GARDNER'S Chemical Synonyms and Trade Names

Eleventh Edition

GARDNER'S Chemical Synonyms and Trade Names

Eleventh Edition

Edited by G W A Milne

Ashgate

Published by
Ashgate Publishing Limited
Gower House
Croft Road
Aldershot
Hants GU11 3HR
England

Ashgate Publishing Company
Old Post Road
Brookfield
Vermont 05036
USA

British Library Cataloguing in Publication Data

Gardner's chemical synonyms and trade names. - 11th ed.
1. Chemicals - Dictionaries 2. Chemical industry - Trademarks - Dictionaries
I. Milne, George W. A. II. Chemical synonyms and trade names
661'.003

ISBN 0 566 08190 3

Library of Congress Cataloging-in-Publication Data

Gardner's chemical synonyms and trade names / edited by George W. A. Milne. -- 11th ed.
p. cm.
Includes index.
ISBN 0-566-08190-3
1. Chemicals--Dictionaries. 2. Chemicals--Trademarks. I. Milne, George W. A., 1937-
TP9.G286 1999
660'.03 -- dc21 98-51143
CIP

Printed in Great Britain by MPG Books Ltd, Bodmin.

Contents

Preface

Through ten previous editions, *Gardner's Chemical Synonyms and Trade Names* has become the best-known and most widely used source of information on chemicals in commerce. The Eleventh Edition reflects the continuing research underlying this reference work and has also seen a major expansion of the information provided for individual chemical compounds.

This edition contains 34,925 entries, many of which have been added for the first time. The main criterion for inclusion of a material in this handbook is its importance as a significant and commercially available chemical. Thus all bulk inorganic chemicals are included, and all major pesticides (herbicides, insecticides, antifungal agents, and so on) and many dyestuffs, surfactants, metals and alloys are described in this edition, which includes the 5,000 highest volume chemicals in the US, as defined by application of the Toxic Substances Control Act. Most drugs are not included and will appear in a separate publication, *Drugs: Synonyms and Properties*, also from Ashgate. Almost all the records describing pure chemicals now carry the appropriate Chemical Abstracts Service (CAS) Registry Number and the associated EINECS (European Inventory of Existing Commercial Chemical Substances) number. Wherever possible, a chemical is thus tagged with the major American and European identification numbers. In addition, all chemicals in this edition which also appear in the Twelfth Edition of the Merck Index have the Merck Index Number provided. Details of the structure of a record are provided in the section *How to Use* Gardner's on page ix.

Entries, whenever possible, contain detailed information on definitions, classification, chemical composition, functions, applications and suppliers. The trade name entries have been obtained directly from chemical manufacturers worldwide and supplemented by a research program into other secondary sources. Verification and correction of the information is a continuous process; this work contains data available at the time of its publication. When a trade name can not be verified or is determined to be obsolete, that fact is noted. When the manufacturer is not verified, that fact is noted. A material whose trade name has no identified manufacturer is also so characterized.

A new feature to this edition is the inclusion of physical properties data for pure chemicals. Properties that have been provided as available include the melting point, boiling point, density or specific gravity, refractive index, optical rotation, ultraviolet absorption, solubility and acute toxicity.

Proprietary Considerations

Every attempt has been made to ensure the accuracy of the information provided in this new edition of *Gardner's*. However, the publishers cannot be held responsible for the accuracy of the information, and users are expected to bear in mind the following information:

> The reporting of a name in *Gardner's* cannot imply definitive legality in establishing proprietary usage. Questions concerning legal ownership of a particular name can be resolved by due legal process.
>
> A manufacturer in some countries may manufacture its product under names different from those cited in *Gardner's*. Similarly, manufacture or marketing of a product may be licensed to a separate company in another country either under the same or a different name.

We trust that readers will find that this new edition contains a wealth of information which is difficult to obtain from any other source. It is the intention of the publishers to produce regularly updated editions and subsets of this compilation at suitable intervals in both printed and digital form. Companies wishing to submit new or updated material for inclusion in future editions should contact George W A Milne (address below).

Acknowledgements

The Editor would like to acknowledge the intensive data entry work performed by Joan Comstock, David Boyer, Teodorina Lessidrenska and Robert Milne, and the skilled programming performed by Dr Ju-yun Li which allowed for accurate formatting and typesetting of this handbook. Finally, I should like to acknowledge the expert assistance and continual support provided by my wife, Kay, without whom none of my deadlines would have been met.

George W A Milne,
Ashgate Publishing,
Old Post Road,
Brookfield, VT 05036, USA.
Telephone: 001-802-276-3162
Fax: 001-802-276-3651
E-mail: gmilne@ashgate.com

How to Use *Gardner's*

Gardner's is divided into four Parts. A brief description of each Part is given below.

a) **PART I** is a dictionary of chemical names and synonyms with extensive cross-referencing from synonyms to the main entry for that chemical. Every entry in this Part now has a unique Gardner's Entry Number (GEN) which is used to refer to, and find, the entry it identifies. Parts II, III and IV make use of the GEN number to make cross-referencing back to Part I easier.

Record Structure

14850 Igepal® Cephene Distilled
122-99-6 7410 204-589-7
$C_8H_{10}O_2$
Phenoxyethanol
2-Phenoxyethanol; Phenoxetol; Phenoxyethyl alcohol; Arosol; Ethylene glycol phenyl ether; 1-Hydroxy-2-phenoxyethane; β-Hydroxyethylphenyl ether; Ethylene glycol mono phenyl ether; Euxyl K 400; Phenyl cellosolve; Phenoxethol; Phenoxyl ethanol; glycol monophenyl ether; phenoxytol; phenylmonoglycol ether; 2-hydroxyethyl phenyl ether; β-phenoxyethyl alcohol; Dowanol EP; Dowanol EPH; Emeressence 1160; Emery 6705; rose ether; Ethanol-2-phenoxy. Surfactant. mp = 14°; bp = 245°; d = 1.1000; insoluble in H_2O, soluble in organic solvents; LD_{50} (rat orl) = 1260 mg/kg. *Rhône-Poulenc France.*

A typical *Gardner's* record is shown above. The first line contains, in bold face, the GEN for the record (14850) and the name of the material. The second line gives, if available, the Chemical Abstracts Service (CAS) Registry Number (122-99-6) for the compound, the corresponding Merck Index Entry Number (7410) and the European Inventory of Existing Commercial Chemical Substances (EINECS) Number (204-589-7). Such numbers always appear in the same position (left, centre or right) enabling the reader to determine which source they belong to. Whenever CAS Registry Numbers are used in the text, they are always enclosed in brackets, for example [122-99-6]. The molecular formula of the compound is provided and the next line carries the chemical name of the compound. This is followed by as many as 50 synonyms, including trade names and other trivial names.

A description of the material, its origins and known uses then follows and, when available, its physical properties are presented. These include melting point, boiling point, density or specific gravity, refractive index, uv absorption, solubility and acute toxicity, usually limited to oral dosage in rats. Finally, the companies who supply, or have supplied, the product are given. A note as to the currency of the information ends the record. The absence of such a note indicates that the record is presumed to be accurate. When a trade name is not verified as available, a note 'name unverified' is attached to the record. When a material which was once available from a given supplier is encountered, it is tagged as 'discontinued'; 'unverified' means that neither the trade name nor the supplier have been confirmed, and 'no supplier' means that the trade name may still be valid, but no supplier can be located.

It should be noted that much of these data are available only for pure compounds. When mixtures of known composition are described, the components are given and each can be examined separately. Properties that are described, however, are those of the mixture.

b) **PART II** is a thesaurus containing all chemical and trade names found in Part I. It is probably the most convenient place to start if only a name is known. This thesaurus will refer the reader to the GEN Number in Part I, which relates to the main entry for that chemical.

c) **PART III** contains two Indexes:

i) **Index 1** enables the reader to locate the GEN for any CAS (Chemical Abstracts Service) Registry Number. In addition, it lists other trade names or synonyms which appear in *Gardner's* and which have the same CAS Number.

ii) **Index 2** enables the reader to locate the GEN for any EINECS (European Inventory of Existing Commercial Chemical Substances) Number. In addition, it lists other trade names or synonyms which appear in *Gardner's* and which have the same EINECS Number.

d) **PART IV** is a directory of chemical manufacturers and suppliers whose products are described in Part I. The entries are in strict alphabetical order by company name. Wherever possible, the postal address, telephone number, fax number and website address are provided.

Glossary of Units

Name	Description
Mass	Unless otherwise specified, mass is expressed in a multiple of grams (g), such as micrograms (μg; = 10^{-6} g), milligrams (mg; = 10^{-3} g), grams (g; = 10^{0} g), kilograms (kg; = 10^{+3} g), etc.
Volume	Volume is expressed in litres (l) or millilitres (ml) unless otherwise specified.
Temperature	When no units are cited, the temperature given is in degrees Celsius (°C).
Melting point	Melting points are cited in degrees Celsius (°C) unless otherwise specified.
Boiling point	When measured at atmospheric pressure, boiling points are cited with no pressure, e.g. bp = 167°. At other pressures, the pressure is also cited, i.e. $bp_{0.01}$ = 167°.
Density	The measurement temperature is given as a superscript; thus a density of 1.123 measured at 25° will appear as d^{25} = 1.123. If the measurement was explicitly referenced to the density of water at 4°, the citation will carry both a superscript and a subscript, as in d^{25}_{4} = 1.123. Specific gravities are denoted by the abbreviation 'sg'.
Refractive index	Denoted by the letter n, refractive indices are usually determined at a temperature which is cited as a superscript, as in n^{25} = 1.5432. The wavelength of the light used in the measurement is cited as a subscript, as in n^{25}_{546} = 1.5432. Most commonly, the sodium D line (wavelength 549 nm) is used and in such cases, the subscript is a D, as in n^{25}_{D} = 1.5432.
Optical rotation	As with refractive indexes, optical rotations (α) are cited with the measurement temperature superscripted, and the measurement wavelength (often the sodium D line) subscripted, as in $[\alpha]^{25}_{D}$ = 105°. When mutarotation can occur, the rotation given is an equilibrium value, measured after some time interval, which is cited, as in $[\alpha]^{25}_{D}$ = 105°(14 hr).
UV absorption	The ultraviolet absorption maxima given by the material are cited in nanometers (nm = millimicrons, mμ) and the absorptivity (E, A, ϵ or log ϵ, all of which are unitless) is also given.
Acute toxicity	Wherever possible the units of toxicity are LD_{50}, i.e. the dose which is lethal to 50% of the test animals. In most cases, acute toxicity is measured with the rat, orally administered, and the result is reported as LD_{50} (rat orl) = 50 mg/kg. Other species (for example, mus = mouse; rbt = rabbit; pgn = pigeon; hmn = human; chd = child; wmn = woman; gpg = guinea pig) are occasionally cited as are other administration routes (sc = subcutaneous; ihl = inhalation; ip = intraperitoneal; iv = intravenous). Chronic toxicity data are not given.

Abbreviations

ABS	acrylonitrile-butadiene-styrene
ACE	acetylcholinesterase
ACN	acrylonitrile
alc.	alcohol
AMP	2-amino-2-methyl-1-propanol
aq.	aqueous
ASA	acrylic-styrene-acrylonitrile
BHA	butylated hydroxyanisole
BHT	butylated hydroxytoluene
BMC	bulk molding compound
bp	boiling point
BP	British Pharmacopeia
BR	butadiene rubbers, polybutadienes
B/S	butadiene/styrene
CAB	cellulose acetate butyrate
CAS	Chemical Abstracts Service
CDA	completely denatured alcohol
CI	Color Index
CMC	carboxymethylcellulose, critical micellar concentration
CNS	central nervous system
CPE	chlorinated polyethylene
CPVC	chlorinated polyvinyl chloride
CR	chloroprene rubber, polychloroprene
cs *or* cSt	centistoke(s)
CTFA	Cosmetic, Toiletries and Fragrance Association
DAP	diallyl phthalate, diammonium phosphate
DB	dichlorophenoxybutyric acid
DEA	diethanolamine, diethanolamide
DEDM	diethylol diethyl
DIBA	diisobutyl adipate
DIDA	diisodecyl adipate
DMC	4,4'-dichloro(methylbenzhydrol)
DMDM	dimethylol dimethyl
DMF	dimethlformamide
DMSO	dimethyl sulfoxide
DNPT	dinitrosopentamethylenetetramine
DOP	dioctyl phthalate
DOT	Department of Transportation
DP acid	diphenolic acid
DPG	diphenylguanidine
DTPA	diethylenetriamine pentaacetic acid
ECTFE	ethylene/chlorotrifluoroethylene copolymer
EDTA	ethylenediamine tetraacetic acid
EINECS	European Inventory of Existing Commercial Chemical Substances
EMC	electromagnetic conductive
EMI	electromagnetic interference
EO	ethylene oxide
EP	extreme pressure
EPDM	ethylene-propylene-diene rubbers
EPM	ethylene-propylene rubbers
EPR	ethylene-propylene rubber
ESCR	environmental stress crack resistance
ETFE	ethylene tetrafluoroethylene

ETU	ethylene thiourea
EVA	ethylene vinyl acetate
F	Fahrenheit
FA	fatty acid
FDA	Food & Drug Administration
FEP	fluorinated ethylene propylene
FFA	free fatty acid
FG	food grade
fp	freezing point
FRP	fiberglass-reinforced plastic(s)
GFRP	glass fiber-reinforced plastic(s)
gran.	granular, granules
GRAS	generally recognized as safe
GRP	glass-reinforced plastics, polyester
HDI	hexamethylene diisocyanate
HDL	high density lipids
HDPE	high density polyethylene
HIPS	high impact polystyrene
HLB	hydrophilic-lipophilic balance
HPLC	high performance liquid chromatography
IC	integrated circuit
ihl	inhalation
IIR	isobutylene-isoprene rubber
IPA	isopropyl alcohol
IPM	isopropyl myristate
IPP	isopropyl palmitate
IR	(synthetic) isoprene rubber, infrared
IU	international units
iv	intravenous
J	joule(s)
KTPP	potassium tripolyphosphate
LDL	low density lipids
LDPE	low density polyethylene
LED	light-emitting diode
LLDPE	linear low density polyethylene
Ltd	Limited
MA	methacrylic acid
MBCA	4,4'-methylene bis(orthochloroaniline)
MBT	mercaptobenzothiazole
MBTS	2-mercaptobenzothiazole disulfide
MCPA	(4-chloro-2-methylphenoxy) acetic acid
MDI	methylene diphenylene diisocyanate
MDM	monomethylol dimethyl
MDPE	medium density polyethylene
MEA	monoethanolamine, monoethanolamide
MEK	methyl ethyl ketone
MIBK	methyl isobutyl ketone
min.	minute(s), mineral, minimum
MIPA	monoisopropylamine, monoisopropylamide
MKP	monopotassium phosphate
MMW-HDPE	medium molecular weight high density polyethylene
MOCA	methylene bis(orthochloroaniline)
mp	melting point
MPK	methyl propyl ketone
MVTR	moisture vapor transmission rate
mw	molecular weight
N	normal
NBR	nitrile-butadiene rubber
NC	nitrocellulose

NCR	nitrile-chloroprene rubber
NEMA	National Electrical Manufacturers Association
N/F	non-flammable
NF	National Formulary
NR	natural (isoprene) rubber
NSF	National Science Foundation
NTA	nitrilotriacetic acid
OEM	original equipment manufacturer
OPP	oriented polypropylene
OTC	over the counter (i.e. non-prescription) drug
o/w	oil-in-water
Pa	Pascal
PAN	polyacrylonitrile
PBT	polybutylene terephthalate
pbw	parts by weight
PC	polycarbonate
PCA	2-pyrrolidine-5-carboxylic acid
PCP	pentachlorophenol
PCTFE	polychlorotrifluoroethylene
PE	polyethylene
PEEK	polyetheretherketone
PEG	polyethylene glycol
PEI	polyetherimide
PEK	polyetherketone
PES	polyether sulfone
PET	polyethylene terephthalate
PFA	perfluoroalkoxy
PG	polypropylene glycol
pH	hydrogen ion concentration as negative logarithm
phr	parts per hundred of rubber or resin
PIB	polyisobutylene
pK	dissociation constant as negative logarithm
PMA	phosphomolybdic acid
PMMA	polymethyl methacrylate
PO	propylene oxide
POE	polyoxyethylene, polyoxyethylated
POM	polyoxymethylene
POP	polyoxypropylene, polyoxypropylated
PP	polypropylene
PPE	polyphenylene ether
PPG	polypropylene glycol
ppm	parts per million
PPO	polyphenylene oxide
PPS	polyphenylene sulfide
PS	polystyrene
PTFE	polytetrafluoroethylene
PTMEG	polytetramethylene ether glycol
PU, PUR	polyurethane
PVA, PVAL	polyvinyl alcohol
PVAc	polyvinyl acetate
PVB	polyvinyl butyral
PVC	polyvinyl chloride
PVDC	polyvinylidene chloride
PVDF	polyvinylidene fluoride
PVE	polyvinyl ethyl ether
PVF	polyvinyl fluoride
PVM	polyvinyl methyl ether
PVP	polyvinyl pyrrolidone
RFI	radio frequency interference

RIM	reaction injection molded (molding)
RTM	resin transfer molding
RTV	room temperature vulcanizing
RV	recreational vehicle
SAN	styrene-acrylonitrile
S/B	styrene/butadiene
SBR	styrene/butadiene rubber
SBS	styrene-butadiene-styrene
SDA	specially denatured alcohol
SE	self emulsifying
SMA	styrene maleic anhydride
SMC	sheet molding compound
SPF	sun protection factor
SR	styrene rubber
SRF	semi-reinforced furnace
TBHQ	*tert*-butylhydroquinone
TDI	toluene diisocyanate
TEA	triethanolamine, triethanolamide
TFE	tetrafluoroethylene
THF	tetrahydrofuran
TIPA	triisopropanolamine
TMC	thick molding compound
TMPTA	trimethylolpropane triacrylate
TPGDA	tripropylene glycol diacrylate
TPO	thermoplastic polyolefin
UF	urea-formaldehyde
UHF	ultra-high frequency
UHMW	ultra high molecular weight
UHMWPE	ultra high molecular weight polyethylene
UL	Underwriter's Laboratory
UPVC	unplasticized polyvinyl chloride
USDA	United States Department of Agriculture
USP	United States Pharmacopeia
uv	ultra-violet
VA	vinyl acetate
VAE	vinyl acetate ethylene
VC	vinyl chloride
VdC, VDC	vinylidenechloride
VHF	very high frequency
VOC	volatile organic compounds
v/v	volume by volume
v/w	volume by weight
w/o	water in oil
w/v	weight by volume
w/w	weight by weight
XLPE	cross-linked polyethylene

PART I

DICTIONARY

Chemical Names and Synonyms

1 **A Metal**
A nickel-iron-copper alloy containing 6-8% copper; used for making audio-frequency transformers. No manufacturer.

2 **A.B.S. 87%**
Alkylbenzene sulfonic acid
Detergent intermediate. *Triantaphyllou SA.*

3 **A.T.S**
Erythromycin
Antibacterial. No manufacturer.

4 **A-0020**
Stryene-butadiene rubber. Styrene/butadiene rubber (100 parts), N-300 carbon black (82.5 parts), high aromatic oil (62.5 parts), black masterbatch. *Goldsmith and Eggleton.*

5 **A-1, A-1 Thiocarbanilide**
102-08-9 3393 203-004-2
$C_{13}H_{12}N_2S$
N,N'-diphenylthiourea
1,3-diphenyl-2-thiourea;N,N-diphenylthiourea; N,N-diphenylsulfourea. Vulcanization accelerator for fast-curing repair stocks; neoprene latex, natural rubber latex and cements; an activator for thiazole accelerators mp = 152-155°; d = 1.32; insoluble in H_2O, soluble in organic solvents; LD_{50} (rat orl) = 50 mg/kg. *Monsanto.*

6 **A-2**
1344-28-1 369 215-691-6
Al_2O_3
Aluminum trioxide
alumina; aluminium oxide; activated alumina; Abradux; Abramant. Activated alumina *La Roche Chem.*

7 **A-17**
106-97-8 1541 203-448-7
C_4H_{10}
n-Butane
butane; n-butane. hydrocarbon propellant bp -0.50°; soluble in H_2O (15%), alcohol 17% (v/v); d = 2.046 (air=1) *Phillips.*

8 **A-31**
75-28-5 200-857-2
C_4H_{10}
Isobutane
2-Methylpropane; Methylpropane. Hydrocarbon propellant bp = -12°; insoluble in H_2O, soluble in organic solvents. *Phillips.*

9 **A-108**
74-98-6 7982 200-827-9
C_3H_8
Propane
n-Propane. Hydrocarbon, used as a propellant. mp = -187.7°; bp = -42.1° *Phillips.*

10 **A-625/641ABS 301K, ABS 500FR-1**
50-70-4 8680 200-061-5
$C_6H_{14}O_6$
Sorbitol
D-glucitol; D-sorbitol; L-Gulitol; orbit; Sorbol; Sorbicolan; Sorbo; Sorbostryl; Nivitin; Cholaxine; Karion, Sionit; Sionon; Sorbilande; Diakarmon. Natrually occurring, synthesized by hydrogenation of glucose; used in manufacture of sorbose, ascobic acid, propylene glycol, synthetic plasticizers and resins, as humectant and softener mp = 93-97°, 110-112°; bp = 105°; n_D^{20}= 1.4600; $[\alpha]_D^{20}$ = -2.0°; soluble in H_2O (83%); moderately soluble in EtOH, alcohols, phenol, Me_2CO, AcOH, DMF, pyridine, acetamide, insoluble in most other solvents. *ICI Americas.*

11 **AA**
118-92-3 442 204-287-5
$C_7H_7NO_2$
Anthranilic acid
o-Aminobenzoic acid; 2-Aminobenzoic acid; Vitamin L1. Antioxidant for fats, greases, lube oils and polyamides. Sludge preventative in furnace and lube oils. Chelating agent and sequestrant. Corrosion inhibitor. Stabilizer of can lacquers, oils and lubricants *PMC Specialities.*

12 **AA Standard, AA USP**
8001-79-4 1946 232-293-8
Castor oil
Ricinus oil; Tangantangan oil; Neoloid. Triglycerides of ricinoleic, oleic, linoleic, palmitic and stearic acids mp = 230°; bp = 313°; d = 0.961; n^D = 1.4780; $[\alpha]_D$ = 5°. *CasChem.*

13 **AA2 Lime Additive**
Lime additive.
Surfactants/antifoams blend for lime or borax coating operations *Crown Tech.*

14 **A-acid**
$C_{10}H_9NO_4S$
2-amino-1-naphthol-5-sulfonic acid.
Intermediate in synthesis of dyestuffs.

15 **AAmlube V**
Inorganic salt; all-purpose sequestering agent for beck dyeing; improves solubility and dispersibility of all dyes.

16 **AAprotect**
137-30-4 10305 205-288-3
$C_6H_{12}N_2S_4Zn$
Bis(dimethylcarbamodithioato-S,S')zinc
Methyl Zimate; Milbam; Corozate; Fuclasin; Fuklasin; Karbam White; Methyl Cymate; Methasan; Zerlate; Zimate; Zirberk; Carbazinc; Cuman; Drupina 90; Fungostop; Hexazir; Mezene; Prodaram; Tricarbamix Z; Triscabol; Vancide MZ96; Zincmate; Ziram; Ziram F4; Ziram W76; Ziramvis; Zirasan 90; Zirex 90; Ziride; Zitox; Fuclasin Ultra; Accelerator L; Methazate; Aceto ZDED; Aceto ZDMD; Methyl Ziram; Mexene; Molurame; Orchard Brand Ziram; Pomarsol Z Forte; Corona Corozate; Cymate; EPTAC 1; Rhodiacid; Soxinal PZ; Soxinol PZ; Tsimat; Vulcacure; Vulcacure ZM; Vulkacite L; Z 75; ZC; Z-C Spray; Zirasan; Aavolex; Aazira; Antene; Amyl Zimate; Ciram; Cuman L; Hermat Zdm; Methyl Zineb; Mycronil; Zarlate; Zirthane; Alcobam Zm;milam; Pomarsolz; Vancide,. Rubber vulcanization accelerator and agricultural fungicide. Used as a bird and animal repellent. mp = 250°; d^{25} = 1.66; insoluble in H_2O, slightly soluble in organic solvents; LD_{50} (rat orl) = 1.4 g/kg. *Universal Crop Protection Ltd.*

17 **Aaterra WP**
2593-15-9 219-991-8
$C_5H_5Cl_3N_2OS$
1,2,4-thiadiazole, 5-ethoxy-3-(trichloromethyl)-
Etridiazole; Banrot; Echlomezol; Ethazol; Dwell; ETCMTB; Ethazol; Ethazole; Ethazole (fungicide).; 5-Ethoxy-3-trichloromethyl-1,2,4-thiadiazol; 5-Ethoxy-3-trichloromethyl-1,2,4-thiadiazole; ETMT; Etridiazol; Etridiazole; Koban;MF-344; Olin Mathieson 2,424; OM 2424. Protective fungicide which is incorporated into soil or compost. *ICI Agrochemicals.*

18 **Aatex**
Natural rubber-based adhesive, used with leather, paper. *Anglo Specialty Adhesives.*

19 **AB**
96-20-8 447 202-488-2
$C_4H_{11}NO$
2-Amino-1-butanol
2-amino-1-hydroxybutane; 1-(Hydroxymethyl)propylamine; 1-Hydroxy-2-butylamine. Pigment dispersant, neutralizing/emulsifying amine; corrosion inhibitor; acid salt catalyst; pH buffer; chemical, pharmaceutical intermediate, solubilizer mp = 2°; bp = 176-178°; d = 0.944; n_D = 1.4520 *Angus.*

20 **AB1000-F**
Thermoset composite molding compound. Extremely high heat resistance electrical grade for compression transfer and injection molding. *Cuyahoga Plastics.*

21 **AB3500, 3500AR**
Polyester. Thermoset, mineral-filled, glass-reinforced. Thermoset for compression, transfer and injection molding. *Cuyahoga Plastics.*

22 **abaca**
Manila hemp; the inner fiber of *Musa textilis.*

23 **Abalon**
Non-metallic floor tiles for use in building *Courtaulds plc.* Discontinued.

24 **Abalyn**
127-25-3 6087 204-832-7
$C_{21}H_{32}O_2$
Methyl abietate
Methyl resinate; Methyl rosinate. Resin with compatibility, surface wetting properties, viscosity. and tack. Used in lacquers, inks, paper coatings, varnishes, adhesives, sealing compounds, plastics, wood preservatives and perfumes. Has Gardner color 6, Gardner Holt viscosity 21 and acid number 6. bp = 360-365°; d^{20} = 1.040; n_D = 1.530; soluble in most organic solvents, insoluble in H_2O. *Hercules.*

25 **aba-odo**
African term for mixture of rubber latexes, probably those from *Funtumia elastica and Ficus vogelii.*

26 **Abate 1-SG, 2-CG, 4-E, 5CG**
3383-96-8 9286 222-191-1
$C_{16}H_{20}O_6P_2S_3$
O,O'-(Thiodi-4,1-phenylene)phosphorothioic acid O,O,O',O'-tetramethyl ester
Temephos; Difenphos; Phosphorothioic acid, O,O'-(thiodi-p-phenylene) O,O,O',O'-tetramethyl ester.; Phosphorothioic acid, O,O'-(thiodi-p-phenylene) O,O,O',O'-tetramethyl ester. Granular and emulsifiable concentrated herbicide. *Amercian Cyanamid/Ag.*

27 **Abatia**
Leaves of *Abatia rugosa.* Used as a black dye.

28 **Abavit B**
Organo-mercury seed dressing. *Murphy Chemical Co Ltd.*

29 **Abavit S**
Organo-mercurial dip. *Murphy Chemical Co Ltd.*

30 **Abbalide**
1,3,4,6,7,8-Hexahydro-4,6,6,7,8,8-hexamethylcyclopenta[s]-2-benzopyran
Musk odor used in fragrances at all price levels *Bush Boake Allen Ltd.*

31 **Abbarome**
Perfumery base. *Bush Boake Allen Ltd.*

32 **Abbavert**
$C_{10}H_{20}O_2$
2-Ethyl hexanal 1,2-ethanediol cyclic acetal
Caproic aldehyde 1,2-ethanediol cyclic acetal. Perfumery base. *Bush Boake Allen Ltd.*

33 **Abbcite No. 2**
An explosive for coal mines. Contains ammonium nitrate, nitroglycerin and dinitrotoluene.

34 **ABC Trieb**
1066-33-7 522 213-911-5
CH_5NO_3
Ammonium bicarbonate
Acid ammonium carbonate; Ammonium hydrogen carbonate. Baking raising agent. In cooling baths, fire extinguishers, manufacture of porous plastics, ceramics, dyes and fertilizer. mp = 107.5; soluble in H_2O 14%, insoluble in alcohol, acetone; pH of 0.1N soln. = 7.8. *BASF AG.*

35 **Abcure S-40-25**
94-36-0 1149 202-327-6
$C_{14}H_{10}O_4$
Benzoyl peroxide
Benzoyl superoxide; Acetoxyl; Acnegel; Benoxyl; Benzagel 10; Benzaknen; Debroxide; Desanden; Lucidol; Nericur; Oxy-5; Oxy-L; PanOxyl; Peroxydex; Persadox; Persa-gel; Sanoxit; Theraderm; Xerac BP 5; Xerac BP 10. Benzoyl peroxide dispersed in an inorganic medium. Alternative to MEK peroxide. Benzoyl peroxide pastes or granules as initiators for catalysis of unsaturated polyester resins; well-suited for spray applications. mp = 103-106°; sparingly soluble in H_2O, EtOH, soluble in C_6H_6, $CHCl_3$, Et_2O *Abco Industries.*

36 **Abelite**
Blasting explosives containing ammonium nitrate and trinitrotoluene.

37 **Abel's reagent**
A 10% solution of normal chromic acid; used in the micro-analysis of carbon steels, for etching.

38 **Abequito**
50435-25-1 6642
$C_9H_8ClNS_2$
4-Chloro-N-1,3-dithietan-2-ylidene-2-methylbenzeneamine
Nimidane; Cyclic methylene (4-chloro-o-tolyl)-dithioimidocarbonate; AC 84633; ENT 29106. Acaricide. mp = 43-46°. *American Cyanamid.*

39 **Abex® 12S**
Surfactant; emulsifier for vinyl acrylics and acrylic polymerization. *Rhône-Poulenc Surf.*

40 **Abex® 23S**
9004-82-4
Sodium laurethyl sulfate
Emulsifier for emulsion polymerization *Rhône-Poulenc Surf.*

41 **Abex® EP-110**
9051-57-4
Ammonium nonoxynol-9 sulfate
Primary emulsifier and stabilizing agent for the preparation of vinyl acetate, vinyl acetate/acrylic, all acrylic, styrene/acrylic and S/B emulsion copolymers; wetting agent, dispersant for agricultural formulations. *Rhône-Poulenc Surf.; Rhône-Poulenc France.*

42 **Abex® JKB**
Proprietary surfactant; emulsifier for high acid polymerization systems *Rhône-Poulenc Surf.; Rhône-Poulenc Franc.*

43 **Abex® VA 50**
Octoxynol-33, sodium laurethyl sulfate
Emulsifier for high solids vinyl acetate emulsion. *Rhône-Poulenc Surf.; Rhône-Poulenc France.*

44 **Abex®LIV/30**
Ammonium alkylaryl ether sulfate
Emulsifier for emulsion polymerization of acrylic, styrene-acrylic, vinyl acetate; detergent, emulsifier, foam stabilizer, wetting agent household and industrial detergents, shampoos, bubble baths *Rhône-Poulence Geronazzo.*

45 **abietic acid**
514-10-3 3 208-178-3
$C_{20}H_{30}O_2$
[1R-(1α,4aβ,4bα,10aα)]-1,2,3,4,4a,4b,5,6,10,10a-decahydro-1,4a-dimethyl-7-(1-methylethyl)-1-phenanthrenecarboxylic acid
sylvic acid; abietinic acid. Used in lacquers, varnishes, soaps and plastics. mp = 173-174°; bp_9 = 250°; $[\alpha]_D^{20}$= -65° (c = 1 EtOH), insoluble in H_2O, soluble in organic solvents; λ_m = 235, 241.5, 250 nm (ε 19500, 22000, 14300); LD_{50} (mus iv) = 180 mg/kg.

46 **Abietic anhydride**
Gum Rosin. Rosin used in manufacturing of soaps, printing inks, paint dryers, wax compositions, artificial amber and medicine.

47 **Abil® AV 8853, 20-1000**
2116-84-9 218-320-6
$C_{15}H_{32}O_3Si_4$
Phenyl trimethicone
Phenyltris(trimethylsiloxy)silane; Tris(trimethylsiloxy)phenylsilane. Emollient providing skin protection; barrier against aqueous media; perfumery ingredient and fixative; provides improved rub and spreadability; faster penetration; non-sticky; prevents aerosol clogging. *Goldschmidt AG.* Discontinued.

48 **Abil® B 8839**
69430-24-6
Cyclomethicone
Conditioner for hair care products, aerosols, sticks, shaving preparations, deodorants, antiperspirants *Goldschmidt AG.* Discontinued.

49 **Abil® B 8851, 8852**
68937-55-3
Dimethicone copolyol
Conditioner for personal care products; emollient for skin and hair care products, aerosol shaving lather; deodorants, antiperspirants, creams and lotions, perfumes and colognes *Goldschmidt AG.* Discontinued.

50 **Abil® B 9950**
102523-96-6
Dimethicone propyl PG-betaine
Silicone surfactant, conditioner; used in hair and skin care products *Goldschmidt; Goldschmidt AG.*

51 **Abil® B 88183, 88184**
68937-55-3
Dimethicone coployol
Dimethylsiloxane-(60% propylene oxide-40% ethylene oxide) block copolymer. Surfactant used as foam former and providing lubricating and gloss properties; refatting agent for skin products; increases slip in shaving creams; also for shampoos, power gels, hand cleaners, aerosols, antiperspirants. *Goldschmidt; Goldschmidt AG.*

52 **Abil® EM-90, 97**
Cetyl dimethicone copolyol
Conditioner, emulsifier for w/o type creams and lotions, roll-ons, sun care products, shampoos and conditioner *Goldschmidt; Goldschmidt AG.*

53 **ABIL® K 4**
69430-24-6
Cyclomethicone
Emollient, conditioner used in hair care products, aerosols, shaving preparations, deodorants, antiperspirants *Goldschmidt AG.* Discontinued.

54 **Abil® OS5**
Polysiloxane polyether copolymer
Abil® OS12, OS13. Silicone surfactant used in personal care products. *Goldschmidt AG.* Discontinued.

55 **ABIL® OSW 12, OSW 13**
69430-24-6
Cyclomethicone, dimethiconol, dimethicone
Glossing and conditioning agent for hair care products *Goldschmidt.*

56 **Abil® S201**
Sodium poly PG-propyl dimethicone thiosulfate
Hair conditioner and setting lotion; improves gloss and sheen of shampoos, conditioners, mousses, gels and styling aids *Goldschmidt.*

57 **Abil® S255**
Oleyl-, palmitoyl-, palmitoleamidopropyl-silkhydroxypropyldimonium chloride. Conditioner and setting lotion; improves gloss and sheen of shampoos and conditioners *Goldschmidt.*

58 **Abil® Wax 2434**
68554-53-0
Stearoxy dimethicone
Siloxanes and Silicones, di-Me, (octadecyloxy)-terminated. Wax improving application and skin care properties of emulsions; spreading and emollient properties for protection against aqueous media; water barrier for creams and lotions. *Goldschmidt; Goldschmidt AG.*

59 **Abil® Wax 2434**
Polysiloxane polyalkyl copolymer
Emollient used in cosmetic creams and lotions, pigmented products and non-aqueous systems. *Goldschmidt.*

60 **Abil® Wax 2440**
Behenoxy dimethicone
Wax improving application and skin care properties of emulsions; spreading and emollient properties for protection against aqueous media; reduces whitening during application of creams and lotions; pigment solubilizer *Goldschmidt; Goldschmidt AG.*

61 **Abil® Wax 9800**
Stearyl dimethicone
Wax improving color, luster and spreadability of pigmented products; spreading, penetrating and emollient properties for skin care products *Goldschmidt; Goldschmidt AG.*

62 **Abil® Wax 9801**
Cetyl dimethicone
Wax providing emolliency and application benefits for antiperspirants; pigment solubilizer; also used in skin care products *Goldschmidt AG.*

63 **Abil® Wax 9809**
68607-75-0
Stearyl methicone
Wax providing water barrier for night creams and protective lotions *Goldschmidt.*

64 **Abil® Wax 9814**
Cetyl dimethicone
Emollient and spreading agent for cosmetic esters and oils; pigment grinding aid and dispersant, especially for titanium dioxide (prevents reagglomeration); used in sunscreens and pigmented products such as pressed powders *Goldschmidt.*

65 **Abil® WE 09**
Polyglyceryl-4 isostearate, cetyl dimethicone copolyol, hexyl laurate
Emulsifier for highly stable w/o creams and lotions; improves UV protection in sunscreens. *Goldschmidt AG.*

66 **Abil® WS 08**
Cetyl dimethicone copolyol, cetyl dimethicone, polyglyceryl-3 oleate, hexyl laurate
Emulsifier for water-oil creams and lotions. *Goldschmidt AG.*

67 **Abil®10-10000**
Dimethicone
Conditioner for skin and hair care, sunscreens, tanning creams or lotions; after care, aerosol preparations. *Goldschmidt AG.* Discontinued.

68 **Abil®-Quat 3270, 3272**
Quaternium-80
Conditioner, antistat for shampoos and hair rinses; also refatting agent for skin cleansers. *Goldschmidt; Goldschmidt AG.*

69 **Abiol**
39236-46-9 254-372-6
$C_{11}H_{16}N_8O_8$
Imidiazolidinyl urea NF
Germall 115; Biopure 100; Imidurea NF; Sept 115; Tristat 1U; Unicide U-13; Imidurea; Imidazolidinyl urea; Methanebis[N,N'-(5-ureido-2,4-diketotetrahydroimidazole)-N,N-dimethylol]; Urea, N,N''-methylenebis[N'-[1-(hydroxymethyl)-2,5-dioxo-4-imidazolidinyl]-. Preservative. *3V-Sigma.*

70 **Abisol**
7631-90-5 8731 231-548-0
HO_3SNa
Sodium hydrogen sulfite
Sodium bisulfite; Sodium acid sulfite. A 40% aqueous solution of sodium bisulfite; used as a disinfectant and preservative. d = 1.48; soluble in 3.5 parts cold H_2O, 2 parts boiling H_2O, 70 parts EtOH; LD_{50} (rat iv) = 115 mg/kg. No manufacturer.

71 **Abitol® E**
26266-77-3 247-574-0
C20H34O
Dihydroabietyl alcohol
1-Phenanthrenemethanol, dodecahydro-1,4a-dimethyl-7-(1-methylethyl)-, [1R-(1α,4aβ,4bα,10aα)]-; Abietyl alcohol, dihydro-; Dihydroabietyl alcohol; Hydroabietyl alcohol; 1-Phenanthrenemethanol, dodecahydro-1,4a-dimethyl-7-(1-methylethyl)-,. Resinous plasticizer and tackifier in plastics, lacquers, inks and adhesives; chemical intermediate. *Hercules.*

72 **Ablefilm® 550**
Glass-supported, toughened epoxy adhesive film; moisture-resistant adhesive film designed for substrate attachment and sealing microelectronic packages. *Ablestik.*

73 **Ablufoam HT**
Compound used as antifoamer for high temperature jet dyeing. *Taiwan Surf.*

74 **Ablufoam SAE**
Silicone emulsion; textile antifoamer effective over a wide pH range. *Taiwan Surf.*

75 **Abluhide DS**
Blend; soaking, degreasing and rewetting agent for leather manufacturing. *Taiwan Surf.*

76 **Abluhide F Series**
Fat liquoring agent for leather manufacturing. *Taiwan Surf.*

77 **Ablumide CDE**
Cocamide/DEA (1:1)
Cocamide DEA foam stabilizer, thickener for shampoos, bubble baths, liquid detergents, toiletries. *Taiwan Surf.*

78 **Ablumide CME**
68140-00-1 268-770-2
Cocamide-MEA (1:1)
Amides, coco, N-(hydroxyethyl); Coconut fatty acid, monoethanolamide; Coconut oil fatty acid ethanolamide; Coconut oil fatty acids, ethanolamine condensate; Coconut oil fatty acids, monoethanolamide; Coconut oil, monoethanolamide; Loramine C212; Loramine C 212. Foam stabilizer, thickener for shampoos, bubble baths, liquid detergents, toiletries. *Taiwan Surf.*

79 **Ablumide LDE**
120-40-1 204-393-1
$C_{16}H_{33}NO_3$
Lauramide-DEA (1:1)
Dodecanamide, N,N-bis(2-hydroxyethyl)-; N,N-Bis(2-hydroxyethyl)dodecanamide; Clindrol 200l; Clindrol 200L; Clindrol 101cg; Clindrol 203cg; Clindrol 210cgn; Clindrol 200 L; Clindrol superamide 100l; Clindrol Superamide 100L; Coco diethanolamide; Coconut oil amide of diethanolamine; Comperlan LD; CO-1214 Natural; Condensate PL; Crillon l.d.e.; Crillon L.D.E.; Diethanolamide lauric acid; Diethanolamine lauric acid amide; Diethanollauramide; N,N-Diethanollauramide; Diethanol lauric acid amide; Clindrol 200 L; LDA; LDE; Ninol AA62; Onyxol 345; Rewomid DLMS; Rewomid DL 203/S; Richamide 6310; Rolamid CD; Standamidd LD; Steinamid DL 203 S; Super Amide L-9A; Super Amide L-9C; Synotol L-60; Unamide J-56,. Foam stabilizer, thickener for shampoos, bubble baths, liquid detergents. Toiletries. *Taiwan Surf.*

80 **Ablumide LME**
142-78-9 205-560-1
$C_{14}H_{29}NO_2$
Lauramide-MEA (1:1)
Dodecanamide, N-(2-hydroxyethyl)-; Amisol LDE; Cocomonoethanolamide; Comperlan LM; Copramyl; Crillon L.M.E.; Cyclomide LM; 2-Dodecanamidoethanol; Lauramide MEA; Lauric acid ethanolamide; Lauridit LM; Laurylamidoethanol; Rewomid L 203; Rolamid CM; Stabilor C.M.H.; Steinamid L 203; Ultrapole H; Vistalan,. Foam stabilizer, thickener for shampoos, bubble baths, liquid detergents, toiletries. *Taiwan Surf.*

81 **Ablumide SDE**
93-82-3 202-280-1
$C_{22}H_{45}NO_3$
Stearamide-DEA (1:1)
Octadecanamide, N,N-bis(2-hydroxyethyl)-; N,N-Bis(hydroxyethyl) octadecanamide; N,N-Bis(2-hydroxyethyl)octadecanamide; N,N-Bis(2-hydroxyethyl)stearamide; N,N-Bis(β-hydroxyethyl)stearamide; Clindrol 868; Clindrol 200-S; Cyclomide SD; Diethanolamine stearic acid amide; Diethanolstearamide; Onyxol 42; Schercomid ST; Stearamide DEA; Stearic acid diethanolamide; Stearic diethanolamide; Stearoyl Diethanolamide; N-Stearoyl diethanolamine; Unamide S,. Thickener, emulsifier for mineral and vegetable oils, microcrystalline wax. *Taiwan Surf.*

82 **Ablumide SME**
111-57-9 203-883-2
$C_{20}H_{41}NO_2$
Stearamide-MEA (1:1)
Octadecanamide, N-(2-hydroxyethyl)-; Clindrol 200-MS; Comperlan HS; Cycloamide SM; N-(2-Hydroxyethyl)octadecanamide; N-(2-Hydroxyethyl)stearamide; Loramine S 280; Marlamid M 18; Monoethanolamine stearic acid amide; Ninol 1301; Onyx Wax EL; Stearamide MEA; Stearamyl;Stearic ethanolamide; Stearic ethylolamide; Stearoylethanolamide; N-Stearoylethanolamine; Stearoyl monoethanolamide; Teric CME7,. Opacifier, thickener for shampoo, cream rinse, bubble bath. *Taiwan Surf.*

83 **Ablumine 08**
Alkyl (98% C_8) benzyl dimethyl quaternary
Leveling agent for acrylic fiber *Taiwan Surf.*

84 **Ablumine 230**
N-Alkyl dimethylammonium chloride
Algicide for industrial cooling towers and swimming pools. *Taiwan Surf.*

85 **Ablumine 280**
122-19-0 204-527-9
$C_{27}H_{50}N.Cl$
Stearyl dimethyl benzyl ammonium chloride
Benzenemethanaminium, N,N-dimethyl-N-octadecyl-, chloride; Ammonium,

benzyldimethyloctadecyl-, chloride; Ammonyx 4; Ammonyx 4002; Ammonyx 485; Ammonyx 490; Ammonyx ca special; Arquad dm18b-90; Baraquat sb-25; Barquat sb-25; Benzyldimethyloctadecylammonium chloride; Carsoquat sdq-25; Carsoquat sdq-85; Dehyquart stc-25; Dimethylbenzyloctadecylammonium chloride; Dimethyl(N-octadecylphenylmethyl)ammonium chloride; Intexan SB-85; Intexsan SB-85; J Soft C 4; Katamine AB; Nissan cation s2-100,. Antistat and hair conditioner. *Taiwan Surf.*

86 Ablumine D10
Didecyl dimethyl ammonium methosulfate
Disinfectant, sanitizer, germicide. *Taiwan Surf.*

87 Ablumine DHT75
Dihydrogenated tallow dimethyl ammonium methosulfate
Antistat, fabric softener suitable for dryer sheets. *Taiwan Surf.*

88 Ablumine DT
Ditallow dimethyl ammonium methosulfate
Household softener with wide dispersion stability. *Taiwan Surf.*

89 Ablumox C-7
61791-14-8
PEG-7 cocamine
Corrosion inhibitor for acid cleaners. *Taiwan Surf.*

90 Ablumox CAPO
68155-09-9 268-938-5
Cocamidopropylamine oxide
Amides, coco, N-[3-(dimethylamino)propyl], N-oxide; Cocamidopropyl)dimethylamine oxide; Cocoamido-3-propyldimethylamine oxide; 3-Cocoamidopropyl dimethylamine oxide; 3-(N,N-Dimethylamino)propyl cocoamido amine oxide; N,N-Dimethyl-N-(3-cocamidopropyl)amine oxide; N,N-Dimethyl-N-[3-(coconut oil alkyl)amidopropyl]amine oxide,. Foamer, wetting agent, foam stabilizer, antistat, detergent, emollient. *Taiwan Surf.*

91 Ablumox LO
1643-20-5 216-700-6
$C_{14}H_{31}NO$
Lauramine oxide
1-Dodecanamine, N,N-dimethyl-, N-oxide; Dimethyldodecylamine oxide; dimethyldodecylamine-N-oxide; Ammonyx LO; Ammonyx AO; Aromox DMMC-W; Conco XAL;DDNO; N,N-dimethyldodecylamine oxide; dodecycldimethylamine oxide; n-dodecyldimethylamine oxide; lauryldimethylamine oxide; Lauramine oxide. Foamer, wetting agent, foam stabilizer, antistat, detergent, emollient. mp = 130-131°; LD_{50} (rat orl)= 1000 mg/kg. *Taiwan Surf.*

92 Ablumox T-15
61791-44-4 263-177-5
PEG-15 tallow amine
Leveling agent for dyeing of nylon with anionic dyes; controls rate of strike and dye migration *Taiwan Surf.*

93 Ablumul AG-306
Blend; emulsifier used with acephate. *Taiwan Surf.*

94 Ablumul EP
Thickener for textile printing. *Taiwan Surf.*

95 Ablunol 200ML
9004-81-3
$(C_2H_4O)_n \cdot C_{12}H_{24}O_2$
PEG 200 laurate
Poly(oxy-1,2-ethanediyl), α-(1-oxododecyl)-ω-hydroxy-; Glycols, polyethylene, monolaurate; Aquafil I; Aquafil II; Atlas G-2127; Atlas G-2129; Cirrasol TCS; Emanon 1112; Empilan AP-100; Ester 14; Ethylan L; Ethylan L 3; G 2129; Hallco CPH 43; Ionet ML-400; Lauric; Lauric acid, ethylene oxide adduct; Lipo-Peg 4-L; Lonzest PEG-4L; Macrogol laurate 600; Newcol 150; Nissan Nonion L 4; Nonex 139; Nonex 27; Nonex 31; Nonex 39; Nonex 55; Nonex 56; Nonex 99; Nonion L 4; Nopalcol 1-L; Nopalcol 6-L; Nopalcol 10-L,. Emulsifier, lubricant, dispersing and leveling agent; defoamer used in cosmetics, textiles, paint, dyestuffs and other industrial uses. *Taiwan Surf.*

96 Ablunol 200MO
9004-96-0
$(C_2H_4O)_nC_{18}H_734O_2$
PEG-200 oleate
Poly(oxy-1,2-ethanediyl), α-(1-oxo-9-octadecenyl)-ω-hydroxy-, (Z)-; Glycols, polyethylene, monooleate; Advawet 33; Agrimul 26B; Akyporox O 50; Atlas g-2142; Atlas g-2144; Atlas G-2142; Atlas G-2144; Cemulsol 1050; Cemulsol d-8; Cemulsol D-8; Cemulsol A; Cemulsol C 105; Chemester 300-OC; Cithrol PO; Colloid 815M; Crodet O 6; E2; Emanon 4115; Emcol h-2a; Emcol H-2A (VAN); Emcol H 31a; Emcol H 31A; Emerest 2646; Emerest 2660; Empilan BP-100; Empilan BQ-100; Emulphor A; Emulphor UN-430; Emulphor vn 430; Emulsifier L-32; Ethofat 0/15,. Emulsifier, lubricant, dispersing and leveling agent used in cosmetic, textile, leather, paint and other industrial uses. *Taiwan Surf.*

97 Ablunol 200MS
9004-99-3
PEG 200 stearate
Emulsifier, thickener, lubricant, softener, defoamer, dispersing and leveling agent used in cosmetic, textile, paint and other industrial uses. *Taiwan Surf.*

98 Ablunol CO 10
61791-12-6
PEG-10 castor oil
Castor oil, ethylene glycol polymer; Castor oil, ethoxylated; AL 846; AL 858; Atlox 1285; Atlox 1300; Castor oil, polyethoxylated; Cremophor El; Crystal Inhibitor No. 5; El-620; Emulphor EL-620; Emulphor EL 719; Emulsifier Component I; Emulsifier Component L; Emulsifier Component M; G-1284; G-1300; Geopan SF365; Industrol CO-36; Nalco L-357; Ninox CZ-1; Nopco polyethoxylated castor oil; Polyoxyethylene castor oil; Retzanol C040; Retzanol CO-40; Retzloff Intermediate No. 123,. Emulsifier for oils, solvents and waxes, lubricant, antistat. *Taiwan Surf.*

99 Ablunol DEGMS
9004-99-3
$(C_2H_4O)_nC_{18}H_{36}O_2$
Poly(oxy-1,2-ethanediyl), α-(1-oxooctadecyl)-ω-hydroxy-
Diethylene glycol stearate; Glycols, polyethylene, monostearate; Akyporox S 100; Arosurf 1855E40; Atlox 5000; Brij 78; Carbowax 1000 monostearate; Carbowax 4000 monostearate; Cerasynt 660; Cithrol 10MS; Cithrol PS; Clearate G; Cremophor A; Crill 20,21,22,23; Emanon 3113; Emcol H 35-A; Emerest 2640; Emery 15393; Empilan CP-100; Empilan CQ-100; Emulphor vt-650; Emunon 3115; Ethofat 60/15; Ethofat 60/20; Ethofat 60/25; Ionet MS-1000; Kessco X-211; Lactine; Lamacit CA; Lipal 15S,. Opacifier, pearlescent for cosmetics, detergents. *Taiwan Surf.*

100 Ablunol EGMS
111-60-4 203-886-9
$C_{20}H_{40}O_3$
Glycol stearate
Ethylene glycol monostearate. Opacifier, pearlescent for cosmetics, detergents. mp = 57-63°. *Taiwan Surf.*

101 Ablunol GML
142-18-7 205-526-6
$C_{15}H_{30}O_4$
Glyceryl laurate
Dodecanoic acid, 2,3-dihydroxypropyl ester; Laurin, 1-mono-2,3-dihydroxypropyl dodecanoate; Glycerin 1-monolaurate; Glycerol 1-dodecanoate; Glycerol 1-laurate; Glycerol 1-monolaurate; Glycerol α-monolaurate; 1-Glyceryl laurate; Glyceryl monododecanoate; Glyceryl monolaurate; Lauric acid 1-monoglyceride; Lauric acid α-monoglyceride; 1-Monolaurin,. Used in mold release agents. *Taiwan Surf.*

102 Ablunol GMO
111-03-5 203-827-7
$C_{21}H_{40}O_4$
Glyceryl oleate
9-Octadecenoic acid (Z)-, 2,3-dihydroxypropyl ester; Olein, 1-mono-; Glycerin 1-monooleate; Aldo HMO; Glyceryl Monooleate; 1-Monooleoylglycerol; 1-Oleoylglycerol; 1-Oleylglycerol; α-Monoolein; Glycerol α-*cis*-9-octadecenate,. Internal lubricant, antistat, antifogging agent for PVC film; mold release agent and rust prevention for compounded oils. *Taiwan Surf.*

103 Ablunol GMS
11099-07-3 234-325-6
Glyceryl stearate
Octadecanoic acid, ester with 1,2,3-propanetriol; Glyceryl stearate; Glyceryl Stearate SE. Emulsifier for hand creams, lotions. cosmetics; textile lubricant softener.

104 Ablunol LA-3
Laureth-3
Emulsifier, dispersant, detergent used in textile processing, cosmetics, metalworking compounds, agriculture and industrial cleaners. *Taiwan Surf.*

105 Ablunol LMO
Surfactant; leveling agent for dyeing polyester fibers with disperse dyes.

106 Ablunol NP4
9016-45-9 6772
$(C_2H_4O)_nC_{15}H_{24}O$
Poly(oxy-1,2-ethanediyl), α-(nonylphenyl)-ω-hydroxy-
Nonoxynol-4; Glycols, polyethylene, mono(nonylphenyl) ether; Agral R; Alfenol 8; Alfenol 18; Alfenol 22; Alfenol 28; Alfenol 710; Alkyl aryl polyoxyethylene; Alkyl phenol ethoxylate; Alphenol 8; Antarox CO; Antarox CO 430; Antarox CO 530; Antarox CO 630; Antarox CO 730; Antarox CO 850; Antarox CO 880; Antarox CO 970; Arkopal N-040; Arkopal N-060; Arkopal N-090; Arkopal N; Arkopal N 100; Arkopal N 110; Arkopal N 150; Arkopal N 300,. Detergent and dispersant used with petroleum oils;

intermediate in manufacture of surfactants and antistats; co-emulsifier for fats, oils and waxes. *Taiwan Surf.*

107 Ablunol OA-6
9004-98-2
$(C_2H_4O)+nC_{18}H_{36}O$
Poly(oxy-1,2-ethanediyl),α-9-octadecenyl-ω-hydroxy-, (Z)-
Oleth-6; Glycols, polyethylene, mono-9-octadecenyl ether, (Z)-; Amerox oe-20; Ameroxol EO 2; Ameroxol OE 10; Ameroxol OE-20; Atlas G-3915; Atlas G-3920; BO 2; BO 7; Brij 92; Brij 93; Brij 96; Brij 97; Brij 98; Brij 99; Brij 92((2)-oleyl); Brij 96((10) oleyl;)Decaethoxy oleyl ether; Dehydol 100; EL-620; EL-719; Emalex 515; Emery 6802; Emulgen 408; Emulgen 420; Emulgen 430; Emulgin 010; Emulgin 05; Emulphor; Emulphor ON-870; Emulphor O,. Emulsifier for mineral oils and cosmetics. *Taiwan Surf.*

108 Ablunol S-20
1338-39-2 8872 215-663-3
$C_{18}H_{34}O_6$
Sorbitan laurate
Sorbitan, monododecanoate; Sorbitan, monolaurate; Anhydrosorbitol monolaurate; Arlacel 20; Armotan ML; Emasol 110; Emsorb 2515; Glycomul L; Glycomul LC; Ionet S-20; Lauric acid, sorbitan ester; Nikkol SL 10; Nonion LP 20R; Radiasurf 7125; Solgen 90; Sorbitan laurate; Sorbitan ML; Sorgen 90; Soruken 90; Span 20,. Emulsifier, emulsion stabilizer; thickener for cosmetic, pharmaceutical, food applications; textile fiber lubricant, softener, anti-fogging agent. *Taiwan Surf.*

109 Ablunol S-40
26266-57-9 247-568-8
$C_{22}H_{42}O_6$
Sorbitan palmitate
Sorbitan, monohexadecanoate; Sorbitan, monopalmitate; Anhydrosorbitol monopalmitate; Aracel 40; Arlacel 40; Crill 2; Emsorb 2510; Glycomul P; Liposorb P; Montane 40; Nikkol sp10; Nissan nonion PP-40; Nissan Nonion PP 40; Nissan nonion PP 40r; Nonion PP 40; Protachem Smp; Rheodol sp-p 10;,. Emulsifier, emulsion stabilizer, thickener for cosmetics, pharmaceutical, food applications; textile fiber lubricant, softener, antifogging agent. *Taiwan Surf.*

110 Ablunol S-60
1338-41-6 8872 215-664-9
$C_{24}H_{46}O_6$
Sorbitan stearate
Sorbitan, monooctadecanoate; Sorbitan, monostearate; Anhydrosorbitol monostearate; Arlacel 60; Armotan MS; Crill 3; Crill-K-3; Drewsorb 60; Durtan 60; Emsorb 2505; D-Glucitol, anhydro-, monooctadecanoate; Glycomul S; Hodag SMS; Ionet S-60; Ionet S 60; Liposorb S-20; Liposorb S; Montane 60; MS-33; Ms 33F; Newcol 60; Nikkol SS-30; Nikkol SS 30; Nissan Nonion SP-60; Nissan nonion sp 60; Nonion sp 60; Nonion sp 60r; Rikemal S 250; Solgen 50;,. Emulsifier, emulsion stabilizer, thickener for cosmetics, pharmaceutical and food applications; textile fiber lubricant, softener, antifogging agent, silicone defoamer emulsions. *Taiwan Surf.*

111 Ablunol S-80
1338-43-8 8872 215-665-4
$C_{24}H_{44}O_6$
Sorbitan oleate
Sorbitan, mono-9-octadecenoate; Sorbitan, monooleate; Anhydrosorbitol monooleate; Arlacel 80; Arlacel A (VAN); Armotan MO; Emsorb 2500; Glycomul O; Ionet S-80; Liposorb o-20; Liposorb O; ML 33F; Ml 55F; MO 55F; Monodehydrosorbitol monooleate; Montan 80; Nikkol SO-10; Nikkol SO-15; Nikkol SO-30; Nikkol SO 10; Nikkol so-15; Nikkol so-30; Nonion op80r; Nonion OP 80R; O 250; Oleic acid, monosorbitan ester; Radiasurf 7155. Emulsifier, emulsion stabilizer, thickener for cosmetics, pharmaceutical and food applications; textile fiber lubricant, softener, antifogging agent, wet processing of synthetic PU leather. *Taiwan Surf.*

112 Ablunol S-85
26266-58-0 8872 247-569-3
$C_{60}H_{108}O_8$
Sorbitan trioleate
Sorbitan, tri-9-octadecenoate, (Z,Z,Z)-; Anhydrosorbitol trioleate; Anhydrosorbitol trioleate; Sorbitan, trioleate; Aracel 85; Arlacel 85; Crill 5; Emasol 430; Emsorb 2503; Glycomul TO; Ionet S-85; Ionet S 85; Liposorb To; Nissan nonion OP-85; Nissan nonion op 85; Nissan nonion op 85r; Nonion OP 85R; Op 85r; Protachem Sto; Rheodol sp 030;,. Emulsifier, emulsion stabilizer, thickener for cosmetics, pharmaceutical and food applications; textile fiber lubricant, softener, antifogging agent. *Taiwan Surf.*

113 Ablunol SA-7
9005-00-9
$(C_2H_4O)_n{\cdot}C_{18}H_{38}O$
Poly(oxy-1,2-ethanediyl), α-octadecyl-ω-hydroxy-
Steareth-7; Glycols, polyethylene, monooctadecyl ether; Avivan SO 6; Berol 08; Brij 72; Brij 76; Brij 78; Ekaline G 80; Emulgen 320P; G 3710; G 3720; Genapol S; Genapol S 020; Genapol S 150; G 3694POE; G-3720-POE; Levenol PW; Lipocol S-20; Marlipal 1850; Myrj 45; Myrj 52; Nonion S-220; Nonion S 220; 1-Octadecanol, monoether with polyethylene glycol; Octadecyl alcohol, ethoxylated; Octadecyloxypoly(ethyleneoxy)ethanol; Oleth-15; OS 20A; Oxanol TsS 21; Polyethylene glycol monooctadecyl ether,. Emulsifier for waxes and cosmetics. *Taiwan Surf.*

114 Ablunol T-20
POE sorbitan laurate
Oil-water emulsifier for waxes and cosmetics. *Taiwan Surf.*

115 Abluphat AP Series
Phosphate ester, sodium salt, free acid
Antistat, penetrant, wetting agent, solubilizer, detergent; for cotton and synthetics processing; high alkaline and acid tolerance. *Taiwan Surf.*

116 Abluphat LP Series
Phosphate ester sodium salt
Antistat for synthetic fibers, penetrant, wetting agent, solubilizer, detergent. *Taiwan Surf.*

117 Ablupol AF
Antifoaming agent for coating color formulation of paper. *Taiwan Surf.*

118 Ablusoft A
Polyamide type surfactant; softener for acrylic fiber, cotton and synthetics. *Taiwan Surf.*

119 Ablusoft C-70
Dioctyl sulfosuccinate
Wetting agent for industrial applications; penetrant/wetting agent for bleaching of cotton, agricultural applications. *Taiwan Surf.*

120 Ablusoft ES
Epoxy-modified silicone; finishing agent imparting durable softness, wrinkle resistance to cotton and synthetic fiber blends. *Taiwan Surf.*

121 Ablusoft PE
Polyethylene emulsion; finishing agent improving tear strength, abrasion resistance and handle of fabrics. *Taiwan Surf.*

122 Ablusol DA
Ethoxylated decyl alcohol sulfosuccinate monoester
Emulsifier for polyacrylate emulsion polymerization. *Taiwan Surf.*

123 Ablusol DBC
26264-06-2 247-557-8
Calcium dodecylbenzene sulfonate
Dodecyl-benzenesulfonic acid calcium salt; Calcium dodecyl benzene sulfonate. Emulsifier for agricultural chemicals, oils and solvents. *Taiwan Surf.*

124 Ablusol DBD
26545-53-9 246-784-2
$C_{18}H_{30}O_3S{\cdot}C_4H_{11}NO_2$
DEA dodecylbenzene sulfonate
Benzenesulfonic acid, dodecyl-, compd. with 2,2'-iminobis[ethanol] (1:1;)Bis(2-hydroxyethyl)ammonium dodecylbenzenesulfonate; BAS 089-00E; Diethanolamine dodecylbenzenesulfonate; Diethanolamine-dodecylbenzenesulfonic acid adduct; Diethanolamine-dodecylbenzenesulfonic acid salt; Diethanolammonium dodecylbenzenesulfonic acid; Dodecylbenzenesulfonic acid diethanolamine salt; Dodecylbenzenesulfonic acid, diethanolamine salt,. Surfactant, emulsifier, wetting agent for bubble baths, shampoos and detergents. *Taiwan Surf.*

125 Ablusol DBM
1331-61-9 215-559-8
$C_{18}H_{30}O_3S.NH_3$
Ammonium dodecylbenzene sulfonate
Benzenesulfonic acid, dodecyl-, ammonium salt; Ammonium dodecylbenzenesulfonate; Ammonium laurylbenzenesulfonate;
Arylan PWS; Conoco SA 597; Conoco SA 597; Dodecylbenzenesulfonic acid ammonium salt,. Surfactant, emulsifier, wetting agent for bubble baths, shampoos and detergents. *Taiwan Surf.*

126 Ablusol DBT
TEA dodecylbenzene sulfonate
Surfactant, emulsifier, wetting agent for bubble baths, shampoos and detergents. *Taiwan Surf.*

127 Ablusol LDE
Lauramide DEA sulfosuccinate monoester
Detergent; foam booster/stabilizer for low irritation shampoos, bubble baths, liquid detergents. *Taiwan Surf.*

128 Ablusol LMS
Lauramide MEA sulfosuccinate monoester
Detergent producing copious lather and extra dry residue for high foaming rug shampoos. *Taiwan Surf.*

129 Ablusol ML
9084-06-4
$(C_{10}H_8O_3S.CH_2O)_x{\cdot}xNa$
Sodium naphthalene sulfonate formaldehyde condensate

Naphthalenesulfonic acid, polymer with formaldehyde, sodium salt; Atlox 4862; Barra Super; Bevaloid 35; Blancol; Blancol Dispersant; Darvan 1; Darvan no. 1; Daxad 11; Daxad 15; Daxad 18; Daxad no. 11; Dispergator NF; Disperser NF; Dispersing agent NF; Dispersol Aca; Flube; Humifen nbl 85; Leukanol NF; Lissatan Ac; Lomar D; Lomar Ls; Lomar Pw; Na-Cemmix; Naphthalenesulfonic acid-formaldehyde condensate sodium salt; NF; Nf-a; Nf (dispersant); Pozzolith 400n; QR 819,. Plasticizer and water reducing agent used in pourable and high-strength concrete. *Taiwan Surf.*

130 Ablusol PM

Alkyl sulfate

Penetrant, scouring agent for mercerizing of cotton. *Taiwan Surf.*

131 Abluter BE

Cocamidopropyl betaine

Detergent for mild cleansing products; antistat, softener, germicide, spreading/wetting agent. *Taiwan Surf.*

132 Abluter GL Series

Glycine derivatives; antistat, softener, germicide, spreading/wetting agent. *Taiwan Surf.*

133 Abluton A

Surfactant blend; penetrant, scouring agent for cotton and synthetics. *Taiwan Surf.*

134 Abluton CAT

Catalyst; durable water repellent for polyamide, cotton and synthetic blends. *Taiwan Surf.*

135 Abluton CMN

Methyl naphthalene

Dye carrier for polyester fiber and its blends. *Taiwan Surf.*

136 Abluton CTP

Trichlorobenzene

Carrier for bleaching polyester fiber and its blends; emulsifiable. *Taiwan Surf.*

137 Abluton MK9

Wax emulsion with zirconium salt; water repelling agent for natural and synthetic fibers. *Taiwan Surf.*

138 Abluton SR

Reactive silicone emulsion; durable water repellent for polyamide, cotton and synthetic blends. *Taiwan Surf.*

139 Abluton T30

108-90-7 2172 203-628-5

C_6H_5Cl

Chlorobenzene

Benzene, chloro-; Benzene chloride; Chlorbenzene; Chlorbenzol; Chlorobenzene; Chlorobenzene, mono-; Monochlorbenzene; Monochlorobenzene; NCI-C54886; MCB; Monochlorbenzene; Phenyl chloride; Un1134,. Carrier for textile dyeing with low stain to wool. mp = -45°; bp = 130-133°; d_{20} = 1.107; n_D^{20} = 1.5240; insoluble in H_2O, freely soluble in C_6H_6, EtOH, $CHCl_3$, Et_2O. *Taiwan Surf.*

140 Abluwax EBS

110-30-5 203-755-6

$C_{38}H_{76}N_2O_2$

Ethylene bis-stearamide

Octadecanamide, N,N'-1,2-ethanediylbis-; Octadecanamide, N,N'-ethylenebis-; Abril wax 10DS; 1,2-Bis(octadecanamido)ethane; Acrawax CT; Acrowax C; Advawachs 280; Advawax; Carlisle Wax 280; Disperse Yellow 60; Ethylenebis[stearamide]; N,N'-Ethylenebis[stearamide]; Lubrol EA; Microtomic 280; Nopcowax 22-DS; Plastflow,. Lubricant for ABS, PS and PVC; defoamer and mold-releasing agent. *Taiwan Surf.*

141 ABM 5C Chlormequat Plus

Mixture of chlormequat and choline chloride; plant growth regulator used in cereals and ornamentals. *ABM Chemicals Ltd.*

142 ABM Chlormequat 40, 72.5

999-81-5 2153 213-666-4

$C_5H_{13}Cl_2N$

Chlormequat chloride

Chlorocholine chloride. Soluble concentrate containing 400 or 725 g/l chlormequat, a plant growth regulator. mp = 241° *ABM Chemicals Ltd.*

143 Abocast

Adhesives, casting epoxies, solventless coatings. *Abatron Inc.*

144 Abocrete

Concrete and masonry patching and resurfacing compound. *Abatron Inc.*

145 Abocure

Catalyst, curing and hardening agent. *Abatron Inc.*

146 Abol

23103-98-2 7651 245-430-1

$C_{11}H_{18}N_4O_2$

2-(Dimethylamino)-5,6-dimethyl-4-pyrimidinyl dimethylcarbamate

2-Dimethylamino-5,6-dimethylpyrimidin-4-yl dimethylcarbamate; Pirimicarb; Pyrimicarbe; ENT 27766; OMS 1330; Aficida; Aphox; Demo; Fernos; Phantom; Pirimor; PP 062; Rapid. Garden aphicide. mp = 90.5°; soluble in H_2O (2.7 g/l); acetone 4.0; EtOH 2.5; xylene 2.9; $CHCl_3$ 3.3 (all g/l, 25°); logP 1.69; LD_{50} (rat orl) = 147 mg/kg, (mus, orl) = 107 mg/kg, (dog, orl) = 100-200 mg/kg. *ICI Chem & Polymers Ltd.*

147 Abopon

A proprietary liquid inorganic resinous product which forms films in a few minutes on drying in air, recommended as an adhesive, a suspending medium for pigments and abrasives, and for sealing of surfaces to be lacquered or painted; a boro-phosphate.

148 Aboseal

Sealants, caulks. *Abatron Inc.*

149 Abracol®

A registered trademark for toluene sulfonic acid esters; used as plasticizers, emulsifying agents. *Bush Boake Allen Ltd.*

150 Abradux

1344-28-1 369 215-691-6

Aluminum oxide

Extra tough and dense types of aluminum oxide; used for abrasive industries, super refractories, sandblasting and for safes. *Lonza AG.*

151 Abramant

1344-28-1 369 215-691-6

Al_2O_3

Aluminum oxide

Alumina. Semi-friable aluminum oxide; used for abrasive industries, super refractories, sandblasting and for safes. *Lonza AG.*

152 Abramax

1344-28-1 369 215-691-6

Al_2O_3

Aluminum oxide

Pure white aluminum oxide; used for abrasive industries, super refractories, sandblasting and for safes. *Lonza AG.*

153 Abrarex

1344-28-1 369 215-691-6

Al_2O_3

Aluminum oxide

Special grades of pure aluminum oxide; used for abrasive industries, super retractories, sandblasting and for safes. *Lonza AG.*

154 Abrasit

1344-28-1 369 215-691-6

Al_2O_3

Aluminum oxide

Alumina. Regular aluminum oxide, finely crystallized; used for abrasive industries, super refractories, sandblasting and for safes. *Lonza AG.*

155 Abrastol

516-18-7 859

$C_{20}H_{14}CaO_8S_2$

Calcium-β-naphthol-γ-sulfonate

asaprol; asaprol-etrasol; calcium 2-hydroxy-1-naphthalene sulfonate; calcium 2-naphthol-1-sulfonate; 2-naphthol-1-sulfonic acid calcium salt; Calcinaphthol. Used as a clarifier for wines. mp = 50° (dec); soluble in H_2O, 0.66 g/ml, EtOH 0.33 g/ml.

156 Abraum salts

Mixed chlorides of magnesium, potassium and sodium

Stassfurt salts; potash salts; stripping salt. The names applied to the upper layers of mixed chlorides of magnesium, potassium, and sodium, overlying the beds of rock-salt at Stassfurt.

157 Abril

Synthetic waxes. *Abril Industrial Waxes.*

158 Abrodil

126-31-8 6051 204-782-6

CH_2INaO_3S

Iodomethanesulfonic acid, sodium salt

Methiodal sodium; Sodium iodomethanesulfonate; Skiodan; Radiographol; Segosin; Diagnorenol. A proprietary preparation of sodium iodomethanesulfonate. Used as diagnostic aid, radio-opaque medium, urographic. Soluble in H_2O (0.7 g/ml), EtOH 0.025 g/ml; slightly soluble in C_6H_6, Et_2O, Me_2CO. *Schering.* Discontinued.

159 Abros

A heat-resisting alloy containing 88% nickel, 10% chromium, and 2% manganese.

160 ABS 124ESG, 236F, 236MA

9003-56-9

Acrylonitrile-butadiene-styrene polymers

ABS copolymers; 124ESG, general purpose molding grade for spools and molded parts requiring high tensile modulus; 236F, automotive grade for injection molding of interior parts; 236MA, medium impact copolymer for profile extrusions. *Mobil/Polystyrene.*

161 ABS 301K, ABS 500FR-1

9003-56-9

Acrylonitrile-butadiene-styrene polymers
301K, ABS copolymer, injection molding, pipefitting grade for DWV fittings, sewer pipe and conduit; 500FR-1, flame-retardant, high impact grade for aircraft, automotive, appliance parts, furniture and smoke detector housings. *Mobil/Polystyrene.*

162 ABS resin
Abbreviation for acrylonitrile butadiene styrene resin.

163 Absaglas®
Flame retardant acrylonitrile-butadiene-styrene.

164 Abson A.B.S 213, A.B.S. 230
A.B.S. 213, A proprietary general-purpose grade of ABS; used in injection molding and extrusion; A.B.S. 230, grade of ABS with higher toughness, used in applications requiring a higher gloss. *BF Goodrich Canada.*

165 Abson A.B.S. 300, A.B.S. 500
A proprietary general-purpose grade of ABS with medium impact resistance and good gloss, used in refrigeration applications; A.B.S. 500, grade with high-medium impact resistance and good strength over temperature range. *BF Goodrich Canada.*

166 ABT-2500®
14807-96-6 9207 238-877-9
Hydrous magnesium silicate
Talc; Soapstone; Steatite; French chalk. Antiblocking agent for polyolefin films. *Pfizer.*

167 Abyssinian gold
Copper-Zinc alloy
Talmi gold; Cuivre poli. A yellow alloy of copper (91%) and zinc (8%). It usually consists of about 91% copper and 8% zinc, but sometimes contains 86% copper, 12% zinc, and 1% tin; employed in the manufacture of cheap jewelry.

168 A-C® 6
9002-88-4 7544
Polyethylene homopolymer
PET; Terylene; Dacron; Mylar. For adhesives, ink, floor finishes, paper coatings, personal care, plastics, rubber, textiles, wax blends. *AlliedSignal.*

169 A-C® 316, A-C®629
Oxidized HDPE homopolymer
Used in ink, floor finishes, personal care, plastics, textiles, wax blends. *AlliedSignal.*

170 A-C® Copolymer 400A
Low molecular weight EVA copolymer; pigment dispersant; PS color concentrates. *AlliedSignal/A-C® Perf. Addit.* Discontinued.

171 A-C® Copolymer 540, 540A, 580, 5120, 5180
9010-77-9
Low molecular weight ethylene-acrylic acid copolymer; plastics lubricant and processing aid, pigment dispersant; internal lubricant PVC, nylon6, nylon color concentrates; alkali-dispersion. Additive for adhesives, floor finishes, personal care, plastics, wax blends. *AlliedSignal/A-C® Perf. Addit.* Discontinued.

172 A-C® Copolymer 580
9010-77-9
Ethylene/acrylic acid copolymer
Alkali-dispersible additive for recyclable hot-melt and aqueous adhesives and coatings. *AlliedSignal/A-C® Perf. Addit.* Discontinued.

173 A-C® Copolymer 580, 5120, 5180
9010-77-9
Low molecular weight ethylene-acrylic acid copolymer
Alkali-dispersible additive for recyclable hot-melt and aqueous adhesives and coatings. *AlliedSignal/A-C® Perf. Addit.* Discontinued.

174 A-C® Copolymer 5180
9010-77-9
Low molecular weight ethylene-acrylic acid copolymer; alkali-dispersible additive for recyclable hot-melt and aqueous adhesives and coatings. *AlliedSignal/A-C® Perf. Addit.* Discontinued.

175 A-C® Polyethylene 6, 6A, 7, 7A, 8, 8A, 9, 9A, 617, 617A
9002-88-4 7728
$(C_2H_4)_n$
Polyethylene wax
Ethene homopolymer; Agilene; Alathon; Alkathene; Courlene; Lupolen; Platilon; Polythene; Pylen; Reevon. Polyethylene wax; processing lubricant, melt index modifier, pigment dispersant, mold release aid; external lubricant PVC, color concentrates, polyolefin flow modifiers; thickener for cosmetic and pharmaceutical gels. *AlliedSignal.* Discontinued.

176 A-C® Polyethylene 316, 316A, 325, 330, 392, 395, 629, 680
$(C_2H_4)_n$
Oxidized polyethylene
Wax for polishes, finishes and emulsions; slip resistance, heel-mark resistance and compatibility. *AlliedSignal.* Discontinued.

177 A-C® Polyethylene 629A
$(C_2H_4)_n$
Oxidised low molecular weight polyethylene
Processing lubricant, mold release aid; PVC lubricant. *AlliedSignal.* Discontinued.

178 A-C® Polyethylene 6702
Oxidized polyethylene
AlliedSignal. Discontinued.

179 Acagine
Lead chromate + bleaching powder
A mixture of lead chromate and bleaching powder; used to purify acetylene.

180 acajou balsam
Cardol
A material obtained from the fruits of *Anacardium* occidentale (mahogany nuts, elephant nuts) or *A. orientale* by the extraction of the powdered nuts. employed in the preparation of indelible inks and colors for die-sink.

181 Acaprin®
532-05-8 3416 208-525-9
$C_{23}H_{26}N_4O_9S_2$
1,3-Di-6-quinolylurea bismethosulfate
N,N'-di-6-quinolinylurea bis methosulfate; 6,6'-diquinolinylurea bis methosulfate; *sym*-di-(6-quinolyl)urea bis methosulfate; bis(6-quinolyl)urea bis methosulfate; SN 5870; Zothelone; Baburan; Pirevan; Pyroplasmin; Atral. Chemotherapeutic against piroplasmosis (babesiasis); used in veterinary medicine. mp = 237° (dec). *Bayer AG.*

182 Acardite 2
$C_{19}H_{14}N_4O$
N-Methyl-N',N'-diphenylurea
Stabilizer improving storage stability of powders and propellants; plasticizer for celluloid. *Lowi.*

183 Acarin
115-32-2 3136 204-082-0
$C_{14}H_9Cl_5O$
4-Chloro-+1a-(4-chlorophenyl)-+1a-(trichloromethyl)benzenemethanol
1,1-bis(*p*-chlorophenyl)-2,2,2-trichloroethanol; di(*p*-chlorophenyl)trichloromethylcarbinol; DTMC; ENT 23648; FW 293; Kelthane; Mitigan. Active ingredient; dicofol, a miticide. mp = 77-78°; bp = 225°; d_{10} = 1.1234; n_D = 1.1234; $[\alpha]_D$ = 100°; soluble in most organics, insoluble in H_2O; λ_m = 226, 258, 266, 276 nm (logε = 4.43, 2.82, 2.85, 2.60); LD_{50} (rat orl) = 1495 mg/kg. *Makhteshim Chemical Works Ltd.*

184 Accelemal
A proprietary rubber vulcanization accelerator. Possibly a thiocarbamide. No manufacturer.

185 Accelerase
53608-75-6 7138 258-659-7
Pancrelipase
Cotazyme; Ilozyme; Ku-zyme HP; Pancrease, Viokase. A concentrate of porcine pancreatic enzymes standardized for lipase content. *Organon Inc.*

186 Accelerator
Range of amine and cobalt-based accelerators for use with unsaturated polyester resins. *Akzo Chemie UK Ltd.*

187 Accelerator 2P
98-77-1 202-698-4
Piperidinium pentamethylene dithiocarbamate
1-Piperidinecarbodithioic acid, piperidine salt; piperidinecarbodithioic acid, piperidinium salt; Pip-Pip, PMP, 522 Rubber accelerator. A proprietary rubber vulcanizing accelerator. *Pacific Anchor Chemical Corp.*

188 Accelerator 4P
Dipentamethylene thiuram disulfide
A proprietary rubber vulcanizing accelerator. *Pacific Anchor Chemical Corp.*

189 Accelerator 100
Schiff's base derivative of an aldehyde, made from both butyraldehyde and acetaldehyde.

190 Accelerator 108
Tetramethyl thiuram disulfide +2/3/2-mercapto-benzthiazole +1/3
A proprietary rubber vulcanizing accelerator. *Naugatuck (US Rubber).*

191 Accelerator A1010
Formaldehyde-aniline
A proprietary rubber vulcanizing accelerator.

192 Accelerator A22
137-97-3 205-309-6
$C_{15}H_{16}N_2S$
Di-o-tolyl thiourea
N,N'-Di-o-tolylthiourea; thiourea, N,N'-bis(2-methylphenyl)-; carbanilide, 2,2'-dimethylthio-; N,N'-bis(2-methylphenyl)thiourea; 2,2'-dimethylthiocarbanilide; DOTT; 1,3-bis(*o*-tolyl)-2-thiourea; USAF EK-1651. A proprietary rubber vulcanizing accelerator. mp = 157-159°. *Lancaster Synthesis Ltd.*

193 **Accelerator A32**
Condensation products of aldehydes and Schiff's base, e.g., butyraldehyde and butylidine aniline; proprietary rubber vulcanizing accelerator.

194 **Accelerator A50**
Adehyde-amine
A proprietary rubber vulcanizing accelerator.

195 **Accelerator A5-10**
Formaldehyde-anilines
A proprietary rubber vulcanizing accelerator.

196 **Accelerator A7**
Ethylidine aniline/acetaldehyde (2:1)
Made from 2 molecules of ethylidene aniline condensed with 1 molecule of acetaldehyde; A proprietary rubber vulcanizing accelerator.

197 **Accelerator BB**
Butyraldehyde *p*-aminodimethylaniline
A proprietary rubber vulcanizing accelerator.

198 **Accelerator BZ Powder**
136-23-2 205-232-8
Zinc dibutyl dithiocarbamate
Nondiscoloring, nonstaining accelerator for ethylene-propylene-diene rubbers, natural rubber and SR latexes; stabilizer and antioxidant in uncured rubber. *Akrochem.*

199 **Accelerator D**
Cyclo amine blended with organo metallic salt; accelerator recommended for use in light-colored compounds, especially shoe soling industry. *Anchor.*

200 **Accelerator DBA**
103-49-1 3058 203-117-7
$C_{14}H_{15}N$
Dibenzylamine
N-(Phenylmethyl)benzenemethanamine. A proprietary rubber vulcanizing accelerator. mp = 68°; bp_{10} = 160-163°; d = 1.026; n_D^{20} = 1.4745; soluble in most organic solvents, insoluble in H_2O.

201 **Accelerator DT**
97-39-2 202-577-6
Di-*o*-tolylguanidine
A French proprietary rubber vulcanizing accelerator.

202 **Accelerator E**
0.4% active cobalt solution in styrene; accelerator. *Scott Bader.* Discontinued.

203 **Accelerator E-A**
Ethylidene-aniline
A proprietary rubber vulcanizing accelerator.

204 **Accelerator EZ Powder**
14324-55-1 238-270-9
Zinc diethyl dithiocarbamate
A nondiscoloring, nonstaining accelerator for natural rubber and SR latexes. *Akrochem.*

205 **Accelerator F-A**
High melting form, formaldehyde-aniline; low melting form methylenedianilide; A proprietary rubber vulcanizing accelerator.

206 **Accelerator G.M.F.**
Quinone dioxime
A proprietary rubber vulcanizing accelerator.

207 **Accelerator Mercapto**
149-30-4 5916 205-736-8
$C_7H_5NS_2$
Mercaptobenzothiazole
2-Benzothiazolethiol; 2(3H)-Benzothiazolethione; MBT; Captax; Dermacid; Mertax; Thiotax. A German proprietary rubber vulcanizing accelerator. mp = 180.2-181.7°; d = 1.42; insoluble in H_2O, solubility (25°) = 2.0 (EtOH), 1.0 (Et_2O), 10.0 (Me_2CO), 1.0 (C_6H_6), <0.2 (CCl_4), ,0.5 (naphtha) g/l.

208 **Accelerator MF**
102-77-2 203-052-4
$C_{11}H_{12}N_2OS_2$
2-Benzothiazyl-N-morpholine disulfide
N-oxydiethylenebenzothiazole-2-sulfenamide; Amax; Morpholine, 4-(2-benzothiazolylthio)-; 2-(morpholinothio)benzothiazole; 2-(Morpholinthio)-benzothiazole; N-oxydiethylene-benzothiazole sulfenamide. An accelerator in natural and synthetic rubbers, e.g., tires and mechanical goods. *Akrochem.*

209 **Accelerator MZ Powder**
137-30-4 10305 205-288-3
$C_6H_{12}N_2S_4Zn$
Zinc dimethyl dithiocarbamate
Ziram; Bis(dimethylcarbamodithioato-S,S')zinc; bis(dimethyldithiocarbamato)zinc; zinc dimethyldithiocarbamate; dimethyldithiocarbamic acid, zinc salt; zinc bis(dimethylthiocarbamoyl) disulfide; Aaprotect; Methyl cymate; Methasan; Zimate; Zirberk; Karbam white; Corozate; Fuclasin; Fuklasin; Zerlate. Nondiscoloring, nonstaining accelerator for natural rubber, isoprene rubber, BR, styrene/butadiene rubber, isobutylene-isoprene rubber, and ethylene-propylene-diene rubbers and natural rubber latex. mp = 250°; d_{25} = 1.66; sol. <0.2 (EtOH), <0.5 (Me_2CO), <0.5 (C_{66}), <0.2 (CCl_4), <0.2 (Et_2O), 0.5 (naphtha) (g/100 ml), insoluble in H_2O; LD_{50} (rat orl) = 1.4 g/kg. *Akrochem.*

210 **Accelerator PTX**
Phenyl-tolyl-xylyl-guanidine
A proprietary rubber vulcanizing accelerator.

211 **Accelerator R2**
Condensation product from methylenedipiperidine and carbon disulfide; a proprietary rubber vulcanizing accelerator.

212 **Accelerator R3**
Zinc salt of dithiocarbaminic acid
Zinc salt of a dithio-carbaminic acid; a proprietary rubber vulcanizing accelerator.

213 **Accelerator R5**
Dithiocarbamate
A proprietary rubber vulcanizing accelerator.

214 **Accelerator VN-2**
Vanadium salt
An accelerator for ketone peroxides, hydroperoxides, and peroxy esters; short gel times and a very high speed of cure can be achieved. *Akzo.*

215 **Accelerator W29**
Diphenyl guanidine/dibenzyl dithiocarbaminic acid
Compound of diphenyl guanidine with dibenzyl dithiocarbaminic acid; a proprietary rubber vulcanizing accelerator.

216 **Accelerator W80**
Diphenyl-guanidine salt of mercaptobenzothiazole
A proprietary rubber vulcanizing accelerator.

217 **Accelerator Z88**
Mercaptobenzothiazole, ammonium salt
Ammonium salt of mercapto-benzothiazole mixed with a softener, A proprietary rubber vulcanizing accelerator.

218 **Accelerator ZBX**
Zinc butyl xanthate
A proprietary rubber vulcanizing accelerator.

219 **Accelerator ZPD**
Zinc pentamethylene dithiocarbamate
A proprietary rubber vulcanizing accelerator.

220 **Accelerene V 1**
p-nitrosodimethylaniline/β-naphthol (1:1)
Consists of equi-molecular proportions of *p*-nitroso-dimethylaniline and β-naphthol; A proprietary rubber vulcanizing accelerator.

221 **Accelguard 80**
Chloride-free set accelerator for concrete and mortar; liquid form. *Feb Ltd.*

222 **Accobetaine CL**
68424-94-2 270-329-4
betaines, coco alkyldimethyl; Dehyton ab 30; Mirataine CDMB; Nissan Anon BF; Quaternary ammonium compounds, (carboxymethyl)coco alkyldimethyl, hydroxides, inner salts; Standapol ab 45; Tego-betain BL 158; Varion Cdg; Velvetex AB 45. Complex coco betaine; detergent, wetting agent, emulsifier, high foaming agent, solubilizer, household and cosmetic uses. *Karlshamns.*

223 **Accobond**
Melamine formaldehyde resins; film resins for process industries. *Cyanamid BV.*

224 **Accomeen C2, C5, C10, C15**
61791-14-8
PEG-10 cocamine
PEG2, PEG5, PEG10 and PEG15 cocamines. Used as emulsifiers, antistats, surfactants. *Karlshamns.*

225 **Accomeen S2, S10, S15**
61791-24-0
PEG-10 soyamine
PEG2, PEG10, PEG15 soyamines. Used as emulsifiers, antistats, surfactants. *Karlshamns.*

226 **Accomeen T2, T5, T15**
61791-44-4 263-177-5
PEG-15 tallow amine
PEG2, PEG5, PEG15 tallow amine. Used as emulsifiers, antistats, surfactants, dispersants. *Karlshamns.*

227 **Accomet C**
Metal precleaning process. *Albright & Wilson Ltd., Phosphates & Speciality Business.*

228 **Accomet, Accomet C**
Paint pretreatment; used for steel strip and coil. *Albright & Wilson Ltd., Phosphates & Specialty Business.*

229 **Accomid 50**
Palm kernerlamide DEA, detergent. *Karlshamns.*

230 **Accomid C**
Cocamide DEA
Cocamide DEA. Detergent, stabilizer, viscosity improver, foam booster for shampoos and dishwashes; biodegradable. *Karlshamns.*

231 **Accomid PK**
68155-12-2
Palm kernelamide DEA
Palm kernalamide DEA (1:1); viscosity builder, foam booster/stabilizer, emulsifier for shampoos, liquid soaps, dish detergents, bubble bath products. *Karlshamns.*

232 **Acconon 200-DL**
9005-02-1
PEG-4 dilaurate
Surfactant used as emulsifier, dispersant, solubilizer, viscosity control agent for cosmetics, pharmaceuticals, and industrial applications. *Karlshamns.*

233 **Acconon 200-MS, 400-MS**
9004-99-3
PEG-4, PEG-8 stearate
Surfactant used as emulsifier, dispersant, solubilizer, viscosity control agent for cosmetics, pharmaceuticals, and industrial applications. *Karlshamns.*

234 **Acconon 400-MO**
9004-96-0
PEG-8 oleate
Emulsifier, dispersant, lubricant, chemical intermediate, solubilizer, viscosity control agent; for cosmetics, pharmaceuticals, food, agricuture, plastics. *Karlshamns.*

235 **Acconon 1300**
9004-94-3
PPG-3-laureth-9
Surfactant used as emulsifier, dispersant, solubilizer, viscosity control agent for cosmetics, pharmaceuticals, and industrial applications. *Karlshamns.*

236 **Acconon CA-5, CA-9, CA-15**
61791-12-6
PEG-15 castor oil
PEG-5, PEG-9, PEG-15 castor oil. Surfactant used as emulsifier, lubricant, dispersant, solubilizer, viscosity control agent for cosmetics, pharmaceuticals, and industrial application. *Karlshamns.*

237 **Acconon CON**
PEG-10 propylene glycol glyceryl laurate; surfactant used as emulsifier, dispersant, solubilizer, viscosity control agent for cosmetics, pharmaceuticals, and industrial applications. *Karlshamns.*

238 **Acconon E**
25231-21-4
PPG-15 stearyl ether
PPG-15 stearyl ether; surfactant used as emulsifier, dispersant, solubilizer, viscosity control agent for cosmetics, pharmaceuticals, and industrial applications. *Karlshamns.*

239 **Acconon ETG**
31694-55-0
Glycereth-26
Glycereth-26; humectant; lubricant for skin care products, creams, lotions, industrial applications. *Karlshamns.*

240 **Acconon TGH**
68783-63-1
PEG-10-PPG-10 glyceryl stearate
PEG-20-PPG-10 glyceryl stearate. Surfactant used as emulsifier, dispersant, solubilizer, viscosity control agent, wetting and foaming agent for cosmetics, pharmaceuticals, and industrial applications; biodegradable. *Karlshamns.*

241 **Acconon W230**
68439-49-6 7737
Ceteareth-20
Surfactant used as emulsifier, dispersant, solubilizer, viscosity control agent for cosmetics, pharmaceuticals, and industrial applications. *Karlshamns.*

242 **Accoquat 2C-75, 2C-75H**
61789-77-3 263-087-6
Dicocodimonium chloride
Emulsifier, coupling agent; used for car spray waxes, dust control oil, spot removal. *Karlshamns.*

243 **Accosize**
Modified alkenylsuccinic anhydride; used in the paper industry. *Cyanamid BV.*

244 **Accosoft 440-75**
Methyl bis (hydrogenated tallowamidoethyl) 2-hydroxyethyl ammonium methyl sulfate
Fabric softener quaternary for textile industry, household and commercial use; good lubricity and scorch resistance; nonyellowing. *Stepan; Stepan Canada.*

245 **Accosoft 540 HC, 550-90 HHV**
Methyl bis (modified tallowamidoethyl) 2-hydroxyethyl ammonium methyl sulfate
Fabric softener quaternary for household products and textile processing. *Stepan; Stepan Canada.*

246 **Accosoft 550L-90, 620-90**
Methyl bis (tallowamidoethyl) 2-hydroxyethyl ammonium methyl sulfate
Fabric softener and antistat with good rewet properties for laundry products, industrial textile processing. *Stepan; Stepan Canada.*

247 **Accosoft 750**
Methyl bis(oleylamidoethyl) 2-hydroxyethyl ammonium methyl sulfate
Fabric softener quaternary for heavy duty liquid laundry detergents. *Stepan; Stepan Canada.*

248 **Accosoft 808HT**
Methyl-1-hydrogenated tallowamidoethyl-2-hydrogenated tallow imidazolinium-methyl sulfate
Fabric softener and antistat for household and industrial applications, dryer products. *Stepan; Stepan Canada.*

249 **Accosperse 20**
9005-64-5 8872
PEG-20 sorbitan laurate
Emulsifier, solubilizer; biodegradable. *Karlshamns.*

250 **Accosperse 60**
9005-67-8 8872
PEG-20 sorbitan stearate
Emulsifier, solubilizer; biodegradable. *Karlshamns.*

251 **Accosperse 80**
9005-65-6 7742
PEG-20 sorbitan oleate
emulsifier, solubilizer; biodegradable. *Karlshamns.*

252 **Accostrength 72**
Acrylamide copolymers (anionic); used in the paper industry. *Cyanamid BV.*

253 **Accostrength, Accostrength 711**
Anionic/cationic polyacrylamide; used in the paper industry. *Cyanamid BV.*

254 **Accostrength, Accostrength 711**
Anionic/cataionic polyacrylamide; used in the paper industry. *Cyanamid BV.*

255 **Accrolon® 9039**
Self-lubricated bearing engineering thermoplastic; bearing material featuring high continuous service temperature excellent wear, abrasion and fatigue resist, outstanding mechanical and excellent chemical resistance. *Accro-Seal.*

256 **Accrolube, Accrolube-FG**
Grease with Teflon® biodegradable food-grade lubricant featuring waterproof props., chemically resistant; for conveyor chains, seamers, filling equipment, sterilizers, pumps, valves, cylinders, mixers, bearings, packaging equipment. *Accro-Seal.*

257 **Accrotan**
Self-basifying chrome tanning material. *British Chrome & Chemicals Ltd.*

258 **Accrox**
1308-38-9 2283 215-160-9
Cr_2O_3
Chromic oxide
Anadonis green; chrome green; chrome ochre; chrome oxide green; chromia; chromium sesquioxide; green cinnabar; green oxide of chromium; green rouge, leaf green; oil green; ultramarine green. Special refractory grades of chromic oxide. mp = 2435°; bp = ca. 3000°; d_{25} = 5.22; insoluble in H_2O, EtOH, Me_2CO, soluble in alkalis, acids. *British Chrome & Chemicals Ltd.*

259 **Acctuf 3045**
9003-07-0 7741
PP impact copolymer; high impact general-purpose grade for extrusion, compression molding. *Amoco Chemical Co.*

260 **Accuglass**
Spin-on Glass; solutions of inorganic polymers (siloxanes or silicates) spun applied to silicon wafers to form insulating or passivating films used in the manufacture of integrated circuits. *AlliedSignal, Planarization & Diffusion Products Division.*

261 **Acculog**
Standard volumetric concentrated solutions used in quantitative analytical chemistry. *Schweizerhall.*

262 **Accurac 39-S**
Cationic polyacrylamide solution; used in the paper industry. *Cyanamid BV.*

263 **Accurac, Accurac 33/35/41**
Polymers based on acrylamide; can be described as nonionic polyacrylamide, anionic polyacrylamide and cationic polyacrylamide; used in the paper industry. *Cyanamid BV.*

264 **Accuspin**
Spin-on dopant; solutions of impurity atoms (boron, phosphorus, arsenic or antimony) spun applied to silicon waters used to dope silicon to form transistors and integrated circuits. *AlliedSignal, Planarization & Diffusion Products Division.*

265 **Accuthane® UR-1100**
One-component urethane adhesive
Patented adhesive for difficult-to-bond substrates such as nylon, SMC/FRP, galvanized and stainless steel; good heat resistance. *H.B. Fuller.*

266 **Accuvette**
Plastic sample vial; used to hold blood cell dilution for counting in semiautomatic analyzers. *Coulter Electronics Ltd.*

267 **Ac-Di-Sol**
croscarmellose sodium
Carboxymethylcellulose sodium that has been internally cross-linked; pharmaceutics aid. *FMC UK.* Discontinued.

268 **Acelan A**
629-70-9 211-103-7
$C_{18}H_{36}O_2$
Cetyl acetate
Acrylated lanolin alcohol; 1-Hexadecanol, acetate; Acetic acid, hexadecyl ester; 1-Acetoxyhexadecane; Cetyl acetate; ENT 1025; Hexadecyl acetate; n-Hexadecyl ethanoate; Palmityl acetate. *Fabriquimica.*

269 **Acelan L**
61788-48-5 262-979-2
Acetylated lanolin. *Fabriquimica.*

270 **Aceloid**
A proprietary cellulose acetate material; used as a molding composition.

271 **Acelon**
Cellulose acetate coated fabric. *May & Baker Ltd.*

272 **Acelose, Aceplus**
A proprietary cellulose acetate material.

273 **Ace-Ite**
A proprietary bituminous or asphalt composition.

274 **Acerado**
Fierroso. Names used for mercurial earths.

275 **Acerdol**
10118-76-0 1735 233-322-7
$CaMn_2O_8$
Calcium permanganate
Used to treat gastro-enteritis and diarrhea. Freely soluble in H_2O, reacts with alcohol.

276 **Ace-Sil**
A proprietary trade name for microporous rubber; used for battery separators and filters.

277 **acesulfame potassium**
55589-62-3 35 259-715-3
$C_4H_5NO_4S$
6-methyl-1,2,3-oxathiazin-4(3H)-one 2,2-dioxide
acetosulfam potassium; Acesulfame-K; Hoechst-095K; Sunette. Sweetener used in foods and cosmetics. dec 225°; d = 1.81; λ_m = 225 nm (ε 10762); soluble in H_2O (360 g/l), organic solvents; LD_{50} (rat orl) = 7431 mg/kg.

278 **Aceta**
A proprietary brand of cellulose rayon; the name is also applied to a French nitrocellulose lacquer.

279 **Acetadeps, Acelan A**
61788-48-5 262-979-2
Acetylated lanolin
Acetylated lanolin; emollient, moisturizer for antiperspirants, baby oils, cleansers, shampoos, hair conditioners, sun-screen; binder for pressed powders. *Westbrook Lanolin, Fabriquimica.*

280 **Acetal**
105-57-7 36 203-310-6
$C_6H_{14}O_2$
1,1-Diethoxyethane
AT-20GF; Acetron®GP; Aceton® NS; Cadco® Acetal; Delrin® 100, 500; Delrin®100ST, 500T; Delrin® 107, 507; Delrin® 150 SA, 550SA; Delrin® 570; Delrin®900; Delrin® AF Blend; Electrafil® J-80/CF/10/TF/10; Thermocomp® KB-1008,. Solvent in synthetic perfumes such as jasmine; used for organic syntheses. Acetal resin, 20% glass fiberreinforced; offers lubricity and chemical and hot water resistance for automotive, hardware, plumbing applications. bp = 102.7°; flash point (closed cup) = 97°F; 100 g H_2O dissolve 5 g acetal; soluble in heptane, methylclohexane, propyl-, isopropyl, butal, isobutyl, alcohols, ethyl acetate; LD_{50} (rat orl) = 4.57 g/kg.

281 **Acetaldehyde**
75-07-0 37 200-836-8
C_2H_4O
acetic aldehyde
ethyl aldehyde; ethanal. Manufacture of acetic acid, acetic anhydride, n-butanol, peracetic acid, pentaerythritol, pyridines, 1,3-butylene glycol, trimethylolpropane; synthetic flavors. mp = -123.5°; bp = 21°; d_{16} = 0.788; miscible with H_2O, EtOH; LD_{50} (rat orl) = 1930 mg/kg. *BP Chem. Ltd.; Eastman; Hoechst-Celanese; Hüls UK; Mitsui Petrochem. Ind.*

282 **Acetaloid**
A proprietary cellulose acetate material in the form of rods, sheet, and molding composition.

283 **Acetamide MEA**
142-26-7 205-530-8
$C_4H_9NO_2$
N-Acetyl ethanolamine
Amidex AME; Amidex AME; Carsamide® AMEA; Foamid AME-70; Foamid AME-75; Foamid AME-100; Hetamide MA; Incromectant AMEA-100, AMEA-70; Lipamide MEAA; Mackamide ™ AME-75, AME-100; Schercomid AME; Upamide ACMEA; Witcamide® CMEA,. Used as antistat, humectant, conditioner for skin and hair products. Solvent, humectant, skin and hair conditioner, intermediate, coupling agent, pigment dispersant, clarifying agent for shampoos, moisturizer.

284 **Acetamin 24**
2016-56-0 217-956-1
$C_{14}H_{31}NO_2$
Cocamine acetate
Dodecanamine, acetate; Dodecylamine acetate; Laurylamine acetate. Surface coating agent for pigments, anticaking agent for fertilizer; emulsifier, dispersant, and softening agent for textiles; mineral flotation reagent. *Kao Corp SA.*

285 **Acetamin 86**
2190-04-7 218-583-7
$C_{20}H_{43}NO_2$
Stearamine acetate
1-Octadecanamine, acetate. Surface coating agent for pigments, anticaking agent for fertilizer; emulsifier, dispersant, and softening agent for textiles; mineral flotation reagent. *Kao Corp SA.*

286 **Acetamin C**
61790-57-6 263-147-1
n-Cocamine acetate
Flotation of minerals, anticaking agents, emulsifier bactericide. *Kao Corp SA.*

287 **Acetamin HT**
61790-59-8 263-149-2
hydrogenated tallow amine acetate
Flotation of minerals, anticaking agents, emulsifier bactericide. *Kao Corp SA.*

288 **Acetamin T**
2190-04-7 218-583-7
$C_{20}H_{43}NO_2$
n-Tallow amine acetate
1-Octadecanamine, acetate. Used for flotation of minerals, anticaking agents, emulsifier bactericide. *Kao Corp SA.*

289 **Acetargol**
A mixture of formic and acetic acids.

290 **Acetest**
A mixture of sodium nitroprusside, aminoacetic acid, disodium phosphate, and lactose, in tablet form; used to test for the presence of ketones in blood and urine. *B. C. Ames.*

291 **Acetex**
A proprietary safety-glass.

292 **Acetic acid**
64-19-7 52 200-580-7
$C_2H_4O_2$
ethanoic acid
aci-jel; Ethylic acid; Vinegar acid; glacial acetic acid; vinegar; Methanecarboxylic acid; Acetic acid glacial; TCLP extraction fluid 2; Shotgun; methane-carboxylic acid. Used in manufacture of acetic anhydride, cellulose acetate, vinyl acetate monomer; acetic esters; production of plastics, pharmaceuticals, dyes, insecticides, photographic chemicals, food additives; solvent reagent. mp = 16.7°; bp = 118°; d_{25} = 1.049; n_D^{20} = 1.3718; miscible with H_2O, EtOH, glycerol, Et_2O; CCl_4; pKa= 4.74; LD_{50} (rat orl) = 3.53 g/kg. *Air Prods & Chem; BASF; BP Chem.; General Chem.; Hoechst Celanese; Janssen Chimica; Quantum/USI.*

293 **acetic anhydride**
108-24-7 53 203-564-8
$C_4H_6O_3$
acetyl oxide
acetic oxide; Acetic acid anhydride; Acetic oxide; Ethanoic anhydride; acetyl oxide; acetyl anhydride; acetyl ether; ethanoic anhydrate. Cellulose acetate fibers and plastics; vinyl acetate; dehydrating and acetylating agent for pharmaceuticals, dyes, perfumes, explosives; aspirin; esterifying agent for food starch. mp = -73°; bp = 140°; d = 1.0870; LD_{50} (rat orl) = 1780 mg/kg.

Ashland; BP Chem. Ltd.; Chisso; CPS; Eastman; Hoechst-Celanese; Schweizerhall; Union Carbide.

294 acetin blue
Solutions of indulines in acetins (acetic esters of glycerol).

295 Acetol
A proprietary cellulose acetate material in the form of flake.

296 Acetol® 1706
Cetyl acetate
acetylated lanolin alcohol. Water repellent; strongly hydrophobic emollient, penetrant, lubricant, and cosolvent used in hair products, creams, lotions, suntan preparations and baby products. *Henkel/Cospha; Henkel Canada.*

297 acetone
67-64-1 64 200-662-2
C_3H_6O
dimethylketone
2-propanone. Solvent for paints, varnishes and lacquers; for cleaning and drying precision equipment; delustrant for cellulose acetate fibers. mp = -94°; bp = 56.5°; d_{25} = 0.788; n_D^{20}= 1.3591; miscible with H_2O, EtOH, DMF, Et_2O; CCl_24, most oils; LD_{50} (rat orl) = 10.5 ml/kg. AlliedSignal; Ashland; BASF; BP Chem. Ltd; Dow; Eastman; Exxon; Mitsui Petrochem.; Montedipe SpA; Shell; Texaco; Union Carbide.

298 acetone bromoform
76-08-4 9741 200-931-4
$C_4H_7Br_3O$
tribromo-*t*-butyl alcohol
1,1,1-tribromo-2-methyl-2-propanol; acetone-bromoform; brombutol; Brometone. Modifier in the polymerization of vinyl chloride.

299 acetonitrile
75-05-8 68 200-835-2
C_2H_3N
methyl cyanide
cyanomethane; ethanenitrile; Cyanomethane; Ethyl nitrile; Methyl cyanide; Ethane nitrile; methanecarbonitrile; AN; ethanonitrile. Solvent for hydrocarbon extraction processes, especially for butadiene; intermediate; catalyst; for separation of fatty acids from vegetable oils; manufacture of synthetic pharmaceuticals. mp = -45°; bp = 81.6°; d_{15} = 0.78745; d_{30} = 0.7138; n_D^{15} = 1.34604; miscible with H_2O, MeOH, organic solvents, immiscible with saturated hydrocarbons; LD_{50} (rat orl) = 3800 mg/kg. *BP Chem. Ltd; Du Pont; R. W. Grief; ICE Ind.; Penta Mfg.*

300 acetophenone
98-86-2 71 202-708-7
C_8H_8O
phenyl methyl ketone
acetylbenzene; 1-phenylethanone; 1-Phenylethanone; Phenyl methyl ketone; Acetylbenzene; Methyl Phenyl Ketone; α-Acetophenone; Hypnone; Benzoyl methide; Acetylbenzol. Perfumery, solvent, intermediate for pharmaceuticals, resins, etc.; flavoring; polymerization catalyst, organic synthesis. mp = 20.5°; bp = 202°; d_{15} = 1.033; n_D^{20}= 1.5339; slightly soluble in H_2O, freely soluble in EtOH, $CHCl_3$, Et_2O, fatty oils, glycerol; LD_{50} (rat orl) = 0.90 g/kg. BP Chem. Ltd.; Enichem Am.; Janssen Chimica; Mitsui Petrochem. Ind.; Mitsui Toatsu Chem.; Montedipe SpA; Penta Mfg.

301 Acetoquat CPB
140-72-7 205-428-3
$C_{21}H_{38}BrN$
Cetyl pyridinium bromide
Hexadecylpyridine bromide; Hexadecylpyridinium bromide; n-Hexadecyl pyridinium bromide monohydrate. Germicide, sanitizing agent. mp = 67-69°. *Aceto.*

302 Acetoquat CPC
123-03-5 2074 204-593-9
$C_{21}H_{38}ClN$
1-Hexadecylpyridinium chloride
Cetyl pyridinium chloride; Ceepryn; Cepacol; Cetamium; Dobendan; Medilave; Merocet; Pristacin; Pyrisept. Germicide, sanitizing agent. mp = 78-83°; slightly soluble in C_6H_6, freely soluble in H_2O, EtOH, $CHCl_3$; LD_{50} (rat orl) = 200 mg/kg. *Aceto.*

303 Acetoquat CTAB
57-09-0 2068 200-311-3
$C_{19}H_{42}BrN$
n-Hexadecyl trimethylammonium bromide
Cetrimonium bromide; cetyl trimethylammonium bromide; N,N,N-trimethyl-1-hexadecanaminium bromide; Bromat; Cetab; Cetavlon; Cetylamine; C.T.A.B.; Lissolamine V; Micol; Quamonium,. Germicide, sanitizing agent. mp = 237-243°; soluble in 10 parts H_2O, freely soluble in EtOH, sparingly soluble in Me_2CO, insoluble in Et_2O, C_6H_6; LD_{50} (mus, iv) = 32.0, 44.0 mg/kg. *Aceto.*

304 Acetosol
79-34-5 9331 204-825-9
$C_2H_2Cl_4$
Tetrachloroethane
1,1,2,2-tetrachloroethane; *sym*-tetrachloroethane; acetylene tetrachloride; Cellon, Bonoform; Westrol. A trade name for tetrachloroethane. mp = -43°; bp = 142-146°; d = 1.596; n_D^{20}= 1.4935; sparingly soluble in H_2O (1gm/350 ml), soluble in organic solvents; LD_{50} (rat orl) = 0.20 ml/kg.

305 AcetOxyl 2.5 and 5
94-36-0 1149 202-327-6
$C_{14}H_{10}O_4$
Dibenzoyl peroxide
benzoyl superoxide; Acnegel; Benoxyl; Benzagel 10; Benzaknen; Debroxide; Desanden; Lucidol; Nericur; Oxy-5; Oxy-L; PanOxyl; Peroxydex; Persadox; Persa-gel; Sanoxit; Theraderm; Xerac BP 5; Xerac BP 10. Contains benzyl peroxide in two strengths (2.5% or 5%) in an aqueous gel base; as an aid in the treatment of *acne vulgaris*. mp = 103-106°; sparingly soluble in H_2O, EtOH, soluble in C_6H_6, $CHCl_3$, Et_2O, 1 gm dissolves in 40 ml CS_2,50 ml olive oil. *Stiefel Laboratories (UK) Ltd.*

306 acetoxyphenylmercury
62-38-4 7453 200-532-5
Phenylmercuric acetate
PMAC; phenylmercury acetate; PMAS; Ceresan Slaked Lime; Gallotox; Liquiphene; Phix; Mersolite; Tag Fungicide; Tag HL-331; Nylmerate; Scutl; Riogen. Herbicide and fungicide. mp = 149°; soluble in 600 parts of H_2O, soluble in EtOH, C_6H_6, Me_2CO; LD_{50} (rat orl) = 22 mg/kg.

307 acetrizoate sodium
129-63-5 79 204-956-1
$C_9H_5I_3NNaO_3$
3-(acetylamino)-2,4,6-triiodobenzoic acid sodium salt
acetrizoic acid sodium salt; Acétiodone; Bronchoselectan; Cystokon; Diaginol; Iodopaque; Pyelokon-R; Salpix; Thixokon; Tri-Abrodil; Triopac; Triurol; Urokon Sodium; Vesamin; Visotrast. Marketed as a sterile 30% solution. A radiopaque medium, used as a diagnostic aid. Free acid: mp = 278-283° (dec); soluble in H_2O (94 g/100 ml), EtOH; LD_{50} (rbt iv) = 5200 mg/kg.

308 Acetron®, Acetron® GP; Acetron® NS
105-57-7 36 203-310-6
$C_6H_{14}O_2$
Acetal
1,1-Diethoxyethane; Acetaldehyde diethyl acetal; Ethylidene diethyl ether; 1,1-Diethoxyacetal; Capsicum annuum l.; Diethyl acetal; Ethylidine diethyl ether. Acetal compound with solid lubricants; for bearing and wear applications, e.g. bearings, bushings, valve seats, seals, wear surfaces, rollers, gears, cams, liners, fooling fixtures, forming dies. mp = -100°; bp = 102.7°; d_{20} = 0.8254; soluble in H_2O (5 g/100 ml), organic solvents; LD_{50} (rat orl) = 4.57 gm/kg. *Polymer Corp.*

309 Acetulan
8028-98-6 80
$C_{18}H_{36}O_2$
Cetyl acetate
Acetol; Acetylated lanolin alcohol; Acetulan®. Acetylated lanolin alcohol; binder for pressed powders; emollient, plasticizer, cosolvent, NV and sebum solvent for personal care products; lubricant for clay, talc, and starch; stabilizer for lanolin; solubilizer in aerosols; penetrant and spreading agent. d = 0.867; miscible with mineral oil, castor oil, vegetable oil, i-PrOH, EtOH, isopropyl palmitate, butyl stearate. *Amerchol Corp.*

310 Acetyl
Anionic dyestuffs (level dyeing); used for wool and wool blends. *Holliday Dyes & Chemicals Ltd.*

311 acetyl chloride
75-36-5 86 200-865-6
C_2H_3ClO
ethanoyl chloride
Acetyl chloride; Acetic acid, chloride; Acetic acid monochloride; Acetic chloride; Ethanoyl chloride; RCRA Waste Number U006; UN1717 (DOT). acetylating agent for organic preparations; dyestuffs; pharmaceuticals. mp = -112°; bp = 52°; d = 1.104; n_D^{20}= 1.3898; miscible with C_6H_6, $CHCl_3$, Et_2O; AcOH, Petroleum ether. *Hoechst-Celanese; Penta Mfg.; Saurefabrik Schweizerhall.*

312 acetyl methyl carbinol
513-86-0 61 208-174-1
$C_4H_8O_2$
3-hydroxy-2-butanone
acetoin; 2,3-butanolone; dimethylketol; γ-hydroxy-β-oxobutane;. Aroma carrier; preparation of flavors and essences. mp = 15°; bp = 148°; d_{17} = 0.9972; $n^{17.3}_D$ = 1.4190; miscible with H_2O, EtOH, sparingly soluble in Et_2O, petroleum ether. *BASF; Penta Mfg.*

313 acetyl tributyl citrate
77-90-7 201-067-0
$C_{20}H_{34}O_8$
2-(acetyloxy)-1,2,3-propane tricarboxylic acid, tributyl ester
Tributyl acetylcitrate; Tributyl 2-(acetyloxy)-1,2,3-propanetricarboxylate;

Tributyl 2-(acetyloxy)-1,2,3-propanetricarboxylic acid; Tributyl citrate acetate; Citric acid, tributyl ester, acetate; 2-Acetoxy-1,2,3-propanetricarboxylic acid tributyl ester; Acetyl Butyl Citrate; Acetylcitric acid, tributyl ester; O-Acetylcitric acid tributyl ester; Acetyl tributyl citrate; Blo-trol; Citroflex A; Citroflex A 4; Pfizer citroflex A-4; Tributyl 2-acetoxy-1,2,3-propanetricarboxylate; Tributyl acetylcitrate; Tributyl O-acetylcitrate. Plasticizer for vinyl resins. *Morflex; Plizer Spec.; Unitex.*

314 acetylacetone
123-54-6 82 204-634-0
$C_5H_8O_2$
2,4-pentanedione
diacetylmethane. Solvent for cellulose acetate; intermediate; chelating agent for metals, paint drier, lubricant additives, pesticides. mp = -23°; bp = 140.5°; d = 0.976; n_D^{20} = 1.4512; soluble in H_2O (12%), miscible with EtOH, C_6H_6, $CHCl_3$, Et_2O, Me_2CO, AcOH; LC_{50} (rat inh) = 4000 ppm/4 hrs. *Aldrich; Penta Mfg.; Union Carbide; Wacker Chem.*

315 Acetylated hydrogenated lard glyceride
8029-91-2
Axol® E 61; Myvacet® 7-00; Tegin® E-61, 61 NSE. Acetylated hydrogenated lard glyceride; food emulsifier, lubricant, solv., plasticizer and coating material for foodstuffs and cosmetics. d = 0.94; mp=37-40°; acid no 3 max; iodine no 5 max; sapon no 316-331; hyd no 80.5-95. *Eastman.*

316 Acetylated hydrogenated tallow glyceride
68990-58-9 273-612-0
Acetylated hydrogenated tallow glyceride
Lamegin® EE. Emulsifier and plasticizer for cosmetic, food, and edible coatings. *Grünau.*

317 acetyl-dinitro-butyl-xylene
81-14-1 201-328-9
$C_{14}H_{18}N_2O_5$
4-t-Butyl-2,6-dimethyl-3,5-dinitroacetophenone
Ethanone, 1-[4-(1,1-dimethylethyl)-2,6-dimethyl-3,5-dinitrophenyl; Ketone musk; ketone moschus; Acetophenone, 4'-tert-butyl-2',6'-dimethyl-3',5'-dinitro-; 1-Acetyl-4-tert-butyl-2,6-dimethyl-3,5-dinitrobenzene; 4-tert-Butyl-2,6-dimethyl-3,5-dinitroacetophenone; 4'-tert-Butyl-2',6'-dimethyl-3',5'-dinitroacetophenone; 1-[4-(1,1-Dimethylethyl)-2,6-dimethyl-3,5-dinitrophenyl]ethanone. An artificial musk perfume. mp = 138-140°.

318 acetylene
74-86-2 91 200-816-9
C_2H_2
ethyne
ethine. An asphyxiant gas; intermediate for manufacture of vinyl chloride, vinylidene chloride, vinyl acetate, acrylates, acrylonitrile, acetaldehyde, perchloroethylene trichloroethylene, 1,4-butanediol, carbon black, welding and cutting metals. mp = -81°; d = 1.165 g/l (0°, 760 mm), d of gas = 0.90 (air=1); soluble in H_2O (1:1 v/v), soluble in AcOH, EtOH, Et_2O, C_6H_6, Me_2CO; LC (rat inh) = 9000 ppm. *Air Prods & Chem; BASF; DMS NV; Union Carbide.*

319 acetylene black
1333-86-4 1856 215-609-9
Shawinigan black
A carbon black made by incomplete combustion of acetylene.

320 Acetylene dichloride
540-59-0 93 208-750-2
$C_2H_2Cl_2$
1,2-dichloroethylene
Dioform; dichloracetylene. Used as a general solvent for organic materials, dye extraction, perfume, lacquers, thermoplastics, organic synthesis. Liquid; gradually decomposes from air, light and moisture forming HCl; d = 1.28; bp = 55°; insoluble in H_2O; soluble in alcohol, ether and most organic solvents; LD_{50} (mus ip) = 2150 mg/kg. *Aldrich.*

321 2-Acetyl pyridine
1122-62-9 214-355-6
C_7H_7NO
Methyl-2-pyridyl ketone
Ethanone, 1-(2-pyridinyl)-; Acetyl pyridine; 2-Acetopyridine. Chemical intermediate. mp = 8-10°; bp = 189-190°; d = 1.082; n_D^{20} = 1.5235; $[\alpha]_D$ = 100°; sol. most organics, insol. H_2O; λ_m = 226, 258, 266, 276 nm (logε = 4.43, 2.82, 2.85, 2.60); LD_{50} (rat orl) = 1495 mg/kg, (rat ip) = 1150 mg/kg. *Aldrich; Penta Mfg; Raschig; Reilly Ind.; Schweizerhall.*

322 Acheson's Deflocculated Graphite
Aquadag. A lubricant obtained by macerating graphite with a solution of tannin for several weeks, forming a permanent emulsion; the graphite with water is called Aquadag; Oildag is prepared by pouring oil over the filtered dag, then freeing the material from moisture. *Acheson Colloids.*

323 Achilles Dipentene
138-86-3 5518 205-341-0
$C_{10}H_{16}$
1-Methyl-4-(1-methylethenyl)cyclohexene
(±)-Limonene; cinene; cajeputene; kautschin. Dipentene, commercial grade; solvent for resins, waxes and oils, perfumery, wetting agent and antiskinning for paint. bp = 170-180°; d = 0.856; n_D^{20} = 1.4750; insoluble in H_2O, miscible with EtOH. *Langley Smith & Co Ltd.*

324 Achilles Pine Oil
Pine oil (mixed isomers of terpene alcohol); disinfectants/cleaners, plasticizer for epoxy resin, solvent and leveler in paint formulations, wetting agent for pigments. *Langley Smith & Co Ltd.*

325 Achilles Tall Oil Fatty Acid
Oleic acid/linoleic acid mixture with a rosin acid content; used in alkyd resins, detergents, disinfectants, soaps and core oils. *Langley Smith & Co Ltd.*

326 Aciculite
12141-46-7 377 235-253-8
Al_2SiO_5
Acicular aluminum silicate
Aluminum oxide silicate;
Pyrax ABB; Silicic acid, aluminum salt (1:2); Aluminum silicate; Aluminum silicate oxide; Kerphalite; Pyrax ABB. High length-to-thickness ratio in naturally occurring mineral; reinforcement for plastic systems, coating, caulks, sealants, and mastics. *Kaopolite.*

327 Acid Aid 5LXS-IH
Lowers surface tension between steel and acid yielding faster pickling rates; protects base metal; excellent detergency properties; reduces HCl acid consumption. *Crown Tech.*

328 Acid Aid X
Accelerator and extender for hydrochloric and sulfuric acids; detergency properties; for metal cleaning, pickling. *Crown Tech.*

329 acid bronze
Alloy of copper, tin, lead and zinc
Alloys containing from 82-88% copper, 8-10% tin, 2-8% lead, and 0-2% zinc. A metal containing 90% copper and 10% aluminum is also known as acid bronze.

330 acid calcium phosphate
7758-23-8 1740 231-837-1
$CaH_4O_8P_2$
Monobasic calcium phosphate
calcium biphosphate; monocalcium orthophosphate; monocalcium phosphate; primary calcium phosphate; calcium superphosphate. Loses H_2O at 100°, decomposes at 200 °; d_{18} = 2.220; soluble in H_2O, HCl, HNO_3, AcOH.

331 Acid Felt Scour
Specially formulated product; pulp and paper felt cleaner for batch and continuous processing. *Hart Chem. Ltd.*

332 Acid Foamer
Quaternary ammonium chloride
Foaming agent, surfactant for strong acids; used in aluminum trailer cleaner, brightener, acid inhibitors and cleaners, chrome plating baths. *Exxon/Tomah.*

333 acid fuchsine
3244-88-0 109 221-816-5
$C_{20}H_{17}N_3Na_2O_9S_3$
2-Amino-5-[(4-amino-3-sulfophenyl)(4-imino-3-sulfo-2,5-cyclohexadien-1-ylidene)methyl]-3-methylbenzenesulfonic acid disodium salt
C.I. Acid Violet 19; C.I. 42685; acid magenta; acid rubin; fuchsin(e) acid; acid roseine; Andrade indicator. Used as a pH indicator and a biological stain. λ_m = 540-545 nm (10 mg/l 0.1N HCl); soluble in H_2O 143 g/l, slightly soluble in EtOH.

334 acid of amber
110-15-6 9038 203-740-4
$C_4H_6O_4$
Succinic acid
Chemical intermediate.

335 acid tar
The waste acid from the washing of crude light oils of coal tar.

336 Acid Thickener
Surfactant; viscosity builder, wetting agent, corrosion inhibitor for acid-based cleaners, e.g., acid bowl cleaners, truck cleaners, building restoration cleaners; perfume solubilizer. *Exxon/Tomah.*

337 Acidan
97593-31-2 307-334-9
Monoglyceride citric acid ester
Emulsifier, surfactant used in food industry. *Grindsted Prods.; Grindsted Prods. Denmark.*

338 Acidan N 12
68990-59-0 273-613-6
Glycerides, tallow mono-, hydrogenated, citrates
Hydrogenated tallow glyceride citrate. Emulsifier and surfactant. *Grindsted Prods.*

339 Acidax
A stearic acid derivative for use in the rubber industry.

340 **Aciderm®**
Dyestuffs; acid dyes for all kinds of leather. *Bayer AG; Bayer plc.*

341 **Acidol®**
Acid dyes for dyeing and printing wool, polyamide, silk. *BASF.*

342 **Acidol® M**
Sulfo-group containing 1:2 metal complex dyes for dyeing and printing wool, polyamide, silk. *BASF.*

343 **acifluorfen**
50594-66-6 111 256-634-5
$C_{14}H_7ClF_3NO_5$
5-[2-chloro-4-(trifluoromethyl)phenoxy]-2-nitrobenzoic acid
(sodium salt) scifluorfen; RH-6201; Blazer. Herbicide. mp = 151-157°; (sodium salt): mp = 124-125°; soluble in H_2O (>25 g/100 ml); LD_{50} (rat orl) = 1300 mg/kg.

344 **Acihib A9**
Amine derivative; controls corrosion in HCl, sulfuric, phosphoric, and other acids; acid restrainer. *ICI Australia.*

345 **Acilan**
Acid wool dyestuffs with good leveling power. *Bayer AG.*

346 **Acillin**
69-53-4 628 200-709-7
$C_{16}H_{19}N_3O_4S$
ampicillin
Antibacterial. *ICN Nutritional Biochemicals Corp.*

347 **Acintene®**
Terpene products including α-pinene, β-pinene middle boilers containing dipentene, technical grade anethole. α-pinene: solvent, terpene resin monomer, intermediate for flavour and fragrance chemicals, intermediate for metal lubricant additive, camphor, camphene manufacture. β-pinene: intermediate for flavor, fragrance chemicals, solvent, pine oil extender, technical grade anethole, anise flavor intermediate. *Arizona.* Discontinued.

348 **Acintol® 736, 2122, D25LR, D30E, DFA, EPG, FA-1, FA-2, R Type3A, LO-3A, SB, SM4**
61790-12-3 263-107-3
Tall oil acid, distilled; surfactant for asphalt emulsifiers, concrete form release and air entraining agents, metalworking fluids, varnishes, printing inks, soaps, cleaners, degreasers. *Arizona.* Discontinued.

349 **Acintol® Liquaros**
Tall of rosin; pigment dispersing agent, fortified paper size, compounding cutting oils and buffing materials, surface coatings, printing ink vehicles and base stock for surfactants. *Arizona.* Discontinued.

350 **Acintol® R Type SFS**
Tall oil rosin
Tall oil rosin; formaldehyde-treated. *Arizona.* Discontinued.

351 **Acintol® R Type SM4**
Tall oil rosin; maleated. *Arizona.* Discontinued.

352 **Aciplex®**
Ion exchange membrane. For food industry. *Asahi Chem. Industry.*

353 **ACL**
9896
Chlorinated *s*-triazinetriones
Machine dishwashing compounds, disinfection of public and private swimming pools, household and industrial cleaners, bleaching agents for both domestic and industrial washing, formulation of industrial bactericides. *Monsanto Co.*

354 **ACL 56, 59, 60, 66,**
2244-21-5 9896 218-828-8
$C_3Cl_2KN_3O_3$
Potassium dichloroisocyanurate dihydrate
1,3-Dichloro-1,3,5-triazine2,4,6-[1H,3H,5H]-trione potassium salt; Triclosene potassium; 3,5-Dichlorotetrahydro-2,4,6-trioxo-s-triazin-1(2H)-yl potassium; potassium troclosene; potassium trichloroisocyanurate; ACL-59. Bleaching compound, sanitizer, disinfectant. *Monsanto Co.*

355 **ACL 85, 90 Plus**
87-90-1 9188 201-782-8
$C_3Cl_3N_3O_3$
1,3,5-Trichloro-1,3,5-triazine2,4,6-[1H,3H,5H]-trione
Symclosene; Trichloroisocyanuric acid; trichloroiminocyanuric acid; ACL-85; Chloreal. Bleaching compound, sanitizer, disinfectant, detergent. mp = 246-247° (dec); pH aq. soln. = 4; soluble in H_2O (0.2%, 25°). *Monsanto Co.*

356 **Aclar Films**
Fluorocarbon thermoplastic; high moisture barrier flexible film for packaging and container liners for pharmaceutical and cosmetic packaging, clean room, military, electronic applications; extruded films in varying widths and thickens; optically clear. *AlliedSignal.*

357 **Aclarat 8678 Granules, Liquid**
Fluorescent whitener for commercial/industrial laundry detergents, rug/upholstery cleaners, fabric softeners, laundry bleach, whitening soap, brightening polymers and plastics; especially for synthetics and wool. *Sandoz.*

358 **Aclon PCTFE**
Homopolymers and copolymers of chlorotrifluoroethylene. *AlliedSignal Inc.*

359 **AClyn®201A**
Calcium acrylate copolymer. Ethylene/calcium acrylate copolymer; low molecular weight ionomers used as processing and performance additives; improves dispersion of additives in plastics; adhesion to variety of substrates. *AlliedSignal.* Discontinued.

360 **AClyn®246A**
magnesium acrylate copolymer. Ethylene/magnesium acrylate copolymer; low molecular weight ionomers used as processing and performance additives; improves dispersion of additives in plastics; adhesion to variety of substrates. *AlliedSignal.* Discontinued.

361 **AClyn®250, 262, 296**
9010-77-9
ethylene/acrylic acid copolymer, Mg ionomer
Low molecular weight ethylene/acrylic acid copolymer, Mg ionomer; alkali-dispersible additive for recyclable hot-melt and aqueous adhesives and coatings. *AlliedSignal/A-C® Perf. Addit.* Discontinued.

362 **Acne-Aid Detergent Soap**
A blend of high molecular weight fatty acids and selected detergents and contains sulfated surfactant blend; as an aid in the management of acne and any condition where greasy skin predominates. *Stiefel Laboratories (UK) Ltd.*

363 **Acnidazil**
A proprietary preparation of miconazole nitrate and benzoyl peroxide; for acne vulgaris. *Janssen Pharmaceutical Ltd.*

364 **Acofor**
61790-12-3 263-107-3
Distilled tall oil fatty acids; latex stabilizer, dispersant (as soap) for pigments and fillers. *Reichhold.* Discontinued.

365 **Aconol X6**
61791-00-2
PEG-6 tallate
Emulsifier for mineral oil. *Hart Chem. Ltd.*

366 **Acorga**
Range of chemicals; used in the extraction and treatment of metals and metalloids. *Acorga Ltd.*

367 **acorn sugar**
488-73-3 8218
$C_6H_{12}O_5$
Quercitol
2-Deoxy-D-*chiro*-inositol, D-1-deoxy-*muco*-inositol; (+)-protoquercitol; 1,2,3,4,5-cyclopentanepentol. Sugar found in acorns. Chemical intermediate. mp = 234-235°; $[\alpha]_D^{20}$ = 24-26°; soluble in H_2O, slightly soluble in hot EtOH, insoluble in cold EtOH, Et_2O.

368 **Acpol**
Polyester resin. *Freeman Chemical Corp.*

369 **acquerite**
$Ag_{12}Hg$
A native alloy of silver and mercury, found in Chile, approximating to the formula $Ag_{12}Hg$.

370 **Acquit®**
For agriculture industry. *DuPont UK.*

371 **Acra-500**
Fatty amine complex; asphalt wetting and antistripping agent. *Exxon/Tomah.*

372 **Acraconc®, Acraconz®**
A range of printing concentrates; dispersion thickeners for printing systems with a low white spirit content or without white spirit. *Bayer AG.*

373 **Acrafil®**
Flame retardant styrene-acrylonitrile.

374 **Acrafloc®**
Range of binders; for flocking of textiles. *Bayer AG; Bayer plc.*

375 **Acralen® A, AFR, ATR, BS**
Aqueous dispersions of polyacrylic resins; suitable for impregnated or coated substrates that must be fast to light; especially suitable as binders for nonwoven fabrics and padding materials. *Bayer AG.*

376 **Acralen® ATR**
Acrylic acid ester
styrene latex. Self-crosslinking binder for nonwoven fabrics, lamination. *Bayer AG.*

377 **Acralen® BS**
Butadiene
styrene; acrylonitrile latex. Self-crosslinking binder for nonwoven fabrics, technical fabrics. *Bayer AG.*

378 Acralux®
Auxiliary to increase the lightfastness of polyamide dyeings. *Bayer AG.*

379 Acramin® Binders
Auxiliary for pigment printing. *Bayer AG.*

380 Acramin® Dyestuffs
Pigment dyestuffs; for the pad dyeing of fabrics made of celluosics, synthetic fibers and their blends. *Bayer AG; Bayer plc.*

381 Acrawax®
A proprietary synthetic wax for lubrication and blending with other waxes to increase their melting point. *Lonza Inc.*

382 Acrawax® B
Amide wax
Amide wax; synthetic wax and plastic lubricants. *Lonza Inc.*

383 Acrawax® C
110-30-5 203-755-6
$C_{38}H_{76}N_2O_2$
N,N'-Ethylene bisstearamide
Octadecanamide, N,N'-1,2-ethanediylbis-; Ethylene bis(stearamide); N,N'-ethylenedi(stearamide). Internal and surface lubricant in resins and plastics; processing aid; plasticizer for resin; flow improver; pigment dispersant; used in hot-melt adhesives and coatings; powdered grade used as lubricant, processing aid, detackifier, mold release, and antiblocking agent. mp = 140-145°; insoluble in H_2O. *Lonza Inc.*

384 Acrex
973-21-7 3339 213-546-1
$C_{14}H_{18}N_2O_7$
Carbonic acid 1,1-methylethyl2-(1-methylpropyl)-4,6-dinitrophenyl ester
Dinobuton; Dessin; Sytasol. A miticide. mp = 56-57°; LD_{50} (rat orl)= 59 mg/kg. *Murphy Chemical Co Ltd.*

385 Acrilan
Acrylic fiber; for sweaters, handcraft yarns and carpets. Density: d_{25} = 1.17; Fiber decomposes as it melts; practically insoluble and unaffected by common solvents. *Monsanto Co.*

386 Acrilester
Two part polyester coating resin which is water-based and catalysed by the addition of equal parts of the two components; high gloss overcoat for a variety of surfaces which exhibits extremely high heat resistance, scuff resistance and smooth durable surface. *ADM Tronics Unlimited Inc.*

387 Acrilev ADK Special
Potassium salt of phosphate ester; caustic stable wetting agent, leveling agent, scouring agent. *Finetex.*

388 Acrilev AM, AM-Special
Phosphate ester, potassium salt
Detergent, wetter, dye leveler used in textiles. *Finetex.*

389 Acrilev OJP-25N
Alkylalkoxylated phosphate ester, sodium salt
Emulsifier, detergent, wetting agent, dispersant. *Finetex.*

390 Acrilpact
Acrylic sealant; sealant in building. *Siliconas Hispania SA.*

391 Acrisint 400, 410, 430
Carbomer. *3V-Sigma.*

392 Acrisorcin
7527-91-5 128 231-389-7
$C_{25}H_{28}N_2O_2$
4-Hexyl-1,3-benzenediol compd with 9-acridinamine(1:1)
Akrinol; Sch-7056; 9-aminoacridinium 4-hexylresorcinolate. Antifungal. Yellow crystals. Name unverified.

393 Acritamer 934, 940, 941
Carbomer; suspending and viscosity agent. *RITA.*

394 Acrodel
An insecticide preparation. *ICI Chem & Polymers Ltd.*

395 Acronal®
Acrylate homo-and copolymers; binders for paper and board coating; binders and coating agents for production of materials based on leather fibers. *BASF AG; BASF plc.*

396 Acronal® 14D
A plasticizer-free acrylic copolymer in the form of a 55% dispersion; adhesive. *BASF plc.*

397 Acronal® 21D, 27D, 30D
Plasticizer free dispersions of thermosetting copolymers of various acrylic esters; adhesive. *BASF plc.*

398 Acronal® 160D
A plasticizer-free acrylic copolymer in the form of a 40% dispersion; adhesive. *BASF plc.*

399 Acronal® 350D
A plasticizer free acrylic copolymer in the form of a 45% dispersion; adhesive. *BASF plc.*

400 Acrosol®
Acrylic acid ester copolymers; cobinders for paper coating. *BASF AG.*

401 Acrosyl
A saponified cresol disinfectant.

402 Acry-Ace
Polymethyl methacrylate
A proprietary molding powder. *Fudow.*

403 Acrycal® MP CP-1000-E
Acrylic resin, impact-modified; extrusion grade with high impact strength; excellent blend stock material; for medical, appliance, industrial, personal accessory, and RV exterior applications. *Continental Polymers.*

404 Acrycal® MP CP-41
Acrylic resin; injection molding resin with lowest heat resistance and maximum flow. *Continental Polymers.*

405 Acrycal® MP CP-924
Acrylic resin, impact-modified; injection molding grade with good impact strength, excellent light transmission; for medical, appliance, industrial, personal accessory, and RV exterior applications. *Continental Polymers.*

406 Acrydur
Reactive acrylic resin systems; for seamless, industrial and commercial floorings. *Ulfcar International A/S.*

407 Acryl
Dyestuffs for acrylic fibers. *BASF plc.*

408 Acryl/bis
Solution or powder of acrylamide/bis-acrylamide. *Am. Research Prods.*

409 Acrylafil G-40/20/FR and G-40/30/FR
Proprietary flame retardant grades of styrene-acrylonitryle, reinforced with glass fiber. *Dart Industries Inc.*

410 Acrylamide
79-06-1 131 201-173-7
C_3H_5NO
2-propenamide
Propenamide; Acrylic amide; Ethylene carboxamide; 2-Propenamide; propenamide; 2-Propeneamide; vinyl amide; propenoic acid, amide. Available as solid, 50% or 30%; used in the process industry. mp = 84.5°; bp_{72} = 103°; Solubility: in g/100ml solvent at 30°=H_2O 215.5; methanol 155; ethanol 86.2; acetone 63.1; ethyl acetate 12.6; chloroform 2.66; benzene 0.346; Human toxicity: highly toxic and irritant. Causes CNS paralysis. *Cyanamid BV.*

411 Acrylates/PVP copolymer
26589-26-4
Luviflex® VBM 35. Film former with excellent hydrocarbon compatibility for weatherproof hairstyles. *BASF AG; BASF plc.* Name unverified.

412 acrylic acid
79-10-7 132 201-177-9
$C_3H_4O_2$
2-propenoic acid
ethylenecarboxylic acid; propenoic acid; acroleic acid; vinylformic acid; glacial acrylic acid; propene acid; acrylate. Corrosive liquid used in the manufacture of plastics. Monomer for polyacrylic and polymethacrylic acids, other acrylic acids, acrylic polymers. Used in manufacture of acrylic fiber; In plastics, surface coatings and adhesives industry. mp = 13°; bp = 139°; d = 1.0510; n_D^{20} = 1.4202; soluble in H_2O, organic solvents; LD_{50} (rat orl) = 2.59 g/kg. *BASF; Hoechst-Celanese; Hüls UK; Penta Mfg.; Rohm & Haas; Union Carbide.*

413 Acrylic acid acrylonitrogens copolimer
136505-00-5, 617-40-7; 136505-01-6
Hypan® SR150H, SA100H. Acrylic acid/acrylonitrogens copolymer; thickener and gellant for aqueous formulations, especially highly concentrated salt solutions, surfactants and drugs; emulsifier. *Kingston Tech.*

414 Acrylic Resin AS
Acrylic copolymer aqueous dispersion; finishing agent giving full supple hand on all fibers. *ICI Surf. UK.*

415 Acrylite®
Acrylic sheet, and molding and extrusion compounds. *Cyro Industries.*

416 Acrylite® H
Proprietary acrylic pellets; used for injection molding or extrusion. *Cyro Industries.*

417 Acrylite® M
Proprietary acrylic pellets providing medium heat resistance; used in injection molding and extrusion where greater flow is required. *Cyro Industries.*

418 Acrylite® FF
Acrylic sheet; clear, lightweight, rigid, dimensionally stable, and weather-resistant thermoplastic for use in skylights, recreational vehicles, boat and motorcycle fairings, signs, displays, boutique items; available in thicknesses from 1.5-12.7 mm. *Cyro Industries.*

419 Acrylite® GP
Cast acrylic sheet; used in industrial plants and building products, e.g., window glazing, skylights, tub enclosures. *Cyro Industries.*

420 Acrylo-40
A 40% solution of acrylamide.
Am. Research Prods.

421 Acryloft, Acryloft Conc.
Modified quaternary ammonium compound
Textile softener. *Rhône-Poulenc.*

422 Acryloid®
Acrylic ester resin
Used for surface coatings. *Rohm & Haas.*

423 Acryloid® 150
Alkyl methacrylate copolymer in solv. refined neutral oil
Pour-point depressant for use in motor oils, gear lubricants, hydraulic fluids, other lubricant applications; dewaxing aid in solvent dewaxing processes for manufacturing of lubricating oils. *Rohm & Haas.*

424 Acryloid® 702
Methacrylate copolymer,
Viscosity index improver and pour point depressant for gasoline and diesel engine oils, hydraulic fluids, industrial lubricants. *Rohm & Haas.*

425 Acryloid® A-101
40% Solids methyl methacrylate polymer in MEK
Thermoplastic resin used for vinyl topcoating and printing inks. *Rohm & Haas.*

426 Acryloid® A-21
Methyl methacrylate polymer in toluene/butanol (90/10)
Thermoplastic resin for automotive finishing, vinyl clear coatings, plastisol primers, specification lacquers, rail car and aircraft finishes. *Rohm & Haas.*

427 Acryloid® A-21LV
Methyl methacrylate polymer in toluene/MEK/butanol (50/40/10)
Thermoplastic resin used in bake-sand-bake finishes, auto refinishes, aircraft and rail care finishes, vinyl clear coatings, plastisol primers. *Rohm & Haas.*

428 Acryloid® A-30
9011-14-7
Methyl methacrylate polymer
Thermoplastic resin used in vinyl topcoating with vinyl copolymers and homopolymers. *Rohm & Haas.*

429 Acryloid® AT-51
Hydroxyl-type acrylic polymer,
Thermosetting resin for applications requiring hardness and chemical, stain, and solvent resistance, e.g., product finishing, light appliance finishing. *Rohm & Haas.*

430 Acryloid® AT-63
Hydroxyl-type polymer
Thermosetting resin with fast cure; for general product finishing, exterior coil coating. *Rohm & Haas.*

431 Acryloid® B-44
9011-14-7
Methyl methacrylate copolymer
Thermoplastic resin for finishes for vacuum-metallized plastics, aluminum sash and chrome plate, luminescent finishes, metal and masonry clear coatings, farm and factory machinery, product finishing. *Rohm & Haas.*

432 Acryloid® B-67
Isobutyl methacrylate polymer in VM&P naphtha
Thermoplastic resin which improves properties of alkyd and oleo-resinous varnishes; used in enamels, aerosols, clear coatings, farm and factory machinery, fluorescent pigment dispersion, plastic top coats, marine finishes. *Rohm & Haas.*

433 Acryloid® B-99
Methyl methacrylate copolymer, xylene, toluene (70/30)
Thermoplastic resin used in pigment grind to improve gloss; used for refinish lacquer maintenance parts, aerosols, farm and factory machinery, pigment dispersions, printing inks, product finishing, auto refinishes, marine finishes. *Rohm & Haas.*

434 Acryloid® F-10
Acrylates copolymer (butyl methacrylate polymer), min. thinner/Aromatic 100 (90/10)
Thermoplastic resin used in clear and pigmented coatings on metal, luminescent pigment vehicles, aerosols, printing inks. *Rohm & Haas.*

435 Acryloid® WR-97
Hydroxyl-type acrylic polymer
Thermosetting solution resin reducible with water or solvent; crosslinkable with urea and melamine resins; pigment dispersant; also for exterior coil coating, product and appliance finishing. *Rohm & Haas.*

436 Acrylonitrile
107-13-1 133 203-466-5
C_3H_3N
Propenenenitrile
Vinyl cyanide; Cyanoethylene; Fumigrain; Ventox; propenenitrile; vinyl cyanide; 2-Propenenitrile; Cyanoethylene; ACN; Fumigrain; propenonitrile; AN; miller's fumigrain; TL 314; VCN; Propenitrile. Monomer for acrylic and modacrylid fibers; in production of ABS and acrylonitrile styrene copolymers. mp = -84°; bp= 77°; d = 0.8060; n_D^{20} = 1.3910; LD_{50}(rat orl) = 78 mg/kg; explosive, flammable and toxic liquid; may polymerize spontaneously, particularly in absence of oxygen; Aldrich; Am, Cyanamid; Asahi Chem Industry Co Ltd; BP Chem. Ltd; DSM NV; Du Pont; Mitsui Toatsu Chem.

437 Acrylonitrile-butadiene rubber
$[C_4H_8 \cdot C_3H_3N]_x$
Acrylonitrile rubber; Acrylonitrile-butadiene copolymer; Nitrile rubber, Heveasyn Nitrile Latex; Hycar 1552; Hycar 1561; Hycar 1571; Hycar 1572; Hycar 1572x64; Nipol DN-601; Perbunan N Latex; Synthetic rubber,. Used for oil well parts, general-purpose oil-resistant applications, gaskets, grommets, o-rings. Density: 0.98; tensile strength = 1000-3000 psi.

438 Acrylonitrile-butadiene-styrene
9003-56-9
Acrylonitrile polymer with 1,3-butadiene
AS-10GF; AS-15CF/000; ABS; Baymod® 50, 90/92, AKU3-2086; Blendex® 131, 336, 338, 467. ABS resin, 10% glass fiberreinforced; thermoplastic with dimensional stability and toughness; for business machine housings, cabinetry, tool housing, appliances, automotive, and construction materials. Combustible; d = 1.04; tensile strength = 6500 psi; flex str. = 10,000 psi. *Aiscondel SA; Bamberger Polymers; BASF; LNP; Mitsui Toatsu Chem; Monsanto; Reichold;.*

439 Acrylonitrile-butadiene-styrene
9003-56-9
Electrafil® J-1200/CF/10; ABS 124ESG; ABS 236F; ABS 236MA; ABS 301K; ABS 500FR-1; AS-10GF; AS-15CF/000; Baymod® A KU3-2086; Blendex® 101; Blendex® 310; Claradex CH-540; Conductomer ABS-22; Cycolac® CKM1; Cycolac® DH; Cycolac® GPM4700; Cycolac® GPX2800; Cycolac® KCS; Cycolac® KJM; Cycolac® X-11; Electrafil® G-1204/SS/3; Electrafil® J-1200/CF/10; EMI-X® PDX-A-88128; Magnum 240; Magnum 275; Magnum 445 HQ; Magnum 788HP; Magnum 2610; Magnum 3661; Magnum 4420; Magnum 9450P;Magnum FG960; Multibase ABS 3075; Novodur® L3FR; Novodur® P2H-AT; Novodur® PMTM; RTP 601; Stat-Kon® AC-1003; Stat-Kon® AS; Stylac® ABS; Styvex 40007 BKL2; Styvex 42023 NAFR; Terluran®;Terlux®;Thermocomp® AF-1004,. ABS, 10% PAN carbon fiber-reinforced; static dissipative and conductive thermoplastic. *Akzo Engineering Plastics.* Name unverified.

440 Acrylweld
Two-part acrylic adhesive. *Hardman.*

441 Acrymul AM 123R
Acrylates copolymer; a proprietary trade name for a self cross-linking acrylic emulsion. *Protex.*

442 Acrysol®
Acrylic and urethane thickeners; used for coatings. *Rohm & Haas; Rohm & Haas UK.*

443 Acrysol® ASE-60
Crosslinked acrylic emulsion copolymer
Used to suspend pigments and fillers in water-based paints, inks, or other coatings, and the abrasive particles in waxes or polishes; viscosity modifier for emulsion and latex compounds; binder; thickener. *Rohm & Haas.*

444 Acrysol® ASE-75
9003-01-4
Polyacrylic acid
Thickener for latexes and emulsions for paints, flocking adhesives. *Rohm & Haas.*

445 Acrysol® ASE-95
Acrylic copolymer emulsion
Thickener for fabric laminants, pigment dispersant in polar solvents, surfactant solutions, paints, inks, waxes, and polishes. Alkali-sol *Rohm & Haas.*

446 Acrysol® G-110
9003-03-6
Ammonium polyacrylate solution.; thickening and stabilizing agent for synthetic latexes; used in coatings, adhesives, dipped, cast, and molded goods, cements for rug backing, spraying, spreading, brushing, and extruding compounds. *Rohm & Haas.*

447 Acrysol® GS
9003-04-7
Sodium polyacrylate
Thickener for natural and synthetic latexes for paints, films, coatings, and adhesives. *Rohm & Haas.*

448 Acrysol® HV-1
9003-04-7
Sodium polyacrylate

All purpose thickener for rubber rug backing, unholstery backing. *Rohm & Haas.*

449 Acrysol® LMW

Polyacrylic acid homo/copolymer; detergent, water treatment and textiles. *Rohm & Haas.*

450 Acrysol® TT-615

Acrylic copolymer emulsion; thickener for coatings, textile printing pastes, and adhesives. Liquid; alkali-soluble. *Rohm & Haas.*

451 Acrysol® WS-24

Acrylic copolymer. *Rohm & Haas.*

452 Acrythane

Paint product available in all colors for automotive/commercial vehicle use; vehicle coating for original and refinishing; also for high quality chemically resistant coating of machinery. *H Marcel Guest Ltd.*

453 ACS 60

37475-88-0 253-519-1
$C_9H_{15}NO_3S$
Ammonium cumenesulfonate
Benzenesulfonic acid, (1-methylethyl)-, ammonium salt. Hydrotrope, solubilizer for personal care applications. *Witco SA.*

454 Acsil

1344-09-8 8824 215-687-4
Na_4O_4Si
Sodium orthosilicate
sodium sesquisilicate; sodium silicate; Water glass; Soluble glass; Silicate of soda; Sodium Orthosilicate; Silicic Acid Sodium Salt; Sodium silicate glass. Used as a preservative agent, for lining reactors and as a binder in abrasive wheels. *Crosfield Chemicals Ltd.*

455 Acsium®

For polymer industry. *DuPont UK.*

456 Actafoam® F-2, R-3

Activator-stabilizers for vinyl foams containing Kempore blowing agents; gas release accelerator. *Uniroyal.*

457 Actan SP

Tannic acid concentrate
Corrosion inhibitor used in surface pretreatment for field applications; base for application of coatings systems. *Troy.*

458 Actellic

29232-93-7 7652 249-528-5
Pirimiphos-methyl
Contact fumigant and organophosphorus insecticide which is available as a dust, liquid or smoke. *ICI Agrochemicals; ICI Chem. & Polymers Ltd.*

459 Actellifog

29232-93-7 7652 249-528-5
Pirimiphos-methyl
Fumigant and insecticide for glasshouse crops. *ICI Agrochemicals; ICI Chem. & Polymers Ltd.*

460 ACter 1450, 1450A

Ethylene/acrylic acid/vinyl acetate copolymer. *AlliedSignal.* Discontinued.

461 Acticarbone

64365-11-3 1856 264-846-4
C
Carbon, activated
Carbon black; Carbon decolorizing. Powdered and granulated activated carbon; used for purification, decolorization, deodorization, separation and recovery in liquid or gas phase, in the chemical, petrochemical, pharmaceutical and food industries (glucose factories, sugar refiners, oil refining, wine treatment); for the treatment of drinking and industrial water, etc. Catalyst supports. *Elf Atochem UK/Ceca.*

462 Acticide 50

Potassium orthophenyl phenate solution
Sterilizing wash concentrate for surfaces prior to re-coating, wet-state protection of water-based compositions, prevention of fungal growth on leather in storage. *Thor Chemicals (UK) Ltd.* Discontinued.

463 Acticide 50X®

Blend of benzalkonium chloride and phenyl phenoxide
Fungicide/algicide wash concentrate for sterilization of surfaces prior to re-coating and for industrial plant cleaning. *Anti-Chem; Thor Chemicals (UK) Ltd.*

464 Acticide APA

Fluorinated sulfonamide-based powder
Insoluble bactericide/fungicide for emulsion paints, textured coatings, renovating plasters, plasterboard jointing adhesives, etc. Available in standard and superfine forms. *Thor Chemicals (UK) Ltd.* Discontinued.

465 Acticide AZ®

Synergistic blend of aliphatic nitrogen and heterocyclic sulfur containing compounds
Wet-state bactericide/fungicide for emulsion paints, adhesives, cellulose solutions, polymer emulsions, fillers, pottery glazes, inks, etc. *Anti-Chem; Thor Chemicals (UK) Ltd.*

466 Acticide CPC

Synergistic mixture of FDA approved chlorinated phenols. Bactericide/fungicide for water-based products in the wet state, e.g., adhesives, carpet backing compounds. *Thor Chemicals (UK) Ltd.* Discontinued.

467 Acticide DDM/S

Dichlorophen sodium salt
Bactericide/fungicide for textiles, cellulose solutions, proteins, adhesives and soaps; bactericide/algicide for water treatment. *Thor Chemicals (UK) Ltd.* Discontinued.

468 Acticide LG®

Blend of chlorinated and nonchlorinated methyl isothiazolones with nonmetal salts stabilizing system
Latex grade biocide for wet state protection of polymer emulsions and other aqueous based formulations. *Anti-Chem; Thor Chemicals (UK) Ltd.*

469 Acticide MPM

A 10% mercury metal glycol solution of phenyl mercury nonane-2-ol
Bactericide/fungicide for wet-state and dry-film protection of interior and exterior aqueous and solvent-based paints, woodstains, textured coatings, adhesive fillers, sealants, etc. *Thor Chemicals (UK) Ltd.* Discontinued.

470 Acticide PMA 100

62-38-4 7453 200-532-5
Phenyl mercury acetate powder
Wet-state bactericide/fungicide for emulsion paints, plasters, wood-pulp, etc. *Thor Chemicals (UK) Ltd.* Discontinued.

471 Acticide PMDDS

A 10% phenyl mercury dodecenyl succinate solvent solution
Bactericide/fungicide for wet-state and dry-film protection of interior and exterior aqueous and solvent-based paints, woodstains, textured coatings, adhesives, fillers, sealants, etc. *Thor Chemicals (UK) Ltd.* Discontinued.

472 Acticrom

Dry powder triazynil-type reactive dyes; used in the textile industry, mainly for dyeing cotton. *Multicrom SA.*

473 Acticryl CL 959

Acrylic carbamate
Radiation-curable monomer. *SNPE Chimie.*

474 Acticulum

Conjugated glycopolypeptides
Cosmetic products to help stimulate cell metabolism; strengthens skin against inflammation. *Active Organics.*

475 Actif 8®

Acidic sodium aluminum phosphate with anhydrous monocalcium phosphate
Leavening agent for baking, cereals. *Rhône-Poulenc Food Ingreds.*

476 Actiflo® 68, 70

8002-43-5 5452 232-307-2
lecithin
Natural lecithin; emulsifier, wetting agent, dispersant; used in the food industry. *Central Soya.*

477 Actigen

Live cell animal extract; used for cosmetic products (skin, body preparations and hair care products). *Active Organics.*

478 Actigen C

9007-34-5 2543 232-697-4
Soluble collagen
Soluble animal collagen, cosmetic protein. *Active Organics.*

479 Actigen E

100085-10-7 309-148-3
Hydrolyzed elastin
Hydrolysed animal elastin, cosmetic protein. *Active Organics.*

480 Actiglow

Hydrolyzed mucopolysaccharides. *Active Organics.* Name unverified.

481 Actiglow C

Hydrolyzed mucopolysaccharides.
Hydrolyzed mucopolysaccharides; skin conditioner for moisturizing lotions, creams, gels, masques, and face packs *Active Organics.*

482 Actilex

Ion exchange materials. *Courtaulds plc.* Discontinued.

483 Actimer FR-803

61368-34-1 262-737-6
Tribromostyrene
Reactive monomer which can be copolymerized with styrene, acrylonitrile and maleic anhydride to impart flame retardancy to HIPS, ABS, SAN, SMA; crosslinking agent providing flame retardancy to thermoset systems. mp = 65-67°; LD_{50} (rat orl) >5 g/kg. *AmeriHaas; Dead Sea Bromine; AmeriBrom.*

484 Actimer FR-1025M
59447-55-1 261-767-7
Pentabromobenzyl acrylate
Flame retardant for engineering thermoplastics (PET, PBT, PC, nylon 6 and 6/6); processing aid; maintains transparency of PC and HIPS resins. mp = 122°; insoluble in H_2O, soluble in organic solvents; LD_{50} (rat orl) >5 g/kg *AmeriHaas; Dead Sea Bromine; AmeriBrom.*

485 Actimer FR-1033
59789-51-4
Tribromophenyl maleimide
Reactive flame retardant for polymeric systems which undergo crosslinking (XLPE, ethylene-propylene-diene rubbers); can be grafted onto unsaturated sites in ABS or copolymerized with styrenics to yield flame retardant systems with improved thermal properties. mp = 138°; insoluble in H_2O, souble in organic solvents; LD_{50} (rat orl) >5 g/kg. *AmeriHaas; Dead Sea Bromine; AmeriBrom.*

486 ActiMoist
9067-32-7 4793
Sodium hyaluronate solution
hyaluronic acid, sodium salt. Cosmetic products for high moisture retention/absorbtion; moisturizer for skin care preparations, foundations, moisturizing creams and lotions, eye shadow, eye liners, mascaras; especially effective for normal, dry, and sensitive skins. *Active Organics.*

487 Actiphyte
Extensive range of botanical extracts; a plant extract in propylene glycol water; for cosmetic products (skin, body preparations and hair care products). *Active Organics.*

488 Actiplex
Custom made mixtures of Actiphyte extracts; actiblend, a combination of actiphytes; for cosmetic products (skin, body preparations and hair care products). *Active Organics.*

489 Actipol® E6
9003-28-5
Activated polybutene; through grafting onto a base polymer, impact resistance, low temperature, flexibility, and water resistance are enhanced in adhesives, sealants, coatings, unsaturated polyesters, electrical compounds, foams, and other applications. *Amoco Chemical Co.*

490 Actipron
Adjuvant containing 97% refined mineral oil; wetting agent for herbicides and fungicides. *Bayer plc.*

491 Actipron
Adjuvant containing 97% refined mineral oil; wetting agent for herbicides and fungicides. *BP Oil Ltd.*

492 Actiron NX 3
90-72-2 202-013-9
$C_{15}H_{27}N_3O$
2,4,6-Tri (dimethylaminomethyl) phenol
DMP-30; tris(dimethylaminomethyl)phenol. Accelerator, catalyst, and hardener for epoxy resins. Slightly soluble in hot H_2O; d = 0.98 g/ml; viscosity = 375 mPa; LD_{50} (rat orl) = 2500 mg/kg. *Protex.*

493 Actisize
Modified starch; used for adhesives and paper. *Roquette (UK) Ltd.*

494 Activ-8®, Activ-8 in Hexylene Glycol
66-71-7 7356 200-629-2
Solution forms containing 1,10-phenanthroline
Drier accelerator and stabilizer used in combination with manganese and/or cobalt in coating systems that cure by oxidative polymerization. mp = 93-94°; d = 0.95; soluble in H_2O (0.3 g/100 ml), more soluble in organic solvents. *R. T. Vanderbilt Co Inc.*

495 Activated alumina
1344-28-1 369 215-691-6
Al_2O_3
Aluminum oxide
Alumina; Tabular alumina; Calcinated alumina; Alumite; Alundum. Colorant; dispersing agent; used in orals bp = 2977°; d = 3.54; insoluble in H_2O, slightly soluble in mineral acids; toxic by inhalation of dust; eye irritant by mechanical abrasion. *Rhône-Poulenc; Aldrich; BA Chem. Ltd; Lonza Sarl; Alcan; Atomergic Chemetals.*

496 Activated sludge
A material obtained by allowing the growth of microorganisms in the sludge deposited by sewage. it is used in the treatment of sewage.

497 Activator 736
Surface treated urea for easy dispersion in elastomers; activator for thiazole, thiuram and dithiocarbamate accelerators; odor reducer when used with nitrosoamine type blowing agents. *Uniroyal.*

498 Activator 1102
Dibutyl ammonium oleate
Accelerator/activator for natural and synthetic rubbers; lubricant. d = 0.87 *Anchor; Air Prods & Chem/Perf. Chems.*

499 Activator STAG
Complex secondary amine, surface treated; relatively nonstaining, nondiscoloring activator for thiazole-type accelerators; primary accelerator for natural rubber; strong secondary accelerator in styrene/butadiene rubber. mp = 130 °; d = 1.26. *Akrochem.*

500 Activax
A proprietary vaccine used in the treatment of fowl pox. *Coopers Animal Health Ltd.*

501 Active 2
68603-42-9 271-657-0
Cocamide DEA
Cocamide Dea; Coconut Oil Diethanolamine; Coconut Acid, Diethanolamide; Coconut Oil Acids Diethanolamide; Clindrol 200CGN; Clindrol 202CGN; Clindrol Superamide 100CG; Comperlan PD; Comperlan KD; Comperlan LS; Conco Emulsifier K; Elromid KD 80; Empilan CDE; Ethylan LD; Ethylan A15; Lauridit KDG; Marlamid D 1218; Monamid 150D; Monamid 150DB; Ninol 2012E; Ninol P 621; Ninol 1281; P and G Amide 72; Purton CFD; Schercomid CDA; Steinamid DC 2129; Steinamid DC 2129e; Varamide A 2; Varamide A 10; Varamide A 83; Witcamide 82; Witcamide 5133; Diethanolamides of the Fatty Acids of Coconut Oil. Coconut diethanolamide, manufactured from coconut oil, is widely used as a surface-active agent in hand gels, hand-washing liquids, shampoos and dish-washing liquids. It has rarely caused allergic contact dermatitis and is used an industrial surfactant. Soluble in H_2O, alcohols, aliphatic and aromatic hydrocarbons; d = 8.34 lb/gal; viscosity = 19.8 poises; pour point = 10°F; pH 9.1 (5%); surface tension = 30 dynes/cm (0.0015%); 100%act.; LD_{50} (rat orl) = 6175 mg/kg. *Blew Chem.*

502 Active 4
Cocamide DEA and DEA-dodecylbenzenesulfonate
Biodegradable high sudsing surfactant used as a base for hand dishwashing detergents, shampoos, bubble baths; wool filling agent; wetting and dispersing agent. Soluble in H_2O; d = 8.7 lb/gal; pour point = 40°F; pH 9 (5%); surface tensin = 30 dynes/cm (0.1%); LD_{50} (rat orl) = 3157 mg/kg. *Blew Chem.*

503 Active 18
93-83-4 202-281-7
$C_{22}H_{43}NO$
Oleamide DEA.
oleic diethanol amide; Oleic acid diethanolamine; 9-Octadecenamide, N,N-bis(2-hydroxyethyl)-, (Z)-; Oleic acid diethanolamide. Surfactant. *Blew Chem.*

504 Activex
Fungicides for garden use. *ICI Garden Products.*

505 Activit
102-08-9 3393 203-004-2
$C_{13}H_{12}N_2S$
Thiocarbanilide
A proprietary rubber vulcanization accelerator.

506 Activol
77-06-5 4426 201-001-0
Gibberellic acid
A hormonal plant growth regulator. *ICI Chem & Polymers Ltd.*

507 Activox
1314-13-2 10279 215-222-5
ZnO
Zinc oxides
Chinese white; C.I. 77947; Zink white. Color additive; used in parenterals, rectals, dental cements; Amorphous powder or crystals; odorless, bitter taste; soluble in dilute acetic or mineral acids, alkalis, insoluble in H_2O, alcohol; d = 5.67; mp = 1975°; pH 6.95 (American process); LD_{50} (rat orl) = 240 mg/kg. *Durham Chemicals Ltd.*

508 Activox B
1314-13-2 10279 215-222-5
ZnO
Zinc oxide
Colloidal zinc oxide. Flame retardant, absorber, activator; pigment in paints. *Harcros.*

509 Acto 450, 500, 630, 632, 636, 639
Alkylaryl sodium sulfonate
Detergent, wetting agent, emulsifier, rust preventative. *Exxon.*

510 ACtone® 1
Low molecular weight proprietary ionomer; pigment wetting agent, dispersion aid to increase color strength; used in thermoplastic and thermoset resins. Free-flowing powder; viscosity = 5500 cps (190°); bulk density = 410 kg/m³; softening point = 101°; acid no. 37. *AlliedSignal/A-C® Perf. Addit.* Discontinued.

511 **ACtone® 2010, 2010P**
Low molecular weight proprietary ionomer resins; pigment wetting agent, dispersion aid for color concentrates and masterbatches for polyester and styrenics. Granular and powdered; melt flow = 1.0 g/10 min. *AlliedSignal/A-C® Perf. Addit.* Discontinued.

512 **ACtone® N**
Low molecular weight proprietary ionomer; pigment wetting agent, dispersion aid, color enhancer for color concs. for nylon and polyester. Granules; viscosity = 30,000 cps (190°); bulk density = 30.21 lb/ft³; acid no. 5; softening point = 100°. *AlliedSignal/A-C® Perf. Addit.* Discontinued.

513 **ACtone® P**
9002-88-4 7728
Low molecular weight branched polyethylene; vehicle for pigment pastes and presscakes; color enhancer. *AlliedSignal/A-C® Perf. Addit.* Discontinued.

514 **Actrabase 31-A**
Soap sulfonate
Primary emulsifier for oil systems, metalworking fluids; base for naphthenic oils. *Actrachem.*

515 **Actrabase 215**
Emulsifier for metalworking fluids; low use level base for paraffinic oils. *Actrachem.*

516 **Actrabase 264**
Emulsifier for metalworking fluids; low use level base for naphthenic oils. *Actrachem.*

517 **Actrabase PS-470**
petroleum sulfonate, sodium salt
Emulsifier and rust inhibitor for cutting and lube oils; dispersant for soluble oil and semi-synthetics for metalworking fluids. Viscous liquid, oil-soluble; 100% concentrate; anionic. *Actrachem.*

518 **Actrabase SS-503**
Semisynthetic concentrate base, emulsifier for metalworking fluids. *Actrachem.*

519 **Actracor 129, 856**
Carboxylic acid amine salt
Corrosion inhibitor for synthetic and semi-synthetic metal working fluids. *Actrachem.*

520 **Actracor 401**
Corrosion inhibitor for water-glycol hydraulic fluids. *Actrachem.*

521 **Actracor 800**
Hydrocarbon
Corrosion inhibitor; metalworking additive. *Actrachem.*

522 **Actracor 1987**
Amine carboxylate
Corrosion inhibitor for metalworking. *Actrachem.*

523 **Actracor M**
Ethanolamine-borate ester
Corrosion inhibitor in cutting oils. *Actrachem.*

524 **Actracor T**
Triethanolamine borate ester
Corrosion inhibitor in cutting oils. *Actrachem.*

525 **Actrafoam A, B, C, S**
Blend of glycols, fatty acids, and nonionic surfactants in a hydrocarbon base
General purpose defoamer (A, B); defoamer for water sewage applications (C, S). d = 7.30 lb/gal; acid no. 7.5; flash pt. (COC) 305 F; may become solid at lower temperatures; store at 70°F for at least 24 hours if product should freeze. *Actrachem.* Discontinued.

526 **Actrafos 104, 109**
Phosphate ester
Coupler for sulfated oils in cleaner formulations; lubricant for synthetic, semi-synthetic and water-based cutting, grinding, and drawing fluids. *Actrachem.* Discontinued.

527 **Actrafos 110, 110A**
Complex aliphatic hydroxyl compound phosphate ester
Pressure additive for cutting and rolling oils; hydrotrope for cleaning compounds; lubricant emulsifier and rust inhibitor; excellent for aluminum; 110A has higher melting point. Soluble in H_2O; 100% concentrate. *Actrachem.*

528 **Actrafos 152A**
Organic phosphate ester
Extreme pressure lubricant for cutting oils; high phosphorus content. Soluble in H_2O; 100% concentrate; anionic. *Actrachem.* Discontinued.

529 **Actrafos SA-216**
Phosphate ester
Lubricant and emulsifier for cutting oils. Soluble in H_2O; 100% concentrate; anionic. *Actrachem.*

530 **Actrafos SN-315**
Phosphate ester
Lubricant, emulsifier for metal working; aluminum corrosion inhibitor. *Actrachem.*

531 **Actrafos SP-407**
Phosphate ester
Low foaming lubricant and rust inhibitor for synthetic metalworking fluids. *Actrachem.*

532 **Actrafos T**
Tridecyl alcohol phosphate ester
Extreme pressure lubricant and release agent for cutting oils. *Actrachem.* Discontinued.

533 **Actrafos TDA**
Organic phosphate ester
Oil-soluble lubricant and release agent. *Actrachem.* Discontinued.

534 **Actralube 21**
Synthetic lubricant with moderate foaming properties. *Actrachem.* Discontinued.

535 **Actralube 100**
Synthetic ester
Lubricant. *Actrachem.* Discontinued.

536 **Actralube 310**
Fatty ester
Lubricity additive for cutting oil. *Actrachem.*

537 **Actralube 1200**
Modified triglyceride
Metal lubricity additive. *Actrachem.*

538 **Actralube 7142**
8002-13-9 8289 232-299-0
Blown rapeseed oil
Metal lubricant additive. *Actrachem.*

539 **Actralube SOS**
Emulsifier, lubricant, metalworking additive; substitute for sperm oil. *Actrachem.* Discontinued.

540 **Actralube Syn-147**
Complex diester
Lubricity additive for synthetic cutting fluids. Soluble in H_2O; 90% concentrate; nonionic. *Actrachem.*

541 **Actralube Syn-153**
Soap
Soap; light duty synthetic lubricant and rust inhibitor for synthetic cutting fluids. Soluble in H_2O; 95% concentrate; anionic. *Actrachem.*

542 **Actramide 176**
Alkanolamide
Cutting fluids. *Actrachem.* Discontinued.

543 **Actramide 202**
2:1 Tall oil fatty acid alkanolamide
Emulsifier for soluble oils, metalworking fluids and emulsion cleaners; corrosion inhibitor. Soluble in H_2O; nonionic. *Actrachem.*

544 **Actramide 410**
Alkanolamide
Secondary emulsifier with lubricating properties. Soluble in H_2O; 100% concentrate; nonionic. *Actrachem.*

545 **Actramide 5264**
Modified 2:1 tall oil fatty acid alkanolamide
Emulsifier, lubricant, rust inhibitor. Soluble in H_2O. *Actrachem.*

546 **Actran Extra**
8016-28-2 232-405-5
Porcine Lard oil
Lard oil for industrial use. *Actrachem.*

547 **Actrasol 167A**
Sulfated caster oil, sodium neutralized
acid etching additive. *Actrachem.*

548 **Actrasol 6092**
61778-68-9
Sulfated rapeseed oil
Lubricant, emulsifier in pigment flushing, cleaners, textiles, paper processing. *Actrachem.*

549 **Actrasol C-50, C-75, C-85**
8002-33-3 232-306-7
Sulfated castor oil, sodium-neutralized
Pigment wetting and dispersing agent; lubricant and emulsifier for metalworking fluids; for pigment flushing, cleaners, textile, and paper processing. Anionic *Actrachem.*

550 **Actrasol CS-75**
Sulfonated soyabean oil, sodium neutralized
Lubricant, emulsifier in pigment flushing, cleaners, textiles, paper processing. *Actrachem.* Discontinued.

551 Actrasol EO
Sulfated glyceryl trioleate, sodium neutralized
Surfactant for shampoos, metalworking. Soluble in H_2O; 75% concentrate. *Actrachem.*

552 Actrasol KAP
Sulfated blend of oils, sodium neutralized
Sperm oil substitute; fat liquor for leather, metalworking. *Actrachem.*

553 Actrasol MY-75
Sulfated methyl ester of soya fatty acid, sodium neutralized
Lubricant and emulsifier in metalworking fluids, water-based drilling muds; oil field defoamer. Soluble in H_2O; 75% concentrate; anionic; biodegradable. *Actrachem.*

554 Actrasol OY-75
Sulfated soyabean oil, sodium neutralized
Lubricant, emulsifier in pigment flushing, cleaners, textiles, paper processing; Soluble in H_2O; 75% concentrate; biodegradable. *Actrachem.*

555 Actrasol PSR
Sulfated ricinoleic acid, potassium neutralized
Aluminum lubricant; lubricant, emulsifier for pigment flushing, cleaners, textiles, paper processing; pigment wetting and dispersing agent. Soluble in H_2O; 75% concentrate; anionic. *Actrachem.*

556 Actrasol SBO
Sulfated butyl oleate, sodium neutralized
Lubricant, emulsifier in pigment flushing, cleaners, textiles, paper processing. *Actrachem.*

557 Actrasol SP
268-372-9
Sulfonated tall oil fatty acid, sodium neutralized wet process phosphoric acid defoamer
Lubricant, emulsifier for pigment flushing, cleaners, textiles, paper processing; wet process phosphoric acid defoamer Soluble in H_2O; 50% concentrate; anionic. *Actrachem.*

558 Actrasol SP 175K
272-349-9
Sulfated tall oil, potassium neutralized
Lubricant, emulsifier for pigment flushing, cleaners, textiles, paper processing; Soluble in H_2O; 50% concentrate; anionic; biodegradable. *Actrachem.*

559 Actrasol SR75
Sulfated oleic acid ammonium neutralized
Mold release agent; lubricant, emulsifier for pigment flushing, cleaners, textiles, paper processing Soluble in H_2O; 75% concentrate; anionic; biodegradable. *Actrachem.*

560 Actrasol SRK 75
Sulfated oleic acid potassium neutralized
Mold release agent; lubricant, emulsifier for pigment flushing, cleaners, textiles, paper processing; Soluble in H_2O; 75% concentrate; biodegradable; anionic. *Actrachem.*

561 Actrasol SS
61790-35-0 273-604-7
Sulfated tall oil
Rust preventative, lubricant, metal polish. Solid; 75% concentrate; anionic. *Actrachem.*

562 Actrel
Aliphatic hydrocarbon with oxygenerated hydrocarbons
Cleaner effective for oils and greases, OEM and maintenance applics.; oil-sol. metalworking oils, quenching oils, electro-discharge machining oils; rust preventative oils, silicone oils; water-sol. Metalworking fluids, waxes, dye penetrants, etc. d = 0.79; flash pt. (CC) 79°; surface tension = 25 dynes/cm; very low acute, oral and inhalation toxicity. *Exxon Int'l.*

563 Actril
Selective weed killer. *May & Baker Ltd.*

564 Actril S
Bromoxynil + dichloroprop + ioxynil + MCPA
Broad spectrum, post emergence contact and translocated herbicide. *Rhône-Poulenc Crop Protection Ltd.*

565 Actrilawn
Selective weedkiller. *May & Baker Ltd.*

566 Actrilawn 10
1689-83-4 216-881-1
Ioxynil
Contact herbicide for use in turf. *Rhône-Poulenc Environmental Prods. Ltd.*

567 Actrol 4DP
PEG diester
Emulsifier and lubricity additive for metalworking fluids. Liquid, 95% concentrate; nonionic. *Actrachem.* Discontinued.

568 Actrol 628
PEG ester
Emulsifier and lubricity additive for metalworking fluids. *Actrachem.*

569 Actrol 6M25P
61791-00-2
PEG 600 tallate
Low foaming surfactant with good lubricating characteristics; adds emulsification, cleaning, hard water stability to soluble oils, cutting and grinding fluids, emulsion cleaners. Nonionic. *Actrachem.* Discontinued.

570 Actylon
Yarns and threads. *Courtaulds plc.* Discontinued.

571 ACuflow AF-1
9010-77-9
Ethylene copolymer [9010-77-9] and aluminum stearate [300-92-5]
Mixture processing additive. *AlliedSignal.* Discontinued.

572 Aculyn 22
Acrylates/steareth-20 methacrylate copolymer:
Thickener for cosmetics and toiletries (hair care products, hand creams, lotions, waterless hand cleaners). Anionic; d = 8.75 lb/gal; viscosity = 20 cps; pH 3.0; LD_{50} (rat orl) > 5g/kg; moderately irritating to eyes; slightly irritating to skin. *Rohm & Haas.*

573 Acumen
Bentazone + MCPA + MCPB
Post-emergence contact and translocated herbicide for undersown cereals. *BASF plc.*

574 Acumer
Water treatment polymers.
Rohm & Haas.

575 Acumer 1000
9003-01-4
$[C_3H_4O_2]_n$
Low molecular weight polyacrylic acid
Poly(acrylic acid). Scale inhibitor in industrial water treatment and oil production. Powder; irritant. *Rohm & Haas.*

576 Acumer 5000
Polymer; scale inhibitor, dispersant for silica and magnesium silicate scale control; for water treatment, geothermal wells, reverse osmosis. *Rohm & Haas.*

577 Acumer A-10
Acidic acrylic emulsion; flocculant, oil/water separator. *Rohm & Haas.*

578 Acumer C-3
Polyamine
Flocculant. *Rohm & Haas.*

579 Acumer QR-1010
Acrylates copolymer; kaolin clay dispersant. *Rohm & Haas.*

580 ACumist A-12, A-18
9002-88-4 7728
Oxidized/micronized polyethylene
Wax additive for adhesives, inks, personal care, rubber applications; suspension aid, flatting and texturing agent, and binder for personal care products. Micronized powder; 12-18μ average particle size; d = 0.99; drop point = 136°; acid no. = 26-40; hardness <0.5 dmm. *AlliedSignal; Perf. Addit.* Discontinued.

581 ACumist B-6, B-12, B-18, C-5, C-12, C-18
9002-88-4 7728
Micronized polyethylene
Wax additive for adhesives, inks, personal care, rubber applications; suspension aid, flatting and texturing agent, and binder for personal care prods.; Micronized powder; 12-9μ average particle size; d = 0.96; drop point = 126°. *AlliedSignal; Perf. Addit.* Discontinued.

582 Acusol® 410N
9003-04-7
Sodium polyacrylate
Detergent polymer, dispersant for cleaners, water treatment, mineral processing. Mw ca.100,000; viscosity = 500-1500 cps; pH 6.5-8.0; 40% solids. *Rohm & Haas.*

583 Acusol® 445
9003-01-4
Polyacrylic acid
Detergent polymer for detergents and cleaners, water treatment, mineral processing, other industrial markets. Mw ca. 4500; pH 3; 48% solids. *Rohm & Haas.*

584 Acusol® 445N
9003-04-7
Sodium polyacrylate
Detergent polymer for detergents and cleaners, water treatment, mineral processing, other industrial markets. Mw ca. 4500; pH 7; 45% solids. *Rohm & Haas.*

585 **Acusol® 445ND**
9003-04-7
Sodium polyacrylate
Detergent polymer for detergents and cleaners, water treatment, mineral processing, other industrial markets. Dry form; mw ca. 4500; 92% solids. *Rohm & Haas.*

586 **Acusol® 460ND**
Maleic acid/olefin copolymer, sodium salt
Detergent polymer for detergents and cleaners, water treatment, mineral processing, other industrial markets. 92% solids. *Rohm & Haas.*

587 **Acusol® 479N**
Sodium acrylic acid/maleic acid copolymer
Detergent polymer for detergents and cleaners, water treatment, mineral processing, other industrial markets. Mw ca. 70,000; pH 7; 40% solids. *Rohm & Haas.*

588 **Acusol® 479ND**
Sodium acrylic acid/maleic acid copolymer
Detergent polymer for detergents and cleaners, water treatment, mineral processing, other industrial markets. Dry form; mw ca. 70,000; 92% solids. *Rohm & Haas.*

589 **Acusol® 480N**
9003-04-7
Modified polyacrylic acid, sodium salt, sodium polyacrylate (45-47%), water (53-55%), and < 0.1 % residual monomers. Detergent polymer for detergents and cleaners, water treatment, mineral processing, other industrial markets. Soluble in H_2O; mw ca. 3500; d = 1.25; viscosity = 400-2000 cps; pH 7; bp = 100°; noncombustible; LD_{50} (rat orl) >5000 mg/kg, inhalation of vapor may cause headache, nausea. *Rohm & Haas.*

590 **Acusol® 810**
Acrylic crosslinked copolymer
Detergent polymer, processing aid, thickener for detergents and cleaners, water treatment, mineral processing, other industrial markets. Alkali-soluble; d = 1.046; viscosity = 200 cps; pH 2.4-3.4; 18% solids; anionic. *Rohm & Haas.*

591 **Acusol® 820**
151066-66-5
Acrylic copolymer emulsion
Detergent polymer, processing aid, thickener for detergents and cleaners, water treatment, mineral processing, other industrial markets. Alkali-sol.; mw ca. 500,000; d = 1.06, 8.75 lb/gal; viscosity = 20cps; pH 2.7; 30% solids; LD_{50} (rat orl) >5 g/kg; moderately irritating to eyes, skin. *Rohm & Haas.*

592 **Acusol® 830**
151066-66-5
Acrylates copolymer
Acrilic crosslinked copolymer. Detergent polymer, thickener and stabilizer for detergents, floor polishes, window cleaners, solv.-based cleaners, paint strippers, lubricant emulsions, car polishes, antifreeze, water treatment, mineral processing, other industrial markets. Alkali-soluble; d = 1.054; viscosity = 10 cps; pH3; 28% solids; anionic. *Rohm & Haas.*

593 **Acusol® 840**
Acrylic copolymer
Detergent polymer, processing aid, thickener for detergents and cleaners, water treatment, mineral processing, other industrial markets. *Rohm & Haas.*

594 **Acusol® 860N**
9003-04-7
Sodium polyacrylate
Detergent polymer, processing aid for detergents, water treatment, mineral processing, other industrial markets. *Rohm & Haas.*

595 **Acylan**
61788-48-5 262-979-2
Acetylated lanolin
Lipid emollient for personal care and pharmaceutical products; forms water-repellent films. Soft solid; bland odor; soluble in mineral oil and soft waxy hydrophobic films; mp = 32-39°; acid no. 2.0 max.; iodine no. 30 max; sapon. no. 100-125; hyd. No. 12 max; 100% act. *Croda Chem. Ltd; Croda Inc.*

596 **Acylglutamate AS-12**
Sodium oleoyl glutamate - sodium cocoyl glutamate. Good detergency for liquid dishwashing detergents, textile detergents, etc.; biodegradable. *Ajinomoto.*

597 **Acylglutamate CS-11**
68187-32-6 269-087-2
Sodium cocoyl glutamate
Amisoft CS-11. Detergent, emollient for personal care products; bacteriostat for facial and body cleansers, shampoo, bar soap, children's and dermatological products Anionic; biodegradable; nonirritating; powder/flakes; pH 5.5 (1%, 40°); 100% concentrate. *Ajinomoto; Ajinomoto USA.*

598 **Acylglutamate CS-21**
Disodium cocoyl glutamate
Detergent, emollient for personal care products; bacteriostatic effect; corrosion inhibitor; biodegradable. *Ajinomoto.*

599 **Acylglutamate CT-12**
68187-29-1 269-084-6
TEA-cocoyl-glutamate
Amisoft CT-12. Detergent, emollient for personal care products; emulsifier for cosmetics, facial and body cleansers, shampoo, bar soap, children's and dermatological products, bacteriostatic effect; biodegradable. pH 5.4; (1%, 40°); *Ajinomoto; Ajinomoto USA.*

600 **Acylglutamate DL-12**
Monolithium salt of N-distilled cocoyl-L-glutamic acid
Foamer for carpet cleansers; easily becomes a powder after drying. *Ajinomoto.*

601 **Acylglutamate GS-11**
Sodium hydrogenated tallow glutamate, sodium cocoyl glutamate
Amisoft GS-11. Detergent, emollient for personal care products; bacteriostatic effect; emulsifier for cosmetics, facial and body cleaners, shampoos, bar soap, children's and dermatological products; biodegradable. Powder/flake; pH 6.6 (1%, 40°); 100% active. *Ajinomoto.*

602 **Acylglutamate GS-21**
Disodium cocoyl/tallowyl glutamate
Basic material for heavy-duty detergents; reduces adverse reactions on human skin in toiletries; capturing agent of heavy metal ions; biodegradable. *Ajinomoto.*

603 **Acylglutamate HS-11**
38517-23-6 253-980-9
Sodium hydrogenated tallow glutamate
Detergent, emollient for personal care products; bacteriostatic effect; biodegradable. pH 6.9 (1%, 40°); 100% concentrate. *Ajinomoto.*

604 **Acylglutamate HS-21**
Disodium stearoyl glutamate
Basic material for heavy-duty detergents; reduces adverse reactions on human skin in toiletries; capturing agent of heavy metal ions; biodegradable. *Ajinomoto.* Name unverified.

605 **Acylglutamate HS-21**
Disodium stearoyl glutamate
basic material for heavy-duty detergents; reduces adverse reactions on human skin in toiletries; capturing agent of heavy metal ions; biodegradable. pH 9.0 (1%, 40°); LD_{50} (mus orl) = 3.5 g/kg. *Ajinomoto.*

606 **Acylglutamate LT-12**
258-636-1
TEA lauroyl glutamate
Acylglutamate LT-12. Detergent, emollient for personal care products; bacteriostatic effect; biodegradable pH 5.2 (1%, 40°); 30% aqueous solution. *Ajinomoto.* Name unverified.

607 **Acylglutamate MS-11**
253-981-4
Sodium myristoyl glutamate
Acylglutamate MS-11. Detergent, emollient for personal care products; bacteriostatic effect; biodegradable. pH 6.1 (1%, 40°); 100% concentrate; anionic. *Ajinomoto.* Name unverified.

608 **AD-700**
Surfactant/coupler/alkaline blend; biodegradable steam cleaner. *Anedco.*

609 **AD-709**
Alkyl sulfonate
Paraffin dispersant. *Anedco.*

610 **AD-710**
Terpenes/amides/surfactants/solvent blend; paraffin dispersant to prevent or remove paraffin. *Anedco.*

611 **AD-713C**
Diamines/amides/surfactant blend; conc. paraffin dispersant to remove or prevent paraffin. *Anedco.*

612 **AD-742C**
Surfactant/coupler blend; biodegradable heavy-duty degreaser. *Anedco.*

613 **AD-749**
Ethoxylated nonyl phenol
Ethoxylated nonyl phenol and surfactant; drilling detergent. Soluble in H_2O. *Anedco.*

614 **AD-763**
Pine oil, surfactants, couplers; for pine oil cleaners. *Anedco.*

615 **ADA**
542-05-2 66 208-797-9
$C_5H_6O_5$
1,3-Acetonedicarboxylic Acid
Acetone Dicarboxylic Acid; β-ketoglutaric Acid; 3-Ketoglutaric acid; 3-Oxoglutaric acid; 3-Oxopentanedioic acid; Acetone-1,3-dicarboxylic acid. Used in organic synthesis, manufacturing. mp = 133°; very soluble in H_2O, EtOH, less soluble in organic solvents.

616 Adal
Alloying elements into aluminum alloys. *Foseco (F.S.) Ltd.*

617 Adalin
$[CH_2CH_2]_x$
Polyethylene emulsion nonionic
Softener for resin finishing; improves sewability, abrasion resistance, crease angles and softness. *Henkel Chemicals Ltd.* Name unverified.

618 Adalox
A proprietary trade name for coated abrasives for sanding metal or plastics. No manufacturer.

619 Ad-aluminum
A proprietary trade name for an alloy of 82% copper, 15% zinc, 2% aluminum, and 1% tin. No manufacturer.

620 Adamac
Specially prepared tar for roads. *Thomas Ness Ltd.* Name unverified.

621 Adamant
1344-28-1 369 215-691-6
Al_2O_3
Aluminum Oxide
Polish and abrasive; absorbent, desiccant; filler for paints and varnishes; ceramic materials, electronics and resistors; dental cements; glass, artificial gems; coating for metals. No manufacturer.

622 Adame
2439-35-2 219-460-0
$C_7H_{13}NO_2$
N,N-Dimethylaminoethyl acrylate
Dimethylaminoethyl acrylate; 2-dimethylaminoethyl acrylate;. Surfactant. Insoluble in H_2O, soluble in organic solvents. *Rhône-Poulenc Surf.*

623 Adamite
Carbon nickel-chromium iron alloy
A proprietary high-carbon nickel-chromium iron alloy; used for dies. No manufacturer.

624 Adamsite
578-94-9 7357 209-433-1
$C_{12}H_9AsClN$
10-Chloro-5,10-dihydrophenarsazine
Phenarsazine Chloride; Chloro-5,10-dihydrophenarsazine; DM; 5-aza-10-arsenaanthracene chloride; 10-chloro-5,10-dihydroarsacridine. A poison gas, war gas; also used in the formulation of wood-treating solutions against marine borers and similar pests. mp = 195°; bp = 410°; insoluble in H_2O, slightly soluble in organic solvents; highly toxic. No manufacturer.

625 Adansonia Fiber
A fiber obtained from the bark of *Adansonia digitata* used for making rope and sacking, and also for special paper. No manufacturer.

626 Adaphax 758
A proprietary factice containing neither sulfur nor chlorine used as a processing aid in the manufacture of polyurethanes and PVC. Enables good electrical properties to be maintained in PVC and tackiness to be reduced; also used in white nitrile mixes. *Hubron Rubber Chemicals.* Name unverified.

627 Adaptinol
Xanthophyl dipalmitate
An antidazzle preparation. *Bayer AG.*

628 Adarola
A proprietary casein plastic. No manufacturer.

629 ADC
Epoxy formulations used in tooling, adhesives, electronic encapsulates, coatings, sealants and ablatives; ultra high temperature laminating resins, fast setting high temperature adhesives, pinhole-free RT laminating resin (requires no vacuum bag) chemically resistant and thermally conductive coatings, encapsulants, sealants, binders, adhesives, specialty coatings for food and chemical industries. *ADC Resins.* Name unverified.

630 Adcora A3
A hard coal tar epoxy coating for tanks and pipes for effluent and for resistance to fumes and weathering. No manufacturer.

631 Adcora P6
C_4H_5Cl
Polychloroprene
Neoprene; Chloroprene, Chloroprene rubber. A proprietary trade name for a neoprene coating with good chemical and abrasion resistance and good flexibility. LD_{50} (rat orl) = 260 mg/kg. *Neoprene.* No manufacturer.

632 Adcora SP
A solventless epoxide-based coating giving a hard finish resembling enamel. No manufacturer.

633 Adcora V
Fluoroelastomer
Viton™ A-35; Viton™ B-50; Viton™ A-35. A Viton (fluoroelastomer) coating with very good heat stability resistant to temperatures up to 260°C. No manufacturer.

634 Adcortyl with Graneodin Cream
Triamcinolone acetonide, neomycin sulfate and gramicidin in cream base. *Bristol-Myers Squibb Pharmaceuticals Ltd.*

635 Adcote
Adhesives and primers for flexible packaging. *Morton Int'l Ltd.*

636 Add It To Oil
Alkylaryl poly (Ethylene oxy)
Natural oil emulsifier for use as an additive for conditioning oil for crop spray use. *Doyle Specialties.* Name unverified.

637 Addabond
Substance for bonding new concrete to old; particularly useful where it is impossible to dry out the substrate. *Addagrip Surface Treatments UK Ltd.* Name unverified.

638 Addacoat
Two part pack; colored material, giving a tile-like finish; applied to walls where cleaning is of paramount importance; used in damp proofing of cellars and basements. *Addagrip Surface Treatments UK Ltd.* Name unverified.

639 Addacol
Organic and inorganic pigments; used for coloring cement, concrete, bricks, etc. *Calder Colours (Ashby) Ltd.*

640 Addaflex
Flexible epoxy compound
For filling cracks and expansion joints in concrete floors. *Addagrip Surface Treatments UK Ltd.* Name unverified.

641 Addaflor
Epoxy top coat
Solvent free colored epoxy top coat; applied to concrete, brick, wood and steel. Oil, acid and chemical resistant; strong enough to hold antislip chippings; easily cleaned. *Addagrip Surface Treatments UK Ltd.* Name unverified.

642 Addagrout
Three-part pack epoxy-based grouting material
Used for bedding machinery, filling anchor points and repairing cracks and depressions in floors. Oil, acid and chemical resistant; strong enough to hold antislip chippings; easily cleaned. *Addagrip Surface Treatments UK Ltd.* Name unverified.

643 Addalevel
Leveling compound for laying between 3 and 6 mm thick; applied to concrete and brick. *Addagrip Surface Treatments UK Ltd.* Name unverified.

644 Addamortar
Epoxy mortar
Three part pack fillers, hardener and resin for filling cracks, holes and undulations, and for screeds; can be power floated. *Addagrip Surface Treatments UK Ltd.* Name unverified.

645 Addapitch
Pitch epoxy material
Pitch epoxy material; suitable for waterproofing and also for holding antislip chipping on ramps and loading bays etc. *Addagrip Surface Treatments UK Ltd.* Name unverified.

646 Addaprime
Two part pack; resin and hardener; solvent-free deep-penetrating epoxy primer. Can also be used as a coating; strong enough to hold antislip chipping; applied to concrete, brick, wood and steel. *Addagrip Surface Treatments UK Ltd.* Name unverified.

647 Addaseal
Polyurethane material of high quality (50% solids); clear or colored; for dust sealing concrete floors; strong enough to hold small particles of antislip aggregate. *Addagrip Surface Treatments UK Ltd.* Name unverified.

648 Addasure
Solvent based epoxy coating; two part pack; used for coating walls and floors. *Addagrip Surface Treatments UK Ltd.* Name unverified.

649 Adder
Adjuvant containing 97% refined mineral oil; wetting agent for herbicides. *Embetec Crop Protection Ltd.*

650 Adder
An asbestos fabric grade of Tufnol industrial laminates. *Tufnol Ltd.*

651 Add-H
Liquid hay additive. *BP Chemicals Ltd.*

652 Addipast
1333-86-4 1856 215-609-9
carbon
Carbon black pigment pastes. *Brockhues AG.*

653 Additin 30
90-30-2 201-983-0
$C_{16}H_{13}N$

N-phenyl-1-naphthylamine
Acetopan; C.I. 44050; Neozone A; phenylnaphthylamine;. Antioxidants for the mineral oil industries; gum inhibitor.; staining antioxidant for rubber technical goods and heavily stressed goods; antiflexing agent for NR and IR; storage stabilizer for petroleum products. mp = 60-62°; bp_{15} = 226°; d = 1.11. *Miles/Polysar Rubber.*

654 Additive-A
8061-52-7
Modified calcium lignosulfonates. Clay conditioners for production of bricks and tiles. Compounds act as plasticizers, lubricants, binders and antiscumming agents. *Miles/Polysar Rubber.*

655 Additol
Range of additives to improve properties such as flow and eliminate film defects etc; used in paints and printing inks. *Hoechst UK.*

656 Additol
Range of additives to improve properties such as flow and eliminate film defects, etc.; used for paints and printing inks. *Resinous Chemicals Ltd.*

657 Add-M
Liquid antimolding compound. *BP Chemicals Ltd.* Discontinued.

658 Adeka Catioace DM, PD Series
Tetraamonium salt type polymer
Tetraamonium salt type polymer; surfactant, antistat for electrification, low foaming, coating applications. *Asahi Denka Kogyo.*

659 Adeka CR-5
Chlorinated rubber; used in printing inks, overprint varnish and paint applications. *Asahi Denka Kogyo.*

660 Adeka ED-505
TGE of trimethylol propane
Adeka Glycilol ED-505. Epoxy resin diluent. d = 1.16; viscosity = 125-200 cps. *Asahi Denka Kogyo.*

661 Adeka EP-4100
Liquid epoxy resin for general applications. *Asahi Denka Kogyo.*

662 Adeka Estol
Fatty ester surfactants; for emulsification, solubilization and dispersing. *Asahi Denka Kogyo.*

663 Adeka GH-200
51258-15-2
PPG-24-glycereth-24
Surfactant. *Asahi Denka Kogyo.* Name unverified.

664 Adeka Hypote
7681-52-9 8773 231-668-3
ClNaO
sodium hypochlorite
Used to bleach fiber, pulp and paper and for the production of bleaching chemicals for use with foods and in water treatment. *Asahi Denka Kogyo.*

665 Adeka Kiku-Lube
Pour point depressants and viscosity index improvers. *Asahi Denka Kogyo.*

666 Adeka Lub E-500
Chlorinated N-paraffin
Extreme pressure agent for metalworking, especially formulation of cutting oil, grinding oil; plasticizer for PVC. Viscous liquid; oil soluble; 50% chlorine. *Asahi Denka Kogyo.*

667 Adeka Optomer KR Series
One-component epoxy resin, UV curing
Primers for materials such as metal/plastic/paper, finishing varnishes, protective varnishes, inks, adhesives, insulating coatings, conductive coatings. *Asahi Denka Kogyo.*

668 Adeka P-400
Polyoxypropylene glycol
Diol for polyurethane production. *Asahi Denka Kogyo.*

669 Adeka PR-3007
Polyether polyol, water-soluble
For water absorbing polyurethane flexible foams, adhesives. Soluble in H_2O. *Asahi Denka Kogyo.*

670 Adeka Sakura-Lube 100
Organo molybdenum additive
Oil-soluble organo-molybdenum additive; for gasoline engine oil, diesel engine oil, gear oil, industrial lubricating oil, greases. *Asahi Denka Kogyo.*

671 Adeka Sole CO
68603-42-9 271-657-0
cocamide-DEA (1:2)
Surfactant providing detergency, stable foaming, antirust; dispersability and thickening properties with only mild irritation. *Asahi Denka Kogyo.*

672 Adeka Sole YA
68140-00-1 268-770-2
cocamide-MEA
Surfactant providing detergency, stable foaming, antirust; dispersability and thickening properties with only mild irritation. *Asahi Denka Kogyo.*

673 Adekacol CS, PS, TS
Aromatic phosphate ester type surfactants
Phosphate ester type surfactants; for hair care products; Stable to acid/alkali at high temperatures; anticorrosive. *Asahi Denka Kogyo.*

674 Adekacol EC Series
Dialkyl sulfosuccinate ester
Emulsifier, detergent; wetting agent; emulsifier dispersant for hair care products. Transparent liquid; 30-70%solids. *Asahi Denka Kogyo.*

675 Adekamine E Series
Mono-or di-alkyl tetraamonium salts
Cationic surfactant, softener, antistat, emulsifier. *Asahi Denka Kogyo.*

676 Adekanol
EO-PO block copolymer
For detergency, emulsification, dispersing and wetting. *Asahi Denka Kogyo.*

677 Adekatol DES, DS, HAN, LS, TR, SAN, YES
Sulfate type anionic surfactants. *Asahi Denka Kogyo.*

678 Adekatol LA, LO, NP, OA, PC, SO Series
Ethoxylated nonionic surfactants. *Asahi Denka Kogyo.*

679 Adenine
73-24-5 150 200-796-1
$C_5H_5N_5$
1H-purin-6-amine
6-aminopurine; 6-amino-9H-purine; 1,6-dihydro-6-iminopurine; 3,6-dihydro-6-iminopurine; Leuco-4. Used in medical and biochemical research. mp = 360-365° (dec); λ_m = 207, 260,5 nm (ε 23200, 13400 pH 7.0); soluble in H_2O (0.05 g/100 ml), less soluble in organic solvents; LD_{50} (rat orl) = 745 mg/kg. *Aldrich; Lonza AG; Penta Mfg.; Schweizerhall; U.S. Biochemical.*

680 adenosine monophosphate
61-19-8 157 200-500-0
$C_{10}H_{14}N_5O_7P$
adenosine-5'-monophosphate
5'-adenylic acid; AMP; muscle adenylic acid; ergadenylic acid; t-adenylic acid; adenosine-5'-phosphoric acid; A-5MP; NSC-20264; Lycedan; Myoston. Nutrient. mp = 200° (dec); $[\alpha]_D^{20}$= -47.5° (c = 2 2% NaOH); soluble in H_2O.

681 Adenosine triphosphate
56-65-5 154 200-283-2
$C_{10}H_{16}N_5OP_3$
ATP
Adenosine, 5·g-(tetrahydrogen) triphosphate; Adetol; Atriphos; Striadine; Triadenyl. Organic compound used in biochemical research. Used to inhibit enzymatic browning of raw edible plant materials, such as sliced potatoes, apples, etc. Soluble in H_2O. *Asahi Chemical Industry Co Ltd; R. W. Greef; Penta Mfg.*

682 Adenotriphos
A proprietary preparation of adenosine triphosphoric acid sodium. *Rona Laboratories.* Unverified.

683 5'-adenylic Acid
61-19-8 157 200-500-0
$C_{10}H_{14}N_5O_7P$
Adenosine phosphate
Adeno; Adenosine 5'-(dihydrogen phosphate); My-B-Den; Lycedan; AMP; Phosaden; Myoston. Nutrient mp = 196-200°; readily soluble in boiling H_2O. Name unverified.

684 Adequan
Polysulfated glycosaminoglycan
For intra-articular injection; for treatment of lameness in horses. *Luitpold-Werk.* Name unverified.

685 ADF-600
Alcohol-based general purpose defoamer for water-based drilling muds. *Anedco.*

686 ADF-610
Silicone-based; defoamer. *Anedco.*

687 Adflex
Compounded polymeric emulsion
Flexing agent for cement based adhesives. *Howlett Adhesives Ltd.* Name unverified.

688 Adheso
A proprietary trade name for a synthetic wax consisting of a modified polymerized terpene. No manufacturer.

689 Adimoll® BO
Benzyloctyl adipate
Monomeric plasticizer used for PVC articles with resistance to low temperatures; calendering, coating, extrusion, injection molding; film, expanded imitation leather; useful in VC copolymers, NC, ethyl cellulose, PS, natural, S/B, N/B, chlorinated, butyl rubber. d = 1.002-1.008; viscosity = 16-17 mPa. S; hardness (Shore) 26; acid no. <0.1. *Bayer AG.*

690 Adimoll® DB
105-99-7 203-350-4
dibutyl adipate
A monomeric plasticizer. *Bayer AG.*

691 Adimoll® DH
110-33-8 203-757-7
$C_{18}H_{34}O_4$
Di-n-hexyl adipate
Monomeric plasticizer. *Bayer AG.*

692 Adimoll® DN
33703-08-1 251-646-7
$C_{24}H_{46}O_4$
diisononyl adipate
Hexanedioic acid, diisononyl ester; Diisononyl adipate. A monomeric plasticizer. *Bayer AG.*

693 Adimoll® DO
103-23-1 203-090-1
Dioctyl adipate
Plasticizer for PVC, applications requiring good low-temperature resistance; rubber plasticizer imparting high elasticity and low-temperature flexibility in conjunction with minimum volatility. d = 0.923-0.926; viscosity = 12-14 mPa.s; acid no (Shore D) 26. *Bayer AG.*

694 Adinol
A range of taurates. Used as anionic surfactants, foaming agents in the cosmetic industry. *Croda Chem. Ltd.*

695 Adinol
Textile auxiliaries based on triethyl citrate. *Fine Dyestuffs & Chemicals Ltd.* Name unverified.

696 Adinol CT
12765-39-8 235-802-1
Sodium N-Cocoyl-N-Methyl Taurate
Sodium Methyl Cocoyl Taurate. Sodium methyl cocoyl taurate in white powder form. An anionic surfactant with high foaming and cleansing capacity; chemically stable, lime soap dispersant, used in cosmetics, toiletries and pharmaceutical preparations. *Croda Chem. Ltd; Croda Inc.*

697 Adinol OT
137-20-2 205-285-7
$C_{21}H_{40}NNaO_4S$
Sodium N-methyl N-oleoyl taurate
(Z)-2-[methyl(1-oxo-9-octadecenyl)amino]ethanesulfonic acid, sodium salt; Sodium N-methyl-N-oleoyltaurate; Sodium methyl oleyl taurate. Sodium N-methyl N-oleoyl taurate in powder, past, gel or liquid form. A biodegradable anionic surfactant with detergent, wetting, emulsifying and foaming properties. Used in the textile and dyeing industries and in the manufacture of leather and paper. *Croda Chem. Ltd.*

698 Adipic acid
124-04-9 161 204-673-3
$C_6H_{10}O_4$
Dicarboxylic acid C_6
Hexanedioic acid; Adi-pure®;. Manufacture of nylon and polyurethane foams; preparation of esters for use as plasticizers and lubricants; food additive (acidulant) for baking powders; flavoring agent; leavening agent, neutralizer, acidulant, pH control agent for baked goods, beverages, mp = 152°; bp = 337°; very soluble in EtOH, Me_2CO; d_4^{25}= 1.360; LD_{50} (mus orl) = 1900 mg/kg. *AlliedSignal; Asahi Chem Industry Co Ltd; Du Pont; Monsanto; Penta MFG.; Rhône-Poulenc; UCB SA.*

699 Adipocere
A wax-like mass left when animal bodies decompose in the earth. It consists of the fatty acid salts of calcium and potassium.

700 Adipon
Fatty alcohol sulfate, anionic
Detergent for scouring and milling of worsted fabrics; softening effect. *Henkel Chemicals Ltd.* Name unverified.

701 Adiprene®
Adiprene® BL-16. A range of polyether-based prepolymers which offer high abrasion resistance, chemical resistance and electrical properties. *Uniroyal.*

702 Adiprene® L-100
Urethane rubber
For products of high hardness, load-bearing capacity, and resistance to abrasion, oils, oxidation, ozone. *Uniroyal.*

703 Adiprene® L-42
Urethane rubber
For products with high flexibility at low temperatures. *Uniroyal.*

704 Adiprene® LW-570
Urethane rubber
For products of extreme hardness and high impact resistance, load-bearing capacity, hydrolytic stability; curable with MDA. *Uniroyal.*

705 Adiprene® M-400
Urethane rubber, non-MBCA
Offers improved processing. *Uniroyal.*

706 Adi-pure®
124-04-9 161 204-673-3
$C_6H_{10}O_4$
adipic acid
High purity adipic acid. For chemical industry, chemical intermediate; used in adhesives, coatings, nylon 66, PU foams/elastomers/fibers, lubricants, textile treatments, cosmetic emollients; pH buffer; food acidulant.

707 Adirondackite
A rubber substitute made from sulfurized oils. Used in the proofing of cloth, and as an insulator. No manufacturer.

708 Adjunct B
7558-79-4 8805 231-448-7
Disodium phosphate
Deposit inhibitor; precipitates soluble calcium hardness from boiler water to prevent scale deposits. *Drew Ind. Div.*

709 Adjust 4
Thixotrope that prevents settling by pigment wetting and controlled flocculation without appreciably increasing apparent viscosity; also improves sag resistance as post-additive. *United Catalysts.*

710 ADK CIZER O-130P
8013-07-8 232-391-0
Soybean oil epoxide
Epoxidized linseed oil; Stabilizer providing higher heat weathering stability than ADK CIZER O-130P; suitable for rigid calendering sheet and rigid bottle formulation. *Asahi Denka Kogyo.*

711 ADK CIZER O-180A
Epoxidized linseed Oil
Stabilizer providing higher heat weathering stability than O-130P; suitable for rigid calendering sheet and rigid bottle formulation. *Asahi Denka Kogyo.*

712 ADK STAB 144
Zinc complex; PVC stabilizer for homogeneous tile flooring; provides excellent heat stability , improves initial color; effective blended with phosphite. *Asahi Denka Kogyo.*

713 ADK STAB 465
51258-15-2
di-n-octyltin mercaptide
PVC stabilizer providing excellent heat stability, color and transparency; suitable for rigid calendering sheet and blow formulation. *Asahi Denka Kogyo.*

714 ADK STAB 466
Octyltin mercaptide
PVC stabilizer providing excellent heat stability, color and transparency; suitable for rigid calendering sheet formulation. *Asahi Denka Kogyo.*

715 ADK STAB 1292
Dibutyltin mercaptide
PVC stabilizer for rigid transparent PVC with highest heat stability; applicable to rigid water pipe requiring NSF approval for high resistance to water extraction. *Asahi Denka Kogyo.*

716 ADK STAB 1413
119-61-9 1129 204-337-6
$C_{13}H_{10}O$
Benzophenone
UV absorber for polyolefins, PVC, etc.; good compatible with polymers. mp = 48°. *Asahi Denka Kogyo.*

717 ADK STAB 1500
Special phosphite
Provides excellent heat, color and weathering stability; for rigid calendering sheet and blow bottle. Viscosity = 10000 cps. *Asahi Denka Kogyo.*

718 ADK STAB 2335
12513-27-8
Zinc borate
Flame retardant, smoke preventer for flooring and cable formulations. *Asahi Denka Kogyo.*

719 ADK STAB AC-122
Ba-Zn
General purpose PVC stabilizer providing good heat stability and color resist. Liquid. *Asahi Denka Kogyo.*

720 ADK STAB AC-133
Ba-Cd-Zn
PVC stabilizer providing excellent initial color, transparency and heat stability; for calendering sheet, leather and plasticized PVC compound by extrusion and injection molding processes. Liquid. *Asahi Denka Kogyo.*

721 ADK STAB AC-169
Ba-Zn
One-package PVC stabilizer providing excellent heat stability and

transparency; for calendar sheets of plasticized PVC. Liquid. *Asahi Denka Kogyo.*

722 ADK STAB AP-536
Ba-Zn
General purpose PVC stabilizer providing good heat stability, color retention. *Asahi Denka Kogyo.*

723 ADK STAB BT-11
77-58-7 3089 201-039-8
dibutyltin dilaurate
PVC stabilizer providing good heat and weathering resistance; improves processability in rigid transparent formulation. Also used as an anthelmintic, against tapeworms in chickens. *Asahi Denka Kogyo.*

724 ADK STAB BT-31
15535-69-0
Dibutyltin maleate
PVC stabilizer providing excellent initial color transparency; for whole rigid PVC products, extrusion, press and blow molding. *Asahi Denka Kogyo.*

725 ADK STAB BT-83
Dibutyltin mercaptide
PVC stabilizer with excellent heat resist. and color stability; for injection molding and rigid PVC extrusion (T-die, contour). Powder. *Asahi Denka Kogyo.*

726 ADK STAB EC-14
Zn-phosphite
Improves initial color with Ba-Zn stabilizers in formulations containing TiO_2 and $CaCO_3$; suitable for plasticized and rigid calendering formulation and paste formulation. *Asahi Denka Kogyo.*

727 ADK STAB FL-21
Na-Zn
Stabilizer for foam PVC at low expansion. Liquid. *Asahi Denka Kogyo.*

728 ADK STAB GR-16
Ca-Zn
PVC stabilizer providing excellent long term heat stability; for extrusion and injection molding by filler loading formulation; effective with ADK STAB 1500. Powder. *Asahi Denka Kogyo.*

729 ADK STAB LA-32
95-14-7 1140 202-394-1
1H-Benzotriazole
UV absorber for PVC, ABS, PS, PU, NMA, etc. mp = 130°. *Asahi Denka Kogyo.*

730 ADK STAB LA-57
Hindered amine. Light stabilizer for polyolefins, PVC, ABS, etc.; superior light stability. mp = 132°. *Asahi Denka Kogyo.*

731 ADK STAB LS-2
15535-69-0
dibutyltin maleate
Lubricant and stabilizer for PVC; provides excellent transparency and gives good properties with tin maleate stabilizer; for extrusion and blow molding. *Asahi Denka Kogyo.*

732 ADK STAB LS-8
123-95-5 1625 204-666-5
$C_{22}H_{44}O_2$
octadecanoic acid butyl ester
butyl stearate. Plastics lubricant. *Asahi Denka Kogyo.*

733 ADK STAB NA-11
85209-91-2 286-344-4
sodium 2,2·g-methylene bis-(4,6-di-*t*-butylphenyl) phosphate
Nucleating agent which upgrades heat deflection temperature, flexibility modulus and impact strength of PP, PET, PBT and polyamides. Gives high transparency at low concentrations and raises crystallization temperatures. *Asahi Denka Kogyo.*

734 ADK STAB OF-14
Ba-Zn
General purpose foam PVC stabilizer for expanded leathercloth and wall paper. *Asahi Denka Kogyo.*

735 ADK STAB OT-1
Di-n-octyltin dilaurate
PVC stabilizer providing excellent external lubricity, heat and process stability with Ca-Zn stabilizers; suitable for rigid sheet and blow formulation. Liquid. *Asahi Denka Kogyo.*

736 ADK STAB OT-9
Di-n-octyltin maleate
PVC stabilizer providing excellent heat stability and weatherability; suitable for rigid extrusion and calendering formulation. Liquid. *Asahi Denka Kogyo.*

737 ADK STAB RUP-9
Ba-Zn
PVC stabilizer for electrical wire using polyester plasticizer; provides excellent heat stability and insulating properties. Powder. *Asahi Denka Kogyo.*

738 ADM
Ammonium dimolybdate
Corrosion inhibitor for vapor phase inhibitor programs. *Climax Performance.* Name unverified.

739 ADM-407, 407C
Alkyl benzyl sulfonic acid
Demulsifier. Anionic. *Anedco.*

740 Adma®
Alkyl dimethyl amines.
Intermediates, chemical and cosmetics products. *Ethyl Corp.*

741 Adma® 8, 10, 12
230-939-3; 214-302-7; 203-943-8
octyldimethylamine [7378-99-6], decyldimethylamine [1120-24-7] and dodecyl dimethylamine [112-18-5]. Intermediate for quaternary ammonium compounds, amine oxides, betaines. *Ethyl Corp.*

742 Adma® 14
112-75-4 204-002-4
$C_{16}H_{35}N$
dimethyl(tetradecyl)amine
Chemical intermediate. *Asahi Denka Kogyo.*

743 Adma® 16
112-69-6 203-997-2
$C_{18}H_{39}N$
N,N-dimethyl-1-Hexadecanamine
Armeen DM 16D; Dimethyl-1-hexadecanamine; N,N-Dimethylhexadecyl amine. Chemical intermediate. *Asahi Denka Kogyo.*

744 Adma® 18
124-28-7 3525 204-694-8
Octadecyl dimethylamine
Dymanthine. Intermediate for manufacturing of quaternary ammonium compounds for biocides, textile and oilfield chemicals, amine oxides, betaines, polyurethane foam catalysts, epoxy curing agents. *Ethyl Corp.*

745 Adma® 246-451
Dodecyl dimethylamine (40%), tetradecyl dimethylamine (50%), hexadecyl dimethylamine (10%)
Intermediate for quaternary ammonium compounds, amine oxides, betaines. Clear Liquid; fatty amine odor; corrosive; f.p. = -13°; amine no. 238; flash pt. (PM) 114°. *Ethyl Corp.*

746 Adma® 1214
dodecyldimethylamine-tetradecyldimethylamine (65:35)
Intermediate for quaternary ammonium compounds, amine oxides. *Asahi Denka Kogyo.*

747 Adma® WC
Octyl dimethylamine (7%), decyl dimethylamine (6%), dodecyl dimethylamine (53%), tetradecyl dimethylamine (19%), hexadecyl dimethylamine (9%), octadecyl dimethylamine (6%)
Intermediate for quaternary ammonium compounds, amine oxides, betaines. Clear Liquid; fatty amine odor; amine no. 249; flash pt. (PM) 102°; f.p. -22°. *Ethyl Corp.*

748 Admerol®
Six common metals and their alloys. Metal building materials; transportable buildings; railway tracks; nonelectric cables and wires; small hardware items; pipes and tubes; safes. *Reichhold.*

749 Admerol® 75-M-70
Varnish, modified oil. Gardner 11 max color; dens. 8.21 lb/gal (solids); viscosity = (G-R)R-V. *Reichhold.*

750 Admex®
Polymeric plasticizers
For plasticizing PVC and other polymers. *Hüls Am.* Discontinued.

751 Admiral® FPS Type 3089 FS
High solids fluidized polymer suspension
Viscosifier for paper coatings; thickener; Tan liquid.; d = 9.6 lb/gal; pH 9.6; 25% active solids.

752 Admiralty brass
Cu-Ca-Zn-Sn
Alloy of copper, calcium, zinc and tin.

753 Admiralty gun metal
Alloys. Some contain from 87-90% copper and from 10-13% tin, while others consist of from 86-88% copper, 6-10% tin, and 2-6% zinc.

754 Admiralty white metal
A bearing alloy containing 86% tin, 8.5% antimony, and 5.5% copper.

755 Admire
105827-78-9 4946
Imidacloprid
Insecticidal seed dressing.

756 Admos Alloys
Brass alloys of varying composition containing small amounts of tin, nickel, lead, and iron.

757 Admox®
ethyl dimethylamine oxides
Ethyl Corp.

758 Admox® 14-85
3332-27-2 222-059-3
Myristamine oxide
Used in soap bars, shaving creams, fabric softeners, hard surface cleaners, laundry detergents, oxygen bleach powders, toothpaste, agriculture, automatic dishwashers, cellulose extraction, gasoline additives and bubble baths. *Ethyl Corp.*

759 Admox® 18-85
2571-88-2 219-919-5
Stearamine oxide
Used in soap bars, shaving creams, fabric softeners, hard surface cleaners, laundry detergents, oxygen bleach powders, toothpaste, agriculture, automatic dishwashers, cellulose extraction, gasoline additives and bubble baths. *Ethyl Corp.*

760 Admox® 1214
Alkyldimethylamine oxides
High foaming material to improve foam profile of anionic surfactants; viscositymodifier, emollient. Liquid; nonionic; 30% concentrate. *Ethyl Corp.*

761 Admul
Mono- and di-glycerides and their acid derivatives. *Quest Int'l.*

762 Adnic
Admiralty nickel
An alloy containing 70% copper, 29% nickel, and 1% tin; it is resistant to corrosion and heat.

763 Adogen® 66
Alkoxylated quaternary
Nylon retarding agent; antistat for fabric, fiber, and yarn finish formulations. *Sherex/Div. of Witco.*

764 Adogen® 137
Dihydrogenated tallow dimethyl ammonium methyl sulfate
Very high melting textile softener, antistat; nonyellowing; minimizes metal corrosion. *Sherex/Div. of Witco.*

765 Adogen® 170
61790-33-8 263-125-1
tallow succinamate
Tallowamine, reacted with maleic anhydride to make tallow succinamate, a high-foaming surfactant used to froth latex carpet backing. *Sherex/Div. of Witco.*

766 Adogen® 185
C_{12-15} ether amine
Reacted with maleic anhydride to make tallow succinamate, a high-foaming surfactant used to froth latex carpet backing. Liquid; 100% solids. *Sherex/Div. of Witco.*

767 Adogen® 412
112-00-5 203-927-0
$C_{15}H_{34}ClN$
Lauryl trimethyl ammonium chloride
Dodecyltrimethylammonium chloride; N,N,N-trimethyl-1-dodecanaminium chloride; Dodecanaminium, N,N,N-trimethyl-, chloride; Trimethyl-1-dodecanaminium chloride; n-Dodecyl trimethylammonium chloride; Lauryl Trimethyl Ammonium Chloride; Laurtrimonium Chloride. Softener for textile, laundry, paper, etc. mp = 235°. *Sherex/Div. of Witco.*

768 Adogen® 417
61790-41-8 263-134-0
soya trimethylammonium chloride
Retarding and leveling agent for textiles. *Sherex/Div. of Witco.*

769 Adogen® 442
61789-80-8 263-090-2
Quaternium-18
Fabric softener concentrate for home and commercial laundries; textile processing. *Sherex/Div. of Witco.*

770 Adogen® 442-P100
61789-80-8 263-090-2
Dihydrogenated tallow dimethyl ammonium chloride
Textile softener producing very slick cationic hand, maximum softness. *Sherex/Div. of Witco.*

771 Adogen® 461
61789-18-2 263-038-9
cocotrimethylammonium chloride
cocotrimonium chloride. An emulsifier and dispersant. Used in corrosion inhibitor formulations for oil field brines and HCl acidizing systems; textile antistat.

772 Adogen® 462
61789-77-3 263-087-6
Dicocodimonium chloride
Arquad® 2C-75. Dicocodimonium chloride in iPrOH and H_2O; antistat, emulsifier, flocculating agent, dispersant used in corrosion inhibitor formulations for oil-field chemicals. *Sherex/Div. of Witco.*

773 Adogen® 470
272-207-6
Ditallowdimonium chloride
Speciality quaternary for nonionic laundry detergent-softeners. *Sherex/Div. of Witco.*

774 Adogen® 471
Tallow trimonium chloride, IPA
Dispersant, antistat, emulsifier, used in corrosion inhibitor formulations for oilfield brines and HCl acidizing systems; textile antistat. Gardner 6 maximum liquid.; m.w. 339; flash pt. 58F (PM). *Sherex/Div. of Witco.*

775 Adogen® 477
Tallow diamine diammonium dichloride
Emulsifier, dispersant; retardant, dyeing assistant. Liquid; 50% concentrate. *Sherex/Div. of Witco.*

776 Adogen® MA-108 SF
124-28-7 3525 204-694-8
$C_{20}H_{43}N$
dimethylstearamine
dymanthine. Neutralizer, conditioner and coemulsifier for personal care products. *Sherex/Div. of Witco.*

777 Adogen® MA-112 SF
dimethyl behenamine
Neutralizer, conditioner and coemulsifier for personal care products. *Sherex/Div. of Witco.*

778 Adogen® S-18 V
7651-02-7 231-609-1
stearamidopropyl dimethylamine
Conditioner, antistat, coemulsifier, plasticizer, neutralizer for personal care products. *Sherex/Div. of Witco.*

779 Adogen® TA-100
107-64-2 203-508-2
$C_{38}H_{80}ClN$
Distearyl dimonium chloride
Di(hardened tallow)dimethylammonium chloride; Distearyldimethylammonium chloride; Aerosurf TA-100; Dimethyl distearyl ammonium chloride; Dimethyl-n-octadecyl-1-octadecanaminium chloride; dimethyldioctadecylammonium chloride; Distearyldimonium Chloride. Textile softener producing very slick cationic hand and maximum softness. Good non-yellowing and high absorbency, can replace wax in sizing formulations. Used as a conditioner, anti-stat and softener. *Sherex/Div. of Witco.*

780 Adogen® TA-101
107-64-2 203-508-2
Distearyl dimonium chloride
Arosurf® TA-100; Arquad® 218-75; Arquad® 218-100; Arquad® 218-75; Blandofen CT; Dehyquart DAM; Adogen® TA-100; Genamin DSAC; Prepagen WK; Sumquat® 6045; Varisoft® TA-100. Textile softener producing very slick cationic hand, maximum softness; good nonyellowing, absorbency; can replace wax in sizing formulations; disperses at lower temperature than TA-100. *Sherex/Div. of Witco.*

781 Adogen® TA-101
107-64-2 203-508-2
Textile softener producing very slick cationic hand and maximum softness. Good non-yellowing and high absorbency, can replace wax in sizing formulations. Disperses at lower temperatures than TA-100. *Sherex/Div. of Witco.*

782 Adol® 52 NF
36653-82-4 2070 253-149-0
$C_{16}H_{34}O$
cetyl alcohol
Cetanol; 1-Hexadecanol; Ethal; Ethol; Palmityl alcohol; Hexadecan-1-ol; Hexadecyl alcohol; Hexadecanol; Alcohol, C16; Atalco C; Cachalot C-50; Cetaffine; Cetal; Cetylol; CO-1670; Crodacol-cas; DYTOL F-11; LorolL 24; Loxanol K; Product 308. Coemulsifier, lubricant, foam control agent, cosolvent; plasticizer, stabilizer, emollient, intermediate for metal lubricants, inks, textiles, emulsions, paper, cosmetics, mineral processing, oil field chemicals and fabric softeners.

783 Adol® 62 NF
112-92-5 8960 204-017-6
$C_{18}H_{38}O$
1-Octadecanol
Stearyl alcohol; Aldol 62; Alfol 18; Atalco S; Cachalot S 43; CO 1895F; Conol 1675; Conol 30F; Crodacol S; 1-Hydroxyoctadecane; Kalcohl 80; Lanol S;

Lorol 28; n-Octadecanol; Octadecyl alcohol; Sipol S; Siponol S; Siponol SC; Stearol; Steraffine; Stenol; Octadecan-1-ol. Emollient, glass frit binders, waxes, emulsion stabilizers, esters, tertiary amines, surfactants, polymers, chemical intermediate; cosmetic formulations.

784 Adol® 64
Ceteary| alcohol
Emollient, emulsion stabilizer, viscosity modifier for skin care products; conditioner imparting velvety feel; opacifier for creams and lotions. *Sherex/Div. of Witco.*

785 Adol® 66
248-470-8
$C_{18}H_{38}O$
isostearyl alcohol
Coemulsifier, lubricant, foam control agent, cosolvent, plasticizer, stabilizer, emollient, intermediate; for metal lubricants, inks, textiles, emulsions, paper, cosmetics, mineral processing, oil field chemicals and fabric softeners.

786 Adol® 85
143-28-2 6968 205-597-3
$C_{18}H_{36}O$
octadec-9-en-1-ol
(Z)-9-octadecen-1-ol; 9-Octadecen-1-ol; oleyl alcohol. Emulsifier, lubricant, foam control agent, cosolvent, plasticizer and emollient. *Sher/Div. of Witco.*

787 Adox 3125
7758-19-2 8743 231-836-6
$ClNaO_2$
sodium chlorite
Antimicrobial for water and waste water treatment. *Int'l Dioxcide.*

788 Adpro AP 2112-GP
9003-07-0 7741
Polypropylene homopolymer
Amoco ® 6400p; Polypropylene; Amoco ® 7234; Amoco® 1246; Amoco® 6114; Amoco ® 5016; Astryn® SD068-4; Astryn® BA16G; Amoco® 7239; Astryn® 63A6-2; Amoco ® 1012; Amoco® 1016; Amoco ® 4018; Astryn® 65F5-4;Astryn® 63F4-2; Amoco® 9119; Adpro AP 8210-HS; Arpro 3313; Amoco ® 7728; Astryn® 78F4-2; Astryn® 73F5-2; Astryn® 734-2,. Features high flow rate, efficient cycle times, UL and FDA approval; for injection molded consumer and pkg. items, medical components. Solid; softens at 155°; insoluble in cold organic solvents, soluble in hot decalin, hot tetralin, boiling tetrachlorethane; good resistance to abrasion; tendency to develop static charges. *Genesis Polymers.*

789 Adpro AP 8210-HS
9003-07-0 7741
Polypropylene
Ultra high impact resistance, high uniform flow rate; for automotive and extreme impact lens; as impact modifier. *Genesis Polymers.*

790 Ad-Pro-MTS
Adhesion promoter for use in adhesives, inks, coatings, and lacquers; improves adhesion especially for difficult substrates, aids gloss. Rit-Chem Soluble in alcohols, esters, ketones, aromatic solvents; insoluble in aliphatic hydrocarbons; softening point = 68-76°. *Genesis Polymers.*

791 Adroit
Cutting oils. *S & D Chemicals Ltd.* Name unverified.

792 Adronal
108-93-0 2794 203-630-6
$C_6H_{12}O$
cyclohexanol

793 Adronal acetate
Cyclohexanol acetate
Hexalin acetate. A resin solvent.

794 Adsee® 775
POE ethers and special resins
Spreader sticker, wetting agent, and penetrant for agricultural spray; surfactant for monosodium methane arsonate formulations. *Witco/Organics.* Discontinued.

795 Adsee® 799
Alkyl POE ether
Agriculture surfactant; soil penetrant. *Witco/Organics.* Discontinued.

796 Adtac™ LV
Aliphatic resin, stabilized with 0.05% antioxidant
Contributes a balance of tack and adhesive properties to elastomer systems; typical softening point grades range from 10-25°C; used in adhesives, in coatings and as a waterproofing agent. Melt viscosity = 62° (10 poises); softening point (R&B) = 5°. Name unverified.

797 Adurol
Monochloro and monobromo hydroquinones
Photographic developers. No manufacturer.

798 Aduvex®
A range of benzophenone derivatives; ultra violet absorbers for the protection of polymers. *Octel Chemicals Ltd.*

799 Advagum
A terpene resin; a proprietary plasticizer. No manufacturer.

800 Advance
Bromoxynil, fluroxypyr and loxynil post-emergence contact herbicide for cereals. Broad-spectrum herbicide. (Sold in UK for Dow Elanco). *ICI Chem & Polymers Ltd.; ICI Agrochemicals.*

801 Advantage
Nonpigment, polymeric colorants for opaque polyolefin applications. Used for molded opaque polyolefin parts, such as lids and closures where the dimensional tolerances are tight or where warpage from nucleation is a problem. *Milliken.* Discontinued.

802 Advantage 52-B
Oil-based defoamer for pulp mill brownstock washing operations or other high-temperature surfactant-stabilized foam systems; improves drainage. *Hercules.*

803 Advantage 70DYX
Silica/Silicone-based aq. Emulsion
Drainage aid for improved drainage in kraft pulpmill brownstock washing operations by removing entrained air and surface foam; also for cold-stock systems, pulpmill bleaching and screening; defoamer for extremely foamy paints. d = 8.32lb/gal; viscosity = 1500-4000 cps; pH 6.5-8.5. *Hercules.*

804 Advantage 1007B Defoamer
Water-based all-purpose defoamer
For solid and surfactant-stabilized aqeous foaming systems, e.g., acid and alkaline paper machines, pulpmill screening, de-inking operations, size press and calendar solutions, mill effluent systems. *Hercules.*

805 Advantage 136 Defoamer
Hydrocarbon oil-based defoamer
For use as drainage aid and foam killer in kraft pulpmill brownstock washing operations. *Hercules.*

806 Advantage CP
Vinyl acetate/butyl maleate/isobornyl acrylate copolymer, ethanol SDA-40B. *Hercules.*

807 Advantage DF 110
Glycol/hydrocarbon oil-based defoamer for papermaking systems. Dispersion in H_2O; d = 7.42 lb/gal; viscosity 30 cps, can be frozen. *Hercules.*

808 Advantage M104 Defoamer
Drainage aid and foam killer for kraft pulpmill brownstock washing systems. *Hercules.*

809 Advantage M1251 Production Aid
Production aid removing entrained air and surface foam in papermaking applications and waste/effluent treatment. Dispersion in H_2O; d = 8.2 lb/gal; viscosity <1000 cps; pH 6-9. *Hercules.*

810 Advantage™ 10 Defoamer
Fatty acid/fatty alcohol-based
Defoamer for acid and alkaline papermaking systems, waste paper deinking systems, coatings; improves drainage and prod. Rates of washing operations. Solid brick, soapy odor; stable for freezing; emulsifiable in H_2O; d = 0.96; mp = 54°; 100% concentrate. *Hercules.*

811 Advantage® 101M
Contains Mekor® volatile oxygen scavenger/metal passivator, DPB-42 antiscalant, sequestrants, organics, and antifoam. Deposit and corrosion inhibitor for treating steam generating systems. *Hercules.*

812 Advantage® 124
DPB-42 antiscalant, sequestrants
Polymeric deposit and corrosion inhibitor for steam generating systems. *Hercules.*

813 Advapak® ML-1325
Ester
Lubricant for injection molding formulations; NSF-accepted for use at 1.3-2.5 parts per hundred in PVC for potable water contact. EZ-FLO beads; d = 0.98; mp = 94 °. *Morton Int'l Specialty Chem.*

814 Advastab® LS-203
Organotin compound
Lubricating stabilizer for high-output PVC pipe production on multiple screw extruders. *Morton Int'l Specialty Chem.*

815 Advastab® TM-181
Methyltin mercaptide
Heat stabilizer for rigid PVC processes (extrusion, calendering, injection and blow molding) and PVC-PVA formulations with prolonged/severe processing temperatures, for chlorinated PVC compds; tin catalyst in PU polymerizations. APHA 70 clear liquid; d = 1.175; 9.8 lb/gal; viscosity = 50 cps. *Morton Int'l Specialty Chem.*

816 Advastab® TM-692
Methyltin mercaptide
Heat stabilizer for most rigid PVC applications for potable water, sewer, irrigation pipe, conduit, and duct, and extrusion applications, profile, foamed profile. d = 0.99; 8.3 lb/gal; viscosity = 60 cps; cloud pt. -7°; pour pt. -18°. *Morton Int'l Specialty Chem.*

817 Advastab® TM-790 Series
Organotin compound.
Heat stabilizer/external lubricant for single and multiscrew extrusion of PVC potable water, sewer and irrigation pipe, conduit, telephone duct. d = 0.89; 7.5 lb/gal; viscosity = 250 cps; pour point = -9.5°. *Morton Int'l Specialty Chem.*

818 Advastab® WS-499
Organotin compound
Heat stabilizer for extrusion of weatherable, rigid PVC articles. *Morton Int'l Specialty Chem.* Name unverified.

819 Advawax® 240
110-31-6 203-756-1
Abrilube 90. Ethylene bisoleamide. A synthetic wax used as a plastics processing lubricant and release agent, antistat, melting point modifier for waxes, industrial asphalts and tar, pigment dispersing agent for resin systems; polyamide-paraffin coupling agent used in adhesive tapes, coatings and food packaging materials. *Abril Industrial Waxes Ltd; Morton Int'l/Speciality Chem.*

820 Advawax® 290
110-30-5 203-755-6
$C_{38}H_{76}N_2O_2$
N,N'-Ethylene bisstearamide
Alkamide® STED; Octadecanamide, N,N'-1,2-ethanediylbis-; Ethylene bis(stearamide); N,N'-ethylenedi(stearamide;)Acrawax® C; Advawax® 290; Armowax EBS; Abuwax EBS; Armowax EBS; Glycowax® 765; Kemamide® W-20; Kemwax; Nopcowax22-DS; Uniwax 1760; Ethylene bis-stearamide,. Synthetic wax used as plastics processing lubricant and release agent; melting point modifier for waxes and resin blends and industrial asphalt and far; pigment dispersing agent for resin systems; paper-making defoamer; used in adhesive tapes, coatings, food packaging materials. Gardners 9 max sm. Beads; insoluble in H_2O and most organic solvents; soluble in Cellosolve, MIBK, benzene, xylene, kerosene, heptane, naphtha,; dens. 4.75 lb/gal; mp = 143-146°; flash pt. (COC) 290°; fire pt. (COC) 310°. *Morton Int'l Specialty Chem.*

821 Advex 91025
9002-86-2 7746 206-625-7
Polyvinyl Chloride

822 Advitagel
Monogliceride/emulsifier blend
Flour and confectionery emulsifier; aerating agent for sponge cakes. Nonionic; paste; HLB 4.0. *Quest Int'l.*

823 Advitamix
Animal feed supplements. *Quest Int'l.* Name unverified.

824 Advitaroma
Butter and meat flavoring. *Quest Int'l.* Name unverified.

825 Advitrol 8-10
Castor/organoclay complex
Antisettling and thickening agents for paints, varnishes, lubricants, adhesives, coatings, putties, and cosmetics; easily dispersed heat activated rheological additive for low to medium. Polarity solvent based coating systems. d = 13.40 lb/gal; bulking value 0.075 gal/lb; 2.5% maximum moisture. Name unverified.

826 Advizor
Chloridazon and lenacil; pre-emergence herbicide for use in sugar beet. *ICI Chem & Polymers Ltd.*

827 AE-1
Ethoxylated lauryl alcohol
Intermediate in manufacturing of surfactants. *Procter & Gamble.*

828 AEI Compound 403/401
Silane crosslinkable polyethylene compound. Flame retardant compound for cable insulation. *AEI Compounds.* Name unverified.

829 AEI Compound 407/424
Chemically crosslinkable lowsmoke, low-toxicity, halogen-free flame-retardant compound; for insulation of LV power cables and sheathing of all types of cables. *AEI Compounds.* Name unverified.

830 AEI Compound 505/401
Crosslinkable ethylene-propylene-diene rubbers
For low and medium voltage cable insulation. *AEI Compounds.* Name unverified.

831 Aeonite
Elaterite.
A nickel silver containing 20% nickel. No manufacturer.

832 AEPD®
115-70-8 459 204-101-2
C5H13NO2
2-amino-2-ethyl-1,3-propanediol
Pigment dispersant, neutralizing amine, corrosion inhibitor, acid-salt catalyst, pH buffer, chemical and pharmaceutical intermediate and solubilizer. *Angus.*

833 AEPD® 85
115-70-8 459 204-101-2
$C_5H_{13}NO_2$
2-amino-2-ethyl-1,3-propanediol
AEPD®. Chemical intermediate, formaldehyde scavenger, acid-salt catalyst for permanent-press resins, corrosion inhibitor. mp = 35-37°; bp_{10} = 152-153°; d = 1.0990; n_D^{20}= 1.490; soluble in H_2O, organic solvents. *Angus.*

834 Aerelle®
For fibers industry. *Du Pont UK.*

835 Aerex®
De-icing and anti-icing fluid for airplanes. *BASF AG.*

836 Aerial cement
A term applied to cements which set in air, the setting being due to desiccation and carbonation.

837 Aerialite
A proprietary synthetic resin. No manufacturer.

838 Aero
Rosin-glycerol varnish and lacquer resins.
A proprietary trade name for rosin-glycerol varnish and lacquer resins. No manufacturer.

839 Aero 301 Xanthate
Sodium *sec*-butyl xanthate
Used in the mining industry. *Cyanamid BV.*

840 Aero 303 Xanthate
Potassium ethyl xanthate
Used in the mining industry. *Cyanamid BV.*

841 Aero 317 Xanthate
Sodium isobutyl xanthate
Used in the mining industry. *Cyanamid BV.*

842 Aero 343 Xanthate
140-93-2 205-443-5
$C_4H_8OS_2$
Proxan sodium
Carbonodithioic acid, O-(1-methylethyl) ester, sodium salt; Aeroxanthate 3443; Good-rite nix; Isopropylxanthic acid, sodium salt; Sodium isopropylxanthate; Sodium isopropylxanthogenate; Sodium o-isopropyl dithiocarbonate; Z 11. Sodium isopropyl xanthate, used in the mining industry. *Cyanamid BV.*

843 Aero 350 Xanthate
Potassium amyl xanthate
Used in the mining industry. *Cyanamid BV.*

844 Aero 3477 Promoter
Sodium diisobutyl dithiophosphate
Used in the mining industry. *Cyanamid BV.*

845 Aero 3501 Promoter
Sodium diisoamyldithiophosphate
Used in the mining industry. *Cyanamid BV.*

846 Aero metal
An aluminum alloy, consisting mainly of aluminum with 2.1-2.9% magnesium, 0.3-1.3% iron, and 0.2-0.6% copper.

847 Aero X
A proprietary rubber vulcanization accelerator. No manufacturer.

848 Aerocol
Polyvinyl acetate adhesives.
Adhesives. *Ciba plc.*

849 Aerodri 100 104
Modified dioctylsulfosuccinate
Used in the mining industry. *Cyanamid BV.*

850 Aerodri 200
Mixture of surfactants
Used in the mining industry. *Cyanamid BV.*

851 Aerodux
Resorcinol/formaldehyde resins
For wood adhesives, glass reinforced plastics, abrasives, and foundry resins. *Ciba plc.*

852 Aerodux
Resorcinol formaldehyde resins.
Dynochem UK Ltd.

853 Aerofloat 208 Promoter
Sodium diethyl and sodium di-sec. butyl dithiophosphate mixture
Used in the mining industry. *Cyanamid BV.*

854 Aerofloat 211 Promoter
Sodium diisopropyl dithiophosphate
Used in the mining industry. *Cyanamid BV.*

855 Aerofloat 238 Promoter
Sodium di-*s*-butyl dithiophosphate
Used in the mining industry. *Cyanamid BV.*

856 Aerofonic
Compressed polyether acoustical foam.
For wall lining. *ScotFoam Corp.* Name unverified.

857 Aerofroth 65
Polypropylene glycol
Used in the mining industry. *Cyanamid BV.*

858 Aerofroth 76
Mixture of higher alcohols
Used in the mining industry. *Cyanamid BV.*

859 Aerofroth 88
104-76-7 3854 203-234-3
$C_8H_{18}O$
2-Ethylhexanol
2-Ethyl-1-hexanol; 2-Ethylhexyl alcohol. Used in the mining industry; mercerizing textiles; solvent for dyes, resin, oils; antifoaming properties bp = 184-185°; soluble in H_2O (1.4 mg/ml), organic solvents; LD_{50} (rat orl) = 12.46 mg/kg. *Cyanamid BV.*

860 Aerofroth 99
2-Ethylhexanol tails
Used in the mining industry. *Cyanamid BV.*

861 Aerolite
Urea/formaldehyde resins.
Ciba plc.

862 Aerolite
Phenol formaldehyde resins.
Novolac resin. Molding materials; bonding agent, reinforcing agent and modifier for nitrile rubber; air drying varnishes. *Dynochem UK Ltd.*

863 Aeromatt
471-34-1 1697 207-439-9
Precipitated calcium carbonate, used in cosmetics. *Rhône-Poulenc Struge Lifford.*

864 Aeromin
An alloy of 91.6% aluminum and 8.4% magnesium.

865 Aeron
Alloys of 95% aluminum, 4% copper, and 1% silicon.

866 Aerophen
Phenol/formaldehyde resins.
Ciba plc.

867 Aerophen
Phenol/formaldehyde resins.
Dynochem UK Ltd.

868 Aerophine 3418A
Sodium diisobutyldithiophosphinate
Used in the mining industry. *Cyanamid BV.*

869 Aeroplex
A proprietary safety-glass. No manufacturer.

870 Aerosil COK 84
Mixture of Aerosil and alumina in 5:1 ratio.
A trade name for a mixture of Aerosil (primarily SiO_2) and alumina in 5:1 ratio. It is suited particularly for thickening aqueous and other polar systems; filler for plastics. d = 2.2; 50 g/l; pH 3.6-4.3; 82-86% SiO_2, 14.8% Al_2O_3. *Bush Beach Ltd.* Name unverified.

871 Aerosil Composition
Mixture of Aerosil with 15% starch
A trade name for a mixture of Aerosil with 15% starch; specially designed for tableting. *Bush Beach Ltd.* Name unverified.

872 Aerosil®
Highly dispersed pyrogenic silica
Highly active filler for natural and synthetic rubber, especially for silicone rubber; as thickening agent for ointments, creams, toothpaste etc.; tableting and dragee production auxiliary; thixotrope. *Degussa, Degussa Ltd.*

873 Aerosil® 130, 150
7631-86-9 8637 231-545-4
SiO_2
Silica
Highly active filler for natural and synthetic rubber; thickening agent; tabletting and dragee production auxiliary; thixotropizing agent for polyester resins; antisetting agent. *Degussa; Degussa AG.*

874 Aerosil® 200
7631-86-9 8637 231-545-4
O_2Si
silicon dioxide
Fumed silica. Anti-caking and free flow agent with high absorption capacity; for adhesives, food, cosmetics, paint, paper, film, pesticides, pharmaceuticals, plastics, silicone rubber, inks, sealants; a thixotrope for greases and mineral oils.

875 Aerosil®
7631-86-9 8637 231-545-4
silicon dioxide
Anti-caking and free flow agent for adhesives, electrical parts, cosmetics, paints, pesticides, pharmaceuticals, plastics and inks. Improves water resistance of greases. *Degussa; Degussa AG.*

876 Aerosil® R972V
Fumed silica
Reinforces and improves storage stability of RTV compounds; yields softer silicone rubbers. *Degussa; Degussa AG.*

877 Aerosol OT
577-11-7 3460 209-406-4
docusate sodium
Wetting agent, lubricant, detergent for dry cleaning, corrosion resistant lubricants, agricultural emulsions, organic solvent system; used when a higher flash is required. *Am. Cyanamid.*

878 Aerosol® 18
14481-60-8 238-479-5
disodium stearyl sulfosuccinamate
Emulsifier, dispersant, foamer, detergent, solubilizer for soaps and surfactants, alkaline cleaner formulations, brick and tile cleaners, emulsion polymerization of vinyl chloride and styrene/butadiene rubbers, emulsifying oils and waxes, household detergents, cleaning paper mill felts, foamer for foamed latexes and plastics; biodegradable. *Am. Cyanamid; Cyanamid BV.*

879 Aerosol® 22
38916-42-6 254-187-0
Tetrasodium dicarboxyethyl stearyl sulfosuccinamate
Emulsifier, dispersant, solubilizer, surfactant, emulsion polymerization of vinyl polymers, polishing waxes, surface tension depressant for writing and drawing inks, demulsifier for water-oil emulsions, cleaning of paper mill felts; industrial, household and metal cleaners. Biodegradable. *Am. Cyanamid; Cyanamid BV.*

880 Aerosol 200
Disodium alkyl amidopolyethoxy sulfosuccinate
Cyanamid BV. Discontinued.

881 Aerosol® DPOS-45
25167-32-2 246-688-8
Disodium mono-and didodecyl diphenyl oxide disulfonate
Emulsifier, dispersant, solubilizer, primary surfactant for emulsion polymerization systems, coupling agent; high electrolyte tolerance, stable in highly acid and alkaline solutions and at elevated temperatures. *Am. Cyanamid; Cyanamid BV.*

882 Aerosol® GPG
577-11-7 3460 209-406-4
docusate sodium
Wetting agent, surface tensile depressant, emulsifier, dispersant; for dust control, industrial cleaners, emulsifying waxes; biodegradable. *Am. Cyanamid, Cyanamid BV; Cyanamid of Great Britain Ltd.*

883 Aerosol® NPES 458
9051-57-4
Ammonium nonoxynol-4 sulfate
High foaming surfactant for emulsion polymerization of acrylic, styrene and vinyl acetate systems, dishwashing detergents, germicides, pesticides, general purpose cleaners, cosmetics and textile wet processing applications. *Am. Cyanamid.*

884 Aerosol® 501
Disodium alkyl sulfosuccinate
Dispersant, emulsifier, wetting agent, foaming agent; used for acrylic and vinyl acetate emulsions; self-cross-linking latexes; textile wetting and foaming applications. Soluble in H_2O; d = 9.66 lb/gal; viscosity = 260 cps; nonirritating or sensitizing to skin. *Am. Cyanamid, Cyanamid BV.*

885 Aerosol® A-102
39354-45-5 255-062-3
Disodium deceth-6 sulfosuccinate
Emulsifier, solubilizer, foamer dispersant, surfactant, wetting agent; used in emulsion polymerization of PVAc/acrylics, textiles, cosmetics, shampoos, wallboard, adhesives. Biodegradable and stable to acid media. *Am. Cyanamid; Cyanamid BV.*

886 Aerosol® A-103
9040-38-4

Disodium nonoxynol-10 sulfosuccinate
Emulsifier, solubilizer, wetting agent, surfactant, surface tension depressant; used in PVAc/acrylic emulsions; textile emulsions, pad-bath additive, textile wetting, cosmetics, shampoos, wallboard and adhesives. *Am. Cyanamid; Cyanamid BV; Cyanamid of Great Britain Ltd.*

887 **Aerosol® A-196-85**
23386-52-9 245-629-3
Dicyclohexyl sodium sulfosuccinate
Dispersant and surfactant. Sole emulsifier for modified S/B; post additive to stabilize latex and promote adhesion. Biodegradable. *Am. Cyanamid; Cyanamid BV.*

888 **Aerosol® A-268**
37294-49-8 253-452-8
Disodium isodecyl sulfosuccinate
Surfactant, sole emulsifier for PVC latexes, vinyl, vinylidene chloride, acrylics; surface tension depressant and solubilizer. *Am. Cyanamid; Cyanamid BV.*

889 **Aerosol® AY-65**
922-80-5 213-085-6
Diamyl sodium sulfosuccinate
Wetting agent, dispersant, surfactant. Used in agriculture, emulsion polymerization, electroplating, ore leaching, cleaning of porcelain, tile, brick and cement. Biodegradable. *Am. Cyanamid; Cyanamid BV.*

890 **Aerosol® C-61**
Ethoxylated alkyl guanidine-amine complex
Antistat, pigment dispersant, flushing agent, wetting agent, settling agent; alkaline, cement, brick, and tile cleaner formulations for crystal growth control, emulsion breaking, alkaline metal and paint brush cleaners; paint removers, textile softener; demulsifying agent; for plastics, paper, textiles, adhesives industries. Tan creamy paste; strong ammoniacal odor; soluble in organic solvents in presence of alcohol; dispersion in H_2O; d = 8 lb/gal; partially biodegradable. *Am. Cyanamid, Cyanamid BV.*

891 **Aerosol® IB-45**
127-39-9 3238 204-839-5
sodium diisobutyl sulfosuccinate
Emulsifier, wetting agent; emulsion polymerization of styrene, butadiene and copolymers; dye and pigment dispersant; for leaching, electroplating; biodegradable Soluble in H_2O; extremely hydrophilic; d = 1.12, 9.3 lb/gal. *Am. Cyanamid; Cyanamid BV.*

892 **Aerosol® IB-45**
127-39-9 3238 204-839-5
$C_{12}H_{21}NaO_7S$
Sulfo-butanedioic acid 1,4-bis(2-methylpropyl)ester sodium salt
Sodium dibutyl sulfosuccinate. Emulsifier, wetting agent. Used in emulsion polymerization of styrene, butadiene and copolymers; dye and pigment dispersant. Used for leaching and electroplating. Biodegradable. *Am. Cyanamid; Cyanamid BV.*

893 **Aerosol® MA-80**
3006-15-3 221-109-1
Dihexyl sodium sulfosuccinate
Dispersant, textile wetting agent, emulsifier, solubilizer and penetrant. Used for emulsion polymerization, battery separators, electroplating, ore leaching and as a germicide. Not as rapidly biodegradable as Aerosol 18 and 22. *Am. Cyanamid; Cyanamid BV.*

894 **Aerosol® NS**
130-14-3 204-976-0
$C_{10}H_7NaO_3S$
Sodium naphthalene sulfonate
Sodium α-naphthalenesulfonate; Sodium 1-Naphthalenesulfonate; Sodium naphthalene sulfonate. Dispersant for pigments, extenders and fillers. Usable in aqueous media over a wide pH range. *Am. Cyanamid.*

895 **Aerosol® OS**
1322-93-6 215-343-3
Diisopropylnaphthalenesulfonic acid, sodium salt
Aerosol OS; Alkanol B; Nekal A; Novonacco; NSAE; Petroll; Sellogen W; Sodium diisopropylnaphthalene sulfonate; Vatsol OS. Emulsifier, dispersant and wetting agent. Used in alkaline cleaning formulations, antigelling agents, automotive radiator cleaners, metal, cement, brick and tile cleaners for crystal growth control; electroplating, filtration, glass cleaning, household detergents, leaching ores and slags, pigment dispersions, soap additives, adjuvant in agricultural chemicals. Slowly biodegradable. *Am. Cyanamid; Cyanamid BV.*

896 **Aerosol® OT-70 PG, OT-S**
577-11-7 3460 209-406-4
Sodium dioctyl sulfosuccinate, propylene glycol/water
Sodium dioctyl sulfosuccinate in propylene glycol/water. Wetting agent, surface tensile depressant, emulsifier, surfactant; for use where high flash required; biodegradable. *Am. Cyanamid, Cyanamid BV.*

897 **Aerosol® OT-75%**
577-11-7 3460 209-406-4
Dioctyl sodium sulfosuccinate
Wetting agent and surface tensile depressant used in textile, rubber, petrol, paper, metal, paint, plastic, and agricultural industries; antistat for cosmetics, dry cleaning detergents, emulsion, plastic, pipelines, and suspension polymerization; emulsifier wax for polish, firefighting, germicide, metal cleaner, mold release agent, dispersant in paints and inks, paper, photography, process aid, rust preventative, soldering flux and wallpaper removal. *Am. Cyanamid, Cyanamid BV.*

898 **Aerosol® OT-MSO**
577-11-7 3460 209-406-4
$C_{20}H_{37}NaO_7S$
sulfosuccinic acid 1,4-bis(2-ethylhexyl) ester
Dioctyl sodium sulfosuccinate in mineral seal oil. Used as a wetting agent, lubricant and detergent for dry cleaning, corrosion resistant lubricants, agricultural emulsions and organic solvent systems. Used when a higher flash is required. *Am. Cyanamid.*

899 **Aerosol® TR-70**
2673-22-5 220-219-7
Ditridecyl sodium sulfosuccinate
Emulsifier and surfactant. Used in emulsion polymerization of vinyl chloride and vinyl acetate, suspension polymerization of vinyl chloride. Dispersant for resins, pigments, polymers and dyes in organic systems. Pigment dispersant in printing inks. Rust preventative. Biodegradable. *Am. Cyanamid; Cyanamid BV.*

900 **Aerothene**
75-09-2 6140 200-838-9
CH_2Cl_2
methylene chloride
Vapor pressure depressant and carrier solvent. *Dow Cheml Co Ltd, UK & Ireland.*

901 **Aerotru 23**
Modified melamine formaldehyde resin
Used in the paper industry. *Cyanamid BV.*

902 **Aerozine A-50**
Rocket propellant. *Olin.*

903 **Aeternol**
A proprietary synthetic resin. No manufacturer.

904 **Aethoxal**
Fattening agent, and dispersing agent; emollient for bath oils, shampoos, skin and personal care products. d^{70} = 0.9340-0.9370; pH 6.5-7.5; nonionic. *Henkel Chemicals Ltd.* Name unverified.

905 **Aethoxal B**
PPG-4-laureth-2
Superfatting agent and emollient for bath oils, skin and personal care products and pharmaceuticals. Biodegradable. *Henkel/Cospha; Henklel KGaA.*

906 **Aethrol**
A plastic of the pyroxylin-cellulose acetate type. No manufacturer.

907 **AF 10 FG**
9006-65-9 3264
$(C_2H_6OSi)_xC_4H_{12}Si$
Polydimethylsiloxane
Dimethicone; Durkex 100DS; SF18-350; AF 9020; AF 30 FG. Silicone antifoam agent used for general food, poultry, and meat processing applications; anticaking agent; Soluble in hydrocarbon solvents, chloroform, ether, H_2O; suspected carcinogen. *Harcros Organics.*

908 **AF 10 IND**
Silicone antifoam for agriculture, cutting oils, drilling muds, effluent, inks, chemicals, detergents and textiles. *Harcros Organics.*

909 **AF 60**
Dimethyl polysiloxane aqueous emulsion
Defoaming agent used in adhesive, ink, latex, soap, starch, and paint manufacturing and other aqeous industrial systems. Sorbic acid odor; soluble in H_2O; d = 1.01, 8.4 lb/gal; viscosity = 1000 cps. *GE Silicones.*

910 **AF 70**
100% silicone compound
Defoamer in petroleum refining, cutting oils, chemical processing, antifoam formulating; food additive for food processing. Soluble in aliphatic and chlorinated hydrocarbons; d = 1.01, 8.4 lb/gal; viscosity = 1500 cps; flash pt. (OC) 315°. *GE Silicones.*

911 **AF 72**
PEG-40 stearate, sorbitan stearate, and silica
Food-grade antifoam agent, surfactant; also for industrial applications such as textile dyeing and finishing, leather finishing, latex processing, soap and detergent manufacturing, adhesive manufacturing, and as a boiler feed water defoamer. Sorbic acid odor; soluble in H_2O; d = 1.01, 8.4 lb/gal; viscosity = 1000 cps; 30% silicone, 44.2% solids. *GE Silicones.*

912 **AF 1025**
Silicone antifoam for delayed coker units. Insoluble in H_2O; d = 0.82, 6.8 lb/gal; flash point (PMCC) = 140ºF. *Harcros Organics.*

913 **AF 8820**
Silicone antifoam
Antifoam for effluent, agriculture, antifreeze, detergent applications; dilutable. Dispersion in H_2O; d = 1.00; 8.3 lb/gal; flash pt. (PMCC) >212ºF; pH 4-5 (1% aq.); nonionic. *Harcros Organics.*

914 **AF 9020**
Dimethicone aqueous emulsion
Defoamer for industrial and food-processing systems including chemical processing (adhesive manufacturing, water-based ink manufacturing, latex processing, soap manufacturing, starch processing, paint additive, alcohol fermentation), waste treatment, petrochemical (resin polymerization, glycol dehydrators, ehtylene oxide and urea production. Dispersion in H_2O with mild agitation; d = 1.01, 8.4 lb/gal; viscosity = 3500 cps; 20% silicone, 28.75% solids. *GE Silicones.*

915 **AF GN-11-P**
Nonsilicone antifoam
Used for fermentation, drilling muds, effluent, adhesives, gas treating. Dispersion in H_2O; d = 1.01, 8.4 lb/gal; flash pt. >200ºF (PMCC); pH 5-7; nonionic. *Harcros Organics.*

916 **AF HL-36**
Nonsilicone antifoam
For fermentation, processing beet sugar and yeast, distillation; Kosher. Supplied as a dispersion in H_2O; d = 1.00, 8.3 lb/gal; flash pt. (PMCC) >300. *Harcros Organics.*

917 **AF HL-52**
Nonsilicone antifoam
For solvents, latex paints, inks, chemical processing, adhesives, paper, paper coatings. Oil-soluble; d = 0.87, 7.25 lb/gal; flash pt. (PMCC) >200ºF. *Harcros Organics.*

918 **Afalon**
330-55-2 5534 206-356-5
linuron
A urea-derived herbicide used for control of weeds in field crops such as potatoes and carrots. *Hoechst UK.*

919 **Afax**
Continuous casting mold flux for all steel grades. *Foseco (F.S.) Ltd.* Discontinued.

920 **A-Fax®**
A range of amorphous polypropylenes
Used in adhesives and sealants, used for asphalt modifications in construction and building industry, in carpet backing, in polyolefin modification, as sound deadening and rubber processing agents.

921 **Afco-Chem CS**
1592-23-0 1750 216-472-8
$C_{36}H_{70}CaO_4$
calcium stearate
Lubricant for metal sintering. Lubricant and stabilizer for resins. Pigment dispersant, mold releasing agent, waterproofing agent and lubricant additive. *Adeka Fine Chem.*

922 **Afco-Chem LIS**
4485-12-5 224-772-5
$C_{18}H_{35}LiO_2$
Lithium stearate
Lubricant for metal sintering. Lubricant and stabilizer for resins. Pigment dispersant, mold releasing agent, waterproofing agent and lubricant additive. mp= 220º; LD_{50} (rat orl) = 15 gm/kg. *Adeka Fine Chem.*

923 **Afco-Chem MGS**
557-04-0 5730 209-150-3
$C_{36}H_{70}MgO_4$
Octadecanoic acid, magnesium salt
magnesium stearate. Lubricant for metal sintering. Lubricant and stabilizer for resins. Pigment dispersant, mold releasing agent, waterproofing agent and lubricant additive. mp = 130-140º. *Adeka Fine Chem.*

924 **Afco-Chem ZNS**
557-05-1 10292 209-151-9
Zinc stearate
Lubricant for metallic sintering; lubricant and stabilizer for resins; pigment dispersant; mold release; waterproofing agent; lubricant additive. *Afco-Chem.*

925 **Afco-Coat**
Inorganic powder used as a precoating agent for stainless steel. Soluble in H_2O. *Adeka Fine Chem.*

926 **Afcolene**
25704-18-1 9028 229-680-9
A proprietary polystyrene.
Atochimie. Unverified.

927 **Afco-Lube Series**
Lubricants for wet wire drawing. *Adeka Fine Chem.*

928 **Afco-Met Series**
Lubricants for dry wire drawing. *Adeka Fine Chem.*

929 **Afenil**
Calcium chloride-urea
No manufacturer.

930 **Afflair® Lustre Pigments**
Mica platelets coated with titanium dioxide and/or iron oxide
Luster pigments for coatings, inks, and plastics. Suitable for high temperature applications *EM Industries.*

931 **Afilan EHS**
22047-49-0 244-754-0
2-Ethylhexyl stearate
Surfactant for textile processing. *Hoechst Celanese/Colorants & Surf.*

932 **Afilan ICS**
25339-09-7 246-868-6
Isocetyl stearate
Surfactant for textile processing. *Hoechst Celanese/Colorants & Surf.*

933 **Afilan PP**
14450-05-6 238-430-8
Pentaerythrityl tetrapelargonate
Surfactant for textile processing. Viscosity = 50 cps. *Hoechst Celanese/Colorants & Surf.*

934 **Afilan TDA**
Ditridecyl adipate
Surfactant for textile processing. *Hoechst Celanese/Colorants & Surf.*

935 **Afilan TMPP**
Trimethylolpropane tripelargonate
Surfactant for textile processing. Viscosity =36 cps. *Hoechst Celanese/Colorants & Surf.*

936 **Aflaban**
Feed preservative based on sorbic acid; growth inhibitor for molds, yeast and bacteria in animal feeds. *Monsanto Co.* Name unverified.

937 **Aflamman CN**
Halogen compound with metal oxide and binder
Flameproofing agent for treatment of technical fabrics from polyester; also suitable for backcoating. Weakly anionic; environmentally nonhazardous; H_2O-resistant. *Thor Chemicals (UK) Ltd.* Discontinued.

938 **Aflammit P**
Organic phosphorus/nitrogen compound
Wash and dry-cleaning resistant flameproofing agent for cotton. *Thor Chemicals (UK) Ltd.* Discontinued.

939 **Aflammit TI**
16919-27-0 240-969-9
F_6K_2Ti
Titanium(IV)Potassium Fluoride
Potassium hexafluorotitanate; Titanate(2-), hexafluoro-, dipotassium, (OC-6-11)-; dipotassium hexafluorotitanate; Dipotassium monotitanium hexafluoride; Dipotassium titanium hexafluoride; Titanium potassium fluoride. Flameproofing agent used in wool processing. *Thor Chemicals (UK) Ltd.* Discontinued.

940 **Aflammit ZAL**
Zirconium acetate solution
Flameproofing agent used in wool processing. *Thor Chemicals (UK) Ltd.* Discontinued.

941 **Aflammit ZR**
16923-95-8 7804 240-985-6
F_6K_2Zr
Potassium hexafluoro zirconate
Potassium zirconium hexafluoride; Dipotassium hexafluorozirconate; Dipotassium zirconium hexafluoride; Potassium fluorozirconate; Potassium hexafluorozirconate. Flameproofing agent used in wool processing. Also used in manufacture of zirconium. Slightly soluble in cold H_2O. *Thor Chemicals (UK) Ltd.* Discontinued.

942 **Aflunox**
Perfluoropolyether fluids.
Heat transfer agents. *PCR.* Discontinued.

943 **Aflux**
A range of fatty acid derivatives partly bound to highly active silica. Used as dispersing agents and internal lubricants in the rubber industry; used for molded and extruded technical articles. d = 1.14 *Bayer plc.*

944 **Afonic**
Embossed sound absorbing foam. *ScotFoam Corp.* Name unverified.

945 AFP®
Solid photopolymer.
Asahi Chem. Industry.

946 AFP 2000
9014-01-1 232-752-2
Enzyme for hydrolysis of proteins under acid conditions; prevents haze in fruit juice. Tan to white powder; free of offensive odor; readily water-soluble. *Solvay Enzymes.*

947 Afranil®
Alcohol and fatty acid derivatives
Grease and foam inhibitor, pulp deaerator for papermaking. *BASF AG; BASF plc.*

948 African phosphates
Mineral phosphates found in Tunis and Algeria. They contain from 55-65% calcium phosphate. Others found at Safaga and Kosseir contain 60-70% calcium phosphate. Used as fertilizers.

949 Afrisect
Insecticide formulation. *Mitchell Cotts Chemicals Ltd.*

950 Afrol
Timber insecticide. *ICI Chem & Polymers Ltd.*

951 Afror Tyne Powder
A nitrated mixture of glycerine and ethylene glycol and ammonium nitrate. A low-freezing explosive. No manufacturer.

952 Afugan
13457-18-6 8146 236-656-1
$C_{14}H_{20}N_3O_5PS$
Pyrazophos; Curamil; HOE 2873. Systemic organophosphorus fungicide. m.p. 38-40°; LD_{50} (rat orl) = 140 mg/kg. *Hoechst UK.*

953 Agalite
Mineral pulp; asbestine pulp. A variety of talc (hydrated magnesium silicate); used in paper manufacture. No manufacturer.

954 Agallol
123-88-6 204-659-7
C_3H_7ClHgO
methoxyethyl mercury chloride
Agallol; Aratan; Aretan 6; Ceresan Universal Nazbeize; Chloro(2-methoxyethyl)mercury; Falisan; Gramisan; Higosan; Agallolat; Agalol; Aretan; Atiran; Cekusil Universal C; Ceresan-Universal Nassbeize; MEMC; Merchlorate; (β-methoxyethyl)mercuric Chloride; Methoxyethyl Mercuric Chloride; 2-methoxyethylmercuric Chloride; β-methoxyethylmercury Chloride; 2-methoxyethylmercury Chloride; Sedresan; Tafasan 6W; Tafasan. Antifungal agent used as a dressing for seed potatoes, flower bulbs and sugar cane cuttings. mp = 65°. Insoluble in H_2O, organic solvents, LD_{50} (rat orl) = 22 mg/kg. *Bayer AG.*

955 Agar (Agar-agar)
9002-18-0 182 232-658-1
Gelose; Bengal isinglass; Chinese insinglass; Layor Carang; Japan agar; Ceylon isinglass. Polysaccharide mixture of agarose and agaropectin, extracted from the agarocytes of algae of the *Gracilaria seaweeds* or *Rhodophyceae*. A phycolloid, also used as a culture medium in microbiology and bacteriology; antistaling agent in baking, confections, meats, poultry, gelation agents in desserts and beverages, protective colloid in foods, pharmaceuticals, laboratory reagents and photographic emulsions. Transparent, odorless, fine powder; insoluble in cold H_2O; slowly soluble in hot H_2O to a viscid solution. A 1% solution forms a stiff jelly on cooling.

956 Agaricic
666-99-9 184 211-566-5
$C_{22}H_{40}O_7$
agaric acid sesquihydrate
Laricic acid; Agaricin; 2-Hydroxy-1,2,3-nonadecanetricarboxylic acid; Laricic acid. A resin acid, obtained by extraction with alcohol of the fruit bodies of *Polyporus officinalis* and *Agaricus albus*. A febrifuge and antiperspirant Crystalline powder, odorless, tasteless; mp = 142°; slightly soluble in cold H_2O, chloroform, or ether; freely soluble in boiling H_2O, alkali, hot glacial acetic acid.

957 Agate ware
Enameled iron-granite. Enameled iron.

958 Agatine
A proprietary phenol-formaldehyde resin. Available in the form of sheet, rods, tubes, etc. No manufacturer.

959 Agavin
Thiosolucin-dihydrostreptomycin preparation
For the veterinary field. *May & Baker Ltd.* Name unverified.

960 Age
Axin
The fat of *Coccus Axin*, growing in Mexico. It consists of the glycerides of lauric and axinic acids.

961 Ageflex
Acrylate and methacrylate monomers.
CPS Chemical Co Inc.

962 Ageflex DEGDMA
2358-84-1 219-099-9
$C_{12}H_{18}O_5$
3-Oxapentane-1,5-diyl dimethacrylate
Diethylene glycol dimethacrylate. Crosslinking for rubber vulcanization, moisture barrier films and coatings, photopolymer printing plates and letterpress inks, conversion coatings and adhesives. d^{20}_{20} = 1.056. *CPS Chemical Co Inc.*

963 Ageflex EOTMPTA
28961-43-5
Ethoxylated trimethylolpropane triacrylate
Monomer having flexibility and fast cure response with other acrylated resins; radiation cured systems offer chemical and abrasion resistance and high gloss to inks, coatings, overprint varnishes. *CPS Chemical Co Inc.*

964 Ageflex 1,3 BGDMA
1,3-Butylene glycol dimethacrylate
Crosslinker for plastisols, hard rubber rolls, cast acrylic sheet/rods, coagent for rubber compounding, impregnant for metal and wood composites, adhesives, glass-reinforced plastics. d = 0.944 *CPS Chemical Co Inc.*

965 Ageflex AGE
106-92-3 203-442-4
$C_6H_{10}O_2$
1-allyloxy-2,3-epoxy-propane
AGE; 1,2-Epoxy-3-Allyloxypropane; Allyl-2,3-epoxypropyl ether; [(2-propenyloxy)methyl]oxirane; 1-allyl-2,3-epoxypropane; Allyl Glycidyl Ether-Ethylene Glycol Prepolymer(18/1). Modifier for elastomer, epoxies, adhesives, fibers; reactive intermediate for coatings, sizing/finishing agent for fiberglass; silane intermediate in electrical coatings. Soluble in methanol, toluene; partly soluble in H_2O; d^{20}_{20} = 0.970; LD_{50} (rat orl) = 922 mg/kg. *CPS Chemical Co Inc.*

966 Ageflex AMA
96-05-9 202-473-0
$C_7H_{10}O_2$
Allyl methacrylate
Sipomer™ AM; 2-Propenoic acid, 2-methyl-, 2-propenyl ester. Silane monomer intermediate; crosslinker offering two-stage polymerization, abrasion and solvent resistance; polymer modifier for high impact plastics, adhesives, acrylic elastomers, photoresists, optical polymers. mp = -65°; bp = 144°; n_D^{20} = 1.4360; d^{20}_{20} = 0.9380; soluble in H_2O (4 g/l), organic solvents; LD_{50} (rat orl) = 430 mg/kg. *CPS Chemical Co Inc; Richman; San Esters.*

967 Ageflex BGE
2426-08-6 219-376-4
$C_7H_{14}O_2$
Butyl glycidyl ether
BGE; Glycidyl butyl ether monomer; Butoxymethyl oxirane; BGE; 1,2-epoxy-3-butoxypropane; Butyl glycidyl ether; n-Butyl 2,3-Epoxypropyl Ether; (butoxymethyl)oxirane; 2,3-epoxypropyl butyl ether; butyl-2,3-epoxypropyl ether; glycidyl butyl ether; 1-butoxy-2,3-epoxypropane; ageflex bge; Glycidyl n-butyl ether; Araldite RD-1. Reactive diluent in epoxy resins, laminating, flooring, electrical casting and encapsulants. d^{25}_{4}= 0.908; bp = 164°; soluble in H_2O (10-20 mg/ml), more soluble in organic solvents; LD_{50} (rat orl) = 2050 mg/kg. *CPS Chemical Co Inc.*

968 Ageflex CHMA
101-43-9 202-943-5
$C_{10}H_{16}O_2$
2-methyl-2-propenoic acid, cyclohexyl ester
Cyclohexyl methacrylate; 2-Propenoic acid, 2-methyl-, cyclohexyl ester; Cyclohexyl methacrylate, monomer. Clear colorless monomeric liquid. Polymer modifier for optical lens coatings, adhesives, floor polishes, vinyl polymerization, anaerobic adhesives. Insoluble in H_2O; d^{20}_{20} = 0.9640; bp = 210°; flash point = 82°; combustible. *CPS Chemical Co Inc.*

969 Ageflex EGDMA
97-90-5 202-617-2
$C_{10}H_{14}O_4$
Ethylene glycol dimethacrylate
MFM-416. Crosslinker and modifier of ABS, acrylic and PVC, ion exchange resins, encapsulation of smokeless powder, glaze coatings, dental polymers, paper processing aids, rubber modifier, adhesives, optical polymers, leather finishing, moisture barrier films; fiberglass-reinforced polyesters, emulsion polymerization. d^{20}_{20} = 1.055; flash point = 68°; moderately toxic by ingestion. *CPS Chemical Co Inc.*

970 Ageflex T4EGDA
17831-71-9 241-789-3
$C_{14}H_{22}O_7$
PEG-4 diacrylate
PEG 200 diacrylate; Photomer™4013; Tetraethylene Glycol Diacrylate; 2-

Propenoic acid oxybis(2,1-ethanediyloxy-2,1-ethanediyl) ester; TTEGDA; Polyethylene glycol 1000 diacrylate. Fast curing monomer providing good adhesion and flexibility, low shrinkage, and good impact strength in inks, coatings, adhesives, photo resists, and rubber products. $bp_{0.3}$ = 120°; d_{20}^{20} =1.11; slightly soluble in H_2O; f.p. <20°; flash pt. >93°. *CPS Chemical Co Inc.*

971 Ageflex FA-1Q75MC
44992-01-0 256-176-6
$C_8H_{16}ClNO_2$
Dimethylaminoethyl acrylate methyl chloride
Ethanaminium, N,N,N-trimethyl-2-[(1-oxo-2-propenyl)oxy]-, chloride; [2-(acryloyloxy)ethyl]trimethylammonium chloride. Quaternary antistatic finish for polyester fibers, flocculant and coagulant for industrial process water treatment, flocculant for mineral recovery, ion exchange resins, adhesives, acid dye receptivity, electrostatic coatings on wodd, retention aids for paper. *CPS Chemical Co Inc.*

972 Ageflex FA-1Q80DMS
13106-44-0 236-029-2
Dimethylaminoethyl acrylate dimethyl sulfate
Quaternary; antistatic finish for polyester fibers, flocculant and coagulant for industrial process water treatment, flocculant for mineral recovery, ion exchange resins, adhesives, acid dye receptivity, electrostatic coatings on wodd, retention aids for paper. *CPS Chemical Co Inc; Aldrich.*

973 Ageflex FA-2Q50DMS
21810-39-9 244-588-9
Diethylaminoethyl acrylate dimethyl sulfate
N,N-Diethylaminoethyl acrylate Q-Salt, methosulfate. Quaternary antistatic finish for polyester fibers, flocculant and coagulant for industrial process water treatment, flocculant for mineral recovery, ion exchange resins, adhesives, acid dye receptivity, electrostatic coatings on wodd, retention aids for paper. *CPS Chemical Co Inc.*

974 Ageflex FA-1
2439-35-2 219-460-0
$C_7H_{13}NO_2$
dimethylaminoethylacrylate
N,N-Dimethylaminoethyl acrylate. Adhesion promoter in UV- and EB-cured coatings for metals, plastic, paper and wood surfaces. Catalyst for epoxy molding and extrusion resins, intermediate for water treatment chemicals, quaternary monomers; silane coupling agents, conductive paper coatings. bp = 64°; insoluble in H_2O, soluble in organic solvents. *CPS Chemical Co Inc.*

975 Ageflex FA-1
2439-35-2 219-460-0
$C_7H_{13}O_2N$
N,N-Dimethylaminoethyl acrylate
Dimethylaminoethyl acrylate. Adhesion promoter in uv and eb cured coatings for metal, plastic, paper, and wood surfaces; catalyst for epoxy molding and extrusion resins; intermediate for water treatment chemicals, quaternary monomers, silane coupling agents, conductive paper coatings. d_{20}^{20} = 0.940; bp_{50} = 94°; insoluble in H_2O, slightly soluble in organic solvents. *CPS Chemical Co Inc.*

976 Ageflex FA-2
2426-54-2 219-378-5
$C_9H_{17}O_2N$
Diethylaminoethyl acrylate
2-Propenoic acid, 2-(diethylamino)ethyl ester; N,N-Diethylaminoethyl acrylate; 2-(Diethylamino)-ethyl acrylate. Industrial and automotive coatings, electronic photo resists, dye additives, lube oil additives; intermediate for water treatment chemicals, silane coupling agents, conductive paper coatings; retention aids for paper manufacturing; flocculant and coagulant. bp_{10} = 81°; d = 0.939; corrosive; severe eye and skin irritant; harmful if swallowed or inhaled. *CPS Chemical Co Inc.*

977 Ageflex FA-6
2499-95-8 219-698-5
$C_9H_{16}O_2$
n-Hexyl acrylate
Acrylic acid hexyl ester; n-Hexyl acrylate. UV-cured inks/coatings, glass coating, viscosity index improver for functional oils, polymer cements and sealants. bp_{24} = 88-90°; d_{20}^{20} = 0.8880; LD_{50} (rat orl) = 26 mg/kg. *CPS Chemical Co Inc.*

978 Ageflex FA-8
29590-42-9 249-707-8
$C_{11}H_{20}O_2$
Isooctyl acrylate
2-Propenoic acid, isooctyl ester; Isoctyl acrylate. Pressure-sensitive adhesives, coatings, caulks and sealants. d_{20}^{20} = 0.880; flash point = 79°. *CPS Chemical Co Inc.*

979 Ageflex FA-10
1330-61-6 215-542-5
$C_{13}H_{24}O_2$
2-Propenoic acid, isodecyl ester
Isodecyl acrylate. Adhesives, coatings, uv-curable reactive diluent in inks and coatings, viscosity index improver. bp_{10} = 121°; d_{20}^{20} = 0.864. *CPS Chemical Co Inc.*

980 Ageflex FA-12
2156-97-0 218-463-4
$C_{15}H_{28}O_2$
Lauryl acrylate
UV-curable reactive diluent in inks and coatings, adhesives, viscosity index improver, finishing aid for leather. d_{20}^{20} = 0.884. *CPS Chemical Co Inc.*

981 Ageflex FM-1Q80DMS
6891-44-7 229-995-1
$C_{10}H_{21}NO_6S$
2-(Methacryloyloxy)ethyltrimethylammonium methyl sulfate
Antistatic finish for polyester fibers, flocculant and coagulant for industrial process water treatment, flocculant for mineral recovery, ion exchange resins, adhesives, acid dye receptivity, electrostatic coatings on wood, retention aids for paper. Soluble in H_2O, d_{20}^{20} = 1.183; cationic. *CPS Chemical Co Inc.*

982 Ageflex FM-1Q80MC
5039-78-1 225-733-5
$C_9H_{18}ClNO_2$
Dimethylaminoethyl methacrylate methyl chloride
Methacryloxyethyltrimethyl ammonium chloride. Quaternary antistatic finish for polyester fibers, flocculant and coagulant for industrial process water treatment, flocculant for mineral recovery, ion exchange resins, adhesives, acid dye receptivity, elec *CPS Chemical Co Inc.*

983 Ageflex FM-1
2867-47-2 220-688-8
$C_8H_{15}NO_2$
Dimethylaminoethyl methacrylate
dimethylaminoethyl methacrylate; N,N-Dimethylaminoethyl methacrylate; 2-Propenoic acid, 2-methyl-, 2-(dimethylamino)ethyl ester; DMAEMA. Detergent and sludge dispersant in lubricants; viscosity index improver; flocculant for waste water treatment; retention aid for paper manufacturing; acid scavenger in PU foams; corrosion inhibitor; resin and rubber modifier; used in acrylic polishes and paints, hair preparation copolymers, sugar and water clarification and adhesives. Very soluble in H_2O; soluble in organic solvents; viscosity = 1.38 cst; mp = -30°; bp = 68.5°; d = 0.9330; n_D^{20}= 1.4391; LD_{50} (rat orl) = 1751 mg/kg. *CPS Chemical Co Inc.*

984 Ageflex FM-4
3775-90-4 223-228-4
$C_{10}H_{19}NO_2$
2-[(1,1-Dimethylethyl)amino]ethyl 2-methyl-2-propenoate
N-(*tert*-Butylamino)ethyl methacrylate; *tertiary*-Butylaminoethyl methacrylate. Automotive dip tanks, coatings, industrial/consumer adhesives and coatings, dye and lube oil additives, intermediate for water treatment chemicals, oil-water separations. d_{20}^{20} = 0.914; f.p.<-60°; flash pt. 96°; *CPS Chemical Co Inc.*

985 Ageflex FM-10
29964-84-9 249-978-2
$C_{14}H_{26}O_2$
Isodecyl methacrylate
Isodecyl 2-methylpropenoate; Isodecyl 2-methyl-2-propenoate. Pressure-sensitive adhesives, coatings for leather, textiles, paper, nonwovens, polymer modifier/stabilizer, viscosity index improver, dispersion for plastic and rubber, floor waxes, potting compounds, sealants, adhesives. d_{20}^{20} = 0.878; bp = 126°; flash pt. 121 °; insoluble in H_2O, soluble in organic solvents; LD_{50} (rat ip) = 2467 mg/kg. *CPS Chemical Co Inc.*

986 Ageflex FM-12
142-90-5 205-570-6
$C_{16}H_{30}O_2$
Lauryl methacrylate
Lauryl methacrylate; Acrylic acid, 2-methyl-, dodecyl ester; Dodecyl 2-methyl-2-propenoate; dodecyl methacrylate; Metazene; Methyl-2-propenoic acid, dodecyl ester; Propenoic acid, 2-methyl-, dodecyl ester; n-Dodecyl methacrylate; n-Lauryl methacrylate. Lube oil additives, coatings for nonwoven fiber, floor waxes, paints, adhesives, varnishes, sealants, caulks, stabilizer for nonaq. dispersions and inks. mp = -7°; bp_4 = 142°; d = 0.8680; insoluble in H_2O, soluble in organic solvents. *CPS Chemical Co Inc.*

987 Ageflex FM-25
2495-25-2 219-671-8
$C_{17}H_{32}O_2$
Tridecyl methacrylate
Lauryl methacrylate. d = 0.88; flash pt. >110°; n_D^{20}= 1.4500. *CPS Chemical Co Inc.* Name unverified.

988 Ageflex FM-68
32360-05-7 228-126-3
$C_{22}H_{42}O_2$
2-Methyl-2-propenoic acid, octadecyl ester

Hexadecyl methacrylate; Octadecyl methacrylate; Stearyl methacrylate. Natural, C_{16-18} methacrylates, with 100 ppm hydroquinone inhibitor; lube oil additive, pour point depressant, paper coatings, textile finishes, paints, varnishes, pressure-sensitive adhesives. d_{20}^{20} = 0.868; f.p. -20°; flash pt. 110° *CPS Chemical Co Inc.*

989 Ageflex FM-1620

32360-05-7 228-126-3

$C_{22}H_{42}O_2$

Octadecyl methacrylate

Stearyl methacrylate. Lube oil additive, pour point depressant. Used in paper coatings, textile finishes, paints, varnishes, pressure-sensitive adhesives. *CPS Chemical Co Inc.* Name unverified.

990 Ageflex HDDA

13048-33-4 235-921-9

$C_{12}H_{18}O_4$

2-Propenoic acid, 1,6-hexanediyl ester

HDDA; HDODA; 2-Propenoic acid 1,6-hexanediyl ester; 1,6-Hexanediol Diacrylate; Hexamethylene diacrylate; HDDA; 1,6-hexamethylene diacrylate. Fast curing monomer providing adhesion to metal and glass, flexibility in inks and coatings, water resistant, good weatherability; reactive diluent for radiation-curable oligomers. d_{20}^{20} = 1.01; f.p. <-20°; flash pt. >93°. *CPS Chemical Co Inc.*

991 Ageflex IBOA

5888-33-5 227-561-6

$C_{13}H_{20}O_2$

Isobornyl acrylate

Monomers when cured providing hardness, low shrinkage, abrasion resistant, heat and water resistant, good weatherability in automotive coatings, electronics, adhesives, and other acrylic polymers. mp = -60°; d_{20}^{20} = 0.986; m.p. -60°; f.p. <-20°; flash pt. 84°; irritant. *CPS Chemical Co Inc; Aldrich.*

992 Ageflex IBOMA

7534-94-3 231-403-1

$C_{14}H_{22}O_2$

Isobornyl methacrylate

Monomers when cured providing hardness, low shrinkage, abrasion resistance, heat and water resistance, good weatherability in automotive coatings, electronics, adhesives, and other acrylic polymers. d_{20}^{20} = 0.983; f.p. <-20°; flash pt. 101°; irritant. *CPS Chemical Co Inc.*

993 Ageflex mDMDAC

7398-69-8 230-993-8

$C_8H_{16}ClN$

Dimethyl diallyl ammonium chloride

N,N-dimethyl-N-2-propenyl-2-propen-1-aminium chloride; diallyl dimethyl ammonium chloride. Monomer for synthesis of homo and copolymers used as coagulant and flocculants for water treatment, mineral processing, demulsifier for petrol, recovery, electrically conductive paper and coatings, wet and dry strength resins, antistatic additives and coatings; cosmetic additives in hair conditioners, biocides, detergent additives, water-soluble polymers and electrographic paper and film. Soluble in H_2O; d = 1.04, 8.7 lb/gal; viscosity = 15 cps; pH 6; cationic; nonhazardous; mildly corrosive. *CPS Chemical Co Inc.*

994 Ageflex MEA

3121-61-7 221-499-3

$C_6H_{10}O_3$

Methoxyethyl acrylate

ethylene glycol monomethyl ether acrylate; glycol monomethyl ether acrylate; 2-methoxyethanol, acrylate; methoxyethyl acrylate; methyl cellosolve acrylate; 2-propenoic acid, 2-methoxyethyl ester; ageflex mea. Solv.-resist. elastomer, polyacrylate rubber, uv-curable reactive diluent, soft contact lenses, PVC impact modifier, fabric coatings, barrier coatings for polyethylene, textile coatings. bp_{17} = 61°; d_{20}^{20} = 1.012; soluble in H_2O, organic solvents; moderately toxic by ingestion and inhalation; LD_{50} (rat orl)= 810 mg/kg. *CPS Chemical Co Inc.*

995 Ageflex NB-50

7398-69-8 230-993-8

$C_8H_{16}ClN$

2-Propen-1-aminium, N,N-dimethyl-N-2-propenyl-, chloride

Diallyl dimethyl ammonium chloride. Diallyl dimethyl ammonium chloride solution in n-butanol. Antistatic finish for polyester fibers, flocculant and coagulant for industrial process water treatment, flocculant for mineral recovery, ion exchange resins, adhesives, acid dye receptivity, electrostatic coatings on wood, retention aids for paper. Nonhazardous; mildly corrosive *CPS Chemical Co Inc.*

996 Ageflex n-HA

2499-95-8 219-698-5

$C_9H_{16}O_2$

n-hexyl acrylate

Acrylic acid, hexyl ester. Monomer for UV-cured inks and coatings, glass coating, viscosity Index improver for functional oils, polymer cements and sealants; polymer modifier. bp_{24} = 88-90°; d = 0.8880; LD_{50} (rat orl) = 26 g/kg. *CPS Chemical Co Ltd.* Name unverified.

997 Ageflex PEA

48145-04-6 256-360-6

$C_{11}H_{12}O_3$

Phenoxyethyl acrylate

2-Propenoic acid, 2-phenoxyethyl ester. UV-curable reactive diluent in inks and coatings, adhesives, viscosity index improver, tile coating. d_{20}^{20} = 1.090; flash pt. 90°. *CPS Chemical Co Inc.*

998 Ageflex TBGE

7665-72-7 219-376-4

$C_7H_{14}O_2$

((1,1-dimethylethoxy)methyl)oxirane

BGE; Glycidyl butyl ether; Butoxymethyl oxirane; *tert*-Butyl Glycidyl Ether; T-BGE; (*Tert*-butoxymethyl)oxirane. Reactive diluent in epoxy resins, corrosion inhibitor in some solvs., modifier for amines, acids and thiols. d_{20}^{20} = 0.910; bp = 164-166°; flash pt. 55°; TWA 25ppm; moderately toxic by ingestion, skin contact, intraperitoneal route; mildly toxic by inhalation *CPS Chemical Co Inc.*

999 Ageflex THFMA

2455-24-5 219-529-5

$C_9H_{14}O_3$

2-Propenoic acid, 2-methyl-, (tetrahydro-2-furanyl)methyl ester

Methacrylic acid, tetrahydrofurfuryl ester; Tetrahydrofurfuryl methacrylate. Anaerobic adhesives and sealants, printed circuit boards, artificial finger nails, modifier for hard rubber rolls, wire and cable coatings, screen printing inks, emulsion polymerization, plastic modifier, EB-curable coatings d = 1.044; bp_4 = 52°; n_D^{20} = 1.4850; flash pt. 90° *CPS Chemical Co Inc; Aldrich; CPS; Monomer-Polymer & Dajac Labs; Rohm Tech.* Name unverified.

1000 Ageflex TM 402, 403, 404, 410, 421, 423, 451, 461, 462

3290-92-4 221-950-4

$C_{18}H_{26}O_6$

2,2-bis(methacryloxymethyl)butyl methacrylate

Trimethylolpropane trimethacrylate; 1,1,1-Trimethylol propane trimethacrylate. Trimethylolpropane trimethacrylate blends; used as a processing aid for extrusion and molding of plastisols and rubber compounds (improves abrasion resistance, adhesion in PVC plastisols, scorch and chemical resistance, elevated temperature stability). d^{20} = 1.06. *CPS Chemical Co Ltd.*

1001 Ageflex TMPTA

15625-89-5 239-701-3

$C_{15}H_{20}O_6$

1,1,1-Trimethylolpropane triacrylate

TMPTA; 2-Ethyl-2-(hydroxymethyl)-1,2,3-propanediol; Trimethylolpropane triacrylate. Crosslinker; uv-cured adhesives, wood fillers, inks, coatings, dry film photo polymer resists, flexographic, offset and screen printing inks, vinyl acrylic latex paint, exterior coatings, highly crosslinked polybutadiene rubber. d_{20}^{20} = 1.108; flash pt. >230°F; LD_{50} (rat orl) = 5190 mg/kg. *CPS Chemical Co Inc.*

1002 Ageflex TMPTMA

3290-92-4 221-950-4

$C_{18}H_{26}O_6$

1,1,1-Trimethylolpropane trimethacrylate

Sipomer™TMPTMA; Trimethylolpropane trimethacrylate. Coagents for wire and cable, hard rubber rolls, polybutadiene and polyethylene, moisture barrier films and coatings, plastisols and vinyl acetate latexes, adhesives, molding compounds, textile products. d = 1.060; bp = 185°; flash pt. >230°F; irritant. *CPS Chemical Co Inc; Aldrich; CPS; U.S. Chems.*

1003 Ageflex TPGDA

68901-05-3 272-647-9

$C_{15}H_{24}O_6$

Tripropylene glycol diacrylate

TRPGDA; PPG-3 diacrylate; Propenoic acid, (1-methyl-1,2-ethanediyl)bis(oxy(methyl-2,1-ethanediyl; ester. Crosslinking monomer for uv-curable inks/coatings, floor tiles, wood coatings and fillers, adhesives, textile finishes, and rubber compounds; thinner for radiation curing systems, inks, etc. bp >120°; flash pt. >100°; d_{20}^{20} = 1.039; irritant. *CPS Chemical Co Inc.*

1004 Ageflex ZDA

14643-87-9 238-692-3

$C_6H_6O_4Zn$

zinc diacrylate

ZDA; Akrochem® ZDA Powd. Crosslinker for molded polybutadiene compounds, conductive and protective coatings, coagent for SBR compounds and reactive pigments; activator for rubber compounding; scorch retarder. *CPS Chemical Co Inc.* Name unverified.

1005 Agefloc A-50

Poly (2, hydroxypropyl-N,N-dimethyl ammonium chloride)

Flocculant and coagulant, dewatering aids in centrifugation, filtration, and flotation of both industrial and municipal waste sludges, potable water treatment. *CPS Chemical Co Inc.*

1006 Agefloc B-50LV
Dimethylamine/epichlorohydrin copolymer
For color removal, low turbidity water clarification, sugar cane/sugar beet processing, latex coagulation, latex waste clarification, liquid/solid separation, sludge dewatering, wire and felt cleaning compounds for paper manufacturing, stabilizer. *CPS Chemical Co Inc.*

1007 Agefloc CF50
Inorganic-aluminum complex; for meat and poultry plant effluents, emulsion breaking, fat and grease separation, oil/water separation, raw water clarification, mining recycle water, replacement for alum or ferric chloride; reduces levels of phosphorus and heavy metals. *CPS Chemical Co Ltd.*

1008 Agefloc PC20HV
7398-69-8 230-993-8
Dimethyl diallyl ammonium chloride
Paper industry retention aid, pigment dispersion, drainage aid, fiber dewatering, stabilizer for sizes, electroconductive polymer, recycling operations, raw and waste water clarification. *CPS Chemical Co Ltd.*

1009 Agefloc WT-20
26062-79-3
Poly(dimethyl diallyl ammonium chloride)
Coagulant for water clarification, potable water treatment, waste water treatment, oil field, flotation enhancement, mining filtration aid. *CPS Chemical Co Ltd.*

1010 Agefloc WT-40
26062-79-3
Polydimethyl diallyl ammonium chloride
2-Propen-1-aminium, N,N-dimethyl-N-2-propenyl-, chloride, homopolymer; N,N-dimethyl-N-2-propenyl-, chloride; Polydiallyldimethylammonium chloride; Ammonium, diallyldimethyl-, chloride, polymers; Cat-floc TL; Propen-1-aminium, N,N-dimethyl-N-2-propenyl-, chloride, homopolymer. Coagulant for water clarification, potable water treatment, waste water treatment, oil field, flotation enhancement, mining filtration aid. *CPS Chemical Co Inc.*

1011 Agenap HMW-H
1338-24-5 215-662-8
Naphthenic acid
Paint dryers, fungicides, metal catalysts, corrosion inhibitors, lubricants, fracturing fluids, cellulose preservatives, solvents, detergents, rubber reclaiming agent. mp = 31°; bp_6 = 160-198°; d = 1.034; slightly soluble in H_2O; LD_{50} (rat orl) = 3000 mg/kg. *CPS Chemical Co Inc.* Name unverified.

1012 Agequat 400
26062-79-3
Polyquaternium-6
For personal care formulations including hair sprays, shampoos, conditioners, mousses and rinses. d = 1.08. *CPS Chemical Co Ltd.*

1013 Agequat C505
7398-69-8 230-993-8
Dimethyl diallyl ammonium chloride
Ageflex mDMDAC. Drainage and retention aid, sludge dewatering. *CPS Chemical Co Ltd.*

1014 Agerite
A full line of phenol and amine rubber antioxidants both primary and secondary; used in all forms of rubber. *BF Goodrich.* Name unverified.

1015 Agerite® Superflex®
Diphenylamine-acetone
Antioxidant used in rubber and mechanical products. Insoluble in H_2O, soluble in organic solvents. *R. T. Vanderbilt Co Inc.*

1016 Agerite® DPPD
74-31-7 3388 200-806-4
$C_{18}H_{16}N_2$
Diphenyl-*p*-phenylene diamine
Agerite DPPD; 1,4-Dianilinobenzene; Diphenyl PPD; DPPD; N,N'-diphenyl-1,4-diaminobenzene; flexamine g; JZF; Nonox DPPD; Diafen; Diafen FF; Altofane DIP; Nonflex H; Permanax 18; Stabilizer DPPD; *p*-bis(phenylamino)benzene; Nocrac DP; Permanax DPPD; DFFD; Antage DP; Ekaland DPPD; Naugard J; N,N'-Diphenyl-p-phenylene diamine. Antioxidant used in rubber and mechanical products; antioxidant for automotive and appliance molded goods, tires, latex; improves environmental flex and stress cracking mp = 144-153°; $bp_{0.5}$ = 220-225°; d = 1.28 g/ml; insoluble in H_2O, soluble in acetone, toluene, chloroform; LD_{50} (rat orl) = 2370 mg/kg. *R. T. Vanderbilt Co Inc.*

1017 Agerite® Hipar®T
65% Dioctylated diphenylamine, 35% diphenyl-*p*-phenylenediamine
An antioxidant for rubber. Soluble in acetone, toluene, chloroform; insoluble in H_2O; mp = 70°. *R. T. Vanderbilt Co Inc.* Discontinued.

1018 Agerite® HP-S
Dioctylated diphenylamine
Diphenyl-*p*-phenylene diamine, 65:35 ratio. Antioxidant for NR, CR, SR, tires, hose, and belting, automotive and appliance molded goods, wire and cable; inhibits oxygen attack, environmental flex and stress cracking; improves heat aging; also used in CR compounds for outdoor service. Powder; soluble in acetone, toluene, chloroform; insoluble in H_2O; d = 1.11 g/ml. *R. T. Vanderbilt Co Inc.* Discontinued.

1019 Agerite® MA
26780-96-1
1,2-dihydro-2,2,4-trimethylquinoline homopolymer
Agerite MA; Antigene RDF; Antioxidant HS; Antioxidant HSL; Flectol H, polymer; Nocrac 224; Nonflex RD; Permanax TQ; Permanax 45; Polnoks R; TDQP. Antioxidant; aging protection to XLPE. mp = 105°; d = 1.03-1.09; insoluble in H_2O, soluble in organic solvents. *R. T. Vanderbilt Co Inc.* Discontinued.

1020 Agerite® Resin D®
26780-96-1
Polymerized 1,2-dihydro-2,2,4-trimethylquinoline
Antioxidant for natural rubber and synthetic rubbers. Insoluble in H_2O, soluble in organic solvents. *R. T. Vanderbilt Co Inc.* Discontinued.

1021 Agerite® Spar
Styrenated phenol
A proprietary antioxidant. *R. T. Vanderbilt Co Inc.* Discontinued.

1022 Agerite® Stalite
68921-45-9 202-965-5
$C_{28}H_{43}N$
Octylated diphenylamines
4-octyl-N-(4-octylphenyl)benzenamine; 4,4'-dioctyldiphenylamine. A proprietary antioxidant. Insoluble in H_2O, soluble in organic solvents. *R. T. Vanderbilt Co Inc.* Discontinued.

1023 Agerite® Stalite® S
68411-46-1 202-965-5
Octylated diphenylamines
Agerite® Stalite. A proprietary antioxidant. mp = 89-103°; insoluble in H_2O; soluble in alcohol, toluene, gasoline. *R. T. Vanderbilt Co Inc.* Discontinued.

1024 Agerite® White
93-46-9 202-249-2
$C_{26}H_{20}N_2$
Sym-di-β-Naphthyl-p-phenylenediamine
N,N'-Di-β-naphthyl-p-phenylenediamine; Agerite white; DBNPD; Aceto DIPP; DNPD; DNPDA; Nonox CL; Tisperse MB-2X. Antioxidant. An antidegradant for latex, nitrile rubber, styrene-butadiene and nitrile-butadiene rubber. mp = 224-230°; d = 1.22-1.28; insoluble in H_2O, EtOH, soluble in organic solvents; LD_{50} (rat orl) = 4500 mg/kg. *R. T. Vanderbilt Co Inc.* Discontinued.

1025 Agesperse 71
Polymeric carboxylic acid, sodium salt
Low foaming dispersant, emulsifier, stabilizer for paper, paints, carpet backcoating, rubber, mining, textiles, ceramic slip, detergents, boiler water compounds, cooling water compounds, adhesives. *CPS Chemical Co Ltd.*

1026 Agesperse 80
Polymeric carboxylic acid, sodium salt
Dispersant, emulsifier, stabilizer for paper, rubber, mining, textiles, ceramic slip, detergents, boiler water compounds, cooling water compounds, adhesives. *CPS Chemical Co Ltd.*

1027 Agestan® 68
Silver amalgam in tablet form. *Bayer AG.*

1028 Agestat 41
7398-69-8 230-993-8
$C_8H_{16}ClN$
Dimethyl diallyl ammonium chloride
Ageflex mDMDAC. Paper industry retention aid, pigment dispersion, drainage aid, fiber dewatering, stabilizer for sizes, electroconductive polymer, recycling operations, raw and waste water clarification. Mildly corrosive; nonhazardous. *CPS Chemical Co Inc.*

1029 Agfa-Gevaert
Imaging systems; for graphic and reprographic systems, X-ray, cinematography, office systems, photography. *Agfa-Gevaert MV.*

1030 Agidex
9032-08-0 232-877-2
Spezyme GA. Glucoamylase (mw = 97,000) an enzyme for conversion of starch dextrins into dextrose. Used in food processing and in manufacture of low-carbohydrate beer. *Glaxo Laboratories.* Name unverified.

1031 Agitan 217
Blend of liquid hydrocarbons, nonionic emulsifiers, and <5% silicone defoamer; for emulsion paints, emulsion polymers, adhesives. Emulsifiable in H_2O, ethylene glycol, isobutanol; d^{20} = 0.91; medium viscosity; flash pt. >140°. *Münzing Chemie GmbH.*

1032 Agitan 281
Blend of liquid hydrocarbons, hydrophobic silica, synthetic copolymers, and

nonionic emulsifiers
Silicone-free defoamer for emulsion paints, emulsion polymers, adhesives, aqueous systems, silicate paints. Soluble in hexanol; emulsifiable in butanol, glycol, glycolethers, isobutanol, MEK, propylene glycol, mineral spirits, H_2O; d^{20} = 0.96; flash pt. >140°; pH 5.0. *Münzing Chemie GmbH.*

1033 Agitan 301
Blend of vegetable oils, modified solids, nonionic emulsifiers, and silicone defoamer
Silicone defoamer; biodegradable, defoamer for emulsion paints, emulsion polymers, synthetic renderings, adhesives. Emulsifiable in H_2O, ethylene glycol, propylene glycol; d^{20} = 0.93; medium viscosity; flash pt. >200°; pH 6.5 (2% in DW, 20°); nonionic. *Münzing Chemie GmbH.*

1034 Agitan E 255
Polysiloxane aqueous emulsion
Defoamer for emulsion paints, gloss emulsion paints, synthetic renderings, adhesives, aqueous systems, glazes, aqueous printing inks, polymerization processes. Soluble in H_2O; emulsifiable in acetone, ethanol, ethylene glycol, IPA, propylene glycol; d_{20} = 1.00; pH 8.5 (2% in H_2O). *Münzing Chemie GmbH.*

1035 Agitan P 800
Blend of liquid hydrocarbons, polyglycols, and amorphous silica
Defoamer for powder systems, powder coatings, synthetic renderings, plasters, fillers, mortars, cements. Apparent d= 320 g/l; pH 7 (1% aqueous suspension). *Münzing Chemie GmbH.*

1036 Agitan VP 725
Blend of modified organo polysiloxanes with nonionic alkoxylated comps.
Silicone compound; defoamer for lacquers, solvent-free systems, printing inks. Soluble in all common laquer solvents; d^{20} = 0.99; medium viscosity; flash point >200°. *Münzing Chemie GmbH.*

1037 Agma
5696
Calcinated magnesite
A magnesium carbonate mineral. *ICI Chem & Polymers Ltd.*

1038 Agnin
Agnolin
Purified wool fat.

1039 Agnowax
Wool wax alcohols. *Croda Chem. Ltd.* Discontinued.

1040 Agomet
Methylmethacrylate adhesives. *Degussa Ltd.*

1041 Agral
Nonionic spreader containing 900 g/l alkylphenol ethoxylate. Wetting, spreading and emulsifying agent for agricultural and horticultural pest control products. *ICI Chem & Polymers Ltd.*

1042 Agramm
Nitrogenous fertilizers. *ICI Chem & Polymers Ltd.* Discontinued.

1043 Agriben
Manure composter for processing liquid and solid manure for agriculture. *Süd-Chemie AG.* Name unverified.

1044 Agricastrol
A range of lubricants and hydraulic fluids for use with farm machinery; for engine oils and hydraulic fluids for tractors of all kinds. *Burmah-Castrol Ltd.* Name unverified.

1045 Agrichem
1702-17-6 2462 216-935-4
Clopyralid
Translocatable herbicide for cereals and established grassland. *Agrichem (International) Ltd.*

1046 Agrichem DB Plus
2,4-DB + MCPA; translocatable herbicide for cereal crops. *Agrichem (International) Ltd.*

1047 Agrichem Flowable Thiram
137-26-8 9510 205-286-2
Thiram
Fungicide with animal repellent properties. *Agrichem (International) Ltd.*

1048 Agrichem MCPA-25, 50
94-74-6 5803 202-360-6
MCPA
Herbicide for cereals and grassland. *Agrichem (International) Ltd.*

1049 Agricol
A proprietary range of alginates
Used for root dipping. *Alginate Industries Ltd.* Name unverified.

1050 Agricorn 500
94-74-6 5803 202-360-6
MCPA
Herbicide for cereals and grassland. *Farmers Crop Chemicals Ltd.*

1051 Agricorn D
1702-17-6 2462 216-935-4
Clopyralid
Translocatable herbicide for cereals and established grassland. *Farmers Crop Chemicals Ltd.*

1052 Agricur®
Systemic nematicide and insecticide with good residual activity against free-living, cyt and root-knot nematodes, soil larvae, banana weevil borers (*Cosmpolites sordidus*) and other sucking and biting pests on the foliage of plants. *Bayer AG.*

1053 Agridin 60
333-41-5 3043 206-373-8
diazinon
An emulsifiable insecticide, active component diazinon; applied as an active insecticide against worms. *Chemical Combine.* Name unverified.

1054 Agrilan® AEC123
Alkoxylate
Coemulsifier, emulsion stabilizer for agricultural toxicant emulsifiable concentrates. d^{40} = 1.040; viscosity = 402 cps; pH 6.9 (1% aqueous solution). *Harcros.*

1055 Agrilan® AEC178
EO/PO copolymer complex
Coemulsifier, emulsion stabilizer for agricultural toxicants. d = 1.043; viscosity = 650 cps.; flash pt. (PMCC)>150°. *Harcros.*

1056 Agrilan® AEC266
577-11-7 3460 209-406-4
Glucoamylase
Emulsifier for production of insecticide emulsifiable concentrates and microemulsion formulations. *Harcros.*

1057 Agrilan® EA14
Surfactant, activity optimizer for agrochemical toxicant formulations. d = 0.928; viscosity = 32 cps; pH 7.0 (1% aq.); flash pt. (PMCC) 46°. *Harcros.*

1058 Agrilan® F513
Polyaromatic ethoxylate phosphate ester
Wetting and dispersing agent for agricultural toxicants. d = 1.130; viscosity = 4300 cps; flash pt. (PMCC) >150°; Ph 6.4 (1% aq.). *Harcros.*

1059 Agrilan® F546
Complex phosphate ester, free acid
Emulsifier for agrochemical flowable/emulsifiable concentrate blends. d = 1.0978; viscosity = 1600 cps; pour point = 14°; pH 6.2 (1% aq.). *Harcros.*

1060 Agrilan® FS101
Epoxidized vegetable oils
Stabilizers for emulsifiable concentrates of agricultural toxicants. d = 0.995; viscosity = 350 cps; pour point = -13°. *Harcros.*

1061 Agrilan® TKA103
Quaternary ammonium compound
Wetting and compatibility agent for quaternary herbicides. Cationic; soluble in H_2O; d = 1.081; viscosity = 400 cs; pH 6.0-9.0 (1% aq.) *Harcros.*

1062 Agrilan® WP101
Aromatic sulfonate
Wetting agent for production of water-dispersed granules and wettable powders of agrochemical toxicants. Soluble in H_2O; pH 9.5 (10%aq.). *Harcros.*

1063 Agrilan® X98
247-557-8
Calcium dodecylbenzene sulfonate
Emulsifier for herbicide and pesticide formulations. Insoluble in H_2O; viscosity = 1080 cs; pH 5-8 (1% aq.) *Harcros.*

1064 Agrilite alloy
A copper-lead alloy containing small amounts of tin.

1065 Agrimer 15L
9003-39-8 7879
poly[1-(2-oxo-1-pyrrolidinyl)ethylene]
Povidone; Polyvidone; P.V.P.; 1-Ethenyl-2-pyrrolidinone polymers; 1-vinyl-2-pyrrolidinone polymers; Plasmosan; Protagent; Vinisil. Preservative and clarifying agent in wines; also used as a pharmaceutic aid. Soluble in H_2O, giving a coloidal solution; also soluble in EtOH, chloroform; insoluble in ether. *ISP.*

1066 Agrimer AL-22
9003-39-8 7879
Alkylated polyvinylpyrrolidone
Povidone. Clarifying agent in wines; pharmaceutic aid *ISP.*

1067 Agrimer VA 6
25086-89-9
$(C_6H_9NO.C_4H_6O_2)_x$
Vinyl pyrrolidone/vinyl acetate copolymer
1-Ethenyl-2-pyrrolidinone, polymer with acetic acid ethenyl ester; PVP/VA copolymer. Film-former used in hairsprays, gels, hair thickeners, tints, and

dyes; suspending agent, dispersant, thickener, stabilizer, adhesion promoter, coatings. *ISP.*

1068 Agrimer VEMA-H-240
9011-16-9
$(C_4H_2O_3.C_3H_6O)_x$
2,5-Furandione, polymer with methoxyethylene
PVM/MA copolymer. Dispersant, coupling, stabilizer, thickener, emulsifier, solubilizer, corrosion inhibitor, film former, antistat, used in paper and textile industries, chemical processing, industrial products, detergents, cosmetics, emulsion polymerization; sequestrant. *ISP; Aldrich; Sigma.*

1069 Agrimul
Anionic/nonionic emulsifiers (Agrimul 26-B, Agrimul 70-A, Agrimul-A-300, Agrimul N-300)
Agrimul 26-B. Agricultural toxicant dispersants. *Harcros.*

1070 Agriphlan 24
1582-09-8 9815 216-428-8
$C_{13}H_{16}F_3N_3O_4$
trifluralin
Trade name for an orange-red liquid containing 240 g/liter trifluralin, a herbicide against cottonweeds, bean, tomato, pear, garlic and sunflower weeds; very efficient against amaranth, bristle-grass, knapweed etc. *Chemical Combine.* Name unverified.

1071 Agrisol PX401
Aromatic alkoxylates
Cosolvent, penetrant, flow promoter for agrochemical toxicants. Clear Liquid.; insoluble in H_2O; d = 1.098; viscosity = 27 cs; pour point = -15°; pH 7.0 (1%aq.); nonionic. *Harcros.*

1072 Agrisorb
61791-44-4 263-177-5
PEG tallowamine ethoxylate
Wetting agent for glyphosate-based herbicides. *ABM Chemicals Ltd.*

1073 Agrispon
Mineral and plant extracts in a water base containing cytokinin, B-vitamin, morphogenic and porphyrin. Activity to aid in increased plant metabolism and soil nutrient availability; used for all agricultural, horticultural and forestry products. *SN Corp/Appropriate Technology Ltd.*

1074 Agrisynth BLO
96-48-0 1632 202-509-5
$C_4H_6O_2$
γ-Butyrolactone
Butyrolactone. Solvent for PAN PS, fluorinated hydrocarbons, cellulose triacetate, shellac; used in paint removers, petrol. processing, hectograph process, specialty inks; intermediate for aliphatic and cyclic compounds; *ISP.*

1075 Agritol
A proprietary ammonium dynamite explosive. No manufacturer.

1076 Agritox
Selective weedkiller. *May & Baker Ltd.*

1077 Agritox 50
94-74-6 5803 202-360-6
MCPA
Selective, systemic, hormone-like herbicide used for post-emergence control of annual and perennial broad-leaved weeds in cereal crops and grassland. *Rhône-Poulenc Crop Protection Ltd.*

1078 Agriwet
Nonionic spreader containing 25% alkylphenol ethoxylate; wetting agent for insecticides, fungicides and foliar feeds. *ABM Chemicals Ltd.*

1079 Agrocide
Insecticide *ICI Chem & Polymers Ltd.*

1080 Agropen
Adjuvant containing 95% emulsifiable vegetable oil. Wetting agent for insecticides. *Ideal Manufacturing Ltd.*

1081 Agrosan GN
62-38-4 7453 200-532-5
phenylmercuric acetate
Organo mercury powder for seed dressing. *ICI Chem & Polymers Ltd.*

1082 Agrosil® LR
Colloidal silicate
Encourages intensive root development, improves irrigation efficiency, improves soils. *BASF AG.* Name unverified.

1083 Agrosol
Liquid mercury seed dressing, a fungicide. *Plant Protection.* Name unverified.

1084 Agrothion
Insecticide. *ICI Chem & Polymers Ltd.*

1085 Agro-Vita
Mineral and plant extracts in a water base containing cytokinin, B-vitamin, morphogenic and porphyrin. Activity to aid in increased plant metabolism and yield; for all agricultural, horticultural and forestry products. *SN Corp/Appropriate Technology Ltd.* Discontinued.

1086 Agroxone 50
94-74-6 5803 202-360-6
MCPA
Herbicide for cereals and grassland. *ICI Agrochemicals.*

1087 Agsol Ex 2
2687-91-4 220-250-6
$C_6H_{11}NO$
N-Ethyl-2-pyrrolidone
NEP; 1-ethyl-2-pyrrolidone. Solvent. bp_{20}= 97°; d = 0.9920; n_D^{20} = 1.4652; LD_{50} (rat orl) = 1350 mg/kg. *ISP.*

1088 Agsol Ex 6C
N-2-Cyclohexyl-2-pyrrolidone.
Solvent. *ISP.*

1089 Agsol Ex 12
2687-96-9 403-730-1
$C_{16}H_{31}NO$
Lauryl pyrrolidone
N-dodecyl-2-pyrrolidone. Conditioner, foam stabilizer, wetting agent; special solvent for commercial cleaning, textile processing, water-borne coatings, inks; replacement for volatile organic compounds. *ISP; Aldrich.*

1090 Agsol Ex BLO
96-48-0 1632 202-509-5
$C_4H_6O_2$
Butyrolactone
Solvent for PAN, PS, fluorinated hydrocarbons, cellulose triacetate, shellac; used in paint removers, petroleum processing, hectograph process, specialty inks; intermediate for aliphatic and cyclic compounds; reaction and diluent solvent for pesticides; used in dyeing of acetate, wetting agent for cellulose acetate films, fibers, solvent welding of plastic films in adhesive applications. *ISP.*

1091 Agsol Ex1
872-50-4 6197 212-828-1
N-Methyl-2-pyrrolidinone
Solvent for resins, paints, acetylene, industrial cleaning, mold cleaning; pigment dispersant; petroleum processing; spinning agent for PVC; intermediate; plastic solvent for the microelectronics industry. *ISP; Aldrich; Allchem Ind; BASF; ARCO; Sigma.*

1092 Agulin
Sheep vaccine *Glaxo Laboratories.* Name unverified.

1093 AH Salt
Hexamethylenediamine adipate
Monomer for production of polyamide 6/6. *BASF AG.*

1094 Aicello
Photo-sensitive diazo film in screen printing technology; stencil film for making photo-stencils in screen printing. *Aicello Chemical Co Ltd.* Name unverified.

1095 Aich metal
Sterro metal; Gedge's metal. An alloy similar to Delta metal, except that it contains iron; usually consists of 60% copper, 38% zinc, and 1.5-2% iron; used for sheathing ships.

1096 AID
Mixture of polyols and salts; used as a detergent; for stabilization of motor fuels and carburetor cleaning. *UOP Inc.* Name unverified.

1097 Air Saltpeter
A mixture of calcium nitrite and nitrate. it is produced by passing air over a series of high intensity alternating arcs, absorbing the gases produced by lime water, and evaporating.
Norwegian saltpeter. *Norwegian Saltpeter.*

1098 Airedale
A range of dyes of various classes; for dyeing of leather. No manufacturer.

1099 Airex
PVC foam (soft and rigid), with a density from 50-400 kg/m³. Used as protective padding and life jackets, as core material in sandwich construction used in boat building, automotive and aviation industries as well as in off-shore oil platforms, used in gymnastic mats, sealings and insulation. *Lonza AG.*

1100 Airflex® 323
24937-78-8
Ethylene/VA copolymer
Base for high-speed paper packaging adhesives; nonwoven binder for wipes, towels, cover stock; formaldehyde-free. *Air Prods & Chem/Polymers.*

1101 Airflex® 456
Vinyl acetate-ethylene-vinyl chloride terpolymer latex
Nonwoven binder for flame retardant fabrics; coating binder and saturant for paper and paperboard applics.; sprayable, excellent water resist., heat

sealable. Dens. 8.8 lb/gal; viscosity = 100-500 cps; pH 5-6.5; anionic *Air Prods & Chem/Polymers.*

1102 **Airflex® 4500**
Ethylene-vinyl chloride copolymer latex
Crosslinkable flexible binder and saturant for paper and paperboard applications; excellent water, alcohol, and grease resistant; inherently flame retardant; nonwoven binder for fiberfill and high loft stock requiring flame retardancy; imparts flexibility and water resistance to caulks, mastics and barrier coats in building applications. d = 9.17 lb/gal; viscosity = 25-150 cps; pH 8.0; anionic. *Air Prods & Chem/Polymers.*

1103 **Airflex® 4514**
Ethylene-vinyl chloride copolymer latex
Coating binder and saturant for paper and paperboard application; inherently flame retardant; binder for flame retardant fabrics, heat sealable nonwovens; imparts flexibility and water resistance to caulks, d = 9.22 lb/gal; viscosity = 25-150 cps. *Air Prods & Chem/Polymers.*

1104 **Airflex® 4530**
Ethylene-vinyl chloride copolymer latex
Coating binder and saturant for paper and paperboard applications; highest stiffness, tensile strength, and flame retardancy in series; excellent water, alcohol, and grease resistance; binder for filters, stiff flame retardant nonwovens; imparts flexibility and water resistance to caulks, mastics and barrier coats in building applications. Dens. 9.37 lb/gal; viscosity = 25-150 cps; pH 8.0; anionic. *Air Prods & Chem/Polymers.*

1105 **Airflex® 4814**
Ethylene-vinyl chloride copolymer latex
Carboxylated version of Airflex 4514; coating binder and saturant for paper and paperboard applications; improved heat and uv stability; also for nonwovens and textile applications. *Air Prods & Chem/Polymers.*

1106 **Airflex® CA-50**
Vinyl acetate-ethylene latex
Proprietary stabilization; for carpetbacking applications; moderately stiff hand; high filler loads; formaldehyde-free. *Air Prods & Chem/Polymers.*

1107 **Airflex® RB-11**
Vinyl acetate-ethylene latex emulsion
PVAL stabilization; for carpetbacking applications; for systems where firmer hand is desired. *Air Prods & Chem/Polymers.*

1108 **Airflex® RB-35**
Ethylene-vinyl chloride emulsion
Surfactant stabilization; for carpetbacking applications; maximum fire retardant characteristics. *Air Prods & Chem/Polymers.*

1109 **Airflex® RB-40**
Ethylene-vinyl chloride emulsion
Surfactant stabilization; for carpetbacking applications; moderately stiff hand, excellent fire retardancy. *Air Prods & Chem/Polymers.*

1110 **Airflex® RB-8**
Vinyl acetate-ethylene emulsion
PVAL stabilization; for carpetbacking applications; good fire retardant characteristics. *Air Prods & Chem/Polymers.*

1111 **Airflex® TL-30**
Vinyl acetate-ethylene latex
PVAL stabilization; for carpetbacking applications. *Air Prods & Chem/Polymers.*

1112 **Airglow**
Bright nickel plating process. *Hanshaw Chemicals.* Unverified.

1113 **Air-hardening steel**
Manganese tool steel containing some tungsten
A trade term applied to a manganese tool steel which hardens when cooled in air.

1114 **Airlift® WB-1210, 1222, 1220, 1270, 1282, 1290**
Waterbased release agents; benefits include elimination of solvent emissions, reduction or elimination of wax build-up on molds, release and surface quality equivalent to solvent-based products. *Air Prods & Chem/Polyurethanes.*

1115 **Airstrip**
Waterproof nonocclusive plaster. *Smith & Nephew Pharmaceuticals Ltd.* Name unverified.

1116 **Airthane® PET-60D; PET-70D; PET-75D; PET-80A; PET-90A; PET-93A; PET-95A**
TDI-PTMEG polyurethane prepolymer
Polymers. *Air Prods & Chem/Polyurethanes.*

1117 **Airthane® PPT-80A; PPT-95A**
TDI-PPG; polyurethane prepolymer. *Air Prods & Chem/Polyurethanes.*

1118 **Airthane® PST-60D**
TDI-polyester polyurethane prepolymer. *Air Prods & Chem/Polyurethanes.*

1119 **Airthane® PST-80A; PST-90A**
TDI-ester polyurethane prepolymer
Polymers. *Air Prods & Chem/Polyurethanes.*

1120 **Airvol®**
Polyvinyl alcohol
Binder, carrier, compounding agent, dispersant, stabilizer, protective colloid in polymerizations; for textiles, paper, adhesives, cement/plaster additive; water and humidity resistance. *Air Prods & Chem/Polymers.*

1121 **Airvol® 103, 107**
9002-89-5 7745
Polyvinyl alcohol, fully hydrolyzed; offers high tensile strength and ease of film formation, excellent adhesive characteristics; super hydrolyzed grades for maximum water and humidity resistance. *Air Prods & Chem/Polymers.*

1122 **Airvol® 125**
9002-89-5 7745
Polyvinyl alcohol, super hydrolyzed
Offers high tensile strength and ease of film formation, excellent adhesive characteristics; super hydrolyzed grades water and humidity resistance. *Air Prods & Chem/Polymers.*

1123 **Airvol® 165**
9002-89-5 7745
Polyvinyl alcohol, super hydrolyzed
Air Prods & Chem/Polymers.

1124 **Airvol® 203, 205**
25213-24-5
Binder, carrier, compounding agent, dispersant, stabilizer, protective colloid in polymerizations; for textiles, paper, adhesives, cement/plaster additive. d = 1.27-1.31; viscosity = 5.2-6.2 cps; pH 4.5-6.5 (4% aq). *Air Prods & Chem/Polymers.*

1125 **Airvol® 205S, 523S, 540S**
25213-24-5
Polyvinyl alcohol, partially hydrolyzed
Polyvinyl alcohol, partially hydrolyzed; binder, carrier, compounding agent, dispersant, stabilizer, protective colloid in polymerization; for textiles, paper, adhesives, cement/plaster additive. *Air Prods & Chem/Polymers.*

1126 **Airvol® 205S, 523S, 540S**
25213-24-5
Binder, carrier, compounding agent, dispersant, stabilizer, protective colloid in polymerizations; for textiles, paper, adhesives, cement/plaster additive. *Air Prods & Chem/Polymers.*

1127 **Airvol® 321, 325, 350**
9002-89-5 7745
Polyvinyl alcohol, fully hydrolyzed
Offers high tensile strength and ease of film formation, excellent adhesive characteristics; super hydrolyzed grades water and humidity resistance. *Air Prods & Chem/Polymers.*

1128 **Airvol® 425, WS 42**
25213-24-5
Polyvinyl alcohol, intermediate hydrolyzed; binder, carrier, compounding agent, dispersant, stabilizer, protective colloid in polymerization; for textiles, paper, adhesives, cement/plaster additive. *Air Prods & Chem/Polymers.*

1129 **Airvol® 425, WS 42**
25213-24-5
Binder, carrier, compounding agent, dispersant, stabilizer, protective colloid in polymerizations; for textiles, paper, adhesives, cement/plaster additive. *Air Prods & Chem/Polymers.*

1130 **Airvol® 523, 540**
25213-24-5
Polyvinyl alcohol, partially hydrolyzed; Airvol 103. Binder, carrier, compounding agent, dispersant, stabilizer, protective colloid in polymerization; for textiles, paper, adhesives, cement/plaster additive. *Air Prods & Chem/Polymers.*

1131 **Airvol® 523, 540**
25213-24-5
Binder, carrier, compounding agent, dispersant, stabilizer, protective colloid in polymerizations; for textiles, paper, adhesives, cement/plaster additive. *Air Prods & Chem/Polymers.*

1132 **Airvol® 705, 723, 740**
25213-24-5
Polyvinyl alcohol, partially hydrolyzed low foam grades. *Air Prods & Chem/Polymers.*

1133 **Airvol® 705, 723, 740**
25213-24-5
Binder, carrier, compounding agent, dispersant, stabilizer, protective colloid in polymerizations. *Air Prods & Chem/Polymers.*

1134 **Airvol® MH-82, MM-14, MM-51, MM-81**
9002-89-5 7745
Polyvinyl alcohol, tackified; derived from fully hydrolyzed grades; tackified grades yield viscous aqueous solutions. Possess tack when applied onto surfaces such as paper, reducing penetration. *Air Prods & Chem/Polymers.*

1135 Airvol® SH-72, SM-73
9002-89-5 7745
Polyvinyl alcohol, tackified
Polyvinyl alcohol, tackified, derived from super hydrolyzed grades. *Air Prods & Chem/Polymers.*

1136 AIT
Pigment dispersions; for in-plant tinting of aqueous coatings. *Pacific Dispersions Inc.* Name unverified.

1137 Aither's Lawn Sand Plus
Dichlorophen-ferrous sulfate
A moss killer/fertilizer mixture for turf. *R. Aitken.*

1138 Aix Oil
Commercial varieties of edible olive oil.
No manufacturer.

1139 Ajax powder
Potassium perchlorate, nitroglycerol, ammonium oxalate, wood meal, and small quantities of collodion cotton, and nitrotoluenes.
An explosive.

1140 Ajicure® MY-24
A proprietary accelerator for latent epoxy resin systems providing high storage stability, longer pot life; curable at lower temperatures. Soluble in H_2O (>0.01 g/100 ml); d = 1.27; pH 8.8 (10% suspension); LD_{50} (mus orl) - 20 g/kg. *Ajinomoto.*

1141 Ajicure® PN-23
A proprietary accelerator for latent epoxy resin systems providing high storage stability, longer pot life; curable at lower temperatures. Soluble in N-methylpyrrolidine (20 g/100 g), DMSO, (<0.01 g/100 g), H_2O; d = 1.28; pH 9.2; LD_{50}: (mus orl) = 1.23 g/kg. *Ajinomoto.*

1142 Ajidew A-100
98-79-3 8185 202-700-3
$C_5H_7NO_3$
L-Pyroglutamic acid
5-oxol-Proline; 5-Oxoproline; L-2-Pyrrolidone-5-carboxylic acid; L-Pyroglutamic acid; 5-Pyrrolidone-2-carboxylic acid; L-5-Pyrrolidone-2-carboxylic acid; 5-Oxopyrrolidine-2-carboxylic acid; Pidolic acid; pyroglutamic acid; PCA. A natural humectant used in cosmetics, soaps, dentifrices, medicinal supplies, tobacco, cellulose film, paper products, fiber products, paints; additive to dyeing agent, softening agent, finishing agent, and antistatic agent; intermediate for synthesis. mp = 181°; soluble in H_2O (10 g/100 ml); $[\alpha]_D^{20}$ = -8.7° (c = 13, H_2O); pH 1.8-2.2. *Ajinomoto.* Name unverified.

1143 Ajidew N-50
28874-51-3 249-277-1
Sodium PCA
Natural humectant used in cosmetics, soaps, dentifrices, medicinal supplies, tobacco, cellulose film, paper products, fiber products; dyeing agent, softening agent, finishing agent, and antistatic agent; intermediate for synthesis; thickener for shampoos. pH 6.8-7.4. *Ajinomoto.* Name unverified.

1144 Ajkaite
Ajkite. A fossil resin found in Hungary. *Ajkite.*

1145 Akarittom fat
A solid fat from *Parinarium laurinum.* mp = 49-50°; iodine value = 214.

1146 Akaustan® A
Flameproofing agent for vegetable and animal fibers - discontinued product. *BASF AG; BASF plc.*

1147 Akbar
A condensation product of formaldehyde and *p*-toluidine. A rubber vulcanization accelerator. No manufacturer.

1148 Akco Resins
A proprietary synthetic phenolic resin
For varnish manufacture. No manufacturer.

1149 Akorex
8016-70-4 232-410-2
Hydrogenated soybean oil
For applications requiring a high stability oil; used as candy centers, color/flavor carriers, in frying, lubricants, spray coatings, vegetable, diary, and anti dusting applications. *Karlshamns.*

1150 Akorol®
Brake fluid components. *BASF AG.*

1151 Akrfax B Light
Sulfur-type vulcanized vegetable oil; extender, processing aid and softener for colored rubber products, especially for dimensional stability in tubing and calendered goods.

1152 Akrochem® 9930 Zinc Oxide Transparent
3486-35-9 10260 222-477-6
CO_3Zn
Zinc carbonate
Carbonic acid, zinc salt (1:1); Zinc carbonate hydroxide. Accelerator-activator for transparent natural and synthetic rubber goods, adhesives. mp = 168°; d = 4.3980. *Akrochem.*

1153 Akrochem® Accelerator CZ-1
N,N dimethyl cyclohexyl ammonium dibutyl dithiocarbamate
Accelerator for natural rubber, styrene/butadiene rubber, or latexes. *Akrochem.*

1154 Akrochem® Accelerator R
103-34-4 3437 203-103-0
$C_8H_{16}N_2O_2S_2$
4,4·g-Dithio dimorpholine
4,4'-Dithiobis[morpholine]; Morpholine, N,N'-disulfide; Dimorpholine N,N'-disulfide; Sulfasan R. Accelerator for uses where a nonblooming or nonstaining sulfur donor is required; used for EV or semi-EV compounds, in synthetic and natural rubbers. Soluble in acetone, benzene, ethyl acetate; d = 1.36. *Akrochem.*

1155 Akrochem® Accelerator VS
Zinc salt of dibutyl dithiophosphoric acid on silica carrier
Nonblooming and nonstaining accelerator for ethylene-propylene-diene rubbers compounds and other sulfur-curable elastomers, especially hose and belt applications. Gray powder; soluble in naphtha, benzene, mineral oils, EtOH; d = 1.42. *Akrochem.*

1156 Akrochem® Accelerator ZIPPAC
Zinc-amine dithiophosphate complex coated with mineral oil
Accelerator; used with thiazoles and thiurams for fast cure rates in ENB-type ethylene-propylene-diene rubberss; nondiscoloring, nonstaining. Gray powder; d = 1.05; mp = 100°. *Akrochem.*

1157 Akrochem® Antioxidant 12
68610-51-5 271-867-2
Cresol dicyclopentadiene butylated reaction product
Nonstaining antioxidant, stabilizer, and antiozonant in polymers, including natural and synthetic polyisoprene, neoprene, nitrile, styrene/butadiene rubber rubber and latex. mp = 115°; d = 1.10. *Akrochem.*

1158 Akrochem® Antioxidant 16
61788-44-1 262-975-0
Alkylated phenols. Nonstaining antioxidant used in styrene/butadiene rubber latex compounding urethanes, ABS, and ethylene-propylene-diene rubbers. Soluble in toluene, ethyl acetate, chloroform, gasoline, hexane; easily emulsifiable in H_2O; d = 1.23. *Akrochem.*

1159 Akrochem® Antioxidant 58
2-Mercapto-4(5)-methyl benzimidazole, Zinc salt
Nonstaining antioxidant for use in rubber compounds to improve heat resistance. White powder; d = 1.75. *Akrochem.*

1160 Akrochem® Antioxidant DQ
26780-96-1
Dihydrotrimethylquinoline polymer
poly(1,2-dihydro-2,2,4-trimethylquinoline); 1,2-dihydro-2,2,4-trimethylquinoline homopolymer; trimethyldihydroquinoline polymer; 2,2,4-trimethyl-1,2-dihydroquinoline polymer; Agerite MA; Antigene RDF; Antioxidant HS; Antioxidant HSL; Flectol H, polymer; Nocrac 224;Nonflex RD; Permanax TQ; Permanax 45; Polnoks R; TDQP. Polymerized; antioxidant in rubber goods requiring resistance to high temperatures; copper and manganese inhibitor; antiozonant. Soluble in acetone, ethyl acetate, alcohol, benzene, methylene chloride, and CCl_4; insoluble in H_2O; m.p. 75°; d = 1.095; LD_{50} (rat orl) = 2250 mg/kg. *Akrochem.*

1161 Akrochem® Antioxidant PANA
90-30-2 201-983-0
$C_{16}H_{13}N$
Phenyl-α-naphthylamine
-(1-Naphthyl)aniline; N-Phenyl-1-Naphthalenamine; N-Phenyl-α-naphthylamine; phenyl-α-naphthylamine; phenyl-1-naphthylamine; Aceto Pan; Additin 30; 1-anilinonaphthalene; C.I. 44050; Neozone A; phenylnaphthylamine; α-phenylnaphthylamine; algerite; 1-Naphthyl phenyl amine. Antioxidant for rubber products; antiflexcracking under dynamic stress. mp = 60-62°; bp_{15} = 226°; d = 1.2; insoluble in H_2O, soluble in benzene, acetone, CCl_4, ethyl acetate, methylene chloride, and ethanol; LD_{50} (rat orl) = 1625 mg/kg. *Akrochem.*

1162 Akrochem® Antioxidant S
Octylated diphenylamine
Antioxidant protecting dynamically stressed rubber products; antiozonant; antiflexcracking. Soluble in acetone, benzene, methylene chloride, and naphtha; insoluble in H_2O; d = 1.00; mp = 88°. *Akrochem.*

1163 Akrochem® Antiozonant MPD-100
Mixed diaryl *p*-phenylene diamine
Antiozonant, antidegradant, antioxidant, and antiflexcracking agent for most diene polymers. d = 1.20; mp = 90-105°. *Akrochem.*

1164 Akrochem® Antiozonant PD-2
793-24-8 212-344-0
N-(1,3-dimethyl butyl)-N'-phenyl-p-phenylene diamine

Antiozonant protecting rubber polymers against heat, oxidation, and flex cracking; copper, manganese inhibitor and styrene/butadiene rubber stabilizer. soluble in benzene, CCl_4, methylene chloride, acetone, ethyl acetate, and ethyl alcohol; insoluble in H_2O; d = 1.10; mp = 45°. *Akrochem.*

1165 Akrochem® Cu.D.D
137-29-1 205-287-8
$C_6H_{12}CuN_2S_4$
Copper dimethyl dithiocarbamate
bis(dimethylcarbamodithioato-S,S') Copper; Copper dimethyldithiocarbamate; Cumate; Copper, bis(dimethylcarbamodithioato-S,S')-, (SP-4-1)-. Accelerator for butyl rubber, ethylene-propylene-diene rubbers rubbers. mp >300°; d = 1.75; slightly soluble in organic solvents. *Akrochem.*

1166 Akrochem® DCBS Granules
Benzothiazyl 1-2-dicyclohexyl sulfenamide
Accelerator for rubber industry. *Akrochem.*

1167 Akrochem® DOTG
97-39-2 202-577-6
$C_{15}H_{17}N_3$
Di-*o*-tolylguanidine
1,3-Di-*o*-tolylguanidine. Accelerator; provides activation for MBTS, MBT, and other thiazole accelerators in natural rubber, styrene/butadiene rubber, NBR, and CR. mp= 167°; d = 1.18; soluble in acetone, ethyl alcohol, ethyl acetate, and methylene chloride. *Akrochem.*

1168 Akrochem® DPG
102-06-7 3383 203-002-1
$C_{13}H_{13}N_3$
Diphenyl guanidine
1,3-Diphenylguanidine; DPG; Melaniline; N,N'-Diphenylguanidine; Nocceler D; Sanceler D; Soxinol D; *sym*-diphenylguanidine; Vulkazit; DPG accelerator; Vulcacid D; Vulkacit D/C. Accelerator-activator for natural rubber, styrene/butadiene rubber, and NBR. Powder; soluble in acetone, methylene chloride, ethyl acetate, ethyl alcohol, and benzene; d = 1.19; mp = 148-150°; LD_{50} (rat orl) = 375 mg/kg. *Akrochem.*

1169 Akrochem® PEG 3350
Polyethylene glycol
Activator for compounding with silica fillers; process aid, lubricant for rubber compounds (natural and synthetic); mold release for foam and mechanical goods. soluble in H_2O; d = 1.20; mp = 54°. *Akrochem.*

1170 Akrochem® Peptizer PTP
133-49-3 205-107-8
C_6HCl_5S
Pentachlorthiophenol
Renacit® 7; pentachlorobenzenethiol; Pentachlorthiophenol. Peptizer for natural rubber, polyisoprene, styrene/butadiene rubber, polybutadiene, NBR, butyl, chloroprene and blends. Mildly toxic by ingestion; severe eye irritant. *Akrochem.*

1171 Akrochem® Plasticizer LN
Naphthenic rubber process oil
Low-viscosity plasticizer. d^{15} = 0.96; d = 7.6 lb/gal; viscosity = 154 SUS (100°F); pour point = -40°F; n_D^{20}= 1.5076. *Akrochem.*

1172 Akrochem® Powder Colors
Organic and inorganic pigments
Colorants for rubber and certain thermoplastic polymers. *Akrochem.*

1173 Akrochem® P.P.D
98-77-1 202-698-4
$C_{11}H_{22}N_2S_2$
Piperidinium pentamethylene dithiocarbamate
Vanax® 552; PIP. Accelerator for latex; in rubber cements, compounds containing white factice and for tank linings; peptizing agent; nondiscoloring and nonstaining. mp = 167° (dec); insoluble in H_2O, soluble in chloroform, acetone, toluene, alcohol; poisonous by ingestion; eye irritant. *Akrochem; R. T. Vanderbilt Co Inc.*

1174 Akrochem® Proaid FILL
Mixture of partially oxidized hydrocarbons; processing aid for injection molding of synthetic and natural rubber. *Akrochem.*

1175 Akrochem® Proaid FLOW
Mixture of fatty acids
Processing aid for extrusion of synthetic and natural rubber. *Akrochem.*

1176 Akrochem® Proaid PEP
Zinc salts of high molecular weight fatty acids
Processing aid for natural and synthetic rubber. *Akrochem.*

1177 Akrochem® Rubbersil RS-150/RS-200
1343-98-2 8634 215-683-2
$SiO_2.xH_2O$, x varies with method of precipitation and extent of drying
hydrated silica
Silica gel. Highly reinforcing filler for use in synthetic and natural rubber compounding, for mechanical rubber goods, tires, adhesives, footwear, soling. Insoluble in H_2O or acids except HF; eye irritant; poison by intravenous route; TLV: TWA 10 mg/m³. *Akrochem.*

1178 Akrochem® TBUT
1634-02-2 216-652-6
$C_{18}H_{36}N_2S_4$
Bis (dibutylthiocarbamoyl) disulfide
Disulfide, bis(dibutylthiocarbamoyl); Butyl Tuads®; Tetrabutyl thiuram disulfide. Accelerator for rubber industry; vulcanizing agent for rubbers Soluble in carbon disulfide, benzene, chloroform, gasoline; insoluble in H_2O; d_{20}^{20} = 1.03-1.06; combustible. *Akrochem.*

1179 Akrochem® TDEC
20941-65-5 244-121-9
$C_{20}H_{40}N_4S_8Te$
Tellurium diethyl dithiocarbamate
TeEDC; Ethyl tellurac; Ethyl Tellurac®; Ethyl Tellurac®Rodform; Perkacit® TDEC; tellurium diethyldithiocarbamate; tetrakis(diethylcarbamodithioato-S,S')tellurium; tellurac; diethyldithiocarbamic acid tellurium salt; tellurium,tetrakis(diethyldithiocarbamate)-; Tellurium (IV) diethyldithiocarbamate. Fast-curing primary or secondary accelerator for use in natural rubber, styrene/butadiene rubber, NBR, ethylene-propylene-diene rubbers, and butyl. Insoluble in H_2O, soluble in benzene, carbon disulfide, alcohol, gasoline; d = 1.44; mp = 108-118°; harmful if inhaled; possible skin and eye irritant; LD_{50}: rat orl >5000 mg/kg. *Akrochem.*

1180 Akrochem® TETD
97-77-8 3428 202-607-8
$C_{10}H_{20}N_2S_4$
disulfiram
TTD; TETD; Bis(diethylthiocarbamyl) disulfide; Disulfiram; Ekaland TETD; Ethyl Tuads® Rodform; Perkacit® TETD. Ultra-accelerator for rubber industry; fungicide. mp = 71-72°; d = 1.30; insoluble in H_2O, soluble in organic solvents; LD_{50} (rat orl) = 8.6 g/kg. *Akrochem; Abbott Labs; Aldrich; Novachem; R. T. Vanderbilt Co Inc.*

1181 Akrochem® Thio No. 1
102-08-9 3393 203-004-2
$C_{13}H_{12}N_2S$
N,N'-diphenylthiourea
sym-Diphenylthiourea; 1,2-Diphenyl-2-thiourea; Thiocarbanilide; A-2; Rhenocure® CA; Stabilizer C. Accelerator for CR latex, natural rubber latex, and cements, ethylene-propylene-diene rubbers sponge compounds; activates thiazole accelerators; essentially nondiscoloring. mp = 153-154°; b.p. dec.; insoluble in H_2O; soluble in EtOH, ether, chloroform; d^{25} = 1.32; combustible; moderate toxic by ingestion and intraperitoneal routes; heated emits highly toxic fumes. *Akrochem.*

1182 Akrochem® TMTD
137-26-8 9510 205-286-2
$C_6H_{12}N_2S_4$
Tetramethylthiuram disulfide
Thiuram disulfide; Ancazide ME; Bis(dimethylthiocarbamyl) disulfide; TMTD; Agrichem Flowable; Akrosperse® D-177; Methyl Tuads°; Methyl Tuads® Rodoform; Naftocit® Thiuram 16; Naftopast® Thiuram 16-P; Perkacit® TMTD; Agrichem Flowable,. Very active, sulfur-bearing, nondiscoloring organic accelerator and activator; for curing systems requiring very low or no sulfur and for butyl and ethylene-propylene-diene rubbers compounds. mp = 155-156°; d^{20} = 1.29; insoluble in H_2O; soluble in organic solvents; LD_{50}: (rat orl) = 640 mg/kg. *Akrochem.*

1183 Akrochem® Z.B.E.D
14726-36-4 238-778-0
$C_{30}H_{28}N_2S_4Zn$
Zinc dibenzyl dithiocarbamate
ZBeDC; Arazete®; Naftocit® ZBEC; Octocure ZBZ-50; Vulkacit ZBEC. Accelerator for rubber, latex dispersions, cements; nondiscoloring and nonstaining. mp = 186°; insoluble in H_2O, acetone, gasoline, soluble in benzene, ethylene dichloride. *Akrochem.*

1184 Akrochem® ZDA Powd
14643-87-9 238-692-3
Zinc diacrylate
Ageflex ZDAZDA. Activator for rubber compounding. *Akrochem.*

1185 Akrochem® Z.P.D
Zinc pentamethylene dithiocarbamate
Ultra-accelerator for dry, natural rubber, especially footwear; nonpigmenting and nonstaining. *Akrochem.* Name unverified.

1186 Akrochlor
Chlorinated paraffin; fire retardant. *Akrochem.*

1187 Akrodye
Dyes. *Akrochem.*

1188 Akrofax 900C
Vulcanized vegetable oil; rubber processing aid; absorbent for mineral oils and other liquid plasticizers on mill and Banbury equipment; speeds

incorporation of fillers; flow promoter; provides unique surface finish to vulcanized rubber goods; improves ozone resistance. *Akrochem.*

1189 Akrofax A

Vulcanized vegetable oil; rubber processing aid; absorbent for mineral oils and other liquid plasticizers on mill and Banbury equipment; speeds incorporation of fillers; flow promoter; provides unique surface finish to vulcanized rubber goods; improves ozone resistance. *Akrochem.*

1190 Akroflex DAZ

A proprietary amine blend used as a combined antioxidant and antiozonant. *Akrochem.* Name unverified.

1191 Akroform® ETU-22 PM

96-45-7 3849 202-506-9
2-Imidazolidinethione; ethylene thiourea
Ethylene thiourea on compatible polymeric binder carrier; accelerator imparting a high state of cure to neoprene compounds; nonstaining and nondiscoloring. *Akrochem.* Name unverified.

1192 Akrol

A proprietary synthetic resin. *Akrochem.* No manufacturer.

1193 Akrolease® E-9410

Silicone polymer; release agent for PU, polyester, or epoxy materials from plastic or metal molds; corrosion inhibitor for metal. *Akrochem.*

1194 Akroplast®

Thermoplastic pigment dispersions; colorants for vinyls, acrylics, polyethylene, cellulose acetate butyrate, rubbers, urethanes, and phthalate pastes. *Akrochem.*

1195 Akrosperse® Color Masterbatches

Dispersions of Akrochem® Powder Colors in various elastomers, most commonly styrene/butadiene rubber and EPR. *Akrochem.*

1196 Akrosperse® Plasticizer Paste Colors

Organic and inorganic pigments; plasticizer pigment dispersions. *Akrochem.*

1197 Akrosperse® Water Paste Colors

Aqueous pigment dispersions; colorants for latex products and adhesives. *Akrochem.*

1198 Akrotak 100

Pentaerythritol ester; tackifier for synthetic and natural rubber, adhesives and hot melts; provides wetting for filler/rubber; speeds up incorporation of mineral fillers. *Akrochem.*

1199 Akrowax PE

9002-88-4 7728
Low molecular weight polyethylene; process aid in natural and synthetic rubbers. *Akrochem.*

1200 Aktiplast

A range of zinc salts of unsaturated acids used as peptizing agents and dispersing agents and aromatic disulfides used as reclaiming agents in the rubber industry; used for molded and extruded articles, production of reclaims. *Bayer plc.*

1201 Aktisil AM

919-30-2 213-048-4
$C_9H_{23}NO_3Si$
γ-Aminopropyltriethoxysilane
3-Aminopropyl-triethoxysilane; 3-Triethoxysilylpropylamine; AMEO; Dynasylan AMEO. Filler for thermosets, thermoplastics. mp = -70°; bp = 217°; d = 0.9420; n_D^{20}= 1.4210; LD_{50} (rat orl) = 1780 mg/kg. *Hoffmann Min.*

1202 Aktisil EM

2530-83-8 219-784-2
$C_9H_{20}O_5Si$
Glycidoxy propyl trimethoxysilane
Glymo; σ-Glycidoxypropyltrimethoxysilane; Dynasylan GLYMO. Filler for thermosets. bp_2 = 120°; d = 1.0700; n_D^{20}= 1.4290; LD_{50} (rat orl) = 23 g/kg. *Hoffmann Min.*

1203 Aktisil MM

4420-74-0 224-588-5
$C_6H_{16}O_3SSi$
γ-Mercaptopropyltrimethoxysilane
Trimethoxysilylpropanethiol; 1-Propanethiol, 3-(trimethoxysilyl)-; γ-Mercaptopropyl trimethoxy silane; Dynasylan MTMO. Filler for sulfur and metal oxide-cured systems. bp = 215°; d = 1.0390; n_D^{20}= 1.4436; LD_{50} (rat orl) = 2940 mg/kg. *Hoffmann Min.*

1204 Aktisil PF 216

40372-72-3 254-896-5
$C_{18}H_{42}O_6S_4Si_2$
3,16-Dioxa-8,9,10,11-tetrathia-4,15-disilaoctacane
Bis(3-(triethoxysilyl)propyl) tetrasulfane; Bis[3-(triethoxysilyl)propyl]tetrasulfide. Filler for sulfur-cured systems. *Hoffmann Min.*

1205 Aktisil VM

1067-53-4 213-934-0
$C_{11}H_{24}O_6Si$
Vinyl-tris(β-methoxyethoxy)silane
Vinyl-tris(2-methoxyethoxy)-silane; 2,5,7,10-Tetraoxa-6-silaundecane, 6-ethenyl-6-(2-methoxyethoxy)-; Tris(2-methoxyethoxy)vinylsilane; Dynasylan VTMOEO. Filler for peroxide-cured systems. mp = -30°; bp = 285°; d = 1.0300; n_D^{20}= 1.4284; reacts with H_2O; LD_{50} (rat orl) = 2960 mg/kg. *Hoffmann Min.*

1206 Akulon

Nylon 6 and 66
Algemene Industriele. Unverified.

1207 Akulon K and M

Proprietary grades of Nylon 6. *Algemene Industriele.* Unverified.

1208 Akulon R2

A grade of Nylon 66. *Algemene Industriele.* Unverified.

1209 Akuloy® J-75/30

Nylon 6 alloy with functionalized polyolefins, glass fiber-reinforced; engineering thermoplastic with improved dimensional stability for automotive, consumer products (power tool housings, furniture components), pump housings, impellers, fans, bearing retainers, gears and fasteners. *DSM.* Discontinued.

1210 Akuloy® J-75/30/HI

Nylon 6 alloy with functionalized polyolefins, glass fiber- reinforced; engineering thermoplastic with improved dimensional stability for automotive, consumer products (power tool housings, furniture components), pump housings, impellers, fans, bearing retainers, gears and fasteners. *DSM.* Discontinued.

1211 Akuloy® NY-75

Nylon 6 alloy with functionalized polyolefins; engineering thermoplastic with improved dimensional stability for automotive, consumer products (power tool housings, furniture components). pump housings, impellers, fans, bearing retainers, gears, fasteners *DSM.* Discontinued.

1212 Akund

A vegetable down of the kapok class, from *Asclepias* species of South Africa.

1213 Akwilox 133

68952-98-7 273-181-9
Brominated soybean oil
Food additive in soft drinks for viscosity adjustment. *Am.Chem. Services.*

1214 Akypo 1690 S

Laureth-5 carboxylic acid
Additive for liquid heavy-duty detergent formulations; thickener for NaOH. *Chem-Y GmbH.*

1215 Akypo AD 100 SPC

Sodium PEG-3 lauramide carboxylate
Surfactant for mild cosmetic products. *Chem-Y GmbH.*

1216 Akypo ITD 30 N

68891-17-8
Sodium trideceth-3 carboxylic acid
Sodium POE(3) Tridecyl Ether Acetate; Sodium Trideceth-3 Carboxylate; Sodium POE(6) Tridecyl Ether Acetate; Sodium Trideceth-6 Carboxylate. Emulsifier for silicone oil; biodegradable. *Chem-Y GmbH.*

1217 Akypo LF 1

Capryleth-6 carboxylic acid
Low foaming surfactant for industrial, institutional and household cleaning; alkaline and acid stable; biodegradable. *Chem-Y GmbH.*

1218 Akypo LF 2

107600-33-9
Capryleth-9 carboxylic acid
Low foaming surfactant for industrial, institutional and household cleaning, cooling tower cleaners, disinfectant cleaners, high-pressure cleaners, metalworking fluids, electroplating, PU foam for orthopedic uses; alkaline and acid stable; biodegradable. *Chem-Y GmbH.*

1219 Akypo LF 3

105391-15-9
Hexeth-4 carboxylic acid
Low foaming surfactant for industrial, institutional and household cleaning; alkaline and acid stable; biodegradable. *Chem-Y GmbH.*

1220 Akypo LF 4

Capryleth-9 carboxylic acid and hexeth-4 carboxylic acid. Low foaming surfactant for industrial, institutional and household cleaning, cutting and drilling oils, drawing and rolling oils, water treatment products, high-pressure cleaners, cooling tower cleaners, electroplating, film developing; solubilizer; alkali and acid stable; biodegradable. *Chem-Y GmbH.*

1221 Akypo LF 4N

Sodium capryleth-9 carboxylate and sodium hexeth-4 carboxylate. Low foaming surfactant for industrial cleaning; alkaline and acid stable; biodegradable *Chem-Y GmbH.*

1222 Akypo LF 5

105391-15-9

Buteth-2 carboxylic acid
Low foaming surfactant for industrial, institutional and household cleaning; alkaline and acid stable; biodegradable. *Chem-Y GmbH.*

1223 Akypo LF 6
Buteth-2 carboxylic acid
Buteth-2 carboxylic acid and capryleth-9 carboxylic acid. Low foaming surfactant for industrial, institutional and household cleaning, metalworking fluids, disinfectant cleaners, high-pressure cleaners, engine cleaners, automatic dishwash, cooling water systems, electroplating, textile pretreatment; alkali and acid stable; biodegradable. *Chem-Y GmbH.*

1224 Akypo MB 1614/1
Capryleth-4 carboxylic acid
Surfactant for industrial cleaning. *Chem-Y GmbH.* Name unverified.

1225 Akypo MB 2528S
105391-15-9
Buteth-2 carboxylic acid
Surfactant for industrial cleaning. *Chem-Y GmbH.* Name unverified.

1226 Akypo NP 70
3115-49-9 221-486-2
$C_{17}H_{26}O_3$
(4-Nonylphenoxy)-acetic acid
Nonoxynol-8 carboxylic acid. Emulsifier. *Chem-Y GmbH.*

1227 Akypo NTS
68987-89-3
Sodium laureth-6 carboxylate
Detergent for carpet and upholstery cleaners especially aerosols; leak detector spray. *Chem-Y GmbH.*

1228 Akypo OCD 10 NV
Sodium deceth-2 carboxylate and sodium capryleth-2 carboxylate. Defoamer for phosphoric acid industry (suitable for nitrate process). *Chem-Y GmbH.*

1229 Akypo OP 190
72160-13-5
Octoxynol-20 carboxylic acid
Octoxynol-9 carboxylic acid. Detergent, emulsifier; used in aqueous solutions, metalworking fluids, electroplating; lime soap dispersant; moderate foam. *Chem-Y GmbH.*

1230 Akypo OP 80
72160-13-5
Octoxynol-9 carboxylic acid
Detergent, emulsifier; used in aqueous solutions, metalworking fluids, emulsion polymerization, film developing baths; lime soap dispersant; moderate foam. *Chem-Y GmbH.*

1231 Akypo RCS 60
68954-89-2
Ceteareth-7 carboxylic acid
Surfactant. *Chem-Y GmbH.*

1232 Akypo RLM 100
Laureth-11 carboxylic acid
Emulsifier for cosmetics applications; foam booster for cleaners, heavy-duty detergent formulations. *Chem-Y GmbH.*

1233 Akypo RLM 100 NV
33939-64-9
Sodium laureth carboxylate series. Surfactant and additive for personal care products. *Chem-Y GmbH.*

1234 Akypo RLM 130
68954-89-2
Laureth-14 carboxylic acid
Emulsifier. *Chem-Y GmbH.*

1235 Akypo RLM 160
27306-90-7
Laureth-17 carboxylic acid
Emulsifier. *Chem-Y GmbH.*

1236 Akypo RLM 38
Laureth-5 carboxylic acid
Surfactant. *Chem-Y GmbH.*

1237 Akypo RLM25
68954-89-2
Laureth-4 carboxylic acid
Emulsifier, dispersant for emulsion and dispersion use; metalworking fluids; lime soap dispersant; moderate foam. *Chem-Y GmbH.*

1238 Akypo RLMQ 38
68954-89-2
Laureth-5 carboxylic acid
Emulsifier, dispersant, additive for personal care products, household and industrial formulas; primary emulsifier for synthetic latex. *Chem-Y GmbH.*

1239 Akypo RO 20
Oleth-3 carboxylic acid
Emulsifier for metalworking fluids; lime soap dispersant; very slow foaming. *Chem-Y GmbH.*

1240 Akypo RO 50
57635-48-0
Oleth-6 carboxylic acid
Emulsifier for cleaning agents and metal cooling liquids; chain lubricant; lime soap dispersant; slow foaming. *Chem-Y GmbH.*

1241 Akypo RO 90
57635-48-0
Oleth-10 carboxylic acid
Emulsifier for metalworking fluids; lime soap dispersant; slow foaming; stable to hard water; biodegradable. *Chem-Y GmbH.*

1242 Akypo RS 100
68954-89-2
Steareth-11 carboxylic acid
Surfactant. *Chem-Y GmbH.*

1243 Akypo RS 60
68954-89-2
Steareth-7 carboxylic acid
Surfactant. *Chem-Y GmbH.*

1244 Akypo RT 60
68954-89-2
Talloweth-7 carboxylic acid
Surfactant. *Chem-Y GmbH.*

1245 Akypo TBP 180
104909-82-2
Butoxynol-19 carboxylic acid
Surfactant for zinc galvanization processes. *Chem-Y GmbH.*

1246 Akypo TBP 40
Butoxynol-5 carboxylic acid
Surfactant for metalworking fluids. *Chem-Y GmbH.*

1247 Akypo TFC-S
Laureth-5 carboxylic acid and sodium octyl sulfate
Biodeg, foaming detergent, thickener, disinfectant for dairy, brewery, sanitary cleaning. *Chem-Y GmbH.*

1248 Akypo TPR
Sodium hexeth-4 carboxylate and trideceth-2
Foam-suppressant surfactant for powder cleaners for dishwashers, steam carpet cleaners, metalworking industry. *Chem-Y GmbH.*

1249 Akypo® Muls 400
90453-59-1
PEG-9 stearamide carboxylic acid
Nontoxic, biodegradable emulsifier for cosmetics, oil-water emulsions. *Chem-Y GmbH.*

1250 Akypo® Soft 100 MgV
99330-44-6
Magnesium laureth-11 carboxylate
Detergent, emulsifier, wetting agent for shampoos, liquid soaps, foam baths, low irritation formulas; biodegradable. *Chem-Y GmbH.*

1251 Akypo® Soft 100 NV
68987-89-3
Sodium laureth 11-carboxylate
Detergent, emulsifier, wetting agent for shampoos, liquid soaps, foam baths, feminine hygiene products, low irritation formulas; PU foam for orthopedic use; biodegradable. *Chem-Y GmbH.*

1252 Akypo® Soft 160 NV
33939-64-9
Surfactant for contact lens cleaning fluids; very low eye irritation. *Chem-Y GmbH.*

1253 Akypo® Soft 45 NV
53610-02-9
Sodium laureth-6 carboxylate
Cosmetics surfactant for baby care products, shampoos, foam baths, medicinal liquid soaps; biodegradable. *Chem-Y GmbH.*

1254 Akypo® Soft KA 250 BV
107628-03-5
Sodium PEG-6 cocamide carboxylate
Economical surfactant for formulation of mild shampoos, foam baths, shower baths, liquid soaps; mild to skin and eyes; biodegradable. *Chem-Y GmbH.*

1255 Akypogene FP 35 T
61790-64-5 263-155-5
TEA cocoate
Cosmetics surfactant for shower, shampoo and bath formulations, liquid hand cleaner. *Chem-Y GmbH.*

1256 Akypogene HM 12
Sodium C12-13 pareth sulfate, sodium PEG-6 cocamide carboxylate, disodium laureth sulfosuccinate and trideceth-2 carboxamide MEA.

Economical surfactant for preparation of mild shampoos, foam baths, shower baths, and liquid soaps; not irritating to optic mucosa. *Chem-Y GmbH.*

1257 Akypogene HM 8
MEA lauryl sulfate, sodium PEG-6 cocamide carboxylate and disodium laureth sulfosuccinate. Economical surfactant for manufacturing of mild shampoos, foam baths, shower baths, liquid soaps; not irritating to optic mucosa. *Chem-Y GmbH.*

1258 Akypogene Jod F
Sodium laureth-11 carboxylate, iodine
Antibacterial surfactant for food, dairy, beverage industries, agriculture and hospitals. *Chem-Y GmbH.*

1259 Akypogene KTS
Ammonium polyacrylate and sodium laureth-6 carboxylate. Base for rug shampoo and upholstery cleaner with antistatic and anticorrosive properties; especially for aerosols. *Chem-Y GmbH.*

1260 Akypogene SO
MEA-PPG-6-laureth-7 carboxylate, nonoxynol-2
Surfactant for metalworking cooling lubricants; emulsifier, lime soap dispersant. *Chem-Y GmbH.*

1261 Akypogene VSM-N
Trideceth-2, sodium dodecylbenzene sulfonate
For manufacturing of liquid scour creams with excellent stability. *Chem-Y GmbH.*

1262 Akypogene WSW-W
Sodium laureth sulfate, MEA laureth-6 carboxylate, cocamide DEA and sodium dodecylbenzene sulfonate
Surfactant for mild wool detergent formulations. *Chem-Y GmbH.*

1263 Akypogene ZA 97 SP
Potassium xylene sulfonate, potassium tallate and potassium cocoate
Surfactant blend for liquid soap. *Chem-Y GmbH.*

1264 Akypomine® BC 50
68130-43-8 268-589-9
Fatty alcohol ether sulfate
Surfactant for mineral industry; flotation agent for barite (selective); collector for flotation of typical salt minerals such as fluorspar, magnesite or scheelite. *Chem-Y GmbH.*

1265 Akypomine® BC/S
Fatty alcohol sulfate
Flotation agent for barite. *Chem-Y GmbH.*

1266 Akypomine® MW 05
39464-66-9
Laureth-7 phosphate
Surfactant for mining industry; fluorspar collector selective for barite. *Chem-Y GmbH.*

1267 Akypomine® P 191
Oleamine hydroxypropyl bistrimonium chloride-polyacrylamide. Filtration auxiliary with flocculant for mining industry. *Chem-Y GmbH.*

1268 Akypopress DB
9002-97-5
Synthetic polymer
Depressant for metal ions. *Chem-Y GmbH.*

1269 Akypoquat 40
Oleoyl PG-trimonium chloride-stearoyl PG-trimonium chloride-behenoyl PG-trimonium chloride-palmitoyl PG-trimonium chloride and trideceth-2. Environmentally safe laundry softener concentrate, textile softener; antistat, rewetting agent; fully biodegradable. *Chem-Y GmbH.*

1270 Akypoquat 129
Isostearoyl PG-trimonium chloride and behenoyl PG-trimonium chloride. Raw material for manufacturing of laundry softeners; molecule is fully biodegradable. *Chem-Y GmbH.* Name unverified.

1271 Akypoquat 131
69537-28-8
Behenoyl PG-trimonium chloride
Raw material for cosmetic hair products, cream rinses; antistat; good wet and dry combing properties; molecule is fully biodegradable. *Chem-Y GmbH.*

1272 Akypoquat 132
Lauroyl PG-trimonium chloride and hexylene glycol. Raw material for cosmetic hair and skin products; molecule is fully biodegradable. *Chem-Y GmbH.*

1273 Akyporox CO 400
61788-85-0
PEG-40 hydrogenated castor oil. Perfume solubilizer, oil-water emulsifier for cosmetic products; eliminates oil bath turbidity. *Chem-Y GmbH.*

1274 Akyporox NP 105
9016-45-9 6772
Nonoxynol-10
Emulsifier. *Chem-Y GmbH.*

1275 Akyporox NP 1200V
9016-45-9 6772
Nonoxynol-120
Emulsifier for emulsion polymerization. *Chem-Y GmbH.*

1276 Akyporox NP 15
27986-36-3 248-762-5
Nonoxynol-1
Emulsifier for emulsion polymerization. *Chem-Y GmbH.*

1277 Akyporox NP 150
9016-45-9 6772
Nonoxynol-15
Emulsifier. *Chem-Y GmbH.*

1278 Akyporox NP 200
9016-45-9 6772
Nonoxynol-20
Emulsifier, wetting agent used in textile products, emulsion polymerization, degreasing baths, electroplating industry. *Chem-Y GmbH.*

1279 Akyporox NP 30
9016-45-9 6772
Nonoxynol-3
For manufacturing of hair dye formulations; emulsifier for emulsion polymerization. *Chem-Y GmbH.*

1280 Akyporox NP 300V
9016-45-9 6772
Nonoxynol-30
Emulsifier for calcium stearate, emulsion polymerization. *Chem-Y GmbH.*

1281 Akyporox NP 40
9016-45-9 6772
Nonoxynol-4
Emulsifier for film developing baths. *Chem-Y GmbH.*

1282 Akyporox NP 90
9016-45-9 6772
Nonoxynol-9
Emulsifier. *Chem-Y GmbH.*

1283 Akyporox NP 95
9016-45-9 6772
Nonoxynol series. Emulsifier for calcium stearate. *Chem-Y GmbH.*

1284 Akyporox OP 100
9002-93-1 6858
Octoxynol-10
Emulsifier; dust suppressant for coal mining industry. *Chem-Y GmbH.*

1285 Akyporox OP 115 SPC
9002-93-1 6858
Octoxynol-12
Emulsifier, wetting agent. *Chem-Y GmbH.*

1286 Akyporox OP 200
9002-93-1 6858
Octoxynol-20
Emulsifier. *Chem-Y GmbH.*

1287 Akyporox OP 250V
9002-93-1 6858
Octoxynol-25
Emulsifier for emulsion polymerization; perfume solubilizer. *Chem-Y GmbH.*

1288 Akyporox OP 400V
9002-93-1 6858
Octoxynol-40
Emulsifier for emulsion polymerization. *Chem-Y GmbH.*

1289 Akyporox RC 200
9004-95-9
$C_{56}H_{114}O_{21}$
Ceteth-20
Brij 58; Polyoxyethylene(20) cetyl ether; polyoxyethylene (10) cetyl alcohol; polyoxyethylene (20) cetyl alcohol; Cetomacrogol 1000; POE (10) cetyl alcohol; POE (20) cetyl alcohol; POE(10) Cetyl Ether; Ceteth-10; POE(15) cetyl Ether; Ceteth-15; Diethylene glycol Cetyl Ether; Ceteth-2; POE(20) Cetyl Ether; Ceteth-20; POE(23) Cetyl Ether; POE(25) Cetyl Ether; Ceteth-25; POE(30) Cetyl Ether; Ceteth-30; POE(40) Cetyl Ether; POE(5.5) Cetyl Ether; Ceteth-6; POE(7) Cetyl Ether. Solubilizer for cosmetic products. *Chem-Y GmbH.*

1290 Akyporox RLM 160
9002-92-0 7717
$[C_{20}H_{42}O_5]_n$
Polidocanol
Polyoxyethylene(23) lauryl ether; Brij 35. Emulsifier for emulsion polymerization. mp = 40-42° *Chem-Y GmbH.*

1291 Akyporox RLM 22
3055-93-4 221-279-7
Laureth-2

Emulsifier for manufacturing of hair dye formulations; additive for manufacturing of snow from spray cans. *Chem-Y GmbH.*

1292 Akyporox RLM 40
5274-68-0 226-097-1
Laureth-4
Emulsifier for manufacturing of hair dye formulations, cosmetic aerosols, oil bath formulations, window cleaners, hand cleaners, heavy-duty detergents. *Chem-Y GmbH.*

1293 Akyporox RLM 80V
Laureth-8
Emulsifier for manufacturing of cosmetic aerosols, heavy-duty detergents, all-purpose cleaners. *Chem-Y GmbH.*

1294 Akyporox RO 90
9004-98-2
Oleth-9
Emulsifier, wetting agent for textile products, metalworking fluids, all-purpose cleaners, hand cleaners, creams and lotions. *Chem-Y GmbH.*

1295 Akyporox RTO 70
9004-98-2
Oleth-7
Emulsifier for emulsion polymerization. *Chem-Y GmbH.*

1296 Akyporox RZO 30
61791-12-6
PEG-3 castor oil; surfactant for metalworking cooling lubricants. *Chem-Y GmbH.*

1297 Akyposal 23 ST 70
Sodium C12-13 pareth sulfate
Detergent; shampoo base. *Chem-Y GmbH.*

1298 Akyposal 100 DAL
107600-36-2
TIPA-laureth sulfate
Emulsifier for bath oils; suited for anhydrous formulations; biodegradable. *Chem-Y GmbH.*

1299 Akyposal 2010 S
Sodium laureth sulfate-Cocamide DEA-glycol distearate. Surfactant, pearling agent, foam stabilizer for cosmetics; biodegradable. *Chem-Y GmbH.*

1300 Akyposal 2010 SD
Sodium PEG-6 cocamide carboxylate, glycol distearate. Surfactant, pearling agent for cosmetics, biodegradable. *Chem-Y GmbH.*

1301 Akyposal 9278 R
9004-82-4
Sodium laureth sulfate
Emulsifier for emulsion polymerization. *Chem-Y GmbH.*

1302 Akyposal ALS 33
2235-54-3 218-793-9
Ammonium lauryl sulfate
Detergent, emulsifier; shampoo base; used in emulsion polymerization. *Chem-Y GmbH.*

1303 Akyposal BA 28
Sodium trideceth sulfate
Detergent for personal care, dishwashing, and textile products. *Chem-Y GmbH.* Name unverified.

1304 Akyposal BD
69011-84-3
Sodium octoxynol-6 sulfate
Emulsifier for emulsion polymerization. *Chem-Y GmbH.*

1305 Akyposal DS 28
68957-18-6 273-328-7
Sodium laureth sulfate
Detergent; shampoo base. *Chem-Y GmbH.*

1306 Akyposal DS 56
68957-18-6 273-328-7
Sodium Laureth Sulfate
Detergent; shampoo base. *Chem-Y GmbH.*

1307 Akyposal EO 20 MW
9004-82-4
Sodium laureth sulfate
Detergent, base for shampoos and bubble baths, emulsifier for emulsion polymerization. *Chem-Y GmbH.*

1308 Akyposal HF 28
Sodium laureth sulfate and magnesium laureth-16 sulfate
Base for mild shampoos and shower products. *Chem-Y GmbH.*

1309 Akyposal MGLS
3097-08-3 221-450-6
Magnesium lauryl sulfate
Detergent for shampoos, foam bath. *Chem-Y GmbH.*

1310 Akyposal MLES 35
68184-04-3
MEA-laureth sulfate
Base, foamer for shampoo and bubble baths. *Chem-Y GmbH.*

1311 Akyposal MLS 30
4722-98-9 225-214-3
MEA-lauryl sulfate
Detergent, shampoo base, shower baths. *Chem-Y GmbH.*

1312 Akyposal MS SPC
9004-82-4
Sodium laureth sulfate
Detergent. *Chem-Y GmbH.*

1313 Akyposal NAF
Sodium dodecylbenzene sulfonate
Emulsifier for emulsion polymerization. *Chem-Y GmbH.*

1314 Akyposal NLS
151-21-3 8782 205-788-1
Sodium lauryl sulfate
Detergent. *Chem-Y GmbH.*

1315 Akyposal NPS 100
9014-90-8
Sodium nonoxynol-10 sulfate
Emulsifier for polymerization. *Chem-Y GmbH.*

1316 Akyposal NPS 60
9014-90-8
Sodium nonoxynol-6 sulfate
Emulsifier for emulsion polymerization. *Chem-Y GmbH.*

1317 Akyposal OP 80
Octoxynol-9 carboxylic acid
Emulsifier for emulsion polymerization. *Chem-Y GmbH.*

1318 Akyposal OPS 85
Sodium octoxynol-9 sulfate
Emulsifier. *Chem-Y GmbH.* Name unverified.

1319 Akyposal RLM 56 S
3088-31-1 221-416-0
Sodium laureth sulfate
Detergent for personal care products, liquid soaps, dishwashing products. *Chem-Y GmbH.*

1320 Akyposal RLM 70
9004-82-4
Fatty alcohol ether sulfate
Detergent; shampoo base. *Chem-Y GmbH.*

1321 Akyposal TIPA 45
661-61-6
TIPA lauryl sulfate
Surfactant for cosmetics, shampoos, shower and foam baths. *Chem-Y GmbH.*

1322 Akyposal TLS42
139-96-8 205-388-7
TEA-lauryl sulfate
Avirol® T 40. Detergent; base for personal care products and car shampoos; foaming agent for agrochemicals, fire extinguishers. *Chem-Y GmbH.*

1323 Akyposept B
Benzylhemiformal
Detergent. *Chem-Y GmbH.*

1324 Akypostat MA 35
38720-61-5
Myreth-5 carboxylic acid
Antistat and antifogging agent for polyester film; anticondensation agent for foil; biodegradable. *Chem-Y GmbH.*

1325 alabaster
$CaSO_4 \cdot 2H_2O$
A form of gypsum, used for ornamental carvings.

1326 alabaster, Oriental
A compact form of marble, $CaCO_3$.

1327 Alachlor
15972-60-8 203 240-110-8
$C_4H_{20}ClNO_2$
2-Chloro-N-(2,6-diethylphenyl)-N-(methoxymethyl)acetamide
Alanex; Alagan; Alazine; 2-chloro-2',6'-diethyl-N-(methoxymethyl)acetanilide; Metachlor; CP-50144; Lasso. Active ingredient: alachlor; pre-emergence and pre-plant incorporated herbicide for the control of most annual grasses and certain broadleaf weeds. mp = 40-41°; $bp_{0.3}$ = 135°; soluble in H_2O, organic solvents; LD_{50} (rat, orl) = 1200 mg/kg.

1328 Alacsan T
Quaternary sulfate
Surfactant. *Rhône-Poulenc Surf. Canada.*

1329 Alagan
15972-60-8 203 240-110-8
Alachlor
Herbicide *Agan Chemical Manufacturers Ltd.*

1330 Alamask
Industrial deodorants. *May & Baker Ltd.* Name unverified.

1331 Alanex
15972-60-8 203 240-110-8
alachlor. Active ingredient: alachlor; pre-emergence and pre-plant incorporated herbicide for the control of most annual grasses and certain broadleaf weeds. *Agan Chemical Manufacturers Ltd.*

1332 Alan-gilan
Cananga oil, a neutral oil from *Cananga odorata.*

1333 α-alanine
56-41-7 205 200-273-8
$C_3H_7NO_2$
2-aminopropionic acid
Ala; L-α-aminopropionic acid; L-2-Aminopropionic Acid; Ala; A; Alanine; 2-Aminopropanoic acid; 2-ammoniopropanoate; L-alanine. (L-form); used in microbiological research, biochemical research, dietary supplement. mp = 314-317° (dec); $[\alpha]_D$ = 14° (c = 6 1N HCl); d = 1.401; soluble in H_2O (158 g/l), cold 80% ethanol; insoluble in ether. *Penta Mfg.; US Biochemical.*

1334 Alar
1596-84-5 2874 216-485-9
Daminozide
Growth regulator. *Murphy Chemical Co Ltd.* Discontinued.

1335 Alargan
An alloy of aluminum and silver, the surface having been dusted with platinum black and hammered or subjected to pressure; a platinum substitute. No manufacturer.

1336 Alathon® H5234
9002-88-4 7728
HDPE copolymer; high flow injection molding resin for frozen food containers, drink cups, housewares. *OxyChem/Alathon.*

1337 Alathon® L5440
9002-88-4 7728
HDPE resin; blow molding resin. *OxyChem/Alathon.*

1338 Alathon® M5560
9002-88-4 7728
HDPE copolymer resin; film-grade resin for water bath or chill roll film; food contact applications. *OxyChem/Alathon.*

1339 Alathon® M6062
9002-88-4 7728
HDPE homopolyer resin; injection molding resin for crates, cases. *OxyChem/Alathon.*

1340 Alazine
15972-60-8 203 240-110-8
alachlor-atrazine
Ready formulated mixture of alachlor and atrazine for use as a selective pre-emergence herbicide. *Agan Chemical Manufacturers Ltd.*

1341 Alba
Petrolatum USP. *Witco Corporation.*

1342 Albacar
471-34-1 1697 207-439-9
calcium carbonate
Highly refined calcite (calcium carbonate). *Pzizer International.*

1343 Albacer
A proprietary synthetic wax for increasing melting point of waxes. No manufacturer.

1344 Albalan
68201-49-0 269-220-4
Lanolin wax; emollient, emulsifier, forms stable water-oil emulsions. *Westbrook Lanolin.*

1345 Albalith
A white, light-resisting lithopone pigment; used in the paint and rubber industries. No manufacturer.

1346 Albanite
A bituminous material found in Albania.

1347 Albanose
A leucite rock.

1348 Albaphos Dental Na 211
10163-15-2 233-433-0
FNa2O3P
Disodium monofluorophosphate
Phosphorofluoridic acid, disodium salt; Sodium fluorophosphate. A proprietary trade name for sodium monofluorophosphate, a fluorine component for toothpastes, the toxic effects of which are only 1/3 of those of sodium fluoride. *Hoechst UK.* Name unverified.

1349 Albar-40
94-75-7 2865 202-361-1
2,4-D
Herbicide. *Makhteshim Chemical Works Ltd.*

1350 albarium
A lime obtained by burning marble; used for stucco.

1351 Albar-M
94-74-6 5803 202-360-6
MCPA
Herbicide *Makhteshim Chemical Works Ltd.*

1352 Albar-Super
94-75-7 2865 202-361-1
2,4-D
Herbicide. *Makhteshim Chemical Works Ltd.*

1353 Albaryt
7727-43-7 1023 231-784-4
BaO_4S
barium sulfate
Finely divided barytes (barium sulfate), used as a pigment. *Sachtleben Chemie GmbH.*

1354 Albata
A nickel-brass or low nickel-silver containing about 8% nickel.

1355 Albatex
Dyeing and printing assistant. *Ciba plc.* Name unverified.

1356 Albatex OR
A proprietary trade name for a nonfoaming polyvalent amide; used as a leveling agent for vat dyes. *Ciba plc.* Name unverified.

1357 Albatra metal
A nickel silver; it contains 57.5% copper, 22.5% zinc, 18.75% nickel, and 1.25% lead.

1358 Albegal
Dyeing and printing assistant. *Ciba plc.* Name unverified.

1359 Albegal CL
A proprietary trade name for an ester of sulfonated fat; used as a leveling agent in wool dyeing. *Ciba plc.* Name unverified.

1360 Alberene
A blue-grey soapstone mined in Virginia.

1361 Alberger® Natural Flake
7647-14-5 8742 231-598-3
ClNa
Sodium chloride
Crystalline salt refined by Alberta process. *Akzo Salt.*

1362 Alberit MP
A proprietary melamine/phenol-formaldehyde thermosetting molding compound. *Canadian Hoechst.* Name unverified.

1363 Alberit PF
A proprietary phenol-formaldehyde resin thermosetting molding compound. *Canadian Hoechst.* Name unverified.

1364 Alberit VP
A proprietary unsaturated polyester thermosetting molding compound used in the production of impact resistant moldings. *Canadian Hoechst.* Name unverified.

1365 Albert
Basic slag used for fertilizing purposes. *Canadian Hoechst.* Name unverified.

1366 Albert MF
A proprietary melamine formaldehyde thermosetting molding compound; used in the manufacture of tracking-resistant moldings for the electrical industry. *Canadian Hoechst.* Name unverified.

1367 Albertat
A proprietary range of chemical fillers, extenders and additives for products containing synthetic resins, such as additives for thickening and preventing setting in paints and varnish. *Vianova Resins.* Discontinued.

1368 Albertol
Rosin modified phenolic resins
Used in printing inks, paints and varnishes. *Hoechst UK.*

1369 Albertol 142-R
Butyl phenol formaldehyde
Vianova Resins. Discontinued.

1370 Albertol 175-A
The aluminum salt of unesterified Albertol IIIL. *Vianova Resins.* Discontinued.

1371 Albertol 237-R
Di-isobutylphenol-formaldehyde
Vianova Resins. Discontinued.

1372 Albertol 326-R (387L)
A rosin modified phenolic resin used in aircraft primers made from 1 part diane (diphenylolpropane), 1 part rosin and 0.1 part paraformaldehyde. *Vianova Resins.* Discontinued.

1373 Albertol 347Q, 369-Q (209-L)
An ester gum-phenolic combination made from xylenol-formaldehyde rosin, pentaerythritol, and glycerogen. *Vianova Resins.* Discontinued.

1374 Albertol IIIL
A phenol-resin condensation product melting at 106-133°; saponification value 15.8; insoluble in alcohol, but soluble in linseed oil; stated to be a good substitute for kauri gum in the manufacture of oil varnishes; rosin glycerine-diane-formaldehyde condensate in the presence of alkali. *Vianova Resins.* Discontinued.

1375 Albigen® A
Textile stripping agent, leveling agent, washing-off agent; prevents re-exhaustion of dyes in washing of prints. *BASF; BASF AG.*

1376 Albiogen
Tetramethylammonium oxalate
No manufacturer.

1377 albion metal
A sheet of metal containing tin and lead; it is formed by pressing together sheets of these metals.

1378 Alboleum, Albolineum
Oil insecticide. *Plant Protection.* Name unverified.

1379 Albolit
A proprietary phenol-formaldehyde synthetic resin. No manufacturer.

1380 Albondur
A Bondur alloy coated on each side with pure aluminum to improve corrosion resistance. No manufacturer.

1381 Albor Die Steel
A proprietary steel containing small amounts of chromium, molybdenum, and carbon. No manufacturer.

1382 Alboresin
A proprietary urea-formaldehyde synthetic resin; molding composition. No manufacturer.

1383 Albral
Flux for use with aluminum bronzes, silicon bronzes and high tensile brasses. *Foseco (F.S.) Ltd.*

1384 Albras Propachlor
1918-16-7 7977 217-638-2
Propachlor
A pre-emergence herbicide for various horticultural crops. *ICI Chem & Polymers Ltd.*

1385 Albrass
1918-16-7 7977 217-638-2
Propachlor
Pre-emergence herbicide for various horticultural crops. *ICI Plant Protection.*

1386 Albrichrome
Textile dyestuffs. *Albright & Wilson Ltd., Phosphates & Specialty Business.* Name unverified.

1387 Albricide
Fungicide
Albright & Wilson Ltd., Phosphates & Specialty Business.

1388 Albrifloc
Flocculating agent. *Albright & Wilson Ltd., Phosphates & Specialty Business.*

1389 Albrightex
Textile optical brighteners. *Albright & Wilson Ltd., Phosphates & Specialty Business.*

1390 Albrilan
Textile dyestuffs. *Albright & Wilson Ltd., Phosphates & Specialty Business.* Name unverified.

1391 Albrilene
Textile dyestuffs. *Albright & Wilson Ltd., Phosphates & Specialty Business.* Name unverified.

1392 Albrilon
Textile dyestuffs. *Albright & Wilson Ltd., Phosphates & Specialty Business.* Name unverified.

1393 Albrilube
Textile lubricant. *Albright & Wilson Ltd., Phosphates & Specialty Business.* Name unverified.

1394 Albrinol
Textile dyestuffs. *Albright & Wilson Ltd., Phosphates & Specialty Business.* Name unverified.

1395 Albrinyl
Textile dyestuffs. *Albright & Wilson Ltd., Phosphates & Specialty Business.* Name unverified.

1396 Albriquest
Sequestering agent. *Albright & Wilson Ltd., Phosphates & Specialty Business.* Name unverified.

1397 Albriscour
Textile scouring agent. *Albright & Wilson Ltd., Phosphates & Specialty Business.* Name unverified.

1398 Albrisolve
Textile dyeing agent. *Albright & Wilson Ltd., Phosphates & Specialty Business.* Name unverified.

1399 Albrisperse
Textile dispersing agent. *Albright & Wilson Ltd., Phosphates & Specialty Business.* Name unverified.

1400 Albritone
Textile leveling agent. *Albright & Wilson Ltd., Phosphates & Specialty Business.* Name unverified.

1401 Albrivap
Boiler scale inhibitor. *Albright & Wilson Ltd., Phosphates & Specialty Business.* Name unverified.

1402 Albumen
9006-50-2
egg albumin; dried egg white; albumin; Egg white Solids Type P-20, P-11, P-18G, P-19, P-21, P-25, P-39; P-110, PF-1; Hentax Type P-1800; Hentex Type P-2100; Sol-U-Tein EA Type PF-1;. Protective colloid and emulsifier in baking; textile dye mordant; adhesives and veneers. Yellow, amorphous lumps, scales, or powder; swells in H_2O, then dissolves gradually; decomposes in moist air. Am. Roland; Atomergic; Chemetals; British Bakels; Dasco Sales; Farbest Brands; Frigova Produce; Igreca; Industrial Protiens; Mitsubishi; Moore Fine Foods; Penta Mfg; Spice King; Alfred L Wolff.

1403 Albumotope I-131
Albumin, iodinated ^{131}I serum; diagnostic aid; radioactive agent. *Bristol-Myers Squibb Co Inc.* Name unverified.

1404 Alburex
Vegetable proteins; used in animal feedstuff. *Roquette (UK) Ltd.*

1405 Albustix
A prepared test strip of tetra-bromphenol blue with a citrate buffer; used to detect protein in urine. *B. C. Ames.* Name unverified.

1406 Albutannin
Protan
Albumen tannate, used in animal feed.

1407 Alcalase® 2,0 T
Proteinase
Enzyme for laundry powder detergents. *Novo Nordisk.* Name unverified.

1408 Alcamine®
A versatile range of cationic softening agents for all textile fibers; includes dyebath stable and durable types. *Allied Colloids Ltd.*

1409 Alcamizer 1
11097-59-9 234-319-3
magnesium aluminum carbonate
Heat stabilizer for PVC that reacts in a unique way with HCl in PVC; offers high heat stability, nontoxicity, and high transparency. *Kyowa Chem. Industry.*

1410 Alcamizer 2
96492-31-8, 11097-59-9 234-319-3
magnesium aluminum carbonate
Heat stabilizer for PVC. *Kyowa Chem. Industry.*

1411 Alcan AA-100
1344-28-1 369 215-691-6
Al_2O_3
aluminum oxide
Activated alumina; in selective absorption processes; as starting material for catalyst. *Alcan.*

1412 Alcan Aluminum Fluoride
7784-18-1 351 232-051-1
$AlF+3$
Aluminum trifluoride
Aluminum fluoride mixed with aluminum oxide and silica for use as an electrolyte in reduction of alumina to aluminum metal; as flux in remelting and refining of aluminum and its alloys; opacifier aid in production of ceramic enamels, glass, and glazes. *Alcan.*

1413 Alcan Aluminum Sulfate Liquid
10043-01-3 381 233-135-0
$Al_2O_{12}S_3$
Aluminum sulfate
Alcan Aluminum Sulfate Liquid; Ant Flip. Used in water purification in pulp and paper mills and water purification plants. *Alcan.*

1414 Alcan C-70, C-71, C-72, C-73, C-75
1344-28-1 369 215-691-6

Al_2O_3
aluminum oxide
alumina. Calcined alumina; ceramic grade alumina for refractory bricks, whitewares. *Alcan.*

1415 Alcan FRF 5, 10, 20, 30, 40, 60, 80, 85
1344-28-1 369 215-691-6
Alumina hydrate
Flame retardant and smoke suppressant properties for plastics and rubber industries. *Alcan.*

1416 Alcan FRF LV2, LV4, LV5, LV6, LV7, LV8, LV9
1344-28-1 369 215-691-6
Alumina hydrate
Flame retardant and smoke suppressant. *Alcan.*

1417 Alcan GB-1S
Wrought, non heat-treatable 99.5% aluminum alloy. *Alcan.* Name unverified.

1418 Alcan GB-2S
Wrought, non heat-treatable commercially pure aluminum alloy. *Alcan.* Name unverified.

1419 Alcan GB-3S
Wrought, non heat-treatable aluminum alloy; stronger and harder than 2S. *Alcan.* Name unverified.

1420 Alcan GB-50S
Wrought, heat-treatable aluminum alloy; forms well in the W (solution heat-treated) condition. *Alcan.* Name unverified.

1421 Alcan GB-99.8%
Wrought, non heat-treatable high purity aluminum alloy. *Alcan.* Name unverified.

1422 Alcan GB-100
Cast, non heat-treatable commercially pure aluminum alloy. *Alcan.* Name unverified.

1423 Alcan GB-160
Cast, non heat-treatable aluminum alloy. *Alcan.* Name unverified.

1424 Alcan GB-350
Cast, heat-treatable aluminum alloy specially impact resisant; good resistance to marine conditions. *Alcan.* Name unverified.

1425 Alcan GB-B116
Cast, heat-treatable aluminum alloy available in four conditions: M (as cast), P (precipitation treated), W (solution treated) and WP (fully heat treated). *Alcan.* Name unverified.

1426 Alcan GB-B320
Cast, non heat-treatable medium strength aluminum alloy. *Alcan.* Name unverified.

1427 Alcan GB-B51S, 65S
Wrought, heat-treatable medium strength aluminum alloys. *Alcan.* Name unverified.

1428 Alcan GB-B535S, 54S, D54S, A56S, M57S
Wrought, non heat-treatable aluminum alloys in which magnesium is the main additive. *Alcan.* Name unverified.

1429 Alcan H-10
1344-28-1 369 215-691-6
Alumina hydrate
For production of low-iron aluminum sulfate, alumina-based catalysts, sodium aluminate and other aluminum salts, coated titanium dioxide pigments, ceramic and glass products, fire-retardant carpet backing. *Alcan.*

1430 Alcan H-10-08
Alumina trihydrate
Filler and extender in plastics, resins, rubber, latex foams, especially where flame retardance is important. *Alcan.*

1431 Alcan Recovered Cryolite
Sodium fluoroaluminate (87-90%), aluminum oxide (2%), sodium sulfate (4%), sodium carbonate (1-1.5%); source of fluorine; as ceramic flux and opacifier aid in the production of vitreous enamels, glass, glazes, abrasives; metallurgical flux in refining of aluminum and its alloys. *Alcan.*

1432 Alcan Superfine 4, 7, 11
Alumina hydrate
Filler for fire retardants/smoke suppressants for plastics, rubber, paints, adhesives, adhesive tapes and in toothpaste, cosmetics, polishes and waxes; in paper coatings. *Alcan.*

1433 Alcaphos 24
Strongly alkaline silicated solid; formulated for heavy duty soak cleaning in steel fabricated metals. *Inviquimica & CIA SCA.* Name unverified.

1434 Alcapsol®
A range of synthetic microencapsulation products for the replacement of gelatin and natural compounds in paper manufacture. *Allied Colloids Ltd.*

1435 Alcement
Alcacement
Fused cement prepared in the electric furnace from bauxite and lime. It contains approximately 40% CaO, 40% Al_2O_3, 10% SiO_2, and 10% Fe_2O_3. No manufacturer.

1436 Alchemie
Epoxy resins, polyurethane resins and RTV silicone rubber systems; used in casting, moldmaking and laminated structures, engineering patterns and toolmaking, electrical potting and encapsulations. *Alchemie Ltd.*

1437 Alchemix
Epoxy and polyurethane resins and RTV silicone rubber systems; used in casting, moldmaking and laminated structures, engineering patterns and toolmaking, electrical potting and encapsulations. *Alchemie Ltd.*

1438 Alchemy
Metal carboxylates; driers for paint or printing ink. *Manchem Ltd.* Name unverified.

1439 Alcian
Dyestuffs, blues, greens and yellow dyes formed by introducing chloromethyl groups into phthalocyanin and its derivatives by means of dichlorodimethyl either in pyridine containing aluminum chloride. *ICI Chem & Polymers Ltd.* Discontinued.

1440 Alclar®
A range of synthetic microencapsulation products for the replacement of gelatin and natural compounds in paper manufacture. *Allied Colloids Ltd.*

1441 Alcoa 2-S
A proprietary trade name for a commercially pure aluminum. No manufacturer.

1442 Alcoa 3-S
A proprietary alloy of aluminum, containing small amounts of copper, iron, silicon, and zinc. No manufacturer.

1443 Alcoa 24-S
A proprietary wrought aluminum alloy containing 93.7% aluminum, 1.5% magnesium, 4.2% copper, 0.6% manganese. No manufacturer.

1444 Alcoa 32-S
A proprietary alloy of aluminum with 12% silicon, 0.8% nickel, 1% magnesium, and 0.8% copper. No manufacturer.

1445 Alcoa 356
A proprietary alloy of aluminum containing 4% silicon, 0.3% magnesium. No manufacturer.

1446 Alcoa 43
A proprietary alloy of aluminum and silicon; contains 5% silicon. No manufacturer.

1447 Alcoa 47
A proprietary alloy of aluminum with 12.5% silicon. No manufacturer.

1448 Alcoa 108
A proprietary aluminum alloy with 3% silicon and 4% copper. No manufacturer.

1449 Alcoa 112
A proprietary aluminum alloy containing 7-8.5% copper, 1-2% zinc, and up to 1.7% of other metals, mostly iron. No manufacturer.

1450 Alcoa 122
A proprietary alloy of aluminum with 10% copper, 0.2% magnesium, and 1.2% iron. No manufacturer.

1451 Alcoa 145
A proprietary aluminum alloy containing 10% zinc, 2.5% copper, and 1.2% iron. No manufacturer.

1452 Alcoa 220-TA
A proprietary aluminum alloy containing aluminum with 10% magnesium. No manufacturer.

1453 Alcoa 515
A proprietary alloy containing aluminum with magnesium, silicon, and iron. No manufacturer.

1454 Alcoa 535
A proprietary alloy of aluminum with 0.25% chromium, 1.25% magnesium, and 0.7% silicon. No manufacturer.

1455 Alcobez
A comprehensive range of compounds to control corrosion, scale build-up and biofouling in boiler and cooling water systems. *Allied Colloids Ltd.*

1456 Alcocare® 1000
Surfactant/humectant blend; provides foam, feel, viscosity and mildness to health care formulations. *Rhône-Poulenc Surf.*

1457 Alcocare® 2011
Chloroxylenol PCMX concentrates; high foaming antimicrobial concentrate; alkali stable; for handwash, surgical scrubs, health care. *Rhône-Poulenc Surf.*

1458 Alcocare® 3020
Iodine concentrates; low foaming antimicrobial concentrate; acid stable; for surgical scrubs, health care. *Rhône-Poulenc Surf.*

1459 Alcodet® 218
9004-83-5

PEG-10 isolauryl thioether
Emulsifier, wetting agent, detergent, carbon soil and grease cleaners, metal cleaning specialties, steel processing, textile scouring, insecticide emulsions, cosmetics, wood pulp and paper industries. *Rhône-Poulenc Surf.; Rhône-Poulenc France.*

1460 Alcodrill HPD-D
Dry, free-flowing, powdered carboxylate copolymers; deflocculant in water based drilling. *Alco Chemical Corp.*

1461 Alcodrill HPD-L
A polycarboxylate copolymer solution; deflocculant for water based drilling. *Alco Chemical Corp.*

1462 Alcofix®
Afterfixing agents for improving the wet-fastness properties of dyes and prints and as a pitch fixative in paper manufacture. *Allied Colloids Ltd.*

1463 Alcoflood®
A complete range of anionic, nonionic, and cationic polymers for enhanced oil recovery in a choice of molecular weights to suit both tight and open, oil or water wet formations; suitable for long-range mobility control or the various short range treatments used for water shut-off. *Allied Colloids Ltd.*

1464 Alcoform
A solution of formaldehyde in one of a variety of alcohols; used for production of butylated resins and methylated resins; methyl alcoforms are used for the production of ion-exchange resins. *Synthite Ltd.*

1465 Alcogas
Fuels containing alcohol and liquid fuels containing alcohol and a hydrocarbon; for chemicals, medicines and pharmaceutical preparations. *Quantum Chemical Corp.*

1466 Alcogum 296-W
A high viscosity sodium polyacrylate thickener; used in adhesives, paint, cement additive, protective colloid. *Alco Chemical Corp.*

1467 Alcogum 310
A high molecular weight inverse-emulsion copolymer (organic phase medical white oil); for adhesives, coatings. *Alco Chemical Corp.* Discontinued.

1468 Alcogum 9639
9003-03-6
Ammonium polyacrylate
Thickener for natural and synthetic latexes; used in dipped molded or cast goods; water solution. *Alco Chemical Corp.*

1469 Alcogum 9710
A sodium polyacrylate thickener; used for coatings, packaging, adhesives, latex thickening. *Alco Chemical Corp.*

1470 Alcogum AN 10
High viscosity sodium polyacrylate thickener for latex systems; used in carpet backing, latex foam, adhesives, dispersants. *Alco Chemical Corp.*

1471 Alcogum L-11
A high efficiency alkali-swellable acrylic emulsion thickener; used for adhesives, latex thickening, paint thickening. *Alco Chemical Corp.*

1472 Alcogum L-15, L-26, L-28, L-31, L-35, L-36
9003-01-4
Polyacrylic acid
Thickener for adhesives, paints, paper coatings, natural and synthetic latexes, alkali reactive emulsion; FDA approved. *Alco Chemical Corp.*

1473 Alcogum L-27
A high efficiency alkali-swellable acrylic emulsion thickener; used for adhesives, latex thickening, dispersion thickening. *Alco Chemical Corp.*

1474 Alcogum L-28, L-29, L-35
A reactive alkali activated acrylic emulsion polymer; used for paper coating, latex compounding. *Alco Chemical Corp.*

1475 Alcogum L-52
A self-crosslinking alkali activated emulsion thickener; used for adhesives, latex thickening. *Alco Chemical Corp.*

1476 Alcogum PT-33
An alkali activated associative emulsion thickener; used for paint, adhesives, cleaners, wall joint compounds. *Alco Chemical Corp.*

1477 Alcogum TSB
A high viscosity sodium polyacrylate thickener; used for textile coatings for upholstery. *Alco Chemical Corp.* Name unverified.

1478 Alcogum VEP-II
A high viscosity sodium polyacrylate thickener; latex adhesive in the tufted carpet industry. *Alco Chemical Corp.*

1479 Alcojet®
Sodium metasilicate-sodium carbonate-POE ester of mixed fatty and resin acids. Detergent used in mechanical washers for stain removal; for cleaning healthcare instruments, laboratory ware, electronic components, pharmaceutical apparatus, industrial parts, etc. *Alconox.*

1480 Alcolec 532
A proprietary preparation of vinyl-based resin in bead form; a carboxylated vinyl copolymer; used in the formulation of flexographic printing inks and paper lacquers. *Am. Lecithin.* Name unverified.

1481 Alcolec® 439-C
8002-43-5 5452 232-307-2
Lecithin
Wetting agent, emulsifier, release agent; for waterbased paints, coatings, textiles. *Am. Lecithin.*

1482 Alcolec® 439-C
8002-43-5 5452 232-307-2
lecithin
Alcolec® 495; Aqualipid 95; Alcolec® S; Alcolec® BS; Alcolec® F-100; Alcolec® 440-WD. Lecithin; oil-water emulsifier, wetting agent for aqueous and oil-base systems; approved for food use. Deoiled lecithin; emulsifier for industrial and food applications; instantizing for milk powder, cake mixes, etc.; choline source. *Am. Lecithin.*

1483 Alcolec® 440-WD
8002-43-5 5452 232-307-2
Lecithin; emulsifier, stabilizer, viscosity control agent, wetting agent, pigment grinding aid and dispersant for waterbased paints, coatings. *Am. Lecithin.*

1484 Alcolec® 495
8002-43-5 5452 232-307-2
Lecithin; oil-water emulsifier, wetting agent for aqueous and oil-based systems; approved for food use. *Am. Lecithin.*

1485 Alcolec® BS
8002-43-5 5452 232-307-2
Single bleached lecithin; commercial lecithin emulsifier, wetting and dispersing agent, stabilizer, release and lubricating agent, foam suppressant, solubilizer for food and industrial applications; choline source. *Am. Lecithin.*

1486 Alcolec® F-100
8002-43-5 5452 232-307-2
Deoiled lecithin; emulsifier for industrial and food applications; instantizing for milk powd., cake mixes, etc.; choline source. *Am. Lecithin.*

1487 Alcolec® S
8002-43-5 5452 232-307-2
Unbleached lecithin; commercial lecithin emulsifier, wetting and dispersing agent, stabilizer, release and lubricating agent, foam suppressant, solubilizer for food and industrial applications; choline source. *Am. Lecithin.*

1488 Alcolec® Z-3
8029-76-3 232-440-6
Hydroxylated lecithin
Hydroxylated lecithin; wetting agent, emulsifier for personal care products, pharmaceuticals, food use; improves dispersability of colors and flavors in aqueous systems, improves wetting of fatty powders. *Am. Lecithin.*

1489 Alcolite
A proprietary product used for denture purposes. No manufacturer.

1490 Alcolube®
Textile lubricants and raising assistants. *Allied Colloids Ltd.*

1491 Alcomer®
Synthetic water-soluble polymers designed to offer a wide range of fluid properties for specialized systems for drilling, completion and workover fluids; used for viscosity shale encapsulation, flocculation thinning, fluid loss, deflocculation, friction reduction, Bentonite extension and gellation in water, brine, acid and cement systems. *Allied Colloids Ltd.*

1492 Alconate L-80
Concentrated form of Petronate L, an anionic surfactant of the petroleum sulfonate type; emulsifier and wetting agent used when low oil content is required. *Witco Chemical Ltd.* Discontinued.

1493 Alconox®
Blend of alkylaryl sulfonates, lauryl alcohol sulfates, phosphates, carbonates; detergent with wetting, sequestering and synergistic agents; for manual cleaning of laboratory and hospital glassware and instruments. *Alconox.*

1494 Alcopal FA
A foaming agent for aqueous systems used for carpet backing. No manufacturer.

1495 Alcophor AC
Tannin compounds, modified; corrosion inhibitor used in petrol-based paint systems. *Henkel KgaA.* Discontinued.

1496 Alcopol
A wide range of nonionic and anionic wetting agents and surfactants. *Allied Colloids Ltd.*

1497 Alcopol AH New
Amine salt of sulfated higher fatty acid ester in the form of a brown fluid oil; powerful wetting, penetrating and emulsifying agent used in pigments; paint; leather; emulsion polymerization; dry cleaning; cutting oils; agricultural chemicals. *Allied Colloids Ltd.*

1498 Alcopol FA
Anionic surfactant in which the anion is a long chain sulfosuccinamate, in the form of a fluid dispersion; powerful surface active agent used in the

production of low density latex foams with good wet stability. *Allied Colloids Ltd.*

1499 Alcopol O

577-11-7 3460 209-406-4

docusate sodium

Range of anionic surfactants in which the anion is dioctyl sulfosuccinate; emulsifiers and powerful wetting agents with applications in industries such as paper, textiles, asbestos, plastics and photographic film, metals, pest control, detergents anddegreasing, dust control, glass cleaning oils, lubricants, paints, pigments, printing inks and a wide variety of proprietary products, e.g. hand cleansers and cosmetics. *Allied Colloids Ltd.*

1500 Alcopol OB

Sodium diisobutyl sulfosuccinate as a water/alcohol solution; powerful wetting agent in the presence of electrolytes. *Allied Colloids Ltd.*

1501 Alcopol OD

Sodium ditridecyl sulfosuccinate in water/alcohol solution; emulsifier for oils, solvents, waxes and polymers. *Allied Colloids Ltd.*

1502 Alcopol OS

Sodium dihexyl sulfosuccinate in water/alcohol solution; emulsifier and powerful wetting agent in the presence of electrolytes; used for oils, solvents, waxes, polymers and windscreen wash concentrate. *Allied Colloids Ltd.*

1503 Alcopol T

Sodium salt of a sulfated higher fatty acid ester as a low viscosity pale yellow liquid; powerful wetting, penetrating and emulsifying agent used in pigments; paints; leather; emulsion polymerization; dry cleaning; cutting oils, agricultural chemicals. *Allied Colloids Ltd.*

1504 Alcoprint®

A comprehensive range of novel polyacrylic thickening agents for pigment, disperse and reactive printing on textiles; can be used in emulsion systems but specially developed for completely aqueous systems. *Allied Colloids Ltd.*

1505 Alcoproof®

Synthetic sizing agents used in papermaking. *Allied Colloids Ltd.*

1506 Alcor 7

A clad combination of stainless steel and aluminum-containing magnetic stainless interlayer; used in the production of high quality cookware and magnetic induction heating stoves. *Pfizer International.* Discontinued.

1507 Alcoseal®

Tail seal adhesive for the tissue industry. *Allied Colloids Ltd.*

1508 Alcoset®

Crosslinking acrylic resins; used as a wash-resistant finish to improve weave locking and fabric stability; fixed in a relatively short time above 150°, no catalyst needed. *Allied Colloids Ltd.*

1509 Alcosist®

Dyeing assistants for wool, acrylic and polyamide fibers. *Allied Colloids Ltd.*

1510 Alcosize®

Acrylic-based blended products for sizing staple yarns, spun synthetics, cotton yarns; excellent performance at low application levels. *Allied Colloids Ltd.*

1511 Alcosol 5

An industrial methylated spirit; solvent. *Sasolchem.*

1512 Alcosol EM

Ethanol/MEK blend; solvent for the paint industry. *Sasolchem.*

1513 Alcosol PI

Ethanol, ethyl acetate blend; solvent for the printing industry. *Sasolchem.*

1514 Alcosperse®

Leveling and dispersing agents for disperse dyeing systems for textiles. *Alco Chemical Corp.*

1515 Alcosperse 104

A polycarboxylate solution polymer; anti-redeposition agent, slurry stabilization. *Alco Chemical Corp.*

1516 Alcosperse 107, 124, 149, 157

9003-04-7

Sodium polyacrylate; dispersant for pigments, high solids slurries, paper coating, paint, textile, mining, and ceramic applications. *Alco Chemical Corp.*

1517 Alcosperse 107-D

A dry powdered polycarboxylate polymer; dispersant in paper, board, coatings and paint. *Alco Chemical Corp.*

1518 Alcosperse 144

An acrylic solution polymer; used for slurry preparation. *Alco Chemical Corp.* Discontinued.

1519 Alcosperse 149-C

An acrylate solution polymer; used for paper coatings, pigment slurries. *Alco Chemical Corp.*

1520 Alcosperse 169

A sodium polyacrylate solution polymer; dispersant in adhesives, paint, kaolin and calcium carbonate. *Alco Chemical Corp.*

1521 Alcosperse 175

A low molecular weight carboxylate copolymer solution; used in laundry detergent, incrustation inhibitor; phosphate replacement. *Alco Chemical Corp.*

1522 Alcosperse 249

Ammonium polyacrylate solution polymer; dispersant in paint, adhesives. *Alco Chemical Corp.*

1523 Alcosperse 602

Polycarboxylate solution polymer; antiredesposition agent; detergents; phosphate replacement. *Alco Chemical Corp.*

1524 Alcostat®

Conductive resins and antistatic agents for general uses including plastics, surface coatings and latex compounds. *Allied Colloids Ltd.*

1525 Alcotabs®

Blend of alkylaryl sulfonates, lauryl alcohol sulfates, phosphates, carbonates; detergent with wetting, sequestering and synergistic agents; for cleaning pipettes and tubes in hospital, clinical, education, R&D, and industrial laboratories. *Alconox.*

1526 Alcotac

A range of organic binders for all mineral agglomeration, including pelletization and briquetting applications. *Allied Colloids Ltd.*

1527 Alcotex®

Polyvinyl alcohol (partially hydrolyzed polyvinyl acetate); used for suspension PVC polymerization, reprographic film/paper, and adhesives. *Harlow Chemical Co Ltd.*

1528 Alcotreat 182

A high molecular weight inverse-emulsion polymer; used for drilling fluids, mineral extraction. *Alco Chemical Corp.* Discontinued.

1529 Alcotreat PC 95

A high molecular weight cationic polymer solution; used for water/oil clarification in oil field applications. *Alco Chemical Corp.* Discontinued.

1530 Alcovar

Fast dyes for spirit and cellulose varnishes. *Morton Int'l Ltd.* Discontinued.

1531 Alcowipe®

67-63-0 5227 200-661-7

isopropyl alcohol

70% w/v isopropyl alcohol wipe; hard surface cleanser. *Seton Healthcare Group plc.*

1532 Alcryn®

Halogenated elhylene interpolymer alloy; thermoplastic elastomer; melt processable synthetic rubber which does not require compounding or vulcanization and can be formed on most plastics equipment; offers excellent weather, ozone, heat and oil resistance. *DuPont; DuPont UK.*

1533 Alcumite

A proprietary corrosion-resisting alloy containing 87.5% copper, 7.5% aluminum, 3.5% iron, and 1.5% nickel. No manufacturer.

1534 Aldamine

75-39-8 38 200-868-2

C_2H_7NO

1-aminoethanol

acetaldehyde-ammonia; Acetaldehyde-ammonia trimer; Hexahydro-2,4,6-trimethyl-*s*-triazine trihydrate. A proprietary trade name for acetaldehyde-ammonia C_2H_7NO; exists as a trimer and is used in the manufacture of plastics and as a pickling inhibitor for steel, and is a rubber vulcanizing accelerator. mp = 96°; bp= 110°; soluble in H_2O. No manufacturer.

1535 Aldehol A

An oxidized kerosene to be used in U.S.A. for denaturing methylated spirit. No manufacturer.

1536 aldehyde C14

Undecalactone

A proprietary flaming material. No manufacturer.

1537 Alder bark

The bark of *Alnus glutionsa*; used for fixing yellow dyes and as a tanning material.

1538 Alderton's solution

A solution of ammonium ichthosulfonate, in glycerol.

1539 Aldo® HMS KFG

11099-07-3 234-325-6

Glyceryl stearate

Glyceryl stearate, high mono; kosher food grade emulsifier, softener. Lipophilic emulsifier; approved food additive; contains75% monoglyceride. *Lonza Inc.*

1540 Aldo® MLD

142-18-7 205-526-6

$C_{15}H_{30}O_4$

Glyceryl laurate

2,3-Dihydroxypropyl dodecanoate. Glyceryl laurate, dispersible; emulsifier for cosmetic, pharmaceutical and industrial use. *Lonza Inc.*

1541 Aldo® MO
111-03-5 203-827-7
Glyceryl oleate
Glyceryl oleate; emulsifier, defoamer. *Lonza Inc.*

1542 Aldo® MR
141-08-2 205-455-0
Glyceryl ricinoleate
Emulsifier, solubilizer for cosmetic, pharmaceutical and industrial applications. *Lonza Inc.*

1543 Aldo® MS FG
11099-07-3 234-325-6
Glyceryl stearate
Emulsifier for general use in foods. *Lonza Inc.*

1544 Aldo® PGHMS KFG
1323-39-3 215-354-3
$C_{21}H_{42}O_3$
Propylene glycol stearate
Propylene glycol monostearate; Propylene Glycol Stearate SE. High mono; kosher food grade emulsifier, whipping agent. *Lonza Inc.*

1545 Aldobond
Blend of synthetic or natural elastomers and resins in a solvent or aqueous medium; adhesive compositions for industrial fabricating operations. *Aldo Products Co Inc.*

1546 Aldocoat
Blend of synthetic or natural elastomers and resins in a solvent of aqueous medium; used for industrial speclalty coatings. *Aldo Products Co Inc.*

1547 Aldogen
A mixture of trioxymethylene and bleaching powder; an antiseptic. No manufacturer.

1548 Aldomax GA-100
9032-08-0 232-877-2
Immobilized amyloglucosidase enzyme; enzyme for saccharification of low viscosity, higher DE (50-95) starch containing streams. *UOP.*

1549 Aldones
Flavor bases. *Bush Boake Allen Ltd.*

1550 Aldosperse
Emulsifiers; for food. No manufacturer.

1551 Aldosperse® 40/60 FG
40% PEG-20 glyceryl stearate, 60% glyceryl stearate (high mono); bakery and food emulsifier; dough strengthener, softener. *Lonza Inc.*

1552 Aldosperse® ML 23
59070-56-3
PEG-23 glyceryl laurate
PEG-23 glyceryl laurate; emulsifier, solubilizer, suspending and dispersing agent used in personal care products, textiles. *Lonza Inc.*

1553 Aldosperse® MO-50
Glycol monooleate and polysorbate 80; emulsifier for frozen desserts; antifog for PVC. *Lonza Inc.*

1554 Aldosperse® MS-20 FG
51158-08-8
PEG-20 glyceryl stearate
PEG-20 glyceryl stearate; bakery and food emulsifier. *Lonza Inc.*

1555 Aldosperse® O-20 KFG
80% Glyceryl stearate, 20% Polysorbate 80; kosher food grade emulsifier; for ice cream. *Lonza Inc.*

1556 Aldosperse® TS-40 KFG
60% Glyceryl stearate, 40% Polysorbate 65; kosher grade food emulsifier; for frozen desserts. *Lonza Inc.*

1557 Aldrey
An aluminum alloy of Swiss origin used for electrical conductors; contains 98.7% aluminum, 0.6% silicon, 0.4% magnesium, and 0.3% iron. No manufacturer.

1558 Aldrin
309-00-2 227 206-215-8
$C_{12}H_8Cl_6$
1,2,3,4,10,10-hexachloro-1α,4α,4aβ,5α,8α,8aβ-hexahydro1,4:5,8-dimethanonaphthalene
HHDN; aldrine; OMS 194; ENT 15949; Aldrex; Aldrite; Aldron; Aldrosol; Algran; Compound 118; Octalene; Seedrin; Aldocit; HHPN; Kortofin; OMS-194; Tatuzinho; Tipula; Aldrin-R. Non-systemic insecticide with contact, stomach and respiratory action. Used for control of soil-dwelling insects, termites and ants and also as a wood preservative. mp = 104-105°; bp_2 = 145°; insoluble in H_2O, moderately soluble in organic solvents; LD_{50} (rat orl) = 38-67 mg/kg.

1559 Aldrin Dust
309-00-2 227 206-215-8
Aldrin
Insecticide. *Murphy Chemical Co Ltd.* Discontinued.

1560 Aldur
A urea-formaldehyde resin. No manufacturer.

1561 Al-dur-ba
A patented alloy of zinc, copper, and aluminum. No manufacturer.

1562 Aldydale
A proprietary phenol-formaldehyde synthetic resin. No manufacturer.

1563 Aldyl®
For polymer industry. *DuPont UK.*

1564 Alecra
Chromium plating processes. *Albright & Wilson Ltd., Phosphates & Speciality Business.*

1565 Alepol
A proprietary preparation of selected sodium salts of hydrocarpus oil acids. No manufacturer.

1566 Alexis
Aluminum etchant *Albright & Wilson Ltd., Phosphates & Speciality Business.*

1567 Alexis Antibloom
Proprietary additive for metal finishing for hot water sealing of anodized aluminum to prevent sealing bloom. *Albright & Wilson Am.* Name unverified.

1568 Alexite
1344-28-1 369 215-691-6
aluminum oxide
A proprietary trade name for an aluminum oxide abrasive. No manufacturer.

1569 alfa
A variety of esparto grass used in the manufacture of paper. It is also the term for a synthetic tannin, a red-brown liquid containing 23% tanning substance, 11% non-tannins, 66% water, and trace of sulfuric acid. *Johnson Matthey plc.*

1570 Alfa®
Proprietary name for extensive range of research chemicals and materials.

1571 Alfacron
35575-96-3 252-626-0
Poultry house insecticide containing azamethiphos. *Ciba-Geigy Agrochemicals.*

1572 Alfacron 10WP
35575-96-3 252-626-0
Azamethiphos
Used for fly control. *Ciba-Gelgy Agrochemicals.*

1573 Alfadex
Pyrethrin
Contact insecticide. *Ciba-Geigy Agrochemicals.*

1574 Alfaprostol
74176-31-1 234 277-746-0
$C_{24}H_{38}O_5$
[1R-[1α(Z),2β(SName unverified.),3α,5α]]-7-[2-(5-Cyclohexyl-3-hydroxy-1-pentynyl)-3,5-dihydroxycyclopentyl]-5-heptenoic acid methyl ester
Alfavet; Ro-22-9000; K-11941; Alphacept. Prostaglandin. Used for estrus control in cows. Name unverified.

1575 Alferium
A proprietary alloy of aluminum with 2.5% copper, 0.62% magnesium, 0.5% manganese, and 0.3% silicon. No manufacturer.

1576 Alferric
10043-01-3 381 233-135-0
Aluminum sulfate
Used in dyeing, tanning *Laporte Industries Ltd.*

1577 Alfol
A proprietary trade name for an aluminum foil in a crumpled condition used for heat insulation. No manufacturer.

1578 Alforder
A proprietary synthetic resin. No manufacturer.

1579 Alfralat
A proprietary name for glyceryl phthalate resins. *Vianova Resins.* Name unverified.

1580 Alfrax B301
1344-28-1 369 215-691-6
aluminum oxide
A commercial grade of bubble aluminum oxide. *Carborundum.* Name unverified.

1581 Alftalat
Alkyd resins (oil modified polyesters); used for air drying decorative paints, air drying and stoving industrial finishes. *Resinous Chemicals Ltd.*

1582 Algae Treat
30% Quatenary ammonia compound; used for algae control in cooling water systems. *Delaware Chemical Corp.* Name unverified.

1583 Algafen
97-23-4 3120 202-567-1
Dichlorophen
A fungicide, bactericide, and algicide used as a moss-killer. *Geeco.*

1584 Algalex 104
A solution of the sodium salt of a chlorinated phenyl derivative of methane containing surface active and antifoam agents; a nontoxic, noncorrosive bactericide for water systems. *Kinnis & Brown.* Name unverified.

1585 Algalith
A proprietary algine plastic. No manufacturer.

1586 Algarobilla
A vegetable tannin material; it consists of the pods of *Caesulpinias brevifolia* of Chile, and contains about 60% tannin.

1587 Algarobillin
A dye product obtained from the carob tree, *Ceratonia siliqua* found in Argentina; it is employed for dyeing cloth khaki.

1588 Algarovilia
A Columbian name for a copal resin obtained there.

1589 Alger metal
An alloy of 90% tin and 10% antimony; a silverywhite alloy used in making jewelry.

1590 Algier's metal
A jeweler's alloy: *(a)* Consists of 90% tin and 10% antimony and is used for the manufacture of forks and spoons; *(b)* contains 94.5% tin, 5% copper, and 0.5% antimony and is used for making hand bells.

1591 Algin
9005-38-3 240
$(C_6H_7O_6Na)_n$
sodium alginate
Sodium polymannuronate; Hydrophilic polysaccharide; Collid 488T; Dariloid® Q, QH; Kelco® HV, LV; Kelcosol®; Kelgin® F, HV, LV, MV, QL, XL; Kelset®; Keltone®, HV, LV; Kelvis®; Kimitsu Algin I-1, I-2, I-3; Manucol DH, DM, DMF, LB; Manugel DJX; Manugel DMB; GHB, GMB; Prime F-25, F-40, F-400, F-600; Proctin BUS; Protanal 686, HF 120M, HFC 60, KC 119, KP, KPM, LF 5/60LF 20, LF 20/40, LF 60; Sobalg FD 100 Range; Sodium Alginate HV NF/FCC, LV, LVC, MV NF/FCC; W-300FG,. Stabilizer in manufacture of ice cream; emulsifying agent in foods and paints. Emulsifier; firming agent; formulation aid; processing aid; surfactant. Cream colored powder; soluble in H_2O; insoluble in alcohol, ether, chloroform; LD_{50} (rat iv) = 1000 mg/kg. *Aldrich; Kelco, Div of Merck.*

1592 Alginade MR, MRE
Alginate blend; ice cream stabilizer. *Keloc Int'l Ltd.*

1593 alginic acid
9005-32-7 241 232-680-1
$(C_6H_8O_6)_n$;
Norgine; Polymannuronic acid; Sazio. Polysaccharide composed of β-d-mannuronic acid residues; suspending, thickening, emulsifying, and stabilizing agent. Very slightly soluble in H_2O; capable of absorbing 200-300 times its weight in H_2O; soluble in alkaline solutions. *Kelco; Mendell; Penta Mfg.; Protan Ltd.*

1594 Alginic acid potassium salt
9005-36-1 241
Potassium alginate
Improved Kelmar®; Stercofuge. A hydrophilic, collodial polysaccharide obtained from seaweed. Used for dietetic and low-sodium foods, dry mixes, dental impression material, surgical impressions. *Kelco.*

1595 Alginoplast®
Alginate impression material for dentistry. *Bayer AG.*

1596 Algiron
Alginoid iron
An iron compound of alginic acid, from seaweed; it contains 11% iron.

1597 Algisium-C
128973-71-7
Methylsilanol mannuronate
Provides cutaneous hydration, lipolytic action, skin regeneration and maintenance, for cosmetic and health products, milks, emulsions, creams, lotions, anti-aging formulations. *Exsymol.*

1598 Algistat
Preparations for water treatment. *BDH Chemicals Ltd.*

1599 Algitox®
Algicide for swimming pools. *Fargro Ltd.* Discontinued.

1600 Algodon
Cotton wool.

1601 Algodon de Seda
The fiber of *Calotropis gigantea* is known in Venezuela by this name.

1602 Algofen
97-23-4 3120 202-567-1
Dichlorophen
Fungicide, bactericide and algicide used as a moss-killer. *Geeco.*

1603 Algoflon®
PTFE resins. *Fluorocarbon Co Ltd; Ausimont; Montedison UK Ltd.*

1604 Algol
Vat dyestuffs; used for textile dyeing and printing. *Hoechst UK.*

1605 Algol®
Vat dyestuffs; used for textile dyeing and printing. *Cassella AG.*

1606 Algon 100
Synergistic blend of a substituted urea and an acid amide; powder; bactericide/fungicide/algicide for wet-state and dry-film protection of interior and exterior water and solvent-based coatings, woodstains, adhesives, fillers, sealants, etc.; recommended *Thor Chemicals (UK) Ltd.* Discontinued.

1607 Algulose
A very pure cellulose obtained from kelp; used in papermaking. No manufacturer.

1608 alibated iron
Iron coated with aluminium to form a protective covering.

1609 Alibi
Herbicide containing bifenox and linuron. *ICI Chem & Polymers Ltd.*

1610 Alicep®
1698-60-8 216-920-2
Chloridazon
For pre- and post-emergence weed control in onions, leeks, chives, and flower bulbs. *BASF AG; BASF plc.* Name unverified.

1611 Ali-Clean
Acid based aluminum cleaner; for cleaning aluminum and magnesium alloy products especially car wheels; removes dirt, brake dust, and oxidation just by brushing on and washing off. Name unverified.

1612 Aliette
15845-66-6
fosetyl
Fungicide. *May & Baker Ltd.*

1613 Aliette
39148-24-8 4278 254-320-2
Fosetyl-aluminum
Fosetyl-aluminum; a systemic phosphonate fungicide for horticultural crops. *Embetec Crop Protection Ltd.*

1614 Aliette Extra
Captan + fosetyl-aluminum + thiabendazole. Fungicide seed dressing for peas. *Embetec Crop Protection Ltd.*

1615 Alimet
Methionine hydroxy analog feed supplement; liquid source of amino acid (methionine) activity for poultry and other animal feeds. Name unverified.

1616 Alipa®
Redesignated Abex or Rhodapex®, an emulsifier. *Rhône-Poulenc Surf.*

1617 Aliso
555-31-7 359 209-090-8
Aluminum isopropoxide
Used for cosmetics, pharmaceuticals. *Rhône-Poulenc UK.*

1618 Alisol®
For agriculture industry. *DuPont UK.*

1619 Alistell
linuron-2,4-DB- MCPA.
Herbicide containing linuron, 2,4-DB and MCPA. *ICI Chem & Polymers Ltd.*

1620 Alistell
linuron-2,4-DB- MCPA.
Emulsifiable concentrate containing 220 g 2,4-DB, 30 g linuron and 30 g MCPA per liter; used to control weeds in undersown cereals and seedling grassland. *Farm Protection Ltd.*

1621 Alizarin
72-48-0 247 200-782-5
$C_{14}H_8O_4$
1,2-dihydroxyanthraquinone
C.I. Mordant Red 11; C.I. Pigment Red 83; C.I. 58000. Acid wool dyes. Also used as an acid-base indicator (pH 5.5 yellow; pH 6.8 red) and as a spot-test reagent for Al, In, Hg, Zn and Zr. mp = 290°; bp = 430°; slightly soluble in H_2O (0.58 mg/l); more soluble in organic solvents.

1622 Alizarine
Anionic dyestuffs (level dyeing); used for wool and wool blends. *Holliday Dyes & Chemicals Ltd.*

1623 Alka
Range of alkali detergents for the food industry; for bottlewashing and tank cleaning. *Harshaw Chemicals Ltd.* Name unverified.

1624 Alkaflo®
Patented phosphate salt mixture; alkalizer for reactive dyes in textile industry. *Sybron.*

1625 Alkafoam D
Alkanolamide/ethoxylated alcohol blend; foaming agent for foam dyed carpets. *Rhône-Poulenc Surf.*

1626 Alkagel
A trade name for alginates made for waterproofing. No manufacturer.

1627 alkali cellulose
Hydrated cellulose. The product of the reaction between cotton and caustic soda. when hydrolyzed by water it gives hydrated cellulose.

1628 Alkali Surfactant NM
Surfactant, wetting and coupling agent used in alkaline formulations for hard surface cleaning, floor strippers, heavy-duty degreasers, steam, soak tank, and household/institutional cleaners; solubilizer for nonionics into high electrolyte cleaners. *Exxon/Tomah.*

1629 Alkalit
A proprietary synthetic resin obtained by heating the sodium salt of phenolphthalein with toluoyl chloride. No manufacturer.

1630 Alkalsite
An explosive containing 25-32% potassium perchlorate, ammonium nitrate, trinitrotoluene, and other constituents. No manufacturer.

1631 Alkamide® 101 CG
68603-42-9 271-657-0
Cocamide DEA
Cocamide DEA; thickener, detergent, emulsifier for lower boiling aliphatic hydrocarbons; for cosmetic and industrial applications; produces emulsions stable in presence of alcohols, glycols, and phenols. *Rhône-Poulenc Surf.*

1632 Alkamide® 200 CGN
Cocamide DEA
Thickener, detergent, emulsifier, foam stabilizer; for hard surface cleaners, floor cleaners, rinsable degreasers, metal cleaners, metalworking compounds; corrosion inhibition characteristics. *Rhône-Poulenc Surf.*

1633 Alkamide® 2104
2:1 Cocamide DEA; detergent, emulsifier, foam booster; base for floor and general purpose cleaners; lubricant for synthetic grinding and cutting fluids. *Rhône-Poulenc Surf. Canada.*

1634 Alkamide® 327
120-40-1 204-393-1
$C_{16}H_{35}NO_2$
Lauric acid diethanolamine
Clindrol 200; N,N-diethanollauric acid amide; lauramide DEA; LDA; LDE; Ninol Aa62; Onyxol 345; Rewomid Dlms; Rewomid DI 203/S; Richamide 6310; Varamid ML-1; Rolamid CD; Standamidd LD; Steinamid DI 203 S; Super Amide L-9A; Super Amide L-9C; Synotol L-60; Unamide J-56. Lauramide DEA; foam and viscosity modifier for personal care products. *Rhône-Poulenc Surf. Canada.*

1635 Alkamide® C-212
68140-00-1 268-770-2
Cocamide MEA (1:1)
Thickener, foam builder and stabilizer for soap or synthetic based washing powders. *Rhône-Poulenc Surf.*

1636 Alkamide® C-5
61791-08-0
PEG-5 cocamide
PEG-5 cocamide; thickener, foam stabilizer, and emulsifier for formulated detergents and cosmetics. *Rhône-Poulenc Surf. Canada.*

1637 Alkamide® CDE
68603-42-9 271-657-0
Cocamide DEA
Aminol KDE. Detergent, emulsifier, stabilizer, thickener, foam stabilizer for personal care and detergent products. *Rhône-Poulenc Surf. Canada.*

1638 Alkamide® CP-1255
27883-12-1 248-710-1
Linoleamide DEA
Linolamide DEA, a detergent and emulsifier. *Rhône-Poulenc Surf.*

1639 Alkamide® DC-212/MP
Coconut poly-diethanolamide
Used where alkaline builders are incorporated into aqueous systems; viscosity builder for liquid potash soaps; conveyor belt lubricant; corrosion inhibitor properties. *Rhône-Poulenc Surf. Canada.*

1640 Alkamide® DIN-295/S
68425-47-8 270-355-6
Linoleamide DEA (1:1); foam booster, emulsifier, viscosity builder, thickener for shampoos, industrial cleaners; conditioning to hair. *Rhône-Poulenc Surf.; Rhône-Poulenc Surf. France.*

1641 Alkamide® DO-280
93-83-4 202-281-7
$C_{22}H_{43}NO$
Oleamide DEA
oleic diethanol amide; Oleic acid diethanolamine; 9-Octadecenamide, N,N-bis(2-hydroxyethyl)-, (Z)-; Oleic acid diethanolamide. 2:1 Oleamide DEA and diethanolamine; emulsifier for solvent oils; corrosion inhibitor. *Rhône-Poulenc Surf.*

1642 Alkamide® DS-280/S
93-82-3 202-280-1
Stearamide DEA (1:1); viscosity builder, thickener, foam booster, dispersant for nonionic and cationic systems, shampoos, bath preparations, industrial cleaners; emulsifier, corrosion inhibitor, lubricant for metalworking fluids; *Rhône-Poulenc Surf.*

1643 Alkamide® HTDE
93-82-3 202-280-1
Stearamide DEA
Detergent, thickener, viscosity builder, emulsifier for kerosene, vegetable and mineral oil, microcrystals, wax. *Rhône-Poulenc Surf. Canada.*

1644 Alkamide® L-203
142-78-9 205-560-1
Lauramide MEA (1:1); foam builder/stabilizer for soap and synthetic washing powds.; viscosity builder. *Rhône-Poulenc Surf.; Rhône-Poulenc Surf. France.*

1645 Alkamide® L7DE
Lauric-myristic DEA (1:1); detergent, foam booster/stabilizer, superfatting and thickening agent for toiletry and cleaning formulations; fortifier for perfumes in soaps. *Rhône-Poulenc Surf. Canada.*

1646 Alkamide® LE
120-40-1 204-393-1
1:1 Lauramide DEA
Amidex LD; Amidex L-9; Alkamide® 327. Viscosity booster, foam stabilizer in shampoo. *Rhône-Poulenc Surf.*

1647 Alkamide® LIPA/C
142-54-1 205-541-8
Lauramide MIPA
Lauramide MIPA; high foaming lubricant for metalworking fluids. *Rhône-Poulenc Surf.*

1648 Alkamide® OIP
111-05-7 203-828-2
Oleamide MIPA
Oleamide MIPA; foam modifier for high-temperature cleaners, especially liquid and powder laundry detergents; lubricant for metalworking fluids. *Rhône-Poulenc Surf. Canada.*

1649 Alkamide® R-280
106-16-1 203-368-2
Ricinoleamide MEA
Ricinoleamide MEA, a detergent and emulsifier. *Rhône-Poulenc Surf.*

1650 Alkamide® S-280
111-57-9 203-883-2
Stearamide MEA (1:1); viscosifier for industrial cleaners; skin protectant in toilet bars, creams, lotions, pastes. *Rhône-Poulenc Surf.*

1651 Alkamide® SDO
68425-47-8 270-355-6
Soyamide DEA
Foam stabilizer, viscosity builder, superfatting agent for toiletries, cutting and solvent oils, textiles, household and industrial cleaners, corrosion inhibitor. *Rhône-Poulenc Surf. Canada.*

1652 Alkamide® STEDA
110-30-5 203-755-6
$C_{38}H_{76}N_2O$
Ethylene bis-stearamide
Ethylene bis(stearamide); N,N'-ethylenedi(stearamide). Additive in pulp and paper defoamer formulations; lubricant, plasticizer, antistat, pigment dispersant for resins and plastics. *Rhône-Poulenc Surf. Canada.*

1653 Alkamide® WRS 1-66
93-83-4 202-281-7
$C_{22}H_{43}NO$
Oleic acid diethanolamine
oleic diethanol amide; 9-Octadecenamide, N,N-bis(2-hydroxyethyl)-, (Z)-; Oleic acid diethanolamide. Oleamide DEA; emulsifier for highly nonpolar aliphatic hydrocarbons and chlorinated aliphatic hydrocarbons; rust inhibitor. *Rhône-Poulenc Surf.*

1654 Alkaminox® C-2
61791-14-8
PEG-2 cocamine; textile scouring, dyeing assistant, softener, antistatic agent;

used as corrosion inhibitor in steam generating and circulating systems; coemulsifier. *Rhône-Poulenc Surf.*

1655 Alkamuls® 14/R
61791-12-6
PEG-60 castor oil; surfactant. *Rhône-Poulenc Surf. France.*

1656 Alkamuls® 400-DO
9005-07-6
PEG-8 dioleate; emulsifier, solubilizer, lubricant, wetting agent for cosmetic, textile, metalworking, and agricutural uses. *Rhône-Poulenc Surf.*

1657 Alkamuls® 400-MO
9004-96-0
PEG-9 oleate; emulsifier for fats, wetting agent, dispersant, lubricant used in dairy industry, cosmetic, metalworking, and industrial applications. *Rhône-Poulenc Surf.*

1658 Alkamuls® 600-DO
9005-07-6
PEG-12 dioleate
PEG-12 dioleate; dispersant, emulsifier for oil-water emulsions; for cosmetic, metalworking, and industrial use. *Rhône-Poulenc Surf.*

1659 Alkamuls® A
9004-96-0
PEG-6 oleate; emulsifier, lubricant for solvent oils, most aliphatic solvents; for lubricating and cutting oils, agricutural formulations. *Rhône-Poulenc Surf.; Rhône-Poulenc Surf. France.*

1660 Alkamuls® AG-900
9016-45-9 6772
Ethoxylate blend; spreading and wetting agents for aqueous pesticide systems. *Rhône-Poulenc Surf. Canada.*

1661 Alkamuls® B
61791-12-6
PEG-33 castor oil; emulsifier, dispersant for textiles, metallurgy, metal degreasing, personal care products; dye leveler, fabric softener. *Rhône-Poulenc France.*

1662 Alkamuls® COH-5
61788-85-0
PEG-5 hydrogenated castor oil; lubricant, softener, antistat, emulsifier, detergent. PEG-40 hydrogenated castor oil; perfume solubilizer, oil-water emulsifier for cosmetic products; eliminates oil bath turbidity. *Rhône-Poulenc Surf.*

1663 Alkamuls® EGDS
627-83-8 211-014-3
$C_{38}H_{74}O_4$
Glycol distearate
Glycol distearate pure; thickener, opacifier, pearlizing agent used in shampoos and cosmetic lotions. mp = 58-65°. *Rhône-Poulenc Surf.*

1664 Alkamuls® EGMS/C
111-60-4 203-886-9
Ethylene glycol monostearate
Viscosity booster, opacifying and pearlescing agent for liquid cosmetic and detergent compounds. *Rhône-Poulenc Surf.*

1665 Alkamuls® EL-620
61791-12-6
PEG-30 castor oil
PEG-30 castor oil; emulsifier, wetting agent, pigment dispersant, antistat, lubricant, solubilizer for industrial/household cleaners, cosmetics, pharmaceuticals, metalworking fluids, leather, pesticides, herbicides, paper industries. *Rhône-Poulenc Surf.; Rhône-Poulenc France.*

1666 Alkamuls® GMR-55LG
Glyceryl mono/dioleate
Coemulsifier, lubricant, softener, emollient, rust preventative additive for mold release agents, synthetic fiber spin finishes, compounded oils; antistat, antifog for PVC film processing. *Rhône-Poulenc Surf. Canada.*

1667 Alkamuls® GMS/C
31566-31-1 4498 250-705-4
glyceryl monostearate
Emulsifier, wetting agent for cosmetic, agricuture, textile industries; coupler used to bind waxes together; emollient and thickener in cosmetic creams. *Rhône-Poulenc Surf.*

1668 Alkamuls® GMS/C
31566-31-1 4498 250-705-4
Glyceryl stearate
Emulsifier, wetting agent for cosmetic, agricuture, textile industries; coupler used to bind waxes together; emollient and thickener in cosmetic creams. *Rhône-Poulenc Surf.*

1669 Alkamuls® L-9
9004-81-3
PEG-9 laurate
PEG-9 laurate; emulsifier, coemulsifier for cosmetic and toiletry preparations.; defoamer, leveling agent for latex paints; dispersant for dyes and pigments. *Rhône-Poulenc Surf.*

1670 Alkamuls® MM/M
3234-85-3 221-787-9
Myristyl myristate
Myristyl myristate; emollient, moisturizer, lubricant, and conditioner for hair and skin care products; viscosity builders and gelling/stiffening agents for makeup and deodorant applications. *Rhône-Poulenc Surf.*

1671 Alkamuls® PSML-20
9005-64-5 8872
Polysorbate 20; emulsifier, solubilizer, antistat, viscosity modifier, lubricant for textiles, cosmetics, pharmaceuticals. *Rhône-Poulenc Surf.*

1672 Alkamuls® PSMO-20
9005-65-6 7742
Polysorbate 20
Polysorbate 80; emulsifier, wetting agent for cosmetic, food, agricutural applications; coemulsifier for aliphatic alcohols, petroleum oils, fats, solvents and waxes. *Rhône-Poulenc Surf.*

1673 Alkamuls® PSMO-5
9005-65-6 7742
Polysorbate 81; emulsifier, solubilizer, antistat, lubricant for paint, food, cosmetic, insecticides, herbicides, fungicides, textiles, cutting oils. *Rhône-Poulenc Surf.*

1674 Alkamuls® PSMS-20
9005-67-8 8872
Polysorbate 60
Polysorbate 60; wetting agent, emulsifier for cosmetic and food applications, textiles, paper coatings; fiber-to-metal lubricant for fibers and yarns. *Rhône-Poulenc Surf.*

1675 Alkamuls® PSTO-20
9005-70-3 8872
Polysorbate 85
Polysorbate 85; emulsifier for cosmetic and food applications; textile and leather lubricant. *Rhône-Poulenc Surf.*

1676 Alkamuls® S-20
1338-39-2 8872 215-663-3
Sorbitan laurate; water-oil emulsifier, lubricant and softener for the textile industry; secondary suspending agent, porosity modifier in PVC suspensions. *Rhône-Poulenc Geronazzo.*

1677 Alkamuls® S-60
1338-41-6 8872 215-664-9
$C_{24}H_{46}O_6$
Sorbitan, monooctadecanoate
Sorbitan Monostearate; Sorbitan stearate. Sorbitan stearate; water-oil emulsifier, lubricant and softener for the textile industry; secondary suspending agent, porosity modifier in PVC suspensions. *Rhône-Poulenc Geronazzo.*

1678 Alkamuls® S-65
26658-19-5 247-891-4
Sorbitan tristearate
Water-oil emulsifier, lubricant and softener for the textile industry; secondary suspending agent, porosity modifier in PVC suspensions. *Rhône-Poulenc Geronazzo.*

1679 Alkamuls® S-65-40
9004-99-3
PEG-40 stearate; emulsifier, self-emulsifying lubricant and softener for synthetic fibers. *Rhône-Poulenc Surf.*

1680 Alkamuls® S-65-8
9004-99-3
$C_{20}H_{40}O_3$
PEG-8 stearate
polyoxyethylene (8) stearic acid (monoester); PEG 400 monostearate; PEG 600 monostearate; polyoxyethylene (40) stearic acid (monester); polyoxyethylene (50) stearic acid (monoester;)Polyethylene glycol monostearate; POE monostearate; POE (4) stearic acid (monoester); POE (8) stearic acid (monoester); POE (40) stearic acid (monester); POE (50) stearic acid (monoester). Emulsifier, self-emulsifying lubricant and softener for synthetic fibers. *Rhône-Poulenc Surf.*

1681 Alkamuls® S-80
1338-43-8 8872 215-665-4
Sorbitan oleate
Sorbitan oleate; water-oil emulsifier for mineral and vegetable oils, in metalworking; oil spill dispersant. *Rhône-Poulenc Geronazzo.*

1682 Alkamuls® S-85
26266-58-0 8872 247-569-3
Sorbitan, tri-9-octadecenoate, (Z,Z,Z)-
Span 85; Sorbitan Trioleate;. Sorbitan trioleate; water-oil emulsifier, lubricant

and softener for the textile industry; secondary suspending agent, porosity modifier in PVC suspensions. *Rhône-Poulenc Geronazzo.*

1683 Alkamuls® SDG

106-11-6 203-363-5

PEG-2 stearate

PEG-2 stearate; emollient, moisturizer, lubricant for skin and hair care systems. *Rhône-Poulenc Surf.*

1684 Alkamuls® SEG

111-60-4 203-886-9

$C_{20}H_{40}O_3$

Ethylene glycol monostearate

Opacifier and pearling agent for shampoos, creams, liquid hand soaps, liquid detergents; emulsion stabilizer, viscosity builder. *Rhône-Poulenc Surf.*

1685 Alkamuls® SML

1338-39-2 8872 215-663-3

$C_{18}H_{34}O_6$

Sorbitan, monododecanoate

Span 20; Sorbitan Monolaurate; Sorbitan laurate. Sorbitan laurate; emulsifier for oils and fats in cosmetic, metalworking and industrial oil products; corrosion inhibitor; antistat for PVC. *Rhône-Poulenc Surf.*

1686 Alkamuls® SMO

1338-43-8 8872 215-665-4

$C_{24}H_{44}O_6$

Sorbitan oleate

Span 80; Sorbitan Monooleate; Sorbitan, mono-9-octadecenoate, (Z)-; Sorbitan oleate. Emulsifier, coupling agent, wetting agent for medicants, petroleum oils, fats, and waxes in the industrial, textile, metalworking, and cosmetic industries; textile and leather lubricant and softener; corrosion inhibitor. *Rhône-Poulenc Surf.*

1687 Alkamuls® SMS

1338-41-6 8872 215-664-9

Sorbitan stearate

Sorbitan stearate; emulsifier and coupling agent; used to prepare silicone defoamer emulsions for industrial applications, paraffin wax emulsions for processing paper coatings; textile process lubricant; internal PVC film lubricant; cosmetics; foods. *Rhône-Poulenc Surf.*

1688 Alkamuls® SS

2778-96-3 220-476-5

Stearyl stearate

Stearyl stearate; emollient, moisturizer, lubricant, and conditioner for hair and skin care products; viscosity builders and gelling/stiffening agents for makeup and deodorant applications. *Rhône-Poulenc Surf.*

1689 Alkamuls® STO

26266-58-0 8872 247-569-3

Sorbitan trioleate

Sorbitan trioleate; emulsifier and coupling agent; used to compound textile and leather softener finishes; in metalworking fluids. *Rhône-Poulenc Surf.*

1690 Alkamuls® STS

26658-19-5 247-891-4

Sorbitan tristearate

Sorbitan tristearate; hydrophobic emulsifier for use as a fiber-to-metal lubricant for synthetic and cotton fibers; cosmetics, foods. *Rhône-Poulenc Surf.*

1691 Alkamuls® T-20

9005-64-5 8872

Polysorbate 20

Polysorbate 20; emulsifier, solubilizer, antistat and lubricant for textile industry; solubilizer for essential oils; raw material for no-tears shampoo. *Rhône-Poulenc Geronazzo.*

1692 Alkamuls® T-60

9005-67-8 8872

Polyoxyethylene Sorbitan Monostearate

Tween(R) 60; Polyoxyethylene (20) sorbitan monostearate; Polysorbate 60; Tween 60; Sorbitan, monooctadecanoate, poly(oxy-1,2-ethanediyl) derivs.; POE (20) sorbitan monostearate. PEG-20 sorbitan stearate; emulsifier, solubilizer, antistat and lubricant for textile industry; solubilizer for essential oils; raw material for no-tears shampoo. *Rhône-Poulenc Geronazzo.*

1693 Alkamuls® T-80

9005-65-6 7742

Polyoxyethylene Sorbitan Monooleate

Tween(r) 80; Tween 80; Sorethytan (20) Mono-oleate; Sorlate; Olothorb; Armotan PMO-20; Capmul POE-O; Drewmulse POE-SMO; Emsorb 6900; Glycosperse 0-20; Glycosperse 0-20 Veg; Glycosperse 0-20X; Liposorb 0-20; Polysorbate 80 B.P.C.; Polysorbate 80; Protasorb O-20; Sorbimacrogol Oleate 300. Polysorbate 80; oil-water emulsifier, solubilizer, textile fiber antistat/lubricant; used in hot and cold rolling formulations. *Rhône-Poulenc Geronazzo.*

1694 Alkamuls® T-85

9005-70-3 8872

Sorbitan, tri-9-octadecenoate, poly(oxy-1,2-ethanediyl) derivs., (Z,Z,Z)-polyoxyethylene (20) sorbitan trioleate; Polysorbate 85; POE (20) sorbitan trioleate. PEG-20 sorbitan trioleate; emulsifier, solubilizer, antistat and lubricant for textile industry; solubilizer for essential oils; raw material for no-tears shampoo. *Rhône-Poulenc Geronazzo.*

1695 Alkanet

517-88-4 253 208-245-7

$C_{16}H_{16}O_5$

(S)-5,8-dihydroxy-2-(1-hydroxy-4-methyl-3-pentenyl)-1,4-naphthalenedione

Anchusin; Alkanna; Alkannin; anchusa acid; Alkanna red; alkanet extract; C.I. Natural Red 20; C.I. 75530. Terms applied to two different plants, *Lawsonia inermis* and *Anchusa tintoria* whose roots are the source of a red dye, anchusine (alkannin), the name is applied to the dye as well as to the plant. mp = 149°; $[\alpha]^{20}_{Cd}$ = -165° (C_6H_6); poorly soluble in H_2O, more soluble in organic solvents; LD_{50} (rat orl) >1 g/kg.

1696 Alkanol®

Surfactants. *DuPont UK.*

1697 Alkanol® 189-S

Sodium alkyl sulfonate

Wetting agent, detergent, penetrant, foamer for textiles, elastomers, plastics, film, metal cleaning and pickling, hard surface cleaning, and chemical manufacturing; effective in acid and alkali media. *DuPont.*

1698 Alkanol® ND

Sodium alkyl diaryl sulfonate

Foaming agent, dyeing assistant, surfactant; for textiles, chemical manufacturing; leveling agent for acid dyes on nylon. *DuPont.*

1699 Alkanol® WXN

Sodium alkylbenzene sulfonate

Wetting, rewetting agent, foaming agent, emulsifier, dyeing assistant; for textiles, paper, chemical manufacturing, alkaline and acid cleaners; leveling agent for acid dyes; stable to acid or alkaline media. *DuPont.*

1700 Alkanol® XC

Sodium alkylnaphthalene sulfonate

Wetting agent, dispersant, penetrant, low foaming; used in bleaching and dyeing of textiles, leather, paper, chemical manufacturing, photography; reduces shrinkage in ceramics manufacturing; dry colors manufacturing. *DuPont.*

1701 Alkanolamine

108-01-0 2900 203-542-8

$C_4H_{11}NO$

N,N-Dimethyl-2-Hydroxyethylamine

β-Dimethylaminoethyl alcohol; Deanol; Dimethyl ethanolamine; 2-dimethylaminoethanol; Dimethylethanolamine; DMAE; dimethylaminoethanol; N,N-Dimethylaminoethanol; N-dimethylaminoethanol; N,N-dimethyl-N-(2-hydroxyethyl)amine; β-hydroxyethyldimethylamine. Solubilizer of synthetic resins for water soluble paints, raw material for ion exchange resins and coagulants. mp = -70°; bp= 139°; d = 0.8870; n_D^{20}= 1.4294; LD_{50} (rat orl).= 2 gm/kg. *Yokkaichi Chemical Co Ltd.* Name unverified.

1702 Alkanox® 24-44

38613-77-3 254-037-4

Tetrakis (2,4-di-t-butylphenyl) 4,4·g-biphenylylene diphosphonite

Processing antioxidant for PE, PP, PC, PS, polyesters; protects against thermo-oxidative degradation during long term aging. *Enichem Synthesis SpA.*

1703 Alkanox® 240

31570-04-4 250-709-6

Trisdibutylphenyl phosphite

High performance antioxidant for stabilization of polymers including PP, HDPE, LDPE, LLDPE, PC, ABS, and polyesters. *Enichem Synthesis SpA.*

1704 Alkanox® 240-3T

693-36-7 211-750-5

$C_{42}H_{82}O_4S$

Tris(2,4-di-t-butylphenyl) phosphite

distearyl thiodipropionate; Dioctadecyl 3,3'-thiodipropionate. Antioxidant for stabilization of PP, PE. *Enichem Synthesis SpA.*

1705 Alkanox® P-24

26741-53-7 247-952-5

Bisdibutylphenyl pentaerythritol diphosphite

Antioxidant for stabilization of ABS, PVC and PC polymers; color stabilizer. *Enichem Synthesis SpA.*

1706 Alkapol PEG 300

25322-68-3 7729

$(C_2H_6O_2)_n$

Ethoxylated 1,2-ethanediol

PEG 1000; Polyoxyethylene 1000; Polyglycol 1000; Polyethylene glycol 400; Poly Ethylene Oxide; Polyethylene Glycol 8000; Carbowax PEG

8000;Carbowax PEG 400; Carbowax 200; Emkapol 200; Gafanol E 200; Pluriol E 200; Polydiol 200; Polyethylene Glycol; PEG; Polyox WSR-301; PEG 200; Macrogol; Polyethyleneglycol 4000; Polyethyleneglycol 1500. PEG-6; intermediate for surfactants; binder/lubricant in pharmaceuticals; plasticizer; paper softener; humectant; solvent; antistat; for cosmetics, textile, plastics processing, dyes and inks. *Rhône-Poulenc Surf.* Name unverified.

1707 Alkaquat® DMB-451-50, DMB-451-80
61789-71-7 263-080-8
Benzalkonium chloride
Wetting agent, emulsifier, biocide, disinfectant for use in beverage industry, dairy industry, food processing, water treatment, paper industry, pest control, preservatives, antidandruff rinses; general disinfection and sanitization for hospitals, and laundries. *Rhône-Poulenc Surf. Canada.*

1708 Alkasil® HNM 1223-15 (70%)
Organo-modified polydimethylsiloxane in aromatic hydrocarbon solvents; for surface modification, waterproofing, release properties on wood, masonry, silica, mineral granules, paper, etc. *Rhône-Poulenc Surf. Canada.*

1709 Alkasil® NE58-50
63148-55-0
Dimethicone copolyol.
Surfactant and mold releasing agent. *Rhône-Poulenc Surf. Canada.*

1710 Alkasit
A proprietary cellulose adhesive. No manufacturer.

1711 Alkasperse 25
A proprietary series of pigment dispersions based on a short oil xylol-thinned alkyd used in the coloring of medium to fast air-drying surface coatings. *Collinda Ltd.*

1712 Alkasperse® A-20
9003-04-7
Sodium polyacrylate
Dispersant for boiler water compounds. *Rhône-Poulenc Surf. Canada.*

1713 Alkasperse® M-10
25086-62-8
sodium polymethacrylate
Polymeric dispersant for clay slurries and drilling muds; antiredeposition agent in phosphate-free detergents; conductivity aid for Electrofax paper. *Rhône-Poulenc Surf.* Name unverified.

1714 Alkastar 83
Organic brightener system; used for alkaline noncyanide zinc electroplating. *Harshaw Chemicals Ltd.*

1715 Alkasurf® NP-4
9016-45-9 6772
Tergitol NP-33
Nonyl Phenol Ethoxylate; Nonylphenyl Polyethyleneglycol Ether, Nonionic; Tergitol TP-9; Antarox; Antarox BI-344; Macrogol Nonylphenyl Ether; Nonoxinol; Nonoxynol; Conco NI; Dowfax 9N; Igepal CO; Makon; Neutronyx 600's; Nonipol NO; Polytergent B; Renex 600's; Solar NP; Triton N; Tergitol Np; T-det-n; Surfionic N; Sterox; Arkopal N-090; Carsonon N-9; Conco Ni-90; Igepal Co-630; Neutronyx 600; Peg-9 Nonyl Phenyl Ether; Protachem 630; Rewpol Hv-9; Tergitol Tp-9 (Non-ionic); Polyoxyethylene (1.5) Nonyl Phenol; Polyoxyethylene (4) Nonylphenol; Tergitol NP-14; Tergitol NP-27; Tergitol NP-35; Tergitol NP-40; Tergitol NPX. Nonoxynol-4; emulsifier, detergent, dispersant, intermediate, stabilizer; plasticizer, antistat for plastics, surfactants, household, industrial, and cosmetic use, fat liquoring, cutting and soluble oils. *Rhône-Poulenc Surf. Canada.*

1716 Alkaterge® C
Oxazolidine
Surfactant, emulsifier, emulsion stabilizer, wetting agent, acid acceptor; pigment grinding and dispersion aid; penetrant for textile and paper industries, metal cleaners; coatings; antifoam for antibiotic fermentation; antioxidant. *Angus.*

1717 Alkaterge® E
68140-98-7 268-820-3
Ethyl hydroxymethyl oleyl oxazoline
Detergent, emulsifier, wetting agent, antifoamer, antioxidant; used in salt, soap, paper, textiles, and metal cleaners; emulsion stabilizer; acid acceptor; pigment grinding and dispersion. *Angus.*

1718 Alkaterge® -T-IV
95706-86-8
Oxazoline derivative, ethoxylated: acid scavenger, offers filming protection to metal surfaces; corrosion inhibitor; oil-water emulsifier; dispersant in aqueous and nonaqueous systems; wetting agent. *Angus.*

1719 Alkateric® A2P-OS
Octyl propionate; low-foaming surfactant for acid and alkaline cleaners; pH stable. *Rhône-Poulenc Surf.* Name unverified.

1720 Alkateric® AP-C
Cocopropionate
Foaming agent for alkaline cleaning compound formulations. *Rhône-Poulenc Surf.* Name unverified.

1721 Alkateric® PB
693-33-4 211-748-4
Cetyl betaine
Mild substantive surfactant for personal care products; conditioner, antistat, emollient; as solubilizer, viscosity builder, foam booster with lauryl sulfates; stable to acid and alkali media. *Rhône-Poulenc Surf.* Name unverified.

1722 Alkathene
Solid polymers of ethylene prepared by subjecting ethylene to extremely high pressures under carefully controlled conditions of temperature. *ICI Chem & Polymers Ltd.*

1723 Alkawet®
Amphoteric surfactant; used as a wetting agent. *Lonza AG.*

1724 Alkawet® AA-60
Alkoxylated alcohol
Low foam wetting agent for continuous dyeing of carpet. *Rhône-Poulenc Surf.*

1725 Alkawet® CF
Proprietary; wetting agent, detergent for industrial use; electrolyte-tolerant, controlled foaming. *Lonza AG.*

1726 Alkazid® Lye DIK, M
For removal of hydrogen sulfide and carbon dioxide from synthesis gas and cracked gas for production of ethylene. *BASF AG.*

1727 Al-kenna
The powdered roots and leaves of *Lawsonia inermis*; used in the East for dyeing the nails, teeth, and hair.

1728 Alkolite
A proprietary phenophthalein resin. No manufacturer.

1729 alkyd resin
Thermosetting coating polymer; vehicles in exterior house paints, marine paints, and baking enamels; elec. components, encapsulation. *Bayer SA; Croda Resins Ltd; DSM GmbH; PPG Industries SA; Reichold Chemie AG; Scott Bader.*

1730 Alkydal®
Phthalate resins modified with oil or fatty acids; for use in the formulation of paints and varnishes. *Bayer AG; Bayer plc.*

1731 Alkylaryl ether sulfate
Agrilan® DG102. Wetting agents for production of water-dispersible granular agrochemical toxicants. d = 1.110; viscosity = 6000 cps; pH 7.8 (1% aq.); anionic.

1732 Alkylate 215
68648-87-3 272-008-4
n-dodecylbenzene
Detergent intermediate used in light-duty detergents, dishwash, laundry, industrial cleaners. *Monsanto/Detergents & Phosphates.*

1733 Alkylate 230
Linear tridecylbenzene
Detergent intermediate used in heavy duty detergents.

1734 Alkylated phenol
61788-44-1 262-975-0
Akrochem® Antioxidant 16; AO47L; AO47P; Montaclere®; Montalere® SPH; Nevastain® 21; Prodox®120; Vulkanox® SP. Nonstaining antioxidant used in styrene/butadiene rubber latex compounding urethanes, ABS, and ethylene-propylene-diene rubbers. Powder; soluble in toluene, ethyl acetate, chloroform, gasoline, hexane; easyly emulsifiable in H_2O; d = 1.23.

1735 Alkynol®
Saturated polyesters free from oil and fatty acids; for use in the formulation of coil coatings and high-grade stoving finishes. *Bayer AG; Bayer plc.*

1736 Allabond Twenty/Twenty Adhesive
25928-94-3
Epoxy adhesive; for production applications; for use with Activator BA-66B; also available in clear and conductive grades. *Bacon.* Name unverified.

1737 Allabond Twenty/Twenty NM
25928-94-3
Epoxy adhesive; nonmagnetic version of Allabond Twenty/twenty Adhesive. *Bacon.* Name unverified.

1738 Allactol
Aluminum lactotartrate
No manufacturer.

1739 Allantoin
97-59-6 255 202-592-8
$C_4H_6N_4O_3$
2,5-dioxo-4-imidazolidinyl urea
glyoxyldiureide; Cordianine; Psoralon; Septalan; 5-ureidohydrantoin. Heterocyclic organic compound; production of animal metabolism, excreted in urine; biochemical research, medicine; soothing agent, skin protectant;

stimulates growth of healthy tissue. Racemic form, monoclinic plates or prisims from H_2O; mp=238°; 1 gram dissolves in 190 ml H_2O, 500 ml alcohol; almost insoluble in ether; pH of saturated H_2O solution = 5.5. 3-V; Atomergic Chemetals; EM Ind; R. W. Greef; Hommel GmbH; ICI Am; Penta Mfg; Schweizerhall; Sutton Lab; Tri-K Ind.

1740 Allegheny 33, 44, 55, 66
Corrosion resisting alloys (formerly Ascoloy 33, 44, 55, 66): they contain iron and chromium : 33 contains 12-16% chromium, and 55 contains 26-30% chromium. No manufacturer.

1741 Allegheny Metal
A proprietary corrosion-resisting alloy containing iron with 17-20% chromium and 7-10% nickel. No manufacturer.

1742 Allenoy
A proprietary molybdenum steel. No manufacturer.

1743 Allen's metal
An alloy of 55.3% copper, 44-6% lead, and 0.1% tin.

1744 allethrin
584-79-2 257 209-542-4
$C_{19}H_{26}O_3$
2-methyl-4-oxo-3-(2-propenyl)-2-cyclopenten-1-yl-2,2-dimethyl-3-(2-methyl-1-propenyl)cyclopropanecarboxylate
allethrine; pallethrine; OMS 468; ENT 17510; Alleviate; Pynamin; Pyresin; Pyrexcel; Pyrocide;. A synthetic insecticide structurally similar to pyrethrin. Gives rapid knockdown and paralyzes insects before killing them. Used for control of flies, mosquitoes, ants and other household and public health insect pests. mp = 4°; $bp_{0.1}$ = 140°; n_D^{20}= 1.5040; insoluble in H_2O, soluble in organic solvents; LD_{50} (rat orl)= 1100 mg/kg. *Fairfield, McLaughlin Gormley King; Sumitomo.*

1745 Allguard
Water repellent coating for concrete, stone and masonry building walls. *Dow Croning.*

1746 Alligator wood
The wood of *Guarea grandifolia*, of West India.

1747 Allisan
99-30-9 202-746-4
dicloran
Horticultural fungicide containing dicloran. *The Boots Co plc.* Discontinued.

1748 All-O
A proprietary liquid soap for use as a rubber lubricant. No manufacturer.

1749 Alloprene
Chlorinated rubber. *ICI Chem & Polymers Ltd.*

1750 Alloxan
50-71-5 290 200-062-0
$C_4H_2N_2O_4$
2,4,5,6(1H,3H)-pyrimidinetetrone
Mesoxalylurea; mesoxalylcarbamide. Used in nutrition experiments. Causes diabetes in experimental animals. dec 256°; soluble in H_2O. No manufacturer.

1751 Alloy 39
An alloy containing aluminum with 3.75-4.25% copper, 1.2-1.7% magnesium, and 1.8-2.3% nickel. No manufacturer.

1752 Alloy 109
An alloy containing 88% aluminum and 12% copper. No manufacturer.

1753 Alloy 122
An alloy containing 88.6% aluminum, 10% copper, 1.2% iron, and 0.25% magnesium. No manufacturer.

1754 Alloy 142
An alloy containing 92.5% aluminum, 4% copper, 2% nickel, and 1.5% magnesium. No manufacturer.

1755 Alloy 145
An aluminum alloy containing 10-11% zinc, 2-3% copper, and 1-1.5% iron. No manufacturer.

1756 Alloy 195
An aluminum alloy containing 4-5% copper, and not more than 1.2% silicon, 1.2% iron, 0.35% magnesium, and 0.35% zinc. No manufacturer.

1757 Alloy 2129
A special nickel-iron alloy having fairly high permeability and excellent mechanical properties. No manufacturer.

1758 Alloy 2L5
An aluminum alloy with 12.5-14.5% zinc and 2.5-3% copper. No manufacturer.

1759 Alloy 2L8
An aluminum alloy containing 11-13% copper. No manufacturer.

1760 Alloy 3L11
An aluminum alloy containing 6-8% copper and tin may be added up to 1%. No manufacturer.

1761 Alloy AM4-4
An aluminum-magnesium alloy; it contains magnesium with 4% aluminum, 0-4% manganese, and 0-15% silicon. No manufacturer.

1762 Alloy AM7-4
An alloy containing magnesium with 7% aluminum, 0-4% manganese, and 0-15% silicon. No manufacturer.

1763 Alloy AMF
An alloy containing 50-60% nickel for low temperature use. No manufacturer.

1764 Alloy AP33
An aluminum alloy containing 4.5% copper and 0-4% titanium. No manufacturer.

1765 Alloy JL
An aluminum alloy containing 4-5% copper, 0-41% iron, and 0-35% silicon. No manufacturer.

1766 Alloy L10
An alloy containing aluminum with 10% copper and 1% tin. No manufacturer.

1767 Alloy L11
An alloy containing aluminum with 7% copper and 1% tin. No manufacturer.

1768 Alloy L5
An aluminum alloy containing, in addition to aluminum, 13% zinc and 2.8% copper. No manufacturer.

1769 Alloy L7
An alloy containing aluminum with 14% copper, and 1% manganese. No manufacturer.

1770 Alloy L8
An alloy containing aluminum with 12% copper. No manufacturer.

1771 Alloy MG7
An alloy consisting mainly of aluminum with magnesium and manganese; it has mechanical properties similar to Duralmin, and is stated to be highly resistant to corrosion. No manufacturer.

1772 Alloy N
An alloy of 91% aluminum, 6% copper, and 3% manganese. No manufacturer.

1773 Alloy NCT3
An alloy containing 44.5% iron, 37.5% chromium, 17.5% nickel, and 0.5% manganese. No manufacturer.

1774 Alloy RR
A series of aluminum alloys containing aluminum with 0.5-5% copper, 0.2-2.5% nickel, 0.05-5% magnesium, 0.6-1.5% iron, 0.05-0.5% titanium, and 0.2-5% silicon. No manufacturer.

1775 Alloy Steel
The term applied to a steel containing one or more elements in addition to carbon;. No manufacturer.

1776 Alloy T
An alloy containing aluminum with 3.8% magnesium, 0.5% iron, 0.5% silicon, and 0.1% copper. No manufacturer.

1777 Alloy W.7.1
1% Silicon, 4.5% manganese, balance nickel with controlled zirconium addition. Unverified.

1778 Alloy W.9
1% Silicon, 4-5% manganese, balance nickel with controlled zirconium addition. Unverified.

1779 Alloy Y
Alloy 24. An alloy of aluminum with 4% copper, 2% nickel, and 1.5% magnesium. No manufacturer.

1780 Alloys Wm
White-bearing metals; Wm5 contains 78.5% lead, 15% antimony, 5% tin, and 1.5% copper, with a specific gravity of 10.1; Wm10 consists of 73.5% lead, 15% antimony, 10% tin, and 1.5% copper; Wm42 contains 42% tin, 41% led, 14% antimony and 3% copper; Wm80 consists of 80% tin, 10% antimony and 10% copper. No manufacturer.

1781 Alluman
An alloy of aluminum with 10-20% tin and 4-6% copper. No manufacturer.

1782 Ally®
74223-64-6 6244
Metsulfuron-methyl
Plant growth regulator, used for control of annual dicotyledons in cereals. *DuPont UK.*

1783 Allyl glycidyl ether
106-92-3 203-442-4
$C_6H_{10}O_2$
1-allyloxy-2,3-epoxy-propane
Sipomer® AGE; Ageflex AGE; AGE; 1,2-Epoxy-3-Allyloxypropane; Allyl-2,3-epoxypropyl ether; [(2-propenyloxy)methyl]oxirane; 1-allyl-2,3-epoxypropane; Allyl Glycidyl Ether-Ethylene Glycol Prepolymer(18/1). Modifier for elastomer, epoxies, adhesives, fibers; reactive intermediate for coatings, sizing/finishing agent for fiberglass; silane intermediate in elec. coatings. mp = -100°; bp =

154°; d^{20}_{20} = 0.970; n^{20}_{D} = 1.4322; slightly soluble in H_2O, more soluble in organic solvents; LD_{50} (rat orl) = 1600 mg/kg.

1784 Allyl methacrylate
96-05-9 202-473-0
$C_7H_{10}O_2$
2-Propenoic acid, 2-methyl-, 2-propenyl ester
Sipomer™ AM; Ageflex AMA. Silane monomer intermediate; crosslinker offering two-stage polymerization, abrasion and solv. resist.; polymer modifier for high impact plastics, adhesives, acrylic elastomers, photoresists, optical polymers. mp = -65°; bp = 144°; d = 0.9380; n^{20}_{D}= 1.4360; soluble in H_2O (4 g/l), organic solvents; LD_{50} (rat orl) = 430 mg/kg. Aldrich; CPS; Monomer-Polymer & Dajac; Polysciences; Rhône-Poulec Spec.; Richman; Rohm Tech; San Esters.

1785 Almag
A proprietary aluminum alloy similar in composition to Alferium. No manufacturer.

1786 Almasilium
An aluminum alloy containing 1% magnesium and 2% silicon. No manufacturer.

1787 Almelec
A proprietary alloy of aluminum containing 0.7% magnesium, 0.5% silicon, and 0.3% iron. No manufacturer.

1788 Almen's reagent
A solution containing 5g tannic acid in 240 ml of 50% alcohol, to which has been added 10 ml of a 25% solution of acetic acid; a precipitate is given with nucleoproteins.

1789 Almo Steel
Proprietary chrome-molybdenum steels. No manufacturer.

1790 Almora
3632-91-5 222-848-2
Magnesium gluconate
Replenisher. Soluble in H_2) (160 g/l); LD_{50} (rat orl) = 9.1 g/kg. *O'Neal, Jones & Feldman Pharmaceuticals.* Name unverified.

1791 Almstab
Stabilizers for use in PVC compositions. *Associated Lead Manufacturers Ltd.* Name unverified.

1792 Alnovol
Nonheat hardening phenol formaldehyde resins; used for printing inks and rubber reinforcement. *Resinous Chemicals Ltd.*

1793 Alnovol
Nonheat hardening phenol formaldehyde resins; novolak types; used in printing inks and rubber reinforcement. *Hoechst UK.*

1794 Alocrom
Chromating pretreatment for aluminum. *ICI Chem & Polymers Ltd.*

1795 Aloe Vera Powd. 200XXXExtract-Microfine
Aloe vera gel
Rapid dissolving ingredient, for cosmetic, health and pharmaceutical industries. *Tri-K Industries.*

1796 Alomite
A trade name applied to a variety of sodalite used as an ornamental stone. No manufacturer.

1797 Alon®
Fumed alumina. No manufacturer.

1798 Alox
A proprietary trade name for a series of methyl esters of higher alcohols. No manufacturer.

1799 Alox® 111
Mold release or parting compound for cast concrete products; rust preventive protecting metal forms. *Alox.* Name unverified.

1800 Alox® 152
Lubricity agent in mineral oil solutions, cutting and metalworking formulations, quench oil additives, for automotive and industrial lubricants. *Alox.* Name unverified.

1801 Alox® 318F
Corrosion inhibitor for use on ferrous and nonferrous metals, especially for bright or highly polished steel surfaces. *Alox.* Name unverified.

1802 Alox® 350
Oxidized petroleum fraction esters; corrosion inhibitor and lubricant for making cuttting and soluble oils. *Alox.* Name unverified.

1803 Alox® 436A
Mold release, lubricant, corrosion inhibitor for concrete molds. *Alox.* Name unverified.

1804 Alox® 488
Oxidized petroleum fractions; upper cylinder lubricant for internal combustion engines; cleans and prevents formation of carbon and lacquer deposits in engines. *Alox.* Name unverified.

1805 Alox® 575
Oxygenated hydrocarbon with barium and a sodium petroleum sulfonate; emulsifier, corrosion inhibitor for cutting and soluble oils. *Alox.* Name unverified.

1806 Alox® 606, 606-55, 606-70
Heavy-duty rust preventive, protective coating for metal surfaces, especially for automotive rustproofing; 606-55 is a total solvent cutback grade with 55% Alox 606 and 45% mineral spirits; 606-70 is a partial solvent cutback grade with 70% Alox 606 and 30% minimum. *Alox.* Name unverified.

1807 Alox® 904
Base for formulation of fingerprint removers or rifle bore cleaners; corrosion preventive. *Alox.* Name unverified.

1808 Alox® 1680
Oxygenated hydrocarbon containing minor amount of phosphatide; corrosion inhibitor; penetrating oil additive. *Alox.* Name unverified.

1809 Alox® 2000
Corrosion inhibitor and penetrant for metal conditioning compounds. *Alox.* Name unverified.

1810 Alox® 2028L
Oxygenated hydrocarbon, calcium soap; nonemulsifiable, nonstaining, water-displacing rust preventive, lubricant. *Alox.* Name unverified.

1811 Alox® 2211Y
Thixotropic rust preventive which deposits film providing long-term protection against humidity and salt fog; can be cutback in solvent or blended 50/50 with naphthenic oil. *Alox.* Name unverified.

1812 Alox® 2301
Oxidized microcrystalline wax; for preparation of wax emulsions for textile sizing, finishing and waterproofing. *Alox.* Name unverified.

1813 Aloxicoll
1327-41-9 356 215-477-2
$Al_2H_5O_5Cl \cdot 2H_2O$
Aluminum chlorohydrate
Astringen; Chlorhydrol; Hyperdrol; Locron; Phosphonorm; Aluminum chloride, basic; Aluminum hydroxychloride; Aluminum chlorhydroxide. Astringent andantihyperphosphatemic. Used in antiperspirants. Soluble in H_2O (< 55% w/w). *Giulini Corp.*

1814 Aloxite
1344-28-1 369 215-691-6
A trademark for abrasive and refractory materials consisting essentially of alumina. No manufacturer.

1815 Alpacca
An alloy of 64% copper, 19% zinc, 14.5% nickel, 2% silver, 0.4% iron, and 0.12% tin; it is a nickel silver. No manufacturer.

1816 Alperox-F
105-74-8 203-326-3
$C_{24}H_{46}O_4$
dodecanoyl peroxide
dilauroyl peroxide; Peroxide, bis(1-oxododecyl); Dilauryl Peroxide. Lauroyl peroxide; initiator for bulk, solution, and suspension polymerization, high-temperature curing of polyester resins, and cure of acrylic syrup. *Elf Atochem.*

1817 Alpex
Cyclized rubber resins; used for chemical resistant coatings. *Resinous Chemicals Ltd.*

1818 Alpex
Cyclized rubber resins; used for chemical resistant coatings. *Hoechst UK.*

1819 Alpfa
Proprietary name for extensive range of research chemicals and materials. *Johnson Matthey plc.*

1820 Alpha Chymar
9004-07-3 2320 232-671-2
Chymotrypsin
Proteolytic enzyme. *Barnes-Hind Inc.* Name unverified.

1821 Alpha Daphnone
α-isomethylionone
Used in perfumery. *Bush Boake Allen Ltd.*

1822 Alphachloralose
Rodenticide. *Rentokil Ltd.*

1823 Alphachroic
Chrome dyestuffs. *J C Bottomley.* Name unverified.

1824 Alphadim® 90AB
67701-33-1 266-952-6
High-purity, molecularly distilled monoglyceride prepared from fully hardened edible fats and glycerin; food additive providing functional improvements in processing and storage stability; stabilizes and disperses fat particles in coffee whiteners; as starch complexing and softening agent in breads. *Am. Ingredients/Patco.*

1825 Alphadim® 90LC
61789-10-4 263-032-6
Dimethyl hydrogenated tallow amine
High-purity, molecularly distilled monoglyceride prepared from edible fats and glycerin; food additive providing a stable emulsion of finely dispersed water droplets for margarine and coffee whiteners. *Am. Ingredients/Patco.*

1826 Alphadim® 90NLK
67701-32-0 266-951-0
High-purity, molecularly distilled mono- and diglycerides of fatty acids prepared from refined sunflower oil and glycerin with BHA and citric acid; Kosher; food additive providing a stable, finely dispersed emulsion in diet margarines. *Am. Ingredients/Patco.*

1827 Alphadim® 90SBK
High-purity, molecularly distilled monoglyceride prepared from fully hardened soybean oil and glycerin; kosher; food additive providing a stable, finely dispersed emulsion in margarines; stabilizes and disperses fat particles in coffee whiteners; starch complexing agent. *Am. Ingredients/Patco.*

1828 Alphamint
Peppermint blend. *Bush Boake Allen Ltd.*

1829 Alphanol
Acid dyestuffs; used for wool dyeing. *Hoechst UK.*

1830 Alphanol
Medium chain length alcohols, forming a plasticizer. *ICI Chem & Polymers Ltd.*

1831 Alphanol®
Acid dyestuffs; used for wool dyeing. *Cassella AG.*

1832 Alpha-Ruvite
1344-28-1 369 215-691-6
Aluminum oxide
Activated alumina; in selective absorption processes; used in manufacture of catalysts.

1833 Alphasol OT
A proprietary trade name for the sodium salt of an alkyl ester of sulfosuccinic acid. A surfactant. No manufacturer.

1834 Alpha-Step® MC-48
Sodium α-sulfomethyl cocoate
Surfactant, foam booster/stabilizer for dishwashing liquids. *Stepan; Stepan Canada.*

1835 Alpha-Step® ML-40
Sodium methyl-2-sulfolaurate and sodium ethyl-2 sulfolaurate; biodegradable surfactant, foaming agent, hydrotrope for dishwashing liquids, fine fabric washes, hard surface cleaners and bubble baths; scouring, leveling, coupling and foaming agent for textiles; metalworking formulations. *Stepan; Stepan Canada; Stepan Europe.*

1836 Alphatex®
12141-46-7 377 235-253-8
Altowhite LL. Metakaolinitic aluminum silicate produced by calcining kaolin clay; paper coating pigment, paper filler. *ECC International Ltd.*

1837 Alphenate
A series of phenolic-alkyd plasticized resins. *Vianova Resins.* Discontinued.

1838 Alphide
A proprietary trade name for a cold molded refractory ceramic. No manufacturer.

1839 Alphogen
$(COOH\ CH_2{\cdot}CH_2CO)_2O_2$
Alphozone
Succinoxate; Succinyl peroxide. An antiseptic. No manufacturer.

1840 Alphol
550-97-0 6501
$C_{17}H_{12}O_3$
α-Naphthol salicylate
An antiseptic and antirheumatic. mp = 83°; insoluble in H_2O, soluble in organic solvents. No manufacturer.

1841 Alphoxat O 105
9004-96-0
PEG-5 oleate
PEG-5 oleate; basic material for textile industry. *Zschimmer & Schwarz.* Discontinued.

1842 Alphoxat S 110
9004-99-3
PEG-10 stearate
PEG-10 stearate; emulsifier and dispersant, greasing agent for textile and chemical technical industries. *Zschimmer & Schwarz.* Discontinued.

1843 Alplate
A proprietary aluminum coated steel. No manufacturer.

1844 Alpolit
Unsaturated polyester resin in styrene; used for glass fiber reinforced laminate, casting and potting. *Resinous Chemicals Ltd.*

1845 Alprokyds
Proprietary drying oil and non-drying oil modified alkyd resins. No manufacturer.

1846 alquifon
Black Lead Ore
Potter's Ore. A mineral. It consists of zinc sulfide; used in pottery to give a green glaze. *Potter's Ore.*

1847 Alreco
Aluminum and aluminum alloy ingot; for aluminum die cast industry. *Reynolds Metal Co.*

1848 Alresat
Maleinized rosin esters; used in nitrocellulose lacquers, printing inks and to improve gloss in air drying paints. *Hoechst AG; Hoechst UK.*

1849 Alresen
Alkyl phenol formaldehyde resins, terpene phenolic resins; used for oil varnishes and adhesives. *Hoechst UK.*

1850 Alromin Ru 1000
Antistatic agent. *Ciba plc.* Name unverified.

1851 Alrosperse® 100
Surfactant blend; surfactant, interfacial tensile depressant, dispersant, deflocculant, solubilizer, emulsifier, corrosion inhibitor, antistat for metal processing, petroleum products, drycleaning, spotting compounds, leather and upholstery cleaners, emulsions, paints and inks. *M. S. Paisner.*

1852 Alscoap AF Series
Surfactants; foaming agent for air drilling. *Toho Chem. Industry.*

1853 Alscoap LN-40, LN-90
151-21-3 8782 205-788-1
Sodium lauryl sulfate
Sodium lauryl sulfate; detergent, shampoo base, toothpaste; polymerization emulsifier for synthetic resins and latex. *Toho Chem. Industry.*

1854 Alsi
A pigment consisting of finely ground aluminum-silicon alloy used to give durable and rust-preventative paints. No manufacturer.

1855 Alsibronz
12001-26-2
Wet ground muscovite mica; used for paint, rubber, plastics, pearlescent pigments. *Franklin Mineral Products Co.*

1856 Alsica Alloys
Aluminum-silicon-copper alloys. No manufacturer.

1857 Alsifer
A proprietary alloy of 40% silicon, 40% iron, and 20% aluminum; a hardener alloy for adding silicon to aluminum alloys. No manufacturer.

1858 Alsifilm
A material made from bentonite; used in place of mica. No manufacturer.

1859 Alsimag
Ceramic materials; used for insulation and for the dielectric of condensers. *3M.* Name unverified.

1860 Alsimag 754
A beryllia ceramic. *3M.* Name unverified.

1861 Alsimag 779
Leachable ceramic cores for precision metal castings. *3M.* Name unverified.

1862 Alsimin
An alloy similar to Alsifer. No manufacturer.

1863 Alsol
333
Aluminum acetotartrate; a germicide, astringent and antiseptic. No manufacturer.

1864 Alstat
Very powerful antistatic agent; for all stages of textile processing. *Altex Chemical Co Ltd.* Name unverified.

1865 Alstromed A 18 LV
n-Octadecyl sodium sulfosuccinamate solution; foaming, frothing and emulsifying agent. *Alco Chemical Corp.* Discontinued.

1866 Alsynates
Metal carboxylates based on C_8-C_{10} branched chain synthetic aliphatic carboxylic acids; driers for paint or printing ink catalyst for unsaturated polyester. *Manchem Ltd.* Name unverified.

1867 Alsynol RS-47
Cyclized rubber resins for coatings. *Daniel Prods.* Discontinued.

1868 Alsystin
64628-44-0 9809 264-980-3
triflumuron
Insect growth regulator. *Bayer AG.*

1869 **Altal**
115-86-6 9872 204-112-2
triphenylphosphate
A proprietary trade name for triphenylphosphate. Incombustible substitute for camphor in cellulose. No manufacturer.

1870 **Altalc 200 USP**
14807-96-6 9207 238-877-9
talc
Artic Mist. Talc; excellent color and purity; for pharmaceutical and cosmetic applications including baby powders, medicated foot powders; glidant, lubricant, pigment carrier. *Cyprus Industrial Minerals.*

1871 **Altan**
Veterinary laxative. *May & Baker Ltd.* Name unverified.

1872 **Altax®**
120-78-5 3435 204-424-9
2,2'-Dithiobis[benzothiazole] and zinc stearate in petroleum process oil; accelerator for natural and synthetic rubbers; primary accelerator and scorch modifying secondary accelerator in natural rubber and styrene/butadiene rubber copolymers; retarder-plasticizer in neoprene (G types); cure modifier in W types. *R. T. Vanderbilt Co Inc.*

1873 **Altene DG**
79-01-6 9769 201-167-4
Trichloroethylene
Degreasing solvent. *Elf Atochem SA.*

1874 **Altolube**
Excellent scrooping agent; for processing of warp knit nylon. *Altex Chemical Co Ltd.* Name unverified.

1875 **Altowhite LL**
12141-46-7 377 235-253-8
Aluminum Silicate
Calcined aluminum silicate, extender pigment; exhibits improved optical properties in paint systems. *Dry Branch Kaolin.*

1876 **Alubrasoft® 12-N**
Softener substantive to cotton, wool, acrylics, nylon, and other synthetics; imparts softness, lubricity, antistatic props. to yarns and fabrics. *PPG/Specialty Chem.* Discontinued.

1877 **Alubrasoft® Super 100**
Fatty polyamide; softener, antistat, conditioner, lubricant; imparts soft hand to synthetic and natural fibers. *PPG/Specialty Chem.* Discontinued.

1878 **Aludone®**
59792-81-3 261-931-8
Aluminum PCA
Aluminum PCA; astringent, antiseptic; peripheral antiperspirant; for spray or stick deodorants, shower gel, hair comb-out balm. *UCIB.*

1879 **Aludur**
An alloy of aluminum and silicon, containing from 5-20% silicon. No manufacturer.

1880 **Alufrit**
1344-28-1 369 215-691-6
Alumina microgrits. *Atomergic Chemetals Corp.*

1881 **Alugan**
1715-40-8 216-996-7
$C_8H_5BrCl_6$
bromocyclene
bromodan. A proprietary preparation of bromocyclene; a veterinary pesticide. No manufacturer.

1882 **Alugel**
21645-51-2 355 244-492-7
Aluminum hydroxide
Antacid. *Giulini Corp.*

1883 **alum**
10043-67-1 373 233-141-3
$AlKO_8S_2$
Potassium aluminum sulfate
Ammonium aluminum sulfate and aluminum sulfate are known by this name, but it is usually applied to potassium aluminum sulfate. An astringent.

1884 **Alumail®**
A range of inorganic smelted products for use in the surface coating of metals; used for special purpose enamels for various uses as well as for electrophorectic or powder electrostatic application. *Bayer AG.*

1885 **Aluman**
An alloy of 88% aluminum, 10% zinc, and 2% copper.

1886 **Alumantine**
Al_2O_3
A proprietary refractory containing 60-65% Al_2O_3. No manufacturer.

1887 **Alumbro**
A trademark for a 2% aluminum-brass alloy having good resistance to corrosion by sea water and marine atmospheres. *ICI plc.* Name unverified.

1888 **Alumedia**
Complex of alkyd resin and aluminum alkoxide; for preparation of high solids paint. *Manchem Ltd.* Name unverified.

1889 **Alumel**
An electrical resistance alloy containing 94% nickel, 1% silicon, 2% aluminum, 2.5% manganese, and 0-5% iron. No manufacturer.

1890 **Alumilite**
A proprietary trade name for chemical coatings applied to aluminum electrically. No manufacturer.

1891 **alumina**
1344-28-1 369 215-691-6
Al_2O_3
Aluminum oxide
An abrasive. Air Prods & Chem; Alcan; Alcoa; Aldrich; Atomergic Chemetals; Lonza Sarl; Nissan Chem. Ind.; Rhone-Poulenc.

1892 **Alumina hydrate**
1333-84-2; 21645-51-2 355 244-492-7
$Al_2O_3 \cdot 3H_2O$
Aluminum trihydroxide
Aluminum hydroxide; Alumina trihydrate; Liquigel®; Martifin; Martinal; Martinal® OL-111 LE; Martinal® ON-4608; Martinal® OS; Mucogel®; Nidrin; R-MA 11®; Rehydragel® Compressed Gel; Theodrex; Almacarb; Almagel; AL terna GEL; Alu-Cap™; Alu-Tab™; Aludrox; Alugel; Amphojel; Colugel; Dialume; F-500, -1000,-3600,etc.; F-1000®; F-100 Dried Gel; F-2000; F-2000 Dried Gel; F-2100 Dried Gel; F-2200 Dried Gel; F-MA 11®; Gastrils; Hydroxal;. Used in production of aluminum, abrasives, refractories, ceramics, electrical insulators; catalysts and catalyst supports, paper, spark plugs, crucibles and lab ware, adsorbent for gases/water vapors, chromatographic analysis, heat-resistant fibers, food additives. Insoluble in H_2O, soluble in alkaline solns or in HCl, H_2SO_4, and other strong acids; absorbs acids, CO_2.

1893 **Aluminac**
A similar alloy to Alpax, for making die castings. No manufacturer.

1894 **Aluminoferric**
10043-01-3 381 233-135-0
Consists of crude aluminum sulfate, and contains some iron sulfate; used as a precipitating agent in sewage and refuse liquids treatment, and also for removing suspended matter from boiler feed water. *Laporte Industries Ltd.*

1895 **Aluminoid, Aluminox**
A trademark for goods of the abrasive and refractory type, the essential constituent being crystalline alumina. No manufacturer.

1896 **alumino-vanadium**
An alloy of aluminum and vanadium, obtained by adding a mixture of vanadium pentoxide and powdered aluminum to liquid aluminum; used as a deoxidizing agent.

1897 **aluminum**
7429-90-5 331 231-072-3
Al
Metallic element; building and construction, corrosion-resistant chemical equipment (desalination plants), die-cast auto parts, electrical industry (power transmission lines), photoengraving plates, permanent magnets, cryogenic technology, machinery, tubes for ointments. mp = 660°; bp = 2327°; d = 2.70. *Alcan GmbH; Norsk Hydro AS.*

1898 **aluminum acetate solution**
8006-13-1 332
Burrow's Solution
Buro-sol Concentrate; Domeboro; Aluminum acetate. Astringent. Colorless liquid; slight odor of acetic acid. *Bayer.* Name unverified.

1899 **aluminum brass**
Alloys of from 59-70% copper, 26-40% zinc, 0.3-5.2% aluminum, and sometimes a little iron.

1900 **aluminum bronze**
There are various alloys under this name. Those containing a high percentage of aluminum are termed light, and have from 83-89% aluminum and 11-17% copper. The other type is called heavy, and contains from 85-95% copper.

1901 **aluminum chlorate**
15477-33-5 347 239-499-7
$AlCl_3O_9$
chloric acid aluminum salt
Mallebrin. Occurs as a hexahydrate and a nonahydrate. Used as an antiseptic and astringent.

1902 **aluminum chloride, anhydrous**
7446-70-0 348 231-208-1
$AlCl_3$
Anhydrol Forte. Ethylbenzene catalyst, dyestuff intermediate, detergent

alkylate, ethyl chloride, pharmaceuticals and organics, butyl rubber, petroleum refining, hydrocarbon resins, nucleating agent for titanium dioxide pigments. White or colorless crystals; strong irritant; freely soluble in many organic solvents. *Aldrich; Asada Chem Industry Co Ltd; Elf Atochem N. Am.; Fluka; Harcros Durahm; Witco/Argus.*

1903 aluminum chlorohydrate
1327-41-9 356 215-477-2
$[Al_2(OH)_5Cl]_x$;
Basic aluminum chloride
Aluminum chlorohydrol; Aluminum chlorohydroxide; Aluminum chloride hydroxide; Astringen; Chlorhydrol; Hyperdrol; Locron; Phosphonorm. Inorganic salt. Commercial antiperspirant and deodorant; water purification; treatment of sewage and plant effluent. Solid; dissolves in H_2O. *Catomance Ltd; Reheis.*

1904 Aluminum Grade Bone Ash, BCP 600
Calcium hydroxyapatite
Used to coat and protect all surfaces contacted by molten nonferrous metals; used extensively in aluminum industries. *Murlin Chemical Inc.*

1905 aluminum hydroxide
21645-51-2 355 244-492-7
H_3AlO_3
aluminum oxide trihydrate
aluminum trihydrate; Alugel; AL terna GEL; Aludrox; Alu-Tab™; Amphojel; Almacarb; aluminum trihydrate; algeldrate; Aldrox; Alkagel I.S.; Alucol; Al-U-Crème; Aludrox; Aludyal; Antidiar; Creamalin; Cremorin; Gelumina; Merlium; Pepsamar; Uracid. Used in dyes, paints, textile finishing. Insoluble in H_2O. Alcan; Alcoa, Atomergic Chemetals; BA Chem. Ltd.; Nyco Minerals; Reheis; Rhône-Poulenc; SeimiChem.;Solem; Vista; Whittaker, Clark & Daniels.

1906 aluminum iron
ferro-aluminum
An alloy of iron and aluminum; used for refining iron, also as a permanent ingredient for increasing the strength. A 15% alloy has been used for crucibles exposed to high temperatures.

1907 aluminum iron brass
An alloy containing 61.1% copper, 35.3% zinc, 1.1% iron, and 2.3% aluminum.

1908 aluminum iron bronze
An alloy containing 85-89% copper, 6-9% aluminum, and 3-7% iron.

1909 aluminum isopropoxide
555-31-7 359 209-090-8
$C_9H_{21}AlO_3$
Aluminum(III)isopropoxide
Aliso; AIP; 2-Propanol aluminum salt; Aluminum isopropylate. Used for cosmetics, pharmaceuticals. mp = 128-133°; $bp_{5.2}$ =125-130°; d = 1.0350; decomposed by H_2O; soluble in organic solvents; LD_{50} (rat orl) = 11.3 g/kg.

1910 aluminum magnesium bronze
An alloy of 89-94% copper, 5-10% aluminum, and 0.5% magnesium.

1911 aluminum manganese
An alloy of aluminum with 2-3% manganese.

1912 aluminum manganese brass
An alloy consisting of 56-3% copper, 40% zinc, 2.7% manganese, and 1% aluminum.

1913 aluminum manganese bronze
An alloy of 89% copper, 9.6% aluminum, and 1-2% manganese.

1914 aluminum nickel
An alloy containing varying amounts of nickel with aluminum. One alloy consists of 76.4% nickel, and 23.6% aluminum.

1915 aluminum nickel bronze
An alloy containing 85% copper, with from 5-10% aluminum and 5-10% nickel.

1916 aluminum nickel zinc
An alloy consisting of 85% aluminum, 10% nickel, and 5% zinc.

1917 aluminum nitrate
7784-27-2 365 236-751-8
AlN_3O_9
Occurs mainly as the nonahydrate. Mordant for textiles, leather tanning, manufacture of incandescent filaments, catalyst in petroleum refining, nucleonics, anticorrosion agent, antiperspirant. mp = 73°; dec 135°; very soluble in H_2O, less soluble in organic solvents; LD_{50} (rat orl) = 4.28 g/kg. *Aldrich; EM Ind.; Hoechst-Celanese; Sherman Chem. Ltd.; Spectrum Chem. Manufacture.*

1918 aluminum nitride
24304-00-5 366 246-140-8
AlN
As semiconductor in electronics, nitriding of steel; steel manufacture. Crystals; hardness 9 to 10 on Moh's scale; mp = 2150-2200°. *Aldrich; Atomergic Chemetals; Carborundum; Dow; Mandoval Ltd.*

1919 aluminum oxide
1344-28-1 369 215-691-6
Al_2O_3
Alumina; Calcined alumina; Alumite; Alcan C-70, C-71, C-72, C-73, C-75; Alexite; α-Ruvite; Alfrax B301; Aluminum Oxide C; Brasivol; C-1; Compalox; D-201; Dirubin; Dural; Dycron; Flame Guard; Italcor; Mafe®; Martipol; Marttisorb; Purdox; Tealox®; Rewagit; S-201; Saffil®; Selexsorb® COS; T-1061; T-64; Versal 150; α-Alumina; Aloxite; Aluminite 37; Activated Alumina; Aluminum Oxide, Acidic; Aluminum Oxide, Activated; Aluminum Oxide, Basic; Alundum; Boileezers; Corundum; Basic Alumina; Alumina acidic; Alumina basic; Alumina neutral; δ-alumina; θ-alumina; Aluminum oxide (fibrous forms); σ-alumina; sapphire; α-Alumina trihydrate; γ-Alumina,. Inorganic compound; production of aluminum, abrasives, refractories, ceramics, elec. insulators, catalysts and catalyst supports, paper, spark plugs, crucibles and lab ware, adsorbent for gases/water vapors, chromatographic analysis, heat-resist. fibers, Manufacture of aluminum, abrasives, refractories, ceramics, electrical insulators, catalyst and catalyst supports, paper, spark plugs, laboratoryware, adsorbent for gases, chromatographic analysis, fluxes, fibers, food additive (dispersant). mp about 2030°; d= 3.97; slowly soluble in alkaline solvents. Alcan; Aldrich; Atomergic Chemetals; BA Chem. Ltd.; Degussa; Ferro/Transelco; Hüls Am.; Lonza; Norton Chem. Process Prods.; Rhône-Poulenc; Vista.

1920 aluminum oxide C
1344-28-1 369 215-691-6
Alumina
Free-flow and anticaking agent; aids in reducing electrostatic charges of powder substances; for electrical industry. *Degussa.*

1921 aluminum PCA
59792-81-3 261-931-8
Pyrrolidone carboxylic acid, aluminum salt
Aludone®; Aluminum, tris(5-oxo-L-prollinato-). Aluminum PCA; astringent, antiseptic; peripheral antiperspirant; for spray or stick deodorants, shower gel, hair comb-out balm. White-cream powder; soluble in H_2O.

1922 aluminum phosphate
7784-30-7 371 232-056-9
AlO_4P
Aluminum orthophosphate
Monoaluminum phosphate; Angelite; Coeruleoactite; Evansite; Lucinite; Metavariscite; Sterretite; Vasheegyite; Wavellite; Zepharovicht; ; Aluphos; Fosfalugel; MALP; Phosphalijel; Phosphalugel; Phosphalutab; Ulcocid,. Acidic solution. Refractory bonding agent, metal processing. Used as cement admixture with calcium sulfate and sodium silicate; as a flux for cermaics; dental cements and for special glasses. mp >1460°; d_{23} = 2.56; practically insoluble in H_2O or acetic acid; very slightly soluble in concentrated hydrochloric acid and nitric acid. *Albright & Wilson; Rasa Ind; Rhône-Poluenc Basic; Superfos Biosector A/S.*

1923 aluminum potassium sulfate
10043-67-1 373 233-141-3
$AlKO_8S_2$
burnt alum; exsiccated alum. Anhydrous or dodecahydrate. Used as an astringent and as a cement hardener. Dodecahydrate: mp = 92°; d = 1.725; soluble in H_2O (10%), insoluble in organic solvents.

1924 aluminum silicate
12141-46-7 377 235-253-8
Alusil; alphatex®; Alusil ET; Altowhite LL; Alusil ET; Aluminum Silicate P820. Calcined aluminum silicate, titanium dioxide extender; exhibits improved optical properties in paint systems.

1925 Aluminum Silicate P820
12141-46-7 377 235-253-8
Precipitated aluminum silicate; titanium dioxide extender in powder coatings, decorative paints. *Degussa.*

1926 Aluminum Silicon Alloy C
A British Chemical Standard alloy; it contains 12.74% silicon, 0.34% iron, 0.005% manganese, 0.020% zinc, 0.006% titanium, and 0.010% copper.

1927 aluminum silver
Silver metal. An alloy consisting of 57% copper, 20% nickel, 20% zinc, and 3% aluminum. The name is also applied to an alloy of 95% aluminum, and 5% silver.

1928 aluminum starch octenyl succinate
9087-61-0
Dry Flo®. Body powders, antiperspirants, feminine hygiene sprays, foot powders. *Nat'l. Starch.*

1929 aluminum stearate
637-12-7 379 211-279-5
$C_{54}H_{105}AlO_6$
Octadecanoic acid aluminum salt
Stearic acid aluminum salt; aluminum tristearate; Aluminum, dihydroxy (octadecanoato-). Aluminum salt of stearic acid; paint, varnish drier, greases,

waterproofing agent, cement additive, lubricants, cutting compounds, flatting agent, cosmetics, pharmaceuticals, and defoaming agent. mp = 117-120°; d= 1.01; practically insoluble in H_2O; when freshly made soluble in alcohol, benzene, turpentine, mineral oils. *Elf Atochem/Wire Mill; Ferro/Grant; Magnesia GmbH; Norac; Synthetic Prods; Witco.*

1930 aluminum sulfate
10043-01-3 381 233-135-0
$Al_2O_{12}S_3$
alum
cake alum; Sulfuric Acid, Aluminum Salt (3:2); Dialuminum Sulfate. Inorganic salt; in pulp and paper mills, water purification plants, leather, textile, gypsum treatment, in fire retardants; deodorizer, decolorizer, food additive. mp = 770°; d = 2.71; LD_{50} (mus orl) = 6207 mg/kg. Alcan; Aldrich; Am. Cyanamid; Asada Chem Industry Co Ltd; Ashland; BA Chem Ltd; Ethyl; General Chem; Rasa Ind; Rhône-Poulenc Basic.

1931 aluminum tin bronze
An alloy of 85% copper, 10% tin, 2-5% aluminum, and 2% zinc.

1932 aluminum-nickel-titanium
An alloy of 97.6% aluminum, 2% nickel, and 0.4% titanium.

1933 Alumite
1344-28-1 369 215-691-6
Alumina microgrits *Atomergic Chemetals Inc.*

1934 Alundum®
1344-28-1 369 215-691-6
A registered trademark for various types of goods, such as grinding wheels, abrasive and refractory grain, refractory articles and cement, porous plates, crucibles, and other articles made from crystalline alumina; or alumina which has been electrically fused and crystallized. No manufacturer.

1935 Aluni
An aluminum-nickel alloy used as an anode for deposition of the alloy coating. No manufacturer.

1936 Alusec
Aluminum organic complexes; rheology modifiers for high solids, air drying paint systems. *Manchem Ltd.* Name unverified.

1937 Alusil
12141-46-7 377 235-253-8
Aluminum Silicate
Aluminum silicate. Used in glass, ceramics, dental cement and semi-precious stones. *Crosfield Chemicals Ltd.*

1938 Alusil ET
12141-46-7 377 235-253-8
Synthetic aluminum silicate of controlled particle size; used an extender in emulsion paint. *Crosfield Chemicals Ltd.*

1939 Alveograf
1306-06-5 3519 215-145-7
Durapatite
calcium phosphate hydroxide. Durapatite; prosthetic aid (artificial bones, teeth), also used as a dietary supplement. *Sterling Drug Inc.* Name unverified.

1940 Alvex
Highly alkaline detergent. *Crosfield Chemicals Ltd.* Discontinued.

1941 Alytol
Bitumen mastic; for roof repair and maintenance. *Vedag GmbH.* Name unverified.

1942 Alzen
An alloy of 66% aluminum and 33% zinc.

1943 amalgam
A name applied to alloys of metals with mercury; it is also used as a term for a native alloy of mercury and silver with a formula varying between AgHG and Ag_2Hg_3.

1944 Amargosite
A trade name for a clay of the bentonite type. No manufacturer.

1945 Amarin
$C_{21}H_{18}N_2$
Triphenyldihydroglyoxaline
No manufacturer.

1946 Amasil®
64-18-6 8912 200-579-1
Formic acid
For ensiling and feed preservation. *BASF AG.* Name unverified.

1947 Amasil® P
4075-81-4 1745 223-795-8
Calcium propionate and calcium formate
Ensiling agent for preservation of feeds. *BASF AG.*

1948 Amatols
Mixtures of trinitrotoluene and ammonium nitrate; an 80/20 amatol contains 80 parts ammonium nitrate and 20 parts trinitrotoluene. Used in high explosives. No manufacturer.

1949 Amax XLP
Low phosphorus copper; oxygen-free copper plus 0.001-0.005% phosphorus; conductivity 98% IACS; copper-silver 99.95%, phosphorus-0.001-0.005%; ideal for applications where a low phosphorus content is beneficial and good conductivity with resistance to embrittlement must be ensured. *Amax Inc.* Name unverified.

1950 Amax® , Amax No 1
102-77-2 203-052-4
$C_{11}H_{12}N_2OS_2$
N-Oxydiethylene 2-benzothiazole-sulfenamide
N-oxydiethylenebenzothiazole-2-sulfenamide; Amax; 2-Benzothiazolyl-N-morpholinosulfide; Morpholine, 4-(2-benzothiazolylthio)-; 2-(morpholinothio)benzothiazole; 2-(Morpholinthio)-benzothiazole; N-oxydiethylene-benzothiazole sulfenamide. Primary and secondary accelerators for rubber; safe at processing temperature and active over a wide curing range; particularly advantageous in styrene/butadiene rubber tires compounded with fine particle furnace blacks. *R. T. Vanderbilt Co Inc.*

1951 Amaze®
25311-71-1 5187 246-814-1
Isofenphos
Soil-applied insecticide used for control of insects in rice crops and pear sucker. *Bayer AG.*

1952 Amazin®
Maneb with zineb; fungicide. *Aceto.*

1953 Ambazyme
Amyloglucosidase. *Hoechst UK.*

1954 amber
A fossil resin formed in certain beds of clay and sand, stated to be derived from *Pinites succinifer*. The following are varieties of amber; Succinite (mp 250-300°), Gedanite (mp 150-180°), Glessite (mp 250-300°), Beck.

1955 Amberglow
A proprietary phenol formaldehyde synthetic resin. No manufacturer.

1956 ambergris
A grey, wax-like product found in the sea. It occurs in certain conditions of the intestines of the sperm whale. The chief constituent is ambrein, $C_{23}H_{40}O$; used in perfumery.

1957 amber-guaiacum resin
A variety of guaiacum resin; it is not an amber.

1958 Ambergum® 721
Cellulose derivative; replaces gum arabic in lithographic printing processes. *Aqualon.*

1959 Amberite
A smokeless powder consisting of 71% nitrocotton, 18.6% barium nitrate, 1.3% potassium nitrate, 1.4% wood meal, and 5.8% petroleum jelly.

1960 Amberlac®
Synthetic resinous material in solid form or solutions; for use in industrial arts and coatings. *Reichhold.*

1961 Amberlac® 13-801
Acrylic monomer-modified alkyd. *Reichhold.*

1962 Amberlite®
Ion-exchange resins. *Rohm & Haas UK.*

1963 Amberlite® IRA-68
Weakly basic acrylic anion exchange resin for industrial water treatment, pharmaceutical, chemicals and food processing industries. *Rohm & Haas.*

1964 Amberlite® IRP-64
Polacrilin
A synthetic ion-exchange resin, supplied in the hydrogen or free acid form; pharmaceutic aid. *Rohm & Haas.* Name unverified.

1965 Amberol
A phenol-formaldehyde resin combined with rosin or other resin; used in the varnish industry. No manufacturer.

1966 Ambersil
Silicone emulsion for shell molding and hot box processes. *Foseco (F.S.) Ltd.*

1967 Ambiflo
Synthetic lubricants which are colorless, viscous liquids made by combining or polymerizing propylene oxide or propylene oxide and ethylene oxide; used in equipment for metal working, heat transfer and as automotive brake fluids, internal combustion engines, gears and bearings. *Dow UK.* Discontinued.

1968 Ambiteric
Ampholytic surfactant. *Hoechst UK.*

1969 Ambiteric D
High molecular weight substituted betaine as a creamy unctuous mass; good alkali stability, wetting, foaming, detergency and solubilizing properties; used

in industrial cleaner formulations; perfume solubilization; antistat. *Hoechst UK.*

1970 Ambitrol
A series of formulated engine coolants made from glycols, deionized water and suitable inhibitors; stationary engines operating for transmission of natural gas and petroleum products, electrical power generated systems, irrigation systems and drilling operations. *Dow UK.* Discontinued.

1971 Amborate
Formoxy methyl isolongifolene
Used in perfumery. *Bush Boake Allen Ltd.*

1972 Amborol
Hydroxy methyl isolongifolene
Used in perfumery. *Bush Boake Allen Ltd.*

1973 Amboryl Acetate
Acetoxymethyl isolongifolene
Used in perfumery. *Bush Boake Allen Ltd.*

1974 Ambra
A phenol-formaldehyde synthetic resin. No manufacturer.

1975 Ambrac Metal
A corrosion resistant nickel silver containing copper, nickel, zinc, and manganese. No manufacturer.

1976 Ambraloys
A proprietary trade name for alloys of copper and aluminum with zincorion. No manufacturer.

1977 Ambrene
Dinitro-*t*-butyl-*m*-cresol-methyl ether
An artificial musk perfume. No manufacturer.

1978 ambrite
A resin found in the lignite of Auckland, New Zealand.

1979 ambroid
Pressed amber. A product consisting of small fragments of amber heated under pressure; adhesive.

1980 Ambrol
A phenol-formaldehyde synthetic resin. No manufacturer.

1981 ambroxan
8α-12-oxido-13, 14, 15, 16 tetra-norlabdane
Fragrance raw material; ambergris type. *Henkel Cospha; Henkel Canada.*

1982 Ambush
52645-53-1 7321 258-067-9
permethrin
Insecticide containing permethrin. *ICI Chem & Polymers Ltd.*

1983 Ambush C
66841-24-5 266-492-6
cypermethrin
An emulsifiable concentrate containing 100 g cypermethrin per liter; a pyrethroid insecticide. *ICI Agrochemicals.*

1984 Amcar CL
Alkyl benzoate ester
Carrier for disperse and cationic dyeing of polyester. *Am. Emulsions.*

1985 Amcar OCP
Chlorinated benzene
Anionically emulsified; carrier for disperse dyeing of polyester. *Am. Emulsions.*

1986 Amcide
7773-06-0 589 231-871-7
$H_6N_2O_3S$
Ammonium sulfamate
Atlacide, AMS, Ammate. An inorganic herbicide to control weeds and grasses in vegetables and ornamentals prior to planting and as a tree-killer. *Battle, Hayward & Bower Ltd.*

1987 Amcron
Oxygen-free copper plus 0.7 to 1.2% chromium; conductivity 82% IACS; chromium 0.7 to 1.2%, copper + silver + chromium 99.95%; principal uses based on its good compressive yield strength, creep resistance and thermal fatigue properties at moderately elevated temperatures *Amax Inc.* Name unverified.

1988 Amdye PH-12
Inorganic salts; replacement for trisodium phosphate in liquid form. *Am. Emulsions.*

1989 AME 4000®
Modified epoxy resin with superior strength/weight characteristics; high performance marine resin for fiberglass power boats and sailboats. *Ashland.*

1990 Ameen
Long chain aliphatic amines. *Akzo Chemie UK Ltd.*

1991 Ameenex 70 WS
Amine salt; concentrate for oilfield down-hole corrosion inhibition. *Chemron.*

1992 Ameenex C-18
Tall oil amido-amine; film-forming corrosion inhibitor; wetting, emulsifying and antistripping agent with asphalt compounds, coal tar pitches; drilling fluid additive; useful in non-metallic mineral flotation. *Chemron.*

1993 Ameenex Polymer
Complex resinous polyamine; corrosion inhibitor intermediate; excellent film persistency; low emulsification tendency; high temperature stability; for oilfield applications. *Chemron.*

1994 Amercell Polymer HM-1500
Nonoxynol hydroxyethylcellulose; nontacky thickener for aqueous solutions, surfactant, emulsion stabilizer, film-former; substantive to skin and hair; for body lotions, moisturizers, hydroalcoholic products, sun lotions, hair conditioners, mousses, liquid makeup. *Amerchol Corp.* Discontinued.

1995 Amerchol Polysorbate
Emulsifying and solubilizing agent in foods; defoamer; used for vegetable oils, vitamins, beet sugar, yeast, cottage cheese. *D F Anstead Ltd.* Name unverified.

1996 Amerchol® 400
lanolin alcohol
lanolin alcohol; cetyl alcohol; lanolin; stearone; Petrolatumlanolin, stearone. Auxiliary emulsifier, emulsion stabilizer for oil-water and water-oil systems, including makeup, pharmaceuticals; emollient, lubricant. *Amerchol; Amerchol Europe.* Discontinued.

1997 Amerchol® BL
Multisterol absorption base of lanolin sterol esters and higher alcohols; nonionic water-oil emulsifier. *Amerchol Corp.* Discontinued.

1998 Amerchol® C
lanolin; Petrolatum; lanolin alcohol; lanolin. Absorption base, auxiliary emulsifier for oil-water systems, conditioner, emollient, moisturizer, stabilizer for cosmetics and pharmaceuticals, textile finishes. *Amerchol; Amerchol Europe.* Discontinued.

1999 Amerchol® CAB
Solid emollient multisterol extract of lanolin alcohols in petrolatum; water-oil emulsifier activity, ideal for pharmaceutical vehicles well tolerated on dry and injured skin. *Amerchol Corp.* Discontinued.

2000 Amerchol® H-9
Absorption base containing cholesterol esters and free sterols; natural emollient and nonionic water-oil emulsifier. *Amerchol; Amerchol Europe.*

2001 Amerchol® L-101
Mineral oil
lanolin alcohol; Mineral oil. Emollient, penetrant, emulsifier, moisturizer, softener, stabilizer for cosmetics, creams, makeup, hair dressing, pharmaceuticals, aerosols, baby products, textile finishes; plasticizer for hair sprays. *Amerchol Corp.* Discontinued.

2002 Amerchol® L-500
Mineral oil
lanolin alcohol; octyldodecanol; Amerchol® 400; Amerchol® L-10; Amerchol Polysorbate1; Amerchol® L-99; Amerchol® RC; Amerchol® C;,. Emulsifier, stabilizer, emollient, moisturizer, conditioner for hair and skin products, creams, lotions, makeup, aerosols, pharmaceutical vehicles, baby toiletries. *Amerchol Corp.*

2003 Amerchol® L-101
lanolin alcohol; Mineral oil; Amerchol® L-99. Emulsifier, stabilizer, conditioner, emollient, moisturizer for creams and lotions, hair and skin preparations, dermatological specialties. *Amerchol; Amerchol Europe.*

2004 Amerchol® RC
Concentrated lipophilic lanolin alcohol fraction with lubricating, nontacky, barrier properties, particularly suited for makeup systems. *Amerchol Corp.* Discontinued.

2005 Amercor® 8730
Blend of volatile amines; corrosion inhibitor providing protection in low, medium and high pressure sections of a steam/condensate system. *Drew Ind. Div.*

2006 Amerfloc Plus® 5270
Amerfloc® 275. High molecular weight, highly anionic fluid polymer; flocculant and coagulant aid for sludge conditioning; for dewatering industrial slurries and water clarification applications in paper industry, food processing, nonpotable water clarification, oily waste treatment. *Drew Ind. Div.*

2007 Amerfloc® 2
1302-42-7 8715 215-100-1
Sodium aluminate
Stabilized solution of sodium aluminate; coagulant for treatment of potable water, water clarification, hot or cold lime softening applications. *Drew Ind. Div.*

2008 Amerfloc® 275
High molecular weight polymer; flocculant and coagulant aid for potable water clarification, water plant sludge dewatering applications. *Drew Ind. Div.*

2009 Amergel® 100
Organics, surfactants, water blend; antifoam for aqueous process systems where oils, solvents, waxes, silicas are undesirable, e.g., pulp and paper, nonwovens, effluent systems. *Drew Ind. Div.*

2010 Amergize
Deposit modifier/combustion improver; a unique blend of oil soluble organometallic compounds. *Ashland Chemical Company.* Name unverified.

2011 Amergy® 5400
Organometallic
Fuel oil treatment; slag modifier preventing deposits and corrosion on boiler firesides. *Drew Ind. Div.*

2012 Ameribond
8061-52-7
Modified calcium lignosulfonates; pelleting aids for animal feeds. *Borregaard Ligno Tech.*

2013 Ameribond 2000
Lignosulfonate
Animal feed binder. *Borregaard Ligno Tech.*

2014 Ameripol Synpol 1009
Hot polymerized pre-crosslinked styrene/butadiene rubber; produces a smooth, nonstringy adhesive that breaks clean when gunned or troweled in place. *Ameripol Synpol.* Discontinued.

2015 Ameripol Synpol 1013/8000
Hot polymerized noncrosslinked styrene/butadiene rubber; used for paper saturation, as barrier coat prior to application of pressure-sensitive adhesive; useful in blends with other elastomer to increase cohesive strength, green strength. *Ameripol Synpol.* Discontinued.

2016 Amerite
A proprietary trade name for rubber derivatives and rubber-like resins in aqueous dispersion. No manufacturer.

2017 Amerlate® LFA
68424-43-1 270-302-7
Lanolin acid
Amerlate® WFA. Emulsifier, stabilizer, emollient for fatty acid systems, aerosol shave creams, cream shampoos, wax systems, household products; pigment dispersant; increases tack and plasticity of wax films. *Amerchol; Amerchol Europe.*

2018 Amerlate® P
63393-93-1 264-119-1
Isopropyl lanolate
Conditioner, penetrant, lubricant, moisturizer, emollient, water-oil emulsifier, stabilizer, opacifier for cosmetics and pharmaceuticals; pigment dispersant; wetting agent and dispersant for solids; plasticizer for wax and pigment systems. *Amerchol; Amerchol Europe.*

2019 Amerlate® WFA
68424-43-1 270-302-7
Lanolin acid
Lanolin acid; emulsifier, stabilizer for emulsions, aerosols, shampoos; stabilizer for conventional soap emulsions; wets and disperses pigments in makeups. *Amerchol; Amerchol Europe.* Discontinued.

2020 Amerol
The methyl ester of saccharin. No manufacturer.

2021 Amerone
Perfumery base. *PPF International Ltd.* Name unverified.

2022 Ameroxol® OE-2
9004-98-2
$C_{58}H_{116}O_{21}$
Polyoxyethylene(20) oleyl ether
Polyoxyl(10)oleyl ether; Brij 99. Oleth-2; solubilizer, emulsifier, dispersant, stabilizer, lipophilic cosolvent for creams and lotions, shampoos, and detergents, fluid and gelled transparent emulsions, fragrance products, and aerosols. *Amerchol; Amerchol Europe.* Discontinued.

2023 Ameroxol® OE-5
9004-98-2
Oleth-5, an emulsifier. *Amerchol Corp.* Discontinued.

2024 Ameroyal
A concentrated liquid blend of polyelectrolyte scale inhibitors and antifoam agents; an evaporator treatment used to prevent scale deposition and foaming in conventional marine evaporators thereby minimizing the need for acid cleaning. *Ashland Chemical Company.* Name unverified.

2025 Amerplex® 605
Amerzine® corrosion inhibitor, Isoquest HT scale inhibitor, modified natural organic components, and antifoam; deposit and corrosion inhibitor for steam generating systems. *Drew Ind. Div.*

2026 Amerscan MDP Kit
5837
methylene bisphosphonic acid ^{99m}Tc complex
Medronic acid ^{99m}Tc complex. Pharmaceutic aid. ^{99m}technetium salt of methylene bisphosphonic acid [1984-15-2]. Used as a radio-imaging diagnostic aid. *Amersham Corp.* Name unverified.

2027 Amerscent 86
Neutralizing/masking agent for odor control in waste water holding areas. *Drew Ind. Div.*

2028 Amerscreen
UV absorber; sunscreen. *D F Anstead Ltd.* Name unverified.

2029 Amersep® MP-3
Sodium dimethyldithiocarbamates
Metals precipitant for use in plating and metal finishing operations. *Drew Ind. Div.*

2030 Amersil® DMC-287, DMC-357
Dimethicone copolyol. An emulsifier. *Amerchol Corp.* Discontinued.

2031 Amersil® L-45 Grades
Dimethicone. *Amerchol Corp.* Discontinued.

2032 Amersil® ME-358
Cyclomethicone and dimethicone copolyol; emulsifier for preparation of water-in-silicone oil for personal care products. *Amerchol Corp.* Discontinued.

2033 Amersil® simethicone
8050-81-5 3264
Simethicone. Ointment base and anti-foaming agent. *Amerchol Corp.* Discontinued.

2034 Amersil® VS-7207
69430-24-6
Cyclomethicone
Ointment base. *Amerchol Corp.* Discontinued.

2035 Amersite® 2
7631-90-5 8731 231-548-0
sodium bisulfite
Sodium bisulfite and selected catalytic agents; corrosion inhibitor for steam generating systems. *Drew Ind. Div.*

2036 Amersite® 2
7631-90-5 8731 231-548-0
Sodium bisulfite
Sodium bisulfite and selected catalytic agents; corrosion inhibitor for steam generating systems. *Drew Ind. Div.*

2037 Amersperse 1200
Scale inhibitor for evaporators, vacuum pans, and juice heating equipment. *Drew Ind. Div.*

2038 Amerstat®
Liquid and solid microbial control agents used for control of slime in the papermaking process for bacterial control in the sugar process and for a preservative in aqueous systems; used in paper mills, sugar mills, preservation of paints, coatings, adhesives, mineral slurries, drilling muds, animal glues, latent metal working fluids and paper coatings. *Drew Ind. Div.*

2039 Amerstat® 233
533-74-4 2892 208-576-7
$C_5H_{10}N_2S_2$
3,5-Dimethyl tetrahydro-2-H,1,3,5-thiadiazone-2-thione
Dazomet. Antimicrobial in industrial water systems, preservative in aqueous systems. Name unverified. *Drew Ind. Div.*

2040 Amerstat® 250
2-Methyl-4-isothiazolin 3-one [2682-20-4] and 5-chloro-2 methyl-4-isothiazolin-3-one [26172-55-4]; paper mill slimicide. *Drew Ind. Div.*

2041 Amerstat® 251
5-Chloro-2-methyl-4-isothiazolin-3-one [26172-55-4] and 2-methyl-4-isothiazolin-3-one [2682-20-4]; antimicrobial, preservative in latex. *Drew Ind. Div.*

2042 Amerstat® 272
Sodium dimethyl dithiocarbamate and disodium ethylene bisdithiocarbamate; paper and sugar mill slimicide. *Drew Ind. Div.*

2043 Amerstat® 282
6317-18-6 228-652-3
$C_3H_2N_2S_2$
Methylene bis(thiocyanate)
Thiocyanic acid, methylene ester; Dithiocyanatomethane; Antiblu 3737; Busan 110; Cytox; Nalco D-1994; Slimicide MC. Antimicrobial in industrial water systems; preservative in water-containing systems. mp = 102-106°; reacts with H_2O, soluble in organic solvents; LD_{50} (rat orl) = 161 mg/kg. *Drew Ind. Div.*

2044 Amerstat® 294
3064-70-8 221-310-4
$C_2Cl_6O_2S$
Bistrichloromethyl sulfone
Hexachlorodimethyl sulfone; sulfonylbis(trichloromethane). Bis(trichloromethyl) sulfone and dispersants; paper mill slimicide. *Drew Ind. Div.*

2045 Amerstat® 300
10222-01-2 233-539-7
$C_3H_2Br_2N_2O$
2,2-Dibromo-3-nitrilo propionamide
2,2-Dibromo-2-cyanoacetamide; DBNPA; Dibromo-2-carbamoylacetonitrile; Dibromo-3-nitrilopropionamide; Slimicide 508; XD-7287L Antimicrobial; XD-1603. Paper mill slimicide, antimicrobial agent for enhanced oil recovery systems; preservative for metal working fluids containing water. *Drew Ind. Div.*

2046 Amertrol
Deposit inhibitors; deposit control agents for boiler deposit control. *Ashland Chemical Company.*

2047 Amerzine®
302-01-2 4809 206-114-9
Hydrazine
Organically catalyzed hydrazine; corrosion inhibitors for preboiler and afterboiler corrosion control, oil field line corrosion control and chromate reduction. *Ashland Chemical Company.*

2048 amesite
A chloritic mineral.

2049 Ametox
7772-98-7 8844 231-867-5
$Na_2S_2O_3$
Sodium thiosulfate
Specially purified and sterilized sodium thiosulfate for use in treatment of metallic poisoning. *May & Baker Ltd.* Name unverified.

2050 Ametrex
834-12-8 411 212-634-7
Ametryn
Active ingredient: ametryn; selective pre-and post-emergence herbicide, also used as an aquatic herbicide and vine desiccant. *Agan Chemical Manufacturers Ltd.*

2051 ametryn
834-12-8 411 212-634-7
$C_9H_{17}N_5S$
N-ethyl-N'-(1-methylethyl)-6-(methylthio)-1,3,5-triazine-2,4-diamine
ametryne; Amephyt; Ametrex; Doruplant; Evik; G 34162; Gesapax; Mebatryne; Evik 80W; Cemerin; 2-Ethylamino-4-isopropylamino-6-methylthio-s-triazine. Selective systemic herbicide used for control of most annual grasses and broad-leaved weeds in pineapples, sugar cane, bananas; citrus fruit, maize, cassava; coffee, tea, cocoa; oil palmsand on non-crop land. An unrestricted, general use pesticide. mp = 84-85°; d_{20} = 1.19; soluble in H_2O (185 mg/l), more soluble in organic solvents; LD_{50} (rat orl) = 1110 mg/kg.

2052 Amfaid
Surface active agents and detergents. *ABM Chemicals Ltd.* Name unverified.

2053 Amfix
High-speed fixer for photographic processing. *May & Baker Ltd.* Name unverified.

2054 Amfix FRL
Formaldehyde-free fixing agent for improving wet fastness properties of cellulosics dyed or printed with direct or reactive dyes. *Am. Emulsions.*

2055 Amgard TBEP
78-51-3 201-122-9
$C_{18}H_{39}O_7P$
tri-(2-butoxyethyl) phosphate
Tris(2-butoxyethyl) phosphate; tributoxyethyl phosphate; KP-140; TBEP; tris(2-butoxyethyl)ester phosphoric acid; 2-butoxy-ethanol phosphate (3:1). Surfactant. bp_4 = 215-228°; d = 1.0060; n_D^{20} = 1.4359; LD_{50} (rat orl) = 3 g/kg. *Surfachem. Ltd.*

2056 Amgard® CPC 452
Red phosphorous-based additive; flame retardant additive for nylon resins; for injection molding, glass-filled, and extrusion resins and electrical applications. *Albright & Wilson Am.*

2057 Amianthus
1322-31-4
Amianth
Mountain flax; asbestos. A white and satiny variety of asbestos. *Mountain flax.*

2058 Amical® 48
20018-09-1 243-468-3
$C_8H_8I_2O_2S$
Diiodomethyl *p*-tolyl sulfone
1-((diiodomethyl)sulfonyl)-4-methylbenzene; Methylphenyl diiodomethyl sulfone; Toluene, 4-(diiodomethylsulfonyl)-; Tolyl diiodomethyl sulfone. Mildewcide, fungicide for latex paints, emulsions, caulks, adhesives and sealants, and in lumber, construction, home improvement, textile, and automotive industries. *Angus.*

2059 Amical® 85
471-34-1 1697 207-439-9
Calcium carbonate
Filler designed for maximum loading in resin systems. *Franklin Industrial Minerals.*

2060 Amical® 101
471-34-1 1697 207-439-9
Calcium carbonate
Filler designed for maximum loading in resin systems. *Franklin Industrial Minerals.*

2061 Amical® Flowable
Amical® 48. Aqueous suspension of diiodomethyl-*p*-tolyl sulfone; preservative, mildewcide, algicide for polymeric systems, especially latex paints, caulks, adhesives, leather. *Angus.*

2062 Amical® SC
471-34-1 1697 207-439-9
calcium carbonate
Stearate surface-coated calcium carbonate; enhanced processability; surface coating is compatible with PVC, polyolefins, silicones, and engineering plastics. *Franklin Industrial Minerals.*

2063 amicarbalide
3459-96-9 414 222-402-7
$C_{15}H_{16}N_6O$
3,3'-(carbonyldiimino)bisbenzenecarboximidamide
3,3·g-Diamidinocarbanilide. Babesiacide (treatment for piroplasmosis) in cattle. Diampron, the isethionate of amicarbalide, is an antiprotozoan for veterinary use.

2064 Amichrome
Premetallized dyes. *ICI Chem & Polymers Ltd.*

2065 Amicon® C-860-4
25928-94-3
Epoxy resin
Conductive epoxy; high Tg die attach adhesive for IC assembly. *Emerson & Cuming Polymer Group.* Name unverified.

2066 Amicon® C-940-4
Conductive polyimide; two-step cure die attach adhesive for assembly of integrated circuits. *Emerson & Cuming Polymer Group.* Name unverified.

2067 Amicon® CT-4042-5
25928-94-3
Two-component conductive epoxy; die attach adhesive for assembly of integrated circuits. *Emerson & Cuming Polymer Group.* Name unverified.

2068 Amicon® ECT-86
25928-94-3
Epoxy tape adhesive; electrically conductive version of TG-86; meets MIL-Std. 883C, Method 5011. *Emerson & Cuming Polymer Group.* Name unverified.

2069 Amicon® ME-868
25928-94-3
Nonconductive epoxy; oxide-filled version of C-868-1; meets MIL-Std. 883C, Method 5011. *Emerson & Cuming Polymer Group.* Name unverified.

2070 Amicon® SC-220
High purity silicone; blob top for high-reliability circuits. *Emerson & Cuming Polymer Group.* Name unverified.

2071 Amicon® SC-2634A/B
Two-component silicone; die coating. *Emerson & Cuming Polymer Group.* Name unverified.

2072 Amicon® SC-3613
One-component silicone; electronic grade silicone; die attach adhesive for IC assembly. *Emerson & Cuming Polymer Group.* Name unverified.

2073 Amicon® TG-86
25928-94-3
Epoxy tape adhesive; meets MIL-Std. 883C, Method 5011. *Emerson & Cuming Polymer Group.* Name unverified.

2074 Amicure® 352
Cycloaliphatic amines
Curing agents. *Air Prods & Chems Inc.*

2075 Amicure® AEP
Cycloaliphatic amines
Curing agents. *Air Prods & Chems Inc.*

2076 Amicure® CL-485
Aliphatic amine tetrol
Crosslinker and reactivity enhancer for PU coatings, adhesives, and sealants. *Air Prods & Chems Inc.*

2077 Amicure® DBU
Tertiary amine accelerator. *Air Prods & Chems Inc.*

2078 **Amicure® PACM**
Cycloaliphatic amines
Curing agents. *Air Prods & Chems Inc.*
2079 **Amicure® SA**
Tertiary amine salts
Curing agents. *Air Prods & Chems Inc.*
2080 **Amicure® TEDA**
Tertiary amine accelerator. *Air Prods & Chems Inc.*
2081 **Amicure® TMR 30**
Tertiary amine accelerator. *Air Prods & Chems Inc.*
2082 **Amidan**
Distilled monoglycerides; emulsifier, dough conditioner, starch complexing agent in bread rolls, bread improvers. *Grindsted Prods.; Grindsted Prods. Denmark.*
2083 **Amide CMA-2**
61791-08-0
PEG-2 coco MEA
Thickener producing stable foam for personal care formulations. *Berol Nobel AB.*
2084 **Amide RMA-2**
PEG-2 rapeseed amide
Thickener, foaming agent for personal care formulations. *Berol Nobel AB.*
2085 **Amidex 1285**
Modified coco diethanolamine (2:1); phosphate-compatible detergent, wetting agent, emulsifier for high-alkaline industrial and specialty cleaning compounds, e.g., degreasers, floor strippers; compatible with high concentration of inorganics in aqueous systems without need for hydrotropes. *Chemron.*
2086 **Amidex AME**
142-26-7 205-530-8
$C_4H_9NO_2$
Acetamide MEA
N-(2-Hydroxyethyl)acetamide; N-acetyl ethanolamine. Acetamide MEA; antistat, humectant, conditioner for skin and hair products. *Chemron.*
2087 **Amidex C**
Modified coco diethanolamide; detergent, thickener, emulsifier, wetting agent, foam stabilizer, viscosity builder; for industrial and household cleaners. *Chemron.*
2088 **Amidex CE**
Cocamide DEA (1:1); detergent, thickener, viscosity builder, foam stabilizer for shampoos, cleaners, bubble baths, industrial cleaners, car shampoos, dishwashes, drycleaning detergents, waterless cleaners, solvent cleaners. *Chemron.*
2089 **Amidex CIPA**
Cocamide MIPA; antidefatting surfactant; for shampoos, skin cleansers, bubble baths. *Chemron.*
2090 **Amidex CME**
68140-00-1 268-770-2
Cocamide MEA; viscosity builder, foam enhancer for personal care products, soap systems, synthetic powder detergents, liquid dishwashing formulations. *Chemron.*
2091 **Amidex CP**
136-26-5 205-234-9
$C_{14}H_{29}NO_3$
Capramide DEA
Bis(2-hydroxyethyl)decanamide; Capric diethanolamide. Capramide DEA; flash foaming detergent, wetting agent for use in pigmented personal care systems. *Chemron.*
2092 **Amidex KD**
Cocamide DEA
Surfactant for ethoxy sulfate systems; yields high stable viscosities at low concentrations; flash foamer, foam stabilizer; for gelled shampoos, bath gels, liquid soaps, facial cleansers. *Chemron.*
2093 **Amidex KME**
68140-00-1 268-770-2
Cocamide MEA
Foam builder, viscosity booster, stabilizer for personal care products, synthetic powdered and liquid detergent systems. *Chemron.*
2094 **Amidex L-9**
120-40-1 204-393-1
Lauramide DEA
Thickener, flash foamer, viscosity enchancer, foam stabilizer/builder; for liquid detergents, household, institutional and industrial cleaning compounds. *Chemron.*
2095 **Amidex LD**
120-40-1 204-393-1
Lauramide DEA
Thickener, viscosity builder, foam booster/stabilizer, detergent, emulsifier, for household, institutional and industrial cleaners, personal care products. *Chemron.*
2096 **Amidex LIPA**
142-54-1 205-541-8
Lauramide MIPA
Mild, low melting, fully active foam booster/stabilizer; for shampoo and detergent systems. *Chemron.*
2097 **Amidex LMMEA**
142-78-9 205-560-1
Lauramide MEA
Viscosity builder, foam booster; for bath products, shampoos, skin cleansers. *Chemron.*
2098 **Amidex LN**
Linoleamide DEA
Thickener, foam builder, emulsifier, conditioner; substantive to hair, bath and skin care products, shampoos, conditioners. *Chemron.*
2099 **Amidex O**
93-83-4 202-281-7
Oleamide DEA
Thickener, emulsifier, lubricant, conditioner; for shampoos, mineral oil emulsions; compatible with hair dye systems. *Chemron.*
2100 **Amidex PK**
Palm kernelamidea DEA
Viscosity builder and foamer for conditioning shampoos, mousses, styling gels. *Chemron.*
2101 **Amidex RC**
40716-42-5 255-051-3
Ricinoleamide DEA
Low foaming surfactant with wetting and softening properties, emulsifier with lubricity; for hair conditioners, shampoos, skin creams and lotions. *Chemron.*
2102 **Amidex S**
68425-47-8 270-355-6
Soyamide DEA
Foamer, viscosity builder, emulsifier with skin feel properties; for shower and facial cleansers, liquid soaps, bath gels. *Chemron.*
2103 **Amidex SME**
111-57-9 203-883-2
Stearamide MEA
Thickener, emulsifier for mineral oil and vegetable oil systems; for toiletry bars, creams, lotions. *Chemron.*
2104 **Amidex TD**
68140-08-9 268-772-3
Tallowamide DEA
Detergent for dry laundry compounds and specialty cleaners. *Chemron.*
2105 **Amido betaine C**
Component of personal care products industrial foamer. *Zohar Detergent Factory.*
2106 **Amidocid®**
25311-71-1 5187 246-814-1
Isofenphos
Soil-applied insecticide used for control of insects in rice crops and pear sucker. *Bayer AG.*
2107 **Amidogene**
An explosive consisting of potassium nitrate, magnesium sulfate, wood charcoal, bran, and sulfur. No manufacturer.
2108 **amidol**
2,4-Diaminophenol
Used as a photographic developer.
2109 **Amidox®**
Ethoxylated alkanolamides
Emulsifiers, detergents, wetting agents. *Stepan.* Name unverified.
2110 **Amidox® C-2**
61791-08-0
PEG-3 cocamide
Emulsifier, detergent, wetting agent for dishwashing detergents, shampoos, emulsions, textile wetting and leveling agent. *Stepan; Stepan Canada.*
2111 **Amidox® L-2**
26635-75-6
PEG-3 lauramide
Emulsifier, detergent, wetting agent for dishwashing detergents, shampoos, emulsions. *Stepan; Stepan Canada.*
2112 **Amidozid**
Soil applied insecticide; used for control of insects in rice crops and pear sucker. *Bayer AG.*
2113 **Amiema MA-OD**
Octyldodecyl N-myristoryl-N-methyl alanate
Oil-phase cosmetics ingredient. *Nihon Emulsion.*

2114 Amiema MA-OL
Oleyl N-myristoyl-N-methyl alanate
Oil-phase cosmetics ingredient. *Nihon Emulsion.*

2115 Amiesite
A proprietary asphalt-rubber product used in road surfacing. No manufacturer.

2116 Amietol
Diethylethanolamine and a range of aminoethanols used in chemical manufacturing. *Imperial Chemical Industries plc.*

2117 Amigan
834-12-8, 886-50-0
Ametryn-terbutryn; Ametrex. Herbicide. *Agan Chemical Manufacturers Ltd.*

2118 Amigel
Sclerotium gum; natural gellifying agent, stable in acid pH range. *Alban Muller.*

2119 Amigen
Protein hydrolysate
Replenisher. *Travenol Laboratories Inc.* Name unverified.

2120 Amihope LL-11
52315-75-0 257-843-4
Lauroyl lysine
Surface modifier, coemulsifier, codispersant; in cosmetics, medical, painting and other fields; filler for ink and paint; chelating agent. *Ajinomoto.* Name unverified.

2121 Amikapron
1197-18-8 9704 214-818-2
tranexamic acid.
A proprietary trade name for tranexamic acid, a hemostatic agent. No manufacturer.

2122 Amilperoxy pivalate
29240-17-3 249-530-6
$C_{10}H_{20}O_3$
t-Amyl peroxypivalate
Aztec® *t*-Amyl peroxypivalate-75 OMS; Esperox® 551M; Lupersol 554-M50, 554-M75; Trigonox® 125-C75; Propaneperoxoic acid, 2,2-dimethyl-, 1,1-dimethylpropyl ester. Initiator used in polymerization of monomers; initiator for bulk, solution, and suspension polymerization. *Atochem.* Discontinued.

2123 Amine
Large range of cationic surfactants composed of primary amines, secondary amines or tertiary amines in liquid, solid or paste form; used in industry and in household and personal care formulations, though mostly in the form of derived quaternaries and various salts. *Keno Gard (UK) Ltd.* Unverified.

2124 Amine 0
Dewatering agent and corrosion inhibitor. *Berol Nobel AB.* Name unverified.

2125 Amine 8 D
111-86-4 203-916-0
$C_8H_{19}N$
n-Octylamine
1-Aminooctane; octyl amine; 1-Octanamine; Monoctylamine. Chemical intermediate. mp = -5 - 1°; bp = 175-177°; d = 0.7820; n_D^{20}= 1.4290; soluble in organic solvents. *Berol Nobel AB.*

2126 Amine 10
2016-57-1 204-690-6
$C_{10}H_{23}N$
Lauramine
n-Decylamine; 1-Aminodecane; Decylamine; 1-Decanamine. Chemical intermediate; ethoxylated end products used in detergent, cosmetic, and agricutural applications. mp = 16°; bp = 211°; d = 0.794. *Berol Nobel AB.*

2127 Amine 12
124-22-1 204-690-6
$C_{12}H_{27}N$
n-dodecylamine
dodecylamine; lauryl amine; 1-Dodecanamine; Lauramine. Chemical intermediate; ethoxylated end products used in detergent, cosmetic, and agricultural applications. *Berol Nobel AB.*

2128 Amine 12-98D
2016-57-1 204-690-6
$C_{10}H_{23}N$
Lauramine
Emulsifier; chemical intermediate; ethoxylated or guanidated end-products used in detergent, cosmetic, and agricultural formulations. *Berol Nobel AB.*

2129 Amine 14D
2016-42-4 217-950-9
$C_{14}H_{31}N$
Myristamine
n-Tetradecylamine; 1-Tetradecanamine; 1-Tetradecylamine; 1-Aminotetradecane; Tetradecylamine. Emulsifier; chemical intermediate; quaternized end products used as bactericides. mp = 40-42°; bp = 291°; insoluble in H_2O, soluble in organic solvents. *Berol Nobel AB.*

2130 Amine 16D
143-27-1 205-569-8
$C_{16}H_{35}N$
Palmitamine
Cetylamine; n-Hexadecylamine; 1-Hexadecanamine; 1-Hexadecylamine; alamine 6; Armeen 16d;
palmitylamine; 1-aminohexadecane. Emulsifier; chemical intermediate; end products such as quaternary ammonium compounds used as bactericides and in shampoo formulations. mp = 38-47°; bp = 330°; insoluble in H_2O, soluble in organic solvents. *Berol Nobel AB.*

2131 Amine 18-90
124-30-1 204-695-3
$C_{18}H_{39}N$
Octadecylamine
Stearylamine; 1-octadecanamine; 1-octadecylamine; octadecylamine; Adogenen 142; Alamine 7; Armeen 1180; n-octadecylamine. Chemical intermediate; ethoxylated end-products used in detergent, cosmetic, and agricutural applications. mp = 50-52°; bp= 349°; insoluble in H_2O (<1 mg/ml), soluble in organic solvents; LD_{50} (rat orl) = 2395 mg/kg. *Berol Nobel AB.*

2132 Amine 2HBG
61789-79-5 263-089-7
N,N-Di(hydrogenated tallow) amine
Surfactant intermediate; for pour point depressant formulations for diesel fuel, paper chemical auxiliary, personal care products. *Berol Nobel AB.*

2133 Amine 2M1214D
N,N-Dimethyl dodecyl tetradecylamine
Chemical intermediate; end products including quaternaries and amine oxides used in disinfectant and cosmetic applications. *Berol Nobel AB.*

2134 Amine 2M1218D
61788-93-0 263-020-0
N,N-dimethyl cocamine
Chemical intermediate; end products including quaternaries and amine oxides used in detergent and cosmetic applications. *Berol Nobel AB.*

2135 Amine 2M12D
67700-98-5 266-922-2
Dimethyl lauramine
Surfactant intermediate; quaternized end products used as bactericides, textile auxiliries. *Berol Nobel AB.*

2136 Amine 2M14-50D
N,N-Dimethyl tetradecylamine
Chemical intermediate; quaternized end products used as bactericides. *Berol Nobel AB.*

2137 Amine 2M14D
68439-70-3 270-414-6
Dimethyl myristamine
Surfactant intermediate; end products including quaternaries and amine oxides for detergent, disinfectant and cosmetic formulations. *Berol Nobel AB.*

2138 Amine 2M16D
68037-93-4 268-217-5
Dimethyl palmitamine
Surfactant intermediate; quaternized end products used in cosmetic formulations. *Berol Nobel AB.*

2139 Amine 2M18D
124-28-7 3525 204-694-8
Dimethyl octadecylamine
Surfactant intermediate; quaternized end products used in cosmetic formulations. *Berol Nobel AB.*

2140 Amine 2M810D
Methyloctyldecylamine
Chemical intermediate for quaternaries used as bactericides. *Berol Nobel AB.*

2141 Amine 2MBGD-M
68814-69-7 272-339-4
N,N-Dimethyl tallowamine
Chemical intermediate; quaternized end products used in detergent applications. *Berol Nobel AB.*

2142 Amine 2MHBGD
61788-95-2 263-022-1
Dimethyl hydrogenated tallowamine
Surfactant intermediate: quaternized end products used in detergent applications. *Berol Nobel AB.*

2143 Amine 2MKKD
61788-93-0 263-020-0
Dimethyl cocoamine
Surfactant intermediate; quaternized end products used as textile auxiliaries; betaines for cosmetic applications. *Berol Nobel AB.*

2144 Amine 2MOLD
68814-69-7 272-339-4
N,N-Dimethyl tallowamine
Chemical intermediate; quaternized end products used in detergent applications. *Berol Nobel AB.*

2145 Amine 740
68911-79-5 272-787-0
Oleotripropylene tetraamine
Chemical intermediate for production of surface active agents. *Berol Nobel AB.*

2146 Amine 760
Tallow tripropylene tetramine
Chemical intermediate; end-products such as acetates used for emulsifiers and dispersants. *Berol Nobel AB.*

2147 Amine 780
97808-04-3 307-919-9
Cocotripropylene tetraamine
Chemical intermediate for production of surface active agents. *Berol Nobel AB.*

2148 Amine Acetate HBG
61790-59-8 263-149-2
Hydrogenated tallowamine acetate
Hydrogenated tallowamine acetate; reagent for pigment flushing, flocculation. *Berol Nobel AB.*

2149 Amine Acetate KK
61790-57-6 263-147-1
Cocamine acetate
Cocamine acetate; reagent for pigment flushing, flocculation. *Berol Nobel AB.*

2150 Amine B11
68037-92-3 268-215-4
Eicosyl docosylamine
Chemical intermediate; end-products such as ethoxylates used for detergent applications, acetates for emulsifiers and dispersants. *Berol Nobel AB.*

2151 Amine BG
61790-33-8 263-125-1
Tallow amine
Emulsifier, corrosion inhibitor; chemical intermediate producing sulfosuccinimides and textile auxiliaries. *Berol Nobel AB.*

2152 Amine C4
109-73-9 1578 203-699-2
$C_4H_{11}N$
n-butylamine
1-Aminobutane; aminobutane; Butyl amine; 1-Butanamine; norralamine; mono-n-butylamine; tutane; Monobutylamine. Chemical intermediate. mp = -49°; bp = 78°, d = 0.7400; n_D^{20} = 1.4010; LD_{50} (rat orl) = 366 mg/kg. *Berol Nobel AB.*

2153 Amine CS-1135®
51200-87-4 257-048-2
$C_5H_{11}NO$
4,4-Dimethyloxazolidine
Dimethyl-1-oxa-3-aza-cyclopentane; Dimethyloxazolidine; Bioban CS 1135. Emulsifying amine, corrosion inhibitor, alkaline pH stabilizer; for metalworking fluids and aqueous systems. A preservative used in cooling fluids and paints. *Angus.*

2154 Amine CS-1246
7747-35-5 231-810-4
Ethyldihydro-1H,3H,5H-oxazolo(3,4-c)oxazole
Chemtan A 60; Ethyldihydro-1H,3H,5H-oxazolo(3,4-c)oxazole; Oxazolidine E; Oxazolo(3,4-c)oxazole, 7a-ethyldihydro-; Zoldine ZE. Catalyst, resin reactant, formaldehyde substitute, crosslinking agent, corrosion inhibitor; raw material for synthesis. *Angus.*

2155 Amine D
1446-61-3 215-899-7
$C_{20}H_{31}N$
1R-1α,2,3,4,4aβ,9,10,10aα-octahydro-1,4a-dimethyl-7-(1-methylethyl)-1-Phenanthrenemethanamine
Dehydroabietylamine; Amine D; Podocarpa-8,11,13-trien-15-amine, 13-isopropyl-; Rosin amine D. Used as asphalt additive, as cationic collectors for calcite, sylrite, mica, feldspar, vermicilulite and phosphate rock concentration operations. n_D^{20} = 1.5460; $[\alpha]_D^{20}$ = +56.10 (c=2.4, pyridine) *Hercules.*

2156 Amine HBG
61788-45-2 262-976-6
Hydrogenated tallow amine
Emulsifier, corrosion inhibitor; chemical intermediate producing ethoxylates for textile auxiliaries and acetates for emulsifiers and dispersants. *Berol Nobel AB.*

2157 Amine HBGD
61788-45-2 262-976-6
Distilled hydrogenated tallow amine; emulsifier, corrosion inhibitor; chemical intermediate producing ethoxylated for textile auxiliaries and acetates for emulsifiers and dispersants. *Berol Nobel AB.*

2158 Amine KK
61788-46-3 262-977-1
Cocamine
Emulsifier, corrosion inhibitor; chemical intermediate producing ethoxylates used as detergents or textile auxiliaries. *Berol Nobel AB.*

2159 Amine M210D
7396-58-9 230-990-1
Methyldidecylamine
Chemical intermediate; quaternized end-products used as bactericides. *Berol Nobel AB.*

2160 Amine M218
4088-22-6 223-819-7
Methyldioctadecylamine
Surfactant intermediate. *Berol Nobel AB.*

2161 Amine M2HBG
61788-63-4 262-991-8
Dihydrogenated tallow methylamine
Surfactant intermediate. *Berol Nobel AB.*

2162 Amine OL
112-90-3 204-015-5
Oleamine
Emulsifier, corrosion inhibitor; chemical intermediate producing sulfosuccinimides for carpetback binding, engine-oil additives, ethoxylates for detergent applications. *Berol Nobel AB.*

2163 Amine2 VT
61789-79-5 263-089-7
Dihydrogenated tallowamine
Armeen® 2HT; Amine 2HBG. Chemical intermediate; end products including pour point depressant formulations for diesel fuel, paper chemical auxiliaries. *Berol Nobel AB.*

2164 Aminitrazole
140-40-9 430 205-414-7
$C_5H_5N_3O_3S$
N-(5-Nitro-2-thiazolyl)acetamide
2-acetamido-5-nitrothiazole; 2-acetylamino-5-nitrothiazole; trichorad; acinitrazole; tritheon; trichloral; Gynofon; Enheptin-A; Pleocide; Acetyl enheptin; Acinitrazol; Acinitrazole; Ametoterina; Aminitrazol; Aminitrozol; Aminitrozole; Aminitrozolum; CL 5,279; Pleocide; Thiazole, 2-acetamido-5-nitro-; Trichlorad; Trichocid; Trichoman; Trichorad; Trichoral,. Antiprotozoal (Trichomonas) and an antihistomonad; used with turkeys. mp = 264-265°; soluble in aqueous alkaline media.

2165 Amino Acid Gelatinization Agent
N-Acyl glutamic acid diamide
Gelatinization agent for oil for solidifying almost all oils ranging from petroleum to vegetable oils. *Ajinomoto.* Name unverified.

2166 Amino Gluten MG
Maize gluten amino acids-sodium chloride. Conditioner for skin creams and lotions and hair conditioners; humectant for cosmetics and pharmaceuticals. *Croda Inc.* Discontinued.

2167 *m*-Aminobenzoic Acid
99-05-8 441 202-724-4
$C_7H_7NO_2$
3-Aminobenzoic acid
Used in chemical synthesis. mp=174°; d = 1.151; soluble in H_2O (5.9 g/l), organic solvents; LD_{50} (mus orl) = 6300 mg/kg. *Penta Mfg.; Schweizerhall; SCM Glidco Organics.*

2168 Amino-Collagen-25,-40
Collagen amino acids
Substantivity agent, penetrant, moisturizer for skin and hair care products, especially conditioners, shampoos, styling and setting products, nutritive skin products. *Maybrook.*

2169 Aminodermin CLR
Sulfur rich amino acid concentrate
Conditioner for structurally damaged hair, oily skin care products. *Dr. Kurt Richter; Henkel/Cospha.*

2170 2-Amino-2-ethyl-1,3-propanediol
115-70-8 459 204-101-2
$C_5H_{13}NO_2$
Aminoethyl propanediol; AEPD®-85; AEPD®. Pigment dispersant, neutralizing amine, corrosion inhibitor, acid-salt catalyst, pH buffer, chemical and pharmaceutical intermediate, solubilizer. mp = 37.5-38.5°; bp_{10} = 152-153°; d = 1.0990; miscible with H_2O; soluble in alcohols; pH 0.1, molar aqueous solution.

2171 Aminoethylaminopropyltrimethoxy silane
1760-24-3 217-164-6
$C_8H_{22}N_2O_3Si$
N-2-aminoethyl-3-aminopropyltrimethoxysilane
CA0700; Dow Corning® Z-6020; Dynasylan® DAMO-T; Dynasylan® DAMO-P; Petrarch® A0700, A0701; Prosil® 3128; Union Carbide® A-1120. Coupling agent, chemical intermediate, blocking agent, release agent, lubricant, primer, reducing agent. Liquid; d = 1.0100; bp:15ks = 146°; n_D^{20} = 1.4450; flash point = 121°; LD_{50} (rat orl) = 7460 mg/kg. *Dow Corning; Hüls Am.*

2172 Aminoethylethanolamine
111-41-1 203-867-5
$C_4H_{12}N_2O$
N-(2-Hydroxyethyl)ethylenediamine
(2-Aminoethyl) ethanolamine; N-aminoethylethanolamine; monoethanolethylenediamine; 2-amino-2'-hydroxydiethylamine. Used in textile finishing compounds (antifuming agents, dyestuffs, cationic surfactants), resins, rubber, insecticides, medicinals. bp_{752} - 238-240°; d = 1.0300; n_D^{20}= 1.4861; LD_{50} (rat orl) = 3 gm/kg. *BASF; Dow; Nippon Nyukazai; Schweizerhall; Union Carbide.*

2173 Aminoethylpiperazine
140-31-8 205-411-0
$C_6H_{15}N_3$
1-(2-aminoethyl) piperazine
2-piperazinoethylamine; AEP; D.E.H. 39. Epoxy curing agent, intermediate for pharmaceuticals, anthelmintics, surface-active agents, synthetic fibers. Soluble in H_2O ; d = 0.9837; mp = -19°; bp = 218-222°; LD_{50} (rat orl) = 2140 mg/kg. *Akzo Nobel; Dow; Fabrichem; Texaco; Tosoh; Union Carbide.*

2174 Aminofoam K
TEA-lauroyl animal keratin amino acids; mild protein surfactant, foaming agent for shampoos, conditioners, facial cleansers. *Croda Inc.; Croda Chem. Ltd.* Discontinued.

2175 Aminofoam W
TEA-lauroyl animal collagen amino acids; mild protein surfactant, detergent, conditioner for skin and hair care cleansing systems, shaving creams. *Croda Inc.; Croda Chem. Ltd.*

2176 Aminogen I
134-32-7 6485 205-138-7
$C_{10}H_9N$
α-Naphthylamine
1-naphthylamine; 1-naphthalenamine; 1-aminonaphthalene; naphthalidine; Fast Garnet B Base; 1-Naphthylamine; naphthalidam; naphthalidine; C.I. azoic diazo component 114; Fast Garnet Base B. Used in manufacture of dyestuffs and as a rubber vulcanization accelerator. mp = 50°, bp = 301°; d = 1.1140; soluble in H_2O (1.7 mg/ml), more soluble in organic solvents; LD_{50} (rat orl) = 779 mg/kg. No manufacturer.

2177 Aminogen II
106-50-3 7439 203-404-7
$C_6H_8N_2$
p-phenylenediamine
orsin; C.I. 76076; Ursol D. A rubber vulcanization accelerator and chemical intermediate. mp = 139-141°; bp = 267°; soluble in H_2O (1 g/100 ml), more soluble in organic solvents; LD_{50} (rat orl) = 80 mg/kg. No manufacturer.

2178 Aminol A-15
Trideceth-2 carboxamide MEA
Biodegradable cosmetics surfactant, thickener, foam stabilizer; excellent dermatological properties. *Chem-Y GmbH.*

2179 Aminol KDE
68603-42-9 271-657-0
Cocamide DEA
Foam booster/stabilizer, superfatting agent for personal care products; solubilizer for perfumes, vegetable oils. *Chem-Y GmbH.*

2180 Aminol N
85536-23-8
PEG-4 rapeseed amide
Biodegradable cosmetics surfactant, thickener; for shower bath, foam bath, shampoo, soap gel, and other surfactant formulations. *Chem-Y GmbH.*

2181 Aminol TEC N
85536-23-8
PEG-4 rapeseedamide
Biodegradable emulsifier for metalworking fluids and lubricants, conveyor chain lubricant, in anticorrosive formulations. *Chem-Y GmbH.*

2182 2-amino-2-methyl-1 propanol
124-68-5 469 204-709-8
$C_4H_{11}NO$
Amino-2,2-dimethylethanol
AMP; AMP-95; AVT-75; 2,2-Diethyl-ethanolamine; Isobutanolamine; 2-Amino-2-Methylpropanol; Amino-2-methyl-1-propanol; Aminoisobutanol; Dimethyl-2-hydroxyethylamine; Hydroxymethyl-2-propylamine; Isobutanol-2-amine. Boiler water treatment chemical, corrosion inhibitor, carbon dioxide absorber. Widely used as a buffer and phosphate acceptor in assay of phosphatases. Suitable as buffer for manual and automated determination of alkaline phosphatase using 4-nitrophenyl phosphate as substrate mp = 31-32°; bp = 164-166°; d = 0.9350; n_D^{20}= 1.4480; LD_{50} (rat orl) = 2900 mg/kg. *Lancaster.*

2183 2-amino-2-methyl-1,3-propanediol
115-69-5 468 204-709-8
$C_4H_{11}NO_2$
1,1-Di(hydroxymethyl)ethylamine
AMPD; Aminomethyl propanediol; 1,3-dihydroxy-2-methyl-2-propylamine. Pigment dispersant, neutralizing amine, corrosion inhibitor, acid-salt catalyst, pH buffer, chemical and pharmaceutical intermediate; solubilizer or emulsifier system component in personal care products. mp = 108-110°; bp_{10} = 151°; soluble in H_2O, organic solvents; LD_{50} (rat orl) = 17 gm/kg.

2184 2-amino-5-nitrothiazole
121-66-4 477 204-490-9
$C_3H_3N_3O_2S$
5-Nitro-2-thiazol-amine
Entramin; aminonitrothiazole; aminonitrothiazolum; Aminzol soluble; 5-nitro-2-aminothiazole; 5-nitro-2-thiazolamine; 5-nitro-2-thiazolylamine; Enheptin; Enheptin premix; Enheptin-T; Nitramin IDO; Nitromin IDOo; Enheptyne. 2-Amino-5 nitrothiazole premix; antihistomonad in turkeys, chickens; for trichomonasis in pigeons Powder; mp = 202°; sparingly soluble in H_2O; almost insoluble in chloroform; soluble in dilute mineral acids. *May & Baker Ltd.* Name unverified.

2185 *p*-aminophenol
123-30-8 482 204-616-2
C_6H_7NO
p-Hydroxyaniline
Azol; 4-Amino-1-hydroxybenzene; Azol; Certinal; Citol; Paranol; Rodinal; Unal; Ursol P; *para*-aminophenol; paramidophenol; Kodelon; Energol; Freedol; Indianol; Kathol; BASF Ursol P base; Benzofur P; C.I. Oxidation Base 6A; Fouramine P; Fourrine 84; PAP; Pelagol Grey P Base; Tertral P Base; Ursol P Base; Zoba Brown P Base; C.I. 76550; Durafur Brown RB; Fourrine P Base; Furro P Base; Nako Brown R; Pelagol P Base; Renal AC; 4-Hydroxyaniline; 4-Aminobenzenol. 4-Aminophenol hydrochloride in solution. Photographic developer, intermediate in manufacture of azo dyes. mp = 188-190°; bp = 284°; insoluble in H_2O, soluble in EtOH; LD_{50} (rat orl) = 375 mg/kg. *Johnsons of Hendon.* Name unverified.

2186 γ-aminopropyltriethoxysilane
919-30-2 213-048-4
$C_9H_{23}NO_3Si$
3-(Triethoxysilyl)-1-propanamine
Aktisil AM; CA750; Dynasylan® AMEO, AMEO-P; Prosil® 220; Union Carbide® A-1100. Filler for thermosets, thermoplastics. Reacts with porous glass to form the aminopropyl derivative of glass, an adsorbent for affinity chromatography. d = 0.9420; bp = 217°; n_D^{20}= 1.4210; LD_{50} (rat orl) = 1780 mg/kg. *Gelest; Hüls Am; PCR.*

2187 Aminopropyltrimethoxysilane
13822-56-5 237-511-5
$NH_2(CH_2)_3Si(OC_2H_5)_3$
3-(Triethoxysilyl)-1-propanamine
CA0880; Dynasylan® AMMO; Union Carbide® A-1110; γ-Aminopropyl triethoxysilane. Coupling agent, chemical intermediate, blocking agent, release agent, lubricant, primer, reducing agent. d = 0.942; bp = 217°. *Aldrich; Fluka; Gelest; Hüls Am.; PCR; Sigma; Union Carbide.*

2188 Amino-Silk SF
Silk amino acids
Substantive protein for elegant skin and hair preparations; penetrant, moisturizer. *Maybrook.*

2189 Aminosol
Protein hydrolysate
Replenisher. *Abbott Laboratories.* Name unverified.

2190 Aminotriazole Bayer
61-82-5 513 200-521-5
Amitrole
Fast-acting herbicide, for control of hard-to-kill grass and broad-leaved weeds; mainly used in mixtures with other compounds (Ustinex products). *Bayer AG.*

2191 Aminotrimethylene phosphonic acid
6419-19-8 229-146-5
Fostex AMP; Chemphonate AMP; Unihib® 305-LC. Scale inhibitor and sequestrant in water treatment. *Henkel; Chemron; Lonza AG.*

2192 Aminox
Reaction product of di-phenylamine and acetone; gives protection against oxygen and heat deterioration; used in tire carcass, heels, soles,

mechanicals, proofing sundries and wire insulation; effective in natural and nitrile rubbers and nylon. *Uniroyal.* Name unverified.

2193 Aminox® Flake, Powd
9003-79-6
Diphenylamineacetone reaction production; antioxidant. *Uniroyal.*

2194 Aminox® Naugard A
Antioxidant protecting polymers from loss of physical properties on extended exposure to heat; for EVA and polyamide hot-melt adhesives, nylon 6. *Uniroyal.*

2195 Aminoxid
Foam stabilizer; wetting; auxiliary; polymerization accelerator; used for shampoos, detergents; aqueous dispersions; photography; electroplating; vulcanizing. *Goldschmidt Ltd.* Name unverified.

2196 Aminoxid WS 35
68155-09-9 268-938-5
Cocamidopropylamine oxide
Tegamine® Oxide WS-35. Detergent, emulsifier, wetting agent, softener, foam stabilizer for detergent preparations, cosmetic and pharmaceutical emulsions. *Goldschmidt; Goldschmidt AG.*

2197 Amipaque
31112-62-6 6240 250-475-5
Metrizamide
Radiopaque medium used as a diagnostic aid. *Sterling Drug Inc.* Name unverified.

2198 Amiter LGOD
82204-94-2 279-917-5
Dioctyldodecyl lauroyl glutamate
Oily surfactant used in personal care products. *Ajinomoto; Ajinomoto USA; Nihon Emulsion.*

2199 Amiter LGOD-2
Dioctyldodeceth-2 lauroyl glutamate
Emulsifier used in cosmetics; oil ingredient. *Ajinomoto; Ajinomoto USA; Nihon Emulsion.*

2200 Amiter LGS-2
Disteareth-2 lauroyl glutamate
Emulsifier used in cosmetics. *Ajinomoto; Ajinomoto USA; Nihon Emulsion.*

2201 Amiter SG-2000
Di(2-octyldodecyl) N-stearoyl-L-glutamate
Amiter SG-OD. Surfactant for cosmetic goods with high affinity to skin or hair. *Nihon Emulsion.*

2202 Amitraz
33089-61-1 510 251-375-4
$C_{19}H_{23}N_3$
N-Methylbis-(2,4-xylyliminomethy1)-amine
Taktic; Triatrix; Triatox; Taktic; Ovasyn; Mitac; Baam; Triazid. An acaricide and insecticide. Used to control mites, bugs, insects and larvae in fruit crops. mp = 86-88°; d^{25}= 1.128; poorly soluble in H_2O (<1 mg/l), more soluble in organic solvents, LD_{50} (rat orl) = 800mg/kg. No manufacturer.

2203 Amjet A-4
Leveling carrier for pressure dyeing of stock yarn and piece goods. *Am. Emulsions.*

2204 Amldex SE
93-82-3 202-280-1
Stearamide DEA
Thickener, emulsifier for personal care products including cold wave neutralizers, vegetable oil emulsions, conditioning shampoos and mousses. *Chemron.*

2205 Amlev ACY-Super
Complex inorganic salt of an organic amine; nonretarding leveling agent for polyacrylic dyeing. *Am. Emulsions.*

2206 Amlev CH641
Surfactant blend; leveling agent for dyeing nylon fiber, yarns, fabric, carpet. *Am. Emulsions.*

2207 Amlev DAS
Quaternary ammonium compound; leveling and retarding agent for Dacro 62 with cationic dyes; also suitable for acrylics. *Am. Emulsions.*

2208 Amlev HBL
Proprietary; low foaming penetrant, dispersant for jet machine dyeing of polyester, polyester/nylon carpet. *Am. Emulsions.*

2209 Amlev MRC
Sulfonated surfactant; acid dye leveler for continuous and batch dye applications. *Am. Emulsions.*

2210 Amlight M-2
Mixture of inorganic compounds; peroxide bleaching stabilizer for cellulosic fibers and their blends. *Am. Emulsions.*

2211 Amlube AEC
Blend; lubricant to prevent crack marks on polyester or nylons; improves backwinding of space-dyed yarns. *Am. Emulsions.*

2212 Ammonal
An explosive consisting of 30% trinitrotoluene, 47% ammonium nitrate, 22% aluminum powder, and 1% charcoal. No manufacturer.

2213 Ammon-Carbonite
An explosive containing ammonium nitrate, flour, nitroglycerin, and collodion wool. No manufacturer.

2214 Ammon-Dynamite
An explosive containing 40% nitroglycerin, 10% wood meal, 10% sodium nitrate, and 40% ammonium nitrate. No manufacturer.

2215 Ammondyne
A coal-mine explosive containing 9-11% of nitroglycerin, 45-51% of ammonium nitrate, 8-10% of sodium nitrate, 17-19% of ammonium oxalate, and 11-13% of wood meal. No manufacturer.

2216 Ammon-Foerdite I
An explosive containing ammonium nitrate, flour, nitroglycerin, collodion wool, glycerin, diphenylamine, and potassium chloride. No manufacturer.

2217 Ammon-Gelatin-Dynamite
A blasting explosive consisting of 50% nitroglycerin, 2.5% collodion cotton, 45% ammonium nitrate, and 2.5% rye meal. No manufacturer.

2218 Ammon-Halalit
An explosive containing nitroglycerin, ammonium nitrate, vegetable meal, nitro-compounds, and potassium perchlorate. No manufacturer.

2219 ammonia
7664-41-7 517 231-635-3
NH_3
ammonia gas
ammonia anhydrous; spirit of Hartshorn. Fertilizers, refrigerant, nitriding of steel, condensation catalyst, neutralizing agent, petroleum industry, latex preservative, explosives. Colorless gas; d (gas) = 0.7714 g/l; mp = -77°; d = 0.5967; bp = -33.35° Air Prods; AlliedSignal; Am. Cyanamid; Asahi Chem Industry Co Ltd; Chevron; General Chem; La Roche Ind; Mitusbishi Toatsu Chem; Monsanto; Nissan Chm Ind; Norsk Hydro A/S; OxyChem; PPG Ind; Unocal.

2220 ammonia dynamites
Explosives usually containing nitroglycerin, wood pulp, ammonium nitrate, sodium nitrate, and calcium or magnesium carbonate. Name unverified.

2221 Ammonia Gelignite
An explosive containing 29.3% nitroglycerin, 0.7% nitro-cotton, and 70% ammonium nitrate. No manufacturer.

2222 ammoniacal Turpethum
$Hg_4N_2O_4S{\cdot}2H_2O$
Hydrated dimercuri-ammonium sulfate $(NHg_2)_2SO_4 \cdot 2H_2O$.

2223 Ammonia-Olein
The trade name for a form of sulfonated castor oil. No manufacturer.

2224 Ammonit C (Anfo-explosives)
Primed from the bottom of the borehole by Gelatine Donarit 1 and detonating fuse. *Dynamit Nobel Wien GmbH.* Name unverified.

2225 Ammonite
An explosive; contains ammonium nitrate, trinitrotoluene, and sodium chloride.

2226 Ammonium acetate
631-61-8 520 211-162-9
$C_2H_7NO_2$
Acetic acid ammonium salt
Reagent in analytical chemistry, drugs, textile dyeing, preserving meats, foam rubbers, vinyl plastics, and explosives. mp = 114°; d = 1.07; freely soluble in alchol; slightly soluble in acetone; 0.5 molar aqueous solution has pH of 7.0. *Aldrich; General Chem; Schaefer Salt & Chem; Verdugt BV.*

2227 Ammonium acrylates acrylonitrogens copolymer
123754-28-9
Hypan® SS201; Ammonium acrylates/acrylonitrogens copolymer. Gellant, emulsifier for cosmetics and related applications. *Kingston Tech.*

2228 Ammonium alginate
9005-34-9
$C_6H_7O_6.NH_4$
ammonium polymannuronate
alginic acid; ammonium salt; Amoloid LV; Amoloid HV; Collatex; Sobag FD 300 Series; Superloid®; Keltose®. Ammonium salt of alginic acid; thickening agent and stabilizer in food products. Filamentous, grainy, granular, or powder; colorless or slightly yellow; slowly soluble in H_2O forming viscous solid; insoluble in alcohol; heated to decomposition emits toxic fumes of NO_x *Kelco int'l.*

2229 Ammonium alum
7784-25-0 335 232-055-3

$AlH_4NO_8S_2$
aluminum ammonium sulfate
ammonium aluminum sulfate; sulfuric acid, aluminum ammonium salt (2:1:1) dodecahydrate. Inorganic salt; mordant in dyeing, water and sewage purification, sizing paper, retanning leather, clarifying agent, food additive, manufacture of lakes and pigments, fur treatment.

2230 Ammonium biborate
Timborised. Preserved timber. *Borax Europe Ltd.*

2231 Ammonium bicarbonate
1066-33-7 522 213-911-5
CH_5NO_3
Ammonium hydrogen carbonate
Acid ammonium carbonate; carbonic acid, monoammonium salt). Inorganic salt; used for production of ammonium salts, dyes; leavening agent for cookies, crackers; fire-extinguishing compounds; pharmaceuticals, degreasing textiles, blowing agent for foam rubber, boiler scale removal, compost treatment. mp = 60° (dec); d = 1.586; soluble in H_2O at 20° = 17.4 g/100 ml; insoluble in alcohol and acetone; *BASF; General Chem; Nissan Chem Ind; Norsk Hydro A/S; Rhône-Poulenc Basic.*

2232 Ammonium bifluoride
1341-49-7 523 215-676-4
F_2H_5N
ammonium acid fluoride
Ammonium hydrogen fluoride; Amminium fluoride; acid ammonium fluoride; Ammonium Fluoride, Acidic; Ammonium hydrogendifluoride; Ammonium fluoride. Ceramics, chemical reagent, etching glass, sterilizer for brewery, dairy, etc.; electroplating processing beryllium; laundry sour. Orthorhombic crystals which readily etch glass; d = 1.5; mp = 124.6°; bp = 235°; soluble in H_2O (630 g/l); *Bayer UK; Hoechst-Celanese; Miles; Solvay GmbH.*

2233 Ammonium bisulfite
10192-30-0 528 233-469-7
H_5NO_3S
Sulfurous acid, monoammonium salt
acid ammonium sulfite. Preservative. mp = 147°; soluble in H_2O (267-620 g/100 ml). *Brotherton Ltd.; General Chem.; Heico Chem.*

2234 Ammonium bromide
12124-97-9 531 235-183-8
BrH_4N
FR-1. Flame retardant for textiles, wood, chipboard, plywood. Used in manufacture photographic films, plates, and papers; in engraving and lithography; as a corrosion inhibitor White odorless crystals, slightly hydroscopic; mp = 452°; mp = 235°; d^{25} = 2.4290; freely soluble in in H_2O, organic solvents; nearly insoluble in ethyl acetate; incompatible with acids, acid salts, salts of Pb, Hg, Ag. *Aldrich; Great Lakes Chem; Johnson Matthey SA.*

2235 Ammonium chloride
12125-02-9 537 235-186-4
ClH_4N
ammonium muriate; Sal ammoniac; Amchlor; Darammon; Salammonite; ammonium chloride fume; Salmiac. In dry batteries; mordant (dyeing and printing); safety explosives; flux for coating sheet and iron with zinc; manufacture of various ammonia compounds, fertilizer, pickling agent; In washing powders Used as a flux and cleanser. Medically as a systemic acidifier and in veterinary medicine as an expectorant and diaphoretic. Colorless, odorless crystal; mp = 340°; d^{25} = 1.5274; soluble in H_2O, methanol, ethanol; LD_{50}=im rats 30 mg/kg; incompatible with Ag, Pb salts; LD_{50} (rat orl) = 1650 mg/kg. *Aldrich; BASF; EM Ind' General Chem; Heico; Hüls Am; Montefluos SpA.*

2236 Ammonium curmene sulfonate
37475-88-0 253-519-1
$C_9H_{15}NO_3S$
Benzenesulfonic acid, (1-methylethyl)-, ammonium salt
Eltesol® AC60; ACS 60; Reworyl® ACS. Surfactant, hydrotrope for agricultural applications. *Albright & Wilson UK; Rewo Chemicals Ltd; Witco SA.*

2237 Ammonium dichromate
7789-09-5 544 232-143-1
$Cr_2H_8N_2O$
Ammonium dichromate (VI)
Dichromic acid, diammonium salt; Chromic acid, diammonium salt; Ammonio; Chromic acid (H2Cr2O7), diammonium salt; ammonium bichromate. Mordant for dyeing, pigments, manufacture of alizarin, chrome alum, catalysts, oil purification, pickling, leather tanning, synthetic perfumes, photography, lithography, pyrotechnics Bright orange-red crystals; flammable; mp = 170°; d = 2.155, 82 lb/f³; dec 180°; very soluble in H_2O. *British Chrome & Chemicals; EM Ind.*

2238 Ammonium dodecylbenzene sulfonate
1331-61-9 215-559-8
$C_{18}H_{29}KO_3S$
Hetsulf 50A; Ablusol DBM; Nansa® AS 40. Wetting agent, emulsifier, dispersant, for light duty detergent formulations. *Heterene; Taiwan Surf; Albright & Wilson UK.* Name unverified.

2239 Ammonium Fluoride
12125-01-8 553 235-185-9
FH_N
Neutral ammonium flouride. Manufacture of fluorides, analytical chemistry, antiseptic in brewing, etching glass, textile mordant, wood preservative, mothproofing agent d = 1.009; soluble in H_2O (100g/100ml); decomposed by hot H_2O into NH_3 and ammonium biflouride; incompatible with quinine salts calcium salts; ingestion produces nausea, vomiting, gastroentroenteritlis, convulsions, death. *Aldrich; Flexchemie BV; General Chem; GE; Hoechst-Celanese; Olin-Hunt.*

2240 Ammonium laureth sulfate
67762-19-0
$(C_2H_4O)_n.C_{12}H_{26}O_4S.H_3N$ (average n = 1-4)
Ammonium lauryl ether sufate
Avirol® AE 3003; Calfoam NEL-60; Carsonal® SES-A; DeSonol AE; Empicol® EAA, EAB, EAC; Texapon EA-1, NA; Ungerol AM3-75; Witcolate AE; Witcolate LES-60A; Zoharpon LAEA 253; Nonasol N4AS; Nurapon AL1, AL 60; Polystep®B-11; Rhodapex® AB-20, EA, EAY; Standapol® EA-1; Steol® CA-460; Sulfochem EA-1, EA-2, EA-3, EA-60, EA-70; Sulfotex OT;. Emulsifier for vinyl acetate copolymers, S/B latexes, vinyl chloride copolymers, acrylate homo-and copolymers. Surfactant; hair and skin detergents; braks up and hols oils and soil. Liquid; d = 1.02. Allchem Ind.; Ashland; Clariant; Great Western; Lonza; Pilot; Rhône-Poulenc Surf & Spec; Sealand; Stepan; Witco/Oleo-Surf.

2241 Ammonium lauroyl sarcosinate
68003-46-3 268-130-2
Hamposyl® AL-30; N-methyl-N-(1-oxododecyl) glycine, ammonium salt. Surfactant for shampoos, skin cleansers, bath gels; sec. emulsifier for emulsion polymerization. *W R Grace/Hampshire.*

2242 Ammonium lauryl sulfate
2235-54-3 218-793-9
$C_{12}H_{26}O_4S{\cdot}H_3N$
Sulfuric acid, monododecyl ester, ammonium salt
Akyposal ALS 33; Avirol ® A; Carsonol® ALS-R; DeSonol A; Lorol NH; Sulfochem ALS; Tensopol N; Texapon ALS; Ungerol AM3-75; Ufarol Am 30; Witcolate 6430; Witcolate AM; Witcolate NH; Zoharpon LAA; Maprofix NH, NHL; Marlinat® DFN 30; Neopon LAM; Nutrapon HA 3841; Nutrapon PP 3563; Octosol ALS-28; Perlankrol® DAF25; Polystep® B-7; Rhodapon® L-22, L-22/C; Sermul EA129; Standapol® A; Stepanol® AM;. Detergent, emulsifier, foaming agent, dispersant, wetting agent; for personal care products, carpet shampoos, firefighting. Anionic detergent; liquid. *Lonza; Sandoz; Stepan; Witco.*

2243 Ammonium lignosulfonate
8061-53-8
Ammonium lignosulfonate
Lignosol TS; Tembind A 002; Wanin AM. Wetting agent, emulsifier, dispersant, tanning extract, slurry water reducer and grinding aid in cement mfg. *Borregaard Ligno Tech; Temfibre.*

2244 Ammonium molybdate
12027-67-7 565 234-722-4
$H_{16}MoN_2O_8$
Hexaammonium molybdate
Ammonium paramolybate. In soil additives, enamel bonding agents, protective and decorative metal coatings, iron and steel alloys, lubricants, petroleum refining catalysts, pigments, corrosion inhibitors, smoke suppressants, production of molybdenum metal. Tetrahydrate; colorless or slightly greenish crystal; LD_{50} (rat orl) = 333 mg/kg. *AAA Molybdenum Prods.; Climax Molybdenum BV; Climax Performance.*

2245 Ammonium nitrate
6484-52-2 567 229-347-8
$H_4N_2O_3$
Ansax; Hero-Prills; Nitram; Old Plantation. Fertilizer, explosives, pyrotechnics, herbicides/insecticides, manufacture of nitrous oxide, absorbent for nitrogen oxides, ingredient of freezing mixtures, oxidizer in solid rocket propellants, nutrient for antibiotics and yeast, catalyst. Transparent, hygroscopic crystals or white granules; mp = 169°; dec at approx. 210° into H_2O and N_2O; d = 1.7250; soluble in H_2O (2 g/ml); pH of 0.1 M solution = 5.43. *Faith, Keyes and Clark; Air Prods; Chevron; La Roche Ind; Norsk Hydro A/S; Unocal.*

2246 Ammonium nonoxynol-4 sulfate
9051-57-4
$(C_2H_4O)_xC_{15}H_{24}O_4S.H_3N$
Ammonium salt of sulfated nonylphenoxy POE ethanol
Aerosol® NPES 458; Abex® EP-110; Polystep® B-1; Rhodapex® CO-415, CO-436; Sulfochem 436. High foaming surfactant for emulsion polymerization of acrylic, styrene and vinyl acetate systems, dishwashing

detergents, germicides, pesticides, general purpose cleaners and cosmetics. Pale yel. clear liq.; alcoholic odor; sol. In H_2O; part. sol. in org. solvents; m.w. 493; dens. 8.9 lb/gal; viscosity = 100 cps.; f.p.<0°; flash pt. (PMCC) 83F; pH 6.5-7.5; *Cytec Ind.; Rhône Poulec Surf & Spec.; Stephan; Chemron.*

2247 Ammonium oxalate

6009-70-7 571

$C_2H_{10}N_2O_5$

Ethanedioic acid diammonium salt monohydrate

Analytical chemistry, safety explosives, manufacture of oxalates, rust and scale removal. Orthohombic odorless crystals or granules; poisonous; mp = 70°; d = 1.50; soluble in H_2O (11.8 g/100 ml); *Brotherton Ltd; General Chem; Heico; Rhône-Poulenc.*

2248 Ammonium pentaborate

Trydil. A safe gel heavy duty hand cleanser *Borax Europe Ltd.*

2249 Ammonium persulfate

7727-54-0 575 231-786-5

$H_8N_2O_8S_2$

Ammonium Peroxydisulfate

Ammonium Peroxodisulfate; Peroxydisulfuric acid, diammonium salt; diammonium peroxodisulfate. Oxidizer, bleaching agent; photography; etchant for printed circuit boards, copper; electroplating; deodorizing oils; aniline dyes; food preservative; depolarizer in batteries; washing infected yeast; manufacture of other persulfates mp = 80° (dec); d = 1.9820; soluble in H_2O (800 g/l); LD_{50} (rat orl) = 689 mg/kg. *Aldrich; Degussa, EM Ind; FMC; Interox Chem Ltd.*

2250 Ammonium phosphate dibasic

7783-28-0 576 231-987-8

$H_9N_2O_4P$

diammonium hydrogen phosphate

secondary ammonium phosphate; Fyrex; Diammonium hydrogenphosphate; Diammonium phosphate; DAP; diammonium hydrogen orthophosphate; Phosphoric Acid, Diammonium Salt; Ammonium hydrogen phosphate. Used for fireproofing textlies, as a soldering flux and in dentrifices, corrosion inhibitors and fertilizers. Flame retardant for wood, paper, textiles fertilizer, plant nutrient sol'ns., feed additive; flux for soldering, purifying sugar; in ammoniacal dentifrices; manufacture of yeast, vinegar, bread improvers; foods, pharmaceuticals. d = 1.6190; soluble in H_2O (0.59 g/ml), insoluble in organic solvents. Albright & Wilson; Aldrich; Chisso; Heico; IMC Fertilizer; La Roche Ind; Monsanto; Oxychem; Rhône-Poulenc Basic.

2251 Ammonium phosphate monobasic

7722-76-1 577 231-764-5

H_6NO_4P

ammonium dihydrogen phosphate

monoammonium phosphate. Used in the manufacture of food products, fertilizer, flame retardants, plant nutrient solutions, manufacture of yeast, vinegar, yeast foods and bread improvers, food additive, analytical chemistry mp= 190°; d = 1.8030; soluble in H_2O (0.4 g/ml), less soluble in organic solvents. Albright & Wilson; Aldrich; Chisso; EniChem SpA; Heico; IMC Fertilizer; Monsanto; Oxychem; Rhône-Poulenc Basic; Showa Denko.

2252 Ammonium polyacrylate

9003-03-6

$(C_3H_4O_2)_x{\cdot}xH_3N$

2-propenoic acid, homopolymer, ammonium salt

Alcogum 9639; Poly(acrylic acid), ammonium salt. Dispersant for paints and coatings; thickening and stabilizing agent for synthetic latices; used in coatings, adhesives, dipped, cast, and molded goods, cements for rug backimg, spraying, spreading, brushing and extruding compounds.

2253 Ammonium stearate

1002-89-7 588 213-695-2

$C_{18}H_{39}NO_2$

Octadecanoic acid, ammonium salt

Ammonium salt of stearic acid; vanishing creams, brushless shaving creams, other cosmetic products, waterproofing of cements, concrete, stucco, paper, textiles. mp = 38-42°; soluble in H_2O, organic solvents except Me_2CO, CCl_4. *Magnesia GmbH; Original Bradforn Soap Works.*

2254 Ammonium sulfamate

7773-06-0 589 231-871-7

$H_6N_2O_3S$

Sulfamic acid, ammonium salt

Necco Fire Retardant 2750, 2578, 2762; Amcide; ammate; Sulfamic Acid, Monoammonium Salt; Ammonium Amidosulfate; Amicide; AMS; ammonia sulfamate; Sepimate; Silvacide; Ammate X; Ammate X-NI; monoammonium sulfamate; Amcide; Ikurin. Flameproofing agent for textiles and paper; weed and brush killer; electroplating; generation of nitrous oxide. mp = 132-135°; bp = 160°; soluble in H_2O (225g/100 ml); pH=5.2; LD_{50} (rat orl) = 2000 mg/kg. *Heico; Nissan Chem Ind; Spartan Flame Retardants.*

2255 Ammonium sulfate

7783-20-2 590 231-984-1

$H_8N_2O_4S$

Sulfuric acid diammonium

Diammonium sulfate; Sulfuric acid, diammonium salt; Ammonium sulfate (2:1); Sulfate, Ammonium. Fertilizers, water treatment, fermentation, fireproofing compositions, viscose rayon, tanning, food additive Orthohombic crystals or white granules; d = 1.7690; mp = 280° (dec); soluble in H_2O (77 g/100 ml); insoluble in alcohol, acetone; pH of 0.1 aqueous solution = 5.5; LD_{50} (rat orl) = 2840 mg/kg. Acurate Chem & Scientific; Aldrich; AlliedSignal; BASF; DSM NV; General Chem; Heico; Nissan Chem Ind; Schaefer Salt & Chem; Showa Denko.

2256 Ammonium thiocyanate

1762-95-4 597 217-175-6

CH_4N_2S

Ammonium rhodanide

Thiocyanic acid, ammonium salt; Ammonium sulfocyanate; Ammonium Sulfocyanide; Ammonium Rhodantate; Ammonium Rhodonide; Ammonium Rhodanide; ATC; Trans-aid. Analytical chemistry; thiourea; fertilizers; photography; in liquid rocket propellants; fabric dyeing; zinc coating; weed killer, defoliant; adhesives; curing resins; pickling iron and steel; electroplating; polymerization catalyst; metals separation. mp = 149°; d = 1.3050; soluble in H_2O (163 g/100 ml); LD_{50} (rat orl) = 750 mg/kg. *Carbo-Tech GmbH; Degussa; Witco/Argus.*

2257 Ammonium thiosulfate

7783-18-8 598 231-982-0

$H_8N_2O_3S_2$

Ammonium hyposulfite

Thiosulfuric acid, diammonium salt; diammonium thiosulfate; thio-sul; ammo hypo; Amthio; Thiosulfuric acid ($H_2S_2O_3$), diammonium salt. Photographic fixing agent; analytical reagent; fungicide; reducing agent; brightener in silver plating baths; cleaning compounds for zinc-base die-cast metals; hair waving preparations; fog screens. mp = 150° (dec); soluble in H_2O (64 g/100 ml), insoluble in alcohol, ether; LD_{50} (rat orl) = 2890 mg/kg. *Blythe, Willam Ltd; Du Pont; General Chem.*

2258 Ammonium tungstate

11120-25-5 600 234-364-9

$H_8N_2O_4W$

ammonium wolframate

ammonium paratungstate; Tungstate, decaammonium; Ammonium tungstate pentahydrate. Preparation of ammonium phosphotungstate and tungsten alloys. mw=283.93; freely soluble in H_2O. *Aldrich; C;imax Molybdenum.*

2259 Ammonium xylene sulfonate

26447-10-9 247-710-9

$C_8H_{13}NO_3S$

Benzenesulfonic acid, dimethyl-, ammonium salt

Eltesol® AX 40; Hartotrope AXS; Naxonate® 4AX; Stepanate® AXS. Hydrotrope, cloud point depressant used in the detergent manufacture; solubilizer, coupler. *Albright & Wilson UK; Hart Chem. Ltd; Ruetgers-Nease; Stepan; Stepan Canada.*

2260 Ammonyx® 4, 4B, 485, 4002

122-19-0 204-527-9

Stearalkonium chloride

Emulsifier, conditioner, softener, emollient for cosmetics. *Stepan; Stepan Canada.*

2261 Ammonyx® CETAC, CETAC-30

112-02-7 203-928-6

Cetrimonium chloride

Emulsifier, conditioner, softener, emollient for cosmetics. *Stepan; Stepan Canada.*

2262 Ammonyx® CO

7128-91-8 230-429-0

palmitamine oxide

Conditioner, detergent, foam stabilizer, viscosity builder used in cosmetic, household, and janitorial products; wetting agent in concentrated electrolyte solutions; textile lubricant, emulsifier, wetter, dye dispersant. *Stepan; Stepan Canada.*

2263 Ammonyx® KP

37139-99-4 253-363-4

Olealkonium chloride

Conditioner, antistat in clear hair rinses. *Stepan; Stepan Canada.*

2264 Ammonyx® LO

1643-20-5 216-700-6

$C_{14}H_{31}NO$

Lauramine oxide

N,N-dimethyldodecylamine-N-oxide; 1-Dodecanamine, N,N-dimethyl-, N-oxide; Dimethyldodecylamine oxide; dimethyldodecylamine-N-oxide; Ammonyx LO; Ammonyx AO; Aromox DMMC-W; Conco XAL; DDNO; N,N-dimethyldodecylamine oxide; dodecycldimethylamine oxide; n-dodecyldimethylamine oxide; lauryldimethylamine oxide; Lauramine oxide. Foamer/foam stabilizer, wetting agent, visc. builder, grease emulsifier for shampoos, bath products, fine fabric cleaners, hard surface cleaners containing acids or bleach, dishwash, shaving creams, lotions; textile lubricant, emulsifier, dye dispersant. *Stepan; Stepan Canada.*

2265 Ammonyx® OAO
14351-50-9 238-311-0
Oleamine oxide
Wetting agent, foam booster/stabilizer, conditioner, viscosity builder for shampoos, bubble baths, hand soaps, conditioners. *Stepan; Stepan Canada.*

2266 Amo Vitrax®
9067-32-7 4793
Hyaluronic Acid Sodium
Surgical aid. *Allergan Inc.*

2267 Amo® Balanced Salt Solution
Balanced salt solution; sterile intraocular surgical irrigation. *Allergan Inc.*

2268 Amoco® 1012
9003-07-0 7741
polypropylene
PP homopolymer resin; extrusion grade resin *Amoco Chemical Co.*

2269 Amoco® 1016
9003-07-0 7741
polypropylene
PP homopolymer resin; general purpose injection molding grade; FDA compliant. *Amoco Chemical Co.*

2270 Amoco® 1246
9003-07-0 7741
polypropylene
PP homopolymer resin; LTHA injection molding grade ; FDA compliant. *Amoco Chemical Co.*

2271 Amoco® 4018
9003-07-0 7741
polypropylene
PP homopolymer resin; general-purpose injection molding/ extrusion grade. *Amoco Chemical Co.*

2272 Amoco® 5016
9003-07-0 7741
polypropylene
PP homopolymer fiber resin; for fiber/film application. *Amoco Chemical Co.*

2273 Amoco® 6114
9003-07-0 7741
polypropylene
PP homopolymer film resin; for oriented film application ; FDA compliant. *Amoco Chemical Co.*

2274 Amoco® 6400p
9003-07-0 7741
polypropylene
PP homopolymer resin; powder grade resin for general purpose applications. *Amoco Chemical Co.*

2275 Amoco® 7234
9003-07-0 7741
polypropylene
PP homopolymer resin; nucleated, antistat; install ; injection molding grade ; FDA compliance. *Amoco Chemical Co.*

2276 Amoco® 7239
9003-07-0 7741
polypropylene
PP homopolymer; nucleated, antistat; injection molding grade with improved clarity, high rigidity and tensile strength properties for packaging, disposable medical , houseware products; FDA compliant. *Amoco Chemical Co.*

2277 Amoco® 7728
9003-07-0 7741
polypropylene
PP homopolymer resin; general-purpose radiation-stable grade. *Amoco Chemical Co.*

2278 Amoco® 8244
9010-79-1
Ethylene/ propylene copolymer; general purpose grade suitable for extrusion blow molding application; improved toughness over Amoco 8217; FDA compliance. *Amoco Chemical Co.*

2279 Amoco® 8410
9010-79-1
Amoco ® 8244. Ethylene/ propylene copolymer; slip and antiblock; for cast film application ; FDA compliant. *Amoco Chemical Co.*

2280 Amoco® 9119
9003-07-0 7741
polypropylene
PP; enhanced grade offering high stiffness, higher teat deflection temperatures, good gloss on finished parts, improved processability; for sheet, extrusion and thermoforming. *Amoco Chemical Co.* Name unverified.

2281 Amoco® BR-310
9003-28-5
Polybutene/polyolefin blend; acts as a moisture barrier and corrosion inhibitor when used to flood the area between the layers of composite metal-plastic sheath in electrical cables; its adhesive quality prevents slippage between the layers. *Amoco Chemical Co.*

2282 Amoco® CI-500
9003-28-5
Polybutene/polyolefin blend; cable filling compound acting as a moisture barrier and corrosion inhibitor in cable core containing paired wires; also to fill the interstitial space in the core. *Amoco Chemical Co.*

2283 Amoco® H-15
9003-28-5
Isobutylene butene copolymer; used as tackifier, strengthener, and extender in adhesives, as plasticizer for rubber , as vehicle and fugitive binder for coatings, as cling additive for LLDPE stretch wrap films, as reactive intermediate for speciality chemicals; as leather impregnant, as vehicle or modifier for caulks, sealants and glazing compounds, and in lubricants, paper treatments and electrical-use compounds. FDA compliant. *Amoco Chemical Co.*

2284 Amoco® H2R
9003-53-6 9028
HIPS; impact-grade resin for use in food containers, plates, trays, thermoformed panels. *Amoco Chemical Co.*

2285 Amoco® H3E
9003-53-6 9028
HIPS; impact-grade resin for use in thin-wall containers for food and dairy. *Amoco Chemical Co.*

2286 Amoco® L-14
9003-28-5
Polybutene
Used as tackifier, strengthener, and extender in adhesives, as plasticizer for rubber, as vehicle and fugitive binder for coating as cling additive for LLDPE stretch wrap films, as reactive intermediate for specialty chemicals as leather impregnant, as vehicle or modifier for caulks, sealants and glazing compounds, in lubricants, paper treatment and electical compounds. FDA compliant. *Amoco Chemical Co.*

2287 Amoco® PIA
1,3-benzene dicarboxylic acid
isophthalic acid. Purified isophthalic acid; high purity product reacted to form esters, amides, salts, acid chlorides and other organic intermediates; In preparation of low color saturated and unsaturated polyesters, engineering resins, PET Copolymers; derivatives used in thermoset composites, paints, coatings, bottle and fiber resins and adhesives. *Amoco Chemical Co.*

2288 Amoco® R1
9003-53-6 9028
Crystal PS resin; for use in foam sheet, food containers, meat and produce trays, jackets for thin-wall glass bottles. *Amoco Chemical Co.*

2289 Amoco® R5
9003-53-6 9028
Polystyrene
Crystal PS resin; for use in injection blow-molded containers. *Amoco Chemical Co.*

2290 Amodel® A-1115HS, A-1145HS
Polyphthalamide
Glassreinforced; semicrystalline thermoplastic resin with outstanding dimensional stability and processing characteristics, high strength and stiffness, high strength and stiffness, high thermal properties, excellent chemical resistance for automotive under-the-hood parts, gears, bearings and industrial parts. *Amoco Chemical Co.*

2291 Amodel® A-1340HS
Polyphthalamide
Mineral/glass-reinforced; thermoplastic resin with lower warp than glass-reinforced grades, higher strength and stiffness than mineral-reinforced grades; used for small engine components, power tools, ignition components. *Amoco Chemical Co.*

2292 Amodel® AF-1115VO, AF-1133VO, AF-1145VO
Polyphthalamide
Glass-reinforced; flame-retarded: semicrystalline thermoplastic resin with high strength and stiffness; used for electrical components, e.g., connectors switches, sockets, sockets, circuit breakers. *Amoco Chemical Co.*

2293 Amodel® ET-1000
Polyphthalamide
Impact-modified; semicrystalline, thermoplastic resin with high toughness and impact, high strength and stiffness, low moisture sensitivity; used for power tools, recreational equipment, clips and fasteners, caster wheels. *Amoco Chemical Co.*

2294 Amollan® A
Oxyethylated fatty amine
Wetting and emulsifying agent for bating and degreasing leather. *BASF AG.*

2295 Amollan® L
Organic esters and fatty acids; leveling agent for application of pigment finishes. *BASF AG.*

2296 Amoloid HV, LV
9005-34-9
Ammonium alginates
Used for textile printing, ceramic binding, can sealant. *Kelco.*

2297 Amonyl 380 BA
Cocamidopropyl betaine
Detergent for shampoos. *Seppic.* Discontinued.

2298 Amonyl 675 SB
68139-30-0 268-761-3
Cocamidopropythydroxysultaine
Surfactant for shampoos. *Seppic.* Discontinued.

2299 Amonyl DM
Quaternium-82
Cosmetic ingredient. *Seppic.* Discontinued.

2300 AMP
124-68-5 469 204-709-8
2-amino-2-methyl-1-propanol
Emulsifier, catalyst; dispersant for pigments and latex paints; corrosion inhibitor; stabilizer; resin solubilizer. *Angus.*

2301 AMP
61-19-8 157 200-500-0
Adenosine monophosphate
Used in biochemical research.

2302 AMP-95
124-68-5 469 204-709-8
2-amino-2-methyl-1-propanol
Emulsifier, catalyst; dispersant for pigments and latex paints corrosion inhibitor; stabilizer; resin solubilizer *Angus.*

2303 Ampco
An aluminum bronze containing from 86-92% copper, 7-11% aluminum, and 1.3% iron. No manufacturer.

2304 AMPD
115-69-5 468 204-709-8
2-Amino-2-ethyl-1,3-propanediol
Pigment dispersant, neutralizing amine, corrosion inhibitor, acid-salt catalyst, pH buffer, chemical and pharmaceutical intermediate; solubilizer or emulsifier system component in personal care products. *Angus.*

2305 Amphenol
A proprietary trade name for polystyrene products. No manufacturer.

2306 Amphionic
Ampholytic surfactant. *Rhône-Poulenc UK.*

2307 Amphionic 25B
High molecular weight amino-acid derivative, supplied as a golden liquid; alkaline cleaning and sanitizing formulations and biocidal soaps; an efficient biocide with a broad spectrum of kill; good stability in the presence of electrolytes; compatibility with other types of surface agent; dispersant. *Rhône-Poulenc UK.*

2308 Amphisol
69331-39-1 273-968-7
DEA-cetyl phosphate
Acid pH emulsifier. *Bernel.*

2309 Amphisol K
potassium cetyl phosphate
Emulsifier; stable over wide pH range. *Bernel.*

2310 Amphobac
Bactericidal amphoterics; used in shampoos and industrial cleaners. *Lonza AG.*

2311 Amphocerin
Mixture of higher molecular fatty alcohol and wax esters; water-in-oil type ointments and creams with good spreading properties. *Henkel Chemicals Ltd.* Name unverified.

2312 Amphocerin K
Cetearyl alcohol
Hydrogenated peanut oil, vegetable oil, mineral oil, petrolatum; cream base for manufacturing of light and smooth creams and ointments of the water-oil type. *Henkel/Cospha; HenkelKGaA.*

2313 Ampholak 7TX
97659-53-5 307-458-3
Tallowamphopolycarboxyglycinate
Medium foaming detergent used in detergent application, nonirritating irritation of anionics; softening agent. *Berol Nobel AB.*

2314 Ampholak 7TX/C
97659-53-5 307-458-3
Stearylamphopolycarboxyglycinate
Used in detergents, shampoos, liquid soaps; conditioner in shampoos; softener; reduces irritation of anionics. *Berol Nobel AB.*

2315 Ampholak 7TX-SD 55
97659-53-5 307-458-3
Tallowamphopolycarboxyglycinate
Detergent, softener for laundry and hard surface cleaners; also for cosmetics. *Berol Nobel AB.*

2316 Ampholak 7TX-T
97659-53-5 307-458-3
Tallowamphopolycarboxyglycinate
Detergent, softener for detergent applications, cosmetics, shampoos, liquid soaps; reduces irritation of anionics; conditioner in shampoos. *Berol Nobel AB.*

2317 Ampholak 7TY
97488-62-5 306-998-7
Tallowamphopolycarboxypropionic acid
Low foaming surfactant for alkaline cleaners, toiletries. *Berol Nobel AB.*

2318 Ampholak CCA
Complex N-alkylamino propionic acid and alkyldimethylbenzyl ammonium chloride; corrosion inhibitor at all pH values against corrosion by H_2 and CO_2 in presence of salt water; algicide, bactericide. *Berol Nobel AB.*

2319 Ampholak MDX-1
Blend of amphoteric surfactants; designed for mild washing up liquids; reduces irritation to skin from anionics; bacteriostatic properties, preservative for formulations. *Berol Nobel AB.*

2320 Ampholak XCE
97659-51-3 307-456-2
Cocoiminodiglycinate
Medium foaming surfactant, hydrotrope, detergent for industrial applications in strong alkaline solutions. *Berol Nobel AB.*

2321 Ampholak XCO-30
68608-65-1 271-793-0
Sodium cocoamphoacetate
Medium foaming surfactant for toiletries, nonirritating shampoos, acid hard surface cleaners. *Berol Nobel AB.*

2322 Ampholak XJO
68608-64-0 371-792-5
Disodium capryloamphodiacetate
Low foaming wetting agent, hydrotrope for high alkaline industrial hard surface cleaners. *Berol Nobel AB.*

2323 Ampholak XO7
97659-53-5 307-458-3
Oleoamphocarboxyglycinate
Medium foaming, multipurpose cleaner component; for nonirritating toiletries, conditioners, liquid soap, as softener. *Berol Nobel AB.*

2324 Ampholak XOO-30P
70024-77-0 274-267-9
Oleoamphocarboxyglycinate
Surfactant for laundry detergents, hard surface cleaners where high viscosity is desirable. *Berol Nobel AB.*

2325 Ampholak XTP
N -Tallowamido-polyamino-polygincate
For detergent applications, cosmetics, shampoos; liquid soaps; conditioner in shampoos; softergent formulations; anti-irritant for anionics.

2326 Ampholak YCA/P
91995-05-0 295-264-9
Cocoiminodipropionate half sodium salt
Cocoiminodipropionate half sodium salt; medium to high foaming surfactant for alkaline cleaners; high stability to alkali. *Berol Nobel AB.*

2327 Ampholak YCE
97659-50-2 307-455-7
Cocoiminodipropionate
Medium foaming surfactant for industrial alkaline cleaners, cosmetic preprations; hydrotrope *Berol Nobel AB.*

2328 Ampholak YCO-40
68919-40-4 272-897-9
Cocoamphocarboxypropionate acid
Mild, stable surfactant for highly alkaline systems, hard surface cleaners, industrial laundry detergents. *Berol Nobel AB.*

2329 Ampholak YJH-40
94441-92-6 305-318-6
Octyl dipropionate
Surfactant for strong alkali cleaners where low foam is required; high caustic stability. *Berol Nobel AB.*

2330 Ampholan B171
Cocoamido propyl betaine in liquid form; foaming and wetting agent, for toiletries, industrial cleaners, cement, gypsum & latex. *Henkel Europe.* Name unverified.

2331 Ampholan®
Complex amphoteric surfactants in high foam formulations; used for froth flotation, fire fighting, toiletries. *Harcros.*

2332 Ampholyt JA 140
Sodium lauroamphoacetate
Mild surfactant for cosmetics, shampoos, detergents, baby care products. *Hüls Am; Hüls AG.*

2333 Ampholyt JB 130
Cocamidopropyl betaine
Surfactant for cosmetics, shampoos, detergents, hair shampoos, foam baths, shower foams, liquid soaps. *Hüls Am; Hüls AG.*

2334 Ampholyte KKDP-60
84812-94-2 284-219-9
Cocaminopropionic acid
Emulsifier, dispersant, corrosion inhibitor. *Berol Nobel AB.*

2335 Ampholyte KKE-70
84812-94-2 284-219-9
Coco alkyl aminopropionic acid
Surfactant for detergents, toiletries, emulsifier, dispersant, corrosion inhibitor *Berol Nobel AB.*

2336 Ampholyte SKKP 70
Amphoteric surfactant with dispersing and corrosion inhibiting properties; for emulsion paints, pigment grinding. *Keno Gard (UK) Ltd.* No manufacturer.

2337 Amphomer® LV-71
Octyl acrylamide/acrylates/butylaminoethyl methacrylate copolymer; hair fixative resin enhancing stiffness, holding and moisture resist. *Nat'l. Starch.*

2338 Amphoram
Alkyl amino acids based on coco and tallow alkyl chains; amphoteric surface active agent used in cosmetics, detergents, paint and pigment industries. *Elf Atochem UK/Ceca.*

2339 Amphosol CA
Derivative of alkyl amido propyl N-dimethyl amino acetic acid, as a clear yellow liquid; for shampoos and bubble baths. *KWR Chemicals Ltd.* Name unverified.

2340 Amphosol DM and DMA
Acetyldimethyl alkylammonium chloride
Sodium salt or in acid form; clear yellow liquid; for bacterial detergent preparation. *KWR Chemicals Ltd.* Name unverified.

2341 Amphosol® CA
Cocamidopropyl betaine
Mild conditioner, detergent, wetting agent, viscosity builder, foam enhancer, base for cosmetics and household and industrial liquid detergents. *KWR Chemicals Ltd.*

2342 Amphosol® CB3
C8-18 alkylamido betaine
Detergent, foaming and wetting agent for household and industrial cleaners. *Stepan Europe.*

2343 Amphoteen 24
66455-29-6 266-368-1
Lauryl betaine
Surfactant for low-irritation shampoos, washing-up liquids, hard surface cleaners, vehicle cleaners. *Berol Nobel AB.*

2344 Amphoteen BCA-30
70851-07-9 274-923-4
Cocoamidopropyl betaine
Foam enhancer, viscosity builder, thickener, mild surfactant for liquid soaps and washing-up liquids. *Berol Nobel AB.*

2345 Amphoteen BCM-30
68424-94-2 270-329-4
Cocobetaine
Surfactant for low irritation shampoos and dishwashing liquids. *Berol Nobel AB.*

2346 Amphotensid 9M
Disodium cocoamphodiacetate and sodium laureth sulfate; detergent for personal care products. *Zschimmer & Schwarz.* Discontinued.

2347 Amphotensid B4 F
61789-40-0 263-058-8
Cocamidopropyl betaine
Surfactant for cosmetics, shampoos, detergents. *Zschimmer & Schwarz.*

2348 Amphoterge®
Substituted imidazoline amphoterics; used in shampoos and industrial cleaners. *Lonza AG.*

2349 Amphoterge® J-2
Disodium capryloamphodiacetate
Wetting agent and detergent for personal care and industrial applications. *Lonza AG.*

2350 Amphoterge® K
68919-41-5
Sodium cocoamphopropionate
Detergents used in shampoos, skin cleansers, dishwashing; salt-free. *Lonza Inc.*

2351 Amphoterge® K-2
Disodium cocoamphodipropionate
Detergents used in shampoos, skin cleansers, dishwashing, heavy duty liquid cleaners. *Lonza AG.*

2352 Amphoterge® KJ-2
68815-55-4 272-383-4
Disodium capryloamphodipropionate
Salt-free version of Amphoterge J-2; wetting agent, detergent for personal care and industrial applications. *Lonza Inc.*

2353 Amphoterge® L Special
Disodium lauroamphodiacetate
Mild shampoo concentrate. *Lonza AG.*

2354 Amphoterge® NX
Coco imidazoline dicarboxylate
Industrial detergent. *Lonza AG.*

2355 Amphoterge® W
Sodium cocoamphoacetate
Surfactant for mild shampoos, skin cleansers, heave duty cleaners, dishwashing preps. *Lonza AG.*

2356 Amphoterge® W-2
Disodium cocoamphodiacetate
Surfactant for nonirritating shampoos and skin cleansers heavy duty liquid cleaners. *Lonza AG.*

2357 amphoteric 300
Sodium eicosyloxypropyliminodipropionate
General surfactant. *Exxon/Tomah.*

2358 Amphoteric 400
Iminopropionate
Partial sodium salt; low foam detergent, coupler for hard surface alkaline or acid detergents, laundry, metal, acid bowl cleaners; defoamer in latex paints; corrosion inhibitor in metalworking lubricants; leather lubricant; stable in acid, alkali and concentrated electrolytes. *Exxon/Tomah.*

2359 Amphoteric L
Coco derivative; detergent, foam stabilizer/booster, wetting agent, mild surfactant for liquid detergents, shampoos, hand soaps, mechanical, foaming systems, dishwash; stable in mildly acid and alkaline media. *Exxon/Tomah.*

2360 Amphoteric N
Sodium C12-15 alkoxypropyl iminodipropionate
High foam wetting agent, coupler for shampoos, detergents; corrosion inhibitor in metalworking lubricants; viscosity builder; fire fighting foams. *Exxon/Tomah.*

2361 Ampicillin
69-53-4 628 200-709-7
$C_{16}H_{19}N_3O_4S$
6-[(Aminophenylacetyl)amino]-3,3-dimethyl-7-oxa-4-thia-1-azabicyclo[3.2.0]heptane-2-carboxylic acid
Acillin; 6-[D(-)-α-aminophenylacetamido] penicillanic acid; D(-)-α-aminobenzylpenicillin; ampicillin A; Ay 6108; BRL 1341; P 50; Adobacillin; Alpen; Amblosin, Amfipen; Amipenix S; Ampi-Bol; Ampicin; Ampicina; Ampilar; Ampimed; Ampipenin; Ampi-Tablinen; Amplisom; Amplital; Ampy-Penyl; Austrapen; Binotal; Bonapicillin; Britacil; Copharcilin; Doktacillin; Grampenil; Guicitrina; Marisilan; Nuvapen; Pen-Bristol; Penbritin; Penbrock; Pénicline; Penstabil; Pentrex; Pentrexyl; Polycillin; Ponecil; Ql Damp; Rosampline, Synpenin; Tokiocillin; Totacillin; Totalciclina; Totapen; Ultrabion; Viccilin. antibacterial. Anhydrous form, mp = 199-202° (dec); $[\alpha]_D^{23}$ = 287.9° (H_2O), soluble in H_2O, DMSO *ICN Nutritional Biochemicals Corp.*

2362 Amplex
1406-65-1 2207 215-800-7
$C_55H_{72}MgN_4O_5$
chlorophyll. The green pigments of plants and green algae contain chlorophyl a and chlorophyl b in a ratio of aprox 3:1; A proprietary preparation of chlorophyll; a deodorant. Used to color soaps, oils, fats, perfumes and sensitizing color film. Source of phytol. soluble in oil *Ashe Chemicals.* Name unverified.

2363 AmpliWax PCR Gems
Specially formulated wax beads to enhance PCR process; replaces mineral oil as vapor barrier in PCR amplifications. *Perkin-Elmer.*

2364 Amprol
121-25-5 631 204-458-4

Amprolium
Coccidiostat. *Merck & Co Inc.* Name unverified.

2365 **Amprolium hydrochloride**
121-25-5 631 204-458-4
$C_{14}H_{19}ClN_4.HCl$
1-[(4-amino-2-propyl-5-pyrimidinyl)methyl]-2-picolinium chloride
Amprol; Corid; Pancoxin. Coccidiostat in veterinary medicine. Soluble in H_2O, methanol, ethanol, dimethylformamide; insol.in isopropanol, butanol, dioxane, acetone, ethyl acetate, acetonitrile, isooctane; pH 2.5-3.0. Name unverified.

2366 **Amron**
A proprietary vinylite base plastic for coatings. No manufacturer.

2367 **Amsco Steel**
A proprietary high manganese steel containing 12-13% manganese and 1.2% carbon. No manufacturer.

2368 **Amsil**
Oxygen free copper plus 8 oz to 30 oz per ton silver; conductivity 100% IACS. principal uses are based on its good creep strength at elevated temperatures and its high softening point. *Amax Inc.* Name unverified.

2369 **Amsoft FA**
Fatty ester
Nonyellowing softener for cotton, knits, yarn and fabric lubricant. *Am. Emulsions.*

2370 **Amsoft MDH-20**
9002-88-4 7728
Polyethylene emulsion; general-purpose nonyellowing hand modifier for improved abrasion and sewability. *Am. Emulsions.*

2371 **Amsol GMS**
Surfactant blend; antigelling agent for concentrated dye solutions. *Am. Emulsions.*

2372 **Amsperse 109**
Dispersing assistant for disperse dyes; prevents agglomeration; compatibilizes dye systems. *Am. Emulsions.*

2373 **Amsulf**
Sulfur copper alloy containing oxygen-free copper and 0.3% sulfur; conductivity 96% IACS. *Amax Inc.* Name unverified.

2374 **Amtel**
Tellurium-copper alloy containing oxygen-free copper and 0.5% tellurium; Conductivity 93% IACS. *Amax Inc.* Name unverified.

2375 **Amterge TC**
Blend of organic solvents, detergents, emulsifiers; scouring and cleaning aid. *Am. Emulsions.*

2376 **Amvis**
An explosive containing ammonium nitrate. *Am. Emulsions.* No manufacturer.

2377 **Amwet DAD**
Ethoxylated alcohol
Detergent, nonrewetting wetting agent, penetrant. *Am. Emulsions.*

2378 **Amwet DOSS**
Dioctyl sulfosuccinate
Fast wetting agent for synthetics, cotton and blends; excellent for space dyeing. *Am. Emulsions.*

2379 **Amwet MS-100**
Solvent-based penetrant, scour, wetting agent for bleaching, alkaline scouring and dyeing operations. *Am. Emulsions.*

2380 **Amyl acetate**
628-63-7 211-047-3
$C_7H_{14}O_2$
n-Amyl acetate
Amylacetic ester; n-Pentyl acetate; 1-Pentanol acetate; Acetic acid amyl ester; Acetic acid pentyl ester; Pentyl acetate; banana oil; pear oil. Solvent for lacquers and paints, extraction of penicillin, photographic film, leather and nail polishes, flavoring agent, printing and finishing fabrics, solvent for phosphors in fluorescent lamps. Colorless, clear liquid; soluble in alchol, ether; slightly soluble in H_2O; d = 0.862-0.866; mp = -0.8°; bp = 149°; flash pt=25°; LD_{50} (rat orl) = 6500 mg/kg. *Aldrich; BP Chem Ltd; Penta Mfg; Pentagon Chemicals Ltd; Union Carbide.*

2381 **Amyl alcohol**
71-41-0 7257 200-752-1
$C_5H_{12}O$
1-Pentanol
2-methyl-1-butanol; 2-pentanol; n-Pentanol; n-Amyl Alcohol; Pentyl alcohol; 1-Pentanol; n-pentyl alcohol; 1-Pentol; Pentan-1-ol; n-butylcarbinol; pentanol-1; pentasol; primary amyl alcohol; primary-n-amyl alcohol; amyl alcohol, normal; n-pentan-1-ol. Eight isomers possible; raw material for pharmaceutical preparation, organic synthesis solvent; flotation agent, organic synthesis, medicine (sedative). mp= -78°; bp = 136-138°; d = 0.8110; n_D^{20}= 1.4093; LD_{50} (rat orl) = 2200 mg/kg. *Ashland; Hoechst-Celanese; MTM Spec. Ltd; Union Carbide; Vista.*

2382 **Amyl mustard oils**
$C_6H_{11}S$.
Amyl thiocarbimides
Name unverified.

2383 **Amyl-m-cresol**
53043-14-4 648
$C_{12}H_{18}O$
6-n-amyl-*m*-cresol
5-methyl-2-pentylphenol. Antiseptic, germicide and mold preventative. mp = 24°; bp_{15} = 137-139°; insoluble in H_2O, soluble in organic solvents.

2384 **Amylase**
Diastase, an enzyme which renders starch soluble, by converting it into maltose.

2385 **Amylit**
A diamalt compound, which is an enzymic product; used for desizing in the textile industry. No manufacturer.

2386 **Amylogen**
A soluble starch. No manufacturer.

2387 **amyloid**
Concentrated sulfuric acid dissolves cellulose, gradually converting it into dextrin, and ultimately into dextrose; if the solution, as soon as it is made, is diluted with water, a gelatinous hydrate is produced ; this substance is known as amyloid.

2388 **Amylopsin**
Pancreatic diastase. No manufacturer.

2389 **Amylozyme**
Amylase starch converting enzyme. *Rhône-Poulenc UK.*

2390 **Amylum**
$(C_{12}H_{20}O_{10})_n$.
Starch
No manufacturer.

2391 **Amyx A-25-S 0040**
122-19-0 204-527-9
Stearalkonium chloride
Conditioner, softener, and emollient for hair rinses, skin creams and lotions; emulsifier. *Clough.*

2392 **Amyx CDO 3599**
68155-09-9 268-938-5
Cocamidopropylamine oxide
Mild high foaming surfactant, foam booster/stabilizer, wetting agent, hair conditioner for personal care, household, and janitorial products. *Clough.*

2393 **Amyx CO 3764**
7128-91-8 230-429-0
Palmitamine oxide
Mild high foaming surfactant, foam booster/stabilizer, conditioner for personal care, household, and janitorial products. *Clough.*

2394 **Amyx LO 3594**
Lauramine oxide
Mild high foaming surfactant, foam booster/stabilizer, wetting agent, grease emulsifier for personal care, household and janitorial products. *Clough.*

2395 **Amyx SO 3734**
2571-88-2 219-919-5
Stearamine oxide
Mild high foaming surfactant, foam booster/stabilizer, conditioner, emulsifier for personal care, household, and janitorial products. *Clough.*

2396 **Amyx ST 3837**
Cetearyl alcohol
PEG-40 hydrogenated castor oil; Stearalkonium chloride. Concentrate for preparation of hair conditioners. *Clough.*

2397 **Amzirc**
Zirconium copper alloys containing oxygen-free copper and 0.13-0.20% zirconium. *Amax Inc.* Name unverified.

2398 **AN-30GF**
Amorphous nylon (modified nylon 12), 30% glass fiber-reinforced; exhibits low moisture absorption and good dimensional stability, improved mechanical, thermal and chemical props. *Compounding Tech.*

2399 **Anadoucissant 88210 T**
Softener for polyamide and cellulose fibers; compatible with electrostatic processes. *Ceca SA.*

2400 **Analar**
Laboratory reagents and chemicals. *British Drug Houses.*

2401 **Analoam**
Soil testing reagent. *Murphy Chemical Co Ltd.* Discontinued.

2402 **Anasite**
An explosive consisting of ammonium perchlorate and myrabolans, usually

with some sodium or potassium nitrate, and a small quantity of agar-agar. No manufacturer.

2403 Anatola
68-26-8 10150 200-683-7
Vitamin A
Antixerophthalmic. *Parke-Davis.* Name unverified.

2404 anatomical alloy
An alloy of 53.5% bismuth, 19% tin, 17% lead, and 10.5% mercury.

2405 Ancaflex
A proprietary trade name for a series of boron trifluoride-based polymers; used as curing agents for epoxy resins to give flexible products. *Pacific Anchor Chemical Corp.* Name unverified.

2406 Ancaflex 70, 150
A proprietary range of polymeric hardeners based on boron trifluoride. *Pacific Anchor Chemical Corp.* Name unverified.

2407 Ancamide 280, 400
A proprietary trade name for a fluid complex polyamide; used as a curing agent for epoxy resins. *Anchor Chemical (UK) Ltd.*

2408 Ancamine LO
An epoxy hardener for use at low temperatures; free from phenolic odor and processing a low irritation index. *Anchor Chemical (UK) Ltd.*

2409 Ancamine LT
A proprietary trade name for an activated aromatic amine; curing agent for epoxy resins. *Anchor Chemical (UK) Ltd.*

2410 Ancamine MCA
Modified cyclo-aliphatic polyamine; used in floorings and coatings. *Anchor Chemical (UK) Ltd.*

2411 Ancaris
4304-40-9 9414 224-318-6
thenium closylate
Thenium closylate; used as an anthelmintic. *The Wellcome Foundation Ltd.* No manufacturer.

2412 Ancatax
Dibenzthiazyl disulfide
A proprietary rubber accelerator. No manufacturer.

2413 Ancazate BU
zinc butyl dithiocarbamate
A self-dispersible zinc butyl dithiocarbamate; used as an accelerator for vulcanization and an antioxidant for rubbers. *Anchor Chemical Group plc.*

2414 Ancazate EPH
Zinc ethyl phenyl dithiocarbamate
Used as an accelerator for vulcanization and an antioxidant for rubbers. *Anchor Chemical Group plc.*

2415 Ancazate ET
14324-55-1 238-270-9
$C_{10}H_{20}N_2S_4Zn$
Zinc diethyl dithiocarbamate
bis(diethylcarbamodithioato-S,S') Zinc; Zinc diethyldithiocarbamate; Ethyl zimate; Zinc, bis(diethylcarbamodithioato-S,S')-, (T-4)-; Bis(diethyldithiocarbamate)zinc complex; Etazin; Ethasan; Ethazate; Nocceler EZ; Soxinol EZ; ZDC; ZDEC; Zinc diethyldithiocarbamate. An accelerator and activator for natural rubber, styrene-butadiene, nitrile-butadiene and butyl rubber. *Anchor Chemical Group plc.*

2416 Ancazate ME
137-30-4 10305 205-288-3
Zinc dimethyl dithiocarbamate
ziram. Used as an accelerator. *Anchor Chemical Group plc.*

2417 Ancazate Q
A proprietary complex of zinc dithiocarbamate used as an accelerator. *Anchor Chemical Group plc.*

2418 Ancazate XX
Butyl dithiocarbamate
Used as an accelerator. *Anchor Chemical Group plc.*

2419 Ancazide ET
97-77-8 3428 202-607-8
Tetraethylthiuram disulfide
A proprietary rubber accelerator. Also an alcohol deterrent. No manufacturer.

2420 Ancazide IS
97-74-5 202-605-7
$C_6H_{12}N_2S_3$
Tetramethyl thiuram monosulfide
Bis(dimethylthiocarbamoyl) sulfide; Tetramethylthiuram monosulfide; UNADS; TMTM; Thiodicarbonic diamide, tetramethyl-; Bis(dimethylthiocarbamyl) sulfide. A proprietary rubber accelerator; an accelerator and activator for natural rubber nitrile-butadiene, and butyl rubber. mp = 106-108°. No manufacturer.

2421 Ancazide ME
137-26-8 9510 205-286-2
Thiram
A proprietary rubber accelerator. No manufacturer.

2422 anchoic acid
$C_9H_{16}O_4$
lepargylic acid
azelaic acid. Used in chemical synthesis.

2423 Anchor
A proprietary trade name for a vanadium tool steel. No manufacturer.

2424 Anchor 1040, 1115, 1170, 1171 and 1222
A proprietary group of boron trifluoride epoxy hardener curing agents; they consist of modified amine complexes of boron trifluoride. *Pacific Anchor Chemical Corp.* Name unverified.

2425 Anchor DLCT, PLCT
Liquid tackifing resins; adhesives. *Anchor Chemical (UK) Ltd.*

2426 Anchoracel
102-08-9 3393 203-004-2
Thiocarbanilide
A proprietary rubber vulcanization accelerator. No manufacturer.

2427 Anchor-bac®
Adhesives. *DuPont UK.*

2428 Anchorite
A light rubber product used in rubber mixings. No manufacturer.

2429 Anchorlube G-771
Animal/vegetable oil in water emulsion; metal cutting and tapping compound for stainless and other hard to work metals or operations on sensitive appliances. *Anchor Chemical Co.*

2430 Anchred
A red oxide used in rubber mixings. No manufacturer.

2431 Ancor® CR-538, CR-539
Proprietary acetylenic alcohol-based; corrosion inhibitor for metals, acid pickling, industrial cleaning, coatings and inks. *Air Prods & Chems Inc.*

2432 Ancor® LB-503, LB-504
Proprietary blends of amine salts; contains no nitrites or nitrates; corrosion inhibitor for metalworking fluids, water-based coatings and inks. *Air Prods & Chems Inc.*

2433 Ancor® OW-1
Secondary acetylenic alcohol
Corrosion inhibitor for mineral acid systems, oil well acidizing, steel pickling. *Air Prods & Chems Inc.*

2434 Ancor® OW-9
Proprietary blend of acetylenic alcohols and diols; corrosion inhibitor for oil well acidizing, steel pickling, industrial cleaning and inks. *Air Prods & Chems Inc.*

2435 Ancrack
Naptalam plus dinitro; pre-emergent herbicide for use on peanuts and soybeans. *Draxel Chemical Co.* Unverified.

2436 andalusite
Al_2SiO_5
A mineral; a silicate of aluminum.

2437 Andaria®
Novelty rayon filament. *Asahi Chem. Industry.*

2438 Anderol®
Synthetic lubricants; for lubrication of compressors, crankcases and other industrial uses. *Hüls Am.* Discontinued.

2439 Anderol® Premium Plus
Synthetic lubricant and coolant; for lubrication and cooling of rotary compressors. *Hüls Am.* Discontinued.

2440 Andersil
High flash point liquid polysilicate; stable liquid binders for coatings and precision casting molds. *Anderson Development Company.* Name unverified.

2441 andersonite
$6[Na_2CaVO_2(CO_3)_2 \cdot 6H_2O]$
A mineral.

2442 Andrez 8000
85% bound styrene/butadiene; used to improve the processing properties and increase the hardness and modulus of many rubber compounds. *Anderson Development Company.* Name unverified.

2443 Androx 3961
Shield
A water displacing fluid meeting DTD 900/4942. No manufacturer.

2444 Andur
Urethane based prepolymers, water emulsion thermoplastics and coatings; for use in the manufacturing of urethane elastomers, varnishes and paints, adhesives, etc. *Anderson Development Company.* Name unverified.

2445 Anedco ADM-407
Long-chain alkylbenzyl sulfonate; surfactant, emulsifier for bad tank bottoms and slop oil in the refinery. *Anedco.*

2446 Anedco ADM-407C
Alkybenzyl sulfonic acid
Surfactant, emulsifier for bad tank bottoms. *Anedco.*

2447 Anedco AF-800
Ammonium salt of an alcohol ether sulfate; fresh-water foamer for air/gas drilling and well clean-out operations. *Anedco.*

2448 Anedco AF-801
Alkyl sulfate salt; brine water foamer for air/gas drilling and well clean-out operations. *Anedco.*

2449 Anedco AW-395
Coco alkyl dimethyl benzyl quaternary amine
Quaternary surfactant; anti-clay swelling agent; corrosion inhibitor for waterfloods, oil and gas wells, pipelines; foamer additive for surfactants, cleaners, water treating. *Anedco.*

2450 Anedco AW-396
Quaternary amine/nonyl phenol ethoxylate blend; surfactant flush aid; anti-clay swelling agent; corrosion inhibitor for waterfloods, oil and gas wells, pipelines; foamer for wetting agents, surfactants, cleaners, water treating. *Anedco.*

2451 Anedco AW-397
Nonylphenol ethoxylate
Surfactant for manufacturing of corrosion inhibitors; emulsifier, detergent, wetting agent, penetrant, antistat, coupling agent. *Anedco.*

2452 Anedco DF-6002
Alcohol-based; defoamer for drilling muds. *Anedco.*

2453 Anedco DF-6031
Silicone emulsion
Effective in acid and alkaline media; for cooling towers, amine scrubbers, glycol dehydrators, water-based drilling muds, cleaning compounds, effluents, cutting oils, abrasive slurries. *Anedco.*

2454 Anedco DF-6130
Dimethyl silicone fluid
Antifoamer for gas-oil separators. *Anedco.*

2455 Angarite
A Russian cast basalt used as an electrical insulator.

2456 Angio-Conray
1225-20-3 5080 214-955-8
Iothalamate acid sodium salt
Diagnostic aid. *Mallinckrodt Inc.* Name unverified.

2457 Angiovist 282
131-49-7 5851 205-024-7
Meglumine Diatrizoate
Diagnostic aid *Berlex Laboratories Inc.* Name unverified.

2458 Anhydrite
7778-18-9 1753 231-900-3
Calcium sulfate, anhydrous
Concrete substitute and soil conditioner. *Pacific Chemical Industries Pty Ltd.* Name unverified.

2459 Anhydrol
1332-58-7 5294 296-473-8
Kaolin
Specially processed calcined kaolin (aluminum silicate); used as a molecular sieve support. *Engelhard.*

2460 Anhydrone
10034-81-8 5715 233-108-3
Cl_2MgO_8
magnesium perchlorate
A proprietary name for a perchlorate of magnesium, a drying agent for gases. d = 2.6000;. No manufacturer.

2461 Anhydrous Lanolin HP-2050
8006-54-0 5371 232-348-6
Lanolin
High purity lanolin for cosmetics applications; emulsifier, emollient, conditioner, lubricant for creams, lip products, hair grooms, makeup, suncare products. *Henkel/Cospha; Henkel Canada.* Name unverified.

2462 Anhydrous Lanolin P80
8006-54-0 5371 232-348-6
Lanolin
Moisturizer for pharmaceuticals and cosmetics (baby creams, cleansers, eye preparations, foundation, lipstick, sunscreen preparations). Emulsifier and emollient. *Westbrook Lanolin.*

2463 Anhydrous Lanolin P9SRA
8006-54-0 5371 232-348-6
Anhydrous lanolin
Emulsifier for pharmaceuticals and cosmetics (baby creams, cleansers, eye preparations, foundation, hypo-allergenic cosmetics, lipstick, sunscreen preparations). *Westbrook Lanolin.*

2464 Anidrisorb
Anhydrosorbitols
Used for pharmaceutical encapsulation. *Roquette (UK) Ltd.*

2465 Anilazine
101-05-3 694 202-910-5
4,6-Dichloro-N-(2-chlorophenyl)-1,3,5-triazin-2-amine
$C_9H_5Cl_3N_4$
Dairene®; Dairin®; Dyrene®; Dyrene; Kemate; Triasyn; Direx; B-622; Bortrysan; Direz; Dyrene 50W Triazine; Zinochlor; Aniyaline; Triasym. Broad spectrum fungicide used for tobacco, potatoes, cereals and ornamentals. Non-systemic foliar fungicide with protective action used to control blights of potatoes and tomatoes and leafspot diseases in many crops. mp = 159-160°; insoluble in H_2O; soluble in toluene, xylene, acetone; LD_{50} (rat orl) >5000 mg/kg. *Bayer AG.*

2466 animal charcoal
Bone charcoal. This term is used for all charcoal produced by the ignition of animal substances with exclusion of air, but more particularly to that obtained from bones. This material contains approximately 10% carbon, and 90% mineral matter, mainly calcium phosphate and is used mainly for absorbing dyes.

2467 animal glycerin
Neatsfoot oil.

2468 animal starch
Glycogen
Found in the blood and liver of mammals.

2469 anime resin
Gum anime. A fossil copal resin from South America.

2470 *p*-anisaldehyde
123-11-5 701 204-602-6
$C_8H_8O_2$
4-methoxybenzaldehyde
aubepine; anisic aldehyde. Used in perfumery. mp = -1°; bp = 248-249°; d = 1.1220; n_D^{20} = 1.5710-1.5740; insoluble in H_2O, soluble in organic solvents; LD_{50} (rat orl) = 1510 mg/kg.

2471 Aniscol
Antiscale paints to reduce oxidation losses and surface decarburization. *Foseco (F.S.) Ltd.*

2472 *m*-anisidine
536-90-3 208-651-4
C_7H_9NO
3-Methoxy-1-aminobenzene
m-Anisidine; *m*-aminoanisole; 3-aminophenol methyl ether; 3-methoxyaniline. Chemical intermediate. mp = 1°; bp = 251°; d = 1.0960; n_D^{20} = 1.5794; soluble in EtOH. *Aldrich; Penta Mfg.; Rhône-Poulenc.*

2473 *o*-anisidine
90-04-0 201-963-1
C_7H_9NO
2-Methoxy-1-aminobenzene
o-Aminoanisole; 2-methoxybenzenamine; *o*-methoxyaniline; 2-Anisidine; *ortho*-Anisidine; 1-amino-2-methoxybenzene; *o*-anisylamine; *o*-methoxyphenylamine; *o*-aminophenol methyl ether. Chemical intermediate. mp = 5-6°; bp = 225°; d = 1.0920; n_D^{20} = 1.5730; soluble in H_2O (14 g/l), more soluble in organic solvents; LD_{50} (rat orl) = 2 g/kg. *Aldrich; Penta Mfg.; Rhône-Poulenc.*

2474 *p*-anisidine
104-94-9 203-254-2
C_7H_9NO
4-Methoxy-1-aminobenzene
para-Anisidine; 4-methoxybenzenamine; *p*-methoxyaniline; *p*-aminoanisole; 1-amino-4-methoxybenzene; *p*-anisylamine; *p*-methoxyphenylamine. Chemical intermediate. mp = 58°; bp = 240-243°; d = 1.0700; soluble in organic solvents; LD_{50} (rat orl) = 1400 mg/kg. *Aldrich; Penta Mfg.; Rhône-Poulenc.*

2475 Anka steel
V2A Steel. Nickel-chromium steels containing from 15-16% chromium and from 7-10% nickel.

2476 aniline black
Aniline black in paste; fine black; oxidation black. A black dyestuff produced on the fiber by the oxidation of aniline salts.

2477 Anlonyx® 12S
56388-43-3 260-143-1
Disodium oleamido PEG-2 sulfosuccinate
Disodium oleamido PEG-2 sulfosuccinate; detergent for personal care products, bubble baths, dishwashing liquids. *Stepan; Stepan Canada.*

2478 annatto
anatto; anotto; arnatto; arnotto. Vegetable dye containing ethyl bixin, obtained from the fleshy covering of the seeds of the ruccu tree, Bixa orellana; $C_{27}H_{34}O_4$ as extract for coloring foods, foodmarking inks. *Haarmann & Reimer/Food Ingred.; Meer; Penta Mfg.;Pfizer; Warner-Jenkinson.*

2479 anode mud
Anode slime. The material which falls to the bottom of the electrolysing vessel during the electrolytic refining of copper; it contains copper (10-25%) gold (0.7-2%) silver (5-40%) tellurium, antimony, and arsenic and is used as a source of these metals.

2480 *p*-anol
539-12-8 715
$C_9H_{10}O$
4-(1-propenyl)phenol
p-propenylphenol; 4-hydroxy-1-propenylbenzene. Intermediate in synthesis of estrogens. mp= 93-94°; bp = 250° (dec); bp_{14} = 138-140°; slightly soluble in H_2O, more soluble in organic solvents.

2481 Anonaid TH
577-11-7 3460 209-406-4
docusate sodium
Sodium dioctyl sulfosuccinate liquid form; wetting and emulsifying agent for the textile, leather and paper industries and for emulsion polymerization. *Rhône-Poulenc UK.*

2482 Anotex
Dyes for anodized metal. *Pointing Ltd.* Name unverified.

2483 Anox® 20
6683-19-8 229-722-6
Pentaerythrityl tetrakis dibutyl hydroxy phenylpropionate
Antioxidant, processing stabilizer for thermoplastic polymers; especially effective against polymer thermo oxidative degradation during long term aging. *Enichem Synthesis SpA.*

2484 Anox® 70
41484-35-9 255-392-8
Thiodiethylene bisdibutyl hydroxy hydrocinnamate
Antioxidant providing processing and end-use stability for LDPE copper wire insulation and cable jacketing, carbon black loaded polyolefins, chem. crosslinked PE, HIPS, ABS, ethylene-propylene-diene rubbers, styrene/butadiene rubber, neoprene, natural and *Enichem Synthesis SpA.*

2485 Anox® IC-14
27676-62-6 248-597-9
Trisdibutyl hydroxybenzyl isocyanurate
Nonvolatile, nondiscoloring antioxidant protecting polymers against high temp. degradation during processing and thermal stability during service life. *Enichem Synthesis SpA.*

2486 Anox® PP 18
2082-79-3 218-216-0
Octadecyl dibutyl hydroxy hydrocinnamate
Antioxidant retarding oxidative degradation during polymerization, processing and in end-use applications; stabilizer for polyolefins, impact styrenics, blocked copolymers, elastomers, adhesives, PVC, PU. *Enichem Synthesis SpA.*

2487 Ansa
Alkyl naphthalene sodium sulfonate.
Albright & Wilson Ltd. Name unverified.

2488 Ansax
6484-52-2 567 229-347-8
ammonium nitrate
Fertilizer prilled ammonium nitrate *L & K Fertilizers Ltd.* Name unverified.

2489 Anscor®
MgO
magnesium oxide. Lightly calcined sea water magnesia. Used in antacids and as a source of magnesium. *Steetley Magnesia Products Ltd.*

2490 anserine
584-85-0 717 209-545-0
$C_{10}H_{16}N_4O_3$
β-alanyl 3-methyl L-histidine
A natural peptide from muscle. It is β-alanyl 3-methyl L-histidine. Used in biochemical research.

2491 Ansol
Solvents and solvent mixtures containing anhydrous alcohol; especially those adpated for use as general organic solvents, solvents for nitrocellulose resins, wood and metal lacquers, paint and varnish removers, cleaners, shellac and floor finishes and alcohol substitutes. *Quantum Chemical Corp.*

2492 Ansol A
A proprietary blend of anhydrous denatured ethyl alcohol with small percentages of esters, other alcohols, and hydrocarbons; a nitrocellulose and resin solvent. No manufacturer.

2493 Ansol B
A preparation similar to A, except that B contains a small amount of n-butyl alcohol in the place of amyl alcohol. ®

2494 Ansol E-121
110-71-4 3274 203-794-9
$C_4H_{10}O_2$
ethylene glycol dimethyl ether
1,2-Dimethoxyethane. A trademark for ethylene glycol dimethyl ether, a general-purpose solvent. *Ansul Co.* Name unverified.

2495 Ansol E-181
143-24-8 9348 205-594-7
$C_{10}H_{22}O_5$
tetraethylene glycol dimethyl ether
tetraglyme. A trademark for tetraethylene glycol dimethyl ether (teraglyme), a solvent. *Ansul Co.* Name unverified.

2496 Anstex AK-25
Special phosphate; antistat for synthetic fibers, plastics. *Toho Chem. Industry.*

2497 Ant Flip
10043-35-3 1364 233-139-2
BH_3O_3
boric acid
A gel bait with boric acid as the active ingredient; for household, hospital, restaurant use for controlling pharoah and other sweet eating ants; workers carry bait back to nest to kill the colony. *Colonial Products Inc.*

2498 Ant Gun
Contains diazinon and pyrethrins; ready-for-use spray for control of ants and household insects. *ICI Garden Products.*

2499 Ant Killer
Insecticide. Name unverified.

2500 Ant Killer Dust
Insecticide.

2501 Antak
Alcohol; contact tobacco sucker control material. *Draxel Chemical Co.* Unverified.

2502 Antara®
Redesignated Lubrhophos® *Rhône-Poulenc Surf.*

2503 Antaron ET-201
PVP/decene copolymer. *Rhône-Poulenc Surf.*

2504 Antarox® 17-R-2
9003-11-6 7721
Meroxapol 172
poloxalene. Defoamer, dispersant, wetting agent, emulsifier, demulsifier, leveling agent, detergent for industrial/household cleaners, fermentation, paper processing, rinse aids, automatic dishwashing, metal cleaning. *Rhône-Poulenc Surf.*

2505 Antarox® 461/P.
EO/PO alkylphenol block polymer
Emulsifier, dispersant used in agricutural industry for preparation of emulsifiable concentrates and toxicant flowable systems. *Rhône-Poulenc Surf.; Rhône-Poulenc Geronazzo.*

2506 Antarox® 497/P
EO/PO alkylphenol block polymer
Emulsifier, dispersant used in agricutural industry for preparation of emulsifiable concentrates and toxicant flowable systems. *Rhône-Poulenc Surf.; Rhône-Poulenc Geronazzo.*

2507 Antarox® B-10
Low molecular weight block copolymer; surfactant. *Rhône-Poulenc France.*

2508 Antarox® BL-214
68603-25-8
ethoxylated and propoxylated octyl/decyl alcohols
Wetting, rewetting agent, detergent for metal cleaning, textile finishing, industrial/household cleaners. *Rhône-Poulenc Surf.; Rhône-Poulenc France.*

2509 Antarox® E-100
9003-11-6 7721
Poloxamer 401
Antarox® 17-R-2; Antarox® L-61; poloxalene. Surfactant *Rhône-Poulenc France.*

2510 Antarox® L-61
9003-11-6 7721
Poloxamer 181
Defoamer, dispersant, wetting agent, emulsifier, demulsifier, leveling agent, detergent, lubricant for household/industrial cleaners, metalworking fluids, agricultural formulations, rinse aids, automatic dishwashing, water treatment. *Rhône-Poulenc Surf.; Rhône-Poulenc France.*

2511 Antarox® LA-EP 15
Modified oxyethylated straight chain alcohol; detergent, dispersant, wetting

agent, emulsifier; for controlled foam applications, machine dishwashing, rinse aid compositions. *Rhône-Poulenc Surf.*

2512 Antarox® PGP 23-7
9003-11-6 7721
Poloxamer 237
Coemulsifier for cosmetics, toiletries, pulp and paper defoamers; dispersant, viscosity control agent. *Rhône-Poulenc Surf. Canada.*

2513 Antec Farm Fluid S
A mixture of organic acids, high molecular weight phenols, low molecular phenols and surfactant; used for the disinfection of all types of livestock buildings. *Antec International Ltd.*

2514 Antec Longlife 250 S
A mixture of organic acids, high molecular weight phenols, synthetic boicide and surfactant; used for the disinfection of all types of livestock buildings. *Antec International Ltd.*

2515 Antec OO-Cide
Two part ammonia release system incorporating a biocide; coccidiacide; for the treatment of all livestock buildings. *Antec International Ltd.*

2516 Antec Virkon S
Oxidizing detergent/disinfectant system with particularly broad spectrum activity against viruses and safety in use; for disinfection of hard surfaces, water, environment and instruments in agriculture, industrial and medical situations. *Antec International Ltd.*

2517 Antelope
7758-16-9 8713 231-835-0
Acid sodium pyrophosphate
Used in baking powders. *Albright & Wilson Ltd.* Name unverified.

2518 Antepan
110-85-0 7617 203-808-3
Piperazine
Anthelmintic, used to control nematodes. *Coopers Animal Health Ltd.* Name unverified.

2519 Anthion
7727-21-1 7825 231-781-8
$K_2O_8S_2$
Potassium persulfate
Potassium peroxydisulfate; Potassium Peroxodisulfate; Peroxydisulfuric acid, dipotassium salt; Dipotassium Peroxydisulfate. Used as a hypo eliminator in photography. No manufacturer.

2520 Anthiphen
97-23-4 3120 202-567-1
$C_{13}H_{10}Cl_2O_2$
Dichlorophen
Algafen; Algofen. Agricultural fungicide, antimicrobial and germicide. *May & Baker Ltd.* Name unverified.

2521 Anthium Dioxcide
10049-04-4 2146 233-162-8
ClO_2
Chlorine Dioxide
chlorine peroxide; Chloroperoxyl; Doxcide 50; Chlorine oxide. Stabilized chlorine dioxide; broad spectrum biocide, preservative; deodorant designed to remove odors caused by residual monomers in resins; works as an oxidizer. *Int'l Dioxcide.*

2522 anthocyanins
The red, blue, and violet coloring matters of plants.

2523 Anthosin®
Based on acid dyes; for paper coloring in papermaking. *BASF AG.*

2524 Anthoxan
4-Isopropyl-5,5-dimethyl-1,3-dioxane
Raw material for herbal fragrances. *Henkel.*

2525 anthracite
A hard coal, containing 85-95% carbon. It burns with little smoke.

2526 anthraflavic acid
84-60-6 201-544-3
$C_{14}H_8O_4$
2,6-Dihydroxyanthraquinone
9,10-Anthracenedione, 2,6-dihydroxy-; 2,6-Dihydroxyanthraquinone. Used as a basis for pigments and dyestuffs. mp > 320°.

2527 anthraflavone
A dyestuff which is prepared by treating 2-methylanthraquinone with condensing agents; used to dye cotton yellow shades.

2528 anthragalanthranol
$C_{14}H_{10}O_4$
Trioxyanthranol
Used to synthesize pigments.

2529 anthranilic acid
118-92-3 442 204-287-5
$C_7H_7NO_2$
o-Aminobenzoic acid
Carboxyaniline; Anthranilic acid; Aminobenzoic acid; 2-Aminobenzoic Acid; Anthranic Acid; 1-amino-2-carboxybenzene; *o*-anthranilic acid; *o*-carboxyaniline; *ortho*-amidobenzoic acid; Vitamin L; Vitamin L1. Used as a dye intermediate. mp = 144-148°; d = 1.4120; soluble in H_2O (5.7 g/l), more soluble in organic solvents; LD_{50} (mus orl) = 1400 mg/kg.

2530 anthranol
1143-38-0 723 214-538-0
$C_{14}H_{10}O_3$
9(10H)-Anthracenone, 1,8-dihydroxy-
anthralin. A smooth soft ointment containing anthralin (in 0.4, 1.0 and 2.0 strengths) w/w in a base containing cetyl alcohol, liquid paraffin, soft white paraffin and sodium sulfate with salicylic acid; for the topical treatment of subacute and chronic psoriasis including psoriasis of the scalp. *Stiefel Laboratories (UK) Ltd.* Name unverified.

2531 anthrapurpurin
$C_{14}H_8O_5$
Isopurpurin, Alizarin GD, Alizarin RF, Alizarin RT, Alizarin RX, Alizarin SC, Alizarin SSA, Alizarin SX, Alizarin SX Extra, Alizarin WG. Trioxyanthraquinone and dyes red shades on alumina; a mordant dyestuff.

2532 Anthraquinone
84-65-1 726 201-549-0
$C_{14}H_8O_2$
9,10-Anthracenedione
9,10-anthraquinone; 9,10-dioxoanthracene; Morkit. Intermediate for dyes and organics, Also used as a discharging auxiliary for textiles and an organic inhibitor, bird repellent for seeds. Light yellow, slender monoclinic prisims by sublimation; orthohombic crystals from H_2SO_4 and H_2O; d = 1.42-1.44; mp = 286°; bp = 377°; insoluble in H_2O. *Buckton Scott Ltd; ICI Am.*

2533 anthrarufin
117-12-4 728 204-275-6
$C_{14}H_8O_4$
1,5-Dihydroxy-anthraquinone.
1,5-dihydroxy-9,10-anthracenedione. Intermediate in dyestuffs manufacture. mp = 280° (dec); slightly soluble in H_2O, moderately soluble in EtOH.

2534 Anthraxolite
A variety of anthracite, a fuel. No manufacturer.

2535 anthropic acid
A mixture of palmitic and stearic acids. Used in chemical manufacturing.

2536 antiar
The milky juice of the upas tree; used as an arrow poison.

2537 Antibacterin
127-65-1 2118 204-854-7
Chloramine T
Antibacterial and topical anti-infective.

2538 Antiblu/Antiboror
Fungicide/insecticide; Used for freshly sawn timber. *Hickson & Welch Ltd.*

2539 Anticor 70
Anticorrosion pigments containing zinc ferrite as the active antirust pigment for the formulation of chromate-free primers. *Bayer AG.*

2540 Anticorodal
An aluminum alloy containing 1% silicon, 0.06% magnesium, and 0.06% manganese.

2541 Antidust 2
557-05-1 10292 209-151-9
zinc stearate
A proprietary preparation similar to Antidust F, but containing a particularly fine zinc stearate in aqueous dispersion. Antiseptic and astringent. *Rhein-Cheme Rheinau.* Name unverified.

2542 Antidust F
A proprietary group of surfactants in combination with polyvalent alcohols; used in a concentrated aqueous solution to prevent undesirable surface tackiness in sheets and extrudates of plastics and rubber materials. *Rhein-Cheme Rheinau.* Name unverified.

2543 Antifoam
Silicone and nonsilicone antifoams for industrial applications. *Bayer plc.*

2544 Antifoam 20WB
Water-based antifoam for screen room, paper machine, bleaching applications, waste water treatment, chem. processing industry, latex, coatings, metal treating, and electroplating. *Stockhausen.*

2545 Antifoam 55
Silicone defoamer; free-rinsing defoamer for garment dyeing. No manufacturer.

2546 Antifoam 6031
Sulfocarboxylic acid/complex fatty blend; defoamer for manufacturing of phos-acid and fertilizer. *Stockhausen.*

2547 Antifoam 7800 New
Higher hydrocarbons and their sulfonic acid derivs.; antifoam for aq. systems

in sugar, fertilizer, phosphoric acid, dyestuffs, paper, leather, plastics, and chemical industries; resistant to acid and weak alkalis. *Miles/Organic Prods.*

2548 Antifoam ET

An emulsified form of Antifoam T. *Bayer AG*. Discontinued.

2549 Antifoam FRS

Blended silicones and emulsifiers; highly effective defoamer and antifoam for ambient and high temp. operations; free rinsing on carpets. *Am. Emulsions.*

2550 Antifoam GEB

Silicone defoamer; free-rinsing defoamer for continuous and beck dyeing of carpets. *Am. Emulsions.*

2551 Antifoam T

126-73-8 9749 204-800-2

Tributyl Phosphate

An antifoaming agent used in the manufacture of paper coatings. *Bayer AG.*

2552 Antifoam TP

Antifoam for pigment printing of textiles. *BASF AG.*

2553 Antifoam VOL

Polysiloxane

Defoamer for use in dyebaths, finish mixes where stability is essential. *Am. Emulsions.*

2554 Antifungin

13703-82-7 5693 237-235-5

B_2MgO_4

magnesium borate

The trade name for magnesium borate, an antiseptic and fungicide. Occurs in a variety of minerals. Slightly soluble in H_2O.

2555 Antigrison

An explosive; it consists of 27% nitro-glycerin, 1% nitro-cotton, and 72% ammonium nitrate.

2556 Antihypo

589-97-9 7821

$C_2K_2O_6$

Potassium percarbonate

Used as a bleaching agent and hypo eliminator in photography. Also in microscopy. Dissolves in H_2O releasing oxygen. Solubility = 6.5 g/100 ml.

2557 Antil® 141 Liquid

PEG-55 propylene glycol oleate

Thickener for aqueous solutions of surfactants, e.g., shampoos, foam baths, shower preparations, liquid soaps; solubilizes essential oils into aqueous surfactant systems; liquid version developed for cold process systems. *Goldschmidt; Goldschmidt AG.*

2558 Antil® 141 Solid

PEG-55 propylene glycol oleate

Thickener for surfactant systems, e.g., shampoos, shower gels; solubilizer for essential oils, fragrances, and perfumes. *Goldschmidt.*

2559 Antil® 208

Carbomer 208

Highly effective thickener for detergent/soap-based products; refatting agent in hair or skin cleansing products. *Goldschmidt.*

2560 Antiluetin

$[SbO(C_4H_4O_6)_2DFKNH_4]\cdot H_2O$

Potassium-ammonium-antimonyl-tartrate

A trypanocide. No manufacturer.

2561 Antilux

Blends of selected paraffins and micro waxes, which protect rubber articles from damage by the sun, ozone and weathering; used for technical molded and extruded articles, articles subjected to dynamic stress, tires, conveyor belts, cable coverings, articles fit for foodstuffs quality. *Bayer plc.*

2562 Antilux AOL

A proprietary blend of paraffinic hydrocarbons used as an antiweathering agent in natural and synthetic rubbers. *Rhein-Chemie Rheinau.* Name unverified.

2563 Antimigrant 157

Antimigrant for continuous dyeing of textiles. *Catawba-Charlab.*

2564 Antimigrant C-45

Sodium alginate

Antimigrant for dyeing. No manufacturer.

2565 Anti-Migrant CAS

Proprietary blend of high molecular weight polymers and sequestrants; antimigrant with excellent lubricity for applications such as pigment padding and continuous disperse dyeing. *Arol Chem. Prods.*

2566 antimonial lead

An alloy of 87% lead and 13% antimony.

2567 antimony

7440-36-0 733 231-146-5

Sb

stibium

Hardening alloy for lead, bearing metal, type metal, solder, collapsible tubes and foil, sheet and pipe, semiconductor technology, pyrotechnics. mp = 630°; bp = 1635°; d = 6.68; LD_{50} (rat orl) = 100 mg/kg. *Aldrich; Amspec Chem.; Atomergic Chemetals.*

2568 antimony pentachloride

7647-18-9 736 231-601-8

Cl_5Sb

antimony (V) chloride

For analytical testing of alkaloids and cesium; dyeing intermediates; as chlorine carrier in organic chlorinations. mp = 2-4°; bp = 140°; Corrosive *Aldrich; Atomergic Chemetals; Hoechst-Celanese.*

2569 antimony pentafluoride

7783-70-2 737 232-021-8

F_5Sb

antimony (V) fluoride

Catalyst and/or source of fluorine in fluorination reactions. Moderaly viscous liquid; poisonous; mp = 8.3°; bp = 141°; $d^{25.8}$ = 3.097. *AlliedSignal; Elf Atochem. N. Am.; Atomergic Chemetals.*

2570 antimony pentasulfide

1315-04-4 738 215-255-5

Sb_2S_5

Antimony (V) sulfide; Antimonial saffron; Golden antimony sulfide; Antimony red. Red pigment, rubber accelerator. Orange-yellow powder; insoluble in H_2O; soluble in conc. HCl, solns of alkali hydroxides or sulfides; LD_{50} (rat orl) = 150 mg Sb/100 g. *Atomergic Chemetals.*

2571 antimony potassium tartrate

$C_8H_4K_2O_{12}Sb_2\cdot 3H_2O$

tartar emetic

potassium antimonyl tartrate; tartrated antimony. Textile and leather mordant; medicine; insecticide. *Aldrich.*

2572 antimony salt

SbF_2

Double salts of antimony fluoride, with alkali sulfates, or with alkali fluorides; used as mordants in dyeing.

2573 antimony trichloride

10025-91-9 233-047-2

$SbCl_3$

Alferric; Aluminoferric; Antimony salts, bronzing iron, mordant, manufacture of lakes, chlorinating agent in organic synthesis, pharmaceuticals, fireproofing textiles, analytical reagent. *Akzo; Aldrich; Hoechst-Celanese; Nihon Kagaku Sangyo.*

2574 antimony trioxide

1309-64-4 752 215-175-0

Sb_2O_3

Antimony oxide

Cooksons; Dechlorane A-O; Thermoguard® L; Thermoguard® UF; Timonox; Timonox Blue Star; Amsperse; Antimony Oxide High Tint, Low Tint, Ultrapure, Very High Tint; AO; Fireshield® H, HPM, HPM-UF, L; Fyraway; Fyrebloc; KR; Octoguard FR-10; Petcat R-9; Thermoguard® HPM, HPM-UF, L, S, UF; Ultrafine® II. Antimony oxide; contains halogens; A proprietary synergistic agent for fire-retardant plastics. mp = 655°; bp = 1425°; slightly soluble in H_2O; soluble in KOH, HCl and sulfuric acid, strong alkalis; d = 5.67; LD_{50} (rat orl) >20 g/kg. Kingsley & Keith Chemical Corp; Aldrich; Alfa Aesar Johnson Matthey; Amspec; Asarco; Ashland; Fluka; Hoechst; HoltraChem; Laurel Ind.; D.N. Lukens; Miljac; Nihon Kagaku Sangyo; Noah; Punda Mercantile; Reade Advanced Materials; Revelli; HM Royal; Sigma;. Name unverified.

2575 antimony trisulfide

1345-44-6 215-713-4

Sb_2S_3

antimony(III) sulfide

Vermilion or yellow pigment, antimony salts, pyrotechnics, matches, percussion caps, camouflage paints, ruby glass. *Atomergic Chemetals, BASF.*

2576 Anti-oxidant 425

88-24-4 201-814-0

$C_{25}H_{36}O_2$

2,2'-methylene-bis(4-ethyl-6-t-butyl) phenol

2,2'-Methylenebis(4-ethyl-6-*tert*-butylphenol). A proprietary preparation of 2,2-methylene-bis(4-ethyl-6-t-butyl) phenol, an anti-oxidant. *Ciba-Geigy Europe.* Name unverified.

2577 Antioxidant 431

Nondiscoloring and nonstaining antioxidant for dry rubber and latex. *Uniroyal.* Name unverified.

2578 Antioxidant 449

A phenolic phosphite antioxidant; a nondiscoloring stabilizer for EDPM polymers; also economical replacement for BHT in compound work. *Uniroyal.* Name unverified.

2579 Antioxidant 451
Alkylated hydroquinone
For synthetic rubbers and plastics; used as a stabilizer for synthetic rubbers, such as polybutadiene and as an antioxidant for uncured adhesives; it functions as an antioxidant in both black and nonblack cured compounds and latex compounds. *Uniroyal.* Name unverified.

2580 Anti-oxidant 2246
119-47-1 204-327-1
$C_{23}H_{32}O_2$
Methylenebisbutyl methylphenol
2,2'-Methylenebis(4-methyl-6-*tert*-butylphenol); 2,2'-methylenebis 6(1,1-dimethylethyl)-4-methyl-phenol; 6,6'-di-*tert*-butyl-2,2'-methylenedi-p-cresol; 2,2'-Methylene-bis(6-*tert*-butyl)-p-cresol. A proprietary preparation of 2,3-methylenebis (4-methyl-6-*t*-butyl) phenol. *Ciba-Geigy Europe.* Name unverified.

2581 Anti-oxidant 4010
$C_{18}H_{22}N_2$
N-phenyl-N'-cyclohexyl-*p*-phenylenediamine
Phenylcyclohexyl PPD; N-Cyclohexyl-N-phenyl-4-phenylenediamine; CPPD; Flexizone GH; N-phenyl-N'-cyclohexyl-*p*-phenylenediamine. A proprietary preparation of N-phenyl-N'-cyclohexyl-*p*-phenylenediamine, an antidegradant in natural rubber, styrene-butadiene, and chloroprene rubber. No manufacturer.

2582 Anti-Oxidant AH
A proprietary aldol-α-naphthylamine resin. *Bayer AG.* Discontinued.

2583 Anti-Oxidant AP
A proprietary aldol α-naphthylamine powder. *Bayer AG.* Discontinued.

2584 Antioxidant BA
Aldol-α-naphthylamine powder, A proprietary antioxidant;. No manufacturer.

2585 Anti-Oxidant DDA
A proprietary derivative of diphenylamine, an anti-oxidant. *Bayer AG. Discontinued.*

2586 Anti-Oxidant DNP
Di-β-naphthyl-*p*-phenylenediamine.
Bayer AG. Discontinued.

2587 Anti-Oxidant DOD
4,4·g-Dioxydiphenyl.
Bayer AG. Discontinued.

2588 Anti-Oxidant EM
A 30% aqueous emulsion of a diphenylamine derivative. *Bayer AG.* Discontinued.

2589 Anti-Oxidant MB
583-39-1 1112 209-502-6
$C_7H_6N_2S$
1,3-dihydro-2H-benzimidazole-2-thione
2-benzimidazolethiol; 2-mercaptobenzimidazole; *o*-Phenylenethiourea; Antioxidant MB; Antiegene MB; AOMB; ASM MB. A proprietary antioxidant; 2-mercaptolbenzimidazole. mp = 301-305°; insoluble in H_2O, soluble in organic solvents; LD_{50} (mus orl) = 750 mg/kg. *Bayer AG.* Discontinued.

2590 Antioxidant PAN
90-30-2 201-983-0
$C_{16}H_{13}N$
Phenyl-α-naphthylamine.
N-Phenyl-1-Naphthylamine; N-(1-Naphthyl)aniline; N-Phenyl-1-Naphthalenamine; N-Phenyl-α-naphthylamine; phenyl-α-naphthylamine; phenyl-1-naphthylamine; acetopan; additin 30; 1-anilinonaphthalene; C.I. 44050; neozone a; phenylnaphthylamine; α-phenylnaphthylamine; algerite; 1-Naphthyl phenyl amine. A proprietary antioxidant. mp = 60-62°; bp_{15} = 226°; LD_{50} (rat orl) = 1625 mg/kg. *Bayer AG.* Discontinued.

2591 Antioxidant PBN
135-88-6 205-223-9
$C_{16}H_{13}N$
Phenyl-β-naphthylamine
N-Phenyl-2-naphthylamine; N-Phenyl-β-naphthylamine; N-Phenyl-2-Naphthalenamine; PBNA; Phenyl-β-naphthylamine; Agerite; PBN; Aceto PBN; anilinonaphthalene; 2-anilinonaphthalene; Antioxidant 116; Antioxidant PBN; N-(2-naphthyl)aniline; 2-naphthylphenylamine; β-naphthylphenylamine; Neozon D; Neozone; Nilox PBNA; nonox d; 2-phenylaminonaphthalene; phenyl-2-naphthylamine; Stabilizator AR; Neosone D; Vulkanox PBN; Nonox DN; N-(2-naphthyl)-N-phenylamine; Stabilizer AR; Nocrac D; Naftam 2; N-β-naphthyl-N-phenylamine. A proprietary antioxidant. mp = 108°; bp = 395°; d = 1.24, insoluble in H_2O, soluble in organic solvents; LD_{50} (rat orl) = 8730 mg/kg. No manufacturer.

2592 Antioxidant RES
Aldol-α-naphthylamine resin
A proprietary antioxidant. No manufacturer.

2593 Anti-Oxidant RR 10 N
A proprietary range of alkylated phenols. *Bayer AG.*

2594 Anti-Oxidant SP
A proprietary styrenated phenol, an antioxidant. *Bayer AG.*

2595 Anti-Oxydant Bayer
128-37-0 1583 204-881-4
$C_{15}H_{24}O$
butylated hydroxytoluene
Deenax; Lowinox®BHT; Nipanox®BHT; BHT; Ralox® BHT food grade; Spectratech® CM 11340, KM 11264; Vanox® PCX; Vulkanox® KB; Antrancine 8; Tenox BHT; Ionol CP; Sustane; Dalpac; Impruvol; Vianol. BHT; for the protection of animal fats, feedstuffs, concentrates and mixed feed from oxidative decomposition and loss of essential nutrients. mp = 70°; bp = 265°; d_4^{20}= 1.048; insoluble in H_2O, soluble in organic solvents; LD_{50} (rat orl) = 890 mg/kg. *Bayer AG.*

2596 Antioxydant NV3
Alkylphenol
Used in tire industry for technical goods. *BASF AG.*

2597 Antioxygene A
A proprietary trade name for phenyl-α-naphthylamine; an antioxidant. *Allied Colloids Ltd.* Name unverified.

2598 Antioxygene AFL
A proprietary trade name for a ketoneamine reaction product. *Allied Colloids Ltd.* Name unverified.

2599 Antioxygene AN
A proprietary trade name for aldol-α-naphthylamine paste; an antioxidant. *Allied Colloids Ltd.* Name unverified.

2600 Antioxygene BN
135-19-3 6471 205-182-7
$C_{10}H_8O$
2-naphthol
A proprietary trade name for β-naphthol; an antioxidant and chemical intermediate. *Allied Colloids Ltd.* Name unverified.

2601 Antioxygene CAS
A proprietary trade name for a mixture of phenyl-α-naphthylarnine and meta-toluylene-di-amine; used as an antioxidant. *Allied Colloids Ltd.* Name unverified.

2602 Antioxygene INC
A proprietary trade name for aldol-α-naphthylamine in powder form; an antioxidant. *Allied Colloids Ltd.* Name unverified.

2603 Antioxygene MC
135-88-6 205-223-9
phenyl-β-naphthylamine
A proprietary trade name for phenyl-β-naphthylamine; an antioxidant. *Allied Colloids Ltd.* Name unverified.

2604 Antioxygene RA
A proprietary trade name for aldol naphthylamines; antioxidants. *Allied Colloids Ltd.* No manufacturer.

2605 Antioxygene RES
A proprietary trade name for an aldol-α-naphthylamine resin; an antioxidant. *Allied Colloids Ltd.* Name unverified.

2606 Antioxygene RM
A proprietary trade name for aldol naphthylamine; an antioxidant. *Allied Colloids Ltd.* Name unverified.

2607 Antioxygene RO
A proprietary trade name for aldol naphthylamine; an antioxidant. *Allied Colloids Ltd.* Name unverified.

2608 Antioxygene STN
A proprietary trade name for phenyl-α-naphthylamine mixed with m-toluylene-diamine and stearic acid; an antioxidant. *Allied Colloids Ltd.* Name unverified.

2609 Antioxygene WBC
A proprietary trade name for an antioxidant. *Allied Colloids Ltd.*

2610 Antiozonant AFD
Nonstaining antiozonant. *Bayer plc.*

2611 Antiprex®
A range of water-soluble polymers developed for the control of scale and deposition formation in aqueous systems; widely used in detergent formulation as builders and in cooling water and evaporation systems. *Allied Colloids Ltd.*

2612 antipyoninum
1330-43-4 8733 215-540-4
sodium borate
Neutral sodium tetraborate, prepared by fusing together borax and boric acid. Used as an antiseptic, detergent and astringent.

2613 Antiquax
Wax polishes; for furniture. *James Briggs & Sons Ltd.* Name unverified.

2614 Antisepsin
103-88-8 1420 203-154-9
C_8H_8BrNO
p-bromoacetanilide
Bromanilide; Asepsin; Bromoantifebrin. Used as an analgesic and antipyretic. mp = 165-166°; d = 1.72°; insoluble in H_2O, soluble in EtOH, organic solvents. No manufacturer.

2615 Antiseptin
A mixture of boric acid zinc iodide, and thymol; an antiseptic dusting powder.

2616 Antisettle
Thixotrope. *Cray Valley Ltd.*

2617 Antispumin ZU
Ethoxylated/propoxylated fatty alcohols
Defoamer for beet sugar industry for flume water, diffusion, liming, and carbonation; also for paper mill effluent. *Stockhausen.*

2618 Antistat 100
Complex blend; nondurable antistat for synthetics; nonyellowing and low foaming. *Am. Emulsions.*

2619 Antistat 7220
Oxyalkylated polyester; antistat. *BASF.*

2620 Antistat BT
Static depressant for natural and synthetic fibers and fabrics. No manufacturer.

2621 Antistat RD2-2351
Quaternized fatty amine; antistat for all textile fibers. *Marlowe-Van Loan.*

2622 Antistatic 812 and 813
100% Active phenolic ethoxylates; antistatic compounds for plastics used in proportions of 5-7%. *Farbwerke Hoechst.* Name unverified.

2623 Antistatic 816
A proprietary 95% active lauric imide; antistatic compound for plastics used in polyethylene, PVC and polystyrene. *Farbwerke Hoechst.* Name unverified.

2624 Antistatin
Antistatic finishing agent for textiles. *BASF plc.* Name unverified.

2625 Antitan
A tannin remover. *S & D Chemicals Ltd.* Name unverified.

2626 Anti-Terra® -202
Alkylammonium salt of a higher molecular weight polycarboxylic acid solution; wetting and dispersing additive to prevent settling and flooding of pigments; gellant for organophilic bentonites; for coating systems. *Byk-Chemie USA.*

2627 Anti-Terra® -P
Phosphoric acid salt of long chain carboxylic acid polyamine amides; wetting and dispersing additive to prevent settling and flooding of pigments in alkyds, alkyd-melamine, chlorinated rubber systems, PVC co-polymers, acid catalyzed paints. *Byk-Chemie USA.*

2628 Antiwick OP
Thickener for pigment print pastes to control wicking and flushing. No manufacturer.

2629 Antlak
333-41-5 3043 206-373-8
diazinon
1.74% Diazinon; ant and crawling insect lacquer (aerosol); used for household crawling insect control on hard surfaces. *Doff Portland Ltd.*

2630 Antoban
A veterinary anthelmintic. *Wellcome Foundation Ltd.*

2631 Anton N
Octylated diphenylamine
A trademark for a general-purpose antioxidant used in many elastomers. *DuPont UK.* Name unverified.

2632 Antox® N
Octylated diphenylamine
A registered trade name for a general purpose antioxidant with only mild discoloring and staining characteristics. *DuPont UK.* Name unverified.

2633 Antozite®
Phenylenediamines
Rubber antiozonants used in natural rubber, isoprene rubber, BR, styrene/butadiene rubber and CR rubbers. *R. T. Vanderbilt Co Inc.*

2634 Antracol®
12071-83-9 8004 235-134-0
Propineb
Fungicide for protective control of potato blight, hop downy mildew, apple scab, leafspot on celery, blackcurrants and gooseberries, downy mildew on grapes and suppression of yellow rust on winter wheat. *Bayer AG.*

2635 Antron®
Fibers. *DuPont UK.*

2636 Antron® Stainmaster®
Fibers. *DuPont UK.*

2637 Anzon
Flame retardant compositions. *Anzon.* Name unverified.

2638 AOR/GR
A proprietary name for technically-comminuted rubber. *Societé Indochine de Plantations d'Hvas (SIPH) Ivory Coast.* Unverified.

2639 Aosoft
Tetrafilcon A
Contact lens material. *Am. Optical.* Name unverified.

2640 Apagallin
2217-44-9 5059 218-715-3
$C_{20}H_8I_4Na_2O_4$
Tetraiodophenol-phthalein
Iodophthalein sodium. An antiseptic and acd-base indicator. No manufacturer.

2641 Apec®
Polyarylate (aromatic polyester carbonate) and copolycarbonate based on bisphenol A and bisphenol TMC (trimethylcyclohexanone); high temperature resistant engineering thermoplastic for the manufacture of injection moldings; used for components exposed to high temperatures in electrical systems for automobiles, lighting engineering; the electrical and electronic industries, medical engineering and domestic appliances. *Bayer plc.*

2642 Apec® DP9-9330
PC; high-heat amorphous thermoplastic with high strength and optical props. for extrusion and injection molding applications including demanding transparent and opaque applications, lenses in auto and industrial lighting, fuse housings, microwave oven doors. *Miles.* Discontinued.

2643 Apec® HT DP9-9350
PC; high-heat amorphous thermoplastic resin suitable for demanding applications such as lenses and reflectors used in high heat-output lighting devices and automotive fuses; injection molding and extrusion grade. *Bayer; Miles.*

2644 Apec® HT KU 1-9350
Amorphous aromatic PC; thermoplastic offering high heat resistance and improved flowability, transparency, and uv-stability for automotive electrical components, lamp housings, lamp reflectors, domestic appliance components. *Bayer AG.* Name unverified.

2645 Apec® KL 1-9306
Polyester carbonate
Amorphous thermoplastic providing heat resist. between 150° and 184° for lighting engineering, electronics, electrical engineering applications. *Bayer AG.* Name unverified.

2646 Apec® KU 1-9309
Polyester carbonate
Flame-retarded grade of polyester carbonate. *Bayer AG.* Name unverified.

2647 Apex 400
A proprietary alloy of aluminum with silicon. No manufacturer.

2648 Apexior
Heat-resistant organic coating for water-side corrosion protection of steam generating equipment and auxiliaries; for steam generating equipment, feed water heaters, evaporators, steam turbines, diesel cylinder liners, condensate tanks. *Dampney Company Inc.* Name unverified.

2649 APG® 225 Glycoside
68515-73-1
C_{8-10} alkyl polysaccharide ether
Caustic-stable wetting agent and coupler for agricutural formulations; biodegradable. *Henkel/Emery.*

2650 APG® 300 CS
Decyl polyglucose
Cosurfactant, auxiliary foaming agent for mild shampoos and other personal care cleansers. *Henkel/Cospha; Henkel Canada.*

2651 APG® 300 Glycoside
113976-90-2
Decyl Polyglucose
Caustic-stable degreaser, emulsifier, dispersing and wetting agent; for general purpose and hard surface cleaners, agricultural formulations; biodegradable. *Henkel/Emery.*

2652 APG® 600 CS
Lauryl polyglucose
Cosurfactant, viscosity modifier and thickener for mild shampoos, other personal care cleansers. *Henkel/Cospha; Henkel Canada.*

2653 APG® 600 Glycoside
110615-47-9
Lauryl glucoside
Caustic-stable detergent active; viscosity modifier for crutcher spray-dryer slurries; biodegradable. *Henkel/Emery.*

2654 Aphox
23103-98-2 7651 245-430-1
Pirimicarb

Granules containing 50% w/w pirimicarb; used for control of aphids. *ICI Chem & Polymers Ltd.*

2655 Aphrogene

Premetallized dyes. *ICI Chem & Polymers Ltd.* Discontinued.

2656 Aphthite

An alloy of 800 parts copper, 25 parts platinum, 10 parts tungsten, and 170 parts gold. No manufacturer.

2657 Aphtite

Zinc and cadmium-containing nickel bronzes; used for high-grade imitation silver products.

2658 Apiezon

A range of high quality oils, waxes and greases prepared by molecular distillation of low volatility hydrocarbon feedstocks. originally developed for high vacuum applications but suitable in several other sectors of industry. *Shell.* Name unverified.

2659 Apifac

Polyglyceryl-2 isostearate

Self-emulsifying base for water-oil cosmetic creams. *Gattefosse; Gattefosse SA.* Discontinued.

2660 Apifil

PEG-8 beeswax

Self-emulsifying base for oil-water emulsions in cosmetics and pharmaceuticals. *Gattefosse; Gattefosse SA.*

2661 Apollo 50C

74115-24-5 2435 277-728-2

Clofentezine

Suspension concentrate containing 500 g clofentezine per liter; an acaricide for use on top fruit. *Schering Agrochemicals Ltd.* Discontinued.

2662 Apolloy

A proprietary copper-iron alloy containing 0.25% copper and 0.08% carbon. No manufacturer.

2663 aposafranine

Diazotized safranine boiled with alcohol, a dyestuff.

2664 Aposet 707

A proprietary trade name for a ketone peroxide catalyst for polyesters. No manufacturer.

2665 Appeel®

Polymer

DuPont UK.

2666 Apperitive Saffron of Iron

Ferric subcarbonate

No manufacturer.

2667 apple acid

617-48-1 5747 210-514-9

$C_4H_6O_5$

malic acid

N-hydroxysuccinic acid. Chelating, buffering and flavoring agent.

2668 Appretan

Washfast finishing agents. *Hoechst UK.*

2669 Appretan Ant

An acrylate-based copolymer; A proprietary dispersant surfactant for finishing woven fabrics. *Alma Paint & Varnish Co.* Unverified.

2670 Appretan CPF

A proprietary polyvinyl acetate dispersion surfactant used for finishing woven, nonwoven and knitted fabrics. *Alma Paint & Varnish Co.* Name unverified.

2671 Appretan GM

A proprietary polyvinyl acetate-based dispersion surfactant used in the finishing of especially lightweight fabrics, knitted fabrics and nonwovens. *Alma Paint & Varnish Co.* Name unverified.

2672 Appretan TN

A vinyl acetate copolymer; A proprietary dispersion surfactant for finishing woven, nonwoven and knitted fabrics. *Alma Paint & Varnish Co.* Name unverified.

2673 Appreteen

Water soluble size. *S & D Chemicals Ltd.* Name unverified.

2674 APR®

Liquid photopolymer, printing plate-making system. *Asahi Chem. Industry.*

2675 Apricot kernel oil PEG-6 esters

97488-91-0 307-030-6

Labrafil M 1944 CS. Apricot kernel oil PEG-6 esters; hydrophilic oil for pharmaceutical and cosmetic formulations. *Gattefosse; Gattefosse SA.*

2676 Apron Combi

Mixture of metalaxyl, thiram and thiabendazole; a protectant fungicide for pea and bean seeds. *Ciba-Geigy Agrochemicals.*

2677 Apron T

Mixture of metalaxyl and thiabendazole; protectant fungicide for grass seed. *Ciba-Geigy Agrochemicals.*

2678 Aqua Gro G Granular

Blended nonionic soil wetting agent; 100% active Aqua-Gro 40% wt. (polyoxyethylene ester of cyclic acids-47%, polyoxyethylene ether of alkylated phenols-47%, silicone antifoam emulsion-6%), vermiculite-60% by weight; a granular wetting agent to aid water penetration and drainage; used by greenhouses, nurseries and interior plantscapers in the manufacture of horticulture growing and potting media. *Aquatrols Corp of Am.* Name unverified.

2679 Aqua magic

Silicones

Stain preventive for fabrics. *Adasco Inc.*

2680 Aqua Mer

A totally aqueous dry film photoresist, which comprises a three-layer sandwich construction of polyolefin/photopolymer/polyester. *Hercules.* Discontinued.

2681 Aqua Paste ™

Inhibited aluminum pigments; aluminum pigments for aqueous paints and coatings used for decorative metallic effects as well asd a wide range of protective coatings applications. *Silberline Mfg Co Inc.*

2682 aqua regia

8007-56-5 6706

Nitrohydrochloric acid

chloronitrous acid; nitromuriatic acid. A mixture of one volume of nitric acid and four volumes of hydrochloric acid; used in metallurgy, testing metals, in dissolving metals such as platinum and gold.

2683 Aqua Thix

Modified polysaccharide thixotrope; thickening agent and protective colloid for water-based coating compositions. *Hüls Am.* Discontinued.

2684 Aquabase

Fatty alcohols/PEG ester blend; base for oil-water emulsions; emulsifier for baby creams, cleansers, day creams/lotions, foundations, night creams, sunscreen preps. *Westbrook Lanolin.*

2685 Aquabase

Waterborne automotive paints. *ICI Chem & Polymers Ltd.*

2686 Aquabloc

Water-based acrylic resins and coatings; for use as a replacement for polyethylene or wax coated paper and board due to its high water barrier and high MVTR barrier properties; used in food packaging, bakery applications, frozen foods; has water, grease, and other product-resistance properties for paper and boards. *ADM Tronics Unlimited Inc.*

2687 Aquabrome

Swimming pool water disinfectant. *Great Lakes Europe.*

2688 Aqua-Chem®

Colorant dispersions; for coloring of emulsion and other water-borne coating compositions. *Hüls Am.*

2689 Aquacoat

Water-based overlacquers. *The Scottish Adhesives Co Ltd.*

2690 Aquadag

7782-42-5 4560 231-955-3

Graphite

Colloidal graphite in water; used as a lubricant for drawing tungsten and molybdenum filament wires, for metal forming operations such as extrusion, as an aid to cutting and for forming opaque coatings for face plates of cathode ray tubes. *Acheson Colloids.*

2691 Aquaflex

Tetrafilcon A

Contact lens material. *UCO Optics Inc.* Name unverified.

2692 Aquaflim

Fluorocarbon

Water and grease proofing for specialty papers and nonwovens. *CNC Int'l L.P.*

2693 Aquafloc

Flocculants *Dearborn Chemicals Ltd.*

2694 Aquaforte

Water-based primer/adhesive with low solids and containing no organic solvents; extrusion primer for film to film, paper and foil, adhesion promoter of inks to foil. *ADM Tronics Unlimited Inc.*

2695 Aquagel

A proprietary trade name for a colloidal bentonite; also a proprietary hydrated silicate of alumina for waterproofing cement. No manufacturer.

2696 Aqua-Gro L Liquid

Blended nonionic soil wetting agent; polyoxyethylene esters of cyclic acids-47%, polyoxyethylene esters of alkylated phenols-47%, silicone antifoam emulsion-6%; provides increased water penetration into and out of soils and horticultural media; used in manufacture of horticultural growing and potting media, on golf courses, turf, lawns and exterior landscapes, drainage of compacted soils and puddles, *poa annua* seedhead inhibition, spreader-activator; adjuvant; hydroseeding; dew, frost control. *Aquatrols Corp of Am.* Name unverified.

2697 Aqua-Gro S Spreadable
Blended nonionic soil wetting agent; 100% active Aqua-Gro 15% wt; (polyoxyethylene ester of cyclic acids-47%, polyoxyethylene ether of alkylated phenols-47%, silicone antifoam emulsion 6%) ground corn cobs-85%; granular soil wetting agent to aid water penetration and drainage, used on golf course, sports turf, lawns and exterior landscapes. *Aquatrols Corp of Am.* Name unverified.

2698 Aqualac
A proprietary shellac. No manufacturer.

2699 Aqualease 2802
High molecular weight silicone polymer; water-based heavy-duty release agent for cast urethane elastomer and integral skin urethane foams. *George Mann.*

2700 Aqualease 6102
Silicone emulsion
Release agent with better corrosion protection and excellent film formation; releases synthetic microcellular and solid elastomers, epoxies and polyester systems. *George Mann.*

2701 Aqualipid 95
8002-43-5 5452 232-307-2
De-oiled lecithin; marine animal feed additive. *Central Soya.*

2702 Aqualite
A proprietary phenol-formaldehyde synthetic resin laminated product and bearing material requiring only water as lubricant. No manufacturer.

2703 Aqualon® Cellulose Gum
9004-32-4 1877
carboxymethylcellulose sodium
Suspending agent for abrasive and polishing agents, and prevents syneresis in toothpaste; rheology control agent in creams and lotions; adhesive and cohesive agent used in denture adhesive and ostomy adhesive products. *Aqualon.*

2704 Aqualon® CMC-T
9004-32-4 1877
carboxymethylcellulose sodium, technical grades
Binder, thickener, stabilizer, suspending agent, film-former, rheology control aid, water-retention aid used for adhesives, aerial-drop fluids, ceramics, coatings, detergents, lithography, paper, textiles, and tobacco. *Aqualon.* Discontinued.

2705 Aqualon® CMC-T
9004-32-4 1877
Carboxymethylcellulose sodium
Aquasorb® A250; Aqualon® Cellulose Gum. Absorbent for urine, blood, and other body fluids; used in feminine hygiene products, medical disposables, disposable diapers. *Aqualon.* Discontinued.

2706 Aqualon® CMHEC-37L
Carboxymethyl hydroxyethyl cellulose
Hydrophilic colloid with good flocculating action on suspended solids and water-binding capacity; possibility for complexing and crosslinking reactions. *Aqualon.* Discontinued.

2707 Aqualon® EHEC
9004-58-4 2014
Ethyl hydroxyethylcellulose
Film-forming polymer for inks and coatings including printing inks, lacquers, varnishes, specialty finishes; available in four viscosity types. *Aqualon.*

2708 Aqualose
Alkoxylated lanolin and lanolin derivatives; water-soluble emollients, oil-water emulsifiers and wetting agents for cosmetics. *Westbrook Lanolin.* Name unverified.

2709 Aqualose L30
61790-81-6
PEG-30 lanolin
PEG-30 lanolin; emollient, emulsifier, plasticizer, solubilizer. *Westbrook Lanolin.*

2710 Aqualose LL100
PPG-40-PEG-60 lanolin oil; emollient, emulsifier, plasticizer, solubilizer used in aqueous/alcoholic preparations of alcohol content. *Westbrook Lanolin.*

2711 Aqualose W20
61791-20-6
Laneth-20
Plasticizer and solubilizer for hydrophobic substances; emollient, emulsifier, solubilizer for aftershaves, antiperspirants, cleansers, foam baths, shampoos, nailcare; carrier for foam baths, shampoos, nailcare. *Westbrook Lanolin.*

2712 Aqualox® 225-100, 225A-100
Amine salt; surfactant, corrosion inhibitor, lubricant, antiwear additive effective in inhibiting the attack of ferrous metals by aqueous solutions; for metalworking formulations; use 225-100 for EP/antiwear use; 225A-100 grade contains no phosphorus. *Alox.* Name unverified.

2713 Aqualox® 232
Amine salts of organic acids; corrosion inhibitor, low foaming surfactant for synthetic metalworking formulations, especially in aqueous solution. *Alox.* Name unverified.

2714 Aqualox® 2268
Low foaming corrosion inhibitor, lubricity and antiwear agent protecting ferrous metals from aqueous solutions and steel against high humidity conditions during extended indoor storage; emulsions deposit very thin waxy films. *Alox.* Name unverified.

2715 Aqualube
A proprietary plasticizer for water soluble materials. No manufacturer.

2716 Aquamet M
A dithiocarbamate, liquid formulation suitable for precipitating metal ions from solution; used in waste water treatment. *Alco Chemical Corp.*

2717 Aquamollin®
Water softening agents. *Cassella AG.*

2718 Aquanol
Liquid blend of polymeric resins and modified siloxane in a petroleum solvent base; used as an internal sealer and an external water repellant for concrete, brick, treated concrete block, stucco, wood, and stone. *Secure Inc.* Discontinued.

2719 Aquapel®
Fatty acid ketene dimer emulsions; sizing agent for paper and paperboard under neutral conditions. *Hercules; Hercules Ltd.* Discontinued.

2720 Aquapel®
Aquapel® 360XC. Reactive sizing emulsion for use against a wide variety of penetrants; for papermaking industry. *Hercules.*

2721 Aquaperle
Textile auxiliary chemicals. *ICI Chem & Polymers Ltd.*

2722 Aquaphil K
Lanolin and lanolin alcohol; emollient, emulsifier with enhanced water-oil emulsion stability. *Westbrook Lanolin.*

2723 Aquaplex
A proprietary trade name for alkyd resin dispersed in an aqueous medium; an emulsion varnish and lacquer resin vehicle for stucco, etc. No manufacturer.

2724 Aquapol
Hydrophilic urethane prepolymer. *Freeman Chemical Corp.* Name unverified.

2725 Aquaprint
Moldable impression material, synthetic resins. *BP Chemicals Ltd.*

2726 Aquaresin
Glyceryl boriborate
A plasticizer. No manufacturer.

2727 Aquarite
Water treatment chemicals *Albright & Wilson Ltd., Phosphates & Speciality Business.*

2728 Aquaseal
Bitumen-based emulsions for waterproofing roofs, asphalt, asbestos-cement, concrete. *Feb Ltd.*

2729 Aquaseal Aquaflex Liquid Felt
Polychloroprene bitumen emulsion with excellent elastomeric properties; for coating asphalt, asbestos cement, roofing felt, concrete, corrugated iron, wood, etc. *Feb Ltd.*

2730 Aquaseal Firmafix
A cold-applied bituminous felt adhesive for bonding roofing felts to metal, timber, concrete screeds, asphalt and built-up felt. *Feb Ltd.*

2731 Aquaseal Reflect
White solar reflective roof coating that prevents heat build-up on roofs and protects roofing membranes from uv attack and environmental pollution. *Feb Ltd.*

2732 Aquaseal Weatherwise Standard
All-weather roofing treatment providing waterproofing to asphalt, asbestos-cement, concrete, corrugated iron, roofing felt, etc. *Feb Ltd.*

2733 Aquasil
Window desiccants. *Laporte Industries Ltd.* Discontinued.

2734 Aquasil®
Inhibited aluminum pigments; used for aqueous paints and coatings used for decorative metallic effects as well as a wide range of protective coatings applications. *Silberline Mfg Co Inc.*

2735 Aquasoft®
Water-based textile ink; direct silk screen for T-shirts, athletic garments, aprons, tote bags, draperies, tablecloths, caps, wallhangings; excellent adhesion to cotton, blends and most synthetic fabrics. *Int'l Coatings Co Inc.*

2736 Aquasol
Aqueous dispersants; used for pigment dispersions. *Tennant-KVK Ltd.*

2737 **Aquasorb® A250**
9004-32-4 1877
Carboxymethylcellulose
Absorbent for urine, blood, and other body fluids; used in feminine hygiene products, medical disposables, disposable diapers. *Aqualon.*

2738 **Aquasperse**
Colorant dispersions; for coloring of emulsion and other water-borne coating compositions. *Hüls Am.*

2739 **Aquastab PA 48**
Dilauryl thiodipropionate
Stabilizer/antioxidant for polymers. *Eastman.*

2740 **Aquastore**
Crosslinked polyacrylamide. *Cyanamid BV.*

2741 **Aquasun**
Sun protection products, to protect skin while promoting a tan. *Richardson-Vicks Inc.* Name unverified.

2742 **Aquatac® 5527**
Water-based resin; freeze/thaw-stable dispersion for removable or permanent label adhesive applications. *Arizona.*

2743 **Aquatac® 6085**
Glyceryl rosinate aqueous dispersion; tackifier for pressure-sensitive adhesives; produces formulations with aggressive tack and peel, excellent shear properties after aging; for waterborne labels, decals, shelf liners, construction adhesives, tapes. *Arizona.*

2744 **Aquatac® 8005**
Water-dispersible resin; designed for freezer-grade label applications and as low softening point modifier for water-based adhesives. *Arizona.* Discontinued.

2745 **Aquatec**
A paraffin wax emulsion sometimes with aluminum acetate; proprietary trade name for a waterproofing material. No manufacturer.

2746 **Aqua-Tein C**
Collagen amino acids-acetamide MEA-propylene glycol. Substantive moisturizer, emollient for hair and skin care products (shampoos, conditioners, ethnic products, nutritive eye creams, face creams and lotions; anti-irritant for anionic shampoos. *Maybrook.*

2747 **Aquathane Series**
Fluorochemical polyurethane aqueous co-polymer; finishes for natural, synthetic and blended constructions; forms tough films with resistance to degradation. *CNC Int'l L.P.*

2748 **Aquathene AQ 120-000**
Ethylene vinylsilane copolymer; wire and cable resin for use in low voltage power cable applications. *Quantum/USI.*

2749 **Aquathene MP**
Structural waterproofing based on 125 micron thick polyethylene film; waterproofs solid floors where mechanical damage is unlikely. *Feb Ltd.*

2750 **Aquathene MV**
Structural waterproofing based on 300 micron black polyvinyl chloride film; provides protection for underground structures and roofing. *Feb Ltd.*

2751 **Aquatherm**
High-density insulation board specifically designed for use as a roof overlay board. *Feb Ltd.*

2752 **Aquatreat AR-225-D**
A free flowing low molecular weight sodium polymethacrylate; dispersant and desludger in water systems, cooling towers, boilers and heat exchangers. *Alco Chemical Corp.*

2753 **Aquatreat AR-232**
A low molecular weight sodium polymethacrylate solution; dispersant and desludger in water systems, cooling towers, boilers and heat exchangers. *Alco Chemical Corp.*

2754 **Aquatreat AR-626**
A low molecular weight acrylate copolymer solution; used for scale prevention in cooling towers, boilers, heat exchangers, oil field applications. *Alco Chemical Corp.*

2755 **Aquatreat AR-648**
A polycarboxylate copolymer solution; used for scale prevention in cooling towers, boilers and heat exchangers. *Alco Chemical Corp.*

2756 **Aquatreat AR-7-H**
A high molecular weight acrylic acid polymer; used for adhesives, lithographics, latex stabilization. *Alco Chemical Corp.*

2757 **Aquatreat AR-900**
A low molecular weight sodium polyacrylate; used for scale prevention in cooling towers and heat exchangers. *Alco Chemical Corp.*

2758 **Aquatreat DNM-30**
Sodium dimethyldithiocarbamate (15-16.6%), nabam (15-16.6%), inert; short-stop in emulsion polymerization of rubber; fungicide and bactericide for use in pulp/paper mills, sugar mills, drilling fluids, petroleum recovery; algicide for use in industrial recirculating water cooling towers, etc. *Alco Chemical Corp.*

2759 **Aquatreat DNM-9, DNM-25, DMN-360**
Dithiocarbamate salts; short-stop in emulsion polymerization of rubber; biocide, fungicide, and algicide used in water treatment, paper, sugar, and petroleum applications. *Alco Chemical Corp.*

2760 **Aquatreat KM**
128-03-0 204-875-1
$C_3H_6KNS_2$
Potassium dimethyl dithiocarbamate
Potassium dimethyldithiocarbamate; BUSAN 85; Carbamic acid, dimethyldithio-, potassium salt, hydrate. Polymerization short stops in the copolymerization of styrene and butadiene. *Alco Chemical Corp.*

2761 **Aquatreat SDM**
128-04-1 204-876-7
$C_3H_6NNaS_2$
Sodium dimethyldithiocarbamate
dimethyl-carbamodithioic acid, sodium salt; Sodium dimethyldithiocarbamate; Aceto SDD 40; Alcobam NM; Brogdex 555; Carbon S; Dibam; Dibam A; DMDK; methyl namate; Sharstop 204; sodium N,N-dimethyldithiocarbamate; Stafresh 615; Steriseal 40; Thiostop N; Vinstop; VulnopolNM; Wing Stop B; SDMDTC; sodium dimethylcarbamodithioate; Dimethyldithiocarbamic acid sodium salt; Freshgard 40. LD_{50} (rat orl) = 1000 mg/kg. *Alco Chemical Corp.*

2762 **Aquatrend®**
Colorant dispersions; for in-plant coloring of latex emulsion and other water-based coating composition. *Hüls Am.*

2763 **Aqua-Trete®**
Alkylalkoxysilane
Weatherproofing agents for concrete terra cotta, brick, stucco and masonry surfaces. *Hüls Am.*

2764 **Aquazym**
α-Amylase produced by submerged fermentation of a selected strain of *Bacillus subtilis;* intended for use in the desizing of textiles. *Novo Nordisk.*

2765 **Aracast**
Heterocyclic epoxide resins. *Ciba plc.* Name unverified.

2766 **Arakote® 3000**
Polyester resin; excellent flow grade. *Ciba-Geigy/Plastics.* Name unverified.

2767 **Araldite®**
Epoxy resin systems; used for casting, encapsulating, laminating, surface coating and as an adhesive. *Ciba plc.*

2768 **Araldite® 2001**
25928-94-3
Two-component epoxy adhesive; used for light engineering structures, boat and vehicle parts, sporting goods; bonds well to glass fiber laminates. *Ciba-Geigy GmbH.* Name unverified.

2769 **Araldite® CY 225**
25928-94-3
Liquid epoxy resin; used for casting systems, electrical insulated components, high strength structural applications. *Ciba-Geigy/Plastics.* Name unverified.

2770 **Araldite® ECN 1235**
25928-94-3
Solid epoxy cresol novolac; for high temperature adhesives, coatings, electrical and laminating applications. *Ciba-Geigy/Plastics.* Name unverified.

2771 **Araldite® GT 6060**
25928-94-3
Solid bisphenol A epoxy; for castings, electrical encapsulating, laminating and adhesive applications. *Ciba-Geigy/Plastics.* Name unverified.

2772 **Araldite® GZ 540 X-90**
25928-94-3
Epoxy solution in xylene; two component epoxy solution for maintenance and architectural coatings. *Ciba-Geigy/Plastics.* Name unverified.

2773 **Araldite® LT 8052**
Solid brominated epoxy; flame retardant epoxy for impregnating, casting applications. *Ciba-Geigy/Plastics.* Name unverified.

2774 **Araldite® LY 8047**
Brominated epoxy liquid; for prepregnated laminating applications. *Ciba-Geigy/Plastics.* Name unverified.

2775 **Araldite® PT 810**
25928-94-3
Epoxy resin; unmodified epoxy with color stability at high temperatures and good weathering; good thermal, adhesive and chemical resistance. *Ciba-Geigy Plastics UK.* Name unverified.

2776 **Araldite® PY 306**
25928-94-3
Bisphenol F epoxy liquid modifier of other resins for lower viscosity, higher solids coatings. *Ciba-Geigy Plastics; Ciba-Geigy Plastics UK.* Name unverified.

2777 Araldite® XD 4955
25928-94-3
Bisphenol F epoxy liquid for civil engineering and coatings requiring higher solids and performance. *Ciba-Geigy Plastics; Ciba-Geigy Plastics UK.* Name unverified.

2778 Araldite® XD 897
25928-94-3
Epoxy resin; stabilizer for chlorinated vinyl resins. *Ciba-Geigy Plastics UK.* Name unverified.

2779 Araldite® XU GY 358
25928-94-3
Epoxy resin; weatherable epoxy for maintenance and marine coatings, automotive refinishing. *Ciba-Geigy/Plastics. Name unverified.*

2780 Aranox®
p-(*p*-toluenesulfonyl amido) diphenylamine
Antioxidant protecting EVA and polyamide hot-melt adhesives, PP, LDPE, LLDPE, HDPE against thermal degradation. *Uniroyal.*

2781 Arassist APH
Organic acid; buffering agent and bath stabilizer in acid dyebaths. *Arol Chem. Prods.*

2782 Arassist HKM
Peroxide stabilizer and sequestrant for use in continuous bleaching of cotton or poly/cotton blends. *Arol Chem. Prods.*

2783 Aratex
68334-28-1 269-820-6
Hydrogenated vegetable oil
Partially hydrogenated vegetable oil (cottonseed, soybean); icing stabilizer; syrups; donut glazes; bakery dry mixes. *Van Den Bergh Foods.*

2784 Aratronic® 5001
25928-94-3
Bisphenol A epoxy liquid; electronic grade. *Ciba-Geigy/Plastics.* Name unverified.

2785 Aratronic® 5040
25928-94-3
Bisphenol F epoxy liquid; electronic grade. *Ciba-Geigy/Plastics.* Name unverified.

2786 Aravite® 3001
Cyanoacrylates
Structural adhesive for manufacturing of loudspeakers, optical instruments, jewelry, toys, rubber seals, domestic appliances, computers. *Ciba-Geigy Plastics UK.* Name unverified.

2787 Arazate®
14726-36-4 238-778-0
$C_{30}H_{28}N_2S_4Zn$
Zinc dibenzyldithiocarbamate
Dibenzyldithiocarbamic acid zinc salt. Activator. mp = 186°. *Uniroyal.*

2788 Arbeflex
Plasticizers. *Robinson Brothers Ltd.* Discontinued.

2789 Arbestab
Antioxidants and uv-stabilizers for polymers; useful in a number of plastics materials but particularly effective in polyolefins. *Robinson Brothers Ltd.*

2790 Arbo
Generic name for a range of putties, mastics, and sealants. *Adshead Ratcliffe & Co Ltd.*

2791 Arbocaulk
An acrylic emulsion-based sealant for gun application; principally used for internal pointing application. *Adshead Ratcliffe & Co Ltd.*

2792 Arbocel
Wood cellulose
For use in bitumen products, adhesives, plastics, and sealants. *ICI Chem & Polymers Ltd.*

2793 Arbocrylic
An acrylic solvent-based sealant for gun application; principally used for sealing external joints in building structures. *Adshead Ratcliffe & Co Ltd.*

2794 Arboflex
A glazing compound based on a blend of vegetable oils, plasticizers and butyl rubber; used for bead glazing aluminum and sealed timber window frames. *Adshead Ratcliffe & Co Ltd.*

2795 Arbofoam
Polyurethane foam packed in an aerosol dispenser; used to seal and insulate gaps around pipes and duct work and as a fixing, gap filling adhesive for doors and windows. *Adshead Ratcliffe & Co Ltd.*

2796 Arbogard
Herbicides. *ICI Chem & Polymers Ltd.*

2797 Arbokol
A range of single and two component polysulfide and epoxy/polysulfide sealants; used for sealants in building joints, floor joints and double glaze unit construction. *Adshead Ratcliffe & Co Ltd.*

2798 Arbolite
A putty composition based on a blend of vegetable oils; used for face glazing steel window frames. *Adshead Ratcliffe & Co Ltd.*

2799 Arbomast
A range of gun-applied sealants based on vegetable oils or butyl rubber low-cost, general-purpose sealants for a range of applications. *Adshead Ratcliffe & Co Ltd.*

2800 Arborsan®
Substitute for creosote; wood preservative. *Lanstar Ltd.*

2801 Arboseal
A range of preformed mastics strips based on butyl rubber and polybutenes; used for making watertight and dustproof seals between components, where the joint is under compression. *Adshead Ratcliffe & Co Ltd.*

2802 Arbosil
A range of RTV silicone-based single component sealants; used for sealing a wide range of industrial and building applications between a variety of substrates. *Adshead Ratcliffe & Co Ltd.*

2803 Arbostrip
Self-adhesive foam strips based on plasticized PVC; compression sealants for draft-proofing and similar applications. *Adshead Ratcliffe & Co Ltd.*

2804 Arbyl
Dispersing and leveling agent; used for dyeing, a wetting agent and detergent for pretreatment, designing and dyeing in the textile industry. *Degussa AG.* Name unverified.

2805 Arbylen
A wetting agent and detergent; used for pretreatment, desizing and dyeing in the textile industry. *Degussa AG.* Name unverified.

2806 ARcare® On/OFF 7810
Medical grade tape; on/off Ma-48 medical grade adhesive coated on one side of flexible nonwoven fabric; intended for mounting medical devices to the skin when good long-term adhesion and/or ease of removal after soaking with water are desired. *Adhesives Research.*

2807 Arcel Moldable Polyethylene Copolymers
9002-88-4 7728
Polyethylene
Resins for molded resilient foam packaging (for computers, military instruments, home entertainment electronics, medical devices), automotive, marine, and fiberglass laminate applications. *Arco Chemical Co.*

2808 Archil
1400-62-0 6994 215-750-6
orchil
orseille; Persio; cudbear; orchellin. A natural coloring matter obtained from *Roccela tinctoria* and other lichens. Also prepared by oxidation of orcinol. The coloring principle is orcein, which, in the presence of air and ammonia, oxidizes to a violet dye.

2809 Arcolloy
A nonmagnetic alloy of iron with 12-16% chromium, and less than 0.12% carbon, 0.5% manganese, 0.025% phosphorus, 0.025% sulfur, and 0.5% silicon. No manufacturer.

2810 Arcoloy
A proprietary copper-silicon casting alloy containing 97.25% copper, 2.63% silicon, 0.12% iron, and 0.01% phosphorus. No manufacturer.

2811 Arconate® Propylene Carbonate
108-32-7 203-572-1
$C_4H_6O_3$
Propylene carbonate
4-methyl-1,3-Dioxolan-2-one; 1,2-Propanediol cyclic carbonate; PC; 1,2-Propylene Carbonate. Solvent with high boiling point, low toxicity, broad range of applications; reactive diluent for woodbinders, urethane foams and coatings, foundry sand binders, in textile and synthetic fiber industry, natural gas treating; lubricant in cosmetics. mp = -55°; bp = 240°; d = 1.1890; n_D^{20} = 1.4210; slightly soluble in H_2O, soluble in organic solvents; LD_{55} (rat orl) = 29 g/kg. *Arco Chemical Co.*

2812 Arcosolv® DPM
34590-94-8 252-104-2
PPG-2 methyl ether
Solvent for coatings, cleaners, inks, agricutural products, cosmetics, chemical intermediate applications. *Arco Chemical Co.*

2813 Arcosolv® DPMA
88917-22-0
Dipropylene glycol methyl ether acetate
Solvent where a slow evaporating nonhydroxylic solvent is required; effective in coatings; as a coalescent in waterborne emulsion systems. *Arco Chemical Co.*

2814 Arcosolv® PM
107-98-2 203-539-1

$C_4H_{10}O_2$
Methoxypropanol
Methoxypropanol, α isomer; Propylene glycol methyl ether; 1-Methoxypropan-2-ol; 1-Methoxy-2-propanol; methoxy ether of propylene glycol; α-propylene glycol monomethyl ether; polypropylene glycol methyl ether; propylene glycol 1-methyl ether; (±)-1-methoxy-2-propanol; Dowanol 33B; Dowanol PM; Dowtherm 209; glycol ether PM; PGME; poly-solve MPM; propasol solvent M; UCAR solvent LM. Solvent for coatings, cleaners, inks, agricutural products, cosmetics, chemical intermediate applications. mp = -97°; bp = 118-119°; d = 0.9220; n_D^{20} = 1.4030; soluble in H_2O, organic solvents; LD_{50} (rat orl) = 5660 mg/kg. *Arco Chemical Co.*

2815 Arcosolv® PMA
108-65-6 203-603-9
$C_6H_{12}O_3$
Propylene glycol methyl ether acetate
Propylene Glycol Monomethyl Ether Acetate; 1-Methoxy-2-propyl Acetate; PGMEA; 1-Methoxy-2-propanol Acetate; Propylene glycol methyl ether acetate; 2-Methoxy-1-methylethyl acetate; 2-(1-Methoxy)propyl acetate. Slow-evaporating solvent with good solvency for many commonly used coating resins, e.g., acrylics, NC, and urethanes; used in lacquers, water-based paints. bp = 150°; d = 0.9690; soluble in H_2O (19 g/100 ml), organic solvents; LD_{50} (rat orl) = 8532 mg/kg. *Arco Chemical Co.*

2816 Arcosolv® TPM
20324-33-8 243-734-9
Triglycol monomethyl ether
Solvent for coatings, cleaners, inks, agricultural products, cosmetics, chemical intermediate applications; slow evaporating. *Arco Chemical Co.*

2817 Arctite Injection Mortar
Expanding cement; used for brickwork; prevents rising damp (DPC); nontoxic. *Arcmann Denmark A/S.*

2818 Arctite Quickbinder
A liquid for treating concrete for severe leaks. *Arcmann Denmark A/S.*

2819 Arctite Slurry 200 B
Waterproofing cement; used for concrete, high water pressure applications; nontoxic. *Arcmann Denmark A/S.*

2820 Arctite Tanking Mortar 500
Cement based waterproofing compounds, nontoxic; used for concrete and brick structures. *Arcmann Denmark A/S.*

2821 Arcton
A range of fluorinated hydrocarbon refrigerants and aerosol propellants (fluorocarbons). *ICI Chem & Polymers Ltd.*

2822 Ardeer Powder
An explosive containing 31-34% nitroglycerin, 11-14% kieselguhr, 47-51% magnesium sulfate, 4-6% potassium nitrate, and 0.5% ammonium or calcium carbonate. No manufacturer.

2823 Ardel® D-100
Polyarylate resin
Engineering plastic for automotive, electrical/electronic, glazing/solar energy, safety equipment, and plumbing applications; features excellent uv resistance, toughness, transparency, good electrical properties. *Amoco Chemical Co.* Discontinued.

2824 Ardenite
A proprietary synthetic resin; a molding composition. No manufacturer.

2825 Ardent
Suspension concentrate containing 40 g diflufenican and 400 g trifluralin per liter; used for control of weeds in winter cereals. *Embetec Crop Protection Ltd.*

2826 Ardmorite
A variety of bentonite, found in the Pierre shales at Ardmore, South Dakota.

2827 Ardux
Modified urea/formaldehyde resin. *Ciba plc.* Name unverified.

2828 arecaidine
499-04-7 814
$C_7H_{11}NO_2$
1,2,5,6-tetrahydro-1-methyl-3-pyridinecarboxylic acid
arecaine; methylguvacine. Isolated from betel nuts. dec 232°; soluble in H_2O, insoluble in organic solvents.

2829 arecoline
63-75-2 815 200-565-5
$C_8H_{13}NO_2$
1,2,5,6-tetrahydro-1-methyl-3-pyridinecarboxylic acid methyl ester
arecaline; arecholine; methyl arecaidin. Used in veterinary medicine as an anthelmintic (Cestodes). bp = 209°; d^{20} = 1.0495; n_D^{20} = 1.4302; LD_{50} (mus sc) = 100 mg/kg.

2830 Arelon
34123-59-6 5237 251-835-4
Isoproturon
Suspension concentrate containing 553 g isoproturon per liter; used for annual weed control in cereals. *Hoechst UK.*

2831 Aremco-Bond
High temperature organic adhesive. *Merlec Co.* Unverified.

2832 Aremco-Cast
High temperature casting material. *Merlec Co.* Unverified.

2833 Aremco-Coat
High temperature coating material. *Merlec Co.* Unverified.

2834 Aremsol A
Cocamidopropyl betaine
Foam booster for conventional shampoos, conditioning shampoos and conditioning rinses. *Ronsheim & Moore.* Name unverified.

2835 Aremsol MA
MEA lauryl sulfate
Base for preparation of liquid shampoos; biodegradable. *Ronsheim & Moore.*

2836 Arenka
A proprietary high-strength yarn manufactured from aramides (aromatic polyamides). *Erika Glanzstoff AG.* Name unverified.

2837 Arenolite
An artificial siliceous-argillaceous-calcareous stone.

2838 Aresenid
7778-39-4 833 231-901-9
AsH_3O_4
Arsenic Acid
Arsenic acid solution; wood preservative and chemical intermediate. *Mechema Chemicals Ltd.* Name unverified.

2839 Aresin
1746-81-2 217-129-5
$C_9H_{11}ClN_2O_2$
N'-(4-chlorophenyl)-N-methoxy-N-methylurea
Monolinuron; Afesin; Arresin; Monorotox. Selective systemic herbicide used to control broad-leaved weeds and some annual grasses in vegetable crops such as potatoes and leeks. mp = 80-83°; soluble in H_2O (735 mg/l), organic solvents; LD_{50} (rat orl) = 1660 mg/kg. *Hoechst UK.* Name unverified.

2840 Areskap
A proprietary trade name for a butylphenyl-phenol sodium sulfonate, a wetting agent. No manufacturer.

2841 Aresket
A proprietary trade name for a wetting agent; stated to be a butyl-diphenyl sodium sulfonate. No manufacturer.

2842 Aresklene
A proprietary trade name for a dibutyl-phenyl-phenol sulfonate; a mold lubricant and emulsifier. No manufacturer.

2843 Aretan
A mercury-based fungicide. *ICI Chem & Polymers Ltd.*

2844 Aretone 270
A proprietary trade name for a fine mica used as a pigment in paints. No manufacturer.

2845 Argentai
An alloy of 85% copper, 10% tin, and 5% cobalt.

2846 Argentalium
An aluminum alloy containing antimony.

2847 argentan
A nickel silver. It consists of 56% copper, 26% nickel, 18% zinc, and 1% iron, and is used as an electrical resistance alloy.

2848 Argidone®
56265-06-6 260-081-5
arginine-PCA
Moisturizing adjuvant for nutritive or generative creams or lotions; activates cell metabolism. *UCIB.*

2849 arginine PCA
56265-06-6 260-081-5
Argidone®; L-Proline; 5-Oxo compound with L-arginine;. Moisturizing adjuvant for nutritive or generative creams or lotions; activates cell metabolism.

2850 Argobase
Water-oil absorption bases containing lanolin and/or lanolin alcohols; emulsifiers for cosmetics and ointments. *Westbrook Lanolin.* Name unverified.

2851 Argobase 125
8027-33-6 232-430-1
Lanolin alcohols extract; emollient, water-oil emulsifier used in personal care products. *Westbrook Lanolin.*

2852 Argobase EU
Sterols and sterol esters lanolin extracts; absorption base. *Westbrook Lanolin.*

2853 Argonol
Liquid lanolin derivatives; fluid emollients and moisturizers for cosmetics. *Westbrook Lanolin.* Name unverified.

2854 Argonol 1SO
Plasticizer for hair sprays. *Westbrook Lanolin.*

2855 Argonol 40
85005-47-6 284-980-7
Isobutylated lanolin oil
Water-oil emulsifier, emollient used in personal care products. *Westbrook Lanolin.*

2856 Argonol 50 Pharmaceutical
Lanolin oil *Westbrook Lanolin.*

2857 Argonol ACE5
Cetyl acetate
acetylated lanolin alcohol. Emollient, moisturizer for baby oils, eye preparations, foam baths, sunscreen preparations, lubricant/glossing aid for hairsprays, lipsticks; binder for lipsticks, pressed powders. *Westbrook Lanolin.*

2858 Argowax
Refined lanolin alcohols; w/o emulsifier in cosmetics and pharmaceuticals. *Westbrook Lanolin.* Name unverified.

2859 Argowax Standard
8027-33-6 232-430-1
Lanolin alcohol BP
Gelling agent, water-oil emulsifier, emollient, moisturizer for baby creams, day creams/lotions, face masks, foundation, sunscreen preparations; plasticizer for hairsprays. *Westbrook Lanolin.*

2860 Argozie
Arguzoid
Argozoil. An alloy of from 54-56% copper, 23-38% zinc, 2-4% tin, 2-3.5% lead, and 13.5-14% nickel. No manufacturer.

2861 Argus DLTDP
123-28-4 204-614-1
$C_{30}H_{58}O_4S$
Dilauryl thiodipropionate
Didodecyl 3,3-thiodipropionate; Didodecyl 3,3'-thiodipropionate. Antioxidant used for polyolefins, thermoplastic elastomers synthetic rubber, antioxidant for cosmetics and pharmaceuticals. *Witco/Argus.* Discontinued.

2862 Argus DMTDP
16545-54-3 240-613-2
Dimyristyl thiodipropionate
Antioxidant for polyolefins and other polymeric systems. *Witco/Argus.* Discontinued.

2863 Argus DSTDP
693-36-7 211-750-5
Distearyl thiodipropionate
Antioxidant for polyolefins and other polymeric systems where long-term heat stability is required; also for pharmaceutical and cosmetic products, oils, greases, and lubricants. *Witco/Argus.* Discontinued.

2864 Argus DTDTDP
10595-72-9 234-206-9
Ditridecyl thiodipropionate
Antioxidant for polyolefins and thermoplastic elastomers especially latexes where a liquid antioxidant/stabilizer is dispersed more effectively. *Witco/Argus.* Discontinued.

2865 Arguzoid
An alloy of 56% copper, 23% zinc, 4% tin, 3.5% lead, and 13.5% nickel. No manufacturer.

2866 Argylene
Powder containing 8% w/w sodium silver thiosulfate; used to prolong flower life in pot plants. *Fargro Ltd.*

2867 Argyrolith
China silver
electroplate. Names given to alloys containing 50-70% copper, 10-20% nickel, and 5-30% zinc. Alfenide and Argentan are similar alloys; they are nickel silvers or German silvers. No manufacturer.

2868 Ariabel
Inorganic and organic cosmetic pigments; for coloring of cosmetic products. *Morton Int'l Ltd.* Discontinued.

2869 Ariagran
High strength water soluble, granular food colors; used for coloring of foodstuffs and pharmaceuticals. *Morton Int'l Ltd.* Discontinued.

2870 Arianor
Semipermanent water soluble hair colors, incorporated in hair products intended to color hair. *Morton Int'l Ltd.* Discontinued.

2871 Ariavit
High strength water soluble powder food colors; for coloring of foodstuffs and pharmaceuticals. *Morton Int'l Ltd.* Discontinued.

2872 Aricel
Melamine-formaldehyde resins, usually 80% solids; cross-linking polymer systems, textile finish, filter papers, nonwoven binder. *Astro Industries Inc.* Name unverified.

2873 Aricyl
Veterinary product; injectable deodorant and tonic. *Bayer AG.*

2874 Aridex
A retexturing and reproofing aid. *Laporte Industries Ltd.* Discontinued.

2875 Aridry B
Methylol stearamide
Water repellent durable to washing and dry cleaning; extender for use with fluorocarbons. *CNC Int'l L.P.*

2876 Arigal PMP
A proprietary trade name for a solution of an organic mercuric compound which, when used together with Arigal C, imparts a mildew resistant and rot-proof finish on cellulosic fibers; fast to water, washing and dry cleaning. *Ciba plc.* Name unverified.

2877 Arigran
Granular food colors of guaranteed purity. *Morton Int'l Ltd.* Discontinued.

2878 Arikrome S
Chrome complex solution
Water repellent for specialty papers and nonwovens; may be used with dyes or pigmented papers to improve resistant to leaching or crocking. *CNC Int'l L.P.*

2879 Arimid®
Fiber. *DuPont UK.*

2880 Aripol
Polyelectrolyte flocculating agent. *Steetley Chemicals Ltd.* Name unverified.

2881 Aristar
Ultrapure reagents and solvents. *BDH Chemicals Ltd.*

2882 Aristoflex
Hair lacquer resins. *Hoechst UK.*

2883 Aristol A
60% mono substituted C_{20-24} benzene; lubricant; lube oil additive; chemical feedstock for sulfonation to produce emulsifiers and corrosion preventatives. *Pilot.*

2884 Aristonate H
78330-12-8
Sodium petroleum sulfonate
Surfactant for formulating drycleaning soaps, cutting oils, textile oils, leather oils, rust preventative and fuel oil compositions; ore floation collectors; emulsifiers for agricutural sprays. *Pilot.*

2885 Arizole® Anethole Extra.
104-46-1 682 203-205-5
$C_{10}H_{12}O$
1-methoxy-4-(1-propenyl)benzene
Anethole; anise camphor; Monasirup; p-Propenylanisole; anise camphor; isoestragole; *p*-methoxy-β-methylstyrene; 1-methoxy-4-propenylbenzene; nauli "gum"; oil of aniseed; 1-(*p*-methoxyphenyl)propene; *p*-1-propenylanisole; *p*-propenylphenyl methyl ether; 1-methoxy-4-(1-propenyl)benzene; Methoxy-4-propenylbenzene; Propenylanisole. Anise flavoring material. (*trans* isomer): mp = 20-21°; $bp_{2.3}$ = 81-81.5°; d_4^{20}= 0.9883; n_D^{20} = 1.56145; insoluble in H_2O, soluble in organic solvents; LD_{50} (rat orl) = 2090 mg/kg. *Arizona.* Discontinued.

2886 Arizole® Pine Oil
Pine oil
Disinfectants and cleaners, odorant, frothing agent in mineral flotation, solvent. *Arizona.* Discontinued.

2887 Arizona 208
Tall oil fatty acid ester; for plasticizers, extenders, surfactants in grinding and cutting oils, specialty lubricant additives, corrosion inhibitors, specialty solvents for printing inks, metalworking, and oil well servicing. *Arizona.* Discontinued.

2888 Arizona DR-22
Disproportionated rosin; emulsifier, detergent, wetting agent; Used to prepare emulsifiers for styrene/butadiene rubber polymerization, as shortstop for solv. polymerizations of rubber, as plasticizer/tackifier; used to make Arizona disproportionated tall oil rosin soaps. *Arizona.* Discontinued.

2889 Arizona DR-24
Disproportionated rosin
Emulsifier for styrene/butadiene rubber production; intermediate for production of disproportionated rosin soaps. *Arizona.* Discontinued.

2890 Arizona DRS-40
61790-50-9 263-142-4

Potassium soap of disproportionated tall oil rosin; emulsifier, detergent, wetting agent; for ABS, styrene/butadiene rubber, other synthetic elastomers. *Arizona.* Discontinued.

2891 Arizona DRS-43
61790-51-0 263-144-5
Sodium rosinate
Sodium salt of disproportionated tall oil rosin; emulsifier, detergent, wetting agent; emulsifier for ABS, styrene/butadiene rubber, synthetic elastomers. *Arizona.* Discontinued.

2892 Arizona DRS-50
61790-50-9 263-142-4
Potassium rosinate
Potassium soap of disproportionated tall oil rosin; polymerization emulsifier for the styrene/butadiene rubber industry. *Arizona.* Discontinued.

2893 Arizona FA-7001
Tall oil dimer acid; for manufacturing of many dimer acid-based derivatives. *Arizona.* Discontinued.

2894 Arklone
76-13-1 200-936-1
Trichlorotrifluoroethane
A cleaning solvent used in aerosols. *ICI Chem & Polymers Ltd.*

2895 Arko metal
An alloy of 80% copper and 20% zinc.

2896 Arkopal
Alkylphenol polyglycol detergent bases. *Hoechst UK.*

2897 Arkopal N
Range of nonionic surfactants of the nonylphenol ethoxylate type in liquid, paste or wax form; wetting, dispersing, foaming, emulsification, detergent and cleaning agent used in domestic and industrial cleaners, disinfectant cleaners, auxiliaries for texttile. Leather and fur dressing, metal working, rubber, electroplating, pesticides, plant protection, building, anti-dusting and many other uses. *Hoechst UK.*

2898 Arkopon T
137-20-2 205-285-7
$C_{21}H_{40}NNaO_4S$
(Z)-2-[methyl(1-oxo-9-octadecenyl)amino]ethanesulfonic acid, sodium salt
Sodium N-methyl-N-oleoyltaurate; Sodium methyl oleyl taurate. Sodium oleoyl methyl tauride in powder form; dispersing and wetting agent used in wettable powders, plant protection and pest control. *Hoechst UK.* Name unverified.

2899 Arlacel® 20
1338-39-2 8872 215-663-3
Sorbitan laurate
Emulsifier for cosmetics, pharmaceuticals. *ICI Spec. Chem.; ICI Surf. Belgium.*

2900 Arlacel® 40
26266-57-9 8872
Sorbitan palmitate
Emulsifier for cosmetics, pharmaceuticals. *ICI Spec. Chem.; ICI Surf. Belgium.*

2901 Arlacel® 60
1338-41-6 8872 215-664-9
Sorbitan stearate
Emulsifier for cosmetics, pharmaceuticals. *ICI Spec. Chem.; ICI Surf. Belgium.*

2902 Arlacel® 80
1338-43-8 8872 215-665-4
Sorbitan oleate
Emulsifier for cosmetics, pharmaceuticals. *ICI Spec. Chem.; ICI Surf. Belgium.*

2903 Arlacel® 83
8007-43-0 8872 232-360-1
Sorbitan sesquioleate
Emulsifier. Cosmetic and pharmaceutical grade of Arlacel® C. *ICI Spec. Chem.; ICI Surf. Belgium.*

2904 Arlacel® 85
26266-58-0 8872 247-569-3
Sorbitan trioleate
Surfactant for cosmetics and pharmaceuticals. *ICI Spec. Chem.; ICI Surf. Belgium.*

2905 Arlacel® 165
Glyceryl stearate
PEG-100 stearate. Surfactant, emulsifier, thickener, opacifier for cosmetics and allied fields; acid-stable; self-emulsifying. *ICI Spec. Chem.; ICI Surf. Belgium.*

2906 Arlacel® 186
Glyceryl oleate, propylene glycol, 0.02% BHA and 0.01% citric acid as preservatives; surfactant, emulsifier, thickener for personal care products; defoamer for oral pharmaceutical products. *ICI Spec. Chem.; ICI Surf. Belgium.*

2907 Arlacide A
56-95-1 2140 200-302-4
Chlorhexidine diacetate
Preservative/bactericide for liquid and powd. preparations. *ICI Spec. Chem.* Name unverified.

2908 Arlagard
Bacteriostat. *ICI Am.*

2909 Arlamol
Blends of specific fatty acid esters; nonionic surfactants. *ICI Am.*

2910 Arlamol® E
25231-21-4
PPG-15 stearyl ether series
Emollient, solvent for personal care products. *ICI Spec. Chem.; ICI Surf. Belgium.*

2911 Arlamol® ISML
Isosorbide laurate
Surfactant. *ICI Australia.*

2912 Arlasolve® 200
Isoceteth-20
Surfactant, emulsifier, solubilizer for cosmetics. *ICI Spec. Chem.*

2913 Arlasolve® DMI
5306-85-4 226-159-8
Dimethyl isosorbide
Surfactant, emollient. *ICI Spec. Chem.*

2914 Arlatone® B
Polysorbate 85 and dinonyl phenol. Emulsifier for cosmetic cleansers, waterless hand cleaners, blooming bath oils. *ICI Spec. Chem.*

2915 Arlatone® G
61788-85-0
PEG hydrogenated castor oil
Surfactant, solubilizer, emollient; formulates clear gels. *ICI Spec. Chem.; ICI Surf. Belgium.*

2916 Arlatone® T
PEG-40 sorbitan peroleate
Emulsifier, solubilizer, antistat, lubricant, spreading agent; used for bath oils, and in the textile industry. *ICI Spec. Chem.; ICI Surf. Belgium.*

2917 Arlin
A proprietary name for polyethylene film. *Polyversion Inc.* Unverified.

2918 Arlinflex
Rigid vinyl. *Arlington Mills Inc.*

2919 Arlon®
Polyether-etherketone
Used for structures requiring toughness and chemical resistance, e.g., valve seats, compressor plates. *Greene, Tweed & Co.*

247-568-8

2920 Arloy®
Blends of Dylark SMA and polycarbonate; engineering resins providing heat resistance, high impact strength and moldability; for automotive instrument panels, seat belt retractors, speaker grilles, surgical appliances, institutional feeding trays, camera components and power tool housings. *Arco Chemical Co.*

2921 Armac®
Series of cationic surfactants which consist of acetate salts of primary amines, and which are soluble; used for pigment flushing, froth flotation and flocculation, particularly in mineral flotation; petroleum processing; leather; paper; pigments and surface coatings, ceramics. *Akzo Chemie UK Ltd.*

2922 Armco
A trademark for ingot iron and stainless steel. No manufacturer.

2923 Armco Ingot Iron
A trademark for a very pure iron, 99.84% pure. No manufacturer.

2924 Armeen
A range of coco, oleyl and stearyl amines; used as chemical intermediates, anticaking agents and in secondary oil recovery. *Harcros Australia.*

2925 Armeen®
Range of cationic surfactants composed mainly of primary amines, in liquid, paste or solid form; degree of surface activity varies with composition; they have the ability to change mineral surfaces from hydrophobic to hydrophilic; used in petroleum; road emulsions, plastics, rubber, textiles, leather, herbicides, fungicides, rodent repellents, mineral flotation, paper, pigments and surface coatings, water and sewage treatment, wax, sealant formulations and cement curing. *Akzo Chemie UK Ltd.*

2926 Armeen® 2-10
1120-49-6 214-312-1
$C_{20}H_{43}N$
Didecylamine

Di-n-decylamine; 1-Decanamine, N-decyl-; Didecylamine. Industrial surfactant. *Akzo.*

2927 Armeen® 2-18
112-99-2 204-020-2
Dioctadecylamine
Industrial surfactant. *Akzo.*

2928 Armeen® 2C
61789-76-2 263-086-0
Dicocamine
Emulsifier, flotation agent, corrosion inhibitor. *Akzo.*

2929 Armeen® 2HT
61789-79-5 263-089-7
Hydrogenated ditallowamine
Emulsifier, flotation agent, corrosion inhibitor. *Akzo.*

2930 Armeen® 2T
68783-24-4 272-191-0
Ditallowamine
Emulsifier, flotation agent, corrosion inhibitor. *Akzo.*

2931 Armeen® 3-12
102-87-4 203-063-4
Tridodecylamine
Chemical intermediate for manufacturing of sol. betaines and quaternary ammonium salts; carrier for manufacturing of citric acid and oil. *Akzo.*

2932 Armeen® 3-16
67701-00-2 266-924-3
Trihexadecylamine
Chemical intermediate for manufacturing of oil-soluble betaines and quaternary ammonium salts. *Akzo.*

2933 Armeen® 12
124-22-1 204-690-6
$C_{12}H_{27}N$
Lauramine
n-Dodecylamine; dodecylamine; lauryl amine; 1-Dodecanamine. Industrial surfactant. *Akzo.*

2934 Armeen® 12D
124-22-1 204-690-6
$C_{12}H_{27}N$
Lauramine (primary amine)
n-Dodecylamine; dodecylamine; lauryl amine; 1-Dodecanamine. Emulsifier, flotation agent, corrosion inhibitor; lubricant for metal treatment. *Akzo; Akzo Chem. BV.*

2935 Armeen® 16
143-27-1 205-596-8
$C_{16}H_{35}N$
Palmitamine
Cetylamine; n-Hexadecylamine; 1-Hexadecanamine; 1-Hexadecylamine; alamine 6; armeen 16d; palmitylamine; 1-amino hexadecane. Industrial surfactant. *Akzo.*

2936 Armeen® 16D
143-27-1 205-596-8
$C_{16}H_{35}N$
palmitamine
Amine 16D; Cetylamine; n-hexadecylamine; 1-hexadecanamine; 1-hexadecylamine; alamine 6; armeen 16d; palmitylamine; 1-amino hexadecane. Emulsifier, flotation agent, corrosion inhibitor. *Akzo; Akzo Chem. BV.*

2937 Armeen® 16D
143-27-1 205-596-8
Palmitamine
Emulsifier, flotation agent, corrosion inhibitor. *Akzo; Akzo Chem. BV.*

2938 Armeen® 18
124-30-1 204-695-3
$C_{18}H_{39}N$
Stearamine
1-Aminooctadecane; Stearylamine; 1-Octadecanamine; 1-Octadecylamine; octadecylamine; Adogenen 142; Alamine 7; Armeen 1180; n-octadecylamine. Emulsifier, flotation agent, corrosion inhibitor, anticaking agent; hard rubber mold release agent. *Akzo; Akzo Chem. BV.*

2939 Armeen® 18D
124-30-1 204-695-3
Stearamine
Distilled stearamine. Emulsifier, flotation agent, corrosion inhibitor, anticaking agent; rubber processing auxiliary; mold release agent for plastics and rubber. *Akzo.*

2940 Armeen® C
61788-46-3 262-977-1
Cocamine
Emulsifier, flotation agent, corrosion inhibitor, stripping agent for paints. *Akzo; Akzo Chem. BV.*

2941 Armeen® CD
61788-46-3 262-977-1
Cocamine (primary amine)
Emulsifier, flotation agent, corrosion inhibitor, stripping agent for paints. *Akzo.*

2942 Armeen® DM12D
112-18-5 203-943-8
Dimethyl lauramine
Intermediate in the manufacture of surfactants. *Akzo.*

2943 Armeen® DM16D
112-69-6 203-997-2
Dimethyl palmitamine
Chemical intermediate, raw material for surfactants. *Akzo.*

2944 Armeen® DM18D
124-28-7 3525 204-694-8
dimanthine
Dimethyl stearamine. Chemical intermediate, raw material for surfactants. *Akzo.*

2945 Armeen® DMCD
61788-93-0 263-020-0
Dimethyl cocamine
Chemical intermediate, raw material for surfactants. *Akzo; Akzo Chem. BV.*

2946 Armeen® DMHTD
61788-95-2 263-022-1
Dimethyl hydrogenated tallow amine
Chemical intermediate, raw material for surfactants. *Akzo; Akzo Chem. BV.*

2947 Armeen® DMOD
28061-69-0 248-811-0
Dimethyloleamine
Surfactant intermediate. *Akzo.*

2948 Armeen® DMSD
61788-91-8 263-017-4
Soyaalkyl dimethylamine
Surfactant intermediate. *Akzo.*

2949 Armeen® DMTD
68814-69-7 272-339-4
Dimethyltallowamine
Surfactant intermediate. *Akzo.*

2950 Armeen® HT
61788-45-2 262-976-6
(Hydrogenated tallow) amine (primary amine)
Emulsifier, flotation agent, corrosion inhibitor, chemical intermediate, anticaking agent. *Akzo; Akzo Chem. BV.*

2951 Armeen® HTD
61788-45-2 262-976-6
Hydrogenated tallowamine
Emulsifier, flotation agent, corrosion inhibitor, chemical intermediate, anticaking agent. *Akzo; Akzo Chem. BV.*

2952 Armeen® L8D
104-75-6 203-233-8
$C_8H_{19}N$
2-Ethylhexylamine
2-Ethylhexylamine; ethylhexylamine; 3-Octylamine; 1-Hexanamine, 2-ethyl-; 1-amino-2-ethylhexane. Distilled 2-Ethylhexylamine, a chemical intermediate for vapor phase corrosion inhibitors. *Akzo.*

2953 Armeen® M2-10D
7396-58-9 230-990-1
Didecyl methylamine
Amine M210D. Chemical intermediate for water-soluble betaines; catalyst for urethane resins. *Akzo.*

2954 Armeen® M2C
61788-62-3 262-990-2
Dicocomethylamine
Chemical intermediate; surfactant; for manufacturing of oil-soluble betaines and quaternary ammonium salts. *Akzo; Akzo Chem. BV.*

2955 Armeen® M2HT
61788-63-4 262-991-8
Dihydrogenated tallow methylamine
Chemical intermediate in the manufacture of oil-soluble betaines and quaternary ammonium salts. *Akzo; Akzo Chem. BV.*

2956 Armeen® OL
112-90-3 204-015-5
Oleamine
Emulsifier, flotation agent, corrosion inhibitor. *Akzo.*

2957 Armeen® OLD
112-90-3 204-015-5

Oleamine
Emulsifier, flotation reagent, corrosion inhibitor. *Akzo.*

2958 Armeen® T
61790-33-8 263-125-1
Tallowamine (primary amine)
Emulsifier, flotation reagent, corrosion inhibitor, dispersant, anticaking agent, chemical intermediate, cosmetics ingredient. *Akzo; Akzo Chem. BV.*

2959 Armeen® TD
61790-33-8 263-125-1
Tallowamine, distilled
Emulsifier, flotation agent, corrosion inhibitor. *Akzo.*

2960 Armenian cement
A jeweler's cement containing gum mastic, isinglass, gum ammoniac, alcohol, and water; it is made by soaking the isinglass in water and mixing it with the spirit containing the gums.

2961 Armid® E
112-84-5 204-009-2
$C_{22}H_{43}NO$
Erucamide
Erucamide; Erucylamide; (Z)-13-Docosenamide. Mold release agent for rubber and plastics; auxiliary for processing rubber. *Akzo.*

2962 Armid® HT
61790-31-6 263-123-0
Hydrogenated tallowamide
Also antifoam in steam generator systems, lubricant additive; auxiliary for rubber processing. *Akzo.*

2963 Armid® HTD
124-30-1 204-695-3
$C_{18}H_{39}N$
Stearylamine
1-Aminooctadecane; Stearylamine; 1-Octadecanamine; 1-Octadecylamine; octadecylamine; adogenen 142; alamine 7; armeen 1180; n-octadecylamine. Processing aid for high viscosity rubber compounds; facilitating flow behavior and improving mold release. mp = 50-52°; bp_{32}= 232°. *Akzo.* Name unverified.

2964 Armid® O
301-02-0 206-103-9
Oleamide
Release agent in cosmetics, penetrant in paper manufacture. *Akzo.*

2965 Armillatox®
Emulsion of polyhydric phenols in soap; home garden lawn treatment and fungicide; controls moss in lawns; reduces the severity of club root; hinders the spread of honey funfus. *Armillatox Ltd.*

2966 Armite
Vulcanized fiber. *Spaulding Fibre Co.* Name unverified.

2967 Armix 146
Formulated product; specialty product designed to enhance the effectiveness of MSMA formulations. *Witco Corporation.*

2968 Armix 176
Formulated product; tank mix which provides a combination of wetting, sticking, spreading and penetration. *Witco Corporation.*

2969 Armofilm
Long chain filming amine emulsion. *Akzo Chemie UK Ltd.*

2970 Armoflo®
Conditioner; hygroscopic salts and fertilizers. *Akzo Chemie UK Ltd.*

2971 Armofog
Anticondensing agent for polyolefins. *Akzo Chemie UK Ltd.*

2972 Armogard
Fuel oil additives. *Akzo Chemie UK Ltd.*

2973 Armogloss
Cationic car wash additive. *Akzo Chemie UK Ltd.* Discontinued.

2974 Armohib®
Acid inhibitors. *Akzo Chemie UK Ltd.*

2975 Armohib® 18, 28
Formulated products containing cationic surfactants consisting of aliphatic nitrogen derived materials; inhibitors used in acid pickling; plant cleaning; oil well acidizing; 18 developed for use with sulfuric, phosphoric, citric and sulfamic acids; 28 for use with hydrochloric acid. *Akzo Chemie UK Ltd.*

2976 Armor-Kote
Emulsified coal tar pitch. *Crowley Chem.*

2977 Armor-ply
A proprietary trade name for metal bonded plywood. No manufacturer.

2978 Armoslip®
Slip and antiblocking agents for polyolefins. *Akzo Chemie UK Ltd.*

2979 Armostat®
Antistatic agent for polyolefins. *Akzo Chemie UK Ltd.*

2980 Armoteric LB
Amphoteric surfactant supplied as a yellowish liquid; for baby shampoos; bubble baths; strong acid, and alkaline cleaning detergents. *Akzo Chemie UK Ltd.*

2981 Armoteric SB
Amphoteric surfactant supplied as a yellowish paste; for baby shampoos; bubble baths; strong acid and alkaline cleaning detergents. *Akzo Chemie UK Ltd.* Name unverified.

2982 Armourcote
Unreinforced fluorocarbon coatings and coatings reinforced with stainless steel, molybdenum and ceramic; for low friction and nonstick surfaces in baking and food processing, and general industrial applications. *Fothergill Tygaflor Ltd.* Name unverified.

2983 Armowax
Synthetic waxes. *Armour Hess Chemicals.* Unverified.

2984 Armowax
Processing aid for highly filled polyolefins. *Akzo Chemie UK Ltd.*

2985 Armowax EBS
110-30-5 203-755-6
$C_{38}H_{76}N_2O_2$
Octadecanamide, N,N'-1,2-ethanediylbis-
Ethylene bis(stearamide); N,N'-ethylenedi(stearamide). A proprietary trade name for N,N'-ethylene bisstearamide. *Armour Hess Chemicals.* Unverified.

2986 Armul 17
Formulated product; emulsifier for paraffinic hydrocarbon crop oils. *Witco/Organics.* Discontinued.

2987 Armul 22, 88
Blended emulsifiers; emulsifiers which, when used in combinations, can be used to formulate a wide variety of pesticide products. *Witco Corporation.*

2988 Arneel DN
A proprietary trade name for the dimerized product of octadecene and octadecadiene nitriles; a vinyl plasticizer. *Armour Pharmaceutical Co.* Name unverified.

2989 Arneel HF
A proprietary trade name for the 18, 20 and 24 carbon atom fatty acid nitriles; vinyl plasticizers. *Armour Pharmaceutical Co.* Name unverified.

2990 Arneel S
A proprietary trade name for a derivative of octadecene and octadecadiene nitriles; a vinyl plasticizer. *Armour Pharmaceutical Co.* Name unverified.

2991 Arneel TOD
A proprietary trade name for a derivative of octadecene and octadecadiene nitriles. A vinyl plasticizer. *Armour Pharmaceutical Co.* Name unverified.

2992 Arneel® OD
Octadecene nitrile
Detergent, wetting agent, rust inhibitor. *Armour Pharmaceutical Co.*

2993 Arnica Oil CLR
825
Arnica extract, soybean oil, tocopherol; emollient, conditioner; protective skin and hair care products. *Dr. Kurt Richter; Henkel/Cospha.*

2994 Arnica Yellow
An azo dyestuff prepared by condensing *p*-nitrotoluenesulfonic acid with *p*-aminophenol, in the presence of aqueous caustic soda; dyes cotton golden-yellow from a salt bath.

2995 Arnite A.K.U
A trade name for a polyethylene glycol terephthalate; injection molding material. *Algemene Industriele.* Unverified.

2996 Arnite G
Thermoplastic polyester grade; for injection molding and extrusion. *Algemene Industriele.* Unverified.

2997 ARO
Asbestos and nonasbestos friction material; friction material for brakes and clutches. *Caramba Chemie GmbH.* Name unverified.

2998 Arobleach HW
Blend of oxygenated inorganics, bleach, and surfactants; for single-bath scouring, bleaching, and dyeing of polyester/cotton blends. *Arol Chem. Prods.*

2999 Arobleach MX
Peroxygen compound; one-bath bleaching and dyeing agent for knit or woven fabrics of cotton or cotton/synthetic blends. *Arol Chem. Prods.*

3000 Arochem
Aroplax
A proprietary trade name for soft oil modified alkyds. No manufacturer.

3001 Aroclean MC-4
Heavy-duty industrial cleaner for metal surfaces, plastic, concrete; solubilizes trimer build-up in textile processing equipment. *Arol Chem. Prods.*

3002 Aroclear
Proprietary blend with alkaline builders; one step clearing agent for reduction

clearing of disperse dyes on polyester; biodeg.; removes excess dye concs. on fabric, anti-redeposition agent on fabric and equip.; stripping agent for overdyed fabric lots. *Arol Chem. Prods.*

3003 Arodet AA-350
Alkylaryl sulfonate with glycol coupling agents; detergent, wetting agent, dyeing assistant, scouring agent, leveler, retarder, dye dispersant, finishing agent, emulsifier. *Arol Chem. Prods.*

3004 Arodet AN-100
Modified nonionic derivative; detergent, wetting agent, scouring agent, dye dispersant. *Arol Chem. Prods.*

3005 Arodet AN-160
Detergent, wetting agent for natural and synthetic fiber processing, textile scouring, cotton desizing, kier boiling; dye dispersant, leveling and penetrating agent; stable to acid and alkalies, hard water, bleaching agents. *Arol Chem. Prods.*

3006 Arodet BLN Special
Blend of long-chain ethoxylates; detergent, wetting agent, penetrant for natural and synthetic fibers; aids dyestuff dispersion; fulling agent for wool; solv. emulsifier. *Arol Chem. Prods.*

3007 Arodet BN-100
Ethoxylated alcohol
Detergent, wetting agent, emulsifier, dyeing assistant, dispersant for dyeing, finishing, textiles, pigments, resins. *Arol Chem. Prods.*

3008 Arodet E-15
Blend; one-bath scouring agent for dyebaths for woolen fabrics; imparts wetting, scouring and leveling without interfering with subsequent dye procedures. *Arol Chem. Prods.*

3009 Arodet HCS
Ethoxylate
General-purpose detergent, wetting agent, scouring agent for synthetic and natural fibers; high cloud point, relatively low foaming, high temperature operating stability. *Arol Chem. Prods.*

3010 Arodet MKD
Blend of nonionic and neutralized phosphate ester surfactants; low temperature textile scouring agent for removal of sizes, waxes, and natural or synthetic oils from fabrics. *Arol Chem. Prods.*

3011 Arodet N-100
Nonylphenol PEG ester
Scouring and soaping off agent for natural and synthetic fibers. *Arol Chem. Prods.*

3012 Arodet TA-8
Phosphate ester
Multipurpose surfactant for textile processing; for desizing, kier boiling, bleaching, wetting and dispersion in jig or beck, after-scouring; emulsifier for polar and nonpolar solvs. used as carriers (trichlorobenzene, butyl benzoate, etc.) *Arol Chem. Prods.*

3013 Arofene
Formaldehyde polymers with phenol and substituted phenols usually supplied in solvent other than water; used for paper impregnation: air, oil and fuel filters; fiber bonding: nonwovens of all types; laminates: rolled and flat stock; adhesives: high performance. *Ashland Chemical Company.* Name unverified.

3014 Arofix F-6
Fixative for acid dyes on nylon when applied as an after-treatment. *Arol Chem. Prods.*

3015 Arofix SRN
Modified formaldehyde resin; fixative for direct dyes where complete removal of inorganic salts from the dyed goods impractical, e.g., in package, beam, or skein work. *Arol Chem. Prods.*

3016 Aroflat®
Alkyd synthetic resin
For manufacture of flat wall paints. *Reichhold.*

3017 Aroflat® 3113-P-30
Flat alkyd *Reichhold.*

3018 Aroflint®
Resin solutions. *Reichhold.*

3019 Aroflint® 202-A6X-60
Polyester resin. *Reichhold.*

3020 Aroflint® 303-X-90
25928-94-3
Epoxy resin. *Reichhold.*

3021 Arofoam
Unsaturated polyester, two component system; used for structural applications such as acrylic tubes and showers with fiberglass reinforcement. *Ashland Chemical Company.* Name unverified.

3022 Arofoam SNI
Ethoxylates blend; micro-foam surfactant for foam dyeing procedures on synthetic fibers; dispersant for dyestuffs; dye leveling agent; antiprecipitant. *Arol Chem. Prods.*

3023 Arofos 200 Conc
Phosphate ester blend; detergent, wetting agent, emulsifier, penetrant, dye leveling agent, dispersant. *Arol Chem. Prods.*

3024 Arofos 326
Polyphosphorylated surfactant
Detergent, emulsifier, wetting agent. *Arol Chem. Prods.*

3025 Aroful BV-50
Amine condensate/ethoxylates blend; fulling detergent for carbonized woolen fabrics under acidic conditions; wetting agent, penetrant. *Arol Chem. Prods.*

3026 Arogrip
Adhesive for marine market. *Ashland Chemical Company.* Name unverified.

3027 Arol Biodet
Biodeg
Scouring agent, rapid wetting agent for textile processing including degreasing, desizing, bleaching, dyeing, and finishing operations. *Arol Chem. Prods.*

3028 Arol Defoamer NA2X
Silicone-stearate blend; defoamer for atmospheric dyeing operations in textile industry. *Arol Chem. Prods.*

3029 Arol Woolbrite
Specialty reducing agent for wool treatment. *Arol Chem. Prods.*

3030 Arolev ADL-30
Long-chain derivative; leveling and retarding agent for wool and acrylic fibers; stable to dilute acids and alkalies, hard water, salts. *Arol Chem. Prods.*

3031 Arolev CDD
Sulfated fatty ester blend; surfactant, fast wetting/rewetting agent, emulsifier, leveling agent for cellulosics dyed with direct dyes, various synthetics and blends; for scouring, solvent scouring, sizing, kier bleaching, etc.; biodegradable. *Arol Chem. Prods.*

3032 Arolev MTR-7
Low foam leveling agent, dye dispersant, lubricant; synergistic with polyester dye carriers; for piece dyeing, yam dyeing, atmospheric or pressure equipment. *Arol Chem. Prods.*

3033 Arolon®
Liquid copolymers for paints. *Reichhold.*

3034 Arolon® 580-W-42
Waterborne dispersion alkyd. *Reichhold.*

3035 Aroltergc 100M
Blend of fatty acid alkanolamides and phosphate esters detergent, emulsifier, wetting agent. *Arol Chem. Prods.*

3036 Arolube MIT-1
Ethoxylates blend; softener, lubricant, crack mark inhibitor for all fibers; for dye bath, bleach bath, scouring bath; promotes leveling; stable under high temp. and pressure. *Arol Chem. Prods.*

3037 Aromabator PC-80
A tamed and stabilized nontoxic chlorine dioxide complex concentrate formulated for use as an additive deodorant without free chlorine release. formulated for use as an airborne spray for industrial applications; effectively arrests malodors caused by viruses, fungi, bacteria and coliform densities; added to coolants, cutting oils, industrial sumps, sludge pits, cooling towers, waste water, marine holding stations, ships bilging areas, animal housing and restroom surfaces; for spray application. *Punati Chemical Corp.* Unverified.

3038 Aromabator PC-88
A tamed and stabilized nontoxic chlorine dioxide complex concentrate formulated for use as an additive deodorant without free chlorine release; as an airborne spray in home and farm application; for use in kitchens, toilets, outhouses, pet and animal housing; can be safely sprayed on fabric and non-fabric syrfaces, on air conditioning and humidifier filters. *Punati Chemical Corp.* Unverified.

3039 Aromaplas
Range of perfumes for plastics. *PPF International Ltd.* Name unverified.

3040 Aromasol
Solvents. *ICI Chem & Polymers Ltd.*

3041 Aromasol 17
A narrow distillation range white spirit substitute with 17% aromatic content. *Sasolchem.*

3042 Aromatic Oil 745
Aromatic plasticizer, used in adhesives, rubber (cements, mechanical and molded goods, tires), caulking compounds. *Neville.*

3043 Aromatic Solvent 150
8030-30-6 7329 232-443-2
150° flash aromatic naphtha with narrow distillation range; high solvency, high flash point for paint and protective coatings, herbicide and pesticide carrier, synthetic resin manufacturing, degreasing applications. *Texaco.*

3044 Aromex
Powdered perfumery compounds. *Bush Boake Allen Ltd.*

3045 Aromix
Solvent-emulsifier concentrates for pesticide formulations. *Plant Protection.* Name unverified.

3046 Aromix
Mixture of heavy aromatics and aliphatics; solvent. *Sasolchem.*

3047 Aromox® C/12-W
61791-47-7 263-180-1
Dihydroxyethyl cocamine oxide
Wetting agent, emulsifier, stabilizer, antistat, foaming agent for detergents, shampoos, cosmetics, textiles, metal plating, petroleum additives, paper, plastics, rubber; gel sensitizer for latex foam; biodegradable. *Akzo; Akzo Chem. BV.*

3048 Aron α
A proprietary cyano-acrylate adhesive. *Toagosie Chemical Co.* Name unverified.

3049 Aroplaz®
Synthetic resins for use in compounding protective and decorative coatings, printing, textile inks, and for general industrial use. *Reichhold.*

3050 Aroplaz® 3667-Z-80
High solids alkyd. *Reichhold.*

3051 Aroplaz® 6820-100
High solids polyester. *Reichhold.*

3052 Aropol
Unsaturated polyester resins, including orthophthalic, isophthalic, and other specialty polymer types; used for fiberglass reinforced polyester applications in construction, transportation, gel coats, marine, consumer, electrical and corrosion resistant markets. *Ashland Chemical Company.* Name unverified.

3053 Aropol 2036
Isophthalic polyester resin; resin for molding and pultrusion applications. *Ashland Chemical Company.*

3054 Aropol 7020
High reactivity isophthalic polyester resin; can be chemically thickened for use in low-shrink and controlled-shrink SMC and BMC applications. *Ashland Chemical Company.*

3055 Aropol 7240T-15
Isophthalic polyester resin; thixotropic, prepromoted, chemical-resistant resin for use in hand lay-up and spray lay-up, fume hoods and ducts, tanks, pipes. *Ashland Chemical Company.*

3056 Aropol 7710
Isophthalic polyester resin; flexible resin with outstanding toughness and tensile strength; for blending with rigid resin; used in potting compounds for electronic components, polyester gaskets, caulking and sealing compounds. *Ashland Chemical Company.*

3057 Aropol 8321
Polyester resin; resilient resin for general purpose matched-die molding, preform molded chairs, tote bins, pultrusion applications. *Ashland Chemical Company.*

3058 Aropol 8420
Polyester resin; nonpromoted, resilient resin for use in matched-die molding, pultrusion, architectural sheeting, gel coats, casting. *Ashland Chemical Company.*

3059 Aropol Phase α
Low profile unsaturated polyester resins; sheet molding compound for compression molding into Class A exterior automotive panels (hoods, roofs, and deck lids) and truck panels (hoods, tilt cabs). *Ashland Chemical Company.* Name unverified.

3060 Aropol Phase II
Low profile unsaturated polyester resins; sheet molding compound for compression molding into Class A exterior automotive panels (hoods, roofs, deck lids) and truck panels (hoods, tilt cabs). *Ashland Chemical Company.* Name unverified.

3061 Aropol WEP
Unsaturated polyester resins which can form water emulsions; for casting of decorative art and other applications. *Ashland Chemical Company.* Name unverified.

3062 Aroquest 100
64-02-8 3557 200-573-9
Edetate sodium
Sequestering agent for textile industry, boiler water treatment (water softening, scale removal and prevention). *Arol Chem. Prods.*

3063 Aroquest 120
Sequestering agent for calcium and iron at neutral and mildly alkaline conditions; for textile scouring, dyeing assistant, industrial cleaning compounds, textile bleaching. *Arol Chem. Prods.*

3064 Aroquest M Special
67-43-6 7266 200-652-8
$C_{14}H_{23}N_3O_{10}$
Diethylene triamine pentaacetic acid adduct
[[(Carboxymethyl)imino]bis(ethylenenitrilo)]-tetra-acetic acid; DTPA; Diethylenetriaminepentaacetic acid; Diethylenetriamine-N,N,N',N",N"-pentaacetic acid; Pentetic acid; N,N-Bis(2-(bis-(carboxymethyl)amino)ethyl)-glycine. Sequestrant for use in peroxide bleach baths. *Arol Chem. Prods.*

3065 Aroquest MLC
Chelating agent for peroxide bleach baths. *Arol Chem. Prods.*

3066 Aroset®
Acrylic solutions and emulsions; crosslinking and thermoplastic resins with outstanding clarity and uv and oxidation resistance; for adhesives. *Arol Chem. Prods.*

3067 Arosoft Base LCS-2
Blend of fatty amides and nonionic softeners; all-purpose softener base for application on nylon, polyester and other synthetics and blends. *Arol Chem. Prods.*

3068 Arosoft GSE-D
Ethoxylated glyceride plus esters; softener and lubricant for cotton and other textile fibers, yarns, and fabrics; imparts full, soft hand with excellent drape. *Arol Chem. Prods.*

3069 Arosoft LC-15
Fatty amide/nonionic softener blend; nonyellowing softener for nylon and other synthetic fibers. *Arol Chem. Prods.*

3070 Arosolve 9D5R
Trichlorobenzene-butyl benzoate. High exhaustion carrier for use on polyester and blends where extremely effective swelling, penetration and leveling are required. *Arol Chem. Prods.*

3071 Arosolve 570-HF
Aromatic solvents/detergent blend; non-red label solvent scouring agent, detergent for textiles; roller cleaner, tar remover. *Arol Chem. Prods.*

3072 Arosolve MN-LF
1321-94-4 215-329-7
Emulsified aromatic naphthas; low foaming polyester dye carrier for use in jets and other high-pressure dyeing equipment; produces level and bright shades on polyester. *Arol Chem. Prods.*

3073 Arosolve RCB
Low-foaming emulsifiers/aromatic petrol. distillate/biphenyl blend; solvent carrier for use on synthetic fabrics and their blends; low foaming for high shear jets. *Arol Chem. Prods.*

3074 Arosolve XNF-1
Solvents/low foaming detergent blend; low foaming pressure jet solvent scour for difficult grease, graphite and oil stains on polyester, nylon and other synthetic and natural fibers; general degreaser and tar remover; stable to most acids and alkalies. *Arol Chem. Prods.*

3075 Arosulf SBO-65
Sulfated fatty acid ester; wetting, rewetting agent, lubricant for textile dyeing operations; dye leveling agent; emulsifier for solvent systems. *Arol Chem. Prods.*

3076 Arosulf SCO-75%
Castor oil derivative; textile processing auxiliary, bleaching, leveling and dyeing assistant; emulsifier, finishing agent; also for industrial waxes, polishes, paints. *Arol Chem. Prods.*

3077 Arosurf® 66-E2
52292-17-8
Isosteareth-2
Emulsifier, emollient for personal care products, cutting oils; oil-water and water-oil systems; coupling agent, emulsion stabilizer, perfume stabilizer. *Sherex/Div. of Witco.*

3078 Arosurf® 66-PE12
PPG-3-isosteareth-9
Low cloud point emulsifier, emollient, dispersant, bath oil spreading agent, perfume solubilizer. *Sherex/Div. of Witco.*

3079 Arosurf® AA 23
Diamine
Asphalt emulsifier for rapid set and mixing grade emulsions. *Sherex/Div. of Witco.*

3080 Arosurf® MG-70
Primary ether amine; flotation reagent for the iron mining industry. *Sherex/Div. of Witco.*

3081 Arosurf® TA-100
107-64-2 203-508-2
Distearyl dimonium chloride
Fabric softener concentrate, conditioner for home and commercial laundry and textile processing. *Sherex/Div. of Witco.*

3082 Arotap®
Acrylic and alkyd copolymers; for photoconductive applications in paper industry. *Ashland Chemical Company.*

3083 Arotech
Acrylamate polymer; for fiberglass reinforced parts requiring superior strength properties for automotive and other applications. *Ashland Chemical Company.* Name unverified.

3084 Arotex
Growth regulator containing 644 g chlormequat and 32.2 g choline chloride per liter; for use on wheat, oats or rye. *ICI Agrochemicals.*

3085 Arothix
Sixteen thixotropic vehicles for wall paints. *Reichhold.*

3086 Arothix 4000-P-40
Flat alkyd. *Reichhold.*

3087 Arotran 50437-8
Polyester resin; for resin transfer molding; used for premium Class A automotive body panels, inner structures. *Ashland Chemical Company.*

3088 Arova 16
54982-83-1 259-423-6
$C_{14}H_{24}O_4$
1,4-Dioxacyclohexadecane-5,16-dione
Ethylene dodecanedioate. Musk perfume. Oriental; Sandalwood; Amber; Woody; Musk. Not Found In Nature. *Hüls AG.*

3089 Arowet 70 E
Sulfonated ester
Wetting/rewetting agent, penetrant for textile processing, desizing, scouring, bleaching, level dyeing and printing, finishing operations. *Arol Chem. Prods.*

3090 Arowet ODA
Biodegradable wetting agent for batch and continous operations on natural and synthetic fibers and blends; relatively low foaming. *Arol Chem. Prods.*

3091 Arowet SC-75
577-11-7 3460 209-406-4
Dioctyl sodium sulfosuccinate
Fast wetting agent, penetrant, and dyeing assistant for mild acidic or alkaline textile processing; dye leveling agent. *Arol Chem. Prods.*

3092 Arozyme TD
9000-92-4 640 232-567-7
α Amylase
Thermo-stable enzyme; used as a textile desizing agent. *Arol Chem. Prods.*

3093 ARP®
Addition agents, wetting agents, other specialty chemicals designed to solve problems of adhesion, leveling, pitting and other conditions affecting quality of surface treatment in metal finishing industry. *Witco/Allied-Kelite.* Discontinued.

3094 Arpak 4322
9002-88-4 7728
Expanded polyethylene beads; produces very flexible closed-cell foam offering high resiliency; foams maintain dimensions and shock absorbence after repeated deformations; for dynamic cushioning applications. *Arco Chemical Co.*

3095 Arpal Non Selex
7775-09-9 8741 231-887-4
sodium chlorate
Powder containing 58.2% w/w sodium chlorate; used for total weed control for paths, drives and noncrop areas. *R. P. Adams Ltd.*

3096 Arpocox
55779-18-5 828 259-817-8
Arprinocid
Coccidiostat. *Merck & Co Inc.* Name unverified.

3097 Arpro 3313
9003-07-0 7741
Expanded PP beads; produces low-density, closed-cell foam with excellent energy absorption characteristics, excellent recoverability from repeated shocks; foams maintain dimensional stability when exposed to temperature extremes. *Arco Chemical Co.*

3098 Arpylen
Polypropylene compounds. *Norsk Hydro Polymers Ltd.* Unverified.

3099 Arquad
Quaternary ammonium salts. *Akzo Chemie UK Ltd.*

3100 Arquad®
Range of cationic surfactants composed of alkyl quaternary ammonium chlorides in mainly liquid form; effective in killing micro-organisms at low concentrations; used in sanitizing foodstuffs, catering, blanket sterilization; algal control; mold inhibition air conditioning, textile softening agents, laundry, dry-cleaning, paper, corrosion inhibition, petroleum, e.g. in drilling; emulsification, e.g. in road-making, metal cleaning, insecticides; antistats; e.g. plastics, rubber, latex, cosmetics, leather. *Akzo Chemie UK Ltd.*

3101 Arquad® 2C-70 Nitrite
71487-01-9 275-532-1
Dicoco nitrite
Dicoco nitrite in methanol/isopropanol; biodegradable. surfactant, dispersant for protective coatings, pigments, inks, textiles, agriculture, acid pickling baths, marine applications, metalworking, electroplating, fuel treatment, emulsion/plastic manufacturing, waste water treatment, mineral processing, paper. *Akzo.*

3102 Arquad® 2C-75
61789-77-3 263-087-6
Dicocodimonium chloride
In aqueous isopropanol. Biodegradable emulsifier, foaming, wetting, dispersing agents, corrosion inhibitor, softener, dyeing aid, antistat for textiles, paper, cosmetics; industrial, agriculture, plastics, petrol. industry, acid pickling baths; bactericide, algicide. *Akzo; Akzo Chem. BV.*

3103 Arquad® 2HT-75
61789-80-8 263-090-2
Quaternium-18
Biodegradable emulsifier, foaming, wetting, dispersing agents, corrosion inhibitor, antistat, bacteriostat for paper softening, household laundry, hair conditioning. *Akzo.*

3104 Arquad® 2T-75
68783-78-8 272-207-6
Ditallow dimonium chloride
Biodegradable surfactant, dispersant for protective coatings, pigments, inks, textiles, agriculture, acid pickling baths, marine applics, metalworking, electroplating, fuel treatment, emulsion/plastic manufacturing, waste water treatment, mineral processing, paper. *Akzo.*

3105 Arquad® 12-37W
112-00-5 203-927-0
Laur-trimonium chloride
Arquad® 12-37W. Emulsifier, corrosion inhibitor, textile softener, antistat, hair conditioner and combing aid emulsifier; biodegradable. *Akzo.*

3106 Arquad® 12-50
112-00-5 203-927-0
$C_{15}H_{34}ClN$
Laurtrimonium chloride
Dodecyltrimethylammonium chloride; N,N,N-trimethyl-1-dodecanaminium chloride; Dodecanaminium, N,N,N-trimethyl-, chloride; Trimethyl-1-dodecanaminium chloride; n-Dodecyl trimethylammonium chloride; Lauryl Trimethyl Ammonium Chloride; Laurtrimonium Chloride. IPA; biodegradable emulsifier, foaming, wetting, dispersing agents, corrosion inhibitor, softener, dyeing aid, antistat for textiles, paper, cosmetics; industrial, agriculture, plastics, petrol. industry, acid pickling baths; bactericide, algicide; gel sensitizer for latex foam. *Akzo.*

3107 Arquad® 16-29
112-02-7 203-928-6
Cetrimonium chloride
Emulsifier, foaming, wetting, dispersion agents, corrosion inhibitor, antistat for textiles, cosmetics, industrial, agriculture; bactericide, algicide. *Akzo.*

3108 Arquad® 16-50
112-02-7 203-928-6
Cetrimonium chloride
Ammonyx® CETAC, CETAC-30; Arquad® 16-29. IPA; emulsifier, foaming, wetting, dispersing agents, corrosion inhibitor, softener, dyeing aid, antistat for textiles, paper, cosmetics; industrial, agriculture, plastics, petrol. industry, acid pickling baths; bactericide, algicide; *Akzo; Akzo Chem. BV.*

3109 Arquad® 18-50
112-03-8 203-929-1
$C_{21}H_{46}ClN$
Stear-trimonium chloride
Octadecyltrimethylammonium chloride; Stearyl Trimethyl Ammonium Chloride; Steartrimonium Chloride; Stearyl Trimethyl Ammoium Chloride. Emulsifier, foaming, wetting, dispersing agents, corrosion inhibitor, softener, dyeing aid, antistat for textiles, paper, cosmetics; industrial, agriculture, plastics, petrol. industry, acid pickling baths; bactericide, algicide; dye leveling agent, viscosity stabilizer in lubricant compounding. Biodegradable. *Akzo; Akzo Chem. BV.*

3110 Arquad® 210-50
7173-51-5 3149 230-525-2
Didecyldimonium chloride
In aqueous EtOH, surfactant, dispersant for protective coatings, pigments, inks, textiles, agriculture, acid pickling baths, marine applics, metalworking, electroplating, fuel treatment, emulsion/plastic manufacturing, waste water treatment, mineral processing, paper. *Akzo; Akzo Chem. BV.*

3111 Arquad® 218-100
107-64-2 203-508-2
Distearyl dimonium chloride
Biodegradable surfactant, dispersant for protective coatings, pigments, inks,

textiles, agricuture, acid pickling baths, marine applics, metalworking, electroplating, fuel treatment, emulsion/plastic manufacturing, waste water treatment, mineral processing, paper. *Akzo.*

3112 Arquad® 218-75
107-64-2 203-508-2
In aqueous isopropanol; biodegradable surfactant, dispersant for protective coatings, pigments, inks, textiles, agriculture, acid pickling baths, marine applics, metalworking, electroplating, fuel treatment, emulsion/plastic manufacturing, waste water treatment, mineral processing, paper. *Akzo.*

3113 Arquad® 316(W)
71060-72-5
Trihexadecylmethyl ammonium chloride
Water; industrial surfactant for pigment dispersing, coatings, inks, paper processing. *Akzo.*

3114 Arquad® C-33W
61789-18-2 263-038-9
Cocotrimonium chloride
Emulsifier, corrosion inhibitor, textile softener, antistat; hair conditioning and combing aid emulsifier; emulsion-break retardant in cosmetics; biodegradable. *Akzo.*

3115 Arquad® C-50
61789-18-2 263-038-9
Cocotrimonium chloride, IPA
Biodegradable emulsifier, foaming, wetting, dispersing agents, corrosion inhibitor, softener, dyeing aid, antistat for textiles, paper, cosmetics; industrial, agriculture, plastics, petroleum industry, acid pickling baths; bactericide, algicide; gel sensitizer for latex foam. *Akzo; Akzo Chem. BV.*

3116 Arquad® DMCB-80
61789-71-7 263-080-8
Cocoalkyl dimethyl benzyl ammonium chloride
Solution in aqueous isopropanol; microbicide for disinfectants, sanitizers, algicides for use in swimming pools, air conditioning cooling towers, bathroom cleaners, petroleum recovery. *Akzo.*

3117 Arquad® DMHTB-75
61789-72-8 263-081-3
Hydrogenated tallowalkonium chloride
Hydrogenated tallow dimethylbenzyl ammonium chloride in aqueous isopropanol; bactericide, disinfectant, softening agent for textiles. *Akzo.*

3118 Arquad® HTL8(W) MS-85
2-Ethylhexyl hydrogenated tallowalkyl methosulfate
Biodegradable surfactant, dispersant for protective coatings, pigments, inks, textiles, agriculture, acid pickling baths, marine applications, metalworking, electroplating, fuel treatment, emulsion/plastic manufacturing, waste water treatment, mineral processing, paper. *Akzo.*

3119 Arquad® M2HTB-80
61789-73-9 263-082-9
Dihydrogenated tallow benzylmonium chloride
In aqueous isopropanol; industrial surfactant for preparation of organophilic clays. *Akzo.*

3120 Arquad® S-50
61790-41-8 263-134-0
Soytrimonium chloride
Adogen® 417. Solution in isopropanol; emulsifier, corrosion inhibitor, textile softener, antistat; hair conditioning and combing aid emulsifier; bitumen emulsions; slime control agent in water systems; biodegradable. *Akzo; Akzo Chem. BV.*

3121 Arquad® T-27W
8030-78-2 232-447-4
Tallow trimonium chloride
Biodegradable emulsifier, foaming, wetting, dispersing agents, corrosion inhibitor, softener, dyeing aid, antistat for textiles, paper, cosmetics; industrial, agriculture, plastics, petrol. industry, acid pickling baths; bactericide, algicide. *Akzo.*

3122 Arquad® T-50
8030-78-2 232-447-4
Tallow trimonium chloride
Arquad® T-27W. Biodegradable emulsifier, foaming, wetting, dispersing agents, corrosion inhibitor, softener, dyeing aid, antistat for textiles, paper, cosmetics; industrial, agriculture, plastics, petrol. industry, acid pickling baths; bactericide, algicide. *Akzo; Akzo Chem. BV.*

3123 Arquard® B-100
68391-01-5 269-919-4
Benzalkonium chloride
In aqueous isopropanol; Antimicrobial for industrial applications, secondary oil recovery, textiles, cosmetics, pharmaceuticals, sanitizers. *Akzo; Akzo Chem. BV; Pentagon Chemicals Ltd.*

3124 Arrconox S.P.
A nonstaining antioxidant. *Rubber Regenerating Co.* Unverified.

3125 Arrcorez 16
A butyl rubber curing resin. *Rubber Regenerating Co.* Unverified.

3126 Arrcorez 17
A tackifying resin. *Rubber Regenerating Co.* Unverified.

3127 Arresin
1746-81-2 217-129-5
monolinuron
Emulsifiable concentrate containing 200 g/l monolinuron; used for control of annual dicotyledons in potatoes, french beans and leeks. *Hoechst UK.*

3128 arrhenal
144-21-8 6020 205-620-7
$CH_3AsNa_2O_3{>}H_2O$
Sodium methylarsinate
arsinyl; new cacodyl; arsinyl; DSMA. Selective contact herbicide.

3129 Arroconox AHT, DNL and DNP
A proprietary range of antioxidants used in the manufacture or processing of rubber. Unverified.

3130 Arrow Tool Steels
Proprietary steels containing 0.9-1.02% chromium, 0.16-0.20% vanadium, 0.5-0.6% manganese, and 0.20-0.30% carbon. No manufacturer.

3131 Arsan600
Moldable resilient resin; produces resilient, low density, closed-cell foam offering low dynamic set; foams maintain shock absorbence after repeated impacts. *Arco Chemical Co.*

3132 Arsenal
81334-34-1 4942
imazapyr
Soluble concentrate containing imazapyr; used for bracken control in noncrop areas. *Chipman Ltd.*

3133 Arsenal XL
A soluble concentrate containing 300g atrazine and 12.5 g imazapyr per liter; used for total weed control in non crop areas. *Chipman Ltd.*

3134 Arsenal®
81334-34-1 4942
$C_{13}H_{15}N_3O_3$
2-[4,5-Dihydro-4-methyl-4-(1-methylethyl)-5-oxo-1H-imidazol-2-yl]-3-pyridinecarboxylic acid
imazapyr. Imazapyr with isopropylamine (1:1) salt; used for bracken control in noncrop areas. *Am. Cyanamid/Ag; Cyanamid of Great Britain Ltd.*

3135 Arsenal® XL
Soluble concentrate containing 300 g atrazine and 12.5g imazapyr per liter; for total weed control in noncrop areas. *Cyanamid of Great Britain Ltd.*

3136 arsenic
7440-38-2 832 231-148-6
As
Fowler's solution; Grey arsenic; Colloidal arsenic. Metallic form: alloying additive for metals, especially lead and copper as shot, battery grids, cable sheaths; high-purity semiconductor grade: manufacture of gallium arsenide for dipoles and other electronic devices; doping agent; solders; medicine. mp = 818°; soluble in H_2O; d = 4.700. *Aldrich; Atomergic Chemetals; Whiting, Peter Ltd.*

3137 arsenic acid
7778-39-4 833 231-901-9
AsH_3O_4
Orthoarsenic Acid
Aresenid; orthoarsenic acid; Dessicant L-10; Hi-Yield Dessicant H-10; Zotox; Desiccant L-10; Hy-Yield H-10; Poly Brand Dessicant; CCA Type C; Chemonite Part A; Crab grass killer; orthoarsenic acid; Dessicant L-10; Hi-Yield Dessicant H-10; Zotox; Desiccant L-10; Hy-Yield H-10; Poly Brand Dessicant; CCA Type C; Chemonite Part A; Crab grass killer. Arsenic acid solution; wood preservative. mp = 35°; bp = 160°; soluble in H_2O (16.7 g/100 ml); LD_{50} (rbt iv) = 6 mg/kg. Name unverified.

3138 arsenic bronze
An alloy of 80% copper, 10% tin, 9.2% lead, and 0.8% arsenic.

3139 arsenic pentasulfide
1303-34-0 838
As_2S_5
diarsenic pentasulfide
Used in paint pigments, light filters and manufacture of other arsenic compounds. Insoluble in H_2O. *Atomergic Chemetals.*

3140 arsenic pentoxide
1303-28-2 839 215-116-9
As_4O_{10}
Arsenic oxide
Arsenic anhydride; Arsenic acid anhydride; arsenic pentoxide; Diarsenic Pentoxide; Fotox; Arsenic (V) Oxide; Arsenic oxide; Arsenic oxide. Arsenates; insecticides; dyeing and printing; weed killer; in colored glass,

metal adhesives. mp = 315°; d = 4.3200; soluble in H_2O, alcohol; LD_{50} (rat orl) = 8 mg/kg. *Atomergic Chemetals; Spectrum Chem. Manufacture.*

3141 arsenic trichloride
7784-34-1 841 232-059-5
$AsCl_3$
arsenic(III) chloride
arsenious chloride; arsenic chloride; arsenous chloride; arsenic trichloride; butter of arsenic; Arsenous Trichloride; Trichloroarsine; Arsenic butter; Arsenious chloride; fuming liquid arsenic. Intermediate for organic arsenicals (pharmaceuticals, insecticides), ceramics. mp = -9°; bp = 130.21°; d = 2.1497; n_D^{20}=1.6006; reacts with H_2O, soluble in organic solvents. *Atomergic Chemetals; Noah Chem.*

3142 arsenic trifluoride
7784-35-2 842 232-060-0
AsF_3
arsenious fluoride
Fluorinating reagent, catalyst, ion implantation source, dopant. mp = -5.95°; bp = 57.8°; d_4^{15} = 2.73; reacts with H_2O, soluble in alcohol, ether, benzene. *Atomergic Chemetals; Elf Atochem N. Am.; Noah Chem.*

3143 arsenic trioxide
1327-53-3 844 215-481-4
As_4O_6
Arsenic (III) oxide
Arsenic oxide;Arsenous trioxide; arsenous acid; arsenous oxide; arsenic sesquioxide; White Arsenic; Diarsenic Trioxide; Crude Arsenic; Arsenic (white); Arsenious oxide; Arsenic (III) trioxide; Arsenous anhydride; arsenite; arsenolite; arsenous acid anhydride; arsenous oxide anhydride; arsodent; claudelite; claudetite; Arsenic oxide (3); Arsenic oxide (As_2O_3); Arsenic sesquioxide (As_2O_3); Arsenicum album; Diarsonic trioxide; Diarsenic oxide. Pigments, ceramic enamels, aniline colors, decolorizing agent in glass, insecticide, rodenticide, herbicide, sheep and cattle dip, hide preservative, wood preservative, preparation of other arsenic compounds. mp = 315°; bp = 465°; soluble in H_2O, dil HCl, alkali hydroxide or carbonate solns; insoluble in EtOH, $CHCl_3$, Et_2O; LD_{50} (rat orl) = 1.46 mg/kg. *Atomergic Chemetals; Noah Chem.; Outokumpu Oy; Transene.*

3144 arsine
7784-42-1 849 232-066-3
AsH_3
Arsenic trihydride
Hydrogen arsenide; Arsine; arsenic trihydride; Arsenic hydride; hydrogen arsenide; arseniuretted hydrogen; arsenous hydride. Organic synthesis, military poison, doping agent for solid-state electronic components. mp = -117°; bp = -62.5°; decomposes when heated at 300°; slightly soluble in H_2O. *Air Prods & Chem; Atomergic Chemetals.*

3145 Arsinette
144-21-8 6020 205-620-7
Arsenate insecticides. *Plant Protection.* Name unverified.

3146 Art Bronze
An alloy of 80-90% copper and 5-8% tin.

3147 Artic
74-87-3 6121 200-817-4
methyl chloride
A proprietary trade name for methyl chloride used in refrigeration. No manufacturer.

3148 Artic Mist
14807-96-6 9207 238-877-9
Talc. Lubricant and mold release agent, filler and extender. *Steetley Minerals Ltd.*

3149 Artisil®
Pigment for coloring oils, waxes, and solvents. *Sandoz.*

3150 Artodan SP 55 Kosher
25383-99-7 246-929-7
Sodium stearoyl lactylate
Food emulsifier, dough conditioner, starch complexing agent, bread improver, freeze/thaw emulsions. *Grindsted Prods.; Grindsted Prods. Denmark.*

3151 Arton F
Transparent polymer with improved flowability for optical discs, optical fibers, lens applications. *Japan Synthetic Rubber.*

3152 Arton G
Transparent polymer with high heat resistance for optical discs, optical fibers, lens applications. *Japan Synthetic Rubber.*

3153 Arubren®
Chlorinated paraffin plasticizer. *Bayer AG; Bayer plc.*

3154 Arvetane
A proprietary adhesive containing polyurethane. *Arveta SA.* Unverified.

3155 Arylan S
Anionic surfactant as pale cream flakes
Primary emulsifier and wetting agent for emulsion polymerization and wettable powders. *Harcros.*

3156 Arylan®
Alkaryl sulfonic acids
Salts and blends with nonionics; for anionic detergency, emulsification, emulsion polymerization. *Harcros.*

3157 Arylan® CA
Calcium dodecylbenzene sulfonate
Emulsifier for degreasers, herbicides, pesticides, waxes, hydrocarbon solvents. *Harcros; Harcros UK.*

3158 Arylan® PWS
26264-05-1 247-556-2
Isopropylamine dodecylbenzene sulfonate
Surfactant and emulsifier for mineral oils, kerosene, waxes, and chlorinated solvents, herbicides and insecticides; for manufacture of emulsion degreasers and kerosene-based hand cleaning gels. *Harcros; Harcros UK.*

3159 Arylan® SBC Acid
Straight chain dodecylbenzene sulfonic acid
Surfactant, detergent, foaming agent, emulsifier for phenolic materials; intermediate for liquid detergents, dishwash, emulsifiers; biodegradable. *Harcros; Harcros UK.*

3160 Arylan® SC15
Sodium dodecylbenzene sulfonate
Biodegradable wetting agent, detergent base, emulsifier for emulsion polymerization. *Harcros; Harcros UK.*

3161 Arylan® SNS
Anionic surfactant supplied as a buff powder; dispersing agent. *Harcros.*

3162 Arylan® SP
Anionic surfactant in acid form, supplied as a brown viscous liquid; low free oil and inorganic content; biodegradable intermediate for detergents, especially liquids. *Harcros.*

3163 Arylan® SX
Anionic surfactant in flake form; base for detergents, wetting agent for detergent powders. *Harcros.*

3164 Arylan® TE/C
Anionic surfactant in liquid form; emulsifier for specialty waxes, chlorinated solvents. *Harcros.*

3165 Arylene M40
577-11-7 3460 209-406-4
Dioctyl sodium sulfosuccinate
Wetting agent, rewetting, dewatering surfactant, filtration aids. *Hart Chem. Ltd.*

3166 Arylmate®
29973-13-5 249-981-9
Ethiofencarb
Insecticide used for control of aphids. *Bayer AG.*

3167 Arylon® LP 401 NC10
Polyarylate resin, unreinforced. *DuPont; DuPont UK.*

3168 AS
9004-70-0 8195
Nitrocellulose with nitrogen content of 11.3 to 11.7%; for coatings on cellophane and in converting operations for paper coatings. *Hercules.* Discontinued.

3169 AS-10GF
9003-56-9
Acrylonitrile-butadiene-styrene
ABS resin, 10% glass fiberreinforced; thermoplastic with dimensional stability and toughness; for business machine housings, cabinetry, tool housing, appliances, automotive, and construction materials. *Compounding Tech.*

3170 AS-15CF/000
9003-56-9
ABS, 15% carbon fiber-reinforced; thermoplastic resin. *Compounding Tech.*

3171 ASA
Alkenyl succinic anhydride
Intermediate for defoamers, demulsifiers, emulsifiers, foam boosters, wetting agents, detergents, dispersants; sizing agent for paper. *Ethyl Corp; Pentagon Chemicals Ltd.* Name unverified.

3172 Asadene®
BR; styrene/butadiene rubber thermoplastic elastomers. *Asahi Chem. Industry.*

3173 Asaflex®
Transparent styrenic resin. *Asahi Chem. Industry.*

3174 Asahi Aji®
527-07-1 8766 208-407-7
Sodium Gluconate
Used in the food industry. *Asahi Chem. Industry.*

3175 Asaprene®
BR; styrene/butadiene rubber thermoplastic elastomers. *Asahi Chem. Industry.*

3176 asaprol
516-18-7 859
$C_{20}H_{14}CaO_8S_2$
Calcium-β-naphthol-γ-sulfonate
asaprol-etrasol; Abrastol; calcium 2-hydroxy-1-naphthalene sulfonate; calcium 2-naphthol-1-sulfonate; 2-naphthol-1-sulfonic acid calcium salt; Calcinaphthol. Used as a clarifier for wines. mp = 50° (dec); soluble in H_2O (0.66 g/ml), EtOH (0.33 g/ml).

3177 Asbesto-Wet
Blend of polyoxyethylene esters of mixed organic acids (47%) and polyoxyethylene ether of alkylated phenols (47%) containing a silicone defoamer (6%) for ease of handling; for dust control and wet removal of asbestos. *Aquatrols Corp of Am.* Name unverified.

3178 Ascinin® P, R, Special
Antiskinning and stabilizing agent used in oil based paints and varnishes. *Bayer AG; Bayer plc.*

3179 Ascorbosilane C
Ascorbyl methylsilanol pectinate
Cosmetic ingredient for anti-aging formulations, after sun bath treatments, superficial burn treatments.

3180 Ascot
Mixture of thiabendazole and thiram; fungicide seed dressing. *Ciba-Geigy Agrochemicals.*

3181 Aseptisil
An alkaline bottle washing detergent. *Staveley Chemicals Ltd. Name unverified.*

3182 Aseptoforms
p-Hydroxybenzoates; used in antiseptics. *R W Greef & Co Inc.*

3183 Ashberry metal
Ashbury Metal
An alloy of 80% tin, 14% antimony, 2% copper, and 1% zinc.

3184 Ashlade 4% At Gran
1912-24-9 902 217-617-8
Atrazine
Herbicide. *Ashlade Formulations Ltd.*

3185 Ashlade 4-60 CCC, 700 CCC
999-81-5 2153 213-666-4
Soluble concentrates containing 460 or 700 g/l chlormequat chloride; plant growth regulator. *ABM Chemicals Ltd.*

3186 Ashlade 5C
Mixture of chlormequat and choline chloride; plant growth regulator. *Ashlade Formulations Ltd.*

3187 Ashlade Adjuvant Oil
Adjuvant containing 99% refined mineral oil; herbicide wetting agent. *Ashlade Formulations Ltd.*

3188 Ashlade Atrazine 50 FL
1912-24-9 902 217-617-8
Atrazine
Atraflow; Atranex; Ashlade 4% At Gran; Atraflow Plus. A residual herbicide. *Ashlade Formulations Ltd.*

3189 Ashlade Blight Fungicide
Mixture of cymoxanil and mancozeb; used to control potato blight. *Ashlade Formulations Ltd.*

3190 Ashlade Cosmic FL
A suspension concentrate containing 40 g carbendazim, 320 g maneb and 90 g tridemorph per liter. systemic fungicide for cereals. *Ashlade Formulations Ltd.*

3191 Ashlade CP
Suspension concentrate containing 86 g chloridazon and 400 g propachlor per liter; a residual herbicide for beet crops. *Ashlade Formulations Ltd.*

3192 Ashlade D-Moss
Mixture of chloroxuron and ferrous sulfate; a lawn sand herbicide to control mosses in turf. *Ashlade Formulations Ltd.*

3193 Ashlade Flotin
76-87-9 9875 200-990-6
triphenyltin hydroxide
Suspension concentrate containing 625 g triphenyltin hydroxide per liter; used for control of potato blight. *Ashlade Formulations Ltd.*

3194 Ashlade Halt
66841-24-5 266-492-6
cypermethrin
Emulsifiable concentrate containing 100 g cypermethrin per liter; a pyrethroid insecticide. *Ashlade Formulations Ltd.*

3195 Ashlade Linuron
330-55-2 5534 206-356-5
Linuron
A residual urea herbicide for the control of weeds in field crops. *Ashlade Formulations Ltd.*

3196 Ashlade M
83601-81-4
Ashlade Mancarb FL. Fungicide for cereals. *Ashlade Formulations Ltd.*

3197 Ashlade Mancarb FL
83601-81-4
A suspension concentrate containing 50 g carbendazim and 320 g maneb per liter; systemic fungicide for cereals. *Ashlade Formulations Ltd.*

3198 Ashlade SMC
Copper Oxychloride, maneb and sulfur; a fungicide for wheat and barley which also stimulates yields. *Ashlade Formulations Ltd.*

3199 Ashlade TCNB
117-18-0 204-178-2
Tecnazene
Granules or dustable powder containing tecnazene; protectant fungicide and potato sprout suppressant. *Ashlade Formulations Ltd.*

3200 Ashland Hi-Sol 10
An aromatic hydrocarbon solvent; with high solvency for paints, varnishes, resins and insecticides. *Ashland Chemical Company.* Name unverified.

3201 Ashland Hi-Sol 15
An aromatic hydrocarbon solvent; used in baked enamels and in chlorinated rubber finishes. *Ashland Chemical Company.* Name unverified.

3202 Ashland Kwik-Dri
An aliphatic hydrocarbon solvent; used in cleaning compounds, waxes, polishes and as a resin solvent. *Ashland Chemical Company.* Name unverified.

3203 Ashland Lacolene
An aliphatic hydrocarbon solvent; used as diluent for lacquer. *Ashland Chemical Company.* Name unverified.

3204 Ashlene®
Thermoplastic engineering resins including polyamide(nylon), polycarbonate, polyester PBT, polyphenytene oxide; used for injection molding and extrusion. *Ashley Polymers.*

3205 Ashlene® 61-2M
32131-17-2
Nylon 66 resin, 40% mineral-reinforced; offers improved dimensional stability, high stiffness, and high heat resistance, improved impact; chrome plateable; for lamp and instrument housings, automotive and marine hardware, range knobs. *Ashley Polymers.*

3206 Ashlene® 520
32131-17-2
Nylon 66, slightly lubricated; general purpose economy injection molding resin. *Ashley Polymers.*

3207 Ashlene® 520-13G
32131-17-2
Nylon 66, 13% glass fiber-reinforced, lubricated; general purpose economy injection molding resin designed for parts requiring excellent thermal and dimensional stability and higher strength than conventional resins; *Ashley Polymers.*

3208 Ashlene® 520MS
32131-17-2
Nylon 66, molybdenum disulfide-modified; self-lubricating economy injection molding resin with excellent wear resist. *Ashley Polymers.*

3209 Ashlene® 525-13G
32131-17-2
Nylon 66, 13% glass fiber-reinforced, lubricated; high impact. *Ashley Polymers.* Name unverified.

3210 Ashlene® 527
32131-17-2
Nylon 66; economy injection molding resin; very high impact resistance. *Ashley Polymers.*

3211 Ashlene® 527LD-13G
32131-17-2
Nylon 66, 13% glass fiber-reinforced, lubricated; very high impact resistance. *Ashley Polymers.*

3212 Ashlene® 528BR-WO
32131-17-2
Nylon 66, flame retardant. *Ashley Polymers.*

3213 Ashlene® 528L-13G
32131-17-2
Nylon 66, 13% glass fiber-reinforced, lubricated; general purpose reinforced resin for parts requiring excellent thermal and dimensional stability and higher

strength than conventional resins; lubricated for improved machine feed and mold release. *Ashley Polymers.*

3214 Ashlene® 541
32131-17-2
Nylon 66; extrusion grade; very high viscosity for thick slab material, large rod stock and large complex profiles, pipes, and slabs. *Ashley Polymers.*

3215 Ashlene® 541S
32131-17-2
Nylon 66, heat-stabilized; extrusion grade; high viscosity for sheet, film, and blown film extrusion. *Ashley Polymers.*

3216 Ashlene® 630-33G
25038-54-4 6832
Nylon 6, 30% glass fiber-reinforced, lubricated; general purpose economy injection molding resin. *Ashley Polymers.*

3217 Ashlene® 735
25038-54-4 6832
Nylon 6 plasticized co-polymer; economy injection molding resin; improved impact resistance. *Ashley Polymers.*

3218 Ashlene® 830L
25038-54-4 6832
Nylon 6, lubricated; for improved machine feed and mold release. *Ashley Polymers.*

3219 Ashlene® 830L
25038-54-4 6832
Nylon 6, lubricated; for improved machine feed and mold release. *Ashley Polymers.*

3220 Ashlene® 840
25038-54-4 6832
Nylon 6; extrusion grade; high viscosity general purpose for film, rigid pipe, large profile, and thick slab extrusions. *Ashley Polymers.*

3221 Ashlene® 858
25038-54-4 6832
Nylon 6
Nylon 6; resin with good surface finish and high strength, rigidity, toughness, and heat, abrasion and chemically resistant; for rotomolding. *Ashley Polymers.*

3222 Ashlene® 870
Amorphous nylon
General-purpose transparent nylon with good dimensional stability and chemical resist. *Ashley Polymers.*

3223 Ashlene® 980L
Nylon 6/12; general-purpose resin with low moisture absorptivity, excellent dimensional stability. *Ashley Polymers.*

3224 Ashlene® 980LS-40G
Nylon 6/12, short glass fiber-rein-forced; general-purpose, heat-stabilized resin with low moisture absorptivity and good dimensional stability. *Ashley Polymers.*

3225 Ashlene® 981S
Nylon 6/12; heat-stabilized, flexible extrusion grade for cable jacketing; low moisture absorptivity and dimensional stability. *Ashley Polymers.*

3226 Askure
Acid catalyst used in conjunction with furan resin binders for the cold set bonding of sand molds and cores. *Foseco (F.S.) Ltd.*

3227 Asmer
Shape-memory resin. *Asahi Chem. Industry.*

3228 A-Sol
68-26-8 10150 200-683-7
Vitamin A
Antixerophthalmic. *The Purdue Frederick Co.* Name unverified.

3229 Asp
An asbestos paper grade of Tufnol industrial laminates. *Tufnol Ltd.*

3230 ASP®
1332-58-7 5294 296-473-8
kaolin
Kaolinite; China clay; Bolus alba; Porcelain clay; Aluminum silicate hydroxide; Kaopectate; Aluminum silicate (hydrated); Aluminum silicate dihydrate. Very fine to coarse particle size hydrous kaolin; used as an extender pigment in coatings, adhesives, caulks, inks, polishes, molecular sieve support, specialty ceramics. *Engelhard.*

3231 Aspac®
Polystyrene foam loose fill. *Asahi Chem. Industry.*

3232 Asparagine
70-47-3 872 218-163-3
$C_4H_8N_2O_3$
(S)-2,4-diamino-4-oxobutanoic acid
L-β-asparagine; α amminosuccinamic acid; Aspartic acid β amide; Altheine; Asparamide; Agedoite; Asn; N. Biochemical research, preparation of culture media, medicine. mp = 234-235°; d_4^{15} = 1.543; $[\alpha]_D^{20}$ = -5.30° (c = 1.41); practically insoluble in methanol, ethanol, ether, benzene; soluble in acids and alkalis *Aldrich; Degussa; Penta Mfg.; Spectrum Chem. Mfg.; Tanabe USA; U.S. Biochemical.*

3233 Aspartame
22839-47-0 874 245-261-3
$C_{14}H_{18}N_2O_5$
3-amino-N-(α-methoxycarbonylphenethyl) succinamic acid
L-aspartyl-L-phenylalanine methyl ester; APM; SC-18862; Canderel; Equal; NutraSweet; Sanecta; Tri-Sweet. A sweetening agent mp = 246-247°; $[\alpha]_D^{22}$= -2.3° (1N HCl). *Searle Chemicals Inc.* No manufacturer.

3234 Aspartic acid
1783-96-6 875 217-234-6
$C_4H_7NO_4$
1-amino-1,2-carboxyethane
L-Aspartic acid; L-Asparagic acid; Asp; D; aminosuccinic acid; (S)-aminobutanedioic acid. Biological and clinical studies, preparation of culture media, organic intermediate, ingredient of aspartame, detergents, fungicides, germicides, metal complexation. L-form is most common, D-form occurs naturally but is less common. mp = 270-271°; soluble in H_2O, more soluble in salt solutions, soluble in acid, alkali, insoluble in EtOH; $[\alpha]_D^{20}$ = 25° (c = 1.97 6N HCl). *Atomergic Chemetals; Penta Mfg.; Schweizerhall; Tanabe USA; U.S. Biochemical.*

3235 Aspect® TPPE
9009-54-5
Polyurethane
Thermoplastic polyester compounds; natural and flame-retarded grades available; offers easy processing, excellent toughness, outstanding mechanical properties. *Phillips.* Name unverified.

3236 asphaltenes
Constituents of bitumen, insoluble in hexane, but soluble in carbon tetrachloride.

3237 Asplit
A range of acid and chemical resisting cements. *Prodorite Ltd.*

3238 Aspon
7558-80-7 8806 231-449-2
sodium phosphate, monobasic
Acid sodium orthophosphate for laundry use. *Albright & Wilson Ltd.* Name unverified.

3239 Aspulum
A mercury derivative of chlorophenol; a seed preservative. No manufacturer.

3240 Aspumit AP
Silicone emulsion; nonionic; defoamer for all textile finishing processes. *Thor Chemicals (UK) Ltd.* Discontinued.

3241 Aspumit SDM
Synergistic blend of deaerating products and surfactants; weakly anionic; deaerating agent and penetration accelerator for all fibers. *Thor Chemicals (UK) Ltd.* Discontinued.

3242 Assaf
Silicone foam control agent. *Rhône-Poulenc UK.*

3243 Asset
An emulsifiable concentrate containing 50 g benazolin, 125g bromoxynil and 62.5 g ioxynil per liter; a post-emergence herbicide for cereal crops and grass. *Schering Agrochemicals Ltd.* Discontinued.

3244 Astacin® Finish PUD
Polyester-polyurethane dispersions; bottoming and top wetting agent for leather and fur industry *BASF AG.*

3245 Asterite
Filled methyl methacrylate dispersion. *ICI Chem & Polymers Ltd.*

3246 Astick
Adhesive promoter for asphaltic shingle; shingle tab adhesive additive. *Chemseco.* Name unverified.

3247 Astingol
A proprietary preparation of dimethyl phthalate and diethyl toluamide; an insect repellant. *Ayrton Saunders plc.* Unverified.

3248 Astix
93-65-2 202-264-4
MCPP
Soluble concentrate containing 600 g/l mecoprop-P (MCPP); used for control of weeds in undersown cereals and grassland. *Rhône-Poulenc Crop Protection Ltd.*

3249 Aston 123
Thermosetting polyamine; durable textile antistat with high resistance to laundering and dry cleaning on all substrates. *Rhône-Poulenc Surf.*

3250 Aston RC
Special cationic surfactant in paste or liquid form; antistat for rugs; reduces resoiling after shampooing. *Millmaster-Onyx UK.* Name unverified.

3251 **Astra®**
Dyestuff; for the paper, printing ink, surface coatings and office supplies industries. *Bayer AG; Bayer plc.*

3252 **Astradur® A and T**
A registered trademark for high impact PVC. No manufacturer.

3253 **Astraflex®**
Dyestuffs; for the printing ink industry. *Bayer AG.*

3254 **Astragal®**
Retarders for dyeing polyacrylonitrile fibers. *Bayer AG; Bayer plc.*

3255 **Astralex**
A proprietary range of chemicals in the bright plating of nickel. *Albright & Wilson Ltd.* Name unverified.

3256 **Astralon®**
A registered trademark for PVC polymers in sheet form. No manufacturer.

3257 **Astrazon®**
Cationic dyestuffs
For dyeing and printing polyacrylonitrile fibers. *Bayer AG; Bayer plc.*

3258 **Astro Floctite**
Acrylic and other polymer blends; adhesive for flocking auto parts, carpets, mats, wall plaques, assorted items. *Astro Industries Inc.* Name unverified.

3259 **Astro Mel**
Melamine-formaldehyde resins, usually 80% solids; water resistant corrugated boxes, abrasive nonwoven pads, textile finish. *Astro Industries Inc.* Name unverified.

3260 **Astrol**
Bromoxynil-ioxynil-isoproturon
A contact herbicide for cereal crops. *Embetec Crop Protection Ltd.*

3261 **Astrolith**
A proprietary trade name for a special lithopone; a pigment. No manufacturer.

3262 **Astrolok**
Sprayable, moisture curing adhesive; adhesive for laminates. *Apollo Chemicals Ltd.*

3263 **Astroplax**
Sodium *p*-glycollylarsanilate
A hydraulic gypsum cement. *May & Baker Ltd.* Name unverified.

3264 **Astroturf**
polyethylene
For doormats and sports turf. *Monsanto Co.* Name unverified.

3265 **Astrowet 0-70-PG**
577-11-7 3460 209-406-4
Sodium dioctyl sulfosuccinate
Solution in propylene glycol. Wetting emulsifying agent; for high flash point applications. *Alco Chemical Corp.*

3266 **Astrowet 0-75**
577-11-7 3460 209-406-4
Dioctyl sodium sulfosuccinate solution; wetting agent and emulsifier. *Alco Chemical Corp.*

3267 **Astryl**
144-87-6 4489 205-643-2
$C_8H_{10}AsNO_5$
glycarsamide
Sodium *p*-glycollylarsanilate. Anthelmintic. *Rhône-Poulenc UK.*

3268 **Astryn® 63A6-2**
9003-07-0 7741
PP, 20% mineral-reinforced; available in homopolymer and copolymer grades; offers high gloss, best surface quality, excellent impact resistance, outstanding heat-aging resistance, high resistance to solvents and chemicals, excellent stress crack resistance; low shrinkage; used for ABS replacement, lawn and garden tools/housings, lawn mower decks, appliance housings. *Himont.*

3269 **Astryn® 63F4-2**
9003-07-0 7741
PP homopolymer, 20% talc-reinforced; aesthetic polymer offering maximum stiffness, excellent heat-aging resistance, high solvent/chemical resistance, excellent stress crack resistance, good dimensional stability, low shrinkage; injection molding grade used for automotive parts, small and large appliances, electrical parts, housewares and utility products. *Himont.* Name unverified.

3270 **Astryn® 65F4-4**
9003-07-0 7741
40% talc-filled; max stiffness and high temperature performance; UL (115 C continuous use); in automotive, appliances, industrial components. *Himont.* Name unverified.

3271 **Astryn® 65F5-4**
9003-07-0 7741
PP, 40% calcium carbonate-filled; best high flex modulus/impact balance, good colorability, surface finish; in housewares, small appliances. *Himont.* Name unverified.

3272 **Astryn® 73F4-2**
9003-07-0 7741
PP copolymer, 20% talc-reinforced; injection molding resin offering outstanding heat-aging resistance, high solvent/chemical resistance, excellent stress crack resistance, low shrinkage, optimum balance of stiffness and impact resistance; used for automotive parts, small and large appliances, electrical parts, housewares, furniture and utility products. *Himont.* Name unverified.

3273 **Astryn® 73F5-2**
9003-07-0 7741
PP copolymer, 20% calcium carbonate-reinforced; high-impact injection molding resin offering good balance of stiffness and impact resistance, outstanding heat-aging resistance, highly resistant to solvents and chemicals, excellent stress crack resistance, used for automotive parts; small and large appliances, electrical parts, housewares, furniture and utility products. *Himont.* Name unverified.

3274 **Astryn® 78F4-2**
9003-07-0 7741
PP copolymer, 20% reinforced; high-impact extrusion grade resin with good surface quality, outstanding heat-aging resistance, highly resistant to solvents and chemicals, excellent stress crack resistance; used for extruded profiles, blow-molded ducts and thermoformed trays. *Himont.* Name unverified.

3275 **Astryn® BA16G**
9003-07-0 7741
PP copolymer, uv-stabilized; extrusion grade offering impact resistance, good dimensional stability, outstanding heat-aging resistance, highly resistant to solvents and chemicals, excellent stress crack resistance; used for weather stripping and extruded profiles. *Himont.* Name unverified.

3276 **Astryn® SD068-4**
9003-07-0 7741
PP, 40% calcium carbonate-reinforced; blow molding resin with high melt strength, high impact resistance, high stiffness, good long-term heat aging, highly resistant to solvents and chemicals, excellent stress crack resistance; used for vacuum-formed wiring channels, extruded profiles and blow molded ducts. *Himont.* Name unverified.

3277 **Asulam**
3337-71-1 222-077-1
Asulox. A soluble concentrate containing 400 g asulam per liter; herbicide for control of docks and bracken. *Embetec Crop Protection Ltd; Rhône-Poulenc Environmental Prods. Ltd.*

3278 **Asulox**
3337-71-1 222-077-1
Asulam
A soluble concentrate containing 400 g asulam per liter; herbicide for control of docks and bracken. *Embetec Crop Protection Ltd; Rhône-Poulenc Environmental Prods. Ltd.*

3279 **Asulox**
3337-71-1 222-077-1
Asulam
Selective weedkiller. *May & Baker Ltd.*

3280 **Asuntol®**
56-72-4 2626 200-285-3
Coumaphos
A veterinary preparation for the control of ectoparasites, including mange mites, on all domestic animals. *Bayer AG.*

3281 **AT 1806M; AT 4030M**
24937-78-8
Ethylene-vinyl acetate copolymer. Used for hot-melt adhesives, coatings, and sealants. *AT Plastics.*

3282 **AT-20GF**
105-57-7 36 203-310-6
Acetal
Acetal resin, 20% glass fiber-reinforced; offers lubricity and chemical and hot water resistance for automotive, hardware, plumbing applications. *Compounding Tech.*

3283 **Atar Phenol**
A natural phenol derived by fractionation from crude tar acids; a colorless, crystalline solid at ambient temperatures, with a distinct cresylic odor; used in the manufacture of phenol-formaldehyde resins and novolacs, in disinfectants, in selective weedkillers, as a preservative and in epoxies such as Bisphenol A. *Sasolchem.* Name unverified.

3284 **Aterite**
A nickel silver; usually contains from 47-68% copper, 17-38% zinc, 10-14% nickel, 1.5-1.9% iron, and 0.16-2.2% manganese.

3285 **Atgard**
62-73-7 3129 200-547-7

Dichlorvos
A proprietary preparation of dichlorvos; an insecticide. No manufacturer.

3286 Atiran
123-88-6 204-659-7
C_3H_7ClHgO
methoxyethyl mercury chloride
Agallol; Aratan; Aretan 6; Ceresan Universal Nazbeize; Chloro(2-methoxyethyl)mercury; Falisan; Gramisan; Higosan; Agallolat; Agalol; Aretan; Atiran; Cekusil Universal C; Ceresan-Universal Nassbeize; MEMC; Merchlorate; (β-methoxyethyl)mercuric Chloride; Methoxyethyl Mercuric Chloride; 2-methoxyethylmercuric Chloride; β-methoxyethylmercury Chloride; 2-methoxyethylmercury Chloride; Sedresan; Tafasan 6W; Tafasan. A potato fungicide. mp = 65°; insoluble in H_2O, organic solvents; LD_{50} (rat orl) = 22 mg/kg. *Plant Protection.* Name unverified.

3287 Atlac®
Synthetic polyester resins. *Reichhold.*

3288 Atlac® 382-05A
Bisphenol A fumarate polyester resin; resin with superior resistance to hydrolysis and chemical attack, resistant to deformation in high temperatures; used in fiberglass reinforced structures, coatings, mortars in pulp/paper, metal treatment, etc. *Reichhold/Reactive Polymers.*

3289 Atlac® 797CT
Neopentyl-chlorendic polyester resin; high thermo-oxidative resistance, provides fire protection in structures at high operating temperatires and in presence of oxidizing materials; for large fabrications, e.g., stack linings and scrubber systems in power and chlor-alkali industries. *Reichhold/Reactive Polymers.*

3290 Atlacide
7775-09-9 8741 231-887-4
sodium chlorate
Powder containing 58.2% w/w sodium chlorate; for total weed control for paths, drives and noncrop areas. *Chipman Ltd.*

3291 Atlacide Extra
Atrazine-sodium chlorate; used for total weed control in non crop areas. *Chipman Ltd.*

3292 Atladox HI
Soluble concentrate containing 240 g 2,4-D and 65 g picloram per liter; used to control weeds in non crop grass and grass verges. *Chipman Ltd.*

3293 Atlas 10 Bronze
A proprietary trade name for an aluminum bronze containing 9.0% aluminum, 7.0% lead with copper. No manufacturer.

3294 Atlas 5C Chlormequat
Soluble concentrate containing 460 g chlormequat and 320 g choline chloride per liter; plant growth regulator. *Atlas Interlates Ltd.*

3295 Atlas 89
A proprietary trade name for an alloy of copper with 9.0% aluminum and 1.0% iron. No manufacturer.

3296 Atlas 90
A proprietary trade name for an aluminum bronze containing 90.0% copper with 10.0% aluminum. No manufacturer.

3297 Atlas Adherbe
Adjuvant containing 83% refined mineral oil; wetting agent for herbicides. *Atlas Interlates Ltd.*

3298 Atlas Adherbe®
Crop chemical enhancer/additive. *Allied Colloids Ltd.*

3299 Atlas Adjuvant Oil
Adjuvant containing 95% refined mineral oil; wetting agent for herbicides. *Atlas Interlates Ltd.*

3300 Atlas Brown
Emulsifiable concentrate containing 150 g chlorpropham and 100 g pentanochlor per liter, a residual herbicide. *Atlas Interlates Ltd.*

3301 Atlas Brown
Herbicide used to protect vegetable crops. *Allied Colloids Ltd.*

3302 Atlas Chlormequat 46, 700
999-81-5 2153 213-666-4
Chlormequat chloride
Plant growth regulator. *Atlas Interlates Ltd.*

3303 Atlas CIPC 40
101-21-3 2240 202-925-7
chlorpropham
Emulsifiable concentrate containing 400 g chlorpropham per liter; a carbamate herbicide. *Atlas Interlates Ltd.*

3304 Atlas D
94-75-7 2865 202-361-1
2,4-D
Translocatable herbicide for cereals and established grassland. *Atlas Interlates Ltd.*

3305 Atlas Defoamer AFC
Hydrophobized silicone
Defoamer for textile and leather industries. *Atlas Refinery.*

3306 Atlas Electrum
Suspension concentrate containing 200 g chloridazon, 30 g chlorpropham, 20 g fenuron and 120 g propham per liter; a residual herbicide for beet crops. *Atlas Interlates Ltd.*

3307 Atlas EM-2
Glycol ester
Fiber lubricant, emulsifier. *Atlas Refinery.*

3308 Atlas EMJ-2
Nonylphenoxyl polyethoxy ethanol
Detergent, dispersant, emulsifier, wetting agent, penetrant; grease dispersant; for leather, textile sizing, bleaching operations, paper industry. *Atlas Refinery.*

3309 Atlas Gold
Suspension concentrate containing 37.5 g chlorpropham 25 g fenuron and 150 g propham per liter; an herbicide for use on beet crops. *Allied Colloids Ltd.*

3310 Atlas Gold
Herbicide used to protect sugar beet. *Allied Colloids Ltd.*

3311 Atlas Indigo
Plant growth regulator used to protect vegetable crops. *Allied Colloids Ltd.*

3312 Atlas Indigo
Mixture of chlorpropam and propham; plant growth regulator to suppress sprout growth in stored potatoes. *Atlas Interlates Ltd.*

3313 Atlas JG1
Raw, refined, and standardized fish oil; leather additive; aids fiber lubrication; masks undesirable finish odors. *Atlas Refinery.*

3314 Atlas Leather Odor
Compounded. masking agent exhibiting a traditional leather aroma. *Atlas Refinery.*

3315 Atlas Libsorb
Nonionic spreader containing 900 g/l alkylalcohol ethoxylate; a wetting agent for herbicides. *Atlas Interlates Ltd.*

3316 Atlas Lignum
Pesticides for forestry and amenity products. *Allied Colloids Ltd.*

3317 Atlas Lignum Granules
Atrazine-dalapon; granular soil acting herbicide for use in forestry plantations. *Atlas Interlates Ltd.*

3318 Atlas Linuron
330-55-2 5534 206-356-5
Linuron
A residual urea herbicide for the control of weeds in field crops. *Atlas Interlates Ltd.*

3319 Atlas Linuron
330-55-2 5534 206-356-5
$C_9H_{10}Cl_2N_2O_2$
N'-(3,4-Dichlorophenyl)-N-methoxy-N-methylurea
N'-(3,4-Dichlorophenyl)-N-methoxy-N-methylurea; Linuron; Methoxydiuron; Arresin; Afalon. A residual urea herbicide for the control of weeds in field crops including potatoes and carrots. mp = 93-94°; soluble in H_2O, acetone, alcohol, benzene, toluene, xylene; LD_{50} (rat orl) = 1500 mg/kg. *Atlas Interlates Ltd.*

3320 Atlas M 130
Methacrylate-based resin filled with 60 or 85% aluminum; for thermoforming large molds, prototype injection and blow molds, soft foam molds, open mold laminating, molds for RIM and other urethane processes, fiber-reinforced processes. *Degussa.*

3321 Atlas MCPA
94-74-6 5803 202-360-6
MCPA
Herbicide for cereals and grassland. *Atlas Interlates Ltd.*

3322 Atlas Orange
1918-16-7 7977 217-638-2
Propachlor
A pre-emergence herbicide for various horticultural crops. *Atlas Interlates Ltd.*

3323 Atlas Pink C
Suspension concentrate containing 25 g chlorpropham, 6 g diuron and 100 g propham per liter; a herbicide for use on outdoor lettuce. *Allied Colloids Ltd.*

3324 Atlas Pink C
Pesticides for vegetables. Herbicide. *Allied Colloids Ltd.*

3325 Atlas Protrum® K
Pesticides for sugar beet. Herbicide. *Allied Colloids Ltd.*

3326 Atlas Red
Suspension concentrate containing 200 g chlorpropham, cresylic acid and 50

g fenuron per liter; an herbicide for use on ornamentals and vegetables. *Atlas Interlates Ltd.*

3327 **Atlas Red**
Pesticides for vegetables. Herbicide. *Allied Colloids Ltd.*

3328 **Atlas Sheriff**
Chlorpyrifos-dimethoate. Mixture of chlorpyrifos [2921-88-2] and dimethoate [60-51-5]; systemic and fumigant insecticide for brassica crops. *Atlas Interlates Ltd.*

3329 **Atlas Sheriff®**
Pesticides for vegetables. Insecticide. *Allied Colloids Ltd.*

3330 **Atlas Silver**
1698-60-8 216-920-2
Chloridazon. Suspension concentrate containing 430 g chloridazon per liter; a pyridazinone herbicide for beet crops. *Atlas Interlates Ltd.*

3331 **Atlas Silver**
1698-60-8 216-920-2
Chloridazon
Suspension concentrate containing 430 g chloridazon per liter; a pyridazinone herbicide for beet crops. *Atlas Interlates Ltd.*

3332 **Atlas Solan**
Pesticides for vegetables. Herbicide. *Allied Colloids Ltd.*

3333 **Atlas Steel**
A proprietary hot die steel containing 9-11% tungsten, 3.25-3.5% chromium, and a little vanadium. No manufacturer.

3334 **Atlas Steward**
58-89-9 5526 200-401-2
lindane
Suspension concentrate containing 560 g γ HCH (lindane) per liter; an organochlorine insecticide. *Allied Colloids Ltd.*

3335 **Atlas Steward®**
Herbicide for cereals. *Allied Colloids Ltd.*

3336 **Atlas Tanked Cod Oil**
8001-69-2 11, 2464 232-289-6
Mainstay; Super D. Cod oil; leather additive providing fullness and mellowness to vegetable tanned sole leather, shoe linings, shoe uppers; ingredient in stuffing compounds and for chamois leather processing. *Atlas Refinery.*

3337 **Atlas Tecgran**
Pesticides for vegetables. Herbicide. *Allied Colloids Ltd.*

3338 **Atlas Total A, Total S**
Pesticides for forestry and amenity products. Herbicide. *Allied Colloids Ltd.*

3339 **Atlas®**
Man-made fibers, monofilament (wire); used for ropes and other industrial uses. *Bayer AG.*

3340 **Atlasbeam 1**
Odorless dehairing assistant for leathers. *Atlas Refinery.*

3341 **Atlasol 103**
142-87-0 205-568-5
$C_{10}H_{21}NaO_4S$
Sodium decyl sulfate
Decyl sodium sulfate; Decylsulfuric acid sodium salt; Sodium n-decyl sulfate; Sulfuric acid, monodecylester, sodium salt. Emulsifier, wetting agent, dispersant, fiber lubricant, synthetic fatliquor; for textile, leather, and general industrial applications. *Atlas Refinery.*

3342 **Atlasol 118-U**
Sulfated neatsfoot oil, hydrocarbon, relatively high boiling solvent; fatliquor for leather, especially white leather. *Atlas Refinery.*

3343 **Atlasol 170**
Emulsified neatsfoot oil, refined coconut oil, and synthetic oils; fatliquor for snow white alum tanned leather. *Atlas Refinery.*

3344 **Atlasol 177**
Sulfonated oil, fatty alcohols, and synthetic lubricants; fatliquor for light, fluffy leathers. *Atlas Refinery.*

3345 **Atlasol 178**
Sulfated neatsfoot oil, sperm oil; lightfast fatliquor for upper leathers. *Atlas Refinery.* Name unverified.

3346 **Atlasol BSC**
Sulfonated sperm oil and chlorinated ester; fatliquor for leather including hair-on skins; chrome and alum stable. *Atlas Refinery.*

3347 **Atlasol CSN**
Modified chlorosulfonated hydrocarbon; fatliquor for leathers, especially for white or pale colored leather where lightfastness and prolonged heat stability are important. *Atlas Refinery.*

3348 **Atlasol KAD**
95-19-2 202-397-8
Heptadecyl hydroxyethylimidazoline
Lightfast fiber lubricant, fatliquor, and softener for textile and leather applications. *Atlas Refinery.*

3349 **Atlasol KMM**
Ricinoleic acid
Triethanolamine salt; leather tanning surfactant, lubricant. *Atlas Refinery.*

3350 **Atlastan AR**
Low molecular weight acrylic polymer; tanning agent. *Atlas Refinery.*

3351 **Atlastan LC**
Aromatic carboxylic acid deriv.; auxiliary synthetic tanning agent for chrome leathers. *Atlas Refinery.*

3352 **Atlavar**
Atrazine + 2,4-D + sodium chlorate
Used for total weed control in non crop areas. *Chipman Ltd.*

3353 **Atlazin**
atrazine-aminotriazole. A liquid formulation containing 250g atrazine [1912-24-9] and 218g aminotriazole [61-82-5] per liter as a suspension concentrate; used for total weed control on industrial sites, paths, kerbs and channels, drives and hard tennis courts, hardstanding and storage areas. *Chipman Ltd.*

3354 **Atlox**
Blends of anionic and ionic surfactants. *ICI Am.*

3355 **Atmer®**
Surfactants. *Imperial Chemical Industries plc.*

3356 **Atmer® 100**
1338-39-2 8872 215-663-3
Sorbitan laurate
Antifog agent for PE and EVA food-wrapping films, antisat, cling additive. *ICI Polymer Additives.*

3357 **Atmer® 103**
Sorbitan ester
Antifog agent for long-lasting properties in LDPE, EVA, and PVC agricultural film; wetting agent for PP and PE films. *ICI Polymer Additives.*

3358 **Atmer® 105**
1338-43-8 8872 215-665-4
sorbitan oleate
Antifog, antistat, cling additive for low-temp. LDPE film applications. *ICI Polymer Additives.*

3359 **Atmer® 106**
26266-58-0 8872 247-569-3
Sorbitan trioleate
Antifog, cling additive for LDPE film. *ICI Polymer Additives.*

3360 **Atmer® 121**
Glyceryl oleate
Antifog, agent and cling additive for PVC food-wrapping film. *ICI Polymer Additives.*

3361 **Atmer® 122**
Glyceryl stearate
Antistat for LDPE, PP, flex, PVC; antifog. *ICI Polymer Additives.*

3362 **Atmer® 1007**
Glyceryl oleate
Antifog, cling agent for LDPE and PVC films. *ICI Polymer Additives.*

3363 **Atmer® 7001**
50% Atmer 129/163 (2:1) in PP; antistat for injection molding PP. *ICI Polymer Additives.*

3364 **Atmer® 8112**
20% in PE; antifog for film. *ICI Polymer Additives.*

3365 **Atmido**
A siliceous earth; used as a filtering medium, also as a rubber filler. No manufacturer.

3366 **Atmos**
Glycerine fatty acid esters. *ICI Am.*

3367 **Atmos 150**
Glyceryl stearate
Antistat for plastics (PP, PS) useful in food packaging. *Witco Corporation.*

3368 **Atmos 300**
Glyceryl oleate
Food emulsifier. *Witco Corporation.*

3369 **Atmos 659K**
Blend of propylene glycol mixed esters, mono/diglycerides and lecithin; food emulsifier for cakes. *Witco/Humko.* Discontinued.

3370 **Atmul 80**
Mono-and diglycerides; food emulsifier. *Witco Corporation.*

3371 **Atmul 124**
Glyceryl monoester
Antistat for PP and PS useful in food packaging. *Witco Corporation.*

3372 **Atmul 2622K**
Glyceryl-lactostearate

Lipophilic food emulsifier for aerating applications. *Witco/Humko.* Discontinued.

3373 Atolex ASL/C
Cationic surfactant in liquid form; lubricating and antistatic agent. *Standard Chemical Company.*

3374 Atolex ASL/C100
Cationic surfactant in the form of a thick liquid; antistatic lubricant. *Standard Chemical Company.*

3375 Atolex DA/25
Naphthalene sulfonate in liquid form; anionic dispersing liquid and leveling assistant particularly for disperse or acid dyes on acrylics or polyester. *Standard Chemical Company.*

3376 Atolex Polythene Emulsions
Full range of cationic surfactants in liquid form; softening and lubricating agents used in additives to resin finishes. *Standard Chemical Company.*

3377 Atolex QE
Cationic surfactant in liquid form; retarding agent. *Standard Chemical Company.*

3378 Atomite
471-34-1 1697 207-439-9
calcium carbonate
Fine ground calcium carbonate, used as a filler. *Thompson, Weinman & Co.* Name unverified.

3379 Atomite®
1317-65-3 5515 215-279-6
Limestone
High brightness, controlled particle size, easy dispersing grade for water and solvent-based coatings, rubber, plastics, caulks, sealants, adhesives, etc. *ECC International Ltd.*

3380 atoquinol
524-34-5 2347 205-067-1
$C_{19}H_{15}NO_2$
Allylphenylcinchoninic ester
cinchophen allyl ester. A powerful uric acid solvent and eliminator. Used experimentally to induce ulcers. mp = 30°; bp_{15} = 260°; insoluble in H_2O, soluble in organic solvents.

3381 ATP Nucleotides
Mixture of collagen amino acids in propylene glycol with adenosine triphosphate; used as a moisturizer. *Croda Inc.* Discontinued.

3382 Atpet
Surfactant with a range of properties including emulsifying, wetting, dispersing, antifoam, antirust. *ICI Am.*

3383 Atpet 300
25322-68-3 7729
Polyethylene glycol
Pharmaceutical aid; ointment base; suppository base; solvent; tablet excipient; tablet and/or capsule lubricant. *ICI Am.* Name unverified.

3384 Atpet 400
25322-68-3 7729
Polyethylene glycol
Ointment base; suppository base; solvent; tablet excipient; tablet and/or capsule lubricant. *ICI Am.*

3385 Atpet 600
25322-68-3 7729
Polyethylene glycol
Pharmaceutical aid; ointment base; suppository base; solvent; tablet excipient; and/or capsule lubricant. *ICI Am.* Name unverified.

3386 Atprime®
Bonding agent for reinforced plastic. *Reichhold.*

3387 Atraflow
1912-24-9 902 217-617-8
atrazine
A soluble concentrate containing 500g atrazine per liter; a residual herbicide. *Rhône-Poulenc Environmental Prods. Ltd.*

3388 Atraflow Plus
A liquid formulation containing 264g atrazine [1912-24-9] and 214g aminotriazole [61-82-5] per liter as a suspension concentrate; for used where total weed control is required including industrial sites, paths, curbs, and channels, drives and hard tennis courts, hardstanding and storage areas. *Burts & Harvey.*

3389 Atraflow Plus
A liquid formulation containing 270g atrazine [1912-24-9] and 160g aminotriazole [61-82-5] per liter as a suspension concentrate; for use where total weed control is required including industrial sites, paths, curbs, and channels, drives and hard tennis courts, hard standing and storage areas. *Rhône-Poulenc Environmental Prods. Ltd.*

3390 Atragan
Herbicide. *Agan Chemical Manufacturers Ltd.*

3391 Atramentum Stone
A mixture of ferric and ferrous sulfates with ferric oxide; used in the manufacture of inks.

3392 Atramet Combi
Active ingredients; atranex plus ametrex, ready formulated mixture of atrazine plus ametryne for use as a selective pre-and post-emergence herbicide. *Agan Chemical Manufacturers Ltd.*

3393 Atranex
1912-24-9 902 217-617-8
2-chloro-4-ethylamino-6-isopropylamino-1,3,5-triazine
atrazine. Active ingredient, atrazine; pre- and post-emergence herbicide. *Agan Chemical Manufacturers Ltd.*

3394 atrazine
1912-24-9 902 217-617-8
$C_8H_{14}ClN_5$
6-chloro-N-ethyl-N'-(1-methylethyl)-1,3,5-triazine-2,4-diamine
Atrex; Atratol; Primatol A; A 361; Aatrex; Aktinit A; G 30027; Gesaprim; Hungazin; Atranex; Fogard; Griffex; Mebazine; Vectal; Atrazines; Extrazine II; Laddock; Aatrex 4l; Aatrex 80W; Atrazine 4l; Atrazine 80W; Griffex 4l; Ortho St. Augustine Weed and Feed; Scotts Bonus Type S; Crisazina; Vectral SC; Attrex; Crisamina; Vectal SC; ATZ; Triazine A 1294; Zeazin; Argezin; Aktikon; Aktikon PK; Aktinit PK; Wonuk; Oleogesaprim; Chromozin; Pitezin; Actinite PK; Gesaprim 50; Akticon; Atrataf; Zeapos; Oleogesaprim 200; Gesaprim 500; Maizina; Aatrex Nine-O; Atazinax; Atrasine; Atratol A; Atred; Cekuzina-T; Crisatrina; Crisazine; Farmco Atrazine; Fenamine; Geigy 30,027; Gesoprim; Inakor; Primatol; Primaze; Radizine; Strazine; Weedex A; Zeaphos; Azinotox 500; Candex; Cyazine; Fenatrol; Radazine; Weedex,. Selective systemic herbicide, absorbed through roots and foliage, inhibits photosynthesis. Used for pre- and post-emergence control of annual grasses and broad-leaved weeds in a variety of crops. mp = 176°; d = 1.38; soluble in H_2O (28 mg/l), more soluble in organic solvents; LD_{50} (rat orl) = 1100, 3080 mg/kg.

3395 Atrinal
52508-35-7 257-976-8
Dikegulac sodium
Soluble concentrate of 200g dikegulac sodium per liter; plant growth regulator for hedges. *Rhône-Poulenc Environmental Prods. Ltd.*

3396 Atrixo
A silicone hand cream. *Smith & Nephew Pharmaceuticals Ltd.* Unverified.

3397 atroxindol
The anhydride of *o*-amino-α-phenylpropionic acid.

3398 Atrust
Rust converter. *Imperial Chemical Industries plc.*

3399 Attaclay
1337-76-4
magnesium aluminum silicate
Fine particle size sorbent attapulgite (magnesium aluminum silicate); used as a chemical conditioning agent (free-flow agent); for prilled ammonium nitrate and urea fertilizers and other bulk granular chemicals. *Engelhard.*

3400 Attacote
1337-76-4
Fine particle size sorbent attapulgite (magnesium aluminum silicate); used as a chemical conditioning agent (free-flow agent); for fire extinguishing chemicals. *Engelhard.*

3401 Attaflow
1337-76-4
Liquid (slurry) attapulgite (hydrous magnesium aluminum silicate); used as a suspending agent for liquid suspension fertilizers, flowable pesticides. *Engelhard.*

3402 Attane
Ultra-low density polyethylene copolymer. *Dow UK.* Discontinued.

3403 Attane 4601, 4802
9002-88-4 7728
Ultra low density. LLDPE; blown film extrusion resin. *Dow Cheml Co Ltd, UK & Ireland.*

3404 Attapulgite
1337-76-4
Palygorskite; Dioctrahedral smectite; Fullers earth; Carrisorb; Diluex®, FG; Economy Flor-Dri; Emcor; Flor-Kleen; Florco®; Florco® X; Florex®; Florex® Ag-Dri 6/30; Florigel® H-Y; Gelsorb B; Minugel; Min-U-Gel® 100, -200,-AR; -CW; _LF; Pharmasorb; QA-555; Refinex; S-60 RVM; Attaclay; Attacote; palygorskite; Attaflow; Attapulgus; Attasorb; Attagel,. A hydrated aluminum-magnesium silicate, chief ingredient in Fullers earth; drilling fluids, decolorizing oils, filter medium, absorbent. *Bromhead & Denison Ltd.*

3405 Attapulgus
1337-76-4
Attasorb; Attagel; attapulgite; Atasorb; Attaflow; Attacote; Attaclay,. Colloidal attapulgite (hydrous magnesium aluminum silicate); used as

suspending agent for oil well drilling clay, particularly salt water formations. *Engelhard.*

3406 Attasorb
1337-76-4
Fine particle size sorbent attapulgite (magnesium aluminum silicate); used as a chemical conditioning agent (free-flow agent) for powdered detergents, agricultural chemicals. *Engelhard.*

3407 Audrey
Automatic dielectric analyzer. *Tetrahedron Association Inc.* Name unverified.

3408 Auel solder
An alloy of 63% tin, 35% zinc, 1.7% copper, and 0.3% aluminum.

3409 Augsburg metal
A brass containing 72% copper, and 28% zinc.

3410 Auracryl
Aqueous color dispersions for use in water-borne industrial coatings. *Engelhard.*

3411 Aurantine
The trade name for osage orange extract (from the bark of a shrub); used in the textile industry for tanning. A residue containing terpenes remaining after the refining of orange oil; used as a perfume in soaps. No manufacturer.

3412 Aurantlol®
Schiff base of hydroxycitronellal and methyl anthranilate; fragrance; sweetly floral. *BASF AG.*

3413 Aurasperse
Aqueous color dispersions for use in water based coatings. *Engelhard.*

3414 Aurocyanase
A colloidal gold and potassium double cyanide.

3415 Auromet55
A proprietary trade name for an alloy containing 76-80% copper 10-12% aluminum, 4-6% iron, and 4-6% nickel. No manufacturer.

3416 Aurum 400, 450, 500
Thermoplastic polyimide resin; high medium and low flow non-filled grades. *Advanced Web Prods.*

3417 Aurum JAF 3040
Thermoplastic polyimide resin, aramid fiber filled; low wear, high PV value grade. *Advanced Web Prods.* Discontinued.

3418 Aurum JCN 6030
Thermoplastic polyimide resin, carbon fiber filled; high flow, high modulus, high strength reinforced grade. *Advanced Web Prods.* Discontinued.

3419 Aurum JCN 3030
Thermoplastic polyimide resin, glass fiber filled; electrically insulated, high modulus reinforced grade. *Advanced Web Prods.*

3420 Aurum JGN 3030
Thermoplastic polyimide resin, carbon fiber filled; high modulus, high strength, low wear grade. *Advanced Web Prods.*

3421 Aurum JNF 3010
Thermoplastic polyimide resin, fluropolymer filled; low friction, reinforced grade. *Advanced Web Prods.* Discontinued.

3422 Aurum JNF 3020
Thermoplastic polyimide resin, fluropolymer filled; low friction, high PV value grade. *Advanced Web Prods.*

3423 Aurum JQF 3025
Thermoplastic polyimide resin, fluropolymer filled; low friction, medium modulus grade. *Advanced Web Prods.* Discontinued.

3424 Aurum JRF 3025
Thermoplastic polyimide resin, graphite filled; low wear grade. *Advanced Web Prods.*

3425 Aurum JRN 3015
Thermoplastic polyimide resin, graphite filled; low wear grade. *Advanced Web Prods.* Discontinued.

3426 Aurum Series
Thermoplastic polyimide resin; super heat-resistant resin for injection and extrusion molding; radiation resistant, good chemical and oil resistance, electrical properties. and weatherability; for mechanical and precision parts, electrical/electronic parts, automotive parts, wire extrusion coating, film, fiber. *Advanced Web Prods.*

3427 austenite
A characteristic constituent of very highly carbonized steel, containing more than 1.1% carbon.

3428 Australian gold
A gold-silver alloy containing 8.33% silver; used for coinage.

3429 Austrapol
Styrene/butadiene and polybutadiene polymers; used for tires, retread, general rubber goods and plastics modification. *Australian Synthetic Rubber Co Ltd.*

3430 Austratex
Styrene-butadiene high solids latex; used for carpet underlay foams, bitumen modification. *Australian Synthetic Rubber Co Ltd.*

3431 Austrostab
A full range of stabilizer systems for PVC, containing stabilizers (Pb, Pb/Ba, Cd, Ca/Zn), lubricants, pigments, fillers, and modifiers; completely ready-for-use formulations for applications such as PVC pipe, cable, profile. *Chemtech (Crop Protection) Ltd.* Name unverified.

3432 Austrox
Melted lead oxide granules
Used as raw material for specialty glasses such as crystal glass and glass for electronic tubes. *Chemson Polymer Additiv GmbH.* Name unverified.

3433 AuSub
Gold substitute inorganic die attach pastes; for electronic applications in computer industry and for military and aerospace uses. *Johnson Matthey plc.*

3434 Autan®
Consumer insect repellent against mosquitoes, gnats and other flying insects. *Bayer AG.*

3435 Auto Command®
For automotive industry. *DuPont UK.*

3436 Autofroth
9009-54-5
Polyurethane foam
Rigid foam systems (forth-in-place) for insulation, flotation, molding. *Olin.* Discontinued.

3437 Autogal
A trademark for a flux used for soldering and welding aluminum; a mixture of the halogen salts of the alkali metals. No manufacturer.

3438 Automate
Liquid dyestuffs for petroleum. *Morton Int'l Ltd.*

3439 automolite
ZnO Al_2O_3
A mineral.

3440 Autopak
9009-54-5
Polyurethane foam; rigid/flexible foam system (pour-in-place) for pre-mold packaging. *Olin.* Discontinued.

3441 Autopoon NI
Cationic surfactants, solvents and solubilizers; water-repellent concentrate for preparations for lacquered surfaces, car finishing. *Zschimmer & Schwarz.* Discontinued.

3442 Autopour
9009-54-5
Polyurethane foam; pressurized rigid foam system (pour-in-place) for insulation, molding applications. *Olin.* Discontinued.

3443 Autopur WK 4121
Surfactant blend; basic material for car shampoo formulations with water-repellent and gloss effects. *Zschimmer & Schwarz.* Discontinued.

3444 Autovisuel®
For automotive industry. *DuPont UK.*

3445 Autumn Kite
Emulsifiable concentrate of 300 g isoproturon [34123-59-6] and 200 g trifluralin [1582-09-8] per liter; used for control of annual grasses in winter wheat and barley. *Schering Agrochemicals Ltd.* Discontinued.

3446 Autumn Lawn Food
Lawn fertilizer. *Fisons plc, Horticultural Div.* Name unverified.

3447 Auxiliary PR-10BT
Antiwicking agent and thickener for textile pigment printing. *Catawba-Charlab.*

3448 Avabond
Adhesives for packaging, woodworking, textiles etc. *ICI Polyurethanes.* Name unverified.

3449 Avadex
2303-16-4 3014 218-961-1
Diallate
Herbicide for wild oats. *Monsanto Co; Monsanto plc.*

3450 Avadex® BW
2303-17-5 9726 218-962-7
Triallate
Herbicide with triallate as active ingredient; for control of wild oats, slender foxtail and bent grass in sugar beet and feed turnips, summer and winter barley, winter rye. *BASF AG.*

3451 Avadyne AV1200/CA100
Two-part urethane prepolymer emulsion adhesive; for lamination of film-to film and film-to metallized film structures. *Pierce & Stevens.*

3452 Avalon
Thermoplastic polyurethanes for injection molding and as adhesives and coatings. *ICI Polyurethanes.* Name unverified.
3453 Avamid
Polyimide prepregnations. *DuPont UK.*
3454 Avamid 150
Avocadamide DEA, avocado oil; biodeg. SE foam stabilizer, viscosity builder, lubricant for conditioning shampoos, hair rinses, creams and lotions; imparts smooth, silky feel to skin and hair. *Mona Industries.*
3455 Avanel® S-30
Sodium C12-15 pareth-3 sulfonate
Biodegradable mild surfactant, emulsifier for personal care, household, institutional and industrial products; stable in presence of hypochlorite and over entire pH range. *PPG/Specialty Chem.* Discontinued.
3456 Avantine
67-63-0 5227 200-661-7
isopropanol
isopropyl alcohol
A brand of isopropyl alcohol; an antiseptic and anesthetic. *Laporte Industries Ltd.*
3457 Avaunt®
For agriculture industry. *DuPont UK.*
3458 Avazyme
Chymotrypsin
Enzyme. *Wallace Laboratories.* Name unverified.
3459 Avecolite
A proprietary phenol-formaldehyde synthetic resin. No manufacturer.
3460 Avenge 2
43222-48-6 3185 256-152-5
Difenzoquat methyl sulfate
Soluble concentrate of 150 g difenzoquat per liter; used for control of wild oats in cereals. *Cyanamid of Great Britain Ltd; Schering Agro-chemicals Ltd.*
3461 Aventox SC
Simazine-trietazine; herbicide used with peas and beans. *Dow Elanco Ltd.*
3462 Aveonal
An alloy of aluminum with 4% copper, 0.05% magnesium, 0.05% manganese, and 0.05% silicon.
3463 Aversin
Paraffin wax emulsion with zirconium salts; nonpermanent water-repellent finish compatible with resins, suitable for all kinds of fibers. *Henkel Chemicals Ltd.* Name unverified.
3464 Avgard
7601-54-9 8808 231-509-8
Trisodium phosphate
Food additive for meat, poultry, and seafood industries. *Rhône-Poulenc Food Ingreds.*
3465 Avgard
Antimisting kerosene. *ICI Chem & Polymers Ltd.* Discontinued.
3466 Avialite
A proprietary trade name for an alloy of copper with about 9.0% aluminum and 1.0% iron. No manufacturer.
3467 Aviamide-6
Policapram
Pharmaceutic aid; a filler. *Avicon Inc.* Name unverified.
3468 Aviashine
A blend of solvents, carriers, abrasives and waxes; aircraft maintenance chemical; provides effective cleaning for paintwork, chrome and other metal surfaces; can be polished to give a durable bright and protective finish. *The Kent Chemical Company Ltd.*
3469 Aviawash
A blend of detergents and surfactants with inhibitors; aircraft maintenance chemical for aircraft exterior cleaning; for cleaning painted and unpainted external surfaces of aircraft and ground equipment. *The Kent Chemical Company Ltd.*
3470 Avicel
Microcrystalline cellulose
Pharmaceutic aid, a filler. *FMC; FMC UK.* Discontinued.
3471 Avicell-RC
A proprietary trade name for a chemically pure colloidal cellulose; forms thixotropic dispersions both mechanically and thermally stable; edible and metabolically alert; used as a thickening agent. *BP Chemicals.*
3472 Aviester
Pegoterate
Pharmaceutic aid. *Avicon Inc.* Name unverified.
3473 Avilon
Metal complex dyes. *Ciba plc.* Name unverified.
3474 Avional D
An alloy of aluminum with 3.9% copper, 0.5% nickel, 0.5% magnesium, 0.55% silicon, and 0.3% iron. No manufacturer.
3475 Avirol
Textile and leather auxiliary. *Hickson & Welch Ltd.*
3476 Avirol® 125E
Sodium alkyl ether sulfate
Emulsifier for vinyl acetate copolymers, S/B latexes, vinyl chloride copolymers, acrylate homo- and copolymers. *Henkel/Functional Prods.*
3477 Avirol® A
2235-54-3 218-793-9
Ammonium lauryl sulfate
Emulsifier for emulsion polymerization; additive for mechanical latex foaming; foaming agent for acrylate dispersions, carpet and upholstery cleaners; air entraining agent for mortars. *Henkel/Functional Prods.*
3478 Avirol® AE 3003
67762-19-0
Ammonium laureth sulfate
Emulsifier for vinyl acetate copolymers, S/B latexes, vinyl chloride copolymers, acrylate homo-and copolymers. *Henkel/Functional Prods.*
3479 Avirol® FES 996
9004-82-4
Sodium laureth sulfate
Emulsifier for vinyl acetate copolymers, S/B latexes, vinyl chloride copolymers, acrylate homo- and copolymers. *Henkel/Functional Prods.*
3480 Avirol® SA 4106
Sodium 2-ethylhexyl sulfate
Wetting agent, stabilizer for plastics, rubber, adhesives, food contact paper; coemulsifier for vinyl chloride, acrylics, vinyl acetate copolymers; biodegradable. *Henkel/Functional Prods.*
3481 Avirol® SA 4108
Sodium N-octyl sulfate
Emulsifier, low foaming surfactant, wetting agent. *Henkel/Functional Prods.*
3482 Avirol® SA 4110
Sodium N-decyl sulfate
Wetting and emulsifying agent for plastics. *Henkel/Functional Prods.*
3483 Avirol® SA 4113
3026-63-9 221-188-2
Sodium tridecyl sulfate
Emulsifier for S/B and vinyl chloride copolymers. *Henkel/Functional Prods.*
3484 Avirol® SE 3002
9004-82-4
Sodium laureth sulfate
Akyposal 9278 R; Akyposal EO 20 MW; Akyposal RLM 70; Akyposal MS SPC; Avirol® FES 996. Emulsifier for vinyl acetate copolymers, S/B latexes, vinyl chloride copolymers, acrylate homo- and copolymers. *Henkel/Functional Prods.*
3485 Avirol® T 40
139-96-8 205-388-7
TEA lauryl sulfate
Emulsifier for emulsion polymerization; additive for mech. latex foaming; foaming agent for acrylate dispersions, carpet and upholstery cleaners; air entraining agent for mortars. *Henkel/Functional Prods.*
3486 Avisol
Neutral soluble sulfathiazole for poultry. *May & Baker Ltd.* Name unverified.
3487 Avisol® G
metiram-cymoxanil. Fungicide for potatoes, vines, and other crops. *BASF AG.* Name unverified.
3488 Avistin®
A range of basic fatty acid condensation products; used for the manufacture of cationic preparing and finishing agents for textiles. *Hüls AG.*
3489 Avistin® PN
141-21-9 205-469-7
Stearamidoethyl ethanolamine
For production of cationic textile auxiliary agents. *Hüls Am; Hüls AG.*
3490 Avitex
Surface active agents. *DuPont UK.*
3491 Avitige®
Fiber. *DuPont UK.*
3492 Avitone®
Dyeing assistants. *DuPont UK.*
3493 Avitone® A
Sodium alkyl sulfonate
Finishing agent, softener, lubricant for improving texture and hand of textiles, leather, and paper, also for elastomers; highly stable to chemicals and oxidation. *DuPont.*

3494 Avitrol
504-24-5 3974 207-987-9
4-aminopyridine
Frampridine. Treated grain baits for pest bird control; classified as a restricted use' pesticide. *Avitrol Corporation.*

3495 Avivage
Combination of sulfated fats with special additives; softener, stabilizer in bleaching liquors. *Henkel Chemicals Ltd.* Name unverified.

3496 Avivan
Finishing agent. *Ciba plc.* Name unverified.

3497 Avocado Oil
8024-32-6 232-428-0
Avocado fatty oil
Avocado Oil CLR; Lipoval A; Super Refined Avocado Oil; Natoil AVO; Nikkol Avacado Oil; Tri-Ol AVO. Emollient. Conditioner for skin and hair care preparations. *Nikko Chem.; Lipo.*

3498 Avocado Oil
8024-32-6 232-428-0
Lipovol A; Alligator Pearl oil; Oils, avocado; Avocado Oil CLR; Natoil AVO; Nikkol Avocado Oil; Super Refined™ Avocado oil; Tri-Ol AVO. Avocado oil; conditioner, glosser, emollient imparting a light, nongreasy, silky afterfeel to skin and hair products; high film gloss and rapid spread; used in personal care products. Yellowish-green oil; sol.in mineral oil, isopropyl esters, ethanol; insoluble in H_2O; d = 0.908-0.925. Lipo; Tri-K Industries; Am. Roland; Arista Industries; Nikko Chem. Co. Ltd.; Croda Inc; Universal Preserv-A-Chem.

3499 Avocado Oil CLR
8024-32-6 232-428-0
Avocado Oil
Emollient and conditioner for skin and hair care preparations. *Dr. Kurt Richter; Henkel/Cospha.*

3500 Avoilefin
Polypropene 25
Pharmaceutic aid. *Avicon Inc.* Name unverified.

3501 Avolan®
Leveling agents for wool; brightening agents to correct faulty dyeings on wool; dispersant for dyeing polyester fibers. *Bayer AG; Bayer plc.*

3502 Avoparcin
37332-99-3 922 253-466-4
Avotan; AV-290; C-254; CL-81588; LL-AV290. Antibacterial. Animal feed glycopeptide produced from *Streptomyes candidus* Soluble in H_2O, DMF, DMSO; λ_m = 280 nm; LD_{50} (rat orl) > 10000 mg/kg. Name unverified.

3503 Avron
Acrylic dispersion. *ICI Chem & Polymers Ltd.*

3504 AVT-75
124-68-5 469 204-709-8
2-amino-2- methyl-1 propanol
Boiler water treatment chemical, corrosion inhibitor, carbon dioxide absorber. *Angus.*

3505 Axall
Herbicide. *May & Baker Ltd.* Name unverified.

3506 Axarel®
For electrical industry. *DuPont UK.*

3507 Axelcon
Thermoplastic lubricant concentrates available in various carrier resins for specific applications; high lubricant concentrations based on the Mold Wiz internal lubricants with a complex polymeric base. *Axel Plastics Research Laboratories Inc.* Discontinued.

3508 Axelglo
Proprietary, polymeric based polishes to maintain and renew fiberglass and metal (painted and unpainted) surface luster; used for molded fiberglass products (yachts, camper tops etc.) and automotive. *Axel Plastics Research Laboratories Inc.*

3509 Axiquel
4171-13-5 10048 224-033-7
Valnoctamide
Tranquilizer. *McNeil Pharmaceuticals.* Name unverified.

3510 Axite
An explosive; smokeless powder which contains guncotton, nitro-glycerine, petroleum jelly, and a little potassium nitrate. No manufacturer.

3511 Axol® C 62
68990-05-6 273-575-0
Citric acid ester of glycerol mono/distearates; food emulsifier. *Goldschmidt AG.* Discontinued.

3512 Axol® E 61
8029-91-2
Acetylated hydrogenated lard glyceride
Acetylated hydrogenated lard glyceride; food emulsifier, lubricant, solvent, plasticizer and coating material for foodstuffs and cosmetics. *Goldschmidt AG.* Discontinued.

3513 Axol® L 61, L62
689990-06-7
Hydrogenated tallow glyceride lactate
Food emulsifier. *Goldschmidt AG.* Discontinued.

3514 Ayrtol
88-04-0 2228 201-793-8
Chloroxylenol
A proprietary preparation of chloroxylenol; a disinfectant. *Ayrton Saunders plc.* Unverified.

3515 AZ
Toothpaste to help fight plaque and cavities. *Richardson-Vicks Inc.* Name unverified.

3516 azadirachtin
11141-17-6 926
$C_{35}H_{44}O_{16}$
Isolated from the seeds of the neem tree. Used experimentally as an insect control agent. mp= 154-158°; $[\alpha]_D$ = -53° (c = 0.5 $CHCl_3$); λ_m =217 nm (ε 9100 MeOH).

3517 azamethiphos
35575-96-3 252-626-0
$C_9H_{10}ClN_2O_5PS$
S-[(6-chloro-2-oxooxazolo[4,5-b]pyridin-3(2H)-yl)methyl] O,O-dimethyl phosphorothioate
OMS 1825; Alfacron; Alfacron 10WP; Alficron; CGA 18809; Dymos; Rubidor; Snip. Insecticide with contact and stomach action; cholinesterase inhibitor, used for control of flies and other insects in animal houses. Used for mosquitoes, tsetse flies, cockroaches etc. mp = 89°; SG_{20}= 1.60; soluble in H_2O(1.1 g/l), more soluble in organic solvents; LD_{50} (rat orl) = 1180 mg/kg.

3518 azaperone
1649-18-9 931 216-715-8
$C_{19}H_{22}FN_3O$
1-(4-fluorophenyl)-4-[4-(2-pyridinyl)-1-piperazinyl]-1-butanone
R-1929; Stresnil; Suicalm. Used in veterinary medicine as a sedative and a tranquillizer. mp = 73-75°.

3519 Azelaic acid
123-99-9 938 204-669-1
$C_9H_{16}O_4$
Nonanedioic acid
Emerox® 1110; 1,7-Heptanedicarboxylic acid; lepargylic acid; Anchoic acid; Skinoren. Intermediate in chemical manufacturing of, for example, detergents; also an antiacne agent. mp = 98-102°; bp_{100} = 286°; soluble in H_2O (2.4g/l), EtOH, Et_2O; LD_{50} (rat orl) >5 gm/kg. *Henkel/Emery.*

3520 Azinphos-ethyl
2642-71-9 220-147-6
$C_{11}H_{16}N_3O_3PS_2$
O,O-diethyl S-[(4-oxo-1,2,3-benzotriazin-3(4H)-yl)methyl] phosphorodithioate
Cotnion-ethyl; Gusathion® A; Cotnion-Ethyl-Methyl; Azinos; Azinugec E; Bay 16259; Bionex; Ethyl Guthion; Gusathion K forte; Gutex; R 1513; Sepizin L; Triazotion; ENT-22014. Insecticide with broad spectrum of activity against biting pests such as caterpillars, beetles and their larvae, as well as sucking pests such as aphids, thrips, leafhoppers, etc. on a wide range of crops; side-effect on mites. mp = 53°; $bp_{0.0013}$ = 111°; d^{20} = 1.284; n_D^{20}= 1.5928; soluble in H_2O (4-5 mg/l), more soluble in organic solvents; LD_{50} (rat orl) = 12.5-17.5 mg/kg. *Bayer AG; Makhteshim Chemical Works Ltd.*

3521 Aziplex
Blended metal chelate. *ABM Chemicals Ltd.* Name unverified.

3522 Azocoll®
53092-90-3
Substrate for organisms such as E. coli and B. subtilis. *Calbiochem Corp.*

3523 azocyclotin
41083-11-8 255-209-1
$C_{20}H_{35}N_3Sn$
1-(tricyclohexylstannyl)-1H-1,2,4-triazole
BAY-BUE 1452; Peropal; tricyclotin. Acaricide used for control of mobile stages of spider mites on pome and stone fruit, grapes, citrus, vegetables, etc. mp = 218°; insoluble in H_2O (< 1 mg/l), soluble in organic solvents; LD_{50} (rat orl) = 99 mg/kg.

3524 Azodicarbonamide
123-77-3 950 204-650-8
$C_2H_4N_4O_2$
1,1-g-azobisformamide
Azodicarboxamide; Genitron; Unifoam; Nitropore; Porofor; Diazenedicarboxamide; 1,1'-azobiscarbamide; Azobiscarbonamide; Azobiscarboxamide; Azodicarbamide; Azodicarboxylic Acid Diamide; Azoformamide; Diazenedicarboxylic Acid Diamide; Celogen® AZ 120, 130,

150, 180, 199; Celosen AZ; CHKHZ 21; CHKHZ21R; Genitron AC; Genitron AC 2; Genitron AC 4; Kempore; Kempore 125; Kempore R 125; Lucel ADA; Pinhole AK 2; Porofore 505; Porofor ADC/R; Ficel® AC2; Genitron; Porofor CHKHZ 21; Porofor CHKHZ 21r; Unifoam AZ; Uniform AZ; Yunihomu AZ; C,C'-azodi(formamide;)Kempore® 60/14FF; Porofor® ADC/E,. Blowing and foaming agent for plastics; bleaching agent in cereal flour. mp = 225° (dec); soluble in hot H_2O; insoluble in cold H_2O; EtOH. *Atochem SA; Gist-Brocades Food Ingreds.; Olin; Schering Berlin Polymers; Uniroyal.*

3525 azodimethyl-2,2'-azobis(2,4-dimethylvaleronitrile)
Initiator for suspension polymerization of vinyl chlorides, solution polymerization of various monomers.

3526 Azoene
Fast red salt; Ponceau fast L salt. A pinkish cream powder for use in automatic SGO-T assays. *British Drug Houses.* Name unverified.

3527 Azofix
Azofix colors
Fine Dyestuffs & Chemicals Ltd. Name unverified.

3528 Azoground
Azoic colors
Fine Dyestuffs & Chemicals Ltd. Name unverified.

3529 Azoguard
A stabilizer for diazo compounds. *Imperial Chemical Industries plc.* Name unverified.

3530 Azolan
61-82-5 513 200-521-5
1,2,4-Triazol-3-ylamine
Active ingredient: aminotriazole; weedkiller with good translocation characteristics for the control of perennial and annual weeds. *Agàn Chemical Manufacturers Ltd.*

3531 Azolith
A proprietary pigment containing 71.0% $BaSO_4$ and 29.0% ZnS. *USV Pharmaceutical Corp.* No manufacturer.

3532 Azomagenta G
A dyestuff obtained by diazotizing sulfanilic acid, and treating the product with S acid (8-Amino-1-naphthol-5-sulfonic acid) [83-64-7]. No manufacturer.

3533 Azoprint
Azoic printing colors. *Fine Dyestuffs & Chemicals Ltd.* Name unverified.

3534 Azorapid
Azoic printing colors. *Fine Dyestuffs & Chemicals Ltd.* Name unverified.

3535 Azosan
A fungicide. *May & Baker Ltd.* Name unverified.

3536 Azostix
A prepared test strip of urease, bromothymol blue and buffers, for the semi-quantitative determination of blood urea levels. *B. C. Ames.* Name unverified.

3537 Azotox
Insecticides. *ICI Chem & Polymers Ltd.* Discontinued.

3538 Azotoz 580
919-86-8 6129 213-052-6
Demeton S methyl
Emulsifiable concentrate containing 580 g demeton-S-methyl per liter; a systemic organophosphorus insecticide. *BritAg Industries Ltd.*

3539 Aztec®
6731-36-8 229-782-3
$C_{17}H_{34}O_4$
1,1-bis(*t*-butylperoxy)-3,3,5-trimethylcyclohexane
1,1-Di-(*tert*-butylperoxy)-3,3,5-trimethyl cyclohexane. 1,1-bis(*t*-butylperoxy)-3,3,5-trimethyl cyclohexane in dibutyl phthalate. Catalyst and initiator. *Catalyst Resources.* Name unverified.

3540 Aztec® *t*-Amyl peroxypivalate-75 OM
29240-17-3 249-530-6
$C_{10}H_{20}O_3$
Amilperoxy pivalate
Propaneperoxoic acid, 2,2-dimethyl-, 1,1-dimethylpropyl ester. Initiator. *Catalyst Resources.* Name unverified.

3541 Aztec® *t*-Butyl Hydroperoxide-70, Aq
75-91-2 1604 200-915-7
$C_4H_{10}O_2$
t-Butyl hydroperoxide
tert-butyl hydroperoxide; TBHP; 2-hydroperoxy-2-methylpropane; T hydro; Cadox TBH;1,1-Dimethylethyl; Perbutyl H; Trigonox AW70; Trigonox A-75; ethyldiethylperoxide; trigonox; Butyl hydroperoxide; Dimethylethyl hydroperoxide; Hydroperoxide,1,1-dimethylethyl; Slimicide DE-488; *tertiary*-Butyl hydroperoxide. Aqueous suspension of *t*-Butyl hydroperoxide; initiator and slimicide. mp = 5°; bp_{15} = 37°; d = 0.9400; n_D^{20} = 1.3840; LD_{50} (rat orl) = 406 mg/kg. *Catalyst Resources. Name unverified.*

3542 Aztec® *t*-butyl peracetate-50 OMS, 60 OMS, 75 OMS
107-71-1 203-514-5
$C_6H_{12}O_3$
t-butyl peroxyacetate
tert-butyl peroxyacetate; *tert*-butyl peracetate; Ethaneperoxoic acid, 1,1-dimethylethyl ester; *tertiary*-Butyl peracetate. *t*-butyl peroxyacetate in odorless mineral spirits. Catalyst and initiator. *Catalyst Resources.* Name unverified.

3543 Aztec® *t*-Butyl Perbenzoate
614-45-9 210-382-2
$C_{11}H_{14}O_3$
tert-butyl peroxybenzoate
tert-butyl perbenzoate; Benzenecarboperoxoic acid, 1,1-dimethylethyl ester; *tertiary*-Butyl perbenzoate. Catalyst and initiator. mp = 8°; bp = 113°; d = 1.021; n_D^{20}= 1.4990; insoluble in H_2O, soluble in organic solvents; LD_{50} (rat orl) = 1012 mg/kg. *Catalyst Resources.* Name unverified.

3544 Aztec® *t*-Butyl Peroctoate
3006-82-4 221-110-7
$C_{12}H_{24}O_3$
Butyl peroctoate
Hexaneperoxoic acid, 2-ethyl-, 1,1-dimethylethyl ester; *tert*-Butyl peroxy-2-ethylhexanoate. Initiator. *Catalyst Resources.* Name unverified.

3545 Aztec® *t*-Butyl Peroctoate-50 OMS
3006-82-4 221-110-7
$C_{12}H_{24}O_3$
tert-butyl peroxy-2-ethylhexanoate
hexaneperoxoic acid, 2-ethyl-, 1,1-dimethylethyl ester. *t*-butyl peroctoate in odorless mineral spirits. *Catalyst Resources.* Name unverified.

3546 Aztec® *t*-Butyl Peroxyisobutyrate-75 OMS
109-13-7 203-650-5
Butylperoxy isobutyrate
Initiator. *Catalyst Resources.* Name unverified.

3547 Aztec® *t*-Butyl Peroxyneodecanoate-50 OMS, 75 OMS
26748-41-4 247-955-1
Butylperoxy neodecanoate
tert-Butyl peroxyneodecanoate. Initiator. *Catalyst Resources.* Name unverified.

3548 Aztec® *t*-Butyl Peroxypivalate-75 OMS
927-07-1 213-147-2
$C_9H_{18}O_3$
t-Butyl peroxypivalate
tert-butyl peroxypivalate; Propaneperoxoic acid, 2,2-dimethyl-, 1,1-dimethylethyl ester. Initiator. mp = -17°; insoluble in H_2O. *Catalyst Resources.* Name unverified.

3549 Aztec® 1,1-Bis(*t*-Butylperoxy)-3,3,5-Trimethyl Cyclohexane
6731-36-8 229-782-3
$C_{17}H_{34}O_4$
Bisbutyl peroxy trimethylcyclohexane
Peroxide, (3,3,5-trimethylcyclohexylidene)bis[(1,1-dimethylethyl); 1,1-Di-(*tert*-butylperoxy)-3,3,5-trimethyl cyclohexane. Initiator. *Catalyst Resources.* Name unverified.

3550 Aztec® 1,1-Bis(*t*-Butylperoxy)Cyclohexane-80 BBP
3006-86-8 221-111-2
Dibutyl peroxy cyclohexane-butylbenzyl phthalate
Initiator. *Catalyst Resources.* Name unverified.

3551 Aztec® 2,5-Di
78-63-7 201-128-1
$C_{16}H_{34}O_4$
2,5-dimethyl-2,5-di(*t*-butylperoxy)hexane
(1,1,4,4-tetramethyl-1,4-butanediyl)bis[(1,1-dimethylethyl) peroxide; 2,5-Dimethyl-2,5-di(*tertiary*-butylperoxy)-hexane. Initiator. *Catalyst Resources.* Name unverified.

3552 Aztec® 2,5-Tri
1068-27-5 213-944-5
$C_{16}H_{30}O_4$
2,5-dimethyl-2,5-di(*t*-butylperoxy)hexyne-3
2,5-Dimethyl-2,5-di-(tert-butylperoxy)hexyne-3; 3-hexyne, 2,5-dimethyl-2, 5-di(t-butylperoxy)-; 2,5-Dimethyl-2,5-di(tertiary-butylperoxy)-3-hexyne. Initiator. *Catalyst Resources.* Name unverified.

3553 Aztec® Benzoyl Peroxide-70-77
94-36-0 1149 202-327-6
$C_{14}H_{10}O_4$
Benzoyl peroxide
Benzoyl peroxide and water. Catalyst and initiator. *Catalyst Resources.* Name unverified.

3554 Aztec® Benzoyl Peroxide-Dry
94-36-0 1149 202-327-6
benzoyl peroxide
Catalyst and initiator. *Catalyst Resources.* Name unverified.

3555 Aztec® CHP-80
80-15-9 201-254-7
$C_9H_{12}O_2$
Cumene hydroperoxide
Cumene Hydroperoxide; Cumyl Hydroperoxide; Hyperiz; Trigorox K 80; α,α-Dimethylbenzylhydroperoxide; 1-methyl-1-phenylethylhydroperoxide; cumenyl hydroperoxide; α-cumyl hydroperoxide; α-cumene hydroperoxide; isopropylbenzene hydroperoxide; Percumyl H; Trigonox K 80. Initiator. mp = -40°; bp = 127° (dec); d = 1.024, insoluble in H_2O, soluble in organic solvents; LD_{50} (rat orl) = 382 mg/kg. *Catalyst Resources.* Name unverified.

3556 Aztec® DCP-R
80-43-3 201-279-3
$C_{18}H_{22}O_2$
dicumyl peroxide
Bis(1-methyl-1-phenylethyl) peroxide; bis(α,α-dimethylbenzyl) peroxide; cumyl peroxide; di-α-cumyl peroxide; Di-cup; diisopropylbenzene peroxide; isopropyl benzene peroxide; Cumene peroxide. Initiator. bp = 130°; insoluble in H_2O. *Catalyst Resources. Name unverified.*

3557 Aztec® Di-*t*-Butyl Peroxide
110-05-4 3515 203-733-6
$C_8H_{18}O_2$
Di-*t*-Butyl Peroxide
tert-Butylperoxide; di-*tert*-butyl peroxide; *tert*-Dibutyl peroxide; TBP; Cadox; DTBP; bis(1,1-dimethylethyl)peroxide; Di-*tertiary*-butyl peroxide. Initiator. mp = -40°; bp = 109-110° (dec explosively); d = 1.334; n_D^{20}= 1.5430; insoluble in H_2O, soluble in organic solvents; LD_{50} (rat orl) = 7710 mg/kg. *Catalyst Resources.* Name unverified.

3558 Aztec® TKB
Mixed peroxide solution in butyl benzyl phthalate. Catalyst and initiator. *Catalyst Resources.* Name unverified.

3559 Azurico
Blue on glaze decoration. *Degussa Ltd.*

3560 B&L 70
Lidofilcon A with 70% water; contact lens material. *Bausch & Lomb, Professional Products Div.* Name unverified.

3561 B.A.R
140-04-5 205-393-4
$C_{24}H_{44}O_4$
Butyl acetyl ricinoleate
9-Octadecenoic acid, 12-(acetyloxy)-, butyl ester, [R-(Z)]-. A vinyl plasticizer. *Harcros.*

3562 B.B.D.C. Standard Alloy
An alloy of 88.5% copper, 10% tin, 1% nickel, 0.25% lead, and 0.25% phosphorus.

3563 B.M. Mixture
A mixture for producing smoke screens; contains 35% zinc, 42% carbon tetrachloride, 9% sodium chlorate, 5% ammonium chloride, and 8% magnesium carbonate.

3564 B.P. Pyro®
7758-16-9 8713 231-835-0
Sodium acid pyrophosphate
Leavening agent for baking, cereals. *Rhône-Poulenc Food Ingreds.*

3565 B.R.V
A coal-tar distillate consisting chiefly of high boiling constituents; used as a rubber softener.

3566 B.S.Sea Water Alloy
An aluminum alloy containing 7.5-9.5% magnesium, 0.2% silicon, and 0.2-0.6% manganese; tensile strength is 45-55 kg/mm^2.

3567 B_1Vac
B_1-type, B_1-strain, Newcastle vaccine; for immunization of poultry. *Intervet Inc.*

3568 B-147
Reclaimed rubber from mechanically defibered passenger tires; used in carcass, sidewall, and undertread of passenger, light truck, and off-road tires, and general purpose mechanical goods. *Midwest Rubber Reclaiming.* Name unverified.

3569 B-182
63394-02-5
polydimethylsiloxane rubber
Silicone elastomer, economical grade with good filler acceptance, low compression set; for extrusion, molding applications. *Wacker Silicones.*

3570 B-618
Reclaimed rubber from modified whole tire, with whiting, clay, and min. rubber added; used in automotive floor mats, semipneumatic tires. *Midwest Rubber Reclaiming.*

3571 B-8880-50%
High molecular weight polyoxyalkylene polymer; dyeing assistant for acrylics; stabilizer for aqueous emulsions. *Ethox.*

3572 BA 27
A non-heat-treatable alloy of aluminum containing 3.5% magnesium and 0.4% manganese.

3573 Babylon
Low density polyethylene homopolymers and ethylene/vinyl acetate copolymers; used for film, extrusion coating and cables. *Bayer AG.* Discontinued.

3574 Bacdip®
Preparation for the control of ectoparasites, especially of multi-host ticks on all domestic animals; veterinary medicine. *Bayer AG.*

3575 Bachite
7440-44-0 1855 231-153-3
C
A specially prepared carbon made from anthracite.

3576 Baclad®
Metal clad products. *Asahi Chem. Industry.*

3577 Bac-N-Fos®
Sodium hexametaphosphate and sodium bicarbonate; food additive for meat, poultry, and seafood industries. *Rhône-Poulenc Food Ingreds.*

3578 Bacote®
68309-95-5 269-682-7
Stabilized ammonium zirconium carbonate aqueous solution.
Paper coating insolubilizers. Clear solution d = 1.37; pH 9.1; 20%ZrO_2 *Magnesium Elektron Ltd.*

3579 Bacterase
Bacterial enzyme or diastase. *Rhône-Poulenc UK.*

3580 Bactiram
Specialty biocides for use in crude oil production, water treatment. *Elf Atochem UK/Ceca.*

3581 Bactistep® MH 80
Dialkyl dimethyl ammonium methoxy sulfate
Sanitizer, germicide, algicide, fungicide, deodorizing agent, antistat for water treatment. *Stepan Europe.*

3582 Bactospeine
Bacillus thuringiensis
Biotrol; Agritrol; Bakthane; Thurcide. *Bacillus thuringiensis;* a bacterial insecticide for control of caterpillars. This organism is considered nonpathogenic. *Fargro Ltd. Discontinued.*

3583 Bactospeine WP
Bacillus thuringiensis; a bacterial insecticide for control of caterpillars. *Koppert (UK) Ltd.*

3584 Bactria
Proprietary range of industrial biocides based on isothiazolinones, formal donors, etc.; used for latex and emulsions, metal working fluids, paints, adhesives, graphic arts, etc. *Reichhold.* Discontinued.

3585 Badin metal
A name used for an alloy of iron with 8-10% aluminum, 18-20% silicon, and 4-6% titanium; used for adding silicon to steel.

3586 Bafixan® Dyes
Disperse dyes for transfer paper printing on textile printing machines. *BASF.*

3587 Bahn metal
A lead base bearing alloy containing copper, 0.7% calcium, 0.6% sodium, and 0.04% lithium; also known as Railway metal.

3588 Bailey solder
An alloy containing 70% tin, 16% zinc, 10% aluminum, and 4% phosphor tin.

3589 Bakelite
Phenolic molding powders. *Bakelite Polymers (UK) Ltd; Rutgerswerke.* Name unverified.

3590 Bakelite A and B
Phenol-formaldehyde resins, soluble in certain solvents. *Bakelite Corporation.* Name unverified.

3591 Bakelite C-9
A proprietary trade name for soft oil modified alkyds. *Bakelite Corporation.* Name unverified.

3592 Bakelite Dilecto
A laminated product consisting of paper or fiber cemented together with a phenol-formaldehyde resin. *Bakelite Corporation.* Name unverified.

3593 Bakelite DQD-3269
ethylene-vinyl acetate copolymer
A proprietary trade name for an ethylene-vinyl acetate copolymer; a plastics material which retains its flexibility and toughness at low temperatures. *Bakelite Corporation.* Name unverified.

3594 Bakelite Micarta
A product similar to Bakelite Dilecto. *Bakelite Corporation.* Name unverified.

3595 Baker P-2C
A proprietary trade name for cellosolve ricinoleate; a vinyl plasticizer. No manufacturer.

3596 Baker P-8
101-34-8 202-935-1
$C_{63}H_{110}O_{12}$
Glyceryl triacetyl ricinoleate
A proprietary trade name for glyceryl triacetyl ricinoleate; a plasticizer for vinyl polymers and GR/S, neoprene GN, ethyl cellulose, and perbunan. d = 0.967; soluble in most organic solvents; combustible. *Aldrich.*

3597 Baker's P and S Liquid
A proprietary mixture of 1% phenol, sodium chloride, liquid paraffin, and water. *Baker Laboratories.* Name unverified.

3598 baker's salt
506-87-6 534 208-058-0
$CH_8N_2O_3$
Ammonium carbonate
Also contains ammonium carbamate. mp = 58° (dec); vaporizes at about 60°; incompatible with acids and acid salts; soluble in four parts H_2O;.

3599 Bakfil
Granular infill materials for use behind Garnex boards in continuous casting tundishes. *Foseco (F.S.) Ltd.*

3600 Baking soda
144-55-8 8726 205-633-8
$NaHCO_3$
sodium bicarbonate
Sodium hydrogen carbonate; Sodium acid carbonate. Used in manufacture of many sodium salts, as a baking additive, in extinguishers and cleaning agents.

3601 BAL
Adhesive; used for fixing ceramic tiles. *Building Adhesives Ltd.*

3602 Balance®
For agriculture industry. *DuPont UK.*

3603 Balanced Salt Solution
A proprietary solution of 0.49% sodium chloride, 0.075% potassium chloride, and 0.036% calcium chloride. Used in medicine and biochemical research. *Alcon Universal.* Name unverified.

3604 Balata
The coagulated milky juice of *Mimusops globosa* (the balata or bullet tree) of South America; used for electrical insulation, for transmission belts, and as soles for shoes.

3605 Ballistite
A trademark for a smokeless powder containing nitroglycerin and collodion cotton. No manufacturer.

3606 Balmex Medicated Lotion
A proprietary protective lotion of lanolin, hexachlorophene, allantoin, balsam Peru and silicone oil in a nonmineral oil base. *Macsil Inc.*

3607 Balmosa®
Menthol, camphor and methyl salicylate. A rubifacient cream. *Pharmax Ltd.*

3608 Bal-nela
An alloy of 28% nickel, 66% copper, with 6% manganese, silicon, iron, and zinc.

3609 Balneol
8012-95-1 232-384-2
Mineral oil; laxative; pharmaceutic aid. *Rowell Laboratories Inc.* Name unverified.

3610 Balsamarome®
Resinoids
Used for perfumery. *Laserson & Sabetay Ets.*

3611 balsams
Resins containing benzoic or cinnamic acids.

3612 Balsan
Proprietary name for a specially purified preparation of balsam Peru. *Macsil Inc.* Name unverified.

3613 BAM
Hot melt adhesive. *Beardow & Adams (Adhesives) Ltd.*

3614 Band-Lock
Hot-melt adhesives. *Reichhold.*

3615 Bandrift
A spray drift retardant suitable for ground and aerial spraying of pesticides and other agrochemicals. *Allied Colloids Ltd.*

3616 Banlene Plus
Soluble concentrate of 18g dicamba, 252g MCPA and 84g mecoprop per liter; used for weed control in cereals and grassland. *Schering Agrochemicals Ltd.* Discontinued.

3617 Banox
Additive to de-icing salts. *Albright & Wilson Ltd., Phosphates & Speciality Business.* Name unverified.

3618 Banox
Food grade antioxidants. *UOP Speciality Products.*

3619 Banox ES
BHT-lecithin-soybean oil-mineral oil. Antioxidant blend for animal feeds. Reddish brown liquid; nutty odor; d = 0.975; viscosity = 67 cps; n_D^{20} = 1.485. *UOP.*

3620 Banwee
Residual herbicide. *Fisons plc, Horticultural Div.* Name unverified.

3621 Banweed-S
A suspension concentrate containing napropamide and simazine; soil-applied residual for control of annual grasses and annual broad-leaved weeds in field and container grown nursery stock. *Fisons plc, Horticultural Div.* Name unverified.

3622 Barabar
Oxidation and corrosion inhibitors. *Exxon Int'l.* Name unverified.

3623 Baragel
Thickening and thixotropic rheological additives; used for grease, underbody coatings. *Exxon Int'l; Rheox Inc.*

3624 Barberite
A corrosion-resisting alloy containing 88.5% copper, 5% nickel, 5% tin, and 1.5% silicon; has a high tensile strength in addition to good corrosion-resisting properties.

3625 Barbouze's Alloy
An alloy of aluminum with 10% tin.

3626 Barchlor
107-05-1 297 203-457-6
C_3H_5Cl
Allyl chloride
3-Chloropropene; Chlorallylene; 3-chloro-1-propene; 1-chloro-2-propene; Chloropropene; 3-chloropropylene; α-chloropropylene; 3-chloroprene; 1-chloro propene-2; 3-chloropropene-1; 3-chloro-1-propylene; 2-propenyl chloride. Used in the synthesis of allyl compounds mp = -134°; bp = 44-45°; d = 0.9390; n_D^{20} = 1.4140; soluble in H_2O (3.6 g/l); more soluble in organic solvents; LD_{50} (rat orl) = 460 mg/kg. *Lonza Ltd.*

3627 *Bardew*
Fungicide. Schering Agrochemicals Ltd. Discontinued.

3628 Bardyne
Isdopher
A biocide with useful hospital and animal health applications. *Lonza AG.*

3629 Barex® 210
Rubber-modified copolymer containing 75% acrylonitrile and 25% methacrylate; high-barrier resins in injection molding and extrusion grades for film, sheet, blow molding and injection molding applications; used for packaging for food, pharmaceuticals, household and industrial chemicals (lighter fluid, polishes, insecticides, detergents), cleaning solvents, waxes, cosmetics and other toiletries. *BP Chemicals Inc.*

3630 Barfoed's Reagent
Copper acetate, 1 part, dissolved in 15 parts water. To 200 ml of this solution, 50 ml of 68% acetic acid are added; used to distinguish glucose and other monosaccharides from disaccharides, such as lactose and maltose.

3631 Bario metal
An alloy containing 90% nickel, 1.22% tungsten, 0.29% silicon, and 4.25% chromium; acid and heat-resistant.

3632 Barisol Super BRM
Complex multicarbon alcohol phosphate potassium salt; textile wetting agent for dye leveling, pectin removal from cottons, scouring of cotton and synthetic blends; dispersant. *Dexter.*

3633 Barium
7440-39-3 991 231-149-1
Ba
Alkaline-earth element; Alloys in vacuum tubes, deoxidizer for copper, lubricant for anode rotors in x-ray tubes, spark-plug alloys. mp = 725°; bp = 1640°; d = 3.51; reacts with H_2O, producing H_2; *Aldrich; Atomergic Chemetals; Degussa; Noah Chem.*

3634 Barium acetate
543-80-6 992 208-849-0
$C_4H_6BaO_4$
Acetic acid, barium salt. Chemical reagent, acetates, textile mordant, catalyst manufacture, paint and varnish driers. d = 2.4680; LD_{50} (rat orl) = 921 mg/kg. *Aldrich; Bemardy Chimie SA; Hoechst Celanese; Mallinckrodt.*

3635 Barium bromide
10553-31-8 995 234-140-0
$BaBr_2$
barium dibromide

Manufacture of bromides, photographic compounds, phosphors. mp = 850°; very soluble in H_2O, soluble in MeOH, insoluble in acetone, ethyl acetate. *Atomergic Chemetals.*

3636 Barium carbonate

513-77-9 996 208-167-3
CBaO3
Carbonic acid, barium salt (1:1)
Durex white. Treatment of brines in chlorine-alkali cells to remove sulfates, rodenticide, production of barium salts, ceramic flux, optical glass, case-hardening baths, ferrites, in radiation-resistant glass for color television tubes. mp = 811°; bp = 1450°; d = 4.430; slightly soluble in H_2O (24 mg/l), LD_{50} (rat orl) = 418 mg/kg. *Benardy Chimie SA; Mallinckrodt; Solvay GmbH.*

3637 Barium chloride

10361-37-2 998 233-788-1
$BaCl_2 \cdot 2H_2O$
Barium dichloride
Ba 0108E. Production of artificial barium sulfate, other barium salts; reagents; lubrication oil additives; boiler compounds; textile dyeing; pigments; manufacture of white leather. mp = 963°; d = 3.86; very soluble in H_2O, insoluble in organic solvents; LD_{50} (rat orl) = 118 mg/kg. *Aldrich; EM Industries; Hoechst Celanese; Mallinckrodt; Sachtleben Chemie GmbH; Solvay GmbH.*

3638 Barium chromate

10294-40-3 999 233-660-5
$BaCrO_4$
barium chromate(VI)
ultramarine yellow; baryta yellow; lemon chrome; C.I. 77103; C.I. Pigment Yellow 31; lemon yellow; permanent yellow; Steinbühl yellow. Safety matches, corrosion inhibitor in metal-joining compounds, pigment for paints, ceramics, fuses, pyrotechnics, metal primers, ignition control devices. Pigment in anti-corrosion pastes. d = 4.50; insoluble in H_2O. *Atomergic Chemetals; BASF AG; Noah Chem.*

3639 Barium nitrate

10022-31-8 1012 233-020-5
BaN_2O_6
barium dinitrate
Nitric Acid, Barium Salt. Pyrotechnics, incendiaries, chemicals (barium peroxide), ceramic glazes, rodenticide, electronics. mp = 592°; d = 3.23; freely soluble in H_2O; LD_{50} (rat orl) = 355 mg/kg. *Aldrich; Berk Chem. Ltd; Hoechst Celanese; Noah Chem.; San Yuan Chem. Co. Ltd.*

3640 Barium nitrite

13465-94-6 1013 236-709-9
$BaN_2O_4 \cdot H_2O$
nitrous acid barium salt monohydrate
Used as a chemical intermediate in diazotization reactions. Also used to prevent corrosion of steel and in explosives. d = 3.187; soluble in H_2O, insoluble in organic solvents.

3641 Barium oxide

1304-28-5 1015 215-127-9
BaO
barium monoxide
barium protoxide; calcined baryta; Barsito; Calcined Barsito. Dehydrating agent for solvents, detergent for lubricating oils. mp = 1920°; bp = 2000°; d = 5.98; soluble in H_2O and dilute acids. *Atomergic Chemetals; Hoechst Celanese; Noah Chem.; Spectrum Chem. Mfg.*

3642 Barium peroxide

1304-29-6 1018 215-128-4
BaO_2
barium binoxide
barium dioxide; barium superoxide. Bleaching, decolorizing glass, thermal welding of aluminum, manufacture of hydrogen peroxide, oxidizing agent, dyeing textiles. Insoluble in H_2O. *Benardy Chimie SA; Hocehst Celanese; Spectrum Chem. Mfg.*

3643 Barium stearate

6865-35-6 229-966-3
$C_{36}H_{70}BaO_4$
Octadecanoic acid, barium salt
Waterproofing agent, lubricant in metalworking, plastics, and rubber; wax compounding; preparation of greases; heat and light stabilizer in plastics. *Adeka Fine Chem; Reagens SpA; Syn. Prods.; Witco Corporation.*

3644 Barium sulfate

7727-43-7 1023 231-784-4
BaO_4S
Synthetic barytes; Bakontal; Barosperse; Barotrast; Blanc fixe macro; Blanc fixe, articifical percipitate; Baridol; Citrobaryum; Neobar; Unibaryt; artifical barite; Barytes, natural; blanc fixe, artificial, precipitated; basofor. Used as an opaque medium in x-ray diagnosis. Weighting mud in oil drilling, paper coating, paints; filler and delustrant for textiles, rubber, plastics, and lithographic inks; radiation shield. Extender pigment. mp = 1580°; d = 4.500; Insoluble in H_2O. *Cyprus Industrial Minerals; J.M. Huber; Mallinckrodt; Sachtlebe Chemie GmbH.*

3645 Barium sulfonate

Rust preventive. *Atomergic Chemetals; Lubrizol; Witco/Sonneborn.*

3646 Barium titanate

12047-27-7 1029 234-975-0
BaO_3Ti
Barium titanate(IV)
Barium titanium oxide; Barium metatitanate. Ferroelectric ceramics used in storage devices, dielectric amplifiers, digital calculators. mp = 1625°; d = 6.0800. *Aldrich; Atomergic Chemetals; Ferro/Transelco; Noah Chem.*

3647 Barkite B

Di(dimethylcyclohexyl)oxalate
A plasticizer for cellulose lacquers. *Laporte Industries Ltd.* Discontinued.

3648 Barlox

A range of alkyl dimethyl amine oxides; used for detergents, shampoos, and textile processing applications. *Lonza AG; Pentagon Chemicals Ltd.*

3649 Barlox® 12

61788-90-7 263-016-9
Cocamine oxide
Amines, coco alkyldimethyl, N-oxides. Detergent; viscosity builder, emollient. pH = 7.0; liquid viscosity = 45 cps. *Lonza Inc.*

3650 Barlox® 14

3332-27-2 222-059-3
Myristamine oxide
Detergent; viscosity builder, emollient. pH = 7.0; liquid viscosity = 60 cps *Lonza Inc.*

3651 Barlox® 16S

7128-91-8 230-429-0
Cetamine oxide
Biodegradable foam stabilizer, viscosity builder, emulsifier, conditioner for personal care and industrial products. *Lonza Inc.*

3652 Barlox® 18S

2571-88-2 219-919-5
Stearyl dimethyl amine oxide
Biodegradable, foam stabilizer, viscosity builder, emulsifier, conditioner for personal care and industrial products. *Lonza Inc.*

3653 Barlox® C

68155-09-9 268-938-5
Cocamidopropylamine oxide
Foam stabilizer and viscosity builder for shampoos, industrial products. pH = 7.0; liquid viscosity = 36 cps. *Lonza Inc.*

3654 Bäropan MC 8046 SP

Ca/Zn. Stabilizer for shoe injection molding of plasticized PVC. *Bärlocher GmbH.*

3655 Bäropan TX 296 KA

Stabilizer for cable compounds. *Bärlocher GmbH.*

3656 Bäropol

Additives for polyolefins and polystyrenes (antioxidants, lubricants, release agents, uv stabilizers, antiblocking agents, antistats, flame retardants, fillers and pigments). *Bärlocher GmbH.*

3657 Baros

A heat-resisting alloy containing 90% nickel and 10% chromium.

3658 Bärostab® CT 901

Ca/Zn product; stabilizer offering very good color properties and excellent transparency for injection molding of plasticized PVC; for extrusion and shoe injection molding. *Bärlocher GmbH.*

3659 Bärostab® KK 47 S

K/Zn
Fast-kicking stabilizer for foam processing in production of wallpaper, artificial leather; highest degree of whiteness. *Bärlocher GmbH.*

3660 Bärostab® L 230

$ZnC_2O_42H_2O$
Zinc octoate
Fast-kicking stabilizer for foam processing in production of sealings for caps; good organoleptic props., excellent color; approved for contact with foodstuffs. Very slightly soluble in H_2O *Bärlocher GmbH.*

3661 Bärostab® NT 1005

Mg/Al/Zn complex; stabilizer for plastisol processing; especially for artificial leather, wallpaper, conveyor belts, nontoxic applications. *Bärlocher GmbH.*

3662 Bärostab® UBZ 632

Barium zinc complex; stabilizer for PVC plastisols, especially topcoats for light pigmented cushion vinyl flooring; features good color; especially suitable in presence of inhibitors such as TMA. *Bärlocher; Vanderbilt.*

3663 Bärostab® UBZ 820 KA

Ba/Zn complex with self-lubrication; stabilizer for transparent cable compounds; imparts excellent heat stability. *Bärlocher GmbH.*

3664 Bärostab® UBZ76BX
Ba/Zn complex with self-lubrication; stabilizer for transparent cable sheathing compounds; hydrolysis-resist.; imparts high transparency and excellent light stability. *Bärlocher GmbH.*

3665 Barquat®
Alkyl dimethyl benzyl quaternary ammonium compounds; for germicidal applications in hospitals, institutions and industrial water treatment, also used as antistatic agents. *Lonza AG.*

3666 Barquat® CME-35
78-21-7 201-094-8
Cetethyl morpholinium ethosulfate
Antistat, combing aid and detangling agent, textile lubricant, odor counteractant. *Lonza Inc.*

3667 Barquat® CT-29
112-02-7 203-928-6
$C_{19}H_{42}BrN$
Cetrimonium chloride
Coagulating agent in manufacturing of antibiotics. *Lonza Inc.*

3668 Barras
A pine resin.

3669 Barrialon® CX
Polyvinylidene chloride
Polyvinylidene chloride coextruded multilayer film. *Asahi Chem. Industry.*

3670 Barrialon® S
Biaxially oriented multilayer film. *Asahi Chem. Industry.*

3671 Barrialon® SF
Polyvinylidene chloride film. *Asahi Chem. Industry.*

3672 Barrier
Suspension concentrate containing 300 g chloridazon, 30 g fenuron and 170g propham per liter; a residual herbicide for beet crops. *Truchem Ltd.*

3673 Barrier-Guard®
Polymer preparations providing moistureproof film to textile fabrics. *Reichhold.*

3674 Barsilowsky's base
Aminoditolyl *p*-toluquinone diimine.

3675 baryta
1304-28-5 1015 215-127-9
BaO
barium monoxide
barium oxide; barium protoxide, calcined baryta. Used as a drying agent mp = 1920°; bp = 2000°; d = 5.7200.

3676 BAS 438
Suspension concentrate containing 250g chlorothalonil and 187g fenpropimorph per liter; a systemic fungicide for winter wheat. *BASF plc.*

3677 BAS 46402F
Systemic fungicide for use in winter wheat and barley. *BASF plc.*

3678 Basacid® Dyes
Anionic dyes
For production of inks, coloring of cleaner, wood preservatives in paint and varnish industry. *BASF AG.*

3679 Basacryl Salt
Levelling agent for textile dyeing. *BASF plc.*

3680 Basacryl/Baflxan
Dyestuffs for textile transfer printing. *BASF plc.* Name unverified.

3681 Basacryl® Dyes
Cationic dyes for dyeing and printing acrylic fibers and cationic dyeable polyester fibers. *BASF.*

3682 Basacryl® Salt NB-KU
Dye retarder for dyeing of acrylic fibers. *BASF.*

3683 Basacryl® Salt TX-412
Resist agent for printing nylon with disperse/acid dyes. *BASF.*

3684 Basammon® Extra 25
Ammonium sulfate nitrate with dicyandiamide as stabilizer, for improved nitrogen utilization and sustained action on agricultural crops. *BASF AG.*

3685 Basantol® Dyes
Anionic dyes for paint and varnish industry. *BASF AG.*

3686 Basazol®
Cationic liquid dyestuff based on basic dyes; coloring for papermaking. *BASF AG; BASF plc.*

3687 Bascal®
Acid donor for textile dyeing. *BASF plc.* Name unverified.

3688 Bascal® S
Mixture of aliphatic dicarboxylic acids; deliming agent and pickling acid; for fur industry. *BASF AG.*

3689 Base 10-L
Sulfurized lard compounds; EP agent for drawing compounds, solvent and cutting oils. *Ferro/Keil.*

3690 Base 36
Sulfurized oleic acid
Can be neutralized with bases to form water-soluble or-dispersible soaps; also used in lubricating oils, cutting and solvent oils. *Ferro/Keil.*

3691 Base 75
Sodium sulfonate-soap derivative.
Oil-water emulsifier for conventional and EP sol. oils and pastes for metalworking industry; imparts stability and wetting properties. *Ferro/Keil.*

3692 Base 104
2-Ethyl hexanol phosphate
Free acid; surfactant base for making low foam wetter, penetrants, antistats, lubricants, rust preventatives. *Clark.*

3693 Base 500-A
Used for preparation of cationic and anionic textile softeners. *CNC Int'l L.P.*

3694 Base 865
Sulfur chlorinated base; extreme pressure additive, detergent, lubricant for high performance fluids, machining and grinding operations. *Ferro/Keil.*

3695 Base 7800
Sulfonate soap
Oil-water emulsifier for soluble oils and pastes for metalworking industry, emulsion cleaners, synthetic and semi-synthetic coolants; hard water tolerance; aids rust protection. *Ferro/Keil.*

3696 Base HS
Sulfur base; high in active sulfur; for use in difficult metalworking applications, cutting oils. *Ferro/Keil.*

3697 Base ML
Methyl lardate
Wetting/oiliness/lubricity agent for metalworking, lubricating, motor, and rolling oils; antiwear additive, process aid, release additive. *Ferro/Keil.*

3698 Base MT
Methyl tallowate
Wetting/oiliness/lubricity agent for metalworking, lubricating, motor, and rolling oils; antiwear additive, process aid, release additive. *Ferro/Keil.*

3699 Base Nacrante 2078
Cocamidopropyl betaine, cocamide DEA and glycol stearate. Foaming base for cosmetic pearlescent preparations. *Seppic.*

3700 Base Nacrante 6030 CP
Sodium coceth-2 sulfate and triethylene glycol distearate. Foaming base for cosmetic pearlescent preparations. *Seppic.*

3701 Base Wax 36-AG
Substituted amide wax concentrate; substantive softener wax for synthetic fibers and fabrics; may be used to prepare stock solution of softeners or softeners themselves. *Eastern Color & Chem.*

3702 Basensol®
Functional block polymers based on propylene oxide/ethylene oxide or polyalkylene glycol ether; sensitizing agents for polymer dispersions. *BASF AG.*

3703 Basex
A proprietary base-exchange material for water softening; approximate composition $Na_2Al_2O_3$. $14SiO_2$; stated to have high softening capacity. No manufacturer.

3704 BASF Reactive Resist Liquid
Auxiliary for resist prints with Basilen P dyes under Primazin dyes. *BASF.*

3705 Basfapon®
127-20-8 11,2806 204-828-5
$C_3H_4Cl_2O_2.Na$
Dalapon-sodium
Propanic acid, 2,2-dichloro-, sodium salt; Antigramigma; Dalacide; Dalaphon; Delapon; Dikopan; Doowpon; Dowpon; Gramevin; Omnidel Spezial; Propinate; Radapon; Sodium α,α-dichloropropionate; sodium Dalapon. Post-emergence systemic herbicide for control of grasses in annual and perennial crops, used on nonagricultural land, in ditches, and pastures. mp = 174-176°; soluble in H_2O (45 g/100 ml); LD_{50} (rat orl) = 9330 mg/kg. *BASF AG.*

3706 Basfoliar® 6-12-6
Liquid foliar fertilizer with phosphate, zinc copper, and other micronutrients; for Indian corn and other crops with high phosphate requirements. *BASF AG.*

3707 Basfoliar® 34
Liquid nitrogenous foliar fertilizer with magnesium and micronutrients; for agricultural crops, vines, fruit, hops, and field vegetables. *BASF AG.*

3708 Basilen®
Reactive dyes for dyeing cellulosic fibers. *BASF plc.*

3709 Basilex
57018-04-9 260-515-3

tolclofos-methyl
A wettable powder containing tolclofos-methyl; protective fungicide for use on all ornamentals and some edible crops against *Rhizoctonia*. *Fisons plc, Horticultural Div.* Name unverified.

3710 Basocoll® CM
9003-08-1
Melamine-formaldehyde condensate; sizing assistant for papermaking. *BASF AG.*

3711 Basoform®
Nonvolatile formaldehyde bonding agent; formaldehyde catcher for cured UF foams. *BASF AG.*

3712 Basogal® C
Nonfoaming leveling agent for vat dyes. *BASF.*

3713 Basojet® PEL 200%
EO condensate/anionic blend; non-foaming dispersing and leveling agent for dyeing of polyester fibers with disperse dyes under high temperature conditions. *BASF.*

3714 Basokol® NB-S
Polymeric organic compound; sequestering agent for dyeing cotton and polyester/cotton; protective colloid; dispersant for dyeing, bleaching, afterscouring; stable to acids, alkalies, and electrolytes. *BASF.*

3715 Basol® WS
Aromatic sulfonic acid condensate; dispersant, protective colloid, dyeing assistant especially for exhaust dyeing of polyester, stabilizes dye dispersions; stable to acids, alkalies, hard water, electrolytes. *BASF.*

3716 Basolan® F
Auxiliary for improving wet fastness of acid dyes on wool, silk, and polyamide fibers. *BASF.*

3717 Basomol®
Synthetic wetting agent containing phosphoric acid for manufacturing of synthetic resin foam in conjunction with Basopor® *BASF AG.*

3718 Basonat®
Isocyanate-based; for adhesive and sealant industry; hardener components for polyols, preparation of prepolymers containing isocyanate groups; polyurethane adhesives and binders for granular materials. *BASF AG.*

3719 Basonyl® Dyes
Cationic dyes for production of inks, carbon paper coatings, for dyeing natural fibers such as jute, sisal, wool; paint and varnish industry for production of daylight fluorescent pigments and gloss paints. *BASF AG.*

3720 Basopal®
Detergent for washing and cleaning processes in textile industry. *BASF AG; BASF plc.*

3721 Basophen® M
Alkyl ester of phosphoric acid with anionic/nonionic emulsifiers; nonfoaming wetting agent, foam suppressant for textile applications, package dyeing of cotton, cotton/polyester, wool, desizing, latex penetration in carpet backing; stable in hard water and liquors containing alkali, acetic or sulfuric acids. *BASF/Fibers.*

3722 Basophen® RA
Sulfosuccinate
Wetting agent for desizing, pretreatment, and bleaching of cellulosic fibers; resistant to chlorine, hard water, weak acids. *BASF/Fibers.*

3723 Basophob®
Aqueous paraffin or polyethylene wax dispersions, solns. of fatty acid derivs.; hydrophobic agents for internal or surface treatment of paints, mortar, concrete and paper. *BASF AG.*

3724 Basoplast®
Alkyl ketene diamide copolymers
Cationic internal and surface sizes for papermaking. *BASF AG; BASF plc.*

3725 Basopon® LN
Alkylphenol ethoxylate
Wetting agent, detergent with dispersing effects for wool and synthetic fibers, desizing and scouring applications; stain removing agent; stable to hard water, alkalis, acids, reducing and oxidizing agents. *BASF/Fibers.*

3726 Basopon® TX-110
Proprietary; nonfoaming scouring agent with emulsifying properties for polyester knit goods. *BASF.*

3727 Basopor®
Urea-formaldehyde resin precondensate; for production of synthetic resin foam (*in-situ* UF foam), for use in building (as thermal insulation) and in mining (e.g., for protection against explosions). *BASF AG.*

3728 Basoset® 162
25928-94-3
Epoxy resin based on epichlorohydrin and an aliphatic polyol; as casting resin, optionally in combination with glass fibers for production of molded articles; for potting/encapsulation in electrical industry; for bonding glass fiber reinforced polyester resin parts to each other, for binding expanded PS. *BASF AG.*

3729 Basosoft®
Brighteners and softeners for textile industry. *BASF AG; BASF plc.*

3730 Basotect®
Open-cell, resilient foam plastic based on melamine resin; for soundproofing and thermal insulation where good fireproofing, high heat stability and low weight are required. *BASF AG.*

3731 Basotol®
Oxidizer for vat dyes; auxiliary for protecting dyes against reduction. *BASF.*

3732 Basotrope® W
Auxiliary for producing white discharges on dyed grounds that are difficult to discharge and for stripping dyed shades. *BASF.*

3733 Basovit® Dyes
Anionic dyes for foodstuffs and cosmetics industry for fiber pen inks, detergents, and seeds. *BASF AG.*

3734 Bastamol®
Dyeing auxiliaries for fixing and deepening anionic dyeings and for leveling cationic dyeings in fur and leather industries. *BASF AG.*

3735 Bastanet
7440-44-0 1855 231-153-3
A decolorizing carbon.

3736 bastose
The cellulose of jute.

3737 Basyntyn®
Synthetic replacement tanning agents and auxiliary tanning agents. *BASF AG; BASF plc.*

3738 Batchite
7440-44-0 1855 231-153-3
An activated carbon produced from coco-nut shells.

3739 Bath metal
A brass containing 83% copper and 17% zinc. Another alloy consists of 55% copper and 45% zinc.

3740 Battal
10605-21-7 1836 234-232-0
carbendazim
Battal FL. Fungicide containing carbendazim. *ICI Chem & Polymers Ltd.*

3741 battery copper
A brass containing 94% copper and 6% zinc.

3742 Baudoin's metal
Complex nickel silvers. One contains 72% copper, 16.6% nickel, 1.8% cobalt, 2.25% tin, and 7.1% zinc, and another 75% copper, 16% nickel, 2.25% zinc, 2.75% tin, 2% cobalt.

3743 Baxton®
For chemical industry. *DuPont UK.*

3744 Bayblend®
Thermoplastic polycarbonate/ABS or polycarbonate/ASA blends; engineering thermoplastics for the manufacture of injection moldings for use in the electrical and automotive industries and for domestic appliances. *Bayer AG; Bayer plc; Miles.*

3745 Bayblend® DP2-1448
PC/ABS blend; bromine/chorine-free flame-retardant thermoplastic with easy processing for business machine housings, portable computer housings, consumer products, personal care items and appliances. *Miles.* Discontinued.

3746 Bayblend® FR 1439
PC/ABS blend, flame retardant; thermoplastic resin offering good impact resist. even at low temp., rigidity and dimensional stability, easy processing; ideal for business machine and automotive markets. *Miles.* Discontinued.

3747 Bayblend® T 44
PC/ABS blend; high-productivity general purpose thermoplastic resin offering good impact resistance, even at low temperatures, rigidity and dimensional stability, easy processing; ideal for business machine and automotive markets. *Bayer; Miles.*

3748 Bayblend® T 45 MN
PC/ABS blend; thermoplastic blend for injection molding, extrusion, blow molding, thermoforming, machining, welding, bonding, screwing, coating, printing, vacuum metallizing, and electroplating; for automotive, electrical, domestic appliances camera bodies, electroplated components and other applications. *Bayer; Miles.*

3749 Bayblend® T 88-2N
PC/ABS blend, 10% glass-reinforced; thermoplastic resin offering good impact resistance even at low temperature, rigidity and dimensional stability, easy processing; ideal for business machine and automotive markets. *Bayer; Miles.*

3750 Baybond®
Hydrous polyurethane dispersion; for glass fiber sizes. *Bayer AG; Bayer plc.*

3751 Bayboran®
Selective reducing agent for aldehydes, ketones, acid chlorides, peroxides, Schiff bases, azides, etc.; used for electrodeless metallizing, purification and stabilization of process streams (odor removal, decolorization). *Bayer AG.*

3752 Baycast
Biomedical preparation; polyurethane supportive dressing. *Bayer AG.*

3753 Baycidal
Insect growth regulator for the control of fly larvae. *Bayer AG.*

3754 Bayclin
Perfumed all-purpose cleaner with disinfecting properties for dissolution or concentrated application. *Bayer AG.*

3755 Bayco®
Monofilament; used for wine and fruit growing. *Bayer AG.*

3756 Baycoll®
Hydroxyl polyethers and hydroxyl polyesters; used in conjunction with Desmodur in the production of reaction adhesives. *Bayer AG.*

3757 Baycoll® 17
Linear saturated hydroxyl polyester; polyol used in two-component reaction adhesives in combination with Desmodur polyisocyanates; suitable for bonding plastic, foam, fabric, and wood substrates; modifier in solvent-based Desmocoll PU adhesives. *Bayer; Miles.* Name unverified.

3758 Baycoll® MD 3040
Linear polyester polyol; used for two-part PU adhesives; as modifiers and building blocks for PU. *Bayer Corp; Polysar.*

3759 Baycryl®
A range of acrylic resins. *Bayer AG.*

3760 Bayderm® A
Solutions of anionic dyestuffs; for spraying, curtain coating and printing as well as for shading the finish. *Bayer AG.*

3761 Bayderm® Colours B-TO
Transparent pigment dispersions with low binder content; used for organic and aqueous finishes and for coating textiles. *Bayer AG.*

3762 Bayderm® Colours C-TO
Covering pigment dispersions with low binder content; used for organic and aqueous finishes and for coating textiles. *Bayer AG.*

3763 Bayderm® KF
Solutions of cationic dyestuffs for brilliant spray dyeings and for shading finishing liquors; used in leather industry. *Bayer AG.*

3764 Bayderm® Lacquers Auxiliaries
Polyurethane-based products for leathers. *Bayer plc.*

3765 Baydur®
Rigid integral skin polyurethane foams; for covers for machinery, housings for computers, sporting goods, furniture, engineering components, automotive interior trim. *Bayer AG.*

3766 Baydur® STR
Polyurethane resin systems. Structural composites. *Miles.* Discontinued.

3767 Bayer 5072
140-56-7 205-419-4
p-Dimethylamino-benzenediazo sodium sulfonate
Fenaminosulf. Fungicide for the prevention of crop damage by soil fungi. *Bayer AG.*

3768 Bayer Base Plates Glass-Clear
Plastic plates; used for the preparation of individual trays, for taking bites and trial fittings and for protective plates for surgical wounds. *Bayer AG.* Discontinued.

3769 Bayer CM
Chlorinated polyethylene
Used for the rubber and cable industries; very good resistance to ageing, ozone, weathering and chemicals. *Bayer AG.*

3770 Bayer Perlon®
Man-made staple fiber and monofilament. *Bayer AG.*

3771 Bayer SBR Latex 200 C
9003-55-8 8534
S/B latex
Nonstaining hard component for blended latex goods. *Bayer AG.*

3772 Bayer UV Absorber 325, 340
UV absorbers for protection of rigid and plasticized PVC, cellulose derivatives, polystyrene, polyacrylate; used in clear finishes to protect substrate against light. *Bayer AG.*

3773 Bayferon®
Interferon inducer for cattle for prophylaxis and treatment of infections due to interferon-sensitive infectious agents and for paramunisation; veterinary medicine. *Bayer AG.*

3774 Bayfill®
Semirigid polyurethane filling foams; for automotive engineering, instrument panels, consoles, arm rests, bumpers. *Bayer AG; Bayer plc.*

3775 Bayfit®
Flexible polyurethane foams; flexible foam for cushioning applications, upholstered furniture, mattresses, automotive engineering, textiles, packaging. *Bayer AG; Bayer plc.*

3776 Bayflex®
Elastic integral skin PU foams, microcellular RIM elastomers; for interior and exterior automotive applications, shoe soles, packaging, engineering components, domestic appliances. *Bayer AG; Bayer plc.*

3777 Bayfol®
Film made from blends based on polycarbonates, other engineering thermoplastics or composites; used for special overlay and decoration film for membrane keyboards, for front facings with integrated membrane switches and for all applications requiring a high level of chemical resistance. *Bayer AG; Miles.*

3778 Bayfol® CR 6-2
PC blend film; technical film with high dynamic strength (especially for membrane switch overlays), increased chemical resistance, scratch resistance. *Bayer AG.*

3779 Bayfolan®
Foliar feed containing macro and micro nutrients; for all agricultural and horticultural crops to help recover from the effects of adverse conditions such as drought, low temperatures or waterlogging. *Bayer AG; Bayer plc.*

3780 Bayfresh®
Perfumed air freshener as aerosol, liquid or on solid base. *Bayer AG.*

3781 Baygal®
A polyurethane based casting resin; used in the electrical industry and for rock consolidation. *Bayer AG.*

3782 Baygal® K30
Branched polyether polyol
Used for producing cast PU resin compounds with high heat resistance for electronics and electrical engineering. *Bayer AG.* Name unverified.

3783 Baygal® K115
Polyether-polyester polyol
Used for producing PU casting compounds for electronics and electrical engineering; improved compatibility with Baymidur isocyanates, reduced sensitivity to moisture. *Bayer AG.* Name unverified.

3784 Baygal® K190
Linear, hydroxyl-bearing polyether polyol; used as flexibilizer in combination with other polyols for PU industry. *Bayer AG.* Name unverified.

3785 Baygal® K390
Branched polyether polyol; used in combination with other polyols as flexibilizer. *Bayer AG.* Name unverified.

3786 Baygard®
Textile finishing product; used for floor covering; soil, oil and water repellent; pile stability. *Bayer AG.*

3787 Baygen®
Lacquers and auxiliaries based on polyurethane reactive lacquers; for cold lacquer finishes. *Bayer AG; Bayer plc.*

3788 Baygenal®
Dyestuffs; for high-grade chrome upper leathers. *Bayer AG.*

3789 Bayglaze
Ready-to-use glazes. *Bayer Ag.*

3790 Baygon®
9003-55-8 8534
$C_{11}H_{15}NO_3$
Propoxur
Phenol, 2-(1-methylethoxy)-,methylcarbamate; Carbamic acid, methyl-,o-isopropoxyphenyl ester; Aprocarb; Arprocarb; Bay 9010 Baygon; Bayer B 5122; Bayer 39007; Blattanex; Blattosep; Bolfo; Boygon; Chemagro 9010; Dalf dust; DDVP; ENT 25,671; Invisi-Gard; Isocarb; IPMC; OMS 33; Propotox; Propoxure; PHC; Sendran; Suncide; Tendex; Unden. Broad spectrum insecticide for the control of household and hygiene pests. mp = 84-87°, 91.5°; d_{20} = 1.12; soluble in H_2O (2 g/l), organic solvents; LD_{50} (rat orl) = 70 mg/kg. *Bayer AG.*

3791 Baygon® MEB Spray
Household insecticide; used for control of flies, moths, wasps and other flying pests. *Bayer AG.*

3792 Bayguard
Waterproofing agents for textiles. *Bayer plc.*

3793 Bayhydrol
Water-thinnable binders for industrial coatings. *Bayer plc.*

3794 Bayhydrol 140 AQ
Aliphatic polyester PU dispersion in water/toluene; adhesion promoter in metal/plastic composite structures, textile and leather coatings, primers for rigid surface coatings. d = 1.04, 8.7 lb/gal; viscosity = 25-500 mPas; pH=6.0-7.5 *Miles.* Discontinued.

3795 Baykanol® AK, HLX, SL
Dyeing auxilliaries; high quality, almost neutral syntans for leveling aniline dyeing; the products facilitate the neutralization of chrome leathers and improve the fullness. *Bayer AG.*

3796 Baykanol® Liquor TN
Light fast special liquor for suede and white leathers; used in leather industry. *Bayer AG.*

3797 Baykisol®
Silica sol
Used for the fining of apple juice, lemon juice, other juices, wines. *Bayer AG.*

3798 Baylan®
Low temperature dyeing auxiliaries for wool. *Bayer AG; Bayer plc.*

3799 Baylectrol®
Chlorine-free electrical insulating fluid for power capacitors. Designed to replace polychlorobiphenyls. *Bayer AG.*

3800 Bayleton® 5
43121-43-3 9723 256-103-8
Triadimefon
Wettable systemic fungicide powder containing 5% w/w triadimefon; used to control powdery mildew on apples, hops, raspberries, strawberries and other cane fruits plus American gooseberry mildew on all varieties of blackcurrants and gooseberries. *Bayer plc.*

3801 Bayleton® BM
A wettable powder systemic fungicide containing 12.5% w/w triadimenol and 25% w/w carbendazim; used to control eyespot, mildew and early attacks of yellow and brown rust on winter wheat and winter barley, rhynchosporium on winter barley and eyespot and mildew on winter rye. *Bayer AG.*

3802 Bayleton® CF
Wettable powder fungicide with contact and systemic properties containing 6.25% w/w triadimefon and 65% w/w captafol; used to control powdery mildew, yellow and brown rust, leaf spot and glume blotch and to reduce the late-season ear disease complex on spring and winter wheat. *Bayer plc.*

3803 Baylith®
Technical grade oxides and zeolites
Uses include intensive driers for use in polyurethane systems, paints and varnishes, plastics and solvents. *Bayer AG.*

3804 Baylube®
Polyether bases for synthetic lubricants. *Bayer AG; Bayer plc.*

3805 Bayluscide®
1420-04-8 6602 215-811-7
Clonitrilide
Ethanolamine salt of niclosamide. Molluscicide for the control of water snails. *Bayer AG.*

3806 Baymat®
Fungicide with good penetrant, protective, curative and eradicative activity for control of scab and blossom wilt on pome and stone fruit, rust, leafspot diseases and mildews on pome and stone fruit, bananas, vegetables, sugar beet, ornamentals. *Bayer AG.*

3807 Baymer®
Rigid polyurethane foams; rigid foams for heating and refrigeration engineering, building industry, insulation, sporting goods, automotive engineering. *Bayer AG; Bayer plc.*

3808 Baymetex®
Metallized textile fabrics, primarily fabrics woven from raw materials such as polyamide, aramid, glass fiber, carbon fiber and polyester/cotton for EMI shielding, reflection of electromagnetic waves and lightning protection of sandwich elements. *Bayer AG.*

3809 Baymicron®
Aqueous microcapsule dispersions for the manufacture of carbonless copying papers giving clear blue and black copies. *Bayer AG.*

3810 Baymid.
Polycarbodiimide foam. *Bayer plc.*

3811 Baymidur®
Polyurethane based casting resin; for use in the electrical industry, for rock consolidation and for the formulation of core sand binders. *Bayer AG.*

3812 Baymidur® K88
Aromatic diisocyanate
For production of PU cast compounds for electronics and electrical engineering. *Bayer AG.* Name unverified.

3813 Baymin®
Flotation chemicals. *Bayer AG; Bayer plc.*

3814 Baymod®
Ethylene-vinyl acetate copolymer.
Bayer AG.

3815 Baymod® A
Acrylo-butadiene-styrene polymer, PVC modifier. *Bayer AG.*

3816 Baymod® PU
Aliphatic polyurethane
Powdered plasticizing polymers for PVC. *Bayer AG.*

3817 Baymoflex A
Acrylo-styrene-acrylate polymers blend. d = 1.07. *Bayer AG.*

3818 Baymoflex A KU3-2069.A
ASA; modifier for semirigid halogen-free sheeting, especially for automotive instrument panels. *Bayer Corp; Polysar.*

3819 Baymol® A, D
Range of nonionic tanning auxilliaries with grease emulsifying and degreasing effect. *Bayer AG; Bayer plc.*

3820 Baymosthrin
Pyrethroid with exceptionally fast knockdown action for control of insects in public health sector. *Bayer AG.*

3821 Baynat®
Rigid polyurethane foams; rigid foams for heating and refrigeration engineering, building industry, insulation, sporting goods, automotive engineering. *Bayer AG.*

3822 Bayofly®
Specific for the control of flies in cattle. *Bayer AG.*

3823 Bayothrin
Pyrethroid with exceptionally fast knockdown action for control of insects in public health sector. *Bayer AG.*

3824 Bayovac
Various vaccines for animals. *Bayer AG.*

3825 Bayplast®
Organic pigments; for the plastics industry. *Bayer AG.*

3826 Baypreg®
Polyols for the Baypreg system (filler and glass fibers containing resin compounds for SMC technology). *Bayer AG.*

3827 Baypren®
Polychloroprene rubber; raw material in the adhesive and rubber industries; used for rubber goods with excellent resistance to weathering and aging, good flame-retardant behavior and insensitivity to many chemicals. *Bayer AG; Bayer plc.*

3828 Baypren® 110
126-99-8 204-818-0
C_4H_5Cl
Polychloroprene
Neoprene; Chloroprene; Baypren Latex B; Baypren Latex GK; Baypren Latex L 200A; Baypren Latex L345; Baypren Latex MKB; Baypren Latex SK; Baypren Latex T; Heveasyn Polychloroprene latex; Neoprene AH; Neoprene Latex 115; Neoprene Latex 622; Neoprene Latex 654; Neoprene Latex 671A; Neoprene Latex 842-A. Polchloroprene rubber; synthetic polymer used for moldings and extrudates, reinforced hoses, roll covers, belting, cable sheathings and insulation, sponge rubber, sheeting, fabric proofings, footwear, food-contact goods, adhesives for footwear, furniture, building industry. d = 0.958. *Bayer; Miles.* Name unverified.

3829 Baypren® 110 VSC
126-99-8 204-818-0
Chloroprene rubber
Chloroprene rubber; modified grade providing reduced tendency to crystallize and harden at low temperatures; offers good resistance to aging, weathering, ozone. *Bayer; Miles.*

3830 Baypren® 216
126-99-8 204-818-0
Chloroprene rubber
Chloroprene rubber; modified grade of Baypren 210 offering superior tensile strength and tear resistance; recommended where improved physical properties, or cost reductions are necessary; applications including cable jackets, conveyor belts, hose covers, seals, bellows, and mechanical goods. *Bayer; Miles.*

3831 Baypren® 310
126-99-8 204-818-0
Chloroprene rubber
Chloroprene rubber elastomer; for adhesive compounding.; used for contact adhesives for bonding high-pressure laminates to wood in the manufacturing of furniture, sole attaching in the footwear industry, cements for building and automotive trades and other operations involving fibrous substrates, plastics, foam materials or metals. *Bayer; Miles.*

3832 Baypren® AT-H, AT-M, AT-S
126-99-8 204-818-0
Chloroprene rubber
Chloroprene rubber elastomer, thiuram modified; for formulating contact adhesives used in bonding of high-pressure laminates to wood in manufacturing of furniture, sole attaching in footwear industry, cements for building and automotive trades and other operations involving fibrous substrates, plastics, foam materials or metals. *Bayer; Miles.* Discontinued.

3833 Baypren® EM1
126-99-8 204-818-0
Chloroprene rubber
Precrosslinked chloroprene rubber; elastomer offering favorable processing chars. in mixing, calendering, extruding, and, to some extent, in injection molding; vulcanizates have good resist. to aging, weathering, and ozone attack but do not exhibit outstanding low temperature properties; applications including extruded profiles, hose, cable and calendered sheeting. *Bayer; Miles.*

3834 Baypren® Latex KA 8348
126-99-8 204-818-0
Polychloroprene
Polychloroprene aqueous colloidal dispersion.; rosin acid soap emulsifier system; for manufacturing of molded foam; vulcanizates have slight to medium crystallization tendency and show little discoloration when exposed to light. *Bayer; Miles.*

3835 Baypren® Latex L 200A
126-99-8 204-818-0
Polychloroprene
Polychloroprene aqueous colloidal dispersion; used in dipped goods, paper and fabric surfactants, and coatings. *Bayer; Miles.* Name unverified.

3836 Baypren® M1
126-99-8 204-818-0
Chloroprene rubber
Chloroprene rubber, mercaptan-modified; for rubber goods of moderate hardness and medium filler loading, especially for applications which do not have to meet stringent low temperature requirements; vulcanizates have good resistance to aging, weathering, and ozone attack, and moderate oil resistance; applications include molded and extruded goods, hose, cable jackets, closed cell sponge and elastomeric linings. *Bayer; Miles.*

3837 Bayprint®
Screen printing pastes for the manufacture of printed circuits. *Bayer AG.*

3838 Bayrena®
Sulfonamide
Preparation for the treatment of bacterial infections; veterinary medicine. *Bayer AG.*

3839 Bayrusil®
Diethchinalphion
Insecticide for the control of biting and sucking pests. *Bayer AG.*

3840 Baysan®
Range of household cleaning products with disinfecting properties. *Bayer AG.*

3841 Bayscrlpt®
Dyes for ink-jet printing. *Bayer AG; Bayer plc.*

3842 Baysical®
Precipitated silicates; inorganic fillers improving the whiteness and opacity of paper; preserves the freeflowing capacity of table salt; as a grinding aid in powders with the tendency to cake and as an extender in emulsion paints. *Bayer plc.*

3843 Baysilex®
Addition curing elastomeric high precision impression for the single phase technique two paste system material; used in dentistry. *Bayer AG.*

3844 Baysilone®
A range of silicone fluids; used in the production of polishes and water repellants, transfer media, dampening fluids, hydraulic fluids, dielectrics, lubricants in the production of plastics and man-made fibers and in the metal industry. *Bayer AG.*

3845 Baysin®
Finish for full grain/corrected grain side leather and splits. *Bayer AG; Bayer plc.*

3846 Baysolvex®
Solvent extractants for nonferrous metals, precious metals and rare earths. *Bayer AG.*

3847 Baysport®
PU elastomer; for building, furniture, shoes, automotive and mechanical engineering, sports surfacing, sporting goods, textiles, electrical industry, domestic appliances. *Bayer AG; Bayer plc.*

3848 Baystal®
Butadiene-styrene copolymers with self-crosslinking groups and possible additional co-monomers; used for special applications in a variety of industries e.g. the textile, paper and rubber industries. *Bayer AG; Bayer plc.*

3849 Baysynthol®
Synthetic sizing agents; used in the paper industry. *Bayer AG; Bayer plc.*

3850 Baytan
fuberidazole and triadimenol. Powder mixture of fuberidazole and triadimenol; seed dressing for cereals. *ICI Agrochemicals.*

3851 Baytan®
123-88-6 204-659-7
Triadimenol-fuberidazole
Dry powder containing 25% w/w triadimenol and 3% w/w fuberidazole; a seed treatment for barley, wheat, oats and rye; controls the important seed and certain soil borne diseases including loose smut, covered smut, foot rot, leaf stripe bunt and early attacks of mildew and rhynchosporium. *Bayer AG; Bayer plc.*

3852 Baytan® IM
fuberidazole, imazalil and triadimenol. Dry powder mixture of fuberidazole, imazalil and triadimenol; seed dressing for barley. *Bayer plc.*

3853 Baytec®
PU elastomer; for building, furniture, shoes, automotive and mechanical engineering, sports surfacing, sporting goods, textiles, electrical industry, domestic appliances. *Bayer AG; Bayer plc.*

3854 Baytherm®
Rigid polyurethane foams; rigid foams for heating and refrigeration engineering, building industry, insulation, sporting goods, automotive engineering. *Bayer AG; Bayer plc.*

3855 Baytigan® AR
Softening, lightfast, anionic polymer retanning material for chrome leather. *Bayer AG.*

3856 Baytroid®
68359-37-5 2826 269-855-7
cyfluthrin
Cyfluthrin, a fast-acting synthetic pyrethroid with a long residual effect and broad spectrum of activity; for use in controlling caterpillars, beetles and their larvae, cutworms, sucking pests on vegetables, pome and stone fruit, grapes, maize, soya, tobacco, cotton and other crops. *Bayer AG.*

3857 Bayvap®
Electric vaporizer tablets to combat gnats and mosquitos in rooms. *Bayer AG.*

3858 Bayvarol
For the control of ectoparasites in bees (*varroa*). *Bayer AG.*

3859 Bazak
Photographic chemicals. *Makhteshim Chemical Works Ltd.*

3860 BB Accelerator
Derivative of dimethyl-*p*-phenylenediamine; a rubber vulcanization accelerator. *BF Goodrich.* Name unverified.

3861 BB10GF/15T
Bayblend, 10% glass fiber, 15% Teflon. *Compounding Tech.*

3862 BBS
68334-28-1 269-820-6
Fats and Glyceridic oils, vegetable, hydrogenated; Partially hydrogenated vegetable oil; Trifat P-52. Partially hydrogenated vegetable oil. mp = 56-58°. *Karlshamns.*

3863 BBTS
95-31-8 202-409-1
$C_{11}H_{14}N_2S_2$
N-*t*-Butyl-2-benzothioazole sulfenamide
Perkacit®TBBS; Santocure®NS; Vanax®NS; Vanax®TBSI; Vulkacit NZ/EG; NTBBTS; Nocceler NS; Vulkacit NZ. Delayed-action accelerator for natural and synthetic rubbers. mp=105°; d = 1.28; practically insoluble in H_2O *Akrochem.*

3864 BCF
353-59-3 206-537-9
bromochlorodifluoromethane
Bromochlorodifluoromethane (Halon 1211). A fire extinguisher. *ICI Chem & Polymers Ltd.*

3865 BE Buffer
Tris-borate-EDTA solution or powder. *Am. Research Prods.* Unverified.

3866 Be Square® 185
63231-60-7 264-038-1
Hard microcrystalline wax consisting of n-paraffinic, branched paraffinic, and naphthenic hydrocarbons; wax used in hot-melt coatings and adhesives, cup and paper coatings, printing inks, plastic modification (as lubricant and processing aid) lacquers, paints, varnishes, binder in ceramics, potting in electronic components, rubber, elastomers, emulsion wax size in papermaking, fabric softener ingredient, in cosmetic hand creams and lipsticks, d = 0.93; Very low solubility in organic solvents *Petrolite.*

3867 Beanfeast
Proteinaceous substances used as food or as ingredients for food. *Courtaulds plc.* Discontinued.

3868 Bear
A cotton fabric grade of Tufnol industrial laminates. *Tufnol Ltd.*

3869 Bearflex® LAO
Extender oil offering light color and low aniline points; for elastomer

compounding, especially nitrile, neoprene, SBR, and natural rubber. *Witco/Golden Bear.* Discontinued.

3870 Bearium
Proprietary alloys of copper with 17.5-28% lead and 10% tin. Bearing metal. No manufacturer.

3871 Beaudouin's reagent
1% furfural in alcohol.

3872 Beaver steel
A proprietary nickel chromium steel containing 1.5% nickel, 0.75% chromium, 0.6% manganese, and 0.55% carbon. No manufacturer.

3873 Beckacite® 110, 115
Phenolic modified rosin ester resin; resins for paste ink applications. *Arizona.* Discontinued.

3874 Beckacite® 425
Maleic modified rosin ester resin; resins for paste ink applications. *Arizona.* Discontinued.

3875 Beckacite® 4900
Modified tall oil rosin; for manufacturing of paper size, intermediate in rosin derivative production, printing ink binders as resins, tackifier resin in sealants and mastics, as starting point rosin for resin esters. *Arizona.* Discontinued.

3876 Beckamine®
Synthetic resins for use in paints, varnishes, enamels, etc. and in industrial arts. *Reichhold.*

3877 Beckamine® 21-500
Urea-formaldehyde resin. d = 8.33 lb/gal; Gardner 1 maximum color. *Reichhold.*

3878 Beckol
A proprietary trade name for synthetic alkyd resin. No manufacturer.

3879 Beckolin®
Synthetic oils for use in industrial arts, paints, varnishes, lacquers, linoleum, oil cloth, inks, etc. *Reichhold.*

3880 Beckolloid
A proprietary synthetic resin plastic. No manufacturer.

3881 Beckopol®
Synthetic resins for use in the industrial arts, paints, varnish, enamel, lacquer, etc. *Reichhold.*

3882 Beckosol
Plastic synthetic resins for paints, varnishes, lacquers, paper/cardboard goods. No manufacturer.

3883 Beckosol 13-400
Waterborne medium oil alkyd. d = 9.40 lb/gal; Gardner 8 maximum color. *Reichhold.*

3884 Becksol
A proprietary trade name for alkyd synthetic resins. No manufacturer.

3885 Becosal
A range of fire extinguishing powders which are also compatible with foam. *Degussa AG.* Name unverified.

3886 Becxopox
Epoxide resins; for paints, adhesives, sealants and encapsulants. *Resinous Chemicals Ltd.* Discontinued.

3887 Bedesol
Synthetic resins. No manufacturer.

3888 beeswax
8006-40-4 (White), 8012-89-3 (Yellow) 11,1031
Cera alba. Purified wax from the honeycomb; food additive, furniture and floor waxes, shoe polishes, leather dressings, anatomical specimens, artificial fruit, textile sizes and finishes, church candles, cosmetic creams, adhesive compounds. mp = 62-65°; d = 0.95-0.96. *British Wax Refining; ICI Spec.; Koster Keunen; Maruzen Fine Chem.; Strahl & Pitsch.*

3889 Beetafil
Foam resins for thermal insulation. *BIP Chemicals Ltd.*

3890 Beetafin
Aqueous polyurethane resins; used in leather and textile industries. *BIP Chemicals Ltd.*

3891 Beetle
Thermosetting molding powders (urea-formaldehyde, polyester), thermoplastic molding materials (nylon 6 and 66, PET and PBT polyesters, polycarbonate, acetal), resins (unsaturated polyester, urea-formaldehyde, melamine-formaldehyde and alkyds for paints textiles and adhesives. *BIP Chemicals Ltd.*

3892 Beetle Resin BT 333
A cyclic reactant; recommended for soft mechanical finishes on cellulosic fabrics, also soft crease resistant and shrink resistant finishes on cellulose/synthetic fiber blends. *BIP Chemicals Ltd.*

3893 Beetle Resin BT 334
A cyclic reactant-modified melamine cross linking agent; recommended for chlorine resistant finishes on cotton fabrics and easy care finishes on cellulose/synthetic fiber blends. *BIP Chemicals Ltd.*

3894 Beetle Resin W69
A modified urea-formaldehyde resin; recommended for glueing Vac-Vac" treated timbers where difficulties may be experienced with standard adhesives. *BIP Chemicals Ltd.*

3895 Behenamide DEA (1:1)
70496-39-8
Incromide BED. High melting pearling and opacifying agent used in stick preparations. *Croda Inc.* Discontinued.

3896 Behenamidopropyl dimethylamine
60270-33-9 262-134-8
N-[3-(Dimethylamino)propyl]docosamide
Chemidex B; Incromine BB; Mackine® 601; Schercodine B; Dimethylaminopropyl behenamide. Emollient, conditioner, lubricant, viscosity builder, and moisturizer for hair care products; intermediate for hair conditioning agent. mp = 70-72°. *Croda Inc.; Croda Surfactants Ltd; Chemron; McIntyre;.*

3897 Behenamine oxide
26483-35-2 247-730-8
N,N-Dimethyl-1-behenamine-N-oxide
Incromine Oxide B-30P; Incromine Oxide B; Incromine Oxide B50; N,N-Dimethyl-1-docosanamine-N-oxide; 1-Docosanamine, N,N-dimethyl-,N-oxide. Softener and conditioner for hair care products; viscosity builder, emulsifier, lubricant, wetting, and foam stabilizer used in personal care products. White paste; pH=6.5-8.0 *Croda Inc.; Croda Surf. Ltd.*

3898 Behentrimonium chloride
17301-53-0 241-327-0
Genamin KDM-F; Varisoft® BT-85. Base, antistat, emulsifier for preparation of hair care products; conditioner. *Hoechst Celanese/Colorants & Surf.*

3899 Behenyl alcohol
661-19-8 8135 211-546-6
$C_{22}H_{46}O$
1-docosanol
alcohol C_{22}; Cachalot BE-22; n-eicosanol. Mixture of fatty alcohols chiefly of n-docosanol; synthetic fibers, lubricants, evaporation retardant on water surfaces. mp = 65-72°; $bp_{0.22}$ = 180° *M. Michel; Schweizerhall; Sherex; Vista.*

3900 Behenyl betaine
84082-44-0 281-991-9
N-(Carboxymethyl)-N,N-dimethyl-1-docosanaiminum hydroxide, inner salt
Incronam B-40; Behenamidopropyl betaine. Foaming surfactant, conditioner, lubricant used in personal care products; excellent slip, conditioning; good wetting and rinse aid. White paste; pH=5.5-7.0. *Croda Inc.; Croda Surf. Ltd.*

3901 Belclene
Water treatment chemicals. *Ciba plc.* Name unverified.

3902 Belco
Cellulose car finish paints and resins. *ICI Chem & Polymers Ltd.*

3903 Beldex
ASS modifiers for PVC. *GE Plastics ABS Ltd.* Name unverified.

3904 Belfasin 320 Crushed
Softener concentrate for use on package dyed yarns. *Henkel/Textiles.*

3905 Belgard
Water treatment chemicals. *Ciba plc.* Name unverified.

3906 Belite
Antifoaming agents. *Ciba plc.* Name unverified.

3907 bell pepper
The fruit of *Capsicum grossum.*

3908 Bellasol
Foam preventives. *Ciba plc.* Name unverified.

3909 Bellauxine
Water treatment chemicals. *Ciba plc.* Name unverified.

3910 Bellclo
Soluble concentrate containing 250g 2,4-DB and 53 g mecoprop per liter; a translocated herbicide. *MTM Agrochemicals Ltd.*

3911 Bellite
Explosives for coal mine consisting of ammonium nitrate and dinitrobenzene with or without sodium chloride.

3912 Belloid
Dispersing agents. *Ciba plc.* Name unverified.

3913 Belmac Straight
94-81-5 202-365-3
MCPB
Soluble concentrate containing 400 g/l MCPB; for control of weeds in undersown cereals and grassland. *MTM Agrochemicals Ltd.*

3914 Belpro
A water-based acrylic copolymer temporary coating and remover (also known as Tempro). *ICI Chem & Polymers Ltd.*

3915 Belro
Wood rosin derivative insoluble in water, soluble in organic solvents, fats and oils; acid number 114; drop softening point of 90°; construction adhesive. *Hercules.*

3916 Belsil ADM 6041E
Amodimethicone emulsion; substantive conditioner for hair conditioners and shampoos. *Wacker-Chemie.*

3917 Belsil DM 0.65
Dimethicone
Used in skin and hair care products and decorative cosmetics; enhances suppleness and gives soft, velvety feel to skin; prevents stickiness and increases water resistance of cosmetic. *Wacker-Chemie.*

3918 Belsil DMC 6031
Dimethicone copolyol
Wetting aid improving surface slip, emollient, moisturizer, fatting agent for cosmetics; stabilizer for foams and emulsions; plasticizer for hair spray resins. *Wacker-Chemie.*

3919 Belsil PDM 20
Phenyl dimethicone
Used in skin and hair care products and decorative cosmetics; give excellent feel to skin; high penetrating and water repellent properties; imparts suppleness and depth of color to hair. *Wacker-Chemie.*

3920 Belsil SDM 6021
68554-53-0
Stearoxydimethicone
Give nongreasy, soft, velvety feel to skin; enhance gloss and color brightness in decorative cosmetics; good spreading properties, protection against aqueous media. *Wacker-Chemie.*

3921 Belsoft
Blend of acid amides and fatty alcohol sulfates, anionic; softener for cellulosic and synthetic fibers, especially suitable for terry cloths and textured polyester. *Henkel Chemicals Ltd.* Name unverified.

3922 Belsol
Wetting and penetrating agents. *J C Bottomley.* Name unverified.

3923 Beltherm
Fuel oil additive. *Ciba plc.* Name unverified.

3924 Belzak AC
$C_7H_{13}NaO_8$
α-Sodium glucoheptonate dihydrate
Sequestering agent for metal cleaning and processing, various industrial cleaning compounds, bottle wash, set retarder in concrete and trace metals for agriculture. dec 161°. *Belzak corporation.*

3925 Belzak BL-50
β sodium glucoheptonate
β sodium glucoheptonate 50% liquid; used for metal cleaning and processing, various industrial cleaning compounds, bottle wash, set retarder in concrete and trace metals for agriculture. *Belzak Corporation.*

3926 Bemal
An alloy of 70% copper, 29% zinc, and traces of lead and iron.

3927 Bemberg®
Cuprammonium rayon; viscose rayon for textile industry. *Asahi Chem. Industry.*

3928 Bemillese®
Nonwoven fabric of cuprammonium rayon. *Asahi Chem. Industry.*

3929 Ben-A-Gel®
1302-78-9 1082 215-108-5
$Al_2O_3 \cdot 4SiO_2 \cdot H_2O$
Bentonite
Wilkinite. Beneficiated bentonite clay; for use as thickening, gelling and emulsifying agent for water systems. *Rheox Inc.*

3930 Benalite
A proprietary trade name for a lignin plastic; cured lignin sheets. No manufacturer.

3931 Benaloid
A proprietary trade name for uncured lignin sheets. No manufacturer.

3932 Benaqua
Rheological additives for aqueous systems. *Rheox Inc.*

3933 Benathix
Thixotropic and thickening additives; used in unsaturated polyester laminating resins. *Rheox Inc.; Steetley Minerals Ltd.*

3934 Benazalox
3813-05-6, 57754-85-5 260-929-4
Benazolin-clopyralid; a herbicide mixture for use in oilseed rape. *Schering Agrochemicals Ltd.* Discontinued.

3935 benazolin
3813-05-6 223-297-0
$C_9H_6ClNO_3S$
4-chloro-2-oxo-3(2H)-benzothiazoleacetic acid
Eunasin; Cresopur; Benzar; BTS-7693; Cornox; BEN-30; Ben-cornox; Benopan; Bensecal; Cornox CWK; EX 10781; Galipan; Grassland weedkiller; Herbazolin; Keropur; Legumex extra; Ley-cornox; Leymin; Metizolin; Tri-cornox special. Selective, systemic growth regulator herbicide Used for control of broad leaved weeds such as black bindweed, chickweed, cleavers and charlock. mp = 193°; soluble in H_2O (600 mg/l), more soluble in organic solvents; LD_{50} (rat orl) > 4800 mg/kg.

3936 Bendalite
A proprietary trade name for a cast styrene synthetic resin. No manufacturer.

3937 Bendiocarb
22781-23-3 1063 245-216-8
$C_{11}H_{13}NO_4$
2,2-Dimethyl-1,3-benzodioxol-4-ol methylcarbamate
Bendiocarb; Fuam; Garvox; Multamat; Multimet; methylcarbamic acid 2,3-(isopropylidenedioxy)phenyl ester; Tattoo; Dycarb; Niomil; Rotate; Sedox; Seedoxin; Turcam; Ficam W; NC 6897FICAM D; Ficam Plus; Ficam ULV; Isopropylidenedioxy)phenyl N-methylcarbamate; Seedox. Insecticide with contact action. mp = 129-130°; soluble in H_2O (40 mg/l), more soluble in hexane.

3938 Benecel® Methylcellulose
Carboxymethylmethylcellulose
Thickener, stabilizer, rheology control agent, filmformer, suspending agent; water-retention aid, binder for food, pharmaceutical, and cosmetic industries. *Aqualon.*

3939 Benedict Plate
A nickel silver containing 57% copper, 28% zinc, and 15% nickel.

3940 Benfluralin
1861-40-1 1067 217-465-2
$C_{13}H_{16}F_3N_3O_4$
N-butyl-N-ethyl-2,6-dinitro-4-trifluoromethylaniline
Benefex; Benefin; EL-110; Balan; Balfin; Quilan;. Active ingredient: benfluralin. A pre-emergence herbicide with a wide range of weed control both of annual grass weeds and broad-leaved weeds. Yellow-orange crystalline solid; mp = 65-66.5°; soluble in most organic solvents. *Agan Chemical Manufacturers Ltd.* Discontinued.

3941 Bennatate
A proprietary trade name for a cellulose acetate plastic. No manufacturer.

3942 benodanil
15310-01-7 239-352-7
$C_{13}H_{10}INO$
2-iodo-N-phenylbenzamide
Apache; BAS 3170F; Benefit; Calirus. Systemic and contact fungicide with protective and curative action. Used for control of rust disease in various crops. mp = 137°; soluble in H_2O (20 mg/l), more soluble in organic solvents; LD_{50} (rat orl) >6400 mg/kg.

3943 Benol
White mineral oil. *Rheox Inc.* Discontinued.

3944 Benomyl
17804-35-2 1073 241-775-7
$C_{14}H_{18}N_4O_3$
[1-[(Butylamino)carbonyl]-1H-benzimidazol-2-yl]carbamic acid methyl ester
Benlate; F-1991;. Fungicide for garden use. (Sold in UK on behalf of Du Pont). Insoluble in H_2O, soluble in $CHCl_3$, LD_{50} (rat orl) > 9590 mg/kg. *ICI Chem & Polymers Ltd.*

3945 Bensapol
A proprietary trade name for a wetting agent consisting of sulfonated oils and a solvent. No manufacturer.

3946 Bentalol
100-51-6 1159 202-859-9
C_7H_8O
benzyl alcohol
Discontinued.

3947 Bentazon
25057-89-0 1080 246-585-8
$C_{10}H_{12}N_2O_3S$
3-(1-Methylethyl)-1H-2,1,3-benzothia-diazin-4(3H)-one 2,2-dioxide
Basagran®; bentazone; BAS 351H; Adagio; Galaxy; Storm; Basagran 4E; Thiadiazinol; Bendioxide. A contact herbicide. mp = 137-139°; soluble in H_2O; LD_{50} (rat orl) = 1100 mg/kg. *BASF plc.*

3948 Bentobrite® 770
Micronized white powder used as general purpose suspension agent and gellant for household and industrial products. *Am. Colloid.*

3949 Bentokol
Foundry coal dust replacement additive. *Foseco (F.S.) Ltd.*

3950 Bentone
Rheological additives; used for paint, ink, grease, caulks, sealants, cosmetics and adhesives. *Rheox Inc.; Steetley Minerals Ltd.*

3951 Bentone Gel
Thixotropic and thickening additives. pregelled bentone additive for cosmetics. *Rheox Inc.; Steetley Minerals Ltd.*

3952 Bentone SD
Super dispersing rheological additives; used for paint, ink, caulks, sealants and adhesives. *Rheox Inc.*

3953 *Bentone-34*
Rheological additives. Rheox Inc. Discontinued.

3954 Bentonite
1302-78-9 1082 215-108-5
$Al_2O_3.4SiO_2.H_2O$
Sodium montmorillonite
Wilkinite; Colloidal clay; soap clay; wilkinite; tonsil L80; Brebent; BentoPharm; Bregel; Ben-A-Gel®. Native hydrated colloidal aluminum silicate clay; oil-well drilling fluids, cement slurries for oil-well casings, thickener, fireproofing, cosmetics, decolorizing agent, filler in ceramics, emulsifier for oils, suspending agent in pharmaceuticals, base for plasters. *Am. Colloid; Dry Branch Kaolin; Norsk Hydro AS; L.A. Salomon; Southern Clay; R. T. Vanderbilt Co Inc.*

3955 BentoPharm
1302-78-9 1082 215-108-5
Bentonite
Pharmaceutical quality; suspension and thickening agent. *Bromhead & Denison Ltd.* Name unverified.

3956 Benvic
Granulated PVC compound. *Laporte Industries Ltd.*

3957 Benzaldehyde
100-52-7 1085 202-860-4
C_6H_5CHO
benzoic aldehyde
Artifical essential oil of almond; Phenylmethanal; almond artificial essential oil; artificial almond oil; benzenecarbonal; benzene carboxaldehyde; oil of bitter almond; Artificial Bitter Almond Oil; Benzene methylal; Benzoyl hydride. Synthetic oil of bitter almond; Chemical intermediate for dyes, flavors, perfumes, aromatic alcohols; solvent for oils, resins, cellulose acetate and nitrate; manufacture of cinnamic acid, benzoic acid; pharmaceuticals; photographic chemicals. mp = -17°; bp = 205°; d = 1.0440; n_D^{20} = 1.5454; insoluble in H_2O, soluble in organic solvents; LD_{50} (rat orl) = 1300 mg/kg. Aceto; Aldrich; DSM BV; R.W. Greeff; Haarmann & Reimer Janssen Chimica; Penta Mfg.; Sina (UK); Spectrum Chem. Mfg.

3958 Benzalkonium chloride
68989-00-4 273-544-1
$C_6H_5CH_2N(CH_3)RCl$, R=C_8H_{17} to $C_{18}H_{37}$
Ablumine 08, 10, 12, 1214, 3500; Alkaquat® DMB-451-50, DMB-451-80; Arquad® DMMCB; Barquat® 50-28, 80-28; BTC® 50NF, 65NF, 835, 8358; Catigene® D80, T50, T80; DeQUAT BC-1214-50; Empigen® BCB50, BAC50/BP, BAC 80, BCB50; Exameen 3500, 3580; Gardiquat 12H, 1450, 1480, SV 480; Heliquat BAC 50; Maquat LC-12S, MC-1412, MC-1416, MC-6025; Noramium DA-50; RP-50; Rhodaquat® RP-80; Sinotex CDB; Uniquat® QAC-50, QAC-80; Variquat® 50MC, 60LC, 80MC, 80ME; Zoharquat 80. Disinfectant for dairy, food processing, restaurant, brewing, and bottling industries; retarder in dyeing of acrylic fibers. White or yellowish powder; very soluble in H_2O, alcohol, acetone; slightyl soluble in benzene; insoluble in ether; mp = 34-37°; LD_{50}=rat orl 400 mg/kg;. Albright & Wilson UK; Akzo nobel; Aldrich; Allchem Ind. Chemron; Fluka; Gresco Mfg; Lonza Ltd; Mason; Rhône-Poulec France; Sigma; Spectrum Quality Prods; Stepan Europe; Witco/Oleo-Surf; Akzo; EM Industries; Lonza AG; Sherex; Stepan; Witco/Humko.

3959 Benzamine®
Diazotizing dyes for cotton and other vegetable fibers, textiles from regenerated cellulose. *Bayer AG.*

3960 Benzanil
A range of direct dyes; for dyeing of cotton and viscose fibers. No manufacturer.

3961 Benzene
71-43-2 1094 200-753-7
C_6H_6
benzol
Cyclohexatriene; Annulene; Benzol. Used in manufacture of ethylbenzene, dodecylbenzene, cyclohexane, phenol, nitrobenzene, chlorobenzene; dyes, medicinals, artificial leather, airplane dopes, lacquers; solvent for waxes, resins, oils. mp = 5.5°; bp = 80°; d = 0.8790; n_D^{20}= 1.5000; highly flammable; LD_{50} (rat orl) = 3.8 ml/kg. BP Chem. Ltd; Exxon; Janssen Chimica; Mitsubishi Petrochem.; Mitsui Petrochem. Ind.; Mobil; OxyChem; Shell.

3962 Benzenesulfonamide
98-10-2 202-637-1
$C_6H_7NO_2S$
Chemical intermediate. mp = 150-152°; soluble in H_2O (4.3 g/l), organic solvents; LD_{50} (rat orl) = 991 mg/kg. *Unitex.*

3963 benzethonium chloride
121-54-0 1103 204-479-9
$C_{27}H_{42}ClNO_2$
N,N-dimethyl-N-[2-[2-[4-(1,1,3,3-tetramethylbutyl)phenoxy]ethoxy]ethyl]benzenemethaminium chloride
Solanine; Benzathonium Chloride; Anti-Germ 77; Antiseptol; Benzethonium Chloride 1622; BZT; Diapp; Disilyn; Hyamine; Hyamine 1622; Phemeride; Phemerol; Phemerol Chloride; Phemersol Chloride; Phemithyn; Polymine D; Quatrachlor; Solamin. A topical anti-infective, used primarily in cosmetics for its antimicrobial and cationic surfactant properties. Bactericide, deodorant, anti-infective; antiseptic (veterinary); preservative for veterinary and pharmaceutical products. mp = 158-163°; soluble in H_2O, organic solvents; LD_{50} (rat orl) = 420 mg/kg. *Lonza Inc; Parke-Davis.*

3964 Benzex
Benzyl cellulose.
Benzyl cellulose. No manufacturer.

3965 Benzhydrol
91-01-0 1121 202-033-8
$C_{13}H_{12}O$
Diphenyl methanol
α-Phenylbenzenemethanol; Diphenylcarbinol; benzohydrol. Used in organic synthesis mp = 65-67°; bp = 297-298°; slightly soluble in H_2O, soluble in organic solvents; LD_{50} (rat orl) = 5 gm/kg.

3966 Benzimidazole
51-17-2 1111 200-081-4
$C_7H_6N_2$
1,3-benzodiazole
Benzoglyoxaline; N,N'-methenyl-*o*-phenylenediamine; 1H-Benzimidazole; benziminazole; 1-azindole; 3-azaindole; *o*-benzimidazole; benzoimidazole; BZI; 1,3-diazaindene. Used in chemical synthesis. mp = 172-174°; soluble in H_2O, organic solvents; LD_{50} (mus orl) = 2910 mg/kg. *Atomergic Chemetals; Janssen Chimica; Penta Mfg.; Schweizerhall.*

3967 Benzo®
Direct dyes; suitable for cotton and rayon goods where no special demands are made on fastness properties. *Bayer AG; Bayer plc.*

3968 Benzo® Cuprol
Direct dyestuffs whose wet fastness and light fastness are considerably improved by an aftertreatment with copper salts. *Bayer AG.*

3969 Benzoflex
Plasticizer. *Velsicol.* Name unverified.

3970 Benzoflex 2-45
120-55-8 204-407-6
$C_{18}H_{18}O_5$
Diethylene glycol dibenzoate.
Ethanol, 2,2'-oxybis-, dibenzoate; Dipropylene glycol dibenzoate/diethylene glycol dibenzoate 1/1 blend; 2,2'-Oxydiethylene dibenzoate. A proprietary trade name for diethylene glycol dibenzoate, a chemical intermediate. bp_7 = 235-237°; d = 1.175; flashpoint >230°F. *Tennessee Corp.*

3971 Benzofloc
Organic and inorganic specialty flocculant and coagulant polymers; used for solids flocculation in industrial and municipal solids, dissolved or collodial form, color, turbidity, algae removal; specialty application. *Benzsay & Harrison Inc.*

3972 Benzoic acid
65-85-0 1122 200-618-2
Nipacide®. Formulated biocides. *Nipa Laboratories Ltd.*

3973 Benzoic acid
65-85-0 1122 200-618-2
$C_7H_6O_2$
benzenecarboxylic acid
Benzenecarboxylic Acid; Carboxybenzene; Diacylic acid; benzene formic acid; benzenemethonic acid; phenyl carboxylic acid; phenylformic acid; Retarded BA; Retardex; Salvo; Tennplas; Oracylic acid. Used in the manufacture of sodium and butyl benzoates, plasticizers, benzoyl chlorides, food preservatives, flavors, perfumes, antifungal agent. mp = 122.4°; bp = 249°; d = 1.321; soluble in H_2O (2.9 g/l), organic solvents; LD_{50} (rat orl) = 1700 mg/kg. Elf Atochem SA; R.W. Greeff; Mallinckrodt; Penta Mfg.; Pentagon Chemicals Ltd; Schaefer Salt; Velsicol.

3974 Benzoic Acid K
Gloss and flow promoter for use in oil and synthetic resin-based topcoats; hardener for rubber soles, heels, floor coverings. *Bayer AG.*

3975 Benzoin
119-53-9 1124 209-441-5

$C_{14}H_{12}O_2$
α-hydroxybenzyl phenyl ketone
α-hydroxy-α-phenylacetophenone; α-hydroxybenzyl phenyl ketone; 2-Hydroxy-1,2-diphenylethanone; hydroxy-2-phenyl acetophenone; Bitter almond-oil camphor. Organic synthesis; intermediate; photopolymerization catalyst. mp = 134-136°; bp_{12} = 194°; slightly soluble in H_2O, more soluble in organic solvents; LD_{50} (rat orl) = 10 gm/kg. *Aldrich; Janssen Chimica; Dr. Madis Labs; Snia (UK); Spectrum Chem. Mfg.*

3976 Benzonitrile
100-47-0 1128 202-855-7
C_6H_5CN
phenyl cyanide
Cyanobenzene. Manufacture of benzoquanamine; intermediate for rubber chemicals; solvent for nitrile rubber, specialty lacquers, resins and polymers, anhydrous metallic salts. mp = -13°; bp = 188-191°; d = 1.0100; n_D^{20}= 1.5271; soluble in H_2O (10 g/l), organic solvents; LD_{50} (mus orl) = 971 mg/kg. *Penta Mfg.; PMC Specialties; Spectrum Chem. Mfg.*

3977 Benzophenone-2
131-55-5 205-028-9
$C_{13}H_{10}O_5$
2,2',4,4'-Tetrahydroxybenzophenone
2,2,4,4-Tetrahydroxybenzophenol; Methanone, bis(2,4-dihydroxyphenyl)-;. Organic benzophenone derivative; used as commercial uv absorber. mp = 200-203°; LD_{50} (rat orl) = 1220 mg/kg. EM Industries; Ferro/Bedford; R.W. Greeff; Haarmann & Reimer; Hoechst Celanese; Quest Int'l.; Sartomer.

3978 Benzophenone-6
131-54-4 1130 205-027-3
$C_{15}H_{14}O_5$
2,2·g-dihydroxy-4,4·g -dimethoxybenzophenone
bis(2-hydroxy-4-methoxyphenyl) methanone; Uvinul D49. Organic benzophenone derivative; uv light absorber, especially in paints, plastics. mp = 133-136°. EM Industries; Ferro/Bedford; R.W. Greeff; Haarmann & Reimer; Hoechst Celanese; Quest Int'l.; Sartomer.

3979 Benzophenone-9
3121-60-6 221-498-8
$C_{15}H_{14}O_8S$
Disodium 2,2·g-dihydroxy-4,4·g-dimethoxy-5,5·g-disulfobenzophenone
Organic benzophenone derivative; uv absorber in cosmetic formulations, textiles and water-based paints. EM Industries; Ferro/Bedford; R.W. Greeff; Haarmann & Reimer; Hoechst Celanese; Quest Int'l.; Sartomer.

3980 benzophenone-11
1341-54-4
Organic benzophenone derivative.; mixture of benzophenone-6 and -2 and other tetrasubstituted benzophenone materials; UV absorber used in NC lacquer, fluorescent paint, inks, and for protecting furniture woods, colored liquid toiletries, cleaninig agents, isocyanate systems and butyrate metal lacquers. EM Industries; Ferro/Bedford; R.W. Greeff; Haarmann & Reimer; Hoechst Celanese; Quest Int'l.; Sartomer.

3981 Benzoresorcinol
131-56-6 1138 205-029-4
$C_{13}H_{10}O_3$
2,4-dihydroxybenzophenone
benzoresorcinol; 4-benzoyl resorcinol. Organic benzophenone derivative; uv absorber in polymers. mp = 144-147°; bp_1 = 194°; insoluble in H_2O, soluble in organic solvents. EM Industries; Ferro/Bedford; R.W. Greeff; Haarmann & Reimer; Hoechst Celanese; Quest Int'l.; Sartomer.

3982 benzotriazole1H-benzotriazole
95-14-7 1140 202-394-1
$C_6H_5N_3$
benzisotriazole
1,2-aminozophenylene; azimidobenzene; Cobratec 99; 1,2,3-Benzotriazole; Azimidobenzene; benzene azimide; 1,2,3-triaza-1H-indene; U-6233; aziminobenzene; 1,2-aminoazophenylene; cobratec 99; 2,3-diazaindole; 1,2,3-triazaindene. Chelating agent and sesquestrant for copper ions; corrosion inhibitor for copper, brass, bronze; used in antifreeze, cleaners, coatings, detergents, functional fluids, metalworking fluids, packaging materials, polishes. mp = 98.5°; bp_{15} = 201-204°. *Atomergic Chemetals; Miles; PMC Specialties; Sandoz.*

3983 Benzoyl peroxide
94-36-0 1149 202-327-6
$C_{14}H_{10}O_4$
dibenzoyl peroxide
Benzoyl Peroxide; Benzac; Benzagel; Acetoxyl; Lucidol; Desanden; Nericur; Oxy-5; PanOxyl; Thermaderm; Xerac; PanOxyl; Thermaderm; Xerac; Novadelox; Acetoxyl; Acnegel; Benzac; Benzaknen; Debroxide; Desanden; Benzagel 10; Benoxyl; Lucidol; Nericur; Oxy-5, Oxy 10; PanOxyl; Peroxydex; Persadox; Persa-gel; sanoxit; Theraderm; Xerac BP 5; Xerac BP 10; Benzoyl peroxide; BPO; trichlorobenzoic acid; TCBA; Tribac; 2,3,6-TBA; aztec bpo; benzaknew; BZF-60; Cadet; cadox bs; dry and clear; epi-clear; fostex; Garox; incidol; loroxide; luperco; luperox fl; nayper b and bo; norox bzp-250; norox bzp-c-35; OXY-5; Oxy-10; oxylite; oxy wash; quinolor compound; superox; Topex; vanoxide; Xerac,. Keratolytic, oxidizing agent. Catalyst, initiator and curing agent. mp = 103-106°; insoluble in H_2O, soluble in organic solvents; explodes when heated. *Galderma Laboratories.*

3984 Benzyl
Acid wool dyestuffs. *Ciba-Geigy.* Name unverified.

3985 benzyl acetate
140-11-4 1158 205-399-7
$C_9H_{10}O_2$
phenylmethyl acetate
Acetic acid phenylmethyl ester; Acetic acid benzyl ester. Artificial jasmine and other perfumes; soap perfume; flavors; solvent and high boiler for cellulose acetate and nitrate, natural and synthetic resins; oils; lacquers; polishes; printing inks; varnish removers. mp = -51°; bp = 213°; LD_{50} (rat orl) = 2490 mg/kg. Haarmann & Reimer; Janssen Chimica; MTM Spec. Ltd; Penta Mfg.; Pentagon Chemicals Ltd; Quest Int'l; Bush Boake Allen Ltd.

3986 benzyl alcohol
100-51-6 1159 202-859-9
C_7H_8O
phenylmethanol
hydroxytoluene; phenylcarbinol; Benzenemethanol; Phenylcarbinol; Phenylmethyl alcohol; Phenylmethanol; α-Hydroxytoluene; Benzoyl alcohol; Hydroxytoluene; Benzenecarbinol; α-toluenol; (hydroxymethyl)benzene. Perfumes; flavors; photographic developer; dyeing nylon, textiles, sheet plastics; solvent for dyestuffs, cellulose esters, casein, waxes; heat-sealing polyethylene films; intermediate for benzyl esters and ethers; bacteriostat; cosmetics; inks. Specially pure benzyl alcohol. Used as a solvent and in chemical synthesis. mp = -15°; bp = 205°; d_4^{20} = 1.04535; n_D^{20}= 1.54035; soluble in H_2O (4 g/100 ml), organic solvents; LD_{50} (rat orl) = 3.1 g/kg. *Givaudan Iberica SA; R.W. Greeff; Janssen Chimica; Penta Mfg.; Quest Int'l; Bush Boake Allen Ltd.*

3987 Benzyl benzoate
120-51-4 1162 204-402-9
$C_{14}H_{12}O_2$
Benzoic acid phenylmethyl ester
Benzylbenzenecarboxylate; Ascabin; Venzoate; Vanzoate; Ascabiol; Benylate. Fixative and solvent for musk in perfumes and flavors; external medicine; plasticizer for nitrocellulose and cellulose acetate; miticide. Also used to treat scabies and pediculosis. mp = 21°; bp = 323-324°; LD_{50} (rat orl) = 1.7 g/kg. Haarmann & Reimer; Janssen Chimica; Penta Mfg.; Pentagon Chemicals Ltd; Schweizerhall; May & Baker Ltd.

3988 benzyl chloroformate
501-53-1 1848 207-925-0
$C_8H_7ClO_2$
benzyl chlorocarbonate
Carbobenzoxy Chloride; Benzyl chloroformate; Chloroformic acid benzyl ester; Benzylcarbonyl chloride. Used to provide the carbobenzoxy group, a protecting group in peptide synthesis. bp_{20} = 103°; d = 1.1950; n_D^{20}= 1.5167. *Janssen Chimica; Pentagon Chemicals Ltd; PPG Industries.*

3989 Benzyl trimethyl ammonium chloride
56-93-9 200-300-3
$C_6H_5CH_2N(CH_3)_3Cl$
trimethylbenzylammonium chloride
benzyl trimethyl ammonium chloride; benztrimonium chloride; TMBAC. Quaternary ammonium salt; dispersant, dye leveler and retarder, emulsifier used in textile industry; solvent for cellulose; gelling inhibitor in polyester resins; intermediate. mp = 239° (dec). *Janssen Chimica; Pentagon Chemicals Ltd; Sherex; Sybron.*

3990 Benzyl Tuex®
10591-85-2
N,N,N',N'-Tetrabenzylthiuram disulfide
Primary or secondary accelerator for natural and synthetic rubbers; for tire compounds, wire insulation, mechanical goods, sponge, footwear, sheeting, hose; activator for sulfenamide and thiazole accelerators; sulfur donor minimizing reversion in natural rubber. *Uniroyal.*

3991 benzylamine
100-46-9 1160 202-854-1
C_7H_9N
benzenemethanamine
aminotoluene; phenylmethylamine. Used in organic synthesis. mp = 10°; bp = 184-185°; d = 0.9810; n_D^{20}= 1.5430; soluble in H_2O, organic solvents.

3992 benzylidene acetone
122-57-6 1172 204-555-1
$C_{10}H_{10}O$
benzalacetone
methyl styryl ketone; *trans*-4-phenyl-3-buten-2-one; Benzylideneacetone. Organic synthesis; perfumery; fixative; flavors. mp = 41.5°. *Penta Mfg.; Raschig; Schweizerhall.*

3993 Benzyltriethyl ammonium chloride
56-37-1 200-270-1
$C_{13}H_{22}ClN$
N,N,N-triethylbenzenemethanaminium chloride
BTEAC. Quaternary ammonium salt; solvent for cellulose; gelling inhibitor in polyester resins; intermediate. mp = 185°; soluble in H_2O (1700 g/l). *Janssen Chimica; Zeeland.*

3994 Bercotox
Cattle dips and sprays. *The Wellcome Foundation Ltd.* Name unverified.

3995 Bergauf
Skin protective soap with 0-48-G skin protective agent; medicated foot-spray with bactericidal and fungicidal properties; eye cleansing cream for intensive and soothing cleansing of the area around eyes; for protection and care of the skin under environmental stress. *Dynamit Nobel Wien GmbH.* Name unverified.

3996 Berger Colorizer - Full Gloss/Vinyl Matte/Vinyl Silk
Gloss; alkyd resin base thinned with white spirit. Matte: V.A. copolymer emulsion, water based. Silk: V.A. copolymer emulsion, water based; 450 colors available, high gloss external and internal application, matt and silk interior application. *Berger Jenson & Nicholson Ltd.* Name unverified.

3997 Berger Cuprinol Woodpaints and Woodstains
Primer, Matte and Gloss: acrylic copolymer emulsion, water thinned; sheen finishes: alkyd based, white spirit thinned; exterior microporous wood paints and stains giving protection and decoration. *Berger Jenson & Nicholson Ltd.* Name unverified.

3998 Berger mixture
A mixture of zinc, potassium chlorate, a chlorinating agent, such as carbon tetrachloride, and a filler, such as kieselguhr; a smoke screen material.

3999 Berkatekt
A range of products designed to protect titanium, steel, and super alloys during heating and during the forging process. *Acheson Colloids.*

4000 Berkstop
A range of coatings to assist in the heat treatment (carburizing and nitriding) of steel. *Acheson Colloids.*

4001 Berlin Brown
A pigment produced by charring Prussian blue; a mixture of ferroso-ferric oxide and charcoal, and is used as an artist's color.

4002 Bernel® Ester 168
Isocetyl octanoate
Emollient with dry, silky feel. *Bernel.*

4003 Bernel® Ester 2014
Octyldodecyl myristate
Rich emollient, pigment disperser, lipstick component. *Bernel.*

4004 Bernel® Ester CO
59130-69-7 261-619-1
Cetyl octanoate
Noncomedogenic emollient; low viscosity ester. *Bernel.*

4005 Bernel® Ester DID
103213-20-3
Diisopropyl dimer dilinoleate
Emollient, film-former. *Bernel.* Discontinued.

4006 Bernel® Ester DISM
Diisostearyl malate; high viscosity emollient for anhydrous systems. *Bernel.*

4007 Bernel® Ester DOM
Dioctyl maleate
Emollient, oxybenzone solubilizer, cleanser and wax solvent; noncomedogenic; imparts shine in hair conditioners. *Bernel.*

4008 Bernel® Ester EHP
29806-73-3 249-862-1
Octyl palmitate
Cost-effective emollient. *Bernel.*

4009 Bernel® Ester NPDC
Neopentyl dicaprate
Light, dry emollient; pigment wetter and binder. *Bernel.*

4010 Bernel® Ester TOC
Trioctyl citrate
Noncomedogenic emollient, pigment wetter. *Bernel.*

4011 Bernel® OPG
59587-44-9 261-819-9
Octyl pelargonate
Dry, nonoily emollient. *Bernel.*

4012 Bernit
A proprietary trade name for a cellulose acetate plastic. No manufacturer.

4013 Berol
Block polymers; machine dishwashing and rinse aids; emulsion polymerization. *Berol Kemi (UK) Ltd.* Name unverified.

4014 Berol 09
68412-54-4
Poly(oxy-1,2-ethanediyl), α-(nonylphenyl)-ω-hydroxy-, branched
Nonyl phenol ethoxylate; Berol 278, 281, 282, 291 and 292. Detergent, wetting agent, emulsifier, dispersant; biodeg. d = 1.05; soluble in ethanol, xylene, trichloroethylene. *Berol Nobel AB.*

4015 Berol 26
Nonionic surfactant of the nonylphenol ethoxylate type, in liquid form; emulsifiers; solvent cleaners. Soluble in ethanol, xylene, trichloroethylene; pH6-7; d = 1.02. *Berol Kemi (UK) Ltd.* Name unverified.

4016 Berol 108
61791-12-6
PEG-40 castor oil
Surfactant, emulsifier for chemical industry; as softener, rewetting agent, pigment dispersant, dye assistant, leveling agent for paints, textiles, leather; lubricant additive and emulsifier in lubricants for plastics, metals, textiles. d = 1.06; flashpoint >100°. *Berol Nobel AB.*

4017 Berol 191
61791-12-6
PEG-200 castor oil
Surfactant, emulsifier for chemical industry; as softener, rewetting agent, pigment dispersant, dye assistant, leveling agent for paint, textile, leather; lubricant additive/emulsifier in lubricants for plastics, metals, textiles. d = 1.06; flashpoint >100°. *Berol Nobel AB.*

4018 Berol 195
61791-12-6
PEG-32 castor oil
Surfactant, emulsifier for chemical industry. d = 1.06; flashpoint >100°. *Berol Nobel AB.*

4019 Berol 198
61791-12-6
PEG-160 castor oil
Surfactant, emulsifier for chemical industry; as softener, rewetting agent, pigment dispersant, dye assistant, leveling agent for paint, textile, leather; lubricant additive/emulsifier in lubricants for plastics, metals, and textiles. *Berol Nobel AB.*

4020 Berol 199
61791-12-6
PEG-32 castor oil
Surfactant, emulsifier for chemical industry; as softener, rewetting agent, pigment dispersant, dye assistant, leveling agent for paint, textile, leather; lubricant additive/emulsifier in lubricants for plastics, metals, textiles. d = 1.06; flashpoint>100°. *Berol Nobel AB.*

4021 Berol 259
Nonylphenol ethoxylate
Nonionic surfactant consisting of nonylphenol ethoxylate in liquid form; foam depressor. *Berol Kemi (UK) Ltd.* Name unverified.

4022 Berol 260
68439-46-3
C9-11 pareth-4
C9-11 pareth-4; surfactant for alkaline systems, hard surface cleaners, industrial, institutional and vehicle cleaners. d = 0.931; pH = 6-8; soluble in EtOH, propylene glycol. *Berol Nobel AB.*

4023 Berol 267
Nonylphenol ethoxylate
Nonionic surfactant of the nonylphenol ethoxylate type, in liquid form; liquid cleaners; wetting agents; emulsifiers. d = 1.04; soluble in ethanol, xylene, trichloroethylene; flashpoint >100°. *Berol Kemi (UK) Ltd.* Name unverified.

4024 Berol 272 and 716
68891-21-4
Dinonylphenol ethoxylate
Dinonylphenol ethoxylate in liquid or paste form; nonionic low foaming detergents. d = 1.04; soluble in H_2O, ethanol, xylene, trichloroethylene; pH=6.0-7.0. *Berol Kemi (UK) Ltd.* Name unverified.

4025 Berol 278, 281, 282, 291 and 292
68412-54-4
Nonylphenol ethoxylate
Nonionic surfactants of the nonylphenol ethoxylate type in liquid or wax form; used for emulsion polymerization. d = 1.09; soluble in H_2O, ethanol. *Berol Kemi (UK) Ltd.* Name unverified.

4026 Berol 302
13127-82-7 236-062-2
PEG-2 oleamine
Emulsifier, dispersant, wetting agent, antistat, anticorrosive for agriculture, leather, textiles, metalworking and plastic industries. d = 0.904; soluble in EtOH. *Berol Nobel AB.*

4027 Berol 307
61791-14-8; 61791-31-9 263-163-9

PEG-2 cocamine
Emulsifier, wetting agent, antistat, anticorrosive for agriculture, leather, textiles, metalworking and plastic industries. d = 0.910; Soluble in EtOH, iPrOH, low aromatic solvents, propylene glycol, xylene. *Berol Nobel AB.*

4028 Berol 381
61791-26-2
PEG-15 tallowamine
Emulsifier, wetting agent, antistat, anticorrosive for agriculture, leather, textiles, metalworking and plastic industries. d = 1.03; soluble in H_2O, EtOH, iPrOH, propylene glycol, xylene; pH=9; Flashpoint >150°. *Berol Nobel AB.*

4029 Berol 386
61791-26-2
PEG-20 tallowamine
Emulsifier, wetting agent, antistat, anticorrosive for agriculture, leather, textiles, metalworking and plastics industries. d = 1.00; soluble in H_2O, EtOH, iPrOH, propylene glycol, xylene; pH=9. *Berol Nobel AB.*

4030 Berol 391
61791-26-2
PEG-5 tallowamine
Emulsifier, wetting agent, antistat, anticorrosive for agriculture, leather, textiles, metalworking and plastics industries. d = 1.04; soluble in H_2O, EtOH, iPrOH, propylene glycol, xylene; pH=9-11. *Berol Nobel AB.*

4031 Berol 397
61791-14-8
PEG-15 cocamine
Emulsifier, wetting agent, antistat, anticorrosive for agriculture, leather, textiles, metalworking and plastic industries. d = 1.043; soluble in H_2O, EtOH, propylene glycol, xylene; corrosive. *Berol Nobel AB.*

4032 Berol 455
61791-44-4 263-177-5
PEG-3 tallow diamine
Emulsifier, wetting agent, antistat, anticorrosive for agriculture, leather, textiles, metalworking and plastic industries. *Berol Nobel AB.*

4033 Berol 456
61791-44-4 263-177-5
PEG-2 tallowamine
Emulsifier, wetting agent, antistat, anticorrosive for agriculture, leather, textiles, metalworking and plastic industries. d = 1.043; soluble in H_2O, EtOH, propylene glycol, xylene; pH=9. *Berol Nobel AB.*

4034 Berol 457
61791-26-2
PEG-5 tallowamine
Emulsifier, wetting agent, antistat, anticorrosive for agriculture, leather, textiles, metalworking and plastic industries. d = 1.03; soluble in H_2O, EtOH, iPrOH, propylene glycol, xylene; pH=10. *Berol Nobel AB.*

4035 Berol 474
68891-38-3
Sodium lauryl sulfate in paste form; anionic surfactant used in emulsion polymerization. d = 1.040; soluble in H_2O (25 g/100 ml), propylene glycol; pH=6.6-8. *Berol Kemi (UK) Ltd.* Name unverified.

4036 Berol 475
Sodium alkyl ether sulfate in liquid or paste form; anionic surfactants used in shampoos and bath preparations and the textile industry. *Berol Kemi (UK) Ltd.* Name unverified.

4037 Berol 484
Monoethanolamine lauryl sulfate in liquid form; anionic surfactant used in shampoos and bath preparations. *Berol Kemi (UK) Ltd.* Name unverified.

4038 Berol 490
Anionic surfactant in liquid form; for emulsion polymerization. *Berol Kemi (UK) Ltd.* Name unverified.

4039 Berol 496
Anionic surfactant in the form of a paste; for powder and liquid detergents. *Berol Kemi (UK) Ltd.* Name unverified.

4040 Berol 513 and 525
Anionic surfactant in acid form, in which the anion is a phosphate ester, supplied as a paste; detergent auxiliary. *Berol Kemi (UK) Ltd.* Name unverified.

4041 Berol 518
Anionic surfactant in acid form in which the anion is a phosphate ester, supplied as a wax; foam regulator. *Berol Kemi (UK) Ltd.* Name unverified.

4042 Berol 521
Potassium alkyl phosphate ester
Corrosion inhibitor, solubilizer of nonionic surfactants in presence of high electrolyte concentrate; for liquid alkaline hard surface cleaners; biodegradable. d = 1.145; soluble in H_2O, propylene glycol; pH=7-8. *Berol Nobel AB.*

4043 Berol 563
Alkyl polyglycol ether ammonium methyl sulfate
Detergent, alkaline degreasing and cleaning agent; hydrotrope for aqueous alkaline cleaners for hard surface cleaning and for acid cleaners. d = 1.110; soluble in H_2O, ethanol, propylene glycol, trichloroethylene; flashpoint >100°; pH=7-9. *Berol Nobel AB; Berol Kemi (UK) Ltd.*

4044 Berol 594
Hydroxy-ethyl 2 alkylimidazoline in liquid form; cationic surfactant used in water displacing acid cleaners in industrial cleaning; corrosion inhibition. *Berol Kemi (UK) Ltd.* Name unverified.

4045 Berol 733
Potassium phosphate ester in liquid form; hydrotrope for nonionics. *Berol Kemi (UK) Ltd.* Name unverified.

4046 Berol 784
Alkylaryl sulfonate/fatty alcohol ethoxylate blend; high foaming surfactant blend for neutral cleaning products. *Berol Nobel AB.*

4047 Berol 822
Calcium alkylaryl sulfonate
Anionic surfactant in liquid form. d = 1.03; soluble in EtOH, trichloroethylene, paraffin oil, xylene, white spirt, fuel oil; flashpoint = 35°; pH=5-8. *Berol Kemi (UK) Ltd.* Name unverified.

4048 Berol 829
61791-12-6
PEG-20 castor oil
Surfactant, emulsifier for chemical industry. d = 1.03; soluble in EtOH, iPrOH, xylene. *Berol Nobel AB.*

4049 Berol WASC
68412-54-4
Nonylphenol ethoxylate
Biodegradable household and industrial detergent, textile washing agent, dispersant for paints, varnishes. d = 1.06; soluble in H_2O, EtOH; LD_{50} (rat orl) >2 g/kg *Berol Nobel AB.*

4050 Berpak
A two-component cold curing polyurethane encapsulant; sold in kits containing molds and electrical accessories; used for jointing of underground mains electrical cable and telecommunications cables. *Berger Elastomers.*

4051 Bersch bearing metal
An alloy of 93% aluminum and 7% nickel.

4052 Berthier's alloy
A copper-nickel alloy containing approximately 32% nickel.

4053 Berylla
1304-56-9 1216 215-133-1
BeO
Beryllium oxide
Used in manufacture of beryllium oxide ceramics, glass in nuclear reactor fuels and moderators; electrically resistive; catalyst for organic reactions mp = 2530°; very sparingly soluble in H_2O.

4054 Beryllium
7440-41-7 1201 231-150-7
Be
Metallic element; Structural material in space technology; moderator in nuclear reactors; source of neutrons; windows for x-ray tubes; in gyroscopes, computer parts, inertial guidance systems; additive in solid-propellant rocket fuels; beryllium-copper alloys. mp = 1287°; bp = 2500°; d = 1.8477. *Atomergic Chemetals; Degussa; Noah Chem.*

4055 beryllium bronze
An alloy of 97.5% copper with 2.5% beryllium.

4056 Besconus
Textile processing aid. *Crosfield Chemicals Ltd.* Discontinued.

4057 Besiege®
For agriculture industry. *DuPont UK.*

4058 Beta Lite® 3503
Hydrocarbon resin; resin for paste ink applications. *Arizona.* Discontinued.

4059 Beta Plus
Partially hydrogenated soybean oil with mono and diglycerides, sodium stearoyl lactylate, ethoxylated mono and diglycerides, TBHQ; kosher; high performance fat for continuous and conventional breads and other yeast-raised goods. *Van Den Bergh Foods.*

4060 betamethasone 21 phosphate
151-73-5 1226 205-797-0
$C_{22}H_{28}FNa_2O_8P$
betamethasone 21 phosphate disodium salt
Betsolan; Bentalan; Betnersol Injectable; Durabetason; Vista-Methasone,. betamethasone 21 phosphate, veterinary preparations. *Glaxo Laboratories.* Name unverified.

4061 betanal E
13684-63-4 7384 237-199-0
phenmedipham
Emulsifiable concentrate containing 114 g/l phenmedipham; used for weed control for beet crops. *ICI Agrochemicals.*

4062 betanal Tandem
26225-79-6, 13684-63-4 237-199-0
ethofumesate-phenmedipham. Emulsifiable concentrate of 100 g ethofumesate and 80 g phenmedipham per liter; used for weed control in beet crops. *Schering Agrochemicals Ltd.* Discontinued.

4063 betapal Concentrate
120-23-0 6484 204-380-0
(2-naphthyloxy) acetic acid
Soluble concentrate containing 16 g/l (2-naphthyloxy)acetic acid; plant growth regulator. *Synchemicals Ltd.*

4064 betaprene® 253
Hydrocarbon resin; resin for paste ink applications; tackifier for adhesives. *Arizona.* Discontinued.

4065 betaprone
57-57-8 8005 200-340-1
$C_3H_4O_2$
Propiolactone
β-Propiolactone; betaprone; 3-hydroxypropionic acid lactone; propanilide; propiolactone; 3-hydroxypropionic acid β-lactone; β-propionolactone; oxetan-2-one; Propanolide; Oxetanone; Propionolactone; NSC-21626. Chemical intermediate; used as a disinfectant. mp = -34.4°; bp = 162° (dec); d_4^{20} = 1.1460; n_D^{20} = 1.4131; soluble in H_2O (35 g/100 ml); LC_{50} (rat ihl) = 25 ppm/6H. *O'Neal, Jones & Feldman Pharmaceuticals.* Name unverified.

4066 betaseal® 43518
Solvent-based silane blend type primer; conditions the surface and promotes adhesion to glass, e.g., windshields of automobiles, trucks, buses, off-road vehicles. *Essex Specialty Prods.*

4067 betaseal® 43520A
Polyurethane-based solvent release type primer; screens out uv rays and promotes adhesion between polyurethane adhesive and glass; for windshield installation and backlite bonding in trucks, buses, off-road vehicles. *Essex Specialty Prods.*

4068 betaseal® 43555
Primer promoting adhesion between betaseal® adhesives and PVC substrates; for bonding automotive vinyl trim to glass and various painted substrates. *Essex Specialty Prods.*

4069 betaseal® 58702
One-component polyurethane adhesive; fast moisture curing adhesive for stationary glass bonding when used with appropriate glass primers. *Essex Specialty Prods.*

4070 betasol Ot-A
A sulfonated ester; A proprietary trade name for a wetting agent. No manufacturer.

4071 betathane
Solid polyurethane elastomers. *Hallam Polymer Engineering Ltd.* Name unverified.

4072 betazole hydrochloride
138-92-1 1230 205-345-2
$C_5H_9N_3 \cdot 2HCl$
1H-pyrazole-3-ethanamine dihydrochloride
Histimin; Histalog. A stimulant of gastric secretion, used as a diagnostic aid. mp = 224-226°; soluble in H_2O, insoluble in EtOH.

4073 Betricing
Partially hydrogenated vegetable oil (soybean, cottonseed), mono and diglcyerides, <0.9% polysorbate 60; kosher; multipurpose shortening for icings, fillings, yeast-raised goods. *Van Den Bergh Foods.*

4074 Betrkake
Partially hydrogenated vegetable oil (soybean, cottonseed), mono and diglcyerides; kosher; high quality emulsified shortening for cakes, icings, and sweet doughs. *Van Den Bergh Foods.*

4075 Betrox
7647-14-5 8742 231-598-3
ClNa
Sodium chloride
Sodium chloride fertilizer. *ICI Chem & Polymers Ltd.*

4076 Better Flowable
1698-60-8 216-920-2
$C_{10}H_8ClN_3O$
chloridazon
Suspension concentrate containing 430 g chloridazon per liter; a pyridazinone herbicide for beet crops. *Sipcam UK Ltd.*

4077 Beutene
Butyraldehyde-aniline
Butyraldehyde-aniline condensation product; fast-curing accelerator most active above 250°F (121°C); moderately nonscorchy at processing temperature; compatible with channel and furnace blacks, reclaimed rubber and acidic softeners; hard clays retard its activity; fast cures in hard rubber, wire insulation and mechanical goods. *Uniroyal.* Name unverified.

4078 Bevacid
Tall oil fatty acids and distilled tall oil; used in surface coatings, soaps, oils and synthetic resins. *Bergvik Sales Ltd.*

4079 Bevaloid
Emulsifier for natural oils; dispersant; foam control agent; used in leather industry; nonaqueous systems. *Rhône-Poulenc UK; Bevaloid Ltd.*

4080 Bevaloid 35 and 36
Anionic surfactant in powder form; Bevaloid 36 is a higher molecular weight version of Bevaloid 35; dispersant for dyestuffs and pigments, and in concrete and resin technology. *Bevaloid Ltd.* Name unverified.

4081 Bevaloid 211
Sodium polycarboxylate
Low molecular weight sodium polycarboxylate, in liquid form. A long term stable dispersant for paint. *Bevaloid Ltd.* Name unverified.

4082 Bevaloid 1299
577-11-7 3460 209-406-4
docusate sodium
Sodium dioctyl sulfosuccinate in liquid form; powerful wetting agent used in textiles and detergent dry cleaning. *Bevaloid Ltd.*

4083 Bevaloid 6423
127-39-9 3238 204-839-5
Diisobutyl Sodium Sulfosuccinate
Sodium di-isobutyl sulfosuccinate in flake form; anionic surfactant having good stability to electrolytes; used in emulsion polymerization and electroplating. *Bevaloid Ltd.*

4084 Bevaloid 6522
Cationic surfactant in liquid form; softener for acrylic fibers. *Bevaloid Ltd.* Name unverified.

4085 Bevaloid 6703
Low molecular weight sodium polycarboxylate in liquid form; dispersant for calcium carbonate, china clay. *Bevaloid Ltd.* Name unverified.

4086 Bevaloid 6744
Sodium polycarboxylate of low molecular weight, in liquid form; dispersant for drilling muds. *Bevaloid Ltd.* Name unverified.

4087 Bevaloid DA 6805
Cationic surfactant in liquid form; leveling agent used in dyeing polyamide with acid dyes. *Bevaloid Ltd.* Name unverified.

4088 Beviros
Tall oil resins; used for surface coatings, synthetic resins and binders. *Bergvik Sales Ltd.*

4089 Bevitack Resins
Esters and derivatives of distilled tall oil resin; components of adhesives and coatings. *Bergvik Sales Ltd.*

4090 Bewoid
Modified resin emulsion. *Tenneco Malros Ltd.* Name unverified.

4091 Bewopac
Modified rosin emulsion. *Tenneco Malros Ltd.* Name unverified.

4092 Bex
Bakery compound. *Albright & Wilson Ltd., Phosphates & Speciality Business.*

4093 Bexfilm
Acetate films, polyester films. *ICI Chem & Polymers Ltd.* Discontinued.

4094 Bexloy®
Automotive engineering resins. *DuPont UK.*

4095 Bexphane
Polypropylene film. *Hercules Ltd.* Discontinued.

4096 Bexton
1918-16-7 7977 217-638-2
Propachlor
Herbicide used for pre-emergence grass and broadleaf weed control in corn and grain sorghum. *Dow UK.* Discontinued.

4097 Beycopon
Specialty blends of anionic surfactants; emulsifiers, industrial detergents, textiles. Anionic. *Elf Atochem UK/Ceca.*

4098 Beycostat
Fatty alcohol phosphate esters; used for detergents, antifoams, emulsifiers, wetting agents for agrochemicals. *Elf Atochem UK/Ceca.*

4099 Beycostat 148 K
7778-77-0 7829 231-913-4
Potassium phosphate, Monobasic
Surfactant. *Ceca SA.*

4100 Beycostat 231
Phosphate ester
A potassium phosphate derivative; antistat, wetting agent. Anionic. *Ceca SA.*

4101 Beycostat 256A
C_8 fatty alcohol phosphate ester

Surfactant, release agent, corrosion inhibitor, antifoam, intermediate. *Ceca SA.*

4102 Beycostat 273 P
7778-77-0 7829 231-913-4
potassium phosphate, monobasic
Surfactant, wetting agent, detergent. *Ceca SA.*

4103 Beycostat 319 A
PEG-6C13 fatty alcohol phosphate ester
Antistat, degreaser. *Ceca SA.*

4104 Beycostat 656 A
PEG-8 alkylphenol phosphate ester
Emulsifier for emulsion polymerization. *Ceca SA.*

4105 Beycostat B 070 A
Phosphate ester
For acid detergent formulations. *Ceca SA.*

4106 Beycostat B 706 A
C_{13} fatty alcohol phosphate ester
Corrosion inhibitor, EP additive; intermediate. Liquid; insoluble in H_2O; pH 2; surface tension = 36 dynes/cm; anionic. *Ceca SA.*

4107 Beycostat B 706 E
Phosphate ester
TEA salt; antistat for textile lubricants. *Ceca SA.*

4108 Beycostat LP 4 A
PEG-4 C12 fatty alcohol phosphate ester
Antistat, degreaser; additive for acid or basic detergents. Liquid; insoluble in H_2O; pH 2; surface tension = 30 dynes/cm; anionic/nonionic. *Ceca SA.*

4109 Beycostat NE
Alkyl ether sulfate and solvent; antistat for PVC, wetting agent. Liquid; pH 7.5; anionic. *Ceca SA.*

4110 Beycostat QA
PEG-4 alkylphenol phosphate ester
Emulsifier, detergent base. Liquid; disperses in H_2O; pH 2; surface tension = 29.5 dynes/cm; anionic/nonionic. *Ceca SA.*

4111 BFP 64K
Mono/diglyceride from hydrogenated soybean oil and glycerin, TBHQ, citric acid; food ingredient, emulsifier, crumb softener; for bread, sweet goods, bakery mixes, shortening, margarine. mp = 48-50°; iodine no. 65-75. *Am. Ingredients/Patco.*

4112 BFP 65
Lard glycerides-TBHQ-citric acid. Mono- and diglycerides from hydrogenated edible fats and glycerin with TBHQ, citric acid; food additive, emulsifier for baked products, mixes, shortenings, icings. HLB 3.2; nonionic; discontinued. *Am. Ingredients/Patco.*

4113 BFP 74
Lard glycerides
Citric acid. Mono- and diglycerides from animal lipid source; food additive, emulsifier for coffee whiteners, whipped toppings, snack food, chewing gum, margarine. Bead, flake; mp = 60-63°; 42% minimum α monoglyceride. *Am. Ingredients/Patco.*

4114 BFP 75
Lard glycerides-citric acid. Mono- and diglycerides from hydrogenated soybean oil and glycerin with citric acid; food additive, emulsifier for coffee whiteners, whipped toppings, snack food, chewing gum, margarine, jelly, frozen dessert, sour cream dips, caramel, nougats. White-cream flake or bead; mp = 140-145°F; HLB 3.2; nonionic. *Am. Ingredients/Patco.*

4115 BH 2,4-D Ester 40
1702-17-6 2462 216-935-4
Clopiralid
Clopyralid formulated with 2,4-D; translocated herbicide for cereals and established grassland. *Rhône-Poulenc Environmental Prods. Ltd.*

4116 BH CMPP/2,4-D
2,4-D-mecoprop. Soluble concentrate containing 116g 2,4-D and 250g mecoprop per liter; used to control weeds in grassland. *Rhône-Poulenc/Agri.*

4117 BH Dalapon
75-99-0 2869 200-923-0
sodium 2,2-dichloropropionate
Water soluble powder containing 85% dalapon; a translocated herbicide. *Rhône-Poulenc Environmental Prods. Ltd.*

4118 BH Dockmaster
Soluble concentrate of 125g dicamba and 125g maleic hydrazide per liter; used to control docks in road verges and noncrop grass. *Rhône-Poulenc Environmental Prods. Ltd.*

4119 BH MCPA 75
94-74-6 5803 202-360-6
MCPA. Herbicide for cereals and grassland. *Rhône-Poulenc Environmental Prods. Ltd.*

4120 BH Prefix D
1194-65-6 3093 214-787-5
Dichlobenil
Granular herbicide containing dichlobenil; used to control weeds in woody crops and noncrop areas. *Rhône-Poulenc Environmental Prods. Ltd.*

4121 BHA
25013-16-5 1582 246-563-8
$C_{11}H_{16}O_2$
butylated hydroxyanisole.
(1,1-dimethylethyl)-4-methoxyphenol; 3-*tert*-butyl-4-hydroxyanisole. Mixture of isomers of tertiary butyl-substituted 4-methoxyphenols; antioxidant for foods, etc. *Eastman; Penta Mfg.; UOP.* Name unverified.

4122 BHT
128-37-0 1583 204-881-4
$C_{15}H_{24}O$
DBPC
butylated hydroxytoluene; 2,6-di-*t*-butyl-*p*-cresol; 2,6-bis (1,1-dimethylethyl)-4-methylphenol. Substituted toluene; antioxidant for foods, animal feed, petrol. products, synthetic rubbers, plastics, soaps; antiskinning agent in paints and inks. *Penta Mfg.; PMC Specialties; Raschig; Uniroyal.*

4123 BI Ammonium Phosphate
Used for other chemicals, water treatment, fire fighting, cosmetic products, food industry, pharmaceutical products. *Rhône-Poulenc NV/CdF Chimie AZF.* Name unverified.

4124 Biactol
Antibacterial face wash to help clear acne blemishes. *Richardson-Vicks Inc.* Name unverified.

4125 Biakmetals
A group of alloys some of which are zinc-copper alloys with small amounts of nickel or manganese or both, and others are aluminum-zinc-copper alloys, or zinc-aluminum alloys.

4126 Biasbeston
A synthetic varnish.

4127 Bicor®
Oriented and nonoriented polypropylene films; for packaging of food and nonfood products and special industrial applications. *Mobil Plastics Europe.* Name unverified.

4128 Bicor® 70 PXS
Oriented PP film, 1 side PVDC coated; used as the inside sealant web in laminations; superior seal strength, wide sealing range, excellent hot tack; lap seals to acrylic coated OPP or to sealable PVDC coated films; aroma barrier and moderate oxygen barrier. *Mobil.* Name unverified.

4129 Bicor® 220 AB, 250 AB, 310 AB, 380 AB, 420 HS
Oriented PP film, 2 side acrylic coated; excellent gloss, aroma barrier, and high speed machinability; designed for printing with solvent- or water-based inks; excellent film for replacing cellophane where oxygen barrier is not required. *Mobil.* Name unverified.

4130 Bicor® 240 B, 306 B, 420 B, 470 B
9003-07-0 7741
PP homopolymer film, unmodified; base sheet designed for lamination where slip is not required; one side treated; nonsealable. *Mobil.* Name unverified.

4131 Bicor® 318 ASB, 252 ASB
Oriented PP film, 1 side PVDC coated, 1 side acrylic coated; film for horizontal, vertical, and overwrap applications; ideal for replacing light gauge cellophane; excellent machinability, gloss, aroma barrier, printability, hot tack, moderate oxygen barrier. *Mobil.* Name unverified.

4132 Bidisin
101-10-0 202-915-2
Cloprop
Used for control of wild oats in cereals, maize and sugar beet. *Bayer AG.*

4133 Bielzite
A resin-asphalt containing sulfur and nitrogen.

4134 Bife
A proprietary mixture of thiamine hydrochloride and ferrous sulfate; used in the treatment of iron deficiency. *Jenkins Laboratories Inc.* Unverified.

4135 bifenox
42576-02-3 1256 255-894-7
$C_{14}H_9Cl_2NO_5$
5-(2,4-dichlorophenoxy)-2-nitrobenzoic acid methyl ester
MC-4379; Modown; Modown 4 Flowable. Selective herbicide used for control of annual broad-leaved weeds and some grasses in cereals, maize, sorghum, soybeans, rice and some other crops. mp = 84-86°; poorly soluble in H_2O (0.43 mg/l), more soluble in organic solvents; LD_{50} (rat orl) > 6400 mg/kg.

4136 Biju®
7787-59-9 1303 232-122-7
Bismuth oxychloride
Colorant and pearlescent for frosted cosmetics; pearl nail enamel because of brilliance and smoothness. *Mearl.* No manufacturer.

4137 BIK
Surface treated urea for easy dispersion in elastomers; activator for thiazole, thiuram, and dithiocarbamate accelerators; odor reducer when used with nitrosoamine type blowing agents. *Uniroyal.* Name unverified.

4138 Bikini Cream
A proprietary suntan cream. *Ayrton Saunders plc.* Unverified.

4139 Bikorit
1344-28-1 369 215-691-6
Al_2O_3
aluminum oxide
Corundum; Alumina. White special fused corundum (crystalline aluminum oxide); used for production of ramming mixes, shape bricks and crucibles for the lining of high temperature furnaces; molding material for precision casting molds. *Hüls UK Ltd.*

4140 Bilevon® -Solution
Veterinary medicine preparation for use against liver flukes (*Fasciola hepactica* and *F. Gigantica*), for subcutaneous injection in cattle. *Bayer AG.*

4141 Bilgen bronze
An alloy of 97% copper, 1.9% tin, 0.52% iron, and 0.24% lead.

4142 Bilimiron
41473-08-9 5077
Iopronic acid
Radiopaque medium used as a diagnostic aid. *Bracco Industria Chimica SpA.* Name unverified.

4143 Bi-Lite®
Bismuth oxychloride/mica; pearlescent pigments for cosmetic eye, face, lip, and body make-up. *Van Dyk.*

4144 Bilivist
1221-56-3 5087 214-945-3
Ipodate sodium
Diagnostic aid. Radiopaque medium used for diagnostic work. *Berlex Laboratories Inc.* Name unverified.

4145 Bilopaque
7246-21-1 9968 230-653-9
Tyropanoate sodium
Diagnostic aid, radiopaque medium, cholecystographic. *Sterling Drug Inc.* Unverified.

4146 Bilston
A basic slag used for fertilizing purposes. *Fisons plc, Horticultural Div.* Name unverified.

4147 Bilt-Cote®
1332-58-7 5294 296-473-8
Kaolin
Kaolinite; China clay; Bolus alba; Porcelain clay; Aluminum silicate hydroxide; Kaopectate; Aluminum silicate (hydrated); Aluminum silicate dihydrate. Kaolin clay; used for agricultural prill conditioning. *R. T. Vanderbilt Co Inc.*

4148 Bilt-Plates®
1332-58-7 5294 296-473-8
Kaolin
Kaolin clay; used as filler when high brightness is not a prerequisite; for paints, cosmetics and pharmaceuticals, pitch control for paper processing. *R. T. Vanderbilt Co Inc.*

4149 Bilt-Rex®
Carboxylated polyethylene resin; used for coatings for paper or paperboard. *R. T. Vanderbilt Co Inc.*

4150 Binab T
Trichoderma viride
Biological fungicide used to control silver leaf in fruit trees. *Henry Doubleday Research Association.*

4151 Bind
Isocyanate-based binders. *Dow UK.* Discontinued.

4152 Bi-nell®
Fibers. *DuPont UK.*

4153 Bio Terge®
α olefin sulfonate
Detergent, shampoo, bubble bath. *Stepan.* Name unverified.

4154 Bioacid
A range of alkylaryl sulfonic acids; detergent for use in the manufacture of liquid and powdered cleaning compounds. *Harcros Australia.*

4155 Bio-add
Fish/meat silage additive. *BP Chemicals Ltd.*

4156 Bioban® BNPD-40
52-51-7 1470 200-143-0
2-Bromo-2-nitropropane-1,3-diol
Broad spectrum antimicrobial, preservative for metalworking fluids. *Angus.*

4157 Bioban® CS-1135
4,4-Dimethyloxazoline (74.7%), 3,4,4-trimethyloxazolidine (2.5%) preservative; antibacterial agent for water-based paints, latexes, emulsions, metalworking fluids; for oilfield water-flooding operations; aids corrosion protection. Liquid; soluble in H_2O, polar and nonpolar solvents; d = 0.98-0.99; 8.2 lb/gal; viscosity = 7.5 cp; flash pt. 120°; pH 10.5-11.5. *Angus.*

4158 Bioban® GK
$C_6H_9N_3O_3$
Hydroxyethyl-s-triazine
Hexahydro-1,3,5-tris(2-hydroxyethyl)-s-triazine; 1,3,5-Triazine-1,3,5-(2H,4H,6H)-triethanol; *s*-Triazine-1,3,5(2H,4H,6H)-triethanol; Busan®1506; Canguard® 454; Nipacide® BK; Triadine® 3; Grotan BK; Grotan B; Kalpur TE; Onyxide 200; Grotan; Miliden X-2;. Hexahydro-1,3,5-tris(hydroxyethyl) triazine is the active component in Gotan BK, a bactericide that releases formaldehyde. A biocide for metalworking fluids, latex paints, emulsions, caulks, and adhesives. Soluble in H_2O, DMSO, EtOH; LD50 (rat orl) = 763 mg/kg; mutagenic. *Angus.*

4159 Bioban® N-95
5-Hydroxymethoxymethyl-1-aza-3,7-dioxabicylco[3.3.0]octane (24.5%), 5-hydroxymethyl-1-aza-3,7-dioxabicyclo[3.3.0]octane (17.7%) and 5-hydroxypoly[methyleneoxyl]methyl-1-aza-3,7-dioxabicyclo[3.3.0] octane (7.8%). Preservative for aqueous metalworking fluids. *Angus.*

4160 Bioban® P-1487®
2224-44-4; 1854-23-5 217-450-0
4-(2-Nitrobutyl) morpholine and 4,4-(2-ethyl-2-nitrotrimethylene) dimorpholine. Preservative, antibacterial and antifungal agent used in metalworking fluids, lubricants and cutting oils. Liquid; oil-soluble; moderately soluble in H_2O. *Angus.*

4161 Biobor
Diesel fuel fungicide *U.S. Borax & Chem.*

4162 BioCare® Polymer HA-24
Polyquaternium-24-hyaluronic acid. Emollient, humectant, conditioner, softener, moisturizer, lubricant for hair and skin; substantive to protein substrates. Opalescent liquid. *Amerchol Corp.*

4163 BioCare® SA
Albumen, hyaluronic acid and dextran sulfate; polymer providing hydration and revitalization to surface skin; lifts wrinkles; substantive to protein substrates; for eye gels, facial treatments, skin toners, makeup, moisturizers, after-sun products cleansing lotions. *Amerchol Corp.*

4164 Biocide
Animal feed preservative. *BP Chemicals Ltd.*

4165 Biocyde
Industrial biocides. *Imperial Chemical Industries plc.*

4166 Biodynes® TRF Ultra-5
Live yeast cell derivative; moisturizer for skin; minimizes dryness and pain of sunburn and wind chapped skin. Odorless liquid. *Brooks Industries.*

4167 Bio-Feed
A range of nonstarch polysaccharide degrading and nutrient hydrolyzing enzyme preparations produced by submerged fermentation; added to animal feeds in order to boost feed digestability and improve feed utilisation. *Novo Nordisk.*

4168 Biomart
Bactericide for use in diesel fuel. *Crowley Chem.*

4169 Biomate
Microbiological growth control agents; biocide for cooling water. *Grace Dearborn Ltd.*

4170 bioMeT 14
10% Diphenylstibine 2-ethyl hexoate solution in dioctylphthalate. Antimicrobial compound for protection of PVC systems. Pale yellow clear liquid; odorless; d = 1.0346; 8.634 lb/gal; viscosity = 66.7 cs; flash pt. 200°. *Elf Atochem.*

4171 bioMeT TBTF
1983-10-4 217-847-9
$C_{12}H_{27}FSn$
Tributyltin fluoride
tributylfluorostannane; Polyflo; KL-990; Amercoat 635; Pro-Line 1077; Sea Hawk Biotin; Vin Clad Super Vinge; Fluorotri-n-butyltin. Antifoulant for marine paints. *Elf Atochem.*

4172 bioMeT TBTO
56-35-9 200-268-0
$C_{24}H_{54}OSn_2$
Bis(tributyltin)oxide
Antimicrobial for paper mill slime control, industrial cooling water, secondary oil recovery, hospital use, textiles, plastics, urethane foam, paper preservation; antifoulant for shipbottom paints; wood preservative. *Elf Atochem.*

4173 Biomin® Cinque
Silicon, zinc, copper, iron and magnesium glyconucleopeptides; five essential minerals bound in a complex matrix of a low molecular weight

peptide/mineral/nucleotide/carbohydrate. Used as a nutrient. *Brooks Industries.*

4174 **Biomin® Marine**
Sea minerals yeast derivative; marine elements in substantive moisturizing form; for skin care cosmetics. *Brooks Industries.*

4175 **Biopal® NR-20**
11096-42-7
Nonoxynol-12 iodine
Biopal NR-20; Nonylphenolpolyethylene glycol-iodine complex . Iodophor used in formulating no rinse sanitizing solutions. *Rhône-Poulenc Surf.*

4176 **Biopal® NR-20W**
Nonylphenoxypoly(ethyleneoxy) ethanol iodine complex; concentrated iodophor suitable for formulating no rinse" sanitizing solutions. *Rhône-Poulenc Surf.* Name unverified.

4177 **Biopal® VRO-20**
Nonylphenoxypoly(ethyleneoxy) ethanol iodine complex; concentrated iodophor used for cleaning, sanitizing and disinfecting hospital, biological laboratory and dairy equipment, breweries, multiple-use eating and drinking utensils in bars, restaurants, etc. *Rhône-Poulenc Surf.* Name unverified.

4178 **Biophos 35**
Phosphoglycoproteins, adenosine triphosphate, magnesium and potassium glycoprotein cosmetic ingredients for moisturizing. *Brooks Industries.*

4179 **Bioplex RNA**
Propylene glycol, hydrolyzed RNA, hydrolyzed DNA; cosmetic ingredients for skin and hair care products. *Brooks Industries.*

4180 **Biopol**
Poly β-hydroxybutyrate (PHB), a biodegradable thermoplastic. *ICI Chem & Polymers Ltd.*

4181 **Bio-Pol® OE**
Sodium C_{8-16} isoalkyl succinyl lactoglobulin sulfonate
Oil-absorbing polymer designed to entrap surface oil of the skin; film-former enhancing skin feel; excellent pigment dispersant and color enhancer. *Brooks Industries.*

4182 **Biopol® TE**
Dermal tissular extract; cosmetic ingredient for moisturizing applications. *Brooks Industries.*

4183 **Bio-Pruf®**
Antimicrobials for plastic products for healthcare, sanitary maintenance, construction, transportation and agriculture markets; protects against cosmetic/hygienic and structural degradation. *Morton Int'l./Plastics Additives.*

4184 **Bioques**
A series of enzyme producing biological cultures used to degrade industrial and municipal organically fouled waste water; both dry and liquid compositions available. *Ques Industries.*

4185 **Bioques Q**
A liquid chemical formulation which can liquify and digest complex fats, oils, grease, cellulose, proteins and starch; for use in toilets, drains and grease traps. *Ques Industries.*

4186 **Bioques Z**
A blend of highly active, broad spectrum bioactive cultures that have the ability to digest and liquefy organic wastes; designed for use in all sewage systems, lagoons, sink drains and traps, and grease traps. *Ques Industries.*

4187 **Biorion 450 Super**
Iron supplement; for food industry. *Asahi Chem. Industry.*

4188 **Biosil**
Cobalt-chrome dental alloy. *Degussa Ltd.*

4189 **Biosint Supra**
Dental chrome investment. *Degussa Ltd.*

4190 **Biosoft C100**
Anionic surfactant as an ivory paste; for detergent paste or liquids. *KWR Chemicals Ltd.* Name unverified.

4191 **Biosoft D**
Anionic surfactant in slurry form; for liquid detergents. *KWR Chemicals Ltd.* Name unverified.

4192 **Biosoft N-300**
Anionic surfactant, in liquid form; for liquid detergents. *KWR Chemicals Ltd.* Name unverified.

4193 **Biosoft S and D-35X**
Anionic surfactant in liquid form; for liquid detergents. *KWR Chemicals Ltd.* Name unverified.

4194 **Biosoft S-100 and JN**
Anionic surfactant in acid form, supplied as a viscous liquid; intermediate in detergent preparations. *KWR Chemicals Ltd.* Name unverified.

4195 **Bio-Soft®**
Alkylbenzene sulfonic acid and alkylbenzene sulfonates; detergents. *Stepan.* Name unverified.

4196 **Bio-Soft® 9283**
Surfactant blend; all-purpose detergent base for industrial/household cleaners. *Stepan Europe.*

4197 **Bio-Soft® E-400**
68131-39-5
C_{12-15} pareth-3
Emulsifier, detergent, wetting and foam stabilizing. *Stepan Canada.*

4198 **Bio-Soft® EA-4**
Linear primary alkoxylate; surfactant, emulsifier, detergent. *Stepan; Stepan Canada.*

4199 **Bio-Soft® FF 400**
26183-52-8
Deceth-2
Surfactant. *Stepan Canada.*

4200 **Bio-Soft® LD-95**
Sodium dodecylbenzene sulfonate, sodium laureth sulfate, lauramide DEA, and urea base for dishwashing, carwashing, and other light duty hard surface detergents; biodegradable. *Stepan; Stepan Canada.*

4201 **Bio-Soft® MT 40**
68952-16-9
TEA-coco hydrolyzed animal protein; very mild surfactant, conditioner, foaming agent for medicated and conditioning shampoos, creams, baby products. *Stepan Europe.*

4202 **Bio-Soft® N-300**
TEA-dodecylbenzene sulfonate
Detergent, wetting and foaming agent for dishwash, carwash detergents, oil hair shampoos. *Stepan; Stepan Canada.*

4203 **Bio-Soft® S-100**
Linear dodecylbenzene sulfonic acid
Emulsifier, detergent intermediate for formulation of built detergents, dishwash, all-purpose cleaners, acid cleaners, degreasers, industrial cleaners; textile scouring, wetting, bleaching, dyeing assistant; high foamer when neutralized; biodegradable. *Stepan; Stepan Canada; Stepan Europe.*

4204 **Bio-Soft® TD400**
24938-91-8
Trideceth-3
Emulsifier. *Stepan Canada.*

4205 **Biosol**
Water soluble coolant used in glass manufacturing. *Specialty Products Co.* Name unverified.

4206 **Biosperse®**
Microbiocidal agents used to control microbiological growth; used for cooling towers, air washers, pasteurizers, cooling water systems, oil field water systems and metal working fluids. *Drew Ind. Div.*

4207 **Biosperse® 240**
10222-01-2 233-539-7
$C_3H_2Br_2N_2O$
2,2-Dibromo-3-nitrilopropionamide
2,2-Dibromo-2-cyanoacetamide; DBNPA; Dibromo-2-carbamoylacetonitrile; Dibromo-3-nitrilopropionamide; Slimicide 508. 2,2-Dibromo-3-nitrilopropionamide and solubilizing agents; antimicrobial for control of bacteria and algae in recirculating cooling water systems, air washer systems, evaporative condensers. *Drew Ind. Div.*

4208 **Biosperse® 250**
5-Chloro-2-methyl-4-isothiazolin-3-one and 2-methyl-4-isothiazolin-3-one. Broad-spectrum antimicrobial for control of bacteria, fungi, and algae in industrial recirculating cooling water systems, oil field aqueous systems; preservative in aqueous metalworking fluids. *Drew Ind. Div.*

4209 **Biostat A.1**
79-57-2 7111 201-212-8
Oxytetracycline
Oxytetracycline for fish preparation. *Pfizer International.*

4210 **Biosulphur Powder**
7704-34-9 9142 231-722-6
S
Sulfur
Brimstone sulfur. Micrograined active sulfur (96.5% S) with protective colloid; products for impure skin, oily hair and dandruff. mp = 115.21°; d = 1.96; insoluble in H_2O, sparingly soluble in alcohol and ether. *Dr. Kurt Richter; Henkel/Cospha.*

4211 **Bio-Surf I-20**
Nonylphenoxypoly (ethyleneoxy) ethanoliodine complex, iodophor concentrate; antimicrobial, germicide, disinfectant, sanitizer for cleaning, sanitizing, disinfecting in hospital, food and beverage plants, breweries, restaurants. *Lonza Inc.*

4212 **Bio-Terge® AS-40**
68439-57-6 270-407-8
Sodium C_{14}-C_{16} olefin sulfonate

Detergent, foaming agent for personal care, commercial and industrial formulations. *Stepan; Stepan Canada.*

4213 Bio-Terge® AS-40 and AS-90F

Sodium α-olefin sulfonate in liquid or flake form; liquid form used for dishwashing detergents and car washing; flake form used in powdered detergents. *KWR Chemicals Ltd.* Name unverified.

4214 Bio-Terge® PAS-8S

Sodium 1-octane sulfonate

Hydrotrope and detergent used in acid, alkaline, high electrolyte or bleach containing cleaners for industrial, institutional and household markets, e.g., acid cleaners, carpet steam cleaners, automatic dishwash; textile penetrant, dye dispersant; metalworking formulations. *Stepan; Stepan Canada; Stepan Europe.*

4215 Biothane System 228

Polyurethane system; for biomedical applications, e.g., potting/encapsulating compounds, adhesives, coatings, and sealants for artificial kidneys, blood oxygenators, blood filters, catheters, industrial filters. *CasChem.* Name unverified.

4216 Biphenyl

92-52-4 3372 202-163-5

$C_{12}H_{10}$

diphenyl

xenene; phenyl benzene; lemonene; phenador-x; PhPh. Organic synthesis; heat transfer agent; fungistat in packaging citrus fruit; plant disease control. mp = 69-72°; bp= 255°; d = 0.9920; insoluble in H_2O, soluble in organic solvents; LD_{50} (rat orl) = 2400 mg/kg. *Aldrich; Monsanto; Sybron.* Name unverified.

4217 Bi-play®

For agriculture industry. *DuPont UK.*

4218 Biquanide

56-03-1 1261 200-251-8

$C_2H_7N_4$

Imidodicarbonimidic diamide

Baquacil; Baquagold; Quanyl-quanidine; Amidinoquanidine; Diquanide. Polymeric biguanide swimming pool sanitizer. mp = 130°; dec = 142°; soluble in H_2O, alcohol, insoluble in organic solvents; *ICI Chem & Polymers Ltd.*

4219 Birgin®

Sprout suppressant for potatoes in storage. *Bayer AG.*

4220 Birlane

470-90-6 2137 207-432-0

chlorfenvinphos

Liquid seed dressing containing chlorfenvinphos for winter wheat. *ICI Chem & Polymers Ltd.* Discontinued.

4221 Birlane

470-90-6 2137 207-432-0

chlorfenvinphos

Liquid seed dressing containing 240g chlorfenvinphos per liter; used for winter wheat. *Shell UK.*

4222 Birmabright

A corrosion-resisting alloy containing 96% aluminum, 3.5% magnesium, and 0.5% manganese.

4223 Birmasil alloy

A special alloy which is a nickel-aluminum-silicon alloy containing up to 3.5% nickel and from 8.13% silicon. It has high tensile strength.

4224 Birmidium

A proprietary trade name for an alloy of aluminum with smaller amounts of copper, nickel, and magnesium; similar to Y-alloy. No manufacturer.

4225 Birmingham platina

A brass containing 47% copper, 53% zinc, and 0.25% iron.

4226 Birox®

Resistor compositions. *DuPont UK.*

4227 Bis(1-methylamyl) Sodium Sulfosuccinate

6001-97-4 1295 227-847-0

$C_{16}H_{29}NaO_7S$

Sulfobutanedioic acid 1,4-bis(1-methylpentyl) ester sodium salt

Sodium dihexyl sulfosuccinate; Dihexyl sodium sulfosuccinate; Aerosol MA; αsol MA; Lankropol® KMA. Emulsifier, wetting agent, especially in solutions of electrolytes; solubilizer for soaps, emulsion polymerization aid. Soluble in H_2O, pine oil, oleic acid, acetone, kerosene, carbon tetrachloride, ethanol, benzene, hot olive oil, glycerol; insoluble in liquid petrolatum. *Harcros.*

4228 Bis(2-ethylhexy) Phthalate

117-81-7 1291 204-211-0

$C_{24}H_{38}O_4$

Dioctyl phthalate

di(2-ethylhexyl phthalate; Bisoflex 81; Bisoflex 82; Octoil; DEHP; DOP; bis(2-Ethylhexyl)phthalate; Diethylhexyl phthalate; Dioctyl Phthalate; 1,2-Benzenedicarboxylic acid bis(2-ethylhexyl) ester; Octoil; Ethyl hexyl phthalate; octyl phthalate; phthalic acid dioctyl ester; BEHP; Bisoflex 81; Bisoflex DOP; compound 889; DAF 68; Ergoplast FDO; Eviplast 80; Eviplast 81; Fleximel; Flexol DOP; Flexol Plasticizer DOP; Good-rite GP 264; Hatcol DOP; Hercoflex 260; Kodaflex DOP; Mollan O; Nuoplaz DOP; Palatinol AH; Pittsburgh PX-138; Platinol AH; Platinol DOP; RC Plasticizer DOP; Reomol DOP; Reomol D 79P; Sicol 150; Staflex DOP; Truflex DOP; Vestinol AH; Vinicizer 80; Witcizer 312; Union Carbide Flexol 380,. A vinyl plasticizer. Used in vacuum pumps. mp = -50°; bp = 384°; d = 0.9810; n_D^{20} = 1.4853; poorly soluble in H_2O (0.1 g/l), more soluble in organic solvents; LD_{50} (rat orl) = 30.6 g/kg. *BP Chemicals Ltd.* Discontinued.

4229 Bis-2

2-Bis-acrylamide solution. *Am. Research Prods.* Unverified.

4230 Bisbutyl peroxy diisopropyl benzene

25155-25-3 246-678-3

$C_{20}H_{34}O_4$

[phenylenebis(1-methylethylidene)bis(1,1-dimethylethyl)]peroxide

Luperox 802; Perkadox® 14; Retilox® F 40 MG; Vul-Cup; Vul-Cup 40KE and R. Crosslinking agent for thermoplastic modification, curing elastomers, and high temperature cure of polyesters; initiator for vinyl polymerization. *Elf Atochem; Akzo; Hercules; Hercules Ltd.*

4231 bishydroxyethyl-N,N-Bis(2-hydroxyethyl) stearyl amine

10213-78-2 233-520-3

N,N-Bis (2-hydroxyethyl) stearyl amine

Chemstat® 273-E. Permanent internal antistat for plastic film and molded products; eliminates electrostatic problems in extrusion, injection and blow molding. *Chemax.* Name unverified.

4232 Bismate®

21260-46-8 244-299-8

$C_9H_{18}BiN_3S_6$

Bismuth dimethyldithiocarbamate

Bismuth, tris(dimethylcarbamodithioato-S,S')-, (OC-6-11)-. Ultra accelerator for NR, IR, BR, and SBR; for high temperature, high speed vulcanization. *R. T. Vanderbilt Co Inc.* Discontinued.

4233 Bismet

21260-46-8 244-299-8

$C_9H_{18}BiN_3S_6$

Bismuth dimethyldithiocarbamate

Bismuth, tris(dimethylcarbamodithioato-S,S')-, (OC-6-11)-. Accelerator for SBR, NR, IR, and BR compounds that are high-temperature cured; nonstaining. *Akrochem.*

4234 Bismica 46

12001-26-2, 7787-59-9 232-122-7

Mica-bismuth oxychloride. Pearlescent pigment. *Presperse.* Discontinued.

4235 Bismuth

7440-69-9 1297 231-177-4

Bi

Used in the synthesis of pharmaceuticals and medicinals, cosmetics, alloys, catalyst in making acrylonitrile, additive, coating selenium. mp = 271°; bp = 1420°; *Asarco; Atomergic Chemetals; Frys Metals Ltd; Noah Chem.*

4236 bismuth bronze

Alloys. a) Consists of 1 part bismuth with 16 parts tin; b) contains 1 part bismuth, 63 parts copper, 21 parts spelter, and 9 parts nickel; resists sea-water.

4237 Bismuth chloride oxide

12001-26-2, 7787-59-9 1303 232-122-7

BiClO

bismuth oxychloride

Bismica 46; Mica, Basic bismuth chloride; bismuthyl chloride; bismuthsubchloride; Pearl white; Chlorbismol. Pearlescent pigment. Used in face powders, as a pigment; manufactuer of artificial pearls; and dry cell cathodes. d = 7.72; melts at low red heat. *Presperse.*

4238 Bismuth dimethyldithiocarbamate

21260-46-8 244-299-8

$C_9H_{18}BiN_3S_6$

Bismuth, tris(dimethylcarbamodithioato-S,S')-, (OC-6-11)-

Bismate®; Bismet. Ultra accelerator for NR, IR, BR, and SBR; for high temperature, high speed vulcanization. *R. T. Vanderbilt Co Inc.*

4239 Bismuth nitrate

10361-46-3 233-792-3

$Bi(NO_3)_3$

Bismuth (III) nitrate, basic

bismuthyl nitrate; bismuthoxy nitrate; bismuth trinitrate. Preparation of other bismuth salts, bismuth luster on tin, luminous paints and enamels, precipitation of alkaloids. d = 2.83; soluble in H_2O containing nitric acid. *Atomergic Chemetals; Mallinckrodt; Nihon Kagaku Sangyo; Noah Chem.*

4240 Bismuth oxide

1304-76-3 1314 215-134-7

Bi_2O_3

bismuth trioxide

bismuth yellow; bismite. Enameling cast iron, ceramic, porcelain colors. Used in disinfectants, magnets, glass, rubber vulcanization, fireproofing papers and polymers. Practically insoluble in H_2O *Atomergic Chemetals; Ferro/Transelco; Mallinckrodt; Nihon Kagaku Sangyo.*

4241 Bismuth oxychloride
7787-59-9 1303 232-122-7
BiClO
Bismuth(III) oxychloride
Bisoxyl; Basic Bismuth Chloride;bismuthyl chloride; bismuth subchloride; Pearl white; blanc d'Espagne; blanc de perle; Chlorbismol. Bismuth oxychloride in liquid suspension; an antisyphilitic; in face powders; as pigment; used in the manufacture of artificial pearls, dry-cell cathodes. Fine powder or crystals; insoluble in H_2O; soluble in HCl, HNO_3. *British Drug Houses; Atomergic Chemetals; Mallinckrodt; Mearl; Van Dyk.* Name unverified.

4242 Bismuth subcarbonate
5892-10-4 1324 227-567-9
CBi_2O_5
bismuth oxycarbonate
bismuth carbonate basic. Bismuth compounds, cosmetics, opacifier in x-ray diagnosis, enamel fluxes, ceramic glazes. Practically insoluble in H_2O and alcohol. *Atomergic Chemetals; Mallinckrodt; Spectrum Chem. Mfg.*

4243 Bismuth subnitrate
1304-85-4 1326 215-136-8
$Bi_5(OH)_9(NO_3)_4O$
Bismuth hydroxide nitrate oxide
bismuth oxynitrate; bismuth nitrate, basic; Bismuthyl nitrate, Bismuth white; Magistry of bismuth; novismuth; paint white; Spanish white. Inorganic salt; Cosmetics, ceramic glazes, enamel fluxes. Odorless, tastless, slightly hydroscopic microcrystalline powder; practically insoluble in H_2O and alcohol. *Atomergic Chemetals; Celtic Chem. Ltd; R.W. Greeff; Mallinckrodt.*

4244 Bismuth subsalicylate
14882-18-9 1327 238-953-1
$Bi(C_7H_5O_3)_3Bi_2O_3$
(2-Hydroxybenzoato-O[1])-oxobismuth
bismuth salicylate basic; basic bismuth salicylate; salicylic acid basic bismuth salt; Bismogenol; Stabisol. Surface-coating plastics and copying paper. Used to impart pearly surface to cellulose-base, polystyrene and phenol-formaldehyde resins. Almost insoluble in H_2O or alcohol. *Atomergic Chemetals; R.W. Greeff; Mallinckrodt; Spectrum Chem. Mfg.*

4245 Bismuth trioxide
1304-76-3 1314 215-134-7
Bi_2O_3
bismuth oxide
bismuth yellow; bismite. Enameling cast iron, ceramic, porcelain colors. *Atomergic Chemetals; Ferro/Transelco; Mallinckrodt; Nihon Kagaku Sangyo.* Name unverified.

4246 Bisoflex
Plasticizers. *BP Chemicals Ltd.*

4247 Bisoflex 799
A trimellitate plasticizer for higher temperature resistant PVC for cable use. *BP Chemicals Ltd.* Discontinued.

4248 Bisoflex 79A
The adipate of mixed C_7-C_9 alcohols; a vinyl plasticizer. *BP Chemicals Ltd.*

4249 Bisoflex 81
117-81-7 1291 204-211-0
bis(2-ethylhexyl) phthalate
A general-purpose plasticizer. *BP Chemicals Ltd.*

4250 Bisoflex 810
C_8-C_{10} aliphatic phthalate
A plasticizer. *BP Chemicals Ltd.* Discontinued.

4251 Bisoflex 819
C_8-C_{10} trimellitates
Plasticizer. *BP Chemicals Ltd.* Discontinued.

4252 Bisoflex 82
117-81-7 1291 204-211-0
bis(2-ethylhexyl) phthalate
A vinyl plasticizer. *BP Chemicals Ltd.*

4253 Bisoflex 88
The phthalate of a C_8 alcohol; a vinyl plasticizer. *BP Chemicals Ltd.*

4254 Bisoflex 8N
Condensation product of 2-ethyl hexyl urethane and formaldehyde; a vinyl plasticizer. *BP Chemicals Ltd.* Discontinued.

4255 Bisoflex BP9
68515-49-1 271-091-4
Diisodecyl phthalate
A plasticizer. *BP Chemicals Ltd.*

4256 Bisoflex DEP
A trademark for polymeric plasticizers. *BP Chemicals Ltd.* Discontinued.

4257 Bisoflex DMP
Dioctyl adipate
A vinyl plasticizer. *BP Chemicals Ltd.*

4258 Bisoflex DNA
Dinonyl adipate
A vinyl plasticizer. *BP Chemicals Ltd.* Discontinued.

4259 Bisoflex DOP
A trademark for an adipic ester plasticizer. *BP Chemicals Ltd.*

4260 Bisoflex DUP
A primary low temperature plasticizer of PVC and synthetic rubber. *BP Chemicals Ltd.*

4261 Bisoflex L711P
ditridecyl phthalate
A trademark for ditridecyl phthalate; a plasticizer. *BP Chemicals Ltd.*

4262 Bisoflex L79
A higher straight chain phthalate plasticizer. *BP Chemicals Ltd.* Discontinued.

4263 Bisoflex L79P
C_6-C_{10} aliphatic alcohol phthalate
A plasticizer. *BP Chemicals Ltd.*

4264 Bisoflex L9
Dialkyl (C_7-C_9) phthalate
Plasticizer. *BP Chemicals Ltd.*

4265 Bisoflex L911
A higher straight chain phthalate plasticizer. *BP Chemicals Ltd.* Discontinued.

4266 Bisoflex L911P
C_6-C_{10} trimellitates
Plasticizer. *BP Chemicals Ltd.*

4267 Bisoflex ODN
A trade mark for an adipic ester plasticizer. *BP Chemicals Ltd.* Discontinued.

4268 Bisol
Solvents, plasticizers, intermediates. *BP Chemicals Ltd.*

4269 Bisolene
Proprietary liquid fuels. *BP Chemicals Ltd.* Discontinued.

4270 Bisolite
Solid fuels, e.g., metaldehyde. *BP Chemicals Ltd.* Discontinued.

4271 Bisolube
Oil additives. *BP Chemicals Ltd.* Discontinued.

4272 Bisomer
Specialty chemicals for use as intermediates in the production of various products in industry, particularly surface coatings, adhesives and sealants. *BP Chemicals Ltd.* Discontinued.

4273 Bisomer 2HEMA
868-77-9 212-782-2
$C_6H_{10}O_3$
2-Hydroxyethyl Methacrylate; Glycol Methacrylate; β-Hydroxyethyl Methacrylate; Mhoromer; heme-a; 2-methyl-2-propenoic acid, 2-hydroxyethyl ester; GMA; Ethylene glycol methacrylate. A monomer which permits the production of polymers with side chain hydroxyl groups suitable for crosslinking and the production of thermosetting acrylic surface coating adhesives. mp= -12°; $bp_{3.5}$ = 67°; d= 1.0340; n_D^{20} = 1.4515; soluble in H_2O, organic solvents; LD_{50} (rat orl) = 5050 mg/kg. *BP Chemicals Ltd.* Discontinued.

4274 Bisomer 2HPMA
27813-02-1 248-666-3
$C_7H_{12}O_3$
2-Hydroxypropyl methacrylate
Propylene glycol monomethacrylate. Monomer for cross-linkable polymers. *BP Chemicals Ltd.* Discontinued.

4275 Bisomer D10M
1330-76-3 215-547-2
Diisooctyl maleate plasticizer. *BP Chemicals Ltd.* Discontinued.

4276 Bisomer DALP
Diallyl phthalate, a plasticizer. *BP Chemicals Ltd.* Discontinued.

4277 Bisomer DAM
Dialkyl maleate (C_7-C_9 alcohols)
Plasticizer. *BP Chemicals Ltd.* Discontinued.

4278 Bisomer DBF
105-75-9 203-327-9
Dibutyl fumarate, a plasticizer. *BP Chemicals Ltd.* Discontinued.

4279 Bisomer DBM
105-76-0 203-328-4
$C_{12}H_{20}O_4$
Dibutyl maleate

Dibutyl Maleate; 2-Butenedioic acid (Z)-, dibutyl ester. Dibutyl maleate, a plasticizer. *BP Chemicals Ltd.* Discontinued.

4280 Bisomer DNM

Dinonyl maleate, a plasticizer. *BP Chemicals Ltd.* Discontinued.

4281 Bisomer DOM

Diethyl hexyl maleate, a plasticizer. *BP Chemicals Ltd.* Discontinued.

4282 Bisoprufe

Preparations for waterproofing cements. *BP Chemicals Ltd.* Discontinued.

4283 Bisphenol A

80-05-7 1338 201-245-8

$C_{15}H_{16}O_2$

4,4'-Isopropylidenediphenol (CTFA)

2,2-bis (4-hydroxyphenol) propane; 4,4'-(1-Methylethylidene) bisphenol; 2,2-bis(4-Hydroxyphenyl)propane; *p,p'*-Isopropylidenediphenol; bis(4-hydroxyphenyl) dimethylmethane; bis(4-hydroxyphenyl)propane; 4,4'-bisphenol a; DIAN; 2,2-(4,4-dihydroxydiphenyl)propane; 4,4'-dihydroxdiphenylpropane; 4,4'-dihydroxydiphenyl-2,2-propane; 4,4'-dihydroxy-2,2-diphenylpropane; dimethylmethylene-*p,p'*-diphenol; β-di-*p*-hydroxyphenylpropane; dimethyl bis(*p*-hydroxyphenyl)methane; diphenylolpropane; 2,2-di(4-phenylol)propane; *p,p'*-isopropylidenebisphenol; 4,4'-dimethylmethylenediphenol; Phenol, 4,4'-(1-methylethylidene)bis-; 2,2-Bis(4,4'-Hydroxyphenyl)propane,. Intermediate in manufacture of epoxy, polycarbonate, phenoxy, polysulfone, polyester resins, flame retardant, rubber chemicals; fungicide. mp = 150-155°; bp_4 = 220°; insoluble in H_2O; soluble in organic solvents; LD_{50} (rat orl) = 3250 mg/kg. *Aristech; Mitsui Petrochem. Ind.; Mitsui Toatsu Chem.; Shell.*

4284 Bistetramethyl piperidinyl sebacate

52829-07-9; 52829-07-0 258-207-9

$C_{36}H_{60}N_2O_6$

Bis (2,2,6,6-tetramethyl-4-piperidinyl) sebacate

Decanedioic acid, bis (2,2,6,6-tetramethyl-4-piperidinyl) ester; Bis (2,2,6,6-tetramethyl-4-piperidinyl) decanedioate; Uvaseb 770;Tinuvin® 770; BLS™ 1770; Lowilite® 77. Light stabilizer (HALS) for polyolefins, PS, HIPS, ABS, SAN, ASA; also suitable for PU, polyamides, acetal. mp = 80-85°; white to yellowish powder; soluble in organic solvents; *Lowi; Ciba-Geigy/Additives; Enichem Synthesis SpA.*

4285 bistrichloromethylsulfone

3064-70-8 221-310-4

$C_2Cl_6O_2S$

sulfonylbis(trichloromethane)

Hexachlorodimethyl sulfone; bis(trichloromethyl) sulfone. Slimicide. Used to control bacteria and fungi in wastewater.

4286 Bistrimethylsilyl acetamide

10416-59-8 233-892-7

$C_8H_{21}NOSi_2$

N,O-Bis(trimethylsilyl)acetamide

N,O-Bis(trimethylsilyl)acetamide; BSA; Bis(trimethylsilyl)acetamide; Ethanimidic acid, N-(trimethylsilyl)-, trimethylsilyl ester; Dynasylan BSA. Coupling agent, chemical intermediate, blocking agent, release agent, lubricant, primer, reducing agent. Liquid, d = 0.832; bp = 71-73°; n_D^{20}= 1.418; flash pt=11°; moderately toxic by intraperitoneal route. *Hüls Am.*

4287 Bi-Tarco

Tar compounds for roads. *Thomas Ness Ltd, North Thames Gas Board.* Name unverified.

4288 Bitertanol

55179-31-2 1342 259-513-5

$C_{20}H_{23}N_3O_2$

β-[1,1'-Biphenyl)-4-yloxy]-α-(1,1-dimethylethyl)-1H-1,2,4-triazol-1-ethanol

BAY KWG 0599; Sibutol; Baycor®. Broad spectrum fungicide. Mixture of diastereoisomers. Systemic foliar fungicide used especially to control scab on apples and black spot on roses. mp = 125-129°; slightly soluble in H_2O (5 mg/l), more soluble in organic solvents; LD_{50} (rat orl) > 5000 mg/kg. *Bayer AG.*

4289 Bitran

Cationic surfactant range, composed of high molecular weight imidazoline compounds and their salts; viscous liquids/brown aqueous solutions; used as adhesion aids, anticorrosives, dispersants, flocculants, dewatering agents; used in road maintenance; metalworking; paints; emulsion cracking; effluent reatment; pigment dispersion. *Rhône-Poulenc UK.*

4290 Bitran H

A proprietary long-chain cyclic polyamine. A surfactant. *Glover (Chemicals) Ltd.* Unverified.

4291 Bitrex®

3734-33-6 2942 223-095-2

Denatonium benzoate

Available as powder and in aqueous ethanol, methanol and N-propanol; a denaturant for use in alcohols; aversive agent in hazardous household, garden and automotive products as an aid to preventing accidental poisoning by ingestion; animal repellent. *Macfarlan Smith Ltd.* Name unverified.

4292 bittern

The mother liquor remaining after the crystallization of sodium chloride from sea-water. It is a source of magnesium, and also contains bromides and iodides.

4293 Bitumastic

A proprietary trade name for a spirit paint made from refined coal-tar pitch, etc. No manufacturer.

4294 Bitumen

8052-42-4 232-490-9

mineral pitch; natural asphalt; compact bitumen; naphthine. A hard pitchy material found at the surface of the Dead Sea, and in the pitch lake at Trinidad; used in hot-melt adhesive, coatings, paints, sealants, roofing and road coatings.

4295 Bitumuls

Asphalt products. *Monsanto (Solaris).* Name unverified.

4296 Bitusize

Asphalt products. *Monsanto (Solaris).* Name unverified.

4297 Bituvar

Anticorrosive paint. *J C Bottomley.* Name unverified.

4298 Bivert

Amine salts of organic acids, aromatic acid, aromatic and aliphatic petroleum distillate; for use with all pesticides (herbicides, insecticides and fungicides) to control evaporation, increase plant coverage and control drift. *Stull Chemical Company.* Name unverified.

4299 BL 3

Semipermanent release agent for application to flexible molds and metal molds; effective for halogenated or peroxide-cured elastomers. *Releasomers.*

4300 BL-60®

Sodium aluminum phosphate

acidic with aluminum sulfate. Leavening agent for baking, cereals. *Rhône-Poulenc Food Ingreds.*

4301 Black 103

1309-37-1 4072 215-168-2

Ferric Oxide

Iron oxides mixed with bismuth oxychloride; inorganic colorant. *Presperse.* Discontinued.

4302 Black and White Bleaching Cream

123-31-9 4853 204-617-8

Hydroquinone

Depigmentor. *Schering Corp.* Discontinued.

4303 Black Grip

Hard polyurethane elastomer resin, MBOCA, silica; for solid fork truck tires, belting (conveyor), friction drive wheels, drum rotators. *Royale Polymers Ltd.* Name unverified.

4304 Black Out® Black

Toluene (77.5%), 2-butanone (10.9%) and carbon black (2.2%). Decorative and protective finishing agent for rubber; deposited films are resistant to ozone, acids, alkalies, and paraffinic hydrocarbons. *R. T. Vanderbilt Co Inc.* Discontinued.

4305 Black Pearls® 1100

1333-86-4 1856 215-609-9

Carbon black

For coloring plastics. *Cabot.*

4306 Black-Out®

Modified elastomeric base material in solvents; finishing agent for decorative and protective flexible finishes on rubber products. *R. T. Vanderbilt Co Inc.* Discontinued.

4307 Blackox

1309-37-1 4072 215-168-2

Fe_3O_4

Ferric Oxide

Foundry grade black iron oxide for core and mold use with particular emphasis on elimination of sub-surface porosity and carbon streaking when phenolic urethane sand binder systems are in use. *DCS Color & Supply Co Inc.* Discontinued.

4308 Bladafum®

3689-24-5 9139 222-995-2

Sulfotep

Insecticide fumigant used for control of greenhouse pests, e.g., aphids, whiteflies, thrips, spider mites, mealybugs, mobile stages of scale insects on vegetables and ornamentals. *Bayer AG.* Name unverified.

4309 Bladan®

56-38-2 7167 200-271-7

Parathion

Spray and dust formulations for control of biting and sucking pests. *Bayer AG.* Name unverified.

4310 Blade

23135-22-0 7052 245-445-3

oxamyl

Granules containing 10% w/w oxamyl; systemic insecticide and nematicide. *Kommer-Brookwick Ltd.*

4311 Blagden Resins

Synthetic resins for the coatings industry; used for industrial and decorative paints and printing inks. *Blagden Chemicals Ltd.* Name unverified.

4312 Blagdenite

Unsaturated polyester resins; used for reinforced plastics, marine, transport and industrial. *Blagden Chemicals Ltd.* Name unverified.

4313 Blanc Fixe Micro

7727-43-7 1023 231-784-4

$BaSO_4$

Barium sulfate

Filler for paints and coatings. *Sachtleben Chemie GmbH.*

4314 Blancol

Optical brightening agents; for whitening of textiles. *Holliday Dyes & Chemicals Ltd.* Discontinued.

4315 Blancol N

The sodium salt of sulfonated naphthaleneformaldehyde condensate; in papermaking disperses pigments, clays and other solids, prevents pitch coagulation, reduces two sidedness, improves sizing; bleaching, dispersing, leveling and neutralizing agent for leather. *ISP.* Name unverified.

4316 Blancorol®

Different type products for retanning; the majority contain chrome oxide. *Bayer AG.*

4317 Blandofen CAZ

Complex polyalkyl amido imidazolinium sulfate as a white-yellow viscous liquid; cold water dispersible cationic surfactant, stable to freeze thaw cycles; used in softeners for fabrics and paper. *ISP.* Name unverified.

4318 Blandofen CT

107-64-2 203-508-2

Distearyl dimethyl ammonium chloride concentrate in isopropanol/water; supplied as a white-yellow soft paste; not cold water dispersible; higher viscosity than other cationics in the range; cationic surfactant used for fabric softeners. *GAF Great Britain.* Name unverified.

4319 Blandofen FA

Complex polyalkyl amido imidazolinium sulfate as a yellow viscous liquid; cationic surfactant, thixotropic at low temperatures, specially developed for bulk deliveries; used in fabric softeners. *ISP.* Name unverified.

4320 Blandol®

White mineral oil. *Witco/Sonneborn.*

4321 Blanket Adhesive H-98

Blanket adhesive for flat bed and rotary screen print machines. *Catawba-Charlab.*

4322 Blankit®

7775-14-6 8771 231-890-0

Sodium dithionite

Stabilized bleaches for textile, leather and fur, pulp and paper industries. *BASF AG; BASF plc.*

4323 Blanko-Blech

An alloy of 80% copper and 20% nickel.

4324 Blankophor®

Fluorescent whitening agents for textiles and detergents; textile finishing agent. *Bayer AG; Bayer plc.*

4325 Blanose

9004-32-4 1877

Carboxymethylcellulose Sodium

Sodium salt of carboxymethyl cellulose; stabilizer and thickener in aqueous systems for food and nonfood uses. *Hercules.* Discontinued.

4326 blasting gelatin

A blasting explosive; a jelly-like mass is obtained when nitrocotton is added to nitroglycerin, and contains from 90-93% nitroglycerin, and 7-10% dry nitrocotton.

4327 Blatt gold

A brass containing 77% copper and 23% zinc.

4328 Blatt silver

An alloy of 91.1% tin, 8.25% zinc, 0.35% lead, and 0.23% iron.

4329 Blattanex®

114-26-1 8022 204-043-8

Propoxur

Emulsifiable concentrate containing 216 g/l propoxur; a carbamate insecticide. *Bayer AG; Bayer plc.*

4330 Blattanex® 20

114-26-1 8022 204-043-8

Propoxur

Emulsifiable concentrate containing 20% propoxur; used for quick knock down and lasting residual control of crawling and flying pests. *Bayer plc.*

4331 Blattanex® Residual Spray

An aerosol containing 0.5% dichlorvos and 2% propoxur; used for quick knock down and lasting residual control of crawling and flying pests. *Bayer plc.*

4332 Blaxon LT

Acid phosphating concentrate for coating of metals with a corrosion-resistant finish. *Eastern Color & Chem.*

4333 Blazer®

62476-59-9 263-560-7

Acifluorfen

For post-emergence control of broad-leaved weeds and suppression of some annual grasses in soybeans. *BASF AG.*

4334 Blazon

Water soluble, biodegradable colorants; for spray pattern indicators for application of herbicides, pesticides in the lawn and turf, forestry and industrial weed control areas. *Milliken.*

4335 BLE

Reaction product of diphenylamine and acetone; used in natural, IR, SBR, BR, neoprene, and nitrile rubbers; particularly recommended in tire treads and carcass to combat the effects of heat and mechanical flexing; a general purpose antioxidant where discoloration and staining are not factors; also available as a 75% active powder on silica. *Uniroyal.* Name unverified.

4336 BLE® 25

Acetone/diphenylamine reaction product; superaging and flex resistant antioxidant for natural and synthetic rubber. *Uniroyal.* Name unverified.

4337 bleach liquor

Bleaching liquid, liquid chloride of lime. Prepared by passing chlorine through milk of lime; a solution of chlorinated lime.

4338 Bleachit® 1A

Auxiliary for bleaching and optical whitening wool and polyamide. *BASF.*

4339 Blenderm

A proprietary preparation of polythene adhesive tape for surgical use. *3M.* Name unverified.

4340 Blendex® 101

9003-56-9

acrylonitrile-butadiene-styrene copolymers

ABS; modifier resin providing good impact strength and improved processing to PVC compounds; used in calendered films, semiflexible PVC film. *GE Specialty.*

4341 Blendex® 310

9003-56-9

acrylonitrile-butadiene-styrene copolymers

ABS; impact modifier for rigid PVC, epoxies, PU, polyesters. *GE Specialty.*

4342 Blendex® 340

26741-53-7 247-952-5

High efficiency impact modifier for PVC siding and profile substrate applications. *GE Specialty.*

4343 Blendex® 586

Poly(α-methylstyrene-styrene acrylonitrile)

High heat modifier resin used to upgrade PVC compounds. *GE Specialty.*

4344 Blendex® 975

ASA copolymer

Weatherable modifier for use in rigid polymer applications; can be alloyed with PVC and other miscible thermoplastics in molded, extruded or calendered applications; provides impact and higher heat distortion properties. *GE Specialty.*

4345 Blendex® HPP 801

Polycarbonate modifier resin; modifier for other polymers; base for color concentrates; reinforcing modifier for recycled polymers and alloys. *GE Specialty.*

4346 Blendmax 322

8002-43-5 5452 232-307-2

Enzyme-modified lecithin; emulsifier with enhanced water dispersibility; for instantizing whole milk powders, emulsifying vegetable and animal fats in milk replacer products; dough conditioner, antistaling agent in baking. *Central Soya.*

4347 Blendur®

High-quality raw materials based on epoxy, acrylic and polyurethane resins; for thermoset applications in electrical/electronics industry, automotive engineering. *Bayer AG.*

4348 Blendur® KU 3-4513

Polyether polyol

Low-viscosity polyol used in combination with isocyanates to produce soft, flexible compounds with improved thermal endurance. *Bayer AG.* Name unverified.

4349 Blex

29232-93-7 7652 249-528-5

Pirimiphos-methyl

Insecticide and contact fumigant. *ICI Agrochemicals; ICI Chem. & Polymers Ltd.*

4350 BLO®

96-48-0 1632 202-509-5

γ-Butyrolactone

Solvent for PAN, PS, fluorinated hydrocarbons, cellulose triacetate, shellac; used in paint removers, petroleum processing, hectograph process, specialty inks; intermediate for aliphatic and cyclic compounds; reaction and diluent solvent for pesticides; used in dyeing of acetate, wetting agent for cellulose acetate films and fibers, solvent welding of plastic films in adhesive applications. *ISP.*

4351 Block-Out A-SF

Smoke-free auxiliary in pigment printing to achieve a discharge effect on dyed grounds. *Catawba-Charlab.*

4352 blood meal

A nitrogenous fertilizer prepared by coagulating blood, drying and grinding the product; contains on average, from 11-14% nitrogen, and 0.75% phosphorus.

4353 Bloodit

A synthetic gum for process engraving. *Johnsons of Hendon.* Name unverified.

4354 blown oils

polymerized oils; oxidized oils; soluble castor oils; base oil. When semi-drying vegetable oils, marine animal oils, and liquid waxes are warmed at from 70-120°, and a current of air blown through them, the oils oxidize to thickened fluids.

4355 Blue Basic Lead Sulphate

A basic lead sulfate containing lead oxide, lead sulfite, lead sulfide, zinc oxide, and carbon, produced from lead ore by volatilization; used in rubber mixing and in priming paint.

4356 Blue Dot

A double-base-type formulation which minimizes charge weight and moisture absorption; graphite glaze enables smooth granule flow; used for ammunition; designed for use in magnum shotshell loads; also for use in magnum handgun loads. *Hercules.* Discontinued.

4357 blue gold

A jeweler's alloy, consisting of 75% gold and 25% iron.

4358 Blue Mold

Penicillium Glaucum, a fungus.

4359 Blue Powder

Zinc dust; zinc fume. A by-product in the smelting of zinc; consists of a mixture of finely divided zinc and zinc oxide. It has the power of absorbing hydrogen; used in the chemical industries; employed to discharge locally the color of dyed cotton goods; also used for the recovery of gold from the cyanide solution of the metal.

4360 Blueminster Resin Emulsions

Dispersions of rosin and rosin derivatives in water; tackifier for water-based adhesives. *Blueminster Ltd.*

4361 BMC 100, 102

Thermoset BMC; low-cost, general-purpose compound with excellent heat distortion resistance, dimensional stability, flow, controlled reactivity; for injection molding of very intricate parts; applications including electrical brush holders high temperature coil bobbins, 102 complies with Ford ESFM3D056A. *BMC.* Name unverified.

4362 BMC 200

Thermoset BMC reinforced with high strand integrity fiberglass; flame-retardant, medium-strength electrical grade formulated to provide proper filling of intricate mold cavities; applications including internal TV parts, electrical connectors, coli bobbins. *BMC.* Name unverified.

4363 BMC 800

Thermoset polyester BMC, 10-20% glass-reinforced; corrosion-and weather-resistant compound for marine, outdoor electrical, corrosive pump parts, and chemical barriers. *BMC.*

4364 BMC 1000

Thermoset polyester BMC, glass-reinforced; food grade compound for molding of dishware for microwave oven use and conventional ovens to 410°F; also for other houseware items. *BMC.* Name unverified.

4365 B-Nine

1596-84-5 2874 216-485-9

Daminozide

Water soluble powder containing 85% daminozide; a plant growth regulator. *Fargro Ltd.*

4366 Bobierre metal

A brass containing 58-66% copper and 34-42% zinc. Ship's sheathing metal.

4367 Bodenstein

Synonym for amber.

4368 Bohemian earth

Veronese earth; Tyrolean earth; Seladon green; terre verte; stone green. Green earths, which are products of the disintegration of minerals, chiefly of the hornblende type. Stone green is a mixture of ground green earth and white clay; used in the manufacture of waterproofing paints.

4369 Bohemian topaz

Synonym for citrine.

4370 Bohnalite B

A proprietary trade name for an aluminum alloy containing 4.5% copper and 0.3% magnesium. No manufacturer.

4371 Bohnalite J

A proprietary trade name for an alloy of aluminum with 10% copper. No manufacturer.

4372 Bohnalite S43

A proprietary trade name for an alloy of aluminum with about 5% silicon. No manufacturer.

4373 Bohnalite S51

A proprietary trade name for an alloy of aluminum with small amounts of magnesium, silicon, and iron. No manufacturer.

4374 Bohnalite U

A proprietary trade name for an alloy of aluminum with 13% silicon. No manufacturer.

4375 Bohnalite Y

A proprietary aluminum alloy containing small quantities of copper, nickel, and magnesium. No manufacturer.

4376 Bohrmittel Hoechst

Alkylsulfamido carboxylic acid

Corrosion inhibitor, lubricant, wetting agent, and emulsifier for metalworking fluids; extreme pressure properties. *Hoechst Celanese/Colorants & Surf.; Hoechst AG.*

4377 boiled oil

Linseed oil, which has been boiled with litharge to render the oil more drying. The term is also applied to linseed oil which has been heated for some time, thereby increasing the tendency to oxidize.

4378 boiler plug alloy

An alloy of 8 parts bismuth, 5-30 parts lead, and 3-24 parts tin.

4379 Boiler-Aid

Various liquid or powder blends of boiler water deposit and corrosion inhibitors; used for steam and hot water boiler treatment. *Schaefer Technologies Inc.*

4380 Boisambrene Forte

Formaldehyde ethyl cyclododecylacetal

Henkel/Cospha.

4381 Bolda

Fungicide containing carbendazim, maneb, and sulfur. *ICI Chem & Polymers Ltd.*

4382 Bolda FL

A suspension concentrate containing 50 g carbendazim, 320 g maneb and 100 g sulfur per liter; systemic fungicide for cereals. *Farm Protection Ltd.*

4383 bolfo®

Preparation for external use on dogs and cats against ecto parasite infestation; veterinary medicine. *Bayer AG.*

4384 Bolstar

35400-43-2 9164 252-545-0

Sulprofos

Insecticide with stomach and contact action mainly against caterpillars of *Heliothis spp.* and *Spodoptera spp.* on cotton and other crops. *Bayer AG.*

4385 bolster silver

A nickel silver. It is an alloy of 65.5% copper, 16% zinc, 18% nickel, and 0.5% lead.

4386 Bombardier

1897-45-6 2219 217-588-1

Chlorothalonil

A fungicide for a wide range of agricultural crops. *Farm Protection Ltd.* Name unverified.

4387 Bond-A-Tint

Pigment dispersions; for coloration of bonded polyurethane carpet underlay. *Pacific Dispersions Inc.* Name unverified.

4388 Bonder/Gardobond

Phosphate systems for iron, steel, zinc, aluminum and its alloys; for corrosion protection, pretreatment prior to painting, assisting cold forming and improving sliding friction properties and running-in characteristics. *Sachtleben Chemie GmbH.*

4389 Bonder/Gardoclean
Alkaline, neutral, acid-passivating cleaners for spray and dip application. *Sachtleben Chemie GmbH.*

4390 Bonderite
Phosphating process for iron, steel, aluminum, and zinc which converts the metal surface into a zinc phosphate coating; corrosion inhibiting and paint bonding treatments. *Brent Chemicals International plc.*

4391 Bonderlube/Gardolube
Water-based systems oils, fatty emulsions; for cold forming solid and hollow ware; drawing, cold extrusion, wall ironing; for steel, stainless steel, aluminum. *Sachtleben Chemie GmbH.*

4392 Bonding Agent 2001
Polymethylene polyphenyl isocyanate solution in dibutyl phthalate; bonding agent for PVC coatings. *Bayer Corp; Polysar.*

4393 Bonding Agent M 3
A methylene donor; used in combination with Bonding Agents R6 in natural, IR, BR, SBR, neoprene, or nitrile rubbers to give improved adhesive bonds to fabrics such as cotton, nylon, polyester, rayon and glass, as well as to metals. *Uniroyal.* Name unverified.

4394 Bonding Agent P 1
4,4'-Methylene-bis-(phenylcarbanilate)
For pre-dip solutioning of polyester tire and industrial cord prior to a secondary treatment with an RFL system; the two dip treatment provides superior adhesion to other known treatment systems. *Uniroyal.* Name unverified.

4395 Bonding Agent R6
A resorcinol donor; used in combination with Bonding Agent M3 in natural rubber, IR, BR, SBR, neoprene or nitrile rubber to give improved adhesive bonds to fabrics such as cotton, rayon, nylon, polyester and glass, as well as to metals. *Uniroyal.* Name unverified.

4396 Bonding Agent TN
70% solution of a polyester containing hydroxyl groups; used to improve adhesion of PVC coatings on polyester and polyamide coatings. *Bayer AG.*

4397 Bondogen
A mixture of an oil soluble sulfonic acid of high molecular weight with a high boiling alcohol and a paraffin oil; used as a peptizing agent and strong plasticizer for all elastomers; functions as a scorch retarder. *King Industries.*

4398 Bond-Plus
Adhesives, industrial coatings for packaging, laminating, paper converting, wood-working, labelling, foil laminating, paper cups; FDA approved adhesives and coatings; used for packaging; case sealing, labelling, lap-gluing; paper converting; foil laminating; paper-to-paper, paper-to-film laminations; paper cups, coatings, woodworking; water-resistant adhesives. *Industrial Adhesives Company.* Name unverified.

4399 Bond-Plus HM
Adhesives, industrial coatings for packaging, laminating, paper converting, wood-working, labeling, foil laminating, paper cups, FDA approved adhesives and coatings; used for packaging; case sealing, labeling, lap-gluing; paper converting; foil laminating paper-to-paper, paper-to-film laminations; paper cups, coatings, woodworking; water-resistant adhesives. *Industrial Adhesives Company.* Name unverified.

4400 BondTint
Family of polymeric, liquid, reactive colorants used to color polyurethane-based adhesives for rebonded carpet underlay. *Milliken.*

4401 Bondur
An aluminum alloy containing 4.2% copper, 0.3-0.6% manganese, and 0.5-0.9% magnesium; a corrosion-resisting alloy. No manufacturer.

4402 Bone oil
8001-85-2 232-294-3
Animal oil; Dippel's oil; oil of Hartshorn; bone tar. A dark brown oil, rich in pyridine bases, obtained by the distillation of bones; used for denaturing spirits.

4403 Bonner L-894
A proprietary trade name for a polyester resin. No manufacturer.

4404 Bonosol
Acrylic resins. *Ernst Jager GmbH.*

4405 Bonotex
Acrylic latexes. *Berol Nobel Ltd.*

4406 Bontex
Range of synthetic blended detergents. *Unilever.* Name unverified.

4407 Bonzi
76738-62-0 7118
Paclobutrazol
Suspension concentrate containing 4 g/l paclobutrazol; plant growth regulator for use on ornamental plants. *ICI Chem & Polymers Ltd.*

4408 Boost
Fish oil concentrates; source of ω-3 fatty acids in fish farm feeds. *Seven Seas Ltd.*

4409 Boral
Aluminum-boro-tartrate
An antiseptic and astringent prescribed in skin diseases, and in inflammation of the ear.

4410 Borascu
Borax decahydrate
General nonselective weedkillers. *Borax Consolidated Ltd.*

4411 Borax
1303-96-4
Sodium tetraborate decahydrate
Dispersant, wetting agent for NR, SR latexes; mold lubricant for general dry rubber molding. *U.S. Borax & Chem.*

4412 Borax Glass
12007-92-0 234-522-7
Sodium pentaborate
Pentabor. Provided as tablets and powder to kill larvae in livestock confinement areas and cockroaches in residences. Rarely, solutions are sprayed as a nonselective herbicide. *Borax Europe Ltd.*

4413 Boraxusta
Calcined borax.

4414 Borcher's metal
Noncorrosible alloys. One alloy contains 64.65 nickel, 32.3% chromium, 1.8% molybdenum, and 0.5% silver; other are stated to contain 30% chromium, 34-35% cobalt, and 34-35% nickel.

4415 Borden
Foundry sand binders. *Borden (UK) Ltd.* Name unverified.

4416 Borderland Black
2032-65-7 6050 217-991-2
methiocarb
Active ingredient; methiocarb; seed protectant to prevent sprout pulling by birds in newly planted corn. *Borderland Products Inc.* Discontinued.

4417 Borester
Organic borates. *Rhône-Poulenc UK.*

4418 Borester 7
Trihexylene glycol diborate
Used for timber preservation. *Manchem Ltd.* Name unverified.

4419 Boresters
A series of esters of boric acid. They are used as a convenient means of introducing boron into organic media such as paints and plastics. No manufacturer.

4420 Borethyl
97-94-9 202-620-9
$C_6H_{15}B$
Triethylboron
triethylborane. Radical reaction initiator. mp = -92°; bp = 95°; d = 0.6750; LD_{50} (rat orl) = 235 mg/kg.

4421 Borfax
Shaped lightweight sideliner lines for steel ingot heading. *Foseco (F.S.) Ltd.* Discontinued.

4422 Boric Acid
10043-35-3 1364 233-139-2
BH_3O_3
Boracic acid
Orthoboric acid; Borofax; sal Sedativus. Waterproofing wood; fireproofing fabrics; preservative; manuf.cements, ceramics, glass, hats, soaps, leather, borates, enamels, crockery; cosmetics; printing, painting, and dyeing; photography; hardening steel. Used for weather- and fire-proofing wood, as an insecticide for beetles and medically as an astringent and antiseptic. mp = 171°; soluble in H_2O, EtOH, glycerol; LD_{50} (rat orl) = 5.14 g/kg. *Janssen Chimica; OxyChem; U.S. Borax & Chem.*

4423 Boric anhydride
1303-86-2 1365 215-125-8
B_2O_3
boric oxide
Diboron trioxide; Boron oxide; Boric Oxide; Boron Trioxide; B-O; Boria; Boric acid, anhydride; Boron oxide (B_2O_3); Boron sesquioxide. Production of boron, heat-resistant glassware, fire-resistant additive for paints, electronics, liquid encapsulation techniques, herbicide. mp = 450°; bp = 1860°; d = 2.46; soluble in H_2O (3 g/100 ml); LD_{50} (mus orl) = 3163 mg/kg. *Atomergic Chemetals; Noah Chem.; Spectrum Chem. Mfg.; U.S. Borax & Chem.*

4424 Boric oxide
Boraxo. Industrial hand cleaners. *Borax Consolidated Ltd.*

4425 Boric oxide

4426 Borite
1344-28-1 369 215-691-6
aluminum oxide
Consists essentially of crystalline alumina; a trademark for goods used as abrasives and refractories. No manufacturer.

4427 Borium
A fused tungsten carbide diamond substitute formed by exposing tungsten at high temperatures to carbon monoxide or hydrocarbon gases.

4428 DL-Borneol
507-70-0 1366 208-080-0
$C_{10}H_{18}O$
endo-1,7,7-Trimethylbicyclo[2.2.1]heptan-2-ol
Bhimsiam camphor; Baros camphor; Borneo camphor; Dryobalanops camphor; Malay camphor; Sumatra camphor; endo-2-bornanol; endo-2-camphanol; endo-2-hydroxycamphane; bornyl alcohol; camphor; Bhimsaim camphor. A terpene from *Dryobalanops camphora;* used in perfumery, in celluloid manufacture, and in medicine as an antiseptic. mp = 208°; bp = 212°; insoluble in H_2O, soluble in organic solvents; LD_{50} (rat orl) = 500 mg/kg.

4429 Bor-Nitrophoska® 13-13-21+0.1B
Complex fertilizer with 13% nitrogen, 13% phosphate, 21% potash, and 0.1% boron; for all agricultural crops requiring boron and horticultural crops which are not sensitive to chloride. *BASF AG.*

4430 Borocil A
1912-24-9 902 217-617-8
Atrazine
A residual herbicide. *ABM Chemicals Ltd.*

4431 Borocil Extra
Atrazine, boromacil and diuron. Atrazine, bromacil and diuron; used for total weed control in non crop areas. *ABM Chemicals Ltd.*

4432 Borocil K
Bromacil and diuron. Bromacil and diuron; used for total weed control in non crop areas. *ABM Chemicals Ltd.*

4433 Boroflow
122-34-9 8681 204-535-2
Simazine
Suspension concentrate containing 500 g/l simazine; a triazine herbicide to control weeds and grasses in cane fruit, roses and some vegetables. *ABM Chemicals Ltd.*

4434 Boroflow A
1912-24-9 902 217-617-8
Atrazine
A liquid formulation containing 500 g atrazine per liter as a suspension concentrate; a residual herbicide. *ABM Chemicals Ltd.*

4435 Boroflow A/ATA
1912-24-9 902 217-617-8
atrazine-aminotriazole. A liquid formulation containing 270 g atrazine and 160 g aminotriazole per liter as a suspension concentrate; used for total weed control on industrial sites, paths, kerbs and channels, drives and hard tennis courts, hardstanding and storage areas. *ABM Chemicals Ltd.*

4436 Boroflow S/ATA
A suspension concentrate containing 160 g aminotriazole and 270 g simazine per liter; used for total weed control in non crop areas and fruit orchards. *ABM Chemicals Ltd.*

4437 Boroflow S/ATA
61-82-5 513 200-521-5
aminotriazole-simazine
A suspension concentrate containing 160 g aminotriazole and 270 g simazine per liter; used for total weed control in non crop areas and fruit orchards. *ABM Chemicals Ltd.*

4438 Boroflux
A mixture of boron suboxide with boric anhydride and magnesia; used to the extent of about 1% to deoxidize copper during purification.

4439 Borogard® ZB
12513-27-8
Zinc borate
Corrosion inhibitor, biocide, fungicide, in-can preservative for paints and coatings. *U.S. Borax & Chem.*

4440 boroglyceride
$C_3H_5BO_3$
boroglycerine
glyceryl borate. Used as a preservative for wines and fruits.

4441 Borolon
1344-28-1 369 215-691-6
aluminum oxide
Consists essentially of crystalline alumina; a trademark for articles made for abrasive and refractory purposes. No manufacturer.

4442 Boron
7440-42-8 1373 231-151-2
B
Nonmetallic element. Used in special-purpose alloys; cementation of iron; neutron absorber in reactor controls; oxygen scavenger for copper and other metals; fibers and filaments in composites; semiconductors; rocket propellant mixtures. mp = 2200°; insoluble in H_2O. *Atomergic Chemetals; Noah Chem.*

4443 Boron nitride
10043-11-5 1376 233-136-6
BN
PBN; pyrolytic boron nitride; BN. Used in manufacture of alloys, in semiconductors, nuclear reactors, lubricants. mp = 3000°. *Atomergic Chemetals; Carborundum Co.; New Metals & Chems. Ltd; Noah Chem.*

4444 Boron tribromide
10294-33-4 1377 233-657-9
BBr_3
boron bromide
boron bromide; Tribromoborane. Catalyst in organic synthesis manufacture of diborane. mp = -46.0°; bp = 91°; d_0 = 2.650; reacts with H_2O. *Aldrich; Atomergic Chemetals; Janssen Chimica; Kerr-McGee.*

4445 Boron trichloride
10294-34-5 1378 233-658-4
BCl_3
boron chloride
trichloroborane. Catalyst in organic synthesis; source of boron compounds; refining of alloys; soldering flux; electrical resistors; extinguishing magnesium fires in heat-treating furnaces; manufacture of diborane. mp = -107°; bp = 12.5°; d_0 = 0.7380 *Air Prods & Chem; Aldrich; Atomergic Chemetals; Kerr-McGee.*

4446 Boron trifluoride
7637-07-2 1379 231-569-5
BF_3
boron fluoride
Lewis acid catalyst in organic synthesis; production of diborane; instruments for measuring neutron intensity; soldering fluxes; gas brazing. bp = -127.1°; d = 0.8700. *Air Prods & Chem; Akzo; Aldrich; AlliedSignal; Atomergic Chemetals.*

4447 borosalicylic acid
A solution containing 4% each boric and salicylic acids. Has been used as an antiseptic.

4448 Borrebond
8061-52-7
Calcium lignosulfonate
Binder, low-cost dispersant. *Borregaard Ligno Tech.*

4449 Borrechel
Lignosulfonate containing chelated trace elements; used as micronutrients. *Borregaard Ligno Tech.*

4450 Borresperse CA/CAF
8061-52-7
Calcium lignosulfonate
Desugared; dispersant for pesticide formulations, concrete admixtures. *Borregaard Ligno Tech.*

4451 Borrewell C
Chrome lignosulfonate
Conditioner in water-based oil well drilling mud systems. *Borregaard Ligno Tech.*

4452 Borrewell FC
Ferrochrome lignosulfonate
Conditioner in water-based oil well drilling mud systems. *Borregaard Ligno Tech.*

4453 Borrewell FE
Iron lignosulfonate
Conditioner in water-based oil well drilling mud systems. *Borregaard Ligno Tech.*

4454 *Bortin 45*
Brucella abortus vaccine for veterinary purposes. *Glaxo Laboratories.* Name unverified.

4455 Borvicote
A proprietary range of emulsions of vinyl homopolymers and copolymers. *Borregaard Ligno Tech.* Name unverified.

4456 Bos MH
123-33-1 5745 204-619-9
Maleic hydrazide
A plant growth regulator for grass and to reduce bud growth in trees, hedges and vegetables. *Bos Chemicals Ltd.*

4457 Boselon
Packaging materials; plastic film made of low density polyethylene or other thermoplastic, containing a volatile corrosion inhibitor effective for rust protection of ferrous metals; used for packaging for metal parts for the

purpose of protection against corrosion. *Aicello Chemical Co Ltd.* Name unverified.

4458 Bostik

Wide variety of formulations based on natural and synthetic materials (rubbers, resins, polymers, etc.); used for adhesives, sealants and coating compounds for a wide variety of bonding, sealing and productive coating applications in all types. *Bostik Ltd.* Name unverified.

4459 Bostik 1100FS

One-part urethane sealant; cures by reaction with atmospheric moisture to form tough, flexible seal; for sealing and bonding applications in general industry, HVAC, marine industry, truck trailer, container construction, log home, RV. *Bostik.*

4460 Botrilex

82-68-8 8264 201-435-0
Quintozene
A horticultural fungicide containing quintozene. *ICI Chem & Polymers Ltd.*

4461 Bouchardt's reagent

A solution of 1 part iodine and 2 parts potassium iodide in 20 parts water. An alkaloid reagent giving a brown precipitate; used in chemical analysis.

4462 Bourbonne's metal

An alloy of 50.48% tin, 48.8% aluminum, 0.25% copper, and 0.33 percent iron.

4463 Bourbouze aluminum solder

An alloy of 47% zinc, 37% tin, 10% aluminum, and 5% copper.

4464 Bourbouze solder

An alloy of 83% tin and 17% aluminum.

4465 Bovinal-20

9048-46-8 8613 232-936-2
Serum albumin
Protein for use in skin and hair care preparations. Also used in biochemical and medical research. *RITA.*

4466 Bovinox

Cattle dip and spray. *The Wellcome Foundation Ltd.* Name unverified.

4467 Bowhill's stain

A microscopic stain; contains 15 ml of a saturated alcoholic solution of orcin, 10 ml of 20% tannic acid solution, and 30 ml water.

4468 Boxite

1344-28-1 369 215-691-6
aluminum oxide
Consists essentially of crystalline alumina; a trademark for articles of the abrasive and refractory class. No manufacturer.

4469 Boxolon

Herbicide containing clopyralid, bromoxynil and mecoprop. *ICI Chem & Polymers Ltd.* Discontinued.

4470 Bozefloc

Polyacrylic flocculants. *Hoechst UK.*

4471 Bozetol

A sulfonated castor oil product; a proprietary trade name for a wetting agent. No manufacturer.

4472 Bozzle

Container for agrochemicals; used with the Electrodyn sprayer. *ICI Chem & Polymers Ltd.*

4473 BP LDPE

Low density polyethylene. *BP Chemicals Ltd.*

4474 BP Mycocide

Liquid preservative based on propionic acid. *BP Chemicals Ltd.* Discontinued.

4475 BP polystrene

Polystyrene. *BP Chemicals Ltd.*

4476 BR Destral

Aminotriazole-bromacil-diuron
Used for total weed control in noncrop areas and on railway tracks. *ABM Chemicals Ltd.*

4477 Brabant PCNB

82-68-8 8264 201-435-0
Quintozene
Dustable powder containing 20% w/w quintozene; fungicide for various agricultural crops. *Bos Chemicals Ltd.*

4478 Bradophen

Quaternary ammonium bactericide. *Ciba plc.* Name unverified.

4479 Bradsil

Silicone-based textile softeners and lubricants. *Hickson & Welch Ltd.*

4480 Bradsyn

Polyethylene softeners for textile finishing. *Hickson & Welch Ltd.*

4481 Braemer's reagent

A tannin reagent. It consists of 1 gram sodium tungstate and 2 g sodium acetate in 10 ml of water.

4482 Brakol

A water treatment alkali. *Laporte Industries Ltd.*

4483 Brasivol

1344-28-1 369 215-691-6
Aluminum Oxide
An abrasive cleaning paste in three grades, each containing synthetic aluminum oxide in a nonirritant soap-detergent base; for the treatment of acne vulgaris. mp = 94-96°; $d^{2\cdot}$ = 1.40; soluble in H_2O (55 mg/l), more soluble in organic solvents; LD_{50} (rat orl) = 3600-5833 mg/kg. *Stiefel Laboratories Inc; Stiefel Laboratories (UK) Ltd.* Name unverified.

4484 Brasoran 50 WP

4658-28-0 225-101-9
$C_7H_{11}N_7S$
4-azido-N-(1-methylethyl)-6-methylthio-1,3,5-triazin-2-amine
Aziprotryne; C 7019; Mesoranil. A selective herbicide used to control a wide range of annual broad-leaved weeds and some grasses in brassicas. *Ciba-Geigy Agrochemicals.*

4485 brass

A copper-zinc alloy of varying proportions. Usually it contains more than 18% zinc, and lead is sometimes added to the extent of 1-2% Ordinary brass contains 67% copper and 33% zinc; used in condenser tube plates, piping, hose nozzles, couplings oil gauges, flow indicators, air and drain cocks, marine equipment, jewelry, etc.

4486 brass, iserlohn

An alloy of 64% copper, 33.5% zinc, and 2.5% tin.

4487 brass, leaded

Alloys of from 71-79% copper, 4.5-9.5% lead, 8.5-23% zinc, and traces to 3% tin.

4488 brass, iron

A brass containing from 1-9% iron.

4489 Bras-sicol

82-68-8 8264 201-435-0
Quintozene
A contact fungicide containing 50% w/w quintozene; used for the control of diseases in turf. *Burts & Harvey.*

4490 Bravo 500

1897-45-6 2219 217-588-1
Chlorothalonil
A fungicide for a wide range of agricultural crops. *BASF plc.* Name unverified.

4491 Bravocarb

A liquid formulation containing 100 g carbendazim and 450 g chlorothalonil per liter as a suspension concentrate; systemic fungicide. *Fermenta ASC Europe Ltd.*

4492 Brazilian Elemi

Elemi from *Icica icariba.*

4493 brazing solder

Alloys of from 35-45% copper, 45-57% zinc, and 8-10% nickel.

4494 Breaker F

Enzyme for hydrolyzing water-soluble polysaccharides such as guar, derivatized guar, and cellulose ethers. *Rhône-Poulenc/Water Soluble Polymers.*

4495 Breaxit

Performance chemicals for oil field. *Exxon Int'l.* Name unverified.

4496 Brebent

1302-78-9 1082 215-108-5
Bentonite
Bentonite clay for binding, suspending and emulsifying sands. *Laporte Industries Ltd.*

4497 Brebond

Bonding clay for foundry sands. *Laporte Industries Ltd.*

4498 Breecht's double salt

Potassium dimagnesium sulfate

4499 Breedervac 1 Plus, II Plus, III Plus, IV plus

For immunization of poultry. *Intervet Inc.*

4500 Breedervac-Reo-Plus, RN Plus

For immunization of poultry. *Intervet Inc.*

4501 Bregel

1302-78-9 1082 215-108-5
Bentonite
Drilling bentonite. *Laporte Industries Ltd.*

4502 Breon

Vinyl materials; nitrile and acrylic rubbers. *British Geon.* Name unverified.

4503 Breox

Polyalkylene glycols and polyethylene glycols; brake fluids, lubricants. *BP Chemicals Ltd.*

4504 **Bresin 2, 2E**
Thermoplastic resins derived from natural materials extracted from pinewood; used in construction adhesives and mastics. *Hercules.* Discontinued.

4505 **Brestan**
fentin acetate-maneb. Mixture of fentin acetate and maneb; used for control of potato blight. *Hoechst UK.*

4506 **Bretol®**
124-03-8 2071 204-672-8
Cetyldimethylethylammonium bromide
Disinfectant and topical anti-infective. *Zeeland.*

4507 **bretonite**
C_3H_5IO
Iodoacetone

4508 **Breviol**
Dyeing auxiliary; dispersing and leveling agent for dyeing blends of polyester/wool and polyester/acrylic; dispersing agent for cationic dyestuffs, dyeing in general. *Henkel Chemicals Ltd.* Name unverified.

4509 **Brightray Alloy B**
A trademark for an electrical resistance alloy of 59% nickel, 16% chromium, 0.3% silicon and the balance iron. *Wiggin Alloys Ltd.* Unverified.

4510 **Brightray Alloy C**
A trademark for an electrical resistance alloy of 19.5% chromium, 1.5% silicon, 0.04% rare earth elements and the balance nickel. *Wiggin Alloys Ltd.* Unverified.

4511 **Brightray Alloy F**
A trademark for an electrical resistance alloy of 37% nickel, 18% chromium, 2.2% silicon, and the balance iron. *Wiggin Alloys Ltd.* Unverified.

4512 **Brightray Alloy H**
A trademark for an electrical resistance alloy of 19.5% chromium, 3.6% aluminum, and the balance nickel. *Wiggin Alloys Ltd.* Unverified.

4513 **Brightray Alloy S**
A trademark for an electrical resistance alloy of 20% chromium, and the balance nickel. *Wiggin Alloys Ltd.* Unverified.

4514 **Brij®**
Polyxyethylene alkyl ethers
Emulsifiers *ICI Am.*

4515 **Brij® 30**
Laureth-4
Oil-water emulsifier for topical cosmetics. *ICI Spec. Chem.; ICI Surf. Belgium.*

4516 **Brij® 30SP**
Laureth-4
Preservatives; oil-water emulsifier for topical cosmetics. *ICI Spec. Chem.*

4517 **Brij® 52**
9004-95-9
Ceteth-2
Ceteth-2 with added antoxidants; a surfactant and emulsifier for topical cosmetics. *ICI Spec. Chem.; ICI Surf. Belgium.*

4518 **Brij® 72**
9005-00-9
Steareth-2
Steareth-2 with preservatives; surfactant, emulsifier especially for topical cosmetic applications. *ICI Spec. Chem.; ICI Surf. Belgium.* Name unverified.

4519 **Brij® 93**
9004-98-2
Oleth-2 with added preservatives; surfactant with low color and odor; disperses emollients, perfume oils, and surfactants; especially for blooming bath oils. *ICI Spec. Chem.*

4520 **Brij® 700 S**
9005-00-9
Steareth-2
Emulsifier, solubilizer. *ICI Spec. Chem.* Name unverified.

4521 **Brij® 721**
9005-00-9
Steareth-2
Cosmetic emulsifier, solubilizer for fragrances. *ICI Spec. Chem.; ICI Surf. Belgium.* Name unverified.

4522 **Briklens**
Laundry detergents. *Laporte Industries Ltd.*

4523 **Briline**
Diazo compounds and coupling agents. *Bridge Pharmaceuticals.* Name unverified.

4524 **Brilliant Indigo**
Dyestuffs for dyeing and printing of cellulosic fibers. *BASF plc.* Name unverified.

4525 **Brimstone Plus**
Mixture of potassium sorbate, sodium metabisulfate and sodium propionate; a broad-spectrum fungicide for field crops. *Mandops (UK) Ltd.*

4526 **Bri-Nylon**
Nylon yarns. *ICI Chem & Polymers Ltd.*

4527 **Briosil**
A nonprecious metal alloy; for dentistry and dental engineering. *Degussa AG.* Name unverified.

4528 **Briotril**
Bromoxynil octanoate and ioxynil octanoate; herbicide used for selective post-emergence weed control. *Agan Chemical Manufacturers Ltd.*

4529 **Briotril Plus**
Bromoxynil octanoate-ioxynil octanoate
An emulsifiable concentrate containing 200 g bromoxynil and 200 g ioxynil per liter; a post-emergence contact herbicide for cereal crops. *Pan Britannica Industries Ltd.*

4530 **Briphos**
Range of anionic surfactants composed of aliphatic or nonylphenol based ethoxylated phosphate ester (mixture of mono- and di-esters) in acid form; viscous liquids or paste; used for agricultural chemicals, cleaners, cosmetics, toiletries, dry cleaning, inks, lubricants, surface coatings. *Albright & Wilson Ltd., Phosphates & Speciality Business.*

4531 **Briquest®**
Phosphonate derivatives. *Albright & Wilson Ltd., Phosphates & Speciality Business.*

4532 **Briquest® 301-30SH**
Sodium nitrilotris(methylene phosphate); sequestrant for scale inhibition and corrosion control. *Albright & Wilson UK.*

4533 **Briquest® 301-50A**
Nitrilotris(methylene phosphonic acid); sequestrant for water treatment, oil-drilling muds, powder detergents, photographic applications. *Albright & Wilson UK.*

4534 **Briquest® 462-23K**
Potassium hexamethylene diamine tetrakis(methylene phosphate); sequestrant for water treatment. *Albright & Wilson UK.*

4535 **Briquest® 543-45AS**
22042-96-2 244-751-4
Diethylenetriamine-pentakis(methylene phosphonic acid)
Sequestrant for peroxide stabilization in pulp bleaching and de-inking, in liquid detergents and oil-field chemicals. *Albright & Wilson UK.*

4536 **Briquest® ADPA-60A**
2809-21-4 3908 220-552-8
$C_2H_8O_7P_2$
Etidronic Acid
(1-Hydroxyethylidene)biphosphonic acid. Used for water treatment and oil-drilling muds, in powder detergents and photographic applications, sequestering agent for calcium carbonate. *Albright & Wilson Am.* Name unverified.

4537 **Brisgo II**
A thermoplastic resin derived from rosin acids; hog carcass dehairing composition. *Hercules.* Discontinued.

4538 **Britannia metal**
An alloy of from 74-91% tin, 5-24% antimony, and 0.15-3.68% copper, sometimes with small quantities of zinc, lead, and bismuth; a Britannia metal containing 90% tin and 10% antimony has a specific gravity of 7.

4539 **Britesil**
Hydrous sodium polysilicate
Used as an alkaline builder in powdered or granular laundry detergents, dishwashing detergents, household cleaners. *PQ Corp.* Name unverified.

4540 **Britesorb**
Hydrous silica
Clarifying and chill proofing agent for beer, clarifying or fining agent for wines, fruit juices etc.; selective absorbent of proteins and metals. *PQ Corp.* Name unverified.

4541 **Britomya**
Ground chalk; whitening fillers. *Croxton & Garry Ltd.*

4542 **Britonite**
An explosive containing nitro-glycerin, potassium nitrate, wood meal, and ammonium oxalate.

4543 **Brittox**
Herbicide. *May & Baker Ltd.* Name unverified.

4544 **Brix**
Cupola flux. *Foseco (F.S.) Ltd.* Discontinued.

4545 **Brixil**
Cupola flux and silicon additive. *Foseco (F.S.) Ltd.* Discontinued.

4546 **Brixmetal**
Alloys of from 60-75% nickel, 15-20% chromium, 5% copper, 1-4% tungsten, 4% silicon, 3% titanium, 2% aluminum, and 1% bismuth; noncorrosive.

4547 **Briz**
A scouring powder. *Unilever.* Name unverified.

4548 **broad salt**
Ground rock salt.

4549 **Broadshot**
Emulsifiable concentrate containing 200 g 2,4-D, 85 g dicamba and 65 g triclopyr per liter; an herbicide to control perennial and woody weeds. *Shell UK.*

4550 **Broenner's acid**
2-Naphthylamine-6-sulfonic acid
Intermediate in manufacture of dyestuffs.

4551 **Brolac Dualcote Acrylic Primer/Undercoat**
Acrylic copolymer emulsion, water thinned, interior primer for soft and hardwoods and an interior and exterior undercoat for Brolac Full Gloss. *Berger Jenson & Nicholson Ltd.* Name unverified.

4552 **Brolac Eggshell Low Odour**
Alkyd resin based, white spirit thinned; for interior application; can be used in kitchens and bathrooms giving satin sheen finish. *Berger Jenson & Nicholson Ltd.* Name unverified.

4553 **Brolac Full Gloss**
Alkyd resin based, white spirit thinned; high gloss protective finish for interior and exterior use by the professional decorator and specifier. *Berger Jenson & Nicholson Ltd.* Name unverified.

4554 **Brolac PEP Vinyl Matte & Vinyl Silk Emulsions**
VA copolymer emulsion based, water thinned; Vinyl Matte is for both exterior and interior application giving a matt finish; Vinyl Silk is for interior application giving a silk sheen. *Berger Jenson & Nicholson Ltd.* Name unverified.

4555 **Brolac Primers, Sealers and Surface Preparation Pr**
Various alkyd/oleo resinous, white spirit thinned; protective primers/sealers for painting surfaces. *Berger Jenson & Nicholson Ltd.* Name unverified.

4556 **Brolac Specialist Coatings**
Various specialty resins; for maintenance and protection of steel, floors, concrete and other surfaces requiring extra resistance and performance. *Berger Jenson & Nicholson Ltd.* Name unverified.

4557 **Brolac Superflat Emulsion**
VA copolymer emulsion, water thinned; for interior application, especially suited to new plasterwork. *Berger Jenson & Nicholson Ltd.* Name unverified.

4558 **Brolac Tartaruga**
VA copolymer emulsion, water thinned; for high build textured wall finish for interior and exterior use. *Berger Jenson & Nicholson Ltd.* Name unverified.

4559 **Brolac Undercoat**
Alkyd resin based, white spirit thinned; used in preparation for all types of Brolac finishes. *Berger Jenson & Nicholson Ltd.* Name unverified.

4560 **Brolac Varnishes**
Alkyd resin base (some P.U. modified), thinned with white spirit; provides surface sheen to soft and hardwoods. *Berger Jenson & Nicholson Ltd.* Name unverified.

4561 **Brolac Weathercoat No. 1-finely textured**
VA copolymer emulsion, water thinned; for masonry exterior coatings. *Berger Jenson & Nicholson Ltd.* Name unverified.

4562 **Brolac Weathercoat No. 2 - smooth**
VA copolymer emulsion, water thinned; for exterior masonry, ideal for airless spray equipment; suitable for buildings subjected to atmospheric pollution. *Berger Jenson & Nicholson Ltd.* Name unverified.

4563 **Brolac Weathercoat No.3**
Pliolite resin based, white spirit thinned; provides smooth finish to masonry; can be applied in cold, wet and damp conditions with no separate sealer requirement. *Berger Jenson & Nicholson Ltd.* Name unverified.

4564 **bromacil**
314-40-9 1402 206-245-1
$C_9H_{13}BrN_2O_2$
5-bromo-6-methyl-3-(1-methylpropyl)-2,4(1H,3H)pyrimidinedione
Borea; Bromax; Croptex Onyx; Du Pont Herbicide 976; Hyvar X; Nalkil; Rokar Xrout; Staa-Free; Uragan; Urox B. Photosynthesis inhibitor, used as a herbicide for total weed and brush control on non-crop land and selective control of annual and perennial weeds and grasses in citrus and pineapple plantations. mp = 158-159°; d^{25} = 1.55; soluble in H_2O (815 mg/l), more soluble in organic solvents; LD_{50} (rat orl) = 5200 mg/kg.

4565 **bromadiolone**
28772-56-7 1403 249-205-9
$C_{30}H_{23}BrO_4$
3-[3-(4'-bromo[1,1'-biphenyl]-4-yl)-3-hydroxy-1-phenylpropyl]-4-hydroxy-2H-1-benzopyran-2-one
Bromone; Canadien 2000; Contrac; Maki; Ratimus; Tamogam; Boldo; LM-637; Super-caid; Super-rozol;. Rodenticide. mp = 200-210°; λ_m = 260 nm($E^{1\%}_{1cm}$ = 538-582 EtOH); soluble in H_2O (19 mg/l), more soluble in organic solvents; LD_{50} (rat orl) = 1.125 mg/kg.

4566 **Bromat®**
57-09-0 2068 200-311-3
Cetyl trimethyl ammonium bromide
Surfactant, emulsifier, germicide. *Zeeland.* Name unverified.

4567 **Bromcresol purple**
115-40-2 1408 204-087-8
$C_{21}H_{16}Br_2O_5S$
Dibromo-*o*-cresolsulfonphthalein
5,5'-dibromo-*o*-cresolsulfonphthalein. An acid-base indicator. mp = 240°; irritant; practically insoluble in H_2O, soluble in alcohol.

4568 **Bromeikon**
860-07-1 9329
$C_{20}H_8Br_4Na_2O_4$
Sodium tetrabromophenolphthalein
Brom-tetragnost; Cholegnostyl; Tetrabrom. Radio-opaque medium, used as a diagnostic aid. Soluble in H_2O, insoluble in organic solvents.

4569 **Bromex**
300-76-5 6445 206-098-3
Naled
Active ingredient: naled; fast-acting agricultural insecticide of short to moderate residual action. *Makhteshim Chemical Works Ltd.*

4570 **Bromicide**
Cooling water biocide. *Great Lakes Europe.*

4571 **Bromidine**
A dry mixture of sodium bisulfate with sodium or potassium bromide and bromate; a disinfectant.

4572 **Bromine**
7726-95-6 1415 231-778-1
Br_2
Used in the manufacture of ethylene dibromide, organic synthesis; in water disinfection; bleaching fibers; medicinals; dyestuffs. mp = -7.25°; bp = 59.47°; soluble in H_2O, alcohol, ether; critical temperature = 315°; critical pressure = 102 atm; Ingestion of solution may cause severe gastroenteritis and/or death; burns and blisters the skin. *Bromine & Chems. Ltd; Ethyl; Great Lakes Chem.; Janssen Chimica; Spectrum Chem. Mfg.*

4573 **Bromine salt**
A mixture made by saturating caustic soda with bromine, draining off the mother liquor, and adding sodium bromate; used in the extraction of gold ores.

4574 **Bromochloro dimethyl hydantoin**
126-06-7 204-766-9
$C_5H_6BrClN_2O_2$
1-Bromo-3-chloro-5,5-dimethyl hydantoin
Dantoin® BCDMH; Dantoin® GSD-550; Halobrom; Quesbrom; Bromo-1-chloro-5,5-dimethyl-2,4-imidazolinedione; Bromo-1-chloro-5,5-dimethylhydantoin; Hydantoin, 3-bromo-1-chloro-5,5-dimethyl-; Imidazolidinedione, 3-bromo-1-chloro-5,5-dimethyl-. Low temperature industrial bleach. *Lonza Inc.*

4575 **Bromocresol green**
76-60-8 1407 200-972-8
$C_{21}H_{14}Br_4O_5S$
Tetrabromo-*m*-cresolsulfonphthalein
An acid indicator pH 3.8-5.4 mp = 225°; λ_m = 168nm.

4576 **Bromocyclene**
1715-40-8 216-996-7
$C_8H_5BrCl_6$
5-bromomethyl-1,2,3,4,7,7-hexachloro-2-norbornene
Bromocyclene; Alugan; Bromodan; 5-(bromomethyl)-1,2,3,4,7,7-hexachloro-norbornene. A veterinary pesticide.

4577 **Bromodan**
1715-40-8 216-996-7
$C_8H_5BrCl_6$
Bromomethyl-1,2,3,4,7,7-hexachloro-2-norbornene
A brominating agent. *Fisons plc, Horticultural Div.* Name unverified.

4578 **1-Bromodecane**
112-29-8 203-955-3
$C_{10}H_{21}Br$
n-Decyl bromide
Decylbromide. Chemical intermediate and solvent. mp = -30°; bp = 238°; d = 1.0660; n_D^{20}= 1.4560 *Ethyl; Great Lakes; Humphrey.*

4579 **Bromoform**
75-25-2 1441 200-854-6
$CHBr_3$
Tribromomethane
Used for gem and mineral testing. Is used for generation of dibromocarbene.

Used medically as sedative, hypnotic and antitussive. mp = 5-8°; bp = 145-150°; d = 2.827; LD 50 (rat orl) = 933 mg/kg. *Geoliquids.* Name unverified.

4580 Bromo-Gas
Methyl bromide with chloropicrin. Used as a fumigant to disinfect cereals and grains. Also used as a war gas. *Great Lakes Europe.*

4581 Bromoklor 50
Halogenated aliphatic liquids containing bromine and chlorine; flame retardant used in plasticized PVC; uses including dispersed molded or coated automotive parts, interior trim, package closures, boots, hand grips, flooring, upholstery carpet backing, furniture and wall covering, and packaging film, plasticizing properties in PVC (not primary). d = 1.36-1.38; viscosity = 270 cps. *Ferro.*

4582 Bromol
118-79-6 9744 204-278-6
$C_6H_3Br_3O$
2,4,6-Tribromophenol
Tribromophenol. Used as a caustic and disinfectant. mp = 87-89°; bp_{746} = 282-290°; LD_{50} (rat orl) = 2 gm/kg.

4583 Bromolaurionite
$Pb_2OBr_2 \cdot H_2O$
A mono-hydrated lead oxydibromide, obtained by heating a solution of lead acetate and sodium bromide for 12 hours.

4584 Bromonitroform
$CBr(NO_2)_3$
Bromotrinitromethane

4585 Bromopicrin
464-10-8 207-348-4
CBr_3NO_2
Tribromonitromethane
nitrotribromomethane; nitrobromoform. Organic synthesis; military poison. mp = 10°; bp = 89-90°; d = 2.79; insoluble in H_2O, soluble in organic solvents; LD_{50} (mus ipr) = 15 mg/kg.

4586 Bromo-purpurin
Bromotrihydroxyanthraquinone

4587 Bromotrill
Bromoxynil octanoate-2,6-dibromo-4-cyanophenyl octanoate
Active ingredients: bromoxynil octanoate and 2,6-dibromo-4-cyanophenyl octanoate; selective postemergence control of a wide range of annual broadleaf weeds in winter and spring cereals and in corn. *Agan Chemical manufacturers Ltd.*

4588 Bromowagnerite
$Ca_3(PO_4)_3\ CaBr_2$
A compound of calcium bromide and calcium phosphate.

4589 Bromox
7758-01-2 7779 231-829-8
Potassium bromate
Potassium bromate dispersion in food grade filler; flour additive. *Diaflex Ltd.* Discontinued.

4590 Bromoxynil
1689-84-5 1465 216-882-7
$C_7H_3Br_2NO$
3,5-dibromo-4-hydroxybenzonitrile
Brominal; Bromotril; Buctril; Certrol B; Litarol; M&B 10064; Merit; Pardner; Sabre; Torch. A selective contact herbicide used for post-emergence control of some annual broad-leaved weeds in cereal crops. mp = 84°; soluble in H_2O (50 mg/l), organic solvents; LD_{50} (rat orl) = 365 mg/kg. .

4591 Bronco
Active ingredients are 2.6 lb of 2-chloro-2·g6·g- diethyl-N-(methoxymethyl) acetanilide and 1.4 lb of the isopropylamine salt of glyphosate; herbicide for no-till farming. *Monsanto Co.* Name unverified.

4592 Bronidox L
30007-47-7 250-001-7
5-Bromo-5-nitro-1,3-dioxane dissolved in 1,2 propylene glycol; preservative for shampoos, foam bath and other surfactant preparations. d = 1.080-1.090; cloud point 15°; flash point 100°; n_D^{20} = 1.435-1.437; corrosive to metals. *Henkel/Cospha; Henkel Canada.*

4593 Bron-Newcavac-M
Bronchitis-Newcastle disease, inactivated vaccine; for immunization of poultry. *Intervet Inc.*

4594 Bronocot
Cotton seed dressings. *ICI Chem & Polymers Ltd.*

4595 Bronopol
52-51-7 1470 200-143-0
$C_3H_6BrNO_4$
2-Bromo-2-nitropropane-1,3-diol
Bronosol; β-bromo-β-nitrotrimethyleneglycol. Broad spectrum antimicrobial agent; preservative for topical cosmetics, pharmaceuticals, toiletries. mp = 120-122°; soluble in H_2O, EtOH, EtOAc; slightly soluble in chloroform, acetone, ether, benzene; LD_{50} (rat orl) = 350 mg/kg. *Angus; Inolex.*

4596 Bronopol-Boots
52-51-7 1470 200-143-0
$C_3H_6BrNO_4$
2-bromo-1-nitro-1,3-propanediol
Bronopol; Bronosol; Lexgard bronopol; Onyxide 500; bronocot; bronidiol. Pharmaceutical and cosmetic preservative. mp = 120-122°; soluble in H_2O, EtOH, EtOAc; slightly soluble in chloroform, acetone, ether, benzene; LD_{50} (rat orl) = 350 mg/kg. *Knoll AG.* Unverified.

4597 Bronotabs
Milk testing preservative. *The Boots Co plc.* Discontinued.

4598 Bronox
Wettable powder containing linuron and tritrazine; used for control of annual dicotyledons and annual grasses in potatoes and nursery stock. *Schering Agrochemicals Ltd.* Discontinued.

4599 Bronze
Alloys usually consisting of copper and tin in varying proportions, often with zinc, and occasionally with lead. The copper varies form about 74-95%, the tin from 1-20%, zinc from 0-17%, and the lead from 0-18%; spark-resistant tools, vacuum dryers, water gauges flow indicators, valves, drain cocks, fine arts.

4600 Bronze A
A British chemical standard. It contains 85.5% copper, 9.96% tin, 1.86% zinc, 1.86% lead, 0.25% phosphorus, 0.24% antimony, 0.07% iron, 0.04% nickel, and 0.06% arsenic.

4601 bronze acetate
A calcium acetate prepared from crude pyroligneous acid and lime. It contains from 60-70% acetate. The name is also applied to an impure variety of lead acetate prepared from the same acid.

4602 bronze bearing metals
Very variable alloys; One type contains from 70-91% tin, 7-26% antimony, and 2-22% copper; while another class contains from 70-86% copper, 4-20% tin, 0-30% zinc, and 0-15% lead.

4603 bronze wire
An alloy of 98.75% copper, 1.2% tin, and 0.05% phosphorus.

4604 bronzing liquids
Volatile liquids which will hold up the metal and some material which will keep the metallic powder from rubbing off after it has been applied. The best one contains pyroxylin dissolved in amyl acetate to which the metallic powder is added.

4605 bronzing solder
An alloy of 50% zinc and 50% copper.

4606 bronzite
$16[(Mg,Fe)SiO_3]$
A pyroxene mineral.

4607 Brookosome® EFA
Phospholipids with ω-6 linoleic acid and ω-3 linolenic acid; moisturizer for skin cosmetics. *Brooks Industries.*

4608 Brookosome® EPO
Phospholipids and evening primrose oil; moisturizer for skin cosmetics. *Brooks Industries.*

4609 Brookosome® Fucus
Phospholipids, fucus extract; skin moisturizer for use in slimming creams; derived from succulent giant kelp (*Fucus vesiculosus*) *Brooks Industries.*

4610 Brookosome® TRF
Live yeast cell derivatives and phospholipids; tissue respiratory factors promoting wound healing and anti-inflammatory effects on skin. *Brooks Industries.*

4611 Brookswax D
Cetearyl alcohol, ceteareth-20; emulsifying wax substantive to hair. *Brooks Industries.*

4612 Brookswax P
Emulsifying wax substantive to hair. *Brooks Industries.*

4613 Brophos 5C10
50643-20-4
PPG-5 ceteth-10 phosphate
Surfactant, emulsifier substantive to hair. *Brooks Industries.* Name unverified.

4614 Brophos OL-3
39464-69-2
Oleth-3 phosphate
Surfactant, emulsifier substantive to hair. *Brooks Industries.*

4615 Brophos OL-3N
58855-63-3

DEA oleth-3 phosphate
Surfactant, emulsifier substantive to hair. *Brooks Industries.*

4616 Broponol
52-51-7 1470 200-143-0
$C_3H_6BrNO_4$
2-Bromo-2-nitropropane-1,3-diol
β-bromo-β-nitrotrimethyleneglycol; Bronosol; Bioban® BNPD-40. Broad spectrum antimicrobial, preservative for metalworking fluids. mp = 120-122°; soluble in H_2O, polar organic solvents; insol.in ligroin; LD50 (rat orl) = 400 mg/kg. *Angus.*

4617 Brown 208
1309-37-1, 7787-59-9 232-122-7
Iron oxides and bismuthoxychloride; inorganic colorant. *Presperse.* Discontinued.

4618 Brown Copp.
1317-39-1 2734 215-270-7
Cuprous Oxide
Brown cuprous oxide, inorganic colorant. *Am. Chemet.*

4619 Brown lead ore
$3[Pb_3(PO_4)_2]\cdot 2PbCl_2$
Linnets
A mixture of lead phosphate and lead chloride.

4620 Brown ore
A variable mixture of hydrated oxides of iron, usually $2Fe_2O_3 \cdot H_2O$.

4621 Brown oxide of tungsten
WO_2
Tungsten dioxide

4622 Brown precipitate
Iodine dissolved in potassium iodide.

4623 Brugre powder
ammonium picrate-potassium nitrate. An explosive; a priming composition, containing 54% ammonium picrate, and 46% potassium nitrate.

4624 Brunner's salt
$HgSK_2S\cdot 5H_{20}$
Obtained by dissolving vermilion (red mercuric sulfide) in potassium monosulfide.

4625 Brunol
2052-14-4 218-142-9
$C_{11}H_{14}O_3$
2-hydroxy-benzoic acid, butyl ester
Benzoic acid, 2-hydroxy-, butyl ester; Butyl salicylate. A proprietary preparation of n-butyl salicylate. Used as a fragrance. *Catomance Group.*

4626 Brunswick black
Asphalt or pitch mixed with turpentine and linseed oil, and heated.

4627 Brunswick blue
Celestial blue
A pigment produced by mixing 50-90% barytes with Prussian blue.

4628 Brush Buster
2,4-D and 2,4-DP; A post-emergence herbicide used to control woody species such as poison oak/ivy and brambles; applied as a foliar spray or straight from the bottle to the cut stump of woody species. *Lawn & Garden Products Inc.*

4629 Brush wire
A brass wire containing 64.25% copper, 35% zinc, and 0.75% tin.

4630 Brush-B-Gon
Brush killer. *Monsanto (Solaris).* Name unverified.

4631 Brushcrete
Two-component cementitious coating providing a hard-wearing, seamless, waterproof membrane which protects and resurfaces concrete masonry and other construction materials. *Feb Ltd.*

4632 Brussels System
A systemic insecticide. *Murphy Chemical Co Ltd.* Discontinued.

4633 BTC® 99
7173-51-5 3149 230-525-2
Didecyl dimonium chloride
Low foaming algicide and slimicide for swimming pool and industrial water treatment. *Stepan; Stepan Canada.*

4634 BSWL 202
10099-76-0 233-246-4
O_3SiPb
Silicic acid (H2SiO3), lead(2+) salt (1:1)
Lead silicate; Lead monosilicate. Basic silicate white lead; white pigment acting as heat stabilizer for chlorinated polyethylene, chlorosulfonated polyethylene, PVC, and polyepichlorohydrin; rust-inhibitive pigment in the automobile industry; used industrial or maintenance paints. mp = 680-730°; d = 6.50. *Eagle-Picher.* Name unverified.

4635 BTC® 818
32426-11-2 251-035-5
$CaCO_3$
Quaternium-24
Disinfectant, sanitizer, and fungicide for hard surfaces; excellent sanitizer in hard water to 800 ppm as $CaCO_3$. Liquid; soluble in H_2O; d = 0.93; flash point = 86°F; cationic. *Stepan; Stepan Canada.*

4636 BTC® 824
139-08-2 205-352-0
Myristalkonium chloride
Antimicrobial for hard surf. disinfection and sanitization; algicide for swimming pools and industrial water treatment. Liquid; d = 0.96; flash point = 120°F; cationic. *Stepan; Stepan Canada.*

4637 BTC® 885 P40
Quaternium-24, benzalkonium chloride; germicide for formulation of disinfectant, sanitizer and fungicidal products for hospitals, nursing homes, public institutions. Powder; cationic. *Stepan; Stepan Canada.*

4638 BTC® 1010-80
7173-51-5 3149 230-525-2
$C_{22}H_{48}ClN$
Quaternium-12
1-Decanaminium, N-decyl-N,N-dimethyl-, chloride; Arquad 10; Bardac 22; Decanaminium, N-decyl-N,N-dimethyl-, chloride; didecyl dimethyl ammonium chloride; Quaternium 12. Fungicide for hard-surface disinfection and sanitization; algicide in swimming pool and industrial water treatment; deodorizer. Soluble in H_2O. *Stepan; Stepan Canada.*

4639 BTC® 2565
139-08-2 205-352-0
Myristalkonium chloride
Algicide and slimicide for swimming pool and industrial water treatment. *Stepan; Stepan Canada.*

4640 BTG alloy
A heat-resisting alloy containing 60% nickel, 12% chromium, 1-4% tungsten, and balance iron.

4641 Bubber Shet
57-13-6 10005 200-315-5
Urea
Prilled urea containing minimum of 46% available nitrogen; used as a fertilizer to supply nitrogen to the crops for better yield per acre. *Dawood Hercules Chemicals Ltd.*

4642 Bubble Breaker® 3056A
Dispersion of reacted silica in a hydrocarbon solvent; defoamer used in latex manufacturing operations, formulation of water-based paints and adhesives, effluent water, asphalt emulsions, PVC monomer stripping. Soluble in mineral spirits, dispersion in H_2O; d = 0.89; flash point >200°F; pH=5.0. *Witco/Organics.* Discontinued.

4643 Bubble Breaker® 748
Silicone-free blend; defoamer for water-based systems, paints and coatings. Soluble in mineral spirits, dispersion in H_2O; d = 0.87; flash point = 200°F; pH = 9.5. *Witco/Organics.* Discontinued.

4644 Bubble Breaker® DMD-1
Complex surfactant; oilfield surfactant, defoamer; personal care formulations. Liquid, soluble in iPrOH; dispersion in H_2O, kerosene, xylene; d = 8.4 lb/gal; viscosity = 600 cps; pour point < 0°F; pH = 10.5. *Witco/Organics.* Discontinued.

4645 Bubblefil
A proprietary trade name for regenerated cellulose. No manufacturer.

4646 Buca
1332-58-7 5294 296-473-8
Kaolin
Very fine particle size ultra pulverized hydrous kaolin (aluminum silicate); used as an extender pigment in coatings, inks, caulks, rubber. *Engelhard.*

4647 Bucarpolate
136-63-0
Piperonylic acid, diethylene glycol monobutyl ether ester
Butoxy(ethoxy)ethyl ester of piperonylic acid; Bucarpolate; Piperonylic acid, diethylene glycol monobutyl ether ester. A pyrethrum synergist. *Bush Boake Allen Ltd.*

4648 Buckland's cement
A label cement consisting of 50% gum arabic, 37.5% starch, and 12.5% white sugar. It is mixed with a little water for use. No manufacturer.

4649 Buckroid
A very tough form of pure vulcanized rubber; used for making mats, and for other purposes.

4650 Buctril
1689-84-5 1465 216-882-7
bromoxynil
Selective weedkiller. *May & Baker Ltd.* Name unverified.

4651 **Budale**
A proprietary preparation of paracetamol, codeine phosphate and butobarbital; a sedative for veterinary use. *Dales Pharmaceuticals Ltd.* Name unverified.

4652 **Budene® 1207**
9003-17-2
Polybutadiene
Nonstaining, solution polymerized; used in tire tread and carcass compounds, v-belts, conveyor belt covering, hose covers and tubes, solid golf balls, footwear, sponge rubber, mechanical goods; imparts abrasion resistance, low temperature properties, resilience and durability. *Goodyear.*

4653 **Budene® 1254**
9003-17-2
Polybutadiene
Extended with 18.5-21.5% petroleum process oil used for treads, belts, retread stocks, mechanical goods; good dynamic properties, enhanced compound processing. *Goodyear.* Name unverified.

4654 **Bufa**
84-74-2 1622 201-557-4
Dibutyl phthalate
A proprietary trade name for dibutyl phthalate, a plasticizer. *Koninklijke Maastrichtsche.*

4655 **bufexamac**
2438-72-4 1497 219-451-1
$C_{12}H_{17}NO_3$
4-butoxy-N-hydroxybenzeneacetamide
CP 1044 J3; Droxarol; Droxaryl; Feximac; Malipuran; Mofenar; Norfemac; Parfenac; Parfenal; para-Butoxyphenylacetohydroxamic acid; Bufexamic acid; Paraderm. A proprietary preparation of bufexamac used as a skin cream. mp = 153-155°; insoluble in H_2O; LD_{50} (rat orl) >4 g/kg. Name unverified.

4656 **Bufferight**
144-55-8 8726 205-633-8
Sodium bicarbonate
An alkalizer. *Kerr-McGee Chemical Corp.* Discontinued.

4657 **Bug Check® BF**
Bacterial and fungal contamination indicator. *Angus.*

4658 **Bug Check® SRB**
Simplified sulfate-reducing bacteria viable counts. *Angus.*

4659 **Bug Gun**
Ready-for-use insecticide spray. *ICI Garden Products.*

4660 **Bu-Gas**
106-97-8 1541 203-448-7
n-Butane
Butane. Fuel and propellant. *Monsanto (Solaris).*

4661 **Bug-Geta**
Slug and snail pellets. *Monsanto (Solaris).* Name unverified.

4662 **Bull metal**
An alloy similar to Delta metal in composition.

4663 **Bulldock**
68359-37-5 2826 269-855-7
β-cyfluthrin
Insecticide with mainly contact action against biting insects on maize, cotton, deciduous fruit, groundnuts, potatoes, vegetables and other crops; also against migratory locusts and grasshoppers. *Bayer AG.*

4664 **Bullion®**
For agriculture industry. *DuPont UK.*

4665 **Bullseye**
A double-base-type formulation minimizing charge weight and moisture absorption; graphite glaze enables smooth granule flow; a smokeless powder used for ammunition. *Hercules.* Discontinued.

4666 **Bullseye**
Water soluble, biodegradable colorants; for spray pattern indicators for application of herbicides, pesticides in the lawn and turf, forestry and industrial weed control areas. *Milliken.*

4667 **Bullseye CDA**
Amitrole, atrazine and diuron; a liquid mixture of herbicides for weed control. *ICI Agrochemicals.*

4668 **Bumal**
Modified rosin emulsion. *Tenneco Malros Ltd.* Name unverified.

4669 **Bumper**
60207-90-1 8003 262-104-4
Propiconazole
Fungicide. *Makhteshim Chemical Works Ltd.*

4670 **Bumyr**
110-36-1 203-759-8
Butyl myristate
Emollient; ingredient in cosmetics. *Amerchol Corp.* Discontinued.

4671 **Buna Hüls AP**
A range of ethylene/propylene rubber; used as modifiers for polyolefins, molded articles and hoses, blend component for improving flowability and green strength. *Hüls AG.* Discontinued.

4672 **Buna Hüls EM**
A range of styrene/butadiene rubbers; emulsion polymerization; used for tires, injection moldings and industrial rubber goods. *Hüls AG.* Discontinued.

4673 **Buna S**
A butadiene-styrene copolymer; A proprietary trade name for a vulcanizable synthetic rubber. No manufacturer.

4674 **Buna SL**
A range of styrene/butadiene rubbers; solution polymerization; used for tire treads, industrial rubber goods, shoe soles. *Bayer AG.* Name unverified.

4675 **Buna SS**
An important series of synthetic products is made by copolymerizing butadiene with 10-30% of another polymerizable substance such as styrene or acetonitrile, e.g., Buna N, etc. Contains a larger percentage of styrene. No manufacturer.

4676 **Buna VI**
A range of vinyl/butadiene rubbers; used for industrial goods and tires. *Bayer AG. Name unverified.*

4677 **Buna® CB**
cis-1,4-polybutadiene polymers; used for tires, conveyor belting, mountings and roll covers and plastics modification. *Bayer AG.*

4678 **Buna® CB 11**
Butadiene rubber, organometallic titanium catalyst; used in tires, conveyor belting, caterpiller tread blocks, footwear soles, transmission belting; for blending with NR in buffers, roll covers, seals, and injection molded goods. *Bayer; Miles; Polysar.*

4679 **Buna® CB 22**
Butadiene rubber, neodymium catalyst type, nonstaining stabilizer; used in blends with other general purpose rubbers (NR and IR); imparts excellent abrasion resist., rebound resilience, dynamic fatigue resist; in NBR and CR, improves the low temp. brittle *Bayer; Miles; Polysar.*

4680 **bunamidine hydrochloride**
1055-55-6 1510 213-890-2
$C_{25}H_{38}N_2O{\cdot}HCl$
N,N-dibutyl-4-(hexyloxy)-1-naphthalenecarboximidamide hydrochloride
Scolaban. Anthelmintic. mp = 214-215°; LD_{50} (rat orl) = 540 mg/kg.

4681 **Bunatex**
A range of styrene/butadiene rubber latex; used as raw materials for foam backing for carpets and textiles, mattresses, upholstery materials for the furniture industry, shoe components, rubberized hair bitumen modifiers. *Hüls AG.*

4682 **Bunte's salt**
1515
Sodium-ethyl thiosulfate
Water soluble salts of certain alkyl or aralkyl thiosulfuric acids of the general formula $RSSO_2ONa$. Have been used as coccodiostats.

4683 **Bur-A-Loy® 3873**
NBR/PVC (60:40), 70 parts DOP plasticizer; highly plasticized; soft roll and printing blanket stocks; excellent PVC dispersions and end-processing. *Mach-1 Compounding.*

4684 **Bur-A-Loy® 3874**
NBR/PVC (60:40), 85 parts DOP; used for low durometer rolls and molded goods; good oil, abrasion, and ozone resist. *Mach-1 Compounding.*

4685 **Bur-A-Loy® 5915**
NBR/PVC (50:50), 15 parts DOP plasticizer; thermoset or thermoplastic elastomer; for heels, soles, mats, and extruded goods where oil, abrasion, and ozone resistance is required. *Mach-1 Compounding.*

4686 **Bur-A-Loy® 7130**
NBR/PVC (70:30), 30 parts DOP plasticizer; for hose, oil seals, and general molded and extruded goods; exhibits excellent oil and fuel resistance. *Mach-1 Compounding.*

4687 **Burco Anionic APS**
Ethoxylated sulfonate
Hypochlorite-stable surfactant, emulsifier for acid and alkaline cleaners, disinfectants, personal care products, household cleaners, tub and tile cleaners, mildew removers, textile scours, dairy cleaners; stable over entire pH range. Liquid; pH = 7.5; flash point > 200°F. *Burlington Chem.* Discontinued.

4688 **Burco CS-LF**
Phosphate ester of EO/PO block polymer, halogen-capped ethoxylated polyether; biodegradable low foaming synergistic surfactant blend, scouring agent for textile processing, dishwash rinse aid, hard surface cleaners, liquid laundry products, steam cleaners; stable to 10% alkali. *Burlington Chem.*

4689 Burco DFE-45
Polyol ester; emulsifier, defoamer for paper, textiles, water treatment, coatings, metalworking applications and low-foaming emulsions. Amber liquid; soluble in oil and solvents; d = 0.94; pH 5. *Burlington Chem.*

4690 Burco LAF-6
Aliphatic alcohol alkoxylate
Biodegradable low foam detergent, wetting agent for rinse aid formulations, automatic dishwash, metal cleaners, textile scouring agents; stable to acids and alkalies. Pale yellow to clear liquid; d = 0.99; cloud point = 37°; flash point >300°F; pH = 6-8. *Burlington Chem.*

4691 Burco NCS-80
Surfactant blend; moderately foaming detergent intermediate for formulating mildly alkaline cleaners for use as vehicle cleaners, floor cleaners, car wash. *Burlington Chem.* Discontinued.

4692 Burco NPS-225
68515-73-1
Alkyl polyglycoside
Dispersant, wetting agent, coupling agent in high electrolyte solutions; for hard surface cleaners (acid or alkali), all-purpose cleaners, metal cleaners; hydrotrope in highly alkaline formulations; biodegradable; stable to acids or alkali. *Burlington Chem.*

4693 Burco TME
Ethoxylated dodecylmercaptan
Detergent for aqueous cleaning systems, wool scouring, hard surface cleaners; oil splitter; replaces chlorinated hydrocarbons in metal degreasing. Yellow, clear liquid; d = 1.07; viscosity = 5.7 cps; HLB = 13.5; cloud point = 160°F; flash point = 127°F; pH=4. *Burlington Chem.*

4694 Burcofac 1060
Organic phosphate ester; wetting agent, detergent, emulsifier, hydrotrope for other surfactants, glycol ethers; for mildly alkaline cleaner formulations, all-purpose cleaners, floor cleaners/wax strippers, carwash, textile scouring; alkaline stable. Clear liquid; d = 1.07; acid no. 205 to pH9.5; pH (1% solution) = 2.0. *Burlington Chem.*

4695 Burcol BP-181
Low-foaming block polymer; defoamer for textiles, paper, metalworking, antifreeze, and other applications; lubricant base; rinse aid formulations. *Burlington Chem.* Discontinued.

4696 Burcop
Soda/Bordeaux fungicide *McKechnie Chemicals Ltd.* Name unverified.

4697 Burcosperse AP Liq
9003-04-7
Low molecular weight sodium polyacrylate solution; chelating agent, antiredeposition agent. *Burlington Chem.*

4698 Burcoterge DG-40
Linear alcohol derivatives; wetting agent, detergent, emulsifier for laundry and hard surface cleaners, microemulsions with solvents, degreasing formulations; excellent performance in hot or cold systems. *Burlington Chem.*

4699 Burcotreat 900-A
9003-01-4
Polyacrylic acid
Chelating agent, antiredeposition agent; free acid version of Burcosperse AP Liquid. *Burlington Chem.*

4700 Burcowet TM-LF
Alkoxylated linear alcohol; defoamer at high temperatures; adds wetting, detergency. *Burlington Chem.*

4701 Burez
Disproportionated rosin and rosin derivatives. *Albright & Wilson Ltd; Tenneco Malros Ltd.*

4702 Burgess 30-P
12141-46-7 377 235-253-8
Aluminum silicate
Thermo-optic; used in PVC compounds; combines ease of incorporation, uniformity of compound color, excellent electrical properties and low specific gravity. *Burgess Pigment.*

4703 Burgess 2211
12141-46-7 377 235-253-8
Aluminum silicate, anhydrous
Surface treated; used in filled nylon applications and polyterephthalate, urethane, PVC, thermoplastic polyester; features low warpage and high impact strength. *Burgess Pigment.*

4704 Burgess KE
12141-46-7 377 235-253-8
Aluminum silicate, anhydrous
Aluminum silicate, anhydrous, surface treated; very pure, high brightness clay for use in EPR, EPT, crosslinked polyethylene and polyester systems; excellent wet and dry, initial and long term electrical characteristics; increases tensile strength and compression set. *Burgess Pigment.*

4705 Burgess solder
An alloy of 76% tin, 21% zinc, and 3% aluminum.

4706 Burgess-Hambuechen solution
A solution containing 275 g ferrous ammonium sulfate and 1000 ml water; used for the electrodeposition of iron, using a current density of 6-10 amps/sq. foot at 30°.

4707 Burgundy Lake
A proprietary trade name for a red lake containing organic colors, aluminum hydroxide, and Blanc fixe. No manufacturer.

4708 Burkeite
Gauslinite
A double salt of sodium carbonate and sulfate.

4709 Burmite
Birmite
Burmese amber of a reddish-brown color.

4710 Burmol®
For stripping of dyeings and for removing discoloration and stains from textiles; for commercial laundries. *BASF AG.*

4711 burnt carmine
A pigment obtained by calcining carmine.

4712 burnt hypo
Black hypo; Eureka compound. A mixture of lead thiosulfate and sulfide, and sulfur; used in the vulcanization of rubber.

4713 burnt iron
Iron which has been heated to a high temperature for a long time. It is brittle.

4714 Burnt magnesia
1309-48-4 5713 215-171-9
Magnesium oxide
Used in refractories, especially for steel furnace linings, polycrystalline ceramic for aircraft windshields; removal of sulfur dioxide from stack gases.

4715 burnt nickel
A term used for a grey pulverulent nickel precipitated by too strong a current during its electrodeposition.

4716 burnt pyrites
Fe_2O_3
Pyrites which have been burnt until 70% of ash is left; consists primarily of iron oxide,.

4717 burnt topaz
When Brazilian topaz is heated, it changes from a cherry-yellow to a rose-pink, and is then known as burnt topaz.

4718 burnt umber
Umber which has been heated, whereby its color is somewhat reddened. Raw umber is a brown earthy variety of ocher, colored by oxides of iron.

4719 Buro-sol Concentrate
8006-13-1 332
Aluminum acetate
Astringent. *Doak Pharmacal Co.*

4720 Burow's solution
A 7.5-8% solution of aluminum acetate. (Liquor *alumini acetatis*, BPC); an astringent and antiseptic.

4721 Bursoline
Sulfonated oil for tanners. *Clayton Aniline Co Ltd.* Name unverified.

4722 Burtolin
123-33-1 5745 204-619-9
Maleic hydrazide
A tree growth inhibitor containing 185 g/liter maleic hydrazide (as the potassium salt); used to control shoots on the trunk and suckers around the base of street trees; it also inhibits the development of buds on the trunk which remain dormant following treatment. *Rhône-Poulenc Environmental Prods. Ltd.*

4723 Bush metal
An alloy of 72% copper, 14% tin, and 14% yellow brass.

4724 Bushwacker
34014-18-1 9255 251-793-7
Tebithiuron
Wettable powder or granules containing tebuthiuron; used for total weed control in noncrop areas. *Rhône-Poulenc Environmental Prods. Ltd.*

4725 but-2-yne-1,4-diol
110-65-6 203-788-6
$C_4H_6O_2$
1,4-butynediol
2-butyne-1,4-diol; 1,4-Dimethoxyacetylene; 1,4-Butynediol; but-2-yne-1,4-diol. Corrosion inhibitor in acid pickles and cleaners. mp = 52-54°; bp = 238°; n_D^{20}= 1.4804; LD_{50} (rat orl) = 105 mg/kg. *BASF; ISP.*

4726 Butac®
Resin acids-amine resin soaps blend; tackifier for IIR, NR, SBR, molding aid

for NBR, SBR; improves pigment dispersions; activates cure slightly. *Whitney & Oettler.* Name unverified.

4727 Butachlor
23184-66-9 1533 245-477-8
$C_{17}H_{26}ClNO_2$
N-(butoxymethyl)-2-chloro-2',6'-diethylacetanilide
2-chloro-2,6-diethyl-N-(butoxymethyl)acetanilide; Butanex; CP-53619; Machete. Active ingredient; selective pre-emergence and early post-emergence weed control in transplanted, direct seeded and upland rice. $bp_{0.5}$ = 196°; d_4^{30}= 1.0695; soluble in H_2O (20 mg/l), organic solvents; LD_{50} (rat orl) = 1740 mg/kg. *Agan Chemical Manufacturers Ltd.* Discontinued.

4728 Butacite®
Polyvinyl butyral. *DuPont UK.*

4729 Butaclor
Polychloroprene rubber. *BP Chemicals Ltd.* Discontinued.

4730 butadiene/styrene copolymer
B/S. Mechanical rubber goods, flooring tile, tire tread, rug backing, adhesives, asphalt modification, injection molded items, medical devices, containers, toys, food containers; blendable modifier. *BASF; Firestone Syn. Rubber & Latex; Goodyear Tire & Rubber; Reichhold/Emulsion Polymers; Shell.*

4731 Butakon A2554
A proprietary butadiene copolymer rubber. *Revertex Ltd.* Discontinued.

4732 Butakon ML 57711
A proprietary butadiene/methyl methacrylate latex. *Revertex Ltd.* Discontinued.

4733 butamisole hydrochloride
54400-62-3 1540
$C_{15}H_{19}N_3OS \cdot HCl$
2-methyl-N-[3-(2,3,5,6-tetrahydroimidazo[2,1-b]thiazol-6-yl)phenyl]propanamide hydrochloride
CL-206214; Styquin. Anthelmintic.

4734 Butane
106-97-8 1541 203-448-7
C_4H_{10}
n-Butane
A-17; Bu-Gas. Butane. Used as producer gas and propellant; raw material for motor fuels, in the manufacture of synthetic rubbers. May be narcotic in high concentrations. A simple asphyxiant. Flammable gas, flash point = -138°; bp = -0.50°; d (gas) = 2.046. *Monsanto (Solaris); Air Prods & Chem; Electrochem Ltd; Phillips 66 Co.* Name unverified.

4735 1,4-butanediol
110-63-4 203-786-5
$C_4H_{10}O_2$
1,4-butylene glycol
tetramethylene glycol; 1,4-Butylene glycol; 1,4-Dihydroxybutane; 1,4-Tetramethylene; Tetramethylene glycol; butanediol; butane-1,4-diol; diol 14b; sucol b; 1,4-tetramethylene glycol; butylene glycol; tetramethylene 1,4-diol; BDO. Intermediate; used in polyurethane formulation in the hard segment as a curative. mp = 16°; bp = 230°; d = 1.017; soluble in H_2O (>10 g/100 ml), organic solvents. *Arco; BASF; DuPont; Hüls UK; ISP.*

4736 Butanex
23184-66-9 1533 245-477-8
$C_{17}H_{26}ClNO_2$
N-(butoxymethyl)-2-chloro-2·g,6·g-diethylacetanilide
Butachlor. Active ingredient, butachlor; selective pre-emergence and early post-emergence weed control in transplanted, direct seeded and upland rice. *Agan Chemical Manufacturers Ltd.*

4737 2-butanone
78-93-3 6149 201-159-0
C_4H_8O
methyl ethyl ketone
MEK; 2-oxobutane. Solvent in nitrocellulose coatings and vinyl films, paint removers, cements, adhesives, organic synthesis; manufacture of smokeless powder, cleaning fluids, priming, catalyst carrier, acrylic coatings. mp = -86°; bp = 80°; d_4^{20}= 0.805; n_D^{15}= 1.3814; soluble in H_2O (27 g/100 ml), organic solvents; LD_{50} (rat orl) = 5.5 g/kg. Elf Atochem N. Am.; BP Chem. Ltd.; Exxon; Hoechst Celanes; Mallinckrodt; Shell; Texaco; Union Carbide.

4738 2-butanone peroxide
1338-23-4 215-661-2
$C_8H_{16}O_4$
methyl ethyl ketone hydroperoxide
MEK peroxide; 2-butanone peroxide; MEKP; ethyl methyl ketone peroxide. Initiator/catalyst for curing of unsaturated polyester resins. bp = 110°; insoluble in H_2O (<5 mg/ml) *Akzo; Elf Atochem; Cook Composites & Polymers; Norac; Witco/Argus.*

4739 Butanox
1338-23-4 215-661-2
Methyl ethyl ketone peroxide
Used for the ambient temperature curing of unsaturated polyester resins. *Akzo Chemie UK Ltd.* Name unverified.

4740 Butarez
Liquid polybutadienes *Phillips.*

4741 Butasan Vulcanization Accelerator
136-23-2 205-232-8
Zinc dibutyldithiocarbamate
Nondiscoloring stabilizer in noncuring applications and in butyl rubber. *Monsanto Co.*

4742 Butazate
136-23-2 205-232-8
Zinc dibutyldithiocarbamate
An ultra accelerator, fast curing from 212°F (100°C) and up; low critical temperature, medium curing range below 250°F (121°C); nonstaining and nondiscoloring; for latex compounding; has strong tendency to precure. *Uniroyal.* Name unverified.

4743 Butazate 50D
136-23-2 205-232-8
Zinc Dibutyldithiocarbamate
A 50% active water dispersion of Butazate: ready-to-use form for latex compounding; used in combination with Ethazate 50D and Oxaf for greater economy and improved physical properties. *Uniroyal.* Name unverified.

4744 Butex
9010-85-9
Reclaimed rubber from butyl inner tubes; used in tire inner liners, inner tubes, and in butyl tapes and sealants. *Midwest Rubber Reclaiming.*

4745 Butex
A proprietary trade name for a synthetic rubber. No manufacturer.

4746 Butisan
67129-08-2 266-583-0
metazachlor
Suspension concentrate containing 500 g/l metazachlor; used for weed control in brassicas and ornamental crops. *Bayer plc.*

4747 Butisan® S
67129-08-2 266-583-0
Metazachlor
Systemic herbicide against annual grasses, broadleaf weeds *BASF AG.*

4748 Butofan®
Butadiene polymer dispersions; paper impregnation and coating; binders for paper, adhesives, and textile coatings; binders for production of materials based on leather fibers. *BASF AG.*

4749 Butofan® D
Polybutadiene dispersion. *BASF plc.*

4750 Butonal®
Butadiene/styrene polymer dispersions; binders for production of adhesives, for treatment of asphalt. *BASF AG.*

4751 Butox
A proprietary preparation of polyisobutylene-isoprene. *Hardman.*

4752 Butoxone
94-82-6 2893 202-366-9
2,4-DB. Herbicide. *ICI Chem & Polymers Ltd.*

4753 2-butoxyethanol
111-76-2 1594 203-905-0
$C_6H_{14}O_2$
Ethylene glycol n-butyl ether
Butyl cellosolve; Dowanol EB; Butyl oxitol; Jeffersol EB; Ektasolve EB; BUCS; gafcol eb; glycol butyl ether; glycol ether eb; glycol ether eb acetate; 3-oxa-1-heptanol; poly-solv eb; 2-n-Butoxyethanol; Ektasolve EB solvent. Solvent for nitrocellulose resins, spray lacquers, co-solvent, gas chromatography. bp_{12} = 67°; d = 0.9020; soluble in H_2O (5 g/100 ml), organic solvents; LD_{50} (rat orl) = 470 mg/kg. *Arco.*

4754 2-butoxyethanol acetate
112-07-2 203-933-3
$C_8H_{16}O_3$
Ethylene glycol butyl ether acetate
Ektasolve™ EB Acetate; Butyl Cellosolve™ Acetate; Ethylene glycol monobutyl ether acetate; Butyl Cellosolve Acetate; Ethylene glycol butyl ether acetate; Ektasolve EB Acetate; Ethylene Glycol Mono-n-butyl Ether Acetate; n-Butyl Cellosolve Acetate; Butyl glycol acetate; 2-Butoxyethanol acetate. Solvent for high-solids coatings d = 7.84 lb/gal (20°); bp = 192.3°; flash point = 165°F. *Arco; Eastman; OxyChem; Union Carbide.*

4755 Butoxyethyl stearate
109-38-6 203-668-3
$C_{24}H_{48}O_3$
2-Butoxyethyl stearate
Dermol BES; KP-23. Butoxyethyl stearate, a solvent. *Alzo.*

4756 Butoxyne 497
A mixture of hydroxyethyl ethers of butynediol; nickel brightener in electroplating; pickling inhibitor used prior to plating copper; corrosion inhibitor for specialty application such as in aerosol cans. *ISP.*

4757 Butter of paraffin
Soft paraffin.

4758 Butter of sulfur
Precipitated sulfur.

4759 Butter yellow
60-11-7 3279 200-455-7
$C_{14}H_{15}N_3$
4-dimethylaminoazobenzene
Oil yellow; butyro flavine; Dimethyl Yellow Analar; Dimethyl Yellow N,N-dimethylaniline; DMAB; Enial Yellow 2G; Fast Oil Yellow B; Fast Yellow; Fat Yellow; Fat Yellow A; Fat Yellow AD OO; Fat Yellow ES; Butter Yellow; Methyl Yellow; Dab; Oil Yellow; C.I. Solvent Yellow 2; Atul Fast Yellow R; Brilliant Fast Oil Yellow; Brilliant Fast Spirit Yellow; Cerasine Yellow GG; C.I. 11020; Dab (Carcinogen); Dimethyl Yellow; Fat Yellow R; Fat Yellow R (8186); Grasal Brilliant Yellow; Oil Yellow II; Oil Yellow 20; Oil Yellow 2625; Oil Yellow 7463; Oil Yellow BB; Oil Yellow D; Oil Yellow DN; Oil Yellow FF; Oil Yellow FN; Oil Yellow G; Oil Yellow 2G; Oil Yellow G-2; Oil Yellow GG; Oil Yellow GR; Oil Yellow N; Oil Yellow Pel; Oleal Yellow 2G; Organol Yellow ADM; Orient Oil Yellow GG; P.D.A.B.; Petrol Yellow Wt; Resinol Yellow GR; Resoform Yellow GGA; Silotras Yellow T2G,. Azo dyestuffs based on dimethylaminoazobenzene were formerly used for coloring butter and oils. Acid-base indicator; red at pH 2.9, yellow at pH 4.0 mp = 111°; bp = 200°; LD_{50} (rat orl) = 200 mg/kg.

4760 Button metal
An alloy of 57% zinc and 43% copper.

4761 Button solder
White solder. Usually contains 50% tin, 30% copper, and 20% brass; or 33% copper, 27% brass, and 40% zinc.

4762 Butvar
Polyvinyl butyral resins; white, free-flowing powders supplied in seven resin types or grade forms with various hydroxyl content whose solutions provide a wide range of viscosities; recommended for upgrading the coating performance of thermosetting phenolics, ureas, melamines and polyesters. *Monsanto Co.*

4763 Butyl benzoate
136-60-7 1587 205-252-7
$C_{11}H_{14}O_2$
benzoic acid butyl ester
Marvanol® Carrier BB. Carrier used in dyeing. mp = -22°; bp = 250°; d = 0.9900; n_D^{20} = 1.4960; insoluble in H_2O, soluble in organic solvents; LD_{50} (rat orl) = 5.14 g/kg. *NovaChem Corp; Penta Mfg.; Pentagon Chemicals Ltd; L5790PMC Specialties; Raschig; Uniroyal.*

4764 Butyl benzyl phthalate
85-68-7 201-622-7
$C_{19}H_{20}O_4$
1,2-Benzenedicarboxylic acid butyl phenylmethyl ester
Santicizer 160; Unimoll® BB; BBP; n-Butyl Benzyl Phthalate; Benzyl butyl phthalate; benzyl n-butyl phthalate; butyl phenylmethyl 1,2-benzenedicarboxylate; santicizer 160; Palatinol BB; Sicol 160; Unimoll BB. Aromatic ester; plasticizer for polyvinyl and cellulosic resins, organic intermediate. mp = -35°; bp = 370; d = 1.100; slightly soluble in H_2O (0.269 mg/100 ml), more soluble in organic solvents; flash point = 230°F; LD_{50} (rat orl) = 2330 mg/kg. *Ashland; Monsanto.*

4765 Butyl bisbutyl peroxy valerate
995-33-5 213-626-6
n-Butyl-4,4-bis (t-butylperoxy) valerate
Lupersol 230; Luperco 230-XL. Initiator for bulk, solution, and suspension polymerization, for high-temperature cure of polyester resins, for curing elastomers, and for cure of acrylic syrup. *Elf Atochem.*

4766 Butyl Carbitol®
112-34-5 1592 203-961-6
$C_8H_{18}O_3$
2-(2-Butoxyethoxy)ethanol
Diethylene glycol monobutyl ether; Butyl digol; Butyl Diicinol; Butoxydiglycol; Butyl Carbitol; Diethylene Glycol Butyl Ether; 2-(2-n-Butoxyethoxy)ethanol; Diethylene Glycol Mono-n-butyl Ether; n-Butyl Carbitol; Diethylene glycol, monobutyl ether; Glycol, monobutyl ether; butoxydiethylene glycol; O-butyl diethylene glycol; butyl dioxitol; diethylene glycol n-butyl ether; diethylene DB; diglycol monobutyl ether; Dowanol DB; Ektasolve DB; glycol ether DB; Jeffersol DB; Poly-solv DB; 3,6-Dioxa-1-decanol; Butoxyethoxy)ethanol; Dowanol OR; 3,6-Dioxadecanol. A lacquer solvent boiling at 230°. mp = -68.1°;bp = 230.4°; soluble in H_2O, organic solvents.

4767 Butyl Cellosolve®
111-76-2 1594 203-905-0
$C_6H_{14}O_2$
2-Butoxyethanol
Ethylene glycol monobutyl ether; Ethylene glycol butyl ether; Ether alcohol; Butoxyethanol; Butyl Icinol; Dowanol™EB; Ektasolve™EB; Butyl Cellosolve™;. Solvent for nitrocellulose resins, spray lacquers, cosolvent, gas chromatography. Liquid; bp = 171-172°; soluble in H_2O, organic solvents; d_{20} = 7.52 lb/gal; LD_{50} (rat orl) = 1.48 g/kg. *Arco; Union Carbide; ICI Australia.*

4768 Butyl cumyl peroxide
3457-61-2 222-389-8
$C_{13}H_{20}O_2$
1,1-dimethylethyl 1-methyl-1-phenylethyl peroxide
Luperco 801-XL; Trigonox® T. *t*-Butyl cumyl peroxide on inert filler; initiator for high-temperature cure of polyester resins, curing elastomers, and polymer modification thermoplastic cross-linking. *Elf Atochem.*

4769 Butyl Di-icinol
112-34-5 1592 203-961-6
2-(2-butoxyethoxy)ethanol
Solvent for use in protective coatings, inks, cleaning products, agricultural chemicals; aids wetting, penetration, and soil removal; coupling solvent. *ICI Australia.*

4770 Butyl Eight®
Dithiocarbamate compound in a solvent; ultra accelerator for vulcanization. *R. T. Vanderbilt Co Inc.*

4771 Butyl Icinol
111-76-2 1594 203-905-0
Butyl Cellosolve
2-Butoxy ethanol; solvent for lacquers, enamels, water-bone systems, inks, industrial cleaners, waterless hand cleaners. *ICI Australia.*

4772 Butyl Kamate®
Potassium dibutyldithiocarbamate
Potassium dibutyldithiocarbamate; accelerator for vulcanization of rubber. *R. T. Vanderbilt Co Inc.* Discontinued.

4773 Butyl lactate
138-22-7 205-316-4
$C_7H_{14}O_3$
n-butyl lactate
Lactic Acid Butyl Ester; Butyl Lactate; 2-Propanoic Acid; Butyl α-Hydroxypropionate; Butyl 2-hydroxypropanoate. Easter; dipped latex products, coatings, specialty finishes, inks, diluents, adhesives, intermediates, photoresist solvents, screen printing of electronic parts, flavors and fragrances, solvents for nitrocellulose, antiskinning agent, dry cleaning fluids. mp = -28°; bp = 185-187°; d = 0.968; n_D^{20}= 1.4220; flash point = 157°F; LD_{50} (rat orl) >5 gm/kg. *CPS; Penta Mfg.; Purac Biochem BV.*

4774 Butyl methacrylate
97-88-1 202-615-1
$C_8H_{14}O_2$
butyl 2-methyl-2-propenoate
2-Methyl-2-Propenoic Acid Butyl Ester; Butyl-2-Methyl-2-Propenate; n-Butyl Methacrylate; BMA; 2-Methyl butylacrylate; butyl 2-methacrylate; butyl 2-methyl-2-propenoate. Ester of n-butyl alcohol and methacrylic acid; monomer for resins, solvent coatings, adhesives, oil additives; emulsions for textiles, leather, paper finishing. bp = 160-163°; d = 0.894; flashpoint = 123°F; Irritant; LD_{50} (rat orl) = 18 g/kg. *Degussa; 101 Spec.; Janssen Chimica; Mitsubishi Gas Chem.; Rohm & Haas.*

4775 Butyl myristate
110-36-1 203-759-8
$C_{18}H_{36}O_2$
n-butyl tetradecanoate
Bumyr; Wickenol®; Crodamol BM; Nikkol BM;. Emollient; cosmetic ingred. Soluble in H_2O, organic solvents; white liquid; practically odorless; d = 0.860; flash point=166°; *Amerchol Corp.* Discontinued.

4776 Butyl Namate®
136-30-1 205-238-0
Sodium di-n-butyl-dithiocarbamate aqueous solution; accelerator for latexes of all elastomers. *R. T. Vanderbilt Co Inc.*

4777 Butyl octyl phthalate
84-78-6 201-562-1
$C_{20}H_{30}O_4$
Used as a plasticizer. *Aristech; BASF.*

4778 Butyl oleate
142-77-8 205-559-6
$C_{22}H_{42}O_2$
butyl 9-octadecenoate
Butyl Oleate C-914; Emerest® 2328; Kemester® 4000; Plasthall® 503; Priolube 1405; Uniflex® BYO; Witcizer 100. Ester of butyl alcohol and oleic acid; plasticizer. *Ferro/Keil; Inolex; NovaChem Corp; Sybron NV; Witco/Humko.*

4779 Butyl Oleate C-914
142-77-8 205-559-6
$C_{22}H_{42}O_2$
n-butyl oleate
9-octadecenoic acid (Z)-, butyl ester; butyl oleate; (Z)-9-octadecenoic acid butyl ester. Plasticizer. *C.P. Hall.*

4780 butyl rubber
A proprietary trade name for a copolymer of isobutylene with a small percentage of diene such as butadiene; an unsaturated synthetic rubber possessing the minimum unsaturation required for vulcanization. Name unverified.

4781 butyl stearate
123-95-5 1625 204-666-5
$C_{22}H_{42}O_2$
butyl octadecanoate
n-butyl octadecanoate; octadecanoic acid butyl ester; ADK STAB LS-8; Butyl Stearate C-895; Emerest® 2325; Kemester® 5510; Kessco® BS; Lexolube® BS-Tech; Priolube 1451; Uniflex® BYS-Tech; Unimate® BYS; Witconol 2326. Ester of butyl alcohol and stearic acid; solvent, spreading and softening agent in plastics, textiles, cosmetics, rubbers. mp = 27°; bp = 343°; flash point = 196°; slightly soluble in H_2O, soluble in alcohol, ether. *Amerchol; Henkel/Organic Prods.; Inolex; Mosselman NV; Penta Mfg.; Stepan; Witco/Humko; C.P. Hall.* Discontinued.

4782 Butyl toluene
98-51-1 202-675-9
$C_{11}H_{16}$
p-t-Butyl toluene
Lowinox® PTBT; *p-tert*-butyl toluene; 1-methyl-4-*tert*-butylbenzene; *p*-methyl-*tert*-butylbenzene; TBT; 1-(1,1-dimethylethyl)-4-methyl-benzene. Antioxidant. mp = -52°; bp = 193°. *Lowi.* Name unverified.

4783 Butyl Zimate®
136-23-2 205-232-8
$C_{18}H_{36}N_2S_4Zn$
Zinc di-n-butyldithiocarbamate
Butasan; Butazate; Butazin; Nocceler BZ; Soxinol BZ; ZBC. Ultra accelerator for EPDM and natural and synthetic latexes; provides fast, flat cures in SBR, nitrile, and neoprene latexes; functions as nondiscoloring antioxidant in noncuring applications and stabilizer in IIR; antioxidant in thermoplastic rubbers and hot melts. *R. T. Vanderbilt Co Inc. Discontinued.*

4784 Butyl-m-xylene
98-19-1 202-647-6
$C_{12}H_{18}$
t-Butyl-*m*-xylene
Lowinox® TBMX; Benzene, 1-(1,1-dimethylethyl)-3,5-dimethyl-; 5-*tert*-butyl-*m*-xylene. Intermediate for perfumes and fragrances. mp = -18°; bp = 205-206°. *Lowi.* Name unverified.

4785 *tert*-Butyl Acetate
540-88-5 1572 208-760-7
$C_6H_{12}O_2$
Acetic acid 1,1-dimethylethyl ester
Acetic acid *t*-butyl ester; Acetic acid *tert*-butyl ester. Ester of *t*-butyl alcohol and acetic acid; a gasoline additive. bp = 97.8°; insoluble in H_2O; soluble in EtOH, ether. *Aldrich; Schweizerhall.*

4786 n-Butyl Acrylate
141-32-2 1574 205-480-7
$C_7H_{12}O_2$
2-Propenoic acid butyl ester
Butyl-2-propenoate; Butyl acrylate; Acrylic acid n-butyl ester. Intermediate in organic synthesis, polymers and copolymers for solvent coatings, adhesives, paints, binders, emulsifiers. Soluble in H_2O; LD_{50} (rat orl) = 3.73 g/kg. *BASF.*

4787 Butylated hydroxyanisole
25013-16-5 1582 246-563-8
$C_{11}H_{16}O_2$
BHA
(1,1-dimethylethyl)-4-methoxyphenol; 3-*tert*-butyl-4-hydroxyanisole; Antracine 12; Embanox; Nipantiox; 1-F; Sustane; Tenox BHA. Mixture of isomers of *tertiary*-butyl-substituted 4-methoxyphenols; antioxidant for foods mp = 48-55°; bp_{733} = 264-270°; insoluble in H_2O, soluble in organic solvents; LD_{50} (rat orl) = 2100 mg/kg. *Eastman; Penta Mfg.; UOP.*

4788 Butylated hydroxytoluene
128-37-0 1583 204-881-4
$C_{15}H_{24}O$
2,6-bis (1,1-dimethylethyl)-4-methylphenol
BHT; 2,6-di-*t*-butyl-*p*-cresol; DBPC; Antracine 8; Tenox BHT; Ionol CP; Sustane; Dalpac; Impruvol; Vianol. Substituted toluene; antioxidant for foods, animal feed, petrol. products, synthetic rubbers, plastics, soaps; antiskinning agent in paints and inks. mp = 70°; insoluble in H_2O; soluble in organic solvents; LD_{50} (mus orl) = 1040 mg/kg. *Penta Mfg.; PMC Specialties; Raschig; Uniroyal.*

4789 Butylbenzene sulfonamide
3622-84-2 222-823-6
$C_{10}H_{15}NO_2S$
n-Butylbenzene sulfonamide
Dellatol® BBS; Plasthal® BSA; Plastomoll® BMB; Uniplex 214. Plasticizer for polymide 6, 66, 11, and 12 and copolymides; also for flexibilizing cellulose derivatives, especially flame-retardant cable coatings based on cellulose acetate and cellulose acetobutyrate. *Bayer AG; Unitex.*

4790 n-Butyl benzoate
136-60-7 1587 205-252-7
$C_{11}H_{14}O_2$
Benzoic acid butyl ester
Hipochem B-3-M; butyl benzoate; Marvanol® Carrier BB. Emulsified butyl benzoate; self-emulsifying carrier for polyester and triacetate fibers. Thick, oily liquid; mp = -22°; insoluble in water; soluble in EtOH or ether; LD_{50} (rat orl) = 5.14 g/kg. *High Point.*

4791 Butylbenzothiazole sulfenamide
95-31-8; 3741-08-8
$C_4H_4NCS(SNH)C_4H_9$
N-*t*-Butyl-*o*-benzothiazole-2-sulfenamide
Akrochem® BBTS; Delac NS; Perkacit® TBBS; Santocure® NS; Santocure® TBSI; Vanax® NS; Vanax® TBSI; Vulkacit NZ/EG; TBBS. A delayed action accelerator very safe at processing temperatures but producing high modulus stocks at curing temperatures; activated by thiurams, dithiocarbamates, aldehyde amines, guanidines and BIK; nondiscoloring and nonstaining to rubber stocks and m Light buff powder or flakes; soluble in most organic solvents; d = 1.29; mp = 104°; combustible; emits very toxic fumes when heated. *Uniroyal; Akrochem. Name unverified.*

4792 *p*-Butylcresol
2409-55-4 219-314-6
$C_{11}H_{16}O$
2-*t*-Butyl-*p*-cresol
Lowinox® MBPC; Cao. Intermediate for antioxidants, uv absorbers. *Lowi; PMS Specialities Group Inc.*

4793 *t*-Butylhydroquinone
1948-33-0 217-752-2
$C_{10}H_{14}O_2$
2-(1,1-Dimethylethyl)-1,4-benzenediol
Eastman® MTBHQ; Sustane® TBHQ; Tenox® TBHQ;TBHQ; Mono-*tert*-butylhydroquinone. Antioxidant for rubber, monomers. Polymerization inhibitor; food additive. White to light tan crystal solid; soluble in organic solvents; slightly soluble in water; mp = 126.5-128.5; flash pt = 171°; LD_{50} (rat orl) = 700 mg/kg. *AC Ind.; Aceto; Allchem Ind.; Charkit; Eastman; Penta Mfg; Schweizerhall; Showa Denko; UOP.*

4794 Butylite
A proprietary butyl rubber sealant. *Polymeric Systems Inc.* Name unverified.

4795 2-butyloctanol
3913-02-8 223-470-0
$C_{12}H_{26}O$
2-butyl-1-octanol
isododecyl alcohol. Used in organic synthesis.

4796 Butylol
Mixture of normal and isobutyl alcohol; solvent. *Sasolchem.*

4797 Butylparaben
94-26-8 1619 202-318-7
$C_{11}H_{14}O_3$
butyl *p*-hydroxybenzoate
4-hydroxybenzoic acid butyl ester; n-butyl *p*-hydroxybenzoate; Lexgard B; Butoben; Butyl Chemosept; Butyl Parasept; Tegosept B. Ester of butyl alcohol and *p*-hydroxybenzoic acid; antifungal for pharmaceuticals; preservative in foods. mp = 68-69°; soluble in H_2O, organic solvents; LD_{50} (mus orl) = 13.2 g/kg. *Inolex; Nipa Labs; Penta Mfg.*

4798 *t*-butyl perbenzoate
614-45-9 210-382-2
$C_{11}H_{14}O_3$
t-butyl peroxybenzoate
Organic peroxide; initiator for polymerization and/or crosslinking of monomers and unsaturated polymers. *Akzo; Elf Atochem N. Am.; Norac; Witco/Argus.* Name unverified.

4799 2,6-di-*t*-butylphenol
128-39-2 204-884-0
$[(CH_3)_3C]_2C_6H_3OH$
2,6-(1,1-Dimethylethyl)phenol
2,6-bis(1,1-dimethylethyl)phenol; 2,6-Di-*tert*-Butylphenol; (2,6 DTBP). Orthoalkylated aromatic; intermediate; antioxidant, stabilizer. mp = 33-37°; bp = 253°; d = 0.9100; insoluble in H_2O, soluble in organic solvents. *Schenectady.*

4800 4-*t*-butyl phenol
98-54-4 1620 202-679-0
$C_{10}H_{14}O$
4-(1,1-Dimethylethyl)phenol
Butylphen; Lowinox® TBMX; Lowinox® PTBT; Terbutol,. Antioxidant; chemical intermediate for synthetic resins, plasticizers, surface active agents. Intermediate in the manf. of varnish and lacquer, as a soap antioxidant, in motor oil additives. mp = 98°; bp = 237°; volitile with steam; practically insoluble in cold H_2O; soluble in alcohol, ether; LD_{50} (rat orl) = 3.25 ml/kg. *Janssen Chimica; PMC Specialties; Schenectady.*

4801 Butyraldehyde
123-72-8 1627 204-646-6
C_4H_8O
n-Butyraldehyde
Butanal; Butyric aldehyde; Butal; Butyl Aldehyde; Butyric Aldehyde; butaldehyde; butalyde; n-butanal; n-butyl aldehyde. Plasticizers, rubber accelerators, solvents and high molecular weight polymers. mp = -99°; bp = 74.8°; d = 0.802; LD_{50} (rat orl) = 5.89 g/kg. *Eastman; Hoechst Celanese; Neste UK; Penta Mfg.; Union Carbide.* Name unverified.

4802 Butyroin
496-77-5 1631 207-830-4
$C_8H_{16}O_2$
5-Hydroxy-4-octanone
5-octanol-4-one. Chemical intermediate. bp = 180-190°; d_4^{16}= 0.91075; n_D^{16}= 1.43455.

4803 Butyrolactone
96-48-0 1632 202-509-5
$C_4H_6O_2$
Dihydro-2(3H)-furanone
BLO®; γ-hydrooxybutyric acid lactone; 1,2-butanolide; 1,4-butanolide; 3-hydroxybutyric acid lactone; γ-butyrolactone; 4-hydroxybutanoic acid lactone. Solvent for PAN, PS, fluorinated hydrocarbons, cellulose triacetate, shellac; used in paint removers, petrol. processing, hectograph process, speciality inks; intermediate for aliphatic and cyclic compounds; reaction and diluent solvent for pesticides. mp = -43.53°; bp = 204°; d_0^{15}= 1.1286; n_D^{25}= 1.4348; flashpoint = 98°; soluble in H_2O, organic solvents; LD_{50} (rat orl) = 19.4 g/kg. *Aldrich; BASF; ISP; Janssen Chimica; Spectrum Chem. Mfg.; UCB SA.*

4804 Butyrone
123-19-3 3408 204-608-9
$C_7H_{14}O$
Di-n-propyl ketone
4-Heptanone; Butyrone; Heptan-4-one; GBL; Dipropyl ketone. Solvent, chemical intermediate. mp = -33°; bp = 145°; d = 0.8170; n_D^{20} = 1.4070; insoluble in H_2O, soluble in organic solvents; LD_{50} (rat orl) = 3.04 g/kg.

4805 Butyronitrile
109-74-0 1633 203-700-6
C_4H_7N
propyl cyanide
Nitrile C_4; Butanenitrile; Butyric acid nitrile. Basic material in industrial, chemical and pharmaceutical intermediates and products, poultry medicines. mp = -112°; bp = 117.5°; sparingly soluble in H_2O; LD_{50} (rat orl) = 0.14 g/kg. *Air Prods & Chem; Eastman; Janssen Chimica; Lonza.*

4806 Bu-White
Disproportionated resin derivatives. *Tenneco Malros Ltd.* Name unverified.

4807 BWF
Extruded acrylic and polycarbonate profiles; for shopfitting, sign work. *Cornelius Chemical Co Ltd.* Name unverified.

4808 BX 310
A range of polypropylene films of different gauge thicknesses and widths; for packaging applications. *Hercules.* Discontinued.

4809 BXA
Diarylamine-ketone-aldehyde reaction product; for CR, NBR, SBR; nonblooming, easily dispersed; protects against heat and oxygen; brown discoloration in stocks exposed to light, may stain light colored materials in contact with cured stocks used in tire treads, carcass, inner tubes, insulated wire, soles, heels and mechanicals. *Uniroyal.* Name unverified.

4810 B-X-A
A reaction product of diarylamine-ketone-aldehyde; a proprietary antioxidant. *Rubber Regenerating Co.* Unverified.

4811 BXA Flake
9003-79-6
Diphenylamine-acetone reaction product; antioxidant. *Uniroyal.*

4812 BXT
Polypropylene film; for wrapping tobacco products. *Hercules.* Discontinued.

4813 BY-59-18
Hydrocarbon resin, modified; used as replacement for rosin derivatives, and in inks. *Neville.*

4814 Byacin
Iodophor in liquid solution; for controlling storage rots in potatoes. *Wheatley Chemical Co Ltd.* Name unverified.

4815 Byakisol 30
Silica solution
Used for binders, surface treatment wine and fruit juice fining, and the textile industry. *Bayer AG.*

4816 Byatran
Iodophor and TBZ as dry granules; for controlling soil borne diseases in growing potato crops. *Wheatley Chemical Co Ltd.* Name unverified.

4817 Byatran
Granules containing nonylphenoxypoly (ethyleneoxy) ethanol-iodine complex and tecnazene; used to control soilborne diseases in growing potato crops. *Dean Agrochemicals Ltd.*

4818 Byco A, C, O
9000-70-8 4388 232-554-6
Gelatin NF
Binders in pharmaceutical tableting; excipient, film former, coating agent; emulsion stabilizer; adjuvant protein in nutritional supplement. *Croda Inc.*

4819 Bygran F
Tecnazene and iodophor as dry granules; for controlling sprouting, dry rot and other storage diseases in potatoes. *Wheatley Chemical Co Ltd.* Name unverified.

4820 Bygran F
Granules containing nonylphenoxypoly (ethyleneoxy) ethanol-iodine complex and thiabendazole; used for controlling sprouting, dry rot and other storage diseases in potatoes. *Dean Agrochemicals Ltd.*

4821 Bygran S
117-18-0 204-178-2
tecnazene
Granules containing 10% w/w tecnazene; used for controlling sprouting and dry rot in stored potatoes. *Dean Agrochemicals Ltd.*

4822 BYK® 020
Solution of a modified polysiloxane copolymer; defoamer for water-reducible coating system, e.g., alkyds, polyester, epoxy esters, acrylic. *Byk-Chemie USA.*

4823 BYK® 024
Polysiloxanes/polymer mixture. Defoamer for water-based systems including polyurethane, acrylate/polyurethane paints, wood varnishes, furniture paints, pigmented dispersion paints, plastic coatings. *Byk-Chemie USA.*

4824 BYK® 045
Emulsion of hydrophobic solids, emulsifiers, foam-destroying Polysiloxanes; defoamer for emulsion paint, paper coating, foil coatings, stains. *Byk-Chemie USA.*

4825 BYK® 151
Solution of an alkylolammonium salt of a polyfunctional polymer; dispersant, wetting agent for inorganic and organic pigments, fillers in aqueous systems; improves color strength development and rheological properties of pigment pastes. *Byk-Chemie USA.*

4826 BYK® 156
9003-03-6
Solution of an ammonium salt of an acrylic acid copolymer; dispersant for gloss and semigloss latex systems; improves gloss and stability on heat aging. *Byk-Chemie USA.*

4827 BYK® 307
Polyether modified dimethyl polysiloxane copolymer; additive to increase surface slip, substrate wetting, scratch and mar resistance for paints, printing inks. *Byk-Chemie USA.*

4828 BYK® A500
Silicone-free polymeric solution; defoamer, air release agent for laminating, spray-up, hand lay-up molding, gel coats, and solvent-free epoxy flooring systems; prevents air entrapment and porosity. *Byk-Chemie USA.*

4829 BYK® -Catalyst 450
Solution of an amine salt of *p*-toluene sulfonic acid; additive to improve curing in acid catalyzable organic systems. *Byk-Chemie USA.*

4830 BYK® ES 80
Solution of an alkylammonium salt of an unsaturated acidic carboxylic acid ester; additive for increasing the conductivity of electrostatically sprayed coatings; used for solvent-based air dry and baking enamels. *Byk-Chemie USA.*

4831 BYK® P104
Solution of higher molecular weight unsaturated polycarboxylic acid; wetting, dispersing, antiflooding and antisettling additive; for coating systems. *Byk-Chemie USA.*

4832 Bykanol® -N
Solution of an alkylolammonium salt of acidic phosphoric acid esters and

ketoxime; antigelling agent and viscosity stabilizer for solvent-based coatings which have a tendency to gel on aging. *Byk-Chemie USA.*

4833 Byketol® OK

Mixture of high boiling aromatics, ketones and esters; additive to counteract surface defects and improve leveling for solvent-based coatings, chlorinated rubber systems, silk screen inks. *Byk-Chemie USA.*

4834 Bykumen®

Solution of a higher molecular weight unsaturated acidic polycarboxylic acid ester; wetting, dispersing additive to improve pigment wetting and prevent setting of pigments; for solvent and solvent-free systems, alkyd trade sales systems, acrylics, polyesters. *Byk-Chemie USA.*

4835 Bynel®

Coextrudable adhesive resins. *DuPont UK.*

4836 C.E. powders

Explosive powders, containing tetryl- or tetranitromethylaniline.

4837 C.P.D

Cadmium pentamethylene dithiocarbamate
A rubber vulcanization accelerator.

4838 C.P.R

Multi-component systems used in the manufacture of rigid foams. *Dow UK.* Discontinued.

4839 C-1

1344-28-1 369 215-691-6
Calcined alumina
For manufacturing of abrasives, fused alumina, high-alumina refractories, technical ceramics, kiln furniture, whitewares, fiberglass, ceramic fiber, electrical insulators and supports. *La Roche Chem.* Discontinued.

4840 C12-14 alkyl dimethylamine oxide

85408-49-7 287-011-6
C12-14 alkyl dimethylamine oxide
Lilaminox M24. Foam booster/stabilizer, antistat, softener for hair shampoos; thickener for household bleaches based on sodium hypochlorite, hard surface cleaners. *Berol Nobel AB.*

4841 C12-16 pareth-1

68551-12-2
Genapol® 26-L-1. Biodegradable detergent intermediate for sulfation for use in cosmetics, shampoos, light duty detergents; emulsifier, prewash spotter, agricultural adjuvant, hydrocarbon-based cleaning systems. *Hoechst Celanese/Colorants & Surf.*

4842 C-2

7758-19-2 8743 231-836-6
Sodium chlorite
Bleach and oxidizer. *Olin.* Discontinued.

4843 C-715u

63394-02-5
Silicone elastomer; features high green strength, high resilience; for extrusion, molding, calendering. *Wacker Silicones.*

4844 C-84

25928-94-3
Epoxy compound; general purpose potting compound for potting and casting applications; inherently high degree of flame resistance; used with 7.3 phr Activator BA-63. *Bacon.*

4845 C-920u

63394-02-5
Silicone elastomer; high temperature grade remaining flexible after 24 h at 371°. *Wacker Silicones.*

4846 CA0397

4369-14-6
$C_9H_{18}O_5Si$
3-Acryloxypropyltrimethoxysilane
3-Acryloxypropyl Trimethoxysilane; 2-Propenoic acid 3-(trimethoxysilyl)propyl ester. Coupling agent, chemical intermediate, blocking agent, release agent, lubricant, primer, reducing agent. bp_3 = 97°. *Hüls Am.*

4847 CA0567

2551-83-9 219-855-8
$C_6H_{14}O_3Si$
Allyltrimethoxy silane
Coupling agent, chemical intermediate, blocking agent, release agent, lubricant, primer, reducing agent. *Hüls Am.*

4848 CA0570

762-72-1 212-104-5
$C_6H_{14}Si$
Allyltrimethyl silane
Silane, trimethyl-2-propenyl-. Coupling agent, chemical intermediate, blocking agent, release agent, lubricant, primer, reducing agent. bp = 84-88°; d = 0.7190; n_D^{20}= 1.4056. *Hüls Am.*

4849 CA0699

3069-29-2 221-336-6
$C_8H_{22}N_2O_2Si$
N-(2-Aminoethyl)-3-aminopropylmethyldimethoxysilane
Dynasylan 1411. Coupling agent, chemical intermediate, blocking agent, release agent, lubricant, primer, reducing agent. *Hüls Am.*

4850 CA0700

1760-24-3 217-164-6
$C_8H_{22}N_2O_3Si$
N-2-Aminoethyl-3-aminopropyltrimethoxysilane
N-[3-(trimethoxysilyl)propyl]-1,2-ethanediamine N-β-(aminoethyl)-γ-aminopropyl trimethoxy-silane; N-[3-(Trimethoxysilyl)propyl]ethylenediamine; [3-(2-Aminoethyl)aminopropyl]trimethoxysilane; DAMO; N-(2-Aminoethyl-3-aminopropyl)trimethoxysilane; Dynasylan DAMO. Coupling agent, chemical intermediate, blocking agent, release agent, lubricant, primer, reducing agent. bp_{15} = 146°; d = 1.0100; n_D^{20}= 1.4450; LD_{50} (rat orl) = 7460 mg/kg. *Hüls Am.*

4851 CA0742

3179-76-8 221-660-8
$C_8H_{21}NO_2Si$
3-Aminopropylmethyldiethoxy silane
1-Propanamine, 3-(diethoxymethylsilyl)-; Dynasylan 1505; Dynasylan 1506; Aminopropylmethyldiethoxysilane; 3-Aminopropylmethyldiethoxysilane. Coupling agent, chemical intermediate, blocking agent, release agent, lubricant, primer, reducing agent. *Hüls Am.*

4852 CA0750

919-30-2 213-048-4
$C_9H_{23}NO_3Si$
3-Aminopropyltriethoxysilane
1-Propanamine, 3-(triethoxysilyl)-; 3-Aminopropyl-triethoxysilane; 3-Triethoxysilylpropylamine; AMEO; Dynasylan AMEO. Coupling agent, chemical intermediate, blocking agent, release agent, lubricant, primer, reducing agent. bp = 217°; d = 0.9420; n_D^{20}= 1.4210; LD_{50} (rat orl) = 1780 mg/kg. *Hüls Am.*

4853 CA0880

13822-56-5 237-511-5
$C_6H_{17}NO_3Si$
3-Aminopropyltrimethoxysilane
1-Propanamine, 3-(trimethoxysilyl)-; Dynasylan AMMO; 3-Aminopropyltrimethoxysilane; γ-Aminopropyl Trimethoxy Silane. Coupling agent, chemical intermediate, blocking agent, release agent, lubricant, primer, reducing agent. *Hüls Am.*

4854 CA0900

107-72-2 203-515-0
$C_5H_{11}Cl_3Si$
Amyltrichlorosilane
amyl trichlorosilane; Trichloropentylsilane; Pentyltrichlorosilane. Coupling agent, chemical intermediate, blocking agent, release agent, lubricant, primer, reducing agent. *Hüls Am.*

4855 CA-25

Calcium aluminate
Refractory bonding agent for use in refractories designed for service above 3000°F. *Alcoa Industrial Chemicals.*

4856 CA-394-60S

9004-35-7 2013
Cellulose acetate
Used where high strength and good resistance to heat, uv light, oils, and greases is required. d = 1.300; n_D^{20} = 1.4750; insoluble in H_2O, soluble in organic solvents. *Eastman.*

4857 CAB-171-15S

9004-36-8
Cellulose acetate butyrate
CAB. Used in cloth coatings for airplanes, wire, leather, and plastics. *Eastman.*

4858 CAB-381-0.1, 381-0.5, 381-2, 381-20

9004-36-8
Cellulose acetate butyrate
Used in lacquers. *Eastman.*

4859 CAB-500-5

9004-36-8
Cellulose acetate butyrate
Used for hot melts and additives for polyurethanes. *Eastman.*

4860 Cabelec®

Compounds capable of conducting electricity; used in the manufacture of electrically conductive and antistatic articles. *Cabot Plastics Ltd.*

4861 Cabelec® 1015

25038-54-4 6832
$[C_6H_{11}NO]_n$, n > 200
Polycaprolactam
Poly[imino(1-oxo-1,6-hexanediyl)]; poly(iminocabonylpentamethylene;)Caprolan; Enkolan; Grilon; Kapron; Mirlon; Perlon; Phrilon; Amilan. Nylon

6/carbon blk. compound; conductive compound for injection molding applications; gives rigidity with permanent electrical conductivity; suggested for packaging and electronic production handling applications; e.g., for handling explosives, electronic mp = 223°; d = 1.084. *Cabot Plastics Ltd.*

4862 Cabelec® 1017
9002-88-4 7728
$[C_2H_4]_n$
Polyethylene
Agilene; Alathon; Alkathene; Courlene; Lupolen; Platilon; Polythene; Pylen; Reevon. Modified polyethylene/carbon black compound; conductive compound for thin film extrusion; for packaging and production handling where freedom from hazard of electrostatic discharge is necessary, e.g., for handling explosive powders, mp = 85-110°, 130-145°; d_4^{20}= 0.922. *Cabot Plastics Ltd.*

4863 Cabelec® 1017
9002-88-4 7728
Modified polyethylene/carbon black compound; conductive compound for thin film extrusion; for packaging and production handling where freedom from hazard of electrostatic discharge is necessary, e.g., for handling explosive powders, pigments, *Cabot Plastics Ltd.*

4864 Cabelec® 3004
9003-07-0 7741
PP copolymer/carbon black compound; permanently static dissipative compound for injection molding; for applications where slower discharge of static electricity is required such as unearthed containers and specialized moldings; *Cabot Plastics Ltd.* Name unverified.

4865 Cabelec® 3172
9002-88-4 7728
HDPE/carbon black compound; conductive compound for molding and sheet extrusion; for production handling applications where freedom from electrostatic discharge hazards is necessary, e.g., in handling of explosive powders, pigments, electronic components. *Cabot Plastics Ltd.* Name unverified.

4866 Cabelec® 3464
9003-07-0 7741
PP copolymer/carbon black compound; conductive compound for extrusion applications including extruded tapes for sacking, fiber/twine for fabric or interweaving into fabric for chair coverings, for flexible pipes and tubing, *Cabot Plastics Ltd.*

4867 Cabelec® 3464, 3004
9003-07-0 7741
Polypropylene
Polypropylene copolymer/carbon black compound; conductive compound for extrusion applications including extruded tapes for sacking, fiber/twine for fabric or interweaving into fabric for chair coverings, for flexible pipes and tubing, and as corrugated *Cabot Plastics Ltd.*

4868 Cab-O-Sil® L-90
7631-86-9 8637 231-545-4
O_2Si
silicon dioxide
7631-86-9. Fumed silica; dispersant, anticaking agent for foods, agricultural products, and powders for cosmetics and coatings industries. *Cabot.*

4869 Cab-O-Sil® TS-530
7631-86-9 8637 231-545-4
Silicon dioxide
Fumed silica, hexamethyldisilazane-surface treated; reinforcing filler for elastomers; free flow agent for toners and powder coatings; antisetting agent in coatings; dry powder carrier for perfumes, pesticides, veterinary products, etc. *Cabot.*

4870 Cab-O-Sperse® A 105
Aqueous dispersion of fumed silica, ammonia stabilized; thickener; rheological control in water-based systems. *Cabot.* Name unverified.

4871 Cabot® PE 6008
30% trimethyl quinoline antioxidant in low density polyethylene carrier; antioxidant masterbatch for polymeric cable applications including cable sheath insulation layers and semiconductive screens; compatible with LDPE, EVA, EBA, LLDPE, HDPE, EPDM, ACN. *Cabot Plastics Ltd.*

4872 Cabot® PE 9006
75% calcium carbonate plus antioxidant in LDPE carrier; antifibrilation masterbatch for use in PP weaving tapes, PP strapping bands, PP string and twine, LDPE shrink film; effective antiblock; compatible with LPDE, LLDPE, HDPE, PP. *Cabot Plastics Ltd.*

4873 Cabot® PE 9007
9002-88-4 7728
LDPE polymer with mildly abrasive fine natural silica; antiblock masterbatch for blown LDPE and LLDPE film extrusion; also as cleaning compound for extruders; compatible with LDPE, LLDPE, HDPE, PP, EVA. *Cabot Plastics Ltd.* Name unverified.

4874 Cabot® PE 9138
Polymer masterbatch with mineral additive; infra-red absorber masterbatch; used to improve thermal barrier properties of polyethylene film in greenhouse and other agricultural applications; optimum addition level is 11-12% in polyethylene. *Cabot Plastics Ltd.*

4875 Cabot® PE 9166
Low density polyethylene
5% erucamide slip agent in LPDE carrier; slip additive masterbatch providing excellent slip and good antiblock effect to LDPE and LLDPE films; slower migration of slip agent is advantageous when printing, sealing, or laminating processes are carried out *Cabot Plastics Ltd.*

4876 Cabuflx
An adhesive for cellulose acetate butyrate. *May & Baker Ltd.* Name unverified.

4877 Cabulite
9004-36-8
Cellulose acetate butyrate film and sheet. *May & Baker Ltd.* Name unverified.

4878 Cachalot® C-50 NF
36653-82-4 2070 253-149-0
$C_{16}H_{34}O$
Cetyl alcohol NF
Cetyl Alcohol; Cetanol; 1-Hexadecanol; Ethal; Ethol; Palmityl alcohol; Hexadecan-1-ol; Hexadecyl alcohol; Hexadecanol; Alcohol, C16; Atalco C; Cachalot C-50; Cetaffine; Cetal; Cetylol; CO-1670; Crodacol-cas; DYTOL F-11; LorolL 24; Loxanol K; Product 308. Emollient used in cosmetics. mp = 48-50°; bp_{15} = 189-190°; d = 0.8180; insoluble in H_2O, soluble in organic solvents; LD_{50} (rat orl) = 5 g/kg. *M. Michel.*

4879 Cachalot® M-43 NF
112-72-1 6418 204-000-3
Myristyl alcohol
Emollient used in cosmetics. *M. Michel.*

4880 Cachalot® O-15
143-28-2 6968 205-597-3
Oleyl alcohol
Conditioner, lubricant for cosmetics; corrosion inhibitor additive to lube oils. *M. Michel.* Discontinued.

4881 Cachalot® S-56
112-92-5 8960 204-017-6
Stearyl alcohol USP
Emollient used in cosmetics. *M. Michel.*

4882 Cacodyl
471-35-2 1639 207-440-4
$As_2C_4H_{12}$
Alkarsin
Tetramethyldiarsine; dicacodyl. Chemical intermediate. mp = -6°; bp = 165°; slightly soluble in H_2O.

4883 Cadaverine
462-94-2 1645 207-329-0
$C_5H_{14}N_2$
Pentamethylenediamine
1,5-Pentanediamine; animal coniine. A base found in ergot and formed by bacterial decomposition; used in the preparation of high polymers; in biomedical research. mp = 9°; bp = 178-180°; d = 0.873; n_D^{20}= 1.4582; pKa_1 = 10.25, pKa_2 = 9.13 ; soluble in H_2O, EtOH, slightly soluble in Et_2O.

4884 Cadco® Acetal
105-57-7 36 203-310-6
Acetal resin; highly resistant to moisture, gasoline and organic solvents; for antifriction parts, bearings, bushings, rollers, cams, dimensionally stable parts for business machines, elec. components, gears, plumbing parts; *Cadillac Plastic & Chem.* Name unverified.

4885 Cadco® Cast Acrylic
Cast acrylic rods, tubes, and blocks; weather/chemical resistance strong material with optical clarity; for bathroom fixtures, chemical retorts, display items, electrical components, food canisters, furniture components, knobs, lamp shades, lighting fixtu *Cadillac Plastic & Chem.*

4886 Cadco® Cast Nylon
25038-54-4 6832
Cast nylon 6; strong, lightweight, self-lubricating thermoplastic resistance to wear, abrasion and chems.; for bearings, bushings, cams, gears, hydraulic seals, insulators, rollers, seals, slide bearings, tooling fixtures, valves, wheels, etc. *Cadillac Plastic & Chem.*

4887 Cadco® Nylon
Nylon
Good abrasion and wear resistance, electrical insulating characteristics, noise dampening properties for unlubricated gears, bearings, and antifriction parts, mechanical parts, auto body parts, electrical parts, high impact parts, food processing parts. *Cadillac Plastic & Chem.*

4888 Cadco® Teflon
PTFE and FEP fluorocarbons; high melting thermoplastic for valve and pump components, gaskets, mechanical components, food processing, chemical equip., tape-wrapped wire, transformers, coils, electronics equipment. *Cadillac Plastic & Chem.*

4889 Cadco® UHMW
9002-88-4 7728
Ultra-high molecular weight polyethylene; high strength, abrasion and impact propereties; for various applications from suction box covers for high-speed paper machines to surfaces on snow skis. *Cadillac Plastic & Chem.* No manufacturer.

4890 Caddy
10108-64-2 1653 233-296-7
$CdCl_2$
Cadmium turf fungicide
cadmium chloride. *W A Cleary.*

4891 Cadet® BPO-70W
94-36-0 1149 202-327-6
Benzoyl peroxide
Initiator for elevated-temperature curing of unsaturated polyester resins in applications such as matched metal die molding, pultrusion, vacuum bag molding, continuous laminating, hotcure casting, injection molding; also for PS, specialized PVC resin, *Akzo.* Name unverified.

4892 Cadmate®
14239-68-0 238-113-4
$C_{10}H_{20}CdN_2S_4$
Cadmium diethyldithiocarbamate
Ethyl cadmate; Cadmium, bis(diethylcarbamodithioato-S,S')-, (T-4)-; Bis(diethyldithiocarbamate)cadmium complex. Activated cadmium diethyldithiocarbamate; accelerator for vulcanization of synthetic rubber. *R. T. Vanderbilt Co Inc.*

4893 cadmium
7440-43-9 1649 231-152-8
Cd
In alloys; electrodeposited and dipped coatings on metals. mp = 321°, bp = 765°; d^{25} = 8.65. *Aldrich; Asarco; Atomergic Chemetals; Pasminco Europe; Zinc Corp. of Am.*

4894 cadmium bronze
A copper alloy containing 0.5-1.2% cadmium; used for telephone and trolley wire.

4895 cadmium chloride
10108-64-2 1653 233-296-7
$CdCl_2$
Caddy; Vi-Cad. Used in photography, dyeing and printing and as a fungicide. Preparation of cadmium sulfide, analytical chemistry, photography, dyeing and calco printing, in electroplating baths and tinning solutions, manufacture of special mirrors, cadmium yellow. mp = 568°; bp = 967°; soluble in H_2O, EtOH, poorly soluble on organic solvents; LD_{50} (rat orl) = 88 mg/kg.

4896 cadmium diethyldithiocarbamate
14239-68-0 238-113-4
Ethyl Cadmate; Penotrane; Cadmate®;. Activated cadmium diethyldithiocarbamate; accelerator for vulcanization of synthetic rubber. Primary accelerator for NR and synthetic rubbers; used with a thiazole; gives heat resistance and low compression set properties. *R. T. Vanderbilt Co Inc; Octel Chemicals Ltd.* Discontinued.

4897 cadmium fluoride
7790-79-6 1655 232-222-0
CdF_2
Cadmium fluoride
Electronic and optical applications, high-temperature dry-film lubricants, starting material for crystals for lasers, phosphors. mp = 1049°; bp = 1748°; d = 6.33; soluble in H_2O 4.3 g/100ml; soluble in HF, mineral acids, insoluble in EtOH, NH_3. *Atomergic Chemetals; Cerac; Noah Chem.*

4898 cadmium hydroxide
21041-95-2 1656 244-168-5
$Cd(OH)_2$
cadmium hydrate
Used in preparation of cadmium salts, in cadmium plating and storage-battery electrodes. *Noah Chem.*

4899 cadmium iodide
7790-80-9 1657 232-223-6
CdI_2
Cadmium iodide
Photography, process engraving and lithography, analytical chemistry, electroplating, lubricants, phosphors, nematocide. mp = 388°; bp = 787°; d = 5.67; soluble in H_2O, Et_2O, EtOH, Me_2CO *Atomergic Chemetals; Cerac; Spectrum Chemical Mfg.*

4900 cadmium lithopone
cadmopone
A pigment analogous to lithopone, in which cadmium replaces zinc. It is made by the precipitation of cadmium sulfate solution with barium sulfide, and contains 38% cadmium sulfide.

4901 cadmium oxide
1306-19-0 1659 215-146-2
CdO
cadmium oxide brown
Cadmium plating baths, electrodes for storage batteries, cadmium salts, catalyst, ceramic glazes, phosphors, nematocide. d = 8.15; insoluble in H_2O, soluble in acids, ammonium salts; LD_{50} (rat orl) = 72 mg/kg. *Amax Inc; Asarco; Atomergic Chemetals; Chemisphere Ltd; Mallinckrodt; Nihon Kagaku Sangyo.*

4902 cadmiumized zinc
Zinc metal placed in a 2% cadmium sulfate solution for five minutes, then well washed; used as a reducing agent.

4903 Cadmopur
A range of cadmium pigments; especially for use in plastics such as polyethylene, polystyrene, polyamides, PVC, etc., as well as in rubber. *Bayer AG.*

4904 Cadon
Engineering thermoplastics, impact-modified styrene-maleic anhydride terpolymers; used in automotive interior/exterior parts, business machines and appliance housings, electrical equipment, electronic parts, plumbing and industrial parts. *Monsanto Co.* Name unverified.

4905 Cadoussant
Blends of cationic surfactants; softening and antistatic agent for synthetic fibers. *Elf Atochem UK/Ceca.*

4906 Cadox® 40E
94-36-0 1149 202-327-6
benzoyl peroxide
Benzoyl peroxide with plasticizer; initiator for ambient-temperature polyester cures. *Akzo.* Name unverified.

4907 Cadox® BFF-50
Benzoyl peroxide with dicyclohexyl phthalate; initiator for elevated-temperature curing of unsaturated polyester resins in applications such as matched metal die molding, pultrusion, vacuum bag molding, continuous laminating, hot-cure casting, injection *Akzo.* Name unverified.

4908 Cadox® BPO-W40
94-36-0 1149 202-327-6
Benzoyl peroxide
Initiator. *Akzo.*

4909 Cadox® BS
94-36-0 1149 202-327-6
Benzoyl peroxide
Cross-linking agent used for curing silicone rubbers. *Akzo.* Name unverified.

4910 Cadox® BTA
94-36-0 1149 202-327-6
Benzoyl peroxide
Mixture of dibenzoyl peroxide and calcium sulfate. Initiator for ambient-temperature polyester cures. *Akzo.* Name unverified.

4911 Cadox® BTW-50
94-36-0 1149 202-327-6
Benzoyl peroxide
Initiator for cure of unsaturated polyester resins *Akzo.* Name unverified.

4912 Cadox® F-85
Ketone peroxide
Initiator for ambient-temperature polyester cures. *Akzo.* Name unverified.

4913 Cadox® HBO-50
1338-23-4 215-661-2
MEK peroxide
Methyl ethyl ketone peroxide. Initiator for cure of unsaturated polyester resins *Akzo.*

4914 Cadox® L-30
1338-23-4 215-661-2
MEK peroxide
Methyl ethyl ketone peroxide. Initiator for cure of unsaturated polyester resins *Akzo.*

4915 Cadox® TDP
94-17-7 202-310-3
2,4-Dichlorobenzoyl peroxide
2,4-Dichlorobenzoyl peroxide in plasticizer; highly reactive initiator used with amine-accelerated polyester resins. *Akzo.*

4916 Cadox® TS-50S
94-17-7 202-310-3
2,4-Dichlorobenzoyl peroxide

2,4-Dichlorobenzoyl peroxide, phthalate-free; cross-linking agent for polymers; co-vulcanizing agent *Akzo.*

4917 CAE

Ethyl N-cocoyl-L-arginate PCA salt; surfactant, foamer, and antistat; preservative; antiseptic; germicide; disinfectant in cosmetics, detergents, dentifrices, medical supplies. *Ajinomoto; Ajinomoto USA.* Name unverified.

4918 Caflon

Foam stabilizer, for liquid detergents; shampoos; bubble baths; hand cleaners. *Ellis & Everard Ltd.*

4919 Caflon MIS

Anionic surfactant as a pale amber clear liquid; emulsifier for hand cleaning gels and degreasers. *Ellis & Everard Ltd.*

4920 Caflon MS30

Monoethanolamine alkyl sulfate in liquid form; for liquid detergents and shampoos. *Ellis & Everard Ltd.*

4921 Caflon NAS 25, SA, SNA

Anionic surfactants as a pale yellow (NAS 25), dark brown (SA, SNA) liquid; high foaming wetting agent for liquid detergents. *Ellis & Everard Ltd.*

4922 Cake Mix 96

Partially hydrogenated soybean oil, propylene glycol mono and diesters of fats and fatty acids, mono and diglycerides, optional lecithin; kosher; high performance fat for single-stage household cake mixes. *Van Den Bergh Foods.*

4923 Cal Plus

10043-52-4 1699 233-140-8
$CaCl_2 \cdot 2H_2O$
Calcium chloride dihydrate
Replenisher. *Mallinckrodt Inc.*

4924 Calac

62-54-4 1683 200-540-9
Calcium acetate
Used in dyeing, tanning and curing skins and as a chemical intermediate. *Mechema Chemicals Ltd.*

4925 Calamide C, O

Cocamide DEA superamide
Foam stabilizer, emulsifier for liquid dishwash, bubble baths, shampoos, all-purpose cleaners; thickener; imparts mildness. *Pilot.*

4926 Calaton

A textile finishing agent. *ICI Chem & Polymers Ltd.*

4927 Calbiosorb™

Absorbent. *Calbiochem Corp.*

4928 Calbrite

7757-93-9 1739 231-826-1
$CaHO_4P$
calcium phosphate, dibasic
calcium monohydrogen phosphate; secondary calcium phosphate; monetite; Caliment. Dentifrice grade of dicalcium phosphate. d = 2.31; insoluble in H_2O, EtOH, soluble in mineral acids *Albright & Wilson Ltd., Phosphates & Speciality Business.*

4929 Calbux

1317-65-3 5515 215-279-6
Calcium carbonate
Agstone; lithographic stone; Solnhofen stone; Portland stone; dolomite, chalk; marble. A proprietary ground limestone. *ICI Chem & Polymers Ltd.* Discontinued.

4930 Calcars

7778-44-1 1686 231-904-5
Calcium arsenate
Insecticide and molluscicide. *Mechema Chemicals Ltd.* Name unverified.

4931 Calcene

A trade name applied to a precipitated calcium carbonate with 2% organic material; prepared for use as a rubber filler. No manufacturer.

4932 Calcet

Contains calcium gluconate and calcium lactate; replenisher. *Mission Pharmacal Co.* Name unverified.

4933 Calcichrome

Cyclo tris-7-(1-azo-8-hydroxynaphthalene 3:6 disulfonic acid)
A mauve crystalline powder; a sensitive and specific reagent for calcium giving a red color with Ca^{2+} ions in alkaline solution. *British Drug Houses.* Name unverified.

4934 Calcisorb

9038-41-9
Each sachet contains sodium cellulose phosphate as a white to beige fibrous powder. *3M Pharmaceuticals.* Discontinued.

4935 calcium

7440-70-2 1682 231-179-5
Ca
Alloying agent for aluminum, copper, lead; reducing agent for Be. mp = 850°; bp = 1440°; d_4^{20} = 1.54. *Atomergic Chemetals; Cerac; Leverton-Clarke Ltd; Noah Chemical; Pfizer.*

4936 calcium acetate

62-54-4 1683 200-540-9
$C_4H_6CaO_4$
Calcium acetate
brown acetate of lime, gray acetate of lime; Calac. Used in manufacture of acetic acid, acetone; in dyeing, tanning, as lubricant, food stabilizer, corrosion inhibitor. Decomposes at 160°; d = 1.50; soluble in H_2O, slightly soluble in MeOH, insoluble in EtOH, Me_2CO, C_6H_6; LD_{50} (rat orl) = 4.28 g/kg. *Mechema Chemicals Ltd.* Name unverified.

4937 calcium acetate

62-54-4 1683 200-540-9
$C_4H_6CaO_4 \cdot H_2O$
vinegar salts
gray acetate of lime, brown acetate of lime; lime acetate. Manufacture of acetone, acetic acid, acetates, mordant in dyeing and printing of textiles, stabilizer in resins, additive to calcium soap lubricants, food additive, corrosion inhibitor. mp = 160° (dec); soluble in H_2O, slightly soluble in EtOH; LD_{50} (rat orl) = 4.28 g/kg *General Chemical; Mallinckrodt; Niacet; Schaefer Salt; Verdugt BV.*

4938 calcium arsenate

7778-44-1 1686 231-904-5
$As_2Ca_3O_8$
Calcium arsenate
Tricalcium arsenate; Pencal; Calcars. Used as insecticide, molluscicide Slightly soluble in H_2O and dilute acids; LD_{50} (rat orl) = 298 mg/kg. *Mechema Chemicals Ltd.* Name unverified.

4939 calcium carbide

75-20-7 1696 200-848-3
C_2Ca
Acetylenogen. Generation of acetylene gas for welding, chloroethylenes, vinyl acetate monomer, acetylene chemicals, reducing agent. mp = 2300°; d = 2.22; soluble in H_2O, MeOH, EtOH, slightly soluble in Me_2CO, insoluble in $CHCl_3$, Et_2O. *Spectrum Chemical Mfg.*

4940 calcium carbonate

1317-65-3 5515 215-279-6
$CCaO_3$
calcium carbonate
carbonic acid, calcium salt; Calcichew; Calcidia; Citrical; Aragonite; Aeromatt; Albacar; Purecal; Chalk; English White; Paris White; Carbonic acid, calcium salt; precipitated calcium carbonate, commercial form; prepared calcium carbonate, native purified form,. Inorganic salt; source of lime; neutralizing agent; opacifying agent in paper; fortification of bread; putty; tooth powders; mp = 825° (dec); d = 2.83; insoluble in H_2O, soluble in dilute acids. BASF; ECC Int'l.; EM Industries; Genstar Stone Prods.; Georgia Marble; J.M. Huber; Mallinckrodt; Nichia Kagaku Kogyo; Pfizer; Whittaker, Clark & Daniels.

4941 calcium carbonate

1317-65-3 5515 215-279-6
Calcium carbonate
Carbital® 50; Calbux; Carbital® 35; Agstone; lithographic stone; Solnhofen stone; Portland stone; dolomite, chalk; marble. Fine, wet-ground filling pigment offering excellent retention in the sheet, excellent optical properties; also for matte or dull coatings. *ECC International Ltd.* Discontinued.

4942 calcium carbonate

1317-65-3 5515 215-279-6
$CaCO_3$
Calcium carbonate
Calcium carbonate, prepared; drop chalk; Chalk, Prepared; Camel-TEX®, CamelWITE®; Carbital®; Camel-CAL®, Camel-CARB®, Camel-FIL, Camel-FINE. Used in medicine, tooth powders, polishing powders, silicate cement.

4943 calcium chloride

10043-52-4 1699 233-140-8
$CaCl_2 \cdot 2H_2O$
calcium chloride
Intergravin-orales; Caloride; calcium chloride, dihydrate; Calcosan; calcium dichloride. Inorganic salt; deicing and dust control of roads, drilling muds, dustproofing, freezeproofing, and thawing coal, coke, stone, sand, ore, concrete conditioning; drying and desiccating agent; sequestrant in foods. mp = 772°; bp >1600°; d_4^{15}= 2.152; soluble in H_2O, EtOH; LD_{50} (mus iv) = 42.2 mg/kg. Akzo Salt; AlliedSignal; EM Industries; Gist-Brocades Food Ingreds.; Kemira Kemi UK; Mallinckrodt; Nichia Kagaku Kogyo; OxyChem; Schaefer Salt.

4944 calcium citrate

813-94-5 1701 212-391-7
$C_{12}H_{10}Ca_3O_{14} \cdot 4H_2O$
2-Hydroxy-1,2,3-propanetricarboxylic acid, calcium salt (2:3)

tricalcium citrate; lime citrate. Dietary supplement, sequestrant, buffer, and firming agent in foods. Soluble in 1050 parts cold H_2O, insoluble in EtOH. *EM Industries; Rit-Chem; Rottapharm SpA.*

4945 calcium cyanide
592-01-8 1704 209-740-0
C_2CaN_2
Calcium cyanide
Cyanolime; Cyanogas. Dissolves in H_2O liberating HCN. Used as a source of HCN to exterminate rodents. Soluble in H_2O, EtOH; LD_{50} (rat orl) = 39 mg/kg. *Mechema Chemicals Ltd.* Name unverified.

4946 calcium disodium versenate
62-33-9 3555 200-529-9
Edetate calcium disodium injection USP; lead chelating agent. *3M Pharmaceuticals.*

4947 calcium fluoride
7789-75-5 1709 232-188-7
CaF_2
calcium fluoride
fluorite, fluorspar. Source of fluorine, flux, ceramics, phosphors, paint pigment, catalyst in wood preservative, spectroscopy, electronics, lasers, high-temperature dry-film lubricants. mp = 1403°; bp = 2500°; d = 3.18; insoluble in H_2O; LD_{50} (gpg orl) > 5 g/kg. *Cerac; GE; Noah Chemical; Solvay GmbH.*

4948 calcium formate
544-17-2 1711 208-863-7
$C_2H_2CaO_4$
Latibon®. As feed additive and as silage additive; binder for fine-ore briquets. Soluble in water; insoluble in alcohol. *Bayer AG.*

4949 calcium hydride
7789-78-8 1715 232-189-2
CaH_2
hydrolete. Reacts with water to produce hydrogen. Powerful reducing agent, used to prepare rare metals from their oxides, as a drying agent or as a source of hydrogen. mp = 816°; d= 1.900.

4950 calcium hydrochlorphosphate
A mixture of calcium chloride and phosphate.

4951 calcium hydrogen phosphate
7757-93-9 1739 231-826-1
$CaHPO_4$ or $CaHPO_4{\cdot}2H_2O$
DCP-O
dicalcium phosphate, anhydrous; calcium hydrogen orthophosphate. Foods, pharmaceuticals, dentifrice, medicine, glass, fertilizer, stabilizer for plastics, dough conditioner, yeast food. d = 2.31; insoluble in H_2O, alcohols. Albright & Wilson; EM Industries; FMC; GE; Janssen Chimica; Mallinckrodt; OxyChem; Rhône-Poulenc Basic.

4952 calcium hydroxide
1305-62-0 1716 215-137-3
CaH_2O_2
calcium hydrate
hydrated lime; slaked lime; caustic lime. Inorganic base; mortar, plaster, cements, calcium salts, disinfectant, food additives, lubricants, pesticides, manufacture of paper pulp, in SBR vulcanization, in water treatment. d = 2.08-2.34; slightly soluble in H_2O; LD_{50} (rat orl) = 7.34 g/kg. *EM Industries; Janssen Chimica; Mallinckrodt; Pfizer; U.S. Gypsum.*

4953 calcium hypochlorite
7778-54-3 1717 231-908-7
$CaCl_2O_2$
calcium oxychloride
Losantin; Induchlor. Algicide, bactericide, deodorant, potable water purification, disinfectant for pools, bleaching agent, oxidizing agent; commercial grade usually 50% or more $Ca(OCl)_2$. *Olin; PPG Industries; Surex Int'l. Ltd.*

4954 calcium iodate
7789-80-2 1719 232-191-3
CaI_2O_6
calcium iodate
Lautarite; calcinol. Deodorant, mouthwashes, food additive, dough conditioner. mp = 740°; bp = 1100°; soluble in H_2O, alcohols, Me_2CO, insoluble in Et_2O, dioxane. *Atomergic Chemetals; Blythe, William Ltd; R.W. Greeff; Mitsui Toatsu; Spectrum Chemical Mfg.*

4955 calcium lactophosphate
A mixture of calcium lactate, calcium acid lactate and calcium acid phosphate. It contains about 2% P_2O_5.

4956 calcium lignosulfonate
8061-52-7
Lignosulfonic acid, sodium salt
Additive-A; Ameribond; Borrebond; Borresperse CA/CAF; Colex G; Darvan® No. 404; Marasperse GFC; Norlig 11DA; Trastan LS; Ultrazine CA; Wafex; Wargotan; Welltex 300 F; Glutrin; Goulac; Lognorit; Lignosite®; Lignosol D; Lignosol SF. Dispersant. Anionic; powder. *R. T. Vanderbilt Co Inc.* Discontinued.

4957 calcium lignosulfonates
8061-52-7
Glutrin; Additive A; Ameribond; Borrebond; Borresperse CA/CAF; Collex G; Darvan® No.404; Norlig 11 DA, A; Trastan LS; Ultrazine CA; Wafex; Wargotan; Welltex 300 F; Goulac; Lignorit; Lignosite®; Lignosol B, SF; Marasperse GFC. Modified calcium lignosulfonates; binders for foundry and refractory brick manufacture. *Borregaard Ligno Tech.*

4958 calcium molybdate
7789-82-4 1728 232-192-9
$CaMoO_4$
Molybdic acid calcium salt
calcium molybdate(VI); Carosella. Alloying agent in production of iron and steel, crystals in optical and electronic applications, phosphors. d = 4.350; insoluble in H_2O, alcohols, soluble in concentrated mineral acids *AAA Molybdenum Prods.; Atomergic Chemetals; Cerac; Noah Chem.*

4959 calcium montanate
68308-22-5 269-637-1
Hostalub® VP Ca W 2. Internal lubricant for PVC, polyamide, thermoplastic PU, other thermoplastics and thermosets. *Hoechst AG.*

4960 calcium nitrate
10124-37-5 1729 233-332-1
CaN_2O_6
Usually crystallizes as the tetrahydrate [13477-34-4] or the trihydrate [15842-29-2]. Used to support combustion in matches, explosives, as a fertilizer and corrosion inhibitor. mp = 560°; very soluble in H_2O, MeOH, EtOH, Me_2CO, insoluble in other organic solvents;.

4961 calcium oleate
142-17-6 1731 205-525-0
$C_{36}H_{66}CaO_4$
9-Octadecenoic acid calcium salt
oleic acid calcium salt. Used as a thickening lubricating grease, waterproofing agent, emulsifier. mp > 140° (dec); insoluble in H_2O and organic solvents.

4962 calcium oxide
1305-78-8 1733 215-138-9
CaO
Lime
quicklime; calxCaO,. Inorganic oxide; refractory, sewage treatment, insecticides, fungicides, manufacture of steel and aluminum; flotation of nonferrous ores; manufacture of glass, paper, Ca salts; in drilling fluids, lubricants; laboratory. mp = 2572°; bp = 2850°; d = 3.32-3.35. *Cerac; GE; Hüls Am.; Mallinckrodt; Pfizer; U.S. Gypsum.*

4963 calcium permanganate
10118-76-0 1735 233-322-7
$CaMn_2O_8$
Acerdol. Used in gastro-enteritis and diarrhea. Freely soluble in H_2O, decomposes in alcohol.

4964 calcium peroxide
1305-79-9 1736 215-139-4
calcium superoxide
calcium dioxide. Seed disinfectant, dentifrices, dough conditioners, antiseptic, bleaching of oils, modification of starches, high-temperature oxidation. Slightly soluble in H_2O. *Chemoxal SA; Eagle-Picher; FMC; Henley; Interox Am.*

4965 calcium petronate 25H, 25C and 300
Calcium petroleum sulfonate
Detergents in lube oil additives, rust inhibitors, emulsifiers. *Witco Corporation.*

4966 calcium phosphate (monobasic)
7758-23-8 1740 231-837-1
$CaH_4O_8P_2$
Phosphoric acid, calcium salt (2:1)
monocalcium phosphate, monohydrate; acid calcium phosphate; calcium biphosphate; monocalcium orthophosphate; primary calcium phosphate; calcium superphosphate; MCP,. Leavening acid in food products, mineral supplement, fertilizer, stabilizer for plastics, to control pH in malt, glass manufacture, firming agent. mp = 200° (dec); d_4^{18}=2.220; moderately soluble in H_2O, soluble in acids. Albright & Wilson; FMC; Kemira Kemi UK; Mallinckrodt; Monsanto; OxyChem; Nichia Kagaku Kogyo; Rhône Poulenc Basic.

4967 calcium phosphate (tribasic)
7758-87-4 1741 231-840-8
$Ca_3O_8P_2$
TCP
tricalcium phosphate; tricalcium orthophosphate; tertiary calcium phosphate; Calcigenol Simple; oxydapatit; voelicherite; whitlockite; bone ash,. Foods,

pharmaceuticals, polystyrene, ceramics, mordant, fertilizers, dentifrices, stabilizer for plastics, in meat tenderizer, anticaking agent, nutrient supplement. mp = 1670°; d = 3.14; insoluble in H_2O, alcohol, soluble in mineral acids. *Albright & Wilson; FMC; Mallinckrodt; Monsanto; Rhône-Poulenc Basic.*

4968 calcium propionate
4075-81-4 1745 223-795-8
$C_6H_{10}CaO_4$
propionic acid, calcium salt
Mycoban; Propionic acid, calcium salt. Mold-inhibitor; food additive; antifungal agent; improves scorch resistance and processibility of butyl rubber. soluble in H_2O, slightly soluble in alcohols, insoluble in organic solvents *Gis-Brocades Food Ingreds.; Niacet; Verdugt BV.*

4969 calcium silicate
1344-95-2 1749 215-710-8
$CaSiO_3$, Ca_2SiO_4, Ca_3SiO_5
silicic acid, calcium salt
Silicic acid, calcium salt; afwillite; akermanite; calcium pectolith; centrallasite; crestmoreite; eaklite; Cecasil; foshagite; foshallasite; gjellebaekite; grammite; gyrolite; hillebrandite; larnite; okenite; parawollastonite; pseudo-wollastonite; riversideite; table spate; tobermorite; wollastonite; xonaltite; xonotlite,. Hydrous or anhydrous silicate with varying proportions of calcium oxide and silica; Common forms: constituent of lime glass, portland cement; reinforcing filler in elastomers and plastics; absorbent for liquids, gases, vapors; anticaking agent, suspending d^{25} = 2.10. *Celite; Crosfield; Degussa; R. T. Vanderbilt Co Inc.*

4970 calcium stearate
1592-23-0 1750 216-472-8
$C_{36}H_{70}CaO_4$
octadecanoic acid, calcium salt
stearic acid, calcium salt; calcium stearate; Calcium octadecanoate; stearic acid, calcium salt. Calcium salt of stearic acid; water repellent; flatting agent in paints, emulsions; release agent for plastic molding powds.; stabilizer for PVC resins; lubricant; in pencils and crayons; food grade as conditioning agent in foods and mp = 147-149°; insoluble in H_2O, EtOH, Et_2O, $CHCl_3$, Me_2CO. Adeka Fine Chemical; Elf Atochem N. Am./Wire Mill; Eka Nobel Ltd; Ferro/Grant; Henkel/Organic Prods.; Mallinckrodt; PPG Industries; R. T. Vanderbilt Co Inc; Witco.

4971 calcium sulfate (anhydrous)
7778-18-9 1753 231-900-3
CaO_4S
calcium sulfate
karstenute; muriacite; anhydrous sulfate of lime; anhydrous gypsum; anhydrite. Inorganic salt; insoluble anhydride: in cement formulations, as paper filler; soluble anhydride: drying agent, desiccant. d = 2.96; soluble in H_2O (0.2%). *U.S. Gypsum.*

4972 calcium sulfate (dihydrate)
10101-41-4 1753 240-991-9
$CaO_4S{\cdot}2H_2O$
calcium sulfate dihdrate p.a.
Calcium sulfate dihydrate NF; Compactrol®. Tablet and capsule filler for pharmaceutical tablets manufacturing by direct compression. d = 2.3200. *Penwest Pharmaceuticals Co.*

4973 calcium sulfonate
61789-86-4 263-093-9
calcium sulfonate
Lubrizol® 2152, 2064; TLA-256; Ircogel® 900; SACI® 200HM, 445a, 450W, 4192, 4194,4215,4253. Pigment dispersant and wetting agent for color concentrates, paints, coatings, inks; pigment flushing aid for organic and inorganic pigments; viscosity stabilizer/reducer in plastisols. *Lubrizol; Ashland; King Ind; Stepan; Witco/Petroleum Spec.; Witco/Sonneborn.*

4974 calcium titanate
12049-50-2 234-988-1
CaO_3Ti
Perovskite. Inorganic compound; electronics. *Atomergic Chemetals; Cerac; Ferro/Transelco; TAM Ceramics.*

4975 calcium zirconate
$CaZrO_3$
Atomergic Chemetals; Cerac; Ferro/Transelco; TAM Ceramics.

4976 Cal-C-Vita
Proprietary preparation of vitamin B_6, vitamin C, vitamin D, and calcium. *Roche Products Ltd.* Name unverified.

4977 Caldiox
Calcium plumbate. *Associated Lead Manufacturers Ltd.* Name unverified.

4978 Caldura
A high temperature (240°) resistant resin, containing aromatic hydrocarbon groups linked by oxygen and methylene bridges. *Associated Electrical Industries (GEC).* Unverified.

4979 Caledon
Vat dyes. *ICI Chem & Polymers Ltd.*

4980 Calendula Oil CLR
1763
calendulin
Marigold; Mary-bud; gold-bloom; holligold; calendulin;. Ester of oleanolic acid, soybean oil, calendula extract, tocopherol; emollient, conditioner; emulsified and oily preparation for skin care. *Dr. Kurt Richter; Henkel/Cospha.*

4981 Calester
α-sulfomethyl laurate
Surfactant for high-quality toilet soaps, laundry detergents, automobile cleaners, spray cleaners, foamers, emulsifiers. *Pilot.*

4982 Calfax 10L-45
36445-71-3 253-040-8
Sodium linear decyl diphenyl oxide disulfonate
Decyl(sulfophenoxy)benzenesulfonic acid, disodium salt. Biodegradable surfactant, solubilizer, dispersant for dye bath leveling, pigment dispersion, heavy-duty cleaners, latex emulsification, agricultural chemicals, phenolic germicides, bottle washing. *Pilot.* Name unverified.

4983 Calfax DB-45
28519-02-0 249-063-8
Sodium alkyl diphenyl oxide disulfonate
Detergent, solubilizer for dye bath and other strongly polar applications, e.g., dip tank cleaners, electroplating baths, heavy-duty cleaners, latex emulsifiers, agricultural chemicals; tolerant of high alkalinity, high acidity, *Pilot.* Name unverified.

4984 Calfoam ES-30
9004-82-4
Sodium laureth sulfate
Detergent, foam stabilizer, flash foamer, wetter for detergent systems, personal care products, wool washing; emulsion polymerization. *Pilot.* Name unverified.

4985 Calfoam NEL-60
67762-19-0
Ammonium lauryl ether sulfate
Flash foamer, foam stabilizer, detergent, emulsifier, wetter for liquid detergents, bubble baths, shampoos, car washing; lime soap dispersant. *Pilot.* Name unverified.

4986 Calfoam SLS-30
151-21-3 8782 205-788-1
Sodium lauryl sulfate
Mild detergent, foamer for personal care products; rug/upholstery shampoos; emulsifier for cosmetics, emulsion polymerization of latex, SBR rubber, polyacrylates, elastomers; foaming agent for foamed rubber. *Pilot.* No manufacturer.

4987 Calgon
8621
$(NaPO_3)_x$
sodium polymetaphosphate
sodium polymetaphosphate; sodium hexametaphosphate; Graham's salt; glassy sodium metaphosphate; Hy-Phos. Water softening compound. mp = 628°; soluble in H_2O. *Albright & Wilson Ltd., Phosphates & Speciality Business.*

4988 Calgon® Type SGL
7440-44-0 1855 231-153-3
Carbon
For purification and decolorization of aqueous and organic liquids. *Calgon Carbon.* No manufacturer.

4989 Calgonite
Dishwashing machine detergents. *Albright & Wilson Ltd., Phosphates & Speciality Business.* Name unverified.

4990 Cal-Grid
A calcium-aluminum alloy used in the production of calcium-containing long-life lead battery plates. *Pfizer International.* Discontinued.

4991 Calibre 200-4
PC resin; FDA compliance; engineering thermoplastic for food contact and medical applications. *Dow Plastics.*

4992 Calibre 302-E
PC resin; excellent clarity and uv/weatherability properties for profile extrusion, sheet and lighting applications. *Dow Plastics.*

4993 Calibre 400-10
PC resin; impact modified grade. *Dow Plastics.*

4994 Calibre 510, 550
PC resin, glass-reinforced; applications are in transportation, electronics, and service parts. *Dow Plastics.* Name unverified.

4995 Calibre 700-4
PC resin; ignition-resistant and clarity grade; for electronics housings, optical parts, transportation, and appliances. *Dow Plastics.*

4996 Calibre 1001CD
PC resin; for the compact disc market; offers proven purity, increased clarity, processing ease, excellent physical properties and dimensional stability. *Dow Plastics.*

4997 Calibre 2060-4
PC resin; formulated to meet sterilization needs of the health care industry, providing exceptional clarity, heat resistance, impact, strength, and processability; 2060 resins are for applications using steam or ethylene oxide sterilization; 2080 resins *Dow Plastics.*

4998 Calibre LG2010
PC resin; opthalmic grade. *Dow Plastics.*

4999 Calibre®
For agriculture industry. *DuPont UK.*

5000 Calibrite
A trademark for series of aluminum pigments for plastics. No manufacturer.

5001 Calido
An alloy of 64% nickel with from 15-25% iron, 8-12% chromium, and 3-8% manganese.

5002 Calido brass
An alloy of 70% copper, 30% zinc, and traces of iron.

5003 Calido-elalco
A heat-resisting alloy containing 60% nickel, 24% iron, and 16% chromium.

5004 Califlux® 90
Aromatic oil; extender oil offering low aniline points; for elastomer compounding, especially nitrile, neoprene, SBR, and natural rubber. *Witco/Golden Bear.* Discontinued.

5005 Calight RPO
Naphthenic process oil; for rubber industry, resin extending, PVC, textiles, caulking compounds, etc. *Calumet Lubricants.*

5006 Caliment
7757-93-9 1739 231-826-1
dicalcium phosphate
Food grade of dicalcium phosphates. *Albright & Wilson Ltd., Phosphates & Speciality Business.*

5007 Calimulse PRS
Isopropylamine dodecylbenzene sulfonate
Biodegradable emulsifier, solubilizer; dry cleaning; degreasers; latex emulsifier; pigment dispersant; agricultural sprays, oil slick emulsifiers. *Pilot.*

5008 Calipharm
7757-93-9 1739 231-826-1
dicalcium phosphate
Pharmaceutical grade of dicalcium phosphate. *Albright & Wilson Ltd., Phosphates & Speciality Business.*

5009 Calirus
15310-01-7 239-352-7
Benodanil
A systemic fungicide. *BASF plc.*

5010 Calite
An alloy of 50% iron, 35% nickel, 10% aluminum, and 5% chromium. It is very resistant to heat, and melts at 2777ºF.

5011 Calixin®
24602-86-6 9793 246-347-3
Tridemorph
Systemic fungicide for control of powdery mildew in cereals, vegetables, etc. *BASF AG; BASF plc.*

5012 Calkleen
Industrial cleaning compound. *The Wellcome Foundation Ltd.* Name unverified.

5013 Callaway 4000 Series
Series of formaldehyde-free cationic polymers; dye fixatives for direct, reactive, sulfur and pigment dyes; minimal effect of lightfastness; for textile use. *Callaway.*

5014 Calloseal, Callotek
Rubber sheet or strip containing expandable graphite for sealing fire doors, pipe joints etc. Expandable, acid treated graphite. *Foseco (F.S.) Ltd.*

5015 Callusolve
1066
Benzalkonium chloride, dibromide
Amber colored paint containing 25% benzalkonium chloride bromine; used for topical use in the treatment of warts, especially multiple or mosaic warts. *Dermal Laboratories Ltd.*

5016 Calofil
471-34-1 1697 207-439-9
Calcium carbonate
Used as an antacid and calcium supplement. *Rhône-Poulenc UK.* Unverified.

5017 Calofort®
471-34-1 1697 207-439-9
$CaCO_3$
Calcium carbonate
Precipitated calcium carbonate, used for sealants, plastics, rubber. *Rhône-Poulenc Sturge Lifford.* Unverified.

5018 Calomel
10112-91-1 5951 233-307-5
Hg_2Cl_2
Mercurous chloride
Fungicide. Used for control of clubroot in brassicas and white rot in onions. *Hortichem Ltd.*

5019 calomic
A nickel-iron-chromium resistance alloy; used for electric heater elements operating up to 900º.

5020 Calopake®
471-34-1 1697 207-439-9
$CaCO_3$
Precipitated calcium carbonate
Used as a filler in paper, paint. *Rhône-Poulenc Sturge Lifford.*

5021 Caloreen
Glucose polymer; nutritional supplement. *Roussel Laboratories Ltd.* Discontinued.

5022 calorite
Alloys. One contains 65% nickel, 15% iron, 12% chromium, and 8% manganese; and another 65% nickel, 23% iron, and 12% chromium.

5023 Caloxol CP2
1305-78-8 1733 215-138-9
CaO
calcium oxide
A proprietary calcium oxide desiccant for rubber. *Rhône-Poulenc Sturge Lifford.*

5024 Calprona K
Mold inhibitor. *BP Chemicals Ltd.*

5025 Calquat
Industrial cleaning compound. *The Wellcome Foundation Ltd.* Name unverified.

5026 Calsan
Sanitary disinfectant. *The Wellcome Foundation Ltd.* Name unverified.

5027 Calsil
Silica/calcium oxide
Precipitated silica with a proportion of calcium oxide; medium-active reinforcing filler with outstanding extrusion characteristics, good processibility for technical rubber articles. *Degussa AG.* Name unverified.

5028 Calsoft AOS-40
Sodium α olefin sulfonate
Surfactant for hand soaps, shampoos, hard surface cleaners, household and industrial cleaners. *Pilot.*

5029 Calsoft F-90
Sodium dodecylbenzene sulfonate
Detergent, emulsifier, wetter for all-purpose and hard surface cleaners, bubble baths, degreasers, laundry powders, textile scouring aids, emulsion polymers, sanitation, emulsion paints, wettable powders, ore flotation, metal pickling. *Pilot.*

5030 Calsoft LAS-99
Dodecylbenzene sulfonic acid, linear
Biodegradable detergent, emulsifier, intermediate for liquid and dry detergents, hard surface cleaners, stripping, wetting, foaming. *Pilot.*

5031 Calsoft T-60
TEA-dodecylbenzene sulfonate
Biodegradable detergent, wetting agent, flash foamer; liquid detergents, wool wash compounds, cosmetics and shampoos, agricultural emulsifiers, industrial cleaners, textile scouring, and car wash compounds. *Pilot.*

5032 Calsol 510
Naphthenic process oil; provides excellent initial color and color stability for thermoplastics, radial and styrene block elastomers. *Calumet Lubricants.*

5033 Calsol 804
Naphthenic process oil; provides excellent color stability in resin extending, PVC, textiles, caulking compounds and other applications. *Calumet Lubricants.*

5034 Calsolene
General industrial wetting agent, emulsifier and emulsion breaker; biodegradable; used in oil and petroleum industry; gas manufacture; control of foam generation. *ICI Chem & Polymers Ltd.*

5035 Calstrip
Industrial cleaning compound. *The Wellcome Foundation Ltd.* Name unverified.

5036 Calsuds 81
Formulated product; concentrate base for liquid detergents, all-purpose and hard surface cleaning, dishwash, car wash, foaming and wetting solutions. *Pilot.*

5037 Calsun Bronze
Copper, aluminum, tin alloy
A proprietary trade name for an aluminum bronze containing copper with 2.5% aluminum and 2% tin. No manufacturer.

5038 Calthane
Two-part solvent-free urethane elastomer systems; for molds for making plastic, concrete and plaster parts. *Cal Polymers.* Name unverified.

5039 Calthane ND 1100
Two-part urethane elastomer system; used in flexible and rigid castings, prosthetics, underwater sports equipment, optical lenses, art objects, mechanical devices, flexible windshields, industrial tubing or sheeting, toys, adhesives; mix ratio: 1:1. *Cal Polymers.* Name unverified.

5040 Calthane NF 0710
Two-part nonfilled urethane elastomer Part A is an MDI-based prepolymer; Part B is a polyester resin; tougher than filled Calthane systems; low mixed viscosity suggests use for machine-dispensed urethane sheets, skate board wheels, mechanical parts; mix *Cal Polymers.* Name unverified.

5041 Calthane NF 1900
Two-part urethane elastomer. Part A is an MDI-based prepolymer, Part B is a wh. resin; ideal for making tough, flexible molds; produces solid castings without degassing; other applications including casting industrial and print rollers, bumper pads, *Cal Polymers.* Name unverified.

5042 Cal-Tint
Colorant dispersions; for coloring of aqueous and nonaqueous coating compositions. *Hüls Am.*

5043 Calurea
urea/calcium nitrate
A nitrogenous fertilizer. It is a mixture of urea and calcium nitrate containing 34% nitrogen.

5044 Cambilene
3206
A selective weedkiller. *Fisons plc, Horticultural Div.* Name unverified.

5045 Cambrelle
Melded fabrics. *Camtex Fabrics Ltd.*

5046 Cambrite
A smokeless powder, containing 22-24% nitroglycerin, 3-4.5% barium nitrate, 26-29% potassium nitrate, 32-35% dried wood meal, 1% calcium carbonate, and 7-9% calcium chloride.

5047 Camel-CAL®, Camel-CARB®, Camel-FIL, Camel-FINE
471-34-1 1697 207-439-9
$CaCO_3$
Calcium carbonate
Filler for water-based coatings and inks, paper, and PVC pipe. Used in interior flat paint and exterior house paints, rubber compounds, putty and caulk, ceramics, adhesives, linoleum, floor tile, and textile coatings.Designed for high filler loading in *Genstar Stone Prods.*

5048 Camelia metal
An alloy of 70.2% copper, 14.7% lead, 10.2% zinc, 4.2% tin, and 0-5% iron.

5049 Camel-TEX®, CamelWITE®
471-34-1 1697 207-439-9
Calcium carbonate
Fineground, general-purpose filler used in interior flat paints, primers, and sealers, polyester fiberglass premixes, preforms, and hand lay-up gel coats, rubber automotive products, household products, tubing, medical products, and closures, putty, caulk *Genstar Stone Prods.*

5050 Cameo®
For agricultural industry. *DuPont UK.*

5051 Camflex
Plastics in the form of sheets. *Courtaulds plc.* Discontinued.

5052 Camie 300
General-purpose spray adhesive with quick adhesion to metal, plastic, wood, paper, cardboard, fabric, leather; for garment, furniture, upholstery, and shipping applications. *Camie-Campbell.*

5053 Camilol
515-69-5 1281 208-205-9
$C_{15}H_{26}O$
α-Bisabolol
α-bisabolol; Bisabolol. Cosmetic ingredient. *Maybrook.*

5054 Camite
12070-12-1 9945 235-123-0
WC
Tungsten carbide
Proprietary trade name for tungsten carbide materials. No manufacturer.

5055 Campbell's CIPC 40%
101-21-3 2240 202-925-7
chlorpropham
Emulsifiable concentrate containing 400 g chlorpropham per liter; a carbamate herbicide. *MTM AgroChemicals Ltd.*

5056 Campbell's DB Straight
94-82-6 2893 202-366-9
2,4-DB
Soluble concentrate containing 300 g 2,4-DB per liter; used to control weeds in lucerne. *MTM AgroChemicals Ltd.*

5057 Campbell's Destox
1702-17-6 2462 216-935-4
clopyralid
Translocated herbicide for cereals and established grassland. *MTM AgroChemicals Ltd.*

5058 Campbell's Dioweed 50
94-75-7 2865 202-361-1
2,4-D
Translocated herbicide for cereals and established grassland. *MTM AgroChemicals Ltd.*

5059 Campbell's DSM
8022-00-2 6129
Demeton S methyl
Emulsifiable concentrate containing 580 g demeton-S-methyl per liter; a systemic organophosphorus insecticide. *MTM AgroChemicals Ltd.*

5060 Campbell's Field Marshal, Grassland Herbicide
Mixture of dicamba, MCPA and mecoprop; used for weed control in cereals and grassland. *MTM AgroChemicals Ltd.*

5061 Campbell's Linuron 45%
330-55-2 5534 206-356-5
Linuron-trifluralin
A residual urea herbicide for the control of weeds in field crops. *MTM AgroChemicals Ltd.* Name unverified.

5062 Campbell's MC Flowable
A suspension concentrate containing 62 g carbendazim [10605-21-7] and 400 g maneb [12427-38-2] per liter. systemic fungicide for cereals. *MTM AgroChemicals Ltd.*

5063 Campbell's MC Flowable
A suspension concentrate containing 62 g carbendazim [10605-21-7] and 400 g maneb [12427-38-2] per liter. systemic fungicide for cereals. *MTM AgroChemicals Ltd.*

5064 Campbell's MCPA 25, 50
94-74-6 5803 202-360-6
MCPA; herbicide for cereals and grassland. *MTM AgroChemicals Ltd.*

5065 Campbell's MCPA 25, 50
94-74-6 5803 202-360-6
MCPA; herbicide for cereals and grassland. *MTM AgroChemicals Ltd.*

5066 Campbell's Nabam Soil Fungicide
142-59-6 6426 205-547-0
nabam
Soluble concentrate containing 320 g/l nabam; used for control of root rot in tomatoes and chrysanthemums. *MTM AgroChemicals Ltd.* Name unverified.

5067 Campbell's New Camppex
Soluble concentrate containing 34 g 2,4-D, 133 g dichlorprop, 53 g MCPA, and 164 g mecoprop per liter; and herbicide for use in cereals. *MTM AgroChemicals Ltd.*

5068 Campbell's Nico-Soap
54-11-5 6611 200-193-3
Nicotine
Alkaloid insecticide. *MTM AgroChemicals Ltd.*

5069 Campbell's Rapier
23950-58-5 8058 245-951-4
Propyzamide
A residual herbicide for agricultural crops. *MTM AgroChemicals Ltd.* No manufacturer.

5070 Campbell's Redipon
120-36-5 3128 204-390-5
Soluble concentrate of 350g/l (Redipon Extra) or 500 g/l (Redipon) dichlorprop; used for control of weeds in barley, wheat and oats. *MTM AgroChemicals Ltd.*

5071 Campbell's Redipon Extra
3068 204-390-5
Soluble concentrate of 350 g dichlorprop and 150 g MCPA per liter; used for control of weeds in cereals and grass. *MTM AgroChemicals Ltd.*

5072 Campbell's Redlegor
2893
2,4-DB [94-82-6] - MCPA [94-74-6]. Translocated herbicide for cereal crops. *MTM AgroChemicals Ltd.*

5073 Campbell's Sugar Beet Herbicide
Emulsifiable concentrate containing 27.5 g chlorpropham, 28.5 g fenuron, and 147 g propham per liter; an herbicide. *MTM AgroChemicals Ltd.*

5074 Campbell's Trifluron
Mixture of linuron [330-55-2] and trifluralin [1582-09-8]; herbicide for winter cereals. *MTM AgroChemicals Ltd.*

5075 Campbell's X-Spor
12427-38-2 5761 235-654-8
Maneb
Suspension concentrate containing 480 g maneb per liter; a dithiocarbamate fungicide to control blight, rusts and mildew. *MTM AgroChemicals Ltd.* No manufacturer.

5076 d-camphorsulfonic acid
3144-16-9 1781 221-554-1
$C_{10}H_{16}O_4S$
7,7-dimethyl-2-oxobicyclo[2.2.1]heptane-1-methanesulfonic acid
CSA; D-Camphor-10-sulfonic acid; Reychler's acid; camsylate; camphostyl. Used for resolution of optical isomers. mp = 193-195° (dec); $[\alpha]_D^{20}$ = 43.5° (c = 4.3 EtOH); soluble in EtOH, poorly soluble in other organic solvents.

5077 Camtex
Melded fabrics. *Camtex Fabrics Ltd.*

5078 Camwood
Cambe wood.

5079 Canacert
Foodstuffs coloring matter meeting Canadian regulations. *Pointing Ltd.* Name unverified.

5080 Canadian Certicol
Food colors meeting Canadian specifications. *Morton Int'l. Ltd.* Discontinued.

5081 Canadium
An alloy of 1 part palladium, 2 parts platinum, and 6 parts nickel; used as a substitute for platinum.

5082 Canagel 75
Semigelatin dynamite with good water resistance, high weight and volume energy and relatively high detonation velocity; used for construction and building industry, mining, explosives. *Hercules.* Discontinued.

5083 Canarin
Persulfocyanogen yellow. A yellow coloring matter obtained by the action of bromine upon potassium or ammonium thiocyanate; used in calico printing.

5084 candelilla wax
8006-44-8 1789 232-347-0
Hentriacontane
Wax from various *Euphorbiaceae* species; manufacture of cosmetics, rubber substitutes, polishes, candles, sealing waxes, varnishes, leather, creams; for waterproofing; electrical insulation; inks; molding compositions; sizing paper; protective coat mp = 68-70°; d = 0.950-0.990; insoluble in H_2O, soluble in organic solvents. *Penta Mfg.; Sevenson Bros.; Strahl & Pitsch.*

5085 Candex®
50-99-7 4467 200-075-1
Glucose
Dextrose with small amounts of higher glucose saccharides; offers sweet, nongritty taste and is easily blended with flavors, lubricants, and other dry additives; excellent flow and compaction properties; for use in chewable tablets, *Penwest Pharmaceuticals Co.*

5086 Canesten®
23593-75-1 2478 245-764-8
Clotrimazole
Broad spectrum antifungal drug. *Bayer AG.*

5087 Canfelzo
1314-13-2 10279 215-222-5
Zinc oxide
Astringent and topical protectant. *Pigment & Chemical Inc.*

5088 Canguard® 327
51200-87-4 257-048-2
$C_5H_{11}NO$
4,4-Dimethyloxazolidine
Dimethyl-1-oxa-3-aza-cyclopentane; Dimethyloxazolidine; Bioban CS 1135. Bioban CS 1135 is a preservative used in cooling fluids and paints. In-can preservative for water-containing systems, e.g., latex paint, resin emulsions, caulks, adhesives; broad spectrum antimicrobial activity; emulsifying capability when used with fatty *Angus.*

5089 Canguard® 409
52-51-7 1470 200-143-0
2-Bromo-2-nitropropane-1,3-diol
Preservative for adhesives, coatings, paints, starch, pigment and extender slurries, latex, antifoam emulsions, and inks. *Angus.* No manufacturer.

5090 Canguard® 454
Hexahydro-1,3,5-tris(2-hydroxyethyl)-s-triazine
In-can preservative for water-containing systems, e.g., latex paint, resin emulsions, adhesives, inks; broad spectrum antimicrobial activity; nonfoaming. *Angus.*

5091 Canilep
Vaccines for dogs. *Glaxo Laboratories.* Name unverified.

5092 Canilep D.D
A combined double-dose vaccine for dogs. *Glaxo Laboratories.* Name unverified.

5093 Canopar
4304-40-9 9414 224-318-6
Thenium closylate
Used in veterinary medicine as an anthelmintic. *Burroughs Wellcome Co.* No manufacturer.

5094 Cantamega 1000, 2000
Fiber based vitamin and mineral supplements; dietary supplements. *Larkhall Laboratories plc.* Name unverified.

5095 Canthaxanthin
514-78-3 1798 208-187-2
$C_{40}H_{52}O_2$
β,β-Carotene-4,4'-dione
4,4'-Dioxo-β-carotene; Canthaxanthin; Carophyll Red; Food Orange 8; C.I. 40850; Orobronze; Roxanthin Red 10; Carotaben Plus; Carophyll. Pigments for animal feedstuffs. mp = 217° (dec); λ_m = 470 nm ($E^{1\%}_{cm}$ = 2250); soluble in $CHCl_3$, oils. *Roche Products Ltd.* Name unverified.

5096 Cantreece®
Fibers. *DuPont UK.*

5097 Canzler Wire
An alloy of 98.8% copper, 1% silver, and 0.2% phosphorus.

5098 Cao
2409-55-4 219-314-6
$C_{11}H_{16}O$
2-*tert*-Butyl-4-methylphenol
Phenol, 2-(1,1-dimethylethyl)-4-methyl-. Butylated *p*-cresol; antioxidants for food, plastic, rubbers, and other general purpose requirements. mp = 51-52°; bp = 244°; insoluble in H_2O, soluble in organic solvents; LD_{50} (rat orl) = 2500 mg/kg. *PMC Specialities Group Inc.* No manufacturer.

5099 caoutchouc
Coagulated latex of various rubber trees and shrubs; used in the manufacture of electrical insulation, toys, and beltings.

5100 C-A-P
Cellulose acetate phthalate
Pharmaceutic aid. *Eastman.*

5101 C-A-P Enteric Coating Polymer
Cellulose acetate phthalate
Eastman. Name unverified.

5102 CA-P20
Clarifying agent concentrate in polypropylene carrier; nucleating agent in PP homopolymer and random copolymers; provides improved clarity, better gloss, improved tensile and flex. strength and dimensional stability. *Polyvel.*

5103 Cap copper
An alloy of 95-97% copper and 3-5% zinc; used for deep drawing and stamping.

5104 Capa
Caprolactone and (Nylon) polymers based on it. *Solvay Interox Ltd.*

5105 Capasal
Brown colored foaming shampoo containing 0.05% w/w salicylic acid BP, 1.0% w/w coconut oil BP, 1.0% w/w distilled coal tar; used for the treatment of dry scaly scalp conditions. *Dermal Laboratories Ltd.*

5106 Capcure Emulsifier 37S
Ethoxylate
Epoxy resin emulsifier. *Henkel/Functional Prods.*

5107 Capella®
For agriculture industry. *DuPont UK.*

5108 Capexco
A coal mine explosive consisting of 32-34% nitroglycerin, 0.5-1.5% nitrocotton, 24-25% sodium nitrate, 30-32% ammonium oxalate, and 8-10% wood meal.

5109 Capital 170
Coconut fatty acids. *Karlshamns.*

5110 Capmul® EMG
51158-08-8
PEG-20 glyceryl stearate
Dough conditioner for yeast-raised baked goods. *Karlshamns.*

5111 Capmul® GDL
27638-00-2 248-586-9
Glyceryl dilaurate
Emulsifier used in fats and oils; for cosmetics, pharmaceuticals. *Karlshamns.*

5112 Capmul® GMO
111-03-5 203-827-7
Glyceryl oleate
Food emulsifier, wetting control agent; dispersant for pigments, solids; defoamer. *Karlshamns.*

5113 Capmul® GMS
31566-31-1 4498 250-705-4
Glyceryl stearate
Octadecanoic acid monoester with propane-1,2,3-triol. Stabilizer; internal lubricant for cosmetics; food emulsifier used in margarine, yeast-raised baked goods. Used in pharmaceutical dispensing. mp = 56-58°; insoluble in H_2O, soluble in hot organic solvents. *Karlshamns.*

5114 Capmul® MCM
Glyceryl caprate/caprylate
Co-solvent and coupler for organic compounds; without emulsifier. *Karlshamns.*

5115 Capmul® O
1338-43-8 8872 215-665-4
Sorbitan oleate
Food emulsifier and dispersant. *Karlshamns.*

5116 Capmul® POE-L
9005-64-5 8872
Polysorbate 20
Food emulsifier and solubilizer for flavors. *Karlshamns.*

5117 Capmul® POE-O
9005-65-6 7742
Polysorbate 80
Food emulsifier for frozen desserts; solubilizer for oils into water systems; used for flavors, fragrances, vitamins, pharmaceuticals, solvent. *Karlshamns.*

5118 Capmul® POE-S
9005-67-8 8872
Polysorbate 60
Food emulsifier; solubilizer for oils into water systems. *Karlshamns.*

5119 Capmul® S
1338-41-6 8872 215-664-9
Sorbitan stearate
Food emulsifier for chocolate and confectionary coatings, shortenings. *Karlshamns.* No manufacturer.

5120 Caposil
A trademark for calcium silicate and asbestos based insulating materials. Caposil 1400 withstands 1400°F (760°) and Caposil HT withstands 1850°F (1000°). *Cape Insulation Cape Asbestos Co.* Name unverified.

5121 Cappagh brown
Euchrome; mineral brown. A natural pigment used in oil-painting, containing hydrated oxide of iron and 27% manganese dioxide.

5122 Capran® 77C
25038-54-4 6832
Nylon 6 film, unoriented; thermoplastic film for general purpose packaging, sterilization packaging; clear extruded films in various widths and thicknesses. *AlliedSignal Engineered Plastics.* No manufacturer.

5123 Capran® Unidraw®
25038-54-4 6832
Nylon 6, monaxially oriented; specifically engineered as a carrier web for molding fiberglass-reinforced panels; exceptional tear and tensile strength. *AlliedSignal Engineered Plastics.*

5124 Capran®, Capran® 77C, Capran® Emblem, Capran®Unidraw®
25038-54-4 6832
Poly(caprolactam). Polyamide 6 film useful for food packaging, composites manufacturing and other industrial applications. Nylon 6 film, unoriented; thermoplastic film for general purpose packaging, sterilization packaging; clear extruded films in various widths *AlliedSignal Inc.*

5125 Capri blue gon
$C_{17}H_{20}N_3OCl$
A dyestuff, it is the zinc double chloride of dimethyldiethyldiamino-toluphenazonium chloride. Dyes tannined cotton greenish-blue.

5126 n-Capric acid
334-48-5 1802 206-376-4
$C_{10}H_{20}O_2$
n-decanoic acid
capric acid; carboxylic acid (C_{10}). Esters for perfumes and flavors; base for wetting agents; intermediate for chemical synthesis. mp = 31.4°; bp = 270°; d_4^{50}= 0.8782; n_D^{40}= 1.4288; insoluble in H_2O, soluble in organic solvents; LD_{50} (mus iv) = 129 mg/kg. *Akzo; Aldrich; Henkel/Emery; Mirachem Srl; Procter & Gamble; Witco/Humko.*

5127 caproic acid
142-62-1 1803 205-550-7
$C_6H_{12}O_2$
n-caproic acid
hexanoic acid. Analytical chemistry, flavors, manufacture of rubber chemicals, varnish driers, resins, pharmaceuticals. bp = 205°; d_4^{20}= 0.9265; n_D^{20} = 1.4163; slightly soluble in H_2O (1.082 g/100g), soluble in organic solvents; LD_{50} (rat orl) = 3.0 g/kg. *Aldrich; Chisso Am.; Janssen Chimica; Penta Mfg.; Schweizerhall.*

5128 Caprol® 10G10S
39529-26-5 254-495-5
Polyglyceryl-10 decastearate
Lubricant for thread finishes, wax additive, crystal modifier; thickener; opacifier. *Karlshamns.*

5129 Caprol® 10G2O
Polyglyceryl-10 dioleate
Oil-in-water emulsifier, humectant, lubricant; for frozen desserts. *Karlshamns.*

5130 Caprol® 10G40
34424-98-1 252-011-7
Polyglyceryl-10 tetraoleate
Food emulsifier, viscosity control, stabilizer. *Karlshamns.*

5131 Caprol® 10G100
11094-60-3 234-316-7
Polyglyceryl-10 decaoleate
Food emulsifier, solubilizer, lubricant, and dispersant. *Karlshamns.*

5132 Caprol® 2G4S
72347-89-8
Polyglyceryl-2 tetrastearate
Food emulsifier; opacifier; wax modifier; thickener. *Karlshamns.*

5133 Caprol® 3GO
9007-48-1
Polyglyceryl-3 oleate
Food emulsifier for frozen desserts, vegetable dairy products, diet spreads; wetting agent for dyes and pigments in cosmetics; defoamer. *Karlshamns.*

5134 Caprol® 3GS
37349-34-1
Polyglyceryl-3 stearate
Food emulsifier, stabilizer and whipping agent used in frozen desserts and fat reduction. *Karlshamns.*

5135 Caprol® 6G2S
Polyglyceryl-6 distearate
Food emulsifier for whipped toppings, frozen desserts, coffee whiteners. *Karlshamns.*

5136 Caprol® 6G20
Polyglyceryl-6 dioleate
Food emulsifier for frozen desserts. *Karlshamns.*

5137 Caprol® ET
Polyglyceryl mixed esters; food emulsifier and crystal inhibitor in oils. *Karlshamns.*

5138 ε-caprolactone monomer
502-44-3 207-938-1
$C_6H_{10}O_2$
6-hexanalactone
2-oxepanone; ε-Caprolactone; 6-Hexanolactone; 6-Hexanolide; 6-Hydroxyhexan-6-olide;
1-oxa-2-oxocycloheptane; hexan-6-olide; 6-caprolactone monomer. Intermediate in adhesives, urethane coatings, elastomers; solvent; diluent for epoxy resins; synthetic fibers; organic synthesis. mp = -1°; bp_{10} = 96-98°; d = 1.0300; n_D^{20}= 1.4630; LD_{50} (rat orl) = 4290 mg/kg. *Union Carbide.*

5139 Capron
Range of type 6 nylon polymers including reinforced grades of nylon 6 and either glass fibers or a blend of glass fibers and selected minerals; used for power tool housings, chain saw housings and other components, housings for lawn and garden equipment, *AlliedSignal Inc.*

5140 Capron 8200
A proprietary nylon used for injection molding applications requiring improved toughness over standard nylons. *AlliedSignal Inc.*

5141 Capron 8202C, 8203C, 8202CQ
A proprietary range of crystalline type nylon molding compounds; for improved cycle time and higher stiffness. *AlliedSignal Inc.*

5142 Capron 8206 S
A proprietary nylon of good flexibility used in extrusion and molding operations. *AlliedSignal Inc.* Discontinued.

5143 Capron 8230G
Nylon with 6% glass fiber; improved stiffness and dimensional stability. *AlliedSignal Inc.*

5144 Capron 8231G, 8323G, 8233G
A range of glass-filled nylons (14%, 25%, and 33% respectively). *AlliedSignal Inc.*

5145 Capron 8253, 8350, 8351, 8352
A proprietary range of copolymer grades of nylon possessing high impact strength; for extrusion and injection molding. *AlliedSignal Inc.*

5146 Capron 8267 G HS
Mineral and glass reinforced nylon for improved mechanical properties and low warpage. *AlliedSignal Inc.*

5147 Capron 8270 HS
A proprietary modified nylon 6 compound used in blow molding. *AlliedSignal Inc.*

5148 Capron 8331G, 8332G, 8333G, 8334G
A family of high impact, glass reinforced materials exhibiting improved dry-as-moled toughness. *AlliedSignal Inc.*

5149 Capron® 8202
25038-54-4 6832
Nylon 6 homopolymer; for consumer and industrial parts, e.g., electric hair brush, chain saw housings, electrical parts; features rigidity, heat resistance, excellent surface appearance. *AlliedSignal Engineered Plastics.*

5150 Capron® 8203C HS
25038-54-4 6832
Nylon 6 homopolymer
Nucleated; Heat-stabilized material for automotive push-pull cable assembly, inner tube; featuring stiffness, wear resistance. *AlliedSignal Engineered Plastics.*

5151 Capron® 8232G HS FR
25038-54-4 6832
Nylon 6, 25% glass-reinforced, flame-retarded; flame-retarded grade for tool housings; features excellent surface appearance, strength, stiffness, impact resistance, creep resistance, heat and chemical resistance *AlliedSignal Engineered Plastics.*

5152 Capron® 8233G HS
25038-54-4 6832
Nylon 6, 33% glass-reinforced; for automotive end fittings; features rigidity, chemical resistance. *AlliedSignal Engineered Plastics.*

5153 Capron® 8253
25038-54-4 6832
Nylon 6 copolymer
For automotive fasteners and clips; features flexibility, toughness, low temperature impact resistance. *AlliedSignal Engineered Plastics.*

5154 Capron® 8259
25038-54-4 6832
Nylon 6 copolymer
For consumer/industrial parts, e.g., fasteners; features flexibility, low temperature impact resistance. *AlliedSignal Engineered Plastics.*

5155 Capron® 8266G HS
25038-54-4 6832
Nylon 6, 40% mineral/glass-reinforced; for automotive grills. *AlliedSignal Engineered Plastics.*

5156 Capron® 8280
25038-54-4 6832
Nylon 6 homopolymer
For consumer/industrial parts, e.g., roto-molded reservoirs, tanks, shapes; processability for large parts, chemical resistance, impact strength, rigidity. *AlliedSignal Engineered Plastics.*

5157 capryl alcohol
111-87-5 6849 203-917-6
$C_8H_{18}O$
1-octanol
octyl alcohol. Solvent and chemical intermediate. *Ethyl; M. Michel; Penta Mfg.; Schweizerhall; Vista.* No manufacturer.

5158 caprylic acid
124-07-2 1808 204-677-5
$C_8H_{16}O_2$
n-octanoic acid
octoic acid. Fatty acid; Used in manufacture of dyes, drugs, perfumes. mp = 16.7°; bp = 239.7°; d_4^{20} = 0.910; n_D^{20} = 1.4280; slightly soluble in H_2O (68 mg/100 g), soluble in organic solvents; LD_{50} (rat orl) = 10,080 mg/kg. *Akzo; Aldrich; Henkel/Emery; Procter & Gamble; Unichema.*

5159 Caprylic imidazoline
37478-68-5 253-521-2
Crodazoline Cy. *Croda Universal Ltd.* Name unverified.

5160 caprylic/capric acid
67762-36-1 267-013-3
Emery® 6358; Industrene® 365. Mixture of caprylic/capric acid; intermediate used in alkyd resins, rubber compounding, water repellents, polishes, soaps, abrasives, cutting oils, candles, crayons, emulsifiers; FG grades as lubricant, release agent, binder, defoamer in foods, *Witco/Humko.* Discontinued.

5161 caprylic/capric acid triglyceride
65381-09-1 265-724-3
Emalex K.T.G; Captex® 300; Labrafac Lipophile WL 1349; Lexol GT 855, GT 865; Mazol® 812; Miglyol® 812; Myritol 318; Neobee® M-5; Tegosoft® 622LD. Oil-phase cosmetic ingredient; emollient. *Nihon Emulsion; Karlshamns.*

5162 capsule metal
An alloy of 92% lead and 8% tin.

5163 captafol
2425-06-1 1814 219-363-3
$C_{10}H_9Cl_4NO_2S$
3a,4,7,7a-tetrahydro-2-[(1,1,2,2-tetrachloroethyl)thio]-1H-isoindole-1,3(2H)-dione
Sanspor; Difolatan. Fungicide, used with potatoes. mp = 160-161°; LD_{50} (rat orl) = 2500-6200 mg/kg.

5164 Captan Granular
133-06-2 1815 205-087-0
Captan as dry granules; used for controlling soil borne fungal diseases in tomato, lettuce and strawberry. *Wheatley Chemical Co Ltd.* No manufacturer.

5165 Captan, Captan-Col, Captan-50, Captan-83P, Captan Granular
133-06-2 1815 205-087-0
$C_9H_8Cl_3NO_2S$
N-trichloromethylthiotetrahydrophthalimide
N-trichloromethylthio-4-cyclohexene-1,2-dicarboximide; N-trichloromethylthio)-4-cyclohexene-1,2-dicarboxamide; N-trichloromethylthio-3a,4,7,7a-tetrahydrophthalimide; N-(trichloromethylmercapto)-Δ^4-tetrahydrophthalimide; ENT 26538; SR-406; Merpan; Orthocide-406; Vancide 89. Organic compound; seed treatment, fungicide, and bacteriostat in plants, plastics, leather, fabrics, and fruit preservation; gas odorant. mp = 178°; d = 1.74; insoluble in H_2O, soluble in organic solvents; LD_{50} (rat orl) = 9000 mg/kg. *Industrias Quimicas del Valles SA; R. T. Vanderbilt Co Inc Co Inc; ICI; Wheatley Chemical Co Ltd; Plant Protection.*

5166 Captax
149-30-4 5916 205-736-8
2-Mercaptobenzothiazole
Primary accelerator for both natural and synthetic rubbers. *R. T. Vanderbilt Co Inc.*

5167 Captex® 200
Propylene glycol dicaprylate/dicaprate
Carrier, coupler, solvent for flavors, fragrance oil, soluble colorants, vitamins, medicinals, cosmetics; emollient for creams, lotions, makeup. *Karlshamns.*

5168 Captex® 300
65381-09-1 265-724-3
Caprylic/capric triglyceride
Solvent for colors and perfumes; emollient, moisturizer in cosmetics, toiletries, pharmaceuticals; plasticizer. *Karlshamns.* No manufacturer.

5169 Captex® 800
56519-71-2
Propylene glycol dioctanoate
Nonoily lubricant imparting rich feel to skin in cosmetics and pharmaceuticals; carrier for essential oils, flavors; vehicle for vitamins, medicinals, nutritional products. *Karlshamns.* No manufacturer.

5170 Captex® 8000
538-23-8 208-686-5
$C_{27}H_{50}O_6$
Caprylic triglyceride
Tricaprylin; Octanoic acid, 1,2,3-propanetriyl ester; Caprylin; Glycerol trioctanoate; caprylic acid triglyceride; glycerin tricaprylate; glyceryl trioctanoate; MCT; panacete 800; RATO; tricaprylyl glycerin; tricapryloylglycerol; tricaprylic glyceride; trioctanoylglycerol. Nonoily lubricant imparting rich feel to the skin; for cosmetics and pharmaceuticals; carrier for essential oils, flavors; vehicle for vitamins, medicinals, nutritional products. mp = 6°; bp = 233°; d = 0.9530; insoluble in H_2O, soluble in organic solvents; LD_{50} (rat orl) = 33300 mg/kg. *Karlshamns.*

5171 Captex® 810B
Caprylic/capric/linoleic triglyceride
Emollient, solvent, carrier, fixative, and extender in pharmaceuticals, nutritional, and cosmetic applications. *Karlshamns.*

5172 Captor
Surfactants for use in oil recovery. *Imperial Chemical Industries plc.*

5173 Caradate, Caradol
A range of polyether polyols and isocyanates used for the production of

flexible and rigid polyurethane foams; may be used in furniture and automotive sealing, in pipes and tanks; effective insulators; also find use in domestic freezers and refrigerators. *Shell; Shell UK.*

5174 Caramba

Rust solvents, penetrating oils, anticorrosives, preservative preparations, anticorrosion lacquers, underbody coatings and cavity preservation compounds for vehicles; cleaning, polishing, scouring and abrasive preparations for metals, motors and machine *Caramba Chemie GmbH.* Name unverified.

5175 Caramba Felgenglanz, Caramba Felgenneu

Detergents for cleansing wheel-rims. *Caramba Chemie GmbH.* Name unverified.

5176 Caramba Lackkrone, Caramba Perlglanz

Cleans, polishes and seals in one operation. Used on paintwork and motor cars. *Caramba Chemie GmbH.* Name unverified.

5177 carbadox

6804-07-5 1825 229-879-0

$C_{11}H_{10}N_4O_4$

(2-quinoxalinylmethylene)hydrazinecarboxylic acid methyl ester N,N'-dioxide

GS-6244; Fortigro; Mecadox. Antimicrobial, used in swine. mp = 239-240°; λ_m = 236, 251, 303, 366, 373 nm (ε 11000, 10900, 36400, 16100, 16200 H_2O); almost insoluble in H_2O.

5178 Carbagel

An activated carbon containing calcium chloride; A proprietary trade name for a drying agent. No manufacturer.

5179 carbanilic ether

101-99-5 7473 202-995-9

$C_9H_{11}NO_2$

Phenylurethane

Phenylurethane; Carbamic acid, phenyl-, ethyl ester; N-Phenyl ethylcarbamate; Ethyl carbanilate; Ethyl N-phenylcarbamate. Chemical intermediate. mp = 49-51°; bp = 238°; d = 1.106; n_D^{20}= 1.5376; slightly soluble in H_2O, soluble in organic solvents.

5180 carbanilide

102-07-8 1829 203-003-7

$C_{13}H_{12}N_2O$

Diphenylurea

N,N'-Diphenylurea; diphenylcarbamide; 1,3-diphenylurea; *sym*-diphenylurea. Organic synthesis. mp = 238°; bp = 260° (dec); slightly soluble in H_2O (0.15 g/l), Me_2CO, EtOH, $CHCl_3$.

5181 Carbaryl

63-25-2 1831 200-555-0

$C_{12}H_{11}NO_2$

1-Naphthyl methylcarbamate

1-Naphthalenol, methylcarbamate; carbamic acid, methyl-, 1-naphthyl ester; α-naphthalenyl methylcarbamate; α-naphthyl N-methylcarbamate; Atoxan; Bercema NMC 50; Caprolin; Carbamic acid, N-methyl-1-naphthyl-; Carbamic acid, N-methyl,1-naphthyl ester; Carbamine; Carbaril; Carbarilo; Carbarilum; Carbatox; Carbatox 75; Carbatox-60; Carbavur; Carpolin; Compound 7744; Denapon; Dicarbam; Dyna-Carbyl; Tomado; Carylderm; Microcarb; Murvin 85; Thinsec; Tornado; Wasp Destroyer. Used as an insecticide. Used in lice infestation. mp = 142°; d = 1.232; soluble in H_2O (40 mg/l), freely soluble in most organic solvents; LD_{50} (rat orl) = 230 mg/kg *Napp Laboratories Ltd; ICI AgroChemicals Professional Products.* Name unverified.

5182 Carbate Flowable

10605-21-7 1836 234-232-0

A liquid formulation containing 500 g carbendazim per liter as a suspension concentrate; systemic fungicide. *Pan Britannica Industries Ltd.*

5183 Carbate Flowable

10605-21-7 1836 234-232-0

A liquid formulation containing 500 g carbendazim per liter as a suspension concentrate; systemic fungicide. *Pan Britannica Industries Ltd.*

5184 Carbathene

A proprietary trade name for an ethylene N-vinyl carbazole copolymer; a molding compound form stable over 240°. *Standard Telecommunications Laboratories.* Unverified.

5185 Carbazole Blue

A dyestuff, obtained by fusing carbazole with oxalic acid.

5186 Carbazole Violet

A dyestuff obtained by fusing 9-ethylcarbazole with oxalic acid.

5187 Carbendazim

10605-21-7 1836 234-232-0

$C_9H_9N_3O_2$

Carbamic acid, 1H-benzimidazol-2-yl, methyl ester

2-Benzimidazolecarbamic acid, methyl ester; Bavistin, Bavistin 3460; Benzimidazole-2-carbamic acid, methyl ester; BAS 3460; BAS 3460F; BAS 67054F; BCM; BMK; Carbendazime; Carbendazol; Carbendazole; Carbendazym; G 665; Kemdazin; Mecarzole; Methyl benzimidazol-2-ylcarbamate; Methyl 1H-benzimidazol-2-ylcarbamate; MBC; BMC; CTR 6669; Hoe 17411; Derosal. Systemic fungicide with protective and curative action. Controls a wide variety of fungal diseases in cerals, fruit and vegetables. Used against Dutch elm disease. mp = 302-307° (dec); d = 1.45; soluble in H_2O (8 mg/l), soluble in organic solvents; LD_{50} (rat orl) 6400 mg/kg.

5188 Carbendazim

10605-21-7 1836 234-232-0

$C_9H_9N_3O_2$

Carbamic acid, 1H-benzimidazol-2-yl, methyl ester

2-Benzimidazolecarbamic acid, methyl ester; Bavistin, Bavistin 3460; Benzimidazole-2-carbamic acid, methyl ester; BAS 3460; BAS 3460F; BAS 67054F; BCM; BMK; Carbendazime; Carbendazol; Carbendazole; Carbendazym; G 665; Kemdazin; Mecarzole; Methyl benzimidazol-2-ylcarbamate; Methyl 1H-benzimidazol-2-ylcarbamate; MBC; BMC; CTR 6669; Hoe 17411; Derosal. Systemic fungicide with protective and curative action. Controls a wide variety of fungal diseases in cerals, fruit and vegetables. Used against Dutch elm disease. mp = 302-307° (dec); d = 1.45; soluble in H_2O (8 mg/l), soluble in organic solvents; LD_{50} (rat orl) 6400 mg/kg.

5189 Carbetamex

16118-49-3 240-286-6

$C_{12}H_{16}N_2O_3$

Propanamide, N-ethyl-2-[[(phenylamino)carbonyl]oxy]-, (R)-

Lactamide, N-ethyl-, carbanilate (ester), D- ; 11,561 RP; Carbanilic acid, (1-ethylcarbamoyl)ethyl ester, D-(-)- ; Carbetamex; Carbetamid; Carbetamide; Carbetamide; Carbethamide; D-N-Ethylacetamide carbanilate; 1-(Ethylcarbamoyl)ethyl phenylcarbamate; D-(-)-1-(Ethylcarbamoyl)ethyl phenylcarbamate; D-N-Ethyllactamide carbanilate; D-N-Ethyllactamide carbanilate (ester); Legurame; (Phenylcarbamoyloxy)-2-N-ethylpropionamide; N-Phenyl-1-(ethylcarbamoyl-1)-ethylcarbamate, D isomer; RP 11561. Herbicide. *Rhône-Poulenc UK; Embetec Crop Protection Ltd.*

5190 Carbilys

Used to render starch insoluble in the manufacture of paper and board, and also in the textile and spinning process industries. *Roquette (UK) Ltd.*

5191 Carbital®

471-34-1 1697 207-439-9

Calcium carbonate

Finely ground natural marble sold as a slurry in water; paper coating pigments; fillers for paper; fillers and extenders for paints. *ECC International Ltd.* Discontinued.

5192 Carbital® 35

471-34-1 1697 207-439-9

Calcium carbonate

Medium-fine wet-ground filler pigment; also for matte or dull coatings. *ECC International Ltd.*

5193 Carbital® 50

471-34-1 1697 207-439-9

$CCaO_3$

Calcium carbonate

limestone [1317-65-3]. Fine, wet-ground filling pigment offering excellent retention in the sheet, excellent optical properties; also for matte or dull coatings. *ECC International Ltd.*

5194 Carbitol®

111-90-0 1847 203-919-7

$C_6H_{14}O_3$

2-(2-Ethoxyethoxy)ethanol

Diethylene glycol monoethyl ether; ethyl digol. A solvent for cellulose nitrate, shellac, copal, rosin, etc.; used in dopes for artificial leather and is added to brushing lacquers. bp = 196°; d_{25}^{25}= 0.9855; n_D^{20}= 1.4273; miscible with all organic solvents; LD_{50} (rat orl) = 8.69 g/kg. *Union Carbide.*

5195 Carbo Alumina

1344-28-1 369 215-691-6

aluminum oxide

Consists mainly of crystalline alumina; a trademark for goods of the abrasive and refractory class. No manufacturer.

5196 Carbo-corundum

1344-28-1 369 215-691-6

aluminum oxide

A trademark for articles made from crystalline alumina; refractories and abrasives. No manufacturer.

5197 Carbodan

1563-66-2 1851 216-353-0

$C_{12}H_{15}NO_3$

7-Benzofurano, 2,3-dihydro-2,2-dimethyl, methylcarbamate

Carbofuran; carbamic acid, methyl, 2,3-dihydro-2,2-dimethyl-7-benzofuranyl ester; BAY 70143; BAY 78537; carbamic acid, methyl, 2,2-dimethyl-2,3-dihydro-7-benzofuranyl ester; Curaterr®; Furadan; Furadan 3G; Niagaral 242; NIA 10242; OMS 864; Yaltox; 2,2-dimethyl-2,3-dihydrobenzoduranyl-7 N-methylcarbamate. Granular soil-applied insecticide for use in sugar beet, tobacco, maize, rice, sugar cane and vegetables. mp = 150-153°; d = 1.180;

soluble in H_2O (320 mg/l), organic solvents; LD_{50} (rat orl) = 5 mg/kg. *Makhteshim Chemical Works Ltd; Bayer AG.*

5198 Carboform
Plastic molding material made from carbon fibers impregnated with resin, in the form of sheets, tape, matting or fabric, and tow for packing and jointing. *Cyanamid Fothergill Ltd.*

5199 Carbofrax
A carborundum refractory containing more than 90% silicon carbide.

5200 carbo-gel
Silica gel impregnated with carbon, for use to absorb organic vapors.

5201 Carbogran, Carbogran E, Carbogran UF
409-21-2 8636 206-991-8
SiC
Silicon carbide
Used in abrasive and electrical industries. UF used in sintered ceramics and for abrasion resistant surfaces. d = 3.23. *Lonza AG.*

5202 Carbokaylene
A proprietary preparation of vegetable charcoal with colloidal aluminum silicate. No manufacturer.

5203 carbol xylene
A microscopic cleaning solution containing 3 parts xylene and 1 part phenol.

5204 Carbolan
Super-milling acid dyes. *ICI Chem & Polymers Ltd.*

5205 Carbolfuchsine
Ziehl's stain. A microscopic stain for bacteria. It contains 5 parts fuchsin, 25 parts phenol, 50 parts alcohol, and 500 parts water.

5206 Carbolic Acid
108-95-2 7390 203-632-7
C_6H_6O
Phenol
In trade, the term is used for pure phenol, the cresols and their mixtures with phenol, and also for crude tar oils. Solvent and chemical intermediate.

5207 Carbolon
A trademark for abrasive articles consisting essentially of silicon carbide. No manufacturer.

5208 Carboloy
A trademark for hard metal compounds consisting of tungsten carbide and cobalt for cutting glass and for high-speed tools. No manufacturer.

5209 Carbomang
A proprietary trade name for a steel containing 1-1.25% manganese, 0.45% chromium, 0.5% tungsten, and 1% carbon. No manufacturer.

5210 Carbomant
409-21-2 8636 206-991-8
SiC
Silicon carbide
Used in abrasive industries and wire sawing. *Lonza AG.*

5211 Carbomer
A suspension agent. A polymer of acrylic acid crosslinked with allyl sucrose.

5212 Carbomix® 1605, 1609, 3651
Cold SBR blk. masterbatch with: 50 phr N550 black; nonstaining masterbatch for high-quality extruded and molded mechanical Goods; 40 phr N1 10 black, 5 phr highly aromatic oil; for applications requiring high tensile strength and low heat buildup. *Copolymer Rubber.*

5213 Carbon
7440-44-0 1855 231-153-3
C
Activated carbon
Calgon® GW 12x40. Activated carbon; filter medium for dechlorination and removal of trace dissolved organics from water supplies. *Calgon Carbon.* Name unverified.

5214 Carbon
7440-44-0 1855 231-153-3
Activated carbon
Calgon® Type BPL 4x10, 6x16, 12x30. Activated carbon; for vapor phase applications. *Calgon Carbon.* Name unverified.

5215 Carbon
7440-44-0 1855 231-153-3
Activated carbon
Calgon® Type RB. Activated carbon; pulverized form for purification applications in chemical, food and pharmaceutical industries. *Calgon Carbon.* Name unverified.

5216 Carbon
7440-44-0 1855 231-153-3
Carbon
Calgon® Type SGL. For purification and decolorization of aqueous and organic liquids. *Calgon Carbon.* Name unverified.

5217 Carbon
7440-44-0 1855 231-153-3
C
Carbon
Carbonado; Black Diamond. A black, compact variety of diamond; used in the steel crowns of rock-drills.

5218 Carbon 4E
A liquid formulation containing 48% trichloropyrester; controls woody weeds and perennial broad-leaved weeds in forestry and noncropped areas. *Burts & Harvey.* Name unverified.

5219 carbon bisulfide
75-15-0 1859 200-843-6
CS_2
Carbon disulfide
Dithiocarbonic anhydride. Solvent. mp = -12°; bp = 46.5°; n_D^{21}= 1.62803.

5220 carbon black
1333-86-4 1856 215-609-9
Continex® LH-10, N-351; Thermal black; channel black; furnace black. Finely divided particles of elemental carbon obtained by incomplete combustion of hydrocarbons (channel or impingement process); tire treads, belt covers, other abrasion-resistant rubber products; uv light *Akrochem; Akzo; Cabot; Degussa; Exxon; R. T. Vanderbilt Co Inc; Witco/Cancarb.*

5221 Carbon Bronze
An alloy containing 75.5% copper, 9.75% tin, and 14.5% lead; a white metal used for bearings.

5222 carbon dioxide
124-38-9 1857 204-696-9
CO_2
carbon dioxide
carbonic acid gas; carbonic anhydride; Cardice. Refrigeration, carbonated beverages, aerosol propelant, chemical intermediate, fire extinguishing, inert atmospheres, municipal water treatment, medicine, oil wells, mining, blowing agent. bp = -78.5° (sublimes); soluble in H_2O (88 ml CO_2 gas/100 ml H_2O) *Air Prods & Chem; Nissan Chemical Ind.; Norsk Hydro AS; Showa Denko; Messer UK Ltd.*

5223 carbon disulfide
75-15-0 1859 200-843-6
CS_2
carbon bisulfide
Carbon bisulfide; Dithiocarbonic Anhydride; Alcohol of sulfur; Carbon bisulfuret; Carbon sulfide; methyl disulfide; weeviltox; sulfocarbonic anhydride. Viscose rayon, cellophane, manufacture of carbon tetrachloride and flotation agents, solvent. mp = -110°; bp = 46°; d = 1.2660; n_D^{20}= 1.6272; soluble in H_2O (300 mg/100 ml); more soluble in organic solvents; LD_{50} (mus orl) = 2780 mg/kg. *Akzo; PPG Industries; Rhône-Poulenc Chemie NV.*

5224 carbon tetrachloride
56-23-5 1864 200-262-8
CCl_4
tetrachloromethane
Perchloromethane; Necatorina; Benzinoform. Chlorinated hydrocarbon; refrigerants, metal degreasing, agricultural fumigant, chlorinating organic compounds, production of semiconductors, solvents. mp = -23°; bp = 76.7°; d_{25}^{25} = 1.589; n_D^{20} = 1.4607; soluble in H_2O (1 ml/2000 ml), soluble in organic solvents; LC_{50} (mus inh) = 9528 ppm. *Ashland; Janssen Chimica; Mitsui Toatsu Chemical; Montefluos SpA; Solvay SA.*

5225 carbon, activated
7440-44-0 1855 231-153-3
C
Activated carbon
Norit A; Norit SA II; charcoal; Active carbon; activated charcoal; Calgon® Type HGR; Carbonado; Calgon® Type SGL ; Calgon® GW 12x40. Clarifying, deodorizing, decolorizing, filtering; activated charcoal in medicine as antidote, adsorptive. *AlliedSignal; Am. Norit; Elf Atochem N. Am.; Calgon Carbon; Ceca SA; United Catalysts; Westvaco.*

5226 Carbondale silver
An alloy of 66% copper, 18% nickel, and 16% zinc.

5227 Carbonet
Nonadherent gauze dressing. *Smith & Nephew Pharmaceuticals Ltd.* Name unverified.

5228 Carbonin
Granular carbonaceous additive for molten iron and steel. *Foseco (F.S.) Ltd.* Discontinued.

5229 Carboplastic
A proprietary fireproof plastic cement, the main constituent of which is carborundum; suitable for repairing furnaces. No manufacturer.

5230 Carbopol® 613, 614
9003-01-4
Polyacrylic acid

Emulsifier for solvent cleaners, emulsion stabilizer, suspending agent, thickener for detergent formulations. *BFGoodrich.* Name unverified.

5231 Carbopol® 910

Carbomer 910, a polymer of acrylic acid, crosslinked with a polyfunctional agent; the viscosity of a neutralized preparation containing 2.5 g of Carbomer 910 in 500 ml water is not less than 2,500 centipoises and not more than 7000 centipoises; *BFGoodrich.* Name unverified.

5232 Carbopol® 934, 934P, 940

Carbomer 934, a polymer of acrylic acid, crosslinked with a polyfunctional agent; the viscosity of a neutralized preparation containing 2.5 g of carbomer 934 in 500 ml water is not less than 3,000 centipoises and not more than 4,000 centipoises; *BFGoodrich.*

5233 Carbopol® 941

Carbomer 941, a polymer of acrylic acid, crosslinked with a polyfunctional agent; a neutralized preparation containing 2.5 g of carbomer 941 in 500 ml water is not less than 4,000 centipoises and not more than 11,000 centipoises; pharmaceutic aid. *BFGoodrich.* Name unverified.

5234 Carbo-Pulbit®

Oral antidiarrhoeic; veterinary medicine. *Bayer AG.*

5235 Carbora

Proprietary name for silicon carbide materials; abrasive. No manufacturer.

5236 Carboraffin

7440-44-0 1855 231-153-3
C
carbon
Powdered activated carbon; used for decolorizing and improving the odor and taste of liquids mainly in the chemical and food industries. *Bayer AG.* Discontinued.

5237 Carborex

409-21-2 8636 206-991-8
SiC
Silicon carbide
Silicon carbide, SiC; for abrasive, refractory, metallurgical and other usages. *Orkla Exolon A/S & Co.* Name unverified.

5238 Carborundum

The name applied to abrasives, the main constituent of which is silicon carbide, SiC. *Harbison-Carborundum Corp.* Unverified.

5239 Carbosal

A range of multipurpose fire extinguishing powders for extinguishing fires of incandescent solid, liquid or gaseous materials. *Degussa AG.* Name unverified.

5240 Carboset 514A

Acrylic resin, IPA; thermoplastic film-forming resin used in protective metal coatings, paints, ceramics, adhesives, textiles, paper, leather, cosmetics, floor polishes, chemical specialties; excellent dispersant, leveling, and binding *BFGoodrich.* Name unverified.

5241 Carboset 531

Acrylic resin, ammonia water; thermoset film-forming resin used in protective metal coatings, paints, ceramics, adhesives, textiles, paper, leather, cosmetics, floor polishes, chemical specialties; excellent dispersant, leveling and binding *BFGoodrich.* Name unverified.

5242 Carbosorb

Soda synthetic silicate, a carbon dioxide absorbent. *BDH Chemicals Ltd.*

5243 Carbostat 2203

Quaternary amine; surfactant for textile processing. *Hoechst Celanese/Colorants & Surf.*

5244 carbostyril

59-31-4 1873 200-420-6
C_9H_7NO
2(1H)-Quinolinone
2-Hydroxyquinoline; 2-quinolinol; 2(1H)-quinolone; *o*-aminocinnamic acid lactam; 2-quinoline; *o*-amino cinnamic acid lactam. mp = 199-200°; slightly soluble in H_2O; soluble in EtOH, Et_2O, HCl.

5245 carbosulfan

55285-14-8 259-565-9
$C_{20}H_{32}N_2O_3S$
2,3-dihydro-2,2-dimethyl-7-benzofuranyl[(dibutylamino)thio]methyl carbamate
OMS 3022; Advantage; FMC 35001; Marshal; Marshall 10G; Posse; Sheriff. Systemic insecticide with contact and stomach action. Used for control of soil-dwelling and foliar insects in a wide variety of crops. bp = 124-128°; d^{20} = 1.056; insoluble in H_2O (0.03 mg/l), soluble in organic solvents; LD_{50} (rat orl) = 250 mg/kg.

5246 Carbotex

A brand of natural rubber and carbon black; used in rubber mixings. No manufacturer.

5247 Carbowax® Compound 20M

25322-68-3 7729
PEG-350
Binder, lubricant for ceramics, powder metallurgy, toilet bowl cleaners. *Union Carbide.* Name unverified.

5248 Carbowax® MPEG 350

9004-74-4
PEG-6 methylether
Surfactant intermediate, lubricant for adhesives, inks, mining, soaps and detergents. *Union Carbide.* Name unverified.

5249 Carbowax® PEG 200

25322-68-3 7729
PEG-4
Intermediate for surfactants, lubricants, urethanes; antistat, humectant, mold release agent, plasticizer for adhesives, inks, lubricants. *Union Carbide.* Name unverified.

5250 Carbowax® PEG 540 Blend

25322-68-3 7729
PEG-6 and PEG-32 (41:59); base for ointments and suppositories; also for adhesives, creams and lotions, deodorant sticks. *Union Carbide.*

5251 Carbowax® PEG 8000

25322-68-3 7729
PEG-150
Antistat, ceramic binder, surfactant intermediate, dye carrier, lubricant, release agent, tablet binder for adhesives, creams and lotions, mining, powder metallurgy, soaps and detergents, tablet coating, toilet bowl cleaners. *Union Carbide.*

5252 Carbowax® Sentry

25322-68-3 7729
Polyethylene glycol
Pharmaceutic aid: ointment base; suppository base; solvent; tablet excipient; tablet and/or capsule lubricant. *Union Carbide.*

5253 carbox metal

An alloy of 84% lead, 14% antimony, 1% iron, and 1% zinc.

5254 Carboxide

A proprietary fumigant containing 1 part ethylene oxide and 8 parts carbon dioxide. No manufacturer.

5255 carboxin

5234-68-4 1874 226-031-1
$C_{12}H_{13}NO_2S$
1,4-Oxathiin-3-carboxamide, 5,6-dihydro-2-methyl-N-phenyl-
1,4-Oxathiin-3-carboxanilide, 5,6-dihydro-2-methyl; Carbathiin; Carboxine; D 735; DCMO; DMOC; F 735; Vitavax; Vitavax 100; Vitavax 735D; Vitavax 75W; 1,4-oxathiin, 2,3-dihydro-5-carboxanilido-6-methyl-; Kemikar; Kisvax; Oxatin. Systemic fungicide; seed treatment for control of smuts, bunts and seedling diseases. mp = 93-95°; soluble in H_2O (175 mg/l, 25°); more soluble in organic solvents; LD_{50} (rat orl) = 3820 mg/kg. *Uniroyal; Kemira; Jin Hung; Diachem.*

5256 Carboxymethyl mercaptosuccinic acid

99-68-3 202-778-9
Evanacid® 3CS. Metal chelate; metal deactivator for the stabilization of glyceride oils. *Evans Chemetics.*

5257 carboxymethylcellulose

9004-32-4 1877
carboxymethyl ether cellulose sodium salt
Cethylose; CMC; Carmethose; Cel-O-Brandt; Glykocellon; Carbose D; Xylo-Mucine; Tylose MGA; Cellolax; Polycell. Used in drilling muds, detergents, as a soil-suspending agent; in resin emulsion paints, adhesives, printing inks, textile sizes and as a protective colloid. A stabilizer in foods and a pharmaceutic aid.

5258 carboxymethylcellulose sodium

9004-32-4 1877
R_nOCH_2COONa
Carboxymethyl ether cellulose sodium salt
CMC; sodium carboxymethylcellulose; sodium cellulose glycolate; Carmethose; Cel-O-Brandt; Cethylose; Glykocellon; Carbose D; Thylose; Xylomucine; Tylose MGA; Cellolax; Polycell; CMC; cellulose gum (CTFA); sodium carboxymethylcellulose; sodium CMC. Sodium salt of the carboxylic acid R-O-CH_2COOH;used in drilling muds; in detergents as soil-suspending agent; in emulsion paints, adhesives, inks, textile sizes; as a protective colloid. *Courtaulds Water Soluble Polymers; J.W.S. Delavau; FMC; Hercules.*

5259 Cardice

124-38-9 1857 204-696-9
CO_2
carbon dioxide
Solid carbon dioxide; used as a refrigerant and a chemical intermediate. *Messer UK Ltd.*

5260 Cardinal

A dyestuff; a mixture of chrysoidine and fuchsine; used for dyeing cotton red.

5261 **Cardinal Red J**
A dyestuff; a British brand of Fast red. No manufacturer.
5262 **Cardio-Green**
Indocyanine green; diagnostic aid. *Hynson Westcott & Dunning.* Name unverified.
5263 **Cardipol® LP**
Oxidized polyethylene wax; for formulating hot-melt adhesives. *Petrolite.*
5264 **Cardis® 36**
Oxidized microcrystalline. wax; used in the formulation of polishes. *Petrolite.*
5265 **Cardolite**
A proprietary cashew nut derivative of the phenol aldehyde polymer class; also proprietary trade name for plasticisers, resins, rubber-like polymers, and solvents. No manufacturer.
5266 **Cardolite® NC-1307**
Terpene-based extender/flexibilizer/diluent for epoxy manufacturing; also for concrete coatings. *Cardolite.* Name unverified.
5267 **Cardolite® NC-507**
501-24-6 207-921-9
$C_{21}H_{36}O$
3-(n-Pentadecyl)phenol
Starting raw material for surfactants, antioxidants, anticorrosives; lubricant additive; cosolvent for insecticides, germicides; resin modifier. mp = 50-53°; bp_1 = 190-195°. *Cardolite.*
5268 **Cardolite® NC-511**
8007-24-7 232-355-4
3-(n-Penta-8-decenyl)phenol
Chemical intermediate. bp_{10} = 223-27°; d = 0.9300; n_D^{20}= 1.5112; insoluble in H_2O, soluble in organic solvents. *Cardolite.*
5269 **Cardox**
An explosive utilizing liquid carbon dioxide.
5270 **Cardura**
A proprietary trade name for nondrying alkyd resins with excellent weathering properties used for surface coatings. *Shell.*
5271 **Cardura E**
Glycidyl ester of the synthetic fatty acid, Versatic 10; versatile intermediate for the production of stoving enamels, nitrocellulose lacquers and urethane paints. *Shell UK.*
5272 **Cariflex**
A range of synthetic rubbers (styrene butadiene, cis-polyisoprene and polybutadiene). *Shell UK.*
5273 **Cariflex Butadiene Rubber (BR)**
Polybutadienes with resiliency and very good abrasion resistance; available in straight and oil-extended versions; used in car tires and in the manufacture of high-impact polystyrenes as well as many industrial applications. *Shell UK.*
5274 **Cariflex Isoprene Rubbers (IR)**
Chemically very similar to natural rubber; available in straight and oil-extended versions; used in tire carcasses, footwear and belting. *Shell UK.*
5275 **Cariflex S**
A proprietary styrene-butadiene rubber. *Shell.* Name unverified.
5276 **Cariflex Styrene-Butadiene Rubbers (SBR)**
A family of general purpose synthetic rubbers available as straight, oil-extended, carbon-black masterbatch or resin/rubber masterbatch; widely used in car tires, footwear, adhesives and a range of industrial and domestic products. *Shell UK.*
5277 **Cariflex Thermoplastic Rubbers (TR)**
Materials possessing the inherent elasticity of rubbers yet processable as thermoplastics; used in the manufacture of adhesives or blended with thermoplastics for improved impact properties; when compounded, used as carpet-backing or injection-molded for *Shell UK.*
5278 **Carina**
A proprietary trade name for polyvinyl chloride *Shell.* Name unverified.
5279 **Carinex, Carinex SB41, SI73**
A series of polystyrenes and toughened polystyrenes. SB41 is an an easy processing polystyrene; SI73 is A proprietary high-impact grade of polystyrene possessing good flow properties in molding. *Shell.* Name unverified.
5280 **Carletti's indicator**
Phenolphthalein which has been reduced by caustic soda and zinc dust.
5281 **Carlona 460, 462, 463**
A proprietary polyethylene used in the production of film; has high impact strength but contains a slip additive. *Shell.* Name unverified.
5282 **Carlona 55-004, 60-010, 60-060, 60-120**
A proprietary high density polyethylene possessing high resistance to stress cracking; used in the blow-molding of containers (55-004), thin-walled bottles (60-010), heavy-duty containers (60-060) and thin sections (60-120). *Shell.* Name unverified.
5283 **Carlona LB 157, LF 456, LF 459**
A proprietary low-density polyethylene used in the production of film with (LF 456, LF 459) good optical properties. *Shell.* Name unverified.
5284 **Carlona P PLZ 532**
A proprietary talc-filled, heat-stabilized polyethylene used in the production of rigid moldings. *Shell.* Name unverified.
5285 **Carlona P PY 61**
A proprietary polypropylene homopolymer used in the production of film, especially stretched tapes for use as plastic string. *Shell.* Name unverified.
5286 **carloon bronze**
An alloy of 75.5% copper, 14.5% lead, and 10% tin.
5287 **carmalum**
A microscopic stain; consists of carminic acid in aqueous alum.
5288 **Carmargo® White**
Calcium bentonite
Filler and binder for technical applications. *Am. Colloid.*
5289 **Carmine**
1390-65-4 215-724-4
$C_{44}H_{43}AlCaO_{32}$
Alum carmine
Alum lake; Cochineal extract; Carmine alum lake. Aluminum lake of the coloring agent, cochineal; cochineal is a natural pigment derived from the dried female insect Coccus cacti; dyes, inks, indicator in chemical analysis, coloring food, medicine. *Aceto; R.W. Greeff; Penta Mfg.; Warner-Jenkinson.*
5290 **Carmine 40**
1390-65-4 215-724-4
Carmine; cosmetic pigment. *RITA.*
5291 **Carmine 224**
Carmine compounded with bismuth oxychloride; natural colorant. *Presperse.* Discontinued.
5292 **carmine lake**
Lac lake, Indian lake. Cochineal carmine prepared by precipitating a decoction of cochineal with alum or stannic chloride, with addition of acid oxalate or tartrate of potassium.
5293 **carmine red**
Obtained by boiling a dilute aqueous solution of carminic acid with a few drops of a mineral acid. It gives colored lakes.
5294 **carminic acid**
1260-17-9 1891 215-023-3
$C_{22}H_{20}O_{13}$
7-α-D-Glucopyrosanosyl-9,10-dihydro-3,5,6,8-tetrahydroxy-1-methyl-9,10-dioxo-2-anthracenecarboxylic acid
C. I. Natural Red 4; C. I. 75470; cochineal. Aluminum lake of the coloring agent, cochineal; cochineal is a natural pigment derived from the dried female insect Coccus cacti; dyes, inks, indicator in chemical analysis, coloring food, medicine. λ_m = 500 nm (ε = 5800); soluble in H_2O, EtOH, insoluble in organic solvents *Aceto; R. W. Greef; Penta Mfg.; Warner-Jenkinson; RITA.*
5295 **carnallite**
7579
$KCl{\cdot}MgCl_2{\cdot}6H_2O$
A double chloride of potassium and magnesium, $MgCl_2{\cdot}KCl{\cdot}6H_2O$, found in the Stassfurt deposits; a source of potassium chloride and potash manures.
5296 **Carnauba**
8015-86-9 1859 232-399-4
Brazil wax. Exudate from leaves of Brazilian wax palm tree *Copernicia prunifera*; greyish, yellowish, or greenish in color, and when freshly purified melts at 85-86 C; shoe polishes, leather finishes, varnishes, waterproofing, furniture and floor *Industrial Waxes Ltd; Penta Mfg.; Stevenson Bros.; Strahl & Pitsch.*
5297 **Carnauba wax**
8015-86-9 1859 232-399-4
Carnauba Spray 200
Brazil wax; ceara wax. Exudate from leaves of Brazilian wax palm tree *Copernicia prunifera*; greyish, yellowish, or greenish in color, and when freshly purified melts at 85-86°; shoe polishes, leather finishes, varnishes, waterproofing, furniture and floor mp = 85-86°; d = 0.990 - 0.999; *Industrial Waxes Ltd; Penta Mfg.; Stevenson Bros.; Strahl & Pitsch; Sherex/Div. Of Witco.*
5298 **carnidazole**
42116-76-7 1897 255-663-0
$C_8H_{12}N_4O_3S$
[2-(2-methyl-5-nitro-1H-imidazoly-1-yl]carbamothioic acid methyl ester
R-25831; Spartrix. Antiprotozoal. mp = 142°.
5299 **Carnot's reagent**
Basic bismuth nitrate (100 g) dissolved in hot, concentrated hydrochloric acid, and diluted to 1 l with 92% alcohol.

5300 **Carnoy's fluid**
A mixture of absolute alcohol and glacial acetic acid; used for fixing animal tissue.

5301 **Caro bronze**
An alloy of 92% copper, 8% tin, and 0.25% phosphorus. A bearing metal.

5302 **Caroat**
A peroxomonosulfate compound with about 4.5% active oxygen; oxidizing agent used for production of cleansers of all types, e.g., denture cleansers, household and toilet cleaners, for detoxicating cyanidic waste water. *Degussa Ltd.*

5303 **carob flour**
5436
Johannisbrotmehl; Arobon; Locust Bean Gum; carob seed gum. Ground kernel endosperms of tree pods of *Ceratonia siliqua L., Leguminosae.* Consists primarily of proteins and carbohydrates. Used as a coffee, chocolate, cocoa substitute.

5304 **Carolid® MN-1**
Methyl naphthalene carrier; carrier for polyester and blends; low odor. *Sybron.*

5305 **Caromax**
A range of close cut aromatic hydrocarbon solvents. *Carless Refining & Marketing Ltd.*

5306 **Caro's acid**
7722-86-3 1900 231-766-6
H_2SO_2
Persulfuric acid, H_2SO_2; used in dye manufacture, oxidizing agent, bleaching.

5307 **Caro's reagent**
Obtained by dissolving ammonium or potassium persulfate in concentrated sulfuric acid; a pasty oxidizing agent; used in testing for alkaloids.

5308 **carotene**
7235-40-7 1902 230-636-6
$C_{40}H_{56}$
β-Carotene
Carotaben; Provatene; Solatene; Provitamin A. Color additive for foods; vitamin A precursor; uv screen. mp = 183°; λ_m = 497, 466 nm; soluble in CS_2, C_6H_6, $CHCl_3$, less soluble in other organic solvents, insoluble in H_2O. *BASF; Hoffmann-La Roche; Penta Mfg.; Schweizerhall; Warner-Jenkinson.*

5309 **Caroubier**
An acid dyestuff, giving crimson shades on wool.

5310 **Carovax**
A proprietary pasteurella vaccine used in veterinary work. *Coopers Animal Health Ltd.* Name unverified.

5311 **Carp**
A cotton fabric grade of Tufnol industrial laminates. *Tufnol Ltd.*

5312 **Carp Brand**
Salt. *ICI Chem & Polymers Ltd.* Discontinued.

5313 **Carpenter 22Cr-13Ni-5Mn**
Nitrogen-strengthened austenitic stainless steel; alloy providing good corrosion resistance and high strength; for valve shafts, pumps and fittings for chemical equipment, fasteners, cables, chains, screens, wire cloth, marine hardware, heat exchanger parts, *Carpenter Tech.*

5314 **Carpenters Wood Glue**
PVA liquid; fast grab wood glue. *Wessex Resins & Adhesives Ltd.*

5315 **carrageenan**
9000-07-1 1914 232-524-2
Carrageen; Carrageenin; Chondrus; Irish moss extract; carrageen. Structural polysaccharide of the red seaweed (*Rhodophyceae*). Sulfated polysaccharide; emulsifier, gelling agent in food products, toothpaste, cosmetics, pharmaceuticals; stabilizing aid in ice cream. *G Fiske & Co Ltd; FMC; Hercules.*

5316 **Carriant Series**
Oil mixture; textile dyeing assistants; carriers for polyester fibers. *Toho Chemical Industry.*

5317 **Carrisorb**
1337-76-4
Attapulgite clay; an absorbent used in treatment and bleaching of oils (mainly mineral). *Bromhead & Denison Ltd.*

5318 **Carrot Oil CLR**
Soybean oil, carrot oil, carrot extract, β-carotene, tocopherol; emollient, conditioner, superfatting agent; emulsified and oily preparations for care of skin and hair. *Dr. Kurt Richter; Henkel/Cospha; Henkel Canada.*

5319 **Carsamide®**
Alkylamines
Surfactants. *Lonza Inc.*

5320 **Carsamide® AMEA**
142-26-7 205-530-8
Acetamide MEA (1:1)
Hair conditioner and antistat. *Lonza Inc.*

5321 **Carsamide® CA**
Cocamide DEA (1:1)
Detergent, dispersant, emulsifier, wetting agent, foam booster, thickener, softener for industrial, cosmetic, and household cleaners; biodeg. *Lonza Inc.*

5322 **Carsamide® CMEA**
68140-00-1 268-770-2
Cocamide MEA (1:1)
Foam booster, stabilizer, viscosity builder for shampoos and detergents. *Lonza Inc.*

5323 **Carsamide® SAL-7**
120-40-1 204-393-1
Lauramide DEA (1:1)
Detergent, emulsifier, foaming agent, foam stabilizer, thickener for shampoos, bath products, household, institutional and industrial detergents. *Lonza Inc.* Discontinued.

5324 **Carset**
Hardener for cold-set sand molds. *Foseco (F.S.) Ltd.*

5325 **Carsil**
Binder for the CO_2 process. *Foseco (F.S.) Ltd.*

5326 **Carsilon**
409-21-2 8636 206-991-8
SiC
Silicon carbide
Used in special refractories. *Lonza AG.*

5327 **Carsofoam® 1618**
Polyethylene glycol ethers of Cetyl and Stearyl alcohols
Cetearyl alcohol; bodying agent, viscosity modifier for cosmetic, personal care, and household products. *Lonza Inc.*

5328 **Carsofoam® BS-I**
PEG-80 sorbitan laurate-sodium trideceth sulfate-PEG-150 distearate-disodium lauroamphodiacetate-cocamidopropylhydroxysultaine-sodium laureth-13 carboxylate. Low irritation concentrate for baby shampoos. *Lonza Inc.*

5329 **Carsofoam® MSP**
TEA-lauryl sulfate-cocamide DEA-cocamidopropyl betaine. Low irritation concentrate for baby shampoos. *Lonza Inc.*

5330 **Carsofoam® T-60-L**
TEA dodecylbenzene sulfonate
High foaming detergent, wetting agent for cosmetic, industrial, and institutional usage. *Lonza Inc.*

5331 **Carsonol®**
Alcohol sulfates
Used for shampoos and bubble baths where high foaming and mildness are desired. *Lonza AG.*

5332 **Carsonol® ALS-R**
2235-54-3 218-793-9
Ammonium lauryl sulfate
Detergent with high foam, good wetting and emulsifying properties used for cosmetics, chemical specialties. *Lonza Inc.* Discontinued.

5333 **Carsonol® AOS**
68439-57-6 270-407-8
Sodium C_{14}-C_{16} olefin sulfonate
Surfactant for shampoos, liquid soaps, industrial cleaners. *Lonza Inc.*

5334 **Carsonol® DLS**
DEA lauryl sulfate
Biodegradable. detergent, foaming agent, wetting agent, emulsifier for personal care products *Lonza Inc.*

5335 **Carsonol® MLS**
Magnesium lauryl sulfate
Detergent with high foam, good wetting and emulsifying properties used for cosmetics, chemical specialties, rug and upholstery formulations. *Lonza Inc.*

5336 **Carsonol® SES-A**
67762-19-0
Ammonium laureth sulfate
Surfactant with excellent foaming in hard and soft water, for cosmetic, household, and industrial uses, shampoos, bubble baths, liquid cleaners. *Lonza Inc.*

5337 **Carsonol® SES-S**
9004-82-4
Sodium laureth sulfate
Surfactant with excellent foaming in hard and soft water, for cosmetic, household, and industrial uses, liquid carwash, laundry detergents. *Lonza Inc.* No manufacturer.

5338 **Carsonol® SHS**
Sodium 2-ethylexyl sulfate
Low foaming detergent, wetting agent, penetrant, emulsifier used in caustic

solutions for peeling of fruits and vegetables; stable to high concentrations of electrolytes. *Lonza Inc.*

5339 Carsonol® SLS Paste B
151-21-3 8782 205-788-1
Sodium lauryl sulfate
Detergent with high foam, good wetting and emulsifying properties used for cosmetics, chemical specialties, shampoo bases, textile scouring. *Lonza Inc.* No manufacturer.

5340 Carsonol® TLS
139-96-8 205-388-7
TEA-lauryl sulfate
Biodegradable detergent with high foam, good wetting and emulsifying properties used for cosmetics, mild shampoos, bubble baths, chemical specialties, emulsion polymerization. *Lonza Inc.* Name unverified.

5341 Carsonon®
Nonionic sulfates and wetting agents; used in industrial and household cleaning products where good detergency is required. *Lonza AG.*

5342 Carsonon® 144-P
63793-60-2
PPG-3 myristyl ether
Lubricant, emollient, solubilizer for cosmetics including silicon systems; aids low temperature stability, antistatic and conditioning effects; contributes to mildness and spreading behavior. *Lonza Inc.* Name unverified.

5343 Carsonon® 169-P
9035-85-2
PPG-10 cetyl ether
Lubricant, emollient, solubilizer for cosmetics including silicon systems; aids low temperature stability, antistatic and conditioning effects; contributes to mildness and spreading behavior. *Lonza Inc.*

5344 Carsonon® N-4
9016-45-9 6772
Nonoxynol-4
Emulsifier, detergent, wetting agent, dispersant for household and industrial uses; intermediate; drycleaning detergent. *Lonza Inc.*

5345 Carsoquat®
Quaternary ammonium compounds; used as creme rinses in personal care applications. *Lonza AG.*

5346 Carsoquat® 816-C
Cetearyl alcohol, PEG-40 castor oil, and stearalkonium chloride; formulated base, cream rinse concentrate. *Lonza Inc.*

5347 Carsoquat® 868 E
Dicetyl dimonium chloride
Hair conditioner, cream rinse, fabric softener. *Lonza Inc.*

5348 Carsoquat® CB
Mixture of Cetyl alcohol, Glyceryl stearate, dicetyl dimonium chloride, cetrimonium chloride
Polysorbate 85, PEG-40 castor oil; cream rinse concentrate. *Lonza Inc.*

5349 Carsoquat® CT-429
112-02-7 203-928-6
Cetyl trimethyl ammonium chloride
Surfactant; conditioner in hair and skin care preparations; antistat. *Lonza Inc.* Name unverified.

5350 Carsoquat® SDQ-25
122-19-0 204-527-9
Stearalkonium chloride
Conditioner, softener, creme hair rinse and conditioner for cosmetics and natural fibers. *Lonza Inc.* Name unverified.

5351 Carsosoft®
Quaternary ammonium compounds; used as fabric softeners. *Lonza AG.*

5352 Carsosoft® CFI-90
Alkylimidazolinium methosulfate
Fabric softener for preparation of clear liquid concentrates *Lonza Inc.*

5353 Carsosoft® S-90
Quaternium-27
Fabric softener and antistat. *Lonza Inc.*

5354 Carsosoft® T-90
130124-24-2
Quaternium-53
Fabric softener, rewetting agent, antistat, and antilint for home and commercial laundries. *Lonza Inc.* Name unverified.

5355 Carspray2, CW
Dicoco quaternary; emulsifier plus glycol ether for car rinses. *Sherex/Div. of Witco.*

5356 Carstab® DLTDP
123-28-4 204-614-1
$C_{30}H_{58}O_4S$
Dilauryl thiodipropionate
Didodecyl 3,3-thiodipropionate. Heat aging stabilizer used in conjunction with primary antioxidants; for PP, HDPE. *Morton Int'l./Specialty Chem.* Name unverified.

5357 Carstab® DSTDP
693-36-7 211-750-5
$C_{42}H_{82}O_4S$
Distearyl thiodipropionate
Long-term heat aging stabilizer in conjunction with primary antioxidants; for PP, HDPE. *Morton Int'l./Specialty Chem.*

5358 Cartaretin F-4
61840-27-5
Adipic acid/dimethylaminohydroxypropyl diethylenetriamine copolymer; polymer substantive to hair; for shampoo systems; imparts lubricity. *Sandoz.*

5359 Carterite
A proprietary trade name for resin-pulp molding compound. No manufacturer.

5360 Cartolac
General-purpose carton lacquers. *The Scottish Adhesives Co Ltd.*

5361 Cartose
Dextrose; replenisher. *Sterling Drug Inc.* Name unverified.

5362 carvacrol
499-75-2 1923 207-889-6
$C_{10}H_{14}O$
2-Methyl-5-(1-methylethyl)phenol
2-Methyl-5-isopropyl-phenol; 2-*p*-cymenol; 2-hydroxy-*p*-cymene; isopropyl-*o*-cresol; isothymol. Fond in oil of origanum; thyme, marjoram, summer savory. Used in perfumes; fungicides, disinfectant, flavoring, organic synthesis. mp = 0°; bp = 237-238°; d_4^{20}= 0.976; n_D^{20}= 1.52295; λ_m = 277.5 nm (log ε 3.262); soluble in H_2O, soluble in organic solvents; LD_{50} (rbt orl) = 100 mg/kg.

5363 carvone
99-49-0 1925 202-759-5
$C_{10}H_{14}O$
2-Methyl-5-(1-methylethenyl)-2-cyclohexene-1-one
p-Mentha-6,8-dien-2-one; 1-methyl-4-isopropenyl-Δ^6-cyclohexen-2-one; carvol; DL - [99-49-0], D-(+) - [2244-16-8], L-(-) - [6485-40-1]. Flavoring, liqueurs, perfumery, soaps. D-(+): bp = 228-230°; d = 0.962; n_D^{20}= 1.4970. L-(-): bp_{760} = 228-230°; d = 0.959; n_D^{20}= 1.4990; DL: bp = 230-231°; d = 0.9645; n_D^{20}= 1.5003. *Penta Mfg.; Schweizerhall.*

5364 Carylderm
63-25-2 1831 200-555-0
Carbaryl
Contact insecticide, used for infestation by lice. *Napp Laboratories Ltd.* No manufacturer.

5365 Casabet, Casabet 655
Coco amido betaine and coco amido sulfo betaine; betaine type surfactants characterized by low toxicity and irritancy and high foam production properties; used in the formulation of household cleaning products, in industrial and institutional detergent *Thomas Swan & Co Ltd.* Discontinued.

5366 Casahib
Amine functional compounds; corrosion inhibitors for crude oil and gas production. *Thomas Swan & Co Ltd.* Discontinued.

5367 Casamer
Epoxy and urethane acrylates; electron beam and ultraviolet curing systems. *Thomas Swan & Co Ltd.* Discontinued.

5368 Casamid
Nonreactive polyamides, dicydiamide, accelerated dicydiamide and substituted dicydiamide; epoxy resin curing agents used in surface coatings, powder coatings, electrical encapsulation and potting, filament winding, flexographic and *Thomas Swan & Co Ltd.*

5369 Casamine
Aminoethyl alkyl imidazolines and hydroxyethyl alkyl imidazolines; antifungal agents, emulsifiers in agricultural sprays, ore flotation, and acid detergent formulations. *Thomas Swan & Co Ltd.*

5370 Casamox
Amine oxides
Nonionic detergent of versatile properties for the formulation of a wide range of cosmetic, detergent and industrial products; lime soap dispersant, detergent thickener and foam booster. *Thomas Swan & Co Ltd.* Discontinued.

5371 Casaquat
Imidazoline-derived quaternary ammonium compounds; widely used in shampoos, hair conditioners, skin preparations, bacteriacidal and algacidal application; especially for the formulation of fabric softeners, both domestic and commercial. *Thomas Swan & Co Ltd.* Discontinued.

5372 Casateric
Fatty acid imidazoline derived carboxylate amphoteric surfactants; used to reduce irritancy of other surfactants used in shampoo, personal care and domestic applications; also used in industrial cleaning, textile processing, ore benefaction, oil-gas *Thomas Swan & Co Ltd.* Discontinued.

5373 Casathane
Prepolymers based on polyether and polyester diols; used for casting systems for the engineering industry. *Thomas Swan & Co Ltd.*

5374 Cascade
Photographic wetting agent. *May & Baker Ltd.* Name unverified.

5375 Cascade
55-63-0 8913 200-240-8
A nitroglycerin dynamite; for construction and building, explosives, mining. *Hercules.* Discontinued.

5376 Cascamite
Precatalyzed urea-formaldehyde powdered resin glue; used for general purpose wood glue; waterproof. *Wessex Resins & Adhesives Ltd.*

5377 Casco
Casein-based adhesives. *Borden (UK) Ltd.* Name unverified.

5378 Cascogel TM
Animal glue based adhesives. *Borden (UK) Ltd.*

5379 Cascomelt
Hot melt adhesives. *Borden (UK) Ltd.*

5380 Cascophen
Phenolic resins. *Borden (UK) Ltd.* Name unverified.

5381 Cascophen Resorcinol Resin RS 216/RXS-8
Liquid resorcinol-formaldehyde resin with powder catalyst; general purpose wood glue; weatherproof. *Wessex Resins & Adhesives Ltd.*

5382 Casco-resin, Casco Resin TM
Urea-formaldehyde resins. *Borden (UK) Ltd.*

5383 Cascorez TM
Vinyl acetate-based adhesives. woodwork adhesive. *Borden (UK) Ltd.*

5384 Cascosel TM
Aluminum foil laminating adhesives. *Borden (UK) Ltd.*

5385 Cascotak TM
Pressure sensitive adhesives. *Borden (UK) Ltd.*

5386 casein
9005-46-3 11, 1892
Milk protein. Mixture of phosphoproteins from cows milk; cheesemaking, plastic items, paper coatings, water dispersed paints, adhesives, textile sizing, foods and feeds, textile fibers, dietetic preparations, binder in foundry sands. *Meggle Marketing GmbH; Nat'l. Casein; U.S. BioChemical; Worthington BioChemical.*

5387 casein glue
Made by stirring casein with 25% distilled water, and 1-4% sodium bicarbonate, adding another 25% distilled water, standing, and adding an antiseptic to prevent mold; can be applied cold.

5388 casein magnesia
A preparation which consists of powdered casein, water, and magnesia; it fixes mineral pigments.

5389 casein paints
Paints formed by the addition of a powder containing casein and alkali, to water; coloring matter and lime are added.

5390 casein silk
Casein dissolved in alkali or zinc chloride solution, and spun into an acid bath.

5391 caseogum
A solution of casein in lime water; used as an adhesive, or for impregnating linen and cotton fabrics.

5392 Cashmilon®
Acrylic staple fiber. *Asahi Chemical Industry.*

5393 Casilan
Calcium caseinate. *Glaxo Laboratories.* Name unverified.

5394 Casoron
1194-65-6 3093 214-787-5
Dichlobenil
Granular herbicide containing dichlobenil; used to control weeds in woody crops and noncrop areas. *Chipman Ltd, ICI AgroChemicals, SynChemicals Ltd.* No manufacturer.

5395 Casoron G
1194-65-6 3093 214-787-5
Dichlobenil
Direct, selective weed killing action in orchards, vineyards, flower beds and parkland areas and along rail tracks, motorways and waterways. *Duphar BV.* Unverified.

5396 Casoron G4
1194-65-6 3093 214-787-5
Dichlobenil
Granule containing dichlobenil; residual herbicide for use among established trees and shrubs, paths, hard surfaces and vacant ground. *Vitax Ltd.* Unverified.

5397 Cassappret®
Textile finishing agents. *Cassella AG.*

5398 Cassastat®
Antistatic finishing agents for textiles. *Cassella AG.*

5399 Cassatan®
Tanning agents. *Cassella AG.*

5400 cassava
A food product. It consists of the starch obtained from the roots of the Manioc, *Manihot utilissima*; used to make tapioca, laundry starch; and adhesives.

5401 cassava meal
Ground cassava root.

5402 Cassella's acid
92-40-0 111,898 202-153-0
$C_{10}H_8O_4S$
7-Hydroxy-2-naphthalenesulfonic acid
2-naphthol-7-sulfonic acid; β-Naphthol-δ-monosulfonic F676; β-naphtholsulfonic acid F; F Acid. Used as an intermediate for azo dyes. mp 89°; soluble in H_2O, EtOH, insoluble in C_6H_6, Et_2O.

5403 Casselmam's green
Cupric sulfate/cupric hydroxide
A pigment made by mixing boiling solutions of copper sulfate and an alkaline acetate; consists of $CuSO_4 \cdot 3Cu(OH)_2 \cdot 4H_2O$.

5404 Cassulfon®
Water-soluble sulfur dyestuffs; used for textile dyeing and printing. *Cassella AG.*

5405 Cassurit®
Melamine resin products; for textile finishing. *Cassella AG.*

5406 cast brass
Brass which is not required to be spun, rolled, drawn or hammered. It is made by melting together the copper and zinc. This usually contains 66% copper with zinc, and the lead varies from 1-3%.

5407 Cast yellow brass
Usually an alloy of 67% copper, 31% zinc, and 2% lead. It melts at 895°.

5408 Castaldo
Natural rubber (*cis*-1,4-polyisoprene compound); for mold making for jewelry casting. *F E Knight Inc.* Name unverified.

5409 Castaloy
A proprietary trade name for high carbon, high chromium steel. No manufacturer.

5410 Castaway Plus
Suspension concentrate containing 60 g γ-HCH (lindane) and 500 g thiophanate-methyl per liter; used for control of earthworms and leather jackets in turf. *Rhône-Poulenc Environmental Prods. Ltd.*

5411 Castellanos powder
A dynamite containing nitro-glycerin, nitrobenzene, fibrous material, and kieselguhr.

5412 Castethane
Two-component systems used in the manufacture of cast polyurethane elastomers. *Dow UK.* Discontinued.

5413 casting copper
7440-50-8 2583 231-159-6
Cu
Copper
An American copper containing 98.5-99.75% copper; used for casting. mp = 1083°; bp = 2595°; d = 8.94. No manufacturer.

5414 castiron D2
A British chemical standard; a grey phosphoric cast iron in the form of fine turnings. It contains 1.31% silicon, 1.075 phosphorus, and 1.64% manganese, also about 2.5% graphitic carbon, 0.8% combined carbon, and 0.03% sulfur.

5415 Casto-Magic
A liquid oil derivative applied to concrete forms to release forms from concrete and improve appearance of concrete. *Rostine Manufacturing & Supply Co.* Name unverified.

5416 Castomer
Two-component high duty polyurethane elastomer systems. *Baxenden Chemicals Ltd.*

5417 castor oil
1323-38-2 11, 1904 215-353-8
triglyceride of ricinoleic, oleic, linoleic, palmitic, stearic acids
oil of Palma Christi; tangantangan oil; Ricinus oil. Fixed oil obtained from seeds of *Ricinus communis*; plasticizer in lacquers and nitrocellulose; polyurethane coatings; hydraulic fluids; industrial lubricants; used in manufacture of Turkey red oil, cosmetics, leather treatment. *Ashland; CasChem; Climax Performance; Fanning; Harcros; Lipo; Norman, Fox.*

5418 Castrol GTX
High performance motor oil; for all types of motor car engines. *Burmah-Castrol Ltd.* Name unverified.

5419 Castrol Turbomax
A heavy duty high performance mineral-based crankcase lubricant for commercial vehicles; used for service, fill and top-up turbocharged and naturally aspirated diesel engines (except Detroit Diesel) and virtually all petrol units. *Burmah-Castrol Ltd.* Name unverified.

5420 Castung 103 G-H
Castor oil, dehydrated; used with phenolics to obtain fast dry coatings with max. alkali resistance, e.g., for sanitary can linings, corrosion resistance coatings, traffic paints, varnishes, ink vehicles, marine finishes; provides effective internal *CasChem.*

5421 Casul® 70 HF
Calcium dodecylbenzene sulfonate
Coemulsifier for oil-water and water-oil formulations, agricultural emulsions. *Harcros Organics.*

5422 Caswell Adhesives
A range of industrial adhesive products. *Caswell & Co Ltd.*

5423 C-A-T Enteric Coating Polymer
Cellulose acetate trimellitate.
Eastman. Name unverified.

5424 Catabond
A proprietary trade name for a phenol-formaldehyde liquid resin used for plywood manufacture where a waterproof bond is required. No manufacturer.

5425 Cata-Chek
For catalysts for polymerization and adhesion of urethanes and other synthetic organic polymers, in Int Class 1. *Ferro.*

5426 Cataflot
Quaternary ammonium compounds; flotation collectors. *Elf Atochem UK/Ceca.*

5427 Catafor
Electrostatic paint and PU foam additive. *Rhône-Poulenc UK.*

5428 Cataid
Surface active agents and detergents. *ABM Chemicals Ltd.* Name unverified.

5429 Catalase L
9001-05-2 1948 232-577-1
Bovine catalase
Caperase; Catalase; Equilase; Optidase. Enzyme which catalyzes the decomposition of hydrogen peroxide to water and oxygen; for milk production *Solvay Enzymes.*

5430 Catalazuli
A proprietary trade name for a phenol-formaldehyde synthetic resin product. No manufacturer.

5431 Catalex
A proprietary trade name for an expanded phenol formaldehyde plastic. No manufacturer.

5432 Catalpo
1332-58-7 5294 296-473-8
Fine particle size ultra pulverized hydrous kaolin (aluminum silicate); used as an extender and reinforcer in coatings, adhesives, rubber. *Engelhard.* Name unverified.

5433 Catalyst 9
Magnesium chloride-based; fast cure catalyst for textile resin finishes. *BASF.*

5434 Catalyst CC
Amine hydrochloride
Catalyst for urea-formaldehyde resins. *CNC Int'l L.P.*

5435 Catalyst RD Liq
Buffered inorganic salt catalyst; fast-acting resin catalyst. *Eastern Color & Chemical.*

5436 Catalyst ZA
Fast-acting accelerator for silicone water repellents. *CNC Int'l L.P.*

5437 Catapol SR
A proprietary range of polyurethane elastomers ranging in hardness form 20-60 Durometer A. *Longfield Chemicals.* Name unverified.

5438 Catarase®
2265
1:5,000 Ophthalmic solution when reconstituted; chymotrypsin; for enzymatic zonulysis prior to intracapsular lens extraction. *Iolab Corp.*

5439 Catavar
A proprietary phenol-formaldehyde surface coating or laminating varnish. No manufacturer.

5440 Catavat Black N-JBB
Vat dyestuffs. *Catawba-Charlab.*

5441 CatCO 600, CatCO 610ST
Catalysts for abatement of carbon monoxide and unburned hydrocarbons; for gas turbines, heaters and boilers. (610ST) for abatement of carbon monoxide and unburned hydrocarbons in the presence of sulfur compounds; for any industrial fossil *Engelhard.*

5442 Catex
Imidazoline derived cationic and amphoteric surfactants; used for textile softeners, antistats, foaming agents and solubilizing detergents. *Thomas Swan & Co Ltd.* Discontinued.

5443 Catigene 4513
Mixture of the chlorides of alkyl dimethyl benzyl ammonium and alkyl dimethyl ethyl benzyl ammonium as a clear yellow liquid; for algicides, bactericides, fungicides, deodorants. *KWR Chemicals Ltd.* Name unverified.

5444 Catigene BR 80 B
7281-04-1 230-698-4
Lauryl dimethyl benzyl ammonium bromide as a clear yellow liquid; for algicides, bactericides, fungicides, deodorants. *KWR Chemicals Ltd.* Name unverified.

5445 Catigene DC/100
139-08-2 205-352-0
Myristyl dimethyl benzyl ammonium chloride
Myristalkonium chloride as a white powder; bactericides for pharmaceuticals. *KWR Chemicals Ltd.* Name unverified.

5446 Catigene Red-brown
A dyestuff obtained by the action of alkali sulfides and sulfur upon aminohydroxy-phenazines.

5447 Catigene SR
Quaternary alkylamine acetate as a white fluid paste; cationic surfactant used in textile softeners. *KWR Chemicals Ltd.* Name unverified.

5448 Catigene T80
Alkyl dimethyl benzyl ammonium chloride as a clear yellow liquid; for bactericides, fungicides, algicides. *KWR Chemicals Ltd.* Name unverified.

5449 Catigene® 1011
7173-51-5 3149 230-525-2
Didecyl dimethyl ammonium chloride
Germicide, algicide, fungicide, deodorizing agent, antistat; also for oilfield production. *Stepan Europe. Name unverified.*

5450 Catigene® 50 USP
n-Alkyl (50% C_{12}, 30% C_{14}, 17% C_{16}, 3% C_{18}) dimethylbenzyl ammonium chloride
Germicide, algicide, fungicide, deodorizing agent, antistat. *Stepan Europe.*

5451 Catigene® 818
Octyl decyl dimethyl ammonium chloride
dioctyl dimethyl ammonium chloride; didecyl dimethyl ammonium chloride. Germicide, algicide, fungicide, deodorizing agent, antistat. *Stepan Europe.*

5452 Catigene® B 50
n-Alkyl (40% C_{12}, 50% C_{14}, 10% C_{16}) dimethylbenzyl ammonium chloride
Germicide, algicide, fungicide, deodorizing agent, antistat. *Stepan Europe.*

5453 Catigene® CA 56
Alkyl amidopropyl trimethyl ammonium methoxysulfate
Antistat additive for carpet latex and bitumen. *Stepan Europe.*

5454 Catigene® CETAC 30
Alkyl trimethylammonium chloride
Antistat, lubricant, emulsifier, softener, germicide, algicide, fungicide, deodorizing agent for household/industrial cleaners, oilfield production. *Stepan Europe.*

5455 Catigene® DC 100
139-08-2 205-352-0
Alkyl (3% C_{12}, 95% C_{14}, 2% C_{16}) dimethylbenzyl ammonium chloride
Germicide, algicide, fungicide, deodorizing agent, antistat. Myristyl dimethyl benzyl ammonium chloride as a white powder; bactericides for pharmaceuticals. *Stepan Europe.*

5456 Catinal CB-50
139-07-1 203-351-5
Lauralkonium chloride
Disinfectant, germicide, antistat *Toho Chem. Industry.* Name unverified.

5457 Catinal HTB-70
57-09-0 2068 200-311-3
Hexadecyl trimethyl ammonium bromide
Base material for hair rinse, antistat, germicide. *Toho Chem. Industry.*

5458 Catinal OB-80E
122-19-0 204-527-9
Stearyl dimethyl benzyl ammonium chloride
Base material for hair rinse. *Toho Chem. Industry.*

5459 Cationaster-C
3327-22-8 222-048-3
3-Chloro-2-hydroxypropyl trimethylammonium chloride (50% aq solution)
Quaternary cationic agent for starch and cellulose. *Yokkaichi Chemical Co Ltd.*

5460 Cationic Collagen Polypeptides
9007-34-5 2543 232-697-4
Cationic collagen polypeptides
Substantive film-former for hair and skin care products (shampoos, conditioners, mousses, shave preparations); Protective colloid effect. *Maybrook.*

5461 Cationic Softener X Concentrate
Alkyl-imidazoline derivative in paste form; softener with good substantivity; used as glass fiber mordant and lubricant. *Millmaster Onyx UK.* Name unverified.

5462 Catisol AO 100
Oleylamine acetate
Emulsifier, wetting agent, antistat, anticorrosive agent, lubricant for textile lubricants for mineral and synthetic fibers. *Stepan Europe; KWR Chemicals Ltd.*

5463 Catomer VA
9003-20-7
Polyvinyl acetate homopolymer emulsion; textile hand builder, stiffener, cationic exhaustible. *Sybron.*

5464 Catosal®
Organic phosphorus preparation; a stimulator of metabolism used in veterinary medicine. *Bayer AG.*

5465 Cat's gold
Cat's silver. Very finely powdered mica is sometimes called by these names. The term cat's gold is also used for mosaic gold (a tin sulfide).

5466 Caucho Blanco
A rubber obtained from different species of the genus *sapium* which belongs to the *euphorbaceae* family and is found in the northern part of South America.

5467 Caust X
Sodium phosphates
Silicates and precipitating agents which are compounded into solid bricks; used as a descalant in the caustic removal sections of bottle washers. *Delaware Chemical Corp.* Name unverified.

5468 Caustic baryta
17194-00-2 1006 241-234-5
$Ba(OH)_2$
Barium hydroxide
Used in glass manufacture, rubber vulcanization, corrosion inhibitors, drilling fluids and lubricants.

5469 Causul
A nickel-copper-chromium cast iron of marked acid and alkaline resistance; used in the manufacture of valves intended particularly for handling sulfuric acid and caustic soda.

5470 Caytur 21 & 22
A trademark for a range of curing agents used in urethane elastomers. *DuPont UK.* Name unverified.

5471 Caytur 4
A proprietary partial complex of zinc chloride with benzothiazyl disulfide; used as a cross-linking agent in sulfur-curable urethane elastomers; formerly known as Id-55. *DuPont UK.* Name unverified.

5472 Cazin
An alloy of cadmium and zinc, containing 82.6% cadmium.

5473 CB2100
3768-58-9 223-200-1
$C_6H_{18}N_2Si$
Bis(dimethylamino)dimethylsilane
Coupling agent, chemical Intermediate, blocking agent, release agent, lubricant, primer, reducing agent. mp = -98°; bp = 128-129°; d = 0.8090; n_D^{20}= 1.4170 *Hüls Am.*

5474 CB2405
126-80-7 204-803-9
$C_{16}H_{34}O_5Si_2$
Bis (glycidoxypropyl) tetramethyldisiloxane
Coupling agent, chemical intermediate, blocking agent, release agent, lubricant, primer, reducing agent. *Hüls Am.*

5475 CB2408
Bis (hydroxyethyl) aminopropyltriethoxy silane
Coupling agent, release agent, lubricant, blocking agent, chemical intermediate. *Hüls Am.*

5476 CB2409.5
16230-35-6 240-354-5
Bis-(N-methylbenzamide)ethoxymethyl silane
Coupling agent, release agent, lubricant, blocking agent, chemical intermediate. *Hüls Am.*

5477 CB2493
Bis (trimethoxysilylpropyl) ethylene diamine
Coupling agent, release agent, lubricant, blocking agent, chemical intermediate. *Hüls Am.*

5478 CB2494
40372-72-3 254-896-5
Bis[3-(triethoxysilyl) propyl]tetrasulfide
3,16-Dioxa-8,9,10,11-tetrathia-4,15-disilaoctacane. Coupling agent, release agent, lubricant, blocking agent, chemical intermediate. *Hüls Am.*

5479 CB2500
10416-59-8 233-892-7
$C_8H_{21}NOSi_2$
Bis(trimethylsilyl) acetamide
N,O-Bis(trimethylsilyl)acetamide; BSA; Ethanimidic acid, N-(trimethylsilyl)-, trimethylsilyl ester; Dynasylan BSA. Coupling agent, chemical intermediate, blocking agent, release agent, lubricant, primer, reducing agent. mp = 24°; bp_{35} = 71-73°; d = 0.8230; n_D^{20}= 1.4170. *Hüls Am.*

5480 CB2785
1000-50-6
$C_6H_{15}ClSi$
n-Butyldimethylchlorosilane
Coupling agent, chemical intermediate, blocking agent, release agent, lubricant, primer, reducing agent. *Hüls Am.*

5481 CB2790
18162-48-6 242-042-4
t-Butyldimethylchloro silane
Coupling agent, chemical intermediate, blocking agent, release agent, lubricant, primer, reducing agent. *Hüls Am.*

5482 CB2805
58479-61-1 261-282-0
t-Butyldiphenylchlorosilane
Coupling agent, chemical intermediate, blocking agent, release agent, lubricant, primer, reducing agent. *Hüls Am.*

5483 CB-4-34
Aromatic plasticizer; used in adhesives, rubber (cements, mechanical and molded goods, tires), caulking compounds. *Neville.*

5484 CBTS
95-33-0 202-411-2
$C_{13}H_{16}N_2S_2$
N-Cyclohexyl-2-benzothiazole sulfenamide
N-Cyclohexyl-2-benzothiazolesulfenamide; Sufenax CB; Rhodifax 16; Accelerator CZ; Durax; N-cyclohexylbenzothiazole-2-sulfenamide. Delayed-action sulfenamide accelerator for natural, reclaimed, and synthetic rubbers; nondiscoloring. An accelerator used in natural rubber and styrene-butadienethiazyl sulfenamide rubber. mp = 93-100°; insoluble in H_2O, soluble in organic solvents. *Akrochem.*

5485 CC-103
471-34-1 1697 207-439-9
Calcium carbonate
Coarse ground filler for polyolefins, carpet backing, caulks, sealants, putties, as mild abrasives in cleaners. *ECC International Ltd.* Discontinued.

5486 CC3005
7787-85-1 232-134-2
$C_3H_7Cl_3Si$
2-Chloroethylmethyldichloro silane
Coupling agent, chemical intermediate, blocking agent, release agent, lubricant, primer, reducing agent. *Hüls Am.*

5487 CC3270
1719-57-9 217-006-6
$C_3H_8Cl_2Si$
Chloromethyldimethylchlorosilane
Chloro(chloromethyl)dimethylsilane; CMDMCS. Coupling agent, chem. intermediate, blocking agent, release agent, lubricant, primer, reducing agent. bp_{752} = 114°; d = 1.0860; n_D^{20}= 1.4373. *Hüls Am.*

5488 CC3275
1558-33-4 216-319-5
$C_2H_5Cl_3Si$
Chloromethylmethyldichlorosilane
Silane, dichloro(chloromethyl)methyl-; Chloromethyl methyldichlorosilane. Coupling agent, chemical intermediate, blocking agent, release agent, lubricant, primer, reducing agent. *Hüls Am.*

5489 CC3285
2344-80-1 219-058-5
$C_4H_{11}ClSi$
Chloromethyltrimethylsilane
Coupling agent, chemical intermediate, blocking agent, release agent, lubricant, primer, reducing agent. bp = 98-99°; d = 0.8790; n_D^{20}= 1.4175. *Hüls Am.*

5490 CC3290
18171-19-2 242-056-0

$C_6H_{15}ClO_2Si$
3-Chloropropylmethyldimethoxysilane
Coupling agent, chemical intermediate, blocking agent, release agent, lubricant, primer, reducing agent. *Hüls Am.*

5491 CC3291
2550-06-3 219-844-8
$C_3H_6Cl_4Si$
Chloropropyltrichlorosilane
Trichloro(3-chloropropyl)silane; 3-Chloropropyltrichlorosilane. Coupling agent, chemical intermediate, blocking agent, release agent, lubricant, primer, reducing agent. *Hüls Am.*

5492 CC3292
5089-70-3 225-805-6
$C_9H_{21}ClO_3Si$
3-Chloropropyltriethoxy silane
Coupling agent, chemical intermediate, blocking agent, release agent, lubricant, primer, reducing agent. *Hüls Am.*

5493 CC3300
2530-87-2 219-787-9
$C_6H_{15}ClO_3Si$
3-Chloropropyltrimethoxysilane
CPTMO. Coupling agent, chemical intermediate, blocking agent, release agent, lubricant, primer, reducing agent. bp = 195-196°; d = 1.0810; n_D^{20} = 1.4196. *Hüls Am.*

5494 CC3433
17932-62-6 241-869-8
2-Cyanoethyltriethoxy silane
Coupling agent, chemical intermediate, blocking agent, release agent, lubricant, primer, reducing agent. *Hüls Am.*

5495 CC3555
1071-27-8 213-990-6
$C_4H_6Cl_3NSi$
3-Cyanopropyltrichlorosilane
(3-Cyanopropyl)trichlorosilane; Butanenitrile, 4-(trichlorosilyl)-. Coupling agent, chemical intermediate, blocking agent, release agent, lubricant, primer, reducing agent. bp_8 = 94°; d = 1.2800; n_D^{20} = 1.4629; LD_{50} (rat orl) = 2830 mg/kg. *Hüls Am.*

5496 CCA Type C Wood Preservative 50-60%
11125-95-4
Chromate copper arsenate in water, wood preservative. *CSI.* Name unverified.

5497 CCC 700
999-81-5 2153 213-666-4
$C_5H_{13}Cl_2N$
Chlormequat
Chlorocholine chloride; 2-Chloro-N,N,N-trimethylethanaminium chloride; (2-chloroethyl)trimethylammonium chloride; choline dichloride; AC 38555; CCC; Cycocel; Cycogan. Plant growth regulator. mp = 241° (dec); soluble in H2O, EtOH, insoluble in organic solvents; LD_{50} (mus orl) = 54 mg/kg. *Farmers Crop Chemicals Ltd.* Discontinued.

5498 CD480
41637-38-1
10 mole ethoxylated bisphenol A dimethacrylate; difunctional monomer providing excellent hardness and good adhesion properties when cured; for use in electronic applications, e.g., dry film resists and solder masks. d = 1.120; n_D^{20} = 1.5320 *Sartomer.*

5499 CD492
53879-54-2
Propoxylated trimethylolpropane triacrylate
Monomer with excellent solvency, good flexibility, improved impact resistance to uv and electron beam curing formulations; for use in wood, metal and vinyl flooring coatings, printing inks, overprint varnishes, solder masks. *Sartomer.*

5500 CD3770
541-02-6 2905 208-764-9
$C_{10}H_{30}O_5Si_5$
Decamethylcylopentasiloxane
Coupling agent, chemical intermediate, blocking agent, release agent, lubricant, primer, reducing agent. *Hüls Am.*

5501 CD3780
141-62-8 2907 205-491-7
$C_{10}H_{30}O_3Si_4$
Decamethyltetrasiloxane
Coupling agent, chemical intermediate, blocking agent, release agent, lubricant, primer, reducing agent. mp = -68°; bp = 194°; d = 0.8540; n_D^{20} = 1.3888. *Hüls Am.*

5502 CD4153
13170-23-5 236-112-3
$C_{12}H_{24}O_6Si$
Di-*t*-butoxydiacetoxysilane
Silicon di-*t*-butoxide diacetate. Coupling agent, chemical intermediate, blocking agent, release agent, lubricant, primer, reducing agent. *Hüls Am.*

5503 CD4368
69304-37-6
$C_{12}H_{28}Cl_2OSi_2$
1,3-Dichlorotetraisopropyl-disiloxane
1,3-Dichloro-1,1,3,3-tetraisopropyldisiloxane; TIPSCl; 1,3-Dichlorotetraisopropyldisiloxane; 1,1,3,3-Tetraisopropyl-1,3-dichlorodisiloxane; 1,3-Dichlorotetraisopropylsiloxane. Coupling agent, chemical intermediate, blocking agent, release agent, lubricant, primer, reducing agent. bp_{15} = 120°; d = 1.0010; n_D^{20} = 1.4543. *Hüls Am.*

5504 CD4450
996-50-9 213-637-6
$C_7H_{19}NSi$
N,N-Diethylaminotrimethyl silane
N,N-Diethyl-1,1,1-trimethylsilylamine; TMSDEA; Trimethylsilyldiethylamine; N-trimethylsilyl diethylamine; N,N-Diethyltrimethylsilylamine; N,N-Diethylaminotrimethylsilane; Diethylaminotrimethylsilane. Coupling agent, chemical intermediate, blocking agent, release agent, lubricant, primer, reducing agent. bp = 125-126°; d = 0.7670; n_D^{20} = 1.4081. *Hüls Am.*

5505 CD5400
2083-91-2 218-222-3
$C_5H_{15}NSi$
Dimethylaminotrimethylsilane
N,N-Dimethyltrimethylsilylamine; N,N-Dimethylaminotrimethylsilane; N-Trimethylsilyldimethylamine; Dimethyl(trimethylsilyl)amine; Dimethylaminotrimethylsilane. Coupling agent, chemical intermediate, blocking agent, release agent, lubricant, primer, reducing agent. bp = 84°; d = 0.7320; n_D^{20} = 1.3925. *Hüls Am.*

5506 CD5430
3,3-Dimethylbutyldimethylchlorosilane
Coupling agent, chemical intermediate, blocking agent, release agent, lubricant, primer, reducing agent. *Hüls Am.*

5507 CD5470
1066-35-9 213-912-0
C_2H_7ClSi
Dimethylchlorosilane
Chlorodimethylsilane; DMCS. Coupling agent, chemical intermediate, blocking agent, release agent, lubricant, primer, reducing agent. mp = -111°; bp = 35°; d = 0.8520; n_D^{20} = 1.3827 *Hüls Am.*

5508 CD5600
78-62-6 201-127-6
$C_6H_{16}O_2Si$
Dimethyldiethoxy silane
diethoxydimethylsilane. Coupling agent, chemical intermediate, blocking agent, release agent, lubricant, primer, reducing agent. bp = 114°; d = 0.8650; n_D^{20} = 1.3811; LD_{50} (rat orl) = 9280 mg/kg. *Hüls Am.*

5509 CD5605
1112-39-6 214-189-4
$C_4H_{12}O_2Si$
Dimethyldimethoxysilane
dimethoxydimethylsilane. Coupling agent, chemical intermediate, blocking agent, release agent, lubricant, primer, reducing agent. *Hüls Am.*

5510 CD5610
71864-46-5
(2,3-Dimethylpropyl) dimethylchlorosilane
Coupling agent, chemical intermediate, blocking agent, release agent, lubricant, primer, reducing agent. *Hüls Am.*

5511 CD5635
14857-34-2 238-921-7
$C_4H_{12}OSi$
Dimethylethoxysilane
ethoxydimethylsilane. Coupling agent, chemical intermediate, blocking agent, release agent, lubricant, primer, reducing agent. *Hüls Am.*

5512 CD5636
18643-08-8 242-472-2
$C_{20}H_{43}ClSi$
Dimethyloctadecylchloro silane
Dimethyl-n-octadecylchlorosilane; Octadecyldimethylchlorosilane. Coupling agent, chemical intermediate, blocking agent, release agent, lubricant, primer, reducing agent. mp = 28-30°; bp = 300°. *Hüls Am.*

5513 CD5950
80-10-4 201-251-0
$C_{12}H_{10}Cl_2Si$
Diphenyldichlorosilane
dichlorodiphenylsilane. Coupling agent, chemical intermediate, blocking agent, release agent, lubricant, primer, reducing agent. mp = -22°; bp = 305°; d = 1.2040; n_D^{20} = 1.5800. *Hüls Am.*

5514 CD6000
2553-19-7 219-860-5
$C_{16}H_{20}O_2Si$
Diphenyldiethoxy silane
Coupling agent, chemical intermediate, blocking agent, release agent, lubricant, primer, reducing agent. *Hüls Am.*

5515 CD6010
6843-66-9 229-929-1
$C_{14}H_{16}O_2Si$
Diphenyldimethoxy silane
Coupling agent, chemical intermediate, blocking agent, release agent, lubricant, primer, reducing agent. *Hüls Am.*

5516 CD6150
947-42-2 213-427-4
$C_{12}H_{12}O_2Si$
Diphenylsilanediol
Dihydroxydiphenylsilane. Coupling agent, chemical intermediate, blocking agent, release agent, lubricant, primer, reducing agent. LD_{50} (mus orl) = 2150 mg/kg. *Hüls Am.*

5517 CD6210
2627-95-4 220-099-6
$C_8H_{18}OSi_2$
Divinyltetramethyldisiloxane
1,3-Diethenyl-1,1,3,3-tetramethyl-disiloxane; 1,3-Divinyltetramethyldisiloxane; Divinyltetramethyldisiloxane; 1,1,3,3-Tetramethyl-1,3-divinyldisiloxane. Coupling agent, chemical intermediate, blocking agent, release agent, lubricant, primer, reducing agent. *Hüls Am.*

5518 CD6220
4484-72-4 224-769-9
$C_{12}H_{25}Cl_3Si$
Dodecyltrichlorosilane
dodecyl trichlorosilane; trichloro(dodecyl)silane; n-Dodecyltrichlorosilane. Coupling agent, chemical intermediate, blocking agent, release agent, lubricant, primer, reducing agent. *Hüls Am.*

5519 CDA Dicotox Extra
1702-17-6 2462 216-935-4
2,4-D; translocated herbicide for cereals and established grassland. *Rhône-Poulenc Environmental Prods. Ltd.*

5520 CDA Mildothane
23564-05-8 9489 245-740-7
Thiophanate-methyl
A systemic insecticide. *Rhône-Poulenc Environmental Prods. Ltd.* Name unverified.

5521 CDA Roval
36734-19-7 5093 253-178-9
Iprodione
A fungicide with protectant activity for use in turf and amenity grasses. *Rhône-Poulenc Environmental Prods. Ltd.* Name unverified.

5522 CDA Simflow Plus
A suspension concentrate containing 100 g aminotriazole [61-82-5] and 300 g simazine [122-34-9] per liter; used for total weed control in non crop areas and fruit orchards. *Rhône-Poulenc Environmental Prods. Ltd.* Name unverified.

5523 CDA Viper
Amitrole-atrazine-diuron; a liquid mixture of herbicides for weed control. *CDA Chemicals Ltd.*

5524 CDB 90
87-90-1 9188 201-782-8
Trichloroisocyanuric acid
Chlorinating agent for automatic dishwashing detergents, scouring cleaners, chlorinated detergents, sanitizers, dry bleaches, wool shrink, cooling tower, and toilet water. *Olin.*

5525 CDB Clearon
2244-21-5 9896 218-828-8
Sodium dichloroisocyanurate dihydrate
Chlorinating agent for automatic dishwashing detergents, scouring cleaners, chlorinated detergents, sanitizers, dry bleaches, wool shrink. *Olin.*

5526 CDS-1801
115047-92-2
Behenyl polyethoxy ethylmethacrylate
Unneutralized surfactant. *Rhône-Poulenc Surf.*

5527 Cérite
A French synthetic resin of the phenol-formaldehyde type.

5528 CE6250
3388-04-3 222-217-1
$C_{11}H_{22}O_4Si$
2-(3,4-Epoxycyclohexyl) ethyltriacetoxysilane
Silane A-186; 1-[2-(Trimethoxysilyl)ethyl]cyclohexane-3,4-epoxide; β-(3,4-Epoxycyclohexyl) ethyl trimethoxy silane; 2-(3,4-Epoxycyclohexyl)ethyltrimethoxysilane. Coupling agent, chemical intermediate, blocking agent, release agent, lubricant, primer, reducing agent. *Hüls Am.*

5529 CE6345
17689-77-9 241-677-4
$C_8H_{14}O_6Si$
Ethyltriacetoxysilane
Coupling agent, chemical intermediate, blocking agent, release agent, lubricant, primer, reducing agent. *Hüls Am.*

5530 CE6350
115-21-9 204-072-6
$C_2H_5Cl_3Si$
Ethyltrichlorosilane
trichloroethylsilane; ethyl silicon trichloride; trichloroethylsilicane. Coupling agent, chem. intermediate, blocking agent, release agent, lubricant, primer, reducing agent. *Hüls Am.*

5531 Ceanel
A proprietary preparation of phenylethyl alcohol, cetrimide and undecenoic acid; scalp antiseptic used as shampoo. *Quinoderm Ltd.*

5532 Cebion®
A proprietary vitamin C preparation. *E Merck.*

5533 Cecagel
Silica gel; for gas drying; catalyst support. *Elf Atochem UK/Ceca.*

5534 Cecaperl
Expanded perlite; used for industrial and cryogenic insulation (tanks and methane tankers). *Elf Atochem UK/Ceca.*

5535 Cecarbon
7440-44-0 1855 231-153-3
Granular activated carbon; used for purification, decolorization, deodorization, separation and recovery in liquid or gas phase, in the chemical, petrochemical, pharmaceutical and food industries (glucose factories, sugar refiners, oil refining, wine *Elf Atochem UK/Ceca.*

5536 Cecasol
10101-39-0 233-250-6
Calcium silicate
Industrial fillers. *Elf Atochem UK/Ceca.*

5537 Ce-Cobalin
Syrup containing vitamins B_{12} and C; provides vitamin supplementation in deficiency states such as self imposed dietary restrictions as in strict vegetarianism or following therapeutic diets for gastro-intestinal ulceration. *Paines & Byrne Ltd.* Name unverified.

5538 Cecolene 1
A proprietary trade name for trichlorethylene. No manufacturer.

5539 Cecolene 2
A proprietary trade name for perchlorethylene. No manufacturer.

5540 Cedepal FA-406
Ammonium deceth sulfate
Nonionic foaming agent, emulsifier, detergent, wetting and foam stabilizing; for gypsum board production. *Stepan; Stepan Canada.*

5541 Cedepal FS-406
Sodium deceth sulfate
Surfactant for high electrolyte applications. *Stepan Canada.*

5542 Cedepal TD-403
Sodium trideceth sulfate
Surfactant, wetting agent, foamer for shampoo, bath products, mild baby products. *Stepan; Stepan Canada.*

5543 Cedepal® Range
Fat carrier compounds; fat components in premixes and cake ready mixes and in yeast raised baked goods. *Grünau.* Discontinued.

5544 Cedephos® FA600
Deceth-4 phosphate
Detergent, emulsifier, wetting agent, hard surface detergent for industrial cleaners, metal cleaners, janitorial products, agricultural, textile wetting, emulsion polymerization, oil emulsification, lubricants, corrosion inhibitor, dedusting agent; *Stepan; Stepan Canada.*

5545 Cederan® P 23
Phosphate single fertilizer (23% phosphate); for all crops and soil types. *BASF AG.*

5546 Ceepree® C200
Blend of vitreous materials with very broad, almost continuous melting range; fire barrier/smoke suppressant additive for polymeric composites, especially used in conjunction with fiber, glass reinforcement or in adhesives, coatings, and mastics. *Brunner Mond.*

5547 Cefadroxil
66592-87-8 1963
$C_{16}H_{17}N_3O_5S.H_2O$

[6R-[6α,7β(RName unverified.)]]-7-[[Amino-(4-hydroxyphenyl)acetyl]amino]-3methyl-8-oxo-5-thia-1-azabicyclo-[4.2.0]oct-2-ene-2-carboxylic acid monohydrate
Duricef; BL-S578, MJF-11567; Baxan; Bidocef; Cefa-Drops; Cefamox; Ceforal; Cephos; Duracef; Kefroxil; Oracefal; Sedral; Ultracef. Antibacterial. Cefadroxil monohydrate, USP is a semisynthetic cephalosporin antibiotic intended for oral administration. It is a white to yellowish-white crystalline powder. It is soluble in water and it is acid-stable. White crystals; mp = 197°. *Mead Johnson & Co.*

5548 Cefkanat®
Mineral single feed for poultry. *BASF AG.*

5549 Cefkaphos®
7758-23-8 1740 231-837-1
Calcium phosphate
Feed phosphate for the mixed feed industry. *BASF AG.*

5550 Cegemett® Range
Mono and diglycerides, partially esterified with lactic and/or citric acid; prevents gelation and fat separation in manufacturing of sausages. *Grünau.*

5551 Cegepal® Range
Fat powders; food additive for dessert and cake mixes, soups. *Grünau.* Discontinued.

5552 Cegepaot® Range
Protein concentrates. Food emulsifier, stabilizer, protein enrichment for meat and sausage industry. *Grünau.*

5553 Cegeskin® Range
Acetic acid esters of mono and diglycerides of edible fatty acids; coating for shelf life and reduction of weight loss for raw and cooked sausages. *Grünau.*

5554 Cegesol® Range
Medium chain triglycerides; dust preventer for spice mixtures and other powder blends in meat/sausage industry. *Grünau.*

5555 Cegesterin® Range
Monoglyceride hydrate dispersions; antistaling effect, formation of starch complexes. *Grünau.*

5556 Ceistran® N66G30-01-4
32131-17-2
Nylon 6
Nylon 6/6, 30% glass long fiber-reinforced; injection molding. *Polymer Composites.* Unverified.

5557 Cekas
A heat-resisting alloy containing 59.7% nickel, 11.25 chromium, 28% iron, and 2% manganese.

5558 Cekol
9004-32-4 1877
$R_nC_2H_2O_3Na$
Sodium carboxymethylcellulose
Carboxymethyl ether cellulose sodium salt; CMC; Sodium carboxymethyl cellulose; sodium cellulose glycolate; Carmethose; Cel-O-Brandt; Cethylose; Glykocellon; Carbose D; Thylose; Xylomucine; Tylose MGA; Cellolax; Polycell. Thickening agent. Used for oil well drilling, detergents, wallpaper adhesives, ceramic glazes, emulsion paints, and paper. mp >300°. *Berol Nobel Ltd; Courtaulds Chemicals, Water Soluble Polymers.*

5559 Celacol
Hydroxypropyl methyl/methyl/methyl/ethyl cellulose; water-soluble polymers with film-forming and water-retaining properties; for plaster and cement products, emulsion paints, foods, pharmaceuticals and cosmetics. *Courtaulds Chemicals, Water Soluble Polymers.*

5560 Celafuse
A proprietary range of polyamide resins. Celafuse 100 is a terpolyamide with good melt flow properties (melting point 103-108 C); Celafuse T is a terpolyamide with a higher viscosity (melting point 115-125 C);Celafuse CP has a melting point of 145-150° *Hoechst Celanese.* Name unverified.

5561 Celanese
Celanese is an English name for cellulose acetate silk; the same type of silk is known in America as Lustron. No manufacturer.

5562 Celanese® Nylon 6/6
Polyamide (PA)
Used for coil bobbins, window winder mechanisms, lighter bodies and valve rocker covers. *Hoechst UK.*

5563 Celanese® Nylon 1000-1
32131-17-2
Celanese® Nylon 1003-1; Celanese® Nylon 1500-1; Celanese® Nylon 7420, 7423; Celanese® Nylon N-186; CTC-3300. Nylon 6/6 resin; general purpose resin for injection molding and extrusion; used for mechanical parts, gears, bearings, hardware. *Hoechst Celanese.* Name unverified.

5564 Celanese® Nylon 1003-1
32131-17-2
Nylon 6/6 resin, heat-stabilized; for high-temperature resistance applications, e.g., under-the-hood automotive parts. *Hoechst Celanese.* Name unverified.

5565 Celanese® Nylon 1500-1
32131-17-2
Nylon 6/6, 33% glass fiber-reinforced; injection molding resin for use in screw injection molding machines; available in surface lubricated grade (Celanese nylon 1503-1) surface-lubricated and heat-stabilized (Celanese nylon 1503-2) with same properties. *Hoechst Celanese.* Name unverified.

5566 Celanese® Nylon 7420, 7423
32131-17-2
Nylon 6/6 resin, 13% glass fiber-reinforced; glass-reinforced version of grade 7020. 7423 is heat stabilized. *Hoechst Celanese.* Name unverified.

5567 Celanese® Nylon N-186
32131-17-2
Nylon 6/6 resin, heat stabilized; high molecular weight extrusion grade. *Hoechst Celanese.* Name unverified.

5568 Celanex®
Thermoplastic polyester (PBT) polybutylene terephthalate; used for domestic equipment housings, plugs and sockets, switches, telecommunications, keyswitches, switchboard components, automotive, distribution housings, exterior door handles and wiper *Hoechst UK.*

5569 Celanex® 1300A, 1600A, 2000K
26062-94-2
Thermoplastic polyester resin (pbt); unmodified, (1300A) super flow, (1600A) extrusion grade, (2000K) unfilled, key cap grade. *Hoechst Celanese.* Name unverified.

5570 Celanex® 3310, 5300, J600
26062-94-2
Thermoplastic polyester resin (pbt); (3310) 30% glass-reinforced; for aggressive end resin (pbt); (5300) 30% fiberglass-filled, improved surface finish, (J600) 40% glass/mineral-filled, Low warpage, improved impact. *Hoechst Celanese.* Name unverified.

5571 Celanex® J600
26062-94-2
Thermoplastic polyester resin (pbt), 40% glass/mineral-filled; low warpage, improved impact. *Hoechst Celanese.* Name unverified.

5572 Celastic
A proprietary trade name for a pyroxylin product. No manufacturer.

5573 Celasyl
Fast dyestuffs for artificial fibers. *J C Bottomley.* Name unverified.

5574 Celatene Colors
Proprietary colors which are amino-anthraquinone derivatives. No manufacturer.

5575 Celatom
7631-86-9 8637 231-545-4
silicon dioxide
Diatomaceous earth, filter powder. *Flexbulk Ltd.* Name unverified.

5576 Celcon® EC90+
Acetal copolymer; semiconductive grade for applications requiring rapid dissipation of static build-up. *Hoechst Celanese.*

5577 Celcon® GB25
Acetal copolymer; 25% glass bead-filled; for low shrinkage and warp resistance In large, flat, and thin walled parts. *Hoechst Celanese.*

5578 Celcon® LW90
Acetal copolymer; low-wear grade for high speed, low-load service against metals; used in bearings, slide plates, bushings, wear surfaces, and conveyor links or plates. *Hoechst Celanese.*

5579 Celcon® M25
Crystalline acetal copolymer; thermoplastic engineering resin used for extrusion and selected injection molding applications in easy-to-fill molds; end-uses include wire coatings, rod, tube, sheet, and slab, and injection molded items requiring extra *Hoechst Celanese.*

5580 Celcon® M270
Crystalline acetal copolymer; high-flow grade thermoplastic engineering resin designed for superior moldability in multi-cavity, intricate, or hard-to-fill mold applications, e.g., combs, marking pen bodies, and housings. *Hoechst Celanese.*

5581 Celcon® U10
Acetal copolymer; high melt strength for extrusion and blow molding applications; used for aerosol containers, industrial tanks and floats, rod, tube, slab, and profiles. *Hoechst Celanese.*

5582 Celcon® UV25Z
Acetal copolymer; uv-stabilized; celcon® m25-based material stabilized for use where uv radiation degradation is a problem; used for automotive interiors and recreational items exposed to sunlight, e.g., knobs, buttons, toys, cams, and levers. *Hoechst Celanese.*

5583 Celestol
A proprietary trade name for an alkyd synthetic resin. No manufacturer.

5584 Celestols
A proprietary trade name for polybasic acid polyhydric alcohol fatty acid type synthetic resins. No manufacturer.

5585 Celestron
A condensation product of phenol and formaldehyde with fillers; used as an insulator.

5586 Celite®
7631-86-9 8637 231-545-4
Siliceous earth; diatomaceous earth; fossil flour; kieselguhr; Celite; Super-Cel;. A registered trademark for a material for separating impurities and other matters from fluids, and used as an aid in filtering, dehydrating, and demulsifying of liquids; also used as a filler in the plastics industry. No manufacturer.

5587 Celite® 110
7631-86-9 8637 231-545-4
Diatomaceous silica, Flux calcined
Extender pigment and flatting agent for solvent- and water-thinned paints; increases toughness and durability; add tooth" for adhesion of subsequent coats, improves sanding properties, control vapor permeability, reduces blistering and peeling; *Celite.*

5588 Celite® 209
7631-86-9 8637 231-545-4
Diatomaceous silica
Functional filler used for agricultural chemicals; grinding aid, and conditioner; low concentrate toxicants (up to 50%) where greater inertness is required; carrier for liquid seed inoculants; improves flow properties in fertilizers. *Celite.*

5589 Celite® 270
7631-86-9 8637 231-545-4
Diatomaceous silica
Calcined; functional filler for rubber industry; processing aid in rubber compounds; semireinforcing filler in mechanical rubber goods. *Celite.*

5590 Celite® R-625
7631-86-9 8637 231-545-4
Diatomite; Filler for use as inert, porous support/carrier for catalysts in industrial processes. *Celite.*

5591 Celite® Snow Floss
7631-86-9 8637 231-545-4
Diatomaceous silica; Functional filler used in polishes and cleaners; suitable for fast-cutting buffing compounds. *Celite.*

5592 Celkate T-21
1343-88-0 215-681-1
Magnesium silicate
Florisil. Synthetic hydrous magnesium silicate; functional filler, chromatographic column material. *Celite.*

5593 Cellacephate
A partial mixed acetate and hydrogen phthalate ester of cellulose; used as an enteric coating.

5594 Cellamine PAD
Semidurable gas-fading inhibitor for printed material *Eastern Color & Chemical.*

5595 Cellanite
A proprietary trade name for a synthetic resin paper product. No manufacturer.

5596 Cellastine
A proprietary trade name for a cellulose-acetate product. No manufacturer.

5597 Cellasto®
Cellular PU elastomers; for casting. *BASF AG.*

5598 Cellestren
Dyestuffs for cellulosic/polyester blends. *BASF plc.* Name unverified.

5599 Cellidor
Cellulose acetate, butyrate and propionate compounds; for the manufacture by injection molding and/or extrusion of spectacle frames. tool handles, seating furniture etc. and the production of fluidized bed coating powders. *Bayer AG.* Discontinued.

5600 Cellit
Cellulose acetate butyrate and propionate; used for photographic films, electrical insulating films as well as block casting. *Bayer AG.* Discontinued.

5601 Cellitazol® STN
Disperse dyes for dyeing and printing acetate and triacetate and synthetic fibers. *BASF AG.*

5602 Celliton® Dyes
Disperse dyes for dyeing and printing acetate, triacetate and synthetic fibers. *BASF AG; BASF plc.*

5603 Cellmore
Polyvinylidene chloride expandable beads. *Asahi Chemical Industry.*

5604 Cellobond
Phenolic products. *BP Chemicals Ltd.* Discontinued.

5605 cellodin
9004-70-0 8195
photoxylin
A substance which is obtained from collodion by precipitating it from its solution in alcohol and ether. It consists of pure nitrocellulose; used in microscopy and in surgery.

5606 Cellokyd®, Cellokyd®-2708
Alkyd resins for use in the manufacture of protective and decorative coatings, (2708) high solids alkyd. *Reichhold.*

5607 Cellolyn
A range of dibasic-modified rosin esters; tackifier for adhesives; used for type C gravure inks, nitrocellulose coatings.

5608 Cellophane
1959
Francephane. Regenerated cellulose film, plain and colored; P: uncoated films; M: nitrocellulose-coated films; X: copolymer-coated films; film for use on high speed automatic packaging machines and for incorporation into laminates. *UCB nv Film Sector.* Discontinued.

5609 Cellosize
9004-62-0 4707
Hydroxyethyl cellulose
Thickening agent and adhesive. *Union Carbide.*

5610 Cellosize® HEC QP Grades
9004-62-0 4707
Hydroxyethylcellulose
Water soluble polymer. *Amerchol Corp.*

5611 Celite® HSC
7631-86-9 8637 231-545-4
Diatomaceous silica, flux calcined; Functional filler used in polishes and cleaners; suitable for fast-cutting buffing compounds. *Celite.*

5612 Cellthane
One and two-component polyurethanes; coatings for industrial and specialty applications; also for elastomers for encapsulation, structural enhancement. *Flexible Prods.*

5613 Celluclast
A cellulase preparation made by submerged fermentation of a selected strain of the fungus *Trichoderma reesei* , can be used in any case where the aim is break-down of cellulose matter for production of fermentable sugar, or reduction of viscosity. *Novo Nordisk.*

5614 Cellucraft
A proprietary trade name for nitro-cellulose spray coating. No manufacturer.

5615 Celluflex M179
A proprietary trade name for alkyl-aryl plasticizer. No manufacturer.

5616 Celluflow C-25
9004-34-6 2012 232-674-9
$(C_6H_{10}O_5)_x$
Cellulose
Cosmetic ingredient with excellent oil absorbency, moisture retention, high lubricity for powders, emulsions and anhydrous systems. Insoluble in H_2O and organic solvents. *Presperse.* Discontinued.

5617 Celluflow TA-25
9004-35-7 2013
Cellulose triacetate
Cosmetic ingred. with excellent oil absorbancy, moisture retention, high lubricity for powds., emulsions and anhyd. systems. mp = 240°; d = 1.30; n_D^{20} = 1.4750. *Presperse.* Discontinued.

5618 Cellufluor
Chemical equivalent for calcofluor white; fungal stain for cytopathology. *Polysciences Inc.*

5619 Cellulase
9012-54-8 232-734-4
Enzyme complex; derived from *Aspergillus niger*, digestive aid in medicine and brewing industry; aids bacteria in the hydrolysis of cellulose. *Schweizerhall.*

5620 Cellulase
9012-54-8 232-734-4
Cellulase
Celluzyme® 2400 T. Enzyme for laundry powder detergents and laundry additives. *Novo Nordisk.* Name unverified.

5621 Cellulase 4000
9012-54-8 232-734-4
Fungal cellulase; enzyme for pharmaceuticals (aids digestion of cellulosics), animal feeds, brewing, fruit juices, essential oils and spices, paper and other waste treatment. *Solvay Enzymes.*

5622 Cellulase AC
9012-54-8 232-734-4
Fungal cellulase; enzyme for extraction of grains, vegetables, fruits, and other plant tissues. *Solvay Enzymes.*

5623 cellulith
A material made by grinding cellulose in water until a jelly is produced and boiling until hard; used as a binding agent for carborundum wheels, also as a packing material.

5624 Celluloid
Composed of a soluble nitrocellulose mixed with camphor, obtained by gelatinising nitrocellulose by means of a solution of camphor in ethyl alcohol; can be molded, and was formerly used extensively for making toys, combs, and other articles. No manufacturer.

5625 celluloid-caoutchouc
A material prepared by dissolving rubber and celluloid in cyclohexanol and mixing the solutions. An elastic substance is produced.

5626 cellulose
9004-34-6 2012 232-674-9
bleached wood pulp; cotton fiber. Natural polysaccharide derived from plant fibers; basic material for textile, paper, for manufacture of nitrocellulose, cellulose acetate, etc.; in chromatography; as ion exchange material; microcrystalline. *Degussa; Eastman; FMC Int'l.; Hercules Ltd; Edw. Mendell.*

5627 cellulose acetate
9004-35-7 2013
Cellulosic
cellulose acetate ester; CA; acetylcellulose; Lanoplast, CA; . Used in manufacture of rubber and celluloid substitutes, nonflammable photographic and cinema films, airplane dopes, varnishes and lacquers; waterproofing and sizing for textiles; coating skins; insulating electrical wires. *Aldrich; Eastman; FMC.*

5628 Cellulose acetate butyrate
9004-36-8
Cellulose acetate butyrate
Cabulite; CAB-171-15S; CAB-500-5; CAB-381-0.1, 381-0.5, 381-2, 381-20. Cellulose acetate butyrate film and sheet. *May & Baker Ltd.* Name unverified.

5629 Cellulose acetate butyrate
9004-36-8
cellulose acetobutyrate
cellulose, acetate butanoate. CAB; butyric acid ester of a partially acetylated cellulose; thermoplastic for automotive, tool, building, furniture industries, domestic appliances, lighting, electrical, radio, and TV industries, optical, photographic, stationery, toy, toiletries, mp = 110-125°; n_D^{20} = 1.4750. *Eastman; FMC; Whitfield Chemical Ltd.*

5630 Cellulose acetate propionate
9004-39-1
cellulose acetate propionate
cellulose acetate propionate ester; cellulose, acetate propanoate; CAP; CAP-482-0.5; CAP-482-20; Dispercap; CAP-482-0.5; CAP-482-20; Tenite® 360-H2; Tenite® Cellulosic Propionate. Propionic acid ester of a partially acetylated cellulose; thermoplastic for automotive, tool, building, furniture industries, domestic appliances, lighting, electrical, radio, and TV industries, optical, photographic, stationery, toy, toiletries, and packaging applications, printing inks and paper and cloth coatings. mp = 188-210°. *Tennant-KVK Ltd; Aldrich; Eastman; SAF Bulk Chems.*

5631 Cellulose acetate propionate ester
9004-39-1
Cellulose acetate propionate ester
For blending with CAP-482-0.5. *Eastman.*

5632 cellulose pitch
The residue obtained from the evaporation of the waste sulfite lye, from the treatment of wood in the sulfite process; used for making briquettes.

5633 cellulose triacetate
9012-09-3
Triacetyl cellulose. Affords protective coatings resistant to most solvents; textile fibers; base for magnetic tape. *Courtaulds Acetate; Eastman; FMC.*

5634 Cellulosine
A proprietary trade name for a bleached celluloid. No manufacturer.

5635 Cellulysin®
Cellulase. *Calbiochem Corp.*

5636 Celluvarno
A proprietary trade name for cellulose nitrate surfacing material. No manufacturer.

5637 Celluzyme®
A cellulytic enzyme preparation produced by submerged fermentation of the fungus *Humicola insolens*; used as color brightening, softening, and removal of particulate soil. *Novo Nordisk.*

5638 Celluzyme® 2400 T
9012-54-8 232-734-4
Cellulase
Enzyme for laundry powder detergents and laundry additives. *Novo Nordisk.*

5639 Celmar®
A trademark for polypropylene/glass fiber reinforced structures. *Hoechst Celanese.* Name unverified.

5640 Celmontite
A coal mine explosive containing 65.5-68.5% ammonium oxalate, 10.5-12.5% trinitrotoluene, and 19.5-21.5% sodium chloride.

5641 Celogen®
A range of nonstaining and nondiscoloring, nontoxic and odorless nitrogen blowing agents for sponge rubber and expanded plastics. *Uniroyal.* Name unverified.

5642 Celogen® AZ 120, 130, 150, 180, 199
123-77-3 950 204-650-8
Azodicarbonamide
Chemical blowing agent for thermoset and thermoplastic polymers; for injection-molding structural foam, extrusion of profiles, sheet, pipe, and wire coatings, and vinyl plastisol, coating, and calendering. *Uniroyal.* Name unverified.

5643 Celogen® OT
80-51-3 201-286-1
$C_{12}H_{14}N_4O_5S_2$
p,p-Oxybisbenzenesulfonyl hydrazide
diphenyloxide-4,4'-disulfohydrazide. Blowing agent for sponge rubber and expanded plastics (LDPE wire/cable, structural foam injection moldings, and rotational casting, flexible PVC structural foam injection molding). *Uniroyal. Name unverified.*

5644 *Celogen® RA*
p-Toluenesulfonyl semicarbazide
Chemical blowing agent for polymers processing at 216-260° range; for injection molding structural foam (HIPS, ABS, PP), extrusion (rigid PVC), plasticized vinyl. *Uniroyal.* Name unverified.

5645 Celogen® TSH
877-66-7 212-895-7
$C_7H_{11}ClN_2O_2S$
p-Toluene sulfonylhydrazide
Chemical blowing agent for thermoset polyester. mp = 202°. *Uniroyal.* Discontinued.

5646 Celogen® XP-100
Sulfonyl hydrazide
Chemical blowing agent. *Uniroyal.*

5647 Celquat® H-100, SC-240
Polyquaternium-4
Substantive polymer providing gloss and antistatic properties to setting lotions, cream rinses, mousses, shampoos, conditioning soaps, skin lotions and creams. *Nat'l. Starch.*

5648 Celsit
An alloy similar in composition to Stellite.

5649 Cel-Soft 2
Softener/lubricant for towel, tissue and specialty creped papers; excellent for dryer and press release. *CNC Int'l L.P.*

5650 Celstran® ACG40-01-4
Acetal copolymer, 40% glass long fiber-reinforced; injection molding. *Polymer Composites.* Unverified.

5651 Celstran® N6G30-01-4, N66G30-01-4
25038-54-4 6832
Poly(caprolactam). Nylon 6, 66, 30% glass long fiber-reinforced; injection molding. d = 1.0840. *Polymer Composites.* Unverified.

5652 Celstran® PBTG30-01-4
26062-94-2
Polybutylene terephthalate, 30% glass long fiber-reinforced; injection molding. *Polymer Composites.* Unverified.

5653 Celstran® PCG30-01-4
Polycarbonate, 30% glass long fiber-reinforced; injection molding. *Polymer Composites.* Unverified.

5654 Celstran® PETG30-01-4
25038-59-9 7730
Polyethylene terephthalate, 30% glass long fiber-reinforced; injection molding. *Polymer Composites.* Unverified.

5655 Celstran® PPG30-01-4
9003-07-0 7741
Polypropylene, 30% glass long fiber-reinforced; injection molding. *Polymer Composites.* Unverified.

5656 Celstran® PPSG30-01-4
9016-75-5

Polyphenylene sulfide, 30% glass long fiber-reinforced; injection molding. *Polymer Composites.* Unverified.

5657 **Celstran® PUG30-01-4**
Polyurethane, 30% glass long fiber-reinforced; injection molding. *Polymer Composites.* Unverified.

5658 **Celstran® SMAG30-01-4**
9011-13-6
Styrene maleic anhydride, 30% glass long fiber-reinforced; injection molding. *Polymer Composites.* Unverified.

5659 **Celtex**
Extruded cellular ceramic filters for cast irons. *Foseco (F.S.) Ltd.* Discontinued.

5660 **Celtite**
An explosive. It consists of 56-59% nitroglycerin, 2-3.5% nitro-cotton, 17-21% potassium nitrate, 8-9% wood meal, and 11-13% ammonium oxalate.

5661 **Celulon**
Series of cellular polyurethane coating materials suitable for application by brush or spray. *Unitex Ltd.*

5662 **cement copper**
Cementation copper; copper precipitate. Copper produced from copper liquors and mine liquors, by means of iron (pig iron, scrap iron, or spongy iron). It is usually contaminated with arsenic, antimony, and iron.

5663 **cement mortar**
A mixture of natural slag or Portland cement, sand, and water. Lime is also added.

5664 **Cement Prodor**
A range of acid resisting cements. *Prodorite Ltd.*

5665 **cement, adamantine**
A mixture of powdered pumice and silver amalgam; used in dentistry. Name unverified.

5666 **cement, American**
A rubber cement made from 10 parts rubber, 6 parts chloroform, and 2 parts mastic.

5667 **cementation steel**
Obtained by heating bars of good malleable iron, packed with nitrogenous matter, or wood charcoal.

5668 **cementite**
Fe_3C
Triferrous carbide
The hardest component of steel.

5669 **cementite, independent**
Cementite in rectilinear lamellae.

5670 **Cemset**
Accelerators for use in cement bonded sand molds and cores. *Foseco (F.S.) Ltd.* Discontinued.

5671 **Ceneg**
84-76-4 201-560-0
$C_{26}H_{42}O_4$
Dinonyl phthalate
1,2-Benzenedicarboxylic acid dinonyl ester; Dinonylphthalate; Di-n-nonyl phthalate; Phthalic acid dinonyl ester. A proprietary trade name for dinonyl phthalate; a vinyl plasticizer *Koninklijke Nederlandsche Gist-En Spiritusfabriek.* Unverified.

5672 **Cenegen® 7**
Alkylaryl sulfonate
Retarding and leveling agent for dyeing nylon with acid and neutral premetallized dyes; good barre coverage. *Crompton & Knowles.*

5673 **Cenegen® CJB**
Alkyl-ethoxy condensate
Leveling, migrating, and retarding agent for dyeing nylon with acid dyes. *Crompton & Knowles.*

5674 **Cenekol® 1141**
Sulfonated phenolic condensate; Fixing and reserving agent for nylon in cellulosic blends and acid dye fixatives; for automotive nylon/rayon blends, nylon/wool blends. *Crompton & Knowles.*

5675 **Cenekol® FT Supra**
Aromatic condensate; surfactant for the textile industry; acid dye fixative on nylon; reserving agent in dyebaths. *Crompton & Knowles.*

5676 **Centa™**
β-Lactamase substrate. *Calbiochem Corp.*

5677 **Centari®**
Acrylic enamels; used for automotive industry. *DuPont UK.*

5678 **Centex**
7775-09-9 8741 231-887-4
Sodium chlorate
Soluble concentrate containing 6.4% w/w sodium chlorate per liter; used for total weed control for paths, drives and noncrop areas. *Chemsearch (UK) Ltd.* Name unverified.

5679 **Centralite**
Dimethyldiphenylurea
Used in explosive powders.

5680 **Centrex® 811**
Weatherable polymer; general-purpose injection molding grade providing good balance of physical properties, gloss, and processing, and excellent color stability. *Monsanto Plastics.*

5681 **Centrex® 833**
Weatherable polymer; extrusion grade; suited for coextrusion over ABS; offers gloss, toughness, property retention. *Monsanto Plastics.*

5682 **Centrifugal Syrup**
A selective weedkiller. *A H Marks & Co Ltd.* Name unverified.

5683 **Centrocap® 162SS, 162US**
8002-43-5 5452 232-307-2
Lecithin
Special grade lecithin; designed for encapsulation where clarity and brilliance are required. d_4^{24}= 1.0305 *Central Soya.*

5684 **Centrol® CA**
8002-43-5 5452 232-307-2
Lecithin
Special grade lecithin; oil-water emulsifier. *Central Soya.* Name unverified.

5685 **Centrolene® A, S**
8029-76-3 232-440-6
Hydroxylated lecithin
Oil-water emulsifiers, increased hydrophilic properties. *Central Soya.* Name unverified.

5686 **Centrolex® C**
8002-43-5 5452 232-307-2
Centrol® 2FSB, 2FUB, 3FSB, 3FUB; Centrophase® HR2B, HR2U. Special grade lecithin; emulsifier, stabilizer, suspending agent for foods. *Central Soya.*

5687 **Centromix® CPS**
8002-43-5 5452 232-307-2
Lecithin-Polysorbate 80; Centrol® 2FSB, 2FUB, 3FSB, 3FUB; Centrophase® HR2B, HR2U. Oil/water emulsifier. *Central Soya.*

5688 **Centrophase® C**
8002-43-5 5452 232-307-2
Lecithin
Centrol® 2FSB, 2FUB, 3FSB, 3FUB; Centrophase® HR2B, HR2U. Special grade lecithin; wetting agents. *Central Soya.*

5689 **Centrophase® HR**
8002-43-5 5452 232-307-2
Lecithin
Centrol® 2FSB, 2FUB, 3FSB, 3FUB; Centrophase® HR2B, HR2U. Heat resistant; multifunctional ingredient; food substance; lubricant and release agent for heated surfaces. *Central Soya.*

5690 **Centrophase® HR2B, HR2U**
8002-43-5 5452 232-307-2
Lecithin
Centrol® 2FSB, 2FUB, 3FSB, 3FUB; Centrophase® HR2B, HR2U. Special grade lecithin; wetting agents. *Central Soya.*

5691 **Centrophil® K**
8002-43-5 5452 232-307-2
Lecithin
Special grade lecithin. *Central Soya.*

5692 **Cenwax® G**
Hydrogenated castor oil; lubricant, wax modifier; used in coatings. *Union Camp.* Name unverified.

5693 **Cenwax® ME**
141-23-1 205-471-8
Methyl hydroxystearate
Lubricant. *Union Camp.*

5694 **Cephreine**
150-84-5 205-775-0
$C_{12}H_{22}O_2$
3,7-Dimethyl-6-octen-1-yl acetate
citronellyl acetate. Pure citronellyl acetate. Used in perfumery. bp = 229°; d = 0.8900; n_D^{20}= 1.4434; LD_{50} (rat orl) = 6800 mg/kg. *Bush Boake Allen Ltd.* Discontinued.

5695 **Cephrol**
106-22-9 2391 203-375-0
$C_{10}H_{20}O$
R-(+)-3,7-Dimethyl-6-octen-1-ol
R-(+)-β-citronellol; 2,6-dimethyl-2-octen-8-ol; citronellol; cephrol. Pure dextro citronellol. bp = 221-222°; d = 0.858; n_D^{20}= 1.4560; $[\alpha]_D^{20}$= 5.22°; slightly soluble in H_2O, soluble in organic solvents. *Bush Boake Allen Ltd.*

5696 **Cerabond 18**
Alumina-based ceramic adhesive; high temperature resistance. Adhesive used to bond selected inorganic substrates including metals. *Master Bond.*

5697 **Cerabrit**
Synthetic waxes. *Abril Industrial Waxes.*

5698 **Cerachem®**
Alumina-silica-zirconia composite; refractory bulk fibers for solving heat problems in furnaces (metal, petrochemical, kilns in ceramic industry, boilers in utility industry.) *Thermal Ceramics.*

5699 **Cerachrome®**
Alumina-silica-chromia composite; refractory bulk fibers for solving heat problems in furnaces (metal, petrochemical, kilns in ceramic industry, boilers in utility industry). *Thermal Ceramics.*

5700 **Ceracolor**
On glaze colors in tubes ready for use for painting on porcelain, bone china and earthenware. *Degussa Ltd.*

5701 **Cerafiber®**
Blend of alumina and silica; refractory bulk fibers for solving heat problems in furnaces (metal, petrochemical, kilns in ceramic industry, boilers in utility industry). *Thermal Ceramics.*

5702 **Ceralan®**
8027-33-6 232-430-1
Lanolin alcohols
Emollient; water-oil emulsifier. *Amerchol Corp.*

5703 **Ceralumin C**
A proprietary alloy of aluminum with 2.5% copper, 1.5% nickel, 1.2% iron, 1.2% silicon, 0.8% magnesium, and 0.15% cerium. No manufacturer.

5704 **Ceramabond**
High temperature ceramic adhesive. *Meclec Ltd.*

5705 **Ceramcel**
Hollow ceramic spheres; for chemical Catalyst supports, polymer and metal matrix composites, automotive emery absorbing structures, plastic fillers, lightweight concrete fillers, thermal insulation, filter bed media, sound attenuation, industrial *Microcel Tech.*

5706 **Ceramer® 67**
Hard isopropyl maleate adduct of polyolefin wax; used in the manufacturing of emulsion coatings and floor polishes. *Petrolite.*

5707 **Ceramitalc**
14807-96-6 9207 238-877-9
talc
Finely ground industrial talc; used in ceramic applications where higher fired strength is required; also used as an auxiliary flux. *R. T. Vanderbilt Co Inc.*

5708 **Ceramite**
A solution of fluorosilicates; used as a disinfectant, as a preservative for wood, and for hardening cements. No manufacturer.

5709 **Ceramol**
Refractory mold and core coatings. *Foseco (F.S.) Ltd.*

5710 **Ceramtex**
Range of colored thermoplastic masterbatches achieving ceramic speckle or fleck effect when molded. *Collinda Ltd.*

5711 **Ceranine PN Base**
Emulsifier/conditioner for skin and hair care products conferring softening, lubricating and antistatic properties *Sandoz.*

5712 **Ceraphyl® 28**
35274-05-6 252-478-7
Cetyl Lactate
Cetyl lactate; emollient binder for pressed powders, lipsticks and hair products *Van Dyk.* Name unverified.

5713 **Ceraphyl® 31**
6283-92-7 228-504-8
Lauryl lactate
Emollient binder for pressed powders, lipsticks, hair products; also antitack agent in antiperspirants. *Van Dyk.* Name unverified.

5714 **Ceraphyl® 41**
C_{12}-C_{15} alcohols lactate; emollient; for sheen on hair; antitack in antiperspirants. *Van Dyk.*

5715 **Ceraphyl® 45**
56235-92-8 260-070-5
Dioctyl maleate
Binder, emollient for cosmetic products applications. *Van Dyk.* Name unverified.

5716 **Ceraphyl® 50**
1323-03-1 215-350-1
Myristyl lactate
Lubricant, emollient for skin products, alcoholic preparations, shaving lotions, colognes, makeup and medicated products *Van Dyk.* Name unverified.

5717 **Ceraphyl® 55**
106436-39-9
Tridecyl neopentanoate
Emollient for creams and lotions, binder for pressed powders. *Van Dyk.* Name unverified.

5718 **Ceraphyl® 60**
51812-80-7 257-440-3
Quatermium-22
Conditioner, emollient, moisturizer, humectant, antistat; highly substantive to skin and hair. *Van Dyk.* Name unverified.

5719 **Ceraphyl® 65**
68953-64-0 273-222-0
Quatermium-26
Emollient; hair conditioner, mild foaming auxiliary emulsifier with antistatic, antitangle properties for shampoos, rinses and other hair products. *Van Dyk.* Name unverified.

5720 **Ceraphyl® 70**
68921-83-5 272-964-2
Quatermium-70
Propylene glycol; antitangle, antistatic ingredient used in all types of hair conditioners; auxiliary emulsifier and emollient for skin creams and lotions. *Van Dyk.* Name unverified.

5721 **Ceraphyl® 85**
87616-36-2 289-325-9
Stearamidopryoyl cetearyl dimonium tosylate, propylene glycol; emollient; hair conditioner; mild foaming auxiliary emulsifier with antistatic, antitangle properties for shampoos, rinses and other hair products *Van Dyk.*

5722 **Ceraphyl® 140**
3687-46-5 222-981-6
Decyl oleate
Emollient; binder for pressed powders; pigment dispersant; co-solvent. *Van Dyk.* Name unverified.

5723 **Ceraphyl® 140-A**
59231-34-4 261-673-6
Isodecyl oleate
Emollient binder for pressed powder; make-up solubilizer; wetting agent for iron oxides; cleansing agent for emulsions. *Van Dyk.* Name unverified.

5724 **Ceraphyl® 230**
6938-94-9 248-299-9
Diisopropyl adipate; emollient; coupler; increased spread of bath oils; reduces oiliness of mineral oil. *Van Dyk.*

5725 **Ceraphyl® 368**
29806-73-3 249-862-1
Octyl palmitate; emollient binder for pressed powders, blushers; gloss agent in lipsticks; antitack for antiperspirants. *Van Dyk.*

5726 **Ceraphyl® 375**
58958-60-4 261-521-9
Isostearyl neopentanoate
Emollient binder for pressed powders, blushers; gloss agent in lipsticks; antitack for antiperspirants. *Van Dyk.* Name unverified.

5727 **Ceraphyl® 424**
3234-85-3 221-787-9
Myristyl myristate
Emollient; increases viscosity of creams and lotions at low concentrations. *Van Dyk.* Name unverified.

5728 **Ceraphyl® 791**
97338-28-8 306-621-6
Isocetyl stearoyl stearate
Pigment dispersant, emollient, lubricant, spreading agent for lipsticks, etc. *Van Dyk.* Name unverified.

5729 **Ceraphyl® 847**
9005-25-8 232-679-6
Octyldodecyl stearoyl stearate
Emollient binder for pigmented sticks and emulsions with dispersing and mold release properties. *Van Dyk.* Name unverified.

5730 **Ceraphyl® GA**
68648-66-8 272-000-0
Maleated soybean oil; emollient for creams and lotions; hair and skin conditioner. *Van Dyk.*

5731 **Ceraphyl® ICA**
36311-34-9 252-964-9
Isocetyl alcohol; emollient for creams and lotions, binder for pressed powders, hair and skin conditioner. *Van Dyk.*

5732 **Ceraphyl® IPL**
22882-95-7 245-289-6
Isopropyl linoleate
Superfatting agent for skin, hair, and cleansing products; skin conditioner leaving luxurious afterfeel on skin; imparts luster and softness in hair care

products; reduces dry afterfeel of liquid soap formulas. *Van Dyk.* Name unverified.

5733 Cerasynt® 303
3179-81-5 221-662-9
Diethylaminoethyl stearate
Viscosity builder in hair dyes; pharmaceutical emulsifier; dispersant, wetting agent. *Van Dyk.*

5734 Cerasynt® 840
9004-99-3
PEG-20 stearate
Emulsifier, viscosity builder, stabilizer for creams, lotions, ointments; superfatting agent in shampoos; vehicle for stick products melting at body temperature. *Van Dyk.*

5735 Cerasynt® D
14351-40-7 238-310-5
Stearamide MEA-stearate
Opacifier, thickener for liquid cream shampoos; auxiliary emulsifier in hydrocarbon aerosol systems such as shave creams. *Van Dyk.* Name unverified.

5736 Cerasynt® M, PA
1323-39-3 215-354-3
Propylene glycol stearate; opacifier for liquid and cream shampoos; secondary emulsifier for cosmetics. (PA) Thickener, pearlescent for liquid and cream shampoos; secondary emulsifier for cosmetics and pharmaceuticals. *Van Dyk.*

5737 Cerasynt® WM
31566-31-1 4498 250-705-4
Glyceryl stearate
Stearyl alcohol, and sodium lauryl sulfate; emulsifier for oil/water creams, lotions, ointments, antiperspirants; electrolyte tolerance and low pH stability. *Van Dyk.*

5738 Cercobin
23564-05-8 9489 245-740-7
thiophanate-methyl
Suspension concentrate containing 500 g/l thiophanate-methyl; a systemic insecticide. *Rhône-Poulenc Crop Protection Ltd.*

5739 Cereclor
Series of secondary plasticizers manufactured from chlorinated waxes; the percentage of chlorine is indicated by the number after the name, e.g., Cereclor 70. *ICI Chem & Polymers Ltd.*

5740 Cereflo
A purified bacterial β-glucanase preparation produced by submerged fermentation of a selected strain of *Bacillus subtilis* ; used as a supplementary glucanase preparation when masking malt or mixtures of malt and barley. *Novo Nordisk.*

5741 Cerelose
50-99-7 4467 200-075-1
glucose
A commercial glucose.

5742 Ceremix
Contains the following enzyme actives: α-amylase, β-glucanase, and proteinase; used in the brewing process when A proportion of the malt is replaced by barley and for the production of malt extract and barley syrups. *Novo Nordisk.*

5743 Cerere
Tricresylmercuroacetate
A mixture of mono and diacetate derivatives of the three cresols, with about 75% of the mono-derivative, and containing about 57% mercury. It accelerates the germination of grain and affords protection against animal and vegetable parasites.

5744 Ceres®
Fat-soluble dyestuffs; used for shoe and floor polishes, office supplies, waxes, oils, fats, fuels, plastics, surface coatings, and printing inks. *Bayer AG.*

5745 Ceresan
123-88-6 204-659-7
C_3H_7ClHgO
Mercury, chloro(2-methoxyethyl)-
β-Methoxyethylmercury chloride; (β-Methoxyethyl)mercuric chloride; Agallol; Agallol '3'; Agallolat; Agalol; Aratan; Aretan 6; Atiran; Baytan; Celmer; Ceresan Universal Nazbeize; Ceresan Universal Wet; Chloro(2-methoxyethyl)mercury; Emisan 6; Falisan; Gramisan; Higosan; Merchlorate; MEMC; Sedresan; Tafasan; Tafasan 6W; Tayssato. Seed dressing for control of fungal diseases on cereals, rice, cotton and vegetables. mp = 65°; insoluble in H_2O, organic solvents; LD_{50} (rat orl) = 22 mg/kg. *Bayer AG.*

5746 ceresin
8001-75-0 2033 232-290-1
White ozokerite wax; earth wax; mineral wax. Waxy mixture of hydrocarbons obtained by purification of ozokerite; candles, sizing; bottles for hydrofluoric acid; shoe and leather polishes; antifouling paints; cosmetics; waterproofing textiles. mp = 61-78°; d = 0.91-0.92; insoluble in H_2O, soluble in 30 parts EtOH, soluble in organic solvents. *Astor Wax; Jonk BV; Stevenson Bros.; Strah & Pitsch.*

5747 Ceresit
Waterproofing compounds for cement, consisting mainly of calcium carbonate, alum, and calcium soap, sometimes with more or less free oil or fat; sold in the form of powder to be mixed with dry cement, or as a paste to be mixed with water. No manufacturer.

5748 Ceresol
62-38-4 7453 200-532-5
Phenylmercury acetate
Organomercury fungicide seed dressing for cereals and fodder beet. *ICI AgroChemicals; ICI Chem. & Polymers Ltd.*

5749 Cerevase
A standardized solution of refined papain or purified papain concentrate; used in the brewing industry for the prevention of colloidal haze caused by repeated chilling and warming of beer. *Pfizer International.*

5750 Cerevax, Cerevax Extra
carboxin-imazalil-thiabendazole. A flowable concentrate containing 360 g carboxin and 20 g thiabendazole; fungicide seed dressing for rye and wheat. Cerevax Extra also contains 25 g imazalil. *ICI AgroChemicals; ICI Chemical & Polymers Ltd.*

5751 Cerex
A copolymer containing carbon, hydrogen, and nitrogen, probably of the acrylonitrile type; a proprietary thermoplastic stated to be resistant to deformation at 100°. No manufacturer.

5752 Ceridust
Micronized polyethylene wax. *Hoechst UK.*

5753 Ceritone
An aromatic flavoring chemical. *PPF International Ltd.* Name unverified.

5754 cerium
7440-45-1 2037 231-154-9
Ce
cerium
A rare-earth element; used in manufacture of cerium salts, cerium-iron alloys, ignition devices, military signaling, illuminant in photography, reducing scavenger, catalyst, alloys for jet engines, solid state devices, rocket propellants, vacuum tubes, mp = 815°; bp = 3257°; d = 6.77. *Aldrich; Cerac; Ferro/Transelco; Rhône-Poulenc Basic.*

5755 cerium sulfate
10294-42-5 11, 1987 237-029-5
$CeO_8S_2 \cdot 4H_2O$
ceric sulfate tetrahydrate
Dyeing and printing textiles, analytical reagent, waterproofing, mildewproofing. *Atomergic Chemetals; Noah Chemical; Rhône-Poulenc Basic.*

5756 Cerone
16672-87-0 3777 240-718-3
$C_2H_6ClO_3P$
2-chloroethyl phosphonic acid
Ethephon; Cerone; Bromeflor; Ethrel; Florel; Prep; Arvest; Chloroethyl)phosphonic acid; Flordimex; 2-Chloroethanephosphonic acid. Growth regulator containing 480 g 2-chlorethyl-phosphonic acid (Ethephon) per liter; used for winter barley. (Sold under license from Union Carbide) *ICI Chem & Polymers Ltd; Embetec Cropt Protection Ltd.*

5757 cerotin
$C_{54}H_{108}O_2$
Ceryl cerolate
Occurs in Chinese wax.

5758 Ceroxin GL
106-14-9 203-366-1
$C_{18}H_{36}O_3$
12-hydroxy stearic acid
A proprietary trade name for 12-hydroxy stearic acid; a lubricant for plastics processing. *J H Little.*

5759 Ceroxin GMO, GMR, GMSI
A proprietary trade name for a partially esterified fatty acid made from naturally occurring saturated or unsaturated fatty acids and a polyhydric alcohol; a lubricant for plastics processing. *J H Little.* Unverified.

5760 Ceroxin TRI
A proprietary trade name for hydroxystearic acid glyceride; a lubricant for PVC which does not cause discoloration and which is particularly suitable for compounds for electrical purposes. *J H Little.* Unverified.

5761 Cerrobase
An alloy of lead and bismuth used as a pattern metal in foundry work.

5762 Cerrobend
An alloy of lead, bismuth, tin, and cadmium used for tube and section bending in foundry work.

5763 Cerromatrix
A proprietary alloy of bismuth, lead, tin, and antimony; expands on cooling. No manufacturer.

5764 Certi-fired
Resistor compositions. *DuPont UK.*

5765 Certincoat
Organotin compound. Patented coating system for application to glass bottles to reduce breakage. *Elf Atochem.*

5766 Certistain
High quality microscopy stains. *BDH Chemicals Ltd.*

5767 Certite
Polyester resin based concrete repair and grouting materials. *SBD Construction Products Ltd.*

5768 Certolake
Water insoluble, aluminum lake food colors; used for coloring of foodstuffs and pharmaceuticals. *Morton Int'l. Ltd.* Discontinued.

5769 Ceruse
1319-46-6 11, 1016 215-290-6
$C_2H_2O_8Pb_3$
Basic lead carbonate
Lead sub-carbonate; Basic lead carbonate; Lead carbonate hydroxide; Lead, bis[carbonato(2-)]dihydroxytri-; Lead(II) carbonate, basic. Basic lead carbonate.

5770 cesium
7440-46-2 2051 231-155-4
Cs
caesium
An alkali-metal (Group I) element; Cs; used in photoelectric cells, vacuum tubes, hydrogenation catalyst, ion propulsion systems, rocket propellant, heat transfer fluid, thermochemical reactions. mp = 28.5°, bp = 706°. *Aldrich; Atomergic Chemetals; Cabot; Cerac.*

5771 cesium bromide
7787-69-1 2052 232-130-0
CsBr
Crystals for infrared spectroscopy, scintillation counters, fluorescent screens. *Atomergic Chemetals; Cabot; Cerac; Noah Chemical.*

5772 cesium carbonate
534-17-8 2053 208-591-9
$C_3Cs_2O_9$
Used in brewing, mineral waters, in specialty glasses, as a polymerization catalyst for ethylene oxide. Soluble in H_2O, organic solvents. *Aldrich; Atomergic Chemetals; Cabot; Cerac.*

5773 cesium chloride
7647-17-8 2054 231-600-2
CsCl
Used in brewing; preparation of cesium compounds; mineral waters; evacuation of radio tubes; in ultracentrifuge separations; fluorescent screens; contrast medium. mp = 646°; bp = 1303°; d = 3.99; LD_{50} (rat ip) = 1.5 g/kg. *Accurate Chemical & Scientific; Atomergic Chemetals; Cabot; Cerac; Janssen Chimica.*

5774 cesium fluoride
13400-13-0 236-487-3
CsF
Used in optics, catalysis, specialty glasses. *Atomergic Chemetals; Cabot; Cerac; Spectrum Chemical Mfg.*

5775 cesium hydroxide
21351-79-1 2055 244-344-1
CsOH
cesium hydrate. Electrolyte in alkaline storage batteries (especially at subzero temperatures), polymerization catalyst for siloxanes. mp = 272°; d = 3.68; LD_{50} (rat ip) = 100 mg/kg. *Aldrich; Atomergic Chemetals; Cabot; Cerac; Noah Chemical.*

5776 cesium iodide
7789-17-5 2056 232-145-2
CsI
Crystals for infrared spectroscopy, scintillation counters, fluorescent screens. mp = 621°, bp = 1280°; d = 4.5; LD_{50} (rat ip) = 1.4 g/kg. *Atomergic Chemetals; Cabot; Cerac; Noah Chemical.*

5777 cesium nitrate
7789-18-6 2057 232-146-8
$CsNO_3$
mp = 414°; d_4^{20} = 3.64-3.68; LD_{50} (rat ip) = 1.2 g/kg. *Cabot; Cerac; Noah Chemical.*

5778 cesium sulfate
10294-54-9 2058 233-662-6
Cs_2O_4S
Brewing, mineral waters, for density gradient in ultracentrifuge separation. mp = 1019°; d = 4.24; soluble in H_2O, insoluble in organic solvents. *Aldrich; Atomergic Chemetals; Cabot; Cerac.*

5779 Cestarsol
900-77-6 815 212-983-5
$C_8H_{13}NO_2$
1,2,5,6-tetrahydro-1-methyl-3-pyridinecarboxylic acid methyl ester
Arecoline-acetarsol. Drocarbil is a formulation of cestarsol with acetarsone and is used in veterinary medicine as an anthelmintic (Cestodes) and a cathartic. *May & Baker Ltd.* Name unverified.

5780 Cetaceum
8892
Cetyl palimitate-cetyl alcohol.
Spermaceti. Waxy substance from the head of he sperm whale.

5781 Cetaffine®
36653-82-4 2070 253-149-0
Cetyl alcohol
Raw material for cosmetics and pharmaceuticals. *Laserson & Sabetay.*

5782 Cetal
An alloy of 87% aluminum and 13% silicon.

5783 Cetal
36653-82-4 2070 253-149-0
Cetyl alcohol NF
Emollient used in emulsions, oils, and makeup; viscosity control in emulsions; auxiliary emulsifier. *Amerchol Corp.*

5784 Cetamoll®
Plasticizers for polyamide-based surface coating resins and for polyamides. *BASF AG.*

5785 Cetaped
Veterinary antiseptic preparation. *ICI Chem & Polymers Ltd.* Discontinued.

5786 Cetats®
138-32-9 2068 205-324-8
$C_{26}H_{49}BrNO_3S$
Cetrimonium *p*-toluene sulfonate
Cetats; cetrimonium tosylate. Surfactant and topical antiseptic. *Zeeland.*

5787 Cetavlex, Cetavlon
57-09-0 2068 200-311-3
$C_{19}H_{42}BrN$
N,N,N-Trimethyl-1-hexadecanaminium bromide
hexadecyltrimethylammonium bromide; cetyltrimethylammonium bromide; Bromat; Cetab; Cetavlon; Cetylamine; C.T.A.B.; Lissolamine V; Micol; Quamonium. A preparation of cetrimide in a cream base; antiseptic skin cream. Cetrimide is a mixture of tetradecyltrimethylammonium, dodecyltrimethylammonium and cetrimonium bromides. mp = 237-243°; soluble in H_2O, EtOH, less soluble in Me_2CO; LD_{50} (rat iv) = 44.0 mg/kg. *ICI Chem & Polymers Ltd.* Name unverified.

5788 Ceteareth-6
68439-49-6 7737
Dehydol PCS 6; Procol CS-6; Prox-onic CSA-1/06. $R(OCH_2CH_2)_nOH$, R represents blend of cetyl and stearyl radicals. Emulsifier; intermediate raw material for mfg. of ethoxysulfates for use in detergents and industrial specialties. Paste, solid; HLB 10.2; Hyd no 106-110. *Pulcra SA.*

5789 Ceteareth-11
68439-49-6 7737
Ceteareth-11
Cremophor® A 11, A 25. Emulsifier for cosmetic and pharmaceutical preparations. *BASF AG.*

5790 Ceteareth-20 BP
68439-49-6 7737
Ceteareth-20 BP
Cetomacrogol 1000 BP. Pharmaceutical oil/water emulsifier, solubilizer; wetting agent for stick formulations; for depilatories, antiperspirants, conditioning rinses. *Croda Inc.; Croda Chemical Ltd.*

5791 Cetearyl octanoate
59130-69-7 261-619-1
Lanol 1688; Cetyl/stearyl 2-ethyl hexanoate; 2-Ethylhexanoic acid, cetyl/stearyl ester; Crodamol CAP; Estalan JB; Hest CSO; Luvitol® EHO; PCL Liq.1002/066240; Schercemol 1688; Tegosoft® Liquid. Cosmetic emulsifier. Seppic; Croda Inc; Croda Surf; Lanaetex Prods.; Heterene; BASF; BASF AG; Dragoco; Scher; Goldschmidt.

5792 Cetec
A proprietary trade name for a cold molding bituminous compound (nonrefractory). No manufacturer.

5793 Cetiol®
3687-45-4 222-980-4
Oleyl oleate
Oily component of strong greasy characteristic, for cosmetic/pharmaceutical

products; carrier for lipid soluble ingredients. *Henkel/Cospha; Henkel Canada.*

5794 Cetiol® 868
22047-49-0 244-754-0
Octyl stearate
Emollient; superfatting oil for oil-water and water-oil emulsions; for cosmetics/pharmaceuticals. *Henkel/Cospha; Henkel Canada.*

5795 Cetiol® 1414E
59686-68-9
Myreth-3-myristate
Self-emulsifying emollient; used in cosmetic preparations such as creams, lotions, lipsticks, and blooming bath oils. *Henkel/Cospha; Henkel Canada.*

5796 Cetiol® A
34316-64-8 251-932-1
Hexyl laurate
Vehicle for lipid-soluble topically active ingredients used in skin lubricants and personal care products; mild emollient. *Henkel/Cospha; Henkel Canada.*

5797 Cetiol® B
105-99-7 203-350-4
Dibutyl adipate
Emollient; oily component for day creams, liquid emulsions. *Henkel/Cospha; Henkel Canada.* No manufacturer.

5798 Cetiol® HE
68201-46-7
PEG-7 glyceryl cocoate
Emollient oil, superfatting agent for aqueous formulations in personal care products; dispersant for biologically active ingredients. *Henkel/Cospha; Henkel KGaA.*

5799 Cetiol® J600
17673-56-2 241-654-9
Oleyl erucate
Emollient; fatty component for cosmetic preparations, jojoba oil substitute. *Henkel/Cospha; Henkel Canada.*

5800 Cetiol® LC
Coco caprylate/caprate; penetrating emollient oil used in personal care products. *Henkel/Cospha; Henkel Canada.*

5801 Cetiol® MM
3234-85-3 221-787-9
Myristyl myristate
Emollient; wax ester with superfatting properties; for skin care and stick preparations. *Henkel/Cospha; Henkel Canada.*

5802 Cetiol® R
Trihydroxy methoxystearin
Emollient; fatty oil for makeup preparations; castor oil substitute. *Henkel/Cospha; Henkel Canada.*

5803 Cetiol® S
$C_{22}H_{44}$
Dioctylcyclohexane
Emollient, superfatting agent; used in cosmetic and pharmaceutical creams and emulsions. *Henkel/Cospha; Henkel Canada.*

5804 Cetiol® SB45
68424-60-2 270-311-6
Shea butter; Emollient, consistency giving agent for oil-water and water-oil creams and emulsions; native fatting agent for creams, lotions, anhydrous creams. *Henkel/Cospha; Henkel Canada.*

5805 Cetiol® SN
Cetearyl isononanoate
Emollient for application in skin care, massage and sun protection preparations; oily component with expressed hydrophobic effect. *Henkel/Cospha; Henkel Canada.*

5806 Cetiol® V
3687-46-5 222-981-6
Decyl oleate
Penetrating emollient, carrier for lipid soluble substances used in personal care products and pharmaceutical topical applications. *Henkel/Cospha; Henkel Canada.*

5807 Cetodan 50-00P Kosher
Acetylated monoglyceride from hydrogenated refined fats; food emulsifier for cake shortenings and fats; edible coating agent for meat products, candy, fruit and nuts. *Grindsted Prods.*

5808 Cetomacrogol 1000
9004-95-9
Polyethylene glycol 1000 monocetyl ether
Emollient.

5809 Cetomacrogol 1000 BP
68439-49-6 7737
Ceteareth-20 BP
Pharmaceutical oil-water emulsifier, solubilizer; wetting agent for stick formulations; for depilatories, antiperspirants, conditioning rinses. *Croda Inc.; Croda Chem. Ltd.*

5810 Cetosan
A mixture of the higher alcohols of spermaceti, mainly cetyl and octodecyl alcohols, with petroleum jelly.

5811 Cetostearyl Alcohol BP, Alcohol NF
Cetearyl alcohol
Croda Inc.

5812 Cetrimide BP
57-09-0 2068 200-311-3
Cetyl trimethyl ammonium bromide
Antistat in hair conditioners; biocidal applications; phase transfer catalysis. *Aceto; Pentagon Chemicals Ltd.*

5813 Cetrimonium bromide
57-09-0 2068 200-311-3
Hexadecyl trimethyl ammonium bromide
Cetrimonium bromide; N,N,N-trimethylammonium-1-hexadecanaminium bromide; hexadecyltrimethyl ammonium bromide; cetyltrimethyl ammonium bromide; Bromat; Cetab; Cetavlon; Cetylamine; Cradocap; C.T.A.B.; Lissolamine V; Micol; Quamonium; Catinal HTB-70. Base material for hair rinse, antistat, germicide. Used for treatment of cradle cap. mp = 237-243°; soluble in H_2O, EtOH, insoluble in Et_2O; LD_{50} (rat iv) = 44 mg/kg. *Aceto; Aldrich; Chemron; Napp Laboratories Ltd; Sherex; Toho Chemical Industry; Zeeland.*

5814 Cetrimonium chloride
112-02-7 203-928-6
Genamin CTAC; Ammonyx® CETAC, CETAC-30; Arguad® 16-29, 16-50; Barquat® CT-29; Carsoquat® CT-429; Chemquat 16-50; Dehyquart A; Genamin CTAC; Incroquat CTC-30; Quartex CTAC; Querton 16Cl-29; Radiaquat 6444; Rhodaquat® M242C/29; Varisoft® 250, 300, 355;. Cosmetic raw material for hair treatment preparations; antistat. Hoechst Celanese/Colorants & Surf; Chemax; Henkel/Cospha; Henkel Functional Prods.; Henkel Chemicals Ltd.

5815 cetyl alcohol
36653-82-4 2070 253-149-0
$C_{16}H_{34}O$
Cetyl alcohol NF
Myristyl alcohol; 1-hexadecanol; ethal; ethol; palmityl alcohol; Cachalot® C-50; n-hexadecyl alcohol. Emollient used in cosmetics. Raw material for cosmetics & pharmaceuticals, as an auxiliary emulsifier and for viscosity control. mp = 47-49°, 54-56°; bp_{10} = 79-81°; d = 0.811; n_D^{79} = 1.4283; insoluble in H_2O, soluble in EtOH, $CHCl_3$, Et_2O. M. Michel; Laserson & Sabetay; Amerchol Corp.; Aarhus Oliefabrik A/S; Chemron; Croda; Ethyl; Lipo; Lonza; Norman, Fox; Procter & Gamble; Stepan; Vista.

5816 cetyl alcohol
36653-82-4 2070 253-149-0
$CH_3(CH_2)_{14}CH_2OH$
palmityl alcohol
C_{16} linear primary alcohol; 1-hexadecanol; Fatty alcohol. Perfumery; emulsifier, emollient, coupling agent; foam stabilizer in detergents; opacifier; thickener; chemical intermediate; cosmetics and pharmaceuticals. Aarhus Oliefabrik A/S; Amerchol; Chemron; Croda; Ethyl; Lipo; Lonza; Michel; Norman, Fox; Procter & Gamble; Sherex; Stepan; Vista.

5817 cetyl alcohol
36653-82-4 2070 253-149-0
$C_{16}H_{34}O$
1-Hexadecanol
Cetyl alcohol; cetyl alcohol NF; ethal; ethol; palmityl alcohol; Crodacol C70. Secondary emulsifier, thickener, opacifier, and structural agent in anhydrous stick systems. Cetyl alcohol NF is a primary structural agent in antiperspirant sticks. mp = 48-50°; $bp_{15}5$ = 189-190°; d = 0.8120; insoluble in H_2O, soluble in EtOH, $CHCl_3$, Et_2O. *Croda Inc.*

5818 cetyl diethanolamine phosphate
90388-14-0 291-394-5
Cetyl diethanolamine phosphate
Crodafos CDP. Emulsifier and stabilizer for oil-water emulsions. *Croda Chem. Ltd.*

5819 cetyl esters
8002-23-1 232-302-5
Dermalcare® SPS; Crodamol SS; Liponate SPS; Ritaceti; Ross Spermaceti Wax Substitute 573; Starfol® WaxCG; W.G.S. Synaceti 116 NF/USP,. Cetyl esters; cosmetic grade emulsifier and emulsion stabilizer for creams, lotions, antiperspirants, creme rinse conditioners, personal care products; substitute for natural spermaceti; acid and alkali stable. *Croda Inc.; Rhône-Poulenc Surf.*

5820 cetyl lactate
35274-05-6 2072 252-478-7
$C_{19}H_{38}O_3$
2-Hydroxypropionic acid hexadecyl ester

lactic acid cetyl ester; lactic acid hexadecyl ester; Cetyl lactate; Ceraphyl 28; Ceraphyl® 28. Non-ionic emollient binder for pressed powders, lipsticks, hair products mp = 41°; bp_1 = 170°; n_D^{40}= 1.4410. *Am. Biorganics; Van Dyk.*

5821 cetyl octanoate
59130-69-7 261-619-1
Emalex CC-168; Bernel® Ester CO; Schercemol CO; Schercemol 1688; Tegosoft® CO;. Lipophilic base for cosmetics *Nihon Emulsion; Bernel; Goldschmidt.*

5822 cetyl palmitate
540-10-3 2073 208-736-6
$C_{32}H_{64}O_2$
Cetyl palmitate
hexadecanoic acid hexadecyl ester, palmitic acid hexadecyl ester; hexadecyl palmitate; Crodamol CP. Emollient for replacing spermaceti wax. Used as a base for ointments and in manufacture of candles and soaps, as a consistency factor in creams, ointmnets, liquid emulsions and fatty make-ups. mp = 55-56°; d^{20} = 0.989; n_D^{20} = 1.4398; insoluble in H_2O, soluble in organic solvents. *Croda Inc.; Sherex; Werner G. Smith; Stepan; Witco/Humco; Henkal Cospha; Henkel Canada.*

5823 cetyl pyridinium chloride
123-03-5 2074 204-593-9
C_21H_38ClN
1-Hexadecylpyridinium chloride
Acetoquat CPC; Ceepryn; Cepacol; Cetamium; Dobendan; Medilave; Merocet; Pristacin; Pyrisept. Germicide, sanitizing agent. Emulsifier; antibacterial, preservative in cough syrups and lozenges; topical anti-infective. mp = 78-83°; slightly soluble in C_6H_6, freely soluble in H_2O, EtOH, $CHCl_3$; LD_{50} (rat, orl) = 200 mg/kg. *Aceto; Schweizerhall; Spectrum Chemical Mfg.; Weiders Farmasoytiske A/S; Zeeland.*

5824 cetyl ricinoleate
10401-55-5 233-864-4
Liponate CRM; Hexadecyl 12-hydroxy-9-octadecenoate; 12-Hydroxy-9-octadecenoic acid ester; Naturechem® CR; Pelemol CR; Protachem CER. Glosser, emollient with dry afterfeel. *Lipo; CasChem; Phoenix; Protameen.* Name unverified.

5825 cetyl stearate
1190-63-2 214-724-1
$C_{34}H_{66}O_2$
n-hexadecyl stearate
Ester of cetyl alcohol and stearic acid;. Emollient. *Koster Keunen; Sherex.*

5826 cetyl trimethyl ammonium chloride
112-02-7 203-928-6
Carsoquat® CT-429. Surfactant; conditioner in hair and skin care preparations; antistat. *Lonza Inc; Pentagon Chemicals Ltd.*

5827 Cevalin
134-03-2 8723 205-126-1
Sodium ascorbate
Vitamin C. *Eli Lilly & Co.*

5828 ceyssatite
SiO_2
silica
A white earth consisting of almost pure silica; an absorbent powder.

5829 CF 1500
65996-61-4 265-995-8
Cellulose fibers; reinforcing filler; contributes excellent mechanical properties in V-belt formulations, etc. *Custom Fibers.* Name unverified.

5830 CF 31,000 C Coarse
9004-34-6 2012 232-674-9
Cellulose fibers; for use as asbestos replacement in industrial applications; relatively coarse fiber with extremely high oil absorption; for use when maximum viscosity increase, sag resistance, and fiber reinforcement is required; suited for solvent and *Custom Fibers.*

5831 CF 42,500T Medium
9004-34-6 2012 232-674-9
Cellulose fibers; asbestos replacement fibers providing increased viscosity and sag resistance, dispersion, and fiber reinforcement for asphalt plastic roof cement, caulks, putties, aluminum roof coating, adhesives. *Custom Fibers.*

5832 CF 70,000WDK, Ex. Superfine
9004-34-6 2012 232-674-9
Cellulosic fibers containing an anionic wetting agent; asbestos replacement fibers providing increased viscosity and sag resistance, dispersion, and fiber reinforcement for textured coatings, adhesives, roof coatings, caulks and sealants, paints, and *Custom Fibers.*

5833 CF1-3510
Fluorosilicone
Oil/solvent resistant material for electronic/aerospace application. *McGhan NuSil.*

5834 C-Flakes
68334-00-9 269-804-9
Hydrogenated cottonseed oil. *Karlshamns.*

5835 CG6710
2897-60-1 220-780-8
$C_{11}H_{24}O_4Si$
(3-Glycidoxypropyl)-methyl-diethoxysilane
Coupling agent, chemical intermediate, blocking agent, release agent, lubricant, primer, reducing agent. *Hüls Am.*

5836 CG6720
2530-83-8 219-784-2
$C_9H_{20}O_5Si$
3-Glycidoxypropyltrimethoxysilane
Glycidoxypropyltrimethoxysilane; Glymo; σ-Glycidoxypropyltrimethoxysilane; Dynasylan GLYMO. Coupling agent, chemical intermediate, blocking agent, release agent, lubricant, primer, and reducing agent. bp_2 = 120°; d = 1.0700; n^{20}_D = 1.4290; LD_{50} (rat orl) = 23 gm/kg. *Hüls Am.*

5837 CG-80
Dairy pipeline cleaner/sterilizer. *Ciba plc.* Name unverified.

5838 CH7250
1009-93-4 213-773-6
1,1,3,3,5,5-Hexamethylcyclotrisilazane
Coupling agent, chemical intermediate, blocking agent, release agent, lubricant, primer, and reducing agent. mp = -10°; bp_{756} = 188°; d = 0.9200. *Hüls Am.*

5839 CH7260
541-05-9 208-765-4
$C_6H_{18}O_3Si_3$
Hexamethylcyclotrisiloxane
Coupling agent, chemical intermediate, blocking agent, release agent, lubricant, primer, and reducing agent. mp = 65-67°; bp = 134°. *Hüls Am.*

5840 CH7280
1450-14-2 215-911-0
$C_6H_{18}Si_2$
Hexamethyldisilane
Coupling agent, chemical intermediate, blocking agent, release agent, lubricant, primer, and reducing agent. mp = 15°; bp = 112-114°; d = 0.7150; n_D^{20}= 1.4221. *Hüls Am.*

5841 CH7310
107-46-0 203-492-7
$C_6H_{18}OSi_2$
Hexamethyldisiloxane
HMDSO. Coupling agent, chemical intermediate, blocking agent, release agent, lubricant, primer, and reducing agent. mp = -59°; bp = 101°; d = 0.7640; n_D^{20}= 1.3775. *Hüls Am.*

5842 CH7332
928-65-4 213-178-1
$C_6H_{13}Cl_3Si$
Hexyltrichlorosilane
Hexyltrichlorosilane; Silane, trichlorohexyl-; n-Hexyltrichlorosilane. Coupling agent, chemical intermediate, blocking agent, release agent, lubricant, primer, and reducing agent. *Hüls Am.*

5843 Chafer 5C Chlormequat
Soluble concentrate containing 640 g chlormequat and 64 g choline chloride per liter; plant growth regulator for use in cereals and ornamentals. *BritAg Industries Ltd.*

5844 Chafer Certrol-E
Bromoxynil, dichlorprop and ioxynil; herbicide mixture for weed control in spring cereals. *DuPont UK.*

5845 Chafer CMPP Super
7085-19-0 5826 230-386-8
Mecoprop
Soluble concentrate containing 570 g/l mecoprop; for control of weeds in cereals and grassland. *BritAg Industries Ltd.*

5846 Chafer MCPA 675
94-74-6 5803 202-360-6
MCPA; herbicide for cereals and grassland. *BritAg Industries Ltd.*

5847 chalcostibnite
$CuSbS_2$
A mineral.

5848 Chaldegal
Foam stabilizers, solubilizers; for detergent formulations; shampoos. *Hoechst UK.* Name unverified.

5849 Chalkone
94-41-7 2078 202-330-2
$C_{15}H_{12}O$
1,3-Diphenyl-2-propen-1-one
1,3-Diphenyl propenone; chalkone; benzylideneacetophenone;

benzalacetophenone; phenyl styryl ketone. mp = 57-58°; bp = 345-34° (dec), bp_{25} = 208°; d_4^{62}= 1.0712; n_D^{62}= 1.6458; insoluble in H_2O, soluble in organic solvents.

5850 Chamotan®
Fish oils; dressing oils for chamois leather. *Seven Seas Ltd.*

5851 Chandor
Emulsifiable concentrate containing 120 g linuron and 240 g trifluralin per liter; herbicide for winter cereals. *DowElanco Ltd.*

5852 Chardot 815 T
Leveling agent for dispersed dyes on polyamide, possessing softening and lubricating properties *Ceca SA.*

5853 Chargemaster® R530
Derived from corn. *Grain Processing.*

5854 Chargepac® 5
Organic and inorganic cationic polymers; coagulant for water and waste water clarification. *Drew Ind. Div.*

5855 Charguard 329
Intumescent flame retardant. *Great Lakes.*

5856 Charisma®
For the agriculture industry. *DuPont UK.*

5857 Chartoon Character®
Polymer. *DuPont UK.*

5858 Chatalier solder
An alloy of 70% tin, 25% zinc, 2% aluminum, and 1.5% phosphorus.

5859 Chatterton's compound
Mixtures of tar, rosin, and gutta-percha; used for cementing gutta-percha to wood and metals.

5860 Chaubert's oil
Consists of 75% oil of turpentine, and 25% oil of hartshorn.

5861 chavicol
501-92-8 2091 207-929-2
$C_9H_{10}O$
4-(2-Propenyl)phenol
p-Allylphenol; γ-(*p*-hydroxyphenyl)-α-propylene;. Occurs in many essential oils such as volatile betel oil. mp = 15.8°; bp = 238°, bp_{16} = 123°; n_D^{18} = 1.5441; insoluble in H_2O, soluble in organic solvents.

5862 Chavosol
A dental antiseptic containing *p*-allylphenol. No manufacturer.

5863 Checkmate
74051-80-2 8620 277-682-3
sethoxydim
Emulsifiable concentrate containing 193 g/l sethoxydim; a cyclohexene-oxime herbicide for annual grasses in field crops. *Embetec Crop Protection Ltd.*

5864 Cheelox® 80
140-01-2 205-391-3
$C_{14}H_{18}N_3Na_5O_{10}$
Diethylenetriaminepentaacetic acid, pentasodium salt
Trilon C; Diethylenetriaminepentaacetic acid, pentasodium salt; Pentasodium diethylenetriaminepentaacetate; Bis(2-(bis(carboxymethyl)amino)ethyl)glycine pentasodium salt; Carboxymethyl)imino)bis(ethylenenitrilo; tetraacetic acid, pentasodium salt; CHEL 330; Detarex PY; Diethylenetrinitrilo)pentaacetic acid, pentasodium salt; Glycine, N,N-bis(2-(bis(carboxymethyl)amino)ethyl)-, pentasodium salt; HAMP-EX 80; Kiresuto P; Pentasodium dtpa; Pentasodium pentetate; Penthanil; Perma Kleer 140; Plexene D; Syntron C; Versenex 80; pentasodium (carboxylatomethyl)iminobis(ethylenenitrilo)tetraacetate. DTPA, pentasodium salt; all purpose chelating agent used in pulp bleaching applications using hydrogen peroxide. *Rhône-Poulenc Surf.*

5865 Cheelox® 100
64-02-8 3557 200-573-9
Tetrasodium EDTA
Sequestrant for metalworking fluids. *Rhône-Poulenc Surf.*

5866 Cheelox® 120
139-89-9 10102 205-381-9
Trisodium HEDTA
Chelating agent used for iron from pH 1.13; sequestrant for magnesium. *Rhône-Poulenc Surf.; Rhône-Poulenc France.* Name unverified.

5867 Cheelox® CG
Substitute for sodium hexametaphosphate; treatment chemical for complexing metal ions in the dye bath without affecting the shade of metal complex dyes. *Rhône-Poulenc Surf.*

5868 Cheelox® HE-24
139-89-9 10102 205-381-9
Trisodium HEDTA
Sequestration agent. *Rhône-Poulenc Surf.*

5869 Cheelox® NTA-Na3
5064-31-3 225-768-6
$C_6H_6NNa_3O_6$
Nitriloacetate trisodium salt
Nitrilotriacetic Acid Trisodium Salt; Glycine, N,N-bis(carboxymethyl)-, trisodium salt;Trisodium nitrilotriacetate; Bis(carboxymethyl)glycine, trisodium salt; NTA, trisodium salt. Nitriloacetate trisodium salt, a sequestrant. *Rhône-Poulenc Surf.*

5870 Cheetah R
66441-23-4 4024 266-362-9
An emulsion containing 60 g fenoxaprop-ethyl per liter; used for grass weed control in wheat. *Hoechst UK.*

5871 Chel
Chelating agents. *Ciba plc.* Name unverified.

5872 Chel DM-41
139-89-9 10102 205-381-9
Trisodium HEDTA
Chelating agent used in bar soaps, photographic developer baths, textiles, and mineral separations. *Ciba-Geigy.*

5873 Chel DTPA
67-43-6 7266 200-652-8
Pentetic acid
Chelating agent used for stabilizing peroxides, biological preparations, cosmetics, textiles, scale removal, and rare earth separations. *Ciba-Geigy.*

5874 Chel DTPA-41
140-01-2 205-391-3
Pentasodium diethylenetriaminepentaacetic acid
Chelating agent for metals. *Ciba-Geigy.* Name unverified.

5875 Chelatex
Water-based adhesives. *Caswell & Co Ltd.*

5876 Chelsea
A range of solvent-based adhesives, dubbins and shoe polishes. *Caswell & Co Ltd.*

5877 Chelsea Melt
Hot melt adhesives. *Caswell & Co Ltd.*

5878 Cheltenham salts
A mixture of 34 parts sodium sulfate, 23 parts magnesium sulfate, and 50 parts sodium chloride.

5879 Chemal 2EH-2
PEG-2 2-ethylhexyl ether
Detergent, wetting agent, emulsifier, dispersant, solubilizer, defoamer for textiles, metal cleaners, industrial and institutional cleaners, household cleaners, hand cleaners, specialties. *Chemax.*

5880 Chemal BP 261
Difunctional block polymers ending in primary hydroxyl groups; defoamer, emulsifier, demulsifier, dispersant, binder, stabilizer, wetting agent, chemical intermediate; for metalworking, cosmetic, paper, textiles, dishwashing detergents, rinse aids, *Chemax.*

5881 Chemal DA-4
26183-52-8
Deceth-4
Detergent, wetting and penetrating agent for textile processing, clay soils, and fire fighting products; emulsifier for polyethylene emulsions; dispersant, solubilizer, defoamer; metal cleaners, industrial, institutional, household and hand cleaners. *Chemax.* Name unverified.

5882 Chemal LA-4
Laureth-4
Oil/water emulsifier, lubricant, detergent, dispersant, solubilizer, defoamer for cosmetic, household, silicone polish, and mold release products. *Chemax.*

5883 Chemal LF 14B, 25B, 40B, LFL-10, -19, -28, -38, -47
Alkoxylated linear alcohol
Low foaming biodegradable surfactant, wetting agent, detergent, defoamer used as rinse aids in mechanical dishwashing, spray metal cleaning formulations and detergent products. *Chemax.*

5884 Chemal OA-20/70CWS
9004-98-2
Oleth-20
Emulsifier, lubricant, solubilizer. *Chemax.*

5885 Chemal OA-4
9004-98-2
Oleth-4
Dispersant, detergent; emulsifier and solubilizer for topical cosmetic applications; stabilizer and anticoagulant for natural and synthetic latexes; emulsifier for waxes used in coating citrus fruit. *Chemax.*

5886 Chemal OA-5
9004-98-2
Oleth-5
Emulsifier, lubricant, and solubilizer. *Chemax.* Discontinued.

5887 Chemal TDA-3
24938-91-8

Trideceth-3
Wetting agent, detergent, emulsifier, dispersant, foam stabilizer; solubilizer, penetrant for scouring and dye leveling in textiles, in cleaning and dishwashing compounds. *Chemax.* Discontinued.

5888 Chematex
Synthetic latex cement for bedding and jointing tiles and bricks. *Prodorite Ltd.*

5889 Chemawinite
Cedarite
A pale yellow Canadian amber.

5890 Chemax AR-497
PEG-15 rosin acid; emulsifier, detergent for acid cleaners, especially for aluminum. *Chemax.*

5891 Chemax CO-16
61791-12-6
PEG-16 castor oil; emulsifier for fiber lubricants; cutting oils and hydraulic fluids; clay and pigment dispersant, rewetting agent, softener, dyeing assistant for paint, paper, textile, and leather industries. *Chemax.*

5892 Chemax CO-5
61791-12-6
PEG-5 castor oil; emulsifier, lubricant for textiles; pigment dispersant in latex paints, paper; essential oils solubilizer. *Chemax.*

5893 Chemax DF-10, DF-10A
Silicone defoamer
Defoamer/antifoam for pulp and paper, textiles, paints, effluent treatment, commercial cleaning processes, adhesives, metalworking. *Chemax.*

5894 Chemax DNP-8, DNP-18
9014-93-1
Nonyl nonoxynol-8, -18. Emulsifiers for nonpolar solvent and oils; detergent for cellulosic and synthetic fibers; dispersant for hard surface cleaners and laundry compounds; solubilizer. *Chemax.*

5895 Chemax DOSS/70
577-11-7 3460 209-406-4
Sodium dioctyl sulfosuccinate
Wetting agent for textile, agricultural, detergent formulations, emulsion polymerization; pigment dispersant in paints and inks; solubilizer for drycleaning solvents. *Chemax.*

5896 Chemax E-200 ML
9004-81-3
PEG-4 laurate
Emulsifier for mineral and cutting oils; dispersant, detergent, lubricant; coemulsifier and defoamer in water-based coating; cosmetics ingredient. *Chemax.*

5897 Chemax E-200 MO
9004-96-0
PEG-5 oleate
Emulsifier for mineral and fatty oils, solvent; degreaser, dispersant, detergent, lubricant; metal, textile, cosmetic, plastisol formulations. *Chemax.*

5898 Chemax E-200 MS
9004-99-3
PEG-5 stearate
Emulsifier for mineral oils and fats used in polishes and metal buffing compounds; dye assistant, lubricant, softener, antistat; for metal lubricants, textiles, cosmetic, plastisol formulations. *Chemax.*

5899 Chemax E-400 ML
9004-81-3
PEG-8 laurate
Emulsifier, dispersant, detergent, lubricant; viscosity control agent in plastisol formulations; welting agent and defoamer in latex paint; cosmetics ingredient. *Chemax.*

5900 Chemax E-400 MO
9004-96-0
PEG-14 oleate
Emulsifier and lubricant for solvents and oils in pesticides and metal cleaners; detergent dispersant; textile, cosmetics, plastisol formulations. *Chemax.*

5901 Chemax E-400 MS
9004-99-3
PEG-9 stearate
Lubricant and softener for synthetic fibers; dye assistant, antistat, emulsifier; for metal lubricants, textiles, cosmetics, plastisols. *Chemax.*

5902 Chemax E-600 ML
9004-81-3
PEG-14 laurate
Emulsifier, dispersant, detergent, lubricant; metal, textile, cosmetics, plastisol formulations. *Chemax.*

5903 Chemax E-600 MO
9004-96-0
PEG-14 oleate
Surfactant used as coemulsifier and lubricant in industrial formulations, cosmetics, metal lubricants, textiles, plastisols; dispersant, detergent. *Chemax.*

5904 Chemax E-600 MS
9004-99-3
PEG-14 stearate
Dye assistant, lubricant, softener, antistat, emulsifier for cosmetic and textile formulations. *Chemax.*

5905 Chemax E-1000 MO
9004-96-0
PEG-20 oleate
Emulsifier for mineral and fatty oils, solvent; degreaser, dispersant, detergent, lubricant. *Chemax.*

5906 Chemax E-1000 MS
9004-99-3
PEG-20 stearate
Emulsifier for cosmetic and textile formulations; dye assistant, lubricant, softener, antistat. *Chemax.*

5907 Chemax HCO-200/50
61788-85-0
PEG-200 hydrogenated castor oil
Emulsifier and lubricant for plastics, metals, textiles, paint, paper, leather industries. *Chemax.*

5908 Chemax HCO-5
61788-85-0
PEG-5 hydrogenated castor oil
Emulsifier, lubricant, softener, dispersant; co-emulsifier for synthetic esters; for plastics, metals, textiles, leather, paint, and paper industries. *Chemax.*

5909 Chemax NP-1.5
9016-45-9 6772
Nonoxynol-1.5
Emulsifier, dispersant, detergent, wetting agent, solubilizer, coupler for textile, metalworking, household, industrial, agricultural, paper, paint, and other industries. *Chemax.*

5910 Chemax NP-4
9016-45-9 6772
Nonxynol-4
Emulsifier, dispersant, detergent, wetting agent, solubilizer, coupler for textile, metalworking, household, industrial, agricultural, paper, paint, and other industries. *Chemax.*

5911 Chemax NP-6
9016-45-9 6772
Nonxynol-6
Emulsifier, dispersant, detergent, wetting agent, solubilizer, coupler for textile, metalworking, household, industrial, agricultural, paper, paint, and other industries. *Chemax.*

5912 Chemax NP-9
9016-45-9 6772
Nonoxynol-9
Emulsifier, dispersant, detergent, wetting agent, solubilizer, coupler for textile, metalworking, household, industrial, agricultural, paper, paint, and other industries. *Chemax.*

5913 Chemax NP-10
9016-45-9 6772
Nonoxynol-10
Emulsifier, dispersant, detergent, wetting agent, solubilizer, coupler for textile, metalworking, household, industrial, agriculture, paper, paint, and other industries. *Chemax.*

5914 Chemax NP-15
9016-45-9 6772
Nonoxynol-15
Surfactant, detergent, wetting and rewetting agent, emulsifier in textile, leather, paper, paint, and metal processing. *Chemax.*

5915 Chemax NP-30
9016-45-9 6772
Nonoxynol-30
Surfactant, detergent, wetting and rewetting agent, emulsifier in textile, leather, paper, paint, and metal processing. *Chemax.*

5916 Chemax NP-40
9016-45-9 6772
Nonoxynol-40
Polymerization emulsifier for vinyl acetate and acrylic emulsions; stabilizer for synthetic lattices; wetting agent in electrolyte solutions. *Chemax.*

5917 Chemax OP-3
9002-93-1 6858
Octoxynol-3
Emulsifier, detergent, stabilizer, dispersant, wetting agent; pesticides and floor finishes. *Chemax.*

5918 Chemax OP-7
9002-93-1 6858
Octoxynol-7
Detergent compounds; industrial metal cleaning, acid and waterless hand cleaners, and floor finishes. *Chemax.*

5919 Chemax OP-30/70
9002-93-1 6858
Octoxynol-30
Emulsifier, dispersant, detergent, wetting agent, solubilizer, coupler for textile, metalworking, household, industrial, agriculture, paper, paint, and other industries. *Chemax.*

5920 Chemax OP-40
9002-93-1 6858
Octoxynol-40
Emulsifier, detergent, stabilizer, wetting agent, dispersant. *Chemax.*

5921 Chemax OP-40/70
9002-93-1 6858
Octoxynol-40
Emulsifier for vinyl acetate and acrylate polymerization. *Chemax.*

5922 Chemax PEG 200 DO
9005-07-6
PEG-4 dioleate
surfactant. Coemulsifier for oils as mold release agent. *Chemax.*

5923 Chemax PEG 400 DO
9005-07-6
PEG-8 dioleate
Emulsifier and solubilizer for solvents, fats, and mineral oils; in lubricant, softener, and defoamer formulations for agriculture, cosmetic, household, leather, metalworking, and textile industries. *Chemax.*

5924 Chemax PEG 600 DO
9005-07-6
PEG-12 dioleate
Emulsifier in lubricant, softener, and defoamer formulations for agricultural, cosmetic, household, leather, metalworking, and textile industries. *Chemax.*

5925 Chemax SBO
Sulfated butyl oleate
Softener, emulsifier, wetting agent in textile and metal working industries. *Chemax.* Name unverified.

5926 Chemax SCO
8002-33-3 232-306-7
Sulfated castor oil; Softener, emulsifier, wetting agent in textile and metal working industries. *Chemax.* Name unverified.

5927 Chemax TO-10
61791-00-2
PEG-10 tallate
Emulsifier, detergent in industrial lubricants and degreasers. *Chemax.*

5928 Chemax TO-16
61791-00-2
PEG-16 tallate
Foam detergent and emulsifier; lubricant, dye assistant. *Chemax.*

5929 Chemax TO-B
61791-00-2
PEG-8 tallate
Emulsifier, lubricant, dye assistant. *Chemax.*

5930 Chembetaine BW
68424-94-2 270-329-4
Coco betaine
Viscosity builder, gelling agent, industrial surfactant; lime soap dispersant; mild to skin; tolerant to hard water; stable in high-electrolyte solutions; cationic in acid and anionic in alkaline media. *Chemron.*

5931 Chembetaine C, CGF
Cocamidopropyl betaine
High foaming mild industrial and personal care surfactant; foam and viscosity builder; lime soap dispersant; foaming agent for water and acid systems; stable over wide pH range. *Chemron.*

5932 Chembetaine CAS
68139-30-0 268-761-3
Cocamidopropylhydroxysultaine
Anti-irritant for other surfactants; esp. for baby shampoos and baby bath products; detergent for heavy-duty industrial alkaline cleaners (steam cleaners, wax remover, hard surface cleaner); wetting agent in acid pickling of metals; lime soap dispersant; *Chemron.*

5933 Chembetaine CB
68424-94-2 270-329-4
Cocobetaine
Foaming surfactant effective in hard and soft water; mild to hair and skin; for shampoos, soaps, conditioners. *Chemron.*

5934 Chembetaine L
Lauramidopropyl betaine
Foam booster, viscosity builder; mild surfactant for shower gels, liquid soaps, skin cleansers, shampoos. *Chemron.*

5935 Chembetaine OL-30
871-37-4 212-806-1
Oleyl betaine
Gentle, substantive viscosity builder for shampoos, conditioners, mousses. *Chemron.*

5936 Chembetaine S
Soyamidopropyl betaine
Conditioner, foamer, viscosity builder for shampoos, conditioners, bath products *Chemron.*

5937 Chembetaine TG
Dihydroxyethyl tallow glycinate
Conditioner providing nonoily feel and resistance to build-up for cream rinses, shampoos, conditioning mousses, combout sprays. *Chemron.*

5938 Chem-Calk® 500
Two-component polyurethane; nonsag, high performance elastomeric joint sealant. *Bostik.*

5939 Chemcaulk
Two-component fluoroelastomer base caulk; corrosion resistance sealant for expansion joint flanges, expansion joint grouting in chem-resistance flooring, expansion joint repair, formed in-place gaskets; recommended for contact with strong acids, *Advanced Polymer Coatings.*

5940 Chemcoat
Two-component fluoroelastomer (Viton®) base coating system; recommended for harsh chemical environments; applied to valves, pipes, fittings, tanks, electronic components, cables; cures at R.T.; excellent adhesion properties. *Advanced Polymer Coatings.*

5941 Chemcogen AC
28519-02-0 249-063-8
Disodium alkyl diphenyl oxide disulfonate. *Rhône-Poulenc Surf.*

5942 Chemdur
Liquid elastomeric urethane membrane for tanking rooms and pits. *Prodorite Ltd.*

5943 Chemeen 18-2
10213-78-2 233-520-3
PEG-2 stearamine
Emulsifier and antistat in textiles, metal buffing, and rubber compounds; lubricant for fiber glass. *Chemax.*

5944 Chemeen C-2
61791-14-8
PEG-2 cocamine
Emulsifier, antistat, dye leveler, wetting agent, lubricant, dispersant; substantive to metals, fibers, and clays. *Chemax.*

5945 Chemeen DT-3
PEG-3 tallow diamine
Emulsifier, textile dyeing assistant, corrosion inhibitor used in preparation of asphalt and agricultural chemical emulsions. *Chemax.*

5946 Chemeen HT-5
61791-26-2
PEG-5 hydrogenated tallow amine
Emulsifier and antistat in textiles, metal buffing and rubber compounds, lubricant for fiber glass. *Chemax.*

5947 Chemeen O-30, O-30/80
PEG-30 oleamine
Emulsifier, antistat, lubricant, and textile dyeing assistant; antiprecipitant in cross dyeing. *Chemax.*

5948 Chemeen S-2
61791-24-0
PEG-2 soya amine
Emulsifier, antistat, lubricant. *Chemax.*

5949 Chemeen T-2
61791-44-4 263-177-5
PEG-2 tallow amine
Antistat for carpet shampoos; emulsifier, lubricant, dispersant, softener, antiprecipitant, leveling and migrating agent in textile dyeing process. *Chemax.*

5950 Chemfac NC-0910
POE alkyl phenol phosphate
Wetting agent, detergent, hydrotrope, emulsifier, rust inhibitor, extreme pressure additive for alkaline detergents, metal cleaners, hard surface cleaners, textile scours, metal and textile lubricants, drycleaning soaps, emulsion polymerization, *Chemax.*

5951 Chemfac PA-080, PB-082
Alcohol phosphate ester
Detergent, emulsifier, wetting agent, lubricant, antistat for alkaline detergents,

metal cleaners, hard surface cleaners, textile scours, metal and textile lubricants, drycleaning soaps, emulsion polymerization, agricultural formulations. *Chemax.*

5952 Chemfac PB-184
39464-69-2
Oleth-4 phosphate
Wetting agent, detergent, hydrotrope, emulsifier, rust inhibitor, EP additive for alkaline detergents, metal cleaners, hard surface cleaners, textile scours, metal and textile lubricants, drycleaning soaps, emulsion polymerization and *Chemax.*

5953 Chemfac PC-006
39464-70-5
Phosphate ester; Wetting agent, lubricant, antistat, detergent, emulsifier; foaming hydrotrope. *Chemax.*

5954 Chemfac PD-600
52019-36-0
POE alkyl ether phosphate
Wetting agent, detergent, hydrotrope, emulsifier, rust inhibitor, EP additive for alkaline detergents, metal cleaners, hard surface cleaners, textile scours, metal and textile lubricants, drycleaning soaps, emulsion polymerization, agriculture formulations. *Chemax.*

5955 Chemfac PN-322, PX-322
Phosphate ester, neutralized; hydrotrope for solubilizing nonionic surfactants in high concentrations of alkali or other electrolytes; wetting agent, detergent, emulsifier, rust inhibitor, EP additive for alkaline detergents, textile scours, emulsion, *Chemax.*

5956 Chemfac RD-1200
Proprietary; environmentally safe industrial defoamer for offshore drilling. *Chemron.*

5957 Chemfax 5AM-100
Hydrocarbon resin polymerized from olefins; used in adhesives, coatings, tackifiers, etc. *Chemfax.*

5958 Chemglaze®, Chemglaze® Z-004
High performance single - and two-pack urethane coatings, both moisture cure and catalyzed cure; used to preserve, protect and beautify all types of substrates including rubber, plastic, concrete, metal and glass. *Lord Corporation (UK) Ltd.*

5959 Chemical 39 Base
141-21-9 205-469-7
Stearamidoethyl ethanolamine
Cosmetic and toiletry base; emulsifying agent for creams and lotions; conditioning additive for hair products; lubricant in skin products *Sandoz.*

5960 Chemical Base 6532
16889-14-8 240-924-3
Stearamidoethyl diethylamine
Emulsifier, emollient, conditioning agent for skin and hair, shaving creams, hydro-alcoholic lotions, skin creams, cream rinse shampoos; substantive to hair. *Sandoz.*

5961 Chemidex B
60270-33-9 262-134-8
Behenamidopropyl dimethylamine
Substantive surfactant for cream rinses, conditioners, shampoos, creams, and lotions. *Chemron.*

5962 Chemidex C
68140-01-2 268-771-8
Cocamidopropyl dimethylamine
Substantive surfactant for cream rinses, conditioners, shampoos, creams, and lotions. *Chemron.*

5963 Chemidex L
3179-80-4 221-661-3
Lauramidopropyl dimethylamine
Substantive surfactant for cream rinses, conditioners, shampoos, creams, and lotions. *Chemron.*

5964 Chemidex M
45267-19-4 256-214-1
Myristamidopropyl dimethylamine
Substantive surfactant for cream rinses, conditioners, shampoos, creams, and lotions. *Chemron.* No manufacturer.

5965 Chemidex O
109-28-4 203-661-5
Oleamidopropyl dimethylamine
Substantive surfactant for cream rinses, conditioners, shampoos, creams, and lotions. *Chemron.* No manufacturer.

5966 Chemidex P
39669-97-1 254-585-4
Palmitamidopropyl dimethylamine
Substantive surfactant for cream rinses, conditioners, shampoos, creams, and lotions. *Chemron.*

5967 Chemidex R
Ricinoleamidopropyl dimethylamine
Substantive surfactant for cream rinses, conditioners, shampoos, creams, and lotions. *Chemron.*

5968 Chemidex S
7651-02-7 231-609-1
Stearamidopropyl dimethylamine
Low irritation emulsifier with substantivity to protein and cellulosic substrates; for cream rinses, conditioners, shampoos, creams, and lotions. *Chemron.*

5969 Chemidex SE
Stearamidoethyl dimethylamine
Substantive surfactant for cream rinses, conditioners, shampoos, creams, and lotions. *Chemron.*

5970 Chemidex SI
67799-04-6 267-101-1
Isostearamidopropyl dimethylamine
Substantive surfactant for cream rinses, conditioners, shampoos, creams, and lotions. *Chemron.*

5971 Chemidex SO
68188-30-7
Soyamidopropyl dimethylamine
Substantive surfactant for cream rinses, conditioners, shampoos, creams, and lotions. *Chemron.* Name unverified.

5972 Chemidex T
68425-50-3 270-356-1
Tallowamidopropyl dimethylamine
Substantive surfactant for cream rinses, conditioners, shampoos, creams, and lotions. *Chemron.* Name unverified.

5973 Chemidex WC
68140-01-2 268-771-8
Cocamidopropyl dimethylamine
Substantive surfactant for cream rinses, conditioners, shampoos, creams, and lotions. *Chemron.*

5974 Chemigum® HR662
Butadiene-acrylonitrile-N-(4-anilinophenyl) methacrylamide terpolymer
NBR, cold polymer. Used in molded and extruded goods such as gaskets and oil seals, oil well cable jackets and hose that require high resistance to oil and heat aging. *Goodyear.* Name unverified.

5975 Chemigum® Latex 260
Nitrile latex
Used as binder for asbestos and cellulose; provides good water resistance; sulfur vulcanizable. *Goodyear.* Name unverified.

5976 Chemigum® N318B, N917
NBR, cold polymer. Nonstaining antioxidants; for extrusions and compression molding, hose tubes; exceptional grn. strength; noncorrosive. *Goodyear.* Name unverified.

5977 Chemigum® N5, N7
NBR, hot polymer. Slightly staining antioxidant; for compounds which must withstand rugged or high-temperature service, e.g., hose, tubes, and belts. *Goodyear.* Name unverified.

5978 Chemigum® NX-775
Carboxylated NBR
Cold polymer. Nonstaining antioxidant; FDA compliance for applications where outstanding abrasion and oil resistance and high strength are required; used for footwear, cable jackets, oil field parts, rollers and roll covers, shaft seals, pump liners, *Goodyear.* Name unverified.

5979 Chemigum® P7-D
nitrile-butadiene rubber. Branched nitrile-butadiene rubber powder partitioned with 9% calcium carbonate; for rubber, adhesives (high gel strength), dry or semiwet process disc-pads. *Goodyear.* Name unverified.

5980 Chemigum® P83
nitrile-butadiene rubber. Nitrile-butadiene rubber powder partitioned with 9% PVC, lightly precrosslinked, stabilized; designed for PVC modification; improves hot-melt stability, plasticizer permanence, and processability for calendered and extruded film, sheet, gaskets and seals, *Goodyear.* Name unverified.

5981 Chemigum® TPE 03050
Nitrile-based thermoplastic elastomer. For automotive weatherstripping, tubing, architectural seals, coextruded glazing gaskets, appliance profiles and moldings, hose and tubing, cable sheathing, conveyor belting, flooring, roofing; modifier for vinyl, TPU and polar TPEs. *Goodyear.*

5982 Chemlease 55
Silicone blend; mold release for urethane elastomers. *Chemlease.* Name unverified.

5983 Chemlease 88
Fluoropolymer blend; mold release for electrical potting, filament winding, phenolic board, adhesive laminate. *Chemlease.* Name unverified.

5984 Chemlease 906E
Silicone water-based emulsion; mold release for rubber and plastics industries. *Chemlease.* Name unverified.

5985 Chemlease 158R
Nonsilicone; water-based release system for epoxy composites, compression molding, RIM urethane molding, rubbers. *Chemlease.* Name unverified.

5986 Chemlease SP 40
General-purpose semi-permanent release system for plastic and rubber, electrical potting, filament winding, phenolic board, roto molding. *Chemlease.* Name unverified.

5987 Chemline
A range of anti-abrasive glass tiles and ceramic tiles for coal and coke bunkers and hoppers. *Prodorite Ltd.*

5988 Chemlock
Rubber to metal heat activated bonding agents and adhesives. *Durham Chemicals Ltd.*

5989 Chemlok® 205
One-coat bonding agent for nitrile-butadiene rubber-based thermoplastic elastomer and primer for cover coat adhesive. *Lord.*

5990 Chemlok® 220
General purpose cover coat bonding agent. *Lord.*

5991 Chemlok® 459
Bonding agent; primer promoting adhesion to thermoplastic elastomers and polyolefins. *Lord.*

5992 Chemlok® 607
Silane-based; one-coat adhesive for bonding elastomers, especially silicone, fluroelastomers, EPDM, nitrile, epichlorohydrin, polyacrylate, hydrogenated nitriles. *Lord.*

5993 Chemlube
A series of diester and polyolester synthetic based oils; for compressor oils, chain oils, gear oils, automotive engine oils, high temperature oils, etc. *Ultrachem Inc.* Name unverified.

5994 Chemoxide CAW
68155-09-9 268-938-5
Cocamidopropylamine oxide
Surfactant for mild, low irritation personal care and industrial applications, e.g., shampoos, facial cleansers, bath products; foam and viscosity builder, emollient over broad pH range. *Chemron.*

5995 Chemoxide L
Lauramidopropyl betaine
Foam booster, viscosity builder; mild surfactant. *Chemron.*

5996 Chemoxide LM-30
Lauramine oxide
viscosity builder, foam enhancer for household and industrial cleaners, personal care products; tolerant to electrolytes for improved hard water performance. *Chemron.*

5997 Chemoxide O
14351-50-9 238-311-0
Oleamine oxide
Thickener, viscosity builder; wetting agent for pigments and dyes; used in hair colorants, gels, permanent waves. *Chemron.*

5998 Chemoxide O1
Oleyl dimethylamine oxide
Foamer, thickener, industrial surfactant; tolerant to electrolytes. *Chemron.*

5999 Chemoxide SAO
25066-20-0 246-598-9
Stearamidopropylamine oxide
Wetting and foaming agent, viscosity builder, conditioner, softener for hair; for shampoos, conditioners, mousses; emulsifier in creams and lotions. *Chemron.*

6000 Chemoxide ST
2571-88-2 219-919-5
Stearamine oxide
Conditioner, softener, viscosity and foam builder for conditioning shampoos, rinses improving comb-out and manageability. *Chemron.*

6002 Chemoxide TAO
68647-77-8 271-972-3
Tallowamidopropylamine oxide
Viscosity builder, foam booster; improves manageability and luster in hair; for use in conditioners, sprays, mousses. *Chemron.*

6003 Chemoxide WC
61788-90-7 263-016-9
Cocamine oxide
Viscosity builder, foam booster, emollient for shampoos, cleansers, bath products *Chemron.* Discontinued.

6004 Chemphonate AMP
6419-19-8 229-146-5
Amino tris(methylene phosphonic acid)
Scale inhibitor for water injection, water disposal and production systems. *Chemron.* Discontinued.

6005 Chemphonate AMP-S
Sodium amino tris(methylene phosphonate)
Scale inhibitor for water systems; dispersant for drilling muds. *Chemron.*

6006 Chemphonate HEDP
1-Hydroxyethyl-1-diphosphonic acid
Temperature-stable scale inhibitor for oilfield application; sequestering agent for calcium carbonate. *Chemron.*

6007 Chemphonate NP
Sodium TEA phosphoric acid ester; Scale inhibitor for oilfield prod. *Chemron.*

6008 Chemphos TC-227
Aromatic phosphate ester. Detergent, wetting agent, emulsifier, coupling agent, surface tension reducer; for alkaline cleaners, heavy-duty all-purpose metalworking detergents, steam cleaning, dairy cleaners, bottle washing compounds, floor strippers. *Chemron.*

6009 Chemphos TC-231S
Sodium nonoxynol-9 phosphate
Emulsifier, solubilizer, antistat, substantivity agent for hair care products, perms, straighteners, depilatories; resistant to hydrolysis. *Chemron.*

6010 Chemphos TC-337
Nonoxynol-20 phosphate
Emulsion polymerization surfactant for vinyl acetate, acrylates, SBR; emulsifier, solubilizer, antistat, substantivity agent for hair care products, perms, straighteners, depilatories; resistant to hydrolysis. *Chemron.*

6011 Chemphos TC-341
39464-64-7
Nonyl nonoxynol-10 phosphate
Emulsifier, solubilizer, antistat, substantivity agent for hair care products, perms, straighteners, depilatories; resistant to hydrolysis. *Chemron.*

6012 Chemphos TC-444
Aliphatic phosphate ester
Coupling agent for nonionic surfactants with liquid alkali detergent systems; surface tension reducer; for oil treating chemicals; compatible with high concentrations of sodium hydroxide and silicate builders. *Chemron.*

6013 Chemphos TDAP
Alkyl acid phosphate
Oil treating surfactant, asphaltine dispersant. *Chemron.*

6014 Chemphos TR-414W
Phosphate ester; lubricant and detergent for water-based drilling fluids. *Chemron.*

6015 Chemphos TR-505
39464-69-2
Oleth-10 phosphate
Emulsifier, solubilizer, antistat, substantivity agent for hair care products, perms, straighteners, depilatories; resistant to hydrolysis. *Chemron.*

6016 Chemphos TR-505D
58855-63-3
DEA-oleth-10 phosphate
Emulsifier, solubilizer, antistat, substantivity agent for hair care products, perms, straighteners, depilatories; resistant to hydrolysis. *Chemron.*

6017 Chemphos TR-510
39464-66-9
Laureth-4 phosphate
Emulsifier, solubilizer, antistat, substantivity agent for hair care products, perms, straighteners, depilatories; resistant to hydrolysis. *Chemron.*

6018 Chemphos TR-510S
42612-52-2
Sodium laureth-4 phosphate
Emulsifier, solubilizer, antistat, substantivity agent for hair care products, perms, straighteners, depilatories; resistant to hydrolysis. *Chemron.*

6019 Chemphos TR-515
39464-69-2
Oleth-3 phosphate
Emulsifier, solubilizer, antistat, substantivity agent for hair care products, perms, straighteners, depilatories; resistant to hydrolysis. *Chemron.*

6020 Chemphos TR-515D
58855-63-3
DEA-oleth-3 phosphate
Emulsifier, solubilizer, antistat, substantivity agent for hair care products, perms, straighteners, depilatories; resistant to hydrolysis. *Chemron.*

6021 Chemphos TR-541
39464-69-2
Oleth-4 phosphate

Emulsifier, solubilizer, antistat, substantivity agent for hair care products, perms, straighteners, depilatories; resistant to hydrolysis. *Chemron.*

6022 Chemprene 50, 75
Isoprenoidal polymer; excellent water, acid, and alkali resistant, excellent oxidative stability. *Chemfax.*

6023 Chemprene R-10
Thermoplastic isoprenoidal polymer; used in compding. of synthetic, natural, and reclaim rubber; softener, tackifier, and reinforcing agent; used in calendering and extruding hard rubber, molded goods, in camelback, tubes, and tire stocks, in rubber *Chemfax.*

6024 Chemquat 12-50
112-00-5 203-927-0
Laurtrimonium chloride
Surfactant, corrosion inhibitor, antistat for plastics, textile dyeing aid, gel sensitizer in latex foam production; viscosity depressant in paper and textile softener formulations. *Chemax.*

6025 Chemquat 16-50
112-02-7 203-928-6
Cetrimonium chloride
Surfactant, corrosion inhibitor, antistat for plastics, textile dyeing aid, gel sensitizer in latex foam production; viscosity depressant in paper and textile softener formulations. *Chemax.*

6026 Chemquat C/33W
61789-18-2 263-038-9
Coco trimethyl ammonium chloride
Surfactant, corrosion inhibitor, antistat for plastics, textile dyeing aid, gel sensitizer in latex foam production; viscosity depressant in paper and textile softener formulations. *Chemax.*

6027 Chemraz®
Perfluoroelastomer (PTFE plus proprietary agents); extremely chemically resistant seals for aggressive chemical environments. *Greene, Tweed & Co.*

6028 Chem-Rez
Phenolic and furfuryl alcohol-based resin systems cured through the application of acid catalyst or with heat; for production of foundry cores and molds. *Ashland Chemical Company.* Name unverified.

6029 Chemsalan NLS 30
151-21-3 8782 205-788-1
Sodium lauryl sulfate
Surfactant. *Chemsal.*

6030 Chemsalan RLM 28
9004-82-4
Sodium laureth sulfate
Biodegradable surfactant. *Chemsal.*

6031 Chemset
Artificial and synthetic resins all adapted for setting; used for adhesives, chemical products for potting, impregnating, flooring, casting, molding, laminating, building, sheathing, coating, enamelling, waterproofing and for the hardening of resins. *R F Bright Enterprises Ltd.*

6032 Chemstat® 106G/90
58767-50-3 261-430-4
Bis (2-hydroxyethyl) octyl methylammonium *p*-toluene sulfonate
Permanent internal antistat for polystyrene and other thermoplastics where high processing temperatures are required; also as external antistat. *Chemax.*

6033 Chemstat® 122
Ethoxylated coco amine
Internal antistat for polyolefins. *Chemax.*

6034 Chemstat® 172
13127-82-7 236-062-2
Ethoxylated oleyl amine
Internal antistat for polyolefins. *Chemax.* Name unverified.

6035 Chemstat® 192/NCP
Ethoxylated stearyl amine
Permanent internal antistat effective in eliminating electrostatic problems in extrusion, injection and blow molding of thermoplastics. *Chemax.*

6036 Chemstat® 273-C
61791-31-9 263-163-9
N,N-Bis (2-hydroxyethyl) coco amine
Permanent internal antistat for plastic film and molded products; eliminates electrostatic problems in extrusion, injection and blow molding. *Chemax.* Name unverified.

6037 Chemstat® 273-E
10213-78-2 233-520-3
N,N-Bis (2-hydroxyethyl) stearyl amine
Permanent internal antistat for plastic film and molded products; eliminates electrostatic problems in extrusion, injection and blow molding. *Chemax.*

6038 Chemstat® 9820A
Bis(2-hydroxyethyl)octyl methyl ammonium p-toluene sulfonate in ABS carrier resin; permanent internal antistat for ABS; recommended use 1-3% of active antistat. *Chemax.* Name unverified.

6039 Chemstat® AF-906
Antifog agent for polyethylene film. *Chemax.*

6040 Chemstat® HTSA1
16260-09-6 240-367-6
Oleyl palmitamide
Slip/antiblock agent, antistat, mold release with high thermal stability; for PP film. *Chemax.*

6041 Chemstat® HTSA3
10094-45-8 233-226-5
Stearyl erucamide
Slip/antiblock agent, mold release with high thermal stability. *Chemax.*

6042 Chemstat® P-400
25322-68-3 7729
Polyethylene glycol
Permanent internal antistat for polyethylene. *Chemax.*

6043 Chemstat® PS-101
68037-49-0 268-213-3
Sodium C_{10}-C_{18} alkyl sulfonate
Permanent internal antistat for thermoplastic film and molded products; temporary external antistat; biodegradable. *Chemax.* Name unverified.

6044 Chemsulf S2EH-Na
Sodium 2-ethylhexyl sulfate
Low foaming surfactant, wetting agent, emulsifier, detergent; electrolyte tolerant. *Chemax.*

6045 Chemsulf SBO/65
Sulfated butyl oleate
Softener, wetting agent, lubricant additive, emulsifier, solubilizer for textile, metalworking industries. *Chemax.*

6046 Chemsulf SCO/75
8002-33-3 232-306-7
Sulfated castor oil; softener, wetting agent, lubricant additive, emulsifier, solubilizer for textile, metalworking industries. *Chemax.* Name unverified.

6047 Chemtac 20, 35
Recommended for harsh chemical environments when additional surface tack is required for laminating and/or joining substrates. *Advanced Polymer Coatings.*

6048 Chemtech Cypermethrin
66841-24-5 266-492-6
Emulsifiable concentrate containing 100 g cypermethrin per liter; a pyrethroid insecticide. *Chemtech (Crop Protection) Ltd.*

6049 Chem-Trete®
Alkylalkoxysilane
Weatherproofing agents for concrete and masonry surfaces. *Hüls Am.*

6050 Chemzoline 1411
Aminoethyl tall oil imidazoline
Corrosion inhibitor base for high-temperature oilfield applications. *Chemron.*

6051 Chemzoline C-22
61791-38-6 263-170-7
Hydroxyethyl coco imidazoline
Emulsifier for oils; corrosion inhibitor for oilfield applications. *Chemron.*

6052 Chemzoline T-11, T-33
Aminoethyl tall oil imidazolines; intermediates for production of film-forming corrosion inhibitors; in automatic car wash rinse aids, oilfield applications; pigment dispersant for paints. *Chemron.*

6053 Chemzoline T-44
61791-39-7 263-171-2
Hydroxyethyl tall oil imidazoline; emulsifier for mineral, vegetable, and animal oils; emulsifier for oil-based drilling muds; corrosion inhibitor for oilfield applications. *Chemron.*

6054 Chenzinsky-Plehn's solution
A microscopic stain. It contains 0.25 gram eosin, 50 g of 70% alcohol, 100 g of a saturated solution of methylene blue, and 100 ml distilled water.

6055 Ches® 500
Nonfat drymilk, xanthan gum, propylene glycol, alginate, glyceryl stearate, sodium glyceryl oleate phosphate; food grade stabilizer; cold mix emulsifier for cosmetics and pharmaceuticals; cold hot emulsion system; unique ambient temperature emulsifier. *CasChem.*

6056 Chia oil
An oil obtained from the Mexican plant *Salvia hispanica.* The raw oil dries slowly, but the boiled oil is a good drying oil.

6057 Chian turpentine
Chio turpentine, Chios turpentine, Cyprian turpentine, Scian turpentine. Names applied to the oleo-resin obtained from the bark of *Pistachi terebinthus*, a tree in the Mediterranean and Asia Minor.

6058 Chicle
11, 2045
A gum obtained from *Achras species, Manilkara zapotilla* and others of Mexico, Belize, and Venezuela; has been used in the manufacture of chewing gum, but has now been considerably superseded by other gums, e.g., Jelutong.

6059 Chierite
A proprietary molding material of urea formaldehyde. *Butese.* Unverified.

6060 Chierol
A proprietary phenolic molding material. *Butese.* Unverified.

6061 Chilcote
Dressings for metal chills. *Foseco (F.S.) Ltd.*

6062 Childion
Emulsifiable concentrate of 166 g dicofol and 58.7 g tetradifon per liter; a contact acaricide. *Hortichem Ltd; ICI AgroChemicals.*

6063 Chili saltpeter
7631-99-4 8792 231-554-3
$NaNO_3$
Chili niter
soda salt peter; Peru saltpeter; cubic saltpeter, soda niter; cubic niter, Nitratine. Sodium nitrate, $NaNO_3$, found as deposits in Chile and Peru; used as an oxidizing agent; in solid rocket propellants, fertilizer, flux, glass manufacture, refrigerant, matches, dynamite, dyes, pharmaceuticals, anaphrodisiac.

6064 Chilisa FE®
57-55-6 8040 200-338-0
Inhibited propylene glycol. *Arco Chemical Co.*

6065 Chiltern Cropspray 11E
Adjuvant containing 99% refined mineral oil; wetting agent for herbicides. *Chiltern Farm Chemicals Ltd.*

6066 Chiltern Cyperkill 10
66841-24-5 266-492-6
An emulsifiable concentrate containing 100 g per liter of cypermethrin, a pyrethroid insecticide. *Chiltern Farm Chemicals Ltd.*

6067 Chiltern Fazor
123-33-1 5745 204-619-9
Maleic hydrazide
A plant growth regulator for grass and to reduce bud growth in trees, hedges and vegetables. *Chiltern Farm Chemicals Ltd.*

6068 Chiltern IPU
34123-59-6 5237 251-835-4
Suspension concentrate containing 500 g isoproturon per liter; used for annual weed control in cereals. *Chiltern Farm Chemicals Ltd.*

6069 Chiltern Kocide 101
20427-59-2 2709 243-815-9
Cupric hydroxide
A protectant fungicide. *Chiltern Farm Chemicals Ltd.*

6070 Chiltern Ole
1897-45-6 2219 217-588-1
Chlorothalonil
A fungicide for a wide range of agricultural crops. *Chiltern Farm Chemicals Ltd.* Name unverified.

6071 Chiltern Pyrazol
1698-60-8 216-920-2
Suspension concentrate containing 430 g chloridazon per liter; a pyridazinone herbicide for beet crops. *Chiltern Farm Chemicals Ltd.*

6072 Chimassorb® 119FL
106990-43-6 401-990-0
$C_{124}H_{234}N_{32}$
1,3,5-Triazine-2,4,6-triamine
N,N',N",N"'-Tetrakis(4,6-bis(butyl-(N-methyl-2,2,6,6-tetramethylpiperidin-4-yl)-amino)triazin-2-yl)-4,7-diazadecane-1,10-diamine. Light and thermal stabilizer for PP fiber applications in automotive, marine and residential carpets, agricultural films, fertilizer bags, thick section pigmented applications, rotational molding applications. *Ciba-Geigy/Additives.* Name unverified.

6073 Chinese bronze
Alloys of from 72.5-74% copper, 15-18.5% lead, 10-14% zinc, and 1-5% tin. One alloy, also known as Chinese bronze, contains 78% copper and 22% tin.

6074 Chinese glue
Shellac dissolved in alcohol; used for joining wood, earthenware, or glass.

6075 Chinese silver
Peru silver. A German silver, containing a little aluminum.

6076 Chinese tallow
Vegetable tallow. A waxy substance obtained from the outer coating of the fruit of *Stillingia sebifera*, of China.

6077 Chinese wax
Vegetable spermaceti; tree wax. Also wrongly called Japanese wax. Insect wax obtained from the insect *Coccus ceriferus*, or *C. pela*, which deposits wax on certain trees. Its chief ingredient is ceryl certate.

6078 Chinese white
Zinc white. The name is sometimes incorrectly applied to barium sulfate.

6079 Chinese white copper
An alloy of 40% copper, 31% nickel, 25% zinc, and 2% iron.

6080 Chinese wood oil
Wood oil; Tung oil. Tung oil obtained from the seeds of *Aleurites* species in China and Japan.

6081 Chisso-rite
A proprietary synthetic resin made by condensing formaldehyde and acid oils from low temperature coal carbonization. No manufacturer.

6082 chitin
1398-61-4 2105 215-744-3
$(C_8H_{13}NO_5)_n$
poly(N-acetyl-1,4-β-D-glucopyranosamine)
A glucosamine polysaccharide. Principal component of exoskeletons. Used in biological research, source of chitosan. *Ajinomoto USA; Amerchol; Atomergic Chemetals; Tri-K Industries.*

6083 Chloracel®
Sodium aluminum chlorhydroxy lactate
Active principle in deodorants. *Reheis Inc.*

6084 chloral iodine
A solution of chloral hydrate (50 g in 20 ml water) saturated with iodine; used for the detection of starch grains.

6085 Chloramine B
127-52-6 2117 204-847-9
$C_6H_5ClNNaO_2S$
N-Chlorobenzenesulfonamide sodium salt
Sodium benzene-sulfochloroamide; (N-Chlorobenzenesulfonamido)sodium; sodium benzensulfochloramine; Neomagnol. Used like Chloramine T. Soluble in 20 parts H_2O, 25 parts of EtOH, insoluble in organic solvents.

6086 Chloramine T
127-65-1 2118 204-854-7
$C_7H_7ClNNaO_2S$
N-Chloro-4-methylbenzenesulfonamide sodium salt
Tochlorine; Sodium *p*-toluenesulfochloramide; (N-chloro-*p*-toluenesulfonamido)sodium; tolamine; chloramine; Aktiven; Chloraseptine; Chlorazene; Chlorazone; Clorina; Euclorina; Gansil; Gyneclorina; Halamid; Mianine; Tochlorine; Tolamine; chloramine-Heyden, Pyrgos; Mianin, Aktivin,. An antiseptic used in medicine; solutions of Chloramine T are also used as detergents and bleaching agents. Soluble in H_2O, insoluble in organic solvents.

6087 chloranil
118-75-2 2121 204-274-4
$C_6Cl_4O_2$
2,3,5,6-Tetrachloro-2,5-cyclohexadiene-1,4-dione
Tetrachlorquinone. Used as an agricultural fungicide, dye intermediate, electrodes for pH measurement, reagent. mp = 288-290°; insoluble in H_2O, slightly soluble in organic solvents; LD_{50} (rat orl) = 4.0 g/kg.

6088 Chlorantine
Dyestuffs fast to light. *Clayton Aniline Co Ltd.* Name unverified.

6089 Chlorasol®
7681-52-9 8773 231-668-3
Sodium hypochlorite solution; used for cleansing and desloughing wounds. *Seton Healthcare Group plc.*

6090 Chlorazol® Dyes
A registered trademark applied to certain dyestuffs. No manufacturer.

6091 Chlorcahücit
A German explosive.

6092 Chlorcosane
63449-39-8 7156 264-150-0
chlorinated paraffin
Cereclor. A liquid chlorinated paraffin wax; the chlorine content varies from 27-50%. Thick oily liquid, d = 1.00-1.07; insoluble in H_2O, EtOH, soluble in C_6H_6, $CHCl_3$, Et_2O, and other non-polar organic solvents.

6093 Chlordispel
Dishwashing detergent. *The Wellcome Foundation Ltd.* Name unverified.

6094 Chlorea
1912-24-9 902 217-617-8
Atrazine
A residual herbicide. *Chipman Ltd.*

6095 Chloresium
2155
Chlorophyllin copper complex obtained from chlorophyll by removing the

methyl and phytyl ester groups with alkali and the magnesium with copper; deodorant. *Rystan Company Inc.* Name unverified.

6096 Chlorez® 700-DF, Chlorez® 760
53449-39-8
Resinous chlorinated paraffin; flame retardant for paints, printing inks, plastics, foams, adhesives, paper and fabric coatings. (760) flame retardant for LDPE, PP, olefins. styrenes, adhesives, wire and cable and other applications. *Dover.*

6097 Chlorez®, Chlorez® 700, Chlorez® 700DD
Chlorinated paraffin; flame retardant in plastics, rubber, coatings etc. (700) flame retardant for LDPE, in coatings, inks, plastics, foams, adhesives, paper, and fabrics. (700DD) flame retardant for use in white coatings; superior color stability. *Dover.*

6098 Chlorfenvinphos
470-90-6 2137 207-432-0
$C_{12}H_{14}Cl_3O_4P$
Phosphoric acid 2-chloro-1-(2,4-dichlorophenyl)ethenyl diethyl ester
2-Chloro-1-(2,4-dichlorophenyl)-vinyl diethyl phosphate; β-2-Chloro-1-(2',4'-dichlorophenyl)vinyl diethyl phosphate; Benzyl alcohol, 2,4-dichloro-α-(chloromethylene)-, diethyl phosphate; Birlan; Birlane; Birlane 10G; C 8949; C-10015; Clofenvinphos; chlorofenvinphos; Clofenvineosum; Clofenvinfos; CVP; SD-7859; Compd. 4072; Dermaton; Sapecron; Steladone; Supona. An insecticide and acaricide with contact and stomach action. Used in seed treatment or soil application for control of root flies, rootworms, fruit flies, Colorado beetles etc. mp = -22 - -16°; $bp_{0.05}$ = 167-170°; soluble in H_2O (145 mg/l), miscible with organic solvents; LD_{50} (rat orl) = 9.6 mg/kg.

6099 Chlorguard
Industrial control gear for effecting automatic closure of valves. *ICI Chem & Polymers Ltd.*

6100 chlorhexidine acetate
56-95-1 2140 200-302-4
$C_{22}H_{30}Cl_2N_{10}{\cdot}2C_2H_4O_2$
N,N-bis(4-chlorophenyl)-3,12-diimino-2,4,11,13-tetraazatetradecanediimidamide diacetate
Chlorasept 2000; Chlorohexidine Diacetate; Arlacide A; Bactigras; Hexamethylenebis(5-(p-chlorophenyl)biguanide) diacetate; Nolvasan; Tetraazatetradecanediimidamide, N,N''-bis(4-chlorophenyl)-3,12-diimino-, diacetate. Antiseptic and disinfectant. mp = 154-155°; soluble in H_2O (1.9 g/100 ml), alcohols; LD_{50} (mus orl) = 2 g/kg.

6101 chlorhexidine digluconate
2140
$C_{34}H_{54}Cl_2N_{10}O_{14}$
N,N'-bis (4-chlorophenyl)-3,12-diimino-2,4,11,13-tetraazatetradecane-diimidamide compound with D-gluconic acid
N,N'-bis (4-chlorophenyl)-3,12-diimino-2,4,11,13-tetraazatetradecane-diimidamide compound with D-gluconic acid; chlorhexidine gluconate; Salt of chlorhexidine and gluconic acid; Bacticlens; Corsodil; Corsodyl; Hibiclens; Hibidil; Hibiscrub; Hibitane; Orahexal; Peridex; pHisomed; Plac Out; Plurexid; Rotersept; Unisept; Hibidil; Hibicare; Hibisol; Chlorhexidine digluconate; Bacticlens; Hibiclens; Hibiscrub; Hibitane;Corsodyl; Plac Out; Peridex; pHisoMed; Plurexid; Rotersept; Unisept; Chlorhexidine gluconate; Hibistat; Orahexal; Hibitane chlorhexidine gluconate. Soluble in H_2O (>50% w/v, 25°); LD_{50} (mus orl) = 1800 mg/kg. *Degussa; Lonza AG; ICI Chemical & Polymers Ltd.*

6102 Chloric acid
7790-93-4 2141 232-233-0
HyPure C. *Olin.*

6103 chloridazon
1698-60-8 216-920-2
$C_{10}H_8ClN_3O$
5-Amino-4-chloro-2-phenyl-3(2H)-pyridazinone
5-Amino-4-chloro-2-phenylpyridazin-3(2H)-one; chloridazone; pyrazon; PAC; PCA; Better; Brek; Curbetan; Gladiator; H 119; Hyzon; Pyramin; Pyrazol; Silver; Starter; Trojan;. Selective systemic herbicide, rapidly absorbed by roots, used for control of annual broad-leaved weeds in sugar beets, fodder beet and beetroot. mp = 205-206° (dec); soluble in H_2O (400 mg/l), more soluble in organic solvents; LD_{50} (rat orl) = 3830 mg/kg. *Sipcam; Agrimont; Wacker; Tripart; BASF; Chiltern; Atlas Interlates, Truchem; Schering.*

6104 chloride of lime
7778-54-3 1717 231-908-7
$CaCl_2O_2$
Calcium hypochlorite

6105 Chlorinated paraffin
63449-39-8 7156 264-150-0
Electrofine® S-70; Chlorocosane; Chlorcosane; Chlorinated Paraffin; Parafin Waxes and hydrocarbon waxes, chlorinated; ADK CIZER E-500; Doverguard® 170,700,152,700-S; Kloro 6001; Rez-O-Sperse® 3; Rez-O-Sperse® A-1; Chlorez® 700-ss, 700, 700-DD, 700-DF; Akrochlor™ L-170, L-57, L-60, L-61, L-40.,. Flame retardant for plastics, textiles, paper and cardboard in conjunction with antimony trioxide; improves hardness, gloss and resist. to acids and bases in paints; solvent; lubricant; plasticizer for PVC in PE sealants. insoluble in H_2O; soluble in EtOH, benzene, chloroform; d = 0.90-1.50; nonflammable; LD_{50} (rat orl) > 4000 mg/kg. Elf Atochem; ACInd.; R.E. Carroll; Chemcentral; Dover; C.P. Hall: Harwick; ICI; Morton Int'l.; Occidental; Punda Mercantile; Sea-Land; StanChem; Tosoh; Tri-Iso; Witco/Polymer Addit.

6106 chlorinated rubber
9006-03-5
Elastomer to which 65% chlorine has been added to give a solid, film-forming resin; used in swimming pool, traffic, marine and masonry paints.

6107 chlorine
7782-50-5 2145 231-959-5
Cl_2
Manufacture of CCl_4, trichloroethylene, chlorinated hydrocarbons, neoprene, PVC, etc.; water purification; shrinkproofing wool; in flame retardant compounds; food processing; bleaching. mp = -101°; bp = -34.05°; soluble in H_2O. Air Prods & Chem; Asahi Chem Industry Co Ltd; Asahi Denka Kogyo; Elf Atochem; BASF; Georgia-Pacific Resins; Olin; OxyChem; PPG Industries; Showa Denko.

6108 chlorine dioxide
10049-04-4 2146 233-162-8
ClO_2
chlorine peroxide
Bleaching wood pulp, fats and oils; biocide; odor control; water purification; oxidizing agent; bactericide, antiseptic. *Drew Ind. Div.; Int'l. Dioxcide.*

6109 Chloritane
7758-19-2 8743 231-836-6
Sodium chlorite
Solvay Interox Ltd. Discontinued.

6110 Chlormequat chloride
999-81-5 2153 213-666-4
$C_5H_{13}Cl_2N$
2-Chloro-N,N,N-trimethylethanaminium chloride
choline chloride; Cycocel®; (2-chloroethyl)trimethylammonium chloride; chlorocholine chloride; choline dichloride; AC 38555; Cycocel; Cycogan; Clifton Chlormequat 46. Improves resistance against lodging of oats, rye, and wheat mp = 245° (dec); soluble in H_2O, EtOH; LD_{50} (mus orl) = 54 mg/kg. *BASF AG; Agan Chemical Manufacturers Ltd.*

6111 Chlormytol
A proprietary preparation of chloramphenicol and prednisolone; used in dermatology as an antibacterial agent. *Parke-Davis.* Name unverified.

6112 chloroacetophenone
532-27-4 2166 208-531-1
C_8H_7ClO
2-Chloro-1-phenylethanone
α-chloroacetophenone; ω-chloroacetophenone; 2-chloroacetophenone; phenacyl chloride; Chemical Mace; CN. Pharmaceutical intermediate, riot-control gas. mp = 54-56°; bp = 243-244°; d^{15} = 1.324; insoluble in H_2O; soluble in organic solvents; LD_{50} (rat orl) = 127 mg/kg. *Janssen Chimica; Penta Mfg.; Schweizerhall.*

6113 *m*-chloroaniline
108-42-9 2169 203-581-0
C_6H_6ClN
m-aminochlorobenzene
3-Chlorobenzenamine. Intermediate for azo dyes and pigments, pharmaceuticals, insecticides, agricultural chemicals. mp = -10.4°; bp = 230.5°; d_4^{22} = 1.2150; n_D^{20} = 1.5931; insoluble in H_2O, soluble in organic solvents. *Du Pont; Janssen Chimica; Schweizerhall.*

6114 *o*-chloroaniline
95-51-2 2169 202-426-4
C_6H_6ClN
o-aminochlorobenzene
2-Chlorobenzenamine. Dye intermediate, standards for colorimetric apparatus, manufacture of petroleum solvents and fungicides. mp = -1.94°; bp_{11} = 95-97°; d = 1.213; n_D^{20} = 1.5880; insoluble in H_2O, soluble in organic solvents. *Du Pont.*

6115 *p*-chloroaniline
106-47-8 2169 203-401-0
C_6H_6ClN
p-aminochlorobenzene
4-Chlorobenzenamine. Dye intermediate, pharmaceuticals, agricultural chemicals. mp = 68-71°; bp_{11}= 113-114°, bp = 232°; d_4^{77}= 1.169; soluble in hot H_2O, organic solvents; LD_{50} (rat orl) = 0.31 g/kg. *Du Pont; Janssen Chimica; Mitsui Toatsu.*

6116 *p*-chloroaniline
106-47-8 2169 203-401-0
$ClC_6H_4NH_2$

p-aminochlorobenzene
Dye intermediate, pharmaceuticals, agricultural chemicals. *Du Pont; Janssen Chimica; Mitsui Toatsu.* Discontinued.

6117 Chloroben
95-50-1 3106 202-425-9
o-Dichlorobenzene
Used in sewage treatment.

6118 chlorobenzaldehyde
35913-09-8
C_7H_5ClO
Intermediate for triphenyl methane and related dyes, organic intermediate. *Hoechst Celanese; Janssen Chimica; Penta Mfg.; Rit-Chem.*

6119 chlorobenzene
108-90-7 2172 203-628-5
C_6H_5Cl
monochlorobenzene
phenyl chloride; benzene chloride. Pesticide intermediate; manufacture of phenol, aniline, DDT; solvent carrier for methylene diisocyanate; solvent for paints; heat transfer medium. mp = -45°; bp = 131-132°; d_4^{20}= 1.107; n_D^{20}= 1.5248; insoluble in H_2O, soluble in organic solvents; LD_{50} (rat orl) = 1110 mg/kg. *Aldrich; Elf Atochem SA; Janssen Chimica; Monsanto; PPG Industries.*

6120 chlorobromhydrin
C_3H_6BrClO
α-Chloro-α-bromo isopropyl alcohol
Chemical intermediate.

6121 chlorobromoform
124-48-1 2186 204-704-0
$CHClBr_2$
Chlorodibromo methane
dibromochloromethane; DBCM; NCI-C55254. Used in organic synthesis. bp = 115-118°; d = 2.420; n_D^{20}= 1.5465; LD_{50} (rat orl) = 370 mg/kg.

6122 2-chloroethylmethyldichlorosilane
7787-85-1 232-134-2
$C_3H_7Cl_3Si$
2-Chloroethylmethyldichlorosilane
CC3005. Coupling agent, chemical intermediate, blocking agent, release agent, lubricant, primer, reducing agent. bp_{744} = 157°; n_D^{20}= 1.4580. *Hüls Am.*

6123 Chlorofin 42
Chlorinated paraffin extender for vinyl plastics containing 40% chlorine. *Hercules.* Discontinued.

6124 Chloroflo® 40, 42
Chlorinated paraffin; lubricant additive. *Dover.*

6125 chloroform
67-66-3 2193 200-663-8
$CHCl_3$
trichloromethane
Fluorocarbon plastics, solvent, analytical chemistry, fumigant, insecticides. mp = -63.5°; bp = 61-62°; d_{20}^{20} = 1.484; n_D^{20}= 1.4476; LD_{50} (rat orl 14 day) = 0.9 - 2.18 ml/kg *Elf Atochem N. Am.; Hüls AG; Mallinckrodt; Mitsui Toatsu Chemical; OxyChem.*

6126 (3-Chloro-2-hydroxy)o-propyltrimethyl ammonium chloride
3327-22-8 222-048-3
$C_6H_{15}Cl_2NO$
CHPTA 65%. Cationizing reagent for natural and synthetic polymers; starch modifier for textiles. mp = 190-193°; d = 1.154; n_D^{20}= 1.4541. *Chem-Y GmbH.*

6127 3-Chloro-2-hydroxypropyl trimethylammonium chloride (50% aq solution)
3327-22-8 222-048-3
Catiomaster-C. Quaternary cationic agent for starch and cellulose. d = 1.154; n_D^{20}= 1.4541. *Yokkaichi Chemical Co Ltd.* Name unverified.

6128 chloromethyldichloromethylsilane
1558-33-4 216-319-5
$C_2H_5Cl_3Si$
Chloromethylmethyldichlorosilane
Chloromethyldichloromethylsilane; (Chloromethyl)methyldichlorosilane; CC3275. Coupling agent, chemical intermediate, blocking agent, release agent, lubricant, primer, reducing agent. bp = 121-122°; d = 1.286; n_D^{20} = 1.4490. *Hüls Am.*

6129 chloromethyldimethylchlorosilane
1719-57-9 217-006-6
$C_3H_8Cl_2Si$
Chloromethyldimethylchlorosilane
(Chloromethyl)dimethylsilane; chloro(chloromethyl)dimethylsilane; CC3270. Coupling agent, chemical intermediate, blocking agent, release agent, lubricant, primer, reducing agent. bp_{752} = 114°; d = 1.0860; n_D^{20}= 1.4373. *Hüls Am.*

6130 chloromethyltrimethyl silane
2344-80-1 219-058-5
$C_4H_{11}ClSi$
Chloromethyltrimethyl silane
CC3285. Coupling agent, chemical intermediate, blocking agent, release agent, lubricant, primer, reducing agent. Reagent for the Peterson olefination and the homologation of ketones and aldehydes via α, β-epoxysilanes. bp = 98-99°; d = 0.8790; n_D^{20}= 1.4175. *Hüls Am.; Janssen Chimica.*

6131 Chloroneb 65W Fungicide
2675-77-6 220-222-3
$C_8H_8Cl_2O_2$
Benzene, 1,4-dichloro-2,5-dimethoxy-
65% Chloroneb wettable powder; seed treatment to suppress seeding blights, soreshin and pre- and post-emergence damp-off caused by *rhizoctonia solani, pythium spp* and *sclerotium* rolfsii on cotton, beans, soybeans and sugar beets. *Kincaid Enterprises Inc.*

6132 Chloroneb Systemic Flowable Fungicide
2675-77-6 220-222-3
$C_8H_8Cl_2O_2$
Benzene, 1,4-dichloro-2,5-dimethoxy-
30% Chloroneb; seed treatment for control of *rhizoctonia solani, pythium spp.* and *sclerotium rolfsii* on cotton, beans, soybeans and sugar beets. *Kincaid Enterprises Inc.*

6133 Chlorophacinone
3691-35-8 2204 223-003-0
$C_{23}H_{15}ClO_3$
2-[(4-Chlorophenyl)phenyl-acetyl]-1H-indene-1,3(2H)-dione
Drat; LM-91; Caid; Liphadione; Quick; RaviacRozol,. An oil formulation containing 2.5 g/liter chlorophacinone, a powerful anticoagulant rodenticide; controls black rats, brown rats, house mice, long-tailed field mice, voles and musk rats. mp = 140°; sparingly soluble in H_2O, soluble in organic solvents; LD_{50} (rat,orl) = 20.5 mg/kg. *Rhône-Poulenc Environmental Prods. Ltd.*

6134 chlorophyll
1406-65-1 2207 215-800-7
Chlorophyll
Leaf green; chromule. The green pgment of plants, leaves and stalks; it is one of two magnesium compounds, chlorophyll a ($C_{55}H_{72}MgN_4O_5$) and chlorophyll b ($C_{55}H_{70}MgN_4O_6$); used as a colorant for soaps, oils, fats, waxes, liquors, confectionery, preserves, chlorophyll a: λ_m = 660, 613, 577, 531, 498, 429, 409 nm; chlorophyll b: λ_m = 642, 593, 565, 545, 453, 427 nm. *Atomergic Chemetals; Biochim Srl; Penta Mfg.; Spectrum Chemical Mfg.*

6135 chloropicrin
76-06-2 2208 200-930-9
CCl_3NO_2
trichloronitromethane
nitrochloroform, acquinite; Larvacide 100; Picfume. Used in organic synthesis, dyestuffs, fumigants, insecticides, rat extermination, tear gas. mp = - 64°; bp_{757} = 112°; d_4^{20}= 1.6558; n_D^{20}= 1.4611; insoluble in H_2O, soluble in organic solvents.

6136 chloroprene
126-99-8 204-818-0
C_4H_5Cl
1,3-chloro-2-butadiene
Chlorobutadiene; β-chlorobutadiene; β-Chloroprene; α-Chloroprene; Neoprene; 2-Chloroprene; 2-chlorobutadiene; 2-chlorobuta-1,3-diene; 2-Chloro-1,3-butadiene. Used as a protective coating and in synthetic rubber manufacture by polymerization. The polymerized product bears the name of Neoprene. mp = -130°; bp = 59°; almost insoluble in H_2O (211 mg/100 ml); d_{22}^{20} = 0.9514; LD_{50} (rat orl) = 450 mg/kg.

6137 3-chloropropylmethyldimethoxysilane
18171-19-2 242-056-0
$C_6H_{15}ClO_2Si$
3-Chloropropylmethyldimethoxysilane
CC3290. Coupling agent, chemical intermediate, blocking agent, release agent, lubricant, primer, reducing agent. *Hüls Am.*

6138 chloropropyltrimethoxysilane
2530-87-2 219-787-9
$C_6H_{21}ClO_3Si$
3-Chloropropyltrimethoxysilane
CC3300; Dow Corning® Z-6076; Dynasylan® CPTMO; Petrarch® C3300; CPTMO. Coupling agent, chemical intermediate, blocking agent, release agent, lubricant, primer, reducing agent. bp = 195-196°; n_D^{20}= 1.4196; d = 1.081; flash point = 78°; irritating to eyes, respiratory sytem, skin. *Dow Corning; Howard Hall; Hüls Am.; Osi Spec; PCR.*

6139 chloropropyltrichlorosilane
2550-06-3 219-844-8
$C_3H_6Cl_4Si$
CC3291. Coupling agent, chemical intermediate, blocking agent, release

agent, lubricant, primer, reducing agent. bp = 181-183°; d = 1.350; n_D^{20} = 1.4666. *Hüls Am.*

6140 3-chloropropyltriethoxysilane
5089-70-3 225-805-6
$C_9H_{21}O_3Si$
CC3292. Coupling agent, chemical intermediate, blocking agent, release agent, lubricant, primer, reducing agent. *Hüls Am.*

6141 3-chloropropyltriethoxysilane
5089-70-3 225-805-6
$C_9H_{21}ClO_3Si$
(3-Chloropropyl)triethoxysilane
Coupling agent, release agent, lubricant, blocking agent, chemical intermediate. *Hüls Am.; Union Carbide.*

6142 3-chloropropyltrimethoxysilane
2530-87-2 219-787-9
$C_6H_{15}O_3Si$
CC3300. Coupling agent, chemical intermediate, blocking agent, release agent, lubricant, primer, reducing agent. *Hüls Am.*

6143 3-chloropropyltrimethoxysilane
2530-87-2 219-787-9
$C_6H_{15}ClO_3Si$
3-(chloropropyl)trimethoxysilane
Coupling agent for epoxies, nylons, urethanes. bp_{750} = 195°; d = 1.077; n_D^{20} = 1.4190. *Dow Corning; Hüls Am.; PCR; Union Carbide.*

6144 2-chloropyridine
109-09-1 203-646-3
C_5H_4ClN
2-Chloropyridine
Intermediate, used in manufacture of antihistamines, germicides, pesticides and agricultural chemicals. mp = -46°; bp = 168-170°; d = 1.2100; n_D^{20} = 1.5320; soluble in H_2O (27 g/l), organic solvents; LD_{50} (mus orl) = 110 mg/kg. *Expansia SA; Olin; Penta Mfg.; Schweizerhall.*

6145 4-chlororesorcinol
95-88-5 202-462-0
$C_6H_5ClO_2$
4-Chloro-1,3-dihydroxybenzene
4-chloro-1,3-benzenediol; 1,3-benzenediol, 4-chloro. Halogenated phenol, used in synthesis. mp = 106.5-108°; bp_{18} = 147°. *Rit-Chem.*

6146 Chloros
7681-52-9 8773 231-668-3
sodium hypochlorite
Disinfectant containing sodium hypochlorite. *ICI Chem & Polymers Ltd.*

6147 Chlorosoda
7681-52-9 8773 231-668-3
sodium hypochlorite
A proprietary form of solidified sodium hypochlorite for use as a bleaching agent; a small proportion of a saturated fatty acid, such as lauric acid, is incorporated. No manufacturer.

6148 N-chlorosuccinimide
128-09-6 2217 204-878-8
$C_4H_4ClNO_2$
1-Chloro-2,5-pyrrolidinedione
2,5-Pyrrolidinedione, 1-chloro-; NCS; Chlorosuccinimide; Pyrrolidinedione, 1-chloro-; Succinchlorimide. Chlorinating agent, disinfectant for swimming pools, bactericide. mp 150-151°; soluble in H_2O (1.4 g/100ml), slightly soluble in organic solvents; MLD (rat orl) = 2.7 g/kg. *Janssen Chimica; Penta Mfg.; Schweizerhall.*

6149 Chlorotex
Reagent for estimating residual chlorine in water. *BDH Chemicals Ltd.* Name unverified.

6150 Chlorothalonil
1897-45-6 2219 217-588-1
$C_8Cl_4N_2$
Chiltern Ole; Forturf; Bravo; Exotherm; *m*-TCPN; Sweep; TCIN; Termil; TPN; Daconil;DAC-2787; Daconil 2787; Exotherm Termil; Chlorothanonil; Bombardier; Farber; Jupital; Ole; Pillarich; Repulse; Taloberg; Tuffcide; Black Leaf Lawn & Garden Fungicide; Bonide; ClortoCaffaro; Clortosip; Dragon Daconil 2787; Ferti-lome; Green Charm Multi-Purpose Fungicide; Green Thumb Lawn & Garden Fungicide; Ortho Multi-Purpose Fungicide Daconil 2787; Pennington's Pride Multi-Purpose Fungicide; Pro-Care Multi-Purpose Fungicide; Rigo's Best Lawn & Garden Fungicide; SA Lawn Ornamental & Vegetable Fungicide; Security Fungi-Gard; Bravo-W-75; Dacobre; Echo 75; Vanox; Evade; BB Chlorothalonil,. A fungicide for a wide range of agricultural crops. mp = 250°; bp = 350°; d = 1.800; insoluble in H_2O (0.6 g/l), more soluble in organic solvents; LD_{50} (rat orl) = 10 gm/kg. *Chiltern Farm Chemicals Ltd; SNIA (UK) Ltd; Fermenta ASC Europe Ltd; Brown Butlin Ltd.*

6151 Chlorothene
71-55-6 9766 200-756-3
$C_2H_3Cl_3$
1,1,1-Trichloroethane
Methyltrichloromethane; Methyl Chloroform; Chlorothene NU; Chlorothene VG; α-trichloroethane; 1,1,1-TCE; Aerothene TT; solvent 111; tri-ethane. A line of inhibited 1,1,1-trichloroethane-based solvents; used in industry. mp = -33°; bp = 74°; d = 1.3376; soluble in H_2O (1.5 g/l); more soluble in organic solvents; LD_{50} (rat orl) = 10300 mg/kg; LC_{50} (rat ihl) = 18000 ppm/4H. *Dow UK.* Discontinued.

6152 2-chlorothiophene
96-43-5 202-505-3
C_4H_3ClS
2-thienyl chloride
Used in organic synthesis. bp = 127-129°; d = 1.285; n_D^{20} = 1.5483. *Janssen Chimica.*

6153 chlorotrfluoromethane
75-72-9 11, 2140 200-894-4
$CClF_3$
Chlorotrifluromethane
trifluoromethyl chloride; trifluorochloromethane; CFC-13; Halocarbon 13; R-13; Trifluoromethyl chloride; CFC; Arcton; Freon; Frigen; Genetron. Refrigerant, dielectric and aerospace chemical, hardening of metals, pharmaceutical processing. As aerosol propellant. Low temperature refrigerant used in low stage of cascade systems to provide evaporator temperatures in the range of - 100°F. mp = -181°; bp = -80°. *Elf Atochem N. Am.; PCR.*

6154 Chlorovis 150A
Chlorinated paraffin
Lubricant additive. *Dover.*

6155 Chlorowax
Chlorinated paraffins. *Occidental Chemical Corp.*

6156 Chlorowax 40, 50, 70, LV
Chlorinated paraffin (50 - 50% chlorine); a vinyl plasticizer. *U.S. Industrial Chem.* Name unverified.

6157 chloroxethose
C_4Cl_6O
Hexachlorodivinyl ether

6158 chloroxuron
1982-47-4 217-843-7
$C_{15}H_{15}ClN_2O_2$
3-[*p*-(*p*-chlorophenoxy)phenyl]-1,1-dimethylurea
chloroxyfenidim; chlorphencarb; C 1983; Gesamoos; Tenoran. Selective herbicide, inhibits photosynthesis. Used for pre- and post-emergence control of annual broad-leaved weeds and some grasses in peas, beans, carrots, celery, onions, leeks, garlic, chives, fennel, parsley, dill, tomatoes, cucurbits, soya beans mp = 151-152°; soluble in H_2O (3.7 mg/l), more soluble in organic solvents; LD_{50} (rat orl) = 3700 mg/kg.

6159 chloroxylenol
88-04-0 2228 201-793-8
C_8H_9ClO
4-chloro-3,5-dimethylphenol
4-Chloro-3-xylenol; 2-Chloro-*m*-xylenol; 2-Chloro-5-hydroxy-*m*-xylene; 2-Chloro-5-hydroxy-1,3-dimethylbenzene; 4-Chloro-3,5-dimethylphenol; Benzytol; Chloroxylenol; Dettol; Husept extra; Nipacide PX; Ottasept; *Para*-chloro-meta-xylenol; PCMX; Chloro-3,5-dimethylphenol; Chloro-3,5-xylenol; Chloro-5-hydroxy-*m*-xylene; Chloro-*m*-xylenol; Desson; Dimethyl-4-chlorophenol; Espadol; Septiderm-hydrochloride; Xylenol, 4-chloro-. Antiseptic and germicide. Used as a mildewcide. mp = 114-116°; bp = 246°; soluble in organic solvents; LD_{50} (rat orl) = 3830 mg/kg.

6160 Chlorozone
127-65-1 2118 204-854-7
Chloramine-T
A bleaching liquor prepared by passing chlorine into caustic soda.

6161 Chlorphenesin
104-29-0 2230 203-192-6
$C_9H_{11}ClO_3$
3-(4-Chlorophenoxy)-1,2-propanediol
Gechophen; *p*-chlorophenyl-α-glyceryl ether; Adermykon; Mycil. A trade name for chlorphenesin B.P. An Antifungal. mp= 77-79°; poorly soluble in H_2O (<1 g/100 ml). *British Drug Houses.* Name unverified.

6162 Chlorpropam
101-21-3 2240 202-925-7
$C_{10}H_{12}ClNO_2$
Carbamic acid, (3-chlorophenyl)-, 1-methylethyl ester
Carbanilic acid, *m*-chloro- isopropyl ester; (3-chlorophenyl)carbamic acid, 1-methylethyl ester; Beet-Kleen; Chloro-IPC; Chloro-IFK; Chloro IPC;

Chloropropham; Chlorpropham; Chlorprophame; ChlorIPC; Cl-IPC; Elbanil; ENT 18,060; Fasco WY-HOE; Furloe; Furloe 3 EC; isopropyl chlorocarbanilate; IPC, CIPC; Croptex Chrome; Campbell's CIPC 40%. Selective systemic herbicide and growth regulator. Used for pre-emergence control of annual grasses and broad-leaved weeds. mp = 40.7-41.1°; bp_2 = 149°; d_{30}= 1.180; soluble in H_2O (89 mg/l), mores soluble in organic solvents; LD_{50} (rat orl) = 1200 mg/kg. *Hortichem Ltd.; Pfizer International.*

6163 Chlorpyrifos

2921-88-2 2242 220-864-4

$C_9H_{11}Cl_3NO_3PS$

Phosphorothioic acid, O,O-diethyl O-(3,5,6-trichloro-2-pyridinyl) ester

Phosphorothioic acid, O,O-diethyl O-(3,5,6-trichloro-2-pyridyl) ester; Brodan; Chloropyrifos; Chloropyriphos; Chloropyripos; Chloroyriphos; Chlorpyrofos; Chlorpyrophos; Crossfire; Detmol U.A.; Dowco 179; Dursban; Dursban F; Dursban 4E; DOWCO; Eradex; ENT 27,311; Empire; Equity; Killmaster; Lentrek; Lock-On; Lorsban; Loxiran; OMS 971; Pageant; Piridane; Pyrinex; Silrifos; Spannit; Stipend; Talon; Trichlorpyrphos; Zidil. Non-systemic insecticide. Crossfire is an emulsifiable concentrate containing 228g chlorpyrifos per liter; an organophosphorus insecticide usd for control of soil insects. mp = 42-43.5°; almost insoluble in H_2O (2 ppm), readily soluble in organic solvents, LD_{50} (rat orl) = 82 mg/kg. *Rhône-Poulenc Crop Protection Ltd; Planters Products; Frowein; Neudorff; Diachem; Makhteshim-Agan; Siapa; Pan Britannica; FCC.*

6164 Chlorpyrifos-methyl

5598-13-0 2242 227-011-5

$C_7H_7Cl_3NO_3PS$

Chlorpyrifos-methyl

Phosphorothioic acid, O,O-dimethyl O-(3,5,6-trichloro-2-pyridinyl) ester; chlormethylfos; Chloropyrifos-methyl; Dowco 214; ENT 27520; Fospirate; M 3196; methyl chlorpyrifos; methyl dursban; Noltran; OMS 1155; Reldan; Reldan F; Trichlormethylfos; Zertell; Cooper Graincote. An organophosphate insecticide for the treatment of pests in stored grain and oilseed rape. mp = 42-43.5°; soluble in H_2O (4 mg/l), more soluble im organic solvents; LD_{50} (rat orl) = 1630-2140 mg/kg *The Wellcome Foundation Ltd.*

6165 Chlor-Tabs

Effervescent chlorine tablets. *PPF International Ltd.* Name unverified.

6166 Chlumin

An aluminum alloy resistant to sea water. it contains chromium and small percentages of magnesium and iron.

6167 Cholebrine

16034-77-8 5030 240-173-1

$C_{12}H_{13}I_3N_2O_3$

Iocetamic acid

Diagnostic aid. *Mallinckrodt Inc.*

6168 cholesterol

57-88-5 2256 200-353-2

$C_{27}H_{46}O$

Cholest-5-en-3-β-ol

cholesterin. Found in all body tissues; used as an emulsifying agent in cosmetics and pharmaceutical products; source of estradiol. mp = 148.5°; bp = 360° (dec); d_4^{18} = 1.052; $[\alpha]_D^{20}$= -31.5° (c = 2, Et_2O); insoluble in H_2O, soluble in organic solvents. *Croda; EM Industries; Schweizerhall; Solvay Duphar BV; U.S. BioChemical; Croda Chemical Ltd.*

6169 Cholestrophane

$C_5H_6N_2O_3$

Dimethylparabanic acid

Preservative.

6170 Choletec

78266-06-5 5812 278-877-6

Mebrofenin

Diagnostic aid. *Bristol-Myers Squibb Co Inc.*

6171 choline chloride

67-48-1 2261 200-655-4

$C_5H_{14}ClNO$

2-Hydroxy-N,N,N-trimethylethanaminium chloride

choline hydrochloride; 2-(hydroxyethyl)trimethylammonium chloride; Biocolina; Hepacholine; Lipotril. Animal feed additive. Freely soluble in H_2O, EtOH; LD_{50} (rat orl) = 6.64 g/kg. *Am. Biorganics; Mitsubishi Gas Chemical; Penta Mfg.; Tanabe USA; UCB SA.*

6172 Cholografin Meglumine

3521-84-4 5043 222-534-5

Iodipamide meglumine

Bis[N-methylglucamine] salt of Cholografin. More soluble in H_2O; used as a 20% solution. Diagnostic aid. *Bristol-Myers Squibb Co Inc.*

6173 CHP

6837-24-7 229-919-7

$C_{10}H_{17}NO$

N-Cyclohexyl-2-pyrrolidone

cyclohexyl-2-pyrrolidone; 2-Pyrrolidinone, 1-cyclohexyl-; 1-Cyclohexyl-2-pyrrolidone. Solvent, reaction intermediate, textile auxiliary, cosmetic ingredient. bp_7 = 154°; d = 1.0070; n_D^{20}= 1.4975; LD_{50} (rat orl) = 370 mg/kg. *ISP.*

6174 CHP-5

80-15-9 201-254-7

α,α-Dimethylbenzyl hydroperoxide

Cumene hydroperoxide solution; initiator for room temperature curing of vinyl ester of other unsaturated polyester resins; especially useful used with vinyl ester resins that are cobalt promoted. *Witco Corporation.*

6175 CHP-158

80-15-9 201-254-7

$C_9H_{12}O_2$

cumene hydroperoxide

Cumyl Hydroperoxide; Hyperiz; Trigorox K 80; α,α-Dimethylbenzylhydroperoxide; 1-methyl-1-phenylethyl hydroperoxide; cumenyl hydroperoxide; α-cumyl hydroperoxide; α-cumene hydroperoxide; isopropylbenzene hydroperoxide; Percumyl H; Trigonox K 80. 80% solution of cumene hydroperoxide; initiator for vinyl monomers and copolymers and the crosslinking of unsat. polyester resins. *Witco Corporation.*

6176 CHPTA 65%

3327-22-8 222-048-3

$C_6H_{15}Cl_2NO$

3-Chloro-2-hydroxypropyltrimethyl ammonium chloride

Cationizing reagent for natural and synthetic polymers; starch modifier for textiles. *Chem-Y GmbH.*

6177 Christolit

A proprietary plastic of the phenol-formaldehyde type. No manufacturer.

6178 Chrogo U42

An alloy of 40% gold, 45% copper, 14% nickel, 1% chromium, and traces of platinum. A dental alloy. No manufacturer.

6179 Chroma-Cal®, Chroma-Chem

Colorant dispersions; for coloring of coating compositions. *Hüls Am.*

6180 Chromaflo®

Pourable dispersions for high pigment loading, controlled viscosity; for pumping and metering. *Plasticolors.*

6181 Chromagan

Nickel-chromium steel for the manufacture of tableware.

6182 Chromagel

Chemical products for use in chromatography and chromtographic systems. *Courtaulds plc.* Discontinued.

6183 Chromagen

Ferrous fumarate USP, ascorbic acid USP, cyanocobalamin USP, desicated stomach substance; phosphorous-free vitamin and mineral dietary supplement; indicated for the treatment of all anemias responsive to oral iron therapy; *Altana Inc.* Unverified.

6184 Chromaguard

Plastics materials in the form of films, sheets, strips or labels incorporating a photochromic image or pattern. *Courtaulds plc.* Discontinued.

6185 Chromalay

Materials for thin-layer chromatography. *May & Baker Ltd.* Name unverified.

6186 chromaline

A chrome mordant made by reducing chromic acid with glycerin; used for printing chrome colors on wool.

6187 Chroma-Lite®

Bonded combinations of colored pigments and bismuth oxychloride on mica; low-dusting powders imparting a subdued satiny luster to pressed powder products. *Van Dyk.*

6188 chromaloy

Nickel-chromium-iron alloys; the specific gravity varies from 8.15-8.35, and the melting point from 1360-1390°.

6189 chromaluminum

An alloy of aluminum, chromium, and other metals. It has a specific gravity of 2.9.

6190 Chroman B

A Rohn alloy containing 64% nickel, 20% iron, 15% chromium, and 1% manganese.

6191 Chroman Co

A Rohn alloy containing 79% nickel, 20% chromium, and 1% manganese.

6192 Chromargans

Nonoxidizing steels, contain chromium; non-magnetic; used for the manufacture of turbine blades, valves, cutlery, etc.

6193 Chromaset DF-100

Cotton fixative for indigo and direct dyes. *Henkel/Textiles.*

6194 Chromasist 1487A

9084-06-4

Sodium naphthalene sulfonate
Dispersing and leveling agent for disperse dyes; especially for high-temperature dyeable polyester. *Henkel/Textiles.*

6195 Chromastral
Pigments for paints and plastics. *ICI Chem & Polymers Ltd.*

6196 chromatized gelatin
Chrome cement; chrome glue. Made by adding 1 part potassium dichromate to 5 parts of a solution (5-10%) of gelatin; a cement for glass.

6197 Chromatogram
Materials for thin layer chromatography. *Kodak Ltd.* Name unverified.

6198 chromax
An electrical resistance alloy containing 75% nickel and 25% chromium.

6199 chromax bronze
An alloy of 15% nickel, 67% copper, 12% zinc, 3% aluminum, and 3% chromium.

6200 Chrombral
Flux for copper/chromium alloys. *Foseco (F.S.) Ltd.*

6201 chrome
acid alizarin
acid anthracene; diamond salicine. Acid chrome colors for wool.

6202 chrome amalgam
An alloy of chromium and mercury obtained by the electrolysis of chromic chloride, using a mercury cathode.

6203 chrome black
12018-10-9 2701 234-634-6
$Cr_2Cu_2O_5$
cupric chromite(III)
Anhydrous cupric chromite. Hydrogenation catalyst.

6204 chrome bronze
1333-82-0 2293 215-607-8
Chromium trioxide
chromic oxide. Crystalline chromium oxide obtained by heating potassium bichromate with sodium chloride or in a stream of hydrogen. It has a metallic sheen, a specific gravity of 5.61, and cuts glass.

6205 Chrome Fast Cyanine B, BN
Dyestuffs; British brands of Palatime chrome blue. No manufacturer.

6206 Chrome Green 106
2283
Chromium hydroxide [1308-38-9]-bismuth oxychloride [7787-59-9]
Inorganic colorant. *Presperse.* Discontinued.

6207 chrome iron
Ferrochrome
An alloy of iron and chromium, usually containing from 62-68% chromium; used in the manufacture of chrome steel.

6208 chrome prune
A mordant dyestuff. It gives claret shades with chrome mordants.

6209 chrome red
18454-12-1 5423 242-339-9
$CrPb_2O_5$
Lead chromate(VI) oxide
Austrian cinnabar; Chinese red; Persian red; Victoria red; Vienna red; derby red; chrome cinnabar; chrome orange; American vermilion; chrome garnet; chrome ruby; chrome carmine, chinese scarlet. Pigments consisting of basic lead chromate,.

6210 chrome steels
These steels represent a range of alloys containing from 0.2-2% carbon and from 0.5-15% chromium. Ball-bearing steel contains 1% carbon and 1% chromium, chrome die steel contains 2% C and 12% Cr.

6211 chrome steels, high
These usually contain from 64-81% iron and 18-35% chromium.

6212 Chrome yellow
7758-97-6 5422 231-846-0
$PbCrO_4$
Lead chromate(VI)
Colgne yellow; King's yellow; Leipzig yellow; Paris yellow; C.I. Pigment Yellow 34; C.I. 77600. Is known as chrome green. It dyes chromed wool green, also used in cotton printing. The name has also been used for various other pigments. mp = 844°; d = 6.3; generally insoluble; LD_{75} (gpg ip) = 156 mg/kg.

6213 Chromeduol
Concentrated chrome sulfate powders; used for tanning, mineral dyeing and electrolytic purposes. *Lancashire Chemical Works Ltd.*

6214 chromels
Nickel/chromium or nickel/chromium/iron alloys. They are used for heating elements, and as molds for glass.

6215 chrome-nickel steel, high
These alloys usually contain 70-75% iron, 17-20% chromium, and 8-10% nickel.

6216 Chrometan
Chrome tanning powders. *British Chrome & Chemicals Ltd.*

6217 Chrome-tin Pink
A pigment obtained by calcining a mixture of stannic oxide and a small amount of chromic oxide.

6218 Chrometrace
10025-73-7 2278 233-038-3
$Cl_3Cr \cdot 6H_2O$
Chromic chloride hexahydrate
dichlorotetraaquochromium chloride dihydrate; Chromic chloride; Chromium (III) chloride; Chromium chloride; Chromium chloride (3). The hexahydrate [10025-73-7] is used as a dietary supplement. mp = 83°; d = 1.760; soluble in H_2O. *Armour Pharmaceutical Co.* Name unverified.

6219 Chromglaserite
potassium sodium chromate
A double salt, $3K_2CrO_4 \cdot Na_2CrO_4$.

6220 chromic acid
7738-94-5 11, 2238 231-801-5
CrO_3
chromium trioxide
chromic anhydride. Chemicals (chromates, oxidizing agents, catalysts), medicine, process engraving, anodizing, ceramic glazes, colored glass, metal cleaning, inks, tanning, paints, textile mordant, etchant for plastics. *Aldrich; Elf Atochem N. Am.; Britsh Chrome & Chemical; OxyChem; Rit-Chem; Spectrum Chemical Mfg.*

6221 chromic chloride
10025-73-7 2278 233-038-3
$CrCl_3$
chromium chloride(III) anhydrous
chromium trichloride; chromic chloride; chromium sesquichloride. Chromium salts, intermediates, textile mordant, chromium plating, preparation of sponge chromium, catalyst for polymerizing olefins, waterproofing. mp = 1152°; d^{25} = 2.87; MLD (mus, iv) = 801 mg/kg. *Atomergic Chemetals; Cerac; Hoechst Celanese.*

6222 chromic fluoride
7788-97-8 2279 232-137-9
CrF_3
chromium(III) fluoride
Chromium trifluoride; Chromium(III) fluoride tetrahydrate. Used in printing, dyeing woolens, coloring marble, treating silk and polishing metals. mp = 1100°; d = 3.8; insoluble in H_2O, EtOH.

6223 chromic oxide
1308-38-9 2283 215-160-9
Cr_2O_3
chromium (III) oxide
chromia; Chrome Green 106; Chromium hydroxide; green cinnabar; chromium sesquioxide; Chrome oxide green; Chromium (III) oxide; Chromium Sesquioxide; Ultramarine Green; Chrome Green; Chrome oxide; Chromic Oxide Sesquioxide; Chromic oxide pigment; Chromium oxide green pigments; Chromium oxide (Cr_2O_3); dichromium trioxide; Chromium (III) oxide hydrate. Metallurgy, green paint pigment, ceramics, catalyst in organic synthesis, green granules in asphalt roofing, component of refractory brick, abrasive. mp = 2435°; bp = 3000°; d^{25} = 5.22 *Atomergic Chemetals; British Chrome & Chemical; Noah Chemical.*

6224 chromic potassium sulfate
10141-00-1 2286 233-401-6
$CrKH_{12}O_{14}S_2{>}H_2O$
Chromic potassium sulfate dodecahydrate
Potassium chromium alum dodecahydrate; potassium chromic sulfate dodecahydrate; potassium disulfato chromate(III) dodecahydrate; chrome alum. Mordant for dyeing fabrics uniformly mp = 89°; d^{25} = 1.83; soluble in 4 parts H_2O.

6225 chromidium
A nickel-chromium cast iron.

6226 Chromiform
A reagent used for the preservation of milk samples. It consists of pastilles containing 0.25 g potassium dichromate and 0.25 g trioxymethylene.

6227 Chromitan®
Basic chromium salts; for tanning and retanning leathers and furs. *BASF AG.*

6228 chromium
7440-47-3 2288 231-157-5
Cr
Metallic element; Cr; alloying and plating element for corrosion resistance, stainless steels, protective coatings, nuclear and high-temperature research,

constituent of inorganic pigments. mp = 1903°; bp = 2642°; d^{20} = 7.14. *Aldrich; Atomergic Chemetals; Cerac; Noah Chemical.*

6229 chromium acetate (ic)
1066-30-4 2275 213-909-4
$C_6H_9CrO_6$
chromic acetate
Chromium Acetate; Chromium Triacetate; Acetic Acid, Chromium (3+) Salt; Chromium (III) acetate; Chromium (III) acetate n-hydrate; Chromium (III) acetate, basic. Textile mordant, tanning, polymerization and oxidation catalyst, emulsion hardener. *Atomergic Chemetals; Noah Chemical; Spectrum Chemical Mfg.*

6230 chromium bronze
The term applied to copper-zinc or copper-tin alloys to which chromium has been added up to 5%.

6231 chromium copper
An alloy of copper and chromium, containing 10% chromium; used in the manufacture of hard steels. Also added to increase elasticity.

6232 chromium green
Guignet's green; chrome emerald green; Mittler's green; permanent green; emerald green; Veridian; chrome green; French Veronese green. Green pigments consisting of hydrated sesquioxide of chromium, mixed with phosphate or borate of chromium.

6233 chromium manganese
An alloy containing 30% chromium and 70% manganese; used in the manufacture of hard steels. Also added to copper to increase elasticity.

6234 chromium molybdenum
An alloy of 50% chromium and 50% molybdenum; used in the manufacture of hard steels.

6235 chromium nickel
An alloy of 10% chromium and 90% nickel, also 50% chromium, and 50% nickel; used in the manufacture of hard steels.

6236 chromium oxide (ic)
1308-38-9 2283 215-160-9
Cr_2O_3
chromic oxide
chromia; chromium(III) oxide; green cinnabar; chromium sesquioxide. Metallurgy, green paint pigment, ceramics, catalyst in organic synthesis, green granules in asphalt roofing, component of refractory brick, abrasive. *Atomergic Chemetals; British Chrome & Chem.; Noah Chem.*

6237 chromium sulfate
15244-38-9 2287
$Cr_2O_{12}S_3 \cdot xH_2O$
chromium(III) sulfate hydrate
chromic sulfate. Peach-colored solid, occurring as several hydrated forms. Used in leather tanning. d = 3.012; MLD (mus iv) = 247 mg/kg.

6238 chromium-molybdenum steel
Alloys containing from 0.06-1.2% carbon, traces to 15% molybdenum, and traces to 6% chromium.

6239 chromium-vanadium steel
An alloy of this type contains 0.3-0.4% carbon, 1-1.5% chromium, and 0.15-0.25% vanadium.

6240 chromium-vanadium-molybdenum steels
Alloys of iron containing 0.1-0.55% carbon, 0.22-1.45% molybdenum 0.8-1.5% chromium, and 0.15-0.45% vanadium.

6241 Chromol
Metachrome dyestuffs. *James Robinson & Co Ltd.* Discontinued.

6242 Chromosal®
Different types of tanning materials containing chrome oxide. *Bayer AG.*

6243 Chromospun
Acetate spun yarn. *Eastman.*

6244 chromotrope acid
148-25-4 2298 205-712-7
$C_{10}H_8O_8S_2$
4,5-Dihydroxynaphthalene-2,7-disulfonic acid
1,8-Dihydroxynaphthalene-3,6-disulfonic acid; Chromogen C; Chromogen I; LL. An acid dye for wool. Soluble in H_2O.

6245 chromous chloride
10049-05-5 2301 233-163-3
$CrCl_2$
chromium (II) chloride anhydrous
chromous chloride. Reducing agent, catalyst, reagent, chromizing. mp = 824°; d_4^{14} = 2.751; soluble in H_2O; LD_{50} (rat orl) = 1.87 g/kg. *Atomergic Chemetals; Cerac; Noah Chemical.*

6246 chromous sulfate
13825-86-0 2305
$CrSO_4 \cdot 5H_2O$
chromous sulfate pentahydrate
Oxygen scavenger, reducing agent, analytical reagent. Soluble in H_2O, EtOH, insoluble in less polar organic solvents.

6247 chromous trioxide
1333-82-0 2293 215-607-8
CrO_3
chromium (VI) oxide
chromium trioxide; chromic acid; chromic anhydride; chrome bronze. Dark red crystals, powerful oxidizing agent; used in chromium plating, copper stripping; aluminum anodizing; photography, corrosion inhibition, hardening microscopical preparations, oxidizing agent in organic chemistry. mp = 197°; d = 2.70; very soluble in H_2O.

6248 Chromovan Steel
A proprietary trade name for a nonsparking tool steel containing 12.5% chromium, 0.8% molybdenum, 1% vanadium, and 1.6% carbon. No manufacturer.

6249 chronin
An alloy of 84% nickel and 15% chromium.

6250 chronite
Heat-resisting alloys containing 63-67% nickel, 13-16% chromium, 12-20% iron, 0-0.4% silicon, 0.1% manganese, and 0-0.8% aluminum.

6251 chrysazol
1,8-Dihydroxyanthracene

6252 Chrysene
218-01-9 2314 205-923-4
$C_{18}H_{12}$
1,2-Benzphenanthrene
Found in coal tar. Used in chemical synthesis. mp = 254°; bp = 448°; d_4^{20}= 1.274; Insoluble in H_2O, slightly soluble in organic solvents. No manufacturer.

6253 Chrysocale
A jeweler's alloy, containing 9 parts copper, 8 parts zinc, and 2 parts lead.

6254 Chrysochalk
Gold-copper
An alloy containing 59-93% copper, 8-39% zinc, and 1.6-1.9% lead. A jeweler's alloy.

6255 Chrysoform
$C_6H_8Br_2I_2N_4$
Dibromodiiodohexamethylene-tetramine
Used as an antiseptic.

6256 Chrysoidine
532-82-1 2317 208-545-8
$C_{12}H_{13}ClN_4$
4-(Phenylazo)-1,3-benzenediamine hydrochloride
4-phenylazo-*m*-phenylenediamine hydrochloride; 2,4-diaminoazobenzene hydrochloride; chrysoidine orange; chrysoidine Y; C.I. 11270;C.I. Basic Orange 2; Chrysoldine crystal. A dyestuff. It consists of the hydrochloride of phenylazo-m-phenylenediamine, with some of the homologues from *o*- and *p*-toluidine. Dyes wool and silk orange. The citrate is used as an antiseptic. mp = 118-118.5°; soluble in H_2O (5.5%, 15°), more polar organic solvents, insoluble in C_6H_6.

6257 Chrysorin
An alloy of 66% copper and 34% zinc.

6258 Chryzoplus, Chryzopon, Chryzosan, Chryzotek
133-32-4 4995 205-101-5
4-Indol-3-ylbutyric acid
A root growth promoter. *Fargro Ltd.*

6259 CHT Activator NB
Hydrogen peroxide activator for low pH bleaching. *Catawba-Charlab.*

6260 CHT Antifoam MI
Mineral oil-based defoaming agent. *Catawba-Charlab.*

6261 CHT Biavin 109
Concentrate gliding, anticreasing and leveling agent for cotton. *Catawba-Charlab.*

6262 CHT Carrier GR-A
Carrier for dyeing of polyester and PES blends. *Catawba-Charlab.*

6263 CHT Contavan ALR
Organic stabilizer for alkaline peroxide bleaching. *Catawba-Charlab.*

6264 CHT Cotoblanc HTD-N
Low-foam alkali-stable scouring auxiliary and chelate. *Catawba-Charlab.*

6265 CHT Defoamer SC
Deaerating agent for transfer printing inks. *Catawba-Charlab.*

6266 CHT Egasol SP
Leveling and penetrating agent for rapid dyeing of polyester. *Catawba-Charlab.*

6267 CHT Felosan TAK-NO
Low foaming stain remover and detergent. *Catawba-Charlab.*

6268 CHT Heptol NWS
Polyphosphate
For textile washing. *Catawba-Charlab.*
6269 CHT Intensol TH-B
Dyestuff solvent and machine cleaner. *Catawba-Charlab.*
6270 CHT Lavotan DS
Wetting, washing and cleaning surfactant. *Catawba-Charlab.*
6271 CHT Lustraffin BA
In-bath yarn lubricant for textile knitting and weaving. *Catawba-Charlab.*
6272 CHT Meropan BRE
Peroxide neutralizer. *Catawba-Charlab.*
6273 CHT Prisulon SNP-113S
Synthetic thickener for transfer paper printing. *Catawba-Charlab.*
6274 CHT Rapidoprint M-4
Leveling agent for transfer paper printing. *Catawba-Charlab.*
6275 CHT Retinol M
Stripping and leveling agent with dyestuff affinity. *Catawba-Charlab.*
6276 CHT Rewin MRT
Nonformaldehyde cationic dye fixative. *Catawba-Charlab.*
6277 CHT Sarabid DLO Conc
Auxiliary with dyestuff affinity for pre-and after-treatment. *Catawba-Charlab.*
6278 CHT Subitol HLF Conc
Low foaming alkali-stable surfactant for continuous bleaching. *Catawba-Charlab.*
6279 CHT Thickener 8300E
Extremely pure synthetic thickener for pigment prints with soft hand; suitable for very fine mesh screens. *Catawba-Charlab.*
6280 CHT Tubingal 220A
Textile softener. *Catawba-Charlab.*
6281 CHT Tubiprint PERL C
Pigment printing paste for pearlescent effects. *Catawba-Charlab.*
6282 CHT Viscavin DMS
Anticreasing agent with leveling and dispersing properties for polyester. *Catawba-Charlab.*
6283 Chymex
37106-97-1 1081 253-349-8
Bentiromide
Diagnostic aid used to examine pancreatic function. *Adria Laboratories Inc.*
6284 Chymopapain
9001-09-6 2319 232-580-8
Discase; BAX-1526; Chymodiactin. Enzyme; used in meat tenderizer. Powder; more soluble in aqueous solutions than in papain. *Travenol Laboratories Inc.* Name unverified.
6285 Chymoral
A proprietary preparation of trypsin and chymotrypsin; digestive enzymes. *Armour Pharmaceutical Co.* Name unverified.
6286 α-Chymotrypsins
9004-07-3 2320 232-671-2
Avazyme; Catarase; Chymar; Chymetin; Chymolase; Enzeon; Kymo-trypure; Quimar; Quimoral; Quimotrase; Zolyse. A proprietary preparation of α-chymotrypsin, a proteolytic enzyme; for treatment of wounds. *Armour Pharmaceutical Co.* Name unverified.
6287 CI-2
25928-94-3
Two-part epoxy compound; coil impregnant for electrical components; also for casting and coating applications. *Bacon.*
6288 CI7810
18395-30-7 242-272-5
$C_6H_{16}O_3Si$
Isobutyltrimethoxysilane
Prosil 178; Isobutyltrimethoxysilane; Dynasylan IBTMO. Coupling agent, chemical intermediate, blocking agent, release agent, lubricant, primer, and reducing agent. *Hüls Am.*
6289 CI7840
24801-88-5 246-467-6
$C_{10}H_{21}NO_4Si$
Isocyanatopropyltriethoxy silane
3-Isocyanatopropyltriethoxysilane; γ-isocyanopropyltriethoxysilane; γ-Isocyanatopropyl triethoxy silane. Coupling agent, chemical intermediate, blocking agent, release agent, lubricant, primer, and reducing agent. *Hüls Am.*
6290 Cialit
The sodium salt of 2-ethyl mercury mercaptobenzoxazol-5-carboxylic acid; an antifouling pigment.
6291 Ciba 1906
500-89-0 207-914-0
$C_{19}H_{25}N_3OS$
thiambutosine
A proprietary preparation containing thiambutosine. *Ciba plc.* Name unverified.
6292 Ciba, Cibanone
Vat dyes. *Ciba plc.* Name unverified.
6293 Cibacet, Cibacrolan; Cibacron; Cibalan
Disperse, reactive dyestuffs, Cibalan used for wool. *Ciba plc.* Name unverified.
6294 Cibacoll™
α-Amylase substrate. *Calbiochem Corp.*
6295 Cibamin
Modified melamine/formaldehyde resins. *Ciba plc.* Name unverified.
6296 Cibanite
A proprietary trade name for an aniline-formaldehyde resin resistant to water, oil, alkalis, and organic solvents. No manufacturer.
6297 Cibanold
A proprietary trade name for urea-formaldehyde synthetic resins. No manufacturer.
6298 Cibaphasol 6042
A proprietary trade name for a fatty acid amide derivative giving improved quality and appearance to continuous dyeings; used as a dyeing auxiliary for polyacrylonitrile fibers. *Ciba plc.* Name unverified.
6299 Cibatex 248
A proprietary trade name for a phenol sulfonic acid derivative which gives improved wet fastness properties of dyeings; used also as a synthetic tanning agent for dyes on nylon 66. *Ciba plc.* Name unverified.
6300 Cibatex PA
A proprietary trade name for a synthetic tanning agent of the phenol sulfonic acid type used as an improver of wet fastness in the dyeing of nylon. *Ciba plc.* Name unverified.
6301 cidrase
Cider yeast.
6302 CIL
Silicon product; corrosion inhibitor for water systems including potable water systems, cannery rinse water, etc.; forms barrier-type protective film on internal metal surfaces handling treated water. *Drew Ind. Div.*
6303 Cilbond
A range of rubber to metal and rubber to plastics adhesives. *Compounding Ingredients Ltd.*
6304 Cilcast
A two-component cold cure polyurethane elastomer. *Compounding Ingredients Ltd.*
6305 Cimet
Stainless alloys containing about 48% chromium with iron.
6306 Cimfix 606, Cimpact 699, 710
14087-96-6 9207 238-877-9
Reinforcement in PP homopolymer; offers unique color and brightness in thermoplastics, good long-term heat resistance in polyolefins. (710) for for high impact, TPO, auto, and appliance applications. *Luzenac Am.* Discontinued.
6307 cinchophen
132-60-5 2347 205-067-1
$C_{16}H_{11}NO_2$
2-Phenyl-4-quinolinecarboxylic acid
2-Phenylcinchoninic acid; Atophan; Quinophan; Phenoquin; Agotan; Artam; Alutyl; Cinconal; Vantyl; Viophan; Atocin; Mylofanol; Rhematan; Rheumin; Tophol; Polyphlogin; Quinofen; Tophosan. Used experimentally in medicine to produce ulcers. Has been used as an analgesic. mp = 213-216°; insoluble in H_2O, poorly soluble in organic solvents.
6308 Cindal
An aluminum alloy containing zinc and small amounts of magnesium and chromium.
6309 Cindumix
Coal tar/bitumen mixture; surface dressing material for roads. *Cindu Chemicals BV.* Name unverified.
6310 cinereine
An induline dyestuff obtained from azoxyaniline, aniline hydrochloride, and *p*-phenylenediamine.
6311 cinnamaldehyde
104-55-2 2356 203-213-9
C_9H_8O
3-Phenyl-2-propenal
trans-3-phenyl-2-propenal; cinnamic aldehyde; phenyl acrolein. Used in flavors and perfumery. mp = -7.5°; bp = 246°; d_{25}^{25} = 1.048-1.052; n_D^{20} = 1.618-1.623; LD_{50} (rat orl) = 2220 mg/kg. *Nipa Labs; Penta Mfg.; Quest Int'l.*
6312 cinnamein
103-41-3 1165 203-109-3

A term applied to benzyl cinnamate. It is, however, also used for the mixture of ester and alcohol in the Balsams of Tolu and Peru, which are not extractable by alkali from an ethereal solution.

6313 cinnamic acid
621-82-9 2358 205-398-1
$C_9H_8O_2$
3-Phenyl-2-propenoic acid
3-phenylpropenoic acid; β-phenylacrylic acid; cinnamylic acid. Medicine (anthelmintic), perfumes, intermediate. mp = 133°; bp = 300°; λ_m = 273 nm (EtOH); soluble in H_2O (0.5 g/l), more soluble in organic solvents. *Aceto; Hüls Am.; Penta Mfg.; Raschig.*

6314 cinnamoyl chloride
102-92-1 2360 203-065-5
C_9H_7ClO
3-Phenyl-2-propenoyl chloride
phenylacrylyl chloride; *trans*-3-phenylacryloyl propenoyl chloride. Reagent for determination of small amounts of water, chemical intermediate. mp = 35-36°; bp_2 = 101°; $d_3^{45.3}$ = 1.1617; $n_D^{42.5}$ = 1.614. *ICI Specialties; Janssen Chimica; Penta Mfg.; Raschig; Schweizerhall.*

6315 cinnamyl alcohol
104-54-1 2362 203-212-3
$C_9H_{10}O$
3-phenyl-2-propen-1-ol
Cinnamic alcohol; Styrone; γ-phenylallyl alcohol; Styrylcarbinol; phenyl-2-propenol; phenylallyl alcohol. Used in perfumery and in deodorants. mp = 33-35°; bp = 249-250°; d_{35}^{35} = 1.0397; n_D^{20} = 1.58190; soluble in H_2O, organic solvents.

6316 Cinquasia
High performance organic pigments. *Ciba plc.*

6317 ciprofloxacin
85721-33-1 2374
$C_{17}H_{18}FN_3O_3$
1-Cyclopropyl-6-fluoro-1,4-dihydro-4-oxo-(1-piperazinyl)-3-qunioline carboxylic acid
Bayo 9867; Ciflox; Ciprobay; Ciproxan; Ciproxin; Velmonit. A synthetic broad spectrum antibacterial agent for oral administration. Ciprofloxacin intercalates with DNA causing blockade of DNA and RNA synthesis. mp = 255-257° (dec) *Bayer AG.*

6318 Cirami No. 1
Beeswax; candelilla wax; shea butter. A natural thickener. *Alban Muller.*

6319 Circacid
Dairy hygiene compound. *The Wellcome Foundation Ltd.* Name unverified.

6320 Circadet
Surface active detergent. *Ciba plc.* Name unverified.

6321 Circaline MK 11
Hypochlorite pipe line cleaner/sterilizer. *Ciba plc.*

6322 Circosan
Dairy hygiene detergent sterilizer. *The Wellcome Foundation Ltd.* Name unverified.

6323 Cire De Lanol CTO
Ceteareth-33-cetearyl alcohol. Emulsifier used in cosmetics. *Seppic.* Discontinued.

6324 Cirol
68334-28-1 269-820-6
Partially hydrogenated vegetable oil (soybean, cottonseed); kosher; domestic oil for nut roasting, margarines, frozen desserts, coffee whiteners. *Van Den Bergh Foods.*

6325 Cirrasol
A textile softening agent, fiber lubricant or antistatic agent. *ICI Chem & Polymers Ltd.*

6326 Cisdene® 1203
9003-17-2
Stereospecific polybutadiene, high *cis*, nonstaining antioxidant; used in mixtures with SBR or NR to enhance abrasion resistance, reduce shrinkage, improve low temperature properties; vulcanized with sulfur and conventional accelerator systems. d = 0.9000; n_D^{20} = 1.5178. *Am. Syn. Rubber.*

6327 Cismollan® BH
Highly bactericidal and fungicidal soaking agent for dried and salted raw hides, outstanding wetting action. *Bayer AG.*

6328 Cithrol
Polyethylene glycol esters; used as nonionic surfactants. *Croda Surf. Ltd.*

6329 Cithrol GMS A/S
Acid-stable glyceryl monostearate; a nonionic surfactant used in the cosmetic and pharmaceutical industries. *Croda Chem. Ltd.*

6330 Citmol 316
Triisocetyl citrate
Emollient for cosmetics, cleansing creams, lipsticks; pigment dispersant. *Bemel; Heterene.*

6331 Citmol 320
Trioctyldodecyl citrate
Noncomedogenic, high viscosity emollient, film-former, castor oil replacement for lipsticks. *Bemel.*

6332 Citobaryum
7727-43-7 1023 231-784-4
Barium sulfate
A proprietary preparation of special barium sulfate. Used in radiology.

6333 Citotray
Light curing plastic for the preparation of custom impression trays; used in dentistry. *Bayer AG.*

6334 Citowett
Nonionic spreader containing 99-100% alkylaryl polyglycol ether; wetting agent for herbicide, fungicide and insecticide sprays. *BASF plc.*

6335 Citraclean
77-92-9 2387 201-069-1
Citric acid
Citric acid metal cleaning process. *John & E Sturge Ltd, Selby.*

6336 citral *(cis and trans)*
5392-40-5 2383 226-394-6
$C_{10}H_{16}O$
3,7-dimethyl-2,6-octadienal
geranial; neral. Mixture of two geometrical isomers, geranial and neral. Perfumes, flavoring agents, intermediate for other fragrances, vitamin A synthesis. $bp_{2.6}$ = 91-93°; d_4^{20}= 0.8869-0.8888; insoluble in H_2O, soluble in organc solvents. *Lucta SA; Penta Mfg.; Schweizerhall; SCM Glidco Organics.*

6337 Citralka
A proprietary preparation containing sodium acid citrate. *Parke-Davis.* Name unverified.

6338 Citranova
Synthetic citrus oils.

6339 Citranox®
Blend of organic acids, anionic and nonionic surfactants, and alkanolamines; phosphate-free biodegradable acid detergent, wetting agent for cleaning dairy equipment, laboratory ware, clean rooms, optical and electronic parts, pharmaceutical apparatus, industrial parts, etc. *Alconox.*

6340 Citrest
A citric acid ester. *Rhône-Poulenc Surf.* Name unverified.

6341 citric acid
77-92-9 2387 201-069-1
$C_6H_8O_7$
2-hydroxy-1,2,3-propanetricarboxylic acid
β-hydroxytricarballylic acid. Preparation of citrates, flavoring extracts, confections, soft drinks; antioxidant in foods; sequestering agent; detergent builder; metal cleaner. mp = 153°; d = 1.665; pK_1 = 3.128, pK_2 = 4.761, pK_3 = 6.396; soluble in H_2O (59.2% w/w 20°), moderately soluble in organic solvents; LD_{50} (rat ip) = 975 mg/kg. *Cargill; R.W. Greef; Haarmann & Reimer; Hoffmann-La Roche; Penta Mfg.*

6342 Citroflex A-2
77-89-4 201-066-5
$C_{14}H_{22}O_8$
Acetyltriethyl citrate
triethyl acetyl citrate; 1,2,3-Propanetricarboxylic acid, 2-(acetyloxy)-, triethyl ester; Triethyl 2-acetoxy-1,2,3-propanetricarboxylate. Plasticizer for cellulosics. *Morflex.*

6343 Citroflex A-4
77-90-7 201-067-0
$C_{20}H_{34}O_8$
Acetyl tri-n-butyl citrate
acetyl tributyl citrate;1,2,3-Propanetricarboxylic acid, 2-(acetyloxy)-, tributyl ester; Tributyl acetylcitrate; Acetyl tri-n-butyl citrate. Plasticizer for PVC, PVDC, especially food films, medical articles. mp = -80°; bp = 172-174°; d = 1.048; insoluble in H_2O, soluble in organic solvents. *Morflex.*

6344 Citroflex A-6
24817-92-3
Acetyl tri-n-hexyl citrate
Vinyl plasticizer for medical applications and other toxicologically sensitive areas. d = 1.005. *Morflex.*

6345 Citroflex A-8
Acetyl tri-2-ethylhexyl citrate
Isooctyl O-acetylcitrate. A vinyl plasticizer. *Pfizer International.*

6346 Citroflex B-6
82469-79-2
n-Butyryl tri-n-hexyl citrate
Vinyl plasticizer for medical applications. *Morflex.*

6347 citronellal
106-23-0 2390 203-376-6
$C_{10}H_{18}O$
3,7-dimethyl-6-octenal
Soap perfumery, manufacture of hydroxycitronellal, insect repellant. bp_1 = 47°; d = 0.848-0.856; n_D^{20} = 1.4460; $[\alpha]_D^{25}$ = 11.50°. *Penta Mfg.; PMC Specialties; Schweizerhall.*

6348 citronellol
106-22-9 2391 203-375-0
$C_{10}H_{20}O$
3,7-dimethyl-6-octen-1-ol
β-Citronellol; 2,6-dimethyl-2-octen-8-ol; citronellol; cephrol; Cephrol. Perfumery (floral odors). bp = 224.5°; d_4^{20}= 0.8550; n_D^{20}= 1.4559; $[\alpha]_D^{20}$= 5.22°; slightly soluble in H_2O, soluble in organic solvents. *Int'l. Flavors & Fragrances; Penta Mfg.; SCM Glidco Organics.*

6349 Citrozone
Vanadium sodium citrochloride.

6350 Cladex®
Metal clad products. *Asahi Chemical Industry.*

6351 Clairolite L
Calcium bentonite
Clay with high brightness and low viscosity; plasticizer and binding agent for white firing ceramic; moisture absorbent, anticaking agent. *Kaopolite.*

6352 Clairsol
Highly de-aromatized aliphatic hydrocarbon solvents. *Carless Refining & Marketing Ltd.*

6353 Clampdown
Silage treatment. *May & Baker Ltd.* Name unverified.

6354 Claradex CH-540
9003-56-9
Poly(acrylonitrile-co-butadiene-co-styrene)
ABS resin; general purpose for injection molding and extrusion. *Shin-A Chemical Mfg.; Syn. Rubber Tech.*

6355 Clar-Apel
A proprietary viscose plastic used as a packing material. No manufacturer.

6356 Clarase® 5,000, 40,000
9000-92-4 640 232-567-7
α-Amylase
Enzyme for hydrolysis of starch and dextrin; for starches, syrups, brewing, fruit juices. *Solvay Enzymes.*

6357 Clarcel
7631-86-9 8637 231-545-4
SiO_2
Silica
Silica powder; silicon dioxide. Diatomaceous earth filter aid; for liquids filtration in chemical, pharmaceutical and food industries. *Elf Atochem UK/Ceca.*

6358 Clarcel Flo
Perlite-based filter aid; for liquids filtration in chemical, pharmaceutical and food industries. *Elf Atochem UK/Ceca.*

6359 Clarex® L
9032-75-1 232-885-6
Pectinase
Enzyme; used in depectinization of fruit juices, grape processing, jams and jellies, berry processing and wine production *Solvay Enzymes.*

6360 Clarfina
Synthetic zeolite; raw material for detergents. *SNIA (UK) Ltd.*

6361 Clarifex 800
Clarifier. *ICI Polymer Additives.*

6362 Clarifloc
Flocculants and coagulants; clarification of water and waste water, sludge treatment, thickening and dewatering. *AlliedSignal Inc/Water Treatment Div.* Name unverified.

6363 Clarifloc
9000-07-1 1914 232-524-2
Polysaccharides extracted from seaweed. *Rhône-Poulenc UK.*

6364 Clarifoil
Films or pellicles made from cellulose acetate or other cellulose derivatives or cellulose; sold in the form of sheets, bands or strips; for wrapping or packing or electrical insulating purposes. *Acordis.*

6365 Clarion®
For the agriculture industry. *DuPont UK.*

6366 Clar+Ion® A410P
10043-01-3 381 233-135-0
Aluminum sulfate and flocculating agent; coagulant and flocculant for industrial and potable water treatment applications; especially where rapid settling is desired; emulsion breaker. *General Chemical.*

6367 Clarit
A particularly pure calcium bentonite for the adsorptive stabilization of wines, fruit juices, vinegar etc. *SüdChemie AG.* Name unverified.

6368 Clarital®
Polyvinyl acetate
Antitranspirant. *Fargro Ltd.*

6369 Clarite
Pretreatment agent. *Ciba plc.* Name unverified.

6370 Clark's patent alloy
An alloy of 75% copper, 7.2% zinc, 14.5% nickel, 1.9% tin, and 1.9% cobalt.

6371 Clar-O-Cel
Cellulose fiber filter aids; for liquids filtration in chemical, pharmaceutical, and food industries. *Elf Atochem UK/Ceca.*

6372 Clarosan 1FG
886-50-0 212-950-5
Granules containing 1% w/w terbutryn; for aquatic weed control. *Ciba-Geigy AgroChemicals.*

6373 Clarstabil
Clarification of wines, beers, etc. *Minas de Gádor SA.* Discontinued.

6374 Clarus metal
A light aluminum alloy used for sheets and tubes.

6375 Clarvin
Clarification of wines, beers, etc. *Minas de Gádor SA.*

6376 Clay Breaker
7778-18-9 1753 231-900-3
calcium sulfate
Mineral gypsum; clay breakdown agent. *Vitax Ltd.*

6377 Claymaster
VR Claymaster. Polystyrene boards of low density colored pink; compressible fill beneath concrete foundations. *Vencel Resil Ltd.*

6378 Claysil
A silicate adhesive. *Crosfield Chemicals Ltd.*

6379 CLD 2
Croscarmellose sodium
Carboxymethylcellulose sodium that has been internally cross-linked; pharmaceutic aid. *Buckeye Cellulose Corp.* Name unverified.

6380 Clean Wiz
Proprietary solvent and aqueous formulations to clean and strip build-up and residue of resin and mold release from metal and composite molds; for fiberglass, polyurethane foam and rubber. *Axel Plastics Research Laboratories Inc.*

6381 Cleanacres CMPP
7085-19-0 5826 230-386-8
Mecoprop
Soluble concentrate containing 600 g/l mecoprop; for control of weeds in cereals and grassland. *Cleanacres Ltd.* Name unverified.

6382 Cleanacres PDR 675
999-81-5 2153 213-666-4
Soluble concentrate containing 675 g/l chlormequat; plant growth regulator. *Cleanacres Ltd.*

6383 Clean-Up
Contains concentrated tar acids in emulsion; general moss-killer and sterilizer for garden use. *ICI Garden Products.*

6384 Cleapact TI-100, TI-100S
Styrene-based copolymers; thermoplastic with excellent balance of transparency, impact strength, processibility; injection molding grade for copy machine trays, printer covers, packaging materials for electrical equipment or food. *Dainippon Ink & Chemical.*

6385 Clear by Design
94-36-0 1149 202-327-6
Benzoyl peroxide
Keratolytic. *Herbert Laboratories.* Name unverified.

6386 Clear Conc. 7174
Synthetic thickener for aqueous pigment printing systems. *Catawba-Charlab.*

6387 Clear Tint®
Nonpigmented, polymeric colorants used to obtain clear color in polyolefins and other plastics, particularly in clarified polypropylene; colorants for housewares, protective packaging, medical devices, blow molded bottles and sheet. *Milliken.*

6388 Clearam
A range of modified starches that require heating in aqueous solution to develop their viscosity; used in the food industry to develop viscosity in a wide number of applications e.g. sauces, gravies, custards etc. *Roquette (UK) Ltd.*

6389 Clearbreak TEB
Polymeric amine
Reverse emulsion breaker for oil treating; may be combined with metal salts or anionic polymers for enhanced performance. *Chemron.*

6390 Clearcol
9007-34-5 2543 232-697-4
Soluble animal collagen; protein for moisturizing and conditioning of facial systems where clarity is important. *Croda Inc.*

6391 Cleargum®
Thin-boiling food modified starch; used for foods, confectionery and coatings. *Roquette (UK) Ltd.*

6392 Clearon
2244-21-5 9896 218-828-8
Troclosene Potassium
Sodium dichloroisocyanurate dihydrate, a topical anti-infective. *Schering AgroChemicals Ltd.* Discontinued.

6393 Clearon®
Terpene tackifying resin. *Aceto.*

6394 Clearpol
11, 231
Alginate for brewing. *Alginate Industries Ltd.* Name unverified.

6395 Clearsite
A proprietary trade name for cellulose acetate used for transparent containers. No manufacturer.

6396 Clearsol®
Blend of surfactants and xylenols; disinfectants used for hospitals. *Coventry Chemicals Ltd.*

6397 Clear-Stat AS401
Nonamine antistatic concentrate for use with polyolefins. *Polycolors.* Name unverified.

6398 Clearway
aminotriazole-simazine. A suspension concentrate containing 100 g aminotriazole [61-82-5] and 300 g simazine [122-34-9] per liter; used for total weed control in non crop areas and fruit orchards. *Rhône-Poulenc Environmental Prods. Ltd.*

6399 Clearys Waterless Hand Cleaner
Concentrate heavy-duty but gentle cleaning agent for removal of pesticide residues, grease, oil, tar, ink, paint, carbon, adhesives from hands, work clothes, cars. *W A Cleary.*

6400 Cleaval
cyanazine-mecoprop. Suspension concentrate containing 60 g cyanazine and 400 g mecoprop per liter; a contact herbicide. *Shell UK.*

6401 Clebrium alloys
Heat-resisting alloys. One contains 76.5% iron, 13% chromium, 3.6% molybdenum, 2.6% carbon, 2% nickel, 1.5% silicon, and 0.75% manganese.

6402 Clelands Reagent
3483-12-3 3441 222-468-7
Dithiothreitol
A white powder, mp = 37°; a reagent for protecting -SH groups in certain biochemical systems. *British Drug Houses.*

6403 Clenecorn
7085-19-0 5826 230-386-8
Mecoprop
Soluble concentrate containing 570 g/l mecoprop; for control of weeds in cereals and grassland. *Farmers Crop Chemicals Ltd.*

6404 Clenesco
A proprietary trade name for anhydrous sodium metasilicate. No manufacturer.

6405 Clerici solution
A molecular mixture of thallium malonate and formate; used for floating mineral specimens to determine the specific gravity.

6406 Clerit
A horticultural fungicide. *ICI Chem & Polymers Ltd.*

6407 Clerite
A horticultural fungicide. *Plant Protection.* Name unverified.

6408 Clermait®
41083-11-8 255-209-1
Azocyclotin
Acaricide used for control of red spider mites. *Bayer AG.*

6409 Cleroxide
A stabilizer for polyvinyl chloride. *Akzo Chemie.* Name unverified.

6410 Cleve's β acid
119-79-9 2409 204-351-2
$C_{10}H_9NO_3S$
α-naphthylamine-6-sulfonic acid
1,6-Cleve's acid; 5-Amino-2-naphthalene sulfonic acid; 1-naphthalene -6-sulfonic acid. Intermediate in synthesis of dyestuffs. Soluble in 1000 parts H_2O, insoluble in EtOH, Et_2O.

6411 Cleve's γ-Acid
134-54-3 205-146-0
2-Naphthalenesulfonic acid, 4-amino-
α-naphthylamine 3-sulfonic acid; 1-Amino-3-naphthalenesulfonic acid; 4-Amino-2-naphthalenesulfonic acid; Cleve's γ-acid; 1-Naphthylamine, 3-sulfo-; 1-Naphthylamine-3-sulfonic acid. Intermediate in synthesis of dyestuffs.

6412 Cleve's ω-Acid or J-acid
119-28-8 2410 204-311-4
$C_{10}H_9NO_3S$
α-naphthylamine 7-sulfonic acid
8-Amino-2-naphthalene sulfonic acid; 1,7-Cleve's acid;. Intermediate in synthesis of dyestuffs. Crystallizes as the monohydrate, soluble in 220 parts H_2O.

6413 Cleve's ω-or δ-acid
α-nitronaphthalene-7-sulfonic acid. Intermediate in synthesis of dyestuffs.

6414 Cliché metal
An alloy of 33% tin, 46% lead, and 21% cadmium. Another alloy called by this name contains 48% tin, 32.5% lead, 9% bismuth, and 10.5% antimony.

6415 Clifton Chlormequat 46
999-81-5 2153 213-666-4
chlormequat
Soluble concentrate containing 460 g/l chlormequat, a plant growth regulator. *Clifton Chemicals Ltd.*

6416 Clifton CMPP Amine 60
7085-19-0 5826 230-386-8
Mecoprop
Soluble concentrate containing 600 g/l mecoprop; for control of weeds in cereals and grassland. *Clifton Chemicals Ltd.*

6417 Clifton Glyphosate Additive
61791-44-4 263-177-5
PEG-2 tallow amine
Wetting and spreading agent containing 850 g/l tallow amine ethoxylate for use with phosphonoglycine herbicides. *Clifton Chemicals Ltd.*

6418 Clifton Wetter
Nonionic adjuvant containing 250 g/l alkyl phenol ethoxylate; wetting and spreading agent for use in herbicides, fungicides, insecticides and foliar feeds. *Clifton Chemicals Ltd.*

6419 Climacel
Protective moisturizer. *Richardson-Vicks Inc.* Name unverified.

6420 Climax
A magnetic alloy containing 30% nickel and 70% iron. Another alloy containing 73% iron, 24.4% nickel, and 2.6% manganese has also been called Climax.

6421 Climax 193
An alloy of 68% iron, 28% nickel, 2% chromium, and 1% manganese.

6422 Clinafarm
73790-28-0
imazalil
A proprietary preparation containing imazalil; used for fumigation. *Janssen Pharmaceutical Ltd.* Name unverified.

6423 Clinifeed 400, Favour, ISO, Protein Rich
Protein, carbohydrate, fat, vitamins, and minerals; liquid feeds. *Roussel Laboratories Ltd.* Discontinued.

6424 Clinistix
A prepared test strip of glucose oxidase, peroxidase and *o-toluidine; used for the detection of glucose in urine. B. C. Ames.* Name unverified.

6425 Clinitest
A prepared test tablet containing copper sulfate, sodium hydroxide, citric acid, and sodium carbonate; used for the detection of reducing substances in urine. *B. C. Ames.* Name unverified.

6426 Clipper
Tree growth regulator. *ICI Chem & Polymers Ltd.*

6427 Clofazimine
2030-63-9 2433 217-980-2
$C_{27}H_{22}Cl_2N_4$
N,5-Bis(4-chlorophenyl)-3,5-dihydro-3-[(1-methylethyl)imino]-2-phenazinamine
Lamprene; G-30320; B-663. A proprietary preparation of clofazimine; an antileprotic, antibacterial and tuberculostatic. mp = 210-212°; soluble in organic solvents; LD_{50} (rat orl) >4 g/kg. *Ciba plc.* Name unverified.

6428 Cloisonné®
Highly lustrous pigments with deep colors produced by combination of light interference and light absorption. *Mearl.*

6429 Clonevac D-78
Gumboro D-78 strain; for immunization of poultry. *Intervet Inc.*

6430 Clonevac-30, 30T
B_1 type, cloned lasota strain, Newcastle vaccine; for immunization of poultry. *Intervet Inc.*

6431 Clonevac-30-Ma5
B_1 type, clone 30 strain Newcastle and Ma5 Mass. type vaccine; for immunization of poultry. *Intervet Inc.*

6432 clonitrilide
1420-04-8 6602 215-811-7
$C_{13}H_8Cl_2N_2O_4$
2',5-dichloro-4'-nitrosalicylanilide compound with 2-aminoethanol (1:1)
Bayluscid; niclosamide ethanolamine; Bayer 73; Bayer 25648; Bayluscide; M 73; SR 73; Mollutox; Niclosamide; Phenasal ethanolamine salt; Yomesan. A molluscicide. mp = 204°; LD_{50} (rat orl) = 500 mg/kg.

6433 Cloparin, Cloparol
Chlorinated paraffins; plasticizers for PVC, paints; additives for lubricating and cutting oils and flame resistant additives. *Collinda Ltd; SNIA (UK) Ltd.*

6434 Cloparten, Cloparten Z
Chlorosulfonated paraffins
Raw material for the preparation of emulsifiable synthetic greasing agents for hides and leather. *Caffaro SpA; SNIA (UK) Ltd.* Name unverified.

6435 Clophen
Synthetic liquid insulants and coolants for the electrical industry; used as high grade, flame-retardant impregnants for capacitors and as liquid coolants for transformers and rectifiers. *Bayer AG.* Discontinued.

6436 Clopyralid
1702-17-6 2462 216-935-4
$C_6H_3Cl_2NO_2$
3,6-Dichloropyridine-2-carboxylic acid
Cirtoxin, Cyronal, Dowco 290; Lontrel; Matrigon; Reclaim; Shield Stinger. Post-emergence control of broad-leaf weeds of *Polygonaceae, Compositae, Leguminosae* and *Umbelliferae.* Good control of creeping thistle, sow thistle, coltsfoot, mayweeds and *Polygonum* spp. mp = 151-152°; soluble in H_2O (9 g/l), soluble in organic solvents; LD_{50} (rat orl) >4300 mg/kg. *DowElanco.*

6437 Clorafin
Chlorinated paraffin
Nonflammable, chemically resistant; good plasticizing properties; plasticizer in inks, additive in cutting oils and drawing compounds, plasticizing resin in chlorinated rubber corrosion-resistant coatings, plasticizer for vinyl resins, waterproofer and flameproofer for textiles. *Hercules.* Discontinued.

6438 Cloran
A trademark for a chlorinated anhydride, a thermal and chemical stabilizer for polymers. *UOP Inc.* Name unverified.

6439 Clortex
Chlorinated rubber; resin for the preparation of paints and varnishes. *Caffaro SpA.* Name unverified.

6440 Clortol
Chlorine releasing sterilant. *Rhône-Poulenc UK.*

6441 Closantel
57808-65-8 2473 260-967-1
$C_{22}H_{14}Cl_2I_2N_2O_2$
N-[5-Chloro-4-[(4-chlorophenyl)cyanomethyl]-2-methylphenyl]-2-hydroxy-3,5-diiodobenzamide
Flukiver; R-31520; Seponver. A proprietary preparation containing closantel; a veterinary anthelmintic (flukicide). mp = 217.8° *Janssen Pharmaceutical Ltd.* Name unverified.

6442 Closyl 30 2089
61791-59-1 263-193-2
Sodium-N-cocoyl sarcosinate
Wetting, foaming detergent used in personal care and household products; corrosion inhibitor for mild steel. *Clough.*

6443 Closyl LA 3584
137-16-6 4379 205-281-5
Sodium lauroyl sarcosinate
Soap-like detergent providing wetting and foaming; for personal care and household products; corrosion inhibitor for mild steel. *Clough.*

6444 Cloth oil
That fraction obtained from the residue of Russian petroleum by distillation.

6445 Cloustonite
A variety of asphalt.

6446 Clout
55635-13-7 259-733-1
Alloxydim-sodium
Herbicide for control of annual weeds. *Embetec Crop Protection Ltd.*

6447 Clovacorn Extra
An emulsifiable concentrate containing 220 g 2,4-DB, 30 g linuron and 30 g MCPA per liter; used to control weeds in undersown cereals and seedling grassland. *Farmers Crop Chemicals Ltd.* Discontinued.

6448 Clovean
A detergent for bakehouse usage. *Laporte Industries Ltd.*

6449 Clovotox
7085-19-0 5826 230-386-8
Mecoprop
Soluble concentrate containing 300 g/l mecoprop; for control of weeds in cereals and grassland. *Rhône-Poulenc Environmental Prods. Ltd.*

6450 Clowes' solution
A solution containing 160 g potassium hydroxide in 200 ml water and 10 g pyrogallol; used in the Hempel pipette for the absorption of oxygen.

6451 CLSP 499
Palm kernel oil, lecithin; kosher; coating fat for pastel and chocolate flavored confectionery coatings. *Van Den Bergh Foods.*

6452 Club
2032-65-7 6050 217-991-2
Methiocarb
Pellets containing 4% w/w methiocarb; snail and slug bait *ICI AgroChemicals.* No manufacturer.

6453 Clysar®
Shrink film. *DuPont UK.*

6454 CM8450
31001-77-1 250-426-8
3-Mercaptopropylmethyldimethoxy silane
Coupling agent, chemical intermediate, blocking agent, release agent, lubricant, primer, and reducing agent. *Hüls Am.*

6455 CM8500
4420-74-0 224-588-5
3-Mercaptopropyltrimethoxysilane
Coupling agent, chemical intermediate, blocking agent, release agent, lubricant, primer, and reducing agent. *Hüls Am.*

6456 CM8550
2530-85-0 219-785-8
3-Methacryloxypropyltrimethoxysilane
Coupling agent, chemical intermediate, blocking agent, release agent, lubricant, primer, and reducing agent. *Hüls Am.*

6457 CM8620
3069-25-8 221-334-5
N-Methylaminopropyltrimethoxysilane
Coupling agent, chemical intermediate, blocking agent, release agent, lubricant, primer, and reducing agent. *Hüls Am.*

6458 CM8645
5578-42-7 226-956-0
Methylcyclohexyldichlorosilane
Coupling agent, chemical intermediate, blocking agent, release agent, lubricant, primer, and reducing agent. *Hüls Am.*

6459 CM8650
17865-32-6 402-140-1
Methylcyclohexyldimethoxysilane
Coupling agent, chemical intermediate, blocking agent, release agent, lubricant, primer, and reducing agent. *Hüls Am.*

6460 CM8750
75-54-7 200-877-1
Methyldichlorosilane
Coupling agent, chemical intermediate, blocking agent, release agent, lubricant, primer, and reducing agent. *Hüls Am.*

6461 CM8930
149-74-6 205-746-2
Methylphenyldichlorosilane
Coupling agent, chemical intermediate, blocking agent, release agent, lubricant, primer, and reducing agent. *Hüls Am.*

6462 CM8980
4253-34-3 224-221-9
Methyltriacetoxysilane
Coupling agent, chemical intermediate, blocking agent, release agent, lubricant, primer, and reducing agent. *Hüls Am.*

6463 CM9000
75-79-6 200-902-6
Methyltrichlorosilane
Coupling agent, chemical intermediate, blocking agent, release agent, lubricant, primer, and reducing agent. *Hüls Am.*

6464 CM9050
2031-67-6 217-983-9
Methyltriethoxysilane
Coupling agent, chemical intermediate, blocking agent, release agent, lubricant, primer, and reducing agent. *Hüls Am.*

6465 CM9100
1185-55-3 214-685-0
Methyltrimethoxysilane

Coupling agent, chemical intermediate, blocking agent, release agent, lubricant, primer, and reducing agent. *Hüls Am.*

6466 CM9160
24589-78-4 246-331-6
N-Methyl-N-trimethylsilyltrifluoroacetamide
Coupling agent, chemical intermediate, blocking agent, release agent, lubricant, primer, and reducing agent. *Hüls Am.*

6467 CM9220
22984-54-9 245-366-4
Methyltris(methylethylketoxime)silane
Coupling agent, chemical intermediate, blocking agent, release agent, lubricant, primer, and reducing agent. *Hüls Am.*

6468 CMD 834
Modified cycloaliphatic amine adduct; room temperature curing agent for epoxy resins; end uses including industrial floor grouting, high build glaze, tank linings, general purpose castings and encapsulations. *Shell.*

6469 CN 103
Epoxy acrylate
For paper clear coatings, wood top coatings, screen inks, litho inks, polyethylene coatings, metal decorative coatings, adhesive papers, wood fillers, solder masks and photoresists. *Sartomer.*

6470 CN 104 A80
Epoxy acrylate/Tripropylene glycol diacrylate
Features excellent flow and gloss; used for paper coatings, wood coatings, inks. *Sartomer.*

6471 CN 104 B80
Epoxy acrylate/HDDA
Offers flexibility for paper coatings, wood coatings, metal coatings, inks. *Sartomer.*

6472 CN 104 C75
Epoxy acrylate/Trimethylolpropane triacrylate
Used for paper coatings, wood coatings, metal coatings, inks. *Sartomer.*

6473 CN 104 D80
Epoxy acrylate/GPTA
Used for paper coatings, wood coatings, metal coatings, inks. *Sartomer.*

6474 CN 104 F50
Epoxy acrylate/Di-isodecyl adipate
Used for paper coatings, wood coatings, metal coatings, inks. *Sartomer.*

6475 CN 111
Epoxidized soybean oil acrylate; offers flexibility, pigment wetting, high adhesion for paper coatings, wood coatings, metal coatings, inks. *Sartomer.*

6476 CN112 C60
Epoxy novolak acrylate/TMPTA
Offers high heat resistance; used for inks, solder resists. *Sartomer.*

6477 CN 300
Polybutadiene diacrylate
Used for optical fibers, pottants and encapsulants, conversion coatings, pressure-sensitive adhesives. *Sartomer.*

6478 CN 953
Aliphatic urethane acrylate
Soft; used for printing inks, polyvinyl chloride floor coatings, wood coatings, anaerobic adhesives, paper, flexible overprint varnishes. *Sartomer.*

6479 CN 960
Urethane acrylate resin
Hard; nonyellowing; provides excellent heat, light, and chemical resistance to coating formulations; recommended for use in overprint varnishes, laminating adhesives, plastic film and foil coatings, screen inks, and rigid plastic and *Sartomer.*

6480 CN 962
Aliphatic urethane acrylate resin
Flexible, nonyellowing; provides excellent falling sand and high impact resistance to coating and adhesive formulations; supplied as base resin or in resin/monomer blends which offer flexibility, clarity, and excellent heat and light stability to uv/eb cured products; used in screen inks, foil coatings, laminating adhesives, pressure-sensitive adhesives and metal coatings requiring extremely flexible properties. *Sartomer.*

6481 CN 966
Aliphatic urethane acrylate resin
Highly flexible, nonyellowing; provides excellent high impact resistance to coating and adhesive formulations; supplied as base resin or as resin/monomer blends; which offer exceptional flexibility, clarity, and excellent heat and light stability to uv/eb cured products. *Sartomer.*

6482 CN 970
Aromatic urethane acrylate resin
Hard; produces hard, fast curing, solv-resistant uv/eb-cured products comparable to those made from acrylated epoxy resins; supplied as base resin or as resin/monomer blends; for use in adhesives, and paper, wood, and metal coatings. *Sartomer.*

6483 CN 974
Aromatic urethane methacrylate
High flexibility; hydrophilic; used for adhesives, paper, wood, ink, metal coatings. *Sartomer.*

6484 CNC Antifoam 30-FG
Silicone emulsion; antifoaming/defoaming agent for paper, textile, water-phase paints and other industrial uses; avail. in food grade versions. *CNC Int'l L.P.*

6485 CNC Antifoam A-107
Filled silicone fluid; defoamer for aqeous and nonaqueous systems; effective over very broad pH and temperature range. *CNC Int'l L.P.*

6486 CNC Defoamer 12, 34, 407
Defoamer for paper applications for difficult, foamy high solids, or high resin content systems. *CNC Int'l L.P.*

6487 CNC Defoamer 69, 97
Water-based nonsilicone defoamer for paper applications. *CNC Int'l L.P.*

6488 CNC Detergent E
Coconut fatty acid amine condensate. Detergent, wetting agent, dyeing assistant; leveling agent for acid dyestuffs on wool; rewetting and softening agent for sanforizing; scouring agent for synthetics; also for laundry processing of garments. *CNC Int'l L.P.*

6489 CNC Dispersant WB Series
Surfactants, leveling agents, dispersants, dyeing assistants for textiles. *CNC Int'l L.P.*

6490 CNC Dispersion PE
Complex phosphate ester of nonionic ethoxylates. Forms self-emulsfiable concentrations when mixed with many solvents; base for textile wetting agents. *CNC Int'l L.P.*

6491 CNC Foam Assist AA, AA-100
High solids foaming agents for textile foam finishing. *CNC Int'l L.P.*

6492 CNC Gel Series
Detergents for textile scouring, wetting, dispersing. *CNC Int'l L.P.*

6493 CNC Inhibitor 30
Dyeing and bleaching assistant providing protection against atmospheric yellowing to Spandex fabrics. *CNC Int'l L.P.*

6494 CNC Leveler JH
Sulfated PEG ester; Low-foaming surfactant, wetting agent, dispersant for textile applications; high compatible with high concentrations of acids. *CNC Int'l L.P.*

6495 CNC PAL 100, 200, 300
Surfactants; anionic, nonionic, and cationic low-foaming biodegradable surfactants for rewetting towels, tissue or corrugating medium, de-inking printed wastes, magazine stocks; wetting and penetrating agent, detergent. *CNC Int'l L.P.*

6496 CNC PAL V-8 Supra
Detergent, emulsifier for textile processing. *CNC Int'l L.P.*

6497 CNC Product PW
Print wash used in conjunction with CNC Fix Concentrate for nylon printed with acid dyes. *CNC Int'l L.P.*

6498 CNC Product ST
Surfactant; print wash for use where high concentrations of residual color and print paste are to be removed from cotton, rayon, or synthetics; wetting and scouring agent. *CNC Int'l L.P.*

6499 CNC Soft C-1
Synthetic softener for use where a soft, luxurious hand is required; all-purpose softener/lubricant for textile processing. *CNC Int'l L.P.*

6500 CNC Sol 72-N Series
Wetting and rewetting agents for textiles; stable to dilute acids and alkalies; not for use with water repellent or fluorochemical finishes. *CNC Int'l L.P.*

6501 CNC Sol BD, CNC Sol XNN 11
Solvent/emulsifier blends; for removal of waxes, greases, oils, and greige-mill dirt from cotton, rayon, and blends. *CNC Int'l L.P.*

6502 CNC Solv 809
Solvent/surfactant blend; for deinking and repulping resin coated mixed waste furnishes; prevents redeposition of agglomerated inks and binders. *CNC Int'l L.P.*

6503 CNC Wet CP Conc
Phosphated long chain alcohol; wetting and dispersing agent, leveler for pigments and dyes, in peroxide bleach baths; dyeing assistant; suitable for cotton and synthetics processing; acid and alkali stable. *CNC Int'l L.P.*

6504 CNComerse IMP
Alkyl phospho-sulfate
Nonfoaming biodegradable mercerizing agent, wetting and penetrating agent. *CNC Int'l L.P.*

6505 CO9745
27668-52-6 248-595-8
Octadecyldimethyl [3-(trimethoxysilyl) propyl] ammonium chloride

Coupling agent, chemical intermediate, blocking agent, release agent, lubricant, primer, reducing agent. *Hüls Am.*

6506 CO9750
112-04-9 203-930-7
Octadecyltrichlorosilane
Coupling agent, chemical intermediate, blocking agent, release agent, lubricant, primer, reducing agent. *Hüls Am.*

6507 CO9800
101205-02-1
Octamethylcyclotetrasilazane
Coupling agent, chemical intermediate, blocking agent, release agent, lubricant, primer, reducing agent. *Hüls Am.*

6508 CO9810
556-67-2 6843 209-136-7
Octamethylcyclotetrasiloxane
Coupling agent, chemical intermediate, blocking agent, release agent, lubricant, primer, reducing agent. *Hüls Am.*

6509 CO9816
107-51-7 6844 203-497-4
Octamethyltrisiloxane
Coupling agent, chemical intermediate, blocking agent, release agent, lubricant, primer, and reducing agent. *Hüls Am.*

6510 CO9817
546-56-5 208-904-9
Octaphenylcyclotetrasiloxane
Coupling agent, chemical intermediate, blocking agent, release agent, lubricant, primer, reducing agent. *Hüls Am.*

6511 CO9819
18162-84-0 242-044-5
Octyldimethylchlorosilane
Coupling agent, chemical intermediate, blocking agent, release agent, lubricant, primer, reducing agent. *Hüls Am.*

6512 CO9830
5283-66-9 226-112-1
Octyltrichlorosilane
Coupling agent, chemical intermediate, blocking agent, release agent, lubricant, primer, reducing agent. *Hüls Am.*

6513 CO9835
2943-75-1 220-941-2
Octyltriethoxysilane
Coupling agent, chemical intermediate, blocking agent, release agent, lubricant, primer, reducing agent. *Hüls Am.*

6514 Coagulant CHA
Cyclohexyl amine acetate
Coagulant used in dipped goods from natural latex. *Bayer AG.*

6515 Coagulant WS
Functional polyorganosiloxane. Heat sensitizer permitting coagulation point of 32-60° in mixes based on natural or synthetic latex. *Bayer; Miles; Polysar.*

6516 Coalatex
oxalic acid and oxalate salts. A good coagulating agent in 10% solution; A proprietary coagulating material for use in coagulating rubber latex. No manufacturer.

6517 Coaldet
An electric blasting cap comprising copper-bronze alloy shell; designed especially for initiating explosives in coal mining. *Hercules.* Discontinued.

6518 Coalite N.T.P
25155-23-1 246-677-8
dimethylphenol phosphate (3:1); tri(dimethylphenyl)phosphate. Nontoxic, trixylenyl phosphate plasticizer. d = 1.155; bp_{10} = 243-265°; insoluble in H_2O, soluble in organic solvents. *Coalite Fuels & Chemicals Ltd.* Name unverified.

6519 Coaltec
Wood preservative. *Coalite Fuels & Chemicals Ltd.*

6520 coarse metal
Matte. An impure mixture of ferrous and cuprous sulfides produced in copper smelting. It usually contains from 20-75% copper, 12-45% iron, and 19-25% sulfur.

6521 Coasol
Diisobutyl adipate/glutarate/succinate; coalescing solvent for emulsion paints/adhesives. *Chemoxy International Ltd.*

6522 Coate®
Metal salts of organic acids; driers for coating compositions. *Hüls Am.*

6523 Coatmaster® K580
Binder chemical derived from corn. *Grain Processing.*

6524 cobalt
7440-48-4 2488 231-158-0
Co
Cobalt
Super cobalt. Oxidizing agent, lamp filaments, in manufacture of cobalt steel; in porcelain, glass, pottery, enamels. *Aldrich; Atomergic Chemetals; Cerac; Noah Chemical.*

6525 Cobalt 254
A paste containing compounds of cobalt; loss-of-dry inhibitor for nonaqueous coating compositions, inhibitor of gel-time drift in unsaturated polyester resins. *Hüls Am.*

6526 cobalt brass
Acid-resisting alloys of 52% copper, 17-25% zinc, and 22-30% cobalt.

6527 cobalt bronze
A phosphate of cobalt and ammonium.

6528 cobalt chloride (ous)
7646-79-9 2498 231-589-4
$CoCl_2$
cobaltous chloride anhydrous
Absorbent for ammonia, gas masks, electroplating, inks, hygrometers, manufacture of vitamin B_{12}, flux for magnesium refining, solid lubricant, dye mordant, catalyst, barometers, laboratory reagent, fertilizer additive. *Celtic Chem. Ltd; Mallinckrodt; Nihon Kagaku Sangyo; Spectrum Chem. Mfg.*

6529 cobalt steel
Alloys of steel with cobalt; used for certain parts of electrical machinery requiring a high permeability. An alloy containing 34.5% cobalt, at high inductions, has a higher permeability than pure iron.

6530 cobalt-chromium-molybdenums steel
Usually an alloy of iron with 3.05% molybdenum, 2.16% chromium, 1.33% cobalt and 0.65% carbon.

6531 cobaltic fluoride
10026-18-3 2491 233-062-4
CoF_3
cobalt trifluoride
cobaltic fluoride. Used in fluorination of hydrocarbons (Fowler process). d = 3.88; reacts with air and H_2O. *Elf Atochem N. Am.; Atomergic Chemetals; Cerac.*

6532 cobaltic oxide
1308-04-9 215-156-7
Co_2O_3
Cobalt (III) oxide
cobalt black; Cobaltic oxide; Cobalt oxide; cobaltic oxide monohydrate. Black solid that is insoluble in water and soluble in sulfuric acid. It is a pigment used in coloring enamels and glazing pottery. *Atomergic Chemetals; Cerac; Noah Chem.*

6533 cobaltic oxide monohydrate
12016-80-7 2492 234-614-7
$CoHO_2 \cdot H_2O$
cobaltic hydroxide
cobalt hydroxide oxide. Oxidation catalyst Insoluble in H_2O.

6534 cobaltic trifluoride
10026-18-3 2491 233-062-4
CoF_3
Cobalt trifluoride
Fluorinating agent. d = 3.88. *Elf Atochem N. Am.; Atomergic Chemetals; Cerac.*

6535 cobaltous acetate tetrahydrate
6147-53-1 11, 2427 213-033-2
$C_4H_6CoO_4 \cdot 4H_2O$
Cobaltous acetate tetrahydrate
Inks, paint and varnish driers, catalyst, anodizing, mineral supplement in feed additives, foam stabilizer. Tetrahydrate; d = 1.705; soluble in H_2O and EtOH *Atomergic Chemetals; Celtic Chemical Ltd; Mallinckrodt; Nihon Kagaku Sangyo; Noah Chemical.*

6536 cobaltous carbonate
513-79-1 2479 235-714-3
$CCoO_3$
cobaltous carbonate basic
cobaltous hydroxide carbonate. Manufacture of cobaltous oxide, cobalt pigments, cobalt salts; intermediate. d= 4.13; insoluble in H_2O, EtOH, EtOAc *Atomergic Chemetals; Celtic Chemical Ltd; Nihon Kagaku Sangyo; Noah Chemical.*

6537 cobaltous chloride
7646-79-9 2498 231-589-4
$CoCl_2$
cobaltous chloride anhydrous
Absorbent for ammonia, gas masks, electroplating, inks, hygrometers, manufacture of vitamin B_{12}, flux for magnesium refining, solid lubricant, dye mordant, catalyst, barometers, laboratory reagent, fertilizer additive. *Celtic Chemical Ltd; Mallinckrodt; Nihon Kagaku Sangyo; Spectrum Chemical Mfg.*

6538 cobaltous naphthenate
Cobaltous naphthenate

Paint and varnish drier, bonding rubber to steel and other metals. *Akzo; Nuodex Espanola SA; Troy.*

6539 cobaltous nitrate

10141-05-6 2505 233-402-1

CoN_2O_6

cobalt(II) nitrate

Oxidizing agent, dangerous fire risk in contact with organic materials. mp = 1493°; d = 8.92. *Atomergic Chemetals; Celtic Chemical Ltd; Mallinckrodt; Nihon Kagaku Sangyo; Noah Chemical.*

6540 cobaltous oxide

1307-96-6 2507 215-154-6

CoO

Cobalt(II) oxide

cobaltous monoxide. Readily absorbs oxygen at room temperature. Used as pigment in ceramics. Pigment, coloring enamels, glazing pottery. mp = 1935°; d = 5.7 - 6.7; LD_{50} (rat orl) = 1.70 g/kg. *Atomergic Chemetals; Cerac; Noah Chemical.*

6541 cobaltous sulfate

10124-43-3 2510 233-334-2

$CoSO_4$

cobalt(II) sulfate

Ceramics, pigments, glazes, in plating baths for cobalt, additive to soils, catalyst, paint and ink drier, storage batteries. *Elf Atochem N. Am.; Atomergic Chemetals; Johnson Matthey plc; Mallinckrodt; Nihon Kagaku Sangyo.*

6542 Cobaltron Steel Alloy

A proprietary alloy containing iron with about 11% chromium, 2.25% cobalt, 1.5% carbon, 1.25% molybdenum, and 0.25% tungsten. No manufacturer.

6543 Cobatope-60

7646-79-9 2498 231-589-4

Cobaltous chloride ^{60}Co

Radioactive agent. *Bristol-Myers Squibb Co Inc.* Name unverified.

6544 Cobex

29091-05-2 249-419-2

dinitramine

Weedkiller containing dinitramine. *ICI Chemical & Polymers Ltd.* Discontinued.

6545 Cobox® L

Ammonical copper polyacrylate

Contact fungicide for control of fungus and bacterial diseases in coffee, cotton, fruits. *BASF AG.*

6546 Cobrol

A photographic developer. *May & Baker Ltd.* Name unverified.

6547 Cocamide DEA

61791-31-9 263-163-9

PEG-2 cocamine; Ethokem C/12; Berol 307; Chemstat® 273-C;Ethomeen® C/12; Norfox® DC. Cationic surfactant adjuvant containing bis-2 hydroxyehyl cocamine; wetting agent for phosphonoglycine herbicides and ammonium sulfate. *Midkem AgroChemicals; Berol Nobel AB;Chemax; Norman, Fox`.*

6548 Cocamide MEA

68140-00-1 268-770-2

Cocamide MEA

Comperlan P 100. Foamer, thickener, viscosity builder for powd. detergents, soaps; dispersant and blending agent for many cosmetic products *Pulcra SA.*

6549 Cocamide MEA (1:1)

68140-00-1 268-770-2

Cocamide MEA (1:1)

Carsamide® CMEA. Foam booster, stabilizer, viscosity builder for shampoos and detergents. *Lonza Inc.*

6550 Cocamide MIPA

68440-05-1 269-793-0

Empilan® CIS; Coconut monoisopropanolamide; Amidex CIPA; Rewomid® IPP 240; Witcamide® PPA. Foam stabilizer, detergent, shampoo additive. *Albright & Wilson UK.*

6551 Cocamidopropyl betaine

61789-40-0 263-058-8

$RCO-NH(CH_2)_3N^+(CH_3)CH_2COO^-$

Cocamidopropyl dimethyl glycine

CADG; Abluter BE, CPB; Amido betaine C, C-45; Amonyl® 380BA, 440 NI; Ampholak BCA-30; Ampholan® 197; Ampholyt™ JB 130; Rewoteric® AMB-14, AMB-14S, AMB-15, AM B45; AM B 50; Ritataine; Rititaine B; Schercotaine CAB, CAB-G; Swanol AM-3130N; ; AMPHOSOL® CA, CG; Amphotensid B4, B4 LS; Caltaine C35; Chembetaine® C, CGF, CL; Dehyton® PK, 3016 B, K, KE, KE 3016, PK 45; Deriphat® BAW; DeTAINE CAPB-35, CAPB-35HV; Tego® betaine C, E, F, F50, L-7, L-7F, L-5351, S, ZF; Velvetex® BA-35,; Emcol® 6748, Coco betaine; Emcol® DG, NA30; Emery® 6744; Empigen ® BS/AU, BS/F, BS/FA, BS/H, BS/P; Euroquat C45, CPB, K, LA, LAC; Foamtaine CAB, CAB-G; Genagen CA 818, CAB; Incronam 30; Lebon 200; Lonzaine® C, CO; Mackam™ 35, 35HP, J, L; Manroteric CAB; Miratiane BD-J, BET-C-30, BET-W, CAB, CAB-A, CAB-O, CB, CBC, CB/M, CBR; Monoteric ADA, CAB, CAB-LC, CAB-XLC, COAB; Naxaine® C, CO, Cocbetaine; Nutrol betaine MD 3863, OL 3798; Proteric CAB, COAB; Ralufon® 414; Rewoteric AMB-13,,. High foaming, conditioning detergent for mild shampoos; solubilizer for lauryl sulfates in concentrated shampoos; thickener. Gardner 4 liquid; soluble in H_2O; disperses in glycerol trioleate; density = 8.8 lb/gal; visc 7cSt (100°F); sp gr = 1.03; clear pt = 0°. Pulcra SA; Henkel /Chems Group; Taiwan Surf; McIntyre; Witco/Oleo Surf.; Lonza; Chemron; Goldschmidt; Henkel/Emery; Huntington Labs; Inolex; McIntyre; Mona; Scher; Sherex.

6552 Cocamidopropyl dimethylamine

68140-01-2 268-771-8

N-[3-Dimethylamino)propyl]coco amides

Chemidex C; Chemidex WC; Incromine CB; Lexamine C-13; Mackine 101; Mazeen® DAPL; Mazeen® SHCFA; Miramine® CODI; Schercodine C. Foaming agent and conditioner used in hair care products; intermediate for hair conditioning agent. mp = 30°; HLB = 10.2; acid no 4 max; pH = 9.5; strong skin and severe eye irritant. *Croda Surfactants Ltd; Chemron; Inolex.*

6553 Cocamidopropyl dimethylamine lactate

68425-42-3

N-[3-(Dimethylamino)propyl]cocoamide lactate

Incromate CDL; Mackalene™ 116. Surfactant, foamer, conditioner for personal care products, conditioners; base for cationic emulsions. Cationic; liquid; pH = 5.0. *Croda Inc.* Discontinued.

6554 Cocamidopropyl dimethylamine propionate

68425-43-4

Emcol® 1655; Foamid 117; Incromate CDP. Cosmetics and toiletry surfactant used as antistat, conditioner, emollient, foaming and substantive agent. *Alzo; Croda Inc.*

6555 Cocamidopropyl hydroxysultaine

68139-30-0 268-761-3

Cocamidopropylhydroxysultaine

Chembetaine CAS. Anti-irritant for other surfactants; especially for baby shampoos and baby bath products; detergent for heavy-duty industrial alkaline cleaners (steam cleaners, wax remover, hard surface cleaner); wetting agent in acid pickling of metals; lime soap dispersant. *Chemron.*

6556 Cocamidopropylamine oxide

68155-09-9 268-938-5

Cocamidopropylamine oxide

Chemoxide CAW. Surfactant for mild, low irritation personal care and industrial applications, e.g., shampoos, facial cleansers, bath products; foam and viscosity builder, emollient over broad pH range. *Chemron.*

6557 Cocamine oxide

61788-90-7 263-016-9

$R-NH-CHCH_2COOHCH_3$, R represents the coconut radical

Coco dimethylamine oxide

Aminoxid A 4080; Aromax® DMC; Aromax® DMC-W; Barlox® 12; Chemoxide® WC; Empigen® OB/AU; Genaminox KC; Mackamine™ CO; Noxamine CA 30; Rhodamox® C; Schercamox DMC; Sochamine OX 30Empigen® 5083. Foam booster/stabilizer and viscosity modifier for shampoos, foam baths, cleaners; improves conditioning in shampoos; solubilizer for liquid bleach products. Gardners 1 max liquid; sp gr = 0.89; HLB 18.6; flash point = 21°; pH = 6-9; prolonged contact may cause severe burns to eyes and severe skin irritation; flammable. *Albright & Wilson UK.*

6558 Coceth-27

61791-13-7

$R(OCH_2CH_2)_nOH$, R represents coconut radical, avg n=27

Dehydol LT 3; Genapol® C-050; Genapol® GC-050. Emulsifier, solubilizer for solvents, oils; base for production of ether sulfates; raw material for low-foaming detergents, dishwashing, cleansing agents, cold cleaners; superfatting agent. Liquid; soluble in oil; HLB 7.3; hyd no. 163-174; pH = 6.0-7.5. *Pulcra SA.*

6559 cochrome

An alloy of 60% cobalt, 12% chromium, 24% iron, and 2% manganese. It has been used in the place of Nichrome for the elements of electrical heating apparatus.

6560 Coclor

7447-39-4 2699 231-210-2

Cl_2Cu

cupric chloride

Industrial cupric chloride (35/36% Cu). Used as a catalyst, deodorant and desulfurizer. *Mechema Chemicals Ltd.*

6561 Coco betaine

68424-94-2 270-329-4

Coco betaine

Chembetaine BW, CB. Viscosity builder, gelling agent, industrial surfactant;

lime soap dispersant; mild to skin; tolerant to hard water; stable in high-electrolyte solutions; cationic in acid and anionic in alkaline media. *Chemron.*

6562 Coco trimethyl ammonium chloride
61789-18-2 263-038-9
Coco trimethyl ammonium chloride
Chemquat C/33W. Surfactant, corrosion inhibitor, antistat for plastics, textile dyeing aid, gel sensitizer in latex foam production; viscosity depressant in paper and textile softener formulations. *Chemax.*

6563 cocoa butter
8002-31-1 11, 9208
Cocoa butter; theobroma oil. The fat extracted from the seeds of *Theobroma cacao*. The seeds contain from 35-45% of the butter. It has an acid value of 0.6-1.3 (oleic acid), a Saponification value of 192-193 and is used in confectionary. mp = 33-34°; d_{25}^{100} = 0.858-0.864; n_D^{40} = 1.4537-1.4578.

6564 Cocodiamine
61791-63-7 263-195-3
N-Coco-1,3-diaminopropane
Diamine KKP; Duomeen® CD; Duomeen® C; Coco propylenediamine; Jet Amine DT; Radiamine 6560. Chemical intermediate; corrosion inhibitor, fuel oil additive, flotation agent; used in metals, textiles, plastics, herbicides; epoxy curing agent. *Akzo; Akzo Chem. BV.*

6565 Cocoloid®
Alginate
Used for chocolate milk, soft-serve frozen desserts, ice cream, custard, variegated syrup and fudge topping. *Kelco.*

6566 coconut butter
8001-31-8 11, 2457 232-282-8
Coconut oil, coconut milk, copra. The sweet watery liquid contained in the coconut is called coconut milk. This disappears and gives place to a soft edible pulp, which hardens in the air, and is sold as copra. Copra contains from 60-70% oil; which is extracted as coconut oil or butter.

6567 Coconut Oils® 76, 92, 110
8001-31-8 11, 2457 232-282-8
Coconut oils, refined bleached, deodorized; base for ice cream coatings, candies, icings; clouding agent for beverages; emollient for creams and lotions; flavor solubilizer; pharmaceutical vehicle. *Stepan/PVO.*

6568 Cocoyl sarcosine
68411-97-2 270-156-4
Hamposyl® C, CZ; Vanseal® CS; N-Cocoyl-N-methyl glycine; N-methyl-N-(1-coconut alkyl) glycine; Crodasinic C. Detergent, wetting and foaming agent, foam stabilizer, emulsifier, anticorrosive agent, conditioner for hair and rug shampoos, cosmetics, skin cleansers; biodegradable. *W R Grace/Hampshire; Chemplex Chems; Croda Chem. Ltd; R. T. Vanderbilt Co Inc.*

6569 cod liver oil
8001-69-2 11, 2464 232-289-6
Oils, cod liver; morrhua oil. Fixed oil expressed from fresh livers of *Gadus morrhua* and other species of codfish; medicine (vitamin A and D content), chamois-leather tanning. *Arista; R.W. Greeff; Penta Mfg.*

6570 Codacide Oil
Adjuvant containing 95% emulsifiable vegetable oil; extender and wetting agent for herbicides, fungicides and insecticides. *Microcide Ltd.*

6571 Code 321
8016-70-4 232-410-2
Partially hydrogenated soybean oil; kosher; high performance fat used for filler, snack frying, prepared foods, spray oil. *Van Den Bergh Foods.*

6572 Codite
A proprietary trade name for a vulcanized fiber and pure cotton cellulose plastic tubing. No manufacturer.

6573 Cofill
A finely ground mixture (50:50) of highly dispersed silica-resorcin; adhesion powder for bonding rubber mixtures to treated and untreated synthetic textiles and metal fabrics. *Degussa AG.* Name unverified.

6574 Co-Gell® A2/B270
Mineral oil, aluminum isostearates/ laurates/ palmitates, isopropyl palmitate; stabilizer for continuous phase of Water-oil emulsion systems and gelling agent for oils and nonaqueous fluids; waterproofing, lubricity properties; improved pigment adhesion and *Rhône-Poulenc Surf.*

6575 Cohedur®
Bonding agents for rubber. *Bayer AG; Bayer plc.*

6576 Cohedur® RK
108-46-3 8323 203-585-2
Resorcinol
Resorcinol component with filler, 1:1 ratio; bonding agent in polychloroprene. *Bayer; Miles; Polysar.* Name unverified.

6577 Coherex®
Concentrate nonvolatile emulsion consisting of 60% natural petroleum resins and 40% wetting solution; dust retardant providing dust control on ballparks, playgrounds, dirt roads, construction sites. *Witco/Golden Bear.* Discontinued.

6578 COHRlastic® 400
63394-02-5
Solid silicone rubber; general purpose molded sheet. *Furon/CHR.* Name unverified.

6579 COHRlastic® 1010
Silicone coated on fiberglass. Thin, tough, dimensionally stable, flexible fabric for applications including belting, vacuum blankets, thermal shielding, and diaphragms; 1000 series has superior electrical properties and good abrasion resistance; general purpose, electrical grade. *Furon/CHR.* Name unverified.

6580 COHRlastic® 1867
63394-02-5
Conductive silicone rubber. Thermally conductive coated fabrics providing thin, cost-effective heat transfer capability; available plain or with pressure-sensitive adhesive 1 side. *Furon/CHR.* Name unverified.

6581 COHRlastic® 3320
63394-02-5
Solid silicone rubber, fiberglass-reinforced (20% for 1/16 in.; 14% 3/32 in., 10% 1/8 in.); dimensionally stable, durable material for press pads, belting, and gasketing; 3320 grade offers lubricating oil resistance, and excellent comprehensive set resistance. *Furon/CHR.* Name unverified.

6582 COHRlastic® 8016
Aluminum mesh impregnated with neoprene rubber; electrically conductive RFI/EMI shield material; the mesh maintains electrical contact, while the elastomer provides an effective seal in gasket applications. *Furon/CHR.* Name unverified.

6583 COHRlastic® 9041
63394-02-5
Solid silicone rubber, phenyl-based; high strength, low temperature grade. *Furon/CHR.* Name unverified.

6584 COHRlastic® F12
Silicone foam sheet; low density, flame retardant foam sheet for aviation, mass transit, automotive, electronics, construction, furnishings industries; applications including fireblocks, thermal barriers, noise and vibration dampeners, insulation, high performance gaskets and seals. Non-toxic on burning. *Furon/CHR.* Name unverified.

6585 COHRlastic® FR17
Silicone rubber foam bonded one side to a reinforcing silicone coated fiberglass fabric; lightweight, thermal insulating/fire blocking composite. *Furon/CHR.*

6586 COHRlastic® R10450
63394-02-5
Silicone closed cell sponge rubber, fiberglass-reinforced; for high performance gasketing, thermal shielding, vibration mounts, press pads; offers the compressbility of sponge and dimensional stability in the X-Y direction; for consistent size and shape of die-cut parts; eliminates outward extrusion under pressure. *Furon/CHR.* Name unverified.

6587 COHRlastic® R10490
Fluorosilicone sponge rubber sheet; for gasket, seal applications where it maintains integrity in presence of jet fuel, engine oil, etc.; military and commercial aircraft industry. *Furon/CHR.* Name unverified.

6588 COHRlastic® TC100
63394-02-5
Conductive silicone rubber; unsupported, thermally conductive grade providing thermal and mechanical protection to electronic devices. *Furon/CHR.* Name unverified.

6589 cohydrol
A colloidal graphite solution.

6590 coinage bronze
An alloy of 95% copper, 4% tin, and 1% zinc.

6591 Coke
50-36-2 2517 200-032-7
The carbonaceous residue from the distillation of coal. Used as a fuel.

6592 Colacryl
Acrylic resin in solid beads or as a solution. *Bonar Polymers Ltd.*

6593 Colamine
141-43-5 3772 205-483-3
Aminoethanol
Chemical intermediate. Name unverified.

6594 Colanyl
Pigments for resin dispersions; used in aqueous based paints. *Hoechst UK.*

6595 Colasta
A proprietary trade name for a phenol-formaldehyde synthetic resin. No manufacturer.

6596 Colastex
A patented mixture of Colas (cold asphalt) and rubber latex for improving and hardening roads. No manufacturer.

6597 Colcar®
A range of carriers for dyeing synthetic fibers and their blends. *Allied Colloids Ltd.*

6598 Colcolor
Carbon black/plastic concentrate; used for pigmenting of paper and cement for the plastics processing industry, pigmenting of PE, PP, EVA, PVC, PS, SAN, and ABS. *Degussa AG.* Name unverified.

6599 cold varnishes
Varnishes obtained by heating linseed oil to 105 C. for four and a half hours, adding manganese borate, linoleate, or resinate, and stirring the mass with compressed air.

6600 Coles solder
An alloy of 82% tin, 11% aluminum, 5% nickel, and 2% manganese.

6601 Col-Evac
144-55-8 8726 205-633-8
Sodium bicarbonate
Replenisher; alkalizer. *O'Neal, Jones & Feldman Pharmaceuticals.*

6602 Colex 900 BP
Acrylic bonding paste. *Colourex Ltd.*

6603 Colex 1000 FR
Fire retardent polyester gel coat; special gel coat resin for manufacture of imitation fuel effects for electric and gas. *Colourex Ltd.*

6604 Colex 2000, 2200
General purpose gel coats. *Colourex Ltd.*

6605 Colex 4300
Casting resin for cold casting. *Colourex Ltd.*

6606 Colfite
A proprietary trade name for a graphited laminate bearing material. No manufacturer.

6607 colfosceril palmitate
63-89-8 2540 200-567-6
$C_{40}H_{80}NO_8P$
(R)-4-hydroxy-N,N,N-trimethyl-10-oxo-7-[(1-oxahexadecyl)oxy]-3,5,9-trioxa-4-phosphapentacosan-1-aminium inner salt 4-oxide
dipalmitoyl phosphatidylcholine; DPPC. Pulmonary surfactant, used to treat infant respiratory distress symptoms. Also used as an aid in the diagnosis of fetal lung immaturity. mp = 234-235°; $[\alpha]_D^{23}$ = 6.6° (c = 4.2 $CHCl_3$-EtOH); soluble in organic solvents.

6608 Collacral®
Thickeners for polymer dispersions and other aqueous systems, paints, adhesives, sealants, for production of nonwovens. *BASF AG.*

6609 Collafix®
A range of acrylic copolymers for use as wallpaper adhesives; all are concentrated liquids; grades are available for both conventional and pre-pasting adhesives. *Allied Colloids Ltd.*

6610 collagen
9007-34-5 2543 232-697-4
Collagen
Collagen fiber; ossein; soluble animal collagen. Fibrous protein derived from connective tissues in animals; clear to hazy, colorless; mw ·e 100,000; as adhesive, in cosmetic industry for face creams, lotions, hair preparations; sausage casing; as fibers in sutures; as a gel in photographic emulsions. *Croda; Hormel; Inolex; Maybrook; RITA; U.S. BioChemical.*

6611 Collagen 15K
2476
Hydrolyzed mixed glycosaminoglycans, hydrolyzed collagen, hydrolyzed dermal proteins; cosmetic ingredient for moisturizing applications. *Brooks Industries.*

6612 Collagen CLR
9007-34-5 2543 232-697-4
Carrier of native soluble collagen in weakly acid hydrophilic medium; product for aging skin, wrinkle and after-sun treatment. *Dr. Kurt Richter; Henkel/Cospha; Henkel Canada.*

6613 Collagen Hydrolyzate Cosmetic 55
9015-54-7 310-296-6
Hydrolyzed collagen
Protective colloid effect, anti-irritant properties; substantive film-former with dye leveling effects for skin and hair care products (shampoos, conditioners, color treatments, shave creams), liquid soaps, dish detergents. *Maybrook.*

6614 Collagen Native Extra 1%
9007-34-5 2543 232-697-4
Soluble collagen
Moisturizer, film-former for skin care products; protective barrier. *Maybrook.*

6615 Collagenase
9001-12-1 2544 232-582-9
Clostridiopeptidase A
3.4.4.19; digests native collagen at about physiological pH. No manufacturer.

6616 Collamino 25
9015-54-7 310-296-6
Collagen amino acids
Cosmetic ingredient. *Brooks Industries.*

6617 Colla-Moist CG
Collagen glycerides
Moisturizing protein complexes for cosmetics. *Brooks Industries.*

6618 Colla-Moist WS
Propylene glycol, hydrolyzed collagen, PPG-12-PEG-65 lanolin oil; moisturizing protein complexes for cosmetics. *Brooks Industries.*

6619 Collasol
9007-34-5 2543 232-697-4
Soluble animal collagen; humectant; hygroscopic film former; conditioner; skin care products *Croda Inc; Croda Chem. Ltd.*

6620 Collatex
9005-34-9
Ammonium alginate
Kelco Int'l. Ltd.

6621 collaurin
4412
A form of colloidal gold.

6622 collene
Colloidal silver.

6623 Collet steel
A manganese steel with small amounts of chromium and 0.75% carbon.

6624 Collex G
8061-52-7
Fermented calcium lignosulfonate; dispersant; low-cost concrete additive. *Borregaard Ligno Tech.*

6625 Collidine
108-75-8 9848 203-613-3
$C_8H_{11}N$
2,4,6-Trimethylpyridine
2,4,6-Collidine; S-collidine; σ-Collidine. Trimethyl- and/or methylethylpyridines; used as chemical intermediate, dehydrohalogenating agent. mp = -43°; bp = 171-172°; d = 0.9170; n_D^{20} = 1.4979; soluble in H_2O (35 g/l), organic solvents; LD_{50} (rat orl) = 400 mg/kg.

6626 collodion
2480
A solution of pyroxylin (nitrocellulose) in ether and alcohol; cements, coating wounds and abrasions, solvent for drugs, corn removers, process engraving, lithography, photography. *Spectrum Chemical Mfg.*

6627 collodion cotton
9004-70-0 8195
Pyroxylin

6628 Colloid 106
Low molecular weight acrylic polymer aqueous solution; scale inhibitor used in water treatment and dishwashing compounds *Rhône-Poulenc.*

6629 Colloid 111, 111D
37199-81-8
Sodium salts of polycarboxylic acid; surfactants, dispersants for agricultural formulations. *Rhône-Poulenc Surf.; Rhône-Poulenc/Water soluble Polymers.*

6630 Colloid 202
9003-04-7
Medium molecular weight sodium polyacrylate solution; builder additive for detergents; antiredeposition aid for laundry applications; dispersion aid for dishwash formulations; process and anticaking aid for dry products; sequestrant for liquid or dry formulations. *Rhône-Poulenc/Water Soluble Polymers.*

6631 Collokit
Technetium ^{99}Tc sulfur colloid; radioactive agent. *Abbott Laboratories.* Name unverified.

6632 Collone
Emulsifying wax. *Rhône-Poulenc UK.*

6633 Colloresin D
A methyl cellulose soluble in cold water. No manufacturer.

6634 Colloresin DK
A proprietary alkyl ether of cellulose soluble in cold water insoluble in hot water; for use as a thickening agent in textile printing. No manufacturer.

6635 Collosol Argentum
7440-22-4 8647 231-131-3
A proprietary preparation of colloidal silver; used as an ocular antiseptic. *Crookes Laboratories.* Name unverified.

6636 Collys
Native and modified starches; used in the board industry. *Roquette (UK) Ltd.*
6637 Colmonoy
A proprietary trade name for a chromium boride; an abrasive. No manufacturer.
6638 Colmonoy No.6
A corrosion-resisting alloy with about 75% nickel base; an essential constituent is chromium boride. No manufacturer.
6639 Colona Steel
A proprietary trade name for a nickel-chromium steel containing some manganese. No manufacturer.
6640 Color Seal
A two-step application of sealers; seals, hardens and dustproofs concrete, providing an attractive, colored, very high gloss, tough abrasion and chemical resistant floor. *Secure Inc.* Discontinued.
6641 Colorado silver
An alloy of 57% copper, 25% nickel, and 18% zinc. It is a nickel silver. *German Silver.*
6642 Colorlok
Inorganic salts; antimigrator for direct dyes on cellulosics. *Marlowe-Van Loan.*
6643 Colormatch®
High pigment loading dispersions for excellent color control and economy of use; for polyester, epoxy, vinyl, urethane, vinyl ester, gel coats. *Plasticolors.*
6644 Color-Max
Pigment/plasticizer dispersions; pourable colors for plastisols and other systems; offered custom matched to standard. *W J Ruscoe Co.*
6645 Colorol 70
Phosphoamino compounds; wetting and dispersing agent, binder for 2-part PU varnishes, epoxy-varnishes in systems based on PVC, PVA. *Lucas Meyer.*
6646 Colorol Rust Binder
Compound based on modified oils and carboxylic ester; rust binder, hardener, corrosion inhibitor; penetrates and wets rust and similar metal oxides allowing water-repellent film to coat and protect; contains xylene. *Lucas Meyer.* Discontinued.
6647 Colorol Standard
Phosphoamino compounds modified with film-formers; wetting and dispersing agent, binder for pigmented air-dry and stoving finishes, anticorrosive paints, alkyd and marine paints, zinc dust paints, bituminous paints, epoxy resins, putty, solventless floor sealers, two-part finishes and inks. *Lucas Meyer.*
6648 Colortrend
31001-77-1 250-426-8
Colorant dispersions; for volumetric machine coloring of coating compositions. *Hüls Am.*
6649 Col-o-tex
A proprietary trade name for lacquer coated fabrics. No manufacturer.
6650 Colour-Chem
Organic pigment powders; for paints, printing inks, plastics and rubber. *Colour-Chem Ltd.* Name unverified.
6651 Col-o-vin
A proprietary trade name for polyvinyl synthetic resin coated fabrics. No manufacturer.
6652 Colsol
Fatty amide; cotton and wool softener and lubricant for the textile industry; finishing agent. *Scher Chemicals Inc.*
6653 Colstar®
For the agriculture industry. *DuPont UK.*
6654 Colturiet
A wide range of epoxy coatings, coal tar, solvent free and solventless coatings; also tank coatings. *Sigma Coatings.* Name unverified.
6655 Colugel
21645-51-2 355 244-492-7
aluminum hydroxide
A proprietary aluminum hydroxide gel; used as an antacid. *Ulmer Pharmacal Co.* Name unverified.
6656 Colvinal®
Vinyl copolymer sizes for application to all continuous filament cellulosic fibers; used with starch or CMC for sizing synthetic staple yarns or blends. *Allied Colloids Ltd.*
6657 Comac Bordeaux Plus
Copper sulfate/lime complex; a protectant fungicide for the control of blight and canker. *McKechnie Chemicals Ltd.*
6658 Comac Macuprax
Wettable powder containing 16.7% weight/weight copper sulfate and cufraneb (outside U.S.); a protectant fungicide against potato blight, canker in fruit trees and mildew in grapes. *McKechine Chemicals Ltd.*
6659 Comac Parasol
20427-59-2 2709 243-815-9
Copper hydroxide
A protectant fungicide. *McKechnie Chemicals Ltd.* Name unverified.
6660 Combined Seed Dressing
Fungicide/insecticide. *Murphy Chemical Co Ltd.* Discontinued.
6661 Combismalt®
Special-purpose enamels for various uses (e.g., regenerative heat exchangers and electronic components); especially for electrophoretic or electrostatic powder application. *Bayer AG.*
6662 Comboloob 0609
Synthetic and hydrocarbon wax system; lubricant for processing under moderate to high temperature and shear conditions; for pipe, window profiles, injection molding. *Astor Wax.* Name unverified.
6663 Combovac-30
B_1 type, lasota strain, Newcastle with Massachusetts and Connecticut types; for immunization of poultry. *Intervet Inc.*
6664 Comet metal
An alloy of 67% iron, 30% nickel, 2.2% chromium, with small quantities of manganese and copper.
6665 Commando
63782-90-1
flamprop-M isopropyl
Emulsifiable concentrate of 200 g flamprop-M-isopropyl per liter; used for control of wild oats in cereal crops. *Shell UK.*
6666 Common Degras
Anhydrous lanolin derivative from crude wool grease; EP and slip aid for wire drawing compounds; in slushing oils, cutting oils and lubricants; long term rust preventive; in leather stuffing greases; waterproofing agent. *Fanning.*
6667 Comodor, Comodor 600
35256-85-0 252-470-3
$C_{15}H_{23}NO$
tebutam
2,2-dimethyl-N-(1-methylethyl)-N-(phenylmethyl)propanamide; Benzyl-N-isopropyltrimethylacetamide; Dimethyl-N-(1-methylethyl)-N-(phenylmethyl)propanamide; Propionamide, N-benzyl-2,2-dimethyl-N-isopropyl-. Emulsifiable concentrate containing 720 g/l or 600 g/l tebutam; weed germination inhibitor. *Farm Protection Ltd.*
6668 Comoflastic®
Fibers. *DuPont UK.*
6669 Comox
Cobalt oxide-molybdenum oxide
Cobalt and molybdenum oxides on alumina. Abrasives. *Laporte Industries Ltd.*
6670 Compactrol®
10101-41-4 1753 240-991-9
CaH_4O_6S
Calcium sulfate dihydrate NF
Alabaster; C.I. 77231; C.I. pigment white 25. Tablet and capsule filler for pharmaceutical tablets manufacturing by direct compression. *Penwest Pharmaceuticals Co.*
6671 Compak
Composite kit of photographic processing chemicals. *May & Baker Ltd.* Name unverified.
6672 Compalox
1344-28-1 369 215-691-6
aluminum oxide
Specialty aluminum oxide product; used for the purification of waste water. *Lonza AG; Lonza Inc.*
6673 Compass
Mixture of iprodione and thiophanate-methyl; systemic fungicide for field corps. *Rhône-Poulenc Crop Protection Ltd.*
6674 Comperlan COD
Cocamide DEA
Foam and viscosity increasing agent with emulsifying properties; conditioner for shampoos and other cosmetic and pharmaceutical applications; solvent. *Pulcra SA.*
6675 Comperlan KD
Cocamide DEA
Thickener, foam booster/stabilizer, superfatting agent for personal care products; hair conditioner. *Henkel; Henkel Canada; Henkel KGaA; Pulcra SA.*
6676 Comperlan LD
120-40-1 204-393-1
Lauramide DEA superamide
Foam booster, stabilizer, detergency and viscosity builder, emulsifier for personal care products, household detergents. *Henkel; Henkel KGaA; Pulcra SA.*

6677 Comperlan P 100
68140-00-1 268-770-2
Cocamide MEA
Foamer, thickener, viscosity builder for powd. detergents, soaps; dispersant and blending agent for many cosmetic products *Pulcra SA*.

6678 Comperlan PKDA
Cocamide DEA (2:1)
Detergent base, foam booster/stabilizer, emulsifier, wetting agent for household detergents. *Pulcra SA*.

6679 Compimide
High temperature bismalemide thermosetting resins. *The Boots Co plc*. Discontinued.

6680 Compitox Extra
7085-19-0 5826 230-386-8
Mecoprop
Soluble concentrate containing 570 g/l mecoprop; for control of weeds in cereals and grassland. *Rhône-Poulenc Crop Protection Ltd*.

6681 Complemix
577-11-7 3460 209-406-4
Dioctyl sodium sulfosuccinate USP/BP
Used in the process industry. *Cyanamid BV*.

6682 Comploment
Vitamin B_6 in a controlled release tablet; for Vitamin B_6 deficiency. *Napp Laboratories Ltd*. Name unverified.

6683 Composibor
Destral BR. Nonselective herbicides; Weedkilling. *Borax Europe Ltd*.

6684 Compound 403/401
9002-88-4 7728
High density polyethylene
Silane crosslinkable flame-retardant polyethylene; for cable insulation applications *AEI Compds*. Name unverified.

6685 Compritol 888
Food emulsifier and additive for tablet manufacturing. *Gattefosse; Gattefosse SA*.

6686 Compron
Veterinary compressed products. *May & Baker Ltd*. Name unverified.

6687 Comtek
Sheetfed offset printing inks; for commercial printing. *AlliedSignal Inc/Sinclair and Valentine Division*. Name unverified.

6688 Com-Trol
2,4-D-mecoprop. Soluble concentrate containing 6.6% w/w 2,4-D and 250 g mecoprop; used to control weeds in grassland. *Certified Laboratories Ltd*.

6689 Conacure
Epoxy and polyurethane curing agents. *Conap*.

6690 Conap® UC-21
Two-component modified urethane resin systems; non-MBOCA, non-TDI; nonmoisture-sensitive; produces tough, high impact-resistance elastoplastics; used as casting or laminating resins; applications including foundry patterns and linings, glass fiber-reinforced plastic tools, holding and checking features, die faces, metal stamping pads, prototype parts, cures at room temperature or at elevated temperatures; mix ratio: 100/9.6. *Conap*.

6691 Conap® UC-28 (formerly Conap DPUC-11898)
Two-component urethane system; surface coat with optimum abrasion resistance; for application to vertical surfaces with min. sag; suitable for fixtures, gauges, patterns, dies, and for repair and redesign. *Conap*.

6692 Conapoxy® FR-1270
25928-94-3
Two-component epoxy; potting and casting system with excellent thermal shock resistance; for encapsulation of strain sensitive devices, electrical/electronic modules, transformers, coils. *Conap*.

6693 Conapoxy® TE-1257/Conacure® EA-08
25928-94-3
Highly filled aluminum epoxy base; produces tough and dimensionally stable castings with excellent heat dissipation; for casting drill jigs, spotting blocks, chuck jaws, dies, vacuum forming models, match plates, molds, core boxes; EA-08 curative for high operating temperature, medium potential life, heat cure; mix ratio: 100/5. *Conap*.

6694 Conastic® AD-20
Two-part polyurethane adhesive; flexible, high strength adhesive for bonding thermoplastics such as PC, Kapton, Hytrel, PVC, ABS, nylon; opaque amber; mix ratio 100/17.5. *Conap*. Name unverified.

6695 Conastic® ST-115
Two-part PU adhesive; dielectrical staking adhesive offering thixotropy, short pot life, good dielectrical properties; ideal for holding PC board components in place, as PU patching compound and water barrier joint sealer; opaque; mix ratio 100/17.5. *Conap*. Name unverified.

6696 Conathane® CE-1155
Two-component PU dielectrical conformal coating; insulating coating providing ultimate protection to aircraft avionics, instrumentation, missiles, spacecraft, fire and smoke detectors, electronic components, coils, and transformers. *Conap*. Name unverified.

6697 Conathane® CE-1163
One-component PU dielectrical conformal coating; insulating coating providing ultimate protection to aircraft avionics, instrumentation, missiles, spacecraft, fire and smoke detectors, electronic components, coils, and transformers. *Conap*. Name unverified.

6698 Conathane® EN-20
Two-part PU resin system; low viscosity, low toxicity, RT curing system for potting, casting, embedding, and encapsulation of electronic circuits, components, and power devices; non-TDI, non-MOCA. *Conap*. Name unverified.

6699 Conathane® EN-2521
Two-component PU potting and encapsulating system; used for modules, strain-sensitive components, transformers, and coils; non-MBOCA, non-TDI; EN-2521 is UL recognized; mix ratio: 20/100. *Conap*. Name unverified.

6700 Conathane® EN-4
Two-component polybutadiene-based liquid urethane casting and potting system; produces tough, stable elastomers with outstanding hydrolytic stability, electrical properties, thermal shock resistance, and low exotherm, shrinkage, and toxicity; applications include cable molding and connections, underwater protection, strain-sensitive, cryogenic and high voltage potting, vibration dampers, sealant bushings and transducer potting; mix ratio: 100/17.5 *Conap*. Name unverified.

6701 Conathane® RN-1501
TDI polyether elastomer; high-performance PTMG polyether prepolymer which produces flexible to semirigid urethanes; uses including rollers, tires, belts, gaskets, seals, wheels, molded parts, die pads, bushings, gears. *Conap*. Name unverified.

6702 Conathane® RN-1558
TDI polyether elastomer; medium-performance polyether elastomer for rolls, seals, casters, bushings, gears, impellers. *Conap*. Name unverified.

6703 Conathane® RN-1570
Aliphatic isocyanate polyether elastomer; high-performance prepolymer yielding tough elastomers with uv and hydrolysis resistance. *Conap*. Name unverified.

6704 Conathane® TU-400
Two-component liquid casting systems; non-MBOCA, non-TDI; produces elastomers of exceptional toughness, high elongation, high tensile and tear strength, and excellent abrasion resistance; cures at room temperature or elevated temperatures. *Conap*.

6705 Conathane® TU-4010
Two-part urethane elastomer system; filled, low-viscosity liquid casting system; filled, low-viscosity liquid casting system for tooling applications. *Conap*.

6706 Conathane® TU-50A
Two-component liquid TDI/MBOCA-based casting systems; tooling resin system producing flexible elastomers of exceptional toughness, high elongation, high tensile and tear strength, and excellent abrasion resistance; applications including cast-in-place linings for metal polishing and finishing equipment, industrial wheels, rollers, metal forming pads, casters, vibration, shock and sound damping pads, drop hammer faces, flexible molds, washers, gaskets, bushings, diaphragms, impellers and flexible couplings. Mix ratio: 100/94. *Conap*. Name unverified.

6707 Conathane® UC-17
Two-component urethane resin systems; non-MBOCA, non-TDI; produces tough, high impact-resistance, dimensionally stable castings; used for prototype parts and limited production runs; mix ratio: 25/100. *Conap*.

6708 Conathane® UC-34
Two-component urethane resin systems; casting system; non-TDI, non-BMOCA; 1:1 mix ratio and low viscosity for ease of handling; fast set and cure characteristics; ideal for casting duplicate masters, jigs, and patterns; mix ratio: 100/100. *Conap*.

6709 Concentrated Dariloid®
Alginate
Used for puddings, pie fillings, custards, cheese dips, cheesecake, egg nog, milk shake, toppings, syrups, baked cream fillings. *Kelco*.

6710 Concentrated Dariloid® KB
9005-37-2
Propylene glycol alginate
Used for sour cream, cottage cheese dressings, cream cheese, egg nog, ice cream fruit, fruit beverages, water ice, sherbert, lowfat novelties. *Kelco*.

6711 Concentrated Dariloid® XL
Alginate

Used for egg nog, liquid dairy applications requiring a low viscosity stabilizer. *Kelco.*

6712 Concurat® L
Anthelmintic against gastrointestinal nematodes and lungworms in cattle, sheep, goats and pigs; veterinary medicine. *Bayer AG.*

6713 Condens-Aid
Blends of amines; used for steam and return line corrosion inhibitors. *Schaefer Technologies Inc.*

6714 condenser foil
An alloy of 90% lead, 9.25% tin, and 0.75% antimony.

6715 Condensol®
Catalysts for textile high-grade finishing. *BASF AG; BASF plc.*

6716 Conditioner Base
Blend of conditioning and stabilizing agents. *Croda Chem. Ltd.*

6717 Conductive Nylon 12
Conductive nylon 12; for static dissipation and high tensile strength applications *Americhem.*

6718 conductivity bronze
A bronze containing copper with 0.8% cadmium and 0.6% tin.

6719 Conducto-Lube®
Pure silver powder and high grade petroleum oil; used in high-speed air blast breakers or virtually any application where a conductive lubricant is needed. *The Cool-Amp Conducto-Lube Co.*

6720 Conductomer ABS-22
9003-56-9
ABS with 22% conductive carbon black; electrical conductive compound for fabricated parts capable of alleviating electrostatic discharge. *Syn. Rubber Tech.*

6721 Conductomer HDC-22HLMI-M
9002-88-4 7728
HDPE, 22% conductive carbon black; electrical conductive compound for fabricated parts capable of alleviating electrostatic discharge. *Syn. Rubber Tech.*

6722 Conducto-Wrap
Copper foil tape with conductive pressure sensitive adhesive; for EMI shielding and electrical splicing. *Custom Coating & Laminating Corp.* Name unverified.

6723 Condux
Electrically conductive elastomers. *Hardman.*

6724 Condy's fluid
A solution of aluminum permanganate with some aluminum sulfate. An oxidizing agent, is used as a disinfectant and deodorizer.

6725 Confidor
105827-78-9
$C_9H_{10}ClN_5O_2$
1-[(6-Chloro-3-pyridinyl)methyl]-4,5-dihydro-N-nitro-1H-imidazol-2-amine
1-(6-Chloro-3-pyridiylmethyl)-N-nitroimidazolidin-2-ylideneamine; Admire; Gaucho; NTN 33 893Imidacloprid,. Systemic insecticide with contact and stomach action; applied as a foliar or soil treatment especially against sucking pests (virus vectors), e.g., aphids, thrips, whiteflies and leafhoppers on rice, potatoes, vegetables, cotton, tobacco, citrus, pome, stone fruit and other crops; also against some biting insects such as rice water weevil, Colorado potato beetle, wireworm, frit fly, beet fly, onion fly and citrus and apple leaf miners. mp = 136.5-144°; soluble in H_2O (0.51 g/l); LD_{50} (rat orl) = 450 mg/kg. *Bayer AG.*

6726 Conifer and Shrub Fertiliser
Powdered fertilizer containing NPK 10:7.5:10 plus 1.85 Mg and trace elements; all-purpose base and top dressing fertilizer. *Vitax Ltd.*

6727 Conlex
Specialized lattices based on acrylic-styrene copolymers; used in floor polishes. *Morton Int'l. Ltd.*

6728 Conn's stain
A microscopic stain. It contains 1 gram rose bengal, 5 g phenol, and 100 ml distilled water.

6729 Conoptic® UC-33
Two-part PU resin system; casting resin system formulated for applications requiring clarity; used for prototype parts, windows, display devices, lens materials. *Conap.*

6730 Conotrane
benzalkonium chloride-dimethicone. A proprietary preparation of benzalkonium chloride and dimethicone; an antiseptic skin barrier cream. *Boehringer Ingelheim Ltd.* Discontinued.

6731 Conrex
Acrylic leveling resin. *Morton Int'l. Ltd.*

6732 Consal
497-19-8 8739 207-838-8
Sodium carbonate hydrated
Chemical intermediate, alkalizer. *Church & Dwight Co.* Name unverified.

6733 Constab
Specialized masterbatches for thermoplastics. *Cornelius Chemical Co Ltd.* Name unverified.

6734 Constantan
An alloy of 60% copper and 40% nickel. It has a specific gravity of 8.9, and melts at 1290°; used as an electrical resistance.

6735 Constantin
Electrical resistance alloys. One contains 54% copper and 46% nickel, and another consists of 54% copper, 44% nickel, 1.3% manganese, and 0.4% iron.

6736 constructal
An aluminum alloy containing 3% of alloying elements, chiefly zinc.

6737 Construction 1200®
Silicone sealant
High strength acetoxy-cure sealant for structural glazing and premium general-purpose applications. *GE Silicones.*

6738 Contact 75
1897-45-6 2219 217-588-1
Chlorothalonil
A fungicide for a wide range of agricultural crops. *Fermenta ASC Europe Ltd.* Name unverified.

6739 Contex
An antifluxing agent for gold-and silversmiths and in the jewelry industry. *Degussa AG.* Name unverified.

6740 Continental® Clay
1332-58-7 5294 296-473-8
Kaolin
Kaolin clay; filler used in agricultural applications. *R. T. Vanderbilt Co Inc.* Discontinued.

6741 Continex® LH-10
1333-86-4 1856 215-609-9
carbon black
Pigment with improved abrasion resistance properties and lower hysteresis; for rubber, plastics, paper and printing ink industries. *Witco/Concarb.* Discontinued.

6742 Continex® LH-10, N-351
1333-86-4 1856 215-609-9
Improved abrasion resistance properties with lower hysteresis; for rubber, plastics, paper, and printing ink industries. *Witco/Concarb.* Discontinued.

6743 Continex® N351
1333-86-4 1856 215-609-9
Carbon black
For rubber, plastics, paper, and printing ink industries. *Witco/Concarb.* Discontinued.

6744 contracid
A corrosion-resisting alloy stable to nitric acid, hydrochloric acid, sulfuric acid, and other reagents. It contains from 50-60% nickel, 15-20% chromium, 0-20% iron, and up to 10% molybdenum or tungsten.

6745 Contractors 1000®
Silicone sealant
General-purpose glazing and sealing product *GE Silicones.*

6746 Contradet
A laundry contra-flow detergent. *Laporte Industries Ltd.*

6747 Contraqua LF
Zirconium wax paraffin emulsion
Cationic; hydrophobic agent for all types of fibers; water-repelling effect stable to gelling temperatures up to 190°. *Thor Chemicals (UK) Ltd.* Discontinued.

6748 Contrastol W
For contrast dyeings. *Bayer plc.* Discontinued.

6749 Control 1-100
9005-25-8 232-679-6
Corn starch
Corn starch; used in prepared foods. *Grain Processing.*

6750 Controx® KS
Tocopherol, citric esters; antioxidant for cosmetics. *Henkel/Cospha; Henkel Canada.* Name unverified.

6751 Controx® Range
Antioxidants for animal and vegetable oils and fats in food industry. *Grünau.*

6752 Cook's alloys
One contains 68.5% antimony and 31.5% zinc, and another consists of 57% antimony and 43% zinc.

6753 Cooksons
1327-33-9 215-474-6

White oxide of antimony. *Associated Lead Manufacturers Ltd.* Name unverified.

6754 Cool-Amp
Sodium chloride-calcium chloride-silver chloride-potassium bitartrate
Powder for silver plating electrical bus bars, connectors, etc. *The Cool-Amp Conducto-Lube Co.*

6755 Coolanol
Series of formulated silicate ester fluids used for a wide range of heat transfer, dielectric and hydraulic fluid applications; for an operating-temperature range of -90ºF to 700ºF; used primarily in aerospace and advanced electronic hardware as a coolant and dielectric. *Monsanto Co.* Name unverified.

6756 Cool-Treet
Various liquid blends of cooling water deposit and corrosion inhibitors; used for cooling water treatment. *Schaefer Technologies Inc.*

6757 Coomassie
Milling and half milling acid dyes. *ICI Chemical & Polymers Ltd.*

6758 Cooper Coopex
52645-53-1 7321 258-067-9
Permethrin
A pyrethroid insecticide. *The Wellcome Foundation Ltd.*

6759 Cooper Graincote
5598-13-0 2242 227-011-5
Chlorpyrifos-methyl
An organophosphate insecticide for the treatment of pests in stored grain and oilseed rape. *The Wellcome Foundation Ltd.*

6760 Coopercote
Insecticidal varnish for paper, board etc. *The Wellcome Foundation Ltd.* Name unverified.

6761 Cooperite
An alloy of 80% nickel, 14% tungsten, and 6% zirconium; used for cutting tools. A modified alloy contains tantalum.

6762 Cooper's gold
An alloy of 19% platinum and 81% copper.

6763 Cooper's pen metal
An alloy of 50% platinum, 37.5% silver, and 12.5% copper.

6764 Cooper's speculum metal
An alloy of 58% copper, 27% tin, 10% platinum, 4% zinc, and 1% arsenic.

6765 Copac
Metal salt complex; accelerator for unsaturated polyesters; drier for nonaqueous coating compositions. *Hüls Am.*

6766 Copac® E
Ammoniacal copper sulfate
For control of bacterial diseases in pears, vegetables, ornamentals *BASF AG.*

6767 copal oils
Oils obtained by the dry distillation of copal; used for the preparation of oil varnishes.

6768 Copaloy
A proprietary platinum-tin-antimony alloy with a small percentage of copper. A bearing metal. No manufacturer.

6769 Copel alloy
An alloy of 55% copper and 45% nickel.

6770 Copene
A proprietary trade name for a polyterpene copolymer resin used in lacquers, paints, and varnishes. No manufacturer.

6771 Coperflex BR-45
1,3-Butadiene polymer; light colored, nonstaining rubber. *Coperbo.*

6772 Coperflex SSBR-B18 4525
1,3-Butadiene/styrene block polymer, solution polymerized with an alkyl-lithium catalyst; light colored, nonstaining rubber. *Coperbo.*

6773 Copernick
An alloy similar to Hypernick; the permeability is constant over a wide range of flux densities. No manufacturer.

6774 Copisil
Color developer for the production of carbonless copying papers, thermoreactive papers and chemical test papers. *Süd-Chemie AG.* Name unverified.

6775 Copo® 1500, Copo® 1712
(1500) Cold SBR, nonpigmented; general purpose, staining, with good physicals and abrasion chars. (1712) Cold SBR, nonpigmented with 37.5 parts-per-hundred oil; general purpose, staining, oil-extended cold SBR. *Copolymer Rubber.*

6776 Copolymer 186
1904
Castor oil, dehydrated; for preparation of quality coating vehicles by cold blending; compatible with phenolic, rosin esters, hydrocarbons and alkyd resins. *CasChem.*

6777 copper
7440-50-8 2583 231-159-6
Cu
Copper
Bronze powder; copper bronze; gold bronze. Electric wiring, switches, plumbing, heating, roofing; chemical and pharmaceutical machinery; alloys; coatings; cooking utensils. mp = 1083°; bp = 2595°; d = 8.94 *Aldrich; Asarco; M&T Harshaw; Noah Chem.*

6778 Copper Antracol®
$C_5H_8CuN_2S_4$
[[(1-methyl-1,2-ethendiyl)bis[carbamodithioato]](2-)copper
Combination product for prevention of fungal infections in fruit, grapes, vegetables and arable crops. *Bayer AG.*

6779 copper dimethyldithiocarbamate
137-29-1 205-287-8
$C_6H_{12}CuN_2S_4$
Copper dimethyldithiocarbamate
Cumate®; bis(dimethylcarbamodithioato-S,S') Copper; Copper dimethyldithiocarbamate; Cumate; Copper, bis(dimethylcarbamodithioato-S,S')-, (SP-4-1)-. Accelerator for vulcanization of rubber. *R. T. Vanderbilt Co Inc.* Discontinued.

6780 Copper Euparen®
3095
copper-dichlorfluanid. Combination product for prevention of fungal infections in fruit, grapes, vegetables and arable crops. *Bayer AG.*

6781 copper gluconate
527-09-3 2706 208-408-2
$C_{12}H_{22}CuO_{14}$
bis(D-gluconato) copper
copper, bis(D-gluconato). Copper salt of gluconic acid; mineral source for pharmaceutical and food products Soluble in H_2O (30 g/100 ml); slightly soluble in EtOH, insoluble in organic solvents; d = 1.710 mg/kg. *Akzo; Glucona; Spectrum Chem. Mfg.*

6782 copper green
malachite. A term applied to the mineral malachite. Used in pigments, pyrotechnics, insecticides, copper salts. An astringent in pomades, antidote for phosphorus poisoning, smut preventive, fungicide for seed treatment, feed additive.

6783 copper naphthenate
1338-02-9 215-657-0
Copper(II)Naphthenate; Naphthenic acids, copper salts; CNC; Copper uversol; Copper-nap-all; Cuprinol; Troysan; WILTZ-65; Wittox-C. Copper salt of petroleum naphthenic acids; wood, canvas and rope preservative; antifouling in paints; insecticide, fungicide. *Akzo; KMZ Chemical Ltd; Troy.*

6784 copper oxychloride
1332-40-7 296-473-8
$CuCl_2 \cdot Cu_3H_6O_6$
Copper oxychloride
dicopper chloride trihydroxide; copper(II) chloride oxide hydrate; Blitox; Ckuper; Cobox; Coprantol; Coptox; Criscobre; Cupravit; Cuprenox; Cuprocaffaro; Cuprokylt; Cuprosan; Cuprossina; Cuprovinol; Cuprox; Fytolan; Kauritil; Neoram; Recop; Viricuivre; Cupravit®; Vitigran; Cuprokylt; Cuprosana; Curenox-50. Copper containing spray used for control of fungal diseases like downy mildews, late blight, early blight, apple scab and various leaf spot diseases on a wide range of crops. Insoluble in H_2O, organic solvents; LD_{50} (rat orl) = 1440 mg/kg. Bayer AG; Cequisa; Ciba-Geigy; All-India Medical; Crystal; Diachem; Caffaro; Universal Crop Protection; Rhône Poluenc; Agrichem; Zeltia; BASF; Sandoz; Hoechst; Industrias Quimicas Del Valles SA;.

6785 copper oxychloride
1332-40-7
FS Dricol 50; Cupravit®; Cuprokylt; Cuprosana; Curenox-50; Headland Inorganic Liquid Copper. A protectant fungicide. Ford Smith & Co Ltd; Bayer AG; Universal Crop Protection Ltd.; Industrias Quimicas Del Valles SA; WBC Technology Ltd.

6786 copper silumin
An alloy of 85.6% aluminum, 13% silicon, 0.8% copper, and 0.6% iron.

6787 copper solder
An alloy of 2 parts lead and 5 parts tin.

6788 copper steel
An alloy of steel with up to 1% copper, usually 0.5%. It resists corrosion.

6789 Copper-Count
Broad spectrum bactericide and fungicide. *Mineral Research & Development Corp.*

6790 **Copper-Lonacol**
Copper-zineb. Combination spray for the control of fungal diseases. *Bayer AG.*

6791 **copper-nickel**
An alloy of 50% copper and 50% nickel; used in the manufacture of nickel copper alloys, is known by this name. The term is also applied to the mineral Niccolite, NiAs.

6792 **Coppertox**
Livestock dips and sprays. *The Wellcome Foundation Ltd.* Name unverified.

6793 **Coppertrace**
7447-39-4 2699 231-210-2
Cupric chloride dihydrate
Supplement, trace mineral. mp = 100°; d = 2.51; soluble in H_2O, EtOH, less souble in organic solvents. *Armour Pharmaceutical Co.*

6794 **Coppesan**
A copper fungicide. *The Boots Co plc.* Discontinued.

6795 **Coprantex**
Dyeing and printing assistant. *Ciba plc.* Name unverified.

6796 **Coprol**
577-11-7 3460 209-406-4
dioctyl sodium sulfosuccinate
A proprietary trade name for dioctyl sodium sulfosuccinate.

6797 **Cops 1**
52556-42-0 258-004-5
Sodium 1-allyloxy-2-hydroxy-propane sulfonate
Detergent and surfactant. *Rhône-Poulenc Surf.*

6798 **Coptal**
Textile auxiliary chemicals. *ICI Chemical & Polymers Ltd.*

6799 **Corafilm, Coravol**
Amine treatments; for steam lines. *Garvey Chemical Corp.* Name unverified.

6800 **Corasole**
Carbon black/plastic concentrate; used for pigmenting of paper and cement for the plastics processing industry, pigmenting of PE, PP, EVA, PVC, PS, SAN, and ABS. *Degussa AG.* Name unverified.

6801 **Corban**
Corrosion inhibitors; used to reduce metal loss from oil and gas well equipment caused by hydrogen sulfide, carbon dioxide and organic acids; the inhibitor coats the metal surfaces with a thin protective film. *Dow UK.* Discontinued.

6802 **Corbel®**
67306-03-0 4035 266-639-4
Fenpropimorph
Systemic fungicide *BASF AG; BASF plc.*

6803 **Corbel® CL, Corbel® Star**
chlorothalonil-fenpropimorph. Suspension concentrate containing 250 g chlorothalonil and 187 g fenpropimorph per liter; a systemic fungicide for winter wheat. *BASF plc.*

6804 **Corbel® Duo**
Fenpropimorph-Carbendazim. Combination of Fenpropimorph and Carbendazim; used as a fungicide. *BASF AG.*

6805 **Corbrite**
Disazo yellow and azo red pigments. *European Colour (Pigments) Ltd.*

6806 **Corcert**
Food colors meeting EEC specifications. *European Colour (Pigments) Ltd.*

6807 **Cordetec 100**
Catalyst made of vanadium pentoxide and titanium dioxide with promoters on a uniform ceramic ring; Used to produce phthalic anhydride from o-xylene reaching yields of 100%. Used in normal and in low energy processes. *Desarrollo Quimico Industrial Spa.* Name unverified.

6808 **Cordite**
Cordite MD
Maximite. An explosive used as powder and filaments. It is a mixture of 65% guncotton, gelatinized by means of acetone, 30% nitroglycerine, and 5% petroleum jelly.

6809 **Cordura®**
Fibers. *DuPont UK.*

6810 **Corephen® 10**
Phenol formaldehyde
For acid-proof stoving finishes. *Bayer AG.*

6811 **Coresize 630**
Hydrocarbon emulsion; sizing agent imparting water resistance to core of gypsum board, especially bathroom sheathing and other high-moisture-resistance boards. *Hercules.* Discontinued.

6812 **Corexit**
Oil slick dispersants. *Exxon UK.* Name unverified.

6813 **Corfast**
Azo red and yellow pigments. *European Colour (Pigments) Ltd.*

6814 **Corfix**
A range of adhesives for all types of resin, silicate and oil-bonded cores. *Foseco (F.S.) Ltd.*

6815 **Corgran**
Granular organic pigments, toners, food colors. *European Colour (Pigments) Ltd.*

6816 **Coriacide**
Leather dyes. *ICI Chemical & Polymers Ltd.*

6817 **Corial Primer**
Aqueous polymer dispersion for leather. *BASF plc.*

6818 **Corial®**
Special organic solvent for leather finishes. *BASF AG; BASF plc.*

6819 **Corialbinder®**
Polymer dispersions; binders for leather finishes. *BASF AG.*

6820 **Corian®**
Polymer. *DuPont UK.*

6821 **corichrome**
79533-80-5
Titanium lactates
Used as mordants and strikers in the leather industry.

6822 **coridine**
n-Propyllutidine

6823 **Corilene**
An aqueous degreasing agent for leather. *ICI Chemical & Polymers Ltd.*

6824 **Corinal**
A trademark for a synthetic tannin prepared by condensing heavy tar oils with formaldehyde and then making the aluminum salt. *National Lead Co.* Name unverified.

6825 **Corindite**
The trade name for an abrasive and refractory of the carborundum type; obtained from bauxite by heating it with anthracite, and contains 69%. Al_2O_3, with SiO_2 and Fe_2O_3. No manufacturer.

6826 **corioflavines**
Red or reddish-brown dyestuffs; used for leather.

6827 **Coripact**
Mineral fiber thermal-insulating boards and strips together with bitumen sheet; for non-ventilated roofs and for all roof coverings. *Vedag GmbH.* Name unverified.

6828 **coriphosphines**
Alkylated aminoacridines used as dyestuffs.

6829 **Coriumine**
Leather dyes. *ICI Chemical & Polymers Ltd.*

6830 **Corlake**
Food color lakes on insoluble substrates. *European Colour (Pigments) Ltd.*

6831 **Corlar®**
Epoxy finish; used for automotive industry. *DuPont UK.*

6832 **Cormix**
Concrete admixtures for the construction industry. *Crosfield Chemicals Ltd.*

6833 **Cormul**
Range of cationic bitumen emulsifiers in liquid form; used in road surfacing: surface dressing tack coats; stone coating; grouting; slurry sealing. *Thomas Swan & Co Ltd.* Discontinued.

6834 **corn oil**
8001-30-7 11, 2528 232-281-2
Maize oil
corn oils; Zea mays oil; Maize oil; Maydol; Mazola. Refined fixed oil obtained from wet milling of corn, *Zea mays*; used in preparing foodstuffs, soap, lubricants, leather finishing, margarine, salad oil, hair dressing; solvent; paints. mp= -18 - -10°; d_{25}^{25} = 0.916-0.921; n_D^{25} = 1.470-1.474; soluble in organic solvents. *Grain Procesing; Karlshamns; Penta Mfg; A.E. Staley Manufacture.*

6835 **Corn oil PEG-6 esters**
85536-08-9 287-489-6
Labrafil M 2125 CS. Hydrophilic oil for pharmaceutical and cosmetic formulations. *Gattefosse; Gattefosse SA.*

6836 **corn starch**
9005-25-8 232-679-6
Obtained from grains of *Zea mays*; carbohydrate polymer consisting primarily of amylose and amylopectin; white powder; source of glucose; filler in baking powder; thickening agent in food products; adhesives, coatings; *Am. Maize Prods.; Cerestar UK; Grain Processing; Nat'l. Starch & Chemical; A.E. Staley Manufacture.*

6837 **corn starch**
9005-25-8 232-679-6
D12F; Mira-Gel® 463; Mira-Set® B; Pure Food Powd. Starch 105-A, 131-C, 142-A; Staley® 7025, 7350 Waxy no. 1 Starch; Argo Brand Corn Starch; Control I-100; D12F; Pure-Dent® B700, B810, B812, B815, B816, B880;

Purity® 21; Supercore® S13F; Stanley® Moulding Starch, Pure Food Powder, Redreid Starch A, B; Sta-Rx®; Stir & Sperse®. Oxidized corn starch. White powder. *ADM; Am. Maize Prods.; Cerestar UK; Grain Processing; Nat'l Starch & Chem; A.E. Stanley.*

6838 **corn syrup**
8029-43-4 232-436-4
Glucose syrup. Mixture of D-glucose, maltose, and maltodextrins; used in the food industry as sweetener, thickener, bodying agent in soft drinks. *Am. Maize Prods.; Gist-Brocades Food ingreds.; A.E. Staley Manufacture.*

6839 **Cornish bronze**
An alloy of 78% copper, 9.5% tin, and 12.5% lead; a white bearing metal.

6840 **Corn-Pro 35**
Hydrolyzed corn protein; cosmetic ingredient for skin and hair care products *Brooks Industries.*

6841 **Cornuite**
A protein-like mineral found in the diatomaceous earth from Neu-Ohe. It is a gelatinous albuminous material with a 3% dry residue.

6842 **Corolox, Corotox, Corowalt**
1344-28-1 369 215-691-6
aluminum oxide
Trademarks for abrasive and refractory materials; the essential constituent is crystalline alumina. No manufacturer.

6843 **Corona**
8006-54-0 5371 232-348-6
Lanolin BP/EP
Used for cosmetic and pharmaceutical preparations. *Croda Chem. Ltd.*

6844 **Coronet**
8006-54-0 5371 232-348-6
Refined lanolin; used as emulsifier, emollient, dispersing, solubilizing, and wetting agent. *Croda Chem. Ltd.*

6845 **Coronium**
An alloy containing 80% copper, 15% zinc, and 5% tin.

6846 **Corox**
1309-48-4 5713 215-171-9
MgO
magnesium oxide.
A proprietary insulating material having a great thermal conductivity and high electrical insulating power; consists essentially of magnesium oxide.

6847 **Corozo**
Vegetable ivory, the seeds of *Phytelephas macrocarpa.*

6848 **Corro-Guard**
Various liquid blends of deposit and corrosion inhibitors for nonpotable water systems; used for recirculating hot and chilled water systems. *Schaefer Technologies Inc.*

6849 **Corrolite®**
Vinyl esters. *Reichhold.*

6850 **Corronel 220**
A trademark for a nickel-molybdenum-vanadium alloy with good resistance to hydrochloric, sulfuric and phosphoric acids under reducing conditions. *Wiggin Alloys Ltd.* Unverified.

6851 **Corronel Alloy 230**
A trademark for an alloy of 35% chromium and 65% nickel; resistant to nitric and nitric/hydrochloric acid mixtures. *Wiggin Alloys Ltd.* Unverified.

6852 **corronil**
An alloy of 70% nickel, 26% copper, and 4% manganese. It is a corrosion resisting alloy.

6853 **Corrosalloy**
A proprietary trade name for stainless steels. No manufacturer.

6854 **corrosiron**
An iron-silicon alloy containing 12% silicon. It is stated to be very resistant to acids.

6855 **Corrosist**
A proprietary trade name for nickel base corrosion resistant alloys; used for valves in chemical plant subject to contact with chlorine. No manufacturer.

6856 **Corry's Moss Remover**
cupric sulfate-sodium borate. Mixture of copper sulfate and sodium borate; used to control moss on paths and hard tennis courts. *SynChemicals Ltd.*

6857 **Corseal**
Refractory moldable products for repairing damaged cores or sealing core joints. *Foseco (F.S.) Ltd.*

6858 **Corten**
A proprietary trade name for a chromium steel containing 0.5% chromium, 0.10% carbon, 0.1% manganese, 0.5% silicon, and 0.3% copper. No manufacturer.

6859 **Corton A1**
hydrocortisone-neomycin. Rubber bitumen-based product designed for use where fuel resistance is important, e.g., airport runways. *Feb Ltd.*

6860 **Cortone**
Organic pigment toners. *European Colour (Pigments) Ltd.*

6861 **Cortymol® LP**
9002-60-2 136 232-659-7
ACTH. Adrenocorticotropic hormone, ACTH. *BASF AG.*

6862 **Corubin**
1344-28-1 369 215-691-6
alumininum oxide
An artificial corundum Al_2O_3. It is the alumina which constitutes the slag formed in the reaction between aluminum and metallic oxides (Thermit); used for polishing purposes, and in the manufacture of fireproof stones.

6863 **Corundite**
1344-28-1 369 215-691-6
alumininum oxide
A trademark used for abrasive and refractory materials the essential constituent of which is crystalline alumina. No manufacturer.

6864 **Corvic**
PVC, emulsion polymers and copolymers. *ICI Chemical & Polymers Ltd.* Discontinued.

6865 **Cosbiol®**
111-01-3 8923 203-825-6
$C_{30}H_{62}$
squalane
Perhydrosqualene. High quality oil for cosmetics and pharmaceuticals. *Laserson & Sabetay Ets.*

6866 **Cosiderm Collagen Masks**
Collagen and soluble collagen; cosmetic ingredients. *Maybrook.*

6867 **Coslettized steel**
A steel whose surface has been rustproofed by dipping in a solution of iron phosphate and phosphoric acid.

6868 **Cosmedia Guar C-261 N**
65497-29-2
Guar hydroxypropyltrimonium chloride
Viscosity builder; personal care product formulating; substantivity provides hair conditioning; stabilizer and thickener for emulsions; provides slip to finished formulations. *Henkel/Cospha; Henkel Canada.*

6869 **Cosmedia Polymer HSP 1180**
Polyacrylamidomethylpropane sulfonic acid
Smooth feel agent for creams, lotions, liquid antiperspirants, shaving creams, soaps, nail polish removers. *Henkel/Cospha; Henkel Canada.*

6870 **Cosmetic Lanolin USP**
8006-54-0 5371 232-348-6
Lanolin
wool fat; oesipos; agnin; alapurin; Agnolin; Lanum; Lanain; Lanalin; Lanesin; Lanichol; Laniol. Superfatting emollient, emulsifier for creams, lotions, lipsticks, make up, sunscreen products mp = 38-40°. *Croda Inc.*

6871 **Cosmetol® X**
8001-79-4 1946 232-293-8
castor oil
castor oil; Ricinus oil; tangantangan oil; oil of Palma Christi; Neoloid. Unrefined castor oil USP, with antioxidant; deodorized specially refined grade of castor oil containing a food grade antioxidant; emollient, pigment wetter, cosolvent, lubricant for cosmetics, makeup, antiperspirant sticks. bp = 313°; d = 0.961; n_D^{20} = 1.4780; $[\alpha]_D^{20}$ = 5° (c = 5, EtOH). *CasChem.*

6872 **Cosmic**
Emulsifying agents. *Abril Industrial Waxes.*

6873 **Cosmic® FL**
Maneb-tridemorph-carbendazim. Broad spectrum cereal fungicide *BASF AG; BASF plc.*

6874 **Cosmica®**
Pigment series where absorption colors are deposited directly on the mica, creating highly intense effects with minimal luster; suitable for most cosmetics. *Mearl.*

6875 **Cosmocil**
Preservatives for cosmetics. *ICI Chemical & Polymers Ltd.*

6876 **Cosmos Alloy**
A proprietary lead base alloy with small amounts of tin and antimony; a bearing metal. No manufacturer.

6877 **Cosmowax**
Nonionic self emulsifying wax base. *Croda Chem. Ltd; Croda Inc.*

6878 **Cosmowax K**
Stearyl alcohol-ceteareth-20
Emulsifier, stabilizer with body, opacity and conditioning properties for personal care products and pharmaceuticals. *Croda Inc.* Discontinued.

6879 **Cosmowax P**
Cetearyl alcohol

ceteareth-20; cetyl/stearyl alcohols. Emulsifier, emulsion stabilizer for lotions, creams. *Croda Inc.*

6880 Cotazym
Pancrelipase
A concentrate of pancreatic enzymes standardized for lipase content; enzyme. *Organon Inc.* Name unverified.

6881 Cothias metal
An alloy of 67% copper and 33% tin; used as a hardener for zinc alloys.

6882 Cotnion-Ethyl
2642-71-9 220-147-6
azinphos-ethyl
Persistent agricultural organo phosphorous insecticide, active ingredient; azinphos-ethyl. *Makhteshim Chemical Works Ltd.*

6883 Cotnion-Ethyl-Methyl
azinphos-methyl-azinphos-ethyl. Active ingredients: azinphos-ethyl [2642-71-9] plus azinphos-methyl [86-50-0]; persistent agricultural organophosphorus insecticide combination. *Makhteshim Chemical Works Ltd.*

6884 Cotnion-Methyl
86-50-0 944 201-676-1
azinphosmethyl
Active ingredient: azinphosmethyl; persistent agricultural organophosphorous insecticide. *Makhteshim Chemical Works Ltd.*

6885 Cotolan Fast
Dyestuffs for the one-bath of union materials (wool/cellulosics). *Bayer plc.* Discontinued.

6886 Cotonerol®
Azo dyestuffs
Used for textile dyeing and printing. *Cassella AG.*

6887 Cotopa
A form of textile-based thermal insulating material composed of acetylated cotton yarn.

6888 Cottestren
Dyestuffs for cellulosic/polyester blends. *BASF plc.*

6889 Cottestren® C Dyes
Mixed dyes for one-bath tone-in-tone dyeing of polyester/cellulose fiber blends. *BASF AG.*

6890 Cottoclarin
Wetting agent; for scouring, bleaching, dyeing and wetting processes. *Henkel Chemicals Ltd.* Name unverified.

6891 cotton black
A dyestuff obtained by the fusion of *o*-, *p*-dinitrodiphenylaminesulfonic acid with sodium polysulfides. Dyes cotton brownish-black.

6892 cotton blue
Many of the direct cotton blues, such as diamine, benzo, and Congo blues, are sold under this name.

6893 cotton wax
The wax found in cotton fiber. It melts at 82-86°, and appears similar to carnauba wax.

6894 Cottonex
2164-17-2 4189 218-500-4
fluometuron
1,1-dimethyl-3-(α,α,α-trifluoro-*m*-tolyl)urea. Active ingredient fluometuron; residual herbicide effective against a wide range of both annual broadleaf weeds and grasses. *Agan Chemical Manufacturers Ltd.*

6895 Cotton-Pro®
7287-19-6 7973 230-711-3
Prometryn
Prometryn suspension; flowable herbicide for selective weed control in cotton and celery crops. *Griffin.*

6896 Couch and Grass Killer
75-99-0 2869 200-923-0
2,2-dichloropropionic acid
Dalapon. Soluble powder containing dalapon; used for control of grasses in crop and noncrop areas. *Vitax Ltd.*

6897 Cougar
Water-soluble polymeric materials for use as thickening agents or as texturizers in chemical products. *Courtaulds plc.* Discontinued.

6898 Couloscope
Coulometric coating thickness gauge. Coulometric electrolytes; for measurement of metallic coating thickness of coatings deposited over metallic and non metallic substrates, or of metal foils. *Fischer Instrumentation (GB) Ltd.* Name unverified.

6899 Coulter Clenz
Cleaning solution; for cleaning of automatic blood cell analyzers. *Coulter Electronics Ltd.*

6900 Coulter Clone
Range of reagents based on monoclonal antibodies; used to identity different cell types. *Coulter Electronics Ltd.*

6901 coumalic acid
500-05-0 2625 207-899-0
$C_6H_4O_4$
2-Oxo-1,2H-pyran-5-carboxylic acid
α-Pyrone-3-carboxylic acid; 2-Pyrone-5-carboxylic acid. Used in chemical synthesis. mp= 203-205° (dec); bp_{120} = 218°; slightly soluble in H_2O, Et_2O, Me_2CO, EtOAc;.

6902 Coumalux®
$C_{12}H_{13}NO_2$
7-Diethylamino-4-methyl-2H-benzopyran-2-one
Optical brightener. *Octel Chemicals Ltd.*

6903 coumarone-indene resin
Thermosetting resin; adhesives, printing inks, floor tile binder; friction tape; paints; varnishes, enamels; in chewing gum. *Allchem Industries; Natrochem; Neville; Spectar Ltd.*

6904 Countdown
41394-05-2 5985 255-349-3
Metamitron
Triazinone herbicide for weed control in beet crops. *Kommer-Brookwick Ltd.*

6905 Country Fresh Disinfectant
1330-85-4 215-551-4
Dodecylbenzyltrimonium chloride
Dodecylbenzyl trimethyl ammonium chloride; Gloquat C; Trimethyl dodecylbenzylammonium chloride. Soluble concentrate of 150 g dodecylbenzyl trimethyl ammonium chloride per liter; used for algae control. *Dimex Ltd.*

6906 Coupler
Photosensitive coupler
Rhône-Poulenc UK.

6907 Coupler SC
clopyralid-cyanazine. Suspension concentrate containing 60 g clopyralid and 350 g cyanazine per liter; a contact herbicide. *Shell UK.*

6908 Coupsil VP 6109
Silica reacted with bis (3-triethoxysilylpropyl) tetrasulfane; reinforcing filler for rubber compounds including NR, IR, SBR, BR, NBR, EPDM, and EPM. *Degussa AG; Struktol.*

6909 Courcel®
Proprietary and blended gums and stabilizers for the food industry; used for pet foods, sauces/salad dressings, meat products, vegetarian products, batters, egg products, water ices. *Courtaulds Chemicals, Water Soluble Polymers.*

6910 Courgel®
Performance water-soluble polymers with gelling properties for industrial applications; used for ceramic tile adhesives, textured building finishes, ceramic glazes, horticultural products. *Courtaulds Chemicals, Water Soluble Polymers.*

6911 Courline
A proprietary trade name for polypropylene monofilament yarns. *Hoechst Celanese.* Name unverified.

6912 Cournova
A proprietary trade name for polypropylene oriented slit film for the manufacture of strings, twines and ropes. *Hoechst Celanese.* Name unverified.

6913 court plaster
Isinglass dissolved in water, alcohol, and glycerin, and painted on taffeta.

6914 Courtaulds
9004-35-7 2013
Cellulose acetate
Courtaulds plc. Discontinued.

6915 Courtek
Chemical products for use in industry; polymeric plastics. *Courtaulds plc.* Discontinued.

6916 Courthene
Plastics materials in the form of sheets (nontextile) for use in manufacture. *Courtaulds plc.* Discontinued.

6917 Courtochrome
Chemical products for use in photochromy; photosensitive paper, plates, films, microfilm and foils. *Courtaulds plc.* Discontinued.

6918 Covar
Aliphatic solvent containing emulsifier. *Exxon Int'l.* Name unverified.

6919 Coveral
Aluminum alloy cleansing flux. *Foseco (F.S.) Ltd.*

6920 Coverite
A wetting agent. *Murphy Chemical Co Ltd.* Name unverified.

6921 Covermark
13463-67-7 9612 236-675-5
O_2Ti
titanium dioxide
A proprietary preparation of titanium dioxide with different pigments in a cream base; a skin masking cream. *Stiefel Laboratories (UK) Ltd.* Name unverified.

6922 Covexin
A proprietary combined sheep vaccine. *Coopers Animal Health Ltd.* Name unverified.

6923 Covon
Glycol-free colorants; formulated for in-plant tinting of high-gloss latex, emulsion and water-based paints. *Hüls Am.*

6924 Cow Gum
Adhesive for mounting of photographs and art work. *Cow Proofings Ltd.* Unverified.

6925 Cowles' aluminum bronze
An alloy containing 89-98.75% copper and 1.25-11% aluminum. One alloy contains small amounts of iron and silicon.

6926 Coyden
Coccidiostat is mixed with chicken feed to prevent coccidiosis (diarrhoea) in poultry. *RMB Animal Health Ltd.*

6927 Cozirc
Chemically combined cobalt-zirconium carboxylate; drier for lead and/or barium free paint or printing ink. *Rhône-Poulenc UK.*

6928 CP Filler
471-34-1 1697 207-439-9
$CCaO_3$
Calcium carbonate
Agricultural limestone [1317-65-3]; Agstone; Bell Mine Pulverized Limestone; Calcium carbonate; Calcium(II) carbonate (1:1); Camadil; Chalk; Chalk, precipitated; Chalk, whiting; Domolite; Franklin; Ground limestone; Limestone; Lithographic stone; Lithograpic Stone; Marble; Marble (osha); Natural calcium carbonate; Portland stone; Sohnhofen stone; VLR Diluent; Aeromatt; Akadama; Albacar; Albacar 5970; Albafil; Albaglos; Albaglos Sf; Allied Whiting; Atomit; Atomite; AX 363; BF 200; Brilliant 15; Brilliant 1500; Britomya M; Britomya S; Calcen CO; Calcene Co; Calcene NC; Calcene TM; Calcicoll; Calcidar 40; Calcilit 8; Calcilit 100; Calcite; Calcium carbonate (1:1); Calcium carbonate (ACN); Calcium carbonate (CaCO3); Calcium carbonate (chips); Calcium carbonate, mud; Calcium carbonate, precipitated,. Competitively priced filler for construction products where color is of secondary importance. *ECC International Ltd.* Discontinued.

6929 CP0110
940-41-0 213-371-0
$C_8H_9Cl_3Si$
2-Phenethyltrichlorosilane
Coupling agent, chemical intermediate, blocking agent, release agent, lubricant, primer, reducing agent. *Hüls Am.*

6930 CP0156
3068-76-6 221-328-2
$C_{12}H_{21}NO_3Si$
N-Phenylaminopropyltrimethoxysilane
Coupling agent, chemical intermediate, blocking agent, release agent, lubricant, primer, reducing agent. *Hüls Am.*

6931 CP0160
768-33-2 212-193-0
$C_8H_{11}ClSi$
Phenyldimethylchlorosilane
Coupling agent, chemical intermediate, blocking agent, release agent, lubricant, primer, reducing agent. n_D^{20} = 1.5090. *Hüls Am.*

6932 CP0280
98-13-5 202-640-8
$C_6H_5Cl_3Si$
Phenyltrichloro silane
Coupling agent, chemical intermediate, blocking agent, release agent, lubricant, primer, reducing agent. bp = 201°; d = 1.3210; n_D^{20} = 1.5230; LD_{50} (rat orl) = 2390 mg/kg. *Hüls Am.*

6933 CP0320
780-69-8 212-305-8
$C_{12}H_{20}O_3Si$
Phenyltriethoxy silane
Coupling agent, chemical intermediate, blocking agent, release agent, lubricant, primer, reducing agent. bp_{10} = 112-113°; d = 0.9960; n_D^{20} = 1.4604; LD_{50} (rat orl) = 2818 mg/kg. *Hüls Am.*

6934 CP0330
2996-92-1 221-066-9
$C_9H_{14}O_3Si$
Phenyltrimethoxysilane
Trimethoxyphenylsilane; Phenylmethoxysilane. Coupling agent, chemical intermediate, blocking agent, release agent, lubricant, primer, reducing agent. *Hüls Am.*

6935 CP0800
141-57-1 205-489-6
$C_3H_7Cl_3Si$
n-Propyltrichlorosilane
trichloropropylsilane; n-Propyltrichlorosilane. Coupling agent, chemical intermediate, blocking agent, release agent, lubricant, primer, reducing agent. *Hüls Am.*

6936 CP0810
1067-25-0 213-926-7
$C_6H_{16}O_3Si$
n-Propyltrimethoxysilane
n-Propyltrimethoxysilane; Dynasylan PTMO. Coupling agent, chemical intermediate, blocking agent, release agent, lubricant, primer, reducing agent. *Hüls Am.*

6937 CP-153-2 (25% in xylene)
Chlorinated polyolefin
Used as primer for polyethylene; excellent adhesion characteristics. *Eastman.*

6938 CP-343-1(100%)
Chlorinated polyolefin
Used as primer for PP, selected plastics, and metals where it promotes adhesion. *Eastman.*

6939 CPE
Chlorinated polyethylene resins. When added to other plastics, CPE can be calendered, injection molded or extruded into tough, chemical resistant, weather resistant products; typical end uses would be pond liners, automotive hose and chemical transfer hose. *Dow Cheml Co Ltd, UK & Ireland.*

6940 C-petroleum naphtha
petroleum benzine
safety oil. That fraction of petroleum distilling at 80-100°, of specific gravity 0.667-0.707.

6941 CPH-43-N
9004-81-3
PEG-12 laurate
Solubilizer and dispersant. *C.P. Hall.*

6942 CPH-52-SE
1323-39-3 215-354-3
Propylene glycol stearate
Dispersant. *C.P. Hall.*

6943 CPH-211-N
9005-07-6
PEG-8 dioleate
Water treatment, lubricant. *C.P. Hall.*

6944 CPH-250-SE
31566-31-1 4498 250-705-4
Glyceryl stearate
glyceryl monostearate; octadecanoic acid monester with propane-1,2,3-triol. Emulsifier. mp = 56-58°. *C.P. Hall.*

6945 CPS 034
9006-65-9 3264
Dimethicone
Fluids for mechanical, heat transfer and dielectric applications. *Hüls Am.*

6946 CPS 076
(N-Trimethoxysilylpropyl)-polyethylenimine
Coupling agent especially for mineral-filled and adhesive bonding applications of high molecular weight thermoplastic polymides and polyesters. *Hüls Am.*

6947 CPS 120
63148-57-2
Polymethylhydrosiloxane
Crosslinker, waterproofing agent. *Hüls Am.*

6948 CPS 130
68607-75-0
Polymethyloctadecylsiloxane
Waterproofing agent. *Hüls Am.*

6949 CPS 140
68440-90-4
Polymethyloctylsiloxane
Waterproofing agent. *Hüls Am.*

6950 CPS 340
67923-14-2 3264
Dimethicone
Silanol-terminated; reactive fluid, room temperature vulcanizing intermediate. *Hüls Am.*

6951 **CPS 925**
68037-87-6
Polyvinylmethylsiloxane
Coupling agent. *Hüls Am.*

6952 **CPS 9120**
Polydiethoxysiloxane
Silicone dioxide source. *Hüls Am.*

6953 **CR-39**
Allyl diglycol carbonate. Monomer. *PPG Industries.*

6954 **Crafol AP-16**
Ethoxylated C_{16} phosphate ester
Surfactant, textile lubricant. *Pulcra SA.*

6955 **Crafol AP-201**
12751-23-4 235-799-7
Lauryl phosphate
Free acid; lubricant, emulsifier for leather and textile auxiliaries. *Pulcra SA.*

6956 **Crafol AP-260**
Sodium laureth (10) phosphate
Emulsifier, wetting agent, antistat for leather, textile auxiliaries, hot baths containing cyanides. *Pulcra SA.*

6957 **Crafol AP-262**
MEA laureth (6) phosphate
Lubricant, antistat for textile auxiliaries. *Henkel KGaA.* Discontinued.

6958 **Crafol AP-31**
Potassium lauryl phosphate
Manufacturing of man-made textile fibers; lubricant, antistat; spinning oil for carpets. *Pulcra SA.*

6959 **Crafol AP-53**
66197-78-2 266-231-6
Nonylphenol ether phosphate, free acid; lubricant, emulsifier, corrosion inhibitor for oil and water-based cutting fluids, hydraulic fluids, rolling oils. *Pulcra SA.*

6960 **Crafol AP-64**
Ethoxylated C_{12-18} phosphate (8 EO) sodium salt
Wetting agent, detergent, humectant. *Pulcra SA.*

6961 **Craig gold**
An alloy of 80% copper, 10% nickel, and 10% zinc. It is a nickel silver *German silver.*

6962 **Cranco**
Lacquers and enamels. *ICI Chemical & Polymers Ltd.* Discontinued.

6963 **Crastine®**
Thermoplastic polyester molding compounds. *Ciba plc.*

6964 **Crastine® S 600, Crastine® SG 625, 665 FR, 653, XB 3035**
26062-94-2
Polybutylene terephthalate polyester; (625, 665FR) 30% glass reinforced; for electronics, electrical, automotive, mechanical engineering, chemical and apparatus engineering, domestic appliances, medical appliances, sporting goods. *Ciba-Geigy GmbH.* Name unverified.

6965 **Crastine® SG 625**
26062-94-2
Polybutylene terephthalate polyester, 30% long glass fiber-reinforced. *Ciba-Geigy GmbH.*

6966 **Crastine® SG 665 FR**
26062-94-2
Polybutylene terephthalate polyester, 30% long glass fiber-reinforced *Ciba-Geigy GmbH.*

6967 **Crastine® SO 653**
26062-94-2
Polybutylene terephthalate polyester, 20% glass bead-reinforced. *Ciba-Geigy GmbH.*

6968 **Crastine® XB 3035**
26062-94-2
Polybutylene terephthalate polyester *Ciba-Geigy GmbH.*

6969 **Crastine® XMB 1068**
25038-59-9 7730
Polyethylene terephthalate polyester, 30% glass fiber-reinforced. *Ciba-Geigy GmbH.* Name unverified.

6970 **Cravenette EFC**
Durable water and oil repellent to be used with and without fluorocarbons; excellent resin/catalyst stability. No manufacturer.

6971 **Craymer**
Dimer acids
Dimer derivatives. *Cray Valley Ltd.*

6972 **Crayvallac**
Thixotropes.
Cray Valley Ltd.

6973 **Cream E45**
A proprietary preparation containing white, soft paraffin, light liquid paraffin and anhydrous lanolin; a dermatological cream. *Crookes Healthcare.*

6974 **Creamtex**
68334-28-1 269-820-6
Partially hydrogenated vegetable oil (soybean, cottonseed); kosher; good quality bakery shortening for use where emulsifier is not required. *Van Den Bergh Foods.*

6975 **Crelan®**
Resins and curing agents for the formulation of electrostatic thermosetting powder coatings. *Bayer AG.*

6976 **Cremalys**
Blends of gelatinized and nongelatinized starches; used in the food industry to develop creamy textures. *Roquette (UK) Ltd.*

6977 **Cremba**
Mineral oil, petrolatum, lanolin alcohol, and lanolin; emollient, moisturizer, and emulsifier for cosmetics and pharmaceuticals, detergent surgical scrubs. *Croda Inc; Croda Chem. Ltd.*

6978 **Cremerol HMG**
Hydroxylated milk glycerides; emollient, auxiliary emulsifier for personal care products. *Amerchol Corp.*

6979 **Cremophor® A 11**
68439-49-6 7737
Ceteareth-11
Emulsifier for cosmetic and pharmaceutical preparations. *BASF AG.*

6980 **Cremophor® A 25**
68439-49-6 7737
Ceteareth-25
Emulsifier for cosmetic and pharmaceutical preparations. *BASF AG.*

6981 **Cremophor® EL**
61791-12-6
PEG-35 castor oil
Solubilizer, emulsifier used for essential oils, pharmaceuticals, cosmetics, veterinary medicine. *BASF; BASF AG.* Name unverified.

6982 **Cremophor® NP 10**
9016-45-9 6772
Nonoxynol-10
Solubilizer for perfumes, essential oils, and flavors. *BASF AG.* Name unverified.

6983 **Cremophor® NP 14**
9016-45-9 6772
Nonoxynol-14
Detergents and surfactants. *BASF AG.* No manufacturer.

6984 **Cremophor® RH 40**
61788-85-0
PEG-40 hydrogenated castor oil
Solubilizer for essential oils and perfumery, emulsifier; for cosmetics and pharmaceuticals. *BASF; BASF AG.* No manufacturer.

6985 **Cremophor® RH 60**
61788-85-0
PEG-60 hydrogenated castor oil
Solubilizer, emulsifier for essential oils and perfumes; for cosmetics. *BASF; BASF AG.*

6986 **Cremophor® S 9**
9004-99-3
PEG-9 stearate
Emulsifier for oil-water type, thickening agent; suspension stabilizer; lubricating and antitack effects; for cosmetics and pharmaceuticals. *BASF AG.*

6987 **Cremophor® WO 7**
61788-85-0
PEG-7 hydrogenated castor oil
Emulsifier for cosmetic water-oil preparations. *BASF AG.*

6988 **Crenette**
Nonwoven glass scrim fabric; for reinforcement of filler media, floor coverings, needle-felt, kraft paper and plastic sheeting. *Fothergill Tygaflor Ltd.* Name unverified.

6989 **Creolin®**
12751-04-1 2639
A coal tar black disinfectant containing 20% coal tar acids conforming to B.S.2462 BA; a disinfectant for public health, industrial and institutional application. d = 1.02-1.04; soluble in organic solvents. *William Pearson Ltd.* Name unverified.

6990 **creosol**
93-51-6 2640 202-252-9
$C_8H_{10}O_2$
2-Methoxy-4-methylphenol
Homocatechol methyl ester; 2-methoxy-*p*-cresol; 4-methylguaiacol; 3-

methoxy-4-hydroxytoluene; 4-hydroxy-3-methoxy-1-methylbenzene; Valspice; Homocatechol monomethyl ether. Constituent of creosote; used as wood preservative, disinfectant. mp = 5.5°; bp_{15} = 104-105°; d = 1.092; n_D^{20} = 1.5370; LD_{50} (rat orl) = 740 mg/kg.

6991 Creosote
8001-58-9 2641 232-287-5
Coal tar creosote. A term used in reference to the mixed phenols obtained from wood tar, coal tar, and other sources; used for wood preservation.

6992 Crepetrol® 190
Polymer, creping and adhesion aid for tissue machines. *Hercules.*

6993 Cresavon
Hospital antiseptic soap. *Laporte Industries Ltd.*

6994 Cresol
Tricresol
cresylol. Crude cresol contains approximately 35% *o*-, 40% *m*-, and 25% *p*-cresol, C_7H_8O; used as an antiseptic.

6995 cresol purple
2303-01-7 218-960-6
$C_{21}H_{18}O_5S$
Phenol, 4,4'-(3H-2,1-benzoxathiol-3-ylidene)bis[3-methyl-, S,S-dioxide
m-cresolsulfonphthalein. Used as an indicator.

6996 cresol red
1733-12-6 2647 217-064-2
$C_{21}H_{18}O_5S$
4,4'-(3H-2,1-Benzoxathiol-3-ylidene)bis(2-methylphenol) S,S-dioxide
o-Cresolsulfonphthalein; α-hydroxy-α,α-bis(4-hydroxy-*m*-tolyl)-*o*-toluenesulfonic acid γ-sultone; *o*-cresolsulfonphthalein. Used as an indicator, range 7.2 (yellow), 8.8 (red), 2-3 (orange). mp = 290° (dec); pK 8.3.

6997 *m*-cresol
108-39-4 2645 203-577-9
C_7H_8O
3-Methylphenol
m-cresylic acid. In disinfectants, fumigants; in photographic developers, explosives; phenolic resins, ore flotation; textile scouring agent; manufacture of coumarin and salicylaldehyde; herbicides, surfactants. mp = 11-12°; bp = 212°; d_4^{20}= 1.034; n_D^{20}= 1.5398; soluble in 40 parts H_2O, very soluble in organic solvents; LD_{50} (rat orl) = 242 mg/kg. *Allchem Industries; Mitsui PetroChemical Ind. Penta Mfg.: Schweizerhall; Spectrum Chemical Mfg.*

6998 *o*-cresol
95-48-7 2645 202-423-8
C_7H_8O
2-methylphenol
o-cresylic acid; *o*-hydroxytoluene. Disinfectant, phenolic resins, tricresyl phosphate, ore flotation, textile scouring, organic intermediate, manufacture of salicylaldehyde, coumarin, and herbicides, surfactant. mp = 30-32°; bp = 191-192°; d = 1.048; n_D^{20} = 1.553; LD_{50} (rat orl) = 121 mg/kg. *Allchem Industries; Crowley Tar Prods.; Penta Mfg.; PMC Specialties.*

6999 *p*-cresol
106-44-5 2645 203-398-6
$CH_3C_6H_4OH$
4-methylphenol
p-cresylic acid; 4-hydroxytoluene. Disinfectant, phenolic resins, ore flotation; textile scouring agent; manufacture of coumarin and salicylaldehyde; herbicides, surfactants, synthetic food flavors. mp = 32-34°; bp = 202°; d = 1.0300; soluble in H_2O (2 g/100 ml), more soluble in organic solvents; LD_{50} (rat orl) = 207 mg/kg *Allchem Industries; Am. Biorganics; Penta Mfg. PMC Specialities; Spectrum Chemical Mfg.*

7000 Cresolox®
Blend of vegetable soaps, coal tar oils and tar acids; disinfectants. *Coventry Chemicals Ltd.*

7001 cresotic acids
$C_8H_8O_3$
Cresol carboxylic acids
These acids have ten possible isomers; dye intermediate; research on plant growth inhibition.

7002 Cressylite
picric acid-trinitrocresol. A mixture of picric acid and trinitrocresol; used in explosives.

7003 Crestalan
Lanolin and isopropyl esters; used in cosmetic and household products. *Croda Chem. Ltd.*

7004 Crester KZ
Polyglycerol esters
Food emulsifier. *Croda Surf. Ltd.*

7005 Crestolan NF, Crestosolve 630
Detergents for textile scouring. *Reilly-White-man.*

7006 Crestomer®
Unsaturated urethane acrylate resins; used for adhesives, binders; tough, impact resistant FRP applications. *Scott Bader.* Discontinued.

7007 Crestomer® 1066A
Tough resin (styrene monomer) with high filler tolerance and good adhesion; cured with benzoyl peroxide catalyst; for casting resin, binder for seamless flooring with good resilience and chemical resistance. *Scott Bader.* Discontinued.

7008 Crestomer® 1080
Tough, flexible base resin in styrene solution; for formulation of high performance, high impact adhesives and filled casting resins; impact modifier for polyester resins. *Scott Bader.* Discontinued.

7009 Crestopene 5X
Wetting agent for textile processing of cellulosics and blends. *Reilly-Whiteman.*

7010 Crestophen®
Phenolic resins, phenolic resin ancillaries; for highly fire resistant FRP applications. *Scott Bader. Discontinued.*

7011 *m*- cresyl acetate
122-46-3 2651 204-546-2
$C_9H_{10}O_2$
m-cresyl acetate
m-Tolyl acetate; acetic acid *m*-tolyl ester; Cresatin. External antiseptic and analgesic. bp_{14} = 100°; d_4^{26}= 1.048; λ_m = 262.5, 269.5 nm (MeOH); insoluble in H_2O, soluble in organic solvents *Merck & Co Ltd.*

7012 Cresylene
Dyestuffs. *Imperial Chemical Industries plc.*

7013 cresylic acid
1319-77-3 2645 215-293-2
$C_{21}H_{24}O_3$
Crysylol; Tricresol; Cresol (All Isomers); Cresylic acid; cresols; Cresol (mixed isomers); Mixed cresols; *m*-,*p*-cresol mixture; Hydroxytoluene; Coal tar acids; Coal tar cresols; Coal tar phenols; Cresylic acid, coal tar acids, coal tar phenols, or coal tar cresols. Commercial mixtures of phenolic materials boiling above the cresol range; phosphate esters, phenolic resins, wire enamel solvent, plasticizers, gasoline additives, laminates, coating for magnet wire, disinfectants, metal cleaning, flotation agents, mp = 1-2°; bp = 88-94°, d = 1.04; soluble in H_2O (1.932 g/l); more soluble in organic solvents; LD_{50} (rat orl) = 1454 mg/kg. *Allchem Industries; Crowley Tar Prods.; PMC Specialties; Spectar Ltd.*

7014 Creto
A cationic surfactant. *Croda Chem. Ltd.* Discontinued.

7015 Crex
Mild laundry alkali. *ICI Chemical & Polymers Ltd.*

7016 Crexathix
Castor oil derivatives; thixotropic agent for paints and printing inks. *Blagden Chemicals Ltd.* Name unverified.

7017 Cri-Line FDA-612, APC-718-75
Fluoroelastomer compounds; (FDA) elastomer for compression, transfer and injection molding, extrusion, calendering; for food contact applications. (APC) Fluoroelastomer compound; elastomer for extruded profile, O-ring cord. *Cri-Tech.*

7018 Cri-Line FDA-715
Fluoroelastomer compound; elastomer for compression, transfer and injection molding, extrusion, calendering; for food contact applications. *Cri-Tech.*

7019 Cri-Line GP-715
Fluoroelastomer copolymer; general purpose elastomer for compression, transfer, injection molding, extrusion, calendering of valve stem seals, custom molded goods, extruded profiles, roll covers. *Cri-Tech.*

7020 Cri-Line HF-618-65
Fluoroelastomer terpolymer; elastomer for compression, transfer and injection molding, extrusion, calendering of fuel line and seals, diesel cylinder linders, diaphragms. *Cri-Tech.*

7021 Cri-Line IF-612
Hexafluoropropylene/vinylidene fluoride/tetrafluoroethylene terpolymer; elastomer for compression, transfer and injection molding, extrusion, calendering of fuel line and seals, expansion joints, shaft seals, valve stem seals. *Cri-Tech.*

7022 Cri-Line LC-508-55
Reprocessed and virgin fluoroelastomer compound; low-cost elastomer for compression, transfer, injection molding, extrusion and calendering of lathe cuts, custom molded goods. *Cri-Tech.*

7023 Cri-Line LC-612-THK
Reprocessed and virgin fluoroelastomer compound; low-cost elastomer for thick cross section curing; for compression, transfer, injection molding, extrusion, calendering of custom molded goods, rubber-to-metal bonded seals. *Cri-Tech.*

7024 Cri-Line LC-708
Reprocessed and virgin fluoroelastomer compound; low-cost elastomer for compression molding and calendering; for calendered sheet packing. *Cri-Tech.*

7025 Cri-Line SP-508
Hexafluoroporpylene/vinylidene fluoride copolymer; elastomer for compression, transfer, injection molding, extrusion and calendering of O-rings, seals, rolls, covers, diaphragms. *Cri-Tech.*

7026 Cri-Line SP-815
Hexafluoroporpylene/vinylidene fluoride copolymer; elastomer for compression and transfer molding of O-rings and seals. *Cri-Tech.*

7027 Criliprint 788FYN
Self-crosslinking acrylic copolymer; durable pigment and textile pad binder. *Sybron.*

7028 Crill 1
1338-39-2 8872 215-663-3
Sorbitan laurate
Emulsifier, pigment dispersant, cosolvent, wetting agent, antifoam, viscosity reducer, mold release, antiblock agent, corrosion inhibitor, lubricant, antistat; used for cosmetics, food and food packaging, insecticides and herbicides, leather treatment, metalworking fluids, oil slick dispersing, paints and inks, pharmaceuticals, plastics, polishes and textiles. *Croda Surf. Ltd.*

7029 Crill 2
26266-57-9 8872 247-568-8
Sorbitan palmitate
Emulsifier, dispersant and wetting agent. *Croda Surf. Ltd.* No manufacturer.

7030 Crill 3
1338-41-6 8872 215-664-9
Sorbitan stearate
Emulsifier, lubricant, antistat; oil-water emulsions; cosmetic and pharmaceutical creams and lotions; polishes; insecticides, herbicides; metal cleaners; buffing compounds; textile lubricants; food applications. *Croda Inc.; Croda Surf. Ltd.* No manufacturer.

7031 Crill 4
1338-43-8 8872 215-665-4
Sorbitan oleate
Emulsifier wetting agent, pigment dispersant, cosmetic and pharmaceutical application; aerosol polishes; insecticidal sprays; inks and surf. coatings; metal working lubricants; cutting oils; textile lubricants; dry cleaning operations; food process antifoam, oil slick dispersant. *Croda Inc.; Croda Surf Ltd.* Name unverified.

7032 Crill 6
71902-01-7 8872 276-171-2
Sorbitan isostearate
Water-oil emulsifier, wetting agent, pigment dispersant for creams, lotions, aerosols. *Croda Inc.; Croda Surf. Ltd.*

7033 Crill 35
26658-19-5 8872 247-891-4
Sorbitan tristearate
Emulsifier, lubricant, antistat for cosmetic, pharmaceutical, food, and industrial applications. *Croda Surf. Ltd.* Name unverified.

7034 Crill 43
8007-43-0 8872 232-360-1
Sorbitan sesquioleate
Water-oil emulsifier, wetting agent, pigment dispersant; for cosmetic, pharmaceutical, food, and industrial applications. *Croda Surf. Ltd.* No manufacturer.

7035 Crill 45
26266-58-0 8872 247-569-3
Sorbitan trioleate
Water-oil emulsifier, wetting agent, pigment dispersant; for cosmetic, pharmaceutical, food, and industrial applications. *Croda Surf. Ltd.* No manufacturer.

7036 Crillet
8872
Polyethoxylated sorbitan esters
Used as nonionic surfactants and food emulsifiers. *Croda Chem. Ltd; Croda Inc.*

7037 Crillet 1
9005-64-5 8872
Polysorbate 20
Solubilizer, emulsifier, dispersant, wetting agent; often combined with a member of the Crill range in emulsification systems; used in cosmetics, food and food packaging, household products, insecticides, herbicides, metalworking fluids, paints and inks, *Croda Inc.; Croda Chem. Ltd.*

7038 Crillet 2
9005-66-7 8872
PEG-20 sorbitan palmitate
Emulsifier, solubilizer, wetting agent for cosmetic, pharmaceutical, food, and industrial applications. *Croda Chem. Ltd.*

7039 Crillet 3
9005-67-8 8872
Polysorbate 60 NF
Oil-water emulsifier for cosmetics and pharmaceuticals, dispersant for insecticides, herbicides, cattle dyes, penetrant, leveling agent, lubricant, antistat. *Croda Inc.; Croda Chem. Ltd.*

7040 Crillet 4
9005-65-6 7742
Polysorbate 80 NF
Emulsifier, dispersant, solubilizer, lubricant, detergent, antistat, wetting agent for cosmetics, pharmaceuticals, polishes, insecticides, leather degreasing, veterinary products. *Croda Inc.; Croda Chem. Ltd.*

7041 Crillet 6
66794-58-9
PEG-20 sorbitan isostearate
Oil-water emulsifier, solubilizer for fragrances and perfumes; for creams, lotions, ointments; improved resistance to oxidation. *Croda Inc.; Croda Chem. Ltd.*

7042 Crillet 11
9005-64-5 8872
PEG-4 sorbitan laurate
Emulsifier, solubilizer, wetting agent for cosmetic, pharmaceutical, food, and industrial applications. *Croda Chem. Ltd.*

7043 Crillet 31
9005-67-8 8872
PEG-4 sorbitan stearate
Emulsifier, solubilizer, wetting agent for cosmetic, pharmaceutical, food, and industrial applications. *Croda Chem. Ltd.*

7044 Crillet 35
9005-71-4 8872
Polysorbate 65
Emulsifier, solubilizer, wetting agent for cosmetic, pharmaceutical, food, and industrial applications. *Croda Chem. Ltd.*

7045 Crillet 41
9005-65-6 7742
PEG-5 sorbitan oleate
Emulsifier, solubilizer, wetting agent for cosmetic, pharmaceutical, food, and industrial applications. *Croda Chem. Ltd.*

7046 Crillet 45
9005-70-3 8872
Polysorbate 85
Emulsifier, solubilizer, wetting agent for cosmetic, pharmaceutical, food, and industrial applications. *Croda Chem. Ltd.*

7047 Crillon
Detergent and foam stabilizing properties; antistatic agents; anticorrosive properties; skin protecting agents; used in shampoos, bubble bath formulations; detergent cleaners; hand cleansers; cutting fluids and soluble cutting oil compositions. *Croda Surf. Ltd.*

7048 Crillon CDY
Cocamide DEA
Solvent and cutting oils. *Croda Chem. Ltd.* Name unverified.

7049 Crillon LDE
120-40-1 204-393-1
Lauramide DEA
Detergent and foam stabilizer; oil-water emulsifiers, antistat; anticorrosive. *Croda Surf. Ltd.*

7050 Crillon LME
142-78-9 205-560-1
Lauramide MEA
Lubrication; skin protection agent. *Croda Chem. Ltd.*

7051 Crillon ODE
93-83-4 202-281-7
Oleamide DEA
Emulsifier, stabilizer, skin protectant, lubricant, anti-irritant used in personal care products; additive for cutting fluids and soluble cutting oils. *Croda Surf. Ltd.*

7052 Crimidesa
7757-82-6 8829 231-820-9
Anhydrous sodium sulfate
Used in manufacture of detergents, glass, dyestuffs, etc. *Bromhead & Denison Ltd.*

7053 Crimidine
535-89-7 2655 208-622-6
$C_7H_{10}ClN_3$
2-chloro-N,N,6-trimethyl-4-pyrimidinamine
Castrix Grains. Rodenticide; used for control of field and house mice. Mode

of action is similar to that of the strychnine alkaloids. Pyridoxine [58-56-0] is an antidote for crimidine poisoning in mice. mp = 87°; insoluble in H_2O, soluble in organic solvents; LD_{50} (mus orl) = 1.8 mg/kg. *Bayer AG.*

7054 Crimson Lake
A cochineal lake containing aluminum salts. No manufacturer.

7055 Cristite
A proprietary trade name for a steel containing 10% chromium, 17% tungsten, 3.5% carbon, and 2.5% molybdenum. No manufacturer.

7056 Criterion
Blend of pyridinium compounds, alcohols, and saccharin; nickel electroplating additive. *Taskem Inc.* Name unverified.

7057 Crllitex H-50
Self-crosslinking acrylic copolymer; hand builder producing soft full hand, durable to washing and dry cleaning. *Sybron.*

7058 Croak
fluometuron-MSMA. Fluometuron plus MSMA; herbicide for post-emergence control of broadleaf and grass weeds in cotton. *Draxel Chemical Company.* Unverified.

7059 Crocell
Hot-dip, thermoplastic strippable coating. *Croda Application Chemicals Ltd.* Discontinued.

7060 Crocidolite
61105-31-5 863
Blue asbestos; cape blue. A fibrous mineral. It is a member of the group of minerals known as soda-amphiboles. It is a hydrated silicate of sodium and iron. Pale blue asbestos, found in South Africa and Western Australia. The name is sometimes applied incorrectly to a variety of quartz known as Tiger's Eye.

7061 crocus
8466
Spanish saffron; French saffron. Saffron, a coloring matter from the dried and powdered flowers of the saffron plant, *Crocus sativus*; used for coloring and flavoring confectionery, also refers to ferric oxide having a bluish tint and used for polishing metals, both known as crocus. Contains about 1% of picrocrocin.

7062 Crocus Martius
1309-37-1 4072 215-168-2
ferric oxide
A hydrated ferric oxide; used as a pigment for pottery and other purposes.

7063 Croda Bath Oil Disperant
Synergistic blend of nonionic surfactants. *Croda Chem. Ltd.*

7064 Croda Fluid
Temporary rust preventative. *Croda Application Chemicals Ltd.* Discontinued.

7065 Crodacel QL
Laurdimonium hydroxyethyl cellulose
Conditioner improving foaming for skin and hair care products *Croda Inc.* Discontinued.

7066 Crodacel QM
Cocodimonium hydroxyethyl cellulose
Conditioner improving foaming and imparting body to skin and hair care products *Croda Inc.*

7067 Crodacel QS
Steardimonium hydroxyethyl cellulose
Conditioner improving foaming and imparting body to skin and hair care products *Croda Inc.*

7068 Crodacol C70
36653-82-4 2070 253-149-0
Cetyl alcohol
Secondary emulsifier, thickener, opacifier, and structural agent in anhydrous stick systems. *Croda Inc.* No manufacturer.

7069 Crodacol C95NF
36653-82-4 2070 253-149-0
Cetyl alcohol NF
Primary structural agent in antiperspirant sticks. *Croda Inc.* No manufacturer.

7070 Crodacol CS50
Cetyl-Stearyl alcohols
Emollient; hair and skin lubricant. *Croda Inc.*

7071 Crodacol S70
112-92-5 8960 204-017-6
Stearyl alcohol
Secondary emulsifier, thickener, opacifier, and structural agent in anhydrous stick systems. *Croda Inc.* No manufacturer.

7072 Crodacol S95NF
112-92-5 8960 204-017-6
Stearyl alcohol NF
Primary structural agent in antiperspirant sticks. *Croda Inc.* No manufacturer.

7073 Crodacreme
Blend of emulsifiers including mono/diglycerides of fatty acids; ice cream emulsifier/stabilizer. *Croda Food Prods. Ltd.*

7074 Crodafos
A range of surfactants composed of alkyl ether phosphates in acid form. Each is a mixture of mon- and diesters and may be converted to salts by neutralization with (e.g.) alkanolamine or metal hydroxide; detergents and coupling agents. (for nonionics) with wetting and emulsifying properties; used in heavy duty clothes washing; hard surface cleaners, floor and industrial cleaners and steam cleaning systems; dry cleaning; dispersible solvent cleaners; cutting and rolling oils, grinding fluids, lubricants; as antistats, lubricants and softeners in synthetic fiber and wool processing; emulsion polymerization; pesticides and fertilizers; waxes, polishes, surface coatings, cosmetics and pharmaceuticals. .

7075 Crodafos 25 D2 Acid
68071-35-2
PEG-2 C_{12-15} ether phosphate
Hair conditioner for shampoo formulations. *Croda Chem. Ltd.* No manufacturer.

7076 Crodafos CAP
111019-03-5
PPG-10 cetyl ether phosphate
Water-oil emulsifier; antistat used in personal care products; modifies pH, thickening, emulsifying and suspending properties; enhances hair conditioning; useful for microemulsion systems. *Croda Inc.* No manufacturer.

7077 Crodafos CDP
90388-14-0 291-394-5
Cetyl diethanolamine phosphate
Emulsifier and stabilizer for oil-water emulsions. *Croda Chem. Ltd.*

7078 Crodafos CS2 Acid
Ceteareth-2 phosphate
Emulsifier and stabilizer for oil-water emulsions. *Croda Chem. Ltd.*

7079 Crodafos N10 Acid
39464-69-2
Oleth-10 phosphate
Oil-water emulsifier, gelling agent for surfactants, preparation of clear mineral oil gels, skin cleansers, clear microemulsion gels. *Croda Inc.*

7080 Crodafos N10 Neutral
58855-63-3
DEA-oleth-10 phosphate
Oil-water emulsifier, gelling agent for surfactants, preparation of clear mineral oil gels, skin cleansers, clear microemulsion gels. *Croda Inc.* Name unverified.

7081 Crodafos N3 Acid
39464-69-2
Oleth-3 phosphate
Surfactant, conditioner, antistat, oil-water emulsifier, gelling agent for surfactants, cosmetics, pharmaceuticals, and toiletries, microemulsion gels; corrosion inhibitor and anti-gelling agent in aerosol antiperspirant systems. *Croda Inc.; Croda Chem. Ltd.* Name unverified.

7082 Crodafos N3 Neutral
58855-63-3
DEA-oleth-3 phosphate
Oil-water emulsifier, gelling agent for surfactants and for preparation of clear mineral oil gels and microemulsion gels; used in hair relaxers. *Croda Inc.* No manufacturer.

7083 Crodafos N5 Acid
39464-69-2
Oleth-5 phosphate
Emulsifier and stabilizer for oil-water emulsions. *Croda Chem. Ltd.*

7084 Crodafos O2 Acid
39464-69-2
Oleth-2 phosphate
Hair conditioners for shampoo formulations. *Croda Chem. Ltd.*

7085 Crodafos O2 TEA
TEA-oleth-2 phosphate
Croda Chem. Ltd. Name unverified.

7086 Crodafos SG
50643-20-4
PPG-5 ceteth-10 phosphate
Oil-water emulsifier, conditioner, wet comb enhancer for shampoos and cream rinses; thickener, gellant, and pH adjuster for shampoos. *Croda Inc.; Croda Chem. Ltd.*

7087 Crodafos T2 Acid
9046-01-9

Trideceth-2 phosphate
Hair conditioners for shampoo formulations. *Croda Chem. Ltd.*

7088 Crodakyd
Water-thinnable resins. *Croda Resins Ltd.*

7089 Crodalan AWS
Polysorbate 80
cetyl acetate; acetylated lanolin alcohol. Emollient, superfatting agent, conditioner, oil-water emulsifier, dispersant, wetting agent, plasticizer, solubilizer used in cosmetics, pharmaceuticals, detergent systems. *Croda Inc.; Croda Chemical Ltd.*

7090 Crodalan C24
Polyethoxylated cholesterol derivatives; used in the cosmetic industry. *Croda Chem. Ltd.*

7091 Crodalan IPL
Isopropyl lanolate
General purpose raw material for cosmetics. *Croda Chem. Ltd.*

7092 Crodalan LA
Cetyl acetate and cetylated lanolin alcohol. Emollient, penetrant, wetting agent, conditioner, plasticizer used in cosmetics, pharmaceuticals. *Croda Inc; Croda Chem. Ltd.*

7093 Crodamer
A proprietary trade name for a range of uv radiation curable resins for lacquers and inks. *Croda Resins Ltd.*

7094 Crodamet Series
Ethoxylated primary amines; emulsifiers, antistats for plastics and fibers. *Croda Chem. Ltd.*

7095 Crodamide
Fatty acid amides; Primary, secondary, and bisamides based on stearic, oleic, erucic and behenic fatty acids; applications mainly in plastics industry as slip and antiblock agents. *Croda Universal Ltd.*

7096 Crodamide E, ER
112-84-5 204-009-2
Erucamide
Internationally used slip and antiblock agent for polyolefins; internal release agent for molded thermoplastic polymers. *Croda Chem. Ltd.* No manufacturer.

7097 Crodamide O, OR
301-02-0 206-103-9
Oleamide
International slip and antiblock agent for polyolefins; internal release agent for moulded thermoplastic polymers. *Croda Chem. Ltd.* No manufacturer.

7098 Crodamide S, SR
124-26-5 204-693-2
Stearamide
International slip and antiblock agent for polyolefins; internal release agent for moulded thermoplastic polymers. *Croda Chem. Ltd.* No manufacturer.

7099 Crodamine 1
Cationic surfactants; primary alkylamines, where the alkyl group may be lauric, palmitic, stearic, coconut, soya, tallow, or oleic in origin; solid, liquid or paste forms; D = distilled. Corrosion inhibitors, anticaking agents. *Croda Universal Ltd.*

7100 Crodamine 1.0, 1.0D
112-90-3 294-015-5
$C_{18}H_{37}N$
(Z)-9-Octadecen-1-amine
Oleylamine; octadecenylamine; 1-Amino-9-octadecene. Emulsifier for herbicides, ore flotation, pigment dispersion; auxiliary for textiles, leather, rubber, plastics, and metal industries. mp = 15-22°; d = 0.8130; n_D^{20} = 1.4578; *Croda Universal Ltd.* No manufacturer.

7101 Crodamine 1.16D
143-27-1 205-596-8
$C_{16}H_{35}N$
n-Hexadecylamine
Cetylamine; 1-Hexadecanamine; 1-Hexadecylamine; alamine 6; Armeen 16d; palmitylamine; 1-amino hexadecane. Primary cetyl amine; emulsifier for herbicides, ore flotation, pigment dispersion; auxiliary for textiles, leather, rubber, plastic, and metal industries. mp = 38-47°; bp = 330°; insoluble in H_2O, soluble in organic solvents. *Croda Universal Ltd.*

7102 Crodamine 1.18D
124-30-1 204-695-3
$C_{18}H_{39}N$
Stearylamine
1-Aminooctadecane; 1-Octadecanamine; 1-Octadecylamine; octadecylamine;adogen 142; alamine 7; armeen 1180; n-octadecylamine. Emulsifier for herbicides, ore flotation, pigment dispersion; auxiliary for textiles, leather, rubber, plastics, and metal industries. mp = 50-52°; bp_{32} = 232°; insoluble in H_2O, soluble in organic solvents; LD_{50} (rat orl) = 2395 mg/kg. *Croda Universal Ltd.*

7103 Crodamine 1.HT
61788-45-2 262-976-6
Hydrogenated tallow amine
Anticaking agent for fertilizers; emulsifier for herbicides, ore flotation, pigment dispersion; auxiliary for textiles, leather, rubber, plastics, and metal industries. *Croda Universal Ltd.*

7104 Crodamine 1.T
61790-33-8 263-125-1
Tallow amine
Emulsifier for herbicides, ore flotation, pigment dispersion; auxiliary for textiles, leather, rubber, plastics, and metal industries. *Croda Universal Ltd.* No manufacturer.

7105 Crodamine 2.C, 2.S and 2.HT
Cationic surfactant consisting of secondary amines R_2NH, in solid form, where the fatty alkyl group, R, may be coconut, soya or hydrogenated tallow in origin respectively; used for corrosion inhibition; emulsifiers for herbicides; mineral flotation; pigment dispersion, antocaking agents, auxiliaries for leather, textiles, rubber, plastics and metal industries. *Croda Universal Ltd.*

7106 Crodamine 3.A16D
112-69-6 203-997-2
$C_{18}H_{39}N$
N,N-Dimethylhexadecylamine
Armeen DM 16D; 1-Hexadecanamine, N,N-dimethyl-; Palmityl dimethylamine. Emulsifier for herbicides, ore flotation, pigment dispersion; auxiliary for textiles, leather, rubber, plastics, and metal industries. *Croda Universal Ltd.* Name unverified.

7107 Crodamine 3.A18D
124-28-7 3525 204-694-8
$C_{20}H_{43}N$
N,N-Dimethyloctadecylamine
1-Octadecanamine, N,N-dimethyl-; Dimethylstearylamine; Armeen DM 18D; Dimethyl-1-octadecanamine; Dimantine; Dymanthine; Stearyl dimethylamine. Emulsifier for herbicides, ore flotation, pigment dispersion; auxiliary for textiles, leather, rubber, plastics, and metal industries. *Croda Universal Ltd.* Name unverified.

7108 Crodamine 3.AED
Dimethyl erucylamine
Emulsifier for herbicides, ore flotation, pigment dispersion; auxiliary for textiles, leather, rubber, plastics, and metal industries; intermediate. *Croda Universal Ltd.*

7109 Crodamine 3.AOD
28061-69-0 248-811-0
Dimethyl oleylamine
Emulsifier for herbicides, ore flotation, pigment dispersion; auxiliary for textiles, leather, rubber, plastics, and metal industries. *Croda Universal Ltd.* Discontinued.

7110 Crodamine 3A
Series of cationic surfactants of the tertiary amine type, where the fatty alkyl groups are lauric, palmitic, stearic and coconut in origin; supplied in liquid or paste form; used for corrosion inhibition; emulsifiers for herbicides and synthetic resins, mineral flotation, pigment dispersion, anticaking agents, e.g. for high analysis NPK fertilizers; auxiliaries for leather, textiles, rubber, plastics and metals industries; catalysts in the production of polyurethane foam. *Croda Universal Ltd.*

7111 Crodamine 3ABD
N,N-Dimethylbehenylamine
An intermediate for the preparation of quaternary ammonium compounds, betaines and amine oxides. Other uses include pharmaceuticals, solvents and anticorrosives. *Croda Universal Ltd.*

7112 Crodamine 3AED
N,N-Dimethylerucylamine
Used as an intermediate, other uses include pharmaceuticals, solvents and anticorrosives. *Croda Universal Ltd.*

7113 Crodamine 3AHRD, 3ARD
N,N-Dimethyl C_{18-22} amines
Used as intermediates, in pharmaceuticals, solvents and anticorrosives. *Croda Universal Ltd.*

7114 Crodamine 3AOD
N,N-Dimethyloleylamine
An intermediate for the preparation of quaternary ammonium compounds, betaines and amine oxides; also in pharmaceuticals, solvents and anticorrosives. *Croda Universal Ltd.*

7115 Crodamol
A range of straight and branched chain mono and dibasic ester; used as cosmetic emollients. *Croda Surf. Ltd.*

7116 Crodamol 1PM
110-27-0 5234 203-751-4
Isopropyl myristate

IPM, perfumery grade; spreading agent, emollient, cosolvent for cosmetic raw materials. *Croda Inc.* Discontinued.

7117 Crodamol BE
18312-32-8 242-201-8
Behenyl erucate
Thickening and opacifying agent, emollient, stabilizer, modifier for lipsticks, powder suspensions. *Croda Inc.* Discontinued.

7118 Crodamol CAP
261-619-1
Cetearyl octanoate
Emollient simulating properties of preen gland oil; provides water repellency in skin care preparations *Croda Inc.*

7119 Crodamol CP
540-10-3 2073 208-736-6
Cetyl palmitate
Emollient for replacing spermaceti wax. *Croda Inc.* Discontinued.

7120 Crodamol CSP
Cetearyl palmitate
Improves feel and body of emulsions; replaces spermaceti wax and beeswax in personal care products *Croda Inc.* Discontinued.

7121 Crodamol GTC/C
Glyceryl tricapryl caprylate
Bakery lubricant, release agent, glazing agent, hypoallergenic baby food formulations, flavor carrier. *Croda Food Prods. Ltd; Croda Inc.*

7122 Crodamol MM
3234-85-3 221-787-9
Myristyl myristate
Emollient, superfatting agent, viscosity builder in emulsions; substitute for spermaceti wax and/or beeswax in cosmetic and pharmaceutical formulations. *Croda Inc.*

7123 Crodamol PMP
PPG-2 myristyl ether propionate
Emollient, lubricant with dry, light greaseless feel; coupling agent; solvent for sunscreen actives; for bath oils, creams, moisturizers, emulsions. *Croda Inc.*

7124 Crodamol PTC
Pentaerythritol tetracaprylate/caprate
Lubricant for creams and lotions; reduces tack in clear gel microemulsions. *Croda Inc.*

7125 Crodamol PTIS
Pentaerythritol tetraisostearate
Lubricant for creams and lotions; reduces tack in clear gel microemulsions. *Croda Inc.*

7126 Crodamol SS
8002-23-1 232-302-5
Cetyl esters
Synthetic spermaceti NF; emollient and viscosity builder for cosmetic and pharmaceutical preparations. *Croda Inc.*

7127 Crodamol W
66009-41-4 266-065-4
Stearyl heptanoate
Nongreasy emollient and water repellent for cosmetics and toiletries, especially stick formulations; melts rapidly on application to skin; synthetic preen gland wax. *Croda Inc.*

7128 Crodapearl
Modified glycol ester used as a nonionic surfactant; pearling agent for cosmetics, perfumes and toiletries. *Croda Surf. Ltd.*

7129 Crodapearl Liq
Sodium laureth sulfate
glycol MIPA stearate. Pearling agent for shampoos, bubble baths, dishwashing liquids. *Croda Inc.*

7130 Crodapearl NI Liquid
Hydroxyethyl stearamide-MIPA
PPG-5 ceteth-20. Pearlescent for detergent systems, lotions, gels, clear rinses, bath products *Croda Inc.*

7131 Crodaplast
A proprietary trade name for a range of hydroxy functional acrylic resins for high-quality automotive and kitchen appliance finishes. *Croda Resins Ltd.*

7132 Crodapol
A propretary trade name for a range of saturated polyester resins for two pack and melamine cure. *Croda Resins Ltd.*

7133 Crodapur
Technical lanolin; plasticizer for tape adhesives, mastics, and putty. *Croda Chem. Ltd.* Name unverified.

7134 Crodarom Avocadin
Avocado oil unsaponifiables; botanical extract for personal care products *Croda Inc.*

7135 Crodarom Calendula O
Soybean oil, calendula extract, tocopherol; botanical extract for personal care products *Croda Inc.*

7136 Crodarom Carrot O
Peanut oil, carrot extract, isopropyl myristate, tocopherol; botanical extract for personal care products *Croda Inc.*

7137 Crodarom Chamomile A
Propylene glycol and matricaria extract; botanical extract for personal care products *Croda Inc.*

7138 Crodarom Chamomile EO, O
Caprylic/Capric triglycerides and matricaria extract; botanical extract for personal care products *Croda Inc.*

7139 Crodarom Nut O
Peanut oil, mineral oil, walnut extract; botanical extract for personal care products *Croda Inc.*

7140 Crodarom St. John's Wort O
Olive oil, hypericum extract; botanical extract for personal care products *Croda Inc.*

7141 Crodascoop
Mono/diglycerides of fatty acids; emulsifier, stabilizer for soft ice cream. *Croda Food Prods. Ltd.*

7142 Crodasinic
N-Acyl sarcosines and their sodium salts; used as anionic surfactants; foaming agents for cosmetics, perfumes and toiletries. *Croda Chem. Ltd; Croda Inc.*

7143 Crodasinic L
97-78-9 202-608-3
$C_{15}H_{29}NO_3$
N-methyl-N-(1-oxododecyl)glycine
n-lauroylsarcosine; Lauroyl sarcosine; N-Cocoyl Sarcosinate; Lauroyl Sarcosinate. Lauroyl sarcosine in acid form, as a white solid; properties include mild detergency, high foaming, bacteriostatic activity, enzyme inhibition, corrosion inhibition, hard water tolerance and stability in mildly acid formulations; anionic surfactant used in cosmetics and toiletries, dentrifices and shampoos, carpet shampoos, emulsion polymerization, metal treatment, food and food packaging, textile and fine fabric detergents. *Croda Chem. Ltd.*

7144 Crodasinic LS30
137-16-6 4379 205-281-5
$C_{15}H_{28}NNaO_3$
Sodium N-lauroyl sarcosinate
Gardol. Foaming, wetting agent and detergent for acidic conditions; corrosion inhibitor; bacteriastat and inhibitor; used in dental care preparations, pharmaceuticals, personal care products, household and industrial applications. *Croda Chem. Ltd.*

7145 Crodasinic LS35
137-16-6 4379 205-281-5
$C_{15}H_{28}NNaO_3$
Sodium lauroyl sarcosinate in the form of a clear liquid; properties include mild detergency, high foaming, bacteriostatic activity, enzyme inhibition, corrosion inhibition, hard water tolerance and stability in mildly acid formulations; *Croda Universal Ltd.*

7146 Crodasinic LT40
16693-53-1 240-736-1
TEA lauroyl sarcosinate
Detergent, foaming agent, wetting agent, dispersant, emulsifier, anticorrosive, foam stabilizing synergist for carpet shampoos, textile and cosmetic detergent systems. *Croda Chem. Ltd.*

7147 Crodasinic OS35
Sodium N-oleoyl sarcosinate
Foaming agent, wetting agent, detergent, lubricant, antistat, corrosion inhibitor, bacteriostat, penetrant for dental, pharmaceutical, shampoos, depilatories, shaving preparations, household and industrial uses. *Croda Chem. Ltd.* Name unverified.

7148 Crodasinic O
110-25-8 203-749-3
N-Oleoyl sarcosine
Corrosion inhibitor in oils, fuels, lubricants, greases, surface coatings; as antifog agent for food packaging polyolefin films. *Croda Chem. Ltd.*

7149 Crodasone W
Hydrolyzed wheat protein polysiloxane copolymer; Substantive film-forming copolymer with lubricity, gloss, conditioning properties for hair and skin care products *Croda Inc.*

7150 Crodasub
Factice
A processing aid for rubber. *Croda Universal Ltd.*

7151 Crodatem
Diacetyl tartaric acid ester. Antistaling agent, starch complexing agent in

biscuits, cake mixes, coffee whitener, gravy mix, toffee emulsifier. *Croda Food Prods. Ltd.*

7152 Crodateric
A range of imidazoline based amphoteric surfactants; properties include dense foaming and high foam stability; stability in acids, alkalies and strong electrolyte solutions; compatability with soaps, quaternary germicides; good detergency; wetting, emulsifying and sequestering properties, hard water solubility, germicidal and fungicidal properties, used for acid and alklai detergents, angricultural sprays, antistats, corrosion inhibitors, toiletries, dry cleaning, emulsions, polishes, and polymers. *Croda Surf. Ltd.*

7153 Crodateric C
Derived from coconut fatty acid; surfactant. *Croda Universal Ltd.*

7154 Crodateric Cy
63451-23-0 264-189-3
Derived from caprylic acid; surfactant. *Croda Universal Ltd.* Name unverified.

7155 Crodateric O, 0.100
32456-28-6
Derived from oleic acid; surfactant. *Croda Universal Ltd.* Name unverified.

7156 Crodateric S
30342-62-2 250-135-6
Derived from stearic acid; surfactant. *Croda Universal Ltd.* Name unverified.

7157 Crodax
Rust preventatives. *Croda Chem. Ltd.*

7158 Crodax DP 100
Calcium soap and oxidate; used in thick film rust preventives providing protection for steel in salt environments; used for underseal, wire rope dressings, waterproofing agents. *Croda Chem. Ltd.* Name unverified.

7159 Crodax DP 50
Complex mixture of organic acids, esters, and lactones; converted to metallic soaps and esters used in solvent or oil-based rust preventives and lubricating products *Croda Chem. Ltd.* Name unverified.

7160 Crodazoline Cy
37478-68-5 253-521-2
Caprylic imidazoline
Surfactant. *Croda Universal Ltd.*

7161 Crodazoline O
95-19-2 202-397-8
Stearic imidazoline
Surfactant, wetting agent, emulsifier, softener, for textiles, asphalt, tar emulsion breaker, paint additive, corrosion inhibitor, lubricating metal processing aids. *Croda Universal Ltd.*

7162 Crodazoline O
95-38-5 428 202-414-9
Oleic imidazoline
Surfactant, wetting agent, emulsifier, softener, for textiles, asphalt, tar emulsion breaker, paint additive, corrosion inhibitor, lubricating metal processing aids. *Croda Universal Ltd.* No manufacturer.

7163 Crodazoline S
95-19-2 202-397-8
Stearic imidazoline
Surfactant, wetting agent, emulsifier, softener, for textiles, asphalt, tar emulsion breaker, paint additive, corrosion inhibitor. *Croda Universal Ltd.*

7164 Croderol G7000
56-81-5 4493 200-289-5
Glycerol
Conditioner, humectant, moisturizing agent in cosmetics and pharmaceuticals, diluent and plasticizer for many polar materials. *Croda Chem. Ltd.*

7165 Crodesta
A range of sucrose esters of fatty acids; used as nonionic surfactants; pigment dispersing agents for paint. *Croda Chem. Ltd; Croda Inc.* Name unverified.

7166 Crodesta DKS F10
27195-16-0 248-317-5
Sucrose mono/di/tri palmitic/stearic acid; emulsifier, wetting agent and dispersant for use in cosmetics, pharmaceuticals. *Croda Surf. Ltd.*

7167 Crodesta DKS F110
25168-73-4 246-705-9
Sucrose mono/di/tri palmitic/stearic acid
Emulsifier, wetting agent, dispersant for use in cosmetics, pharmaceuticals. *Croda Surf. Ltd.*

7168 Crodesta F-10
27195-16-0 248-317-5
Sucrose distearate
Dispersant, emulsifier, wetting agent, solubilizer, emollient, detergent in cosmetics, toiletries, pharmaceuticals (suntan and baby lotions). *Croda Inc.*

7169 Crodesta F-160
25168-73-4 246-705-9
Sucrose monostearate
Sucrose stearate; dispersant, emulsifier, wetting agent, solubilizer, detergent, foaming agent in cosmetics, toiletries, ingestible pharmaceuticals; thickener and suspending agent. *Croda Inc.* No manufacturer.

7170 Crodesta SL-40
Sucrose cocoate
Dispersant, emulsifier, wetting agent, solubilizer, mild detergent, high foaming emollient in cosmetics, toiletries, pharmaceuticals. *Croda Inc; Croda Surf. Ltd.*

7171 Crodet C10
61791-29-5
PEG-10 coconut fatty acid
Surfactant for cosmetic and industrial applications. *Croda Chem. Ltd.* Name unverified.

7172 Crodet L4
9004-81-3
PEG-4 laurate
Oil-water emulsifier for cosmetics and pharmaceutical creams, lotions and ointments, industrial applications, wetting agent, solubilizer for perfumes or aqueous alcoholic preparations; dispersant; plasticizer for hair setting sprays. *Croda Chem. Ltd.* Name unverified.

7173 Crodet O4
9004-96-0
PEG-4 oleate
Surfactant for cosmetic and industrial applications. *Croda Chem. Ltd.*

7174 Crodet S4
9004-99-3
PEG-4 stearate
Oil-water emulsifier for cosmetics and pharmaceutical creams, lotions and ointments, industrial applications; wetting agent, solubilizer for perfumes or aqueous alcoholic preparations, dispersant. *Croda Chem. Ltd.*

7175 Crodex A
Cetostearyl alcohol and sodium lauryl sulfate
Emulsifying wax BP for pharmaceuticals and cosmetic uses. *Croda Chem. Ltd.*

7176 Crodex C
Cetostearyl alcohol and Cetrimide BP
Emulsifying wax BPC, bactericides, for pharmaceuticals and cosmetics, hair conditioning rinses. *Croda Chem. Ltd.*

7177 Crodex N
Cetostearyl alcohol and ceteth-20
Emulsifying wax BP, wetting agent, penetrant, emulsifier for most emollient materials in cosmetics and pharmaceuticals. *Croda Chem. Ltd.*

7178 Crodinhib
Amine borates
Used as lubricants and plasticizers; corrosion inhibitors for paints and oils. *Croda Chem. Ltd.*

7179 Crodinhib RT70, RT70S
Borate derivs
Corrosion inhibitor used in metal working fluids and oils. *Croda Chem. Ltd.* Name unverified.

7180 Crodol
A barrier cream. *Croda Chem. Ltd.* Name unverified.

7181 Croduret 10
61788-85-0
PEG-10 hydrogenated castor oil
PEG-10 hydrogenated castor oil; emulsifier, solubilizer, emollient, superfatting agent, detergent used for cosmetics, textiles, metalworking fluids, emulsion polymerization, insecticides, herbicides, household detergents. *Croda Chem. Ltd.*

7182 Crodyne BY19
9000-70-8 4388 232-554-6
gelatin
Pharmaceutical gelatin NF; protective colloid, moisturizer, conditioner for skin and hair care products; humectant, thickener for pharmaceutical and food applications. *Croda Inc; Croda Chem. Ltd.*

7183 Croian
Lanolin fatty acid esters; used in the cosmetic industry. *Croda Chem. Ltd.* Name unverified.

7184 Crolactil CSL
5793-94-2 11, 1711 227-335-7
$C_{48}H_{86}CaO_{12}$
Calcium 2-stearoyl lactylate
stearic acid, ester with lactate of lactic acid, calcium salt; calcium stelate; Verv-Ca. Water-oil emulsifier, emollient for skin care and treatment products; ingredients in food products *Croda Surf. Ltd.*

7185 Crolactil SISL
66988-04-3 266-533-8
Sodium isostearoyl lactylate
Oil-water emulsifier, emollient for skin care and treatment products; ingredients in food products *Croda Surf. Ltd.*

7186 Crolactil SSL
25383-99-7 246-929-7
Sodium stearoyl lactylates
Oil-water emulsifier, emollient for skin care and treatment products; ingredients in food products *Croda Chem. Ltd.*

7187 Crolastin
Hydrolyzed animal elastin; moisturizer and conditioner for skin care products and cleansers. *Croda Inc; Croda Chem. Ltd.*

7188 Crolastin 10 Powder
Partially hydrolyzed elastin; conditioner for skin care cosmetics. *Croda Inc.* Discontinued.

7189 Crolec 4135
8002-43-5 5452 232-307-2
Modified lecithin. Conditioner, superfatting, emulsifier for cosmetics and pharmaceuticals. *Croda Chem. Ltd.*

7190 Croloy
A proprietary trade name for high chromium steel containing molybdenum and vanadium. No manufacturer.

7191 Cromal
An alloy of aluminum with 2-4% chromium and smaller amounts of nickel and manganese. It melts at 700 C., and is specially suitable for castings.

7192 Cromalit
Concentrated chrome alums; tanning and drilling mud additive. *Lancashire Chemical Works Ltd.*

7193 Cromalit 150
Concentrated basic potash chrome alum powder. *Lancashire Chemical Works Ltd.*

7194 Cromaloy II
An alloy of 80% nickel, 15% chromium, and 5% iron.

7195 Cromaloy III
An alloy of 85% nickel and 15% chromium.

7196 Cromaloy IV
An alloy of 80% nickel and 20% chromium.

7197 Cromeen
Substituted alkylamine derived lanolin acids
Mild multifunctional surfactant with high foaming, detergency, and emulsifying properties; for shampoos, detergents, and hand cleansing preparations, aerosol skin and shaving foams. *Croda Chem. Ltd.*

7198 Cromo Steel
A proprietary trade name for a chrome-molybdenum steel. No manufacturer.

7199 Cromoist CS
9007-28-7 2270 232-696-9
Chondroitin sulfate
Hydrolyzed animal protein; moisturizer, conditioner for face and body creams and lotions. *Croda Inc.* Discontinued.

7200 Cromoist HYA
9004-61-9 4793 232-678-0
hyaluronic acid
Hydrolyzed animal protein, moisturizer, conditioner for skin care products, facial creams. *Croda Inc.*

7201 Cromoist O25
Hydrolyzed whole oats; skin care ingred. improving feel properties of creams and lotions imparting soft, cushiony feel on skin; moisturizer. *Croda Inc.*

7202 Cromophtal
High performance organic pigments for plastics; a registered trade name. *Ciba plc.*

7203 Cromophytal C-20, M-20
Pigment preparations. *Ciba plc.* Name unverified.

7204 Cromul 0685
Complex ethoxy (6) fatty alcohol ether; emulsifier and opacifier for cosmetic and pharmaceutical creams and lotions. *Croda Chem. Ltd.*

7205 Cronectin
Hydrolyzed fibronectin
Conditioner for skin and hair care products *Croda Chem. Ltd.* Name unverified.

7206 Croneton®
29973-13-5 249-981-9
Ethiofencarb
Systemic insecticide used for control of aphids. *Bayer AG.* No manufacturer.

7207 Cronite
An alloy of nickel and chromium. No. 1 contains 85% nickel, and 15% chromium.

7208 Cropepsol, Cropeptone
9015-54-7 310-296-6
Hydrolyzed collagen
Conditioner for skin and hair conditioner. *Croda Chem. Ltd.* Name unverified.

7209 Cropeptide W
Hydrolyzed wheat protein; wheat oligosaccharides; film-forming conditioning protein for controlling moisture and strengthening hair. *Croda Inc.*

7210 Cropol 60
577-11-7 3460 209-406-4
docusate sodium
Sodium dioctyl sulfosuccinate as an ethanol solution; emulsifier and powerful wetting agent used in a wide range of manufacturing industries, especially textiles. *Croda Chem. Ltd.* No manufacturer.

7211 Cropotex®
37893-02-0 253-703-1
$C_{17}H_{10}F_6N_4S$
N-(3-Phenyl-4,5-bis((trifluoromethyl)imino)-2-thiazolidinylidene)benzenamine
Bay SLJ 0312; Benzenamine, N-(3-phenyl-4,5-bis((trifluoromethyl)imino)-2-thiazolidinylidene)-; Cropotex; Flubenzimine; SLJ 0312. Contact acaricide for control of spider mites on pome and stone fruit, plums, damsons, citrus fruit and citrus rust mites. *Bayer AG.*

7212 Croptex Amber
1918-16-7 7977 217-638-2
Propachlor
A pre-emergence herbicide for various horticultural crops. *Hortichem Ltd.*

7213 Croptex Bronze
2307-68-8 8851 218-988-9
Solan
Emulsifiable concentrate containing 400 g/1 pentanochlor; used to control weeds in horticultural crops. *Hortichem Ltd.*

7214 Croptex Chrome
chlorpropham-fenuron. Suspension concentrate containing 80 g chlorpropham and 15 g fenuron per liter, an herbicide for use on ornamentals and vegetables. *Hortichem Ltd.*

7215 Croptex Fungex
33113-08-5 251-381-7
$C_2H_4CuN_2O_6$
Cupric ammonium carbonate
Copper carbonate, basic; copper ammonium carbonate; Carbonic acid, ammonium copper salt; Copper count N. Cupric ammonium carbonate; a protectant fungicide. *Hortichem Ltd.*

7216 Croptex Pewter
cetrimide-chlorpropham. A suspension concentrate containing 80 g cetrimide (hexadecyltrimethylammonium bromide) and 80 g chlorpropham per liter; soil acting herbicide for lettuce. *Hortichem Ltd.*

7217 Cropure Apricot Kernel
Apricot kernel oil; emollient; lubricant and softener in nail oils; conditioner for hair care products. *Croda Inc.*

7218 Cropure Avocado
8024-32-6 232-428-0
Avocado oil; emollient; provides elegant skin feel and promotes spreading in creams, lotions, bath oils; emollient for hair care products. *Croda Inc.*

7219 Cropure Babassu
Babassu oil; emollient for sunscreen products. *Croda Inc.*

7220 Cropure EPO
Evening primrose oil; contains essential fatty acids for skin care. *Croda Inc.*

7221 Cropure Orange Roughy
Orange roughy oil; emollient, softener, spreading agent in skin care products. *Croda Inc.*

7222 Cropure Wheat Germ
8006-95-9
Wheat germ oil; emollient; contains essential fatty acids vital to health of skin; used in facial creams; moisturizer in hair care products. *Croda Inc.*

7223 Croquat HH
68915-25-3
Cocodimonium hydroxypropyl hydrolyzed hair keratin; substantive conditioning protein for hair shampoos and conditioners. *Croda Inc.*

7224 Croquat L
Laurdimonium hydrolyzed animal protein; substantive protein, conditioner for clear rinses, shampoos, conditioners. *Croda Inc.*

7225 Croquat M
Cocodimonium hydrolyzed animal protein; substantive protein, conditioner for shampoos, perms, hair relaxers. *Croda Inc.*

7226 Croquat S
Steardimonium hydrolyzed animal protein; substantive protein, conditioner for cream rinses. *Croda Inc.*

7227 **Croquat WKP**
68915-25-3
Cocodimonium hydrolyzed animal keratin; permanent conditioning protein for cream rinses, shampoos, conditioners, perms, nail care products *Croda Inc.*

7228 **Croscolor**
Dyeing auxiliaries. *Croda Universal Ltd.*

7229 **Croscour**
Scouring agents for textiles. *Croda Universal Ltd.*

7230 **Crosdurn**
Durable finishes. *Crosfield Chemicals Ltd.* Discontinued.

7231 **Crosil**
Textile processing system; for wool shrink proofing. *Croda Universal Ltd.*

7232 **Crosilk**
Silk amino acids
Croda Chem. Ltd, Croda Inc. No manufacturer.

7233 **Crosilk 10,000**
96690-41-4 306-235-8
Hydrolyzed silk; protein conditioner providing manageability, gloss, texture in hair care products, moisturizing and protection in skin care products. *Croda Inc.*

7234 **Crosilk Liq**
977077-71-6
Silk amino acids; conditioner, humectant for skin and hair care products *Croda Inc.*

7235 **Crosilk Powder**
9009-99-8
Silk protein for solid make-up, hair products where it absorbs oil, improves leveling, enhances spreading, gives elasticity and lubricity, modifies application properties and provides a silky, lustrous appearance; improves pigment binding and stability. *Croda Inc.*

7236 **Crosilkquat**
Cocodimonium silk amino acids; substantive conditioner and moisturizer with excellent foaming properties for skin and hair conditioners, shampoos, styling mousses, perms and relaxers, night creams and lotions; effective over wide pH range. *Croda Inc.*

7237 **Croslube**
Yarn lubricants. *Croda Universal Ltd, Crosfield Chemicals Ltd.*

7238 **Crosoft**
Softening agents for fiber softeners and finishes. *Croda Universal Ltd, Crosfield Chemicals Ltd.*

7239 **Crossfire**
2921-88-2 2242 220-864-4
Chlorpyrifos
Emulsifiable concentrate containing 228g chlorpyrifos per liter; an organophosphorus insecticide. *Rhône-Poulenc Crop Protection Ltd.*

7240 **Crosterene**
Stearic, oleic, behenic, erucic and palmitic fatty acids. *Croda Universal Ltd; Croda Chemical Ltd.*

7241 **Crosultaine C-50**
68139-30-0 268-761-3
Cocamidopropyl hydroxysultaine
Foam booster/stabilizer effective over wide range of pH and water hardness; for shampoos, baby shampoos; lime soap dispersant. *Croda Inc.*

7242 **Crosultaine E-30**
Erucamidopropyl hydroxysultaine
Viscosity booster and conditioner giving silky feel, comb and static control at levels less than 2% active; improves creaminess and lubricity of lather; lime soap dispersant. *Croda Inc.*

7243 **Crosultaine T-30**
Tallowamidopropyl hydroxysultaine
Foam and viscosity booster; improves wet combing and hair condition; lime soap dispersant. *Croda Inc.*

7244 **Crotein A, C, O**
9015-54-7 310-296-6
Hydrolysed collagen *Croda Chem. Ltd.*

7245 **Crotein AD Anhyd**
AMP isostearoyl hydrolyzed animal protein; conditioners for hair preparations, alcohol products, aerosol hair sprays, skin tonics, setting lotions, aftershaves. *Croda Inc.*

7246 **Crotein ADW**
AMP isostearoyl hydrolyzed wheat protein; conditioning protein, film modifier for alcoholic and hydroalcoholic lotions, hair care products, skin tonics, aftershave, nail preparations, quick-breaking foam aerosols; plasticizer for resins in hair products *Croda Inc.*

7247 **Crotein ASC**
68951-89-3
Ethyl ester of hydrolyzed animal protein; protein conditioner and film modifier for alcoholic and hydroalcoholic lotions and hair care products *Croda Inc.* Discontinued.

7248 **Crotein ASK**
69430-36-0 274-001-1
Hydrolyzed animal keratin; conditioner, film modifier for hair setting/conditioning systems. *Croda Inc.*

7249 **Crotein CAA**
9015-54-7 310-296-6
Collagen amino acids *Croda Chem. Ltd; Croda Inc.*

7250 **Crotein CAA/SF**
9015-54-7 310-296-6
Animal collagen amino acids; moisturizer for skin creams and lotions. *Croda Inc.*

7251 **Crotein HKP**
68238-35-7 269-409-1
Keratin amino acids *Croda Chem. Ltd; Croda Inc.*

7252 **Crotein HKP Powd**
Hair keratin amino acids with added sodium chloride Substantive conditioner and moisturizer for shampoos, cream rinses. *Croda Inc.*

7253 **Crotein HWE**
Hydrolysed whole eggs. *Croda Chem. Ltd.*

7254 **Crotein IP**
Isostearoyl hydrolyzed animal protein; conditioner for use in solvent-based nail polish and removers. *Croda Inc.*

7255 **Crotein K, WKP**
69430-36-0 274-001-1
Hydrolyzed animal keratin; conditioners for hair products *Croda Inc.*

7256 **Crotein O**
9015-54-7 310-296-6
Refined collagen hydrolysates; conditioning agent, foam booster and stabilizer, dye leveling agent in hair products, substantive to hair and skin. *Croda Inc.* Discontinued.

7257 **Crotein Q**
11174-62-0
Steartrimonium hydrolyzed animal protein; substantive conditioner, body and gloss agent for hair care preparations. *Croda Inc; Croda Chem. Ltd.* Discontinued.

7258 **Crotein SPA**
9015-54-7 310-296-6
Hydrolyzed animal protein; conditioner for hair care preparations; foam stabilizer/booster in shampoo; peptizing aid in shampoos; dye leveling aid in hair dyes and bleaches; also for skin treatment, in depilatories. *Croda Inc.*

7259 **Crothix**
Polyol alkoxy ester; mild thickener for aqueous systems, shampoos, auxiliary emulsifier and bodying agent for creams and lotions. *Croda Inc.*

7260 **Crotonic Acid**
107-93-7 2664 223-077-4
$C_4H_6O_2$
β-methylacrylic acid
trans-2-butenoic acid; β-methacrylic acid; α-crotonic acid. Synthesis of resins, polymers, plasticizers, drugs. mp = 71-73°; bp = 180-181°; d = 1.0270; n_D^{80} = 1.4228; soluble in H_2O (54.6 g/l), more soluble in organic solvents; LD_{50} (rat orl) = 1.0 g/kg. *Aldrich; Allchem Industries; Atomergic Chemetals; Chisso Am.; Eastman.*

7261 **Crotorite**
A cupro-manganese alloy. It contains 68% copper, 30% manganese, and 2% iron.

7262 **Cro-tung**
A proprietary trade name for a chromium-tungsten steel. No manufacturer.

7263 **Crovol A40**
PEG-20 almond glycerides
Coemulsifier, superfatting agent, emollient, counter-irritant, wetting aid, solubilizer, dispersant for skin and hair care products, soaps, bath oils, astringents, antiperspirants; plasticizer for styling mousses and aqueous *Croda Inc.; Croda Chemical Ltd.*

7264 **Crovol M40**
PEG-20 corn glycerides
Coemulsifier, superfatting agent, emollient, counter-irritant, wetting aid, solubilizer, dispersant for skin and hair care products, soaps, bath oils, astringents, antiperspirants; plasticizer for styling mousses, aqueous aerosols; fragrance solubilizer. *Croda Inc.; Croda Chemical Ltd.*

7265 **Crovol PK40**
PEG-12 palm kernel glycerides
Coemulsifier, superfatting agent, emollient, counter-irritant, wetting aid, solubilizer, dispersant for skin and hair care products, soaps, bath oils, astringents, antiperspirants; plasticizer for styling mousses, aqueous

aerosols; fragrance solubilizer. *Croda Inc.; Croda Chemical Ltd.* Discontinued.

7266 Crow

A cotton fabric grade of Tufnol industrial laminates. *Tufnol Ltd.*

7267 Crown Acid Aid X

Accelerator and extender for hydrochloric and sulfuric acid; increases pickling action of acid; excellent detergency properties. *Crown Tech.*

7268 Crown Anti-Foam

Silicone emulsion; antifoam for dehydrating, evaporating, fermentation processes, abrasive slurries, commercial cleaners, adhesives, latex emulsions, cutting oils, insecticides and pesticides; effective at concentrations as low as 5 ppm; stable to high shear and high pH. *Crown Tech.*

7269 Crown Foamer 20

Foamer, wetting agent producing a detergent-type foam for pickling solutions; effective in sulfuric and hydrochloric acid baths. *Crown Tech.*

7270 Crown L-1011 Acid Inhibitor

Complex amine and thioureas; Inhibitor for sulfuric acid; used in metalworking industry in continuous and batch pickling operations. *Crown Tech.*

7271 Crown L-60B Acid Inhibitor

Complex amine and thioureas

HCl inhibitor protecting metals through absorption to specific corrosion sites; for pickling operations, cleaning industrial boilers, heat exchangers and condensers, in copper cleaners and toilet bowl cleaners. *Crown Tech.*

7272 Crown solder

An alloy of 63% tin, 18% zinc, 13% aluminum, 1% lead, 3% copper, and 2% antimony.

7273 Cru Fax P9N

Antistat for paper manufacturing. *Crucible.*

7274 Cruverlite

A proprietary trade name for a luminous plastic molding powder. No manufacturer.

7275 Crylene

acetaldehyde-aniline condensation product

A proprietary rubber vulcanization accelerator. No manufacturer.

7276 CryOfine® Butyl

9010-85-9

Fine particle rubber. Produced by cryogenically grinding butyl scrap in a liquid nitrogen environment; blends easily into butyl and halobutyl compounds for improved processing, lower cost, reduced blisters, and increased air flow during curing; used in molded and mechanical goods, sealants, membranes, tubes, dynamic parts, diaphragms and tire inner liners. *Midwest Elastomers.*

7277 CryOfine® EPDM

Fine particle rubber filler

Prepared by cryogenically grinding cured EPDM flash, trim, and other scrap; reduces cost of EPDM compounds; used in hose, molded and extruded goods, sponge, roofing, and thermoplastic elastomers. *Midwest Elastomers.*

7278 Cryofluorane

76-14-2 2672 200-937-7

$C_2Cl_2F_4$

1,2-Dichloro-1,1,2,2-tetrafluoroethane

Genetron® 114; Dichlorotetrafluoroethane; Freon 114; Frigen 114; Arcton 114; Dichloro-1,1,2,2-tetrafluoroethane; Propellant 114; 1,2-Dichloro-1,1,2,2-tetrafluoroethane; R-114; 1,2-Dichlorotetrafluoroethane; CFC-114; *sym*-Dichlorotetrafluoroethane; Refrigerant 114; Halon 242; Freon 114; Halocarbon 114; Refrigerant R114; Fluorocarbon 114; cryofluorane. Used with centrifugal compressors for higher capacities or for lower evaporator temperature process applications; also for foam applications. mp = -94°; bp = 4°. *AlliedSignal.*

7279 cryolite

15096-52-3 2673 239-148-8

AlF_6Na_3

sodium aluminum fluoride

Kryolith; ice spar. Found in deposits in Greenland and Russia. Used in the aluminum industry as a source of the metal. Also used as an insecticide.

7280 Cryoseal

Modified natural rubber latex; water-based cold seal adhesives; for confectionery packaging, applied by gravure printing. *The Scottish Adhesives Co Ltd.*

7281 Cryptone

Proprietary trade names for white pigments containing lithopone (a mixture of ZnS, $BaSO_4$ and ZnO) and TiO_2. No manufacturer.

7282 Cryptonol

134-31-6 4890 205-137-1

8-Hydroxyquinoline sulfate

Used for control of soil borne diseases in ornamentals. *Fargro Ltd.*

7283 Crystal

Sodium, potassium and lithium silicates. *Croda Universal Ltd.*

7284 Crystal

1344-09-8 8824 215-687-4

Sodium silicates

Used in abrasives and refractories. *Crosfield Chemicals Ltd.* No manufacturer.

7285 Crystal 1000

63148-62-9

Release blend in water/ethanol; semi-permanent mold release providing high slip for injection, compression and transfer molding of urethane parts, natural, silicone and other synthetic elastomer parts. *TSE Industries.*

7286 crystal glass

lead silicate, potassium silicate

A glass composed of lead and potassium silicates.

7287 Crystal Inhibitor 5

Blend; retards formation of crystals in 2,4-D amine formulation dilutions. *Harcros Organics.*

7288 Crystal Polystyrene

Polymerization product of styrene monomer, with additives; general commodity plastic for extrusion and injection. *Lin-Pac International.* Name unverified.

7289 Crystal® O, Crystal® Crown

8001-79-4 1946 232-293-8

Refined castor oil USP; emollient, pigment wetter, cosolvent, lubricant for cosmetics, makeup, antiperspirant sticks. *CasChem.*

7290 Crystalite

Organic brighteners; for cyanide zinc electroplating. *Engelhard Technologies Ltd.*

7291 Crystalor BC-1

Polymethylpentene resin, 30% glass-filled; injection molding resin with outstanding impact properties, high heat distortion temperature, light weight, good chemical resistance, excellent electrical properties; for automotive, appliances, *Phillips.* Name unverified.

7292 Crystalor DC-6

25068-26-2

Polymethylpentene resin; high impact properties and clarity, light weight, good elongation; for automotive, appliances, medical, housewares, electrical/electronic and packaging applications. *Phillips.* Name unverified.

7293 Crystamet 1020

6834-92-0 8788 229-912-9

Sodium metasilicate, pentahydrate

Alkaline silicate used for formulating specialty detergents and contributing buffering capacity and corrosion inhibition of soft metals and ceramic glazes. *Crosfield.*

7294 Crystex®, Crystex® Regular

7704-34-9 9142 231-722-6

sulfur, insoluble

Insoluble sulfur (polymeric sulfur and sulfur); prevents crystallization of sulfur on uncured rubber surfaces (sulfur bloom); detackifier; retards bin scorch; minimizes sulfur migration *Akzo.*

7295 Crystic® 2-414PA

o-Phthalic polyester resin; for building, land transport, boat building applications; low exothermicity; thixotropic. *Scott Bader.* Discontinued.

7296 Crystic® 39PA

Isophthalic/HET polyester resin; fire retardant gelcoat for building, land transport applications. *Scott Bader.* Discontinued.

7297 Crystic® 199

Isophthalic polyester resin; heat resistance resin for high temperature use and electrical applications, tanks and chemical plant. *Scott Bader.* Discontinued.

7298 Crystic® 471PALV

o-phthalic polyester resin; preaccelerated, rapid-hardening resin with fast wetting of reinforcement and short mold release time, for contact molding applications, catalyst injection spray equipment; for automotive, marine, industrial applications. *Scott Bader.* Discontinued.

7299 Crystic® 581PA

Orthophthalic polyester resin; tough resin for formulation of body fillers in building and land transport industries; thixotropic; gel time 5 minutes at 20°. *Scott Bader.* Discontinued.

7300 Crystic® Fireguard

Unsaturated polyester gelcoat; fire retardant. *Scott Bader.* Discontinued.

7301 Crystic® Fireguard 75PA

DBNPG polyester resin; thixotropic, intumescent surface coating resin giving fire protection to GRP laminates; fire-retardant coating for timber and other slightly porous surfaces. *Scott Bader.* Discontinued.

7302 Crystic® Impel

Unsaturated polyester molding granules. *Scott Bader.* Discontinued.

7303 Crystic® Impreg
Sheet molding compound; viscosity controlled by temperature. *Scott Bader.* Discontinued.

7304 Crystic® Prefil F
1309-64-4 752 215-175-0
Antimony trioxide
Antimony trioxide and chlorinated organic compound dispersed in polyester resin; additive to facilitate the manufacturing of reduced fire hazard moldings. *Scott Bader.* Discontinued.

7305 Crystic® Pregel 17
Silica in a general purpose polyester resin dispersion; resin additive to confer thixotropic properties to general purpose polyester resins for laminating or gelcoat applications. *Scott Bader.* Discontinued.

7306 Crystolon®
A registered trade name for various types of goods such as grinding wheels, abrasives, and refractory grain, etc., made from silicon carbide. No manufacturer.

7307 CS1590
3-(N-Styrylmethyl-2-aminoethylamino) propyltrimethoxy silane hydrochloride
Coupling agent, chemical intermediate, blocking agent, release agent, lubricant, primer, reducing agent. *Hüls Am.*

7308 CSC 2-aminobutane
13952-84-6 1579 237-732-7
2-Aminobutane
sec-butylamine. Fungicide for stored potatoes. *Chemical Spraying Co Ltd.*

7309 CSE-6000 Series
Cold set epoxies; for vessel linings and heavy-duty maintenance applications; high resistance to chemicals. *Heresite Protective Coatings.*

7310 CSI
Crofilcon A
Contact lens material. *Syntex Ophthalmics Inc.* Name unverified.

7311 CT1750
4766-57-8 225-305-8
$C_{16}H_{36}O_4Si$
Tetra-n-butoxysilane
Tetrabutyl silicate; Tetrabutyl orthosilicate; Tetrabutoxysilane. Coupling agent, chemical intermediate, blocking agent, release agent, lubricant, primer, reducing agent. *Hüls Am.*

7312 CT1800
10026-04-7 233-054-0
Cl_4Si
Tetrachlorosilane
silicon tetrachloride; SIC-L(TM); silicon chloride; Tetrachlorosilicon; Silicon(IV) chloride. Coupling agent, chemical intermediate, blocking agent, release agent, lubricant, primer, reducing agent. mp = -70°; bp = 58°; d = 1.4830; n_D^{20}= 1.4130; LC_{50} (rat inh) = 8000 ppm/4H. *Hüls Am.*

7313 CT2015
13528-93-3 236-871-0
$C_6H_{16}Cl_2Si_2$
1,1,4,4-Tetramethyldichlorodisilethylene
1,2-Bis(Chlorodimethylsilyl)-ethane. Coupling agent, chemical intermediate, blocking agent, release agent, lubricant, primer, reducing agent. *Hüls Am.*

7314 CT2030
3277-26-7 221-906-4
$C_4H_{14}OSi_2$
Tetramethyldisiloxane
Coupling agent, chemical intermediate, blocking agent, release agent, lubricant, primer, reducing agent. bp = 71°; d = 0.7600; n_D^{20}= 1.371; LD_{50} (mus orl) = 3 g/kg. *Hüls Am.*

7315 CT2050
75-76-3 200-899-1
$C_4H_{12}Si$
Tetramethyl silane
Tetramethyl silicane; TMS. Coupling agent, chemical intermediate, blocking agent, release agent, lubricant, primer, reducing agent. mp= -99°; bp = 26-28°; d = 0.6480. *Hüls Am.*

7316 CT2090
682-01-9 211-659-0
$C_{12}H_{28}O_4Si$
Tetra-n-propoxysilane
Tetrapropoxysilane; Tetrapropyl orthosilicate; Tetrapropyl silicate; Dynasil P. Coupling agent, chemical intermediate, blocking agent, release agent, lubricant, primer, reducing agent. bp_5 = 94°; d = 0.9160; n_D^{20}= 1.4013. *Hüls Am.*

7317 CT2500
998-30-1 213-650-7
$C_6H_{16}O_3Si$
Triethoxysilane
Coupling agent, chemical intermediate, blocking agent, release agent, lubricant, primer, reducing agent. *Hüls Am.*

7318 CT2507
23779-32-0 245-876-7
$C_7H_{18}N_2O_4Si$
N-(Triethoxysilylpropyl) urea
Dynasylan 2201; Ureidopropyltriethoxysilane; N-(Triethoxysilylpropyl)urea. Coupling agent, chemical intermediate, blocking agent, release agent, lubricant, primer, reducing agent. *Hüls Am.*

7319 CT2520
994-30-9 213-615-6
$C_6H_{15}ClSi$
Triethylchlorosilane
Coupling agent, chemical intermediate, blocking agent, release agent, lubricant, primer, reducing agent. bp = 142-144°; d = 0.8980; n_D^{20}= 1.4301. *Hüls Am.*

7320 CT2523
617-86-7 210-535-3
$C_6H_{16}Si$
Triethylsilane
Coupling agent, chemical intermediate, blocking agent, release agent, lubricant, primer, reducing agent. bp = 107-108°; d = 0.7280; n_D^{20}= 1.4130. *Hüls Am.*

7321 CT2902
50975-76-3 256-873-5
1-trimethoxysilyl-2-(chloromethyl) phenylethane
Coupling agent, chemical intermediate, blocking agent, release agent, lubricant, primer, reducing agent. *Hüls Am.*

7322 CT2910
35141-30-1 252-390-9
Trimethoxysilylpropyldiethylenetriamine
(3-Trimethoxysilylpropyl)diethylenetriamine; Dynasylan TRIAMO. Coupling agent, chemical intermediate, blocking agent, release agent, lubricant, primer, reducing agent. *Hüls Am.*

7323 CT2925
N-Trimethoxysilylpropyl-N,N,N-trimethyl ammonium chloride
Coupling agent, chemical intermediate, blocking agent, release agent, lubricant, primer, reducing agent. *Hüls Am.*

7324 CT2928
2857-97-8 220-672-0
C_3H_9BrSi
Trimethylbromosilane
Bromotrimethylsilane; Trimethylsilyl bromide. Coupling agent, chemical intermediate, blocking agent, release agent, lubricant, primer, reducing agent. mp = -43°; bp = 79°; d = 1.1600; n_D^{20}= 1.4245. *Hüls Am.*

7325 CT2950
75-77-4 200-900-5
C_3H_9ClSi
Trimethylchlorosilane
Chlorotrimethylsilane; chlorotrimethylsilicane; trimethylsilyl chloride; monochlorotrimethylsilicon; silylium, trimethyl-, chloride; tl 1163; CSI. Coupling agent, chemical intermediate, blocking agent, release agent, lubricant, primer, reducing agent. mp = -40°; bp = 57°; d = 0.8560; n_D^{20} = 1.3870; LD_{50} (rat orl) = 4.8 g/kg. *Hüls Am.*

7326 CT2970
1825-62-3 217-370-6
$C_5H_{14}OSi$
Trimethylethoxy silane
ethoxytrimethylsilane. Coupling agent, chemical intermediate, blocking agent, release agent, lubricant, primer, reducing agent. *Hüls Am.*

7327 CT3250
13435-12-6 236-565-7
$C_5H_{13}NOSi$
Trimethylsilylacetamide
N-(Trimethylsilyl)acetamide. Coupling agent, chemical intermediate, blocking agent, release agent, lubricant, primer, reducing agent. mp = 40-46°; bp_{18} = 84°. *Hüls Am.*

7328 CT3254
2754-27-0 220-404-2
$C_5H_{12}O_2Si$
Trimethylsilylacetate
Trimethylacetoxysilane; Acetoxytrimethylsilane; O-trimethylsilylacetate. Coupling agent, chemical intermediate, blocking agent, release agent, lubricant, primer, reducing agent. *Hüls Am.*

7329 CT3600
18156-74-6 242-040-3
$C_6H_{12}N2Si$
Trimethylsilyl imidazole

N-(Trimethylsilyl)imidazole; TSIM; 1-(Trimethylsilyl)imidazole; 1H-Imidazole, 1-(trimethylsilyl)-. Coupling agent, chemical intermediate, blocking agent, release agent, lubricant, primer, reducing agent. bp_{14} = 93-94°; d = 09560; n_D^{20}= 1.4751 *Hüls Am.*

7330 CT3610
16029-98-4 240-171-0
C_3H_9ISi
Trimethylsilyl iodide
Iodotrimethylsilane; Trimethyliodosilane. Coupling agent, chemical intermediate, blocking agent, release agent, lubricant, primer, reducing agent. bp = 106°; d = 1.4060; n_D^{20}= 1.4770. *Hüls Am.*

7331 CT3795
27607-77-8 248-565-4
$C_4H_9F_3O_3SSi$
Trimethylsilyl trifluoromethane sulfonate
Trimethylsilyl triflate; Trifluoromethanesulfonic acid trimethylsilyl ester. Coupling agent, chemical intermediate, blocking agent, release agent, lubricant, primer, reducing agent. bp_{80} = 77°; d = 1.1500. *Hüls Am.*

7332 CT-690, CT-700
Polyurethane elastomeric resins. *Unichem.* Name unverified.

7333 CTC-3300
32131-17-2
Nylon 6/6, 10%, 33% glass fiber-reinforced. *Compounding Tech.* Name unverified.

7334 CTW
Epoxy resin based cements, screeds and coatings. *Prodorite Ltd.*

7335 CTX-308
30% polybutylene terephthalate/50% acrylonitrile-butadiene-stryene, 20% glass fiber-reinforced. *Compounding Tech.* Name unverified.

7336 CTX-312
25038-54-4 6832
Nylon 6, 10% glass fiber-reinforced, impact-modified. *Compounding Tech.* Name unverified.

7337 CTXC-020
Polyarylsulfone, carbon fiber-reinforced; for ESD applications. *Compounding Tech.* Name unverified.

7338 Cuba black
Dianil Black
A dyestuff. Dyes cotton black from an alkaline bath.

7339 Cuba orange
A dyestuff prepared by the action of sodium sulfite upon diazo-naphthalene-sulfonic acid. Dyes wool orange.

7340 cubanite
$CuFe_2S_4$
A mineral.

7341 Cuclat
Kerosene sweeting catalyst. *Mechema Chemicals Ltd.* Name unverified.

7342 Cudbear
2618
Crottle
A lichen, *Ochrolechia tatratrea* L. *Lecanoraceae.* Has red-purple color and is used as a pigmenting agent. No manufacturer.

7343 Cufenlum
An alloy of 22% nickel, 72% copper, and 6% iron.

7344 cufraneb
11096-18-7
ethylenebis(dithiocarbamate) complexed with 8.15% Zn, 8.05% Mn, 5.5% Cu, 1.0% Fe. cufranebe. Foliar fungicide; also behaves as an acaricide. Used together with cupric sulfate to control late blight in tomatoes and potatoes. LD_{50} (rat orl)= 2700 mg/kg.

7345 Cuite
Natural silk freed from the silk gum or sericin. Raw silk consists of about 66% fibroin, forming the real silk substance, and silk gum or sericin. The silk gum is removed by treating with hot neutral soap solution.

7346 Culminal® Hydroxypropylmethylcellulose
9004-65-3 4889
Methylhydroxypropylcellulose
Aqualon.

7347 Culminal® Methylcellulose
9004-67-5 6120
Methylcellulose
Aqualon.

7348 Cultar
76738-62-0 7118
paclobutrazol
Suspension concentrate containing 250 g/l paclobutrazol; plant growth regulator. *ICI Chem & Polymers Ltd.*

7349 Cumal
122-03-2 2687 204-516-9
p-Isopropylbenzaldehyde
Intermediate for pharmaceuticals, perfumes. *Mitsubishi Gas.*

7350 Cumar®, Cumar® P-10, R-1
Coumarone-indene resin; resin with excellent resistance to alkalies, dilute acids and moisture; for use in adhesives, aluminum paints, varnishes, rotogravure inks, rubber compounds *Neville Chemical Co.*

7351 Cumate®
137-29-1 205-287-8
$C_6H_{12}CuN_2S_4$
Copper dimethyldithiocarbamate
bis(dimethylcarbamodithioato-S,S') Copper; Copper dimethyldithiocarbamate; Cumate; Copper, bis(dimethylcarbamodithioato-S,S')-, (SP-4-1)-; Compound-4018; Wolfen. Accelerator for high-speed vulcanization of SBR, IIR, EPDM. *R. T. Vanderbilt Co Inc.* Discontinued.

7352 cumene
98-82-8 2683 202-704-5
C_9H_{12}
isopropylbenzene
(methylethyl)benzene; cumol. Production of phenol, acetone and α-methylstyrene; solvent. mp = -96°; bp = 152-154°; d = 0.8640; n_D^{20}= 1.4917; LD_{50} (rat orl) = 2.91 g/kg. *Ashland; Chevron; Georgia Gulf; Hüls AG; Mitsubishi PetroChemical; Mitsui PetroChemical Ind.*

7353 Cumene sulfonic acid
28631-63-2 249-112-3
Eltesol® CA 65; Reworyl® C. Catalyst for foundry resins; descaling agent for metal cleaning; anti-stress additive and plating aid in electroplating bath; curing aid in the plastics industry; raw material in the manufacture of dyes and pigments; detergents industry. *Albright & Wilson UK; Rewo Chemicals Ltd.*

7354 cuminaldehyde
122-03-2 2687 204-516-9
$C_{10}H_{12}O$
p-Isopropylbenzaldehyde
cuminaldehyde; 4-(1-methylethyl)benzaldehyde; Cumal. Intermediate for pharmaceuticals, perfumes. bp = 235-236°; d^{20} = 0.978; n_D^{20} = 1.5301; insoluble in H_2O, soluble in organic solvents; LD_{50} (rat orl) = 1390 mg/kg. *Mitsubishi Gas.*

7355 cumyl phenol
599-64-4 209-968-0
$C_{15}H_{16}O$
4-cumylphenol
4-(2-phenylisopropyl) phenol. Intermediate for resins, insecticides, lubricants. mp = 70-73°; bp = 335°. *Hüls AG; ICI Specialties; PMC Specialties; Schenectady.*

7356 Cunilate 2174-NO
Antimicrobial for solvent systems, adhesives, latexes, FDA-approved preservatives. *Morton Int'l/Plastics Additives.*

7357 Cuniloy
An alloy of nickel, manganese, copper, and small quantities of lead.

7358 Cuniphen 2778-1
Antimicrobial for in-can preservation; FDA approved. *Morton Int'l./Plastics Additives.*

7359 Cunitex
Bird and animal repellant. *May & Baker Ltd.* Name unverified.

7360 Cupalit
A softening agent and lubricant; used by the textile industry for the after treatment of all types of fibers. *Degussa AG.* Name unverified.

7361 cuperatin
Copper albuminate.

7362 Cupertine
maneb-cupric sulfate. Bordeaux mixture 84%, maneb 8%; wettable powder used as protective fungicide for foliage application to ornamental and crop plants. *Industrias Quimicas Del Valles SA.*

7363 Cupertine Folpet
folpet-cupric sulfate. Bordeaux mixture 80%, folpet 10%; wettable powder used as protective fungicide for foliage application to ornamental and crop plants. *Industrias Quimicas Del Valles SA.*

7364 Cupertine Super
Bordeaux mixture 90%, cymoxanil 3%; wettable powder used as protective and curative fungicide for foliage application to ornamental and crop plants. *Industrias Quimicas Del Valles SA.*

7365 cupferron
135-20-6 2688 205-183-2
$C_6H_9N_3O_2$
N-Hydroxy-N-nitrosobenzenamine ammonium salt
copperone; Nitrosophenylhydroxylamine ammonium salt. The ammonium

salt is used as a precipitating agent for copper, in the determination of copper; used as an analytical reagent, especially for the separation and precipitation of metals, e.g., copper, iron, vanadium. mp = 163-164°; soluble in H_2O, EtOH.

7366 Cupolloy
Ferro-molybdenum briquettes for cupola. *Foseco (F.S.) Ltd.*

7367 cuprammonium silk
cuprate silk
Artificial silks made by dissolving cotton in a cuprammonium solution (copper hydrate dissolved in ammonia), and precipitating it in a fine thread.

7368 cupranium
A name for certain brass and bronze alloys. The bronze contains tantalum and vanadium.

7369 Cuprase
A form of colloidal copper; a proprietary trade name for a vulcanizing accelerator containing copper. No manufacturer.

7370 Cupravit®
1332-40-7
Copper oxychloride
Copper oxychloride; Dacobre DG; Agrizan; Agrizan-15; Basic copper chloride; Basic cupric chloride; Blitox; Blitox 50; Chemocin; Colloidox; Copper chloride (CuCl2), mixt. with copper oxide (CuO), hydrate; Copper chloride oxide; Copper chloride oxide, hydrate; Copper chloride, mixed with copper oxide, hydrate; Coppesan; Coprantol; Coprex; Coprosan blue; Cupral 45; Cupramar; Cupravit; Cupravit forte; Cupric oxide chloride; Copper(II) Oxychloride; Cupricol; Cupritox; Cuprox; Cuproxol; Faligruen; FYCOL 8; Fytolan; Kauritil; Kupricol; Kuprikol; Miedzian; Miedzian 5; Miedzian 50; Oleocuprit; Oxicob; Oxivor; Oxycur; Parrycop; Tamraghol; Vitigran; Vitigran blue,. Copper containing spray used for control of fungal diseases like downy mildews, late blight, early blight, apple scab and various leaf spot diseases on a wide range of crops. *Bayer AG.*

7371 Cuprex
Copper alloy general purpose flux. *Foseco (F.S.) Ltd.*

7372 cupric acetate
142-71-2 2690 205-553-3
$C_4CuH_6O_4$
cupric acetate
crystallized verdigris; neutralized verdigris. Pesticide, catalyst, fungicide, pigments, manufacture of Paris green. mp = 115°; d = 1.882; soluble in H_2O, slightly soluble in organic solvents; LD_{50} (rat orl) = 0.71 g/kg. *Celtic Chemical Ltd; Mallinckrodt; Nihon Kagaku Sangyo.*

7373 cupric acetate, basic
52503-64-7 2691 257-974-7
cupric subacetate
complex with different cupric acetate:cupric hydroxide:water ratios; verdigris, green verdigris. Used as pigments, insecticides, fungicides, mold-preventatives.

7374 cupric acetoarsenite
12002-03-8 2692
$C_4H_6As_6Cu_4O_{16}$
(acetato)trimetaarsenitodicopper
copper acetate arsenite; emerald green; French green; imperial green; mineral green; Mitis green; parrot green; Schweinfurt green; Vienna green; C.I. Pigment Green 21; C.I. 77410. Used as a pigment, particularly in ships and submarines, and as a wood preservative and insecticide. Insoluble in H_2O, unstable in acid or base and with H_2S; LD_{50} (rat orl) = 100 mg/kg.

7375 cupric arsenate
7778-41-8
$As_2Cu_3O_8$
Copper(II)arsenate
Cupar; Copper arsenate. Protective fungicide. *Mechema Chemicals Ltd.* Name unverified.

7376 cupric arsenite
10290-12-7 2693 233-644-8
$CuHAsO_3$
arsonic acid copper(2+) salt (1:1)
arsenious acid coppper(2+) salt(1:1); Scheele's green. Used as a pigment, wood preservative, insecticide, fungicide, and rodenticide. Insoluble in H_2O, organic solvents.

7377 cupric bromide
7789-45-9 2695 232-167-2
Br_2Cu
cupric bromide
Photography (intensifier), organic synthesis (brominating agent), battery electrolyte, wood preservative. mp = 498°; bp = 900°; d^{20}= 4.710. *Aldrich; Atomergic Chemetals; Cerac; Hoechst Celanese; Mallinckrodt.*

7378 cupric carbonate
12069-69-1 2697 235-113-6
$CuCO_3 \cdot Cu(OH)_2$
Basic Cupric carbonate
malachite; Copper, [μ-[carbonato(2-)-O:O']]dihydroxydi-; (carbonato(2-; dihydroxydicopper; copper carbonate hydroxide; cupric subcarbonate; Copper(II)hydroxide carbonate; Basic cupric carbonate; Bremen blue; Bremen green; Carbonato(2-)-O:O'; dihydroxydicopper; Basic copper carbonate ($Cu_2(OH)_2CO_3$). Used in pigments, pyrotechnics, insecticides, copper salts. An astringent in pomades, antidote for phosphorus poisoning, smut preventive, fungicide for seed treatment, feed additive. mp = 200° (dec); d = 4.000; LD_{50} (rat orl) = 1350 mg/kg. *Am. Chemet; Boliden Intertrade; Nihon Kagaku Sangyo.*

7379 cupric chloride dihydrate
7447-39-4 2699 231-210-2
Cl_2Cu
Copper(II) chloride dihydrate
Coppertrace. Supplement, trace mineral. mp = 100°; d = 2.51; freely soluble in H_2O, EtOH, less soluble in Me_2CO, EtOAc. *Armour Pharmaceutical Co.* Name unverified.

7380 cupric formate
544-19-4 2705 208-865-8
$C_2H_2CuO_4$
cupric formate
Tubercuprose; Copper formate; Cufor. Antibacterial agent. Soluble in H_2O, insoluble in organic solvents *Mechema Chemicals Ltd.* Name unverified.

7381 cupric hydroxide
20427-59-2 2709 243-815-9
CuH_2O_2
Copper hydroxide (ic)
copper oxide hydrated; copper hydrate; hydrated cupric oxide; Kocide; Copper hydroxide; Comac parasol; Copper dihydroxide; Copper hydrate; Criscobre; Cudrox; Cuidrox; Cupravit blue; Kocide 101. Copper salts, mordant, paper staining; pesticide, fungicide, catalyst. d = 3.37. *Am. Chemet; Cuproquim; Faesy & Besthoff; Griffin.*

7382 cupric hydroxide
20427-59-2 2709 243-815-9
CuH_2O_2
Copper hydroxide
cupric hydroxide; copper hydrate, hydrated cupric oxide; Chiltern Kocide 101. A protectant fungicide. Also used as a pigment, catalyst. d = 3.37; insoluble in H_2O. *Chiltern Farm Chemicals Ltd.*

7383 cupric hydroxide
20427-59-2 2709 243-815-9
$C_4O_2H_2$
cupric hydroxide
hydrated copper oxide; Comac Parasol; copper hydrate. Copper salts, mordant, paper staining; pesticide, fungicide, catalyst. d = 3.37; insoluble in H_2O. *Am. Chemet; Cuproquim; Faesy & Besthoff; Griffin; McKechnie Chemicals Ltd.*

7384 cupric nitrate
3251-23-8 2710 221-838-5
$CuN_2O_6 \cdot 3H_2O$
cupric nitrate trihydrate
Light-sensitive papers, analytical reagent, textile dyeing mordant, nitrating agent, insecticide, coloring copper black, electroplating, paints, varnishes, enamels, pharmaceuticals, catalyst. mp = 255-256°; sublimes 150-225°; (trihydrate) mp = 114.5°; d = 2.05; LD_{50} (rat orl) = 0.94 g/kg. *Blythe, Williams Ltd; Mallinckrodt.*

7385 cupric oxalate
814-91-5 2712 212-411-4
C_2CuO_4
Ethanedioic acid copper salt
Crow Chex; copper oxalate. Active ingredient: copper oxalate; seed protectant to prevent sprout pulling by birds in newly planted corn. mp = 310° (dec); insoluble in H_2O, organic solvents. *Borderland Products Inc.* Discontinued.

7386 cupric oxide
1317-38-0 2713 215-269-1
CuO
cupric oxide
copper oxide, black copper oxide. Ceramic colorant, reagent in analytical chemistry, insecticide, catalyst, purification of hydrogen, batteries and electrodes, electroplating, solvent, desulfurizing oils, rayon, metallurgical and welding fluxes, antifouling paints, phosphors. d^{14}= 6.315; insoluble in H_2O, EtOH. *Aldrich; Am. Chemet; Cerac; Chemisphere Ltd; Nihon Kagaku Sangyo; Noah Chem.*

7387 cupric sulfate
7758-98-7 2722 231-847-6
$CuSO_4 \cdot 5H_2O$
copper sulfate pentahydrate

cupric sulfate, copper sulfate; hydrocyanite; blue vitriol; blue stone; blue copperas,. CSP; soil additive, pesticides, feed additive, germicides, textile mordant, leather, pigments, batteries, electroplated coatings, copper salts, analytical reagent, medicine, wood preservative, lithography, ore flotation, petroleum, rubber, steel $d^{15.6}_4$ = 2.886; very soluble in H_2O, EtOH; LD_{50} (rat orl) = 960 mg/kg. *Allchem Industries; Farleyway Chemical Ltd.*

7388 cupric sulfate, basic
1332-14-5 2723 215-568-7
$Cu_4H_6O_{10}S$
copper hydroxide sulfate
Caldo Bordeles Valles; brochantite; langite. Bordeaux mixture plus adjuvants; wettable powder used as protective fungicide for foliage application to ornamental and crop plants. *Industrias Quimicas Del Valles SA.* Discontinued.

7389 Cupridan
1317-39-1 2734 215-270-7
CuO
Copper oxide
Fungicide. *Makhteshim Chemical Works Ltd.*

7390 Cuprinol
A copper naphthenate or sodium pentachlorphenate preparation used as a preservative for timber and fabrics. No manufacturer.

7391 Cuprit
Copper alloy cleansing flux. *Foseco (F.S.) Ltd.*

7392 cupro-aluminums
Alloys of aluminum and copper containing 1-20% aluminum. They are also wrongly called aluminum bronzes.

7393 Cuprodine
A medium for coating steel with copper. *ICI Chemical & Polymers Ltd.* Discontinued.

7394 Cuproid
7758-89-6 2730 231-842-9
cuprous chloride
Cuprous chloride. Catalyst in organic reactions, decolorizing and desulfurizing agent. *Mechema Chemicals Ltd.*

7395 Cuprokylt
1332-40-7
Copper oxychloride
Copper oxychloride 87% w/w wettable powder formulation; a protectant fungicide for agriculture and horticulture. *Universal Crop Protection Ltd.*

7396 Cupron
A proprietary trade name for a copper nickel alloy with a low temperature coefficient of resistance. No manufacturer.

7397 Cuprophenyl
An after coppering direct dye. *Ciba plc.* Name unverified.

7398 cupror
An aluminum bronze. It contains 94.2% copper and 5.8% aluminum.

7399 Cuprosana
1332-40-7
Copper oxychloride
A dustable powder containing 10% w/w copper oxychloride; a protectant fungicide for hops. *Universal Crop Protection Ltd.*

7400 cupro-steel
An alloy of steel with copper up to 4%. Occasionally used for printing rollers and projectiles.

7401 cupro-titanium
Usually an alloy of copper with 10% titanium; used as a deoxidizer in making brass and bronze castings.

7402 cuprous bromide
7787-70-4 2729 232-131-6
CuBr
copper(I) bromide
copper bromide. Catalyst in organic reactions. mp = 504°; bp = 1345°; d_4^{25}= 4.72; slightly soluble in H_2O. *Aldrich; Atomergic Chemetals; Cerac; Hoechst Celanese.*

7403 cuprous chloride
7758-89-6 2730 231-842-9
ClCu
cuprous chloride
Copper chloride; Cu-lyt. Catalyst, preservative and fungicide, desulfurizing and decolorizing agent for petroleum industry, absorbent for carbon monoxide. mp = 430°; d_4^{25} = 4.14; slightly soluble in H_2O. Atomergic Chemetals; Cerac; Hoechst Celanese; Mallinckrodt; Nihon Kagaku Sangyo; Zinder SpA; Mechema Chemicals Ltd.

7404 cuprous iodide
7681-65-4 2732 231-674-6
CuI
copper(I) iodide
copper iodide; hydro-giene; marshite. Feed additive, in table salt as source of dietary iodine, catalyst, cloud seeding. mp = 588-606°; bp= 1290°; d_4^{25}= 5.63; insoluble in all solvents. Aldrich; Atomergic Chemetals; Blythe, William Ltd; Cerac; R.W. Greeff; Mitsui Toatsu; Nihon Kagaku Sangyo.

7405 cuprous oxide
1317-39-1 2734 215-270-7
Cu_2O
Cuprous oxide
Cupridan; Copper(I) oxide; Red Copper Oxide; C.I. 77402; Perenex; Yellow Cuprocide; Copper-Sandoz; Caocobre; Cuprox; Copper oxide. Fungicide. mp = 1232°; bp = 1800°; d = 6.000; LD_{50} (rat orl) = 0.47 g/kg. *Makhteshim Chemical Works Ltd.*

7406 cuprous potassium cyanide
13682-73-0 2735 237-192-2
C_2CuKN_2
potassium cyanocuprate
potassium copper cyanide; cuprous potassium cyanide; potassium dicyanocuprate(I); potassiocuprous cyanide; cuprocyan. A double cyanide of potassium and copper. d = 2.38; insoluble in H_2O.

7407 cuprous thiocyanate
1111-67-7 2739 214-183-1
CCuNS
Copper(I) thiocyanate
Cusyd;. Used as marine anti-fouling agent. d = 2.85; insoluble in H_2O, organic solvents. *Mechema Chemicals Ltd.* Name unverified.

7408 cupro-vanadium
An alloy usually containing from 10-15% vanadium, 60-70% copper, 10-15% aluminum, and 2-3% nickel.

7409 Cuprox
1317-39-1 2734 215-270-7
OCu_2
Copper(I) oxide
Cuprous oxide. Fungicide. *Makhteshim Chemical Works Ltd.*

7410 Curbetan®
For the agriculture industry. *DuPont UK.*

7411 Curafos® STP
7758-29-4 8846 231-838-7
Sodium tripolyphosphate
Food additive for meat, poultry, and seafood industries, processed foods, confections. *Rhône-Poulenc Food Ingreds.*

7412 Curaseal
Natural and synthetic rubber latex based adhesives and coatings; cohesive, pressure-sensitive and synthetic resin emulsions; used for numerous adhesive bonding applications. *Testworth Laboratories Inc.*

7413 Curasol
Soil erosion inhibitor. *Hoechst UK.*

7414 Curaterr®
1563-66-2 1851 216-353-0
Carbofuran
Granular soil-applied insecticide for use in sugar beet, tobacco, maize, rice, sugar cane and vegetables. *Bayer AG.*

7415 Curatin
1229-29-4 3492 214-966-8
Doxepin hydrochloride
Veterinary antipruritic. *Pfizer Inc.*

7416 Curb
7784-25-0 335 232-055-3
Aluminum ammonium sulfate
Bird and animal repellent. *Sphere Laboratories (London) Ltd.* Discontinued.

7417 Curbeton 0550
Desugared hardwood calcium/magnesium lignosulfonate
Water-reducing and strength-increasing concrete additive. *Borregaard Ligno Tech.*

7418 Curene
Catalysts and light and heat stabilizers; catalysts and chain extenders for urethane polymers. *Anderson Development Company.* Name unverified.

7419 Curenox-50
1332-40-7
Copper oxychloride
Copper oxychloride formulated to 50% metallic copper contents; wettable powder used as protective fungicide for foliage application to ornamental and crop plants. *Industrias Quimicas Del Valles SA.*

7420 Cure-Rite® 18
Thiocarbamyl sulfenamide
Nonstaining primary accelerator for EPDM, SBR, nitrile, natural and butyl rubbers. *Akrochem.*

7421 Curgon
A proprietary trade name for naphthenate driers. No manufacturer.

7422 Curithane
Reactive polyamines used as a hardener and curing agent for polyisocyanurate foams; rigid PU catalysts. *Dow Cheml Co Ltd, UK & Ireland.*

7423 Curithane 103
Polymethylene polyaniline
Amine curing agent for PU; intermediate in manufacturing of polyamides, polyimides, coatings, and plastics. *Dow Plastics.* Name unverified.

7424 Curodex
Rubber odorants and deodorants. *Bush Boake Allen Ltd.*

7425 Curolac
Precatalysed and two-pack curing lacquers. *The Scottish Adhesives Co Ltd.*

7426 Curtexil 100S
Hardwood calcium/magnesium lignosulfonate; general binder and dispersant. *Borregaard Ligno Tech.*

7427 Curzate® M
cymoxanil-mancozeb. Mixture of cymoxanil and mancozeb; used to control potato blight. *DuPont UK.*

7428 Cusamon
Copper sulfate monohydrate
Mechema Chemicals Ltd. Name unverified.

7429 Cusatrib
1332-14-5 2723 215-568-7
$Cu_4H_6O_{10}S$
Tribasic copper sulfate
brochantite; langite. Fungicide for plants, seed treatment. *Mechema Chemicals Ltd.* Name unverified.

7430 Cusiloy A
A proprietary trade name for an alloy containing 95.5% copper, 3% silicon, 1% iron, and 0.5% tin. No manufacturer.

7431 Custom Age 625 Plus® Alloy
Nickel-base alloy; precipitation-hardenable alloy with good corrosion resistance; used for deep sour gas wells, refineries, chemical process industry, high-temperature nuclear water. *Carpenter Tech.*

7432 Cut Aid
71-55-6 9766 200-756-3
1,1,1-trichlorethane
ASTM S-315 Oil, mask odor No 3; excellent cutting aid for general machine operations; exceptional for disc sanding, filing and punch press operations. *Doyle Specialties.* Name unverified.

7433 Cutavit Richter
Complex of vitamins A and E and essential fatty acids in lipophilic medium; product for dry skin and hair. *Dr. Kurt Richter; Henkel/Cospha.*

7434 Cutina®
Self-emulsifying raw material; creams and emulsions. *Henkel Chemicals Ltd.* Name unverified.

7435 Cutina® BW
Glyceryl hydroxystearate-cetyl palmitate-trihydroxystearin
Microcrystalline wax for use as beeswax substitute; viscosity agent for personal care products. *Henkel/Cospha; Henkel Canada.*

7436 Cutina® CBS
Glyceryl stearate-cetearyl alcohol-cetyl palmitate-cocoglycerides
Cream base for manufacturing of creams and lotions of the oil-water type; viscosity agent and stabilizer. *Henkel/Cospha; Henkel KGaA.*

7437 Cutina® CP
540-10-3 2073 208-736-6
Cetyl palmitate
Synthetic spermaceti; consistency factor for creams, ointments, liquid emulsions, fatty make-ups, and sticks. *Henkel/Cospha; Henkel Canada.* No manufacturer.

7438 Cutina® E24
51158-08-8
PEG-20 glyceryl stearate
Oil-water emulsifier for mild creams and emulsions for baby and childrens preparations, sun preparations. *Henkel Canada; Henkel/Emery/Cospha; Henkel KGaA.*

7439 Cutina® FS 25, FS 45
Fatty acid mixture; consistency factor after saponification; oil-water emulsifier for emulsions and ointments. *Henkel KGaA.*

7440 Cutina® GMS, Cutina® KD16, Cutina® MD-A
31566-31-1 4498 250-705-4
Glyceryl stearate
Nonself-emulsifying cream base for oil-water and Water-oil emulsions; viscosity agent for creams, ointments, sticks; beads. *Henkel/Cospha; Henkel Canada.*

7441 Cutlanego
An alloy of 50% bismuth and 50% tin; used for tempering steel tools.

7442 Cutlass
18467-77-1 3243 242-348-8
Dikegulac
Contains dikegulac; growth regulator for hedges. *ICI Garden Products.*

7443 Cutonic
Micronutrient foliar sprays. *McKechnie Chemicals Ltd.* Name unverified.

7444 Cuvan®
Proprietary formulations; corrosion inhibitors and metal deactivators for industrial and automotive oils and metalworking fluids. *R. T. Vanderbilt Co Inc.*

7445 CV-1142
One-part silicone adhesive/sealant; RTV non-corrosive low-temperature adhesive/sealant. *McGhan NuSil.*

7446 CV-2640
Two-part silicone adhesive/sealant; electrically conductive RTV RF/EMI shielding/grounding adhesive/sealant, potting compound. *McGhan NuSil.*

7447 CV-4720
1719-58-0 217-007-1
C_4H_9ClSi
Vinyldimethylchlorosilane
Coupling agent, chemical intermediate, blocking agent, release agent, lubricant, primer, reducing agent. *Hüls Am.*

7448 CV-4772
124-70-9 204-710-3
$C_3H_6Cl_2Si$
Vinylmethyldichlorosilane
Methylvinyldichlorosilane. Coupling agent, chemical intermediate, blocking agent, release agent, lubricant, primer, reducing agent. bp = 92°; d = 1.08; n_D^{20} = 1.4300. *Hüls Am.*

7449 CV-4800
4130-08-9 223-943-1
$C_8H_{12}O_6Si$
Vinyltriacetoxysilane
Ethenyltriacetate-silanetriol; Triacetoxy(vinyl)silane. Coupling agent, chemical intermediate, blocking agent, release agent, lubricant, primer, reducing agent. mp = 7°; bp_{13} = 112°; d = 1.1600; n_D^{20} = 1.4200. *Hüls Am.*

7450 CV-4900
75-94-5 200-917-8
$C_2H_3Cl_3Si$
Vinyltrichlorosilane
trichlorovinylsilane; Silane, trichloroethenyl-; VTC. Coupling agent, chemical intermediate, blocking agent, release agent, lubricant, primer, reducing agent. bp = 90°; d = 1.2700; n_D^{20} = 1.4362; LD_{50} (rat orl) = 1280 mg/kg. *Hüls Am.*

7451 CV-4910
78-08-0 201-081-7
$C_8H_{18}O_3Si$
Vinyltriethoxysilane
Triethoxyvinylsilane; vinyltriethoxysilane; Silane, ethenyltriethoxy-; Ethenyltriethyloxy-silane; VTEO; Dynasylan VTEO. Coupling agent, chemical intermediate, blocking agent, release agent, lubricant, primer, reducing agent. bp = 160-161°; d = 0.9030; n_D^{20} = 1.3978; LD_{50} (rat orl) = 22500 mg/kg. *Hüls Am.*

7452 CV-4917
2768-02-7 220-449-8
$C_5H_{12}O_3Si$
Vinyltrimethoxysilane
Silane, ethenyltrimethoxy-; Trimethoxyvinylsilane; VTMO; Dynasylan VTMO. Coupling agent, chemical intermediate, blocking agent, release agent, lubricant, primer, reducing agent. bp = 123°; d = 1.1300; n_D^{20} = 1.3915; LD_{50} (rat orl)= 11300 mg/kg. *Hüls Am.*

7453 CV-5000
1067-53-4 213-934-0
$C_{11}H_{24}O_6Si$
Vinyltris(methoxyethoxy)silane
2,5,7,10-Tetraoxa-6-silaundecane, 6-ethenyl-6-(2-methoxyethoxy)-; Tris(2-methoxyethoxy)vinylsilane; Dynasylan VTMOEO. Coupling agent, chemical intermediate, blocking agent, release agent, lubricant, primer, reducing agent. mp = -30°; bp = 285°; d = 1.0300; n_D^{20} = 1.4284; LD_{50} (rat orl) = 2960 mg/kg. *Hüls Am.*

7454 CV-5050
2224-33-1 218-747-8
Vinyl tris (methylethylketoxime) silane
Coupling agent, chemical intermediate, blocking agent, release agent, lubricant, primer, reducing agent. *Hüls Am.*

7455 CV-5100
5356-84-3 226-342-2
$C_{11}H_{30}O_3Si_4$
Vinyl tris(trimethylsiloxy) silane

Coupling agent, chemical intermediate, blocking agent, release agent, lubricant, primer, reducing agent. *Hüls Am.*

7456 CW-35
Chlorinated paraffin; offers good stability and resistance to thermal decomp.; used in metalworking applications providing corrosion inhibition; extreme pressure agent for soluble and cutting oils, some drawing compounds. *Ferro/Keil.*

7457 CW-79
Lidofilcon B material containing 79% water; contact lens material. *Bausch & Lomb, Professional Products Div.* Name unverified.

7458 CWT 102
Blend; corrosion inhibitor for open cooling water systems. *Drew Ind. Div.*

7459 CXL 400, 78T
General purpose adhesive for concrete or steel. *Colebrand Ltd.*

7460 Cyanacryl® 35
Acrylic elastomer. *Am. Cyanamid.*

7461 Cyanamer P35-P70
Acrylamide copolymers (anionic). *Cyanamid BV.*

7462 Cyanamer P-80
Sulfonated polycarboxylic acid polymer. *Cyanamid BV.*

7463 Cyanaprene®, 2070, 2160, 2175, 2180
TDI-polyether urethane prepolymer *Am. Cyanamid; Air Prods & Chem.*

7464 Cyanaprene® 1050, 1080, 1090
TDI-polyether urethane prepolymer; used for solid tires, coated materials, rolls, molded mechanical goods; high abrasion resistance, good resilience, hydrolytic stability, and load bearing capability, oil and fuel resistance; vulcanizable with diamines. *Am. Cyanamid; Air Prods & Chem.*

7465 Cyanaprene® A-7QM, A-75QM
TDI-polyester urethane prepolymer; used in solid tires, coated materials, rolls, molded mechanical goods; high abrasion resistance, tear strength oil and fuel resistance, good load bearing characteristics, hydrolytic stability; vulcanizable with diamines. *Am. Cyanamid; Air Prods & Chem.*

7466 Cyanaprene® A-8, A-8QM, A-9
TDI-polyester urethane prepolymer; used in solid tires, coated materials, heels, soles, rubber coved rolls, molded mechanical goods, metal forming pads; high abrasion resistance, tear strength oil and fuel resistance; vulcanizable with diamines. *Am. Cyanamid; Air Prods & Chem.*

7467 Cyanaprene® A-85
TDI-polyester urethane prepolymer; castable urethane used in casters, lift truck and press-on tires; high abrasion resistance, cutting and flex fatigue, solvent resistance, exceptional thermal stability; vulcanizable with diamines. *Am. Cyanamid; Air Prods & Chem.*

7468 Cyanaprene® A-9QM, D-55, D-5QM
TDI-polyester urethane prepolymer. *Am. Cyanamid; Air Prods & Chem.*

7469 Cyanaprene® D-6
TDI-polyester urethane prepolymer; castable urethane used in top lifts, pellet and caster wheels, molded mechanical goods, rubber-covered rolls; high abrasion resistance, tear strength, oil and fuel resistance; vulcanizable with diamines. *Am. Cyanamid; Air Prods & Chem.*

7470 Cyanaprene® D-7
TDI-polyester urethane prepolymer; Castable urethane used in rubber-covered rolls, bearing blocks, drilling jig fixtures, cutting pads; high abrasion resistance, oil and fuel resistance; vulcanizable with diamines. *Am. Cyanamid; Air Prods & Chem.*

7471 Cyanatrol
Modified polyacrylamide in water-in-oil emulsion. *Cyanamid BV.*

7472 cyanazine
21725-46-2 2755 244-544-9
$C_9H_{13}ClN_6$
Propanenitrile, 2-[[4-chloro-6-(ethylamino)-1,3,5-triazin-2-yl]amino]-2-methyl Propionitrile, 2-[[4-chloro-6-(ethylamino)-1,3,5-triazin-2-yl]amino]-2-methyl; s-triazine, 2-chloro-4-(ethylamino)-6-(1-cyano-1-methyl)(ethylamino)-; Bladex; Bladex; 80WP; Cyanazine SD 15418; DW 3418; Fortol; Fortrol; Payze; SD 15418; WL 19805; Match. Selective systemic herbicide, absorbed by roots and foliage. Used for control of annual grass and broad-leaved weeds. mp = 167.5-169°; soluble in H_2O (171 mg/l), more soluble in organic solvents; LD_{50} (rat orl) = 149 mg/kg. *DuPont; Shell.*

7473 Cyanex
Modified phosphine derivatives. *Cyanamid BV.*

7474 Cyanine
Anionic dyestuffs (level dyeing); for wool and wool blends. *Holliday Dyes & Chemicals Ltd.*

7475 Cyanine
Leitch's blue; quinoline blue. A blue pigment consisting of a mixture of cobalt blue and Prussian blue. A quinoline dyestuff, is also known as cyanine or quinoline blue; used as a panchromatic sensitizer, and also dyes silk.

7476 Cyanine Fast
Acid dyes. *Holliday Dyes & Chemicals Ltd.* Discontinued.

7477 Cyanine Moderns
Dyestuffs, the condensation products of gallo-cyanines and allyldiamines.

7478 2-cyanoethyltriethoxysilane
17932-62-6 241-869-8
2-Cyanoethyltriethoxysilane
CC3433. Coupling agent, chemical intermediate, blocking agent, release agent, lubricant, primer, reducing agent. *Hüls Am.*

7479 cyanogen bromide
506-68-3 2763 208-051-2
CBrN
bromine cyanide
Organic synthesis, parasiticide, fumigating compositions, rat extermination, cyaniding reagent in gold extraction processes. mp = 52°; bp = 61-62°; d_4^{20}= 2.015; soluble in H_2O, organic solvents. *Aldrich; Atomergic Chemetals; Eastman; Janssen Chimica.*

7480 Cyanolime
592-01-8 1704 209-740-0
C_2CaN_2
Calcium cyanide
Fumigant and rodenticide. Also used in stainless steel manufacture and extraction of precious metals. Soluble (reacts) in H_2O, EtOH; LD_{50} (rat orl) = 39 mg/kg. *Mechema Chemicals Ltd.* Name unverified.

7481 Cyanolit-Hitemp
A proprietary range of high-temperature cyanoacrylate adhesives. *Unichem.* Name unverified.

7482 3-cyanopropyltrichlorosilane
1071-27-8 213-990-6
$C_4H_6Cl_3NSi$
Trichloro-3-cyanopropylsilane
4-(Trichlorosilyl)butyronitrile; CC3555. Coupling agent, chemical intermediate, blocking agent, release agent, lubricant, primer, reducing agent. bp_8 = 93-94°; d = 1.30; n_D^{20}= 1.4630. *Hüls Am.*

7483 Cyanosin A
Cyanosin; Spirit Soluble. The alkali salt of tetrabromodichlorofluoresceine methyl ether; used in silk dyeing.

7484 Cyanosin B
A dyestuff. The sodium salt of tetrabromotetrachlorofluoresceine ethyl ether. Dyes wool bluish-red.

7485 Cyanox® 425
88-24-4 201-814-0
$C_{25}H_{36}O_2$
2,2'-Methylenebis (4-ethyl-6-*tert*-butylphenol)
Antioxidant for impact molding resins; stabilizes acrylics and ABS; end-uses include extruded and molded products. *Am. Cyanamid.*

7486 Cyanox® 711
10595-72-9 234-206-9
Ditridecyl thiodipropionate
Secondary antioxidant for stabilizing polyolefins, ABS, petrol. lubricants, SBR latex compositions. *Am. Cyanamid.*

7487 Cyanox® 1212
Mixed lauryl-stearyl thiodipropionate; secondary antioxidant for stabilizing polyolefins; applications including pipe, hot-melt adhesives, and molded olefin products. *Am. Cyanamid.*

7488 Cyanox® 1790
1,3,5-Tris(4-tert-butyl-3-hydroxy-2,6-dimethylbenzyl)-1,3,5-triazine-2,4,6-(1H,3H,5H)-trione
Antioxidant for use in polyolefin pipe, film, household appliances, olefin and urethane fibers, styrenics and polyesters. *Am. Cyanamid.*

7489 Cyanox® 2246
119-47-1 204-327-1
$C_{23}H_{32}O_2$
2,2'-Methylenebis (4-methyl-6-*tert*-butylphenol)
2,2'-methylenebis 6-(1,1-dimethylethyl)-4-methyl-phenol; 6,6'-di-*tert*-butyl-2,2'-methylenedi-*p*-cresol; 2,2'-Methylene-bis(6-*tert*-butyl)-*para*-cresol. Antioxidant preventing thermal oxidation of ABS, polyethylene, PP, and EVA; oxidation inhibitor for fats, oils, and paraffin wax; polymerization inhibitor in chemical processes; for ABS, hot-melt adhesives, latex carpet backing, and speciality olefin applications. *Am. Cyanamid. Name unverified.*

7490 Cyanox® 2777
1,3,5-tris (4-*t*-butyl-3-hydroxy-2, 6-dimethylbenzyl)-1,3,5-triazine-2,4,6-(1H, 3H,5H) trione-tris (2,4-di-*t*-butylphenyl) phosphite (1:2). Antioxidant, stabilizer for polymers, especially polyolefins in high temperature processing. *Am. Cyanamid.*

7491 Cyanox® LTDP
123-28-4 204-614-1
$C_{30}H_{58}O_4S$

Dilauryl thiodipropionate
Dodecyl-3,3'-thiodipropionate; Didodecyl 3,3'-thiodipropionate. Secondary antioxidant in ABS, PP, and polyethylene; used in food packaging materials, automotive, appliance, battery casing, pipe; stabilization of oils, lubricants, sealants, and adhesives. mp = 40-42°; d = 0.915. *Am. Cyanamid.*

7492 Cyanox® MTDP
16545-54-3 240-613-2
Dimyristyl thiodipropionate
Secondary antioxidant for protection of polyolefins. *Am. Cyanamid.*

7493 Cyanox® STDP
693-36-7 211-750-5
Distearyl thiodipropionate
Secondary antioxidant used in polyolefins and other polymers; used in automotive, appliance, container film, sealant, and adhesive applications. *Am. Cyanamid.*

7494 cyanuric acid
108-80-5 2767 203-618-0
$C_3H_3N_3O_3$
1,3,5-Triazine2,4,6(1H,3H,5H)-trione
2,4,6-trihydroxy-1,3,5-triazine; cyanuric acid, anhydride; tricyanic acid; trihydroxycyanidine; tricarbimide; tricyanide. Intermediate for chlorinated bleaches, selective herbicide, whitening agents. d_4^{20}= 2.500; soluble in H_2O (1 g/200 ml), soluble in organic solvents. *Allchem Industries; Monsanto; Orkem UK Ltd; 3-V.*

7495 Cyasorb® UV 9
131-57-7 7088 205-031-5
$C_{14}H_{12}O_3$
2-Hydroxy-4-methoxybenzophenone
2-Hydroxy-4-methoxyphenyl)phenylmethanone; 4-Methoxy-2-hydroxybenzophenone; Benzophenone 3; Escalol 567; Eusolex 4360; MOB; Oxybenzone; Uvinul M-40; Spectra-Sorb UV-9; 2-benzoyl-5-methoxyphenol; Cyasorb UV 9; Syntase 62; UF 3; Advastab 45; Anuvex; Chimassorb 90; Cyasorb UV 9 light absorber; MOD; Ongrostab HMB; Sunscreen UV 15; Uvinul 9; Uvistat 24; HMB. Light stabilizer and uv absorber for plastics and coatings; especially for flexible and rigid PVC, unsaturated polyesters, and acrylics; used in outdoor sheeting and glazing applications, molded products, adhesives. mp = 63-65°; bp_5 = 150-160°; LD_{50} (rat orl) = 7400 mg/kg. *Am. Cyanamid.*

7496 Cyasorb® UV 24
131-53-3 3357 205-026-8
$C_{14}H_{12}O_4$
2,2-g-Dihydroxy-4-methoxybenzophenone
dioxybenzone, benzophenone 8; Spectro-Sorb UV 24. Light stabilizer and uv absorber for coatings and plastics, e.g., alkyds, phenolics, PU coatings; stabilizer for polyester film and PVC formulations. mp = 68°; insoluble in H_2O, soluble in organic solvents. *Am. Cyanamid.*

7497 Cyasorb® UV 531
1843-05-6 6838 217-421-2
$C_{21}H_{26}O_3$
2-Hydroxy-4-n-octoxybenzophenone
Methanone, [2-hydroxy-4-(octyloxy)phenyl]phenyl-; 2-Hydroxy-4-(octyloxy)benzophenone; Octabenzone. Light stabilizer and uv absorber for plastics and coatings, e.g., polyethylene, PP, PVC, and EVA; uses include pipe, storage tanks, and auto, marine, garden products, auto refinish and industrial coatings, adhesives and sealants. *Am. Cyanamid.*

7498 Cyasorb® UV 1084
2,2'-Thiobis (4-*tert*-octylphenolato)-n-butylamine nickel II
Light and heat stabilizer for polyolefins, e.g., PP fiber, LDPE agricultural films and pool liners, and molded products *Am. Cyanamid.*

7499 Cyasorb® UV 2098
2-Hydroxy-4-acryloyloxyethoxy benzophenone
Light stabilizer, uv absorber which may be chemically bonded with monomers or polymers. *Am. Cyanamid.*

7500 Cyasorb® UV 2126
Polymer of 4-(2-acryloyloxyethoxy)-2-hydroxy benzophenone
Light stabilizer and uv absorber for films and plastics in automotive, greenhouse, home siding, and solar applications. *Am. Cyanamid.*

7501 Cyasorb® UV 2908
3,5-Di-*t*-butyl-4-hydroxybenzoic acid n-hexadecyl ester
Light stabilizer, free radical scavenger for polyolefins, especially pigmented opaque formulations; antioxidant; for pipe, crates, drums, auto, marine, garden, recreational products *Am. Cyanamid.*

7502 Cyasorb® UV 3346
Oligomeric hindered amine; stabilizer for polymers alone or in combination with UV absorbers. *Am. Cyanamid.*

7503 Cyasorb® UV 5411
3147-75-9 221-573-5
$C_{20}H_{27}N_3O$
2-(2-Hydroxy-5-*t*-octylphenyl)-benzotriazole
Octrizole. Light stabilizer and uv absorber for polymeric systems including polyester, PVC, styrenics, acrylics, PC, and polyvinyl butyral; end-uses including molding, sheet, and glazing materials for window, marine, and auto applications; also in coatings, photoproducts, sealants and elastomeric materials. *Am. Cyanamid.*

7504 Cyastat® 609
N,N-Bis (2-hydroxyethyl)-N-(3-dodecyloxy-2-hydroxypropyl) methyl ammonium methosulfate
Antistatic agent with good heat stability; for PVC phonograph records, specialty packaging. *Am. Cyanamid.*

7505 Cyastat® LS
(3-Lauramidopropyl) trimethylammonium methyl sulfate
Antistatic agent for polymeric materials, i.e., coatings, PVC, PS, polyolefins, ABS. *Am. Cyanamid.*

7506 Cyastat® SN
Stearamidopropyl dimethyl-2-hydroxyethyl ammonium nitrate
Antistatic agent for polymers; used for plastics, surface coatings, paper, glass, and other materials; dispersant in coatings. *Am. Cyanamid.*

7507 Cyastat® SP
Stearamidopropyl dimethyl-2-hydroxyethyl ammonium dihydrogen phosphate
Antistatic agent with surface-active properties; for plastics, waxes, textiles, and glass; emulsifier, settling, dispersing, and rewetting agent. *Am. Cyanamid.*

7508 Cybond WD-4517 and WD-4521
Proprietary two-component polyurethane adhesives of the solvent type. *Unichem.* Name unverified.

7509 Cyclamic acid
100-88-9 2770 202-898-1
$C_6H_{13}NO_3S$
Cyclohexylsulfamic acid
hexamic acid; cyclohexanesulfamic acid. Used as a non-nutritive sweetener, acidulant. mp = 169-170°; sparingly soluble in H_2O; LD_{50} (rat orl) = 15.25 g/kg.

7510 Cyclanon®
4266
Gardinol type wetting agents for textiles. *BASF plc.*

7511 Cyclanon® R
4266
Gardinol type after-treatment agent for polyester dyeings and prints. *BASF AG.*

7512 Cyclatex
A proprietary cyclized rubber. *Hubron Rubber Chemicals.* Name unverified.

7513 Cyclite
A proprietary cyclized rubber. *Durham Raw Materials.* Name unverified.

7514 Cyclo
Surfactant for cosmetics, toiletries, pharmaceutical, processing, agricultural and other industries. *Baxenden Chemicals Ltd.*

7515 Cyclochem
A fatty acid ester and emulsifying and wetting agent. *Rhône-Poulenc Surf.* Name unverified.

7516 cyclodextrins
7585-39-9 2787 231-493-2
cycloamylose
β-cycloamylose; cycloglucan; Schardinger dextrin; cyclomaltoheptaose. There are 3 cyclodextrins: α-cyclodextrin [10016-20-3], β-cyclodextrin [7585-39-9] and γ-cyclodextrin [17465-86-0]. They are cyclic polysaccharides comprised of six to eight glucopyranose units and are natural clathrates. *Am. Maize Prods.; Janssen Chimica; Pfanstiehl Labs; U.S. BioChemical.*

7517 Cyclo-Flo
Dispersant blend; fuel oil additive designed to remove and prevent formation of sludge and carbon aceous deposit when using heavy residual fuel oil. *Taiwan Surf.*

7518 Cyclofor
A paint additive. *Rhône-Poulenc Surf.* Name unverified.

7519 Cyclogol
Low foaming detergent, rinse aid; for machine dishwashing, soluble bath oils. *Witco Chemical Ltd.* Discontinued.

7520 cyclohexane
110-82-7 2792 203-806-2
C_6H_{12}
hexahydrobenzene
hexamethylene; hexanaphthene. Manufacture of nylon; solvent for cellulose ethers, fats, oils, waxes; paint and varnish remover; glass substitutes; in analytic chemistry; chemical intermediate; in fungicidal formulations. mp = 6-7°; bp = 80-81°; d = 0.779; n_D^{20}= 1.4260. *Exxon BV; Phillips 66; Texaco.*

7521 cyclohexanol
108-93-0 2794 203-630-6
$C_6H_{12}O$
Cyclohexanol
hexalin; hexahydrophenol. Used in soap making and manufacture of phenolic insecticides; source of adipic acid for nylon, textile finishing; solvent for alkyd and phenolic resins and rubber. bp = 160-161°; d = 0.951; n_D^{20} = 1.4650; LD_{50} (rat orl) = 2.06 g/kg *AlliedSignal; BASF; UCB SA.*

7522 cyclohexanone
108-94-1 2795 203-631-1
$C_6H_{10}O$
Cyclohexanone
pimelic ketone; ketohexamethylene; Hytrol O; Anone; Nadone. Used as paint and varnish remover; solvent for cellulose acetate, nitrocellulose, natural resins, vinyl resins, rubber, waxes, fats; in production of adipic acid for nylon, cyclohexanone resins. mp = -32°; bp = 156°; d_4^{25} = 0.9421; n_D^{20} = 1.4507; soluble in H_2O (150 g/l), more soluble in organic solvents; LD_{50} (rat orl) = 1.53 g/kg. *Aldrich; AlliedSignal; BASF; DSM BV; Union Carbide.*

7523 cyclohexyl benzothiazole sulfenamide
95-33-0 202-411-2
$C_6H_4SNCSNHC_6H_{11}$
N-Cyclohexyl-2-benzothiazolesulfenamide
Durax®;Durax® Rodform; Delac S; Akrochem® CBTS; Ekaland CBS; Perkacit ® CBS; Santocure®; Vulkacit CZ/EG; Vulkacit CZ/EGC; Vulkacit DZ/EGC. An all-purpose delayed action accelerator which combines superior scorch safety with shorter curing cycles; used in tire tread, carcass, camelback and mechanical goods. mp = 93-100°; d = 1.27; insoluble in H_2O, soluble in benzene, chloroform. *Uniroyal.* Name unverified.

7524 cyclohexyl chloride
542-18-7 2801 208-806-6
$C_6H_{11}Cl$
chlorocyclohexane
Solvent and chemical intermediate. mp = -44°; bp = 141-143°; d = 1.000; n_D^{20} = 1.4620. *Hüls AG; Janssen Chimica.*

7525 cyclohexylamine
108-91-8 2798 203-629-0
$C_6H_{13}N$
aminocyclohexane
cyclohexanamine; hexahydroaniline. Boiler water treatment, rubber accelerator, intermediate in organic synthesis. bp = 133-134°; d = 0.868; n_D^{20} = 1.4585; LD_{50} (rat orl) = 0.61 mg/kg. *Air Prods & Chem; Elf Atochem N. Am.; BASF; PMC Specialties.*

7526 Cyclolac
ABS polymers; for injection molding, sheet extrusions. *GE Plastics ABS Ltd.* Name unverified.

7527 Cyclolube® 62, NN-1
Naphthenic oil distillate; process oils offering light color, low pour points for general-purpose compounding of elastomers such as neoprene, SBR, isoprene, EDPM, butyl, and natural rubber. *Witco/Golden Bear.* Discontinued.

7528 Cyclolube® 85
Low aromatic naphthenic oil; process oil offering excellent color and heat stability for rubber compounding, especially EPDM and butyl; also for SBR, isoprene, and natural rubber. *Witco/Golden Bear.* Discontinued.

7529 Cyclomatic
Blend of anionic detergents; used for degreasing of synthetic fibers under pressure. *Atochem UK/Ceca.* Discontinued.

7530 Cyclomatic Dur
Wetting agent for use in hydrosulfite bleaching solutions for textiles. *Ceca SA.*

7531 Cyclomide
A fatty acid alkanolamide. *Rhône-Poulenc Surf.* Name unverified.

7532 Cyclonette
Emulsifiable wax composition. *Rhône-Poulenc Surf.* Name unverified.

7533 Cyclonox
Cyclohexanone peroxide-based range of catalysts; for the ambient temperature curing of unsaturated polyester resins. *Akzo Chemie UK Ltd.*

7534 Cyclophos
A phosphate ester. *Rhône-Poulenc Surf.* Name unverified.

7535 Cyclopol
A detergent composition. *Rhône-Poulenc Surf.* Name unverified.

7536 Cyclops metal
A nickel-chromium-iron alloy, containing 18% nickel, 18% chromium, and the rest iron; resists corrosion.

7537 Cyclorans
They contain a high-boling alcohol emulsified with a potassium olein soap and are used as wetting-out agents for textiles. No manufacturer.

7538 Cyclorubbers
Thermoplastic products made by heating a mixture of rubber sheet with ·e 10% of its weight of an organic sulfonyl chloride or an organic sulfonic acid. These products resemble gutta-percha or can be made to resemble shellac according to treatment.

7539 Cycloryl
A liquid surface active agent. *Rhône-Poulenc Surf.* Name unverified.

7540 Cycloryl 580 and 585N
Sodium lauryl sulfate, conforming to BP specification; powder or needle form; emulsification polymerization; emulsification. *Witco Chemical Ltd.* Discontinued.

7541 Cycloteric
An amphoteric surface active agent. *Rhône-Poulenc Surf.* Name unverified.

7542 Cycloton® SCS
122-19-0 204-527-9
$C_{27}H_{50}ClN$
N,N-dimethyl-N-octadecylbenzenemethanaminium chloride
dimethyl octadecylbenzylammonium chloride; benzyl dimethyloctadecylammonium chloride; benzyl dimethylstearylammonium chloride; benzyl stearyldimethylammonium chloride; dimethyl benzyloctadecylammonium chloride; N-octadecyl-N-benzyl-N,N-dimethylammonium chloride; octadecyldimethylbenzylammonium chloride; stearalkonium chloride; stearyldimethylbenzylammonium chloride; tallow benzyldimethyl ammonium chloride; Ammonyx 4; Ammonyx 485; Ammonyx 490; Ammonyx 4002; Ammonyx ca Special; 2b; Barquat Sb-25; Carsoquat Sdq-25; Carsoquat Sdq-85; Intexan Sb-85; Intexsan Sb-85; J Soft C 4; Katamine AB; Orthosan Mb; Quaternol 1; Stebac; Stedbac; Triton X-40; Triton X-400; Varisoft Sdc; Arquad Dm18b-90; Dehyquart Stc-25; Nissan Cation S2-100; Benzenemethaminium, N-octadecyl-n,n-dimethyl, Chloride; Dimethyl-n-octadecylbenzenemethanaminium Chloride,. Stearalkonium chloride and additives. Insoluble in H_2O, soluble in EtOH, organic solvents; LD_{50} (rat orl) = 1250 mg/kg. *Rhône-Poulenc Surf.*

7543 Cyclovertal
3,6-Dimethyl-3-cyclohexene-1-carbaldehyde
General fragrance raw material, green, fruity notes. *Henkel/Cospha.*

7544 Cycloxydim
101205-02-1
$C_{17}H_{27}NO_3S$
2-(1-(ethoxyimino)butyl)-3-hydroxy-5-(tetrahydro-2H-thiopyran-3-yl)-2-cyclohexene-1-one
Laser; Stratos; Focus; BAS-517H. Selective herbicide, absorbed primarily by leaves. Inhibits mitosis. Used for post-emergence control of annual and perennial grasses in broad-leaved crops, such as beans and potatoes. mp = 36°; soluble in H_2O (88 mg/l), freely soluble in organic solvents; LD_{50} (rat orl)= 3940 mg/kg.

7545 Cycogan
999-81-5 2153 213-666-4
$C_5H_{13}Cl_2N$
2-chloroethyltrimethyl-ammonium chloride
chlormequat. Active ingredient: chlormequat chloride, a versatile plant growth regulant widely used for the prevention of lodging in wheat. *Agan Chemical Manufacturers Ltd.*

7546 Cycolac®
ABS thermoplastic compounds. *GE Plastics Ltd.*

7547 Cycolac® CKM1
9003-56-9
ABS resin, flame retardant; UL-recognized grade offering uv stability, high flow, excellent finished part esthetics; for business machines, computer equipment and electrical applications. *GE Plastics.*

7548 Cycolac® DH
9003-56-9
ABS resin; medium-impact, high heat, modulus, and tensile grade for injection molding applications. *GE Plastics.*

7549 Cycolac® GPM4700
9003-56-9
ABS resin; general purpose grade with enhanced flow and more economical processing. *GE Plastics.*

7550 Cycolac® GPX2800
9003-56-9
ABS resin; extrusion grade resin for bathtub surrounds, recreational vehicle parts, high-abuse products (luggage, automotive components). *GE Plastics.*

7551 Cycolac® KCS
9003-56-9
ABS resin, flame-retardant; extrusion grade resin with superior thermal stability; ideal for thermoformed applications. *GE Plastics.*

7552 Cycolac® KJM
9003-56-9
Polyacrylonitrile-co-butadiene

ABS resin, flame-retardant; injection-molding polymer for many applications previously designed for flame retardant structural foam. *GE Plastics.*

7553 Cycolac® X-11
9003-56-9
ABS resin; high heat injection molding grade used in building products and interior automotive trim. *GE Plastics.*

7554 Cycolin® GCM1900, GCM2900
Chemically resistant G series; excellent processability, impact strength and heat resistance *GE Plastics.*

7555 Cycoloy® C1110
PC/ABS alloy; general purpose injection molding grade formulated to maintain impact and ductility below -30°; provides high heat resistance and impact strength and still processes well into long flow lengths and complicated parts. *GE Plastics.*

7556 Cycoloy® C2800
PC/ABS alloy; formulated for flame retardance without chlorinated or brominated additives; superior flow for thin-wall sections. *GE Plastics.*

7557 Cycoloy® MC8100
PC/ABS alloy; designed for extrusion blow molding. *GE Plastics.*

7558 Cycom
Resin impregnated glass, carbon or aramid fabrics; advanced composite molded components for military and aerospace applications. *Fothergill Tygaflor Ltd.* Name unverified.

7559 Cycom® MCG Fiber
Nickel-coated graphite fiber; for integrally heated tooling applications. *Am. Cyanamid.*

7560 Cydril
Cationic and anionic modified polyacrylamides. *Cyanamid BV.*

7561 Cy-Ex
Polyacrylate; used for oil field applications. *Cyanamid BV.*

7562 Cyfloc 6000
Modified polyamine; used for oil field applications. *Cyanamid BV.*

7563 cyfluthrin
68359-37-5 2826 269-855-7
$C_{22}H_{18}Cl_2FNO_3$
cyano(4-fluoro-3-phenoxyphenyl)methyl 3-(2,2-dichloroethenyl)-2,2-dimethyl cyclopropanecarboxylate
Baythroid; BAY-FCR 1272; FCR 1272; Sofac; Baythroid 2; Tempo 2; BAY-V1 1704; Baythroid H; Cyfoxylate; Eulan SP; Solfac. Non-systemic insecticide with contact and stomach action. Used for control of chewing and sucking insects on oilseed rape, cereals, ornamentals, maize, cotton etc. n_D^{23} = 1.5511; almost insoluble in H_2O; LD_{50} (rat orl) = 500-800 mg/kg. (1R,3R,αR) form: $[\alpha]_D^{20}$ = -15°; (1R,3R,αS form): $[+1a]_D^{20}$ = 24.5°; mp = 50-52°; (1R,3S,αS form): $[\alpha]_D^{20}$ = -2°; mp = 68-69°. *Bayer.*

7564 Cyglas®
Thermoset polyester composites; molding compounds for industrial applications requiring electrical insulation, high strength, rigidity at elevated temperatures and corrosion resistance *Am. Cyanamid.* Name unverified.

7565 Cygna
Self-emulsifiable mineral oil based and water soluble synlube based products; fiber processing aids and lubricants. *Thomas Swan & Co Ltd.* Discontinued.

7566 Cy-Guard
Modified polyacrylamide; used for oil field applications. *Cynamid BV.*

7567 λ-cyhalothrin
91465-08-6
$C_{23}H_{19}ClF_3NO_3$
[1α(S),3α(Z)]-(±)-cyano-(3-phenoxyphenyl)methyl 3-(2-chloro-3,3,3-trifluoro-1-propenyl)-2,2-dimethylcyclopropanecarboxylate
Charge; Commodore; Excalibur; Hallmark; Icon; Karate; Matador; PP 321; Saber; Sentinel. Non-systemic insecticide with contact and stomach action. Used for control of a wide variety of insects. mp = 48°; almost insoluble in H_2O (0.005 mg/l), very soluble in organic solvents; LD_{50} (rat orl) = 79 mg/kg.

7568 cyhexatin
13121-70-5 2829 236-049-1
$C_{18}H_{34}OSn$
tricyclohexylhydroxystannane
tricyclohexylstannoltricyclohexyltin hydroxide; Acarstin; Aracnol F; ENT-27395; Dowco 213; Mitacid; Plictran; TCTH; Tetran; Triran. Non-systemic acaricide with contact action. Used for control of mites on fruit crops. mp = 195-198°; insoluble in H_2O (<1 mg/l) , soluble in organic solvents; LD_{50} (rat orl) = 540 mg/kg.

7569 Cykelin
Drying oils for use in paints, varnishes, lacquers and enamels. *Reichhold.*

7570 Cylence
Parasiticide for the control of lice in cattle and sheep; also known as Synticol. *Bayer AG.*

7571 Cylink ISOBU-M-AMD
N-(Isobutoxymethyl)acrylamide
Used in the process industry. *Cyanamid BV.*

7572 Cylink M.B.A
N,N'-Methylenebisacrylamide
Used in the process industry. *Cyanamid BV.*

7573 Cylink NBMA
N-(Butoxymethyl)acrylamide
Used in the process industry. *Cyanamid BV.*

7574 Cylink NM-AMD
N-Methylolacrylmide
Used in the process industry. *Cyanamid BV.*

7575 Cylok® GM
Cyanoacrylate adhesive; for high tensile strength bonds of metals and rubber to metal; bonds sl. dirty or oily metals. *Lord.*

7576 Cylok® R
Cyanoacrylate adhesive; splice adhesive for buttand-mitre cut joints and O-ring repairs; adheres to natural and synthetic rubbers. *Lord.*

7577 Cymag
143-33-9 8750 205-599-4
Sodium cyanide
Poisonous gassing compound containing sodium cyanide; a rodent, rabbit and insect exterminator. *ICI Chem & Polymers Ltd.*

7578 Cymbal metal
A brass containing 78% copper and 22% zinc.

7579 Cymbilide
Cyclamen aldehyde. *May & Baker Ltd.* Name unverified.

7580 Cymbush
52315-07-8 2836 257-842-9
Cypermethrin
Insecticide containing cypermethrin. *ICI Chem & Polymers Ltd.*

7581 Cymogran
Phenylalanine-free casein hydrolysate. *Allen & Hanburys Ltd.* Name unverified.

7582 cymoxanil
57966-95-7 261-043-0
$C_7H_{10}N_4O_3$
2-Cyano-N-[(ethylamino)carbonyl]-2-(methoxyimino)acetamide
1-(2-Cyano-2-methoxyiminoacetyl)-3-ethylurea; Curzate; DPX-3217;. Foliar fungicide with protective and curative action. Used for control of Peronosporates, particularly *Peronospora, Phytophthora* and *Plasmopara* spp. mp = 160-161°; d_{25} = 1.31; soluble in H_2O (1 g/l), freely soluble in organic solvents; LD_{50} (rat orl) = 1196 mg/kg. *DuPont.*

7583 Cymperator
52315-07-8 2836 257-842-9
Cypermethrin
Insecticide containing cypermethrin. *ICI Chem & Polymers Ltd.*

7584 Cymyl orange
An indicator. It is an azo dye obtained by combining diazotized sulfonated amino-cymene with dimethyl-aniline.

7585 Cynorex
A bright cyanide copper plating process. *Hanshaw Chemicals.* Unverified.

7586 Cypan
Hydrolyzed polyacrylonitrile
Used for oil field applications. *Cyanamid BV.*

7587 Cyperkill
52315-07-8 2836 257-842-9
Cypermethrin
An emulsifiable concentrate containing 50 g cypermethrin per liter; a pyrethroid insecticide. *Mitchell Cotts Chemicals Ltd.*

7588 cypermethrin
52315-07-8 2836 257-842-9
$C_{22}H_{19}Cl_2NO_3$
Cyclopropanecarboxylic acid, 3-(2,2-dichloro ethenyl)-2,2-dimethyl-, cyano (3-phenoxyphenyl) methyl ester
Barricade, Cymbush; FMC 30980; FMC 45806; Imperator; Kafil; PP 383; Ripcord; WL 43467; Aimcocyper; Ammo; Arrivo; Basathrin; CCN52; Cymperator; Cynoff; Cyper; Cypercopal; Cyperguard; Cyperkill; Cyperscet; Cypertox; Demon; Fenom; Flectron; Fligene CI; Folcord; Halt; LE 79600; NRDC 149; Nurelle; Polytrin; Prevail; Ralothrin; Sherpa; Siperin; Stockade; Toppel; Ustaad; WL 43467. Non-systemic insecticide with contact and stomach action. Used to control wide range of insects, e.g. *lepidoptera, coleoptera, diptera* and *hemiptera.* mp = 60-80°; d^{20}= 1.25; insoluble in H_2O, soluble in organic solvents; LD_{50} (rat orl) = 70 mg/kg.

7589 Cypersect
52315-07-8 2836 257-842-9
Cypermethrin

Emulsifiable concentrate containing 100 g cypermethrin per liter; a pyrethroid insecticide. *Barclay Chemicals UK.*

7590 Cypertox

52315-07-8 2836 257-842-9

Cypermethrin

100 g/l Cypermethrin; broadspectrum insecticide with many crop uses. *Farmers Crop Chemicals Ltd.*

7591 Cypro Promoters

Polyamines; used in paper industry. *Cyanamid BV.*

7592 Cyquest

Modified polyacrylamides; antiprecipitant for mining applications. *Cyanamid BV.*

7593 Cyracure

Ultra-violet curing agents. *Union Carbide; Union Carbide (UK) Ltd.*

7594 Cyrene

A resin used in adhesives, coatings, and molding compounds.

7595 Cyrez 963

Hexamethoxymethylamine

Adhesion promoter; used in rubber industry. *Cyanamid BV.*

7596 Cyrez 963/4 Powders

Hexamethoxymethylamine on silica/silicate carrier; adhesion promoter; used in rubber industry. *Cyanamid BV.*

7597 Cyrolite® G-20

Acrylic multipolymer; compound for extrusion, blow molding, vacuum forming, and injection molding applications where transparent material with good chemical resistance is required; ideal for food and medical packaging; FDA approved for food confact use. *Cyro industries.*

7598 Cyrolite® G-20 Hiflo®

Acrylic multipolymer; molding/extrusion compound suited for hard-to-fill molds, small intricate moldings, as well as large-area molding; FDA approved for food contact use. *Cyro Industries.*

7599 Cyrolon® UVP

PC sheet; weather-resistance sheet for glazing applications where long-term impact resistance, high service temperatures, and optical clarity are required. *Cyro Industries.*

7600 Cyrolon® ZX

PC sheet; continuously manufacturing lightweight, rigid sheet for applications requiring excellent impact resistance or high service temperatures; high optical quality, dimensional stability; used for safety glazing, machine guards, panels for vending machines, signs, RV windscreens and windows, displays. *Cyro Industries.*

7601 cysteine hydrochloride

52-89-1 2850 200-157-7

$C_3H_7NO_2S \cdot HCl$

L-cysteine hydrochloride anhydrous

Essential amino acid. Used as a dough conditioner. mp = 175-178° (dec); $[\alpha]_D^{25}$ = 5° (5N HCl); soluble in H_2O, organic solvents. Bretagne Chimie Fine SA; Degussa; EM Industries; R.W. Greeff; Nippon Rikagakuyakuhin; Penta Mfg.; Tanabe USA; U.S. BioChemical.

7602 L-cysteine

52-90-4 2850 200-158-2

$C_3H_7NO_2S$

(+)-2-amino-3-mercaptopropionic acid

α-amino-lb-thiolpropionic acid. A nonessential amino acid; biochemical and nutrition research, reducing agent in bread doughs. *Diamalt GmbH; Nippon Rikagakuyakuhin; Showa Denko.* Name unverified.

7603 cystine

56-89-3 2851 200-296-3

$C_6H_{12}N_2O_4S_2$

di(α-amino-β-thiolpropionic acid

β,β'-dithiobisalanine; 3,3'-dithiobis(2-aminopropionic acid); dicysteine; β,β'-dithiodialanine; α-diamino-lb-dithiolactic acid; β,β'-diamino-β,β'-dicarboxydiethyl disulfide; bis(β-amino-β-carboxyethyl) disulfide; gelucystine (L form). Biochemical and nutrition research, nutrient and dietary supplement. mp = 260-261° (dec); $[\alpha]_D^{20}$ = -223.4° (1.0N HCl); soluble in H_2O, insoluble in EtOH. Aldrich; Am. Biorganics; Bretagne Chimie Fine SA; Degussa; R.W. Greeff; Tanabe USA; U.S. BioChemical.

7604 Cy-Temp

Modified polyacrylamide; used for oil field applications. *Cyanamid BV.*

7605 Cytop

Fluoropolymer

Transparent, low refractive index, durable to uv light, good coating properties; as clad material for optics, electronics, chemical industry applications. *Asahi Glass.* Name unverified.

7606 Cytro-Lane

950-10-7 5900 213-447-3

Mephosfolan

Emulsifiable concentrate containing 250 g/l mephosfolan; used for control of damsonhop aphid in hops. *Cyanamid of Great Britain Ltd.* Name unverified.

7607 D S M

8022-00-2 6129

demeton-S-methyl

Demetox. Preparation containing 50% demeton-S-methyl; insecticide. *L W Vass (Agricultural) Ltd.* Discontinued.

7608 D.D.D.

Dimethylamine dimethyldithiocarbamate

Used as an accelerator for rubber vulcanization.

7609 D.E.H

Epoxy curing agents (hardeners) specifically designed to provide a desirable range of properties and handling characteristics for the hardening of epoxy resins. *Dow Cheml Co Ltd, UK & Ireland.*

7610 D.E.H. 20

111-40-0 203-865-4

$C_4H_{13}N_3$

Diethylenetriamine

2,2'-Diaminodiethylamine; DETA; N-(2-aminoethyl)-1,2-Ethanediamine; aminoethylethanediamine; 3-azapentane-1,5-diamine; bis(2-aminoethyl)amine; bis(beta-aminoethyl)amine; 2,2'-iminobisethylamine; N-(2-aminoethyl)ethylenediamine; β,β'-diaminodiethylamine; Diethylenetriamine. Aliphatic polyamine curing agent for epoxy resins; used for civil engineering, adhesives, grouts, casting and electric encapsulation; D.E.H. 20 is a general purpose curing agent. mp = -33°; bp = 206-209°; d = 0.9510; n_D^{20} = 1.4826; insoluble in H_2O, soluble in organic solvents; LD_{50} (rat orl) = 1080 mg/kg. *Dow Chemical.*

7611 D.E.H. 40

Accelerated dicyandiamide; epoxy curing agent for powder coating formulations. *Dow Chemical.*

7612 D.E.H. 52

111-40-0 203-865-4

Diethylenetriamine adduct

Adducted aliphatic polyamine curing agent for epoxy resins. *Dow Chemical.*

7613 D.E.N

Epoxy novolac resins and solutions designed to provide high temperature service for epoxy-type applications; used for coatings and adhesives for abrasives. *Dow Cheml Co Ltd, UK & Ireland.*

7614 D.E.N. 431

Epoxy-novolac resin; used in high-performance adhesives, structural and electrical laminates, potting and molding compounds, coatings and castings for elevated temperature service, and filament wound pipe; used primarily in electrical applications and as an additive to conventional epoxies; low viscosity. *Dow Chemical.*

7615 D.E.R

Epoxy resins; a range of solid, liquid, flexible and brominated thermosetting polymers; features adhesion, hardness, flexibility, toughness, dimensional stability, clarity and chemical resistance. *Dow Cheml Co Ltd, UK & Ireland.*

7616 D.E.R. 317

25928-94-3

Bisphenol-A epoxy resin; ideal for adhesive, casting, potting, encapsulation, and wet lay-up applications; after cure, yields highly crosslinked thermoset polymers; a high-viscosity, fast-reacting resin for adhesive applications. *Dow Chemical.*

7617 D.E.R. 362

25928-94-3

Bisphenol A epoxy resin; low viscosity liquid epoxy resin for applications requiring high filler loading, excellent wet-out, and high performance; crystallization resistance; no diluents or solvents; used for adhesives, flooring, encapsulants, coatings, composites. *Dow Chemical.*

7618 D.E.R. 383

25928-94-3

Epoxy resin; liquid epoxy resin for reduced viscosity and extended pot life of catalyzed formulations; for filament winding, potting and encapsulation, etc.; yields higher filler loadings, better process control and handling, and reduced resin/hardener waste. *Dow Chemical.*

7619 D.E.R. 511-A80

Brominated epoxy resin. *Dow Chemical.*

7620 D.E.R. 642U

25928-94-3

Epoxy resin; for powder coatings. *Dow Chemical.*

7621 D.E.R. 732

25928-94-3

D.E.R. 642U; D.E.R.362; D.E.R 317; D.E.R 383. Epichlorohydrin-polyglycol reaction product; flexible epoxy resin; used as additives to base epoxy systems; increases flexibility in adhesives, construction and civil engineering applications (aggregates, seamless floors, machine grouts), industrial maintenance coatings. *Dow Chemical.*

7622 D.N.T
121-14-2 204-450-0
$C_7H_6N_2O_4$
2,4-Dinitrotoluene
Dinitrotoluene; DNT; 2,4-DNT; Dinitrotoluol; 1-Methyl-2,4-Dinitrobenzene; 2,4-dinitrotoluol. Used in synthesis of toluidines, dyes and explosives. mp = 66-68°; bp = 300°; insoluble in H_2O, solble in organic solvents; LD_{50} (rat orl) = 268 mg/kg.

7623 D.O.T.T
$C_{15}H_{16}N_2S$
Di-*o*-tolylthiourea
Used as a metal pickling inhibitor.

7624 D.P.G
102-06-7 3383 203-002-1
Diphenylguanidine
A rubber vulcanization accelerator. No manufacturer.

7625 D.S.H.C
128952-18-1
Dimethylsilanol hyaluronate
Provides skin regeneration, strong hydrating action for cosmetic and health products such as milks, emulsions, creams, lotions. *Exsymol.*

7626 D.T.S.
130-17-6 204-979-7
$C_{14}H_{12}N_2O_3S_2$
Dehydrothio-*p*-toluidinesulfonic acid.
2-(4-aminophenyl)-6-methylbenzothiazole sulfonic acid; 2-(*p*-aminophenyl)-6-methyl-7-benzothiazolesulfonic acid. Chemical intermediate. mp > 300°; soluble in DMSO, insoluble in H_2O, organic solvents; LD_{50} (mus iv) = 178 mg/kg.

7627 D.X.L.
High boiling tar acids. *Coalite Fuels & Chemicals Ltd.* Name unverified.

7628 D-129
Propylene/hexene copolymer adhesives, and sealants raw materials. *Eastman.*

7629 D12F
9005-25-8 232-679-6
Oxidized corn starch. *Grain Processing.*

7630 D-151
Propylene/ethylene copolymer, adhesives, and sealants raw materials. *Eastman.*

7631 D1-SEA 210 Silicone
63394-02-5
Two-component silicone elastomeric adhesive; cures fast to a tough, durable, resilient silicone rubber at room temperature with primerless adhesion to many substrates; for deep sections, automotive headlamps, as gasketing material for clothes iron steam chambers. *GE Silicones.*

7632 D-201
1344-28-1 369 215-691-6
Activated alumina; abrasion-resistant desiccant and drying agent for use in pressure swing and other packaged dryers; also for drying organic liquids and gases in industrial processes, and air in air conditioning systems; desiccant/adsorbent in reconditioning oils. *La Roche Chem.*

7633 D2T2
Dye diffusion thermal transfer process for printing. *Imperial Chemical Industries plc.*

7634 D-300 Conc
Electrolyte used to produce hard and durable architectural and engineering anodized surfaces. *Pilot.*

7635 D3770
541-02-6 2905 208-764-9
$C_{10}H_{30}O_5Si_5$
Decamethylcyclopentasiloxane
As cleaning, polishing, and damping media; offers low toxicity, inertness. *Hüls Am.*

7636 D3780
141-62-8 2907 205-491-7
$C_{10}H_{30}O_3Si_4$
Decamethyltetrasiloxane
As cleaning, polishing and damping media; offers low toxicity, inertness. mp = -68°; bp = 194°; d = 0.8540; n_D^{20} = 1.3888. *Hüls Am.*

7637 D6219.5
Dodecamethylpentasiloxane
As cleaning, polishing and damping media; offers low toxicity, inertness. *Hüls Am.*

7638 D-7040
Silicone fluid; high purity diffusion pump fluid. *McGhan NuSil.*

7639 DAB
91-95-2 202-110-6
$C_{12}H_{14}N_4$
3,3'-Diaminobenzidine
3,3',4,4'-Biphenyltetramine. A proprietary intermediate for various high-temperature plastics, used to make polypyrones and polyquinoxalines. mp = 172-174°; soluble in H_2O (0.55 g/l), more soluble in organic solvents; LD_{50} (mus orl) = 1834 mg/kg, carcinogen. *Upjohn Ltd.* Name unverified.

7640 Dabco® 33-LV
280-57-9 9801 205-999-9
33% Triethylenediamine in dipropylene glycol; amine-based catalyst for PU high resiliency slabstock and molded foam, coating. *Air Prods & Chem/Polyurethanes.*

7641 Dabco® 120
Metal-based catalyst for PU flexible molded foam. *Air Prods & Chem/Polyurethanes.*

7642 Dabco® 1027, 1028
Catalysts for improved processing in microcellular polyurethane foams. *Air Prods & Chem/Polyurethanes.*

7643 Dabco® 7928
Amine-based catalyst for flexible slabstock PU foam; optimized for conventional equipment. *Air Prods & Chem/Polyurethanes.*

7644 Dabco® 8154
Amine-based catalyst for improved processing of flexible slabstock PU foam; gives delayed action. *Air Prods & Chem/Polyurethanes.*

7645 Dabco® 8264
Amine-based catalyst for improved processing of flexible slabstock PU foam. *Air Prods & Chem/Polyurethanes.*

7646 Dabco® B-16
103-83-3 203-149-1
$C_9H_{13}N$
N,N-dimethylbenzenemethanamine
N,N-dimethylbenzylamine; BDMA; benzyl dimethylamine; N-benzyl-N,N-dimethylamine; N-benzyldimethylamine; Dimethyl benzylamine. Amine-based catalyst for flexible slabstock PU foam. mp = -75°; bp_{765} = 183-194°; d = 0.9000; n_D^{20} = 1.5011; insoluble in H_2O, soluble in organic solvents; LD_{50} (rat orl) = 265 mg/kg. *Air Prods & Chem/Polyurethanes.*

7647 Dabco® BDO
110-63-4 203-786-5
$C_4H_{10}O_2$
1,4-Butanediol
1,4-Butylene glycol; 1,4-Dihydroxybutane; 1,4-Tetramethylene; Tetramethylene glycol; butanediol; butane-1,4-diol; diol 14b; Sucol b; 1,4-tetramethylene glycol; butylene glycol; tetramethylene 1,4-diol. Curative, chain extender; provides reactive H-source in prepolymer production; used to provide hard segments in PU. mp = 16°; bp = 230°; d = 1.0170; n_D^{20} = 1.4452; soluble in H_2O, organic solvents; LD_{50} (rat orl) = 1525 mg/kg. *Air Prods & Chem/Polyurethanes.*

7648 Dabco® BL-11
70% bis(dimethylaminoethyl) ether, 30% dipropylene glycol. Amine-based catalyst for flexible slabstock PU foam. *Air Prods & Chem/Polyurethanes.*

7649 Dabco® BL-17
Amine-based catalyst for improved processing of flexible slabstock PU foam; gives delayed action version of Dabco® BL-11. Liquid; soluble in H_2O; d = 1.04; vapor pressure = 0.61 mm Hg; hyd no 476; flash point = 65°. *Air Prods & Chem/Polyurethanes.*

7650 Dabco® BL-22
Blowing catalyst for polyurethane; designed for use as co-catalyst with Dabco 33-LV, etc. Liquid; miscible with H_2O; d = 0.8304; viscosity = 255 cps; vapor pressure = 5 mm Hg; fp <-78°; bp = 204°; hyd no 0; flash point = 70°; pH=11.3. *Air Prods & Chem/Polyurethanes.*

7651 Dabco® BLV
Amine-based catalyst for flexible slabstock PU foam; optimized preblend. Optimized blend of 3:1 blend of Dabco 33-LV and Dabco Bl-11. *Air Prods & Chem/Polyurethanes.*

7652 Dabco® Crystalline
280-57-9 9801 205-999-9
Triethylenediamine
Catalyst for PU coatings. *Air Prods & Chem/Polyurethanes.*

7653 Dabco® CS90
71-55-6 9766 200-756-3
Trichloroethane
Catalyst for chlorinated solvent-blown foams *Air Prods & Chem/Polyurethanes.*

7654 Dabco® DC198
Silicone glycol copolymer; high-potency surfactant used in production of flexible slabstock polyurethane foam. *Air Prods & Chem/Polyurethanes.*

7655 Dabco® DC5043
Silicone glycol copolymer; surfactant for production of molded and slabstock

high-resiliency polyurethane foam; nonhydrolyzable; broad processing latitude. *Air Prods & Chem/Polyurethanes.*

7656 Dabco® DC5125

Silicone glycol copolymer; wide-processing-latitude surfactant for production of conventional and flame-retarded flexible polyurethane slabstock foam systems. *Air Prods & Chem/Polyurethanes.*

7657 Dabco® DC5160

Silicone glycol copolymer; surfactant used in production of conventional and flame-retarded slabstock polyurethane foam. *Air Prods & Chem/Polyurethanes.*

7658 Dabco® DC5164

Silicone product; surfactant for production of difficult-to-stabilize high-resiliency molded PU foam formulations. Clear to slightly hazy liquid; insoluble in H_2O; d = 1.005; viscosity = 305 cps; vapor pressure 20.8 mm Hg; fp = -53°; hyd no 20; flash point = 92°. *Air Prods & Chem/Polyurethanes.*

7659 Dabco® DC5169

Silicone product; surfactant for production of high-resiliency MDI-based PU molded foams. Insoluble in H_2O; d = 0.968; viscosity = 29.2 cps; vapor pressure 8.6 mm Hg; fp = -54°; hyd no 0; flash point >113°. *Air Prods & Chem/Polyurethanes.*

7660 Dabco® DC5180

Surfactant for flexible slabstock PU foam. *Air Prods & Chem/Polyurethanes.*

7661 Dabco® DC5258

Surfactant; provides for maximum airflow in polyurethane foam without sacrificing surf. stability. Liquid; insoluble in H_2O; d = 1.020; viscosity = 289 cps; vapor pressure 8.9 mm Hg; fp = -55°; hyd no 46; flash point = 92°. *Air Prods & Chem/Polyurethanes.*

7662 Dabco® DC5365

Silicone surfactant; low-to-medium efficiency stabilizer for TDI/MDI flexible mold foams. Liquid; insoluble in H_2O; d = 1.000; viscosity = 290 cps; vapor pressure = 7.7 mm Hg; fp = -60°; hyd no 170; flash point >113°. *Air Prods & Chem/Polyurethanes.*

7663 Dabco® DC5425

Silicone surfactant; for flexible molded foams; medium-efficiency stabilizer for water-blown TDI-based molded foams. Liquid; insoluble in H_2O; d = 0.980; viscosity = 57.1 cps; vapor pressure 6.0 mm Hg; fp= -65°; hyd no 14; flash point = 78°. *Air Prods & Chem/Polyurethanes.* Name unverified.

7664 Dabco® DC5450

Silicone surfactant; surfactant for waterblown MDI-based PU foams; provides enhanced bulk stability. Liquid; insoluble in H_2O; d = 1.019; viscosity = 302 cps; vapor pressure = 9.0 mm Hg; fp = -35°; hyd no 85; flash point >113°. *Air Prods & Chem/Polyurethanes.*

7665 Dabco® DC5885

Silicone surfactant; for flexible molded foams; med.-efficiency stabilizer for water-blown TDI-based molded foams. Liquid; insoluble in H_2O; d = 0.984; viscosity = 68.5 cps; vapor pressure <5.0 mm Hg; fp = -63°; hyd no 123; flash point = 74°. *Air Prods & Chem/Polyurethanes.* Name unverified.

7666 Dabco® DC5890, DC5895

Surfactant for flexible slabstock PU foam processing. Liquid; insoluble in H_2O; d = 1.0; viscosity = 46.8 cps; vapor pressure = 5 mm Hg; fp = -69°; flash point = 156°. *Air Prods & Chem/Polyurethanes.*

7667 Dabco® DEOA-LF

111-42-2 3156 203-868-0

Diethanolamine

Crosslinker for flexible slabstock PU foam processing. *Air Prods & Chem/Polyurethanes.*

7668 Dabco® DM9534, DM9793

Metal-based catalyst for flexible slabstock PU foam. *Air Prods & Chem/Polyurethanes.*

7669 Dabco® DMEA

108-01-0 2900 203-542-8

$C_4H_{11}NO$

N,N-Dimethyl-2-Hydroxyethylamine

β-Dimethylaminoethyl alcohol; Deanol; 2-dimethylaminoethanol; Dimethylethanolamine; DMAE; dimethylaminoethanol; N,N-Dimethylaminoethanol; N-dimethylaminoethanol; N,N-dimethyl-N-(2-hydroxyethyl)amine; β-hydroxyethyldimethylamine. Amine-based catalyst for flexible slabstock PU foam. *See Deanol Air Prods & Chem/Polyurethanes.*

7670 Dabco® MC

Catalyst for improved processing of flexible slabstock PU foam; for methylene chloride systems. Liquid; soluble in H_2O; d = 1.023; viscosity = 63.5 cps; vapor pressure 1.0 mm Hg; fp = -66.7°; hyd no 757; flash point = 84°. *Air Prods & Chem/Polyurethanes.*

7671 Dabco® NCM, NEM, NMM

100-74-3; 109-02-4 202-885-0; 203-640-0

N-ethyl morpholine and N-methyl morpholine; amine-based catalyst for flexible slabstock PU foam. *Air Prods & Chem/Polyurethanes.*

7672 Dabco® T-10, T-11

Metal-based catalyst for flexible slabstock PU foam. Liquid; insoluble in H_2O; d = 1.02; viscosity = 102 cps; flash point = 93.3°. *Air Prods & Chem/Polyurethanes.*

7673 Dabco® T-12

77-58-7 3089 201-039-8

$C_{32}H_{64}O_4Sn$

Dibutyltin dilaurate

Di-n-butyltin Dilaurate; dibutylbis[(1-oxododecyl)oxy]stannane; Dibutyltin didodecanoate; bis(lauroyloxy)di(n-butyl)stannane; butyl norate; DBTL; dibutylbis(lauroyloxy) tin; lauric acid, dibutylstannylene derivative; laudran; lauric acid, dibutyltin derivative; stabilizer D-22; bis(dodecanoyloxy)di-n-butylstannane; tin dibutyl dilaurate; tinostat; Di-n-butyldilauryltin. Metal-based catalyst for PU flexible molded foam, coatings. mp = 22-24°; d = 1.066; n_D^{20} = 1.4683; insoluble in H_2O, soluble in organic solvents; LD_{50} (rat orl) = 175 mg/kg. *Air Prods & Chem/Polyurethanes.*

7674 Dabco® T-95

Metal-based catalyst for flexible slabstock PU foam. Liquid; slightly soluble in H_2O; sg = 1.10; viscosity = 101cps; flash pt > 130°. *Air Prods & Chem/Polyurethanes.*

7675 Dabco® TL

Catalyst for improved processing of flexible slabstock PU foam; for methylene chloride systems. Liquid; soluble in H_2O; d = 0.988; viscosity = 40 cps; vapor pressure 20.1 mm Hg. *Air Prods & Chem/Polyurethanes.*

7676 Dabco® X2-5357, X2-5367

Surfactants for reduced CFC and non-CFC blown rigid polyurethane foam systems. Gardner 3 color; d = 1.04; viscosity = 400 cps; flash pt > 61°; hyd no 31. *Air Prods & Chem/Polyurethanes.*

7677 Dabco® X-542, X-543

Amine-based catalyst for PU flexible molded foam; provides fast cure. Liquid; soluble in H_2O; d = 1.048; vapor pressure = 8 mm Hg; hyd no 321; flash point = 88.3°. *Air Prods & Chem/Polyurethanes.*

7678 Dabco® XDM™

Amine-based catalyst for PU flexible molded foam; for improved surface cure. Liquid; soluble in H_2O; d = 0.942; viscosity = 2.7 cps; vapor pressure = 3 mm Hg; flash point = 76°. *Air Prods & Chem/Polyurethanes.*

7679 Dabco® XF-C10-40

Amine-based catalyst for PU flexible molded foam; provides fast cure. *Air Prods & Chem/Polyurethanes.*

7680 Dabco® XF-F2002, XF-F2003

Amine catalyst; for all-water-blown methylene -*p*-phenyl di-isocyanate cold cure seating process in polyurethane production. *Air Prods & Chem/Polyurethanes.*

7681 Daconil Turf

1897-45-6 2219 217-588-1

Chlorothalonil

A fungicide for a wide range of agricultural crops. *ICI Agrochemicals Professional Products.*

7682 Dacospin 1735-A

PEG-30 castor oil glycerides; surfactant, lubricant for PP carpet backing giving excellent fiber-to-fiber lubricity and plasticizing properties. *Henkel/Textiles.*

7683 Dacospin LA-704

PEG-7/PPG-4 tridecyl alcohol; surfactant for textile use. *Henkel/Textiles.*

7684 Dacron®

A polyethylene terephthalate fiber having high strength and low water absorption. *DuPont UK.*

7685 Dacthal

1861-32-1 2896 217-464-7

$C_{10}H_6Cl_4O_4$

Dimethyl tetrachloroterephthalate

DCPA. Wettable powder containing 75% w/w dimethyl tetrachloroterephthalate (DCPA); an herbicide for use on ornamentals, turf, fruit and vegetables. *Fermenta ASC Europe Ltd.*

7686 Dag 137

7782-42-5 4560 231-955-3

Colloidal graphite in water; lubricant additive for dry gear, dry chain lubes, aerosols, machine oils, assembly lubes, thread lubes. *Acheson Colloids.*

7687 Dag 154

7782-42-5 4560 231-955-3

Colloidal graphite in anhydrous isopropanol; lubricant additive for dry gear and chain lubes, aerosols, machine oils, assembly and thread lubes. *Acheson Colloids.*

7688 Dag 155

Colloidal graphite in trichloroethane; lubricant additive for dry gear and chain lubes, aerosol, machine oils, assembly and thread lubes. *Acheson Colloids.*

7689 Dag 197

7782-42-5 4560 231-955-3

Colloidal graphite in polyglycol synthetic oil; lubricant additive for chain lubes. *Acheson Colloids.*

7690 Dag 243
Graphite/MoS_2 in 500 solvent refined paraffinic petroleum oil; lubricant additive for aerosols, thread lubes. *Acheson Colloids.*

7691 Dag 2412
Colloidal graphite in mineral spirits; lubricant additive for assembly and thread lubes. *Acheson Colloids.*

7692 Dagenite
A proprietary trade name for a bituminous asbestos-filled thermoplastic for accumulator cases. No manufacturer.

7693 Dagger
81405-85-8
imazamethabenz-methyl
Suspension concentrate containing 300 g imazamethabenz-methyl per liter; used for control of grass weeds in winter cereals. *Cyanamid of Great Britain Ltd.*

7694 Dahl's acid II
85-74-5 6488 201-629-5
$C_{10}H_9NO_6S_2$
α-naphthylamine-4,6-disulfonic acid
Intermediate in synthesis of dyestuffs. Soluble in H_2O, EtOH.

7695 Dahl's acid III
85-75-6 6489 201-630-0
$C_{10}H_9NO_6S_2$
α-naphthylamine-4,7-disulfonic acid.
Intermediate in synthesis of dyestuffs. Soluble in H_2O, insoluble in EtOH.

7696 Dahmenite A
An explosive containing ammonium nitrate, naphthalene, and potassium bichromate.

7697 Daintex
A proprietary trade name for a wetting agent; contains miscible terpene alcohols. No manufacturer.

7698 Dairene®
101-05-3 694 202-910-5
Anilazine
Fungicide for the prevention of leaf and fruit diseases in vegetables, tobacco, coffee, cereals, berry fruit, and ornamental plants. *Bayer AG.*

7699 Dairos
A dairy detergent. *Laporte Industries Ltd.*

7700 Dairozon
A dairy sterilization agent. *Laporte Industries Ltd.*

7701 Dairy Fly Spray
pyrethrin
Synergized pyrethrins; insecticides. *Ciba plc.*

7702 Dairy Flyspray
Mixture of lindane, resmethrin and tetrametramethrin; controls flies in livestock houses. *Deosan Ltd.*

7703 Dakamballi starch
A starch prepared from the fruit of *Aldina insignis,* a tree of British Guiana.

7704 Dakin's solution
7681-52-9 8773 231-668-3
sodium hypochlorite-sodium perborate. A mixture of hypochlorite and perborate of sodium, with small amounts of hypochlorous and boric acids; used as an antiseptic for wound treatments.

7705 dalapon
75-99-0 2869 200-923-0
$C_3H_4Cl_2O_2$
2,2-dichloropropanoic acid
Basfapon/N; BH dalapon; Basinex; Crisapon; Davpon; Gramevin; Kenapon; Uropon; Dalapon; Dowpon Proprop; Revenge; Unipon; Alatex; S95; Dalapon 85; Basfapon. Selective systemic herbicide absorbed by roots and leaves. Used for control of annual and perennial grasses on non-crop land and also orchards and vineyards. bp_{18} = 98-100°; d = 1.4014; n_D^{20} = 1.4544; LD_{50} (rat orl) = 7126 mg/kg.

7706 Dalfratex
Flexible silica textiles. *The Chemical & Insulating Co Ltd.*

7707 Dalpad
Coalescing agent assisting film formation of certain latexes; maintains that characteristic during formulation storage and resists tendency (of films) towards water sensitivity. *Dow Cheml Co Ltd, UK & Ireland.*

7708 Daltocel
Polyesters or polyethers for flexible foams. *ICI Chem & Polymers Ltd.*

7709 Daltoflex
Polyurethane flexible foams. *ICI Chem & Polymers Ltd.*

7710 Daltofoam
Rigid polyurethane foams. *Imperial Chemical Industries plc.*

7711 Daltogard
A polyurethane foam additive. *ICI Chem & Polymers Ltd.*

7712 Daltolac
Polyurethane rigid foams. *ICI Chem & Polymers Ltd.*

7713 Daltomold
Trademark for a range of plastic molding compounds. No manufacturer.

7714 Daltoped
Polyurethane systems for shoe soling. *ICI Chem & Polymers Ltd.*

7715 Daltorez
Polyester resins for adding to polyurethane foams and elastomers. *ICI Chem & Polymers Ltd.*

7716 Daltorol
Polyester for printers rollers. *ICI Chem & Polymers Ltd.* Discontinued.

7717 Dama® 810
Dioctyl/octyldecyl/didecyl methyl amines; Intermediate for manufacture of quaternary ammonium compounds for biocides, textile chemicals, oil field chemicals, amine oxides, betaines, polyurethane foam catalysis, epoxy curing agent; in fabric softeners, disinfectants, laundry detergents. Clear liquid; fatty amine odor; d = 0.801; fp = -38.6°; flash pt (PM) = 166°; corrosive. *Ethyl Corp; Albemarle.*

7718 Dama® 1010
7396-58-9 230-990-1
Didecyl methylamine
Intermediate for manufacture of quaternary ammonium compounds for biocides, textile chemicals, oil field chemicals, amine oxides, betaines, polyurethane foam catalysis, epoxy curing agent; in fabric softeners, disinfectants and laundry detergents. *Ethyl Corp.*

7719 Damascenized steel
Steel made by repeatedly welding, drawing out, and doubling up a bar made of a mixture of steel and iron, the surface of which has been treated with an acid. The steel is with a black coating of carbon, and the iron retains its metallic luster.

7720 Damascus bronze
Damar bronze
An alloy of 76% copper, 10.5% tin, and 12.5% lead; a white bearing metal.

7721 Damfin
62610-77-9 6004
Methacrifos
Emulsifiable concentrate containing 950 g/l methacrifos; insecticide and acaricide used for pest control in stored grain. *Ciba-Geigy Agrochemicals.*

7722 Damiana
The dried leaves of a Mexican plant, *Turnera diffusa*; a tonic.

7723 daminozide
1596-84-5 2874 216-485-9
$C_6H_{12}N_2O_3$
butanedioic acid mono(2,2-dimethylhydrazide)
Dazide; DMASA; Alar; Aminozide; B-NINE; SADH; B 995; Dimas; Alar-85; Kylar; dimethylaminosuccinamic acid; NINE. Plant growth regulator, absorbed by leaves with translocation throughout the plant. Used on apples to restrict vegetative growth. mp = 157-164°; soluble in H_2O (100 g/l), less soluble in organic solvents; LD_{50} (rat orl) = 8400 mg/kg.

7724 Dammar Resin
9000-16-2 2872 232-528-4
Gum Damar; Dammar; resin Damar. A resin obtained from *Hopea, Shorea* and *Balanocarpus* spp, mainly from Malaysia; used in varnishes, cellulosic lacquers, paper. Insoluble in H_2O; soluble in EtOH, chloroform, ether, carbon disulfide, oil rosemary; the melting point usually varies from 90-200°, acid value from 33-72, and the ash from 0.04-0.52%.

7725 Damoil
Phytonomic oil
Dormant and summer spray oil for use as a contact insecticide. *Draxel Chemical Company.* Unverified.

7726 Damox
Dialkymethylamine oxides
Ethyl Corp.

7727 Danex
52-68-6 9753 200-149-3
trichlorphon
An organophosphate agricultural insecticide with a very broad range of activity. *Makhteshim Chemical Works Ltd.*

7728 Danol
17230-88-5 2875 241-270-1
Danazol
A proprietary preparation of danazol; used as a suppressant of gonadotropins. *Winthrop Laboratories.* Name unverified.

7729 Danol diols
Propoxylated, ethoxylated esters *Witco/Organics.* Discontinued.

7730 Dantocol® DHE
26850-24-8 248-052-5
DEDM hydantoin
Diethylol diethyl (DEDM) hydantoin; resin crosslinker in coatings and polymers; intermediate for epoxies, urethane resins, and antistatic lubricants for the textile and plastics industries. *Lonza Inc.*

7731 Dantogard®
Monomethylol dimethyl (MDM) hydantoin and dimethylol dimethyl (DMDM) hydantoin; industrial preservative. *Lonza Inc.*

7732 Dantoin® DCDMH
118-52-5 3115 204-258-7
1,3-Dichloro-5,5-dimethyl hydantoin
Intermediate for custom chemical synthesis, laundry bleach formulations, and automatic dishwashing compounds. *Lonza Inc.*

7733 Dantoin® GSD-550
126-06-7 204-766-9
$C_5H_6BrClN_2O_2$
1-Bromo, 3-chloro-5,5-dimethyl hydantoin
BCDMH; Bromo-1-chloro-5,5-dimethyl-2,4-imidazolinedione; Bromo-1-chloro-5,5-dimethylhydantoin; Hydantoin, 3-bromo-1-chloro-5,5-dimethyl-; Imidazolidinedione, 3-bromo-1-chloro-5,5-dimethyl-. Used for water treatment. *Lonza Inc.*

7734 Dantoin® MDMH
116-25-6 4877 204-132-1
$C_6H_{10}N_2O_3$
1-(hydroxymethyl)-5,5-dimethylhydantoin
Imidazolidinedione, 1-(hydroxymethyl)-5,5-dimethyl-; MDM hydantoin; Monomethylol-5,5-dimethylhydantoin. MDM hydantoin; intermediate for cosmetics and other applications; preservation and gelation agent. *Lonza Inc.*

7735 Daotan
Polyurethane modified alkyd resins; used for air drying paints and varnishes. *Resinous Chemicals Ltd.*

7736 Daphnetin
486-35-1 2881 207-632-8
$C_9H_6O_4$
7,8-dihydroxy-2H-1-benzopyran-2-one
1,8-Dihydroxycoumarin; 7,8-Dihydroxycoumarin. Used in manufacture of dyes and medicines. mp= 256° (dec); soluble in H_2O, less soluble in organic solvents.

7737 Dapon 35
131-17-9 205-016-3
A trademark for diallyl phthalate; a molding material. *FMC UK.* Discontinued.

7738 Dapon M
1087-21-4 214-122-9
$C_{14}H_{14}O_4$
isophthalic acid, di-(2-propenyl) ester
A trademark for diallyl isophthalate; a molding material. *FMC UK.* Discontinued.

7739 Daponite Sheet
131-17-9 205-016-3
Diallyl phthalate resin; for decorative impregnated paper. *Sumitomo Bakelite Co Ltd.* Name unverified.

7740 Dappol
Boiler and cooling water treatment. *Laporte Industries Ltd.*

7741 Dapro 5005
Special drier, chelating catalyst replacing cobalt drier in modified polyurethanes and alkyds. *Elementis Specialties.*

7742 Dapro Defoamer NA 1621
Blend of hydrocarbon liquids and metallic soaps modified with silicone polymer; defoamer for nonaqueous clear urethane wood finishes. d = 0.94; 7.8 lb/gal; flash point = 37°. *Elementis Specialties.*

7743 Dapro DF 880
Dispersion of metallic salt of fatty acid; foam suppressor for aqueous coatings, inks and adhesives. d = 0.86; 7.2 lb/gal; flash point = 48°. *Elementis Specialties.*

7744 Dapro DF 900
Dispersion of olefinic solids; foam suppressor for aqueous coatings, inks and adhesives. d =0.84; 7.0 lb/gal; flash point >149°. *Elementis Specialties.*

7745 Dapro DF 1181
Silicone-modified dispersion of olefinic solids; foam suppressor for aqueous coatings, inks, and adhesives. d = 0.85; 7.1 lb/gal; flash point >149°. *Elementis Specialties.*

7746 Dapro S-65
Surfactant blend; interfacial tension modifier for minimizing crawling, fish-eyes and cratering in solvent-based coating systems. d =0.98; 8.2 lb/gal; flash point = 47°. *Elementis Specialties.*

7747 Dapro U-99
Surfactant blend; interfacial tension modifier for minimizing crawling, fish-eyes and cratering in water and solvent-based coating systems and inks. d =1.01; 8.4 lb/gal; visc (G-H) A3; flash point = 51°. *Elementis Specialties.*

7748 Dapro W-77
Surfactant blend; interfacial tension modifier for minimizing crawling, fish-eyes and cratering in water-based coating systems. d =1.03; 8.6 lb/gal; visc (G-H) A3; flash point = 51°. *Elementis Specialties.*

7749 Dapro-7
Viscosity stabilizer, antigelling agent for solvent-thinned coatings. Amber liquid;d = 0.90; 7.5 lb/gal; flash point = 27°. *Elementis Specialties.*

7750 Dapsetyn
chloramphenicol-dapsone
Dapsyvet. A chloramphenicol/dapsone veterinary preparation. *Allen & Hanburys Ltd.* Name unverified.

7751 Dapsone
80-08-0 2885 201-248-4
$C_{12}H_{12}N_2O_2S$
4,4'-Sulfonylbisbenzeneamine
4,4·g-sulfonyldianiline; bis(4-aminophenyl) sulfone; 4,4·g-diaminodiphenyl sulfone; 4,4·g-DDS; DDS; Diaphenylsulfone; DADPS; 1358F; Avlosulfon; Croysulfone; Diphenasone; Disulone; Dimitone; Eporal; Novophone; Sulfona-Mae; Sulphadione; Udolac; DADPS; Diaphenylsulfone; Diphone; Dumitone; Bis(4-aminophenyl)sulfone; 4,4'-Diaminodiphenyl sulfone; Acedapsone; Araldite HT; Metabolite C; Sulfanona-mae; Sulfonyldianiline; Sumicure S; Dapsone; DDS; Sulfona; Dapsonum; DSS; Dubronax; 1358F; F 1358; ICI; Sulfona-mae; Sulfone UCB; Sulfon-mere; Tarimyl; WR 448. Curing agent for epoxy resins; antibacterial. mp = 175-176°; soluble in EtOH, methanol, acetone, hydrochloric acid; insoluble in H_2O; LD_{50} (rat orl) = 1 g/kg. *Crown Metro.*

7752 Dapsyvet
chloramphenicol-dapsone
A chloramphenicol-dapsone veterinary preparation. *Allen & Hanburys Ltd.* Name unverified.

7753 Daracide
Slimicide for use in the paper and pulp industry. *Grace Dearbom Ltd.*

7754 Daraclean
Cleaning materials. *Grace Dearborn Ltd.*

7755 Daradefoam
Defoamers; used in paper and pulp industry. *Grace Dearborn Ltd.*

7756 Darafloc
Flocculants. *Grace Dearborn Ltd.*

7757 Daran® 229
PVDC emulsion; high adhesion coating (paper and board); low heat seal tempratures, high barriers, non-blocking and excellent slip characteristics. *W R Grace/Organics.*

7758 Daran® 8350
PVDC copolymer emulsion; for use as a water-borne laminating adhesive. *W R Grace/Organics.* Name unverified.

7759 Darasperse
Dispersants; used in paper and pulp industry. *Grace Dearborn Ltd.*

7760 Daraspray
Dispersants; used in paper and pulp industry. *Grace Dearborn Ltd.*

7761 Daratak® 89L
Adhesive emulsion for formulation of remoistenable adhesives for envelopes, labels, stamps, tapes; also for textile warp sizing and adhesives for arts and crafts. *W R Grace/Organics.*

7762 Daratak® MX
Acrylic vinyl acetate terpolymer; adhesive polymer with fast grab and set; provides adhesion of vinyl films to wood as good as EVAs with tougher bonds. *W R Grace/Organics.*

7763 Daratak® RP2000
9003-20-7
Carboxylated polyvinyl acetate emulsion; adhesive base for paper/paper, paper/foil laminations, bottle labels, coated paperboard. *W R Grace/Organics.*

7764 Daratak® SP1011
9003-20-7
PVAc emulsion; highly water-resistant, borax-tolerant adhesive emulsion; can be formulated with borax to give fire-retardant adhesives and paints; also used in heat-sealing applications and in aluminum foil adhesives. *See Polyacrylic acid* *W R Grace/Organics.*

7765 Daratak® XB-3631
Vinylidene chloride-based adhesive emulsion; for use on difficult-to-adhere substrates including films, plastics, and vinyls. *W R Grace/Organics.*

7766 Darathane® WB-4000
Waterborne urethane prepolymer; dispersant; emulsifier; latex

polymerization; cross linkable, emulsifier-free polymers on curing; waste water treatment. *W R Grace/Organics.*

7767 **D'Arcet's alloy**
a) Consists of 50% bismuth, 25% lead, and 25% tin, melting-point 93 C; b) contains 50 parts bismuth, 25 parts lead, 25 parts tin, and 250 parts mercury.

7768 **Darco**
7440-44-0 1855 231-153-3
Activated carbon, powdered and granular; for water treatment, glucose and sugar purification, chemical manufacture, gas and air treatment. A decolorizing and refining carbon; a substitute for bone charcoal. *NORIT Americas Inc.*

7769 **Darex® 110L**
Nitrile latex
Copolymer latex for paper and felt saturation; high delamination resistance and edge tear; used in tape and sandpaper bases; heat curable. *W R Grace/Organics.*

7770 **Darex® 165L**
9003-54-7
SAN latex; water-based latex for bonding and impregnation of fibers, webs and films; forms stiff films by itself; produces flexible films by plasticizing with softer latices or compatible plasticizers e.g., the phthalyl glycollates. *W R Grace/Organics.*

7771 **Darex® 5281L**
SB latex; low emulsifier latex providing clear, smooth, high gloss, water-resistant coatings with good resistance to blocking; dries rapidly, with little heat; gross water holdout after application (paper and paperboard substrates). *W R Grace/Organics.* Name unverified.

7772 **Darex® 550L**
Vinylidene chloride emulsion; for use in flame retardant adhesives; designed for adhesion to nonporous substrates (e.g., film-to-paper, foil-to-paper applications). *W R Grace/Organics.*

7773 **Darex® 636L**
S/B latex; nonfilm forming latex used as box toe fabric saturant. *W R Grace/Organics.*

7774 **Dariloid® 100**
11138-66-2 10191 234-394-2
Xanthan gum; used for sherbert, water ice, cottage cheese dressings and sour cream. *Kelco.*

7775 **Dariloid® Q**
9005-38-3 240
Alginate
Used for sour cream-based chip dip, cheese dip, cheese sauce, pie filling. *Kelco.*

7776 **Dariloid® QH**
9005-38-3 240
Alginate
Used for puddings, cheesecake mix, whipped toppings, bakery fillings, instant egg nog, milk shake, dessert dry mix. *Kelco.*

7777 **Darilold®**
9005-38-3 240
Alginates; dairy stabilizers and stabilizer/emulsifiers; thickener for food preparations and milk solutions at pasteurization temperatures *Kelco.*

7778 **Darmex**
High technology lubricant; used for all types of industrial/automotive equipment. *Darmex Corp.*

7779 **Darmex Plus**
High technology lubricant; used for all types of industrial/automotive equipment. *Darmex Corp.*

7780 **Darmycel Agarifume Smoke**
52645-53-1 7321 258-067-9
Permethrin
A pyrethroid insecticide. *Darmycel UK.*

7781 **Dartex®**
Fibers. *DuPont UK.*

7782 **Darvan®**
Nonionic dispersing agents and surfactants; used for dispersing materials in rubber, paint, ceramics, plastics, cosmetics, pharmaceuticals, agriculture and house hold products. *R. T. Vanderbilt Co Inc.* Discontinued.

7783 **Darvan® No. 1**
9003-24-1 2886
Sodium naphthalene formaldehyde sulfonate
Naphthalenesulfonic acid, polymer with formaldehyde, sodium salt. Latex dispersant. *R. T. Vanderbilt Co Inc.*

7784 **Darvan® No. 2**
8061-51-6
Sodium lignosulfonate
Dispersing and emulsifying agent for rubber industry, especially for zinc oxide, clays, and sulfur. *R. T. Vanderbilt Co Inc.* Discontinued.

7785 **Darvan® No. 404**
8061-52-7
Calcium lignosulfonate
Dispersant. *R. T. Vanderbilt Co Inc.* Discontinued.

7786 **Darvan® SMO**
Sodium salt of sulfated methyl oleate; dispersant; improves smoothness and gloss of dipped CR latex films. *R. T. Vanderbilt Co Inc.*

7787 **Dasanit®**
115-90-2 4042 204-114-3
fensulfothion
Granular insecticide; used for treatment of biting insects and nematodes. *Bayer AG.*

7788 **Dastar**
57-88-5 2256 200-353-2
$C_{27}H_{46}O$
Cholesterol
Croda Chem. Ltd. Discontinued.

7789 **Datac**
Industrial adhesives for packaging, paper-converting and woodworking. *Datac Adhesives Ltd.*

7790 **Datagel**
Activated datem gel. *Croda Chem. Ltd.* Discontinued.

7791 **Datamuls**
Diacetyl tartaric acid esters; increases volume, fermenting stability and gas retention; improved crumb structure; prolonged shelf-life; improved dough compatability with machines; for bread products. *Thomas Goldschmidt Ltd.*

7792 **Datem**
Diacetyl tartaric esters of monoglycerides; edible emulsifiers for use in lipsticks and similar products. *Croda Chem. Ltd.* Discontinued.

7793 **Daubond DC-9200-A/DC-9200-B**
Modified polyurethane latex system (polyurethane and polyisocyanate respectively); for vacuum forming operations; good adhesion to vinylbacked foam and various substrates in manufacture of dashboards, door panels for automotive industry. *Daubert.*

7794 **Daubond DC-9300**
126-99-8 204-818-0
Polychloroprene latex
Adhesive for vacuum forming operations; good adhesion to shoddy and ABS substrates that are vacuum formed to foam-backed vinyl used on automotive dashboards, door panels, etc. *Daubert.*

7795 **Daudelin solder**
An alloy of 65.6% tin, 12.2% zinc, 1% aluminum, 17.4% lead, 3.1% copper, and 0.4% phosphorus.

7796 **Davey's gray**
A pigment prepared from siliceous earths; used principally in mixtures with other colors to reduce tones.

7797 **Davis metal**
An alloy of 67% copper, 29% nickel, 2% iron, and 1.5% manganese.

7798 **Dawson bronze**
An alloy of 83.9% copper, 15.9% tin, and traces of antimony, lead, iron, arsenic, and zinc.

7799 **Daxad® 11**
9084-06-4
Low molecular weight naphthalene sulfonate formaldehyde condensate, sodium salt; dispersant for pigments in aqueous media; used in agricultural chemicals, mastics, caulks, sealants, pigment slurries and dispersions. *W R Grace/Organics.*

7800 **Daxad® 19L-40**
High molecular weight naphthalene sulfonate formaldehyde condensate, sodium/potassium salt, in solution; cold weather-stable dispersant, water-reducing agent, and viscosity-reducer for high-solids slurries like concrete, cement, gypsum, lime, coal. *W R Grace/Organics.* Name unverified.

7801 **Daxad® 23**
Sodium salts of polymerized substituted benzenoid alkyl sulfonic acids; dispersant for agricultural chemicals, concrete admixtures, dyes, high-strength cements, linoleum pastes, pigment slurries and dispersions, and water treatment chemicals. *W R Grace/Organics.*

7802 **Daxad® 30-30**
Sodium polymethacrylate solution; dispersant, especially for pigments in aqueous solutions; used in paint formulations, emulsion polymerization, water treatment (as a scale control agent for boiler systems), in large particle suspensions. *W R Grace/Organics.*

7803 **Daxad® 37LN10-35**
9003-04-7
Sodium polyacrylate solution; dispersant for rapid dispersion of solid materials (especially clays) in aqueous systems; dispersant, deflocculant for pigment manufacture, paper coating, latex paints, oil well drilling muds;

antideposition agent in cleaning formulas; stable over a wide range of pH. *W R Grace/Organics.*

7804 Daxad® CP-2
Cationic polyelectrolyte
Dispersant for fillers, pigments, other additives; precipitates wetting agents, detergent soaps, emulsifiers, and dispersants out of industrial waters; coagulates latexes; flocculates kaolin clay dispersions; breaks oil-water and water-oil emulsions. *W R Grace/Organics.* Name unverified.

7805 Dazide
1596-84-5 2874 216-485-9
daminozide
Water soluble powder containing 85% daminozide; a plant growth regulator. *Fine Agrochemicals Ltd.*

7806 Dazomet
533-74-4 2892 208-576-7
$C_5H_{10}N_2S_2$
tetrahydro-3,5-dimethyl-2H-1,3,5-thiadiazine-2-thione
Basimid G; Basamid; Basamid P; Basamid-puder; Carbothialdine; Crag; Crag 974; Crag 85W; Dimethylformocarbothialdine; Fennosan B 100; Micofume; Mylone; Mylone 85; N 521; Nalcon 243; Nefusan; Prezervit; Stauffer N 521; Tiazon; Troysan 142; UCC 974. Granules containing 98-99% dazomet; soil fumigant, which acts by decomposition to methyl isothiocyanate. Used as a slimicide and soil sterilant, controlling nematodes, and soil fungi. mp = 104-105°, d = 1.30; insoluble in H_2O, soluble in organic solvents; LD_{50} (rat orl) = 320 mg/kg. *DowElanco Ltd.*

7807 DB Oil
Castor oil
For applications requiring maximum purity with minimum acidity and moisture, e.g., sonar fluid, liquid dielectrics, urethane reactions. *CasChem.*

7808 DB-1 Defoamer
Organo-modified silicone polymer, silica, and waxes; defoamer for emulsifiable oils (metalworking lubricant solvent oils), paints, water treatment, coolants, agricultural sprays, rolling oils, coatings. *Genesee Polymers.*

7809 DB-19 Antifoam Compd.
Silicone compound with water carrier; antifoam for aqueous systems, hot or cold foaming systems, alkaline systems, industrial processing, chemical processing, cleaning products, paints, paper and latex processing, metalworking. *Genesee Polymers.* Discontinued.

7810 2,4-DB
94-82-6 2893 202-366-9
$C_{10}H_{10}Cl_2O_3$
4-(2,4-dichlorophenoxy)butanoic acid
Embutone; Venceweed; Legumex; DB; Butirex; Butormone; Butoxone; Butyrac; M&B 2878. Selective systemic hormone type herbicide. Used for post-emergence control of many annual and perennial broad-leaved weeds in lucerne, clovers, cereals, grassland, forest legumes, soybeans and ground nuts. mp = 117-119°; soluble in H_2O (46 mg/l), more soluble in organic solvents; LD_{50} (rat orl) = 370-700 mg/kg.

7811 DBA Accelerator
Dibenzylamine and mono-benzylamine blend; a nonstaining and nondiscoloring activator for CPB, other xanthates or carbon disulfide; used in natural, synthetic and nitrile rubbers. *Uniroyal.* Name unverified.

7812 DBP
84-74-2 1622 201-557-4
Dibutyl phthalate
A plasticizer for vinyl and other plastics.

7813 DBPC
128-37-0 1583 204-881-4
Di-*t*-butyl-*para*-cresol
A proprietary antioxidant. *Koppers Co Inc.* Discontinued.

7814 DC 150
Chemical catalyst system; for decorative bright chromium plating (using low chromic acid concentrations). *Harshaw Chemicals Ltd.* Name unverified.

7815 DC Cristobalite
Cristobalite containing investment material for precious metals; dental preparation. *Bayer AG.*

7816 DC700
Chemical catalyst system; for decorative bright chromium plating. *Harshaw Chemicals Ltd.* Name unverified.

7817 DCI-3
Corrosion inhibitor. *DuPont UK.*

7818 DCP
131-15-7 205-014-2
Dicapryl phthalate
A plasticizer for vinyl plastics and cellulosic resins.

7819 DDBS 100
Alkylbenzene sulfonic acid
Contains branched alkyl chains; detergent intermediate. *Zohar Detergent Factory.*

7820 DDT
50-29-3 2898 200-024-3
$C_{14}H_9Cl_5$
1,1'-(2,2,2,-Trichloroethylidene)bis[4-chlorobenzene]
Dichlorodiphenyltrichlorethane; Chlorophenothane; Clofenotane; dicophane; Pentachlorin; Agritan; Gesapon; Gesarex; Gesarol; Guesapon; Neocid; Chlorophenothane; Clofenotane; Pentachlorin; Agritan; Gesapon; Gesarex; Gesarol; Geusapon; *p,p'*-DDT; Anofex; Didigam; Didimac; Estonate; Genitox; Gyron; Ixodex; Neocid; Santobane;Zeidane; Zerdane; Arkotine; Azotox; Bosan Supra; Bovidermol; Chlorophenotoxum; Citox; Dedelo; Deoval; Detox; Detoxan; Dibovan; Dnsbp; Dykol; Gesafid; Guesapon; Guesarol; Havero-extra; Hildit; Ivoran; Kopsol; Micro DDT 75; Mutoxin; OMS 16; Parachlorocidum; Peb1; Pentech; PPZeidan; R50; Rukseam; Tech DDT,. A powerful insecticide. Polyclorinated, nondegradable pesticide. One of the Dirty Dozen pesticides. mp = 108.5-109°; λ_m =236 nm; insoluble in H_2O; soluble in acetone (58 mg/100ml), benzene (78 mg/100ml), carbon tetrachloride (45 mg/100ml); LD_{50} (rat orl) = 113 mg/kg. Name unverified.

7821 De De Tane
50-29-3 2898 200-024-3
DDT
DDT products. *Murphy Chemical Co Ltd.* Discontinued.

7822 De Han salt
$SbF_3(NH_4)_2SO_4$
antimony trifluoride-ammonium sulfate
A double salt of antimony trifluoride and ammonium sulfate; a mordant.

7823 De Rossi's stain
A microscopic stain. It consists of two solutions: a) tannic acid 25 g, distilled water 100 ml; b) fuchsin 0.25 gram, phenol 5 g, alcohol 10g, and distilled water 100 g.

7824 De-Acidite
An anion exchange material. *The Permutit Co.* Name unverified.

7825 dead silver
Frosted silver. Silver, whitened by heating in air, and immersed in dilute sulfuric acid.

7826 Deadline
Range of pesticides for public health use. *Rentokil Ltd.*

7827 DEA-lauryl sulfate
68585-44-4 205-577-4
Diethanolamine lauryl sulfate
Empicol® 0031/T; Carsonal® DLS; Empicol® DA; Nutrapon DE 3796; Standapol® Conc. 7021; Sulfochem DLS; Supralate EP; Texapon® DEA; Unipol DEA; Witcolate™ DLS-35; Zoharpon LAD. Detergent used in personal care products Soluble in H_2O; d = 1.05, 8.49 lb/gal.; pH = 7.5-8.5. *Albright & Wilson UK; Chemron; Lonza.*

7828 Deanase D.C.
2320
δ-chymotrypsin
A proprietary preparation of δ-chymotrypsin; a digestive enzyme. *Consolidated Chemicals Ltd.* Name unverified.

7829 deanol
108-01-0 2900 203-542-8
$C_4H_{11}NO$
N,N-Dimethyl-2-Hydroxyethylamine
β-Dimethylaminoethyl alcohol; Deanol; 2-dimethylaminoethanol; Dimethylethanolamine; DMAE; dimethylaminoethanol; N,N-Dimethylaminoethanol; N-dimethylaminoethanol; N,N-dimethyl-N-(2-hydroxyethyl)amine; β-hydroxyethyldimethylamine. Amine-based catalyst for flexible slabstock PU foam. mp = -70°; bp = 139°; d = 0.8870; n_D^{20} = 1.4294; miscible with EtOH, Et_2O, LD_{50} (rat orl) = 2 gm/kg. *Air Prods & Chem; Elf Atochem N. Am.; BASF; Nippon Nyukazai; Texaco; Union Carbide.*

7830 Deanox
1309-37-1 4072 215-168-2
Iron oxides. Used as pigments. *Deanshanger Oxides Ltd.* Name unverified.

7831 Dearcide
Biocides used for cooling waters. *Grace Dearborn Ltd.*

7832 DEAS Base
Fatty alkanolamide; base for textile softeners; finished products yield a dry, hard, slightly waxy hand. *Clark.*

7833 Deasol
High boiling aromatic solvent used for fuel treatment. *Grace Dearborn Ltd.*

7834 Debenal®
Geriatric drug for dogs and cats; veterinary medicine. *Bayer AG.*

7835 Debron 711
9016-75-5
Phenylene Sulfide Resin

Ryton. A proprietary coating compound manufactured from polyphenylene sulfide. d = 1.36. *de Beers Laboratories Inc.* Unverified.

7836 Debut®

For the agriculture industry. *DuPont UK.*

7837 Decabromodiphenyl oxide

1163-19-5 214-604-9

$C_{12}Br_{10}O$

1,1'-Oxybis (2,3,4,5,6-pentabromobenzene)

DECA; Decabromobiphenyl ether; Decabromobiphenyl oxide; Decabromophenyl ether; Pentabromophenyl ether; FR-1210; Great Lakes DE-83™; Octaguard FR-01; Themoguard® 505; Decabromodiphenyl ether; Pentabromodiphenyl ether; DPBPO; FR300BA; FRP 53; Berkflam B 10E; Bromkal 82-ODE; BR 55N; FR 300; DE 83R; Bromkal 83-10DE; pentabromophenyl ether; Saytex 102; Saytex 102E; Tardex 100; DBDPO. Flame retardant used in thermoplastics and fibers, including HIPS, glass-reinforced thermoplastic polyester molding resins, LDPE extrusion coatings, PP (homo and copolymers), ABS, nylon, PBT, PET, PU, SBR latex, textiles, rubber. mp=300-310°; d = 3.00; LD_{50} (rat orl) > 5000 mg/kg; experimental carcinogen with teratogenic and reproductive effects. AmeriHaas; Albemarle; Aldrich; Allchem Ind.; AmeriBrom; Dead Sea Bromine; Elf Atochem N. Am; Fluka; Great Lakes.

7838 Decadex

Water-based elastomeric weatherproofing compound applied by brush or spray; for roof refurbishment, anticarbonation/all application external coating in range of colors. *Liquid Plastics Ltd.*

7839 Decalex

A photographic developer. *May & Baker Ltd.* Name unverified.

7840 Decalin®

91-17-8 2903 202-046-9

$C_{10}H_{18}$

Decahydronaphthalene

Perhydronaphthalene; Naphthalane; Naphthane; Dec; DeKalin. A paint and resin solvent; used as turpentine substitute. mp = -43°; bp = 194°; d = 0.8960; insoluble in H_2O; very soluble in organic solvents; LD_{50} (rat orl) = 4.2 g/kg. *DuPont.* Unverified.

7841 Decaltal®

Nonswelling organic acids and their salts; deliming and masking agents in chrome tanning. *BASF AG.*

7842 decamphorized oil of turpentine

Oxidized oil of turpentine. The residue from the manufacture of camphor; consists mainly of dipentene.

7843 Decanox-F

14156-10-6

$C_{10}H_{20}O_3$

Decanoyl peroxide

peroxydecanoic acid. Initiator for bulk, solution, and suspension polymerization, curing elastomers, and high-temp. cure of polyester resins. *Elf Atochem.*

7844 Decanoyl chloride

112-13-0 203-938-0

$C_{10}H_{19}ClO$

capric acid chloride

Intermediate, polymerization initiator. bp = 94-96°; bp_5 = 94-96°; d = 0.919; n_D^{20} = 1.4410; reacts with H_2O, soluble in organic solvents. *Elf Atochem N. Am.; Janssen Chimica.*

7845 Decap

67-68-5 3308 200-664-3

DMSO

DMSO-based solvent for depotting, deflashing, decapsulation of epoxy castings, transfer moldings; used hot (150°); nonselective. *Dynaloy.*

7846 Deccox

18507-89-6 2910 236-948-9

Decoquinate

Coccidiostat for sheep. *RMB Animal Health Ltd; May & Baker Ltd.*

7847 Decelox

1314-13-2 10279 215-222-5

zinc oxide

Zinc oxide, modified to give slow rubber cure. *Durham Chemicals Ltd.* Name unverified.

7848 Deceth-6

26183-52-8

$C_{22}H_{46}O_7$

3,6,9,12,15,18-Hexaoxaoctacosan-1-ol

Bio-soft® FF 400; Chemal DA-4; Desonic® DA-4; Desonic® DA-6;Ethal DA-4; Genapol® DA-040; Iconol DA-4; Iconol DA-6; Iconal DA-9; Marlipal® 1012/4; Oxetal D 104; Prox-onic DA-1/04; Rhodasurf® DA-4; Trycol® 5950; Surfonic® DA-6; Synthrapol KB; PEG-6 decyl ether. Wetting agent for built scour systems. Detergent; penetrant; emulsifier for textiles, clay soils, fire fighting; surfactant. d = 1.0014; viscosity = 109 cps; HLB 12.5; hyd no 132; pour point =12.8°; cloud point =42°. *Rhône-Poulec Surf & Spec.; Huntsman.*

7849 Dechlorane A-O

1309-64-4 752 215-175-0

O_3Sb_2

Antimony oxide

Antimony oxide; contains halogens; a proprietary synergistic agent for fire-retardant plastics. *Kingsley & Keith Chemical Corp.* Name unverified.

7850 Dechlorane® Plus 25, Plus 35, Plus 515

13560-89-9 2908 236-948-9

$C_{18}H_{12}Cl_{12}$

1,2,3,4,7,8,9,10,13,13,14,14-dodecachloro-1,4,4a,5,6,6a,7,10,10a,11,12,12a-dodecahydro-1,4:7,10-dimethanodibenzo[a,e]cyclooctene

Chlorine-containing cycloaliphatic compound. Used as a flame retardant in polymer systems (thermoplastics, thermosets and elastomers); usually combined with antimony oxide as a synergist. soluble in *o*-dichlorobenzene; mp >325°; LD_{50} (rat orl) = 25 g/kg. *OxyChem.*

7851 Decimate

chlorthal-dimethyl-propachlor

Suspension concentrate containing 225 g chlorthal-dimethyl and 216 g propachlor per liter; an herbicide for use on brassicas and onions. *Fermenta ASC Europe Ltd.*

7852 Decis

52820-00-5

deltamethrin

Emulsifiable concentrate containing 25 g deltamethrin per liter; a pyrethroid insecticide. *Hoechst UK.*

7853 Decisquick

deltamethrin-heptenophos

Emulsifiable concentrate containing 25 g deltamethrin and 400 g heptenophos per liter; a systemic aphicide. *Hoechst UK.*

7854 Deck Seal-PD

Polysiloxane and methylmethacrylate in an aromatic and aliphatic solvent vehicle system; sealer for parking decks, bridge decks, ramps, stadiums etc. *Nova Chemical Inc.* Name unverified.

7855 Declar®

Thermoplastic sheet with Tedlar® PVF film surface finish; formulated to meet aircraft industry flammability requirements for interior parts. *DuPont; Du Pont (Name unverified.UK) Ltd; Sheffield Plastics.* Name unverified.

7856 Deco Board P, Deco Poly

Polyester resin; for decorative laminates. *Sumitomo Bakelite Co Ltd.* Name unverified.

7857 Decoart

Phenolic resin; for decorative laminates. *Sumitomo Bakelite Co Ltd.* Name unverified.

7858 Decol

A range of alkyl aryl sulfonates; all purpose liquid detergent concentrates, detergent bases for powdered cleansers, foaming agents for plaster board production, concrete foaming agents, wetting agents in textile processing, raw wool scouring detergent, *Harcros Australia.*

7859 Decola Back sheet

Phenolic resin; for back sheet. *Sumitomo Bakelite Co Ltd.*

7860 Decola Excel, New Marine, F, FG, MA, MF

Melamine resin; for decorative laminates. *Sumitomo Bakelite Co Ltd.* Name unverified.

7861 Decola PFC

Melamine resin; for postform. *Sumitomo Bakelite Co Ltd.* Name unverified.

7862 Decolamide

A range of alkanolamides; used as foam boost additives in detergents and shampoos, superfatting agents, opacifiers, thickeners, demulsifiers and emulsifiers. *Harcros Australia.*

7863 Deconyl

A proprietary trade name for a weatherproof nylon coating. *Plastic Coatings Ltd.* Name unverified.

7864 Decopress

Antithrombotic stocking. *Bayer AG.*

7865 Decothane® SP

An elastomeric high-build single-pack polyurethane roofing compound; forms a seamless, durable, uv resistant, waterproof barrier that is immediately resistant to rain and ponded water; applied by brush roller or airless spray; for refurbishment of all roofs - flat or pitched. *Liquid Plastics Ltd.*

7866 Decrolin®

24887-06-7 246-515-6

Zinc formaldehyde sulfoxylate

Zinc formosul. Reducing agent used after catalytic bleaching of fur skins. *BASF AG; BASF plc.*

7867 decyl alcohol
68526-85-2 271-234-0
$C_{10}H_{22}O$
n-decyl alcohol
Exxal® 10; decanol. Solvent, chemical intermediate *Exxon.* Name unverified.

7868 decyl bromide
112-29-8 203-955-3
$C_{10}H_{21}Br$
1-bromodecane
n-Decyl bromide. Chemical intermediate, solvent. mp = -30°; bp = 238°; d = 1.0660; n_D^{20} = 1.4560; insoluble in H_2O, soluble in organic solvents. *Ethyl; Great Lakes; Humphrey.*

7869 decyl oleate
3687-46-5 222-981-6
$C_{28}H_{54}O_2$
9-Octadecenoic acid(Z), decyl ester
Ceraphyl® 140. Emollient; binder for pressed powders; pigment dispersant; co-solvent. *Van Dyk.* Name unverified.

7870 Dedevap®
62-73-7 3129 200-547-7
dichlorvos
DDVP. Insecticide usd for spray control of sucking, biting and mining insects in greenhouses. *Bayer AG.*

7871 Dedico 5981
Dehydrated castor oil fatty acid; for production of fast drying alkyd resins and epoxy esters including styrenated and maleinized resins; cured coatings show excellent mechanical and chemical properties; for automotive lacquers, car repair lacquers, can coatings. *Unichema.*

7872 DEDM hydantoin
26850-24-8 248-052-5
$C_9H_{16}N_2O_4$
Di-(2-hydroxyethyl)-5,5-dimethyl hydantoin; 1,3-Bis(2-hydroxyethyl)-5,5-dimethyl-2,4-imidazolidinedione; Diethylol dimethyl hydantoin; Dantocol® DHE. Resin crosslinker in coatings and polymers; intermediate for epoxies, urethane resins, and antistatic lubricants for the textile and plastics industries. mp = 63°; pH 6.5 (5%). *Lonza Inc.*

7873 DEDM hydantoin dilaurate
$C_{33}H_{64}N_2O_6$
Dantoest® DHE DL; Di-(2-hydroxyethyl)-5,5-dimethyl hydantoin dilaurate; Diethylol dimethyl hydantoin dilaurate. Fiber lubricant; intermediate for epoxies, urethane resins, and antistatic lubricants for the textiles industry. *Lonza Inc.*

7874 Deenax
128-37-0 1583 204-881-4
$C_{15}H_{24}O$
2,6-di-*t*-butyl 4-methylphenol
BHT. A proprietary antioxidant. *Exxon.* Name unverified.

7875 Deep Feed
Liquid fertilizer. *Fisons plc, Horticultural Div.* Name unverified.

7876 Def®
78-48-8 201-120-8
$C_{12}H_{27}OPS_3$
S,S,S-tributyl phosphorotrithioate
Tribufos; Tributylphosphorotrithioate;Tribufos; S,S,S-Tributyltrithiophosphate; DEF; De-Green; E-Z-Off D; Easy off-D; Butyl phosphorotrithioate; S,S,S-tributyl phosphorotrithioate; Phosphorotrithioic acid S,S,S-tributyl ester; Deleaf defoliant; Folex 6EC; FOS-FALL "A"; Ortho phosphate defoliant; Tribuphos. Plant growth regulator which acts as a defoliant. Used on cotton to facilitate harvesting. mp < -25°; $bp_{0.3}$ = 150°; d^{20} = 1.057; n_D^{25} = 1.532; poorly soluble in H_2O (2.3 mg/l), more soluble in organic solvents; LD_{50} (rat orl) = 233 mg/kg. *Bayer AG.*

7877 Defirust
A proprietary trade name for a rustless iron containing 12-15% chromium and 0.1% carbon. No manufacturer.

7878 Deflavit® ZA
Stripping agent for wool and nylon dyeings. *BASF AG.*

7879 Defoamer 1713
Proprietary blend; paper machine defoamer for fine paper. *Hart Chem. Ltd.*

7880 Defoamer A 50
Tall oil acid, octoxynol-200; defoamer for phosphate mineral processing for the sulfuric acid process in manufacture of artificial fertilizer; biodegradable. *Chem-Y GmbH.*

7881 Defoamer B 90
Octeth-3 carboxylic acid
Defoamer for phosphate mineral processing for the sulfuric acid process in manufacture of artificial fertilizer and for the regeneration of silver in electroplating baths. *Chem-Y GmbH.*

7882 Defoamer C5B
Nonsilicone mineral oil-based; defoamer for effluents in pulp and paper industry; effective over wide pH range. *Hart Chem. Ltd.*

7883 Defoamer DF-160-L
Defoamer for synthetic latex coating systems, especially acrylic emulsions and binders; also controls foaming of sodium lauryl sulfate solutions. *Henkel/Textiles.*

7884 Defoamer KCE/S
Proprietary blend; water-based defoamer for screen room, bleachery, and effluent applications in pulp and paper industry. *Hart Chem. Ltd.*

7885 Defoamer NXZ
Defoamer for use with latex coating and latex printing; also in atmospheric dyeing, printing and textile finishing processes. *Henkel/Textiles.*

7886 Defoamer S
Silicone-based defoamer, multipurpose defoaming agent. *Hart Chem. Ltd.*

7887 Defoamer SF
Perfluoro alkyl phosphinate/phosphonate
Surfactant for agricultural formulations. *Hoechst Celanese/Colorants & Surf.*

7888 Defoamer WB Series
Proprietary blend; water-based defoamers for screen room and linerboard applications in pulp/paper industry. *Hart Chem. Ltd.*

7889 Defol
7775-09-9 8741 231-887-4
Sodium chlorate
Cotton defoliant, dessicant for corn, grain sorghum, sunflowers, safflowers, rice, soybeans, chili peppers and guar beans. *Draxel Chemical Company.* Unverified.

7890 Defolia
Defoliant for hops. *Murphy Chemical Co Ltd.* Discontinued.

7891 Defomax
Nonionic surfactant/wax emulsions; antifoaming agents for rubbers and plastics. Liquid/paste *Toho Chem. Industry.*

7892 Degadur
Methacrylic resins; self-leveling coatings, mortar, as floor and wall coverings, road construction, ship deck coatings. *Degussa Ltd.*

7893 Degalan
Polymethyl methacrylate compounds for injection molding and extrusion; used in production of light-and weather-resistant sheets, tubes and specially shaped profiles as well as finished parts such as car light assemblies, graduated dials, writing and drawing equipment, household utensils, watch crystals, optical lenses and lamp covers. *Degussa Ltd.*

7894 Degalex
Aqueous pure acrylic emulsions; used as binder for the production of high grade emulsion paints and synthetic resin plasters. *Degussa.* Discontinued.

7895 Degament
Cold-curing, low-viscosity methacrylate resins for precast elements (concrete/polymer/composite); used for coverings on buildings, walk on elements, pictures, sanitary components, stationary sport installations, machine mountings, stone bondings and marble agglomerates. *Degussa Ltd.*

7896 Degapas
Aqueous solutions of acrylic acid and methacrylic acid polymers (S series) and their sodium salts (N series); used for the dispersion of organic solid materials; as a builder in phosphate free or low phosphate cleaners; as thickening agents; for the complexing of multivalent metal cations and the sizing of polyamide fibers. *Degussa Ltd.*

7897 Degaplast
Acrylic resins and foam resins; used in production of cast arm and leg prostheses, apparatus, support corsets, night splints, bed casts; as filler resin for last tip extensions and corrections. *Dequssa AG.*

7898 Degaroute
Acrylic road marking. *Degussa Ltd.*

7899 Degaser
A degassing agent for aluminum alloys. *Foseco (F.S.) Ltd.*

7900 Deglas
Extruded acrylic sheet; used to cover lighting fixtures, for light domes, illuminated signs, wash basins and bathtubs, roof canopies, etc. *Degussa Ltd.*

7901 Degopol
Components for polyurethane shoe soling systems. *ICI Chem & Polymers Ltd.* Discontinued.

7902 Degopur
Polyurethane systems. *ICI Chem & Polymers Ltd.* Discontinued.

7903 Degras
Wool grease. *Croda Inc.*

7904 Degras Special
8020-84-6 232-418-6

Lanolin from crude wool grease; EP and slip agent for wire drawing compounds, slushing oils, cutting oils, lubricants; long term rust preventative; in leather stuffing greases; waterproofing agent. *Fanning.*

7905 Degreez
Methyl oleate, methyl stearate, methyl palmitate, methyl laurate, methyl myristate; degreasing agent for removing oil stains from metal, textiles, etc.; biodegradable. *Alzo.* Discontinued.

7906 Degressal® SD 20
Polypropoxylate
Defoamer for cleaners and detergents. *BASF AG.*

7907 Degressal® SNC
Modified phosphoric acid monoester; defoamer for detergents and cleaners. *BASF AG.*

7908 Degubond
A precious metal alloy; for dentistry and dental engineering. *Degussa AG.* Name unverified.

7909 Degucast
A precious metal alloy; used for dentistry and dental engineering. *Degussa Ltd.*

7910 Degudent
A range of precious metal alloys; used for dentistry and dental engineering. *Degussa Ltd.*

7911 Deguflex
Dental impression material. *Degussa Ltd.*

7912 Deguform
Dental silicone duplicating material. *Degussa Ltd.*

7913 Degulor
A range of precious metal alloys; used for dentistry and dental engineering. *Degussa Ltd.*

7914 Deguphos
Precious metal heterogeneous catalysts. *Degussa Ltd.*

7915 Degupress
Dental acrylic denture base material. *Degussa Ltd.*

7916 Degusorb
7440-44-0 1855 231-153-3
carbon
A range of activated carbons; used for decolorizing, cleaning and deodorizing in the chemical, pharmaceutical, beverage and food industry. *Degussa AG.*

7917 Degutron
Dental casting machine. *Degussa Ltd.*

7918 Deguvest
Plaster free material for the dental casting process. *Degussa Ltd.*

7919 Dehesive®
Coating, separating and impregnating media for paper, leather, cardboard, asbestos, synthetic plastics, metal and textiles; all containing synthetic resins and silicones. *Wacker-Chemie GmbH.*

7920 Dehesive® 920
Silicone
Solvent-free silicone system with high curing rate for high-quality release coatings. *Wacker Silicones.*

7921 Dehscofix 904
TEA salt of substituted phenol ethoxylated phosphate ester; emulsifier, dispersant for agricultural chemicals, leather processing. *Albright & Wilson Am.*

7922 Dehscofix 911
Naphthalene sulfonic acid, formaldehyde condensate; surfactant for leather processing. *Albright & Wilson Am.*

7923 Dehscofix 912
9084-06-4
Sodium naphthaleneformaldehyde sulfonate
Surfactant for leather processing. *Albright & Wilson Am.*

7924 Dehscofix 916
1322-93-6 215-343-3
Sodium diisopropyl naphthalene sulfonate
Wetting agents for agricultural chemicals, leather processing. *Albright & Wilson Am.*

7925 Dehscofix 917
25417-20-3 246-960-6
Sodium di-n-butyl naphthalene sulfonate
Wetting agents for agricultural chemicals, leather processing. *Albright & Wilson Am.*

7926 Dehscofix 918
120-18-3 6464 204-375-3
β-Naphthalene sulfonic acid
Surfactant used in leather processing. mp = 124°; d = 1.4400. *Albright & Wilson Am.*

7927 Dehscofix 923
26264-58-4 247-564-6
Sodium dimethyl naphthalene-formaldehyde sulfonate
Wetting agents for agricultural chemicals, leather processing. *Albright & Wilson Am.*

7928 Dehscofix 929
Ammonium naphthalene-formaldehyde sulfonate
Surfactant for leather processing. *Albright & Wilson Am.*

7929 Dehscofix CO Series
Castor oil ethoxylates; emulsifier for agricultural chemicals *Albright & Wilson Am.*

7930 Dehscotex
Formulated auxiliaries for textiles and leather. *Albright & Wilson Am.*

7931 Dehscotex BA Series
Formulated surfactant blends; bleaching agent for textile processing. *Albright & Wilson Am.*

7932 Dehscotex DT 809
Formulated surfactant; detergent for leather processing. *Albright & Wilson Am.*

7933 Dehscotex DT Series
Formulated surfactant blends; washing/scouring detergent for textile processing. *Albright & Wilson Am.*

7934 Dehscotex DY Series
Formulated surfactant blends; dye bath dispersant and leveling agent for textile processing. *Albright & Wilson Am.*

7935 Dehscotex FW Series
Formulated surfactant blends; felting auxiliary for textile processing *Albright & Wilson Am.*

7936 Dehscotex MC Series
Formulated surfactant blends; wetting agent for textile mercerizing and causticizing. *Albright & Wilson Am.*

7937 Dehscotex SN Series
Formulated surfactants; softener for textile processing. *Albright & Wilson Am.*

7938 Dehscotex VP-PF
Nonsilicone surfactant blends; fiber antistat for textile processing. *Albright & Wilson Am.*

7939 Dehscotex WA Series
Formulated surfactant blends; wetting agent for textile processing. *Albright & Wilson Am.*

7940 Dehscoxid 700 Series
Synthetic alcohol ethoxylates
Wetting agents for agricultural chemicals, textiles, leather. *Albright & Wilson Am.*

7941 Dehscoxid 730/740 Series
C_{13} alcohol ethoxylates
Surfactant. *Albright & Wilson UK.*

7942 Dehybor
1330-43-4 8733 215-540-4
sodium borate
Anhydrous Borax. Borax from which the water of crystallization has been removed by heat (anhydrous borax); it is widely used in glass, vitreous enamel, ceramic glaze and metallurgical industries where borax has to be melted. mp = 741°; d = 2.3670. *Borax Europe Ltd.*

7943 Dehydag Wax 14
112-72-1 6418 204-000-3
Myristyl alcohol
Consistency factor for cosmetic and pharmaceutical oil-water and water-oil creams, ointments, emulsions, liniments, and sticks. *Henkel.* Name unverified.

7944 Dehydag Wax 16
36653-82-4 2070 253-149-0
Cetyl alcohol
Consistency factor for cosmetic and pharmaceutical oil-water and water-oil creams, ointments, emulsions, liniments, and sticks. *Henkel.* Name unverified.

7945 Dehydag Wax 18
112-92-5 8960 204-017-6
Stearyl alcohol
Consistency factor for cosmetic and pharmaceutical oil-water and water-oil creams, ointments, emulsions, liniments, and sticks. *Henkel.* Name unverified.

7946 Dehydag Wax 22 (Lanette)
661-19-8 8135 211-546-6
Behenyl alcohol
Consistency factor for cosmetic and pharmaceutical oil-water and water-oil creams, ointments, emulsions, liniments, and sticks. *Henkel.* Name unverified.

7947 **Dehydag Wax E**
59186-41-3
Sodium cetearyl sulfate
Oil-water emulsifier used in personal care products and powder cleaners. *Henkel.* Name unverified.

7948 **Dehydag Wax N**
cetearyl alcohol-sodium cetearyl sulfate. Cetearyl alcohol and sodium cetearyl sulfate; manufacture of oil-water creams and liquid emulsions; solid. *Henkel.* Name unverified.

7949 **Dehydag Wax O**
Cetearyl alcohol
Consistency modifier used in personal care products and pharmaceuticals. *Henkel.* Name unverified.

7950 **Dehydag Wax SX**
cetearyl alcohol-sodium lauryl sulfate. Cetearyl alcohol, sodium lauryl sulfate (90:10 ratio); surfactant used in manufacture of oil-water emulsions for personal care products; self-emulsifying base. *Henkel.* Name unverified.

7951 **Dehydag Wax W**
cetearyl alcohol-sodium lauryl sulfate. Cetearyl alcohol and sodium lauryl sulfate (90:10 ratio); base for manufacture of ointments, creams and liniments; granular. *Henkel.* Name unverified.

7952 **Dehydol LS 2**
Laureth-2l
Emulsifier, solubilizer for solvents, oils, bases for production of sulfates; raw material for dishwashing, cleansing agent and cold cleaners; in bath oils, waterless hand cleaners. *Henkel Canada; Henkel KGaA; Pulcra SA.* Discontinued.

7953 **Dehydol LT 3**
61791-13-7
Coceth-27
Emulsifier, solubilizer for solvents, oils; base for production of ether sulfates; raw material for low-foaming detergents, dishwashing, cleansing agents, cold cleaners; superfatting agent. *Pulcra SA.*

7954 **Dehydol PCS 6**
68439-49-6 7737
Ceteareth-6
Emulsifier; intermediate raw material for manufacture of ethoxysulfates for use in detergents and industrial specialties. *Pulcra SA.*

7955 **Dehydol PID 6**
34938-91-8
Laureth-6
Emulsifier, wetting agent for industrial, cosmetic, pharmaceutical applications, in high electrolyte concentrations. *Pulcra SA.*

7956 **Dehydol PLS 1**
Ethoxylated C_{12}-C_{14} fatty alcohol (1 EO)
Surfactant. Liquid; HLB 3.7; hyd no. 232-236. *Pulcra SA; Henkel.*

7957 **Dehydol PTA 7**
61791-28-4
Talloweth-7
Surfactant. *Pulcra SA.*

7958 **Dehydol TA 11**
61791-28-4
Talloweth-11
Raw material for detergents, toilet cubes; emulsifier for technical applications *Pulcra SA.*

7959 **Dehydran 520**
Fatty acid ester, defoamer for polymerization, especially during monomer stripping in manufacture of suspension PVC; biodegradable. Brown, to clear liquid; miscible with H_2O; d = 0.98; 8.1 lb/gal; viscosity = 100 cps; flash point > 93°; pH = 9; LD_{50} (rat orl) > 5g/kg. *Henkel/Functional Prods.*

7960 **Dehydran 1019**
Nonsilicone defoamer; defoamer for production and processing of polymer dispersions. White cloudy liquid; disperses in H_2O; d = 0.86; 7.20 lb/gal; flash point = 165°. *Henkel/Functional Prods.*

7961 **Dehydran 1293**
Modified polysiloxane; defoamer for water based industrial coatings, printing inks, wood lacquers. H_2O-white clear liquid; disperses in H_2O; d = 0.90; 7.6 lb/gal; viscosity = 15 cps; flash point = 65°; VOC 90.7%; causes skin, eye and respiratory irritation. *Henkel/Functional Prods.*

7962 **Dehydran P 12**
Nonsilicone defoamer; for monomer recovery in production of synthetic rubber; antifoaming agent for synthetic polymer dispersions, water-based paints, adhesives and glues. *Henkel/Functional Prods.*

7963 **Dehydrite®**
10034-81-8 5715 233-108-3
Cl_2MgO_8
magnesium perchlorate trihydrate
anhydrone. A registered trade name for magnesium perchlorate trihydrate; a drying agent for gases. dec > 250°; d = 2.6000; soluble in H_2O, (exothermic), EtOH. No manufacturer.

7964 **Dehydroacetic acid**
520-45-6 2919 208-293-9
$C_8H_8O_4$
3-acetyl-6-methyl-2H-pyran-2,4(3H)-dione, ion(1-)
2H-pyran-2,4(3H)-dione, 3-acetyl-6-methyl-; 3-acetyl-4-hydroxy-6-methyl-2-pyrone cyclic ketone; DHA; DHAA; Methylacetopyronone. DHA; fungicide, bactericide, plasticizer, chemical intermediate, medicated toothpastes. mp = 109-111°; bp = 269.9°; insoluble in H_2O, soluble in organic solvents; LD_{50} (rat orl) = 570 mg/kg.

7965 **Dehydrophen 65**
Alkylaryl polyglycol ether
Detergent, wetting, dishwashing agent. *Henkel KGaA.* Discontinued.

7966 **Dehydrophen PNP 4**
9016-45-9, 2311-27-5
Nonoxynol-6
Detergent, dispersant, emulsifier, wetting agent for solvent cleaners; emulsifier/stabilizer for paint additives; as deicing fluid; sludge dispersant in petroleum products *Pulcra SA.*

7967 **Dehydrophen POP 4**
9002-93-1 6858
Octoxynol-4
Emulsifier, detergent, dispersant; used in wax-based washing formulations, emulsion cleaners. *Pulcra SA.*

7968 **Dehymuis SSO**
8007-43-0 8872 232-360-1
Sorbitan sesquioleate
Water-oil emulsifier and coemulsifier for waxes and oils. *Henkel; Henkel KGaA.*

7969 **Dehymuls**
Mixture of higher molecular esters mainly mixed ester of penthaerithrityl fatty acid ester; water-oil type ointment and creams. *Henkel Chemicals Ltd.* Name unverified.

7970 **Dehymuls E**
Sorbitan sesquioleate, pentaerythritol cocoate, stearyl citrate, beeswax, aluminum stearate; water-oil emulsifier; suitable for cosmetics. Yellowish, waxy solid; drop point = 45-60°; HLB 6.0; iodine no 20-30; saponification no 160-170. *Henkel/Cospna; Henkel Canada; Henkel KGaA.*

7971 **Dehymuls F**
Microcrystalline wax, pentaerythritol cocoate, stearyl citrate, glyceryl oleate, aluminum stearate, propylene glycol; water-oil emulsifier for creams and emulsions. White to slightly yellow wax; mp = 60-75°; acid no 4-10; iodine no 15-20; saponification no 120-140. *Henkel/Cospha; Henkel Canada; Henkel KGaA.*

7972 **Dehymuls HRE 7**
61788-85-0
PEG-7 hydrogenated castor oil; emulsifier for water-oil emulsions for personal care products Pale yellow, cloudy viscous liquid; acid no <1; sapon no 125-140; hyd no 110-130. *Henkel KGaA.*

7973 **Dehymuls K**
Petrolatum, decyl oleate, sorbitan sesquioleate, pentaerythritol cocoate, stearyl citrate, beeswax, mineral oil, ceresin, aluminum stearate; SE base for manufacture of cosmetic and pharmaceutical preparations of the water-oil type. White to slight yellow soft waxy solid; mp = 35-50°; iodine no 18-23; saponification no.= 75-85. *Henkel; Henkel KGaA.*

7974 **Dehymuls SML**
1338-39-2 8872 215-663-3
Sorbitan laurate
Water-oil emulsifier and coemulsifier for cosmetic and pharmaceutical applications *Henkel; Henkel KGaA.*

7975 **Dehymuls SMO**
1338-43-8 8872 215-665-4
Sorbitan oleate
Emulsifier and coemulsifier for ointments and creams of the water-oil type. *Henkel; Henkel KGaA.*

7976 **Dehymuls SMS**
1338-41-6 8872 215-664-9
Sorbitan stearate
Emulsifier for water-oil emulsions; used in the cosmetic and pharmaceutical industry. *Henkel; Henkel KGaA.*

7977 **Dehypon Conc**
Ethoxylated, propoxylated fatty alcohols and anion; wetting agent, emulsifier and dispersant; for carpet cleaners; high detergency, low foam. *Henkel; Henkel Canada; Henkel KGaA.*

7978 **Dehypon LS-24**
Ethoxylated, propoxylated lauryl alcohol

Detergent and wetting agent; low foaming. *Henkel Canada; Henkel KGaA.* Discontinued.

7979 Dehypon LT 054
109075-72-1
Fatty alcohol polyglycol ether
Surfactant for bottle washing. *Henkel KGaA.*

7980 Dehyquart A
112-02-7 203-928-6
Cetrimonium chloride
Emulsifier for emulsion polymerization; softener, conditioner, bactericide, fungicide, and odor inhibitor in personal care product; antistat for hair and fibers. *Henkel/Cospha; Henkel/Functional Prods.; Henkel Chemicals Ltd.*

7981 Dehyquart AU-36
Bis(acyloxyethyl) hydroxyethyl methylammonium methosulfate
15% solution in isopropanol; highly biodegradable surfactant; raw material for production of softeners; good rewetting properties. Semi solid paste at 20°; yellowish liquid at 45°; drop point = 37-42°; flash point = 26-30°; pH = 2-3; highly biodegradable. *Pulcra SA.*

7982 Dehyquart C Crystals
104-74-5 203-232-2
$C_{17}H_{30}ClN$
Lauryl pyridinium chloride
LPC; Lauryl pyridinium chloride; n-Lauryl pyridinium chloride; 1-dodecylpyridinium chloride. Emulsifier in creams and lotions; hair conditioners, skin creams; antistat for hair and fiber; bactericide, fungicide, corrosion inhibitor, sequestrant; conditioner used in personal care products mp = 66-70°. *Henkel/Cospha; Henkel/Functional Prods.; Henkel Chemicals Ltd.*

7983 Dehyquart CDB
Cetyl dimethyl benzylammonium chloride in liquid form; cationic emulsifier used in hair cosmetics and antistatics. *Henkel Chemicals Ltd.* Name unverified.

7984 Dehyquart D
17342-21-1 241-364-2
Lauryl pyridinium bisulfate
Germicide, wetting agent with anticorrosive effect, emulsion breaker; acid stable. *Henkel/Cospha; Henkel/Functional Prods.*

7985 Dehyquart DAM
107-64-2 203-508-2
Distearyl dimethyl ammonium chloride
Emulsifier for plastics industry; conditioning component for hair care preparations; antistat. *Henkel/Functional Prods.; Henkel Chemicals Ltd.*

7986 Dehyquart LDB
139-07-1 203-351-5
$C_{21}H_{38}ClN$
Lauralkonium chloride
Benzenemethanaminium, N-dodecyl-N,N-dimethyl-, chloride; C_{12}-alkylbenzyldimethylammonium chloride; Dimethyl benzyl lauryl ammonium chloride; Dodecyl dimethyl benzyl ammonium chloride; Lauryl dimethyl benzyl ammonium chloride; benzyl dimethyl dodecyl ammonium chloride; Benzododecinium chloride. Bactericide and fungicide for disinfectants; emulsifier; external antistat for plastics. *Henkel/Functional Prods.*

7987 Dehyquart LT
112-00-5 203-927-0
Laurtrimonium chloride
Emulsifier for plastics industry, wetting agent, antistat, bactericide, demulsifier, deodorant, conditioning component for hair care products. *Henkel/Functional Prods.; Henkel Chemicals Ltd.*

7988 Dehyquart SP
58069-11-7
Quaternium-52
Emulsifier, conditioning, softening and antistatic agent used in personal care products; metal corrosion inhibitor. *Henkel/Cospha; Henkel Canada; Henkel Chemicals Ltd.*

7989 Dehyton® AB-30
Fatty amine derivative with betaine structure, in liquid form; for liquid shampoos, especially baby and special shampoos. *Henkel Chemicals Ltd.* Name unverified.

7990 Dehyton® G
68647-53-0 272-043-5
Cocoamphocarboxyglycinate
Surfactant for mild and conditioning shampoos, bath products, baby shampoos, skin cleansers, foam baths. *Henkel Canada; Henkel KGaA; Pulcra SA.* Discontinued.

7991 Dehyton® K
Betaine in liquid form; additive for shampoos. *Henkel Chemicals Ltd.* Name unverified.

7992 Dehyton® KE
Cocamidopropyl betaine
Surfactant base. *Pulcra SA.*

7993 Dehyton® PAB-30
683-10-3 211-669-5
Lauryl betaine
High foaming detergent for mild shampoos; solubilizer for lauryl sulfates in concentrated shampoos; thickener. *Pulcra SA.*

7994 Dehyton® PG
68650-39-5 272-043-5
Disodium cocoamphodiacetate
Foamer, wetting agent for shampoos, foam baths, cosmetics requiring high foaming, mildness. *Pulcra SA.*

7995 Dehyton® PK
61789-40-0 263-058-8
Cocamidopropyl betaine
High foaming, conditioning detergent for mild shampoos; solubilizer for lauryl sulfates in concentrated shampoos; thickener. *Pulcra SA.*

7996 Dehyton® PLG
14350-96-0 238-305-8
Sodium lauroamphoacetate
Mild, high foaming surfactant for shampoos, foam baths, cosmetics. *Pulcra SA.*

7997 Dekol®
Protective colloids for use in dyeing cellulose fibers and mixtures with vat, sulfur, and naphthol dyes. *BASF AG; BASF plc.*

7998 Dekol® N
Lignin sulfonate
Protection and glossing agent for furs. *BASF AG.*

7999 Dekryll
534-52-1 3331 208-601-1
4,6-dinitrocresol.
A proprietary preparation of 4,6-dinitrocresol, used as a herbicide and insecticide. No manufacturer.

8000 Delac MOR
102-77-2 203-052-4
$C_{11}H_{12}N_2OS_2$
N-Oxydiethylene-benzo-thiazole-2-sulfenamide
Amax; 2-Benzothiazolyl-N-morpholinosulfide; Morpholine, 4-(2-benzothiazolylthio)-; 2-(morpholinothio)benzothiazole; 2-(Morpholinthio)-benzothiazole; N-oxydiethylene-benzothiazole sulfenamide. The most delayed action accelerator offered by Uniroyal; activated by thiurams, dithiocarbamates, BIK, guanidines and aldehyde-amines; nondiscoloring and nonstaining to rubber stocks and materials in contact with them; used in tire treads, carcass, mechanicals and wire jackets. *Uniroyal.* Name unverified.

8001 Delac NS
95-31-8 202-409-1
$C_{11}H_{14}N_2S_2$
N-*t*-Butyl-*o*-benzothiazole-2-sulfenamide
N-*t*-Butyl-2-Benzothiazolesulfenamide; NTBBTS; Nocceler NS; Santocure NS; Vulkacit NZ. A delayed action accelerator very safe at processing temperatures but producing high modulus stocks at curing temperatures; activated by thiurams, dithiocarbamates, aldehyde amines, guanidines and BIK; nondiscoloring and nonstaining to rubber stocks and materials in contact with them; used in tire treads, carcass, mechanicals and wire jackets. mp = 105°; d = 1.25-1.31. *Uniroval.* Name unverified.

8002 Delac S
95-33-0 202-411-2
$C_{13}H_{16}N_2S_2$
N-Cyclohexyl-2-benzothiazolesulfenamide
Sufenax CB; Rhodifax 16; Accelerator CZ; Durax; N-cyclohexylbenzothiazole-2-sulfenamide; CBTS; CBS; Durax; N-Cyclohexyl-2-benzothiazyl sulphenamide; N-Cyclohexylbenzothiazyl sulphenamide; N-Cyclohexyl-2-benzothiazole sulfenamide; Santocure. Accelerator used in natural rubber and styrene-butadienethiazyl sulfenamide rubber. An all-purpose delayed action accelerator which combines superior scorch safety with shorter curing cycles; used in tire tread, carcass, camelback and mechanical goods. mp = 93-100°, insoluble in H_2O. *Uniroyal.* Name unverified.

8003 Delafila
Slate powder. *Delabole Slate.*

8004 Delaglas® A
PMMA extruded sheet. *Asahi Chem. Industry.*

8005 Delalot's alloy
An alloy containing 80% copper, 18% zinc, 2% manganese, and 1% calcium phosphate.

8006 Delan-Col
3347-22-6 3433 222-098-6

Dithianone
Suspension concentrate containing 600 g dithianon per liter; fungicide used for control of scab in fruit apples and pears. *ICI Chem & Polymers Ltd.*

8007 Delanium
A proprietary trade name for carbon and graphite materials highly resistant to all chemicals except some oxidizing agents. *Powell Duffryn Quarries Ltd.* Name unverified.

8008 Delaphos
7779-90-0 10284 231-944-3
$O_8P_2Zn_3$
Zinc orthophosphate
Heucophos™ZPO; Heucophos™ZPZ; Zinc Phosphate ZP-10. Anticorrosive pigment for paints, particularly primer formulations; in dental cements; flame retardant for plastics. White powder; mp = 900°; d^{15} = 3.998; insoluble in H_2O, EtOH; soluble in dilute mineral acids, AcOH, ammonium hydroxide, alkali hydroxide solutions. Pasminco Europe Ltd/ISC Alloys Div; BASF; Calgon; Colores Hispania SA; Elf Atochem N. Am.; Halox Pigments; Hammond Lead Prods.; Heucotech Ltd.; Landers-Segal Color; Lohmann; McGean-Rohco; Min. Pigments; Nat'l Chem.; Pasminco Europe; Sino-Am. Pigment Sys.

8009 Delaprism®
PMMA prismatic sheet. *Asahi Chem. Industry.*

8010 Delaville
7440-66-6 10255 231-175-3
Zn
zinc
Zinc dust (minimum 95% metallic). Used in zinc-rich anticorrosive paints and as a reducing agent in chemical processes. *Pasminco Europe Ltd/ISC Alloys Div.*

8011 Delchowyte
A decolorizing carbon, prepared from peat.

8012 Delegol®, Delegol-T
Disinfectant on a phenol base; veterinary medicine. *Bayer AG.*

8013 Delf® Clene
2-Butoxyethanol and surfactants; general-purpose water-based cleaner for the furniture industry. *D L Forster Ltd.*

8014 Delf® Drape
Isopropyl alcohol and surfactants; crease and wrinkle remover for curtain and furniture fabrics. *D L Forster Ltd.*

8015 Delf® Fabric Protector
71-55-6 9766 200-756-3
1,1,1-Trichloroethane-surfactants-butane-propane. Stain protector for general fabrics. *D L Forster Ltd.*

8016 Delf® HD Aerosol Adhesive
Synthetic rubber and resin; universal heavy duty adhesive. *D L Forster Ltd.* Name unverified.

8017 Delf® MP Aerosol Adhesive
Synthetic rubber and resin in blend of chlorinated and ketonic hydrocarbon solvents; general purpose adhesive in aerosol form for use in the furniture Industry. *D L Forster Ltd.*

8018 Delf® Silicone Aerosol
Silicone oil-butane-propane. Universal release and lubricating agent; reduces friction on a wide range of surfaces. *DL Forster Ltd.*

8019 Delfloc® 50
Cationic polymer solution; retention aid and flocculant for the paper industry. *Hercules.*

8020 Delft blue
A pigment; a mixture of indigo and ultramarine.

8021 Delhi rustless iron
A corrosion resisting alloy containing 18% chromium, 1.5% silicon, and not more than 0.08% carbon.

8022 Delial®
Suntan products combining sun protection with skin care. *Bayer AG.*

8023 Delicron
Elastomeric precision impression material; for double-mix technique. *Bayer AG.* Discontinued.

8024 Delight
Spray dried light density granular detergent builder containing disilicate. *Kemira Kemi Ltd.* Discontinued.

8025 Delios®
C_8-C_{10} triglycerides
Solvent for flavors and oil-soluble food additives; surface treating agent for dried fruits, confectionery, dietary foodstuffs. *Grünau.*

8026 Dellatol® BBS
3622-84-2 222-823-6
$C_{10}H_{15}NO_2S$
n-Butylbenzenesulfonamide
N-butyl-benzenesulfonamide; N-n-butylbenzene sulfonamide; 4-n-butylbenzene sulfonamide. Plasticizer for polymide 6, 66, 11, and 12 and copolymides; also for flexibilizing cellulose derivatives, especially flame-retardant cable coatings based on cellulose acetate and cellulose acetobutyrate. *Bayer AG.*

8027 Delnet
A nonwoven fabric made foam high density polyethylene, polypropylene and polypropylene copolymers; nonadherent facing for surgical dressings, feminine hygiene products, disposable press cloth, breathable laminate, fusible adhesive, reinforcement for paper, plastic foam and other non-woven fabrics. *Hercules.* Discontinued.

8028 Delo
Lubricating oil. *Monsanto (Solaris).* Name unverified.

8029 Deloxan
Organo functional polysiloxanes. *Degussa Ltd.*

8030 Deloxil
bromoxynil-oxynil. An emulsifiable concentrate containing 190 g bromoxynil and 190 g ioxynil per liter; a post-emergence contact herbicide for cereal crops. *Hoechst UK.*

8031 Delpet®
PMMA resin. *Asahi Chem. Industry.*

8032 Delrin®
A proprietary trade name for a stiff strong engineering plastic of the acetal resin type; excellent fatigue resistance; used as a replacement for die cast parts in gears, bearings and housings. *DuPont UK.*

8033 Delrin® 100, 500
105-57-7 36 203-310-6
Acetal homopolymer resin
Acetal homopolymer resin; engineering thermoplastic with high strength and stiffness; also offers high fatigue endurance, natural lubricity, corrosion resistance, and resilience where recovery and toughness under wide range of temperature and humidity conditions are required; 100 grade is the most viscous in the series, used in machinery, agricultural equipment, interior automotive door handles, clock mechanisms, ballcock valves, videocassettes and other molding applications. *DuPont.*

8034 Delrin® 100AF, 500AF
105-57-7 36 203-310-6
Acetal homopolymer resin
Acetal homopolymer resin with Teflon fibers; low friction/low wear grade. *DuPont.*

8035 Delrin® 100ST, 500T
105-57-7 36 203-310-6
Acetal resin
Acetal resin; toughened grades. *DuPont.*

8036 Delrin® 107, 507
105-57-7 36 203-310-6
Acetal resin
Acetal resin; uv-stabilized grades for improved weatherability. *DuPont.*

8037 Delrin® 150A, 550SA
105-57-7 36 203-310-6
Acetal resin
Acetal resin; low die deposit grades for improved extrusion. *DuPont.*

8038 Delrin® 570
105-57-7 36 203-310-6
Acetal resin
Acetal homopolymer resin, glass fiber-filled; high stiffness grade; *DuPont.*

8039 Delrin® 900
105-57-7 36 203-310-6
Acetal resin
Acetal homopolymer resin; high flow grade; *DuPont.*

8040 Delrin® AF Blend
105-57-7 36 203-310-6
Acetal resin
Acetal homopolymer; thermoplastic for use in moving parts where low friction and long wear are important; for bushings, gears, slides, housing, guides, low friction components, electrical components. *Dupont; Polymer Corp.*

8041 Delsene® 50 DF
83601-81-4
$C_{13}H_{18}ClN_3O_2$
Carbamic acid, (5-butyl-1H-benzimidazol-2-yl)-, methyl ester, monohydrochloride
Systemic fungicide. *DuPont UK.*

8042 Delsene® M Flowable
83601-81-4, 12427-38-2
carbendazim-maneb. A suspension concentrate containing 50 g carbendazim and 320 g maneb per liter; systemic fungicide for cereals. *DuPont UK.*

8043 Delta
Two-piece can inks; for two-piece cans, side wall printing of pilfer proof closures and fast cure sheet fed metal containers. *Coates Coatings Ltd.*

8044 Delta acid
119-28-8 2410 204-311-4
$C_{10}H_9NO_3S$
β-Naphthylamine-7-sulfonic acid
F-acid; Cassella's Acid F; 1,7-Cleve's acid. Used as an intermediate for azo dyes. Soluble in H_2O (4.5 mg/ml).

8045 Delta Metals®
The trademark for a variety of metals, metallic alloys, and metal articles. No manufacturer.

8046 Deltacast
A range of specially designed release agents for the foundry industry, most usually the pressure die casting of aluminum. *Acheson Colloids.*

8047 Deltaforge
A range of lubricating dispersions of graphite or other lubricating pigments in water; designed especially for lubrication of dies or for coating of billets in the metal forging process. *Acheson Colloids.*

8048 Deltaglaze
A range of products designed to protect titanium, steel and super alloys during heating and during the forging process. *Acheson Colloids.*

8049 Delta-Therm® NDT-300
Rigid polyurethane systems; for dispensing froth foams through pressurized equipment for water heaters. *Flexible Prods.*

8050 Delta-Therm® NDT-402
Rigid polyurethane systems; for dispensing froth foams through pressurized equipment for sandwich panels, thin wall refrigeration, coolers/freezers trucks. *Flexible Prods.*

8051 Deltyl
A proprietary trade name for a fatty acid ester which serves as a plasticizer. No manufacturer.

8052 Delweve
Nonwoven fabric made from polypropylene by the film extrusion-embossing-orientation process; excellent backing material, skirt and cambric liners for furniture, reinforcement for films, plastic foam, needlepunched nonwovens and wall coverings and bale coverings. *Hercules.* Discontinued.

8053 Demavet
67-68-5 3308 200-664-3
dimethyl sulfoxide
Dimethyl sulfoxide in aqueous solution. *Ciba plc.* Name unverified.

8054 Demelloy
Soft solder and brazing alloys for semiconductor industry. *Degussa Ltd.*

8055 demeton-S-methyl
919-86-8 6129 213-052-6
$C_6H_{15}O_3PS_2$
S-[2-(ethylthio)ethyl] O,O-dimethyl phosphorothioate
methyl demeton; methylmercaptofostiol; Bay 18436; Bay 25/154; DSM; Duratox; Metasystox 55; Metasystox I; Mifatox; Persyst. Systemic insecticide and acaricide with contact and stomach action. Used for control of aphids and other insects in a wide variety of crops. bp_1= 118°; d^{20} = 1.207; n_D^{20}= 1.5065; soluble in H_2O (3.3 g/l), more soluble in organic solvents; LD_{50} (rat orl) = 40-106 mg/kg.

8056 Demetox
919-86-8 6129 213-052-6
demeton-S-methyl
Insecticide containing demeton-s-methyl. *ICI Chem & Polymers Ltd.*

8057 Demix® 7730
Catalytically disproportionated tall oil product; emulsifier for SBR production. *Arizona.* Discontinued.

8058 Demulsifier 3837
Polymer; demulsifier for metalworking applications; for use at alkaline pH and without the use of any other chemical. *Hoechst Celanese/Colorants & Surf.*

8059 DenClen
Denture cleanser (powder, tablets and liquid), for cleaning dentures. *Richardson-Vicks Inc.* Name unverified.

8060 Denesive
Coating, separating and impregnating media for paper, leather, cardboard, asbestos, synthetic plastics, metal and textiles; all containing synthetic resins and silicones. *Wacker-Chemie GmbH.* Name unverified.

8061 De-NOx Catalyst MDN-100
Nitric acid plant exhaust gas; applicable temperatures 300-500°. *Mitsubishi Kasei.* Name unverified.

8062 Denquel
Sensitive teeth toothpaste, to relieve the discomfort of sensitive teeth. *Richardson-Vicks Inc.* Name unverified.

8063 Densil
Industrial biocides. *ICI Chem & Polymers Ltd.*

8064 Densites
Mining explosives; contain ammonium nitrate, sodium or potassium nitrate, and trinitrotoluene.

8065 Densodrin®
Combination of special natural and synthetic oils; lightfast waterproof agents; also for vegetable-tanned, synthetic-tanned or retanned leather. *BASF AG; BASF plc.*

8066 Dentomat
Dental amalgamator. *Degussa Ltd.*

8067 Dentplus Special
7757-93-9 1739 231-826-1
A proprietary trade name for dicalcium phosphate dihydrate; used as a thickening agent, cleaning agent, and carrier in toothpaste. *Hoechst UK.* Name unverified.

8068 Denzox
1314-13-2 10279 215-222-5
Dense zinc oxide
Used in speciality ceramics. *Eagle Zinc.*

8069 Deodorant 4761-C, OS
Mixture of fragrance materials; heat-resistant deodorizer for residual odors in the finished product *Andrea Aromatics.* Name unverified.

8070 Deoxidine
Metal treating compositions. *ICI Chem & Polymers Ltd.* Discontinued.

8071 Deoxidizing Tubes
Pure copper tubes containing a variety of deoxidizing agents for the treatment of copper and nickel alloys. *Foseco (F.S) Ltd.*

8072 Deoxiphos 600
Liquid, phosphoric acid based cleaner; deoxidizer for aluminum and steel. *Invequimica & CIA SCA.* Name unverified.

8073 Deoxylyte
Metal treating compositions. *ICI Chem & Polymers Ltd.* Discontinued.

8074 Depat
O,O-Diethyl phosphoroamidothioate
An intermediate in the manufacture of phosphorus pesticides. *A/S Cheminova.* Name unverified.

8075 Dephosphex
Highly basic oxidizing supplementary flux for phosphorus removal from arc melted steel. *Foseco (F.S) Ltd.* Discontinued.

8076 Depsoline
Textile auxiliary chemicals. *ICI Chem & Polymers Ltd.*

8077 Dequalinium Chloride
522-51-0 2959 208-330-9
$C_{30}H_{40}Cl_2N_4$
1,1'-(1,10-Decanediyl)-bis-[4-amino-2-methylquinolinium chloride]
Dequadin; 1,1'-decamethylenebis[4-aminoquinaldinium chloride]; BAQD-10; decamine; Dekamin; Decatylen; Dekadin; Dequadin Chloride; Dequafungan; Dequavet; Dequavagyn; Eriosept; Evasol; Grocreme; Labosept; Optipest; Phylletten; Polycidine; Sorot. A proprietary preparation containing dequalinium chloride; antibacterial. mp = 326°; soluble in H_2O. *Crookes Healthcare.*

8078 Dequaspon
A proprietary preparation of gelatin sponge impregnated with dequadin; dental packing for hemorrhage. *Allen & Hanburys Ltd.* Name unverified.

8079 Dequest
Phosphonates; used in scale prevention and corrosion inhibition for industrial water treatment and to control metal ions in aqueous systems. *Monsanto Co.*

8080 Derakane®
Vinyl ester resins formulated for the reinforced plastics industry; used to make articles by molding, spray-up and filament winding, in sheet molding compounds and in coatings. *Dow Cheml Co Ltd, UK & Ireland.*

8081 Derakane® 411-35
Epoxy-based vinyl ester resin; thermoset; provides superior toughness and corrosion resistance; used in chemical processing industry, pulp and paper mills; FDA compliant. *Dow Chemical.*

8082 Derakane® 411C-50
Vinyl ester resin; resin transfer molding grade. *Dow Chemical.*

8083 Derakane® 470-45
A proprietary polyvinyl ester resin possessing good chemical resistance to chlorinated solvents. *Dow Cheml Co Ltd, UK & Ireland.*

8084 Derakane® 510-40
A proprietary polyvinyl ester resin used in fire-retardant laminates; contains 20% bromine. *Dow Cheml Co Ltd, UK & Ireland.*

8085 Derakane® 510B-700PAT
Vinyl ester resin; formulated for use in the military marine market for the production of high-quality lightweight navy craft. *Dow Chemical.*

8086 Derakane® 8084
Elastomer-modified vinyl ester resin; offers increased adhesive strength, superior resistance to abrasion and mechanical stress, exceptional toughness, chemical resistance; for FRP applications. *Dow Chemical.*

8087 Deraspan
Insulating panels using Styrofoam brand plastic form as the core material and a variety of outside layers such as aluminum, plywood, etc; for cold storage warehouses, dairy coolers, cold rooms for food processing, etc. *Dow UK.* Discontinued.

8088 Derifil
Chlorophyllin copper complex, obtained by removing the methyl and phytyl ester groups with alkali and replacing the magnesium with copper; a deodorant. *Rystan Company Inc.* Name unverified.

8089 Deriphat 154
61791-56-8 263-190-6
Disodium N-tallow-β iminodipropionate
Detergent, solubilizer for personal care products, hard surface cleaning, textiles, emulsion polymerization; good substantivity. *Henkel/Cospha; Henkel Canada.*

8090 Deriphat 160
3655-00-3 222-899-0
Disodium N-lauryl β-iminodipropionate
Detergent, solubilizer, primary emulsifier used in organic and inorganic compounds; emulsion polymerization and stabilization; wetting agent; mild surfactant for hair and skin products. *Henkel/Cospha; Henkel/Functional Prods.; Henkel Canada.*

8091 Deriphat 160C
26256-79-1 247-552-0
Sodium-N-lauryl β-iminodipropionate
Detergent, solubilizer, stabilizer; used in petroleum processing, emulsion polymerization, foaming cleaners, personal care products. *Henkel/Cospha; Henkel/Functional Prods.; Henkel Canada.*

8092 Derma
Synthetic organic dyestuffs for use in the leather industry. *Sandoz Products Ltd.*

8093 Dermaffine®
143-28-2 6968 205-597-3
Oleyl alcohol
Raw material for cosmetics and pharmaceuticals. *Laserson & Sabetay Ets.*

8094 Dermafill
Impregnating agents in aqueous solution; for leather. *Colour-Chem Ltd.* Name unverified.

8095 Dermalac
Nitrocellulose lacquers and emulsions; for leather. *Colour-Chem Ltd.* Name unverified.

8096 Dermalcare® 1673
Sodium laureth sulfate-disodium laureth sulfosuccinate-laureth-6 carboxylic acid-cocamidopropyl betaine-ammonium chloride. High performance base for preparation of ultra-mild face and skin cleanser formulations. *Rhône-Poulenc Surf.; Rhône-Poulenc France.*

8097 Dermalcare® GMS/SE
Glyceryl stearate SE; self-emulsifying emulsifier for creams and lotions. *Rhône-Poulenc Surf.*

8098 Dermalcare® GMS-165
Glyceryl stearate, PEG 100 stearate; emulsifier for oil-water creams and lotions; high electrolyte tolerance. *Rhône-Poulenc Surf.*

8099 Dermalcare® NI
Cetearyl alcohol, ceteareth-20; broad tolerance emulsifier for oil-water systems, viscosity builder for cosmetic creams, lotions, ointments. *Rhône-Poulenc Surf.; Rhône-Poulenc UK; Rhône-Poulenc France.*

8100 Dermalcare® POL
Cetearyl alcohol, ceteth-20, and glycol stearate; lubricant self-emulsifying wax, emulsifier for lotions and creams; effective over broad pH range. *Rhône-Poulenc Surf.*

8101 Dermalcare® SPS
8002-23-1 232-302-5
Cetyl esters; cosmetic grade emulsifier and emulsion stabilizer for creams, lotions, antiperspirants, creme rinse conditioners, personal care products; substitute for natural spermaceti; acid and alkali stable. *Rhône-Poulenc Surf.*

8102 Dermalex
squalene-hexachlorophene-allantoin
A proprietary preparation of squalene, hexachlorophene and allantoin; a protective skin lotion. *The Dermalex Co Ltd.* Name unverified.

8103 Dermane
111-01-3 8923 203-825-6
squalane
Synthetic squalane. Used in cosmetics and as a lubricant. *Universal Preserv-A-Chem Inc.* Unverified.

8104 Dermasome® A
lecithin-allantoin
Lecithin and allantion; cosmetic ingredient encapsulated in lipid spheres; for skin care products *Microfluidics.* Discontinued.

8105 Dermasome® E
lecithin-tocopheryl acetate
Lecithin and tocopheryl acetate; cosmetic ingredient encapsulated in lipid spheres; for skin care products *Microfluidics.* Discontinued.

8106 Dermasome® H
9067-32-7 4793
lecithin and sodium hyaluronate
Lecithin and sodium hyaluronate; cosmetic ingredient encapsulated in lipid spheres; for skin care products *Microfluidics.* Discontinued.

8107 Dermasome® MT
8002-43-5 5452 232-307-2
lecithin
Lecithin; cosmetic ingredient encapsulated in lipid spheres; for skin care products *Microfluidics.* Discontinued.

8108 Dermasome® P
panthenol-lecithin
Panthenol and lecithin; cosmetic ingredient encapsulated in lipid spheres; for skin care products *Microfluidics.* Discontinued.

8109 Dermasome® S
squalane-lecithin
Squalane and lecithin; cosmetic ingredient encapsulated in lipid spheres; for skin care products *Microfluidics.* Discontinued.

8110 Dermasome® SC
lecithin-soluble collagen
Lecithin and soluble collagen; cosmetic ingredient encapsulated in lipid spheres; for skin care products. *Microfluidics.* Discontinued.

8111 Dermasome® SOD
lecithin-superoxide dismulase
Lecithin and superoxide dismulase; cosmetic ingredient encapsulated in lipid spheres; for skin care products *Microfluidics.* Discontinued.

8112 Dermasome® TRF
8002-43-5 5452 232-307-2
lecithin
Tissue/skin respiratory factors and lecithin; cosmetic ingredient encapsulated in lipid spheres; for skin care products. *Microfluidics.* Discontinued.

8113 Dermasome® U
urea-lecithin
Urea and lecithin; cosmetic ingredient encapsulated in lipid spheres; for skin care products *Microfluidics.* Discontinued.

8114 Dermasome® V
Aloe vera gel-lecithin
Aloe vera gel and lecithin; cosmetic ingredient encapsulated in lipid spheres; for skin care products. *Microfluidics.* Discontinued.

8115 Dermasulph
A proprietary preparation of polythionates. Sulfur-containing skin ointment. *Crookes Laboratories.* Name unverified.

8116 Dermoblock OS
118-60-5 204-263-4
$C_{15}H_{22}O_3$
Octyl salicylate
Ethylhexyl salicylate; Sunarome O; Sunarome WMO; 2-Ethylhexyl salicylate. UV light absorber for sunscreens. *Alzo.* Discontinued.

8117 Dermol 89
71566-49-9 275-637-2
Octyl isononanoate
Emollient for skin care and make-up products; partial replacement for silicone oils; antitackiness aid in antiperspirants; resin plasticizer for hair sprays. *Alzo.*

8118 Dermol 105
60209-82-7 262-108-6
Isodecyl neopentanoate
Dry emollient, low viscosity oil; sun protection factor booster. *Bernel.*

8119 Dermol 108
34962-91-9 252-302-9
Isodecyl octanoate
Emollient for cosmetics; antitackiness aid. *Alzo.*

8120 Dermol 185
58958-60-4 261-521-9
Isostearyl neopentanoate
Emollient, pigment binder, freeze/thaw stabilizer. *Bernel.*

8121 Dermol 334
Isodecyl octanoate, octyl isononanoate, diethylene glycol diisononanoate,

diethylene glycol dioctanoate; Carbopol dispersing aid; dry feel emollient. *Alzo.* Discontinued.

8122 Dermol 489
72269-52-4, 106-01-4
Diethylene glycol dioctanoate, diethylene glycol diisononanoate; emollient for cream and lotion formulations; dispersant for Carbopol powders. *Alzo.* Discontinued.

8123 Dermol DISD
127358-81-0
diisostearyl dimer dilinoleate
Emollient for creams, lotions, and makeup; anti-irritant in formulations. *Alzo.*

8124 Dermol G-76
125804-12-8
Glycereth-7 benzoate
Emollient for creams, lotions, bath products, liquid soaps, hydro-alcoholic solutions; softener and moisturizer for skin. *Alzo.*

8125 Dermol GL-7A
57569-76-3
Glycereth-7 triacetate
Emollient, solubilizer. *Alzo.*

8126 Dermol ICSA
138208-68-1
Isocetyl salicylate
Emollient; solvent for benzophenone in sunscreen formulations. *Alzo.*

8127 Dermol L45
125804-13-9
Glycereth-4,5 lactate
Emollient, humectant for hydro/alcoholic formulations, aftershave, body splashes. *Alzo.*

8128 Dermol MO
Glycereth-7 diisononanoate, diethylene glycol dioctanoate, diethylene glycol diisononanoate; emollient for skin care formulations and lip products; replacement for mineral oil. *Alzo.* Discontinued.

8129 Dermol OO
29710-25-6 249-793-7
Octyl oxystearate
Prevents defatting of skin from harsh surfactants and detergents; for cosmetic and cleansing formulas; binder in pressed powders. *Alzo.*

8130 Dermolan GLH
138314-11-1
Glycereth-7,5-hydroxystearate
Emollient in creams and lotions; self emulsifier; viscosity builder for shampoos; conditioner for shampoos. *Alzo.* Discontinued.

8131 Dermoplast
benzocaine-benzethonium chloride-menthol-hydroxyquinalone benzoate-methyl paraben
A proprietary preparation of benzocaine, benzethonium chloride, menthol, hydroxyquinalone benzoate, and methyl paraben in the form of an aerosol; used as a soothing skin spray. *Wyeth Laboratories.* Discontinued.

8132 Dermoxyl®
94-36-0 1149 202-327-6
Benzoyl peroxide
For treatment of acne. *ICN Pharmaceuticals Inc.*

8133 Derosal WDG
83601-81-4
Carbendazim
Systemic fungicide. *Hoechst UK.*

8134 Derris Dust
83-79-4 8427 201-501-9
$C_{23}H_{22}O_6$
rotenone
Contains rotenone; garden insecticide. *ICI Garden Products.*

8135 Derussole
1333-86-4 1856 215-609-9
carbon
Diablack® A. Carbon black dispersions; for simple and dust-free dyeing of paints, lacquers, paper, cardboard, plastics, synthetic fibers, printing inks, and mineral binders. *Degussa AG.* Name unverified.

8136 Desamidocollagen
9007-34-5 2543 232-697-4
Soluble collagen; moisturizer and film-former for toiletries. *Henkel/Cospha; Henkel Canada.* Name unverified.

8137 Desavin
Plasticizer for use in coating materials based on unsaponifiable binders as well as in plastics dispersions and film coatings. *Bayer AG.*

8138 Descale
Acid detergent for milkstone removal. *Ciba plc.*

8139 Desicchlora
13465-95-7 236-710-4
$BaCl_2O_8$
barium perchlorate
Perchlorate of barium; Perchloric acid, barium salt. A proprietary name for a perchlorate of barium, a drying agent to replace calcium chloride, sulfuric acid, and potassium hydroxide; absorbs 20% of its weight of water. mp = 505°; d = 3.200; soluble in H_2O. No manufacturer.

8140 Desimpal
An active oxygen compound; used in the textile industry during peroxide bleaching in order to prevent precipitation of hardening silicate when stabilizing with water glass. *Degussa AG.* Name unverified.

8141 Desmalkyd®
Oil-modified polyurethanes; for the formulation of paints and varnishes as well as printing inks. *Bayer AG.*

8142 desmetryn
1014-69-3 213-800-1
$C_8H_{15}N_5S$
N-methyl-N'-(1-methylethyl)-6-(methylthio)-1,3,5-triazine-2,4-diamine
Semeron; Isopropylamino)-4-(methylamino)-6-(methylthio)-s-triazine; Methylamino)-4-(isopropylamino)-6-(methythio)-s-triazine; desmetryne; G 34360; Topusyn; Methylamino-4-methylthio-6-isopropylamino-1,3,5-triazine; Methylthio-4-isopropylamino-6-methylamino-s-triazine; Topusyn; Triazine, 2-(isopropylamino)-4-(methylamino)-6-(methylthio)-; N-methyl-N'-(1-methylethyl)-6-(methylthio)-; 2-Methylthio-4-methylamino-6-isopropyl-1,3,5-triazine. Selective systemic herbicide, inhibits photosynthesis. Used for post-emergence control of broad-leaved weeds and some grasses in brassicas, herbs, onions, leeks and conifer seed beds. mp = 84-86°; soluble in H_2O (580 mg/l), more soluble in organic solvents; LD_{50} (rat orl) = 1390 mg/kg.

8143 Desmocap®
Polyurethane coating raw material; for automotive engineering, mechanical and plant engineering, industrial coatings, building, wood and furniture finishing, coatings for plastics and paper. *Bayer AG; Bayer plc.*

8144 Desmocast®
Polyols for elastomers. *Bayer AG.*

8145 Desmocoll®
High molecular hydroxyl polyurethanes; used for solution based adhesives with particular suitability for high grade footwear sole bonding and production of plastics laminates. *Bayer AG; Bayer plc.*

8146 Desmocoll® 110
High molecular weight linear polyester urethane; for use in adhesives; used primarily in contact bonding with heat activation in the footwear, laminating and packaging industries. *Bayer; Miles.*

8147 Desmocoll® 526
Linear polyester PU based on MDI; elastomeric PU for use in adhesives for bonding plastics, rubber, leather, wood, textiles, paper; suitable for contact bonding at room temperature *Bayer; Miles.*

8148 Desmoderm®
One-component polyurethane finishes; for polyurethane coated materials. *Bayer AG.*

8149 Desmoderm® Foil
Microporous polyurethane foil. *Bayer plc.* Discontinued.

8150 Desmodur®
Crosslinking agent for adhesives based on Baycoll, Baypren, Desmocoll, and Dispercoll and PUR coatings; isocyanates for foams and elastomers. *Bayer AG; Bayer plc.*

8151 Desmodur® HL
TDI/HDI-based polyisocyanate; used for fast curing two-component coatings for furniture and metal; improved light stability. *Bayer AG.*

8152 Desmodur® IL
TDI-based polyisocyanate; used for extremely fast-curing two-component coatings for wood and metal. *Bayer AG.*

8153 Desmodur® KA-8331
Polyisocyanate polyether PU based on MDI; solvent-free, fast-curing moisture-cure adhesive for bonding textiles, leather, wood, metal, and plastics. *Bayer; Miles.* Name unverified.

8154 Desmodur® L
TDI-based polyisocyanate; for fast-drying topcoats and primers for metal, plastic and several substrates; recommended for interior use. *Bayer AG.*

8155 Desmodur® L75A
Adduct of toluene diisocyanate and polyol; cross-linking agent that improves resistance to heat, greases, oils, plasticizers, organic solvents, and the adhesion to many substrates; primarily used in adhesives. *Bayer Corp; Polysar.*

8156 Desmodur® R, RF
Bonding agent. *Bayer plc.*

8157 Desmodur® RE
2422-91-5 219-351-8
$C_{22}H_{13}N_3O_3$
triphenylmethane-4,4',4-triisocyanate
triphenylmethane triisocyanate; benzene, 1,1',1''-methylidynetris[4-isocyanato-. Triphenylmethane 4,4·g,4·h-triisocyanate in ethyl acetate; room temperature crosslinking agent for adhesives based on Desmocoll and Baycoll polymers, natural rubber, and synthetic rubbers. Triphenylmethane triisocyanate is also used as a plasticizer. *Bayer Corp; Polysar.*

8158 Desmodur® RFE
Tris (*p*-isocyanatophenyl)-thiophosphate in ethyl acetate; room temperature crosslinking agent for adhesives based on Desmocoll and Baycoll polymers, natural and synthetic rubbers. *Bayer Corp; Polysar.*

8159 Desmodur® SL
66% solution of a polyisocyanate; for use in combination with Leguval to increase adhesion on a wide variety of materials and to produce hard, tack-free surfaces. *Bayer AG.*

8160 Desmodur® VKS-2, VKS-4, VKS-18
101-68-8 202-966-0
Polymethylene polyphenyl isocyanate
Room temperature crosslinking agent for adhesives based on Desmocoll and Baycoll polymers, used with natural and synthetic rubbers. *Bayer Corp; Polysar.*

8161 Desmodur® VL
MDI-based polyisocyanate; used for high solids and solvent-free coatings, sealants and related products, cationic electrodeposition. *Bayer AG.*

8162 Desmoflex®
PU elastomer; for building, furniture, shoes, automotive and mech. engineering, sports surfacing, sporting goods, textiles, electrical industry, domestic appliances. *Bayer AG.*

8163 Desmolac®
Polyurethane coating raw material; for automotive engineering, mechanical and plant engineering, industrial coatings, building, wood and furniture finishing, coatings for plastics and paper. *Bayer AG; Bayer plc.*

8164 Desmopan®
Thermoplastic polyurethane elastomer with various levels of shore hardness; for processing by injection molding and extrusion. *Bayer AG; Bayer plc.*

8165 Desmopan® 150
9009-54-5
Thermoplastic ester polyurethane; injection molding grades with high mechanical strength, good flow, short cycles, easy mold release; used for articles subjected to wear, e.g., castors, shoe heels. *Bayer AG.*

8166 Desmopan® 385
9009-54-5
Thermoplastic ester polyurethane; injection molding and extrusion grade with high mechanical strength, increased resistance to hydrolysis, improved low temperature flexibility; used for extruded articles and heavy duty engineering components. *Bayer AG.* Name unverified.

8167 Desmopan® 585
9009-54-5
Thermoplastic ester PU; soft extrusion and injection molding grades with good resist to microbial attack; used for parts where exposure to high mechanical loads is combined with risk of attack from microorganisms, e.g., sports shoe soles, cable sheathing, animal marking tags; 585 is an extremely wear-resistant grade. *Bayer AG.* Name unverified.

8168 Desmopan® 786
9009-54-5
Desmopan® 385; Desmopan® KA 8333; Desmopan® 585; Desmopan® 150. Thermoplastic ether carbonate PU; exhibits very good resistance to hydrolysis and microbial attack; provide dependable performance in harshest outdoor environments; 786 for injection molding and extrusion; used for fire hoses. *Bayer AG.* Name unverified.

8169 Desmopan® KA 8333
9009-54-5
Thermoplastic ether PU; injection molding grades with very good resistance to hydrolysis and microbial attack, very good low-temperature flexibility; used where impact resistance and flexibility at very low temperatures is required, e.g., cable, ski boots. *Bayer AG.* Name unverified.

8170 Desmophen®
Flexible and rigid polyurethane foams; flexible foam for cushioning applications, upholstered furniture, mattresses, automotive engineering, textiles, packaging; rigid foam for heating and refrigeration engineering, building industry, sporting goods, automotive engineering, also for industrial coatings. *Bayer AG; Bayer plc; Miles.*

8171 Desmorapid®
Catalyst for adhesives based on Baycoll, Desmodur and Desmocoll. *Bayer AG.*

8172 Desmotherm®
Polyurethane coating raw material; for automotive engineering, mech. and plant engineering, industrial coatings, building, wood and furniture finishing, coatings for plastics and paper. *Bayer AG; Bayer plc.*

8173 Desodora
Skin protective soap with substances against bacterial attack and fungal infections of the skin; for protection and cleansing of the skin under environmental stress. *Dynamit Nobel Wien GmbH.* Name unverified.

8174 Desogrip
Trade name for resin based adhesive in polychloroprene digression; used for attaching soles and uppers for shoe manufacturing. *Seal Ltd.*

8175 Desomeen
Ethoxylated fatty amines. *Witco/Organics.* Discontinued.

8176 DeSonate 50-S
25155-30-0 8757 246-680-4
Sodium dodecylbenzene sulfonate
53% active; detergent component. *Witco/Organics.* Discontinued.

8177 DeSonate 60-S
25155-30-0 8757 246-680-4
Sodium dodecylbenzene sulfonate
60% active; detergent component. *Witco/Organics.* Discontinued.

8178 DeSonate AOS
sodium α olefin sulfonate
High foaming detergent component for personal care products, light duty detergents and industrial cleaners. *Witco/Organics.* Discontinued.

8179 DeSonate SA
Dodecylbenzene sulfonic acid
Component for manufacture of many detergents. *Witco/Organics.* Discontinued.

8180 DeSonate SA-H
Alkylbenzene sulfonic acid
Emulsifier, chemical intermediate (non biodegradable). *Witco/Organics.* Discontinued.

8181 Desonic® 30C
Ethoxylated castor oil; emulsifier, lubricant, dye leveler, antistatic agent and dispersant for various textile applications and the pulp and paper industry. *Witco/Organics.* Discontinued.

8182 Desonic® DA-4
26183-52-8
Deceth-4
Wetting and rewetting agent for industrial applications. *Witco Corporation.*

8183 Desonic® DA-6
26183-52-8
Deceth-6
Wetting agent for built scour systems. *Witco Corporation.*

8184 Desonic® 20N
Ethoxylated nonylphenols
1-100 moles ethylene oxide, emulsifiers, wetting agents, detergents and dispersants. *Witco Corporation.*

8185 Desonic® S Series
Octylphenol ethoxylates
Various degrees of ethoxy; emulsifiers, wetting agents, detergents and dispersants. *Witco Corporation.*

8186 DeSonol A
2235-54-3 218-793-9
Ammonium lauryl sulfate
Detergent and shampoo component. *Witco/Organics.* Discontinued.

8187 DeSonol AE
67762-19-0
Ammonium laureth (3) sulfate
Component for detergents, shampoos and personal care products. *Witco/Organics.* Discontinued.

8188 DeSonol S
Sodium lauryl sulfate
Detergent and shampoo component. *Witco/Organics.* Discontinued.

8189 DeSonol SE
Sodium laureth (3) sulfate
Component for detergents, shampoos and personal care products. *Witco/Organics.* Discontinued.

8190 DeSonol SE-2
Sodium laureth (2) sulfate
Component for detergents, shampoos and personal care products. *Witco/Organics.* Discontinued.

8191 DeSonol T
139-96-8 205-388-7
TEA lauryl sulfate
Mild shampoo component. *Witco/Organics.* Discontinued.

8192 Desophos®
Phosphate esters. Detergents. *Witco/Organics.* Discontinued.

8193 Desoplas
Trade name for resin based adhesive in polyurethene dispersion; used for attaching soles and uppers for shoe manufacturing. *Seal Ltd.*

8194 DeSotan® SMO
1338-43-8 8872 215-665-4
Sorbitan monooleate
Lipophilic emulsifier, fiber lubricant and softener. *Witco/Organics.* Discontinued.

8195 DeSotan® SMO-20
9005-65-6 7742
Polysorbate 80
Hydrophilic emulsifier and wetting agent. *Witco/Organics.* Discontinued.

8196 DeSotan® SMT
Sorbitan monotallate
Lipophilic emulsifier, fiber lubricant and softener. *Witco/Organics.* Discontinued.

8197 DeSotan® SMT-20
PEG-20 sorbitan tallate; hydrophilic emulsifier and wetting agent. *Witco/Organics.* Discontinued.

8198 Destral
2,4-D-dalapon-diuron
Mixture of 2,4-D, dalapon and diuron; used for total weed control in non crop areas. *ABM Chemicals Ltd.*

8199 Desulfex
Fluxes for removal of sulfur from molten iron or steel. *Foseco (F.S.) Ltd.*

8200 Detac
Denaturing materials for use in water washed spray booths. *Brent Chemicals International plc.*

8201 Detaclad®
For the chemical industry. *DuPont UK.*

8202 Detarex
Chelating agents of the aminopoly/carboxylic acid type; sequestering agents for trace metal control. *W R Grace Ltd.*

8203 Detarol
Ethylene diamine tetraacetic acid salts; chelating agents for trace metal control. *W R Grace Ltd.*

8204 Deterflo A 210
Modified ethoxylated alcohol; low foaming degreasing sol'ns. for surfaces requiring strongly acid or alkaline detergents, at low temperature and under pressure. Stable to acids and bases; biodegradable; nonionic; liquid; pH 3. *Ceca SA.*

8205 Detergent8®
Blend of an alkanolamine, glycol ethers, and an alkoxylated fatty alcohol biodegradable detergent, wetting agent for cleaning circuit boards, electronic parts, phosphate-sensitive labware, delicate industrial parts, nuclear reactor cavities. *Alconox.*

8206 Detergyl
Textile auxiliary chemicals. *ICI Chem & Polymers Ltd.*

8207 Deterpal
Specialty blends of anionic surfactants; used for industrial detergents, paints. *Elf Atochem UK/Ceca.*

8208 Deterpal 832
Alkyl ether sulfate/solvent blend; degreaser, emulsifier. *Ceca SA.*

8209 Deterpal LC
Alkylsulfate/solvent blend; dispersant and antisettling agent for alkyd paints. *Ceca SA.*

8210 Dethlac
62-73-7 3129 200-547-7
dichlorvos
An insecticidal lacquer, containing dichlorvos; a slow release fly killer. *Gerhardt Pharmaceuticals.* Name unverified.

8211 Dethmor
81-81-2 10174 201-377-6
Warfarin
An anti-coagulant used as a rodenticide. *Gerhardt Pharmaceuticals.* Name unverified.

8212 Det-O-Jet®
potassium hydroxide-sodium silicate-sodium hypochlorite
Highly alkaline detergent containing potassium hydroxide, silicate of soda, and sodium hypochlorite; low sudsing biodegradable detergent with good wetting, emulsifying, and penetrating properties, for ultrasonic and mechanical washers, hospital and lab ware, optical and electronic components, pharmaceutical apparatus, industrial parts. *Alconox.*

8213 Dettol
A proprietary trade name for a germicide containing chloroxylenols and terpineol; very powerful in action and yet non-poisonous. No manufacturer.

8214 Deva
A range of precious metal alloys; used for dentistry and dental engineering. *Degussa.*

8215 Devarda's alloy
An alloy of 45% aluminum, 50% copper, and 5% zinc.

8216 Devrinol
15299-99-7 6503 239-333-3
napropamide
Suspension concentrate containing 450 g/l napropamide; amide herbicide for oilseed, rape and fruit. *Embetec Crop Protection Ltd.*

8217 Devrinol T
15299-99-7 6503 239-333-3
napropamide-trifluralin. Suspension concentrate containing 140 g napropamide and 140 g trifluralin per liter; amide herbicide for oilseed rape and fruit. *Embetec Crop Protection Ltd.*

8218 Deward Steel
A proprietary nonshrinking steel containing 1.55% manganese, 0.3% molybdenum, and 0.9% carbon. No manufacturer.

8219 Dewitt Deadline
Magnesium photoengraving plate; manufactured in different gauges and sizes to meet requirements of the newspaper and printing industry. *Dow UK.* Discontinued.

8220 DEWT L
Blend of chromate and an organic; corrosion control aid for closed recirculating water systems. *Drew Ind. Div.*

8221 Dexamist Ear Spray
dexamethasone-neomycin sulfate
Nonpressurized pump action spray containing dexamethasone, neomycin sulfate and glacial acetic acid; for treatment of *otitis externa* in dogs. *Stafford-Miller.* Unverified.

8222 Dexel
Plastic molding materials, cellulose derivatives or synthetic resins; for industrial use. *Courtaulds Fibres Ltd.*

8223 Dexil
Core breakdown agent. *Foseco (F.S) Ltd.*

8224 Dexine 521
A polyisobutylene material with a good resistance to chemical attack from oxidizing liquors up to 110°; used for lining and covering metal tanks. *Dexine Rubber Co Ltd.* Name unverified.

8225 Dexine 656
A natural rubber based compound with good abrasion resistance used for lining metal tanks. *Dexine Rubber Co Ltd.* Name unverified.

8226 Dexine 687
A natural rubber based material with good resistance to chemicals especially sodium hypochlorite. *Dexine Rubber Co Ltd.* Name unverified.

8227 Dexine 759
A Hypalon-based lining and covering material with very good resistance to chemical attack. It can be used with sulfuric acid at concentrations up to 95% *Dexine Rubber Co Ltd.* Name unverified.

8228 Dexine 779
A polyurethane lining and covering material with a very good resistance to abrasion. *Dexine Rubber Co Ltd.* Name unverified.

8229 Dexlar®
Staple cement; for the automotive industry. *DuPont UK.*

8230 Dexon
Polyglycolic acid
Surgical aid. *Davis & Geck.* Name unverified.

8231 Dexonite
Ebonite. A trade name for a proprietary hard rubber molded material for electrical insulation; a proprietary trade name for ebonite. *Dexine Rubber Co Ltd.* Name unverified.

8232 Dexoplas
A proprietary trade name for a butadiene-styrene plastics material used for constructing corrosion resistant fittings. *Dexine Rubber Co Ltd.* Name unverified.

8233 Dextran
9004-54-0 2989
$(C_6H_{10}O_5)_n$
Dextraven; Gentran; Hemodex; Intradex; Macrose; Onkotin; Plavolex; Polyglucin; Promit. Polymers of glucose with chain-like structures; blood plasma substitute or extender, confections, lacquers, oil-well drilling muds, filtration gel, food additive. Accurate Chem. & Scientific; Am. Biorganics; Pharmacia AB; Schweizerhall; Spectrum Chem. Mfg.; U.S Biochemical.

8234 Dextran
A proprietary trade name for α-D(1→6) polyglucose or polyanhydroglucose. Used as a blood plasma substitute or extender, in confections, lacquers, oil-well drilling muds, filtration gel, food additive. *Fisons plc, Pharmaceuticals Div.* Name unverified.

8235 dextrin
9004-53-9
dextrine
starch gum; amylin; gommelin; artificial gum; vegetable gum. Gum produced by incomplete hydrolysis of starch; thickener, adhesives, sizing paper and textiles, substitute for natural gums, food industry, printing inks, in penicillin manufacture, fuel in pyrotechnic devices. *Am. Maize Prods.; Avebe BV; Grain Processing; Nat'l. Starch & Chem.; A.E. Staley Manufacture.*

8236 Dextroform
A condensation product of dextrin and formaldehyde; used as an antiseptic.

8237 Dextrol OC-20
51811-79-1
Complex organic phosphate ester free acid detergent, wetting agent, emulsifier; for pesticides, emulsion polymerization; corrosion inhibitor; dedusting agent for alkaline powders. Liquid; soluble in H_2O; anionic. *Dexter.*

8238 Dextrol OC-70
Phosphated aliphatic ethoxylate; detergent, wetting agent, dispersant, emulsifier; for alkyd paints, solvent-based coatings, colorant systems, pesticides, PVA and acrylic polymerization; corrosion inhibitor; dispersant for magnetic oxide in aromatic solvents. d = 1.06; 8.83 lb/gal; viscosity = 800 cps; flash point > 300°F; pH 1.5; anionic. *Dexter.*

8239 Dextrone X
4685-14-7 7165 225-141-7
paraquat
Soluble concentrate containing 200 g/l paraquat; a pre-emergence bipyridilium herbicide to control weeds in field crops and ornamentals. *ICI Agrochemicals Professional Products; Chipman Ltd.*

8240 Dextrostix®
A proprietary test strip impregnated with a buttered mixture of glucose oxidase, peroxidase, and a chromogen system; used to estimate blood glucose. *Bayer AG.*

8241 Dextrozyme
A balanced mixture of glucoamylase and pullulanase obtained from selected strains of *Aspergillus niger* and *Bacillus acidopullulyticus* by submerged fermentation; used in the starch syrup industry for the saccharification of liquified starch. *Novo Nordisk.*

8242 Dexuron
diuron-paraquat
Suspension concentrate containing 300 g diuron and 100 g paraquat per liter; used to control weeds around trees and shrubs. *Chipman Ltd.*

8243 DI-43
Polymeric lubricity additive for draw and iron can body lubricants. *Ferro/Keil.*

8244 Diabase Developer®
Azoic bases used in combination with various Diathol Grounder prepares; prior to coupling the bases must be diazotized to form diazo salts. *Mitsubishi Kasei.* Name unverified.

8245 Diablack® A
1333-86-4 1856 215-609-9
Carbon black; for rubber industry. *Mitsubishi Kasei.* Name unverified.

8246 Diacelliton Dye®
Disperse dyes for acetate and nylon. *Mitsubishi Kasei.* Name unverified.

8247 Diacetazotol
83-63-6 3004 201-490-0
$C_{18}H_{19}N_3O_4$
N-Acetyl-N-[(2-methyl-4-[(2-methyl-phenyl)azo]phenyl]acetamide
Dimazon; Diacetylaminoazotoluene; 4"-(o-tolylazo)-o-diacetotoluidide; N,N-diacetyl-o-tolyazo-o-toluidine; 2,3'-dimethyl-4'-(diacetylamino)azobenzene; Dermagan; Diacetotoluide; Pellidol. A red dye used in ointment or as a dusting powder. Crystallizes in two modifications: brick-red needles, mp = 65°, or stout, red prisms, mp = 75°; insoluble in H_2O; soluble in organic solvents.

8248 Diacetin
25395-31-7 3005 246-941-2
$C_{14}H_{24}O_{10}$
Glycerol diacetate
1,2,3-Propanetriol diacetate. Auxiliary for use in foundries. mp = -30°; bp = 280°; d = 1.18; LD_{50} (mus orl) = 8500 mg/kg. *Bayer AG.*

8249 Diacetone alcohol
123-42-2 3008 204-626-7
$C_6H_{12}O_2$
4-hydroxy-4-methyl-2-pentanone
4-Hydroxy-4-methyl-2-pentanone; Diacetone alcohol; 4-Hydroxy-2-keto-4-methylpentane;
2-methyl-3-pentanol-4-one; Acetonyldimethylcarbinol; Diketone Alcohol; 4-Methyl-4-Hydroxy-2-pentanone; Tyranton; Diacetone; Hydroxy-4-methyl-2-pentanone; 2-Hydroxy-2-methyl-4-pentanone; 4-hydroxy-4-methylpentan-2-one. Solvent for nitrocellulose, cellulose acetate, oils, resins, waxes, fats, dyes, tars, lacquers, dopes, coatings, wood preservatives, rayon, artificial leather, metal cleaning; laboratory reagent; hydraulic fluids; textile stripping agent. mp = -43°; bp = 166°; d = 0.9310; n_D^{20}= 1.4233; soluble in H_2O, organic solvents; LD_{50} (rat orl) = 4.0 g/kg. *Allchem Industries; Elf Atochem N. Am.; BP Chem. Ltd; Hoechst Celanese; Shell; Union Carbide.*

8250 Diacetyl
431-03-8 3010 207-069-8
$C_4H_6O_2$
2,3-butanedione
biacetyl; dimethyl diketone; 2,3-diketobutane; diacetyl; 2,3-butadione; 2,3-butanedione; dimethyl diketone; dimethylglyoxal; Butanedione; butadione; 2,3-diketobutane; 2,3-dioxobutane; butane-2,3-dione. Aroma carrier in food products. mp = -2 - -4°; bp = 88°; soluble in H_2O, organic solvents; d = 0.9810; n_D^{20}= 1.3951; LD_{50} (rat orl) = 1580 mg/kg. *Aldrich; Penta Mfg.*

8251 Diacid Dye®
Acid colors for wool, silk, and nylon; produces bright shades. *Mitsubishi Kasei.* Name unverified.

8252 Diaclear® MA-3000-4L
Super high molecular weight polyacrylamide; flocculant for industrial waste water treatment and sedimentation, filtration, and centrifugal processes. *Mitsubishi Kasei.* Name unverified.

8253 Diaclear® MK-166
Polyacrylamide
Flocculant for industrial waste water treatment, dewatering and filtration of suspended organic sludge. *Mitsubishi Kasei.* Name unverified.

8254 Diaclear® MO-3000 H
Super high molecular weight polyacrylamide; flocculant for enhanced oil recovery. *Mitsubishi Kasei.* Name unverified.

8255 Diacotton Dye®
Direct dyes for vegetable fiber and viscose rayon. *Mitsubishi Kasei.* Name unverified.

8256 Diacron Dye
Reactive hot-type dyes, especially suitable for printing and continuous dyeing method. *Mitsubishi Kasei.* Name unverified.

8257 Diacryl
Range of methacrylate monomers used in adhesive and coating resins. *Akzo Chemie UK Ltd.*

8258 Diacryl Dye®
Colors for dyeing polyacrylonitrile fiber. *Mitsubishi Kasei.* Name unverified.

8259 Diacryl P Series®
Colors for dyeing cationic dyeable polyester. *Mitsubishi Kasei.* Name unverified.

8260 Diacupro Dye®
Direct dyes whose fastness to light and washing can be improved considerably after treatment with copper salt. *Mitsubishi Kasei.* Name unverified.

8261 Diadavin®
Cleaning and stain removing agent; for textile finishing. *Bayer AG; Bayer plc.*

8262 Diadem Chrome
Acid mordant wool dyestuffs. *Akzo Chemie.* Name unverified.

8263 Diadol 13
Higher alcohol; surfactant for shampoo, light duty liquid detergents. *Mitsubishi Kasei.* Name unverified.

8264 Diadol 18G
248-470-8
Isostearyl alcohol
Base for oily cosmetic preparations; superfatting agent. *Mitsubishi Kasel.*

8265 Diadur
Zirconia-alloyed aluminum oxide
Used for abrasive industries, super refractories, sandblasting and for safes. *Lonza AG.*

8266 Diaflex
Intraperitoneal dialysis solutions. *Knoll AG.* Unverified.

8267 Diaformer Z-AT
Methacryloyl ethyl betaine/methacrylate copolymer; polymer for hair care products (aerosol or pump hairspray, spritz, mousse); good compatibility with propellants. *Mitsubishi Petrochem.; Sandoz.*

8268 Diahold A-503
Methacrylates/acrylates copolymer amine salt; fixative and film-forming polymer for aerosol and pump hairsprays, setting lotions, spritzes; forms

hard, glossy films; excellent hair holding performance. *Mitsubishi Petrochem.; Sandoz.*

8269 Diahope® -006
Activated carbon; for waste water treatment, water purification, solvent. purification, chems. purification. *Mitsubishi Kasei.* Name unverified.

8270 Diahope® -S60
Activated carbon; for sugar refining, water works, food additives purification. *Mitsubishi Kasei.* Name unverified.

8271 Diaion® CR10
Chelating resin for recovery of metal, waste water treatment, purification of brine for NaOH, membrane process, etc. *Mitsubishi Kasei.* Name unverified.

8272 Diaion® HP 10
Synthetic adsorbent for vitamins, antibiotics, enzymes, steroids, fatty acids, perfumes, other bioactive substances; removal of phenol and surfactants; decolorization applications *Mitsubishi Kasei.* Name unverified.

8273 Diaion® SA10A
Strongly basic anion exchange resin for demineralization of water, catalyst, separation of amino acid, recovery of metal. *Mitsubishi Kasei.* Name unverified.

8274 Diaion® SK1B
Strongly acidic cation exchange resin for softening/demineralization of water, separation and recovery of metal, refining of chems., sugar, dextrose, and amino acid, dehydration of organic solvent, catalyst. *Mitsubishi Kasei.* Name unverified.

8275 Diaion® WK10
Weakly acidic cation exchange resin (methacrylic type) for recovery of metal, dealkalization, iron removal, refining of cane sugar, prep. of antibiotics, medicines, amino acids. *Mitsubishi Kasei.* Name unverified.

8276 Diaion® WK20
Weakly acidic cation exchange resin (acrylic type) for water treatment, recovery of metal, dealkalization, iron removal, prep. of antibiotics, medicines, amino acids. *Mitsubishi Kasei.* Name unverified.

8277 Diak
Curing agent. *DuPont UK.*

8278 Diak No. 4
13253-82-2 236-239-4
4,4-Methylenebis(cyclohexylamine)carbamate
Curing agent. *R. T. Vanderbilt Co Inc.* Discontinued.

8279 Diakon
Acrylic molding and extrusion powders. *ICI Chem & Polymers Ltd.*

8280 Dialead®
Coal tar pitch carbon fiber; for thermoset plastics, cements, metal, rubber applications *Mitsubishi Kasei.* Name unverified.

8281 Dialen 6
α olefin
Comonomer for polyolefin; chemical intermediate. *Mitsubishi Kasei.* Name unverified.

8282 diallyl maleate
999-21-3 213-658-0
$C_{10}H_{12}O_4$
Polymers and copolymers, insecticides. *Aceto; Ashland.*

8283 diallyl phthalate
131-17-9 205-016-3
$C_{14}H_{14}O_4$
1,2-Benzenedicarboxylic acid, di-2-propenyl ester
Daponite Sheet; Allyl Phthalate; diallylester phthalic acid; diallyl ester *o*-phthalic acid; Dapon R; Dapon 35; DAP. DAP; primary plasticizer; intermediate. mp = -70°; bp_5 = 165-167°; d = 1.1200; n_D^{20} = 1.5194; slightly soluble in H_2O, more soluble in organic solvents; LD_{50} (rat orl) = 656 mg/kg. *Allchem Industries; Arco; BP Chem. Ltd; C.P. Hall; OxyChem; Rogers.*

8284 Dialose
7491-09-0 3460 231-308-5
Docusate potassium
Stool softener. *Stuart Pharmaceuticals.* Discontinued.

8285 Dialose Plus
Contains casanthranol and docusate potassium (casanthranol is a purified mixture of the anthranol glycosides derived from *Cascara sagrada*); laxative and stool-softener. *Stuart Pharmaceuticals.* Discontinued.

8286 Dialpha
Fungal α-amylaze dispersion in food grade filler; flour additive. *Diaflex Ltd.* Discontinued.

8287 Dialume
21645-51-2 355 244-492-7
Aluminum hydroxide
Antacid *Armour Pharmaceutical Co.* Name unverified.

8288 Dialuminous Dye®
Direct dyes with good to excellent lightfastness. *Mitsubishi Kasei.* Name unverified.

8289 dialyl
Methylamine and lithium citrate.

8290 Diamalt
Malt extract. *Abri Industrial Waxes.* Discontinued.

8291 Diamex
Malt extract. *ABM Chemicals Ltd.* Name unverified.

8292 Diamin(e)®
Substansive dyestuffs; used for textile dyeing and printing. *Cassella AG.*

8293 Diamine
Range of cationic surfactants of the diamine type, in liquid, solid or paste form; widespread use in industry and in household and personal care formulations, though mostly in the form of derived quaternaries and various salts. *Hoechst AG.* Name unverified.

8294 Diamine B11
90640-45-2 292-564-1
N-C16-22 alkyl propylene diamine
Emulsifier, corrosion inhibitor; chemical intermediate producing ethoxylates for surfactants, acetates for emulsifiers and dispersants. *Berol Nobel AB.*

8295 Diamine BG
61791-55-7 263-189-0
n-Tallow-propylene diamine
Duomeen® T. Emulsifier, corrosion inhibitor; chemical intermediate, surfactant producing ethoxylates for detergent applications, and hydrochlorides/acetates for emulsifiers and dispersants. *Berol Nobel AB.*

8296 Diamine HBG
68603-64-5 271-696-6
Hydrogenated tallow propylene diamine; emulsifier, corrosion inhibitor, surfactant; chemical intermediate producing ethoxylates for detergent applications, hydrochlorides/acetates for emulsifiers, dispersants. *Berol Nobel AB.*

8297 Diamine KKP
61791-63-7 263-195-3
n-Coco propylene diamine
Emulsifier, corrosion inhibitor, surfactant; chemical intermediate producing ethoxylates for detergent applications, hydrochlorides/acetates for emulsifiers, dispersants, and alginates. *Berol Nobel AB.*

8298 Diamine OL
7123-62-8 230-528-9
n-Oleyl propylene diamine
Emulsifier, corrosion inhibitor, surfactant; chemical intermediate producing ethoxylates for detergent applications, hydrochlorides/acetates for emulsifiers and dispersants. *Berol Nobel AB.*

8299 Diamine® H Extra
1,6-Hexanediamine
Intermediate for production of AH salt. *BASF AG.*

8300 Diaminocyclohexane
694-83-7 211-776-7
$C_6H_4N_2$
1,2-cyclohexanediamine
1,2-diaminocyclohexane; DCH-99. High-quality polyamine; epoxy curing agent; chelating agent for oilfield, textile, water treatment, detergent fields; herbicide intermediate; polyamide resins for adhesives, films, plastics, inks and corrosion inhibitors. Colorless clear liquid; miscible with H_2O; sg = 0.94 (20°); mp = 2°(*cis*), 15° (*trans*); bp = 183°; flash point = 75°; LD_{50} (rat orl) = 2300 mg/kg; corrosive; combustible. *Aldrich; Du Pont; Milliken; Pacific Anchor.*

8301 1,2-diaminocyclohexane
694-83-7 211-776-7
$C_6H_{14}N_2$
1,2-cyclohexanediamine
cyclohexanediamine;1,2-Cyclohexanediamine; 1,2-Diaminocyclohexane; DACH. Chemical intermediate. bp_{18} = 92-93°; d = 0.9310; n_D^{20} = 1.4864; soluble in H_2O, organic solvents. *Aldrich; Du Pont; Milliken; Pacific Anchor.*

8302 4,4'-diaminodiphenyl sulfone
80-08-0 2885 201-248-4
$C_{12}H_{12}N_2O_2S$
4,4'-DDS
dapsone; 4,4'-sulfonyldianiline; bis(4-aminophenyl) sulfone. Curing agent. *Crown Metro.*

8303 3,3'-diaminodiphenylsulfone
599-61-1 209-967-5
$C_{12}H_{12}N_2O_2S$
3,3'-diaminodiphenyl sulfone; 3,3'-sulfonyldianiline; 3-Aminophenyl sulphone. mp = 171-172°; irritant. *BASF; Mitsui Toatsu; Schweizerhall.*

8304 Diaminogen®
A range of dyestuffs. *Cassella AG.* Name unverified.

8305 2,6-diaminopyridine
141-86-6 205-507-2
$C_5H_7N_2$
2,6-pyridinediamine
Aromatic amine; used as a chemical intermediate. *Cilag AG; Janssen Chimica; Reilly Industries; Schweizerhall.*

8306 diaminopyridine
141-86-6 205-507-2
$C_5H_7N_3$
2,6-diaminopyridine; 2,6-pyridinediamine. Aromatic amine mp = 118-121°; irritant. *Cilag AG; Janssen Chimica; Reilly Industries; Schweizerhall.*

8307 Diamite Epoxy Brushkote
Two-component solvent-release, epoxy compound used as a heavy-duty coating for protecting floors, walls and equipment against attrition and chemical attack. *Metalcrete Mfg Co.*

8308 Diamite Epoxy Flooring
100% solids epoxy; used for repairing and surfacing floors subject to heavy traffic. *Metalcrete Mfg Co.*

8309 Diamond cement
A cement containing 8 parts isinglass, 1 part gum ammoniacum, 1 part galbanum, and 4 parts alcohol; used for mending china and glass.

8310 Diamond D 31
8016-70-4 232-410-2
Partially hydrogenated soybean oil; kosher; all-purpose deep fat frying filler fat for cookies, whipped toppings. *Van Den Bergh Foods.*

8311 Diamond D21
Partially hydrogenated vegetable oil (soybean, cottonseed), mono/diglycerides, glycerol-lacto ester of fatty acids, lecithin; kosher; shortening for cakes. *Van Den Bergh Foods.*

8312 Diamond Fiber
A proprietary trade name for a vulcanized fiber; a laminated acid-treated cotton cellulose. No manufacturer.

8313 Diamond Quality®
Unrefined castor oil USP; emollient, pigment wetter, cosolvent, lubricant for cosmetics, makeup, antiperspirant sticks. *CasChem.*

8314 Diamondite
An alloy of 95.65% tungsten with 3.91% carbon.

8315 Diamonine
Textile auxiliary chemicals. *ICI Chem & Polymers Ltd.*

8316 Diamox
59-66-5 50 200-440-5
A proprietary preparation of acetazolamide; a diuretic. *Lederle Laboratories.*

8317 Diampron
NN-Di-(3-amidinophenyl)urea diiesthionate
Rhône-Poulenc Rorer Ltd.

8318 Dianal
Thermoplastic acrylic resins. *British Traders & Shippers Ltd.*

8319 Diane
Diphenylolpropane used as the phenolic reactant in resin manufacture. *Vianova Resins.* Discontinued.

8320 Dianette
2 mg Cyproterone acetate, 35 µg ethinylestradiol *Schering Health Care Ltd.* Discontinued.

8321 Dianix Dye®
Disperse dyes with good fastness for dyeing synthetic fibers, especially polyester; excellent dispersing properties *Mitsubishi Kasei.* Name unverified.

8322 Dianol®
Raw materials for the manufacture of saturated and unsaturated polyester resins, glassmat binder resins, alkyd resins, microcellular and elastomeric polyurethanes. *Akzo Chemie UK Ltd.*

8323 Dianol® 220
32492-61-8, 901-44-0
Ethoxylated bisphenol A diol; reactive modifier for sat. and unsat. polyesters, vinyl esters, and PU resin formulations. *Akzo.*

8324 Dianol® FSD
Modified amine condensate in alcohol; provides fulling qualities and scouring efficiency for processing woolen and worsted fabrics. *Rhône-Poulenc Surf.*

8325 Dianthine
A trademark for a range of dyes. *National Lead Co.* Name unverified.

8326 Diaparene
25155-18-4 6103 246-675-7
Methylbenzethonium chloride
Anti-infective, topical. *Sterling Drug Inc.* Name unverified.

8327 Diapen
1538-09-6 216-260-5
A proprietary preparation of penicillin G benzathine; an antibiotic. *Pfizer International.* Discontinued.

8328 Diaphan oil
A mixture of methylhexalin and sodium oleate; used in the preparation of transparent soaps.

8329 Diapol® WMB® 1808
SBR carbon black masterbatch; for rubber industry. *Mitsubishi Kasei.* Name unverified.

8330 Diaresin Dye®
Solvent dyes for coloring thermoplastics and thermosets; provides high brilliancy and transparency with good fastness properties *Mitsubishi Kasei.* Name unverified.

8331 diastase
9000-92-4 640 232-567-7
Diastatic enzyme. An enzyme that converts starch into sugar; enzyme in textile industry used for starch desizing; degrades starch, dextrins, and oligosaccharides to maltose. *Spectrum Chem. Mfg.; U.S. Biochemical.*

8332 diastatic enzyme

8333 Diastix®
A proprietary test strip impregnated with glucose oxidase and peroxidase, plus potassium iodide; used to detect glycosuria. *Bayer AG.*

8334 Diathol Grounder®
Azoic coupling component used on cotton and rayon as prepare prior to combining with Diabase Developer to form insolutions dye; also combined directly with diazotized bases to produce organic pigments. *Mitsubishi Kasei.* Name unverified.

8335 diatomaceous earth
7631-86-9 8637 231-545-4
Kieselguhr; diatomite; infusorial earth. Mineral material consisting chiefly of the siliceous frustules and fragments of various species of diatoms; Filtration, clarifying, decolorizing; insulation absorbent; mild abrasive; catalyst carrier; anticaking agents for dust, soil. *Celite; L.A. Salomon; Spectrum Chem. Mfg.*

8336 Diatomaceous silica
Discelite. A proprietary trade name for a diatomaceous silica filler. No manufacturer.

8337 diatrizoate sodium
737-31-5 3040 212-004-1
$C_{11}H_8I_3N_2NaO_4$
3,5-bis(acetylamino)-2,4,6-triiodobenzoic acid sodium salt
Hypaque; Hypaque Sodium; Sodium amidotrizoate; Triognost. Radio-opaque medium used as a diagnostic acid. mp = 261-262°; soluble in H_2O (60 g/100 ml), less soluble in organic solvents; LD_{50} (rat iv) = 14.7 g/kg.

8338 diaveridine
5355-16-8 3041 226-333-3
$C_{13}H_{16}N_4O_2$
5-[(3,4-dimethoxyphenyl)methyl]-2,4-pyrimidine diamine
2,4-diamino-5-veratrylpyrimidine. Used with sulfaquonoxaline as Darvisul, a veterinary antiprotozoal agent. mp = 233°.

8339 Diavite
Vitamin mixture dispersed in food grade filler; flour additive. *Diaflex Ltd.* Discontinued.

8340 Diax
A proprietary product used as a diastatic ferment. No manufacturer.

8341 Diazabicycloundecene
6674-22-2 229-713-7
$C_9H_{16}N_2$
2,3,4,6,7,8,9,10-Octahydropyrimido[1,2-a]azepine
DBU; 1,8-diazabicyclo[5.4.0]undec-7-ene. Can be used as a biotin substitute as cofactor in a number of enzymatic carboxylation reactions. $bp_{0.6}$ = 80-83°; d = 1.018; n_D^{20}= 1.5219. *Air Prods & Chem; BASF; Fluka; Schweizerhall.*

8342 Diazamine
Direct dyes. *ICI Chem & Polymers Ltd.* Discontinued.

8343 Diazine
Direct dyestuffs for cotton and artificial silk. *J C Bottomley.* Name unverified.

8344 Diazinon Liquid
333-41-5 3043 206-373-8
$C_{12}H_{21}N_2O_3PS$
O,O-diethyl-O-(6-methyl-2-(1-methylethyl)-4-pyrimidinyl)phosphorothioate
Spectracide; Dimpylate; Basudin; Knox Out; Dianon; Gardentox; Kayazinon; G-24480; Basudin, Neocidol; Dipofene; Diazitol; Ag-500; Antigal; Dacutox; Dassitox; Dazzel; Diagran; Diaterr-fos; Diazajet; Diazide; Diazol; Diethyl Dimpylatum; Dizinon; Drawizon; Dyzol; Exodin; Fezudin; Flytrol; Galesan; Kayazol; Knox Out 2FM; Neocidol; Nipsan; Nucidol; Sarolex. Insecticide. mp = 120°; bp = 306°; d = 1.117; slightly soluble in H_2O (0.004 g/100 ml), more soluble in organic solvents; LD_{50} (rat orl) = 76 mg/kg. *DowElanco Ltd.*

8345 Diazitol, Diazitol Liquid
333-41-5 3043 206-373-8

diazinon
Organophosphorus insecticide. *Ciba plc.* Name unverified.

8346 Diazol
Direct dyes. *ICI Chem & Polymers Ltd.*

8347 Diazol
333-41-5 3043 206-373-8
diazinon
Active ingredient: diazinon; organophosphorus agricultural insecticide with acaricidal properties. *Makhteshim Chemical Works Ltd.*

8348 Diazolidinyl urea
78491-02-8 278-928-2
$C_6H_{10}N_4O_7$
Urea, N-(1,3-bis(hydroxymethyl)-2,5-dioxo-4-imidazolidinyl)-N,N'-bis(hydroxymethyl)-
Germall® II; Diazolidinylurea. Broad-spectrum antimicrobial preservative for cosmetics and toiletries. *Sutton Labs.*

8349 Diazone
Fast dyestuffs for cotton. *J C Bottomley.* Name unverified.

8350 Diazopon SS-837
Polyoxyethylated alkyl phenol (nonionic), a water-soluble surfactant with dispersing and solubilizing properties; anticrock agent for naphthol dyeings; soaping agent for naphthol-dyed stock, yarn and fabrics; leuco-vat-ester dyeing assistant for use in the acid bath to inhibit pigment agglomeration on the fiber. *ISP; Rhóne-Poulenc Surf.* Name unverified.

8351 diazoresorcin
550-82-3 8309 208-987-1
$C_{12}H_7NO_4$
azoresorcin
resazoin; 7-hydroxy-3H-phenoxazin-3-one 10-oxide; Resazurin. Used as an acid-base indicator and in enzymology - a reductase detector.

8352 Diazyme® L-200
9032-08-0 232-877-2
Glucoamylase
Distillase® L-200. Enzyme for hydrolysis of starch dextrins to glucose; used in brewing and fermentation applications *Solvay Enzymes.*

8353 Dibenzo GMF
120-52-5 204-403-4
$C_{20}H_{14}N_2O_4$
Dibenzoyl-*p*-quinone-dioxime
2,5-Cyclohexadiene-1,4-dione, bis(O-benzoyloxime). A non-sulfur vulcanizing agent for natural, SBR, butyl and EPDM rubber; used to impart heat resistance to tyre curing bags, gaskets and wire insulation, also used as a coagent with peroxide curatives. *Uniroyal.*

8354 Dibenzosuberone
1210-35-1 214-912-3
$C_{15}H_{12}O$
10,11-dihydro-5H-dibenzo[a,d] cyclohepten-5-one
Dibenzocycloheptadienone; 10,11-Dihydro-5H-dibenzo[a,d]cyclohepten-5-one; Dibenzosuberone; Dibenzocycloheptadienone; 10,11-Dihydrodibenzo[a,d]cyclohepten-5-one. Used in chemical synthesis. mp = 32-34°; $bp_{0.3}$ = 148°; d = 1.156; n_D^{20} = 1.6332; LD_{50} (mus orl) = 2.1 g/kg. *Lonza; Penta Mfg.; Sandoz; Schweizerhall.*

8355 Dibexin
A proprietary preparation of Vitamin B. *Parke Davis.* Name unverified.

8356 Dibrom
300-76-5 206-098-3
naled
An insecticide. *Monsanto (Solaris).* Name unverified.

8357 1,4-dibromobutane
110-52-1 203-775-5
$C_4H_8Br_2$
tetramethylene dibromide
1,4-butylene bromide. Used in chemical synthesis. mp = -20°; bp = 197°; d = 1.8080; n_D^{20} = 1.5186; insoluble in H_2O, soluble in organic solvents. *Humphrey; Janssen Chimica.*

8358 dibromoneopentyl glycol
3296-90-0 221-967-7
$C_5H_{10}Br_2O_2$
2,2-Bis(bromomethyl)-1,3-propanediol
DBNPG; Dibromopentaerythritol; Pentaerythritol dibromide; Emery® 9336; FR-522; FR-521. Flame retardant for unsaturated polyesters, rigid PU and foams; flame retardant intermediate. mp=109°; d = 2.23; LD_{50} (rat orl) = 3450 mg/kg. *AmeriHaas; Albemarle; Aldrich; AmeriBrom; Dead Sea Bromine;.*

8359 2,4-dibromophenol
615-58-7 210-436-5
$C_6H_4Br_2O$
DBP; Emery® 9331; FR-612. Flame retardant for epoxy resins, phenolic resins, polyester resins; flame retardant intermediate. mp=33-36°; d = 1.04; LD_{50} (mus orl) = 282 mg/kg. *AmeriHaas; Aldrich; AmeriBrom; Dead Sea Bromine; Esprit; Fluka; Great Lakes.*

8360 2,3-dibromo-1-propanol
96-13-9 202-480-9
$C_3H_6Br_2O$
2,3-dibromo-1-propanol
dibromo-1-propanol; 2,3-Dibromopropanol; glycerol 1,2-dibromohydrin; β-dibromohydrin; dibromopropanol; allyl alcohol dibromide. Intermediate in preparation of flame retardants, insecticides, and pharmaceuticals. bp_{10} = 95-97°; d = 2.1200; n_D^{20}= 1.5599. *Aldrich; ICI Am.; Schweizerhall.*

8361 dibromopropanol
96-13-9 202-480-9
$C_3H_6Br_2O$
2,3-dibromo-1-propanol
2,3-Dibromo-1-Propanol; dibromo-1-propanol; 2,3-Dibromopropanol; glycerol 1,2-dibromohydrin; β-dibromohydrin; dibromopropanol; allyl alcohol dibromide. Intermediate in preparation of flame retardants, insecticides, and pharmaceuticals. bp_{10} = 94-97°; d = 2.1200; n_D^{20}= 1.5599; *Aldrich; ICI Am.; Schweizerhall.*

8362 Di-*t*-butoxydiacetoxysilane
13170-23-5 236-112-3
CD4153; Dynasylan® BDAC. Coupling agent, chemical intermediate, blocking agent, release agent, lubricant, primer, reducing agent. bp = 102°; d = 1.0196; n_D^{20}= 1.4040; flash point = 95°. *Hüls Am.*

8363 Dibutoxyethyl phthalate
117-83-9 204-213-1
$C_{20}H_{30}O_6$
n-butyl glycol phthalate
ethylibutoxy phthalate; DBEP; Plasthall® 200; Staflex DBEP. Plasticizer for PVC, polyvinyl acetate, and other resins. bp = 270°; d = 1.06; fp = -55°; flash point = 407°F. *Ashland; C.P. Hall; Unitex.*

8364 dibutyl fumarate
105-75-9 203-327-9
$C_{12}H_{20}O_4$
2-Butenedioic acid (E)-, dibutyl ester
Di-n-butyl fumarate; Staflex DBF; Bisomer DBM. Monomeric plasticizers copolymers, intermediate. *Penta Mfg; AC Ind.; Monomer-Polymer&Dajac; Unitex.*

8365 dibutyl maleate
105-76-0 203-328-4
$C_{12}H_{20}O_4$
2-Butenedioic acid (Z)-, dibutyl ester
DBM; Di-n-butyl Maleate; Dibutyl Maleate; 2-butenedioic acid, dibutyl ester. Intermediate in the manufacture of copolymers and plasticizers, bp_4 = 129°; d = 0.9930; LD_{50} (rat orl) = 3730 mg/kg. *Aristech; Pentagon Chemicals Ltd; Penta Mfg.; Unitex.*

8366 dibutyl phthalate
84-74-2 1622 201-557-4
$C_{16}H_{22}O_4$
1,2-benzenedicarboxylic acid, dibutyl ester
dibutyl-1,2-benzene dicarboxylate; butyl phthalate; DBP; Di-n-Butyl Phthalate; n-Butylphthalate; 1,2-Benzenedicarboxylic acid dibutyl ester; Phthalic acid dibutyl ester; *o*-benzenedicarboxylic acid, dibutyl ester; benzene-*o*-dicarboxylic acid di-n-butyl ester; dibutyl 1,2-benzenedicarboxylate; Celluflex DPB; Elaol; Hexaplas M/B; Palatinol C; Polycizer DBP; PX 104; Staflex DBP; Witcizer 300; Araldite 502; Benzenedicarboxylic acid, dibutyl ester; Dibutyl-*o*-Phthalate. Diester of butyl alcohol and phthalic acid; plasticizer in nitrocellulose, lacquers, elastomers, explosives, nail polishes; solvent for perfumes, oils; perfume fixative; textile lubricating agent. mp = -35°; bp = 340°; d = 1.0430; n_D^{20}= 1.4910; LD_{50} (rat orl) = 8 g/kg. Aldrich; Aristech; BP Chem. Ltd; Chisso Am.; Daihachi Chem. Ind.; Eastman; C.P. Hall; Mitsubishi Gas; Unitex.

8367 Di-*t*-butyl dicarbonate
24424-99-5 246-240-1
$C_{10}H_{18}O_5$
BOC anhydride
Di-*t*-butyl pyrocarbonate; Dicarbonic acid, bis(1,1-dimethylethyl) ester; Di-*tert*-butyl pyrocarbonate. Plasticizer. mp = 23°; $bp_{0.5}$ - 56-57°; n_D^{20} = 1.4075; *Aldrich; Fluka; Schweizerhall.*

8368 Dibutylhydroquinone
88-58-4 201-841-8
$C_{14}H_{22}O_2$
2,5-Di-*t*-butylhydroquinone
Eastman® DTBHQ; 2,5-Bis(1,1-dimethylethyl)-1,4-benzenediol. Antioxidant for rubber, polyesters. Polymerization inhibitor, stabilizer against UV deterioration of rubber. mp=215-220°; insoluble in H_2O, soluble in organic solvents; flash point = 216°. *Eastman; Fluka.*

8369 di-*t*-butyl peroxide
110-05-4 3515 203-733-6

$C_8H_{18}O_2$
t-butyl peroxide bis(1,1-di-methylethyl)peroxide
DTBP; *tert*-Butylperoxide; *tert*-Dibutyl peroxide; TBP; Cadox; DTBP; bis(1,1-dimethylethyl)peroxide; Di-*tertiary*-butyl peroxide. Polymerization catalyst for resins (e.g., olefins, styrene, styrenated alkyds, silicones); ignition accelerator for diesel fuel; organic synthesis; intermediate. *Akzo; Elf Atochem N. Am.; Witco/Argus.*

8370 2,6-dibutylphenol
128-39-2 204-884-0
$C_{14}H_{22}O$
2,6-di-*t*-butylphenol
Isonox® 103; 2DTBP; 2,6-Bis(*t*-butyl)phenol; 2,6 DTBP. Orthoalkylated aromatic; intermediate; antioxidant, stabilizer for foods, aviation and other gasolines. mp = 37°; bp = 253°; insoluble in H_2O, soluble in organic solvents; d^{20} = 0.914; flash point = 118°; LD_{50} (mus orl) = 120 mg/kg. *Allchem Industries; Ethyl; Penta Mfg.; PMC Spcialties; Schenectady.*

8371 dibutyltin laurate
77-58-7 3089 201-039-8
$C_{32}H_{64}O_4Sn$
dibutylbis[(1-oxodecyl)oxy]stannane
Butynorate; Davainex; Tinostat; ADK STAB BT-11. PVC stabilizer providing good heat and weathering resistance; improves processability in rigid transparent formulation. Also used as an anthelmintic, against tapeworms in chickens. mp = 22-24°; n_D^{20} = 1.4683; insoluble in H_2O, MeOH, soluble in organic solvents. *Asahi Denka Kogyo; Air Prods & Chem; Elf Atochem N. Am.; Ferro/Bedford; KMZ Chem. Ltd; Witco.*

8372 Dibutyltin maleate
78-04-6 201-077-5
$C_{12}H_{20}O_4Sn$
2,2-Dibutyl-1,3,2-dioxastannepin-4,7-dione
Dibutyltin maleate; Di-n-Butyl(maleate)tin; DBM; ADK STAB BT-31; AK STAB BT-52; ADK STAB BT-53A; ADK STAB LS-2; Stanclere® TM; Therm-Chek® 837. DBM Stabilizer for PVC resins; condensation catalyst. mp = 110°; insoluble in H_2O; soluble in organic solvents; flash point = 400°F; *Elf Atochem N. Am.; Ferro/Bedford; Gelest.*

8373 Dicalite 14, 14B, and 14W
Proprietary trade names for diatomaceous silica fillers; used for heat insulating and as a filler for plastics, etc. No manufacturer.

8374 Dicamba
1918-00-9 3090 217-635-6
$C_8H_6Cl_2O_3$
3,6-dichloro-2-methoxybenzoic acid
3,6-dichloro-*o*-anisic acid; MDBA; Banvel; Fallowmaster; Mediben; Metambane; Tracker; Trooper; Velsicol 58-CS-11. Selective systemic herbicide which acts as an auxin-like growth regulator. Used for control of annual and perennial broad-leaved weeds and brush sp-ecies in cereals, maize, sorghum, sugar cane, asparagus, seed pastures, turf, pastures and rangeland. mp = 114-116°; d^{25} = 1.57; soluble in H_2O (6.5 g/l), more soluble in organic solvents; LD_{50} (rat orl) = 1707 mg/kg. *Diachem; Sandoz.*

8375 Dicapryl phthalate
131-15-7 205-014-2
$C_{24}H_{38}O_4$
Di-(2-octyl)phthalate
DCP; Uniflex® DCP. A plasticizer for vinyl plastics and cellulosic resins. bp = 227-234°; insoluble in H_2O; d = 0.965; fp = -60°; flash point = 295°F. *Union Camp.*

8376 Dicestal
97-23-4 3120 202-567-1
Dichlorophen
Agricultiural fungicide and antimicrobial. *May & Baker Ltd.* Name unverified.

8377 Dichan 100
3129-91-7 221-515-9
$C_{12}H_{24}N_2O_2$
Dicyclohexylamine nitrite
Dicyclohexyl ammonium nitrite; Cyclohexanamine, N-cyclohexyl-, nitrite; dicyclohexylamine nitrite; Dechan; dicyclohexylaminonitrite; dodecahydrophenylamine nitrite; NDA; N-cyclohexylcyclohexanamine nitrite;. Vapor-phase corrosion inhibitor for ferrous metals; designed for items enclosed in packaging, e.g., hot water heating systems, nuclear reactor heat-exchange units, gas recovery systems, jet aircraft engine compressors, internal combustion engines, welding electrodes and double-walled pipes. soluble in H_2O, EtOH, less soluble in organic solvents; LD_{50} (rat orl) = 284 mg/kg. *Olin.* Discontinued.

8378 Dichevrol
Dielectric oil *Monsanto (Solaris).* Name unverified.

8379 Dichlofluanid
1085-98-9 3095 214-118-7
$C_9H_{11}Cl_2FN_2O_2S_2$
1,1-Dichloro-N-[(dimethylamino)-sulfonyl]-1-fluoro-N-phenylmethanesulfenamide
Euparen®; Bayer 47531; KUE 13032c; Elvaron® ; Euparen (e); Euparen®. Fungicide with specific action against *Botrylis.* mp = 105-105.6°; insoluble in H_2O, soluble in organic solvents; LD_{50} (mus orl) = 1250 mg/kg. *Bayer AG.*

8380 Dichlofuanide
N-Dimethylamino-N'-phenyl-N'-(fluorodichlormethylthio) sulfamide
A proprietary paint fungicide. *Bayer AG.*

8381 Dichlone
117-80-6 3096 204-210-5
$C_{10}H_4Cl_2O_2$
2,3-Dichloro-1,4-naphthalenedione
Dichloronaphthoquinone; USR-604; Phygon; Phygon Paste; Phygon XL; Quintar; USR 604; 2,3-dichloro-1,4-naphthoquinone. A fungicide used as a seed dressing, insecticide, organic catalyst. mp = 193°; insoluble in H_2O; soluble in organic solvents; LD_{50} (rat orl) = 1300 mg/kg.

8382 Dichloramine T
473-34-7 3098 207-462-4
$C_7H_7Cl_2NO_2S$
N, N-Dichloro-*p*-toluenesulfonamide
A disinfectant, germicide and antibacterial. mp = 83°; insoluble in H_2O, EtOH, soluble in organic solvents.

8383 Dichloramine-M
Methyldiphenylmethyl-dichloramine
A disinfectant, germicide and antibacterial.

8384 dichloro-1,3,5-triazinetrione, potassium salt
2244-21-5 9896 218-828-8
$C_3Cl_2KN_3O_3$
potassium dichloro-*s*-triazine-2,4,6-trione
potassium dichloroisocyanurate; triazine 2,4,6(1H,3H,5H)trione, 1,3-dichloro, potassium salt; 1,3-dichloro-s-triazine-2,4,6(1H,3H,5H)-trione, potassium salt; dichloro-*s*-triazin-2,4,6(1H,3H,5H)trione potassium; isocyanuric acid, dichloropotassium salt; potassium troclosene; 3,5-dichlorotetrahydro-2,4,6-trioxo-s-triazin-1(2H)yl-potassium; Potassium dichlorocyanurate; triclosene potassium; ACL 59. Active ingredient in dry bleaches, water and sewage treatment, etc. Topical anti-infective. mp = 250°; soluble in H_2O (10-50 mg/ml). *Biachem Ltd; ICI Spec.; Monsanto; Nissan Chem. Ind.; 3-V.*

8385 dichloro-1,3,5-triazinetrione, sodium salt
2893-78-9 220-767-7
$C_3Cl_2N_3NaO_3$
sodium dichloro-s-triazine-2,4,6-trione
sodium dichloro-s-triazine-2,4,6-trione; sodium dichloro isocyanurate; sodium dichlorocyanurate; 1,3-dichloro-1,3,5-triazine-2,4,6(1H,3H,5H)-trione sodium salt; 1-sodium-3,5-dichloro-s-triazine-2,4,6-trione; dichloro-s-triazine-2,4,6-(1H,3H,5H)-trione sodium salt; monosodium dichloroisocyanurate; sodium salt of dichloro-s-triazinetrione; triazine, 2,4,6(1H,3H,5H)-trione, 1,3-dichloro-, sodium salt; troclosene sodium. Active ingredient in dry bleaches, water and sewage treatment, etc. Topical anti-infective. mp = 250°; soluble in H_2O.

8386 *p*-dichlorobenzene
106-46-7 3107 203-400-5
$C_6H_4Cl_2$
1,4-dichlorobenzene
Paracide; dichloricide; DCB; Dichlorocide; PDCB; PDB; Para-zene; Paramoth; *p*-chlorophenyl chloride; *p*-dichlorobenzol; Evola; globol; PARA; Paradow; paranuggets; persia-perazol; santochlor. Used as an insecticidal fumigant and as a room deodorant. Intermediate in plastics manufacture. mp = 53-56°; bp = 174°; d = 1.2410; n_D^{60} = 1.5285; insoluble in H_2O, soluble in organic solvents; LD_{50} (rat orl) = 500 mg/kg.

8387 2,4-dichlorobenzyl alcohol
1777-82-8 3110 217-210-5
$C_7H_6Cl_2O$
2,4-dichlorobenzenemethanol
Dybenal. Antiseptic. mp = 55-58°.

8388 dichlorodifluoromethane
75-71-8 3114 200-893-9
CFC-12; Refrigerant 12; Freon-12; Halon 122; Halocarbon 12; Refrigerant R12; Fluorocarbon 12; Difluorodichloromethane; Arcton 12; Frigen 12; Genetron 12; Halon; Isotron 2; R-12; algofrene type 2; arcton 6; electro-cf 12; F 12; FC 12; eskimon 12; freon f-12; Ledon 12; Propellent 12. Used as a refrigerant and aerosol propellant. Has no anesthetic activity. mp = -158°; bp = -30°; insoluble in H_2O, soluble in organic solvents.

8389 1,3-Dichloro-5,5-dimethyl hydantoin
118-52-5 3115 204-258-7
$C_5H_6Cl_2N_2O_2$
1,3-Dichloro-5,5-dimethyl-2,4-imidazolidinedione
Dantoin® DCDMH; Dactin; Halane; Omchlor; DDH; Dantochlor. Intermediate for custom chemical synthesis, laundry bleach formulations, and automatic dishwashing compounds; disinfectant, industrial deodorant; stabilizer for vinyl chloride polymers; polymerization catalyst. mp = 132°; soluble in H_2O,

soluble in chlorinated and highly polar solvents; LD_{50} (rat orl) = 542 mg/kg. *Lonza Inc.*

8390 Dichloroditane
2051-90-3 218-134-5
$C_{13}H_{10}Cl_2$
p-dichlorodiphenylmethane
A proprietary trade name for *p*-dichloro diphenyl methane, a chemical intermediate. *Bakelite Corporation.* Name unverified.

8391 Dichlorofluoroethane
1717-00-6
$C_2H_3Cl_2F$
1,1-Dichloro-1-fluoroethane
Genetron® 141b; HCFC-141b; dichlorofluoroethane; Freon 141b. Blowing agent (replacement for CFCs) in foam applications, rigid board, flexible foam. *AlliedSignal.*

8392 Dichlorophen
97-23-4 3120 202-567-1
$C_{13}H_{10}Cl_2O_2$
2,2'-Methylenebis[4-chlorophenol]
Acticide DDM; Dichlorophene; G-4; Anthipen; Dicestal; Didroxane; Diphenthane-70; Parabis; Plath-Lyse; Preventol G-D; Teniathane; Teniatol; Wespuril. Bactericide/fungicide for textiles, cellulose solutions, proteins, adhesives and soaps; bactericide/algicide for water treatment. mp = 177-178°; practically insoluble in H_2O; soluble in organic solvents; LD_{50} (rat orl) = 1506 mg/kg. *Thor Chemicals (UK) Ltd.* Discontinued.

8393 Dichlorotrifluoroethane
306-83-2 206-190-3
$C_2HCl_2F_3$
2,2-dichloro-1,1,1-trifluoroethane
Genetron® 123; HCFC-123; dichlorotrifluoroethane; Freon 123; 1,1-Dichloro-2,2,2-trifluoroethane; Fluorocarbon 123; FC-123. Centrifugal refrigerant with low operating pressures. *AlliedSignal.*

8394 dichlorvos
62-73-7 3129 200-547-7
$C_4H_7Cl_2O_4P$
Dedevap®
DDVP; Dichlorophos; Equigand; No-pest Strip; SD 1750; Astrobot; Atgard; Canogard; Dedevap; Dichlorman; Divipan; Equigard; Equigel; Estrosol; Herkol; Nogos; Nuvan; Apavap; Atgard C; Atgard V; Bay-19149; Benfos; Bibesol; Brevinyl; Brevinyl E50; Chlorvinphos; Deriban; Derribante; Devikol; Duo-kill; Duravos; Estrosel; Fecama; Fly-Die; Fly Fighter; Herkal; Krecalvin; Mafu; Mafu Strip; Marvex; Mopari; Nerkol; Nogos 50; Nogos G; No-pest; Nuva; Nuvan 100ec; Oko; OMS 14; Phosvit; Szklarniak; Task; Tenac; Task Tabs; Tetravos; UDVF; Unifos; Unifos 50 EC; Vaponite; Vapora II; Verdican; Verdipor; Vinylofos; Vinylophos; Bayer 19149; Fekama; Insectigas D; Nefrafos; Nogos 50 EC; Novotox; Nuvan 7; Panaplate; Winylophos; Cekusan; Cypona; Delevap; Derriban; Equiguard; Prentox; Verdisol,. DDVP (dichlorvos); an insecticide used for spray control of sucking, biting and mining insects in greenhouses. mp = 84°; bp_{20} = 140°; d_4^{25}= 1.415; soluble in H_2O (10-20 mg/ml), more soluble in organic solvents; LD_{50} (rat orl) = 25 mg/kg. *Bayer AG.*

8395 dicloran
99-30-9 202-746-4
$C_6H_4Cl_2N_2O_2$
2,6-dichloro-4-nitrobenzeneamine
DCNA; Botran; Botran 75W; Dichloran; Dicloron; Allisan; Ditranil; Kiwi Lustr 277; Resisan; RD-6584; AL-50; U-2069; CDNA; CNA; Resissan; Dichloro-4-nitroaniline; Fumite Dicloran. Protective fungicide used for control of *Botrytis Monilinia, Rhizopus, Sclerotinia* and *Sclerotium* spp. in fruits and vegetables. mp = 195°; soluble in H_2O (6.3 mg/l), more soluble in organic solvents; LD_{50} (rat orl) = 4040 mg/kg. *Octavius Hunt Ltd.*

8396 Dicloxacillin sodium
13412-64-1 3134
$C_{19}H_{16}Cl_2N_3NaO_5S.H_2O$
[2S-(2α,5α,6β)]-6-[[[3-(2,6-dichlorophenyl)-5-methyl-4-isoxazolyl]carbonyl]amino]-3,3-dimethyl-7-oxo-4-thia-1-azabicyclo[3.2.0]heptane-2-carboxylic acid
Sodium dicloxacillin monohydrate; Dycill; Dicloxin; P-1011; Brispen; Constaphyl; Dichlor-Stapenor; Dynapen; Noxaben; Pathocil; Pen-Stint; Stampen; Syntarpen; Veracillin. Antibacterial. dec 222-225°; $[\alpha]_D^{20}$ = 127.2° (H_2O); soluble in H_2O, organic solvents; LD_{50} (rat orl) >5 g/kg. *Bristol laboratories.* Name unverified.

8397 Dicofen
122-14-5 4017 204-524-2
fenitrothion
Emulsifiable concentrate of 500 g fenitrothion per liter; an organophosphorus insecticide. *Pan Britannica Industries Ltd.*

8398 Dicofol
115-32-2 3136 204-082-0
$C_{14}H_9Cl_5O$
4-Chloro-α-(4-chlorophenyl)-α-(trichloromethyl)benzenemethanol
1,1-bis(*p*-chlorophenyl)-2,2,2-trichloroethanol; DTMC; ENT 23648; FW 293; Kelthane; Mitigan; Acarin. Active ingredient; dicofol; an acaricide and miticide. mp = 77-78°; bp = 225°; d^{10} = 1.1234; n_D = 1.1234; $[\alpha]_D$ = 100°; , insoluble in H_2O, soluble in organic solvents; λ_m = 226, 258, 266, 276 nm (logε = 4.43, 2.82, 2.85, 2.60); LD_{50} (rat, orl) = 1495 mg/kg. *Makhteshim Chemical Works Ltd.* Discontinued.

8399 Dicontal® New
Broad spectrum insecticide; used with grape crops. *Bayer AG.*

8400 Dicosal
A range of fire extinguishing powders to extinguish metal fires including burning uranium. *Degussa AG.* Name unverified.

8401 Dicotox
A selective weedkiller. *May & Baker Ltd.*

8402 Dicotox Extra
94-75-7 2865 202-361-1
2,4-D
2,4-D; translocated herbicide for cereals and established grassland. *RhônePoulenc Environmental Prods. Ltd.*

8403 Dicrodamine
Series of cationic surfactants in the form of fatty alkyl propylene diamines or diamine salts, in which the alkyl group is coconut, tallow, hydrogenated tallow or oleic in origin; intermediates for di-quaternaries and polyethoxylates; dispersants for pigments; corrosion inhibitors, drawing aids for copper wire and tubing; flexible hardeners for epoxy resins; auxiliaries for oil and petroleum industries; bitumen emulsions. *Croda Chem. Ltd.* Discontinued.

8404 Dicron 45Sc
Organophosphorus insecticide *Ciba plc.* Name unverified.

8405 Dicrylan
Finishing agent. *Ciba plc.* Name unverified.

8406 Dicrylan 270
A proprietary trade name for an aqueous emulsion of an acrylic resin for increasing the abrasion resistance of crease-resistant fabrics. *Ciba plc.* Name unverified.

8407 Dicumyl peroxide
80-43-3 201-279-3
$C_{18}H_{22}O_2$
di-α-cumyl peroxide
cumene peroxide; dicumyl peroxide; Bis(1-methyl-1-phenylethyl) peroxide; bis(α,α-dimethylbenzyl) peroxide; cumyl peroxide; di-α-cumyl peroxide; Dicup; diisopropylbenzene peroxide; isopropyl benzene peroxide; Cumene peroxide; diisopropylbenzene peroxide. Polymerization catalyst and vulcanizing agent; crosslinking agent for olefinic polymers. bp = 130°; insoluble in H_2O, soluble in organic solvents; LD_{50} (rat orl) = 4100 mg/kg. *Akzo; Elf Atochem; Hercules; Mitsui Petrochem. Ind.; R. T. Vanderbilt Co Inc.*

8408 Di-Cup
80-43-3 201-279-3
$C_{18}H_{22}O_2$
dicumyl peroxide
A range of dicumyl peroxide preparations; vulcanizing and polymerization agent. *Hercules.* Discontinued.

8409 Dicurane 500 FW
15545-48-9 239-592-2
chlorotoluron
Suspension concentrate containing 500 g chlorotoluron per liter; a contact urea herbicide for cereal crops. *Ciba-Geigy Agrochemicals.*

8410 Dicurane Duo 495FW
bifenox-chlorotoluron
A liquid formulation containing 106 g bifenox and 389 g chlorotoluron per liter as a suspension concentrate; a residual herbicide for the control of weeds in winter wheat. *Ciba-Geigy Agrochemicals.*

8411 Dicyandiamide
461-58-5 3142 207-312-8
$C_2H_4N_4$
Cyanoguanidine
Fertilizers; nitrocellulose stabilizer; organic synthesis, especially of melamine, barbituric acid, and guanidine salts; explosives; catalyst for epoxy resin; pharmaceuticals; fireproofing compounds; stabilizer in detergents; modifier for starch production. mp = 209.5°; (eutectic with cyanamide at 35.6°); soluble in H_2O (32 g/l), EtOH, insoluble in organic solvents. *Allchem Industries; Andrulex Trading Ltd.*

8412 dicyclohexyl carbodiimide
538-75-0 3146 208-704-1
$C_{13}H_{22}N_2$
N,N'-Methanetetraylbiscyclohexanamine
N,N·g-dicyclohexylcarbodiimide; DCC; DCCI; Carbodicyclohexylimide. DCC; chemical intermediate, coupling agent in peptide synthesis. mp = 35-36°;

bp$_{11}$ = 154-156°; reacts with H_2O; contact allergen. *Janssen Chimica; Schweizerhall.*

8413 dicyclohexyl phthalate

84-61-7 201-545-9

$C_{20}H_{26}O_4$

1,2-Benzenedicarboxylic acid, dicyclohexyl ester

KP 201; Morflex 150; Unimoll® 66M; Uniplex 250; DCHP. DCHP; plasticizer for nitrocellulose, ethylcellulose, chlorinated rubber, PVAc, PVC, and other polymers. mp = 62-65°; insoluble in H_2O, soluble in organic solvents; d^{25}_{25} = 1.20; flash point = 405°; nonvolatile. *Morflex; Novachem; Schweizerhall; Unitex.*

8414 dicyclohexylamine nitrite

3129-91-7 221-515-9

$C_{12}H_{24}N_2O_2$

Cyclohexanamine, N-cyclohexyl-, nitrite

Dichan 100; Dicyclohexyl ammonium nitrite; dicyclohexylamine nitrite; dechan; dicyclohexylaminonitrite; dodecahydrophenylamine nitrite; NDA; N-cyclohexylcyclohexanamine nitrite; Naphthalene dialdehyde. Vapor-phase corrosion inhibitor for ferrous metals; designed for items enclosed in packaging, e.g., hot water heating systems, nuclear reactor heat-exchange units, gas recovery systems, jet aircraft engine compressors, internal combustion engines, welding electrodes and double-walled pipes. mp = 182-183°; oxidizer; soluble in H_2O, EtOH, less soluble in organic solvents; LD_{50} (rat orl) = 284 mg/kg. *Olin.* Discontinued.

8415 didecyl methylamine

7396-58-9 230-990-1

$C_{21}H_{45}N$

Dama® 1010; Amine M210D; Armeen® M2-10D; Radiamine 6310; Methyl decyl-1-amino decane; N-Methyldidecylamine. Intermediate in the manufacture of quaternary ammonium compounds for biocides, textile chemicals, oil field chemicals, amine oxides, betaines, polyurethane foam catalysis, epoxy curing agent; in fabric softeners, disinfectants and laundry detergents. d = 0.807; fp = -6.3°; flash point (PM) >93°; corrosive. *Ethyl Corp; Fluka; Albemarle.*

8416 Didi-Col

An insecticide. *ICI Chem & Polymers Ltd.*

8417 Didigram

An oil insecticide. *Plant Protection.* Name unverified.

8418 Didimac

An insecticide. *ICI Chem & Polymers Ltd.*

8419 Didin Fluid

Nitrification inhibitor. *Omex Agriculture Ltd.*

8420 Didpex® A40 and N40

Ammonium or sodium salt of polymeric carboxylate as a pale yellow liquid; nonfoaming pigment dispersing agent used in emulsion paints, sometimes in combination with polyphosphate dispersants. *Allied Colloids Ltd.*

8421 Didronel

7414-83-7 3908 231-025-7

Etidronate disodium

Regulator; pharmaceutic aid. *Norwich Eaton Pharmaceuticals Inc.*

8422 die-casting alloys

These are usually zinc-base alloys containing 86% zinc, 7-10% tin, 4-7% copper, 0.5-1% aluminum. Some alloys have a tin base, and a typical one contains 90% tin, 4% copper, and 6% antimony.

8423 Di-el

A material used as an insulating material for high voltage engineering.

8424 Dieline

Sym-dichlorethylene, CHCl·cCHCl; used as a solvent for cellulose acetate, rubber, and oils.

8425 Dielmoth

A moth proofing agent for wool. *Shell UK.* Name unverified.

8426 Diene®

Polybutadiene rubber; for high impact polystyrene and tires. *Firestone Syn. Rubber.*

8427 Diene® 35AC, 55AC

9003-17-2

Stereospecific polybutadiene rubber; used as the backbone rubber for graft polymers; impact modifier for PS and ABS to produce beverage cups, plates, dinnerware, football helmets, coolers, packaging, toys, furniture, automotive grills, dashboard components, consoles, headlight and interior light fixtures. *Firestone Syn. Rubber.*

8428 Diene® 70AC

9003-17-2

polybutadiene

Polybutadiene rubber, alkyl lithium polymerized; features excellent color and low gel; imparts superior toughness and lower gloss when used as a grafted impact modifier for thermoplastic resins. *Firestone Syn. Rubber.*

8429 dienochlor

2227-17-0 3154 218-763-5

$C_{10}Cl_{10}$

1,1',2,2',3,3',4,4',5,5'-decachlorobi-2,4-cyclopentadien-1-yl

HRS-16; Pentac. Acaricide used to control mites on ornamental plants. mp = 121-122°; λ_m = 330 nm (ε 2950).

8430 Dienol

7439-96-5 5762 231-105-1

manganese

Colloidal manganese, a catalyst.

8431 Diepoxy

An epoxy mold material; used for dental molds. *Kemtron International Inc.*

8432 Dieselect

Lube oils. *Monsanto (Solaris).* Name unverified.

8433 Diesterex N

A proprietary trade name for a rubber vulcanizing accelerator stated to be 60% of the dinitrophenyl ester of mercaptobenzthiazole and 40% of the acetate of diphenylguanidine. No manufacturer.

8434 Diethanolamine

111-42-2 3156 203-868-0

$C_4H_{11}NO_2$

2,2'-Iminobisethanol

2,2·g-iminobisethanol; Aliphatic amine; di(2-hydroxyethyl) amine; 2,2·g-iminodiethanol; Diethylolamine; bis(hydroxyethyl)amine. DEA; gas scrubbing; rubber chemicals intermediate; manufacture of surfactants for textiles, herbicides, petroleum demulsifier; emulsifier and dispersant in agriculture., cosmetics, pharmaceuticals; textile lubricants; humectant; softening agent; in organic synthesis. mp= 28°; bp = 268.8°; soluble in H_2O, organic solvents; viscosity at 30°=351.9 cps; flash point = 300°F; LD_{50} (rat orl) = 12.76 g/kg. *Hüls AG.*

8435 Diethoxol

111-90-0 1847 203-919-7

$C_6H_{14}O_3$

2-(ethoxyethoxy)ethanol

carbitol; 3,6-dioxa-1-octanol; APV; carbitol cellosolve; dioxitol; Dowanol; Dowanol DE; losungsmittel APV; solvolsol; transcutol; Poly-solv DE ethanol, 2,2'-oxybis-, monoethyl ether; ethyl digol; 2-(2-Ethoxyethoxy)ethanol; Diethylene glycol ethyl ether; Diethylene Glycol Monoethyl Ether; Ethyl Carbitol; diglycol monoethyl ether; ethoxy diglycol; ethyl diethylene glycol; ethylene diglycol monoethyl ether; monoethyl ether of diethylene glycol; Ektasolve DE; 1-hydroxy-3,6-dioxaoctane; O-ethyldigol; 3,6-dioxaoctan-1-ol; 2-(β-ethoxyethoxy)ethanol; diglycol; Ethoxyethoxy)ethanol. Monoethyl ether of diethylene glycol. Solvent for lacquers and varnishes. mp = -76°; bp = 197°; d = 0.9990; n_D^{20}= 1.4270; soluble in H_2O, organic solvents; LD_{50} (rat orl) = 5500 mg/kg. *ICI Chem & Polymers Ltd.* Discontinued.

8436 diethyl ketone

96-22-0 3170 202-490-3

$C_5H_{10}O$

3-pentanone

Dimethylacetone; Propione; Methacetone; DEK; Ethyl Ketone; pentan-3-one. Medicine, organic synthesis. mp = -42°; bp = 101.5°; soluble in H_2O (4 g/100 ml); organic solvents; LD_{50} (rat orl) = 2.1 g/kg. *BASF; Janssen Chimica; Penta Mfg.; Union Carbide.*

8437 diethyl phthalate

84-66-2 201-550-6

$C_{12}H_{14}O_4$

1,2-Benzenedicarboxylic acid diethyl ester

DEP; phthalic acid; diethyl ester; diethyl 1,2-benzenedicarboxylate; ethyl phthalate; Phthalol; diethyl 1,2-benzenedicarboxylate; Kodaflex® DEP. Solvent for nitrocellulose and cellulose acetate; plasticizer, wetting agent, insectidal preps.; in perfumery as solvent and fixative; plasticizer in solid rocket propellants. bp = 298°; n_D^{20}= 1.500-1.505; insoluble in H_2O, soluble in organic solvents; d = 1.120; fp = -40.5°; flash pt = 325° F; viscosity = 31.3 cps; LD_{50} (rat orl) = 8600 mg/kg. Aldrich; Allan; Allchem Ind.; BerjeBP Chem. Ltd; Daihachi Chem. Ind.; Eastman; Hüls Am.; Morflex; Penta Mfg.; Unitex.

8438 diethyl toluamide

134-62-3 2912 205-149-7

$C_{12}H_{17}NO$

N,N-Diethyl-3-methylbenzamide

N,N-diethyl-*m*-toluamide; *m*-toluic acid diethylamide; DEET; M-Det; *m*-DETA; ENT-20218; Autan; *m*-Delphene; Detamide; Dieltamid; Flypel; Metadelphene; Off; Repel. Insect repellant, resin solvent, film formers. bp$_{19}$ = 160°; insoluble in H_2O; soluble in organic solvents; LD_{50} (rat orl) > 2 g/kg. *Du Pont; Honywill & Stein Ltd; Morflex.*

8439 diethylamine

109-89-7 3160 203-716-3

$C_4H_{11}N$

N-Ethylethanamine

DEA; diethamine; N,N-diethylamine. Used in rubber chemicals, textile specialties, selective solvent, dyes, flotation agents, resins, pesticides, polymerization inhibitor, pharmaceuticals, petroleum chemicals, electroplating, corrosion inhibitors. mp = -50°; bp = 55.5°; flash pt < 20° F; soluble in H_2O, EtOH; LD_{50} (rat orl) = 540 mg/kg. *Air Prods & Chem; Allchem Industries; Elf Atochem; BASF; Union Carbide.*

8440 2-diethylaminoethanol
100-37-8 3161 202-845-2
$C_6H_{15}NO$
2-hydroxytriethylamine
diethylethanolamine; DEAE; 2-diethylaminoethyl alcohol; N,N-diethyl-2-hydroxyethylamine; N,N-diethylethanolamine; N,N-diethylaminoethanol; β-diethylaminoethanol; N,N-diethyl-N-(β-hydroxyethyl)amine; β-hydroxytriethylamine. Used in making medicines, pharmaceuticals, pesticides, and other chemicals. mp = -70°; bp = 161°; d = 0.8840; n_D^{20} = 1.4410; soluble in H_2O, organic solvents; LD_{50} (rat orl) = 1100 mg/kg.

8441 diethyldiphenyldichloroethane
72-56-0 200-785-1
$C_{18}H_{20}Cl_2$
1,1-bis(*p*-ethylphenyl)-2,2-dichloroethane
perthane; diethyldiphenyl dichloroethane; ethylan; *p,p'*-ethyl DDD; Q-137. An organochlorine insecticide marketed under the trade name Perthane, has a lower toxicity to both insects and mammals than its structural analogs, DDT and DDD and is of moderate persistence in the environment. First marketed in 1950 for use against houseflies and cloth moths, it has since been used on vegetables, pears, and livestock. Not a carcinogen. mp = 60-61°; insoluble in H_2O, soluble in organic solvents; LD_{50} (rat orl) = 6600 mg/kg.

8442 diethylene glycol
111-46-6 3168 203-872-2
$C_4H_{10}O_3$
2,2'-Oxybisethanol
diglycol; D.E.H. 52; D.E.H. 20; dihydroxydiethyl ether; 2,2'-oxydiethanol. DEG; Aliphatic diol production of polyurethane and unsaturated polyester resins, triethylene glycol; textile softener; solvent for nitrocellulose, dyes and oils; dehydration of natural gas, elasticizers, and surfactants; humectant for tobacco, casein, and synthetic sponges. mp = -6.5°; bp = 244-245°; flash point = 290° F; soluble in H_2O, organic solvents; LD_{50} (rat orl) = 20.76 g/kg. BASF; BP Chem. Ltd; Du Pont; Eastman; Hoechst Celanese; Mitsui Petrochem. Ind.; Mobil; Olin; OxyChem; Shell; Texaco; Union Carbide.

8443 diethylene glycol butyl ether acetate
124-17-4 204-685-9
$C_{10}H_{20}O_4$
2-(2-Butoxyethoxy) ethyl acetate
Butoxyethanol acetate; Diglycol monobutyl ether acetate; Butyl carbitol acetate; Butyldiglycol acetate; Butyl Carbitol® Acetate; Dba; DB Acetate; Eastman® DB Acetate. Solvent. d = 0.977; n_D^{20} = 1.4300; soluble in H_2O, organic solvents; fp=-32.2°; flash point = 102°; LD_{50} (rat orl) = 6500 mg/kg. Arco; Eastman; OxyChem; Aldrich; Allchem Ind.; Ashland; Eastman; Fluka; Occidental; Spectrum Chem Mfg.

8444 diethylene glycol dimethacrylate
2358-84-1 219-099-9
$C_{12}H_{18}O_5$
3-Oxapentane-1,5-diyl dimethacrylate
Ageflex DEGDMA; MFM-418. Crosslinker for rubber vulcanization, moisture barrier films and coatings, photopolymer printing plates and letterpress inks, conversion coatings and adhesives. bp_2 = 134°; d = 1.082; flash point = 230°. *CPS; Sartomer; Aldrich; CPS; Monomer-Polymer Name unverified. Dajac; Polysciences; Rohm Tech.*

8445 diethylene glycol propyl ether
6881-94-3 229-985-7
$C_7H_{16}O_3$
2-(2-Propoxyethoxy) ethanol
Diethylene glycol monopropyl ether; DP; Ektasolve® DP. Evaporating, water-miscible solvent used in solution and water-dilutable coatings; active for many coating materials including NC, acrylic copolymers, epoxy resins, chlorinated rubber, and alkyd resins; strong coupling agent with some resin systems in water-dilutable coatings. bp = 202°; soluble in H_2O, organic solvents; d = 0.963; fp < -90°; flash point = 93°; causes eye irritation. *Ashland; Eastman; Great Western.*

8446 diethylenetriamine
111-40-0 203-865-4
$C_7H_{16}N_3$
2,2-iminodiethylamine
aminoethylethandiamine; DETA; bis(2-aminoethyl)amine; D.E.H. 20; D.E.H. 52; D.E.H. 58; DETA; Texacure EA-20. DETA; solvent for sulfur, acid gases, various resins, dyes; saponification agent for acidic materials; fuel component. bp = 206.7°; soluble in H_2O and hydrocarbons; d = 0.9542; fp = -39°; strongly alkaline. *Allchem Industries; Janssen Chimica; Nayler Chem. Ltd; Tosoh; Union Carbide.*

8447 diethylin
The diethyl ether of glycerol, a solvent.

8448 Di-Farmon
dicamba-mecoprop
Soluble concentrate of 21 g dicamba and 319g mecoprop per liter; used for weed control in cereals and grassland. *ICI Chem & Polymers Ltd; Farm Protection Ltd.*

8449 difenacoum
56073-07-5 259-978-4
$C_{31}H_{24}O_3$
biphenyl-4-yl-1,2,3,4-tetrahydro-1-naphthyl)-4-hydroxy-1(2H)-benzopyran-2-one
WBA 8107; Ratak; Neosorexa PP580. Anti-coagulant, used as a rodenticide.

8450 Diflubenzuron
35367-38-5 3188 252-529-3
$C_{14}H_9ClF_2N_2O_2$
N-[[(4-chlorophenyl)amno]carbonyl]-2,6-difluorobenzamide
Dimilin; Micromite; Difluron; DU 112307; Largon; PDD 6040-I; PH-60-40; TH-6040; ENT-29054; OMS-1804; Duphacid;. Insecticide. mp = 210-230°; insoluble in H_2O, soluble in polar organic solvents; LD_{50} (rat orl) > 10 g/kg.

8451 Diflufenican
83164-33-4
$C_{19}H_{11}F_5N_2O_2$
N-(2,4-difluorophenyl)-2-[3-(trifluoromethyl)phenoxy]-3-pyridinecarboxamide
diflufenicanil, M&B 38544; Kwarc; DFF. Selective contact and residual herbicide, used for control of broad-leaved weeds and some grasses, particulalry *Galium, Veronica* and *Viola* spp. in cereal crops. mp = 161-162°; almost insoluble in H_2O (0.05 mg/l), readily soluble in organic solvents; LD_{50} (rat orl) >2000 mg/kg.

8452 Difolatan
02425-06-1 219-363-3
captafol
Captafol fungicide. *Monsanto (Solaris).* Name unverified.

8453 diformyl
107-22-2 4519 203-474-9
$C_2H_2O_2$
Glyoxal
Used in textiles, glues, biocides and in organic synthesis.

8454 Diglycolamine® Agent (DGA®)
929-06-6 213-195-4
$C_4H_{11}NO_2$
2-(2-aminoethoxy) ethanol
aminoethoxyethanol. Solvent for removal of CO_2 or H_2S from gases, for recovery of aromatics from refinery streams, for preparation of foam stabilizers, wetting agents, emulsifiers, and condensation polymers. mp = -12.5°; bp = 218-224°; d = 1.0500; n_D^{20} = 1.0500; soluble in H_2O, organic solvents; LD_{50} (rat orl) = 5660 mg/kg. *Texaco.*

8455 diglyme
111-96-6 3208 203-924-4
$C_6H_{14}O_3$
1,1'-Oxybis[2-methoxyethane]
Dricoid® 200; Dariloid® 100; diethylene glycol dimethyl ether; Diglyme; Diethylene glycol dimethyl ether; Dimethyl Carbitol; 2-Methoxyethyl ether; diglycol methyl ether; 2,2'-oxybisethanol dimethyl ether; poly solv; 1,1'-oxybis(2-methoxy)ethane; 2,5,8-Trioxanonane. The dimethyl ether of diethylene glycol; a solvent for polystyrene PVC/PVA copolymer and polymethyl methacrylate. mp = -68°; bp = 162°; flash point = 158°; miscible with H_2O, organic solvents. *ICI Chem & Polymers Ltd.* Discontinued.

8456 Dihalo
Swimming pool water disinfectant. *Great Lakes Europe.*

8457 dihydroisophorone
873-94-9 212-855-9
$C_9H_{16}O$
3,5,5-triethylcyclohexanone
A high boiling point ketone solvent for surface coatings. bp = 188-194°; fp = 68°; d = 0.889.

8458 dihydrojasmone
1128-08-1 214-434-5
$C_{11}H_{18}O$
2-Pentyl-3-methyl-2-cyclopenten-1-one.
Jasmine, fresh, fruity odor. Used in floral & citrus perfumes. bp_{12} = 120-121°; d = 0.916. *Quest Int'l. UK Ltd.*

8459 Dihydroxyacetone
96-26-4 3225 202-494-5
$C_3H_6O_3$
1,3-Dihydroxy-2-propanone
1,3-dihydroxy-2-propanone; 1,3-dihydroxydimethyl ketone; aliphatic ketone; Protosol; Ketochromin. Used as cosmetic stain for suntan lotion and as a

reagent in chemical synthesis; humectant, plasticizer, fungicides. mp = 75-80°; soluble in H_2O (1 g/ml), EtOH. *EM Industries; Gist-Brocades Food Ingreds.; Janssen Chimica; Penta Mfg.; Spectrum Chem. Mfg.*

8460 diisobutyl adipate
141-04-8 205-450-3
$C_{14}H_{26}O_4$
bis(2-methylpropyl) hexanedioate
diisobutyl hexanedioate; DIBA; Isobutyl adipate; hexanedioic acid, diisobutyl ester; DBE-IB. Used as a plasticizer. bp=270-280°; d = 0.950; insoluble in H_2O, soluble in most organic solvents; fp = -20°; LD_{50} (rat ip) = 5950 mg/kg. *Aceto; Aldrich.*

8461 diisobutyl ketone
108-83-8 203-620-1
$C_9H_{18}O$
2,5-dimethyl-4-heptanone
Isovalerone; DIBK; Dibutyl ketone; Isobutyl ketone; Valerone. Solvent for nitrocellulose, rubber, synthetic resins; lacquers, coatings, organic synthesis, roll-coating inks, stains. mp = -46°; bp = 165°; d = 0.8076; n_D^{20}= 1.43130; insoluble in H_2O, soluble in organic solvents; flash point = 49°; LD_{50} (rat orl) = 4300 mg/kg. *Allchem Industries; Eastman; Hüls AG; Union Carbide.*

8462 diisodecyl phthalate
68515-49-1 271-091-4
$C_{28}H_{46}O_4$
1,2-Benzenedicarboxylic acid, diisodecyl ester
DIDP; Diplast® R; Jayflex® DIDP; Nuoplaz® DIDP; Palatinol® DIDP; Plasthall® DIDP-E; PX-120. Used as a plasticizer. bp= 250-257°; d = 0.966; insoluble in H_2O, soluble in organic solvents; fp = -50°; viscosity = 108 cps; flash point = 450°F. BASF; C.P. Hall; OxyChem; Allchem Ind.; Aristech; Ashland; BASF; Coyne; Exxon; C.P. Hall; Harwick; Hatco; Hoechst AG.

8463 diisononyl phthalate
68515-48-0 271-090-9
$C_{26}H_{42}O_4$
Plasticizer. *BASF; Chemisphere Ltd; Chisso Am.; C.P. Hall.*

8464 diisooctyl hydrogen phosphite
Doverphos® 298. Antioxidant and extreme pressure additive for lubricants. *Dover.*

8465 diisooctyl phosphite
36116-84-4 252-287-4
$C_{16}H_{35}PO_3$
Doverphos® DIOP. Stabilizer for PVC, PP; lubricant additive. bp = 129°; d = 0.925-0.933; flash point = 146°. *Dover.*

8466 diisopropanolamine
110-97-4 203-820-9
$C_6H_{15}NO_2$
1,1'-iminobis-2-propanol
Aliphatic amine; DIPA; DIPA Commercial Grade; DIPA Low Freeze Grade 85; DIPA Low Freeze Grade 90; DIPA NF; Bis(2-hydroxypropyl)amine; Di(2-Hydroxy-n-propyl) amine; 1,1'-imino-2-propanol; DIPA; dipropyl-2,2'-dihydroxy-amine; 1,1'-Iminodi-2-propanol; 1,1'-iminodipropan-2-ol. DIPA; emulsifying agents for polishes, textile specialties, leather compounds, insecticides, cutting oils, aqueous paints. bp = 249°; d = 0.992, 8.27 lb/gal; n_D^{20}= 1.4595; viscosity = 870 cps; fp = 44°; flash point = 276°F. *BASF.*

8467 1,3-diisopropyl benzene
99-62-7 202-773-1
$C_{12}H_{18}$
m-diisopropylbenzene
1,3-diisopropyl benzene; 3-isopropylcumene. Solvent; chemical intermediate. mp = -63°; bp = 203°; d = 0.8560; n_D^{20}= 1.4890; LD_{50} (rat orl) = 7400 mg/kg. *Koch.*

8468 1,4-diisopropylbenzene
100-18-5 202-826-9
$C_{12}H_{18}$
p-diisopropylbenzene
4-isopropylcumene. Solvent; chemical intermediate. mp = -17°; bp = 203°; d = 0.8570; n_D^{20}= 1.4889; LD_{50} (mus orl) = 3400 mg/kg. *Koch.*

8469 diisopropylamine
108-18-9 3240 203-558-5
$[(CH_3)_2CH]_2NH$
N-(1-Methylethyl)-2-propamine
DIPA; N-(1-methylethyl)-2-propanamine; N,N-diisopropyl amine. Intermediate, catalyst. bp = 84°; d^{22} = 0.722; flash point = 21°F; soluble in H_2O, EtOH; LD_{50} (rat orl) = 770 mg/kg. *Air Prods & Chem; Aldrich; Elf Atochem N. Am.; BASF; Union Carbide.*

8470 diisostearyl dimer dilinoleate
127358-81-0
Dermol DISD. Emollient for creams, lotions, and makeup; anti-irritant in formulations. *Alzo.* Discontinued.

8471 diisotridecyl hydrogen phosphite
Doverphos® 269; Bis-isotridecyl hydrogen phosphite. Antioxidant and extreme pressure additive for lubricants. *Dover.*

8472 Dikar
Fungicide and miticide. *Rohm & Haas.*

8473 Dilasoft
Softening and filling agent with hydrophilic properties; for synthetic fibers and blends with cellulosics. *Sandoz Products Ltd.* Name unverified.

8474 Dilatin NA Liquid
Based on unchlorinated aromatic hydrocarbons; A proprietary preparation used in the dyeing of 100% polyester goods. *Sandoz.* Name unverified.

8475 dilauryl phosphite
21302-09-0 244-325-8
$C_{24}H_{51}O_3P$
Doverphos® 271L; Doverphos® 274; Duraphos™ AP-230; Dilauryl hydrogen phosphite; Didocyl phosphite; Didodecyl phosphite; Di-n-dodecyl phosphite; Weston® DLP. Antioxidant and EP additive for lubricants; catalyst in polymerization of unsaturated compounds; stabilizer. d = 0.898-0.906; flash point = 138°; LD_{50} (rat orl) >10 g/kg. *Dover; Albright & Wilson Am.; GE Spec.; Witco/Polymer Addit.*

8476 dilauryl thiodipropionate
123-28-4 204-614-1
$C_{30}H_{58}O_4S$
didodecyl 3,3·g-thiodipropionate
Thiodipropionic acid; Dilauryl ester; Thiodipropionic acid dilauryl ester; Proponic acid, 3,3'-thiodipionate. Diester of lauryl alcohol and 3,3·g-thiodipropionic acid; antioxidant, additive for high-pressure lubricants and greases, plasticizer and softening agent, antioxidant for edible fats and oils. mp = 40°; bp = 240°; insoluble in H_2O; soluble in organic solvents; d = 0.975; acid no <1; flash point = 110°; LD_{50} (rat orl) >10.3 g/kg. *Am. Cyanamid; Evans Chemetics; Morton Int'l.; Witco/Argus.*

8477 Dilectene
A proprietary trade name for an aniline-formaldehyde synthetic resin. No manufacturer.

8478 Dilecto
A proprietary trade name for a phenol-formaldehyde synthetic resin laminated product. No manufacturer.

8479 Dilexo
Copolymer dispersions for paints, adhesives and sealants. *RWE-DEA Chemicals UK Ltd.*

8480 Dilinoleic acid
6144-28-1
$C_{36}H_{64}O_4$
9,12-octadecadienoic acid, dimer
Dimer acid; Empol® 1016; Empol® 1020; Empol® 1022, 1004, 1026; Industrene® D; Pripol 1017, 1022, 1025. C_{36} dicarboxylic acid formed by the catalytic dimerization of linoleic acid; lubricant, corrosion inhibitor, mildness additive in household detergents, plastics, and protective coatings. *Arizona; Henkel/Emery; Sherex; Union Camp; Witco/Humko;.* Discontinued.

8481 Dillex
Gripe mixture. *Knoll AG.* Unverified.

8482 Diluex®, FG
1337-76-4
Attapulgite clay
Absorbent carrier for pesticides in dusts and wettable powders. *Floridin.* Name unverified.

8483 Dilver
An alloy containing 42% nickel; used in filament lamps, as it has the same coefficient of expansion as glass.

8484 Dimacide
Acid dyes. *ICI Chem & Polymers Ltd.* Discontinued.

8485 Dimanin
Quaternary ammonium compound; used for disinfection. *Bayer AG.*

8486 Dimanin A Special
Used for control of algae, molds and bacteria in greenhouses and gardens. *Bayer AG.*

8487 Dimazon
83-63-6 3004 201-490-0
$C_{18}H_{19}N_3O_4$
Diacetylaminoazotoluene
A red dye used in ointment or as a dusting powder.

8488 Dimdac
Dimethyldiallylammonium chloride
Synthetic Chemicals Ltd.

8489 Dimension
Range of nylon/PPE alloys including reinforced grades with glass and talc;

applications include wheel covers, wheel hubs, water meter components, resonators, mirror housings and electrical components. *Allied Colloids Ltd.*

8490 Dimension D-9300 BK

Modified nylon 6/PPE alloy; blow molding grade with ultra-high melt viscosity, load-bearing at elevated temperatures, good impact resistance, used for automotive applications. *AlliedSignal Engineered Plastics.*

8491 Dimension Master®

High stability films; for the electrical industry. *DuPont UK.*

8492 Dimension®

For the medical industry. *DuPont UK.*

8493 dimethicone

9006-65-9 3264

dimethyl polysiloxane

A polydimethylsiloxane; used in ointments and topical drug ingredients.

8494 dimethicone copolyol

67674-67-3; 68937-55-3; 68938-54-5

Silicone glycol copolymer

Dabco® DC193, DC 197, DC198, DC5043, DC5098, DC5103, DC5125, DC5160; Dow Corning ®28; Dow Corning® 29, 54, Q2-5211 Superwetting Agent, Q2-5220 Surfactant, Q4-3667 Fluid; Silwet® L-7200, L-7210, L-7230, L-7622. Clear to light straw colored liquid; soluble in H_2O; d = 1.03-1.08; viscosity = 250 cps; fp = -7-12°. *Air Products/Perf. Chems; Dow Corning; Osi Spec.*

8495 dimethirimol

5221-53-4 3266 226-021-7

$C_{11}H_{19}N_3O$

5-butyl-2-(dimethylamino)-6-methyl-4(1H)pyrimidinone

dimethyrimol; Milcurb. Systemic fungicide with protective and curative action. Used as soil application for control of powderyt mildews in curcubits, tobacco, capsicum, tomatoes andsome ornamentals. mp= 102°; soluble in H_2O (1.2 g/l), more soluble in organic solvents; LD_{50} (rat orl) = 2350 mg/kg.

8496 Dimethoate

60-51-5 3269 200-480-3

$C_5H_{12}NO_3PS_2$

O,O-Dimethyl-S-(N-methylcarbamoylmethyl) phosphorodithioate

Cygon; Perfekthion; Rogor; Roxion; Fosfamid; Phosphamide; Defend; Fostion M M; American Cyanamid 12880; Rebelate. An insecticide and acaricide with contact and plant systemic activity suitable for protection against a broad range of insects and mites. mp = 52-53°; $bp_{0.05}$ = 107°; d = 1.281; soluble in H_2O (2.5 g/10 ml), more soluble in organic solvents; LD_{50} (rat orl) = 60 mg/kg. *A/S Cheminova.* Name unverified.

8497 Dimethoate Bayer

60-51-5 3269 200-480-3

Insecticide with limited systemic action for control of aphids, spider mites, whiteflies, mealy bugs, scale insects, fruit flies, leaf-eating caterpillars and leafminers on pome and stone fruit, grapes, sugar beet, cotton, citrus fruit, tobacco and vegetables. *Bayer AG.*

8498 dimethoxane

828-00-2 3273 212-579-9

$C_8H_{14}O_4$

2,6-dimethyl-1,3-dioxan-4-ol acetate

Acetomethoxane; 2,6-dimethyl-*m*-dioxan-4-yl ester acetic acid; 2,6-dimethyl-*m*-dioxan-4-yl acetate; 6-acetoxy-2,4-dimethyl-*m*-dioxane; 2,6-dimethyl-*m*-dioxan-4-ol acetate; acetic acid, ester with 2,6-dimethyl-*m*-dioxan-4-ol; DDOA; g1v gard dxn; Acetoxy-2,4-dimethyl-*m*-dioxane; Dimethyl-1,3-dioxan-4-ol acetate; Dimethyl-*m*-dioxan-4-ol acetate; Dimethyl-*m*-dioxan-4-yl acetate; Dioxan-4-ol, 2,6-dimethyl-, acetate. Preservative and gasoline additive. bp_6 = 74-75°; n_D^{20} = 1.4310; d_4^{20} = 1.0655; soluble in H_2O, organic solvents; LD_{50} (rat orl) = 1930 mg/kg.

8499 dimethoxy Tetrahydrofuran

696-59-3 211-797-1

$C_6H_{12}O_3$

dimethoxytetrahydrofuran; 2,5-dimethoxy tetrahydrofuran; Protectol®DMT;. Biocide. bp = 143-146°; d_4^{20} = 1.023; flash point < 50°F. *Chemie Linz UK Ltd; Fluka; Aldrich.*

8500 dimethoxyphenol

91-10-1 202-041-1

$C_8H_{10}O_3$

2,6-Dimethoxy phenol

pyrogallol 1,3-dimethyl ether; Syringol; Pyrogallol dimethyl ether; 2-Hydroxy-1,3-dimethoxybenzene. mp = 53-56°; bp = 261°; soluble in H_2O (2 g/100 ml), more soluble in organic solvents; LD_{50} (rat orl) = 550 mg/kg. *Janssen Chimica; Penta Mfg.; Schweizerhall; Spectrum Chem. Mfg.*

8501 dimethoxypropane

201-056-0

$C_5H_{12}O_2$

dimethoxypropane; acetone dimethyl acetal; 2,2-Dimethoxypropane; DMP. Protein precipitant; Chemical intermediate for pharmaceuticals; tissue dehydrating agent. mp = -47°; bp = 81-83°; d_4^{20} = 0.849; soluble in H_2O, organic solvents; flash point = -6.7° (CC); fire hazard when exposed to heat; irritant. *Aldrich; Janssen Chimica; Penta Mfg.; Schering Berlin Polymers; Ube Ind.; Sherex.*

8502 dimethyl adipate

627-93-0 211-020-6

$C_8H_{14}O_4$

dimethyl hexanedioate

methyl adipate. Diester of methyl alcohol and adipic acid. Solvent and plasticizer. mp = 8°; bp_{14} = 109-110°; d = 1.0630; n_D^{20} = 1.4285. *DuPont; Morflex; UCB SA.*

8503 dimethyl anthranilate

85-91-6 212-308-4

$C_9H_{11}NO_2$

2-Methylamino methyl benzoate

N-methyl methyl anthranilate; Dimethyl anthranilate; Methyl N-methyl anthranilate; Methyl O-methyl amino benzoate;. Used in perfumes, flavoring, drugs. *Am. Bio-Synthetics; Bell Flavors & Fragrances; Penta Mfg.*

8504 dimethyl azelate

1732-10-1 217-060-0

$C_{11}H_{20}O_4$

nonanedioic acid dimethyl ester

Emery® 2914. Synthetic lubricant. *Henkel/Emery.*

8505 dimethyl ether

115-10-6 6148 204-065-8

C_2H_6O

methane, oxybis-

oxybismethane; methyl ether. Organic compound.; solvent, motor fuel and aerosol propellant. mp = -142°; bp = -25°; d^{25} = 1.91855.

8506 dimethyl formamide

68-12-2 3292 200-679-5

C_3H_7NO; $HCON(CH_3)_2$

N,N-Dimethylformamide

Dynasolve 100; DMF; DMFA; N-Formyldimethylamine; U-4224. Solvent in vinyl resins and acetylene, butadiene, acid gases; polyacrylic fibers; catalyst in carboxylation reactions, organic synthesis; carrier for gases; reagent. mp = -61°; bp = 152.8°; soluble in H_2O, organic solvents; flash point = 57.7°; LD_{50} (rat orl) = 2800 mg/kg. Air Prods & Chem; Ashland; Baychem;Brown; Browning; Chemcentral; Coyne; Aceto; Air Prods & Chem; J.T. Baker; Aldrich; BASF; DuPont; ICI Spec.; Mallinckrodt; Mitsubishi Gas; Nissan Chem. Ind.; Mitsubishi Gas; Monomer-Polymer &Dajac Labs; UCBSA; Van Waters&Rogers;.

8507 dimethyl lauramine

112-18-5 203-943-8

$C_{14}H_{31}N$

Lauryl dimethylamine

Dodecyldimethylamine; Adma®12; Amine 2M12D; Armeen® DM12D; Barlene® 12S; Empigen®ABE; Kemamine® T-6902; Nissan Tert. Amine BB. Liquid cationic detergent; corrosion inhibitor; acid stable emulsifier. cationic detergent; moderately toxic by ingestion; severe skin irritant. *Ethyl; Lonza; Mason; Sherex; Albemarle; Aldrich; Fluka.*

8508 dimethyl phthalate

131-11-3 3304 205-011-6

$C_{10}H_{10}O_4$

1,2-Benzenedicarboxylic acid dimethyl ester

DMP; phthalic acid dimethyl ester; dimethyl 1,2-benzene-dicarboxylate; DMP; Palatinol M; Fermine; Avolin; Mipax; Palatinol M; Solvanom; Solvarone; DMF (insect repellent); NTM; phthalic acid methyl ester; Repeftal; Unimoll DMPalatinol M; Fermine; Avolin; Mipax,. Diester of methyl alcohol and phthalic acid; solvent for resin, plasticizer for cellulose acetate and nitrocellulose lacquers; plastics, rubber; coating agents; safety glass. mp = 2°; bp = 282°; d = 1.1900; n_D^{20} = 1.5145; λ_m = 277 nm ($E^{1\%}_{1cm}$ 57.7 EtOH); poorly soluble in H_2O (430 mg/100 ml), soluble in organic solvents; LD_{50} (rat orl) = 8.2 g/kg. *Allchem Industries; BASF; Daihachi Chem. Ind.; Eastman; Hüls Am.; Morflex; UCB SA; Unitex.*

8509 dimethyl succinate

106-65-0 203-419-9

$C_6H_{10}O_4$

dimethyl butanedioate

methyl succinate; DBE-4; Santosol™ DMS. Light and heat stabilizer for polyolefins, ABS polymer systems, flexible PVC, food packaging; chemical intermediate. mp = 16-19°; bp = 190-193°; d_4^{20} = 1.119; flash point = 90°. *DuPont; Penta Mfg.; Fluka; Sigma; Schweizerhall; Aldrich; Ashland; Chemie Linz N. Am.;.*

8510 dimethyl sulfide

75-18-3 6204 200-846-2

C_2H_6S

thiobismethane

methylthiomethane; 2-thiapropane; 2-thiopropane; methyl sulfide; DMS;

methyl thioether; thiopropane. Used with *t*-butyl mercaptan as an additive to natural gas for leak detection. mp = -83°; bp = 36°; insoluble in H=72O, soluble in organic solvents.

8511 dimethyl sulfoxide
67-68-5 3308 200-664-3
C_2H_6OS
Sulfinylbismethane
Domoso; Decap; Demavet; Sulfinylbis (methane); Methyl sulfinylmethane; DMSO; Herpid; methyl sulfoxide; SQ-9453; DMS-70; DMS-90; Deltan; Demasorb; Demeso; Dolicur; Dromisol; Gamasol 90; Hyadur; Kemsol; Rimso-50; Sclerosol; Somipront; Syntexan. Solvent for polymerization; analytical reagent; industrial cleaners, pesticides, paint stripping, hydraulic fluids, medicine (anti-inflammatory), veterinary medicine, plant pathology and nutrition, pharmaceuticals, spinning synthetic fibers. mp = 18.4°; bp = 189°; soluble in H_2O; insoluble in organic solvents; d = 1.101; flash point = 95°; LD_{50} (rat, orl) = 17,500 mg/kg. Aldrich; Atochem N. Am.; Gaylord; Fluka; Allchem Ind.; Elf Atochem; Howard Hall; Itochu Spec.; Monomer-Polymer & Dajac Labs; Research Organics; Spectrum Chem. Mfg.; Toray Fine Chems.

8512 dimethyl tetrachloroterephthalate
1861-32-1 2896 217-464-7
$C_{10}H_6Cl_4O_4$
2,3,5,6-Tetrachloro-1,4-benzenedicarboxylic acid dimethyl ester
Dacthal; DCPA; Chlorothal; Chlorthal-dimethyl; Chlorthal-methyl; Chlorthal-Dimethyl; Dacthalor; Dimethyl; Dimethyl ester of tetrachloroterephthalic acid; 2,3,5,6-tetrachloro-1,4-benzenedicarboxylate; DAC 4; DAC-893; DCP; DCPA; Fatal; Terephthalic acid, 2,3,5,6-tetrachloro-,dimethyl ester; Rid; Terechloroterephthalic acid dimethyl ester. Wettable powder containing 75% weight/weight DCPA; an herbicide for use on ornamentals, turf, fruit and vegetables. mp = 155-156°; soluble in H_2O, organic solvents; LD_{50} (rat orl) >3000 mg/kg. *Fermenta ASC Europe Ltd.*

8513 dimethylamine hydrochloride
506-59-2 3278 208-046-5
C_2H_8ClN
Methanamine, N-methyl-, hydrochloride
Dimethylamine hydrochloride; Dimethylammonium chloride; Hydrochloric acid dimethylamine. Used as a vulcanization accelerator and in manfacture of soaps. mp = 160°; LD_{50} (rat orl) = 1070 mg/kg. *Janssen Chimica; Penta Mfg.; Schweizerhall.*

8514 4-(dimethylamino)benzoic acid
619-84-1 3282 210-615-8
$C_9H_{11}NO_2$
Solarchem® O. Esters used as ultraviolet screens. mp = 242-243.5°; soluble in alcohol, HCl, KOH solutions, ether; insoluble in acetic acid. *Van Dyk; Lipo; CasChem.*

8515 dimethylaminobenzaldehyde
100-10-7 3280 202-819-0
$C_9H_{11}NO$
4-dimethylaminobenzaldehyde
p-dimethylaminobenzaldehyde; Dimethylaminobenzenecarbonal; Ehrlich's Reagent; *p*-formyl-N,N-dimethylaniline. Dyes, medicine, reagent. mp = 73-75°; bp_{17} = 176-177°; soluble in H_2O (300 mg/l), more soluble in organic solvents. *Aceto; BASF; Penta Mfg.; Schweizerhall.*

8516 dimethylaminoethyl chloride hydrochloride
4584-46-7 224-970-1
$(CH_3)_2NCH_2CH_2Cl \cdot HCl$
1-chloro-2-dimethylaminoethane hydrochloride
DMC; 2-chloro-N,N-dimethylethylamine hydrochloride; N-(2-chloroethyl)dimethylamine hydrochloride. Used in the manufacture of antihistamines and other pharmaceuticals; organic intermediate for introduction of β-dimethylaminoethyl radical. mp = 205-208°; soluble in H_2O (2000 g/l); irritant. *ICI Spec.; Janssen Chimica; Lonza; Aldrich; Schweizerhall.*

8517 2-dimethylamino-2-methyl-1-propanol
7005-47-2 230-279-6
$C_6H_{15}NO$
DMAMP-80. Amine solubilizer for resins in aqueous coatings; emulsifier for waxes; vapor-phase corrosion inhibitor; urethane catalyst; titanate solubilizer raw material for synthesis. *Angus.*

8518 dimethyldibutyl peroxyhexane
78-63-7 201-128-1
$C_{16}H_{34}O_4$
2,5-dimethyl-2,5-bis(*t*-butylperoxy)hexane
Luperco 101-P20; (1,1,4,4-tetramethyl-1,4-butanediyl)bis[(1,1-dimethylethyl) peroxide; 2,5-Dimethyl-2,5-di(tertiary-butylperoxy)-hexane. 20% dispersion of 2,5-dimethyl-2,5-di (*t*-butylperoxy) hexane on a PP powder carrier; crosslinking agent for elastomers and thermoplastic resins. mp = 8°; bp = 50-52°; insoluble in H_2O; LD_{50} (mus ip) = 1700 mg/kg. *Atochem.* Discontinued.

8519 N,N-dimethyldodecylamine
112-18-5 203-943-8
$C_{14}H_{31}N$
dimethyl lauramine
A chemical intermediate. mp = -20°; d = 0.7870. *Asahi Denka Kogyo.*

8520 dimethylethanolamine
108-01-0 2900 203-542-8
$C_4H_{11}NO$
2-dimethylaminoethanol
deanol; DMAE. Used in synthesis of dyestuffs, pharmaceuticals and textile auxiliaries. *Air Prods & Chem; Elf Atochem N. Am.; BASF; Nippon Nyukazai; Texaco; Union Carbide.*

8521 dimethylethylamine
598-56-1 209-940-8
$C_4H_{11}N$
N-ethyldimethylamine
N,N-dimethylethylamine; Dimethylethylamine; Ethyldimethylamine; N,N-Dimethylethanamine; N-Ethyldimethylamine. Used in chemical synthesis. mp = -140°; bp = 36°; d = 0.675.

8522 N,N-dimethylethylamine
598-56-1 209-940-8
$C_4H_{10}N$
N-ethyldimethylamine
Chemical intermediate.

8523 dimethylglyoxime
95-45-4 3295 202-420-1
$C_4H_8N_2O_2$
2,3-Butanedionedioxime
butane dioxime; dimethyl glyoxime; diacetyldioxime; Chugaev's reagent. Analytical chemistry, especially as reagent for nickel, biochemical research. mp = 238-240°; insoluble in H_2O, soluble in organic solvents; *Atomergic Chemetals; Pfaltz & Bauer.*

8524 N,N-dimethylhexadecylamine
112-69-6 203-997-2
$C_{18}H_{39}N$
N,N-dimethyl-1-Hexadecanamine
Adma® 16; Armeen DM 16D; hexadecyldimethylamine; Dimethyl-1-hexadecanamine; N,N-Dimethylhexadecylamine. *Asahi Denka Kogyo.*

8525 dimethylmethyl phosphonate
756-79-6 212-052-3
$C_3H_9O_3P$
Dimethyl methylphosphonate
Fyrol® DMMP; DMMP; Dimethyl methanephosphonate; methylphosphonic acid dimethyl ester. Flame retardant for applications where high phosphorus content, good solvency, and low viscosity are desired; lowers viscosity of epoxy resins and unsaturated polyesters filled with hydrated alumina oxide bp = 181°; d = 1.145; soluble in H_2O, organic solvents; LD_{50} (rat orl) > 5000 mg/kg. *Akzo.*

8526 dimethyl-phenylene-green
$C_{16}H_{20}N_3Cl$
The tetramethyl derivative of Phenylene blue, dyes silk and other fabrics.

8527 N,N-dimethyltetradecylamine
112-75-4 204-002-4
$C_{16}H_{35}N$
dimethyl(tetradecyl)amine
Chemical intermediate. *Asahi Denka Kogyo.*

8528 Dimetridazole
551-92-8 3315 209-001-2
$C_5H_7N_3O_2$
1,2-Dimethyl-5-nitro-1H-imidazole
Emtryl; RP-8595; Unizole. A veterinary antiprotozoan (Histomonas). mp = 138-139°; sparingly soluble in H_2O, ether, soluble in EtOH. *RMB Animal Health Ltd.*

8529 Dimilin
35367-38-5 3188 252-529-3
Diflubenzuron
Insecticides to counteract a number of harmful organisms occurring in agricultural, horticultural and forestry circles, fungus growth, sub-tropical cultures (weevils, cotton worm, many varieties of fruit insects, etc.); blocks development from the larvae to the adult insect stage; does not harm the environment. *ICI Agrochemicals; Duphar BV.*

8530 Dimodan LS Kosher
Sunflower oil distilled monoglyceride, unsaturated; food emulsifier; water-oil emulsifier for low-calorie spreads, icing shortenings and cake shortening. *Grindsted Prods.; Grindsted Prods. Denmark.*

8531 Dimodan O Kosher
Partially hydrogenated soybean oil glyceride; food emulsifier for margarine, icing shortenings, coffee whiteners. *Grindsted Prods.; Grindsted Prods. Denmark.*

8532 Dimodan PM

Hydrogenated lard or tallow distilled monoglyceride, saturated; food emulsifier; food emulsifier; starch complexing agent; for margarine, cake shortenings, confectionery coatings; softener for bread; peanut butter stabilizer. *Grindsted Prods.; Grindsted Prods. Denmark.*

8533 Dimodan PV, PV 300 Kosher

Hydrogenated soybean oil distilled monoglyceride, unsaturated; food emulsifier, starch complexing agent, antisticking agent; crumb softener for bread; aerating agent in cake mixes and frozen desserts. *Grindsted Prods.; Grindsted Prods. Denmark.*

8534 Dimodan PVP Kosher

97593-29-8 307-332-8

Hydrogenated palm oil distilled monoglyceride; food emulsifier, starch complexing agent, aerating agent; for margarine, cake shortening, etc.; crumb softener for bread; peanut butter stabilizer. *Grindsted Prods.; Grindsted Prods. Denmark.*

8535 Dimodan S

97593-29-8 307-332-8

Lard glyceride. Lard distilled monoglyceride, unsaturated; food emulsifier for margarine, cake shortenings, icing shortenings, coffee whiteners. *Grindsted Prods.*

8536 Dimul DDM K

Monoglycerides; food emulsifier. *Witco Corporation.*

8537 Dimycin

streptomycin, dihydrostreptomycin

Streptomycin and dihydrostreptomycin for veterinary purposes. *Glaxo Laboratories.* Name unverified.

8538 Dinamene

Selective weedkiller. *Murphy Chemical Co Ltd.* Discontinued.

8539 Dingler's green

A pigment, a chromium phosphate.

8540 Dinitra

51-28-5 3333 200-087-7

$C_6H_4N_2O_5$

2,4-Dinitrophenol

α-dinitrophenol; aldifen; Fenoxyl Carbon N; 2,4-DNP; Solfo Black B; Solfo Black BB; tertrosulfur black pb; Dinofan; Maroxol-50; Solfo Black 2B Supra; Solfo Black G; Solfo Black SB; tertrosulfur pbr; nitro kleenup; fenoxyl; Dinitrophenols. Used in the manufacture of dyes, as a wood preservative and insecticide and as an acid-base indicator (pH 2,6 colorless, pH 4.4 yellow). mp = 106-108°; d = 1.683; soluble in H_2O (0.28 g/100 ml). No manufacturer.

8541 Dinitrobenzoic acid

99-34-3 3328 202-751-1

$C_7H_4N_2O_6$

3,5-dinitrobenzoic acid

Identification of alcohols; chromatographic determination of the essential oil constituents. mp = 205-207°; soluble in H_2O (2 g/100 ml), EtOH and glacial acetic acid, less soluble in ether, carbon disulfide and benzene; sublimes. *Exchem Organics; Nobel; Schweizerhall.*

8542 4,6-dinitrocresol

534-52-1 3331 208-601-1

$C_7H_6N_2O_5$

3,5-Dinitro-2-Hydroxytoluene

Dinitro-*o*-cresol; DNOC; DNC; Nitrador; Dinitrocresol; Antinonnin; Detal; Dinitrol; Elgetol; K III; K IV; Ditrosol; Prokarbol; Effusan; Lipan; Selinon; Dekrysil; Antinonin; dinitrosol; Elgetox; 4,6-DNOC; Elgetol 30; Sinox. Selective herbicide and insecticide. mp = 83-85°; slightly soluble in H_2O, soluble in organic solvents; LD_{50} (rat orl) = 7 mg/kg.

8543 Dinobuton

973-21-7 213-546-1

$C_{14}H_{18}N_2O_7$

Carbonic acid 1-methylethyl 2-(1-methylpropyl)-4,6-dinitrophenyl ester

Acrex; Dessin; Sytasol; Isopropyl 2,4-dinitro-6-*sec*-butylphenyl carbonate. Miticide. mp = 56-57°; LD_{50} (rat orl) = 59 mg/kg. *Murphy Chemical Co Ltd.*

8544 Dinonyl phenol

1323-65-5 215-356-4

$C_{24}H_{42}O$

Mixture of dialkyl substituted phenols; used as a solvent. Colorless liquid; insoluble in H_2O; soluble in organic solvents *Allchem Industries; Texaco; Huntsman;.*

8545 Dinoram

Alkyl propylene diamines based on coco, tallow, and oleyl amines; bitumen emulsifiers and adhesion agents, production of cationic emulsions of oils and waxes, chemical synthesis intermediates, treatment of pigments and filers for paints, epoxy curing agent, bactericides. *Elf Atochem UK/Ceca.*

8546 Dinoramox

Ethoxylated alkyl propylene diamines; emulsifiers, corrosion inhibitors, wetting agents. *Elf Atochem UK/Ceca.*

8547 Dioctyl

577-11-7 3460 209-406-4

dioctyl sodium sulfosuccinate

A proprietary trade name for dioctyl sodium sulfosuccinate. No manufacturer.

8548 dioctyl adipate

103-23-1 203-090-1

$C_{22}H_{42}O_4$

bis(2-ethylhexyl) hexanedioate; hexanedioic acid bis(2-ethylhexyl) ester

Good-rite® GP-223; Adimoll® DO; Jayflex® DOA; Kodaflex® DOA; Monoplex® DOA; Palatinol® DOA; Plasthall® DOA; Plastomoll® DOA; Polycizer 332; PX-238; Uniflex® DOA; Wickenol® 158; Witcizer 412; BEHA; DEHA; DOA; Adipol 2EH; Bisoflex DOA; Effemoll DOA; Effomoll DOA; Ergoplast Addo; Flexol A 26; Flexol Plasticizer 10-A; Flexol Plasticizer A-26; Kemester 5652; Kodaflex Doa; Mollan S; Monoplex DOA; Plastomoll DOA; PX-238; Reomol DOA; Rucoflex Plasticizer Doa; Sicol 250; Staflex DOA; Truflex DOA; Uniflex DOA; Vestinol OA; Wickenol 158; Witamol 320. Vinyl plasticizer. mp = -67°; bp = 417°; d = 0.9268; n_{20D} = 1.4472; insoluble in H_2O; flash point = 196°; LD_{50} (rat orl) = 9110 mg/kg. BF Goodrich; Eastman; C.P.Hall; BASF; BASF plc; Harwick Standard Chemical Co; Union Camp; CasChem; Chisso; Hüls AG; Inolex; Monsanto. Name unverified.

8549 dioctyl phthalate

117-81-7 1291 204-211-0

$C_{24}H_{38}O_4$

di(2-ethylhexyl) phthalate

di-*sec*-octyl phthalate; 1,2-benzenedicarboxylic acid; Kodaflex DOP; Mollan O; Nuoplaz DOP; Palatinol AH; Pittsburgh PX-138; Platinol AH; Platinol DOP; RC Plasticizer DOP; Reomol DOP; Reomol D 79P; DEHP; DOP; Octoil; BEHP; Bisoflex 81; Bisoflex DOP; Compound 889; DAF 68; Ergoplast FDO; Eviplast 80; Eviplast 81; Fleximel; Flexol DOP; Flexol Plasticizer DOP; Good-rite GP 264; Hatcol DOP; Hercoflex 260; Sicol 150; Staflex DOP; Truflex DOP; Vestinol AH; Vinicizer 80; Witcizer 312; Union Carbide Flexol 380. Diester of 2-ethylhexyl alcohol and phthalic acid; plasticizer for many resins and elastomers. mp = -50°; bp = 384°; d = 0.9810; n_D^{20} = 1.4853; soluble in H_2O (100 mg/l), more soluble in organic solvents; LD_{50} (rat orl) = 30600 mg/kg. Aristech; BASF; Chemisphere Ltd; Chisso; Daihachi Chem. Ind.; Eastman; C.P. Hall; Hüls AG; Mitsubishi Gas; UCB SA.

8550 dioctyl sodium sulfosuccinate

1369-66-3 3460 209-406-4

$C_{20}H_{37}NaO_7S$

Sodium dioctyl sulfosuccinate; Empimin® OP70; DSS; NaDOSS;Sodium 1,4-bis(2-ethylexyl) sulfosuccinate; Dioctyl sulfosodiumsuccinate; Sodium di(2-ethylhexyl)sulfosuccinate; Bis (2-ethylhexyl)-S-sodium sulfosuccinate; 2-Ethylhexyl sulfosuccinate sodium; Docusate sodium; Ablusol C-78; Aerosol® GPG; Arylene M60; Atlas WA-100; Calgene DOSS-70; Complemix® 100; Coptal WA OSN; Denwet CM; Discol DFW; Disponyl® SUS 87 Special; Emcol® 4560, 4500; Empimin® OP70; Gemtex SC-40; Geropon® DOS; Mackanate™DOS-75;,. Emulsifier, dispersant, wetting agent used for emulsion polymerization, oil slicks, textiles, agrochemicals. Wax-like solid; slowly soluble in H_2O; freely soluble in alcohol, glycerol, CCl_4; LD_{50} (rat orl) = 1900 mg/kg. Albright & Wilson UK; Am. Cyanamid.; Alco; Brotherton Ltd; Calgene; Cytec Ind.; Eastern Color & Chem.; EM Ind.; Finetex; Fluka; Hart Prods.; Henkel/Chems.; Hodag; Macintyre; Mona Ind.; Sherex; Sigma; Spectrum Quality Prods.; Venchem; Witco/Oleo-Surf.

8551 Diofan®

Vinylidene chloride polymer dispersions; for production of chemically resistant and heat sealable packaging materials impervious to water vapor, gas and aromas; binders for textile coatings; moisture-resistant coatings for building materials. *BASF AG; BASF plc.*

8552 Diofan® D

Polyvinylidene chloride dispersion. *BASF plc.* Name unverified.

8553 Dioform

540-59-0 93 208-750-2

$C_2H_2Cl_2$

1,2-dichloroethylene

acetylene dichloride; dichloracetylene; 1,2-Dichloroethylene; *sym*-Dichloroethylene; Dichloroethylene; acetylene dichloride; dioform; 1,2-Dichloroethylene (mixture); Dichloroethylenes; 1,2-Dichloroethene (mixed isomers;)*cis & trans* 1,2-dichloroethylene; *cis,trans*-1,2-dichloroethylene; 1,2-Dichloroethylene (*cis & trans*). Used as a general solvent for organic materials, dye extraction, perfume, lacquers, thermoplastics, organic synthesis. mp = -57°; bp = 48-60°; d = 1.265; n_D^{20} = 1.4463; insoluble in H_2O, soluble in organic solvents; LD_{50} (rat orl) = 770 mg/kg.

8554 Dioleyl hydrogen phosphite

64051-29-2 264-626-8

Doverphos® 253; DOHP; Duraphos™AP-240. Antioxidant and extreme pressure additive for lubricants. *Dover.*

8555 Diolpate®
Saturated polyester; used for PVC and rubber compounding, surface coating industry. *Hyperlast.*

8556 Dioltech 311
Molybdate-based; corrosion and scale control agent for open recirculating cooling water systems. *Drew Ind. Div.*

8557 Di-On
Herbicide. *Agan Chemical Manufactures Ltd.* Discontinued.

8558 Dion®
Polyester resins. *Reichhold.*

8559 Dion® Cor-Res
Synthetic resins. *Reichhold.*

8560 Dion® FR 6308
Polyester resin; flame retardant for corrosion applications in filament winding and compression molding/SMC/BMC process; used for molded parts (industrial grating, computer housings, etc.); high physical strengths, MIL-R-7575 compliance; good fume handling corrosion resistance,high heat resistance, ambient or elelvated temperature gel and cure. *Reichhold.*

8561 Dion® FR6657
Polyester resin; flame retardant, promoted resin for transportation and construction applications; only for use with alumina trihydrate. *Reichhold.*

8562 Dion® VER
Vinyl ester resins. *Reichhold.*

8563 Dion® VER 9100 NP
Bisphenol epoxy vinyl ester resin; thermoset with good corrosion resistance, inherent strength and impact resistance; for filament winding centrifugal casting processes. *Reichhold/Reactive Polymers.*

8564 Dional 11,113
Fluorinated hydrocarbon; solvent for dry cleaning. *Hoechst Celanese.*

8565 Dionil®
A range of fatty acid amide polyglycol ethers; special detergent, soil suspending/leveling and protective colloid action; some of the range have an additional superfatting effect. *Hüls AG.*

8566 Dionil® OC
26027-37-2
PEG-3 oleamide
Detergent for light and heavy-duty detergents, dishwashing agents, cosmetic prepations; component in textile auxiliaries; refatting agent. *Hüls Am; Hüls AG.*

8567 Dionil® SD
Fatty acid amide polyglycol ether; superfatting and preparation agent. *Hüls Am; Hüls AG.*

8568 Dion-Iso®
Common metals and their alloys; for metal building materials, transportable buildings, railway tracks, nonelectric cables and wires, small hardware items, pipes and tubes, safes. *Reichhold.*

8569 Dionosil
587-61-1 8048 209-603-5
Propyliodone
Propyliodone suspensions or powders. *Glaxo Laboratories.* Name unverified.

8570 Diorez®
Saturated polyester polyols; used for polyurethane systems, microcellular elastomers, thermoplastic polyurethane fabrics and surface coatings. *Hyperlast.*

8571 Diorez® SC
Polyester, acrylics, epoxies; surface coatings intermediates. *Hyperlast.*

8572 Diosal
133-91-5 3234 205-124-0
$C_7H_4I_2O_3$
2-hydroxy-3,5-diiodobenzoic acid
3,5-diiodosalicylic acid. An intermediate in the manufacture of thyroxine. mp = 235-236°.

8573 Diox DR 22
13463-67-7 9612 236-675-5
titanium dioxide
A trade name for a rutile (titanium dioxide) type white pigment with a blue tone and good dispersability. *U.S. Industrial Chem.* Name unverified.

8574 Dioxane
123-91-1 3353 204-661-8
$C_4H_8O_2$
1,4-dioxane
1,4-diethylene dioxide; diethylene oxide ether; diethylene ether; diethylene oxide; diokan; ethylene glycol ethylene ether; glycol ethylene ether8; Diox; DuPont Zonyl FSO-100 Fluorinated Surfactants; 1,4-Dioxane; 1,4-diethylene dioxide; diethylene ether; 1,4-Dioxacyclohexane; Dioxane; Dioxyethylene ether; Tetrahydro-*p*-dioxin; Tetrahydro-1,4-dioxin; 1,4-Diethyleneoxide; Diethylene dioxide. Stabilizer for chlorinated hydrocarbons; solvent for adhesives, dyes, cellulose, lacquer, wax, pharmaceuticals, and coatings. mp = 11°; bp = 101°; d = 1.0350; .80°; viscosity = 0.0120 poise; LD_{50} (rat orl) = 5.1 gm/kg. *Ashland; BASF; CPS; Ferro/Grant; Mallinckrodt; Union Carbide.*

8575 Dioxine
$C_{10}H_7NO_3$
nitrosodihydroxynaphthalene
Gambine R. A dyestuff; dyes iron mordanted fabrics, green, and chrome mordanted materials, brown.

8576 Dioxitol
111-90-0 1847 203-919-7
Diethylene glycol monoethyl ether
A colorless, slightly hygroscopic liquid with mild odor; used as a solvent in paints, lacquers, textile printing inks, and stains; it is effective as a metal degreasing agent and is used in production of safety glass; it is also used as a coupling agent for cutting oils and emulsifiable oils, in the production of plasticizersand as an extraction agent for essences and perfumes. Colorless liquid; miscible with H_2O; d = 0.986-0.990. *Shell UK.*

8577 dioxogen
7722-84-1 4839 231-765-0
hydrogen peroxide
A 3% solution of hydrogen peroxide, H_2O_2; used as disinfectant.

8578 Dipalmitoyl hydroxy proline
41672-81-5 255-490-0
Dipalmitoyl hydroxyproline
Lipacide DPHP. Mild surfactant with excellent lathering and wetting properties, substantivity to hair and skin; for frequent-use shampoos, bath gels, soaps, shaving creams; used in cosmetic preparations for maintenance of the skin's physiological balance. *Rhône-Poulenc; R.T. Vandervilt.*

8579 Dipanol
Terpene rubber reclaimimg oil. *Crowley Chem.*

8580 Dipel®
Bacillus thuringiensis; a bacterial insecticide for control of caterpillars. *English Woodlands Ltd.*

8581 Dipentaerythritol
126-58-9 204-794-1
$C_{10}H_{22}O_7$
2,2-(oxybis(methylene)bis(2-(hydroxymethyl; -1,3-propanediol
Hercules® Tech Di-PE; Dipentek; Dipentek; 1,3-Propanediol, 2,2'-[oxybis(methylene)]bis[2-(hydroxymethyl)-; 2,2,6,6,-Tetra(hydroxymethyl)-4-oxaheptane-1,7-diol. Used as intermediate in manufacture of alkyds and drying oils, paints and coatings. mp = 212-220°; d = 1.33. *Allchem Industries; Honeywill & Stein Ltd.*

8582 Dipentaerythrityl hexacaprylate/hexacaprate
68130-24-5 268-581-5
Dipentaerythrityl hexacaprylate/hexacaprate
Liponate DPC-6. Nontacky emollient for treatment products. *Lipo.*

8583 Dipentek
A technical grade of dipentaerythritol. No manufacturer.

8584 Dipentene
138-86-3 5518 205-341-0
$C_{10}H_{16}$
cinene
limonene, cajeputene. Solvent for oleoresinous products, rosin, ester gum, etc.; rubber compounding. and reclaiming; dispersant for oils, resins and combinations, pigments, driers, paints, enamels, lacquers: general wetting; printing inks, perfumes, flavors, waxes, polishes. bp = 170-180° *Arizona; Hercules; Penta Mfg.; SCM Glidco Organics; Veitsiluofo Oy.* Discontinued.

8585 Dipentene No.122
Terpene liquid; solvent and antiskinning agent in paints. *Hercules.* Discontinued.

8586 Di-Petronate Series
Range of diluted sodium petroleum sulfonates; emulsifiers, dispersing and wetting agents for use when lower viscosity and easier handling than the Petronate Series is required. *Witco Chemical Ltd.*

8587 Dipex
A proprietary mold lubricant for rubber; a water-soluble sodium sulfonate obtained from petroleum-acid sludges. No manufacturer.

8588 Diphasol
Dyeing and printing assistant. *Ciba plc.* Name unverified.

8589 Diphen 60-B
A phenol-urea formaldehyde resin. No manufacturer.

8590 Diphenal
The sodium salt of diaminodihydroxybiphenyl; used as a photographic developer.

8591 Diphenyl
Direct dyes. *Ciba plc.* Name unverified.

8592 diphenyl
92-52-4 3372 202-163-5
$C_6H_5C_6H_5$
1,1'-biphenyl
bibenzene; phenylbenzene; biphenyl. Used as a heat transfer agent, fungistat for agricultural use, in organic synthesis. *Aldrich; Coalite Chem. Div; Koch; Monsanto; Sybron.*

8593 diphenyl acetonitrile
86-29-3 201-662-5
$C_{14}H_{11}N$
Diphenylacetonitrile
α-phenylbenzeneacetonitrile; Dipan; Diphenatrile; benzhydryl cyanide. Preparation of diphenylacetic acid, synthesis of antispasmodics, herbicide. mp = 71-73°; bp_{16} = 181°; LD_{50} (rat orl) = 3500 mg/kg. *Andeno BV; R.W. Greef; Janssen Chimica; Schweizerhall.*

8594 diphenyl isodecyl phosphite
26544-23-0 247-777-4
$C_{22}H_{31}O_3P$
Doverphos® 8; DPDP; Weston® DPDP. Chelating agent with metal carboxylates as polymer additives, especially for chlorinated polymers such as PVC and chlorinated PE; improves color, heat and light stability. bp = 190°; d = 1.022-1.032; flash point = 154°. *Dover; Akzo Nobel;.*

8595 diphenyl isooctyl phosphite
26401-27-4 247-658-7
$(C_6H_5O)_2POC_8H_{17}$
Doverphos® DPIOP; DPIOP; Weston®ODPP. Color stabilizer for ABC, PC; stabilizer for PVC; antioxidant bp = 190°; d = 1.040-1.047. *Dover; Aldrich.*

8596 diphenyl isotridecyl phosphite
Doverphos® 75. Color stabilizer for ABC, PC; stabilizer for PVC; antioxidant *Dover.*

8597 diphenyl octyl phosphate
1241-94-7 214-987-2
$C_{20}H_{27}O_4P$
Diphenyl-2-ethylhexyl phosphate
Disflamoll® DPO; Diphenyloctyl phosphate; Phosflex® 362; santicizer 141; 2-Ethylhexyl diphenyl ester phosphoric acid; 2-Ethyl-1-hexanol ester with diphenyl phosphate; 2-Ethylhexyl diphenyl phosphate; Octyl diphenyl phosphate. Flame retardant plasticizer for type PVC applications, dip, rotationally, extruded and injection molded parts, mechanical foam. d = 1.088-1.093; flash point = 224° (COC); LD_{50} (mus ip) = 930 mg/kg. *Bayer AG; Miles; Polysar; Akzo Nobel; Ashland; Harwick; Monsanto.*

8598 diphenyl oxide
101-84-8 7442 202-981-2
$C_{12}H_{10}O$
Diphenyl ether
Phenyl ether; 1,1'-oxybisbenzene; phenoxybenzene. Perfumery, soaps; heat-transfer medium; chemical intermediate for halogenation, acylation, alkylation, etc. mp = 27-28°; bp = 259°; d = 1.0730; n_D^{20}= 1.5795; LD_{50} (rat orl) = 3370 mg/kg. *Monsanto; Penta Mfg.*

8599 diphenyl phosphite
4712-55-4 225-202-8
$C_{12}H_{11}O_3P$
Phosphonic acid, diphenyl ester
Doverphos® DPP; Doverphos® 213; Weston® DPP; DPP. Used in the synthesis of organophosphorous compounds; color stabilizer for unsaturated polyesters; stabilizer for PVC; antioxidant for PP. mp = 12°; d_{15}^{25} = 1.221; flash point = 176°. *Dover; Aldrich, Fluka, Janssen Chimica; Spectrum Chem, Mfg.*

8600 diphenylacetonitrile
86-29-3 201-662-5
$C_{14}H_{11}N$
diphenatrile
α-phenyl-benzeneacetonitrile; Dipan; benzhydryl cyanide. Preparation of diphenylacetic acid, synthesis of antispasmodics, herbicide. mp = 71-73°; bp_{16} = 181°; slightly soluble in H_2O, more soluble in organic solvents; LD_{50} (rat orl) = 3500 mg/kg. *Andeno BV; R.W. Greef; Janssen Chimica; Schweizerhall.*

8601 diphenylcresyl phosphate
26444-49-5 247-693-8
$C_{19}H_{17}O_4P$
Cresyl diphenyl phosphate
Disflamoll® DPK; Phosphoric acid methylphenyl diphenyl ester; Tolyl diphenyl phosphate; Diphenyl cresol phosphate; Phosflex® CDP; Methylphenyldiphenyl phosphate. Flame retardant plasticizer for plasticized PVC products; used in air ducts, tarpaulins, driving and conveyor belts, imitation leather, coatings, hoses and extruded goods, cable sheathing and insulation, soles and injection molded items. d = 1.204-1.208; fp = -38°; flash point = 233-237° (COC); LD_{50} (rat orl) = 6400 mg/kg. *Bayer AG; Miles; Polysar; FMC; Velsicol.*

8602 diphenyloxazole
92-71-7 202-181-3
$C_{15}H_{11}NO$
2,5-diphenyloxazole; DPO. Primary fluor used as scintillation counters or in wavelength shifters. *Du Pont Medical Prods.; Packard Instrument BV; Penta Mfg.; Spectrum Chem. Mfg.*

8603 N,N'-diphenyl-*p*-phenylenediamine
74-31-7 3388 200-806-4
$C_{18}H_{16}N_2$
N,N-Diphenyl-1,4-benzenediamine
N,N'-Diphenyl-*p*-phenylenediamine; Diphenyl phenylene diamine; 1,4-dianilinobenzene; DPPD; Agerite® DPPD; Naugard® J; Permanax DPPD; JZF; Agerite DPPD; Diphenyl PPD; DPPD; Flexamine G; JZF; Nonox DPPD; Diafen; diafen FF; Altofane DIP; Nonflex H; Permanax 18; Stabilizer DPPD; Nocrac DP; Permanax DPPD; DFFD; Antage DP; Ekaland DPPD; Naugard J. An antioxidant for use in rubber, polyethylene, petroleum and vegetable oils and animal fats and oils; in natural rubber it protects against copper and manganese; protection against outdoor flexing and static weather cracking; polymerization inhibitor mp = 146-148°; $bp_{0.5}$= 220-225°; d = 1.20; insoluble in H_2O, soluble in organic solvents; LD_{50} (rat orl) = 2370 mg/kg. *Uniroyal; R. T. Vanderbilt Co Inc; Akzo.* Name unverified.

8604 diphenylthiourea
102-08-9 3393 203-004-2
$C_{13}H_{12}N_2S$
sym-Diphenylthiourea
N,N'-diphenylthiourea; sulfocarbanilide; thiocarbanilide; DPTU; S-diphenylthiocarbamide; 1,3-diphenylthiourea; 1,3-diphenyl-2-thiourea; N,N'-diphenylthiocaramide. Used as a vulcanizing accelelrator and in manufacture of sulfur dyes. mp = 152-155°; d = 1.32; insoluble in H_2O, soluble in organic solvents; MLD (rbt orl) = 1.5 g/kg.

8605 Diphone
A range of sulfone chemicals; for use in continuous tin-plating processes, and as monomers for engineering plastics. No manufacturer.

8606 Diphyl® T Diphyl® T
Heat transfer media for indirect heating and cooling in the chemical, petrochemical, plastics, man-made fibers, fat and soap industries. *Bayer AG; Bayer plc.*

8607 Dipicrylamine
131-73-7 3396 205-037-8
$C_{12}H_5N_7O_{12}$
2,4,6-Trinitro-N(2,4,6-trinitrophenyl)benzenamine
hexyl; hexite; hexil; hexanitrodiphenylamine; 2,4,6,2',4',6'-hexanitrodiphenylamine. Used as a booster explosive, analysis for potassium. mp = 238°; insoluble in H_2O, organic solvents.

8608 Dipolymer
Coumarone-indene resin.

8609 Dipping metal
A jeweler's alloy containing 48 parts of copper and 15 parts of zinc.

8610 Diprane®
Polyester polyurethane systems; used for mining, quarrying, construction and engineering. *Hyperlast.*

8611 Dipropylene glycol
110-98-5 203-821-4
$C_6H_{14}O_3$
Methyl-2(methyl-2)oxybispropanol;
1,1'-Oxybis-2-propanol; Di-1,2-propylene glycol; DPG; Adeka Dipropylene Glycol. Mixture of diols; solvent; polyester and alkyd resins, reinforced plastics, plastics, plasticizers, solvents. bp = 233°; d = 1.023; soluble in H_2O, organic solvents; flash point = 280°F; irritant, mildly toxic by ingestion. Aldrich; Olin; Texaco; Allchem Ind.; Arco; Ashland; Berje; Brown; Coyne; Great Western; Olin; PMC Specialties.

8612 Diprosin A-100
Disproportionated resin; polymerization emulsifier for synthetic rubbers and plastics. *Toho Chem. Industry.*

8613 Diprosin K-80
61790-50-9 263-142-4
Disproportionated rosin potassium soap; polymerization emulsifier for synthetic rubbers and plastics. *Toho Chem. Industry.*

8614 Diprosin N-70
61790-51-0 263-144-5
Disproportionated rosin sodium soap; polymerization emulsifier for synthetic rubbers and plastics. *Toho Chem. Industry.*

8615 Dipsal
7491-14-7
Dipropylene glycol salicylate
UV absorbent for sunscreens; suitable as inhibitor for UV degradation of polymers and dyestuffs; does not deteriorate in contact with perspiration; emollient for toiletries, alcohol lotions, vegetable or mineral-type products,

and pharmaceutical specialties; suitable for hair applications; useful to reduce deterioration and discoloration of polymers. *Scher.*

8616 Dipterex®
52-68-6 9753 200-149-3
Trichlorfon
Primarily a stomach poison insecticidal spray used for control of mangold fly, fruit fly and pests of maize, alfalfa, and cotton. *Bayer AG.*

8617 Dipterex® 80
52-68-6 9753 200-149-3
trichlorfon
Organophosphorus insecticide as a soluble powder containing 80% w/w trichlorfon; used to control mangolf fly on beet, cabbage white, leaf minor, and other caterpillars on brassicas. *Bayer plc.*

8618 1,3-di-6-quinolylurea
532-05-8 3416 208-525-9
$C_{23}H_{26}N_4O_9S_2$
1,3-Di-6-quinolylurea bismethosulfate
N,N'-di-6-quinolinylurea bis methosulfate; 6,6'-diquinolinylurea bis methosulfate; *sym*-di-(6-quinolyl)urea bis methosulfate; Acaprin®; bis(6-quinolyl)urea bis methosulfate; SN 5870; Zothelone; Baburan; Pirevan; Pyroplasmin; Atral,. Chemotherapeutic against piroplasmosis (babesiasis); veterinary medicine. mp = 237° (dec) *Bayer AG.*

8619 Diresul
Synthetic organic dyestuffs for use in the textile industry. *Sandoz Products Ltd.*

8620 Direx® 4L
330-54-1 3447 206-354-4
Diuron
Diuron suspension; a flowable herbicide for control of many weeds and grasses in a variety of crops. *Griffin.*

8621 Diroval
7779-90-0 10284 231-944-3
Zinc phosphate
Used in dental cements. *James M. Brown Ltd.*

8622 Dirubin
1344-28-1 369 215-691-6
aluminum oxide
Refined corundum and crystalline aluminum oxide; for production of ramming mixes, shape bricks, and crucibles for the lining of high temperature furnaces; molding material for precision casting molds and the casting of aggressive steels; raw material for the electroslag remelting process, separating agent for the annealing process. *Hoechst AG.*

8623 Disadine
25655-41-8 7880
povidone-iodine
Antiseptic for topical use, presented as a dry powder spray. *ICI Chem & Polymers Ltd.*

8624 Discase
9001-09-6 2319 232-580-8
Chymopapain
Enzyme. *Travenol Laboratories Inc.* Name unverified.

8625 Discharge Agent DP
Reducing agent for discharge and discharge resist printing on polyester, acetate, triacetate, and their blends with polyamide. *BASF.*

8626 Disco 727
Nonionic wax
For use with cationic fixatives in textiles. *Callaway.*

8627 Discodye 1148
Natural organic compound; retarding agent for cotton and cotton/polyester blends. *Callaway.*

8628 Discofix DBA
Fixative for improvement of home laundry, fastness of fiber reactive, sulfur and indigo dyes. *Callaway.*

8629 Discol 715
Cationic water-soluble polymer; for indigo dye fixation in textiles. *Callaway.*

8630 Discol 1457
Silicone defoamer
For textile dyeing and finishing. *Callaway.*

8631 Discol DFW
577-11-7 3460 209-406-4
docusate sodium
Dioctyl sodium sulfosuccinate
Fast wetting agent for denim finishing and continuous carpet dyeing. *Callaway.*

8632 Discolite
$NaHSO_2 \cdot CH_2O \cdot 2H_2O$
A reducing agent for stripping in dyeing. No manufacturer.

8633 Discoloc 70-A
Cationic resinous polymer dye; fixing agent and color control aid; for indigo dye fixation. *Callaway.*

8634 Discolube 473-A
Cationic lubricant improving beaming efficiency after indigo dyeing. *Callaway.*

8635 Discopen 216
Nonrewetting penetrant for rainwear-type finishes; decomposes at temperatures > 250°F. *Callaway.*

8636 Discosoft 1043-S
Fatty blend; softener and sanforizing lubricant that does not promote color bleeding in finish bath; excellent resistance to scorching and yellowing. *Callaway.*

8637 Discoterge 326-D
Crypto-anionic surfactant; surfactant for use in desizing, alkaline scouring, and peroxide bleaching. *Callaway.*

8638 Discozone MAC
Concentrated antiozonant/softener for textiles; easy handling in automated feed systems. *Callaway.*

8639 Disfico
A trade name for a vulcanized fiber used for electrical insulation. No manufacturer.

8640 Disflamoll®
Phosphate-based plasticizers. *Bayer plc.*

8641 Disflamoll® DPK, TPK
26444-49-5 247-693-8
$C_{19}H_{17}O_4P$
Diphenylcresyl phosphate
Phosphoric acid methylphenyl diphenyl ester; cresyl diphenyl phosphate; diphenyl tolyl ester phosphoric acid; cresol diphenyl phosphate; diphenyl cresol phosphate; diphenyl cresyl phosphate; diphenyl tolyl phosphate; Disflamoll DPK; Kronitex CDP; methyl phenyl diphenyl phosphate; monocresyl diphenylphosphate; Phosflex 112; Santicizer 140; tolyl diphenyl phosphate; cresyl phenyl phosphate. Flame retardant plasticizer for plasticized PVC products; used in air ducts, tarpaulins, driving and conveyor belts, imitation leather, coatings, hoses and extruded goods, cable sheathing and insulation, soles and injection molded items. *Bayer AG; Miles; Polysar.*

8642 Disflamoll® TP
115-86-6 9872 204-112-2
Triphenyl phosphate
Dymel®. Flame retardant; gelatinizing and plasticizing agent for collodion cotton; plasticizer without gelatinizing properties for acetyl cellulose; reduces flammability of NC and acetyl cellulose-based plastic compounds and lacquer films; used in manufacture of photographic film materials and surface coatings. mp = 49-50°; bp_{11} = 245°; insoluble in H_2O, soluble in organic solvents; LD_{50} (rat orl) = 3500 mg/kg. *Bayer AG; Miles; Polysar.*

8643 disodium cocoamphodiacetate
68647-53-0 272-043-5
Cocoamphocarboxyglycinate
Abluter DCM-2; Afoteric™ 2C; Ampholak XCO-30; Amphotensid GB 2009; Amphoterge® W-2; Chimin IMB; Dehyton® G; Dehyton® PG; Dehyton®W; Mackam™ 2C; Manroteric CDX38, CLV, CSH-32; Miranol® 2CIB; Miranol® C2M Conc. NP; Proteric CDX-38; Rewoteric® AM 2C NM; Rewoteric® AM 2C W; Schercoteric MS-2; Sochamine A 7527; Surfax AC 50; Surfax ACl; Unibetaine 2C; Velvetex® CDC; Zoharteric D; Zoharteric D-SF 70%. Surfactant for mild and conditioning shampoos, bath products, baby shampoos, skin cleansers, foam baths. Liquid; pH = 11; soluble in H_2O, EtOH; poorly soluble in organic solvents; d = 1.140; viscosity = 400 mPa; pour point = -18°; surface tension = 34 mN/m. *Henkel Canada; Henkel KGaA; Pulcra SA; Taiwan Surf; McIntyre; Berol Nobel AB.*

8644 disodium EDTA
139-33-3 3556 205-358-3
$C_{10}H_{16}N_2O_8 \cdot 2Na$
disodium edetate
disodium dihydrogen ethylene diamine tetraacetetate; ethylenediaminetetraacetic acid; disodium salt; edetate disodium,. Food preservative, chelating and sequestering agent; anticoagulant; pharmaceutic aid. *W.R. Grace/Hampshire; R.W. Greeff.*

8645 disodium inosinate
4691-65-0 225-146-4
$C_{10}H_{13}N_4O_8P \cdot 2Na$
sodium inosinate
IMP sodium; Disodium IMP; Sodium 5-iosinate; Disodium 5'-inosinate; Inosine 5'-disodium phosphate; Gluxor® 1626; Luxor® 1639. Disodium IMP; A 5'-nucleotide derived from seaweed or dried fish; Flavor potentiator in foods. soluble in H_2O; slightly soluble in alcohol; insoluble in ether; LD_{50} (rat orl) = 15,900 mg/kg. *Penta Mfg.; Schweizerhall.*

8646 disodium laneth-5 sulfosuccinate
68890-92-6
$C_{14}H_{26}O_8.2Na$
Sulfobutanedioic acid, 4-isodecyl ester, sodium salt
Incrosul LAFS; Rewolan® 5. Mild, low foaming, conditioning surfactant with good emulsifying properties used in personal care products Liquid; pH = 6.5-7.5. *Croda Inc.; Witco Surf. GmbH.*

8647 disodium lauramido MEA-sulfosuccinate
25882-44-4 247-310-4
Geropon® SBL-203; Incrosul LMS; Mackanate LM-40; Rewopol® SBL 203; Varsulf® SBL-203. Improves flash foam of anionic systems; produces brittle, tack-free residue for carpet shampoos. *Rhône-Poulenc Surf.; Croda Inc.; McIntyre; Rewo GmbH; Sherex/Div. Of Witco.*

8648 disodium lauryl ethoxy sulfosuccinate
37354-45-5
Empicol® SDD. Mild raw material for toiletries and detergents. *Albright & Wilson UK.*

8649 disodium lauryl sulfosuccinate
36409-57-1; 13192-12-6; 19040-44-9;
$(C_2H_4O)_xC_{16}H_{30}O_8.2Na$
Sulfobutanedioic acid 4-[2-[2-(2-(dodecyloxy)ethoxy]ethoxy]ethyl]ester, disodium salt
Ablusol LAE; Disponil® SUS 65; Elfanol® 616; Emcol® 4403; Empicol® SDD; Euranaat LS3; Empicol® SLL; Foampol LPS; Geronol® ACR/4; Geropon® ACR/4, SBFA-30; Grillosol SB3/12; Mackanate™ SL3; Rewopol® SBFA 30; Schercopol LPS; Setacin 103 Spezial; Standapol® SH-124-3; STEPAN-MILD® SL3; Texapon® SB-3; Texapon®SB-3KC; Thorowet ML-3 0532; Varsulf® SBFA30; Zoharpon SE. Emusifier for emulsion polymerization. Mild raw material for toiletries and detergents. Emulsifier, detergent, wetting agent for shampoos, bubble bath base and skin cleanser. Liquid; d = 1.11; cloud point < 0°; pH = 5.5-6.5; LD_{50} (rat orl) >2000 mg/kg. *Albright & Wilson UK; Ceca SA; Chemron; Henkel; Tiawan Surf; Stephan Europe.*

8650 disodium lauryliminodipropionate
3655-00-3 222-899-0
Disodium N-lauryl β-iminodipropionate
Deriphat 160; Disodium 3,3'-(dodecylimino) dipropionate; N-(2-Carboxyethyl)-N-dodecyl-β-alanine, disodium salt; Deriphat® 160; Monateric 1188M,. Detergent, solubilizer, primary emulsifier used in organic and inorganic compounds; emulsion polymerization and stabilization; wetting agent; mild surfactant for hair and skin products whitepowder; d = 2.0 lb/gal. *Henkel/Cospha; Henkel/Functional Prods.; Henkel Canada.*

8651 disodium oleamido MIPA sulfosuccinate
43154-85-4 256-120-0
Emcol® 416L; Mackanate OP; Sole Terge 8. Dispersant, wetting, foam booster/stabilizer, detergent, conditioner, and emulsifying agent for bubble bath, shampoos, cleansers for cosmetics and toiletries. *McIntyre; Calgene.*

8652 disodium phosphate (anhydrous)
7558-79-4 8805 231-448-7
HNa_2O_4P
Sodium phosphate, dibasic
DSP-O; disodium hydrogen phosphate; disodium orthophosphate; DSP; Phosphate of soda; Secondary sodium phosphate. Inorganic salt of phosphoric acid, disodium salt controls pH in mildly alkaline solutions; food products, water treatment, animal feed, textiles, pharmaceuticals, chemicals, fertilizers, detergents. Hygroscopic powder; on exposure to air absorbs 2-7 moles H_2O; soluble in H_2O (12 g/100 ml), insoluble in alcohol; pH of 1% aq soln = 9.1. *Albright & Wilson; Monsanto; Rhône Poulenc Basic; U.S. Biochemical; Whiting, Peter Ltd.*

8653 disodium phosphate, dihydrate
10028-24-7 8805 231-448-7
$Na_2HPO_4.2H_2O$
sodium phosphate, dibasic dihydrate
DSP-2; disodium hydrogen phosphate, dihydrate; Sodium phosphate; Sorenson's phosphate; Sorensen's sodium phosphate. Controls pH in mildly alkaline solutions. Heptahydrate, crystals or granular powder; d = 1.7; soluble in H_2O (25 g/100 ml), insoluble in alcohol; pH = 9.5; LD_{50} (rat orl) = 12.93 g/kg. *Albright & Wilson; Monsanto; Rhône Poulenc Basic; U.S. Biochemical.*

8654 disodium ricinoleamido MEA-sulfosuccinate
40754-60-7
Geropon® SBR-3. Skin protecting anti-irritant surfactant. *Rhône-Poulenc Surf.*

8655 disodium tallowiminodipropionate
61791-56-8 263-190-6
Disodium N-tallow-β iminodipropionate
Deriphat 154; N-(2-Carboxyethyl)-N-(tallow acyl)-β-alanine; Mirataine® T2C-30; Mirataine® T2C-35%; Monateric TDB-35. Detergent, solubilizer for personal care products, hard surface cleaning, textiles, emulsion polymerization; good substantivity. Soluble in H_2O; d = 8.746 lb/gal; biodegradable. *Henkel/Cospha; Henkel Canada.*

8656 Disolite
A disinfectant used in the mushroom industry. *Coventry Chemicals Ltd.*

8657 Dispargen
A form of colloidal mercury.

8658 Disparit B
79-01-6 9769 201-167-4
C_2HCL_3
Trichlorethylene
A disinfecting cleaning compound. No manufacturer.

8659 Disperbyk®
Solution of an alkylolammonium salt of a higher molecular weight polycarboxylic acid; wetting, dispersing additive to prevent settling and flooding of pigments; for solvent and aqueous systems, stains, wood preservatives, anticorrosive primers, wash primers, nitrocellulose primers, fillers, antifouling paints, emulsion paints; an emulsifier. *Byk-Chemie USA.*

236-[illegible]; 248-0[illegible]
8660 Disperbyk® -181
Solution of an alkanolammonium salt of a polyfunctional polymer; wetting additive for emulsion paints based on polymethacrylates and copolymers, vinyl esters, styrene copolymers, water-reducible paint systems, solvent-based paint systems; improves color development and gloss. *Byk-Chemie USA.*

8661 Dispercab
9004-36-8
Cellulose acetate butyrate dispersions. *Tennant-KVK Ltd.*

8662 Dispercap
9004-39-1
Cellulose acetate propionate dispersions. *Tennant-KVK Ltd.*

8663 Dispercel
9004-70-0 8195
Nitrocellulose dispersions *Tennant-KVK Ltd.*

8664 Dispercoll C-74
Poly-2-chlorobutadiene aqueous dispersion; CR latex for formulation of water-based contact cements, laminating and mastic adhesives for the automotive, construction, furniture, footwear, and packaging industries. *Bayer; Miles.*

8665 Dispercryl
Acrylic copolymer dispersions. *Tennant-KVK Ltd.*

8666 Disperfin
1309-37-1 4072 215-168-2
ferric oxide
Transparent iron oxide pigment pastes. *Brockhues AG.*

8667 Disperkyd
Alkyd copolymer dispersions. *Tennant-KVK Ltd.*

8668 Dispermid
Polymide copolymer dispersions. *Tennant-KVK Ltd.*

8669 Dispersant 1084
Proprietary blend; pitch dispersant for pulp/paper industry. *Hart Chem. Ltd.*

8670 Dispersant LF-88
Low foaming pulp machine pitch dispersant. *Hart Chem. Ltd.*

8671 Disperse-Ayd
Dispersing agents; for paints, inks, etc. *Cornelius Chemical Co Ltd.* Name unverified.

8672 Disperse-Ayd 1
Proprietary reaction product of high molecular weight surfactants with a long oil alkyd; broad spectrum dispersant, wetting agent, stabilizing aid for solvent-thinned coatings; wets and deflocculates most organic and inorganic pigments. Gardner 12 max color; d = 0.97; 8.1 lb/gal; flash point = 40°. *Elementis Specialties.*

8673 Disperse-Ayd 6
Synergistic blend of wetting and dispersing agents; wetting, dispersing, and deflocculating agents for use with carbon black pigments in solvent-thinned coatings. Gardner 15 max color; d = 1.01; 8.4 lb/gal; flash point = 27°. *Elementis Specialties.*

8674 Disperse-Ayd 15
Modified thermoplastic acrylic in propylene glycol monomethyl ether acetate; pigment dispersing vehicle for high performance solvent-thinned coatings. Gardner 8 max color; d =1.04; 8.7 lb/gal; flash point = 43°. *Elementis Specialties.*

8675 Disperse-Ayd W-22
Surfactant blend; pigment dispersant for water-thinned coatings. Gardner 7 max color; d = 1.06; 8.8 lb/gal; flash point = 107°; pH = 8.5-9.5; surface tension = 36.4 dynes/cm. *Elementis Specialties.*

8676 **Dispersite**
A proprietary trade name for a dispersion of rubber, rubber-like and film-forming resins in water. No manufacturer.

8677 **Dispersogen A**
9084-06-4
Sodium naphthalene formaldehyde sulfonate
Dispersing and dyeing auxiliary; emulsifier, wetting agent, adjuvant for agricultural formulations. *Hoechst Celanese/Colorants & Surf.; Hoechst AG.*

8678 **Dispersogen SL**
Sodium polynaphthalene sulfonate and sodium C_{12-14} alkyl sulfate; surfactant for agricultural formulations. *Hoechst Celanese/Colorants & Surf.*

8679 **Dispersol**
Condensation product of formaldehyde and sodium naphthalene sulfonate; dispersible powders and aqueous dispersions e.g. for dyestuffs, pigments, pest control. *ICI Chem & Polymers Ltd.*

8680 **Dispersol**
Various liquid blends of dispersants for cooling water systems; used for open recirculating water systems. *Schaefer Technologies Inc.*

8681 **Dispersol 103, 105**
Pitch dispersant for neutral to alkaline systems in paper/pulp processing; keeps particles suspended, prevents agglomeration; dispersant for clay slurries, pigments and coatings. *CNC Int'l L.P.*

8682 **Disperstat**
Antistatic agent. *Stephenson Thompson Textile Chemicals.*

8683 **Dispersyd**
Alkyd copolymer dispersions. *Tennant-KVK Ltd.*

8684 **Dispervyn**
Vinyl copolymer dispersions. *Tennant-KVK Ltd.*

8685 **Dispex®**
Polymeric dispersing agents; effective on minerals and inorganic pigments in aqueous systems; very widely used in aqueous paints, adhesives and ceramic production. *Allied Colloids Ltd.*

8686 **Dispex® G40 and GA40**
Sodium salt of carboxylated polymer in the form of a pale yellow liquid; pigment dispersant for emulsion paints, especially sheen and gloss water based paints. *Allied Colloids Ltd.*

8687 **Displasol DP**
Quatenary ammonium compound blend; displaces acid dyes on nylon producing novel styling effects. *Am. Emulsions.*

8688 **Disponil AAP 307**
Alkylaryl polyglycol ether; coemulsifier for polyacrylates, acrylate-vinyl acetate copolymers, other applications; dispersant for emulsion paints. Liquid; HLB 17.3. *Henkel/Functional Prods.*

8689 **Disponil AEP 5300**
Ether phosphate, acid ester; emulsifier for rosin, vinyl acetate and acrylate systems. *Henkel/Functional Prods.*

8690 **Disponil AES 13**
Sodium alkylaryl ether sulfate; emulsifier for vinyl acetate homopolymers, acrylate homo- and copolymers, styrene acrylate copolymers, vinyl acetate-acrylate copolymer, VAE copolymers, PVDC latexes, vinyl chloride homo and copolymer latexes. *Henkel/Functional Prods.*

8691 **Disponil FES 32**
9004-82-4
Sodium laureth sulfate
Emulsifier for vinyl acetate copolymers, s/b latexes, vinyl chloride copolymers, acrylate homo-and copolymers. *Henkel/Functional Prods.*

8692 **Disponil FES 92E**
Sodium laureth-12 sulfate
Surfactant for low-irritation shampoos, emulsion polymerization. *Pulcra SA.*

8693 **Disponil G 200**
Isoeicosanol
Surfactant; liquid fatty alcohol. *Henkel/Functional Prods.*

8694 **Disponil O 5**
9004-95-9
Cetoleth-5
Emulsifier for emulsion polymerization; wetting agent. *Henkel/Functional Prods.*

8695 **Disponil SML 100 F1**
1338-39-2 8872 215-663-3
Sorbitan laurate
Surfactant for polymerization. *Henkel/Functional Prods.*

8696 **Disponil SML 104 F1**
9005-64-5 8872
Polysorbate 21
Surfactant for polymerization. *Henkel/Functional Prods.*

8697 **Disponil SML 120 F1**
9005-64-5 8872
Polysorbate 20
Surfactant for polymerization. *Henkel/Functional Prods.*

8698 **Disponil SMO 100 F1**
1338-43-8 8872 215-665-4
Sorbitan oleate
Surfactant for polymerization. *Henkel/Functional Prods.*

8699 **Disponil SMO 120 F1**
9005-65-6 7742
Polysorbate 80
Surfactant for polymerization. *Henkel/Functional Prods.*

8700 **Disponil SMP 100 F1**
26266-57-9 8872 247-568-8
Sorbitan palmitate
Surfactant for polymerization. *Henkel/Functional Prods.*

8701 **Disponil SMP 120 F1**
9005-66-7 8872
Polysorbate 40
Surfactant for polymerization. *Henkel/Functional Prods.*

8702 **Disponil SMS 100 F1**
1338-41-6 8872 215-664-9
Sorbitan stearate; surfactant for polymerization. *Henkel/Functional Prods.*

8703 **Disponil SMS 120 F1**
9005-67-8 8872
Polysorbate 60
Surfactant for polymerization. *Henkel/Functional Prods.*

8704 **Disponil SSO 100 F1**
8007-43-0 8872 232-360-1
Sorbitan sesquioleate
Surfactant for polymerization. *Henkel/Functional Prods.*

8705 **Disponil STO 100 F1**
26266-58-0 8872 247-569-3
Sorbitan trioleate
Surfactant for polymerization. *Henkel/Functional Prods.*

8706 **Disponil STO 120 F1**
9005-70-3 8872
Polysorbate 85
Surfactant for polymerization. *Henkel/Functional Prods.*

8707 **Disponil STS 100 F1**
26658-19-5 8872 247-891-4
Surfactant for polymerization. *Henkel/Functional Prods.*

8708 **Disponil STS 120 F1**
9005-71-4 8872
Polysorbate 65
Surfactant for polymerization. *Henkel/Functional Prods.*

8709 **Disponil SUS IC 8**
577-11-7 3460 209-406-4
Dioctyl sodium sulfosuccinate
Wetting agent, coemulsifier for plastics industry. *Henkel/Functional Prods.*

8710 **Dispray**
Disinfectant used for rapid disinfections of the skin before operations, injections or venepuncture. *ICI Chem & Polymers Ltd.*

8711 **Dissolvine**
Sequestering and chelating agents. *Akzo Chemie UK Ltd.*

8712 **Distearyl pentaerythritol diphosphite**
3806-34-6 223-276-6
$C_{41}H_{82}O_6P_2$
Doverphos® S-680; Doverphos® S-686, S-687; Weston® 618F; Weston ® 618; Weston® 619; Mark® 5060. Color stabilizer and melt flow aid for polymer processing; antioxidant. mp = 40-70°; d = 0.920-0.935. *Dover; Aldrich; GE Spec.*

8713 **Distearyl thiodipropionate**
693-36-7 211-750-5
$C_{42}H_{82}O_4S$
3,3·g-Thiobispropanoic acid
3,3·g-dioctadecyl thiodipropionate; thiodipropionic acid; Thiodipropionic acid, distearyl ester; dioctadecyl ester; DSTDP; PAG DSTDP; PAG DXTDP; Vanox® DSTDP; distearyl ester; Argus DSTDP; Carstab® DSTDP; Cyanox® STDP; Evanstab® 18; Lankromark® DSTDP; Lowinox® DSTDP,. Diester of stearyl alcohol and 3,3·g thiodipropionic acid; antioxidant, plasticizer, softening agent; antioxidant for cosmetics, pharmaceuticals; stabilizer for plastics and elastomers; mp = 58-62°; bp = 360°; insoluble in H_2O, soluble in organic solvents; LD_{50} (rat orl) > 2500 mg/kg. Am. Cyanamid;Aldrich;Sigma; Evans Chemetics;Cytec Ind.; Hampshire; Hoechst; ICI Am.; Morton Int'l; Witco/Arqus.

8714 **Distearyldimonium chloride**
107-64-2 203-508-2
Distearyl dimethyl ammonium chloride
Dehyquart DAM; Adogen® TA-100, TA-101. Emulsifier for plastics industry;

conditioning component for hair care preparations; antistat. *Henkel/Functional Prods.; Henkel Chemicals Ltd; Fluka.*

8715 Distec
Fatty acids and glycerides. *Akzo Chemie UK Ltd.* Discontinued.

8716 Distillase® L-200
9032-08-0 232-877-2
Glucoamylase
Industrial grade enzyme for fuel alcohol applications *Solvay Enzymes.*

8717 Distillex DS1
71-55-6 9766 200-756-3
1,1,1-trichloroethane
Recovered 1,1,1-trichloroethane. *Distillex Ltd.*

8718 Distillex DS2
79-01-6 9769 201-167-4
1,1,1-trichloroethylene
Recovered trichloroethylene. *Distillex Ltd.*

8719 Distillex DS3
75-09-2 6140 200-838-9
dichloromethane
Recovered dichloromethane. *Distillex Ltd.*

8720 Distillex DS4
127-18-4 9332 204-825-9
tetrachloroethylene
Recovered tetrachloroethylene. *Distillex Ltd.*

8721 Distillex DS5
76-13-1 200-936-1
Trichlorotrifluoroethane
Recovered 1,1,2-trichloro-1,2,2-trifluoroethane. Solvent used in microelectronics industry. *Distillex Ltd.*

8722 Distillex DS6
75-69-4 9770 200-892-3
trichlorofluoromethane
Recovered trichlorofluoromethane. *Distillex Ltd.*

8723 Distillex DS7
Recovered 75% tetrachloroethylene and 25% n-butanol v/v. *Distillex Ltd.*

8724 Distoline
112-80-1 6965 204-007-1
oleic acid
A proprietary trade name for commercial oleic acid obtained from vegetable oils; plasticizer, softener, in rubbers;. No manufacturer.

8725 Disulfiram
97-77-8 3428 202-607-8
$C_{10}H_{20}N_2S_4$
Tetraethylthioperoxydicarbonic diamide
Tetraethylthiuram disulfide; Ethyl Tuads®; Akrochem® TETD; Ancazide ET; Etyl Tuex; Ethylthiurad; Perkait® TETD; teturamin; TTD; Cronetal; Abstensil; Stopetyl; Contralin; Antadix; Antietanol; Exhoran; Antabuse; Etabus; Abstinyl; Esperal; Tetradine; Noxal; Tetraetil. Accelerator and vulcanizing agent for rubber. mp = 70°; insoluble in H_2O, soluble in organic solvents; LD_{50} (rat orl) = 8.6 g/kg. *R. T. Vanderbilt Co; Uniroyal; Monsanto Co.; Mitchell Cotts Chemicals Ltd.* Discontinued.

8726 Disyston® FE-10
298-04-4 3429 206-054-3
disulfoton
Granular systemic insecticide containing 10% w/w disulfoton on fuller's earth granules; used to control aphids and certain aphid-borne virus diseases on potatoes, carrots, celery, marrows, parsley, french and runner beans and Brussels sprouts. *Bayer plc.*

8727 Ditensamine C, O and S
Cationic surfactants in the form of alkyl propylene diamines in which the alkyl group is coconut, oleic and tallow respectively; liquid or solid forms; synthesis intermediate, bitumen emulsions, corrosion inhibition. *Tensia SA.* Name unverified.

8728 Dithane
8018-01-7 5756
mancozeb
Wettable powder or water dispersible granules containing mancozeb; protective fungicide for fruit, field crops and roses. *Pan Britannica Industries Ltd.*

8729 Dithianone
3347-22-6 3433 222-098-6
$C_{14}H_4N_2O_2S_2$
5, 10, Dihydro-5,10-dioxonaphtho[2-3-b]-1,4-dithiin-2,3-dicarbonitrile
Delan; Delan-col; DTA; MV 119A; Naphtho(2,3-b)-p-dithiin-2,3-dicarbonitrile, 5,10-dihydro-5,10-dioxo-; Stauffer MV-119a; Thynon. Suspension concentrate containing 600 g dithianon per liter; fungicide used for control of scab in fruit apples and pears. mp = 225°; insoluble in H_2O, soluble in organic solvents; LD_{50} (rat orl) = 638 mg/kg. *ICI Chem & Polymers Ltd.*

8730 1,4-Dithiothreitol
3483-12-3 3441 240-263-0
$C_4H_{10}O_2S_2$
(RName unverified.,RName unverified.)-1,4-Dimercapto-2,3-butanediol
1,4-dithio-L-threitol; Cleland regent; Cleland's reagent; Dithiothreitol; L-DTT. Reducing agent for proteins and enzymes, used in biochemical research. mp = 42-43°; soluble in H_2O, organic solvents. *Aldrich; Bio-Rad Labs; Biosynth AG; Schweizerhall; U.S. Biochemical.*

8731 dithizone
102-08-9 3393 203-004-2
Diphenylthiocarbazone
sym-diphenylthiourea. Used for the detection of heavy metals.

8732 Ditolyguanidine
97-39-2 202-577-6
$C_{15}H_{17}N_3$
1,3-Di-*o*-tolylguanidine
D.O.T.G; Di-*o*-tolylguanidine; Acrochem® DOTG; Anchor® DOTG; Ekaland DOTG; Perkacit® DOTG; Vanax® DOTG. A rubber vulcanization accelerator.

8733 Ditrimethylolpropane tetraacrylate
Curing agent. *Sartomer.*

8734 Diurex
330-54-1 3447 206-354-4
3-(3,4-dichlorophenyl)-1,1-dimethylurea
Active ingredient: diuron; residual herbicide effective against a wide range of both broadleaf weeds and annual grasses. *Agan Chemical Manufacturers Ltd.*

8735 Diurol
azolan-diurex
Active ingredients: azolan plus diurex; multlpurpose herbicidal mixture which eradicates a wide spectrum of established weeds while preventing further weed germination for extended periods. *Agan Chemical Manufacturers Ltd.* Discontinued.

8736 diuron
330-54-1 3447 206-354-4
$C_9H_{10}Cl_2N_2O$
3-(3,4-dichlorophenyl)-1,1-dimethylurea
DMU; DCMU; Diurex; Aguron; M Velpar; Karmex; Urox D; Direx 4L; Direx 80W; Diuron 4L; Diuron 80; Karmex 80W; Karmex DL; Cekiuron; Crisuron; Dailon; Di-on; Diater; Unidron; Vonduron; Xarmex, Krovar; Drexel diuron 4L; Dynex. Pre-emergent herbicide, sugar cane flowering suppressant. Inhibits photosynthesis. Used for total control of weeds and mosses in non-crop areas and selectiver control of germinating grass and broad-leaved weeds in many crops. mp = 158-159°; soluble in H_2O (42 mg/l), more soluble in organic solvents; LD_{50} (rat orl) = 3400 mg/kg. *Griffin; Pacific Anchor; Rhône-Poulenc Agrochimie SA; Rhône-Poulenc Environmental Prods. Ltd.*

8737 Diuron Bayer
330-54-1 206-354-4
Diuron; herbicide effective against emerging and young broad-leaved and grass weeds as well as mosses; suitable for both selective (sugar cane, cotton) and total weed control (orchards, vineyards). *Bayer AG.*

8738 Divergan® F, R
Crosslinked PVP for brewing industry; stabilizing agents for drinks (beer, wine, clear fruit juices). *BASF AG; BASF plc.*

8739 Diver's liquid
A liquid formed by absorbing ammonia in solid ammonium nitrate. It is capable of dissolving ammonium nitrate.

8740 Divipan
62-73-7 3129 200-547-7
dichlorvos
Active ingredient: dichlorvos; one of the most useful fast-acting agricultural insecticides-acaricides. *Makhteshim Chemical Works Ltd.*

8741 Diwatex 30, 40
Sulfonated kraft lignin derivatives; dispersant for dyestuffs and pesticides. *Borregaard Ligno Tech.*

8742 Dixie 5 and Dixie Special 102
Proprietary carbon black pigments. No manufacturer.

8743 Dixie Clay®
1332-58-7 5294 296-473-8
Hydrated aluminum silicate
kaolin clay. Filler, extender or reinforcing pigment for paint, paper, rubber, ceramics, plastics, and specialities. *R. T. Vanderbilt Co Inc.*

8744 Dizene
95-50-1 *3106* *202-425-9*
o-dichlorobenzene
Emulsifiable *o*-dichlorobenzene. *PPG Industries.*

8745 Diziktol
333-41-5 3043 206-373-8

Diazinon
Insecticide. *Makhteshim Chemical Works Ltd.*

8746 DK-Ester
Sugar esters; food emulsifier. *Grünau.* Discontinued.

8747 DLG-10, 20
557-05-1 10292 209-151-9
zinc stearate, dispersible
Sanding sealer aid in lacquers. *Hüls Am.* Discontinued.

8748 DLPA 375
150-30-1 7425 205-756-7
DL-phenylalanine
Tablets of DL-phenylalanine 375 mg in a natural basis; used as a dietary supplement. *Larkhall Laboratories plc.* Name unverified.

8749 DLS Base
Preneutralized emulsifiable amidoamine condensate; softener base imparting a dry, soft hand to cotton and other fabrics. *Clark.*

8750 D-Lube
Fuel additive. *Kalon Chemicals Ltd.*

8751 DM hydantoin
77-71-4 201-051-3
$C_5H_8N_2O_2$
5,5-dimethyl hydantoin
5,5-dimethyl-2,4-imidazolidinedione. Intermediate for textiles and other applications. *Great Lakes; Janssen Chimica; Lonza.*

8752 DM-2
Mold repair/potting compound and general purpose adhesive; epoxide resin based mastic use a heat resistant encapsulating material; used for encapsulation of electric components where solder connections are affected by heat, repair of mold porosity in rotational plastic molds. *Dynamold Inc.*

8753 DMAMP-80
7005-47-2 230-279-6
2-Dimethylamino-2-methyl-1-propanol
Amine solubilizer for resins in aqueous coatings; emulsifier for waxes; vapor-phase corrosion inhibitor; urethane catalyst; titanate solubilizer, raw material for synthesis. *Angus.*

8754 DMDM hydantoin
6440-58-0 229-222-8
$C_7H_{12}N_2O_4$
1,3-bis (hydroxymethyl)-5,5-dimethyl-2,4-imidazolidinedione
Glydant®; DMDM hydantoin; Dantoin DMDMH; Dimethyloldimethyl hydantoin; DMDMH; Glydant; Glydant Plus; Mackgard DM; Mackstat® DM; Nipaguard DMDMH; Dantion DMDMH 55; Dantoguard;. A preservative and formaldehyde donor; broad spectrum antimicrobial for cosmetics and toiletries; effective against Gram-positive and Gram-negative bacteria, fungi, and yeast. *Lonza Inc; McIntyre; Nipa Labs.*

8755 DMI-689
A two-part urethane mixture, fast curing, 1:1 mix; hard copy replication system useful for replication of parts, measurement of dimensional accuracy, moldmaking and prototype tooling. *Dynamold Inc.*

8756 DM-Nitrophen™
Reagent. *Calbiochem Corp.*

8757 DMP
4744-10-9 225-258-3
$C_5H_{12}O_2$
2,2-Dimethoxypropane
Propane, 1,1-dimethoxy-; Propionaldehyde dimethyl acetal. Chemical intermediate for pharmaceuticals; dehydrating agent. *Schering Berlin Polymers.* Discontinued.

8758 DMP
Catalyst systems for polyurethane and epoxy resins. *Rohm & Haas.*

8759 DMR-503
A two-part silicone elastomer, fast curing, 1:1 mix, hand mixable; flexible putty useful for replication of parts, measurement of dimensional accuracy, moldmaking and prototype tooling. *Dynamold Inc.*

8760 DMR-504
A two-part silicone elastomer, fast curing, 1:1 mix, hand mixable; pourable flexible system useful for replication of parts, measurement of dimensional accuracy, moldmaking and prototype tooling. *Dynamold Inc.*

8761 DMS-4-828
Epoxy resin based mastic; moldable liquid shim materials used as a spacer between engines or skin of aircraft or ships; surface conforming structural epoxies used to fill gaps between metal parts and between structural members and materials such as graphite. *Dynamold Inc.*

8762 DMSO
67-68-5 3308 200-664-3
C_2H_6OS
Dimethyl sulfoxide
Methyl Sulfoxide; Dimethylsulfoxide; Sulfinylbismethane; Dimethyl Sulfur Oxide; Methylsulfinylmethane; A 10846; Deltan; Demeso; Demasorb; Demavet; Demsodrox; Dermasorb; Dimexide; Dipirartril-tropico; DMS-70; DMS-90; Dolicur; Doligur; Domoso; Dromisol; Durasorb; Gamasol 90; Hyadur; Infiltrina; M 176; Rimso-50; Somipront; Sq 9453; Syntexan; Topsym. Aprotic solvent used as reaction medium in the manufacture of pesticides, pharmaceuticals, dyes, inks; solvent in polymers, electronics, refining; chemical intermediate for pesticides. mp = 18°; bp = 189°; d = 1.1010; n_D^{20} = 1.4748; soluble in H_2O, organic solvents; LD_{50} (rat orl) = 14500 mg/kg. *Elf Atochem N. Am./Fine Chems.*

8763 DO-160
Chlorinated olefin. *Dover.*

8764 Dobane (Detergent Alkylate)
Linear alkylbenzenes which yield light colored, biodegradable sulfonates; used for general detergent applications, from household powders to light duty liquids. *Shell.* Name unverified.

8765 Dobanic Acids JN and 83
Dark, viscous liquids which on neutralization give light colored sulfonates particularly suitable for the production of high performance, liquid detergents. *Shell.* Name unverified.

8766 Dobanol
Detergent alcohol. *Mitsubishi Petrochem.* Name unverified.

8767 Dobanol
Colorless high purity liquids used as base materials for the manufacture of alcohol sulfates, alcohol ethoxylates, and alcohol ethoxysulfates; for production of detergents, wetting agents, dispersants, and emulsifiers. *Shell.* Name unverified.

8768 Dobanol ethoxylates
Intermediates for the production of shampoo components, toiletry products, dishwashing liquids, and washing powders. *Shell.* Name unverified.

8769 Dobanol ethoxysulfates
Aqueous or aqueous/ethanol solutions for various applications, such as components for toiletry products, light-duty liquid detergents, and high-performance liquid detergents.

8770 Dobanox
Surfactant/emulsifier. *Shell UK.* Name unverified.

8771 Dobatex
An anionic detergent. *Shell UK.* Name unverified.

8772 Dobbin's reagent
Prepared by adding mercuric chloride solution to a solution of potassium iodide until a permanent precipitate is obtained. the solution is filtered and 1 gram of ammonium chloride added, then dilute caustic soda until a precipitate is formed; used for detection of traces of caustic alkali in soap.

8773 Dobell solution
An aqueous solution containing 1.5% sodium borate, 1.5% sodium bicarbonate, and 0.3% phenol and glycerin. an alkaline antiseptic.

8774 Dock-Ban
dicamba-MCPA-mecoprop
Soluble concentrate of 19.5 g dicamba, 245 g MCPA and 86.5 g mecoprop per liter; used for weed control in cereals and grassland. *Quadrangle Agrochemicals.*

8775 Docklene
dicamba-MCPA-mecoprop
Soluble concentrate of 336 g mecoprop, 84 g dicamba and 84 g MCPA per liter; used for weed control in cereals and grassland. *Schering Agrochemicals Ltd.* Discontinued.

8776 Doctor metal
An alloy of 88% copper, 9.5% zinc, and 2.5% tin.

8777 1-dodecanol
112-53-8 3464 203-982-0
$C_{12}H_{26}O$
Emery® 3326; Dodecyl alcohol; lauryl alcohol;. Chemical intermediate for detergent manufacture. Insoluble in H_2O; soluble in EtOH, ether. *Henkel/Emery.*

8778 dodecene-1
112-41-4 203-968-4
$C_{12}H_{24}$
α-dodecylene
tetrapropylene; Neodene® 6/12; Neodene® 12; Neodene® 1012. Intermediate for surfactants and specialty industrial chemicals, flavors, perfumes, medicine, oils, dyes, resins. mp = -35°; bp = 213°; d = 0.7580; n_D^{20} = 1.4294; insoluble in H_2O, soluble in organic solvents. *Monsanto (Solaris); Ethyl; Shell.*

8779 dodecylbenzene sulfonic acid
27176-87-0 248-289-4
$C_{18}H_{30}O_3S$
Laurylbenzenesulfonic Acid
DDBSA. Substituted aromatic acid; animonic detergent. mp = 10°; bp = 315°; d = 1.2000; n_D^{20} = 1.5064; LD_{50} (rat orl) = 650 mg/kg.

8780 4-dodecyloxy-2-hydroxybenzophenone
$C_{25}H_{34}O_3$
Eastman® Inhibitor DOBP; UV-Chek® AM-320. Industrial grade uv absorber/stabilizer for polyethylene and PP; suitable for unsaturated polyesters, PS, PVC, CAB, NC, urethane, and acrylic surface coatings; also used as a screening agent in protecting rubber, fluorescent pigments, polishes, and papers from the degrading effects of uv light. mp = 52°; soluble in polar and nonpolar organic solvents. *Eastman.* Name unverified.

8781 dodecylthioethanol
1462-55-1 215-969-7
$C_{14}H_{30}S$
2-(Dodecylthio)ethanol
DV-1936. *Rhône Poulenc Surf.*

8782 dodemorph-acetate
31717-87-0 250-778-2
$C_{20}H_{39}NO_3$
4-cyclodecyl-2,6-dimethylmorpholine acetate
4-cyclododecyl-2,6-dimethylmorpholine acetate; Meltatox; Mehltaumittel; Milban; Morpholine, N-cyclododecyl-2,6-dimethyl-, acetate. Systemic fungicide with protective and curative action. Used for control of powdery mildews on roses. mp = 63-64°; sparingly soluble in H_2O (<100 mg/l), more soluble in organic solvents; LD_{50} (rat orl) = 3944 mg/kg.

8783 Dodicor 2565
Quaternary arylammonium chloride
Metalworking surfactant; corrosion inhibitor for oil and gas industry; maximum protection for zinc in acid cleaners. *Hoechst Celanese/Colorants & Surf.*

8784 Dodigen
Range of cationic surfactants of the quaternary ammonium chloride type, in liquid, paste or solid form; antistatic agents, fabric conditioner/softener, fiber finishers, water repellant and dewatering agents; wetting agents for oils, dispersants for pigments, flushing, foaming agents, spinning bath, viscous additives, corrosion inhibitors, flotation chemicals, abti-caking, anchoring and wetting agents, surface coatings, lacquers, adhesives, disinfectants, auxiliaries for leather, rubber and metals. *Hoechst UK.* Name unverified.

8785 Dodine FL, WP
2439-10-3 3468 219-459-5
$C_{15}H_{33}N_3O_2$
dodecylguanidine monoacetate
Dodine; doguadine; AC 5223; Carpene; CL 7521; Curitan; Cyprex; Efuzin; Melprex; Radspor; Venturol. A foliar fungicide with some protective nd curative action; used for the control of scab in apples and pears. mp = 136°; soluble in H_2O (630 mg/l), EtOH, insoluble in organic solvents; LD_{50} (rat orl) = 100 mg/kg. *Truchem Ltd.*

8786 Dodoxynol series
9014-92-0
Dodoxynol-6
Igepal® RC-520; RC-620; Prox-onic DDP-09; Rexol 65/4; T-Det® DD-5; Teric DD5. A low foaming rewetting agent suitable for paper towels, tissues, and semichemical corrugating media. *Rhône-Poulenc Surf.; Protex; Hart Chem. Ltd.; Harcros; ICI Australia.*

8787 Doff
Range of insecticides and herbicides; for horticultural/household use. *Doff Portland Ltd.*

8788 Dog Off
8208
Quassia
Quassia; animal repellent for outdoor crops. *Fieldspray.*

8789 Dolan, Dolanit
Acrylic fiber. *Hoechst UK.*

8790 Dolasol TF
Nonionic surfactants and inorganic builders; industrial cleaner. *Zschimmer & Schwarz.* Discontinued.

8791 Doler Brass
A proprietary alloy; it is a silicon brass. No manufacturer.

8792 Dolofil
Dolomite. *Steetley Minerals Ltd.*

8793 Dolomol
A white insoluble powder, consisting mainly of magnesium stearate, with small amounts of magnesium oleate and palmitate; used as a dusting powder for skin.

8794 Domeboro
8006-13-1 332
Aluminum acetate
Astringent. *Bayer.* Name unverified.

8795 Domestos
7681-52-9 8773 231-668-3
sodium hypochlorite
Stabilized sodium hypochlorite, a bleach and oxidizer. *Unilever.* Name unverified.

8796 Dominate
A wettable powder containing a mixed culture of micro-organisms (*Anthrobacter, Aspergillus terreus, Bacillis subtilis, Bacillis thuringiensis, Bacteroides, Nocardia,* and *Pseudomonas* spp.); used to suppress growth of pathogenic soil fungi; applied to the soil; used for a wide variety of crops. *Westbridge Research Group.* Unverified.

8797 Donarit1, Donarit2, Donarit3
Powdery ammon dynamites with the addition of aromatic nitro-compounds and explosive oil; particularly suitable for medium-hard rock; used in quarries, in agriculture and forestry under dry conditions. *Dynamit Nobel Wien GmbH.* Name unverified.

8798 Donarite
An explosive containing 80% ammonium nitrate, 12% trinitrotoluene, 4% flour, 3.8% nitroglycerin, and 0.2% collodion wool.

8799 Dontalol®
Pharmaceutical preparation; mouthwash concentration. *Bayer AG.*

8800 Donut Pyro®
7758-16-9 8713 231-835-0
Sodium acid pyrophosphate
Leavening agent for baking, cereals. *Rhône-Poulenc Food Ingreds.*

8801 Doom
A microbial insecticide in powder form containing viable spores of *Bacillus popilliae*, a specific pathogen which infects and kills Japanese beetle grubs; ready-to-use; for control of Japanese beetle grubs; only one application is needed as the living spores are self-perpetuating. *Fairfax Biological Laboratory Inc.*

8802 Dope
The name given to various solutions of cellulose or cellulose compounds in acetone, amyl alcohol, amyl acetate, and other solvents; used for painting aeroplane wings, and other purposes.

8803 Dorcolor®
Spin-dyed man-made fibers. *Bayer AG.*

8804 Dore silver
A silver containing small amounts of gold.

8805 Dorin
triadimenol-tridemorph
Dorindan. Emulsifiable concentrate containing 125 g triadimenol and 375 g tridemorph per liter; used for control of mildew and rust in cereals. *Bayer plc.*

8806 Dorindan
triadimenol-tridemorph
Emulsifiable concentrate containing 125 g triadimenol and 375 g tridemorph per liter; used for control of mildew and rust in cereals. *Bayer plc.*

8807 Dorix®
Polyamide staple fiber; used for nonwovens, felts, insulation, coverings and mats. *Bayer AG.*

8808 Dorlastan®
Man-made fibers, filament yarn. *Bayer AG.*

8809 Dormakil
Fungicide (sold in UK for DowElanco). *ICI Chem & Polymers Ltd.*

8810 Dormone
94-75-7 2865 202-361-1
2,4-D
A selective weed killer containing 465 g/l 2,4-D as the diethanolamine salt; may be used for the control of broad leafed weeds on amenity areas, golf courses, playing fields, etc. *Burts and Harvey; Rhône-Poulenc Environmental Prods. Ltd.*

8811 Dorox
Aluminum alcoholates. *RWE-DEA Chemicals UK Ltd.*

8812 Dosaflo
19937-59-8 243-433-2
metoxuron
Suspension concentrate containing 500 g/l metoxuron; residual urea herbicide for the control of weeds in cereals and carrots. *ICI Chem. & Polymers Ltd; Farm Protection Ltd.*

8813 D.O.T.G
97-39-2 202-577-6
Di-o-tolylguanidine
A rubber vulcanization accelerator.

8814 Double Bond
Two component adhesive. *Stag Polymers & Sealants Ltd.*

8815 double nickel salt
$Ni(NH_4)_2 \cdot (SO_4)_2 \cdot 6H_2O$
Nickel ammonium sulfate
Used in the plating trade.

8816 Double Shield
Marine antifouling; does not contain any tin compounds *Llewellyn Ryland Ltd.*

8817 Double White
1312-76-1 7838 215-199-1
potassium silicate
A proprietary trade name for a general purpose potassium silicate cement for acid conditions, e.g., as a bedding and jointing material for tiles. *Haworth (ARC) Ltd.* Unverified.

8818 Double/Bubble
Package for two-part epoxies. *Hardman.*

8819 Doublet
Herbicide. *May & Baker Ltd.* Name unverified.

8820 Doublet Twitchell reagent
The barium salt of the sulfonated mixture of naphthalene and fatty acid; used in the decomposition of fats.

8821 Doucil
Chemical for use in industry. *Crosfield Chemicals Ltd.*

8822 Doverchlor 10
Chlorinated olefin, a solvent and fire retardant. *Dover.*

8823 Doverguard® 700
53449-39-8
Resinous chlorinated paraffin; flame retardant for plastics, rubbers, adhesives, paints, fabric coatings. d = 1.6; softening point = 95-100°. *Dover.* Discontinued.

8824 Doverguard® 8133
Resinous bromochlorinated paraffin; flame retardant for plastics, rubbers, adhesives, paints, fabric coatings. d^{50} = 2.3; softening point = 130-300°. *Dover.*

8825 Doverguard® 8207-A
Liq. bromochlorinated paraffin; flame retardant for plastics, rubbers, adhesives, paints, fabric coatings. d = 1.42; viscosity = 22 poise. *Dover.*

8826 Doverguard® 8410
Liq. brominated paraffin; flame retardant for plastics, rubbers, adhesives, paints, fabric coatings. d^{50} = 1.52; viscosity = 0.5 poise. *Dover.*

8827 Doverhos® 251
Distearyl hydrogen phosphite
Detergent. *Dover.* Discontinued.

8828 Doverlub 8136
Heavy-duty extreme-pressure soluble oil base providing excellent emulsion stability, good EP and antiwear properties, good rust protection and lubricity. *Dover.*

8829 Doverlub 8506
Chlorinated methyl ester; lubricity and EP agent for lubricants including synthetic coolant, cutting, drawing, and gear oils. *Dover.*

8830 Doverlub 8527
Chlorinated fatty acid; lubricity and EP agent for lubricants including synthetic coolant, cutting, drawing, and gear oils. *Dover.*

8831 Doverlub 8531
Chlorinated lard oil; lubricity and EP agent for lubricants including synthetic coolant, cutting, drawing, and gear oils. *Dover.*

8832 Doverlub 8621
Mixture of fatty compounds and chlorinated paraffin; lubricity and EP agent for lubricants including synthetic coolant, cutting, drawing, and gear oils. *Dover.*

8833 Doverphos®
Liquid and solid phosphites performing as antioxidants; provides process and service life stability in various polymers and intermediates in vinyl stabilizers. *Dover.*

8834 Doverphos®4
26523-78-4 247-759-6
Trisnonylphenyl phosphite
Doverphos® 4-HR; tris(nonylphenol)phosphite; Phenol, nonyl-, phosphite (3:1); Tris(nonylphenyl) phosphite. Heat stabilizer for PVC, ABS, polyolefins, some rubber products. *Dover.*

8835 Doverphos® 6
25448-25-3 246-998-3
Triisodecyl phosphite
Chelating agent with metal carboxylates as polymer additives, especially for chlorinated polymers such as PVC and chlorinated PE; improves color, heat and light stability; antioxidant and EP additive for lubricants. *Dover.*

8836 Doverphos® 7
25550-98-5 247-098-3
Phenyl diisodecyl phosphite
Chelating agent with metal carboxylates as polymer additives, especially for chlorinated polymers such as PVC and chlorinated PE; improves color, heat and light stability. *Dover.*

8837 Doverphos® 8
26544-23-0 247-777-4
Diphenyl isodecyl phosphite
Chelating agent with metal carboxylates as polymer additives, especially for chlorinated polymers such as PVC and chlorinated PE; improves color, heat and light stability. *Dover.*

8838 Doverphos® 10
101-02-0 202-908-4
$C_{18}H_{15}O_3P$
Triphenyl phosphite
Doverphos® 10; Triphenyl phosphite; Phosphorous acid, triphenyl ester; EFED. Improves color stability in polyesters, polyurethanes, and alkyd resins; also aids curing and hardening in epoxies. mp = 22-24°; bp = 360°; d = 1.1840; n_D^{20}= 1.5903; soluble in organic solvents; LD_{50} (rat orl) = 444 mg/kg. *Dover.*

8839 Doverphos® 10-HR
101-02-0 202-908-4
Triphenyl phosphite
Contains 0.5% triisopropanolamine; improves color stability in polyesters polyurethanes, and alkyd resins; also aids curing and hardening in epoxies. *Dover.*

8840 Doverphos® 11
80584-85-6 279-498-9
Tetraphenyl dipropylene glycol diphosphite
Stabilizer in polymers. *Dover.*

8841 Doverphos® 12
80584-86-7 279-499-4
Poly(dipropylene glycol)phenyl phosphite
Chelating agent with metal carboxylates as polymer additives, especially for chlorinated polymers such as PVC and chlorinated PE; improves color, heat, and light stability. *Dover.*

8842 Doverphos® 53
3076-63-9 221-356-5
$C_{36}H_{75}O_3P$
Trilauryl phosphite
Antioxidant and EP additive for lubricants. *Dover.*

8843 Doverphos® 213
4712-55-4 225-202-8
$C_{12}H_{11}O_3P$
Diphenyl phosphite
Phosphonic acid, diphenyl ester. Chelating agent. mp = 12°; bp_{26} = 218-219°; d = 1.2230; n_D^{20}= 1.5575; insoluble in H_2O. *Dover.*

8844 Doverphos® 274
Dilauryl hydrogen phosphite
Surfactant. *Dover.*

8845 Doverphos® 4-HR
26523-78-4 247-759-6
Trisnonylphenyl phosphite, 0.75% triisopropanolamine; hydrolysis-resistant antioxidant for elastomer manufacture *Dover.*

8846 Doverphos® DIOP
36116-84-4 252-287-4
Diisooctyl phosphite
Chelating agent. *Dover.*

8847 Doverphos® DPGDP
80584-85-6 279-498-9
Tetraphenyl dipropylene glycol diphosphite
Chelating agent. *Dover.*

8848 Doverphos® DPIOP
26401-27-4 247-658-7
diphenyl isooctyl phosphite
Chelating agent. *Dover.*

8849 Doverphos® DPP
4712-55-4 225-202-8
$C_{12}H_{11}O_3P$
Diphenyl phosphite
Phosphonic acid, diphenyl ester. Chelating agent. mp = 12°; bp_{26} = 218-219°; d = 1.2230; n_D^{20}= 1.5575; insoluble in H_2O. *Dover.*

8850 Doverphos® S-680
3806-34-6 223-276-6
Distearylpentaerythritol diphosphite
Color stabilizer and melt flow aid for polymer processing. *Dover.*

8851 Doverphos® TIOP
Triisooctyl phosphite
Chelating agent. *Dover.*

8852 Doverphos® TLP
3076-63-9 221-356-5
$C_{36}H_{75}O_3P$

Trilauryl phosphite
Dover.

8853 Dow 276-V2
An α-methyl styrene derivative; a vinyl plasticizer. *Dow UK.* Discontinued.

8854 Dow Corning® 1-2531 Release Coating
Silicone resin coating; provides durable release coating for bakers pans, waffle irons; FDA approved for food contact use. *Dow Corning.*

8855 Dow Corning® 7 Compound
Silicone compound; lubricant, release agent for mold break-in; preservative and lubricant for rubber. *Dow Corning.*

8856 Dow Corning® 24 Emulsion
Silicone emulsion; food grade lubricant for manufacture of paper/paperboard in contact with food; FDA approved; water-dilutable. *Dow Coming.*

8857 Dow Corning® 190 Surfactant
Dimethicone copolyol
Silicone surfactant, surface tensile depressant, wetting agent, emulsifier, foam builder, humectant, softener, used for producing flexible slab stock urethane foam; ingredient in personal care products; plasticizer for hair resins. *Dow Corning.*

8858 Dow Corning® 197 Surfactant
Silicone surfactant
For polyurethane foam industry. *Dow Corning.*

8859 Dow Corning® 200 Fluid
Dimethicone
Foam control agent for nonaqueous systems, distillation, resin manufacture, asphalt, oil refining, gas-oil separation. *Dow Corning.*

8860 Dow Corning® 203 Fluid
Silicone fluid; release film providing internal release and lubrication. *Dow Corning.*

8861 Dow Corning® 344 Fluid, 345 Fluid
69430-24-6
Cyclomethicone
Lubricant, spreading agent, detackifier for skin cleansers. *Dow Corning.*

8862 Dow Corning® 556 Fluid
Phenyldimethicone
Lubricant, emollient for skin care products *Dow Corning.*

8863 Dow Corning® 929
Amodimethicone, tallowtrimonium chloride and nonoxynol-10; imparts wet and dry combing ease to hair care formulations; also car polish ingredient. *Dow Corning.*

8864 Dow Corning® 1500 Compd.
Silica filled polydimethyl siloxane; defoamer for aqueous or nonaqueous systems, food processing, rendering, glycol scrubbing, cutting oils. *Dow Corning.*

8865 Dow Corning® 3225C Formulation Aid
Silicone glycol copolymer; surfactant for preparing water-oil volatile silicone emulsions used in personal care products. *Dow Corning.*

8866 Dow Corning® 7224
Amino functional silicone; hair conditioning ingredient. *Dow Corning.*

8867 Dow Corning® ACH-303
1327-41-9 356 215-477-2
Aluminum Chlorhydrate
Active ingredient in antiperspirant and deodorant formulations. *Dow Corning.*

8868 Dow Corning® ACH7-308
Aluminum sesquichlorhydrate
Active ingredients in antiperspirant formulations. *Dow Corning.*

8869 Dow Corning® Antifoam 1410
Silicone emulsion; foam control agent for inks, textile starching/sizing, cutting oils, resin manufacture, gas processing, adhesives/coatings, waste water treatment, pesticide/fertilizer industries; for extreme pH conditions. *Dow Corning.*

8870 Dow Corning® Antifoam A
8050-81-5 3264
Simethicone
Foam control agent for distillation, resin sizes, textile latex backing, paper, asphalt, lubricants, detergents, pesticides, edible oils, soaps, shampoos; also available in food grade. *Dow Corning.*

8871 Dow Corning® Antifoam C
8050-81-5 3264
Simethicone
Food grade foam control agent for food industry, paper coatings, pesticides, herbicides, fertilizers. *Dow Corning.*

8872 Dow Corning® AZG-368
Aluminum/zirconium chlorohydrate
Active ingredient for all forms of topical antiperspirants. *Dow Corning.*

8873 Dow Corning® AZG-370
Aluminum/zirconium glycine
Active ingredient for all forms of topical antiperspirants. *Dow Corning.*

8874 Dow Corning® FF-400
Silicone glycol copolymer; lubricant, heat stabilizer, antistat, fiber finish for textiles threads. *Dow Corning.*

8875 Dow Corning® FS-1265 Fluid
Dimethyl silicone fluid; foam control agent for nonaqueous. systems, aromatic/chlorinated solvs., gas-oil separation, dry cleaning, metal cleaning and degreasing, oil refining. *Dow Corning.*

8876 Dow Corning® Q1-6106
Coupling agent designed to promote adhesion of two dissimilar materials; in reinforced plastic systems, provides coupling of most reinforcing agents, e.g., glass, Kevlar to most polar organic polymers and engineering plastics, e.g., epoxies, urethanes, acrylics, ploysulfones, PPs, melamines, polyimides, PC and the thermoplastic polyesters; adhesion promoter for plastics and metals. *Dow Corning.*

8877 Dow Corning® QF1-3593A
Dimethicone trimethylsiloxysilicate
Emollient with water repellency and low slip for skin care products *Dow Corning.*

8878 Dow Corning® Z-6020
1760-24-3 217-164-6
N-2-aminoethyl-3-aminopropyl trimethoxy silane
Coupling agent used for epoxies, phenolics, melamines, nylons, PVC, acrylics, polyolefins, polyurethanes, nitrile rubbers. *Dow Corning.*

8879 Dow Corning® Z-6030
2530-85-0 219-785-8
$C_{10}H_{20}O_5Si$
3-Methacryloxypropyl trimethoxysilane
Silane A-174; 3-(Trimethoxysilyl)propyl methacrylate; 3-Methacryloxypropyltrimethoxysilane; MEMO; Dynasylan MEMO; γ-Methacryloxypropyl trimethoxysilane; Methacryloxypropyltrimethoxy silane. Coupling agent used for free-radical cross-linked polyester, rubber, polyolefins, styrenics, acrylics. bp = 190°; d = 1.0450; n_D^{20} = 1.4313. *Dow Corning.*

8880 Dow Corning® Z-6032
N-[2-(Vinylbenzylamino)-ethyl]-3-aminopropyl trimethoxysilane
Coupling agent used for most thermoset and thermoplastic resins. *Dow Corning.*

8881 Dow Corning® Z-6040
2530-83-8 219-784-2
3-Glycidoxy-propyl trimethoxysilane
Coupling agent used for epoxies, urethane, acrylic, and polysulfide sealants. *Dow Corning.*

8882 Dow Corning® Z-6075
4130-08-9 223-943-1
$C_8H_{12}O_6Si$
Vinyltriacetoxy silane
Ethenyltriacetate-silanetriol; Vinyltriacetoxy-silane; Triacetoxy(vinyl)silane. Coupling agent used for polyesters, polyolefins, EPDM, EPM (peroxide cured). mp = 7°; bp_{13} = 112°; d = 1.1600; n_D^{20}= 1.4200; *Dow Corning.*

8883 Dow Corning® Z-6076
2530-87-2 219-787-9
3-Chloropropyl trimethoxy silane
Coupling agent used for epoxy, styrenics, nylon. *Dow Corning.*

8884 Dow DBR
Dibenzoyl resorcinol
An ultraviolet absorber for plastics. *Dow UK. Discontinued.*

8885 Dow Plasticizer No. 5
Diphenyl mono *o*-xylenyl phosphate
A vinyl plasticizer. *Dow Cheml Co Ltd, UK & Ireland.*

8886 Dow Plasticizer No. 55
diphenyl mono *o*-xylenyl phosphate
Technical grade diphenyl mono *o*-xylenyl phosphate; a vinyl plasticizer. *Dow Cheml Co Ltd, UK & Ireland.*

8887 Dow Shield
1702-17-6 2462 216-935-4
clopyralid
Benazalox; Format. Soluble concentrate containing 200 g clopyralid per liter; a foliar herbicide for use on brassicas and field vegetables. *DowElanco Ltd.*

8888 Dow V9
α methyl styrene derivative; a vinyl plasticizer. *Dow UK.* Discontinued.

8889 Dowanol®
A line of glycol ethers and glycol ether acetates used as solvents in a variety of unrelated industrial applications. *Dow Cheml Co Ltd, UK & Ireland.*

8890 Dowanol® DB
112-34-5 1592 203-961-6
Diethylene glycol n-butyl ether
Solvent, coupling agent providing improved surface wetting, soil penetration in household, commercial and industrial cleaning products *Dow Chemical.*

8891 Dowanol® DM
111-77-3 6116 203-906-6
Methoxydiglycol
Solvent, coupling agent providing improved surface wetting, soil penetration in household, commercial and industrial cleaning products *Dow Chemical.*

8892 Dowanol® DPMA
PPG-2 methyl ether acetate
Solvent, coupling agent providing improved surface wetting, soil penetration in household, commercial and industrial cleaning products *Dow Chemical.*

8893 Dowanol® DPnB
29911-28-2 249-951-5
Dipropylene glycol n-butyl ether
Solvent, coupling agent providing improved surface wetting, soil penetration in household, commercial and industrial cleaning products. *Dow Chemical.*

8894 Dowanol® EB
111-76-2 1594 203-905-0
Ethylene glycol n-butyl ether
butyl cellosolve. Solvent, coupling agent providing improved surface wetting, soil penetration in household, commercial and industrial cleaning products *Dow Chemical.*

8895 Dowanol® EPh
Ethylene glycol phenyl ether
Solvent, coupling agent providing improved surface wetting, soil penetration in household, commercial and industrial cleaning products *Dow Chemical.*

8896 Dowanol® PM
107-98-2 203-539-1
$C_4H_{10}O_2$
Propylene glycol methyl ether
Methoxypropanol; Propylene Glycol Monomethyl Ether; Methoxypropanol, α Isomer; Propylene Glycol Methyl Ether; 1-methoxypropan-2-ol; 1-methoxy-2-propanol; Methoxy Ether of Propylene Glycol; α-Propylene Glycol Monomethyl Ether; Polypropylene Glycol Methyl Ether; Propylene Glycol 1-methyl Ether; (±)-1-methoxy-2-propanol; Dowanol 33b; Dowanol PM; Dowtherm 209; Glycol Ether PM; PGME; Poly-solve MPM; Propasol Solvent M; UCAR Solvent IM. Solvent, coupling agent providing improved surface wetting, soil penetration in household, commercial and industrial cleaning products. mp = -97°; bp = 118-119°; d = 0.9220; n_D^{20}= 1.4030; soluble in H_2O, organic solvents; LD_{50} (rat orl) = 5660 mg/kg. *Dow Chemical.*

8897 Dowanol® PMA
108-65-6 203-603-9
$C_6H_{12}O_3$
Propylene glycol methyl ether acetate
Propylene Glycol Monomethyl Ether Acetate;-Methoxy-2-propyl Acetate; PGMEA; 1-Methoxy-2-propanol Acetate; Propylene glycol methyl ether acetate; 2-Methoxy-1-methylethyl acetate; 2-(1-Methoxy)propyl acetate; 1-methoxy-2-acetoxypropane; 2-Methoxy-1-methylethyl acetate; 2-(1-Methoxy)propyl acetate; 1-methoxy-2-acetoxypropane. Solvent, coupling agent providing improved surface wetting, soil penetration in household, commercial and industrial cleaning products mp = -67°; bp = 150°; d = 0.9690; soluble in H_2O, organic solvents; LD_{50} (rat orl) = 8532 mg/kg. *Dow Chemical.*

8898 Dowanol® PnB
Propylene glycol n-butyl ether
Solvent, coupling agent providing improved surface wetting, soil penetration in household, commercial and industrial cleaning products *Dow Chemical.*

8899 Dowanol® PPM
Dipropylene glycol methyl ether
Solvent, coupling agent providing improved surface wetting, soil penetration in household, commercial and industrial cleaning products *Dow Chemical.*

8900 Dowanol® TPM
PPG-3 methyl ether
Solvent, coupling agent providing improved surface wetting, soil penetration in household, commercial and industrial cleaning products. *Dow Chemical.*

8901 Dowclene
Industrial solvents, primarily for metal and suede. *Dow Cheml Co Ltd, UK & Ireland.*

8902 Dowco 179
2921-88-2 2242 220-864-4
chlorpyrifos
A proprietary preparation of chlorpyrifos; an insecticide. *Dow Cheml Co Ltd, UK & Ireland.*

8903 Dowel
Ion exchange resins; used for water softening and recovering waste or undesirable materials from process streams. *Dow UK.* Discontinued.

8904 Dowex M-31
Catalyst for production of motor fuel oxygenates (MTBE, ETBE, and TAME). *Dow Chemical.*

8905 Dowex Monosphere
Ion exchange resins. *Dow Cheml Co Ltd, UK & Ireland.*

8906 Dowfax
A family of disulfonated anionic surfactants used in a variety of end-use applications such as cleaning products and bleaches, latex production and agricultural products. *Dow Cheml Co Ltd, UK & Ireland.*

8907 Dowfax 2A0
Dodecyl diphenyloxide disulfonic acid
For use in acidic systems or for preparation of various salts. *Dow; Dow Europe.*

8908 Dowfax 2A1
Sodium dodecylated oxydibenzene disulfonate
Light colored free-flowing powder; anionic surfactant with high solubility, stability, coupling ability and surface activity in strong aqueous solutions of acids, alkalies and salts; moderate sudsing agent; used in metal cleaning including soak tank, steam and electrolytic systems; textiles; shampoos and cosmetics; emulsion polymerization; pulp and paper, mining and food processing industries. *Dow Cheml Co Ltd, UK & Ireland.*

8909 Dowfax 2A1, 2EP
Sodium dodecyl diphenyloxide disulfonate
Detergent, emulsifier, wetting agent, solubilizer, dispersant, spreading agent, penetrant for detergent formulation emulsion polymerization, agriculture, electroplating, ore flotation, drilling muds; leveling agent for acid dyeing of nylon, dyeing assistant, emulsifier for dye carriers. *Dow; Dow Europe.*

8910 Dowfax 30C05, 30C10, 50C15
106392-12-5
EO/PO block copolymer; low foaming surfactant, defoamer base. *Dow Europe.*

8911 Dowfax 3B2
36445-71-3 253-040-8
Sodium n-decyl diphenyloxide disulfonate
Detergent, emulsifier, wetting agent, solubilizer, dispersant, spreading agent, penetrant for detergent formulation, emulsion polymerization, agriculture, electroplating, ore flotation, drilling muds; leveling agent for acid dyeing of nylon, dyeing assistant, emulsifier for dye carriers, biodegradable. *Dow; Dow Europe.*

8912 Dowfax 3BO
N-Decyl diphenyloxide disulfonate
For formulating cleaning products and agricultural products where salts other than NaCl are required. *Dow Chemical.*

8913 Dowfax 8174
Alkylated disulfonated diphenyl oxide
Detergent, emulsifier, wetting agent, solubilizer, dispersant, spreading agent, penetrant for detergent formulation, emulsion polymerization, agriculture, electroplating, ore flotation, drilling muds; leveling agent for acid dyeing of nylon, dyeing assistants, emulsifier for dye carriers, biodegradable. *Dow Chemical.*

8914 Dowfax 8390
Sodium n-hexadecyl diphenyloxide disulfonate
Detergent, emulsifier, wetting agent, solubilizer, dispersant, spreading agent, penetrant for detergent formulation, emulsion polymerization, agriculture, agriculture, electroplating, ore flotation, drilling muds; leveling agent for acid dyeing of nylon, dyeing assistant, emulsifier for dye carriers; biodegradable. *Dow; Dow Europe.*

8915 Dowfax 9N2, 9N3 and 9N4
Nonionic surfactants of the nonylphenol ethoxylate type in the form of an almost colorless liquid; emulsifier, wetting agent, dry cleaning soap, antifoam; used in chemical intermediates; pesticides; metal cleaners; latex paints; dry cleaning formulations. *Dow Cheml Co Ltd, UK & Ireland.*

8916 Dowfax 9N5, 9N6 and 9N7
Nonionic surfactants of the nonylphenol ethoxylate type in the form of an almost colorless liquid; emulsifier, wetting agent and dry cleaning soap; used in pesticides; wax and polish; metal cleaners; metal working compounds; dry cleaning formulations. *Dow Cheml Co Ltd, UK & Ireland.*

8917 Dowfax 9N8, 9N9 and 9N10
Nonionic surfactant of the nonylphenol ethoxylate type in liquid form; detergent, emulsifier, wetting agent and penetrant; used in household, industrial, and institutional cleaners and specialties; textiles; pesticides; latex paints; pulp and paper; leather. *Dow Cheml Co Ltd, UK & Ireland.*

8918 Dowfax 9N12, 9N14/15 and 9N12W
Nonionic surfactant of the nonyl phenol ethoxylate type in liquid or semi-solid form; used for general detergency and wetting, involving elevated

temperatures; used in light duty liquid detergents; cleaning specialties; soak tank and metal cleaners. *Dow Cheml Co Ltd, UK & Ireland.*

8919 Dowfax XDS 8292.00
Sodium hexyl diphenyloxide disulfonate
Lowest foaming and highest charge density product in Dowfax series; high solubilizing capabilities in acids, alkalies, and electrolytes; for cleaning, latex manufacture, paints, adhesives, mineral and metal processing, textile applications. *Dow Chemical.*

8920 Dowfax XDS 8390.00
Sodium n-hexadecyl diphenyloxide disulfonate
Surfactant for emulsion polymerization. *Dow; Dow Europe.*

8921 Dowflake
10043-52-4 1699 233-140-8
Calcium chloride
Dow Cheml Co Ltd, UK & Ireland.

8922 Dowfrost
57-55-6 8040 200-338-0
$C_3H_8O_2$
1,2-Propanediol
Propylene glycol. Inhibited heat transfer fluid; used as a coolant in the manufacture of beer, wine, milk and other liquids; also used to freeze poultry and fish. *Dow Cheml Co Ltd, UK & Ireland.*

8923 Dowfroth
Flotation frothers; low-viscosity water-soluble liquids which produce highly selective foams; used by the mining industry in the recovery of minerals for ores. *Dow Cheml Co Ltd, UK & Ireland.*

8924 Dowgard
Coolant/antifreeze; designed to protect automobile radiators against against overheating in summer and freezing in winter and contains additives to protect against foaming and corrosion. *Dow UK.* Discontinued.

8925 Dowicide
Phenolic-based antimicrobials used as active ingredients in disinfectant formulations, and also as preservatives in a variety of applications such as metal working fluids, adhesives and cosmetic preparations. *Dow Cheml Co Ltd, UK & Ireland.*

8926 Dowicide 1
90-43-7 7458 201-993-5
o-phenylphenol
A proprietary trade name for *o*-phenylphenol; an antiseptic and fungicide. *Dow Chemical.*

8927 Dowicide 2
95-95-4 9772 202-467-8
2,4,5-trichlorophenol
A proprietary trade name for 2,4,5-trichlorophenol; an antiseptic and fungicide. *Dow Chemical.*

8928 Dowicide 3
85-97-2 7434 201-644-7
$C_{12}H_9ClO$
3-chlorobiphenyl-2-ol
Chloro-*o*-phenylphenol; Chloro-6-phenylphenol; 2-Chloro-6-phenylphenol; 6-Chloro-2-phenylphenol. A proprietary trade name for chloro-*o*-phenylphenol; an antiseptic and fungicide. *Dow Chemical.*

8929 Dowicide 5
brom-*p*-phenylphenol
A proprietary trade name for brom-*p*-phenylphenol; a germicide. *Dow Chemical.*

8930 Dowicide 6
25167-83-3
$C_6H_2Cl_4O$
2,3,4,5-tetrachlorophenol
tetrachlorophenol. A proprietary trade name for tetrachlorophenol; an antiseptic. *Dow Chemical.*

8931 Dowicide 7
87-86-5 7242 201-778-6
pentachlorophenol
A proprietary trade name for pentachlorophenol; an antiseptic and fungicide. *Dow Chemical.*

8932 Dowicide A
132-27-4 205-055-6
$C_{12}H_9NaO$
2-Biphenylol, Sodium Salt
Sodium *ortho*-phenylphenate; Dowicide; 2-Phenylphenol Sodium Salt; Sodium *o*-phenylphenoxide; [1,1'-Biphenyl]-2-ol, sodium salt; OPP-NA; bactrol; D.C.S.; Dorvicide A; Dowicide A; Dowizid; 2-hydroxydiphenyl sodium; Mil-du-rid; natriphene; orphenol; *o*-phenylphenol, sodium derivative; preventolon; (2-biphenylyloxy)sodium; sodium *o*-phenylphenolate; SOPP; Stopmold B; Topane; Biphenyl)-2-ol, sodium salt; Biphenylol, sodium salt; Hydroxydiphenyl, sodium salt; Phenylphenol, sodium salt. A proprietary trade name for sodium-*o*-phenylphenate; an antiseptic and germicide. mp = 55-57°; bp = 275°; d_4^{25}= 1.2130; soluble in H_2O (>10 g/100ml), organic solvents; LD_{50} (rat orl) = 2000 mg/kg. *Dow Chemical.*

8933 Dowicide B
136-32-3 205-239-6
$C_6H_2Cl_3NaO$
sodium 2,4,5-trichlorphenate
Dowicide B; Phenol, 2,4,5-trichloro-, sodium salt; Sodium 2,4,5-trichlorophenate; Sodium trichlorophenate; Trichlorophenol, sodium salt. A proprietary trade name for sodium 2,4,5-trichlorphenate; an antiseptic and germicide. *Dow Chemical.*

8934 Dowicide C
sodium-chloro-*o*-phenylphenate
A proprietary trade name for sodium-chloro-*o*-phenylphenate; an antiseptic and germicide. *Dow Chemical.*

8935 Dowicide F
sodium tetrachlorophenate
A proprietary trade name for sodium tetrachlorophenate; an antiseptic and germicide. *Dow Chemical.*

8936 Dowicil®
Preservatives which provide microbial protection for various applications such as cosmetic and personal care formulations, household products, paints, adhesives, metal working fluids and latex emulsions. *Dow Cheml Co Ltd, UK & Ireland.*

8937 Dowicil® 75
4080-31-3 2168 223-805-0
$C_9H_{16}Cl_2N_4$
1-(3-chloroallyl)-3,5,7-triaza-1-azonia-adamantane chloride
1-(3-Chloroallyl)-3,5,7-triaza-1-azoniaadamantane chloride; 3,5,7-Triaza-1-azoniatricyclo[3.3.1.13,7]decane, 1-(3-chloro-2-propenyl)-, chloride; *cis*-N-(3-chloroallyl)hexaminium chloride; Dowicil 100; Dowco 184; Dowicide 184; Cinartc 200; Dowicide Q; Dowicil 75; Hexamethylenetetramine chloroallyl chloride; Triaza-1-azoniaadamantane, 1-(3-chloroallyl)-, chloride; Triaza-1-azoniatricyclo[3.3.1.1(3,7)]decane, 1-(3-chloro-2-propenyl)-, chloride. 1-(3-chloroallyl)-3,5,7-triaza-1-azonia-adamantane chloride (active ingredient) and sodium bicarbonate (stabilizer); preservative for aqueous end products such as adhesives, latex emulsions, paints, metal-cutting fluids, drilling muds, biodegradable detergents, and paper coatings; antimicrobial activity. Very soluble in H_2O, less soluble in organic solvents; LD_{50} (rat orl) = 500 mg/kg. *Dow Chemical.*

8938 Dowlex
Linear low density polyethylene resins used for cast and blown films, injection molding, blow molding, and extrusion coating. *Dow Cheml Co Ltd, UK & Ireland.*

8939 Dowlex 2032 2035, 2042, 2500, 3010
9002-88-4 7728
LLDPE resin; general purpose liner resin with improved thermal stability for blown film extrusion (thin gauge, medium stiffness film applications); FDA compliance. 2035 is for cast film extrusion, 2042 for blown film extrusion, 2500 for injection molding and 3010 is for extrusion coating. *Dow Chemical.*

8940 Dowmetal alloys
Aircraft alloys containing magnesium with small amounts of aluminum and manganese, sometimes with the addition of small quantities of copper, cadmium, and zinc. They have low specific gravity. Name unverified.

8941 Downright
Latex additives used to improve low and high temperature performance and durability in asphalt concrete applications. *Dow UK.* Discontinued.

8942 Dowper
127-18-4 9332 204-825-9
tetrachloroethylene
Inhibited tetrachloroethylene; solvent for dry cleaning. *Dow Cheml Co Ltd, UK & Ireland.*

8943 Dowpon
127-20-8 11,2806 204-828-5
$C_3H_3Cl_2NaO_2$
2,2-dichloropropanoic acid, sodium salt
Dalapon sodium salt; Dowpon; Radapon; Dichloropropionic acid, sodium salt; Basfapon B. Herbicide; for controlling grass species; used primarily in sugar cane, sugar beets, orchards and also in noncrop applications such as railroads and rubber plantations. *Dow UK.* Discontinued.

8944 Dowtherm 209
Heat transfer agent used as a temperature controlling liquid (coolant) for diesel-powered vehicles. *Dow Cheml Co Ltd, UK & Ireland.*

8945 Dowtherm A
A proprietary trade name for a product consisting of a mixture containing biphenyl ether; used for heating industrial machinery to high temperatures (e.g. 200°) in place of steam. Name unverified.

8946 Doxycycline hydrochloride
24390-14-5 3496
$C_{22}H_{25}ClN_2O_8 \cdot 1/2H_2O \cdot 1/2C_2H_6O$
[4S-(4α,4aα,5α,5aα,6α,12aα)]-4-(dimethylamino)-1,4,4a,5,5a,6,11,12a-octahydro-3,5,10,12,12a-pentahydroxy-6-methyl-1,11-dioxo-2-naphthacenecarboxamide monohydrate

8947 DP. 250
A polyester vinyl plasticizer. No manufacturer.

8948 DP/4137-16
A proprietary fast-curing polyester resin with low viscosity; a fire retardant. *Synthetic Resins Ltd.* Name unverified.

8949 DPNR
9006-04-6 232-689-0
Deproteinized NR; used for products with high electrical resistivity, low affinity for water; engineering grades for off-shore and underwater use, brake pads, vibration mounts, pharmaceutical products *H.A.Astlett.*

8950 DPPG
41395-83-9 255-350-9
Propylene glycol dipelargonate
Emollient and oily rancidless additive for cosmetic and pharmaceutical preparations. *Gattefosse SA.*

8951 DPR® 40, 75, 400
cis-1,4-Polyisoprene of low molecular weight derived from virgin synthetic polyisoprene; used for potting, sealants, caulk, adhesives, and flexible rubber molds and ebonite stocks; cures to soft rubber; two-part compounds can be formulated for self-curing *Hardman.*

8952 DPTT
120-54-7 204-406-0
Dipentamethylene thiuram hexasulfide
Very active sulfur-bearing accelerator imparting heat resistance to sulfurless compounds; primary accelerator for Hypalon, butyl, and EPDM; vulcanizing agent for heat-resistant latex; nondiscoloring, nonstaining. *Akrochem.*

8953 Dracyl
10808803 9667 203-625-9
C_7H_8
Toluene
Solvent.

8954 Dragendorf's reagent
Potassium iodobismuthate
Kraut's reagent. Used for testing alkaloids.

8955 Dragon
Insecticide *ICI Chem & Polymers Ltd.*

8956 Dragon gum
9000-65-1 4609 232-552-5
Gum tragacanth.' Dried gummy exudate from *Astragalus gummifer* Labil, found in Asia Minor (Syria and Iran). Contains polysaccharides in a water-soluble fraction. Used in pharmaceutical compounding, in making emulsions and as a stabilizer, thickener and texturizer in foods.

8957 Dragonmat
Insecticide vaporizing device. *ICI Chem & Polymers Ltd.*

8958 Dragon's blood
A red resin. The two varieties are Palm Dragon's Blood, obtained from the rattan palm, *Daemonorops draco* of Sumatra and Borneo, and Socotra Dragon's Blood, from *Dracoena cinnabari* of Socotra, and the West Indies; used as a pigment for the preparation of red lakes and varnishes.

8959 Drainaid GL-73
Drainage aid for acidified pulp in paper industry. *Hart Chem. Ltd.*

8960 Drakeol 5
8042-47-5 232-455-8
White mineral oil, petroleum
Emollient; Light mineral oil USP d = 0.845 - 0.905. *Penreco.*

8961 Drakeol 7
8042-47-5 232-455-8
White mineral oil, petroleum
Light mineral oil USP; carrier, base ingredient in ointments, lotions, baby oils, sun tan lotions, makeup; solvent and emollient in creams, waterless hand cleaners; protective coating for foods; pigment dispersant, lubricant for plastics coning and finishing oil base for nylon and rayon production d = 0.845 - 0.905. *Penreco.*

8962 Drakeol 10
8042-47-5 232-455-8
White mineral oil, petroleum
Light mineral oil USP; plasticizer, lubricant for plastics. d = 0.845 - 0.905. *Penreco.*

8963 Drakeol 19
8042-47-5 232-455-8
White mineral oil, petroleum
Mineral oil USP; primary plasticizer for ethyl cellulose; lubricant for textile/paper. d = 0.845 - 0.905. *Penreco.*

8964 Dralon®
Polyacrylic staple fiber and tow; used for outdoor textiles and acid proof protective clothing. *Bayer AG.*

8965 Dralon® T
Polyacrylic staple fiber; used for industrial felts and filter media. *Bayer AG.*

8966 Drapex
White mineral oil.
Pennzoil Products Co.

8967 Drapex®
Epoxy plasticizer. *Argus Chemical Corporation.*

8968 Drapex® 4.4
Octyl epoxy tallate
Plasticizer for vinyl compounds; compatible with primary plasticizers; used with Argus Ba/Cd stabilizers to synergistically improve heat and light stability; resistant to extraction and migration; food packaging use. *Witco Corporation.*

8969 Drapex® 6.8
8013-07-8 232-391-0
Epoxidized soybean oil
Plasticizer for vinyl compounds; compatible with primary plasticizers; used with Argus Ba/Cd stabilizers to synergistically improve heat and light stability; resistant to extraction and migration; food packaging use. *Witco Corporation.*

8970 Drapex® 10.4
Epoxidized linseed oil ; plasticizer for vinyl compounds; compatible with primary plasticizers; used with Argus Ba/Cd stabilizers to synergistically improve heat and light stability; resistant to extraction and migration; food packaging use. *Witco Corporation.*

8971 Drapex® P-1
Polyester plasticizer
Plasticizer for use in PVC compounds requiring durability, resistant to migration, high dielectric properties, e.g., for electrical tape compounds, refrigerator gasketing, wall covering, automotive interiors, oil-resistant wire insulation. *Witco Corporation.*

8972 Drat
3691-35-8 2204 223-003-0
Chlorophacinone
Rozol. An oil formulation containing 2.5 g/liter chlorophacinone, a powerful anticoagulent rodenticide; controls black rats, brown rats, house mice, long-tailed field mice, voles and musk rats. *Rhône-Poulenc Environmental Prods. Ltd.*

8973 Draza®
2032-65-7 6050 217-991-2
Pellet formulation containing 4% methiocarb, a molluscicide used to control slugs and snails in any crop; reduces populations of leatherjackets; there is some evidence that cutworms, earwigs, and millipedes are controlled. *Bayer AG; Bayer plc.*

8974 drazoxolan
5707-69-7 3499 227-197-8
$C_{10}H_8ClN_3O_2$
3-methyl-4-[(2-chlorophenyl)hydrazone]-4,5-isoxazoledione
PP-781; Ganocide; Mil-Col; Saisan. Fungicide. mp = 168°; insoluble in H_2O, soluble in organic solvents; LD_{50} (rat orl) = 126 mg/kg.

8975 Dreadnought powder
An explosive containing 73-77% ammonium nitrate, 14-17% sodium nitrate, 4-6% ammonium chloride, and 3-5% trinitrotoluene.

8976 Dreft
151-21-3 8782 205-788-1
sodium lauryl sulfate
A proprietary trade name for a washing material consisting of sodium lauryl sulfate. No manufacturer.

8977 Drene
151-21-3 8782 205-788-1
sodium lauryl sulfate
A proprietary trade name for a shampoo containing sodium lauryl sulfate. No manufacturer.

8978 Dresden Thick Oil
A thick turpentine or oleo-resin; similar to Venice turpentine; used as a vehicle for colors for painting. No manufacturer.

8979 Dresinate
Potassium and sodium soaps of rosin, modified rosin and tall oil derivatives; improves latex stability in rubber latices; used in formulating soluble cutting oils and drawing compounds; as modifiers of heavy-duty metal cleaners and other industrial cleaners. *Hercules.*

8980 Dresinol
A range of resin dispersions; used to modify polymer properties in adhesives, in the production of pigments and resinated colors in the graphics and inks

industry, in paint polymer emulsions and to help wet paint pigments. *Hercules.*

8981 **Dress All**
Partially hydrogenated soybean oil formulated with artificial flavor, TBHQ and artificial color; kosher; dressing oil for seafood, vegetables, frying, other prepared foods. *Van Den Bergh Foods.*

8982 **Drewamine**
Corrosion inhibitor for control of corrosion in the afterboiler section of steam generating systems by neutralizing the acidity of the condensate. *Drew Ind. Div.*

8983 **Drewbrom**
Aqueous solution of bromide ion; precursor for production of biocide used as disinfectant, sanitizer, bactericide, slimicide, and algicide in recirculating cooling water systems, once-through cooling water, and waste water treatment systems. *Drew Ind. Div.*

8984 **Drewchlor®**
Algicide and precursor for generation of chlorine dioxide for treating industrial recirculating cooling water systems including cooling towers, air washers and evaporator condensers. *Drew Ind. Div.*

8985 **Drewclean® 26**
Alkaline cleaner; cleaner for ion exchange resins and filter media fouled by oil, grease, other organics, clay or silt. *Drew Ind. Div.*

8986 **Drewcor® 2130**
Blended neutralizing amines and mekor® volatile oxygen scavenger/metal passivator; corrosion inhibitor for protection against low pH and oxygen induced corrosion in condensate and steam lines. *Drew Ind. Div.*

8987 **Drewfax® 412**
Blend of silicone derivatives and glycol ethers; surfactant reducing surface tension, wetting agent, flow and leveling agent, slip and mar aid for solvent and water based coatings and inks based on alkyds, epoxy, urethanes, acrylics. *Drew Ind. Div.*

8988 **Drewfax® 680**
Silicone surfactant
Surfactant, antifoam, flow and leveling agent, air release agent for solvent-based coatings and inks based on alkyd melamine, epoxy, urethanes, acrylics; especially suited for wood coatings. *Drew Ind. Div.*

8989 **Drewfax® 818**
Blend of silicone copolymers and glycol ether; surfactant, wetting agent, flow and leveling agent, slip and mar aid, anticrater for water-based coatings and inks based on acrylics, alkyds. *Drew Ind. Div.*

8990 **Drewfax® S-600**
Silicone/silica derivs
Surfactant, antifoam, flow and leveling agent for solvent- and water-based coatings and inks based on alkyds, epoxy, urethanes, acrylics; especially suited for clear lacquers, pigmented coatings, uv coatings. *Drew Ind. Div.*

8991 **Drewfax® S-700**
577-11-7 3460 209-406-4
Sodium dioctyl sulfosuccinate
Surfactant, anticrater, dispersant, flow and leveling agent for high-solids and acrylic solvent-based inks and coatings. *Drew Ind. Div.*

8992 **Drewfex® 0007**
577-11-7 3460 209-406-4
Sodium dioctyl sulfosuccinate
Wetting, penetrating, surface tension reducing agent, dispersant for industrial coatings, adhesives, inks, pigments, textile, cosmetic, paper, metal, paint, rubber, plastics, petroleum and agricultural industries. *Drew Ind. Div.* Name unverified.

8993 **Drewfloc® 2270**
High molecular weight emulsion polymer; flocculant and coagulant aid for dewatering industrial slurries, water clarification, and sludge conditioning applications. *Drew Ind. Div.*

8994 **Drewgard® 120**
Conc. polyphosphate
Corrosion and deposit inhibitor in once-through and open recirculating cooling water systems and potable water distribution systems. *Drew Ind. Div.*

8995 **Drewgard® 189E**
Isoquest-LT polymer, organic phosphorus and zinc; corrosion and deposit inhibitor for open recirculating cooling water systems. *Drew Ind. Div.*

8996 **Drewmulse® 10K**
Glyceryl mono-shortening from soya oil; lipophilic emulsifier used in food industry as dispersing aid, antistaling agent, antistick agent. *Stepan/PVO.*

8997 **Drewmulse® 200K**
Glyceryl stearate;. Emulsifier, emollient, antistat, stabilizer, viscosity builder, opacifier for creams, lotions, hair conditioners. Nonionic. *Stepan Europe; Stepan/PVO.*

8998 **Drewmulse® 900K**
Glyceryl stearate; emulsifier for food industry as dispersing aid, antistaling agent, antistick agent. Nonionic; mp = 59°. *Stepan/PVO.*

8999 **Drewplast® 017**
Glycerol mono and diesters of fatty acid; food-grade plastics additive; antistat for LDPE, lubricant for rigid and flexible PVC; good thermal properties, stable to oxidative degradation. *Stepan/PVO; Harwick.*

9000 **Drewplast® 030**
Glycerol mono and diesters of fatty acids; food-grade plastics additive; antifog for PVC sheet and film; good thermal properties, stable to oxidative degradation. *Stepan/PVO; Harwick.*

9001 **Drewplast® 051**
Glycerol mono and diesters of fatty acids; food-grade plastics additive; antifog for PVC sheet and film; cling agent for LDPE; lubricant for rigid and flexible PVC; good thermal properties, stable to oxidative degradation. *Stepan/PVO; Harwick.*

9002 **Drewplex**
Blends of natural and synthetic sludge conditioning agents; used for boiler water sludge conditioning, boiler water treatment. *Ashland Chemical Company.*

9003 **Drewplus® L-108**
Blend of mineral oils, emulsifiers, silica derivatives; defoamer for latex/rubber applications especially monomer stripping, acrylic, PVAc, NBR, SBR, PVC. Straw-colored opaque liquid; emulsifible in H_2O; d = 0.89; 7.58 lb/gal; viscosity = 800 cps; flash point = 93°. *Drew Ind. Div.*

9004 **Drewplus® L-123**
Nonsilicone defoamer; defoamer for latexes. Opaque liquid; d = 0.92; 7.68 lb/gal; flash pt >300°F. *Drew Ind. Div.*

9005 **Drewplus® L-162**
Silicone defoamer; defoamer for latexes. *Drew Ind. Div.*

9006 **Drewplus® L-407**
Modified polysiloxane copolymer emulsion; foam control agent for architectural paints and coatings (interior/exterior gloss and semigloss paints, waterborne industrial coatings, polymer emulsions, water-thinnable gravure and flexographic inks). *Drew Ind. Div.*

9007 **Drewplus® L-464**
Blend of silica and organic solids; surfactant, defoamer, air release agent effective in water-reducible coatings such as polyurethanes, interior and exterior paints. *Drew Ind. Div.*

9008 **Drewplus® L-768**
Blend of silicone fluid, silica derivatives and surfactants; defoamer for industrial/chemical processes, food/fermentation and agricultural applications, aqueous and some nonaqueous systems. *Drew Ind. Div.*

9009 **Drewplus® L-790**
Blend of mineral oils, silica derivatives, and emulsifiers; defoamer for food/fermentation and agricultural applications. *Drew Ind. Div.*

9010 **Drewplus® L-813**
Blend of dimethylpolysiloxane, silica, and emulsifiers; foam control agent for industrial food processing (starch slurries, fermentation, calcium chloride brines, adhesives, glues, vegetable and fruit processing, sugar, instant coffee and fruit juices). *Drew Ind. Div.*

9011 **Drewplus® M-111**
Blend of silica derivatives and surfactants; defoamer for gypsum, starch, cement, pigments, joint compound and adhesives that require a dry defoamer. *Drew Ind. Div.*

9012 **Drewplus® Y-200**
Nonsilicone defoamer; defoamer for paints/coatings, inks, adhesives. *Drew Ind. Div.*

9013 **Drewplus® Y-381**
Blend of organics and hydrocarbons; defoamer/antifoam for latex paints (acrylic, vinyl acrylic, S/B), coatings, ink, adhesives. *Drew Ind. Div.*

9014 **Drewpol® 10-4-O**
34424-98-1 252-011-7
Polyglyceryl-10 tetraoleate
Emulsifier, emollient, antistat, stabilizer, viscosity builder, opacifier for creams, makeup, lotions, hair conditioners; solubilizer for vitamins, flavors, medicaments. *Stepan; Stepan Canada; Stepan Europe.*

9015 **Drewpol® 3-1-O**
9007-48-1
Polyglyceryl-3 monooleate
Emulsifier for creams, makeup, lotions, conditioners; food emulsifier. *Stepan; Stepan/PVO.*

9016 **Drewpone® 60K**
9005-67-8 8872
Food emulsifier, foaming agent, dough conditioner. *Stepan/PVO.*

9017 **Drewpone® 65K**
9005-71-4 8872
Polysorbate 65
Food emulsifier, foaming agent, dough conditioner. *Stepan/PVO.*

9018 Drewpone® 80K
9005-65-6 7742
polysorbate 80
Food emulsifier, foaming agent, dough conditioner. *Stepan/PVO.*

9019 Drewsorb® 60K
1338-41-6 8872 215-664-9
Sorbitan stearate; food emulsifier. *Stepan/PVO.*

9020 Drewsperse® 611
9003-04-7
Sodium polyacrylate
Pigment and paint dispersant. *Drew Ind. Div.* Name unverified.

9021 Drewsperse® S-825
High molecular weight polymer; wetting agent, dispersant, film-former for water-based acrylic, alkyd, styrene acrylic, epoxy, alkyd-melamine, or polyurethane coatings and inks; maximizes color development; stabilizes the dispersion against flocculation and settling. *Drew Ind. Div.*

9022 Drewtrol® 6955S
Polyphosphate, synthetic polymers, natural organic compounds, sequestrants, alkali and antifoam; corrosion and deposit inhibitor for boiler water treatment. *Drew Ind. Div.*

9023 Drexar 530
2163-80-6 6020 218-495-9
Monosodium methylarsonate
MSMA (Monosodium methylarsonate); selective herbicide for post emergent weed control on lawns and ornamental turf. *Drexel Chemical Company.* Name unverified.

9024 Dri Film® DF1040
Polymethyl hydrosiloxane fluid; reactive silicone fluid forming water-repellent fluids with heat or heat/catalyst; used in textiles, particle treatment, magnesium oxide, and Calrod® units. *GE Silicones.*

9025 Dricoid® 200
11138-66-2 10191 234-394-2
Xanthan gum
Xanthan gum; Used for ice cream, lowfat frozen dessert, frozen yogurt, ice milk and novelties. *Kelco.*

9026 Dricold
124-38-9 1857 204-696-9
carbon dioxide
Solid carbon dioxide pellets or slices; used as refrigerant. *ICI Chem & Polymers Ltd.*

9027 drierite
7778-18-9 1753 231-900-3
calcium sulfate
Anhydrous calcium sulfate; used for drying gases.

9028 Driers
Those substances which are added during the process of boiling linseed oil, to accelerate its drying properties; they absorb the oxygen from the air and transfer it to the oil, thereby aiding its oxidation; the term is used for the oxides of lead, manganese, and cobalt. More recently, the oxalate, acetate and borate of manganese have been used. Currently, fatty acid salts such as lead and manganese linoleate and the metal resinates are used.

9029 Drift Proof
Non-phytotoxic drift control agent, spreadersticker and pesticide deposit builder. *W A Cleary.*

9030 Driftol
Drift retardant. *Monsanto (Solaris).* Name unverified.

9031 Drikalite®
471-34-1 1697 207-439-9
Calcium carbonate
Extender pigment. *ECC International Ltd.*

9032 Drikold
124-38-9 1857 204-696-9
carbon dioxide
Sold carbon dioxide blocks or slices; refrigerating agent. *ICI Chem & Polymers Ltd.*

9033 Drimarene®
Specialty dye for aqueous media, wood stains. *Sandoz.*

9034 Drimax®
Filter cake dewatering aids offering improvement in filter performance with substantial reductions in final cake moisture contents. *Allied Colloids Ltd.*

9035 Drimix®
Powder dispersion of liquids, dry liquid concentrates; free flowing power of liquid additives widely used in elastomer compounding. *Kenrich Petrochemicals.*

9036 Dri-Sil
Silicone water repellents. *Midland Silicones.* Unverified.

9037 Drisorb
Desiccant. *Production Chemicals Ltd.*

9038 Drisoy®
Modified soybean oils; used in paints. *Reichhold.*

9039 Dritan
Vegetable tannins. *Hodgson Chemicals Ltd.*

9040 Driton
Colloidal silica; antislip agent; prevents fiber slippage, dry hand; used only under alkaline conditions. *Sybron.*

9041 Drittel silver
An alloy of 67% aluminum and 33% silver.

9042 Drivanil
A range of alkylene oxide addition products; used as a base for synthetic lubricants, base components for brake fluids, heat transfer medium, and a viscosity modifier for hydraulic fluids containing no mineral oil. *Hüls AG.*

9043 Driverit
75-09-2 6140 200-838-9
methylene chloride
Stabilized methylene chloride; solvent for metal degreasing; also suitable for the treatment of aluminum. *Hüls AG.*

9044 Driverol MPL
Partial phosphate ester; corrosion inhibitor for the mineral oil industry. *Hüls AG.*

9045 Driverol OMM
An amide/anhydride mixture; corrosion inhibitor. *Hühs AG.*

9046 Driveron
1634-04-4 6111 216-653-1
Methyl *tert*-butyl ether
An antiknock agent for motor fuels. *Hüls AG.*

9047 Drivolan
A range of dodecanedioic acid esters; base components for synthetic and semisynthetic lubricants. *Hüls AG.*

9048 Drivosol
Aerosol propellant. *Hüls AG.*

9049 Droncit®
55268-74-1 7896 259-559-6
Praziquantel
Praziquantel, an anthelmintic used to control tapeworm in dogs and cats, also against *Echinococcus* spp. *Bayer AG.*

9050 Drontal
Multispectrum anthelmintic for dogs and cats. *Bayer AG.*

9051 Drontal Plus
Multispectrum anthelmintic for dogs. *Bayer AG.*

9052 Drossa
Protective hand lotion; for practice and laboratory. *Bayer AG.* Discontinued.

9053 Drott
A proprietary pyroxylin plastic. No manufacturer.

9054 Droxol 200
25322-68-3 7729
PEG-200; chemical intermediate for textiles, lubricants, printing, solvents, cleaning formulations, humectants, viscosity modifiers. *Henkel/Textiles.*

9055 Droxychrome
A photographic color developer. *May & Baker Ltd.* Name unverified.

9056 Dry Flo®
9087-61-0
Aluminum starch octenyl succinate
Used in body powders, antiperspirants, feminine hygiene sprays and foot powders. *Nat'l. Starch.*

9057 dry ice
124-38-9 1857 204-696-9
carbon dioxide
Solid carbon dioxide. Its specific gravity is 1.56.

9058 Dry Lightning
Inhibited sulfamic acid (dry); removes water-formed deposits. *Garvey Chemical Corp.* Name unverified.

9059 Dry Pexol® 200
Fortified pale rosin; dry size used with alum in paper and paperboard to produce resistance to water and aqueous solutions *Hercules.* Discontinued.

9060 Dry Pexol® 243
Fortified dark rosin; dry size for lower brightness paper grades. *Hercules.* Discontinued.

9061 Dry Seed TRIGGRR
A dry powder containing trace minerals; used to enhance germination, maturation, and crop yields; applied to seed prior to planting; used for a wide variety of crops including corn, sorghum, wheat and vegetables. *Westbridge Research Group.* Unverified.

9062 Dry Size XL20C
Pale rosin, paraffin wax, ratio 80:20; dry size used with alum to impart high level of resistance to water and aqu solutions in specialized applications like molded pulp products; high levels can be used without brightness, foam, or other operating problems on foaming machine; food packaging, paper and paperboard. *Hercules*. Discontinued.

9063 Dry-Blend® NCG Fiber
Nickel-coated graphite fiber in ABS/PC, ABS or PC; for conductive plastics. *Am. Cyanamid.*

9064 Drymax®
Drier accelerator; for accelerating the drying of nonaqueous coatings. *Hüls Am.*

9065 Drymet® 59
6834-92-0 8788 229-912-9
Anhydrous sodium metasilicate; alkaline silicate used for formulating specialty detergents and contributing buffering capacity and corrosion inhibition of soft metals and ceramic glazes. *Crosfield.*

9066 Dryspersion®
Powder dispersions. *Kenrich Petrochemicals.*

9067 Drytech
Absorbant for aqueous liquids. *Dow Cheml Co Ltd, UK & Ireland.*

9068 D-steel
A steel containing 1.1-1.4% manganese, 0.33% carbon and 0.12% silicon.

9069 DSX 1514
Urethane associative thickener; rheology modifier for latex paints and adhesives; maximize high sheer viscosity. *Henkel.* Name unverified.

9070 D-Trans allethrin
257
allethrins
D-trans allethrin based intermediates. Insecticides. *McLaughlin Gormley King Co.*

9071 Du Pont Adjuvant
Accelerators for rubber vulcanization; No. 1: *p*-nitroso-dimethylaniline; No. 4: aniline; No. 5: formaldehyde-aniline; No.6: methylene dianilide; No. 8: anhydro-formaldehyde-*p*-toluidine; No. 11: triphenylguanidine; No. 12: diphenylguanidine. *DuPont UK.*

9072 Du Pont Enrich®
For the agriculture industry. *DuPont UK.*

9073 Du Pont Linuron 50, 4L
330-55-2 5534 206-356-5
Linuron
A residual urea herbicide for the control of weeds in field crops. *DuPont UK.*

9074 Du Pont Pakwrap®
Polymer. *DuPont UK.*

9075 Du Pont Permissible No. 1
A trademark for an explosive containing nitroglycerin, ammonium nitrate, wood pulp, and sodium chloride. *Du Pont.* Name unverified.

9076 Dualite M6001AE, M6017AE
Hollow composite microspheres (shell; PVDC/acrylonitrile copolymer; coating; calcium carbonate); low-density filler for use in plastics, coatings, adhesives, BMC, SMC, rubber compounding, paper manufacture, etc.; M6001AE for moderate exposure to non-aggressive solvents; M6017AE for prolonged exposure to aggressive solvents. *Pierce & Stevens.*

9077 Duallor
A range of precious metal alloys; for dentistry and dental engineering. *Degussa AG.* Name unverified.

9078 Duasyn
Water soluble dyes for paper, toiletries and cleansers. *Hoechst AG.*

9079 Dubbin
Mixtures of waxes and tallow with coloring matter, sometimes with the addition of rosin; used to render leather waterproof, and to preserve it. No manufacturer.

9080 Dubox®
For the electrical industry. *DuPont UK.*

9081 Duco
A proprietary trade name for pyroxylin lacquers, containing cellulose nitrate. No manufacturer.

9082 Ducobee-Hy
13422-51-0 4854 236-533-2
Hydroxocobalamin
Vitamin B_{12}. *Sterling Drug Inc.* Name unverified.

9083 Dudley metal
An alloy of 98% tin, 1.6% copper, and 0.25% lead.

9084 Duet®
For the agriculture industry. *DuPont UK.*

9085 Duette
Paint system with two tone decorative effect for walls and ceilings. *ICI Chem & Polymers Ltd.*

9086 Dufox
A potato fungicide. *Murphy Chemical Co Ltd.* Discontinued.

9087 DUK-880
lenacil-phenmedipham
Lenacil and phenmedipham; used for control of annual dicotyledons in sugar beet. *DuPont UK.*

9088 Dukatalon
diquat-paraquat
Mixture of diquat and paraquat; herbicide. *Makhteshim Chemical Works Ltd.*

9089 Duke's metal
A heat-resisting alloy, containing 81% iron, 12% chromium, 4% cobalt, 1.5% carbon, and small quantities of manganese, tungsten, and silicon.

9090 Dulceta
Textile auxiliary chemicals. *ICI Chem & Polymers Ltd.* Discontinued.

9091 Dulcin
3517
$C_9H_{12}N_2O_2$
(4-Ethoxyphenyl)urea
4-ethoxyphenylurea; Sucrol; Sucrene; Valzin; Mono-*p*-phenetol-carbamide; dulcine; *p*-phenetocarbamide; *p*-phenetylurea. Used as a sweetening substance; 200 times sweeter than cane sugar. Use of this material in foods is prohibited by the USA FDA. mp = 173-174°; soluble in H_2O, EtOH; TDLo (wmn orl) = 600 mg/kg, LDLo (chd orl) = 400 mg/kg.

9092 Dulenza
A proprietary viscose silk. No manufacturer.

9093 Dullray
A heat-resisting alloy containing 60% iron, 34% nickel, and 5% chromium.

9094 Dulux
A range of various interior/exterior paints and paint-related products. *ICI Chem & Polymers Ltd.*

9095 Dumacene C13, NP707, NP7710 and NPX10
Alkylphenol ethoxylate nonionic surfactant; for textile scouring; wool washing. *Tensia SA.* Name unverified.

9096 Dumet
A copper-clad nickel-iron alloy.

9097 Dunclad CE
A proprietary trade name for a laminate of Penton and a synthetic rubber; offers a highly corrosion resistant lining at temperatures up to 125-130°. *Dunlop Rubber.* Name unverified.

9098 Dunclad VN
A laminate of unplasticized PVC and synthetic rubber designed to enable metal tanks to be lined using special adhesives, giving a highly corrosive resistant surface to prevent attack from oxidizing and other acids up to 85°. *Dunlop Rubber.* Name unverified.

9099 Dunlop 6593
A high grade neoprene compound giving high chemical and abrasion resistance at elevated temperatures. *Dunlop Rubber.* Name unverified.

9100 Dunlop Grade 6167
A first quality butyl rubber compound which can be used up to 110°. in corrosive conditions. *Dunlop Rubber.* Name unverified.

9101 Dunlop PL
A laminate of polypropylene and synthetic rubber for tank lining. *Dunlop Rubber.* Name unverified.

9102 Dunnite
An American explosive. The main constituent is picric acid.

9103 Dunova®
Polyacrylic staple fiber; water absorbent used for sports wear. *Bayer AG.*

9104 Duocrome®
Irridescent colors producing dual-color effects. *Mearl.*

9105 Duofol T
Sulfated ester
Wetting, rewetting agent, dispersant, penetrant, leveling agent, finishing agents for textile wet processing of cotton piece goods; stable to mild alkalies and organic acids. *Hart Chem. Ltd.*

9106 Duolite
Ion exchange resins. *Rohm & Haas UK.*

9107 Duomac®
Series of cationic surfactants which consist of the acetate salts of alkyl propylene diamines, and which are water-soluble; used in pigment flushing, froth flotation and flocculation, particularly in mineral flotation; petroleum processing; leather; paper; pigments and surface coatings; ceramics. *Akzo Chemie UK Ltd.* Discontinued.

9108 Duomat
Dental amalgamator. *Degussa Ltd.*

9109 Duomatic
Injector/drencher gun. *May & Baker Ltd.* Name unverified.

9110 Duomeen
A range of coco, oleyl and stearyl diamines; used as waterproofing agents, bitumen emulsifiers, bitumen adhesion additives, anticorrosives and as agricultural sprays. *Harcros Australia.*

9111 Duomeen®
Range of cationic surfactants of the alkyl propylene diamine type, possessing both primary and secondary amine groups; they form strongly bonded films on the surfaces of metal, textiles, plastics, etc.; used in petroleum; road emulsions; plastics; rubber; textiles; leather; herbicides; fungicides; rodent repellents, mineral flotation; paper; pigments and surface coatings; water and sewage treatment; wax; sealant formulations; cement curing; metal working; car underseals; carbon paper and typewriter ribbon. *Akzo Chemie UK Ltd.*

9112 Duomeen® C
61791-63-7 263-195-3
N-Coco-1,3-diaminopropane
Chemical intermediate; corrosion inhibitor, fuel oil additive, flotation agent; used in metals, textiles, plastics, herbicides; epoxy curing agent. *Akzo; Akzo Chem. BV.*

9113 Duomeen® CD
61791-63-7 263-195-3
Coco-1,3-diaminopropane
Intermediate in manufacture of industrial surfactants. *AKzo.*

9114 Duomeen® OL
7173-62-8 230-528-9
N-Oleyl-1,3-diaminopropane
Industrial surfactant. *Akzo.*

9115 Duomeen® OTM
68715-87-7 272-103-0
N,N,N'-Trimethyl-N'-9-octadecenyl-1,3-diaminopropane
Industrial surfactant. *Akzo.*

9116 Duomeen® T
61791-55-7 263-189-0
Tallow-1,3-diaminopropane
Bitumen emulsifier, corrosion inhibitor, oil additive, antisettling agent; textile finishing agent, dispersant for inorganic pigments in paints, larvacidal oil additive. *Akzo; Akzo Chem. BV.*

9117 Duomeen® TDO
61791-53-5 263-186-4
N-Tallow-1,3-propanediamine dioleate
Bitumen emulsifier, corrosion inhibitor, oil additive, antisettling agent; also in metal treatment as film and boundary lubricant, metal drawing additive; dispersant in paint industry. *Akzo; Akzo Chem. BV.*

9118 Duomeen® TTM
68783-25-5 272-192-6
N,N,N'-Trimethyl-N'-tallow-1,3-diaminopropane
Industrial surfactant. *Akzo.*

9119 Duoquad®
Diamine quaternary ammonium chloride, a surfactant. *Akzo Chemie UK Ltd.*

9120 Duoquad® O-50
68310-73-6 269-730-7
N,N,N',N',N'-pentamethyl-N-octadecenyl-1,3-diammonium dichloride
In aqueous isopropanol; industrial surfactant. *Akzo.*

9121 Duoquad® T-50
68607-29-4 271-762-1
N,N,N',N',N'-pentamethyl-n-tallow-1,3-propanediammonium dichlorides
In aqueous isopropanol; detergent, corrosion inhibitor, metal cleaner; emulsifier for secondary oil recovery. *Akzo.*

9122 Duoteric
Surfactant blend. *Rhône-Poulenc UK.*

9123 Duothane
Two-component hot cure polyurethane adhesive. *Compounding Ingredients Ltd.*

9124 Duovac-C
B_1 type, B_1 strain, Newcastle and Connecticut type; for immunization of poultry. *Intervet Inc.* Discontinued.

9125 Duovac-M
B_1 type, B_1 strain, Newcastle and mild Massachusetts type; for immunization of poultry. *Intervet Inc.*

9126 Duovac-Ma5
B_1 Type, B_1 strain Newcastle and Ma_5 Massachusetts type vaccine; for immunization of poultry. *Intervet Inc.*

9127 Dupical
4-Tricyclo(5.2.1.0. 2,6)-decylidene-8)-butanal. *Quest Int'l. UK Ltd.*

9128 Duplosan New System CMPP
93-65-2 202-264-4
mecoprop-P
Soluble concentrate containing 600 g/l mecoprop-P; for control of weeds in undersown cereals and grassland. *Rhône Poulenc Crop Protection Ltd.*

9129 Duplosan®
Translocated herbicide for cereals and grassland. *BASF plc.*

9130 Duplosan® CMPP
93-65-2 202-264-4
Mecoprop-P
Selective herbicide for control of broadleaf weeds in cereals, meadows, pastures *BASF AG.*

9131 Duplosan® DP
28631-35-8 249-110-2
Dichlorprop-P
Herbicide for control of broadleaf weeds in cereals *BASF AG.*

9132 Duponol
Surfactants. *DuPont UK.* Name unverified.

9133 Dupranin CR
Quaternary ammonium compound; leveling agent for dyeing with cationic dyes (retarder). Cationic, clear liquid. *Thor Chemicals (UK) Ltd.* Discontinued.

9134 Dupranin W
Alkyl amino polyglycol ether
Leveling agent for acid, 1:1 metallic complex and after-chroming dyestuffs, recleaning agent for dyed and printed goods. Nonionic; yellow liquid. *Thor Chemicals (UK) Ltd.* Discontinued.

9135 Duprene
A proprietary trade name for a synthetic rubber made by the polymerization of chloroprene; resistant to heat, oil, ozone, and most other chemicals. No manufacturer.

9136 Durabond
Calcium/magnesium lignosulfonate
Pellet binder in animal feeds; contributes some nutritive value. *Borregaard Ligno Tech.*

9137 Durabond 650, 655
Durable synthetic resin stiffener and hand modifier for finishing formulations. *Eastern Color & Chem.*

9138 Duracore
High strength aluminum honeycomb cores; for military and aerospace sandwich panel applications.

9139 Durad
Lubricant additives. *FMC.*

9140 Duradene® 706
Solution-polymerized B/S copolymer, nonstaining; used for mechanical goods. *Firestone synthetic Rubber.*

9141 Duradene® 711
Solution SBR, staining; used where abrasion resistance is important and low oil levels are required; suitable for tire treads, conveyor belting, molded goods. *Firestone Syn. Rubber.*

9142 Duradene® 750
Solution-polymerized B/S copolymer, 37.5 parts per hundred of aromatic oil, staining; for tire tread service and body sidewall compounding. *Firestone Syn. Rubber.*

9143 Duradene® 755
Solution-polymerized B/S copolymer, 20 parts per hundred of naphthenic oil, slightly staining; for tire stocks. *Firestone Syn. Rubber.*

9144 Duradiene®
Styrene/butadiene solution SBR; for tires, molded goods and adhesives. *Firestone Syn. Rubber.*

9145 Duraflex® 8410
9003-28-5
Polybutylene
Base polymer or modifier in adhesive and sealant formulations; modifier for atactic PP, olefin polymers, EPDM, elastomers; for packaging, transportation, construction, nonrigid bonding adhesives/sealants; improved hot melt performance. *Shell.*

9146 Durafoam
High yield lightweight grout for void filling. *Foseco (F.S.) Ltd.*

9147 Duraform
A proprietary trade name for asbestos reinforced thermoplastics; they have greater stiffness, lower coefficient of expansion, higher heat distortion point, lower creep, higher tensile and flexural strengths than the basic resins. No manufacturer.

9148 Duraguard
Polyester epoxy and epoxy polyester-based coating powders; RAL color range; used for electrostatic spray applications for decorative finishes, including wireworking, sheet metal fabrication and domestic wirework. *Plascoat Systems Ltd.*

9149 **Dural**
1344-28-1 369 215-691-6
aluminum oxide
Macro-crystalline regular aluminum oxide; used for abrasive industries, super refractories, sandblasting and for safes. *Lonza AG.*

9150 **Duralac**
Barium chromate jointing compound conforming to specification DTD 369A; for sealing of joints between dissimilar metals, protection of metals in contact with wood, synthetic resin compositions, leather, rubber, fabrics etc. *Liewellyn Ryland Ltd.*

9151 **Duralam®**
For the electrical industry. *DuPont UK.*

9152 **Duralcon**
Antistatic textile cohesive agent. *Stephenson Thompson Textile Chemicals.*

9153 **Duralit**
Dental stone. *Degussa Ltd.*

9154 **Duralium**
An alloy of aluminum with from 3.5-5.5% copper and small amounts of magnesium and manganese.

9155 **Duralkan K Concentrate**
Amine-amide-formaldehyde condensation product; improves the wet-fastness of dyeings and prints with direct and reactive dyes; fiber-affinitive dyestuff fixing agent. Cationic; clear yellowish liquid. *Thor Chemicals (UK) Ltd.* Discontinued.

9156 **Duralloy**
Dental silver alloys and mercury for mixing amalgams; for dentistry and dental engineering. *Degussa Ltd.*

9157 **Duralon**
A furan resin; vinyl plasticizer *U.S. Stoneware Co.* Name unverified.

9158 **Duraloy**
Alloyed thermoplastics; used for vehicle bumpers, high impact applications. *Hoechst-Celanese.* Name unverified.

9159 **Duralum**
An alloy of 79% aluminum, 10% copper, zinc, and tin.

9160 **Duralumin**
Alloys of aluminum with from 3-5.5% Copper, 0.5-1% manganese, 0.5% magnesium, and small quantities of silicon and iron. The alloy, containing 95.5% aluminum, 3% copper, 1% manganese, and 0.5% magnesium; resistant to sea water and dilute acids.

9161 **Duramite®**
471-34-1 1697 207-439-9
Calcium carbonate
Pigment with unique particle size distribution, easy dispersion; for mix-in application in carpet backing, roofing compounds, spackles, coatings. *ECC International Ltd.*

9162 **Durana metal**
An alloy of 65% copper, 30% zinc, 2% tin, 1.5% iron, and 1.5% aluminum, has a golden yellow color.

9163 **Duranalium**
An aluminum alloy similar in composition and properties to hydronalium and B.S. sea-water alloy.

9164 **Durance's metal**
Dewrance metal. A bearing metal, consisting of about 33% tin, 23% copper, and 45% antimony.

9165 **Durand's metal**
An alloy of 66.6% aluminum, and 33.3% zinc.

9166 **Duranic**
An alloy containing aluminum with from 2.5% nickel and 1.5-2.5% manganese.

9167 **Duranit**
A range of styrene/butadiene copolymers; used as reinforcing resins in rubber mixes and as reinforcing dispersion for natural and synthetic latex. *Hüls AG.*

9168 **Duranite**
A proprietary trade name for a fast-baking synthetic enamel. No manufacturer.

9169 **Duranox**
amitrole-atrazine-diuron
Amitrole, atrazine and diuron; a liquid mixture of herbicides for weed control. *Agri-Technics Ltd.*

9170 **Duranthrene® Dyes**
A trade name for certain British dyestuffs. No manufacturer.

9171 **Duraplex®**
Synthetic resins, common metals and their alloys. *Reichhold.*

9172 **Duraplus® 1**
Emulsion polymer, emulsion for ultra-high wear floor polishes with exceptional buffability. *Rohm & Haas.*

9173 **DuraSoft**
Phemfilcon A; contact lens material. *Wesley Jessen.* Name unverified.

9174 **Durasol Acid Blue B**
The British brand of Alizarin saphirol B; a dyestuff. No manufacturer.

9175 **Durastat® AS-5760**
Highly loaded static dissipative concentrate for LDPE, LLDPE. *PPG/Polymer Prods.*

9176 **Durastat® AS-5814-2**
Antistat concentrate for dry foods which can be let-down in HDPE in extrusion process; prevents dust buildup during extrusion and molding. *PPG/Polymer Prods.*

9177 **Durastat® AS-5903-3**
Highly loaded flame retardant and static dissipative film extrusion concentrate which can be letdown in polyethylene at 8.20%. *PPG/Polymer Prods.*

9178 **Durastic**
A bitumen compound stated to be composed of high grade bitumen freed from organic acids and used as a protective coating. No manufacturer.

9179 **Durastrength 200**
Acrylic impact modifier; impact modifier for PVC, exterior durable building products. *Elf Atochem.*

9180 **Duratex**
68334-00-9 269-804-9
Duromel B108; 68334-00-9. Partially hydrogenated cottonseed oil; lubricant for tablets; tablet release; candy dusting and coating for hydroscopic materials. *Van Den Bergh Foods.*

9181 **Durax®**
95-33-0 202-411-2
$C_{13}H_{16}N_2S_2$
N-Cyclohexyl-2-benzothiazolesulfenamide
Sufenax CB; Rhodifax 16; Accelerator CZ; Durax. Primary delayed action accelerator used in both natural and synthetic rubbers; safe at processing temperatures, fast at curing temperatures. *Goodyear; R. T. Vanderbilt Co Inc.*

9182 **Durazol**
Direct dyestuffs. *ICI Chem & Polymers Ltd.* Discontinued.

9183 **Durbar Bronze**
A proprietary trade name for an alloy of copper with 24% lead and 4% tin. No manufacturer.

9184 **Durbar Hard Bronze**
A proprietary trade name for an alloy of copper with 10% tin and 20% lead. No manufacturer.

9185 **Durcoton**
A proprietary phenol-formaldehyde resin impregnated textile. No manufacturer.

9186 **Durecol**
A proprietary glycero-phthalic acid synthetic resin. No manufacturer.

9187 **Durehete 900**
A proprietary trade name for steel containing 1% chromium and 1/2% molybdenum. It is suitable for studs and bolts for service at temperatures up to 900°F (482°C). *Samuel Fox & Co Ltd.* Unverified.

9188 **Durehete 950**
A proprietary trade name for a steel containing 1% chromium, 1/2% molybdenum and 1/4% vanadium for studs and bolts for service at temperatures up to 950°F (510°C). *Samuel Fox & Co Ltd.* Unverified.

9189 **Durehete 1050**
A proprietary trade name for a steel containing 1% chromium, 1% molybdenum and 3/4% vanadium for bolting materials capable of operating at metal temperatures up to 1050°F. *Samuel Fox & Co Ltd.* Unverified.

9190 **Durel**
Polyarylate
Used for microwave cookers, vehicle light lens and traffic light lens; customized product. *Hoechst AG.*

9191 **Durelast**
Moisture cure polyurethane adhesive. *MacPherson Polymers.*

9192 **Durelast®**
Polyurethane prepolymer
Used for adhesives, sealants and surface coatings. *Hyperlast.*

9193 **Dur-Em® 117**
Glyceryl stearate. Textile lubricant and finishing agent; emulsifier for cosmetic and pharmaceutical creams and lotions, foods; lubricant for thermoplastics; dispersant for inorganic pigments. *Van Den Bergh Foods.*

9194 **Dur-Em® GMO**
Glyceryl oleate
Solubilizer, dispersant, lubricant, wetting aid, penetrant for foods, personal care products, dry cleaning bases, paints, and insecticides. *Van Den Bergh Foods.*

9195 Durene
95-93-2 3520 202-465-7
$C_{10}H_{14}$
1,2,4,5-Tetramethylbenzene
Durol. 1,2,4,5-tetramethylbenzene, used in organic synthesis, plasticizers, polymers, fibers. mp = 80-82°; bp = 196-197°; d = 0.8380; insoluble in H_2O, soluble in organic solvents; LD_{50} (rat orl) = 6989 mg/kg.

9196 Durethan®
Polyamide 6 and 6/6, copolyamide and amorphous polyamide; engineering thermoplastic for the manufacture of injection moldings; some grades also for film production; high stiffness and hardness, good impact strength; glass fiber reinforced, glass sphere and mineral-filled grades; polymer and elastomer modified formulations; used for electrical engineering/electronics, power tools, domestic appliances, automotive industry, mechanical engineering, furniture, sporting goods, toys, semi-finished products. *Bayer AG.*

9197 Durethan® A 30 S
32131-17-2
Nylon 66
Engineering plastic (electrical, mechanical, precision, automotive), household appliances, building construction, furniture manufacture, and packaging applications; general-purpose grade for injection moldings (fast cycles, easy release). *Bayer AG.*

9198 Durethan® B 30 S, B 31 SK
25038-54-4 6832
Nylon 6
B 30 S and B 31 SK are used for injection moldings with fast cycles; B 31 SK has a higher impact strength. *Bayer; Miles.*

9199 Durethan® B 35 F, B 38 F, B 40 F
25038-54-4 6832
Nylon 6
Extrusion grades for use in flat and blown film. *Bayer AG.* Name unverified.

9200 Durethan® BKV
Glass-filled polyamide 6. *Bayer AG.* Name unverified.

9201 Durethan® BKV 115
25038-54-4 6832
Nylon 6, 15% glass-reinforced. *Miles.* Discontinued.

9202 Durethan® BKV 30 H
25038-54-4 6832
Nylon 6 with 30% glass fiber and heat stabilizer; used for injection moldings exposed to elevated temperatures; stiffness, and hardness. *Bayer; Miles.*

9203 Durethan® BM 30 X
25038-54-4 6832
Nylon 6, 30% glass fiber/mineral-reinforced; engineering plastic (electrical, mechanical, precision, automotive), household appliances, building construction, furniture manufacture, and packaging applications; general-purpose grade for injection moldings (fast cycles, early release). *Bayer; Miles.*

9204 Durethan® KL1-2402/30
25038-54-4 6832
Nylon 6, impact-modified, 30% mineral-reinforced; injection molding grade for applications requiring impact strength, stiffness, heat resistance and dimensional stability, e.g., automotive fans, shrouds, mirror housings, power tools, furniture components, appliances, impeller blades, motor gears. *Miles.* Discontinued.

9205 Durethan® RM KU 2-2501/30
25038-54-4 6832
Nylon 6, 30% glass-reinforced, reduced moisture; reduced moisture resin for electrical housings, automotive parts, consumer products, office chair seats, lawn/garden equipment, power tool housings, hydraulic cylinder components. *Miles.* Discontinued.

9206 Durethane
A proprietary range of polyurethane thermoplastics used in injection molding. *Bayer AG.* Name unverified.

9207 Durex
A proprietary trade name for an alloy of 83% copper, 10% tin, and 4-5% carbon; also A proprietary trade name for phenol-formaldehyde synthetic resin. No manufacturer.

9208 Durex white
513-77-9 996 208-167-3
$BaCO_3$
Barium carbonate
Barium carbonate, used in treatment of brines in chlorine-alkali cells to remove sulfates, rodenticide, barium salts, ceramic flux, optical glass, case-hardening baths.

9209 Durez® 115
Two-stage phenolic resin; general-purpose thermoset resin for applications including automotive transmission parts and braking systems; for compression, transfer, and injection molding applications. *OxyChem/Durez.*

9210 Durez® 123
Phenolic resin; impact-grade resin; used in compression, transfer, and injection molding; high shock resistance. *OxyChem/Durez.*

9211 Durez® 152
Phenolic resin; Electrical, asbestos-free resin for molded products requiring heat resistance and dimensional stability; used in compression, transfer, and injection molding. *OxyChem/Durez.*

9212 Durez® 18420
Phenolic resin; specialty molding material used in compression molding. *OxyChem/Durez.*

9213 Durez® 24150
Alkyd resin; molding resin for arc-track resistance (retains electrical properties at elevated temperatures); used in transfer and injection moldings. *OxyChem/Durez.*

9214 Durez® 25000
Two-step phenolic molding compound; general purpose molding material for injection molding; good balance of mechanical and electrical properties. *OxyChem/Durez.*

9215 Durez® 32633
Two-step phenolic molding compound, glass-filled; molding compound for applications requiring high strength, dimensional stability, and heat resistance; for small motor and gear housings, brush holders, commutators, underhood automotive parts. *OxyChem/Durez.*

9216 Durfax® 60
9005-67-8 8872
Polysorbate 60
Food emulsifier; personal care products; preshave beard lubricant and softener products *Van Den Bergh Foods.*

9217 Durfax® 65
9005-71-4 8872
Polysorbate 65
Food emulsifier; pesticide dispersant. *Van Den Bergh Foods.*

9218 Durfax® 80
9005-65-6 7742
Polysorbate 80
Food emulsifier; personal care products; antifog agent in plastics and aerosol furniture polish. *Van Den Bergh Foods.*

9219 Durferrit
Carburizing, nitriding, annealing, hardening, tempering and heat transfer salts; used for heat treatment of metal in the machine, machine tool, motor vehicle, aircraft and other metal working industries. *Degussa Ltd.* Discontinued.

9220 Durham's stain
A microscopic stain. It contains a saturated solution of stannous chloride and a 15% solution of tannic acid in equal parts, with a few drops of an alcoholic solution of methylene blue.

9221 Durichlor
A hydrochloric acid resisting alloy containing 81% iron, 14.5% silicon, 3.5% molybdenum, and 1% nickel.

9222 Duridine
$C_{10}H_{15}N$
Tetramethylphenylamine

9223 Durifan AR30, BK 30
Silicic acid dispersions; antislipping agents, especially for linings made of natural and synthetic fibers. *Thor Chemicals (UK) Ltd.* Discontinued.

9224 Durimet Alloys
Proprietary alloys for acid resistance. Alloy A is stated to contain iron with 25% nickel, no chromium, and 5% silicon; B contains iron with more nickel than A, and with chromium content about one-third of the nickel, and 5% silicon; alloy D contains iron with 15% nickel and smaller amounts of chromium and silicon. No manufacturer.

9225 Durine
50-00-0 4262 200-001-8
formaldehyde
A formalin preparation used as an antimicrobial and antiseptic..

9226 Duriron
An acid resisting alloy, which is a silicon-iron alloy. It contains 15.5% silicon, 82% iron, 0.66% manganese, 0.83% carbon, and 0.57% phosphorus.

9227 Durisol
A trademark for products of wood or paper impregnated with synthetic resin. No manufacturer.

9228 Durkex 500
68334-28-1 269-820-6
Partially hydrogenated vegetable oil (cottonseed, soybean); kosher; used where high stability is needed; for coating/spraying (dried fruit, crackers), frying/roasting (nuts), antidusting (gravies, soups), color/flavor carrier,

lubricant, moisture barrier, mineral oil replacement, ingredient in prepared foods. *Van Den Bergh Foods.*

9229 Durkex Durola
Partially hydrogenated canola oil (kosher); for coating/spraying (crackers, snacks); food ingredients where enhanced nutritional profile is desired. *Van Den Bergh Foods.*

9230 Durko
68334-28-1 269-820-6
Partially hydrogenated vegetable oil (soybean, cottonseed), mono and diglcyerides; kosher; shortening used in yeast-raised sweet goods. *Van Den Bergh Foods.*

9231 Durlac® 100W
Glyceryl stearate lactate; emulsifier for food; confectionery gloss enhancer; starch gelling agent in industrial processes. *Van Den Bergh Foods.*

9232 Durlite F
68334-28-1 269-820-6
Partially hydrogenated vegetable oil (soybean, cottonseed); kosher; multipurpose shortening for vegetable dairy, bakery processed food. *Van Den Bergh Foods.*

9233 Durlite Gold MBN II
Partially hydrogenated vegetable oil (soybean, cottonseed), lecithin, TBHQ, β carotene, natural flavors; kosher; shortening system for bakery items, replacement for cholesterol-and tropical oil-containing shortening systems. *Van Den Bergh Foods.*

9234 Dur-Lo®
Mono-and diglycerides with BHA and citric acid; food emulsifier for fat-reduced foods; fat replacement or reduction in sour dressings and other vegetable dairy systems, bakery cake mixes. *Van Den Bergh Foods.*

9235 Duro cement
A cement used in the manufacture of acid towers. It contains 96% silica and 4% sodium silicate.

9236 Durocide
Biocides. *Durham Chemicals Ltd.*

9237 Durodi Steel
A proprietary trade name for a nickel-chromium-molybdenum steel. No manufacturer.

9238 Durofer
Nontoxic carburizing process. *Degussa Ltd.*

9239 Duroftal
A proprietary synthetic resin. No manufacturer.

9240 Duroglass
A proprietary borosilicate resistance glass for chemical use. No manufacturer.

9241 Duroil
Self-emulsifiable oil. *Stephenson Thompson Textile Chemicals.*

9242 Durola Select
Partially hydrogenated canola oil; kosher; shortening for cookies, crackers, biscuits; filler fat for cream centers, whipped toppings. *Van Den Bergh Foods.*

9243 Durolastik
Waterproofing, roofing materials. *Weatherguard/Marbleloid Products Inc.* Name unverified.

9244 Duroloy®
Self-lubricating ball bearings with thermoplastic polyimide retainers; for photodiode manufacturing equipment. *Rogers.*

9245 Durolube
Textile yarn lubricant. *Stephenson Thompson Textile Chemicals.*

9246 Duromel
68334-00-9 269-804-9
Partially hydrogenated cottonseed oil; kosher; high performance fat for vegetable dairy, confectionery, bakery. *Van Den Bergh Foods.*

9247 Duromel B108
68334-00-9 269-804-9
Partially hydrogenated cottonseed oil; kosher; coating fat for bakery coatings. *Van Den Bergh Foods.*

9248 Duronze
A proprietary trade name for a high-silicon copper alloy. No manufacturer.

9249 Durophen
A trade name for a series of plasticized phenolic resins widely used in baking finishes. *Vianova Resins.* Discontinued.

9250 Durophen 127-B
An ammonia condensed phenol formaldehyde resin melting at 55°. *Vianova Resins.* Discontinued.

9251 Durophen 170W
A butylated diene formaldehyde condensate heated with trimethylene glycol maleate (65% solids). *Bush Beach Ltd.* Name unverified.

9252 Durophen 218V
A butylated diene formaldehyde castor oil resin. *Vianova Resins.* Discontinued.

9253 Durophen 287W
A butylated phenol urea formaldehyde resin sold at 58% solids. *Vianova Resins.* Discontinued.

9254 Durophen 309W
A butylated xylenol formaldehyde resin containing butyl glyceryl adipate. *Vianova Resins.* Discontinued.

9255 Durophen 330V
A butylated diene formaldehyde resin cooked with glyceryl phthalate and synthetic fatty acids. *Vianova Resins.* Discontinued.

9256 Duroplaz 610, 810, 911
Proprietary trade names for phthalate esters of straight chain alcohols. Used as plasticizers. No manufacturer.

9257 Duroprene®
A registered trade name for a product obtained by the exhaustive chlorination of natural rubber; can be molded, and is soluble in benzene, coal-tar naphtha, and carbon tetrachloride; it is resistant to chemical action and is used in paints and varnishes. No manufacturer.

9258 Duroseal
637-12-7 379 211-279-5
aluminum stearate
Modified aluminum stearate, for waterproofing. *Durham Chemicals Ltd.*

9259 Durosehl
Epoxy and polyurethane resin systems; for casting and laminated structures, engineering patterns and toolmaking, electrical encapsulations. *Solochart Ltd.* Name unverified.

9260 Durosil
Precipitated silica; low surface area reinforcing silica for rubber compounds. *Degussa.*

9261 Duroslip
Antistatic textile fibers lubricant. *Stephenson Thompson Textile Chemicals.*

9262 Durosoft
Fiber softener. *Stephenson Thompson Textile Chemicals.*

9263 Durosol
Self emulsifiable oil. *Stephenson Thompson Textile Chemicals.*

9264 Durostabe
Stabilizers for PVC. *Durham Chemicals Ltd.* Name unverified.

9265 Duroterm
Investment material; for casting precious metal alloys; used in dentistry. *Bayer AG.*

9266 Durotex
Yarn strengthening agent. *Stephenson Thompson Textile Chemicals.*

9267 Durotex 7603
Antimicrobial for latex carpet backing, dry film preservative. *Morton Int'l./Plastics Additives.*

9268 Durotint
Textile fugitive tints. *Stephenson Thompson Textile Chemicals.*

9269 Durowynd
Hydrophilic wetting agent. *Stephenson Thompson Textile Chemicals.*

9270 Durox
A mullite made by fusing kyanite and alumina; A proprietary trade name for an ammonium dynamite; an explosive. No manufacturer.

9271 Duroxyn
Epoxide resins esterified with fatty acids; for chemical resistant paints. *Resinous chemicals Ltd.* Discontinued.

9272 Dursban®
2921-88-2 2242 220-864-4
Chlorpyrifos
Chlorpyrifos; an organophosphate insecticide for control of ticks, mosquitoes, and other insects. *DowElanco Ltd.*

9273 Dursban® 14G
2921-88-2 2242 220-864-4
chlorpyrifos
Chlorpyrifos-based insecticide used for control of white grubs in sugar cane *Incitec Ltd.*

9274 Dursban® 2E
2921-88-2 2242 220-864-4
Chlorpyrifos
Insecticide containing 2 lb/gal of chlorpyrifos; for broad spectrum control of chinch bugs, sod web worms, ants, and earwigs. *Dow; W.A. Cleary.*

9275 Durtan® 60
1338-41-6 8872 215-664-9
Sorbitan stearate; food emulsifier, gloss enhancer for chocolate coatings; dispersant for inorganics used in thermoplastics. *Van Den Bergh Foods.*

9276 Dustallay
Wetting system for reducing dust in mines. *Foseco (F.S.) Ltd.*

9277 Dustex
Lignosulfonate
Biodegradable, environmentally safe dust binder and road stabilizer. *Borregaard Ligno Tech.*

9278 Dutch camphor
Obtained from the wood of the Japanese camphor laurel, *Cinnamomun camphora.*

9279 Dutch metal
An alloy of 80% copper and 20% zinc.

9280 Dutch pink
A pigment. It is a yellow color made by absorbing quercitron on barytes or alumina.

9281 Dutch varnish
A solution of rosin in turpentine.

9282 Dutch white
A pigment consisting of 1 part white lead with 3 parts heavy spar.

9283 Du-Ter
76-87-9 9875 200-990-6
fentin hydroxide
Fungicide containing fentin hydroxide; for prevention of potato blight and disease control in sugar beet. *ICI Chem. & Polymers Ltd; Chiltern Farm Chemicals Ltd.* Name unverified.

9284 Du-Ter®
76-87-9 9875 200-990-6
Triphenyltin hydroxide
Flowable fungicide for pecans, potatoes, sugarbeets; restricted use. *Griffin.*

9285 Duthane
A proprietary polyester-based polyurethane elastomer cross-linked with diols. No manufacturer.

9286 Dutral
A proprietary trade name for an ethylene-propylene synthetic rubber copolymer suitable for tank linings, seals, hose, cables. *Shell.* Name unverified.

9287 Dutral-Co
A proprietary range of ethylene-propylene copolymers. *Montedison UK Ltd.* Name unverified.

9288 Dutral-Ter
A proprietary range of ethylene-propylene-diene terpolymers (EPDM). *Montedison UK Ltd.* Name unverified.

9289 Dutrex
Hydrocarbon oils. *Shell UK.* Name unverified.

9290 Dutrex 20, 25
Extender-plasticizer oil; vinyl resin additive and plasticizer *Shell.*

9291 Dutrex Process and Extender Oils
Aromatic extracts produced during the refining of lubricating oils; by suitable selection and blending, a range of Dutrex grades with varying viscosities are obtained possessing good solvent characteristics and excellent polymer compatibility. *Shell.* Name unverified.

9292 Duxalid
A proprietary trade name for a synthetic resin. No manufacturer.

9293 Duxlte
An explosive containing nitro-glycerin, collodion cotton, sodium nitrate, wood meal, and ammonium oxalate; also the name applied to a resin from lignite.

9294 Duxol
A proprietary trade name for a synthetic resin. No manufacturer.

9295 DV-1801
115047-92-2
Behenyl polyethoxyethyl methacrylate
Surfactant. *Rhône-Poulenc Surf.*

9296 DV-1936
1462-55-1 215-969-7
2-(Dodecylthio)ethanol
Surfactant. *Rhône Poulenc Surf.*

9297 DV-2301
6281-42-1 228-491-9
$C_5H_{11}N_3O$
1-(2-Aminoethyl)-2-imidazole
Aminoethyl ethylene urea, a surfactant. *Rhône-Poulenc Surf.*

9298 D-Visor
Deodorant for industrial processes and products. *CPL Group Ltd.*

9299 D-Wax
Wax inhibitors and removers. *Baker Performance Chemicals Ltd.*

9300 DY 023
2210-79-9 218-645-3
$C_{10}H_{12}O_2$
1-(2-methylphenoxy)-2,3-epoxypropane
o-cresyl glycidyl ether; ((2-methylphenoxy)methyl)oxirane; cresol, o-epoxypropyl ether; o-tolyl epoxypropyl ether; o-tolyl glycidyl ether; Cresyl glycidyl ether. Diluent for epoxy. bp = 259°; d = 1.09 g/ml; insoluble in H_2O, soluble in organic solvents. *Ciba-Geigy/Plastics.* Name unverified.

9301 Dyafac PEG 6DO
9005-07-6
PEG-12 dioleate
Surfactant for textile use. *Henkel/Textiles.*

9302 Dyamul
A range of dye bath auxiliaries, including detergents, buffers, solvents, lubricants and combinations of these; assistants in textile processing and dyeing. No manufacturer.

9303 Dyapol
Dispersants, levelers, and retardants for disperse dyeing. No manufacturer.

9304 Dybin®
For the chemical industry. *DuPont UK.*

9305 Dycastal
Shrinkage prevention in aluminum diecastings. *Foseco (F.S.) Ltd.*

9306 Dy-Chek
Four step process for detecting surface flaws in metallic and nonmetallic components. *Foseco (F.S.) Ltd.*

9307 Dycill
13412-64-1 3134
Dicloxacillin sodium
Antibacterial. *SmithKline Beecham.*

9308 Dycote
Foundry die coatings. *Foseco (F.S.) Ltd.*

9309 Dycron
1344-28-1 369 215-691-6
Crystalline aluminum oxide
For lapping of gearings and hydraulics, motors, electronics, glass etc. *Dynamit Nobel Wien GmbH.* Name unverified.

9310 Dye Retarder 1
61789-77-3 263-087-6
Dicoco dimethyl ammonium chloride
Retardant for acrylics; oil and solvent emulsifier. *See Dicocodimonium chloride Sherex/Div. of Witco.*

9311 Dyeset® 100 Conc
Resin polymer; dye fixative for direct dyes for exhaust last rinse application for textile finish formulation. *Sybron.*

9312 Dyetone®
7789-38-0 8736 232-160-4
sodium bromate
Aqueous solution of sodium bromate; dye oxidant for vat and sulfur dyes. *Olin.* Discontinued.

9313 Dyeweld SUPR
Fixative for acid dyes; for exhaust application after dyeing. *Sybron.*

9314 Dyflor 2000
Polyvinylidene fluoride
PVDF. For injection, extrusion and blow molding, manufacture of chemical apparatus, mechanical engineering, cable industry, electronics, manufacture of medical equipment. *Dynamit Nobel Wien GmbH.*

9315 Dyflor L90
A proprietary polyvinyl fluoride; processed in dispersion form. No manufacturer.

9316 Dyfonate
944-22-9 4261 213-408-0
Fonofos
Fonofos; soil and seed insecticide. *Farm Protection Ltd.*

9317 Dykor 204
24937-79-9
Polyvinylidene fluoride
Primer coating containing mica to enhance adhesion at metal/coating interface; protective coating for lining chemical processing vessels, acid etching in manufacture of electronic components. bleaching operations in pulp/paper mills, corrosion protection. *Whitford.*

9318 Dylark® 132
9011-13-6
SMA-based engineering copolymer; resin offering excellent clarity. *Arco Chemical Co.*

9319 Dylene
9003-53-6 9028
A proprietary polystyrene. *Arco Chemical Co.*

9320 Dylite® D195B
9003-53-6 9028

Expandable PS; for dynamic cushioning applications, insulation, foam drink cups, packaging. *Arco Chemical Co.*

9321 Dylite® R2595B EPS
9003-53-6 9028
Dylite® D195B. Expandable PS with 25% recycled content; resin for cushioning, insulation, protective packaging applications. *Arco Chemical Co.*

9322 Dylon
Colloidal graphite in aqueous or solvent carriers; high temperature parting agent, release compounds, with good electrical and thermal conductivity. *Dylon Industries Inc.* Name unverified.

9323 Dylonite
Resin impregnated impervious graphite; for heat exchangers, tubes and pumps with excellent resistance to high temperature corrosive chemicals. *Dylon Industries Inc.* Name unverified.

9324 Dylox®
52-68-6 9753 200-149-3
Trichlorfon
Insecticidal spray containing Trichlorfon; used for control of mangold fly, fruit fly and pests of maize, alfalfa, and cotton. *Bayer AG.*

9325 Dylux®
DuPont UK.

9326 Dymacryl
100% Acrylic copolymers, color stable inorganic pigments and surface penetrating agents; available in 10 colors plus clear; beautifies and protects above-grade masonry surfaces against damage caused by water absorption; used on architectural concrete, precast and poured concrete, glass fiber-reinforced concrete, brick, natural stone, stucco and unglazed tile. *Dampney Company Inc.* Name unverified.

9327 Dymanthine
124-28-7 3525 204-694-8
$C_{20}H_{43}N$
N,N-Dimethyl-1-octadecanamine
Barlene® 18S; Stearyl dimethyl amine; Dimantine; 1-Octadecanamine, N,N-dimethyl-; Dimethylstearylamine; N,N-Dimethyloctadecylamine; Armeen DM 18D; Dimethyl-1-octadecanamine; Dimantine. Chemical intermediate; personal care additive. Also used as an anthelmintic. Hydrochloride salt: GS-1339; NSC-5547; Thelmesan. mp = 23°; d = 0.800. *Lonza Inc.*

9328 Dymax® Light-Weld® Adhesives
Aerobic adhesives which cure by exposure to ultraviolet light; used for medical devices, electronic assembly glass and plastics. *Dymax Corp.*

9329 Dymax® Light-Welder™
Ultraviolet light sources including flood lights, focus beam lights, spot lights and a combination spot light/dispenser. *Dymax Corp.*

9330 Dymax® Multi-Cure® Adhesives
Aerobic adhesives which cure by exposure to ultraviolet activator or heat; used for magnet, metal, fiber optics, assembly adhesives and coatings for electronics. *Dymax Corp.*

9331 Dymel®
115-10-6 6148 204-065-8
methyl ether
Methyl ether used as a propellant. *DuPont UK.*

9332 dymerex
Partially dimerized rosin acids; used to reinforce specialty adhesive products, in specialty protective coatings. *Hercules.*

9333 Dymetrol®
Fiber/polymer. *DuPont UK.*

9334 Dymsol® 38C
A proprietary anionic, biodegradable polymerization emulsifier for improving the processing characteristics of SBR, nitrile rubber and neoprene. *Henkel Inc.* Name unverified.

9335 Dymsol® 2031
Sulfonated fatty acid, sodium salt; primary or secondary emulsifier for emulsion polymerization. *Henkel/Functional Prods.; Henkel-Nopco.*

9336 Dymsol® PA
Sulfated ester; primary emulsifier for emulsion polymerization. *Henkel/Functional Prods.*

9337 Dynacal
1305-78-8 1733 215-138-9
Fused calcium oxide
Additive for the melting of high purity metallurgical products; raw material for the electroslag remelting process; auxiliary material for metallurgical desulfurization processes. *Hüls AG.* Discontinued.

9338 Dynacast
Fusion cased bricks; for the lining of slab pusher furnaces, billet pusher furnaces, ingot pusher furnaces, forging furnaces, rocker bar furnaces, roller hearth furnaces, soaking pit furnaces, tundishes of continuous casting plants, steel degassing plants and high temperature furnaces. *Hüls AG.* Discontinued.

9339 Dynacerin®
17673-56-2 241-654-9
Oleyl erucate
Jojoba oil substitute, liquid wax ester; oily component for emulsions with emollient properties in pharmaceuticals and cosmetics. For ointments, creams, liquid emulsions, external suspension, skin lotions, bubble baths and shampoos. *Hüls AG.*

9340 Dynacerin® 660
17673-56-2 241-654-9
Oleyl erucate
Jojoba oil substitute; emollient for cosmetic preparations. *Hüls Am.*

9341 Dynacet®
91723-32-9 294-538-5
Acetylated monoglycerides of edible fatty acids; food emulsifier; coating agent for shortenings, toppings, meat and sausages. *Hüls Am; Hüls AG.*

9342 Dynacoll
Reactive polyesters; adhesive raw material for the formulation of reactive hot melts. *Hüls AG.*

9343 Dynaflex
Two-part polyurethane sealant; used as a security sealant for applications in institutional and correctional complex installations wherever the sealant may be exposed to physical abuse. *Pecora Corporation.*

9344 Dynaflock
1302-42-7 8715 215-100-1
Sodium aluminate
Caustic product for paper industry water treatment, ceramic industry and building industry. *Hüls AG.*

9345 Dynaflush
Nonchlorinated, nonflammable, noncarcinogenic, non-ozone depleting solvent for flushing and cleaning residues from equipment; used in the urethane foam industry. *Dynaloy.*

9346 Dyna-Form
50-00-0 4262 200-001-8
Formaldehyde
Soil sterilant and fumigant for glass houses. *Fargro Ltd.*

9347 Dynaglaze
Polishes and valeting products for cars. *Spectra Brands plc.*

9348 Dynagrout
1302-42-7 8715 215-100-1
Sodium aluminate
Reactive agent for forming injection gels (grout), for the mining and building industry. *Hüls AG.*

9349 Dynagunit
Alkali aluminate formulations; liquid concrete setting accelerator for the mining and building industry. *Hüls AG.*

9350 Dynamag
1309-48-4 5713 215-171-9
Electromagnesia (magnesium oxide)
Raw material for ramming mixes and shape bricks for the lining of high temperature furnaces; raw material for welding powder and the coating of electrodes. *Hüls AG.*

9351 Dynamar® Brand Specialities
A line of speciality chemicals used in the manufacture/processing of elastomer and plastics compositions. *3M.*

9352 Dynamar® FC
Fluorochemical; mold release agent for hot molding operations; especially useful with silicone, ethylene acrylic and fluorosilicone elastomers. *3M.*

9353 Dynamar® PPA-790
Polymer processing additive providing enhanced release and flow in molding and extrusion of elastomer-based compounds *3M.*

9354 Dynamask
Dry film solder mask. *Morton Int'l. Ltd.*

9355 Dynamine
102-06-7 3383 203-002-1
$C_{13}H_{13}N_3$
diphenylguanidine
1,3-Diphenylguanidine; DPG; Melaniline; N,N'-Diphenylguanidine; Nocceler D; Sanceler D; Soxinol D; *sym*-diphenylguanidine; Vulkazit; DPG Accelerator; Vulcacid D; Vulkacit D/C. Diphenylguanidine; a rubber vulcanization accelerator. mp = 148-150°; insoluble in H_2O, soluble in organic solvents; LD_{50} (rat orl) = 375 mg/kg.

9356 dynamite
Kieselguhr dynamite
Principal ingredient is nitroglycerin, absorbed in Kieselguhr; industrial explosive detonated by blasting caps.

9357 dynamite acid
7697-37-2 6671 231-714-2

nitric acid
Concentrated nitric acid; used for making 96% mixed acids (34% nitric acid + 62% sulfuric acid).

9358 dynamite glycerin
56-81-5 4493 200-289-5
glycerol
Glycerin of specific gravity 1.263, containing 98-99%. It contains no lime, sulfuric acid, chlorine, or arsenic.

9359 Dynamites
Explosives first patented by Nobel, which consist of nitroglycerine rendered shock-stable by absorption onto Kieselguhr or some other absorbent.

9360 Dynamullit
12141-46-7 377 235-253-8
aluminum silicate
Fused mullite. For production of shape bricks for high-charged zones in heating and melting furnaces; molding material for precision casting molds. *Hüls AG.*

9361 Dynamyte
88-85-7 3341 201-861-7
Dinoseb
Herbicide for use on beans, small grains, forage, cereal crops. *Draxel Chemical Company.* Unverified.

9362 Dynapol® H
Low-molecular, saturated polyesters; used for the production of high-quality baking enamels; industrial enamels, enamels for household appliances and metallic finishes for vehicles. *Dynamit Nobel Wien GmbH; Hüls UK Ltd.*

9363 Dynapol® L
High molecular copolyesters; raw material for coil coating, can coating, metal decorating. *Dynamit Nobel Wien GmbH.* Name unverified.

9364 Dynapol® L
High-molecular, copolyesters, raw material for coil coating, can coating, metal decorating. *Hüls UK Ltd.*

9365 Dynapol® L 205
Thermoplastic polyester; for corrosion resistant primers and stamping enamels. *Hüls Am.*

9366 Dynapol® LH
Medium and low-molecular, saturated polyesters; particularly suitable for coatings coming in contact with foodstuffs; raw material for coil coating, metal decorating and adhesives. *Dynamit Nobel Wien GmbH.* Name unverified.

9367 Dynapol® LH
Low-molecular, saturated polyesters; particularly suitable for coatings coming into contact with foodstuffs; raw material for coil coating, metal decorating, and adhesives. *Hüls UK Ltd.*

9368 Dynapol® P
Thermoplastic copolymers; suitable where protection against corrosion and good resistance to aggressive chemicals are required. *Dynamit Nobel Wien GmbH.* Name unverified.

9369 Dynapol® P
Thermoplastic copolymers; used where protection against corrosion and good resistance to aggressive chemicals are required. *Hüls UK Ltd.*

9370 Dynapol® S
Thermoplastic, linear, saturated copolyesters; adhesive raw materials for hot melts. *Dynamit Nobel Wien GmbH.* Name unverified.

9371 Dynapol® S
Thermoplastic, linear saturated copolyesters; adhesive raw material for hot melts. *Hüls UK Ltd.*

9372 Dynapor
Phenolic foaming resins; flame inhibiting and rotproof insulating material for ceilings in flat roof construction, for wall and facade paneling or pipe sheatings. *Hüls AG.*

9373 Dynasan®
Fatty acid esters; for use as lubricants and retarding agents in tablets, crystallization accelerator in suppositories and chocolate, consistency regulators in ointments, creams and lotions, basic material for fatty powders. *Hüls AG.*

9374 Dynasan® 110
621-71-6 210-702-0
Tricaprin
Emollient, lubricant, consistency regulator for cosmetic applications such as sticks, creams, lotions, powds.; for prep. of margarine, confectionery, milk products, fruit diets. *Hüls Am.*

9375 Dynasan® 112
538-24-9 208-687-0
Trilaurin
Consistency regulator for cosmetic creams, lotions; lubricant, powder base, stick component *Hüls Am.*

9376 Dynasan® 114
555-45-3 9854 209-099-7
Trimyristin
Binder, lubricant for tablets and compressed confectioneries; consistency regulator for creams and lotions; lubricant, powder base, stick component. *Hüls Am.*

9377 Dynasan® 116
555-44-2 9865 209-098-1
Tripalmitin
Lubricant, consistency regulator in cosmetic powders, cakes, and production of tablets; powder and makeup base; stick component. *Hüls Am.*

9378 Dynasan® 118
555-43-1 9885 209-097-6
Tristearin
Crystallization accelerator in chocolate; lubricant in cosmetic powders, cakes, and production of tablets. *Hüls Am; Hüls AG.*

9379 Dynasan® P60
Hydrogenated palm oil; consistency regulator for creams, lotions, makeup, decorative cosmetics; stabilizer for hindering oiling out of an emulsion. *Hüls Am.*

9380 Dynasil®
Silicic acid esters; binders for the foundry industry and for inorganic zinc dust coatings. *Hüls AG.*

9381 Dynasil® 40
Polydiethoxysiloxane
Silicone dioxide source. *Hüls Am.*

9382 Dynasil® A
78-10-4 3895 201-083-8
Tetraethoxysilane
Coupling agent, chemical intermediate, blocking agent, release agent, lubricant, primer, reducing agent. *Hüls Am.*

9383 Dynasil® CA
18407-94-8 242-287-7
Tetrakis (2-ethoxyethoxy)silane
Coupling agent, chemical intermediate, blocking agent, release agent, lubricant, primer, reducing agent *Hüls Am.*

9384 Dynasil® CM
Tetrakis (2-methoxyethoxy) silane
Coupling agent, chemical intermediate, blocking agent, release agent, lubricant, primer, reducing agent *Hüls Am.*

9385 Dynasil® M
681-84-5 211-656-4
Tetramethoxysilane
Coupling agent, chemical intermediate, blocking agent, release agent, lubricant, primer, reducing agent *Hüls Am.*

9386 Dynasolve 100
68-12-2 3292 200-679-5
N,N-dimethylformamide
Dimethylformamide-based solvent; non-selective solvent. *Dynaloy.*

9387 Dynasolve 150
109-99-9 9351 203-726-8
tetrahydrofuran
Tetrahydrofuran-based solvent; flammable solvent for room temperature use only. *Dynaloy.*

9388 Dynasolve 165
Blend of solvents and organic chemicals; room temperature decapsulation solvent for dissolving cast epoxy resins. *Dynaloy.*

9389 Dynasolve 699
96-48-0 1632 202-509-5
γ-butyrolactone
A σ-butyrolactone-based solvent designed for use at room temperature for thin acrylic coatings; used hot (120-150°) for acrylic adhesives and epoxy powder coatings. *Dynaloy.*

9390 Dynasolve M-30
Nonchlorinated, nonflammable solvent for general cleaning operations; especially for curing of uncured polymers such as epoxies, urethanes and silicones. *Dynaloy.*

9391 Dynasperse A
8061-51-6
Sodium lignosulfonate
Primary dye dispersant, good heat stability and milling properties, low staining and very low dye reduction. *Borregaard Ligno Tech.*

9392 Dynaspinell
magnesium aluminate
Fused spinel; raw material for the production of pyrometer sheaths, crucibles, refractory bricks, ramming mixes. *Hüls AG.*

9393 Dynastite
An explosive containing 94% potassium chlorate, and 6% barium nitrate, dipped in nitro-toluene.

9394 Dynasylan®
Organo-functional silane adhesion promoters, alkyl silanes; bonding agent between inorganic surfaces and organic polymers; binders for the foundry industry. *Dynamit Nobel Wien GmbH.* Name unverified.

9395 Dynasylan®
Organo-functional silane adhesion promoters, alkyl silanes; bonding agent between inorganic surfaces and organic polymers; binders for the foundry industry. *Hüls UK Ltd.*

9396 Dynasylan® AMEO, Dynasylan® AMEO-P
919-30-2 213-048-4
Aminopropyltriethoxysilane
Coupling agent, chemical intermediate, blocking agent, release agent, lubricant, primer, reducing agent. Organo-silicon compounds for bonding between organic and inorganic components of a system. *Hüls Am.*

9397 Dynasylan® AMMO
13822-56-5 237-511-5
$C_6H_{17}NO_3Si$
Aminopropyltrimethoxysilane
1-Propanamine, 3-(trimethoxysilyl)-; Dynasylan AMMO; 3-Aminopropyltrimethoxysilane; γ-Aminopropyl Trimethoxy Silane. Coupling agent, chemical intermediate, blocking agent, release agent, lubricant, primer, reducing agent. *Hüls Am.*

9398 Dynasylan® BDAC
13170-23-5 236-112-3
$C_{12}H_{24}O_6Si$
Di-*t*-butoxydiacetoxysilane
Silicon di-*t*-butoxide diacetate; Di-*t*-butoxydiacetoxysilane. Coupling agent, chemical intermediate, blocking agent, release agent, lubricant, primer, reducing agent. *Hüls Am.*

9399 Dynasylan® BSA
10416-59-8 233-892-7
$C_8H_{21}NOSi_2$
Bis(trimethylsilyl)acetamide
N,O-Bis(trimethylsilyl)acetamide; BSA; Bis(trimethylsilyl)acetamide; Ethanimidic acid, N-(trimethylsilyl)-, trimethylsilyl ester; Dynasylan BSA. Coupling agent, chemical intermediate, blocking agent, release agent, lubricant, primer, reducing agent. mp = 24°; bp_{35} = 71-73°; d = 0.8230; n_D^{20}= 1.4170; reacts with H_2O. *Hüls Am.*

9400 Dynasylan® BSM
Alkyl-alkoxy-silane mixtures; impregnating material for buildings. *Dynamit Nobel Wien GmbH.* Name unverified.

9401 Dynasylan® BSM
Alkyl-alkoxy-silane mixtures; impregnating material for buildings. *Hüls UK Ltd.*

9402 Dynasylan® CPTEO
5089-70-3 225-805-6
3-Chloropropyltriethoxysilane
Coupling agent, chemical intermediate, blocking agent, release agent, lubricant, primer, reducing agent. *Hüls Am.*

9403 Dynasylan® CPTMO
2530-87-2 219-787-9
$C_6H_{15}ClO_3Si$
3-Chloropropyltrimethoxysilane
(3-Chloropropyl)trimethoxy-silane; CPTMO. Coupling agent, chemical intermediate, blocking agent, release agent, lubricant, primer, reducing agent. bp = 195-196°; d = 1.0810; n_D^{20}= 1.4196; reacts with H_2O. *Hüls Am.*

9404 Dynasylan® DAMO
1760-24-3 217-164-6
$C_8H_{22}N_2O_3Si$
N-2-aminoethyl-3-aminopropyltrimethoxysilane
N-[3-(trimethoxysilyl)propyl]-1,2-ethanediamine; N-β-(aminoethyl)-γ-aminopropyl trimethoxy-silane; N-[3-(Trimethoxysilyl)propyl]ethylenediamine; [3-(2-Aminoethyl)aminopropyl]trimethoxysilane; DAMO; N-(2-Aminoethyl-3-aminopropyl)trimethoxysilane; Dynasylan DAMO. Coupling agent, chemical intermediate, blocking agent, release agent, lubricant, primer, reducing agent. bp_{15} = 146°; d = 1.0100; n_D^{20}= 1.4450; LD_{50} (rat orl) = 7460 mg/kg. *Hüls Am.*

9405 Dynasylan® DAMO-P
1760-24-3 217-164-6
Aminoethylamino propyltrimethoxy silane
Coupler for epoxy, phenolic, melamine, nylons, PVC, acrylics, polyolefins, PU and nitrile rubber. *Hüls Am.*

9406 Dynasylan® DAMO-T
1760-24-3 217-164-6
Aminoethylamino propyltrimethoxy silane, tech. grade
Coupler for epoxy, phenolic, melamine, nylons, PVC, acrylics, polyolefins, PU and nitrile rubber. *Hüls Am.*

9407 Dynasylan® ETAC
17689-77-9 241-677-4
$C_8H_{14}O_6Si$
Ethyltriacetoxysilane
Coupling agent, chemical intermediate, blocking agent, release agent, lubricant, primer, reducing agent. *Hüls Am.*

9408 Dynasylan® GLYMO
2530-83-8 219-784-2
3-Glycidoxypropyltrimethoxysilane
Coupling agent, chemical intermediate, blocking agent, release agent, lubricant, primer, reducing agent. *Hüls Am.*

9409 Dynasylan® HMDS
999-97-3 4725 213-668-5
hexamethyldisilazane
Coupling agent, chemical intermediate, blocking agent, release agent, lubricant, primer, reducing agent. *Hüls Am.*

9410 Dynasylan® IBTMO
18395-30-7 242-272-5
$C_6H_{16}O_3Si$
Isobutyltrimethoxysilane
Prosil 178. Coupling agent, chemical intermediate, blocking agent, release agent, lubricant, primer, reducing agent. *Hüls Am.*

9411 Dynasylan® IMEO
58068-97-6 261-093-3
N-[3-(Triethoxysilyl)-propyl]-4,5-dihydroimidazole
Coupling agent, chemical intermediate, blocking agent, release agent, lubricant, primer, reducing agent. *Hüls Am.*

9412 Dynasylan® MEMO
2530-85-0 219-785-8
3-Methacryloxypropyltrimethoxysilane
Coupling agent, chemical intermediate, blocking agent, release agent, lubricant, primer, reducing agent. *Hüls Am.*

9413 Dynasylan® MTES
2031-67-6 217-983-9
$C_7H_{18}O_3Si$
Methyltriethoxysilane
Triethoxymethylsilane; Methyl triethoxysilane; MTES; Dynasylan MTES. Coupling agent, chemical intermediate, blocking agent, release agent, lubricant, primer, reducing agent. bp = 141-143°; d = 0.8950; n_D^{20}= 1.3845; LD_{50} (rat orl) = 7.67 g/kg. *Hüls Am.*

9414 Dynasylan® MTMO
4420-74-0 224-588-5
3-mercaptopropyltrimethoxysilane
Coupling agent, chemical intermediate, blocking agent, release agent, lubricant, primer, reducing agent. *Hüls Am.*

9415 Dynasylan® MTMS
1185-55-3 214-685-0
$C_4H_{12}O_3Si$
Methyltrimethoxysilane
trimethoxy(methyl)silane. Coupling agent, chemical intermediate, blocking agent, release agent, lubricant, primer, reducing agent. bp = 102°; d = 0.9550; n_D^{20}= 1.3703; LD_{50} (rat orl) = 12500 mg/kg *Hüls Am.*

9416 Dynasylan® OCTEO
2943-75-1 220-941-2
$C_{14}H_{32}O_3Si$
Octyltriethoxysilane
Triethoxyoctylsilane; n-Octyltriethoxysilane. Coupling agent, chemical intermediate, blocking agent, release agent, lubricant, primer, reducing agent. *Hüls Am.*

9417 Dynasylan® TCS
10025-78-2 9776 233-042-5
Trichlorosilane
Coupling agent, chemical intermediate, blocking agent, release agent, lubricant, primer, reducing agent. *Hüls Am.*

9418 Dynasylan® TRIAMO
35141-30-1 252-390-9
Trimethoxysilylpropyldiethylene triamine
Coupling agent, chemical intermediate, blocking agent, release agent, lubricant, primer, reducing agent. *Hüls Am.*

9419 Dynasylan® VTC
75-94-5 200-917-8
Vinyltrichlorosilane
Coupling agent, chemical intermediate, blocking agent, release agent, lubricant, primer, reducing agent. *Hüls Am.*

9420 Dynasylan® VTEO
78-08-0 201-081-7

Vinyl triethoxysilane
Coupling agent, chemical intermediate, blocking agent, release agent, lubricant, primer, reducing agent. *Hüls Am.*

9421 **Dynasylan® VTMO**
2768-02-7 220-449-8
Vinyltrimethoxysilane
Coupling agent, chemical intermediate, blocking agent, release agent, lubricant, primer, reducing agent. *Hüls Am.*

9422 **Dynasylan® VTMOEO**
1067-53-4 213-934-0
Vinyl tris (methoxyethoxy) silane
Coupling agent, chemical intermediate, blocking agent, release agent, lubricant, primer, reducing agent. *Hüls Am.*

9423 **Dynat W**
A proprietary brand of mechanically comminuted rubber from Malaysia. No manufacturer.

9424 **Dynatex GTZ**
9006-04-6 232-689-0
Low ammonia natural isoprene rubber latex; suitable for all latex applications where excellent stability and good preservation parameters are desired. *Guthrie Latex.*

9425 **Dynatherm**
1309-48-4 5713 215-171-9
Electromagnesia
Fused magnesia for heating elements. *Hüls AG.*

9426 **Dynatred**
Polyurethane sealant
Used for sealing horizontal joints in parking decks, plazas, warehouse floors or other areas subject to heavy foot and vehicular traffic, particularly where slope exceeds 1%. *Pecora Corporation.*

9427 **Dynatrol**
Polyurethane sealant
Used in general construction for caulking vertical expansion joints in walls and sealing around door and window frames. *Pecora Corporation.*

9428 **Dynaweld**
Urethane polymeric sealant adhesive; used for adhering pre-formed sections of Urexpan NR-200 to concrete or epoxy substrates and to each other. *Pecora Corporation.*

9429 **Dynazirkon**
1314-23-4 10313 215-227-2
Zirconium oxide
Raw material for the production of high refractory, slag resistant crucibles and moldings for the melting of high temperature alloys; raw material for high wear resisting casting nozzles in continuous casting plants. *Hüls AG.* Discontinued.

9430 **Dyne**
Iodophor/acid pipe line cleaner and sterilizer. *Ciba plc.*

9431 **Dynemate 200**
Hypochlorite/acid pipe line cleaner and sterilizer. *Ciba plc.*

9432 **Dynobel**
An explosive containing potassium perchlorate, nitroglycerin, ammonium oxalate, wood meal, and a little collodion cotton. No manufacturer.

9433 **Dynomel**
Melancil and urea melamine resins; for decorative laminates and surfacing of wood-based panels. *Dynochem UK Ltd.*

9434 **Dyphene**
Alkyl phenolic resins; used in adhesives, rubber, elastomers, bonding, abrasives and friction materials. *P M C Specialities Group Inc.*

9435 **Dypur**
Insulating pouring/feeding bush containing a ceramic foam filter. *Foseco (F.S.) Ltd.*

9436 **Dyquex®**
Sodium lignosulfonate
Anionic dye dispersant extender. *Georgia-Pacific.* Name unverified.

9437 **Dyrene®**
101-05-3 694 202-910-5
Anilazine
Broad spectrum fungicide used for tobacco, potatoes, cereals and ornamentals. *Bayer AG.*

9438 **Dysoid**
A bearing bronze containing 62% copper, 18% lead, 10% tin, and 10% zinc.

9439 **Dytek® A**
15520-10-2 239-556-6
$C_6H_{16}N_2$
2-Methylpentamethylenediamine
2-methylpentamethylenediamine; 1,5-pentanediamine, 2-methyl-; MPMD; Methylpentamethylenediamine; 2-methyl-1,5-diaminopentane. Epoxy curing agent; also used in polyurethanes, wet strength resins, scale and corrosion inhibitors, motor oil and gasoline additives, polyamide plastics, films, adhesives, and inks. mp = -50 - -60°; bp = 193°; d^{25} = 0.86; LD_{50} (rat orl) = 1690 mg/kg. *DuPont.* Name unverified.

9440 **Dytel®**
Leak detective dye. *DuPont UK.*

9441 **Dytherm®**
Expandable copolymer; for rigid foams providing excellent resistance to temperatures up to 250°F; end uses include automotive door liners, instrument panels, solar panels, cores for composite structures. *Arco Chemical Co.*

9442 **Dytron® XL**
Thermoplastic rubber; for electrical applications, e.g., wire and cable. *Advanced Elastomer Systems.*

9443 **Dyvax®**
Polymer. *DuPont UK.*

9444 **DZ910**
Dissociated zircon consisting of spheres of free silica with embedded zirconia dendrites; used for manufacture of ceramic pigments and refractory compositions. *Ferro.*

9445 **E alloy**
An alloy of 76% aluminum, 20% zinc, 2.5% copper, 0.2% iron, 0.5% manganese, 0.5% magnesium, and 0.2% silicon.

9446 **E.B Golden Glitter, Neutral Glitter**
Pearlescent material and aqueous acrylic resinous binders; for decorative textile printing. *Eastern Color & Chem.*

9447 **E.C.A**
Cresylic acids
Intermediates. *Murphy Chemical Co Ltd.* Discontinued.

9448 **E-3810**
25928-94-3
Epoxy; low-cost, general purpose encapsulant for applications requiring minimum thermal shock resistance, electrical performance. *ICI Fiberite.*

9449 **E-3824**
25928-94-3
Glass-filled epoxy; low pressure molding compound for encapsulation in electronics industry; good thermal shock resistance and wet electrical properties. *ICI Fiberite.*

9450 **E45 Cream**
A proprietary preparation of white soft paraffin, light liquid paraffin and wool fat used as a skin cream. *Crookes Healthcare.* Discontinued.

9451 **E-9405**
25928-94-3
Glass fiber-filled epoxy; molding compound for encapsulation (delicate electrical components, large coils); thermal shock, moldability, retains electrical properties under humid conditions. *ICI Fiberite.*

9452 **Earex®**
Arachis oil, almond oil, rectified camphor oil (1:1:1); used for softening and removal of ear wax. *Seton Healthcare Group plc.*

9453 **Early Impact**
carbendazim-flusilazole
A suspension concentrate containing 150 g carbendazim and 94 g flusilazole per liter; systemic fungicide for use on cereals. *DuPont UK.*

9454 **earth archil**
Archil contaminated with mineral matter; used for the preparation of litmus.

9455 **Ease Release 200 Series**
General purpose release agent for release of most types of casting and molding systems, e.g., PU elastomers, PU foam, epoxy resin, polyester, RTV silicones, rubber, and thermoplastic polymers; effective on aluminum, chrome; RTV silicone, epoxy, polyester, rubber and steel molds. *George Mann.*

9456 **Ease Release 2040 Series**
Semi-permanent release coating providing long lasting release for urethane, epoxy, rubber, aluminum, and steel molds, molding presses, coating machinery; effective with urethanes, rubbers, epoxies, hot-melt adhesives, polyesters, and other polymers and gums. *George Mann.*

9457 **Ease Release 2191**
Silicone compound blend; release agent for casting wax models and molds, injection molding, assembly lube; provides multiple releases. *George Mann.*

9458 **Easigel**
Rheological additive for use in organic liquid systems of widely differing type and polarity. *Akzo Chemie UK Ltd.*

9459 **Easisperse**
Easy dispersing pigments; used for printing inks and paints. *Manox Ltd.* Name unverified.

9460 **East India Gum**
Bombay gum. A variety of gum arabic, pale amber or pinkish in color.

9461 East Indian Balsam of Copaiba
8030-55-5 232-444-8
A name given to Gurjun balsam (the oleo-resin from the stems of *Dipterocarpus* sp.).

9462 Eastbond
Hot melt adhesives. *Eastman.*

9463 Eastman® 910
A proprietary trade name for a cyanoacrylate adhesive which sets with the application of pressure. Variants are: 910 EM for vinyls; 910 FS for quicker setting; 910 MHT for applications involving high temperatures. *Eastman Chemical Int'l. AG.* Name unverified.

9464 Eastman® AQ-38S
Polymer; adhesives and sealants raw materials. *Eastman.*

9465 Eastman® C-11 Ketone
Solvent. *Eastman.* Name unverified.

9466 Eastman® DTBHQ
88-58-4 201-841-8
$C_{14}H_{22}O_2$
2,5-Di-*t*-butylhydroquinone
2,5-Di-*tert*-butyl hydroquinone. Antioxidant for rubber, polyesters. mp = 215-220°; poorly soluble in H_2O, organic solvents. *Eastman.*

9467 Eastman® HQMME
150-76-5 205-769-8
$C_7H_8O_2$
Hydroquinone monomethyl ether
p-Hydroxyanisole; Hydroquinone monomethyl ether; 4-Methoxyphenol; Mono Methyl Ether Hydroquinone; *p*-methoxyphenol; Mequinol (INN); hydroxyanisole; mequinol. Antioxidant for monomers. mp = 55-57°; bp = 243°; d^{20} = 1.55; insoluble in H_2O, soluble in organic solvents. *Eastman.* Discontinued.

9468 Eastman® Inhibitor OPS
A proprietary trade name for *p*-octylphenyl salicylate; a uv light inhibitor suitable for polyolefins. *Eastman.* Name unverified.

9469 Eastman® Inhibitor Poly TDP 2000
63123-11-5
Thiodipropionate polyester. *Eastman.* Name unverified.

9470 Eastman® Inhibitor RMB
136-36-7 205-241-7
$C_{13}H_{10}O_3$
Resorcinol monobenzoate
1,3-Benzenediol, monobenzoate. Industrial grade uv absorber/stabilizer for cellulose plastics and PVC formulations; uv absorber. mp = 133-135°; LD_{50} (rat orl) = 1600 mg/kg. *Eastman.*

9471 Eastman® MTBHQ
1948-33-0 217-752-2
$C_{10}H_{14}O_2$
Mono-*tert*-butylhydroquinone
tert-Butylhydroquinone; TBHQ; 1,4-Benzenediol, 2-(1,1-dimethylethyl)-; 2-*tert*-Butylhydroquinone; Mono-*tert*-butylhydroquinone. Antioxidant for rubber, monomers. mp = 127-129°; bp= 273°; LD_{50} (rat orl)= 700 mg/kg. *Eastman.*

9472 Eastman® P4C5B-030
9003-07-0 7741
Polypropylene homopolymer
Eastman. Name unverified.

9473 Eastman® Poly TDP 2000
63123-11-5
Thiodipropionate polyester; secondary antioxidant used with phenolic primary antioxidants in polypropylene and other polyolefins. *Eastman.*

9474 Eastman® Yellow
The sodium salt of glucose phenylosazone *p*-carboxylic acid; a yellow coloring matter used as a corrective filter in photography. *Eastman.*

9475 Eastobrite® OB-1
1533-45-5 216-245-3
$C_{28}H_{18}N_2O_2$
2,2'-(1,2-Ethenediyldi-4,1-phenylene)bisbenzoxazole
Optical brightener; fluorescent whitening agent for use in linear polyester, PET, nylon fibers. *Eastman.*

9476 Eastoflex B1020
Propylene/butene copolymer; adhesives and sealants raw materials. *Eastman.*

9477 Eastoflex E1003
Propylene/ethylene copolymer; adhesives and sealants raw materials. *Eastman.*

9478 Eastoflex P1010
Amorphous polyolefin homopolymer; adhesives and sealants raw materials. *Eastman.*

9479 Eastotac H-100
Hydrocarbon tackifying resin; tackifier for adhesives and sealants applications. *Eastman.*

9480 Eastozone 32
N, N' -Dimethyl-N, N' -di-(1 methylpropyl)-p-phenylenediamine
A proprietary antioxidant. *Eastman.* Name unverified.

9481 Easy Cleen
Mild alkaline degreasing agent containing a blend of emulsifiers, solvents and solubilizing agents; general-purpose cleaner. *Vitax Ltd.*

9482 Easy-Flo
Fluxes for silver alloy brazing. *Johnson Matthey plc.*

9483 Easypoxy
Two-component epoxy adhesives *Conap.*

9484 eau de Brouts
Petitgrain water.

9485 eau de Goudron
Tar water.

9486 EB1500-1AR
Polyester, thermoset; flame retardant, abrasion-resistant electrical encapsulation grade general-purpose resin; for compression, transfer, and injection molding. *Cuyahoga Plastics.*

9487 EB3000-2
Polyester, thermoset, glass-reinforced; flame retardant thermoset for compression, transfer, and injection molding. *Cuyahoga Plastics.*

9488 Ebal
4748-78-1 225-268-8
$C_9H_{10}O$
p-Ethyl-benzaldehyde
4-Ethylbenzaldehyde. Additive for resins; intermediate for pharmaceuticals, fragrances. bp = 221°; d = 0.9790. *Mitsubishi Gas.*

9489 Ebecryl®
Acrylic pre-polymers which are radiation curable. *UCB Chemical Sector.*

9490 Ebecryl® 110
Oxyethylated phenol acrylate monomer
Low viscosity, low odor monomer used in uv light and electron beam cure products *UCB Radcure.* Discontinued.

9491 Ebecryl® 150
Bisphenol A derivative diacrylate monomer; for applications where high reactivity and low volatility are required; especially useful in inks and varnishes. *UCB Radcure.* Discontinued.

9492 Ebecryl® 600
Bisphenol A epoxy diacrylate; low color grade featuring very fast cure, high gloss, excellent solvent resistance; for overprint varnishes, lithographic and screen inks, coatings for paper, paperboard, wood chipboard, and rigid plastics, laminating adhesives, paper upgrading. *UCB Radcure.* Discontinued.

9493 Ebecryl® 629
Epoxy novolac acrylate in 33% monomer blend; provides heat and solvent resistance to solder resists, marking inks, adhesion on metallized substrates, low shrinkage coatings, heat-resistant applications. *UCB Radcure.* Discontinued.

9494 Ebecryl® 1360
Silicone hexaacrylate
Additive contributing slip, substrate wetting and flow properties to overprint varnishes, clear coatings on paper, plastics and metals. *UCB Radcure.* Discontinued.

9495 Ebert and Merz's α-acid
92-41-1 6462 202-154-6
$C_{10}H_8O_6S_2$
2,7-Naphthalenedisulfonic acid
Used as an intermediate for dyes.

9496 Ebert and Merz's β-acid
581-75-9 6461 209-471-9
$C_{10}H_8O_6S_2$
2,6-Naphthalenedisulfonic acid
Chemical intermediate. Soluble in H_2O, EtOH, less soluble in organic solvents.

9497 Ebner's fluid
A mixture of 2.5 ml hydrochloric acid, 2.5 g sodium chloride, 100 ml water, and 500 ml alcohol; used for decolorizing in bacteriological work.

9498 Ebonestos
A trade name for a series of proprietary molded products for electrical and heat insulation. No manufacturer.

9499 Ebonite
Vulcanite; Hardened rubber. A material prepared by vulcanizing rubber with up to 75% sulfur or metallic sulfides, with the addition of chalk, gypsum, or

other filling and coloring substances; used as an insulating material in the manufacture of electrical parts, filler for low-cost rubber parts.

9500 Ebonized monel
A monel metal with a fine finish produced by an oxidizing process.

9501 Ebontex
A proprietary trade name for an emulsified asphalt used for waterproofing tanks. No manufacturer.

9502 Ebony black
A blackish-brown dyestuff mixed with a blue dyestuff; used for dyeing cotton from a bath containing sodium sulfate and sodium carbonate, and half-wool.

9503 Eborex
A proprietary preparation containing about 65-70% sodium fluosilicate; it is a light fluosilicate for use as an insecticide. No manufacturer.

9504 E-BR® 8405
9003-17-2
Emulsion polybutadiene; nonstaining; used for tire carcass, sidewalls, mechanical goods; exhibits high tack, good dynamic and low temperature properties. *Ameripol Synpol.* Discontinued.

9505 E-BR® 8471
9003-17-2
Carbon black masterbatch emulsion polybutadiene, 62.5 parts high aromatic oil, 82.5 parts N-339 carbon blk., antiozonant; used for conveyor and power belting, extruded goods, retread rubber. *Ameripol Synpol.* Discontinued.

9506 Ebrok
A proprietary trade name for a bituminous plastic. No manufacturer.

9507 EC-25®, -25K
Propylene glycol mono-and diesters of fats and fatty acids, mono-and diglycerides, partially hydrogenated soybean oil, lecithin, BHA, citric acid; food emulsifier for cakes, mixes. *Van Den Bergh Foods.*

9508 Eca
Oil additive packages. *Exxon UK.* Name unverified.

9509 Ecco Defoamer Heavy
Silicone defoamer; antifoam for a wide variety of systems. *Eastern Color & Chem.*

9510 Ecco Defoamer NSD
Nonsilicone self-emulsifying hydrocarbon; antifoam for dyeing processes. *Eastern Color & Chem.*

9511 Ecco Defoamer S
Petroleum/silicone; defoamer for mills waste water. *Eastern Color & Chem.*

9512 Ecco Fast Binder 1500
Pigment binder for machine and rotary screen printing to give soft, durable prints with wet and dry crock resistance. *Eastern Color & Chem.*

9513 Ecco MP® 2004
97-23-4 3120 202-567-1
Dichlorophen
Textile fungicide, mildewproofing agent; compatible with aluminum/wax type water-repellents. *Eastern Color & Chem.*

9514 Ecco Resin 234
Complexed acrylic emulsion compound; pigment binder for variety of fabric substrates; nonyellowing; gives maximum softness of hand. *Eastern Color & Chem.*

9515 Ecco Rez 3070
Cyclic methylated high solids resin; stiffener for nylon and synthetics; requires catalysts. *Eastern Color & Chem.*

9516 Ecco Rez M-300-7
9003-08-1
High solids melamine formaldehyde resin; produces stiffness and excellent durability to fabrics in resin finishing baths; requires catalyst. *Eastern Color & Chem.*

9517 Ecco White® FW-5
Distyryl-phenyl
Optical brightener for nylon and blends. *Eastern Color & Chem.*

9518 Eccoblanc W-55-Q
Single-package pigment white printing compound containing its own binding system; useful in discharge printing; produces excellent whites with good durability to laundering and drycleaning; for screen printing and roller printing. *Eastern Color & Chem.*

9519 Eccobond®
High strength, high performance adhesives for industrial and electronic use; thermally conductive, fire retardant, and clear formulations; electrical conductive adhesives also available. *Emerson & Cuming Polymer Group.*

9520 Eccobond® 114
A proprietary one part filled epoxy adhesive. *Emerson & Cuming Polymer Group.*

9521 Eccobond® Adhesive Special 2
Resin/solvent adhesive material; solvent system with initial tack and somewhat rapid evaporation; for textiles and paperboard. *Eastern Color & Chem.*

9522 Eccobond® Paste 99
A proprietary one part thixotropic epoxy adhesive of high thermal conductivity; used in cones heat sink applications. *Emerson & Cuming.*

9523 Eccobond® SF40
A proprietary low density two-component epoxy-based adhesive and rigid filler. *Emerson & Cuming Polymer Group.*

9524 Eccobrite RB
588-59-0 8972 209-621-3
Stilbene
Stilbene; whitening agent for cotton and acetates. *Eastern Color & Chem.*

9525 Eccoclean CR-46
Self-emulsifying solvent cleaner; added solvency for extra cleaning power; useful on difficult cleaning problems. *Eastern Color & Chem.*

9526 Eccocoat®
Coatings for electronics, dip coats for resistors and capacitors, conformal coatings for circuit boards, spray or brush coating for general purpose applications *Emerson & Cuming Polymer Group.*

9527 Eccodye Colors
Complete range of colors for oil-phase print systems containing alkyd resins as binding agents. *Eastern Color & Chem.*

9528 Eccofix 101-40
Resinous substantive dye fixative for direct and fiber reactive dyestuffs. *Eastern Color & Chem.*

9529 Eccoflo® HiK
Free-flowing and pour-in-place powders with adjusted surface current suppressants; for dielectric applications. *Emerson & Cuming.*

9530 Eccofloat
A proprietary polyester-resin-bound syntactic foam used to fill voids in submarine hulls. *Emerson & Cuming.* Name unverified.

9531 Eccofloat EG35
A proprietary epoxy resin-bound syntactic foam material used in deep-sea applications. *Emerson & Cuming.* Discontinued.

9532 Eccofloat Encapsulant 1421
A proprietary epoxy-resin-bound encapsulant used to protect under-sea components. *Emerson & Cuming.* Discontinued.

9533 Eccofloat HG452
A proprietary polyester-resin-bound low density float material for use in deep-sea applications. *Emerson & Cuming.* Discontinued.

9534 Eccofloat PC61
A proprietary polyester-resin-bound castable material for use in deep-sea applications. *Emerson & Cuming.* Discontinued.

9535 Eccofloat PP22 and 24
Proprietary grades of polyester-bound syntactic foam which can be packed *in situ* to fill voids and to make buoys. *Emerson & Cuming.* Discontinued.

9536 Eccofloat SP 12, 20
A proprietary polyester-bound low density syntactic foam for use where buoyancy is required in harbor and off-shore applications. *Emerson & Cuming.* Discontinued.

9537 Eccofloat SS40
A proprietary polyurethane rubber-bound material used in the making of deep-sea diving suits. *Emerson & Cuming.* Discontinued.

9538 Eccofloat UG 36
A proprietary polyurethane-bound semiflexible non compressible material. *Emerson & Cuming.* Discontinued.

9539 Eccofloat US 35
A proprietary polyurethane - bound material - flexible, compressible and usable down to about 1000 ft depth of water. *Emerson & Cuming.* Discontinued.

9540 Eccofoam® FPH
Rigid, high temperature resistant polyurethane foam-in-place resin for dielectrical applications. *Emerson & Cuming Polymer Group.* Discontinued.

9541 Eccofoam® PP
A proprietary group of hydrocarbon resin closed-cell foams used in high-frequency electrical applications. *Emerson & Cuming Polymer Group.* Name unverified.

9542 Eccoful DL Conc
Modified amide ethoxylate; detergent, scouring and fulling agent for wool and wool blends; softener. *Eastern Color & Chem.*

9543 Eccogel F
Proteinaceous compound; gives stiffness and hand to suit desired finishes on various tapes and narrow fabrics. *Eastern Color & Chem.*

9544 Eccolube L-54
Self-dispersing lubricant for narrow fabrics and a yarn lubricant for braiding and weaving; resists yellowing. *Eastern Color & Chem.*

9545 Eccopel 10
Semidurable solvent phase water repellent for use in a complete solvent system; used in drycleaning establishments for the reprocessing of outerwear garments to produce water repellency. *Eastern Color & Chem.*

9546 Eccopuff
Printing system providing a raised, three-dimensional type of print for novelty printing; applicable to paper, textiles, nonwoven, and decorative tiles in building applications. *Eastern Color & Chem.*

9547 Eccoro®
Imidazolines; corrosion resistance of metals. *Eastern Color & Chem.*

9548 Eccoscour CB
Sodium alkylaryl sulfonate
Detergent, wetting agent, and emulsifier for textile scouring and dyeing applications. *Eastern Color & Chem.*

9549 Eccoscour D-7
Sulfate alkyl phenol ethoxylate
Surfactant and scouring agent for textile fabric and yarns; prevents redeposition of soils, dyestuffs, or pigments. *Eastern Color & Chem.*

9550 Eccoseal®
Epoxy impregnant and casting resins. *Emerson & Cuming Polymer Group.* Name unverified.

9551 Eccoshield®
Adhesives, coatings, caulks, sealants, tapes, and gaskets for EMI/RFI shielding applications and static dissipation and electrical ground planning. *Emerson & Cuming Polymer Group.*

9552 Eccosil®
Silicone adhesives and casting resins. *Emerson & Cuming Polymer Group.*

9553 Eccosil® 1776
RTV silicone product; one-part general purpose adhesive, caulk, sealant; easy-to-use, nonflowing flexible, moisture-resistant electrical grade for glass ceramics, metal, plastics. *Emerson & Cuming Polymer Group.* Name unverified.

9554 Eccosil® 2CN
RTV silicone product; industrial grade for water clear repairable potting, embedment of circuits. *Emerson & Cuming Polymer Group.* Name unverified.

9555 Eccosoft C-2000
Fatty material; substantive softener for use on cottons and cellulosic materials, blends of synthetics and cellulosics; shows minimal yellowing. *Eastern Color & Chem.*

9556 Eccosol 150
Self-emulsifying mineral oil lubricant for fine wire drawing. *Eastern Color & Chem.*

9557 Eccosolv C-14
Self-emulsifying solvent cleaner for cleaning and degreasing of machine parts. *Eastern Color & Chem.*

9558 Eccosorb® 269E
A proprietary epoxy coating; brushed onto surfaces, it increases their electrical loss in the S to K bands of the high frequency range. *Emerson & Cuming Polymer Group.*

9559 Eccosorb® AN
Lightweight, flexible foam sheet, broadband microwave absorber. *Emerson & Cuming Polymer Group.*

9560 Eccosorb® Coating 268E
A proprietary epoxy coating; brushed onto surfaces, it increases their electrical loss in the L-band of the high frequency range. *Emerson & Cuming Polymer Group.*

9561 Eccosorb® MF
A proprietary range of magnetically-loaded epoxy resins. *Emerson & Cuming Polymer Group.*

9562 Eccospheres
Trade name for small hollow glass or silica spheres of diameter ranging from 10-250 microns; used as a loading material for plastics to impart lightness and reduced permittivity which it does by virtue of the large airspace. *Emerson & Cuming.* Name unverified.

9563 Eccostat
Antistat for textiles. *Eastern Color & Chem.*

9564 Eccostat C
Antistatic agent, softener for synthetic and natural fibers and blends. *Eastern Color & Chem.*

9565 Eccoterge 200
PEG 400 ester
Wetting agent, emulsifier, dispersant for solvents in aqueous systems. *Eastern Color & Chem.*

9566 Eccoterge ASB
Amine alkylaryl sulfonate
Detergent, wetting agent, emulsifier, scouring agent for textile applications. *Eastern Color & Chem.*

9567 Eccoterge MV Conc
Fatty amino condensate; detergent, emulsifier, wetting and scouring agent for textile use. *Eastern Color & Chem.*

9568 Eccotherm® TC-11
General-purpose silicone grease with moderate thermal conductivity; for circuit assembly. *Emerson & Cuming Polymer Group.*

9569 Eccowax UL-100
Substituted fatty amide condensate; wax concentrate for preparation of stock softener solution or softener products at 20-27% solids; requires use of acetic acid in order to produce a cationic softener system; for cotton, synthetics and cotton/synthetic blends. *Eastern Color & Chem.*

9570 Eccowet® W-50
Sodium aliphatic ester sulfonate; wetting agent, penetrant, dispersant, solubilizer, emulsifier, detergent for textiles, metal processing, disinfectants, paints, pigments, wallpaper, rubber cements, adhesives, drycleaning detergents, topical pharmaceuticals and cosmetics. *Eastern Color & Chem.*

9571 Ecdel® 9965
Copolyester ether elastomer; offers clarity, toughness, and chemical resistance for flexible packaging applications including medical applications. *Eastman.* Name unverified.

9572 Ecdel® 9967
Copolyester ether elastomer; offers clarity, toughness, and chemical resistance for flexible packaging applications including medical applications. *Eastman.* Name unverified.

9573 Ecepox® PB1 and PB2
Epoxidized esters used as plasticizers and stabilizers for PVC compounds. Registered trade names. No manufacturer.

9574 Echappe silk
A name for floss or waste silk.

9575 Echicaoutchin
A low-grade gutta-like material from *Alstonia scholaris.*

9576 Echo
Thiadiazole derivatives used as crosslinking agents; used in the vulcanization of halogen-containing polymers such as polyepichlorohydrins and chlorinated polyethylene; typical applications are extruded and molded hose and tubing. *Hercules.* Discontinued.

9577 Echo
53780-34-0 5846 258-767-4
mefluidide
Soluble concentrate containing 240 g/l mefluidide; grass growth suppressant. *ICI Agrochemicals Professional Prods.*

9578 echurin
picric acid-nitro-flavin
A mixture of picric acid and nitro-flavin.

9579 Eclipse
Granular or powder organic based fertilizers; a steady release, lower nitrogen fertilizer for horticultural crops, parks and gardens. *Humber Fertilizers plc.*

9580 Ecobinder®
Polymer emulsion; nonsolvent-containing binder for emulsion paint. *Scott Bader.* Discontinued.

9581 Ecobond
Environmentally friendly phosphating processes. *Brent Chemicals International plc.*

9582 Ecolac
A proprietary trade name for an air drying lacquer and adhesive for plastics. No manufacturer.

9583 Ecolo
Compounded plastic material. *Mitsubishi Petrochem.* Name unverified.

9584 Ecolotec
Resin binder for sand cores, hardened with carbon dioxide gas. *Foseco (F.S.) Ltd.*

9585 Econocat
Catalysts. *Courtaulds Engineering Ltd.*

9586 Econogel
High performance rheological additive for paints, inks, sealants, mastics, etc.; a range of metal-based drying agents. *Akzo Chemie UK Ltd.*

9587 Economy Flor-Dri
1337-76-4
Attapulgite clay; absorbent *Floridin.* Name unverified.

9588 Econopred
52-21-1 7901 200-134-1
Prednisolone acetate
Glucocorticoid. *Alcon Laboratories Inc.* Name unverified.

9589 Ecoro
A proprietary packaging material of polypropylene loaded with calcium

carbonate to ease disposal by incineration. *Mitsubishi Petrochem.* Name unverified.

9590 Ecosyl
Concentrated biological silage additive. *ICI Chem & Polymers Ltd.*

9591 Ecothene EC 101
9002-88-4 7728
Polyethylene
HDPE homopolymer; post-consumer homopolymer usable straight, as a blend, or in coextruded structures; for blow molding, thermoforming, injection molding, and film extrusion applications. *Quantum/USI.*

9592 ECP-170
Emulsifiable chlorinated paraffin; designed for metal displacement applications which leave residual films removable with water/alkali cleaners. *Ferro/Keil.*

9593 ecru silk
Silk which has lost about 3-4% of its weight of sericin or silk gum.

9594 Edaplan LA 400
Silicone-free leveling agent for lacquers and varnishes. *Munzing Chemie GmbH.*

9595 Edaplan LA 411
Silicone compound; leveling agent for lacquers, varnishes, aqueous coatings, powder coatings. *Munzing Chemie GmbH.*

9596 Edaplan VP LA 420
Silicone-free polymer; deaerator and leveling agent for coil coating systems. *Munzing Chemie GmbH.*

9597 Edasil
Natural ion exchanger for improvement and conditioning of the soil. *Süd-Chemie AG.* Name unverified.

9598 Edecrin®
3761
Ethacrynic acid and sodium ethacrynate; potent diuretic for the relief of certain edemas. *Merck & Co Inc.*

9599 Edelfeka
A nickel-containing silver-copper-cadmium alloy.

9600 Edelresanol
A proprietary synthetic resin. No manufacturer.

9601 Edelwit
1305-62-0 1716 215-137-3
$Ca(OH)_2$
Hydrated lime
Used in chemical industries, drinking water treatment, waste water treatment. *BV Nekami.* Name unverified.

9602 Edenol 74
A proprietary alkyl-epoxy stearate type plasticizer for plastisols. *Henkel Chemicals Ltd, Hental & Cie.*

9603 Edenol 302
Propylene glycol dicaprylate/dicaprate;. Emollient oil and cosolvent for personal care products, bath oils, aerosols, antiperspirants; stable vehicle for pigmented cosmetics. *Henkel/Cospha; Henkel Canada.*

9604 Edenol B316
Epoxidized linseed oil; plasticizer used in PVC. *Henkel Chemicals Ltd, Hental & Cie.*

9605 Edenol B35
A proprietary alkyl-epoxy stearate type plasticizer for plastisols. *Henkel Chemicals Ltd, Hental & Cie.*

9606 Edenol D72
A proprietary alkyl-epoxy stearate type plasticizer for plastisols. *Henkel Chemicals Ltd, Hental & Cie.*

9607 Edenol D82
8013-07-8 232-391-0
An epoxidized soybean oil; plasticizer used in both rigid and plasticized PVC. *Henkel Chemicals Ltd, Hental & Cie.*

9608 Edenol HS 235
A proprietary alkyl-epoxy stearate type plasticizer for plastisols. *Henkel Chemicals Ltd, Hental & Cie.*

9609 Edenor ITS
Isotridecyl stearate *Henkel/Cospha.*

9610 Edenor PTO
19321-40-5 242-960-5
Pentaerythritol tetraoleate
Plasticizer. *Henkel/Cospha.*

9611 Eder's solution
mercuric chloride-ammonium oxalate
A solution of mercuric chloride and ammonium oxalate used in photometric determinations.

9612 Edeta®
Complexing agents for cosmetics industry. *BASF AG.*

9613 edetate disodium
139-33-3 3556 205-358-3
$C_{10}H_{16}N_2O_8 \cdot 2Na$
N,N'-1,2-Ethanediylbis[N-(carboxymethyl)glycine] disodium salt
disodium dihydrogen ethylene diamine tetraacetetate; ethylenediaminetetraacetic acid; disodium salt; disodium EDTA; disodium edetate; Tetracemate disodium; Disodium edetate; Chelaplex III; Endrate Disodium; Sequestrene NA 2; Sodium Versenate; Titriplex III; Versene disodium salt,. Food preservative, chelating and sequestering agent; anticoagulant; pharmaceutical aid. Soluble in H_2O; LD_{50} (rat orl) = 2 g/kg. *W.R. Grace/Hampshire; R.W. Greeff.*

9614 edetate trisodium
150-38-9 3558 205-758-8
$C_{10}H_{13}N_2Na_3O_8$
N,N'-1,2-Ethanediylbis[N-(carboxymethyl)glycine]trisodium salt
Hamp-Ene® Na3 Liq; Sequestrene® NA3; Limclair; Versene-9; Trisodium EDTA; EDTA trisodium; (ethylenedinitrilo)tetraacetic acid trisodium salt; ethylenediaminetetraacetic acid trisodium salt; trisodium ethylenediaminetetraacetate; trisodium edetate; edetic acid trisodium salt,. Chelating agent. mp > 300°; soluble in H_2O. *W R Grace/Hampshire.*

9615 edetic acid
60-00-4 3559 200-449-4
$C_170H_{16}O_8$
N,N'-1,2-ethanediylbis [N-(carboxymethyl glycine]
ethylene diamine tetraacetic acid; edathamil; versene acid. EDTA (CTFA); substituted diamine; chelating agent; stabilizer; antioxidant in foods; pharmaceutics aid. *Allchem Industries; Allied Colloids; W.R. Grace/Hampshire; Protex SA; Showa Denko.*

9616 Edicol
Coloring matter for use in foodstuffs, pharmaceuticals, and cosmetics. (Sold under licence from ICI). *ICI Chem & Polymers Ltd.*

9617 Edifas
Vegetable derivatives; used in food manufacturing as binding, stabilizing and emulsifying agents. *Imperial Chemical Industries plc.*

9618 Edifenphos
17109-49-8 3560 241-178-1
$C_{14}H_{15}O_2PS_2$
O-ethyl S,S-diphenyl phosphorodithioate
EDDP; Bay 78418; Blastoff; Hinosan. Foliar fungicide ith protective and curative action. Used for control of rice blast. Also for blight diseases, stem rot and *Fusarium* leaf spot in rice. $bp_{0.01}$ = 154°; d^{20} = 1.23; $n_D^{2?}$ = 1.6112; soluble in H_2O (56 mg/l), more soluble in organic solvents; LD_{50} (rat orl) = 212-240 mg/kg.

9619 Edimet
Polymethyl methacrylate
No manufacturer.

9620 Edinol
It contains *p*-aminosaligenin, acetone sulfite, potassium hydroxide, and potassium bromide; a photographic developer.

9621 Edistir®
A proprietary range of polystyrene molding granules. *Montedison UK Ltd.* Name unverified.

9622 Edistir® FA
9003-53-6 9028
Polystyrene
PS; general purpose, high flow grade for injection molding. *Montedipe Srl.*

9623 Edistir® N 1280, N 1281
9003-53-6 9028
PS; high thermal resistance, and high mechanical properties; for extrusion; also suitable for injection molding. *Montedipe Srl.*

9624 Edistir® RC
9003-53-6 9028
HIPS; glossy grade with balanced properties for injection molding, extrusion. *Montedipe Srl.*

9625 Edistir® RK
9003-53-6 9028
HIPS; flame-retardant grade for injection molding, extrusion. *Montedipe Srl.*

9626 Edistir® RKV
9003-53-6 9028
Semiexpandable HIPS; flame-retardant grade for injection molding. *Montedipe Srl.*

9627 Edistir® RV 8
9003-53-6 9028
Semiexpandable HIPS; medium-high impact, high flow grade for injection molding. *Montedipe Srl.*

9628 Edistir® SR 550, SRL 550
9003-53-6 9028

HIPS; high impact, improved flow grade for injection molding, thin sheet extrusion. *Montedipe Srl.*

9629 Edistir® UT/1
9003-53-6 9028
HIPS; very high impact, thermal resistant grade for injection molding. *Montedipe Srl.*

9630 Edistir® UT/SF
9003-53-6 9028
HIPS; high impact, thermal resist. grade for injection molding. *Montedipe Srl.*

9631 Edit®
For the agriculture industry. *DuPont UK.*

9632 Edolan®
Resist agent for wool/polyamide by the exhaust process. *Bayer AG.*

9633 Edunine
Textile auxiliary chemicals. *ICI Chem & Polymers Ltd.*

9634 Edward's speculum
A zinc and arsenic bearing bronze containing 63.3% copper, 32.2% tin, 2.9% zinc, and 1.6% arsenic.

9635 EE Acetate
111-15-9 3798 203-839-2
$C_6H_{12}O_3$
ethylene glycol monoethyl ether acetate
2-ethoxyethanol acetate; 2-ethoxyethyl acetate; Cellosolve acetate; Ethylene glycol monoethylether acetate; Ethylglycol acetate; Ethyl Cellosolve Acetate; 2-Ethoxyethanol acetate; Glycol, monoethyl ether acetate; 2EEA; Ethylene glycol monethyl ether acetate; Ethoxyethyl acetate; Ethoxyethanol acetate; 2-Ethoxy etheracetate; ethylene glycol ethyl ether acetate; CSAC; Ektasolve EE acetate solvent; 2-ethoxyethanol, ester with acetic acid; β-ethoxyethyl acetate; glycol ether EE acetate; oxytol acetate; Poly-solv EE acetate; Ethyl acetyl glycolate. Ethylene glycol monoethyl ether acetate. Used in automobile lacquers. mp = -61°; bp = 156°; d = 0.9750; n_D^{20}= 1.4040; soluble in H_2O (230 g/l), organic solvents; LD_{50} (rat orl) = 2700 mg/kg. *Eastman.* Name unverified.

9636 EE Solvent
110-80-5 3797 203-804-1
$C_4H_{10}O_2$
Ethylene glycol monoethyl ether.
2-ethoxyethanol; Cellosolve; Oxitol. Solvent. mp = -70°; bp = 135°; d_{20}^{20} = 0.931; n_D^{25}= 1.406; soluble in H_2O, organic solvents; LD_{50} (rat orl) = 3 g/kg. *Eastman.* Name unverified.

9637 Eel antifriction metal
An alloy of 75% lead, 15% antimony, 6% tin, 1.5% cadmium, 0.5% arsenic, and 0.1% phosphorus.

9638 Efetaal
1-Ethoxy-1-phenylethoxyethane
Use in perfumery. *Quest Int'l. UK Ltd.*

9639 Effesay
Sulfonated alcohols and detergents. *ABM Chemicals Ltd.* Name unverified.

9640 Efica
Additives for the coatings industry. *Stoller Chemicals Ltd.*

9641 Efweko
1333-86-4 1856 215-609-9
Carbon black chips in nitrocellulose or plastics; used for dyeing lacquers, gravure and flexographic printing inks. *Degussa AG.* Name unverified.

9642 Egalex
Liquid dyeing assistant; for textile industry. *Tensia SA.* Name unverified.

9643 Egalisal
A fiber protection agent; used in the textile industry for dyeing wool. *Degussa AG.* Name unverified.

9644 Egalon Colours, Auxiliaries, Thinners
Pigments and auxiliaries for nitrocellulose finishing. *Bayer plc.*

9645 Egg oil
8001-17-0 232-271-8
EmCon E-5. Emollient, moisturizer for hair and skin care products; water-oil emulsifier; superfatting agent, humectant, mold release agent; occlusive agent. *Fanning.*

9646 Eglantine
A name which has been applied to both isobutyl benzoate and to isobutyl-phenol acetate; used in perfumery.

9647 EGO-4
Antisettling additive. *United Catalysts.*

9648 Egyptianized clay
Clay rendered more plastic by the addition of tannin.

9649 EHIDA Kit
63245-28-3 264-041-8
$C_{16}H_{22}N_2O_5$
Etifenin
N-(2,6-Diethylphenylcarbamoylmethyl)iminodiacetic acid. Diagnostic aid. mp = 186-189°; soluble in H_2O. *Amersham Corp.* Name unverified.

9650 Ehrhard's metal
An alloy of 89% zinc, 4% copper, 4% tin, and 3% lead.

9651 Ehrlich-Biondi stain
A microscopic stain containing 100 ml of a saturated solution of Orange G, 30 ml of a saturated solution of acid fuchsin, and 50 ml of a saturated solution of methyl green.

9652 Ehrlich's diazo reagent
For indole: 4 g *p*-dimethylaminobenzaldehyde in 380 ml alcohol and 80 ml concentrated hydrochloric acid. One volume of the solution to be tested is used with 1 volume of the reagent, a positive color being red.

9653 Ehrlich's hematoxylin
A microscopic stain. It consists of 30 grains hematoxylin, 100 ml absolute alcohol, 100 ml glycerin, 30 grains ammonium alum, and 100 ml distilled water.

9654 Eisler's bronze
A bronze containing 5.9% tin; used for art castings.

9655 Ekaline
Aliphatic polyglycol ether; versatile dispersing, leveling, and scouring agent. *Sandoz Prods. Ltd.* Name unverified.

9656 Ekanda rubber
A rubber obtained from the shrub, *Raphionacme utilis* in Angola.

9657 Ekatin
640-15-3 211-362-6
thiometon
Insecticide containing thiometon for aphid control. *ICI Chem & Polymers Ltd.* Discontinued.

9658 E-Kote 3042
A trade name for a silver filled air drying epoxy coating material soluble in isobutyl ketone. *Allied Prods. Corporation.* Unverified.

9659 Ektapro® EEP Solvent
763-69-9 212-112-9
Ethyl 3-ethoxypropionate
High-performance retarder solvent for formulating enamels, lacquers, topcoats, and primers; urethane-grade quality; polymerization solvent. bp = 170°; d = 0.9500; n_D^{20}= 1.4064 *Eastman.*

9660 Ektar® FB PG003
PP, 30% glass fiber-reinforced; for electrical/electronic applications *Eastman.*

9661 Ektasolve®
Glycol ethers and esters. Solvents. *Eastman.*

9662 Ektasolve® DB
112-34-5 1592 203-961-6
$C_8H_{18}O_3$
Butoxydiglycol
butyl carbitol; ethylene glycol monobutyl ether; butyl digol; butyl diicinol; 2-(2-butoxyethoxy)ethanol. Solvent mp = -68°; bp = 230°; d_{20}^{20} = 0.9536; n_D^{27} = 1.4258; soluble in H_2O, organic solvents; LD_{50} (rat orl) = 6.56 g/kg. *Eastman.* Name unverified.

9663 Ektasolve® DB Acetate
Diethylene glycol butyl ether acetate; solvent. *Eastman.* Name unverified.

9664 Ektasolve® DE, DE-HG
111-90-0 1847 203-919-7
$C_6H_{14}O_3$
2-(2-ethoxyethoxy)ethanol
carbitol; diethylene glycol monoethyl ether; ethyl digol. Solvent. bp = 196°; d_4^{25}= 0.9855; n:20_D = 1.4273; soluble in H_2O, organic solvents; LD_{50} (rat orl) = 8.69 g/kg. *Eastman.*

9665 Ektasolve® DE Acetate
Diethylene glycol ethyl ether acetate; solvent *Eastman.* Name unverified.

9666 Ektasolve® DM
111-77-3 6116 203-906-6
$C_5H_{12}O_3$
Diethylene glycol monomethyl ether
methyl carbitol; 2-(methoxyethoxy)ethanol; methyl digol;. Evaporating solvent used in brushing lacquers and dye stains; useful in wood stains, printing inks, and dye pastes for textiles; coalescing aid for PVAc latex paints; used in stamp pad and stencil inks; dilutent for hydraulic brake fluids. mp < -84°; bp = 193°; n_D^{27}= 1.4264; soluble in H_2O, organic solvents; LD_{50} (rat orl) = 9.21 g/kg. *Eastman.*

9667 Ektasolve® DP
6881-94-3 229-985-7
Diethylene glycol monopropyl ether
Evaporating, water-miscible solvent used in solution and water-dilutable coatings; active for many coating materials including NC, acrylic copolymers, epoxy resins, chlorinated rubber, and alkyd resins; strong coupling agent with some resin systems in water-dilutable coatings. *Eastman.* Name unverified.

9668 Ektasolve® EB
111-76-2 1594 203-905-0
$C_6H_{14}O_2$
Butoxyethanol
butyl cellosolve; 2-butoxyethanol; ethylene glycol monobutyl ether. Solvent for alkyd, phenolic, maleic, and cellulose nitrate resins; excellent retarder for lacquers, improving gloss and flow-out, blush resistance, and reducing the formation of orange peel; useful in formulating hot-spray, brushing, flow-coat, and aerosol lacquers. bp = 171-172°; d_4^{20} = 0.9012; n_D^{20} = 1.4196; soluble in H_2O, organic solvents; LD_{50} (rat orl) = 1.48 g/kg. *Eastman.*

9669 Ektasolve® EB Acetate
112-07-2 203-933-3
Butoxyethanol acetate
High-boiling; useful as coalescing aid for latex paints; used in multicolor lacquers, lacquer emulsions; retarder in high-low lacquer thinners, printing inks, and epoxy coatings; used as solvent in silk-screen, stamp pad, and stencil inks and component of varnish removers. d = 0.9410 *Eastman.*

9670 Ektasolve® EEH
Ethylene/diethylene glycol 2-ethythexyl ether
Solvent *Eastman.* Name unverified.

9671 Ektasolve® EP
2807-30-9 220-548-6
$C_5H_{12}O_2$
Ethylene glycol monopropyl ether
2-Propoxyethanol; propoxyethanol; ethylene glycol monopropyl ether; Propyl cellosolve; 3-Oxa-1-hexanol. Slow evaporating solvent used in coatings; useful in waterborne coating systems; coupling solvent for resin/water systems; controls viscosity of waterborne resins; effective for NC, acrylic, epoxy, polyamide, and alkyd resins; retarder in coating systems. mp = -70°; bp = 149°; d = 0.9130; soluble in H_2O, organic solvents; LD_{50} (rat orl) = 3089 mg/kg. *Eastman.*

9672 Ektasolve® PM Acetate
108-65-6 203-603-9
Propylene glycol methyl ether acetate
Retarder solvent. bp = 150°; d = 0.9690 *Eastman.*

9673 Ektogan
zinc oxide-zinc peroxide
zinc perhydrol; zinconal; ektogen. A preparation of zinc oxide, containing from 40-60% zinc peroxide; an antiseptic used for dressing wounds and burns, also as an astyptic.

9674 Elacid CLR
Elastin partial hydrolysate with elastin specific protein structures in weakly acid hydrophilic medium; products for aging and inelastic skin. *Dr. Kurt Richter; Henkel/Cospha.*

9675 Elacid Richter
Oil-water hair conditioner concentrate for damaged hair; antistat. *Dr. Kurt Richter; Henkel/Cospha.*

9676 Elaol
Stated to be dibutyl phthalate; a proprietary plasticizer. No manufacturer.

9677 Elaol 1
A trade name for a plasticizer made from C_4 - C_6 paraffin fatty acids and hexanetriol. No manufacturer.

9678 Elaol 2
A trade name for a plasticizer made from C_6 - C_9 paraffin fatty acids and hexanetriol. No manufacturer.

9679 Elaol 3
A trade name for a plasticizer made from C_4 - C_6 paraffin fatty acids and pentaerythri-l. No manufacturer.

9680 Elaol 4
A trade name for a plasticizer made from C_6 - C_9 paraffin fatty acids and pentaerythri-l. No manufacturer.

9681 Elaol VI
Flame retardant hydraulic fluid; for use as a safety precaution (fire and consequential damage) in mining. *Bayer AG.*

9682 Elargol
A silver finish for mica and plastics. *Octel Chemicals Ltd.* Discontinued.

9683 Elastalloy® 6713
Contains Kraton® polymer. *GLS Plastics.*

9684 Elastan®
Polyol-isocyanate formulations; systems for sports ground surfaces, play grounds, tennis courts, artificial lawns. *BASF AG.*

9685 Elas-Tein AS-20
Elastin, ethyl ester in ethanol; cosmetic ingredient. *Maybrook.*

9686 Elasti-glass
A proprietary trade name for a vinyl copolymer used for belts, braces, raincoats, tobacco pouches, etc. No manufacturer.

9687 Elastinhydrolysate, Liquid
Hydrolyzed animal elastin; filmformer; moisturizer; enhances system stability. *Henkel/Cospha; Henkel Canada.* Name unverified.

9688 Elastite
A sulfurized oil rubber substitute; also a proprietary flooring block made from asphalt, fiber, and fillers. No manufacturer.

9689 Elastoblend 8480
Emulsion polybutadiene/SBR (40/60), 55 parts N-234 carbon black, 20 parts high aromatic oil; masterbatch for high grade truck retreads for both precure and mold cure. *Ameripol Synpol.* Discontinued.

9690 Elastobond
Elastomeric bonding bitumen. *Feb Ltd.*

9691 Elastocarb Tech Light, Heavy
546-93-0 208-915-9
$CMgO_3$
Magnesium carbonate
Magnesite;Magnesium Carbonate Hydroxide; Magnesium Carbonate Basic; Carbonic acid, magnesium salt (1:1). Inorganic filler providing flame retardancy and smoke suppression to elastomers, plastics, and thermosets incl. EPDM, PP, PE, PVC; used in wire and cable compounds, conduit/tubing, film and sheet. *Morton Int'l.*

9692 Elastocarb UF
546-93-0 208-915-9
Magnesium carbonate
Precipitated ultra-fine grade filler for use in conduit, wire and cable compounds, aircraft interior components, other transportation uses; reduces smoke dens. and increases physical properties. *Morton Int'l./Plastics Additives.*

9693 Elastocell®
Microcellular polyurethane suspension components. *BASF.*

9694 Elastocoat®
PU; casting and coating systems. *BASF AG.*

9695 Elastoflex®
Soft PU; foam systems. *BASF AG.*

9696 Elastofoam® I
Polyol-isocyanate formulations; for polyurethane flexible integral skin foam systems for decorative parts subject to heavy duty. *BASF AG.*

9697 Elastoid 1300
High build elastomeric coatings based on multiphase synthetic rubber copolymers; tough, flexible, corrosion and weather resistant, abrasion and impact resistant; for heavy-duty protection in aggressive chemical, industrial and marine environments; extremely low water vapor transmission rate; used on metal, concrete, foam insulation; unaffected by immersion in H_2O. *Dampney Company Inc.* Name unverified.

9698 Elastolac
A proprietary trade name for a shellac derivative; water and alcohol soluble. No manufacturer.

9699 Elastolit® D
Polyol-isocyanate formulations; systems for rigid integral skin foams for production of desks, chairs, cabinets, window profiles, cable guides, molded parts, housing. *BASF AG.*

9700 Elastolith
A synthetic resin. No manufacturer.

9701 Elastollan® 1154D
Thermoplastic polyether PU elastomer; offers excellent low temperature properties, hydrolysis and fungus resistant; for injection and blow molding and extrusion. *BASF.* Name unverified.

9702 Elastollan® C-59D
Thermoplastic polyester PU elastomer; exhibits very good hydrolytic stability, excellent oil, fuel, and solvent resistance; for injection and blow molding and extrusion. *BASF.* Name unverified.

9703 Elastollan® S-60D
Thermoplastic polyester PU elastomer; exhibits good hydrolytic stability, good oil, fuel, and solvent resistance; for injection and blow molding and extrusion. *BASF.* Name unverified.

9704 Elastomag® 100
1309-48-4 5713 215-171-9
Magnesium oxide
Chemical thickener for polyester resins; anticaking agent; used in synthetic rubber compounding, adhesives, fuel oil additives, and as acid acceptor for specialty plastics. *Akrochem.*

9705 Elastopal®
PU elastomers; for casting. *BASF AG.*

9706 Elastopan®
PU; Shoe foam systems. *BASF AG.*

9707 Elastopor®
Hard PU; foam systems. *BASF AG.*

9708 Elastopreg®
Thermoplastic fiber composite; Semifinished products of glass mat-reinforced thermoplastics. *Elastogran.*

9709 Elastorid®
Wood dust-filled PP sheets; carrier for inner trim of passenger cars. *BASF AG.*

9710 Elastosil®
RTV-1 silicone rubbers; adhesives, sealing, coating, sealants in the electrical and electronics industry, seals in the automobile industry. *Wacker-Chemie GmbH.*

9711 Elastosil® LR 3001
63394-02-5
Liquid silicone rubber; used for anode caps; flame retardant. *Wacker Silicones.*

9712 Elastosil® LR 3003/20
63394-02-5
Liquid silicone rubber; features high mechanical strength; for precision molded components, o-rings, gaskets; approved for health care, food contact applications. *Wacker Silicones.*

9713 elaterite
Elastic bitumen
mineral caoutchouc; Helenite. A fossil resin, resembling asphaltum, found in some of the lead mines in Derbyshire. It contains 6-7% mineral matter, and is slightly soluble in ether.

9714 Elaterite, Artificial
A proprietary product made from liquid bitumen and vegetable oils, then treatment with heat and pressure with sulfur chloride, saltpeter, and sulfur; used for waterproofing and insulation. No manufacturer.

9715 Elbasol
Solvent dyestuffs; for coloration of non-aqueous solvents. *Holliday Dyes & Chemicals Ltd.*

9716 Elbelan
2:1 Metal complex dyestuffs; for wool, nylon and blends containing one or both of these fibers. *Holliday Dyes & Chemicals Ltd.* Discontinued.

9717 Elbelene
Dyes for polypropylene. *Holliday Dyes & Chemicals Ltd.* Discontinued.

9718 Elbenyl
Anionic dyestuffs; dyes specially selected for their suitability for dyeing nylon. *Holliday Dyes & Chemicals Ltd.*

9719 Elbeplast
Mainly azo and anthraquinone-based solvent dyes; for coloration of plastics. *Holliday Dyes & Chemicals Ltd.*

9720 Elbestret
For desulfurization of gas. *Holliday Dyes & Chemicals Ltd.*

9721 Elcema® F150, G250, P100
9004-34-6   2012   232-674-9
Cellulose NF
Anticaking agent, tabletting aid for pharmaceutical industry. *Degussa.*

9722 Elcomet
A proprietary trade name for a steel containing chromium, silicon, copper, and nickel. No manufacturer.

9723 Eldopaque
123-31-9   4853   204-617-8
Hydroquinone
Depigmentor. *ICN Pharmaceuticals Inc.*

9724 Electrafil® F-1700/CF/10/A
25134-01-4
PPO, modified, 10% PAN carbon fiber-reinforced; static dissipative and conductive thermoplastic. *Akzo Engineering Plastics.* Name unverified.

9725 Electrafil® F-4/CN/40
Nylon, nickel-coated carbon fiber filled; static dissipative and conductive thermoplastic. *Akzo Engineering Plastics.* Name unverified.

9726 Electrafil® G-1/SS/5
Nylon, stainless steel fiber filled; static dissipative and conductive thermoplastic. *Akzo Engineering Plastics.* Name unverified.

9727 Electrafil® G-50/SS/10
PC, 10% stainless steel fiber filled; static dissipative and conductive thermoplastic. *Akzo Engineering Plastics.* Name unverified.

9728 Electrafil® G-1204/SS/3
9003-56-9
ABS, 3% stainless steel fiber filled; static dissipative and conductive thermoplastic. *Akzo Engineering Plastics.* Name unverified.

9729 Electrafil® G-1704/SS/5
25134-01-4
Polyphenylene oxide
Polyphenylene oxide, modified, 5% stainless steel fiber filled; static dissipative and conductive thermoplastic. *Akzo Engineering Plastics.*

9730 Electrafil® G-1854/SS/7
26062-94-2
PBT, 7% stainless steel fiber filled; static dissipative and conductive thermoplastic. *Akzo Engineering Plastics.* Name unverified.

9731 Electrafil® J-1/30/CF/7/H
32131-17-2
Nylon 66, 30% PAN carbon fiber-reinforced; static dissipative and conductive thermoplastic. *Akzo Engineering Plastics.* Name unverified.

9732 Electrafil® J-2/CF/30
9008-66-6
Nylon 6/10, 30% PAN carbon fiber-reinforced; static dissipative and conductive thermoplastic. *Akzo Engineering Plastics.* Name unverified.

9733 Electrafil® J-3/CF/30
25038-54-4   6832
Nylon 6, 30% PAN carbon fiber-reinforced; static dissipative and conductive thermoplastic. *Akzo Engineering Plastics.* Name unverified.

9734 Electrafil® J-4/CF/30
Nylon 6/12, 30% PAN carbon fiber-reinforced; static dissipative and conductive thermoplastic *Akzo Engineering Plastics.* Name unverified.

9735 Electrafil® J-7/20/EC
Nylon, carbon black filled; static dissipative and conductive thermoplastic. *Akzo Engineering Plastics.* Name unverified.

9736 Electrafil® J-30/CF/20
9003-53-6   9028
PS, 20% PAN carbon fiber-reinforced; static dissipative and conductive thermoplastic *Akzo Engineering Plastics.* Name unverified.

9737 Electrafil® J-50/20/CF/10
PC, 10% PAN carbon fiber-reinforced; static dissipative and conductive thermoplastic. *Akzo Engineering Plastics.* Name unverified.

9738 Electrafil® J-60/CF/30
9003-07-0   7741
PP, 30% PAN carbon fiber-reinforced; static dissipative and conductive thermoplastic. *Akzo Engineering Plastics.* Name unverified.

9739 Electrafil® J-80/CF/10/TF/10
105-57-7   36   203-310-6
Acetal, 10% PAN carbon fiber-reinforced, 10% PTFE; static dissipative and conductive thermoplastic. *Akzo Engineering Plastics.* Name unverified.

9740 Electrafil® J-100/CF/30
9009-54-5
Polyurethane, thermoplastic, 30% PAN carbon fiber-reinforced; static dissipative and conductive thermoplastic. *Akzo Engineering Plastics.* Name unverified.

9741 Electrafil® J-1100/CF/30
PES, 30% PAN carbon fiber-reinforced; static dissipative and conductive thermoplastic. *Akzo Engineering Plastics.* Name unverified.

9742 Electrafil® J-1105/CF/30
PEEK, 30% PAN carbon fiber-reinforced; static dissipative and conductive thermoplastic. *Akzo Engineering Plastics.* Name unverified.

9743 Electrafil® J-1106/CF/30
61128-46-9
Polyetherimide
30% polyacrylonitrile, carbon fiber-reinforced; static dissipative and conductive thermoplastic. *Akzo Engineering Plastics.* Name unverified.

9744 Electrafil® J-1200/CF/10
9003-56-9
ABS, 10% PAN carbon fiber-reinforced; static dissipative and conductive thermoplastic. *Akzo Engineering Plastics.*

9745 Electrafil® J-1300/CF/30/TF/15
9016-75-5
PPS, 30% PAN carbon fiber-reinforced, 15% PTFE; static dissipative and conductive thermoplastic *Akzo Engineering Plastics.* Name unverified.

9746 Electrafil® J-1400/CF/20
ETFE, 20% PAN carbon fiber-reinforced; static dissipative and conductive thermoplastic *Akzo Engineering Plastics.* Name unverified.

9747 Electrafil® J-1500/CF/20
25135-51-3
Polysulfone
20% PAN carbon fiber-reinforced; static dissipative and conductive thermoplastic. *Akzo Engineering Plastics.*

9748 Electrafil® J-1700/CF/10
25134-01-4
PPO, modified, 10% PAN carbon fiber-reinforced; static dissipative and conductive thermoplastic. *Akzo Engineering Plastics.* Name unverified.

9749 Electrafil® J-1701/CF/10/FR
PPE, modified, 10% PAN carbon fiber-reinforced; flame retardant static dissipative and conductive thermoplastic *Akzo Engineering Plastics.* Name unverified.

9750 Electrafil® J-1800/CF/30
25038-59-9 7730
PET, 30% PAN carbon fiber-reinforced; static dissipative and conductive thermoplastic. *Akzo Engineering Plastics.* Name unverified.

9751 Electrafil® J-1850/CF/30
26062-94-2
PBT, 30% PAN carbon fiber-reinforced; static dissipative and conductive thermoplastic. *Akzo Engineering Plastics.* Name unverified.

9752 Electrafil® JM-61/CF/10
9003-07-0 7741
PP, 10% PAN carbon fiber-reinforced; static dissipative and conductive thermoplastic. *Akzo Engineering Plastics.* Name unverified.

9753 Electrafil® M-1526/EC
Nylon, carbon black filled; static dissipative and conductive thermoplastic. *Akzo Engineering Plastics.* Name unverified.

9754 Electrafil® PC-50/EC
PC, carbon black filled; static dissipative and conductive thermoplastic. *Akzo Engineering Plastics.* Name unverified.

9755 Electrafil® PE-90/EC
9002-88-4 7728
HDPE, carbon black filled; static dissipative and conductive thermoplastic. *Akzo Engineering Plastics.* Name unverified.

9756 Electrafil® PP-60/CC/20/EC
9003-07-0 7741
PP, carbon black and calcium carbonate filled; static dissipative and conductive thermoplastic. *Akzo Engineering Plastics.* Name unverified.

9757 Electrafil® TR-1900/EC
9002-84-0 7743 204-126-9
TPE, carbon black filled; static dissipative and conductive thermoplastic. *Akzo Engineering Plastics.* Name unverified.

9758 Electran
Reagents for electrophoresis. *BDH Chemicals Ltd.*

9759 Electrathane
Antistatic castable polyurethane; passive static discharge, charged transfer roller (photo copiers). *PEI Precision Elastomers Inc.* Unverified.

9760 Electraurol
A form of colloidal gold.

9761 Electric bronze
An alloy of 87% copper, 7% tin, 3% zinc, and 3% lead.

9762 Electricidal
Electro-collodial iridium.

9763 Electriridol
Colloidal iridium.

9764 Electrisil
A proprietary silicone rubber composition used for insulating conductors. *GE Silicones.* Name unverified.

9765 Electrisil 758
A proprietary flame retardant silicone rubber compound used to insulate high-voltage cables. *GE Silicones.* Name unverified.

9766 Electrisil 9025
A proprietary silicone rubber compound used in applications where radiation and high temperatures may be encountered. *GE Silicones.* Name unverified.

9767 Electrit
A trademark for goods of the abrasive and refractory class, the essential constituent of which is crystalline alumina. No manufacturer.

9768 Electroclear
Clear lacquers used for metal protection. *ICI Chem & Polymers Ltd.*

9769 Electrocuprol
A form of colloidal copper.

9770 Electrodag®
Dispersions of conducting pigment in resin; used for silk screen printable conducting inks for both flexible and rigid circuitry; shielding coatings; conducting coatings; heat generating coatings and inks; conducting impregnants. *Acheson Colloids.*

9771 Electrodag® 112
Graphite pigment, acrylic binder in water; EMC shielding coating for plastics; protects sensitive electronic equipment *Acheson Colloids.*

9772 Electrodag® 415
Silver pigment, PVC binder in SB-1 or MIBK; EMC shielding coating for plastics; protects sensitive electronic equipment. *Acheson Colloids.*

9773 Electrodag® 415C
Silver pigment, acrylic binder in SB-1; EMC shielding coating for plastics; protects sensitive electronic equipment. *Acheson Colloids.*

9774 Electrodag® 4371
Copper pigment, acrylic binder in SB-1, SB-8, or SB-10; EMC shielding coating for plastics; protects sensitive electronic equipment. *Acheson Colloids.*

9775 Electrodag® 438
Copper/silver pigment, acrylic binder in SB-1, SB-8, or SB-10; EMC shielding coating for plastics; protects sensitive electronic equipment. *Acheson Colloids.*

9776 Electrodag® 439
Nickel pigment, acrylic binder in SB-1, SB-8, or SB-10; EMC shielding coating for plastics; protects sensitive electronic equipment. *Acheson Colloids.*

9777 Electrodag® 442
Nickel pigment, polyester binder in MEK; xylol (1:1); EMC shielding coating for steel frames; protects sensitive electronic equipment. *Acheson Colloids.*

9778 Electrodag® 550
Nickel pigment, acrylic binder in SB-8 or SB-10; EMC shielding coating for plastics; protects sensitive electronic equipment. *Acheson Colloids.*

9779 Electrodag® 24501
Silver pigment, urethane binder in PM acetate; EMC shielding coating for composites/aluminum; protects sensitive electronic equipment. *Acheson Colloids.*

9780 Electrodyn
Sprayers. *ICI Chem & Polymers Ltd.*

9781 electro-filtros
A diaphragm material. It consists of grains of pure crystalline silica cemented together with a fused siliceous binding substance; used in electrolytic processes.

9782 Electrofine® S-70
63449-39-8 7156 264-150-0
Paraffin wax, chlorinated
Chlorowax 500c; Chlorowax 40; Paraffin waxes and Hydrocarbon waxes chlorinated; Chlorinated wax; Chlorinated waxes. Flame retardant for plastics, textiles, paper and cardboard in conjunction with antimony trioxide; improves hardness, gloss and resist. to acids and bases in paints; solvent; lubricant; plasticizer for PVC in PE sealants. d = 0.900-1.50; insoluble in H_2O, slightly soluble in alcohol; soluble in benzene, chloroform; LD_{50} (rat orl) >4000 mg/kg. *Elf Atochem.*

9783 electro-granodized iron and steel
A process for forming a rust-preventing coat on iron and steel, An alternating current plates a continuous coating of zinc phosphate on the surface to be protected.

9784 Electrolyilc chlorogen (E.C.)
A chlorinated soda prepared by the electrolysis of brine.

9785 Electromartiol
A form of colloidal iron.

9786 Electromate
Zinc plating specialty chemicals. *Stowlin Ltd.*

9787 Electromercurol
A form of colloidal mercury.

9788 Electronite
A safety explosive containing 75% ammonium nitrate, 5% barium nitrate, with wood meal and starch.

9789 Electronite No. 2
An explosive consisting of 95% ammonium nitrate, and 5% wood meal and starch.

9790 Electropalladiol
A form of colloidal palladium.

9791 Electroplatinol
A form of colloidal platinum.

9792 Electrorhodiol
A form of colloidal rhodium.

9793 Electrorubin
A trademark for abrasive and refractory materials; the essential constituent is crystalline alumina. No manufacturer.

9794 Electrose
A proprietary trade name for a shellac plastic. No manufacturer.

9795 Electroselenium
A form of colloidal selenium.

9796 Electrotype metal
An alloy of 93% lead, 4% antimony, and 3% tin.

9797 Electro-Wash®
General-purpose cleaning agent, degreaser for removal of encrusted dirt, grease, oxidation and contaminants; for use with electrical interfaces, contacts and connectors. *Chemtronics.*

9798 Electrox
1314-13-2 10279 215-222-5
zinc oxide
Photoconductive zinc oxides. *Durham Chemicals Ltd.*

9799 Electrozone
A similar preparation to Chloros (sodium hypochlorite solution); a disinfectant.

9800 **Electrundum**
A trademark for materials of the abrasive type and consisting essentially of alumina. No manufacturer.

9801 **Elefac I-205**
Octyldodecyl neopentanoate
Dry emollient, SPF booster, pigment wetter and binder. *Bernel.*

9802 **Elektra**
Organic brightener system; acid copper electroplating (decorative). *Harshaw Chemicals Ltd.*

9803 **Elektron®**
Alloys, flues, and hardeners. *Magnesium Elektron Ltd.*

9804 **Elektron®**
A trademark used in connection with certain magnesium alloys containing up to about 10% of various alloying constituents, such as aluminum, zinc and manganese; used in cast and wrought forms for aero engines and other purposes. No manufacturer.

9805 **Elemite**
A proprietary trade name for a wetting agent and detergent; a combination of sulfonated oils and solvents. No manufacturer.

9806 **Elephant bronze**
An alloy of 85% copper, 10.5% tin, 2.75% zinc, 1.5% lead, and 0.1-0.2% phosphorus.

9807 **Elephant-S bronze**
An alloy of 80.5% copper, 10.2% tin, 9% antimony, and 0.1-0.3% phosphorus.

9808 **Eleudron-Solution**
Sulfonamide
Used for coccidiosis in poultry; veterinary medicine. *Bayer AG.*

9809 **Elexar®**
A proprietary range of thermoplastic rubbers designed for use in the cable industry. *Shell.* Name unverified.

9810 **Elexar® 8421**
Thermoplastic elastomer; UL recognized, 105 C flexible cord material for insulation, jacketing, and molding applications; excellent abrasion and impact resistance, heat aging characteristics. *Shell.*

9811 **Elfacos®**
A range of nonionic emulsifiers and stabilizers used in water-in-oil and oil-in water cosmetic formulations. *Akzo Chemie UK Ltd.*

9812 **Elfan® 200**
Sodium lauryl sulfate
Sodium lauryl sulfate (C_{12}) as a fine white powder; anionic surfactant used in toothpastes. *Akzo Chemie Nederland BV.*

9813 **Elfan® 240 and 240S**
Sodium lauryl sulfate, either natural (240) or based on a synthetic fatty alcohol (240S); supplied as a transparent to white paste; detergent and emulsifier used in shampoos, light duty detergents and cleaning pastes. *Akzo Chemie Nederland BV.*

9814 **Elfan® 240M and 240M/S**
Monoethanolamine lauryl sulfate, either natural (240M) or based on a synthetic fatty alcohol (240M/S); supplied as a clear, yellowish, medium viscous liquid; detergent and emulsifier for shampoos and bubble baths. *Akzo Chemie Nederland BV.*

9815 **Elfan® 240T and 240T/S**
139-96-8 205-388-7
Triethanolamine lauryl sulfate, either natural (240T) or based on a synthetic fatty alcohol (240T/S); supplied as a yellowish clear liquid; detergent for shampoos and bubble baths. *Akzo Chemie Nederland BV.*

9816 **Elfan® 280**
Sodium coconut fatty alcohol sulfate (C_{12}-C_{18})
Supplied as a white powder or paste; detergent raw material for light duty detergents, all-purpose washing agents and hand cleansers. *Akzo Chemie Nederland BV.*

9817 **Elfan® 680**
Sodium oleyl-cetyl alcohol sulfate
Supplied as a yellowish-brown paste; detergent for heavy and light duty detergent powders, washing and cleaning pastes. *Akzo Chemie Nederland BV.*

9818 **Elfan® A432**
Amphoteric surfactant supplied as a clear yellowish liquid; used for baby shampoos; bubble baths; strong acid and alkaline cleaning detergents. *Akzo Chemie Nederland BV.*

9819 **Elfan® KT550**
Anionic surfactant in which the cation is sodium and the anion is composed of 50% coconut/50% tallow fatty alcohol sulfate; supplied as a white paste; detergent raw material for heavy and light duty detergents, and washing powders. *Akzo Chemie Nederland BV.*

9820 **Elfan® NS 242, 243S, 252 S**
Anionic surfactant in liquid or paste form; used for shampoos, bubble baths, dishwashing liquids, light duty liquids, washing pastes, car shampoos. *Akzo Chemie UK Ltd.*

9821 **Elfan® NS 243 S Mg**
Anionic surfactant with magnesium as counter ion giving excellent mildness qualities; used for shampoos, bubble baths, special care and baby products. *Akzo Chemie UK Ltd.*

9822 **Elfan® NS 682 KS**
Anionic surfactant supplied as a yellowish paste; used for shampoos, bubble baths, dishwashing liquids, light duty liquids, washing pastes, car shampoos. *Akzo Chemie UK Ltd.*

9823 **Elfan® OS 46**
Sodium α-olefine sulfonate (C_{14}/C_{16})
Supplied as a yellowish liquid; anionic surfactant for shampoos, bubble baths, dishwashing detergents, liquid and paste-form cleaners. *Akzo Chemie UK Ltd.*

9824 **Elfan® WA Series**
Dodecylbenzene sulfonate as triethanolamine or sodium salt or in acid form; supplied as liquid, paste or power, anionic surfactants used in heavy and light duty, all-purpose and dishwashing detergents, scouring and other powder formulations. *Akzo Chemie Nederland BV.*

9825 **Elfanol® 510**
Sodium sulfosuccinic acid monoester of a fatty acid alkylolamide; creamy colored paste; for washing and cleaning pastes; hand cleansers. *Akzo Chemie UK Ltd.*

9826 **Elfanol® 616**
Sodium sulfosuccinic acid monoester of an ethoxylated fatty alcohol; yellowish viscous liquid; used for shampoos, bubble baths, baby baths, liquid hand cleansers. *Akzo Chemie UK Ltd.*

9827 **Elfanol® 850**
Sodium sulfosuccinic acid monoester of an ethoxylated fatty acid; yellow-brown, nearly clear liquid; used for baby baths, shampoos, bubble baths, liquid hand cleansers. *Akzo Chemie UK Ltd.*

9828 **Elfanol® 883**
577-11-7 3460 209-406-4
Docusate Sodium
Sodium dioctylester of sulfosuccinic acid; colorless to slight yellow liquid; wetting agent for technical processes. *Akzo Chemie UK Ltd.*

9829 **Elfapur® N50**
Nonylphenol ethoxylate nonionic surfactant in the form of a clear, nearly colorless liquid; used for dishwashing detergents for automatic machines; emulsifiers for fats and mineral oils. *Akzo Chemie Nederland BV.*

9830 **Elfapur® N70**
Nonylphenol ethoxylate nonionic surfactant in the form of a clear, nearly colorless liquid; low foaming dishwashing detergents for automatic machines, industrial and solvent cleansers, fat-dissolving pastes, all purpose washing pastes. *Akzo Chemie Nederland BV.*

9831 **Elfapur® N90, N120 and N150**
Nonylphenol ethoxylate nonionic surfactants in liquid or paste form; wide range of detergents, cleansers, car shampoos, fat-dissolving pastes, all purpose washing pastes. *Akzo Chemie Nederland BV.*

9832 **Elftex® 675**
1333-86-4 1856 215-609-9
Carbon black; for coloring plastics; very good uv protection. *Cabot; Cabot Carbon Ltd.*

9833 **Elfugin**
Wash-fast antistatic agent; for synthetic fibers. *Sandoz Prods. Ltd.* Name unverified.

9834 **Elhuyarite**
A red allophane mineral.

9835 **Elianite I**
An acid-resisting alloy, containing 82% iron, 15% silicon, and 0.6% manganese.

9836 **Elianite II**
An acid-resisting alloy, consisting of 81% iron, 15% silicon, 0.5% manganese, 2.2% nickel, 0.8% carbon, and 0.06% phosphorus.

9837 **Eliminal**
Aluminum-removing flux for copper alloys. *Foseco (F.S.) Ltd.*

9838 **Elimite®**
52645-53-1 7321 258-067-9
Permethrin
Scabicide. *Allergan Inc.*

9839 **Elintaal**
1-Ethoxy-1-(3,7-dimethyl-1,6-octadienyloxy)-ethane
Used in perfumery. *Quest Int'l. UK Ltd.*

9840 Elinvar
A nickel steel containing 36% nickel, 46% iron, 12% chromium, 4% tungsten, and 1-2% manganese.; used for the more delicate parts of watches.

9841 Elite Fast
Anionic dyestuffs (neutral dyeing); for dyeing shades of wool with good wash fastness. *Holliday Dyes & Chemicals Ltd.*

9842 Eljon
Diazo yellow and orange pigment; azo yellow and red pigments; toners; used for printing inks, paints, plastics and artists colors. *European Colour (Pigments) Ltd.*

9843 Elkalub
Oils, oil additives, and greases. No manufacturer.

9844 Elkem Microsilica
Raw and processed amorphous silica (condensed silica fume); for refractories, polymers, insulation, fluid cracking catalysts and a range of chemical and mineral uses. *Elkem Chemicals Inc.* Name unverified.

9845 Elkonite
A copper-tungsten alloy used for making welding dies. It has a Brinnell hardness of 225, a compression strength of 208,000 lb per sq. in, and is not annealed at red heat.

9846 Ellagitannin
A variety of tannin found in divi-divi, knoppern, and myrobaians.

9847 ellagite
A mineral. It is a variety of natrolite.

9848 Elliott's Lawn Sand
10028-22-5 4079 233-072-9
$Fe_2O_{12}S_3$
Ferric sulfate
Iron (III) sulfate; ferric persulfate; ferric sesquisulfate; ferric tersulfate; Sulfuric Acid, Iron(3+) Salt (3:2); Iron Tersulfate; Ferric sulfate monohydrate. Used for moss control in turf. *Thomas Elliott Ltd.* Name unverified.

9849 Elliott's Moss Killer
10028-22-5 4079 233-072-9
Ferric sulfate
Used for moss control in turf. *Thomas Elliott Ltd.*

9850 Elmarid
An alloy of 89% tungsten, 4.5% cobalt, 5.9% carbon, and 0.4% iron.

9851 Elner's German silver
A nickel silver containing 57.4% copper, 26.6% zinc, 13% nickel, and 3% iron.

9852 Elocril
18181-70-9 5053 242-069-1
Iodofenphos
An organophosphorus insecticide. *Ciba-Geigy Agrochemicals.*

9853 Elotex
Redispersible homo or copolymer powders (powdered emulsions); for building products. *Ebnother AG.* Name unverified.

9854 Eloxal
A proprietary trade name for an anodized aluminum. No manufacturer.

9855 ELP-3
Epoxy terminated polysulfide polymer; modifier for epoxy resins; used in concrete adhesives, chemically resistant, linings or coatings, bonding to metallic substrates with oily finishes. *Morton Int'l./Polymer Systems.*

9856 Elsner's reagent
A basic zinc chloride solution obtained by dissolving 500 g zinc chloride and 20 g zinc oxide in 425 ml water and warming. A solvent for silk.

9857 Eltaga®
Fiber. *DuPont UK.*

9858 Eltesol®
Aromatic sulfonates for laundry detergents. *Albright & Wilson Ltd.*

9859 Eltesol® 4009, 4018
Xylene sulfonic acid modified with methanol and sulfuric acid; catalysts for curing cold-setting phenol-formaldehyde and phenol-furane resins used in the foundry industry as binders for sand in the production of molds and cores. *Albright & Wilson UK.*

9860 Eltesol® 4402, 4403, FDA 55/8
Alkylaryl sulfonate anionic in acid form; used widely in the foundry industry for curing cold setting resins. *Albright & Wilson Ltd, Detergents Div, Marchon.*

9861 Eltesol® 5400 Series
Phenol sulfonic acid condensates; surfactant for leather processing. *Albright & Wilson Am.*

9862 Eltesol® 7200 Series
Dihydroxy diphenyl sulfonates
Surfactant for leather processing. *Albright & Wilson Am.*

9863 Eltesol® AC60
37475-88-0 253-519-1
$C_9H_{15}NO_3S$
Ammonium cumene sulfonate
Benzenesulfonic acid, (1-methylethyl)-, ammonium salt. Surfactant, hydrotrope for agricultural applications. *Albright & Wilson UK.*

9864 Eltesol® ACS 60
Alkylaryl sulfonate anionic
Pale yellow liquid; for manufacture of liquid detergent formulations. *Albright & Wilson Ltd, Detergents Div, Marchon.*

9865 Eltesol® AX 40
26447-10-9 247-710-9
Ammonium xylene sulfonate
Hydrotrope, cloud point depressant used in detergent manufacture; solubilizer, coupler. *Albright & Wilson UK.* Name unverified.

9866 Eltesol® CA 65
28631-63-2 249-112-3
Cumene sulfonic acid
Catalyst for foundry resins; descaling agent for metal cleaning; anti-stress additive and plating aid in electroplating bath; curing aid in the plastics industry; raw material in the manufacture of dyes and pigments; detergents industry. *Albright & Wilson UK.*

9867 Eltesol® MGX
Magnesium xylene sulfonate
Hydrotrope for liquid and spray-dried detergent formulations. *Albright & Wilson UK.* Name unverified.

9868 Eltesol® PSA 65
1333-39-7 215-587-0
$C_6H_6O_4S$
m-hydroxybenzenesulfonic acid
Sulfocarbolic acid; Phenol sulfonic acid. Catalyst for foundry resins; descaling agent for metal cleaning; anti-stress additive and plating aid in electroplating bath; curing aid in the plastics industry; raw material in the manufacture of dyes and pigments; detergents industry; pharmaceutical chemicals. *Albright & Wilson UK.*

9869 Eltesol® PT 93
Potassium toluene sulfonate
Hydrotrope, cloud point depressant used in the manufacture of detergents; solubilizer, coupler. *Albright & Wilson UK.*

9870 Eltesol® PX 40, PX 93
Potassium xylene sulfonate
Hydrotrope, solubilizer, coupling agent, and viscosity modifier in liquid. formulations; cloud point depressant in detergent formulations. *Albright & Wilson UK.*

9871 Eltesol® SC 93
32073-22-6 250-913-5
Sodium cumene sulfonate
Hydrotrope for hard surface cleaners. *Albright & Wilson UK.*

9872 Eltesol® ST 40
657-84-1 9671 211-522-5
$C_7H_7NaO_3S$
sodium toluenesulfonate
Sodium *p*-Toluenesulfonate; Benzenesulfonic acid, 4-methyl-, sodium salt. Hydrotrope, solubilizer, coupling agent, and viscosity modifier in liquid formulations; cloud point depressant in detergent formulations. *Albright & Wilson UK.*

9873 Eltesol® SX 30
1300-72-7 215-090-9
$C_8H_9NaO_3S$
Sodium xylene sulfonate
sodium xylenesulfonate; xylenesulfonic acid, sodium salt; dimethylbenzenesulfonic acid, sodium salt; sodium dimethylbenzenesulfonate; Conco SXS; Cyclophil SXS30; Eltesol SX 30; hydrotrope; Naxonate; Naxonate G; Stepanate X; Surco SXS; Ultrawet 40SX; Calsoft SXS 96; Alkatrope SX-40; Carsosulf SXS; Eltesol SX93; Reworyl NXS40; Richonate SXS; Witconate SXS; Sodium *m*-xylene sulfonate. Hydrotrope, cloud point depressant used in detergent manufacture; solubilizer, coupler. mp = 27°; bp = 157°; d = 1.023; soluble in H_2O (>10 g/100 ml), less soluble in organic solvents; LD_{50} (rat orl) = 1000 mg/kg. *Albright & Wilson UK.*

9874 Eltesol® TA 65
65% Toluene sulfonic acid and 1.4% sulfonic acid aqueous solution; curing agent for resins in foundry cores, plastics, coatings; intermediate; catalyst in foundry and chemical industries; hardening agent in plastics; activator for nicotine insecticides. *Albright & Wilson Uk.*

9875 Eltesol® TPA
Formulated product; tin-plating additive. *Albright & Wilson UK.*

9876 Eltesol® TSX
104-15-4 9671 203-180-0
$C_8H_{18}O$
p-Toluene sulfonic acid monohydrate BP

Catalyst for organic synthesis, synthetic resins, manufacture of *p*-cresol, toluene derivatives, pharmaceutical products, dyestuffs; chemical intermediate. *Albright & Wilson UK.*

9877 **Eltesol® XA**
25321-41-9 246-839-8
Xylene sulfonic acid aqueous solution; catalyst in foundry and chemical industries; hydrotrope for agricultural formulations. *Albright & Wilson UK.*

9878 **Eltex**
High density polyethylene. *Laporte Industries Ltd.* Discontinued.

9879 **Eltex P**
Polypropylene. *Laporte Industries Ltd.* Discontinued.

9880 **Eludril Mouthwash**
Antibacterial/antifungal mouthwash. 0.1% chlorhexidine, 0.1% chlorbutol, 0.5% chloroform; dilute 10 ml: half glass warm water; used for gingivitis, apthous/dental ulcers and mouth and throat infection; does not stain teeth or composites. *Concept Pharmaceuticals Ltd.* Name unverified.

9881 **Eludril Spray**
Aerosol spray containing chlorhexidine 0.05% and amethecaine 0.015%; antibacterial/antifungal for apthous/dental ulceration and for mouth and throat infections. *Concept Pharmaceuticals Ltd.* Name unverified.

9882 **Elugent™**
Detergent. *Calbiochem Corp.*

9883 **Elvace®**
Vinyl acetate and ethylene emulsion copolymers and terpolymers. *Reichhold.*

9884 **Elvace® 1870**
VAE copolymer emulsion; adhesive base with improved plasticizer response, quick tack, and adhesion compared with homopolymer. *Reichhold.*

9885 **Elvacite®**
Acrylic resins. *DuPont UK.*

9886 **Elvaloy®**
Resin modifiers. *DuPont UK.*

9887 **Elvaloy® EP-4043, HP441**
Ethylene/acrylate/carbon monoxide terpolymer. *DuPont.*

9888 **Elvamide®**
Nylon multipolymer resins. *DuPont UK.*

9889 **Elvanol®**
Polyvinyl alcohol resins. *DuPont UK.*

9890 **Elvanol® 20-25**
9002-89-5 7745
Fully hydrolyzed polyvinyl alcohol; for use in adhesive, textile and paper applications. *DuPont.*

9891 **Elvanol® 71-30**
9002-89-5 7745
Fully hydrolyzed polyvinyl alcohol; film-forming binder used in adhesives, paper, paperboard sizing and coatings, textiles, films, building products, hoses, gaskets, emulsification in emulsions and latexes, additive for concrete, cement, and food. *DuPont.*

9892 **Elvanol® 90-50**
9002-89-5 7745
Fully hydrolyzed polyvinyl alcohol; provides high film strength and binding power in low viscosity systems; used in paper and paperboard coating and sizing, adhesives; pigment binder; food additive application. *DuPont.*

9893 **Elvaron®**
1085-98-9 3095 214-118-7
Dichlofluanid
A wettable powder containing 50% w/w dichlofluanid; fungicide used to control botrytis on strawberries, raspberries, loganberries, blackberries, blackcurrants, redcurrants, gooseberries, outdoor grapes, tomatoes under cover, tulips and peonies; also controls cane spot and stamen blight; reduces raspberry mildew and gives some reduction of spur blight on raspberries, mildew and blackspot on roses, downy mildew on strawberries and leaf spot on blackcurrants and gooseberries. *Bayer AG; Bayer plc.*

9894 **Elvax®**
EVA polymer resins. *DuPont UK.*

9895 **Elvax® 40-W**
EVA copolymer, antiblock; resin useful as base polymers in hot melt and solvent-thinned coatings and adhesives; modifiers to improve properties such as processability, thermoplasticity, adhesion, and abrasion resistance; food packaging applications. *DuPont.*

9896 **Elvax® 260**
EVA copolymer resin; high-viscosity resin, toughness, flexibility, adhesion, and barrier properties for coating and adhesive applications, and flocked or fabric-laminated counters. *DuPont.*

9897 **Elvax® 310**
EVA copolymer resin; wax-compatible resin; enhances properties of hot melt blends containing microcryst. waxes; used in heat-sealable barrier coatings and hot melt adhesives; also for blending with waxes and modifying resins for formulations; giving fiber- or film-tearing bonds between paper substrates and non-porous packing materials such as aluminum foil, PP and Mylar polyester films, K cellophane. *DuPont.*

9898 **Elvax® 550,560**
EVA copolymer resin; high molecular weight and high melt viscosity grades used in paper roll wrap and carpet seaming tape. *DuPont.*

9899 **Elvax® 4260**
EVA/acid terpolymer; high molecular weight resin used in hot melt systems requiring improved adhesion to polar, nonporous substrates; in coatings, provides superior hot tack, improved grease resistance, and optimum barrier properties. *DuPont.*

9900 **Elvax® 4320**
EVA/acid terpolymer; intermediate molecular weight resin higher in viscosity than Elvax 4310, intermediate in performance between Elvax 4310 and 4355; combinable with Elvax 4355 or 4260 to optimize performance at a desired viscosity level. *DuPont.*

9901 **Elvax® D**
Proprietary dispersions of ionomers and vinyl resins. *DuPont UK.* Name unverified.

9902 **Elverite**
A proprietary trade name for the charcoal iron used for crushing mills. No manufacturer.

9903 **Elveron®, Elvon®**
Polymers. *DuPont UK.*

9904 **Em-1**
Lard oil; additive improving wetting and lubricity in industrial applications. *Ferro/Keil.*

9905 **EM-550**
Fatty compound; Lubricity and detergency additive for compounding water-sol. machining fluids; excellent film strength. *Ferro/Keil.*

9906 **EM-600**
61791-00-2
PEG 600 monotallate
Surfactant, emulsifier, wetting agent, detergent for industrial applications, solvent in cutting oils and drawing compounds. *Ferro/Keil.*

9907 **EM-980**
Mixed fatty acid diethanolamide containing excess diethanolamine; solubilizer, viscosity builder, detergency and lubricity aid used in synthetic and semi-synthetic metalworking fluids, floor cleaners, paint strippers, buffing compounds. *Ferro/Keil.*

9908 **EMA**
Ethylene-maleic anhydride copolymers resins; dispersing agents, film formers and chemical intermediates for use in capsule walls, liquid detergents and drilling muds, thickening agents in textile print pastes and cosmetics. *Monsanto Co.*

9909 **Ema Resins**
EMA resins. *Monsanto plc.*

9910 **Emac SP 2205**
Ethylene methyl acrylate copolymer
Excellent heat stability; for coatings and laminations, compounding (impact modification, compatibilizer), as tie layer for LDPE, HDPE, PET, PP, PVDC, EVA, PVC, PC in films. *Monsanto (Solaris).*

9911 **Emaillit**
Bitumin on a solvent base; primer. *Vedag GmbH.* Name unverified.

9912 **Emalex 103**
9004-95-9
Ceteth-3
Emulsifier, cleaner, dispersant for cosmetics, especially creams and lotions; especially suited for creamy hair conditioners *Nihon Emulsion.*

9913 **Emalex 1605**
9004-95-9
Ceteth-5
Polyoxyethylene Isocetyl Ether. Emulsifier, dispersant, solubilizer, cleaner, wetting agent for cosmetics, creams, milky lotions, skin lotions. *Nihon Emulsion.*

9914 **Emalex 1805**
52292-17-8
Isosteareth-5
Emulsifier, dispersant, solubilizer, cleaner, wetting agent for cosmetics, creams, milky lotions, skin lotions. *Nihon Emulsion.*

9915 **Emalex 200 di-IS**
PEG-4 diisostearate
Oil-phase cosmetic ingredient. *Nihon Emulsion.*

9916 **Emalex 200 di-L**
9005-02-1
PEG-4 dilaurate

Oil-phase cosmetic ingredient, emulsifier for creams, milky lotions, hair conditioners; cleaner, superfattening agent, thickener, reforming agent. *Nihon Emulsion.*

9917 **Emalex 200 di-O**
9005-07-6
PEG-4 dioleate
Oil-phase ingredient, emulsifier, dispersant with good spreadability for cosmetics, creams, milky lotions, foundations. *Nihon Emulsion.*

9918 **Emalex 200 di-S**
9005-08-7
PEG-4 distearate
Pearling agent, hydrophobic component, reforming agent, emulsifier, thickener for cosmetics. *Nihon Emulsion.*

9919 **Emalex 218**
9004-96-0
PEG-3 oleat
Surfactant for cosmetics; produces stable emulsions. *Nihon Emulsion.* Discontinued.

9920 **Emalex 300 di-IS**
PEG-6 diisostearate. Emulsifier for cosmetic emulsions, dispersant, reforming agent. *Nihon Emulsion.*

9921 **Emalex 400 di-IS**
PEG-8 diisostearate
Emulsifier for cosmetic emulsions, dispersant, reforming agent. *Nihon Emulsion.*

9922 **Emalex 400A**
9004-99-3
PEG-3 stearate
Oil-phase cosmetic ingredient, pearlescent, emulsifier, dispersant, emulsion stabilizer, thickener. *Nihon Emulsion.*

9923 **Emalex 508**
9004-98-2
Oleth-8
Emulsifier, dispersant, solubilizer for cosmetics; suitable for hair tonics and hair care products as solubilizer for perfumes. *Nihon Emulsion.*

9924 **Emalex 600 di-IS**
PEG-12 diisostearate
Emulsifier for cosmetic emulsions, dispersant, reforming agent. *Nihon Emulsion.*

9925 **Emalex 640**
9005-00-9
Steareth-40
Emulsifier, dispersant, thickener for cosmetics, creams, milky lotions. *Nihon Emulsion.*

9926 **Emalex 709**
Laureth-9
Emulsifier, penetrant, wetting agent, cleaner, dispersant for cosmetics, creams, milky lotions; paint and itch relieving effect for ointments. *Nihon Emulsion.*

9927 **Emalex 805**
9004-99-3
PEG-5 stearate
Oil-phase cosmetic ingredient, pearlescent, emulsifier, dispersant, emulsion stabilizer; thickener for cleansing foam. *Nihon Emulsion.*

9928 **Emalex 2505**
PEG-5 decylpentadecyl ether
Emulsifier, solubilizer, cleaner, thickener for cosmetics, creams, milky lotions, skin lotions. *Nihon Emulsion.* Discontinued.

9929 **Emalex 6300 Di-ST**
9005-08-7
PEG-150 distearate
Thickener, stabilizer for shampoos, hair conditioners, cleansing foams. *Nihon Emulsion.*

9930 **Emalex 6300 M-ST**
9004-99-3
PEG-150 stearate
Oilphase cosmetic ingredient, thickener for shampoos and hair conditioners. *Nihon Emulsion.*

9931 **Emalex BHA-30**
Beheneth-30
Emulsifier, dispersant, solubilizer for cosmetics; produces moist, spreading emulsion. *Nihon Emulsion.*

9932 **Emalex C-20**
61791-12-6
PEG-20 castor oil; emulsifier, solubilizer, dispersant in cosmetics, medical pharmaceuticals. *Nihon Emulsion.*

9933 **Emalex CC-10**
Cetyl caprate
Lipophilic base for cosmetics *Nihon Emulsion.*

9934 **Emalex CC-16**
540-10-3 2073 208-736-6
Cetyl palmitate
Lipophilic base for cosmetics *Nihon Emulsion.*

9935 **Emalex CC-168**
59130-69-7 261-619-1
Cetyl octanoate
Lipophilic base for cosmetics *Nihon Emulsion.*

9936 **Emalex CC-18**
2778-96-3 220-476-5
Stearyl stearate
Lipophilic base for cosmetics *Nihon Emulsion.*

9937 **Emalex CS-5**
Choleth-5
Emulsifier, solubilizer, dispersant, thickener for cosmetics, creams, milky lotions, skin lotions; emollient for hair care products; gloss aid for creams and milky lotions. *Nihon Emulsion.*

9938 **Emalex CWS-3**
PEG-3 cetyl ether stearate
SE emulsifying cosmetic ingredient; hydrophobic component and reforming agent for cosmetics and industrial areas. *Nihon Emulsion.*

9939 **Emalex DEG-di-IS**
PEG-2 diisostearate
Oil-phase ingredient for cosmetic emulsions. *Nihon Emulsion.*

9940 **Emalex DEG-di-L**
9005-02-1
PEG-2 dilaurate
Oilphase cosmetic ingredient, emulsifier for creams, milky lotions, hair conditioners; cleaner, superfattening agent, thickener, reforming agent. *Nihon Emulsion.*

9941 **Emalex DEG-di-O**
PEG-2 dioleate
Oil-phase ingredient, emulsifier, dispersant with good spreadability for cosmetics, creams, milky lotions, foundations. *Nihon Emulsion.*

9942 **Emalex DEG-m-S**
9004-99-3
PEG-2 stearate
Oilphase cosmetic ingredient, pearlescent, emulsifier, dispersant, emulsion stabilizer, thickener. *Nihon Emulsion.*

9943 **Emalex DISG-2**
67938-21-0 267-821-6
Diglyceryl diisostearate
Emulsifier for cosmetics and foods. *Nihon Emulsion.*

9944 **Emalex DISG-3**
Polyglyceryl-3 diisostearate
Emulsifier for cosmetics and foods. *Nihon Emulsion.*

9945 **Emalex DSG-2**
Polyglyceryl-2 distearate
Emulsifier for cosmetics and foods. *Nihon Emulsion.*

9946 **Emalex EG-2854-IS**
PEG-4 sorbitol triisostearate
Oilphase base for emulsions; emulsifier for water-oil emulsions. *Nihon Emulsion.*

9947 **Emalex EG-2854-O**
PEG-4 sorbitol tetraoleate
Oil-phase base for emulsions; emulsifier for water-oil emulsions. *Nihon Emulsion.*

9948 **Emalex EG-2854-S**
PEG-4 sorbitol tristearate
Oil-phase base for emulsions. *Nihon Emulsion.*

9949 **Emalex EG-di-L**
624-04-4 4507 210-827-0
$C_{26}H_{50}O_4$
dodecanoic acid 1,2-ethanediyl ester
Ethylene glycon dilaurate; ethylene dilaurate. Oil-phase cosmetic ingredient, emulsifier for creams, milky lotions, hair conditioners; cleaner, superfattening agent, thickener, reforming agent. Plasticizer in lacquers and varnishes. mp = 50-52°; bp_{20} = 188°; insoluble in EtOH, Et_2O. *Nihon Emulsion.*

9950 **Emalex EG-di-O**
928-24-5 213-170-8
Ethylene glycol dioleate
Oil-phase ingredient, emulsifier, dispersant, plasticizer with good spreadability for cosmetics, creams, milky lotions, foundations. *Nihon Emulsion.*

9951 Emalex EG-di-S
627-83-8 211-014-3
Ethylene glycol distearate
Pearling agent, hydrophobic component, reforming agent, emulsifier, thickener for cosmetics. *Nihon Emulsion.*

9952 Emalex EGS-A
111-60-4 203-886-9
Ethylene glycol monostearate
Oil-phase cosmetic ingredient, pearlescent, emulsifier, dispersant, emulsion stabilizer, thickener; clouding agent for shampoos/hair conditioners. *Nihon Emulsion.*

9953 Emalex ET-2020
9005-64-5 8872
Polysorbate 20
Emulsifier, solubilizer for cosmetics, medical pharmaceuticals. *Nihon Emulsion.*

9954 Emalex ET-8020
9005-65-6 7742
Polysorbate 80
Emulsifier, solubilizer for cosmetics, medical pharmaceuticals. *Nihon Emulsion.*

9955 Emalex ET-8040
PEG-40 sorbitan oleate
Emulsifier, solubilizer for cosmetics and medical pharmaceuticals. *Nihon Emulsion.*

9956 Emalex GM-5
51158-08-8
PEG-5 glyceryl stearate
Emulsifier, solubilizer, thickener for cosmetics, creams, milky lotions, hair conditioners, facial cleansers. *Nihon Emulsion.*

9957 Emalex GMS-55FD
Glyceryl monostearate
SE; surfactant for food, cosmetic and medical pharmaceutical applications; stabilizer, superfatting agent, reforming agent for cleansing foams. *Nihon Emulsion.*

9958 Emalex GMS-A
Glyceryl monostearate
Surfactant for food, cosmetic and medical pharmaceutical applications. *Nihon Emulsion.*

9959 Emalex GMS-ASE
Glyceryl monostearate
SE; surfactant for food, cosmetic and medical pharmaceutical applications. *Nihon Emulsion.*

9960 Emalex GWIS-115
68958-58-7
PEG-15 glyceryl monoisostearate
Emulsion stabilizer, dispersant, emulsifier, solubilizer for cosmetics. *Nihon Emulsion.*

9961 Emalex GWIS-303
PEG-3 glyceryl triisostearate
Oil-phase cosmetic ingredient for creams and milky lotions. *Nihon Emulsion.*

9962 Emalex GWO-303
PEG-3 glyceryl trioleate
Oil-phase cosmetics ingredient. *Nihon Emulsion.*

9963 Emalex GWS-204
PEG-4 glyceryl distearate
Oil-phase cosmetics ingredient. *Nihon Emulsion.*

9964 Emalex GWS-303
PEG-3 glyceryl tristearate
Oil-phase cosmetics ingredient; for lipsticks, ointments. *Nihon Emulsion.*

9965 Emalex HC-5
61788-85-0
PEG-5 hydrogenated castor oil; oil-phase ingredient, water-oil emulsifier, dispersant for cosmetics and medical products. *Nihon Emulsion.*

9966 Emalex J.J O-V
61789-91-1
Jojoba Oil
Cosmetics ingredient. *Nihon Emulsion.* Discontinued.

9967 Emalex K.T.G
65381-09-1 265-724-3
Caprylic/capric acid triglyceride
Oil-phase cosmetic ingredient; emollient. *Nihon Emulsion.*

9968 Emalex LWIS-2
PEG-2 lauryl ether isostearate
SE emulsifying cosmetic ingredient; oily base for creams, milky lotions, hair conditioners. *Nihon Emulsion.*

9969 Emalex LWS-3
PEG-3 lauryl ether stearate
SE emulsifying cosmetic ingredient; medical ointment base producing good emollient and spreading properties. *Nihon Emulsion.*

9970 Emalex MSG-2
9009-32-9
Diglyceryl monostearate
Emulsifier for cosmetics and foods. *Nihon Emulsion.*

9971 Emalex MTS-30E
2116-84-9 218-320-6
$C_{15}H_{32}O_3Si_4$
Phenyl trimethicone
Phenyltris(trimethylsiloxy)silane; Tris(trimethylsiloxy)phenylsilane. Oil-phase cosmetic ingredient for alcoholic milky-white lotions. *Nihon Emulsion.*

9972 Emalex N-83
Cocamide DEA
Thickener or bubbling agent in hair shampoos, body shampoos, facial cleansing foams. *Nihon Emulsion.*

9973 Emalex NN-15
7545-23-5 231-426-7
Myristamide DEA
Thickener or bubbling agent in hair shampoos, body shampoos, facial cleansing foams. *Nihon Emulsion.*

9974 Emalex NN-7
120-40-1 204-393-1
Lauramide DEA
Cleaner effective in hard water. *Nihon Emulsion.*

9975 Emalex NP-2
9016-45-9 6772
Nonoxynol-2
Wetting agent, emulsifier, cleaner, dispersant, foaming agent, solubilizer for cosmetics. *Nihon Emulsion.*

9976 Emalex O.T.G
7360-38-5 230-896-0
$C_{27}H_{50}O_6$
Glyceryl trioctanoate
Tricaprylin; Octanoic acid, 1,2,3-propanetriyl ester; Caprylin; Glycerol trioctanoate; caprylic acid triglyceride; glycerin tricaprylate; glyceryl trioctanoate; MCT; Panacete 800; RATO; tricaprylyl glycerin; tricapryloylglycerol; tricaprylic glyceride; trioctanoylglycerol. Oil-phase cosmetic ingredient; emollient. mp = 6°; bp = 233°; d = 0.9530; insoluble in H_2O, soluble in organic solvents; LD_{50} (rat orl) = 33300 mg/kg. *Nihon Emulsion.*

9977 Emalex OD-5
Octyldodeceth-5
Emulsifier, dispersant, solubilizer, cleaner, wetting agent for cosmetics, creams, milky lotions. *Nihon Emulsion.*

9978 Emalex OE-6
9004-96-0
PEG-6 oleate
Cosmetics surfactant. *Nihon Emulsion.*

9979 Emalex OP-25
Octoxynol-25
Emulsifier, wetting agent, cleaner, dispersant, foaming agent, solubilizer for cosmetics. *Nihon Emulsion.* Discontinued.

9980 Emalex PEIS-3
56002-14-3
PEG-3 isostearate
Emulsifier, dispersant, solubilizer for cosmetics; emulsion stabilizer, skin fitness reformer for milky lotions, hair conditioners. *Nihon Emulsion.*

9981 Emalex PG-di-IS
Propylene glycol diisostearate
Oil-phase cosmetic ingredient. *Nihon Emulsion.*

9982 Emalex PG-di-L
22788-19-8 245-217-3
Propylene glycol dilaurate
Cosmetic ingredient. *Nihon Emulsion.*

9983 Emalex PG-di-O
105-62-4 203-315-3
Propylene glycol dioleate
Cosmetic ingredient. *Nihon Emulsion.*

9984 Emalex PG-di-S
6182-11-2 228-229-3
Propylene glycol distearate
Oil-phase cosmetic ingredient. *Nihon Emulsion.*

9985 Emalex PGML
142-55-2 205-542-3
Propylene glycol monolaurate

Surfactant for food and cosmetics; emulsion stabilizer for creams, milky lotions, hair conditioners. *Nihon Emulsion.*

9986 Emalex PGMS
1323-39-3 215-354-3
Propylene glycol monostearate
Surfactant for food and cosmetics. *Nihon Emulsion.*

9987 Emalex PGO
1330-80-9 215-549-3
Propylene glycol monooleate
Surfactant for food and cosmetics. *Nihon Emulsion.*

9988 Emalex RWIS-105
PEG-5 hydrogenated castor oil isostearate; oil-phase component, emulsifier for water-oil emulsions, dispersant for hydrophobic components. *Nihon Emulsion.*

9989 Emalex RWIS-305
PEG-5 hydrogenated castor oil triisostearate; solubilizer for oil-phase ingredients. in cosmetics; emulsifier for water-oil emulsions; reforming agent, dispersant for hydrophobic components. *Nihon Emulsion.*

9990 Emalex RWL-120
PEG-20 hydrogenated castor oil laurate; emulsifier, dispersant, reforming agent for cosmetic creams, milky lotions, hair conditioners. *Nihon Emulsion.*

9991 Emalex SG-37
Caprylic/capric/stearic triglyceride
Oil-phase cosmetic ingredient; excellent waxing material with high emulsifying ability. *Nihon Emulsion.*

9992 Emalex SPE-100S
1338-41-6 8872 215-664-9
Sorbitan stearate
Oil-phase cosmetic ingredient, surfactant; for creams, milky lotions, hair conditioners. *Nihon Emulsion.*

9993 Emalex SPE-150S
Sorbitan sesquistearate
Oil-phase cosmetic ingredient, surfactant; for creams, milky lotions, hair conditioners. *Nihon Emulsion.*

9994 Emalex SPIS-100
54392-26-6 276-171-2
Sorbitan isostearate
Oil-phase cosmetic ingredient, surfactant; for creams, milky lotions, hair conditioners. *Nihon Emulsion.*

9995 Emalex SPIS-150
Sorbitan sesquiisostearate
Oil-phase cosmetic ingredient, surfactant; for creams, milky lotions, hair conditioners. *Nihon Emulsion.*

9996 Emalex SPO-100
1338-43-8 8872 215-665-4
Sorbitan oleate
Oil-phase cosmetic ingredient, surfactant; for creams, milky lotions, hair conditioners. *Nihon Emulsion.*

9997 Emalex SPO-150
8007-43-0 8872 232-360-1
Sorbitan sesquioleate
Oil-phase cosmetic ingredient, surfactant; for creams, milky lotions, hair conditioners. *Nihon Emulsion.*

9998 Emalex SWS-4
PEG-4 stearyl ether stearate
SE emulsifying cosmetic ingredient; hydrophobic component and reforming agent for cosmetics and industrial areas. *Nihon Emulsion.*

9999 Emalex TEG-di-IS
PEG-3 diisostearate
Cosmetic ingredient. *Nihon Emulsion.*

10000 Emalex TEG-di-L
PEG-3 dilaurate
Oil-phase cosmetic ingredient, emulsifier for creams, milky lotions, hair conditioners; cleaner, superfattening agent, thickener, reforming agent. *Nihon Emulsion.*

10001 Emalex TPIS-303
PEG-3 trimethylolpropane triisostearate
Oil-phase cosmetics ingredient, emulsifier, solubilizer; offers smooth, clear appearance producing transparent cosmetic products. *Nihon Emulsion.*

10002 Emalex TPM-303
PEG-3 trimethylolpropane trimyristate
Emulsifier, solubilizer, oil-phase ingredient for cosmetics; reforming agent for creams, milky lotions, hair conditioners. *Nihon Emulsion.*

10003 Emalex TPS-203
PEG-3 trimethylolpropane distearate
Oil-phase cosmetics ingredient; for hair creams, lipsticks, ointments, cold creams. *Nihon Emulsion.*

10004 Emalex TPS-303
PEG-3 trimethylolpropane tristearate
Oil-phase cosmetics ingredient; for hair creams, ointments, cold creams; superfatting agent for creamy hair conditioners, cleansing foams. *Nihon Emulsion.*

10005 Emaline
Conventional and high build bituminous coatings. *Sigma Coatings.* Name unverified.

10006 Emaweld®
Electromagnetic welding systems for assembling thermoplastic components. *Ashland.*

10007 Embacel
Kieselguhr for gas chromatography. *May & Baker Ltd.* Name unverified.

10008 Embacide
Sheep dip. *May & Baker Ltd.* Name unverified.

10009 Embacoid
A cement for cine films. *May & Baker Ltd.* Name unverified.

10010 Embadot
Photographic developer. *May & Baker Ltd.* Name unverified.

10011 Embaflx
Photographic fixer. *May & Baker Ltd.* Name unverified.

10012 Embafume
74-83-9 6108 200-813-2
CH_3Br
methyl bromide
Used as a fumigant. *May & Baker Ltd.* No manufacturer.

10013 Embalith
Photographic developer. *May & Baker Ltd.* Name unverified.

10014 Embanox
A food grade antioxidant. *May & Baker Ltd.*

10015 Embanox®
Antioxidant for food applications. *Rhône-Poulenc Food Ingreds.*

10016 Embaphase
Stationary phases for gas chromatography. *May & Baker Ltd.* Name unverified.

10017 Embark
53780-34-0 5846 258-767-4
mefluidide
Soluble concentrate containing 240 g/l mefluidide; grass growth suppressant. *Gordon International Corp.*

10018 Embaspeed
Rapid working photographic developer. *May & Baker Ltd.* Name unverified.

10019 Embatex
A cellulose acetate coated fabric. *May & Baker Ltd.* Name unverified.

10020 Embathion
An insecticide. *May & Baker Ltd.* Name unverified.

10021 Embatype
Photographic developer. *May & Baker Ltd.* Name unverified.

10022 Embazin
9109
sulfaquinoxaline sodium
A proprietary preparation of sulfaquinoxaline sodium; a veterinary coccidiostat. *May & Baker Ltd.* Name unverified.

10023 Embedyne
Chlorodyne
May & Baker Ltd. Name unverified.

10024 Embequin
83-73-8 5063 201-497-9
iodoquinol
Diiodohydroxyquinoline; an anti-amebic agent. *May & Baker Ltd.* Name unverified.

10025 Embesafe
Plastic coated glass bottles. *May & Baker Ltd.* Name unverified.

10026 Embesol
Photographic developer. *May & Baker Ltd.* Name unverified.

10027 Emblet® M
Biaxially oriented polyester film, one side chemically treated. *SNIA (UK) Ltd.*

10028 Embond 55
Polymer reinforced resin/alcohol; needle-punched carpet flooring adhesive. *Marley Floors Ltd.*

10029 Embond 66
Tackified synthetic rubber latex; carpet flooring adhesive. *Marley Floors Ltd.*

10030 Embond 125
Tackified acrylic emulsion; wood and cork flooring adhesive. *Marley Floors Ltd.*

10031 Embond 168
Tackified acrylic emulsion; flooring adhesive. *Marley Floors Ltd.*

10032 Embond 169
Tackified acrylic/EVA copolymer emulsion; flooring adhesive. *Marley Floors Ltd.*

10033 Embond 212
Tackified synthetic rubber latex; flooring adhesive. *Marley Adhesives Ltd.*

10034 Embond 401
Two part solvent-free epoxide base; flooring adhesive. *Marley Floors Ltd.*

10035 Embond 560
Neoprene solution in flammable organic solvents; contact adhesive. *Marley Floors Ltd.*

10036 Embond Surface Tackifier
Modified acrylic emulsion; flooring adhesive. *Marley Floors Ltd.*

10037 embrithite
$Pb_3Sb_2S_6$
A mineral.

10038 Embrol
Photographic developer. *May & Baker Ltd.* Name unverified.

10039 Embutox
94-82-6 2893 202-366-9
2,4-DB
A selective weedkiller. Contains 2,4-DB. *May & Baker Ltd.* Name unverified.

10040 Emcast 1510, 1511
25928-94-3
One-part epoxy; uv-curable, semirigid epoxy for electronic encapsulation, dipping, and adhering; cures rapidly to tough, hard, resistant material with good impact. *Electronic Materials.*

10041 Emcast 1550, 1551
25928-94-3
One-part epoxy; uv-curable, flexible epoxy for electronic encapsulation, dipping, and adhering; cures rapidly to tough, resistant materials with good impact. *Electronic Materials.*

10042 Emcocel® 90M
9004-34-6 2012 232-674-9
Microcrystalline, cellulose NF/BP; tablet binder, disintegrant for pharmaceuticals; features low frability, inherent lubricity, enhanced compression of other excipients. *Penwest Pharmaceuticals Co.*

10043 Emcol
Surfactant for cosmetics, toiletries, pharmaceutical, processing, agricultural and other industries. *Baxenden Chemicals Ltd.*

10044 Emcol® 4
122-19-0 204-527-9
Stearalkonium chloride
Antistat, substantive conditioner, emollient. *Witco Corporation.*

10045 Emcol® 416L
43154-85-4 256-120-0
Disodium oleamido MIPA sulfosuccinate
Dispersant, wetting, foam booster/stabilizer, detergent, conditioner, and emulsifying agent for bubble bath, shampoos, cleansers for cosmetics and toiletries. *Witco/Organics; Witco SA.* Discontinued.

10046 Emcol® 1655
68425-43-4
Cocamidopropyl dimethylamine propionate
Cosmetics and toiletry surfactant used as antistat, conditioner, emollient, foaming and substantive agent. *Witco/Organics.* Discontinued.

10047 Emcol® 3780
55819-53-9 259-837-7
Stearamidopropyl dimethylamine lactate
Cosmetics and toiletry surfactant used as antistat, conditioner, emollient, foaming and substantive agent. *Witco/Organics.* Discontinued.

10048 Emcol®
Sodium alkanolamide sulfosuccinate in liquid or soft paste form; for hair and carpet shampoos; 4161-L has low irritancy and is suitable for baby products; 4150 is used in emulsion polymerization and ore flotation. *Witco Chemical Ltd.*

10049 Emcol® 4300
39354-47-5
Disodium C_{12}-C_{15} pareth sulfosuccinate
Dispersant, wetting, foaming, detergent, emulsifying agent for bubble bath, shampoo, cosmetics and toiletries; emulsifier for acrylic, vinyl acetate, vinyl acrylic polymerization. *Witco/Organics; Witco Chemical Ltd.* Discontinued.

10050 Emcol® 4350
Triethanolamine fatty alcohol ethoxylate sulfosuccinate in liquid form; used for low irritancy hair shampoos and foam baths. *Witco Chemical Ltd.* Discontinued.

10051 Emcol® 4500
577-11-7 3460 209-406-4
Sodium dioctyl sulfosuccinate
Dispersant, detergent, wetting, foaming, emulsifying agent; for cosmetics, toiletries, textiles, industrial processing slurries. *Witco Corporation.*

10052 Emcol® 4600
Sodium bistridecyl sulfosuccinate in liquid form; dispersing agent; used for dry cleaning. *Witco Chemical Ltd.* Discontinued.

10053 Emcol® 4776
Sodium dihexyl sulfosuccinate in liquid form; used for dry cleaning. *Witco Chemical Ltd.* Discontinued.

10054 Emcol® 4910
Sodium lauryl/propoxy sulfosuccinate
Emulsifier for acrylic, vinyl acrylic polymerization. *Witco/Organics.* Discontinued.

10055 Emcol® 5430
Cocamidopropyl betaine; detergent, foam booster/stabilizer, viscosity modifier, wetting agent. *Witco/Organics.* Discontinued.

10056 Emcol® CC-42
9076-43-1
PPG-40 diethylmonium chloride
Pigment dispersant, particle suspension aid, emulsifier, solvent, conditioner, antistat, lubricant, corrosion inhibitor for toiletries, cosmetics, germicides, synthetic fibers and plastics, textiles, industrial processes; ore flotation additive. *Witco Corporation; Witco Chemical Ltd.*

10057 Emcol® CC-55
Polypropoxy quaternary ammonium acetate
General nonirritant quaternary cationic surfactant used especially where anionic computability is required. *Witco Corporation.*

10058 Emcol® DOSS
577-11-7 3460 209-406-4
Sodium dioctyl sulfosuccinate
Emulsifier for acrylonitrile polymerization; improves surface wetting. *Witco Corporation.*

10059 Emcol® E-607L
6272-74-8 5379 228-464-1
Lapyrium chloride
Emollient, emulsifier, conditioner, foamer, cleanser, substantive agent, deodorant for cosmetics, toiletries, industrial applications; hair conditioner. *Witco/Organics; Witco SA.* Discontinued.

10060 Emcol® E-607S
Steapyrium chloride
Emollient, emulsifier, conditioner, foamer, cleanser, substantive agent, deodorant for cosmetics, toiletries, industrial applications; hair conditioner. *Witco/Organics; Witco SA.* Discontinued.

10061 Emcol® ISML
Isostearamidopropyl morpholine lactate
Cosmetics and toiletries surfactant used as antistat, conditioner, emollient, foaming and substantive agent; nonirritating base for cream rinses and conditioning shampoos. *Witco Corporation.*

10062 Emcol® K8300
43154-85-4 256-120-0
Disodium oleamido-MIPA sulfosuccinate
Dispersant, particle suspension aid, wetting agent, foam booster/stabilizer, detergent; emulsifier in emulsion polymerization of acrylic, styrene acrylic, vinyl acrylic. *Witco Corporation.*

10063 Emcol® LO
Lauramine oxide
Foam booster/stabilizer, viscosity modifier. *Witco Corporation.*

10064 Emcompress®
7789-77-7 1739
$CaH_5O_6P \cdot 2H_2O$
Dibasic calcium phosphate dihydrate
Calcium monohydrogen phosphate dihydrate; Dicalcium phosphate dihydrate. Dibasic calcium phosphate dihydrate USP/BP; excipient for production of pharmaceutical tablets by direct compression process. *Penwest Pharmaceuticals Co.*

10065 EmCon E-5
8001-17-0 232-271-8
Egg oil; emollient, moisturizer for hair and skin care products; water-oil emulsifier; superfatting agent, humectant, mold release agent; occlusive agent. *Fanning.*

10066 EmCon Limnanthes Alba
Meadowfoam seed oil; skin/hair conditioner, occlusive agent for shampoos, makeup, face, body and hand creams and lotions, baby products, lipsticks, cleansing products *Fanning.*

10067 EmCon TEA TREE
68647-73-4

Melaleuca alternifolia oil; Melasol; Tea tree oil; Ti tree oil; Ti-trol. Fragrance component, antimicrobial. *Fanning.*

10068 EmCon W
8006-95-9
Wheat germ oil; skin/hair conditioner, occlusive solvent; for hair conditioners, shampoos, lipsticks, cleansers, moisturizing creams and lotions, skin care products. *Fanning.* No manufacturer.

10069 Emcor
1337-76-4
Ultra-short fiber reinforcement based upon attapulgite (magnesium aluminum silicate) used in nonasbestos friction compounds; used for nonasbestos disc pads, drum linings, truck and railroad blocks, friction papers, gasketing. *Engelhard.*

10070 Emcosoy®
68513-95-1
Soy polysaccharides
Tablet disintegrant for direct compression preparation. *Penwest Pharmaceuticals Co.*

10071 Emdex®
50-99-7 4467 200-075-1
Dextrose (95%), isomaltose (2%), gentiobiose (2%), maltose (1%), maltotriose (< 0.1%), panose (< 0.5%); vehicle for direct compression of pharmaceutical tablets. *Penwest Pharmaceuticals Co.*

10072 Emdite
A proprietary trade name for a 50% w/w aqueous solution of ethyl-ammonium ethyl-dithiocarbonate, an alternative to hydrogen sulfide in qualitative inorganic analysis. *British Drug Houses.*

10073 Emdithene
A range of PU resins specifically formulated for the protection of light electrical and electronic components, but offering better retention of flexibility than normal PU types under thermal cycling conditions at higher temperatures. *Robnorganic Systems Ltd.* Name unverified.

10074 Emerald
Extender, wetting agent and antitranspirant for use in pesticides. *Intracrop Ltd.*

10075 Emerald bronze
An alloy of 50% copper, 49.7% zinc, and 0.3% aluminum.

10076 Emercide® 1199
Liquid preservative system for cosmetics. *Henkel/Cospha; Henkel Canada.*

10077 Emeressence® 1150
105-95-3 203-347-8
$C_{15}H_{26}O_4$
Ethylene brassylate
Ethylene brassylate; Astratone; Ethylene undecane dicarboxylate; Musk T; 1,4-Dioxacycloheptadecane-5,17-dione. Musk chemical for fragrance of odor masking applications. *Henkel/Emery.* Name unverified.

10078 Emeressence® 1151
Ethylene dodecanedioate
Musk chemical for fragrance or odor masking applications. *Henkel/Emery.* Name unverified.

10079 Emeressence® 1160 Rose Ether
122-99-6 7410 204-589-7
Phenoxyethanol
Cosmetic preservative; effective against gram negative microorganisms. *Henkel/Emery.*

10080 Emeressence® 1174 Fir Balsam
Aromatic with sweet balsamic odor reminiscent of Canadian fir. *Henkel/Emery.* Name unverified.

10081 Emerest® 2301
112-62-9 6965 203-992-5
Methyl oleate
Base for industrial lubricants; mold release agent, defoamer, flotation agent, plasticizer for cellulose plastics, needle lubricants; when sulfated is useful as wetting, rewetting, and dye leveling agent in textile and leather industries. *Henkel/Emery.*

10082 Emerest® 2302
Propyl oleate
Base for industrial lubricants; mold release agent, defoamer, flotation agent, plasticizer for cellulose plastics, needle lubricants; when sulfated is useful as wetting, rewetting, and dye leveling agent in textile and leather industries. *Henkel/Emery; Henkel/Textile.*

10083 Emerest® 2308
31556-45-3 250-696-7
Tridecyl stearate
Lubricant used in sewing thread manufacture and fiber finish applications where high heat stability is desired. *Henkel/Emery; Henkel/Textile.*

10084 Emerest® 2310
Isopropyl isostearate
Low viscosity emollient, lubricant for bath oils, creams, lotions, shampoos; binder for pressed powder. *Henkel/Cospha; Henkel Canada.*

10085 Emerest® 2314
110-27-0 5234 203-751-4
Isopropyl myristate
Cosmetic emollient; sewing thread lubricant. *Henkel/Cospha; Henkel Canada.*

10086 Emerest® 2316
142-91-6 205-571-1
$C_{19}H_{38}O_2$
Isopropyl palmitate
Lubricant used for synthetic fibers in applications where low friction is essential; emollient in cosmetic formulations; high purity. *Henkel/Cospha; Henkel Canada.*

10087 Emerest® 2324
646-13-9 5165 211-466-1
$C_{22}H_{44}O_2$
Isobutyl stearate
Lubricant for textile applications. *Henkel/Emery; Henkel/Textile.*

10088 Emerest® 2325
123-95-5 1625 204-666-5
Butyl stearate
Emollient in creams and lotions; dye solubilizer in lipsticks; lubricant. *Henkel/Cospha; Henkel Canada.*

10089 Emerest® 2328
142-77-8 205-559-6
Butyl oleate
Lubricant. *Henkel/Emery.*

10090 Emerest® 2350
Glycol stearate
Emulsifier, opacifying and pearlescing agent, thickener, stabilizer used in liquid. cosmetic and detergent compounds *Henkel/Cospha; Henkel Canada.*

10091 Emerest® 2355
627-83-8 211-014-3
Glycol distearate
Emulsifier, opacifier, pearlescent, thickener, stabilizer used in liquid. detergent and cosmetic products *Henkel/Cospha; Henkel Canada.*

10092 Emerest® 2380
1323-39-3 215-354-3
Propylene glycol stearate
Auxiliary emulsifier, opacifier, pearlescent; for lotions, makeup, textile processing. *Henkel/Cospha; Henkel/Textile; Henkel Canada.*

10093 Emerest® 2384
68171-38-0 269-027-5
Propylene glycol isostearate
Solubilizer for fragrances in low alcohol or oil preparations; emollient for personal care products *Henkel/Cospha; Henkel Canada.*

10094 Emerest® 2388
41395-83-9 255-350-9
Propylene glycol dipelargonate
Lubricant, low viscosity emollient for preshaves, bath oils, creams and lotions, textile processing. *Henkel/Cospha; Henkel/Textile; Henkel Canada.*

10095 Emerest® 2400
Glyceryl stearate
Emulsifier for hand creams, cosmetics, textiles, industrial lubricants, polishes, agriculture; lubricant softener for textiles; opacifier and pearling agent. *Henkel/Cospha; Henkel Canada.*

10096 Emerest® 2410
Glyceryl isostearate
Emollient, lubricant, pearling agent, and water-oil emulsifier for creams and lotions, textile applications; excellent oxidation and color stability. *Henkel/Cospha; Henkel/Textile; Henkel Canada.*

10097 Emerest® 2419
25637-84-7 247-144-2
Glyceryl dioleate
glycerol dioleate. Emulsifier, lubricant, rust preventive additive, mold release agent, solvent for dyes and pigments; used in leathers, lubricants and softeners. *Henkel/Cospha; Henkel Canada.*

10098 Emerest® 2421
37220-82-9 253-407-2
Glyceryl monooleate
Emulsifier for cosmetics and industrial applications, in mold release agents, anti-icing fuel additive, rust preventative; in textiles as a lubricant component in synthetic fiber spin finishes; vehicle for agricultural insecticides. *Henkel/Emery; Henkel/Cospha; Henkel Canada.*

10099 Emerest® 2423
122-32-7 204-534-7
Triolein; glyceryl trioleate

Lubricant, water-oil emulsifier for metals, leather, textiles, cosmetics, pharmaceuticals; called synthetic olive oil; sulfated form used as softener in leather and textile industries. *Henkel/Cospha; Henkel/Textile; Henkel Canada.*

10100 Emerest® 2452
Polyglyceryl-3 diisostearate
Emulsifier, solubilizer, dye and pigment wetter; emollient; thickener; solvent; for creams, lotions, lip products. *Henkel/Cospha; Henkel Canada.*

10101 Emerest® 2485
14450-05-6 238-430-8
Pentaerythritol tetrapelargonate
Primary lubricant base or modifier in lubricant formulations used in metal working and synthetic fiber processing. *Henkel/Emery.*

10102 Emerest® 2610
9004-99-3
PEG-20 stearate
Emulsifier for glyceryl stearate in nonionic textile lubricants and softeners; thickener; antigellant in starch solutions. *Henkel/Emery; Henkel/Textile.*

10103 Emerest® 2617
9004-96-0
PEG-150 oleate
Strongly hydrophilic emulsifier, stabilizer, lubricant; for agricultural formulations. *Henkel/Emery; Henkel/Cospha; Henkel Canada.*

10104 Emerest® 2620
9004-81-3
PEG-4 laurate
Emulsifier, coupling agent, defoamer in water base coatings, viscosity control additive; viscosity depressant in vinyl plastisols; agriculture, textiles. *Henkel/Emery; Henkel/Cospha; Henkel/Textile; Henkel Canada.*

10105 Emerest® 2622
9005-02-1
PEG-4 dilaurate
Coemulsifer and lubricant in SE textile and industrial oils, agriculture, mold release agent, viscosity control agent. *Henkel/Emery; Henkel/Cospha; Henkel Canada.*

10106 Emerest® 2624
9004-96-0
PEG-4 oleate
Lubricant component in textile processing; softener/lubricant for leather during tanning; emulsifier for mineral oils, fatty oils, and solvents. for cutting oils, solvents in metal cleaners and degreasers, water-oil emulsifier for consumer pesticide aerosols. *Henkel/Emery; Henkel/Cospha; Henkel/Textile; Henkel Canada.*

10107 Emerest® 2625
56002-14-3
PEG-4 isostearate
Emulsifier; component in fiber lubricants, processing aids, and concentrated liquid fabric softeners. *Henkel/Cospha; Henkel Canada.*

10108 Emerest® 2630
9004-81-3
PEG-6 laurate
Hydrophilic emulsifier; lubricant component and scrooping agent for textile fibers and yarns; viscosity control agent for plastisols; agricultural formulations. *Henkel/Cospha; Henkel Canada.*

10109 Emerest® 2634
PEG-6 pelargonate
Emulsifier, wetting agent, emollient, textile softener, lubricant, defoamer, stabilizer, viscosity control agent, pigment wetting, mold release agent; agricultural formulations, textile processing. *Henkel/Emery; Henkel/Cospha; Henkel/Textile; Henkel Canada.*

10110 Emerest® 2636
9004-99-3
PEG-6 stearate
Waxy emulsifier for oils and fats in industrial lubricants, agricultural formulations; softener and lubricant in textiles and leather. *Henkel/Emery.*

10111 Emerest® 2647
PEG-8 sesquioleate
Emulsifier, wetting agent, emollient, textile softener, lubricant, defoamer, stabilizer, viscosity control agent, pigment wetting, mold release agent, agricultural formulations. *Henkel/Emery; Henkel/Cospha; Henkel/Textile; Henkel Canada.*

10112 Emerest® 2660
9004-96-0
PEG-12 oleate
Dye leveling agent in textiles; emulsifier in specialty lubricants, agricultural formulations; detergent; acid washing of printed circuit boards. *Henkel/Emery; Henkel/Cospha; Henkel/Textile; Henkel Canada.*

10113 Emerest® 2704
9005-02-1
PEG-4 dilaurate
Emulsifier, lubricant, dispersant for bath oils; viscosity control agent for creams and lotions; for cosmetic and industrial applications. *Henkel/Cospha; Henkel Canada.*

10114 Emerlube® 5919
Ethoxylated vegetable oil; lubricant for PP yarns, carpet backings. *Henkel/Emery; Henkel/Textile.*

10115 Emerox® 1110
123-99-9 938 204-669-1
Azelaic acid
Intermediate in manufacture of detergents. *Henkel/Emery.*

10116 Emersist 7210
Acid dye leveling agent for nylon; optimum leveling and enchanced color yield. *Henkel/Textiles.*

10117 Emersoft 7700
Softener concentrate for resin finishing of cotton and polyester/cotton blends; also as plasticizer for starch-finished goods. *Henkel/Textiles.*

10118 Emersol® 110
57-11-4 8959 200-313-4
Stearic acid
Detergent intermediate; opacifier in cosmetics, soaps, emulsifiers, chemical specialties. *Henkel/Emery.*

10119 Emersol® 143
57-10-3 7128 200-312-9
Palmitic acid
Detergent intermediate; opacifier in cosmetics, soaps, emulsifiers, chemical specialties. *Henkel/Emery.*

10120 Emersol® 210
112-80-1 6965 204-007-1
Oleic acid
Detergent intermediate for personal care, emollient, household and industrial applications. *Henkel/Emery.*

10121 Emersol® 315
60-33-3 5529 200-470-9
Linoleic acid
Detergent intermediate for personal care, emollient, household and industrial applications. *Henkel/Emery.*

10122 Emersol® 871
2724-58-5 220-336-3
Isostearic acid
Detergent intermediate for personal care, emollient, household and industrial applications. *Henkel/Emery.*

10123 Emersol® 6333 NF
112-80-1 6965 204-007-1
Low-linoleic content oleic acid USP/NF; food grade fatty acid; also as binder, defoamer and lubricant for pesticides; detergent intermediate in personal care, emollients, household/industrial detergents. *Henkel/Emery.*

10124 Emersol® 6349
57-11-4 8959 200-313-4
Stearic acid
Food grade fatty acid; also as binder, defoamer and lubricant for pesticides; detergent intermediate in personal care, emollients, household/industrial detergents. *Henkel/Emery.*

10125 Emersol® 7021
112-80-1 6965 204-007-1
Oleic acid
Food grade kosher fatty acid; detergent intermediate in personal care, emollients, household/industrial detergents. *Henkel/Emery.*

10126 Emerstat® 6660
Antistat for fiber lubricants. *Henkel/Emery.*

10127 Emerwax® 1251
Synthetic ester wax; for lipsticks, makeup, nail care products, etc. *Henkel/Cospha; Henkel Canada.*

10128 Emerwax® 1253
Beeswax substitute; viscosity agent for cosmetic formulations incl. sticks, cold creams, makeup; beeswax substitute. *Henkel/Cospha; Henkel Canada.*

10129 Emerwax® 1266
Cetearyl alcohol and ceteareth-20; type emulsifying wax; oil-water emulsifier for pharmaceuticals, creams, lotions, antiperspirants, hair care products, depilatories. *Henkel/Cospha; Henkel Canada.*

10130 Emery® 400
57-11-4 8959 200-313-4
Stearic acid
Detergent intermediate. *Henkel/Emery.*

10131 Emery® 515
Tallow/coconut fatty acid blend; detergent intermediate. *Henkel/Emery.*

10132 Emery® 610
Soya fatty acid; detergent intermediate. *Henkel/Emery.*

10133 Emery® 621
Coconut fatty acid; detergent intermediate in personal care, emollients, household/industrial detergents. *Henkel/Emery.*

10134 Emery® 650
143-07-7 5396 205-582-1
Lauric acid
Detergent surfactant. *Henkel/Emery.* Name unverified.

10135 Emery® 654
544-63-8 6416 208-875-2
Myristic acid
Detergent surfactant. *Henkel/Emery.* Name unverified.

10136 Emery® 657
124-07-2 1808 204-677-5
Caprylic acid
Detergent surfactant. *Henkel/Emery.*

10137 Emery® 912
56-81-5 4493 200-289-5
glycerol
CP/USP glycerin; skin softener, solubilizer, viscosity modifier, flavor enhancer, moisturizer, solvent, humectant, thickener, and solubilizer in cosmetics, drug vehicles, food applications, glass, ceramics, agriculture, and adhesives. *Henkel/Emery.*

10138 Emery® 1202
112-05-0 7198 203-931-2
$C_9H_{18}O_2$
Pelargonic acid
n-Nonanoic Acid; nonoic acid; nonanoic acid; 1-Octanecarboxylic acid; Nonylic acid. Intermediate in manufacture of detergents. *Henkel/Emery.*

10139 Emery® 1650
Anhydrous lanolin USP; emulsifier, emollient, conditioner, lubricant for cosmetics, sun care products textiles. *Henkel/Cospha; Henkel/Textile; Henkel Canada.*

10140 Emery® 1730
Mineral oil/fraction of lanolin alcohols and sterols blend; liquid. absorption base, penetrant, water-oil emulsifier, emollient, stabilizer for creams, lotions, makeup, shampoos; provides hypoallergenic skin penetration. *Henkel/Cospha; Henkel Canada.*

10141 Emery® 2203
Methyl tallowate
Detergent intermediate; solvent for pesticides and herbicides. *Henkel/Emery.*

10142 Emery® 2204
Hydrogenated methyl tallowate; detergent intermediate. *Henkel/Emery.*

10143 Emery® 2209
67762-39-4 267-017-5
Methyl caprylate-caprate
Detergent intermediate; solvent for pesticides and herbicides. *Henkel/Emery.*

10144 Emery® 2214
124-10-7 204-680-1
$C_{15}H_{30}O_2$
Methyl myristate
methyl tetradecanoate. Detergent intermediate; solvent for pesticides and herbicides. *Henkel/Emery.*

10145 Emery® 2216
112-39-0 203-966-3
$C_{17}H_{34}O_2$
Methyl palmitate
methyl hexadecanoate. Detergent intermediate; solvent for pesticides and herbicides. mp = 28°, bp_{30} = 212°; d = 0.8520. *Henkel/Emery.*

10146 Emery® 2218
112-61-8 8959 203-990-4
$C_{19}H_{38}O_2$
Methyl stearate
methyl octadecanoate; methyl n-octadecanoate; n-octadecanoic acid methyl ester; kemester 9718; methyl (Z)-9-octadecenoate. Detergent intermediate; solvent for pesticides and herbicides. mp = 37-39°; bp_{15} = 215°. *Henkel/Emery.*

10147 Emery® 2219
112-62-9 6965 203-992-5
Methyl oleate
Detergent intermediate; solvent for pesticides and herbicides. *Henkel/Emery.*

10148 Emery® 2224
Methyl soyate
Solvent for pesticides and herbicides. *Henkel/Emery.*

10149 Emery® 2230
Methyl sunflowerate
Solvent for pesticides and herbicides. *Henkel/Emery.*

10150 Emery® 2231
Dist. methyl canolate
Solvent for pesticides and herbicides. *Henkel/Emery.*

10151 Emery® 2232
Methyl canolate
Solvent for pesticides and herbicides. *Henkel/Emery.*

10152 Emery® 2253
61788-59-8 262-988-1
Methyl coconate
Detergent intermediate; solvent for pesticides and herbicides. *Henkel/Emery.*

10153 Emery® 2255
Methyl palm kernelate
Detergent intermediate. *Henkel/Emery.*

10154 Emery® 2270
Methyl laurate
Detergent intermediate; solvent for pesticides and herbicides. *Henkel/Emery.*

10155 Emery® 2301
112-62-9 6965 203-992-5
$C_{19}H_{36}O_2$
Methyl oleate
(Z)-9-octadecenoic acid methyl ester; oleic acid methyl ester; methyl 9-octadecenoate; methyl cis-9-octadecenoate; 9-Octadecenoic acid (Z)-, methyl ester. Solvent for pesticides and herbicides. bp_9 = 186°; d = 0.8700. *Henkel/Emery.*

10156 Emery® 2895 Foamaster Soap L
8052-48-0 232-491-4
Sodium tallowate
Defoamer for dry agricultural formulations. *Henkel/Emery.* Name unverified.

10157 Emery® 2900
Dimer ester; synthetic lubricant basestock for industrial applications. *Henkel/Emery.*

10158 Emery® 2914
1732-10-1 938 217-060-0
$C_{11}H_{20}O_4$
Dimethyl azelate
nonanedioic acid dimethyl ester; Dimethyl azeleate. Synthetic lubricant. *Henkel/Emery.* Name unverified.

10159 Emery® 2957
Diester; synthetic lubricant basestock. *Henkel/Emery.*

10160 Emery® 3304
661-19-8 8135 211-546-6
Behenyl alcohol (5-15% C_{18}, 5-20% C_{20}, +zw70% C_{22}); chemical intermediate for detergent manufacture. *Henkel/Emery.*

10161 Emery® 3310
Oleyl/cetyl alcohol (25-33% C_{16}, 60-70% C_{18}); chemical intermediate for detergent manufacture. *Henkel/Emery.*

10162 Emery® 3312
143-28-2 6968 205-597-3
Oleyl alcohol (2-8% C_{16}, 87-93% C_{18}); chemical intermediate for detergent manufacture. *Henkel/Emery.*

10163 Emery® 3317
143-28-2 6968 205-597-3
Oleyl alcohol (2-10% C_{16}, 87-95% C_{18}); chemical intermediate for detergent manufacture. *Henkel/Emery.*

10164 Emery® 3320
Tallow alcohol (25-35% C_{16}, 60-67% C_{18}); chemical intermediate for detergent manufacture. *Henkel/Emery.*

10165 Emery® 3321
Hexyl alcohol
Chemical intermediate for detergent manufacture. *Henkel/Emery.*

10166 Emery® 3322
111-87-5 6849 203-917-6
Octyl alcohol
Chemical intermediate for detergent manufacture. *Henkel/Emery.*

10167 Emery® 3323
Decyl alcohol
Chemical intermediate for detergent manufacture. *Henkel/Emery.*

10168 Emery® 3324
111-87-5 6849 203-917-6
Octyl alcohol
Chemical intermediate for detergent manufacture. *Henkel/Emery.*

10169 Emery® 3326
112-53-8 3464 203-982-0
$C_{12}H_{26}O$

Dodecan-1-ol
1-Dodecanol; Dodecyl alcohol; Lauryl alcohol; dodecanol; Alcohol, C12; Hydroxydodecane; Undecyl carbinol. Lauryl alcohol (44-50% C12, 14-20% C14, 8-10% C16, 8-12% C18); chemical intermediate for detergent manufacture. mp= 24-27°; bp = 260-262°; d = 0.8200; LD_{50} (rat orl) = 12800 mg/kg. *Henkel/Emery.* Name unverified.

10170 Emery® 3332
112-53-8 3464 203-982-0
$C_{12}H_{26}O$
Lauryl alcohol
Chemical intermediate for detergent manufacture. *Henkel/Emery.*

10171 Emery® 3334
112-72-1 6418 204-000-3
Myristyl alcohol
Chemical intermediate for detergent manufacture. *Henkel/Emery.*

10172 Emery® 3336
36653-82-4 2070 253-149-0
Cetyl alcohol
Chemical intermediate for detergent manufacture. *Henkel/Emery.*

10173 Emery® 3343
112-92-5 8960 204-017-6
Stearyl alcohol
Chemical intermediate for detergent manufacture. *Henkel/Emery.*

10174 Emery® 3357
112-53-8 3464 203-982-0
Lauryl alcohol (40-48% C_8, 51-59% C_{10}); chemical intermediate for detergent manufacture. *Henkel/Emery.*

10175 Emery® 5353 Lomar PW
Condensed naphthalene sulfonate; dispersant for agricultural formulations. *Henkel/Emery.*

10176 Emery® 5366 (Lomar PWA Llq.)
Ammonium naphthalene sulfonate
Dispersant for agricultural formulations. *Henkel/Emery.*

10177 Emery® 5370 Sellogen W
9084-06-4
Sodium alkyl naphthalene sulfonate. Wetting agent for agricultural formulations. *Henkel/Emery.*

10178 Emery® 6220
EO/PO block polymer; surfactant for agricultural formulations. *Henkel/Emery.*

10179 Emery® 6221 Monolan 2500
EO/PO block polymer; surfactant for agricultural formulations. *Henkel/Emery.*

10180 Emery® 6686
25322-68-3 7729
PEG 600
Chemical intermediate for coatings, adhesives, lubricants, metalworking, paper manufacture, petroleum production, ceramics, printing, electronics, solvents., cleaners, latex paints, mold release agent, rubber. *Henkel/Emery.*

10181 Emery® 6701
Trimethylolpropane tripelargonate
Lubricant with good heat stability for textile processing. *Henkel/Emery; Henkel/Textile.*

10182 Emery® 6717
Internal-external lubricant for manufacture of glass fibers. *Henkel/Textiles.*

10183 Emery® 6744
Cocamidopropyl betaine; surfactant, foamer for personal care products, shampoos, baby preparations, skin cleansers; also for textile applications. *Henkel/Cospha; Henkel/Textile; Henkel Canada.*

10184 Emery® 6750 Nopcosperse AD-6 Liq.
Dispersant for agricultural dry flowables, wettable powders, and aqueous suspensions. *Henkel/Emery.*

10185 Emery® HP-2050
Anhydrous lanolin; emulsifier, emollient, conditioner, moisturizer, pigment dispersant for personal care products, pharmaceuticals. *Henkel/Cospha; Henkel Canada.*

10186 Emery's L-110
123-99-9 938 204-669-1
azelaic acid
A proprietary form of azelaic acid; used as a softener for alkyd resins. No manufacturer.

10187 Emery's L-114
112-05-0 7198 203-931-2
pelargonic acid
A proprietary mixture of low molecular weight aliphatic acids in which pelargonic acid, C_6H_7COOH, predominates; used in the oil modification of alkyd resins. No manufacturer.

10188 Emgard® 2033
Blend; emulsifier for methyl oleate, agricultural formulations. *Henkel/Emery.*

10189 Emgard® 2063
Formulated product; spreading and penetrating agent for agricultural herbicide formulations. *Henkel/Emery.*

10190 EMI-24
931-36-2 213-234-5
$C_6H_{10}N_2$
2-ethyl-4-methylimidazole
A curing agent for epoxy resins used in low proportions thus improving chemical resistance. mp = 36-42°; bp = 292-295°; d = 0.9750; n_D^{20} = 1.4995; soluble in H_2O (180 g/l), organic solvents. No manufacturer.

10191 Emid® 6500
68140-00-1 268-770-2
Cocamide MEA (1:1); thickener, foam stabilizer for shampoos, hair coloring products, liquid detergents, and rug cleaners. *Henkel/Cospha; Henkel Canada.*

10192 Emid® 6515
Cocamide DEA (1:1); emulsifier, foam booster and stabilizer; inhibits redeposition of soils; thickener, superfatting agent for shampoos, bubble baths, cleansers, liquid. detergents; antiredeposition agent for soils on textiles. *Henkel/Cospha; Henkel/Textile; Henkel Canada.*

10193 Emid® 6519
120-40-1 204-393-1
Lauramide DEA (1:1)
Foam booster, stabilizer, viscosity modifier for personal care products, soaps, bath additives. *Henkel/Cospha; Henkel Canada.*

10194 Emid® 6545
93-83-4 202-281-7
Oleamide DEA (1:1)
Foam suppressant in dye carrier and solvent emulsions; emulsifier for mineral oils for antistatic fiber processing aids and yarn lubricants. *Henkel/Emery; Henkel/Textile.*

10195 Emid® 6590
Lauramide DEA and propylene glycol; foam booster, viscosity modifier for personal care products, bath additives, liquid. soaps. *Henkel/Cospha; Henkel Canada.*

10196 EMI-X®
Thermoplastic compounds having electrical conduction properties. *LNP; ICI Chemicals & Polymers Ltd.*

10197 EMI-X® DC-1008
PC with 40% carbon fiber reinforcement; highly conductive composite effectively shielding electromagnetic and/or radio frequency interference; used in avionics housings, business machine enclosures, other electronic devices. *LNP.*

10198 EMI-X® OC-1008
9016-75-5
PPS with 40% carbon fiber reinforcement; highly conductive composite effectively shielding electromagnetic and/or radio frequency interference; used in avionics housings, business machine enclosures, other electronic devices. *LNP.*

10199 EMI-X® PC-1008
25038-54-4 6832
Nylon 6 with 40% carbon fiber reinforcement; highly conductive composite effectively shielding electromagnetic and/or radio frequency interference; used in avionics housings, business machine enclosures, other electronic devices. *LNP.*

10200 EMI-X® PDX-83393
PC with 25% nickel; conductive attenuating composite effectively shielding electromagnetic and/or radio frequency interference; used in avionics housings, business machine enclosures, other electronic devices. *LNP.*

10201 EMI-X® PDX-A-88128
9003-56-9
ABS composite; attenuating composite for EMI/RFI shielding applications. *LNP.*

10202 EMI-X® PDX-D-87815
PC composite; attenuating composite for EMI/RFI shielding applications. *LNP.*

10203 EMI-X® PDX-O-91074
9016-75-5
PPS composite; attenuating composite for EMI/RFI shielding applications. *LNP.*

10204 EMI-X® PDX-P-90305
25038-54-4 6832
Nylon 6 composite; attenuating composite for EMI/RFI shielding applications. *LNP.*

10205 EMI-X® PDX-R-89496
32131-17-2
Nylon 6/6 composite; attenuating composite for EMI/RFI shielding applications. *LNP.*

10206 EMI-X® PDX-W-88341
26062-94-2
PBT composite; attenuating composite for EMI/RFI shielding applications. *LNP.*

10207 EMI-X® RC-1008
32131-17-2
Nylon 6/6 with 40% carbon fiber reinforcement; highly conductive composite effectively shielding electromagnetic and/or radio frequency interference; used in avionics housings, business machine enclosures, other electronic devices. *LNP.*

10208 EMI-X® WC-1008
Thermoplastic polyester with 40% carbon fiber reinforcement; highly conductive composite effectively shielding electromagnetic and/or radio frequency interference; used in avionics housings, business machine enclosures, other electronic devices. *LNP.*

10209 Emka Catalyst P-35
Organic amine catalyst giving excellent bath stability for thermosetting resins in textile industry. *Emkay.*

10210 Emka DDBSA
Dodecylbenzene sulfonic acid
Base for compounding detergents and emulsifiers for textile processing. *Emkay.*

10211 Emka Defoam AA
Solvent type defoamer for textile processing. *Emkay.*

10212 Emka Defoam BC, NC
Nonsilicone; foam and froth control for industrial wastes, textile and paper processing. *Emkay.*

10213 Emka Defoam DP
Silicone emulsion; defoamer; may be cut 50-50 with water. *Emkay.*

10214 Emka Graphite Remover
Effective cleaner for graphited fabrics. *Emkay.*

10215 Emka Transfer Remover
Cleaner for redying nylon hosiery. *Emkay.*

10216 Emkabase CA
Emulsifier for mineral oil; mixed with 90 parts mineral oil to yield a water-dispersible substantive product for textile processing. *Emkay.*

10217 Emkabase ODC-2
Alkylaryl sulfonate blend; emulsifier for orthodichlorobenzene. *Emkay.*

10218 Emkabond UR
Fiber reactive laminating adhesive for bonding urethane foam to fabric. *Emkay.*

10219 Emkacide GS-2
Disinfectant, cleaner, deodorizer, and fungicide with softening properties. *Emkay.*

10220 Emkadixol
Lubricants. *ICI Chem & Polymers Ltd.*

10221 Emkafix RXC
Quatenary ammonium salt; dispersant which improves fixation of direct dyes. *Emkay.*

10222 Emkafol D
Highly sulfonated fatty acid ester; wetting and rewetting agent with excellent leveling properties for textile processing; also activates enzymes for starch removal. *Emkay.*

10223 Emkafume FA
Nondurable type gas fading inhibitor for acetate colors. *Emkay.*

10224 Emkagen 49
Amino condensate and alkylaryl sulfonate blend; detergent for print washing in textile applications. *Emkay.*

10225 Emkagen 49AM
Ammoniated detergent based on long chain alkylaryl sulfonate; detergent, wetting agent, emulsifier for heavy duty cleaning and scouring, industrial cleaning; resistant to hard water. *Emkay.*

10226 Emkagen BT
Amino-condensate blend; detergent for wool scouring. *Emkay.*

10227 Emkal BNS
Butylnaphthalene sodium sulfonate
Dispersant, detergent, dyeing assistant, scouring, and wetting agent, emulsifier used in textile industry. *Emkay.*

10228 Emkal BNS Acid
Butylnaphthalene sulfonic acid
Wetting, dispersing agent, detergent, dye assistant, scouring agent, emulsifier. *Emkay.*

10229 Emkal NNS, NNS Acid
Nonylnaphthalene sodium sulfonate
Dispersant, detergent, dyeing assistant, scouring, and wetting agent, emulsifier used in textile industry. *Emkay.*

10230 Emkal NOBS
Sodium nonylbenzene sulfonate
Dispersant, detergent, dyeing assistant, scouring, and wetting agent, emulsifier used in textile industry. *Emkay.*

10231 Emkalane WL
Amino condensate; scouring agent for wool and wool blends. *Emkay.*

10232 Emkalar Base C50L
Emulsifier; base for liquid dye carriers by mixing with ortho-phenyl phenol. *Emkay.*

10233 Emkalite BAC
Anionic sulfonate blend; wetting agent, partial stripping agent for cationic dyes from acrylic fibers; helps produce level redyes. *Emkay.*

10234 Emkalon KLA
Polyethoxylated condensate; lanolin base softener. *Emkay.*

10235 Emkalon TN
Polyethoxylate
Lubricant, softener; highly compatible. *Emkay.*

10236 Emkalube F-11
Bright oil which forms an emulsion in water; used as a trough oil in twisting and as a needle lubricant (textile industry). *Emkay.*

10237 Emkane Acid
Alkylaryl sulfonic acid
Detergent, dyeing assistant, scouring and wetting agent; emulsifier used in the textile industry; biodegradable. *Emkay.*

10238 Emkane HAD
Alkylaryl sulfonate
Detergent, dyeing assistant, scouring and wetting agent, especially for cleansing stubborn soils; emulsifier used in the textile industry; biodegradable. *Emkay.*

10239 Emkane HAX
Amine-neutralized alkylaryl sulfonate; double-strength detergent, dyeing assistant, scouring and wetting agent; emulsifier used in the textile industry; biodegradable. *Emkay.*

10240 Emkanet B
Stiff resin finish for veils and nets (textiles). *Emkay.*

10241 Emkanol NC, NCD 25, 35, 45, 55
Sulfonates
Penetrants, wetting agents for mercerizing caustic solutions *Emkay.*

10242 Emkanyl 85
Leveling agent for dyeing of nylon hosiery. *Emkay.*

10243 Emkanyl BRX
Nylon dyeing assistant for acid dyes; prevents barre in woven fabrics; leveling agent. *Emkay.*

10244 Emkapel DE
Durable type water repellent for textile industry. *Emkay.*

10245 Emkapene AV, AVX
Low-foaming penetrant, emulsifier, scum preventatives for use in sulfur and vat dyeing; compatible with sodium sulfide. *Emkay.*

10246 Emkapene RW
Mixture of sulfonated oils and pine oil; leveling agent and dye assistant. *Emkay.*

10247 Emkaperm
Permanent stiffening finish which withstands several washings (textiles). *Emkay.*

10248 Emkapol PO-18
143-18-0 7818 205-590-5
Potassium oleate
Detergent, soap, emulsifier; stabilizer for natural latex; biodegradable. *Emkay.* Name unverified.

10249 Emkapon 4S, DS, SS, TS
Amide sulfonates
General textile detergents and emulsifiers. *Emkay.*

10250 Emkapon BC
Wool scouring agent containing optical bleach. *Emkay.*

10251 Emkapon DAC, DAC-50
Detergent/emulsified solvent blend; detergent for cleaning obstinate soils, grease, stains, wax, water repellents; recommended as Dacron scour. *Emkay.*

10252 Emkapon Jel 500 Conc
Fatty methyl taurate
Dispersant and surfactant for textile processing. *Emkay.*

10253 Emkapon ML
Modified alkylaryl sulfonate; scouring and dyeing agent for synthetics. *Emkay.*

10254 Emkapruf ABR, FL
Concentrated flame resistant finishes for textiles. *Emkay.*

10255 Emkarate
Lubricant. *ICI Chem & Polymers Ltd.*

10256 Emkaron GA-1
Sulfated fatty acid ester; low-foaming dyeing assistant, dispersant, wetting agent for textiles; resistant to acid and alkaline media. *Emkay.*

10257 Emkarox
Synthetic lubricants. *Imperial Chemical Industries plc.*

10258 Emkasan QA-50
Quaternary ammonium compound; algicide, germicide, disinfectant, and deodorant for textile and industrial use. *Emkay.*

10259 Emkasene 800
139-89-9 10102 205-381-9
Versen-Ol®
N-(2-hydroxyethyl)ethylene dinitrilo triacetic acid
Trisodium hydroxyethyl ethylene diamine triacetate aqueous solution; chelating agent for sequestering calcium, magnesium, and ferric ions in solution. mp = 288-290° (dec) *Emkay.*

10260 Emkaset
Modified formaldehyde resin, thermosetting; stiff, crease-resistant, shrink-resistant; for textiles. *Emkay.*

10261 Emkasize CF
Sizing agent for cotton and jute. *Emkay.*

10262 Emkasol DE
Modified oleyl-lauryl alcohol sulfate; dispersant, wetting agent, detergent, dye bath stabilizer. *Emkay.*

10263 Emkasorb
Used in finishing of nylon and other synthetics to increase the absorbency. *Emkay.*

10264 Emkastat MLT
Antistat lubricant, soil releasing agent; hard wax base which protects delicate fabrics. *Emkay.*

10265 Emkatan K
Synthetic softener and lubricant used as a finish for cotton and rayon knit goods and as a lubricant on dyed skeins to improve winding and eliminate chafing and abrasions; good antistat. *Emkay.*

10266 Emkatard
Dispersant for vat colors (textiles). *Emkay.*

10267 Emkaterge B
Synthetic detergents with alkyl terpenes; wetting agent, base product for compounding graphite remover. *Emkay.*

10268 Emkatex 11, 21
POE fatty acid
Emulsifier for water kerosene formula. *Emkay.*

10269 Emkatex 49-P
Synthetic detergents; color dispersant, dyebath stabilizer. *Emkay.*

10270 Emkatex AA
Alkylaryl sulfonate
Detergent, wetting agent, penetrant for textile applications. *Emkay.*

10271 Emkatex DX, DXP
Phosphated high molecular weight alcohol; dispersant, leveling agent, penetrant, scouring agent, and textile dye bath stabilizer; for kier, bleaching processes; also increases absorbency of flame proofing finishes. *Emkay.*

10272 Emkatex NE
Synthetic detergents, emulsifiers, dispersants; textile auxiliary. *Emkay.*

10273 Emkatint BRN
Optical bleach and brightener for nylon, arnel, and orlon. *Emkay.*

10274 Emkatol M
Graphite remover for nylon lace. *Emkay.*

10275 Emkawate AS
Weighter finish with nonslip properties; minimizes fraying and ravelling (textiles). *Emkay.*

10276 Emkazyme
Mixture of amylolytic enzymes and catalysts, solubilizing resins, starches, etc.; for textile applications. *Emkay.*

10277 Emol Keleet
A purified fuller's earth.

10278 Empal
94-74-6 5803 202-360-6
MCPA
Aqueous solution containing 25% w/v MCPA as mixed sodium and potassium salts; for use as an agricultural herbicide. *Universal Crop Protection Ltd.*

10279 EM-PB
Lard oil; additive improving wetting and lubricity in industrial applications. *Ferro/Keil.*

10280 Empee® FR 42 LM
9002-88-4 7728
HDPE, flame retardant; used in wire and cable markets or type. PE applications requiring flame retardance. *Monmouth Plastics.*

10281 Empee® PE-112
9002-88-4 7728
LDPE, flame-retarded; UL-recognized grade for injection molded parts, certain electrical applications. *Monmouth Plastics.*

10282 Empee® PE-113
9002-88-4 7728
LDPE, flame-retarded; for wire and cable insulation. *Monmouth Plastics.*

10283 Empee® PO Conc. 61
Imparts flame retardancy to low and high density polythylene. *Monmouth Plastics.*

10284 Empee® PP Conc.33
9003-07-0 7741
PP, flame retardant; imparts flame retardancy to most PP homopolymers; for use in fiber and film applications. *Monmouth Plastics.*

10285 Empee® PP-301
9003-07-0 7741
PP, flame-retarded; UL-recognized resin for injection molded parts, extruded sheet, shapes, certain electrical applications. *Monmouth Plastics.*

10286 Empee® PP-459
9003-07-0 7741
PP homopolymer, flame-retarded; high impact, UL-recognized resin for injection molding. *Monmouth Plastics.*

10287 Empee® PP-560
9003-07-0 7741
PP copolymer, flame-retarded; for extrusion, wire insulation. *Monmouth Plastics.* Name unverified.

10288 Empee® PS-921
9003-53-6 9028
PS, flame-retarded; UL-recognized resin for injection molded transparent crystal parts. *Monmouth Plastics.*

10289 Emperor alloy
203-066-0
A nickel-chromium alloy. It will resist a temperature of 1750-1800°.

10290 Emperor brass
An aluminum bronze. It consists of 60% copper, 20% aluminum, and 20% zinc.

10291 Empetal
4-(4-Methyl-3-penten-1-yl)-3-cyclohexen-1-carboxyaldehyde
Flavorant. *Quest Int'l. UK Ltd.*

10292 Emphos
Series of anionic surfactants of the phosphate ester type showing a range of properties e.g. surface tension lowering, wetting, foaming, according to composition; antistats; dry cleaning; emulsion polymerization; industrial alkaline cleaners; papermaking; pesticides; textile processing; extreme pressure lubricants; corrosion inhibitors; release agents; moisture barrier agents; oil well fluids; pigment dispersion. *Witco Corporation.*

10293 Emphos CS-136
Nonoxynol-6 phosphate
Lubricant; antistat; emulsifier for cutting fluids, PVAc, and acrylic film formation; detergent for hard surfaces, metal cleaners, and dry cleaning systems; waterless hand cleaner component. *Witco Corporation.*

10294 Emphos CS-1361
Sodium nonoxynol-9 phosphate
Antistat; emulsifier for transparent gels; detergent for hard surfaces, metal cleaners, and drycleaning systems; particle dispersant for aqueous systems; coupling agent; wetting agent. *Witco Corporation.*

10295 Emphos PS-21A
Alcohol ethoxylate phosphate ester; detergent base, emulsifier, foaming and wetting agent, lubricant, dispersant; used for detergent industry and industrial surfactants. *Witco Corporation.*

10296 Emphos TS-230
Phenol ethoxylate phosphate ester; acid form; lubricant; improves lubricity and load-bearing of water-based lubricants; industrial corrosion inhibitor, defoamer, metal processing surfactant for synthetic oils; low-foaming hydrotrope for alkaline cleaners. *Witco Corporation.*

10297 Empicol
An aliphatic alcohol sulfate and aliphatic alcohol alkyl ether sulfate; wetting agent. *Albright & Wilson Ltd.*

10298 Empicol® 0031/T
68585-44-4 271-556-1
DEA-lauryl sulfate
Detergent used in personal care products. *Albright & Wilson UK.* Name unverified.

10299 Empicol® 0045
Sodium lauryl sulfate

Raw material and foaming agent for toothpaste, shampoos, foam baths. *Albright & Wilson UK.*

10300 Empicol® 0045V
Sodium lauryl sulfate
Surfactant for toiletries. *Albright & Wilson Am.*

10301 Empicol® 0303
151-21-3 8782 205-788-1
Sodium lauryl sulfate BP
Surfactant in toothpaste and pharmaceutical preparations; emulsion polymerization. *Albright & Wilson UK.*

10302 Empicol® 0585/A
126-92-1 204-812-8
Sodium ethylhexyl sulfate
Surfactant for toiletries; low foam degreasing agent for textiles. *Albright & Wilson Am.*

10303 Empicol® 0758
142-87-0 205-568-5
Sodium decyl sulfate
Surfactant for toiletries. *Albright & Wilson Am.*

10304 Empicol® 0775
Sodium lauryl/tallow sulfate
Surfactant for toiletries. *Albright & Wilson Am.*

10305 Empicol® 9060X
Formulated product; pearling/opacifying concentrate for toiletries. *Albright & Wilson Am.*

10306 Empicol® AL30
68081-96-9 218-793-9
Ammonium lauryl sulfate
Detergent for shampoos, carpet shampoos, leather processing. *Albright & Wilson/Australia.* Name unverified.

10307 Empicol® ALL
Ammonium lauryl triethoxy sulfate
Shampoo ingredient. *Albright & Wilson Australia.*

10308 Empicol® BSD 52
Sodium/magnesium laureth sulfate
Surfactant for toiletries. *Albright & Wilson Am.*

10309 Empicol® CHC 30
Sodium cetyl oleyl sulfate
Multipurpose textile detergent. *Albright & Wilson UK.* Name unverified.

10310 Empicol® DA
68585-44-4 271-556-1
DEA-lauryl sulfate
Ingredient in personal care products, automobile cleaners. *Albright & Wilson Australia.*

10311 Empicol® DLS
DEA-lauryl sulfate
Detergent used in liquid and lotion shampoos. *Albright & Wilson UK.* Name unverified.

10312 Empicol® EAA
67762-19-0
Ammonium laureth sulfate
Surfactant for toiletries. *Albright & Wilson Am.*

10313 Empicol® EAB
67762-19-0
Ammonium laureth sulfate
Detergent and base used in the cosmetic industry. *Albright & Wilson UK.*

10314 Empicol® EAC
67762-19-0
Ammonium laureth sulfate
Surfactant for toiletries. *Albright & Wilson Am.*

10315 Empicol® EGB, EGC
67702-21-4 221-450-6
Magnesium laureth sulfate
Detergent and toiletry raw material. *Albright & Wilson UK.* Name unverified.

10316 Empicol® EL
68184-04-3
MEA-lauryl sulfate
Detergent used in the manufacture of liquid shampoos. *Albright & Wilson UK.* Name unverified.

10317 Empicol® EMB
977067-77-8
MEA laureth sulfate
Detergent raw material for shampoos. *Albright & Wilson UK.*

10318 Empicol® ESA
68585-34-2
Sodium laureth sulfate
Detergent and toiletry raw material. *Albright & Wilson UK.*

10319 Empicol® ESB
9004-82-4
Sodium laureth sulfate
Shampoo ingredient; mild detergent for textile and leather processing; dispersant/emulsifier for emulsion and suspension polymerization. *Albright & Wilson Australia.*

10320 Empicol® ESC/AU
68585-34-2
Sodium laureth sulfate
Surfactant used as a base in personal care products *Albright & Wilson Australia.*

10321 Empicol® ETB
270028-82-6
TEA-laureth sulfate
Detergent and toiletry raw material. *Albright & Wilson UK.*

10322 Empicol® HL25
2044-56-6 218-058-2
Lithium lauryl sulfate
Active shampoo detergent ingredient. *Albright & Wilson UK.*

10323 Empicol® L Series
Alkyl sulfate
Dispersant/emulsifier for emulsion and suspension polymerization, agricultural formulations; foaming agent for carpet backing; air entraining agent for construction; flotation aid. *Albright & Wilson UK.*

10324 Empicol® LM
68955-19-1 273-257-1
Sodium lauryl sulfate
Detergent, foaming agent for toothpaste, shampoos, shaving products. *Albright & Wilson UK.*

10325 Empicol® LMV/T
68955-19-1 273-257-1
Sodium lauryl sulfate
Detergent for cream shampoos, shaving products. *Albright & Wilson UK.*

10326 Empicol® LQ33/T
4722-98-9 225-214-3
MEA-lauryl sulfate
Surfactant in the manufacture of personal care products; emulsifier in the manufacture of rubber latexes and for resins; bactericidal detergents. *Albright & Wilson UK.*

10327 Empicol® LS30
151-21-3 8782 205-788-1
Sodium lauryl sulfate
Wetting, dispersing, emulsifying, and foaming agent for industrial processes, detergent/cleaner formulations. *Albright & Wilson UK.* Name unverified.

10328 Empicol® LX
68585-47-4
Sodium lauryl sulfate
Emulsifier in the manufacture of plastics, resins, and synthetic rubbers; foaming agent for rubber foams, personal care products, pharmaceuticals, and carpet and upholstery shampoos; lubricant in manufacture of molded rubber goods. *Albright & Wilson; Albright & Wilson UK.*

10329 Empicol® LXV
151-21-3 8782 205-788-1
Sodium lauryl sulfate
Emulsifier in the manufacture of plastics, resins, and synthetic rubbers; foaming agent for rubber foams, personal care products, toothpaste, and carpet and upholstery shampoos; lubricant in manufacture of molded rubber goods. *Albright & Wilson; Albright & Wilson UK.*

10330 Empicol® LY28/S
151-21-3 8782 205-788-1
Sodium lauryl sulfate
Coprecipitant in manufacture of photographic film; emulsion polymerization in the plastic and rubber industries; foaming agent in carpet processes. *Albright & Wilson UK.* Name unverified.

10331 Empicol® LZ
68955-19-1 273-257-1
Sodium lauryl sulfate
Detergent, wetting and foaming agent in personal care products, pharmaceuticals; emulsifier in manufacture of rubbers, plastics, and resins by emulsion polymerization; foaming agent in manufacture of foam rubber goods; lubricant used in plastic goods. *Albright & Wilson UK.*

10332 Empicol® LZG 30
151-21-3 8782 205-788-1
Sodium lauryl sulfate
Surfactant, foaming and wetting agent used in industrial processes; emulsifier in manufacture of synthetic rubbers, plastics, and resins by emulsion polymerization. *Albright & Wilson UK.* Name unverified.

10333 Empicol® LZP
151-21-3 8782 205-788-1
Sodium lauryl sulfate
Detergent, wetting and foaming agent in personal care products, pharmaceuticals; emulsifier in manufacture of rubbers, plastics, and resins by emulsion polymerization; foaming agent in manufacture of foam rubber goods; lubricant used in plastic goods. *Albright & Wilson UK.* Name unverified.

10334 Empicol® LZV
68955-19-1 273-257-1
Sodium lauryl sulfate
Detergent and foaming agent for shampoos, toothpaste. *Albright & Wilson UK.*

10335 Empicol® MD
Sodium lauryl ethoxy sulfate as a pale straw liquid; foaming agent with good hard water performance and low viscosity and irritancy; used in baby and medical shampoos, foam baths, hand cleaners and fine fabric washing. *Albright & Wilson Ltd, Detergents Div, Marchon.* Name unverified.

10336 Empicol® ML 26/F
68081-97-0 268-365-0
Magnesium lauryl sulfate
Detergent used in the manufacture of shampoos and toothpaste. *Albright & Wilson UK.*

10337 Empicol® SDD
37354-45-5
Disodium lauryl ethoxy sulfosuccinate
Mild raw material for toiletries and detergents. *Albright & Wilson UK.* Name unverified.

10338 Empicol® SLL
36409-57-1 253-019-3
Disodium lauryl sulfosuccinate
Mild raw material for toiletries and detergents. *Albright & Wilson UK.*

10339 Empicol® STT
Disodium cetearyl sulfosuccinate
Mild raw material for toiletries and detergents. *Albright & Wilson UK.*

10340 Empicol® TA40
68908-44-1 272-675-1
TEA-lauryl sulfate
Ingredient for personal care products. *Albright & Wilson Australia.*

10341 Empicol® TAS30
68955-20-4 273-258-7
Sodium tallow sulfate
Collector in the beneficiation of minerals by ore flotation; detergent active raw material; cofoaming agent for latex foam compounds *Albright & Wilson UK.*

10342 Empicol® TL40
139-96-8 205-388-7
TEA-lauryl sulfate
Detergent used in liquid and lotion shampoos; foam booster for fire fighting. *Albright & Wilson UK.*

10343 Empicol® XC35
Blend; pearlescent base for shampoo, bubble bath, liquid soap; formulations should contain a preservative. *Albright & Wilson UK.*

10344 Empicol® XM 17
Formulated product; concentrated detergent for carpet shampoo. *Albright & Wilson UK.*

10345 Empicol® XPA
Formulated blend; pearlized and opacified concentrate for toiletries. *Albright & Wilson UK.*

10346 Empicryl®
Pigment dispersant; for a wide variety of water based paints. *Albright & Wilson Ltd, Detergents Div.* Name unverified.

10347 Empicryl® 6045
Alkyl methacrylate polymer in hydrocarbon oil; viscosity index improver for formulation of high viscosity index hydraulic fluids; emulsifier, dispersant for emulsion polymerization; reactive diluent for adhesives and coatings. *Albright & Wilson UK.*

10348 Empicryl® APD
Maleic anhydride/diisobutylene copolymer, disodium salt, aqueous solution; pigment dispersant for water based paints, pigment stabilizer. *Albright & Wilson UK.* Name unverified.

10349 Empicryl® DH122
Polyalkyl methacrylate copolymer in hydrocarbon oil; viscosity index improver, pour point depressant, and low temperature sludge dispersant for formulation of multigrade crankcase oils. *Albright & Wilson UK.* Name unverified.

10350 Empicryl® PPT38
Alkyl methacrylate copolymer in hydrocarbon oil; pour point depressant for formulation of lubricants including multigrade crankcase oils in conjugation with hydrocarbon-based polymeric viscosity index improvers. *Albright & Wilson UK.* Name unverified.

10351 Empicryl® PT1334
Alkyl methacrylate polymer in hydrocarbon oil; viscosity index improver/pour point depressant for formulation of multigrade crankcase oils. *Albright & Wilson UK.* Name unverified.

10352 Empicyl
Lubricating oil additives. *Albright & Wilson Ltd.*

10353 Empigen
Tertiary amine and imidazoline derivates. *Albright & Wilson Ltd, Detergents Div.*

10354 Empigen® 5083
61788-90-7 263-016-9
Coco dimethylamine oxide
Foam booster/stabilizer and viscosity modifier for shampoos, foam baths, cleaners; improves conditioning in shampoos; solubilizer for liquid bleach products *Albright & Wilson UK.*

10355 Empigen® 5089
112-00-5 203-927-0
Laurtrimonium chloride
Bactericide for disinfectant and sanitizer formulations for household, institutional, agriculture, food processing applications, antiseptic detergents in pharmaceuticals; algicide for swimming pools; wood preservatives. *Albright & Wilson UK.*

10356 Empigen® 5107
Alkyl dimethylamine betaine; foaming agent/stabilizer, antistat, solubilizer for shampoos, foam baths, latex foam compounds, oil production, alkaline industrial hard surface cleaners; wetting and coupling agent for cleaners, traffic film removers. *Albright & Wilson UK.*

10357 Empigen® 5509
Cocamino hydroxy sulfobetaine
Foam booster, conditioner for shampoos, skin cleaners. *Albright & Wilson UK.*

10358 Empigen® AB
67700-98-5 266-922-2
Dimethyl lauramine
Intermediate for manufacture of high quality derivatives. such as quaternary ammonium compounds, betaines, amine oxides; catalyst for PU foam, resin curing agent, corrosion inhibitor and flotation aid. *Albright & Wilson UK.*

10359 Empigen® AF
C_{12-16} alkyl dimethyl amine; intermediate for amine derivatives; resin curing agent; corrosion inhibitor; ore flotation chemicals. *Albright & Wilson UK.*

10360 Empigen® AG
C_{12-16} alkyl dimethyl amine; intermediate for amine derivatives; resin curing agent; corrosion inhibitor; ore flotation chemicals. *Albright & Wilson UK.*

10361 Empigen® AH
112-75-4 204-002-4
Dimethyl myristamine
Intermediate for manufacture of high quality derivatives. such as quaternary ammonium compounds, betaines, amine oxides; catalyst for PU foam, resin curing agent, corrosion inhibitor, and flotation aid. *Albright & Wilson UK.*

10362 Empigen® BAC50
Benzalkonium chloride NF
Bactericide for disinfectant and sanitizer formulations for household, institutional, agriculture, food processing applications, antiseptic detergents in pharmaceuticals; algicide for swimming pools; wood preservatives; masonry biocides. *Albright & Wilson UK.*

10363 Empigen® BB
66455-29-6 266-368-1
Lauryl betaine; foam booster/stabilizer, emulsifier, dispersant, wetting agent, thickening agent, conditioner used for shampoos, detergents, latex foam compounds for carpet backing, industrial applications; formulation of film removers; stable over wide pH range. *Albright & Wilson Am.; Albright & Wilson UK.*

10364 Empigen® BCB50
68989-00-4 273-544-1
Benzalkonium chloride
Disinfectant for dairy, food processing, restaurant, brewing, and bottling industries; retarder in dyeing of acrylic fibers. *Albright & Wilson UK.*

10365 Empigen® BS
Cocamidopropyl dimethylamine betaine; foaming agent/stabilizer and antistat for shampoos, foam baths, latex foam compounds, oil production; wetting and coupling agent for cleaners, traffic film removers. *Albright & Wilson UK.*

10366 Empigen® BS/H
Cocamidopropyl dimethylamine betaine; foaming agent/stabilizer and antistat for shampoos, foam baths, latex foam compounds, oil production; wetting and coupling agent for cleaners, traffic film removers. *Albright & Wilson UK.*

10367 Empigen® CDL60
68608-66-2 271-794-6
Disodium lauroamphodiacetate
Detergent for nonirritating shampoos, skin cleansing, personal care products *Albright & Wilson UK.*

10368 Empigen® CDR10
Coco imidazoline betaine. Coconut imidazoline betaine; surfactant used in personal care products, household and industrial cleaners, textile processing. *Albright & Wilson UK.*

10369 Empigen® CDR40
68334-21-4 269-819-0
Sodium cocoamphoacetate; Cocoamphoacetate; Cocoamphoglycinate; Afoteric™ 1C; Ampholak XCO-40; Amphoterge® W; Emcol®CMCD; Emery®5412; Empigen® CDR60; Kelisema Collagen-IMZ Complex; Mackam™; 1C; Miranol®CM Conc. NP; Monateric CM-36S; Proteric CM-36S; Schercoteric MS; Standapol+sdo CIM-40; Unibetaine GC-88; Unipol CIM-40; Velvetex® GC-88; Zoharteric M,. Mild detergent for shampoos, skin cleansers, personal care products, textile processing. Albright & Wilson UK; McIntyre; Berol Nobel AB; Lonza;Witco/H-I-P; Henkel/Organic Products; Albright & Wilson UK; Kelisema Srl; Rhône-Poulenc Surf. & Spec.; Mona Industries; Protameen; Scher; Universal Preserv-A-Chem;Zohar Detergent Factory.

10370 Empigen® CHB40
Alkyl trimethyl ammonium bromide
Bactericide. *Albright & Wilson UK.*

10371 Empigen® CM
Tallow trimethyl ammonium methosulfate
Antistat and conditioning agent for personal care products; emulsifier and antistat for industrial use; retarder for dyeing of acrylic fibers. *Albright & Wilson UK.*

10372 Empigen® FKC75L
Alkyl diamido amine lactate
Raw material used in fabric softener formulations. *Albright & Wilson UK.*

10373 Empigen® FKH75L
Dialkyl diamido amine lactate
Surfactant. *Albright & Wilson UK.*

10374 Empigen® FRC75S
Tallow alkyl imidazoline methosulfate
Textile conditioning agent, antistat, and lubricant for fibers; raw material for fabric softener formulations. *Albright & Wilson UK.*

10375 Empigen® OB
1643-20-5 216-700-6
$C_{14}H_{31}NO$
Lauramine oxide
N,N-dimethyldodecylamine-N-oxide; 1-Dodecanamine, N,N-dimethyl-, N-oxide; Dimethyldodecylamine oxide; dimethyldodecylamine-N-oxide; Ammonyx LO; Ammonyx AO; Aromox DMMC-W; Conco XAL; DDNO; N,N-dimethyldodecylamine oxide; dodecycldimethylamine oxide; n-dodecyldimethylamine oxide; lauryldimethylamine oxide; Lauramine oxide. Coactive, detergent, antistat, foam booster/stabilizer and viscosity modifier for personal care products, surgical scrubs, fire fighting foam concentrates, foamed rubbers, bleach additive; solubilizer. mp = 130-131°; LD_{50} (rat orl) = 1000 mg/kg. *Albright & Wilson; Albright & Wilson UK.*

10376 Empigen® OC
Alkyl dimethyl amine oxide
Surfactant. *Albright & Wilson Am.*

10377 Empigen® OH25
3332-27-2 222-059-3
n-Myristyl dimethyl amine oxide
Foam booster/stabilizer and viscosity modifier for shampoos, foam baths, detergents; improves conditioning in shampoos; solubilizer for liquid. bleach products. *Albright & Wilson; Albright & Wilson UK.*

10378 Empigen® OS/A
68155-09-9 268-938-5
Cocamidopropyl dimethyl amine oxide
Foam booster/stabilizer and viscosity modifier for shampoos, foam baths, detergents; improves conditioning in shampoos. *Albright & Wilson; Albright & Wilson UK.*

10379 Empigen® OY
59355-61-2
PEG-3 lauramine oxide
Detergent, antistat, foam booster/stabilizer for foamed rubbers, fire fighting, bleach additive. *Albright & Wilson; Albright & Wilson UK.*

10380 Empigen® XDR302
Sodium cocoamphoacetate with sodium lauryl sulfate; used in personal care products *Albright & Wilson UK.*

10381 Empilan®
Fatty acid esters, emulsifiers and foam stabilizers. *Albright & Wilson Ltd.* Name unverified.

10382 Empilan® 0004
Castor oil EO/PO condensate; emulsifier and lubricant for formulation of cutting oils, grinding fluids, and textile lubricants; manufacture of brake fluids. *Albright & Wilson UK.*

10383 Empilan® 2020
Lauryl alcohol ethoxylate
Stabilizer for latexes produced by emulsion polymerization. *Albright & Wilson UK.* Name unverified.

10384 Empilan® 2502
8051-30-7 232-483-0
Cocamide DEA (1:1)
Foam booster/stabilizer. *Albright & Wilson UK.*

10385 empilan® 7132
Fatty amine alkoxylate
Surfactant. *Albright & Wilson UK.*

10386 Empilan® AM Series
Amine ethoxylates
Emulsifier for agriculture emulsifiable and suspension concentrates; antistat for plastics. *Albright & Wilson UK.*

10387 Empilan® BD
Blend of quaternary ammonium germicide with nonionic detergent; bactericidal detergent used in washing of dishes, pans, glassware, dishcloths, etc. *Albright & Wilson Ltd, Detergents Div, Marchon.* Name unverified.

10388 Empilan® BQ 100
9004-96-0
PEG-8 oleate
Emulsifier for insecticides, spindle oils, industrial wetting agent, scouring agent, antifoam, PVC antistat; in laundry works, glue manufacture, paper coating, hand cleaning jellies, turbine oil additive. *Albright & Wilson UK.*

10389 Empilan® CDE
Cocamide DEA (1:1)
Foam boosting/stabilizing agent, solubilizer, detergent for use in personal care and detergent products; antistat in plastics; softener for leather processing. *Albright & Wilson UK.*

10390 Empilan® CIS
68440-05-1 270-431-9
Cocamide MIPA
Foam stabilizer, detergent, shampoo additive. *Albright & Wilson UK.*

10391 Empilan® CME
68140-00-1 268-770-2
Cocamide MEA
Detergent, foam booster/stabilizer in detergent systems; stabilizer for hair and carpet shampoos; base for manufacture of ethoxylated alkylolamides; viscosity modifier. *Albright & Wilson UK.*

10392 Empilan® EGMS
Glycol stearate
Opacifier/pearling and emulsifying/stabilizing agent in shampoos; emollient. *Albright & Wilson UK.*

10393 Empilan® GMS LSE32
Glyceryl stearate SE
Food grade emulsifier. *Albright & Wilson UK.*

10394 Empilan® GMS NSE32
Glyceryl stearate
Food grade emulsifier. *Albright & Wilson UK.*

10395 Empilan® K Series
Alkyl ethoxylates
Emulsifier, wetting agent for agricultural emulsifiable and suspension concentrates, spinning oils in textile processing; emulsifier, dispersant for emulsion polymerization; rubber compounding aid; antistat for plastics. *Albright & Wilson UK.*

10396 Empilan® KA10/80
68439-45-2
C_{10-12} alcohol ethoxylate; scouring and wetting agent for textiles, emulsifier, dye leveling and dispersing agent, detergent, in metal processing, cutting oils, paper industry, paints, insecticides and pesticides; biodegradable. *Albright & Wilson UK.*

10397 Empilan® KB 2
68002-97-1
Laureth-2
Emulsifier, foam booster, superfatting agent, used in detergents and emulsifying systems; mortar plasticizer. *Albright & Wilson UK.*

10398 Empilan® KCA Series
C_{12-13} alcohol ethoxylates; surfactant. *Albright & Wilson UK.*

10399 Empilan® KCB Series
C_{11} alcohol ethoxylates; surfactant. *Albright & Wilson UK.*

10400 Empilan® KCL Series
$C_{12\text{-}15}$ alcohol ethoxylates; surfactant. *Albright & Wilson UK.*

10401 Empilan® KCMP 0703/F
68551-13-3
Fatty alcohol alkoxylate; emulsifier, detergent, wetting agents for industrial and domestic applications. *Albright & Wilson UK.*

10402 Empilan® KCP Series
$C_{14\text{-}15}$ alcohol ethoxylates; surfactant. *Albright & Wilson UK.*

10403 Empilan® KCXSeries
$C_{13\text{-}15}$ alcohol ethoxylates; surfactant. *Albright & Wilson UK.*

10404 Empilan® KI Series
C_{13} alcohol ethoxylates; wetting agent for leather processing. *Albright & Wilson UK.*

10405 Empilan® KL 6
Cetoleth-6
Emulsifier for mineral oils in cosmetics, detergent and wetting agents in textile industry. *Albright & Wilson UK.*

10406 Empilan® KM 11
68439-49-6 7737
Ceteareth-11
Emulsifier, foam control agent in synthetic heavy duty detergents, soap additive; dispersant for textile processing. *Albright & Wilson UK.*

10407 Empilan® KS Series
$C_{9\text{-}11}$ alcohol ethoxylates; surfactant. *Albright & Wilson UK.*

10408 Empilan® LDE
120-40-1 204-393-1
Lauramide DEA (1:1)
Foam booster/stabilizer, solubilizer, thickener, detergent used in shampoos and liquid detergent formulations; antistat for plastics. *Albright & Wilson UK.*

10409 Empilan® LIS
142-54-1 205-541-8
Lauramide MIPA
Foam booster/stabilizer for liquid and powder detergents, shampoos; viscosity modifier, base for manufacture of ethoxylated derivatives. *Albright & Wilson UK.*

10410 Empilan® LME
142-78-9 205-560-1
Lauramide MEA
Foam booster/stabilizer in detergent systems; stabilizer for hair and carpet shampoos; base for manufacture of ethoxylated alkylolamides; viscosity modifier. *Albright & Wilson UK.*

10411 Empilan® LP10
61791-08-0
Ethoxylated cocamide
Foam and detergent booster. *Albright & Wilson UK.*

10412 Empilan® MAA
61791-08-0
PEG-6 cocamide
Liquid detergent additive as foam stabilizer, detergent booster, solubilizer. *Albright & Wilson UK.*

10413 Empilan® NP9
9016-45-9 6772
Nonoxynol-9
Wetting agent, detergent, emulsifier, solubilizer; for agricultural emulsifiable and suspension concentrates, leather processing, emulsion and suspension polymerization; mortar plasticizer; plastics antistat. *Albright & Wilson UK.*

10414 Empilan® OPE9.5
Octyl phenol ethoxylate
Emulsifier for agricultural emulsifiable and suspension concentrates. *Albright & Wilson UK.*

10415 Empilan® P7061
EO/PO condensate; wetting agent for agricultural emulsifiable and suspension concentrates; compounding aid for pharmaceuticals. *Albright & Wilson UK.*

10416 Empilan® SM Series
Sorbitan ester ethoxylates; emulsifier for agricultural emulsifiable concentrates. *Albright & Wilson UK.*

10417 Empimin®
Alcohol ethoxy sulfates
Surfactants. *Albright & Wilson Ltd, Detergents Div.* Name unverified.

10418 Empimin® 3060
Surfactant blend; high expansion fire fighting foam concentrate. *Albright & Wilson UK.*

10419 Empimin® 3116
Sulfosuccinate blend; flotation aid for phosphate rock. *Albright & Wilson UK.*

10420 Empimin® BMA
Formulated surfactant; surfactant used as air entraining agent for cementitious mixes, foaming agent for concrete, mortar plasticizer. *Albright & Wilson UK.*

10421 Empimin® BMB
Formulated surfactant; plasticizer/air entraining agent for construction industry. *Albright & Wilson UK.*

10422 Empimin® BMC
Formulated product; air entraining agent for mortar/cement. *Albright & Wilson UK.*

10423 Empimin® KSN27
9004-82-4
Sodium laureth sulfate (3 EO)
Detergent raw material for high-quality liquid detergents, textile and leather processing; emulsifier, dispersant for emulsion polymerization; biodegradable. *Albright & Wilson UK.*

10424 Empimin® LAM30/AU
75422-21-0
Ammonium alkyl ether sulfate
Wetting and foaming agent used in manufacture of plasterboard, geological drilling operations. *Albright & Wilson Australia.*

10425 Empimin® LR28
151-21-3 8782 205-788-1
Sodium lauryl sulfate
Foaming agent in rubber latex systems; emulsifier for emulsion polymerization. *Albright & Wilson UK.*

10426 Empimin® LSM30
Sodium alkyl ether sulfate
Foaming agent for slurries in plasterboard manufacture; foam-boosting additive in latexes for carpet backing; wetting agent; ingredient in alkaline liquid cleaners. *Albright & Wilson UK.*

10427 Empimin® MA
3006-15-3 221-109-1
Sodium dihexyl sulfosuccinate
Emulsifier, dispersant, wetting agent for emulsion polymerization, agricultural emulsifiable concentrates. *Albright & Wilson UK.*

10428 Empimin® MH
Disodium cocoyl sulfosuccinamate
Foaming agent for carpet backing. *Albright & Wilson UK.*

10429 Empimin® MHH
Disodium N-lauryl sulfosuccinamate as a pale cream liquid; foaming agent for rubber latexes e.g. in carpet manufacture. *Albright & Wilson Ltd, Detergents Div, Marchon.* Name unverified.

10430 Empimin® MK/B
Disodium cetyl stearyl sulfosuccinamate
Foaming agent for carpet backing. *Albright & Wilson UK.*

10431 Empimin® MKK
14481-60-8 238-479-5
Disodium N-stearyl sulfosuccinamate as a pale cream soft paste or spray dried powder; foaming agent for rubber latexes e.g., in carpet manufacture. *Albright & Wilson Ltd, Detergents Div, Marchon.* Name unverified.

10432 Empimin® MSS
Diammonium N-lauryl sulfosuccinamate as a pale cream liquid; foaming agent for rubber latexes e.g. in carpet manufacture. *Albright & Wilson Ltd, Detergents Div, Marchon.* Name unverified.

10433 Empimin® MTT
Disodium N-oleyl sulfosuccinamate as a pale cream liquid; foaming agent for rubber latexes e.g., in carpet manufacture. *Albright & Wilson Ltd, Detergents Div, Marchon.* Name unverified.

10434 Empimin® OP70
577-11-7 3460 209-406-4
$C_{20}H_{37}NaO_7S$
Sodium dioctyl sulfosuccinate
Di(2-ethylhexyl)sulfosuccinic acid, sodium salt; Aerosol OT; Sulfo-butanedioic acid 1,4-bis(2-ethylhexyl)ester sodium salt; Dioctyl sodium sulfosuccinate; docusate sodium; Bis(2-ethylhexyl) sulfosuccinate, sodium salt; Colace; Bis(2-ethylhexyl)sodium sulfosuccinate. Emulsifier, dispersant, wetting agent used for emulsion polymerization, oil slicks, textiles, agrochemicals. *Albright & Wilson UK.*

10435 Empimin® OT
577-11-7 3460 209-406-4
Sodium dioctyl sulfosuccinate
Dispersant, emulsifier, detergent, wetting agent for oil-water emulsions, agrochemicals, emulsion polymerization, filler and extender dispersions, leather processing. *Albright & Wilson UK.*

10436 Empimin® SDS
84501-49-5 282-968-6
Sodium decyl sulfate

Emulsifier, dispersant, detergent, and wetting agent for industrial and institutional cleansers, manufacture of pigments, alkaline cleansers; dust suppression. *Albright & Wilson Australia.*

10437 Empimin® SQ25
68585-34-2
Sodium alkyl triethoxysulfate
Dispersant, wetting and foaming agent in institutional, household, and industrial cleaners. *Albright & Wilson Australia.*

10438 Empiphos
A synthetic detergent comprising aliphatic phosphate. *Albright & Wilson Ltd.* Name unverified.

10439 Empiphos DF Series
Phosphate esters; emulsifier for agricultural emulsifiable concentrates. *Albright & Wilson UK.*

10440 Empiquaternary
An alkyl dimethylbenzyl ammonium chloride, a surfactant. *Albright & Wilson Ltd.* Name unverified.

10441 Empiwax
Oil/water emulsifier; pharmaceutical and toilet preparations, ointments; penicillin cream bases. *Albright & Wilson Ltd, Detergents Div.*

10442 Empiwax SK
Cetearyl alcohol and sodium lauryl sulfate; SE wax as oil-water emulsifier for pharmaceutical and toilet preparations and ointments. *Albright & Wilson UK.*

10443 Emplex
25383-99-7 246-929-7
Sodium stearoyl lactylate
Starch and protein complexing agent for bakery products; emulsifier, conditioning agent, softener for processed foods. *Am. Ingredients/Patco.*

10444 Empol® 1004
Dimer acid, hydrogenated; surfactant for industrial applications, lubricants. *Henkel/Emery.*

10445 Empol® 1010
Dimer acid
Polymer grade; surfactant for industrial applications, lubricants. *Henkel/Emery.*

10446 Empol® 1040
Trimer acid
Surfactant for industrial applications, lubricants. *Henkel/Emeryl.*

10447 Empol® 1061
Dimer acid
Surfactant for industrial applications, lubricants. *Henkel/Emery.*

10448 Emralon
Dispersions of PTFE in resin; used for dry film lubricants and parting agents for rigid and flexible substrates. *Acheson Colloids.*

10449 Emralon 304
9002-84-0 7743 204-126-9
Colloidal PTFE in mixture of alcohol/esters/aromatics; lubricant additives for dry gear and chain lubes, aerosols, assembly and thread lubes. *Acheson Colloids.*

10450 Emralon 8301-01
9002-84-0 7743 204-126-9
Colloidal PTFE in water; lubricant additive for dry gear and chain lubes, aerosols, assembly and thread lubes. *Acheson Colloids.*

10451 EMS 209
Microsilica with a proprietary coating and treatment; incorporated in extruded PVC pipe to increase elastic modulus and pipe stiffness, increase impact resistance and to increase volumetric output from the extruder. *Elkem Chemicals Inc.* Name unverified.

10452 Emsac Concrete Additive
Microsilica (condensed silica fume) based dry or slurried products formulated with dispensing agents such as those commonly used in the concrete industry meeting the requirements of ASTM C494; used to improve the strength and durability of ordinary Portland cement concretes, grouts and mortars. *Elkem Chemicals Inc.* Name unverified.

10453 Emsodur
Cubic grit produced to close dimensional tolerances (0.5-3 mm); highly resistant to abrasion; for deflashing of thermoset molded parts, rubber parts, deflashing of small electric components, tool cleaning, varnish removal from sensitive surfaces, other surface treatments. *EMS-Am. Grilon; EMS-Grilon (UK) Ltd.*

10454 Emsodur Micro
Special deflashing grit with a cube size of 200-400 microns; for deflashing of the smallest electronic components. *EMS-Am. Grilon.*

10455 Emsorb® 2500
1338-43-8 8872 215-665-4
Sorbitan oleate; coupler, emulsifier, lubricant, and softener for textile fibers, leather, cosmetics, agriculture, household products; for formulating petroleum oils and waxes, natural fats and waxes, and alkyl esters. *Henkel/Emery; Henkel/Textile.*

10456 Emsorb® 2502
8007-43-0 8872 232-360-1
Sorbitan sesquioleate
Coupler, coemulsifier for oil-water systems; emulsifier for water-oil systems; household aerosols, cosmetics, agriculture, industrial and textile oils. *Henkel/Emery; Henkel/Textile.*

10457 Emsorb® 2503
26266-58-0 8872 247-569-3
Sorbitan trioleate
Coupler, emulsifier, lubricant, and softener for textile fibers and leather, cosmetics, agriculture, household products; for formulating petroleum oils and waxes, natural fats and waxes, and alkyl esters; coemulsifier for mineral oil. *Henkel/Emery; Henkel/Textile.*

10458 Emsorb® 2505
1338-41-6 8872 215-664-9
Sorbitan stearate
Coupler, hydrophobic emulsifier; coemulsifier for industrial oils, household products, agriculture, and cosmetics; textile lubricant; paper and textile processing. *Henkel/Emery.*

10459 Emsorb® 2507
26658-19-5 8872 247-891-4
Sorbitan tristearate
Coupler, emulsifier, lubricant, and softener for textile fibers and leather, cosmetics, agriculture, household products; for formulating petroleum oils and waxes, natural fats and waxes, and alkyl esters; coemulsifier for mineral oil. *Henkel/Emery.*

10460 Emsorb® 2510
26266-57-9 8872 247-568-8
Sorbitan palmitate
Coupler, emulsifier for cosmetic, agriculture and household products; fiber-to-metal lubricant. *Henkel/Emery.*

10461 Emsorb® 2515
1338-39-2 8872 215-663-3
Sorbitan laurate
Coupler, emulsifier; used in household specialties, industrial oils, agriculture, cosmetics, and emulsion polymerization; antifoam properties. *Henkel/Emery; Henkel/Textile.*

10462 Emsorb® 2516
71902-01-7 8872 276-171-2
Sorbitan isostearate
Auxiliary emulsifier, solubilizer, corrosion inhibitor in lubricants, metal protectants and cleaners, emulsion polymerization. *Henkel/Emery.*

10463 Emsorb® 2518
68238-87-9 269-410-7
Sorbitan diisostearate
Auxiliary emulsifier, solubilizer, corrosion inhibitor in lubricants, metal protectants and cleaners, emulsion polymerization. *Henkel/Emery.* Name unverified.

10464 Emsorb® 2720
9005-64-5 8872
Polysorbate 20
Oil-water emulsifier, solubilizer, viscosity modifier; used in creams, lotions, shampoos, conditioners, liquid soaps. *Henkel/Cospha; Henkel Canada.*

10465 Emsorb® 2721
9005-64-5 8872
PEG-80 sorbitan laurate
Wetting agent, dispersant, mild cleanser for baby products; anti-irritant. *Henkel/Cospha; Henkel Canada.*

10466 Emsorb® 2722
9005-65-6 7742
Polysorbate 80
Emulsifier, coemulsifier for cosmetics; dispersant for pigments in makeup; solubilizer for oils, flavors, fragrances. *Henkel/Cospha; Henkel Canada.*

10467 Emsorb® 2726
PEG-40 sorbitan diisostearate
Emulsifier, solubilizer for flavors in mouthwashes, lipstick, perfume, for perfumes, germicides, and other polar substances in aqueous systems. *Henkel/Cospha; Henkel Canada.*

10468 Emsorb® 2728
9005-67-8 8872
Polysorbate 60
Oil-water emulsifier for cosmetics, hair straighteners, shaving products, sun care products; binder in powders; with Emsorb 2505 to stabilize wax emulsions. *Henkel/Cospha; Henkel Canada.*

10469 Emsorb® 6900
9005-65-6 7742

PEG-20 sorbitan oleate
Dispersant for pigments in coatings; solubilizer for oils and fragrances; hydrophilic emulsifier; coemulsifier for petroleum oils, fats, solvents., and waxes in cosmetics, household products, industrial lubricants, and textile dye carriers; emulsifier for tobacco sucker control concentrates. *Henkel/Emery; Henkel/Textile.*

10470 Emsorb® 6901
9005-65-6 7742
PEG-5 sorbitan oleate
Oil-water emulsifier and lubricant in industrial lubricants, textile lubricants and softeners, metal treatment, paints, emulsion polymerization, agriculture; color dispersant in plastics. *Henkel/Emery; Henkel/Textile.*

10471 Emsorb® 6903
9005-70-3 8872
PEG-20 sorbitan trioleate
Oil-water emulsifier for petroleum oils, fats, waxes, and alkyl esters; lubricant for metals, textiles, leather; in soluble oils for metal processing and finishing; glass fiber lubricants; automotive lubricant additives; agricultural formulations. *Henkel/Emery; Henkel/Textile.*

10472 Emsorb® 6906
9005-67-8 8872
PEG-4 sorbitan stearate
Water-oil emulsifier used in household formulations; fiber-to-metal lubricant for synthetic and cellulosic fibers and yarns. *Henkel/Emery.*

10473 Emsorb® 6908
PEG-16 sorbitan tristearate
Oil-water emulsifier, lubricant, softener; for textile processing and finishing compounds, agricultural formulations. *Henkel/Emery.*

10474 Emsorb® 6909
9005-67-8 8872
PEG-4 sorbitan stearate
Emulsifier for hydraulic fluids, metal treatment, emulsion polymerization, paints; color dispersant for plastics. *Henkel/Emery.*

10475 Emsorb® 6913
9005-70-3 8872
PEG-20 sorbitan trioleate
Emulsifier for agricultural formulation. *Henkel/Emery.*

10476 Emsorb® 6915
9005-64-5 8872
PEG-20 sorbitan laurate
Oil-water emulsifier and solubilizer of petroleum oils, solvents, and fats; used in cosmetic creams and lotions; viscosity modifier in shampoos; emulsifier for dye carriers, antistatic scrooping agent in primary spin finishes, and fiber processing aid in textile industry; agricultural formulations. *Henkel/Emery; Henkel/Textile.*

10477 Emsorb® 6917
PEG-16 sorbitan trioleate
Oil-water emulsifier, lubricant for metals, agriculture, textiles, leather, glass fiber, automotive additives. *Henkel/Emery.*

10478 Emthox® 2730
PEG-75 cocoa butter glycerides; emollient, humectant, emulsifier for personal care products. *Henkel/Cospha; Henkel Canada.*

10479 Emthox® 5882
Laureth-4; dispersant for bath oil; emulsifier for creams and lotions; wetting agent; for eye make-up, deodorants, hair coloring products. *Henkel/Cospha; Henkel Canada.*

10480 Emthox® 5885
68439-49-6 7737
Ceteareth-20; emulsifier, solubilizer for cosmetics, conditioners, depilatories, hair straighteners, sun care products. *Henkel/Cospha; Henkel Canada.*

10481 Emthox® 5967
Laureth-12; emollient, thickener in shampoos; emulsifier in creams and lotions. *Henkel/Cospha; Henkel Canada.*

10482 Emtryl
551-92-8 3315 209-001-2
Dimetridazole
A veterinary antiprotozoan. *RMB Animal Health Ltd.*

10483 EMU® Powd. 120 FD
Styrene-based copolymer; binder for production of paints and impregnation/coating materials for paper; for modifying cement-containing dry mortar. *BASF AG.*

10484 Emulamid TO-21
68092-28-4 268-452-3
Tall oil fatty alkanolamide; emulsifier, lubricant, corrosion inhibitor, wetting agent for water-extendable metalworking coolants. *Mayco Oil & Chem.*

10485 Emulan
Mink oil
Emollient with high spreading coefficient for skin and hair care formulations requiring good oxidative stability. *Emulan.*

10486 Emulan®
Emulsifiers; used for impregnation, lubrication, polishing and cleaning; anticorrosive emulsifier. *BASF plc.*

10487 Emulan® A
9004-96-0
Oleic acid oxyethylate
Oil-water emulsifier for mineral oils and metal polishing emulsions. *BASF AG.*

10488 Emulan® AF
Fatty alcohol oxyethylate
Emulsifier for paraffin wax and mineral oil emulsions. *BASF AG.*

10489 Emulan® EL
Castor oil oxyethylate
Emulsifier for nonaqueous solvents. in water processing, light chemicals industry, for emulsifiable concentrations. for crop protection. *BASF AG.*

10490 Emulan® OC
Fatty alcohol oxyethylate
Oil-water emulsifier for dry bright emulsions, waxes; dispersant, stabilizer for emulsions. *BASF AG.*

10491 Emulan® PO
Alkylphenol oxyethylate
Emulsifier for cleaners, formwork release oils, drilling and cutting oils; flotation agent. *BASF AG.*

10492 Emulan® PO
Nonionic surfactant of the alkylphenol ethoxylate type as a colorless to pale yellow clear oil; used for emulsification, mainly in combination with other emulsifiers; used in cold cleaners and other solvent based cleaners; drilling oil. *BASF AG.* Name unverified.

10493 Emulcid
Hexahydro-1,3,5-tris (2-hydroxyethyl)-s-triazine
Liquid; wet-state bactericide/fungicide for metalworking fluids, detergents, cooling waters, and for industrial plant cleaning. *Thor Chemicals (UK) Ltd.* Discontinued.

10494 Emuldan HV 40 Kosher, HV 52 Kosher
Hydrogenated vegetable oil monodiglycerides; food emulsifier and stabilizer. *Grindsted Prods.*

10495 Emulgade 1000 NI
Cetearyl alcohol and ceteareth-20; emulsifying agent, SE oil-water base for creams, lotions. *Henkel/Cospha; Henkel Canada; Henkel KGaA.*

10496 Emulgade C
Emulsifying wax NF; self-emulsifying raw material for emulsions, hair conditioners, creams, lotions. *Henkel/Cospha; Henkel Canada.*

10497 Emulgade EO-10
Oleth-25 and cetyl alcohol; self-emulsifying base for preparation of creams and emulsions. *Pulcra SA.*

10498 Emulgade F
Cetearyl alcohol, PEG-40 castor oil, and sodium cetearyl sulfate; SE raw material for cosmetic creams and fluid emulsions; emulsifier for cosmetics, pharmaceuticals. *Henkel/Cospha; Henkel Canada.*

10499 Emulgator E 2149
Steareth-7
stearyl alcohol. Emulsifier and stabilizer for cosmetic and pharmaceutical oil-water emulsions. *Goldschmidt; Goldschmidt AG.*

10500 Emulgator E 2155
Stearyl alcohol, steareth-7, steareth-10; emulsifier and stabilizer for cosmetic and pharmaceutical oil-water emulsions. *Goldschmidt; Goldschmidt AG.*

10501 Emulgator U4
9016-45-9 6772
Nonoxynol-4; surfactant, emulsifier, wetting agent for liquid and powder detergents, industrial preparations, textile, leather, pulp/paper, dry-cleaning, metal cleaners. *Unger Fabrikker AS.* Discontinued.

10502 Emulgeant 710
Polyethoxy ether
Emulsifier for aliphatic and aromatic solvents, paraffin; in degreasers. *Ceca SA.*

10503 emulgen
A jelly-like mass used for the rapid emulsification of oils and resins. It contains tragacanth, gum arabic, pittoporad, glycerin, alcohol, and water.

10504 Emulphopal HC
Proprietary; waterless hand cleaner base. *Stepan Canada.*

10505 Emulpon
Emulsifiers, dispersants, solubilizers for oils, solvents and waxes; used for textile, agriculture and cosmetic industries. *Witco Corporation.*

10506 Emulsamin
A proprietary trade name for menthol diurethane, a wetting agent and detergent. No manufacturer.

10507 Emulsene
Emulsifying agents. *Bush Boake Allen Ltd.*

10508 Emulsifier 4
Dialkyl quaternary ammonium chloride in water/isopropanol; emulsifier for nonpolar hydrophobes, e.g., mineral seal oil, waxes, silicones; used for auto spraywax, carnauba spraywax, mop treatment emulsions, stainless steel cleaners, vinyl dressings. *Exxon/Tomah.*

10509 Emulsifier 632/90%
Modified alkyl phenol ethoxylate; low foam surfactant, dispersant for fats, oils, and waxes. *Ethox.*

10510 Emulsifier K 30 40%
Sodium alkane sulfonates based on n-paraffin; emulsifier for emulsion polymerization; effective over wide pH range; antistat; biodegradable. *Bayer AG; Miles/Organic Prods.*

10511 Emulsifier L.W.
Cyclohexylammonium oleate, a surfactant. *Laporte Industries Ltd.*

10512 Emulsifier WHC
Formulated emulsifier, emulsifier and detergent used in waterless hand cleaner formulations and degreasers. *Stepan; Stepan Canada.*

10513 Emulsil
Food grade silicone products; antifoams and release agents. *Siliconas Hispania SA.*

10514 Emulsilac S
25383-99-7 246-929-7
Sodium stearoyl lactylate
Emulsifier, dough conditioner, freeze/thaw stabilizer for food industry. *Witco Corporation.*

10515 Emulsin
A ferment; decomposes the glucoside, amygdalin, into grape sugar, benzaldehyde, and hydrocyanic acid.

10516 Emulsion C-340
8015-86-9 1859 232-399-4
Carnauba wax emulsion; film-former for most substrates including fabric, metal, wood, leather, and painted surfaces; for spraywaxes, wood water repellent, protective coatings. *Exxon/Tomah.*

10517 Emulsi-Phos
Food grade phosphate; emulsifier specially designed for use in process cheese, cheese food and cheese spreads. *Monsanto Co.* Name unverified.

10518 Emulsogen
Emulsifiers for oils and waxes. *Hoechst UK.*

10519 Emulsogen 2144
Ethoxylated rosin; low foaming emulsifier, wetting agent for metalworking fluids, high acidic conditions. *Hoechst Celanese/Colorants & Surf.*

10520 Emulsogen CP 136
Calcium dodecylbenzene sulfonate/castor oil ethoxylate; surfactant for agricultural formulations. *Hoechst Celanese/Colorants & Surf.*

10521 Emulsogen EL-050
61791-12-6
PEG-5 castor oil; emulsifier additive for chlorinated paraffins, triglycerides; surfactant for textile processing. *Hoechst Celanese/Colorants & Surf.*

10522 Emulsogen HEL-050
61788-85-0
PEG-5 hydrogenated castor oil; surfactant for textile processing. *Hoechst Celanese/Colorants & Surf.*

10523 Emulsogen IC
Calcium dodecylbenzene sulfonate/oleyl alcohol ethoxylate
Surfactant for agricultural formulations. *Hoechst Celanese/Colorants & Surf.*

10524 Emulsogen M
Fatty alcohol ethoxylate; mineral oil emulsifier, wetting agent; for metalworking fluids. *Hoechst Celanese/Colorants & Surf.*

10525 Emulsynt GDL
27638-00-2 248-586-9
Glyceryl dilaurate
Emulsifier, thickener, and emollient for creams and lotions. *Van Dyk.* Name unverified.

10526 Emultex
Liquid emulsifier; for sizing oils. *Tensia SA.* Name unverified.

10527 Emultex 307, 328
A proprietary range of unplasticized vinyl acetate homopolymer emulsions stabilized with polyvinyl alcohol and used in the manufacture of adhesives. *Harlow Chemical Co Ltd.*

10528 Emultex AC431
A proprietary vinyl acetate-acrylic ester copolymer emulsion; used for general-purpose emulsion paints. *Harlow Chemical Co Ltd.*

10529 Emultex®
Vinyl acetate homo and copolymer dispersions; used for adhesives, surface coating and textiles. *Harlow Chemical Co Ltd.*

10530 Emulvin®
Emulsifying agents. *Bayer AG; Bayer plc.*

10531 Emulvin® W
Aromatic polyglycol ether
Emulsifier, stabilizer, wetting agent for latexes. *Bayer Corp; Polysar.*

10532 Emulzome
8012-95-1 232-384-2
Hydrogenated polyisobutene, stearyl heptanoate, mineral oil; for cosmetic creams and milks. *Exsymol.*

10533 Emvelop®
68334-00-9 269-804-9
Hydrogenated cottonseed oil
Hydrogenated vegetable oil; for production of sustained-release tablet formulations. *Mendell.* Discontinued.

10534 Emzyaml No. 1
Phytoenzymatic complex, anti-free radical protector with anti-solar and anti-bacterial activity; helps prevent skin from aging. *Alban Muller.*

10535 Enathene® EA 705-009
Ethylene n-butyl acrylate copolymer; for adhesives for case and carton sealing, bookbinding, laminates, sealants, caulks, textile coatings. *Quantum/USI.*

10536 Enathene® EA 720-009
Ethylene n-butyl acrylate copolymer; for extrusion coating; excellent heat seal response at low temps.; good thermal stability at high processing temps.; for coatings and laminations with oriented PP, PET, papers, clay-coated board and nylon substrates for speciality packaging. *Quantum/USI.*

10537 Enbucrilate
Butyl 2-cyanoacrylate
histoacryl. A surgical tissue adhesive. No manufacturer.

10538 Encapsulated MgO
1309-48-4 5713 215-171-9
Magnesium oxide
Olin. Discontinued.

10539 Encapsulation
Thermosetting material; for molded products. *Sumitomo Bakelite Co Ltd.* Name unverified.

10540 Encelac®
Organic and inorganic pigment formulations in ester-soluble nitrocellulose and dibutyl phthalate; for pigmenting nitrocellulose lacquers. *BASF AG.*

10541 Encem Steel
A proprietary trade name for a nickel-chromium-molybdenum steel. No manufacturer.

10542 Enceprint®
Organic and inorganic pigment formulations in a low viscosity alcoholic nitrocellulose containing plasticizers; for pigmenting of flexographic and gravure inks. *BASF AG.*

10543 Encore
isoproturon-pendimethalin
Suspension concentrate containing 125 g isoproturon and 250 g pendimethalin per liter; used for annual weed control in winter wheat, rye and barley. *Cyanamid of Great Britain Ltd.*

10544 Endanil
Nylon dyes and pigments for plastics. *ICI Chem & Polymers Ltd.*

10545 Endcor
A wide range of high performance, corrosion-resistant coating systems. Types include: acrylic, alkyd, chlorinated rubber, epoxy, pretreatments, primers, polyurethane, vinyl and zinc rich; general industrial maintenance for metal and masonry surfaces in a wide variety of exposure conditions, including weathering and chemical attack, immersion service. *Dampney Company Inc.* Name unverified.

10546 Endegal
Polyester dyes and pigments for plastics. *ICI Chem & Polymers Ltd.* Discontinued.

10547 Endermol
A compound ointment vehicle, containing hydrocarbons of the paraffin series, and stearic acid amide. No manufacturer.

10548 Endobil
31127-82-9 250-478-1
$C_{26}H_{26}I_6N_2O_{10}$
Iodoxamic acid
Diagnostic aid. *Bracco Industria Chimica SpA.* Name unverified.

10549 Endocrocine
$C_{16}H_{10}O_7$

The orange-yellow coloring matter isolated from *Nephioniopsis endocrocea*, a lichen growing in Japan. It is a pentahydroxyanthraquinone.

10550 **Endomirabil**
Iodoxamic acid
Diagnostic aid. *Bracco Industria Chimica SpA*. Name unverified.

10551 **endotryptase**
A proteolytic enzyme.

10552 **Endox**
3090
dicamba-mecoprop
Soluble concentrate of 112 g dicamba [1918-00-9] and 265 g mecoprop 7085-19-0] per liter; herbicide used for weed control in cereals and grassland. *Farmers Crop Chemicals Ltd*. Discontinued.

10553 **Endrate**
139-33-3 3556 205-358-3
Edetate disodium
Chelating agent; pharmaceutical aid. *Abbott Laboratories*. Name unverified.

10554 **Endspray**
fentin hydroxide-metoxuron
Mixture of fentin hydroxide and metoxuron; used for control of potato blight and haulm dessicant. *Pan Britannica Industries Ltd*.

10555 **Endura**
Epoxy and polyurethane primers. *Feb Ltd*.

10556 **Endura®**
Flame-retardant concentrates. *PPG/Polymer Prods*.

10557 **Enduracrete**
A trowel-applied polyurethane screed with good impact and abrasion resistance; especially suited to hygienic demands; resistant to steam cleaning. *Feb Ltd*.

10558 **Enduraflex**
Two-component pitch-modified polyurethane pouring grade sealant for industrial floors. *Feb Ltd*.

10559 **Enduraflor**
Self-smoothing epoxy system. *Feb Ltd*.

10560 **Enduragloss**
Purified rosin ester gum/2-octyldodecyl myristate; neutral oil improving adhesion, gloss, conditioning in lipsticks, foundation, eye shadow stick, hair creams. *U.S. Cosmetics*.

10561 **Enduraguard EP**
A pitch epoxy coating incorporating nonslip aggregate; produces high strength, extremely hard wearing coating. *Feb Ltd*.

10562 **Endurakote**
Nonstainting water-dispersible epoxy coating. *Feb Ltd*.

10563 **Enduralay**
A trowel-applied nontainting epoxy system suitable for use in food processing, brewing, and beverage industries. *Feb Ltd*.

10564 **Enduralith**
Premixed polymer-modified cementitious floor topping suitable for medium to heavy duty floor areas. *Feb Ltd*.

10565 **Enduratop**
A high strength polymer modified overlay for areas requiring resistance to impact and abrasion. *Feb Ltd*.

10566 **En-Dur-Lon**
Rubber base coating, contains no abrasive; antislip or decorative coating. *W J Ruscoe Co*. Discontinued.

10567 **Enduro Alloys**
Proprietary corrosion resisting alloys of iron with chromium, or with nickel and chromium; Enduro a contains iron with from 16.5-18.5% chromium; Enduro KA2 contains iron with 17-20% chromium and 7-10% nickel; Enduro S has iron with 12.5-14.5% chromium. No manufacturer.

10568 **Endurol**
Vat dye colors. *James Robinson & Co Ltd*.

10569 **Enelchem Products**
Lead pigments, lead oxides, lead silicates, calcium carbonates, etc. *Rheox Inc*. Discontinued.

10570 **Enerade® 3045**
Oxyalkylated polyol
Emulsion breaker base. *Rhône-Poulenc Oil Field Chem*.

10571 **Enerade® 7101, 7102**
Oxyalkylated nonyl phenolic resin; emulsion breaker base. *Rhône-Poulenc Oil Field Chem*.

10572 **Energex**
A broad range carbohydrase preparation produced by submerged fermentation of a selected strain of the aspergillus niger group; added to animal feed in order to boost feed digestability and improve feed utilization. *Novo Nordisk*.

10573 **Enervite®**
Fish oils, fish liver oils, cod liver oil B. vet C; dietary supplement for animals; conditioning oil for animals. *Seven Seas Ltd*.

10574 **English bearing metal**
Antifriction and fitting metal. It contains usually 53% tin, 33% lead, 10.5% antimony, and 2.5% copper.

10575 **English metal**
A jeweler's alloy containing 88 parts tin, 2 parts copper, 2 parts brass, 2 parts nickel, 1 part bismuth, 8 parts antimony, and 2 parts tungsten.

10576 **English white bearing metal**
An antifriction metal, containing 77% tin, 15% antimony, and 8% copper.

10577 **engobe**
A fusible mixture of clay, telspar, and silica; used for the manufacture of glazes on pottery.

10578 **Enhance**
Floor polishes. *Evans Vanodine International Ltd*.

10579 **Enhance**
Nonionic wetting agent containing 90% alkylphenol ethoxylate; spreader for use in herbicide, fungicide and insecticide sprays. *Midkem Agrochemicals*.

10580 **Enhance®**
For the agriculture industry. *DuPont UK*.

10581 **Enide**
957-51-7 3364 213-482-4
Diphenamid
Used for weed control in horticultural crops. *ICI Agrochemicals*.

10582 **Enilconazole**
35554-44-0 3622 252-615-0
$C_{14}H_{14}Cl_2N_2O$
1-[2-(2,4-Dichlorophenyl)-2-(2-propenyloxy)ethyl]-1H-imidazole
Imaverol; Imazalil; R-23979; Clinafarm. A proprietary preparation containing enilconazole; used to treat veterinary dermatomycoses; disinfectant; agricultural fungicide. Soluble in organic solvents; slightly soluble in H_2O. *Janssen Pharmaceutical Ltd*. Name unverified.

10583 **Enisyl**
657-27-2 5667 211-519-9
Lysine hydrochloride
Amino acid. Used in biochemical research. *Person & Covey Inc*. Discontinued.

10584 **Enmag**
Magnesium ammonium phosphate fertilizer. *Scottish Agricultural Industries plc*.

10585 **ENR 25**
9006-04-6 232-689-0
Polyisoprene
Natural rubber; modified by epoxidation reaction to achieve greater oil resist., gas permeability, and damping characteristics; used for specialty tires, milking inflations, hoses, footwear. *Guthrie Latex*.

10586 **ENSA-6**
Ethoxylated naphthol sulfonic acid
Brightening and dispersing agent for tin plating. *Hart Chem. Ltd*.

10587 **Ensecote S**
A proprietary trade name for a modified epoxy resin coating material for high temperature stoving or spraying. No manufacturer.

10588 **Enso DTO 10-30**
Distilled tall oils with a rosin acids content of 10-30%; used for alkyd resins, liquid soaps, emulator, latex. *Enso-Gutzeit OY*. Discontinued.

10589 **Enso Rosin**
Tall oil rosin; used for paper sizes. *Enso Gutzelt OY*. Discontinued.

10590 **Ensol2**
Tall oil fatty acids with 2% rosin acids; used for alkyd resins, liquid soaps. *Enso-Gutzeit OY*. Discontinued.

10591 **Ensoline**
Self-emulsifying mineral oils; lubricant for synthetic fibers. *Atochem UK/Ceca*. Discontinued.

10592 **Ensoline 203 AS**
Antistatic oiling of all fibers. *Ceca SA*.

10593 **Entamide**
579-38-4 3246 209-439-4
diloxanide
A proprietary trade name for diloxanide; an amebicide. *Knoll AG*. Discontinued.

10594 **Entarex**
NTA salts; chelating agents for trace metal control. *W R Grace Ltd*.

10595 **Entramin**
121-66-4 477 204-490-9
2-Amino-5-nitrothiazole

2-Amino-5 nitrothiazole premix. Used as an anti-histomonad in turkeys and chickens and to treat trichomoniasis in pigeons. *May & Baker Ltd.*

10596 Entrox
1314-13-2 10279 215-222-5
zinc oxide
Zinc oxide, coated to improve incorporation into rubber. *Durham Chemicals Ltd.* Name unverified.

10597 Enusin Colours
Pigment colors based on casein binders, for aniline and semi-aniline finishing. *Bayer plc.*

10598 Envex® 1001
Polyimide, thermoplastic; for ultrasonic transducers, probes, tips; flow meter gears. *Rogers.*

10599 Envirez
Unsaturated polyester resins formulated for low smoke, reduced smoke toxicity and low volatility combined with fire retardancy; used for transportation (subways, aircraft etc.) and construction materials requiring both fire retardancy and reduction. *Ashland Chemical Company.* Name unverified.

10600 Envirocats
A range of catalysts used in alkylation, oxidation, acetylation, and sulfonylation processes. *Contract Chemicals (Knowsley) Ltd.*

10601 Envirosafe
Cleaner and degreaser. *Kalon Chemicals Ltd.*

10602 Enviroseal
Water based water repellent; used for concrete water proofing. *PCR.* Discontinued.

10603 Enviroset
Binder system for fuel agglomerates. *Foseco (F.S.) Ltd.*

10604 Enzeon
Chymotrypsin
Enzyme. *Sterling Drug Inc.* Name unverified.

10605 Enzypan
A proprietary preparation of pepsin, pancreatin, and bile; digestive enzyme supplement. *Norgine Ltd.*

10606 Eosin
15086-94-9 239-138-3
$C_{20}H_6O_5Br_4Na_2$
Eosin A; eosin B; eosin C; eosin A extra; eosin DH; eosin GGF; eosin G Extra; eosin 3J; eosin JJS; eosin G; eosin KS; eosin yellowish; water soluble eosin; acid eosin; eosin 4J extra. The alkali salts of tetrabromo-fluoresceine, dyes wool and silk yellowish red; used as a microscopic stain and a fluorescent tracer dye; red writing ink; cosmetic products and a colorant for motor fuel.

10607 Eosin YS
A British brand of Eosin. No manufacturer.

10608 Eosolate
Silver acetylguaiacoltrisulfonate
An antiseptic.

10609 EP Lead
Specially-prepared lead naphthenate; used in formulating extreme pressure gear oils and greases. *Hüls Am.*

10610 EP Pastes
Aqueous dispersions of organic and inorganic pigments. *Reckitts Colours Ltd.*

10611 EP-1
298-06-6 206-055-9
$C_4H_{11}O_2PS_2$
O,O-Diethyl phosphorodithioic acid
O-O-diethylphosphorodithioic acid; O,O-diethylphosphorodithioate; Diethylphosphorodithioate; Phosphorodithioic acid, O,O'-diethyl ester. Used as an intermediate for organophosphorus insecticides. *A/S Cheminova.* Name unverified.

10612 EP-2
O,O-Diethyl phosphorochlorodithioate
Mainly used in the production of organophosphorus insecticides. *A/S Cheminova.* Name unverified.

10613 Epal+20·2
68911-61-5 272-778-1
C_{18-32} linear and branched alcohols (66%), C_{24-40} hydrocarbons (40%); biodegradable detergent/ emulsifier intermediate. *Ethyl Corp.*

10614 Epal®
Linear primary alcohols. *Ethyl Corp.*

10615 Epal® 6
111-27-3 4732 203-852-3
Hexanol
Intermediate in the manufacture of detergents and emulsifiers. *Ethyl Corp.*

10616 Epal® 8
111-87-5 6849 203-917-6
Octanol
Detergent and emulsifier intermediate. *Ethyl Corp.*

10617 Epal® 10
112-30-1 2911 203-956-9
Decanol
Detergent and emulsifier intermediate. *Ethyl Corp.*

10618 Epal® 12
112-53-8 3464 203-982-0
Dodecanol
Detergent and emulsifier intermediate. *Ethyl Corp.*

10619 Epal® 14
112-72-1 6418 204-000-3
Tetradecanol
A biodegradable detergent/emulsifier intermediate. *Ethyl Corp.*

10620 Epal® 108
Octanol, decanol, hexanol (55:41:4); a biodegradable detergent/emulsifier intermediate. *Ethyl Corp.*

10621 Epal® 610
C_8 alcohol (42%), decyl alcohol (54%), hexanol (4%); biodegradable detergent/emulsifier intermediate. *Ethyl Corp.*

10622 Epal® 1012
Decyl alcohol (75%) and dodecanol (23%); a biodegradable detergent/emulsifier intermediate. *Ethyl Corp.*

10623 Epal® 1214
Dodecanol (66%), tetradecanol (27%), hexadecanol (6%); a biodegradable detergent/emulsifier intermediate. *Ethyl Corp.*

10624 Epal® 1412
Tetradecanol (58%), dodecanol (40%); intermediate for surfactants, plasticizers, lubricant additives, thioesters, specialty chemicals. *Ethyl Corp.*

10625 Epal® 1618
Hexadecanol (47%), octadecanol (50%); biodegradable detergent/emulsifier intermediate. *Ethyl Corp.*

10626 Epal® 16NF
36653-82-4 2070 253-149-0
Hexadecanol NF
USP grade biodegradable detergent/emulsifier intermediate. *Ethyl Corp.*

10627 Epal® 18NF
112-92-5 8960 204-017-6
Octadecanol NF
USP grade biodegradable detergent/emulsifier intermediate. *Ethyl Corp.*

10628 Ephos
7758-87-4 1741 231-840-8
calcium phosphate, tribasic
A basic phosphate, containing 60-65% tricalcium phosphate.

10629 Ephynal
59-02-9 10159 200-412-2
Vitamin E
A proprietary preparation of vitamin E; an essential vitamin. *Roche Products Ltd.* Name unverified.

10630 Epibond® 1217-A/B
Quick-set adhesive paste for aircraft industry; rapid cure; excellent adhesion to steel, Al, wood, plastics. *Ciba-Gelgy/Furane.*

10631 Epibond® 1544-A/B
Self-extinguishing adhesive for thermoplastics in aircraft industry; low flow; bonds dissimilar substrates. *Ciba-Geigy/Furance.*

10632 Epichlorhydrin
106-89-8 3648 203-439-8
C_3H_5ClO
Chloromethyloxirane
chloropropylene oxide; chloromethyl oxirane; dl-α-epichlorohydrin; 1-chloro-2,3-epoxypropane; γ-chloropropylene-oxide,. Major raw material for epoxy and phenoxy resins, manufacture of glycerol, curing propylene-based rubbers; solvent for cellulose esters and ethers; high wet-strength resins for paper industry.

10633 Epichlorohydrin elastomer
$(CHCH_2ClCH_2O)_x$
chloropropylene oxide elastomer
Epichlorohydrin elastomer; elastomer used for fuel pump diaphragms, pipe gaskets, hose for fuel, oil, and gas, vibration isolators, motor mounts, rolls, adhesives, sponge goods, air conditioning hose and seals. Ciba-Geigy; Conap; Hardman; Key Polymer; Morton Int'l.; Reichhold; Rhône-Poulenc/Perf. Resins; Sartomer; Shell; Union Carbide.

10634 Epiclon
Anydyride epoxy curing agents. *Anchor Chemical Group plc.*

10635 EPICS
Multiparameter computer and laser instrument; used for counting and sorting types of biological cells. *Coulter Electronics Ltd.*

10636 Epi-Cure 8515
A proprietary amidoamine curing agent for epoxy resins. *Alma Paint & Varnish Co.* Unverified.

10637 Epi-Cure® 87
Modified aliphatic amine adduct; curing agent for applications where relatively short cure periods at room temperature required; end uses include tooling gel coats, laminating compounds, adhesives. *Shell.*

10638 Epidermin in Oil
Polyvalent tissue complex in oil medium; product for aging and sensitive skin. *Dr. Kurt Richter; Henkel/Cospha.*

10639 Epidermin Water-Soluble
Polyvalent tissue complex, hydro-alcohol solubilized; product for aging and sensitive skin. *Dr. Kurt Richter; Henkel/Cospha.*

10640 Epiflex
A proprietary trade name for an epoxy resin expansion jointing material. *Haworth (ARC) Ltd.* Unverified.

10641 Epiglaubite
An impure calcium phosphate.

10642 Epihydrin
75-56-9 8041 200-879-2
C_3H_6O
Propylene oxide
Antifreeze, food emulsifier and preservative.

10643 Epikem
Nonstaining epoxy adhesive and grouting system used in the food processing and brewing industries. *Feb Ltd.*

10644 Epikote
A series of epoxy resins whose characteristics may be modified by hardeners and other additives. *Shell.*

10645 Epikote DX-209-B-80, DX-210-B-80
A proprietary 80% solution of epoxy resin in methyl ethyl ketone; used with Epikure 3400 as a curing agent. *Shell.* Name unverified.

10646 Epikote DX-231-B-91
A proprietary solution containing 91% epoxy resin in methyl ethyl ketone for use in work involving carbon film; it is cured with Epikure 3400. *Shell.* Name unverified.

10647 Epikure 3400
A proprietary curing agent for epoxy resins. *Shell.* Name unverified.

10648 Epilink
A range of water-based, solvent-free and solvent-based curing agents for use with epoxy resins. *Akzo Chemie NV.*

10649 Epilok
Epoxy curing agents. *Akzo Chemie.* Name unverified.

10650 Epilon
A proprietary trade name for an epoxy resin cement. *Haworth (ARC) Ltd.* Unverified.

10651 Epiphassol
A viscous oily liquid containing naphthene-sulfonic acids. It is a similar preparation to Kontakt, and is used for cleaning cotton fabrics.

10652 Epiphen
Epoxy resin. *Monomer-Polymer & Dajac Laboratories, Inc.*

10653 Epires
Epoxy resin. *Akzo Chemie UK Ltd.*

10654 Epirez 501
2426-08-6 219-376-4
Butyl glycidyl ether
A proprietary reactive diluent for epoxy resin systems. *Alma Paint & Varnish Co.* Unverified.

10655 Epirez 502
A proprietary aliphatic diepoxide. *Alma Paint & Varnish Co.* Unverified.

10656 Epirez 520C
A proprietary epoxy resin of the bisphenol A type. *Alma Paint & Varnish Co.* Unverified.

10657 Episol
15599-36-7 4622
$C_{19}H_{21}ClN_2OS$
5-chloro-2-[*p*-(diethylaminoethoxy)phenyl]benzothiazole
Halethazole. An antiseptic and antifungal agent. mp = 93-94°. *Crookes Laboratories.*

10658 Episol
Dyes for leather. *James Robinson & Co Ltd.*

10659 Epitar, Epitate
Epoxy resin additives. *Midland-Yorkshire Tar Distillers.* Unverified.

10660 Epitone
7- and 8-acetyl-5-isopropyl-2-methylbicyclo[2.2.2]oct-2-ene. Flavorants. *Quest Int'l. UK Ltd.*

10661 Epivax
A proprietary vaccine used against canine distemper. *Coopers Animal Health Ltd.* Name unverified.

10662 Eplink
Epoxy curing agents. *Akzo Chemie UK Ltd.*

10663 EPlstatic® 100
Tribasic lead sulfate; head stabilizer for flexible and rigid PVC compounds. *Eagle-Picher.* Name unverified.

10664 EPlstatic® 110
Basic lead silicosulfate; heat stabilizer for use in low-temp. flexible PVC. *Eagle-Picher.* Name unverified.

10665 EPlthal 120
Basic lead phthalate
A heat stabilizer in 90° and 105° PVC wire and cable compounds *Eagle-Picher.*

10666 EPM rubber
9010-79-1
Ethylene-propylene copolymer; EPR rubber for injection molded and extruded goods (electrical components, wire insulation, o-rings, brake components); modifying PP and other plastics. *Hüls AG.* Discontinued.

10667 Epocap®
Two-part epoxy compounds. *Harcros UK, Hardman.*

10668 Epocap® 16129 A/B
25928-94-3
Flexible, low viscosity, unfilled system with excellent crack resistance and moisture protection for stress sensitive components; excellent penetration for cost effective sand potting. *Hardman.*

10669 Epocap® 16358 A/B
25928-94-3
Epoxy; rigid, high heat temperature distortion, filled, flame retardant system with excellent thermal conductivity, very low exothermicity; excellent high temperature electrical properties. *Hardman.*

10670 Epocast 1610
Syntactic lightweight honeycomb structure reinforcement compound *Ciba-Geigy Plastics UK.* Name unverified.

10671 Epocrete
Two-part epoxy concrete materials. *Hardman.*

10672 Epocure
Epoxy curing agents. *Hardman.*

10673 Epodil
Glycidyl ether reactive dilutents for epoxy resins. *Anchor Chemical Group plc.*

10674 Epodur
High solids and solventless epoxy maintenance coatings for long-term corrosion protection under extreme exposure conditions; used for buried pipelines, marine and offshore equipment, heavy construction, chemical process equipment, water and sewage works equipment, structural sheet, tank linings. *Dampney Company Inc.* Name unverified.

10675 Epok
Elastomeric sealant coating. *BP Chemicals Ltd.* Discontinued.

10676 Epolast
Two-part epoxy compounds. *Hardman.*

10677 Epolene®
Polyolefin waxes. *Eastman.*

10678 Epolene® C-10
9002-88-4 7728
Polyethylene wax
Nonemulsifiable low-density wax used in hot melt adhesives and coatings for papers and packaging materials, as paraffin modifiers, in slush and cast molding, rubber compounding, and in rigid and flexible vinyl compounds; high gloss coatings, low water-vapor-transmissions rates, grease- and blocking-resistance. *Eastman.*

10679 Epolene® C-13
9002-88-4 7728
Polyolefin resin, low molecular weight; wax designed for use with Epolene waxes or blends containing lower molecular weight materials; used as petrol.-wax modifiers to increase blend visc., improve grease resistance, blocking temp., scuff resistance; and gloss of paraffin; additive for inks and as compounding resins for hot melt adhesives. *Eastman.*

10680 Epolene® C-16
Polyolefin resin; nonemulsifiable wax used as hot melt coatings for paper (glossy barrier coatings, readily heat sealable to paper products, metal foils, and plastic films); in petrol.-wax coatings, and in wax-copolymer coatings for

increased scuff resistance, gloss stabilization and hot tack; tolerates high levels of inorganic fillers without drastic increases in melt viscosity. *Eastman.*

10681 Epolene® E-14
Oxidized polyethylene homopolymer resin; low-density and low softening point emulsifiable wax, imparts slip resistance to floor polish films. *Eastman.* Name unverified.

10682 Epoleon® N-7C
Glycine betaine, organic and salt of organic acids, amine compounds, essential oils; odor neutralizing agent for restaurant and household garbage, wastewater treatment, sludge treatment, landfill applications, scrubber systems. *Epoleon Corp. of Am.*

10683 Epolene® N-15
9003-07-0 7741
PP homopolymer resin; high melting point and hardness; modifier for petroleum. waxes to increase resistance to blocking, scuffing, and abrasion, compound resin for hot melt adhesives. *Eastman.*

10684 Epolene® N-20, N-21
9002-88-4 7728
Polyethylene homopolymer wax; nonemulsifiable wax with improved resistance to solvents and oils, and good hardness; used in cosmetics, hot-melt adhesives, dispersing aids for color concentrations., cable filling composition, slip additives for printing inks; modifiers for hot-melt highway marking. *Eastman.* Name unverified.

10685 Epoleon® N-100
Glycine betaine, sodium citrate, sodium dihydrogen phosphate, PEG lauryl ether; odor neutralizing agent for scrubber systems, waste water treatment. *Epoleon Corp. of Am.*

10686 Epolite 1301
25928-94-3
Two-part epoxy resin; general purpose, chip-resistance, epoxy surface coat (fast tack free time); used in models, masters, stretch dies, checking and drill fixtures, etc.; nonsag properties; cures readily under high humidity conditions; mix ratio: 100 parts resin/10 parts hardener by weight. *Hexcel.*

10687 Epolite 1302
25928-94-3
Two-part epoxy resin and hardener; abrasion-and impact-resistance epoxy, metallic black surface coat with rapid cure; for automotive gates, router and trim fixtures, foundry patterns and core boxes, check and inspection fixtures, draw and stretch dies; mix ratio: 100 parts resin/7 parts hardener by weight. *Hexcel.*

10688 Epolite 2300
Two-part resin laminating system; laminating system, max, wettability, low visc., high mechanical strength; used in checking and inspection fixtures and gauges, Keller models, spotting racks, router and trim, assembly and drill fixtures; foundry patterns, core boxes and stretch dies and models; mix ratio: 8 parts resin/1 part of hardener by weight. *Hexcel.*

10689 Epolite 2315
25928-94-3
Two-part epoxy system; room temperature curing laminating resin. *Hexcel.*

10690 Epolite 3300
25928-94-3
Two-part epoxy compound; mass casting epoxy compounds, variable hardness, high impact resistance; used in hammer forms, stretch dies, drop hammer, close tolerance masters or models; mix ratio: 100 parts resin/10-30 parts hardener by weight. *Hexcel.*

10691 Epolite 5363
25928-94-3
Two-part epoxy system; adhesive for thin film bonding, encapsulating. *Hexcel.*

10692 Epomarine
Two-part epoxy compounds for splash zone and for underwater application. *Hardman.*

10693 Epon®
Epoxy resins. *Shell.*

10694 Epon® 8280
A proprietary liquid epoxy resin for use in filled compounds. *Shell.* Name unverified.

10695 Epon® Resin DPL-1911
25928-94-3
Rubber modified liquid epoxy resin; used for composites exhibiting enhanced fracture toughness. *Shell.*

10696 Eponc®
A range of epoxy resins; used for varnish industry, electrotechnical field and building industry. *AMC SPREA S.p.A.*

10697 Epophen
Epoxide resins. *Borden (UK) Ltd.* Name unverified.

10698 Eposet
Paint brush bristle-setting two-component epoxies. *Hardman.*

10699 Eposet
Two-part epoxy adhesives. *Harcros.*

10700 Eposolve 299-R
Versatile clean-up solvent (non-red label) for cured and uncured epoxies. *Hardman.*

10701 Epotal® 181 D
Ethylene polymer dispersion; additive for other dispersions; increases blocking resistance of adhesives; improves slip in paper finishing. *BASF AG.*

10702 Epo-Tek® E-3081
Single-component, electrical conductive die attach adhesive; for bonding large chips and substrates. *Epoxy Tech.*

10703 Epotuf®
Epoxy resins and hardeners. *Reichhold.*

10704 Epotuf® 38-690
Water-reducible epoxy ester. *Reichhold.*

10705 Epotuf® Hardener 37-612
Polyamide; epoxy hardener; general purpose hardener, good resiliency and adhesion; used in electrical potting, encapsulating, and casting, flooring, surfacing, coatings, and adhesives. *Reichhold.*

10706 Epotuf® Hardener 37-621
Polyamide; epoxy hardener; solution form; used in coatings. *Reichhold.*

10707 Epoweld®
Two-part epoxy compounds; structural adhesives. *Harcros UK, Hardman.*

10708 Epoweld® 19157
25928-94-3
Two-part epoxy adhesive; for ski industry applications; excellent cohesive strength and good peel strength. *Hardman.*

10709 Epoxidized X-70 and X-75
Polysiloxane resin in an aliphatic vehicle system; water repellant for masonry. *Nova Chemical Inc.* Name unverified.

10710 Epoxol 7-4
8013-07-8 232-391-0
Epoxidized soybean oil; auxiliary plasticizer, acid scavenger, stabilizer for PVC compounds; food packaging materials. *Am. Chem. Services.*

10711 Epoxol 8-2B
Epoxidized butyl esters of linseed oil fatty acids; auxiliary plasticizer with low viscosity and low volatility, heat and light stabilization; acid scavenger; stabilizer for PVC compounds; food packaging materials. *Am. Chem. Services.*

10712 Epoxol 9-5
Epoxidized linseed oils; stabilizing plasticizer; food packaging materials. *Am. Chem. Services.*

10713 Epoxol 80, 130
A proprietary trade name for vinyl plasticizers manufactured from soya bean. *Swift & Co.* Unverified.

10714 Epoxol G-5
Linseed oil epoxy resin. No manufacturer.

10715 Epox-S
A proprietary trade name for an epoxidized triester vinyl plasticizer. *Ruco Div.* Name unverified.

10716 Epoxy Adhesive
High strength general purpose epoxy adhesive. *Sealocrete Ltd.*

10717 Epoxy Plus
Epoxy resins used for repairing concrete, grouting and bonding. *SBD Construction Prods. Ltd.*

10718 Epoxy Putty Pack (EP-3/EHP-12)
Epoxy putty resin and hardener; for filling and fairing; can be used underwater. *Wessex Resins & Adhesives Ltd.*

10719 epoxy resin
25928-94-3
Derived from epichorohydrin and diethylene glycol; surface coatings, adhesive, casting metal-forming tools and dies; encapsulation of electrical parts. Asahi Chem Industry Co Ltd; Ciba-Geigy; Conap; Hardman; Key Polymer; Morton Int'l Reichhold; Rhône-Poulenc/Perf. Resins; Sartomer; Shell; Union Carbide.

10720 Epoxyprene 50
Epoxidized natural rubber. *Guthrie Latex.*

10721 Eprolin
59-02-9 10159 200-412-2
Vitamin E
Vitamin E supplement. *Eli Lilly & Co.* Discontinued.

10722 Eprylac
Acrylic adhesives, sealants and coatings. *BP Chemicals Ltd.* Discontinued.

10723 Epsilan-M
59-02-9 10159 200-412-2
Vitamin E
Vitamin E supplement. *Adria Laboratories Inc.* Name unverified.

10724 **EPsyn® 40-A**
EPDM terpolymer; general purpose rubber for med.-loaded extruded and molded compounds requiring short cure times; suggested for wire and cable covering, injection molding, sponge; excellent ozone and heat resist. in vulcanizates; sulfur-vulcanizable; very fast curing with conventional EPDM acceleration systems. *Copolymer Rubber.*

10725 **EPsyn® 55**
EPDM terpolymer; used as additive in SBR, nitrile, neoprene, and other highly sat. rubbers for improved ozone and heat resist.; sulfur-vulcanizable; ultra fast curing with conventional EPDM acceleration systems. *Copolymer Rubber.*

10726 **EPsyn® 5508**
EPDM terpolymer; used in wire and cable, sponge, molded applications, where dimensional stability is important; accepts very high loading of fillers and extenders; aids processing in blends with other EPDMs; thermoplastic EPDM yielding high green strength; sulfur-vulcanizable; very fast curing with conventional EPDM acceleration systems. *Copolymer Rubber.*

10727 **EPsyn® P-557**
EPDM terpolymer, 50 parts per hundred paraffinic oil; general purpose rubber which attains high tensiles and low compression set; sulfur-curable; very fast curing with conventional EPDM accelerator systems. *Copolymer Rubber.*

10728 **Eptam 6E**
759-94-4 212-073-8
Emulsifiable concentrate of 720 g EPTC per liter; used for pre-planting weed control in potatoes. *Farm Protection Ltd.*

10729 **EPTC**
759-94-4 212-073-8
C9H19NOS
S-ethyl dipropylcarbamothioate
Alirox; Eptam; Genep; Niptan; R-1608; Witox. Selective systemic herbicide, used to control annual and perennial grasses in many crops. bp_{20} = 127°; d^{30} = 0.9546; n_D^{30} = 1.4750; soluble in H_2O (375 mg/l), more soluble in organc solvents; LD_{50} (rat orl) = 1630 mg/kg.

10730 **Epurite**
A mixture of bleaching powder, iron sulfate, and copper sulfate; used for the production of oxygen, which gas is obtained by action of water.

10731 **Equal**
22839-47-0 874 245-261-3
Aspartame
NutraSweet. Sweetener. *G D Searle & Co;.* Name unverified.

10732 **equalized guano**
Natural guanos, blended or mixed with ammonium salts, to obtain proportions of nitrogen and phoshorus; a fertilizer.

10733 **Equiben**
26097-80-3 1774 247-459-5
Cambendazole
Anthelmintic. *Merck & Co Inc.*

10734 **Equionic**
Sanitizer/detergent compounds. *Glover (Chemicals) Ltd.* Unverified.

10735 **equisetic acid**
499-12-7 119 207-877-0
$C_6H_6O_3$
Citridic acid
Aconitic acid. A plasticizer for buna rubber.

10736 **Equivurm Plus**
31431-39-7 5807 250-635-4
Mebendazole
A proprietary preparation containing mebendazole; veterinary antihelmintic (horses). *Janssen Pharmaceutical Ltd.* Name unverified.

10737 **Eqvalan**
70288-86-7 5264 274-536-0
Ivermectin
Ivermectin (a mixture of ivermectin component B1a and ivermectin component B1b); antiparasitic. *Merck & Co Inc.* No manufacturer.

10738 **Era 147**
A proprietary trade name for a steel containing 0.22% carbon, 0.20% silicon, 0.04% sulfur, 0.04% phosphorus, 0.50% manganese, 5.00% chromium and 0.50% molybdenum; used for forging steel pressure vessels for service with hydrogen. *Hadfield, George & Co.* Unverified.

10739 **Era 164**
A proprietary trade name for a steel containing 0.20% carbon, 0.25% silicon, 0.04% sulfur, 0.04% phosphorus and 1.5% manganese; used in the manufacture of forged steel pressure vessels for use with intermediate pressures. *Hadfield, George & Co.* Unverified.

10740 **Era CR1**
A proprietary trade name for a steel containing 0.06% carbon, 0.30% silicon, 0.04% sulfur, 0.04% phosphorus, 1.0% manganese, 18.50% chromium and 9.00% nickel; used in the manufacture of forged steel pressure vessels with good corrosion resistance. *Hadfield, George & Co.* Unverified.

10741 **Era CR15 (CB)**
A proprietary trade name for a steel containing 0.06% carbon, 0.50% silicon, 0.04% phosphorus, 1.00% manganese, 19.00% chromium, 10.00% nickel and 0.6% niobium. *Hadfield, George & Co.* Unverified.

10742 **Era metal**
A steel containing 21% chromium and 7% nickel.

10743 **Era® Dyes**
These are registered trade names for certain British dyestuffs. No manufacturer.

10744 **Eraclene**
A registered trademark for low density polyethylene. *Anic Agricoltura Spa.* Unverified.

10745 **Eranol**
A form of colloidal iodine. An antiseptic.

10746 **Erbium**
7440-52-0 3679 231-160-1
Er
Rare-earth element; used in nuclear controls, special alloys, room-temperature laser. d = 9.066; mp = 1529°; bp = 2868°. *Atomergic Chemetals; Cerac; Noah Chem.; Rhône-Poulenc Basic.*

10747 **Erbium oxide**
12061-16-4 235-045-7
Er_2O_3
erbia
Phosphor activator, infrared-absorbing glass. d = 8.64; changing into crystals on heating at 1300°; soluble in acids, H_2O. *Atomergic Chemetals; Cerac; Noah Chem.; Rhône-Poulenc Basic.*

10748 **ercerhinol**
A colloidal silver.

10749 **Ercusol®**
Aqueous acrylic polymer dispersion. *Bayer AG; Bayer plc.*

10750 **Erdmann's reagent**
Made by adding 40 ml of concentrated sulfuric acid to 20 drops of a solution containing 10 drops of nitric acid (specific gravity 1.153) and 20 ml of water; used in testing for alkaloids.

10751 **Erganol**
Dibenzyl ether
A softening agent for cellulose esters.

10752 **Ergol**
$C_{14}H_{12}O_2$
Stated to be benzyl benzoate, A proprietary plasticizer; used as a softening agent for cellulose esters. No manufacturer.

10753 **Eribate**
6381-77-7 5141 228-973-9
$C_6H_7NaO_6$
Sodium erythorbate
Erythorbic Acid Monosodium Salt; Isoascorbate; D-erythro-Hex-2-enonic acid, γ-lactone, monosodium salt; isoascorbic acid, sodium salt; neo-cebitate; Mercate 20; Isona; sodium D-isoascorbate; Monosodium erythorbate; Araboascorbic acid, monosodium salt, D-; Hex-2-enonic acid, γ-lactone, monosodium salt; Sodium D-araboascorbate. Antioxidant and antimicrobial agent used in foods. *PMP Fermentation Prods. Inc.*

10754 **Ericon**
A proprietary trade name for a phenol-formaldehyde synthetic resin. No manufacturer.

10755 **Erinofort**
A proprietary cellulose acetate plastic. No manufacturer.

10756 **Erinoid**
A proprietary trade name for casein-formaldehyde synthetic resin insulating material. *Mobil. Name unverified.*

10757 **Erio**
Acid dyes. *Ciba plc.* Name unverified.

10758 **Eriochrome**
Mordant dyes. *Ciba plc.* Name unverified.

10759 **Erioclarite**
Pretreatment agents. *Ciba plc.* Name unverified.

10760 **Erional**
Dyeing and printing assistant. *Ciba plc.* Name unverified.

10761 **Erional**
Acid dyes for polyamide fabrics. *Ciba plc.* Name unverified.

10762 **Eriopon**
Surface active agents. *Ciba plc.* Name unverified.

10763 **Erkantol®**
Wetting agents and padding auxilliaries; used for textile finishing. *Bayer AG; Bayer plc.*

10764 Erlickl's solution
A hardening agent used in microscopy; consists of potassium dichromate 2.5 parts, calcium sulfate 1 part and water 100 parts.

10765 Ermite
A proprietary trade name for a synthetic resin. No manufacturer.

10766 Ertalon® LFX
Internally lubricated cast nylon; for unlubricated moving parts applications. *Polymer Corp.*

10767 Ertalon® PETP
PETP; used for bearing and wear applications requiring higher dimensional stability. *Polymer Corp.*

10768 Ertalyte®
25038-59-9 7730
PET polyester; thermoplastic with excellent wear resist., low coeff. of friction, superior dimensional stability; for bearings, bushings, seals, spacers, rollers, guides, insulators, food contact parts, pump components, valve parts. *Polymer Corp.*

10769 Ertilen
Ethical veterinary antibiotics. *Ciba plc.* Name unverified.

10770 erucamide
112-84-5 204-009-2
$C_{22}H_{43}NO$
erucic acid amide
13-docosenamide; *cis*-13-docosenamide; Erucylamide; (Z)-13-Docosenamide. Aliphatic amide; foam stabilizer; solvent for waxes and resins, emulsions; slip/antiblock agent for polyethylene. mp = 79°; insoluble in H_2O, soluble in organic solvents. *Akzo; Chemax; Croda; Syn. Prods.; Witco/Humko.*

10771 erucic acid
112-86-7 3713 204-011-3
$C_{22}H_{42}O_2$
(Z)-13-docosenoic acid
Major fatty acid in mustard, rapeseed and wallflower seed. Used as a chemical intermediate. mp = 33°; bp = 381°; d_4^{55} = 0.860; n_D^{45} = 1.4534; insoluble in H_2O, soluble in organic solvents.

10772 Ervamine
A proprietary trade name for melamine-formaldehyde. No manufacturer.

10773 Ervamix
A proprietary trade name for a fibrous glass reinforced polyester molding compound. No manufacturer.

10774 Ervol
White mineral oil. *Witco Corporation.*

10775 Erythorbic acid
89-65-6 5141 201-928-0
$C_6H_8+O_6$
D-erythro-hex-2-enonic acid
isoascorbic acid-γ-lactone; isovitamin C. Isomer of ascorbic acid; antioxidant (industrial, food, brewing), reducing agent in photography. mp = 169-172°; $[\alpha]_D^{20}$ = -16.80° (c = 2 H_2O) *Ashland; Spice King.*

10776 Erythrin
Spirit eosin; methyl eosin; primrose soluble in alcohol; erythrin methyl eosin. A dyestuff; the potassium salt of tetrabromofluorescein methyl ether; dyes silk bluish-red, with a reddish fluorescence.

10777 Erythrosiderite
$2KCl \cdot FeCl_3 \cdot H_2O$
Potassium iron chloride; a mineral.

10778 erythrosin
16423-68-0 240-474-8
$C_{20}H_6O_5I_4Na_2$
Erythrosin B; erythrosin D; pyrosin B; iodeosin B; eosin Blush; Rose B; dianthine B; primrose soluble; soluble primrose. The sodium or potassium salt of tetraiodofluorescein, Dyes silk and wool bluish-red; used for paper staining.

10779 Erythrosin A
The sodium or potassium salt of triiodofluorescein. A dyestuff.

10780 Erythrosin G
$C_{20}H_8O_5I_2Na_2$
dianthine G
pyrosin G; iodeosin G. The sodium or potassium salt of diiodofluoresceine. Dyes wool yellowish-red, with yellowish-red fluorescence.

10781 ES-7CF/000
Polyethersulfone, 7% carbon fiber-reinforced; thermoplastic resin. *Compounding Tech.* Name unverified.

10782 Esaflon
2551-62-4 9146 219-854-2
F_6S
sulfur hexafluoride
SF_6 gas *Montedison UK Ltd.* Name unverified.

10783 Esbenite
A material made from cellulose, powdered mica, and magnesium silicate.

10784 Esbenite
Nylon composite with 50% high strength glass and ceramic fibers; thermoplastic molding composite used as replacements for die cast metal parts in business machines, home appliances, automotive applications. *Thermofil.*

10785 Escacure® KB1
Benzyldimethyl ketal
Photoinitiator for white coatings, inks photopolymers, electronic photoresists, polyester-styrene wood filler composites; optimum absorption is at 250-350 nm. *Sartomer.*

10786 Escaid
Specialty hydrocarbon for mineral applications. Metal extraction solvent. *Exxon Int'l.* Name unverified.

10787 Escalol® 507
21245-02-3 3282 244-289-3
$C_{17}H_{27}NO_2$
4-(Dimethylamino)benzoic Acid
Padimate O; Octyl dimethyl PABA. Topical sunscreen. *Van Dyk.* Name unverified.

10788 Escalol® 557
5466-77-3 6864 226-775-7
Octyl methoxycinnamate
Sunscreen. *Van Dyk.*

10789 Escalol® 567
131-57-7 7088 205-031-5
Benzophenone-3
Sunscreen. *Van Dyk.*

10790 Escalol® 587
118-60-5 204-263-4
Octyl salicylate
Sunscreen. *Van Dyk.*

10791 Escane
Detergent intermediates. *Exxon UK.* Name unverified.

10792 Eschel
A fine-grained light colored small, a blue pigment.

10793 Eschka mixture
A mixture of 2 parts by weight pure calcined magnesia and 1 part pure anhydrous sodium carbonate; used for the determination of sulfur in coal by heating the coal with the mixture, then adding hydrochloric acid and barium chloride, when barium; sulfate is precipitated.

10794 Esco Extract
A synthetic tannin prepared from sulfonated heavy tar oils, by condensation with formaldehyde, and then forming the chromium salt. No manufacturer.

10795 Escomer
Polyethylene wax. *Exxon Int'l.*

10796 Escopol® R-020
68527-24-2
Reactive polymer. *Exxon.* Name unverified.

10797 Escor
Polyethylene copolymer; low density polyethylene. *Exxon Int'l.*

10798 Escorene
Linear low density PE, high density PE; specialty plastics. *Exxon Int'l.*

10799 Escorez
Hydrocarbon resins from petroleum feedstocks. *Exxon UK.*

10800 Escorto
A proprietary brand of artificial silk. No manufacturer.

10801 Escoweld
Liquid epoxy adhesives and grouts. *Exxon UK.* Name unverified.

10802 Esdeform
Stain removers. *S & D Chemicals Ltd.* Name unverified.

10803 Esdesol
A paint stripper. *S & D Chemicals Ltd.* Name unverified.

10804 Esdogen
Wetting agent and solvent soaps. *S & D Chemicals Ltd.* Name unverified.

10805 E-Series® Electronics Cleaner/Degreaser 2000
General-purpose cleaning agent, penetrant removing oil, grease, silicone, flux, and adhesive; for PCBs, motors and interfaces. *Chemtronics.*

10806 E-Series® Freez-It®
Refrigerant for locating thermal intermittent electronic components during soldering or testing; replaces CFC12 refrigerants; freezes adhesives for easy removal. *Chemtronics.*

10807 E-Series® Freez-It® Antistat
High-pressure circuit refrigerant for static-sensitive devices; also for heat sink protection of components during soldering and testing. *Chemtronics.*

10808 E-Series® Ultrajet®
Ultra-pure precision duster for removal of dust, lint, and metallic oxide particles. *Chemtronics.*

10809 Eshalit
A bakelite (phenol-formaldehyde resin). No manufacturer.

10810 Esi-Cryl 1E10N
9002-88-4 7728
Polyethylene emulsion (based on Epolene E-10); used for floor finishes, general coatings, inks, lubricants, textiles. *Emulsion Systems.*

10811 Esi-Cryl 20/20
Zinc-free polymer; produces excellent maintenance floor finishes with high performance and durability without the need of metal crosslinking for removability. *Emulsion Systems.*

10812 Esi-Cryl 246
Metal crosslinked acrylic copolymer; used in detergent-resist., high gloss, long-term durability floor finishes. *Emulsion Systems.*

10813 Esi-Cryl 325N
9002-88-4 7728
Polyethylene emulsion (based on AC-325); used for floor finishes, general coatings, inks, lubricants, textiles. *Emulsion Systems.*

10814 Esi-Cryl 1540A
9002-88-4 7728
Polyethylene emulsion (based on AC-540 copolymer); used for adhesives, floor finishes, and general coatings. *Emulsion Systems.*

10815 Esi-Cryl Respond I
62180-77-2
Acrylic copolymer; used for floor finish formulations exhibiting superior gloss and durability. *Emulsion Systems.*

10816 Esi-Cryl40
Polyethylene wax emulsion; provides excellent blk. heel mark and scuff resist., high gloss, color stability to floor finish formulations. *Emulsion Systems.*

10817 Esi-Det 21M
Modified alkanolamide; detergent for hard surface cleaners; visc. builder for aqueous systems; base for multipurpose cleaners, strippers, degreasers; high tolerance to alkaline builders, solvents. *Emulsion Systems.*

10818 Esi-Det CDA
1:1 Cocamide DEA; detergent, foam stabilizer, thickener for personal care products, light duty dishwash and household cleaners, industrial products; 100% biodegradable. *Emulsion Systems.*

10819 Esi-Det EP-20
Complex phosphate ester of POE ethanol; detergent for wide range of products; high compatibility. *Emulsion Systems.*

10820 Esi-Terge 10
Cocamide DEA; foam stabilizer, thickener for household, cosmetic, industrial products *Emulsion Systems.*

10821 Esi-Terge 320
Phosphated nonylphenoxy polyethoxy ethanol
Detergent, viscosity aid, coupling and wetting agent for synthetic hard surface cleaners with high caustic concentration; high alkali stable. *Emulsion Systems.*

10822 Esi-Graph 743
Acrylic quad polymer; graphics arts emulsion for gift wrap printing, general coatings. *Emulsion Systems.*

10823 Esi-Graph 745
Styrenated acrylic emulsion; graphics arts emulsion for general coatings, as extending vehicle. *Emulsion Systems.*

10824 Esi-Graph 1045
Acrylic copolymer emulsion; graphics arts emulsion for single vehicle corrugated news ink. *Emulsion Systems.*

10825 Esi-Terge 40% Coconut Oil Soap
8001-31-8 232-282-8
Coconut oils® 76,92,110; Konut; Super Refined Coconut oil. Cochin oil soap; detergent for liquid hand soap; mild cleaning and coupling. *Emulsion Systems; van Den Bergh Foods.*

10826 Esi-Terge B-15
2:1 Cocamide DEA; detergent for mild all-purpose cleaners, foam builder, and thickener, wetting agent; for household and industrial cleaners. *Emulsion Systems.*

10827 Esi-Terge HA-20
Modified amine condensate; self-coupling detergent for cleaners, detergent base; viscosity builder; for low and high alkali cleaners. *Emulsion Systems.*

10828 Esi-Terge L-75
Blend of surfactants, chelating and coupling agents; base for high foaming cleaners for dishwashing, carwashing, all-purpose cleaning, butyl degreasers. *Emulsion Systems.*

10829 Esi-Terge N-100
PEG ether surfactant; rapid wetter for formulating mild acid cleaners as well as natural and mildly alkaline products. *Emulsion Systems.*

10830 Esi-Terge S-10
1:1 Cocamide DEA; detergent, emulsifier, foam stabilizer, thickener, for liquid dishwashing and car washing detergents, household, industrial, and cosmetic products. *Emulsion Systems.*

10831 Esi-Terge SXS
1300-72-7 215-090-9
Sodium xylene sulfonate
Solubilizer for light-and heavy-duty cleaners. *Emulsion Systems.*

10832 Esi-Terge T-60
TEA dodecylbenzene sulfonate
Detergent, wetting agent, foam stabilizer, for car and dishwashing detergent, synthetic hand soap, household cleaners. *Emulsion Systems.*

10833 Eskimo
Lubes. *Monsanto (Solaris).* Name unverified.

10834 Esmaillite
A dope consisting of a mixture of cellulose acetate and volatile solvent, usually ethyl formate.

10835 Esophotrast
7727-43-7 1023 231-784-4
Barium sulfate
Diagnostic aid. *Armour Pharmaceutical Co.* Name unverified.

10836 Esperal® 115
80-43-3 201-279-3
Dicumyl peroxide
For medium-temperature applications as a polymerization and crosslinking agent. *Witco Corporation.*

10837 Esperal® 120
78-63-7 201-128-1
2,5-dimethyl-2,5-di(*t*-butylperoxy)hexane
Initiator in crosslinking of polymers and elastomers; melt flow modifier for PP; vulcanization agent for silicone rubber, fluoro elastomers, EPDM, EVA; crosslinking agent for LDPE. *Witco/Argus.* Discontinued.

10838 Esperal® 230
1068-27-5 213-944-5
2,5-Dimethyl-2,5-di(t-butylperoxy)hexyne-3
Initiator in crosslinking of polyolefins and other polymers; crosslinking agent for HDPE and LLDPE at temperatures above 180°. *Witco/Argus.* Discontinued.

10839 Esperase® 16.0L
Proteinase
Enzyme; defoamer for built liquid detergents. *Novo Nordisk.* Name unverified.

10840 Espercarb® 438M-60
19910-65-7 243-424-3
Dibutylperoxy dicarbonate; Trigonox® SBP,ADC; Lupersol 225. 60% solution di-*sec*-butyl peroxydicarbonate in odorless mineral spirits; initiator, crosslinking agent for polymers. *Witco Corporation; Elf Atochem; Akzo.*

10841 Espercarb® 840
16111-62-9 240-282-4
Di-2-ethylhexyl peroxydicarbonate
Diethylhexyl peroxy dicarbonate; Trigonox® EHP; Lupersol 223;. Initiator. *Witco Corporation; Elf Atochem; Akzo.*

10842 Esperfoam® FR
MEK peroxide; Methyl ethyl ketone peroxide; Butanox; Cadox® HBO-50, L-30; Hi-Point® 90; HPC-9; Lupersol DDM-9, DSW; Quickset® Extra; Sprayset® MEKP. MEK peroxide [1338-23-4] and acetylacetone peroxide 37187-22-7]; initiator for rapid cures of polyester resins; DOT org. peroxide label is not required. *Akzo Chemie UK Ltd; Akzo;.*

10843 Esperox®
Organic peroxide catalyst. *Witco Corporation.*

10844 Esperox® 10
614-45-9 210-382-2
t-Butyl peroxybenzoate
Initiator for polymerization of ethylene and styrene, and for high temp. molding of polyesters. *Witco Corporation.*

10845 Esperox® 12MD
107-71-1 203-514-5
t-butyl peroxy acetate
50% solution of *t*-butyl peroxy acetate in odorless mineral spirits; initiator for polymerization of styrene, ethylene, acrylates, etc. *Witco Corporation.*

10846 Esperox® 13M
23474-91-1 245-679-6
t-butyl peroxycrotonate

Butylperoxy crotonate. 75% solution of *t*-butyl peroxycrotonate in odorless mineral spirits; initiator for polymerization application. *Witco/Argus.* Discontinued.

10847 Esperox® 28
3006-82-4 221-110-7
t-Butyl peroxy 2-ethyl hexanoate
Initiator recommended for polymerization of ethylene plus use in medium temperature molding of polyester resin systems. *Witco Corporation.*

10848 Esperox® 31M
927-07-1 213-147-2
t-Butyl peroxypivalate
Aztec® t-Butyl Peroxypivalate-75 OMS; Lupersol 11; Trigonox® 25-C75;. *t*-Butyl peroxypivalate in mineral spirit diluent; initiator used in polymerization of ethylenically unsaturated monomers. *Witco Corporation; Elf Atochem; Catalyst Resources; Akzo.*

10849 Esperox® 33M
26748-41-4 247-955-1
t-Butyl peroxyneodecanoate
t-Butyl peroxyneodecanoate in mineral spirit diluent; efficient and reactive initiator for polymerization of ethylenically unsaturated monomers. *Witco Corporation.*

10850 *Esperox® 41-25A*
1931-62-0 *217-691-1*
t-Butyl peroxy maleate
25% t-Butyl peroxy maleate dispersion; for polymerization of various ethylenically unsaturated resins and monomers *Witco Corporation.*

10851 *Esperox® 497M*
22313-62-8 *244-906-6*
t-butyl peroxy 2-methylbenzoate
Trigonox® 97-C75. *t*-butyl peroxy 2-methylbenzoate solution in mineral spirit diluent; initiator for polymerization of ethylene and styrene, high temperature molding of polyester resin systems, and vulcanization of silicon rubber. *Witco/Argus; Akzo.*

10852 Esperox® 545M
68299-16-1 269-597-5
t-amyl peroxyneodecanoate
Lupersol 546-M75; Trigonox® 123-C75. *t*-amyl peroxyneodecanoate in mineral spirit dilutent; efficient and reactive initiator for polymerization of ethylenically unsaturated monomer. *Witco/Argus; Elf Atochem; Akzo.*

10853 Esperox® 551M
29240-17-3 249-530-6
t-amyl peroxypivalate
t-amyl peroxypivalate in mineral spirit diluent; initiator used in polymerization of ethylenically unsaturated monomers. *Witco Corporation.*

10854 Esperox® 570
686-31-7 211-687-3
t-amyl peroxy 2-ethylhexanoate
Amil peroxyethylhexanoate; Lupersol 575; Trigonox® 121. 95% *t*-amyl peroxy 2-ethylhexanoate; for polymerization of ethylene, acrylates, and unsaturated polyester resins *Witco Corporation;Elf Atochem; Akzo.*

10855 Esperox® 740M
130097-36-8
cumyl peroxyneoheptanoate
75% solution of cumyl peroxyneoheptanoate in odorless mineral spirits; initiator for polymerization of vinyl chloride. *Witco Corporation.*

10856 Esperox® 747M
t-amyl peroxyneoheptanoate
75% solution of *t*-amyl peroxyneoheptanoate in odorless mineral spirits; initiator for polymerization of ethylenically unsaturated monomers. *Witco Corporation.*

10857 Esperox® 750M
26748-38-9 247-954-6
t-butyl peroxyneoheptanoate
75% solution of *t*-butyl peroxyneoheptanoate in odorless mineral spirits; initiator for polymerization of ethylenically unsaturated monomers. *Witco Corporation.*

10858 Esperox® 939M
26748-47-0 247-956-7
Cumylperoxy neodecanoate
Lupersol 188-M75, 288-M75; Trigonox® 99-B75. Cumyl peroxyneodecanoate in mineral spirit diluent; efficient and reactive initiator for polymerization of vinyl chloride. *Witco/Argus; Akzo;Elf Atochem.*

10859 Esperox® 5100
4511-39-1 224-831-5
t-amyl peroxybenzoate
Initiator for polymerization of ethylene, styrene, acrylates, and curing of unsaturated polyester resins. *Witco Corporation.*

10860 Esperox® C-496
34443-12-4 252-029-5
t-butyl peroxy-2-ethylhexyl carbonate
Initiator for polymerization of vinyl monomers, styrene, acrylates, and unsaturated polyester resins. *Witco Corporation.*

10861 Esrakon
Tablets for use in start up of electro slag refining process for steel billets and slabs. *Foseco(F.S.) Ltd.*

10862 Essar (W)
A proprietary trade name for a general purpose acid and alkali resistant furan resin cement for bedding in acid resisting tiles. No manufacturer.

10863 Essence of Bigarade
Oil of bitter orange peel.

10864 Essential Oil of Bitter Almonds
C_7H_6O
Benzaldehyde
Benzaldehyde.

10865 Essex powder
An explosive, containing 22-24% nitroglycerin, 0.5-1.5% collodion cotton, 33-35% potassium nitrate, 33-35% wheat flour, and 5-7% ammonium chloride.

10866 Esshete 1250
A steel containing 16% chromium, 10% nickel and 6% manganese; an austenitic creep resisting steel. *Samuel Fox & Co Ltd.* Unverified.

10867 Esshete CML
A proprietary trade name for a steel containing 1% chromium and 1% molybdenum possessing high creep strength and corrosion resistance; suitable for operating up to 1000 C. *Samuel Fox & Co Ltd.* Unverified.

10868 Esshete CRM2
A steel containing 2 1/4% chromium and 1% molybdenum and superior properties to CML; can be used up to 1100 CF. *Samuel Fox & Co Ltd.* Unverified.

10869 Esshete CRM5
A steel containing 5% chromium and 1% molybdenum suitable for tubes exposed to high temperature steam. *Samuel Fox & Co Ltd.* Unverified.

10870 Esskol®
Vegetable drying oils used in preparation of paints, varnishes. *Reichhold.*

10871 Estabex
Epoxidized soya bean oil and epoxy ester. *Akzo Chemie UK Ltd.*

10872 Estabex 2307, 2349
Epoxidized soya bean oils. No manufacturer.

10873 Estabex 2386
Epoxidized monoesters. No manufacturer.

10874 Estaflex
77-90-7 201-067-0
Acetyl tributyl citrate
Citroflex A-4; Uniplex 84. Plasticizer. *Akzo Chemie UK Ltd; Unitex; Morflex.*

10875 Estaloc 61000 Series
Mixture of thermoplastic polyurethane, reinforcement, polymeric alloys, additives; reinforced engineering thermoplastic. *BFGoodrich/Spec. Polymers.* Name unverified.

10876 Estalol®
Dibasic ester; for paper, foundry, electronics, textile industries. *Aceto.*

10877 Estane®
A proprietary range of thermoplastic polyurethane molding and extrusion compounds. *BFGoodrich.* Name unverified.

10878 Estane® 5701 F1
9009-54-5
Polyester-urethane polymer, calcium stearate-dusted; thermoplastic polymer, flexibility, chemical and abrasion resistance, wide compatibility; used in binders, coated fabrics, adhesive blending resins. *BFGoodrich/Spec. Polymers.* Name unverified.

10879 Estane® 58092
9009-54-5
Polyester-urethane compound; firm, flexible, low-blocking, thermoplastic compound, highest chemical and abrasion resistance and properties; used in injection molding and extrusion of film, sheet, tubing, and shapes. *BFGoodrich/Spec. Polymers.* Name unverified.

10880 Estane® 58300
9009-54-5
Polyether base urethane compound; thermoplastic compound, resistance to hydrolysis, improved fungus resistance, low-temperature flexibility; used in extrusion and injection molding for hose and wire jacketing, molded junctions. *BFGoodrich/Spec. Polymers.* Name unverified.

10881 Estasol
Dimethyl esters of adipic, glutaric, and succinic acids; low toxicity compound solvent; general purpose solvent coil coatings, foundry resins. *Chemoxy International Ltd.*

10882 Ester Copal
An ester gum obtained by the interaction between glycerin and copal.

10883 Estergel
110-27-0 5234 203-751-4
Isopropyl myristate
Pharmaceutic aid. *Merck & Co Inc.* Name unverified.

10884 Esterkem
Three-component product based on modified unsaturated polyester resin with peroxide catalyst; adhesive and grout for ceramic tiles and paviors in heavy-duty chemical-resistant application. *Feb Ltd.*

10885 Estermone
2,4-D-dicamba
Mixture of 2,4-D and dicamba; used to control weeds in turf. *Synchemicals Ltd.*

10886 Esterol
A proprietary brand of benzyl succinate; also a proprietary trade name for alkyd synthetic varnish and lacquer resins. No manufacturer.

10887 Esterolane
Disperse dyes. *ICI Chem & Polymers Ltd.* Discontinued.

10888 Esterox
A fast, air-drying, modified epoxy sealer and dustproofer; for protecting interior concrete floors. *Secure Inc.* Name unverified.

10889 Esterpol
A proprietary synthetic resin. No manufacturer.

10890 Estol
Textile lubricants. *Crosfield Chemicals Ltd.* Discontinued.

10891 Estol 1407
Glyceryl oleate
Antistat for polyethylene and polypropylene; antifog agent for LDPE and PP; for lubricants, cosmetic emollients. *Unichema.*

10892 Estol 1468
Glyceryl stearate
Glyceryl stearate; antistat for polyethylene and polypropylene; antifog agent for LDPE; for lubricants, cosmetic emollients. *Unichema.*

10893 Estol 1476
646-13-9 5165 211-466-1
Isobutyl stearate
Isobutyl stearate; lubricant for PVC processing; cosmetic emollients. *Unichema.*

10894 Estol 1481
Cetostearyl stearate; lubricant for PVC processing; cosmetic emollients. *Unichema.*

10895 Estolan
Polyester polyols
Used for textile lamination foams, packaging foams, microcellular formulated systems, adhesives and flexible coatings. *Harcros.*

10896 Estoral
53370-45-9 5885
$C_{30}H_{57}BO_3$
Menthyl borate
Decomposes in solution into its components. Source of menthol. Insoluble in H_2O, EtOH, soluble in organic solvents.

10897 Estragole
140-67-0 3748 205-427-8
$C_{10}H_{12}O$
1-methoxy-4-(2-propenyl)benzene
p-propenylanisole; esdragol; chavicol methyl ether; 4-allylanisole; tarragon; 4-allyl-1-methoxybenzene; chavicol methyl ether; esdragon; isoanethole; *p*-methoxyallylbenzene; methyl chavicol; Chavicyl methyl ether. Used as a perfume and flavorant. bp = 216°; d_4^{21} = 0.9645; n_D^{20} = 1.5195; soluble in organic solvent; LD_{50} (rat orl) = 1820 mg/kg.

10898 Estron
Acetate tow and yarn. *Eastman.*

10899 Etadurin 31
Amine copolymer aqueous solution; thermosetting wet strength resin for paper manufacture *Akzo.*

10900 Etapuron FT
Quaternary compound; softening agent for flame resistant finishes; compatible with most of the salt-based flame retardants. Cationic; yellow liquid. *Thor Chemicals (UK) Ltd.* Discontinued.

10901 Etard's reagent
Anhydrous chromium oxychloride
An oxidizing agent.

10902 Eteleen
Trigallic acetal

10903 EternaBrite
Leafing aluminum pigments; metallic printing inks (silver and gold). *Silberline Mfg Co.*

10904 Eternite
A slate-like mass made from 6 parts Portland cement and 1 part asbestos fiber; used for roofing.

10905 Ethacol
Pyrocatechol monoethyl ether
Chemical intermediate.

10906 Ethacure® 100
Diethyl toluene diamine
High-performance curing agent for epoxy resins; used in filament winding, electrical encapsulation, prepregs, tooling, potting and casting, laminating, coating, molding, and adhesive applications. *Ethyl Corp.*

10907 Ethacure® 300
Curing agent for polyurethane and epoxy resins. *Ethyl Corp.*

10908 Ethafoam
Polyethylene foam used principally in cushioning and packaging applications. *Dow Cheml Co Ltd, UK & Ireland.*

10909 Etha-Keratin ISO
AMP-isostearoyl hydrolyzed keratin
Cosmetics ingredient. *Brooks Industries.*

10910 ethal
36653-82-4 2070 253-149-0
$C_{16}H_{33}OH$
Cetyl alcohol
Used in manufacture of surfactants.

10911 Ethal 326
Laureth-3
Emulsifier for textile and industrial applications. *Ethox.*

10912 Ethal 368
PEG-3 $C_{16\text{-}18}$ alcohol
Emulsifier, lubricant. *Ethox.*

10913 Ethal 926
Laureth-9
Detergent, emulsifier. *Ethox.*

10914 Ethal 3328
PEG-3 C12-18 alcohol
Low HLB emulsifier and coupler. *Ethox.*

10915 Ethal BPA-6
PEG-6 bisphenol A
Monomer for polyester and urethane coatings. *Ethox.*

10916 Ethal CSA-25
PEG-25 C16-18 alcohol
Hydrophilic emulsifier, detergent for mild acidic or alkaline solutions and hot aqueous systems. *Ethox.*

10917 Ethal DA-4
26183-52-8
PEG-4 decyl alcohol
Wetting agent for aqueous solutions; intermediate for manufacture of anionic surfactants. *Ethox.*

10918 Ethal DDP-7
Multipurpose emulsifier and detergent; excellent greaser cutter. *Ethox.*

10919 Ethal DNP-8
9014-93-1
PEG-8 dinonyl phenol
Nonyl nonoxinol series; Chemax; Chemax DNP-18, DNP-8; Hetoxide DNP-4; Igepal® DM-430, DM-530, DM-710, DM-730, DM-970; Macol® DNP-5; Marlophen® DNP 16; Prox-onic DNP-08; Sellig DN 10 100, DN 22 100; T-Det® D-150; Trycol® 6985,. Low foam emulsifier for solvent systems; coupling agent for introducing water into nonaqueous systems. *Ethox.*

10920 Ethal EH-2
PEG-2 2-ethylhexanol
Intermediate for anionic surfactants; component of low foam wetting systems. *Ethox.*

10921 Ethal LA-4
Laureth-4
General-purpose emulsifier, lubricant. *Ethox.*

10922 Ethal NP-1.5
9016-45-9 6772
Nonoxynol-1.5
Emulsifier, intermediate. *Ethox.*

10923 Ethal NP-6
9016-45-9 6772
Nonoxynol-6
Coupling agent for surfactant systems, for coupling water into oil systems. *Ethox.*

10924 Ethal OA-10
9004-98-2
Oleth-10
Moderate HLB emulsifier for esters and oils. *Ethox.*

10925 Ethal TDA-18
24938-91-8
PEG-18 tridecyl alcohol
High-foaming detergent for elevated temperature applications. *Ethox.*

10926 Ethal TDA-3
24938-91-8
PEG-3 tridecyl alcohol
Surfactant, coupling agent. *Ethox.*

10927 Ethal TDA-6
24938-91-8
PEG-6 tridecyl alcohol
Wetting agent, detergent, foamer, dispersant, emulsifier. *Ethox.*

10928 Ethana®
71-55-6 9766 200-756-3
1,1,1-Trichloroethane
Solvent. *Asahi Chem. Industry.*

10929 Ethanite
A proprietary plastic made from ethylene dichloride and calcium polysulfide. No manufacturer.

10930 ethanolamine
141-43-5 3772 205-483-3
C_2H_7ON
2-aminoethanol
MEA; 2-aminoethyl alcohol; monoethanolamine. Scrubbing acid gases, especially in synthesis of ammonia; nonionic detergents for dry cleaning wool treatment, emulsion paints, polishes, agricultural sprays; chemical intermediate; pharmaceuticals; corrosion inhibitor; rubber accelerator. *BP Chem. Ltd; OxyChem; Texaco; Union Carbide.*

10931 Ethanox® 323
Nonylphenol disulfide
Nonylphenol disulfide; antioxidant for disproportion and bleaching rosins and rosin esters. *Ethyl Corp.*

10932 Ethanox® 330
1,3,5-Trimethyl-2,4,6-tris(3,5-di-*tert*-butyl-4-hydroxybenzyl)benzene
Antioxidant and stabilizer for plastic, resin, rubber, and wax; food industry. *Ethyl Corp.*

10933 Ethanox® 398
118337-09-0
2,2'-Ethylidenebis(4,6-di-*t*-butylphenyl) fluorophosphonite
Ethylidenebis dibutylphenyl fluorophosphonite. Antioxidant for PP, LLDPE. *Ethyl Corp.*

10934 Ethanox® 702
4,4 -Methylenebis(2,6-di-*tert*-butylphenol)
Antioxidant for rubber, plastic, resin, adhesive, petroleum oil, and wax. *Ethyl Corp.*

10935 Ethanox® 703
2,6-Di-*tert*-butyl-α-dimethylamino-*p*-cresol
Antioxidant for rubber, plastic, resin, adhesive, petroleum oil, and wax. *Ethyl Corp.*

10936 Etha-Soy ISO
AMP-isostearoyl hydrolyzed soy protein; cosmetic ingredient for skin and hair care products *Brooks Industries.*

10937 Ethavan
121-32-4 3904 204-464-7
$C_9H_{10}O_3$
Ethyl vanillin
3-Ethoxy-4-hydroxybenzaldehyde; 3-Ethoxy-4-hydroxybenzadehyde; Bourbonal; Ethyl proto-catechualdehyde-3-ethyl ether; Ethyl protal. Used by flavor manufacturers to replace part of the vanillin to give bouquet to the finished flavor or fragrance. mp = 76-78°; bp = 285°; LD_{50} (rat orl) = 1590 mg/kg. *Monsanto Co.* Name unverified.

10938 Ethazate
14324-55-1 238-270-9
Zinc diethyldithiocarbamate
Fast cures at low temperature; medium precure rate activated by Oxaf; used for latex compounding; in latex foam dipped goods and fabric coatings. *Uniroyal.* Name unverified.

10939 Ethazate 50D
A 50% active water dispersion of Ethazate; ready-to-use form for latex compounding. *Uniroyal.* Name unverified.

10940 Ethenediyl diphenylene bisbenzoxazole
1533-45-5 216-245-3
$C_{28}H_{18}N_2O_2$
2,2'-(1,2-Ethenediyldi-4,1-phenylene)bisbenzoxazole
Eastobrite® OB-1. Optical brightener; fluorescent whitening agent for use in linear polyester, PET, nylon fibers. d = 1.39. *Eastman.*

10941 ether, methylated
Ether prepared from methylated spirit. A solvent.

10942 Etheramine 13
3-(Isotridecyloxy) 1-propaneamine
Corrosion inhibitor, chemical intermediate. *Berol Nobel AB.*

10943 Etherdiamine 13
Branched aliphatic ether diamine; corrosion inhibitor, chemical intermediate. *Berol Nobel AB.*

10944 Ethereal Oil of Bitter Almonds
100-52-7 1085 202-860-4
C_7H_6O
Benzaldehyde
Benzoic aldehyde; Artificial essential oil of almond; Phenylmethanal; almond artificial essential oil; artificial almond oil; benzenecarbonal; benzene carboxaldehyde; oil of bitter almond; Artificial Bitter Almond Oil; Benzene methylal; Benzoyl hydride. Used in manufactures of dyes, as a solvent and in flavors. mp = -26°; bp = 178-185°; d = 1.0440; n_D^{20} = 1.5454; LD_{50} (rat orl) = 1300 mg/kg.

10945 Ethfac 1018
Aliphatic phosphate ester; antistat and softener with excellent lubricating properties when neutralized. *Ethox.*

10946 Ethfac 104
Aliphatic phosphate ester; low foaming penetrant and emulsifier. *Ethox.*

10947 Ethfac 142W
Aliphatic phosphate ester; emulsifier, lubricant additive for process fluids. *Ethox.*

10948 Ethfac 391
Aliphatic phosphate ester; detergent for textile scouring; effective dispersant at elevated temperatures. *Ethox.*

10949 Ethfac NP-110
Aromatic phosphate ester; hydrophilic detergent and emulsifier, especially for liquid. detergent formulations. *Ethox.*

10950 Ethfac PB-1
Aliphatic phosphate ester; hydrophilic surfactant with high caustic stability; for highly alkaline cleaners. *Ethox.*

10951 Ethfac PD-6
Aliphatic phosphate ester; higher tolerance to alkali than Ethfac 161; useful as component of alkaline cleaners. *Ethox.*

10952 Ethfac PP-16
Aromatic phosphate ester; low foaming hydrotrope for coupling nonionics into alkaline systems. *Ethox.*

10953 Ethidium
Animal health insecticide. *Schering Agrochemicals Ltd.* Discontinued.

10954 Ethiodan
99-79-6 5074 202-787-8
Ethyl 4-iodophenylundec-10-enoate
iophendylate. X-ray contrast medium for myclography. *British Drug Houses.*

10955 Ethion
563-12-2 3782 209-242-3
$C_9H_{22}O_4P_2S_4$
O,O,O',O'-Tetraethyl-S,S'-methylene di(phosphorodithioate)
Ethanox; Ethiol; FMC 1240; Hylemox; Rhodiacide; Rhodocide; RP-Thion; Vegfru-Fosmite. Has both acaricidal and insecticidal properties; its acaricidal action is widely used in the abatement of cattle ticks; as a non-systemic insecticide it is used on citrus, deciduous fruits, tea, cotton and ornamental plants. mp = -12 - -15°; $bp_{0.3}$ = 164-165°; slightly soluble in H_2O, more soluble in organic solvents; LD_{50} (rat orl) = 208 mg/kg. *A/S Cheminova.* Name unverified.

10956 Ethocel Standard Premium
9004-57-3 3828
Ethyl cellulose NF; binders for pharmaceutical tabletting. *Dow Chemical.*

10957 Ethoduomeen
A range of fatty diamine ethoxylates; used as wax emulsions, polishes, fuel additives, antistatic additives, dye assistants, viscose spinning additives, algicides, bactericides, and disinfectants. *Harcros Australia.*

10958 Ethoduomeen® T/13
61790-85-0
PEG-3 tallow aminopropylamine
Emulsifier used in making of bitumen emulsions; dispersant for waxes; for textiles, asphalt, agricultural emulsions; wetting agent, corrosion inhibitor. *Akzo; Akzo Chem. BV.*

10959 Ethoduomeen® T/20
61790-85-0
PEG-10 tallow aminopropylamine

Emulsifier, dispersant, wetting agent used in coating preparation on paperboard; corrosion inhibitor. *Akzo; Akzo Chem. BV.*

10960 Ethoduomeen® T/25
61790-85-0
PEG-15 tallow aminopropylamine
Corrosion inhibitor in water treatment chemicals used in secondary oil recovery. *Akzo; Akzo Chem. BV.*

10961 Ethoduoquad® T/15-50
N,N,N', N', N'-Penta(2-hydroxyethyl)-N-tallowalkyl-1,3-propane diammonium diacetate
In aqueous isopropyl alcohol, an industrial surface active agent for agriculture, textiles, protective coatings, inks, pigment dispersions, acid pickling baths, metal-working, electroplating, and plastics manufacture. *Akzo.*

10962 Ethofat®
Ethoxylated fatty acids; used for mineral flotation; metalworking; paper; leather; pigments and surface coatings; plastic foams; textiles. *Akzo Chemie UK Ltd.* Discontinued.

10963 Ethofat® 18/14
9004-99-3
PEG-4 stearate
Industrial surfactant. *Akzo.*

10964 Ethofat® 242/25
61791-00-2
PEG-15 tallate
Emulsifier, detergent, dispersant. *Akzo.*

10965 Ethofat® O/20
9004-96-0
PEG-10 oleate
Emulsifier, detergent, dispersant. *Akzo.*

10966 Ethofoil
A proprietary trade name for ethyl cellulose film. No manufacturer.

10967 ethofumesate
26225-79-6 3788 247-525-3
$C_{13}H_{18}O_5S$
(±)-2-ethoxy-2,3-dihydro-3,3-dimethyl-5-benzofuranylmethanesulfonate
Norton; Nortran; Tramat; Nortron; Prograss; Nortranese; NC-8438. Herbicide. mp = 69-71°; soluble in H_2O (110 mg/l), more soluble in organic solvents; LD_{50} (rat orl) = 6400 mg/kg.

10968 Ethohexadiol
94-96-2 3790 202-377-9
$C_8H_{18}O_2$
2-ethyl-1,3-hexanediol; ethyl hexylene glycol; Ethyl hexanediol; octylene glycol; Rutgers 612; 6-12. Insect repellent, cosmetics, vehicle and solvent in printing inks, medicine, chelating agent for boric acid. *Hüls Am.; Union Carbide.*

10969 Ethokem
Cationic surfactant containing 870 g/l polyethanoxy alkylamine; wetting agent for phosphonoglycine herbicides. *Midkem Agrochemicals.*

10970 Ethokem C/12
61791-31-9 263-163-9
Cationic surfactant adjuvant containing bis-2-hydroxyethyl cocamine; wetting agent for phosphonoglycine herbicides and ammonium sulfate. *Midkem Agrochemicals.*

10971 Ethomeen®
A range of fatty amine ethoxylates; used as wax emulsions, polishes, fuel additives, antistatic additives, dye assistants, viscose spinning additives, algicides, bactericides and disinfectants. *Akzo.*

10972 Ethomeen® 18/12
10213-78-2 233-520-3
PEG-2 stearamine
Emulsifier, dispersant used in textile processing. *Akzo.* Name unverified.

10973 Ethomeen® 18/15
26635-92-7
PEG-5 stearamine
Emulsifier, dispersant for textile processing. *Akzo.*

10974 Ethomeen® 18/20
26635-92-7
PEG-10 stearamine
Emulsifier, dispersant used in textile processing. *Akzo.*

10975 Ethomeen® 18/25
26635-92-7
PEG-15 stearamine
Emulsifier, dispersant used in textile processing. *Akzo.*

10976 Ethomeen® 18/60
26635-92-7
PEG-50 stearamine
Emulsifier, dispersant used in textile processing; prevents premature coagulation of latex rubber *Azko.*

10977 Ethomeen® C/12
61791-14-8
PEG-2 cocamine
Emulsifier, dispersant used in textile processing. *Akzo; Akzo Chem. BV.*

10978 Ethomeen® C/15
61791-14-8
PEG-5 cocamine
Emulsifier, dispersant for textile processing, dyeing assistant, desizing assistant, softener, antistat. *Akzo; Akzo Chem. BV.*

10979 Ethomeen® C/20
61791-14-8
PEG-10 cocamine
Emulsifier, dispersant used in textile processing. *Azko.*

10980 Ethomeen® C/25
61791-14-8
PEG-15 cocamine
Emulsifier, dispersant used in textile processing. *Akzo; Akzo Chem. BV.*

10981 Ethomeen® S/12
61791-24-0
PEG-2 soyamine
Emulsifier, dispersant used in textile processing. *Akzo; Akzo Chem. BV.*

10982 Ethomeen® S/15
61791-24-0
PEG-5 soyamine
Emulsifier, dispersant for textile processing. *Akzo; Akzo Chem. BV.*

10983 Ethomeen® S/20
61791-24-0
PEG-10 soyamine
Emulsifier, dispersant used in textile processing. *Azko.* Name unverified.

10984 Ethomeen® S/25
61791-24-0
PEG-15 soyamine
Emulsifier, dispersant used in textile processing. *Azko; Akzo Chem. BV.*

10985 Ethomeen® T/12
61791-44-4 263-177-5
PEG-2 tallow amine
Emulsifier, dispersant used in textile processing. *Azko; Akzo Chem. BV.*

10986 Ethomeen® T/15
61791-26-2
PEG-5 tallow amine
Emulsifier, dispersant for textile processing. *Azko; Akzo Chem. BV.*

10987 Ethomeen® T/25
61791-26-2
PEG-15 tallow amine
Emulsifier, dispersant used in textile processing. *Akzo; Akzo Chem. BV.*

10988 Ethomid®
A range of ethoxylated amides; used as lubricants. *Harcros Australia.*

10989 Ethomid® HT/60
68155-24-8
PEG-50 hydrogenated tallow amide
Surfactant, emulsifier, secondary stabilizer for emulsion systems; dispersant, detergent. *Akzo; Akzo Chem. BV.*

10990 Ethomid® HT/60
68155-24-8
PEG-50 hydrogenated tallow amide
Surfactant, emulsifier, secondary stabilizer for emulsion systems; dispersant, detergent. *Akzo; Akzo Chem. BV.* Name unverified.

10991 Ethomid® O/17
26027-37-2
PEG-7 oleamide
PEG oleamide series. Emulsifier, dispersant, detergent. *Akzo.*

10992 Ethomulsion
A proprietary trade name for ethyl cellulose lacquer emulsion. No manufacturer.

10993 ethopabate
59-06-3 3791 200-414-3
$C_{12}H_{15}NO_4$
4-acetamido-2-methoxybenzoic acid methyl ester
ethyl pabate; pancoxin. Coccidiostat, used in poultry, often as Amprol Plus, a combination with amprolium mp = 148-149°; λ_m = 298, 267 nm ($A^{1\%}_{cm}$ 805, 365); slightly soluble in H_2O, more soluble in organic solvents.

10994 Ethoquad®
Ethoxylated quaternary ammonium salts; used as electroplating bath additives and as bacteriostats. *Harcros Australia.*

10995 Ethoquad® 18/12
PEG-2 stearmonium chloride and IPA; antistat, emulsifier, dyeing assistant,

leveling agent, antifoam used in textile industry, as electroplating bath additives. *Azko.*

10996 Ethoquad® 18/25
28724-32-5
PEG-15 stearmonium chloride
Antistat, emulsifier, dyeing assistant, leveling agent, antifoam used in textile industry, as electroplating bath additives. *Azko.*

10997 Ethoquad® C/12
70750-47-9 274-846-6
PEG-2 cocomonium chloride and IPA; antistat, emulsifier, dyeing assistant, electroplating bath additive. *Azko.*

10998 Ethoquad® C/12 Nitrate
71487-00-8
PEG-2 cocomethyl ammonium nitrate
In isopropanol; industrial surfactant for agricultural, textiles, protective coatings, inks, pigment dispersions, acid pickling baths, metalworking, electroplating, plastics manufacture. *Azko.*

10999 Ethoquad® C/25
61791-10-4
PEG-15 cocomonium chloride
Variquat® K 1215; Tomah Q-C-15. Antistat, emulsifier, dyeing assistant, leveling agent, antifoam used in textile industry, as electroplating bath additives. *Azko;Exxon/Tomah; Sherex/Div. Of Witko.*

11000 Ethoquad® CB/12
61789-68-2 263-078-7
PEG-2 cocobenzonium chloride
In isopropanol; industrial surfactant for agriculture, textiles, protective coatings, inks, pigment dispersions, acid pickling baths, metalworking, electroplating, plastics manufacture. *Akzo.*

11001 Ethoquad® O/12
18448-65-2 242-332-0
PEG-2 oleamonium chloride and IPA; antistat, emulsifier, dyeing assistant, electroplating bath additive. *Azko.*

11002 Ethoquad® T/12
67784-77-4 267-052-6
PEG-2 tallowalkyl methyl ammonium chloride
In ethanol. Industrial surfactant for agriculture, textiles, protective coatings, inks, pigment dispersions, acid pickling baths, metalworking, electroplating, plastics manufacture *Azko.*

11003 Ethoquad® T/13-50
PEG-3 tallow alkyl ammonium acetate
In aqueous isopropanol; industrial surfactant for agriculture, textiles, protective coatings, inks, pigment dispersions, acid pickling baths, metalworking, electroplating, plastics manufacture *Akzo.*

11004 Ethosperse®
Polyoxyethylene ether
Lonza Inc.

11005 Ethosperse® CA-2
9004-95-9
Ceteth-2
Oil-water emulsifier, thickener, stabilizer for hair care products, antiperspirants. *Lonza Inc.*

11006 Ethosperse® G-26
31694-55-0
Glycereth-26
Emulsifier, humectant for cosmetic, pharmaceutical and industrial uses. *Lonza Inc.*

11007 Ethotal
Emulsifier dispersant, solubilizer; used for textiles; agricultural; cosmetics. *Witco Chemical Ltd.* Discontinued.

11008 Ethox 25-R-8
Block polymer, 75% EO; dispersant, emulsifier and intermediate for esters. *Ethox.*

11009 Ethox 1122
31621-91-7
PEG-9 pelargonate
High cohesion lubricant for synthetic fiber production and processing. *Ethox.*

11010 Ethox 1212
Ethoxylated coconut glyceride; lubricant for textiles and metals. *Ethox.*

11011 Ethox 1358
Aliphatic phosphate ester; potassium salt as antistat and lubricant. *Ethox.*

11012 Ethox 1372
Polyoxyalkylene fatty amine
Low foam dye leveler for acid dyes. *Ethox.*

11013 Ethox 2156
Short chain triglyceride ester; lubricant for textiles and metals. *Ethox.*

11014 Ethox 2423
PPG-25-laureth-25
Low foam detergent, wetting and rewetting agent for textiles, emulsifier. *Ethox.*

11015 Ethox 2610
PPG 1025 ditallate
Oil-in-oil dispersant and lubricant; for silicone and nonsilicone defoamers; low foam additive for various formulations. *Ethox.*

11016 Ethox 2659
Ethoxylated styrenated phenol; dispersant for pigments and organic emulsions; higher HLB than Ethox 2938. *Ethox.*

11017 Ethox 2684
Aliphatic phosphate ester; coupling agent for nonionic surfactants in alkali systems. *Ethox.*

11018 Ethox 3113
PPG-2 bisphenol A
Monomer for polyester resins. *Ethox.*

11019 Ethox CAM-2
61791-14-8
PEG-2 cocamine
Emulsifier, dispersant, textile dyeing assistant, lubricant; intermediate for amphoterics. *Ethox.*

11020 Ethox CO-5
61791-12-6
PEG-5 castor oil; emulsifier, lubricant; dispersant for pigments and clays. *Ethox.*

11021 Ethox COA
Cocamide DEA
Foaming agent, detergent, and dispersant. *Ethox.*

11022 Ethox DL-5
9005-02-1
PEG 200 dilaurate
Lubricant and emulsifier for low foaming application. *Ethox.*

11023 Ethox DO-9
9005-07-6
PEG-8 dioleate
Emulsifier for oils and solvents, used for industrial lubricants. *Ethox.*

11024 Ethox DT-15
PEG-15 tallow diamine
Retarder for acid dyes; dispersant for use in acidic solutions. *Ethox.*

11025 Ethox DTO-9A
61791-01-3
PEG-8 ditallate
Emulsifier for oils and solvents; used for industrial lubricants. *Ethox.*

11026 Ethox HCO-16
61788-85-0
PEG-16 hydrogenated castor oil; heat-stable emulsifier for natural and synthetic oils, lubricant. *Ethox.*

11027 Ethox HO-50
57171-56-9
PEG-50 sorbitol hexaoleate
Emulsifier and lubricant for heat-stable systems. *Ethox.* Name unverified.

11028 Ethox L-61
Block polymer, 10% EO; low foam detergent, emulsifier, lubricant; intermediate for esters and polyesters. *Ethox.*

11029 Ethox LF-1226
Polyoxyalkylene glycol ether
Detergent for removal and dispersion of oils and waxes; intermediate for anionic systems. *Ethox.*

11030 Ethox MA-8
PEG-8 monomerate
Cost-effective emulsifier, lubricant. *Ethox.*

11031 Ethox MI-9
56002-14-3
PEG-9 isostearate
Emulsifier and lubricant. *Ethox.*

11032 Ethox ML-5
9004-81-3
PEG-5 laurate
Lubricant, emulsifier, detergent, softener, coupling agent; viscosity control agent for plastisol resins. *Ethox.*

11033 Ethox MO-9
9004-96-0
PEG-9 oleate
Lubricant, emulsifier for natural and synthetic oils, detergent, softener. *Ethox.*

11034 Ethox MS-8
9004-99-3

PEG-8 stearate
Lubricant, wax and oil emulsifier, detergent, softener; for aqueous processing. *Ethox.*

11035 Ethox OAM-308
PEG-30 oleyl amine
Emulsifier, dispersant, textile dyeing assistant and lubricant; leveling agent for dyeing nylon; antiprecipitant for cross dyeing. *Ethox.*

11036 Ethox PPG 1025 DTO
PPG-20 ditallate
Oil-in-oil dispersant and lubricant. *Ethox.*

11037 Ethox SAM-10
26635-92-7
PEG-10 stearamine
Moderate HLB emulsifier; component of corrosion inhibitors. *Ethox.*

11038 Ethox SO-9
PEG-8 sesquioleate
All-purpose oil emulsifier. *Ethox.*

11039 Ethox TAM-2
61791-44-4 263-177-5
PEG-2 tallow amine
Emulsifier, dispersant, textile dyeing assistant, lubricant; used for preparation of high pressure wax dispersions. *Ethox.*

11040 Ethox TO-8
61791-00-2
PEG-8 tallate
Lubricant, emulsifier, detergent, softener, dispersant, *Ethox.*

11041 Ethoxol
A proprietary glycol ether. A solvent. *ICI Chem & Polymers Ltd.*

11042 Ethoxy carbonyl phenylethyl phenylformamidine
65816-20-8 265-932-4
N-(*p*-ethoxycarbonylphenyl)-N'-ethyl-N'-phenylformamidine
Givsorb® UV-2. Industrial UV absorber, light stabilizer, and antioxidant for PU, PVC, polyolefins, ABS, nylon, and acetal resins; photostable, broad spectrum screening agent for protection against the adverse effects of both UV-B and UV-A radiation; used in surface coats. *Givaudan; Givaudan SA.* Name unverified.

11043 Ethoxy carbonyl phenylmethyl phenylformamidine
57834-33-0 260-976-0
N^2-(4-Ethoxy-carbonylphenyl)-N'-methyl-N'-phenylformamidine
Givsorb® UV-1Givsorb® UV-1. Industrial UV absorber. *Givaudan; Givaudan SA.* Name unverified.

11044 Ethoxydiglycol
111-90-0 1847 203-919-7
$C_6H_{14}O_3$
Diethylene glycol monoethyl ether
DEGEE; Diethylene glycol ethyl ether; Diglycol monoethyl ether; Carbitol; Ethanol, 2-(2-ethoxyethoxy); Ethylene diglycol monoethyl ether; 2-(2-Ethoxyethoxy) ethanol; Carbitol®; Carbitol® Low Gravity; Dioxitol-Low Gravity; Ektasolve® DE; Eastman® DE; Ethyl Di-Icinol; Poly-Solv® DE; Solvent APV Spec.; Transcutol. Solvent sg = 0.990; mp = -76°; bp = 195-202°; fp = 96.1°; soluble in H_2O and organic solvents; LD_{50} (rat orl) = 5500 mg/kg; moderately toxic by ingestion. Aldrich; Allchem Ind.; Ashland; Eastman; Fluka; Great Western; Occidental; Oxiteno; Sigma; Spectrum Chem. Mfg.; Union Carbide. Name unverified.

11045 Ethoxyethanol
110-80-5 3797 203-804-1
$C_4H_{10}O_2$
ethylene glycol ethyl ether
2-ethoxyethanol; Cellosolve. EE; solvent for nitrocellulose, lacquer, lacquer thinners, dyeing and printing textiles, varnish removers, cleaning solutions. *Arco; Ashland; OxyChem; Union Carbide.*

11046 Ethoxyethanol acetate
111-15-9 3798 203-839-2
$C_4H_{10}O_4$
ethylene glycol ethyl ether acetate
2-ethoxyethanol acetate; 2-ethoxyethyl acetate; EEA. Ester of ethoxyethanol and acetic acid; solvent; automobile lacquers; retards evaporation and imparts high gloss. mp = -61°; bp = 156°; d = 0.9750; n_D^{20} = 1.4040 *Allchem Industries; Arco; OxyChem; Union Carbide.*

11047 Ethoxylan® 1685
61790-81-6
PEG-75 lanolin
Emollient, emulsifier, dispersant, foam stabilizer, resin plasticizer for cosmetic and pharmaceutical preparations, textile processing. *Henkel/Cospha; Henkel/Textile; Henkel Canada.*

11048 Ethoxyol® 1707
Polysorbate 80
cetyl acetate; acetylated lanolin alcohol. Self-emulsifying emollient with lubricating and penetrating properties; auxiliary emulsifier, solubilizer, pigment wetting agent for cosmetics, liquid. soaps. *Henkel/Cospha; Henkel Canada.*

11049 ethoxyquin
91-53-2 3800 202-075-7
$C_{14}H_{19}NO$
6-ethoxy-1,2-dihydro-2,2,4-trimethylquinoline
EMQ, Santoflex; Santoquin. Antioxidant in feed and food; antidegradation agent for rubber. bp_2 = 123-125°; d_{25}^{25} = 1.029-1.031; n_D^{25} = 1.569-1.672; LD_{50} (rat orl) = 1925 mg/kg.

11050 Ethrel
16672-87-0 3777 240-718-3
A plant growth regulator containing ethephon. (Sold in UK on behalf of Amchem Prods. Inc.) *ICI Chem & Polymers Ltd.*

11051 Ethrel C
16672-87-0 3777 240-718-3
Ethephon
Soluble concentrate containing 480 g 2-chlorethyl phosphonic acid per liter; plant growth regulator. *ICI Agrochemicals.*

11052 Ethrel-E
16672-87-0 3777 240-718-3
Ethephon
Growth regulator. *A H Marks & Co Ltd.* Name unverified.

11053 Ethrel-R
16672-87-0 3777 240-718-3
Ethephon
Defoliant. *A H Marks & Co Ltd.* Name unverified.

11054 Ethsorbox L-20
9005-64-5 8872
PEG-20 sorbitan laurate
Emulsifier for oils, solvents, and fats; lubricant for cotton and rayon. *Ethox.*

11055 Ethsorbox O-20
9005-65-6 7742
PEG-20 sorbitan oleate
Emulsifier for industrial and textile lubricants. *Ethox.*

11056 Ethsorbox S-20
9005-67-8 8872
PEG-20 sorbitan stearate
Waxy lubricant, softener. *Ethox.*

11057 Ethsorbox TO-20
9005-70-3 8872
PEG-20 sorbitan trioleate
Lubricant and emulsifier for industrial process fluids. *Ethox.*

11058 Ethsorbox TS-20
9005-71-4 8872
PEG-20 sorbitan tristearate
Lubricant, softener for textile goods; emulsifier for fats and oils. *Ethox.*

11059 Ethulon
Ethyl cellulose film for tracing and other industrial purposes. *May & Baker Ltd.* Name unverified.

11060 Ethulose
A proprietary preparation of alcohol and levulose; a parenteral source of calories. *Geistlich Sohne AG. Name unverified.*

11061 ethyl *p*-hydroxybenzoate
120-47-8 3883 204-399-4
$C_9H_{10}O_3$
4-hydroxybenzoic acid ethyl ester
ethylparaben; ethyl-4-hydroxybenzoate; 4-hydroxybenzoic acid ethyl ester; Nipagin A; Ethyl Parasept; Solbrol A; *para*-hydroxybenzoic acid ethyl ester. Preservative for pharmaceuticals. mp = 116-118°; bp = 297°; soluble in H_2O (0.07 g/100ml), more soluble in organic solvents.

11062 ethyl 3-ethoxypropionate
763-69-9 212-112-9
$C_7H_{14}O_3$
Ethylethoxy propionate
Solvent; intermediate for manufacture of vitamin B_1. mp = -75°; bp = 170°; d = 0.9500; n_D^{20} = 1.4064; soluble in H_2O (1.6 g/100 ml), more soluble in organic solvents; LD_{50} (rat orl) = 5 gm/kg. *Eastman; Union Carbide.*

11063 ethyl acetate
141-78-6 3803 205-500-4
$C_4H_8O_2$
acetic ether
acetic acid, ethyl ester; Ethyl acetic ester; acetoxyethane; Acetic ether; vinegar naphtha; acetidin; Acetic ester. Ester of ethyl alcohol and acetic acid; general solvent in coatings and plastics, organic synthesis, smokeless powders, artificial leather, photographic films and plates, pharmaceuticals, synthetic fruit essences; cleaning textiles. mp = -84°; bp = 77°; d = 0.9000; n_D^{20} = 1.3716; soluble in H_2O (80 g/l), organic solvents; LD_{50} (rat orl) = 5620

mg/kg. Allchem Industries; BP Chem. Ltd; Chisso; Eastman; Hoechst Celanese; Lonza AG; Mallinckrodt; Monsanto; Union Carbide.

11064 ethyl alcohol
64-17-5 3806 200-578-6
C_2H_6O
ethanol
absolute alcohol, ethyl hydrate, ethyl hydroxide; Anhydrol; alcohol; methylcarbinol; Denatured alcohol; Algrain; Cologne spirit; fermentation alcohol; grain alcohol; Jaysol; Jaysol s; molasses alcohol; potato alcohol; spirit; spirits of wine; Tecsol; alcohol dehydrated; ethanol 200 proof; Cologne spirits (alcohol); sd alcohol 23-hydrogen; Synasol. In alcoholic beverages. Used as a solvent, in pharmaceuticals, perfumery and organic synthesis. Also as an additive in gasolines; and an antiseptic. mp = -114°; bp = 78°; d^{25}_{25} = 0.931; soluble in H_2O, organic solvents. Eastman; Georgia-Pacific Resins; Gist-Brocades SpA; Grain Processing; Quantum/USI; Union Carbide; Vista.

11065 ethyl cadmate
14239-68-0 238-113-4
$C_{10}H_{20}CdN_2S_4$
Cadmium, bis(diethylcarbamodithioato-S,S')-, (T-4)-
Cadmium diethyldithiocarbamate; Ethyl cadmate; Bis(diethyldithiocarbamate)cadmium complex. Activated cadmium diethyldithiocarbamate; primary accelerator for NR and synthetic rubbers; used with a thiazole; gives heat resistance and low compression set properties. *R. T. Vanderbilt Co Inc.*

11066 ethyl cellulose
9004-57-3 3828
Cellulose ethyl ether
ethocel; Ethocel Standard Premium; Hercules® K, N, T. Ethyl ether of cellulose; hot-melt adhesives and coatings for cables, etc; extrusion wire insulation, protective coatings, pigment-grinding bases, food additive; tablet binder in pharmaceuticals. *Aqualon; Colorcon; FMC; Hercules.*

11067 ethyl chloride
75-00-3 3829 200-830-5
C_2H_5Cl
monochloroethane
hydrochloric ether; Muriatic Ether; Chloroethane; aethylis;aethylis chloridum; chloridum; chloryl; ether chloratus; ether hydrochloric; ether muriatic; chlorethyl; kelene; chelen; anodynon; chloryl anesthetic; narcotile; chlorene; cloretilo; dublofix. Used as a refrigerant, and solvent and as a chemical feedstock. mp = -139°; bp = 12°; d^0_4 = 0.9214; soluble in H_2O (0.574 g/100 ml), soluble in organic solvents.

11068 ethyl di-Icinol
111-90-0 1847 203-919-7
$C_6H_{14}O_3$
2-(2-ethoxyethoxy)ethanol
Ethoxydiglycol; Diethylene glycol ethyl ether; Diethylene Glycol Monoethyl Ether; Ethyl Carbitol; diglycol monoethyl ether; ethoxy diglycol; ethyl diethylene glycol; ethylene diglycol monoethyl ether; monoethyl ether of diethylene glycol; 3,6-dioxa-1-octanol; APV; carbitol cellosolve; dioxitol; Dowanol; Dowanol DE; Losungsmittel APV; Solvolsol; Transcutol; Poly-solv DE; ethanol, 2,2'-oxybis-, monoethyl ether; ethyl digol; Ektasolve DE; 1-hydroxy-3,6-dioxaoctane; O-ethyldigol; 2-(ethoxyethoxy)ethanol; 3,6-dioxaoctan-1-ol; 2-(β-ethoxyethoxy)ethanol; diglycol; Carbitol; Ethoxyethoxy)ethanol. Solvent for use in protective coatings, inks, cleaning products, agricultural chems.; aids wetting, penetration, and soil removal; coupling solvent bp = 196°; d^{20}_{20} = 1.0273; n^{20}_D = 1.4273; soluble in H_2O, organic solvents; LD_{50} (rat orl) = 11 g/kg. *ICI Australia.*

11069 ethyl ether
60-29-7 3852 200-467-2
$CH_3CH_2OCH_2CH_3$
diethyl ether
ether; Diethyl ether; Ethyl ether; 1,1'-Oxybisethane; ethyl oxide; Diethyl Oxide; Ethoxyethane; sulfuric ether; pronarcol. Organic synthesis; smokeless powders; industrial solvent; analytical chemistry; anesthetic; extractant. mp = -116°; bp = 34.6°; d= 0.7134; n^{20}_D= 1.3490-1.3520; slightly soluble in H_2O (6.9 g/100 ml), soluble in organic solvents. *Aldrich; Exxon; Hüls AG; Mallinckrodt; Quantum/USI; Spectrum Chem. Mfg.*

11070 ethyl ether Anhydrous A.C.S.
60-29-7 3852 200-467-2
Ethyl ether
Reagent grade. *Quantum/USI.* Name unverified.

11071 ethyl ether USP/ACS
60-29-7 3852 200-467-2
Ethyl ether
Laboratory reagent, surface antiseptic and cleaning agent, in liniments, as analgesic, expectorant; denaturant for alcohol formulations; chem. reaction media. *Quantum/USI.* Name unverified.

11072 Ethyl Green
$C_27H_{35}N_3C1_3BrZn$
Methyl Green
A dyestuff; zinc double chloride of ethylhexamethyl-*p*-rosaniline bromide, Dyes wool mordanted with sodium thiosulfate and sulfuric acid or zinc acetate, and silk and cotton mordanted with tannin a bluish-green.

11073 ethyl hexanediol
94-96-2 3790 202-377-9
$C_8H_{18}O_2$
2-Ethyl-1,3-hexanediol
ethyl hexylene glycol; ethyl hexanediol; Rutgers 6-12; Carbide 6-12; 3-hydroxymethyl-n-heptan-4-ol; EH diol; 2-ethylhexane-diol-1,3; 2-Ethyl-1,2-Hexanediol; Ethyl-1,3-Hexane Diol-2; 2-ethyl hexanediol; Ethyl hexylene glycol; ethohexadiol; 2-ethyl-1,3-hexylene glycol; 2-ethyl-2-propyl-1,3-propanediol; octylene glycol; 6-12 insert repellent; EHD; Ethyl-1,3-hexanediol; Ethyl-3-propyl-1,3-propanediol; Hydroxymethyl-n-heptan-4-ol; Insect repellant; Repellent 6-12; 2-Ethylhexane-1,3-diol. Insect repellent, cosmetics, vehicle and solvent in printing inks, medicine, chelating agent for boric acid. mp = -40; bp = 241-249°; d = 0.9330; n^{20}_D= 1.4497; soluble in H_2O, ethanol, isopropanol, propylene glycol, castor oil; LD_{50} (rat orl) = 1400 mg/kg. *Hüls Am.; Union Carbide.* Name unverified.

11074 ethyl icinol
110-80-5 3797 203-804-1
$C_4H_{10}O_2$
2-Ethoxyethanol
Glycol ethyl ether; Cellosolve; Dowanol EE; Ethylene glycol monoethyl ether; Glycol monoethyl ether; 2EE; Ethoxyethanol; EGEE; ethylene glycol ethyl ether; Cellosolve solvent; glycol ether EE; hydroxy ether; oxitol; Poly-solv EE; Ektasolve EE; Jeffersol EE; Ethyl Cellosolve; Ethyl Glycol. Solvent for use in protective coatings, inks, cleaning products, agricultural chemicals; aids wetting, penetration, and soil removal; coupling solvent mp = -90°; bp = 135°; d = 0.931; n^{20}_D= 1.4068; soluble in H_2O, organic solvents; LD_{50} (rat orl) = 2125 mg/kg. *ICI Australia.*

11075 ethyl isothiocyanate
542-85-8 3861 208-831-2
C_3H_5NS
ethyl mustard oil; isothiocyanatoethane. Used in chemical synthesis and as a military poison gas. mp = -6°; bp = 130-132°; d^{18}_4= 1.003; n^{18}_D = 1.5142; insoluble in H_2O, soluble in organic solvents.

11076 ethyl lactate
97-64-3 3863 211-694-1
$C_5H_{10}O_3$
Ethyl L-(-)-lactate
lactic acid ethyl ester; Lactic acid ethyl ester; Propanoic acid, 2-hydroxy-, ethyl ester; Ethyl L-lactate; Eusolvan. Ester; dipped latex products, coatings, specialty finishes, inks, diluents, adhesives, intermediates, photoresist solvents., screen printing of electronic parts, flavors and fragrances, solvent resins. mp = -26°; bp = 154°; d = 1.0420; n^{20}_D= 1.4130; slightly soluble in H_2O. Soluble in organic solvents; *CPS; Farleyway Chem. Ltd; Penta Mfg.*

11077 ethyl methacrylate
97-63-2 202-597-5
$C_6H_{10}O_2$
ethyl-2-methyl-2-propenoate
ethyl-α-methyl acrylate; 2-Methyl-2-Propenoic Acid Ethyl Ester; Rhoplex AC-33; Methacrylic Acid Ethyl Ester; EMA; ethyl α-methylacrylate; ethyl 2-methacrylate; ethyl 2-methyl-2-propenoate. Ester of ethyl alcohol and methacrylic acid; polymers; chemical intermediates. mp = -75°; bp = 118-119°; d = 0.9170; n^{20}_D= 1.4128; soluble in H_2O (4 g/l), more soluble in organic solvents; LD_{50} (rat orl) = 14800 mg/kg. *Rohm & Haas.*

11078 ethyl namate®
148-18-5 3443 205-710-6
Sodium diethyldithiocarbamate
Immunomodulator and chelating agent. Used in copper analysis. *R. T. Vanderbilt Co Inc.* Discontinued.

11079 ethyl parathion
56-38-2 7167 200-271-7
parathion
O,O-Diethyl-O-(4-nitrophenyl) phosphorothioate. Used as an insecticide for the protection of field crops, vegetables and fruit. *A/S Cheminova.* Name unverified.

11080 ethyl safranate
Ethyl dehydrocyclogeranate.
Flavorant. *Quest Int'l UK Ltd.*

11081 ethyl salicylate
118-61-6 3894 204-265-5
$C_9H_{10}O_3$
2-hydroxybenzoic acid ethyl ester
salicylic acid ethyl ester; salicylic ether; sal ethyl. Used in perfumery; has been used in veterinary medicine as a counter-irritant. mp = 1°; bp = 231-234°; d^{20}_0= 1.131; n^{20}_D= 1.5226; slightly soluble in H_2O, soluble in organic solvents.

11082 ethyl selenac
5456-28-0 226-713-9
$C_{10}H_{20}N_2S_4Se$
Selenium diethyl dithiocarbamate
Rubber accelerator for NR, SBR and IIR; vulcanizing agent; effective in low sulfur and sulfurless heat-resistant compounds. *R. T. Vanderbilt Co Inc.* Discontinued.

11083 ethyl silicate
78-10-4 3895 201-083-8
$(C_2H_5)_4SiO_4$
ethyl orthosilicate
tetraethyl orthosilicate. Intermediate for manufacture of ethyl silicate products; produces binders; chemical and heat-resistant paints, cements, weatherproofing; protective coatings. *Akzo; Aldrich; R.W. Greeff; Hüls Am.; PCR; Wacker Silicones.*

11084 ethyl tellurac®
20941-65-5 244-121-9
$C_{20}H_{40}N_4S_8Te$
Tellurium diethyldithiocarbamate
Tellurium, tetrakis(diethylcarbamodithioato-S,S')-, (DD-8-1,1,1",1",1',1',1"',1"')-; tellurium diethyldithiocarbamate; tetrakis(diethylcarbamodithioato-S,S')tellurium; tellurac; diethyldithiocarbamic acid tellurium salt; tellurium,tetrakis(diethyldithiocarbamate)-; Tellurium(IV) diethyldithiocarbamate. Ultra accelerator for NR, SBR, NBR, EPDM; used with thiazole modifiers; produces high modulus vulcanization; particularly active in IIR compounds. mp = 108-118°; d = 1.44; insoluble in H_2O, soluble in organic solvents. *R. T. Vanderbilt Co Inc.* Discontinued.

11085 ethyl tuads®
97-77-8 3428 202-607-8
Tetraethylthiuram disulfide
Accelerator and vulcanizing agent for rubber. mp = 71-72°; LD_{50} (rat orl) = 500 mg/kg. *R. T. Vanderbilt Co Inc.* Discontinued.

11086 ethyl tuex
97-77-8 3428 202-607-8
Tetraethylthiuram disulfide
A nondiscoloring and nonstaining accelerator; sharp curing range with normal to high sulfur; used in natural, butyl and nitrile rubbers, steam hose and calendered air-cured sheeting. mp = 71-72°; LD_{50} (rat orl) = 500 mg/kg. *Uniroyal.* Name unverified.

11087 ethyl vanillin
121-32-4 3904 204-464-7
$C_9H_{10}O_3$
3-ethoxy-4-hydroxybenzadehyde
3-ethoxy-4-hydroxy benzaldehyde; Ethavan; Rhodiarome; Vanbeenol; Ethavan; Rhodiarome; Vanbeenol; ethylprotocatechuic aldehyde; bourbonal; ethylprotal; vanillal; Ethovan. Substituted phenolic; flavoring agent. mp = 76-78°; LD_{50} (rat orl) = 1590 mg/kg. *Boehringer Mannheim GmbH; Penta Mfg.; Schweizerhall.*

11088 ethyl zimate®
14324-55-1 238-270-9
Zinc diethyldithiocarbamate
Ultra accelerator; primary accelerator in NR and SBR; requires a thiazole modifier for safe processing and wide cure range; nondiscoloring in light colored stocks; stabilizer in thermoplastic rubbers and hot melts; 1:1 combination (dry weight) with Zetax. *R. T. Vanderbilt Co Inc. Discontinued.*

11089 ethyl-*p*-toluenesulfonate
80-40-0 3903 201-276-7
$C_9H_{12}O_3S$
4-methylbenzenesulfonic acid ethyl ester
Used in ethylation reactions. mp = 33°; bp_{15} = 173°; d = 1.17; n_D^{20} = 1.5110; insoluble in H_2O, soluble in organic solvents.

11090 Ethylan
Nonionic ethoxylates of fatty alcohols, acids, amines, phenols and alkylolamides; used for detergency, low foam wetting, latex stabilization, emulsification, cosolvency and petroleum recovery. *Harcros.*

11091 Ethylan® 172
Cetoleth-3
Emulsifier for mineral oils and hydrophobic waxes. *Harcros.*

11092 Ethylan® 44
9016-45-9 6772
Nonoxynol-4
Emulsifier for paraffin hydrocarbons, silicones, mineral oil, alkyd resins, etc.; formulation of degreasers, hand cleaning gels. *Harcros.*

11093 Ethylan® A10
9004-96-0
PEG 1000 oleate
Surfactant, emulsifier, wetting agent, detergent for areas where low toxicity is important, e.g., oil slick dispersants. *Harcros.*

11094 Ethylan® A2
9004-96-0
PEG-4 oleate
Emulsifier for kerosene and mineral oil, antifoam agent. *Harcros.*

11095 Ethylan® BD10
Low foam biodegradable surfactant, wetting agent for rinse aids, machine dishwashing, metal cleaning, bottle washing. *Harcros.*

11096 Ethylan® BV
9016-45-9 6772
Nonoxynol-14
Foam stabilizer and booster, solubilizer, emulsifier used in pesticides, perfumes, emulsion polymerization. *Harcros.*

11097 Ethylan® CD109
Isodeceth-9
Wetting agent, emulsifier, detergent; sudsing agent for liquid detergents; scouring and dye leveling agent; coemulsifier for aromatic solvents, waxes, toxicants; detergent sanitizer; essential oil solubilizer. *Harcros.*

11098 Ethylan® CD123
C_{12} fatty alcohol ethoxylate; emulsifier, wetting agent, detergent; coemulsifier for mineral oils and waxes, alkyd resins, paraffin hydrocarbons; biodegradable. *Harcros.*

11099 Ethylan® CD913
Synthetic lower fraction primary alcohol EO condensate (2.9 EO); detergent, oil-water and water-oil emulsifier, wetting agent, solubilizer; emulsifier for mineral oils and waxes, alkyd resins, paraffinic hydrocarbons. *Harcros.*

11100 Ethylan® CDP2
Linear middle fraction fatty alcohol ethoxylate (2 EO); surfactant, intermediate for toiletry grade sulfates and other specialty surfactants; coemulsifier for mineral oils, waxes, alkyd resins, paraffinic hydrocarbons. *Harcros.*

11101 Ethylan® D252
Synthetic primary alcohol ethoxylate (2EO); oil-water and water-oil emulsifier for mineral oils and waxes, alkyd resins, paraffin hydrocarbons. *Harcros.*

11102 Ethylan® FO30
9004-96-0
PEG 1000 oleate
Surfactant, emulsifier, wetting agent, detergent for areas where low toxicity is important, e.g., oil slick dispersants. *Harcros.*

11103 Ethylan® FO30
Fish oil ethoxylate; surfactant, emulsifier, wetting agent, detergent for areas where low toxicity is important, e.g., oil slick dispersant; also in antifoams. *Harcros.*

11104 Ethylan® GEL2
9005-64-5 8872
Polysorbate 20
Water-oil emulsifier, solubilizer especially with sorbitan esters; used in cosmetics, agriculture, perfumes, fiber and textile lubricants, textile antistats, polymer additives, suspension and emulsion polymerization. *Harcros.*

11105 Ethylan® GEO8
9005-65-6 7742
Polysorbate 80
Emulsifier for cosmetics, pharmaceuticals, agrochem. formulations, textile lubricants, plastic additives, emulsion and suspension polymerization; solubilizer for perfume, flavors, essential oils. *Harcros.*

11106 Ethylan® GEO81
9005-65-6 7742
Polysorbate 81
Emulsifier for cosmetics, agriculture, plastic additives, textile fiber lubricants and softeners, suspension and emulsion polymerization; solubilizer for perfume, flavors, essential oils. *Harcros.*

11107 Ethylan® GEP4
9005-66-7 8872
Polysorbate 40
Water-oil emulsifier, solubilizer especially with sorbitan esters; used in cosmetics, agriculture, perfumes, fiber and textile lubricants, textile antistats, polymer additives. *Harcros.*

11108 Ethylan® GES6
9005-67-8 8872
Polysorbate 60, 61; Accosperse 60; Alkamuls® PSMS-20, T-60; Capmul® POE-S; Crillet 3, 31; Disponil SMS 120 F1; Drewpone® 60K; Durfax® 60; Emsorb® 2728, 6906, 6909; Ethsorbox S-20; Eumulgin SMS 20; Fanwax P; Glycosperse® S-20 FG, S-20 KFG; Hetsorb S-20, S-4; Hodag PSMS-20, SVS-18; Liposorb S-4, S-20; Montanox 60 DF; Norfox® Sorbo T-60; Polycon T60K; Prox-onic SMS-020; Radiamuls® Sorb 2147; Radiasurf® 7147; Ritabate 60; Sorbax PMS-20; T-Maz® 60, 61; Tandem 5K, 8; Tween® 60, 60K, 61,. General purpose, low toxicity emulsifier for cosmetics and agrochemicals; textile lubricant; plastics additive; emulsion and suspension polymerization; solubilizer for perfume, flavors, essential oils. *Harcros.*

11109 Ethylan® GL20
1338-39-2 8872 215-663-3
Sorbitan laurate
Emulsifier for cosmetics, pharmaceuticals, agriculture, plastic antifog, textile fiber lubricant/softener, suspension and emulsion polymerization. *Harcros.*

11110 Ethylan® GO80
1338-43-8 8872 215-665-4
Sorbitan oleate
Emulsifier for cosmetics, pharmaceuticals, agriculture, plastic antifog, textile fiber lubricant/softener, suspension and emulsion polymerization. *Harcros.*

11111 Ethylan® GPS85
9005-70-3 8872
Polysorbate 85
Emulsifier for cosmetics, agriculture, plastic additives, textile fiber lubricants and softeners, suspension and emulsion polymerization; solubilizer for perfume, flavors, essential oils. *Harcros.*

11112 Ethylan® GS60
1338-41-6 8872 215-664-9
Sorbitan stearate
Emulsifier for cosmetics, pharmaceuticals, agriculture, plastic antifog, textile fiber lubricant/softener, suspension and emulsion polymerization. *Harcros.*

11113 Ethylan® GT85
26266-58-0 8872 247-569-3
Sorbitan trioleate
Emulsifier for cosmetics, pharmaceuticals, agriculture, plastic antifog, textile fiber lubricant/softener, suspension and emulsion polymerzation. *Harcros.*

11114 Ethylan® LD
Cocamide DEA
Foam stabilizer, emulsifier for hand cleaning gels, hard surface cleaners, shampoos; plastics antistat. *Harcros.*

11115 Ethylan® NP1
9016-45-9 6772
Nonoxynol-1.5
Defoamer, oil emulsifier. *Harcros.*

11116 Ethylan® TN-10
61791-14-8
PEG-10 cocamine
Wetting agent for acid or alkaline metal cleaners, stripping of surface coatings; oil emulsifier with anticorrosive properties, antistat for synthetic fibers with PS; for cosmetics manufacture. *Harcros.*

11117 Ethylan® TT-15
61791-44-4 263-177-5
PEG-15 tallow amine
Wetting agent for metal cleaning and stripping of surface coatings, fiber antistat, used in cosmetics. *Harcros.*

11118 ethyl-benzaldehyde
4748-78-1 225-268-8
$C_9H_{10}O$
4-ethylbenzaldehyde
Ebal; *p*-Ethyl-benzaldehyde. Additive for resins; intermediate for pharmaceuticals, fragrances. bp = 221°; d = 0.9790. *Mitsubishi Gas.*

11119 ethylbutylcarbinol
589-82-2 209-661-1
$C_7H_{16}O$
sec-Heptyl alcohol
3-heptanol. Chemical intermediate and solvent. mp = -70°; bp_{20} = 66°; d = 0.8180; n_D^{20}= 1.4214; slightly soluble in H_2O, more soluble in organic solvents; LD_{50} (rat orl) = 1870 mg/kg.

11120 ethylcyanine
Lepidinquinoline ethylcyanine bromide
A dyestuff; a similar dye to Cyanine, using ethyl bromide, instead of amyl iodide; used as a substitute for cyanine in dyeing.

11121 ethylene brassylate
105-95-3 203-347-8
$C_{15}H_{26}O_4$
1,4-Dioxacycloheptadecane-5,17-dione
Emeressence® 1150; Astratone; Ethylene undecane dicarboxylate; Musk T. Musk chemical for fragrance of odor masking applications. *Henkel/Emery.* Name unverified.

11122 ethylene dichloride
107-06-2 3843 203-458-1
$C_2H_4Cl_2$
1,2-dichloroethane
Dutch oil; symdichloroethane; ethylene chloride; EDC; Dutch liquid; Brocide; ethane, 1,2-dichloro. Halogenated aliphatic hydrocarbon; production of vinyl chloride, trichloroethylene, vinylidene chloride, trichloroethane; lead scavenger in gasoline; paint, varnish remover; metal degreasing; soaps; wetting/penetrating agents; organic synthesis; ore flotation; solvent; fumigant. mp = -35°; bp = 82.5-84.5°; d = 1.2410; insoluble in H_2O, soluble in organic solvents; LD_{50} (rat orl) = 670 mg/kg. Albright & Wilson Am.; Ashland; BASF; BP Chem. Ltd; Ethyl; Georgia Gulf; Norsk Hydro AS; Oxychem; PPG Industries.

11123 ethylene glycol
107-21-1 3844 203-473-3
$C_2H_6O_2$
1,2-Ethanediol
1,2-ethanediol; ethylene alcohol; Aliphatic diol; glycol; Ilexan E,. Antifreeze in cooling and heating systems; in hydraulic brake fluids; industrial humectant; solvent in paints, plastics, inks; softening agent for cellophane; stabilizer; in explosives, alkyd resins, elastomers, synthetic fibers and waxes; asphalt. bp = 195°; d = 1.1200; soluble in H_2O, organic solvents; LD_{50} (rat orl) = 4700 mg/kg. Ashland; BASF; Eastman; Hoechst Celanese; Mitsui Petrochem. Ind.; Mitsui Toatsu Chem.; Mobil; Olin; OxyChem; Shell; Texaco; Union Carbide.

11124 ethylene glycol diacetate
111-55-7 3845 203-881-1
$C_6H_{10}O_4$
glycol diacetate
ethylene diacetate; 1,2-diacetoxyethane. Ester; extraction solvent, foundry resins, perfume fixative, solvent for coatings. mp = -31°; bp = 190-191°; d = 1.104; n_D^{20} = 1.415; soluble in H_2O (143 mg/ml), more soluble in organic solvents; LD_{50} (rat orl) = 6.86 g/kg. *Chemoxy Int'l. plc; CPS; Eastman.*

11125 ethylene glycol dimethacrylate
97-90-5 202-617-2
$C_{10}H_{14}O_4$
1,2-ethanediol dimethacrylate
1,2-bis(methacryloyoxy) ethane; Glycol Dimethacrylate; Ethylene Glycol Dimethacrylate; Ethylene Methacrylate; EGDMA; 2-Propenoic acid, 2-methyl-, 1,2-ethanediyl ester; ethylene dimethacrylate; Ageflex EGDMA; Perkalink® 401. Crosslinker and modifier for ABS, acrylic sheet and rods, PVC, ion exchange resins, glaze coatings, dental polymers, paper processing aids, rubber modifier, adhesives, optical polymers, leather finishing, moisture barrier films. bp = 85°; d = 1.0510; soluble in H_2O, organic solvents; LD_{50} (rat orl) = 3300 mg/kg. *Akzo; CPS; Sartomer.*

11126 ethylene glycol distearate VA
627-83-8 211-014-3
Glycol distearate
Opacifier and pearling agent for hair care *Goldschmidt.* Discontinued.

11127 ethylene glycol methyl ether
109-86-4 6118 203-713-7
$C_3H_8O_2$
Methyl Cellosolve
Solvent. *Arco; Ashland; OxyChem; Union Carbide.*

11128 ethylene glycol monobutyl ether

11129 ethylene glycol monostearate VA
Glycol stearate
Stabilizer and thickener for creams and lotions; pearlizing and opacifying agent. *Goldschmidt.* Discontinued.

11130 ethylene glycol propyl ether
2807-30-9 220-548-6
$C_5H_{12}O_2$
2-Propoxyethanol
Ethylene glycol monopropyl ether; PropylCellosolve; Ethanol, 2-propoxy; Ektasolve® EP; Eastman® EP; EP; Propyl Cellosolve®. Slow evaporating solvent used in coatings; useful in waterborne coating systems; coupling solvent for resin/water systems; controls viscosity of waterborne resins; effective for NC, acrylic, epoxy, polyamide, and alkyd resins; retarder in coating systems. d = 0.913; bp = 149.5°; fp = -90°; flash point = 51°; soluble in H_2O; LD_{50} = rat orl 3089 mg/kg; moderately toxic by ingestion and skin contact. *Aldrich; Ashland; Eastman.*

11131 ethylene oxide
75-21-8 3848 200-849-9
C_2H_4O
oxirane
epoxyethane; anprolene; dimethylene oxide; 1,2-epoxyethane; ETO; oxacyclopropane; Dihydrooxirene; Epoxyethane; Oxane; E.O.; Qazi-ketcham; Ethene oxide; α,β-oxidoethane; oxidoethane; amprolene; anproline; Ethox; FEMA No. 2433; Merpol; oxyfume; oxyfume 12; T-GAS. Manufacture of ethylene glycol and higher glycols, surfactants, acrylonitrile, ethanolamines; petroleum demulsifier; fumigant; rocket propellant; industrial sterilant (medical plastic tubing); fungicide. mp = -111°; bp = 11°; d_4^{10} = 0.882, soluble in H_2O, organic solvents; LD_{50} (rat orl) = 72mg/kg. BASF; Hoechst Celanese; Hüls AG; Mitsubishi Petrochem.; Mitsui Petrochem.; Mitsui Petrochem. Ind.; Mitsui Toatsu Chem.; OxyChem; Shell; Union Carbide.

11132 ethylene thiourea
96-45-7 3849 202-506-9

$C_3H_6N_2S$
2-Imidazolidinethione
imidazoline-2-thiol; 2-mercaptoimidazoline; ETU 2-imidazolidinethione; imidazoline-2-thiol; Akrochem ETU-22; NA-22; Robac 22; Sanceller 22; Vulkacit NPV/C,. Electroplating baths; intermediate for antioxidants, insecticides, fungicides, synthetic resins, vulcanization accelerators, dyes. mp = 203-204°; soluble in H_2O, methane, ethanol, ethylene glycol, pyridine, insoluble in acetone, ether, chloroform, benzene; LD_{50} (rat orl) = 1832 mg/kg. *Faesy & Besthoff; Ore & Chem. Corp.*

11133 ethylenediamine
107-15-3 3841 203-468-6
$C_2H_8N_2$
1,2-Ethanediamine
Ethylene diamine; 1,2-diaminoethane. Fungicide, manufacture of chelating agents, dimethylolethylene-urea resins, chemical intermediate, solvent, emulsifier, textile lubricants, antifreeze inhibitor. LD_{50} (rat orl) = 1.16 g/kg. *Aldrich; Allchem Industries; BASF; Texaco; Union Carbide.*

11134 Ethylflo 162
Poly-α-olefin
Synthetic lubricant for automotive crankcase oils, hydraulic fluids, gear and transmission fluids, compressor lubricants, metalworking lubricants, and personal care items. *Ethyl Corp.*

11135 Ethylflo 180
Poly-α-olefin
Synthetic lubricant for automotive crankcase oils, hydraulic fluids, gear and transmission fluids, compressor lubricants, metalworking lubricants, and personal care items. *Ethyl Corp.*

11136 2-ethylhexanol
104-76-7 3854 203-234-3
$C_8H_{18}O$
2-ethyl-1-hexanol
2-EH; 2-Ethylhexyl alcohol; Octyl alcohol; Alcohol C_8; Surfynol® 104A. Plasticizer for PVC resins; defoaming agent, wetting agent, organic synthesis, solvent mix for nitrocellulose; penetrant for plasticizing inks, etc. d = 0.83; bp = 183.5°; fp = 81.1°; ref index = 1.4300; miscible withmost organic solvents; slightly soluble in H_2O; LD_{50} (rat orl) = 12.46 ml/kg; moderately toxic by ingestion. *Aristech; Ashland; BASF; BP Chem. Ltd; Eastman; Shell.*

11137 2-ethylhexoic acid
149-57-5 205-743-6
$C_8H_{16}O_2$
butylethylacetic acid
2-ethylhexoic acid; 2-ethylhexanoic acid; Butyl ethyl acetic acid; Ethylhexanoic acid; α-ethylcaproic acid; hexonic acid, 2-ethyl-; 2-Ethylcapronic acid. Paint and varnish drying agents (metallic salts); esters as plasticizers. mp = -59°; bp = 228°; d = 0.9030; n_D^{20} = 1.4250; soluble in H_2O (2 g/l), organic solvents; LD_{50} (rat orl) = 3 g/kg. *Aldrich; Ashland; BASF; Eastman; Neste UK; Union Carbide.* Name unverified.

11138 2-ethylhexyl acetate
103-09-3 6860 203-079-1
$C_{10}H_{20}O_2$
octyl acetate
Ethyl Hexyl Acetate; Octyl acetate; Ethyl(2)-Hexyl Acetate. Solvent for nitrocellulose, resins, lacquers, baking finishes. bp = 197°; d = 0.8700; n_D^{20} = 1.4190-1.4210; LD_{50} (rat orl) = 3 g/kg. *Eastman; Hüls AG; MTM Speciality Chem. Ltd; Penta Mfg.*

11139 2-ethyl-4-methyl imidazole
931-36-2 213-234-5
$C_6H_{10}N_2$
Ethylmethyl imidazole. Curing epoxy resin systems. mp = 36-42°; bp = 292-295°; d = 0.9750; n_D^{20} = 1.4995; soluble in H_2O (180 g/l), organic solvents. *Schweizerhall.*

11140 N-ethyl morpholine
100-74-3 202-885-0
$C_6H_{13}NO$
4-ethylmorpholine
HEM; ethylmorpholine. Amine-based catalyst for flexible slabstock PU foam. mp = -63°; bp = 139°; d = 0.9050; n_D^{20} = 1.4410; soluble in H_2O, organic solvents; LD_{50} (rat orl) = 1780 mg/kg.

11141 4,4'-(2-Ethyl-2-nitrotrimethylene)-dimorpholine
1854-23-5 217-450-0
$C_{13}H_{25}N_3O_4$
Ethyl-(2-nitrotrimethylene)dimorpholine
Morpholine, 4,4'-(2-ethyl-2-nitro-1,3-propanediyl)bis-; Bioban P 1487. Bioban P 1487 is a preservative composed of 4,4-(2-Ethyl-2-nitro-trimethylene) dimorpholine (30%) and 4-(2-Nitrobutyl)morpholine (70%).

11142 ethylol
64-17-5 3806 200-578-6
C_2H_6O
ethyl alcohol
Essentially ethyl alcohol containing approximately 6% isopropyl alcohol; denatured with 3% butyl alcohol and 1% lead-free petrol; solvent in paint and lacquers, printing inks, foundries, dyes, industrial detergents, explosives, polishes, degreasers, rust romovers, manufacture of xanthates and esters. *Sasolchem.* Name unverified.

11143 Ethylol Colored
Mineralized methylated spirits; solvent. *Sasolchem.*

11144 ethylthiurad
97-77-8 3428 202-607-8
Tetraethylthiuram disulfide
Sulfur-bearing accelerator and vulcanizing agent; cure modifier for neoprene. *Monsanto Co.* Name unverified.

11145 ethyltoluene sulfonamide
80-39-7 201-275-1
$C_9H_{13}NO_2S$
N-ethyl-*p*-toluenesulfonamide
ethyl toluenesulfonamide; Unipex 108. Mixture of isomers of aromatic amides; plasticizer for shellac, cellulose acetate, and protein materials; used in NC lacquers, cellulose acetate compositions, PVAc emulsion adhesives, synthetic polyamides; food packaging applications. *ICI Spec.; Rit-Chem.*

11146 ethyltriacetoxysilane
17689-77-9 241-677-4
CE6345; Dynasylan® ETAC. Coupling agent, chemical intermediate, blocking agent, release agent, lubricant, primer, reducing agent. d = 1.14; bp = 107°; n_D^{20} = 1.412; fp = 8°. *Hüls Am.*

11147 ethyltrichlorosilane
115-21-9 204-072-6
$C_2H_5SiCl_3$
trichloroethylsilane
CE6350. Intermediate for silicones. *Hüls Am.; PCR.*

11148 Etingal® A
Phosphoric acid ester
Antifoam for papermaking. *BASF AG.*

11149 Etingal® L
Ethyl ether derivitive of fatty acid; foam controller for papermaking *BASF AG.*

11150 Etingal® S
Phosphoric acid ester
Foam breaker for papermaking, leather, furs. *BASF AG.*

11151 Etocas
Emulsifiers and solubilizers for oils, solvents and waxes; used for cosmetics; metalworking fluids; textiles; insecticides, herbicides; household products. *Croda Chem. Ltd.*

11152 Etophen 102
9016-45-9 6772
Nonoxynol-2
Detergent, dispersant, emulsifier, wetting agent for household and industrial detergents, textiles, paper, leather, and ceramic industries. *Zschimmer & Schwarz.* Discontinued.

11153 Etophen 114
9016-45-9 6772
Nonoxynol-14
Washing and cleansing agent. *Zschimmer & Schwarz.* Discontinued.

11154 E-Toplex
59-02-9 10159 200-412-2
Vitamin E
Vitamin E supplement. *USV Pharmaceutical Corp.* Unverified.

11155 etrimfos
38260-54-7 3936 253-855-9
$C_{10}H_{17}N_2O_4PS$
O-(6-ethoxy-2-ethyl-4-pyrimidinyl) O,O-dimethyl phosphorothioate
Ekamet; Satisfar; Dimethyl-O-(2-ethyl-4-ethoxy-pyrimidinyl-6)-thionophosphate; Ekamet G; Ekamet ULV; Etrimphos;. Organophosphorus insecticide with stomach and contact action. mp = -3°; d = 1.195; n_D^{20} = 1.5068; soluble in H_2O (40 mg/l), more soluble in organic solvents; LD_{50} (rat orl) = 1800 mg/kg.

11156 Etrofolan®
2631-40-5 220-114-6
$C_{11}H_{15}NO_2$
2-Isopropyl-phenyl-N-methylcarbamate
isoprocarb; Etrofolan; MIPC; 2-(1-methylethyl)phenyl methylcarbamate; Etrolan; Hytox; Mipcin; 2-(1-Methylethyl)-phenol methyl-carbamate; N-Methyl-2-isopropylphenylcarbam; Methylcarbamic acid, *o*-cumenyl ester; Cumenyl N-methylcarbamate; KHE 0145;. Fast-acting insecticide effective against leafhoppers, aphids, codling moths, capsids and bugs in rice, cocoa, vegetables, cereals, hops, coffee, potatoes, sugarcane, deciduous fruits and other crops. mp = 88-93°; insoluble in H_2O, soluble in organic solvents; LD_{50} (rat orl) = 485 mg/kg. *Bayer AG.*

11157 Etronite
The trade name for a synthetic resin-paper product; used for electrical insulation. No manufacturer.

11158 ETU
96-45-7 3849 202-506-9
2-imidazolidinethione
Accelerator in synthetic rubber production.

11159 Eubeco
Vitamin B complex for injection. *Allen & Hanburys Ltd.* Name unverified.

11160 Eucopine
A pine disinfectant containing substituted phenols as germicides. *William Pearson Ltd.* Name unverified.

11161 Euderm®
Finely dispersed acrylic resin; used in the leather industry for tightening the grain and as a base coat in pigment finishing. *Bayer AG.*

11162 Euka-drya
A Dutch cellulose rayon.

11163 Eukanol®
Covering pigment dispersions with casein binder for finishing; used in leather industry. *Bayer AG; Bayer plc.*

11164 Eukesol® Binder S
Alcoholic polymer solution; binder and thickener for applying pigment finishes in leather and fur industry. *BASF AG.*

11165 Eukesolar® Dyes
Metal complex dyes for the leather and fur industry. *BASF AG.*

11166 Eukinase
A powder obtained by the desiccation of the pancreatic juice of swine, and contains enzymes, mainly trypsin.

11167 Eulan® 33
General purpose textile finishing agent, fast to washing and processing; for the permanent protection of wool, hair, feathers and brush filaments against moths, carpet beetles (*Anthrenus*) and black carpet beetles (*Attagenus*). *Bayer AG; Bayer plc.*

11168 Eulan® BLS
Mothproofing agent; used for protection of carpets and furnishing fabrics against moths and carpet beetles. *Bayer AG; Bayer plc.*

11169 Eulan® WA
Mothproofing agent; auxiliary for the protection of goods made from wool or wool blends against moths and carpet beetles (*Anthrenus*) where demands made on fastness to washing are not exacting, e.g., carpets and furnishing fabrics. *Bayer AG; Bayer plc.*

11170 Eulysin® WP
pH regulator for use when dyeing wool or nylon with acid and premetallized acid dyes. *BASF.*

11171 Eumulgin 05
9004-98-2
Oleth-5
Emulsifier, solubilizer for pesticides, perfumes, cosmetics, floor polishes, hair dressings, creams, lotions, mineral oil, terpenes; dispersant improving color acceptance of emulsion paints. *Henkel/Cospha; Henkel Canada; Henkel KGaA.*

11172 Eumulgin B2
68439-49-6 7737
Ceteareth-20
Emulsifier for ointments, creams, low viscosity emulsions, cosmetics, pharmaceuticals. *Henkel/Cospha; Henkel Canada; Henkel KGaA; Pulcra SA.*

11173 Eumulgin C4
61791-08-0
PEG-5 cocamide
PEG cocamide series. Foaming agent for detergents; emulsifier. *Henkel; Henkel KGaA.*

11174 Eumulgin EP .5.2L
68920-66-1
Ethoxylated oleyl/cetyl alcohol
Emulsifier component of pronounced cold behavior. *Henkel KGaA.*

11175 Eumulgin EP2
Cetoleth-2
Emulsifier for mineral oil, hydrocarbons, fats, metalworking oils, textile auxiliaries. *Henkel/Cospha; Henkel Canada; Henkel KGaA; Pulcra SA.*

11176 Eumulgin HRE 40
61788-85-0
PEG-40 hydrogenated castor oil; solubilizer and oil-water emulsifier for cosmetic and pharmaceutical preparations *Henkel KGaA.*

11177 Eumulgin L
PPG-2-ceteareth-9
Emulsifier, solubilizer for aqueous or hydroalcoholic media; for skin and hair care preparations *Henkel/Cospha; Henkel KGaA.*

11178 Eumulgin M8
9004-98-2
Oleth-10 and oleth-5; emulsifier, solubilizer for pesticides and cosmetics. *Henkel/Cospha; Henkel KGaA.*

11179 Eumulgin PA 10
61791-44-4 263-177-5
PEG-10 tallowamine
Surfactant. *Pulcra SA.*

11180 Eumulgin PA 12
61791-14-8
PEG-12 cocamine
Surfactant. *Pulcra SA.*

11181 Eumulgin PA 30
PEG-30 oleamine
Surfactant. *Pulcra SA.*

11182 Eumulgin PC 2
61791-08-0
PEG-2 cocamide
Detergent, foam stabilizer, solubilizer for liquid. detergent systems. *Pulcra SA.*

11183 Eumulgin PK 23
61791-29-5
PEG-23 cocoate
PEG cocoate series. Surfactant. *Pulcra SA.*

11184 Eumulgin PLT 4
9004-96-0
Ethoxylated oleic acid
PEG oleate series. Emulsifier for paraffinic waxes and compounds, mineral oils; dyeing assistant, lubricant, antistat, emulsifier for textile industry. *Pulcra SA.*

11185 Eumulgin PPG 40
PPG-40
Surfactant. *Pulcra SA.*

11186 Eumulgin PRT 36
61791-12-6
PEG-36 castor oil; detergent, emulsifier, dispersant, solubilizer for concentrated pesticides, metal, leather, cosmetics, toiletries, pharmaceuticals, textile, and polymer industries. *Pulcra SA.*

11187 Eumulgin PST 5
9004-99-3
PEG-5.2 stearate
Surfactant. *Pulcra SA.*

11188 Eumulgin PWM2
9004-98-2
Oleth-2
Water-oil emulsifier, dispersant, lipophilic cosolvent.; fragrance grade; broad pH and electrolyte tolerance. *Pulcra SA.*

11189 Eumulgin RO 40
61791-12-6
PEG-40 castor oil; oil-water emulsifier, solubilizer for perfume oils; for personal care creams and lotions. *Henkel/Cospha; Henkel Canada; Henkel KGaA.*

11190 Eumulgin SML 20
9005-64-5 8872
Polysorbate 20
Solubilizer and emulsifier for cosmetics and pharmaceuticals. *Henkel; Henkel KGaA; Pulcra SA.*

11191 Eumulgin SMO 20
9005-65-6 7742
Polysorbate 80
Solubilizer and emulsifier for cosmetics and pharmaceuticals. *Henkel KGaA.*

11192 Eumulgin SMS 20
9005-67-8 8872
Polysorbate 60
Solubilizer and emulsifier for cosmetics and pharmaceuticals. *Henkel KGaA.*

11193 Eumulgin WM5
9004-98-2
Oleth-5
Raw material for foam-controlled laundry detergents. *Pulcra SA.*

11194 Eunatrol
143-19-1 6965 205-591-0
$C_{18}H_{33}NaO_2$
sodium oleate
Eunatrol. Pure sodium oleate; used for ore flotation, waterproofing textiles and emulsification of oil/water systems. soluble in H_2O (10 g/100 ml), less soluble in EtOH, organic solvents.

11195 euosmite
A resin found in the lignite in Bavaria.

11196 **Eupad**
An antiseptic consisting of bleaching powder and boric acid in equal parts. The solution prepared from Eupad is known as Eusol.

11197 **Euparen®**
1085-98-9 3095 214-118-7
Dichlofluanid
Fungicide with specific action against *Botrylis*. *Bayer AG*. No manufacturer.

11198 **Euparen® M**
731-27-1 211-986-9
$C_{10}H_{13}Cl_2FN_2O_2S_2$
1,1-dichloro-N-((dimethylamino)sulfonyl)-1-fluoro-N-(4-methylphenyl)methanesulfonamide
Tolylfluanid; BAY 49854; BAY 5712a; Dichlofluanid-methyl. Broad spectrum fungicide; effective against *Botrytis*. *Bayer AG*. Name unverified.

11199 **Euperlan®**
Mixture of fatty alcohol ether sulfates with special additives; for production of pearly sheen shampoos. *Henkel Chemicals Ltd.* Name unverified.

11200 **Euperlan® MPK 850**
Sodium laureth sulfate, magnesium laureth sulfate, sodium laureth-8 sulfate, magnesium laureth-8 sulfate, sodium oleth sulfate, magnesium oleth sulfate, glycol stearate, PEG-3 distearate, cocamide MEA, laureth-10; fatty alcohol ether sulfates with pearly gloss concentrate for baby shampoos and bath products. *Henkel/Cospha; Henkel Canada.*

11201 **Euperlan® PK 3000**
Glycol distearate, laureth-4, cocamidopropyl betaine; surfactant for pearlescent shampoos, bubble bath creams; brilliant, dense pearly gloss. *Henkel/Cospha; Henkel Canada.*

11202 **Euperlan® PK 810**
Glycol distearate, sodium laureth sulfate, cocamide MEA, laureth-9; pearlescent base for lotion shampoos and bath products *Henkel/Cospha; Henkel Canada.*

11203 **Euperlan® PK 900**
PEG-3 distearate
sodium laureth sulfate. Pearlescent base for lotion shampoos, bath products *Henkel/Cospha; Henkel Canada.*

11204 **Eupolen®**
Formulations of organic and inorganic pigments with low molecular weight polyethylene; used for mass dyeing of polyolefins and other polymers. *BASF AG.*

11205 **Eurecryl**
Cyanoacrylates
Rapid curing, solvent free adhesives. *Schering Industrial Chemicals Ltd.* Discontinued.

11206 **Euredur**
Epoxy resin curing agents. *Schering Industrial Chemicals Ltd.* Discontinued.

11207 **Eureka 102**
8002-33-3 232-306-7
Sulfated castor oil
Actrasol C-50, C-75, C-85; Chemax SCO, SCO/75; Eureka 102; Laurel R-50; Marvanol® SCO (50%); Nopcocastor; Nopcosulf CA-60, -70, CA; Standapol® SCO. Emulsifier, detergent; grinding aid in pigment dispersement.; plasticizer in finish coatings; topping oil for suede leather. *Atlas Refinery.*

11208 **Eureka 392**
61790-35-0 263-127-2
Actrasol SS; Eureka 392. Sulfated tall oil fatty acid; emulsifier, carrier for refined oils, base for solvent fat liquiduor systems; for detergent formulations; penetrant, fiber lubricant for textile and leather processing. *Atlas Refinery.*

11209 **Eureka 400-R**
Sulfonated fish oil; emulsifier, lubricant, softener, fat liquiduor used in leather processing, fibers. *Atlas Refinery.*

11210 **Eureka 800**
Sulfonated natural oil, fatliquiduor for leathers; nonyellowing softener. *Atlas Refinery.*

11211 **Eureka 800-R**
Sulfonated animal oil; fiber lubricant; fat liquiduor for leather tanning. *Atlas Refinery.*

11212 **Eureka 1014-M**
Sulfated neatsfoot oil, sulfated synthetic sperm oil; fat liquiduor for finer upper leathers; imparts whiteness, softness and fullness. *Atlas Refinery.*

11213 **Eureka 1067-A**
Fish oil, sulfated fish oil, hydrocarbon oils; economical fat liquiduor for split leather. *Atlas Refinery.*

11214 **Eureka Alloy**
A trademark for a copper-nickel alloy used for electrical resistance wires. No manufacturer.

11215 **Eureka E-2**
Fatty acid diamine condensate derivitive; emulsifier for raw oils, fiber lubricant, fabric softener. *Atlas Refinery.*

11216 **Eurelon**
Thermoplastic polyamides
Binders for flexo and rotogravure inks, overprint varnishes and thixotropic agents for alkyd resins. *Sherex Chemical Co Inc.* Discontinued.

11217 **Euremelt**
Polyamides and ethylene vinyl acetate hot melt adhesives. *Schering Industrial Chemicals Ltd.* Discontinued.

11218 **Eurepox**
Epoxy resins. Raw material base for paints, surface coatings, resin mortars, two component adhesives, casting resins and laminates. *Schering Industrial Chemicals Ltd.* Discontinued.

11219 **Euresyst**
Epoxy resin systems; binders for printing inks and injection resins for the building industry. *Schering Industrial Chemical Ltd.* Discontinued.

11220 **Euretek®**
Vinyl plastisol adhesion promoter. *Sherex/Div of Witco.*

11221 **Euretek® 540**
Adhesion promoter for PVC plastisol; for noncritical, conventional bake applications. *Schering Berlin Polymers.* Discontinued.

11222 **European Elastin 10**
73049-73-7
Hydrolyzed elastin
Substantive protein adding to skin elasticity; protective colloid effect; film-former; moisturizer; for skin nourishing creams, shave creams, sun tan products, hair care and treatment products. *Maybrook.*

11223 **Europolymer**
Thermoplastic polyurethanes for injection molding, and as adhesives and coatings. *ICI Polyurethanes.* Name unverified.

11224 **Europrene**
A registered trademark for general purpose butadiene styrene copolymers coded as follows: CIS 1-4 cis-polybutadiene; SS high styrene copolymers; N butadiene acrylonitrile copolymers; AR acrylic; SOL butadiene styrene. *EniChem Elastomers Ltd.*

11225 **Europrene AR**
Polyacrylic rubbers. *EniChem Elastomers Ltd.*

11226 **Europrene CIS**
Polybutadiene rubbers. *EniChem Elastomers Ltd.*

11227 **Europrene Lattice**
Synthetic latexes. *EniChem Elastomers Ltd.*

11228 **Europrene N**
Nitrile rubbers. *EniChem Elastomers Ltd.*

11229 **Europrene NEOCIS**
Polybutadiene rubbers. *EniChem Elastomers Ltd.*

11230 **Europrene SOL**
Solution butadiene styrene rubbers. *EniChem Elastomers Ltd.*

11231 **Europrene SOL T**
Thermoplastic rubbers. *EniChem Elastomers Ltd.*

11232 **Eurotex**
Polyaminoamide resins, used as adhesion promoters for PVC plastisols. *Schering Industrial Chemicals Ltd.* Discontinued.

11233 **Eurylon**
Maize starch, high in amylose; used for food, textiles, board and paper industries. *Roquette (UK) Ltd.*

11234 **Eusapon®**
Wetting agents for leather and fur processing. *BASF AG.*

11235 **Eusin®**
Brilliant, transparent pigment dispersions based on casein binders for aniline and aniline effect finishing; used in the leather industry. *Bayer AG; Bayer plc.*

11236 **Eusol**
hypochlorous acid-calcium biborate-calcium chloride.
A solution containing 0.54% hypochlorous acid, 1.28% calcium biborate, and 0.17% calcium chloride. It is made by shaking Eupad in water, and filtering.

11237 **Eusolvan**
A proprietary solvent stated to consist of ethyl lactate. No manufacturer.

11238 **Eutannin**
A mixture of gallic acid and milk sugar. An intestinal astringent.

11239 **Eutanol G**
5333-42-6 226-242-9
$C_{20}H_{42}O$
Octyldodecanol
2-Octyl-1-dodecanol; 2-octyldodecan-1-ol. Lubricant, emollient for cosmetics and pharmaceuticals; for oil-soluble active ingredients; pigment dispersant. *Henkel/Cospha; Henkel Canada.* Name unverified.

11240 Eutanol G16
36311-34-9 252-964-9
Isocetyl alcohol
Emollient for personal care products; carrier for oil-soluble active ingredients; dispersant for pigments. *Henkel/Cospha; Henkel Canada.*

11241 Euthylen®
Formulations of organic and inorganic pigments in PE with low molecular weight constituents; for coloring polyolefins. *BASF AG.*

11242 Euvinyl® C
Formulations of organic and inorganic pigments in a VC copolymer; for mass coloring of rigid PVC, plasticized PVC, PVC pastes and VC copolymers. *BASF AG.*

11243 Euviprint®
Formulations of organic and inorganic pigments in a VC copolymer; for pigmenting gravure inks and surface coatings for plasticized PVC and aluminum foil and other packaging materials. *BASF AG.*

11244 Euxyl
A range of preservatives used in cosmetics and toiletries. *Sterling-Winthrop Group Ltd.*

11245 EVA
Ethylene vinyl acetate. A flexible polythene-like polymer for molding and extrusion; adheres well to metals.

11246 Eval
Ethylene vinyl alcohol resins. *Quantum Chemical Corp.*

11247 Eval® E105
Ethylene vinyl alcohol copolymer; thermoplastic barrier resin used in packaging materials, gas fill packaging, packaging for oily foods, edible oils, mineral oils, agricultural pesticides, organic solvents; suitable for film extrusion, sheet coextrusion, blow molding and tube profile coextrusion, extrusion coating and laminating procedures; EP-E105 grade is used for sheets and films. *Eval Co. of Am.*

11248 Eval® E151
Ethylene vinyl alcohol copolymer resin; melt phase forming, solid phase pressure forming, blow molding, tubes. *Eval Co. of Am.*

11249 Eval® G115
Ethylene vinyl alcohol copolymer resin; cast film, speciality applications. *Eval Co. of Am.*

11250 Eval® K102
Ethylene vinyl alcohol copolymer resin; solid phase pressure forming applications. *Eval Co. of Am.*

11251 Evanacid® 3CS
99-68-3 202-778-9
Carboxymethyl mercaptosuccinic acid
Metal chelator; metal deactivator used to stabilize glyceride oils. *Evans Chemetics.*

11252 Evangard® 18MP
31778-15-1 250-801-6
Octadecyl 3-mercaptopropionate
Evans Chemetics.

11253 Evanol
Proprietary; cosmetic cream base for depilatories, hair relaxing creams, and curl kits. *Evans Chemetics.*

11254 Evans' cement
A metallic cement made by adding cadmium amalgam (74% mercury) to mercury.

11255 Evanstab® 12
123-28-4 204-614-1
Dilauryl thiodipropionate
Antioxidant for polyethylene, PP, and polyolefins, ABS; stabilizer for polyolefins, oils and fats, food application; plasticizer for rubber products; lubricating oil additive; synthetic lubricant; chemical preservative in fats and oils. *Evans Chemetics.*

11256 Evanstab® 13
10595-72-9 234-206-9
Ditridecyl thiodipropionate
Secondary antioxidant in ABS, polyolefins, and other polymer systems. *Evans Chemetics.*

11257 Evanstab® 14
16545-54-3 240-613-2
Dimyristyl thiodipropionate
Secondary antioxidant for polyolefins. *Evans Chemetics.*

11258 Evanstab® 18
693-36-7 211-750-5
$C_{42}H_{82}O_4S$
Distearyl thiodipropionate
Dioctadecyl 3,3'-thiodipropionate. Secondary antioxidant for use in polyolefins; used in food-pkg. materials and edible fats and oils. *Evans Chemetics.*

11259 Evasperse
Ethylene vinyl acetate dispersions. *Runnymede Dispersions Ltd.*

11260 Evatane
Ethylene-vinyl acetate copolymers; used for hot melt adhesives. *Elf Atochem UK Ltd; Elf Atochem Inc.*

11261 Event
Adjuvant containing 90% alkyl and aryl ether phosphates; wetting agent for pesticide and liquid fertilizer mixtures. *Ideal Manufacturing Ltd.*

11262 Everbrite
A proprietary trade name for an alloy of 60% copper, 30% nickel, 3% iron, 3% silicon, and 3% chromium. No manufacturer.

11263 Eveready Prestone
The trademarks applied to ethylene glycol antifreeze. No manufacturer.

11264 Everflex® 515L
Vinyl emulsion in anionic/nonionic emulsifier system; pigment binder for use in surface coatings, e.g., interior flat wall paints and primer sealers. *W R Grace/Organics.*

11265 Everflex® E
Vinyl acrylic emulsion; general purpose paint/coatings emulsion for interior and exterior paints with well-balanced properties; latex, interior flat, primer sealers, exterior masonry. *W R Grace/Organics.* Name unverified.

11266 Everflex® SP-1084
9003-20-7
Polyvinyl acetate; Catomer VA; Daratak® RP2000, SP1011; Liquid Latex; Quabond® 210; Rovace® 571, 2113; Ucar® Latex 130; Vinac® 1000 DEV, XX-210, XX-220, XX-230, XX-240;. Vinyl acetate copolymer emulsion; high molecular weight emulsion used as concrete additive to improve adhesion and compressive strength. *W R Grace/Organics.*

11267 Evergreen
Lawn fertilizer combined with selective weedkiller. *Fisons plc, Horticultural Div.* Name unverified.

11268 Everitt's salt
$K_2FeFe(CN)_6$
Potassium ferrous ferro-cyanide

11269 Everlastic
A proprietary trade name for a textile incorporating Duprene. No manufacturer.

11270 Everlube
Family of oven cured dry film lubricant coatings containing solid lubricants and suitable binder system; used for fasteners, slides, pins, clips and numerous applications requiring dry lubrication and corrosion protection particularly under high loads. *E/M Corporation.*

11271 Everseal
A bituminastic liquid applied as a corrosion-resisting material.

11272 Eversoft Plastex
A low-freezing explosive containing a nitrated mixture of glycerin and ethylene glycol, ammonium nitrate, sodium chloride, wood meal, trinitrotoluene, and nitro-cotton.

11273 Eversoft Sea Mex
A low-freezing explosive containing a nitrated mixture of glycerin and ethylene glycol, ammonium nitrate, sodium chloride, and wheat flour.

11274 Eversoft Tees Powder
A low-freezing explosive consisting of a nitrated mixture of glycerin and ethylene glycol, ammonium nitrate, wood meal, and sodium chloride.

11275 Eversun
Sun protection products, to protect skin while promoting a tan. *Richardson-Vicks Inc.* Name unverified.

11276 Evolve
Hydrocarbon or oxyhydrocarbon and aqueous-based solvents and equipment for industrial cleaning processes. *ICI Chem & Polymers Ltd.*

11277 Evo-stik 873 Super
A proprietary synthetic rubber latex adhesive having a high solids content. *Evode Plastics Ltd.*

11278 Ewer-Pick acid
525-37-1 6460
$C_{10}H_8O_6S_2$
1,6-Naphthalene-disulfonic acid
Chemical intermediate. Soluble in H_2O.

11279 Ewo
7727-43-7 1023 231-784-4
barium sulfate
Micronized white barytes. Pigment and filler. Also used in medicine as a contrast medium. *Sachtleben Chemie GmbH.*

11280 EW-POL 8021
Aryl polyglycol ether

Surface-active plasticizer, thickener for PVAc adhesive dispersions. *Henkel/Functional Prods.*

11281 Exact-S®
75-18-3 6204 200-846-2
Dimethyl sulfide
Coking suppressor for ethylene production and for steel mill furnace walls; odorant for natural gas; presulfiding agent for catalysts in refinery processes. *Gaylord Chem.*

11282 Exaltone
541-91-3 6392 208-795-8
muscone
A trade name for muscone (cyclopentadecanone), the perfuming principle of natural musk. No manufacturer.

11283 Exam
Exem. Anionic, nonionic and special anionic and nonionic blends of emulsifier; emulsifier for agricultural pesticides. *Makhteshim Chemical Works Ltd.* Discontinued.

11284 Exameen 824 3724
139-08-2 205-352-0
Myristalkonium chloride
BTC® 824, 2565; Catigene® DC 100; FMB 65-15 Quat, 65-28 Quat; JAQ Powdered Quat. Germicidal quaternary for hard surface disinfection and sanitization; algaecide and slimicide for swimming pool and industrial water treatment. *Clough.*

11285 Exameen 2125 M 3704
Myristalkonium chloridequaternium-14
Germicidal quaternary for bacteriological control in disinfectant and sanitizer formulations for hospitals, nursing homes, public institutions, and industry. *Clough.*

11286 Exameen 3500 3714
1086
Benzalkonium chloride
Germicidal quaternary for hard surface disinfection and sanitization; algicide and slimicide in swimming pool and industrial water treatment. *Clough.*

11287 EXC-33
Fluorochemical copolymer; release agent for rubber, plastic, and metal industries. *Releasomers.*

11288 Excelite
An American trade name for a thermosetting fibrous plastic. No manufacturer.

11289 Excellerex
A proprietary rubber vulcanization accelerator; it is an aniline derivative. No manufacturer.

11290 Excello
An electrical resistance alloy containing 85% nickel, 14% chromium, 0.5% iron, and 0.5% manganese; also the name for a carbon black used in rubber mixings.

11291 Excelo
A proprietary trade name for a hot die steel containing 2.5% tungsten, 1.5% chromium, 0.35% vanadium, and 0.55% carbon. No manufacturer.

11292 Excelon
A proprietary trade name for acrylic denture material. No manufacturer.

11293 Excelsior
A proprietary trade name for carbon black. No manufacturer.

11294 Exchem GO-1
27247-96-7 248-363-6
$C_8H_{17}NO_2$
2-Ethylhexyl nitrate
Cetane improver. *Exchem Organics.*

11295 Exell
61791-44-4 263-177-5
Wetting agent containing 64% polythoxylated tallow amine; spreader for use with phosphonoglycine and contact herbicides. *Truchem Ltd.*

11296 Exgraphite
Expandable acid treated graphite. *Foseco (F.S.) Ltd.* Discontinued.

11297 Exkin
Oxime antiskinning agents; for prevention of skin formation in solvent-based coating compositions. *Hüls Am.*

11298 Exl-die Steel
A proprietary trade name for a non-shrinking steel containing 1.15% manganese, 0.5% chromium, 0.5% tungsten, and 0.9% carbon. No manufacturer.

11299 Exobloc BF-1000
Modified closed-cell silicone foam; fireresistant, nontoxic foam for critical electrical component insulation, vibration damping/gap filler; HVAC line insulation; high and low temperature gasketing. *Bisco Prods.*

11300 Exocerol® OM
Blowing system based on citric acid and azodicarbonamide; chemical blowing agent for foaming of thermoplastics in injection molding. *Boehringer Ingelheim; Henley.*

11301 Exolit® IFR-23
Ammonium polyphosphate and synergists; halogen-free flame retardant for PE, PP, EVA, elastomers. *Hoechst Celanese/Spec. Chem.*

11302 Exolon
409-21-2 8636 206-991-8
silicon carbide
A trademark for abrasive articles consisting essentially of silicon carbide. No manufacturer.

11303 Exoryl
Water based acrylic adhesives. *Harcros.*

11304 EXP-28
Silicone fluid; paint and protective coatings additive improving performance properties. *Genesee Polymers.*

11305 EXP-49
78-62-6 201-127-6
$C_6H_{16}O_2Si$
Dimethyldiethoxy silane
CD5600. Intermediate for blocking hydroxyl and amino groups in organic synthesis reactions; also for preparing hydrophobic and release materials and enhancing flow of powders. bp = 114°; d = 0.8650; n_D^{20}= 1.3811; LD_{50} (rat orl) = 9280 mg/kg. *Genesee Polymers.*

11306 EXP-51
1825-62-3 217-370-6
$C_5H_{14}OSi$
Trimethylethoxy silane
CT2970. Intermediate useful for blocking hydroxyl or amino groups in order to perform reactions on multifunctional organic compounds or polymers; also for deactivating glass surfaces used in gas chromatographic applications. *Genesee Polymers.*

11307 EXP-58
Silicone wax copolymer; lubricant, mold release agent and internal lubricant for plastics. *Genesee Polymers.*

11308 Expancel® 091
Wet, unexpanded microspheres used as blowing agents. *Expancel.*

11309 Expancel® 551 DE
Dry, expanded microspheres used as resilient lightweight fillers. *Expancel.*

11310 expanded graphite
A substance prepared by covering flake graphitic with an oxidizing agent, to produce a film of graphitic acid, then heating strongly to cause the particles to become distended.

11311 Expandex® 175
5-Phenyltetrazole, barium salt; high temperature chemical blowing agent for engineering resins. *Uniroyal.*

11312 Expandex® 5PT
5- Phenyltetrazole, a blowing agent for foaming plastics and elastomers at elevated temperatures. *Uniroyal.*

11313 expanding solder
An alloy of 37.5% lead, 6.75% bismuth, and 56.25% antimony. It expands on cooling, and is used for fixing metal into holes.

11314 Explosive D
131-74-8 581 205-038-3
$C_6H_6N_4O_7$
2,4,6-trinitrophenol ammonium salt
ammonium carbazoate; ammonium picronitrate; ammonium picrate. Used in explsoives, fireworks and rocket propellants. d = 1.72; soluble in H_2O (1 g/00 ml), less soluble in organic solvents.

11315 explosive gum
55-63-0 6704 200-240-8
nitroglycerin
Nitroglycerin gelatinized with collodion cotton. It contains 96% of the former, and 4% of the latter compound.

11316 Explotab®
9063-38-1
Sodium starch glycolate. Sodium starch glycolate NF/BP; tablet disintegrant for formulations prepared by direct compression or wet granulation techniques. *Penwest Pharmaceuticals Co.*

11317 Exprol
Photographic developer. *May & Baker Ltd.* Name unverified.

11318 Exsyproteines 2%
91080-18-1 293-509-4
Hydrolyzed animal elastin; cosmetic ingredient for moisturizers, oily skin treatments, anti-aging creams. *Exsymol.*

11319 Extend®
For the medical industry. *DuPont UK.*

11320 Extendopel
Hydrophobic complex; fluorocarbon extender with no effect on oil repellency; improves water repellency. *Sybron.*

11321 Extendospheres® Metalite® Zinc.
Zinc clad hollow microspheres used as filler to build high-mil coatings at less weight, and in zinc-rich coatings for structural steel, marine, automobile underbodies, tank linings, metal coils, etc. *PQ Corp.*

11322 Extendospheres® SG
Hollow microspheres used as low specific gravity extender in resin matrices and concrete applications to reduce weight, add thermal insulation and increase impact strength. *PQ Corp.*

11323 Extiat®
9015-54-7 310-296-6
Hydrolyzed collagen; Collamino 25; Collagen Hydrolyzate Cosmetic 55; Cropepsol; Cropeptone; Crotein A,C,O; Crotein CAA; Crotein CAA/CF, O, SPA; Peptein® 2000®; Polypeptide 10, 37; Protogest; Travamin; Hydrocoll AL-50, AL-55, EN-40, EN-55, EN-55X, EN-SD; Lexein® X-250HP; Nutrilan® FPK, H, M, I-50, L; Parenamine. Hydrolyzed animal protein; for food industry. *Asahi Chem. Industry.*

11324 Extir
A proprietary range of expandable polystyrene beads. *Montedison UK Ltd.* Name unverified.

11325 Extol
A proprietary trade name for a sulfated compound with solvents used as a detergent. No manufacturer.

11326 Extra Bond
Polyvinyl acetate liquid; used for wood and general builders glue. *Wessex Resins & Adhesives Ltd.*

11327 extra white metal
An alloy of 50% copper, 30% nickel, and 20% zinc. It is a nickel silver (German silver).

11328 Extrakt 52
Surfactant blend; for bath additives, hair shampoos, liquid. body cleaners. *Zschimmer & Schwarz.* Discontinued.

11329 Extrox
1314-13-2 10279 215-222-5
zinc oxide
Coated zinc oxide. A pigment. *Harcros.*

11330 Extrudoil®
Lubricants which reduce friction between workpiece and tool or die in extruding or cold forming operations. *Witco/Allied-Kelite.* Discontinued.

11331 Extrusil
10101-39-0 233-250-6
Calcium silicate
Cecasil; Keical-Ace; Micro-Cel® A, T-38; Microcal; Microcal ET; Paratemp; Silasorb; Silene. Extender for rubber compounds. *Degussa.*

11332 Extrusion-Plus
Nonchlorinated and nonsulfonated oil; amber colored liquid; extensively used by automotive fastener and aerospace fastener manufacturers. *Rustlan Chemical Co.* Unverified.

11333 Exxal®
Industrial alcohols. Used as solvents and intermediates. *Exxon UK.*

11334 Exxal® 6
111-27-3 4732 203-852-3
Hexyl alcohol.
Solvent and chemical intermediate. *Exxon.* Name unverified.

11335 Exxal® 7
70914-20-4
Isoheptyl alcohol.
Solvent and chemical intermediate. *Exxon.* Name unverified.

11336 Exxal® 8
26952-21-6 5211 248-133-5
Isooctyl alcohol.
Solvent and chemical intermediate. *Exxon.* Name unverified.

11337 Exxal® 9
68526-84-1 271-233-5
Isononyl alcohol
Solvent and chemical intermediate. *Exxon.* Name unverified.

11338 Exxal® 10
112-30-1 2911 203-956-9
$C_{10}H_{22}O$
Decyl alcohol
1-Decanol; Decyl alcohol; n-Decyl Alcohol; decanol; Decan-1-ol; Capric alcohol; Alcohol C-10; Nonyl acarbinol. Solvent. *Exxon.*

11339 Exxal® 12
112-53-8 3464 203-982-0
Dodecyl alcohol
Exxon. Name unverified.

11340 Exxal® 13
68526-86-3 271-235-6
Tridecyl alcohol
Solvent and chemical intermediate. *Exxon.* Name unverified.

11341 Exxal® 16
36653-82-4 2070 253-149-0
Hexadecanol
Cetyl alcohol; Isocetyl alcohol; Ceraphyl ICA; Eutanol G16; Michel XO-150-16. Solvent and chemical intermediate. *Exxon.* Name unverified.

11342 Exxal® 18
Octyldecanol
Solvent and chemical intermediate. *Exxon.* Name unverified.

11343 Exxal® 20
5333-42-6 226-242-9
Octyldodecanol
Solvent and chemical intermediate. *Exxon.* Name unverified.

11344 Exxal® 26
70693-05-9
Undecylpentadecanol
Undecyl pentadecanol. Solvent and chemical intermediate. *Exxon.* Name unverified.

11345 Exxal® L1315
67762-41-8 267-019-6
Linear C_{13}-C_{15} alcohol
Solvent and chemical intermediate. *Exxon.* Name unverified.

11346 Exxate®
Acetates of higher oxo-alcohols. solvents for high solids paint formulations. *Exxon Int'l.; Exxon UK.*

11347 Exxate® 600
Oxo-hexyl acetate
Solvent and chemical intermediate. *Exxon.* Name unverified.

11348 Exxate® 800
108419-32-5
C_8 alkyl acetate
Solvent and chemical intermediate. *Exxon.* Name unverified.

11349 Exxate® 900
108419-33-6
C_9 alkyl acetate
Solvent and chemical intermediate. *Exxon.* Name unverified.

11350 Exxate® 1000
108419-34-7
C_{10} alkyl acetate
Solvent. *Exxon.* Name unverified.

11351 Exxate® 1300
108419-35-8
C_{13} alkyl acetate
Solvent. *Exxon.* Name unverified.

11352 Exxelor
Polymer modifiers. *Exxon UK.*

11353 Exxon® Bromo XP-50
Brominated copolymer of isobutylene and paramethylstyrene; elastomer with high damping properties, good environmental and aging resistance, ozone and uv stability; suggested for tire applications (sidewalls, treads, innerliners, carcasses, bladders), mechanical goods (gaskets, diaphragms), hoses, belting, adhesives. *Exxon.*

11354 Exxon® Butyl 065
Isobutylene-isoprene elastomer, 0.05-0.20% zinc dibutyl dithiocarbamate (stabilizer); used for flat belts, coated materials, rubber-covered rolls, hose, mats, molded and extruded products, o-rings, shock and vibration products, inner tubes, water barrier applications, electrical goods; products exhibit shock absorption, weather resistance, flex-, tear- and abrasion resistance, impermeability, chemical resistance, excellent electrical properties; vulcanized with sulfur, resin or GMF. *Exxon.* Name unverified.

11355 Exxon® Butyl 077
9010-85-9
Isobutylene-isoprene elastomer
Isobutylene isoprene copolymer; Butex; CryOfine® Butyl; Exxon® Butyl 077; Kalar® 5214, 5263; Kalene® 800; Polysar Butyl 100; Polysar XL 30102. For use in applications requiring FDA compliance. *Exxon.* Name unverified.

11356 Exxon® Butyl 268
Isobutylene-isoprene elastomer, 0.05-0.20% zinc dibutyl dithicarbamate (stabilizer); used for inner tubes and extrusions; products exhibit impermeability to gases, ozone weathering, heat, chemical, tear, and abrasion resistance, good electrical properties, low compression set;

excellent clor retention; vulcanized with sulfur, resin or GMF. *Exxon.* Name unverified.

11357 Exx-Print
Solvents for offset printing inks. *Exxon UK.*

11358 Exxsol®
High quality dearomatized aliphatic solvent. *Exxon Int'l.*

11359 Exxsol® D-40, D-60, D-80, D-110, D-130
64742-47-8 265-149-8
C isoparafin series; Isopar® M. Dearomatized aliphatic solvent. *Exxon.* Name unverified.

11360 Exxsol® Heptane
142-82-5 4694 205-563-8
C_7H_{16}
n-heptane
Heptane; solvent. *Exxon.*

11361 Exxsol® Hexane
110-54-3 4729 203-777-6
C_6H_{14}
n-hexane
Hexane; solvent. *Exxon.*

11362 Exxsol® Isopentane
78-78-4 201-142-8
C_5H_{12}
isopentane
Isopentane, dearomatized solvent. *Exxon.* No manufacturer.

11363 Exxtraflex
Polyolefin film. *Exxon Int'l.* Name unverified.

11364 Exzyme
Protein digesting enzyme powder. *PMP Fermentation Prods. Inc.*

11365 Eymid
Polyamides. *Ethyl SA.*

11366 Eymyd® Prepreg
Fluorine-containing thermoplastic polyimides prepregged onto unidirectional fibers or woven fabric; structural parts for aircraft, engines and rockets for use up to 700°F. *Ethyl Corp.*

11367 Eymyd® Resin L-20N
Polyamic acid solution in 1-methyl-2-pyrrolidinone; forms a thermoplastic fluorinated polyimide after solvent evaporation and curing; produces coatings with excellent thermal oxidative stability, adhesion, frictional wear, and electrical properties; applications include erosion barrier coating, corrosion-resistant coating, thermal protective coating, tool coating, adhesives and fiber sizes. *Ethyl Corp.* Name unverified.

11368 Eypel®
Polyphosphazene polymers. *Ethyl Corp.*

11369 Eypel® A Acoustic Barrier Sheet
Poly(aryloxyphosphazene) elastomer; fire-resistant acoustic mass sheeting. *Ethyl Corp.*

11370 Eypel® F
Polyfluoroalkoxyphosphazene compound with incorporated peroxide curative; fluoroelastomer for use where low temperature flexibility, flex fatigue resistance, and chemical resistance are required, e.g., o-ring, other sealing applications; broad temperautre use (65-350°F), excellent fatigue and solvent resistance. *Ethyl Corp.*

11371 E-Z Mix
Series of dry liquid concentrates; for rubber and vinyl. *C P Hall.* Name unverified.

11372 EZ Mold Lubricant
Glycol surfactant; mold release lubricant for natural and synthetic rubber compounds *TSE Industries.*

11373 EZA®
Zeolite A; detergent builder; solvent; anticaking agent for detergents and desiccants. *Ethyl Corp.*

11374 F.A.S.T. Lube System
Airless spray lubrication systems for the metalforming industry. *Franklin Oil Corporation (Ohio).*

11375 F.E.P
A proprietary name for Teflon 100. *DuPont UK.* Name unverified.

11376 F-238
31717-87-0 250-778-2
dodemorph
Emulsifiable concentrate of 400g dodemorph per liter; used for control of mildew in ornamental nursery stock. *BASF plc.*

11377 F-309
Iodophor concentrate. *Evans Vanodine International Ltd.*

11378 F-310
Gasoline additive. *Monsanto (Solaris).*

11379 F-500, -3600, etc
21645-51-2 355 244-492-7
Aluminum Hydroxide
Aluminum hydroxide compressed gels; antacid actives. *Reheis.*

11380 F-1000 Dried Gel
21645-51-2 355 244-492-7
H_3AlO_3
Aluminum Hydroxide
USP grade aluminum hydroxide; antacid powder. *Reheis.*

11381 F-1000®
21645-51-2 355 244-492-7
H_3AlO_3
Aluminum hydroxide
Antacid active with good resuspending props. *Reheis.*

11382 F-2000
21645-51-2 355 244-492-7
H_3AlO_3
Aluminum Hydroxide
Aluminum hydroxide compressed gel; antacid. *Reheis.*

11383 F-2000 Dried Gel
21645-51-2 355 244-492-7
H_3AlO_3
Aluminum Hydroxide
USP grade aluminum hydroxide; antacid powder. *Reheis.*

11384 F-2000 FR
Brominated epoxy; exhibits excellent performance in engineering plastics (PBT, PET, nylon), styrenics (ABS, PC), and thermosets (epoxy, phenolic, unsat. polyesters); high thermal stability and thermal aging. *AmeriHaas; Dead Sea Bromine.*

11385 F-2001
Brominated epoxy oligomer; flame retardant additive for thermosets (epoxy, phenolic, unsaturated polyesters); good uv stability. *AmeriHaas; Dead Sea Bromine.*

11386 F-2100 Dried Gel
21645-51-2 355 244-492-7
H_3AlO_3
Aluminum Hydroxide
USP grade aluminum hydroxide; antacid powder. *Reheis.*

11387 F-2200
Brominated epoxy resin; flame retardant resin for potting, wet lay-up, and pre-preg laminates, adhesives, molding compounds, coatings. *AmeriHaas; Dead Sea Bromine.*

11388 F-2200 Dried Gel
21645-51-2 355 244-492-7
H_3AlO_3
Aluminum Hydroxide
USP grade aluminum hydroxide; antacid powder. *Reheis.*

11389 F-2300
Medium molecular weight brominated epoxy oligomer; flame retardant additive for use in thermoplastics (PBT, PC/ABS, ABS, HIPS); high thermal and uv stability. *AmeriHaas; Dead Sea Bromine.*

11390 F-2300H
High molecular weight brominated epoxy polymer; flame retardant additive for use in thermoplastics (PBT, PET), thermoplastic elastomers, alloys (PC/ABS), etc.; high thermal and uv stability, good melt flow. *AmeriHaas; Dead Sea Bromine.*

11391 F-2400
High molecular weight brominated polymer; flame retardant for use in thermoplastics (PBT, PET, polyamides, PU), alloys (PC/ABS), styrenics (ABS, HIPS); high thermal stability and thermal aging, high uv stability. *AmeriHaas; Dead Sea Bromine.*

11392 F-2400E
High molecular weight brominated polymer; flame retardant for use in thermoplastics (PBT, PET, PU), alloys (PC/ABS), styrenics (ABS, HIPS), etc.; high thermal and uv stability, thermal aging. *AmeriHaas; Dead Sea Bromine.*

11393 FA-1
25928-94-3
Epoxy resin adhesive, gyro-grade; for bonding and sealing aluminum and other metals; mix ratio: 100 pbw adhesive/3.2 pbw of Activator BA-4. *Bacon.*

11394 FA-14
25928-94-3
Two-part modified epoxy compound; adhesive coil for impregnation of electronic components with high percentage of fine wires where complete penetration and freedom from voids is important; for casting and coating applications; able to penetrate cracks and crevices; bonding for components made from fused beryllium oxide when low surface tensile compounds are needed; suitable for gyro use in contact with poly(bromotrofluoroethylene) oil. *Bacon.*

11395 FA-8
25928-94-3
Epoxy resin adhesive, gyro-grade; for bonding and sealing beryllium and other metals; mix ratio: 100 pbw adhesive/13.5 pbw Activator BA-5. *Bacon.*

11396 Fabelnyl
Polyamide injection molding compounds. *Rhône-Poulenc UK.*

11397 Fabrene®
Fibers. *DuPont UK.*

11398 Fabrethane
A proprietary one-component foamable polyurethane, 100% solids. *W R Grace & Co.* Name unverified.

11399 Fabrex
Preformed insulating refractory shapes for various applications. *Foseco (F.S.) Ltd.*

11400 Fabrifil
A proprietary trade name for a macerated cotton fabric filler. No manufacturer.

11401 Fabriglide
Fabric lubricant. *Dow Corning.*

11402 Fabrikoid
A trademark for a fabric coated with pyroxylin. No manufacturer.

11403 Fabroil
A proprietary trade name for a synthetic resin. No manufacturer.

11404 Fabrolite
A synthetic resin of the phenol-formaldehyde type. *Associated Electrical Industries (GEC).* Unverified.

11405 Facet®
84087-01-4 402-780-1
Quinclorac
For post-emergence control of barnyard grass and some other weeds in rice. mp = 274°. *BASF AG.*

11406 Facteka
A proprietary trade name for a rubber substitute. No manufacturer.

11407 factice
A polymerization product of natural fatty oils and sulfur or sulfur chloride, a processing promoter for an economical processing of rubber; used for extruded articles. *Rhein-Chemie Rheinau.* Name unverified.

11408 factis
A term applied to rubber substitutes prepared from oils.

11409 Factoprene NS, Z
Proprietary hard factices. *Hubron Rubber Chemicals.* Name unverified.

11410 Faexin extract
The fatty acids of yeast.

11411 Fagacid
A product derived from beech wood tar. It is an antiseptic agent used in the preparation of soaps and plasters.

11412 Fahlun diamonds
Tin brilliants. Lead-tin alloys containing about 40% lead; used for theater jewelery.

11413 Fahralloy
A proprietary trade name for chromium-nickel-iron alloys. No manufacturer.

11414 Fairey Metal
A proprietary trade name for an alloy of aluminum with copper and smaller amounts of magnesium. No manufacturer.

11415 Fairprene
A proprietary trade name for a chloroprene polymer for fabric coating. No manufacturer.

11416 Fairy Ring Destroyer
26644-46-2 247-872-0
triforine
Emulsifiable concentrate containing 190 g/l triforine; a systemic fungicide. *Synchemicals Ltd.*

11417 Faktex
A proprietary trade name for a yellow rubber substitute which can be dispersed in water for addition to latex. No manufacturer.

11418 Faktogel
Range of factices for use in the rubber industry. *Bayer AG.*

11419 FAL Cypermethrin 10
66841-24-5 2836 266-492-6
Emulsifiable concentrate containing 100 g cypermethrin per liter; a pyrethroid insecticide. *Fine Agrochemicals Ltd.*

11420 Falkaloid, Falkyd
A proprietary trade name for a soft oil modified alkyd resin. No manufacturer.

11421 FAM 30
An iodophor, antiseptic, disinfectant. *Evans Vanodine International Ltd.*

11422 Famid
6988-21-2 230-253-4
dioxacarb
Carbamate insecticide. *Ciba plc.* Name unverified.

11423 Famodan MS Kosher
1338-41-6 8872 215-664-9
Sorbitan stearate
Food emulsifier for fat crystal modification and bloom retarders; for cocoa butter substitutes, compound coatings, imitation dairy systems. *Grindsted Prods.*

11424 Famodan TS Kosher
26658-19-5 8872 247-891-4
Sorbitan tristearate; fat crystal modifier preventing fat bloom in cocoa butter substitutes and compound coatings; improves texture of frostings and icings. *Grindsted Prods.*

11425 Famosan
Agricultural disinfectant. *The Wellcome Foundation Ltd.* Name unverified.

11426 Famous
8016-70-4 232-410-2
Partially hydrogenated soybean oil, stabilized. *Karlshamns.*

11427 Fanal
Specialty colorants. *BASF plc.*

11428 Fancol Acel
61788-48-5 262-979-2
Acelan L; Acetadeps; Acylan; Modulan®; Ritacetyl®. Acetylated lanolin; emollient, superfatting agent, lipophilic spreading agent for oils, creams, lotions, hair grooms, ointments, pharmaceuticals. *Fanning; Fabriquimica; Westbrook Lanolin; Croda Chem. Ltd.; Croda Inc.; Amerchol Corp.; RITA.*

11429 Fancol ALA
Cetyl acetate and acetylated lanolin alcohol; cosmetic, toiletry conditioner, emollient, pigment binder, spreading agent, oil coupler, solvent, glossing and plasticizing agent for hairsprays. *Fanning.*

11430 Fancol ALA-10
Polysorbate 80, cetyl acetate, acetylated lanolin alcohol. skin/hair conditioner, secondary emulsifier, pigment wetter, solubilizer, emollient, humectant, superfatting agent for antiperspirants, lotions and creams, sunscreens. *Fanning.*

11431 Fancol C
Petrolatum, lanolin, lanolin alcohol; water-oil emulsifier, moisturizing lubricant, emollient, fatting agent, water absorbent. *Fanning.*

11432 Fancol CA
36653-82-4 2070 253-149-0
Cetyl alcohol
Surfactant, emulsifier, emulsion stabilizer, opacifier, skin conditioner/emollient, viscosity booster, foam booster; for makeup, hair conditioners, cleansers, moisturizing creams and lotions. *Fanning.*

11433 Fancol CAB
Petrolatum, lanolin, lanolin alcohol; water-oil emulsifier, moisturizing lubricant, emollient, fatting agent, water absorbent. *Fanning.*

11434 Fancol CB
8002-31-1
Cocoa butter; skin conditioner, occlusive solvent, skin protectant for OTC drug products, makeup, moisturizing creams and lotions, suntan preparations. *Fanning.*

11435 Fancol CH
57-88-5 2256 200-353-2
Cholesterol
Film-former with lubricating, protective and anti-irritant properties, aids cell regeneration, emulsifier for water-oil formulations, precursor for production of vitamin D; for cosmetics, pharmaceuticals, hair/skin care products. *Fanning.*

11436 Fancol CH-24
Choleth-24-ceteth-24
Choleth-24; Forlan C-24. Surfactant, emulsifier for hair dyes, bubble bath, moisturizing creams and lotions, cleansing products. *Fanning; RITA.*

11437 Fancol CO-30
61791-12-6
PEG-30 castor oil
Surfactant, emulsifier for cleansing products, hair conditioners, wave sets, face, body and hand creams, and lotions. *Fanning.*

11438 Fancol DL
81-13-0 2988 201-327-3
DL-Panthenol
Conditioning agent for hair care products, makeup, moisturizing creams and lotions. *Fanning.*

11439 Fancol Gingko Extract
Ginko extract; biological extract for use in lotions, milks, creams, shampoos, bubble baths, soaps, gels. *Fanning.*

11440 Fancol HCO-25
61788-85-0
PEG-25 hydrogenated castor oil
Surfactant, emulsifier for bath oils, tablets and salts, aftershave lotions, skin fresheners. *Fanning.*

11441 Fancol HL
8031-44-5 232-452-1
Hydrogenated lanolin
Ivarlan HL; Lipolan; Satulan; Super-Sat. Emollient, moisturizer, lubricant, plasticizer, chemical intermediate, humectant, mold release agent for pharmaceuticals, cosmetics, industrial applications. *Fanning; Lipo; Brooks industries; Croda Inc.; RITA.*

11442 Fancol HL-20
68648-27-1
PEG-20 hydrogenated lanolin
Solubilizer, superfatting agent, gelling agent for cosmetics, pharmaceuticals, makeup, nail polish, night creams, microemulsions. *Fanning.*

11443 Fancol HON
8028-66-8
Honey; biological additive, flavoring, skin conditioner, humectant for shampoos, face, body and hand creams and lotions, bath products, hair conditioners, cleansing products, moisturizing creams and lotions. *Fanning.*

11444 Fancol Karite Butter
68424-60-2 270-311-6
Shea butter
Ointment base, anti-irritant for skin, skin conditioner, occlusive agent, solvent for suntan preparations, body lotions, winter sports products, wrinkle creams, soaps, shave foams, shampoos, balsams. *Fanning.*

11445 Fancol Karite Extract
68424-59-9 270-310-0
Shea butter extract; emollient with excellent spreadability for suntan preparations, skin toners, lipsticks, eye liners, ointments, suppositories. *Fanning.*

11446 Fancol LA
8027-33-6 232-430-1
Lanolin alcohol
Emollient, thickener, emulsifier, stabilizer, plasticizer, superfatting agent, dye dispersant, chemical intermediate, lubricant, humectant, mold release agent, conditioner for cosmetics, pharmaceuticals, soaps industrial applications. *Fanning.*

11447 Fancol LA-5
61791-20-6
Laneth-5; emollient, emulsifier, moisturizer, spreading agent, coupler for shampoos, skin care products, cosmetics. *Fanning.*

11448 Fancol LAO
Mineral oil and lanolin alcohol; conditioner, surfactant, stabilizer, moisturizer, humectant, penetrant, emollient, plasticizer, and primary emulsifier for use in cosmetics and pharmaceuticals; plasticizer in aerosol formulas. *Fanning.*

11449 Fancol Menthol
89-78-1 5882 201-939-0
Menthol
Denaturant, flavoring agent, fragrance component for mouthwashes, aftershaves, cleansing products, face, body and hand creams/lotions, skin care products. *Fanning.*

11450 Fancol OA-95
143-28-2 6968 205-597-3
Oleyl alcohol
Plasticizer, emulsion stabilizer, antifoam and coupling agent, aerosol lubricant, petroleum additive, pigment dispersant; rust preventive; detergent, release agent, cosolvent, softener, tackifier; spreading agent used for metalworking, petrochemicals, pulp and paper, paints and coatings, plastics, polymers, food applications, pharmaceuticals, cosmetics and chemical intermediates. *Fanning.*

11451 Fancol SA
112-92-5 8960 204-017-6
Stearyl alcohol
Emulsion stabilizer, opacifier, skin conditioner, emollient, viscosity booster, foam booster for makeup, hair conditioners, moisturizing creams and lotions, cleansers. *Fanning.*

11452 Fancol SA-15
25231-21-4
PPG-15 stearyl ether
Skin conditioner, emollient for bath oils, tablets and salts, fragrance products, lipsticks, deodorants, cleansers. *Fanning.*

11453 Fancol TOIN
97-59-6 255 202-592-8
Allantoin
Skin conditioner for makeup, cleansers, face, body and hand creams and lotions, moisturizing creams, skin care products. *Fanning.*

11454 Fancol WGFA
68484-43-1
Wool grease fatty acid; emulsifier with high adhesion to various substrates, oil binding properties for use in lubricating greases, polishes, anti-corrosion compounds. *Fanning.*

11455 Fancor D
Blend of animal fats and petroleum-derived materials; long term rust preventive; EP and slip agent for wire drawing compounds, slushing oils, cutting oils, lubricants; crystallization inhibitor for waxes; dispersant; in leather stuffing greases. *Fanning.*

11456 Fancor IPL
63393-93-1 264-119-1
Isopropyl lanolate
Hydrophilic emollient, moisturizer, water-oil auxiliary emulsifier, stabilizer and opacifier, wetting and dispersing agent for pigment and talc, aids slip and gloss of stick cosmetics, plasticizes wax systems, mold release, superfatting agent; binder for pressed powders. *Fanning.*

11457 Fancor Lanwax
68201-49-0 269-220-4
Natural lanolin wax ester; plasticizer in wax crayons; water repellent, humectant, conditioner, corrosion inhibitor, emollient, lubricant for cosmetics, toiletries, pharmaceuticals, industrial applications; extender and crystallization inhibitor for natural waxes in industrial applications; emulsifier. *Fanning.*

11458 Fancor LFA
68424-43-1 270-302-7
Lanolin fatty acids
Emollient, stabilizer, emulsifier, corrosion inhibitor for personal care and pharmaceutical products; used in industrial leather treating, coatings, polishes, corrosion inhibitors, lubricants. *Fanning.*

11459 Fancorsil A
69430-24-6
Cyclomethicone and dimethicone; hair/skin conditioner, emollient, solvent; used in hair conditioners, makeup, moisturizing creams and lotions. *Fanning.*

11460 Fancorsil P
Modified silicone hydrocarbon; skin/hair conditioning agent, emollient; excellent afterfeel and lubricity. *Fanning.*

11461 Fancorsil SLA
Dimethicone copolyol adipate
Skin/hair conditioning agent, emollient; provides soft and velvety afterfeel. *Fanning.*

11462 Fancoscour PO, VC
Detergent for textile scouring. *Reilly-Whiteman.*

11463 Fanfare
isoproturon-isoxaben
Suspension concentrate containing 450 g isoproturon and 19 g isoxaben per liter; used for annual weed control in cereals. *Ciba-Geigy Agrochemicals.*

11464 Fanghidi Sclofani
A yellow powder of volcanic origin, consisting chiefly of sulfur, with small quantities of iron, calcium, and manganese.

11465 Fantan
Phenylcinchonoylurethane

11466 Fantasit
A proprietary casein plastic. No manufacturer.

11467 Fanwax G
Stearyl alcohol-ceteareth-20. Oil-water self-emulsifying wax for cosmetics and toiletries, skin/hair lotions, antiperspirants, depilatories, creme rinses, opacified hair dyes, bleaches. *Fanning.*

11468 Fanwax P
8005-44-5, 9005-67-8
Cetearyl alcohol-polysorbate 60. Oil-water self-emulsifying wax for lotions, creams, hair and skin care products, creme rinses, antiperspirants. *Fanning.*

11469 FAR Mark I through X
Elastomeric coatings made up of asphalt, neoprene, acrylic and Kraton rubber imbedded in polyester fabric; used for cold-applied liquid or fully adhered membrane type roof systems. *Flex-Shield.* Name unverified.

11470 Far-Go/Avadex BW
2303-17-5 9726 218-962-7
Trichloroallyl diisopropylthiocarbamate
Triallate. Herbicide for wild oats. *Monsanto Co.* Name unverified.

11471 Fargro Chlormequat
999-81-5 2153 213-666-4
Chlormequat Chloride
Soluble concentrate containing 460 g/l chlormequat; plant growth regulator. *Fargro Ltd.*

11472 farina
Flour, or potato starch.

11473 farine
A term applied in the West Indies to a product obtained by grating fresh cassava root, draining away the juice from the wet pulp, and then heating the residue. It is also known as Cassava meal.

11474 Faringets
A proprietary preparation of myristyl benzalkonium iodine chloride, an antimicrobial and disinfectant. *Bayer AG.*

11475 farinose
Starch cellulose, or the outer covering of the starch granule.

11476 Farlite
A proprietary trade name for a phenol-formaldehyde synthetic resin laminated product. No manufacturer.

11477 Farmacel, Farmacel 645
999-81-5 2153 213-666-4
Chlormequat Chloride
Soluble concentrate containing 460 g/l or 645 g/l chlormequat; plant growth regulator for wheat and oats. *Farm Protection Ltd.*

11478 Farmacel
999-81-5 2153 213-666-4
Chlormequat Chloride
Growth regulator for wheat and oats containing chlormequaternary. *ICI Chem & Polymers Ltd.*

11479 Farmaneb
Fungicide for prevention of potato blight. *ICI Chem & Polymers Ltd.*

11480 Farmatin
76-87-9 9875 200-990-6
Fentin hydroxide
Triphenyltin hydroxide; Quadrangle Super-Tin 4L; Super Tin® 4L; Ashlade Flotin; Du-Ter®. Used for control of potato blight. *Farm Protection Ltd; Ashlade Formulations Ltd.;Griffin; Quadrangle Agrochemicals;.*

11481 Farmon
Range of liquid herbicides of different formulations. *ICI Chem & Polymers Ltd.*

11482 Farmon 2,4-D
94-75-7 2865 202-361-1
2,4-D
2,4-D; a herbicide for use on cereals and established grassland. *Farm Protection Ltd.*

11483 Farmon 2,4-DB Plus
2,4-DB-MCPA
2,4-DB and MCPA; a herbicide for use with cereal crops. *Farm Protection Ltd.*

11484 Farmon 2,4-DP·CPA
dichlorprop-MCPA
Soluble concentrate of 358 g dichlorprop and 177 g MCPA per liter; used for control of weeds in cereals and grass. *Farm Protection Ltd.*

11485 Farmon Blue
Nonionic wetting agent containing 900 g/l alkylphenol ethoxylate; spreader for use in herbicide, fungicide, and insecticide sprays. *Farm Protection Ltd.*

11486 Farmon Condox
dicamba-mecoprop
Soluble concentrate of 112 g dicamba and 265 g mecoprop per liter; used for weed control in cereals and grassland. *Farm Protection Ltd.*

11487 Farmon MCPA 50
94-74-6 5803 202-360-6
MCPA
MCPA; herbicide for cereals and grassland. *Farm Protection Ltd.*

11488 Farmon MCPB Plus
MCPA-MCPB
Soluble concentrate containing 41 g MCPA and 244 g MCPB per liter; for control of weeds in undersown cereals and grassland. *Farm Protection Ltd.*

11489 Farmon Mini Slug Pellets
108-62-3 5983 202-945-6
metaldehyde
Pellets containing 6% w/w metaldehyde; snail and slug bait. *Farm Protection Ltd.*

11490 Farmon PDQ
diquat-paraquat
Soluble concentrate of 80 g diquat and 120 g paraquat per liter; used for weed control in field crops. *Farm Protection Ltd.*

11491 Farmon TCA
76-03-9 9756 200-927-2
trichloroacetic acid
trichloroacetic acid (TCA); for control of weeds in field crops. *Farm Protection Ltd.*

11492 Farnesol
4602-84-0 3978 225-004-1
$C_{15}H_{26}O$
3,7,11-trimethyl-2,6,10-dodecatrien-1-ol
Stirrup-A/WF; Stirrup-CRW; Stirrup-H; Stirrup-HB; Stirrup-TPW; Trimethyl-2,6,10-dodecatriene-1-ol. A sesquiterpene alcohol prepared from nerolidol; a perfume. bp_4 = 149°; d = 0.8880; n_D^{20} = 1.4870-1.4920; insoluble in H_2O, soluble in organic solvents; LD_{50} (rat orl)= 6 g/kg.

11493 Farrant's medium
A microscopic medium. It consists of a mixture of equal parts of glycerl and arsenious acid, to which is added powdered gum arabic, allowed to stand and filtered.

11494 Farronic
A heat-resisting nickel-copper alloy.

11495 Fascat® 2000
Esterification catalyst; effective where catalyst removal is required; source of bivalent tin. *Elf Atochem.*

11496 Fascat® 2004
7772-99-8 8939 231-868-0
Anhydrous stannous chloride
Esterification catalyst for manufacture of plasticizers and polyesters. *Elf Atochem.*

11497 Fascat® 4400
7646-78-8 8929 231-588-9
Anhydrous tin tetrachloride
Strong Lewis acid catalyst for variety of reactions including acetylation and alkylation of aromatic compounds, chloromethylation, reaction of epichlorohydrin with alcohols to form glycidyl ethers. *Elf Atochem.*

11498 Fascol
70-30-4 4716 200-733-8
Hexachlorophene
Anthelmintic (flukicide). *The Wellcome Foundation Ltd.* Name unverified.

11499 Fasinex
Cattle and sheep flukicide. *Ciba plc.*

11500 Fastac
67375-30-8
Alphacypermethrin
Contact insecticide. *Shell UK.*

11501 Fasteeth
Denture adhesive powder, for securing dentures. *Richardson-Vicks Inc.* Name unverified.

11502 Fasteeth Extra Hold
To secure hard-to-hold lower dentures. *Richardson-Vicks Inc.* Name unverified.

11503 Fastex
A proprietary trade name for a specially stabilized and purified rubber latex supplied in concentrations of 40 and 60%. No manufacturer.

11504 Fastusol®
Substantive dyes for paper coloring. *BASF AG.*

11505 Fatal Flip
Trade name for total release fogger insecticide; for use by both professional and consumer trade to control many flying and crawling insects. *Colonial Products Inc.*

11506 Fatsco
7778-43-0 8720 231-902-4
sodium arsenate
Sodium arsenate 3%; ant poison for sweet eating ants, kills roaches, moles, moles, mice, woodchucks etc. *Fatsco.* Unverified.

11507 Faversham powder No. 2
An explosive containing ammonium nitrate, potassium nitrate, trinitrotoluene, and ammonium chloride.

11508 Favlerite No. 1
An explosive consisting of 88% ammonium nitrate and 12% dinitronaphthalene.

11509 Favlerite No. 2
An explosive containing 90% of Favlerite No. 1 and 10% ammonium chloride.

11510 Favour
metalaxyl-thiram
Mixture of metalaxyl and thiram; protectant fungicide against downy mildew in lettuce. *Ciba-Geigy Agrochemicals.*

11511 Fax
Edible fats. *Marfleet Refining Co.* Name unverified.

11512 Faxola
Cooking oil. *Marfleet Refining Co.* Name unverified.

11513 FB 48
1330-43-4 8733 215-540-4

sodium borate
Sodium borate. A fertilizer. *U.S. Borax & Chem.*

11514 FBC CMPP
7085-19-0 5826 230-386-8
mecoprop
Mecoprop, a selective herbicide. *Schering Agrochemicals Ltd.* Discontinued.

11515 FBC Fly Dip
Insecticide. *Schering Agrochemicals Ltd.* Discontinued.

11516 FBC MCPA
94-74-6 5803 202-360-6
MCPA
MCPA, a selective herbicide. *Schering Agrochemicals Ltd.* Discontinued.

11517 FBC Pirimicarb 50
23103-98-2 7651 245-430-1
Pirimicarb
Pirimicarb, an insecticide. *Schering Agrochemicals Ltd.* Discontinued.

11518 FBC Protectant Fungicide
Fungicide *Schering Agrochemicals Ltd.* Discontinued.

11519 FBC Slug Destroyer
Molluscicide. *Schering Agrochemicals Ltd.* Discontinued.

11520 FBC Winter Dip
Insecticide. *Schering Agrochemicals Ltd.* Discontinued.

11521 FC 113
Cleaning solvent. *ICI Chem & Polymers Ltd.*

11522 feathered tin
Granulated tin.

11523 Feb Brickclean
Cleaner and degreasing solution for the rapid removal of cement mortar stains, grime, oil, grease, algae, moss, and wax from brickwork, concrete, asbestos, terrazzo, and ceramics. *Feb Ltd.*

11524 Feb Hybit
High solids protective coating based on emulsified bitumen; dries to tough water- and vapor-proof film; used for general waterproofing and dampproofing of concrete, roofing felt, metal protection, insulation, etc. *Feb Ltd.*

11525 Feb Hyseal Slurry
A surface brush or trowel-applied waterproofing cement slurry; completely watertight even under conditions of high hydrostatic pressures. *Feb Ltd.*

11526 Feb Oxide
Coloring pigments for cement; contains no inert fillers or earth colors. *Feb Ltd.*

11527 Feb Supercrete
Two-component multi-use polyester repair and patching compound for concrete. *Feb Ltd.*

11528 Feb Sylane
Based on alkyl alkoxy silane; water repellent for treatment of concrete. *Feb Ltd.*

11529 Feb Wintamix
Mortar admixture for winter use. *Feb Ltd.*

11530 Febclean
Hand cleaner to remove grease, fresh paint, mastic, ink, glue, bitumen, rubber cement, tile adhesive and food stains without the need for water. *Feb Ltd.*

11531 Febclear Super
Surface coating providing resistance to oil, dusting, abrasions and splash contact from acid and alkali and common chemicals. *Feb Ltd.*

11532 Febco
Concrete floor hardening and dust proofing liquid based on sodium silicate. *Feb Ltd.*

11533 Febcrete AEA
Resin-based air entraining agent for concrete. *Feb Ltd.*

11534 Febcure
Curing compounds. *Feb Ltd.*

11535 Febdura Standard
Premixed material for concrete repair. *Feb Ltd.*

11536 Febexpan
Pre-mixed shrinkage compensating grouting compounds. *Feb Ltd.*

11537 Febface
Single-place polyurethane mold coating providing an abrasion-resistant protective coating to formwork; unaffected by alkalis, petrol, oil and grease; provides impervious barrier to water and damp. *Feb Ltd.*

11538 Febfast PG
A fast setting, high early strength epoxy resin pouring grade mortar. *Feb Ltd.*

11539 Febflex One Coat
One-coat solvent-based bitumen roofing compound; for coating and renovating slate, galvanized iron, asbestos, cement, lead, zinc, asphalt, and timber. *Feb Ltd.*

11540 Febflor
Powder floor smoothing compound for preparing substrates for final floor covering. *Feb Ltd.*

11541 Febflow Accelerating
Concrete plasticizer, water reducing admixture with set accelerating properties for protection of concrete. *Feb Ltd.*

11542 Febflow Retarding
Non-air entraining, water reducing, set retarding admixture for concrete. *Feb Ltd.*

11543 Febfoam
Liquid foaming agent producing aerated concrete or cement. *Feb Ltd.*

11544 Febglaze
A lightly filled pigmented, solventless, epoxy resin-based compound; for surface application to achieve chemical resistant coatings for concrete storage tanks, etc. *Feb Ltd.*

11545 Febgrout
A water-reducing, expanding admixture for cement and mortar grouts. *Feb Ltd.*

11546 Febguard Bonding Agent
Adhesion promoter between concrete substrates and repair mortars. *Feb Ltd.*

11547 Febkol Elastomer 110 and 122
Proprietary brands of polysulfide sealant. *Feb Ltd.* Name unverified.

11548 Febkol Plastomer 555
A proprietary polysulfide/epoxy sealant and adhesive. *Feb Ltd.* Name unverified.

11549 Febmast GP
General-purpose pre-mixed trowel-applied sealing and bedding compound. *Feb Ltd.*

11550 Febmix Admix
Mortar plasticizer. *Feb Ltd.*

11551 Febol Standard
Shutter-applied surface retarder for concrete. *Feb Ltd.*

11552 Febond PVA
An integral PVA bonding admixture; multi-use adhesive, sealer, primer and bonding agent for bonding common building materials to themselves or to each other. *Feb Ltd.*

11553 Febond SBR
A styrene-butadiene synthetic rubber latex bonding compound for concrete and mortar; for water resistant renderings and general purpose floor screeding. *Feb Ltd.*

11554 Febox
Multi-use epoxy patching mortar. *Feb Ltd.*

11555 Febpitch
Epoxy providing a waterproof membrane for use in concrete ducts, vats, tanks, retaining walls. *Feb Ltd.*

11556 Febplast Ready Mixed
A premixed decorative ceiling and wall finish; also as caulking and taping compound. *Feb Ltd.*

11557 Febplate
A proprietary range of epoxy mortars. *Feb Ltd.*

11558 Febproof
An integral liquid waterproofing admixture for use in mortar and concrete. *Feb Ltd.*

11559 Febrail No. 1
Epoxy grout specially formulated for British Rail for fixing rail shoulders into precast concrete railway sleepers. *Feb Ltd.*

11560 Febrok
Penetrates the concrete surface and reacts with free lime to form insoluble pore blocking compounds; results in a dust-free, durable and abrasive resistant surface. *Feb Ltd.*

11561 Febseal
Building sealant and caulking compound. *Feb Ltd.*

11562 Febset NF
Two-part, multipurpose epoxy material for bedding and repair work for concrete. *Feb Ltd.*

11563 Febsilicon
Colorless waterproofer containing silicone; external surface-applied waterproofer for permeable materials such as brickwork, cement rendering, concrete block, asbestos sheeting. *Feb Ltd.*

11564 Febspeed
Calcium chloride-based liquid rapid hardener; accelerates setting times of concrete and mortar to enable paths, floors, drives to be open for early use. *Feb Ltd.*

11565 Febstik
Contact adhesive for bonding plastic laminates, aluminum sheet, chipboard, plywood, plaster board and natural stone. *Feb Ltd.*

11566 Febstrike
Release agent for application to mold and shutter faces prior to casting concrete; approved for use with potable water. *Feb Ltd.*

11567 Febtex
Cold mix textured finish for internal use on plasterboard, cement rendering, sealed plaster, concrete. *Feb Ltd.*

11568 Febtile
Nonslip ceramic tile adhesive. *Feb Ltd.*

11569 Febtite Liquid
Waterproofing admixture for waterproofing cement, mortar. *Feb Ltd.*

11570 Febtone
Coloring agent that also plasticizes and improves water and frost resistance of concrete and mortar. *Feb Ltd.*

11571 Febweld
A permanent high bond strength epoxy adhesive for internal or external bonding of renderings, granolithic toppings and concrete to concrete; used in repair of chloride contaminated concrete. *Feb Ltd.*

11572 Fecap
ferrous fumarate-folic acid
A proprietary preparation of ferrous fumarate and folic acid; an iron supplement. *Pfizer International.*

11573 Fecraloy
A proprietary trade name for an alloy of iron, chromium, aluminum and yttrium under development for use in sintered form as a catalyst to assist in the control of atmospheric pollution by reducing the emission of carbon monoxide and other fumes from atomic energy power stations. *Atomic Energy Establishment.* Name unverified.

11574 Feculose
The name given to various commercial starch esters.

11575 Fedralite
A proprietary trade name for a Vinsol resin-treated laminated paper. No manufacturer.

11576 Feedercalc
Micro computer programs to facilitate rapid reliable calculation of optimum feed requirements for all types of iron and steel castings. *Foseco (F.S.) Ltd.*

11577 Feedercalc
Micro computer programs to facilitate rapid reliable calculation of optimum feed requirements for all types of iron and steel castings. *Foseco (F.S.) Ltd.*

11578 Feedex
Moldable exothermic feeding compound. *Foseco (F.S.) Ltd.*

11579 Feedmate
Boiler water treatment. *Grace Dearborn Ltd.*

11580 Feedol
Feeding compounds for nonferrous metals. *Foseco (F.S.) Ltd.*

11581 Fehlings solution
An alkaline solution of potassio-tartrate of copper. It is prepared in two solutions. a) Consisting of copper sulfate, and b) a solution of Rochelle salt (potassium sodium tartrate) and caustic soda; used for the identification, determination of sugars.

11582 Fehling's solution, neutral
A solution made by adding 25 ml of a solution containing 2 g copper sulfate (7.86 g $CuSO_4 \cdot 7H_2O$) per liter to 25 ml of a solution containing 3.292 g sodium carbonate and 20 g Rochelle salt per liter. It is stated to be a sensitive reagent for the detection of sugars.

11583 Fekta® RT
Algicide for swimming pools. *Goldschmidt.*

11584 Felamine
cholic acid-hexamine
A proprietary preparation of cholic acid and hexamine. An antibacterial and antiseptic. *Sandoz.* Name unverified.

11585 Feliniffa P
Feline panieucopenia vaccine. *RMB Animal Health Ltd.*

11586 Feliniffa RC
Feline rhinotracheitis and calicivirus vaccine. *RMB Animal Health Ltd.*

11587 Felixite
An explosive. A 42-grain powder, containing metallic nitrates, nitro-hydrocarbons, and 3% petroleum jelly.

11588 Felsinosima
A trade name for cultures of *Bacillus felsineus.* No manufacturer.

11589 Felspar
$(K_2O \cdot 3SIO_2) \cdot 2(Al_2O_3 \cdot 3SIO_2)$
Potassium Felspar
Orthoclase. A potassium aluminum silicate, used in the manufacture of porcelain, as a building material, and as a fertilizer.

11590 Felton 3T
Blend; detergent for felt of paper manufacturing machines. *Toho Chem. Industry.*

11591 Felvinone
7-acetyl-5-isopropyl-2-methylbicyclo (2.2.2)oct-2-ene and 8-acetyl-5-isopropyl-2-methylbicyclo (2.2.2)oct-2-ene. *Quest Int'l. UK Ltd.*

11592 Felzodox
1314-13-2 10279 215-222-5
Zinc oxide
Pigment and vulcanization accelerator. *Pigment & Chemical Inc.*

11593 Fenafix
A proprietary trade name for a series of modified vinyl-pyrrolidone resins; used to modify the properties of other vinyl films and also to improve the adhesion of difficult surfaces. *Fine Dyestuffs & Chemicals Ltd.* Name unverified.

11594 fenbendazole
43210-67-9 4000 256-145-7
$C_{15}H_{13}N_3O_2S$
[5-(phenylthio)-1H-benzimidazol-2-yl]carbamic acid methyl ester
HOE-881v; Panacur; Safe-Gard. Anthelmintic (nematodes). mp = 233° (dec); insoluble in H_2O, organic solvents, freely soluble in DMSO.

11595 Fenitex
122-14-5 4017 204-524-2
Fenitrothion
Insecticide. *Makhteshim Chemical Works Ltd.*

11596 Fenitrothion EC
122-14-5 4017 204-524-2
Fenitrothion
Organophosphorus insecticide. *Ciba plc.* Name unverified.

11597 Fennite
A fungicide. *Fisons plc, Horticultural Div.* Name unverified.

11598 Fenocil
bromacil-pentachlorophenol
Bromacil and pentachlorophenol; used for total weed control in noncrop areas. *Chemsearch (UK) Ltd.*

11599 Fenoil®
Range of chemicals for exploration, production and enhanced recovery of oil and natural gas. *Bayer AG; Bayer plc.*

11600 Fenolac
Cyclized rubber used in adhesives and bonding agents. No manufacturer.

11601 Fenolite
An Italian synthetic resin material for use in electrical insulation.

11602 Fenopon AC-78
Anionic surfactant in which the cation is sodium and the anion is a coconut ester of isethionate; powder form; low salt content detergent with good foaming, lathering, and dispersing properties; used in detergent bars, dentifrices, shampoos, bubble baths and other cosmetics. *ISP.* Name unverified.

11603 Fenopon CD
ammonium ethoxy sulfate
Supplied in liquid form; high foaming surfactant for high electrolyte aqueous systems; air entraining properties used in the concrete industry; foaming agent for light weight cements; frothing agent for gypsum wallboard. *GAF Great Britain.* Name unverified.

11604 Fenopon CN-42
Sodium N-cyclohexyl-N-palmitoyl taurate
Supplied in paste form; low foaming detergent, dispersing agent, stabilizer; used in mechanical diswashing detergents; industrial cleaners; synthetic rubber emulsions. *ISP.* Name unverified.

11605 Fenopon CO
Sodium or ammonium nonylphenol ethoxy sulfate in liquid form; high foaming detergent with wetting, dispersing, emulsifying, antistatic and lime soap dispersion properties; used for scrub soaps; car washes; rug and hair shampoos; vinyl polymerization; petroleum wax, plastics and synthetic fibers. *ISP.* Name unverified.

11606 Fenopon EP
Ammonium nonylphenol ethoxy sulfate
Supplied in liquid form; versatile primary emulsifier and stabilizing agent used in emulsion copolymers. *GAF Great Britain.* Name unverified.

11607 Fenopon SE
Sodium alkylaryl polyether sulfonate
Supplied in liquid form; detergent base with good foaming and rinsability; used for cosmetic and pharmaceutical products and emulsion polymerization. *ISP.* Name unverified.

11608 Fenopon T-33 and T-43
137-20-2 205-285-7
Sodium methyl oleoyl taurate

Supplied as a clear liquid or a slurry; anionic surfactant used in textiles, rug shampoos, cleaning and detergent formulations. *GAF Great Britain.* Name unverified.

11609 Fenopon T-51
137-20-2 205-285-7
Sodium N-methyl N-oleyl taurate
Supplied in the form of a readily soluble gel; used for textile processing; latex emulsion stabilizer. *GAF Great Britain.* Name unverified.

11610 Fenopon T-77
137-20-2 205-285-7
Sodium N-methyl N-oleyl taurate
Supplied in powder form; wetting and dispersing agent used in dry blending; industrial and herbicidal wetting powders; textiles; rug shampoos; cleaning and detergent formulations. *GAF Great Britain.* Name unverified.

11611 Fenopon TC-42
Sodium N-coconut acid N-methyl taurate
Supplied in the form of a slurry; chemically stable detergent with foaming, lathering and dispersing properties; used in detergent bars; shampoos; bubble baths; cosmetics. *ISP.* Name unverified.

11612 Fenopon TK32
61791-41-1 263-173-3
Sodium N-methyl N-tall oil acid taurate
Supplied in liquid form; dispersing and suspending agent used as a precipitation inhibitor for salts of barium, calcium and strontium, and for scale prevention in oil well tubing and flow lines. *ISP.* Name unverified.

11613 Fenopon TN-74
Sodium N-methyl N-palmitoyl taurate
Supplied in powder form; dispersing and suspending agent used for dry blending into detergent formulations and in herbicides and insecticides. *GAF Great Britain.* Name unverified.

11614 Fenotec
Ester-phenolic resin binders for cold-set bonding of sand cores. *Foseco (F.S.) Ltd.*

11615 fenpropathrin
39515-41-8 4033 254-485-0
$C_{22}H_{23}NO_3$
cyano(3-phenoxyphenyl)methyl 2,2,3,3-tetramethylcyclopropanecarboxylate
fenpropathrine; Danitol; Herald; Kilumal; Meothrin; Ortho Danitol; Rody; S-3206; Tame. Acaricide and insecticide with contact, stomach and repellentaction. Used for control of mites and insects in fruit and vegetable crops. mp = 45-50°; d^{25} = 1.15; almost insoluble in H_2O (0.33 mg/l), more soluble in organic solvents; LD_{50} (rat orl) = 71 mg/kg.

11616 fenpropidin
67306-00-7 4034
$C_{19}H_{31}N$
1-[3-[4-(1,1-dimethylethyl)phenyl]-2-methyl propyl] piperidine
Patrol; Ro-12-3049. Inhibitor of ergosterol biosynthesis. Used as a fungicide. $bp_{0.2}$ = 117°; insoluble in H_2O.

11617 Fenton's metal
A bearing metal. It contains about 80% zinc, 15% tin, and 5% copper.

11618 Fenton's reagent
Hydrogen peroxide and a ferrous salt; used for the oxidation of polyhydric alcohols.

11619 Fentro
Insecticide. *Murphy Chemical Co Ltd.* Discontinued.

11620 fenvalerate
51630-58-1 4051 257-326-3
$C_{25}H_{22}ClNO_3$
4-chloro-α-(1-methylethyl)benzeneacetic acid cyano(3-phenoxyphenyl)methyl ester
S-5602; SD-43775; WL-43775; Belmark; Pydrin; Pyridin; Sumicidin; Tirade. Synthetic pyethroid insecticide lacking the cyclopropane ring. d^{23} = 1.17; n_D^{20}= 1.5533; insoluble in H_2O, soluble in organic solvents; LD_{50} (rat orl) = 451 mg/kg.

11621 Fenyrane
2,4-Dimethyl-6-phenyldihydropyrane
Quest Int'l. UK Ltd.

11622 Ferad
Bismuth and boron additives for malleable cast iron. *Foseco (F.S.) Ltd.* Discontinued.

11623 Feraloy
A nickel-steel-chromium alloy, having a specific gravity of 8.15, and melting at 1480 C.

11624 Fergapol
Artificial and synthetic resins; for adhesives, paints, mastics and sealants. *Ferguson & Menzies Ltd.* Name unverified.

11625 Fergatac
Artificial and synthetic resins; for adhesives, paints, mastics and sealants. *Ferguson & Menzies Ltd.* Name unverified.

11626 Ferlosa
A proprietary range of synthetic pulp for the paper and cement industry. *Montedison UK Ltd.* Name unverified.

11627 Fermenticide
An antiseptic compound. *Bush Boake Allen Ltd.*

11628 Fermenzyme® L-200
9032-08-0 232-877-2
Glucoamylase from *Aspergillus niger;* enzyme for fermentation of fuel alcohol. *Solvay Enzymes.*

11629 Fermet alloy
An alloy of 74.5% iron, 18% nickel, 2.2% manganese, 0.7% tungsten, 0.3% copper, and 0.35% carbon.

11630 Fermin
68-19-9 10152 200-680-0
Vitamin B_{12}
A proprietary preparation of cyanocobalamin in an oral form; a hematinic. *Albion Group Ltd.* Unverified.

11631 Fermine
131-11-3 3304 205-011-6
dimethyl phthalate
A proprietary trade name for dimethyl phthalate, a chemical intermediate and a plasticizer. No manufacturer.

11632 Fernacol
A fungicide. *ICI Chem & Polymers Ltd.*

11633 Fernasan
A nonmercurial seed dressing. *ICI Chem & Polymers Ltd.*

11634 Fernasul
A lime and sulfur fungicide. *Plant Protection.* Name unverified.

11635 Fernesta
A selective weed killer. *Plant Protection.* Name unverified.

11636 Fernex
23505-41-1 7652 245-704-0
pirimiphos-ethyl
Granules containing 10% w/w pirimiphos-ethyl; an organophosphorus insecticide to control scarid and phorid flies. *Fargro Ltd.* Discontinued.

11637 Fernex
23505-41-1 7652 245-704-0
pirimiphos-ethyl
Insecticide containing pirimiphos-ethyl. *ICI Chem & Polymers Ltd.*

11638 Fernide
137-26-8 9510 205-286-2
thiram
A thiram-based foliage fungicide. *ICI Chem & Polymers Ltd.*

11639 Fernimine
94-75-7 2865 202-361-1
2,4-D
A selective weed killer which has 2,4-D as the active principle. *ICI Chem & Polymers Ltd.*

11640 Fernol
Textile flame retardants; for textile finishing. No manufacturer.

11641 Fernox
Water treatment chemicals including cleansers, descalers, corrosion proofers and inhibitors, and antifreezes. No manufacturer.

11642 Fernoxone
94-75-7 2865 202-361-1
2,4-D
A selective weed killer based on 2,4-D. *ICI Chem & Polymers Ltd.*

11643 Ferox-Celotex
A proprietary trade name for Celotex which has been treated to resist attack by fungi and termites. No manufacturer.

11644 Ferozon
A disinfectant. No manufacturer.

11645 Ferquatac
Artificial and synthetic resins, emulsions of artificial and synthetic resins; tackifiers, and modifiers for water-based adhesives, coatings, mastics, and sealants. *Ferguson & Menzies Ltd.* Name unverified.

11646 Ferralium® Alloy 255
Super stainless steel alloy; high tensile strength. *Haynes Int'l.*

11647 Ferrax
ethirimol-flutriafol-thiabendazole
Flowable concentrate containing 400g ethirimol, 30g flutriafol and 10g thiabendazole per liter; fungicide seed treatment for barley. *Bayer AG; Bayer plc; Dow Elanco Ltd;.*

11648 ferric chloride
7705-08-0 4061 231-729-4
$FeCl_3$
iron(III) chloride
ferric chloride anhydrous; Flores martis; ferric trichloride; iron sesquichloride. Treatment of sewage and industrial wastes; etching agent, mordant, disinfectant, pigment, feed additive. d = 2.90; mp = 300°; bp = 316°; readiliy absorbs water; readily soluble in water, alcohol, ether and acetone; LD_{50} (rat orl) = 450 mg/kg. *Asahi Denka Kogyo; BASF; Eka Nobel AB; Mallinckrodt; Penta Mfg.; Rasa Ind.*

11649 ferric hydroxide
20344-49-4 4068 243-746-4
$FeHO_2$
ferric hydroxide oxide
hydrated ferric oxide. Red-brown powder or crystals. Used as a catlayst, pigment and in water purification. d = 3.4-3.9; insoluble in H_2O, EtOH, soluble in mineral acids.

11650 ferric nitrate
10421-48-4 4071 233-899-5
FEN_3O_9
iron(III) nitrate
Dyeing, tanning, analytical chemistry. d = 1.68; mp = 47°; dec below 100°; d^{21} = 1.68; freely soluble in water, alcohol, acetone; LD_{50} (rats orl) = 3.25 g/kg. *Aldrich; General Chem.; Hoechst Celanese; Mallinckrodt; Sherman Chem. Ltd.*

11651 ferric oxide
1309-37-1 4072 215-168-2
Fe_2O_3
ferric oxide red
Iron(III) oxide; iron oxide; red iron trioxide; ferrosoferric oxide. Metallurgy, gas purification, paint and rubber pigment, in thermite, polishing compounds, mordant, laboratory reagent, catalyst, feed additive, electronic pigments for TV, permanent magnets, memory cores for computers, magnetic tapes. *BASF; Kerr-McGee; Miles.*

11652 ferric sulfate
10028-22-5 4079 233-072-9
$Fe_2O_{12}S_3$
Iron (III) sulfate
ferric persulfate; ferric sesquisulfate; ferric tersulfate; Iron (III) sulfate; ferric persulfate; ferric sesquisulfate; ferric tersulfate; Sulfuric Acid, Iron(3+) Salt (3:2); Iron Tersulfate; Ferric sulfate monohydrate; Hart Lawn Sand; Hart Moss Killer; Maxicrop Moss Killer & Conditioner; Taylors Lawn Sand; Vitax Micro Gran; Vitax Turf Tonic; Walkover Moss Killer; Elliott's Lawn Sand; Elliott's Moss Killer; Green-up Mossfree; Greenmaster Autumn; Greenmaster Mosskiller,. Used for moss control in turf; polymerization catalyst; coagulant in water purification and sewage treatment; in etching aluminium. d = 3.097; soluble in water, alcohol, insoluble in acetone, ethyl acetate. *Thomas Elliott Ltd; Vitax Ltd; Maxwell Hart Ltd; Fisons plc; Horticulture Div.*

11653 Ferri-Darotin
A sodium iron(III) ethylenediamine tetracetic acid trihydrate and ammonium iron ethylenediamine tetracetic acid composition; used as a bleaching bath component for the development of color pictures, as an agent to act against chloros. *Degussa AG.* Name unverified.

11654 Ferrikalite
An artificial potassium ferric sulfate. No manufacturer.

11655 Ferriplex
Iron chelate. *Rhône-Poulenc UK.*

11656 Ferriplus
Iron chelate *Rhône-Poulenc UK.*

11657 Ferrisul
A trade name for ferric sulfate. No manufacturer.

11658 ferrite
Nearly pure iron; phosphorus and sulfur may be present in minute quantities, but the carbon content is not more than 0.05%.

11659 ferro-aluminum
It contains up to 20% aluminum with iron, and is used in the preparation of iron and steel.

11660 ferroaluminum silicate
12178-41-5
Ferrosil 14. Filler for abrasion-resistant plastic systems, caulks, sealants, polishes, abrasive compounds; extender pigment for primers and other coatings. *Kaopolite.*

11661 ferro-argentan
An alloy resembling silver and containing 70% copper, 20% nickel, 5.5% zinc, and 4.5% cadmium.

11662 ferro-boron
An alloy of iron and boron, containing from 20-25% boron. It is added to steel.

11663 ferro-carbon-titanium
An alloy of iron and titanium, containing carbon; used for making steel.

11664 ferrochlor
ferric chloride-calcium hypochlorite
A mixture of ferric chloride and calcium hypochlorite; used to clarify water.

11665 ferro-chromium
A British Chemical Standard Alloy; low carbon No.203 contains 69% and over chromium, 0.08% carbon, and 0.01% sulfur, while the high carbon alloy contains 71.4% and over chromium, 5.09% carbon, and 0.02%.

11666 ferro-cobalt
An alloy of 70% cobalt with iron; used for adding cobalt to steel.

11667 ferrocobaltite
A mineral. It is a cobaltite containing iron.

11668 Ferrocrete
A brand of rapid hardening Portland cement of high strength. No manufacturer.

11669 ferro-cupralium
An alloy of 75-80.5% copper, 11-12% aluminum, and 2-13% iron.

11670 Ferro-Cure
For rubber compounding additives, such as antioxidants, vulcanizing agents and the like in International Class 1. *Ferro.* Discontinued.

11671 ferrodur
A substance containing Nitrolim (calcium cyanamide); used for case hardening and tempering iron and steel.

11672 Ferrofloc
7758-94-3 4091 231-843-4
Ferrous chloride
Water treatment chemical developed from ferrous chloride. *Rheox Inc.* Discontinued.

11673 Ferrogen
Ladle cleaning flux. *Foseco (F.S.) Ltd.*

11674 ferro-magnesite
Obtained by burning magnesite with iron ore; used as a lining for furnaces.

11675 ferro-manganese
An alloy of manganese with iron and carbon, usually made in the blast furnace. High-grade alloys contain about 78% manganese and 8% carbon. These alloys vary from 50-80% manganese, 10-42% iron, 2% silicon, and 5-8% carbon.

11676 ferro-molybdenum
An alloy of iron with 80% molybdenum; used in the place of molybdenum in the manufacture of hard steel.

11677 ferron
An alloy of 50% iron, 35% nickel, and 15% chromium. Also a building material prepared from the pickling liquor from steel mills. It consists of precipitated iron oxide.

11678 ferro-nickel
nickel iron
An alloy of iron and nickel, usually containing 25% nickel.

11679 ferronite
A solid solution of about 0.27% carbon in β-iron.

11680 Ferrophos Pigment
A refractory ferro-alloy developed for use in high performance specialty coatings; its primary use is in zinc-rich coatings where it works with zinc dust to provide good corrosion resistance and weldability characteristics; for use in weldable coil coatings and conductive paints for both EMI shielding and electrostatic needs. *Occidental Chemical Corp.*

11681 ferro-phosphorus
An alloy of iron and phosphorus; used in steel making for thin castings.

11682 Ferrosil 14
12178-41-5
Ferroaluminum silicate
Filler for abrasion-resistant plastic systems, caulks, sealants, polishes, abrasive compounds; extender pigment for primers and other coatings. *Kaopolite.*

11683 ferro-silicon
Alloys of iron and silicon made in the arc type electric furnace. They are graded upon the silicon content. The ordinary grades containing 25, 45-50, 75, and 95% silicon are used in steel works.

11684 ferro-silicon-aluminum
An alloy of iron, silicon, and aluminum, containing up to 15% silicon.

11685 Ferrotone
Mildly acidic scale remover. *Solvay Interox Ltd.* Discontinued.

11686 Ferrotubes
An iron degassing and scavenging agent. *Foseco (F.S.) Ltd.* Discontinued.

11687 ferro-tungsten
An alloy of iron and tungsten. It usually contains from 65-85% tungsten, and from 1-2% carbon; used in the steel industry.

11688 ferro-uranium
Alloys of iron and uranium, containing from 30-50% uranium; used in steel making.

11689 ferrous chloride
7758-94-3 4091 231-843-4
Cl_2Fe
Iron(II) chloride
Ferrofloc; Ferro 66; Iron Protochloride; Iron chloride; iron dichloride. Water treatment chemical. Used in metallurgy; as reducing agent; in pharmaceutical preparations; as mordant in dyeing. mp = 674°; bp = 1023°; d^{25} = 2.90; freely soluble in water, alcohol, acetone, slightly soluble in benzene, practicallly insoluble in ether; LD_{50} (rat orl)= 900 mg/kg. *Rheox Inc.* Discontinued.

11690 ferrous sulfate
7782-63-0 4105 233-072-9
$FeO_4S \cdot 7H_2O$
ferrous sulfate heptahydrate
copperas; green vitriol; iron vitriol; Geosol; Feospan; Fesofor; Fesotyme; Fero-Gradumet; Fer-in-Sol; Haemofort; Ironate; Mol-Iron; Presfersul; Sulferrous. Used in manufacture of iron and iron compounds, also as a hematinic. d = 1.897; soluble in H_2O, insoluble in organic solvents; LD_{50} (mus orl) = 1.52 g/kg.

11691 ferro-vanadium
An alloy of iron and vanadium, containing from 20-40% vanadium. It is added to steel and iron.

11692 ferro-vanadium (No. 205)
A British Chemical Standard alloy. Vanadium 52.2%. (standard). It also contains carbon 0.16%. sulfur 0.03%. (not standard), silicon 1.18%. phosphorus 0.05%.

11693 Ferrox
Trade name for yellow iron oxides used as paint pigments; consists of 98-99%. $Fe(OH)_3$, with added calcium sulfate. No manufacturer.

11694 Ferroxide
1309-37-1 4072 215-168-2
Fe_2O_3
ferric oxide
ferric sesquioxide; jeweller's rouge; Red iron oxide; C.I. 77491; Rouge; Iron oxide fume; Iron (III) oxide; Ferric oxide red; Iron oxide; Iron oxide (Fe2O3); diiron trioxide; English iron oxide red; Iron oxide red; Pigment red 101; Iron(III) oxide dihydrate. Synthetic iron oxide pigments; for surface coatings, plastics and cement coloring. *Mercian Minerals & Colours Ltd.* Unverified.

11695 ferroxyl reagent
A gelatin or agar-agar jelly containing phenolphthalein and potassium ferricyanide. When a piece of iron is placed in the jelly, colors are formed at the ends of the metal after a time. Iron ions give a color of Turnbull's blue with the potassium; ferricyanide and hydroxyl ions, a pink color with the phenolphthalein; used to test corrosion in iron.

11696 Ferrozell
A proprietary trade name for a synthetic resin. No manufacturer.

11697 ferro-zirconium
A 20% zirconium alloy with iron; used to remove nitrogen and oxides from steel.

11698 Ferrozoid
Vestalin
Alloys. They are usually 28% nickel steels and are used as electrical resistances.

11699 ferrozone
A saccharated iron and vanadium compound.

11700 ferrugo
20344-49-4 4068 243-746-4
$FeHO_2$
Ferric hydroxide
Used as a pigment, catalyst, in water purification.

11701 ferrul
An alloy of 54.6% copper, 40% zinc, 5% lead, and 0.4% aluminum.

11702 ferrum
The Latin name for iron.

11703 Ferrux
Feeding compound for ferrous metals. *Foseco (F.S.) Ltd.*

11704 Ferry Alloy
A trademark for an electrical resistance alloy of 54% copper and the balance nickel, a material with a very low temperature coefficient of resistance. *Wiggin Alloys Ltd.* Unverified.

11705 Ferry metal
Alloys used for electrical resistance and containing 40-45% nickel and 55-60% copper. The name appears to be also applied to a bearing alloy and to a solder containing lead with 2% barium, 1% copper, and 0.25% mercury.

11706 Fertibor
Braxo. Industrial hand cleanser. *Borax Europe Ltd.*

11707 Fertilox
1305-79-9 1736 215-139-4
calcium peroxide
An antiseptic. Used in agriculture and horticulture. *Solvay Interox Ltd.* Discontinued.

11708 Festoform
A solid preparation of formaldehyde, obtained by mixing an aqueous solution of formaldehyde with a soda soap solution; an antiseptic disinfectant and deodorizer.

11709 Fetrilon® Combi
Easily soluble micronutrient mixed fertilizer containing 9% magnesium; for all agricultural crops, vines, fruit, and hops. *BASF AG; BASF plc.*

11710 Fettel
dicamba-mecoprop-triclopyr
Emulsifiable concentrate of 78g dicamba, 130g mecoprop, and 72g triclopyr per liter; used for weed control in cereals and grassland. *Farm Protection Ltd.*

11711 Feuille Morte
A jeweler's alloy, containing 70% gold and 30% silver.

11712 FFA-5, FFA-9
25928-94-3
Epoxy resin adhesive; offers strong bond; mix ratio: 100 pbw adhesive/150 pbw Activator BA-15; FFA-9 is fast setting, R.T. curing adhesive; bond is more brittle than FFA-5; mix ratio: 100 pbw adhesive/100 pbw Activator BA-11. *Bacon.*

11713 Fiba-Bond Cl, W
Modified urea-formaldehyde resins; wet strength resins for paper industry; for controlled sizing and wet strength at lower use levels. *CNC Int'l L.P.*

11714 Fiber Pare
A short length olefin fiber; asphalt reinforcement for highway paving, patching, seal coating, crack sealing, curb mix designs. *Hercules.* Discontinued.

11715 Fiberfil® J-1/30
32131-17-2
Nylon 6/6, glass fiber-filled; reinforced engineering thermoplastic for demanding components in automotive, consumer products, mechanical, and electrical/electronic Industries. *DSM.* Discontinued.

11716 Fiberfil® J-17/30/VO
32131-17-2
Nylon 6/6, glass filed; flame retardant thermoplastic for electrical/electronic, mechanical and consumer/industrial applications. *DSM.* Discontinued.

11717 Fiberfil® J-4/35
Nylon 6/12, glass fiber-filled; reinforced engineering thermoplastic for demanding components in automotive, consumer products, mechanical, and electrical/electronic industries. *DSM.* Discontinued.

11718 Fiberfil® J-60/30/E8
9003-07-0 7741
Polypropylene, glass fiber-filled; reinforced engineering thermoplastic for automotive, consumer, and mechanical components where price/performance ratio is important and good chemical and moisture resistance are critical. *DSM.* Discontinued.

11719 Fiberfil® J-60/30/FR
9003-07-0 7741
Polypropylene, glass-filled; flame retardant thermoplastic for electrical/electronic, mechanical and consumer/industrial applications. *DSM.* Discontinued.

11720 Fiberfil® J-7/33
25038-54-4 6832
Nylon 6, glass-filled; toughened nylon for automotive and consumer products. *DSM.* Discontinued.

11721 Fiberfil® J-7/33/IT
25038-54-4 6832
Nylon 6, glass-filled; incrementally toughened nylon for automotive and consumer products. *DSM.* Discontinued.

11722 Fiberfil® NY-16/MF/40
32131-17-2
Nylon 6/6, mineral-filled; reinforced engineering thermoplastic for demanding components in automotive, consumer products, mechanical, and electrical/electronic industries. *DSM.* Discontinued.

11723 Fiberfil® NY-7
25038-54-4 6832
Nylon 6; toughened nylon for automotive and consumer products. *DSM.* Discontinued.

11724 Fiberfil® NY-7/VO
25038-54-4 6832

Nylon 6; high impact flame retardant thermoplastic for electrical/electronic, mechanical and consumer/industrial applications. *DSM.* Discontinued.

11725 Fiberfil® PP-60/TC/40
9003-07-0 7741
Polypropylene, talc-filled; reinforced engineering thermoplastic for automotive, consumer, and mechanical components where price/performance ratio is important and good chemical and moisture resistance are critical. *DSM.* Discontinued.

11726 Fiberfrax® 6000 RPS
1332-58-7 5294 296-473-8
Kaolin ceramic fiber; reinforcement and filler for phenolic, epoxy, nylon, melamine and polyurethane systems. *Carborundum.* Name unverified.

11727 Fiberglas® 101C
Chopped glass strand; reinforcement for bulk molding compounds. *Owens-Corning Fiberglas.* Name unverified.

11728 Fiberil® M-1492
9003-07-0 7741
Polypropylene, glass and mineral-filled; reinforced engineering thermoplastic for automotive, consumer, and mech. components where price/performance ratio is important and good chemical and moisture resistance are critical. *DSM.* Discontinued.

11729 Fiberite
Thermosetting composites. *ICI Chem & Polymers Ltd.*

11730 Fiberite 944
Polyimide resin
For missile fins, high temp. aerospace structures; self-extinguishing, low smoke generation; service temperature 316°. *ICI Fiberite.*

11731 Fiberite 986
Bismaleimide resin
For high temperature structural aircraft parts; low odor; service temp. 218°. *ICI Fiberite.*

11732 Fiberite 6070
Phenolic resin
For aircraft interiors; low smoke, self-extinguishing; designed for press molding; service temp. 177°. *ICI Fiberite.*

11733 Fiberite 7669
25928-94-3
Epoxy resin
For aerospace structures; excellent tack and drape; service temp. 177°. *ICI Fiberite.*

11734 Fiberite 7701
25928-94-3
Epoxy resin
For structural aircraft exteriors; self-extinguishing; solvent resistant; self-adhesive to honeycomb core; press mold and vacuum bag processability; service temperature 93°. *ICI Fiberite.*

11735 Fiberite 9002
Polyester resin
Polyester; for aircraft ducting, radomes; self-extinguishing; vacuum bag, press, and autoclave processability; service temperature 163°. *ICI Fiberite.*

11736 Fiberkal
1332-58-7 5294 296-473-8
Fibrous calcined kaolin; reinforcement for plastic systems; asbestos replacement in some applications; filtration aid; texture paints. *Kaopolite.*

11737 Fiberlac
A proprietary trade name for cellulose nitrate lacquer. No manufacturer.

11738 Fiberloc® 803GR10
Glass-reinforced vinyl composite; for engineered applications including appliances, business equipment, construction, electrical equipment, heating, ventilating and air conditioning, marine products, plumbing and water treatment, windows; high strength, stiffness, dimensional stability. *BFGoodrich/Geon Vinyl.*

11739 Fiberloid
A proprietary trade name for a cellulose nitrate plastic resistant to oils. No manufacturer.

11740 Fiberlon
A proprietary trade name for a phenol-formaldehyde synthetic resin resistant to oils. No manufacturer.

11741 Fiberod
Long, continuous, parallel reinforcing fibers set in thermoplastic resin systems; available in rod, bar, tape, and bidirectional fabric foam; thermoplastic molding compounds for compression, transfer and injection molding; used for automotive, appliances, equipment, sporting and other applications. *Polymer Composites.* Unverified.

11742 Fiberoid
An American grade of vulcanized fiber used for electrical insulation; also the name for a celluloid. No manufacturer.

11743 Fiberstos
A silica-asbestos product used as an insulator.

11744 Fiberstran® G-1/50
32131-17-2
Nylon 6/6, 50% long glass-reinforced; long fiber reinforced thermoplastic with high impact strength for demanding metal replacement applications in automotive, consumer products, mechanical and electrical/electronic industries. *DSM.* Discontinued.

11745 Fiberstran® G-3/50
25038-54-4 6832
Nylon 6, 50% long glass-reinforced; long fiber reinforced thermoplastic with high impact strength for demanding metal replacement applications in automotive, consumer products, mechanical and electrical/electronic industries. *DSM.* Discontinued.

11746 Fiberstran® G-4/45
Nylon 6/12, 45% long glass-reinforced; long fiber reinforced thermoplastic with high impact strength for demanding metal replacement applications in automotive, consumer products, mechanical and electrical/electronic industries. *DSM.* Discontinued.

11747 Fibestos
A proprietary trade name for a cellulose acetate plastic. No manufacturer.

11748 Fibra-Cel®
9004-34-6 2012 232-674-9
Cellulose
Cellulose fibers; functional fillers, reinforcement, thickening aid, processing aid, conditioning agent for the rubber industry (shoe soles, belts, tires, gaskets), thermoset and thermoplastic resins (building products, pet food, asphalt, latex paints). *Celite.*

11749 Fibral
Polishing pads and lenses. *Carl Freudenberg.* Name unverified.

11750 Fibralda
A proprietary trade name for untwisted cellulose acetate fibers. No manufacturer.

11751 Fibro
A proprietary artificial silk product. It is a staple fiber made by the viscose process. No manufacturer.

11752 Fibroc
A laminated product consisting of fiber impregnated with a phenol-formaldehyde resin. No manufacturer.

11753 Fibron
A trademark for a surfacing material for resurfacing and treatment of floors resulting in a plastic finish. No manufacturer.

11754 Fibro-Silk Powd
9009-99-8
Crosilk Powder; Ritasilk Powder; Silk powder. Silk protein pigment. *Brooks Industries.*

11755 Fibrotex
A proprietary trade name for a roofing cement consisting of asbestos mixed with oil and gum. No manufacturer.

11756 Fibrox
A silicon oxycarbide. It is a thermal insulator.

11757 Fibrox 030 SC
Mineral wool fiber; mineral fiber for bitumen, paper, gaskets, paints, friction products, plastics, rubber, coatings, caulks and other applications. *Industrial Fibers.*

11758 Fibrox 300
Mineral wool fiber; filler/reinforcer for bitumen, paper, gaskets, paints, friction products, plastics, rubber, coatings, caulks, and other applications. *Industrial Fibers.*

11759 Fibrredux
Fiber reinforced laminating resins and adhesives. *Ciba plc.* Name unverified.

11760 Ficam
22781-23-3 1063 245-216-8
Bendiocarb
Insecticide used in public health applications. *Cambridge Animal & Public Health Ltd.*

11761 Ficel®
Blowing agents; azo polymerization initiators. *Sherex/Div of Witco.*

11762 Ficel® AC2
123-77-3 950 204-650-8
Azodicarbonamide
Chemical blowing agent for sponge rubber, cushion vinyl floor covering, expanded vinyl coated fabrics, profiles/pipe, sealants. *Schering Berlin Polymers.* Discontinued.

11763 Ficel® AZDN-LF
78-67-1 949 201-132-3
$C_8H_{12}N_4$

2,2-Azodiisobutyronitrile
AIBN; Porofor 57. Polymerization initiator for a wide range of monomers. A blowing agent for elastomers and plastic. It is an initiator for free radical reactions. Used as a foaming agent and an inhibitor in plastic materials. mp = 102-103°; insoluble in H_2O. *Schering Berlin Polymers.* Discontinued.

11764 Fi-Chlor
Chlorocyanurate used for rendering wool shrink resistant. *Fisons plc, Horticultural Div.*

11765 fichtelite
2221-95-6 4120
$C_{19}H_{34}$
[1S-(1α,4aα,4bβ,7β,8aα,10aβ)-tetradecahydro-1,4a-dimethyl-7-(1-methylethyl)phenanthrene
18-norabietane. A hydrocarbon, $C_{19}H_{34}$, found in fossil coniferous resins. mp = 45-46°; bp_{43} = 235-236°; d_4^{22}= 0.9380; n_D^{20}= 1.5052; $[\alpha]_D$ = 19°.

11766 Fi-Clor
87-90-1 9188 201-782-8
symclosene
Chlorinated cyanuric acid. Used as a topical anti-infective, chlorinating agent, disinfectant and industrial deodorant. *Schering Agrochemicals Ltd.* Discontinued.

11767 Ficote
Range of coated fertilizers. *Fisons plc, Horticultural Div.* Name unverified.

11768 Fi-Cryl
Acrylic copolymer solutions and dispersions. *Fisons plc, Horticultural Div.* Name unverified.

11769 Fielder®
For the agriculture industry. *DuPont UK.*

11770 Fi-Gard
Rubber and plastics additives. *Fisons plc, Horticultural Div.* Name unverified.

11771 Filamid
Polymer-soluble dyes for spin coloration of polyamide. *Ciba plc.*

11772 Filastic
A proprietary rubber textile yarn in which the rubber latex impregnation takes place during spinning; the yarns contain 50% of rubber. No manufacturer.

11773 Filcryl
A proprietary acrylic polymer for dental fillings. No manufacturer.

11774 file bronze
An alloy of 64.4% copper, 18% tin, 10% zinc, and 7.6% lead.

11775 Filester
Dyes for PET bottles. *Ciba plc.*

11776 Filex
25606-41-1 247-125-9
Propamocarb hydrochloride
An aqueous concentrate containing propamocarb hydrochloride; a protective fungicide for use on all ornamentals and some edible crops against *Pythium, Peronospora* and *Phytophthora. Fisons plc.*

11777 Fi-Line
Dyeline chemicals for photocopying. *Fisons plc.* Name unverified.

11778 Fillite
A smokeless explosive. It is Ballistite, drawn out into cords with the aid of a solvent.

11779 Fillite
A proprietary inert silicate in the form of spheres; and as glass, it is used as filling material for plastics. *Fillite (Runcom) Ltd.* Name unverified.

11780 Fillite Hollow Microspheres
Free flowing hollow alumina silica microspheres; used for concretes and various refactory cements, ceramic fillers, thermoplastics, undercoating filler material and in molded compounds. *Fillite USA Inc.* Name unverified.

11781 Fillite Solid Microspheres - PFA
Solid alumina silica microspheres; used for concretes and various refractory cements, paints and pigment industry, ceramic fillers and in other molded compounds. *Horn's Crop Service Center.* Name unverified.

11782 Fillmaster
VR Fillmaster. Expanded polystyrene blockform for lightweight landfill civil engineering use; loading bearing fill for roads and embankments. *Vencel Resil Ltd.*

11783 Fillpak
Polyurethane foams; for sealing, as filler. *James Briggs & Sons Ltd.* Name unverified.

11784 Film Plus
Corrosion inhibitors. *Baker Performance Chemicals Ltd.*

11785 Filmex
Alcohol solvent; used in processing of ink, fabrics, latex, lacquers, pharmaceuticals, and chemical specialties, and in other industrial applications. *Quantum Chemical Corp.*

11786 Filmite
White oil preparations. *Murphy Chemical Co Ltd.* Discontinued.

11787 Filmon®
Biaxially oriented nylon 6 film; heat shrinkable. *SNIA (UK) Ltd.*

11788 Filofin
Pigment dispersions for polypropylene fibers. *Ciba plc.*

11789 Filon
Glass fiber reinforced polyester (GRP) sheeting; used for roofing and cladding. *BIP Chemicals Ltd.* Discontinued.

11790 Filpro
Protein hydrolysates; suitable for any savory application: soups, frozen meals, snacks, convenience meals, meat pies and sausages. *PPF International Ltd.* Name unverified.

11791 Filt-char
A proprietary brand of bone charcoal; used as a filtering medium. No manufacturer.

11792 Filter-cel
A proprietary preparation of infusorial earth; used in filtering. No manufacturer.

11793 Filtracarb
Carbonaceous materials for use as filtration media. *Coal Products Ltd.*

11794 Filtracite
Carbonaceous materials for use as filtration media. *Coal Products Ltd.*

11795 Filtram
Laminated filter drains. *ICI Chem & Polymers Ltd.* Discontinued.

11796 Filtrasorb® 100, 200
Activated carbon; for removal of dissolved organic chemicals from municipal and industrial water. *Calgon Carbon.* Name unverified.

11797 Filtrez
Rosin resin; used in printing inks, coatings and adhesives. *Monsanto Co.* Name unverified.

11798 Filtrol
A trademark for a decolorizing substance consisting of fine silica with a little aluminum silicate. No manufacturer.

11799 Filtros
Gum rosin; used in paints, varnishes and lacquers. *Monsanto Co.* Name unverified.

11800 Filtrosol A
118-56-9 4776 204-260-8
Homosalate
Ultraviolet screen. *Norda Inc.* Unverified.

11801 Final Flip
Trade name for a pelleted rodenticide, single feed type, indoor or outdoor use, place pack size; contains a chronic toxicant as the active material. *Colonial Products Inc.*

11802 Final Touch
General degreasers. *Evans Vanodine International Ltd.*

11803 Finaplix
10161-34-9 9716 233-432-5
Trenbolone acetate
Anabolic (veterinary). *Roussel UCLAF, Fine Chemicals.* Name unverified.

11804 fine gold
A jeweler's alloy, containing 75% gold and 25% silver.

11805 fine silver
99.9% pure silver.

11806 Fine-Clad®
Synthetic resins for use in protective coatings. *Reichhold.*

11807 Finesse®
Weed killer. *DuPont UK.*

11808 Finestol
A phloroglucinol dye coupler. *Fisons plc.* Name unverified.

11809 Finex-25
1314-13-2 10279 215-222-5
Zinc oxide
Provide excellent feel, covering power and skin adherance; uv absorber; effective as sun protection factor booster. *Presperse.*

11810 Finex-25-020
1314-13-2 10279 215-222-5
Zinc oxide
methicone. Provides excellent feel, covering power and skin adherance; uv absorbed; effective as sun protection factor booster. *Presperse.* Discontinued.

11811 finings
The term applied to isinglass dissolved in an acid such as tartaric acid; used to clarify beer.

11812 Finish®
For the agriculture industry. *DuPont UK.*

11813 Finistrol ESJ Concentrate
Polyethylene emulsion; nonionic; light brown viscous liquid; fat-free softener for cellulose fibers and synthetics, preferably polyamide and its blends. *Thor Chemicals (UK) Ltd.* Discontinued.

11814 Finistrol GZ
Polysiloxane emulsion; nonionic; thin white liquid; softener for use alone or in combination with synthetic resins or reactants; improves sewability of resin-finished materials. *Thor Chemicals (UK) Ltd.* Discontinued.

11815 Finitron
N-Ethyl perfluorooctanesulfonamide
Insecticide. *Griffin.*

11816 Finizym
A fungal β-glucanase preparation produced by submerged fermentation of a selected strain of Aspergillus niger; used during fermentation and storage of the beer to prevent filtration difficulties and to prevent precipitation of β-glucans. *Novo Nordisk.*

11817 Finntitan
13463-67-7 9612 236-675-5
Titanium dioxide
Used in paints, inks, rubbers, cosmetics, ceramics etc. *Cornelius Chemical Co Ltd.* Name unverified.

11818 Fintex 572
Quaternary ammonium compound in liquid form; cationic surfactant used in fabric conditioning. *Berol Kemi (UK) Ltd.* Name unverified.

11819 Fiolax
A trade name for a resistant glass. No manufacturer.

11820 Fiorivert
1-Ethoxy-1-phenylethoxyethane
Quest Int'l. UK Ltd.

11821 fir wool oil
The oil of Scotch fir leaves.

11822 Fire PRF$_2$ 1000 FM
Two-component preaccelerated, precatalyzed phenol resorcinol-formaldehyde resin; flame-retardant, low-smoke thermoset resin for fabricating factory mutual-approved fume and smoke exhaust duct systems; mix ratio 45.7/34.3/20 (Part A/B/filler). *Indspec.* Discontinued.

11823 Fire Retardant FR-8
Salt-type all-purpose fire retardant for textile fabrics and cellulosics. *Emco Services.*

11824 fire-armour
Heat-resisting alloys containing 60-61% nickel, 18-20% chromium, 10-20% iron, 0-1.8% manganese, and 0.5% carbon.

11825 Firebrake ZB
Boroxo. A heavy-duty soap powder hand-cleanser, based on borax. *Borax Europe Ltd.*

11826 Firebrake® ZB
12513-27-8
zinc borate
Flame retardant synergistic with antimony oxide or alumina trihydrate in most halogenated polyesters and vinyl esters; delays onset of oxidation; used in chlorinated polyesters PVC plastisols, and epoxy systems; smoke suppressant; char promoter; improved electrical properties in polyester andnylon; promotes adhesion between plastics and metals; translucent, low toxicity and non-hydroscopic. *U.S. Borax & Chem.*

11827 Firecheck
Water-based elastomeric fire retardant protective coating, applied by brush or spray; for asbestos encapsulation, fire retardant weatherproofing, combustible substrates. *Liquid Plastics Ltd.*

11828 fireclay
Clay containing a considerable amount of free silica.

11829 Firecol
An alginate-containing fire-fighting suspension. *Alginate Industries Ltd.* Name unverified.

11830 Firecrete
A proprietary trade name for a calcined high alumina clay used as a refractory in furnaces. No manufacturer.

11831 fire-damp
74-82-8 6019 200-812-7
methane
A gas, consisting primarily of methane, often found in coal mines.

11832 FireGuard 910
Plenum compound; low smoke, low flame compound for plastics industry. *Teknor Apex.*

11833 FireMaster®
Fire protection products available in bulk, as blanket, board, putty, etc. for cable tray, duct, plastic pipe, and structural steel fire protection. *Thermal Ceramics.*

11834 Firemaster® 642
Proprietary flame retardant. *Great Lakes.*

11835 Firemaster® HP-36
Halogenated phosphate ester; flame retardant. *Great Lakes.*

11836 Firesaife
7722-76-1 577 231-764-5
ammonium phosphate, monobasic
Ammonium phosphate fire retardant. *Scottish Agricultural Industries plc.*

11837 FireShield® H
1309-64-4 752 215-175-0
antimony trioxide
General purpose flame-retardant synergist for plastics, rubber, paper, and paint; polymerization catalyst for PET resins and fibers; also for electronics, glass, ceramics, petroleum refining, and chemical manufacturing. *Laurel Industries.*

11838 FireShield® HPM
1309-64-4 752 215-175-0
antimony trioxide
Catalyst, chemical intermediate and flame retardant; suitable for sensitive electronic applications. *Laurel Industries.*

11839 FireShield® L
1309-64-4 752 215-175-0
antimony trioxide
General purpose flame-retardant synergist for plastics, rubber, paper, and paint; polymerization catalyst for PET resins and fibers; also for electronics, glass, ceramics, petroleum refining, and chemical manufacturing. *Laurel Industries.*

11840 Firit
A foundry refractory coating. *Foseco (F.S.) Ltd.*

11841 Firmadent
Denture adhesive (powder, cream and liquid), for securing dentures. *Richardson-Vicks Inc.* Name unverified.

11842 Firnagral
Iranolin
A mineral drying oil extracted from aromatic petroleum residues; used to replace up to 30% of the linseed oil in putty. It is stated to give the putty a harder and smoother surface.

11843 Firnis
Linseed oil and driers.

11844 First Choice Electroless Palladium
Electroless plating solution for the autocatalytic chemical deposition of palladium metal; for plating of electronic components for its low contact resistance and high wear properties and for the metalization of ceramics, plastics, and other nonmetallics. *Callery Chemical Company.* Name unverified.

11845 Firthite
A proprietary trade name for a material consisting of a mixture of tungsten and other carbides. Abrasives. No manufacturer.

11846 Fischer-Langbein solution
A solution containing 450 g ferrous chloride, 500 g calcium chloride, and 750 ml water; used for the electro-deposition of iron, with a current density of up to 120 amps/ft^2. A temperature of 60-70° is used.

11847 Fischer's reagent
A test solution for sugars. It consists of 2 parts phenylhydrazine hydrochloride and 3 parts sodium acetate in 20 parts of water.

11848 Fischer's salt
13782-01-9 2493 237-435-2
$CoK_3N{+}6O_{12}$
Potassium cobaltic nitrite
Fischer's yellow; C.I. Pigment Yellow40; C.I. 77357. Used for the detection and determination of potassium. Also used as a pigment in painting glass, porcelain and rubber.

11849 fisetin
$C_{15}H_{10}O_6$
Tetrahydroxymethylanthraquinone
A yellow coloring from the wood of *Quebracho colorado* etc.

11850 Fisons 18-15, MCPB
94-81-5 202-365-3
MCPB
MCPB, a selective weedkiller. *Fisons plc, Horticultural Div.* Name unverified.

11851 Fisons P.C.P
A weedkiller. *Fisons plc, Horticultural Div.* Name unverified.

11852 Fi-Vi
A lightweight expanded PVC. *Fisons plc.* Name unverified.

11853 **Fixanal**
Analytical chemicals accurately weighed and sealed, ready for rapid volumetric solution.
11854 **Fixaplus**
X-ray fixer. *May & Baker Ltd.* Name unverified.
11855 **Fixapret®**
Textile finishing aids for anti-crease, anti-shrink, and easy care finishing. *BASF AG; BASF plc.*
11856 **Fixat**
Inorganic molding sand binders for foundries, also in connection with lustrous carbon formers to obtain better castings. *Sud-Chemie AG.* Name unverified.
11857 **Fixatek**
X-ray fixer. *May & Baker Ltd.* Name unverified.
11858 **fixed white**
7727-43-7 1023 231-784-4
$BaSO_4$
barium sulfate
Commercial barium sulfate, a white pigment and radiopaque medium.
11859 **Fixegal**
Dispersing and leveling agent; used for dyeing in the textile industry. *Degussa.* Name unverified.
11860 **Fixin**
18917-91-4 360 242-670-9
aluminum lactate
Used in foam fire extinguishers and in dental impression materials.
11861 **Fixinvar**
An alloy having the same properties as Elinvar, but having greater stability.
11862 **Fixodent**
Denture adhesive cream for securing dentures. *Richardson-Vicks Inc.* Name unverified.
11863 **Fixogene**
Textile auxiliary chemicals. *ICI Chem & Polymers Ltd.*
11864 **Fixol**
Adhesives. *Associated Adhesives.* Unverified.
11865 **Fix-Sol**
A concentrated fixing and hardening solution. *Johnsons of Hendon.* Name unverified.
11866 **FK 140**
Precipitated silica; for silicone rubber industry. *Degussa.*
11867 **FK 300DS**
Silica, precipitated; for paper and film industry. *Degussa.*
11868 **FK 500LS**
Precipitated silica; thickener for cosmetic creams and lotions; also in thermal insulation, paper, films, pesticides, pharmaceuticals. *Degussa.*
11869 **FL7P**
Liquid fertilizer. *Fisons plc, Horticultural Div.* Name unverified.
11870 **flake litharge**
Litharge, a pigment made by the oxidation of lead.
11871 **flake white**
1304-85-4 1326 215-136-8
Cremnitz; cremnitz white; kremnitz; crems white; blanc d'argent; silver white; London white; Nottingham white; Lead whites. Carbonates of lead containing varying quantities of hydrated oxide of lead. Flake white is a variety of chamber white lead, obtained in flaky pieces by heating lead plates. A basic bismuth is also known as flake white and the term flake or pearl white is sometimes used for bismuth oxychloride.
11872 **Flamarret®**
1309-48-4 5713 215-171-9
magnesium oxide
Magnesium oxide for fire protection. *Steetley Magnesia Products Ltd.*
11873 **Flamco**
Mold and core dressings. *Foseco (F.S) Ltd.*
11874 **Flame guard**
1344-28-1 369 215-691-6
aluminum oxide trihydrate
Fire retardant. *Alcoa Industrial Chemicals.*
11875 **Flame Out 44**
decabromodiphenyl oxide-antimony trioxide
Decabromodiphenyl oxide-antimony trioxide 2:1 aqueous dispersion; a general-purpose flame retardant for textile applications. *Emco Services.*
11876 **Flame Out CO**
Low-cost flameproofing agent for use on cottons; nonyellowing; imparts a soft hand. *Emco Services.*
11877 **Flamegard® 908**
Methylolthiourea-based compound; flame retardant for nylon. *Sybron.*
11878 **Flamenco®**
Pigments of exceptional brilliance displaying twin colors; for cosmetics. *Mearl.*
11879 **Flamenol**
A proprietary trade name for a polyvinyl chloride synthetic resin. No manufacturer.
11880 **Flaming**
A decolorizing agent for sugar juices.
11881 **Flammastik**
Incombustible cable coatings applied by spray guns or spatula. *Degussa AG.* Name unverified.
11882 **Flammentin ASN**
Ammonium-phosphorus compound; clear solution; nonwater resistant flameproofing agent for textiles made from cellulose and wool; especially suitable for re-impregnation for decoration textile materials. *Thor Chemicals (UK) Ltd.* Discontinued.
11883 **Flammentin PS**
Reactive phosphorus-nitrogen compound; thermo-stable flameproofing agent for textiles of cellulose; especially for carrier fabrics to be coated and exposed to higher temperatures. *Thor Chemicals (UK) Ltd.* Discontinued.
11884 **Flammex**
Flame retardant. *The Wellcome Foundation Ltd.* Name unverified.
11885 **Flammocite**
A safety explosive. It contains 44% ammonium nitrate, 16% sodium chloride, 14% sodium nitrate, 10% trinitro-toluene, 6% nitro-glycerin, 5% ammonium sulfate, and 5% cellulose.
11886 **Flamolin**
Flame retarded polymers. *Quantum Chemical Corp.*
11887 **flamprop-isopropyl**
52756-22-6 258-154-1
$C_{19}H_{19}ClFNO_3$
Isopropyl-N-benzoyl-N-(3-chloro-4-fluorophenyl)alanine
flamprop-isopropyl; Suffix BW; WL 29762; Barnon; Flufenprop-isopropyl;. Herbicide.
11888 **Flamprop-M-isopropyl**
52756-22-6 258-154-1
Commando; Gunner; Power Flame; Power Flamprop. Emulsifiable concentrate of 200 g flamprop-M-isopropyl per liter; used for control of wild oats in cereal crops. *Quadrangle Agrochemicals; Shell UK; Kommer-Brookwick Ltd.*
11889 **Flamtard H**
12027-96-2
Zinc hydroxystannate
Flame retardant for PVC, polychloroprene, chlorosulfonated polyethylene, other halopolymers. *Alcan.*
11890 **Flamtard S**
12036-37-2
O_3SnZn
Zinc stannate
Flame retardant for PVC, polychloroprene, chlorosulfonated polyethylene, polypropylene, other halopolymers. *Alcan.*
11891 **Flandrac**
A decolorizing agent for sugar juices.
11892 **Flat-Ayd**
Dispersed flatting agents; used for paints, inks etc. *Cornelius Chemical Co Ltd.* Name unverified.
11893 **Flat-Ayd® Bases**
Dispersions of flatting agents in variety of vehicles; predispepersed flatting bases for gloss control of coatings. *Elementis Specialties.*
11894 **Flatting Agent OK412**
Surface-treated precipitated silica; flatting agent for clear and pigmented solvent- and water-based coatings, printing inks; antisettling agent. *Degussa.*
11895 **Flavaxin**
83-88-5 8367 201-507-1
Riboflavin
Vitamin B_2. *Sterling Drug Inc.* Name unverified.
11896 **Flavazol**
$C_{13}H_{12}N_2O_2$
A dyestuff, prepared from *p*-toluidine and salicylic acid. It dyes yellow on chrome mordanted wool.
11897 **Flaveosine**
A dyestuff obtained by condensing *m*-acetaminodimethylaniline with phthalic anhydride. It dyes tannined cotton and wool reddish yellow, and silk golden yellow.
11898 **Flavinduline**
Induline yellow. A dyestuff. Dyes tannined cotton yellow.

11899 flavine
Three materials are known by this name; a) Diaminobenzophenone; b) a grade of quercitron bark extract; and c) diaminomethylacridinium chloride.

11900 Flavocents
Concentrated flavor compositions. *PPF International Ltd.* Name unverified.

11901 flavoline
2-Phenyl-4-methylquinoline

11902 Flav-O-Lok
Free-flowing nonhygroscopic flavor powders; used in formulations in food, beverages and pharmaceuticals where liquid flavors cause difficulties. *Hercules.* Discontinued.

11903 flavone
525-82-6 4132 208-383-8
$C_{15}H_{10}O_2$
2-phenylbenzopyran-4-one
2-phenylchromone. One of a group of flavonoid plant pigments existing as colorless needles, insoluble in water and melting at about 100°. The flavones produce ivory and yellow colors in plants and flowers. mp = 97-99°; insoluble in H_2O; LD_{50} (mus orl) = 2500 mg/kg.

11904 flavopurpurin
$C_{14}H_8O_5$
Alizarin No. 10CA; alizarin FA; alizarin GB; alizarin GI; Alizarin RG; alizarin SDG; alizarin X; alizarin VCA; alizarin CAF; alizarin DCA; alizarin JCA; alizarin VAR. Trihydroxy-anthraquinone. Red dyes cotton mordanted with alumina.

11905 Flavotint
Food color and flavor compounds. *PPF International Ltd.* Name unverified.

11906 Flav-R-Keep FP-51
7758-29-4 8846 231-838-7
Sodium tripolyphosphate
Lemon juice solids, flavorings; food additive for meat, poultry, and seafood industries. *Rhône-Poulenc Food ingreds.*

11907 flax wax
A wax associated with flax fiber and with the cortical tissues. The air-dried cortex contains as much as 10% by weight of wax. It is removed by extraction with a volatile solvent.

11908 FLC-2
Aliphatic hydrocarbon resin; high softening point resin used as a fluid loss control additive. *Shell.*

11909 Flea Flip
A line of flea control products; insecticide for household, pet, outdoor use in the control of fleas, ticks and other insects; includes aerosols, concentrates and ready-to-use liquids. *Colonial Products Inc.*

11910 Flea-B-Gon
Flea killer. *Monsanto (Solaris).* Name unverified.

11911 Flectol H
26780-96-1
Polymerized 1,2-dihydro-2,2,4-trimethyl-quinoline
Antioxidant; resists effect of heat deterioration and normal aging in dry rubber and latex. *Monsanto Co.*

11912 Flectol ODP
Octylated diphenylamine
Antioxidant protecting rubber against heat deterioration and normal aging; for neoprene. *Monsanto Co.*

11913 Flectol Pastilles
26780-96-1
Polymerized 1,2-dlhydro-2,2,4-trimethylquinoline
Antioxidant used in tires, belts, hose retread rubber and general mechanicals. *Monsanto Co.* Name unverified.

11914 Fletcher's alloy
An alloy of 95.5% aluminum, 3% copper, 1% tin, 0.5% antimony, and 0.5% phosphor tin.

11915 Fletcher's bearing alloys
Aluminum base alloys. One contains 92% aluminum, 7.5% copper, and 0.25% tin, and another 90% aluminum, 7% copper, and 1% zinc.

11916 Fleur
Intensive decorating colors with flux mask for enamel and earthenware. *Degussa Ltd.*

11917 Fleurelle®
Fibers. *DuPont UK.*

11918 Flex Carbon
1333-86-4 1856 215-609-9
A proprietary carbon black. No manufacturer.

11919 Flex®
For the medical industry. *DuPont UK.*

11920 Flexade Regular
Lubricant, corrosion inhibitor; easily soluble in water; biodegradable; used for metal treatment. *Carboxyl Chemicals Ltd.* Name unverified.

11921 Flexalyn
Diethylene glycol diabietate
A proprietary trade name for a plasticizer. No manufacturer.

11922 Flexamine
A superflexing antioxidant containing 35% JZ; for use in heavy service truck tread and SBR treads, camelback, wire insulation, neoprene belting and molded soles; offers protection against copper and manganese. *Uniroyal.* Name unverified.

11923 Flexan® 130
Sodium polystyrene sulfonate
Hair fixative for setting lotions, conditioners, blow drying aid; gloss and antistatic chars. *Nat'l. Starch.*

11924 Flexane®
Two-part urethane; for making flexible molds, cast parts and nonmarring holding fixtures, forming abrasion resistant and noise reduction linings, encapsulating parts. *ITW Devcon.*

11925 Flexane® 80
Two-part urethane elastomer; tough rubber compound for repairs for lining and protection applications, tooling and molding applications. *ITW Devcon.*

11926 Flexane® 94
Two-part urethane elastomer; tough rubber compound for repairs for lining and protection applications, tooling and molding applications. *ITW Devcon.*

11927 Flexbond
Vinyl acetate copolymer emulsions. *Air Prods & Chems. Inc.*

11928 Flexchlor
Chlorinated paraffins; solvents. *Witco.* Discontinued.

11929 Flexcote
Rubber/vinyl lacquer; used for decoration and protection of rubber articles. *W J Ruscoe Co.* Discontinued.

11930 Flexcrete
Polymer-modified cementitious mortars (range) for repair and maintenance of (reinforced) concrete; used for high-rise housing, PRC housing, any damaged or defective concrete structures. *Liquid Plastics Ltd.*

11931 Flexcryl
Acrylic emulsions. *Air Prods & Chems. Inc.*

11932 Flexel® 1010
Thermoplastic compound; for wire and cable applications. *BFGoodrich/Geon Vinyl.*

11933 Flexible Fyrex®
monoammonium phosphate-diammonium phosphate
Mixture of crystalline diammonium phosphate and monoammonium phosphate, penetrating and softening agents; inorganic flame retardant; improved surface wetting; improved hand in treated fabrics *Akzo.*

11934 Flexidor
82558-50-7 5256
Isoxaben
Suspension concentrate containing 500 g isoxaben per liter; used for control of annual dicotyledons in cereals, grass and fruit. *DowElanco Ltd.*

11935 Flexin
1343-98-2 8634 215-683-2
Silicic Acid
Selected compound of silicic acid, cationic; antislip agent for fabrics and knitted goods. *Henkel Chemicals Ltd.* Name unverified.

11936 Flexipol® FP-100(M)
Flexible polyurethane systems; foams for acoustic, packaging, toys, padding cushioning applications. *Flexible Prods.*

11937 Flexipol® FSF-106
Flexible polyurethane systems; foams for toys, seats and cushions, padding cushioning and integral skin applications. *Flexible Prods.*

11938 Flexipol® NDTP-311-1.8
Rigid polyurethane systems; for dispensing froth foams through low or high pressure pour machines, spray equipment or hand mix; for flotation and walk-in refrigeration applications. *Flexible Prods.*

11939 Flexipol® NP-311-2
Rigid polyurethane systems; for dispensing froth foams through low or high pressure pour machines, spray equipment or hand mix; for carving and decorative molding applications. *Flexible Prods.*

11940 Flexipol® NS-322-2
Rigid polyurethane systems; for dispensing froth foams through low or high pressure pour machines, spray equipment or hand mix; for flotation and spray insulation applications. *Flexible Prods.*

11941 Flexobond 329
Two-part urethane system; non-TDI system curing to a tough rubber solid with good adhesion to many substrates; used as a fairing compound,

adhesive, or encapsulant; for cure at ambient or elevated temperatures. *Bacon.*

11942 Flexobond 423
Two-part urethane system; non-TDI system curing to a hard solid with high clarity and low odor; used as an adhesive, encapsulant, or casting compound; for cure at ambient or elevated temperatures. *Bacon.*

11943 Flexocel
Twin component flexible polyurethane foam systems. *Baxenden Chemicals Ltd.*

11944 Flexol
A proprietary trade name for vinyl plasticizers. No manufacturer.

11945 Flexol Plasticizer 3GH
Triethylene glycol-di-2-ethyl butyrate
A proprietary trade name for a plasticizer. No manufacturer.

11946 Flexol Plasticizer 3GO
Triethylene glycol-di-2-ethyl hexoate
A proprietary trade name for a plasticizer. No manufacturer.

11947 Flexomer DFDA-1137 Natural 7
Ethylene copolymer; resin with good resilience, toughness, flexibility, chem. resist., stress crack resist. for blow molding, injection blow molding, profile and sheet extrusion, thermoforming, small part injection molding applications where high flexibility is desired, e.g. hose and tubing, squeeze bottles and tubes; blending resins with high pressure or LLDPE to improve their properties. FDA compliant. *Union Carbide.*

11948 Flexonyl
Pigments for aqueous flexographic inks. *Hoechst AG.*

11949 Flexoresin
Proprietary brands of glycol and glyceryl phthalates, which are used as plasticizers; also polymerized terpenes. No manufacturer.

11950 Flexricin® 9
26402-31-3 247-669-7
Propylene glycol ricinoleate
Wetting agent, dye solvent, wax plasticizer, stabilizer for textile, household, and cosmetic applications, rewetting dried skins. *CasChem.*

11951 Flexricin® 13
141-08-2 205-455-0
Glyceryl ricinoleate
Wetting agent, wax plasticizer, and mold release agent for rubber polymers, antifoam agent, household, and cosmetic applications, rewetting dried skins. *CasChem.*

11952 Flexricin® 15
106-17-2 203-369-8
Glycol ricinoleate
Wetting agent, plasticizer, textile, household, and cosmetic applications, rewetting dried skins; chemical intermediate. *CasChem.* Name unverified.

11953 Flexricin® 17
Pentaerythritol ricinoleate
Plasticizer; chemical intermediate. *CasChem.*

11954 Flexricin® 100
141-22-0 8378 205-470-2
Ricinoleic acid
Lubricant for textile, metalworking compounds; corrosion inhibitor intermediate; intermediate for water soluble and dispersible lubricant. *CasChem.*

11955 Flexricin® 115
N(β-Hydroxyethyl) ricinoleamide
Lubricant/antistat for plastics, metals; mold release, antiblocking agent for textile coatings; slip agent for varnishes and lacquers; also for electrical potting compounds, crayons, wax blends, high-temperature greases. *CasChem.* Name unverified.

11956 Flexricin® 185
N,N'-Ethylene bis-ricinoleamide
Lubricant/antistat for plastics, metals; mold release, antiblocking agent for textile coatings; slip agent for varnishes and lacquers; also for electrical potting compounds, crayons, wax blends, high-temperature greases. *CasChem.* Name unverified.

11957 Flexricin® P-1
141-24-2 205-472-3
Methyl ricinoleate
Low temperature lubricant plasticizer for rubber, phenolic and epoxy resins. *CasChem.* Name unverified.

11958 Flexricin® P-3
Butyl ricinoleate
General purpose plasticizer, lubricant for nitrocellulose. *CasChem.* Name unverified.

11959 Flexricin® P-4
140-03-4 205-392-9
Methyl acetyl ricinoleate
All purpose plasticizer, lubricant for vinyls and lacquers. *CasChem.* Name unverified.

11960 Flexricin® P-6
Butyl acetyl ricinoleate
Lubricity additive for textile finishes; plasticizer. *CasChem.* Name unverified.

11961 Flexricin® P-8
101-34-8 202-935-1
Glyceryl (triacetyl ricinoleate)
Plasticizer for vinyl wire jacketing and semirigid vinyls; emollient; stabilizer for anhydrous pigmented systems. *CasChem.* Name unverified.

11962 Flex-Shield
Elastomeric coatings made up of asphalt, neoprene, acrylic and Kraton rubber imbedded in polyester fabric; used for roof coating, maintenance cleaning products, degreaser, waxes, and wall coatings. *Flex-Shield.* Name unverified.

11963 Flexsol 43
Deltafilcon A; contact lens material. *Alcon Laboratories Inc.* Name unverified.

11964 Flexthane 610 EXP, 611 EXP, 620 EXP
Urethane hybrid polymer, aqueous for coating and adhesive applications including graphic arts polymers, textile and vinyl coatings, leather finishes, plastic coatings, industrial coatings, film lamination. *Air Prods & Chem/Polymers.* Name unverified.

11965 Flextron
A nontoxic, nonbiodegradable two-component, cold-curing polyurethane sealant; for expansion and construction joints in water retaining structures; sold in three viscosity grades, suitable for pouring or pressure gunning. *Berger Elastomers.* Name unverified.

11966 Flexzone
A range of antizonants which offer high protection against flexing, ozone, heat, and oxygen in rubber products. *Uniroyal.* Name unverified.

11967 Flexzone 3C
101-72-4 202-969-7
$C_{15}H_{18}N_2$
N-Isopropyl-N'-phenyl-*p*-phenylene diamine
N-isopropyl-N'-phenyl-4-phenylenediamine; Akrochem Antioxidant Pd1; Anto H; Flexone 3c; IPPD; Isopropyl 0 PPD; Permanex IPPD; Santoflex IP; Cyzone; Elastozone 34; Nonox ZA; Santoflex 36; Cyzone IP; Vulkanox 4010 NA. A proprietary antioxidant. *Uniroyal. Name unverified.*

11968 Flexzone 6-H
N-Phenyl-N'-cyclohexyl-*p*-phenylene diamine
A proprietary antioxidant. *Uniroyal.* Name unverified.

11969 Flicker Flake™
Aluminum pigments; used for coatings requiring a glitter or high sparkle effect. *Silberline Mfg Co.*

11970 Flint
linuron-trifluralin
Emulsifiable concentrate containing 120 g linuron and 240 g trifluralin per liter; herbicide for winter cereals. *Ashlade Formulations Ltd.*

11971 flint
7631-86-9 8637 231-545-4
SiO_2
silicon dioxide
A form of silica, used as an abrasive.

11972 flint alloy
A heat and corrosion-resisting alloy containing 83% iron, 12.5% chromium, 3% carbon, and 0.5% silicon.

11973 flint glass
lead glass crystal. A glass composed of lead and potassium silicates; used for hollow-ware, superior bottles, and optical work.

11974 flint Metal
A proprietary trade name for an alloy of iron with 4-4.5% nickel, 1.25-1.75% chromium, and 3-3.5% carbon. No manufacturer.

11975 flintcast
A white iron made in the electric furnace. It resists abrasion.

11976 flinty zinc ore
Flinty calamine.

11977 Fliselina
Nonwoven textile for the apparel industry. *Carl Freudenberg.* Name unverified.

11978 Flit
A proprietary insecticide. No manufacturer.

11979 Flixapret
Synthetic resins for textile finishing. *BASF plc.* Name unverified.

11980 Flo Chem Extra
A water-soluble fluorochemical compound, used as a stain repellent and dry soil resistant finish for fabrics for aqueous application. *Emco Services.*

11981 Flo-Aid
Anticaking agent; used to improve flow properties of fertilizers and explosives. *Crowley Chem.*

11982 float tin
CAS siterite, occurring in the soil, and formed by the disintegration of tin rocks.

11983 floatstone
Porous opal.

11984 Flocculant T-9
Complex; cohesion and sedimentation agent for TiO_2 manufacture. *Toho Chem. Industry.*

11985 Flochel
Poly-electrolyte flocculating agent. *Steetley Chemicals Ltd.* Name unverified.

11986 Flock-Lok® 850
Flexible polyurethane flock adhesive; adhesive featuring excellent adhesion of flock to elastomers and plastics such as ABS; wide range of cure conditions; solvent-resistance; moisture-curable. *Lord.*

11987 Flock-Lok® 851
Flexible polyurethane flock adhesive; moisturing curing adhesive with excellent adhesion of flock to elastomers such as SBR and plastics such as ABS. *Lord.*

11988 Floclean 103
A proprietary preparation of polycarboxylic, alkyl sulfonic and organic acids; for removal of metallic salts from cellulosic reverse osmosis membranes. *Pfizer International.* Discontinued.

11989 Flo-Con
Textile chemical. *Albright & Wilson Ltd., Phosphates & Speciality Business.* Name unverified.

11990 Flocon 100
35% aqueous solution of a proprietary polyacrylate polymer; a scale control agent in desalination. *Pfizer International.*

11991 Flocsil
7631-86-9 8637 231-545-4
silicon dioxide
Activated silica. *Crosfield Chemicals Ltd.*

11992 Floex
A proprietary trade name for a wetting agent for paint, etc.; a condensation product of higher fatty alcohols. No manufacturer.

11993 Flogel
Slurry explosives; for a wide range of surface blasting. *Hercules.* Discontinued.

11994 Flolys®
50-99-7 4467 200-075-1
Glucose
Glucose syrup; sweetening agent for general food use. *Roquette (UK) Ltd.*

11995 Flomac
Granular desulfurizing agent for blast furnace iron and steel. *Foseco (F.S.) Ltd.* Discontinued.

11996 Fl-Mo 5BMP
Free acid of a complex organic phosphate ester; emulsifier for phosphated and chlorinated pesticides; coupling agent for agricultural and industrial applications. *Witco/Organics.* Discontinued.

11997 Fl-Mo 80/20
Modified alcohol ethoxylate
Agricultural adjuvant. *Witco/Organics.* Discontinued.

11998 Fl-Mo 1082
Formulated product; emulsifier for petroleum based agricultural sprays. *Witco/Organics.* Discontinued.

11999 Fl-Mo 1093
Formulated product; emulsifier for vegetable oil based agricultural sprays. *Witco/Organics.* Discontinued.

12000 Fl-Mo DEL, DEH
Formulated products; matched pair of emulsifiers for a broad range of pesticides. *Witco Corporation.*

12001 Fl-Mo Lowfoam
Modified alcohol ethoxylate; surfactant for the formulations of agricultural adjuvants. *Witco Corporation.*

12002 Fl-Mo Suspend
Formulated product; compatibility agent for use in agricultural formulations. *Witco Corporation.*

12003 Flor Sherry
Dry yeast for wine production. *Ciba plc.* Name unverified.

12004 Florafoam
Rigid foam (urethane and phenolic) for floral arrangements. *Baxenden Chemicals Ltd.*

12005 Floramat
Ethyl-2-*t*-butylcyclohexylcarbonate
Fragrance raw material, for floral notes. *Henkel/Cospha.*

12006 Florane
2-Heptyl tetrahydrofuran
Quest Int'l. UK Ltd.

12007 Floranid® N 32
Special slow-release nitrogenous fertilizer with 32% nitrogen for intensive horticultural crops, ornamentals, and lawns. *BASF AG.*

12008 Floranit
Wetting agent; for mercerizing and caustic lye padding. *Henkel Chemicals Ltd.* Name unverified.

12009 Floratex
Melded fabrics. *ICI Chem & Polymers Ltd.*

12010 Florco®
1337-76-4
Attapulgite clay; floor absorbent for oil, grease, water, and other liqs.; offers lighter dens.; antislip agent on floors; used by automotive, steel, transportation, commercial, pet, food and beverage, institutional, and amusements industries. *Floridin.* Name unverified.

12011 Florco® -X
1337-76-4
Attapulgite clay; 8/30 mesh; absorbs oil and water spills; pet absorbent. *Floridin.* Name unverified.

12012 Florel Fruit Eliminator
Etephon
Fungicide used to eliminate messy fruit set from ornamental olives, carobs, apples and crab apples; applied at bloom; also used to control mistletoe in conifers and deciduous trees. *Lawn & Garden Products Inc.*

12013 Florence lake
Vienna Lake; Paris Lake. Lakes produced from cochineal, by precipitating alkaline solutions of cochineal with alum, or with a mixture of alum and tin salts.

12014 Florentine brown
Vandyck red; hatchette brown. Copper ferrocyanide, a pigment.

12015 Florex®
1337-76-4
Granular attapulgite clay; agrochemical carrier, also used in oil refining. *Whitecourt Ltd.*

12016 Florex® Ag-Dri 6/30, LVM 8/16, RVM 8/16
1337-76-4
Attapulgite clay; absorbent and adsorbent *Floridin.* Name unverified.

12017 Floricin
florizine; derizine; dericin. A substance produced by heating castor oil to 300 C, and distilling 10% of it. The residue is a product called floricin, which solidifies at -20°. It is also made by heating castor oil with formaldehyde. It is miscible with ceresin and petroleum jelly and is used as a vehicle for menthol and oil of eucalyptus; unlike castor oil, it is insoluble in EtOH.

12018 Florida phosphates
Mineral phosphates
There are two types, hard rock phosphates, containing 80% calcium phosphate, and soft clay phosphates, containing 40-60% calcium phosphate; used as fertilizers.

12019 Florigel® H-Y
1337-76-4
Attapulgite clay; absorbent and adsorbent *Floridin.* Name unverified.

12020 Florisil
Chromatography absorbent. *Whitecourt Ltd.*

12021 Florite
A proprietary trade name for a carefully prepared and screened bauxite. No manufacturer.

12022 Flor-Kleen
1337-76-4
Attapulgite clay; absorbent. *Floridin.* Name unverified.

12023 Florocyclene
Tricyclodecenyl propionate
Quest Int'l UK Ltd.

12024 Florosal
Concrete additive. *Steetley Chemicals Ltd.* Name unverified.

12025 Florox
94-36-0 1149 202-327-6
Benzoyl Peroxide
Benzoyl peroxide dispersion in food grade filler, flour additive. *Diaflex Ltd.* Discontinued.

12026 Flosol
17125-80-3 1005 241-189-1
BaF_6Si

barium hexafluorosilicate
Silicate(2-), hexafluoro-, barium (1:1); barium fluosilicate. A proprietary trade name for colloidal barium silicofluoride, used as a horticultural pesticide. d_4^{21}= 4.29; soluble in H_2O (0.0235 g/100 ml). No manufacturer.

12027 Flotox
7704-34-9 9142 213-722-6
sulfur
Sulfur, used as a pesticide. *Monsanto (Solaris)*. Name unverified.

12028 flour of sulfur
7704-34-9 9142 213-722-6
sulfur
Powdered sulfur. It has been powdered by grinding, but is not so finely powdered as flowers of sulfur.

12029 Flovan
Proofing agents. *Ciba plc*. Name unverified.

12030 flowers of antimony
1309-64-4 752 215-175-0
antimony trioxide
Formed when antimony burns in air. It is primarily antimony trioxide, Sb_4O_6; used in medicinal preparations, and as a white pigment.

12031 flowers of Benjamin
65-85-0 1122 200-618-2
$C_7H_6O_2$
benzoic acid
Flowers of Benzoin. Chemical intermediate.

12032 flowers of bismuth
1304-76-3 1314 215-134-7
Bi_2O_3
Bismuth oxide
Obtained by burning bismuth metal at a red heat. An astringent, also used in disinfectants, magnets, glass, vulcanization, catalysts and fireproofing.

12033 flowers of tin
18282-10-5 8933 242-159-0
SnO_2
Stannic oxide
A polishing powder.

12034 Flo-Guard
7631-86-9 8637 231-545-4
silicon dioxide
Synthetic, amorphous, precipitated silica *PPG Industries*.

12035 flow-powder
A mixture of white lead and a salt which gives off chlorine on heating; used for the production of flow blues (cobalt blues) on ceramics. Cobalt chloride is formed, which, being volatile, gives blues of varying intensity.

12036 Floxan SC-5211
Flocculant. *Henkel/Cospha*.

12037 fluates
This term is used for fluosilicates. It is also the name for waterproofing compounds consisting of solutions of sodium silcate, or silicofluoride, and other silicofluorides, such as those of zinc, magnesium, and aluminum.

12038 Fluazifop-butyl
69806-50-4 4152 274-125-6
$C_{19}H_{20}F_3NO_4$
2-[4-[[5-(trifluoromethyl)-2-pyridinyl]oxy]phenoxy]propanoic acid butyl ester
Fusilade; PP-009; TF-1169; butyl 2-[4-(5-trifluoromethyl-2-pyridyloxy)phenoxy]propionate. Emulsifiable concentrate of 125 g fluazifop-p-butyl per liter; used for grass weed control for broad-leaved crops. bp = 167°. *ICI Chem & Polymers Ltd.*

12039 Flubendazole
31430-15-6 4154 250-624-4
$C_{16}H_{12}FN_3O_3$
[5-(4-Fluorobenzoyl)-1H-benzimidazol-2-yl]carbamic acid methyl ester
Flubenol; R-17889; Flumoxal; Flumoxane; Fluvermal; 5-(*p*-fluorobenzoyl)-2-benzimidazolecarbamic acid methyl ester. A proprietary preparation containing flubendazole; veterinary antihelmintic. mp = 260°; LD_{50} (rat orl) > 2560 mg/kg. *Janssen Pharmaceutical Ltd.* Name unverified.

12040 Fludor solder
An aluminum solder containing 56.5% tin, 40% zinc, 3% lead, 0.2% antimony, and 0.1% copper.

12041 Fluf® 10-0-10
Urea-formaldehyde suspension; fertilizer providing nitrogen and potassium to golf course turf. *W A Cleary*.

12042 Flugne 113
Trifluorotrichloroethane
A proprietary noninflammable solvent of low toxicity for cleaning precision equipment. *Rhône-Poulenc NV*. Name unverified.

12043 Fluid EEZ 1000
Soybean oil with glycerol lactate esters of fatty acids, propylene glycol, mono- and diglycerides, TBHQ; kosher; fluid shortening for high quality cakes made without hydrogenated shortening; added emulsifiers for improved performance. *Van Den Bergh Foods*.

12044 Fluid Flex
Soybean oil, fatty acid glyceryl lactates, mono and diglycerides, TBHQ; kosher; fluid shortening for high quality cakes made without hydrogenated shortening; increases pliability and shelf life of soft tortillas. *Van Den Bergh Foods*.

12045 Fluidiram
Fatty amine blends and aminated clays; anticaking agents for fertilizers. *Atochem UK/Ceca*. Discontinued.

12046 Fluilan
8006-54-0 5371 232-348-6
Lanolin oil; water-oil emulsifier; dispersant for pigments; conditioning agent; emollient, penetrant, superfatting agent for lipsticks, baby oils, brilliantines, cleansing lotions; plasticizer for hair spray resins; moisturizer in water-oil emulsions; also for soaps, shampoos, dishwashing liquids and germicidal skin cleansers. *Croda Inc.; Croda Chem. Ltd.*

12047 Fluilan AWS
68458-58-8
PPG-12-PEG-65 lanolin oil; emollient, solubilizer; plasticizer and film modifier for hair sprays. *Croda Inc.*

12048 Fluisil® S55K
A low temperature silicone lubricant; used for refrigeration machinery. *Bayer AG*.

12049 Fluisol
A range of sulfated castor oils; for various applications in many industries. No manufacturer.

12050 Fluitex
Thinned (or fluidized) starches; used in textiles to strengthen and finish and in the paper industry. *Roquette (UK) Ltd.*

12051 Flukiver
57808-65-8 2473 260-967-1
Closantel
A proprietary preparation containing closantel; veterinary flukicide and anthelmintic. *Janssen Pharmaceutical Ltd.* Name unverified.

12052 Fluolite
Textile auxiliaries. *ICI Chem & Polymers Ltd.*

12053 Fluon®
A trademark for polytetrafluoroethylene (PTFE); a hard plastics material with a very water repellant surface which has a working temperature range from 200° to 2280°; has a low high frequency power factor and permittivity and outstanding nonstick properties; it is used in heat-resistant glands, packings and bearings. *ICI Chem & Polymers Ltd.*

12054 Fluon® AD1, AD1L, AD1H
9002-84-0 7743 204-126-9
PTFE resin; general purpose aqueous dispersion polymer for impregnation of packing yarns and glasscloth. *ICI Fluoropolymers*.

12055 Fluon® CDI
9002-84-0 7743 204-126-9
PTFE resin; general purpose coagulated dispersion polymer for thick-walled tubing and unsintered tape. *ICI Fluoropolymers*.

12056 Fluon® G170
9002-84-0 7743 204-126-9
PTFE resin; general purpose molding powder. *ICI Fluoropolymers*.

12057 Fluorad Surfactants
Fluorinated surfactants; used as wetting agents in paint and polishes. *3M UK plc*.

12058 Fluorad® FC-118
3825-26-1 223-320-4
$C_8H_4F_{15}NO_2$
Ammonium perfluorooctanoate
ammonium perfluoroctanate; Ammonium pentadecafluorooctanoate; Pentadecafluoro-1-octanoic acid, ammonium salt; Perfluorooctanoic acid, ammonium salt; Pentadecafluorooctanoic acid, ammonium salt. Surfactant for emulsion polymerization of fluorinated monomers. *3M/Industrial Chem. Prods.*

12059 Fluorad® FC-24
1493-13-6 216-087-5
CHF_3O_3S
Trifluoromethanesulfonic acid
Catalyst and reactant increasing yields in polymerization of epoxies, styrenes, THF, in alkylation and acylation reactions; improves octane rating; used with nitric acid for higher yields of pharmaceuticals, explosives, dyes, and intermediates. mp = -40°; bp = 162°; d = 1.6960; n_D^{20}= 1.3270. *3M.*

12060 Fluorad® FC-26
335-67-1 206-397-9
$C_8HF_{15}O_2$

Perfluorooctanoic acid
Pentadecafluorooctanoic acid; Perfluoroheptanecarboxylic acid. Intermediate for preparation of monomers and surfactants. mp = 59-60°; bp = 192°; soluble in H_2O (3.4 g/l), organic solvents. *3M/Industrial Chem. Prods.*

12061 Fluorad® FC-722
Fluorochemical polymer in inert fluorocarbon solvent blend; conformal coating producing thin, transparent films possessing antimigration properties; as nonwetting barrier coating, as protective coating for electrical contacts and electronic components. *3M.*

12062 Fluorad® FC-724
Fluorochemical acrylate polymer in fluorinated inert vehicle; surface modifier producing transparent films with excellent antiwetting and antisticking properties against oils, silicones and photoresist solutions in the manufacture of semiconductor devices; antimigration barrier; release coating. *3M.*

12063 Fluorad® FC-740
Fluoroaliphatic polymeric ester; oil well stimulation surfactant; foams hydrocarbon liquids; active in low polarity organic solvents. *3M/Industrial Chem. Prods.*

12064 Fluorad® FC-93
Ammonium perfluoroalkyl sulfonate
Wetting agent in etching solutions in semiconductor devices; foaming agent, leveling agent, corrosion inhibitor. *3M/Industrial Chem. Prods.*

12065 Fluorad® FC-95
Potassium perfluoroalkyl sulfonate
Wetting and foaming agents for coatings, etchants, plating baths, cleaning systems; corrosion inhibitor, leveling agent. *3M/Industrial Chem. Prods.*

12066 Fluorad® FX-13
2-(N-Ethylperfluorooctanesulfonamido) ethyl acrylate
Monomer for polymerization reactions to provide oil and water repellency, lubricity and release to polymers; used to manufacture textile treating resins, paper sizes, surfactants, and inert fluids. *3M.*

12067 Fluorad® FX-8
307-35-7 206-200-6
$C_8F_{18}O_2S$
Perfluorooctanesulfonyl fluoride
Perfluorooctane sulfonyl fluoride; Perfluoro-1-octanesulfonyl fluoride; Perfluorooctanesulfonyl fluoride. Intermediate for preparation of monomers and surfactants for textile treatment, paper sizes, inert fluids. mp = -1°; bp = 154-155°; d = 1.8240. *3M/Industrial Chem. Prods.*

12068 fluoram
1341-49-7 523 215-676-4
F_2H_5N
Ammonium bifluoride.
Used in manufacture of Mg and Mg alloys and for cleaning metal food handling machinery.

12069 Fluor-Amps
518-47-8 4194 208-253-0
Fluorescein sodium
Fluorescein sodium 5ml amps 10 and 20%; diagnostic aid in the determination of circulation time, examination of opthalmic vasculature and differentiation of malignant and healthy tissue, visualization of gall bladder and bile duct before surgery. *SAS Pharmaceuticals Ltd.* Unverified.

12070 Fluoranar
Coil coating compositions. *ICI Chem & Polymers Ltd.* Discontinued.

12071 Fluoraz
Modified structure of tetrafluoroethylene and propylene copolymers; used in producing sealing materials for chemical, petrochemical, and other applications where temperature and chemical resistance are of paramount importance. *Greene, Tweed & Co.*

12072 fluorchrome
7788-97-8 2279 232-137-9
$CrF_3 \cdot 4H_2O$
Chromium fluoride
A mordant.

12073 Fluorel® FC-2120
Fluoroelastomer; incorporated cure gumstock developed for automotive fuel line hose; for extrusion, injection and transfer molding applications. *3M.*

12074 Fluorel® FC-2144
Fluoroelastomer; designed for molding of complex shapes requiring good compression set resistance; capable of higher filler loading at equivalent durometer; suitable for compression molding. *3M.*

12075 Fluorel® FC-2173
Specialty masterbatch of incorporated cure fluoroelastomer gum; best-flowing incorporated cure polymer with improved release over 2170 (no sacrifice in physical properties); used in molded goods, o-rings, hose. *3M.*

12076 Fluorel® FC-2211
Fluoroelastomer; used for modifying the viscosity of other Fluorel grades to improve flow, sprayability, etc.; lowest viscosity in series. *3M.*

12077 Fluorel® FT-2481
Fluoroelastomer terpolymer; gumstock with excellent heat and chemical resistance; developed for shaft seals, molded goods, diaphragms, and extrusions where improved chemical resistance is required; vulcanizable with amine and dihydroxy cure systems. *3M.*

12078 Fluorel® FX-9038
Fluoroelastomer; high fluorine, incorporated cure fluoroelastomer offering improved processability, mold release, and compression set; designed for molded goods applications requiring very high chemical resistance. *3M.*

12079 Fluoresbrite
Fluorescent monodisperse carboxylated microspheres (polymer beads containing fluorescent dye); an identification tag and a size reference for agglutination tests, flow cytometry, instrument calibration, gel filtration, light scattering and phagocytosis. *Polysciences Inc.*

12080 Fluorescein
2321-07-5 4194 219-031-8
$C_{20}H_{12}O_5$
3',6'-Dihydroxyspiro[isobenzofuran-1(3H),9'-[9H]xanthen]-3-one
diresorcinolphthalein; resorcinolphthalein; 3',6'-fluorandiol; D & C Yellow No. 7; C. I. Solvent Yellow 94; C.I. 45350:1. Dyeing seawater for spotting purposes, tracer to locate impurities in wells, dyeing silk and wool, diagnostic aid in ophthalmology, indicator and reagent for bromine. mp = 314-316°; insoluble in water, benzene, chloroform, ether, soluble in hot alcohol, glacial acetic acid, alkali hydroxides or carbonates. *EM Industries; R.W. Greeff; Hilton Davis; Kraeber GmbH; U.S. Biochemical.*

12081 Fluorescein disodium salt
518-47-8 4194 208-253-0
$C_{20}H_{10}Na_2O_5$
3',6'-Dihydroxyspiro[isobenzofuran-1(3H),9'-[9H]xanthen]-3-one disodium
Fluor-I-Strip;Fluorescein sodium; soluble fluorescein; resorcinol phthalein sodium; uranin(e); uranine yellow; D & C yellow No. 8; C.I. Acid Yellow 73; C.I. 45350; Fluorescite. Diagnostic aid. Soluble in water, slighlty soluble in alcohols; LD_{50} (rat orl) = 6721 mg/kg. *Wyeth Laboratories.*

12082 Fluoresceine
2321-07-5 4194 219-031-8
$C_{20}H_{12}O_5$
3',6'-Dihydroxyspiro[isobenzofuran-1(3H),9'-[9H]xanthen]-3-one
diresorcinolphthalein; resorcinolphthalein. It is obtained by heating phthalic anhydride with resorcinol, with green fluorescence; Its alkali salts dye silk and wool yellow, as does fluoresceine itself; it is used also as a tracer to locate impurities in wells, diagnostic aid in ophthalmology, indicator and reagent for bromine.

12083 Fluorescent
Dyestuffs that fluoresce in daylight and/or uv light; for textiles, resin pigments, and various solvents. *Holliday Dyes & Chemicals Ltd.*

12084 Fluorescent Blue
$C_{12}H_6Br_4N_2O_3$
Resorcin blue; Iris blue. A dyestuff. It is the ammonium salt of tetrabromoresorufin. Dyes silk and wool blue, with brownish fluorescence.

12085 Fluorescent Red 5B
A proprietary organic fluorescent-red dye used for coloring polystyrene polymethyl-methacrylate and unplasticized PVC. *Farbwerke Hoechst.* Name unverified.

12086 Fluorescite
518-47-8 4194 208-253-0
Fluorescein sodium
Diagnostic aid. *Alcon Laboratories Inc.*

12087 Fluorfolpet
719-96-0 211-952-3
$C_9H_4Cl_2FNO_2$
N-(Fluordichloromethylthio) phthalimid
N-(Dichlorofluoromethylthio)phthalimide;
Dichlorofluoromethyl)thio)phthalimide; Fluorofolpet; Isoindole-1,3(2H)-dione, 2-((dichlorofluoromethyl)thio)-; Preventol A3. Fungicide applied as a paint. *Bayer AG.*

12088 Fluorine
7782-41-4 4199 231-954-8
F
Nonmetallic halogen element; F; Production of metallic and other fluorides, fluorocarbons, fluoridating compounds for drinking water and toothpaste. mp = -219°; bp = -188.13°; d (liquid) = 1.5127; LD_{50} (rat ihl 1hr) = 185 ppm. *Air Prods & Chem; AlliedSignal; Solvay GmbH.*

12089 Fluorinert Liquids
Perfluorinated liquids; used in the electronics industry as a vapor phase soldering medium; in direct testing of electronic components and in direct cooling. *3M UK plc.*

12090 Fluorinse
7681-49-4 8762 231-667-8

Sodium fluoride
Dental caries prophylactic. *orl-B Laboratories Inc.* Name unverified.

12091 Fluor-I-Strip
518-47-8 4194 208-253-0
Fluorescein sodium
Diagnostic aid. *Wyeth Laboratories.*

12092 Fluorl
7681-49-4 8762 231-667-8
Sodium fluoride
Dental caries prophylactic. *orl-B Laboratories Inc.* Name unverified.

12093 Fluorobenzene
462-06-6 4208 207-321-7
C_6H_5F
phenyl fluoride
monofluorobenzene; Fluorobenzenes. Intermediate in the manufacture of insecticides, larvicides and pharmaceuticals, identification reagent for plastic or resin polymers. mp = -42°; bp = 85°; d = 1.0240; n_D^{20} = 1.4653; insoluble in H_2O, miscible with alcohol, ether; LD_{50} (rat orl) = 4399 mg/kg. *Hoechst Celanese; ICI Am.; Schweizerhall.*

12094 Fluorocomp® FC-101
9002-84-0 7743 204-126-9
PTFE resin, 5% glass fiber-filled; long wearing, mechanical stable composite for chemical, electrical, and mechanical applications; for self-lubricating bearings, compressor rings, gaskets, seals, valve seats and liners, thin-wall rings. *ICI Fluoropolymers.*

12095 Fluorocomp® FC-144
9002-84-0 7743 204-126-9
PTFE resin, 40% bronze-filled; long-wearing, self-lubricating filled composite for most mechanical and some chemical applications; for bearings, rings gaskets, seals, valve seats, and liners. *ICI Fluoropolymers.*

12096 Fluorocomp® FC-174
9002-84-0 7743 204-126-9
PTFE resin, 15% glass fiber-filled, 5% MoS_2; long wearing, mechanical stable composite for chemical, electrical, and mechanical applications; for self-lubricating bearings, compressor rings, gaskets, seals, valve seats, and liners, thin-wall rings. *ICI Fluoropolymers.*

12097 Fluorocomp® FC-182
9002-84-0 7743 204-126-9
PTFE resin, 55% bronze-filled, 5% MoS_2; long-wearing, self-lubricating filled composite for most mechanical and some chemical applications; for bearings, rings gaskets, seals, valve seats and liners. *ICI Fluoropolymers.*

12098 Fluoroether Grease 834
Solvent and chemical resistant grease with superior high temperature properties. *Wm F Nye.*

12099 Fluoroform
75-46-7 4211 200-872-4
CHF_3
Trifluoromethane
propellant 23; fluoroform; freon 23; Halocarbon 23; refrigerant 23; R-23; Carbon trifluoride. Refrigerant, intermediate in organic synthesis, direct coolant for infrared detector cells, blowing agent for infrared foams. mp = -155°; bp = -82°; d = 1.935.

12100 Fluoroglide
Fluoropolymer powders used as dry lubricants and additives. *Imperial Chemical Industries plc.*

12101 fluorol
7681-49-4 8762 231-667-8
NaF
Sodium fluoride

12102 Fluorol® Dyes
Polycyclic dyes with intensive fluorescence; for coloring carburetor fuels, oils and lubricating greases. *BASF AG.*

12103 Fluorolene®
9002-84-0 7743 204-126-9
Polytetrafluoroethylene
LNP. Name unverified.

12104 Fluorolube® GR-290, GR-362, GR-470, GR-544, GR-660
9002-83-9
Polychlorotrifluoroethylene
Lubricating grease for use in chlorine and oxygen systems, metalworking, nuclear service, drilling, heat transfer media, damping fluids, plasticizers. *OxyChem.*

12105 Fluorolux
Coil coating compositions. *ICI Chem & Polymers Ltd.* Discontinued.

12106 Fluoromelt®
9002-84-0 7743 204-126-9
Melt processable fluorocompounds based on FEP, ECTFE, PFA, ETFE, and PVDF; color concs. for wire and cable applications. *ICI Fluoropolymers.*

12107 Fluoromelt® FP-F-FMX1
Foam concentrate for blending with FEP base resin and subsequent extrusion onto wire. *ICI Fluoropolymers.*

12108 fluoromide
41205-21-4
$C_{10}H_4Cl_2FNO_2$
2,3-dichloro-N-4-fluorophenylmaleimide
MK-23; Sparticide. Foliar fungicide with protective action. Used for control of scab, *Alternaria* leaf spot and powdery mildew in apples, scab of citrus fruit and coffee berry disease. mp = 240-242°; soluble in H_2O (5.9 mg/l), slightly more soluble in organic solvents; LD_{50} (rat orl) > 15000 mg/kg.

12109 Fluorophenol
371-41-5 206-736-0
C_6H_5FO
p-fluorophenol
4-fluorophenol. Fungicide, intermediate for pharmaceuticals. mp = 46-48°; bp = 185-188°. *ICI Am.; PCR; Schweizerhall.*

12110 Fluorosint® 500
9002-84-0 7743 204-126-9
PTFE with mica filler; low coefficient of thermal expansion, good electrical properties, high chemical resistance and dimensional stability; for electrical/electronic components, bearings, bushings, wear plates, thrust buttons. *Polymer Corp.*

12111 Fluorotex
Fluorochemical; oil and water repellent for outerwear market. *CNC Int'l L.P.*

12112 FluorSave
Reagent. *Calbiochem Corp.*

12113 fluorspar
CaF_2
fluor; fluorite; Derbyshire spar. A mineral. It is calcium fluoride, (CaF_2); principal source of fluorine and its compounds by way of hydrogen fluoride, flux in open hearth steel furnaces and in metal smelting, ceramics, wood preservatives, optical equipment.

12114 Fluowet 40 M
Fluoroaliphatic ethoxylate
Surfactant for metalworking fluids; intermediate products; wetting agent for mordant baths, aqueous and organic solvent systems. *Hoechst Celanese/Colorants & Surf.; Hoechst AG.*

12115 Fluowet OL
Ammonium salt of perfluorosulfate ester; surfactant for textile processing. *Hoechst Celanese/Colorants & Surf.*

12116 Fluphenazine
76674-21-0 4226
$C_{22}H_{26}F_3N_3OS$
4-[3-[2-(Trifluoromethyl)-10H-phenothiazin-10-yl]propyl]-1-piperazineethanol
Impact; Impact Excel; S-94; SQ-4918; 10-[3'-[4-(β-hydroxyethyl)-1-piperazinyl]propyl]-3-trifluoromethylphenothiazine. Fungicide. $bp_{0.5}$ = 268-274°. *ICI Chem & Polymers Ltd.*

12117 fluroxypyr
69377-81-7 4238
$C_7H_5Cl_2FN_2O_3$
[(4-amino-3,5-dichloro-6-fluoro-2-pyridinyl)oxy]acetic acid
EF-689; Dowco 433; Starane. Selective systemic herbicide. Used for post-emergence control of broad-leaved weeds such as *Galium aparine* and *Stella media* and some deep-rooted perennial weeds. mp = 232-233°; soluble in H_2O (91 mg/l), more soluble in organic solvents; LD_{50} (rat orl) = 2405 mg/kg.

12118 fluroxypyr 1-methylheptyl ester
81406-37-3 4238 279-752-9
$C_{15}H_{21}Cl_2FN_2O_3$
[(4-amino-3,5-dichloro-6-fluoro-2-pyridinyl)oxy]acetic acid 1-methylheptyl ester
fluroxypyr-meptyl; Dowco 433; Starane. Selective systemic herbicide absorbed by leaves and roots. Used for control of many broad-leaved weeds in wheat and barley. mp = 56-57°; insoluble in H_2O (0.9 mg/l), freely soluble in organic solvents; LD_{50} (rat orl)= 2405 mg/kg.

12119 flusilazole
85509-19-9 4240
$C_{16}H_{15}F_2N_3Si$
1-[[bis-(4-fluorophenyl)methylsilyl]methyl]-1h-1,2,4-triazole
fluzilazol; DPX-H6573; Olymp; Punch; NuStar; Sanction. Foliar, systemic fungicide with protective and curative action. Used to control *Ascomycetes, Basdiomycetes* and *Deuteromycetes* in cereals, apples, vines and sugar beet. mp = 55°; soluble in H_2O (54 mg/l), more soluble in organic solvents; LD_{50} (rat orl) = 1110 mg/kg.

12120 Flutec
Range of extremely inert, temperature stable, nontoxic, noninflammable

liquids exhibiting excellent electrical insulating properties and good heat transfer characteristics. *Rhône-Poulenc UK.*

12121 Flutec PP1
355-42-0 206-585-0
C_6F_{14}
perfluoro-n-hexane
tetradecafluorohexane; Perfluoro-compound FC-72; Fluorinert FC72; n-Perfluorohexane; Perfluoro-n-Hexane; FLUTEC PP1. A fluorinated hydrocarbon, used in heat transfer. mp = -90°; bp = 56°; d = 1.6800; n_D^{20} = 1.2510. No manufacturer.

12122 Flutec PP2
355-02-2 206-573-5
C_7F_{14}
perfluoro-methylcyclohexane
(Trifluoromethyl)undecafluorocyclohexane; perfluoro-(methylcyclohexane). A fluorinated hydrocarbon used in heat transfer. mp = -37°; bp = 76°; d = 1.7880. No manufacturer.

12123 Flutec PP3
335-27-3 206-386-9
C_8F_{16}
perfluoro-1,3-dimethylcyclohexane
FLUTEC PP3102. A fluorinated hydrocarbon used in heat transfer. mp = -55°; bp = 101-102°; d = 1.8280. No manufacturer.

12124 Flutec PP9
306-92-3 206-191-9
$C_{11}F_{20}$
perfluoro-1-methyldecalin
FLUTEC PP9. A fluorinated hydrocarbon. No manufacturer.

12125 Flux-Off®
Defluxing agent removing activated and nonactivated rosin flux, and dirt, grease, and molding compounds; for use on PC boards, electrical assemblies and other electronic components. *Chemtronics.*

12126 Fluxol
A hardwood pitch prepared from the distillation of hardwood; used as a rubber softener.

12127 FM 1132
Two-stage phenolic resin with macerated fabric as reinforcement; molding compound, good shock resistance and dielectric properties for impact, compression, and transfer molding. *ICI Fiberite.*

12128 FM 3510
Two-stage cotton fabric-filled phenolic resin; impact molding resin, good torque strength and molding surface. *ICI Fiberite.*

12129 FM 21288
One-stage glass roving-reinforced phenolic compound; high impact molding material. *ICI Fiberite.*

12130 F-MA 11®
Aluminum hydroxide-magnesium carbonate
Antacid which minimizes constipative or laxative effects. *Reheis.*

12131 FMB 210-8, 210-15
7173-51-5 3149 230-525-2
Didecyldimonium chloride
For formulation of disinfectants, sanitizers, fungicides, water treatment microbicides, swimming pool algicides, mildewcides. *Huntington Lab.*

12132 FMB 302-8
68424-95-3 270-331-5
Dicapryl/dicaprylyl dimonium chloride
FMB 302-8 Quat. For formulation of disinfectants, sanitizers, and fungicides. *Huntington Lab.*

12133 FMB 500-15 U.S.P
Benzalkonium chloride
Preservative in OTC drug products. *Huntington Lab.*

12134 FMB 504-5
Dicapryl/dicaprylyl dimonium chloride-myristalkonium chloride
For formulation of disinfectants, sanitizers, fungicides, water treatment microbicides, mildewcides. *Huntington Lab.*

12135 FMB 65-15, 65-28
139-08-2 205-352-0
Myristalkonium chloride
For formulation of mildewcides and swimming pool algicides. *Huntington Lab.*

12136 FMB 1210-5, 1210-8
Dicetyldimonium chloride-myristalkonium chloride
For formulation of disinfectants, sanitizers, fungicides, water treatment microbicides, swimming pool algicides. *Huntington Lab.*

12137 FMB 3328-5, 3328-8
Myristalkonium chloride-quaternium-14
For formulation of disinfectants, sanitizers, fungicides, water treatment microbicides, swimming pool algicides. *Huntington Lab.*

12138 FMB 6075-5, 6075-8
N-Alkyl dimethyl benzyl ammonium chloride and n-alkyl dimethyl ethylbenzyl ammonium chloride; for formulation of disinfectants, sanitizers, fungicides, water treatment microbicides, and swimming pool algicides. *Huntington Lab.*

12139 foam tannin
Tannin extracted from sumach, or galls, by means of ether.

12140 Foam Tint
Pigment dispersions; for coloration of polyurethane foam. *Pacific Dispersions Inc.* Name unverified.

12141 Foamacure
Defoaming agents; for aqueous coating compositions. *Hüls Am.*

12142 Foamaster
Antifoams/defoamers. *Henkel Chemicals Ltd.*

12143 Foamaster 340
Nonsilicone defoamer; for pressure dyeing applications in textile industry. *Henkel/Textiles.*

12144 Foamaster 371-S
Silicone emulsion; defoamer for pressure and atmospheric processes in textile industry. *Henkel/Textiles.*

12145 Foamaster A
Polymerized alcohol; defoamer for use in degassing and monomer stripping operations involving synthetic latexes. *Henkel/Functional Prods.*

12146 Foamaster AP
Low silicone hydrophobic silica; defoamer for use in degassing and monomer stripping operations involving synthetic latexes. *Henkel/Functional Prods.*

12147 Foamaster NXZ
Low silicone, fatty acid base; defoamer for synthetic latexes, paints, adhesives, blade and roll coatings, emulsifiable latex stripping. *Henkel/Functional Prods.*

12148 Foamaster Soap L
8052-48-0 232-491-4
Sodium tallowate
Dry defoamer for wettable powder and dry flowable pesticide formulations. *Henkel/Emery.*

12149 Foamaster VC
Low silicone hydrophobic silica; defoamer for use in degassing and monomer stripping operations involving synthetic latexes. *Henkel/Functional Prods.*

12150 Foam-Coll 4C
68920-65-0
Potassium coco-hydrolyzed collagen
Lamepon S; May-Tein C; Maypon 4C; Monteine LCK-32. Foaming protein, cosmetic ingredient; mild surfactant. *Brooks Industries; Henkel/Cospha; Henkel Canada; Henkel KGaA; Maybrook; Inolex; Seppic.*

12151 Foam-Coll 4CT
68952-16-9
TEA coco-hydrolyzed collagen
Bio-Soft® MT-40; Lamepon ST40; May-Tein CT; Maypon 4CT; Monteine LCT. Foaming protein, cosmetic ingredient; mild surfactant. Brooks Industries; Stepan Europe; Henkel/Cospha; Henkel Canada; Henkel KGaA; Maybrook; Inolex; Seppic.

12152 Foamer CD
Proprietary blend; foamer for air mist drilling, general detergents, gypsum board production, ether sulfates. *Harcros Organics.*

12153 Foamex AD-50, AD-100, AD-300, J-275
Silicone defoamers. *Rhône-Poulenc Surf.*

12154 FoamFlush™
N-Methyl-2-pyrrolidone
butyrolactone. Other ingredients; urethane remover for urethane foam manufacture. *ISP.*

12155 Foamid 117
68425-43-4
Cocamidopropyl dimethyl-amine propionate, water
Surfactant, softener, emollient for creams, lotions, fingernail polish removers. *Alzo.* Discontinued.

12156 Foamkill® 30 Series
Organic and organo-silicone concentrate; defoamer for food/pharmaceutical applications. *Crucible.* Name unverified.

12157 Foamkill® 30HP
Organo-silicone emulsion; antifoam/defoamer for waste treatment, pulp/paper, severe foaming situations. *Crucible.*

12158 Foamkill® 400A
Organo-silicone emulsion; defoamer for pulp/paper applications including adhesives backings, latex paints and coatings, water-reducible inks, floor and ceiling coatings. *Crucible.* Name unverified.

12159 Foamkill® 608
Nonsilicone; defoamer for paper coatings and adhesives and formulations

sensitive to fish-eyeing, janitorial supply houses; readily emulsifiable. *Crucible.*

12160 Foamkill® 614
General purpose antifoam/defoamer containing very small amount of silicone; for inks, adhesives, and coatings; suitable for SBU, PVA, PVE, polyethylene, and miscellaneous copolymers. *Crucible.*

12161 Foamkill® 618 Series
Organic and organo-silicone concentrate; defoamer for food/pharmaceutical applications including paper coatings and adhesives in contact with food. *Crucible.* Name unverified.

12162 Foamkill® 634C
Nonsilicone; defoamer for food applications, canning trade, pasteurizer defoaming. *Crucible.*

12163 Foamkill® 639
Nonsilicone; defoamer for aqueous systems, pulp/paper applications, acrylic, PVC, PVA, PVPC, and other coatings, paints, inks, antifoam formulating. *Crucible.*

12164 Foamkill® 639JOH
Organo-silicone
Defoamer for aqueous paint, ink, and coatings systems; excellent for vinyl acrylic paints and other severe foaming applications; may be used in grind or let-down. *Crucible.*

12165 Foamkill® 639Q
Organo-silicone
Defoamer for aqueous systems, paints, inks, adhesives and coatings, especially for highfoaming acrylic or Joncryl resin systems; may be used in grind or let-down; high stability. *Crucible.*

12166 Foamkill® 649
Nonsilicone; highly compatible defoamer for aqueous systems, adhesives, coatings, especially acrylic systems, paints, inks, drawing and cutting fluids, paper applications; very little tendency to cause fish-eyeing. *Crucible.*

12167 Foamkill® 654NS
Nonsilicone; defoamer for drawing and cutting fluids; emulsifiable and compatible with synthetic or oil-based media. *Crucible.*

12168 Foamkill® 663J
Organo-silicone
Defoamer for food/pharmaceutical applications including paper coatings and adhesives in contact with food, pulp/paper applications, water cooling towers; lubricant for fine papers; water-dispersion. *Crucible.*

12169 Foamkill® 80J Series
Silicone compound; defoamer for food/pharmaceutical applications *Crucible.* Name unverified.

12170 Foamkill® 810F
Dimethicone
Defoamer for food/pharmaceutical applications incl. general aqueous systems, paper coatings, and adhesives in contact with food, egg washing, cleaning/sanitizing solutions, cosmetics, pulp/paper applications. *Crucible.*

12171 Foamkill® 836B
Silicone emulsion; highly compatible defoamer for chemical compounding, synthetic or waterbased metalworking lubricants, aqueous systems, inks, janitorial applications, paints, coatings, paper industry, waste treatment, severe foaming applications; dilutable with H_2O. *Crucible.*

12172 Foamkill® CMP
Bis-stearamide
Efficient defoamer for paper reclaiming, caustic treatment, waste treatment, textile jet, or atmospheric dyeing. *Crucible.*

12173 Foamkill® EFT
Silicone emulsion; defoamer for waste treatment; good flash knockdown and staying power; for intermittent or continuous feed. *Crucible.*

12174 Foamkill® MS
Silicone emulsion; defoamer for waste treatment, atmospheric or pressure dyeing of textiles; for intermittent or continuous feed. *Crucible.*

12175 Foam-Kon 20
Thermoplastic structural foam concentrate; for high pressure foam systems; used in injection molding compounds (polyethylene, PP, ABS, SAN, PS, acetal, etc.). *LNP.*

12176 Foamole A
Linoleamide DEA (1:1)
Hair conditioner for shampoos, hair dyes; viscosity booster, foam stabilizer for shampoos, liquid soaps. *Van Dyk.*

12177 Foamole B
68953-11-7 273-187-1
Minkamidopropyl dimethylamine
Superfatting agent, conditioner for hair care products. *Van Dyk.*

12178 Foamole M
68140-00-1 268-770-2
Cocamide MEA (1:1)
Foam booster/stabilizer, thickener, emulsifier for creams and lotions, shampoos, bubble baths, other detergents. *Van Dyk.*

12179 Foamosul
Sodium formaldehyde sulferylate
Textile printing/emulsion polymerisation. *RV Chemicals Ltd.*

12180 Foamquat IAES
67633-63-0 266-778-0
Isostearyl dimethylamidopropyl ethonium ethosulfate
M-Quat® 522; Schercoquat IAS. Detergent. *Alzo; PPG/Specialty Chem.; Scher.*

12181 Foam-Soy C
Sodium cocoyl hydrolyzed soy protein; foaming cosmetic ingredient for skin and hair care products; mild surfactant. *Brooks Industries.*

12182 Foam-Wheat C
Sodium cocoyl hydrolyzed wheat protein; mild foaming protein for use in shampoos; gentle cleanser in facial makeup removers. *Brooks Industries.*

12183 Focal
Flowable fungicide *Schering Agrochemicals Ltd.* Discontinued.

12184 Focus®
101205-02-1
Cycloxydim
Post-emergence graminicide against annual and perennial grasses; selective in broadleaf crops, e.g., sugar beet, cotton, soybean, vegetables, onions *BASF AG.*

12185 Fodel®
For the electrical industry. *DuPont UK.*

12186 Foerdite
An explosive consisting of 25% nitroglycerin, 1.5% collodion wool, 5% nitrotoluene, 4% dextrin, 3% glycerin, 37% ammonium nitrate, and 24% potassium chloride.

12187 foil lead
An alloy of 86.5% lead, 12.5% tin, and 1% copper.

12188 Foilcote
Overprint lacquers for aluminum, foil, carton boards, labels etc. *The Scottish Adhesives Co Ltd.*

12189 Foilgrip
Aluminum foil to paper laminating adhesives. *The Scottish Adhesives Co Ltd.*

12190 Folex-P
Foliar phosphate-based fertilizer. *Omex Agriculture Ltd.*

12191 Foliac Super Red
A proprietary trade name for a graphite jointing compound. No manufacturer.

12192 Folia-Feed
Foliar nutrient. *DowElanco Ltd.*

12193 Foliar 36 Extra
Foliar feed. *BASF plc.*

12194 Foliar Nitrophoska
Foliar feed. *BASF plc.*

12195 Foliar TRIGGRR
A liquid containing plant growth regulators (cytokinin and gibberellic acid) and trace minerals; used to increase crop yields and quality; applied to the foliage; used for a wide variety of crops including cotton, soybeans, wheat, fruits and vegetables. *Westbridge Research Group.* Unverified.

12196 folic acid
59-30-3 4253 200-419-0
$C_{19}H_{19}N_7O_6$
N-(*p*-(((2-amino-4-hydroxy-6-pteridinyl)methyl)amino)benzoyl)-L-glutamic acid
Vitamin M; Pteroylglutamic Acid; folacin; folcysteine; L-pteroylglutamic acid; pteglu; pteroyl-L-glutamic acid; pteroylmonoglutamic acid; pteroyl-L-monoglutamic acid; Vitamin B11; Vitamin BC; PGA; liver lactobacillus casei factor; pteroylglutamic acid; cytofol; foliamin; folipac; folsaure; foluite; incafolic; millafol; folettes; acifolic; folsav; folacid; Vitamin B; folbal. Used as a vitamin (hematopoietic) and nutritional factor. dec 160°; mp = 250°; pKa = 5.40; soluble in H_2O (1.6 mg/l), less soluble in organic solvents; $[\alpha]_D^{20}$ = 21° (c = 0.5 0.1N NaOH); LD_{50} (mus orl) = 10 g/kg. *Mission Pharmacal Co; Lederle Laboratories USA.*

12197 Folicote Transpiration Minimizer
Refined wax, emulsifiers, preservatives, minimum 50% solids; FDA approved for use on edible crops; reduces water loss from plant foliage, winter protection, transplanting and transporting plants, christmas trees, wreaths, agricultural crops such as potatoes, corn, tobacco, transplants, stone and citrus fruits. *Aquatrols Corp of Am.* Name unverified.

12198 Folicur
107534-96-3 9253 403-640-2
Tebuconazole
Fungicide with systemic properties and broad spectrum activity against rusts, leaf spot diseases, e.g., *Septoria* spp., powdery mildew and several Fusarium

species on cereals; whitemold, Phoma and various leaf spot diseases on oilseed rape. *Bayer AG.*

12199 Folidol® E605
56-38-2 7167 200-271-7
Parathion
Insecticide and acaricide. *Bayer AG.* Name unverified.

12200 Folidol® M
298-00-0 6183 206-050-1
Parathion-methyl
Insecticide with contact, stomach, and breathing poison action. *Bayer AG.*

12201 Folimat®
1113-02-6 214-197-8
Omethoate
Systemic acaricide and insecticide. *Bayer AG; Bayer plc.*

12202 Folin-Dennis solution
A solution prepared by adding slowly 400 ml of a 0.7268% solution of silver nitrate to a solution containing 10 g mercuric cyanide and 180 g caustic soda, in 1,200 ml water; used for the determination of acetone.

12203 Folin-McEllroy sugar reagents
Qualitative. 100 g sodium pyrophosphate, $Na_4P_2O_7 \cdot 10H_2O$, 30 g crystalline disodium mono hydrogen phosphate, Na_2HPO_4, and 50 g dry sodium carbonate in 900 ml water. Dissolve and add 13 g copper sulfate dissolved in 200 ml water; Quantitative. A a) acidified copper sulfate solution with 60 g in 900 ml H_2O, add 5 ml concentrated sulfuric acid and make up to 1 liter. b) phosphate-carbonate-thiocyanate dry mixture.

12204 Folin's uranium acetate mixture
Reagent for assay of uric acid in urine. It consists of 500 g ammonium sulfate, 5 g uranium acetate, and 6 ml glacial acetic acid dissolved in 650 ml water and made up to 1 liter.

12205 Folin's Uric Acid Reagent
Add to 160 ml water, 50 ml syrupy phosphoric acid. Heat to 85°, and add 100 g sodium tungstate. Boil for 1 hour under reflux. Place 25 g lithium carbonate in a beaker, add 50 ml syrupy phosphoric acid and 200 ml water. Boil 10 minutes, cool and add first solution, mix and dilute to 1000 ml.

12206 Folio 575FW
chlorothalonil-metalaxyl
Suspension concentrate containing 500 g chlorothalonil and 75 g metalaxyl per liter, a systemic fungicide for field crops. *Ciba-Geigy Agrochemicals.*

12207 Folithion®
122-14-5 4017 204-524-2
Fenitrothion
Broad spectrum insecticide for controlling biting and sucking insect pests. *Bayer AG.*

12208 Folosan
82-68-8 8264 201-435-0
Quintozene
Horticultural fungicides. *Plant Protection.* Name unverified.

12209 Folpan
133-07-3 4255 205-088-6
folpet
Active ingredient; folpet; agricultural fungicide. *Makhteshim Chemical Works Ltd.*

12210 Folpet
133-07-3 4255 205-088-6
$C_9H_4Cl_3NO_2S$
2-[(Trichloromethyl)thio]-1H-isoindole-1,3(2H)-dione
Folpan; N-(trichloromethylthio)phthalimide; N-(trichloromethylmercapto)-phthalimide; Phaltan; Phalton; Folpel; Folpex; Trichloromethyl-(thio)phthalimide. Agricultural fungicide. Used for control of downy mildews, powdery mildews, leaf spot diseases, scab, etc. in fruit and vegetable crops. mp = 177°; insoluble in H_2O, slightly soluble in organic solvents; LD_{50} (rat orl) > 10000 mg/kg. *Makhteshim Chemical Works Ltd.* Discontinued.

12211 Folvite
59-30-3 4253 200-419-0
Folic acid
Vitamin. *Lederle Laboratories USA.* Name unverified.

12212 Fomac
A fungicide for use with rubber. *ICI Chem & Polymers Ltd.* Discontinued.

12213 Fomblin
A proprietary range of perfluoropolyether fluids. Used as diffusion pump oils. *Montedison UK Ltd.* Name unverified.

12214 Fome-Cor
Light-weight, rigid board made of extruded polystyrene foam securely bonded between two layers of tough kraft linerboard; used in manufactured housing, automotive and graphic arts. *Monsanto Co.* Name unverified.

12215 Fomescol
Nonionic surfactant. *Rhône-Poulenc UK.*

12216 fomitine
A liquid extract from the fungi *Fomes cinn'amomeus.*

12217 Fomox®
Intumescent one- and two-component compounds for use as flame retardants in civil engineering, marine engineering, automotive engineering, and industrial construction. *Bayer AG.*

12218 Fomrez
Polyesters for flexible PU foam and PU elastomers. *Baxenden Chemicals Ltd.*

12219 Fomrez® 50
Trimethylol propane branched adipate polyester; used for flexible cellular PU foam for textile and industrial applications, dispersing agents, coatings and adhesives, elastomers. *Witco/Organics.* Discontinued.

12220 Fomrez® 4393
Polyfunctional specialty polyether; used as semirigid foam crosslinkers, rigid foam additives, precursors for reactive diluents and chain extenders in urethane synthesis. *Witco/Organics.* Discontinued.

12221 Fomrez® A1228
Primary polyether triol based on glycerin; used in the manufacture of one-shot, high-resilient molded and slab foam, RIM, microcellular, and shoe sole systems. *Witco/Organics.* Discontinued.

12222 Fomrez® ED400
Polyether diol; used to produce PU coatings, PU prepolymers for elastomers, adhesives, caulks, sealants, and dispersing agents. *Witco/Organics.* Discontinued.

12223 Fomrez® EPD28
Polyether primary diol; faster reaction than conventional prepolymer grade diols; allows production of high vol. intricate elastomer parts where rapid mold turnover is required; also suggested as partial or total replacement for the polytetramethylene glycols. *Witco/Organics.* Discontinued.

12224 Fomrez® ET190
Polyether triol; used in the manufacture of PU adhesives, coatings, foam, and elastomers; also used as crosslinkers and pigment dispersing vehicles. *Witco/Organics.* Discontinued.

12225 Fongarid
57646-30-7 260-875-1
furalaxyl
Fungicide for ornamentals. *Ciba-Geigy Agrochemicals.*

12226 Fonoline® White, Yellow
Petrolatum USP; soft, low melting point for consumer use as petroleum jelly, ointments, industrial applications; as emollient, protective coating, binder, carrier, lubricant, moisture barrier, plasticizer, protective agent, softener. *Witco/Sonneborn.* Name unverified.

12227 Fontaine's powder
An explosive consisting of potassium picrate and potassium chlorate.

12228 foots
foots oil. Matter deposited by oils on standing.

12229 Foraflon® 1000 HD
24937-79-9
Polyvinylidene fluoride
PVDF semicrystalline homopolymer; thermoplastic for general use for extrusion and injection molding applications in the chemical engineering, packaging, and electrical wiring fields and for protection of other materials, e.g., paints, plastic films. *Elf Atochem.*

12230 Foraperle®
Leather finishing agents. *Elf Atochem SA.*

12231 Forbes metal
An alloy of 53.5% zinc and 46.5% copper.

12232 Forbest 1000B
Fatty acid/paraffin oil blend; defoamer for solvent and aqueous systems; used for solvent-based air-dry and stoving finishes, aqueous systems, dip-coatings, emulsion paints, varnishes. *Lucas Meyer.*

12233 Forbest 13
Polyester/fatty alkyls compound; wetting and dispersing agent for water-based varnishes, highly pigmented slurries; binding agent. *Lucas Meyer.*

12234 Forbest 50
Modified aryl-alkyl silicones; antifloating and antisilking agent for solvent systems, curtain coats. *Lucas Meyer.*

12235 Forbest 62B
Based on hydrocarbon polymers; slip agent to improve smoothness and abrasion resistance; recommended for pigmented air-dry and stoving enamels, clear coatings. *Lucas Meyer.*

12236 Forbest 410
9002-88-4 7728
Polyethylene
Polyethylene compound with reactive groups; leafing and stabilizing agent for

leafing aluminum pigments; recommended for air-dry and stoving paints. *Lucas Meyer.*

12237 Forbest 780

Carboxylic reaction product; catalyst for gel formation in offset and heat-set printing inks, hydrocarbon-based structural varnishes. *Lucas Meyer.*

12238 Forbest MW 23

63231-60-7 264-038-1

Microcrystalline wax finely dispersed in C_{17} fatty acid; increases slip and abrasion resistance in clear and pigmented stoving enamels, e.g., industrial paints, coil-coating finishes, printing inks (flexo, letterpress). *Lucas Meyer.*

12239 Forbest WP

High molecular weight synthetic compound; wetting, dispersing, and antisettling agent for all air-dry and stoving solvent-based paints and systems with polar solvs. *Lucas Meyer.*

12240 Force

A malt preparation.

12241 Forcite

A trademark for various types of explosives. No manufacturer.

12242 Fordath Resins

A proprietary range of phenolic and urea resins used in foundry work. *Fordath Engineering Co.* Unverified.

12243 Forest Bark

Chipped, ground or composted bark for use as a soil conditioner, planting aid or mulch. *ICI Garden Products.*

12244 For-Ester

94-75-7 2865 202-361-1

2,4-D

2,4-D; herbicide for cereals and grass. *Synchemicals Ltd.*

12245 Forex®

9002-86-2 7746 206-625-7

Polyvinylchloride

PVC foam panels, slightly expanded with d 700 kg/m^3, and 450 kg/m^3; possible applications include fair stands and exhibition booths, screen printing, signs and displays, lightweight construction in automotive and aviation industries and wall cov *Airex AG.*

12246 Foral 85, 105 and AX

A range of hydrogenated rosin and rosin esters; tackifiers and polymer-modifying resins in adhesives and in hot-melt-applied decorative, pressure sensitive and heat-sealable coatings. *Hercules.*

12247 Forlan

Petrolatum

lanolin alcohol; lanolin. Absorption base, emollient. *RITA.*

12248 Forlan 200

Petrolatum-lanolin alcohol

Absorption base for personal care products; enhance stability of emulsions, dispersions, and suspensions; epidermal emollient, moisturizer, lubricant. *RITA.*

12249 Forlan C-24

27321-96-6

Choleth-24

Emulsifier, emulsion stabilizer, emollient, moisturizer, solubilizer, visc. modifier, pigment dispersant, plasticizer for cosmetics, pharmaceuticals. *RITA.*

12250 Forlan L

Synthetic lanolin; an ingredient that qualitatively approximates the composition of lanolin; enhances emulsion stability in both oil-water and water-oil systems, assists in the wetting and dispersion of pigments in facial makeup, lipstick and eye shadow preparations. *RITA.* Name unverified.

12251 Forlan LM

Synthetic lanolin; An ingredient that qualitatively approximates the composition of lanolin. It enhances emulsion stability in both o/w and w/o systems, assists in the wetting and dispersion of pigments in facial makeup and lipstick. *RITA.* Name unverified.

12252 Forlanit P

Sodium laureth phosphate

Wetting agent, dispersant, flotation auxiliary for metal cleaners, galvanic baths, hot copper baths containing cyanides. *Henkel/Cospha; Henkel/Functional Prods.; Henkel Canada; Henkel KGaA.*

12253 Forlay

Selective weed killers. *Murphy Chemical Co Ltd.* Discontinued.

12254 Forlkyl

Perfluoroalkyl iodides and derivatives. *Atochem UK Ltd.* Discontinued.

12255 Formac 40

A solution polymer based on acrolein and formaldehyde with an active ingredient content of 40% by weight; used to control and eliminate algae, fungi and bacteria from industrial water. *Degussa Ltd.*

12256 Formacel®

For the chemical industry. *DuPont UK.*

12257 formagen

A dental cement consisting of two parts : a) A liquid containing creosote, phenol, olive oil, and alcoholic formalin, and b) a powder consisting of aluminum silicate, magnesium and zinc carbonates and lime.

12258 formaldehyde

50-00-0 4262 200-001-8

CH_2O

methanal

formalin; formic aldehyde; oxymethylene; methyl aldehyde; aldehyde C_1; oxomethane; methylene oxide; Hercules® 37M6-8. Used in urea and melamine resins, polyacetal resins, phenolic resins, fertilizers, preservatives, reducing agent, corrosive inhibitor. mp = -92°; soluble in water, alcohol, ether. Aqualon; Du Pont; Farleyway Chem. Ltd; Georgia-Pacific Resins; Hoechst Celanese; Mallinckrodt; Mitsubishi Gas; Monsanto.

12259 formalin

formol

formol-chlorl. A 40% aqueous solution of formaldehyde which may contains 15% methyl alcohol to prevent the separation of polymerized compounds.

12260 Formalite

A trade name for phenol-formaldehyde resin molded material for use in electrical insulation. No manufacturer.

12261 formamide

75-12-7 4264 200-842-0

CH_3NO

amide C_1

Methanamide; Carbamaldehyde; Formimidic Acid. Solvent, softener, intermediate in organic synthesis. mp = 2-3°; bp = 210°; d = 1.1300; n_D^{20} = 1.4475; soluble in H_2O, organic solvents; LD_{50} (rat orl) = 5570 mg/kg. *Aldrich; BASF; Fluka; Penta Mfg.*

12262 formammidinesulfinic acid

1758-73-2 217-157-8

$CH_4N_2O_2S$

thiourea dioxide

Aminoiminomethanesulfinic acid; Formamidinesulfinic acid. mp = 126°; soluble in H_2O (30 g/l).

12263 forman

$C_{10}H_{19}O{\cdot}CH_2Cl$

Chloromethyl menthyl ether

12264 Formanek's indicator

Alizarin green used as an indicator. It gives a violet color with pH 0.3, pink with pH.0, yellow with pH 12, and brown with pH 14.

12265 formaniline

Anhydro-formaldehyde aniline; a rubber vulcanization accelerator.

12266 Formapex

A proprietary phenol-formaldehyde resin varnish. No manufacturer.

12267 Format

1702-17-6 2462 216-935-4

Clopyralid

Herbicide containing clopyralid. *ICI Chem & Polymers Ltd.*

12268 Formax

141-53-7 8765 205-488-0

$CHNaO_2$

Sodium Formate

Formic acid, sodium salt; Salachlor; Sodium formate, hydrated; Sodium formate, hydrate. Sodium formate; a caustic and astringent. mp = 253°; bp = 360°; d = 1.92; soluble in H_2O (0.77 g/ml), EtOH; LD_{50} (mus orl) = 11200 mg/kg. *May & Baker Ltd.* Name unverified.

12269 Formel NF

A mixture of Isceon chlorofluorocarbons and chloromethanes; totally nonflammable; dielectric fluid for use in distribution transformers. *Rhône-Poulenc UK.*

12270 Formex

A proprietary trade name for an enamelled wire; the enamel is flexible and heat resisting up to about 185°; contains a mixture of polyvinyl acetal and a thermosetting phenol-formaldehyde synthetic resin. No manufacturer.

12271 formic acid

64-18-6 8912 200-579-1

CH_2O_2

hydrogen carboxylic acid

methanoic acid. Organic acid; reducing agent, dyeing and finishing of textiles, leather treatment, chemicals, manufacture of fumigants, insecticides, refrigerants, solvs. for perfumes, lacquers; electroplating; silvering glass; ore flotation. *BASF; BP Chem. Ltd; Hoechst Celanese; Mallinckrodt; Norsk Hydro AS.*

12272 Formica

A proprietary trade name for phenolic and urea resins. No manufacturer.

12273 Formkote
Dry film lubricant coating containing graphite and suitable binder system; high temperature titanium metal forming lubricant. *E/M Corporation.*

12274 Formodac
Chemical products for use as additives in the manufacture of concrete. *BP Chemicals Ltd.*

12275 Formol 55
Urea formaldehyde precondensate; for further condensation with urea, melamine, phenol, or furfuryl alcohol for adhesives and foundry binders. *BASF AG.*

12276 formolide
An antiseptic consisting of an aqueous solution of 15% alcohol, 2% boric acid, 4% sodium benzoate, and 1% formaldehyde.

12277 formolites
Phenol-formaldehyde resins.

12278 Formon®
Solder and braze compositions; for the electrical industry. *DuPont UK.*

12279 formose
A mixture of sugars obtained from formaldehyde by polymerization.

12280 Formrez
Polyesters for flexible PU foam and PU elastomers. *Baxenden Chemicals Ltd.* Discontinued.

12281 Formrez®
A registered trade name for a polyester-based polyurethane elastomer cross-linked with a diamine. *Witco Corporation.*

12282 Formrez® 11
Linear poly (diethylene adipate) polyester; used in PU industry for manufacture of prepolymers (isocyanate or hydroxyl-terminated), thermoplastic elastomers, coating and adhesives, dispersing agents, microcellular shoe sole systems, millable gums, cast elastomers. *Witco Corporation.*

12283 Formrez® L49-28
Primary polyether triol based on glycerin; for manufacture of one-shot urethane elastomers, RIM, microcellular, and shoe sole systems. *Witco/Organics.* Discontinued.

12284 Formrez® T-279, T-280
Polyether triol based on trimethylol propane; used as urethane crosslinker, precursors for reactive diluents, epoxy and melamine cure diluents. *Witco/Organics.* Discontinued.

12285 Formula 90
Squash court plaster. *Prodorite Ltd.*

12286 Formula 111
Heavy-duty solvent for strong cleaning/degreasing on interfaces, metal contacts, electro-mechanical assemblies, PCB connectors. *Chemtronics.*

12287 Formula 405
7915
Pregnenolone succinate
Nonhormonal sterol derivative. Used in synthesis of pharmaceuticals. *Doak Pharmacal Co.* Unverified.

12288 Formula AC
Self-emulsifying mineral oil lubricant for wire drawing; gives superior antitarnish properties. *Eastern Color & Chem.*

12289 Formula S
Squash court plaster. *Prodorite Ltd.*

12290 Formusol® SA
Special acetaldehyde sulfoxylate. *RV Chemical Ltd.* Discontinued.

12291 Formvar
Polyvinyl formal resins; suitable for formulating structural adhesives, wash primers, can and drum linings and wood and knot sealers. *Monsanto Co.*

12292 Fornax
Finishing agents. *Ciba plc.* Name unverified.

12293 fornitrol
A reagent used for the estimation of nitric acid, nitro-compounds, and nitrates.

12294 Forociben Premix
Sulfonamide animal feed additive. *Ciba plc.* Name unverified.

12295 Foron
Synthetic organic dyestuffs for use in textiles. *Sandoz Products Ltd.*

12296 Fortafix
A range of high temperature resistant adhesive cements and sealing compounds; used for bonding, sealing or insulating inorganic materials where heat resistance up to 1600° is required. *Fortafix Ltd.* Name unverified.

12297 Fortex
Flavors for confectionery, foodstuffs, and beverages. *Bush Boake Allen Ltd.*

12298 Fortex®
63231-60-7 264-038-1
paraffin wax
Hard microcrystalline wax consisting of n-paraffinic, branched paraffinic, and naphthenic hydrocarbons; used in hot-melt coatings and adhesives, paper coatings, printing inks, plastic modification (as lubricant and processing aid), lacquers, paints, and varnishes, binder in ceramics, in electrical/electronic components, investment casting, rubber, eleastomers, in emulsion wax size, as a fabric softener ingredient, in emulsions and latex coatings, hand creams and lipsticks. *Petrolite.*

12299 Fortiflex®
9002-88-4 7728
polyethylene
High density polyethylene(HDPE); used for packaging, blow molded and injection molded containers, film for a wide variety of materials including foodstuffs; automotive and industrial parts; extruded pipe for potable water and natural gas transport; sheet industrial parts. *Solvay Polymers.*

12300 Fortiflex® A60-70-99, A60-70-119
9002-88-4 7728
Polyethylene
HDPE; blow molding grade with excellent stiffness and processability on reciprocating screw equipment for milk, water, juice, and fountain syrup botties. *Solvay Polymers.*

12301 Fortiflex® B45-06R-09
9002-88-4 7728
Polyethylene
HDPE copolymer; high molecular weight ammunition grade for shot shell tubes. *Solvay Polymers.*

12302 Fortiflex® G36-24-149
9002-88-4 7728
Polyethylene
HDPE; sheet extrusion and thermoforming grade with excellent ESCR and thermal performance; for geomembranes. *Solvay Polymers.*

12303 Fortiflex® G38-70C
9002-88-4 7728
Polyethylene
HDPE; bulk concentrate for use in NSF pipe grade resins. *Solvay Polymers.*

12304 Fortiflex® J36-25-142
9002-88-4 7728
Polyethylene
HDPE; film grade with high impact strength and tear; for blown film merchandise bags. *Solvay Polymers.*

12305 Fortiflex® K36-55-122
9002-88-4 7728
Polyethylene
HDPE; pipe grade; natural base resin suitable for pipe when blended with color concentrate. *Solvay Polymers.*

12306 Fortiflex® T50-200
9002-88-4 7728
Polyethylene
HDPE; injection molding grade with superior toughness and ESCR; for industrial parts. *Solvay Polymers.*

12307 Fortiflex® XF-855
9002-88-4 7728
Polyethylene
HDPE copolymer containing 25% recycled polyethyene; blow molding grade for applications requiring moderate ESCR. *Solvay Polymers.*

12308 Fortiflex® XF-855
9002-88-4 7728
Polyethylene
HDPE copolymer containing 25% recycled polyethyene; blow molding grade for applications requiring moderate environmental stress crack resistance. *Solvay Polymers.*

12309 Fortilene®
9003-07-0 7741
Polypropylene
Used for fiber applications including nonwovens and slit-tape; packaging, particularly injection molded containers, caps and closures, and blow molded bottles, film, industrial and automotive parts. *Solvay Polymers.*

12310 Fortilene® 1001
9003-07-0 7741
Polypropylene
PP homopolymer; extrusion grade resin with excellent thermal stability for sheet, strapping, profile extrusion. *Solvay Polymers.*

12311 Fortilene® 1602
9003-07-0 7741
Polypropylene
PP homopolymer; medium flow resin with good thermal stability for conventional injection molding of caps and closures, all purpose molding. *Solvay Polymers.*

12312 Fortilene® 1802
9003-07-0 7741
Polypropylene
PP homopolymer, antistat; controlled rheology resin with high flow, good impact for injection molding of caps and closures, thin-wall containers. *Solvay Polymers.*

12313 Fortilene® 2104
9003-07-0 7741
Polypropylene
PP homopolymer; film resin with good stretching and gauge control for bioriented film applications. *Solvay Polymers.*

12314 Fortilene® 3151
9003-07-0 7741
Polypropylene
PP homopolymer; fiber/filament grade for slit film, carpet backing applications. *Solvay Polymers.*

12315 Fortilene® 4104, 4109
9003-07-0 7741
Polypropylene
PP copolymer, antistat; blow molding grade with good clarity for hot-fill containers. *Solvay Polymers.*

12316 Fortilene® 4209
9003-07-0 7741
Polypropylene
PP copolymer; film grade with good processability, excellent clarity for biaxially oriented film. *Solvay Polymers.*

12317 Fortilene® 5801
9003-07-0 7741
Polypropylene
PP copolymer; injection molding resin with good impact for specialty closures, medical applications. *Solvay Polymers.*

12318 Fortilene® 9000
9003-07-0 7741
Polypropylene
PP homopolymer; unstabilized sphere for compounding, automotive applications. *Solvay Polymers.*

12319 Fortimax
Reinforcing fillers for synthetic elastomers. *ICI Chem & Polymers Ltd.*

12320 Fortisan
A proprietary trade name for a synthetic fiber made from regenerated cellulose. No manufacturer.

12321 Fortress
Heat-resistant glass-fiber tissues. *ICI Chem & Polymers Ltd.*

12322 Fortrex
An activated clay for reinforcing natural and synthetic rubber. *Croxton & Garry Ltd.* Name unverified.

12323 Fortrol
21725-46-2 2755 244-544-9
Cyanazine
Suspension concentrate containing 500 g cyanazine per liter; a triazine herbicide. *Shell UK.*

12324 Fortron® 0205B4
9016-75-5
Polyphenylene sulfide
PPS resin; high temperature polymer with excellent thermal and electrical properties, good solvent resistance; used for connectors, switches, coil bobbins, vapor phase soldering, automotive applications (engine compartment components, fuel line systems), applications for high temperature and corrosive environments (caps, gears, fittings, industrial parts); 0205B4 is a low viscosity unfilled powder. *Hoechst Celanese/Engineering Plastics; Hoechst AG.*

12325 Fosalsil
A proprietary trade name for a natural diatomaceous material made into bricks or used as a cement. No manufacturer.

12326 Foscast
Range of preformed refractory shapes. *Foseco (F.S) Ltd.*

12327 Foset
Acid-catalysed phenolic resin systems for cold-set bonding of sand cores. *Foseco (F.S.) Ltd.* Discontinued.

12328 Fosfamide CPD-170
Complex alkanolamide phosphate ester; emulsifier for preparation of metalworking fluids and greases; imparts lubricity, anticorrosion and mild EP properties; grease additive. *Henkel/Cospha; Henkel Canada.*

12329 Fosfamide N
Fatty amido phosphate complex; detergent, foaming and wetting agent for textile, all-purpose, institutional, and metal cleaners. *Henkel/Cospha; Henkel Canada.*

12330 Fosferno
56-38-2 7167 200-271-7
Parathion
Parathion insecticide. *ICI Chem & Polymers Ltd.*

12331 Fosfil
High alumina refractory material. *Foseco (F.S.) Ltd.*

12332 Foshell
Liquid phenol-formaldehyde novalac resins for coating sand by the warm process. *Foseco (F.S.) Ltd.* Discontinued.

12333 Fosoil
Oil-based binders for cold-set bonding of sand cores. *Foseco (F.S.) Ltd.* Discontinued.

12334 Fostap
Preformed highly insulating launder liners. *Foseco (F.S.) Ltd.*

12335 Fosterge LF
Alkylphenoxy POE acid phosphate
Intermediate for foaming cleaners and solvent emulsion cleaners; emulsifier for pesticides; corrosion inhibitor. *Henkel/Cospha; Henkel Canada.*

12336 Fosterge R
Mono and dialkyl phosphoric acid; intermediate for emulsifying, penetrating, and anticorrosion compounds. *Henkel/Cospha; Henkel Canada.*

12337 Fostex AMP
6419-19-8 229-146-5
$C_3H_{12}NO_9P_3$
Aminotrimethylene phosphonic acid
Aminotris(methanephosphonic acid); Aminotris(methylphosphonic acid); ATMP; Briquest 301-500; Budex 5130; Dequest 2000; Dequest 2001; Dowell L 37; Ferrofos 509; Masquol P 320; Mayoquest 1320; Nitrilotrimethanephosphonic acid; Nitrilotris(methylphosphonic acid); NTF; NTMP; NTPA; Sequion 20H45; Sequion OA; Tris(phosphonomethyl)amine; Turpinal MD2; Nitrilotris(methylene)triphosphonic acid. Scale inhibitor and sequestrant in water treatment. mp = -14°; bp = 105°; d = 1.33; LD_{50} (rat orl) = 2100 mg/kg. *Henkel.*

12338 Fostex P
2809-21-4 3908 220-552-8
Etidronic acid
Scale inhibitor and sequestrant for water treatment. *Henkel.* Name unverified.

12339 Fotofax
1314-13-2 10279 215-222-5
Zinc Oxide
Zinc oxide, high purity; for varistors, reprographics. *Manchem Ltd.* Name unverified.

12340 fotosensin
A condensation product of phthalic acid and resorcinol containing small proportions of copper and iron, Small quantities increase the root and stem growth of plants.

12341 Foulagan
Wool fulling agent. *Ceca SA.*

12342 Foulon
Paper dyes. *ICI Chem & Polymers Ltd.* Discontinued.

12343 Foundrinier wire
A brass containing from 80-85% copper, and 15-20% zinc.

12344 Foundrox
1309-37-1 4072 215-168-2
Fe_2O_3
Ferric Oxide
Foundry grade red iron oxide for core and mold use in eliminating veining and other casting expansion defects. *DCS Color & Supply Co Inc.* Name unverified.

12345 foundry clay
A clay containing from 80-90% silica and 15-18% aluminia.

12346 Foundry pattern metal
An alloy of 75% zinc and 25% tin.

12347 Four Sure Liquid
Foot-rot treatment compound for sheep. *Charles Tennant & Co Ltd.*

12348 Fouramine
Leather dyes. *ICI Chem & Polymers Ltd.*

12349 Foxstar
bifenox-isoproturon-mecoprop
A liquid formulation containing 107 g bifenox, 286 g isoproturon and 143 g mecoprop per liter as a suspension concentrate; a post-emergence herbicide for the control of weeds in winter cereals. *Rhône-Poulenc Crop Protection Ltd.*

12350 FP-Vac
Fowl pox vaccine; for immunization of poultry. *Intervet Inc.*

12351 FR-11
12124-97-9 531 235-183-8
Ammonium bromide
Flame retardant for paper, chipboard, a nonwashable textiles. *AmeriHaas; Dead Sea Bromine.*

12352 FR-20
1309-42-8 5706 215-170-3
Magnesium hydroxide
Flame retardant and smoke suppressant for ABS, PP, PS, rubbers. *AmeriHaas; Dead Sea Bromine.*

12353 FR 28
Flame retardant for cellulose. *U.S. Borax & Chem.*

12354 FR-513
36483-57-5 253-057-0
Tribromoneopentyl alcohol
Flame retardant for flexible and rigid PU; flame retardant intermediate. *AmeriHaas; Dead Sea Bromine.*

12355 FR-522
3296-90-0 221-967-7
Dibromoneopentyl glycol
Flame retardant for unsaturated polyesters, rigid PU and foams; flame retardant for intermediate. *AmeriHaas; Dead Sea Bromine.*

12356 FR-612
615-58-7 210-436-5
2,4-dibromophenol
Flame retardant for epoxy resins, phenolic resins, polyester resins; flame retardant intermediate. *AmeriHaas; Dead Sea Bromine.*

12357 FR-613
118-79-6 9744 204-278-6
2,4,6-Tribromophenol
Reactive flame retardant used mainly as an intermediate for polymeric flame retardant. *AmeriHaas; Dead Sea Bromine.*

12358 FR-705
87-83-2 201-774-4
Pentabromotoluene
Flame retardant.

12359 FR-913
3278-89-5 221-913-2
Tribromophenyl allyl ether
Aromatic flame retardant for expandable PS; synergist with hexabromocyclododecane. *AmeriHaas; Dead Sea Bromine.*

12360 FR-1025
594477-57-3
Poly(pentabromobenzyl) acrylate
Polymeric flame retardant for engineering thermoplastics, PET, PBT, nylon, PP and PS. *AmeriHaas; Dead Sea Bromine.*

12361 FR-1034
109678-33-3
Tetrabromodipentaerythritol
Flame retardant for PP extruded fibers; processing aid for ABS and HIPS. *AmeriHaas; Dead Sea Bromine.*

12362 FR-1205
32534-81-9 251-084-2
$C_{12}H_5Br_5O$
Pentabromodiphenyl oxide
Flame retardant for use in laminates (both epoxy and phenolic), unsat. polyesters, synthetic fibers, and flexible PU foams; suitable for textiles. *AmeriHaas; Dead Sea Bromine.*

12363 FR-1206
25637-99-4 247-148-4
$C_{12}H_{18}Br_6$
Hexabromocyclododecane
Fire retardant for wide range of plastics, textiles, adhesives, and coatings; especially for styrene-based systems. *AmeriHaas; Dead Sea Bromine.*

12364 FR-1208
32536-52-0 251-087-9
Octabromodiphenyl oxide
Flame retardant for thermoplastics, e.g., ABS, HIPS, LDPE, PP random copolymer; recommended for injection moldings. *AmeriHaas; Dead Sea Bromine.*

12365 FR-1210
1163-19-5 214-604-9
Decabromodiphenyl oxide
Flame retardant used in thermoplastics and fibers, incl. HIPS, glass-reinforced thermoplastic polyester molding resins, LDPE extrusion coatings, PP (homo and copolymers), ABS, nylon, PBT, PET, PU, SBR latex, textiles, rubber. *AmeriHaas; Dead Sea Bromine.*

12366 FR-1360
tribromoneopentyl alcohol
A proprietary flame-retardant material comprising tribromoneopentyl alcohol. Used in the production of flexible polyurethane foam. *Dow Cheml Co Ltd, UK & Ireland.*

12367 FR-1524
79-94-7 201-236-9
tetrabromobisphenol-A
Reactive flame retardant used in the manufacture of epoxy, PC, ABS, phenolic, PS, and polyester resins, rubber; flame retardant intermediate. *AmeriHaas; Dead Sea Bromine.*

12368 FR-2124
tetrabromobisphenol-A allyl ether
Flame retardant for expandable PS. *AmeriHaas; Dead Sea Bromine.*

12369 Fracton
A refractory dressing. *Foseco (F.S.) Ltd.*

12370 Fractorite
An explosive containing 90% ammonium nitrate, 4% resin, 4% dextrin, and 2% potassium dichromate.

12371 Fractorite B
An explosive consisting of 75% ammonium nitrate, 2.8% dinitronaphthalene, 2.2% ammonium oxalate, and 20% ammonium chloride.

12372 Fragarol
β-naphthyl butyl ether
Fragasol. The butyl ether of β-naphthol; a synthetic perfume.

12373 Fragaroma
Perfumery products. *May & Baker Ltd.* Name unverified.

12374 Fraissite
620-05-3 210-623-1
C_7H_7I
Benzyl iodide
Iodomethylbenzene. Used in chemical synthesis.

12375 Francolor
Dyes, pigments and chemical auxiliaries. *ICI Chem & Polymers Ltd.*

12376 frankincense
6969
Olibanum; gum thus;. Olibanum, a gum resin from *Boswellia Carterii* Birdwood. Consists of terpenes (pinene, dipentene), resins, gims anda bitter principle, bassorin.

12377 Frankonite
silitonite
tonsil. Bleaching earths obtained from deposits in Germany.

12378 Franocide
1642-54-2 3165 216-696-6
Diethylcarbamazine Citrate
A proprietary preparation of diethylcarbamazine citrate used in veterinary medicine. *Coopers Animal Health Ltd.* Name unverified.

12379 Frantin
A proprietary anthelmintic given to unweaned lambs. *Coopers Animal Health Ltd.* Name unverified.

12380 Frary metal
A calcium-barium-lead alloy containing up to 2% barium and 1% calcium; used as a bearing metal. It melts at 445°.

12381 Fraude's reagent
7601-90-3 7296 231-512-4
$HClO_4$
Perchloric acid
Oxidizing agent and bleach.

12382 Fredo
Calcium hydrosulfite

12383 Freeflo
Industrial cleaning compound. *The Wellcome Foundation Ltd.* Name unverified.

12384 Freeman's nonpoisonous white lead
A pigment. It is a mixture of white lead, zinc white, baryta white, and magnesium carbonate.

12385 Freeteem
Fluxes for uphill teemed killed steel ingots. *Foseco (F.S.) Ltd.*

12386 Freez-Gard® Formula FP-88E
Glassy sodium hexametaphosphate-sodium chloride-sodium erythorbate. Food additive for meat, poultry, and seafood industries. *Rhône-Poulenc Food Ingreds.*

12387 Fre-Flex
Focofilcon A; 2-hydroxyethyl methacrylate polymer with methacrylic acid; contains 55% water; contact lens material. *Optech Inc.* Name unverified.

12388 Frekote® 33
Releasing interface; aerosols and bulk. *Dexter/Frekote.* Discontinued.

12389 Frekote® 44-NC
Nonchlorofluorocarbon version of Frekote 44 *Dexter/Frekote.*

12390 Frekote® 800
Release agent for natural and synthetic organic rubber compounds; for use on molds at temperatures above 150°. *Dexter/Frekote.*

12391 Frekote® EXITT®
Silicone-based; release agent for urethane elastomers, compression-molded resins, organic rubbers, EPDM rubber, and specialty molded resins; aerosol and bulk. *Dexter/Frekote.*

12392 Frekote® No. 1-NC
Fluorocarbon release agent for acrylics, epoxies, nylons, phenolics, polycarbonates, polypropylenes, polystyrenes, and rubber; applications including epoxy potting and encapsulating, prepreg fabricating, filament winding; dry lubricant for plastic gears, belts, threaded connectors, rubber, glass, wood, metals; available in aerosol and bulk. *Dexter/Frekote.*

12393 Frelen
Crosslinked polyethylene foam. *Carl Freudenberg.* Name unverified.

12394 Fremy's salt
7789-29-9 7771 232-156-2
F_2HK
Potassium hydrogen fluoride
potassium bifluoride; Potassium acid fluoride; Potassium hydrogen difluoride. Used as an electrolyte and in manufacture of frosted glass. Also as a flux for silvr solders and as an organic catalyst. mp = 238°; d = 2.37; soluble in H_2O (39 g/100 ml).

12395 French cement
A mucilage of gum arabic mixed with powdered starch; used by naturalists, artificial flower makers, and confectioners.

12396 French chalk
A variety of Steatite or Soapstone. It is a hydrated silicate of magnesium, and is used for marking cloth, removing grease from silk, as a filler, and for other purposes.

12397 French polish
A polish for wood. It consists of shellac dissolved in alcohol.

12398 French turpentine
Bordeaux turpentine
The oleo-resin from *Pinus maritima*. It contains from 15-20% essential oil and 70-80% rosin.

12399 Frenokone
A selective weedkiller. *Plant Protection.* Name unverified.

12400 Freon®
Fluorocarbons useful as refrigerants. *DuPont UK.*

12401 Frescile
3-Methyl dodecanonitrile
Quest Int'l. UK Ltd.

12402 Fresh Pak
Extra-thin polyethylene HD-film; for packaging or incorporation into laminates. *UCB nv Film Sector.* Discontinued.

12403 Freudenberg Megulastik
Mechanical dampeners. *Carl Freudenberg.* Name unverified.

12404 Freund's acid
1-Naphthylamine-3,6-disulfonic acid
Chemical intermediate.

12405 Frick's alloys
Nickel silvers containing from 50-69% copper, 18-39% zinc, and 5-31% nickel.

12406 friedelite
Mn_2SiO_4
Manganese silicate, a mineral.

12407 Friedländer's stain
A microscopic stain. it contains 50 g of a saturated alcoholic solution of gentian violet, 10 g of glacial acetic acid, and 100 ml water.

12408 Frigate
61791-44-4 263-177-5
Wetting agent containing 800 g/l tallow amine ethoxylate; spreader for use with phosphonoglycine herbicides. *Fermenta ASC Europe Ltd.*

12409 Frigen
Fluorinated hydrocarbon refrigerant. *Hoechst UK.*

12410 Frigesa® D, F, IC
Hydrocolloids
Thickener and stabilizer for desserts, ice cream, dressings. *Grünau.*

12411 Frigid-Go® 2815, 2816
Molybdenum-containing low-temperature multipurpose synthetic grease. *Henkel/Emery.*

12412 Frimulsion
Hydrocolloid blends used to improve stability, appearance, texture, and mouthfeel of food; for food and beverage industries. *Hercules.* Discontinued.

12413 Frishmuth's aluminum solder
Alloys. One contains 67% tin, 27% lead, and 3% aluminum. Another consists of 94% zinc, 4% aluminum, and 2% copper.

12414 Frishout solder
An alloy of 46% tin, 23% zinc, 15% aluminum, 8% copper, and 9% silver.

12415 Fritzsche's reagent
$C_{14}H_6N_2O_6$
Dinitroanthraquinone

12416 Frodingham G.G.B.S
Ground granulated blastfurnace slag (GGBFS) for blending with Portland cement in the concrete mixer to produce Portland blastfurnace cement concrete; used for all classes and types of concrete to give long term durability; provides high chemical and sulfate resistance, low heat properties and reduces the risk of alkali aggregate reaction when required. *Frodingham Cement Co Ltd.*

12417 Frohde's reagent
An alkaloid reagent. It consists of 0.5 gram of sodium molybdate dissolved in 100 ml of concentrated sulfuric acid.

12418 Frost-Off
Antifreeze liquid; deicing fluid for automobile windshields and door locks. *Merix.*

12419 Frother 4171
A terpene-alcohol based flotation reagent; produces a more brittle froth than typical pine oil; for flotation of nonmetallic minerals. *Hercules.* Discontinued.

12420 fructose
57-48-7 4295 200-333-3
$C_6H_{12}O_6$
levulose
D-fructose, sugar occurring in fruit and honey; Used as a sweetener. *Am. Maize Prods.; Corn Prods; Laevosan GmbH; Pfantsiehl Labs; A. E. Staley Manufacture.*

12421 Fruit Tree Grease
Mineral grease; banding grease to protect trees from wingless moths and other crawling insects. *Vitax Ltd.*

12422 Fruitonile
2-Methyl decanitrile. *Quest Int'l. UK Ltd.*

12423 Fruvit®
12071-83-9 8004 235-134-0
Propineb
Fungicide for control of downy mildews. *Bayer AG.* Name unverified.

12424 FS Bordeaux Powder
Copper sulfate/lime complex; a protectant fungicide for the control of blight and canker. *Ford Smith & Co Ltd.*

12425 FS Derris
83-79-4 8427 201-501-9
Rotenone
Emulsifiable concentrate or dustable powder containing rotenone; a contact insecticide for fruit, vegetables and greenhouse crops. *Ford Smith & Co Ltd.*

12426 FS Dricol 50
1332-40-7
Copper oxychloride
A protectant fungicide. *Ford Smith & Co Ltd.* Discontinued.

12427 FS Thiram 15% Dust
137-26-8 9510 205-286-2
Thiram
Fungicide with animal repellent properties. *Ford Smith & Co Ltd.*

12428 fuberidazole
3878-19-1 223-404-0
$C_{11}H_8N_2O$
2-(2-furanyl)-1H-benzimidazole
Bay 33172; Voronit; W VII/117;. Systemic fungicide used as a seed treatment for control of *Fusarium* spp. in cereals. mp = 284-288° (dec); soluble in H_2O (78 mg/l), more soluble in organic solvents; LD_{50} (rat orl) = 500 mg/kg. *Bayer AG.*

12429 Fubol
mancozeb-metalaxyl
Wettable powder containing mancozeb and metalaxyl; protectant fungicide for field crops and fruit. *Ciba-Geigy Agrochemicals.*

12430 Fudow
A proprietary range of phenolic molding materials. *Fudow.* Name unverified.

12431 Fudowlite U
A proprietary range of urea formaldehyde molding materials. *Fudow.* Name unverified.

12432 Fuelsaver®
4-(2-Nitrobutyl)morpholine-4,4'-(2-ethyl-2-nitrotrimethylene) dimorpholine
Fuelsaver F-15-Fuel Additive. Bioban P 1487 is a preservative composed of 4,4-(2-Ethyl-2-nitro-trimethylene) dimorpholine [1854-23-5] (30%) and 4-(2-Nitrobutyl)morpholine [2224-44-4] (70%). Preservative for diesel and other hydrocarbon fuels. *Angus.*

12433 Fulacolor
Reactive clay used for carbonless copy paper. *Laporte Industries Ltd.*

12434 Fulbent
1302-78-9 1082 215-108-5
Bentonite clay for binding, suspending and emulsifying. *Laporte Industries Ltd.*

12435 Fulbond
1302-78-9 1082 215-108-5
Foundry bentonite
Minas de Gador SA.

12436 Fulcat Catalysts
1318-93-0 6341 215-288-5
Montmorillonite
Acid treated montmorillonite clays; catalysts for a range of organic reactions notably the alkylation of phenols. *Laporte Industries Ltd.*

12437 Fulgurite
An explosive containing 60% nitroglycerin and 40% wheaten flour and magnesium carbonate.

12438 fuligo
Soot. Source of organic compounds.

12439 Fullasorb
Absorbent earth granules. *Laporte Industries Ltd.*

12440 Fullerite
A proprietary trade name for a slate powder used as a rubber filler. No manufacturer.

12441 fuller's earth
A term applied to a sandy loam or argillaceous earth found in Surrey and Kent, UK. It consists of aluminum and magnesium hydrosilicates. A deodorizer; used for clarifying oils, and used in cosmetics, rubber compounds, carrier for catalysts, filtering medium.

12442 Fulmargin
A solution of colloidal silver.

12443 Fulmenit
An explosive consisting of 86.5% ammonium nitrate, 5.5% trinitrotoluene, 2.5% paraffin oil, 1.5% charcoal, and 4% guncotton.

12444 fulminating gold
An explosive compound having the formula $2AuN_2H_3 \cdot H_2O$ prepared by the action of concentrated ammonia on gold hydroxide.

12445 fulminating platinum
Explosive compounds formed by acting upon ammonium platinochloride with potassium hydroxide.

12446 fulminating silver
A compound of nitrogen and silver prepared by the action of ammonia on precipitated silver oxide. It is explosive.

12447 Fulmont
Used for refining of vegetable, animal and mineral oils, fats and waxes. *Minas de Gador SA.*

12448 Fulmont Activated Bleaching Earths
1318-93-0 6341 215-288-5
Montmorillonite
K-10 Bentonite clay. Acid activated montmorillonite clays; color and impurity removal from edible oils, re-refining or purification of mineral oils. *Laporte Industries Ltd.* Discontinued.

12449 Fulton 404®
Acetal, 20% PTFE lubricant; thermoplastic compound for moving parts requiring excellent frictional properties, resistance to wear, and ease of fabrication; used in bearings and sliding-contact parts, e.g., gears, valve seats, stem packing, thrust washers, etc., in textile, printing, automotive, appliance and furniture industries. *LNP.*

12450 Fulton White
A proprietary brand of lithopone, a mixture of zinc sulfide, barium sulfate and zinc oxide. Used as a white pigment, but now largely replaced by titanium dioxide. No manufacturer.

12451 Fulvite
An artificial titanium monoxide.

12452 Fumarate 6-18
Diesters of fumaric acid and C_6-C_{18} alcohols; basic monomers for specialty polymers. *Henkel/Functional Prods.*

12453 Fumaric acid
110-17-8 4309 203-743-0
$C_4H_4O_4$
(E)-2-Butenedioic acid
trans-1,2-ethylenedicarboxylic acid; allomaleic acid; boletic acid. Modifier for polyester, alkyd and phenolic resins; paper sizing resins; plasticizers, rosin esters and adducts, alkyd resin coatings, upgrading natural drying oils, food additive, acidulant, flavoring agent, mordant, organic synthesis, inks. d = 1.625; sublimes at 200°; mp = 287°; pK_1 (25°) = 3.03; soluble in water, alcohol, acetone, ether; almost insol.in olive oil, chloroform, benzene, xylene, molten camphor, liquid ammonia. *Chemie Linz UK; Haarmann & Reimer; Lonza; Mitsubishi Gas; Monsanto; Schaefer Salt.*

12454 Fume-Control
Fume suppressant for hydrochloric and sulfuric acids. *Crown Tech.*

12455 Fumexol
Pretreatment agents. *Ciba plc.* Name unverified.

12456 fuming nitric acid
7697-37-2 6671 231-714-2
nitric acid
Nitric acid containing some of the lower oxides of nitrogen.

12457 fuming sulfuric acid
8014-95-7 9147 231-639-5
H_2SO_4
oleum;
Nordhausen acid; pyrosulfuric acid. Consists of sulfur trioxide dissolved in sulfuric acid. The commonest fuming acid contains 55% sulfuric acid, and 45% sulfur trioxide; used as a sulfating and sulfonating agent, dehydrating agent in nitrations, dyes, and explosives. d = 1.84; bp = 290°; dec 340°; mp = 10°; miscible with H_2O and alcohol; LD_{50} (rat orl) = 2.14 g/kg.

12458 Fumite
General purpose insecticide in smoke form for greenhouse use. *ICI Garden Products.*

12459 Fumite Dicloran
99-30-9 202-746-4
$C_6H_4Cl_2N_2O_2$
2,6-Dichloro-4-nitroaniline
Botran; dicloran; DCNA; Botran 75W; Dichloran; Dicloron; Allisan; ditranil; Kiwi Lustr 277; Resisan; rd-6584; AL-50; U-2069; bortran; CDNA; CNA; Resissan; Dichloro-4-nitroaniline; 2-amino-N-ethyl-N-phenyl-benzamine. Smoke fungicide (active ingredient dicloran); for use in enclosed areas against botrytis and rhizoctonia on protected crops. mp = 190-192°; LD_{50} (rat orl) = 2400 mg/kg. *Octavius Hunt Ltd.*

12460 Fumite Dicofol
115-32-2 3136 204-082-0
Dicofol
Smoke acaricide (active ingredient dicofol); for use in enclosed areas against red spider and other mites on protected crops. *Octavius Hunt Ltd.* Discontinued.

12461 Fumite Lindane
58-89-9 5526 200-401-2
indane
Smoke insecticide (active ingredient lindane); broad spectrum insecticide for use in (enclosed) poultry, mushroom farms, ships' holds, public buildings, protected crops, against a wide range of pests. *Octavius Hunt Ltd.*

12462 Fumite Permethrin
52645-53-1 7321 258-067-9
Permethrin
Smoke insecticide (active ingredient permethrin); for use in enclosed areas against whitefly and other pests of protected crops, cockroaches on stored produce, domestic insect pests. *Octavius Hunt Ltd.*

12463 Fumite Pirimiphos Methyl Smoke
29232-93-7 7652 249-528-5
Pirimiphos-methyl
Smoke insecticide (active ingredient pirimiphos methyl); for use in enclosed areas against whitefly and other pests of protected crops, pests in grain silos etc. *Octavius Hunt Ltd.*

12464 Fumite Pirimiphos-Methyl Smoke
29232-93-7 7652 249-528-5
Pirimiphos-methyl
Contact fumigant and organophosphorus insecticide for greenhouse crops. *Octavius Hunt Ltd.*

12465 Fumite Propoxur
114-26-1 8022 204-043-8
Propoxur
Smoke insecticide (active ingredient propoxur); for use in enclosed areas against whitefly and other pests of protected crops. *Octavius Hunt Ltd.*

12466 Fumite Ronilan
50471-44-8 10122 256-599-6
Vinclozolin
Smoke fungicide (active ingredient vinclozolin); for use in enclosed areas against botrytis on protected crops. *Octavius Hunt Ltd.*

12467 Fumite TCNB
117-18-0 204-178-2
Tecnazene
Smoke fungicide (active ingredient tecnazene); for use in enclosed areas against botrytis and mildew on protected crops. *Octavius Hunt Ltd.* Discontinued.

12468 Fumite TCNB Smoke
117-18-0 204-178-2
Tecnazene
Protectant fungicide and potato sprout suppressant. *Octavius Hunt Ltd.*

12469 Fumite Tecnalin
lindane-tecnazene
Smoke insecticide/fungicide (active ingredients lindane/tecnazene); for use in enclosed areas against botrytis, whitefly and other insect pests on protected crops. *Octavius Hunt Ltd.* Discontinued.

12470 Fumyl-O-Gas
methyl bromide-amyl acetate
Methyl bromide with amyl acetate; soil fumigant against soil-borne diseases. *Brian Jones & Associates Ltd.*

12471 Fungaflor
35554-44-0 3522 252-615-0
Imazalil
Used for control of mildew in greenhouse plants. *Hortichem Ltd.*

12472 Fungal Lactase 100,000
Fungal lactase
Enzyme for hydrolyzing lactose in dairy products (milk, whey, cheese, yogurt), pharmaceuticals (digestive aids). *Solvay Enzymes.*

12473 Fungal Protease Conc
Protease
Enzyme for hydrolysis of peptide bonds. Used for baking (improves grain, texture, loaf volume), in meat tenderizer formulations; hydrolyzes and modifies plant and animal protein under acid conditions. *Solvay Enzymes.*

12474 Fungalysin™
Glucanase. *Calbiochem Corp.*

12475 Fungamyl
A purified fungal α-amylase preparation produced from a selected strain of *Aspergillus oryzae*; used in starch processing, brewing, alcohol production and baking industries. *Novo Nordisk.*

12476 Fungex
A copper-containing fungicide. *Murphy Chemical Co Ltd.* Discontinued.

12477 Fungi-Fluor
Stain solution and counterstain in an eight ounce kit for rapid identification of fungi; used for fungalidentification in clinical cell cultures and tissue biopsy samples. *Polysciences Inc.*

12478 Funginex
26644-46-2 247-872-0
triforine
Emulsifiable concentrate containing 66 g/l triforine; a systemic insecticide. *Synchemicals Ltd.*

12479 Fungisterol
53260-54-1 4313 209-127-8
$C_{28}H_{44}O$
5α-Ergosta-6,8,22E-trien-3β-ol
mp = 147.5°; $[\alpha]_D^{25}$= -21.9° (chloroform).

12480 Fungitex 656
A proprietary trade name for a solution of an organo-mercuric complex; used as a durable mildew proofing agent for textiles. *Ciba plc.* Name unverified.

12481 Fungitrol
Fungicides
Used for nonaqueous coating compositions, caulking compounds, and other substrates. *Hüls Am.*

12482 Fungitrol Tinox
Bis(tri-n-butyl)oxide
Wood preservative and antifouling coating compositions. *Hüls Am.*

12483 Fungo®
97-23-4 3120 202-567-1
Dichlorophen
15% Dichlorophen; fungicide, bactericide and algicide used as a moss, lichen, and mold killer. *Dax Products Ltd.*

12484 Fungus Fighter
Fungicide. *May & Baker Ltd.* Name unverified.

12485 Furac No. 3
The lead salt of dithiofuroic acid; a proprietary rubber vulcanization accelerator. No manufacturer.

12486 Furakem S
Two-component product based on furane resin; used for bedding, jointing files and paviors in highly corrosive environments; for operating temperatures to 180°; resists most commonly used nonoxidizing acids, alkalis and up to 50% concentrated sulfuric acid. *Feb Ltd.*

12487 Furalaxyl
57646-30-7 260-875-1
$C_{17}H_{19}NO_4$
methyl N-(2,6-dimethylphenyl)-N-(2-furanylcarbonyl)-DL-alanine
Fongarid; furalaxyl; Fonganil; CGA-38140. Systemic fungicide with protective and curative action for ornamentals. Used for control of soil diseases caused by *Phytophthora* and *Pythium* spp. mp = 70°, 84°; d^{20} = 1.22; soluble in H_2O (230 mg/l), more soluble in organic solvents; LD_{50} (rat orl) = 940 mg/kg. *Ciba-Geigy Agrochemicals.*

12488 Furanculine
Dried yeast.

12489 Fura-Tone® NC-1012
Furan resin; cured to hard infusible state resist, to solvents and chemicals; suggested for resin transfer molding. *Cardolite.*

12490 Furbac
95-33-0 202-411-2
N-Cyclohexyl-2-benzthiazyl sulfenamide
A proprietary accelerator. No manufacturer.

12491 furfural
98-01-1 4324 202-627-7
$C_5H_4O_2$
2-furaldehyde
2-furancarboxaldehyde; artificial oil of ants; Cyclic aldehyde. Chemical intermediate for manufacture of derivatives (furan, THF); solvent for petroleum lube, nitrocellulose; wetting agent; in manufacture of furfuralphenol plastics; vulcanization accelerator; insecticide, fungicide, germicide; reagent in analytical chemistry. *Aldrich; Allchem Industries; Great Lakes Sarl; QO.*

12492 furfural resins
Artificial resins obtained by the condensation of furfurldehyde with phenols, cresols, or other similar bodies.

12493 furfuryl alcohol
98-00-0 4325 202-626-1
$C_5H_6O_2$
2-furanmethanol
2-furylcarbinol; furfuralcohol; Furfuryl Alcohol; 2-furanmethanol; 2-hydroxymethylfuran; 2-Furylmethanol; 2-Furylcarbinol; furfural alcohol; furfuralcohol; furyl alcohol; 2-furancarbinol; α-furylcarbinol; FA. Wetting agent, furan polymers, solvent for dyes and resins, flavoring. mp = -37°; bp = 162°; d_4^{25}= 1.1563; soluble in H_2O (9 g/100 ml), organic solvents; LD_{50} (rat orl) = 127 mg/kg. *Aldrich; Allchem Industries; Great Lakes Sarl; QO.*

12494 Furlong®
For the agriculture industry. *DuPont UK.*

12495 furnace-calamine
Masses consisting mainly of zinc oxide, formed during the smelting of zinciferous iron ores.

12496 Furnex
1333-86-4 1856 215-609-9
Carbon black
Sevalco Ltd.

12497 2-Furoic acid
88-14-2 4329 201-803-0
$C_5H_4O_3$
2-Furancarboxylic acid
pyromucic acid; α-furoic acid; furoica; Furoic acid; furan-2-carboxylic acid. Preservative, bactericide, furoates for perfume and flavoring, fumigant, textile processing, chemical intermediate. mp = 133-134°; sublimes at 130-140°; bp = 230-232°; soluble in alcohol, ether, water. *R.W. Greeff; Penta Mfg.; QO.*

12498 Furotec
Furane resin binders; used for cold-set boning of sand cores. *Foseco (F.S.) Ltd.*

12499 Furoxone
67-45-8 4320 200-653-3
Furazolidone
Anti-infective, topical; antiprotozoal *Norwich Eaton Pharmaceuticals Inc.* Name unverified.

12500 Fursatil CS12
A proprietary cold-setting resin based on urea/furane. *Fordath Engineering Co.* Unverified.

12501 Fursatil CS15
A faster-setting, lower strength variant of Fursatil CS 12. *Fordath Engineering Co.* Unverified.

12502 Fursatil CS25
A fast-setting, medium strength variant of Fursatil CS12 having a low fume level. *Fordath Engineering Co.* Unverified.

12503 Fursatil CS30
A proprietary phenol/furane cold-setting resin. *Fordath Engineering Co.* Unverified.

12504 Fursatil CS40
A proprietary plasticized urea cold-setting resin. *Fordath Engineering Co.* Unverified.

12505 Fursatil CS60, CS65
A proprietary phenol/urea cold-setting resin. *Fordath Engineering Co.* Unverified.

12506 Fursatil CS71
A proprietary modified phenolic cold-setting resin. *Fordath Engineering Co.* Unverified.

12507 Fursatil CS81
A proprietary modified urea formaldehyde cold-setting resin. *Fordath Engineering Co.* Unverified.

12508 Furuculin
A yeast preparation.

12509 Fusabond® MB-110D
9002-88-4 7728
Polyethylene
Chemically modified LLDPE; anhydride-modified resin for use as compatibilizers in blends and alloys, polymeric coupling agents in reinforced or recycled PE or PP, adhesives and sealants. *DuPont.*

12510 Fusabond® MC-197D
24937-78-8
Chemically modified ethylene/vinyl acetate; anhydride-modified resin for use as compatibilizers in blends and alloys, polymeric coupling agents in reinforced or recycled PE or PP, adhesives and sealants. *DuPont.*

12511 Fusabond® MZ-109D
9003-07-0 7741
Polyethylene
Chemically modified polypropylene hompolymer; anhydride-modified resin for use as compatibilizers in blends and alloys, polymeric coupling agents in reinforced or recycled PE or PP, adhesives and sealants. *DuPont.*

12512 Fusabond® MZ-203D
9003-07-0 7741
Polyethylene
Chemically modified polypropylene impact copolymer; anhydride-modified resin for use as compatibilizers in blends and alloys, polymeric coupling agents in reinforced or recycled PE or PP, adhesives and sealants. *DuPont.*

12513 Fusarex
117-18-0 204-278-2
Tecnazene
Granules or dustable powder containing tecnazene; protectant fungicide and potato sprout suppressant. *ICI Chem & Polymers Ltd.*

12514 Fusariol
A mercury-formaldehyde preparation. A seed preservative.

12515 Fuscochlorin
A dark-green pigment from algae.

12516 Fuscorhodin
A dark-red pigment from algae.

12517 fused cement
electro-fused cement. Terms applied to an aluminous cement with a high alumina content.

12518 fusel oil
fermentation amyl alcohol; potato oil; grain oil; Marc brandy oil. A by-product in alcoholic fermentation, especially in the preparation of potato spirit, and in the rectification of alcohol. It consists mainly of mixed C_2-C_5 alcohols; used in the manufacture of chemicals, explosives, as an intermediate, pharmaceuticals, synthetic rubber, varnishes, lacquers, solvent for resins and waxes, perfumery.

12519 Fusilade
69806-50-4 4152 274-125-6
fluazifop-butyl
Emulsifiable concentrate of 125 g fluazifop-butyl per liter; used for grass weed control for broad-leaved crops. *ICI Chem & Polymers Ltd.*

12520 Fusion®
For the agriculture industry *DuPont UK.*

12521 Fussolon
A proprietary range of fluorocarbon and similar resins. *Daikin Kogyo Co.* Unverified.

12522 Fustic
$C_{13}H_{10}O_6$
old fustic; Brazil wood; yellow wood; Cuba wood. The chips or extract from *Morus tinctoria*; natural dyestuff, the dyeing agents being morin, and maclurin (pentahydroxybenzophenone); chiefly used for dying wool yellow.

12523 fustin
The diazobenzene compound of maclurin (obtained from fustic). A dyestuff.

12524 Futura Flex
High performance protective and waterproofing coatings for harsh environments. *Baxenden Chemicals Ltd.*

12525 Futura Thane
High performance protective and waterproofing coatings for harsh environments. *Baxenden Chemicals Ltd.*

12526 Futurit
A proprietary plastic of the phenol-formaldehyde type. No manufacturer.

12527 FW 18
1333-86-4 1856 215-609-9
Channel-type carbon black; for high-quality film calendered systems. *Degussa.*

12528 FW 200 Beads and Powd
1333-86-4 1856 215-609-9
Channel-type carbon black; for paints and coatings. *Degussa.*

12529 FX Pastes
Acrylic dispersions of organic pigments. *Reckitts Colours Ltd.*

12530 FX-512
UV-activated epoxy curative for coating formulations. *3M.*

12531 Fyarestor
Flame retardant. *Witco.* Discontinued.

12532 Fydulan, Fydumas, Fydusit
1194-65-6 3093 214-787-5
Dichlobenil
Direct, selective weed killing action in orchards, vineyards, flower beds, and parkland areas and along rail tracks, motorways and waterways. *Duphar BV.* Name unverified.

12533 Fyfanon
121-75-5 5740 204-497-7
$C_{10}H_{19}PO_6S_2$
O,O-Dimethyl-S-(1,2-di(ethoxycarbonyl)ethyl) phosphorodithioate
Malathion. A minimally toxic insecticide effective against insect pests on livestock, stored crops, agriculture, home and garden. *A/S Cheminova.* Name unverified.

12534 Fyran J2K
Sulfamate-based compound; flame retardant for treatment of cellulosic fabrics; minimal effect on hand; buffered for nonyellowing. No manufacturer.

12535 Fyrex®
diammonium phosphate-monoammonium phosphate
Mixture of crystalline diammonium phosphate and monoammonium phosphate; inorganic flame retardant; produces essentially neutral solution. *Akzo.*

12536 Fyrol® 6
Diethyl N,N-bis(2-hydroxyethyl) aminomethyl phosphonate
Reactive flame retardant for rigid urethane foams; incorporated into foam structure by reacting as a polyol; foams exhibit good dimensional stability; for spray, froth, or pour-in-place applications. *Akzo.*

12537 Fyrol® 38
Tri(β,β'-dichloroisopropyl)phosphate
Scorch stabilized flame retardant for flexible and rigid polyurethane foams; reduces discoloration caused by high exotherm following processing of flexible polyether urethane foam *Akzo.*

12538 Fyrol® CEF
Tri(β-chloroethyl)phosphate
Flame retardant well suited for transparent or pastel-shaded plastics and coatings. *Akzo.*

12539 Fyrol® DMMP
756-79-6 212-052-3
$C_3H_9O_3P$
Dimethyl methylphosphonate
DMMP; Dimethyl methanephosphonate; methyl phosphonic acid dimethyl ester. Flame retardant for applications where high phosphorus content, good solvency, and low viscosity are desired; lowers viscosity of epoxy resins and unsaturated polyesters filled with hydrated alumina oxide. bp = 181°; d = 1.145; soluble in H_2O (>10 g/100 ml), organic solvents; LD_{50} (rat orl) >5000 mg/kg. *Akzo.*

12540 Fyrol® FR-2
13674-87-8 4342 237-159-2
$C_9H_{15}Cl_6O_4P$
Tri(β,β'-dichloroisopropyl)phosphate
TCPP; PF-38; Emulsion 212. Flame retardant for flexible urethane foams bp_5 = 236-237°; n_D^{20}= 1.5022; soluble in H_2O (100 mg/l); LD_{50} (rat orl) = 1.85 g/kg. *Akzo.*

12541 Fyrol® PBR
32534-81-9 251-084-2

$C_{12}H_5Br_5O$
Pentabromo diphenyloxide. Flame retardant additive for flexible polyurethane foams; low propensity for discoloration caused by high exotherm in the processing of flexible polyether urethane foam Insoluble in H_2O, soluble in organic solvents. *Akzo.*

12542 Fyrol® PCF
Tri(β-chloroisopropyl)phosphate
Flame retardant. *Akzo.*

12543 Fyrquel® 150
Triaryl phosphate ester
Fire-resistant hydraulic fluid for industrial applications including air compressors, glass, and metal furnace hydraulics, gas and steam turbine bearing lubrication, vacuum pumps. *Akzo.*

12544 Fyrquel® EHC
68937-41-7 273-066-3
Triaryl phosphate
Triisopropylated Phenyl Phosphate; Durad 110 hydraulic fluid; Phenol, isopropylated, phosphate (3:1); Kronitex 50 triaryl phosphate. Fire-resistant electro-hydraulic control fluid. *Akzo.*

12545 Fytospore
cymoxanil-mancozeb
Mixture of cymoxanil and mancozeb; fungicide for the control of potato blight. *ICI Chem & Polymers Ltd.*

12546 Fyzol 11E
Adjuvant containing 99% highly refined mineral oil; wetting agent for contact carbamate herbicides. *Schering Agrochemicals Ltd.* Discontinued.

12547 G Resin
A proprietary trade name for cumarone-indene resins. No manufacturer.

12548 G Varnish
A proprietary trade name for varnish and lacquer resins made from glycerol and phthalic anhydride. No manufacturer.

12549 G-623
Silicone compound; general-purpose water-repellent coatings or sealants to prevent galvanic corrosion; lubricant for plastics and rubber. *GE Silicones.*

12550 G-635
Methyl phenyl silicone compound; its sealing and dielectric properties make it useful for high voltage insulators, connectors, automobile and aircraft ignition systems and other electronic equipment. *GE Silicones.*

12551 G-2162
PEG-25 propylene glycol stearate
Surfactant. *ICI Am.; ICI Spec. Chem.*

12552 G-3300
Alkylaryl sulfonate anionic surfactant in the form of a reddish-brown liquid; emulsifier, dispersant used for pigments in paints. *Atlas Chemical Industries (UK) Ltd.* Discontinued.

12553 G-4252
9005-66-7 8872
PEG-80 sorbitan palmitate
Surfactant. *ICI Specialty Chemicals.*

12554 G-4280
9005-64-5 8872
PEG-80 sorbitan laurate
Surfactant. *ICI Specialty Chemicals.*

12555 Gaardocyclene
Tricyclodecenyl isobutyrate
Quest Int'l UK Ltd.

12556 Gabbett's stain
A microscopic stain. It contains 2 g methylene blue, 25 ml sulfuric acid, with water up to 100 ml.

12557 Gabbro
A coarse, crystalline rock, composed mainly of lime-soda felspar.

12558 Gabian oil
An inflammable mineral naphtha.

12559 Gabraster
Polyester resins.

12560 Gabrosa
Carboxymethyl cellulose for drilling muds. *Montedison UK Ltd.* Name unverified.

12561 G-acid
2-Naphthol-6- or 8-sulfonic acid; used as an azo dye intermediate.

12562 Gadalan brands
Special catalysts for textile wash-and-wear finishes. *Thor Chemicals (UK) Ltd.* Discontinued.

12563 Gadorgel
1302-78-9 1082 215-108-5
Bentonite
For drilling fluid and civil engineering to achieve high quality thixotropic muds. *Minas de Gádor SA.*

12564 Gadose
A grease prepared from cod-liver oil and lanolin; used as a basis for ointments.

12565 Gad's cement
A mason's cement, consisting of 3 parts clay and 1 part ferric oxide.

12566 Gaduol
An extract containing the alcohol-soluble constituents of cod-liver oil.

12567 Gafac®
Redesignated Rhodafac® *Rhône-Poulenc Surf.*

12568 Gafamide CDD-518
Coconut oil diethanolamine condensate; adjuvant for nonionics, alkylaryl sulfonates and sulfated fatty alcohols; used in liquid manual-dishwashing formulations, drycleaning detergents, heavy-duty household detergents and industrial cleaners. *ISP.* Name unverified.

12569 Gafen LB-400, LE-500 and LS-500
Organic phosphate esters in free acid form; supplied as liquids; extreme pressure additives for metal working fluids, with lubricant, emulsifier and rust inhibition properties; used in lubricating and rolling oils, cutting and hydraulic fluids. *ISP.* Name unverified.

12570 Gafen LE-700, LP-700 and LK-500
Organic phosphate esters in free acid form; supplied as liquids; lubricity additives used in water-based cutting fluids with high concentration of inorganic rust inhibitors. *GAF Great Britain.* Name unverified.

12571 Gafen LM-400
Organic phosphate ester in free acid form, supplied as a liquid; oil-soluble, water-dispersible lubricant for use as a rust inhibitor in aqueous cutting fluids, rolling oils, and hydraulic fluids. *GAF Great Britain.* Name unverified.

12572 Gafen LM-600
Organic phosphate ester in free acid form, supplied as a liquid; oil-soluble emulsifier and metal lubricant used in rolling oils, cutting fluids, and hydraulic fluids. *ISP.* Name unverified.

12573 Gaffix® VC-713
Vinylcaprolactam/PVP/dimethylaminoethyl methacrylate copolymer, ethanol; film-forming, fixative resin for use in mousses, gels, glazes, lotions and hairsprays. *ISP.*

12574 Gafgard 233 and 233E
Clear amber liquid blends of reactive monomers; radiation curable coatings formulated to provide abrasion resistant surfaces to plastics, paper, wood and other substrates. *ISP.*

12575 Gafgard 238
A general purpose, nonyellowing aliphatic urethane based oligomer; used in tough scuff resistant coatings for flexible substrates, including plastics, textiles and leather. *ISP.*

12576 Gafgard 245
Clear amber liquid blend of aliphatic oligomers and monomers; radiation curable coatings designed for high speed application/curing. *ISP.*

12577 Gafgard 277
Clear amber liquid blend of aliphatic oligomers and monomers; radiation curable coatings designed for flooring use. *ISP.*

12578 Gafgard 280
Clear amber liquid blend of aliphatic oligomers and monomers; radiation curable coating designed for curtain coating. *ISP.*

12579 Gafite
A range of thermoplastic polyester molding compounds; resins which can replace metals as well as other plastics in numerous applications, e.g., under-the-hood parts, electrical/electronic components, appliance parts, hardware, pumps, and hydraulic controls. *ISP.*

12580 Gafite LW
A range of thermoplastic polyester molding compounds; used for automotive exterior parts such as rear end panels, cowl vents, fender extensions and headlight housings. *ISP.*

12581 Gaflex
Thermoplastic polyester elastomers; uses include industrial tubing, fuel lines, hydraulic hoses, flexible couplings, fasteners, gaskets, seals, boots and bellows for mechanical drives, wire and cable jacketing, noise dampening devices and pump parts. *ISP.*

12582 Gafoam AD
Ammonium salt of ethoxylate sulfate; foaming agent primarily used as an air-drilling surfactant for oil and gas wells; also used in well clean out and as a mobility control agent for carbon dioxide. *ISP.*

12583 Gafquat® 734
53633-54-8
Polyquaternium-11
Film-forming substantive polymer, conditioner for formulation of hair conditioners, rinses, sprays, shampoos, dyes, semipermanents, deodorants,

antiperspirants, shaving preparations, antiseptics, toilet soaps, skin creams, sunburn remedies. *ISP*.

12584 Gafquat® HS-100
Polyquaternium-28
Conditioning resin for personal care products; substantive, film-forming properties; for shampoos, conditioners, permanent waves, glazes, moisturizing creams. *ISP*.

12585 Gafstat AD-510 and AE-610
Free acids of complex phosphate esters, in liquid form; anionic surfactants in which the hydrophobe is aromatic; internal antistatic agents for plastics, with heat stability and low toxicity. Compatible with PVC, polyolefins, polystyrene and many other plastics. *GAF Great Britain*. Name unverified.

12586 Gafstat AS-610 and AS-710
Free acids of complex phosphate esters, in liquid form; anionic surfactants in which the hydrophobe is aliphatic; internal antistatic agents for plastics, with heat stability and low toxicity; compatible with PVC, polyolefins, polystyrene, and many other plastics. *GAF Great Britain*. Name unverified.

12587 Gaftuf
Impact modified polybutylene terephthalate compounds; used for hand tool housings, shrouds subjected to severe abuse, gasoline and brake clips etc. *ISP*.

12588 gagat
A variety of soft coal.

12589 Gala
A proprietary casein plastic. No manufacturer.

12590 galactan
$(C_6H_{10}O_5)_n$
Gelose
A gum, from agar-agar.

12591 Galactasol® Guar Derivs
39421-75-5
Hydroxypropyl guar
Gums used to increase viscosity in water and most brines; high viscosity grades also as flocculants; some in series as stabilizers, suspending aids. *Aqualon*.

12592 galactitol
608-66-2 4350 210-165-2
$C_6H_{14}O_6$
dulcitol
dulcite; dulcose; euonymit; melampyrite; melampyrum; melampyrin. Occurs in Madagascar manna. mp = 188-189°; bp_1 = 275-280°; d^{20} = 1.47; soluble in H_2O (3 g/100 ml), less soluble in organic solvents.

12593 galactosidase
9031-11-2 232-864-1
lactase
β-galactosidase. Derived from *E. coli*; enzyme-linked immunoassay, biochemical research.

12594 Galag
Magnesium impregnated metallurgical coke for iron desulfurization. *Foseco (F.S.) Ltd*. Discontinued.

12595 galagum
A mixture of modified polysaccharides. It gives a colloidal solution when boiled in water. It is a protective colloid, and is used in making baker's and flavoring emulsions, and in cosmetic and hair lotions.

12596 Galahad A
A proprietary trade name for a steel containing 0.10% carbon, 0.50% silicon, 0.04% sulfur, 0.04% phosphorus, 0.60% manganese, 13.00% chromium, optional 1.50% nickel, and 0.25% molybdenum; used in the manufacture of forged steel pressure vessels with good *Hadfield, George & Co*. Unverified.

12597 Galalith
The trade name for a polymeric material, obtained by the action of formaldehyde upon casein. No manufacturer.

12598 Galam butter
Shea butter.

12599 Galaxy®
Bentazon
acifluorfen. For post-emergence control of annual broadleaf weeds in soybeans and peanuts *BASF AG*.

12600 Galben M
Benalaxyl-mancozeb
Benalaxyl-mancozeb; used for control of blight in potatoes. *DowElanco Ltd*.

12601 Galden
A proprietary range of perfluoropolyether fluids. *Montedison UK Ltd*. Name unverified.

12602 Galenite
A pigment. It is a basic sulfate of lead. The name is sometimes applied to Galena.

12603 Galettame silk
Ricotti silk
Neri silk; Basinetto silk. The residue of the silk cocoon after reeling.

12604 Galicar
An anti-friction bearing metal containing 83% tin.

12605 Galipot
The resin from *Pinus maritima*.

12606 Gallal
$Al_4(C_7H_2O_5)_3$
Aluminum subgallate
An antiseptic and astringent.

12607 Gallant
69806-50-4 4152 274-125-6
haloxyfop
Herbicides based primarily on haloxyfop. *Dow UK*. Discontinued.

12608 Gallatite
Lactite; Lactoform; Cornalith; Ingalite; Lactorite; Sicalite; Proteolite. Casein preparations of a similar type to Galalith.

12609 Gallery One
Universal color dispersions for use in architectural coatings. *Engelhard*.

12610 Gallic acid
149-91-7 4363 205-749-9
$C_7H_6O_5$
3,4,5-trihydroxybenzoic acid
Photography, writing ink, dyeing; manufacture of pyrogallol, tannins, paper; tanning agent; pharmaceuticals, engraving, lithography; analytical reagent. Sublimes at 210° giving stable form with mp = 258-265°; 1 gram dissoles in 87 ml water; partially insoluble in benzene; LD_{50} (rbt orl) = 5.0 g/kg. *Fuji Chem. Ind.; Mallinckrodt; Penta Mfg.; U.S. Biochemical.*

12611 Gallicin
99-24-1 4363 202-741-7
$C_8H_8O_5$
methyl gallate
Gallic acid methyl ester, used by oculists as an antiseptic in conjunctivitis.

12612 Gallion
Nonionic wetting agent containing ethylene oxide condensates; spreading agent for translocated phenoxy herbicides. *Intracrop Ltd*.

12613 Gallipoli oil
An olive oil used in the textile industries.

12614 Gallisin
That portion of commercial starch syrup which resists fermentation.

12615 Gallium
7440-55-3 4364 231-163-8
Ga
Compounds used as semiconductors. Metallic element mp = 29.78°; bp = 2400°. *Aldrich; Atomergic Chemetals; Cerac; Eagle-Picher; Int'l. Gallium GmbH; Rhône-Poulenc Basic.*

12616 Gallobromol
$C_7Br_2H_4O_5$
Dibromogallic acid

12617 Gallstone
A yellow pigment obtained from the gall bladder of oxen.

12618 Galorn
A proprietary trade name for a casein plastic. No manufacturer.

12619 Galoryl
Range of various surfactants. *CFPI*.

12620 Galt glass
Polyester resin bonded glass fiber moldings with a special surface giving good resistance to weather and chemicals.

12621 Galvanit
A plating powder consisting of a mixture of the salt of the metal to be deposited (silver for silver plating), and a more electro-positive metal.

12622 galvanized iron
(Zinced iron). Iron coated with metallic zinc.

12623 Galvano Lac
A mixture of celluloid varnish with powdered metal.

12624 Galvoline
Magnesium ribbon anode. *Dow UK*. Discontinued.

12625 Galvomag
High current magnesium anode. *Dow UK*. Discontinued.

12626 Galvorod
Standard magnesium anode. *Dow UK*. Discontinued.

12627 Gamanase
A hemicellulase prepared by submerged fermentation of *Aspergillus niger*, particularly suited for applications where a rapid break-down of galactomannans to less viscous products is required. *Novo Nordisk*.

12628 Gamboge
Gummi gutta
A gum resin, the product of *Garcinia morella* of Thailand. It is a yellow pigment used for water colors, also for coloring spirit and other varnishes.

12629 Gamboge butter
A fat obtained from *Garcinia morelia*.

12630 Gamma acid
2-Amino-8-naphthol-6-sulfonic acid
Used as an azo dye intermediate.

12631 Gamma-BHC Dust
58-89-9 5526 200-401-2
lindane
Insecticide. *Murphy Chemical Co Ltd.* Discontinued.

12632 Gamma-Col
58-89-9 5526 200-401-2
lindane
Suspension concentrate containing 800 g γ-HCH per liter; an organochlorine insecticide.

12633 Gamma-HCH Dust
58-89-9 5526 200-401-2
lindane
Garden insecticide. *Murphy Chemical Co Ltd.* Discontinued.

12634 Gammalex
captan-lindane
Captan·2 γ-HCH; insecticide and fungicide seed dressing for brassicas and oilseed rape. *ICI Agrochemicals; ICI Chem. & Polymers Ltd.*

12635 Gammalin
58-89-9 5526 200-401-2
lindane
Insecticides. *ICI Agrochemicals; ICI Chem. & Polymers Ltd.*

12636 Gammasan
58-89-9 5526 200-401-2
lindane
γ-HCH. Liquid seed dressing. *ICI Agrochemicals; ICI Chem. & Polymers Ltd.*

12637 Gammatox
Livestock dips and sprays. *The Wellcome Foundation Ltd.* Name unverified.

12638 Gammatrol
Nucleonic instruments for process control and measurement. *ICI Chem & Polymers Ltd.*

12639 Gammexane
58-89-9 5526 200-401-2
lindane
Insecticides. *ICI Chem & Polymers Ltd.*

12640 Ganex® Et-201
PVP/decene copolymer. *ISP.*

12641 Ganex® P-904
Butylated PVP; used in cosmetics and toiletries as moisture barrier, adhesive, protective colloid, and microencapsulating resin; as dispersant for pigments; as solubilizer for dyes; in petroleum industry as sludge and detergent dispersant; protective colloid in coatings; suspending aid in polymerization, dyeing assistant, antiredepositioin agent in dry cleaning; especially as a dispersant in aqueous agricultural chemicals or in pigmented skin care products. *ISP.*

12642 Ganex® V-216
32440-50-9
PVP/hexadecene copolymer
Polyvinylpyrrolidone/hexadecene copolymer; used in cosmetics and toiletries as moisture barrier, adhesive, protective colloid, and microencapsulating resin; as dispersant for pigments; as solubilizer for dyes; in petroleum industry as sludge and detergent dispersant. *ISP.*

12643 Ganex® V-220
28211-18-9
PVP/eicosene copolymer
Polyvinylpyrrolidone/eicosene copolymer; used in cosmetics and toiletries as moisture barrier, adhesive, protective colloid, and microencapsulating resin; as dispersant for pigments; as solubilizer for dyes; in petroleum industry as sludge and detergent dispersant. *ISP.*

12644 Ganex® WP-660
Tricontanyl PVP
Waterproofing polymer for personal care products; for sunscreens, skin care products, cosmetics, makeup, baby care products. *ISP.*

12645 gangue
The earthy portion of an ore which leaves the metal when reduced.

12646 Ganicin
Zinc-rich coatings. *DuPont UK.*

12647 Ganister
A rock mineral with a composition corresponding to a pure silica with about 1/10 of its weight of clay; used in the manufacture of siliceous fire-bricks, and for lining furnaces.

12648 Ganocide
5707-69-7 3499 227-197-8
Drazoxolon
Fungicides containing drazoxolon. *ICI Chem & Polymers Ltd.*

12649 Gant
A barrier cream. *Croda Chem. Ltd.* Name unverified.

12650 Gantrez®
A range of poly(methylvinyl ether/maleic anhydride) copolymers; gelling agents, thickeners, stabilizers, explosive stabilizers, anticorrosion coatings, and suspending aids. *ISP.*

12651 Gantrez® AN-119
52229-50-2
PVM/MA copolymer; thickener, dispersant, stabilizer, gelling agent, coupler, protective colloid, suspending aid used in emulsion polymerization; adhesives, household detergents, liquid hand cleaners, acid bowl cleaners; produces clear films of high tensile and cohesive strength. *ISP.*

12652 Gantrez® AN-8194
Stearylvinyl ether/maleic anhydride copolymer, toluene; hydrocarbon-soluble copolymer that forms waxy films with good water resistance and thermal properties; compatible with silicone release agents and modifies release level for pressure-sensitive labels. *ISP.*

12653 Gantrez® B-773
Poly(vinyl isobutyl ether), hexane
Tacky polymer with excellent adhesion to plastic, metal, and coated surfaces; plasticizer and leveling agent for surface coatings. *ISP.*

12654 Gantrez® ES-225
Ethyl ester of PVM/MA copolymer, ethanol; copolymer forming clear, glossy films with substantivity and moisture resistance; used in hairsprays, mousses, gels, and lotions, coatings, polishes; emulsion stabilizer in creams and lotions. *ISP.*

12655 Gantrez® M-154
9003-09-2
Polyvinyl methyl ether
Polymer functioning as tackifier, binder, and plasticizer; used in printing inks, textile sizes and finishes, latex modification. *ISP.*

12656 Gantrez® S-95
9011-16-9
PVM/MA copolymer, hydrolyzed low molecular weight polymer; water-soluble polyelectrolyte similar to Gantrez AN series; chelating agent. *ISP.*

12657 Gantrez® SP-215
Ethyl ester of PVM/MA copolymer; hair fixative for stiffer, harder holding products; up to 10% can be formulated in pump hair sprays. *ISP.*

12658 Gantrez® V-215
Methylvinyl ether/maleic anhydride copolymer, ethyl ester, ethanol SDA-40B. *ISP.*

12659 Garbacryl
Lubricating oil additives. *Rhône-Poulenc UK.*

12660 Garbritol
Textile auxiliary. *Hickson & Welch Ltd.* Discontinued.

12661 Garcrete
Refractory castable. *Foseco (F.S.) Ltd.*

12662 Gardamide
N-Methyl-N-phenyl-2-methyl butyramide
Quest Int'l. UK Ltd.

12663 Garden Lime
1317-65-3 5515 215-279-6
Limestone. Ground limestone; used to increase pH of acid soils. *Vitax Ltd.*

12664 Gardeniol
$C_{10}H_{12}O_2$
Phenylmethylcarbinyl acetate
A perfume.

12665 Gardinol
Sodium salts of sulfonated higher fatty alcohols; wetting agents. *Hickson & Welch Ltd.*

12666 Gardinox
Textile auxiliaries. *Hickson & Welch Ltd.*

12667 Gardlite
A proprietary trade name for a synthetic resin, produced from toluenesulfonamide and formaldehyde. No manufacturer.

12668 Gardol®
137-16-6 4379 205-281-5
Sodium lauroyl sarcosinate
Closyl LA 3584; Crodasinic LS30, LS35. Soap-like detergent providing wetting and foaming; for personal care and household products; corrosion inhibitor for mild steel. *Clough.* Name unverified.

12669 Gardoprim A 500FW
atrazine-terbuthylazine
A suspension concentrate containing 100 g atrazine and 400 g terbuthylazine per liter; used for total weed control in forestry plantations. *Ciba-Geigy Agrochemicals.*

12670 Garganine
A madder extract.

12671 Garj
A bituminous sandstone. It contains from 6-17% bitumen.

12672 Garlon
69633-04-1
$C_8H_7Cl_3O_5S$
Ethanol, 2-(2,4,5-trichlorophenoxy)-, hydrogen sulfate
(Trichlorophenoxy)ethyl hydrogen sulfate; triclopyr. Herbicides based primarily on triclopyr. *Dow UK.* Discontinued.

12673 Garlon 2
69633-04-1
Emulsifiable concentrate containing 240 g/l triclopyr, herbicide to control perennial and woody weeds. *ICI Chem & Polymers Ltd.*

12674 Garlon 4
69633-04-1
Emulsifiable concentrate containing 480 g/l triclopyr; herbicide to control perennial and woody weeds. *Chipman Ltd.*

12675 Garnex
Refractory insulating expendable tundish linings. *Foseco (F.S.) Ltd.*

12676 Garoflam
Flame retardant blends; for plastics and rubber. *Croxton & Garry Ltd.* Name unverified.

12677 Garomix
magnesium oxide-zinc oxide
Magnesium oxide and zinc oxide blend; for rubber goods. *Croxton & Garry Ltd.* Name unverified.

12678 Garosorb
Desiccants; for plastics and rubbers. *Croxton & Garry Ltd.* Name unverified.

12679 Garospers
Dispersed rubber chemicals; used for rubber goods. *Croxton & Garry Ltd.* Name unverified.

12680 Garozinc
1314-13-2 10279 215-222-5
Zinc oxides
Filler and pigment; used for rubber goods. *Croxton & Garry Ltd.* Name unverified.

12681 Garpak
Rammable refractory. *Foseco (F.S.) Ltd.*

12682 Garseal
Refractory air-setting mortar. *Foseco (F.S.) Ltd.*

12683 Gartop
Refractory insulating powder or boards for covering the steel in continuous casting tundishes. *Foseco (F.S.) Ltd.*

12684 Gartube
Insulating shrouds fitted on the bottom of teeming ladles. *Foseco (F.S.) Ltd.*

12685 Garvox 3G
22781-23-3 1063 245-216-8
Bendiocarb
A contact, systemic insecticide. *Schering Agrochemicals Ltd.* Discontinued.

12686 gas black
Satin gloss black; hydrocarbon black; hydrocarbon gas black; silicate of carbon; jet black; ebony black. A carbon black made by the incomplete combustion of natural gas; used in rubber mixings.

12687 gas oil
The name applied to all mineral oils intended for the preparation of gas, such as the light oils of brown coal tar, and shale oil; used as an absorption oil and in the manufacture of ethylene.

12688 Gasbinda
1344-09-8 8825 215-687-4
Sodium Silicate
Sodium silicate binders for sand cores and molds using a CO_2 hardening process. *Foseco (F.S.) Ltd.*

12689 Gasil
Micronized silica gel. *Crosfield Chemicals Ltd.*

12690 Gasil EBC, EBN
Proprietary compounds of silica used in the paint, resin and plastics industries, e.g., in the matting of electron beam-cured coatings. *Crosfield Chemicals Ltd.*

12691 Gaskoid
A proprietary trade name for a rubber jointing resistant to oil and petroleum. No manufacturer.

12692 Gastex
A proprietary gas black used in rubber mixings. No manufacturer.

12693 Gastratox 6G Slug Pellets
108-62-3 5983 202-945-6
metaldehyde
Pellets containing 6% w/w metaldehyde; snail and slug bait. mp = 246°; sublimes at 112°; practically insoluble in H_2O, soluble in benzene, chloroform; LD_{50} (rat orl) = 3100 mg/kg. *Truchem Ltd.*

12694 Gastro Caloreen
A proprietary preparation of a polyglucose polymer used as a high-calorie food supplement. *Scientific Hospital Supplies.* Name unverified.

12695 Gastrografin
5851
Sodium and meglumine salts of diatrizoate; x-ray contrast media. *Schering Health Care Ltd.* Discontinued.

12696 Gat 15
Rousselot special gelatin powder; developed as a substitute for gum arabic. *SKW Biosystems Ltd.*

12697 Gatodan 415
Propylene glycol stearate and distilled monoglycerides; food emulsifier for cake and sponge improvers. *Grindsted Prods.; Grindsted Prods. Denmark.*

12698 Gaucho
105827-78-9 4946
Imidacloprid
Insecticidal seed dressing with excellent root systemic properties for control of sucking pests (virus vectors), e.g., aphids, thrips, leafhoppers, white files and some biting insects such as rice water weevil, Colorado potato beetle, wireworms. *Baver AG.*

12699 Gauduin's fluid
A mixture of finely powdered cryolite, and a solution of phosphoric acid in alcohol. A soldering fluid.

12700 gauging metal
An alloy similar to Delta metal, but containing iron.

12701 Gaultheria oil
119-36-8 6200 204-317-7
methyl salicylate
Oil of wintergreen; used in medicine, flavoring compounds and perfumery.

12702 Gaultheric acid
119-36-8 6200 204-317-7
$C_8H_8O_3$
Methyl salicylate
Perfume and flavorant.

12703 Gazelle
Acid phosphates for the baking trade. *Albright & Wilson Ltd.* Name unverified.

12704 GBL
96-48-0 1632 202-509-5
γ-Butyrolactone
Spinning and coagulating solvent for textiles, in nail polish removers, detergents, sunscreens, household cleaners, paint removers, agricultural use, polymers, petroleum industry. *Arco Chemical Co.*

12705 GDL
90-80-2 4465 202-016-5
$C_6H_{10}O_6$
Gluconolactone
Glucono delta lactone; D-(+)-Glucono-1,5-lactone; Glucono δ-lactone; D-Gluconic acid, δ-lactone; δ-Gluconolactone; Gluconic lactone; Gluconic acid lactone; Glucarolactone; D-Aldonolactone; D-threo-Aldono-1,5-lactone; D-Glucono-δ-lactone. Can behave as a chelator and is used in metal cleaning materials. mp = 153° (dec); $[\alpha]_D^{20}$ = 61.7° (c = 1); soluble in H_2O (59 g/100 ml), less soluble in organic solvents. *Pfizer International.*

12706 GE 2557
A proprietary trade name for a polyester vinyl plasticizer. *GE Plastics.* Name unverified.

12707 Geax
A trade name for a synthetic resin-paper product used as an electrical insulation. No manufacturer.

12708 Geblitol
7775-14-6 8771 231-890-0
Sodium hydrosulfite
Used as a disinfectant.

12709 Gecet F100
Expandable engineering resin; foam bead material for use in sporting goods (water sports, boat components), automotive (upper instrument panels), furniture, building and construction, medical (sterilizable insulating containers and lab specimen shippers), shipping, materials handling (functional pallets, cores in composite structures) applications: F100 offers widest range of

mechanical properties, moderate heat resistance and end-use temperature of 220ºF. *GE Plastics; Huntsman.*

12710 Gedanite

Friable amber; soft amber. A resin found on the shores of the Baltic. It is a variety of amber, and melts at 150-180º. It is also called soft amber. A variety of amber low in succinic acid.

12711 Gedeflex®

A registered trademark for dibutyl, butylbenzyl and dioctyl phthalates; plasticizers. No manufacturer.

12712 Gedelite®

A registered trademark for phenolic molding powders and resins. No manufacturer.

12713 gedrite

A yellow resin found with Prussian amber. It is also the name for a mineral.

12714 Geko

Inorganic molding sand binders for foundries, also in connection with lustrous carbon formers to obtain better castings. *Süd-Chemie AG.* Name unverified.

12715 Gel Flo

A slurry explosive; used in iron ore mining. *Hercules.* Discontinued.

12716 Gel II

7681-49-4 8762 231-667-8

Sodium fluoride

Dental caries prophylactic. *orl-B Laboratories Inc.* Name unverified.

12717 Gel Power

Slurry explosives; underground non coal-type mines. *Hercules.* Discontinued.

12718 gel rubber

A term used for the residue of rubber left undissolved when raw rubber is treated with a solvent.

12719 Gelamite D

A nitroglycerin dynamite; used for construction and building, explosives, mining, petroleum and related industries. *Hercules.* Discontinued.

12720 Gelaprime F

A nitroglycerin dynamite; used for construction and building, explosives, mining, petroleum and related industries. *Hercules.* Discontinued.

12721 Gelatase

Bacterial proteolytic enzymes. *ABM Chemicals Ltd.* Name unverified.

12722 gelatin

9000-70-8 4388 232-554-6

Gelfoam; Puragel; Hydrocoll G-40; Hydrocoll AG-SD, PGA, PGB; Hydrocoll™ G-40, G-55; Colla -Gel AC; Cosmetic Gelatin; HiPure Liq. Gelatin, Cosmetic Grade; Byco A,C,O; Crodyne BY 19; Pharmgel; Pronel Capsules; Rousselot Gelatine; SPG Gelatine. Cosmetics ingredient; stabilizer; thickener in food; tablet binder; inhibitor of crystallization in bacteriology, for preparing cultures; suspending agent in pharmaceuticals. Soluble in hot water, glycerol, acetic acid; insoluble in organic solvents. *Brooks Industries; Maybrook; Hormel; Norland Prods.;.*

12723 gelatin carbonite

An explosive consisting of 25% nitroglycerin, 0.7% collodion wool, 7% gelatin, 25% sodium chloride, and 42% ammonium nitrate.

12724 gelatin, vegetable

Agar-agar.

12725 Gelatinastralite

An explosive consisting of ammonium nitrate, with some sodium nitrate, up to 20% dinitrochlorhydrin, and maximum amounts of 5% nitroglycerin, and 1% collodion cotton.

12726 Gelatine Donarit 1,2,3

Gelatinous ammon dynamites

ammon gelatins; ammon gelignites. Main ingredients are explosive-oil, ammonium nitrate, aromatic nitrocompounds and cellulose nitrate; suitable for use at the surface as well as underground; very good water resistance. *Dynamit Nobel Wien GmbH.* Name unverified.

12727 Gelatine Donarit 2 E

Gelatinous explosive; main ingredients are explosive-oil, ammonium nitrate, nitrocellulose, etc. *Dynamit Nobel Wien GmbH.* Name unverified.

12728 Gelatine Donarit S

An explosive for seismic prospecting which, on account of special sensitizers, can achieve best results even under high pressure; supplied in plastic tubes which can be coupled by means of screw threads to form longer charging units. *Dynamit Nobel Wien GmbH.* Name unverified.

12729 gelato-glycerin

Glycerin jelly.

12730 Gelcharg

High viscosity, low cost polymer for gelling watergel explosives; thickener, water-blocking agent. *Hercules.* Discontinued.

12731 Gelcotar®

Tar BP; strong coal tar solution in an aqueous gel base; used to treat psoriasis and eczema. *Quinoderm Ltd.*

12732 Gelcotar® Liquid

Strong coal tar solution and cade oil; a shampoo for treatment of psoriasis of the scalp, seborrhoeic dermatitis and dandruff. *Quinoderm Ltd.*

12733 Geleol

31566-31-1 4498 250-705-4

Glyceryl stearate

Emulsifier. *Gattefosse; Gattefosse SA.*

12734 Gelflex

Dimefilcon A

2-hydroxyethyl methacrylate polymer with methyl methacrylate and ethylenebis dimethacrylate; contact lens material. *Dow Corning Ophthalmics Inc.* Name unverified.

12735 Gelkyd

Thixotropic alkyds

Cray Valley Ltd.

12736 Gellan gum

71010-52-1 4389 275-117-5

Gelrite®; Native gellam gum; High acyl gellan gum; Polysaccharide S-60; PS-60; Deacylated gellan gum; Kelcogel; Deacylated PS-60. Gellan gum; used for microbiological applications. Extracellular polysaccharide obtained by aerobic fermentation of *Pseudomonas elodea*; readily soluble in water; forms thermoreversible gels. Readily soluble in water, partially soluble in cold water; bulk density ca. 50 lb/cu ft; LD_{50} (rat orl) > 5000mg/kg. *Kelco.*

12737 Gelline

A colloidal iodine gel.

12738 Gelobel

A proprietary trade name for a gelatin dynamite; an explosive. No manufacturer.

12739 Gelosine

A mucilaginous material extracted from a Japanese algae. It is soluble in alcohol and water.

12740 Gelot 64®

Glyceryl stearate and PEG-75 stearate; SE base for oil-water cosmetic and pharmaceutical emulsions. *Gattefosse; Gattefosse SA.*

12741 Geloxite

An explosive consisting of 54-64 parts nitroglycerin, 4-5 parts nitro-cotton, 13-22 parts potassium nitrate, 12-15 parts ammonium oxalate, 0-1 part red ocher, and 4-7 parts wood meal.

12742 Geloy® BG10

Acrylic-styrene-acrylonitrile resin; blow molding grade for applications requiring prolonged outdoor exposure. *GE Plastics.*

12743 Geloy® GY1020

Weatherable polymer. *GE Plastics.*

12744 Geloy® GY1220

ASA/PVC alloy; weatherable polymer. *GE Plastics.*

12745 Geloy® XP1001

Acrylic-styrene-acrylonitrile terpolymer resin; injection molding resin with excellent weatherability; applications include interior and exterior automotive/truck/RV trim and parts, providing color stability and property retention without painting. *GE Plastics.*

12746 Gelrite®

71010-52-1 4389 275-117-5

Gellan gum

Gellan gum; used for microbiological research applications. *Kelco.*

12747 Gelsorb B

1337-76-4

Attapulgite. Attapulgite clay. *Bromhead & Denison Ltd.*

12748 Gelucire 35/10

Saturated polyglycolized glycerides; excipient for hard gelatin capsules. *Gattefosse SA.*

12749 Gelva

Polyvinyl acetate resin emulsions/solutions; used to make pressure sensitive coatings and laminating adhesives for difficult-to-bond surfaces; used to formulate surface coatings and for specialty paper coatings. *Monsanto Co.*

12750 Gelvatol

Powdered resins. *Monsanto plc.*

12751 Gemex

Surfactants. *Akzo.* Name unverified.

12752 Gemglo

A proprietary trade name for styrene and methacrylate polymers. No manufacturer.

12753 Gemlite

A proprietary trade name for a urea formaldehyde synthetic resin. No manufacturer.

12754 Gemme
Crude turpentine. Oil and solvent.
12755 Gemstone
A proprietary trade name for a phenol-formaldehyde cast resin. No manufacturer.
12756 Gemstone M.1.2
A proprietary trade name for a phenolic laminating resin. No manufacturer.
12757 Gemtone®
Rich, lustrous pigments deriving color from both light interference and light absorption. *Mearl.*
12758 Genagen C-100
61791-29-5
PEG-10 cocoate
Surfactant for textile processing. *Hoechst Celanese/Colorants & Surf.*
12759 Genagen CA-050
61791-08-0
PEG-5 cocamide
Cleansing skin protective component for detergents and cleaning agents. *Hoechst Celanese/Colorants & Surf.; Hoechst AG.*
12760 Genagen O-090
9004-96-0
PEG-9 oleate
Surfactant for textile processing. *Hoechst Celanese/Colorants & Surf.*
12761 Genagen P-070
9004-94-8
PEG-7 palmitate
Surfactant for textile processing. *Hoechst Celanese/Colorants & Surf.*
12762 Genagen PL-090
31621-91-7
PEG-9 pelargonate
Surfactant for textile processing. *Hoechst Celanese/Colorants & Surf.*
12763 Genagen S-080
9004-99-3
PEG-8 stearate
Surfactant for textile processing. *Hoechst Celanese/Colorants & Surf.*
12764 Genagen TA-080
61791-00-2
PEG-8 tallate
Surfactant for textile processing. *Hoechst Celanese/Colorants & Surf.*
12765 Genal P4300-CM
A proprietary heat resistant but asbestos-free phenolic molding compound. *GE Plastics.* Name unverified.
12766 Genamid®
Polyamide resins, liquid reactable. *Cray Valley Ltd.*
12767 Genamid® 151
Amidoamine resin; epoxy curing agent offering superior wetting, excellent chemical resistance, better internal plasticizer, fast cure time; used for high solids coatings, castings, laminates, and adhesives. *Henkel.* Name unverified.
12768 Genamin
Large range of cationic surfactants which may be primary, secondary or tertiary amines, diamines or amine salts; solid, liquid or paste forms; antistatic agents, fabric conditioner and softener, fiber finishers, water-repellent agents. *Hoechst UK.* Name unverified.
12769 Genamin C Grades
Ethoxylated cocamine (2-25 EO)
Surfactant for agricultural formulations. *Hoechst Celanese/Colorants & Surf.*
12770 Genamin CTAC
112-02-7 203-928-6
Cetrimonium chloride
Cosmetic raw material for hair treatment preparations; antistat. *Hoechst Celanese/Colorants & Surf.*
12771 Genamin DSAC
107-64-2 203-508-2
Distearyldimonium chloride
Adogen® TA-100, TA-101; Arosurf® TA-100; Arquad® 218-75, 218-100; Blandofen CT; Dehyquart DAM; Genamin DSAC; Prepagen WK. Cosmetic raw material for hair and skin treatment preparations; good wet combing, skin softening properties. *Hoechst Celanese/Colorants & Surf.*
12772 Genamin KDM-F
17301-53-0 241-327-0
Behentrimonium chloride
Base, antistat, emulsifier for preparation of hair care products; conditioner. *Hoechst Celanese/Colorants & Surf.*
12773 Genamin KSE
Distearyldimonium chloride
cetyl alcohol; ceteareth-15; ceteareth-3; PEG-3 distearate. SE base material, conditioner for hair treatment preparations. *Hoechst Celanese/Colorants & Surf.*
12774 Genamin KSL
PEG-5 stearyl ammonium lactate
Gloss aid for hair care products. *Hoechst Celanese/Colorants & Surf.*
12775 Genamin T-020
61791-44-4 263-177-5
PEG-2 tallow amine
Surfactant for agricultural formulations, textile processing. *Hoechst Celanese/Colorants & Surf.*
12776 Genamin TA Grades
61790-33-8 263-125-1
Tallow fatty acid amine
Surfactant for agricultural formulations. *Hoechst Celanese/Colorants & Surf.*
12777 Genamine C-020
Coconut fatty amine oxethylate; raw material for mineral oil additives, insecticides, pesticides, cosmetic bases, adhesives. *Hoechst Celanese.*
12778 Genaminox CS
61788-90-7 263-016-9
Coco dimethyl amine oxide
Foaming agent and stabilizer, thickener for personal care products; hair conditioner. *Hoechst AG.*
12779 Genaminox KC
61788-90-7 263-016-9
Cocamine oxide
Foam booster/stabilizer over wide pH range, thickener for shampoos, bath products; surfactant for textile processing. *Hoechst Celanese/Colorants & Surf.; Hoechst AG.*
12780 Genapol®
General surfactant properties, low foaming power and ability to reduce the foam of other surfactants; stable to acids, alkalies and most metal salts; drain aid in machine dish and bottle washing. *Hoechst UK.* Name unverified.
12781 Genapol® 24-L-3
68439-50-9
C_{12}-C_{14} pareth-2.9
Biodegradable detergent intermediate for sulfation for use in cosmetics, shampoos, light duty detergents; emulsifier, prewash spotter, agricultural adjuvant, hydrocarbon-based cleaning systems. *Hoechst Celanese/Colorants & Surf.*
12782 Genapol® 26-L-1
68551-12-2
C_{12}-C_{16} pareth-1
Biodegradable detergent intermediate for sulfation for use in cosmetics, shampoos, light duty detergents; emulsifier, prewash spotter, agricultural adjuvant, hydrocarbon-based cleaning systems. *Hoechst Celanese/Colorants & Surf.*
12783 Genapol® 42-L-3
68439-50-9
C_{12}-C_{14} pareth-3
C_{12}-C_{14} pareth-3;. Biodegradable detergent intermediate for sulfation for use in cosmetics, shampoos, light duty detergents; emulsifier, prewash spotter, agricultural adjuvant, hydrocarbon-based cleaning systems. *Hoechst Celanese/Colorants & Surf.*
12784 Genapol® 2299
Coconut alcohol ethoxylate
Surfactant for textile processing. *Hoechst Celanese/Colorants & Surf.*
12785 Genapol® AMS
TEA-PEG-3 cocamide sulfate
Detergent, foaming agent used in top-grade cosmetics cleansers; lime soap dispersant; biodegradable. *Hoechst Celanese/Colorants & Surf.; Hoechst AG.*
12786 Genapol® ARO
9004-82-4
Sodium laureth sulfate
Raw material for cosmetics, detergents, and cleaning agents. *Hoechst Celanese/Colorants & Surf.; Hoechst AG.*
12787 Genapol® C-050
61791-13-7
Coceth-5
Raw material for manufacture of textile, leather, paper auxiliaries, detergents, emulsifiers, cosmetics, agricultural uses. *Hoechst Celanese/Colorants & Surf.*
12788 Genapol® DA-040
26183-52-8
Deceth-4
Surfactant for textile processing. *Hoechst Celanese/Colorants & Surf.*
12789 Genapol® GC-050
61791-13-7

Coceth-5
Surfactant for textile processing. *Hoechst Celanese/Colorants & Surf.*

12790 Genapol® LRO Liq., Paste
9004-82-4
Sodium laureth sulfate
Detergent, foaming agent used in cosmetic products, personal care products, agricultural, lime soap dispersant; biodegradable. *Hoechst Celanese/Colorants & Surf.; Hoechst AG.*

12791 Genapol® O-020
9004-98-2
Oleth-2
Raw material for manufacture of textile, leather, paper auxiliaries, detergents, emulsifiers, cosmetics, agricultural uses. *Hoechst Celanese/Colorants & Surf.*

12792 Genapol® PF 10
EO/PO block copolymer; surfactant for agricultural formulations, textiles. *Hoechst Celanese/Colorants & Surf.*

12793 Genapol® PGM Conc
Sodium laureth sulfate, glycol distearate and cocamide MEA; pearl-luster concentrate used in shampoos, bubble baths, cosmetics, liquid soaps, detergents; biodegradable. *Hoechst Celanese/Colorants & Surf.; Hoechst AG.*

12794 Genapol® PL 120
EO/PO fatty alcohol adduct; low foaming surfactant. *Hoechst AG.*

12795 Genapol® PMs
627-83-8 211-014-3
Glycol distearate
Pearlescent agent for cosmetic washing agents and shampoo. *Hoechst Celanese/Colorants & Surf.; Hoechst AG.*

12796 Genapol® S-020
Stearyl alcohol polyglycol ether
Raw material for manufacture of textile, leather, paper auxiliaries, detergents, emulsifiers, cosmetics. *Hoechst Celanese/Colorants & Surf.* Name unverified.

12797 Genapol® T Grades
Tallow alcohol polyglycol ether (8-25 EO); detergent base and basic material for cosmetic and specialty chemical industries, agricultural formulations; biodegradable. *Hoechst Celanese/Colorants & Surf.; Hoechst AG.*

12798 Genapol® TS Powd
9005-08-7
PEG-3 distearate
Detergent, pearlescent, opacifier for shampoos, bubble baths, shower preparations. *Hoechst Celanese/Colorants & Surf.; Hoechst AG.*

12799 Genapol® TSM
PEG-3 distearate-sodium laureth sulfate-glycol distearate. Detergent, opacifier for shampoos, bubble baths, shower preparations. *Hoechst Celanese/Colorants & Surf.; Hoechst AG.*

12800 Genapol® UD-030
PEG-3 fatty alcohol
Surfactant for oil and gas industry. *Hoechst Celanese/Colorants & Surf.*

12801 Genapol® V 2908
C_{12}-C_{15} synthetic oxo-alcohol EO/PO block copolymer; surfactant for agricultural formulations. *Hoechst Celanese/Colorants & Surf.*

12802 Genapol® X-040
24938-91-8
Trideceth-4
Surfactant for textile processing. *Hoechst Celanese/Colorants & Surf.*

12803 Genapol® ZRO Liq., Paste
9004-82-4
Sodium laureth sulfate
Raw material with good foaming and cleansing for cosmetics, detergents, cleaning agents. *Hoechst Celanese/Colorants & Surf.; Hoechst AG.*

12804 Genasco
A proprietary trade name for a bituminous softener. No manufacturer.

12805 Genatosan Skin Bar
A soap-free detergent bar. *Fisons plc, Pharmaceuticals Div.* Name unverified.

12806 Gen-che
Preparation of zinc, manganese, copper and sulfur compounds; chelated micronutrient for vegetable, tree and field crops. *Draxel Chemical Company.* Unverified.

12807 Genclor
Chlorinated polyvinyl chloride; adhesive. *ICI Chem & Polymers Ltd.*

12808 Gendriv 162
Fiber recovery and retention aid for paper machine applications. *Hercules.* Discontinued.

12809 Gendriv 492S
Formation aid for low shear or fast draining paper applications *Hercules.* Discontinued.

12810 Genelit
A spongy bronze-like bearing metal prepared from a very finely ground mixture of copper, tin, and graphite. When the mixture is heated the graphite burns away, the copper and tin melt together, leaving behind a porous mass which is able to absorb large quantities of lubricating oil.

12811 Generol® 122
Soya sterol
Emollient, auxiliary or primary water-oil emulsifier, emulsion stabilizer, viscosity modifier, solubilizer for cosmetics. *Henkel/Cospha; Henkel Canada; Henkel KGaA.*

12812 Generol® 122E5
PEG-5 soya sterol
Emollient, primary and secondary emulsifier, conditioner, stabilizer and consistency modifier in oil-water emulsions; substantive to hair and in shampoo. *Henkel/Cospha; Henkel Canada; Henkel KGaA.*

12813 Generon
A unit used to separate components. *Dow UK.* Discontinued.

12814 Genesolv A Solvent
75-69-4 9770 200-892-3
Trichlorofluoromethane
Cleaning solvent for metal, plastic and glass; extractant. *AlliedSignal Inc.*

12815 Genesolv D Solvent
76-13-1 200-936-1
Trichlorotrifluoroethane
Cleaning solvent for electronic, electrical and other high value assemblies; carrier for specialty coatings; drying agent; dielectric fluid. *AlliedSignal Inc.*

12816 Genetron Dry Refrigerants
Chlorofluorocarbon refrigerant fluids; used for refrigeration and air conditioning equipment, expansion agents for plastic foam applications, aerosol propellants. *Norplex.* Name unverified.

12817 Genetron® 11
75-69-4 9770 200-892-3
Trichlorofluoromethane
Centrifugal refrigerant; secondary coolant in low temperature systems; used in thermal insulation construction projects. *AlliedSignal.*

12818 Genetron® 12
75-71-8 3114 200-893-9
Dichlorodifluoromethane
Refrigerant for reciprocating and rotary type equip., household to industrial applications, centrifugal applications, sterilant gas, blowing agents, aerosols. *AlliedSignal.*

12819 Genetron® 13
75-72-9 11, 2140 200-894-4
$CClF_3$
Chlorotrifluoromethane
Chlorotrifluoromethane; CFC-13; Halocarbon 13; R-13; Trifluoromethyl chloride. Low temperature refrigerant used in low stage of cascade systems to provide evaporator temperatures in range of -100°F. *AlliedSignal.*

12820 Genetron® 22
Hydrochlorofluorocarbon 22
Refrigerant for residential, commercial and industrial applications; blowing agent in aerosols; intermediate to produce fluoropolymers. *AlliedSignal.*

12821 Genetron® 113
76-13-1 200-936-1
Trichlorotrifluoroethane
Used in low capacity centrifugal chiller packaged units; operates with very low system pressures, high gas volumes. *AlliedSignal.*

12822 Genetron® 114
76-14-2 2672 200-937-7
Dichlorotetrafluoroethane
Refrigerant used with centrifugal compressors for higher capacities or for lower evaporator temperature process applications; also for foam applications. *AlliedSignal.*

12823 Genetron® 123
306-83-2 206-190-3
$C_2HCl_2F_3$
Dichlorotrifluoroethane
2,2-dichloro-1,1,1-trifluoroethane; HCFC-123; dichlorotrifluoroethane; Freon 123; 1,1-Dichloro-2,2,2-trifluoroethane; Fluorocarbon 123; FC-123. Centrifugal refrigerant with low operating pressures. *AlliedSignal.*

12824 Genetron® 134a
Tetrafluoroethane
Refrigerant; CRC subsitute for wide range of air conditioning and refrigeration systems in residential, commercial and industrial applications including automotive. *AlliedSignal.*

12825 Genetron® 500
Dichlorodifluoromethane (73.8%), hydrofluorocarbon 152a (26.2%) azeotrope; slightly higher vapor pressures; a refrigerant which provides higher capacities from same compressor displacement. *AlliedSignal.*

12826 Genetron® 502
Hydrochlorofluorocarbon 22 (48.8%), chlorofluorocarbon 115 (51.2%); for low evaporation temperature applications *AlliedSignal.*

12827 Genetron® 503
Hydrofluorocarbon 23 (40.1%), chlorotrifluoromethane (59.9%); used in low stage of cascade type systems where it provides gains in compressor capacity and in low temperature capability. *AlliedSignal.*

12828 Genetron® HFC 125
354-33-6 206-557-8
C_2HF_5
Pentafluoroethane
1,1,1,2,2-Pentafluoroethane; R-125; Fc-125. For use in low temperature refrigeration applications bp = -48°. *AlliedSignal.*

12829 Genetron® HFC23
75-46-7 4211 200-872-4
CHF_3
Trifluoromethane
fluoroform; freon 23; Halocarbon 23; R-23; Carbon trifluoride. Low temperature refrigerant replacing CFC-13 and R-503 in low stage of cascade systems. mp = -155°; bp = -82°. *AlliedSignal.*

12830 Genetron® Refrigerant 32/125
difluoromethane-pentafluoroethane
Azeotropic mixture of difluoromethane and pentafluoroethane; substitute for air conditioning and refrigeration applications *AlliedSignal.*

12831 Genetron® Refrigerant 125/143a
HFC-125 (45%) and HFC-143A (55%); for low temperature refrigeration applications *AlliedSignal.*

12832 Genie®
For the agriculture industry. *DuPont UK.*

12833 Genisol
Medicated shampoo. *Fisons plc, Pharmaceuticals Div.* Name unverified.

12834 Genitron
123-77-3 950 204-650-8
Azodicarbonamide
Blowing agents and azo-initiators. *Schering Industrial Products Ltd.* Discontinued.

12835 Genkiene
Industrial solvents. *ICI Chem & Polymers Ltd.*

12836 Genoa oil
Fine olive oil.

12837 Genochrome
A stabilized photographic color developer. *May & Baker Ltd.* Name unverified.

12838 Genomoll P
115-96-8 204-118-5
$C_6H_{12}Cl_3O_4P$
tris(β-chloroethyl) phosphate
Tris(2-chloroethyl)phosphate; Celluflex CEF; 2-chloroethanolphosphate; Phosphoric Acid, Tris(2-chloroethyl)ester; Fyrol CEF; Niax Flame Retardant 3CF; Tris(chloroethyl) Phosphate; Tris(2-chloroethyl) Orthophosphate; 3CF; Niax 3CF; Disflamoll TCA; Genomoll P; TCEP; Fyrol CF; 2-chloro-ethanol Phosphate (3:1). Trichlorethyl phosphate; a plasticizer mp = -15°; bp = 330°; d = 1.39; n_D^{20} = 1.4721; soluble in H_2O (7 g/l) LD_{50} (rat orl) = 1230 mg/kg. *Hoechst UK.*

12839 Genotherm
Flexible PVC film. *Hoechst UK.*

12840 Genoxide®
7722-84-1 4839 231-765-0
hydrogen peroxide
A registered trade name for a special quality of hydrogen peroxide for medical purposes. *Laporte Industries Ltd.* Discontinued.

12841 Gensil
Silicone antifoaming agents. *Bevaloid Ltd.* Name unverified.

12842 Genster
A trademark for a manufacture of carbon. No manufacturer.

12843 Genthane SR. (GS 338)
A proprietary trade name for a polyurethane based molding compound with a temperature range from -60° to +160°. No manufacturer.

12844 gentheivite
A mineral. It is (Zn, Fe, Mn)$_8$Be$_6$Si$_6$O$_{24}$S$_2$.

12845 gentian violet
548-62-9 4401 208-953-6
$C_{25}H_{30}ClN_3$
N-[4-[bis[4-(dimethylamino)phenyl]methylene]-2,5-cyclohexadien-1-ylidene]-N-methylmethanaminium chloride
Basic Violet 3; C.I. 42555; hexamethylpararosaniline chloride; aniline violet; crystal violet; Adergon; Axuris; Badil; Gentiaverm; Meroxylan; Meroxyl; Pyoktanin; Vianin; Viocid; Brilliant Violet 5B; Methyl Violet 10BNS; aizen crystal violet; aniline violet pyoktanine; basic violet BN; bismuth violet; brilliant violet 58; calcozine violet C; calcozine violet 6BN; crystal violet O; crystal violet 5BO; crystal violet 6B; crystal violet 6BO; crystal violet 10B; crystal violet AO; crystal violet AON; crystal violet base; crystal violet BPC; crystal violet FN; crystal violet HI2; crystal violet SS; gentersal; hectograph violet SR; hecto violet R; hidaco crystal violet; meroxylanwander; methyl violet 5BNO; methyl violet 5BO; methyl violet 10B; methyl violet 10BD; methyl violet 10BK; methyl violet 10BN; methyl violet 10BO,. A dye for wood, paper and silk; used in inks and as a biological stain. mp = 215° (dec); soluble in H_2O, $CHCl_3$, insoluble in Et_2O; LD_{50} (rat orl) = 1.0 g/kg.

12846 Gentiannie
$C_{14}H_{14}N_3SCl$
The zinc double chloride of dimethyldiaminophenazthionium chloride. Dyes mordanted cotton bluish-violet.

12847 gentisin
437-50-3 4405 207-114-1
$C_{14}H_{10}O_5$
1,3,7-trihydroxyxanthone-3-methyl ether
gentianic acid. The yellow pigment of *Gentiana lutea.* It is 1,3,7-trihydroxyxanthone-3-methyl ether. mp = 266-267°; λ_m = 260, 275, 315, 410 nm (log ε 4.35, 4.30, 4.10, 3.70 MeOH); poorly soluble in H_2O, organic solvents.

12848 Genu®
A range of carrageenan and pectin powders; emulsifiers, stabilizers, thickeners and gelling agents. *Hercules.*

12849 Genuzan
A proprietary gum blend; thickener and suspender in food and beverage industries, for personal care products and cosmetics, and in the pharmaceutical and medical industry. *Hercules.* Discontinued.

12850 Geolast®
Thermoplastic rubber; high oil resistance rubber. *Advanced Elastomer Systems.*

12851 Geolite
Polyurethane intermediate. *Union Carbide (UK) Ltd.*

12852 Geolith
Grout for combatting spontaneous combustion in coal mines. *Foseco (F.S.) Ltd.*

12853 Geon® 8700A
9002-86-2 7746 206-625-7
Polyvinyl chloride
Rigid vinyl compound; extrusion compound for interior building products; profile applications *BFGoodrich/Geon Vinyl.*

12854 Geon® 8720
9002-86-2 7746 206-625-7
Polyvinyl chloride
Vinyl compound; wire and cable sheath; coaxial cable, noncontaminating jacket. *BFGoodrich/Geon Vinyl.*

12855 Geon® 8812, 8813
9002-86-2 7746 206-625-7
Polyvinyl chloride
Vinyl compound; general purpose flexible molding and extrusion compound. *BFGoodrich/Geon Vinyl.*

12856 Geon® 8896
9002-86-2 7746 206-625-7
Polyvinyl chloride
Vinyl compound; wire and cable insulation; SPT cord. *BFGoodrich/Geon Vinyl.*

12857 Geon® 83457
9002-86-2 7746 206-625-7
Polyvinyl chloride
Vinyl compound; general purpose flexible molding and extrusion compound. *BFGoodrich/Geon Vinyl.*

12858 Geon® 83718
9002-86-2 7746 206-625-7
Polyvinyl chloride
Vinyl compound; flexible molding and extrusion compound; weatherstrip. *BFGoodrich/Geon Vinyl.*

12859 Geon® 86100, 86101, and 86103
9002-86-2 7746 206-625-7
Polyvinyl chloride
Vinyl compound; low-gloss grade for coextrusion capstock. *BFGoodrich/Geon Vinyl.*

12860 Geon® 87239, 87241
9002-86-2 7746 206-625-7
Polyvinyl chloride
Rigid vinyl compound; tin-stabilized; injection molding compound. *BFGoodrich/Geon Vinyl.* Name unverified.

12861 Geon® 87396
9002-86-2 7746 206-625-7
Polyvinyl chloride
Rigid PVC compound; high gloss, high impact general-purpose bottle compound for extrusion blow molding; for opaque bottles, nonfood contact applications such as packaging for automotive and marine additives, shampoos, and charcoal lighter fluids. *BFGoodrich/Geon Vinyl.* Name unverified.

12862 Geon® 87420
9002-86-2 7746 206-625-7
Polyvinyl chloride
Vinyl compound; medical grade injection molding compound with high flow, normal impact; FDA compliant. *BFGoodrich/Geon Vinyl.* Name unverified.

12863 Geon® HTX 92190
9002-86-2 7746 206-625-7
Polyvinyl chloride
PVC compound; interior grade, high heat deflection, low expansion, rigid cube profile extrusion compound. *BFGoodrich/Geon Vinyl.*

12864 Geon® HTX-6110
Vinyl-based alloy; engineering thermoplastic for elevated temperature performance, inherent flame retardance, excellent chemical and corrosion resistance, toughness, electrical performance, surface appearance, processability. *BFGoodrich/Geon Vinyl.* Name unverified.

12865 Geon® W015
9002-86-2 7746 206-625-7
Polyvinyl chloride
Vinyl compound; for wire and cable applications. *BFGoodrich/Geon Vinyl.*

12866 Georgia Gulf 3131 Clear 02
9002-86-2 7746 206-625-7
Polyvinyl chloride
Vinyl compound rigid profile extrusion compound for profiles, electronic pkg., integrated circuit magazine tubes; high impact, good clarity. *Georgia Gulf.*

12867 Georgia Gulf 5006
9002-86-2 7746 206-625-7
Polyvinyl chloride
Vinyl compound rigid profile extrusion compound for window profiles, siding and sheet; excellent weatherability. *Georgia Gulf.*

12868 Georgia Gulf 5006 General
9002-86-2 7746 206-625-7
Polyvinyl chloride
Vinyl compound rigid profile extrusion compound for general purpose applications incl. profile, sheet and elec.; high gloss and tensile strength. *Georgia Gulf.*

12869 Georgia Gulf 9105
9002-86-2 7746 206-625-7
Polyvinyl chloride
Vinyl compound general purpose blow molding grade for pkg. of cosmetics, toiletries and household products; excellent combination of impact strength, clarity and chem. resist. *Georgia Gulf.*

12870 Georgia Gulf 9151
9002-86-2 7746 206-625-7
Polyvinyl chloride
Vinyl compound chem. resist. grade for blow molding bottles for chem. active products like pine oil cleaners; good impact strength. *Georgia Gulf.*

12871 Georgia Gulf 9175J
9002-86-2 7746 206-625-7
Polyvinyl chloride
Vinyl compound general purpose injection blow molding grade for pkgs. for hair dyes and shampoos; excellent clarity, flow and stability. *Georgia Gulf.*

12872 Georgia Gulf 9202
9002-86-2 7746 206-625-7
Polyvinyl chloride
Vinyl compound food grade for mfg. of large returnable water bottles; excellent processing, low odor and taste. *Georgia Gulf.*

12873 Georgia Gulf CL-7049
9002-86-2 7746 206-625-7
Polyvinyl chloride
Engineering vinyl; clear grade with med. flow for injection molding for FDA applications; excellent impact strength. *Georgia Gulf.*

12874 Georgia Gulf EH-71L
9002-86-2 7746 206-625-7
Polyvinyl chloride
PVC disp. resin; high molecular weight, low visc. grade for coating and molding applications. *Georgia Gulf.*

12875 Georgia Gulf EX-240
9002-86-2 7746 206-625-7
Polyvinyl chloride
PVC disp. resin; high performance grade with elastomer-like props. for low temp. and fat fusion applications. *Georgia Gulf.*

12876 Georgia Gulf HH-1900
9002-86-2 7746 206-625-7
Polyvinyl chloride
Engineering vinyl alloy; high heat resistant molding material with high flow. *Georgia Gulf.*

12877 Georgia Gulf HM-7054
9002-86-2 7746 206-625-7
Polyvinyl chloride
Engineering vinyl; special purpose molding material with high modulus and very good surface hardness; used for load bearing applications. *Georgia Gulf.*

12878 Georgia Gulf SP-7107
9002-86-2 7746 206-625-7
Polyvinyl chloride
Engineering vinyl; special purpose molding material with high flow for applications requiring excellent thermal stability and very good mech. props. *Georgia Gulf.*

12879 Georgia Gulf UV-7160
9002-86-2 7746 206-625-7
Polyvinyl chloride
Engineering vinyl; uv-resistant, high flow super tough blend for enclosures. *Georgia Gulf.*

12880 Geoseal
Soil grouting resins. *Borden (UK) Ltd.* Name unverified.

12881 Geostone®
Class IV dental stone; for high quality models and dyes. *Bayer AG.*

12882 Geostop
Accelerated anhydrite for mines roadway stoppings. *Foseco (F.S.) Ltd.* Discontinued.

12883 Gepel
Reactive silane-modified siloxane; protective penetrant. *GE Silicones.*

12884 Geracryl
Patterned extruded acrylic sheet; for glazing, shower screens etc. *Cornelius Chemical Co Ltd.* Name unverified.

12885 Geraniol
106-24-1 4411 203-377-1
$C_{10}H_{18}O$
trans-3,7-dimethyl-2,6-octadien-1-ol
(E)-3,7-Dimethyl-2,6-octadien-1-ol; Lemonol; geranyl alcohol; guaniol. A terpene alcohol; used in perfumery, as a constituent of synthetic fragrances and with synthetic linalool. d = 0.8894; mp = -15°; bp_{757} = 229-230°; d = 0.8770; n_D^{20} = 1.4770; LD_{50} (rat orl) = 3600 mg/kg. *Penta Mfg.; Schweizerhall; SCM Glidco Organics.*

12886 geraniol acetate
16409-44-2 4411 240-458-0
Geranyl acetate

12887 geranium crystals
101-84-8 7442 202-981-2
$C_6H_5{\cdot}O{\cdot}C_6H_5$
Diphenyl ether
Used in perfumery.

12888 geranium oil
8000-46-2
Pale yellow or greenish liquid, having geraniol as its main constituent; used in perfumes and cosmetics.

12889 Geranyl acetate
16409-44-2 4411 240-458-0
$C_{14}H_{20}O_2$
geraniol acetate
Perfumery, flavoring. *Firmenich; Int'l. Flavors & Fragrances; Penta Mfg.; SCM Glidco Organics.*

12890 Gerhardt's caustic
This consists of litharge boiled with potassium hydroxide until it is dissolved, and water added.

12891 Germaben® II
Diazolidinyl urea-propylene glycol-methylparaben-propylparaben
Diazolidinyl urea (30%), propylene glycol (56%), methylparaben (11%), and propylparaben (3%); broad-spectrum antimicrobial preservative for cosmetic products. *Sutton Labs.*

12892 Germalgene
79-01-6 9769 201-167-4

C_2HCl_3
Trichlorethylene
Solvent and cleaner.

12893 Germall® 115
39236-46-9 254-372-6
$C_{11}H_{16}N_8O_8$
Imidazolidinyl urea
Biopure 100; Germall 115; Imidurea NF; Sept 115; Tristat 1U; Unicide U-13; Imidurea. Imidazolidinyl urea; antimicrobial preservative for cosmetics. *Sutton Labs.*

12894 Germall® II
78491-02-8 278-928-2
$C_6H_{10}N_4O_7$
Diazolidinyl urea
A broad-spectrum preservative used in cosmetics and pharmaceutical preparations. It is especially active against gram-negative bacteria and is often combined with parabens. Diazolidinylurea is the newest and most active member of the imidazolidinyl urea group. It has been used since 1982. Diazolidinylurea may be a stronger sensitizer than imidazolidinyl urea. *Sutton Labs.*

12895 German silver solder
Usually consists of 5 parts German silver, and 4 parts zinc.

12896 German yeast
Dried yeast.

12897 Germanium
7440-56-4 4415 231-164-3
Ge
Nonmetallic element; used in electronics: manufacture of Ge diodes, transistors, solid state electronic devices, semiconducting applications, brazing alloys, phosphors, gold and beryllium alloys, infrared-transmitting glass; dental alloys. mp = 937°; bp = 2700°; poor electrical conductor. *Atomergic Chemetals; Cabot; Cerac; Eagle-Picher; New Metals & Chems. Ltd; Noah Chem.*

12898 Germ-i-Tol
Benzalkonium chloride
Pharmaceutic aid. *Hexcel Chemical Products.* Name unverified.

12899 Germul
Vinyl acetate and vinyl acrylic emulsions; base for adhesives. *Elf Atochem UK/Ceca.*

12900 Germul A 735
Crosslinking acrylic resin; textile hand modifier. *Ceca SA.*

12901 Geronol
Blend of nonionic and anionic surfactants; pesticide emulsifiers. *Geronazzo S.p.A.*

12902 Geronol AG-100/200 Series
Complex anionic-nonionic ethoxylate blends; emulsifiers for toxicants, especially organophosphate insecticides and nonsaponifiable herbicides. *Rhône-Poulenc Surf.*

12903 Geropon® ACR/4
Disodium laureth sulfosuccinate
Emulsifier for emulsion polymerization, detergent base, foamer, foam stabilizer, dispersant. *Rhône-Poulenc France.*

12904 Geropon® AS-200
Sodium cocoyl isethionate-coconut acid-stearic acid. Detergent, wetting agent, dispersant, suspending agent for textile wet processing. industrial/household detergents, cosmetics, agricultural pesticides, leather, rubber, etc.; readily biodegradable. *Rhône-Poulenc Surf.*

12905 Geropon® CYA/60
577-11-7 3460 209-406-4
Sodium dioctyl sulfosuccinate
Wetting agent, surface tension reducer, viscosity depressant for emulsion polymerization of PVC. *Rhône-Poulenc Geronazzo.*

12906 Geropon® CYA/DEP
127-39-9 3238 204-839-5
Sodium diisooctyl sulfosuccinate
Wetting agent for textile industry, pesticides, emulsion polymerization, printing inks, water paints; stable in acid media. *Rhône-Poulenc Geronazzo.*

12907 Geropon® DOS
577-11-7 3460 209-406-4
Sodium dioctyl sulfosuccinate
Wetting agent, emulsifier, demulsifier for textile wet processing, specialty cleaners, dewatering agent for flotation concentrates, oil spill clean-up blends; limited stability in alkaline or acidic media. *Rhône-Poulenc France.*

12908 Geropon® LSS
36409-57-1 253-019-3
Disodium lauryl sulfosuccinate
Low-irritant detergent for shampoos and bubble baths. *Rhône-Poulenc Surf.; Rhône-Poulenc France.*

12909 Geropon® MLS/A
1561-92-8 216-341-5
$C_4H_8O_3S$
Sodium 2-methylprop-2-ene-1-sulfonate
Sodium methallyl sulfonate. Dye improver reactive co-monomer for acrylic fibers polymerization; reactive emulsifier or coemulsifier in latex emulsion polymerization. *Rhône-Poulenc Geronazzo.*

12910 Geropon® S-1585
Disodium octylphenoxy sulfosuccinate
Surfactant. *Rhône-Poulenc France.*

12911 Geropon® SBFA-30
68815-56-5
Disodium laureth sulfosuccinate
Skin protecting anti-irritant surfactant. *Rhône-Poulenc Surf.*

12912 Geropon® SBL-203
25882-44-4 247-310-4
Disodium lauramido MEA-sulfosuccinate
Detergent. Improves flash foam of anionic systems; produces brittle, tack-free residue for carpet shampoos. *Rhône-Poulenc Surf.*

12913 Geropon® SBR-3
40754-60-7
Disodium ricinoleamido MEA-sulfosuccinate
Skin protecting anti-irritant surfactant. *Rhône-Poulenc Surf.*

12914 Geropon® SS-L7DE
Sodium lauramido DEA sulfosuccinate
Mild detergent, foam booster/stabilizer for liquid dish detergent and toiletry preparations. *Rhône-Poulenc Surf.*

12915 Geropon® T-22/A
137-20-2 205-285-7
Sodium methyl oleoyl taurate
Surfactant, dispersant for agricultural formulations. *Rhône-Poulenc Surf.*

12916 Geropon® T-33
137-20-2 205-285-7
Sodium N-methyl-N-oleoyl taurate
Detergent, dispersant, wetting agent for textile and general-purpose applications; dye assistant; for bleaching, wetting, finishing of textiles; in industrial detergents, rug shampoos, bottle washing compounds, metal cleaners, paper industry. *Rhône-Poulenc Surf.; Rhône-Poulenc France.*

12917 Geropon® TC-42
61791-42-2 263-174-9
Sodium N-methyl-N-cocoyl taurate
Foamer, dispersant, detergent for detergent bars, shampoos, bubble baths, cosmetics; chemically stable. *Rhône-Poulenc Surf.; Rhône-Poulenc France.*

12918 Geropon® TK-32
61791-41-1 263-173-3
Sodium N-methyl-N-tallowyl taurate
Detergent, suspending agent, dispersant; precipitation inhibitor for organic and inorganic salts of Ba, Ca, Sr; for petroleum industry. *Rhône-Poulenc Surf.*

12919 Geropon® WS-25, WS-25-I
63217-13-0 264-016-1
Sodium dinonyl sulfosuccinate
Rewetting agent for textile finishing, in application of resins, softeners, starches; wetting and dispersing agent for latex paints. *Rhône-Poulenc Surf.; Rhône-Poulenc France.*

12920 Geropon® X2152
Ammonium dioctyl sulfosuccinate
Wetting agent, emulsifier, demulsifier, stabilizer in water treatment and petroleum processing; stable at neutral pH. *Rhône-Poulenc France.*

12921 Gesagard
7287-19-6 7973 230-711-3
Prometryn
Wettable powder containing 50% w/w prometryn; a contact triazine herbicide. *Ciba Geigy Agrochemicals.*

12922 Gesaprim 500FW
1912-24-9 902 217-617-8
atrazine
A liquid formulation containing 500 g atrazine per liter as a suspension concentrate; a residual herbicide. *Ciba-Geigy Agrochemicals.*

12923 Gesatop
122-34-9 8681 204-535-2
Simazine
Suspension concentrate or wettable powder containing simazine; a triazine herbicide to control weeds and grasses in cane fruit, roses and some vegetables. *Ciba-Geigy Agrochemicals.*

12924 Gesilit
Safety explosives. No. 1 contains 30.75% nitroglycerin jelly, 5.25% dinitrotoluene, 7% sodium chloride, 18% sodium nitrate, and 39% dextrin. No.

2 consists of 30.75% nitroglycerin jelly, 5.25% dinitrotoluene, 22% ammonium nitrate, 21% sodium chloride, 21% dextrin.

12925 Gesteins-tremonit V
An explosive containing ammonium nitrate, nitroglycerin, vegetable meal, potassium perchlorate, and nitro-compounds.

12926 Gesteins-Westfalit B and C
Explosives. They are ammonals containing dinitrobenzene, and dinitrotoluene respectively.

12927 Gesterol 50
57-83-0 7956 200-350-6
Progesterone
Progestin. Synthetic intermediate. *O'Neal, Jones & Feldman Pharmaceuticals.* Name unverified.

12928 Gestinal®
Amniotic fluid substitute and lubricant; veterinary medicine. *Bayer AG.*

12929 Getren® 4/200
Silicone-free release agent for foundry and steel industry, plastics include unsaturated polyester, epoxy, and PU resins; lubricant for tire production. *Goldschmidt.*

12930 Getren® FD 575
Release agent for the rubber industry; mainly for production of radiator hoses. *Goldschmidt.*

12931 Gevral
A proprietary preparation of multivitamins and minerals. *Lederle Laboratories.* Name unverified.

12932 GFS®
Xanthan gum/galactomannan blend; gum for use in food preparations; stabilizer, suspending agent, thickener; provides rheological control to water- and milk based foods. *Kelco.*

12933 GH5
Granular fertilizer. *Fisons plc, Horticultural Div.* Name unverified.

12934 gibberellic acid
77-06-5 4426 201-001-0
$C_{19}H_{22}O_6$
(1α,2β,4aα,4bβ,10β)-2,4a,7-trihydroxy-1-methyl-8-methylgibb-3-ene-1,10-dicarboxylic acid 1,4a-lactone
gibberellin X; gibberellin A_3. Used in agriculture and horticulture as a plant growth regulator, malting of barley with improved enzymatic characteristics. mp = 233-235°; $[\alpha]_D^{19}$ = 86° (c = 2.12); soluble in H_2O, polar organic solvents, insoluble in non-polar organic solvents; *Am Biorganics; Atomergic Chemetals; U.S. Biochemical.*

12935 Giemsa's stain
A microscopic stain for white blood corpuscles. It contains eosin, glycerin, and methanol.

12936 Gilalgin
Alginate; dental impression material. *Giulini Corp.*

12937 Gildent
Stump material for dentures. *Giulini Corp.*

12938 gilding metal
A jeweler's alloy of 90% copper, and 10% zinc. Another alloy contains 70% copper, 17.5% brass, and 12.5% tin.

12939 gilding solutions
These generally consist of solutions of gold chloride and potassium carbonate in water; used for the electro-deposition of gold.

12940 Gilsonite
Naturally occurring bitumen; for oil well cements and drilling fluids, printing inks, foundry sand additive, explosives, asphalt pavement sealer, bituminous paints. *Mercian Minerals & Colours Ltd.* Unverified.

12941 gilsonite
An asphaltic material or solidified hydrocarbon; acid, alkali and waterproof coatings; black varnishes, lacquers, baking enamels; wire-insulation compounds; linoleum, floor tile; paving; insulation; diluent in low grade rubber compounds; possible source of gasoline and fuel oil. *R. E. Carroll.*

12942 Gilsonite and Design
Uintaite resin. *Monsanto (Solaris).* Name unverified.

12943 Gilstone
Synthetic gypsum. *Giulini Corp.*

12944 Gilumag
1309-42-8 5706 215-170-3
Magnesium hydroxide.
Giulini Corp.

12945 ginal
A purifier for sugars. It contains the sodium alginate (from seaweed).

12946 Gingelly oil
8008-74-0 8614 232-370-6
Gingili oil
teal oil; teel oil; til oil, beni oil; benne oil, beniseed oil. Sesame oil obtained from the seeds of *Sesamum indicum* and of *S orientate*; used in the manufacture of margarine and soap and as a burning oil. mp = -5°; d = 0.9160; n_D^{20}= 1.4720-1.4760; LD_{50} (rat orl) = 18900 mg/kg.

12947 Ginsene
Ethylcyclohex-3-ene carboxylate
Used in perfumery. *Bush Boake Allen Ltd.*

12948 Gin-shi-bui-chi
A Japanese alloy of 30-50% silver with copper.

12949 Gippon
Antiseptic paint. *J C Bottomley.* Name unverified.

12950 Girard's reagent T
123-46-6 4436 204-629-3
$C_5H_{14}ClN_3O$
2-hydrazino-N,N,N-trimethyl-2-oxoethanaminium chloride
Betaine hydrazide hydrochloride. Reacts with ketones to form water-soluble hydrazones. Used in the isolation of ketosteroids. Girard's reagent P [1126-58-5], used in the same way, is 1-(2-hydrazino-2-oxoethyl)pyridinium chloride. mp = 192°; soluble in H_2O, polar organic solvents, insoluble in non-polar organic solvents.

12951 Givgard DXN
828-00-2 3273 212-579-9
6-Acetoxy-2,4 dimethyl-*m*-dioxane
dimethoxane. Nonformaldehyde bactericidal and fungicidal agent for industrial use; used for all kinds of industrial water-based systems such as emulsions, suspensions and dispersions. *Givaudan SA.* Name unverified.

12952 Givsorb® UV-1
57834-33-0 260-976-0
N^2-(4-Ethoxy-carbonylphenyl)-N'-methyl-N'-phenylformamidine
Industrial UV absorber. *Givaudan; Givaudan SA.* Name unverified.

12953 Gladiator
1698-60-8 216-920-2
Chloridazon
Suspension concentrate containing 430 g chloridazon per liter; a pyridazinone herbicide for beet crops. *Tripart Farm chemicals Ltd.*

12954 Glagerite
A Bavarian white clay.

12955 Glanzan PHN Conc
Wax emulsion; nonionic; luster finish for yarns, woven and knitted fabrics made of cotton, regenerated cellulose as well as their blends. *Thor Chemicals (UK) Ltd.* Discontinued.

12956 Glascol®
A range of acrylic polymers in water; these resins are film formers, have pigment binding properties and dry to form flexible, clear films; uses include paper and board coatings, printing inks, overprint varnishes artists colors. *Allied Colloids Ltd.*

12957 Glascol® HN2
A proprietary aqueous acrylic copolymer, the sodium salt of which gives hard and brittle films soluble in water. *Allied Colloids Ltd.* Name unverified.

12958 Glascol® HN4
A proprietary aqueous acrylic copolymer the sodium salt of which gives soft and flexible films soluble in water. *Allied Colloids Ltd.* Name unverified.

12959 Glascol® PA6
A proprietary acrylic copolymer supplied in the form of low viscosity aqueous solutions; the ammonium salt gives tacky, pressure-sensitive films resistant to water. *Allied Colloids Ltd.* Name unverified.

12960 Glascol® PA8
A proprietary acrylic copolymer supplied in the form of low viscosity aqueous solutions; the ammonium salt gives soft, tacky, pressure-sensitive films resistant to water. *Allied Colloids Ltd.* Name unverified.

12961 Glascol® PN 8
A proprietary acrylic copolymer supplied in the form of low viscosity aqueous solutions; the sodium salt gives soft, tacky, pressure-sensitive films soluble in water. *Allied Colloids Ltd.* Name unverified.

12962 Glasdag
A range of lubricating greases, oils and coatings based on graphite as the lubricating pigment; designed especially for the manufacture of glass containers. *Acheson Colloids.*

12963 Glaser's salt
potassium sulfate-potassium sulfite.
A mixture of potassium sulfate and potassium sulfite.

12964 Glasgro
Range of granular fertilizers. *Fisons plc, Horticultural Div.* Name unverified.

12965 Glaskyd®
Thermoset polyester molding compounds for industrial applications requiring electrical insulation, high strength, rigidity at elevated temperatures and corrosion resistance. *Am. Cyanamid.* Name unverified.

12966 Glass Guard
Viscous liquid applied to glass and metal for protection against acid-based cleaners; used for historical glass, aluminum window castings etc. *Nova Chemical Inc.* Name unverified.

12967 Glass H
50813-16-6 8814 256-779-4
$Na_6O_{18}P6$
Sodium hexametaphosphate
Sodium hexametaphosphate; Metaphosphoric acid, hexasodium salt; Glassy sodium; Hexasodium metaphosphate; Metaphosphoric acid ($H_6P_6O_{18}$), hexasodium salt; Graham's Salt, Sodium Polymetaphosphate. Water softener and detergent. d = 2.181; mp = 628°; d = 2.181; soluble in H_2, insoluble in organic solvents; LD_{50} (rat orl) = 6200 mg/kg. *FMC.* Discontinued.

12968 glass silk
Glass wool.

12969 Glass Sponge
A patented sponge-like product obtained by mixing glass wool with salt, heating, then dissolving out the salt. No manufacturer.

12970 Glassclad® 18
t-butanol-diacetone alcohol
20% solution in *t*-butanol and diacetone alcohol; hydrophobic treatment and lubricant for glass and ceramics. *Hüls Am.*

12971 glassite
1345-25-1 4100 215-721-8
ferrous oxide
Magnetic oxide of iron from precipitation of ferrous sulfate with caustic soda. Also called black rouge; used for buffing.

12972 glass-maker's soap
1313-13-9 5770 215-202-6
MnO_2
Manganese dioxide
Oxidizing agent.

12973 Glassona
Cellulose-bonded fiberglass bandage. *Smith & Nephew Pharmaceuticals Ltd.* Name unverified.

12974 Glauber's salt
7727-73-3 8829
$Na_2O_4S{\cdot}10H_2O$
Sodium sulfate decahydrate
Mirabilite. Used in solar heat storage, air conditioning and in freezing mixtures.

12975 glauconic acids
Bluish-violet dyestuffs, obtained by the successive action of pyroracemic acid and formaldehyde on aromatic primary amines.

12976 Glaucosil
7631-86-9 8637 231-545-4
silicon dioxide
The siliceous residue obtained by extracting green sand with mineral acids. It consists of practically pure silica; used as an absorbent for gases.

12977 Glauramine
A proprietary solution of specially purified auramine; a powerful antiseptic. No manufacturer.

12978 Glaurin
141-20-8 3169 205-468-1
diethylene glycol monolaurate
A proprietary trade name for a plasticizer. No manufacturer.

12979 Glazamine
Textile auxiliary chemicals. *ICI Chem & Polymers Ltd.*

12980 Glaze N Seal Waterbase Clear Concrete and Brick Sealer
Waterbase sealer; an acrylic emulsion for application to interior concrete, stone and masonry type surfaces; nonflammable. *Glessner Corporation Inc (GGI Products) DBA.* Name unverified.

12981 Glaze 'N Seal Concrete and Masonry Sealer
Solvent base sealer; a clear nonyellowing acrylic for application to exterior and interior concrete, stone and masonry type surfaces. *Glessner Corporation Inc (GGI Products) DBA.* Name unverified.

12982 glazier's salt
7778-80-5 7845 231-915-5
K_2SO_4
Potassium sulfate

12983 Glean® TP
Bromoxynil-chlorsulfuron-ioxynil
Herbicide mixture for weed control in cereals. *DuPont UK.*

12984 Glendion
Polyether and polyester polyols. *Montedison UK Ltd.* Name unverified.

12985 Glessite
A variety of amber melting at 250-300°.

12986 Glievor bearing metals
One alloy contains 76% lead, 14% antimony, 8% tin, and 2% iron, and another consists of 73% zinc, 9% antimony, 7% tin, 5% lead, and 4% copper.

12987 glimmer
glist. Mica.

12988 Glissofluid® A 10, A 13
Aliphatic dicarboxylic acid ester; component for synthetic lubricants. *BASF AG; BASF plc.*

12989 Glissopal®
Intermediates and components for production of lubricating oil additives and lubricating oils. *BASF AG.*

12990 Glissosafe®
Hydraulic fluids for mining. *BASF AG; BASF plc.*

12991 Glissoviscal® B
Polyisobutylenes
Thickener and viscosity improver for lubricating oils. *BASF AG; BASF plc.*

12992 Glitzi
Household pad. *Carl Freudenberg.* Name unverified.

12993 Glizarin Binder
Textile finishing cross-linking agents. *BASF plc.* Name unverified.

12994 Glissolube®
Polyalkyleneoxide derivatives; synthetic components for high-performance lubricants. *BASF AG; BASF plc.*

12995 GLOB®
Urethane polymer with sika powders; urethane/carbide mixture. *TSE Industries.*

12996 Globe Granite
A stone similar to Ward's stone.

12997 Glocure
Benzoin ether.
Used in organic synthesis. *Rhône-Poulenc UK.*

12998 Glofoam
Synthetic wax *Rhône-Poulenc UK.*

12999 Glokem
Synthetic wax. *ABM Chemicals Ltd.* Name unverified.

13000 Glokill
Heterocyclic biocide. *Rhône-Poulenc UK.*

13001 Glokill 77
Nonsurface active biocide. *Rhône-Poulenc UK.*

13002 Glokill PQ
Polymeric quaternary ammonium compound in the form of a pale yellow aqueous solution; slow acting nonfoaming biocide for water treatment where foam is a problem. *Rhône-Poulenc UK.*

13003 Glomeen
Chlorine releasing sterilant. *ABM Chemicals Ltd.* Name unverified.

13004 Glo-Mold®
Elastomeric compound; mold cleaning compounds for preventive maintenance. *Glo-Mold.* Name unverified.

13005 Glonoine oil
55-63-0 6704 200-240-8
nitroglycerin
Explosive.

13006 Glopol
Dispersing agent for aqueous systems. *Rhône-Poulenc UK.*

13007 Glopol 461
Polymeric quaternary ammonium salt as a pale yellow slightly viscous liquid; cationic surfactant used as resin to impart electro-conductive properties to paper and ceramics. *Rhône-Poulenc UK.*

13008 Gloquat 1032
Dialkyl quaternary ammonium salt as a brown viscous liquid; cationic surfactant with hydrophobe, anticorrosive, dispersant and electrostatic properties; used in pigment dispersion, paints and coatings, car washes. *Rhône-Poulenc UK.*

13009 Gloquaternary
Alkylaryl trimethyl ammonium chloride
Horticultural algicide and moss-killer. *Flowering Plants Ltd.* Discontinued.

13010 Gloquaternary
Alkyaryl trimethyl ammonium chloride
Horticultural algicide and moss-killer. *Fargro Ltd.*

13011 Gloquaternary
Cationic surfactant, biocide. *Rhône-Poulenc UK.*

13012 Gloria®
White mineral oil. *Witco/Sonneborn.*

13013 glossite
1345-25-1 4100 215-721-8
ferrous oxide
An abrasive, the active material of which is stated to be black oxide of iron.

13014 Glossova
Floor polishes. *Evans Vanodine International Ltd.* Discontinued.

13015 Glowtein
Poultry feed additive, colorant. *Mitchell Cotts Chemicals Ltd.*

13016 Gloy
A trademark for an adhesive said to be a mixture of dextrin and starch, with magnesium chloride. No manufacturer.

13017 Gluadin® AGP
Hydrolyzed wheat protein; pleasant skin feel for emulsions and surfactant preparations. *Henkel/Cospha; Henkel Canada.* Name unverified.

13018 glucal
$C_6H_{10}O_4$
A reducing compound, obtained by the reduction of β-aceto-bromo-glucose with zinc dust and acetic acid.

13019 Glucam® E-10
68239-42-9
Methyl gluceth-10
Humectant for personal care products; freezing point depressant; emollient in aqueous and hydroalcoholic products; moisturizer; foam modifier in detergent and shampoo systems; solvent and solubilizer for topical pharmaceuticals; used in emulsions, toilet articles, adds gloss, conditioning. *Amerchol Corp.*

13020 Glucam® E-20
68239-43-0
Methyl gluceth-20
Humectant for personal care products; freezing point depressant; emollient in aqueous and hydroalcoholic products; moisturizer; foam modifier in detergent and shampoo systems; solvent and solubilizer for topical pharmaceuticals. *Amerchol Corp.*

13021 Glucam® E-20 Distearate
98073-10-0
Methyl gluceth-20 distearate
Auxiliary oil-waterw emulsifier, moisturizer, emollient and lubricant for cosmetics and pharmaceuticals; conditioner. *Amerchol Corp.*

13022 Glucam® P-10
61849-72-7
PPG-10 methyl glucose ether
Humectant for personal care products; freezing point depressant; emollient in aqueous and hydroalcoholic products; moisturizer; foam modifier in detergent and shampoo systems; solvent and solubilizer for topical pharmaceuticals. *Amerchol Corp.*

13023 Glucam® P-20
61849-72-7
PPG-20 methyl glucose ether
Humectant for personal care products; freezing point depressant; emollient in aqueous and hydroalcoholic products; moisturizer; foam modifier in detergent and shampoo systems; solvent and solubilizer for topical pharmaceuticals. *Amerchol Corp.*

13024 Glucam® P-20 Distearate
PPG-20 methyl glucose ether distearate
Skin moisturizer, conditioner, and emollient for cosmetics and pharmaceuticals; binder and plasticizer for pressed powders. *Amerchol Corp.*

13025 Glucamate® DOE-120
PEG-120 methyl glucoside dioleate
Thickener, emulsifier, solubilizer for shampoos. *Amerchol Corp.*

13026 Glucamate® SSE-20
68389-70-8
PEG-20 methyl glucose sesquistearate
Oil-water emulsifier, solubilizer used with Glucate SS; effective at low concentrations. *Amerchol; Amerchol Europe.*

13027 Glucanal
Proprietary preparations of silver and anthraquinone glucosides. No manufacturer.

13028 Glucanex
A β-glucanase preparation produced by a selected strain of *Trichoderma*; can be used in all cases where the aim is to improve the clarification and the filtrability of wines made from botrytized grapes. *Novo Nordisk.*

13029 Glucanex® L-300
β-Glucanase
Enzyme for hydrolysis of β-glucans. for beer breweries. *Solvay Enzymes.*

13030 glucase
Maltase, an enzyme which converts maltose into glucose.

13031 Glucate® DO
Methyl glucose dioleate
Water-oil emulsifier, auxiliary emulsifier for oil-water systems; conditioner, emollient, lubricant, plasticizer, and pigment dispersant. *Amerchol; Amerchol Europe.*

13032 Glucate® IS
Methyl glucose sesquiisostearate
Primary water-oil emulsifier for personal care products. *Amerchol Corp.* Discontinued.

13033 Glucate® SS
68936-95-8 273-049-0
Methyl glucose sesquistearate
Water-oil emulsifier used with Glucamate SSE-20 to provide viscosity stability, mildness. *Amerchol; Amerchol Europe.*

13034 Glucidex
A range of dried glucose syrups or maltodextrins; used in various food applications to provide, for example, bulk with or without sweetness. *Roquette (UK) Ltd.*

13035 Glucina
1304-56-9 1216 215-133-1
BeO
Beryllium oxide
Used as a catalyst and in the manufacture of ceramics.

13036 Gluckauf
A German safety explosive containing ammoniuna nitrate, wood meal, dinitrobenzene, and copper oxalate.

13037 Glucodin
glucose-ascorbic acid
A proprietary preparation of glucose and ascorbic acid; a food supplement. *Farley Health Products.* Unverified.

13038 Gluconal® CA A, CAM
299-28-5 1712 206-075-8
Calcium gluconate anhydrous
Pharmaceutical/food grade mineral source for human and veterinary pharmaceutical preparations, dietary supplements, fortified foods and animal feed. *Akzo Chemie.*

13039 Gluconal® CA M B
5743-34-0 1692 227-264-1
Calcium borogluconate
Pharmaceutical/food grade mineral source for human and veterinary pharmaceutical preparations, dietary supplements, fortified foods and animal feed. *Akzo Chemie.*

13040 Gluconal® CO
Cobalt gluconate
Pharmaceutical/food grade mineral source for human and veterinary pharmaceutical preparations, dietary supplements, fortified foods and animal feed. *Akzo Chemie.*

13041 Gluconal® CU
527-09-3 2706 208-408-2
$C_{12}H_{22}CuO_{14}$
Copper gluconate
cupric gluconate. Pharmaceutical/food grade mineral source for human and veterinary pharmaceutical preparations, dietary supplements, fortified foods and animal feed. Also used as an oral deodorant. Soluble in H_2O (500 g/l), insoluble in organic solvents. *Akzo Chemie.*

13042 Gluconal® FE
299-29-6 4095 206-076-3
Ferrous gluconate
Pharmaceutical/food grade mineral source for human and veterinary pharmaceutical preparations, dietary supplements, fortified foods and animal feed. *Akzo Chemie.*

13043 Gluconal® K
299-27-4 7796 206-074-2
$C_6H_{11}KO_7$
Potassium gluconate
gluconic acid potassium salt; Gluconsan K; Kalimozan; Kaon; Katorin; Potasorl; Potassuril; K-IAO; Tumil-K. Pharmaceutical/food grade mineral source for human and veterinary pharmaceutical preparations, dietary supplements, fortified foods and animal feed. dec 180°; freely soluble in H_2O, insoluble in organic solvents. *Akzo Chemie.*

13044 Gluconal® MG
Magnesium gluconate
Pharmaceutical/food grade mineral source for human and veterinary pharmaceutical preparations, dietary supplements, fortified foods and animal feed. *Akzo Chemie.*

13045 Gluconal® MN
Manganese gluconate
Pharmaceutical/food grade mineral source for human and veterinary

pharmaceutical preparations, dietary supplements, fortified foods and animal feed. *Akzo Chemie.*

13046 Gluconal® NA
527-07-1 8766 208-407-7
Sodium gluconate
Pharmaceutical/food grade mineral source for human and veterinary pharmaceutical preparations, dietary supplements, fortified foods and animal feed. *Akzo Chemie.*

13047 Gluconal® ZN
4468-02-4 224-736-9
$C_{12}H_{22}O_{14}Zn$
Zinc gluconate
Pharmaceutical/food grade mineral source for human and veterinary pharmaceutical preparations, dietary supplements, fortified foods and animal feed. Freely soluble in H_2O (100 g/l); insoluble in organic solvents; LD_{50} (rat orl) > 5,000 mg/kg. *Akzo Chemie.*

13048 Gluconic Acid
526-95-4 4464 208-401-4
$C_6H_{12}O_7$
D-Gluconic acid
glycogenic acid; dextronic acid; maltonic acid; glyconic acid; pentahydroxycaproic acid. Pharmaceuticals and food products, cleaning and pickling metals, sequestrant, cleansers, catalyst in textile printing. mp = 131°; soluble in water, slightly soluble in alcohol, insoluble in ether and most organic solvents. *Akzo; Am. Biorganics; Faesy & Besthoff; Glucona; PMP Fermentation Prods.*

13049 Glucopon 225
68515-73-1
C_8, C_{10} alkyl polyglycoside
Surfactant, detergent, wetting agent, surface tensile reducer, hydrotrope, dispersant for laundry detergents, liquid cleaners, hard surface cleaners, institutional and industrial cleaners; biodegradable. *Henkel/Emery.*

13050 Glucopon 425
68515-73-1
C_{18}-C_{16} alkyl polyglycoside
Surfactant, detergent, wetting agent, surface tensile reducer, hydrotrope, dispersant for laundry detergents, liquid cleaners, hard surface cleaners, institutional and industrial cleaners; biodegradable. *Henkel/Emery.*

13051 Glucopon 600
110615-47-9
C_{12}-C_{16} alkyl polyglycoside
Surfactant, detergent, wetting agent, surface tensile reducer, hydrotrope, dispersant for laundry detergents, liquid cleaners, hard surface cleaners, institutional and industrial cleaners; biodegradable. *Henkel/Emery.*

13052 glucose
50-99-7 4467 200-075-1
$C_6H_{12}O_6$
α-D-glucose
Dextrose; grape sugar; corn sugar; blood sugar; Dextropur; Dextrosol; Glucolin. Sugar obtained from the hydrolysis of starch; (anhydrous), (hydrous); Confectionery, foods, medicine, brewing, baking, canning. mp = 83°; $[\alpha]_D$ = 102.0°, mutarotates to 47.9° (H_2O); soluble in H_2O (1 g/ml), less soluble in organic solvents; LD_{50} (rbt iv) = 35 g/kg. *Am. Biorganics; Amerchol NV; Avebe BV; Corn Prods.; Mallinckrodt; Mendell; U.S. Biochemical.*

13053 glucose or sugar vinegar
A vinegar prepared by the conversion of starch substance into sugar, by the action of dilute acids, followed by fermentation and acetification.

13054 glucose syrup
Dextrin-maltose. A partially hydrolyzed starch employed in brewing and in confectionery.

13055 Glucostix®
Reagent for blood sugar determination. *Bayer AG.*

13056 glucotannin
A tannin found in Chinese rhubarb. It is 1-galloyl-β-glucose.

13057 Glucox
Glucose oxidase, an enzyme. *John & E Sturge Ltd, Selby.* Name unverified.

13058 Glucquat® 125
Lauryl methyl gluceth-10 hydroxypropyldimonium chloride
Substantive conditioner for hair and skin care products; humectant, moisturizer. *Amerchol Corp.*

13059 Gluma
Dentin bonding agent for composite filling materials; used in dentistry. *Bayer AG.*

13060 Glurub
A proprietary rubber-glue compound for rubber stiffening. No manufacturer.

13061 Glutalys®
Maize gluten; used for animal feed. *Roquette (UK) Ltd.*

13062 glutaraldehyde
111-30-8 4480 203-856-5
$C_5H_8O_2$
1,3-diformylpropane
pentanedial; glutaral; glutaric dialdehyde; Cidex; glutarol; Novaruca; Verucasep; 1,5-Pentanedial; Alhydex; Dioxopentane; Sporicidin; Ucarcide; 1,5-pentanedione; potentiated acid glutaraldehyde; Sonacide; Pentane-1,5-dial; Aldesan; Coldcide-25 microbiocide; Glutaralum; Hospex. Intermediate, fixative for tissues, for crosslinking protein and polyhydroxy materials, tanning of soft leathers. bp= 187-189°; n_D^{25}= 1.4338; d= 1.1310; soluble in H_2O; LD_{50} (rat orl) = 134 mg/kg. *Allchem Industries; BASF; Transol Chem. UK Ltd; Union Carbide.*

13063 Glutarom® Range
Amino acid salt mixture on a carrier; natural flavor enhancer. *Grünau.*

13064 Glutolin
9004-67-5 6120
Methyl cellulose
Cellulose methyl ether; Methocel MC; Cellothyl; Syncelose; Bagolax; Celevac; Cellucon; Cologel; Cellumeth; Hydrolose, Nicel; Tylose. Used as a substitute for water-soluble gums, as a suspending agent and as a laxative. Soluble in cold (not hot) water.

13065 Glutrin
8061-52-7
Calcium lignosulfonates
Modified calcium lignosulfonates; binders for foundry and refractory brick manufacture. *Borregaard Ligno Tech.*

13066 Glyakol
Diglyceryl ether tetracetate
Solvent.

13067 glycarbin
Glyceryl carbonate

13068 Glycene
A proprietary trade name for an alkyd synthetic resin used for dentures. No manufacturer.

13069 Glycereth series
31694-55-0
Ethosperse® G-26; Glycereth-26; Acconon ETG; Hetoxide G-7; Liponic EG-1. Emulsifier, humectant for cosmetic, pharmaceutical and industrial uses. *Lonza Inc; Lipo; Karlshamns; Heterene.*

13070 Glyceria wax
A wax formed in the stem of cane grass, *Glyceria ranirgera*, of Australia. It melts at 82°.

13071 glycerin
56-81-5 4493 200-289-5
$C_3H_8O_3$
Polyhydric alcohol
glycerol; 1,2,3-propanetriol; glycerine; glycyl alcohol. Used in manufacture of alkyd resins, dynamite, ester gums, pharmaceuticals, cosmetics, perfumery, lubricants, softener, bacteriostat, penetrant, solvent, humectant, plasticizer, emollient; antifreeze; in production of antibiotics. Alba Int'l.; Asahi Denka Kogyo; Farleyway Chem. Ltd; Fina; Henkel/Emery; Lonza; Procter & Gamble; Unichema; Witco/Humko.

13072 glycerin-formal
A condensation product of glycerol and formaldehyde; used as a solvent for lacquer, preservatives against molds and bacteria.

13073 Glycero-ester
A trade name for ester gum. No manufacturer.

13074 Glycerogen
A mixture of polyhydric alcohols obtained by inversion of sugar to hexose, then reduction with hydrogen and vacuum distilled. The final product is 40% glycerine, 40% propylene glycol and 20% hexyl alcohols.

13075 Glycero-piperaz
Basic piperazine glycerophosphate.

13076 Glycerox
Water soluble emollients; used for cosmetics and toiletries. *Croda Chem. Ltd, Croda Inc.*

13077 glyceryl caprylate
26402-26-6 247-668-1
Imwitor® 988; Imwitor® 308; Imwitor® 988. Surfactant for pharmaceutical, cosmetic and nutritional fields; as emulsifier, solubilizer, dispersion aid, plasticizer, lubricant, consistency regulator, skin and mucous membrane protectant, refatting agent, penetrant, carrier, adsorption promoter. Soluble in water/ethanol, acetone, ether, heptane; acid no = 3 max; iodine no = 3 max; sapon no = 275-300°. *Hüls AG.*

13078 glyceryl cocoate
61789-05-7 263-027-9
Imwitor® 928. Surfactant for pharmaceutical, cosmetic, and nutritional fields;

as emulsifier, solubilizer, dispersion aid, plasticizer, lubricant, consistency regulator, skin and mucous membrane protectant, refatting agent, penetrant, carrier, adsorption promoter. mp = 33-37°; acid no = 2 max; iodine no = 3 max; sapon no = 200-220. *Hüls AG.*

13079 glyceryl dilaurate
27638-00-2 248-586-9
Capmul® GDL; Emulsynt GDL; Kemester® GDL; Kessco® GDL; Lexemul® GDL. Emulsifier, thickener, and emollient for creams and lotions. *Van Dyk; Karlshamns;Witco/Humko;Stepan, Stepan Canada; Inolex.*

13080 glyceryl dioleate
25637-84-7 247-144-2
Emerest® 2419; Priolube 1409. Emulsifier, lubricant, rust preventive additive, mold release agent, solvent for dyes and pigments; used in leathers, lubricants and softeners. *Henkel/Cospha; Henkel Canada; Unichema.*

13081 glyceryl lanolate
97404-50-7 306-817-1
Glyceryl lanolate
Lanesta G. Emollient, emulsifier; forms stable water-oil emulsions. *Westbrook Lanolin.*

13082 glyceryl mono/distearate
Emulsifier for personal care products, food applications.

13083 glyceryl monolaurate
142-18-7 205-526-6
$C_{15}H_{30}O_4$
2,3-Dihydroxypropyl dodecanoate
glyceryl laurate; dodecanoic acid, monoester with 1,2,3-propanetriol; dodecanoic acid, 2,3-dihydroxypropyl ester. Monoester of glycerin and lauric acid; emulsifier, dispersant for food products, oils, waxes, solvents; antifoaming agent; drycleaning soap base. *Grindsted; Henkel/Emery; Lonza; Protameen.*

13084 glyceryl monooleate
37220-82-9 203-827-7
$C_{21}H_{40}O_4$
9-Octadecenoic acid (Z)-, ester with 1,2,3-propanetriol
glyceryl oleate; monoolein; 9-octadecenoic acid, monoester with 1,2,3-propanetriol. Monoester of glycerin and oleic acid; Emulsifier, coemulsifier, stabilizer, wetting agent, lubricant, and antistat; used in cosmetic, pharmaceutical, industrial, and food applications. Calgene; Ferro/Keil; Grindsted; Henkel/Emery; ICI Spec.; Inolex; Karlshamns; Lonza; Mona; Patco; Stepan/PVO; Witco/Humko.

13085 glyceryl monoricinoleate
141-08-2 205-455-0
$C_{52}H_{102}O_9$
12-hydroxy-9-octadecenoic acid, monoester with 1,2,3-propanetriol
glyceryl ricinoleate; monoricinolein. Monoester of glycerin and ricinoleic acid; Emulsifying agent. *CasChem; Lonza.*

13086 glyceryl montanate
68476-38-0 270-679-8
Hostalub® WE4. Lubricant and release agent for PVC and other polymers. *Hoechst AG; Hoechst Celanese.*

13087 glyceryl stearate
123-94-4 250-705-4
$C_{21}H_{42}O_4$
monostearin
glyceryl monostearate; glyceryl stearate; Octadecanoic acid, monoester with 1,2,3-propanetriol. Monoester of glycerin and stearic acid; Nonionic sec. o/w emulsifier for creams and lotions; visc. booster for emulsions. Eastman; Goldschmidt; Grindsted; Hart Prod. Corp.; Henkel/Emery; ICI Spec.; Inolex; Karl-shamns; Lanaetex; Lipo; Lonza; MTM Spec. Chem. Ltd; Patco; Van Dyk; Witco/Humko.

13088 glyceryl stearate citrate
91744-38-6 294-600-1
$C_{27}H_{48}O_{10}$
2-Hydroxy-1,2,3-propanetricarboxylic acid, monoester with 1,2,3-propanetriol monooctadecanoate
Imwitor® 370. Food emulsifier; oil-water emulsifier for very polar oils, fats and liquid wax esters in cosmetics. *Hüls Am; Hüls AG.*

13089 glyceryl triacetyl ricinoleate
101-34-8 202-935-1
$C_{63}H_{110}O_{12}$
9-Octadecenoic acid, 12-(acetyloxy)-,1,2,3-propanetriol ester
Flexricin® P-8; Naturechem® GTR. Plasticizer for vinyl wire jacketing and semi-rigid vinyls; emollient; stabilizer for anhydrated pigmented systems. Triester of glycerin and acetyl ricinoleic acid. d = 0.967; iodine no 76; saponification no 300; insoluble in H_2O. *CasChem.* Name unverified.

13090 glycidoxypropyl methyl diethoxysilane
2897-60-1 220-780-8
GP-137; 3-Glycidoxypropylmethyldiethoxy silane; CG6710. Coupling agent between inorganic fillers and epoxy, melamine, phenolic and urethane resins, PS, acrylic sealants, butyl rubber. *Genesee Polymers.* Discontinued.

13091 glycidyl acrylate
106-90-1 203-440-3
$C_6H_8O_3$
acrylic acid 2,3-epoxypropyl ester
Polyfunctional monomer; in coatings to improve adhesion to substrate and solvent resistance. *Estron.*

13092 glycidyl decanoate
26761-45-5 247-979-2
$C_{13}H_{24}O_3$
2,3-epoxypropyl neodecanoate
Glydexx N-10; Neodecanoic acid oxiranylmethyl ester; glycidyl neodecanoate; cardura E-10 (monoglycidyl ester). Polyfunctional monomer; in coatings to improve adhesion to substrate and solvent resistance. bp = 260°; d = 0.97; insoluble in H_2O; LD_{50} (rat orl) > 9.59 g/kg. *Exxon.* Name unverified.

13093 glycidyl ether
2,3-Epoxypropyl (n-butyl, allyl) ether. A reactve diluent for epoxy resin and raw material for curing agents of epoxy resins. *Yokkaichi Chemical Co Ltd.* Name unverified.

13094 glycidyl methacrylate 2,3-epoxypropyl methacrylate
106-91-2 203-441-9
$C_7H_{10}O_3$
methacrylic acid 2,3-epoxypropyl ester
Glycidyl Methacrylate 2,3-epoxypropyl methacrylate; CP-105; glycidyl α-methylacrylate; 1-propanol, 2,3-epoxy-, methacrylate; SR-379. Polyfunctional monomer; in hydrogels for contact lenses and membranes, molding and casting compounds, impregnating paper, concrete, wood, coatings, printing inks, adhesives, sealants, elastomers. bp = 189°; d = 1.042; n_D^{20} = 1.4494; soluble in H_2O, organic solvents; LD_{50} (rat orl) = 597 mg/kg. *Estron; Mitsubishi Gas; Sartomer.*

13095 glycine
56-40-6 4500 200-272-2
$C_2H_5O_2N$
Amino acid
aminoacetic acid; glycocoll; Gly; G salt; Aminoacetic Acid; Glycocoll; glycine, free base; Athenon; Iconyl; Monazol; Aminoethanoic acid; glycosthene; glycine--iron sulfate (1:1). Organic synthesis, nutrient, biochemical research, buffering agent, chicken-feed additive, reduces bitter taste of saccharin, retards rancidity in animal and vegetable fats. dec 233°; d = 1.1607; soluble in H_2O (25 g/100 ml), less soluble in organic solvents. *Aldrich; Allchem Industries; Degussa; W.R. Grace/Hampshire; U.S.Biochemical.*

13096 glyco metal
An alloy of 70-74% lead, 14-16% antimony, and 8-12% tin.

13097 glycobrom
The glyceryl ester of dibromohydrocinnamic acid.

13098 Glycoderm
Sphingolipid liposomes with glycosaminoglycans; preparation for dry and cracked skin; for restoration of the lipid barrier of the stratum corneum. *Dr. Kurt Richter; Henkel/Cospha.*

13099 glycol
107-21-1 3844 203-473-3
$C_2H_6O_2$
ethylene glycol
1,2-ethanediol; ethylene alcohol. Antifreeze in cooling and heating systems; in hydraulic brake fluids; industrial humectant; solvent in paints, plastics, inks; softening agent for cellophane; stabilizer; in explosives, alkyd resins, elastomers, synthetic fibers and waxes; asphalt. mp = -13°; bp = 198°; d_4^{20}= 1.1135; n_D^{25}= 1.43063; soluble in H_2O, polar organic solvents; LD_{50} (rat orl) = 8.5 g/kg. Ashland; BASF; Eastman; Hoechst Celanese; Mitsui Petrochem. Ind.; Mitsui Toatsu Chem.; Mobil; Olin; OxyChem; Shell; Texaco; Union Carbide.

13100 glycol dilaurate
624-04-4 4507 210-827-0
$C_{26}H_{50}O_4$
dodecanoic acid 1,2-ethanediyl ester
Emalex EG-di-L; Ethylene glycol dilaurate; Kemester® EGDL. Oil-phase cosmetic ingredient, emulsifier for creams, milky lotions, hair conditioners; cleaner, superfattening agent, thickener, reforming agent. mp = 50-52°; bp_{20} = 188°; insoluble in H_2O, EtOH, Et_2O. *Nihon Emulsion; Witco/Humko.*

13101 glycol dimercaptoacetate
123-81-9 204-653-4
$C_6H_{10}O_4S_2$
ethylene glycol bisthioglycolate
Glycol dimercaptoacetate; Acetic acid, mercapto-, 1,2-ethanediyl ester. Crosslinking agent for rubbers, accelerator in curing epoxy resins. bp_1 = 137-139°; d = 1.3170; n_D^{20}= 1.5185; soluble in H_2O (20 g/l), organic solvents; LD_{50} = 330 mg/kg. *Evans Chemetics; Jansse Chimica.*

13102 glycol distearate
627-83-8 211-014-3
$C_{38}H_{74}O_4$
EGDS
ethylene glycol distearate; octadecanoic acid, 1,2-ethanediyl ester. Diester of ethylene glycol and stearic acid; pearlescent and opacifier; thickener, intermediate, lubricant, emulsifier, emollient; for emulsion shampoos and foam baths. mp = 58-65°.

13103 glycol palmitate
4219-49-2 224-160-8
Glycol palmitate
Lanol P; Ethylene glycol monopalmitate; Glycol monopalmitate; 2-Hydroxyethyl hexadecanoate. Cosmetic emulsifier. *Seppic.* Discontinued.

13104 glycol ricinoleate
106-17-2 203-369-8
$C_{20}H_{38}O_4$
Ethylene glycol monoricinoleate
Flexricin® 15; Cithrol EGMR N/E; Glycol monoricinoleate; 2-Hydroxyethyl 12-hydroxy-9-octadecenoate. Wetting agent, plasticizer, textile, household, and cosmetic applications, rewetting dried skins; chemical intermediate. d = 0.965; miscible with most organic solvents, insoluble in water. *CasChem.* Name unverified.

13105 glycolic acid
79-14-1 4508 201-180-5
$C_2H_4O_3$
hydroxyacetic acid
Hydroxyethanoic acid. Leather dyeing and tanning; textile dyeing; cleaning, polishing and soldering compounds; copper pickling; adhesives; electroplating; petroleum demulsifier; chelating agent for iron; chemical milling; pH control. mp = 80°; soluble in H_2O, organic solvents; LD_{50} (rat orl) = 1.95 g/kg. *Du Pont; R.W. Greeff; Hoechst Celanese.*

13106 Glycolube® 100
Polyol ester
Internal lubricant for thermoplastics. *Lonza Inc.*

13107 Glycolube® 140
Polyol ester; thermoplastic antistat. *Lonza Inc.*

13108 Glycolube® P
Ester wax; external lubricant and antistat. *Lonza Inc.*

13109 Glycolube® VL
8002-74-2 7155 232-315-6
Paraffin
Synthetic wax; internal and external lubricant, release agent for thermoplastics; also used for textiles. *Lonza Inc.*

13110 Glycomul®
8872
Sorbitan esters. Emulsifiers, stabilizers and thickeners. *Lonza Inc.*

13111 Glycomul® L
1338-39-2 8872 215-663-3
Sorbitan laurate
Emulsifier for edible, cosmetic, industrial, pharmaceutical uses; antistat, antifog for PVC. *Lonza Inc.*

13112 Glycomul® O
1338-43-8 8872 215-665-4
Sorbitan oleate
Emulsifier for cosmetic, pharmaceutical, and industrial applications. *Lonza Inc.*

13113 Glycomul® P
26266-57-9 8872 247-568-8
Sorbitan palmitate; emulsifier for cosmetic, pharmaceutical, and industrial applications. *Lonza Inc.*

13114 Glycomul® S FG
1338-41-6 8872 215-664-9
Sorbitan stearate; emulsifier for food, cosmetic, household and industrial use. *Lonza Inc.*

13115 Glycomul® SOC
8007-43-0 8872 232-360-1
Sorbitan sesquioleate
Emulsifier for cosmetic, pharmaceutical, and industrial applications. *Lonza Inc.*

13116 Glycomul® TO
26266-58-0 8872 247-569-3
Sorbitan trioleate; emulsifier for cosmetic, pharmaceutical, and industrial applications. *Lonza Inc.*

13117 Glycomul® TS KFG
26658-19-5 8872 247-891-4
Sorbitan tristearate; emulsifier for food, cosmetic, household and industrial applications. *Lonza Inc.*

13118 Glycon® G 100, G 300
56-81-5 4493 200-289-5
glycerol
Glycerin; humectant, bodying agent, moisture control agent for toothpaste, cosmetics, sugarless confections, controlled moisture foods and industrial applications. *Lonza Inc.*

13119 Glycon® P-45
57-10-3 7128 200-312-9
palmitic acid
45% palmitic acid; lubricant, defoamer, and component of other food additives. *Lonza Inc.*

13120 Glycon® S-65
61790-38-3 263-130-9
Hydrogenated tallow fatty acid
Lubricant, defoamer, and component of other food additives. *Lonza Inc.*

13121 Glycon® S-90
57-11-4 8959 200-313-4
stearic acid
90% Stearic acid; lubricant, defoamer, and component of other food additives. *Lonza Inc.*

13122 Glycon® TP
57-11-4 8959 200-313-4
stearic acid
Triple pressed; lubricant, defoamer, and component of other food additives. *Lonza Inc.*

13123 Glyconol®
Amide wax; synthetic wax for textiles and other applications; gelling and viscosity. modifier for hot melts, wax blends, oil modifiers, metalworking lubricants. *Lonza Inc.*

13124 Glyconyl
122-87-2 4885 204-580-8
$C_8H_9NO_3$
N-(4-Hydroxyphenyl)glycine
glycin; *p*-hydroxyphenylglycine. A photographic developer, the active constituent of which is *p*-hydroxyphenylglycine. mp = 248°; slightly soluble in H_2O.

13125 glycosal
The mono-salicylic ester of glycerol.

13126 Glycosperse®
8872
Polyethoxylated sorbitan esters. *Lonza Inc.*

13127 Glycosperse® HTO-40
8872
PEG-40 sorbitan hexatallate
Emulsifier for food, cosmetic, household or industrial applications. *Lonza Inc.*

13128 Glycosperse® L-10
9005-64-5 8872
PEG-10 sorbitan laurate
Emulsifier for food, cosmetic, household or industrial applications. *Lonza Inc.*

13129 Glycosperse® L-20
9005-64-5 8872
Polysorbate 20
Emulsifier for food, cosmetic, pharmaceutical, and industrial uses; flavor solubilizer and dispersant. *Lonza Inc.*

13130 Glycosperse® O-20 FG, O-20 KFG
9005-65-6 7742
Polysorbate 80
Emulsifier for ice cream, frozen desserts; solubilizer and dispersant for shortenings; adjuvant for herbicides and plant growth regulators. *Lonza Inc.*

13131 Glycosperse® O-5
9005-65-6 7742
Polysorbate 81
Flavor solubilizer and dispersant; emulsifier for cosmetic, pharmaceutical, and industrial use. *Lonza Inc.*

13132 Glycosperse® P-20
9005-66-7 8872
Polysorbate 40
Emulsifier for food, cosmetic, pharmaceutical, and industrial uses; flavor solubilizer and dispersant. *Lonza Inc.*

13133 Glycosperse® S-20 FG, S-20 KFG
9005-67-8 8872
Polysorbate 60
Emulsifier for chocolate and confectionery coatings, icings, toppings, cakes, cream fillings, shortenings, desserts, dough conditioners. *Lonza Inc.*

13134 Glycosperse® TS-20 FG, TS-20 KFG
9005-71-4 8872
Polysorbate 65

Emulsifier for ice cream, frozen desserts, cakes, whipped toppings, nondairy creamers, icings, fillings. *Lonza Inc.*

13135 Glycosterine
A glycol glyceryl stearate; used to replace beeswax in certain polishes. No manufacturer.

13136 glycothymoline
An antiseptic, usually containing thymol, eucalyptol, menthol, borates, bicarbonates, benzoates, and glycerin.

13137 Glycowax® 765
110-30-5 203-755-6
$C_{38}H_{76}N_2O_2$
N,N'- Ethylene bisstearamide
Octadecanamide, N,N'-1,2-ethanediylbis-; Ethylene bis(stearamide); N,N'-ethylenedi(stearamide). Synthetic wax for kraft brownstock pulp and paper defoaming, plastics processing, powdered metal lubrication, water treatment defoaming, other industrial applications. *Lonza Inc.*

13138 Glycozone
glyceric acid-glycerol
A proprietary preparation containing 5% glyceric acid and 90% glycerol; an antiseptic. No manufacturer.

13139 N-glycylglycine
556-50-3 4513 209-127-8
$C_4H_8N_2O_3$
diglycine. Used in peptide synthesis. dec 262-264°; soluble in H_2O, EtOH, less soluble in organic solvents. *Am. Biorganics; Penta Mfg.; Wchweizerhall; Spectrum Chem. Mfg.*

13140 Glydant®
6440-58-0 229-222-8
$C_7H_{12}N_2O_4$
1,3-bis (hydroxymethyl)-5,5-dimethyl-2,4-imidazolidine dione
Dantoin DMDMH; Dimethyloldimethyl hydantoin; DMDMH; Glydant; Mackgard DM; Nipaguard DMDMH; Dantion DMDMH 55; Dantoguard;. DMDM hydantoin in water; preservative, broad spectrum antimicrobial for cosmetics and toiletries; effective against Gram-positive and Gram-negative bacteria, fungi, and yeast. A preservative and formaldehyde donor. *Lonza Inc.*

13141 Glydexx N-10
26761-45-5 247-979-2
$C_{13}H_{24}O_3$
Glycidyl decanoate
Neodecanoic acid oxiranylmethyl ester; glycidyl neodecanoate; Cardura E-10 (monoglycidyl ester); 2,3-epoxypropyl neodecanoate. bp = 260°; $d^{12.3}$ = 0.967; insoluble in H_2O, soluble in organic solvents; LD_{50} (rat orl) = 9.59 g/kg. *Exxon.* Name unverified.

13142 Glydus
Herbicide. *Agan Chemical Manufacturers Ltd.*

13143 Glyecin
Ethylthiodiglycol
A solvent used in treatment of wool for dyeing.

13144 Glyezin®
Dye solvents and fixing assistants for textile finishing. *BASF AG; BASF plc.*

13145 Glymin®
Leak detector fluid for double-walled storage tanks. *BASF AG.*

13146 glyoxal 40%
107-22-2 4519 203-474-9
$C_2H_2O_2$
Glyoxal
Ethane-1,2-dione; Oxalaldehyde; Ethanedial; biformal; biformyl; ethandial; 1,2-ethanedione; glyoxylaldehyde; OXAL; aerotex glyoxal 40; ODIX. Used in textile, paper and oil field industries. mp = 15°; bp_{776} = 51°; $n^{20.5}_D$ = 1.3826; LD_{50} (rat orl) = 2020 mg/kg. *Cyanamid BV.*

13147 Glyphogan
1071-83-6 4522 213-997-4
Glyphosate
Herbicide. The monoisopropylamine salt is in Round-up. *Agan Chemical Manufacturers Ltd.*

13148 glyphosate
1071-83-6 4522 213-997-4
$C_3H_8NO_5P$
N-(phosphonomethyl)glycine
MON-0573; Bronco; Landmaster; Ranger; Pondmaster; Rattler 4AS; Roundup 2.5; glialka; MON 2139; MON 6000; phosphonomethyliminoacetic acid; N-phosphomethylglycine; Sonic; Spasor; Sting; tumbleweed. Broad spectrum translocatable herbicide. mp = 230° (dec); soluble in H_2O (12 g/l), insoluble in most organic solvents; LD_{50} (rat orl) = 4873 mg/kg.

13149 Glyprosol 20
Yeast glycoproteins; natural skin smoothing protein. *Brooks Industries.*

13150 Glyptal 2557,2559
Proprietary trade names for polyester plasticizers. *GE Plastics.* Name unverified.

13151 glyptal resins
Resinous products obtained by the interaction of glycerol and organic acids. A resin of this type is made by reacting upon glycerol with oleic acid and phthalic anhydride; used in paints, varnishes, and lacquers.

13152 Glyrol
56-81-5 4493 200-289-5
Glycerin
Pharmaceutical aid. *Lolab Corp.*

13153 Glysantin®
Antifreeze for combustion engines. *BASF AG; BASF plc.*

13154 Glytex
isoxaben-methabenzthiazuron
Mixture of isoxaben and methabenzthiazuron; soil-acting herbicide for winter cereals. *Bayer plc.*

13155 Glytex® 203
Polyol ester; heat-stable nylon and polyester lubricant, emulsifier, antistat. *Lonza Inc.*

13156 Glytex® 513
Ester ethoxylate
Nylon and polyester tire cord finish emulsifier, lubricant, antistat. *Lonza Inc.*

13157 Glytex® 663
Hydantoin ester
Nylon and polyester industrial yarn lubricant, emulsifier, antistat. *Lonza Inc.*

13158 Glytex® 1085
Alcohol ester; textile spin finish lubricant, emulsifier, antistat. *Lonza Inc.*

13159 Glythermin®
Ethylene glycol-based antifreeze thermal liquids. *BASF AG.*

13160 GMD
Pigment dispersions; for coloration of aqueous and solvent-based trade sale paints. *Pacific Dispersions Inc. Name unverified.*

13161 GMF
p-Quinonedioxime
A rapid and economical vulcanizing agent when used in conjunction with red lead; gives fast-curing high-modulus stocks; recommended for use in butyl curing bags, wire insulation and where a fast curing high-modulus stock is desired. *Uniroyal.* Name unverified.

13162 GMS Base
11099-07-3 4498 234-325-6
Glyceryl stearate
Octadecanoic acid, ester with 1,2,3-propanetriol; Glyceryl stearate; monostearin. Base for nonionic textile softeners. *Clark.*

13163 GMS/SE Base
Glyceryl stearate SE
Base for nonionic textile softeners. *Clark.*

13164 Gobapur Acide Pur
Sulfuric acid and oleum. *Rhône Poulenc NV/CdF Chimie AZF.* Name unverified.

13165 Gofrativ
7704-34-9 9142 231-722-6
Sulfur
Fungicide *Makhteshim Chemical Works Ltd.*

13166 Gofravik
7704-34-9 9142 231-722-6
Sulfur
Fungicide *Makhteshim Chemical Works Ltd.*

13167 Gohi Iron
A proprietary trade name for iron containing manganese, sulfur, phosphorus, copper (total less than 0.125%.). No manufacturer.

13168 Gohsefimer
Modified polyvinyl alcohol. *British Traders & Shippers Ltd.*

13169 Gohsenol
Polyvinyl alcohol
British Traders & Shippers Ltd.

13170 Gohseran
Modified polyvinyl alcohol. *British Traders & Shippers Ltd.*

13171 Gold
7440-57-5 4526 231-165-9
Au
Metallic element; Au; infrared reflectors; electrical contact alloys; brazing alloys; laboratoryware; decorative arts; electronics; dental alloys; jewelry; colloidal dispersions for coloring glass, as nucleating agent, for specialized medical treatments. mp = 1064.76°; bp = 2700°; d = 19.3. *Cerac; Degussa; Koch Chem. Ltd; Noah Chem.*

13172 Gold Guard
Contact cleaner/lubricant for removal of oxides, dust and contaminants from precious metals and connectors. *Chemtronics.*

13173 gold size
Consists of a mixture of 1 part yellow ocher, 2 parts copal varnish, 3 parts linseed oil, 4 parts turpentine, and 5 parts boiled oil.

13174 Gold solders
Various alloys of gold, silver, and copper, sometimes with zinc. An ordinary gold solder contains 43% gold, 30% silver, 20% copper, and 7% zinc; consists of 66% gold, 22% copper.

13175 gold trichloride
13453-07-1 4541 236-623-1
$AuCl_3$
auric chloride; gold chloride. Laboratory reagent. d = 3.9; bp = 229°; sublimes at 180° (760mm); soluble in water, alcohol, ether; LD_{50} (mus sc) = 1.5 g/kg. *Degussa; Métaux Précieux SA; Spectrum Chem. Mfg.*

13176 golden acorn
Nutmeg.

13177 golden antimony sulfide
1315-04-4 738 215-255-5
S_5Sb_2
Golden sulfuret of antimony; golden sulfide of antimony; antimonic sulfide; antimonial saffron; antimony red. Antimony pentasulfide. Used as a pigment; for vulcanizing and coloring rubber and in matches and fireworks.

13178 Golden Bear 102
Naphthenic distillate
Cheap oil used in high-loss systems; motor oils for cars with excessive oil consumption. *Witco/Golden Bear.* Discontinued.

13179 Golden Bear 2013-10
Solvent-refined naphthenic oil; used in industrial, metalworking, ink and similar formulations, for blending approved railroad and marine engine oils, motor oils designed for older automobiles where premium quality is not necessary. *Witco/Golden Bear.* Discontinued.

13180 Golden Bear 4013-10
Solvent-refined naphthenic oil; premium grade featuring extra low pour points, outstanding oxidation stability to produce base stock for manufacture of refrigeration oils, white oils, transformer oils, high-performance motor oils either by themselves or in combination with paraffinic oils. *Witco/Golden Bear.* Discontinued.

13181 Golden Dawn
Refined lanolins of pharmaceutical/cosmetic quality; emollients and water-oil emulsifiers. *Westbrook Lanolin.* Name unverified.

13182 Golden Fleece
Hypo-allergenic and super-refined lanolins; emollients and water-oil emulsifiers in cosmetics and pharmaceuticals. *Westbrook Lanolin.* Name unverified.

13183 Golden Hermetite
Golden gel, nonhardening; a high technology gasket jointing compound with excellent chemical and temperature resistance; high tack stabilizes gasket during assembly; clean to use. *Hermetite Products Ltd.* Name unverified.

13184 golden syrup
Drip Syrup. This is the product remaining when raw or brown sugar is dissolved, the solution clarified with animal charcoal and the white sugar crystallized from it.

13185 Golden Wax
Mold release liquid wax for plastic molding. *Specialty Products Co.* Name unverified.

13186 Golden-Pea-Pro EN-15
Hydrolyzed golden-pea protein; cosmetic ingredient for skin and hair care products *Brooks Industries.*

13187 goldenseal
Hydrastis Candensis; used as a source of yellow dye.

13188 Goliath
13684-63-4 7384 237-199-0
phenmedipham
Emulsifiable concentrate containing 114 g/l phenmedipham; for weed control for beet crops. *ABM Chemicals Ltd.*

13189 Golpanol®
Electroplating additives. *BASF plc.*

13190 Golpanol® MBS
Oxidant and demetallizer for industrial cleaners. *BASF AG.*

13191 Goltix®
41394-05-2 5985 255-349-3
Metamitron
A water dispersible granular formulation containing 70% w/w metamitron; used to control annual weeds in sugar beet grown on mineral and organic soils and red beet, fodder beet and mangolds grown on mineral soils. *Bayer AG; Bayer plc.*

13192 Gonacrine
8048-52-0 125
3,6-diamino-10-methylacridinium chloride mixed with 3,6-acridinediamine
Acriflavine; Neutral Acriflavine; euflavine; Trypaflavine; neutroflavine. Preparation of 2,8-diamino-10-methylacridinium chloride and diamino acridine. Used as a topical anti-infective. Soluble in H_2O (30 g/100ml), less soluble in organic solvents. *May & Baker Ltd.* Name unverified.

13193 gong metal
A brass containing 78% copper and 22% zinc.

13194 Goniosol
9004-65-3 4889
Hydroxypropyl methylcellulose
Pharmaceutic aid. *Cooper Vision Inc.* Unverified.

13195 Gooch and Eddy reagent
A reagent used for precipitating magnesium as magnesium carbonate. It contains 180 parts concentrated ammonia, 800 parts water, and 900 parts absolute alcohol the solution being saturated with ammonium carbonate.

13196 Good Gulf
Gasoline. *Monsanto (Solaris).* Name unverified.

13197 GoodLife
A powder organic based fertilizer comprising a composted organic base and chemical N, P & K to form the analysis; four types sold: all-purpose Fertilizer, flower garden fertilizer, vegetable fertilizer and lawn weed and feed fertilizer. *Humber Fertilizers plc.*

13198 Good-rite Polyacrylates
Homopolymers and copolymers of acrylic acid and their salts in both liquid and dry powder forms; for water treatment, soap and detergents and dispersants. *BFGoodrich UK.* Name unverified.

13199 Good-rite® 2528X10
1337-81-1
Vinyl pyridine latex
Used for adhesives in tire cord chafer fabric and industrial rubber goods; produces wickproof chafer fabric in one pass; very stable high solids RFL dips; excellent adhesion retention. *BF Goodrich.* Name unverified.

13200 Good-rite® 3150
1,1',1-(1,3,5-Triazine-2,4,6-triyltris ((cyclohexylimino)-2,1-ethanediyl)tris(3,3,5,5-tetra-methylpiperazinone)
Hindered amine light stabilizer. *BF Goodrich.* Name unverified.

13201 Good-rite® GP-223
103-23-1 203-090-1
Dioctyl adipate
Plasticizer used with vinyl polymers. *BF Goodrich.* Name unverified.

13202 Good-rite® GP-235
Octyldecyl adipate
Vinyl plasticizer. *BF Goodrich.* Name unverified.

13203 Good-rite® GP-236
didecyl adipate
Vinyl plasticizer. *BF Goodrich.* Name unverified.

13204 Good-rite® GP-261
A phthalate vinyl plasticizer. *BF Goodrich.* Name unverified.

13205 Good-rite® GP-265
119-07-3 204-295-9
$C_{26}H_{42}O_4$
Octyldecyl phthalate
n-octyl n-decyl phthalate; 1,2-benzenedicarboxylic acid, decyl octyl ester; decyl octyl phthalate; n-decyl n-octyl phthalate; dinopol 235; octyl decyl phthalate; polycizer 532; polycizer 562; staflex 500. Plasticizer used with vinyl polymers. d_{20}^{20} = 0.980; mp = -50°; bp_4 = 239°; n_D^{25}= 1.4880; LD_{50} (rat orl) = 45 g/kg. *BF Goodrich.* Name unverified.

13206 Good-rite® GP-266
Didecyl phthalate
A vinyl plasticizer. *BF Goodrich.* Name unverified.

13207 Good-rite® K-702
9003-01-4
Polyacrylic acid aqueous solution; detergent assistant, soap builder, particulate soil dispersant, sequestrant for calcium, magnesium, iron; scale inhibitor; for laundry, dishwash, consumer/institutional cleaning products *BF Goodrich.*

13208 Good-rite® K-705BD
9008-04-7
Sodium polyacrylate
Detergent assistant, soap builder, particulate soil dispersant, sequesterant for calcium, magnesium, iron; scale inhibitor; for laundry, dishwash, consumer/institutional cleaning products. *BFGoodrich.*

13209 Good-rite® K-752
9003-01-4

Polyacrylic acid aqueous solution; detergent assistant, soap builder, particulate soil dispersant, sequesterant for calcium, magnesium, iron; scale inhibitor; for laundry, dishwash, consumer/institutional cleaning products *BF Goodrich/Spec. Polymers.*

13210 Goodyear LPR-6632
SBR latex
For textile applications, carpet lamination and precoat, fabric backcoating; as sprayable binders, tape base saturants, binders for adhesive and coating applications *Goodyear.* Name unverified.

13211 Gopmann solder
An alloy of 49.1% tin, 20.3% zinc, and 26% lead.

13212 Gordon superflex D
A trade name for a graft of stereospecific rubber and polystrene for high impact. Izod 1.35-1.65; elongation 27%; tensile strength 3800 psi; modulus 291,000 psi (flex). *PBI - Gordon Corp.* Discontinued.

13213 Gossamer®
Polymer. *DuPont UK.*

13214 Gougeon Laminating Epoxy
Liquid epoxy resin and hardener; laminating resin specifically for use with glass cloth, carbon fibers, aramid and hybrids. *Wessex Resins & Adhesives Ltd.*

13215 Goulac
8061-52-7
Calcium lignosulfonate
Binder for foundry, ceramic, and refractory products *Borregaard Ligno Tech.*

13216 Goulard powder
301-04-2 5411 206-104-4
Lead acetate
Used in dyeing and as an astringent and, in veterinary medicine, a sedative.

13217 GP-4
Amine functional silicone fluid; intermediate for synthesis of silicone/organic copolymers used in textiles, coatings car polishes; also in lubricant, coating and mold release formulations. *Genesee Polymers.*

13218 GP-4-E
Silicone emulsion; provides durability and detergent resistance in car polishes, vinyl conditioners. *Genesee Polymers.*

13219 GP-66 Miracle Cleaner
Nonionic surfactant, synthetic detergent, wetting agents, emulsifiers, builders etc; a USDA A-1 certified cleaning compound used to clean forklifts, diesel engines, conveyors, robots, concrete floors, ovens, whitewalls, vinyl tops. *GP66 Chemical Corporation.* Name unverified.

13220 GP-71-SS
Mercapto-modified dimethyl silicone copolymer fluid; plastic and rubber release agent; internal lubricant and release agent for sulfur and peroxide cure rubber; coreactant in vinyl polymerization; synthesis of organic/silicone copolymers; heat stabilizer; in corrosion inhibitor coatings and inks. *Genesee Polymers.*

13221 GP-137
2897-60-1 220-780-8
$C_{11}H_{24}O_4Si$
3-Glycidoxypropylmethyldiethoxy silane
Coupling agent between inorganic fillers and epoxy, melamine, phenolic and urethane resins, PS, acrylic sealants, butyl rubber. *Genesee Polymers.* Discontinued.

13222 GP-165
Silicone resin emulsion; cures to solvent-resistant coating; as binder in aqueous paints intended for application to substrates exposed to high temperatures; component of waterbased masonry water-repellent formulations. *Genesee Polymers.* Discontinued.

13223 GP-180
Proprietary organosilicone resin concentrate in 1,1,1-trichlorethane solvent; used as a dry film release coating for molding of plastics or rubber. *Genesee Polymers.* Discontinued.

13224 GP-187
Methyl silicone resin solution in isopropanol; cures to a soft pliable consistency at ambient conditions, to a clear, hard film at elevated temperatures; used in masonry water repellents, leather and fabric treating agents, mold release formulations; as binder for high temperature application paints and coatings. *Genesee Polymers.*

13225 GP-209
Dimethyl silicone EO/PO block copolymer; emulsifier, wetting agent, pigment dispersant, leveling agent, profoaming additive for PU foams, hard surface cleaners, polishes, cosmetic formulations; inverse solution suggests use as defoamer for hot aqueous surfactant solutions. *Genesee Polymers.*

13226 GP-210
Silicone emulsion; defoamer for hot and cold foaming systems, industrial applications, commercial cleaning compounds, latex stripping, adhesives, cutting oils, leather treating, sewage treatment, chemical processing, paints/coatings. *Genesee Polymers.*

13227 GP-217
Dimethylpolysiloxane EO block copolymer; wetting agent, emulsifier for water-based coatings, inks, polishes, hard surface cleaners; dispersant for clays, pigments; thread lubricant; leveling and flow control agent; profoaming additive in aqueous systems. *Genesee Polymers.*

13228 GP-227
Organo-modified dimethylsilicone polymer; surfactant, water-oil emulsifier for emulsions, auto and furniture polishes, vinyl conditioners; gloss aid; adds detergent resistance. *Genesee Polymers.*

13229 GP-262
Nonsilicone; defoamer for aqueous systems, metalworking fluids, cleaner formulations, paints, coatings, paper/paperboard, waste water treatment; alkali resistant. *Genesee Polymers.*

13230 GP-310-I
Silicone aqueous emulsion; dilutable antifoam for industrial processing in hot, cold, alkaline or aqueous systems; for chemical processing, cleaning products, paints, paper and latex processing. *Genesee Polymers.*

13231 GP-7000
Methyl alkyl dimethyl silicone fluid; surface tensile reducer for nonaqueous solvent and oil systems; wetting and leveling agent in inks, coatings, plastisols; antifoam, mold release agent; internal release agent for plastics and rubber; textile lubricant. *Genesee Polymers.*

13232 G-P-D
Colorant dispersions; for coloring of nonaqueous coating compositions. *Hüls Am.*

13233 GP-II
Porofocon B
Contact lens material. *Barnes-Hind Inc.* Name unverified.

13234 GR Acid
α-Naphtholdisulfonic acid
Intermediate in dyestuffs manufacture.

13235 Graessorb
Sunscreen agents and ultra violet filters. *Nipa Laboratories Ltd.*

13236 Grafene
A proprietary trade name for a lubricant containing graphite and oils. No manufacturer.

13237 Grafil
High tensile carbon fibers for industrial use. *Courtaulds plc.* Discontinued.

13238 Grafita
A proprietary trade name for a lubricant containing graphite and grease. No manufacturer.

13239 Grafitix (Anti-graffiti)
Nonflammable solvent mixture manufactured in aerosol spray; used for removing of paint, felt, applied on stone, cement, bricks, wood, metal and fabrics. *S F C.* Name unverified.

13240 Grafitix (Baent)
Nonflammable solvent mixture water miscible, containing special wetting agent, specially designed for building trade applications; it strips all surfaces covered with paint, rough coat and plastic facing coatings. *S F C.* Name unverified.

13241 Grafitix (Ravalement)
Aqueous product, basic reaction, low viscosity, completely soluble in water; used for cleaning and stripping the outside surfaces of buildings mainly in stone or cement soiled by atmospheric pollution. *S F C.* Name unverified.

13242 Grafix
Graphic arts fixer. *May & Baker Ltd.* Name unverified.

13243 grahamite
An asphaltic substance found in Mexico and Cuba. It is usually associated with mineral matter. It has a specific gravity of 1.17, melts at 175-230°, contains up to 45% mineral matter, and is very soluble in carbon disulfide.

13244 grain alcohol
64-17-5 3806 200-578-6
C_2H_5OH
Ethyl alcohol

13245 Grain Store Smoke
Insecticide. *DowElanco Ltd.*

13246 grains D'Ambrette
Musk seeds.

13247 grains of paradise
Guinea Grains. Seeds of *Amomum melegueta.*

13248 Gramazine
Herbicide. *ICI Chem & Polymers Ltd.*

13249 Graminon® Plus
bentazon-isoproturon-dichlorprop. For post-emergence control of grasses and broadleaf weeds in winter wheat and winter barley *BASF AG.*

13250 Gramixel
paraquat-diuron
Herbicide containing paraquat [1910-42-5] and diuron [330-54-1]. *ICI Chem & Polymers Ltd.*

13251 Gramonol
paraquat-monolinuron
Weedkiller containing paraquat [1910-42-5] and monolinuron [330-55-2]. *ICI Chem & Polymers Ltd.*

13252 Gramonol 5
paraquat-monolinuron
Suspension concentrate containing 154 g monolinuron [330-55-2] and 110 g paraquat [1910-42-5] per liter; for control of annual dicotyledons in cereals. *Hoechst UK.*

13253 Gramonol Five
paraquat [1910-42-5]-monolinuron [330-55-2]
Suspension concentrate containing 154 g monolinuron and 110 g paraquat per liter; for control of annual dicotyledons in cereals. *ICI Agrochemicals.*

13254 Gramoxone
1910-42-5 7165 217-615-7
Paraquat
Herbicide. *Schering Agrochemicals Ltd.* Discontinued.

13255 Gramoxone
1910-42-5 7165 217-615-7
Paraquat
Paraquat weedkiller preparations. *ICI Chem & Polymers Ltd.*

13256 Gramoxone X
1910-42-5 7165 217-615-7
Paraquat
Soluble concentrate containing 200 g/l paraquat; a pre-emergence bipyridinium herbicide to control weeds in field crops and ornamentals. *ICI Agrochemicals; Schering Agriculture.*

13257 Gramp's solder
An alloy of 60.4% tin, 36.1% zinc, 3% copper, 0.25% lead, and 0.18% antimony. An aluminum solder.

13258 Gram's iodine stain
This consists of 1 g iodine, 2 g potassium iodide, and 300 ml distilled water.

13259 Gram's stain
A microscopic stain. It contains gentian violet.

13260 Gramuron
paraquat-diuron
Herbicide containing paraquat [1910-42-5] and diuron [330-54-1]. *ICI Chem & Polymers Ltd.*

13261 Graney bronze
An alloy of 76.5% copper, 9.2% tin and 15.2% lead.

13262 Granodine
Zinc phosphating solution for metal treatment. *ICI Chem & Polymers Ltd.*

13263 granodized steel
Steel which has been treated with zinc phosphate to give the surface resistance to corrosion.

13264 granol
Carbonized granulated peat.

13265 Granolube
Metal treating compositions. *ICI Chem & Polymers Ltd.* Discontinued.

13266 Granosan®
For the agriculture industry. *DuPont UK.*

13267 Granstock
Animal feed additive. *ICI Chem & Polymers Ltd.*

13268 Granubor
Calbor. Borated animal bone and borate containing flux. *Borax Europe Ltd.*

13269 Granuform
30525-89-4 7158 264-150-0
Paraformaldehyde
Free-flowing, dust-free, small white beads of 90.5 ±1% paraformaldehyde; used in the plastics industry for production of phenolic, urea, melamine and coating resins; in the chemical and pharmaceutical industries for chloromethylation processes; as disinfectant. *Degussa Ltd.*

13270 Granular Weedkiller
7775-09-9 8741 231-887-4
sodium chlorate
Granules containing 30% w/w sodium chlorate; total weed control for paths, drives and noncrop areas. *Dimex Ltd.*

13271 Granulite® BF 6/16
Ground rubber from truck and passenger retread buffings, free of foreign material, fabric, and steel; used in pneumatic tires, auotmotive components, construction and paving materials, industrial rubber goods, building materials, sporting goods, plastic materials. *Baker Rubber.*

13272 Granulite® TR-10
Ground rubber from truck and passenger tread rubber, free of foreign material, fabric, and steel; used in pneumatic tires, auotmotive components, construction and paving materials, industrial rubber goods, building materials, sporting goods, plastic materials. *Baker Rubber.*

13273 Granulite® WTP-10
Ground rubber from whole passenger and/or truck tire or equivalent with fabric removed; free of foreign material; used in pneumatic tires, auotmotive components, construction and paving materials, industrial rubber goods, building materials, sporting goods, plastic materials. *Baker Rubber.*

13274 granulose
The inner part of the starch granule is known by this name.

13275 Grape seed oil
8024-22-4
Oils, grape seed; Nikkol Grapeseed Oil; Super Refined™ Grapeseed oil; Lipovol G. Grape seed oil; conditioner, glosser, emollient imparting a light, nongreasy, silky afterfeel to skin and hair products; high film gloss and rapid spread; used in personal care products. *Lipo; Nikko Chem. Co. Ltd; Croda Inc.*

13276 Graphalloy, Silver
A trademark for a molded graphite impregnated with silver used in electrical brushes and similar appliances. No manufacturer.

13277 graphite
7782-42-5 4560 231-955-3
black lead
plumbago; mineral carbon. The allotropic form of carbon, occuring naturally in Madagascar, Ceylon, Mexico, Korea, Austria, USSR; can be synthetically produced by heating petroleum coke to +zw 3000° in an electric resistance furnace; reinforcing agent for prepregging, filament winding, lubricant additive for greases, engine oils, etc.; fillers, paints, coatings and self-lubricating bearings. *Cerac; Lonza; Sigri GmbH; Ucar Carbon.*

13278 graphite metal
An anti-friction and fitting metal. It contains 15% tin, 68% lead, and 17% antimony.

13279 graphitic carbon
The black shiny flakes of carbon present in pig iron.

13280 graphitic temper carbon
The black amorphous carbon present in certain varieties of iron.

13281 graphitites
Graphites which swell on moistening them with strong sulfuric acid, and then heating them to redness. They are not true graphites.

13282 Graphitol
Synthetic organic pigment colors; used for printing inks and paints. *Sandoz Products Ltd.*

13283 Graphsize
Polyurethane based paper surface size. *Akzo.*

13284 Graphtol®
Pigment for coloring bar soaps, gels, powders and solids. *Sandoz.*

13285 Grappier cements
Hydraulic limes are slaked and passed through sieves. The hard lumps left on the sieve consist of unchanged limestone and calcium silicates. These are finely ground, and are then known as grappier cements. Le Farge cement belongs to this class.

13286 Graslam
Herbicide *May & Baker Ltd.* Name unverified.

13287 Grass Greenizit
Permanent grass colorant restoring green color to dormant or discolored grass. *W A Cleary.*

13288 Grasselerator 101
A trademark for a rubber vulcanization accelerator; it is aldehyde ammonia. No manufacturer.

13289 Grasselerator 102
100-97-0 6036 202-905-8
$C_6H_{12}N_4$
Hexamethylenetetramine
Methenamine; HMT; HMTA; hexamine; Aminoform; Ammoform; Cystamin; Cystogen; Formin; Uritone; Urotropin. A trademark for a rubber vulcanization accelerator. No manufacturer.

13290 Grasselerator 508
Butylideneaniline
A proprietary rubber vulcanization accelerator. No manufacturer.

13291 Grasselerator 552
A trademark for a rubber vulcanization accelerator. No manufacturer.

13292 Grasselerator 833
A trademark for a liquid aldehyde amine condensation product; it is a low temperature rubber vulcanization accelerator, and has antioxidant properties. No manufacturer.

13293 Grasshopper
Compound fertilizer 8:1.5:1.5 plus 2,4-D, dicamba and ferrous sulfate; lawn fertilizer. *ICI Garden Products.*

13294 Gravidox
58-56-0 8166 200-386-2
Pyridoxine hydrochloride
Vitamin B_6. *Eli Lilly & Co.* Discontinued.

13295 Gravulac
Gravure overlacquers. *The Scottish Adhesives Co Ltd.*

13296 Grazon 90
clopyralid-triclopyr
Emulsifiable concentrate containing 60 g clopyralid and 240 g triclopyr per liter; used for treatment of perennial weeds in grassland. *Dow Elanco Ltd.*

13297 Great Lakes BA-50
Bis(2-hydroxyethyl ether) of tetrabromobisphenol A
Difunctional alcohol providing flame retardance; for unsaturated polyester and epoxy thermoset resins; used for laminates for electronic circuit boards; corrosion resistance of systems useful in materials *Great Lakes.*

13298 Great Lakes BA-59P
79-94-7 201-236-9
Tetrabromobisphenol A
Flame retardant for thermoplastic and thermoset resin systems, epoxy systems; also used as reactive flame retardant for PC and additive for styrenic thermoplastics. *Great Lakes.*

13299 Great Lakes BE-51
Bis (allyl ether) of tetrabromobisphenol A
Flame retardant used in EPS and foamed PS. *Great Lakes.*

13300 Great Lakes CD-75P
25637-99-4 247-148-4
Hexabromocyclododecane
Flame retardant used in thermoplastic and thermosetting polymers; textile treatments, latex binders, adhesives, unsaturated polyesters, and coatings. *Great Lakes.*

13301 Great Lakes DBS
Dibromostyrene
Flame retardant. *Great Lakes.*

13302 Great Lakes DE-71
32534-81-9 251-084-2
Pentabromodiphenyl oxide; Pentabromo diphenyloxide. High viscosity flame retardant for thermosetting and thermoplastic resin systems; used for unsaturated polyester, rigid and flexible urethane foams, epoxies, laminates, adhesives, and coatings. *Great Lakes.*

13303 Great Lakes DE-79
32536-52-0 251-087-9
Octabromodiphenyl oxide; Octabromodiphenyloxide. Flame retardant for ABS, nylon, polycarbonate, and polyester thermoplastic polymers; additive for unsat. polyesters and epoxy thermoset resins. *Great Lakes.*

13304 Great Lakes DE-83R
1163-19-5 214-604-9
Decabromo biphenyloxide; Decabromobiphenyl oxide. Halogenated flame retardant for thermoplastic, elastomeric, and thermoset polymer systems incl. HIPS, PBT, ABS, nylons, PP, LDPE, EPDM, unsat. polyesters, and epoxy resins. *Great Lakes.*

13305 Great Lakes FF 680
Bis (tribromophenoxy) ethane
Flame retardant for application where thermal stability at high processing temperatures is important; for thermoplastic and thermoset systems, light-stable applications. *Great Lakes.*

13306 Great Lakes FR-756
sodium tetrabromophthalate
Disodium salt of tetrabromophthalate; flame retardant. *Great Lakes.*

13307 Great Lakes PDBS-10
Poly(dibromostyrene)
Flame retardant. *Great Lakes.*

13308 Great Lakes PDBS-80
Poly(dibromostyrene)
Flame retardant. *Great Lakes.*

13309 Great Lakes PE-68
Bis (2,3-dibromopropyl ether) of tetrabromobisphenol A. Flame retardant used in PP, polyethylene, polybutylenes, and polyolefin copolymers; effective at low loading levels. *Great Lakes.*

13310 Great Lakes PH-73
118-79-6 9744 204-278-6
2,4,6-Tribromophenol
Reactive intermediate for phenol-based reactions; flame retardant, antifungal agent, or chemical intermediate. *Great Lakes.*

13311 Great Lakes PHE-65
3278-89-5 221-913-2
Tribromophenol allyl ether; Tribromo phenyl allyl ether. Flame retardant. *Great Lakes.*

13312 Great Lakes PHT4
632-79-1 211-185-4
$C_8Br_4O_3$
Tetrabromophthalic anhydride
4,5,6,7-tetrabromo-1,3-isobenzofurandione; 3,4,5,6-tetrabromophthalic anhydride. Flame retardant in production of unsaturated polyester resins and rigid PU polyols; co-hardener for epoxy resins; cost efficient additive for latex emulsions; derivatives used as flame retardants in diverse applications (wire coating, and wool, etc.). mp = 279-280°. *Great Lakes.*

13313 Great Lakes PHT4-Diol
20566-35-2 243-885-0
Tetrabromophthalatediol
Reactive intermediate used to produce flame retardant rigid urethane foam; can replace chlorinated polyols; for PU elastomers, coatings, adhesives, and fibers. *Great Lakes.*

13314 Great Lakes PO-64P
Poly-dibromophenylene oxide
Flame retardant which melts into most polymers to optimize physical properties; especially for crystalline polymers (polyesters, polyamides); enhances flow into thin wall sections; permits higher regrind loading levels. *Great Lakes.*

13315 Great Lakes SP75
Stabilized hexabromocyclododecane; flame retardant. *Great Lakes.*

13316 green acid
Green sulfonate
green sulfonic acid. Crude mixtures of sulfonic acids from refining petroleum sludge.

13317 Green Dot
A smokeless powder, double-base-type formulation minimizes charge weight and moisture absorption. *Hercules.* Discontinued.

13318 green earth
Veronese Green
Veronese Earth. A natural pigment. It contains ferrous iron, silica, magnesia, alumina, and lime; used as an absorbent for basic dyestuffs.

13319 green gold
Alloys of 75% gold, with from 11-25% silver, and 4-12% cadmium.

13320 Green Magic
Various granular fertilizer blends; fertilizers for lawns, gardens and flowers. *Horn's Crop Service Center.* Name unverified.

13321 green mordant
7772-98-7 8844 231-867-5
Sodium thiosulfate
Used as a mordant for fixing aniline greens on fiber.

13322 green ocher
Usually a mixture of silica, clay, and ferrous hydroxide; used as a base for cheap lakes.

13323 green oils
A fraction of oil obtained from shale by treating the distillate with sulfuric acid and sodium hydroxide and then again distilling.

13324 Green Sulphur
7704-34-9 9142 231-722-6
Sulfur
Sulfur powder and green dye; garden fungicide. *Vitax Ltd.*

13325 Green Tape
Co-fired ceramic multilayer materials. *DuPont UK.*

13326 green ultramarine
A pigment produced by heating kaolin, silica, sodium carbonate, sulfur, coal, and rosin, washing and grinding; used in water-colors, and is known as Lime green. When heated again with sulfur, it gives blue ultramarine.

13327 Green Up
2,4-D-dicamba
Mixture of 2,4-D and dicamba; used to control weeds in turf. *Synchemicals Ltd.*

13328 Green Up Autumn Liquid Lawn Feed
Liquid concentrate containing NPK 3:6:6 and 1% Fe; autumn feed for lawn areas. *Vitax Ltd.*

13329 Green Up Feed and Weed Plus Mosskiller
Dry powder containing NPK 8:4:4 plus 2,4-D, mecoprop and ferrous sulfate; combined fertilizer, weed and mosskiller. *Vitax Ltd.*

13330 Green Up Lawn Feed and Weed
Liquid concentrate containing NPK 14.5:3:3 and 2,4-D and dicamba; combined feed and weed for turf. *Vitax Ltd.*

13331 **Green Up Lawn Feedn Weed Plus Moss Killer**
dichlorophen-mecoprop-dichlorprop-dicamba-benazolin
Liquid concentrate containing NK 11:4 plus dichlorophen, mecoprop, dichlorprop, dicamba and benazolin; combined feed, weed and mosskiller. *Vitax Ltd.*

13332 **Green Up Lawn Spot Weedkiller**
2,4-D-mecoprop
Trigger spray pack containing 2,4-D [1702-17-6] and mecoprop [7085-19-0]; selective spot weedkiller for lawn areas. *Vitax Ltd.*

13333 **Green Up Liquid Lawn Feed**
Liquid concentrate containing NPK 17:3.5:3.5; liquid fertilizer for turf. *Vitax Ltd.*

13334 **Green Up Mossfree**
Soluble powder containing ferrous sulfate heptahydrate. *Vitax Ltd.*

13335 **Green Up Weedfree Lawn Weedkiller**
2,4-D-dicamba
Emulsifiable concentrate containing 2,4-D [1702-17-6] and dicamba [1918-00-9]; selective herbicide for use on turf. *Vitax Ltd.*

13336 **Green Up Weedfree Spot Weedkiller for Lawns**
2,4-D-dicamba
Aerosol containing 2,4-D [1702-17-6] and dicamba [1918-00-9]; selective spot weedkiller for lawns. *Vitax Ltd.*

13337 **green vitriol**
7720-78-7 4105 231-753-5
ferrous sulfate
For moss control.

13338 **green wood spirits**
acetone alcohol

13339 **Greenfly & Blackfly Killer**
Systemic insecticide. *Fisons plc, Horticultural Div.* Name unverified.

13340 **Greenkeeper**
Range of turf fertilizers. *Fisons plc, Horticultural Div.* Name unverified.

13341 **Greenkeeper Mosskiller**
Turf fertilizer with iron sulfate. *Fisons plc, Horticultural Div.* Name unverified.

13342 **Greenmaster Autumn**
10028-22-5 4079 233-072-9
Ferric sulfate
Used for moss control in turf. *Fisons plc.*

13343 **Greenmaster Extra**
MCPA-mecoprop
Granules containing MCPA and mecoprop; for control of weeds in amenity and roadside grass. *Fisons plc.*

13344 **Greenmaster Mosskiller**
10028-22-5 4079 233-072-9
Ferric sulfate
Used for moss control in turf. *Fisons plc.*

13345 **Greenol**
Liquid, iron-based plant nutrient. *Monsanto (Solaris).* Name unverified.

13346 **Green-up Mossfree**
7720-78-7 4105 231-753-5
Ferrous sulfate
Used for moss control in turf. *Synchemicals Ltd.*

13347 **Grefco**
A proprietary trade name for a chrome ore cement used as a refractory. No manufacturer.

13348 **Greggio**
7704-34-9 9142 231-722-6
Crude Sicilian sulfur
Fungicide.

13349 **Grenacher's alum carmine**
An aqueous solution containing 1-5% common or ammonia alum, boiled with 0.5-1% carmine, and filtered. A microscopic stain.

13350 **Grenacher's borax carmine**
A microscopic stain. It contains 3 g carmine, 4 g borax, and 100 ml distilled water. After dissolving by heat, 100 ml of 70% alcohol added, and the solution filtered.

13351 **grey acetate**
62-54-4 1683 200-540-9
calcium acetate
Crude acetate of lime, prepared with distilled pyroligneous acid. It contains from 80-82% calcium acetate, and 20% water.

13352 **grey antimony**
7440-36-0 733 231-146-5
antimony
Trigonal antimony obtained by allowing molten antimony to cool in a crucible.

13353 **grey cast iron**
A cast iron containing much of its carbon in the uncombined state. A typical one contains 94% iron, 3.5% carbon, and 2.5% silicon. It has a specific gravity of 7.0 and melts at 1230 C.

13354 **grey forge pig**
A pig iron usually containing less silicon than other grey irons.

13355 **grey gold**
An alloy of gold and iron, sometimes with silver. One alloy contains 86% gold, 8.5% silver, and 5.5% iron, and an other 83% gold and 17% iron.

13356 **grey mixture**
A mixture of 7 parts meal powder, with 100 parts saltpeter and sulfur, used in fireworks.

13357 **grey tin**
A form of tin obtained by exposing the metal to low temperatures. It reverts back to the ordinary form when heated.

13358 **Grid®**
For the agriculture industry. *DuPont UK.*

13359 **Grignard's reagent**
Magnesium reacts with alkyl and aryl halides, in the presence of ether, forming reactive compounds of the type RMgX. Used in organic synthesis.

13360 **Grilamid®**
Nylon 12 (polyamide 12)
Granules for injection molding and extrusion; for precision engineering, components for optical and medical appliances, electrical/electronic industry, sports equipment, automotive industry; extrusion types for manufacture of tubes, sausage skins, films, cable sheathing. *EMS-Chemie AG.*

13361 **Grilamid® ELY**
Elastomeric polyamide 12; used for sports equipment. *EMS-Chemie AG.*

13362 **Grilamid® ELY20NZ**
Thermoplastic nylon 12 elastomer; general-purpose elastomer for injection molding and extrusion; used in ski boots, trekking boots, and applications requiring high flexibility. *EMS-Am. Grilon.*

13363 **Grilamid® L16**
Nylon 12 resin; extrusion grade *EMS-Am. Grilon.*

13364 **Grilamid® L16G**
Nylon 12 resin, with internal lubricant; high-flow injection molding resin for thin-section and long-flow-path components. *EMS-Am. Grilon.* Name unverified.

13365 **Grilamid® L25**
Nylon 12 resin; general-purpose film extrusion grade without additives or stabilizers. *EMS-Am. Grilon.*

13366 **Grilamid® TR55**
Nylon 12 resin; transparent resin with superior chemical resistance, strength, stiffness, and toughness for fuel, air, or water filter bodies, domestic appliance parts, faucet and shower handles, electrical/electronic components, eyeglass frames, automotive parts; agricultural and food processing components, medical equipment, fiber optic cable sheathing, instrument gauge lenses. *EMS-Am. Grilon; EMS-Chemie AG.*

13367 **Grilamid® TR55LX**
Nylon 12 resin; high flow grade for injection molding and extrusion; used for eyeglass frames, etc. *EMS-Am. Grilon.*

13368 **Grilbond®**
Blocked isocyanates; bonding agents for polyester-reinforced rubber goods in tire and belt industry. *EMS-Chemie AG.*

13369 **Grilbond® PVC Bonding Agents**
Modified polyurethanes and polyamines; bonding agents in PVC plastisols for underbody protection in the automotive industry. *EMS-Chemie AG.*

13370 **Grilene Swiss Polyester**
Polyethylene terephthalate fibers; used for nonwovens and sieves. *EMS-Chemie AG.* Discontinued.

13371 **Grilesta®**
Saturated polyester resins in granular form; used for polyester/epoxy and polyester TGIC powder coatings, particularly for short curing time or low curing temperatures; special products for transparent systems. *EMS-Chemie AG.*

13372 **Grilesta® P 7205**
Hybrid polyester resin; 50/50 hybrid with excellent flow, normal curing. *EMS-Am. Grilon.* Name unverified.

13373 **Grilesta® P 7304**
Polyester resin; for outdoor durable systems; 93/7 TGIC-PES, normal curing. *EMS-Am. Grilon.* Name unverified.

13374 **Grilesta® P 7401**
Hybrid polyester resin; for powder coatings; 70/30-80/20 hybrid, TMA-free, fast curing. *EMS-Am. Grilon.* Name unverified.

13375 **Grillocin HY-77**
Zinc ricinoleate-triethanolamine-zinc rosinate-isostearic acid-dipropylene

glycol-Sodium lactate-abietic acid-tocopherol. Absorbs malodors from solutions and surfaces. *RITA.*

13376 Grilloten LSE87
25339-99-5 246-873-3
Sucrose laurate
Ryoto Sugar Ester LWA-1570. Oil-water emulsifier, solubilizer, foam booster, counter-irritant; stable over a wide pH range. *RITA.*

13377 Grilloten LSE87K
Sucrose cocoate
Oil-water emulsifier, solubilizer, foam booster, counter-irritant; stable over a wide pH range. *RITA.*

13378 Grilloten PSE141G
25168-73-4 246-705-9
Sucrose stearate
Saccharose monostearate; Crodesta DKS F110; Crodesta F-160; Ryoto Sugar Ester S-1170; Sucro Ester 7; Sucro Ester 15. Oil-water emulsifier, solubilizer; stable over a wide pH range. *RITA; Croda Inc.; Mitsubishi Kasei; Gattefosse SA.*

13379 Grilloten ZT40, ZT80, PSE 141G, LSE 87, LSE 87K
Sucrose ricinoleate
Mild solvent-free surfactants; nontoxic, nonsensitizing and anti-irritant; both water and oil soluble; moisturizing and emulsifying properties do not affect formulation. *RITA.* Name unverified.

13380 Grilon®
25038-54-4 6832
Nylon 6 (polyamide 6), granular form; used for injection molding, extrusion and blow molding for automotive, machinery, electrical/electronic, building industries, medical instruments, packaging films; also as staple fibers, monofilaments. *EMS-Chemie AG.*

13381 Grilon® A23GM
25038-54-4 6832
Nylon 6 resin, internally lubricated; general-purpose injection molding resin for thin-section or long-flow-path components. *EMS-Am. Grilon.*

13382 Grilon® BT
Polyamide alloy, granular form; used for extrusion and injection molding of technical parts for machinery, mining, electrical/electronic, automotive and building industries. *EMS-Chemie AG.*

13383 Grilon® BT40Z
25038-54-4 6832
Nylon 6 resin; impact modified grade for injection molding, extrusion, and cast and blown film applications. *EMS-Am. Grilon.*

13384 Grilon® C
Copolyamide, granular form; used for coextrusion of laminated blown films with high gloss and transparency; also for flexible cable sheathing and monofilament, masterbatch base material. *EMS-Chemie AG.*

13385 Grilon® CA6E
Nylon 6/12 copolymer resin; for shrink wrap film applications. *EMS-Am. Grilon.*

13386 Grilon® CF6S
Nylon 6/12 copolymer resin; film grade intended for shrink wrapping applications, including wrapping of food. *EMS-Am. Grilon.*

13387 Grilon® ELX
Elastomeric polyamide 6, granular form; used in ski, mountaineering and hiking boots, hydraulic and pneumatic tubing. *EMS-Chemie AG.*

13388 Grilon® ELX23NZ
Thermoplastic nylon 6 elastomer
General-purpose elastomer for injection molding and extrusion; used where cold impact resistance and high flexibility are needed. *EMS-Am. Grilon.*

13389 Grilon® PV-15H
25038-54-4 6832
Nylon 6 resin, 15% glass fiber reinforced, stabilized; injection-molding grade offering intermediate rigidity and strength characteristics. *EMS-Am. Grilon.*

13390 Grilon® PVN-15H
25038-54-4 6832
Nylon 6 resin, 15% glass fiber reinforced, stabilized; impact-modified injection-molding grade offering high toughness. *EMS-Am. Grilon.*

13391 Grilon® R47HW
25038-54-4 6832
Nylon 6 resin, plasticized, stabilized; higher molecular weight resin for injection molding and extrusion; used for tubes, rods, and profiles, flexible tubing for automotive and industrial applications, and for the manufacture of mandrels. *EMS-Am. Grilon.*

13392 Grilon® T
Nylon 6/6, granular form; used for injection molding in electrical/electronic, automotive, building and machinery industries. *EMS-Chemie AG.*

13393 Grilon® T300GM
32131-17-2
Nylon 6/6; unreinforced injection molding resin offering fast cycling, easy processability, excellent chemical, solvent, and fuel resistance, wear and abrasion performance, electrical insulating properties; used for electrical/electronic components, automotive connectors, furniture and window components, buttons, switches, housing for household appliances and power tool components. *EMS-Am. Grilon.*

13394 Grilon® TV-15H
32131-17-2
Nylon 6/6, 15% glass fiber-reinforced; heat-stabilized injection molding resin. *EMS-Am. Grilon.*

13395 Grilon® XE3106
Thermoplastic nylon 6 elastomer; very low hardness, highly flexible elastomer for injection molding and extrusion. *EMS-Am. Grilon.*

13396 Grilon® XE3222
Nylon 6-6/9 copolymer resin; for coextrusion for blown and cast film with good clarity and flexibility. *EMS-Am. Grilon.*

13397 Grilon® XE3303
Nylon 6/6-6/10 copolymer resin; for food packaging and medical applications. *EMS-Am. Grilon.*

13398 Grilonit®
Epoxy resins and hardeners, solid and liquid, unmodified and modified; used in civil engineering, general surface protection, flooring systems, mortars, sealers, priming coats, crack injection systems, structural adhesives and coatings in the building industry. *EMS-Chemie AG.*

13399 Grilonit® Reactive Diluents
Liquid glycidyl ethers, mono and difunctional; auxiliary agents to facilitate processing, enhance wetting properties and increase flexibility in epoxy resin systems. *EMS-Chemie AG.*

13400 Grilpet®
Polyethylene terephthalate polyester, granular form; used for extrusion and injection molding of technical parts for electrical/electronic and automotive industries, sports and leisure articles, domestic appliances. *EMS-Chemie AG.*

13401 Grilpet® EV-30
25038-59-9 7730
PET polyester resin, 30% glass fiber-reinforced; fast cycling resin for injection molding applications; produces moldings with high dimensional stability, low moisture absorption, high rigidity; suitable for electrical/electronic, automotive, or structural applications. *EMS-Am. Grilon.* Name unverified.

13402 Grilpet® XE3060
26062-94-2
Polybutylene terephthalate
Celenex® 1300A, 1600A, 2000K, 3310, 5300, J600; Celstran® PBTG30-01-4; Crastine®; Electrafil®; EMI-X®; Grilpet®XE3060; Mitsubishi Kasei PBT 5008, PBT 5010F1; PBT-1100, 1100G15, 1300; PDX-84369; Pocan®; PS-30GM/000; RTP 1001; Stat-Kon® W; Thermocomp® WC-1006; Ultradur® B 2550, B 4300 G10, B 4500; Valox®; Vandar® 2100; Vybex 22008 BKFR, 22028 BKFR. PBT polyester resin; high molecular weight extrudable resin with good mechanical and thermal properties, outstanding hydrolysis resistance; ideal for the outer layer of fiber optic loose buffer dual-tubes. EMS-Am. Grilon; Mitsubishi Kasei; Polymer Composites; Ciba Geigy GmbH; Akzo Engineering Plastics; LNP; ICI Chemicals & Polymers Ltd.; Bay Resins; Bayer AG; Bayer plc.; Miles; Compounding Tech.; GE Plastics; Hoechst Celanese; Ferro/Engineering Thermopl. Name unverified.

13403 Griltex®
Copolyamide and copolyester granules and powders; for coating of fusible interlinings; excellent resistance to laundering and dry-cleaning; short fusing cycles at moderate temperatures; manufacture of hot-melt adhesive films, monofilaments; structural hot melts for automotive, textile and machinery industries. *EMS-Chemie AG.*

13404 Grime Go
Solvent based hand cleaner liquid and paste. *Momar Industrial Services Ltd.*

13405 Grimm aluminum solder
An alloy of 69% tin, 29% lead, 1.5% zinc, and 0.75% silver.

13406 grinding oils
Drying oils such as linseed, etc.; used for grinding pigments for paints.

13407 grindstone
A sandstone consisting almost entirely of quartz.

13408 Grindtek AML 60
Acetylated palm kernel glycerides; lubricant and plasticizer for plastics and coatings; cosolvent for polar additives in low polarity systems. *Grindsted Prods.* Name unverified.

13409 Grindtek FAL 1
25383-99-7 246-929-7
Sodium stearoyl lactylate
Oil-water emulsifier. *Grindsted Prods.* Name unverified.

13410 Grindtek ML 90
142-18-7 205-526-6
$C_{15}H_{30}O_4$

Glyceryl laurate
2,3-Dihydroxypropyl dodecanoate. Component in water-oil and oil-water creams, lubricant, antistat, antifogging agent in plastics; antimicrobial effects reported. *Grindsted Prods.*

13411 Grindtek MM 90
Glyceryl myristate
Component in water-oil and oil-water creams, lubricant, antistat, antifogging agent in plastics; antimicrobial effects reported. *Grindsted Prods.*

13412 Grindtek MSP 32-6
Glyceryl stearate SE
Component in water-oil and oil-water creams, lubricant, antistat, antifogging agent in plastics; antimicrobial effects reported. *Grindsted Prods.* Name unverified.

13413 Grindtek PGE 25
9007-48-1
Polyglyceryl-3 oleate
Oil-water emulsifier, antifogging agent for plastics. *Grindsted Prods.* Name unverified.

13414 Grindtek PGMS 90
1323-39-3 215-354-3
Propylene glycol stearate
Water-oil emulsifier. *Grindsted Prods.* Name unverified.

13415 Grindtek SMS
1338-41-6 8872 215-664-9
Sorbitan stearate
Water-oil and oil-water emulsifier. *Grindsted Prods.* Name unverified.

13416 Grindtek STS
26658-19-5 8872 247-891-4
Sorbitan tristearate
Water-oil and oil-water emulsifier. *Grindsted Prods.* Name unverified.

13417 Griplet®
For the electrical industry. *DuPont UK.*

13418 Grisounites
Coal mine explosives. They contain nitroglycerin, collodion cotton, ammonium nitrate, and sometimes magnesium sulfate in the place of ammonium nitrate. Others contain ammonium nitrate and nitro- naphthalene.

13419 Grisoutite
An explosive consisting of 53% nitroglycerin, 14.5% kieselguhr, and 32.5% magnesium sulfate.

13420 Grivory®
Partial aromatic polyamides, granular form; used for injection molding, blow molding and extrusion for automotive, leisure and sports equipment, electrical and telecommunications technology, packaging industry. *EMS-Chemie AG.*

13421 Grivory® XE3215
Amorphous, unreinforced engineering thermoplastic; impact-resistant resin for injection molding, extrusion, and blow molding; features high rigidity, dimensional stability, gloss, and easy colorability. *EMS-Am. Grilon.*

13422 Grivory® XE3290
Amorphous nylon, 35% glass fiber-reinforced; for injection molding applications; good dimensional stability. *EMS-Am. Grilon.*

13423 Grodex
Seed germination indicator. *May & Baker Ltd.* Name unverified.

13424 GRO-HY
Nitrogen, phosphorus, potash plus trace elements as a slow-release fertilizer tablet; fertilizer for trees, shrubs and bushes. *Envhy Ltd.* Unverified.

13425 Gross solder
An alloy of tin, zinc, aluminum, lead and phosphorus.

13426 Grossmann reagent
591-01-5 3143 209-697-8
$C_4H_{14}N_8O_6S$
(aminoiminomethyl)urea sulfate (2:1)
Guanylurea Sulfate; Urea, (aminoiminomethyl)-, sulfate (2:1); N-Guanylurea sulfate. An ammoniacal solution of dicyandiamidine sulfate, a reagent for nickel, cobalt. mp = 190°; d = 1.61; soluble in H_2O (5-30%), EtOH, insoluble in organic solvents.

13427 Grossman's alloy
An alloy containing 87% aluminum, 8% copper, and 5% tin.

13428 Grotan
Range of biocides for the oil and metal working industries. *Sterlin-Winthrop Group Ltd.*

13429 Groundhog
aminotriazole-diquat-paraquat-simazine
Used for total weed control in non crop areas. *ICI Agrochemicals Professional Products; ICI Chem. & Polymers Ltd.*

13430 ground-nut oil
A non-drying oil obtained from the groundnut. It consists chiefly of the glycerides of oleic and linoleic acids. It is edible, and when hydrogenated is used in the manufacture of margarine. Lower qualities are employed in soap-making.

13431 grout
A mixture of cement and water. Sometimes including sand.

13432 Groutcide 75
1897-45-6 2219 217-588-1
Chlorothalonil
Fungicide for Portland cement grout. *Henkel.*

13433 Growmore
Granular fertilizer containing NPK 7:7:7; general-purpose fertilizer. *Vitax Ltd.*

13434 Gruber solder
An alloy of 60% tin, 25% zinc, 2% aluminum, 10% copper, and 3% cadmium.

13435 Grudekok
Lignite char
Lignite from which about 30-40% water has been expelled; used in the manufacture of briquettes for fuel.

13436 gru-gru fat
The fat from the seeds of *Acrocomia sclerocarpa*, of the palm family.

13437 GTI
Lubricants for hot glass handling in the glass container industry. *Darmex Corp.*

13438 guaiacum resin
A resin obtained from the wood of *Guaiacum officinale* and *Guaiacum sanctum.*

13439 Guaic
Guaiacol resin.

13440 Gualol
C_5H_8O
Tiglic aldehyde.

13441 Guanidine nitrate
506-93-4 4591 208-060-1
$CH_6N_4O_3$
guanidinium nitrate
Guanidine, mononitrate. Used in manufacture of explosives, disinfectants, photographic chemicals. mp = 213-215°; soluble in H_2O (13 g/100 ml), EtOH; LD_{50} (rat orl) = 730 mg/kg. *Dajac Labs; R.W. Greeff; Spectrum Chem. Mfg.*

13442 Guanidine thiocyanate
593-84-0 209-812-1
$C_2H_6N_4S$
guanidine hydrothiocyanate
guanidinium thiocyanate; guanidinium rhodanide; Thiocyanic acid, compd. with guanidine (1:1); Guanidine isothiocyanate; Guanidinium Isothiocyanate. Potent protein denaturant used in isolation of intact DNA, RNA. mp= 117°. *Dajac Labs; Eastman; Fulka; U.S. Biochemical.*

13443 guanine
73-40-5 4593 200-799-8
$C_5H_5N_5O$
2-amino-1,7-dihydro-6H-purin-6-one
2-aminohypoxanthine; 2-Amino-6-hydroxypurine. Naturally occuring purine; used in biochemical research and in cosmetics. dec above 360°; λ_m = 246, 275 nm (ε = 10700, 8100); soluble in ammonia water, aqueous KOH solns, dilute acids; sparingly soluble in alcohol, ether; almost insoluble in water. *R.W. Greeff; Henley; Janssen Chimica; Mearl; Penta Mfg.*

13444 guano
Bird Manure
Consists of deposits of excrement and skeletons of birds and animals; a fertilizer rich in phosphorus and nitrogen.

13445 guar gum
9000-30-0 4601 232-536-8
guar flour
gum cyamopsis; Guaran. Natural polysaccharide derived from the ground endosperms of *Cyamopsis tetragonolobus*; paper coating, cosmetics, pharmaceuticals; thickener, emulsifier in food products; as protective colloid, stabilizer; binding agent in tablets; flocculant in mining industry, coagulant aid in water treatment. Soluble in cold H_2O; $[\alpha]_D^{25}$ = 53°. *Aqualon; Hercules; Rhône-Poulenc; Stan Chem Int'l. Ltd.*

13446 Guara
The ground fruits of a species of *Caesalpinia*, from Central and South America; used as a tanning material.

13447 Guardar
Acetic acid 2%, petroleum distillates 1%, polybutenes 1%, inert ingredients 96%
A liquid adjuvent for reducing pesticide use and for enhancing the sticking of foliar-applied materials to leaf surfaces and seeds. *SN Corp/Appropriate Technology Ltd.*

13448 Guardian
Chemicals and equipment for treating water, effluent and sewage. *ICI Chem & Polymers Ltd.*

13449 Guardsep
Chemicals and equipment for treating water, effluent and sewage. *ICI Chem & Polymers Ltd.*

13450 Guardsman
7784-25-0 335 232-055-3
aluminum ammonium sulfate
Aluminum ammonium sulfate; bird and animal repellent. *Sphere Laboratories (London) Ltd.*

13451 Guardsman
Chemicals and equipment for treating water, effluent and sewage. *ICI Chem & Polymers Ltd.*

13452 Guartec CAP
Water-soluble polymer for strength enhancement in paper. *Hercules.* Discontinued.

13453 Guartec CIP
Water-soluble polymer for strength enhancement in paper. *Hercules.* Discontinued.

13454 Guayale
Durango
A rubber obtained from *Parthenium argentatum.*

13455 Guettier metal
An alloy of 62% copper, 32% zinc, and 6% tin.

13456 Guignet's Green
1308-14-1 2281 215-158-8
chromic hydroxide
Chromium hydroxide or hydrated chromium oxide. *Reading, Green & Marvell Ltd.*

13457 Guillaume Alloy
A proprietary trade name for an alloy of 66% iron and 34% nickel; has a low coefficient of expansion. No manufacturer.

13458 Guillaume Metal
An alloy of 64% copper and 36% bismuth. No manufacturer.

13459 Gulf Lite
Charcoal starter, patio torch fuel. *Monsanto (Solaris).* Name unverified.

13460 Gulf Lubcote
Lube oil and anticorrosive. *Monsanto (Solaris).* Name unverified.

13461 Gulf No-Rust
Lube oil and anticorrosive. *Monsanto (Solaris).* Name unverified.

13462 Gulfad-C
Chemical additive for cements used in oil and gas wells. *Monsanto (Solaris).* Name unverified.

13463 Gulfco
Lube oils and greases. *Monsanto (Solaris).* Name unverified.

13464 Gulfcrest
Greases and motor fuel. *Monsanto (Solaris).* Name unverified.

13465 Gulfcrown
Greases. *Monsanto (Solaris).* Name unverified.

13466 Gulfcut
Cutting fluids. *Monsanto (Solaris).* Name unverified.

13467 Gulfgem
Greases. *Monsanto (Solaris).* Name unverified.

13468 Gulfknit
Lube oils. *Monsanto (Solaris).* Name unverified.

13469 Gulfleet
Lube. *Monsanto (Solaris).* Name unverified.

13470 Gulflex
Greases. *Monsanto (Solaris).* Name unverified.

13471 Gulflube
Motor oils. *Monsanto (Solaris).* Name unverified.

13472 Gulfpride
Motor oils. *Monsanto (Solaris).* Name unverified.

13473 Gulfspin
Lube oil. *Monsanto (Solaris).* Name unverified.

13474 Gulftene
Alpha olefins. *Monsanto (Solaris).* Name unverified.

13475 Gulftex
Lubes. *Monsanto (Solaris).* Name unverified.

13476 Gulftow
Lubes. *Monsanto (Solaris).* Name unverified.

13477 Gulftronic
Electrostatic precipitator. *Monsanto (Solaris).* Name unverified.

13478 Gulfwax
Paraffin wax. *Monsanto (Solaris).*

13479 gum Amritsar
An acacia gum from *Acacia modesta*; used in calico printing.

13480 gum arabic
9000-01-5 232-519-5
acacia gum
arabic gum; Arabin; calcium gummate; Acacia; Gum acacia; acacia syrup; Indian gum; Acacia arabica gum africa; Senegal gum; Gum senegal; Acacia senegal l. willd gum west africa. Gum acacia from *Acacia senegal* and other species; used in pharmaceuticals, adhesives, inks, textile printing, cosmetics, thickening agent, colloidal stabilizer, food products, binding agent in tablets, emulsifier. d = 1.35; soluble in H_2O (5-10 g/100 ml). *Dr. Madis Labs; Meer; Penta Mfg.; TIC Gums.*

13481 gum benguela
A semi-fossil copal used in varnishes.

13482 gum benjamin
9000-05-9 4608 232-523-7
gum benzoin
Resin benzoin; resin benjamin;. Balsamic resin from *Styrax benzoin*; for fumigating pastilles; contains free and combined benzoic and cinnamic acid. Used in perfumery and cosmetics and as a topical protectat and antiseptic. Also a source of aromatic carboxylic acids.

13483 gum catechu
4371
Gambir; Catechu; pale catechu; Gambir catechu; terra japonica. Extract from leaves and twigs of *Uncaria gambier.* Contains catechol [154-23-4], catechutannic acid, quercetin [117-39-5], catechu-red; gambir-fluorescein, oil and wax.

13484 gum Cowrie
2872
Gum dammar; gum damar; Damar; Dammar; resin damar. Exudate from *Shorea, Dipterocarpaceae.* Used in plasters, varnishes and lacquers.

13485 Gum D
A gelatin dynamite. It contains 69.5% nitroglycerin.

13486 Gum E
A gelatin dynamite. It contains 49% nitroglycerin.

13487 gum Lini
A gum made from linseed by treatment with water, then treating the mass with 90% alcohol. The gum is soluble in water and is used as a substitute for gum arabic.

13488 Gum MB
A gelatin dynamite. It contains 74% nitroglycerin.

13489 gum thus
6969
French pine resin
Common frankincense, the crude turpentine from French pine-trees.

13490 gum tragacanth
9000-65-1 4609 232-552-5
A gum obtained from shrubs of the *Astragalus* family. Used in pharmaceutical compounding and dispensing.

13491 Gun metal
An alloy containing from 89-91% copper, 8-11% tin, and 1-2% zinc. The alloy, containing 90% copper, 8% tin, and 2% zinc, has a specific gravity of 8.8, and melts at 1010°.

13492 guncotton
nitro-cotton
pyroxylin; collodion cotton. Nitro-cellulose, an explosive made by acting upon cotton with nitric and sulfuric acids.

13493 Gunner
52756-22-6 258-154-1
$C_{19}H_{19}ClFNO_3$
1-methylethyl-N-benzoyl-N-(3-chloro-4-fluorophenyl)-DL-alanine
flamprop-isopropyl; Flamprop-M-isopropyl; Suffix BW; WL 29762; Barnon; Flufenprop-isopropyl; Isopropyl N-benzoyl-N-(3-chloro-4-fluorophenyl)alanine. Emulsifiable concentrate of 200 g flamprop-M-isopropyl per liter; used for control of wild oats in cereal crops. *Quadrangle Agrochemicals.*

13494 Gunning's reagent
A 10% iodine solution in alcohol; used for the detection of acetone in urine.

13495 gunpowder
black powder
A mixture of saltpeter. carbon, and sulfur, in varying proportions. An average material contains 75% potassium nitrate, 10% sulfur, and 15% carbon. An explosive.

13496 Guntapite
A monolithic refractory material used for lining and repairing linings of steelmaking vessels. *Pfizer International.* Discontinued.

13497 Gurjun balsam or oil
wood oil
The oleo-resin from the stems of *Dipterocarpus* species.

13498 Gurley's metal
An alloy of 86.5% copper, 5.4% zinc, 5.4% tin, and 2.7% lead.

13499 Gurney's bronze
An alloy of 76% copper, 15% lead, and 9% tin.

13500 Gurr
Biological stains and reagents. *BDH Chemicals Ltd.*

13501 Gusathion®
86-50-0 944 201-676-1
Azinphos-methyl
Insecticide with broad spectrum of activity, for control of biting insects, especialy lepidopterous larvae, and sucking pests such as aphids, suckers, scale insects, etc. on a wide range of crops in all climatic zones; side-effect on mites. *Bayer AG.*

13502 Gusathion® A
2642-71-9 220-147-6
$C_{11}H_{16}N_3O_3PS_2$
O,O-diethyl S-[(4-oxo-1,2,3-benzotriazin-3(4H)-yl) methyl] phosphorodithioate
Azinphos-ethyl; Azinos; Ethyl Guthion; Acifon; Bionex; Cotnion-ethyl; Crysthion. Insecticide with broad spectrum of activity against biting pests such as caterpillars, beetles and their larvae, as well as sucking pests such as aphids, thrips, leafhoppers, etc. on a wide range of crops; side-effect on mites. *Bayer AG.*

13503 Gusathion® MS
azinphos-methyl-demeton-S-methyl sulfone
A wettable powder containing 25% w/w Azinphos-methyl and 7.5% w/w demeton-S-methyl sulfone; used to control a wide range of biting and sucking pests on agricultural and horticultural crops. *Bayer plc.* Discontinued.

13504 Gusto
13684-63-4 7384 237-199-0
phenmedipham
Emulsifiable concentrate containing 118 g/l phenmedipham; for weed control for beet crops. *Farm Protection Ltd.*

13505 Guthion®
86-50-0 944 201-676-1
azinphos-methyl
Insecticide with broad spectrum of activity, for control of biting insects, especialy lepidopterous larvae, and sucking pests such as aphids, suckers, scale insects, etc. on a wide range of crops in all climatic zones; side-effect on mites. *Bayer AG.*

13506 Guthrie's eutectic alloy
An alloy of 47% bismuth, 20% tin, 20% lead, and 13% cadmium.

13507 gutta-percha
The coagulated latex of species of *Palaquium and Payena*, of Malaysia, Sumatra, and Borneo. The material is plastic when hot and can be molded. It was formerly used extensively for electrical insulaters, chemical containers etc.

13508 gutta-shea
Karite gum
A product resembling gutta-percha obtained from an African tree, *Bassia Parkii.* The gutta is separated from the fat (Shea butter).

13509 Gutta-Siak
A low grade gutta-percha gum from *Payena Leerii* and other *Payena* species, of Siak, Sumatra. It is a mixed product with a high resin content.

13510 Gutta-sundik
A gutta-percha from *Payena* species.

13511 gutta-susu
Assam white
gutta-gerip; gutta-Singarip; Borneo rubber. A wild rubber mainly obtained from *Willoughbeia firma* of Borneo.

13512 guvacine
498-96-4 4612
$C_6H_9NO_2$
1,2,5,6-tetrahydropyridine-3-carboxylic acid
1,2,5,6-tetrahydronicotinic acid. A pyridine alkaloid derived from betel nuts. Its methyl ester is also called guvacine. mp = 295° (dec); soluble in H_2O, insoluble in organic solvents.

13513 Gyne-Lotrimin
23593-75-1 2478 245-764-8
Clotrimazole
Antifungal. *Schering-Plough Health Care Products Inc.* Discontinued.

13514 gynocardic acid
chaulmoogric acid-hydnocarpic acid
A term used for the acids contained in the oil expressed from the seeds of *Gynocardia odorata.* The main constituents are chaulmoogric acid and hydnocarpic acid.

13515 Gynol
9016-45-9 6772
Nonoxynol 9
Spermaticide. *Ortho Pharmaceutical Corp.*

13516 gypsite
A deposit consisting of small grains of gypsum, disseminated through an earthy mass; used for the production of wall plastics.

13517 Gypsona
Plaster of Paris bandage. Used in medicine. *Smith & Nephew Pharmaceuticals Ltd.* Name unverified.

13518 Gypsum
13397-24-5 1753
$CaSO_4 \cdot 2H_2O$
calcium sulfate
alabaster; selenite; Light spar; satin spar; Lenzit; annaline; terra alba; satinite. A mineral used in plasters, in Portland cement, paints, and as a filler for paper and cotton.

13519 Gypsum-F
Source of calcium and sulfur in fertilizers for use on golf courses, ball parks, cemeteries, nurseries, industrial lawns, home lawns, and other turf applications; increases permeability of soils to lower sodium content. *W A Cleary.*

13520 Gyrane
2-butyl-4,6-dimethyldihydropyran
Flavorant and perfume. *Quest Int'l. UK Ltd.*

13521 H Chrome Green 105
chromium hydroxide-bismuth oxychloride oxide
Chromium hydroxide [12001-99-9], bismuthoxychloride [7787-59-9]; inorganic colorant. *Presperse.* Discontinued.

13522 H.A. Solvent
622-45-7 210-736-6
$C_8H_{14}O_2$
Cyclohexyl acetate
A proprietary solvent for cellulose acetate and nitrate, rosin, rubber, oils, and metallic resinates. bp = 172-173°; d = 0.966. No manufacturer.

13523 H.M.T.D.
Hexamethylenetriperoxydiamine
A detonating explosive.

13524 H.T
7758-23-8 1740 231-837-1
calcium phosphate, monobasic
Monocalcium phosphate monohydrate; leavening agent. *Monsanto Co.* Name unverified.

13525 H.T.S
A salt mixture used as a heat transfer medium. It contains approximately 40% sodium nitrite, 7% sodium nitrate, and 50% potassium nitrate by weight. The temperature limits are 290-1000°F.

13526 H₂old EP-1
9003-39-8 7879 294-352-4
Povidone
Vinylpyrrolidone terpolymer. *ISP.*

13527 H-30
1344-28-1 369 215-691-6
Hydrated aluminum oxide
For production of petroleum hydrotreating and cracking catalysts, aluminum sulfate, aluminum chloride, sodium aluminate, aluminum phosphate, TiO_2, zeolites for detergent industry; chemical heat sink (retards burning of host compounds). *La Roche Chem.*

13528 H7250
1009-93-4 213-773-6
Hexamethylcyclotrisilazane
Difunctional blocking agent; reagent for cyclosilylation. *Hüls Am.*

13529 H7301
999-97-3 4725 213-668-5
Hexamethyldisilazane
Adhesion promoter for photoresists including positive and negative photoresists on SiO_2 substrates; highly purified for electronic applications. *Hüls Am.*

13530 H7310
107-46-0 203-492-7
Hexamethyldisiloxane
As cleaning, polishing and damping media; offers low toxicity, inertness. *Hüls Am.*

13531 HA 819
Polyoxyethylene nonyl phenol
A water soluble material of medium chain length; used as a detergent. *Honeywell Atlas.* Unverified.

13532 H-acid
1,8-Diaminonaphthol-3,6-disulfonic acid
Used as an azo dye intermediate.

13533 Hadranol
1702-17-6 2462 216-935-4
Clopyralid
Plant growth regulator. *Makhteshim Chemical Works Ltd.*

13534 Hagafilm, Hagatreat
Water treatment chemicals. *Albright & Wilson Ltd., Phosphates & Speciality Business.* Name unverified.

13535 Hager's reagent
88-89-1 7562 201-865-9
picric acid
Picric acid (1 gram) dissolved in 100 ml water.

13536 Hagevap
Treatment for sea water evaporators. *Albright & Wilson Ltd.*

13537 Hahnmann's mercury
15829-53-5 5955 239-934-0
Hg_2O
mercurous oxide
Black oxide of mercury.

13538 Haine's solution
Copper sulfate 8.314 g, is dissolved in 400 ml water, 40 ml glycerol, and 500 ml of 5% potassium hydroxide added. A test for sugar.

13539 Hair Complex 20/70n
Placenta extract, B vitamins, sulfurcontaining amino acids in water-alcohol medium; aqueous-alcoholic lotions for regenerative hair care, oily scalps and dandruff. *Dr. Kurt Richter; Henkel/Cospha.*

13540 Hair Complex Aquosum
Herbs and B vitamins in water-alcohol medium; treatment of scalps with dandruff or greasiness; general hair protection. *Dr. Kurt Richter; Henkel/Cospha.*

13541 Halar E-CTFE Film
Copolymer of ethylene and chlorotrifluoroethylene; useful as release films. *AlliedSignal Inc.*

13542 Halar® 300
ethylene/chlorotrifluoroethylene copolymer (ECTFE). ECTFE; melt processable fluoropolymer offering excellent chemical resistance, good electrical properties, broad use temperature range, flame retardance, abrasion resistant, and resistant. to cobalt radiation; suitable for extrusion, injection molding, blow molding, rotomolding, and for application by fluidized bed or electrostatic coating processes; 300 grade has intermediate melt viscosity, for compression and blow molding, extrusion (wire coating, tubing and film). *Ausimont; Fluorocarbon Co. Ltd.*

13543 Halbase 10
12202-17-4 235-380-9
Lead sulfate, tribasic; PVC stabilizer. *Halstab.*

13544 Halberland metal
A brass containing 87% copper and 13% zinc.

13545 Halcarb 20
1319-46-6 11, 1016 215-290-6
CO_3Pb
lead carbonate
Carbonic acid, lead(2+) salt (1:1); Lead (II) carbonate. Lead carbonate, basic; PVC stabilizer. *Halstab.*

13546 half-stuff
Refined wood cellulose obtained as sulfite pulp and in the form of thick sheets; used either alone or mixed with esparto pulp in the manufacture of paper.

13547 Haliborange
Orange flavored and colored chewable sugar coated tablets containing vitamin A 750 ug, vitamin D 5 ug, Vitamin C 25 mg; used as a vitamin supplement for adults and children. *Evans Medical.* Name unverified.

13548 Hallco® C-491
25395-31-7 3005 246-941-2
Glyceryl diacetate
C.P. Hall.

13549 Hallco® C-918
26446-35-5 6325 247-704-6
Glyceryl acetate
C.P. Hall.

13550 Hallcomid®
Series of dimethyl amides; grinding aids, dispersing agents, insecticides, bio-agents. *C P Hall.* Name unverified.

13551 Hallcomid® M-18-OL
N,N-Dimethyl oleamide
Solubilizer, solvent, dispersant, wetting agent. *C.P. Hall.*

13552 Hallcomid® M-8-10
N,N-Dimethyl caprylamidecapramide
Mutual solvent for polar and nonpolar ingredient. in cosmetics, cleaners, corrosion inhibitor. *C.P. Hall.*

13553 Hallcote®
Rubber lubricants. *C.P. Hall.*

13554 Hallcote® 573
471-34-1 1697 207-439-9
Calcium carbonate
Rubber dips and coatings. *C.P. Hall.*

13555 Hallcote® CSD
1592-23-0 1750 216-472-8
calcium stearate
Calcium stearate-based slab dip; rubber dips and coatings. *C.P. Hall.*

13556 Hallcote® ZS 5050
557-05-1 10292 209-151-9
zinc stearate
Zinc stearate-based slab dip; Rubber dips and coatings. *C.P. Hall.*

13557 Hallmark
91465-08-6
λ-cyhalothrin
Emulsifiable concentrate of 50 g λ-cyhalothrin per liter; a pyrethroid insecticide. *ICI Agrochemicals.*

13558 Hal-Lub-D
12578-12-0 5442 214-005-2
lead stearate
Lead stearate, dibasic; PVC stabilizer, lubricant. *Halstab.*

13559 Hal-Lub-N
1072-35-1 5442 214-005-2
lead stearate
Lead stearate, normal; PVC stabilizer, lubricant. *Halstab.*

13560 Halobrom
126-06-7 204-766-9
$C_5H_6BrClN_2O_2$
1-Bromo-3-chloro-5,5-dimethylhydantoin
BCDMH; Bromo-1-chloro-5,5-dimethyl-2,4-imidazolinedione; Bromo-1-chloro-5,5-dimethylhydantoin; Hydantoin, 3-bromo-1-chloro-5,5-dimethyl-; Imidazolidinedione, 3-bromo-1-chloro-5,5-dimethyl-. Broad spectrum biocide for control of algae, bacterial and fungal slimes in swimming pools and industrial water systems; nonflammable. *Dead Sea Bromine.*

13561 Haloflex
Vinyl acrylic-based polymer resin products. *ICI Resins BV.*

13562 halofuginone hydrobromide
64924-67-0 4627
$C_{16}H_{17}BrClN_3O_3 \cdot HBr$
trans-(±)-7-bromo-6-chloro-3-[3-(3-hydroxy-2-piperidinyl)-2-oxopropyl]-4(3H)-quniazolinone hydrobromide
RU-19110; Stenorol. Antiprotozoal (coccidiostat). mp = 247° (dec).

13563 Halon
A proprietary polytetrafluoroethylene. *AlliedSignal Inc.* Discontinued.

13564 Halowax 1014
1335-87-1 215-641-3
$C_{10}H_2Cl_6$
hexachlornaphthalene
Halowax 1014. A proprietary trade name for hexachlornaphthalene. Used to insulate electrical equipment, in flameproofing and water proofing, and as a lubricant additive. mp = 137°; bp = 343-387°. *Bakelite Corporation.* Name unverified.

13565 Halowax 4000 B-2
A proprietary trade name for a chlorinated hydrocarbon vinyl plasticizer. *Union Carbide (UK) Ltd.* Name unverified.

13566 Haloxil
haloxon-oxyclozanide
Haloxon/oxyclozanide anthelmintic. *The Wellcome Foundation Ltd.* Name unverified.

13567 Haloxon
321-55-1 4636 206-289-1
$C_{14}H_{14}Cl_3O_6P$
Phosphoric acid bis(2-chloroethyl) 3-chloro-4-methyl-2-oxo-2H-1-benzopyran-7-yl ester
7-[[bis(2-chloroethoxy)phosphinyl]oxy]-3-chloro-4-methyl-2H-1-benzopyran-2-one; Loxon; 3-chloro-7-hydroxy-4-methylcoumarin bis(2-chloroethyl) phosphate; 3-chloro-7-hydroxy-4-methyl-2H-1-benzopyran-2-one bis(chloro ethyl) phosphate; 3-chloro-4-methylumbelliferone di(2-chloroethyl) phosphate; galoxone; helmirone; 96-H-60; Galloxon; Luxon. A proprietary preparation of haloxon; a veterinary anthelmintic. Used in combination with oxyclozanide. mp = 91°; LD_{50} (rat orl) = 900 mg/kg. No manufacturer.

13568 Halphen reagent
A solution of sulfur (1% in carbon disulfide; used to test for cotton-seed oil. To 1 ml of oil add 1 ml of reagent and 1 ml of amyl alcohol, and heat. A red color is given with cotton-seed oil.

13569 Halphos
12141-20-7 235-252-2
HO_3PPb
Lead phosphite dibasic
Lead phosphite. Lead phosphite, dibasic; PVC heat/light stabilizer *Halstab.*

13570 Halso® 99
25168-05-2 246-698-2
C_7H_7Cl
Monochlorotoluene
o-chlorotoluene. Extender, diluent or substitute for other organic solvents; for dye carrier, fuel oil additive, sludge solvent, in paint thinners, metal parts cleaners, adhesives. *OxyChem.*

13571 Halstab 30
Complex lead salt; heat stabilizer for PVC. *Halstab.*

13572 Halstab P-1
Stabilizer/lubricant for PVC. *Halstab.*

13573 Haltex 300
1344-28-1 369 215-691-6
Alumina trihydrate
Smoke suppressive and flame retardant filler for plastics; used for carpet backings, tub and shower stalls, wire and cable insulation, elec. uses, vinyl coated fabrics, rubber products. *Hitox.*

13574 Halthal
69011-06-9 273-688-5
C8H4PbO4
Lead phthalate, dibasic; PVC heat stabilizer. *Halstab.*

13575 Halumin
An aluminum alloy with 1.48% copper, 2% nickel, 2.3% manganese, 0.47% iron, and 0.09% silicon. It is specially resistant to corroding agents.

13576 Halycitrol
Vitamin emulsion with halibut liver oil; each ml is standardized to contain not less than 276 µg (920 IU) Vitamin A and 1.9 µg (76 IU) Vitamin D. *LAB Ltd.* Unverified.

13577 Hamamell tannin
A tannin originally isolated from *Hamamelis virginica.*

13578 hambergite
$BeOH{\cdot}BeBO_3$
A mineral.

13579 Hamburg white
A pigment consisting of 1 part white lead and 2 parts heavy spar.

13580 Hamilton metal
A brass containing 67% copper and 33% zinc. An alloy of 90% zinc with small quantities of copper, lead, antimony, and phosphor tin, is also known as Hamilton metal.

13581 hammer slag
A basic silicate of iron produced and used in the puddling process for iron.

13582 Hammonia metal
An alloy of 64.5% tin, 32.2% zinc, and 3.2% copper.

13583 Hamonite
An activated carbon, made from peat. Used as an absorbent.

13584 Hampamide B
4862-18-4 225-469-0
Nitrilotriacetamide
Solution of nitriloacetamide, a chelator. *Evans Chemetics.*

13585 Hampden Steel
A proprietary trade name for a chromium tool steel containing 12.5% chromium, 0.25% nickel, 0.25% manganese, and 2.1% carbon. No manufacturer.

13586 Hamp-Ene® 100
64-02-8 3557 200-573-9
edetate sodium
Tetrasodium EDTA; general purpose chelating agent. *W R Grace/Hampshire.*

13587 Hamp-Ene® Acid
60-00-4 3559 200-449-4
edetic acid
EDTA; chelating agent; used where sodium ion is undesirable. *W R Grace/Hampshire.*

13588 Hamp-Ene® Na2
139-33-3 3556 205-358-3
edetate disodium
Disodium EDTA (dihydrate); chelating agent. *W R Grace/Hampshire.*

13589 Hamp-Ene® Na3 Liq
150-38-9 3558 205-758-8
edetate trisodium
Trisodium EDTA; chelating agent. *W R Grace/Hampshire.*

13590 Hamp-Ene® Na4
64-02-8 3557 200-573-9
edetate tetrasodium
Tetrasodium EDTA (dihydrate); chelating agent. *W R Grace/Hampshire.*

13591 Hamp-Ex® 80
140-01-2 205-391-3
Pentasodium pentetate
Chelating agent for alkaline earth and heavy metal ions; peroxide bleaching. *W R Grace/Hampshire.*

13592 Hamp-Ex® Acid
67-43-6 7266 200-652-8
Pentetic acid
Chelating agent. *W R Grace/Hampshire.* Name unverified.

13593 Hampfoam 35
61791-59-1 263-193-2
Sodium cocoyl sarcosinate
Surfactant used in alkaline industrial formulations, textile applications; forms stable small bubbles. *W R Grace/Organics; Chemplex Chems.*

13594 Hamp-Ol® 120
139-89-9 10102 205-381-9
Trisodium HEDTA
General purpose chelating agent for control of iron at pH 6.5-9.5 as well as Ca and Mg. *W R Grace/Hampshire.*

13595 Hamp-Ol® Acid
HEDTA; chelating agent. *W R Grace/Hampshire.* Name unverified.

13596 Hamp-Ol® Crystals
139-89-9 10102 205-381-9
Trisodium HEDTA (dihydrate); chelating agent. *W R Grace/Hampshire.*

13597 Hamposyl® AL-30
68003-46-3 268-130-2
Ammonium lauroyl sarcosinate
Surfactant for shampoos, skin cleansers, bath gels; sec. emulsifier for emulsion polymerization. *W R Grace/Hampshire.*

13598 Hamposyl® C
68411-97-2 270-156-4
Cocoyl sarcosine
Detergent, wetting and foaming agent, foam stabilizer, emulsifier, anticorrosive agent, conditioner for hair and rug shampoos, cosmetics, skin cleansers; biodegradable *W R Grace/Hampshire; Chemplex Chems.*

13599 Hamposyl® C-30
61791-59-1 263-193-2
Sodium cocoyl sarcosinate
Detergent, wetting and foaming agent, foam stabilizer, emulsifier, anticorrosive agent, conditioner for hair and rug shampoos, cosmetics, skin cleansers; biodegradable. *W R Grace/Hampshire; Chemplex Chems.*

13600 Hamposyl® L
97-78-9 202-608-3
Lauroyl sarcosine
Detergent, wetting and foaming agent, foam stabilizer, emulsifier, anticorrosive agent, conditioner for hair and rug shampoos, cosmetics, skin cleansers; biodegradable. W R Grace/Hampshire; Chemplex Chems.; R. T. Vanderbilt Co Inc; Croda Chem Ltd.;Nikko Chem Co. Ltd; Seppic; Ciba-Geigy AG;.

13601 Hamposyl® L-30
137-16-6 4379 205-281-5
Sodium lauroyl sarcosinate
Detergent, wetting and foaming agent, foam stabilizer, emulsifier, anticorrosive agent, conditioner for hair and rug shampoos, cosmetics, skin cleansers; biodegradable. *W R Grace/Hampshire; Chemplex Chems.*

13602 Hamposyl® M
52558-73-3 258-007-1
Myristoyl sarcosine
Detergent, wetting and foaming agent, foam stabilizer, emulsifier, anticorrosive agent, conditioner for hair and rug shampoos, cosmetics, skin cleansers; biodegradable. *W R Grace/Hampshire; Chemplex Chems.*

13603 Hamposyl® M-30
30364-51-3 250-151-3
Sodium myristoyl sarcosinate
Detergent, wetting and foaming agent, foam stabilizer, emulsifier, anticorrosive agent, conditioner for hair and rug shampoos, cosmetics, skin cleansers; biodegradable. *W R Grace/Hampshire.*

13604 Hamposyl® O
110-25-8 203-749-3
Oleoyl sarcosine

Detergent, wetting and foaming agent, foam stabilizer, emulsifier, corrosion inhibitor, mold release agent, conditioner for hair and rug shampoos, cosmetics, skin cleansers; ceramic dispersant; biodegradable. *W R Grace/Hampshire; Chemplex Chems.*

13605 Hamposyl® S
142-48-3 205-539-7
Stearoyl sarcosine
Detergent, wetting and foaming agent, foam stabilizer, emulsifier, anticorrosive agent, conditioner for hair and rug shampoos, cosmetics, skin cleansers; biodegradable. *W R Grace/Hampshire.*

13606 Hamposyl® TL-40
TEA lauroyl sarcosinate
Provides mild, high lathering properties to skin cleansers. *W R Grace/Hampshire.*

13607 Hamposyl® TOC-30
TEA-oleoyl sarcosinate, TEA-cocoyl sarcosinate. Biodegadable surfactant for hair and rug shampoos. *W R Grace/Hampshire.*

13608 Hampshire® DEG
139-41-3 205-360-4
Sodium dihydroxyethylglycinate
Chelating agent used for control of iron only in alkaline sol's. *W R Grace/Hampshire.*

13609 Hampshire® EDG
Ethanoldiglycine disodium salt; chelating agent used for the control of iron and chelates of calcium. *W R Grace/Hampshire.*

13610 Hampshire® NTA 150
5064-31-3 225-768-6
Trisodium NTA
Chelating agents used in laundry detergents and specialty cleaning products, water treatment, textiles, metal finishing, pulp and paper processing , and petroleum industry; nonphosphate detergent builder. *W R Grace/Hampshire.*

13611 Hampshire® NTA Acid
139-13-9 6675 205-355-7
Nitrilotriacetic acid
General purpose chelating agent. *W R Grace/Hampshire.*

13612 Hampshire® NTA Na3
5064-31-3 225-768-6
Trisodium NTA monohydrate
Chelating agent; detergent builder for low or nonphosphate formulations. *W R Grace/Hampshire.*

13613 HAN® 857
64742-06-9 265-105-8
Heavy naphtha. Aromatic solvent. *Exxon.* Name unverified.

13614 Handi-Wrap
Plastic film for use as a household wrapping material. *Dow UK.* Discontinued.

13615 Handyfoam
One-part polyurethane foam in an aerosol; highly adhesive, expanding, filling and sealing material. *Feb Ltd.*

13616 Hansa
Pigment powders; used for paints and inks. *Thomas Goldschmidt Ltd.*

13617 Hansa oil
A proprietary trade name for polymerized marine animal oil for soap manufacture. No manufacturer.

13618 Hanus' Iodine bromide solution
Iodine bromide (10 g), dissolved in 500 ml of glacial acetic acid; used for determining the iodine value of fats and oils.

13619 Harco Foamstopper®
Blends of mineral oil and surfactants; used for adhesives and surface coatings. *Harlow Chemical Co Ltd.*

13620 hard aluminum
An alloy of 77% aluminum, 11% zinc, and 11% magnesium. Another alloy contains copper in the place of zinc.

13621 Hard Cure
Balanced blend of sodium, potassium and meta silicates combined with surface tension reducing agents; for application to freshly placed concrete following finishing. *Secure Inc.* Discontinued.

13622 hard jatoba
A Brazilian copal resin.

13623 hard lead
Antimony lead
An alloy of lead with 10-30% antimony; used as type metal.

13624 hard metal
The name usually applied to a tin-copper alloy, containing 1 part tin, and 2 parts copper.

13625 hard platinum
Platinum containing from about 5-30% iridium.

13626 hard solder
An alloy of 86% copper, 9% zinc, and 4% tin.

13627 hard zinc
An alloy of 92% zinc, 5% Iron, and 3% lead.

13628 Hardcote
Dressing for cores and molds. *Foseco (F.S.) Ltd.*

13629 hardened rosins
Metallic resinates prepared by heating rosin with metallic oxides, usually calcium, magnesium oxide and zinc oxide.

13630 hardenite
A collective name for austenite and martensite of eutectoid composition.

13631 hard-finish plaster
Plaster made from oven-burnt gypsum dipped in alum solution, and again calcined.

13632 hard-head
The name by which the impurities obtained from the refining of tin are known.

13633 Hardite
Heat-resisting alloys containing 55-65% nickel, 15-18% chromium, 1-4% silicon, 1-2% manganese, the balance being iron.

13634 Hardite X
A heat-resisting alloy. It contains 82-86% nickel, 10-13% chromium, and 2% manganese.

13635 Hardset
Heat-curing hard rubber compound. *Hardman.*

13636 Hargus Steel
A proprietary trade name for a die steel containing 1.0% manganese, 0.35% nickel, and 1.0% carbon. No manufacturer.

13637 Harle's solution
7784-46-5 8721 232-070-5
sodium arsenite
A solution of sodium arsenite. Used as a topical acaricide.

13638 Harmomang A and B
Iron alloys containing carbon, manganese and molybdenum. Proprietary alloys.

13639 Harmonia bronze
An alloy of 57% copper, 40% zinc, 1.8% Iron, and 0.4% lead.

13640 Harmony
Lube oil, hydraulic fluid. *Monsanto (Solaris).* Name unverified.

13641 Harmony®
metsulfuron-methyl-thifensulfuron-methyl
Mixture of metsulfuron-methyl and thifensulfuron-methyl; for control of annual dicotyledons in cereals. *Du Pont Uk.*

13642 Haro® Chem P28G
1072-35-1 5442 214-005-2
lead stearate
Normal lead stearate; stabilizer for rigid and plasticized PVC applications. *Harcros.*

13643 Haro® Chem PDF
Dibasic lead phosphite
Stabilizer for rigid and plasticized PVC applications offers heat and light stability and good electrical properties. *Harcros.*

13644 Haro® Chem PTS-E
7446-14-2 5444 231-198-9
Tribasic lead sulfate
Stabilizer for rigid and plasticized PVC applications offers heat stability and good electrical properties. *Harcros.*

13645 Haro® Mix CE-701
7439-92-1 5410 231-100-4
lead
Lead one-pack system; one-pack lead heat stabilizer systems for rigid PVC applications, UPVC pressure and nonpressure pipes. *Harcros.*

13646 Haro® Mix CK-711
7439-92-1 5410 231-100-4
lead
Lead one-pack system; stabilizer/lubricant for extrusion of UPVC pressure or nonpressure pipes *Harcros.*

13647 Haro® Mix MH-204
7439-92-1 5410 231-100-4
lead
Lead one-pack system; stabilizer/lubricant for extrusion of UPVC pressure pipes. *Harcros.*

13648 Haro® Mix YK-110
barium-lead
Barium/cadmium/lead one-pack; stabilizer/lubricant for UPVC window profile extrusion. *Harcros.*

13649 Haro® Mix ZC-028, ZC-029, ZC-030
calcium-zinc

Calcium/zinc one-pack; nontoxic one-pack heat and light stabilizer for rigid PVC applications (profiles). *Harcros.*

13650 Haro® Mix ZT-514
7440-31-5 9587 231-141-8
tin
Tin one-pack; stabilizer/lubricant for extrusion of UPVC pipes *Harcros.*

13651 Haro® Wax L333
Lubricant blend for rigid and plasticized PVC applications, including food contact applications *Harcros.*

13652 Harrier
mecoprop-3,6-dichloropicolinic acid-ioxynil
Weedkiller containing mecoprop, 3,6-dichloropicolinic acid and ioxynil as potassium salts. *ICI Chem & Polymers Ltd.*

13653 Harrington bronze
Harlington bronze
An alloy of 55.75% copper, 42.5% zinc, 1% tin, and 0.75% iron. A white bearing metal.

13654 Harringtonite
An Irish mineral containing 41.4% SiO_2, 30.2% $Al+2O_2$, 11.2% CaO, 5.2% Na_2O, and 12.5% H_2O.

13655 Harris' hematoxyiin stain
A microscopic stain. It contains 1 g haematoxylin, 10 ml alcohol, 20 g alum, 0.5 g mercuric oxide, and 200 ml water.

13656 Harrison's indicator
A small amount of starch boiled with a few mls. of water, adding to it 100 ml of a freshly prepared 10% potassium iodide solution. Used for detection of free iodine.

13657 Hart Lawn Sand
10028-22-5 4079 233-072-9
Ferrous sulfate
Used for moss control in turf. *Maxwell Hart Ltd.*

13658 Hart Moss Killer
10028-22-5 4079 233-072-9
Ferrous sulfate
Used for moss control in turf. *Maxwell Hart Ltd.*

13659 Hartamide 9137
93-83-4 202-281-7
Oleamide DEA; coupling agent, emulsion stabilizer, lubricant and antistat. *Hart Chem. Ltd.*

13660 Hartamide AD
Cocamide DEA (2:1); foam booster, solubilizer, emulsifier, coupling agent for detergent formulations. *Hart Prods. Corp.*

13661 Hartamide LDA
120-40-1 204-393-1
Lauramide DEA; detergent, foam stabilizer and thickener for liquid and powder detergent systems, shampoos, bubble baths. *Hart Chem. Ltd.*

13662 Hartamide LMEA
142-78-9 205-560-1
Lauramide MEA; foam stabilizer for spray-dried powd. detergents and bubble bath preparations; visc. modifier for detergents, shampoos, bubble baths. *Hart Chem. Ltd.*

13663 Hartamide OD
Cocamide DEA; detergent, foam stabilizer and visc. regulator for liq. and powd. detergent systems, shampoos, bubble baths. *Hart Chem. Ltd.*

13664 Hartamine Series
Fatty acid imidazolines
Corrosion inhibitors for gas and oil industry. *Hart Chem. Ltd.*

13665 Hartasist 16
Silicone-based; defoamer for oil and gas industry. *Hart Chem. Ltd.*

13666 Hartasist 46
Specialty surfactant; cold flow improver for oil deposit; heavy-duty cleaner. *Hart Chem. Ltd.*

13667 Hartasperse DI-4900 Series
De-inking agent for wash de-inking systems; increases pulp brightness. *Hart Chem. Ltd.*

13668 Hartbreak Series
Phenolic resins, polyol ester, alkylaryl sulfonate; demulsifiers for water-oil emulsions in oil and gas industry. *Hart Chem. Ltd.*

13669 Hartenol LAS-30
151-21-3 8782 205-788-1
Sodium lauryl sulfate
Detergent, wetting agent, emulsifier, foaming agent for light duty household detergents; biodegradable. *Hart Prods. Corp.*

13670 Hartenol LES 60
9004-82-4
Sodium laureth sulfate
Detergent, wetting agent, dispersant, emulsifier, foaming agent for light duty household detergents; biodeg. *Hart Prods. Corp.*

13671 hartin
A white resin found in the lignite in Austria.

13672 hartite
A similar resin to Harlin, and found with it.

13673 Hartofix 2X
Cationic resin; fixative for direct and reactive dyes; improves fastness of most substantive colors to water, soaps, and detergents. *Hart Chem. Ltd.*

13674 Hartofol 40
Sodium dodecylbenzene sulfonate
High foaming biodegradable surfactant for heavy-duty liquid detergents, textile processing, industrial cleaners, shampoos, automotive cleaners. *Hart Prods. Corp.*

13675 Hartofol 60T
TEA dodecylbenzene sulfonate
Surfactant for cosmetics, shampoos, liquid detergents, wool wash formulations; biodegradable; mild, high purity. *Hart Prods. Corp.*

13676 Hartolan
8027-33-6 232-430-1
Lanolin alcohol
Spreading agent, dispersant, stabilizer, plasticizer, w/o emulsifier, conditioner, superfatting agent, moisturizer, and emollient for cosmetic and pharmaceutical systems. *Croda Inc.*

13677 Hartolite
8027-33-6 232-430-1
Lanolin alcohols fraction; water-oil emulsifier, emollient, skin conditioner, moisturizing agent. *Croda Chem. Ltd.*

13678 Hartolon 5683
9002-88-4 7728
Polyethylene and selected esters; nonyellowing softener for 100% cotton toweling; provides static control and absorbency. *Hart Chem. Ltd.*

13679 Hartolon NA
Saturated esters and alkyl condensate; softener, lubricant, antistat used in textiles. *Hart Chem. Ltd.*

13680 Hartomer 4900
Sodium vinyl sulfonate
Monomer for latex emulsion polymerization systems. *Hart Chem. Ltd.*

13681 Hartomer GP 2164
Phosphate ester
Surfactant for water-based adhesives and latex emulsion polymerization. *Hart Chem. Ltd.*

13682 Hartomer LD 31
Polyacrylate
Antiredeposition agent, dispersant, processing aid. *Hart Chem. Ltd.*

13683 Hartomul PE-30
Blend of polyethylene and polyether derivatives; softener for use with resin finishes on lightweight fabrics; mold release agent for brake linings, urethane foam. *Hart Chem. Ltd.*

13684 Hartonyl L531
Ethoxylated fatty amine
Leveling aid for dyeing polyamides; migrating leveling agent and dispersant for acid and disperse dye stuffs; increases contrast and clarity between nylon fibers of different affinities; effective in continuous and winch dyeing of carpets. *Hart Chem. Ltd.*

13685 Hartopol 25R2
Polyoxyalkylene glycol
Low foaming surfactant for rinse aids and windshield washer fluids. *Hart Chem. Ltd.*

13686 Hartopol L42
EO/PO block copolymer; low foaming rinse aid, defoamer, emulsifier, detergent, dispersant for many industries. *Hart Chem. Ltd.*

13687 Hartopol LF-1
Polyoxalkylene glycols
Detergent, emulsifier, defoamer for low-foam dishwash, rinse aids, general household and industrial use. *Hart Chem. Ltd.*

13688 Hartopol P65
Polyoxyalkylene glycol
Dispersant, demulsifier, defoamer, emulsifier, detergent for many industries. *Hart Chem. Ltd.*

13689 Hartosoft 171
Esters and quaternary amines; nonyellowing softener with antistatic and lubricating properties for velours and toweling. *Hart Chem. Ltd.*

13690 Hartosoft S5793
Amino functional silicone emulsion; softener providing lubricity to cotton and cotton blended knits and woven fabrics. *Hart Chem. Ltd.*

13691 **Hartotrope AXS**
26447-10-9 247-710-9
Ammonium xylene sulfonate
Detergent, solubilizer and cloud point depressant for light duty and built liquid detergent systems. *Hart Chem. Ltd.*

13692 **Hartotrope KTS 44**
30526-22-8 250-228-1
$C_7H_7KO_3S$
Potassium toluene sulfonate
Hydrotrope for high activity or built liquid detergent systems; for use where sodium ion undesirable. *Hart Chem. Ltd.*

13693 **Hartotrope STS-40, Powd**
12068-03-0 235-088-1
Sodium toluene sulfonate
Detergent, solubilizer and cloud point depressant for light duty and built liquid detergent systems. *Hart Chem. Ltd.*

13694 **Hartotrope SXS 40, Powd**
1300-72-7 215-090-9
Sodium xylene sulfonate
Detergent, solubilizer and cloud point depressant for light duty and built liquid detergent systems. *Hart Chem. Ltd.*

13695 **Hartowet 5917**
Blend ; wetting aid for fabrics; contains no phosphates. *Hart Chem. Ltd.*

13696 **Harvesan**
Mercurial seed dressing. *The Boots Co plc.* Discontinued.

13697 **Harvestra®**
For the agriculture industry. *DuPont UK.*

13698 **Harvite**
A proprietary trade name for a shellac compound. No manufacturer.

13699 **Hascrome**
A proprietary alloy. It is a manganese-chromium-iron welding rod.

13700 **Hastelloy®**
Nickel-base alloys; excellent corrosion resist. with moderate wear resist. *Haynes Int'l.*

13701 **Hayem's solution**
A solution of 5 g sodium sulfate, 1 g sodium chloride, and 0.5 g mercuric chloride in 200 ml water; used in the examination of blood corpuscles.

13702 **Haylite No. 1**
An English explosive. It contains 25-27% nitroglycerin, 0.5-1.5% collodion cotton, 19-21% potassium nitrate, 19-21% barium nitrate, 12-14% wood meal, 6-8% mineral jelly, and 10-12% ammonium oxal.

13703 **Haylite No. 3**
An explosive. It consists of 9.5% nitroglycerin, 60% ammonium nitrate, 5% wood meal, 19.5% sodium chloride, and 5% ammonium oxalate.

13704 **Haynes alloy No. 25**
A proprietary trade name for an alloy of cobalt, nickel, tungsten, and chromium. Possesses exceptional mechanical properties up to 1800ºF. No manufacturer.

13705 **Haynes metals**
Alloys of from 10-75% iron, 20-30% chromium, and 5-25% cobalt. A harder alloy contains 45% cobalt, 40% tungsten, and 15% chromium, and a softer one, 62% cobalt, 28% tungsten, and 10% chromium; they are noncorrosive.

13706 **Haynes® 242**
Ni-Mo-Cr alloy ; age-hardenable alloy with high temp. strength, low thermal expansion chars., good oxidation resist. for service to 760 C; resist. to high-temp. fluorine and fluoride environments; for fluoropolymer and fluoroelastomer production. *Haynes Int'l.*

13707 **Haysite**
Thermoset polyester resin, fillers and glass; for electrical insulation, corrosion resistant; sold as custom molded parts, sheets, molding compound, SMC-BMC and pultruded shapes; MIL spec, NEMA and UL recognized materials. *Haysite Reinforced Plastics.*

13708 **Haysite 14100**
Electrical grade bulk molding compound for compression, transfer and injection molding. *Haysite Reinforced Plastics.*

13709 **Haysite 24500**
Corrosion-resistant bulk molding compound for compression, transfer and injection molding. *Haysite Reinforced Plastics.*

13710 **Haysite 42000**
Electrical grade sheet molding compound for compression molded parts requiring high strength and good electrical properties. *Haysite Reinforced Plastics.*

13711 **Haysite EHC-P**
Fiberglass-reinforced polyester; general purpose economy pultrusion grade offering 155º performance; for transformer and electrical applications. *Haysite Reinforced Plastics.*

13712 **HB-40**
Partially hydrogenated terphenyl
Plasticizer for vinyl sheeting, films and fabric or paper coatings, and for vinyl protective coatings and adhesives. *Monsanto Co.* Name unverified.

13713 **HBR**
A compressible closed-cell polyethylene foam rod; placed in a joint before applying a sealant and to assist the sealant in assuming the proper configuration; also serves as a bond breaker to prevent three-side adhesion of sealant to joint substance. *Hercules.* Discontinued.

13714 **HC 200 Concentrate**
8148
Pyrethins
Contact insecticide; consists of a mixture of pyethrin I [121-21-1] and pyethrin II[121-29-9]. *Certified Laboratories Ltd.*

13715 **HC-913**
Acrylate polymer in ethyl acetate solvent; paint primer concentrate used as primer over galvanized metal, aluminum and plastics. *3M.*

13716 **HD-Echelon 110/130**
Oleyl linoleyl alcohol (5-10% C_{16}, 90-95% C_{18}); intermediate in surfactant manufacture. *Henkel/Emery.*

13717 **HD-Echelon 45/50**
Oleyl/cetyl alcohol (25-33% C_{16}, 60-70% C_{18}); intermediate in surfactant manufacture. *Henkel/Emery.*

13718 **HD-Echelon 90/95**
143-28-2 6968 205-597-3
Oleyl alcohol (2-10% C_{16}, 87-95% C_{18}); emollient, superfatting agent, carrier for cosmetics; intermediate in surfactant manufacture. *Henkel/Emery.*

13719 **HD-Eutanol**
143-28-2 6968 205-597-3
Oleyl alcohol
Ultra pure grade used as solubilizer for dyes and waxes; superfatting agent, emollient in cosmetic emulsions, creams, alcoholic lotions. *Henkel/Cospha; Henkel Canada.*

13720 **HDPE 04352N**
9002-88-4 7728
Polyethylene
HDPE; injection molding resin. *Dow Plastics.*

13721 **HDPE 25053-P**
9002-88-4 7728
Polyethylene
HDPE copolymer resin; high molecular weight resin with outstanding hot-melt strength, high flexibility modulus, superior ESCR, and toughness; used for blow molding and sheet vacuum forming applications including gas tanks, drums, large industrial parts; FDA compliant. *Dow Plastics.*

13722 **HDPE 32060C**
9002-88-4 7728
Polyethylene
HDPE; blow molding resin. *Dow Plastics.*

13723 **HDPE 35053**
9002-88-4 7728
Polyethylene
HDPE; corrugated tubing resin. *Dow Plastics.*

13724 **HDPE IP-10**
9002-88-4 7728
Polyethylene
HDPE; improved processing resin designed for reusable cases. *Dow Plastics.*

13725 **Hégor**
Shampoo, to wash and condition hair. *Richardson Vicks Inc.* Name unverified.

13726 **Head and Shoulders**
13463-41-7 8178 236-671-3
Pyrithione zinc
Antibacterial; antifungal; antiseborrheic. *Procter & Gamble.* Name unverified.

13727 **Headland Charge**
7085-19-0 5826 230-386-8
Mecoprop
Soluble concentrate containing 570 g/l mecoprop; for control of weeds in cereals and grassland. *SBC Technology Ltd.* Unverified.

13728 **Headland Dephend**
13684-63-4 7384 237-199-0
phenmedipham
Emulsifiable concentrate containing 114 g/l phenmedipham; for weed control for beet crops. *SBC Technology Ltd.* Unverified.

13729 **Headland Dual**
carbendazim-maneb

A suspension concentrate containing 62 g carbendazim and 400 g maneb per liter; systemic fungicide for use with cereals. *WBC Technology Ltd.*

13730 Headland Guard
Mixture of organo copolymers and surfactants; spreader and sticking agent for agricultural chemicals. *WBC Technology Ltd.*

13731 Headland Inorganic Liquid Copper
1332-40-7
copper oxychloride
Suspension concentrate containing 435 g copper oxychloride per liter; a protectant fungicide. *WBC Technology Ltd.*

13732 Headland Intake
Mixture of organic acids and surfactants; wetting additive for a wide range of herbicide, fungicide, dessicant and growth regulator sprays. *WBC Technology Ltd.*

13733 Headland Relay
dicamba, MCPA, mecoprop
Mixture of dicamba, MCPA and mecoprop; used for weed control in cereals and grassland. *WBC Technology Ltd.*

13734 Headland Spirit
12427-38-2 5761 235-654-8
Maneb
Suspension concentrate containing 480 g maneb per liter; a dithiocarbamate fungicide to control blight, rusts and mildew. *WBC Technology Ltd.*

13735 Headland Swift
999-81-5 2153 213-666-4
chlormequat chloride
Soluble concentrate containing 750 g/l chlormequat; plant growth regulator. *WBC Technology Ltd.*

13736 Heavithane
A proprietary polyester-based polyurethane elastomer crosslinked by diols. *HMC Wheels.* Unverified.

13737 Hebron Pabracr
cetrimide-chlorpropham
A suspension concentrate containing 80 g cetrimide and 80 g chlorpropham per liter. soil acting herbicide for lettuce. *Atlas Interlates Ltd.*

13738 Heckel's solution
sodium sulfite-benzoic acid
A solution containing sodium sulfite, benzoic acid, and water.

13739 Hecla 35
A proprietary trade name for alloy steel containing 0.15% carbon, 0.25% silicon, 0.04% sulfur, 0.04% phosphorus and 0.70% manganese; used to manufacture forged steel pressure vessels. *Hadfield, George & Co.* Unverified.

13740 Hecla 115
Alloy steel containing 0.35% carbon, 0.25% silicon, 0.04% sulfur, 0.04% phosphorus, 0.7% manganese and 1.00% nickel; used for manufacturing forged steel pressure vessels. *Hadfield, George & Co.* Unverified.

13741 Hecla 135, 138
An alloy steel containing 0.60% carbon, 0.30% silicon, 0.30% manganese, 2.00% chromium, 2.00% nickel and 0.45% molybdenum; used for the manufacture of forged high tensile steel pressure vessels. *Hadfield, George & Co.* Unverified.

13742 Hecla 138H
An alloy steel containing 0.40% carbon, 0.30% silicon, 0.04% sulfur, 0.04% phosphorus, 0.60% manganese, 0.70% chromium, 2.70% nickel, 0.50% molybdenum and 0.25% vanadium (optional); used for the manufacture of forged high tensile steel pressure vessels. *Hadfield, George & Co.* Unverified.

13743 Hecla 155
An alloy steel containing 0.12% carbon, 0.20% silicon, 0.04% sulfur, 0.04% phosphorus, 0.40% manganese, 2.30% chromium, 1.00% molybdenum; used for the manufacture of forged steel pressure vessels for use at higher temperatures. *Hadfield, George & Co.* Unverified.

13744 Hecla 174
An alloy steel containing 0.30% carbon, 0.30% silicon, 0.04% sulfur, 0.04% phosphorus, 0.50% manganese, 5.00% chromium, 0.90% vanadium; used for the manufacture of forged steel pressure vessels for hydrogen service and high tensile purposes. *Hadfield, George & Co.* Unverified.

13745 Hecla 180
An alloy steel containing 0.40% carbon, 0.30% silicon, 0.04% sulfur, 0.04% phosphorus, 0.60% manganese, 1.00% chromium, 3.2% nickel, 0.50% molybdenum and 0.25% vanadium; used for the manufacture of forged high tensile steel pressure vessels. *Hadfield George & Co.* Unverified.

13746 Hecla 306
An alloy steel containing 0.30% carbon, 0.20% silicon, 0.04% sulfur, 0.04% phosphorus, 0.50% manganese, 3.00% chromium, 0.50% molybdenum and 0.20% vanadium; used for the manufacture of forged steel pressure vessels for hydrogen service. *Hadfield, George & Co.* Unverified.

13747 Hecla 307
An alloy steel containing 0.18% carbon, 0,20% silicon, 0.04% sulfur, 0.04% phosphorus, 0.50% manganese, 3.00% chromium, 0.50% molybdenum and 0.20% vanadium; used for the manufacture of forged steel pressure vessels for hydrogen service. *Hadfield, George & Co.* Unverified.

13748 Hecla powder
An explosive similar in composition to Giant powder.

13749 Hectabrite® AW
12173-47-6 4653 235-340-0
Hectorite USP/NF; Montmorillonite (smectite) clay mineral-layered, with swelling and intercalation properties. Used as an emulsifier, thickener, suspension agent in pharmaceutical, cosmetic and personal care products. *Am. Colloid.*

13750 Hectalite® 200
12173-47-6 4653 235-340-0
Hectorite USP/NF; viscosifier, suspension agent, binder for pharmaceutical, cosmetic and personal care prods. *Am. Colloid.*

13751 Hector bases
Basic substances obtained by the oxidation of thioureas with hydrogen peroxide; good vulcanization accelerators.

13752 Hectorite
12173-47-6 4653 235-340-0
Hectabrite® AW; Hectalite® 200; Bentone® EW, MA;. Hectorite USP/NF; Montmorillonite (smectite) clay mineral-layered, with swelling and intercalation properties. Used as an emulsifier, thickener, suspension agent in pharmaceutical, cosmetic and personal care products. *Am. Colloid; Rheox.*

13753 hectorite laponite
Sodium magnesium lithium fluoro silicate
A white, iron free suspending and gelling agent for aqueous systems.

13754 hedgehog crystals
Crystals of ammonium urate found in urinary deposits.

13755 Hedonal®
2,4-D MCPA-dichlorprop-MCPP
Range of herbicides containing 2,4-D MCPA, dichlorprop, MCPP either alone or in combinations; growth regulator herbicide used for control of weeds in cereals. *Bayer AG.*

13756 Hegolit 3
A patented product which is a preparation of higher aliphatic alcohols, with a melting-point of 50°; a plasticizer. No manufacturer.

13757 Heidenhain's chrome hematoxyiin
A microscopic stain. It is produced by staining the object in 0.33% solution of haematoxylin in water, then soaking in 0.5% solution of potassium chromate.

13758 Heiloy
A proprietary trade name for stainless steels used for dairy utensils. No manufacturer.

13759 Helarion
9002-91-9 5983 202-945-6
metaldehyde
A pelleted bait containing 6% w/w metaldehyde; used for control of slugs and snails. *Fisons plc.*

13760 Heliane
Vat dyes. *ICI Chem & Polymers Ltd.* Discontinued.

13761 Helio®
Pigments; for surface coatings, printing inks, wallpaper, colored coated paper and lake producing industries. *Bayer AG.*

13762 Heliochrysin
Sun gold
The sodium salt of tetranitro-α-naphthol.

13763 Heliofil
Pigment dyestuffs in paste form for the dope dyeing of viscous continuous filaments and stable fibers. *Bayer AG.*

13764 Heliogen®
Phthalocyanine pigments; for superfast letterpress, offset, flexographic and gravure printing inks, paints, lacquers, for coloring plastics, for production of artists chalks, paints and crayons. *BASF AG; BASF plc.*

13765 Heliolac
A nitrocellulose lacquer.

13766 Heliophan
118-56-9 4776 204-260-8
Homosalate
Ultraviolet screen. *R W Greeff & Co Inc.* Name unverified.

13767 Helizarin®
Pigment formulations for printing and dyeing textiles. *BASF AG; BASF plc.*

13768 Helothion
35400-43-2 9164 252-545-0
Sulprofos
Insecticide with stomach and contact action mainly against caterpillars of

Heliothis spp. and *Spodoptera* spp. Used on cotton and other crops. *Bayer AG.*

13769 Heloxy® 7

Glycidyl ether of a mixture of C_8-C_{10} alcohols; viscosity reducing modifier for epoxy formulations used in flooring, casting, tooling, laminating, potting, coatings, etc. *Shell.*

13770 Heloxy® 61

2426-08-6 219-376-4

Butyl glycidyl ether

epoxy modifier. Maximum viscosity reduction, minimum loss of properties; increases impregnation of resin systems, and level of filler loading; used in electrical equipment, laminating, casting, tooling, flooring, and coatings. *Shell.* Name unverified.

13771 Heloxy® 62

2210-79-9 218-645-3

o-Cresyl glycidyl ether

Reactive diluent; increases level of filler loading in epoxyresins; used in flooring, low viscosity casting, laminating, and decoupage. *Shell.* Name unverified.

13772 Heloxy® 64

Nonyl phenyl glycidyl ether

Epoxy functional modifier *Shell.* Name unverified.

13773 Heloxy® 65

3101-60-8 221-453-2

p-tert-Butyl phenyl glycidyl ether

Moderate reactive diluent, used in casting, tooling, laminating, and in flooring. *Shell.* Name unverified.

13774 Heloxy® 67

2425-79-8 219-371-7

Butanediol diglycidyl ether

Diglycidyl ether of 1,4 butanediol; diluent; used in casting, laminating, tooling, potting, and electrical applications. bp = 266°; d = 1.1000; n_D^{20} = 1.4611. *Shell.* Name unverified.

13775 Hemachates

Agates marked with red jasper.

13776 Hema-Combistix

A proprietary test strip comprised of four separate tests: (1) methyl red and bromothymol blue for pH; (2) a buffered mixture of glucose oxidase, peroxidase, *o*-toluidine, and a red dye for glucose; (3) buffered tetrabromophenol blue for glucose; (4) *o*-toluidine for blood. Used to test urine. *B. C. Ames.* Name unverified.

13777 Hemalum

A microscopic stain. One gram haematoxylin or its ammonium salt is dissolved in 50 ml of 90% alcohol, added to a solution of 50 g of alum in 1,000 ml water, and filtered.

13778 Hemastix

A proprietary test-strip containing *o*-tolidine and an organic peroxide; used to detect the presence of blood in urine. *B. C. Ames.* Name unverified.

13779 Hematest

A proprietary test tablet of *o*-toluidine and an organic peroxide; used to detect blood in feces and urine. *B. C. Ames.* Name unverified.

13780 Hematin

15489-90-4 4668 239-518-9

$C_{34}H_{33}FeN_4O_5$

(SP-5-13)-[7,12-diethenyl-3,8,13,17-tetramethyl-21H,23H-porphine-2,8-dipropanoato(4-)-N^{21},N^{22},N^{23},N^{24}]hydroxyferrate(2-) dihydrogen

Used in biochemical research. λ_m = 580 nm; LD_{50} (rat iv) = 43.2 mg/kg.

13781 Hemicellulase Conc

Fungal hemicellulase; enzyme for petroleum industry (breaker with guar gum), coffee (hydrolyzes coffee gums), flavors (extraction of essential oils and plant extracts). *Solvay Enzymes.*

13782 Hemicelluloses

These are contained in plant cell walls. They are reserve celluloses and are readily converted into hexoses and pentoses by acids and by the enzyme cytase.

13783 Hemoterge

Rinse and reference solution; used for cleaning and blanking of automatic hemoglobinometers. *Coulter Electronics Ltd.*

13784 Hemoxone

dichlorprop-MCPA

Soluble concentrate of 392 g dichlorprop and 210g MCPA per liter; used for control of weeds in cereals and grass. *ICI Chem & Polymers Ltd.*

13785 Hempel's solution

A solution made by mixing a solution of 120 g potassium hydroxide in 80 ml water with a solution of 5 g pyrogallol in 15 ml water; used for the determination of oxygen.

13786 HEP

N-(2-Hydroxyethyl)-2 pyrrolidone

Solvent reaction intermediate, textile auxiliaries, cosmetic ingredient. *ISP.* Name unverified.

13787 hepar sulfur

1312-73-8 7846 215-197-0

potassium sulfide.

13788 Heparin, Sodium Salt

9005-49-6 4685 232-681-7

Liquemin; Heparin sodium; Heprinar; Hepsal; Lipo-Hepin; Monoparin; Panheprin; Pularin; Minihep; Thrombo-Hepin; Thrombophob; Unihep; Lipo-Hepinette; Longheparin;. Proprietary preparation of heparin; anticoagulant. $[\alpha]_D^{25}$ = 47°; soluble in water, saline solutions; insoluble in alcohol, acetone, benzene, chloroform, ether. *Roche Products Ltd.* Name unverified.

13789 Hepatolite

65717-97-7 3422

$C_{18}H_{25}N_2O_5{}^{99m}Tc$

N-[2-[[2,6-bis(1-methylethyl)phenyl]amino]2-oxoethyl]-N-(carboxymethyl)glycine

Technetium ^{99m}Tc salt of Disofenin. Diagnostic aid, radioactive agent used to measure liver function. *DuPont-NEN Medical Products.* Name unverified.

13790 n-heptane

142-82-5 4694 205-563-8

C_7H_{16}

Aliphatic hydrocarbon

Exxsol® Heptane; alkane C_7. Heptane; solvent. Hydrocarbon from petroleum. d = 0.684; mp = -90.7°; bp = 98.4°; fp = 30° F; n_D^{20} = 1.3855; Insoluble in water, soluble in alcohol, chloroform, ether. *Ashland; Exxon; Humphrey; Phillips 66; Texaco.* Name unverified.

13791 heptanoic acid

111-14-8 4695 203-838-7

$C_{14}H_{26}O_3$

enanthic acid; oenanthic acid; oenanthylic acid; carboxylic acid C_7; enanthic anhydride; n-heptoic acid; heptoic acid; n-heptylic acid;. Organic synthesis, production of lubricants for aircraft and brake fluids. d = 0.9181; mp = -7.5°; bp_{760} = 223.01°; specific heat = 0.54 cal/g; n_D^{20} = 1.42162; pKa (25°) = 4.4; soluble in water, ethanol, ether, DMF, DMSO; LD_{50} (mus iv) = 1200 mg/kg. *Elf Atochem N. Am.; Exxon UK; Hoechst Celanese; Penta Mfg.*

13792 1-heptanol

111-70-6 4696 203-897-9

$C_7H_{16}O$

n-Heptyl alcohol

alcohol C7; enanthyl alcohol; enanthic alcohol; heptylalcohol; heptyl alcohol; 1-hydroxyheptane. Organic intermediate, solvent, cosmetic formulations. mp = -34.6°; bp = 175.8°; d = 0.8187; n_D^{25}= 1.4224; soluble in water; miscible with alcohol, ether. *Elf Atochem N. Am.; Penta Mfg.; Suchema AG.*

13793 Hepteen Base

Heptaldehyde-aniline reaction product; fastcuring, high temperature accelerator with maximum processing safety; works well with basic furnace blacks used in natural rubber pure gum stocks. *Uniroyal.* Name unverified.

13794 heptene

Heptaldehyde-aniline

A proprietary rubber vulcanization accelerator. No manufacturer.

13795 heptenophos

23560-59-0 4699 245-737-0

$C_9H_{12}ClO_4P$

7-chlorobicyclo[3.2.0]hepta-2,6-dien-6-yl dimethyl phosphate

Ragadan; Heptenophos; Hostaquick; OMS 1845; Hoe 02982. Systemic insecticide with contact, stomach and respiratory action. Used for control of sucking insects, particularly aphids, in a variety of crops. $bp_{0.1}$ = 64°; d^{20} = 1.294; soluble in H_2O (2.2 g/l), more soluble in organic solvents; LD_{50} (rat orl) = 121 mg/kg.

13796 Heptokill

Insecidal formulation. *Mitchell Cotts Chemicals Ltd.*

13797 heptyl alcohol

111-70-6 4696 203-897-9

$C_7H_{16}O$

1-heptanol

alcohol C_7; enanthyl alcohol; enanthic alcohol; 1-hydroxyheptane. Organic intermediate, solvent, cosmetic formulations. mp = -35°; bp = 176°; d_4^{25} = 0.8187; n_D^{25}= 1.4224; soluble in H_2O (1 g/l), more soluble in organic solvents. *Elf Atochem N. Am.; Penta Mfg.; Suchema AG.*

13798 Her

111-46-6 3168 203-872-2

diethyleneglycol

Chemical resorcinol di(β-hydroxyethyl) ether; curative for urethane polymers. *Anderson Development Company.* Name unverified.

13799 Herapath's salt

7631-46-1 8246 231-544-9

$C_{80}H_{104}I_6N_8O_{20}S_3$

Quinine iodo-sulfate
Used in the manufacture of polarizing glasses and plastics.

13800 Herbatox®
bentazon-isoproturon-dichlorprop
Bentazon, isoproturon, dichlorprop; for postemergence control of grasses and broadleaf weeds in winter cereals and spring wheat. *BASF AG.*

13801 Herbazin 50
122-34-9 8681 204-535-2
Simazine
A wettable powder containing simazine; Long term maintenance of weed-free pathways, bare ground and other areas requiring total weed control. *Fisons plc, Horticultural Div.* Name unverified.

13802 Herbazin Plus
simazine-aminotriazole
A wettable powder containing simazine and aminotriazole; a quick acting herbicide for control of existing weeds with long term persistence. *Fisons plc, Horticultural Div.*

13803 Herbazin Plus SC
amitrole-simazine
A suspension concentrate containing 180 g aminotriazole [61-82-5] and 300 g simazine [122-34-9] per liter; used for total weed control in non crop areas and fruit orchards. *Fisons plc.*

13804 Herbazin Special
amitrole-atrazine-2,4-D
Used for total weed control in non crop areas. *Fisons plc.*

13805 Herbazin Total
1912-24-9 902 217-617-8
atrazine
A granule containing atrazine; a persistent herbicide for the control of grasses and many annual and perennial broad-leaved weeds. *Fisons plc, Horticultural Div.* Name unverified.

13806 Herbohn bronze
An alloy of 71% copper, 26% tin, and 3% zinc.

13807 Herborane
54546-26-8 259-210-8
$C_{11}H_{22}O_2$
2-butyl-4,4,6-trimethyl-1,3-dioxane.
A flavorant. *Quest Int'l. UK Ltd.*

13808 Herbrak®
Herbicide for controlling the grass and broad leaved weeds occurring in sugar and fodder beet; predrilling, pre-emergence or post-emergence application. *Bayer AG.*

13809 Hercat 627
Dispersed rosin size; sizing agent for paper machine wet end processing. *Hercules.* Discontinued.

13810 Herclor
Epichlorhydrin elastomer. *Hercules Ltd.* Discontinued.

13811 Herco®
Herco® Pine Oil. Pine oil; various grades for household and industrial cleaners; disinfectant, antifoam agent, textile specialties. *Hercules.*

13812 Hercobind DS
An oleoresin emulsion; resistant to rain and wind erosion; binds and agglomerates fine minerals for dust control. *Hercules.* Discontinued.

13813 Hercobond® 339
Water-soluble polymer for strength enhancement in paper. *Hercules.*

13814 Hercoflat®
Paint texturing/flatting agents. *Hercules Ltd.* Discontinued.

13815 Hercoflat® Texturing Pigments and Flatting Agent
9003-07-0 7741
Polypropylene
Special type of PP; easily dispersed in coating vehicles for textured, nonglare finishes; used in most finishes, maintains original texture in fully cured, baked systems. *Aqualon.* Discontinued.

13816 Hercoflav
Formulated flavors for foods. *Hercules Ltd.* Discontinued.

13817 Hercoflex® 600
Nonvolatile, low molecular weight plasticizers. *Hercules.*

13818 Hercoflex® 707
Nonvolatile, low molecular weight plasticizers. *Hercules.*

13819 Hercoflex® 707A
Nonvolatile, low molecular weight plasticizers. *Hercules.*

13820 Hercoflex® 900
A polymeric plasticizer; contributes to clarity, green tack, permanence and low-temperature flexibility in polyvinyl acetate adhesives. *Hercules.*

13821 Hercoflex® Plasticizer
Pentaerythrito ester
Plasticizer for PVC with outstanding heat resistance; wire insulation for high-temperature service and government specification cable construction and high-quality plastisol formulations. *Aqualon.*

13822 Hercofloc
A series of high molecular weight synthetic water soluble polymers; flocculant polymers with many uses in water management. *Hercules.* Discontinued.

13823 Hercofroth
A series of frothers; range includes pine oil, modified terpene alcohols, aliphatic-terpene alcohol, polypropylene glycol water-soluble based frothers; in flotation, frothers stabilize the air bubbles containing the unwetted particles into a froth that is easily removed from the liquid. *Hercules.* Discontinued.

13824 Hercol 2
55-63-0 6704 200-240-8
Nitroglycerin dynamite
Used in construction and building, explosives, mining, petroleum and related industries. *Hercules.* Discontinued.

13825 Hercol 2X
55-63-0 6704 200-240-8
Nitroglycerin dynamite
Used in construction and building, explosives, mining, petroleum and related industries. *Hercules.* Discontinued.

13826 Hercolube®
A range of synthetic esters (polyol type) used as lubricants; functional fluids where exposure to both high and low temperature is a primary consideration. *Hercules.*

13827 Hercolube® Synthetic Ester
Polyol esters; ready-to-use completely compound lubricating oils; wide-temperature variance gear oils designed to reduce energy consumption and maintenance in automotive and industrial uses; combine low channel point and superior thermal stability, low volatility at elevated temperatures and a high flash point. *Aqualon.*

13828 Hercolyn®
Hydrogenated methyl abietate. *Hercules Ltd.*

13829 Hercolyn® D
The hydrogenated methyl ester of rosin; tackifying resin for adhesives, plasticizing and softening agent for chewing gum, used in flexographic, type T gravure and screen-process inks, as a plasticizing resin in cellulose-based coatings. *Hercules.*

13830 Hercomix
A blasting agent containing ammonium nitrate and fuel oil; suitable for use under dry borehole conditions. *Hercules.* Discontinued.

13831 Hercon® 2
55-63-0 6704 200-240-8
Nitroglycerin dynamite
Used for construction and building, explosives, mining, petroleum and related industries. *Hercules.*

13832 Hercon® 2X
55-63-0 6704 200-240-8
Nitroglycerin dynamite
Used for construction and building, explosives, mining, petroleum and related industries. *Hercules.* Discontinued.

13833 Hercon® 32
Cationic cellulose-reactive sizing emulsion for application in gypsum paper. *Hercules.* Discontinued.

13834 Hercon® 40, 48
Cationic cellulose reactive sizing emulsions. *Hercules.* Discontinued.

13835 Herco-Prills
6484-52-2 567 229-347-8
Ammonium nitrate prills containing a minimum of 33.5% nitrogen; used alone or for bulk blending of mixed fertilizers. *Hercules.* Discontinued.

13836 Hercoprime
Adhesion promoters for powder coatings. *Hercules Ltd.* Discontinued.

13837 Hercopruf
Glycol-based freeze conditioning agents; prevents ice binding of wet minerals and ice build-up on conveyor belts. *Hercules.* Discontinued.

13838 Hercose AP
A trademark for cellulose acetate propionate; used in lacquer manufacture. No manufacturer.

13839 Hercose C
A trademark for cellulose acetobutyrate for use in lacquer manufacture. No manufacturer.

13840 Hercosett®
Polyamide epichlorohydrin resin; for shrinkproofing of wool. *Hercules.*

13841 Hercosett® 125
A water-soluble cationic resin containing reactive polyamide epichlorohydrin; imparts shrink resistance to wool. *Hercules.* Discontinued.

13842 Hercosol
A trademark for a solvent made from pine oil. No manufacturer.

13843 Hercosol TP-S
65% solids solution of dark pine-tree-derived resin in a mixed-terpene hydrocarbon solvent; designed for rubber reclaimers who need an additional solvent or a second reclaiming solvent in their process. *Hercules.* Discontinued.

13844 Hercosplit WR
55-63-0 6704 200-240-8
A nitroglycerin dynamite; for construction and building industry, explosives, mining, petroleum and related industries. *Hercules.* Discontinued.

13845 Hercotac®
Hercotac® AD. A low molecular weight modified aromatic hydrocarbon resin; used mainly in pressure sensitive systems containing natural rubber and in hot-melts based on ethylene vinyl acetate copolymer. *Hercules.* Discontinued.

13846 Hercotac® LA
A low molecular weight modified aliphatic hydrocarbon resin; used mainly in pressure sensitive systems containing natural rubber, and in hot melts based on ethylene vinyl acetate copolymer. *Hercules.* Discontinued.

13847 Hercotuf
9003-07-0 7741
Polypropylene
Powdered polypropylene. *Hercules Ltd.* Discontinued.

13848 Hercules metal
A bronze containing 85.5% copper, 10% tin, 2.5% aluminum, and 2% zinc. Another alloy contains 54% copper, 36% zinc, 7.5% iron, and 2.5% aluminum. The term is also used for an alloy of copper, nickel, and aluminum.

13849 Hercules® 4 Defoamer
High-efficiency defoamer for acid and alkaline papermaking systems, deinking systems. *Hercules.* Discontinued.

13850 Hercules® 37M6-8
50-00-0 4262 200-001-8
Formaldehyde solution; for manufacture of synthetic resins by reaction with phenols, urea, melamines for molded goods, electrical insulation, binders, plywood adhesives, varnishes, wet-strength resins for paper and textiles; chemical intermediate. mp = -92°; soluble in water, alcohol, ether. *Hercules.*

13851 Hercules® 247
Release agent for tissue and towelling machines; improves wet and dry creeping. *Hercules.*

13852 Hercules® 356 Defoamer
Defoamer for wet-end use in acid or alkaline papermaking systems. *Hercules.*

13853 Hercules® 752 Size
Nonfortified dark paste rosin; dry size used with alum in paperboard and building products to produce high level of resistance to water and aqueous solutions; primarily for unbleached kraft southern pine pulps. *Hercules.* Discontinued.

13854 Hercules® 1098
Rosin soap emulsion; air entrainment aid for flotation save-alls. *Hercules.*

13855 Hercules® 2051GS Defoamer
Hydrocarbon oil-based defoamer for use as drainage aid and foam killer in kraft pulpmill brownstock washing systems. *Hercules.*

13856 Hercules® AR 150
PEG-15 rosinate; low foaming surfactant, detergent, emulsifier, wetting agent, suspending agent, dispersant for industrial cleaners. *Hercules.* Discontinued.

13857 Hercules® AR 160
PEG-16 rosinate; low foaming surfactant, detergent, emulsifier, wetting agent, suspending agent, dispersant for industrial cleaners and food related areas. *Hercules.* Discontinued.

13858 Hercules® AS
9004-70-0 8195
Nitrocellulose
Film-former; intermediate type used in place of Hercules SS in some applications. *Aqualon.*

13859 Hercules® Improved Tech. PE
115-77-5 7245 204-104-9
Pentaerythritol
Used in production of alkyd resins, rosin esters, urethane resins, drying oils, synthetic lubricants, plasticizers, intumescent paints, plastics, stabilizers for plastics, explosives. *Hercules.* Discontinued.

13860 Hercules® K
9004-57-3 3828
Ethylcellulose
For formulation of lacquers, varnishes, inks, foils, adhesives, plastics and for food contact, animal feed, and pharmaceutical goods; good toughness, flexibility at low temperatures; low flammability. *Hercules.* Discontinued.

13861 Hercules® Mono-PE
115-77-5 7245 204-104-9
Pentaerythritol
Used in production of alkyd resins, rosin esters, urethane resins, drying oils, synthetic lubricants, plasticizers, intumescent paints, plastics, stabilizers for plastics, explosives. *Hercules.* Discontinued.

13862 Hercules® N
9004-57-3 3828
Ethylcellulose
For formulation of lacquers, varnishes, inks, foils, adhesives, plastics and for food contact, animal feed, and pharmaceutical goods; good toughness, flexibility at low temperatures; low flammability. *Hercules.* Discontinued.

13863 Hercules® RES A-2338
Synthetic resin aqueous dispersion; tackifier for acrylic-latex polymers used in label, tape, and construction adhesive applications; gives good oxidative and uv light stability in latexes. *Hercules.* Discontinued.

13864 Hercules® RS
9004-70-0 8195
Nitrocellulose
Film-former. *Aqualon.*

13865 Hercules® SS
9004-70-0 8195
Nitrocellulose
Film-former. *Aqualon.*

13866 Hercules® T
9004-57-3 3828
Ethylcellulose
For formulation of lacquers, varnishes, inks, foils, adhesives, plastics and for food contact, animal feed, and pharmaceutical goods; good toughness, flexibility at low temperatures; low flammability. *Hercules.* Discontinued.

13867 Hercules® Type AS4
PAN-based carbon fiber; continuous, high strength fiber which has been surface treated and sized to improve props.; for use in weaving, prepregging, filament winding, pultrusion and molding compounds. *Hercules.* Discontinued.

13868 Hercules® X Dry Size
Nonfortified pale dry rosin size; dry size used with alum in paperboard and building prods. to produce high level of resistance to water and aqueous solutions. *Hercules.* Discontinued.

13869 Herculine FR
Surface explosive; surface charge for geophysical prospecting. *Hercules.* Discontinued.

13870 Herculite
Molding plaster for molds used in making aluminum alloy castings. *Foseco (F.S.) Ltd.* Discontinued.

13871 Herculon
Polypropyleneolefin fiber; for floor coverings, furniture and fixtures and textiles. *Hercules.* Discontinued.

13872 Herculoy
A patented alloy; it is a silicon bronze containing tin; high tensile strength; resists corrosion. Herculoy 418 contains copper with about 3% silicon and 0.5% tin. No manufacturer.

13873 Hercures®
Aryl and alkylaryl resins; used for adhesives, hot melts, coatings, inks. *Hercules; Hercules Ltd.* Discontinued.

13874 Heresite
A proprietary trade name for a phenol-formaldehyde synthetic resin molding compound. *Heresite.* Name unverified.

13875 Herkules
An alloy of 50% silver and 50% copper; used for fuse wire.

13876 Hermann's fluid
Platino-Aceto-Osmic Acid
A fixing agent used in microscopy. It contains 15 parts of a 1% platinum chloride solution, 4 parts of a 2% osmium tetroxide solution, and 1 part of glacial acetic acid.

13877 Hermes
metoxuron-simazine
Suspension concentrate containing 480 g metoxuron and 30 g simazine per liter; post-emergence herbicide for winter cereals. *Atlas Interlates Ltd.*

13878 Hermite fluid
It contains magnesium oxide and hypochlorous acid, with from 4-5% available chlorine. A disinfectant.

13879 Herolith
Bakelite
A synthetic resin of the phenol-formaldehyde type. No manufacturer.

13880 Heron
A paper-based grade of Tufnol industrial laminates. *Tufnol Ltd.*

13881 Herox®
Polymer. *DuPont UK.*

13882 Herrifex DS
mecoprop
A liquid containing 587.5 g per liter mecoprop as the potassium salt; used to control cleavers, common chickweed and a wide range of other broad-leaved weeds in cereals, sports turf, grass seed crops and apple and pear orchards. *Bayer plc.* Discontinued.

13883 Herrisol
dicamba-MCPA-mecoprop
A liquid containing 35.4% w/w dicamba (Banvel D), MCPA and mecoprop; used to control common chickweed, cleavers, knotgrass, redshank, scentless mayweed, corn spurrey and a wide range of other broadleaved weeds in cereals, grass crops and orchards. *Bayer plc.*

13884 Herschel's crystals
$H_4Ca_3O_4S_3{\cdot}8H_2O$
Hydrated calcium-tetrahydroxytrisulfide.

13885 Hest MS
17661-50-6 241-640-2
Myristyl stearate
Bodying agent, emollient for creams and lotions; pearlescent in anionic systems; spermaceti replacement. *Heterene.*

13886 Hesthsulphid
1313-82-2 8830 215-211-5
A proprietary brand of sodium sulfide; used as a depilatory in leather manufacture. No manufacturer.

13887 Hetamine 5L-25
55819-53-9 259-837-7
Stearamidopropyl dimethylamine lactate
Antistat, conditioner for hair; compatible, with most anionic surfactants. *Heterene.*

13888 Hetan SL
1338-39-2 8872 215-663-3
Sorbitan laurate
Lipophilic emulsifier. *Heterene.*

13889 Hetan SO
1338-43-8 8872 215-665-4
Sorbitan oleate
Lipophilic emulsifier. *Heterene.*

13890 Hetan SS
1338-41-6 8872 215-664-9
Sorbitan stearate
Lipophilic emulsifier. *Heterene.*

13891 Hetastarch
9004-62-0 4707
Starch 2-hydroxyethylether
hydroxyethylcellulose; hydroxyethyl cellulose; Cellulose, 2-hydroxyethyl ether; H.E. cellulose; hydroxyethyl starch; HES; 6-H.E.S.; Hespan; Hespander; Hestar; Hestat; Hestsol; Plasmasteril; Volex. Modified cellulose polymer containing hydroxyethyl side chains; thickener, suspending agent; stabilizer for vinyl polymerization; retards evaporation of water in mortars and cements; binder in ceramic glazes. *Amerchol NV; Aqualon; Union Carbide.*

13892 Hetester 412
2778-96-3 220-476-5
Stearyl stearate
Emollient for stick cosmetics. *Heterene; Bernel.*

13893 Hetester 3236S
Myristyleicosyl stearate.
Bernel. Name unverified.

13894 Hetester FAO
C_{12-15} alcohols octanoate
Unique skin feel ingredient, wetting agent; used in personal care products, hair and skin care products, makeup, antiperspirants; emollient replacement for isopropyl myristate. *Heterene; Bernel.*

13895 Hetester HCA
27233-00-7 248-351-0
Glyceryl triacetyl hydroxystearate
Gloss agent, film-former for lip oils, skin oils, lipsticks, eye makeup; high viscosity and adherence. *Heterene; Bernel.*

13896 Hetester HCP
PPG-3 hydrogenated castor oil; emollient for lip and cosmetic products. *Heterene; Bernel.*

13897 Hetester HSS
97338-28-8 306-621-6
Isocetyl stearoyl stearate
Emollient with unusual skin feel, pigment dispersant for stick products, emulsion systems; binding oil for pressed powders. *Heterene; Bernel.*

13898 Hetester ISS
Isostearyl stearoyl stearate
Emollient for use in stick products, pigmented emulsion-type systems; binding oil for use in pressed powders. *Heterene; Bernel.*

13899 Hetester MS
17661-50-6 241-640-2
Myristyl stearate; bodying agent in creams and lotions; in certain anionic systems, imparts pearling effects; replacement for spermaceti. *Heterene; Bernel.*

13900 Hetester PCA
Propylene glycol ceteth-3 acetate
Emulsifier, pigment, wetter, antichalking agent, emollient used in personal care products, antiperspirants. *Heterene; Bernel.*

13901 Hetester PCP
Propylene glycol ceteth-3 propionate
Emulsifier, pigment, wetter, antichalking agent, emollient used in personal care products, antiperspirants. *Bernel.* Name unverified.

13902 Hetester PHA
Propylene glycol isoceteth-3 acetate
Emulsifier, pigment, wetter, antichalking agent, emollient used in personal care products, antiperspirants. *Heterene; Bernel.*

13903 Hetester PMA
Propylene glycol myristyl ether acetate
Emollient, solvent, and plasticizer for anhydrous oil systems, emulsions. *Heterene; Bernel.*

13904 Hetester SSS
Stearyl stearoyl stearate
Emollient used in cosmetic stick formulations, emulsion systems. *Heterene; Bernel.*

13905 Hetester TICC
Triisocetyl citrate
Oily liquid emollient useful in stick and pigmented emulsion-type products, skin care products, cleansing creams, makeup: pigment dispersing properties. *Heterene; Bernel.*

13906 Hetlan AC
61788-49-6 262-980-8
Acetylated lanolin alcohols; emollient for creams and lotions. *Heterene.*

13907 Hetoxamate 200 DL
9005-02-1
PEG-4 dilaurate
Thickener, foam booster, foam stabilizer, emulsifier. *Heterene.*

13908 Hetoxamate 400 DS
9005-08-7
PEG-8 distearate
Thickener, foam booster, foam stabilizer, emulsifier. *Heterene.*

13909 Hetoxamate FA-5
61791-00-2
PEG-5 tallate
Detergent, emulsifier, lubricant, softener, coupling agent for cosmetics, textiles, leather, metal cleaning. *Heterene.*

13910 Hetoxamate LA-5
9004-81-3
PEG-5 laurate
Detergent, emulsifier, lubricant, softener, coupling agent for cosmetics, textiles, leather, metal cleaning. *Heterene.*

13911 Hetoxamate MO-2
9004-96-0
PEG-2 oleate
Detergent, emulsifier for personal care products; softener for leather. *Heterene.*

13912 Hetoxamate SA-5
9004-99-3
PEG-5 stearate
Detergent, emulsifier, lubricant, softener, coupling agent for cosmetics, textiles, leather, metal cleaning. *Heterene.*

13913 Hetoxamide C-4
61791-08-0
PEG-5 cocamide
Detergent, emulsifier, lubricant, softener. *Heterene.*

13914 Hetoxamine 0-2
PEG-2 oleamine
Emulsifier, softener, antistat, water repellent, desizing agent in agriculture, waxes, oils, textile/leather, metal cleaning. *Heterene.*

13915 Hetoxamine C-2
61791-14-8
PEG-2 cocamine
Emulsifier, softener, antistat, water repellent, desizing agent in agriculture, waxes, oils, textile/leather, metal cleaning. *Heterene.*

13916 Hetoxamine S-2
61791-24-0
PEG-2 soyamine
Emulsifier, softener, antistat, water repellent, desizing agent in agriculture, waxes, oils, textile/leather, metal cleaning. *Heterene.* Name unverified.

13917 Hetoxamine ST-5
26635-92-7
PEG-5 stearamine
Emulsifier, softener, antistat, water repellent, desizing agent in agriculture, waxes, oils, textile/leather, metal cleaning. *Heterene.*

13918 Hetoxamine T-2
61791-44-4 263-177-5
PEG-2 tallow amine
Emulsifier, softener, antistat, water repellent, desizing agent in agriculture, waxes, oils, textile/leather, metal cleaning. *Heterene.*

13919 Hetoxide BN-13
35545-57-4
PEG-13 β-naphthol ether
Emollient, emulsifier, viscosity control agent, lubricant, pigment dispersant, perfume solubilizer, used in cosmetics, household, textile industry, metal treating and plating; also as a chemical intermediate. *Heterene.*

13920 Hetoxide BY-1.8
PEG-1.8 butynediol
Metal plating emulsifier. *Heterene.* Name unverified.

13921 Hetoxide C-2
61791-12-6
PEG-2 castor oil; emollient, emulsifier, solubilizer, pigment dispersant, detergent used in cosmetics, household, textile industry. *Heterene.*

13922 Hetoxide C-200-50%
61791-12-6
PEG-200 castor oil; lubricant, emulsifier, solubilizer. *Heterene.*

13923 Hetoxide DNP-4
9014-93-1
Nonylnonoxynol-4
Intermediate, emulsifier. *Heterene.*

13924 Hetoxide G-7
31694-55-0
Glycereth-7
Emulsifier. *Heterene.*

13925 Hetoxide HC-16
61788-85-0
PEG-16 hydrogenated castor oil; emollient, emulsifier, solubilizer, pigment dispersant, detergent used in cosmetics, household, textile industry. *Heterene.*

13926 Hetoxide MPC
m, p-Cresol hydrophobe
Solvent for lacquers and coatings. *Heterene.*

13927 Hetoxide NP-4
9016-45-9 6772
Nonoxynol-4
Detergent and emulsifier. *Heterene.*

13928 Hetoxide P-3
PEG-3 phenyl ether. *Heterene.*

13929 Hetoxol 15 CSA
68439-49-6 7737
Ceteareth-15
Detergent, emulsifier, leveling agent, intermediate for cosmetics, textiles, scouring agents, dyes, household products, silicone emulsifiers surfactants. *Heterene.*

13930 Hetoxol 916P
PEG-6 PPG-2.5 C9-C11 alcohols ether
Surfactant for industrial applications. *Heterene.*

13931 Hetoxol C-24
Intermediate, emulsifier, wetting agent solubilizer, coupling agent. *Heterene.*

13932 Hetoxol CA-2
9004-95-9
Ceteth-2
Detergent, emulsifier, leveling agent, intermediate, used for cosmetics, household formulations, silicone emulsification, textile processing. *Heterene.*

13933 Hetoxol CAWS
PPG-5, ceteth-20
Intermediate, emulsifier, wetting agent, solubilizer, coupling agent. *Heterene.*

13934 Hetoxol CD-4
PEG-4 2-ethyl hexyl ether
Intermediate, emulsifier, wetting agent, solubilizer, coupling agent. *Heterene.*

13935 Hetoxol CS
Cetearyl alcohol, surfactant. *Heterene.*

13936 Hetoxol CS-4
68439-49-6 7737
Ceteareth-4
Detergent, emulsifier, leveling agent, intermediate for personal care prods., wax, oils, textiles, scouring agents, dyes, household prods., silicone emulsification, surfactants. *Heterene.*

13937 Hetoxol D
Cetearyl alcohol and ceteareth-20; intermediate, emulsifier, wetting agent, solubilizer, coupling agent; for cosmetics, paper, textile industries. *Heterene.*

13938 Hetoxol G
Stearyl alcohol and ceteareth-20; intermediate, emulsifier, wetting agent, solubilizer, coupling agent. *Heterene.*

13939 Hetoxol IS-2
52292-17-8
Isosteareth-2
Surfactant. *Heterene.*

13940 Hetoxol J
Cetearyl alcohol and ceteareth-20; intermediate, emulsifier, wetting agent, solubilizer, coupling agent. *Heterene.*

13941 Hetoxol L-4
5274-68-0 226-097-1
Laureth-4
Intermediate, emulsifier, wetting agent, solubilizer, coupling agent, *Heterene.*

13942 Hetoxol LS-9
Laureth-9, steareth-9
Detergent, emulsifier, leveling agent, intermediate, used for cosmetics, household formulations, silicone emulsification, textile processing. *Heterene.*

13943 Hetoxol M-3
27306-79-2
Myreth-3
Emulsifier and pigment dispersant in makeup; intermediate, wetting agent, solubilizer, coupling agent. *Heterene.*

13944 Hetoxol MP-3
63793-60-2
PPG-3 myristyl ether
Emulsifier and pigment dispersant in makeup; intermediate, wetting agent, solubilizer, coupling agent. *Heterene.*

13945 Hetoxol OA-3 Special
9004-98-2
Oleth-3
Emulsifier and pigment dispersant for cosmetic applications. *Heterene.*

13946 Hetoxol OL-2
9004-92-2
Oleth-2
Intermediate, emulsifier, wetting agent, solubilizer, coupling agent. *Heterene.*

13947 Hetoxol P
Emulsifying wax NF; emulsion base. *Heterene.*

13948 Hetoxol PLA
68439-53-2
PPG-30 lanolin ether; oily emollient, intermediate, emulsifier, wetting agent, solubilizer, coupling agent. *Heterene.*

13949 Hetoxol SP-15
25231-21-4
PPG-15 stearyl ether; oily emollient material in cosmetics. *Heterene.*

13950 Hetoxol STA-2
9005-00-9
Steareth-2; intermediate, emulsifier, wetting agent, solubilizer, coupling agent. *Heterene.*

13951 Hetoxol TD-3
24938-91-8
Trideceth-3
Detergent, emulsifier, leveling agent, intermediate, used for cosmetics, household formulations, silicone emulsification, textile processing. *Heterene.*

13952 Hetoxol TDEP-15
PEG-10 PPG-15 tridecyl ether; surfactant for industrial use, automatic dishwashing formulations. *Heterene.*

13953 Hetphos OA-3
39464-69-2
Oleth-3 phosphate
Emulsifier and detergent. *Heterene.*

13954 Hetphos SG
50643-20-4
Emulsifier and detergent. *Heterene.*

13955 Hetquat S-20
122-19-0 204-527-9
Emulsifier and detergent. *Heterene.*

13956 Hetron®
Chlorendic, bisphenol, furan and vinyl ester resins for chemical resistant applications: halogenated resins for fire retardant applications; used for fiberglass reinforced parts for the chemical resistant building construction, electrical and transportation markets. *Ashland Chemical Company.* Name unverified.

13957 Hetron® 92
Unsaturated polyester resin, unpromoted; for flame retardant reinforced thermosetting equipment manufactured by BMC/SMC molding or pultrusion processes; as halogen concentrated resin for blending with non-flame retardant resins. *Ashland.*

13958 Hetron® 99P
Isophthalic polyester resin, halogenated, promoted. *Ashland.*

13959 Hetron® 197-3
Polyester resin, halogenated, unpromoted, thermoset; corrosion and heat resistant, flame retardant resin for filament wound and hand lay-up tanks, pipes, pumps, stacks, scrubbers, and other equipment, handling corrosive gases, vapors or liquids. *Ashland.*

13960 Hetron® 197AT
Unsaturated, styrene-containing polyester resin; Class I fire retardant, chemical- and heat-resistant, for high-temperature applications such as stacks and ducts, handling corrosive gases and vapors. *Ashland.*

13961 Hetron® 692
Polyester resin, halogenated; flame retardant resin for corrosion-resistant fiberglass-reinforced plastic equipment including press molded equipment. *Ashland.*

13962 Hetron® 700 DMA
Accelerated bisphenol A fumarate polyester resin; low-viscosity, chemical-resistant resin, good wet-out, excellent gel time, stability, and craze resistance; for corrosion control applications. *Ashland.*

13963 Hetron® 800
Furan resin; thermoset for corrosion-resistant fiberglass-reinforced plastic equipment including pipes, tanks, fume hoods, ducts, linings, and flake glass coatings where fire retardancy is not required. *Ashland.*

13964 Hetron® 922
Vinyl ester resin; used for hand lay-up, sprayup, and filament winding, flake glass, filled lining and coating compounds. *Ashland.*

13965 Hetron® FR 991
Vinyl ester resin; flame retardant, corrosion resistant resin for filament wound, hand lay-up and sprayup boats, tanks, pipes, duct, stacks, scrubbers, linings. *Ashland.*

13966 Hetsorb L-10
9005-64-5 8872
PEG-10 sorbitan laurate
Detergent, emulsifier, lubricant for cosmetics; corrosion inhibitor. *Heterene.*

13967 Hetsorb L-20
9005-64-5 8872
Polysorbate 20
Detergent, emulsifier, lubricant for cosmetics; corrosion inhibitor. *Heterene.*

13968 Hetsorb L-4
9005-64-5 8872
Polysorbate 21
Emulsifier, lubricant, thickener, corrosion inhibitor. *Heterene.*

13969 Hetsorb L-80-72%
9005-64-5 8872
PEG-80 sorbitan laurate
Surfactant for shampoos, used as counterirritant. *Heterene.*

13970 Hetsorb O-20
9005-65-6 7742
Polysorbate 80
Emulsifier, lubricant for cosmetics; corrosion inhibitor. *Heterene.*

13971 Hetsorb O-5
9005-65-6 7742
Polysorbate 81
Food emulsifier. *Heterene.*

13972 Hetsorb P-20
9005-66-7 8872
Polysorbate 40
General purpose emulsifier. *Heterene.*

13973 Hetsorb S-20
9005-67-8 8872
Polysorbate 60
Emulsifier for vitamins. *Heterene.*

13974 Hetsorb S-4
9005-67-8 8872
sorbitan stearate
Emulsifier. *Heterene.*

13975 Hetsorb TO-20
9005-70-3 8872
Polysorbate 85
General purpose emulsifier, thickener, lubricant, corrosion inhibitor. *Heterene.*

13976 Hetsorb TS-20
9005-71-4 8872
Polysorbate 65
General purpose emulsifier, thickener, lubricant, corrosion inhibitor. *Heterene.*

13977 Hetsulf 40, 40X
Sodium dodecylbenzene sulfonate
Wetting agent, emulsifier, dispersant, intermediate, detergent, liquid formulation syndet. *Heterene.* Name unverified.

13978 Hetsulf 50A
1331-61-9 215-559-8
Ammonium dodecylbenzene sulfonate
Wetting agent, emulsifier, dispersant, for light duty detergent formulations. *Heterene.* Name unverified.

13979 Hetsulf 60S
Sodium dodecylbenzene sulfonate
Wetting agent, emulsifier, dispersant, base for formulated prods. *Heterene.* Name unverified.

13980 Hetsulf 60T
Amine dodecylbenzene sulfonate
Wetting agent, emulsifier, dispersant, for cosmetic, bath and shampoos uses. *Heterene.* Name unverified.

13981 Hetsulf Acid
Sodium dodecylbenzene sulfonate
Wetting agent, emulsifier, dispersant, intermediate, base for neutralized surfactant. *Heterene.* Name unverified.

13982 Hetsulf IPA
MIPA-dodecylbenzenesulfonate
Wetting agent, emulsifier, dispersant. *Heterene.* Name unverified.

13983 Heusler alloy
An alloy of 66-68% copper, 18-22% manganese, 10-11% aluminum, and 0-4% lead.

13984 Heveacrumb
A proprietary compressed rubber crumbled in oil *RRI Malaya.* Unverified.

13985 Heveatex
The trade name to denote a series of preserved, concentrated, or processed Hevea rubber latexes. No manufacturer.

13986 Hevikote
Two peak pitch epoxy protective coating. *Thomas Ness Ltd.*

13987 Hexa-Betalin
58-56-0 8166 200-386-2
Pyridoxine hydrochloride
Vitamin B_6 *Eli Lilly & Co.* Discontinued.

13988 Hexabromocyclododecane
25637-99-4 247-148-4
$C_{12}H_{18}Br_6$
1,2,5,6,9,10-Hexabromocyclododecane
FR-1206; Great Lakes CD-75P™; Great Lakes SP-75™; Saytex® 60006L; Saytex® BCT-610; Saytex® HBCD-LM. Fire retardant for wide range of plastics, textiles, adhesives, and coatings; esp. for styrene-based systems. mp = 188-191; SG = 2.36; soluble in common solvents; LD_{50} (rat orl) >10,000 mg/kg, (rbt dermal) >10,000 mg/kg. *AmeriHaas; Dead Sea Bromine.*

13989 Hexacal
Fire retardant additives for rigid foams. *ICI Chem & Polymers Ltd.*

13990 Hexacarb
Black dye for leather. *Pointing Ltd.* Name unverified.

13991 Hexacert
USA certified foodstuff colorant. *Pointing Ltd.* Name unverified.

13992 hexachlorophene
70-30-4 4716 200-733-8
$C_{13}H_6Cl_6O_2$
2,2'-methylenebis[3,4,6-trichlorophenol]
AT-7; Bilevon; Dermadex; Exofene; G-11; Hexosan; pHisohex; Surgi-Cen; Surofene; bis(2-hydroxy-3,5,6-trichlorophenyl)methane; 2,2'-dihydroxy-3,5,6,3',5',6'-hexachlorodiphenylmethane; hexophene; methane, bis(2,3,5-trichloro-6-hydroxyphenyl); HCP; Compound G-11; Nabac; Acigena; Almederm; B32; Bivelon; Cotofilm; Dihydroxy-3,3'5,5',6,6'-hexachlorodiphenylmethane; Eleven; Fomac; Fostril; Gamophene; Germa-medica; Hexabalm; Hexafen; Hexide; Methylenebis(3,4,6-trichlorophenol); Neosept V; Phisodan; Ritosept; Septisol; Septofen; Steral; Steraskin; Tersaseptic; Trichlorophene; Turgex,. Used in manufacture of germicidal soaps. mp = 164-165°; insoluble in H_2O, soluble in organic solvents; LD_{50} (rat orl) = 66 mg/kg.

13993 Hexacide
Insect and vermin killer. *Pointing Ltd.* Name unverified.

13994 Hexacol
Foodstuff colorants of guaranteed specification. *Pointing Ltd.* Name unverified.

13995 Hexaderm
Fast dyes for leather. *Pointing Ltd.* Name unverified.

13996 Hexafoam
Isocyanate foams. *ICI Chem & Polymers Ltd.* Discontinued.

13997 Hexalan
Fast to light colors for wool. *Pointing Ltd.* Name unverified.

13998 Hexalin
108-93-0 2794 203-630-6
Cyclohexanol
Chemical intermediate. *DuPont.* Name unverified.

13999 Hexallac
Dyes for cellulose lacquers. *Pointing Ltd.* Name unverified.

14000 hexamethylamine
100-97-0 6036 202-905-8
$C_6H_{12}N_4$
hexamethylene-tetramine
aminoform; methenamine. HMTA; used in the curing of phenol formaldehyde and resorcinol formaldehyde, resins, rubber to textile adhesives, protein modifier, organic synthesis, pharmaceuticals.

14001 Hexamethylcyclotrisilazane
1009-93-4 213-773-6
$C_6H_{21}N_3Si_3$
1,1,3,3,5,5-Hexamethylcyclotrisilazane
Hexamethylcyclotrisilazane. Difunctional blocking agent; reagent for cyclosilylation. mp= -10°; bp_{756} = 188°; d = 0.9200. *Hüls Am.*

14002 Hexamethyldisilane
1450-14-2 215-911-0
$C_6H_{18}Si_2$
Starting material for preparation of trimethylsilyl alkali compounds. mp = 15°; bp = 112-114°; d = 0.7150; n_D^{20}= 1.4221. *R.W. Greeff.*

14003 hexamethyldisilazane
999-97-3 4725 213-668-5
$C_6H_{19}NSi_2$
bis(trimethylsilyl) amine
HMDS; 1,1,1,3,3,3-Hexamethyldisilazane; hexamethyl disilizane; 1,1,1-Trimethyl-N-(trimethylsilyl)silanamine; Dynasylan HMDS. HMDS; silica reaction product; chemical intermediate, chromatographic packings, silylating agent. Used for the preparation of volatile derivatives of a wide range of biologically active compounds for glc analysis. bp = 125°; d = 0.7650; n_D^{20}= 1.4071; LD_{50} (rat orl) = 850 mg/kg. *Aldrich; Dow Corning; Hüls Am.; Janssen Chimica; PCR.*

14004 Hexamethyldisiloxane
107-46-0 203-492-7
$C_6H_{18}OSi_2$
H7250; H7310; HMDSO. As cleaning, polishing and damping media; offers low toxicity, inertness. mp = -59°; bp = 101°; d = 0.7640; n_D^{20}= 1.3775. *Hüls Am.*

14005 Hexamethylene Glycol
629-11-8 4726 211-074-0
$C_6H_{14}O_2$
1,6-hexanediol
1,6-dihydroxyhexane. Intermediate in the production of nylon; manufacture of hexamethylenediamine, polyesters, polyesters, polyurethanes; in gasoline refining; as plasticizer. mp = 42.8°; bp = 208°; n_D^{20}= 1.4579; soluble in water, alcohol, sparingly soluble in hot ether; LD_{50}(rat orl) = 3.73 g/kg.

14006 hexamethylene tetramine
100-97-0 6036 202-905-8
$C_6H_{12}N_4$
methenamine
aminoform; urotropine; hexamine; HMTA. Used for curing of phenolformaldehyde and resorcinolformaldehyde resins, adhesives, fungicide, antibacterial. *Allchem Industries; Dajac Labs; R.W. Greeff; Mitsubishi Gas; OxyChem/Durez.*

14007 hexamic Acid
100-88-9 2770 202-898-1
Cyclamic acid
Sweetener. *Abbott Laboratories.* Name unverified.

14008 hexamine
100-97-0 6036 202-905-8
$C_6H_{12}N_4$
Hexamethylenetetramine
Cystamin; cystogen; metramine; urotropine; formin; naphthamine; xametrin; vesaloin; urisol; uritone; hex; H.M.T.; formaldehyde-ammonia 6:4; carin; ammonioformaldehyde; vesalvine. Used in curing of phenolformaldehyde and resorcinolformaldehyde resins.

14009 n-hexane
110-54-3 4729 203-777-6
C_6C_{14}
Exxsol® Hexane; alkane C_6. Hexane; solvent. Aliphatic compound.; solvent, alcohol denaturant, paint diluent, polymerization reaction medium; filling for thermometers. d = 0.660; mp = -100° to -95°; bp = 69°; n_D^{20}= 1.375; insoluble in water; miscible with alcohol, chloroform, ether; LD_{50} (rat orl) = 32 mg/kg. *Exxon; Ashland; BP Chem. Ltd; Exxon; Humphrey; Mitsui Petrochem. Ind.; Phillips 66; Texaco.* Name unverified.

14010 hexanhexol
69-65-8 5788 200-711-8
$C_6H_{14}O_6$
Mannitol
Used in organic synthesis.

14011 1-hexanol
111-27-3 4732 203-852-3
$C_6H_{14}O$
n-Hexyl alcohol;
Epal® 6; Hexanol; amylcarbinol; pentylcarbinol; 1-hydroxyhexane; Hexyl alcohol. Detergent and emulsifier intermediate. d = 0.8153; mp = -51.6°; bp = 157°; n_D^{20} = 1.4162; fp (closed cup) = 145° F; slightly soluble in water; miscible with alcohol, ether; LD_{50} (rat orl) = 4.59 g/kg. *Ethyl Corp.*

14012 Hexapar
Detergent preparations. *Pointing Ltd.* Name unverified.

14013 Hexaphos
50813-16-6 8814 256-779-4
Sodium hexametaphosphate
FMC. Discontinued.

14014 Hexaplant Richter
Water, alcohol, fennel extract, hops extract, balm mint extract, mistletoe extract, matricaria extract, yarrow extract; polyvalent herbal extract in aqueous alcohol; emollient for aqueous and hydroalcoholic herbal cosmetics, skin and hair products, emulsified preparations. *Dr. Kurt Richter; Henkel/Cospha.*

14015 Hexaplus
Plasticizers. *ICI Chem & Polymers Ltd.* Discontinued.

14016 Hexaryl D60L
Triethanolamine alkylaryl sulfonate in liquid form; anionic surfactant. *Witco Chemical Ltd.* Discontinued.

14017 Hexasol
Alcohol-soluble dyes. *Pointing Ltd.* Name unverified.

14018 Hexatype
Dyes for doubletone printing inks. *Pointing Ltd.* Name unverified.

14019 Hexavibex
58-56-0 8166 200-386-2
Pyridoxine hydrochloride
Vitamin B_6. *Parke-Davis.* Name unverified.

14020 hexazinone
51235-04-2 4734 257-074-4
$C_{12}H_{20}N_4O_2$
3-cyclohexy-6-(dimethylamino)-1-methyl-1,3,5-triazine-2,4(1H,3H)-dione
DPX 3674; Velpar. Non-selective contact herbicide. Inhibits photosynthesis. Used for control of annual, biennial, perennial weeds and woody plants in non-crop areas and coniferous plantations. mp = 115-117°; d = 1.25; soluble in H_2O (33 g/l), more soluble in organic solvents; LD_{50} (rat orl) = 1590 mg/kg.

14021 hexazinone
51235-04-2 4734 257-074-4
$C_{12}H_{20}N_4O_2$
3-cyclohexy-6-(dimethylamino)-1-methyl-1,3,5-triazine-2,4(1H,3H)-dione
DPX 3674; Velpar. Non-selective contact herbicide. Inhibits photosynthesis. Used for control of annual, biennial, perennial weeds and woody plants in non-crop areas and coniferous plantations. mp = 115-117°; d = 1.25; soluble in H_2O (33 g/l), more soluble in organic solvents; LD_{50} (rat orl) = 1590 mg/kg.

14022 Hexcel 164M
Nonexpanding two-part PU compound (TDI-and MOCA-free); universal cable plugging compound; forms open sheath moisture blocks and pressure dams in paper/pulp and plastic insulated telecommunications cables. *Hexcel.*

14023 Hexcel 174 Rapid-Dam.
Nonexpanding, two-part polyether PU compound (TDI-free); noncorrosive, cable plugging compound; for low cost moisture blocks and pressure dams in paper/pulp and plastic (PIC) insulated telecommunications cables. *Hexcel.*

14024 Hexcel 195 RE Hexagel
Re-enterable encapsulant; remains permanently soft and tacky; blocks out moisture, adhesion to cable and closures. *Hexcel.*

14025 Hexcelcure 160
Propoxylated amine
Epoxy curative. *Zeeland.*

14026 Hexcelcure 169
Cyanoethylated amine
Epoxy curative. *Zeeland.*

14027 hexedine
5980-31-4 4736
$C_{22}H_{45}N_3$
2,6-bis(2-ethylhexyl)hexahydro-7a-methyl-1H-imidazo[1,5-c]imidazole
Sterisol. Antibacterial. $bp_{0.025}$ = 131°; n_D^{22}= 1.4660.

14028 Hexela
Cellulose acetate and nylon dyes. *Pointing Ltd.* Name unverified.

14029 1-hexene
592-41-6 209-753-1
C_6H_{12}
C_6 linear alpha olefin
butyl ethylene; hexylene. Intermediate for surfactants and specialty industrial chemicals (flavors, perfumes, dyes, resins); polymer modifier. mp = -140°; bp = 63°; d = 0.6780; n_D^{20}= 1.3877. *Monsanto (Solaris); Ethyl; Hüls Am.; Phillips 66; Shell.*

14030 Hexetidine
141-94-6 4741 205-513-5
Hexetidine
Antimicrobial, antifungal agent for orl hygiene products. *Angus.*

14031 Hexil
131-73-7 3396 205-037-8
$C_{12}H_5N_7O_{12}$
2,4,6-trinitro-N-(2,4,6-trinitrophenyl)benzenamine
dipicrylamine; hexyl; hexite; hexanitrodiphenylamine. Used as a booster explosive, in gravimetric analysis for potassium. mp = 238°; insoluble in H_2O, organic solvents.

14032 Hexnitrol
Leather stains. *Pointing Ltd.* Name unverified.

14033 Hexo
Industrial detergent. *Crosfield Chemicals Ltd.* Discontinued.

14034 Hexoll
Oil and varnish dyes. *Pointing Ltd.* Name unverified.

14035 Hexomax
Immobilized invertase enzyme. *UOP Speciality Products.*

14036 Hexoran
A sodium alkylaryl sulfonate; for scouring, wetting-out level bleaching, dyeing etc on all fibers. *Roehm Ltd.* Name unverified.

14037 Hexoran A15
Sodium alkylaryl sulfonate
Anionic surfactant in liquid form; used for scouring, wetting out, level bleaching, dyeing, etc.; for all fibers. *Roehm Ltd.* Name unverified.

14038 Hexsotate
hexamethylene-tetramine sodio-acetate

14039 hexyl alcohol
111-27-3 4732 203-852-3
$C_6H_{14}O$
1-hexanol
n-hexanol; pentylcarbinol; amylcarbinol. Aliphatic alcohol; pharmaceuticals (antiseptics, perfume esters), solvent, plasticizer. *Ashland; Ethyl; Penta Mfg.; Vista.*

14040 Hexyl Jasmat®
Acetyl ethyloctanoate
Floral, herbal, green, jasmine-like fragrance *BASF AG.*

14041 hexylene glycol
107-41-5 4748 203-489-0
$C_4H_{14}O_2$
2-Methyl-2,4-pentanediol
α,α,α-trimethyltrimethyleneglycol; pinakon. Aliphatic alcohol used in hydraulic brake fluids, printing inks, as a coupling agent and penetrant for textiles, cosmetics; ice inhibitor in carburetors; fuel, lubricant additive d = 0.924; bp = 198°; fp = ca. 93°; n_D^{20} = 1.4276; soluble in water, alcohol, ether; LD_{50} (rat orl) = 4.70 g/kg. *Ashland; BP Chem. Ltd; Mitsui Petro-chem. Ind.; Penta Mfg.; Shell; Union Carbide.*

14042 hexylene glycol
107-41-5 4748 203-489-0
$C_6H_{14}O_2$
2-methyl-2,4-pentanediol
α,α,α'-trimethyltrimethylene glycol; pinakon; Diolane. Used in cosmetics and hydraulic fluids. bp = 198°; d_{15}^{15} = 0.924; n_D^{20}= 1.4276; soluble in H_2O, organic solvents; LD_{50} (rat orl)= 4.7 g/kg.

14043 Heyn's reagent
The double chloride of copper and ammonia; used to reveal ferrite in the micro-analysis of carbon steels.

14044 HFC
Compounded plastic material. *Mitsubishi Petrochem.* Name unverified.

14045 Hi Temp EC-1000
Emulsion of aliphatic oils, salts and soaps with water; die casting and mold release agent, drilling and tapping fluid/coolant. *Hi Temp Lubricants Inc.* Name unverified.

14046 Hi Temp EC-4000
Emulsion of aliphatic oils, salts and soaps with water; die casting mold release agent. *Hi Temp Lubricants Inc.* Name unverified.

14047 Hi Temp EC-5000
Emulsion of aliphatic oils, salts and soaps with water; die casting mold release agent, die casting plunger lubricant. *Hi Temp lubricants Inc.* Name unverified.

14048 Hibbo
A proprietary aluminum bronze containing iron. No manufacturer.

14049 Hibiclens
18472-51-0 2140 242-354-0
Chlorhexidine gluconate
An antimicrobial. *Imperial Chemical Industries plc.*

14050 Hibidil
55-56-1 2140 200-238-7
Chlorhexidine
An antiseptic and bactericidal solution containing chlorhexidine. *ICI Chem & Polymers Ltd.*

14051 Hibiscrub
55-56-1 2140 200-238-7
Chlorhexidine
A proprietary preparation of chlorhexidine; an antiseptic. *ICI Chem & Polymers Ltd.*

14052 Hibisol
55-56-1 2140 200-238-7
Chlorhexidine
Antiseptic hand rub containing chlorhexidine. *ICI Chem & Polymers Ltd.*

14053 Hibispray
55-56-1 2140 200-238-7
Chlorhexidine
Antiseptic spray containing chlorhexidine. *ICI Chem & Polymers Ltd.*

14054 Hibitane
18472-51-0 2140 242-354-0
Chlorhexidine digluconate
A proprietary preparation of Chlorhexidine digluconate; an antiseptic. *ICI Chem & Polymers Ltd.*

14055 Hi-bor
Destral. Nonselective herbicides; For weedkilling. *Borax Europe Ltd.*

14056 Hibosol
High boiling point solvents. *BP Chemicals Ltd.* Discontinued.

14057 Hibudine
A proprietary trade name for a synthetic rubber, probably derived from butadiene. No manufacturer.

14058 Hi-Build
Paint filler. *ICI Chem & Polymers Ltd.*

14059 Hi-Carbolon®
Carbon fiber and carbon fiber prepreg. *Asahi Chem. Industry.*

14060 Hi-Care® 1000
65497-29-2
Guar hydroxypropyltrimonium chloride
For skin care products. *Rhône-Poulenc Surf.*

14061 Hi-Cat
Cationic starch; used in the paper industry, wet end additive. *Roquette (UK) Ltd.*

14062 Hickstor
117-18-0 204-278-2
Tecnazene
Tecnazene; used to prevent fungus in stored potatoes. *Hickson & Welch Ltd.*

14063 Hickstor 6 ·2 MBC
carbendazim-tecnazene
Carbendazim and tecnazene; Protectant fungicide and sprout suppressant for stored potatoes. *Hickson & Welch Ltd.*

14064 Hicond-2000
Electrically conductive olefin; high melt strength compound suitable for profile sheet extrusion, thermoforming and blow molding. *United Composites.*

14065 Hicore 90
A nickel-chromium-molybdenum case-hardening steel for heavy motor vehicles and other gears.

14066 HiD 9301
9002-88-4 7728
Polyethylene
HDPE resin; pressure pipe resin; excellent ESCR. *Monsanto (Solaris).*

14067 HiD 9602
9002-88-4 7728
Polyethylene
HDPE homopolymer resin; sheet resin used for toys, housewares, and other stiff sheet and thermoformed parts. *Monsanto (Solaris).*

14068 HiD 9632
9002-88-4 7728
Polyethylene
HDPE copolymer resin; film resin used for notion and millinery bags, coextrusions, insulation bags, shipping bags, multiwall liners. *Monsanto (Solaris).*

14069 HiD 9650
9002-88-4 7728
Polyethylene
HDPE copolymer resin; film resin for high strength and barrier films; used for notion and millinery bags, barrier packaging, coextrusions. *Monsanto (Solaris).*

14070 Hidosin
Disinfectant. *Thomas Goldschmidt Ltd.*

14071 Hiduminium
A registered trade name for a range of aluminum alloys. *High duty Alloys (Properties) Ltd.* Name unverified.

14072 Hi-Ex Foam
Fire fighting foam concentrate. *Henkel/Cospha.*

14073 HiFax AB 6023
9003-07-0 7741
Polypropylene
PP resin; acoustical barrier injection molding grade offering sound deadening, excellent processability, flexibility; used for automotive isolators, fender seals, door seals. *Himont.*

14074 HiFax CA 45A
9003-07-0 7741
Polypropylene
PP resin; semirigid, medium modulus injection molding grade with superior processability, excellent low temperature performance; used for automotive bumper fascias, valence panels, air dams, body side moldings, wheel flares, rub strips. *Himont.*

14075 HiFax CB 17AC
9003-07-0 7741
Polypropylene
PP resin; rigid, high modulus extrusion/blow molding grade with good dimensional stability, high heat resistance; used for thermoformed truck bumpers, blow molded truck wind skirts, roofing, and automotive blow molded bumper beams, extruded roof rack strips,blow molded rigid air ducts. *Himont.*

14076 HiFax ETA 3011
9003-07-0 7741
Polypropylene
PP resin; low modulus, soft grade for injection molding applications; excellent low temperature performance for automotive mud flaps, isolators, map pockets, bumper and caps, and cup holders. *Himont.*

14077 HiFax ETA 3095
9003-07-0 7741
Polypropylene
PP resin; semirigid, medium modulus injection molding grade with superior processability, excellent low temperature performance; used for automotive bumper fascias, valence panels, air dams, body side moldings, wheel flares, rub strips. *Himont.*

14078 HiFax ETA 5012
9003-07-0 7741
Polypropylene
PP resin; soft, low modulus grades for extrusion/blow molding; offers excellent low temperature performance, high heat oxidative stability; used for roofing and automotive air ducts, intake/outtake air hoses, seat back headrest covers. *Himont.*

14079 HiFax RTA 3263E
9003-07-0 7741
Polypropylene
PP resin; extruded roof rack strips, blow molded rigid air ducts; d = 1.10 g/cc; melt flow 0.4 dg/min; tensile strength 22.8 MPa; tensile elong. 330%; Izod impact 748 J/m notch; tear strength 112 kN/m; Shore hardness D63; distort. temp. 131 F(264 psi). *Himont.*

14080 Hi-Fibre
Asbestos substitute. *Hill Brothers Chemical Co.*

14081 HiGel
637-12-7 379 211-279-5
Aluminum stearate
Nonaqueous coating compositions. *Hüls Am.* Discontinued.

14082 High Temperature Deodorant 4896, OS
Mixture of fragrance materials; masks odors released in the plant during high temperature processing, (i.e. 400ºF). *Andrea Aromatics.* Name unverified.

14083 high tensile brass
An alloy of 76% copper, 22% zinc, and 2% aluminum.

14084 High Trees Mixture B
Nonionic spreader; wetting agent for phosphonoglycine herbicide sprays. *Service Chemicals Ltd.*

14085 Highlink 40®, 80®
aliphatic dialdehyde
H_2S scavenger for oil and gas industry. *Hoechst Celanese.*

14086 Hightensite
A trade name for a proprietary hard rubber composition for use in electrical insulation. No manufacturer.

14087 HiGlass BJ44A
9003-07-0 7741
Polypropylene
PP copolymer, glassreinforced; high melt flow, high impact grade for underhood automotive, fan shrouds, headlamp retainers, lawn tractor grilles and other applications. *Himont.*

14088 HiGlass PF062-2
9003-07-0 7741
Polypropylene
PP homopolymer glassreinforced; high melt flow series with 20 and 50% glass reinforcement; features superior resistance to warpage, excellent stiffness and high temperature performance, and a wide processing window; used for appliances, automotive parts, electrical/electronic connectors. *Himont.*

14089 HiGlass PF072-1
9003-07-0 7741
Polypropylene
PP homopolymer, glassreinforced; standard melt flow series with 10-40% glass reinforcement; features superior resist. to warpage, excellent stiffness and high temperature performance, wide processing window, improved impact and creep performance; used for dishwasher components, pump housings and components, plumbing parts and athletic equipment. *Himont.*

14090 HiGlass SB 224-2
9003-07-0 7741
Polypropylene
PP copolymer, 20% glass-reinforced standard melt flow series with 20-50% glass reinforcement; features excellent cold temperature impact resistance; uses include wheel well housings, housings for lawn and garden equipment, splice closures. *Himont.*

14091 Hi-Gloss 1
Solvent base sealer; an acrylic sealer with high solids for application to exterior and interior concrete, stone and masonry type surfaces for commercial use only. *Glessner Corporation Inc (GGI Products) DBA.* Name unverified.

14092 Hi-heet
An American synthetic resin molded product for electrical insulation. No manufacturer.

14093 Hiirogane
A blood-red colored metal prepared either by the treatment of copper with a solution of copper sulfate and verdigris, or by heating a copper alloy with a paste containing a salt of copper, borax, and water.

14094 Hills-McCanna Alloy No.45
An alloy containing 88% copper, 10.5% aluminum, and 1.5% iron.

14095 Hinge
10605-21-7 1836 234-232-0
Carbendazim
A liquid formulation containing 500 g carbendazim per liter as a suspension concentrate; systemic fungicide. *Quadrangle Agrochemicals.*

14096 Hinochloa®
73250-68-7 277-328-8
Mefenacet
Herbicide effective against grasses (especially against *Echinochloa crus-galli* and some broad-leaved weeds in transplanted paddy rice; mainly used in combinations with other compounds. *Bayer AG.*

14097 Hinosan®
17109-49-8 3560 241-178-1
Edifenphos
Fungicide especially effective against *Pyricularia oryzae on rice; also for control of other rice diseases. Bayer AG.*

14098 Hiotrol
For commercial and residential agents for the control and reduction of odors associated with latrines, toilets, lavatories, locker and shower rooms, kennels, livestock pens, industrial malodors, sewage wastes, garbage, land fills and lagoons. *Ferro.* Discontinued.

14099 Hipec®
Semiconductor protective coating. *Dow Corning.*

14100 Hipernick®
A registered trademark for an alloy of nickel and iron in equal parts; used for making cores of audiofrequency transformers. No manufacturer.

14101 Hipersil®
A registered trademark for high permeability silicon steel; used in high-frequency communications equipment. No manufacturer.

14102 Hipersolv
High performance solvents for HPLC. *BDH Chemicals Ltd.*

14103 Hi-Pflex® 100
471-34-1 1697 207-439-9
Calcium carbonate
Surface-treated filler improving impact strength and flexibility modulus in polymers; high-performance reinforcing agent for plastics; applications incl. PVC (pipe, wire and cable), PP (interior and exterior auto parts, toys, pallets, corrugated boxes), HDPE (pipe and other rigid applications, wire and cable and other flexible applications. *Pfizer.*

14104 Hi-pHase® 35
Dispersed rosin size; sizing agent for paper/paperboard. *Hercules.*

14105 Hi-pHorm 67
Rosin-based; surface treatment agent *Hercules.* Discontinued.

14106 Hipochem B-3-M
136-60-7 1587 205-252-7
Butyl Benzoate
Emulsified butyl benzoate; self-emulsifying carrier for polyester and triacetate fibers. *High Point.*

14107 Hipochem Carrier 761
Emulsified aromatic hydrocarbon; dye carrier for poly/wool blends. *High Point.*

14108 Hipochem Compatibilizer WMC
Surfactant blend; compatibilizer for cationic/disperse dyes. *High Point.*

14109 Hipochem EK-18
577-11-7 3460 209-406-4
Sodium dioctyl sulfosuccinate
Wetting, rewetting, desizing, bleaching, scouring for textiles. *High Point.*

14110 Hipochem GM
120-82-1 9760 204-428-0
1,2,4-trichlorobenzene
Emulsified 1,2,4-trichlorobenzene; dyeing assistant; carrier for polyester in removal of trimers. *High Point.*

14111 Hipochem Jet Dye T
Emulsified trichlorobenzene; dyeing assistant; carrier for jet dyeing machine; aids in removal of trimers. *High Point.*

14112 Hipochem Jet Scour
Modified ethoxylated alcohol; low foaming scour for all fibers; good wetting; excellent removal of mineral oil, butyl stearate, greases. *High Point.*

14113 Hipochem M-51
Nonsurface active quaternary ammonium compound; dyeing assistant; migrating agent for dyeing basic dyeable fibers. *High Point.*

14114 Hipochem Migrator J
56-93-9 200-300-3
Benzyl trimethyl ammonium chloride
Migrating agent for basic dyes. *High Point.*

14115 Hipochem MS-BW
Emulsified aromatic and aliphatic solvents; low foaming detergent, solvent scour for removal of oils, grease, waxes, and fatty materials in textile operations. *High Point.*

14116 Hipochem MTD
Complex phosphate ester; wetting agent, leveling agent in caustic boil-off, bleach baths, other wet processing of piece goods, hosiery, knit goods; dispersant, antiredeposition aid. *High Point.*

14117 Hipochem PDO
Optically brightened detergent powder for improving wash effectiveness on denim. *High Point.*

14118 Hipochem SRC
Self-emulsifying solvent systems; detergent, removes dyes, oil, grease, pigment, binders from pads and machinery. *High Point.*

14119 Hipofix 491
Methyl methylol resin; fixing agent for direct dyes on celluosics. *High Point.*

14120 Hi-Point® 90
1338-23-4 215-661-2
MEK peroxide in dimethyl phthalate; catalyst/initiator for R.T. cures of polyester. *Witco Corporation.*

14121 Hipolon New
Modified cationic surfactant; detergent, dye retarder, leveler. *High Point.*

14122 Hiposcour® 3-80
Modified ethoxylate; scouring agent for synthetics and cellulosics; detergent with nonredeposition props. for mineral oils, motor oils, grease, butyl stearate stains; biodegradable. *High Point.*

14123 Hiposcour® BFS
Complex Varsol/surfactant blend; detergent scour for removal of oil and grease from cotton and blends; wetting agent; stable over broad pH range, to caustic; biodegradable. *High Point.*

14124 Hipowet IBS
Complex modified ethoxylate; wetting and scouring agent, detergent, suspending agent; caustic stable; stable with enzymes. *High Point.*

14125 Hippuran-^{131}I
133-17-5 5055 205-097-5
$C_9H_7INNaO_3$
N-(2-iodobenzoyl)glycine-^{131}I monosodium salt
o-Iodohippurate sodium-^{131}I; Hippodin; Jodairol; Hippuran I-131; Hipputope. Diagnostic aids; The unlabelled form provides a radiopaque medium, the radioactive form serves as a radioactive imaging agent. *Mallinckrodt Inc.* Name unverified.

14126 Hippuric acid
495-69-2 4752 207-806-3
$C_9H_9NO_3$
benzaminoacetic acid
benzoylaminoacetic acid; benzoylglycocoll; benzoylglycin. Used in organic synthesis and medicine. mp = 187-188°; soluble in H_2O (4 g/l), less soluble in organic solvents. *Penta Mfg.; Schweizerhall; U.S. Biochemical.*

14127 Hippuryl Amide
$C_9H_8O_2N_3$
N-Benzoyl-glycinamide
A white powder; a substrate in studies of papain action. *British Drug Houses.* Name unverified.

14128 Hipputope
133-17-5 5055 205-097-5
N-(2-iodobenzoyl)glycine-^{131}I monosodium salt
Iodohippurate sodium I 131; diagnostic acid; radioactive agent. *Bristol-Myers Squibb Co Inc.* Name unverified.

14129 Hirathiol
A compound used as a substitute for ichthyol, a topical anti-infective.

14130 Hi-Selon
Polyvinyl alcohol film. *British Traders & Shippers Ltd.*

14131 Hi-Sil
7631-86-9 8637 231-545-4
silicon dioxide
Synthetic, amorphous, precipitated silica. An abrasive. *PPG Industries.*

14132 Hismanal
68844-77-9 891 272-441-9
1-[(4-fluorophenyl)methyl]-N-[1-[2-(4-methoxyphenyl)ethyl]4-piperidinyl]-1H-benzimidazol-2-amine
R-43512; Astemisan; Histamen; Histaminos; Histazol; Kelp; Laridal; Metodik; Novo-Nastizol A; Paralegin; Retolen; Waruzol. A proprietary preparation containing astemizole; a nonsedative antihistaminic used for treatment of allergic rhinitis and conjunctivitis (hay fever), and other allergic reactions. mp = 149°; insoluble in H_2O, soluble in organic solvents; λ_m= 219, 249, 286 nm (ε 27250, 6480; 8634 EtOH). *Janssen Pharmaceutical Ltd.* Name unverified.

14133 Hi-Sol®
Aromatic hydrocarbon solvents; used in coatings. *Ashland Chemical Company.* Name unverified.

14134 Hi-Sol® 10,15,70
8030-30-6 7329 232-443-2
Naphtha
Aromatic solvent. *Ashland.*

14135 Hispor 45WP
Carbendazim-maneb
Carbendazim and maneb; systemic fungicide for winter cereals. *Ciba Geigy Agrochemicals.*

14136 Histazarin
2,3-Dihydroxyanthraquinone

14137 Histo-Acryl
A proprietary preparation of enbucrilate; a surgical tissue adhesive. No manufacturer.

14138 Hitac 300
9003-07-0 7741

polypropylene
High viscosity amorphous polypropylene. *Crowley Chem.*

14139 HiTEC®
Poly-alpha-olefins *Ethyl Corp; Ethyl SA.*

14140 HiTEC® 300
Lubricant additive for automotive and industrial gear oils. *Ethyl Corp.*

14141 HiTEC® 800
Lubricant additive for diesel engine oils. *Ethyl Corp.*

14142 HiTEC® 2900
Lubricant additive for antiwear hydraulic oils. *Ethyl Corp.*

14143 HiTEC® 4000
Performance additive for cleaner burning distillate fuel. *Ethyl Corp.*

14144 HiTEC® 4400
Performance additive for gasolines with enhanced detergency for optimum engine performance. *Ethyl Corp.*

14145 HiTEC® 4700
Performance additive for gasoline and jet fuel with enhanced oxidation inhibition; lubricant additive for lubricants with improved oxidation inhibition. *Ethyl Corp.*

14146 Hitenso
A proprietary trade name for a cadmium bronze. No manufacturer.

14147 HiTint
Pigment dispersions; for coloration of solvent-based coatings. *Pacific Dispersions Inc.* Name unverified.

14148 Hitox®
13463-67-7 9612 236-675-5
Titanium dioxide
Buff-colored pigment developed as alternative to white titanium dioxide; used in alkyds, acrylic urethanes, high solids systems, water reducibles, water bases, powder coatings, inks, and adhesives. *Hitox.*

14149 Hi-Zex
9002-88-4 7728
polyethylene
A trademark for Japanese high d polyethylene. *Mitsui Co Ltd.* Name unverified.

14150 H-K Mastitis
5321-32-4 4706 226-182-3
Hetacillin potassium
Antibacterial. *Bristol Laboratories.* Name unverified.

14151 HM-0230
Thermoplastic polyamide adhesive; offers good penetration of fibrous or porous substrates and rapid setting; as folding cement in shoe manufacture. *H.B. fuller.* Name unverified.

14152 HM-0652
Thermoplastic polyamide adhesive; offers rapid setting and high temperature resistant bonds with fibrous or porous substrates; for manufacture of pleated air filters for automotives. *H.B. Fuller.* Name unverified.

14153 HM-0814
Polyamide hot-melt adhesive; for fiber/foil tube winding operations; excellent adhesion to metals, foils, coated paper and other dense surfaces. *H.B. Fuller.* Name unverified.

14154 HM-6300
Thermoplastic polyester hot-melt adhesive; for demanding production assembly applications; produces flexible bonds over broad temperature range with sufficient open time to bond large components or complex shapes. *H.B. Fuller.* Name unverified.

14155 Hobane
bromoxynil [1689-84-5]-ioxynil [16849-83-4]
Bromoxynil and ioxynil; herbicide. *ICI Chem & Polymers Ltd.*

14156 Hobane
bromoxynil [1689-84-5]-ioxynil [16849-83-4]
An emulsifiable concentrate containing 240 g bromoxynil [1689-84-5] and 160 g ioxynil [1689-33-4] per liter; a post-emergence contact herbicide for cereal crops. *Farm Protection Ltd.*

14157 Hodag 20-L
9004-81-3
PEG-4 laurate
Emulsifier, wetting agent, plasticizer for cosmetic, pharmaceutical and other uses. *Calgene.*

14158 Hodag 22-L
9005-02-1
PEG-4 dilaurate
Emulsifier, wetting agent, plasticizer for cosmetic, pharmaceutical and other uses. *Calgene.*

14159 Hodag 40-O
9004-96-0
PEG-8 oleate
Emulsifier, wetting agent, plasticizer for cosmetics, pharmaceuticals, other uses. *Calgene.*

14160 Hodag 40-R
9004-97-1
PEG-8 ricinoleate
Emulsifier, wetting agent, plasticizer for general cosmetic, pharmaceutical, and other uses. *Calgene.*

14161 Hodag 42-0
9005-07-6
PEG-8 dioleate
Emulsifier, wetting agent, plasticizer for cosmetic, pharmaceutical and other uses. *Calgene.*

14162 Hodag 150-S
9004-99-3
PEG-6-32 stearate
Emulsifier, wetting agent, plasticizer for cosmetic, pharmaceutical and other uses. *Calgene.*

14163 Hodag 602-S
9005-08-7
PEG-150 disearate
Thickener and auxiliary emulsifier for creams and lotions, esp. amphoteric type shampoos. *Calgene.*

14164 Hodag Antifoam CO-350
Silicone antifoam; antifoam for nonaqueous foaming systems; for fermentation, chemical processing, food, pharmaceutical, paint, adhesives, paper coatings, metalworking, lubricants, textile processing, petroleum, pulp and paper, cleaning compounds. *Calgene.*

14165 Hodag Antifoam F-1
8050-81-5 3264
Simethicone
Antifoam for aqueous and nonaqueous foaming systems; for fermentation, chemical processing, food, pharmaceutical, paint, adhesives, paper coatings, metalworking, lubricants, textile processing, petroleum, pulp and paper, cleaning compounds. *Calgene.*

14166 Hodag Antifoam FD-82
Silicone emulsion; antifoam for aqueous foaming problems; for fermentation, chemical processing, food, pharmaceutical, paint, adhesives, paper coatings, metalworking, lubricants, textile processing, petroleum, pulp and paper, cleaning compounds. *Calgene.*

14167 Hodag C-100-L
136-99-2 205-271-0
1-Hydroxyethyl-2-lauric imidazoline
Intermediate for quaternary ammonium compounds; strongly absorbed on textiles, paper and many metal surfs.; for agricultural, asphalt, cleaners, corrosion inhibitors, demulsifiers, flotation, metalworking, paints, pigment grinding, inks, textiles, wax emulsions. *Calgene.*

14168 Hodag C-100-O
1-Hydroxyethyl-2-oleic imidazoline
Intermediate for quaternary ammonium compounds; strongly absorbed on textiles, paper and many metal surfaces; for agricultural, asphalt, cleaners, corrosion inhibitors, demulsifiers, flotation, metalworking, paints, pigment grinding, inks, textiles, and wax emulsions. *Calgene.*

14169 Hodag C-100-S
95-19-2 202-397-8
1-Hydroxyethyl-2-stearic imidazoline
Intermediate for quaternary ammonium compounds; strongly absorbed on textiles, paper and many metal surfaces; for agricultural, asphalt, cleaners, corrosion inhibitors, demulsifiers, flotation, metalworking, paints, pigment grinding, inks, textiles, wax emulsions. *Calgene.*

14170 Hodag C-100-T
61791-39-7 263-171-2
1-Hydroxyethyl-2-tall oil imidazoline
Intermediate for quaternary ammonium compounds; strongly absorbed on textiles, paper and many metal surfaces; for agricultural, asphalt, cleaners, corrosion inhibitors, demulsifiers, flotation, metalworking, paints, pigment grinding, inks textiles and wax emulsions. *Calgene.*

14171 Hodag CC-22
Propylene glycol dicaprylate/dicaprate
Surfactant for food, cosmetics, and pharmaceutical industries; vehicle/diluent/carrier for vitamins, drugs, flavors, color, fragrance; emollient for makeup, bath and skin oils. *Calgene.*

14172 Hodag CSA-101
Cetearyl alcohol-ceteth-20-glycol stearate. Emulsifier, opacifier, stabilizer for food, drug, and cosmetic industries. *Calgene.*

14173 Hodag CSA-80
31394-71-5
PPG-26 oleate
Nongreasy spreading agent, skin moisturizer, emollient, vehicle for skin and

hair care products, bath oils; carrier and dispersant for additives (hormones, vitamins, essential oils, germicides, etc.); foam depressing properties. *Calgene.*

14174 Hodag DGL
9004-81-3
PEG-2 laurate
Emulsifier. *Calgene.*

14175 Hodag DGO
9004-96-0
PEG-2 oleate
Lubricant. *Calgene.*

14176 Hodag DGS
9004-99-3
PEG-2 stearate
Emulsifier, opacifier, thickener for cosmetics, oil-water emulsions; lubricant for stamping and drawing; protective coating for hygroscopic materials (tablets); opacifier/lubricant for paper industry; antitack agent. *Calgene.*

14177 Hodag DGS-C
PEG-2 stearate SE
Self-emulsifying emulsifier, opacifier, thickener for cosmetics, oil-water emulsions; lubricant for stamping and drawing; protective coating for hygroscopic materials (tablets); opacifier/lubricant for paper industry; antitack agent. *Calgene.*

14178 Hodag DOSS-70
577-11-7 3460 209-406-4
Dioctyl sodium sulfosuccinate
Wetting agent, surface tension reducer. *Calgene.*

14179 Hodag EGMS
Glycol stearate
Wetting agent, surface tension reducer. *Calgene.*

14180 Hodag GML
142-18-7 205-526-6
Glyceryl laurate
Emulsifier, opacifier, stabilizer for food, drug, and cosmetic industries. *Calgene.*

14181 Hodag GMO
Glyceryl oleate
Emulsifier, opacifier, stabilizer for food, drug, and cosmetic industries. *Calgene.*

14182 Hodag GMP
Glyceryl palmitate
Surfactant. *Calgene.*

14183 Hodag GMR
141-08-2 205-455-0
Glyceryl ricinoleate
Emulsifier, opacifier, stabilizer for food, drug, and cosmetic industries. *Calgene.*

14184 Hodag GMS
Glyceryl stearate
Emulsifier, opacifier, stabilizer for food, drug, and cosmetic industries. *Calgene.*

14185 Hodag GTO
Glyceryl trioleate
Emulsifier, opacifier, stabilizer for food, drug, and cosmetic industries. *Calgene.*

14186 Hodag Nonionic 1017-R
9003-11-6 7721
Meroxapol 171
Surfactant. *Calgene.*

14187 Hodag Nonionic 1035-L
9003-11-6 7721
Poloxamer 105
Detergent, antifoam, wetting agent, emulsifier, antistat, demulsifier, viscosity modifier, deduster, gelation aid, metalworking lubricants, dispersants. *Calgene.*

14188 Hodag Nonionic 1044-L
9003-11-6 7721
Poloxamer 124
Detergent, antifoam, wetting agent, emulsifier, antistat, demulsifier, viscosity modifier, deduster, gelation aid, metalworking lubricants, dispersants. *Calgene.*

14189 Hodag Nonionic 1064-L
9003-11-6 7721
Poloxamer 184
Detergent, antifoam, wetting agent, emulsifier, antistat, demulsifier, viscosity modifier, deduster, gelation aid, metalworking lubricants, dispersants. *Calgene.*

14190 Hodag Nonionic 1088-F
9003-11-6 7721
Poloxamer 238
Detergent, antifoam, wetting agent, emulsifier, antistat, demulsifier, viscosity modifier, deduster, gelation aid, metalworking lubricants, dispersants. *Calgene.*

14191 Hodag Nonionic 2017-R
9003-11-6 7721
Meroxapol 172
Detergent, antifoam, dispersant, demulsifier, dishwashing rinse and viscosity control agent. *Calgene.*

14192 Hodag Nonionic E-5
9016-45-9 6772
Nonoxynol-5
Detergent and wetting agent for cosmetics, insecticides and other formulations. *Calgene.*

14193 Hodag Nonionic GR-8
61791-12-6
PEG-8 castor oil; surfactant. *Calgene.*

14194 Hodag Nonionic GRH-25
61788-85-0
PEG-25 hydrogenated castor oil; surfactant. *Calgene.*

14195 Hodag Nonionic ID-5
Isodeceth-5
Surfactant. *Calgene.*

14196 Hodag Nonionic L-4
Laureth-4
Surfactant. *Calgene.*

14197 Hodag Nonionic S-2
9005-00-9
Steareth-2
Surfactant. *Calgene.*

14198 Hodag Nonionic TD-15
24938-91-8
Trideceth-15
Surfactant. *Calgene.*

14199 Hodag PB-285
9003-13-8
PPG-15
Surfactant. *Calgene.*

14200 Hodag PE-005
Alkylaryl phosphate ester
Emulsifier, EP lube additive, antistat, corrosion inhibitor, surfactant. *Calgene.*

14201 Hodag PE-005-K
Potassium phosphate ester
Solubilizer for nonionic surfactants in built liquid concentrates, especially low-foam nonionics; also raises cloud point of rinse-aid formulations; stable over wide pH range; nondiscoloring on solid caustic. *Calgene.*

14202 Hodag PEG 200
25322-68-3 7729
PEG-4
Cosmetics and pharmaceuticals formulation; plasticizer for adhesives; inks; resins and coatings. *Calgene.*

14203 Hodag PEG 300
25322-68-3 7729
PEG-6
Cosmetics and pharmaceutical formulation; latex coagulating bath; plasticizer for adhesives, spray-on bandages; resins and coatings. *Calgene.*

14204 Hodag PEG 400
25322-68-3 7729
PEG-8
Cosmetic and pharmaceutical formulation; humectant, coupler for lotions; release agent for rubber; latex coagulating bath; in PVAc paints. *Calgene.*

14205 Hodag PEG 540
25322-68-3 7729
PEG-6
PEG-32. Cosmetic, pharmaceutical, and suppository formulation; humectant, plasticizer in adhesives; base for metal polishes; lubricant for paper sizes; inks; in alkyd resins and coatings. *Calgene.*

14206 Hodag PEG 600
25322-68-3 7729
PEG-12
Cosmetic and pharmaceutical formulation; resins and coatings. *Calgene.*

14207 Hodag PEG 1000
25322-68-3 7729
PEG-20
Cosmetic and pharmaceutical formulation; resins and coatings; imparts

dimensional stability to paper wet strength resins, improves coatings gloss. *Calgene.*

14208 Hodag PEG 1450
25322-68-3 7729
PEG-32
Cosmetic and pharmaceutical formulation; resins and coatings. *Calgene.*

14209 Hodag PEG 3350
25322-68-3 7729
PEG-75
Cosmetic and pharmaceutical formulation; resins and coatings; hu-mectant, plasticizer for adhesives; antistat for rubber conveyor belt; in shoe polish; lubricant for paper sizing; printing inks; tablet binder, lubricant. *Calgene.*

14210 Hodag PEG 8000
25322-68-3 7729
PEG-150; cosmetic and pharmaceutical formulation; resins and coatings; antistat for rubber conveyor belting; in shoe polish; lubricant for paper size; printing inks; release agent for rubber; tablet binder/lubricant. *Calgene.*

14211 Hodag PGL
Triglyceryl monolaurate, emulsifier. *Calgene.*

14212 Hodag PGL-101
Decaglyceryl monolaurate
Emulsifier. *Calgene.*

14213 Hodag PGML
Propylene glycol laurate
Surfactant for food industry. *Calgene.*

14214 Hodag PGMP
Propylene glycol palmitate
Surfactant. *Calgene.*

14215 Hodag PGMS
1323-39-3 215-354-3
Propylene glycol stearate
Surfactant for food industry. *Calgene.*

14216 Hodag PGO-101
9007-48-1
Decaglyceryl monooleate
Surfactant. *Calgene.*

14217 Hodag PGO-1010 (formerly Hodag SVO-10107)
11094-60-3 234-316-7
Decaglycerol decaoleate
Surfactant. *Calgene.*

14218 Hodag PGO-102
Decaglyceryl dioleate
Surfactant. *Calgene.*

14219 Hodag PGO-103
Decaglyceryl trioleate
Surfactant. *Calgene.*

14220 Hodag PGO-104 (formerly Hodag SVO-1047)
34424-98-1 252-011-7
Decaglycrol tetraoleate
Surfactant. *Calgene.*

14221 Hodag PGO-108
Decaglycryl octaoleate
Surfactant. *Calgene.*

14222 Hodag PGO-61
9007-48-1
Hexaglyceryl monooleate
Emulsifier. *Calgene.*

14223 Hodag PGO-62
Hexaglyceryl dioleate
Emulsifier. *Calgene.*

14224 Hodag PGS
1323-39-3 215-354-3
Propylene glycol stearate
Emulsifier for food processing. *Calgene.*

14225 Hodag PGS-101
37349-34-1
Decaglyceryl monostearate
Surfactant. *Calgene.*

14226 Hodag PGS-1010
39529-26-5 254-495-5
Decaglyceryl decastearate
Surfactant. *Calgene.*

14227 Hodag PGS-102
Decaglyceryl distearate
Surfactant. *Calgene.*

14228 Hodag PGS-103
Decaglyceryl tristearate
Emulsifier. *Calgene.*

14229 Hodag PGS-104
Decaglyceryl tetrastearate
Emulsifier. *Calgene.*

14230 Hodag PGS-108
Decaglyceryl octastearate
Emulsifier. *Calgene.*

14231 Hodag PGS-61
37349-34-1
Hexaglyceryl monostearate
Surfactant. *Calgene.*

14232 Hodag PGS-62
Hexaglyceryl distearate
Surfactant. *Calgene.*

14233 Hodag PGSH
Triglyceryl monoshortening
Surfactant. *Calgene.*

14234 Hodag PGSH-61
Hexaglyceryl monoshortening
Surfactant. *Calgene.*

14235 Hodag PGSH-62
Hexaglyceryl dishortening
Surfactant. *Calgene.*

14236 Hodag Polyglycol 5035
PPG-28-buteth-35
Surfactant. *Calgene.*

14237 Hodag PSML-20
9005-64-5 8872
Polysorbate 20
Foodgrade emulsifier. *Calgene.*

14238 Hodag PSMO-20
9005-65-6 7742
Polysorbate 80
Emulsifier for food processing, industrial applications. *Calgene.*

14239 Hodag PSMP-20
9005-66-7 8872
Polysorbate 40; emulsifier for food processing, industrial applications. *Calgene.*

14240 Hodag PSMS-20
9005-67-8 8872
Polysorbate 60; emulsifier for food processing, industrial applications. *Calgene.*

14241 Hodag PSTS-20
9005-71-4 8872
Polysorbate 65; emulsifier for food processing, industrial applications. *Calgene.*

14242 Hodag SML
1338-39-2 8872 215-663-3
Sorbitan laurate; emulsifier, oil additive, corrosion inhibitor; foodgrade emulsifier. *Calgene.*

14243 Hodag SMO
1338-43-8 8872 215-665-4
Sorbitan oleate; emulsifier, oil additive, corrosion inhibitor; food grade emulsifier. *Calgene.*

14244 Hodag SMP
26266-57-9 8872 247-568-8
Sorbitan palmitate; emulsifier, oil additive, corrosion inhibitor; food-grade emulsifier. *Calgene.*

14245 Hodag SMS
1338-41-6 8872 215-664-9
Sorbitan stearate; emulsifier for food processing. *Calgene.*

14246 Hodag SSO
8007-43-0 8872 232-360-1
Sorbitan sesquioleate, emulsifier. *Calgene.*

14247 Hodag STO
26266-58-0 8872 247-569-3
Sorbitan trioleate; emulsifier, oil additive, corrosion inhibitor; food-grade emulsifier. *Calgene.*

14248 Hodag STS
26658-19-5 8872 247-891-4
Sorbitan tristearate; emulsifier, oil additive, corrosion inhibitor; food-grade emulsifier. *Calgene.*

14249 Hodag SVO-1047
34424-98-1 252-011-7
Decaglyceryl tetraoleate; food emulsifier. *Calgene.*

14250 Hodag SVO-629
Hexaglyceryl distearate
Food emulsifier. *Calgene.*

14251 Hodag SVO-9
9005-65-6 7742
PEG-20 sorbitan oleate; emulsifier; kosher grade. *Calgene.*

14252 Hodag SVS-18
9005-67-8 8872
PEG-20 sorbitan stearate; food emulsifier; kosher grade. *Calgene.*

14253 Hoe S 1816
EO/PO block copolymer; surfactant for agric. formulations. *Hoechst Celanese/Colorants & Surf.*

14254 Hoe S 1984 (TP 2279)
Calcium dodecylbenzene sulfonate/nonylphenol ethoxylate; surfactant for agric. formulations. *Hoechst Celanese/Colorants & Surf.*

14255 Hoe S 2650
Dilaureth-4 dimonium chloride; surfactant for conditioners, shampoos. *Hoechst Celanese/Colorants & Surf.*

14256 Hoe S 2713
Sodium dodecylbenzene sulfonate
Surfactant for agric. formulations. *Hoechst Celanese/Colorants & Surf.*

14257 Hoe S 2721
Polyglyceryl-2 sesquiisostearate
W/o emulsifier for personal care prods,; for prods. with high temp. stability. *Hoechst Celanese/Colorants & Surf.*

14258 Hoe S 2749
TEA dodecylbenzene sulfonate
Surfactant for agric. formulations. *Hoechst Celanese/Colorants & Surf.*

14259 Hoe S 3618
EO/PO block copolymer bis-mono-phosphate ester
Surfactant for agric. formulations. *Hoechst Celanese/Colorants & Surf.*

14260 Hoe S 3680
8013-07-8 232-391-0
Epoxidized soybean oil; surfactant for agric. formulations. *Hoechst Celanese/Colorants & Surf.*

14261 Hoechst New Blue
The calcium salt of the di-and trisulfonic acids of trimethyltriphenylpararosanilinetrisulfonic acid; a dyestuff; dyes wool blue from an acid bath. No manufacturer.

14262 Hoechst Wax C
Amide wax based on bis-stearoyl ethylenediamine; lubricant and release agent for polyolefins; improves processing in engineering thermoplastics such as polyamides and acetal. *Hoechst AG.*

14263 Hoechst Wax E
Glycol/butylene glycol montanate
Internal/external lubricant and release agent for PVC and other polymers; for sheet, film and bottle applications, esp. for tin stabilized systems. *Hoechst AG; Hoechst Celanese.*

14264 Hoechst Wax OP
Butylene glycol montanate, calcium montanate; internal/external lubricant and release agent for PVC and other polymers; maintains clarity of finished product. *Hoechst AG; Hoechst Celanese.*

14265 Hoechst Wax PE 190
9002-88-4 7728
Polyethylene wax
For mfg. of transparent PVC articles with high surface gloss. *Hoechst AG; Hoechst Celanese.*

14266 Hoechst Wax PE 520
9002-88-4 7728
Polyethylene wax
Carrier for pigment concs. for polyolefins; external lubricant for PVC. *Hoechst AG; Hoechst Celanese.*

14267 Hoechst Wax PED 191
9002-88-4 7728
Oxidized polyethylene wax
Lubricant for PVC; processing aid for rigid PVC compounds; reduces sticking tin-stabilized PVC melts to hot machine parts. *Hoechst AG.*

14268 Hoechst Wax PED 521
9002-88-4 7728
Polyethylene wax
Lubricant for PVC, reduces sticking of tin-stabilized PVC melts to hot machine parts. *Hoechst AG.*

14269 Hoechst Wax PP 230
9003-07-0 7741
Polypropylene wax
External lubricant for PVC, carrier for pigments. *Hoechst AG.*

14270 Hoechst Wax S
68476-03-9 270-664-6
Montan acid wax
Lubricant and release agent for PVC and other polymers. *Hoechst AG; Hoechst Celanese.*

14271 Hoegrass
51338-27-3 3133 257-141-8
Diclofop-Methyl
Emulsifiable concentrate containing 378g diclofop-methyl per liter; used for control of weeds in grass. *Hoechst UK.*

14272 Hoenle's cement
A cement, consisting of 2 parts shellac, and 1 part Venice turpentine.

14273 Holcote
Thixotropic ready-for-use water based coatings for sand molds and cores. *Foseco (F.S) Ltd.*

14274 Holdfast D
dicamba-paclobutrazol
Soluble concentrate of 25 g dicamba and 250 g paclobutrazol per liter; used for weed control in cereals and grassland. *ICI Agrochemicals.*

14275 Holfos bronze
An alloy of copper with 11-12% tin, 0.25% lead, and 0.1-0.2% phosphorus.

14276 Holite
A proprietary trade name for a synthetic resin for molding and laminating. No manufacturer.

14277 Hollofil®
Fibers. *DuPont UK.*

14278 holmium
7440-60-0 4763 231-169-0
Ho
Metallic element; getter in vacuum tubes, research in electrochemistry, spectroscopy. mp = 1474°; bp = 2700°. *Atomergic Chemetals; Cerac; Noah Chem.; Rhône-Poulenc Basic.*

14279 holmium oxide
12055-62-8 235-015-3
Ho_2O_3
Holmia
Refractories, special catalyst. *Atomergic Chemetals; Cerac; Noah Chem.; Rhône-Poulenc Basic.*

14280 Holtox
atrazine-cyanazine
A suspension concentrate containing 250 g atrazine and 250 g cyanazine per liter; a residual herbicide. *Shell UK.*

14281 Homac
A proprietary synthetic resin. No manufacturer.

14282 Homagenets Aorl
68-26-8 10150 200-683-7
Vitamin A
Vitamin A; vitamin (antixerophthalmic). *SmithKline Beecham.* Name unverified.

14283 Homberg's metal
A fusible alloy, containing 3 parts bismuth, 3 parts tin, and 3 parts lead. It has a melting point of 122 C.

14284 Homberg's phosphorus
10043-52-4 1699 233-140-8
$CaCl_2$
calcium chloride
Anhydrous calcium chloride, which, when fused, and exposed to the sun, becomes phosphorescent in the dark.

14285 Homberg's salt
10043-35-3 1364 233-139-2
H_3BO_3
Boric acid
Astringent and antiseptic.

14286 Hombifine®
Chemical products for industrial purposes, especially pigments for paints and colors, fibers, cosmetics and catalysts. *Sachtleben Chemie GmbH.*

14287 Hombisorp®
Chemical adsorbents for purifying liquids and gases. *Sachtleben Chemie GmbH.*

14288 Hombitan®
13463-67-7 9612 236-675-5
titanium dioxide
Titanium dioxide, TiO_2; white inorganic pigment for paints and coatings. *Sachtleben Chemie GmbH.*

14289 Hombitec®
Chemical products for industrial purposes, especially pigments for paints and colors, fibers, cosmetics and catalysts. *Sachtleben Chemie GmbH.*

14290 **Homo Size 7A**
Blend; neutral sizing agent for paper manufacture (petroleum resin base). *Toho Chem. Industry.*

14291 **Homokol**
A sensitize for silver bromide plates. It is a mixture of quinoline red with an isocyanine dye.

14292 **Homosalate**
118-56-9 4776 204-260-8
$C_{16}H_{22}O_3$
2-hydroxybenzoic acid 3,3,5-trimethylcyclohexyl ester
salicylic acid 3,3,5-trimethylcyclohexyl ester; homosalate; homomenthyl salicylate; heliophan; Kemester® HMS. Used as an ultraviolet screen in Coppertone products. bpss4 = 161-165°; n^{20} = 1.516 to 1.518; d^{25}_{25} = 1.045.; insoluble in H_2O, soluble in organic solvents. *Norda Inc; Witco/Humko; RW Greeff & Co Inc.*

14293 **Hondostab**
Lead stabilizers. *British Traders & Shippers Ltd.*

14294 **Hondurite**
A proprietary molded composition, with cotton and a vulcanized binder, for electrical insulation. No manufacturer.

14295 **Honey**
8028-66-8
Fancol HON. Honey; biological additive, flavoring, skin conditioner, humectant for shampoos, face, body and hand creams and lotions, bath products, hair conditioners, cleansing products, moisturizing creams and lotions. *Fanning.*

14296 **Honey Wax**
Mold release based wax for plastic molding. *Specialty Products Co.* Name unverified.

14297 **Honeycat**
Catalysts for the control of air pollution; used as off-highway exhaust purification catalysts, industrial air pollution control catalysts. *Johnson Matthey plc.*

14298 **Hopcalite I**
A mixture of 50% manganese dioxide, 30% copper oxide, 15% cobalt oxide, and 5% silver oxide; used as a catalyst to oxidize carbon monoxide.

14299 **Hopkin's lactic acid reagent**
110-02-1 9490 203-729-4
Thiophene

14300 **Hopkin's-Cole reagent**
To a liter of a saturated solution of oxalic acid, 60 g of sodium amalgam are added, and the mixture allowed to stand. It is then filtered and diluted with from 2-3 volumes of water; used for the detection of proteins.

14301 **Hopkin's-Cole Tyrosine C reagent**
This contains mercuric sulfate dissolved in a solution of sulfuric acid.

14302 **Hopp II**
A modified aqueous hop extract; standardized to 35% reduced iso-alpha acids; for addition to malt beverages after fermentation to standardize bitterness. *Pfizer International.*

14303 **Hopper salt**
7647-14-5 8742 231-598-3
sodium chloride
Sodium chloride, which has been caused to crystalize in large hollow cubes which float when alum is added to the bath.

14304 **Hoppit**
8208
Bitter wood; Bitter ash. Quassia; animal repellent for outdoor crops. Has been used as an anthelmintic. *Fieldspray.*

14305 **Horco X**
A proprietary trade name for a thermosetting polyvinyl butyral synthetic resin. No manufacturer.

14306 **Hordaflex, Hordalub**
Chlorinated paraffins. *Hoechst UK.*

14307 **Hordamer**
Polyethylene primary dispersions. *Hoechst UK.*

14308 **Horizon/Horizont**
107534-96-3 9253 403-640-2
Tebuconazole
Fungicide with systemic properties/broad spectrum activity against rusts, leaf spot diseases, e.g., *Septoria spp.*, powdery mildew and several *Fusarium* species on cereals; whitemold, Phoma and various leaf spot diseases on oilseed. *Bayer AG.*

14309 **Horizon®**
For the agriculture industry. *DuPont UK.*

14310 **Hormone Rooting Powder**
133-06-2 1815 205-087-0
captan
Captan and a hormonal compound. Fungicide. *Murphy Chemical Co Ltd.* Discontinued.

14311 **Horn O'Plenty**
Various granular fertilizer blends; farm and garden fertilizers. *Horn's Crop Service Center.* Name unverified.

14312 **Horna**
Lead chromate yellow and molybdate red pigments. *Ciba plc.*

14313 **Hortag Aquasulf**
7704-34-9 9142 231-722-6
sulfur
Suspension concentrate containing 900 g/l sulfur; a protectant fungicide. *Avon Packers Ltd.*

14314 **Hortag Carbotec**
carbendazim, tecnazene
Carbendazim and tecnazene; protectant fungicide and sprout suppressant for stored potatoes. *Avon Packers Ltd.*

14315 **Hortag Tecnacarb Dust**
carbendazim-tecnazene
Carbendazim and tecnazene; protectant fungicide and sprout suppressant for stored potatoes. *Avon Packers Ltd.*

14316 **Hortag Tecnazene Plus**
tecnazene-thiabendazole
Dustable powder containing 6% w/w tecnazene [117-18-0] and 1.8% thiabendazole [148-79-8]; a protectant fungicide and potato sprout suppressant. *Avon Packers Ltd.*

14317 **Hortag Thiram**
137-26-8 9510 205-286-2
Thiram
Fungicide with animal repellent properties which is available as a dust, wettable powder or suspension. *Avon Packers Ltd.*

14318 **Hortichem Spraying Oil**
Emulsifiable concentrate containing 710 g/l petroleum oil; an insecticide and acaricide. *Hortichem Ltd.*

14319 **Hortus**
Fertilizers. *Scottish Agricultural Industries plc.*

14320 **Hostacerin CG**
Cetearyl alcohol, triceteareth-4 phosphate, PEG-6 oleamide, sodium C_{14-17} secondary alkane sulfonate; self-emulsifying base material for manufacture of oil-water creams. *Hoechst Celanese/Colorants & Surf.; Hoechst AG.*

14321 **Hostacerin DGO**
Polyglyceryl-2 sesquioleate
Emulsifier for cosmetics and pharmaceuticals. *Hoechst Celanese/Colorants & Surf.; Hoechst AG.*

14322 **Hostacerin DGS**
Polyglyceryl-2-PEG-4 stearate
Coemulsifier and thickener for cosmetic oil-water emulsions. *Hoechst Celanese/Colorants & Surf.; Hoechst AG.*

14323 **Hostacerin O-20**
9004-98-2
Oleth-20; surfactant. *Hoechst Celanese/Colorants & Surf.*

14324 **Hostacerin T-3**
68439-49-6 7737
Ceteareth-3; emulsifier, superfatting agent, base for ointments, creams, liquid emulsions, shampoo additive. *Hoechst Celanese/Colorants & Surf.; Hoechst AG.*

14325 **Hostacerin WO**
Polyglyceryl-2 sesquiisostearate, beeswax, mineral oil, magnesium stearate, and aluminum stearate; emulsifier conc. for cosmetic water-oil emulsions. *Hoechst Celanese/Colorants & Surf.; Hoechst AG.*

14326 **Hostacor**
Corrosion inhibitors. *Hoechst UK.*

14327 **Hostacor 2098**
Complex carboxylic acid; wetting agent, corrosion inhibitor in aqueous systems, synthetic metalworking and cleaning formulations. *Hoechst Celanese/Colorants & Surf.*

14328 **Hostacor 2125**
Carboxylic acid complex; corrosion inhibitor for synthetics on ferrous and nonferrous metals. *Hoechst Celanese/Colorants & Surf.*

14329 **Hostacor BBM**
Boric acid/carboxylic acid amine condensate; emulsifier, wetting agent, and corrosion inhibitor for semi-synthetic metalworking fluids. *Hoechst Celanese/Colorants & Surf.*

14330 **Hostacor H Liq. N**
Arylsulfonamidocarboxylic acid
Chemical intermediate for manufacture of corrosion inhibitors for aqueous systems, metalworking fluids. *Hoechst Celanese/Colorants & Surf.*

14331 Hostacor TP 2445
Boric acid/carboxylic acid amine condensate; high boron content corrosion inhibitor with surface active properties; wetting agent. *Hoechst Celanese/Colorants & Surf.*

14332 Hostadrill Brands
Vinylamide/vinylsulfonic acid copolymers; filtration agents. *Hoechst AG.*

14333 Hostadur
A trade name for partially crystalline thermoplastic polyester based on ethylene terephthalate used for construction of rigid components, e.g., gears. *Hoechst UK.* Name unverified.

14334 Hostaflam
Flame retardants for polymers. *Hoechst UK.*

14335 Hostaflex
Vinyl acetate modifying copolymer. *Hoechst UK.*

14336 Hostaflon®
Polytetrafluoroethylene resins and dispersions. *Hoechst AG.*

14337 Hostaflon® C2
A proprietary polychlorotrifluoroethylene. *Hoechst UK.* Name unverified.

14338 Hostaflon® ET
A proprietary ethylene tetrafluoroethylene copolymer. *Hoechst UK.* Name unverified.

14339 Hostaflon® TF 1101
9002-84-0 7743 204-126-9
PTFE; ram extrusion grade. *Hoechst Celanese.* Name unverified.

14340 Hostaflon® TF 1620
9002-84-0 7743 204-126-9
PTFE suspension polymer; compression molding grade for isostatic compression molding; produces skived film with excellent electrical properties. *Hoechst AG.* Name unverified.

14341 Hostaflon® TF 2071
9002-84-0 7743 204-126-9
PTFE emulsion polymer; paste powder for extrusion applications; d = 450 g/l (bulk), 2.1 g/cc (sintered tube); Sintered tube: tensile strength 28 MPa; tensile elong. 300% (break). *Hoechst AG.* Name unverified.

14342 Hostaflon® TF 5032
9002-84-0 7743 204-126-9
Unmodified PTFE dispersion polymer; good film-forming and penetration properties; for coatings on prepared metal or ceramic surfaces, and impregnation of absorbent materials (e.g., asbestos, glass fibers, carbon, sintered metal); nonstick coating on cookware, impregnation of glass fiber fabrics, production of PTFE-impregnated asbestos packimg. *Hoechst AG.* Name unverified.

14343 Hostaflon® TF 5537
9002-84-0 7743 204-126-9
Modified PTFE dispersion polymer; pigmented, aqueous dispersion for topcoats. *Hoechst AG.* Name unverified.

14344 Hostaflot L Grades
Aliphatic dithiophosphates; flotation collector for sulfide minerals. *Hoechst AG.*

14345 Hostaform
Acetal copolymer molding resins. *Hoechst AG.*

14346 Hostaform C 2521
Acetal copolymer resin; standard grade for extrusion; also for injection molding of thick-walled parts. *Hoechst AG.* Name unverified.

14347 Hostaform C 9021 ELS
Acetal copolymer resin, carbon bulk-modified; injection molding grade with very low surface resistance; for applications in explosion-proof areas. *Hoechst AG.* Name unverified.

14348 Hostaform C 9021 K
Acetal copolymer resin, mineral additive modified; low-friction low-wear injection molding grade modified with mineral additive for moldings with low dry sliding wear. *Hoechst AG.* Name unverified.

14349 Hostaform S 27076
Acetal copolymer resin; high impact resistant, very flexible, vibration absorbing injection molding grade. *Hoechst AG.* Name unverified.

14350 Hostalen®
A trademark for a high density polythene. *Hoechst UK.* Name unverified.

14351 Hostalen® EP 4450
9002-88-4 7728
UHMW polyethylene
Flow enhanced polymer with high bearing properties. *Hoechst Celanese.* Name unverified.

14352 Hostalen® G
9002-88-4 7728
polyethylene
High density polyethylene. *Hoechst UK.*

14353 Hostalen® GB 6950
9002-88-4 7728
Polyethylene
HDPE copolymer with a narrow molecular weight distribution; high melt flow index, fast injection speeds, reduced dens., improved impact strength and ESCR behavior; used in paint pails, domestic ware, etc. *Hoechst AG.* Name unverified.

14354 Hostalen® GM 5010 T2
9002-88-4 7728
Polyethylene
HDPE, HMW, carbon bulk; NSF-approved pipe grade for potable water applications. *Hoechst Celanese.* Name unverified.

14355 Hostalen® GM 7745 HP
9002-88-4 7728
Polyethylene
HDPE, HMW; blow molding grade for storage tank liners, lawn and garden spray tanks, automated refuse carts, intermediate bulk containers; superior ESCR. *Hoechst Celanese.* Name unverified.

14356 Hostalen® GUR 5121
9002-88-4 7728
UHMW polyethylene
Enhanced flow injection molding grade with excellent abrasion and chem. resist., high impact strength. *Hoechst Celanese.*

14357 Hostalen® OO
9003-07-0 7741
polypropylene
A trademark for isotactic polypropylene; a thermoplastic moderately rigid molding material. *Hoechst UK.* Name unverified.

14358 Hostalen® PP 927
9003-07-0 7741
polypropylene
PP homopolymer; stabilized extrusion grade resin; good mechanical properties at moderate draw ratios (10/1); for extrusion of bristles, ribbons, and general purpose applications. *Hoechst Celanese UK.* Name unverified.

14359 Hostalit
Polyvinyl chloride polymers and copolymers. *Hoechst AG.*

14360 Hostalit Z
A blend of PVC and chlorinated polyolefin; a thermoplastic used for manufacturing pipes. *Hoechst AG.*

14361 Hostalub®
Lubricants for polymers. *Hoechst UK.*

14362 Hostalub® CAF 484 SB
Calcium stearate, synthetic paraffin wax, and oxidized polyethylene wax; lubricant for high line-speed extrusion of vinyl siding. *Hoechst Celanese.*

14363 Hostalub® FA 1
Amide wax; internal/external lubricant for PVC and thermosets; carrier for pigments. *Hoechst AG.*

14364 Hostalub® H 4
Modified hydrocarbon wax; external lubricant for PVC pipe and profile extrusion and for plasticized PVC. *Hoechst AG.*

14365 Hostalub® VP Ca W 2
68308-22-5 269-637-1
Calcium montanate; internal lubricant for PVC, polyamide, thermoplastic PU, other thermoplastics and thermosets. *Hoechst AG.*

14366 Hostalub® WE4
68476-38-0 270-679-8
Glyceryl montanate
Lubricant and release agent for PVC and other polymers. *Hoechst AG; Hoechst Celanese.*

14367 Hostalub® XL 165
Blend of paraffinic wax components; external lubricant for rigid PVC extrusion, wire and cable applications. *Hoechst Celanese.*

14368 Hostalux KCB
Benzoxazole type; optical brightener, whitening agent used in all types of polymer processing; used in plastic films, press molding, and injection molding material fibers and bristles, paints and lacquers. *Hoechst Celanese.*

14369 Hostamer Brands
Vinylamide/vinylsulfonic acid copolymers; polymers for enhanced oil recovery, especially at high temperatures and salinity. *Hoechst AG.*

14370 Hostamid
A proprietary trade name for a group of nylons. *Hoechst UK.* Name unverified.

14371 Hostamont
6339
Lignite waxes. Montan waxes. *Hoechst AG.*

14372 Hostanox®
Antioxidants for polymers. *Hoechst UK.*

14373 Hostanox® 03
Benzene propanoic acid
Antioxidant for HDPE, PP, PA, POM. *Hoechst Celanese.* Name unverified.

14374 Hostanox® OSP 1
Phenol, 4,4'-thiobis 2-(1,1-dimethylethyl) phosphite
Antioxidant for HDPE, PP, PMMA, PA. *Hoechst Celanese.* Name unverified.

14375 Hostanox® PAR 24
31570-04-4 250-709-6
$C_{42}H_{63}O_3P$
Tris-(2,4-di-*t*-butylphenyl) phosphite
Phenol, 2,4-bis(1,1-dimethylethyl)-, phosphite (3:1); tris(2,4-di-*tert*-butylphenyl) phosphite. Antioxidant for LLDPE, HDPE, PP. *Hoechst Celanese.* Name unverified.

14376 Hostanox® SE 10
Dioctadecyl disulfide
Antioxidant for HDPE, PP, PMMA, PA. *Hoechst Celanese.* Name unverified.

14377 Hostapal 2345
Nonylphenol polyglycol ether carboxylate
Surfactant for agricultural formulations. *Hoechst Celanese/Colorants & Surf.*

14378 Hostapal N-040
9016-45-9 6772
Nonoxynol-4
Detergent, wetting agent, emulsifier for general industrial applications, agricultural, textiles. *Hoechst Celanese/Colorants & Surf.*

14379 Hostaperm
Pigments powders for prints and inks. *Hoechst UK.*

14380 Hostaphan
polyethylene terephthalate
Polyethylene terephthalate films. *Hoechst UK.*

14381 Hostaphane
polyethylene terephthalate
A proprietary trade name for polyethylene terephthalate film. *Hoechst UK.* Name unverified.

14382 Hostaphat
Phosphoric acid ester emulsifier. *Hoechst UK.*

14383 Hostaphat 2122
2-Ethylhexyl phosphate
Surfactant for textile processing. *Hoechst Celanese/Colorants & Surf.*

14384 Hostaphat AR K
Nonylphenol ethoxylate phosphate ester, potassium salt; surfactant for agric'ultural formulations. *Hoechst Celanese/Colorants & Surf.*

14385 Hostaphat HI
Hexyl phosphate
Surfactant for textile processing. *Hoechst Celanese/Colorants & Surf.*

14386 Hostaphat KL 340N
Trilaureth-4 phosphate
Oil-water emulsifier for cosmetic purposes. *Hoechst Celanese/Colorants & Surf. Hoechst AG.*

14387 Hostaphat KO 300
Trioleyl phosphate
Emulsifier for cosmetic w/o emulsions. *Hoechst Celanese/Colorants & Surf.; Hoechst AG.*

14388 Hostaphat KO 380
Trioleth-8 phosphate
Emulsifier for oil-water emulsions for cosmetics industry. *Hoechst Celanese/Colorants & Surf.; Hoechst AG.*

14389 Hostaphat KW 340 N
Triceteareth-4 phosphate
Emulsifier for creamy cosmetic emulsions based on hydrocarbons, fatty alcohols and fatty acids. *Hoechst Celanese/Colorants & Surf.; Hoechst AG.*

14390 Hostapon CAS
Mixture of sulfation products of fatty alcohol derivatives and fatty acid condensation products, supplied as a clear yellowish liquid; optimal foaming agent with very good skin compatability; used in cosmetics, e.g. cleansing agents for intimate hygiene, baby shampoos and foam baths. *Hoechst AG.* Name unverified.

14391 Hostapon CT Paste
Sodium methyl cocoyl taurate
Detergent used in shampoos; good foaming, skin compat. *Hoechst Celanese/Colorants & Surf.; Hoechst AG.*

14392 Hostapon IDC
Sodium N-palmityl N-cyclohexyl taurine
Surfactant for textile processing. *Hoechst Celanese/Colorants & Surf.*

14393 Hostapon KA Powd
Sodium cocoyl isethionate
Detergent base for cosmetic industry, detergent bars; foamer, dispersant. *Hoechst Celanese/Colorants & Surf.; Hoechst AG.*

14394 Hostapon KTW New
70609-66-4 274-695-6
Sodium lauroyl taurate
Detergent base for cosmetics, toothpastes. *Hoechst Celanese/Colorants & Surf.; Hoechst AG.*

14395 Hostapon SO
137-20-2 205-285-7
Sodium methyl oleoyl taurate
Detergent, foamer, dispersant for cosmetic industry. *Hoechst Celanese/Colorants & Surf.*

14396 Hostapon STT Paste
149-39-3 205-738-9
Sodium methyl stearoyl taurate
Detergent for high quality cream shampoos. *Hoechst Celanese/Colorants & Surf.; Hoechst AG.*

14397 Hostapon T Powd
137-20-2 205-285-7
Sodium methyl oleoyl taurate
Cleaner for textile, leather, household prods., agricultural, cosmetics, etc. *Hoechst Celanese/Colorants & Surf.; Hoechst AG.*

14398 Hostapon TF
Anionic surfactant in which the cation is sodium and the anion is composed of unsaturated fatty acids condensed with methyl taurine; supplied as a clear, yellowish viscous liquid; foaming, washing and lime soap dispersing agent used in clear liquid and cream shampoos, foam baths, toothpastes. *Hoechst UK.* Name unverified.

14399 Hostapor
Polystyrene foaming grades. *Hoechst UK.* Name unverified.

14400 Hostapren
Chlorinated polyethylene. *Hoechst AG.*

14401 Hostaprime® HC 5
Furandione propene polymer; coupling agent for PP, PA. *Hoechst Celanese.* Name unverified.

14402 Hostaprint
Pigment preparations for plastics printing. *Hoechst UK.*

14403 Hostapur DOS Hi conc.
Dialkyl sodium sulfosuccinate
Surfactant for textile processing. *Hoechst Celanese/Colorants & Surf.*

14404 Hostapur DTC
Isotridecyl alcohol polyglycol ether carboxylate
Surfactant for agricultural formulations. *Hoechst Celanese/Colorants & Surf.*

14405 Hostapur OS
Sodium olefin sulfonate with an alkylene radical (C_{15}-C_{18}) as a liquid or free-flowing powder; used for powdered detergents of all kinds, light duty detergents and cleaning agents, textile and leather auxiliaries, and upholstery and carpets. *Hoechst UK.* Name unverified.

14406 Hostapur SAS
Series of three anionic surfactants of the alkane sulfonate type, where the chain length is C_{13}-C_{18} and the cation is sodium; liquid, paste or flake form; wetting and foaming agent used in detergents, washing-up liquids, cleaning agents of all kinds, textiles ad leather auxiliaries. *Hoechst UK.* Name unverified.

14407 Hostaquick
23560-59-0 4699 245-737-0
$C_9H_{12}ClO_4P$
7-chlorobicyclo[3.2.0]hepta-2,6-dien-6-yl dimethyl phosphate
Ragadan; Heptenophos; Hostaquick. Emulsifiable concentrate containing 550 g heptenophos per liter; used for control of insects in greenhouses and a wide range of vegetable crops. *Hoechst UK.*

14408 Hostarex Grades
Secondary and tertiary amines; anion/cation exchangers. *Hoechst AG.*

14409 Hostastat®
Polymer antistatic agents. *Hoechst AG.*

14410 Hostastat® FA 14
Ethoxylated alkylamine, coco derivative; antistat for polyolefins, polystyrene, ABS. *Hoechst Celanese.* Name unverified.

14411 Hostastat® System E1956
Laurylamide in LDPE carrier; antistat concentrate for PE, PP. *Hoechst Celanese.* Name unverified.

14412 Hostatec
Polyether ketone. Solvent. *Hoechst AG.*

14413 Hostathion
24017-47-8 9736 245-986-5
triazophos
Emulsifiable concentrate containing 420 g/l triazophos; an organophosphorus insecticide. *Hoechst UK.*

14414 Hostatint
Pigment preparations for tinting paint. *Hoechst UK.*

14415 Hostatron®
Foaming agents for polymers. *Hoechst UK.*

14416 Hostatron® System P1941
Carbonic acid-based endothermic blowing agents in proprietary carriers; blowing agent for PS, ABS, PE, PP. *Hoechst Celanese.* Name unverified.

14417 Hostavin®
UV stabilizers for polymers. *Hoechst UK.*

14418 Hostavin® ARO 8
1843-05-6 6838 217-421-2
Benzophenone-12
octabenzone. UV absorber for plastics, esp. LDPE, HDPE, PP, polyisobutylene, cellulosics, PC, EVA copolymers, plasticized PVC. *Hoechst Celanese.* Name unverified.

14419 Hostavin® N 30
Oligomeric HALS
UV stabilizer for PP, PE, elastomers, POM, PU. *Hoechst Celanese.* Name unverified.

14420 Hostawet TDC
Ether carboxylate
Emulsifier, wetting agent for metal cleaning formulations. *Hoechst Celanese/Colorants & Surf.*

14421 Hostiren
Polystyrene resins. *Hoechst UK.* Name unverified.

14422 Hot Melt Wetnes Indicator®
Dry adhesive particles for manufacturing use; to be melted and applied to textile fabrics. *Reichhold.*

14423 Hotbac®
Polymer. *DuPont UK.*

14424 Hotspur
Herbicide for broad leafed weeds in cereals. *ICI Chem & Polymers Ltd.*

14425 Hotspur
clopyralid-fluroxypyr-ioxynil
Mixture of clopyralid, fluroxypyr and ioxynil; used to control broad leaf weeds in cereals. *Farm Protection Ltd.*

14426 houillite
Anthracite; a fossil fuel.

14427 House Plant Leaf Shine
Aerosol containing leafshine material; house plant spray. *Vitax Ltd.*

14428 House Plant Liquid Feed
Liquid concentrate containing NPK 5:2:2; house plant feed. *Vitax Ltd.*

14429 House Plant Pest Killer
Aerosol containing pyrethrum and resmethrin; house plant insecticide. *Vitax Ltd.*

14430 Houseplant Long Lasting Feed
Coated fertilizer. *Fisons plc, Horticultural Div.* Name unverified.

14431 Howard's silver
628-86-4 211-057-8
Fulminic acid, mercury(2+) salt
Mercury fulminate; Mercuric oxycyanide. Explodes on impact and is used in percussion caps.

14432 Howflex
Plasticizer. *Laporte Industries Ltd.* Discontinued.

14433 Howstik
Rubber/resin solution in various solvents; contact adhesives for insulation, building, footwear, foam adhesives. *Howlett Adhesives Ltd.*

14434 Howtex
Ready-mixed water-based ceramic tile adhesives; for ceramic tiling, building, woodworking, laminating. *Howlett Adhesives Ltd.*

14435 Howtol
Cyclic alcohols. *Laporte Industries Ltd.* Discontinued.

14436 Hoyle's metals
Bearing metals usually containing about 46% tin, 12% antimony, and 42% lead.

14437 Hoyt Metal
A proprietary trade name for an antimonial lead containing 6-10% antimony. No manufacturer.

14438 HPC-9
80-15-9 201-254-7
$C_9H_{12}O_2$
Cumene hydroperoxide
Cumene hydroperoxide/MEK peroxide solution; facilitates R.T. curing of vinyl ester of other unsaturated polyester resins. *Witco Corporation.*

14439 HSB 1900
High styrene resin/SBR fluxed blend (100:100); used in shoe soles and household goods; gives easy processing, high durometer compounds; vulcanized with standard curing systems. *Mach-1 Compounding.*

14440 H-scale
A proprietary trade name for a synthetic pearl essence. No manufacturer.

14441 HSZ-320NAA
Crystalline aluminosilicate
Zeolite for catalyst and adsorbent applications. *Tosoh.*

14442 HT Non-Ionic Wetter
Spreader containing 90% alkylphenol ethylene oxide condensate; wetting agent for use in horticultural sprays. *Service Chemicals Ltd.*

14443 HTH
7778-54-3 1717 231-908-7
Calcium hypochlorite
Dry chlorinating agent. *Olin.*

14444 HT-Proteolytic Conc
Bacterial protease
Enzyme for hydrolysis of proteins over neutral and alkaline pH range; for baking, proteins. *Solvay Enzymes.*

14445 HTSA 1
16260-09-6 240-367-6
Oleyl palmitamide
Release agent providing slip, antiblocking to thermoplastics including PP film, nylon. *Hexcel.*

14446 HTSA 3
10094-45-8 233-226-5
Stearyl erucamide
Release agent providing slip, antiblocking to thermoplastics including PP film, nylon. *Hexcel.*

14447 Hünefeld solution
This contains 25 ml alcohol, 5 ml chloroform, 1.5 ml glacial acetic acid, and 15 ml turpentine; used for the detection of blood.

14448 Hubel's reagent
a) Iodine (50 g), dissolved in 1 liter of 95% alcohol. b) Mercuric chloride (60 g), dissolved in 1 liter of alcohol; used for obtaining the iodine value of fats and oils.

14449 Huber 40C
1332-58-7 5294 296-473-8
Kaolin clay; functional filler with medium brightness and particle size, moderate oil and water demand; for paint and plastics applications. *J.M. Huber.*

14450 Huber 65A
1332-58-7 5294 296-473-8
Kaolin clay; air floated functional filler with medium brightness, medium particle size; for adhesives, paints, plastics. *J.M. Huber.*

14451 Huber 95
1332-58-7 5294 296-473-8
Kaolin clay; water washed functional filler with medium brightness, fine particle size; for adhesives, paints, plastics and inks. *J.M. Huber.*

14452 Huberbrite 1
7727-43-7 1023 231-784-4
barium sulfate
Pigment grade with high assay, high brightness, fine whiteness. *J.M. Huber.*

14453 Hubercarb® Q 6-20
471-34-1 1697 207-439-9
calcium carbonate
Filler/extender for plastics, caulks/sealants, rubber, adhesives, glass, ceramics, paper, cleansers, paints/coatings, pesticides, asphalt, drilling mud, rice polishing, environmental cleanup. *J.M. Huber.*

14454 Hubercarb® W 2
471-34-1 1697 207-439-9
calcium carbonate
Filler and extender with low moisture pickup; ideal for moisture-sensitive applications such as SMC, BMC, TMC polyethylene film and single-component urethane caulks and for paints and coatings, rubber, adhesives, ceramics, paper, nonabrasive cleaners. *J.M. Huber.*

14455 Huber's reagent
ammonium molybdate-potassium ferrocyanide
A solution of ammonium molybdate and potassium ferrocyanide; used for the detection of free mineral acids.

14456 Hudroson
Metal cleaning and pretreatment systems. *Nickerson Chemicals Ltd.*

14457 Hugel A
A proprietary solution of the sodium salt of an acrylic copolymer; used for thickening natural and synthetic rubber latices. *Hubron Rubber Chemicals.* Name unverified.

14458 Hugel AH
A proprietary viscous solution of an acrylic copolymer. *Hubron Rubber Chemicals.* Name unverified.

14459 Hugel B
A proprietary trade name for emulsions of acrylic copolymer containing free carboxyl groups. *Hubron Rubber Chemicals.* Name unverified.

14460 Hugel BC 10
A proprietary viscous solution, colorless and odorless, of the sodium salt of an acrylic polymer. *Hubron Rubber Chemicals.* Name unverified.

14461 Hugel CH14
The sodium salt of a proprietary acrylic polymer; very viscous; takes the form of a short nonstringy yellow gel. *Hubron Rubber Chemicals.* Name unverified.

14462 Hulot's solder
An alloy of 37.5% tin, 37.5% lead, and 25% zinc amalgam; used for soldering aluminum bronze.

14463 Humber
Granular or powder organic based fertilizers comprising organic base and chemical N, P and K to form the analysis; a steady release, lower nitrogen fertilizer for agricultural crops and grasslands. *Humber Fertilizers plc.*

14464 Humectant SD-35
667-83-4 211-569-1
Panthenyl ethyl ether
Humectant. *Presperse.*

14465 Humectol®
Wetting, dispersion and leveling agents. *Cassella AG.*

14466 Humifen
Sulfonated aliphatic polyesters; wetting agent for dyeing; yarn textile finishing; dry cleaning detergents; glass cleaners; wallpaper removers; battery separators. *GAF Great Britain.* Name unverified.

14467 Humifen BA-77
Anionic surfactant in powder form; wetting, dispersing and antistatic agent for paints, printing inks, latex stabilization, leather, textile processing and agricultural chemistry. *GAF Great Britain.* Name unverified.

14468 Humifen BX-78
Anionic surfactant in powder form; wetting, penetrating, dispersing and emulsification agent; used for cotton, rayon, dyestuffs, leather, agricultural chemicals, insecticides, paper, rubber latex and polymerization. *ISP.* Name unverified.

14469 Huntsman 201
9003-53-6 9028 203-066-0
Polystyrene
Crystalline PS. *Huntsman.*

14470 Huntsman 240
9003-53-6 9028 203-066-0
HIPS, rubber-modified; shock resistance at ambient and low temperatures, gloss characteristics, good thermal properties; used for refrigerator and appliance parts. *Huntsman.*

14471 Huntsman 312
9003-53-6 9028 203-066-0
Polystyrene
Medium impact PS. *Huntsman.*

14472 Huntsman 331
9003-53-6 9028 203-066-0
Polystyrene
High impact PS. *Huntsman.*

14473 Huntsman 351
9003-53-6 9028 203-066-0
Polystyrene
High impact PS, ignition-retardant. *Huntsman.*

14474 Huntsman 474
9003-53-6 9028 203-066-0
Polystyrene
PS, rubber-modified; high-heat and impact polymer, high thermal properties, shock resistance at ambient and low temperatures, moldability; for small appliances, consoles, automotive parts (interior parts). *Huntsman.*

14475 Huntsman 765
9003-53-6 9028 203-066-0
Polystyrene
High impact PS; color-stable grade. *Huntsman.*

14476 Huppert's reagent
10043-52-4 1699 233-140-8
calcium chloride
A 10% aqueous solution of calcium chloride used for the detection of biliary pigments in urine.

14477 Huron
An aluminum alloy containing from 3.5-6.6% copper and small amount of manganese, magnesium, and chromium. The name is also used for a chromium steel containing 12.5% chromium, 1.0% vanadium, and 0.2% carbon. It is used for dies.

14478 Hurr nut
Myrobalans.

14479 Husman metal
An alloy of 74% tin, 11% antimony, 10.6% lead, 4% copper, 0.22% iron, and 0.18% zinc.

14480 HVA-2
3006-93-7 221-112-8
$C_{14}H_8N_2O_4$
N,N'-*m*-Phenyledimaleimide
N,N'-*meta*-Phenylenedimaleimide; N,N'-1,3-Phenylene bismaleimide. A free radical regulator used as an auxiliary in the curing of Hypalon. *DuPont UK.* Name unverified.

14481 HVP
Hydrolyzed vegetable proteins; natural flavors for all types of processed foods to improve or impart a characteristic flavor. *Hercules.* Discontinued.

14482 Hyalase
Hyaluronidase-based spreading agent. *Fisons plc, Pharmaceuticals Div.* Name unverified.

14483 hyaluronic acid sodium salt
9067-32-7 4793
sodium hyaluronate
ARTZ; Connettivina; Equron; Healon; Healonid; Hyacid; Hyalgan; Hyalovet; Hyonate; Ial; Opegan; Provisc; Synacid. Natural mucopolysaccharide formed by bonding N-acetyl-D-glucosamine with glucuronic acid. Moisturizer for skin care products. $[\alpha]_D^{25}$= -74° (c = 0.25 H_2O). *Am. Biorganics; Croda; Solabia; Worthington Biochemical.*

14484 Hyamine® 10X
25155-18-4 6103 204-479-9
Methylbenzethonium chloride
Germicide, disinfectant, sanitizer in restaurant and pharmaceutical uses. *Lonza Inc.*

14485 Hyamine® 1622 50%
121-54-0 1103 204-479-9
Benzethonium chloride
Germicide, disinfectant, sanitizer in restaurant and pharmaceutical uses; antistat, bacteriostat on fabrics; preservative for starch, glue, casein; cocatalyst for curing polyesters. *Lonza Inc.*

14486 Hybaite
7775-14-6 8771 231-890-0
Sodium hydrosulfite
Sodium dithionite/alkali buffers/clorites; for bleaching of paper pulps. *RV Chemicals Ltd.*

14487 Hyban
dicamba-mecoprop
Soluble concentrate of 18.7 g dicamba and 300 g mecoprop per liter; used for weed control in cereals and grassland. *Agrichem (International Ltd.*

14488 Hyb-lum
An alloy of aluminum in which the alloying elements, consisting of about 2% are mainly nickel and metals of the chromium group; used for reflectors of therapeutic lamps.

14489 Hybon® 2011
Continuous filament single-end glass strand rovings with proprietary sizing; reinforcement for polyester and vinyl ester resin systems in pultrusion applications. *PPG/Fiber Glass.*

14490 Hybri-Chem 100
Polyester/polyurethane hybrid system; for open mold casting or lamination; low exothermicity; high heat distortion temperature. *Hybri-Chem.* Name unverified.

14491 Hy-Brite
Metal cleaner. *FMC.* Discontinued.

14492 Hybrite®
7775-14-6 8771 231-890-0
Sodium hydrosulfite
Sodium dithionite/alkali/buffers/chelates; used for bleaching of paper pulps. *RV Chemicals Ltd.*

14493 Hycal
A proprietary preparation of dextrose and related compounds; used as a high-calorie diet supplement. *SmithKline Beecham.* Name unverified.

14494 Hycar® 1203X17
A proprietary 70:30 blend of Hycar medium-high acrylonitrile rubber and Geon PVC resin. *BF Goodrich.* Name unverified.

14495 Hycar® 1204X5
A proprietary 100:70:120 pre-fluxed blend of Hycar medium-high acrylonitrile rubber, Geon PVC resin and a phthalate plasticizer. *BF Goodrich.* Name unverified.

14496 Hycar® 1204X9
A proprietary 100:70:100 pre-fluxed blend of Hycar medium-high acrylonitrile rubber, Geon PVC resin and a phthalate plasticizer, protected by nonstaining stabilizers. *BF Goodrich.* Name unverified.

14497 Hycar® 1205X3
A proprietary 50:50:60 pre-fluxed blend of Hycar medium-high acrylonitrile rubber, Geon PVC resin and a phthalate plasticizer. *BF Goodrich.* Name unverified.

14498 Hycar® 1273
A proprietary 70:30 blend of carboxymodified Hycar acrylonitrile rubber and Geon PVC resin, possessing good resistance to abrasion. *BF Goodrich.* Name unverified.

14499 Hycar® 1300X8
Reactive liquid polymers. *BF Goodrich/Spec. Polymers.* Name unverified.

14500 Hycar® 1402 H82
A proprietary acrylonitrile-bytadiene copolymer in powder form, having a medium-high content of acrylonitrile. *BF Goodrich UK.* Name unverified.

14501 Hycar® 1402 H83
A proprietary acrylonitrile-butadiene copolymer in powder form, having a high content of acrylonitrile. *BF Goodrich UK.* Name unverified.

14502 Hycar® 1403 H84
A proprietary acrylonitrile-butadiene copolymer having a medium acrylonitrile content; supplied in powdered form and used when good behavior at low temperatures is required. *BF Goodrich UK.* Name unverified.

14503 Hycar® 1552
Nitrile latex
Used for adhesives, carpet backcoating, paper coatings and saturation, pigment binding, textile coatings; exhibits fast penetration. *BF Goodrich/Spec. Polymers.* Name unverified.

14504 Hycar® 1577
Butadiene-acrylonitrile-styrene terpolymer latex; used for waterproof and greaseproof coatings, leather finishes, paints, paper coatings and saturation. *BF Goodrich/Spec. Polymers.* Name unverified.

14505 Hycar® 2100
A proprietary range of polyacrylic-solution polymers used as pressure-sensitive adhesives. *BF Goodrich UK.* Name unverified.

14506 Hycar® 2550H33
A proprietary reinforced styrene-butadiene copolymer rubber used in the manufacture of foam rubber. *BF Goodrich UK.* Name unverified.

14507 Hycar® 2550H5
A proprietary aqueous, anionic dispersion of a cold-polymerized styrene-butadiene copolymer. *BFGoodrich UK.* Name unverified.

14508 Hycar® 2550H55
A proprietary aqueous, anionic dispersion of a styrene-butadiene copolymer used in the foambacking of carpets. *BF Goodrich UK.* Name unverified.

14509 Hycar® 2570H28 and 2570H29
A proprietary group of aqueous anionic dispersions self-reactive styrenebutadiene copolymers; used in the making of carpet backings. *BF Goodrich UK.* Name unverified.

14510 Hycar® 2570X5
A proprietary aqueous, anionic dispersion of a carboxy-modified styrene-butadiene copolymer reactive to heat; used for leather finishes and adhesives. *BF Goodrich UK.* Name unverified.

14511 Hycar® 26345
Acrylic latex, heat-reactive; used for upholstery, drapery, mattress ticking coatings; foamable; durable. *BF Goodrich/Spec. Polymers.* Name unverified.

14512 Hycar® 2671H49
A proprietary aqueous anionic dispersion of a heat-reactive carboxy-modified acrylic polymer used in the making of surgical rubber materials. *BF Goodrich UK.* Name unverified.

14513 Hycar® 4021
A proprietary copolymer of ethyl acrylate having a small percentage of 2-chloroethyl vinyl ether. *BF Goodrich.* Name unverified.

14514 Hycar® 4032
A proprietary polyacrylic rubber used to make rubber seals and gaskets. *BF Goodrich UK.* Name unverified.

14515 Hycar® 4043
A proprietary acrylic rubber which remains flexible at -40°, but which also gives good resistance to oil at high temperatures. *BF Goodrich.* Name unverified.

14516 Hycar® 4201
A proprietary copolymer of ethyl acrylate having a small percentage of 2-chlorovinyl ether. *BF Goodrich.* Name unverified.

14517 Hycar® ATBN, CTB, CTBN, VTBN
Reactive liquid polymer which cures at room or elevated temperatures to solid elastomer; improves crack resistance of epoxy, polyester systems, enhances fatigue performance of FRP; good electrical and wetting properties; used as modifier for epoxy structural adhesives; improves impact resistance of epoxy coatings; used in potting, encapsulation, cable fillers, moisture blocks and castable elastomers for roto-molding. *BF Goodrich/Spec. Polymers.* Name unverified.

14518 Hycar® Reactive Liquid Polymer
Reactive Liquid Polymers (RLP) are homopolymers of butadiene or copolymers of butadiene/acrylonitrile; reactive groups are in both terminal positions of the polymer chain and, optionally, may have additional reactive groups pendent on the chain; three types (carboxyl, acrylated vinyl, sec. amine); available commercially; CT polymers in epoxy structural adhesives, encapsulants, coatings, composites, solid propellant binder; AT in epoxy adhesives, maintenance coatings, encapsulation, geographical cable fillers, VT in polyester BMC *BF Goodrich UK.* Name unverified.

14519 Hycathane
A proprietary polyurethane elastomer. *FPT Industries.* Name unverified.

14520 Hycel M Series
Isocyanate/resin system; for rigid PU foam for thermal insulation, structural material, packing materials, etc. *Toho Chem. Industry.*

14521 Hycol
A coal tar black disinfectant composed of 30% coal tar acids, hydrocarbon oils and vegetable oil soap. *William Pearson Ltd.* Name unverified.

14522 Hycolln
A general purpose hospital disinfectant containing 16% substituted phenolic germicides in a detergent base. *William Pearson Ltd.* Name unverified.

14523 Hycon
Photographic developer. *May & Baker Ltd.* Name unverified.

14524 Hycote
Spray paints; for matched colors and touch up paints. *James Briggs & Sons Ltd.* Name unverified.

14525 Hydagen® B
515-69-5 1281 208-205-9
$C_{15}H_{26}O$
(R*,R*)-α,4-dimethyl-α-(4-methyl-3-pentenyl)-3-cyclohexene-1-methanol
α-Bisabolol; Camilol; Dragosantol. Anti-inflammatory agent for emulsions, oils, lotions, and oral hygiene preparations. bp_{12} = 155-157°; d_4^{23}= 0.9223; n_D^{23} = 1.4917; insoluble in H_2O, soluble in organic solvents. *Henkel/Cospha; Henkel Canada.*

14526 Hydagen® C.A.T
77-93-0 201-070-7
$C_{12}H_{20}O_7$
Triethyl citrate
Citric acid triethyl ester; Ethyl citrate. Non-microbiocidal deodorant active agent. bp_1 = 127°; d = 1.137. *Henkel/Cospha; Henkel Canada.*

14527 Hydagen® DEO
triethyl citrate-BHT
Triethyl citrate and BHT; nonmicrobiocidal active ingredient for deodorant systems and personal care products. *Henkel/Cospha; Henkel Canada.*

14528 Hydan
118-52-5 3115 204-258-7
1,3-Dichloro-5,5-dimethylhydantoin
1,3-Dichloro-5,5-dimethylhydantoin. Disinfectantand industrial deodorant. *Rhône-Poulenc UK.*

14529 Hydan®
For the agriculture industry. *DuPont UK.*

14530 Hydecat
Range of catalysts and absorbents used for purification. *ICI Chem & Polymers Ltd.*

14531 Hydex® 100 Gran.206
50-70-4 8873 200-061-5
Sorbitol
Humectant, bodying agent, moisture control agent for toothpaste, cosmetics, sugarless confections, controlled moisture foods, industrial applications. *Lonza Inc; Lonza AG.*

14532 Hydon
bromacil-picloram
Bromacil and picloram; used for total weed control in non crop areas. *Chipman Ltd.*

14533 Hydraffin
7440-44-0 1855 231-153-3
Carbon
Granulated carbon; used for treatment of drinking, process, and waste water. *Bayer AG.*

14534 Hydra-guard
lindane-thiram
Mixture of lindane and thiram; seed dressing for brassica crops. *Agrichem (International) Ltd.*

14535 Hydral® 710
21645-51-2 355 244-492-7
Hydrated alumina
Fire retardant, smoke suppressant improving arc track resistance in thermosets, thermoplastics elastomerics; in fillers and coating pigments in fine printing papers for increased brightness; for vinyl compounding; as reinforcing pigment in adhesives; fine mild abrasive in waxes and polishes; filler for cosmetic powders, lotions; polishing agent in dentrifices. *Alcoa Industrial Chemicals.*

14536 Hydralin
108-93-0 2794 203-630-6
cyclohexanol
Chemical intermediate; also used as a solvent.

14537 Hydrangea Colourant
aluminum sulfate-ferrous sulfate
Soluble powder containing aluminum sulfate hexahydrate [10043-01-3] and ferrous ammonium sulfate hexahydrate [10045-89-3]; hydrangea colorant. *Vitax Ltd.*

14538 Hydraphthal
A preparation containing 90% tetralin, 5% ammonium oleate, and 3% water. A wetting-out and scouring agent.

14539 hydrargaphen
14235-86-0 4805 238-107-1
$C_{33}H_{24}Hg_2O_6S_2$
phenylmercuric dinaphthylmethane disulfonate
phenyl mercuric Fixtan; Conotrane; P.M.F.; Versatrane; Hydraphen; Penotrane; Fibrotan; Septotan. Bactericide and fungicide for the treatment of wool, hides, leather, textiles, timber and wood-pulp, paints and adhesives. Amorphous powder, insoluble in H_2O, LD_{100} (mus orl) = 80 mg/kg.

14540 Hydrasal®
Zirconium, magnesium, rare-earth and other chemicals, powders and alloys; for use in industry, including metalworking. *Magnesium Elektron Ltd.*

14541 hydraulan®
Brake fluids. *BASF AG; BASF plc.*

14542 hydraulic bronze
An alloy of 83% copper, 5% lead, 5% zinc, 5% tin, and 2% nickel. Another alloy contains 83% copper, 10.8% tin, 6% zinc, and 0.1% lead.

14543 hydraulic cements or mortars
$3CaO{\cdot}SiO_2$ $3CaO{\cdot}Al_2O_3$
These are prepared by calcining mixtures of calcium carbonate with from 10-30% clay. Tricalcium silicate, and tricalcium aluminate, are formed.

14544 hydraulic limes
Limes containing from 15-30% claylike matter (aluminum silicate). They are made by burning impure limestones at a low temperature. They slake in water, but show hydraulic properties.

14545 hydrazine
302-01-2 4809 206-114-9
H_4N_2
hydrazine
diamine; diamide; hydrazine base. Chemical intermediate and reducing agent. mp = 1°; bp = 114°; miscible with H_2O; LD_{50} (rat orl) = 60 mg/kg. *Aldrich; Elf Atochem SA; Fairmount; Miles; Olin; Spectrum Chem. Mfg.*

14546 Hydrazine sulfate
10034-93-2 4811 233-110-4
$NH_2NH_2{\cdot}H_2SO_4$
diamine sulfate
diamidogen sulfate. Chemical intermediate, condensation reactions, catalyst for making acetate fibers; analysis of minerals, slags, fluxes; determination of arsenic in metals; separation of polonium from tellurium; fungicide, germicide. *Fairmount; Janssen Chimica; Mallinckrodt; Otsuka Chem.; Spectrum Chem. Mfg.*

14547 Hydrenol D
Tallow alcohol (25-35% C16, 60-67% C18); wetting agent, emulsifier, emollient, consistency giving agent for skin creams and lotions; intermediate for surfactant manufacture. *Henkel/Cospha; Henkel KGaA.*

14548 Hydresol®
Condensates of naphthalene sulfonic or phenolsulfonic acid and formaldehyde; superplasticizers for mineral binders, for improving workability of mineral-bonded mortar and concrete. *BASF AG.*

14549 Hydrex®
Polyester resins for coatings, casting and laminate application. *Reichhold.*

14550 Hydrholac
Nitrocellulose lacquer; used for leather finishing. *Rohm & Haas UK.*

14551 Hydrin® C
Epichlorohydrin/ethylene oxide copolymer; elastomer used for fuel pump diaphragms, pipe gaskets, hose for fuel, oil, and gas, vibration isolators, motor mounts, rolls, adhesives, sponge goods, boots, seals, o-rings, bladders; excellent resistance to ozone, oils and fuels, and low temperature flexibility; vulcanized with ETU, TETA, HMDAC, Thiate E. *Zeon.*

14552 Hydrin® C-CG
Epichlorohydrin/ethylene oxide copolymer; cement grade of Hydrin C. *Zeon.*

14553 Hydrin® T
Ethylene oxide/epichlorohydrin/allyl glycidyl ether terpolymer; elastomer used for fuel pump diaphragms, pipe gaskets, hose for fuel, oil, and gas, vibration isolators, motor mounts, rolls, adhesives, sponge goods, boots, seals, o-rings, bladders; blendable with SBR and NBR; to imporve low temperature performance, fuel oil and ozone resistance; vulcanized with sulfur, peroxide or ETU. *Zeon.*

14554 Hydrine
9004-99-3
PEG-2 stearate; consistency stabilizer for ointments, cream lotions; opacifier in shampoos, liquid soaps. *Gattefosse; Gattefosse SA.*

14555 Hydriodic acid
10034-85-2 4817 233-109-9
HI
Hydroiodic acid
Preparation of iodine salts, organic preparations, analytical reagent, disinfectant, pharmaceuticals. d = 1.70; bp = 127°; pH = 1.0 (0.1 molar solution). *Janssen Chimica.*

14556 Hydrisan
Flocculating aid for water treatment. *Goldschmidt.*

14557 Hydro
Fertilizers, chemicals, gases and plastics. *Norsk Hydro AS.*

14558 Hydro Paste®
Water dispersible aluminum pigment; used for aqueous paints and coatings used for decorative metallic effects as well as a wide range of protective coatings applications. *Silberline Mfg Co Inc.*

14559 Hydroace Series
Emulsifier for emulsion type fuel oils. *Toho Chem. Industry.*

14560 Hydroba
Jojoba oil derivatives, cosmetics, and toiletries additive. *A & E Connock (Perfumery & Cosmetics) Ltd.*

14561 Hydroblok
Cross-linked modified polyacrylamide. *Cyanamid BV.*

14562 Hydrobol
Water repellent finishes. *Ciba plc.* Name unverified.

14563 Hydroboracite
Firebrake®. Flame retardant for plastics and coatings *Borax Europe Ltd.*

14564 Hydrobromic acid
10035-10-6 4819 233-113-0
$HBr{\cdot}C_2H_4O_2$
Hydrogen bromide in acetic acid. Used in analytical chemistry, as a solvent for ore minerals, in manufacture of inorganic and some alkyl bromides, alkylation catalyst. *Allchem Industries; Associated Octel Co Ltd; EM Industries; Ethyl; Great Lakes.*

14565 hydrobuna
A name applied to hydrogenated rubber.

14566 hydrocarbon cement
A cement made by mixing heated pitch or tar with from one to four times its volume of calcium or magnesium sulfate; used for paving or building purposes.

14567 hydrocarbon-aldehyde resins
Resins obtained by the interaction of hydrocarbons such as naphthalene with formaldehyde in the presence of sulfuric acid.

14568 Hydrocell YP-30
Hydrolyzed yeast protein; cosmetic ingredient for skin and hair care products. *Brooks Industries.*

14569 Hydrocerol Compound
Blowing agent. *Henley Chemicals Inc.*

14570 Hydrocerol® BIH
Blowing agent for foaming of thermoplastics in injection molding and extrusion; high gas yield at lower temperatures. *Boehringer Ingelheim; Henley.*

14571 Hydrocerol® CF 70
Blowing and nucleating agents for plastics; for injection molding and extrusion for fine-celled foam. *Boehringer Ingelheim; Henley.*

14572 Hydrocerol® LC
Citric acid esters; blowing agent for foaming of thermoplastics at higher processing temperatures; especially suitable for PC. *Boehringer Ingelheim; Henley.*

14573 Hydrocerol® TAF 50
Nucleating agent for extrusion processes for direct gassing PS and polyolefins. *Boehringer Ingelheim; Henley.*

14574 Hydro-Chem
Driers for water based coatings. *Mooney Chemicals Inc.*

14575 hydrochloric acid
7647-01-0 4821 231-595-7
HCl
hydrogen chloride
Muriatic acid; Chlorohydric acid; Hydrochloride; Spirits of salts; Hydrogen chloride (acid); Hydrogen chloride; Marine acid. Used as a chemical intermediate and reactant. mp = -114°; bp = -85°; azeotrope with H_2O (20% HCl): bp = 109°; d_{15}^{15}= 1.096.

14576 hydrochloric acid
7647-01-0 4821 231-595-7
Muriatic acid. Acidizing of petroleum wells, boiler scale removal, chemical intermediate, ore reduction, food processing, pickling and metal cleaning, industrial; acidizing, general cleaning, alcohol denaturant, laboratory reagent; pharmaceutic aid (acidifier). d = 1.05; n_D^{18}= 1.34168. AlliedSignal; Asahi Chem Industry Co Ltd; Asahi Denka Kogyo; Elf Atochem; Dover; Du Pont; ICI SA; Miles; Nissan Chem. Ind.; OxyChem; PPG Industries; Rasa Ind.; Showa Denko; Vista; Witco/Argus.

14577 hydrocinnamic acid
501-52-0 4825 207-924-5
$C_9H_{10}O_2$
3-phenylpropionic acid
β-Phenylpropionic Acid; Benzenepropanoic Acid; Hydrocinnamic acid; 3-Phenylpropanoic acid. Fixative for perfumes, flavoring. mp = 47-48°; bp = 280°; soluble in H_2O (5.8 g/l), more soluble in organic solvents. *R.W. Greeff; Janssen Chimica; Penta Mfg.*

14578 Hydrocol®
Drainage aids for pulp, paper and board manufacture. *Allied Colloids Ltd.*

14579 Hydrocol®
Aqueous dispersion of iron oxides, ochers and composite pigments; used for coloration of cementitious products (mortars, blocks, bricks, slabs, panels, reconstructed stone, split blocks, etc.). *W Hawley & Son Ltd.*

14580 Hydrocoll AL-50, AL-55, EN-40, EN-55, EN-55-X, EN-SD, EN-SD-1M, EN-SD-10M
9015-54-7 310-296-6
Hydrolyzed animal protein; cosmetics ingredient. *Brooks Industries.*

14581 Hydrocoll G-40
9000-70-8 4388 232-554-6
Gelatin; cosmetics ingredient. *Brooks Industries.*

14582 Hydrocoll G-55
Hydrolyzed gelatin; cosmetic ingredient. *Brooks Industries.*

14583 Hydro-Cure
Drying catalyst for water based paints. *Mooney Chemicals Inc.*

14584 hydrocyanic acid
74-90-8 4836 200-821-6
CHN
prussic acid; formonitrile; Cyclon; HCN. Colorless or pale blue liquid or gas with a bitter almond odor detectable at 1 to 5 ppm. Present in apricot and peach pits in low concentrations, extremely toxic, used to exterminate rodents and insects. mp = -13°; bp = 26°; soluble in H_2O, LC_{50} (rat ihl 5 min) = 544 ppm.

14585 Hydrodarco
7440-44-0 1855 231-153-3
carbon
Activated carbon; used for water purification, purification of chemicals/pharmaceuticals, air purification and solvent recovery. *Am. Norit.* Name unverified.

14586 Hydroferrox®
1309-37-1 4072 215-168-2
ferric oxide
Aqueous dispersion of Bayferrox iron oxides; used for coloration of cementitious products (mortars, blocks, bricks, slabs, panels, reconstructed stone, split blocks, etc.). *W Hawtey & Son Ltd.*

14587 hydrofluosilicic acid
16961-83-4 4220 241-034-8
H_2F_6Si
fluorosilicic acid
hexafluorosilicic acid; hydrogen hexafluorosilicate; hexafluosilicic acid; silicofluoric acid. Stable in 60-70% aqueous solution. Etches glass. Used as 2% solution to sterilise equipment and in electroplating.

14588 Hydrofol Acid 1655
57-11-4 8959 200-313-4
Stearic acid
Stearic acid; pharmaceutic aid. *Sherex/Div of Witco.*

14589 hydrogen Fluoride
7664-39-3 4831 231-634-8
FH
Hydrofluoric acid gas
anhydrous hydrofluoric acid. Catalyst; used in aluminum production, fluorocarbons, pickling stainless steel, etching glass, acidizing oil wells, gasoline production, processing uranium. mp = -83.55°; LD_{50} (rat ihl 1h) = 1278 ppm; soluble in water, alcohol, and many organic solvents, slightly soluble in ether. *AlliedSignal; Du Pont; Farleyway Chem. Ltd; General Chem,; Hoechst Celanese; Seimi Chem.*

14590 hydrogen peroxide
7722-84-1 4839 231-765-0
H_2O_2
dihydrogen dioxide
high-strength hydrogen peroxide; peroxide; hydrogen dioxide; Albone; Inhibine; Perhydrol; Peroxan; Oxydol; Hydroperoxide; Hioxy; t-stuff; superoxol; H2O2. Bleaching and deodorizing textiles, wood pulp, hair, fur, etc.; plasticizers; refining and cleaning metals; viscosity control for starch and cellulose derivatives. mp = -11°; bp = 150°; d = 1.4067 *Elf Atochem N. Am.; Degussa; Du Pont; Farteyway Chem. Ltd; FMC; Mallinckrodt; Mitsubishi Gas.*

14591 hydrogen rubeanide
79-40-3 8436 201-203-9
$C_2H_4N_2S_2$
dithiooxamide
ethanedithioamide; rubeanic acid. The amide of dithiooxalic acid, a chelating agent, especially for copper, cobalt and nickel. mp = 200° (dec); slightly soluble in H_2O, EtOH, insoluble in organic solvents.

14592 hydrogenated castor oil
8001-78-3 232-292-2
Opalwax; castorwax; castor oil, hydrogenated. End product of controlled hydrogenation of castor oil; used in water-repellent coatings, candles, polishes, ointments, cosmetics; impregnant for paper, wood, cloth; as lubricant, mold release in manufacture of formed plastics and rubber goods. *Akzo; Arista; Sothern Clay Prods.*

14593 hydrogenated cottonseed oil
68334-00-9 269-804-9
Emvelop®. Hydrogenated vegetable oil; for production of sustained-release tablet formulations. *Mendell.* Discontinued.

14594 hydrogenated palm/palm kernel oil PEG-6 esters
Labrafil M 2130 BS; Hydrogenated palm/palm kernel oil PEG-6 complex. Hydrophilic wax for pharmaceutical and cosmetic formulations. *Gattefosse; Gattefosse SA.*

14595 hydrogenated polyisobutene
61693-08-1
Hydrogenated polyisobutene
Luvitol HP. Emollient oil component for cosmetics and pharmaceuticals. *BASF.* Name unverified.

14596 hydrogenated soybean oil
8016-70-4 232-410-2
Lipovol HS; Akorex; Code 321; Diamond D 31; Famous; S-Flakes; Witarix® 440; Soybean oil hydrogenated; Sterotex; Witarix® 440; Akofame; Akofil; Akoleno; Durkex; Bunge Biscuit Flakes; Superb;. Fat, used for filler, snack frying, spray oil. *Lipo; Karlshamns; Karlshamns Food Ingredients; Bunge Foods; Van Den Bergh Foods; Hüls AG;.*

14597 hydrogenated stripped coconut acid
68938-15-8 273-118-5
Hystrene® 5012; Industrene® 223. Hydrogenated stripped coconut acid; chemical intermediate, emulsifier; used for personal care products, soaps, lubricants, waxes, textile auxiliary, pharmaceuticals. *Witco/Humko.* Discontinued.

14598 hydrogenated tallow acid
61790-38-3 263-130-9
Hydrogenated tallow fatty acid
Glycon® S-65; Petrac® PHTA; Prifac 9428. Lubricant, defoamer, and component of other food additives. *Lonza Inc; Syn. Prods.; Unichema.*

14599 hydrogenated tallow glyceride citrate
68990-59-0 273-613-6
Hydrogenated tallow glyceride citrate
Lamegin® ZE 30, 60; Tegin® C-62 SE; Tegin® C-63 SE. Emulsifier for cosmetic, margarine and meat industry; food additive. *Grünau; Goldschmidt.*

14600 hydrogenated tallow glyceride lactate
68990-06-7 273-576-6
Hydrogenated tallow glyceride lactate
Lamegin® GLP 10, 20; Tegin® L61, L 62. Emulsifier and plasticizer for cosmetics, foods, and drugs. *Grünau; Goldschmidt.*

14601 hydrogenated vegetable oil
68334-28-1 269-820-6
End product of controlled hydrogenation of vegetable oil; emulsifier for food processing; binder, lubricant in pharmaceutical tableting, pressed powders; cocoa butter replacement in cosmetics and pharmaceuticals; emollient, wax.

Arista; Jojoba Growers & Processors; Karlshamns; Lipo; A.E. Staley Manufacture.

14602 hydrogenite
A mixture of 5 parts ferro-silicon, 90-95 parts silicon, 12 parts sodium hydroxide, and 4 parts slaked lime. When ignited it yields hydrogen.

14603 Hydrokeratin AL-30
69430-36-0 274-001-1
Hydrolyzed keratin
Cosmetics ingredient. *Brooks Industries.*

14604 Hydrokote® 95
68334-28-1 269-820-6
Hydrogenated vegetable oil; specialty base used as replacement for cocoa butter in cosmetic and pharmaceutical applications. *Karlshamns.*

14605 Hydrol 100
Partially hydrogenated coconut oil; kosher; oil for whipped topping, coffee whitener, vegetable dairy systems, biscuit and crackers. *Van Den Bergh Foods.*

14606 Hydrolact
Lactase. *John & E Sturge Ltd, Selby.* Name unverified.

14607 Hydrolactin 2500
Hydrolyzed milk protein; substantive skin and hair care conditioner, moisturizer; for shampoos, conditioner rinses, setting and waving lotions, moisturizing creams and lotions, night creams and lotion. *Croda Inc.*

14608 Hydrolactol 70
Glyceryl stearate, propylene glycol stearate, glyceryl isostearate, propylene glycol isostearate, oleth-25, ceteth-25; self-emulsifying base for fluid, semifluid lotions and oil-water creams, mineral pigment formulations. *Gattefosse.*

14609 Hydrolete
7789-78-8 1715 232-189-2
CaH_2
Calcium hydride
Reacts with water to produce hydrogen. Drying agent and reducing agent.

14610 Hydrolin
7775-14-6 8771 231-890-0
sodium hydrosulfite
Sodium hydrosulfite solution; for bleaching of clay, groundwood, and thermomechanical pulp. *Olin.*

14611 hydrolith
7789-78-8 1715 232-189-2
calcium hydride
A 90% calcium hydride which yields hydrogen on contact with water.

14612 hydrolyzed elastin
73049-73-7
European Elastin 10. Substantive protein adding to skin elasticity; protective colloid effect; film-former; moisturizer; for skin nourishing creams, shave creams, sun tan products, hair care and treatment products *Maybrook.*

14613 hydrolyzed vegetable protein
HVP; Hydrosoy 2000; Solu-Veg EN-35. Hydrolyzed vegetable proteins; natural flavors for all types of processed foods to improve or impart a characteristic flavor. *Croda Inc; Brooks Industries.*

14614 hydrolyzed wheat protein
70084-87-6
Hydrotriticum 2000. Ñubstantive conditioning agent, moisturizer for hair waving systems, shampoos, cream rinses, skin care prods. *Croda Inc.*

14615 Hydro-Marc
Etafilcon A; contact lens material. *Vistakon Inc.* Name unverified.

14616 Hydromilk EN-20
Hydrolyzed milk protein; complete food containing all essential amino acids. *Brooks Industries.*

14617 Hydromol Cream
Arachis oil, isopropyl myristate, liquid paraffin, sodium pyrrolidone carboxylate, sodium lactate; an emollient cream for treatment of dry skin conditions. *Quinoderm Ltd.*

14618 Hydromol Emollient
Light liquid paraffin, isopropyl myristate; an emollient bath additive for dry skin conditions. *Quinoderm Ltd.*

14619 Hydron Blue
A dark-blue vat dye, obtained by reducing nitrosophenol with carbazole to form an indophenol, which is heated with sodium sulfide, and subsequently with sulfur; also used as a trademark for a range of other dyestuffs. *Cassella AG.* Discontinued.

14620 Hydron®
Sulfur vat dyestuffs; used for textile dyeing and printing. *Cassella AG.*

14621 Hydronal
Pure (+)-hydroxy citronellaldehyde. Used in perfumery. *Bush Boake Allen Ltd.*

14622 Hydronalium
An aluminum alloy containing from 7-9% magnesium and small amounts of silicon and manganese. It is resistant to sea-water, soap, and soda.

14623 hydronaphthol
135-19-3 6471 205-182-7
$C_{10}H_8O$
β-naphthol
Chemical intermediate.

14624 hydrone
An alloy of 35% sodium and 65% lead, which generates hydrogen by action of water.

14625 Hydronyx
7775-14-6 8771 231-890-0
sodium hydrosulfite
A proprietary trade name for sodium hydrosulfite, a reducing agent used in dyeing. No manufacturer.

14626 Hydropalat 535
Amine neutralized polyester; suspending agent for water-reducible paints. *Henkel KgaA.* Discontinued.

14627 Hydropalat® A
A registered trade name for diethylhydrophthalate, a solvent. No manufacturer.

14628 Hydropalat® B
A registered name for dibutylhydrophthalate, a solvent. No manufacturer.

14629 Hydrophilol ISO
68171-38-0 269-027-5
Propylene glycol isostearate, a surfactant. *Gattefosse SA.*

14630 Hydrophobol
Proofing agent. *Ciba plc.* Name unverified.

14631 Hydropur®
Water reworking and treatment agents, flocculating agents. *Cassella AG.*

14632 hydroquinone
123-31-9 4853 204-617-8
$C_6H_6O_2$
1,4-benzenediol
p-dihydroxybenzene; hydroquinol. Photographic developer (not for color film); dye intermediate, inhibitor; stabilizer in paints and varnishes; motor fuels and oils; antioxidant for fats and oils. mp = 172-175°; bp = 285°; d = 1.3280; LD_{50} (rat orl) = 320 mg/kg. *Aldrich; Allchem Industries; Eastman; Goodyear Tire & Rubber; Kraeber GmbH.*

14633 hydroquinone monomethyl ether
150-76-5 205-769-8
$C_7H_8O_2$
4-Methoxyphenol
MEHQ; *p*-Hydroxyanisole; Eastman® HQMME. Antioxidant for monomers. Used in manufacture of pharmaceuticals, plasticizers, dyestuffs; stabilizer for chlorinated hydrocarbons and ethylcellulose; UV inhibitor; inhibitor for arcylic and vinyl monomers and acrylonitrile. d = 1.55; mp = 52.5°; bp = 243°; soluble in benzene, acetone, ethyl acetate, alcohol. *Alemark; Alfa; Arenol; ChemDesign; Eastman; Fluka; Kincaid Enterprises; Penta Mfg.; Speciality Chem. Prods.*

14634 Hydro-resin A
A proprietary trade name for a water soluble resin. No manufacturer.

14635 Hydros
7775-14-6 8771 231-890-0
Sodium hydrosulfite
Specially prepared for use as a reducing agent in vat color dyeing. *Albright & Wilson Ltd.*

14636 Hydros® 1
sodium hydrosulfite-sodium pyrophosphate.
Sodium hydrosulfite and sodium pyrophosphate. A reducing agent and bleach. *RV Chemicals Ltd.*

14637 Hydros® F
7775-14-6 8771 231-890-0
Sodium hydrosulfite
sodium dithionite; sodium metabisulfite. Used for the textile coloration/mineral bleaching/general bleaching. *RV Chemicals Ltd.*

14638 Hydrosil®
Organofunctional silane coupling agents; for surface treatment of fillers, substrate primers, integral blend components of thermoplastic and thermoset polymers. *Hüls Am.*

14639 hydrosol
An aqueous colloidal silver solution.

14640 Hydrosol
Solubilized sulfur dyestuffs; used for textile dyeing and printing. *Hoechst AG.*

14641 Hydrosol®
Sulfur dyestuffs; used for textile dyeing and printing. *Cassella AG.*

14642 Hydrosoy 2000
977059-33-8
Hydrolyzed soy protein; conditioner for cosmetic products. *Croda Inc.*

14643 hydrosulfite
7775-14-6 8771 231-890-0
sodium hydrosulfite
Rongalite; hyraldite; decroline; redo; sodium dithionite; sodium hydrosulfite; blanchite; blankit. Used in dyeing, and for decolorizing sugar syrups.; oxygen scavenger for synthetic rubbers.

14644 Hydrosulfite A
A 10% solution of hydrosulfite NF, or hyraldite (sodium hydrosulfite); used for testing dyed fabrics.

14645 Hydrosulfite AWC
7775-14-6 8771 231-890-0
Sodium hydrosulfite
Henkel/Cospha.

14646 Hydrosulfite BASF
7775-14-6 8771 231-890-0
$Na_2S_2O_4$
A 90% sodium hydrosulfite. No manufacturer.

14647 Hydrosulfite NF
Rongalite Conc.; Brittalite; Formosul. A condensation product of formaldehyde and sodium hydrosulfite; consists of a mixture of formaldehyde sodium bisulfite, $NaHSO_3{\cdot}CH_2O$, and formaldehyde-sodium sulfoxylate, $NaHSO_2{\cdot}CH_2O$; used as a dye discharger in calico printing.

14648 Hydrosulphit®
7775-14-6 8771 231-890-0
Sodium hydrosulfite
Sodium dithionite. Reducing bleaches for wood pulps and wood-based old paper. *BASF AG.*

14649 Hydrotek
Pigment dispersions; for coloration of aqueous coatings. *Pacific Dispersions Inc.* Name unverified.

14650 Hydrotriticum 2000
70084-87-6
Hydrolyzed wheat protein; substantive conditioning agent, moisturizer for hair waving systems, shampoos, cream rinses, skin care products. *Croda Inc.*

14651 Hydrotriticum Powd
Hydrolyzed whole wheat protein. *Croda Inc.*

14652 Hydrotriticum QL
Laurdimonium hydroxypropyl hydrolyzed wheat protein. Film-forming conditioner for hair and skin care products, e.g., waving systems, activated conditioner treatments, shampoos, styling mousses, hair coloring, wrinkle remover creams and lotions, liquid soap, facial scrubs, skin cleansers and bath products. *Croda Inc.*

14653 Hydrotriticum QM
Cocodimonium hydroxypropyl hydrolyzed wheat protein; film-forming conditioner for hair and skin care prods., e.g., waving systems, activated conditioner treatments, shampoos, styling mousses, hair coloring, wrinkle remover creams and lotions, liquid soap, facial scrubs, skin cleanser and bath products. *Croda Inc.*

14654 Hydrotriticum QS
Steardimonium hydroxypropyl hydrolyzed wheat protein; film-forming conditioner for hair and skin care prods., e.g., waving systems, activated conditioner treatments, shampoos, styling mousses, hair coloring, wrinkle remover creams and lotions, liq. soap, liquid soap, facial scrubs, skin cleanser and bath products. *Croda Inc.*

14655 Hydrotriticum WAA
Wheat amino acids; substantive moisturizer, humectant for hair and skin care products, cleansers, antiwrinkle preparations, sun screens; leaves soft and conditioned afterfeel. *Croda Inc.*

14656 Hydroxal
21645-51-2 355 244-492-7
aluminum hydroxide
A proprietary oral antacid consisting of a suspension of aluminum hydroxide. *Blue Line Chemical Co. Unverified.*

14657 o-hydroxyacetophenone
118-93-4 204-288-0
$C_8H_8O_2$
2'-hydroxyacetophenone
2-Hydroxyacetophenone; 1-(2-Hydroxyphenyl)ethanone; 2-Acetylphenol. Used in organic synthesis. mp = 4-6°; bp_{717} = 213°; d = 1.1310; n_D^{20} = 1.5584. *R.W. Greeff; Hoechst Celanese; Janssen Chimica; Penta Mfg.*

14658 p-hydroxyacetophenone
99-93-4 202-802-8
$C_8H_8O_2$
4'-hydroxyacetophenone
4-hydroxyacetophenone; 1-(4-hydroxyphenyl)ethanone; 4-Acetylphenol. Used in organic synthesis. mp = 106-108°; bp_3 = 147-148°; d = 1.1090. *R.W. Greeff; Hoechst Celanese; Janssen Chimica; Schweizerhall.*

14659 1-hydroxybenzotriazole
2592-95-2 219-989-7
$C_6H_5N_3O{\cdot}H_2O$
1-Hydroxy-1H-benzotriazole
HOBt; N-Hydroxybenzotriazole. Widely used additive to decrease racemization during dicyclohexylcarbodiimide-catalyzed peptide coupling. mp = 157-158°. *Aldrich; Janssen Chimica; Schweizerhall.*

14660 hydroxybutylmethylphenylchlorobenzotriazole
3896-11-5 223-445-4
$C_{19}H_{22}ClN_3O$
2-(2'-Hydroxy-3'-*t*-butyl-5'-methylphenyl)-5-chlorobenzotriazole
Lowitite® 26. UV stabilizer for polyolefins, polyester resins and coatings. *Lowi.*

14661 4-hydroxycoumarin
1076-38-6 214-060-2
$C_9H_6O_3$
4-hydroxy-2H-1-benzopyran-2-one
4-Coumaryl alcohol. Chemical intermediate. mp = 213-214°. *R.W. Greeff; Janssen Chimica; Penta Mfg.*

14662 hydroxyethyl methacrylate
868-77-9 212-782-2
$C_6H_{10}O_3$
2-hydroxyethyl methacrylate; Bisomer 2HEMA; Sipomer® HEM-D. HEMA; acrylic resins, binder for nonwoven fabrics, enamels. *BP Chemicals Ltd.; Mitsubishi Gas; Rohm & Haas; Rhône-Poulenc Surf.*

14663 hydroxylated lanolin
68424-66-8 270-315-8
Hydroxylan; Ivarlan OH; Ohlan OH; Ritahydrox. Emulsifier, conditioner, emollient. *Fanning; Brooks Industries; Amerchol; Amerchol Europe; RITA.*

14664 hydroxymimetite
An artificially made basic lead arsenate.

14665 hydroxyprolisilane C
12897-74-0
Methylsilanol aspartate hydroxyprolinate
Provides collagen restructuring for cosmetic and health products including stretch mark prevention creams, anti-aging formulations, acne preventives, eye contour creams, etc. *Exsymol.*

14666 hydroxypropyl guar
39421-75-5
Hydroxypropyl ether guar gum
Galactasol® Guar Derivs; Hydroxypropyl guar gum; Guar gum; Chesguar HP4; Chesguar HP4R; Chesguar HP6; Jaguar® HP8, HP60, HP120, HP-11, HP-200. Gums used to increase viscosity in water and most brines; high viscosity grades also as flocculants; some in series as stabilizers, suspending aids. Develops viscosity rapidly when pH is adjusted below 7.0; disperses in cold water; visc = 3600-4200 mPa.s. *Aqualon; Chesman Chem Ltd.*

14667 hydroxypropyl methacrylate
27813-02-1 248-666-3
$C_7H_{12}O_3$
Propylene glycol monomethacrylate
Monomer for acrylic resins, nonwoven fabric binders, detergent lube oil additives. *BP Chem.; Dajac Labs; Rohm & Haas.*

14668 hydroxypropyl methylcellulose
9004-65-3 4889
Hydroxypropyl methylcellulose
Lacril®. Ocular lubricant. *Allergan Inc.*

14669 1-hydroxy-2-pyridine
142-08-5 205-520-3
C_5H_5NO
2-hydroxypyridine
2-pyridinol; 2-pyridone. Educt for the preparation of 1-alkenyl-2-pyridones; useful in Diels-Alder reactions.

14670 8-hydroxyquinoline
148-24-3 4890 205-711-1
C_9H_7NO
8-Quinolinol
Oxine; Hydroxy quinoline; oxyquinoline; hydroxybenzopyridine; oxybenzopyridine; phenopyridine; oxychinolin. Precipitating and separating metals, preparation of fungicides, chelating agent, disinfectant. mp = 76°; bp = 267°; almost insoluble in water, ether; soluble in alcohol, acetone, chloroform, benzene, aqueous mineral acids; LD_{50} (mus ip) = 48 mg/kg. *Penta Mfg.; Spectrum Chem. Mfg.; Superol BV; Tanabe USA.*

14671 hydroxysuccinic acid
6915-15-7 5747 230-022-8
$C_4H_6O_5$
malic acid

apple acid. Manufacture of esters and salts, wines; chelating agent, food acidulant, flavoring. mp = 131-132°; soluble in H_2O (55 g/100 ml), organic solvents. *Allchem Industries; Haarmann & Reimer; Janssen Chimica; Schweizerhall.*

14672 hydrozone
7722-84-1 4839 231-765-0
hydrogen peroxide
Pyrozone; Glycozone. Trade names for hydrogen peroxide, an oxidizer used as an antiseptic in dental practice.

14673 Hyflo NS, S
9006-04-6 232-689-0
Powdered NR, dusted with inert silica; used for adhesives, pharmaceutical and other tapes; Mooney viscosity NS = nonstabilized, S = stabilized. *H.A. Astlett.*

14674 Hyflo Super-Cel
7631-86-9 8637 231-545-4
Diatomaceous earth; filter aids for industrial use (beer, dry cleaning, pharmaceuticals, chemicals, water, industrial waste). *Celite.*

14675 Hyflux
Soldering fluxes. *ABM Chemicals Ltd.* Name unverified.

14676 Hyflux M
13775-80-9 237-412-7
BrH_5N_2
Hydrazine hydrobromide
hydrazine monohydrobromide. Reducing agent. mp = 87-92°. *ABM Chemicals Ltd.* Name unverified.

14677 Hy-glo Steel
A proprietary trade name for a stainless steel containing 17.0% chromium and 0.6% carbon. No manufacturer.

14678 Hygrass
dicamba-mecoprop
Soluble concentrate of 18.7 g dicamba and 300 g mecoprop per liter; used for weed control in cereals and grassland. *Agrichem (International Ltd.*

14679 hygrol
Colloidal mercury.

14680 Hygroplex HHG
Natural moisturizing factor in hydrophilic medium; for emulsified aqueous and hydroalcoholic skin and hair moisturizing cosmetics. *Dr. Kurt Richter; Henkel/Cospha.*

14681 Hyjet
Hydraulic fluid. *Monsanto (Solaris).* Name unverified.

14682 hylastic
A high-manganese steel containing 1.6-1.8% manganese and 0.35% carbon.

14683 Hylene®
A proprietary trade name for an organic diisocyanate used in the manufacture of polyurethane foam having a range of rigidities. *DuPont UK.*

14684 Hylite Color-Max
Fluorescent pigment/plasticizer dispersions; pourable colors for plastisols and other systems; offered custom matched to standard. *W J Ruscoe Co.*

14685 Hymec
7085-19-0 5826 230-386-8
mecoprop
Soluble concentrate containing 570 g/l mecoprop; for control of weeds in cereals and grassland. *Agrichem (International) Ltd.*

14686 Hymod
Copper powder for sintering iron. *Steetley Chemicals Ltd.* Name unverified.

14687 Hymolon CWC
Cocamide DEA
Fulling agent for wool and household cleaners. *Hart Chem. Ltd.*

14688 Hymono
Distilled monoglycerides; for various applications in the food industry e.g., margarines, shortenings, bakery and dairy. *Quest Int'l.*

14689 Hymush
148-79-8 9426 205-725-8
thiabendazole
Wettable powder containing 60% w/w thiabendazole; a systemic fungicide. *Agrichem (International) Ltd.*

14690 Hyonic
Wetting agents. *Henkel Chemicals Ltd.* Name unverified.

14691 Hyonic GL 400
Polyethoxy alkyl phenol
Surfactant, coemulsifier for specialty polymerizations. *Henkel/Functional Prods.*

14692 Hyonic NP-40
9016-45-9 6772
Nonoxynol-4
Emulsifier for solvent cleaning compounds, waterless hand cleaners, agricultural emulsifiable concentrates, silicone products, detergent for petroleum oils; viscosity reducer for plastisols; intermediate for production of anionic surfactants; metal cleaners; food contact applications. *Henkel/Functional Prods.; Henkel/Organic Prods.*

14693 Hyonic OP-55
Scouring for textile desizing; emulsifier for waxes and oils preventing redeposition. *Henkel/Textiles.*

14694 Hyonic OP-7
9002-93-1 6858
Octoxynol-7
Surfactant for textile use. *Henkel/Textiles.*

14695 Hyonic PE-100
9016-45-9 6772
Nonoxynol-10
Detergent, wetter, emulsifier, base, penetrant for household and industrial cleaners, paper toweling; latex stabilizer. *Henkel.* Name unverified.

14696 Hypacel
Bleaching agent for textiles. *RV Chemicals Ltd.* Discontinued.

14697 Hypalon®
Chlorosulfonated polyethylene; synthetic elastomer; resistant to weather, ozone, oil, solvents, chemicals, heat, flame, and abrasion; used in jacketing and insulation for wire/cable, soles and heels, automotive components, coated fabrics, white tire sidewalls, sheet roofing, liners and covers for reservoirs and waste containment ponds, protective, decorative coatings, hose, linings for chemical processing equipment, industrial rolls, seals, gaskets and diaphragms. *DuPont.* Name unverified.

14698 Hypalon® CP 826
Chlorinated, maleic anhydride-grafted PP resin; adhesion promoter for PP and blends; adhesion modifier for coatings, adhesives and inks. *DuPont.*

14699 Hypan® QT 100
Acrylic multiblock copolymer; substantive to skin and hair; thickener; carrier for active substances; provides protective coating; primary emulsifier. *Kingston Tech.*

14700 Hypan® SA100H
61788-40-7
Acrylic acid/acrylonitrogens copolymer. Emulsifier and gellant for neutral pH; able to form conjugates with certain drugs for controlled delivery formulations. *Kingston Tech.*

14701 Hypan® SR150H
136505-00-5
Acrylic acid/acrylonitrogens copolymer. Thickener and gellant for aqueous formulations, especially highly concentrated salt solutions, surfactants and drugs; emulsifier. *Kingston Tech.*

14702 Hypan® SS201
123754-28-9
Ammonium acrylates/acrylonitrogens copolymer. Gellant, emulsifier for cosmetics and related applications. *Kingston Tech.*

14703 Hypan® SS500V
Tromethamine acrylates/acrylonitrogens copolymer. Gellant, emulsifier for cosmetics and related applications. *Kingston Tech.*

14704 Hypan® SS500W
TEA-acrylates/acrylonitrogens copolymer. Gellant, emulsifier for cosmetics and related applications. *Kingston Tech.*

14705 Hypaque Meglumine
131-49-7 5851 205-024-7
Diatrizoate meglumine
Diagnostic aid. *Sterling Drug Inc.* Name unverified.

14706 Hypax
Oxidized wax/printing inks; anticorrosives. *Chemoxy international Ltd.* Discontinued.

14707 Hyper+Ion 1050
Coagulant and flocculant for potable water treatment, municipal wastewater treatment, industrial water and waster treatment, industrial applications. *General Chem.*

14708 Hyper+Ion 2050A
Coagulant and flocculant for potable water treatment, municipal wastewater treatment, industrial water and wastewater treatment, industrial applications. *General Chem.*

14709 Hyperiast®
Polyether polyurethane systems; used for oil and marine, electrical, engineering, and automotive applications. *Hyperlast.*

14710 Hyperkil Bait
Caciferol
A rodenticide. *Antec international Ltd.*

14711 Hypermer
Dispersants. *Imperial Chemical Industries plc.*

14712 Hyperol
hydrogen peroxide-urea
Proprietary name for a compound of hydrogen peroxide and urea, $CO(NH_2)_2 \cdot H_2O_2$; contains 35% hydrogen peroxide, which is obtained by dissolving in water or ether; one gram in 10 ml = a 10-volume strength solution of hydrogen peroxide. No manufacturer.

14713 Hypersal
Functional additives for decorative laminates. *Hoechst AG.*

14714 Hypersal®
Auxiliary agents used in impregnation with melamine and urea formaldehyde resins. *Cassella AG.*

14715 Hypersol
Hyperdispersants for pigment dispersions. *Tennant-KVK Ltd.*

14716 Hypnodil
6232
$C_{13}H_{14}N_2O_2 \cdot HCl$
1-(1-phenylethyl)-1H-imidazole-5-carboxylic acid methyl ester hydrochloride
R-7315. A proprietary preparation containing metomidate hydrochloride; veterinary hypnotic used with pigs. The free base is metomidate [5377-20-8]. mp = 173-174°; LD_{50} (rat iv) = 50 mg/kg. *Janssen Pharmaceutical Ltd.* Name unverified.

14717 Hypnorm
fentanyl citrate-fluanisone
A proprietary preparation containing fentanyl citrate and fluanisone; veterinary anesthetic (dogs, rabbits and guinea pigs). *Janssen Pharmaceutical Ltd.* Name unverified.

14718 Hypochlorous acid
7790-92-3 4912 232-232-5
ClHO
HyPure A. For bleaching, in organic synthesis; disinfectant In aqueous solution, decomposes slowly to chlorine, oxygen and perchloric acid. *Olin.*

14719 Hypol® FHP 2000
PU prepolymer derived from TDI; foamable hydrophilic prepolymer offering water activation, high additive loading, controllable flexibility and texture, high water absorbence, low temperature exothermicities, inherent flame retardancy. *W R Grace/Organics.*

14720 Hyporit
7778-54-3 1717 231-908-7
calcium hypochlorite
The trade name for calcium hypochlorite, containing 80% available chlorine; used as an antiseptic. No manufacturer.

14721 hypoxanthine
68-94-0 4917 200-697-3
$C_5H_4N_4O$
1,7-dihydro-6H-purin-6-one
6-hydroxypurine; H; 6-oxopurine; purine-6-ol; sarcine; sarkin. Used in biochemical research. mp > 300°; soluble in H_2O (700 mg/l), poorly soluble in organic solvents.

14722 Hypromellose
A partial mixed methyl and hydroxypropyl ether of cellulose used as a surface-active agent.

14723 Hyprone
dicamba-MCPA-mecoprop
Soluble concentrate of 18.7 g dicamba, 100 g MCPA and 194 g mecoprop per liter, used for weed control in cereals and grassland. *Agrichem (International) Ltd.*

14724 HyPure A
7790-92-3 4912 232-232-5
HClO
Hypochlorous acid
For bleaching, organic synthesis. *Olin.*

14725 HyPure C
7790-93-4 2141 232-233-0
Chloric acid.
Oxidizing agent. *Olin.*

14726 HyPure K
7778-66-7 231-909-2
OClK
Potassium hypochlorite
For liquid bleach, water treatment, hard surface cleaners. *Olin.*

14727 HyPure L
13840-33-0 237-558-1
lithium hypochlorite
For liquid bleach, water treatment, hard surface cleaners. *Olin.*

14728 HyPure N
7681-52-9 8773 231-668-3
ClNaO
Sodium hypochlorite
For liquid bleach, water treatment, hard surface cleaners. *Olin.*

14729 Hyquat 70, 75
999-81-5 2153 213-666-4
Chlormequat
Plant growth regulator. *Agrichem (International) Ltd.*

14730 Hyraldite C Ext
149-44-0 8764 205-739-4
$NaHSO_2 \cdot CH_2O \cdot 2H_2O$
Sodium formaldehyde sulfoxylate
Used as a reducing agent in calico printing. No manufacturer.

14731 Hysa
Derivative of Hybis polybutene; used in surfactant manufacture and as a corrosion inhibiting agent. *BP Chemicals Ltd.* Discontinued.

14732 Hysede
lindane-thiabendazole-thiram
Mixture of lindane, thiabendazole and thiram; seed dressing for brassica crops. *Agrichem (International) Ltd.*

14733 Hysol® 1C Epoxi-Patch Kit
25928-94-3
Two-part epoxy system; general-purpose room temperature-curing adhesive/sealant for bonding metal, wood and most plastics; used in contact with meat during processing; easily sanded, cut, tapped, and machined; nontoxic when cured. *Dexter/Hysol.*

14734 Hysol® 2000
Vinyl acetate-based; hot-melt adhesive with unique combination of tackifiers allowing good adhesion of styrene-based plastics. *Dexter/Hysol.*

14735 Hysol® 342
24937-78-8
EVA-based; hot-melt adhesive with fast set time and excellent hot tack; for packaging applications. *Dexter/Hysol.*

14736 Hysol® 6C Epoxi-Patch Kit
25928-94-3
Aluminum-filled two-part epoxy system; epoxy kit used for bonding, sealing, and repairing appliance condensers, radiators, other aluminum substrates. *Dexter/Hysol.*

14737 Hysol® 7804
Polyamide; tough, elastomeric hot-melt adhesive with excellent impact resistance at low temperatures; for demanding applications where substrates are exposed to temperature extremes; bonds to metals, plastics, wood, leather, fabric, nonwoven fabric, films and foils. *Dexter/Hysol.*

14738 Hysol® 1942
24937-78-8
EVA-based adhesive; medium setting adhesive for general purpose applications; excellent adhesion to wood and many plastics (nylon, PC, PVC, PS, ABS, acrylic). *Dexter/Hysol.*

14739 Hysol® EA9460
Two-part epoxy; outstanding combination of peel and shear strength, good adhesion to metals, ceramics, plastics, and FRP without primers; for structural bonding applications. *Dexter/Hysol.*

14740 Hysol® EE0067/HD3561 and Hysol EE0067/HD3615
Resin/hardener system with nonabrasive filler; low viscosity, low cost bulk casting system with nonabrasive filler for general purpose applications; mix ratios; 100/20 and 100/14 by weight respectively. *Dexter/Hysol.*

14741 Hysol® EO1016
25928-94-3
One-component epoxy system; casting compound for electrical insulation applications, thermal shock resistance and flameout properties; extremely stable at room temperature; not for casting masses larger than 50 g; heat curing. *Dexter/Hysol.*

14742 Hysol® ES4228
Thixotropic version of Hysol ES4128; for conformal coating or blobbing semiconductor chips; for use in potting; solid state watch circuits and chip carriers; mix ratio: 100/100 by weight. *Dexter/Hysol.*

14743 Hysol® MBI-02
A proprietary epoxy resin molding powder modified for use as a load-bearing material. *Dexter/Hysol.* Name unverified.

14744 Hysol® MG1 Series
25928-94-3
One-component epoxy molding powder; short flow compression grade; fast gel times, good dimensional stability and good release; not for molding delicate inserts; in thin wall cases, bobbins, insulators. *Dexter/Hysol.*

14745 Hysol® OSO100
General purpose LED lamp encapsulant, color stability and moisture resistance; mix ratio: 100/100 by weight. *Dexter/Hysol.*

14746 Hysol® PC18
One-component urethane; clear printed circuit coating for brush dip or spray

application; tough abrasion resistance, and environmental protection. *Dexter/Hysol.*

14747 Hysol® Polyshot 1X

Hot melt adhesive for use on particle board, softwoods, and foam substrates. *Dexter/Hysol.*

14748 Hysol® XC7-W529

A proprietary flexible, one-component epoxy casting and potting compound possessing good thermal shock properties. *Dexter/Hysol.* Name unverified.

14749 Hysorb

7758-29-4 8846 231-838-7
sodium tripolyphosphate
Sodium tripolyphosphate, a sequestrant used as a water softener. *FMC.* Discontinued.

14750 Hyspray

61791-44-4 263-177-5
Cationic surfactant containing 800 g/l polyethoxylated tallow amine; wetting and spreading agent for phosphonoglycine herbicide sprays. *Fine Agrochemicals Ltd.*

14751 Hystar®

Hydrogenated starch; for use in toiletries and special dietary food. *Lonza Ltd.*

14752 Hystar® 3375

Hydrogenated starch hydrolyzate; humectant, bodying agent, moisture control agent for toothpaste, cosmetics, sugarless confections, controlled moisture foods and industrial applications. *Lonza Inc.*

14753 Hystar® TPF

Hydrogenated starch hydrolyzate; lubricant; maintains optimal moisture control in industrial, pet food, tobacco, teat dip, and oral hygiene products. *Lonza Inc.*

14754 Hystor 10

117-18-0 204-278-2
Tecnazene
Granules containing 10% w/w tecnazene; protectant fungicide and potato sprout suppressant. *Agrichem (International) Ltd.*

14755 Hystrene® 1835

67701-05-7 262-978-7
Coconut acid; Coco fatty acids; Coconut oils acids; Industrene® 325, 328; Kartacid C 60; C-108; C-110; Emery® 621, 622, 626, 627; C-108; C 70; Prifrac 5901; Radiacid® 631. Mixted tallow and coconut acids (CTFA); chemical intermediate, emulsifier; used for personal care products, soaps, waxes, textile auxiliary, and pharmaceuticals. *Witco Corporation.*

14756 Hystrene® 3022

Hydrogenated menhaden acid; chemical intermediate, emulsifier; used for personal care products, waxes, greases, textile auxiliary, pharmaceuticals. *Witco Corporation.*

14757 Hystrene® 3675

75% Dimer acid; corrosion inhibitor, intermediate; derivatives used as synthetic lube components, corrosion inhibitors for petroleum processing, as extenders and crosslinking agents for high polymeric systems; mildness additive in detergents. *Witco Corporation.*

14758 Hystrene® 4516

57-11-4 8959 200-313-4
Stearic acid
Lubricant, textile auxiliary, emulsifier, plasticizer, intermediate, used in cosmetics, shampoos, pharmaceuticals. *Witco Corporation.*

14759 Hystrene® 5012

68938-15-8 273-118-5
Hydrogenated stripped coconut acid; chemical intermediate, emulsifier; used for personal care products, soaps, lubricants, waxes, textile auxiliary, pharmaceuticals. *Witco Corporation.*

14760 Hystrene® 5016

57-11-4 8959 200-313-4
Stearic acid, triple pressed; food grade acids used as lubricants, release agents, binders, and defoamers, and in components for producing other food grade additives. *Witco Corporation.*

14761 Hystrene® 5460

68937-90-6
Trilinoleic acid
Corrosion inhibitor, lubricant; intermediate for manufacture of soaps, emulsions, creams, lotions, ethoxylates, buffing compounds, lubricants. *Witco Corporation.*

14762 Hystrene® 7018 FG

57-11-4 8959 200-313-4
Stearic acid
Food grade acids used as lubricants, release agents, binders, and defoamers, and in components for producing other food grade additives. *Witco Corporation.*

14763 Hystrene® 8016

57-10-3 7128 200-312-9
Palmitic acid
80% Palmitic acid; intermediate for manufacture of soaps, emulsions, creams, lotions, ethoxylates, buffing compounds, lubricants. *Witco Corporation.*

14764 Hystrene® 9014

544-63-8 6416 208-875-2
Myristic acid (90%)
Lubricant, textile auxiliary, emulsifier, plasticizer, intermediate, used in cosmetics, shampoos, pharmaceuticals. *Witco Corporation.*

14765 Hystrene® 9022

112-85-6 1051 204-010-8
Behenic acid (90%)
Lubricant, textile auxiliary, emulsifier, plasticizer, intermediate, used in cosmetics, shampoos, pharmaceuticals. *Witco Corporation.*

14766 Hystrene® 9512

143-07-7 5396 205-582-1
Lauric acid (95%)
Lubricant, textile auxiliary, emulsifier, plasticizer, intermediate, used in cosmetics, shampoos, pharmaceuticals. *Witco Corporation.*

14767 Hystrene® 9514

544-63-8 6416 208-875-2
Myristic acid (95%)
Chemical intermediate, emulsifier; used for personal care prods., waxes, textile auxiliary, pharmaceuticals. *Witco Corporation.*

14768 Hystrene® 9718 NF FG

57-11-4 8959 200-313-4
Stearic acid NF (92%)
Food grade acids used as lubricants, release agents, binders, and defoamers; intermediate for producing food grade emulsifiers. *Witco Corporation.*

14769 HyStretch V-29

Heat-reactive, carboxylated saturated acrylic elastomer; sprayable, foamable, high compression recovery, low tack, durable, soft, inherent heat and light stability; can be heat sensitized; self-crosslinking at 275+F; accepts fillers, pigments, thickeners and cross-linkers; used for textile, non-woven, adhesive bindings, coatings and finishes. *BFGoorich.* Name unverified.

14770 HyStretch V-43

Acrylates copolymer; heat-reactive, carboxylated sat. acrylic elastomer; sprayable, foamable, high compression and stretch recovery, low tack, durable, soft, low blocking; inherent heat and light stability; can be heat sensitized; self-crosslinking at 275ºF; accepts fillers, pigments, thickeners and cross-linkers; used for textile, non-woven, adhesive bindings, coatings and finishes. *BFGoodrich.* Name unverified.

14771 Hystrene® 5522

112-85-6 1051 204-010-8
Behenic acid
Chemical intermediate, emulsifier; used for personal care products, soaps, waxes, textile auxiliary, pharmaceuticals. *Witco/Humko.* Discontinued.

14772 Hysward

dicamba-MCPA-mecoprop
Mixture of dicamba, MCPA and mecoprop; used for weed control in cereals and grassland. *Agrichem (International) Ltd.*

14773 Hytak

Hot-melt adhesives for packaging and product assembly formulated from a variety of polymers, resins, waxes, plasticizers and antioxidants; used for carton sealing, bottle and can labelling, manufacture of disposable sanitary products, assembling of components in woodworking and in automitive and general industries. *Hytak Ltd.* Name unverified.

14774 Hytane

34123-59-6 5237 251-835-4
Isoproturon
Suspension concentrate containing 500 g isoproturon per liter; used for annual weed control in cereals. *Ciba-Geigy Agrochemicals.*

14775 Hytec

117-18-0 204-178-2
Tecnazene
Dustable powder containing tecnazene; protectant fungicide and potato sprout suppressant. *Agrichem (International) Ltd.*

14776 Hytec Super

tecnazene-thiabendazole
Dustable powder containing 6% w/w tecnazene [117-18-0] and 1.8% thiabendazole [148-79-8]; a protectant fungicide and potato sprout suppressant. *Agrichem (International) Ltd.*

14777 Hytemco

A proprietary trade name for an iron-nickel resistance alloy. No manufacturer.

14778 HyTemp 4051

Acrylic ester copolymer; elastomer for hose, tubing, cable jacket, belting, and mechanical goods; offers service temperature of 0-400ºF; excellent oil resistance. *Zeon.*

14779 HyTemp 4052
Acrylic ester copolymer; elastomer for hose, tubing, cable jacket, belting, and mechanical goods; service temperature -30 to 375ºF. *Zeon.*

14780 HyTemp NPC-50
Quaternary ammonium compound; nonpost cure agent used with all HyTemp 4050 series elastomers. *Zeon.*

14781 Hyten M Steel
A proprietary trade name for a nickel-chromium-molybdenum steel. No manufacturer.

14782 Hyten®
Fibers. *DuPont UK.*

14783 Hy-ten-sl
A proprietary trade name for an alloy of 66% copper, 19% zinc, 10% aluminum, and 5% manganese. No manufacturer.

14784 Hytex
Printing ink alkyds. *Croda Resins Ltd.*

14785 Hythane
Polyurethane alkyds. *Croda Resins Ltd.*

14786 Hytherm
Heat resistant dyes for plastics. *Morton Int'l. Ltd.*

14787 Hytin
fentin acetate-aneb
Mixture of fentin acetate and maneb; used for control of potato blight. *Agrichem (International) Ltd.*

14788 Hy-TL
thiabendazole-thiram
Mixture of thiabendazole [148-79-8] and thiram [137-26-8]; fungicide seed dressing. *Agrichem (International) Ltd.*

14789 Hytox
Germicidal detergent. *Unilever.* Name unverified.

14790 Hytrel®
A group of polyester elastomers used as thermoplastic rubbers; they are graded for hardness as follows; Hytrel 4055, 92A; Hytrel 5550, 55D Hytrel 6350, 63D. *DuPont UK.*

14791 Hytrol
aminotriazole-2,4-D-diuron-simazine
Aminotriazole, 2,4-D, diuron and simazine; used for total weed control in non crop areas. *Farmers Crop Chemicals Ltd.* Discontinued.

14792 Hytrol
2,4-D-diuron-amitrole-simazine
2,4-D, diuron, amitrole, simazine; total weedkiller with more than one seasons resistance; for use on garden paths and other noncrop areas. *Agrichem (International) Ltd.*

14793 Hyvar X
314-40-9 1402 206-245-1
Bromacil
A wettable powder containing 80% w/w bromacil; used for control of weeds in cane fruit and noncrop areas. *Selectokil Ltd.*

14794 Hyvar® X
314-40-9 1402 206-245-1
Bromacil
A wettable powder containing 80% w/w bromacil; used for control of weeds in cane fruit and noncrop areas. *DuPont UK.*

14795 Hy-Vic
thiabendazole-thiram
Mixture of thiabendazole [148-79-8] and thiram [137-26-8]; fungicide seed dressing. *Agrichem (International) Ltd.*

14796 Hy-Vin
Homo and copolymeric polyvinyl chloride compounds (S-PVC/E-PVC); for production of plastic articles. *Norsk Hydro AS.*

14797 Hyvis
Polybutenes
BP Chemicals Ltd.

14798 Hyzod® AC-1000
PC sheet; features low flame spreadability, low smoke generation, low toxicity, high heat resistance, excellent dimensional stability; for aircraft seat parts, tray tables, lavatories, cargo liners, window reveals, partitions, moldings. *Sheffield Plastics.* Name unverified.

14799 I.P.S.
67-63-0 5227 200-661-7
Isopropyl alcohol
Solvent. *Foseco (F.S.) Ltd.*

14800 I T Talc
14807-96-6 9207 238-877-9
Hydrous magnesium calcium silicate
Industrial talc. Used as fillers and extenders for paint, ceramics, plastics, rubber etc. *R. T. Vanderbilt Co Inc.* Discontinued.

14801 Iachiol
7775-41-9 8657 231-895-8
AgF
Silver fluoride
Used as a chemical intermediate and an anti-infective agent.

14802 IA-IA alloy
An alloy of 60% copper and 40% nickel; used for electrical resistances.

14803 Ibbal
p-Isobutylbenzaldehyde
Intermediate for pharmaceuticals; fragrance. *Mitsubishi Gas.*

14804 Ibdu®
Slow release nitrogen fertilizer. *Mitsubishi Kasei.* Name unverified.

14805 Ibex
7758-23-8 1740 231-837-1
calcium phosphate, monobasic
Acid calcium phosphate for aerating flour. *Albright & Wilson Ltd.*

14806 Icdal
Terephthalic acid resins; for insulation of heat-resistant electrical conductors. *Dynamit Nobel Wien GmbH.* Name unverified.

14807 ice colors
Colors formed on a fiber by treating it with a phenol, and then with a diazotized amine, in the presence of ice. They are also known as ingrain colors.

14808 Ice Melt
10043-52-4 1699 233-140-8
$CaCl_2$
Calcium chloride pellets
Used for snow and ice melting, dust control and tire weighting. *Standard Tar Products Co Inc.*

14809 Ice No. 12K
9005-71-4 8872
80/20 blend of mono-and diglycerides and polysorbate 65; emulsifier used with stabilizers in frozen desserts. *Van Den Bergh Foods.*

14810 Ice No. 2
Glyceryl stearate and polysorbate 80; stabilizer and emulsifier for the food industry; lubricant for textiles and plastics; fabric softener. *Van Den Bergh Foods.*

14811 Iceberg®
12141-46-7 377 235-253-8
Aluminum silicate, anhydrous
Pigment used in paints, paper, board, rubber and plastics. *Burgess Pigment.*

14812 Icecap® K
12141-46-7 377 235-253-8
Aluminum silicate, anhydrous
Extender for TiO_2 in coatings where high Hegman grind is required; used in wire and cable, molded and extruded rubber and plastic products, paper coatings, paints; good electricals, low compression set, low water absorption. *Burgess Pigment.*

14813 Iceland spar
471-34-1 1697 207-439-9
Crystalline calcium carbonate
Used as a chemical intermediate.

14814 Icinol
Polyalkylene glycols
Solvents. *Imperial Chemical Industries plc.*

14815 Icinol DPM
34590-94-8 252-104-2
$C_7H_{16}O_3$
(2-Methoxy-methylethoxy) propanol
Dipropylene Glycol Methyl Ether; dipropylene glycol monomethyl ether; Dowanol 50b; Oxybispropanol, Methyl Ether; Dowanol DPM; bis-(2-Methoxypropyl) ether; Propanol, 1(or 2)-(2-methoxymethylethoxy)-; Arcosolve DPM; Methoxymethylethoxy)propanol; PPG-2 methyl ether; (2-methoxymethylethoxy)propanol. Solvent for use in protective coatings, inks, cleaning products, agricultural chemicals; aids wetting, penetration, and soil removal; coupling solvent. *ICI Australia.*

14816 Icinol PM
107-98-2 203-539-1
$C_4H_{10}O_2$
1-Methoxy 2-propanol
Propylene Glycol Monomethyl Ether; Methoxypropanol, α isomer; Propylene glycol methyl ether; 1-Methoxypropan-2-ol; 1-Methoxy-2-propanol; methoxy ether of propylene glycol; α-propylene glycol monomethyl ether; polypropylene glycol methyl ether; propylene glycol 1-methyl ether; (±)-1-methoxy-2-propanol; Dowanol 33B; Dowanol PM; Dowtherm 209; glycol ether PM; PGME; Poly-solve MPM; propasol solvent M; UCAR solvent LM.

Solvent for use in protective coatings, inks, cleaning products, agricultural chemicals; aids wetting, penetration, and soil removal; coupling solvent. mp = -97°; bp = 118-119°; d = 0.9220; n_D^{20}= 1.4030; LD_{50} (rat orl) = 5660 mg/kg. *ICI Australia.*

14817 Icomeen® T-2
61791-44-4 263-177-5
PEG-2 tallow amine
Wetting agent, penetrant, emulsifier, stabilizer, dispersant, antistat, lubricant, solubilizer. *BASF.*

14818 Icomeen® T-15
61791-44-4 263-177-5
PEG-15 tallow amine
Surfactant. *BASF.*

14819 Iconol DA-4
26183-52-8
Deceth-4
Wetting agent, detergent, emulsifier, dispersant, solubilizer for textile scouring and dyeing, industrial and institutional cleaners, household cleaning products and specialities. *BASF.*

14820 Iconol DA-6
26183-52-8
Deceth-6
Wetting agent, detergent, emulsifier, dispersant, solubilizer for textile scouring and dyeing, industrial and institutional cleaners, household cleaning products and specialities. *BASF.*

14821 Iconol DA-9
26183-52-8
Deceth-9
Wetting agent, detergent, emulsifier, dispersant, solubilizer for textile scouring and dyeing, industrial and institutional cleaners, household cleaning products and specialities. *BASF.*

14822 Iconol NP-100
9016-45-9 6772
Nonoxynol-100
Wetting agent, penetrant, detergent, cleaning agent, emulsifier, latex stabilizer, dispersant for industrial, institutional and household cleaning products; emulsifier for emulsion polymerization, asphalt emulsions. *BASF.*

14823 Iconol NP-30
9016-45-9 6772
Nonoxynol-30
Wetting agent, penetrant, detergent, cleaning agent, emulsifier, latex stabilizer, dispersant for industrial, institutional and household cleaning products; emulsifier for emulsion polymerization, asphalt emulsions. *BASF.*

14824 Iconol NP-40
9016-45-9 6772
Nonoxynol-40
Wetting agent, penetrant, detergent, cleaning agent, emulsifier, latex stabilizer, dispersant for industrial, institutional and household cleaning products; emulsifier for emulsion polymerization, asphalt emulsions. *BASF.*

14825 Iconol NP-50
9016-45-9 6772
Nonoxynol-50
Wetting agent, penetrant, detergent, cleaning agent, emulsifier, stabilizer, dispersant, coemulsifier. *BASF.*

14826 Iconol NP-70
9016-45-9 6772
Nonoxynol-70
Wetting agent, penetrant, detergent, cleaning agent, emulsifier, latex stabilizer, dispersant for industrial, institutional and household cleaning products; emulsifier for emulsion polymerization, asphalt emulsions. *BASF.*

14827 Iconol OP-10
9002-93-1 6858
Octoxynol-10
Wetting agent, penetrant, detergent, cleaning agent, emulsifier, latex stabilizer, dispersant for industrial, institutional and household cleaning products; emulsifier for emulsion polymerization, asphalt emulsions. *BASF.*

14828 Iconol OP-30
9002-93-1 6858
Octoxynol-30
Emulsifier, wetting agent, dispersant, synthetic latex stabilizer, detergent in formulating industrial, institutional, and household cleaning products; primary emulsifier for acrylic and vinyl emulsion polymerization and for asphalt emulsion systems. *BASF.*

14829 Iconol OP-40
9002-93-1 6858
Octoxynol-40
Wetting agent, penetrant, detergent, cleaning agent, emulsifier, latex stabilizer, dispersant for industrial, institutional and household cleaning products; emulsifier for emulsion polymerization, asphalt emulsions. *BASF.*

14830 Iconol TDA-3
24938-91-8
Trideceth-3
Wetting agent, detergent, emulsifier, dispersant, solubilizer for textile scouring and dyeing, industrial and institutional cleaners, household cleaning products and specialities. *BASF.*

14831 Iconol TDA-6
24938-91-8
Trideceth-6
Wetting agent, detergent, emulsifier, dispersant, solubilizer for textile scouring and dyeing, industrial and institutional cleaners, household cleaning products and specialities. *BASF.*

14832 Iconol TDA-8
24938-91-8
Trideceth-8
Wetting agent, detergent, emulsifier, dispersant, solubilizer for textile scouring and dyeing, industrial and institutional cleaners, household cleaning products and specialities. *BASF.*

14833 Iconol TDA-9
24938-91-8
Trideceth-9
Wetting agent, degreaser, emulsifier, detergent. *BASF.*

14834 Iconol TDA-10
24938-91-8
Trideceth-10
Wetting agent, detergent, emulsifier, dispersant, solubilizer for textile scouring and dyeing, industrial and institutional cleaners, household cleaning products and specialities. *BASF.*

14835 Iconol WA-1
Alkoxylated phenolic
Surfactant; pigment dispersant for aqueous systems. *BASF.*

14836 Iconol WA-4
Alkoxylated phenolic
Surfactant; dispersant. *BASF.*

14837 Iconsim®
For the electrical Industry. *DuPont UK.*

14838 ICR
Catalyst. *Monsanto (Solaris).* Name unverified.

14839 Ideal alloy
An alloy of 53.5% copper, 45% nickel, 0.66% iron, and 0.45% manganese.

14840 Iditol
A proprietary shellac substitute. No manufacturer.

14841 Idryl
206-44-0 205-912-4
$C_{16}H_{10}$
Fluoranthrene
1,2-Benzacenaphthene; FA; Idryl; benzo[j,k]fluorene; 1,2-(1,8-naphthalenediyl)benzene; 1,2-(1,8-naphthalene)benzene. Use in chemical research. mp = 109-111°; bp = 384°; insoluble in H_2O, soluble in organic solvents; LD_{50} (rat orl) = 2 gm/kg.

14842 IE-40-A
PU elastomer
Tough, abrasion-resistant engineering elastomer for R.T. hand batch processing; for rollers, pallets, prototypes, molds, fixtures, wheels, bumpers. *Innovative Engineering of Mich.*

14843 Igelit PCU
polyvinyl chloride
A proprietary manufacture of polyvinylchloride. No manufacturer.

14844 Igepal® 131
α-Naphthol + 8-9 EO
Surfactant. *Rhône-Poulenc France.*

14845 Igepal® BPA-6
Bisphenol A ethoxylate
Monomer for coatings; reactive diluent; useful in alkyds and polyurethanes. *Rhône-Poulenc Surf.*

14846 Igepal® CA-210
9002-93-1 6858
Octoxynol-1.5
Emulsifier for solvent cleaners, drycleaning, pesticides, floor polish; defoamer. *Rhône-Poulenc Surf.; Rhône-Poulenc France.*

14847 Igepal® CA-620 and CA-630
A range of nonionic surfactants; may be used in all phases of detergent compounding and aqueous processing in textile and paper industries, in industrial metal cleaners, acid cleaners and floor cleaners, detergent-sanitizers and waterless hand cleaners. *Rhône-Poulenc Surf.* Name unverified.

14848 Igepal® CA-720
Oxtoxynol-12.5
A nonionic surfactant used as a hard surface detergent with aqueous solubility at high temperatures; used in hot spray, soak and steamcleaning systems, electrolytic cleaning and metal pickling operations. *Rhône-Poulenc Surf.* Name unverified.

14849 Igepal® CA-897
9002-93-1 6858
Octoxynol-40
Emulsifier for fats and waxes, stabilizer for vinyl acetate and acrylate polymerization; dyeing assistant. *Rhône-Poulenc Surf.; Rhône-Poulenc France.*

14850 Igepal® Cephene Distilled
122-99-6 7410 204-589-7
$C_8H_{10}O_2$
Phenoxyethanol
2-Phenoxyethanol; Phenoxetol; Phenoxyethyl alcohol; Arosol; Ethylene glycol phenyl ether; 1-Hydroxy-2-phenoxyethane; β-Hydroxyethyl phenyl ether; Ethylene glycol mono phenyl ether; Euxyl K 400; Phenyl cellosolve; Phenoxethol; Phenoxyl ethanol; glycol monophenyl ether; phenoxytol; phenylmonoglycol ether; 2-hydroxyethyl phenyl ether; β-phenoxyethyl alcohol; Dowanol EP; Dowanol EPH; Emeressence 1160; Emery 6705; rose ether; Ethanol-2-phenoxy. Surfactant. mp = 14°; bp = 245°; d = 1.1000; insoluble in H_2O, soluble in organic solvents; LD_{50} (rat orl) = 1260 mg/kg. *Rhône-Poulenc France.*

14851 Igepal® CO-210
9016-45-9 6772
Nonoxynol-2 (1.5 EO)
Foam and emulsion stabilizer, detergent, coemulsifier, intermediate, defoamer, dispersant; for metalworking; biodegradable. *Rhône-Poulenc Surf.; Rhône-Poulenc France.*

14852 Igepal® CO-430
9016-45-9 6772
Nonoxynol-4
An oil soluble surfactant used as a coemulsifier with CO-850 and CO-880; plasticizer and antistatic agent for polyvinyl acetate; freeze-thaw stabilizer for latex emulsions, intermediate for the synthesis of high foaming, water soluble sulfate esters. *Rhône-Poulenc Surf.*

14853 Igepal® CO-520
9016-45-9 6772
Nonoxynol-5
An oil-soluble surfactant and intermediate for anionic surfactants; de-icing fluid for jet aircraft fuels and automotive gasoline, added to home storage tanks to inhibit rusting, dispersant for petroleum oils. *Rhône-Poulenc Surf.*

14854 Igepal® CO-530
9016-45-9 6772
Nonoxynol-6
An oil-soluble surfact antand intermediate for anionic surfactants; uses are similar to those given for CO-520 and it is also used as an emulsifier for silicones, agricultural compounds, and mineral oils. *Rhône-Poulenc Surf.*

14855 Igepal® CO-610
9016-45-9 6772
Nonoxynol-8
A water soluble surfactant widely used for detergency, wetting and emulsification applicable where low foaming is particularly important. *Rhône-Poulenc Surf.*

14856 Igepal® CO-630
9016-45-9 6772
Nonoxynol-9
Nonionic surfactant; water soluble detergents, wetting agents and emulsifiers. *Rhône-Poulenc Surf.*

14857 Igepal® CO-660
9016-45-9 6772
Nonoxynol-10
Nonionic surfactant, water soluble detergents, wetting agents and emulsifiers. *Rhône-Poulenc Surf.*

14858 Igepal® CO-710
9016-45-9 6772
Nonoxynol-10.5
Nonionic surfactant, water soluble detergents, wetting agents and emulsifiers. *Rhône-Poulenc Surf.*

14859 Igepal® CO-720
9016-45-9 6772
Nonoxynol-12
Nonionic surfactant, water soluble detergents, wetting agents and emulsifiers. *Rhône-Poulenc Surf.*

14860 Igepal® CO-730
9016-45-9 6772
Nonoxynol-15
Nonionic surfactant, water soluble detergents, wetting agents and emulsifiers. *Rhône-Poulenc Surf.*

14861 Igepal® CO-850
9016-45-9 6772
Nonoxynol-20
A water-soluble detergent and wetting agent, emulsifier for fats, oils, waxes, solvents, demulsifier for crude petroleum oil emulsions. *Rhône-Poulenc Surf.*

14862 Igepal® CO-880
9016-45-9 6772
Nonoxynol-30
Nonionic surfactant used as a detergent for the high temperature scouring of textiles in pressure equipment. *Rhône-Poulenc Surf.*

14863 Igepal® CO-887
9016-45-9 6772
An aqueous solution of CO-880; surfactant for emulsion polymerization and post-additive stabilization. *Rhône-Poulenc Surf.*

14864 Igepal® CO-890
9016-45-9 6772
Nonoxynol-40
A highly water-soluble emulsifier and stabilizer useful in concentrated electrolyte solutions, synthetic latexes, floor waxes and polishes. *Rhône-Poulenc Surf.*

14865 Igepal® CO-970
9016-45-9 6772
Nonoxynol-50
A very highly soluble surfactant effective at high temperatures and in concentrated electrolyte solutions, stabilizer for synthetic latexes, emulsifier and stabilizer in the preparation of floor waxes and polishes. *Rhône-Poulenc Surf.*

14866 Igepal® CO-997
9016-45-9 6772
Nonoxynol-100
Emulsifier, stabilizer, wetting agent, dyeing assistant for plastics, latexes, floor polishes, etc. *Rhône-Poulenc Surf.; Rhône-Poulenc France.*

14867 Igepal® CTA-639W
Water-soluble surfactant with exceptional wetting and emulsifying properties; used in latex paints as a wetting agent, with titanium oxide and as an emulsifier for oil or alkyd additives. *Rhône-Poulenc Surf.*

14868 Igepal® DM-430
9014-93-1
Nonyl nonoxynol-7
Emulsifier for agricultural, emulsion polymerization, leather industries. *Rhône-Poulenc Surf.; Rhône-Poulenc France.*

14869 Igepal® DM-530
9014-93-1
Nonyl nonoxynol-9
An emulsifier used in acid-based cleaners, aerosol, cosmetic, insecticide and wax emulsions, textile finishing oils, dry-cleaning soaps, inks, lacquers and paints. *Rhône-Poulenc Surf.*

14870 Igepal® DM-710
9014-93-1
Nonyl nonoxynol-15
A water-soluble surfactant for detergency and emulsification, especially where low foaming is desired; detergent for washing paper mill felts. *Rhône-Poulenc Surf.*

14871 Igepal® DM-730
9014-93-1
Nonyl nonoxynol-24
Highly water-soluble surfactant used in emulsion polymerization, latex stabilization and emulsifiable pesticide formulations. *Rhône-Poulenc Surf.*

14872 Igepal® DM-970
9014-93-1
Nonyl nonoxynol-150
Unique high-melting flake or solid nonionic surfactant used in household detergents and in industrial detergent formulations. *Rhône-Poulenc Surf.*

14873 Igepal® OD-410
122-99-6 7410 204-589-7
Phenoxydiglycol
A solvent for various resins (vinyl, phenolic, polyester, alkyd, nitrocellulose, cellulose acetate), ingredient of metal cleaners, paint strippers and other cleaning compounds and as an ink vehicle. *Rhône-Poulenc Surf.*

14874 Igepal® RC-520
9014-92-0
Dodoxynol-6
A low foaming rewetting agent suitable for paper towels, tissues, and semichemical corrugating media. *Rhône-Poulenc Surf.*

14875 Igepal® RC-620
9014-92-0
Dodoxynol-10
An all-purpose detergent and wetting agent; detergent in acid cleaning, detergent-sanitizing and grease cutting formulations, as an emulsifier and wetter in agricultural compounds, rug shampoos, and whitewall tire cleaners. *Rhône-Poulenc Surf.*

14876 Igepon®
Redesignated Geropon®. *Rhône-Poulenc Surf.*

14877 Igetaleim MA
Melamine resin; used for adhesives. *Sumitomo Bakelite Co Ltd.* Name unverified.

14878 Igetaleim UA
Urea resin; used for adhesives. *Sumitomo Bakelite Co Ltd.* Name unverified.

14879 Igewsky's reagent
A solution of 5% picric acid in absolute alcohol; used for etching in the micro-analysis of carbon steels.

14880 Iglodine
A proprietary preparation of phenol and iodine; an antiseptic. *Ayrton Saunders plc.* Unverified.

14881 Ignicide
Phosphate base; for flameproof plywood and hardboard. *Stanley Smith & Co Plastics Ltd.* Discontinued.

14882 Iguafen
Dispersing and soaping agent; solubilizing properties; retarder; antiprecipitant; used for textiles particularly acrylics and wool. *GAF Great Britain.* Name unverified.

14883 Ilexan E
107-21-1 3844 203-473-3
Ethylene glycol
Inhibited ethylene glycol; leak-indicating liquid, antifreeze agent and solar fluid. *Hüls AG.*

14884 Ilexan HT
25322-68-3 7729 200-849-9
Polyethylene glycol
Antifreeze agent stable at high temperatures; used in solar collectors. *Hüls AG.*

14885 Ilexan P
57-55-6 8040 200-338-0
Inhibited 1,2-propylene glycol
Antifreeze agent and solar fluid for heating drinking water. *Hüls AG.*

14886 Ilexan S
A mixture of alkylbenzenes; heat transfer agent. *Hüls AG.*

14887 Illium
An acid-resisting alloy containing 60.65% nickel, 21.07% chromium, 6.42% copper, 4.67% molybdenum, 2.13% tungsten, 1.04% silicon, 1.09% aluminum, 0.98% manganese, 0.76% iron, and 1.19% carbon and boron. *Stainless Foundry and Engineering Co.* Name unverified.

14888 Ilmenite
titanicIron; titaniferous iron; titaniferous iron ore. A mineral; it consists of about 52% titanic oxide, TiO_2, and 48% ferrous oxide, FeO.

14889 Ilozyme
Pancrelipase
A concentrate of pancreatic enzymes standardized for lipase content. *Adria Laboratories Inc.* Name unverified.

14890 Imacol
Low foaming lubricant; for prevention of creases and abrasions when wet-finishing piece goods of synthetic fibers and blends with natural fibers. *Sandoz Products Ltd.* Name unverified.

14891 Imaverol
35554-44-0 3622 252-615-0
Enilconazole
A proprietary preparation containing enilconazole; used for treatment of veterinary dermatomycoses. *Janssen Pharmaceutical Ltd.* Name unverified.

14892 imazalil
35554-44-0 3622 252-615-0
$C_{14}H_{14}Cl_2N_2O$
(±)-1-[2-(2,4-dichlorophenyl)-2-(2-propenyloxy)ethyl]-1H-imidazole
enilconazole; chloramizol; Bromazil; Deccozil; Fecundal; Florasan; Freshguard; Fungaflor; Fungazil; R 23979; Clinafarm; Magnate; Rappor Plus; Fungaflor. Ergosterol biosynthesis inhibitor, used as a systemic fungicide with protective and curative action. Used to control fungal diseases in fruit, vegetables and ornamentals. mp = 50°; SG_{23} = 1.243; n_D^{20} = 1.5643; soluble in H_2O (1.4 g/l), more soluble in organic solvents; LD_{50} (rat orl) = 320 mg/kg. *Hortichem Ltd; Janssen Pharmaceutical Ltd; Makhteshim Chemicak Works Ltd; Dow; Elanco Ltd.*

14893 imazapyr
81334-34-1 4942
$C_{13}H_{15}N_3O_3$
2-[4,5-dihydro-4-methyl-4-(1-methylethyl)-5-oxo-1H-imidazol-2-yl]-3-pyridine carboxylic acid
AC 252,925; Arsenal; Assault; Chopper; CL 252,925; Contain; Pivot. Non-selective systemic herbicide. Inhibits acetohydroxy acid synthase. Used for control of annual and perennial grass and broad-leaved weeds in non-crop areas. mp = 169-173°; soluble in H_2O (10-15 g/l), organic solvents; LD_{50} (rat orl) >5000 mg/kg.

14894 Imicure EMI-24, 24S
Imidazole accelerators. *Air Prods, & Chems. Inc.*

14895 Imidazole
288-32-4 4948 206-019-2
$C_3H_4N_2$
1,3-diaza-2,4-cyclopentadiene
Glyoxaline; glyoxalin; 1,3-diazole; iminazole; miazole; pyrro[*b*]monazole. Biological control of pests, especially fabric-feeding insects; contact insecticide in an oil spray. mp = 90-91°; soluble in H_2O, organic solvents; LD_{50} (mus orl) = 1880 mg/kg. *Aldrich; BASF; Janssen Chimica; Penta Mfg.*

14896 Imidazolidinyl urea
39236-46-9 254-372-6
Germall® 115; Abiol. Antimicrobial preservative for cosmetics. *Sutton Labs; 3V-Sigma.*

14897 Imidazoline 18
Series of organic nitrogenous bases; oleic or stearic acid reacted with various short chain amines to produce alkyl imidazolines; for dewatering of metal and other surfaces, anticorrosives, cationic emulsions for oils, pigment dispersing aids, antistatic agents. *Lake-land Laboratories Ltd.*

14898 Imidrol
Cationic emulsifiers based on fatty imidazoline derivatives; used for the production of oil-based drilling muds and for lubrication in water-based muds. *Allied Colloids Ltd.*

14899 Imiodid
A substance obtained by heating *p*-ethoxyphenyl-succinimide with a solution containing potassium iodide and iodine; a powerful antiseptic.

14900 Imlar®
Vinyl coating; for the automotive industry. *DuPont Uk.*

14901 Immadium
An alloy of manganese bronze containing aluminum.

14902 Immedial®
Sulfur dyestuffs; used for textile dyeing and printing. *Cassella AG.*

14903 Immergan® A
An aliphatic sulfochloride, a lightfast oil tanning agent for chamois and auxiliary tanning agent for very soft leathers. *BASF AG.*

14904 Immetal
Diiodoerucic acid isobutyl ester.

14905 Immuno-bed
Plastic embedding kit; for light microscopy immunohistochemistry procedures. *Polysciences Inc.*

14906 Imogen
Sodium diaminonaphtholsulfonate
Chemical intermediate.

14907 Imogen sulfite
A photographic developer.

14908 Impact
76674-21-0 4246
flutriafol
Fungicide containing flutriafol. *ICI Chem & Polymers Ltd.*

14909 Impact Excel
chlorothalonil-flutriafol
Suspension concentrate containing 300 g chlorothalonil [1897-45-6] and 47 g flutriafol [76674-21-0] per liter; a systemic fungicide. *ICI Agrochemicals.*

14910 Impact®
For the chemical industry. *DuPont UK.*

14911 Impad
High density impact pads for continous casting tundishes. *Foseco (F.S.) Ltd.*

14912 imperatorin
482-44-0 4960 207-581-1
$C_{16}H_{14}O_4$
9-[(3-Methylbut-2-enyl)oxy]-7H-furo[3,2-g][1]benzopyran-7-one
Ammidin; Imperatorin; Marmelosin; Pentosalen; 8-Isopentenyloxypsorlene. Chemical intermediate. mp = 102°; poorly soluble in H_2O, soluble in organic solvents; λ_m = 302, 265, 250 nm (log ε 3.95, 4.00, 4.24).

14913 Imperial metal
An alloy of 80% copper and 20% nickel.

14914 Impet® 330
25038-59-9 7730
PET polyester, 30% glass reinforced; injection molding resin for high-performance applications requiring toughness, rigidity, and exceptional dimensional stability; used for automotive applications (distributor housings, rotors, headlamp bezels, structural body parts), consumer electronics/appliances (motor housings, coffee makers, hairdryers), electrical and electronics (connectors, terminal blocks, bobbins), furniture. *Hoechst Celanese.* Name unverified.

14915 Implenal®
Mixture of salts of dicarboxylic acids; finishing agent for chrome tanning, furs. *BASF AG.*

14916 Imposil
Veterinary iron-dextran complex. *Fisons plc, Pharmaceuticals Div.* Name unverified.

14917 Impra®
Organic wood preservatives and wood stainers; applied by brushing, spraying and dipping. *Weyl GmbH.*

14918 Impra® Concentrates
Water-dilutable concentrates for protection of freshly sawn timber against sap stain and of logs against pinhole borers; applied by dipping and spraying. *Weyl GmbH.*

14919 Impra-biolan
Ecological wood stainer for indoor and outdoor use; applied by brushing and spraying. *Weyl GmbH.*

14920 Impra-color
Wood stainer protecting against fungi and insect attack; for outdoor use; applied by brushing, spraying, and dipping. *Weyl GmbH.*

14921 Impra-elan
Fast-drying satin gloss joinery stainer; applied by brushing and spraying. *Weyl GmbH.*

14922 Imprafix®
Polyurethane products for coating and laminating textiles; fast to washing and to solvents; also suitable as binders for pigment printing. *Bayer AG.*

14923 Impralan®
Water-dilutable ecological stainer for indoor treatment of wood; applied by spraying and pouring. *Weyl GmbH.*

14924 Impraleum
A coaltar distillate for fencing; applied by brushing, spraying and dipping. *Weyl GmbH.* Discontinued.

14925 Impralit
Wood preservative salts and fire retardant salts; for dipping, vacuum-pressure method. *Weyl GmbH.*

14926 Impranil®
Polyurethane products; used for coating and laminating textiles; fast to washing and to solvents. *Bayer AG.*

14927 Impregnant
Acrylic polymers; used for electrical moldings. *Monomer-Polymer & Dajac Laboratories, Inc.*

14928 Impression
High temperature on-glaze colors. *Degussa Ltd.*

14929 Impressional
Alginate impression material; used in dentistry. *Bayer AG.*

14930 Imprez
Petroleum resins. *ICI Chem & Polymers Ltd.*

14931 Impriment Black 7101-A
Pigment dyestuffs. *Catawba-Charlab.*

14932 Imprimus®
Concentrates of pigment pastes for sheet offset printing inks. *BASF AG.*

14933 Improved Kelmar®
9005-36-1 241
Potassium alginate
Used for dietetic and low-sodium foods, dry mixes, dental impression material, surgical impressions. *Kelco.*

14934 Impsonite
An asphaltic substance found in Arkansas and other places; it can be fused with difficulty, and is only slightly soluble in carbon disulfide.

14935 Imron®
Urethane finish; for the automotive industry. *DuPont UK.*

14936 Imsil® A-10
7631-86-9 8637 232-545-4
Silica, amorphous
Filler/extender in solvent- based or latex paints; thickener, carrier, free flow agent; powder coating; paint flatting agent; used in protective and wire and cable coatings, PU elastomers, epoxy, phenolic, buffing, polishing and polyester compounds, electrical resistors, refractory products, insulations, antiblock agents, toothpastes, agricultural compositions, cosmetics, metal castings, injection thermoset moldings. *Unimin.*

14937 Imsol
Solvents. *ICI Chem & Polymers Ltd.* Discontinued.

14938 Imwitor®
Emulsifier for the pharmaceutical, cosmetic and food industries. *Dynamit Nobel Wien GmbH.*

14939 Imwitor® 191
31566-31-1 4498 250-705-4
Glyceryl stearate
Coemulsifier, dispersant for personal care products; emulsifier in oil-water and water-oil emulsions; lubricants and binders used in the pharmaceutical industry; suspending agent, stabilizer, thickener; food emulsifier. *Hüls Am; Hüls AG.*

14940 Imwitor® 308
26402-26-6 247-668-1
Glyceryl caprylate
Solubilizer for pharmaceutical drugs; carrier/vehicle for drugs in capsules; coemulsifier for lipophilic materials. *Hüls Am.*

14941 Imwitor® 312
142-18-7 205-526-6
Glyceryl laurate
Coemulsifier for oil-water emulsions; solubilizer, carrier for lipophilic drugs; superfatting agent for bath products; bacteriostatic effect. *Hüls Am; Hüls AG.*

14942 Imwitor® 369
91744-38-6 294-600-1
Monoglyceride citric ester
Food emulsifier. *Hüls Am; Hüls AG.*

14943 Imwitor® 370
91744-38-6 294-600-1
Glyceryl stearate citrate
Food emulsifier; oil-water emulsifier for very polar oils, fats and liquid wax esters in cosmetics. *Hüls Am; Hüls AG.*

14944 Imwitor® 375
Glyceryl citrate/lactate/linoleate/oleate. Oil-water emulsifier for very polar oils and fats; for oil baths, cream baths. *Hüls Am; Hüls AG.*

14945 Imwitor® 742
26402-26-6 247-668-1
Caprylic/capric glycerides. Plasticizer for hard fats, solvent for lipophilic ingredients; emollient; coemulsifier, solubilizer, carrier for lipophilic drugs; dispersant, absorption promoter; bacteriostatic effect. *Hüls Am; Hüls AG.*

14946 Imwitor® 780 K
66085-00-5 266-124-4
Isostearyl diglyceryl succinate
Water-oil emulsifier for polar oils and fats, cosmetic and pharmaceutical preparations. *Hüls Am; Hüls AG.*

14947 Imwitor® 914
Glyceryl myristate
Coemulsifier for oil-water emulsions; solubilizer, carrier for lipophilic drugs; dispersant, consistency regulator in creams, lotions, anhydrous formulations. *Hüls Am; Hüls AG.*

14948 Imwitor® 928
61789-05-7 263-027-9
Glyceryl cocoate
Surfactant for pharmaceutical, cosmetic, and nutritional fields; as emulsifier, solubilizer, dispersion aid, plasticizer, lubricant, consistency regulator, skin and mucous membrane protectant, refatting agent, penetrant, carrier, adsorption promoter. *Hüls AG.*

14949 Imwitor® 960
Glyceryl stearate SE
Emulsion base for cosmetics; suspending agent in creams, lotions, emulsions, cosmetics. *Hüls Am; Hüls AG.*

14950 Inalium
An aluminum alloy containing 2% cadmium, 0.8% magnesium, and 0.4% silicon.

14951 Inco Chrome Nickel
Proprietary nickel alloys containing 12-14% chromium and 6-7% iron; they resist corrosion by the organic acids met within foodstuffs; the alloy containing 80% nickel, 14% chromium, and 6% iron melts at 1390 c. *Inco Alloys Int'l. Inc.*

14952 Incoblend
Flexible vinyl compound based on intrinsically conductive polymer (polyaniline); for applications requiring high electrical conductivity, such as EMI shielding. *AlliedSignal; Americhem; Zipperling Kessler.*

14953 Incoloy Alloy 800
A trademark for an alloy of 20% chromium, 32% nickel and 48% iron; resistant to hydrogen/hydrogen sulfide corrosion. *Inco Alloys Int'l. Inc; Wiggin Alloys Ltd.* Name unverified.

14954 **Incoloy Alloy 825**
A trademark for an alloy of 40% nickel, 21% chromium, 3% molybdenum, 2% copper, 1% titanium and the balance iron; resistant to corrosion in hot oxidizing acid conditions. *Inco Alloys Int'l. Inc; Wiggin Alloys Ltd.* Name unverified.

14955 **Incoloy Alloy 901**
A trademark for an alloy of 12.5% chromium, 5.7% molybdenum, 2.9% titanium, 42% nickel and the balance iron. *Inco Alloys Int'l. Inc; Wiggin Alloys Ltd.* Name unverified.

14956 **Incoloy Alloy DS**
A trademark for an alloy of 18% chromium, 2.3% silicon, 37% nickel and the balance iron. *Inco Alloys Int'l. Inc; Wiggin Alloys Ltd.* Name unverified.

14957 **Incomparable**
Soldering fluxes. *Reade Metals & Minerals Corp.* Unverified.

14958 **Inconel**
A proprietary trade name for an alloy resistant to corrosion and heat, containing 80% nickel, 14% chromium, and 6% iron. *Inco Alloys Int'l. Inc.* Name unverified.

14959 **Inconel Alloy 600**
A trademark for an alloy of 16% chromium, 7% iron and 77% nickel; good oxidation resistance at high temperatures. *Inco Alloys Int'l. Inc; Wiggin Alloys Ltd.* Name unverified.

14960 **Inconel Alloy 700**
A trademark for an alloy of 15% chromium, 29% cobalt, 3% molybdenum, 2.25% titanium, 3.3% aluminum, and the balance nickel. *Inco Alloys Int'l. Inc; Wiggin Alloys Ltd.* Name unverified.

14961 **Inconel Alloy 718**
A trademark for an alloy of 19% chromium, 3% molybdenum, 0.8% titanium, 5% niobium, 53% nickel, and the balance iron. *Inco Alloys Int'l. Inc; Wiggin Alloys Ltd.* Name unverified.

14962 **Inconel Alloy X-750**
A trademark for an alloy of 15% chromium, 2.5% titanium, 0.9% aluminum, 0.9% niobium, 7% iron, and the balance nickel. *Inco Alloys Int'l. Inc; Wiggin Alloys Ltd.* Name unverified.

14963 **Incrocas 30**
61791-12-6
PEG-30 castor oil; emulsifier, solubilizer, emollient, superfatting agent, lubricant for personal care products, detergents, metalworking fluids, insecticides, herbicides, household products; lubricant, antistat, softener, dye leveling agent for textiles; also lime soap dispersant in alkaline scouring systems. *Croda Inc.* Name unverified.

14964 **Incrodet TD7-C**
Trideceth-7 carboxylic acid
Mild surfactant for shampoos, bath gels, cleansers; neutralizer for trace caustic or thioglycolate residues present after hair permanents or relaxers; in neutralizing shampoos which follow ethnic hair straighteners; lime soap dispersant; emulsifier; stable at low and high pH. *Croda Inc.*

14965 **Incromate ALL**
Almondamidopropyl dimethylamine lactate
Conditioner providing slip for wet comb in hair products. *Croda Inc.* Discontinued.

14966 **Incromate BAL**
Babassamidopropyl dimethylamine lactate
Conditioner with good foam for improved dry comb. *Croda Inc.* Discontinued.

14967 **Incromate CDP**
68425-43-4
Cocamidopropyl dimethylamine propionate
Moderate foaming conditioner for clear rinses and shampoos; good detangling. *Croda Inc.*

14968 **Incromate IDL**
55852-15-8
Isostearamidopropyl dimethylamine lactate
Substantive conditioner improving slip, wet comb and manageability of hair, feel in hand creams and lotions. *Croda Inc.* Discontinued.

14969 **Incromate ISML**
Isostearamidopropyl morpholine lactate
Substantive surfactant, conditioner, viscosity builder for personal care products and cationic emulsions; improves slip, wet comb, and manageability in hair, feel in hand creams and lotions. *Croda Inc.*

14970 **Incromate Mink L**
Minkamidopropyl dimethylamine lactate
Foamer, conditioner for hair and skin care products. *Croda Inc.* Discontinued.

14971 **Incromate ODL**
Oleamidopropyl dimethylamine lactate
A cationic salt used in clear rinses, conditioners, conditioning shampoos, compatible in anionic systems. *Croda Chem. Ltd.* Name unverified.

14972 **Incromate OLL**
Olivamidopropyl dimethylamine lactate
Conditioner and foamer for hair and skin care products. *Croda Inc.* Discontinued.

14973 **Incromate SDL**
55819-53-9 259-837-7
Stearamidopropyl dimethylamine lactate
Viscosity builder, opacifier, softener for hair shampoos, conditioners, fabrics; emulsifier for hand creams, cleansers, lotions; raw material. *Croda Inc.*

14974 **Incromate SEL**
Sesamidopropyl dimethylamine lactate
Conditioner and foamer for good slip, wet combing, and detangling. *Croda Inc.* Discontinued.

14975 **Incromate WGL**
Wheat germamidopropyl dimethylamine lactate; conditioner with excellent slip, wet comb, and dry feel. *Croda Inc.* Discontinued.

14976 **Incromectant AMEA-100**
142-26-7 205-530-8
Acetamide MEA
Clarifying detangling agent for shampoos, conditioners, cream rinses; humectant in creams and lotions. *Croda Inc.*

14977 **Incromectant AQ**
Acetamidopropyl trimonium chloride
Antistat for shampoos and conditioners; humectant; plasticizer for hair conditioning/setting polymers. *Croda Inc.*

14978 **Incromectant LAMEA**
Acetamide MEA and lactamide MEA; humectant, moisturizing agent for hair and skin care products. *Croda Inc.*

14979 **Incromectant LMEA-70**
Lactamide MEA
Conditioning agents in cream rinses, humectants in creams and lotions. *Croda Surf. Ltd.*

14980 **Incromectant LQ**
Lactamidopropyl trimonium chloride
Antistat for shampoos and conditioners; humectant; plasticizer for hair conditioning/setting polymers. *Croda Inc.*

14981 **Incromide ALd**
Almondamide DEA (1:1)
Conditioner, viscosity builder, foam stabilizer; for shampoos, bubble baths, soaps, bath products. *Croda Inc.* Discontinued.

14982 **Incromide BAD**
Babassuamide DEA (1:1)
Viscosity builder, foam stabilizer, emulsifier; for shampoos, bubble baths, soaps, bath products. *Croda Inc.* Discontinued.

14983 **Incromide BED**
70496-39-8
Behenamide DEA (1:1)
High melting pearling and opacifying agent used in stick preparations. *Croda Inc.* Discontinued.

14984 **Incromide BEM**
94109-05-4 302-442-2
Behenamide MEA (1:1)
Surfactant structural wax for antiperspirant and other stick products. *Croda Inc.* Discontinued.

14985 **Incromide CA**
Cocamide DEA
Surfactant, foam stabilizer, emulsifier and thickener used in household, cosmetic, and industrial formulations. *Croda Inc.*

14986 **Incromide CM**
68140-00-1 268-770-2
Cocamide MEA
Foam stabilizer, thickener, opacifier for cosmetic, industrial, household formulations. *Croda Inc.* Discontinued.

14987 **Incromide L-90**
120-40-1 204-393-1
Lauramide DEA
Foam stabilizer, thickener, detergent, and foaming agent in household, industrial and institutional cleaning compositions, car washes, rug and upholstery cleaners, and personal care products. *Croda Inc.*

14988 **Incromide LA**
Linoleamide DEA
Superfatting agent and thickener for personal care and household products; useful in increasing viscosity of various sulfate and ether sulfate dilutions; conditioner and lubricant. *Croda Inc.*

14989 **Incromide LCL**
142-78-9 205-560-1
Lauramide MEA
Surfactant. *Croda Inc.* Discontinued.

14990 Incromide LLT
120-40-1 204-393-1
Lauramide DEA
Surfactant. *Croda Inc.* Discontinued.

14991 Incromide LM-70
120-40-1 204-393-1
Lauramide DEA
Foam stabilizer, thickener, detergent, and foaming agent in household, industrial and institutional cleaning compositions, car washes, rug and upholstery cleaners, and personal care products. *Croda Inc.* Discontinued.

14992 Incromide Mink D
Minkamide DEA
Viscosity builder, foam stabilizer for shampoos, bubble baths, liquid soaps, bath products. *Croda Inc.* Discontinued.

14993 Incromide OD
93-83-4 202-281-7
Oleamide DEA (1:1)
Low color viscosity builder with good solubility in anionic surfactants; for shampoos, bubble baths, soaps, bath products. *Croda Inc.* Discontinued.

14994 Incromide OLD
Olivamide DEA (1:1)
Viscosity builder, foam stabilizer; for shampoos, bubble baths, soaps, bath products. *Croda Inc.* Discontinued.

14995 Incromide SED
Sesamide DEA (1:1)
Viscosity builder and foam stabilizer; for shampoos, bubble baths, soaps, bath products. *Croda Inc.* Discontinued.

14996 Incromide WGD
Wheat germamide DEA (1:1)
Viscosity builder and foam stabilizer; for shampoos, bubble baths, soaps, bath products. *Croda Inc.* Discontinued.

14997 Incromine BB
60270-33-9 262-134-8
Behenamidopropyl dimethylamine
Emollient conditioner, lubricant, viscosity builder, and moisturizer for hair care products; intermediate for hair conditioning agent. *Croda Inc.; Croda Surfactants Ltd.*

14998 Incromine CB
68140-01-2 268-771-8
Cocamidopropyl dimethylamine
Foaming agent and conditioner used in hair care products; intermediate for hair conditioning agent. *Croda Surfactants Ltd.*

14999 Incromine IB
Isostearamidopropyl dimethylamine
Used as an intermediate for hair conditioners and shampoo rinses. *Croda Surf. Ltd.*

15000 Incromine OPB
109-28-4 203-661-5
Oleamidopropyl dimethylamine
Emollient conditioner, lubricant and moisturizer for hair care products; intermediate for hair conditioning agent. *Croda Surfactants Ltd.*

15001 Incromine Oxide AL
Almondamidopropylamine oxide
Viscosity builder, cationic conditioner for hair care products. *Croda Inc.* Discontinued.

15002 Incromine Oxide BA
Babassuamidopropylamine oxide
Foamer, foam stabilizer; for hair care products, facial cleaners. *Croda Inc.* Discontinued.

15003 Incromine Oxide C
68155-09-9 268-938-5
Cocamidopropyl amine oxide, aqueous solution; viscosity builder, foam stabilizer, conditioner used in personal care products. *Croda Inc.; Croda Surf. Ltd.*

15004 Incromine Oxide I
Isostearamidopropylamine oxide
Foam stabilizer, thickener, lubricant, and viscosity builder used in cosmetic, household, and janitorial products; wetting agent for concentrated electrolyte solutions. *Croda Inc.; Croda Surf. Ltd.*

15005 Incromine Oxide ISMO
Isostearamidopropyl morpholine oxide
Foam stabilizer, thickener, lubricant, and viscosity builder used in cosmetic, household, and janitorial products; wetting agent for concentrated electrolyte solutions *Croda Inc.* Discontinued.

15006 Incromine Oxide L-40
Lauramine oxide
Foaming agent, foam stabilizer, degreaser; for shampoos and light-duty liquids. *Croda Inc.* Discontinued.

15007 Incromine Oxide M
3332-27-2 222-059-3
Myristamine oxide
Surfactant, emulsifier, emollient, conditioner, viscosity builder, foam booster used in personal care products. *Croda Inc.; Croda Surf. Ltd.*

15008 Incromine Oxide Mink
Minkamidopropylamine oxide
Foaming agent, viscosity builder. *Croda Inc.* Discontinued.

15009 Incromine Oxide OD-50
14351-50-9 238-311-0
Oleyl dimethylamine oxide
Foam stabilizer, viscosity builder for cosmetic and household products. *Croda Surf. Ltd.*

15010 Incromine Oxide OL
Olivamidopropylamine oxide
Thickener for clear systems. *Croda Inc.* Discontinued.

15011 Incromine Oxide S
2571-88-2 219-919-5
Stearamine oxide
Conditioner, emulsifier, viscosity builder for personal care products. *Croda Inc.; Croda Surf. Ltd.*

15012 Incromine Oxide SE
Sesamidopropylamine oxide
Viscosity builder, conditioner for hair care products. *Croda Inc.* Discontinued.

15013 Incromine Oxide WG
Wheat germamidopropylamine oxide
Viscosity builder, conditioner for hair care products. *Croda Inc.* Discontinued.

15014 Incromine SB
7651-02-7 231-609-1
Stearamidopropyl dimethylamine
Viscosity builder, conditioner for hair care products; intermediate for hair conditioning agent for shampoo rinses. *Croda Inc; Croda Surfactants Ltd.*

15015 Incronam I-30
Isostearamidopropyl betaine
Foaming surfactant, conditioner, lubricant, viscosity builder used in personal care products. *Croda Surf. Ltd.*

15016 Incronam 30
61789-40-0 263-058-8
Cocamidopropyl betaine
Surfactant, emulsifier, coupling agent, viscosity builder, foam detergent for personal care products, chemical specialities, rug and upholstery shampoos, dishwashing compounds. *Croda Inc.; Croda Surf. Ltd.*

15017 Incronam AL-30
Almondamidopropyl betaine
Foam booster/stabilizer for skin products *Croda Inc.* Discontinued.

15018 Incronam B-40
84082-44-0 281-991-9
Behenyl betaine
Foaming surfactant, conditioner, lubricant used in personal care products; excellent slip, conditioning; good wetting and rinse aid. *Croda Inc.; Croda Surf. Ltd.*

15019 Incronam BA-30
Babassamidopropyl betaine
Foam booster/stabilizer for shampoos. *Croda Inc.* Discontinued.

15020 Incronam CD-30
68424-94-2 270-329-4
Cocobetaine
Mild, high foaming surfactant for detergent systems. *Croda Surf. Ltd.*

15021 Incronam Mink 30
Minkamidopropyl betaine
Conditioner; foam booster/stabilizer, viscosity builder. *Croda Inc.* Discontinued.

15022 Incronam OD-50
871-37-4 212-806-1
Oleyl betaine
High foaming surfactant, viscosity builder for conditioning shampoos, clear rinses, cleansing creams and cosmetic lotions. *Croda Surf. Ltd.*

15023 Incronam OL-30
Olivamidopropyl betaine
Foam booster/stabilizer for skin care products *Croda Inc.* Discontinued.

15024 Incronam SE-30
Sesamidopropyl betaine
Foam booster/stabilizer, conditioner with good slip. *Croda Inc.* Discontinued.

15025 Incropol CS-20
68439-49-6 7737

Ceteareth-20
Surfactant, emulsifier, lubricant, detergent for industrial and household products; coupling agent, antistat, fiber lubricant and solubilizer for personal care products. *Croda Inc.*

15026 Incropol L-23
Laureth-23
Detergent, wetting agent, emulsifier, coupling agent, antistat, fiber lubricant, solubilizer for cosmetic formulations. *Croda Inc.* Discontinued.

15027 Incroquat 100
Methyl bis (hydrogenated tallow amidoethyl)2-hydroxyethyl ammonium chloride
Nonfoaming, strongly substantive conditioner, antistat for hair care products. *Croda Inc.* Discontinued.

15028 Incroquat 248
Quaternium-72
Economical hair conditioner. *Croda Inc.* Discontinued.

15029 Incroquat AL-85
Almondamidopropalkonium chloride
Conditioner, foamer for hair care products *Croda Inc.* Discontinued.

15030 Incroquat B65C
Behenalkonium chloride
cetyl alcohol. Substantive conditioner for hair care products *Croda Inc.*

15031 Incroquat BA-85
Babassamidopropalkonium chloride
Conditioner, foamer, antistat for hair care products *Croda Inc.*

15032 Incroquat Behenyl BDQ/P
propylene glycol-behenalkonium chloride
Propylene glycol and behenalkonium chloride; oil-water emulsifier and conditioner. *Croda Inc.; Croda Universal Ltd.*

15033 Incroquat Behenyl TMC
Cetearyl alcohol-behentrimonium chloride
Surfactant. *Croda Inc.* Name unverified.

15034 Incroquat Behenyl TMC/P
propylene glycol-behenalkonium chloride
Propylene glycol and behenalkonium chloride; oil-water emulsifier and conditioner. *Croda Inc.; Croda Universal Ltd.*

15035 Incroquat Behenyl TMS
Behenalkonium methosulfate
cetearyl alcohol. Self-emulsifying wax, conditioner, softener, emollient, oil-water emulsifier used in hair and skin care products. *Croda Inc.*

15036 Incroquat CR Conc
Cetearyl alcohol-EG-40 castor oil-stearalkonium chloride. Self-emulsifying wax, conditioner, softener, emollient, oil-water emulsifier used in cream rinses, conditioners. *Croda Inc.*

15037 Incroquat CTC-30
112-02-7 203-928-6
Cetrimonium chloride
Conditioner, antistat for hair care products. *Croda Inc.*

15038 Incroquat DBM-90
Dibehenyldimonium methosulfate
Conditioner with good wetting and slip for hair care products *Croda Inc.* Discontinued.

15039 Incroquat I-85
67633-59-4 266-777-5
Isostearaminopropalkonium chloride
Conditioner with good foaming, slip and detangling for hair products *Croda Inc.* Discontinued.

15040 Incroquat Mink-85
Minkamidopropalkonium chloride
Conditioner, foamer, excellent slip for cosmetics, hair care products *Croda Inc.* Discontinued.

15041 Incroquat O-50
37139-99-4 253-363-4
Olealkonium chloride
Conditioner with good slip for hair care products. *Croda Inc.*

15042 Incroquat OL-85
Olivamidopropalkonium chloride
Conditioner with good foaming and slip for hair products. *Croda Inc.* Discontinued.

15043 Incroquat S-75CG
Quaternium-27
Low foaming conditioner with dry feel on hair, good bodying. *Croda Inc.* Discontinued.

15044 Incroquat SDQ-25
122-19-0 204-527-9
Stearalkonium chloride
Surfactant used as ingredient in personal care products, textile and paper; dispersant for pigments and dyestuffs; antistat for fibers and synthetics; hair conditioner. *Croda Inc.; Croda Universal Ltd.*

15045 Incroquat SE-85
Sesamidopropalkonium chloride
Conditioner, antistat with good foam and dry feel for hair care products. *Croda Inc.* Discontinued.

15046 Incroquat WG-85
Wheat germamidopropalkonium chloride
Conditioner with good foam and slip for hair products. *Croda Inc.*

15047 Incrosoft 100
Methyl bis (hydrogenated tallow amido ethyl)2-hydroxyethyl ammonium chloride
Fabric softener with good hand and antistatic properties; for home, commercial or industrial laundry applications, textile and finishing operations. *Croda Inc.*

15048 Incrosoft 248
Quaternium-72
Softener for dryer sheets. *Croda Inc.*

15049 Incrosoft CF1-75
A range of alkyl imidazolinium methosulfates; fabric softeners for preparation of clear detergent liquid concentrates. *Croda Surf. Ltd.* Discontinued.

15050 Incrosoft S-75
86088-85-9 289-151-3
Quaternium-27
Softener base, lubricant, antistat and rewetting agent for fabrics and synthetics. *Croda Inc.*

15051 Incrosoft T-90
130124-24-2
$RCONH(CH_2)_2N(CH_2CH_2O)_nHCH_3(CH_2)_2NHCRO]^+CH_3OSO_3^-$, RCO-=tallow acid radical, avg. n=3
Ditallow diamido methosulfate
Carsoft® T-90; Incrosoft T-75; Incrosoft T-90; Incrosoft T-90HV. Fabric softener, lubricant and antistat for home and commercial laundry prods. Yellow liquid; disperses in water; Gardner 7 max; pH=4-7. *Croda Inc.*

15052 Incrosperse
Polymeric dispersants. *Croda Resins Ltd.*

15053 Incrosul LAFS
68890-92-6
Disodium laneth-5 sulfosuccinate
Mild, low foaming, conditioning surfactant with good emulsifying properties used in personal care products *Croda Inc.* Discontinued.

15054 Incrosul LMA
Diammonium lauramido-MEA sulfosuccinate
Mild foaming surfactant for cosmetic products. *Croda Inc.* Discontinued.

15055 Incrosul LMS
25882-44-4 247-310-4
Disodium lauramido MEA-sulfosuccinate
Mild foaming agent and cleanser used in personal care products and carpet shampoos; lime soap dispersant. *Croda Inc.* Discontinued.

15056 Incrosul LS
Disodium lauryl sulfosuccinate
Mild foaming and conditioning surfactant, viscosity modifier for cosmetic products. *Croda Inc.* Discontinued.

15057 Incrosul LSA
Diammonium lauryl sulfosuccinate
Mild, high foaming surfactant used in personal care products; lime soap dispersant. *Croda Inc.* Discontinued.

15058 Incrosul LTS
Disodium laureth sulfosuccinate
Mild high foaming surfactant used in shampoos and general cleansing preparations; low eye irritation for baby shampoos. *Croda Inc.* Discontinued.

15059 Incrosul OMS
Disodium oleamido MEA-sulfosuccinate
Mild high foaming surfactant used in personal care products; good conditioner. *Croda Inc.* Discontinued.

15060 Incrosul OTS
Disodium oleth-3 sulfosuccinate
Mild, conditioning surfactant, foamer for shampoos, baby shampoos, bath gels, mild skin cleansers; anti-irritant for anionic surfactants. *Croda Inc.*

15061 Incrosul TS
Disodium tridecyl sulfosuccinate
Mild high foaming surfactant used in personal care products. *Croda Inc.* Discontinued.

15062 Indalca
A thickening agent derived from natural gum; thickener for fabric-printing dyes. *Hercules.* Discontinued.

15063 2-indanone
615-13-4 210-410-3
C_9H_8O
indanone; β-hydrindone. Chemical intermediate. mp= 54-56°; insoluble in H_2O, soluble in organic solvents. *Aldrich; Penta Mfg.; Schweizerhall.*

15064 Indanthren®
A range of vat dyes, having unsurpassed overall fastness properties on cotton and other vegetable fibers as well as on textiles made from regenerated cellulose. *Bayer AG; BASF plc.*

15065 Indazin
A proprietary trade name for a range of products used in the dyeing of textiles. *Cassella AG.* Name unverified.

15066 India gum
Persian gum
A gum resembling gum arabic.

15067 Indian Ocher
1309-37-1 4072 215-168-2
ferric oxide
A native ferric oxide of North America.

15068 Indian Red
Venetian red; Venetian bole; rouge; colcothar; red bole; bole; Armenian bole; English red; angel red; chemical red; Pompeian red; Berlin red; iron minium; iron red; Persian red; raddle; reddle; red rudd; red ocher; red chalk; red earth; iron saffron; Spanish oxide; Turkey red oxide. Red pigments consisting mainly of ferric oxide with varying amounts of natural agrillaceous compounds. Some are natural products, burnt or unburnt, but the name is also applied to products made by heating ferric sulfate. Used as pigments in paints, rubbers and plastics, and as polishing agents.

15069 Indian Yellow
Piuri; Purree; Pioury. A pigment used in India, obtained from the urine of cows fed on mango leaves. The coloring principal is the magnesium or calcium salt of euxanthic acid, $C_{19}H_{16}O_{11}Mg{\cdot}5H_2O$; used as a permanent water and oil color.

15070 indican
487-94-5 4975
$C_8H_6KNO_4S$
indol-3-yl potassium sulfate
urinary indican; potassium indoxyl sulfate. Chemical intermediate. mp = 179-180° (dec); soluble in H_2O, insoluble in organic solvents.

15071 indicator, universal
A mixture of indicators, usually methyl red, α-naptholphthalein, phenolphthalein, bromothymol blue, and cresol red. The color indicates the pH value when added to the solution.

15072 indicolite
A mineral from Brazil. It is a blue tourmaline.

15073 Indigal
Substrate for β-galactosidase. *U.S. Biochemical.*

15074 Indigo
482-89-3 4977 207-586-9
$C_{16}H_{10}N_2O_2$
2-(1,3-Dihydro-3-oxo-2H-indol-2-ylidene)-1,2-dihydro-3H-indol-3-one
Indigotine; Indigotin; Indigo blue; D & C Blue No 6; C.I. Pigment Blue 66; C.I, Vat Blue 1; C.I. 73000. Natural indigo is obtained by steeping the leaves of indigo bearing plants in water then oxidizing the extract. Synthetic indigo is prepared by several methods; used for cotton, wool, silk, by steeping the material in a vat containing the leuco compound. Sublimes at approximately 300°; dec 390°; practically insoluble in water, alcohol, ether and diluted acids; dissolves in nonpolar solvents with red and in polar solvents with blue color.

15075 Indigo Carmine
860-22-0 4978 212-728-8
$C_{16}H_8N_2Na_2O_8S_2$
Indigotin disulfonate sodium
Acid Blue 74; C.I. Acid Blue 74; Food Blue 1; FD&C Blue No. 2; Acid Blue 74; C.I. Acid Blue 74; C.I. 73015; indigotine; soluble indigo blue. Aid in diagnosis of kidney function. Also used as a dyestuff. Soluble in H_2O (1 g/100 ml); slightly soluble in EtOH, insoluble in organic solvents; *Hynson Westcott & Dunning.* Name unverified.

15076 Indilitans
An alloy of 36% nickel, 0.06% carbon, 0.68% manganese, 0.09% silicon, and remainder iron. The alloy has a low coefficient of thermal expansion.

15077 Indio
Decorating colors for vitreous enamel. *Degussa Ltd.*

15078 indirubin
$C_{16}H_{10}N_2O_2$
Indigo red
A red coloring matter associated with indigo in amounts usually between 1% and 5%. It is produced from the decomposition of indican.

15079 Indisin®
A registered trademark currently awaiting reallocation by its proprietors. *Cassella AG.* Name unverified.

15080 indium
7440-74-6 4980 231-180-0
In
Metallic element,; automobile bearings, electronic and semiconductor devices, brazing and soldering alloys, reactor control rods, electroplated coatings on aircraft bearings. d = 7.3; mp = 155°; bp = 2000°; specific heat = 0.0568 cal/g/°. *Aldrich; Atomergic Chemetals; Cerac; Noah Chem.*

15081 indium oxide
1312-43-2 4984 215-193-9
In_2O_3
Indium oxide. Manufacture of special glasses. d = 7.18; volatilizes at 850°; insoluble in water; soluble in hot mineral acids. *Atomergic Chemetals; Cerac; Noah Chem.*

15082 Indocarbon
Black sulfur dyestuffs; used for textile dyeing and printing. *Cassella AG.*

15083 Indofast®
High-quality organic pigments for the surface coatings industry, especially automotive coatings. *Bayer AG.*

15084 Indoil CPD 142 and CPD 143
A proprietary trade name for petroleum type vinyl plasticizers. *Standard Oil Co.* Name unverified.

15085 indoline
A basic navy blue dye.

15086 4-indol-3-ylbutyric acid
133-32-4 4995 205-101-5
$C_{12}H_{13}NO_2$
4-(3-indole)butyric acid
indolebutyric acid; 3-indolebutyric acid; 1H-indole-3-butanoic acid; Hormodin; Seradix.; hormodin; IBA. Used for promotion and acceleration of root formation in plant clippings. mp = 123-125°; insoluble in H_2O, soluble in organic solvents; LD_{50} (mus ip) = 100 mg/kg. *Biosynth AG; Penta Mfg.; Pfaltz & Bauer; Spectrum Chem. Mfg.*

15087 Indonex VG
A proprietary trade name for aromatic hydrocarbon vinyl plasticizer. *Standard Oil Co.* Name unverified.

15088 Indopol
Polybutenes
Amoco Chemical Co.

15089 Induclor
7778-54-3 1717 231-908-7
Calcium hypochlorite
Bleach and oxidizer. *PPG Industries.*

15090 Indulin® 201
Tall oil fatty acid; emulsifier for soap solutions for high float emulsions. *Westvaco.*

15091 Indulin® AQS
Tall oil amino polycarboxylic acid; emulsifier for asphalt emulsions. *Westvaco.*

15092 Indulin® MQK
C_{21} dicarboxylic amido alkyl amine
Asphalt emulsifier for quick set slurry seal applications. *Westvaco.*

15093 Indulin® SA-L
Sodium lignate
Emulsifier, stabilizer, retarder for asphalt emulsions. *Westvaco.*

15094 Indulin® W-1
110152-58-4
Amine derivative of pine lignin; emulsifier, stabilizer in asphalt emulsions, retarding agent in cement; also for oilfield chemicals. *Westvaco.*

15095 Indulin® XD-70
9016-45-9 6772
Nonylphenol ethoxylate
Coemulsifier for anionic or cationic slow-setting asphalt emulsions. *Westvaco.*

15096 Induline
A mixture of aryl-amino-azines, made by heating together aminoazobenzene, aniline, and aniline hydrochloride; used in making inks.

15097 Indur
A proprietary trade name for a phenol-formaldehyde synthetic resin used for molding. No manufacturer.

15098 Indurite
An explosive containing 40% guncotton, freed from lower nitrates, and 60% nitro-benzene. The name is also applied to a molding powder of the phenol-formaldehyde condensation product type.

15099 Indusoil
A proprietary product; it is a refined Talleol (Swedish for 'pine oil). No manufacturer.

15100 Industrene® 104
112-80-1 6865 204-007-1
Oleic acid
Low titer; chemical intermediate. *Witco Corporation.*

15101 Industrene® 126
Soya acid
Chemical intermediate. *Witco Corporation.*

15102 Industrene® 143
61790-37-2 263-129-3
Tallow acid
Intermediate used in alkyd resins, rubber compounding, water repellents, polishes, soaps, abrasives, cutting oils, candles, crayons, emulsifiers; FG grades as lubricant, release agent, binder, defoamer in foods, intermediate for food emulsifiers. *Witco Corporation.*

15103 Industrene® 223
68938-15-8 273-118-5
Hydrogenated coconut acid
Chemical intermediate, emulsifier; used for personal care products, waxes, textile auxiliary, pharmaceuticals. *Witco Corporation.*

15104 Industrene® 224
oleic acid-linoleic acid
Chemical intermediate. *Witco Corporation.*

15105 Industrene® 365
67762-36-1 267-013-3
Caprylic acid-capric acid. Mixture of caprylic acid and capric acid; intermediate used in alkyd resins, rubber compounding, water repellents, polishes, soaps, abrasives, cutting oils, candles, crayons, emulsifiers; FG grades as lubricant, release agent, binder, defoamer in foods, intermediate. *Witco Corporation.*

15106 Industrene® 4516
57-10-3 7128 200-312-9
palmitic acid
45% Palmitic acid; intermediate for manufacture of soaps, emulsions, creams, lotions, ethoxylates, buffing compounds, lubricants. *Witco Corporation.*

15107 Industrene® 4518
57-11-4 8959 200-313-4
Stearic acid
Single pressed stearic acid; intermediate used in alkyd resins, rubber compounding, water repellents, polishes, soaps, abrasives, cutting oils, candles, crayons, emulsifiers; FG grades as lubricant, release agent, binder, defoamer in foods. *Witco Corporation.*

15108 Industrene® 5016
57-11-4 8959 200-313-4
Stearic acid
Double pressed stearic acid; intermediate used in alkyd resins, rubber compounding, water repellents, polishes, soaps, abrasives, cutting oils, candles, crayons, emulsifiers. *Witco Corporation.*

15109 Industrene® 7018 FG
57-11-4 8959 200-313-4
Stearic acid
70% Stearic acid; lubricant, release agent, binder, defoamer for foods; intermediate for food-grade emulsifiers. FG grades as lubricant, release agent, binder, defoamer in foods, intermediate for food emulsifiers. *Witco Corporation.*

15110 Industrene® D
40% Dimer acid; intermediate for lubricants, corrosion inhibitors for petroleum industry, extenders and cross-linking agents for high polymeric systems, in hotmelt adhesives, epoxy curing agents. *Witco Corporation.*

15111 Industrene® R
57-11-4 8959 200-313-4
Stearic acid
Hydrogenated rubber grade stearic acid; intermediate used in alkyd resins, rubber compounding, water repellents, polishes, soaps, abrasives, cutting oils, candles, crayons, emulsifiers. *Witco/Humko.* Discontinued.

15112 Industrial Dyne
Iodophor/acid pipeline cleaner and sterilizer. *Ciba plc.* Name unverified.

15113 Industrial Dynemate
Hypochlorite pipeline cleaner/sterilizer. *Ciba plc.* Name unverified.

15114 Industrial gum
Tragasol; locust bean gum; Carob bean gum. An ingredient of mucilages; it is also used as a protective colloid.

15115 Industrol® DW-5
Modified oxyethylated alcohol; biodegradable, chlorine-stable, low-foaming detergent, dispersant, wetting agent, emulsifier for home/commercial machine dishwash, spray cleaners; defoamer for protein soils. *BASF.*

15116 Industrol® N3
9003-11-6 7721
Poloxalene
EO/PO block copolymer; detergent, dispersant, wetting agent for low to high temperature rinse aids where low foam and sheeting are important. *BASF.*

15117 Industrol® TFA-8
Fatty acid ethoxylate (8 EO)
Surfactant. *BASF.*

15118 Industrol® TO-16
61791-00-2
PEG-16 tallate
Surfactant. *BASF.*

15119 Inertex
Protective powder for molten magnesium *Foseco (F.S.) Ltd.*

15120 Infacare
Baby bath, additive for bath. *Richardson-Vicks Inc.* Name unverified.

15121 Infasoft
Shampoo, to wash and condition hair. *Richardson-Vicks Inc.* Name unverified.

15122 Infavina
Vitamin B complex; tonic. *Merck & Co Inc.* Discontinued.

15123 InFilm
Kaolin clay or metakaolinitic aluminum silicate produced by calcining kaolin clay; sold in powder form or as a slurry in ethylene glycol; antiblocking agents for plastic films. *ECC International Ltd.*

15124 infusible white precipitate
Dimercuridiammonium chloride

15125 Infusorial earth
61790-53-2 212-293-4
Celite; fossil flour; kieselguhr; Tripolite; mountain flour. The siliceous remains of diatoms; used as a non-conducting material for boilers, an absorbent for liquids and liquid manures, in the preparation of dynamite, extender in paints, rubbers.

15126 Ingotol
Nonferrous chill and ingot dressing. *Foseco (F.S.) Ltd.*

15127 Inhibitor
Gel time/pot life extenders for use with unsaturated polyester resins. *Akzo Chemie UK Ltd.*

15128 Inhibitor 60S
Corrosion inhibitor for acid cleaning, pickling, industrial cleaning, consumer products *Exxon/Tomah.*

15129 Inhibitor RT 212
Fatty acid alkanolamide
Anticorrosion additive for metal cleaning and metal working; lubricant in drilling and cutting oils. *Zschimmer & Schwarz.* Discontinued.

15130 Inipol
Fatty diamines and derivatives; cationic surfactant for organic solvent media for paint and pigment dispersion, corrosion inhibitors, lubricity enhancers. *Elf Atochem UK/Ceca.*

15131 Initial®
For the agriculture industry. *DuPont UK.*

15132 Initiator BK
Silyl ether of benzopinacol in triethylphosphate (30%) and toluene (5%); initiator for radical polymerization processes with vinyl monomers, acrylic compounds or olefins; also suitable for BMC and SMC polyester products *Bayer Corp; Polysar.*

15133 Injacom
A proprietary vitamin injection for animals. *Roche Products Ltd.* Name unverified.

15134 Inklurit®
Urea-formaldehyde precondensation product; water-soluble, in conjunction with emulsifiers and hardening agent for enclosing and manufacture of water-immiscible substances, e.g., liquid petroleum products *BASF AG.*

15135 Inkovar 335
Modified hydrocarbon resins; designed specifically for compatibility with cellulosic polymers. *Hercules.* Discontinued.

15136 Inkovar 617
Modified hydrocarbon resins; designed for low energy heatset ink vehicles and flushed colors. *Hercules.* Discontinued.

15137 Inkrustin®
Fluxes for wet and dry galvanizing, tinplating and leadcoating; as a soldering machine flux in the manufacture of tin cans. *BASF AG.*

15138 Innovex
9002-88-4 7728

Polyethylene
Polyethylene. *BP Chemicals Ltd.*

15139 Inochrome
Premetallized dyes. *ICI Chem & Polymers Ltd.*

15140 Inoculin
Inoculating compound for cast iron. *Foseco (F.S.) Ltd.*

15141 Inoderme
Leather dyes. *ICI Chem & Polymers Ltd.*

15142 Inopak
Prepacked mold inoculants for use in production of grey and ductile iron castings. *Foseco (F.S.) Ltd.*

15143 Inotab
Tableted mold inoculants for use in production of grey and ductile iron castings. *Foseco (F.S.) Ltd.* Discontinued.

15144 Insect Spray for House Plants
Contact fertilizer aerosol. *Fisons plc, Horticultural Div.* Name unverified.

15145 Insektigun
bioallethrin-permethrin
Bioallethrin and permethrin; used for control of flies in agricultural premises. *Spraydex Ltd.*

15146 Insol-U & RM
Urea-formaldehyde concentrate. *Georgia-Pacific.*

15147 Instant Gasket
RTV silicone gasketing material; replaces cork, felt paper, asbestos, rubber and metal gaskets; gasket made *in situ* squeezed from the tube using unique applicator nozzle. *Hermetite Products Ltd.* Name unverified.

15148 Instant Ocean
A dry, granular mixture of salts for preparation of synthetic seawater; used in aquariums for marine animals, culture medium and corrosion testing. *Aquarium Systems Inc.*

15149 Instrument bronze
An alloy of 82% copper, 13% tin, and 5% zinc.

15150 Insulgard
Polycarborate sheet; for conservatory and glass-house glazing. *Insulgard Ltd.*

15151 Insullac
A copal spirit varnish with the resin acids neutralized; an insulating material.

15152 Insulmag®
1309-48-4 5713 215-171-9
magnesium oxide
Magnesium oxide for electrical cable filling. *Steetley Magnesia Products Ltd.*

15153 Insural
Insulating refractory shapes for use with aluminum alloys. *Foseco (F.S.) Ltd.*

15154 Insurok
A proprietary trade name for a phenol-formaldehyde synthetic resin used for molding compounds and laminated products. No manufacturer.

15155 Integuard
Plastics materials in the form of films, sheets, strips or labels; tamper-evident plastic films for sealing and packaging applications. *Acordis.*

15156 Intercept
9016-45-9 6772
Nonoxynol 9
Spermaticide. *Ortho Pharmaceutical Corp.* Name unverified.

15157 Intercide
Biocide/fungicide for PVC. *AKZO Chemie UK Ltd.*

15158 Interferon
A protein formed by the interaction of animals cells with viruses. It is capable of conferring resistance to virus infection on animal cells.

15159 Interox H48
Magnesium monoperoxyphthalate hexahydrate low temperature bleach. *Solvay Interox Ltd.* Discontinued.

15160 Intersept®
Synergistic blend of substituted ammonium salts of alkylated phosphoric acids admixed with a free alkylated phosphoric acid; broad-spectrum biostat. *Interface Research.* Name unverified.

15161 Interstab®
Lead and mixed metal heat stabilizers for PVC. *Akzo Chemie UK Ltd.*

15162 Interstab® BZ-4828A
Liquid metal soap stabilizer for flexible PVC *AKZO.*

15163 Interstab® BZ-4836
Barium-zinc-phosphite complex; liquid metal soap heat and light stabilizer for flexible PVC *AKZO.*

15164 Interstab® FR930
Fire retardant for polypropylene and polystyrene; lubricants for PVC. *AKZO Chemie UK Ltd.*

15165 Interstab® LT-4805R
Barium-zinc
Heat and light stabilizer for flexible PVC *AKZO.*

15166 Interwood®
Machines, tools, parts and fittings for wood-working. *Interwood Ltd.*

15167 Intimate Contact
Silicone release film comprising super clear polyester coated with silicone release; release film for optical applications. *Custom Coating & Laminating Corp.* Name unverified.

15168 Intob
Mold inoculants for iron castings. *Foseco (F.S.) Ltd.* Discontinued.

15169 Intracarrier® ATM
Blend of esters and aromatic hydrocarbon derivatives with nonionic emulsifier; biodegradable dyeing accelerant for disperse dyeing of synthetic fibers and their blends; carrier for polyester and triacetate; moderate foaming; high flash point. *Crompton & Knowles.*

15170 Intrafomil® AK
Nonsilicone defoamer for textile operations, especially jet equipment. *Crompton & Knowles.*

15171 Intral®
Mineral iron sulfide (pyrite)
48-50% S, 44-45% Fe; used for manufacture of amber glass. *Metallgesellschaft AG.*

15172 Intralan® Salt HA
Condensed sodium alkylaryl sulfonate; low foaming dispersing and leveling agent for disperse dyes and dyeing of synthetic fibers; especially for beam, package and jet dyeing. *Crompton & Knowles.*

15173 Intralan® Salt N
Ethylene oxide condensate; dye assistant, leveling agent, penetrant for textiles. *Crompton & Knowles.* Name unverified.

15174 Intraphasol COP
Hydrocarbons, Solubilized, aliphatic and sulfonic acid salts; emulsifier, wetting agent assistant for dyeing of fabrics and carpets. *Crompton & Knowles.* Name unverified.

15175 Intraquest® TA Solution
64-02-8 3557 200-573-9
Tetrasodium EDTA
Sequestering agent for scouring, bleaching, dyeing, and other wet finishing textile operations; softens water by complexing calcium, magnesium, and divalent heavy metal ions over broad pH range; detergent builder for scouring. *Crompton & Knowles.*

15176 Intrasoft® OCN
Fatty acid condensate; softener, antistat for textile fibers, especially acrylic, polyamide, cellulosics; aqueous solution are stable to dilute acids, salts, hard water, nonionics and cationics. *Crompton & Knowles.*

15177 Intrassist® LA-LF
Quaternary compound; nonretarding, low-foaming leveling agent for Sevron® cationic dyes on acrylic fibers. *Crompton & Knowles.*

15178 Intratex® A
Complex amino condensate; nonfoaming leveling agent for dyeing wool. *Crompton & Knowles.* Name unverified.

15179 Intratex® CA-2
Aliphatic ethoxylates
Compatibilizer. *Crompton & Knowles.*

15180 Intratex® DD
Surfactant blend; leveling agent for disperse, direct and acid dyes with penetrating, emulsifying, and compatibilizing properties; for cotton and blends with polyester and nylon. *Crompton & Knowles.*

15181 Intratex® N
Phenolic condensate; dyebath auxiliary, leveling and aftertreating agent for acid dyes on nylon carpet and apparel. *Crompton & Knowles.*

15182 Intravon® JU
Ethylene oxide condensate; detergent, emulsifier, dispersant, wetting agent, penetrant for textile, household and cosmetic applications; antiprecipitant for dyeing acrylic blends with anionic/cationic dyes in a one-bath method. *Crompton & Knowles.*

15183 Intravon® SOL-N
Aliphatic solvent/surfactant blend; low odor solvent-based scouring detergent for all fibers, in one-bath dyeing/scouring processes; oil solubilizer; good wetting and penetration, moderate foamer; stable to alkali. *Crompton & Knowles.*

15184 Intrex
Functional surfactant blend. *Rhône-Poulenc UK.*

15185 Intrex DW81
Corrosion inhibitor. *Rhône-Poulenc UK.*

15186 Intrex HA70
Organic phosphoric acid derivative as an amber liquid; hydrotroping agent

used in the solubilizing of other surface active agents in highly alkaline cleaner bases; base for acid cleaners for metals, etc. *Rhône-Poulenc UK.*

15187 Invaderm
Dyeing auxiliaries. *Ciba plc.* Name unverified.

15188 Invaderm C9B
A sodium aryl disalphonate in powder form used as an anionic level dyeing assistant for leather. *Ciba plc.* Name unverified.

15189 Invadine
A proprietary trade name for a wetting agent containing a sodium alkylphenylene sulfonate. No manufacturer.

15190 Invalon
Dyeing and printing assistant. *Ciba plc.* Name unverified.

15191 Invar
An alloy of 36% nickel and 64% steel (0.2% carbon in steel). It has very little expansion on heating, and is used for delicate instruments for measuring.

15192 invariant
An alloy of 47% nickel and 53% iron. Has similar properties to Permalloy.

15193 Invaro Steel
A proprietary trade name for a tool steel containing 1.15% manganese, 0.5% tungsten, and 0.9% carbon. No manufacturer.

15194 Invasol
Fatliquors. *Ciba plc.* Name unverified.

15195 Invephos 20
Liquid chemical containing calcium to be diluted with water; used to produce a microcrystalline, corrosion resisting, paint bording, zinc phosphate coating on iron, steel, and zinc. *Invequimica & CIA SCA.* Name unverified.

15196 Invephos 21C
Liquid chemical containing calcium to be diluted with water; corrosion preventing chemical used in pretreatment for electropainting, produces microcrystalline coatings on iron and steel. *Invequimica & CIA SCA.* Name unverified.

15197 Inveres EVH
Polyvinyl acetate homopolymer; sizing agent, adhesives for wood and paper; available in a wide variety of viscosities. *Invequimica & CIA SCA.* Name unverified.

15198 Inveres K-82
Highly plasticized homopolymer of polyvinyl acetate; binder for carpet backing. *Invequimica & CIA SCA.* Name unverified.

15199 Inversol 140
Complex polyglycol ester
Inversely soluble lubricity agent for metalworking liquids; reduces surface tensile and coefficient of friction. *Ferro/Keil.*

15200 Invert Sugar
8013-17-0 5026 232-393-1
Nulomoline; Calorose; Invesol; Insubeta; Travert. A mixture of molecular proportions of glucose and fructose; obtained in the hydrolysis of cane sugar by acids; used to improve wines, also in the manufacture of liqueurs, fruit preserves and honey substitutes; humectant.

15201 Invicta Duo 495FW
bifenox-isoproturon
A liquid formulation containing 160 g bifenox and 400 g isoproturon per liter as a suspension concentrate; a residual herbicide for the control of weeds in winter cereals. *Farm Protection Ltd.*

15202 Invicta Duo 495FW
bifenox-isoproturon
A liquid formulation containing 160 g bifenox and 400 g isoproturon per liter as a suspension concentrate; a residual herbicide for the control of weeds in winter cereals. *Farmers Crop Chemicals Ltd.* Discontinued.

15203 Iodal
A substance prepared by the action of iodine upon a mixture of alcohol and nitric acid.

15204 Iodanisol
529-28-2 5049 208-456-4
o-Iodoanisole
An antiseptic based upon *o*-iodoanisole.

15205 Iodesin
16423-68-0 3734 240-474-8
$C_{20}H_8I_4O_5$
Erythrosine
Biological stain and color additive.

15206 Iodex
A proprietary ointment containing 4% organically-combined iodine in a petroleum jelly base. *SmithKline Beecham.* Name unverified.

15207 Iodin
Iodized arachis oil.

15208 Iodine
7553-56-2 5034 231-442-4
I_2
Nonmetallic hologen element; dyes, alkylation and condensation catalyst, iodides, iodates, antiseptics, germicides, x-ray contrast media, food and feed additive, stabilizers, photographic film, water treatment, pharmaceuticals, medicinal soaps. mp = 113.60°; bp = 185.24°; soluble in water, organic solvents. *Aldrich; Andeno BV; Atomergic Chemetals; Cerac; Mallinckrodt.*

15209 iodized salt
Common salt, NaCl, containing a very small amount of sodium iodide. Goiter in certain districts is associated with lack of iodide in the water supply.

15210 iodobenzene
591-50-4 5050 209-719-6
C_6H_5I
phenyl iodide
Chemical intermediate. d = 1.8384; mp = -30°; bp = 188-189°; insoluble in H_2O; miscible with alcohol, chloroform, ether. *R.W. Greeff.*

15211 Iodobio 45
TEA hydro-iodide
Slendering product for cosmetics. *Alban Muller.*

15212 iodoeosin
16423-68-0 3734 240-474-8
erythrosine
Erythrosin or Pyrosin; used as an indicator.

15213 Iodofenphos
18181-70-9 5053 242-069-1
$C_8H_8Cl_2IO_3PS$
Phosphorothioic acid O-(2,5-dichloro-4-iodophenyl) O,O-dimethyl ester
Elocril; Nuvanol-N; jodfenphos; Alfacron; Iodofenfos; Nuvanol N; C-9491; ENT-27408; OMS 1211; Elocril; Waspex; C-9491; Nuvanol N. An organophosphorus insecticide. mp = 72-73°; soluble in acetone, xylene, alcohol, water; LD_{50} (rat orl) = 2100 mg/kg. *Ciba-Geigy Agrochemicals.*

15214 iodoform
75-47-8 5054 200-874-5
CHI_3
Triiodomethane
Used as a topical anti-infective and disinfectant. mp = 120°; d = 4.100; slightly soluble in H_2O, more soluble in organic solvents; LD_{50} (rat orl) = 355 mg/kg.

15215 iodol
87-58-1 5062 201-754-5
C_4HI_4N
Tetraiodopyrrole
Used externally as a disinfectant for superficial lesions. mp = 162-164° (dec); soluble in H_2O (2.04 g/100 ml), more soluble in organic solvents.

15216 Iodosorb
1646
Cadexomer iodine powder. A modified starch polymer containing about 0.9% (w/w) iodine. Used in venous ulcers. A vulnerary. *ICI Chem & Polymers Ltd.* Discontinued.

15217 Iodotope I-125
7790-26-3 8778
^{125}IK
Sodium iodide ^{125}I
Oriodide; Iodotope; Radiocaps-125; Theriodide-125. Diagnostic aid (thyroid function); radioactive agent. *Bristol-Myers Squibb Co Inc.* Name unverified.

15218 Iodotope I-131
7790-26-3 8778
^{131}IK
Sodium iodide ^{131}I
Oriodide; Iodotope; Radiocaps-131; Theriodide-131. Antineoplastic; diagnostic aid; radioactive agent. *Bristol-Myers Squibb Co Inc.*

15219 Iodotope Therapeutic
7790-26-3 8778
^{131}IK
Sodium iodide ^{131}I
Oriodide; Iodotope; Radiocaps-131; Theriodide-131. Antineoplastic; diagnostic aid; radioactive agent. *Bristol-Myers Squibb Co Inc.*

15220 Iodoval
$(CH_3)_2CH{\cdot}CHI{\cdot}CO{\cdot}NH{\cdot}CO{\cdot}NH_2$
Monoiodoisovalerylurea

15221 Iodozol
554-71-2 8881 209-069-3
$C_6H_4I_2O_4S$
4-hydroxy-3,5-diiodo-benzenesulfonic acid
sozoiodolic acid; Diiodo-*p*-phenolsulfonic acid; Optojod. Used in a test for albumin. Also a diagnostic aid; used as a radiopaque medium in retrograde pyelography.

15222 Iodron
Bulk milk tank sanitizer. *The Wellcome Foundation Ltd.* Name unverified.

15223 iohydrin
534-08-7 5081 208-586-1
$C_3H_6I_2O$
1,3-diiodoisopropyl alcohol
ilothion; iopropane; 1,3-diiodo-2-hydroxypropane; α-diiodohydrin. Chemical intermediate. d = 2.4-2.5; soluble in H_2O (12.5 mg/ml); more soluble in organic solvents.

15224 Ionac ECP-88
A proprietary acrylic polymer used in the formulation of coatings applied by electrostatic spray. *Ionac Chemical Co.* Name unverified.

15225 Ionex
Lube oil additives. *Shell UK.* Name unverified.

15226 Ionol
128-37-0 1583 204-881-4
BHT
A proprietary trade name for butylated hydroxytoluene, an antioxidant possessing a symmetrical structure thus giving a low power loss at high frequencies when used in dielectrics. *Shell UK.*

15227 Ionol CP
Antioxidants based on butylated hydroxytoluene (BHT). *Shell UK.*

15228 Ionol CPA-Feed
Antioxidants based on butylated hydroxytoluene (BHT). *Shell UK.*

15229 Ionol K65
Antioxidants based on butylated hydroxytoluene (BHT). *Shell UK.*

15230 Ionomer resins
A name given to thermoplastic resins containing both covalent and ionic bonds; carboxyl groups are located along the polymer chain by copolymerization to provide the anionic portion of the ionic cross links; metal ions constitute the cationic part of the links. Sodium, potassium, magnesium and zinc are xamples of the ions used.

15231 Ionox
Range of antioxidants for polymeric systems, thermoplastic rubber and fatty acid distillates. *Shell UK.*

15232 Ionpure
Amorphous water-soluble, inorganic preservative for cosmetics and plastic packaging. *U.S. Cosmetics.*

15233 Iopamidol
60166-93-0 5071 262-093-6
$C_{17}H_{22}I_3N_3O_8$
(S)-N,N'-Bis[2-hydroxy-1-(hydroxymethyl)ethyl]-5-[(2-hydroxy-1-oxopropyl)amino]-2,4,6-triiodo-1,3-benzenedicarboxamide
Isovue; iomapidol; B-15000; SQ-13396; Iopamiro; Iopamiron; Isovie; Jopamiro; Niopam; Solutrast. Radiopaque medium used as a diagnostic aid. dec at about 300°; $[\alpha]_D^{20}$ = -2.01°; soluble in H_2O, methanol, insoluble in chloroform; LD_{50} (rat iv) = 28.2 g/kg. *Bristol-Myers Squibb Co Inc.*

15234 Iophendylate
99-79-6 5074 202-787-8
$C_{19}H_{29}IO_2$
4-Iodo-ι-methylbenzenedecanoic acid ethyl ester
Ethyl 4-iodophenylundec-10-enoate; Ethiodan; ethyl 10-(*p*-iodophenyl)undecylate; ethyl 10-(*p*-iodophenyl)hendecanoate. Diagnostic aid - X-ray contrast medium for myelography. d = 1.240-1.263; n^{20}_D = 1.5230-1.5280; slightly soluble in water; soluble in alcohol, benzene, chloroform, ether; LD_{50} (rat, ip) = 19 g/kg. *British Drug Houses.* Name unverified.

15235 Iopydone
5579-93-1 5079 226-969-1
$C_5H_3I_2NO$
3,5-Diiodo-4(1H)-pyridinone
Hytrast; 3,5-Diiodo-4-pyridone. Diagnostic aid. A radiopaque medium used in bronchography. mp = 321° (dec); insoluble in H_2O, organic solvents;.

15236 Iosan
Iodophor teat dip and udder cream; used for in dairies for control of mastitis. *Ciba plc.*

15237 Iosol
552-22-7 209-007-5
C20H24I2O2
Hypoiodous acid, 2,2'-dimethyl-5,5'-bis(1-methylethyl)[1,1'-biphenyl]-4,4'-diyl ester
Thymol iodide; Dithymol diiodide; Iodothymol.

15238 Iotect
Iodine indicator *May & Baker Ltd.* Name unverified.

15239 Iotek 7010
Ionomer resin, zinc cation; thermoplastic for molding and extrusion of sporting goods, footwear, automotive parts, foams (buoys, ski lift seat cushions), perfume bottle stoppers, ignition tubes for explosives, impact modifiers. *Exxon.*

15240 Iotek 8000
Ionomer resin, sodium cation; thermoplastic for molding and extrusion of sporting goods, footwear, automotive parts, foams (buoys, ski lift seat cushions), perfume bottle stoppers, ignition tubes for explosives, impact modifiers. *Exxon.*

15241 Iotox
ioxynil-mecoprop
Soluble concentrate containing 72 g ioxynil and 214 g mecoprop per liter; used for weed control in turf. *Rhône-Poulenc Environmental Prods. Ltd.*

15242 Iotril
ioxynil octanoate
4-cyano-2,6-diiodophenyl octanoate. Active ingredient; selective post-emergence herbicide which controls a wide range of annual broadleaf weeds in cereals, onions, leeks, and sugar cane. *Agan Chemical Manufactures Ltd.*

15243 Iotrilex
Ioxynil octanoate
Herbicide. *Agan Chemical Manufactures Ltd.*

15244 iotrolan
79770-24-4 5082
$C_{37}H_{48}I_6N_6O_{18}$
5,5'-[(1,3-dioxo-1,3-propanediyl)bis(methylimino)]bis[N,N'-bis[2,3-dihydroxy-1-(hydroxymethyl)propyl]-2,4,6-triiodo-1,3-benzenedicarboxamide
Iotrol; DL-3117; SH-437; ZK-39482; Isovist. Radiopaque medium used as a diagnostic aid. Soluble in H_2O (> 850 g/l); LD_{50} (rat iv) = 27.1 g/kg.

15245 Ipecac
ipecacuanha
The dried root of *Uragoga ipecacuanha*; used in medicine as an emetic; source of the alkaloid emetine.

15246 Iphaneine
102-13-6 203-007-9
$C_{12}H_{16}O_2$
2-Methyl propyl phenyl acetate
Isobutyl phenylacetate; Isobutyl α-toluate. Pure isobutyl phenyl acetate. Used in perfumery as a sweet musk chocolate amber scent. d = 0.988; *Bush Boake Allen Ltd.*

15247 ipomic acid
111-20-6 8558 203-845-5
$C_{12}H_{16}O_2$
Sebacic acid
Chemical intermediate.

15248 Iporka
Foamable urea resins. *BASF plc.* Name unverified.

15249 Ipsilene
A disinfecting gas made by heating ethyl chloride and iodoform under pressure.

15250 Ipso
isoproturon-isoxaben
Suspension concentrate containing 450 g isoproturon and 19 g isoxaben per liter; used for annual weed control in cereals. *DowElanco Ltd.*

15251 Irabond
Primer and adhesive systems for irathane coating and linings; for use on a variety of substrates to enhance the performance of irathane. *Irathane International Ltd.* Unverified.

15252 Iragcet
Solvent soluble dyes. *Ciba plc.* Name unverified.

15253 Irasolve
Solvents for use in cleaning/dissolving irathane; for assistance in surface preparation and cleaning of equipment. *Irathane International Ltd.* Unverified.

15254 Irathane
Elastomeric polyurethane coating and lining materials; for protection from severe abrasion and corrosion problems in mines, power plants, processing and offshore. *Irathane International Ltd.* Unverified.

15255 Ircogel
Metallo-organic complexes high in calcium content, having a particle size of 0.01 micron; thixotropic agents for PVC plastosols and organosols for cold dipping and general use to prepare nondrip compounds. *Stoller Chemicals Ltd.*

15256 Ircogel® 900
Organic calcium compound.
Thixotropic, antisag, and flow control agent for use in coating plastisols and organosols, cloth coating plastisols, PVC sealants, polysulfide sealants, polymercaptan sealants; exhibits wetting or dispersant effect on fillers. *Lubrizol.*

15257 Ircogel® 905
Organic thixotrope for urethane and high solids, solvent-based coatings. *Lubrizol.*

15258 Ircosperse 2170
Imidazoline in xylene dispersant. *Lubrizol.*

15259 iretol
$C_7H_8O_4$
1,2,3-Trihydroxy-5-methoxybenzene
Used in perfumery.

15260 Irgaclarol
Wetting and scouring agents. *Ciba plc.* Name unverified.

15261 Irgacure
UV curing agents. *Ciba plc.* Name unverified.

15262 Irgaderm
Dyes for leather finishing. *Ciba plc.* Name unverified.

15263 Irgaferm BC Champagne
Dry yeasts for wine production. *Ciba plc.* Name unverified.

15264 Irgafin
Predispersed pigments for plastics polymers. *Ciba plc.* Name unverified.

15265 Irgafiner
Predispersed pigments for polyolefins. *Ciba plc.* Name unverified.

15266 Irgafos
Co-stabilizers for plastics. *Ciba plc.* Name unverified.

15267 Irgalan
Metal complex dyes. *Ciba plc.* Name unverified.

15268 Irgalevone
Dyeing and printing assistants. *Ciba plc.* Name unverified.

15269 Irgalite
Pigment dispersions for emulsion paints and other organic pigments. *Ciba plc.*

15270 Irgalite Blue GST
Beta form copper phthalocyanine blue pigment (CI Pigment Blue 15) with excellent texture, dispersibility, gloss, and high strength for letter press and lithographic inks. *Ciba plc.* Name unverified.

15271 Irgalite C-20
Pigment preparations. *Ciba plc.* Name unverified.

15272 Irgalite Dispersed
Pigment plasticizer dispersions. *Ciba plc.* Name unverified.

15273 Irgalite M-20
Pigment preparations. *Ciba plc.* Name unverified.

15274 Irgalite MPS
Multipurpose pigment stainers for paints. *Ciba plc.* Name unverified.

15275 Irgalite PDS
Predispersed pigment powders for paints. *Ciba plc.* Name unverified.

15276 Irgalite PR
Predispersed pigment powders for rotogravure. *Ciba plc.* Name unverified.

15277 Irgalite Yellow BGW
C.I. Pigment Yellow 13. Metaxylidide bis-arylamide yellow pigment with excellent dispersibility and improved gloss and transparency for letterpress and lithographic inks. *Ciba plc.* Name unverified.

15278 Irgalite Yellow F4G
A proprietary monoazo pigment of the arylamide type; greenish hue which makes it suitable for use in letterpress, offset litho, flexographic, and gravure printing inks. *Ciba plc.* Name unverified.

15279 Irgalon
Textile pretreatment agent. *Ciba plc.* Name unverified.

15280 Irganol
Acid dyes for wool. *Ciba plc.* Name unverified.

15281 Irganox
A trademark for a range of speciality antioxidants of the hindered phenol type developed initially for the stabilization of polyolefins for use at high frequencies. *Ciba plc.* Name unverified.

15282 Irgapadol
Dyeing and printing assistants. *Ciba plc.* Name unverified.

15283 Irgaphor
Rubber masterbatch pigments. *Ciba plc.*

15284 Irgaplastol M-20
Pigment preparations. *Ciba plc.* Name unverified.

15285 Irgapyrol
Flame proofing agents. *Ciba plc.* Name unverified.

15286 Irgarol
Paint additives. *Ciba plc.* Name unverified.

15287 Irgasan
Bacteriostats. *Ciba plc.* Name unverified.

15288 Irgasol®
Dyeing and printing assistant. *Ciba plc.* Name unverified.

15289 Irgasperse
Pigment dispersions for nonaqueous decorative paints. *Ciba plc.*

15290 Irgastab®
Stabilizers for polymers. *Ciba-Geigy/Additives; Ciba plc.* Name unverified.

15291 Irgatan
Synthetic tanning agents for leather. *Ciba plc.* Name unverified.

15292 Irgatron
Premetallized dyes for polyamides. *Ciba plc.* Name unverified.

15293 Irgawax
Plastics lubricants. *Ciba plc.* Name unverified.

15294 Irgazin
A range of organic pigments derived from isoindolinone and dioxazine; typical colors are Yellow 2GLT, Yellow 3RLT, Orange RLT, Red 2BJT and Violet BLT. *Ciba plc.*

15295 Irgazin C-20, M-20
Pigment preparations. *Ciba plc.* Name unverified.

15296 Irgoferm CM Montrachet
Dry yeasts for wine production. *Ciba plc.* Name unverified.

15297 iridin
$C_{24}H_{26}O_{13}$
7-(glucosyloxy)-3',5-dihydroxy-4',5',6-trimethoxyisoflavone
irisin. The powdered extract of iris; a pigment. mp = 217°; soluble in H_2O (2 g/l), more soluble in organic solvents.

15298 Iridite®
Chromate coatings; corrosion protective coating for plated and unplated metals; base coat for painting of nonferrous metals; approved for governmental and industrial specifications finishing of zinc plate, cadmium plate, hotdipped galvanized steel, zinc die-cast alloys. *Witco/Allied-Kelite.* Discontinued.

15299 iridium steel
A German steel containing 4% cobalt, 16% tungsten, 3.5% chromium, 0.67% vanadium, 0.8% molybdenum, and 0.6% carbon.

15300 Iridosmine
Osmiridium
An alloy of iridium and osmium containing 40-77% iridium, and 20-50% osmium. If there is more iridium, the alloy is called Nevyanskite, and Siserskite if the content of osmium is high.

15301 Irigenin
548-76-5 5103 208-958-3
$C_{18}H_{16}O_8$
3',5,7-trihydroxy-4',5',6,-trimethoxyisoflavone
The aglycone of iridin. A pigment. mp = 185°; λ_m = 267 nm; insoluble in H_2O, soluble in organic solvents.

15302 Irilac®
Clear coatings providing protection against corrosion, fingerprinting, tarnish, and abrasion on all metals. *Witco/Allied-Kelite.* Discontinued.

15303 Irish pearl moss
Carageen moss, a gelatinous seaweed, *Chondrus crispus.* It contains carrageenin allied to pectin and is employed as a substitute for isinglass, as a size for thickening colors in calico printing, and for stiffening silk.

15304 Irish Peat Wax
Montana wax
montanin wax. Waxes extracted from Irish peat are sold under these names. They resemble montan wax.

15305 Irisol®
Spirit soluble dyestuffs; for the surface coatings, printing ink, office supplies, and plastics industries. *Bayer AG.*

15306 Irlux®
For the chemical industry. *DuPont UK.*

15307 iron
7439-89-6 5106 231-096-4
Fe
Used to form steels by alloying with other elements such as C, Ni, Mn, Cr. Radioisotopes of iron are used in biological research and in medicine. mp = 1535°; bp = 3000°; d = 7.8600; electrical resistivity = 9.71 microhm-cm.

15308 Iron A
A British chemical standard. It is a cast iron containing 0.734% combined carbon, 1.989% silicon, 0.047% sulfur, 0.049% phosphorus, 0.688% manganese, 0.042% arsenic, 0.052% titanium, and 2.387% graphitic carbon.

15309 Iron B
A British chemical standard. It is a cast iron containing 0.39% combined carbon, 0.031% sulfur, 0.026% phosphorus, 0.031% arsenic, 0.108% titanium, and 2.67% graphitic carbon.

15310 iron black
Antimony precipitated as a fine powder, by action of zinc upon an acid solution of an antimony salt; imparts the appearance of polished steel to paper mache and plaster of Paris.

15311 Iron D
Phosphoric D
A British chemical standard iron alloy containing 11.8% phosphorus.

15312 Iron D2
A British chemical standard. Grey phosphoric cast iron in the form of fine turnings. It contains 1.31% silicon, 1.07% phosphorus, and 1.64% manganese. The approximate quantities of other elements are 2.5% graphitic carbon, 0.8% combined carbon, and 0.03% sulfur.

15313 iron flint
An opaque variety of quartz containing iron.

15314 iron froth
A spongy variety of hematite.

15315 Iron G
A standard cast iron containing 1.82% graphite carbon, 0.86% combined carbon, 1.3% silicon, 0.41% manganese, 0.125% sulfur, and 0.45% phosphorus.

15316 Iron L
nickel-chromium-copper-Austenitic Iron L
A British Chemical Standard containing 3.06% total carbon, 2.26% silicon, 1.01% manganese, 0.119% phosphorus, 13.45% nickel, 3.96% chromium, 4.73% copper, 0.031% sulfur, the remainder being iron.

15317 Iron Man
Vitamin-fortified tonic. *Richardson-Vicks Inc.* Name unverified.

15318 Iron ore A, hematite type
A standard iron ore containing 58.19% iron, 0.056% phosphorus, 8.14% silica, and 0.066% sulfur; used as a standard for checking analyses of iron ore.

15319 iron putty
ferric oxide-linseed oil
A mixture of ferric oxide and boiled linseed oil; used for joints in iron pipes.

15320 iron sulfate (ic)
10028-22-5 4079 233-072-9
$Fe_2(SO_4)_3$
Ferric sulfate
ferric trisulfate; iron persulfate; iron tersulfate. Pigments, reagent, etching aluminum, disinfectant, textile dyeing and printing, flocculant in water and sewage purification, soil conditioner, polymerization catalyst, metal pickling, chelated iron products, chemical intermediate. *Boliden Intertrade; Faesy & Besthoff; Rhône-Poulenc.*

15321 iron sulfate (ous)
7720-78-7 4105 231-753-5
FeO_4S
ferrous sulfate anhydrous
copperas; green vitriol; ferrous sulfate; sal chalybis; iron vitriol. Iron oxide pigment, other iron salts, ferrites, water and sewage treatment, catalyst especially for synthetic ammonia, fertilizer, feed additive, flour enrichment, reducing agent, herbicide, wood preservative, process engraving. *EM Industries; J.M. Huber; Mallinckrodt.*

15322 Iron vitriol
7782-63-0 4105 233-072-9
$FeSO_4{\cdot}7H_2O$
ferrous sulfate heptahydrate
Green vitriol; copperas; green copperas; Ferrous sulfate. As a chemical intermediate, in electroplating, as a pesticide and a hematinic.

15323 Ironac
An acid-resistant alloy of iron and silicon. It contains 13% silicon, 84% iron, 0.77% manganese, 1.08% carbon, and 0.78% phosphorus.

15324 iron-andradite
Synonym for Skiagite.

15325 iron-leucite
$KFeSi_3O_8$
A mineral.

15326 iron-ore cement
Cements in which a large proportion of the alumina is replaced by ferric oxide.

15327 iron-orthoclase
Synonym for Ferriorthoclase.

15328 Irox
A proprietary trade name for a synthetic yellow iron oxide. No manufacturer.

15329 Isanol®
60% isobutanol [78-83-1] and 40% n-butanol [71-36-3]; solvent for production of adhesives and surface coating resins. *BASF AG.*

15330 Isarit
A bleaching earth.

15331 Isatin
91-56-5 5116 202-077-8
$C_8H_5NO_2$
1H-Indole-2,3-dione
Indole-2,3-dione; 2,3-Indolinedione. Intermediate for production of pharmaceuticals and dyes, stabilizer for plastics industry. mp = 193-195°; insoluble in H_2O, soluble in organic solvents. *BASF AG.*

15332 Isceon
A range of halogen derivations of aliphatic hydrocarbons used as refrigerants and propellants. *ISC Chemicals Ltd.*

15333 isinglass
Sodium silicate solution
Used for the clarification of liquids such as beer and wine.

15334 Iso Isostearyle WL 3196
41669-30-1 255-485-3
Isostearyl isostearate
Cosmetics ingredient. *Gattefosse SA.*

15335 iso soap
A solid sulfonic derivative of castor oil, soluble in hot water; used as a bleaching, washing, and dressing agent in the textile industries.

15336 isoamyl acetate
123-92-2 5125 204-662-3
$C_7H_{14}O_2$
isoamyl acetate
Isopentyl Acetate; 3-Methylbutyl acetate; Isoamyl ethanoate; Isoamyl Acetate; amyl acetate ester; 3-methyl-1-butanol acetate; Banana Oil; 3-Methyl-1-Butyl Acetate; 3-methyl butyl ester acetic acid; isopentyl ester acetic acid; isopentyl alcohol, acetate; pear oil; β-methyl butyl acetate; Amyl acetate, common. Provides a pear flavor. Used as a flavorant in foods and beverages and as a solvent. mp = -78°; bp_{756} = 142°; d = 0.8760; n_D^{20} = 1.4000; slightly soluble in H_2O, more soluble in organic solvents; LD_{50} (rat orl) = 16600 mg/kg. *Aldrich; Penta Mfg.; Spectrum Chem. Mfg.*

15337 isoanthraflavic acid
2,7-Dihydroxyanthraquinone

15338 isoascorbic acid
89-65-6 5141 201-928-0
$C_6H_8O_6$
D-erythro-hex-2-enonic acid γ-lactone
D-araboascorbic acid; isovitamin C; saccharosonic acid; glucosaccharonic acid; D-*erythro*-3-ketohexonic acid lactone; erycorbin; Mercate; erythorbic acid; erythorbic acid. Isomer of ascorbic acid; antioxidant (industrial, food, brewing), reducing agent in photography. dec 174°; $[\alpha]_D^{20}$= -16.6°; soluble in water, alcohol, pyridine, moderately soluble in acetone, slightly soluble in glycerol. *Ashland; Spice King.*

15339 *Isobond*
Isocyanate binders. Dow UK. Discontinued.

15340 Isobrite®
Zinc brightener systems providing full bright plate with greater tolerance for high temperatures. *Witco/Allied-Kelite.* Discontinued.

15341 Isobu-M-AMD
1669-59-3
$C_8H_{15}NO_2$
N-(Isobutoxymethyl)acrylamide
Monomer. $bp_{0.3}$ = 99-100°. *Cyanamid BV.* Name unverified.

15342 isobutad
A mineral rubber or bitumen.

15343 isobutane
75-28-5 200-857-2
C_4H_{10}
2-methylpropane
propane, 2-methyl-isobutane. Hydrocarbon gas; Aerosol propellant. bp = -12°; insoluble in H_2O. *Air Prods & Chem; Phillips 66.*

15344 isobutanol
78-83-1 5146 201-148-0
$C_4H_{10}O$
2-methylpropan-1-ol
isobutyl alcohol; isobutanol; IBA; isopropyl carbinol; 2-methylpropyl alcohol; i-butyl alcohol; butanol-iso; fermentation butyl alcohol; 1-hydroxymethylpropane; isopropyl carbitol. Used as a solvent and chemical feedstock. mp = -108°; bp = 108°; d = 0.8030; n_D^{20} = 1.3960; soluble in H = 72o (5 g/100 ml), more soluble in organic solvents; LD_{50} (rat orl) = 2.46 g/kg. *BASF; CPS; Eastman; Hoechst Celanese; Neste UK; Shell; Union Carbide.*

15345 isobutyl acetate
110-19-0 5145 203-745-1
$C_6H_{12}O_2$
Solvent for nitrocellulose; in thinners, sealants, topcoat lacquers; perfumery; flavoring agent. d = 0.871; mp = -99°; bp = 118°; fp (closed cup) = 64°F; n_D^{18} = 1.3907; soluble in H_2O, alcohol. *BASF; Eastman; Hoechst Celanese; Janssen Chimica; Union Carbide.*

15346 isobutyl formate
542-55-2 5157 208-818-1
$C_5H_{10}O_2$
2-methyl-1-propyl formate
tetryl formate; 2-methylpropyl formate. Industrial solvent. mp = -95°; bp = 98°; d = 0.8850; n_D^{20} = 1.3854; soluble in H_2O (1 g/100 ml), more soluble in organic solvents.

15347 Isobutyl Niclate®
15317-78-9 239-354-8
Nickel diisobutyldithiocarbamate
Antioxidant/antiozonant for protection in epichlorohydrins. *R. T. Vanderbilt Co Inc.* Discontinued.

15348 isobutyric acid
79-31-2 5170 201-195-7
$C_3H_7O_2$
2-methylpropanoic acid
isopropylformic acid; dimethylacetic acid. Organic acid; manufacture of esters for solvents, flavors, perfume bases, disinfecting agent, varnish, deliming hides, tanning agent. *Eastman; Hoechst Celanese; Hüls Am.*

15349 isocetyl stearoyl stearate
97338-28-8 306-621-6
Hetester HSS; Ceraphyl 791. Emollient with unusual skin feel, pigment dispersant for stick products, emulsion systems; binding oil for pressed powds. *Heterene; Bernel; Van Dyk.*

15350 Isoclad
Water-based, elastic anticorrosive cladding for all ferrous and nonferrous metals; used for any corrosive environment including industrially polluted areas. *Liquid Plastics Ltd.*

15351 Isocon
Refined, semirefined and polymeric isocyanates; curing components for use with Propocon polyether systems. *Harcros.*

15352 Iso-Cornox
Selective weedkillers. *The Boots Co plc.* Discontinued.

15353 Iso-Cornox
7085-19-0 5826 230-386-8
Mecoprop
Soluble concentrate containing 570 g/l mecoprop; herbicide for control of weeds in cereals and grassland. *Schering Agrochemicals Ltd.* Discontinued.

15354 Isocracking
Catalyst. *Monsanto (Solaris).* Name unverified.

15355 Isocreme
Compounded lanolin-derived sterols. *Croda Chem. Ltd.* Discontinued.

15356 Isocure
A tertiary amine vapor-cured phenolic urethane resin system; liquid resins and catalysts used as binders for foundry core and mold production. When mixed with sand, these resins and catalysts act as an adhesive for foundry sands; these resin-bonded sand articles are used in the production of ferrous and nonferrous castings. *Ashland Chemical Company.* Name unverified.

15357 isocyanuric acid
108-80-5 2767 203-618-0
$C_3H_3N_3O_3$
s-triazine-2,4,6-triol
cyanuric acid; 2,4,6-trihydroxy-1,3,5-triazine. Ketone isomer of cyanuric acid; used to stabilize chlorine solutions in swimming pools; bleaches, sanitizers. *Allchem Industries; Nissan Chem. Ind.; Schaefer Salt & Chem.; 3-V.*

15358 Isodamp® C-1002
Thermoplastic; highly damped thermoplastic material for military tank ammunition racks. *E-A-R.*

15359 isodecyl methacrylate
29964-84-9 249-978-2
$C_{14}H_{26}O_2$
Isodecyl 2-methylpropenoate
isodecyl methacrylate; Isodecyl 2-methyl-2-propenoate; Ageflex FM-10. Pressure-sensitive adhesives, coatings for leather, textiles, paper, nonwoven fiber, polymer modifier and stabilizer, viscosity index improver, dispersion for plastics and rubber, floor waxes, potting compounds, sealants. *CPS; Rohm & Haas; Sartomer.*

15360 Isoderm®
Comprehensive range of nitrocellulose seasons for leather, aqueous or organic systems. *Bayer AG.*

15361 Isodurindine
$C_{10}H_{15}N$
Tetramethylaniline
Solvent.

15362 isoflupredone acetate
338-98-7 5190 206-423-9
$C_{23}H_{29}FO_6$
9-fluoro-11β,17,21-trihydroxypregna-1,4-diene-3,20-dione 21-acetate
U-6013; Predef. Veterinary anti-inflammatory. mp = 244-246°; $[\alpha]^{23}_D$ = 108° (c = 0.611 EtOH); λ_m = 240 nm (ε = 16250 EtOH).

15363 Isoflurane
26675-46-7 5191 247-897-7
$C_3H_2ClF_5O$
2-Chloro-2-(difluoromethoxy)-1,1,1-trifluoroethane
Forane; 1-chloro-2,2,2-trifluoroethyl ether; compd 469; Aerrane; Forene. An inhalation anesthetic; solvent and dispersant for fluorinated materials bp = 48.5°; d = 1.45. *Atochem SA.* Discontinued.

15364 Isofoam
Twin component rigid polyurethane foam systems. *Baxenden Chemicals Ltd.*

15365 isoform oxiosol
p-Iodoxyanisol

15366 Isol
107-41-5 4748 203-489-0
hexylene glycol
An oil forming a permanent emulsion with hot or cold water; used for oiling textiles.

15367 Isol R
Antiskinning agent for paints and varnishes. *Akzo Chemie UK Ltd.*

15368 Isolan® GI 34
91824-88-3
Polyglyceryl-4 isostearate
Water-oil emulsifier for vegetable oils; low odor, easy to use; forms very stable emulsions; for personal care industry. *Goldschmidt.*

15369 Isolan® GO 33
9007-48-1
Polyglyceryl-3 oleate
Water-oil emulsifier for vegetable oils and natural triglycerides; for personal care products industry. *Goldschmidt.*

15370 Isolan®, Isolan® K
1:2 Metal complex dyestuffs; used for dyeing wool and polyamide fibers. *Bayer AG.*

15371 Isolantite
A proprietary trade name for a ceramic material made from talc with binders. No manufacturer.

15372 Isolene® 40, 75, 400
cis-1,4-Polyisoprene of low molecular weight derived form natural rubber, polymer used in adhesives, sealants, caulks, and lubricants. *Hardman.*

15373 Isolit
Dental separating agent. *Degussa Ltd.*

15374 Isoloss® LS
Polyurethane elastomer, cellular; high-density foam for difficult mechanical energy control problems; good shock absorption and vibration isolation performance; for gaskets, motor mounts, cushion pads, bumpers, springs, laminates, pressure pads, athletic paddings. *E-A-R.*

15375 Isomol
Thixotropic, ready for use spirit based coatings for all types of iron, steel, and nonferrous castings. *Foseco (F.S.) Ltd.*

15376 Isonal
Shading dyestuffs complementing the Isolan range. *Bayer plc.*

15377 Isonaphthol
101-68-8 202-966-0
$C_{15}H_{10}N_2O_2$
4,4'-diphenylmethane diisocyanate
Methylene diphenyl diisocyanate; MBI; PMDI; Caradate 30; Desmodur 44; Hylene M50; Isonate 125M; Isonate 125MF; Nocconate 300. Intermediate for the synthesis of polyethanes. mp = 37-39°; bp_5= 194°; d = 1.1900; LD_{50} (mus orl) = 2200 mg/kg.

15378 Isonate® 125M
101-68-8 202-966-0
MDI; processing aid, intermediate for the production of cast, RIM, and thermoplastic PU elastomers, adhesives, binders, coatings, and sealants. *Dow Chemical.*

15379 Isonate® 2125M
101-68-8 202-966-0
MDI
For PU industry as intermediate in manufacture of adhesives, binders, coatings, and sealants. *Dow Chemical.*

15380 isonicotinic acid
55-22-1 5204 200-228-2
$C_6H_5NO_2$
4-picolinic acid
pyridine-4-carboxylic acid; σ-picolinic acid. Chemical intermediate. mp = 319°; soluble in H_2O (0.52 g/100 ml), less soluble in organic solvents. *Raschig; Reilly Industries.*

15381 Isonol
Polyols used in the manufacture of polyurethane products. *Dow UK.* Discontinued.

15382 Isonox® 103
128-39-2 204-884-0
2,6-Di-*t*-butylphenol
Antioxidant for control of formation of gums and peroxides in fuels and oils; stabilizer for aviation turbine fuels. *Schenectady.*

15383 Isonox® 129
35958-30-6 252-816-3
2,2'-ethylidenebis (4,5-di-*tert.*-butylphenol)
Ethylidenebisdibutylphenol. Antioxidant and thermal stabilizer for polymers; food packaging application; used in PP, polyethylene, PVC, PS, ABS, hydrocarbon resins, EVA-modified compounds. *Schenectady.*

15384 Isonox® 132
17540-75-9 241-533-0
$C_{18}H_{30}O$
2,6-Di-*tert*-butyl-4-*sec*-butylphenol
Di-*t*-butyl-4-butylphenol. Antioxidant used in polyols and rubber systems. *Schenectady.*

15385 isooctyl alcohol
26952-21-6 5211 248-133-5
$C_8H_{18}O$
isooctanol
2-ethyl-1-hexanol; alcohol C_8. Intermediate in manufacture of plasticizers; intermediate for nonionic detergents and surfactants, hydraulic fluids; resin, solvent, emulsifier, antifoaming agent. *CDF Chimie Nederland BV.*

15386 isooctyl thioglycolate
25103-09-7 246-613-9
$C_{10}H_{20}O_2S$
isooctyl mercaptoacetate
Antioxidants, fungicides, oil additives, plasticizers, insecticides, stabilizers, polymerization modifiers, stabilizer for tin-sulfur compounds, stripping agent for polysulfide rubber. *Bock, Bruno Chemische Fabrik KG.*

15387 Isopar®
High purity isoparaffinic solvent. *Exxon Int'l.*

15388 Isopar® C
64742-48-9 265-150-3
C_7-C_8 isoparaffin
Solvent. *Exxon.* Name unverified.

15389 Isopar® E
64742-48-9 265-150-3
C_8-C_9 isoparaffin
Solvent *Exxon.* Name unverified.

15390 Isopar® G
64742-48-9 265-150-3
C_{10}-C_{11} isoparaffin
Solvent. *Exxon.* Name unverified.

15391 Isopar® L
64742-48-9 265-150-3
C_{11}-C_{13} isoparaffin
Solvent. *Exxon.* Name unverified.

15392 Isopar® M
64742-47-8 265-149-8
C_{13}-C_{14} isoparaffin
Solvent. *Exxon.* Name unverified.

15393 Isopaste
Foundry core paste that cures fast without heat *Ashland Chemical Company.* Name unverified.

15394 isopentane
78-78-4 201-142-8
C_5H_{12}
2-Methylbutane
Exxsol® Isopentane; ethyl dimethyl methane. Isopentane, dearomatized; solvent. bp = 30°F; d = 0.620; fp =-70°. *Exxon; Phillips.* Name unverified.

15395 isophorone
78-59-1 5213 201-126-0
$C_9H_{14}O$
3,5,5-trimethyl-2-cyclohexen-1-one
3,3,5-trimethyl-2-cyclohexen-1-one. In solvent mixtures for printing inks and finishes, for polyvinyl and nitrocellulose resins, pesticides, stoving lacquers. mp = -8°; bp = 215°; d_4^{20}= 0.9613; n_D^{20}= 1.4778; soluble in H_2O (12 g/l), organic solvents; LD_{50} (rat orl)= 2700 mg/kg. *Allchem Industries; Elf Atochem; BP Chem. Ltd; Hüls AG; Union Carbide.*

15396 Isoplac
A proprietary insulation. No manufacturer.

15397 Iso-Planotox
Selective weedkiller. *May & Baker Ltd.* Name unverified.

15398 Isoplast
Thermoplastic engineering resins. *Dow Cheml Co Ltd, UK & Ireland.*

15399 Isoplast 101
9009-54-5
Thermoplastic PU; amorphous engineering resin with crystalline properties; for extrusion and injection molding; offers high impact, low moisture sensitivity, excellent chemical and solvent resistance, high abrasion resistance; opaque; specific gravity 1.19; melt flow 8g/10 min (5 kg, 224°); tensile strength 48 Mpa (break); tensile elongation 6% (yield); 160% (break); flex strength 68 Mpa; Izod impact 1175 J/m notch; Rockwell hardness > R100. *Dow Plastics.*

15400 Isoplast 101LGF40 Nat, 101LGF60 Blk
9009-54-5
Thermoplastic PU, long glass fiber-reinforced; amorphous engineering resin with crystalline properties; offers high mod., strength, impact, excellent chem. and solv. resistance *Dow Plastics.*

15401 Isoplast 302
9009-54-5
Thermoplastic PU; amorphous engineering resin with crystalline properties *Dow Plastics.*

15402 Isopoxy
Insulating varnish. *Schenectady-Midland Ltd.*

15403 Isoprep®
Cleaners for use on ferrous and nonferrous metals and on nonconductive substrates. *Witco/Allied-Kelite.* Discontinued.

15404 isoprocarb
2631-40-5 220-114-6
$C_{11}H_{15}NO_2$
2-(1-methylethyl)phenyl methylcarbamate
isoprocarbe;MIPC; OMS 32; ENT 25670; Bay 105807; Etrofolan; Hytox; Mipcin. Insecticide with contact and stomach action. Used for control of leafhoppers, planthoppers, aphids, capsids, bugs and other insects in rice, cocoa, vegetables and other crops. mp = 88-93°; insoluble in H_2O, soluble in organic solvents; LD_{50} (rat orl) = 485 mg/kg.

15405 isopropanol
67-63-0 5227 200-661-7
C_3H_8O
isopropyl alcohol
Isopropanol; IPA; dimethyl carbinol; 2-propanol; Sec-propanol; Rubbing Alcohol; Dimethylcarbinol; sec-Propyl; Alcohol; Propan-2-ol; i-Propanol; 2-hydroxypropane; alcojel; alcosolve; avantin; chromar; combi-schutz; hartosol; imsol a; isohol; lutosol; petrohol; n-propan-2-ol; propol; spectrar; sterisol hand disinfectant; takineocol; alcosolve 2; (-)-2,3-o-Isopropyl alcohol; DuPont Zonyl FSP Fluorinated Surfactants; DuPont Zonyl FSJ Fluorinated Surfactants; DuPont Zonyl FSA Fluorinated Surfactants; DuPont Zonyl FSN Fluorinated Surfactants. Used in antifreeze, as a solvent and as an antiseptic. mp = -88°; bp = 82°; d = 0.7850; n_D^{20} = 1.3774; soluble in H_2O, organic solvents; LD_{50} (rat orl) = 5.8 g/kg.

15406 isopropanolamine
78-91-1 488 201-156-4
C_3H_9ON
2-amino-1-propanol
1-amino-2-propanol; monoisopropanolamine; β-propanolamine; 2-aminopropyl alcohol; 2-hydroxypropylamineMIPA; Aliphatic amine,. Solubilizer, neutralizer, emulsifying agent; plasticizers, insecticides. bp = 173-176°; soluble in H_2O, organic solvents. *Ashland; BASF; Mitsui Toatsu Chem.*

15407 isopropyl chloroformate
108-23-6 203-563-2
$C_4H_7ClO_2$
Carbonochloridic acid 1-methylethyl ester
Chemical intermediate for free-radical polymerization initiators, organic synthesis. *Elf Atochem N. Am.; BASF; PPG Industries.*

15408 isopropyl mercaptan
75-33-2 200-861-4
C_3H_8S
2-propanethiol
Unpleasant stench oder. Used as a leak detecting additive to natural gas. mp = -131°; bp = 57-60°; d = 0.8200; n_D^{20}= 1.4255.

15409 isopropyl myristate
110-27-0 5234 203-751-4
$C_{17}H_{34}O_2$
IPM
1-methylethyl tetradecanoate; tetradecanoic acid, 1-methylethyl estermyristic acid isopropyl ester,. Cosmetic creams, topical medicinals. *Amerchol; Goldschmidt; Henkel/Emery; Inolex; Lanaetex; Stepan; Unichema.* Discontinued.

15410 isopropyl palmitate
142-91-6 205-571-1
$C_{19}H_{38}O_2$
1-methylethyl ester;1-methylethyl hexandecanoate
hexadecanoic acid;isopropyl n-hexadecanoate; Kessco® IPP;Emerest® 2316; Lexol IPP; Liponate IPP; Propal; Radia® 7200; Tegosoft® P; Unimate® IPP; Wikenol® 111,. Lubricant used for synthetic fibers in applications where low friction is essential; emollient in cosmetic formulations; high purity. Henkel/Cospha; Henkel Canada; Stepan; Lipo; Inolex; Amerchol Corp.; Fina Chemicals; Goldschmidt; Union Camp; Amerchol; Goldschmidt; Henkel/Emery; Inolex; Stepan; Unichema; CasChem.

15411 isopropyl phenylmethyl carbamate
2631-40-5 220-114-6
2-Isopropyl-phenyl-N-methylcarbamate
Etrofolan®. Insecticide effective against leafhoppers and bugs. *Bayer AG.*

15412 Isopropylan® 33
Isopropyl palmitate, lanolin oil; binder in talc and pearl powder systems; plasticizer, emollient, and moisturizer. *Amerchol Corp. Discontinued.*

15413 N-Isopropyl-N'-phenyl-p-phenylene diamine
101-72-4 202-969-7
$C_{15}H_{18}N_2$
p-Isopropylaminodiphenylamine
N-Phenyl-N'-isopropyl-p-phenyl-enediamine; Flexzone 3C; Dusantox IPPD; Permanax™ IPPD; Santoflex® IP, IPPD; Vanox® 3C; Vulkanox® 410NA. A proprietary antioxidant. d = 1.04; fp = 72-76°; insoluble in H_2O; LD_{50} = (rat orl) 555 mg/kg, (mus orl) 1122 mg/kg. *Uniroyal; Akrochem; Bayer/Fiber, Orgs., Rubbers; Elf Atochem N. Am.* Name unverified.

15414 Isopto Atropine
55-48-1 907 200-235-0
atropine sulfate
Atropine sulfate solution for ocular use; anticholinergic. *Alcon Laboratories Inc.*

15415 Isopto Cetamide
6209-17-2 9067
sulfacetamide sodium
Sulfacetamide sodium solution; an ocular antiseptic, antibacterial. *Alcon Laboratories Inc.*

15416 Isorate
Urethanes. *Dow Cheml Co Ltd, UK & Ireland.*

15417 Isoset®
High performance, two-component adhesive systems for woodworking, panels, beams, plywood, laminating, coating, contact and pressure-sensitive adhesives, fiber bonding. *Ashland.*

15418 Isoset® WD3-A322 Emulsion Resin
Proprietary, reactive, water-based polymer for crosslinking with CX-hardener at ambient temperature; structural wood laminations, Type 1 waterproof/no-creep performance. *Ashland Chemical Company.* Name unverified.

15419 Isoset® WD3-CM402 Emulsion Resin
Proprietary, reactive, water-based polymer for crosslinking with CX-hardener at ambient temperature; structural sandwich composite, metal or plastic faces to porous, e.g. wood cores. *Ashland Chemical Company.* Name unverified.

15420 isostearamidopropyl dimethylamine lactate
55852-15-8
Propanic acid, 2-hydroxy-, compound with N-[3-(dimethylamino)propyl]-16-methylheptadecanamide
Afalene™ 416; Emcol® 6613; Mackalene™ 416; Incromate IDL. Substantive conditioner improving slip, wet comb and manageability of hair, feel in hand creams and lotions. pH=7.0. *Croda Inc.* Discontinued.

15421 isostearic acid
2724-58-5 220-336-3
$C_{18}H_{36}O_2$
heptadecanoic acid
16-methyl-Isooctadecanoic acid; 16-methylheptadecanoic acid; Emersol® 871; Proto-Lan IP; Prisorine 3508; Imwitor® 780K,. Mixture of branched chain 18 carbon aliphatic acids; cosmetics, chemicals, dispersant, softener in rubber compounds, food packaging, suppositories, ointments. SG = 0.96-0.98; viscosity = 700-900; mPa·s; HLB=3.7; acid no = 3 max; iodine no = 10 max; sapon no = 240-260; LD_{50} (rat orl) >5 g/kg. *Hüls Am.; Henkel/Emery; Nissan Chem. Ind.; Unichema; Union Camp.*

15422 isostearyl isostearate
41669-30-1 255-485-3
Iso Isotearyle WL 3196; Schercemol 1818; Lipacide DPHP. Cosmetics ingredient. *Gattefosse SA; Scher; Rhône-Poulenc; R. T. Vanderbilt Co Inc.*

15423 isotachiol
Ag_2SiF_6
Silver silicofluoride

15424 Isotagetone
2,7-Dimethyloct-5-en-4-one
Used in perfumery. *Bush Boake Allen Ltd.*

15425 Isotagetone 50
2,7-Dimethyloct-5-en-4-one in isopropyl myristate; tagette, chamomile odor for use in fragrances. *Bush Boake Allen Ltd.*

15426 Isoterge
Detergent solutions; used for cleaning of automatic blood cell analyzers. *Coulter Electronics Ltd.* Discontinued.

15427 Isothan Q-75
93-23-2 202-230-9
$C_{21}H_{32}BrN$
Lauryl isoquinolinium bromide
Dodecylisoquinolinium bromide; Isoquinolinium, 2-dodecyl-, bromide; Isothan Q-15; Lauryl isoquinolinium bromide. Anti-infective. *Onyx Chemical Co.* Name unverified.

15428 Isoton
Diluents based on normal saline; used for blood cell counting and sizing; analysis medium for electrical sensing zone particle size analyzers. *Coulter Electronics Ltd.*

15429 Isotox
Insecticide seed treater. *Monsanto (Solaris).* Name unverified.

15430 Issolin
A phenol-formaldehyde resin, which is soluble in alcohol. No manufacturer.

15431 Italcor
1344-28-1 369 215-691-6
Aluminum oxide
An abrasive. *Winchem Ltd.*

15432 ITP®
Thermosetting resin compositions. *Reichhold.*

15433 itrol
126-45-4 8654 204-786-8
$C_6H_5O_7Ag_3$
Silver citrate
Anti-infective dusting powder.

15434 *t*-butyl mercaptan
75-66-1 1613 200-890-2
$C_4H_{10}S$
2-Methyl-2-propanethiol
tert-butanethiol. Has skunk-like odor. Used as additive to natural gas for leak detection. mp = 1°; bp = 62-65°; d_4^{25}= 0.7943; n_D^{25}= 1.4198; slightly soluble in H_2O, soluble in organic solvents.

15435 Iupital® F10
Acetal copolymer; outstanding thermal and color stability, resistant to dimensional change, fatigue, wear, and corrosive environments. *Mitsubishi Gas.* Name unverified.

15436 Ivaleur
A proprietary trade name for pyroxylin (cellulose nitrate). No manufacturer.

15437 Ivarbase 98
Polysorbate 80, cetyl acetate, acetylated lanolin alcohol; cosmetics ingred. *Brooks Industries.*

15438 Ivarbase 101
Mineral oil, lanolin alcohol. easy-to-use oil-water emulsifier for cosmetics applications. *Brooks Industries.*

15439 Ivarbase 3210
Cetyl acetate, acetylated lanolin alcohol. light greaseless emollient for cosmetics. *Brooks Industries.*

15440 Ivarbase 3230
Mineral oil, PEG-30 lanolin, cetyl alcohol; absorb. base, cosmetics ingred.; makes very stable milks and creams. *Brooks Industries.*

15441 Ivarbase 3240
Isopropyl lanolate, lanolin oil, oleyl alcohol; absorp. base, cosmetics ingred. *Brooks Industries.*

15442 Ivarbase 3250
Isopropyl palmitate, lanolin oil; rich light emollient for cosmetics use. *Brooks Industries.*

15443 Ivarlan 3100
Lanolin oil; cosmetic ingredient. *Brooks Industries.*

15444 Ivarlan 3310
8027-33-6 232-430-1
Lanolin alcohol
Cosmetics ingredient; distilled grade. *Brooks Industries.*

15445 Ivarlan 3400
61790-81-6
PEG-75 lanolin
Cosmetics ingredient. *Brooks Industries.*

15446 Ivarlan 3406
61790-81-6
PEG-60 lanolin
Cosmetics ingredient. *Brooks Industries.*

15447 Ivarlan 3420
68458-88-8
PPG-12-PEG-50 lanolin
Cosmetics ingredient. *Brooks Industries.*

15448 Ivarlan 3450
68648-27-1
PEG-20 hydrogenated lanolin
Cosmetics ingredient. *Brooks Industries.*

15449 Ivarlan AWS
68458-58-8
PPG-12-PEG-65 lanolin oil; cosmetics ingredient. *Brooks Industries.*

15450 Ivarlan C-24
Choleth-24, ceteth-24; cosmetics ingredient. *Brooks Industries.*

15451 Ivarlan HL
8031-44-5 232-452-1
Hydrogenated lanolin; cosmetics ingredient. *Brooks Industries.*

15452 Ivarlan Light
Lanolin USP; cosmetic ingredient. *Brooks Industries.*

15453 Ivarlan OH
68424-66-8 270-315-8
Hydroxylated lanolin; cosmetic ingredient. *Brooks Industries.*

15454 Ivermectin
70288-86-7 5264 274-536-0
Eqvalan; Ivomec; 22,23-Dihydroabamectin; 22,23-dihydroavermectin B_1; MK-933; Cardomec; Cardotek-30; Heartgard 30; Mectizan; Zimecterin. Amixture of ivermectin component B1a and ivermectin component B1b; anthelmintic, insecticide andacaricide. λ_m = 238, 245 nm (MeOH); $[\alpha]_D^{25}$= 71.5° (c = 0.755 $CHCl_3$); soluble in H_2O, methyl ethyl ketone, propylene glycol, polyethlene glycol. *Merck & Co Inc.* Name unverified.

15455 ivoride
A casein product used as an electrical insulation.

15456 Ivorin-Profalon
Herbicide for beans and potatoes. *Hoechst UK.* Name unverified.

15457 Ivosit
Selective contact herbicide. *Hoechst UK.* Name unverified.

15458 Iwox
Oxidized microcrystalline wax; emulsion-based wax polish. *Industrial Waxes Ltd.*

15459 Ixan
Polyvinylidene chloride. *Laporte Industries Ltd.* Discontinued.

15460 Ixan®
Copolymers of vinylidene chloride and vinyl chloride or methylacrylate base; used for coextruded film, coatings on plastic film, coatings on paper. *Solvay Polymers.*

15461 Ixef® 1022
Polyarylamide
Characterized by shock-absorbing properties, low creep, high density and modulus; used for parts subjected to heavy vibrations in automotive (body parts, under-the-hood parts), electromechanical and electronic; (electric motors, alternators, machine tools, high fidelity equipment housings). *Solvay Polymers.*

15462 Ixol
Halogenated polyols. Solvents. *Laporte Industries Ltd.* Discontinued.

15463 Ixolite
A resin found in Austria.

15464 Ixper
Inorganic peroxides. *Solvay Interox Ltd.*

15465 Izal
A distillate from coke residues; it is a proprietary disinfectant. No manufacturer.

15466 J Slip NS-77
7631-86-9 8637 231-545-4
O_2Si
silicon dioxide
Modified silica; antislip agent; prevents fiber slippage; very effective on polyester; can be used over a wide pH range. *Sybron.*

15467 J Wet 19A
Nonrewetting wetting agent for flame retardant textile finishes. *Sybron.*

15468 J-13
14807-96-6 9207 238-877-9
Jet-milled talc; cosmetic raw material. *U.S. Cosmetics.*

15469 Jabclad
Molded expanded polystyrene panels; for insulation panel applied externally on masonry walls for subsequent rendering. *Vencel Resil Ltd.*

15470 Jabdec
Roof insulation board. *Vencel Resil Ltd.*

15471 Jabdie
Laminate of 12 mm fiberboard with expanded polystyrene; used for insulation of flat roofs under felt and mastic asphalt weatherproofing. *Vencel Resil Ltd.* Discontinued.

15472 Jablina Insulating Panels
Laminate of expanded polystyrene with aluminum/paper facings; used for insulation lining for factory buildings. *Vencel Resil Ltd.*

15473 Jabilite Thermodek
Dry roof screed system. *Vencel Resil Ltd.*

15474 Jablite
Expandable polystyrene. *Vencel Resil Ltd.*

15475 Jablite Cavity
Expanded polystyrene boards; used for partially filling the cavity in masonry wall construction. *Vencel Resil Ltd.*

15476 Jablite Flooring
Expanded polystyrene boards; used for underfloor thermal insulation. *Vencel Resil Ltd.*

15477 Jablite Insulation Board
Expanded polystyrene boards; for insulation of walls and floors, flat roof insulation. *Vencel Resil Ltd.*

15478 Jablite Thermacel
Expanded polystyrene beads; for insulation of masonry cavity walls, infill for 'bean bags'. *Vencel Resil Ltd.*

15479 Jablite Thermoclik
Laminate of bitumen roofing felt with expanded polystyrene; for insulation of flat roofs, used under felt and mastic asphalt weatherproofing. *Vencel Resil Ltd.* Discontinued.

15480 Jablite WallLok
Expanded polystyrene tongue and grooved panels for partial cavity full in house building. *Vencel Resil Ltd.*

15481 Jacana metal
An alloy of 70% lead, 20% antimony, and 10% tin.

15482 J-acid
6-Amino-1-naphthol-3-sulfonic acid
Used as an azo dye intermediate.

15483 Jacoby metal
An alloy of 85% tin, 10% antimony, and 5% copper.

15484 Jacquemart's reagent
An aqueous solution of mercuric nitrate with nitric acid; used as a test for ethyl alcohol.

15485 Jacutin®
58-89-9 5526 200-401-2
lindane
Insecticide preparation based upon Lindane. *E Merck.*

15486 Jaffamine®
A series of polyoxyalkylene-derived di-and triamines; epoxy curing agents, polymer flexibilizers, and specialty polyamides. *Texaco.*

15487 Jagalux
UV and electron-beam curing resins. *Ernst Jager GmbH.*

15488 Jagalyd
Alkyd, epoxy ester synthetic resins. *Emst Jager GmbH.*

15489 Jagapol
Polyester resins. *Emst Jager GmbH.*

15490 JagDril CC
Semi-synthetic polymeric viscosifier for completion, workover, and low solids drilling operations. *Rhône-Poulenc/Water Soluble Polymers.*

15491 Jagotex
Acrylic resins. *Emst Jager GmbH.*

15492 Jagotex Esi-Cryl
Acrylic resins for floor polishes, etc. *Ernst Jager GmbH.*

15493 Jaguar
benazolin-bromoxynil-ioxynil-mecoprop
An emulsifiable concentrate containing 22.2 g benazolin, 55.6 g bromoxynil, 27.8 g ioxynil and 413 g mecoprop per liter; a post-emergence herbicide for cereal crops and grass. *Schering Agrochemicals Ltd.* Discontinued.

15494 Jaguar® 413
39421-75-5
Hydroxypropyl guar gum; thickener for alcohol solutions. *Rhône-Poulenc/Water Soluble Polymers.*

15495 Jaguar® C
9000-30-0 4601 232-536-8

Guar gum. Used as a flocculant and coagulant, an ointment base and filler. *Rhône-Poulenc/Water Soluble Polymers.*

15496 Jaguar® C-13S, C-14S
65497-29-2
Guar hydroxypropyl trimonium chloride; Cationic Guar C-261; Chesguar C10, C10R,C17,C20, C20R; Guar hydroxypropyl trimonium chloride; Guar gum, 2-hydroxy-3-(trimethylammonio) propyl ether, chloride; Guar hydroxypropyl trimethyl ammonium chloride; Cosmedia Guar® C-14-S; N-Hance® 3000; Rhaballgum CG-M; Uniquart COSM GUAR. Thickener and conditioner for hair and skin care products, shampoos, creme rinses, lotions, creams. *Rhône-Poulenc Surf.*

15497 Jaguar® C-13S, C-14S
65497-29-2
Guar hydroxypropyltrimonium chloride
Thickener and conditioner for hair and skin care products, shampoos, creme rinses, lotions, creams. *Rhône-Poulenc Surf.*

15498 Jaguar® C-162
71329-50-5
Hydroxypropyl guar hydroxypropyl trimonium chloride. Conditioner for hair and skin care products, conditioning shampoos, bath gels, liquid soaps. *Rhône-Poulenc Surf.*

15499 Jaguar® C-162
71329-50-5
Hydroxypropyl guar hydroxypropyltrimonium chloride
Conditioner for hair and skin care products, conditioning shampoos, bath gels, liquid soaps. *Rhône-Poulenc Surf.*

15500 Jaguar® Guar Gum
9000-30-0 4601 232-536-8
Guar gum; hydrocolloid for food applications (baking, cereal, dairy/cheese, processed foods, beverages). *Rhône-Poulenc/Water Soluble Polymers.*

15501 Jaguar® HP 60
39421-75-5
Hydroxypropyl guar
Hydroxypropyl guar gum. Polymer providing thickening, lubricating properties, viscosity, suspension, and slip to aqueous or hydroalcoholic cosmetic systems (shampoos, creme rinses, lotions, creams). *Rhône-Poulenc/Water Soluble Polymers.*

15502 Jaguar® HP 8
39421-75-5
Hydroxypropyl guar
Polymer providing thickening, lubricating properties, viscosity, suspension, and slip to aqueous or hydroalcoholic cosmetic systems (shampoos, creme rinses, lotions, creams). *Rhône-Poulenc Surf.*

15503 Jaguar® HP-11
39421-75-5
Hydroxypropyl guar
Rhône-Poulenc Surf.

15504 jalcase
A steel with a high resistance to wear but with a soft core. It has forging properties.

15505 janthone
A synthetic perfume obtained by condensing citral or lippial with mesityl oxide. It has a violet odor.

15506 Janus
linuron-trifluralin
Liquid mixture of linuron [330-55-2] and trifluralin [1982-09-8]; herbicide for winter cereals. *Atlas Interlates Ltd.*

15507 Jaon
299-27-4 7796 206-074-2
Potassium gluconate
Replenisher. *Adria Laboratories Inc.* Name unverified.

15508 Japan camphor
76-22-2 1779 200-945-0
$C_{10}H_{16}O$
Laurel Camphor
Ordinary camphor, which separates from the essential oil of *Laurus camphora.*

15509 Japan sago
The starch from *Cycas revoluta.*

15510 Japan tallow
Sumach wax; vegetable wax; Japan wax. Japan wax, derived from *Rhus vernicifera* and *Rhus sylvestric*; used for candles, floor waxes, polishes, substitute for beeswax, food packaging.

15511 Japan varnishes
These are obtained by blending asphalt varnishes with dark colored copal or amber varnishes.

15512 Japanese acid clay
Kambara earth. A clay with the formula $Al_2O_3{\cdot}6SiO_2{\cdot}xH_2O$ (x > 6). It has powerful adsorptive and decolorizing properties and a strong dehydrating action.

15513 Japanese bell metal
An alloy of 60.5% copper, 18.5% tin, 12% lead, 6% zinc, and 3% iron.

15514 Japanese bronze
An alloy of from 81-83% copper, 10% lead, 4.6% tin, and 0-1.8% zinc.

15515 Japanese Silver
An alloy of 50% aluminum and 50% silver.

15516 Japidermic
A microbial insecticide in powder form containing viable spores of *Bacillus popilllae*, a specific pathogen which infects and kills Japanese beetle grubs; ready to-use; for control of Japanese beetle grubs; only one application is needed as the living spores are self-perpetuating. *Fairfax Biological Laboratory Inc.*

15517 JAQ Powdered Quaternary
139-08-2 205-352-0
$C_{22}H_{40}ClN$
Myristalkonium chloride
N,N-Dimethyl-N-tetradecylbenzene-methanaminium chloride; Myristyl dimethyl benzyl ammonium chloride; Tetradecyl dimethyl benzyl ammonium chloride; Arquad® DM14B-90; Barquat® MX-50, MX-80; BTC® 824; FMB 451-8 Quat; Kemmamine® Q-7903B; Zephiramine; Benzenemethanaminium, N,N-dimethyl-N-tetradecyl-, chloride; C_{14} benzyl dimethyl ammonium chloride; C_{14} dimethyl benzyl ammonium chloride; Benzenemethaminium, N-tetradecyl-N,N-dimethyl, chloride; Benzyl dimethyl tetradecyl ammonium chloride; C_{14}-alkylbenzyldimethylammonium chloride; Roccal MC-14; Tetradecyl dimethyl benzyl ammonium chloride; Miristalkonium chloride. For formulation of disinfectants, sanitizers, and swimming pool algicides. Soluble in H_2O. *Huntington Lab.*

15518 jara jara
93-04-9 6076 202-213-6
$C_{11}H_{10}O$
2-Methoxynaphthalene
β-Naphthyl methyl ether; nerolin old; yara-yara. Used in perfumery. mp = 73-75°; bp = 274°; insoluble in H_2O, soluble in organic solvents.

15519 Jarcal
10043-52-4 1699 233-140-8
$CaCl_2$
Calcium Chloride
Calcium Chloride, dihydrate; Unichem CALCHLOR. Food grade calcium chloride. *Jarchem Industries Inc.*

15520 Jarfix 391
Cellulose fixing agent; cationic auxiliary used in the Jarofast system of cationic dyeing for cellulosic fibers. *James Robinson & Co Ltd.*

15521 Jargonelle pear essence
123-92-2 5125 204-662-3
$C_7H_{14}O_2$
isoamyl acetate
A solution of isoamyl acetate in ethyl alcohol; used for flavoring confectionery.

15522 Jarofast
System name for the batchwise application of solubilized sulfur dyes by a cationic dyeing system, using a cationic auxiliary (Jarofix 391); the system is used in the garment dyeing industry for cellulosic fibers. *James Robinson & Co Ltd.*

15523 Jarosol
Solubilized sulfur dyestuffs; used in the Jarofast system of cationic dyeing for cellulosic fibers. *James Robinson & Co Ltd.*

15524 Jarozyme 491
Cationic reduction enzyme; enzyme used in the Jarofast system of cationic dyeing for cellulosic fibers to achieve the washdown effect. *James Robinson & Co Ltd.*

15525 Jarytherm
Heat transfer fluids. *Elf Atochem SA.*

15526 Jascitile
3-Methyloctanitrile
Quest Int'l. UK Ltd.

15527 Jasilyn
4-Acetoxy-3-pentyltetrahydropyran
Quest Int'l. UK Ltd.

15528 Jasmacyclat
61699-38-5 262-912-7
$C_{10}H_{18}O_3$
Methylcyclooctylcarbonate
Fragrance raw material for floral notes. *Henkel/Cospha.*

15529 Jasmacyclene
2500-83-6 219-700-4
$C_{12}H_{16}O_2$
Tricyclo decenyl acetate

Greenyl acetate; Tricyclodecenyl acetate; Jasmacyclene; Verdyl acetate; Hexahydro-4,7-methanoinden-5(or 6)-yl acetate; Cyclacet; Herbyl acetate. A perfumery specialty. *Quest Int'l. UK Ltd.*

15530 Jasmal
That fraction of the essential oil of jasmine flowers distilling at 100°C.

15531 Jasmatone
2-n-Hexyl cyclopentanone.
Used in perfumery. *Quest Int'l. UK Ltd.*

15532 Jasmolide
A perfumery chemical. *PPF International Ltd.* Name unverified.

15533 Jasmopyrane
18871-14-2 242-640-5
$C_{13}H_{24}O_4$
4-Acetoxy-3-pentyltetrahydropyran
Jasmin acetate; Pentyltetrahydropyranyl acetate. Used in perfumery (fresh, sweet, oily, jasmin, watery). *Quest Int'l. UK Ltd.*

15534 Jasmorange®
2-Methyl-3-(4-methylphenyl) propanal
Fragrance (fruity, balsamic, green, florl, aldehydic). *BASF AG.*

15535 Jatex
A proprietary brand of pure concentrated rubber latex, 60%. No manufacturer.

15536 Jatob duro
A hard copal obtained from Ceara and Northern Bahia, Brazil; used in varnishes.

15537 Jatob lagrima
Trapoc resin. A rather soft copal from the Jatoba tree. Used in spirit varnish.

15538 Jatob resin
Brazilian copals from *Hymenoea courbaril* and *Hymenoea parvifolia*. There are hard and soft qualities. The soft is called jatob, tean, and trapoc. Used in varnishes.

15539 Java wax
Sumatra wax; Gondang wax; Kondang wax; Getah wax. A wax obtained from the bark of the gondang (wild fig) tree, *Ficus variegata* .

15540 Javelin
diflufenican-isoproturon
Suspension concentrate containing 62.5 g diflufenican and 500 g isoproturon per liter; used for control of weeds in winter cereals. *Rhône-Poulenc Crop Protection Ltd.*

15541 Jaydalene
A proprietary soldering paste consisting of orthophosphoric acid with a base which vaporizes without decomposition, e.g., aniline, etc. No manufacturer.

15542 Jayflex®
Plasticizers. *Exxon Int'l.*

15543 Jayflex® 77
71888-89-6 276-158-1
Diisoheptyl phthalate
Plasticizer d = 0.995; fp.= 390°F (TCC). *Exxon.* Name unverified.

15544 Jayflex® 210
64742-53-6 265-156-6
hydrotreated light naphthenic distillate; Distillates (petroleum), hydrotreated light naphthenic. Secondary plasticizer. *Exxon.* Name unverified.

15545 Jayflex® 215
64742-14-9 265-114-7
Petroleum distillates. Secondary plasticizer. *Exxon.* Name unverified.

15546 Jayflex® 911
Dinonyl undecyl linear phthalate; plasticizer. *Exxon.* Name unverified.

15547 Jayflex® 3209
A proprietary ester of adipic acid. A plasticizer *Exxon.* Name unverified.

15548 Jayflex® 7911
Diheptyl, nonyl, undecyl linear phthalates; plasticizer. *Exxon.* Name unverified.

15549 Jayflex® DHP
68515-50-4 201-559-5
Dihexyl phthalate
Di(2-ethylbutyl)phthalate; 1,2-Benzenedicarboxylic acid, dihexyl ester. Plasticizer. bp_{735} = 350°; d = 1.008; fp = -50°; flash point = 380°F (TCC). *Exxon.* Name unverified.

15550 Jayflex® DIDP
68515-49-1 271-091-4
$C_{24}H_{38}O_4$
Diisodecyl phthalate
Plasthall® DIOP; Staflex DIOP. Plasticizer mp = -50°; bp = 370°; insoluble in H_2O; d_{20}^{20} = 0.980-0.983; LD_{50} (rat orl) = 22 g/kg. *Exxon.* Name unverified.

15551 Jayflex® DINA
33703-08-1 251-646-7
Diisononyl adipate
Plasticizer *Exxon.* Name unverified.

15552 Jayflex® DINP
68515-48-0 271-090-9
Diisononyl phthalate
DINP; 1,2-Benzenedicarboxylic acid, diisonyl ester; Diplast® N; Palatinol® N; Plasthall® DINP; PX-139. Plasticizer d = 0.973; flash point (TCC) = 415°F. *Exxon.* Name unverified.

15553 Jayflex® DIOP
68515-48-0 271-090-9
Diisononyl phthalate. Plasticizer *Exxon.* Name unverified.

15554 Jayflex® DOA
103-23-1 203-090-1
Dioctyl adipate
Plasticizer *Exxon.* Name unverified.

15555 Jayflex® DTDP
119-06-2 204-294-3
$C_{34}H_{58}O_4$
Ditridecyl phthalate
Nuoplaz® DTDP; PX-126; Staflex DTDP; Undecyl dodecyl phthalate; DTDP; 1,2-Benzenedicarboxylic acid, ditridecyl ester. Plasticizer. d = 0.951 (20/20°); bp > 285° (5mm); fp = 470°F (OC). *Exxon.* Name unverified.

15556 Jayflex® DUP
3648-20-2 222-884-9
$C_{30}H_{50}O_4$
Diundecyl phthalate
Jayflex® L11P; PX-111; Diundecyl linear phthalate; 1,2-Benzenedicarboxylic acid, diundecyl ester; Santicizer 711. Plasticizer. d = 0.954; fp = 2-7°; fp = 460°F (TCC); insoluble in H_2O, soluble in organic solvents. *Exxon.* Name unverified.

15557 Jayflex® L9P
Linear phthalate; high permanence plasticizer for extrusion and molding film, sheet and coated fabric for automotive, weather stripping, pool liners, membranes, tarps, specialty wire and cable applications. *Exxon.*

15558 Jayflex® TINTM
53894-23-8 258-847-9
Triisononyl trimellitate
Plasthall® TIOTM; PX-339; Staflex TIOTM. Plasticizer d = 0.979; pour point = -40°; flash point = 465°F (TCC); n_D^{20} = 1.484. *Exxon; Aristech; Unitex.* Name unverified.

15559 Jayflex® TOTM
89-04-3 201-877-4
$C_{33}H_{54}O_6$
Trioctyl trimellitate
TOTM; Tri (2-ethylhexyl) trimellitate; ADK CIZER C-8; Diplast® TM; Diplast® TM8; Kodaflex® TOTM; Nuoplaz® 6959; Nuoplaz® TOTM; Palatinol® TOTM; Plasthall® TOTM; PX-338; Staflex TOTM; Uniflex® TOTM. Plasticizer d_{20}^{20} = 0.989; bp = 414°; f.p. - 38°; fp = 263°. *Exxon.* Name unverified.

15560 Jayflex® UDP
68515-47-9 271-089-3
Undecyl dodecyl phthalate; plasticizer *Exxon. Name unverified.*

15561 Jazz®
Fiber. *DuPont UK.*

15562 JB-4
Plastic embedding kit; used for light microscopy. *Polysciences Inc.*

15563 J-Black-20
Silicone color masterbatch. *Dow Corning STI.*

15564 Jectoflo
Lime based fluxes for desulfurisation of steel by deep ladle injection. *Foseco (F.S.) Ltd.*

15565 Jectomag
Magnesium based powders for sulfur removal from blast furnace iron. *Foseco (F.S.) Ltd.*

15566 Jectothane
A proprietary polyester-based polyurethane thermoplastic injection-molding compound. No manufacturer.

15567 Jeffamine® BuD-2000
Urea condensate of POP polyamine; epoxy modifier; nonreactive additive used in concentrations of 5-20 parts per hundred to provide enhancement of metal-to-metal adhesion. thermal shock properties; results in increased elongation, higher impact and tensile strength, and lowered modulus, while heat deflection values are only slightly affected; reactive with formaldehyde to produce polymeric materials. *Texaco.*

15568 Jeffamine® D-2000
POP diamine; epoxy curing agent and modifier usable alone or in combination. *Texaco.*

15569 Jeffamine® DU-700
Urea condensate of POP polyamine; epoxy curing agent usable alone or in combination. *Texaco.*

15570 Jeffox
A series of poly(oxyethylene) glycols and poly(oxypropylene) glycols and triols; flexibilizers, humectants, and intermediates. *Texaco.*

15571 Jel-O-Mer®
Liquid alkyd resins and solutions. *Reichhold.*

15572 Jel-O-Mer® 46-902
Flow control agent. *Reichhold.*

15573 Jelonet
Paraffin gauze dressing. *Smith & Nephew Pharmaceuticals Ltd.* Name unverified.

15574 Jelutong
Pontianac; Fluvia; Gambia. A resinous latex yielded by *Dyera costulata*. It contains from 19-24% of rubber and 75-80% of resin, and is used for mixing with rubber and for other purposes.

15575 Jenner's stain
A microscopic stain for white blood corpuscles. It consists of a) a solution of water-soluble, yellowish eosin, 0.5 g, in 100 ml methyl alcohol, and b) a solution of methyl blue, 0.5 g, in 100 ml methyl alcohol. For use 25 ml of a) are mixed with 20 ml of b.

15576 Jer Dri WRN
Wax/Zirconium salt complex; semidurable water repellent; effective on all fibers with excellent high temperature resistance. *Sybron.*

15577 Jerotex P
Methylated urea-formaldehyde resin; high concentration thermoset resin used as low cost stiffening agent for synthetic fabrics. *Sybron.*

15578 Jersey lily white
A pigment; a lithopone.

15579 Jessate
Ethyl-2-hexylacetoacetate
Quest Int'l. UK Ltd.

15580 Jesuit's balsam
8001-61-4 2580 232-288-0
Balsam copaiba; balsam capivi; Copaifera langsdorffi oil; South american spp. of copaifera l. oil; Copaiba oil. Copaiba, an oleo-resin from South American species of *Copaifera (Copaiba) Leguminosae.* Contains illuric acd, metacopaivic acid, copaivic acid and oxycopaivic acid. Used in varnishes and varnish removers and in perfumery (pepper; spice; violet; woody bases; balsam). The oily liquid is pale yellow to yellow green or bluish. The odor is similar to the balsam but much milder, sweeter, almost creamy balsamic. d = 0.930-0.995; insoluble in H_2O, soluble in organic solvents.

15581 Jet
A mineral that is a fossilized wood, and falls between lignite and coal; used for ornaments.

15582 Jet Amine DC
61791-63-7 263-195-3
Amines, n-cocoalkyltrimethylenedi-; Cocoalkyltrimethylenediamine; Coop turbex; Onyxide; Turbex. Cocodiamine; emulsifier, fuel oil and gasoline additives, corrosion inhibitors, mineral flotation. *Jetco.*

15583 Jet Amine DE 810
Octadecyl ether diamine
Emulsifier, corrosion inhibitor. *Jetco.*

15584 Jet Amine DE-13
22023-23-0 244-726-8
Tridecyl ether diamine
Emulsifier, corrosion inhibitor. *Jetco.*

15585 Jet Amine DMCD
61788-93-0 263-020-0
Dimethyl cocoamine
Intermediate for quaternaries, surfactants, agriculture, and detergent formulations. *Jetco.*

15586 Jet Amine DMOD
14727-68-5 238-781-7
Dimethyl oleamine
Dimethyl oleylamine; Oleyl dimethylamine; Armeen® DMOD; Crodamine 3.AOD; Kemamine® T-9892D. Intermediate for quaternaries, surfactants, agriculture, and detergent formulations. *Jetco.*

15587 Jet Amine DMSD
61788-91-8 263-017-4
Dimethyl soya amine
Intermediate for quaternaries, surfactants, agriculture, and detergent formulations. *Jetco.*

15588 Jet Amine DMTD
68814-69-7 272-339-4
Dimethyl tallowamine
Tallow dimethylamine; Tallow alkyl dimethylamine; Armeen® DMTD; Kemamine® T-9742D; Noram DMSD. Intermediate for quaternaries, surfactants, agriculture, and detergent formulations. *Jetco.*

15589 Jet Amine DO
7173-62-8 230-528-9
Oleyl propylene diamine
Emulsifier, fuel oil and gasoline additives, corrosion inhibitors, mineral flotation. *Jetco.*

15590 Jet Amine DT
61791-55-7 263-189-0
Tallow diamine
Amines, n-tallow alkyltrimethylenedi-; N-Tallowalkyl-1,3-propanediamine; N-(Tallowalkyl)trimethylenediamine; N-Tallow-1,3-propylenediamine. Emulsifier, fuel oil and gasoline additives, corrosion inhibitors, mineral flotation. *Jetco.*

15591 Jet Amine M2C
61788-62-3 262-990-2
Methyl dicocamine
Dicocomethylamine. Intermediate for quaternaries, surfactants, agriculture, and detergent formulations. *Jetco.*

15592 Jet Amine PC
61788-46-3 262-977-1
Cocamine
Corrosion inhibitor, flotation agent, emulsifier, mold release agent, lube oil additive, fertilizer anticaking agent, fabric finishing. *Jetco.*

15593 Jet Amine PE 1214
68511-41-1 270-939-0
Dodecyl/tetradecyl ether amine
Emulsifier, corrosion inhibitor. *Jetco.*

15594 Jet Amine PHT
61788-45-2 262-976-6
Hydrogenated tallow amine
Corrosion inhibitor, flotation agent, emulsifier, mold release agent, lube oil additive, fertilizer anticaking agent. *Jetco.*

15595 Jet Amine PO
112-90-3 204-015-5
$C_{18}H_{37}N$
Oleamine
octadecenylamine; (Z)-9-Octadecen-1-amine; 1-Amino-9-octadecene. Corrosion inhibitor, flotation agent, emulsifier, mold release agent, lub oil additive, fertilizer anticaking agent. mp = 15-22°; d = 0.8130; n_D^{20} = 1.4578; insoluble in H_2O, soluble in organic solvents. *Jetco.*

15596 Jet Amine PS
61790-18-9 263-112-0
Soyamine
Corrosion inhibitor, flotation agent, emulsifier, mold release agent, lube oil additive, fertilizer anticaking agent. *Jetco.*

15597 Jet Amine PT
61790-33-8 263-125-1
Tallowamine
Corrosion inhibitor, flotation agent, emulsifier, mold release agent, lube oil additive, fertilizer anticaking agent. *Jetco.*

15598 Jet Amine TET
68911-79-5 272-787-0
Oleotripropylene tetraamine
Tallow tetramine; gasoline detergent, corrosion inhibitor, in petroleum products, dispersion agents for mineral pigments in organic vehicles, asphalt emulsifier. *Jetco.*

15599 Jet Amine TP
Tallow pentamine
Gasoline detergent, corrosion inhibitor, in petroleum products, dispersion agents for mineral pigments in organic vehicles, asphalt emulsifier. *Jetco.*

15600 Jet Amine TRT
61791-57-9 263-191-1
Tallow triamine
Gasoline detergent, corrosion inhibitor, in petroleum products, dispersion agents for mineral pigments in organic vehicles, asphalt emulsifier. *Jetco.*

15601 Jet Jel®
Flocculant for settling and clarifying reserve mud pits; biodegradable. *Rhône-Poulenc/Water Soluble Polymers.*

15602 Jet Quat 2C-75
61789-77-3 263-087-6
Dicoco dimethyl ammonium chloride; bactericide, textile softener, asphalt emulsifier, petroleum processing. *Jetco.*

15603 Jet Quat C-50
61789-18-2 263-038-9
Coco trimethyl ammonium chloride

Cocotrimonium chloride. Bactericide, textile softener, asphalt emulsifier, petrol. processing; home and personal care products. *Jetco.*

15604 Jet Quat DT-50
68607-29-4 271-762-1
Pentamethyl tallow propane diammonium dichloride
Pentamethyl tallow propane diammonium dichloride in isopropanol; Methyl quaternary of tallow diamine; bactericide, textile softener, asphalt emulsifier, petroleum processing. *Jetco.*

15605 Jet Quat S-50
61790-41-8 263-134-0
Soya trimethyl ammonium chloride
Soytrimonium chloride. Bactericide, textile softener, asphalt emulsifier, petroleum processing. *Jetco.*

15606 Jet Quat T-50
8030-78-2 232-447-4
Tallow trimethyl ammonium chloride
Tallow trimonium chloride. Bactericide, textile softener, asphalt emulsifier, petroleum processing. *Jetco.*

15607 Jetfill 700C
14807-96-6 9207 238-877-9
Talc; a filler and extender. *Luzenac Am.*

15608 Jet-Flex® 101
Weatherable engineering plastic; for camper tops, spas, extrusion profile; stabilized for long-term outdoor weathering; can be coextruded with ABC substrates for thermoforming applications. *Multibase.*

15609 Jet-Lube J-75®
Lead-free drill steel lubricant for percussion rock drilling, blast hole drilling, road construction, logging, mining, coal drilling, pneumatic and tract drilling. *Jet-Lube.*

15610 Jeunite
Copper-based fungicide. *Murphy Chemical Co Ltd.* Discontinued.

15611 Jeweller's rouge
1309-37-1 4072 215-168-2
Fe_2O_3
ferric oxide
Red iron oxide; C.I. 77491; Rouge; Iron oxide fume; Iron (III) oxide; Ferric oxide red; Iron oxide; Iron oxide (Fe2O3); diiron trioxide; English iron oxide red; Iron oxide red; Pigment red 101; Iron(III) oxide dihydrate. The finest calcined ferric oxide or hematite. Used as an abrasive. mp = 1538°; d = 5.24; insoluble in H_2O, organic solvents.

15612 Jeyes disinfectant
A disinifectant containing creosote, rosin, caustic soda and water. It forms emulsions with water.

15613 JL 43155AS
Silver pigment, epoxy/phenolic binder in MEK:toluene (1:1); EMC shielding coating for metals; protects sensitive electronic equipment. *Acheson Colloids.*

15614 JL 43176
Silver pigment, epoxy/amine binder; EMC shielding coating for metals; protects sensitive electronic equipment. *Acheson Colloids.*

15615 Jodomiron
440-58-4 5031 207-125-1
Iodamide, N-Methyl-D-glucamine salt
Diagnostic aid. *Bracco Industria Chimica SpA.* Name unverified.

15616 Jogral
61791-44-4 263-177-5
Cationic surfactant containing 800 g/l tallow amine ethoxylate; wetting agent for phosphonoglycine herbicide sprays. *Ideal Manufacturing Ltd.*

15617 jojoba oil
61789-91-1 5279
Oils, jojoba; Oil of Jojoba. A liquid wax ester from the seeds of the Jojoba desert shrub *Simmondsia chinensis*; emollient, conditioner, and lubricant for cosmetics and toiletries. *Arista; R.W. Greeff; Jojoba Growers & Processors; Lipo.*

15618 Jojobeads
Jojoba oil derivatives; used for cosmetics & toiletries. *A & E Connock (Perfumery & Cosmetics) Ltd.*

15619 Jonylon
Nylon 6 and 66 molding compounds. *BIP Chemicals Ltd.*

15620 Jordapon® CI-50 Disp
Sodium cocoyl isethionate
Mild foaming surfactant for shampoos, bubble baths, creams, and lotions. *PPG/Specialty Chem.* Discontinued.

15621 Jordaquat® 350
Benzalkonium chloride
Used for products requiring bacteriostatic, germicidal, and algicidal activity; also for static elimination at low use levels; used to disinfect hard surfaces, sanitize food contact surface and fabrics, algae control in water systems. *PPG/Specialty Chem.* Discontinued.

15622 JR Surfacer
Water soluble epoxy; for tool/pattern making applications and filler. *J R Technology Ltd.*

15623 JR-228, JR-228-1
Two-component copolymers, one of which is reacted with a high molecular weight epoxide; semiflexible adhesive systems; superior bonds to thermoplastics, elastomers, glass, metals (outstanding adhesion to PC, polyesters, nylon, ABS, PVC, and acrylics). *Bacon.*

15624 J-Red-10
Silicone color masterbatch. *Dow Corning STI.*

15625 J-Red-12FS
Fluorosilicone color masterbatch. *Dow Corning STI.*

15626 J-Soft 111E
Fatty amide
Softener producing luxurious soft hand with no effect on crocking. *Sybron.*

15627 JSR-10
A proprietary ABS material possessing high impact strength. *Japan Syn. Rubber.* Name unverified.

15628 JSR-12
A proprietary ABS material possessing high impact strength. *Japan Syn. Rubber.* Name unverified.

15629 JSR-21
A proprietary ABS resin capable of giving good surface finish in molding operations. *Japan Syn. Rubber.* Name unverified.

15630 Jubilee®
For the agriculture industry. *Du Pont UK.*

15631 Juglone
481-39-0 5282 207-567-5
$C_{10}H_6O_3$
5-Hydroxynaphthoquinone
5-Hydroxy-1,4-naphthoquinone; 5-Hydroxy-*p*-naphthoquinone; 5-Hydroxy-1,4-naphthalenedione; 8-Hydroxy-1,4-naphthalenedione; 8-hydroxy-1,4-naphthoquinone; C.I. 75500; C.I. Natural Brown 7; nucin; regianin. Used as a pigment and a chemical intermediate. mp = 161-163°; λ_m = 420 nm (log ε 3.56 MeOH); soluble in hot H_2O, organic solvents; LD_{50} (rat orl) = 112 mg/kg.

15632 Julin's chloride
118-74-1 4714 204-273-9
C_6Cl_6
Hexachlorobenzene
perchlorobenzene; HCB; hexa c.b.; Anticarie; Bunt-cure; Bunt-no-more; Ceku C.B.; No Bunt; pentachlorophenyl chloride; phenyl perchloryl; No Bunt 40; Julian's carbon chloride; No Bunt 80; Sanocide; Smut-Go; amatin; co-op hexa; Granox NM; Snieciotox. Different from benzenehexachloride (lindane). Used in organic synthesis and formerly in agriculture as a fungicide. mp = 227-229°; bp = 332°; insoluble in H_2O, soluble in organic solvents; LD_{50} (rat orl) = 10 g/kg.

15633 Jupital
1897-45-6 2219 217-588-1
Chlorothalonil
A fungicide for a wide range of agricultural crops. *Fermenta ASC Europe Ltd.*

15634 Justice®
For the agriculture industry. *DuPont UK.*

15635 justite
A substance approximating to the formula, $(Ca{\cdot}Mg{\cdot}Fe{\cdot}Zn{\cdot}Mn{\cdot})_3Si_2O_7$, found in furnace slag. The name was formerly applied to the mineral Koenite.

15636 jutahycica
jutahy
A copal from Brazil. It is obtained from the roots of *Hymenoea courbaril* and *Hymenoea Parvifolia.*

15637 jutahycica resins
paragum; resina animé. Brazilian copal resins from *Hymenoea courbaril* and *Hymenoea parvifolia.*

15638 JZF
74-31-7 3388 200-806-4
$C_{18}H_{16}N_2$
N,N'-Diphenyl-*p*-phenylenediamine
Agerite DPPD; Diphenyl PPD; DPPD; Flexamine G; Nonox DPPD; Diafen; Diafen FF; Altofane DIP; Nonflex H; Permanax 18; Stabilizer DPPD; Nocrac DP; Permanax DPPD; DFFD; Antage DP; Ekaland DPPD; Naugard J. An antioxidant for use in rubber, polyethylene, petroleum and vegetable oils; in natural rubber, it protects against copper and manganese and gives protection against outdoor flexing and static weather cracking; protects against thermal oxidation in polyethylene, inhibits gum formation and degradation at elevated temperatures in petroleum oils. mp = 146-148°; $bp_{0.5}$ = 220-225°; insoluble in H_2O, soluble in organic solvents; LD_{50} (rat orl) = 2370 mg/kg. *Uniroyal.* Name unverified.

15639 K 129
1318-93-0 6341 215-288-5
$R^{+}_{0.33}(Al,Mg)_2Si_4O_{10}(OH)_2 \cdot nH_2O$ where R = Na^+, K^+, Mg^{2+}, Ca^{2+}
White montmorillonite
Binder and plasticizer for ceramic formulations; ion exchange builder in detergents; thixotropic agent for liq. soaps; flocculant for water treatment. Industrial chromatographic techniques Bulk d = 300-370 g/l. *Kaopolite.*

15640 K 129-H
1302-78-9 1082 215-108-5
$Al_2O_3 \cdot 4SiO_2 \cdot H_2O$
White bentonite
Wilkinite. Economical thickener and suspending agent for liq. abrasive cleaning compounds, water treatment; retention aid for paper. *Kaopolite.*

15641 K de Krizia
A fragrance from Milan's foremost name in fashion design. *Richardson-Vicks Inc.* Name unverified.

15642 K. Tab
7447-40-7 7783 231-211-8
KCl
Potassium chloride
Chloropotassuril; Diffu-K; Enseal; Kaleorid; Kalitabs; Kalium-Duriles; Kaon-Cl; Kaskay; Kayback; Kay-Cee-L; K-Contin; Klor-Con; K-Norm; K-Tab; Lento-Kalium; Micro K; Nu-K; Peter-Kal; PfiKlor; Rekawan; Repone K; Slow-K; Span-K. Replenisher. d = 1.98; mp = 773°. *Abbott Laboratories.* Name unverified.

15643 K.A. alloy
An aluminum alloy resembling duralumin.

15644 K.L.X.
68334-28-1 269-820-6
Partially hydrogenated vegetable oil (cottonseed, soybean); icing stabilizer; adds solids to shortening; fat encapsulatin. *Van Den Bergh Foods.*

15645 K.S. magnet steel
A cobalt steel containing 35% cobalt. It is suitable for short magnets.

15646 K.S. powder
A 42-grain powder; an explosive.

15647 K154
Aluminium-based clear masonry waterproofing solution; treats masonry, slates, tiles and all porous substrates. *Liquid Plastics Ltd.*

15648 K285
Absorbable dusting powder. *The Boots Co plc.* Discontinued.

15649 kabaite
A mineral wax of the oxokerite type.

15650 Kabikinase
9002-01-1 8981 232-647-1
Streptokinase
Streptococcal fibrinolysin; plasminokinase; Streptase. A proprietary preparation of streptokinase; a fibrinolytic agent. *KabiVitrum AB.* Name unverified.

15651 kachin
120-80-9 8183 204-427-5
$C_6H_6O_2$
Pyrocatechol
1,2-Benzenediol; pyrocatechin; catechol; 1,2-dihydroxybenzene. A photographic developer. Its active constituent is pyrocatechol. d = 1.344; mp = 104-106°; bp_{760} = 245.5°; bp_{400} = 221.5°; bp_{200} = 197.7°; bp_{100} = 176°: bp_{60} = 161.7°; bp_{40} = 150.6°; bp_{20} = 134°; bp_{10} = 118.3°; bp_5 = 104°.

15652 K-acid
$C_{10}H_9NO_7S_2$
5-Amino-4-hydroxynaphthalene-1,7-disulfonic acid
Used as an azo dye intermediate.

15653 Kadel® E-1000
Polyketone
High performance thermoplastic for high temperature applications in chemical processing, aviation/aerospace composites, advanced electrical/electronic uses, oil drilling, self-lubricating sleeve bearings, antifriction parts, seals, backup rings, extruded shapes for use in hot corrosive media, wire and cable coating, films, structural parts, excellent solvent and hydrolytic resistance, low smoke and toxicity. *Amoco Chemical Co.*

15654 kaempferol
520-18-3 5288 208-287-6
$C_{15}H_{10}O_6$
3,5,7-Trihydroxy-2-(4-hydroxyphenyl)-4H-1-benzopyran-4-one
3,4',5,7-tetrahydroxyflavone; nimbecetin; pelargidenolon 1497; populnetin; rhamnolutein; robigenin; swartziol; trifolitin. The coloring matter of the blue flowers of *Delphinium consolida.* It is a trihydroxyflavonol. mp = 276-278°; λ_m = 265, 365 nm.

15655 Kafil
52645-53-1 7321 258-067-9
$C_{21}H_{20}Cl_2O_3$
Permethrin
3-(2,2-Dichloroethenyl)-2,2-dimethylcyclopropanecarboxylic acid (3-phenoxyphenyl)methyl ester; 3-(phenoxyphenyl)methyl (±)-*cis,trans*-3-(2,2-dichloroethenyl)-2,2-dimethylcyclopropanecarboxylate; *m*-phenoxybenzyl(±)-*cis,trans*-3-(2,2-dichlorovinyl)-2,2-dimethylcyclopropanecarboxylate: FMC-33297; NIA-33297; NRDC-143; PP-557; SBP-1513; S-3151; Ambush; Corsair; Dragnet; Ectiban; Eksmin; Nix; Pulvex; Pounce; Pynosect; Ridect PourOn. Insecticide containing permethrin. mp around 35°; $bp_{0.005}$ = 220°; d^{20} = 1.190-1.272. *ICI Chem & Polymers Ltd.*

15656 Kagolin 5.8FG
Amitrole + atrazine + diuron; a granular mixture of herbicides for weed control. *Ciba-Geigy Agrochemicals.*

15657 kainite
1318-72-5
$K_2Mg(SO_4)_2.MgCl_2.6H_2O$
A salt found in the Stassfurt deposits, consisting mainly of potassium magnesium sulfate and magnesium chloride. The crude material consists of a mixture of kainite and rock salt, and contains 23% of potassium; used in chemicals and fertilizers.

15658 Kairoline
N-Ethyltetrahydroquinoline

15659 Kaiserling solution
A solution used for preserving tissue. It contains 3 g potassium acetate, 1 g potassium nitrate, 75 ml water, and 30 ml formaldehyde.

15660 Kaladex
PEN film. *Imperial Chemical Industries plc.*

15661 Kalammon
A fertilizer containing 17% nitrogen and 30% calcium carbonate.

15662 Kalar® 5214
9010-85-9
$[-C(CH_3)_2CH_2-]_x-[CH_2CH=C(CH_3)CH_2-]_y$
Poly(isobutylene-co-isoprene)
butyl rubber. Cross-linked butyl composition; produces nonsagging butyl-based sealants, e.g., automotive windshield tape, hot melt sealant; as base for butyl mastics Viscosity (Mooney, ML 1+8, 100°) = 42-52. *Hardman.*

15663 Kalar® 5263
9010-85-9
$[-C(CH_3)_2CH_2-]_x-[CH_2CH=C(CH_3)CH_2-]_y$
Poly(isobutylene-co-isoprene)
More highly cross-linked butyl rubber composition; used in elastic sealants requiring more resistance to flow or sag; as green strength enhancer for uncured butyl rubber compositions. Viscosity (Mooney, ML 1+8, 100°) = 42-52. *Hardman.*

15664 Kalbord
Insulating feeder head liners supplied in the form of flexible boards. *Foseco (F.S.) Ltd.*

15665 Kalcrete
Refractory castable for use in foundry ladles. *Foseco (F.S.) Ltd.*

15666 Kalene® 800
9010-85-9
$[-C(CH_3)_2CH_2-]_x-[CH_2CH=C(CH_3)CH_2-]_y$
Poly(isobutylene-co-isoprene)
Low molecular weight flowable butyl rubber derived from virgin butyl rubber; polymer used in sealants, coatings, electrical encapsulating compounds, and conformal coatings; gray; often used in solv. systems, e.g. toluene and Varsol 18. Viscosity (Mooney, ML 1+8, 100°) = 42-52. *Hardman.*

15667 Kaleoilris
A proprietary filling compound. No manufacturer.

15668 Kalex 220 Crystal
64-02-8 3557 200-573-9
$C_{10}H_{12}N_2Na_4O_8 \cdot 2H_2O$
Edetate sodium
N,N'-1,2-Ethanediylbis[N-(carboxymethyl)glycine] tetrasodium salt; (ethylenedintrilo)tetraacetic acid tetrasodium salt; ethylenediaminetetraacetic acid tetrasodium salt; sodium edetate; tetrasodium ethylenediaminetetraacetate; ethylenebis(iminodiacetic acid) tetrasodium salt; tetrasodium ethylenbis(iminodiacetate);EDTA tetrasodium; edetic acid tetrasodium salt; tetracemate tetrasodium; tetrasodium edetate; tetracemin; Endrate Tetrasodium; Questex; Versene; Sequestrene; Tetrine; Trilon B; Komplexon; Nullapon; Aquamollin; Complexone; Distol 8 Irgalon; Calsol; Syntes 12a; Tyclarosol; Nervanaid B. Tetrasodium EDTA; chelating agent. mp > 300°; apparent d = 6.9 lb/gallon. *Hart Chem. Ltd.*

15669 Kalex Acids
60-00-4 3559 200-449-4
$C_{10}H_{16}N_2O_8$
Edetic Acid

N,N'-1,2-Ethanediylbis[N-(carboxymethyl)glycine]; (ethylenedintrilo)tetraacetic acid; ethylenediaminetetraacetic acid; edathamil; EDTA; Havidote; Versene Acid. EDTA; sequestrant for preparation of amine or alkali metal salts; used where sodium ion is undesirable. dec 220°. *Hart Chem. Ltd.*

15670 Kalex HMP

10124-56-8 8814 233-343-1

$(NaPO_3)_x$

Sodium hexametaphosphate

Sodium polymetaphosphate; Graham's salt; glassy sodium metaphosphate; Hy-Phos. Sequestering agent for calcium and magnesium ions for certain sensitive textile application. mp = 628°. *Hart Chem. Ltd.*

15671 Kalex Liq. 50%

64-02-8 3557 200-573-9

$C_{10}H_{12}N_2Na_4O_8$

Edetate sodium

N,N'-1,2-Ethanediylbis[N-(carboxymethyl)glycine] tetrasodium salt; (ethylenedintrilo)tetraacetic acid tetrasodium salt; ethylenediaminetetraacetic acid tetrasodium salt; sodium edetate; tetrasodium ethylenediaminetetraacetate; ethylenebis(iminodiacetic acid) tetrasodium salt; tetrasodium ethylenbis(iminodiacetate);EDTA tetrasodium; edetic acid tetrasodium salt; tetracemate tetrasodium; tetrasodium edetate; tetracemin; Endrate Tetrasodium; Questex; Versene; Sequestrene; Tetrine; Trilon B; Komplexon; Nullapon; Aquamollin; Complexone; Distol 8 Irgalon; Calsol; Syntes 12a; Tyclarosol; Nervanaid B. Tetrasodium EDTA; general purpose chelating agent; complexes Ca, Mg. other common metals over wide pH range; for pulp/paper processing. *Hart Chem. Ltd.*

15672 Kalex OH

139-89-9 10102 205-381-9

$C_{10}H_{15}N_2Na_3O_7 \cdot xH_2O$

N-(2-Hydroxyethyl)ethylenediaminetriacetic acid trisodium salt hydrate

N-[2-[Bis(carboxymethyl)amino]ethyl]-N-(2-hydroxyethyl)glycine trisodium salt; N-(carboxymethyl)-N'-(2-hydroxyethyl)-N,N'-ethylenediglycine trisodium salt; N-hydroxyethylethylenediaminetriacetic acid trisodium salt; trisodium N-hydroxyethylethylenediaminetriacetate; Versen-Ol®. Trisodium HEDTA; sequesters Ca, Mg, iron

pH 8.0-10.5. *Hart Chem. Ltd.*

15673 Kalex Penta

140-01-2 205-391-3

Pentasodium DTPA; sequestrant used when slightly higher chelate stability is required and when other strong complexing agents are present; for pulp and paper, textile bleaching operations. *Hart Chem. Ltd.*

15674 Kalex® 13361

Two-component filled urethane compound; potting and encapsulating compound designed to withstand high humidity and temperature exposure, excellent thermal shock resistance *Hardman.*

15675 Kalex® 15036

Two-part urethane adhesive; extra fast setting, semiflexible, high shear strength adhesive with good peel strength. *Hardman.* Name unverified.

15676 Kalex® 20171

Two-component urethane system; room temperature curing, low viscosity electrical potting and encapsulation compound with very good electrical properties, excellent hydrolytic stability, low elevated temperature wt. loss and low water absorp. *Hardman.*

15677 Kaliammon saltpeter

7815

A potassium ammonium nitrate prepared by mixing equivalent molecular proportions of solid potassium chloride and ammonium nitrate in the presence of a little water; a fertilizer.

15678 Kalidone®

4810-50-8 225-373-9

Potassium PCA

Moisturizing agent for dermatological soap, shampoo, after-sun lotion, shower gel, nutritive and regenerative creams, hair comb-out balm. *UCIB.*

15679 Kalif

A proprietary copper-lead bearing alloy; melts at 952 C, with tensile strength of 10,000 psi at 21 C. No manufacturer.

15680 Kalipol

Polyphosphate solution *Albright & Wilson Ltd., Phosphates & Speciality Business.*

15681 Kalipol 18

A proprietary potassium polyphosphate solution used in the manufacture of liquid detergents. *Albright & Wilson Ltd.*

15682 kalkeisenollvin

Synonym for iron monticellite.

15683 Kalkor

Disposable refractory plugs for plug-bottom ingot molds. *Foseco (F.S.) Ltd.*

15684 Kalle's acid

486-54-4 6487

$C_{10}H_9NO_6S_2$

1-Naphthylamine-2,7-disulfonic acid

1-Amino-2,7-napthalenedisulfonic acid.

15685 Kallodent®

Trade name for a methyl methacrylate thermoplastic material used for molding dentures. No manufacturer.

15686 Kallodoc

Acrylic powder for artificial teeth and eyes. *Imperial Chemical Industries plc.*

15687 Kallte

A proprietary form of chalk prepared by a special process whereby it has a very small particle size, and the particles are coated with a calcium soap; a rubber filler. No manufacturer.

15688 Kalluzoto

A fertilizer containing nitrogen, potassium, and organic matter manufactured from residual molasses.

15689 Kalmex

Thermit compound for feeding ingots and castings; also used in continuous casting tundishes to facilitate start up of casting. *Foseco (F.S.) Ltd.*

15690 Kalmin

Insulating riser sleeves and shapes. *Foseco (F.S.) Ltd.*

15691 Kalminex

Exothermic sleeves with highly insulating residual structures for lining feeder heads on iron and steel castings. *Foseco (F.S.) Ltd.*

15692 Kalorex

Lightweight exothermic sideliner tiles for steel ingots. *Foseco (F.S.) Ltd.*

15693 Kalpack

Refractory ramming compound for securing nozzles in foundry ladles. *Foseco (F.S.) Ltd.*

15694 Kalpad

Moldable and preformed insulating materials to assist in feeding of steel and iron casting sections. *Foseco (F.S.) Ltd.*

15695 Kalpur

Insulating feeding sleeve with ceramic foam filter. *Foseco (F.S.) Ltd.*

15696 Kalrez®

Fluoroelastomer parts. *DuPont UK.*

15697 Kalseal

Refractory air-setting mortar used in foundry ladles. *Foseco (F.S.) Ltd.*

15698 Kalsert

System for applying feeder sleeves by insertion into pre-formed cavities in molds of cores. *Foseco (F.S.) Ltd.*

15699 Kaltas

Highly insulating disposable ladle lining. *Foseco (F.S.) Ltd.*

15700 Kaltop

Exothermic antipiping compounds in board form. *Foseco (F.S.) Ltd.*

15701 Kalvan

471-34-1 1697 207-439-9

$CaCO_3$

Calcium carbonate

Carbonic acid calcium salt; Cacit; Calcichew; Calcidia; Citrical. A proprietary trade name for calcium carbonate for use in rubber to give wear resistance; has an ultra fine particle size. mp = 825° (dec); d = 2.83. No manufacturer.

15702 Kalzana

A proprietary calcium-sodium lactate. No manufacturer.

15703 kalzose

9000-71-9 1934 232-555-1

Casein

A casein preparation containing calcium.

15704 Kamax

Imidized methyl methacrylate molding powder. *Rohm & Haas UK.*

15705 Kamax T-150

Acrylic-imide copolymer; amorphous thermoplastic; injection molding resin featuring high heat resistance, excellent optics, outdoor durability, excellent stiffness, easy processing; used for lighting, automotive, optical, packaging, medical, and appliance applications. *AtoHaas.*

15706 Kamela

5292

kamala

Kamila; kameela; spoonwood. A dyestuff obtained from the seeds or fruits of *Mallotus phillpenis* or *Rottlera tinctoria*; used in India as an anthelmintic (Cestodes), and for dyeing silk orange.

15707 Kamillosan

23089-26-1 1281 245-423-3

$C_{15}H_{26}O$

α-(-)-Bisabolol

[S-(R+,R+)]-α-bisabolol; levomenol; 6-methyl-2-(4-methyl-3-cyclohexen-1-yl)-5-hepten-2-ol; 1-methyl-4-(1,5-dimethyl-1-hydroxyhex-4(5)-enyl) cyclohexen-1. Pharmaceutical preparation containing the active principle of

Matricaria chamomilla. Anti-inflammatory. bp_{12} = 153°; d^{20} = 0.9211; n_D^{20} = 1.4936; $[\alpha]_D$ = -55.7°. *Degussa AG*. Name unverified.

15708 Kamoran
134
Actaplanin
A-4696. A complex of glycopeptide-type antibiotics; growth stimulant. *Eli Lilly & Co.*

15709 Kane-Ace
MBS impact modifier for PVC. *European Vinyls Corporation Ltd.*

15710 Kane's salt
A salt prepared by dissolving mercuric nitrate in a boiling solution of ammonium nitrate.

15711 Kanfotrex
A proprietary preparation of kanamycin, amphymycin and hydrocortisone used in dermatology as an antibacterial agent. *Bristol-Myers Squibb Co Inc.*

15712 Kanigen
Chemical nickel plate. *Albright & Wilson Ltd.* Name unverified.

15713 Kankerex
21908-53-2 5936 244-654-7
HgO
Mercuric oxide
For control of canker in apples and pears. dec 500°; d = 11.14. *Universal Crop Protection Ltd.*

15714 Kannasyn
A proprietary preparation containing kanamycin (as the sulfate); an antibiotic. No manufacturer.

15715 Kanten
A variety of agar-agar from red tegusa seaweed of Japan.

15716 Kanthal Alloy
A trademark for an iron alloy containing aluminum, cobalt, and chromium, and having a high degree of resistance to heat, a low electrical conductivity, and good hot and cold working properties. No manufacturer.

15717 Kantmelt
Nonmelting industrial grease. *Specialty Products Co.* Name unverified.

15718 Kantrex
25389-94-0 5293 246-933-9
$C_{18}H_{36}N_4O_{14}S$
Kanamycin A sulfate
Cantrex; Cristalomicina; Enterokanacin; Kamycin; Kamynex; Kanabristol; Kanacedin; Kanamytrex; Kanasig; Kanatrol; Kanicin; Kannasyn; Kantrox; Klebcil; Otokalixin; Resistomycin; Ophtalmokalixan; Kantrexil; Kano; Kanescin; Kanaqua. A proprietary preparation containing kanamycin sulfate; an antibiotic. *Bristol-Myers Squibb Co Inc.*

15719 Kantrexil
A proprietary preparation containing kanamycin sulfate, pectin, bismuth carbonate and activated attapulgite; an antidiarrheal. *Bristol-Myers Squibb Co Inc.*

15720 Kantrim
25389-94-0 5293 246-933-9
Kanamycin sulfate; antibacterial. *Bristol Laboratories.* Name unverified.

15721 Kantstik Q Powd
Refined synthetic wax ester; internal lubricant for injection molding; improves plastic flow in hard-to-reach mold areas; valuable in heavily pigmented molding resin mix; used in butyrate, PP, nylon, glassfilled nylon, PC, SAN, styrene, polyethylene, ABS, polyethylene, ABS, high-impacy PS, PVC. *Specialty Prods.*

15722 Kaokote
Partially hydrogenated vegetable oil (cottonseed, soybean), sorbitan stearate, polysorbate 60; kosher; coating fat for no-tempering low-liquor type coatings and centers. *Van Den Bergh Foods.*

15723 Kaokote F
68334-28-1 269-820-6
Partially hydrogenated vegetable oil (cottonseed, soybean); kosher; high performance fat for center fat, vegetable dairy systems. *Van Den Bergh Foods.*

15724 Kaola
68334-28-1 269-820-6
Partially hydrogenated vegetable oil (soybean, palm kernel); kosher; oil for vegetable dairy systems, candies, ice cream bar coatings, nut roasting. *Van Den Bergh Foods.*

15725 kaolin
1332-58-7 5294 296-473-8
$Al_2O_3 \cdot 2SiO_2 \cdot 2H_2O$
kaolin
bolus alba; China clay; porcelin clay; white bole; argilla. (Bolus alba; china clay) Native hydrated aluminum silicate; filler and coatings for paper, rubber, refractories, ceramics; in anticaking preps., paint; adsorbent for clarification of liqs. Burgess Pigment; Dry Branch Kaolin; ECC Int'l.; J.M. Huber; Kapolite; Southeastern Clay; R. T. Vanderbilt Co Inc.

15726 kaolinase
A purified kaolin.

15727 Kaomax
Partially hydrogenated soybean oil, sorbitan tristearate; kosher; high performance fat for no-tempering confectioners coating, vegetable dairy systems; steep melting shortening, oil migration inhibitor. *Van Den Bergh Foods.*

15728 Kaomel
68334-28-1 269-820-6
Partially hydrogenated vegetable oil (cottonseed, soybean); kosher; high performance fat for no-tempering confectioners coatings, vegetable dairy systems; steep melting shortening, oil migration inhibitor. *Van Den Bergh Foods.*

15729 Kaon-Cl
7447-40-7 7783 231-211-8
KCl
Potassium chloride
Chloropotassuril; Diffu-K; Enseal; Kaleorid; Kalitabs; Kalium-Duriles; Kaon-Cl; Kaskay; Kayback; Kay-Cee-L; K-Contin; Klor-Con; K-Norm; K-Tab; Lento-Kalium; Micro K; Nu-K; Peter-Kal; PfiKlor; Rekawan; Repone K; Slow-K; Span-K. Replenisher. d= 1.98; mp = 773°. *Adria Laboratories Inc.* Name unverified.

15730 Kaopolite® 1152
12141-46-7 377 235-253-8
Al_2O_5Si
Aluminum silicate
Aluminum silicate, anhydrous; mild polishing properties for auto polishes, plastics; gentle abrasive that speeds cleaning; antiblocking agent for plastic film. *Kaopolite.*

15731 Kaopolite® AB
12141-46-7 377 235-253-8
Al_2O_5Si
Aluminum silicate
Hydrated aluminum silicate; ultra-fine particle, soft silicate to moderate the polishing action of other abrasives; for auto polishes. *Kaopolite.*

15732 Kaopolite® SF
12141-46-7 377 235-253-8
Al_2O_5Si
Aluminum silicate
Aluminum silicate, anhydrous; gentle abrasive for auto polish, metals, plastics, household, dentifrices; antiblocking agent in plastic film. *Kaopolite.*

15733 Kaoprem-E
68334-28-1 269-820-6
Partially hydrogenated soybean oil with sorbitan tristearate; kosher; coating fat for no-tempering confectioners coating. *Van Den Bergh Foods.*

15734 Kaorich Beads
68334-28-1 269-820-6
Vegetable oil (cottonseed, soybean), hydrogenated; emulsifier for food processing. *Van Den Bergh Foods.*

15735 Kaorich Gold
68334-28-1 269-820-6
Partially hydrogenated vegetable oil (soybean, cottonseed), artificial butter flavor, artificial color (β carotene); kosher; shortening system for breading mixes and other dry mixes. *Van Den Bergh Foods.*

15736 Kaowool®
1332-58-7 5294 296-473-8
Kaolin clay-based; refractory bulk fibers for solving heat problems in furnaces (metal, petrochem., kilns in ceramic industry, boilers in utility industry). *Thermal Ceramics.*

15737 kapak
A material made from the mineral rubber elaterite; used in rubber mixings.

15738 Kapazang oil
A fat obtained from the seeds of *Hodgsonia heteroclita.*

15739 Kapex
Exothermic shapes to cover the surface of blind risers for iron and steel castings. *Foseco (F.S.) Ltd.*

15740 Kapitol
Captan + nuarimol; systemic fungicide for apple and pear trees. *DowElanco Ltd.*

15741 kapok
A cotton-like down produced in the seed-pods of the kapok tree; used in upholstery, life jackets, insulation, and pillows.

15742 Kapsol
A proprietary trade name for a methoxy-ethyloleate; used as a plasticizer. No manufacturer.

15743 Kapsovit
Multi-vitamin capsules. *Allen & Hanburys Ltd.* Name unverified.

15744 Kapton®
A proprietary trade name for polyimide resin in the form of film. *DuPont UK.*

15745 Kara Lube Al-Conc
Formulated product; fiber lubricant and dyestuff dispersant; leveling and migrating agent for direct dyeing of polyester/cotton blends; retarder/leveler for dyeing modacrylic fiber. *Rhône-Poulenc Surf.*

15746 Karafac 78
Conc. surfactant; neutralized (potassium salt) version of Karaphos XFA. *Clark.*

15747 Karakane
An alloy of 62.5% copper, 25% tin, 9.4% zinc, and 3.1% iron.

15748 Karalube DKL
Fatty derivatives/surfactant blend; lubricant for jet machine scouring and dyeing of synthetic fabrics. *Rhône-Poulenc Surf.*

15749 Karamate
8018-01-7 5756
$[C_4H_6N_2S_4Mn]_x[Zn^{2+}]_y$ where x:y = 10:1
Mancozeb
[[1,2-Ethanediylbis(carbamodithioato)](2-)]manganese mixt. With [[1,2-ethanediylbis(carbamodithioato)](2-)]zinc; ethylenebis(dithiocarbamic acid) manganese zinc complex; manzeb; manganese ethylenbis(dithiocarbamate) (polymeric) complex with zinc salt; zinc manganese ethylenebisdithiocarbamate; Dithane M-45; Manzate; Manzin; Nemispor; Penncozeb; Vondozeb. Wettable powder or water dispersible granules containing mancozeb; protectant fungicide for fruit, field crops, and roses. *Rohm & Haas UK.*

15750 Karamide 121
Cocamide DEA (1:1); heavy-duty detergent, thickener, foam stabilizer in cleaning compounds, shampoos, textile scours; as lubricant, antistat. *Clark.*

15751 Karamide CO9A
Cocamide DEA (2:1); detergent, base for floor cleaners, all-purpose cleaners. *Clark.*

15752 Karamide ST-DEA
93-82-3 202-280-1
Stearamide DEA; thickener, emulsifier for vegetable oil, mineral oil, microcrystalline wax. *Clark.*

15753 Karaphos HSPE
Blended phosphate ester, free acid; hydrotrope and wetter in heavy-duty detergents; high caustic-stable wetter and penetrant in textile formulations; as antistat; as rust preventative in metal cleaners; very low foaming; excellent alkali stability. *Clark.*

15754 Karaphos SWPE
Blended phosphate ester, free acid; fast wetting agent for detergents, cleaners, textile scours, penetrants, metal prep. compounds, antistats. *Clark.*

15755 Karasoft YB-11
Conc. self-emulsifying softener base; produces soft, dry, fluffy hand when applied to cotton, blends, and synthetics. *Clark.*

15756 Karasurf AS-26
Ammonium 2-ethyl hexanol sulfate
Wetter, penetrant, hydrotrope, solubilizer for industrial cleaners, fire fighting foams, etc. *Clark.*

15757 Karate
68085-85-8 2827 268-450-2
$C_{23}H_{19}ClF_3NO_3$
λ-cyhalothrin
3-(2-Chloro-3,3,3-trifluoro-1-propenyl)-2,2-dimethylcycloproanecarboxylic acid cyano(3-phenoxyphenyl)methyl ester; Grenade. Pyrethroid insecticide. *ICI Chem & Polymers Ltd.*

15758 Karate
3691-35-8 2204 223-003-0
$C_{23}H_{15}ClO_3$
Chloophacinone
2-[(4-Chlorophenyl)phenylacetyl]-1H-indene-1,3(2H)-dione; 2-[(p-chlorophenyl)phenylacetyl]-1,3-indandione; LM-91; Caid; Drat; Liphadione; Quick; Raviac; Rozol. An oil formulation containing 2.5% of chlorophacinone, an anti-coagulant rodenticide; a bait to control black rats, brown rats, house mice and voles. mp = 140°. *Lever Industrial Ltd.*

15759 Karathane
39300-45-3 3340 254-408-0
$C_{18}H_{24}N_2O_6$
Dinocap
DNOCP; CR-1639; ENT-24727; Arathane; Mildex; Isocothane; Crotothane. Emulsifiable concentrate of 350 g dinocap per liter; used to control mildew in fruit, chrysanthemums and roses. $bp_{0.05}$ = 138-140°. *Rohm & Haas UK.*

15760 karaya gum
9000-36-6 5296 232-539-4
Sterculia gum
India tragacanth; gum karaya; kadaya; katilo; kullo; kuteera; mucara. A hydrophilic polysaccharide from the genus *Sterculia*; used in pharmaceuticals, textile coatings, ice cream and other food products, adhesives, protective colloids, stabilizers, thickeners, emulsifiers. *Meer; Penta Mfg.; Rhône-Poulenc Perf. Resins; TIC Gums.*

15761 Karaya Paste
A proprietary preparation of sterculia in isopropyl alcohol; used as a dressing around colostomies. *Abbott laboratories.* Name unverified.

15762 Karbolite
A Russian artificial resin made from phenols, formaldehyde, and naphtholsulfonic acid. No manufacturer.

15763 Karboresin
Hydrocarbon-modified printing ink resins. *Hoechst AG.*

15764 Karbos
A carbonaceous decolorizer and filtering medium.

15765 Karetnja
880
Asphalt
Asphaltum; mineral pitch; Judean pitch; bitumen. A bituminous insulation. It consists mainly of asphalt, with an aluminum stearate binder.

15766 Karlex 12006 BKFR, 12018 BKFR
Polycarbonate, 20% glass fiber-reinforced; flame retardant thermoplast *Ferro/Engineering Thermoplastics.* Name unverified.

15767 Karlex 40002 NA
PC/PET alloy; high heat deflection temperature *Ferro/Engineering Thermoplastics.* Name unverified.

15768 Karlex 40003 NA
PC blend; high-impact. *Ferro/Engineering Thermoplastics.* Name unverified.

15769 Karma metal
An alloy of 80% nickel and 20% chromium. It is a heat-resisting alloy and melts at 1415°.

15770 Karmarsch metal
An alloy containing 88.8% tin, 7.4% antimony, and 3.7% copper.

15771 Karmex
330-54-1 3447 206-354-4
$C_9H_{10}Cl_2N_2O$
Diuron
N'-(3,4-Dichlorophenyl)-N,N-dimethylurea; 1,1-dimethyl-3-(3,4-dichlorophenyl)urea; Diurex; Urox D. Diuron; residual urea herbicide. mp = 158-159°. *Rohm & Haas UK.*

15772 Karmex®
330-54-1 3447 206-354-4
$C_9H_{10}Cl_2N_2O$
Diuron
N'-(3,4-Dichlorophenyl)-N,N-dimethylurea; 1,1-dimethyl-3-(3,4-dichlorophenyl)urea; Diurex; Urox D. Diuron weedkiller. mp = 158-159°. *DuPont UK.*

15773 Karolith
A casein preparation similar to Galalith; used for the manufacture of buttons and other objects.

15774 Karox AO-30
61788-90-7 263-016-9
Cocamine oxide
High foaming, mild surfactant, wetter, emulsifier, and coupling agent with excellent alkali tolerance. *Clark.*

15775 Karox LO
Dimethyl lauramine oxide
High foaming, mild detergent, coupling agent, wetter, emulsifier with good alkali and bleach stability. *Clark.*

15776 Karvol
A proprietary preparation containing various volatile oils; a decongestant. *Crookes Healthcare.*

15777 Kasal®
10102-71-3 378 233-277-3
$AlNa(SO_4)_2$
Aluminum sodium sulfate
Sodium aluminum phosphate, basic; food additive for baking, cereals, dairy and cheese. *Rhône-Poulenc Food Ingreds.*

15778 Kasil
1312-76-1 7838 215-199-1
$K_2Si_2O_5$ to $K_2Si_3O_7$
Potassium silicate
Soluble potash glass; soluble potash water glass. Potassium silicates supplied in varying SiO_2:K_2O ratios; alkaline builder for liquid detergents and cleaners, binder for phosphors in cathode ray tubes. *PQ Corp.* Name unverified.

15779 Kastone®
For the chemical industry. *DuPont UK.*

15780 Katadolon
56995-20-1 4227 260-503-8
$C_{15}H_{17}FN_4O_2 \cdot C_4H_4O_4$
Flupirtine maleate
Flupirtine; capsules and suppositories; analgesic. mp = 175.5-176°. *Degussa AG.* Name unverified.

15781 Katalabu gum
A gum of Nigeria; from *Acacia sieberiana* ; an adhesive.

15782 Katalon
85-00-7 3415 201-579-4
$C_{12}H_{12}Br_2N_2$
Diquat dibromide
6,7-Dihydrodipyrido[1,2-a:2',1'-c]pyrazinediium dibromide; 1,1'-ethylene-2,2'-dipyridylium dibromide; FB/2; Aquacide; Reglone. Diquat; herbicide. mp < 320° dec; Also reported as mp = 335-340°; λ_m = 308.31 nm (ε 18000). *Makhteshim Chemical Works Ltd.*

15783 Katanol
A trademark for a range of mordants used in the dyeing of textiles. *Cassella AG.* Name unverified.

15784 Katapol®
Redesignated Rhodameen® *Rhône-Poulenc Surf.*

15785 Katapone VV-328
12125-02-9 537 235-186-4
NH_4Cl
Quaternary ammonium chloride
Ammonium chloride; ammonium uriate; sal ammoniac; salmiac; Amchlor; Darammon. Water soluble, acid-corrosion inhibitor for steel, copper, and aluminum. d^{25} = 1.5274; sublimes without melting; soluble in water; pH of 1% aqueous solution at 25° = 5.5. *ISP.* Name unverified.

15786 katarsit
10257-55-3 1755 233-596-8
$CaSO_3$
Calcium sulfite
A calcium sulfite pellet for use as a dechlorinating agent for water.

15787 Katavel oil
The oil from *Hydnocarpus wightiana.*

15788 Katchung oil
8002-03-7 7191 232-296-4
Peanut oil
Arachis oil; groundnut oil; earthnut oil. Peanut oil. d^{15}_{15} = 0.917-0.921; d^{25}_{25}.=.0.910-0.915; solidifies at -5°.

15789 Katemul IG-70
118777-77-8
Isostearamidopropyl dimethylamine glycolate
Emulsifier, conditioner and softener for skin and hair products. *Scher.*

15790 Katemul IGU-70
129541-36-2
Isostearamidopropyl dimethylamine gluconate
Emulsifier, conditioner and softener for skin and hair products. *Scher.*

15791 Katharin
56-23-5 1864 200-262-8
CCl_4
Carbon tetrachloride
Tetrachloromethane; perchloromethane; Necatorina; Benzinoform. A proprietary trade name for carbon tetrachloride; used as a grease remover. d^{25}_{25} = 1.589; bp = 76.7°; mp = -23°; n_D^{20}= 1.4607. No manufacturer.

15792 Kathon®
26530-20-1 6853 247-761-7
$C_{11}H_{19}NOS$
Octhilinone
2-Octyl-3(2H)-isothiazolone; 2-octyl-4-isothiazolin-3-one; RH-893. Biocide; preservative for coatings, floor polish and adhesives. $bp_{0.01}$ = 120°; λ_m = 280 nm (log ε 3.88 MeOH). *Rohm & Haas.*

15793 Kathon® 886
5-Chloro-2-methyl isothiazolones
Used as biocides and preservatives in a wide range of industrial applications. *Rohm & Haas.*

15794 Kathon® 893
26530-20-1 6853 247-761-7
$C_{11}H_{19}NOS$
N-Octyl isothiazolones
2-Octyl-3(2H)-isothiazolone; 2-octyl-4-isothiazolin-3-one; RH-893. Used as biocides in paint, leather, textiles, paper and plastics. $bp_{0.01}$ = 120°; λ_m = 280 nm (log ε 3.88 MeOH). *Rohm & Haas.*

15795 Kathro
57-88-5 2256 200-353-2
$C_{27}H_{46}O$
Cholesterol
Cholest-5-en-3β-ol; cholesterin. Semi-refined cholesterol. (anhydr) mp = 148.5°; $bp_{0.5}$ = 233°; bp_{760} = 360°; d = 1.03; d^{19}_{19} = 1.052; $[\alpha]_D^{20}$= -31.5° (c = 2 in Et_2O); $[\alpha]_D^{20}$= -39.5° (c = 2 $CHCl_3$). *Croda Chem. Ltd.* Discontinued.

15796 Katigen®
A registered trademark currently awaiting reallocation by its proprietors to cover a range of dyestuffs. *Cassella AG.* Name unverified.

15797 Katioran® AF
Fatty acid hydroxylalkylamide and ethoxylated fatty alcohol; emulsifier and thickener for hair care products; makes hair soft and manageable and prevents electrostatic charges. *BASF AG.*

15798 Katorin
299-27-4 7796 206-074-2
$C_6H_{11}KO_7$
Potassium gluconate
Gluconic acid potassium salt; Gluconsan K; Kalimozan; Kaon; Katorin; Potasorl; Potassuril; K-IAO; Tumil-K. A proprietary preparation of potassium gluconate; a potassium supplement. Dec 180° *The Boots Co plc.* Discontinued.

15799 Kauk Catalyst
A proprietary trade name for a spherical catalyst of 5 mm diameter consisting of potassium salts and vanadium on a porous silica carrier. V_2O_5 content 6.5%; used for converting SO_2 into SO_3. No manufacturer.

15800 Kauramin®
Melamine-formaldehyde based; glues for product of weather-resistance chipboard and plywood; impregnating resin for decorative and overlay papers. *BASF AG.*

15801 Kauranat®
MDI-based; for production of weather-resistance chipboard. *BASF AG.*

15802 Kauresin®
Phenol-formaldehyde based; glues for production of weather-resistance plywood. *BASF AG.*

15803 Kaurit®
Urea-formaldehyde based; glues for production of veneered board, furniture; impregnating resins for decorative papers. *BASF AG.*

15804 Kauropal®
Assistants for glue and impregnating resins. *BASF AG.*

15805 kava-kava resin
5298
Kava
Ava-ava; kawa. A mixture of resins and resin acids from the dried roots of *Piper methysticum.*

15806 Kawasaki Hakkinko
A proprietary Japanese steel containing 0.19% carbon, 1.8% silicon, 1.0% manganese, 17.0% nickel, 25.0% chromium, and 0.2% molybdenum; offers resistance to hydrogen embrittlement. No manufacturer.

15807 Kay Ciel
7447-40-7 7783 231-211-8
KCl
Potassium chloride
Chloropotassuril; Diffu-K; Enseal; Kaleorid; Kalitabs; Kalium-Duriles; Kaon-Cl; Kaskay; Kayback; Kay-Cee-L; K-Contin; Klor-Con; K-Norm; ; K-Tab; Lento-Kalium; Micro K; Nu-K; Peter-Kal; PfiKlor; Rekawan; Repone K; Slow K; Span-K. Replenisher. d =1.98; mp = 773°. *Berlex Laboratories Inc.* Name unverified.

15808 Kayamer
Phosphate acrylate/methacrylate monomers. *British Traders & Shippers Ltd.*

15809 Kayarad
Monomeric/oligomeric acrylate monomers. *British Traders & Shippers Ltd.*

15810 Kaydol
8012-95-1 7327 232-384-2
Mineral oil
Liquid paraffin; liquid petrolatum; paraffin oil; Clearteck; Drakeol; Hevyteck; Kremol; white mineral oil; Alboline; Nujol; Paroleine; Saxol; Adepsine oil;Glymol. White mineral oil. d = 0.838. *Witco Corporation.*

15811 Kaydox
106-46-7 3107 203-400-5
$C_6H_4Cl_2$
1,4-Dichlorobenzene paste
p-Dichlorbenzene; Paracide; PDB; paradichlorobenzene; Para-zene; Di-chloricide; Paramoth. Insecticide Sublimes; mp = 53.5°; bp 174.12°; n_D^{60} = 1.5285; LD_{50} (rat orl) = 500 mg/kg. *Murphy Chemical Co Ltd.* Discontinued.

15812 Kayexalate
9003-59-2 8815

Sodium polystyrene sulfonate
Resonium A. Ion-exchange resin. *Sterling Drug Inc.* Name unverified.

15813 Kaylene
12141-46-7 377 235-253-8
Al_2O_5Si
Aluminum silicate
A proprietary preparation of colloidal aluminum silicate. No manufacturer.

15814 Kaylene-ol
12141-46-7 377 235-253-8
Al_2O_5Si
Aluminum silicate
A proprietary preparation of colloidal aluminum silicate with liquid paraffin. No manufacturer.

15815 Kaynitro
Concentrated nitrogen/potash fertilizer. *ICI Chem & Polymers Ltd.* Discontinued.

15816 Kayphobe-ABO
1332-58-7 5294 296-473-8
$H_2A_12Si_2O_8 \cdot H_2O$
Kaolin
Bolus alba; China clay; porcelain clay; white bole; argilla. Kaolin. *Kaopolite.*

15817 K-Bond
Aluminum tripolyphosphate dihydrate or metaphosphate; hardener for waterglass; used in inorganic coatings (interior and exterior heat resistance and nonflammable). *Bromhead & Denison Ltd.*

15818 K-Contin
7447-40-7 7783 231-211-8
KCl
Potassium chloride
Chloropotassuril; Diffu-K; Enseal; Kaleorid; Kalitabs; Kalium-Duriles; Kaon-Cl; Kaskay; Kayback; Kay-Cee-L; K-Contin; Klor-Con; K-Norm; ; K-Tab; Lento-Kalium; Micro K; Nu-K; Peter-Kal; PfiKlor; Rekawan; Repone K; Slow K; Span-K. A proprietary preparation of potassium chloride in a controlled release tablet; used as a potassium supplement in diuretic therapy. d = 1.98; mp = 773°. *Napp Laboratories Ltd.* Name unverified.

15819 K-Cop
Copper-ammonium complex; agricultural fungicide. *Griffin.*

15820 Keene's Cement
The name for a number of different plasters prepared by various manufacturers; usually obtained from plaster of Paris, dipped into a solution of alum or aluminum sulfate, and recalcining. No manufacturer.

15821 Keffekilite
A fuller's earth.

15822 Keical-Ace
10101-39-0 233-250-6
Ultra lightweight calcium silicate insulation; insulation for reactor, heat exchanger, vessel, tank, pipe. *Mitsubishi Kasei.* Name unverified.

15823 Kekuna oil
Bakoly oil
candlenut oil.

15824 Kelacid®
9005-32-7 241 232-680-1
Alginic acid
Norgine; polymannuronic acid; Sazio. Used as gelling agent, emulsifier and stabilizer in food, pharmaceutical, and industrial applications; stabilizer in paper and textile industry. *Kelco.*

15825 Kelburon
An elastomer-modified polypropylene used for automotive applications (bumpers, dashboards). *DSM NV.* Discontinued.

15826 Kelco-Crete™
Welan gum; for cementitious applications. *Kelco.*

15827 Kelco-Gel® Gellan Gum
71010-52-1 4389 275-117-5
Gellan Gum
Native gellan gum; high acyl gellan gum; polysaccharide S-60; PS-60. Purified gellan gum; high molecular weight anionic polysaccharide; gelling agent for use in foods, pet foods, personal care products, industrial applications. *Kelco.*

15828 Kelcoloid® D
9005-37-2
Propylene glycol alginate
Used as gelling agent, emulsifier and stabilizer in food, pharmaceutical, and industrial applications; stabilizer in paper and textile industry. *Kelco.*

15829 Kelcoloid® DH, DSF
9005-37-2
Propylene glycol alginate
Propylene glycol alginate; used for emulsions and low pH systems. *Kelco.*

15830 Kelcoloid® HV, LV
9005-37-2
Propylene glycol alginate
Propylene glycol alginates; used for emulsions and low pH systems. *Kelco.*

15831 Kelcoloid® O, S
9005-37-2
Propylene glycol alginate
Propylene glycol alginate; beef foam stabilizer. *Kelco.*

15832 Kelcosol®
9005-38-3 240
Algin
Alginic acid sodium salt; sodium alginate; sodium polymannuronate; Alto; Alman; Alloid; Allose; Kelgin; Protanal. Algin; used as gelling agent, emulsifier and stabilizer in food, pharmaceutical, and industrial applications; stabilizer in paper and textile industry. *Kelco.*

15833 Keldax®
Ethylene interpolymer resin. *DuPont UK.*

15834 Keleastoi
A proprietary trade name for a ricinoleate type of vinyl plasticizer. *Spencer Kellogg & Sons.* Name unverified.

15835 Kelecin®
8002-43-5 5452 232-307-2
Lecithin
Phosphatidylcholine; Lecithol; Vitellin; Kelecin; Granulestin. Industrial lecithin; emulsifying and dispersion agent, paint, mastics, feed and rubber. d_4^{24}= 1.0305. *Reichhold.*

15836 Kel-F 3700 Elastomer
A proprietary synthetic rubber resistant to high temperatures. *3M.*

15837 Kel-F 81 Plastic
Family of thermoplastic extrusion and molding materials that provide excellent chemical resistance over a broad temperature range; polychlorotrifluoroethylene. *3M.*

15838 Kelfizina
152-47-6 9078 205-804-7
$C_{11}H_{12}N_4O_3S$
Sulfalene
4-Amino-N-(3-methoxypyrazinyl)benzenesulfonamide; N¹-(3-methoxy-2-pyrazinyl)sulfanilamide; 2-(*p*-aminobenzenesulfonamido)-3-methoxypyrazine; 3-methoxy-2-sulfanilamidopyrazine; sulfamethopyrazine; 2-sulfanilamido-3-methoxypyrazine; sulfapyrazinemethoxyine; Farmitalia 204/122; Dalysep; Kelfizina; Kelfizine W; Longum; Policydal; Vetkelfizina; sulfamethoxypyrazine. Sulfalene; antibacterial. mp = 176°. *Abbott Laboratories.* Name unverified.

15839 Kelflo®
11138-66-2 10191 234-394-2
Xanthan Gum
Polysaccharide B-1459; Keltrol F; Kelzan. Xanthan gum; liquid feed supplements. *Kelco.*

15840 Kelfo®
Xanthan gum/limestone blend; gum for use in animal feed; thickenr, stabilizer. *Kelco.*

15841 Kelgin
9005-38-3 240
Algin
Alginic acid sodium salt; sodium alginate; sodium polymannuronate; Alto; Alman; Alloid; Allose; Kelgin; Protanal. Sodium alginate, food grade. *Kelco Int'l. Ltd.*

15842 Kelgin® F
9005-38-3 240
Algin
Alginic acid sodium salt; sodium alginate; sodium polymannuronate; Alto; Alman; Alloid; Allose; Kelgin; Protanal. Algin, refined; used as gelling agent, emulsifier and stabilizer in food, pharmaceutical, and industrial applications; stabilizer in paper and textile industry. *Kelco.*

15843 kelgin® HV, LV, MV
9005-38-3 240
Algin
Alginic acid sodium salt; sodium alginate; sodium polymannuronate; Alto; Alman; Alloid; Allose; Kelgin; Protanal. Sodium alginates; used for paper coating and sizing, textile printing, dyeing and sizing, alkaline carpet dyeing and wallpaper adhesives. *Kelco.*

15844 Kelgin® QH, QL, QM
9005-38-3 240
Algin
Alginic acid sodium salt; sodium alginate; sodium polymannuronate; Alto; Alman; Alloid; Allose; Kelgin; Protanal. Sodium alginates; dispersible products for paper coating and textile applications. *Kelco.*

15845 Kelgin® XL
9005-38-3 240

Algin
Alginic acid sodium salt; sodium alginate; sodium polymannuronate; Alto; Alman; Alloid; Allose; Kelgin; Protanal. Sodium alginates; used for paper coating and sizing, textile printing, dyeing and sizing, alkaline carpet dyeing and wallpaper adhesives. *Kelco.*

15846 Kelgum
9005-38-3 240
Algin
Alginic acid sodium salt; sodium alginate; sodium polymannuronate; Alto; Alman; Alloid; Allose; Kelgin; Protanal. A linseed oil rubber substitute.

15847 Kelgum®
11138-66-2 10191 234-394-2
Xanthan Gum
Polysaccharide B-1459; Kelrol F; Kelzan. Xanthan gum; used for desserts, gelled confectioneries, gels. *Kelco.*

15848 Kelig
8061-51-6
Lignosulfonic acid, sodium salt
Modified sodium lignosulfonates; metal complexing agents for micronutrient formulations, industrial cleaners and water treatment formulations. *Borregaard Ligno Tech.*

15849 Kelig 32
Lignosulfonate and modified sugar acids; sequestrant, dispersant for the lubricant removal from zinc phosphate coatings, oil well cement retardant; used for cooling water treatment compounds, alkaline cleaners, scale control. *Borregaard Ligno Tech.*

15850 Kelig FS
Lignosulfonate
Carrier for micronutrient metals in liq. formulations; inhibits leaf burn. *Borregaard Ligno Tech.*

15851 Kellin®
Drying oils. *Reichhold.*

15852 Kellite
A proprietary trade name for a synthetic resin. No manufacturer.

15853 Kel-Lite™ CM
11138-66-2 10191 234-394-2
Xanthan Gum
Polysaccharide B-1459; Keltrol F; Kelzan. Xanthan gum; used for cakes, cookies, muffins, waffles, pancakes. *Kelco.*

15854 Kellox®
Oxidized fish oils. *Reichhold.*

15855 Kelly's paint
A benzoated collodion containing tincture of benzoin, glycerin, and collodion; used for painting on abrasions of the skin.

15856 Kelmar®
9005-36-1 241
Alginic acid, potassium salt
Stercofuge. Potassium alginate; gellant, emulsifier, and stabilizer in food and indust. application; gum, bodying agent for creams and lotions, dental impression materials; used for water holding in foods and industry. *Kelco.*

15857 Kelmer®
Industrial oils and greases; lubricants; dust absorbents; wetting and binding compositions; fuels; illuminants; candle wicks. *Reichhold.*

15858 kelp
(Varec). Seaweed or the ash from seaweed; a source of iodine.

15859 kelp salt
7447-40-7 7783 231-211-8
KCl
Potassium chloride
A mixture of potassium chloride, with some alkaline sulfates and carbonates, formed in the preparation of potassium chloride from kelp.

15860 Kelpak
Seaweed concentrate; soil improver. *Omex Agriculture Ltd.*

15861 kelpchar
A decolorizing carbon obtained from seaweed by carbonizing in two stages and extracting with water and dilute hydrochloric acid.

15862 Kelpol®
Paints, varnishes lacquers; preservatives against rust and deterioration of wood; colorants; mordants; raw natural resins; metals in foil and powder form for painters, decorators, printers, and artists. *Reichhold.*

15863 kelpol® 835-M-50
Vinyl toluene monomer-modified alkyd. *Reichhold.*

15864 Kelpoxy®
Epoxy resins; for use in adhesives and coatings. *Reichhold.*

15865 Kelpoxy® G202-100
Elastomer modified epoxy conc. *Reichhold.*

15866 Kelprox
Thermoplastic EPDM-rubber; for automotive, appliances, industrial worms. *DSM NV.* Discontinued.

15867 Kelrinal
CM rubber; used as a high-grade vulcanized rubber in the wire and cable industry and in the automotive industry. *DSM NV.* Discontinued.

15868 Kelset®
Alginate; used as gelling agent, emulsifier and stabilizer in food, pharmaceutical, and industrial applications; stabilizer in paper and textile industry; self-gelling gum. *Kelco.*

15869 Kelsol®
Modified vegetable oil used as medium for tinting color in paints. *Reichhold.*

15870 Kelsol® 3922-G-80
Waterborne long oil alkyd. *Reichhold.*

15871 Kelstar
Pigment dispersions; for coloration of nonpolyurethane plastics. *Pacific Dispersions Inc.* Name unverified.

15872 Keltan
Range of EPDM terpolymers having differing Mooney viscosities; used in the automotive industry for wires and cables, in the building industry and in domestic appliances and technical products. *DSM NV.* Discontinued.

15873 Keltan 312
EPDM, nonstaining stabilizer; improved processability; very fast curing. *Copolymer Rubber.*

15874 Keltan 512
EPDM, nonstaining stabilizer; suitable for continuous curing; very fast cure rate. *Copolymer Rubber.*

15875 Keltan 4703
EPDM terpolymer, nonstaining stabilizer; high loading capacity, excellent mixing and extrusion behavior; very suitable for sponge articles, profiles; ultra fast curing. *Copolymer Rubber.*

15876 Keltan 4802
EPDM terpolymer, nonstaining stabilizer; good collapse resistance, very good mechanical and elastomeric properties; for solid profiles, radiator hoses, sealing rings, door seals for washing machines; very fast curing. *Copolymer Rubber.*

15877 Keltan TP
Thermoplastic rubber; used in the automotive, medical and pharmaceutical industries, for wires and cables, in household equipment and shoes. *DSM NV.* Discontinued.

15878 Keltex®
9005-38-3 240
Algin
Alginic acid sodium salt; sodium alginate; sodium polymannuronate; Alto; Alman; Alloid; Allose; Kelgin; Protanal. Sodium alginates; used for textile printing, silver recovery. *Kelco.*

15879 Keltex® HV
9005-38-3 240
Algin
Alginic acid sodium salt; sodium alginate; sodium polymannuronate; Alto; Alman; Alloid; Allose; Kelgin; Protanal. Alginate; print paste thickener. *Kelco.*

15880 Keltex® S
9005-38-3 240
Algin
Alginic acid sodium salt; sodium alginate; sodium polymannuronate; Alto; Alman; Alloid; Allose; Kelgin; Protanal. Specially designed sodium alginate; buffered product for textile printing. *Kelco.*

15881 Kelthane
115-32-2 3136 204-082-0
$C_{14}H_9Cl_5O$
Dicofol
4-Chloro-α-(4-chlorophenyl)-α-(trichloromethyl)benzenemethanol; 4,4'-dichloro-α-(trichloromethyl)-benzhydrol; 1,1-bis(*p*-chlorophenyl)-2,2,2-trichloroethanol; di(*p*-chlorophenyl)trichloromethylcarbinol; DTMC; ENT-23648; FW-293; Acarin; Mitigan. Dicofol; an acaricide. mp = 77-78°; λ_m (ethanol) = 226, 258, 266, 276 nm (4.43, 2,82, 2,85, 2.60). *Rohm & Haas UK.*

15882 Kelthix
Thixotropic resins; used for paint.

15883 Keltone®
9005-38-3 240
Algin
Alginic acid sodium salt; sodium alginate; sodium polymannuronate; Alto; Alman; Alloid; Allose; Kelgin; Protanal. Sodium alginate; used for hot and cold water dessert gels, dental impression material. *Kelco.*

15884 Keltone® HV, LV
9005-38-3 240

Algin
Alginic acid sodium salt; sodium alginate; sodium polymannuronate; Alto; Alman; Alloid; Allose; Kelgin; Protanal. Sodium alginate; used for puddings, pie fillings, bakery fillings, chiffons, sauces and dessert gels. *Kelco.*

15885 Keltose®
Calcium alginate and ammonium alginate; gellant, binder, emulsifier, and stabilizer in food and industrial application. *Kelco.*

15886 Keltrol®
11138-66-2 10191 234-394-2
Xanthan gum
Polysaccharide B-1459; Keltrol F; Kelzan. Xanthan gum, food grade; used for bakery fillings, flavor emulsions, canned foods, dry mixes, frozen foods, juice drinks, pourable and spoonable dressings, relishes, gravies, syrups, baked goods, batters, puddings, toothpaste, shampoo, pharmaceuticals. *Kelco.*

15887 Keltrol®
Drying oils used in paints, varnishes, lacquers, and enamels. *Reichhold.*

15888 Keltrol® 1001-M-60
Vinyl toluene monomer-modified alkyd. *Reichhold.*

15889 Keltrol® F
11138-66-2 10191 234-394-2
Xanthan gum
Polysaccharide B-1459; Kelzan. Food grade xanthan gum; stabilizer for foods; thickener and emulsion stabilizer in creams and lotions; binder in toothpaste; suspending agent for fruit pulp. *Kelco.*

15890 Kelzan®
11138-66-2 10191 234-394-2
Xanthan gum
Polysaccharide B-1459; Keltrol F. Xanthan gum; used for textured paint, carpet printing, cleaners and polishes, water based lubricants, coatings, ceramic glazes, pigment and dye suspensions, agricultural products, metal working products, foam dyeing/printing, coal slurries; other systems *Kelco.*

15891 Kelzan® AR
11138-66-2 10191 234-394-2
Xanthan gum
Polysaccharide B-1459; Keltrol F. For industrial uses in highly alkaline systems. *Kelco.*

15892 Kelzan® D, M, XC Polymer
11138-66-2 10191 234-394-2
Xanthan gum
Polysaccharide B-1459; Keltrol F. Industrial grade xanthan gum; Foam stabilizer, flocculant suspending, gelling agent, rheology modifier, lubricant for industrial applications incl. abrasives, adhesives, herbicides, fertilizers, ceramics, cleaners, emulsions, gels, mining, thixotropic paints, paper, pigments, viscosifier for drilling fluids. Standard industrial grade, D grade is for use with galactomannens; M grade for use where lower salt content is required, XC polymer is used as an additivre to oil well drilling mud Dissolves readily in water with stirring to give highly viscous solutions at low concentrations; aqueous solutions are highly psuedoplastic. *Kelco/Oil Field Products.*

15893 Kelzan® S
11138-66-2 10191 234-394-2
Xanthan gum
Polysaccharide B-1459; Keltrol F. Dispersible xanthan gum product; gum providing suspension of solids slurries and rheological control for aqueous systems. *Kelco.*

15894 Kemamide® B
3061-75-4 1051 221-304-1
Behenamide
Lubricant, slip, antiblock, and mold release agent for plastics, crayons, petrol. products, asphalts, inks, metals, textiles; mold release agent for thermoplastic resins in injection molding; defoamer and water repellent in industrial/household application, corrosion inhibitor, pigment grinding aid and dyestuff dispersant in paints, enamels, varnishes and lacquers; intermediate for textile emulsifiers and softeners and foam stabilizer in household detergents. *Witco Corporation.*

15895 Kemamide® E
112-84-5 204-009-2
Erucamide
Lubricant, slip, antiblock, and mold release agent for plastics, crayons, petrol. products, asphalts, inks, metals, textiles; mold release agent for thermoplastic resins in injection molding; defoamer and water repellent in industrial/household applicatio *Witco Corporation.*

15896 Kemamide® E-180
10094-45-8 233-226-5
Stearyl erucamide
Lubricant additive, friction modifier for high-temperature. plastics applications. *Witco Corporation.*

15897 Kemamide® E-221
10094-45-8 233-226-5
Erucyl erucamide
Lubricant additive, friction modifier for high-temperature plastics applications. *Witco Corporation.*

15898 Kemamide® O
301-02-0 206-103-9
Oleamide, technical; lubricant, slip, antiblock, and mold release agent for plastics, crayons, petrol. products, asphalts, inks, metals, textiles; mold release agent for thermoplastic resins in injection molding; defoamer and water repellent in industrial/household application. *Witco Corporation.*

15899 Kemamide® P-181
16260-09-6 240-367-6
Oleyl palmitamide
Lubricant additive, friction modifier for high-temperature plastics applications. *Witco Corporation.*

15900 Kemamide® S
124-26-5 204-693-2
$C_{18}H_{37}NO$
Octadecanamide
Stearamide. Lubricant, slip, antiblock, and mold release agent for plastics, crayons, petrol. products, asphalts, inks, metals, textiles; mold release agent for thermoplastic resins in injection molding; defoamer and water repellent in industrial/household application. mp = 98-102°; bp_{12} = 250°. *Witco Corporation.*

15901 Kemamide® S-180
stearyl stearamide
Lubricant additive, friction modifier for high-temperature plastics applications. *Witco Corporation.*

15902 Kemamide® S-221
Erucyl stearamide
Lubricant additive, friction modifier for high-temperature plastics applications. *Witco Corporation.*

15903 Kemamide® W-20
110-31-6 203-756-1
$C_{38}H_{72}N_2O_2$
N,N'-Ethylenebisoleamide
Ethylene dioleamide. Lubricant, slip, antiblock, and mold release agent for plastics, crayons, petrol. products, asphalts, inks, metals, textiles; mold release agent for thermoplastic resins in injection molding; defoamer and water repellent in industrial/household application, lubricant in ABS, PS, polyethylene; PP, PVC, nylon, nylon, cellulose acetate, PNAc and phenolic resins; defoamer in paper industry and fabric dyeing; latex systems; metal processing and asphalts. mp = 115-118° *Witco Corporation.*

15904 Kemamide® W-39
110-30-5 203-755-6
$C_{38}H_{76}N_2O_2$
N,N'-Ethylenebisstearamide
Ethylene distearamide. Lubricant, slip, antiblock, and mold release agent for plastics, crayons, petrol. products, asphalts, inks, metals, textiles; mold release agent for thermoplastic resins in injection molding; defoamer and water repellent in industrial/household applications. mp =144-146° *Witco Corporation.*

15905 Kemamine® AS-650
Nitrogen derivs
Antistat for polyolefins, styrenics, and other plastics, esp. film applications; lubricity aid, mold release aid, pigment dispersant. *Witco Corporation.*

15906 Kemamine® BQ-2802C
16841-14-8 240-865-3
Behenalkonium chloride
Antistat, textile softener, dyeing aid, corrosion inhibitor, emulsifier; used in personal care products, e.g., creams, lotions, shampoos, hair conditioners. *Witco/Humko.* Discontinued.

15907 Kemamine® BQ-2982B
90730-68-0
Erucalkonium chloride
Witco/Humko. Discontinued.

15908 Kemamine® BQ-9701C
61789-73-9 263-082-9
Dihydrogenated tallow benzylmonium chloride. *Witco Corporation.*

15909 Kemamine® BQ-9702C
61789-72-8 263-081-3
Dimethyl hydrogenated tallow benzyl ammonium chloride; germicide, sanitizer, slimicide, antistat, textile softener, dyeing aid, corrosion inhibitor, emulsifier; also for personal care products. *Witco Corporation.*

15910 Kemamine® BQ-9742C
61789-75-1 263-085-5
Tallowalkonium chloride

Antistat, textile softening agent, dyeing aid, corrosion inhibitor, emulsifier, leveling agent, shampoo and cream rinse conditioner. *Witco Corporation.*

15911 Kemamine® D-190
Arachidyl-behenyl 1,3-propylenediamine
Gasoline detergent, bactericide, corrosion inhibitor in petrol. production, epoxy hardener, dispersant, asphalt emulsifier. *Witco Corporation.*

15912 Kemamine® D-650
N-Coconut 1,3-propylenediamine
Gasoline detergent, bactericide, corrosion inhibitor in petrol. production, epoxy hardener, dispersant, asphalt emulsifier. *Witco/Humko.* Discontinued.

15913 Kemamine® D-970
68603-64-5 271-696-6
N-hydrogenated tallow 1,3-propylenediamine
Gasoline detergent, bactericide, corrosion inhibitor in petrol. production, epoxy hardener, dispersant, asphalt emulsifier. *Witco/Humko.* Discontinued.

15914 Kemamine® D-974
68439-73-6 270-416-7
Tallowaminopropylamine
Gasoline detergent, bactericide, corrosion inhibitor in petrol. production, epoxy hardener, dispersant, asphalt emulsifier. *Witco/Humko.* Discontinued.

15915 Kemamine® D-989
7173-62-8 230-528-9
N-Oleyl 1,3-propylenediamine
Gasoline detergent, bactericide, corrosion inhibitor in petrol. production, epoxy hardener, dispersant, asphalt emulsifier. *Witco/Humko.* Discontinued.

15916 Kemamine® D-999
Soyaminopropylamine
Gasoline detergent, bactericide, corrosion inhibitor in petrol. production, epoxy hardener, dispersant, asphalt emulsifier. *Witco Corporation.*

15917 Kemamine® P-150, P-150D
50% Arachidyl-behenyl primary amine (P-150Ddistilled); emulsifier, flotation agent, dispersing and flushing agent, intermediate, used in metalworking oils, as fuel oil additive; mold release for rubber and plastics; lubricant and spinning aid in metalworking oils. *Witco/Humko.* Discontinued.

15918 Kemamine® P-650D
61788-46-3 262-977-1
Cocamine, distilled; emulsifier, flotation agent, dispersing and flushing agent, intermediate, used in metalworking oils, as fuel oil additive; mold release for rubber and plastics; lubricant and spinning aid in metalworking oils. *Witco Corporation.*

15919 Kemamine® P-880D
143-27-1 205-596-8
$C_{16}H_{35}N$
n-Hexadecylamine
1-Aminohexadecane; cetylamine. Palmityl primary amine (technical and distilled respectively); emulsifier, flotation agent, dispersing and flushing agent, intermediate, used in metalworking oils, as fuel oil additive; mold release for rubber and plastics; lubricant and spinning aid in metalworking oil mp = 42-44°; fp = 140°. *Witco Corporation.*

15920 Kemamine® P-970
61788-45-2 262-976-6
Hydrogenated tallow amine (tech.)
Emulsifier, flotation agent, dispersing and flushing agent, intermediate, used in metalworking oils, as fuel oil additive; mold release for rubber and plastics; lubricant and spinning aid in metalworking oils. *Witco Corporation.*

15921 Kemamine® P-989D
112-90-3 204-015-5
$C_{18}H_{37}N$
Oleylamine
Oleamine, distilled; emulsifier, flotation agent, dispersing and flushing agent, intermediate, used in metalworking oils, as fuel oil additive; mold release for rubber and plastics; lubricant and spinning aid in metalworking oils. n_D^{20} = 1.4596; d = 0.813; fp = 154°. *Witco Corporation.*

15922 Kemamine® P-990, P-990D
124-30-1 204-695-3
$C_{18}H_{39}N$
1-Octadecylamine
1-Aminooctadecane; stearylamine. Stearamine (technical and distilled respectively); emulsifier, flotation agent, dispersing and flushing agent, intermediate, used in metalworking oils, as fuel oil additive; mold release for rubber and plastics; lubricant and spinning aid in metalworking oils. mp = 50-52°; bp = 349°; fp > 110°. *Witco Corporation.*

15923 Kemamine® Q-2802C
26597-36-4
Dibehenyldimonium chloride
Antistat, textile softener, dyeing aid, corrosion inhibitor, emulsifier; used in personal care products, e.g., creams, lotions, shampoos, hair conditioners. *Witco/Humko.* Discontinued.

15924 Kemamine® Q-6502C
61789-77-3 263-087-6
Dimethyl dicoconut ammonium chloride
Germicide, sanitizer, slimicide, antistat, textile softener, dyeing aid, corrosion inhibitor, emulsifier; also for personal care products. *Witco/Humko.* Discontinued.

15925 Kemamine® Q-9702C
61789-80-8 263-090-2
Quaternium-18
Antistat, textile softener, dyeing aid, corrosion inhibitor, emulsifier; used in personal care products, e.g., creams, lotions, shampoos, hair conditioners. *Witco Corporation.*

15926 Kemamine® Q-9743C
Tallowtrimonium chloride
Antistat, textile softening agent, dyeing aid, corrosion inhibitor, emulsifier. *Witco Corporation.*

15927 Kemamine® S-970
61789-79-5 263-089-7
Hydrogenated ditallow amine
Industrial use, grease, corrosion inhibitor. *Witco Corporation.*

15928 Kemamine® T-1902D
distilled dimethyl-90% arachidylbehenyl tert. amine; chemical intermediate for quaternary ammonium derivatives used for cosmetics and textiles; acid scavenger in petrol. products; epoxy hardener, catalyst in manufacture of flexible PU foams. *Witco/Humko.* Discontinued.

15929 Kemamine® T-6501
61788-62-3 262-990-2
Methyl dicoconut tert, amine
Chemical intermediate for quaternary ammonium derivatives used for cosmetics and textiles; acid scavenger in petrol. products; epoxy hardener, catalyst in manufacture of flexible PU foams. *Witco/Humko.* Discontinued.

15930 Kemamine® T-6502D
61788-93-0 263-020-0
distilled dimethyl cocamine; chemical intermediate for quaternary ammonium derivatives used for cosmetics and textiles; acid scavenger in petrol. products; epoxy hardener, catalyst in manufacture of flexible PU foams. *Witco Corporation.*

15931 Kemamine® T-9701
61788-63-4 262-991-8
Dihydrogenated tallow methylamine
Chemical intermediate for quaternary ammonium derivatives used for cosmetics and textiles; acid scavenger in petrol. products; epoxy hardener. catalyst in manufacture of flexible PU foams; corrosion inhibitor; gasoline additive. *Witco Corporation.*

15932 Kemamine® T-9742D
68391-07-1 269-926-2
Dimethyl hydrogenated tallow amine
Intermediate for manufacture of quaternary ammonium chlorides; corrosion inhibitor; gasoline additive. *Witco/Humko.* Discontinued.

15933 Kemamine® T-9902
124-28-7 3525 204-694-8
$C_{20}H_{43}N$
Dimethyl stearamine
Dymanthine; N,N-dimethyl-1-octadecanamine; N,N-dimethyloctadecylamine; dimantine. Intermediate for quats; used in cosmetics and textiles; acid scavenger in petrol, products. mp = 22.89°. *Witco Corporation.*

15934 Kematal
Acetal copolymer (POM); used for domestic equipment. i.e., jug kettle bodies, components for washing machines, food preparation equipment, plumbing (taps, cistern valves, and basins) *Hoechst AG.*

15935 Kemester® 104
112-62-9 6965 203-992-5
$C_{19}H_{36}O_2$
Methyl oleate
Methyl cis-9-octadecanoate; oleic acid methyl ester. Emulsifier, emollient for cosmetics; lubricant for leather; carrier for agricultural spray products. bp_2 = 168-170°; d = 0.876; n_D^{20} = 1.4520. *Witco Corporation.*

15936 Kemester® 143
Methyl tallowate
Intermediate in production of superamides, in metalworking lubricants, as solv.; lubricant, plasticizer for cosmetics, leather, rubber products. *Witco Corporation.*

15937 Kemester® 226
Methyl soyate
Intermediate in production of superamides, in metalworking lubricants, as specialized solv.; opacifier, viscosity control agent. *Witco Corporation.*

15938 Kemester® 1000
122-32-7 9861 204-534-7

$C_{57}H_{104}O_6$
Triolein
9-Octadecenoic acid 1,2,3-propanetriyl ester; olein; glyceryl trioleate. Emollient used in cosmetics, textiles, leather, metalworking lubricants; base for sulfation. d_{4}^{15}= 0.915; mp = -4 to -5°; bp_{15} = 215-240°; n_D^{20}= 1.4676; n_D^{60}= 1.4561. *Witco Corporation.*

15939 Kemester® 1418
17661-50-6 241-640-2
Myristyl stearate
For cosmetics/pharmaceuticals. *Witco/Humko.* Discontinued.

15940 Kemester® 2000
111-03-5 203-827-7
Glycerol oleate
Emollient, emulsifier, stabilizer, wetting agent for textiles, personal care products. *Witco Corporation.*

15941 Kemester® 2050
1120-28-1 214-304-8
$C_{21}H_{42}O_2$
Methyl eicosenate
methyl aracidate. Intermediate in production of alkanolamides, in metalworking lubricants, as specialized solvs.; foam depressant and nutrient in fermentation. mp = 46-51°; bp_{10} = 215-216°; fp > 110°. *Witco Corporation.*

15942 Kemester® 3681
Dioctyl dilinoleate
Lubricant for crankcase, turbine, compressor, gear, and metalworking formulations; also cosmetic emollient, pearling and bodying agent. *Witco Corporation.*

15943 Kemester® 4000
142-77-8 205-559-6
Butyl oleate
Emollient, wetting agent; plasticizer for textiles, leathers, elastomers, personal care products. *Witco/Humko.* Discontinued.

15944 Kemester® 4516
112-61-8 8959 203-990-4
$C_{19}H_{38}O_2$
Methyl stearate
Methyl n-octadecanoate; stearic acid methyl ester. Intermediate in production of alkanolamides, in metalworking lubricants, as specialized solvs.; foam depressant and nutrient in fermentation. mp = 38-39° also 44-46°; bp_{15} = 215°. *Witco Corporation.*

15945 Kemester® 5221SE
9004-99-3
PEG-2 stearate SE
Emollient, emulsifier, plasticizer, lubricant for cosmetics, rubber, textiles. *Witco Corporation.*

15946 Kemester® 5415
646-13-9 5165 211-466-1
$C_{22}H_{44}O_2$
Isobutyl stearate
Emollient, lubricant for textiles, metalworking fluids, personal care products. mp about 20°. *Witco Corporation.*

15947 Kemester® 5500
555-43-1 9885 209-097-6
$C_{57}H_{110}O_6$
Glyceryl tristearate
Tristearin; octadecanoic acid 1,2,3-propanetriyl ester; stearin; glyceryl tristearate. Emollient, emulsifier; stabilizer, plasticizer, lubricant for cosmetic, paper, textile, and industrial uses. $d^{80}{}_4$ = 0.862; mp about 55°; n_D^{80}= 1.4385. *Witco Corporation.*

15948 Kemester® 5510
123-95-5 1625 204-666-5
$C_{22}H_{44}O_2$
Butyl stearate
Octadecanoic acid butyl ester. Emollient for cosmetics; lubricant, plasticizer. mp = 27° also 16°; bp = 343°; fp = 196°; d_{25}^{25} = 0.855-0.875. *Witco Corporation.*

15949 Kemester® 5654
Ditridecyl adipate
Lubricant for crankcase, turbine, compressor, gear, and metalworking formulations; also cosmetic emollient, pearling and bodying agent. *Witco Corporation.*

15950 Kemester® 5721
31556-45-3 250-696-7
Tridecyl stearate
Emollient, dye carrier, textile lubricant, cosmetics/pharmaceuticals. *Witco Corporation.*

15951 Kemester® 5822
25339-09-7 246-868-6
Isocetyl stearate
Emollient, plasticizer for cosmetics. *Witco Corporation.*

15952 Kemester® 6000
555-43-1 9885 209-097-6
$C_{57}H_{110}O_6$
Glyceryl tristearate
Tristearin; octadecanoic acid 1,2,3-propanetriyl ester; stearin; glyceryl tristearate. Cosmetic and industrial emulsifier, emollient; plasticizer for elastomers. $d^{80}{}_4$ = 0.862; mp about 55°; n_D^{80}= 1.4385. *Witco Corporation.*

15953 Kemester® 9022
929-77-1 213-207-8
$C_{23}H_{46}O_2$
Methyl docosanoate
Methyl behenate. Intermediate in production of superamides, in metalworking lubricants, as solv. mp = 54-56°; fp > 110°. *Witco Corporation.*

15954 Kemester® BE
18312-32-8 242-201-8
Behenyl erucate
Industrial lubricant, cosmetic emollient. *Witco/Humko.* Discontinued.

15955 Kemester® CP
540-10-3 2073 208-736-6
$C_{32}H_{64}O_2$
Cetyl palmitate
Hexadecanoic acid hexadecyl ester; palmitic acid hexadecyl ester; hexadecyl palmitate. Industrial lubricant, cosmetic emollient. n_D^{70} = 1.4398. *Witco Corporation.*

15956 Kemester® DMP
131-11-3 3304 205-011-6
$C_{10}H_{10}O_4$
Dimethyl phthalate
1,2-Benzenedicarboxylic acid dimethyl ester; phthalic acid dimethyl ester; methyl phthalate; dimethyl 1,2-benzenedicarboxylate; DMP; Palatinol M; Fermine; Avolin; Mipax. Emollient used in cosmetics, textiles, metalworking lubricants. n_D^{20}= 1.5168; d_{15}^{15} = 1.196; d_{20}^{20} = 1.1940; d_{25}^{25} = 1.189; fp = 146°; mp = 5.5°; bp_{760} = 283.7°; bp_{400} = 257.8°; bp_{200} = 232.7°; bp_{100} = 210.0°. *Witco Corporation.*

15957 Kemester® EE
27640-89-7 248-587-4
Erucyl erucate
Industrial lubricant, cosmetic emollient. *Witco/Humko.* Discontinued.

15958 Kemester® EGDL
624-04-4 4507 210-827-0
$C_{26}H_{50}O_4$
Glycol dilaurate
Dodecanoic acid, 1,2-ethanediyl ester; ethylene dilaurate. For cosmetics/pharmaceuticals. mp = 50-52°; bp_{20} = 188°. *Witco/Humko.* Discontinued.

15959 Kemester® EGDS
627-83-8 211-011-3
Glycol distearate
Intermediate in production of superamides, in metalworking lubricants, specialized solv.; industrial lubricant; opacifier, pearling additive, thickener for cosmetics and pharmaceuticals. *Witco Corporation.*

15960 Kemester® EGMS
Glycol stearate
Intermediate in production of superamides, in metalworking lubricants, specialized solv.; industrial lubricant; opacifier, pearling additive, thickener for cosmetics and pharmaceuticals. *Witco/Humko.* Discontinued.

15961 Kemester® GDL
27638-00-2 248-586-9
Glyceryl dilaurate
For cosmetics/pharmaceuticals. *Witco/Humko.* Discontinued.

15962 Kemester® GMS (Powd.)
555-43-1 9885 209-097-6
$C_{57}H_{110}O_6$
Glyceryl tristearate
Tristearin; octadecanoic acid 1,2,3-propanetriyl ester; stearin; glyceryl tristearate. Industrial lubricant; cosmetic emollient. $d^{80}{}_4$ = 0.862; mp about 55°; n_D^{80}= 1.4385. *Witco/Humko.* Discontinued.

15963 Kemester® HMS
118-56-9 4776 204-260-8
$C_{16}H_{22}O_3$
Homosalate
2-Hydroxybenzoic acid 3,3,5-trimethylcyclohexyl ester; salicylic acid 3,3,5-trimethylcyclohexyl ester; 3,3,5-trimethylcyclohexyl salicylate; homomenthyl salicylate; Heliophan. Uv absorber; sunscreen agent; used in cosmetic skin preps. d_{25}^{25} = 1.045; bpss4 = 161-165°; n^{20} = 1.516 to 1.518. *Witco Corporation.*

15964 Kemester® JO
Erucyl arachidate
Emollient for cosmetics. *Witco/Humko.* Discontinued.

15965 Kemester® MM
3234-85-3 221-787-9
Myristyl myristate
Lubricant for crankcase, turbine, compressor, gear, and metalworking formulations; also cosmetic emollient, pearling and bodying agent. *Witco/Humko.* Discontinued.

15966 Kemester® S20
1338-39-2 8872 215-663-3
$C_{18}H_{34}O_6$
Sorbitan laurate
Sorbitan monolaurate; Alkamuls SML; Arlacel 20; Emsorb 2515; Glycomul L; Span 20. For cosmetics/pharmaceuticals. d = 1.032; n_D^{20}= 1.4740; fp > 110°. *Witco Corporation.*

15967 Kemester® S40
26266-57-9 8872 247-568-8
Sorbitan palmitate
Sorbitan monopalmitate; Span 40. For cosmetics/pharmaceuticals. mp = 46-47°; fp > 110°. *Witco Corporation.*

15968 Kemester® S60
1338-41-6 8872 215-664-9
$C_{24}H_{46}O_6$
Sorbitan stearate
Sorbitan monostearate; Alkamuls SMS; Arlacel 60; Glycomul S; Span 60. For cosmetics/pharmaceuticals. mp = 49-65°; acid value = 5-11; saponification value = 140-157; hydroxyl value = 230-260; soluble in alcohols. *Witco Corporation.*

15969 Kemester® S65
26658-19-5 8872 247-891-4
Sorbitan tristearate
For cosmetics/pharmaceuticals. *Witco Corporation.*

15970 Kemester® S80
1338-43-8 8872 215-665-4
$C_{24}H_{44}O_6$
Sorbitan oleate
Sorbitan monooleate; Alkamuls SMO; Arlacel 80; Capmul O; Emsorb 2500; Glycomul O; Span 80. For cosmetics/pharmaceuticals. d = 0.986; fp > 110°; n_D^{20}= 1.4800. *Witco/Humko.* Discontinued.

15971 Kemester® S85
26266-58-0 8872 247-569-3
Sorbitan trioleate
Span 85. For cosmetics/pharmaceuticals. d = 0.956; fp > 110°; n_D^{20}= 1.4760. *Witco Corporation.*

15972 Kemester® THFO
Tetrahydrofurfuryl oleate
Emollient used in cosmetics, textiles, metalworking lubricants. *Witco Corporation.*

15973 Kemflorseal
Epoxy die material and epoxy floor and wall product; used for dental molds, floor sealer, wall decorating and stain resistance. *Kemtron International Inc.*

15974 Kemgo
A proprietary trade name for inks for use with heat. No manufacturer.

15975 Kemick
Heat-resisting paint. *ICI Chem & Polymers Ltd.* Discontinued.

15976 Kemira Phlogopite Mica
12001-26-2
Mica; filler for plastics, surface coatings, sound-deadening compounds etc. *Comelius Chemical Co Ltd.* Name unverified.

15977 Kemite
A ceramic material. Labstone is the name given to the material when used for laboratory bench tops.

15978 Kemix
Paste form dispersions of activators and accelerators used in elastomer cures. *Kenrich Petrochemicals.*

15979 Kemlet metal
An alloy consisting mainly of zinc, with aluminum and copper.

15980 Kemlex 10007 NAL1
Polyacetal
PTFE-lubricated. *Ferro/Engineering Thermoplastics.*

15981 Kemmat
A range of glycol esters; used as emulsifiers for cosmetic and pharmaceutical creams and lotions, thickeners for shampoos and emulsifiers for mineral oils. *Harcros Australia.*

15982 Kemmest
A range of glycol esters; used as emulsifiers in the cosmetic and pharmaceutical industry. *Harcros Australia.*

15983 Kemopol
Propylene oxide condensates of ethylene oxide/glycol adducts; used as a detergent base for machine dishwashing and cattle antibloat. *Harcros Australia.*

15984 Kemotan
A range of polysorbates; used as emulsifiers in the food and cosmetic industry. *Harcros Australia.*

15985 Kemp
The shorter fibers of mohair.

15986 Kempol
A proprietary trade name for vulcanizable vegetable oil polymers. No manufacturer.

15987 Kempore® 60/14FF
123-77-3 950 204-650-8
$C_2H_4N_4O_2$
Azodicarbonamide
Diazenedicarboxamide; 1,1'-azobisformamide; azodicarboxamide; azobiscarbonamide; azobiscarboxamide; 1,1'azobiscarbamide. Azodicarbonamide based; blowing agent for dynamic foaming processes, e.g., injection molding, extrusion, and calendering. mp = 225° (dec); *Uniroyal.*

15988 Kempoxy
An epoxy flooring material; floor resurfacer, for chemical resistance floors. *Kemtron International Inc.*

15989 Kempy wool
Wool prepared from sheep that have been badly fed or subject to exposure. It dyes badly..

15990 Kemset Epoxy
Three-pack epoxy mortar based on nonsolvented resin and silica aggregate; high compressive, flexural and tensile strengths. *Feb Ltd.*

15991 Kemstrene® 96.0%
56-81-5 4493 200-289-5
$C_3H_8O_3$
Glycerol
1,2,3-Propanetriol; glycerin; glycerine; trihydroxypropane; incorporation factor, IFP; Bulbold; Cristal; Glyceol; Ophthalgan. Refined glycerin USP; humectant, solv.; use in cosmetics, liq. soaps, confections, inks, and lubricants; intermediate used in polyester and PU formulations. mp = 17.8°; bp_{760} = 290.0° (dec); bp_{200} = 240.0°; bp_{20} = 182.2°; fp 160°; d = 1.260; n_D^{20}= 1.4740; d_{20}^{20} = 1.26362. *Witco Corporation.*

15992 Kemtop
Epoxy die material and epoxy floor and wall product; used for dental molds, floor sealer, wall decorating and stain resistance. *Kemtron International Inc.*

15993 Kemwax
110-30-5 203-755-6
$C_{38}H_{74}N_2O_2$
N,N'-Ethylene bisstearamide
Abrilube 84. Used as plastics lubricants and slip agents in plastics. mp = 144-146° *Abril Industrial Waxes Ltd; Harcros Australia.*

15994 Kena® Formula FP-28, FP-85
Food additive for meat, poultry, and seafood industries. *Rhône-Poulenc Food Ingreds.*

15995 Kencolor®
Dispersions of pigments in silicone oils and gums for silicone elastomer compositions; pigment and catalyst dispersions used in the coloring and curing of silicone elastomers. *Kenrich Petrochemicals.*

15996 Kendex 0220
Petroleum hydrocarbon crude petrolatum wax (black); used as a fuel in fire logs or for further refining to petrolatum. *Witco.* Discontinued.

15997 Kendex 0834
Petroleum hydrocarbon heavy petroleum resins; used as a viscosity modifier in lubrication applications, plasticizer extender. *Witco.* Discontinued.

15998 Kendex 0842
Petroleum hydrocarbon-cylinder stock; a base oil for lubrication formulations or plasticizer/extender. *Witco.* Discontinued.

15999 Kendex 0847
Petroleum hydrocarbon 150 bright stock; base oil in the formulation of motor oil or lubricants; a plasticizer/extender in the rubber industry. *Witco.* Discontinued.

16000 Kendex 0866
Petroleum hydrocarbon-extract
A plasticizer/extender; can be used in adhesives. *Witco.* Discontinued.

16001 Kendex 0898
Petroleum hydrocarbon-intermediate petroleum resins; viscosity modifier for

lubricants; plasticizer/extender; additive stock for quench oils. *Witco.* Discontinued.

16002 Kendex OCTG
7328
Petroleum
Crude oil; mineral oil; rock oil; coal oil; seneca oil. Petroleum; corrosion inhibitor; specialty anticorrosion formulation for ferrous metals for tubular goods and other machined parts. *Witco.* Discontinued.

16003 Kendurit
Skin protective soap with granules; for cleansing of dirty hands. *Dynamit Nobel Wien GmbH.* Name unverified.

16004 Kenflex®
The condensation products of alkyl naphthalenes; vinyl plasticizer. *Kenrich Petrochemicals.*

16005 Kenflex® A
A dimethyl naphthalene derivative; elastomer and vinyl plasticizer and softener. *Kenrich Petrochemicals.*

16006 Kenlastic®
Elastomeric dispersion forms of activators and accelerators used in elastomer compounding; provides faster Banbury incorporation times; eliminates dust, *Kenrich Petrochemicals.*

16007 Ken-Mag®
1309-48-4 5713 215-171-9
MgO
Magnesium oxide
Magnesia; calcined magnesia; magnesia usta; Magcal; Maglite. For the polymer industry. mp = 2800° *Kenrich Petrochemicals.*

16008 Kenmix®
Paste dispersions of activators and accelerators; used in elastomer cures. *Kenrich Petrochemicals.*

16009 Kenplast®
Distilled aromatic and cumylphenol derived plasticizers; nonreactive and reactive diluents in epoxies, primary plasticizer in urethanes. *Kenrich Petrochemicals.*

16010 Kenplast® ES-2
Cumyl-phenyl acetate
Plasticizer for urethanes; high flash pt. reactive diluent for epoxy, reducing odor and irritation; ideal for polyamide-cured epoxy floorings and coal tar pipe coatings; comonomer and impact modifier for phenolics; improves adhesion of vinyl plastisols coating metals and polar plastics. *Kenrich Petrochemicals.* Name unverified.

16011 Kenplast® ESB
Cumyl-phenyl benzoate
Primary plasticizer for PVC and PVC/nitrile; process aid for semirigid PVC extrudates; solvates conductive polyester and acrylic inks to improve impact. *Kenrich Petrochemicals.* Name unverified.

16012 Kenplast® ESI
Biscumylphenyl trimellitate
Plasticizer, process aid for extrusion of PVC, urethanes; lubricant and process aid for filled PS, PC; reactive diluent and flow promoter for epoxy powd. coatings; prevents oxidative depolymerization in nylons. *Kenrich Petrochemicals.* Name unverified.

16013 Kenplast® ESN
Cumylphenyl neodecanoate
Primary plasticizer for PVC, PVC/nitrile; process aid for semirigid PVC extrusions; solvates and impact modifies conductive polyester and acrylic inks. *Kenrich Petrochemicals.* Name unverified.

16014 Ken-React®
Titanate, zirconate and aluminate coupling agents; typical composition; LICA 01 - titanium IV neoalkenolato, tris neodecanolato-O, NZ 01 - zirconium IV neoalkenolato, tris neodecanolato-O, KA 322 - diisopropyl (oleyl) aceto acetyl aluminate; coupling agents, adhesion promoters and composite improvers. *Kenrich Petrochemicals.*

16015 Ken-React® 7 (KR 7)
Isopropyl dimethacryl isostearoyl titanate (monoalkoxy)
IPA *Kenrich Petrochemicals.*

16016 Ken-React® 9S (KR 9S)
Isopropyl tridodecylbenzenesulfonyl titante (monoalkoxy)
IPA *Kenrich Petrochemicals.*

16017 Ken-React® 12 (KR 12)
Isopropyl tri(dioctylphosphato) titanate (monoalkoxy)
IPA *Kenrich Petrochemicals.*

16018 Ken-React® 26S (KR 26S)
Isopropyl 4-aminobenzenesulfonyl di(dodecylbenzenesulfonyl) titanate (monoalkoxy)
IPA *Kenrich Petrochemicals.*

16019 Ken-React® 33DS (KR 33DS)
Alkoxy trimethacryl titanate (monoalkoxy)
IPA *Kenrich Petrochemicals.*

16020 Ken-React® 38S (KR 38S)
Isopropyl tri(dioctylpyrophosphato) titanate (monoalkoxy)
IPA *Kenrich Petrochemicals.*

16021 Ken-React® 39DS (KR 39DS)
Alkoxy triacryl titanate (monoalkoxy)
IPA *Kenrich Petrochemicals.*

16022 Ken-React® 41B (KR 41B)
Tetraisopropyl di (dioctylphosphito) titanate (coordinate type), IPA *Kenrich Petrochemicals.*

16023 Ken-React® 44 (KR 44)
Isopropyl tri(N ethylaminoethylamino) titanate (monoalkoxy)
IPA *Kenrich Petrochemicals.*

16024 Ken-React® 46B (KR 46B)
Tetraoctyloxytitanium di(ditridecylphosphite)(coordinate type), IPA; *Kenrich Petrochemicals.*

16025 Ken-React® 55 (KR 55)
Tetra (2, diallyoxymethyl-1 butoxy titanium di (di-tridecyl) phosphite (coordinate type), IPA; *Kenrich Petrochemicals.*

16026 Ken-React® 133DS (KR 133DS)
Titanium dimethacrylate oxyacetate (oxyacetate chelate type), IPA; *Kenrich Petrochemicals.*

16027 Ken-React® 134S (KR 134S)
Titanium di(cumylphenylate) oxyacetate (oxyacetate chelate type), IPA; *Kenrich Petrochemicals.*

16028 Ken-React® 138D (KR 138D)
KR 138S and 2-dimethylamino methyl propanol (quat type), IPA; *Kenrich Petrochemicals.*

16029 Ken-React® 138S (KR 138S)
Titanium di (dioctylpyrophosphate) oxyacetate (oxyacetate chelate type), IPA; *Kenrich Petrochemicals.*

16030 Ken-React® 158FS (KR 158FS)
Titanium di (butyl, octyl pyrophosphate) di (dioctyl, hydrogen phosphite) oxyacetate (oxyacetate chelate type), IPA; *Kenrich Petrochemicals.*

16031 Ken-React® 212 (KR 212)
Di (dioctylphosphato) ethylene titanate (A, B ethylene chelate type), IPA; *Kenrich Petrochemicals.*

16032 Ken-React® 238S (KR 238S)
Di (dioctylpyrophosphato) ethylene titanate (A, B ethylene chelate type); *Kenrich Petrochemicals.*

16033 Ken-React® 262ES (KR 262ES)
Di (butyl, methyl pyrophosphato) ethylene titanate di (dioctyl, hydrogen phosphite) (A, B ethylene chelate type), IPA; *Kenrich Petrochemicals.*

16034 Ken-React® OPP2 (KR OPP2)
Dicyclo (dioctyl) pyrophosphato, titanate (cycloheteroatom type), IPA; *Kenrich Petrochemicals.*

16035 Ken-React® OPPR (KR OPPR)
Dicyclo (dioctyl) pyrophosphato dioctyl titanate (cycloheteroatom type), IPA; *Kenrich Petrochemicals.*

16036 Kensol 10
7329
Petroleum naptha
Petrolum benzin; naphtha; benzin. Light naptha used as fuel in sporting and camping equipment or as a solvent. bp = 35-80°; d = 0.625-0.660; fp = -40° *Witco.* Discontinued.

16037 Kensol 13
7329
Petroleum naptha
Petrolum benzin; naphtha; benzin. Light petroleum naptha used in oil well deparaffinizing and as a solvent in blending asphalt. bp = 35-80°; d = 0.625-0.660; fp = -40° *Witco.* Discontinued.

16038 Kensol 30
6288
Regular mineral spirits
Stoddard solvent; Texsolve S; Varsol 1. Petroleum hydrocarbon; regular mineral spirits or Stoddard solvent; wide range of applications as a commercial solvent and as a parts cleaner or diluent. fp = 38°; $d^{15.6}_{15.6}$ = 0.754-0.820 *Witco.* Discontinued.

16039 Kensol 48T
Petroleum hydrocarbon; low odor petroleum distillate fraction; used as a solvent in various commercial formulations or as an oil to roll aluminum foil. *Witco.* Discontinued.

16040 Kensol 50T
Petroleum hydrocarbon; low odor middle petroleum distillate; oil used in

aluminum rolling applications and as a diluent in commercial formulations. *Witco.* Discontinued.

16041 Kensol 51

Petroleum hydrocarbon; middle distillate or a light mineral seal oil; used as a lubricant in the rolling of aluminum and as a diluent in commercial formulations. *Witco.* Discontinued.

16042 Kensol 53

Petroleum hydrocarbon; mineral seal oil; formulation of commercial products, coke absorber oil, compounding base stock, cutting oil. *Witco.* Discontinued.

16043 Kensol 61

Petroleum hydrocarbon; mineral seal oil; oil for compounding various commercial products, hydraulic oil base stock, light oil for diluents. *Witco.* Discontinued.

16044 Kensol 80

7329

Petrolum naptha

Petroleum benzin; naphtha; benzin; petrolum naphtha. Petroleum hydrocarbon V M & P type naphtha; solvent in paint or as a solvent in commercial products. fp = -40°; d = 0.625-0.660; bp 35-80° *Witco.* Discontinued.

16045 Kensol KM Metal Cleaner

Liquid metal polish; water-based suspension containing ammonia, oxalic acid and methanol among other ingredients; used primarily as an allmetal cleaner for commercial metal maintenance on architectural and ornamental metalwork as well as household use on brass, chrome, stainless steel etc. *Kensol Corporation.*

16046 Kensol KV Rust Retarder

Hydrocarbon mixture, blend of petroleum products; primarily used as a rust preventative on polished steel and stainless steel on bank security equipment. *Kensol Corporation.*

16047 Kensol KX Oxide Resistor

Hydrocarbon mixture, blend of petroleum products; used primarily as an all-metal preservative for commercial metal maintenance on architectural and ornamental metalwork as well as household use to protect against rust and tarnish. *Kensol Corporation.*

16048 Kentish rag

A siliceous limestone; used as an adulterant of Portland cement.

16049 Kentite

An explosive containing from 32-35% ammonium nitrate, 32-35% potassium nitrate, 16-18% ammonium chloride, and 14-16% trinitrotoluene.

16050 Ken-Zinc®

1314-13-2 10279 215-222-5

ZnO

Zinc oxide

Flowers of zinc; Philosopher's wool; zinc white; C.I. Pigment White 4; C.I. 77947. Zinc oxide dispersion; for use as an activator in elastomer compounding. n_D = 2.0042, 2.0203 *Kenrich Petrochemicals.*

16051 Kephos

Nonaqueous phosphating metal pretreatment. *ICI Chem & Polymers Ltd.*

16052 Keramine H

Liquid protein ampules, to strengthen hair. *Richardson-Vicks Inc.* Name unverified.

16053 Keramino 25

68238-35-7 5302 269-409-1

Keratin amino acids

Cosmetics ingred. *Brooks Industries.*

16054 keramyl

16961-83-4 4220 241-034-8

H_2SiF_6

Fluorosilicic acid

Hexafluorosilicic acid; hydrogen hexafluorosilicate; hydrosilicofluoric acid; hydrofluosilicic acid; silicofluoric acid; fluorosilicic acid. A solution of hydrofluosilicic acid $d^{17.5}_{17.5}$ (30% soln) = 1.2742; 60-70% solution forms crystaline dihydrate around 19°.

16055 keraphen

2217-44-9 5059 218-715-3

$C_{20}H_8I_4Na_2O_4$

Iodophthalein sodium

Soluble iodophthalein; tetraiodophenolphthalein sodium; T.I.P.P.S.; tetraiodophthalein sodium; tetiothalein sodium; Iodeikon; Cholepulvis; Shadocol; Bilitrast; Iodognost; Stipolac; Tetraiode; Foriod; Iodtetragnost; Antinosin; Cholumbrin; Iodorayorl; Opacin; Photobiline; Piliophen; Videophel; Radiotetrane; Nosophene sodium; Iodophene sodium. Radiopaque medium, used as a diagnostic aid in cholecystography.

16056 Kera-Quat WKP

68915-25-3

Cocodimonium hydroxypropyl hydrolyzed keratin

Substantive protein, moisturizer for hair and skin care products (shampoos, conditioners, styling products, liq. hand soaps, creams, lotions). *Maybrook.*

16057 Kerasol

5302

Soluble keratin; proteinic conditioner for hair and nail care products. *Croda Inc.*

16058 Kerasol

386-17-4 5059 206-857-9

$C_{20}H_{10}I_4O_4$

Tetraiodophenolphthalein

Iodophthalein; tetraiodophthalein; Iodophene; Nosophen. dec about 200°.

16059 Kera-Tein 1000

69430-36-0 274-001-1

Hydrolyzed keratin

Moisturizer for skin and hair care products; protective colloid effect; provides Sistine to hair; minimizing damage from harsh treatments. *Maybrook.*

16060 Kera-Tein AA

68238-35-7 269-409-1

Keratin amino acids

Substantive/protective protein, penetrant, moisturizer for skin and hair care products; source of cystine. *Maybrook.*

16061 Kera-Tein H

Keratin amino acids and sodium chloride; cosmetics ingred. *Maybrook.*

16062 Keratite

A name applied to a vulcanite.

16063 Keratol

A cellulose waterproofing compound.

16064 Kerb 50W

23950-58-5 8058 245-951-4

$C_{12}H_{11}Cl2_NO$

Propyzamide

3,5-Dichloro-N-(1,1-dimethyl-2-propynyl)benzamide; pronamid; RH-315; Kerb. A residual herbicide in a wettable powder from for a wide range of agricultural crops. mp = 155-156°; vapor pressure at 25° = 8.5x10^{-5} mm Hg. *Rohm & Haas UK.*

16065 Kerb 50W

23950-58-5 8058 245-951-4

$C_{12}H_{11}Cl2_NO$

Propyzamide

3,5-Dichloro-N-(1,1-dimethyl-2-propynyl)benzamide; pronamid; RH-315; Kerb. A residual herbicide in a wettable powder from for a wide range of agricultural crops. mp = 155-156°; vapor pressure at 25° = 8.5x10^{-5} mm Hg. *Pan Britannica Industries Ltd.*

16066 Kerb Propyzamide 50

23950-58-5 8058 245-951-4

$C_{12}H_{11}Cl2_NO$

Propyzamide

3,5-Dichloro-N-(1,1-dimethyl-2-propynyl)benzamide; pronamid; RH-315; Kerb. A residual herbicide in a wettable powder form for a wide range of agricultural crops. mp = 155-156°; vapor pressure at 25° = 8.5x10^{-5} mm Hg. *Kommer-Brookwick Ltd.*

16067 Kerecid

54-42-2 4934 200-207-8

$C_9H_{11}IN_2O_5$

Idoxuridine

2'-Deoxy-5-iodouridine; 1-(2-deoxy-β-D-ribofuranosyl)-5-iodouracil; 5-iodo-2'-deoxyuridine; IDU; IDUR; IUDR; Dendrid; Emanil; Herpes-Gel; Herplex; Idexur; Idoxene; Idulea; Iduridin; Ophthalmadine; Stoxil; Virudox. A proprietary preparation of idoxuridine; an ocular antiseptic. mp = 160°(dec) also 190-195°, 240° and > 175°; λ_m = 288nm (log ε 3.87); $[\alpha]^{25}_{D}$ = .4° (c = 0.108 in water). *SmithKline Beecham.* Name unverified.

16068 Kericompost

Compost for houseplants. *ICI Garden Products.*

16069 Kerigrow

Compound fertilizer 6:4:4; liquid fertilizer for houseplants. *ICI Garden Products.*

16070 Keriguards

Pellets containing fertilizer and insecticide in combination. *ICI Chem & Polymers Ltd.*

16071 Kerimid 500

A proprietary polyamide-imide polymer in solution form. *Rhône-Poulenc NV.*

16072 Kerimid 501

A proprietary polyamide-imide polymer in film form. *Rhône-Poulenc NV.*

16073 Kerimid 502

A proprietary polyamide-imide polymer in the form of a green paste, comprising a thermosetting polymer and an aluminum powder filler. *Rhône-Poulenc NV.*

16074 Kerimid 503
A proprietary polyamide-imide polymer in film form, colored green. *Rhône-Poulenc NV.*

16075 Kerimid 601
A proprietary thermosetting polyimide used in the manufacture of glass-fiber laminates. *Rhône-Poulenc NV.*

16076 Keriroot
Hormone rooting powder. *ICI Chem & Polymers Ltd.*

16077 Kerishine
Leaf glosser for houseplants. *ICI Garden Products.*

16078 Kerispikes
Compound fertilizer 6:4:4; food spikes for houseplants. *ICI Garden Products.*

16079 Kerispray
Contains pirimiphos-methyl with synergized pyrethrins; pesticide spray for houseplants. *ICI Garden Products.*

16080 Keristicks
Capillary sticks for houseplants. *ICI Chem & Polymers Ltd.*

16081 Kern's hydraulic bronze
An alloy of 78% copper, 12% tin, and 10% zinc.

16082 Kerobit®
Mineral oil fraction antioxidants. *BASF plc.*

16083 Kerobit® BPD
101-96-2 202-992-2
N,N'-Di-*s*-butyl-*p*-phenylene diamine
Antioxidant for crude oil distillates. *BASF AG.* Name unverified.

16084 Kerofluid®
Additive for prevention of ice formation in carburetors and jet fuels. *BASF AG; BASF plc.*

16085 Keroflux® 5323, 5486, H, M
Low molecular weight wax; cold flow improver for middle distillates. *BASF AG; BASF plc.*

16086 kerogen
8032-30-2
The organic matter contained in shale. It amounts to 20-27% and on distillation gives water, ammonia, gas, and oil; after fractionation and refinement the oil can yields 18% gasoline 30% kerosene, 27% gas oil, 15% light lube oil, 10% heavy lube oil.

16087 Kerokorr®
Corrosion inhibitors for fuels. *BASF AG.*

16088 Kerol
Disinfectant fluid. *The Wellcome Foundation Ltd.* Name unverified.

16089 Kerolite
Space heater fuel. *Monsanto (Solaris).* Name unverified.

16090 Keromask
A proprietary cosmetic preparation containing titanium oxide and ochre pigments. *Innoxa (England) Ltd.* Name unverified.

16091 Keromet
Mineral oil fraction metal deactivator. *BASF plc.*

16092 Keromet MD 60, MD 80
94-91-7 202-374-2
N,N'-Disalicylidene-1,2-propane diamine
Chelating agent and metal deactivator for fuels and lubricating oils. *BASF AG.* Name unverified.

16093 Keronyx
A proprietary trade name for a casein plastic material used for the manufacture of combs, etc. No manufacturer.

16094 Keropon®
Antifouling agents for crude oil processing plants; stabilizers for storage of petrol. distillates. *BASF AG.*

16095 Keropur®
Additives for cleaning and maintaining the inlet systems of internal combustion engines. *BASF AG; BASF plc.*

16096 kerosene
8003-20-6 5305
C_{10} to C_{16} hydrocarbons
Kerosene
Kerosine. (Paraffin oil, astral oil, coal oil) A refined distillate of petroleum, 150-300 C; used as an illuminating oil, acetylating agent, starting point for production of acetic anhydride and acetate esters. d about 0.80; bp = 175-325°; fp = 65-85°.

16097 Kerosol 200
A low odor hydrocarbon fraction; aliphatic solvent. *Sasolchem.*

16098 Kerostat® 5009
Salts of amino carboxylic acid; antistatic additive for fuels. *BASF AG.* Name unverified.

16099 Kessco®
Esters; emollients, lubricants, emulsifiers. *Stepan.*

16100 Kessco® 653
540-10-3 2073 208-736-6
$C_{32}H_{64}O_2$
Cetyl palmitate
Hexadecanoic acid hexadecyl ester; palmitic acid hexadecyl ester; hexadecyl palmitate. Synthetic spermaceti wax, thickener, viscosity booster for pharmaceutical and cosmetic products; water-oil emulsifier, lubricant for metalworking fluids. n_D^{70}= 1.4398. *Stepan; stepan Canada.*

16101 Kessco® 874
Pentaerythritol tetracaprylate/caprate
High temperature stable lubricant base for textile and metalworking applications. *Stepan; Stepan Canada.*

16102 Kessco® 887
68956-08-1 273-299-0
Trimethylolpropane tricaprylate/caprate
High temperature stable lubricant base for textile and metalworking applications. *Stepan; Stepan Canada.*

16103 Kessco® BS
123-95-5 1625 204-666-5
$C_{22}H_{44}O_2$
Butyl stearate
Octadecanoic acid butyl ester. Biodegradable replacement for mineral oil; used as lubricants in textile spin finish, coning oils, carding, dye bath. mp = 27° also 16°; bp = 343°; fp = 196°. *Stepan; Stepan Canada.*

16104 Kessco® EGAS
Glyceryl stearate, stearamide AMP; pearlescent, bodying agent for shampoos, liq. hand soaps; imparts soft, smooth skin feel to formulations. *Stepan; Stepan Canada.*

16105 Kessco® EGDS
627-83-8 211-014-3
Glycol distearate
Pearlescent, emollient, emulsifier. *Stepan; Stepan Canada.*

16106 Kessco® EGMS
Glycol stearate
Pearlescent, bodying agent, emulsion stabilizer. *Stepan; Stepan Canada.*

16107 Kessco® GDL
27638-00-2 248-586-9
Glyceryl dilaurate
Surfactant for free-flowing lotions. *Stepan; Stepan Canada.*

16108 Kessco® GDS 386F
1323-83-7 215-359-0
Glyceryl distearate
Emulsifier, opacifier, bodying agent. *Stepan; Stepan Canada.*

16109 Kessco® GMC-8
Glyceryl caprylate/caprate
Solubilizer and emulsifier for vitamins, flavors, and medicaments. *Stepan; Stepan Canada.*

16110 Kessco® GML
142-18-7 205-526-6
Glyceryl laurate
Primary emulsifier for water-oil emulsions. *Stepan; Stepan Canada.*

16111 Kessco® GMO
111-03-5 203-827-7
Glyceryl oleate
water-oil emulsifier, emollient, spreading agent, pigment dispersant; lubricant for textiles, metalworking compounds; sperm oil replacement. *Stepan; Stepan Canada.*

16112 Kessco® GMS
Glyceryl stearate
Emulsifier, opacifier, bodying agent; lubricant for textiles, metalworking compounds. *Stepan; Stepan Canada.*

16113 Kessco® IBS
646-13-9 5165 211-466-1
$C_{22}H_{44}O_2$
Isobutyl stearate
Lubricant for textiles, metalworking compounds; slip aid, wetting agent for pigmented lipsticks, bath oils, nail polish and removers, skin cleaners, creams, lotions. mp about 20°. *Stepan; Stepan Canada.*

16114 Kessco® ICS
25339-09-7 246-868-6
Isocetyl stearate
Emollient for makeup formulations. *Stepan; Stepan Canada.*

16115 Kessco® IPM
110-27-0 5234 203-751-4
$C_{17}H_{34}O_2$
Isopropyl myristate

Tetradecanoic acid 1-methylethyl ester. Biodeg. replacement for mineral oil; used as lubricants in textile spin finish, coning oils, carding, dye bath; emollient, solubilizer, vehicle for makeup, shaving preps., bath oils, hair preps. n_D^{25}= 1.432-1.434 *Stepan; Stepan Canada.*

16116 Kessco® IPP
142-91-6 205-571-1
$C_{19}H_{38}O_2$
Isopropyl palmitate
Hexadecanoic acid isopropyl ester; palmitic acid isopropyl ester. Biodeg. replacement for mineral oil; used as lubricants in textile spin finish, coning oils, carding, dye bath; emollient, solubilizer, vehicle for makeup, shaving preps., bath oils, hair preps. mp = 11-13°; fp > 110°; d = 0.852; n_D^{20}= 1.4380. *Stepan; Stepan Canada.*

16117 Kessco® Octyl Isononanoate
71566-49-9 275-637-2
Octyl isononanoate
Emollient with dry, nonoily skin feel; for creams, lotions, makeup, lipstick, antiperspirants. *Stepan.*

16118 Kessco® Octyl Palmitate
29806-73-3 249-862-1
Octyl palmitate
Biodeg. replacement for mineral oil; used as lubricants in textile spin finish, coning oils, carding, dye bath; gloss aid, emollient for makeup, hair grooms, creams, lotions; binder for pressed powds. *Stepan; Stepan Canada.*

16119 Kessco® PEG 200 DL
9005-02-1
PEG-4 dilaurate; lubricant, emulsifier, softener for textile and metalworking applications. *Stepan; Stepan Canada.*

16120 Kessco® PEG 200 DO
9005-07-6
PEG-4 dioleate; surfactants for cosmetics, pharmaceuticals, food, agriculture, plastic, and other industries; thickener, solubilizer. *Stepan; Stepan Canada.*

16121 Kessco® PEG 200 DS
9005-08-7
PEG-4 distearate
Poly(ethylene glycol-400) distearate. Surfactants for cosmetics, pharmaceuticals, food, agriculture, plastic, and other industries; thickener, solubilizer. mp = 35-37°; fp > 110°. *Stepan; Stepan Canada.*

16122 Kessco® PEG 200 ML
9004-81-3
PEG-4 laurate
Surfactants for cosmetics, pharmaceuticals, food, agriculture, plastic, and other industries; thickener, solubilizer. *Stepan; Stepan Canada.*

16123 Kessco® PEG 200 MO
9004-96-0
PEG-4 oleate
Surfactants for cosmetics, pharmaceuticals, food, agriculture, plastic, and other industries; thickener, solubilizer. *Stepan; Stepan Canada.*

16124 Kessco® PEG 200 MS
9004-99-3
PEG-4 stearate
Surfactants for cosmetics, pharmaceuticals, food, agriculture, plastic, and other industries; thickener, solubilizer. *Stepan; Stepan Canada.*

16125 Kessco® PGML
1323-39-3 215-354-3
Propylene glycol laurate
Emollient, emulsifier. *Stepan; Stepan Canada.*

16126 Kessco® PGMS
1323-39-3 215-354-3
Propylene glycol stearate
Emulsifier with m.p. near body temperature *Stepan; Stepan Canada.*

16127 Kester
Fatty acid esters. *Croda Chem. Ltd.* Discontinued.

16128 Kester Wax® K 48
136097-97-7
Ester Wax. *Koster Keunen.*

16129 Ketjenblack® EC-310 NW
1333-86-4 1856 215-609-9
Carbon black aqueous dispersed electroconductive carbon black for rubber compounding. *Akzo.*

16130 Ketjencat
Fluid cracking catalyst. *Akzo.*

16131 Ketjenflex
Toluene sulfonamides. *Akzo.*

16132 Ketjenlube® 115
Butanol ester of α-olefin dicarboxylic acid copolymer; lubricant for load carrying applications; provides wear reduction. *Akzo.*

16133 Ketjensept
127-65-1 2118 204-854-7
$C_7H_7ClNNaO_2S$
Chloramine T
N-Chloro-4-methylbenzenesulfonamide sodium salt; (N-chloro-*p*-toluenesulfonamido)sodium; sodium *p*-toluenesulfonchloramide; chloramine; Aktiven; Chloraseptine; Chlorazene; Chlorazone; Euclorina; Gansil; Halamid; Mianine; Tochlorine; Tolamine. Antibacterial. mp = 167-170°; *Akzo.*

16134 Ketjensil® SM 405
1344-00-9 215-684-8
Sodium-aluminum silicate, precipitated; reinforcing filler for silicone rubber applications; filler in disp. paints for partial replacement of TiO_2. *Akzo.* Name unverified.

16135 keto resins
Artificial resins obtained by the polymerization of aldehyde ketone condensation products.

16136 Keto-Diastix
A proprietary test-strip used to detect ketones and glucose in urine. *B. C. Ames.* Name unverified.

16137 α-ketoglutaric acid
328-50-7 5314 206-330-3
$C_5H_6O_5$
2-oxopentanedioic acid
2-Oxoglutaric acid; 2-ketoglutaric acid; α-ketoglutaric acid; 2-oxopentanedioic acid; 2-oxo-1,5-dipentanedioic acid. mp = 113-115°. *Am. Biorganics; Penta Mfg.; Schweizerhall; U.S. Biochemical.*

16138 β-ketoglutaric acid
542-05-2 66 208-797-9
$C_5H_6O_5$
1,3-Acetonedicarboxylic acid
3-Ketoglutaric acid; 3-oxoglutaric acid; 3-oxopentanedioic acid. Organic synthesis. mp = 134° dec., also 138° dec. *ADA; acetonedicarboxylic acid, dajac Labs; Janssen Chimica; Lonza; Penta Mfg.*

16139 Ketomax
Immobilized glucose isomerase. *UOP Speciality Products.*

16140 ketone base
90-94-8 6265 202-027-5
$C_{17}H_{20}N_2O$
Tetramethyldiaminobenzophenone
Michler's Ketone; 4,4'-bis(dimethylamino)benzophenone; bis[4-(dimethylamino)phenyl]methanone. An intermediate for dyes. mp = 172-176°; bp > 360° dec.

16141 ketone musk
81-14-1 201-328-9
$C_{14}H_{18}N_2O_5$
Dinitro-tert-butylxylyl methyl ketone
4-tert-Butyl-2,6-dimethyl-3,5-dinitroactophenone; Ketone Moschus. An artificial musk perfume. mp = 138-140°.

16142 Ketonone
A proprietary trade name for benzoic acid derivatives used as plasticisers for cellulose acetate and cellulose nitrate. No manufacturer.

16143 Ketonone B
A proprietary trade name for butylbenzoyl benzoate; a plasticizer. No manufacturer.

16144 Ketonone E
A proprietary trade name for ethyl *o*-benzoyl benzoate; a plasticizer. No manufacturer.

16145 Ketonone M
A proprietary trade name for methyl *o*-benzoyl benzoate; a plasticizer. No manufacturer.

16146 Ketonone M.O
A proprietary trade name for methylethyl benzoyl benzoate; a plasticizer. No manufacturer.

16147 Ketostix
A proprietary test strip of buffered sodium nitroprusside and glycine; used for the detection of ketones in urine, serum or milk. *B. C. Ames.* Name unverified.

16148 Ketovite
A complete vitamin supplement for restricted or synthetic diets; Ketovite tablets used in conjunction with Ketovite liquids will provide a complete vitamin supplement for use in conditions such as phenylketonuria, disorders of carbohydrate or amino acid metabolism. *Paines & Byme Ltd.* Name unverified.

16149 Ketrax
14769-73-4 5486 238-836-5
$C_{11}H_{12}N_2S$
Levamisole
(S)-2,3,5,6-Tetrahydro-6-phenylimidazo[2,1-b]thiazole; (-)-6-phenyl-2,3,5,6-

tetrahydroimidazo[2,1-b]thiazole; Levovermax; Totalon. Anthelmintic containing levamisole. mp = 60-61.5°; $[\alpha]_D^{25}$ = -81.5° (c = 10 in chloroform) *ICI Chem & Polymers Ltd.*

16150 Kevlar® 29, 49
Aramid; high strength, lightweight, flexible material; used for aircraft/aerospace, boat hulls, prosthetics, footwear, sporting goods, ropes and cables, fiber optics, bullet-resistant vests, fabrics, brakes and other friction products, in radial tires, as a reinforcement for mechanical rubber goods; available as continuous filament yarn, staple, engineered short fiber, pulp, spun yarn, needlepunched felt, paper, woven fabrics, cord or narrow webbing. *DuPont.*

16151 Kevlar®
Aramid fiber. *DuPont UK.*

16152 key alloy
A nickel-silver containing 60-65% copper, 20-26% zinc, 12% nickel, 1-2% lead, and 0-0.4% iron.

16153 Keycide® X-10
56-35-9 200-268-0
$C_{24}H_{54}OSn_2$
Bis(tri-n-butyltin) oxide
Hexabutyl distannoxane; Tri-n-butyltin oxide. Tributyltin oxide, stabilized; antimildew additive, antimicrobial for PVAc latex paints for packaged stability and mildew-resistant applied films. mp = -45°; bp_2 = 179-180°; fp > 110°; d = 1.172; n_D^{20}= 1.4680. *Witco/Organics.* Discontinued.

16154 Keydime
Ketene dimer emulsion. *Tenneco Malros Ltd.* Name unverified.

16155 Keykote®
Iron, zinc, or manganese phosphate coatings; protective coating for ferrous and nonferrous metals. *Witco/Allied-Kelite.* Discontinued.

16156 Keystone
Adhesives. *Associated Adhesives.* Unverified.

16157 Keytrol
A total weedkiller containing aminotriazole, atrazine and 2,4-D in a wettable formulation; provides broad spectrum control of grassy and broad-leaf weed species, including deep-rooted perennials. *Burts & Harvey.* Name unverified.

16158 KF Polymer® C-1000
24937-79-9
Poly(vinylidene fluoride)
PVDF; fluid-bed powd. coating. *Kureha Chem. Industry.*

16159 KF Polymer® T-1300
24937-79-9
Poly(vinylidene fluoride)
PVDF; for extrusion, compression and blow molding of pipe and bottles. *Kureha Chem. Industry.* Name unverified.

16160 KF Polymer® T-850
24937-79-9
Poly(vinylidene fluoride)
PVDF; for injection molding and extrusion of pipe, sheet, plate, film, filament, wire, insulation. *Kureha Chem. Industry.* Name unverified.

16161 KF Polymer® U-1000
24937-79-9
Poly(vinylidene fluoride)
PVDF; powd. coating for anticorrosive coating applications. *Kureha Chem. Industry.* Name unverified.

16162 KF Polymer® W-1000
24937-79-9
Poly(vinylidene fluoride)
PVDF; extrusion molding for color masterbatch. *Kureha Chem. Industry.* Name unverified.

16163 Khaki Yellow C
Sulfur dyestuffs. No manufacturer.

16164 Khakl
A coloring matter produced on the fiber. The material is dipped in chrome alum, ferrous sulfate, and pyrolignite of iron, and then passed through a solution of sodium silicate.

16165 khari salt
A native salt of India consisting chiefly of sodium sulfate; used for curing skins.

16166 Kidnamin
A proprietary preparation of essential aminoacids; used as a dietary supplement. *KabiVitrum Ltd.* Unverified.

16167 kidney cotton
Peruvian cotton.

16168 kiel compound
An insulating material containing rubber, sulfur and mineral oil. It sometimes also contains pumice and beeswax.

16169 kien oll
Turpentine oil obtained by the dry distillation of resinous wood.

16170 Kienmeyer's amalgam
An amalgam consisting of 2 parts mercury, 1 part tin, and 1 part zinc; used as a coating for frictional electrical machines.

16171 Kieralon®
Textile scouring agents. *BASF plc.*

16172 Kieralon® C
Detergent blend; detergent for scouring; used in neutral and high strength caustic. *BASF/Fibers.*

16173 Kieralon® ED
Blend; low-foaming detergent, emulsifier, dispersant for scouring cotton, synthetic fibers, and blends. *BASF.*

16174 Kieralon® NB-ED
Nonionic/anionic formulation; low foaming detergent with emulsifying and dispersing properties; for desizing, scouring, caustic boil-off, bleaching processes. *BASF/Fibers.*

16175 Kieralon® TX-199
Emulsion; low foaming wetting/scouring additive for pretreatment, bleaching of all fabrics; for use in high turbulence equipment *BASF/Fibers.*

16176 Kieralon® TX-410 Conc
Polyoxyethylated blend; afterscouring agent, wetting agent, detergent for dyed and printed fabrics; stable to hard water, moderate concs. of acids and alkalies. *BASF/Fibers.*

16177 Kieselguhr dynamite
Gurdynamite
dynamite. Ordinary dynamite, consisting of nitroglycerin absorbed by kieselguhr.

16178 Kil
Insecticides. *Fisons plc, Horticultural Div.* Name unverified.

16179 Kilfoam
Silicone, emulsifiers, and stabilizer emulsion; defoamer for textile and aqueous industrial processing, dye and finish baths; emulsifier for paints. *Arol Chem. Prods.*

16180 Kilianite
A proprietary synthetic resin product. No manufacturer.

16181 killed spirits
A solution of zinc chloride. Prepared by dissolving zinc in commercial hydrochloric acid until action ceases.

16182 Killgerm® Py-Kill W
7696-12-0 9362 231-711-6
$C_{19}H_{25}NO_4$
Tetramethrin
2,2-Dimethyl-3-(2-methyl-1-propenyl)cyclopropanecarboxylic acid (1,3,4,5,6,7-hexahydro-1,3-dioxo-2H-isoindol-2-yl)methyl ester; 2,2-dimethyl-3-(2-methylpropenyl)cyclopropanecarboxylic acid ester with N-(hydroxymethyl)-1-cyclohexene-1,2-dicarboximide; N-(3,4,5,6-tetrahydrophthalimide)methyl-*cis,trans*-chrysanthemate; N-(chrysanthemoxymethyl)-1-cyclohexene-1,2-dicarboximide; phthalthrin; FMC-9260; SP-1103; Neo-Pynamin. Emulsifiable concentrate containing 22 g/l tetramethrin; for control of files in livestock houses. mp = 65-80°; d_{20}^{20} = 1.108; $n^{21.5}_D$ = 1.5175. *Killgerm Chemicals Ltd.*

16183 Killgerm® Ratak Cut Wheat Rat Bait
56073-07-5 259-978-4
Difenacoum; a ready-to-use anticoagulant rodenticide. *Killgerm Chemicals Ltd.*

16184 Killgerm® Sewarin P
81-81-2 10174 201-377-6
$C_{19}H_{16}O_4$
Warfarin
4-Hydroxy-3-(3-oxo-1-phenylbutyl)-2H-1-benzopyran-2-one; 3-(α-acetonylbenzyl)-4-hydroxycoumarin; 1-(4'-hydroxy-3'-coumarinyl)-1-phenyl-3-butanone; 3-α-phenyl-β-acetylethyl-4-hydroxycoumarin; compound 42; WARF compound 42; Co-Rax; Rodex. 0.025% Warfarin on pin-head oatmeal, with sugar and mold inhibitor; rat and mouse killer. mp = 161°; λ_m (water, pH 10) = 308 nm (ε 13610). *Killgerm Chemicals Ltd.*

16185 Killgerm® Sol Odamask H
Mixture of odor absorbents, fixatives and odor masking agents; industrial deodorant in sewage works, on refuse tips, maggot farms, etc. *Killgerm Chemicals Ltd.*

16186 Killgerm® ULV 400
8148
Pyrethin
Contact insecticide. *Killgerm Chemical Ltd.*

16187 Killgerm® ULV 500
Mixture containing phenothrin and tetramethrin; for control of flying insects in agricultural premises. *Killgerm Chemicals Ltd.*

16188 Kilmet
Selective weedkillers. *May & Baker Ltd.* Name unverified.
16189 Kilnet
Selective weedkiller. *May & Baker Ltd.* Name unverified.
16190 Kinel 5502
A proprietary polyimide casting, potting, and encapsulating resin. *Rhône-Poulenc NV.*
16191 Kinel 5514
A proprietary polyimide molding composition reinforced with glass fiber. *Rhône-Poulenc NV.*
16192 Kinel 5517
A proprietary free-sintering self-lubricated, heatresistant polyimide molding powder. *Rhône-Poulenc NV.*
16193 Kinetite
A mixture of the jelly formed by dissolving guncotton in nitrobenzene with potassium chlorate or potassium nitrate and sulfur; an explosive.
16194 Kingston bronze
An alloy of 85% copper, 12% zinc, 2.5% tin, and 0.05% iron.
16195 Kinite
A proprietary trade name for steel containing 12.5-14.5% chromium, 1.5% carbon, 1.1 cent molybdenum, 0.7% cobalt, 0.55% silicon, 0.5% manganese, and 0.4% nickel. No manufacturer.
16196 kino
5325
Kino
Resin kino; gum kino. The dried juice obtained from incisions in the trunk of *Pterocarpus marsupium.* It resembles catechu, and is used in dyeing and medicine.
16197 kino, Australian
The dried exudate from *Eucalyptus species.*
16198 kino, Bengal
Kino, Madras; Dhak gum. Butea gum, from *Butea frondosa.*
16199 Kirnol® Range
Fatty acid mono and diglycerides; general purpose food emulsifiers. *Grünau.*
16200 kish
Crystalline graphite formed in blast fumace slag during iron smelting.
16201 Kite
A paper based grade of Tufnol industrial laminates. *Tufnol Ltd.*
16202 Kiton
Acid wool dyestuffs. *Clayton Aniline Co Ltd.* Name unverified.
16203 Kittool fiber
A fiber obtained from the leaves of a Ceylon palm, *Caryota urens*; used in the manufacture of brushes.
16204 Kival
Insecticide. *May & Baker Ltd.* Name unverified.
16205 Klea
CFC ozone-benign replacement refrigerant. *ICI Chem & Polymers Ltd.*
16206 Klearfac® 870
Block copolymer phosphate ester; surfactant. *BASF.*
16207 Klearfac® AA270
Alcohol alkoxylate phosphate ester, free acid; emulsifier, solubilizer, dedusting agent, hydrotrope, metal cleaner. *BASF.*
16208 Klearol
7327
Petroleum, Liquid
Liquid paraffin; mineral oil; white mineral oil; paraffin oil; Clearteck; Drakeol; Hevyteck; Kremol; Kaydol; Alboline; Nujol. White mineral oil. *Witco.* Discontinued.
16209 Klebcil
25389-94-0 5293 246-933-9
Kanamycin A sulfate
Cantrex; Cristalomicina; Enterokanacin; Kamycin; Kamynex; Kanabristol; Kanacedin; Kanamytrex; Kanasig; Kanatrol; Kanicin; Kannasyn; Kantrex; Kantrox; Otokalixin; Resistomycin; Ophtalmokalixan; Kantrexil; Kano; Kanescin; Kanaqua. Antibacterial. dec above 250°. *SmithKline Beecham.* Name unverified.
16210 Kleen-Dent
Tooth polishing agent. *Reheis Inc.*
16211 Kleenite
Denture cleanser powder; for cleaning dentures. *Richardson-Vicks Inc.* Name unverified.
16212 Kleenmold
Graphite lubricants for use in the glass industry, principally as mold releases. *Specialty Products Co.* Name unverified.
16213 Kleenup
Weed and grass Killer. *Monsanto (Solaris).* Name unverified.
16214 Kleerox® HCS
Org. stabilizer for hydrogen peroxide bleach liquors in textile industry. *Rhône-Poulenc Surf.*
16215 Klee's salt
$KHC_2O_4 \cdot H_2O$
Acid potassium oxalate
16216 Klegecell R30
9002-86-2 7746 206-625-7
High performance PVC foam; rigid closed-cell foam with tridimensional grid structure, high thermal stability; structural sandwich core material providing low weight, insulation and stuffiness. *Polimex.*
16217 Kleinenberg's fat mixture
A solution of cacao butter and spermaceti in castor oil; used as an embedding material in microscopy.
16218 Kleinenberg's fixative
Used in microscopy. It consists of 100 ml of a saturated aqueous solution of picric acid, 3 ml of sulfuric acid, and 300 mlc of water.
16219 Klein's reagent
$2(Cd_2H_2W_8O) \cdot {}_7(WO_3)B_2O_3 \cdot H_2O$
A saturated solution of cadmium borotungstate, used for the separation of minerals.
16220 Klenal
Industrial cleaner. *Specialty Products Co.* Name unverified.
16221 Klenenberg's stain
A microscopic stain. It consists of a saturated solution of alum and calcium chloride in alcohol (70%) diluted with 6 times its volume of alcohol (70%) to which is added an alcoholic solution of hematoxylin until the color is violet blue.
16222 Kleptose®
Betacyclodextrin
Encapsulating agent for pharmaceutical and chemical industries. *Roquette (UK) Ltd.*
16223 Klerat
Rodenticide. *ICI Chem & Polymers Ltd.*
16224 Klere-Seal®
Alkylalkoxysilane solution; reacts with minerals in concrete and masonry to form a chemical bond preventing the intrusion of water and water-borne salts into the substrates. *Pecora Corporation.*
16225 Kloben®
555-37-3 6523 209-096-0
$C_{12}H_{16}Cl_2N_2O$
Neburon
N-Butyl-N'-(3,4-dichlorophenyl)-N-methylurea; 3-(3,4-dichlorophenyl)-1-methyl-1-n-butylurea; Kloben Neburon. For the agriculture industry. mp = 101.5-103°. *DuPont UK.*
16226 K-Lor
7447-40-7 7783 231-211-8
KCl
Potassium chloride
Chloropotassuril; Diffu-K; Enseal; Kaleorid; Kalitabs; Kalium-Duriles; Kaon-Cl; Kaskay; Kayback; Kay-Cee-L; K-Contin; Klor-Con; K-Norm; K-Tab; Lento-Kalium; Micro K; Nu-K; Peter-Kal; PfiKlor; Rekawan; Repone K; Slow-K; Span-K. Replenisher. d = 1.98; mp = 773°. *Abbott Laboratories.* Name unverified.
16227 Kloro 6001
Chlorinated paraffin
Very good heat stability and resistance to hydrolysis; used for sol. oils, synthetics, and semi-synthetics. *Ferro/Keil.*
16228 Kluberlubrication
Specialty lubricant and grease. *Carl Freudenberg.* Name unverified.
16229 Klucel®
9004-64-2 4888
Hydroxypropylcellulose
Cellulose 2-hydroxypropyl ether; oxypropylated cellulose; Lacrisert. Emulsion stabilizer, emulsification aid, whipping aid, suspending agent, thermoplastic resin and thickener. Softens at 130°. *Hercules.*
16230 Klucel® E, G, H, J, L, M
9004-64-2 4888
Hydroxypropylcellulose
Cellulose 2-hydroxypropyl ether; oxypropylated cellulose; Lacrisert. Hydroxypropylcellulose, standard grades; surface active thickener, stabilizer, film-former, suspending agent, protective colloid for coatings, adhesives, extrusions, moldings, paper, paint removers, encapsulations, inks. Softens at 130°. *Aqualon.*
16231 Klucel® EF
9004-64-2 4888
Hydroxypropylcellulose
Cellulose 2-hydroxypropyl ether; oxypropylated cellulose; Lacrisert.

Hydroxypropylcellulose, premium grade; stabilizer, film-former, suspending agent, protective colloid; esp. for nonaerosol hairspray formulations. Softens at 130°. *Aqualon.*

16232 Klucine
A proprietary waterproofing compound. No manufacturer.

16233 K-Lyte
298-14-6 7770 206-059-0
$CHKO_3$
Potassium bicarbonate
Potassium hydrogen carbonate; potassium acid carbonate; Kafylox; K-Lyte. Pharmaceutic necessity. *Mead Johnson & Co.*

16234 K-Lyte/C1
7447-40-7 7783 231-211-8
KCl
Potassium chloride
Chloropotassuril; Diffu-K; Enseal; Kaleorid; Kalitabs; Kalium-Duriles; Kaon-Cl; Kaskay; Kayback; Kay-Cee-L; K-Contin; Klor-Con; K-Norm; K-Tab; Lento-Kalium; Micro K; Nu-K; Peter-Kal; PfiKlor; Rekawan; Repone K; Slow-K; Span-K. Replenisher. d = 1.98; mp = 773°. *Mead Johnson & Co.*

16235 KM Ammonium Metavanadate
7803-55-6 605 232-261-3
NH_4VO_3
Ammonium Vanadate (V)
Ammonium metavanadate. *Kerr-McGee Chemical Corp.*

16236 KM Ammonium Perchlorate
7790-98-9 574 232-235-1
NH_4ClO_4
Ammonium perchlorate
Depending on grade; manufactured in ordnance and industrial grades. d = 1.95; decomposes when heated. *Kerr-McGee Chemical Corp.* Discontinued.

16237 KM Fly Ash
Damp bulk form. *Kerr-McGee Chemical Corp.* Discontinued.

16238 KM Manganese Dioxide
1313-13-9 5770 215-202-6
MnO_2
Manganese(IV) oxide
Manganese dioxide; manganese binoxide; manganese peroxide; manganese superoxide; black manganese oxide. AB and SB battery active grades 90% minimum MnO_2 for use in Leclanche, alkaline and zinc chloride dry cell batteries. mp = 1650° *Kerr-McGee Chemical Corp.*

16239 KM Muriate of Potash
White agricultural grade.; 62% minimum K_2O in granular, coarse and standard grades. *Kerr-McGee Chemical Corp.* Discontinued.

16240 KM Pebble Lime
1305-78-8 1733 215-138-9
CaO
Calcium oxide
Lime; burnt lime; calx; quicklime. 90% minimum available CaO, in mill run, coarse, medium, fine and crushed grades. mp = 2572°; bp = 2850°; d = 3.32-3.35. *Kerr-McGee Chemical Corp.* Discontinued.

16241 KM Phosphate Rock
68, 70, 72 and 75% BPL (bone phosphate of lime expressed as $Ca_3(PO_4)_2$) grades. *Kerr-McGee Chemical Corp.* Discontinued.

16242 KM Potassium Chloride
7447-40-7 7783 231-211-8
KCl
Potassium chloride
Chloropotassuril; Diffu-K; Enseal; Kaleorid; Kalitabs; Kalium-Duriles; Kaon-Cl; Kaskay; Kayback; Kay-Cee-L; K-Contin; Klor-Con; K-Norm; K-Tab; Lento-Kalium; Micro K; Nu-K; Peter-Kal; PfiKlor; Rekawan; Repone K; Slow-K; Span-K. High purity white industrial grade; 98.3% KCl (62% minimum K_2O equivalent). d = 1.98; mp = 773°. *Kerr-McGee Chemical Corp.* Discontinued.

16243 KM Potassium Perchlorate
7778-74-7 7822 231-912-9
$KClO_4$
Potassium perchlorate
Peroidin; Perchloracap. 99.7% $KClO_4$, industrial and military grades. mp = 610° dec; d = 2.52. *Kerr-McGee Chemical Corp.* Discontinued.

16244 KM Sodium Chlorate
7775-09-9 8741 231-887-4
$NaClO_3$
Sodium chlorate
Atlacide; Defol; Dervan. Technical grade; 99.5% minimum in bulk; drummed may contain 0.25-0.5% anticaking agent; can be supplied in dry bulk, solution and with salt added to custom specifications. mp = 248° dec; d = 2.5. *Kerr-McGee Chemical Corp.*

16245 KM Sodium Perchlorate
7601-89-0 8798 231-511-9
$NaClO_4$
Sodium perchlorate
Irenat. Aqueous solution; 60-64%. dec about 130°; d = 2.02. *Kerr-McGee Chemical Corp.* Discontinued.

16246 KM Vanadium Pentoxide
1314-62-1 10056 215-239-8
V_2O_5
Vanadium(V) oxide
Vanadium pentoxide; vanadic anhydride. Fused flake and fine granular, 98% minimum V_2O_5. d = 3.35; mp = 690°; looses oxygen reversibly from 700-1125°. *Kerr-McGee Chemical Corp.*

16247 KMC
Diisopropyl naphthalene
High boiling solvent. *Collinda Ltd.*

16248 KMD-50
N,N'-Disalicylidene-1,2-propanediamine
Metal deactivator for gasoline, distillate fuels, and other petrol. products. *Ferro/Keil.*

16249 K-Monel
A proprietary alloy containing nickel and copper in approximately the same ratio as in monel metal with the addition of 4% aluminum. No manufacturer.

16250 Knapp's solution
Mercurous chloride (10.8 grains) are treated with potassium cyanide solution until the addition of caustic soda causes no precipitate. Caustic soda solution (100 ml of specific gravity 1.145), added, and the whole diluted to 1 liter; used for the estimation of glucose.

16251 Knauerit 2
A plaster-shooting (mud-capping) explosive of greatest brisance and high velocity of detonation. Developed for high performance and used without stemming for pop shots; from the cartridge of Knaurerit 2 slices of adequate thickness are cut to be closely pressed against the boulder. *Dynamit Nobel Wien GmbH.* Name unverified.

16252 Knauerit S
An explosive plaster charge of best adhesive strength which can be formed by handto fit the shape of the underground; advantageous to use Knaurerit S for demolition work, e.g., for cutting off iron constructions, bridge girders, rails, etc. *Dynamit Nobel Wien GmbH.* Name unverified.

16253 Knave
Granular mixture of disulfoton and quinalphos; an organophosphorus insecticide. *Hortichem Ltd.*

16254 Kneiss alloy
An alloy of 42% lead, 40% zinc, 15% tin, and 3% copper; used for machine bearings; another alloy contains 50% zinc, 25% tin, and 25% lead.

16255 Knit-Soft 30 NCPM
Blended softener to be pad or exhaust applied; excellent napping lubricant. No manufacturer.

16256 Knittex
Finishing agent. *Ciba plc.* Name unverified.

16257 Knot Out
82558-50-7 5256
$C_{18}H_{24}N_2O_4$
Isoxaben
N-[3-(1-Ethyl-1-methylpropyl)-5-isoxazolyl]-2,6-dimethoxybenzamide; benzamizole; EL-107; NA-8318; Flexidor; Gallery. Suspension concentrate containing 125 g isoxaben per liter; used for control of annual dicotyledons in cereals, grass and fruit. mp = 176-179°. *Synchemicals Ltd.*

16258 kochenite
A fossil resin resembling amber.

16259 Kochlin's Bearing Bronze
An alloy of 90% copper and 10% tin. No manufacturer.

16260 Koch's acid
1-Naphthylamine-3,6,8 trisulfonic acid
Used as an azo dye intermediate.

16261 Kocide® 20/20
Copper hydroxide plus nutritional zinc; wettable powd. agricultural fungicide, bactericide, and nutritional. *Griffin.*

16262 Kocide® Copper Sulfate Pentahydrate Crystals
7758-99-8 2722 231-847-6
$CuSO_4 \cdot H_2O$
Cupric sulfate, pentahydrate
Copper(II) sulfate pentahydrate; cupric sulphate, pentahydrate; copper(II) sulphate pentahydrate; bluestone; blue vitriol; Roman vitriol; Salzburg vitriol. Fungicide to control plant diseases, in fertilizers to correct copper deficiencies in soils. Becomes anhydrous by 250°; $d^{15.6}_{4}$ = 2.286. *Griffin.*

16263 Kodabond® Copolyester 5116
Terephthalate-based copolyester; heat-seal layer in pkg. applications; applied as coextrusion or an extrusion coating; for coating on paper, polyesters, copolyesters, PC, polyvinylidene chloride. *Eastman.*

16264 Kodaflex® DBP
84-74-2 1622 201-557-4
$C_{16}H_{22}O_4$
Dibutyl phthalate
Di-n-butyl phthalate; phthalic acid di-n-butyl ester; n-butyl phthalate; 1,2-benzenedicarboxylic acid dibutyl ester; phthalic acid dibutyl ester; dibutyl phthalate; DBP. Plasticizer used in coatings industry as primary plasticizer-sol. for nitrocellulose lacquers; for rubbers and CAB, ethyl cellulose, PVAc, and synthetic resins; solv. for oil-sol. dyes, insecticides, peroxides, and org. compounds; antifoamer and fiber lub n_D^{20}= 1.4900; fp = 171°. *Eastman.*

16265 Kodaflex® DBS
109-43-3 203-672-5
Dibutyl sebacate
Plasticizer mp = -12°; d = 0.9400; fp = 167°. *Eastman.*

16266 Kodaflex® DEP
84-66-2 201-550-6
$C_{12}H_{14}O_4$
Diethyl phthalate
Phthalic acid diethyl ester; ethyl phthalate; Neantine; Palatinol A. Plasticizer; wetting agent in grinding pigments; pigment-disp. medium in cellulose acetate solutions and plastics, and solv. for natural resins and polymers; PVC products due to relatively high volatility. mp = -3°; bp = 298-299°; fp = 160°; d = 1.118; n_D^{20}= 1.5020. *Eastman.*

16267 Kodaflex® DIBP
84-69-5 201-553-2
$C_{16}H_{22}O_4$
Diisobutyl phthalate
Phthalic acid diisobutyl ester. Plasticizer mp = -64°; bp = 327°; fp = 185°; d = 1.039; n_D^{20}= 1.4900. *Eastman.*

16268 Kodaflex® DIDA
27178-16-1 248-299-9
Diisodecyl adipate
Plasticizer. *Eastman.*

16269 Kodaflex® DIDP
68515-49-1 271-091-4
Diisodecyl phthalate
Plasticizer. *Eastman.*

16270 Kodaflex® DMEP
117-82-8 204-212-6
Di-(2-methoxyethyl) phthalate
Plasticizer. *Eastman.*

16271 Kodaflex® DMP
131-11-3 3304 205-011-6
$C_{10}H_{10}O_4$
Dimethyl phthalate
Phthalic acid dimethyl ester; 1,2-benzenedicarboxylic acid dimethyl ester; methyl phthalate; dimethyl 1,2-benzendicarboxylate; DMP; Palatinol M; Fermine; Avolin; Mipax. Plasticizer with high solv. power for cellulose acetate extrusion compounds; compatible with ethyl cellulose, CAB, PS, PVAc, polyvinyl butyral, and PVC; used in NC-based printing inks. mp = 5.5°; d_{20}^{20} = 1.1940; d_{25}^{25} = 1.189; fp = 146°; bp_{760} = 283.7°; bp_{400} = 257.8°; bp_{200} = 232.7°; bp_{100} = 210.0°; bp_{40} = 182.8°; bp_5 = 131.8°; n_D^{20}= 1.5168. *Eastman.*

16272 Kodaflex® DOA
103-23-1 203-090-1
$C_{22}H_{42}O_4$
Dioctyl adipate
Di-(2-ethylhexyl)adipate; adipic acid bis(2-ethylhexyl) ester. Plasticizer providing flexibility at low temperatures to vinyl products; used in unfilled garden hose, clear sheeting, electrical insulation. bp = 166-168°; fp >110°; d = 0.990; n_D^{20}= 1.4470 *Eastman.*

16273 Kodaflex® DOP
117-81-7 1291 204-211-0
$C_{24}H_{38}O_4$
Dioctyl phthalate
Bis(2-ethylhexyl) phthalate; phthalic acid bis(2-ethylhexyl)ester; 1,2-benzenedicarboxylic acid bis(2-ethylhexyl)ester; di(2-ethylhexyl) phthalate; Octoil. All-purpose plasticizer used with PVC resins incl. film and sheeting for upholstery, clothing, food pkg., paper coatings, molded vinyl products, electrical wire insulation; compatible with PS, methylmethacrylate, chlorinated rubber, NC, and CAB; low odor, relatively low toxicity and low volatility. bp = 385-386°; fp = 207°; d = 0.985; n_D^{20}= 1.4860. *Eastman.*

16274 Kodaflex® DOTP
6422-86-2 229-176-9
$C_{24}H_{38}O_4$
Dioctyl terephthalate
bis(2-ethylhexyl) terephthalate. Primary plasticizer used with PVC resins, in PVC plastisols, rubber; application incl. wire coatings, automotive and furniture upholstery; compatible with acrylics, CAB, cellulose nitrate, polyvinyl butyral, styrene, oxidizing alkyds, nitrile rubber. mp = 30-34°; bp = 400°; n_D^{20}= 1.4900; d = 0.980; fp > 110°. *Eastman.*

16275 Kodaflex® DOZ
103-24-2 203-091-7
Dioctyl azelate
Plasticizer. *Eastman.*

16276 Kodaflex® HS-3
Butyl octyl phthalate; Primary plasticizer for polyvinyl homopolymer and copolymer resins, vinyl plastisols, expanded vinyl foams, and rotational molding and dip coating. *Eastman.*

16277 Kodaflex® HS-4
60% Dioctyl phthalate, 5% dibutyl phthalate; high-solvating primary plasticizer; used in formulating vinyl plastisols having processing chars. required to be mechanically frothed; for PVC formulations, rotational molding, slush molding. dip coating and filament coating applications. *Eastman.*

16278 Kodaflex® OIDP
Octyl isodecyl phthalate; plasticizer. *Eastman.*

16279 Kodaflex® TEG-EH
94-28-0 202-319-2
Triethylene glycol di-2-ethylhexanoate
Plasticizer *Eastman.*

16280 Kodaflex® TOTM
3319-31-1 222-020-0
$C_{33}H_{54}O_6$
Trioctyl trimellitate
tris(2-ethylhexyl) trimellitate. Primary plasticizer used in vinyl film and vinyl-coated fabrics. bp = 414°; n_D^{20}= 1.4850; d = 0.984; fp > 110°. *Eastman.*

16281 Kodaflex® Triacetin
102-76-1 9721 203-051-9
$C_9H_{14}O_6$
Glyceryl triacetate
Triacetin; 1,2,3-propanetriol triacetate; triacetyl glycerine; Enzactin; Fungacetin. Low-toxicity plasticizer for vinyl compounds; used in adhesives, resinous and polymeric coatings, paper, and paperboard for food contact; water-insol. hydroxyethyl cellulose films. mp = 2-3°; bp = 257-260°; fp = 138°; d = 1.55; n_D^{20}= 1.4310. *Eastman.*

16282 Kodaflex® TXIB
6846-50-0 229-934-9
$C_{16}H_{30}O_4$
2,2,4-Trimethyl-1,3-pentanediol diisobutyrate
Primary plasticizer used in surface coatings, vinyl floorings, molding, and vinyl products; compatible with film-forming vehicles used in lacquers for wood, paper, and metals; primary plasticizer for PVC plastisols for rotocasting and slush molding; used in PNC organosols processed by extrusion and injection molding. mp = -70°; bp = 280°; n_D^{20}= 1.4340; d = 0.941; fp > 110°. *Eastman.*

16283 Kodaloid
9004-70-0 8195
Pyroxylin
Cellulose nitrate; nitrocellulose; collodion cotton; soluble gun cotton; collodion wool; colloxylin; xyloidin; celloidin; Parlodion. A proprietary trade name for a cellulose nitrate. It is made in the form of sheets. fp = 4°; ignites at 160°-170°. No manufacturer.

16284 Kodapak®
9004-35-7 2013
Cellulose acetate
Transparent cellulose acetate film; used for making packets. *Eastman.*

16285 Kodapak® 5214A
25038-59-9 7730
$(-OCH_2CH_2O_2CC_6H_4-4-CO-)_n$
Poly(ethylene terephthalate)
PET polyester hydropolymer; light weight material resistant to breaking, bursting and shattering; improved barrier properties; for bottles of carbonated soft drinks fruit juices, foods, etc. *Eastman.*

16286 Kodapak® PET Copolyester 13339
25038-59-9 7730
$(-OCH_2CH_2O_2CC_6H_4-4-CO-)_n$
Poly(ethylene terephthalate)
PET polyester; for use in packaging and other applications. *Eastman.* Name unverified.

16287 Kodar®
Copolyester thermoplastic *Eastman.*

16288 Kodar® A150 Copolyester
Copolyester

Produces extruded sheet with optical clarity, toughness at low temps., and high tear strength and elongation. *Eastman.*

16289 Kodar® PETG Copolyester 6763
25038-59-9 7730
(-$OCH_2CH_2O_2CC_6H_4$-4-CO-)$_n$
Poly(ethylene terephthalate)
Amorphous glycol-modified PET; offers sparkling clarity in film and sheet form, easy thermoformability, toughness, sterilizability with ethylene oxide or gamma rays for medical applications; used for medical containers, thermoformed food containers and lids, blister packaging for cosmetics, pharmaceuticals, heavy hardware and electronic parts. FDA compliant. *Eastman.*

16290 Kodar® Thermx Copolyester 6761
Crystallizable copolyester; for thermoformed dual ovenable trays providing superior temperature resistance; FDA compliance. *Eastman.*

16291 Kodel
5329
Polyester yarn and fiber. mp = 290-295°; sp gr = 1.22 *Eastman.*

16292 Kodofil, Kodolite, Kodosoff
Polyester fiberfill. *Eastman.*

16293 Koenig solder
An alloy of 60% tin, 30% aluminum, and 10% antimony.

16294 Koerzit
An alloy for permanent magnets containing 1.1% carbon, 3.5% manganese, 36% cobalt, 4.8% chromium, the remainder being iron.

16295 Koerzit, I, II, III
Proprietary cobalt steels containing 10, 20, and 30% cobalt respectively. No manufacturer.

16296 Kogasin III
Nonaromatic hydrocarbon oil, approximately C15-C20. *Sasolchem.*

16297 Kohacool L-400
Alkylether sulfosuccinate
Wetting and foaming agent for nonirritating hair shampoos, bubble baths, hair conditioners and lotions. *Toho Chem. Industry.*

16298 Koka Seki
A variety of pumice stone found in the Nujima Islands, near Tokio; used as a building material.

16299 Koken
A proprietary synthetic resin. No manufacturer.

16300 koko
Celastrus buxifolia ; used in Natal as a sumac substitute for tanning.

16301 kokowal
A variety of rouge used by the Maori.

16302 kokum butter
A fat obtained from the seeds of *Garcinia indica* or *G. Purpurea.* It is composed of stearine, myristicine, and oleine.

16303 kola nut
5331
Kola
Cola; Soudan coffee; Bissy nuts; gooroo nuts; guru nuts. The seeds of *Cola acuminate* and *C. vera.*

16304 Kolax
An explosive of the same type as carbonite.

16305 Kolene
Metal cleaning salts. *Degussa Ltd.*

16306 kol-kol gum
A gum of Nigeria; from *Acacia senegal.*

16307 Kollercast
Moldable synthetic resins; for industrial purposes, e.g., flooring. *Scott Bader.* Discontinued.

16308 Kollercure®
Epoxy curing agents. *Scott Bader.* Discontinued.

16309 Kollerdur® L 90
Polyurethane elastomer, IPDI based; evaporation curing high performance coatings for flexible substrates, e.g., foams, fabrics, leather, plastics, and clear overprint coatings. *Scott Bader.* Discontinued.

16310 Kollerdur® MO118
One-pack polyurethane resin; moisture curing systems to produce primers, sealers, and coatings or formulated to produce aggregate-filled floors, roofing systems. *Scott Bader.* Discontinued.

16311 Kollerdure® M0122
PU resin in methoxy propyl acetate/xylene; aliphatic type; one-pack moisture cure coating for high performance industrial finishes; superior hardness, color retention. *Scott Bader.* Discontinued.

16312 Kollermox®
Epoxy resins. *Scott Bader.* Discontinued.

16313 Kollidon® CL
Crospovidone
Tablet-disintegrating agent, suspension stabilizer *BASF AG; BASF plc.*

16314 kolm
A variety of bituminous coal found in Sweden. The ash contains from 1-3% of uranium oxide, U_{38}.

16315 Kombat
Carbendazim + mancozeb
Systemic fungicide for cereals. *Hoechst UK.*

16316 Kombé arrow poison
9018
Strophanthus, the seed of *Strophanthus hispidus.*

16317 Komeen®
Copper-ethylenediamine complex; aquatic herbicide for hydrilla control in golf course, ornamental, and fish ponds, potable water reservoirs, fresh water lakes, fish hatcheries. *Griffin.*

16318 Kommoid
A sulfurized corn oil rubber substitute.

16319 Kompak
Granular product for use on cast iron to produce a compacted graphite structure. *Foseco (F.S.) Ltd.* Discontinued.

16320 Kompolite
Flooring materials. *Weatherguard/Marbleloid Products Inc.* Name unverified.

16321 kon oil
5675
Kusum oil
Macassar oil; Kon oil; Paka oil. Macassar oil obtained from the seeds of *Schleichera trijuga.* It has a saponification value of 215-230, an iodine value of 48-69, and an acid value of 6-35.

16322 Konakion
84-80-0 7536 201-564-2
$C_{31}H_{46}O_2$
Phytomenadione
Vitamin K; [R-[RName unverified.,RName unverified.-(E)]]-2-methyl-3-(3,7,11,15-tetramethyl-2-hexadecenyl)-1,4-napthalenedione; 2-methyl-3-phytyl-1,4-napthoquinone; 3-phytylmenadione; phytonadione; AquaMephyton; Mephyton; Mono-Kay; Veda-K_1; Veta-K_1. A preparation of Vitamin K. mp = -20°; n_D^{20} = 1.5252; $[\alpha]_D^{20}$ = -0.28° (dioxane). *Roche Laboratories; Roche Products Ltd.*

16323 Konator
Dental equipment. *Degussa Ltd.*

16324 Konel
Proprietary nickel-cobalt-iron alloys containing about 2.5% titanium; they are high temperature resisting alloys and possess high tensile strength at elevated temperatures. No manufacturer.

16325 Konforme® AR
Acrylic resin; conformal coating providing insulation against high voltage arcing and corona shorts. *Chemtronics.*

16326 konilite
A silica in powder form.

16327 Konker®
10605-21-7 1836 234-232-0
$C_9H_9N_3O_2$
Carbendazim
1H-Benzimidazol-2-ylcarbamic acid methyl ester; 2-benzimidazolecarbamic acid methyl ester; 2-(methoxycarbonylamino)benzimidazole) methyl 2-benzimidazolecarbamate; carbendazole; BMC; MBC; BCM; Vinclozolin; BAS-3460; BAS-67054; CTR-6669; HOE-17411; Bavistin; Derosal. Fungicide with systemic activity and contact effect for use in rape, sunflower, strawberries, vegetables mp = 302-307° dec; pKa = 4.48 *BASF AG.*

16328 konnan bark
Obtained from *Cassia fistula* of Southern India; a tanning material.

16329 Konservan SN
Organic tin compound; nonionic dispersion; preservative for textiles. *Anti-Chem; Thor Chemicals (UK) Ltd.*

16330 Konstrastin
Basic zirconium basic acetate.

16331 Konstructal
An aluminum alloy containing 1% copper or 8% zinc.

16332 Kontakt
A purified form of the Twitchell reagent; used for the hydrolysis of fatty glycerides.

16333 Kontrastin
1314-23-4 10313 215-227-2
ZrO_2
Zirconium(IV) oxide

Zirconium oxide; Zirconia; zirconium dioxide; zirconic anhydride. d = 5.85; mp = 2680°; bp = 4300°.

16334 Konut
8001-31-8 232-282-8
Coconut oil
Coconut oil; kosher; oil for ice cream, coatings, nut roasting, corn popping. mp = 21-25°; n_D^{20}= 1.4560; d = 0.903; fp > 110°. *Van Den Bergh Foods.*

16335 konzentrole
A term used for essential oils free from terpenes and sesquiterpenes; used for flavoring.

16336 Koolkat
Furane resin cold-set binders for sand cores. *Foseco (F.S.) Ltd.*

16337 Kopol®
Synthetic resins, soluble plastic resins. *Reichhold.* Discontinued.

16338 Koppert Moss Killer
A soap concentrate; use to kill mosses in turf. *Koppert (UK) Ltd.*

16339 Koppeschaar solution
A bromine solution of N/10 strength.

16340 Kopr-Kote®
Copper flake graphite; anti-seize compound protecting metal parts for seizure, galling, rust, corrosion, and heat-freeze. *Jet-Lube.*

16341 Korad A
Acrylic film. *Rohm & Haas.* Name unverified.

16342 Koraid PSM
Alumino-silicate, modified; suspension aid for pigments and abrasive particles in water-based systems without increasing viscosity; used for paints, polishes, inks, agricultural, and pharmaceutical formulations. *Kaopolite.*

16343 Korantin
Additives for oil/water emulsions. *BASF plc.*

16344 Koraton
A proprietary trade name for a synthetic resin. No manufacturer.

16345 Koreforte®
Reinforcing resins for natural and synthetic rubbers. *BASF AG.*

16346 Koreon
2287
A basic chromium sulfate, $Cr(OH)SO_4$; used in the tanning industry.

16347 Koresin®
Tackifier for natural and synthetic rubbers. *BASF AG; BASF plc.*

16348 Korestab®
Antioxidant for natural and synthetic rubbers. *BASF AG.*

16349 Koretack®
Tackifier for natural and synthetic rubbers. *BASF AG; BASF plc.*

16350 Korever®
Vulcanization resins. *BASF AG.*

16351 Korlan
299-84-3 8415 206-082-6
$C_8H_8Cl_3O_3PS$
Ronnel
Phosphorthioic acid; O,O-dimethyl O-(2,4,5-trichlorophenyl)ester; fenchlorphos; dimethyl trichlorophenyl thiophosphate; Trolene; Etrolene; Nankor; Viozene; Ectorl. Active ingredient: ronnel; insecticide used on cattle for the control of ticks, files, maggots, and lice. mp = 41°. *Dow UK.* Discontinued.

16352 Korlite
Natural zeolite silicate mineral; absorbent for org. compounds, odors, sulfur dioxide gases and ammonia; builder in laundry detergents, cleaning compounds; decolorizer for org. liqs.; specialty filler and water treatments for lead and heavy metal removal. *Kaopolite.*

16353 Koro
A proprietary trade name for an alloy of 98% copper and 2% nickel. No manufacturer.

16354 Korogel
9002-86-2 7746 206-625-7
$(-CH_2CHCl-)_n$
Polyvinyl chloride
Chlorethene homopolymer; chloroethylene polymer; PVC; Geon; Breon; Welvic; Movyl; Tevilon; Marvinol; Rhovyl; Fibravyl; Thermovyl; Isovyl; Retractyl; Crinovyl; Envilon; Nip. A proprietary name for a soft Koroseal. d = 1.406. No manufacturer.

16355 Koron
Refractory coatings for use on ingot molds, bottom plates, and slag pots. *Foseco (F.S.) Ltd.*

16356 Koronit
A German explosive.

16357 Koroplate
9002-86-2 7746 206-625-7
$(-CH_2CHCl-)_n$
Polyvinyl chloride
Chlorethene homopolymer; chloroethylene polymer; PVC; Geon; Breon; Welvic; Movyl; Tevilon; Marvinol; Rhovyl; Fibravyl; Thermovyl; Isovyl; Retractyl; Crinovyl; Envilon; Nip. A proprietary synthetic paint in which Koroseal is the base; extremely resistant to acid fumes. d = 1.406. No manufacturer.

16358 Koroseal
A proprietary trade name for a rubber-like thermoplastic varying in hardness, from soft jellies to hard rubber; obtained by treating highly polymerized vinyl chloride with plasticizers at hugh temperatures and cooling; it can be worked like rubber when hot but requires higher temperatures; resistant to light, water, oils and most other chemicals, used for impregnating and coating paper, fabrics and metals for the manufacture o resistant tubing and cable sheathing. No manufacturer.

16359 Korpad
Antisplash pads to minimise splash defects during direct teeming of steel ingots. *Foseco (F.S.) Ltd.*

16360 Korspray
Foundry solvent (I.P.S.) *Foseco (F.S.) Ltd.*

16361 Kortaid
Saturated fatty acids. *Akzo Chemie UK Ltd.*

16362 Korthix
1302-78-9 1082 215-108-5
$Al_2O_3{\cdot}4SiO_2{\cdot}H_2O$
Bentonite
Wilkinite. Bentonite, refined; thixotropic agent for water-based paints, inks, polishes, adhesives, and for household products. *Kaopolite.*

16363 Korthix H-NF
1302-78-9 1082 215-108-5
$Al_2O_3{\cdot}4SiO_2{\cdot}H_2O$
Bentonite
Wilkinite. Bentonite; bacteria-controlled grade for cosmetics and pharmaceuticals. *Kaopolite.*

16364 korung oil
Kagoo oil
Pongam oil, obtained from the fruts of *Pongamia glabra* .

16365 Kosmos Black, 3XB, BB, and F4
A proprietary gas black used in rubber mixings; also used as a black pigment. No manufacturer.

16366 Kosmos®
Tin catalysts. *Thomas Goldschmidt Ltd.*

16367 Kosmos® 10
Stannous octoate, dioctyl phthalate; tin-org. catalyst for the manufacture of polyether PU foam; accelerates the gel reaction and intensifies the activation of the blowing reaction. *Goldschmidt AG.* Discontinued.

16368 Kosmos® 21
Dialdyl-tin-mercaptide
Catalyst for PU polymerization, for molded PU foams, e.g., microcellular foams, rigid foams, high resilient foams, and RIM. *Goldschmidt AG.* Discontinued.

16369 Kostil
A proprietary range of styrene acrylonitrile molding granules. *Montedison UK Ltd.* Name unverified.

16370 Kotamite®
471-34-1 1697 207-439-9
$CaCO_3$
Calcium carbonate
Carbonic acid calcium salt; Cacit; Calcichew; Calcidia; Citrical. Coated pigment with easy dispersion in plastic compounds, e.g., polyolefins, rigid and flexible PVC; for wire and cable insulation compounds, improved impact properties in PP. mp = 825° (dec); d = 2.83. *ECC International Ltd.* Discontinued.

16371 Kotebond
Etch primers and finishes. *Brent Chemicals International plc.*

16372 Kovar® Alloy
A registered trademark for an alloy of iron with 23-30% nickel, 17-30% cobalt and 0.6-0.8% manganese; used for glass to metal seals. No manufacturer.

16373 Kowet 12
26264-06-2 247-557-8
Calcium dodecylbenzene sulfonate
Rhône-Poulenc Surf.

16374 Kowet 3300
26264-05-1 247-556-2
Isopropylamine dodecylbenzene sulfonate
Rhône-Poulenc Surf.

16375 KP-2
Designed to impart detergency and lubricity while minimizing the buildup of hard film deposits; plasticizer for dry rust inhibitor films. *Ferro/Keil.*

16376 KP-23
109-38-6 203-668-3
A proprietary trade name for a plasticizer consisting of butoxyethyl stearate. No manufacturer.

16377 KP 90
A proprietary trade name for a vinyl plasticizer of the epoxy type. No manufacturer.

16378 KP-140®
78-51-3 201-122-9
$C_{18}H_{39}O_7P$
Tributoxyethyl phosphate
Tris(2-butoxyethyl) phosphate. Plasticizer; leveling agent in floor polish formulations allowing uniform coverage, eliminates high and low spots in gloss, and preventing streaking, crazing, powd., and film contracting; flame retardant for plastics or synthetic rubber of lower flammability; imparts low temperature flexibility to plastics and acrylonitrile rubbers, reduces viscosity in plastisols. bp_4 = 215-228°; n_D^{20}= 1.43590; d = 1.006; fp > 112°. *FMC.*

16379 KP 201
84-61-7 201-545-9
$C_{20}H_{26}O_4$
Dicyclohexyl phthalate
A proprietary trade name for a vinyl plasticizer. mp = 64-66°. No manufacturer.

16380 KP 555
Bis(dimethylbenzyl) ether
A proprietary trade name for a vinyl plasticizer. No manufacturer.

16381 K-Pool
Copper-TEA complex; algicide for swimming pools. *Griffin.*

16382 KR01 K-Resin Polymer
9003-55-8 8534
$[-CH_2CH(C_6H_5)]_x-(CH_2CH=CHCH_2)+_y$
Poly(styrene-co-butadiene)
SBR Rubber; sBR Rubber; styrene-butadiene rubber; GR-S; Government Rubber Styrene. Styrene-butadiene copolymer; resin for injection molding and extrusion processing; for crystal clear and warp resistant parts; in housings, blister packs, extruded tubes, molded boxes with integral hinges, toys, and where impact PS, oriented PS sheet, cellulosics and rigid PVC have been used. d = 0.965. *Phillips.*

16383 KR04 K-Resin Polymer
9003-55-8 8534
$[-CH_2CH(C_6H_5)]_x-(CH_2CH=CHCH_2)+_y$
Poly(styrene-co-butadiene)
SBR Rubber; sBR Rubber; styrene-butadiene rubber; GR-S; Government Rubber Styrene. Styrene-butadiene copolymer, antiblock; extrusion grade resin; clarity, toughness, rigidity in extruded parts, and detail on fast production cycles; for blending with general purpose PS; thermoformed blister packs, disposable containers, and where impact PS, oriented PS sheet, polyesters, cellulosics and rigid PVC have been used. Resin can be tinted or colored in a variety of transparent and opaque shades. d = 0.965. *Phillips.*

16384 kraft paper
A paper produced by the sulfate pulp process. It is strong and inexpensive.

16385 Kraft's metal
A fusible alloy containing 5 parts bismuth, 3 parts lead, and 1 part tin. mp = 104°.

16386 Kraton®
9003-55-8 8534
$[-CH_2CH(C_6H_5)]_x-(CH_2CH=CHCH_2)+_y$
Poly(styrene-co-butadiene)
SBR Rubber; styrene-butadiene rubber; GR-S; Government Rubber Styrene. Thermoplastic rubbers; used for footwear and adhesives. d = 0.965. *Shell.*

16387 Kraton® D 1101
9003-55-8 8534
$[-CH_2CH(C_6H_5)]_x-(CH_2CH=CHCH_2)+_y$
Poly(styrene-co-butadiene)
SBR Rubber; styrene-butadiene rubber; GR-S; Government Rubber Styrene. Linear styrene-butadiene-styrene block copolymer; thermoplastic rubber requiring no vulcanization; for formulating adhesives for the building/construction trade, hot-melt adhesives; FDA compliance. d = 0.965. *Shell.*

16388 Kraton® D 1107
Linear styrene-isoprene-styrene block copolymer; thermoplastic rubber; FDA compliance. *Shell.*

16389 Kraton® D 1116
9003-55-8 8534
$[-CH_2CH(C_6H_5)]_x-(CH_2CH=CHCH_2)+_y$
Poly(styrene-co-butadiene)
SBR Rubber; styrene-butadiene rubber; GR-S; Government Rubber Styrene. Branched styrene-butadiene copolymer; thermoplastic rubber for use in solv. -based construction mastic adhesives; high strength. d = 0.965. *Shell.*

16390 Kraton® D 1320X
Branched styrene-isoprene multiarm copolymer; radiation crosslinkable thermoplastic rubber providing good tack in hot-melt pressure-sensitive adhesives even after cure. *Shell.*

16391 Kraton® D 2103
9003-55-8 8534
$[-CH_2CH(C_6H_5)]_x-(CH_2CH=CHCH_2)+_y$
Poly(styrene-co-butadiene)
SBR Rubber; styrene-butadiene rubber; GR-S; Government Rubber Styrene. Thermoplastic elastomer (S-B-S block copolymer); injection molding and extrusion compound for medical/food, sporting goods, and misc. molded items; 2103 has FDA compliance. d = 0.965. *Shell.*

16392 Kraton® FG 1901X
Maleic anhydride-functionalized styrene-ethylene-butylene-styrene block copolymer; thermoplastic rubber with excellent thermal, oxidative, and uv stability for toughened engineering thermoplastics, coextruded tie layers; FDA compliance. *Shell.*

16393 Kraton® G 1650
Linear styrene-ethylene-butylene-styrene block copolymer; thermoplastic rubber for use in adhesives, sealants, and coatings that must withstand weathering and high processing temps.; FDA compliance. *Shell.*

16394 Kraton® G 1701X
Styrene-ethylene-propylene diblock copolymer; thermoplastic rubber. *Shell.*

16395 Kraton® G 1726X
Linear styrene-ethylene-butylene-styrene block copolymer (70% S-EB diblock); thermoplastic rubber for use in adhesives, sealants, and applications where resistance to degradation is necessary and where softness and processability are more important than tensile strength; blendable with Kraton G 1652 to give desired properties. *Shell.*

16396 Kraton® G 2701
Thermoplastic rubber compound (S-EB-S block copolymer blend); for use in extruded film, blowmolded containers, injection molded products, general rubber products; also to improve processing behavior of other thermoplastics rubbers and as impact modifier for high-melt flow PP homopolymers. FDA compliant. *Shell.*

16397 Kraton® G 7430
Thermoplastic elastomer (S-EB-B block copolymer blend); injection molding and extrusion compound for automotive parts, sporting goods, and misc. molded items. *Shell.*

16398 Kraton® RP 6404
S-I-S block copolymer; provides low viscosity, color-stable adhesives with high adhesion to various substrates used in hot melt assembly. *Shell.*

16399 Kremser White
1319-46-6 1034 215-290-6
$(PbCO_3)_2 \cdot Pb(OH)_2$
Basic lead carbonate
C.I.; Pigment White 1; C.I. 77597; lead subcarbonate; white lead; flake lead ceruse; cerussa; bleiweiss. The purest form of white lead. dec at 400°.

16400 Krenite
25954-13-6 247-363-3
Soluble concentrate of 480 g fosamine-ammonium per liter; used for control of woody weeds in noncrop and forestry areas. Applied to unwanted brush in late summer or autumn prevents bud break leading to death of treated plants the following spring. *Selectokil Ltd; Burts & Harvey; DuPont UK.*

16401 Kresamin
Cresamol
A mixture of 25% tricresol with ethylenediamine. A powerful antiseptic.

16402 Kresatin
122-46-3 2651 204-546-2
$C_9H_{10}O_2$
m-Cresol acetate
m-Tolyl acetate; acetic acid 3-methylphenyl ester; *m*-cresol acetic acid ester; acetic acid *m*-cresol ester; acetylmetacresol; metacresol acetate; Cresatin; Cresatin-Sulzberger; Cresatin Metacresylacetate; Metacresylacetate-Sulzberger; Kresatin; acetic acid *m*-tolyl ester; *m*-cresyl acetate. bp_{760} = 212°; bp_{14} = 100°; d_4^{26}= 1.048; λ_m = 262.5, 269.5 nm (MeOH).

16403 Kreside
Cresylic creosote *Sasolchem.*

16404 K-Resin Polymer KR01
9003-55-8 8534
$[-CH_2CH(C_6H_5)]_x-(CH_2CH=CHCH_2)_y$
Poly(styrene-co-butadiene)
SBR Rubber; styrene-butadiene rubber; GR-S; Government Rubber Styrene. Transparent and shatter resistant styrene-butadiene block copolymer containing at least 70 weight percent polymerized styrene; an injection molding grade for use where higher stiffness and warpage resistance is

required; used for dust covers, point-of-purchase displays, molded boxes, and containers, lids and office supplies. d = 0.965. *Phillips.*

16405 K-Resin Polymer KR03
9003-55-8 8534
[-$CH_2CH(C_6H_5)]_x$-$(CH_2CH=CHCH_2)_y$
Poly(styrene-co-butadiene)
SBR Rubber; styrene-butadiene rubber; GR-S; Government Rubber Styrene. Transparent and shatter resistant styrene-butadiene block copolymer containing at least 70 weight percent polymerized styrene; an injection molding grade for use where higher impact resistance is required; used for overcaps, molded boxes and containers, toys, medical devices and tool handles. d = 0.965. *Phillips.* Name unverified.

16406 K-Resin Polymer KR04
9003-55-8 8534
[-$CH_2CH(C_6H_5)]_x$-$(CH_2CH=CHCH_2)_y$
Poly(styrene-co-butadiene)
SBR Rubber; styrene-butadiene rubber; GR-S; Government Rubber Styrene. Transparent and shatter resistant styrene-butadiene block copolymer containing at least 70 weight percent polymerized styrene; a resin for blending with general purpose polystyrene for sheet extrusion and thermoforming applications such as disposable cups and containers, blister packages and portion packaging. d = 0.965. *Phillips.*

16407 K-Resin Polymer KR05
9003-55-8 8534
[-$CH_2CH(C_6H_5)]_x$-$(CH_2CH=CHCH_2)_y$
Poly(styrene-co-butadiene)
SBR Rubber; styrene-butadiene rubber; GR-S; Government Rubber Styrene. Transparent and shatter resistant styrene-butadiene block copolymer containing at least 70 weight percent polymerized styrene; a resin for nonblended sheet extrusion and thermoforming, for blow molding (both extrusion and injection) and for profile extrusion; uses include blister packages, bottles, jars, medical devices and extruded tubes and profiles. d = 0.965. *Phillips.* Name unverified.

16408 K-Resin Polymer KR10
9003-55-8 8534
[-$CH_2CH(C_6H_5)]_x$-$(CH_2CH=CHCH_2)_y$
Poly(styrene-co-butadiene)
SBR Rubber; styrene-butadiene rubber; GR-S; Government Rubber Styrene. Transparent and shatter resistant styrene-butadiene block copolymer containing at least 70 weight percent polymerized styrene; a resin for blown or cast film production; uses include medical packaging, shrink wrap, overwrap, skin packaging, produce wrap, windows for envelopes and boxes and twist wrap. d = 0.965. *Phillips.* Name unverified.

16409 Kresival
A German preparation. It contains the water-soluble calcium salts of the sulfonic acids of the cresols.

16410 Kresopolin
Kresolin
Preparations of crude carbolic acid; disinfectants.

16411 Kricinol 35
7492-30-0 231-314-8
Potassium ricinoleate
Mold release. *Actrachem.*

16412 Kriegr-o-dip
A proprietary trade name for liquid dyes for plastics. S-standard chemical dye; A-for cellulose acetate. W-powder dye for use in hot water. V-for polystyrene. No manufacturer.

16413 Kristalex
α-Methylstyrene copolymer hydrocarbon resins; used for hot-melt product assembly adhesives and light colored caulking compounds. *Hercules.*

16414 Kristel Gold II
Partially hydrogenated vegetable oil (soybean, cottonseed), lecithin, artificial color and flavor; kosher; shortening system for danish pastries, sweet rolls, coffee cakes; higher melt pt. for high shear applications. *Van Den Bergh Foods.*

16415 Krist-o-kleer
A proprietary trade name for a plasticizer containing 50% dextrose and 50% levulose. No manufacturer.

16416 Kristol
White oils. *Carless Refining & Marketing Ltd.*

16417 Krokoloy
A proprietary trade name for a steel containing 14% chromium with some cobalt. No manufacturer.

16418 Kroma Red
1309-37-1 4072 215-168-2
Fe_2O_3
Iron(III) oxide
Ferric oxide; Ferric sesquioxide; jeweller's rouge. A precipitated red iron oxide for color pigment use. *Pfizer International.* Discontinued.

16419 Kromaplast
Blended dry pigments; used in plastics. *Ampacet Corporation.*

16420 Kromax
An electrical resistance alloy of 80% nickel and 20% chromium.

16421 Kromore
An alloy of nickel with 15% chromium; used for the heating elements in wire-wound electric furnaces. It has a specific resistance of 98 micro-ohms/cm at 0°.

16422 Kromosperse
Pigment dispersing agent; for use in nonaqueous and aqueous coating compositions. *Hüls Am.*

16423 Kronagold
Family of lubricating oils; gear oils, hydraulic oils, compressor oils, etc. *E/M Corporation.*

16424 Kronaplate
Family of lubricating greases; bearings, gears, cams, slides, etc. *E/M Corporation.*

16425 Krona-Syn
Synthetic lubricating fluids; compressor fluids, high temperature oven chain lubricants. *E/M Corporation.*

16426 Kronitex®
Flame retardant plasticizers. *FMC.* Discontinued.

16427 Kronitex® 25
68937-41-7 273-066-3
Triaryl phosphate
Flame retardant plasticizer for PC/ABS, engineering resins, PVC, phenolic laminates, cellulosics. *FMC.* Discontinued.

16428 Kronitex® 50
68937-41-7 273-066-3
Triaryl phosphate
Flame retardant plasticizer for PVC, flexible polyurethanes, synthetic rubber, belting. *FMC.* Discontinued.

16429 Kronitex® 100
68937-41-7 273-066-3
Triaryl phosphate
Flame retardant plasticizer for PVC; aids fusion; in plastisols, viscosity stability; compatibilizing plasticizer; catalyst carrier and pigment vehicle for PU; processing aid in rubber belting and mechanical goods; flame retardant and processing aid in engineering *FMC.* Discontinued.

16430 Kronitex® 200
68937-41-7 273-066-3
Triaryl phosphate
Flame retardant plasticizer for cellulose polymers, PVC. *FMC.* Discontinued.

16431 Kronitex® 1840
68937-41-7 273-066-3
Triaryl phosphate
Thixotropic flame retardant gel for use in coating of fiberglass and other media requiring high loading and low volatility. *FMC.* Discontinued.

16432 Kronitex® 3600
Alkylaryl phosphate ester; plasticizer with improved low temperature flexibility, high flame retardance, and low smoke evolution; ideal for vinyl film and sheeting, wire and cable insulation, coated fabrics, plastisols. *FMC.* Discontinued.

16433 Kronitex® PB-460
Brominated aromatic phosphate ester
Flame retardant for engineering thermoplastics incl. PC, modified PPO, PBT, PET, ABS, and blends. *FMC.* Discontinued.

16434 Kronitex® TBP
126-73-8 9749 204-800-2
$C_{12}H_{27}O_4P$
Tri-n-butyl phosphate
Phosphoric acid tri-n-butyl ester; tributyl phosphate; phosphoric acid tributyl ester. Primary plasticizer and solv.; antifoam for paints, pigment dispersant. d_{25}^{25} = 0.976; bp = 289°; bp_{27} = 177-178°; mp < -80°; fp = 146°; n_D^{25} = 1.4215. *FMC.* Discontinued.

16435 Kronitex® TCP
68952-35-2 273-168-8
Tricresyl phosphate
General purpose flame retardant for vinyl compounds, low air, oil, and water loss; processing aid by improving flux rate of compounds containing slow-solvating plasticizers; rapid gellation and fusion rate make it useful in plastisols; plasticizer for NC lacquers and coatings; plasticizer and processing aid for rubbers; flame retardant sheeting. *FMC.*

16436 Kronitex® TOF
Trioctyl phosphate

Low temperature plasticizer for use in PVC, rubber, paints, coatings; highly efficient solvent. *FMC.* Discontinued.

16437 Kronitex® TPP
115-86-6 9872 204-112-2
$C_{18}H_{15}O_4P$
Triphenyl phosphate
Phosphoric acid triphenyl ester. Flame retardant plasticizer for engineering resins, cellulosics. mp = 79-81°; fp = 181°; bp_{11} = 245°. *FMC.* Discontinued.

16438 Kronitex® TXP
25155-23-1 246-677-8
Trixylenyl phosphate
Flame retardant with better milling action in filled PVC compounds; for superior electrical compounds (wire and cable application). *FMC.*

16439 Kronos®
13463-67-7 9612 236-675-5
TiO_2
Titanium dioxide
Unitane; C.I. Pigment White 6; C.I. 77891. Titanium dioxide pigments; used for paint, paper, glass, ceramics, plastics and ink. mp = 1855°; d (rutile) = 4.23, (anatase) = 3.90, (brookite) = 4.13. *Rheox Inc.* Discontinued.

16440 Kronos®
13463-67-7 9612 236-675-5
TiO_2
Titanium dioxide
Unitane; C.I. Pigment White 6; C.I. 77891. Titanium dioxide pigment; used to impart opacity and brightness; for paints, inks, plastics, paper etc. mp = 1855°; d (rutile) = 4.23, (anatase) = 3.90, (brookite) = 4.13. *NL Chemicals (UK) Ltd.* Name unverified.

16441 Kronos® 1000
13463-67-7 9612 236-675-5
TiO_2
Titanium dioxide; anatase; Titanox
Unitane; C.I. Pigment White 6; C.I. 77891. Pigment for fibers, paper and coatings. mp = 1855°; d = 4.23; insoluble in water, HCL, HNO_3 or diluted H_2SO_4, soluble in hot concentrated H_2SO_4. *Kronos.* Name unverified.

16442 Kronos® 2020
13463-67-7 9612 236-675-5
TiO_2
Titanium dioxide; rutile; Titanox 2020
Unitane; C.I. Pigment White 6; C.I. 77891. Pigment for coatings. mp = 1855°; d = 4.23; insoluble in water, HCL, HNO_3 or diluted H_2SO_4, soluble in hot concentrated H_2SO_4. *Kronos.* Name unverified.

16443 Kronos® 2073
13463-67-7 9612 236-675-5
TiO_2
Titanium dioxide; rutile; Titanox 2073
Unitane; C.I. Pigment White 6; C.I. 77891. Pigment primarily for plastics. mp = 1855°; d = 4.23. *Kronos.* Name unverified.

16444 Kronos® 3020
13463-67-7 9612 236-675-5
TiO_2
Titanium dioxide; Titanox 3020
Unitane; C.I. Pigment White 6; C.I. 77891. Nonpigmentary product for ceramics, glass/glass fibers, glazes, vitreous enamels, welding rods. mp = 1855°; d (rutile) = 4.23, (anatase) = 3.90, (brookite) = 4.13. *Kronos.* Name unverified.

16445 Krovar
A wettable powder containing 40% w/w bromacil and 40% w/w diuron; used for total weed control in noncrop areas. *Selectokil Ltd.*

16446 Krovar®
A wettable powder containing 40% w/w bromacil and 40% w/w diuron; used for total weed control in noncrop areas. *DuPont UK.*

16447 Kruppin
An electrical resistance alloy containing 28% nickel and the rest iron.

16448 Kryalith
15096-52-3 2673 239-148-8
$3NaF \cdot AlF_3$
Cryolite
ice spar; sodium aluminum fluoride. A proprietary trade name for a synthetic cryolite. d = 2.95; mp = 1000°. No manufacturer.

16449 Krylene® 606
9003-55-8 8534
$[-CH_2CH(C_6H_5)]_x-(CH_2CH=CHCH_2)+_y$
Poly(styrene-co-butadiene)
SBR Rubber; styrene-butadiene rubber; GR-S; Government Rubber Styrene. A registered trademark for a cold polymerized, alum coagulated non-staining butadiene-styrene rubber. d = 0.965. No manufacturer.

16450 Krylene® 608
9003-55-8 8534
$[-CH_2CH(C_6H_5)]_x-(CH_2CH=CHCH_2)+_y$
Poly(styrene-co-butadiene)
SBR Rubber; styrene-butadiene rubber; GR-S; Government Rubber Styrene. A registered trademark for a cold polymerized styrene butadiene rubber. d = 0.965. No manufacturer.

16451 Krynac® 19.65
9003-18-3
$[-CH_2CH(CN)]_x-(CH_2CH=CHCH_2)+_y$
Poly(acrylonitrile-co-butadiene)
Butadiene-acrylonitrile copolymer, nonstaining, cold polymerized; NBR used for low temperature oil well specialties, belt covers, and idler rolls for low-temperature service, o-rings, seals, gaskets, hydraulic hose; vulcanized with sulfur, sulfur donor, or peroxide. d = 0.980. *Bayer Corp; Polysar.*

16452 Krynac® 20H35
9003-18-3
$[-CH_2CH(CN)]_x-(CH_2CH=CHCH_2)+_y$
Poly(acrylonitrile-co-butadiene)
Butadiene-acrylonitrile copolymer, nonstaining, hot polymerized; excellent low temperature flexibility; vulcanized with sulfur, sulfur donor, or peroxide. d = 0.980. *Bayer Corp; Polysar.*

16453 Krynac® 27.50
A proprietary acrylonitrile rubber. *Bayer Corp; Polysar.* Name unverified.

16454 Krynac® 34.140
9003-18-3
$[-CH_2CH(CN)]_x-(CH_2CH=CHCH_2)+_y$
Poly(acrylonitrile-co-butadiene)
Butadiene-acrylonitrile copolymer, nonstaining, cold polymerized; NBR used for plasticizer masterbatches, as viscosity modifier, for o-rings, lip seals, gaskets; vulcanized with sulfur, sulfur donor, or peroxide. d = 0.980. *Bayer Corp; Polysar.*

16455 Krynac® 34.35, 34.50.
9003-18-3
$[-CH_2CH(CN)]_x-(CH_2CH=CHCH_2)+_y$
Poly(acrylonitrile-co-butadiene)
A proprietary cold-polymerized gel-free oil-resistant butadiene/acrylonitrile rubber. d = 0.980. *Bayer Corp; Polysar.* Name unverified.

16456 Krynac® 34.60 SP
A proprietary nitrile rubber capable of withstanding temperatures up to 135°. *Bayer Corp; Polysar.* Name unverified.

16457 Krynac® 34.80
9003-18-3
$[-CH_2CH(CN)]_x-(CH_2CH=CHCH_2)+_y$
Poly(acrylonitrile-co-butadiene)
A proprietary cold-polymerised gel-free oilresistant butadiene/acrylontrile rubber. d = 0.980. *Bayer Corp; Polysar.* Name unverified.

16458 Krynac® 823X2
9003-18-3
$[-CH_2CH(CN)]_x-(CH_2CH=CHCH_2)+_y$
Poly(acrylonitrile-co-butadiene)
A registered trademark for a copolymer of acrylonitrile and butadiene containing a medium level of bound acrylonitrile. d = 0.980. *Bayer Corp; Polysar.* Name unverified.

16459 Krynac® 833
A proprietary isoprene acrylonitrile rubber containing 31.0% bound acrylonitrile; Mooney viscosity is 70. *Bayer Corp; Polysar.* Name unverified.

16460 Krynac® 843
9003-18-3
$[-CH_2CH(CN)]_x-(CH_2CH=CHCH_2)+_y$
Poly(acrylonitrile-co-butadiene)
Modified butadiene-acrylonitrile copolymer, 50 phr DOP, nonstaining, cold polymerized; NBR used for mechanical goods, plasticizer masterbatch, soft roll covers; vulcanized with sulfur, sulfur donor, or peroxide. d = 0.980. *Bayer Corp; Polysar.*

16461 Krynac® 850
A trademark for a vinyl-modified nitrile rubber. *Bayer Corp; Polysar.* Name unverified.

16462 Krynac® 881 and 882
Proprietary names for synthetic rubbers of the ethylacrylate type. *Bayer Corp; Polysar.* Name unverified.

16463 Krynac® 882X1
A registered trademark for a low temperature resistant acrylic rubber for oil seals. *Bayer Corp; Polysar.* Name unverified.

16464 Krynac® NV 850
NBR/PVC fluxed blend (50:50), nonstaining, cold polymerized; used for flame-resistance belting, footwear, fire hose covers, cable jackets, cellular

products; excellent resistance to ozone, weathering, oils, and fuels; vulcanized with sulfur, sulfur donor, or peroxide. *Bayer Corp; Polysar.*

16465 Krynac® PXL 34.17
9003-18-3
[-$CH_2CH(CN)$]$_x$-(CH_2CH=$CHCH_2$)$_y$
Poly(acrylonitrile-co-butadiene)
Crosslinked butadiene-acrylonitrile copolymer, nonstaining, hot polymerized; impact modifier for PVC. d = 0.980. *Bayer Corp; Polysar.*

16466 Krynac® X 1.46
Lightly carboxylated NBR-org. acid terpolymer, nonstaining, cold polymerized; used for spinning cots, roll covers, automotive seals, mechanical goods, industrial footwear; offers high abrasion resistance, tensile and tear strength; vulcanized with sulfur, sulfur sulfur donor or peroxide in combination with a metal oxide. *Bayer Corp; Polysar.*

16467 Krystalex
Hydrogenated hydrocarbon resins; for adhesives and hot melts. *Hercules.*

16468 Krystallazurin
14283-05-7 9323 238-177-3
$CuH_{12}N_4O_4S$
Tetraamminecopper sulfate
Cuprammonium sulfate; ammonium cupric sulfate; cupric sulfate, ammoniated. A fungicide consisting of ammoniacal copper sulfate. d_4^{20} = 1.81.

16469 Krystallos
7631-86-9 8637 231-545-4
SiO_2
Silicon dioxide
Silica; silicic anhydride. Quartz. d = 2.2 (amorphous); 2.65 (quartz).

16470 Krystaltite Film
PVC film useful as a shrink film for consumer applications. *AlliedSignal Inc.*

16471 Krytox®
A range of fluorinated greases used as lubricants in aircraft and missiles. *DuPont UK.*

16472 Krytox® GPL 206
Synthetic lubricant
Miller-Stephenson. Name unverified.

16473 KS-052P
Thermoplastic olefin resin; for air quenched blown film extrusion of biohazard bags, medical pkg., trash bags; excellent puncture and low temperature impact resistance *Himont.*

16474 K-Slag
Potassium basic slag *Fisons plc, Pharmaceuticals Div.* Name unverified.

16475 KT-012P
Polyolefin resin; for air quenched blown and cast film processes for pkg. and shrink wrap; excellent heat sealing, clarity and gloss, good moisture barrier properties *Himont.*

16476 K-Tea
Copper-TEA complex; algicide for use in golf course, ornamental, and fish ponds, potable water reservoirs, fresh water lakes, and fish hatcheries. *Griffin.*

16477 Küttner silk
An artificial silk prepared by the viscose process.

16478 Kuhne phosphor bronze
An alloy of 78% copper, 10.6% tin, 10.45% lead, 0.57% phosphorus, and 0.26% nickel.

16479 Kukident
Denture cleanser for cleaning dentures; also denture adhesive for securing dentures. *Richardson-Vicks Inc.* Name unverified.

16480 Kukkersite
An oil shale of Estonia, of specific gravity 1.2-1.4. It contains about 55% volatile matter, and when distilled at 500°. yields from 70-80 gallons per ton of oil of specific gravity 0.92-0.93.

16481 Kumulan®
Nitrothal-isopropyl, sulfur
For control of powdery mildew in apples. *BASF AG.*

16482 Kumulus® DF, FL
7704-34-9 9142 231-722-6
S
Sulfur
Sulphur; brimstone. For control of diseases and spider mites in fruit, vines, vegetable, ornamentals, and agricultural crops. mp = 117-120°; bp = 444°; d = 2.060. *BASF AG.*

16483 Kumulus® S
7704-34-9 9142 231-722-6
S
Sulfur
Sulphur; brimstone. Wettable powder containing 80% w/w sulfur per liter; a protectant fungicide/foliar feed. mp = 117-120°; bp = 444°; d = 2.060. *BASF plc; BASF AG.*

16484 Kunstharz HW
An ammonia condensed phenol formaldehyde resin melting at 55°. *Vilanova Resins.* Discontinued.

16485 Kunststein
An artificial stone made from magnesite.

16486 kuoxam
A cellulose solvent prepared by dissolving 50 grams of copper sulfate in 300 ml water and adding ammonia until all the copper hydroxide is precipitated. The precipitate after filtration is dissolved in 25% ammonia solution.

16487 Kupferdermasan
A salicyl-copper soap preparation containing 2% copper; a bactericide.

16488 Kuracap
2-Mercaptobenzthiazole + dibenzthiazole disulfide; A proprietary accelerator. *Nipa Laboratories Ltd.* Name unverified.

16489 Kurade
Accelerator for rubber. *Akzo Chemie.* Name unverified.

16490 Kurchi
The root bark of *Holarrhena antidysenteriea*; a febrifuge.

16491 Kuro Bishi®
Coke; general use for carbide, ferro-alloy, copper and nickel refining. *Mitsubishi Kasei.* Name unverified.

16492 Kurofan
Polyvinylidene chloride
BASF plc.

16493 Kurofan D
Polyvinylidene chloride dispersion. *BASF plc.* Name unverified.

16494 Kurom 1
A jewellery alloy of copper with tin and cobalt.

16495 Kuromoji oil
An oil from *Lindera* species; used in Japan for perfuming soaps and oils and contains α-phellandrene, nerolidol, linalol, and geraniol.

16496 Kuron
93-72-1 8679 202-271-2
$C_9H_7Cl_3O_3$
2-(2,4,5-Trichlorophenoxy)propionic acid
Silvex; fenoprop; 2,4,5-TC. Herbicide containing silvex as the active ingredient; herbicide used in ponds and other still water for the control of aquatic weeds, as well as control of brush on rangeland; also used industrially on railroads or under power lines for the control of weeds and brush. mp = 177-170°, also 181.6°. *Dow UK.* Discontinued.

16497 Kurrodur
A proprietary trade name for an alloy of copper with 0.75% nickel and 0.5% silicon. No manufacturer.

16498 Kurrol salts
Alkaline metaphosphates insoluble in water, but soluble in pyrophosphate solutions. They are produced by heating sodium trimetaphosphate or ethyl sodium phosphate.

16499 K-Van
13769-43-2 237-388-8
KVO_3
Potassium metavanadate.
mp = 520°; d = 2.840. *Kerr-McGee Chemical Corp.*

16500 K-White
Aluminium tripolyphosphate
Nontoxic white anticorrosive pigment used for anticorrosive coatings. *Bromhead & Denison Ltd.*

16501 Kwik Dri
Aliphatic hydrocarbon solvent; used in coatings, in fabric drycleaning, and in cold degreasing. *Ashland Chemical Company.*

16502 Kwikfill
Polyester two-part resin filler; for mending damaged body panels, filling dents, etc. *Hermetite Products Ltd.* Name unverified.

16503 Kwlk-Green
Nitrogen, sulfur, iron and zinc; used on turf, shrubs, trees, and potted plants to promote deep rich green foliage. *Lawn & Garden Products Inc.*

16504 Kymene®
Cationic wet-strength resins, including polyamide, polyamine, epoxide, and urea-formaldehyde resins; imparts strength to wet paper and paperboard; used primarily as internal additives in papermaking processes, but also as cationizing agents for starch added internally and as insolubilizing agents for starch in size press and pigmented coatings. *Hercules; Hercules Ltd.*

16505 Kymene® 109
Wet-strength resin for use in bleached or unbleached pulps and secondary fiber systems in paper industry. *Hercules.*

16506 Kymene® 435
Cationic, urea-formaldehyde resin; high-solids wet strength resin for paper and paperboard under acid pH conditions. *Hercules.*

16507 Kymene® 557H
Cationic, wet-strength resin; high-efficiency under acid or alkaline papermaking conditions; wet-strength resin, retention aid, and starch-cationizing agent, in tissue, toweling, and sanitary wadding grades, to wet-strength corrugating media, in linerboar *Hercules.*

16508 Kymene® 557LX
Environmentally compatible wet-strength resin for paper/paperboard applications. *Hercules.*

16509 Kynar® 301 F
24937-79-9
$(-CH_2CF_2-)+_n$
Poly(vinylidene fluoride)
PVDF, crystalline high molecular weight; for solv.-based coatings; produces films with high resistance to gamma radiation and transparency to uv radiation. Tm = 165-172°. *Elf Atochem N. Am./Plastics.*

16510 Kynar® 460
24937-79-9
$(-CH_2CF_2-)+_n$
Poly(vinylidene fluoride)
PVDF; tough engineering resin with high abrasion resistance and stability in harsh thermal, chem., and uv environments; readily melt processable in molding and extrusion; used for coatings incl. corrosion-resistance coatings for chem. process equipment, long life decorative finishes on building panes, flim, filter cloth, intrumentation, and control equipment linings, membranes, static mixers, pipes and fittings, pumps, stock shapes, valves and electronic and electrical jacketing. Tm = 165-172°. *Elf Atochem N. Am./Plastics.*

16511 Kynar® 700 Series
24937-79-9
$(-CH_2CF_2-)+_n$
Poly(vinylidene fluoride)
700 series available in a wide range of viscosities. Tm = 165-172°. *Elf Atochem N. Am./Plastics.*

16512 Kynar® 7200, 7201
Vinylidene fluoride-tetrafluoroethylene copolymer; easily melt processed by extrusion or molding into film, sheet, tube, monofilament, cable jackets, etc.; esp. useful for fiber optic applications. *Elf Atochem N. Am./Plastics.*

16513 Kynar® Flex® 2800, 2801
24937-79-9
$(-CH_2CF_2-)+_n$
Poly(vinylidene fluoride)
PVDF copolymer; used in wire and cable construction and other uses requiring high flexibility and improved impact resistance Tm = 165-172°. *Elf Atochem N. Am./Plastics.*

16514 Kynar® Flex® 2850
9011-17-0
$(-CH_2CF_2-)+_x[-CF_2CF(CF_3)-]_y$
Poly(vinylidene fluoride-co-hexafluoropropylene)
Polyvinylidene fluoride-hexafluoropropylene copolymer; for various applications esp. wire and cable jacketing. Tm = 140-145°. *Elf Atochem N. Am./Plastics.*

16515 Kynar® Flex® 2900
24937-79-9
$(-CH_2CF_2-)+_n$
Poly(vinylidene fluoride)
Polyvinylidene fluoride copolymer; for wire and cable applications requiring extremely low smoke emission and low flame spread. Tm = 165-172°. *Elf Atochem N. Am./Plastics.*

16516 Kynite
An explosive containing 24-26% nitroglycerin, 2-3% wood pulp, 32-321/2% starch, 31-34% barium nitrate, and 0-0.5% calcium carbonate.

16517 Kynol
A highly cross-linked amorphous phenolic polymer. It resists temperatures up to 2500°.

16518 Kyolox BAT
Alkyl polyglycol ether
Nonionic; colorless viscous liquid; washing, wetting and dispersing agent for all fibers and temperature ranges; used for kier boiling, bleaching, milling, and for after-soaping of dyed and printed materials. *Thor Chemicals (UK) Ltd.* Discontinued.

16519 Kypfarin
81-81-2 10174 201-377-6
$C_{19}H_{16}O_4$
Warfarin
4-Hydroxy-3-(3-oxo-1-phenylbutyl)-2H-1-benzopyran-2-one; 3-(α-acetonylbenzyl)-4-hydroxycoumarin; 1-(4'-hydroxy-3'-coumarinyl)-1-phenyl-3-butanone; 3-α-phenyl-β-acetylethyl-4-hydroxycoumarin; compound 42; WARF compound 42; Co-Rax; Rodex. Warfarin. mp = 161°; λ_m (water, pH 10) = 308 nm (ε 13610). *Mechema Chemicals Ltd.* Name unverified.

16520 Kyrock
A rock asphalt consisting of sand with about 7% bitumen.

16521 Kysite
A proprietary trade name for a phenol-formaldehyde synthetic resin with a fiber filler. No manufacturer.

16522 Kytamer® PC
1398-61-4 2105 215-744-3
Chitan
Chitosan PCA. Water-sol. polymer. *Amerchol Corp.*

16523 KZ 55
Tetra (2,2 diallyloxymethyl) butyl, di(ditridecyl) phosphito zirconate
IPA; coupling agent. *Kenrich Petrochemicals.*

16524 KZ OPPR
Cyclo(dioctyl) pyrophosphato dioctyl zirconate, methyl naphthalene
Coupling agent. *Kenrich Petrochemicals.*

16525 KZ TPP
Cyclo dineopentyl (diallyl) pyrophosphato dineopentyl (diallyl) zirconate
IPA; coupling agent. *Kenrich Petrochemicals.*

16526 KZ TPPJ
Cycloneopentyl, cyclo (dimethylaminoethyl) pyrophosphato zirconate
Di mesyl salt; coupling agent. *Kenrich Petrochemicals.*

16527 K-Zinc
1314-13-2 10279 215-222-5
ZnO
Zinc oxide
Flowers of zinc; philosopher's wool; zinc white; C.I. Pigment White 4; C.I. 77947. Zinc oxide formulation; flowable nutrient for rice seed dressing. d = 5.67 also d^{20}_4 = 5.607; n_D = 2.0041, 2.0203. *Griffin.*

16528 L.A.S
85536-07-8 287-488-0
PEG-8 caprylic/capric glycerides
Nontoxic excipient for creams, lotions; surfactant for microemulsions. *Gattefosse SA.*

16529 L-55R® Acid Neutralizer
Aluminum-magnesium hydroxy carbonate
Used for acid neutralization in the production of polyolefin resins. *Reheis Inc.*

16530 L-66
Sulfur chlorinated base; EP agent for use in threading and tapping operations, cutting and grinding oils. *Ferro/Keil.*

16531 Labitan
Degussa AG. Name unverified.

16532 Laboprin
A proprietary preparation of aspirin [50-78-2] and lysine [56-78-1]; an analgesic. *LAB Ltd.* Unverified.

16533 Labosept
522-51-0 2959 208-330-9
A proprietary preparation of dequalinium chloride; an antibacterial lozenge taken orally. *LAB Ltd.* Unverified.

16534 Labrafac Hydro WL 1219
Caprylic/capric triglycerides PEG-4 esters; hydrophilic oil for pharmaceutical and cosmetic formulations. *Gattefosse; Gattefosse SA.*

16535 Labrafac Lipophile WL 1349
65381-09-1 265-724-3
Caprylic/capric triglyceride. *Gattefosse.*

16536 Labrafil ISO
Isostearic ethoxylated glycerides. Hydrophilic oil. *Gattefosse SA.*

16537 Labrafil M 1944 CS
97488-91-0 307-030-6
Apricot kernel oil PEG-6 esters. Apricot kernel oil PEG-6 esters; hydrophilic oil for pharmaceutical and cosmetic formulations. *Gattefosse; Gattefosse SA.*

16538 Labrafil M 1969 CS
Peanut oil PEG-6 esters. Hydrophilic oil for pharmaceutical and cosmetic formulations. *Gattefosse; Gattefosse SA.*

16539 Labrafil M 1980 CS
103819-46-1
Olive oil PEG-6 esters. Hydrophilic oil for pharmaceutical and cosmetic formulations. *Gattefosse; Gattefosse SA.*

16540 Labrafil M 2125 CS
85536-08-9 287-489-6
Corn oil PEG-6 esters. Hydrophilic oil for pharmaceutical and cosmetic formulations. *Gattefosse; Gattefosse SA.*

16541 Labrafil M 2130 BS
Hydrogenated palm/palm kernel oil PEG-6 esters; Hydrogenated palm/palm

kernel oil PEG-6 complex. Hydrophilic wax for pharmaceutical and cosmetic formulations. *Gattefosse; Gattefosse SA.*

16542 Labrafil M 2735 CS
Triolein PEG-6 esters; Triolein PEG-6 complex. Hydrophilic hydrogenated oil; excipient for pharmaceutical and cosmetic formulations. *Gattefosse; Gattefosse SA.*

16543 Labrasol
85536-07-8 287-488-0
PEG-8 caprylic/capric glyceride; PEG 400 caprylate/caprate glycerides; L.A.S. Hydrophilic oil; excipient, solubilizer for pharmaceutical and cosmetic formulations; surfactant for microemulsions; wetting agent; penetration enhancer. *Gattefosse; Gattefosse SA.*

16544 LABS 100/H.V.
25496-01-9 247-036-5
Tridecylbenzene sulfonic acid
Nansa® TDB. Higher viscosity detergent intermediate. *Zohar Detergent Factory.*

16545 LABS-100
Linear alkylbenzene sulfonic acid
Detergent intermediate. *Zohar Detergent Factory.*

16546 Labstix®
A proprietary test-strip used for the detection of pH, protein, glucose, ketones, and blood in urine. *Bayer AG.*

16547 lac
(Gum lac, lacca, button lac, sheet lac, shellac). The resinous excretion of the lac insect, *Laccifer lacca* , cultivated in India, Burma and Siam. The insects living on the twigs become surrounded with the lac.

16548 lac, Japanese
The lac obtained from *Rhus vernicifera* . It is a natural varnish or lacquer, and contains 85%, of urushic acid.

16549 Lacanite
A proprietary trade name for a shellac compound. No manufacturer.

16550 Laccain
A phenol-formaldehyde resin made with the aid of hydroxy acids, such as tartaric acid.

16551 lac-dye
Lack-lack, dyer's lac. The coloring matter derived from lac. Stick-lac contains 10% of coloring matter; used for dyeing wool mordanted with aluminum or tin salts.

16552 Lackmoid
The blue coloring matter obtained by heating resorcinol with sodium nitrite; used as an indicator in alkalimetry.

16553 lacmus
Chemically pure litmus.

16554 Lacolene
Aliphatic hydrocarbon solvent; used in coatings, adhesives and printing inks. *Ashland Chemical Company.*

16555 Lacorene
A proprietary trade name for a polystyrene molding resin. No manufacturer.

16556 Lacqran
A proprietary trade name for an ABS molding resin. No manufacturer.

16557 Lacqrene 550
A proprietary polystyrene used in extrusion and injection molding. *Aquitaine-Organico.* Unverified.

16558 Lacqrene 635, 811, 835 and 836
Proprietary polystyrenes used to produce extrusions of differing tensile strengths. *Aquitaine-Organico.* Unverified.

16559 Lacqrene 740
An impact and heat resistant polystyrene suitable for use at 90°. *Aquitaine-Organico.* Unverified.

16560 Lacqrene E
Antistatic polystyrene. *Aquitaine-Organico.* Unverified.

16561 Lacqsan 125 and 125L
Proprietary copolymers of styrene and acrylonitrile used in extrusion and injection molding. *Aquitaine-Organico.* Unverified.

16562 Lacqsan E
Antistatic styrene acrylonitrile. No manufacturer.

16563 Lacqtene 1070 MN20, 1200 MN26
A proprietary LDPE used in injection molding. *Aquitaine-Organico.* Unverified.

16564 lacquer
Shellac dissolved in alcohol, and colored with saffron, annatto, or dragon's blood.

16565 Lacril®
9004-65-3 4889
Hydroxypropyl methylcellulose
Ocular lubricant. *Allergan Inc.*

16566 Lacri-Lube® NP, S.O.P®
White petrolatum, mineral oil; ocular lubricant. *Allergan Inc.*

16567 Lacrinite
A proprietary casein-phenolformaldehyde product. No manufacturer.

16568 Lactamide MEA
5422-34-4 226-546-1
2-Hydroxy-N-(2-hydroxyethyl)propanamide
Incromectant LMEA; Lipamide LMEA; Mackamide™ LME; Naetex-LAM; Parapel® LAM-100; Schercomid LME; Upamide LACAMEA; Lactic acid monoethanolamide; Monoethanolamine lactic acid amide. Clarifying detangling agent for shampoos, conditioners, and cream rinses; humectant in creams and lotions. Soluble in water; pH = 5.0. *Croda Inc.; Lipo; McIntyre.* Discontinued.

16569 lacteol
lactigen
lactilloids; lactobacilline; lactone. Preparations of lactic acid bacilli.

16570 lactic acid
50-21-5 5349 200-018-0
$C_3H_6O_3$
2-hydroxypropanoic acid
2-hydroxypropionic acid; milk acid; Acetonic acid; Ethylidenelactic acid; 2-Hydroxypropionic acid; 1-Hydroxyethane 1-carboxylic acid; α-Hydroxypropanoic acid; Purac®,. Organic acid; cultured dairy products, as acidulant, chemicals (salts, plasticizers, adhesives, pharmaceuticals), mordant in wool dyeing, food additive, manufacture of lactates. d = 1.249; mp = 18°; bp = 122° (15mm); LD_{50} (rat orl) = 3730 mg/kg; miscible with water, glycerol, furfural. Penta Mfg.; Pfanstiehl Labs; Purac Biochem BV; AB Tech; ADM; Ashland; Dinoval; Ellis & Everard; Honeywill & Stein; Lohmann; Mitsubishi; Siber Hegner; Van Waters & Rogers.

16571 L-lactic acid
79-33-4 5350 201-196-2
$C_3H_6O_3$
(S)-2-hydroxypropanoic acid
d-lactic acid; sarcolactic acid; paralactic acid;. mp = 53°; $[\alpha]^{21}_{546}$ = 2.6° (c = 2.5); soluble in H_2O.

16572 D-lactic acid
10326-41-7 5348 233-713-2
$C_3H_6O_3$
(R)-2-hydroxypropanoic acid
l-lactic acid; D(-)-lactic acid. mp = 53°; $[\alpha]^{21}_{546}$ = -2.6° (c = 8); soluble in H_2O.

16573 DL-lactic acid
598-82-3 5349 209-954-4
$C_3H_6O_3$
2-hydroxypropanoic acid
lactovagan; Tonsillosan;. Occurs in sour milk. Used as an acidulant and, in veterinary medicine, as an internal antiseptic. mp = 17°; bp_{20} = 122°; n_D^{20} = 1.4259; soluble in H_2O, EtOH, less soluble in organic solvents; LD_{50} (rat orl) = 3.73 g/kg.

16574 Lacticol 336
Specialty algin blend; milk-soluble stabilizer for milk-based systems. *Kelco Int'l Ltd.*

16575 Lactil®
Sodium lactate, sodium PCA, hydrolyzed animal protein, fructose, urea, niacinamide, inositol, sodium benzoate, lactic acid; humectant, moisturing agent for creams and lotions. *Goldschmidt; Goldschmidt AG.*

16576 lactilloids
Lacteol.

16577 Lactimon®
Solution of a partial amide and alkylammonium salt of a higher molecular weight unsaturated polycarboxylic acid and a polysiloxane copolymer; wetting and dispersing additive to prevent setting of pigments; for solvent or solvent-free coating systems. *Byk-Chemie USA.*

16578 Lactin
63-42-3 5356 200-559-2
Lactose
A sugar.

16579 Lactinium
A German preparation of neutral aluminum lactate; used an astringent and disinfectant.

16580 Lactitis
A casein preparation containing borax and lead acetate. It is an artificial ivory.

16581 Lactodan B 30
Glyceryl stearate lactate
Food emulsifier, improves aeration and foam stabilization. *Grindsted Prods.* Name unverified.

16582 Lactoid
A casein preparation.

16583 Lactol
$CH_3CHOH{\cdot}COO{\cdot}C_{10}H_7$
Lactonaphthol
The lactic acid ester of β-naphthol, an intestinal astringent.

16584 Lactolin
Antimonin
The double salts of antimony lactate with alkalis, alkaline earths, and zinc salts. A convenient means for the transport of lactic acid. Also used as a substitute for tartar emetic in dyeing.

16585 Lactolith
A casein preparation.

16586 Lactoloid
A proprietary casein product. No manufacturer.

16587 Lactoprene
A patented vulcanizable synthetic rubber made from emulsified methyl or ethyl acrylate copolymerized with small quantities of a polyfunctional monomer such as butadiene, isoprene or allyl maleate; the copolymer is compounded with sulfur and accelerator an. No manufacturer.

16588 Lactosan
Dairy hygiene detergent sterilizer. *The Wellcome Foundation Ltd.* Name unverified.

16589 α-D-Lactose
5989-81-1 5356 200-559-2
$C_{12}H_{22}O_{11}$
4-O-β-D-Galactopyranosyl-D-glucose
4-O-β-d-galactopyranosyl; Milk sugar; d-glucose; 4-(β-D-galactosido)-D-glucose; Saccharum lactis; Disaccharide,. Pharmacy; infant foods; baking and confectionary; manufacture of penicillin, yeast; adsorbent in chromatography. d = 1.23; mp = 201-202°; $[\alpha]_D^{20}$= +96°; soluble in water, alcohol, insoluble in chlorform, ether. *Dajac Labs; Penta Mfg.; Schweizerhall; Simonis BV.*

16590 Lactose molasses
Molasses obtained from the preparation of milk sugar.

16591 Lactozym
A β-galactosidase (lactase) preparation produced by submerged fermentation of a selected strain of the yeast *Kluyveromyces fragilis*; for sweet milk products, production of low lactose milk for persons suffering from lactose malabsorption. *Novo Nordisk.*

16592 Ladalrod
Aluminum and aluminum alloy rod; used in alloying steel. *Reynolds Metal Co.*

16593 Laddok®
Bentazon
Atrazine. Herbicide for selective postemergence control of broadleaf weeds. *BASF AG.*

16594 Ladelloy
Ferro-alloy ladle additions. *Foseco (F.S.) Ltd.*

16595 Laevuflex
A proprietary preparation of levulose; used as a parenteral calorie supplement. *Geistlich Sohne AG.* Name unverified.

16596 Lafil WL 3254
Polyglyceryl isostearostearate. *Gattefosse SA.*

16597 Lakeland AMA LF
Salt-free low-foam version of Lakeland AMA; low foam cleaners. *Lakeland Laboratories Ltd.*

16598 Lakeland C
Series of cationic emulsions of polyethylene wax and other polymers; used in textiles as softeners and lubricants. *Lakeland Laboratories Ltd.*

16599 Lakeland CAB
Amido betaine based on coconut alkyl chain; viscosity modifier, solubilizer and foaming aid in cosmetic and toiletry formulations. *Lakeland Laboratories Ltd.*

16600 Lakeland CTA/N
Amino betaine based on coconut alkyl chain; amphoteric for production of mild shampoos, foam baths, hand cleaners; high foaming and detergency properties. *Lakeland Laboratories Ltd.*

16601 Lakeland N
Series of nonionic emulsions of polyethylene wax and other polymers; used in floor polish for nonslip, buffability, water resistance; in textiles as needle lubricants; in paper for scuff resistance, gloss, hardness, and water resistance. *Lakeland Laboratories Ltd.*

16602 Lakeland PA, PAE, PPE
Series of anionic surfactant phosphate esters in which the bases are nonylphenol ethoxylates, alcohol ethoxylates and alcohols; wetting agents, detergents, emulsifiers and lubricants which are stable to acid, alkaline and electrolyte conditions. *Lakeland Laboratories Ltd.*

16603 Lakewax
Series of nonionic emulsions of polyethylene wax and other polymers; used in paints and printing inks for scuff or rub resistance, slip properties, matt finish, water resistance. *Lakeland Laboratories Ltd.*

16604 Lalicopharsol
Mild emollient and emulsifier; used for fine cosmetics; pharmaceuticals; nonirritative barrier creams. *Solaver SA.* Name unverified.

16605 Lalitecsol
Mild emollient and emulsifier; used for fine cosmetics; nonirritative barrier creams. *Solaver SA.* Name unverified.

16606 Lamalgin
A thickening agent used for textile printing. *Degussa AG.* Name unverified.

16607 Lambrex
Emulsion explosives (slurries), in which none of the ingredients is classified as an explosive; cap sensitivity is obtained by the mixing process; characterized by greatest handling safety, and can be supplied in cartridges or can be mixed in a pump truck *Dynamit Nobel Wien GmbH.* Name unverified.

16608 Lambrit (Anfo-explosives)
Primed from the bottom of the borehole by Gelatine Donarit 1 and detonating fuse. *Dynamit Nobel Wien GmbH.* Name unverified.

16609 Lamecreme LPM
Hydrogenated palm oil glycerides, cetyl alcohol, TEA- isostearyl hydrolyzed collagen; self-emulsifying raw material for emulsions, hair conditioners, creams, lotions. *Henkel/Cospha; Henkel Canada.*

16610 Lamecreme SA 7
POE stearoylether
Emulsifier for cosmetics. *Grünau.* Discontinued.

16611 Lamefin
A softening agent and lubricant; used by the textile industry for the after treatment of all types of fibers. *Degussa AG.* Name unverified.

16612 Lamefix
Printing oils and fixation accelerators used for textile printing. *Degussa AG.* Name unverified.

16613 Lamefix 680
Fatty acid polyglycol ether and PEG; accelerator for HT-fixation of polyester and triacetate. *Grünau.* Discontinued.

16614 Lameform TGI
Polyglyceryl-3 diisostearate
Emulsifier, emollient for water-oil emulsions with good stability. *Henkel/Cospha; Henkel KGaA.*

16615 Lamefrost® Range
Emulsifier/stabilizer blends; emulsifier and stabilizer for ice cream. *Grünau.*

16616 Lamegin® DW 8000 HW
Diacetyl tartaric acid ester of mono/diglycerides; dough conditioner for bread and rolls. *Grünau; Henkel/Functional Prods.*

16617 Lamegin® DW Range
Diacetyl tartaric acid ester of mono/diglycerides; dough conditioner for bread and rolls. *Grünau.*

16618 Lamegin® EE
68990-58-9 273-612-0
Acetylated hydrogenated tallow glyceride
Emulsifier and plasticizer for cosmetic, food, and edible coatings. *Grünau.*

16619 Lamegin® GLP 20
68990-06-7 273-576-6
Hydrogenated tallow glyceride lactate
Emulsifier and plasticizer for cosmetics, foods, and drugs. *Grünau.*

16620 Lamegin® ZE 30, 60
68990-59-0 273-613-6
Hydrogenated tallow glyceride citrate
Emulsifier for cosmetic, margarine and meat industry. *Grünau.*

16621 Lamegum
A thickening agent used for textile printing. *Degussa AG.* Name unverified.

16622 Lamemul® Range
Mono and diglycerides; food emulsifiers for baking additives, antistaling effect. *Grünau.*

16623 Lamephan
A softening agent and lubricant; used by the textile industry for the after treatment of all types of fibers. *Degussa AG.* Name unverified.

16624 Lamepon
Dispersing and leveling agent; used for dyeing and stabilizer for peroxide bleaching in the textile industry. *Degussa AG.* Name unverified.

16625 Lamepon PA-TR
68918-77-4
TEA-abietoyl hydrolyzed animal protein; mild detergent for shampoos, cleansers; skin compatible. *Henkel/Cospha; Henkel Canada; Henkel KGaA.*

16626 Lamepon S
68920-65-0
Potassium coco-hydrolyzed animal protein; mild detergent for personal care products, cleansers. *Henkel/Cospha; Henkel Canada; Henkel KGaA.*

16627 Lamepon ST40
68952-16-9
TEA-coco-hydrolyzed animal protein; mild detergent for shampoos, cleansers. *Henkel/Cospha; Henkel Canada; Henkel KGaA.*

16628 Lamepon UD
68951-92-8
Potassium undecylenoyl hydrolyzed animal protein; detergent for hair preparations, shampoos, skin cleansers for damaged skin, antidandruff shampoos. *Henkel/Cospha; Henkel Canada; Henkel KGaA.*

16629 Lameprint
A thickening agent used for textile printing. *Degussa AG.* Name unverified.

16630 Lamequat L
Lauryldimonium hydroxypropyl hydrolyzed collagen; conditioning component for hair and body care preparations *Henkel/Cospha; Henkel Canada.*

16631 Lamequick® Range
Fat/emulsifier compounds; whipping base for toppings, desserts, other aerated foods. *Grünau.*

16632 Lamesoft LMG
Glyceryl laurate
TEA-coco-hydrolyzed collagen; refatting agent and thickener for foam bath and shampoos. *Henkel/Cospha; Henkel Canada.*

16633 Lamesorb® Range
Sorbitan esters of fatty acids/polysorbates; food emulsifier for oil-water and water-oil emulsions. *Grünau.*

16634 Lamex 173/FR
A self extinguishing type of the above resins complying with BS 476 Part I (Class II). *Croda Resins Ltd.* Discontinued.

16635 Lamex 185
A trade name for a flexible amine preaccelerated polyester resin used for motor car body repairs. *Croda Resins Ltd.* Discontinued.

16636 Lamex 186
A trade name for a rigid amine preaccelerated polyester resin used for motor car body repairs. *Croda Resins Ltd.* Discontinued.

16637 Lamicoid
A proprietary trade name for a phenol-formaldehyde synthetic resin with a mica filler used for laminated products. No manufacturer.

16638 Lamictal Tablets
84507-84-1
A proprietary formulation containing lamotrigine; anti-epileptic. *The Wellcome Foundation Ltd.*

16639 Laminac EPX-176
A self extinguishing (flame resistant) polyester resin; suitable for manufacture of reinforced plastics in the transportation industry. No manufacturer.

16640 Lamitex
A proprietary trade name for a hard vulcanized fiber. No manufacturer.

16641 Lamol
Rust preventative. *Croda Chem. Ltd.* Discontinued.

16642 Lampronol
Leather dyes and finishes, pigments for plastics, dyes for printing inks. *ICI Chem & Polymers Ltd.*

16643 Lamy butter
Kanja butter, Kanga butter, Sierra Leone butter. The fat obtained from the seeds of *Pentadesma butyracea* of West Africa.

16644 Lanacron
Wool dyestuffs. *Clayton Aniline Co Ltd.* Name unverified.

16645 Lanaire
Scouring preparations. *Crosfield Chemicals Ltd.*

16646 Lanamine®
10525-14-1 234-077-9
Mixed isopropanolamines myristate; high foaming mild detergent for shampoos, shaving soaps. *Amerchol Corp.*

16647 Lanaperl
Acid dyestuffs. *Hoechst AG.*

16648 Lanapex
Textile auxiliary chemicals. *ICI Chem & Polymers Ltd.*

16649 Lan-Aqua-Sol 100
8039-09-6
PEG-75 lanolin
Emulsifier for cosmetic and pharmaceutical emulsions; emollient, superfatting agent, conditioner for skin and hair care products, household detergents; solubilizer, wetting agent, dispersing aid. *Fanning.*

16650 Lanasol
Reactive dyes for wool and silk. *Clayton Aniline Co Ltd.* Name unverified.

16651 Lanbritol
Self-emulsifying waxes. *Hickson & Welch Ltd.*

16652 Lancare
Household detergents and toiletries. *Harcros.*

16653 Lancer
37924-13-3 7300 253-718-3
Flamprop-methyl
Herbicide for wild oat control. *ICI Chem & Polymers Ltd.*

16654 Lancosol
Mild emollient and emulsifier; used for fine cosmetics; nonirritative barrier creams. *Solaver SA.* Name unverified.

16655 Landemul
Demulsifier components for oil field production chemicals. *Harcros.*

16656 Landromil
Ticlatone
Antibacterial; antifungal. *Dorsey Pharmaceuticals.* Name unverified.

16657 Lanesta G
97404-50-7 306-817-1
Glyceryl lanolate
Emollient, emulsifier; forms stable water-oil emulsions. *Westbrook Lanolin.*

16658 Lanesta S
63393-93-1 264-119-1
Isopropyl lanolate
Nonsticky emollient rapidly absorbed by skin; lubricant; for lipsticks, nailcare, sunscreen preparations, toilet soap; binder for pressed powds. *Westbrook Lanolin.*

16659 Lanestren
Dyestuffs for wool/polyester fiber blends. *BASF plc.* Name unverified.

16660 Laneto 27
61790-81-6
PEG-27 lanolin
Emulsifier, emollient, conditioner, moisturizer, stabilizer and solubilizer in makeup, lipstick, skin care products, bath products, shampoos, soap, shave creams, ointments, sun preparations, veterinary products. *RITA.*

16661 Lanette 14
112-72-1 6418 204-000-3
Myristryl alcohol
Emollient, consistency agent for cosmetic and pharmaceutical oil-water and water-oil creams, emulsions, sticks. *Henkel/Cospha; Henkel Canada.*

16662 Lanette 14
112-72-1 6418 204-000-3
Myristyl alcohol
Emollient, consistency agent for cosmetic and pharmaceutical oil-water and water-oil creams, emulsions, sticks. *Henkel/Cospha; Henkel Canada.*

16663 Lanette 16
36653-82-4 2070 253-149-0
Cetyl alcohol
Emollient, consistency agent for cosmetic and pharmaceutical oil-water and water-oil creams, emulsions, sticks. *Henkel/Cospha; Henkel Canada.*

16664 Lanette 18 DEO
112-92-5 8960 204-017-6
Stearyl alcohol
Emollient, consistency agent for cosmetic and pharmaceutical oil-water and water-oil creams, emulsions, sticks. *Henkel/Cospha; Henkel Canada.*

16665 Lanette E
59186-41-3
Sodium cetearyl sulfate
Emulsifier and wetting agent for oil-water emulsions, personal care products. *Henkel/Cospha; Henkel/Functional Prods.; Henkel Canada; Henkel KGaA.*

16666 Lanette N
Cetearyl alcohol and sodium cetearyl sulfate; emulsifying raw material for oil-water creams and emulsions. *Henkel/Cospha; Henkel Canada.*

16667 Lanette O
Cetearyl alcohol
Emollient, base, consistency factor for ointments, creams, emulsions. *Henkel/Cospha; Henkel KGaA.*

16668 Lanette SX
Cetearyl alcohol (90%) and sodium lauryl sulfate (10%); emulsifier; self-emulsifying base for manufacture of oil-water ointment, creams, and emulsions. *Henkel/Cospha; Henkel Canada.*

16669 Lanette Wax
A proprietary trade name for a mixture of cetyl and stearyl alcohols. No manufacturer.

16670 Lanette Wax Ester
A proprietary trade name for palmitic acid ester of cetyl and stearyl alcohols used in emulsions. No manufacturer.

16671 Lanette Wax SX
A proprietary trade name for an emulsified mixture of cetyl and stearyl alcohols. No manufacturer.

16672 Lanexol AWS
68458-88-8
PPG-12-PEG-50 lanolin
Emollient, conditioner, superfatting agent, foam stabilizer, and lubricant for alcoholic and aqueous compositions, plasticizer for hair sprays, oil-water emulsifier, solubilizer. *Croda Inc.*

16673 Lanfrax® 1776
68201-49-0 269-220-4
USP lanolin wax fraction; emulsifier, emollient; waxing agent, oil-water auxiliary emulsifier; slip reducing agent in floor finishing compounds; water-oil emulsion stabilizer and thickener. *Henkel/Cospha; Henkel Canada.*

16674 Langford Clay
1332-58-7 5294 296-473-8
Hydrated aluminum silicate
kaolin clay. Low cost reinforcer and inert filler for paint, paper, rubber, ceramics, plastics, and specialties. *R. T. Vanderbilt Co Inc.* Discontinued.

16675 Lanital
A proprietary trade name for a casein textile fiber made by dissolving casein in a dilute alkaline solution and extruding the viscous compound in the form of thin filaments; these are treated with acid and rendered insoluble by means of formaldehyde. No manufacturer.

16676 Lankrocell® D15L
Blend of organic surfactants; mechanical foam promoter for PVC for carpet and floor backing applications. *Harcros.*

16677 Lankroflex®
Range of epoxidized fatty acid esters and oils; plasticizers for polymers. *Harcros.*

16678 Lankroflex® ED3
Octyl epoxy stearate; plasticizer. *Harcros.*

16679 Lankroflex® ED6
A proprietary epoxy plasticizer. *Harcros.*

16680 Lankroflex® GE
8013-07-8 232-391-0
Epoxidized soya bean oil; plasticizer/stabilizer for epoxy. *Harcros.*

16681 Lankrol
Sulfated oils. *Harcros.*

16682 Lankrolan
Textile auxiliaries; shrink proofing agent for wool. *Harcros.*

16683 Lankroline
Sulfated oils and pigment finishes. *Harcros.*

16684 Lankrolyte
Sequestering agents. *Harcros.*

16685 Lankromark®
Range of mixed metal carboxylates, organotin compounds and organophosphites; stabilizers and antioxidants for polymers. *Harcros.*

16686 Lankromark® BM271
Butyltin carboxylate
Transparent heat and light stabilizer for rigid and plasticized PVC applications *Harcros.*

16687 Lankromark® BT050
Butyl thiotin
Transparent heat stabilizer for rigid and plasticized PVC applications. *Harcros.*

16688 Lankromark® BT120A
Dibutyltin mercaptide
Stabilizer for PVC processing; excellent heat stability and clarity; for rigid sheeting, protiles, injection molding. *Harcros.*

16689 Lankromark® DLTDP
123-28-4 204-614-1
Dilauryl thiodipropionate
Stabilizer for polyolefins and other polymers; synergist with phenolic antioxidant or organophosphite. *Harcros.*

16690 Lankromark® DP6404Z
Barium/Zinc
Self-lubricating stabilizer for suspension PVC for demanding semirigid formulations. *Harcros.*

16691 Lankromark® DSTDP
693-36-7 211-750-5
Distearyl thiodipropionate
Stabilizer for polyolefins and other polymers; synergist with phenolic antioxidant or organophosphite. *Harcros.*

16692 Lankromark® LC475
Cadmium-containing; stabilizer for suspension PVC resins; high efficiency, self-lubricating. *Harcros.*

16693 Lankromark® LC68
Barium/cadmium
Nonlubricating stabilizers for suspension PVC resins; LC310 grade also for emulsion resins. *Harcros.*

16694 Lankromark® LC90
Cadmium/zinc
Stabilizer/activator for chemically blown PVC; fast action. *Harcros.*

16695 Lankromark® LE109
26523-78-4 247-759-6
Tris (nonyl phenyl) phosphite
Stabilizer for nontoxic rigid and flexible PVC; antioxidant for ABS/MBS, polyolefins, SBR. *Harcros.*

16696 Lankromark® LE230
Substituted benzophenone
Stabilizer for polyolefins, ABS, MBS, and SBR. *Harcros.*

16697 Lankromark® LE285
1843-05-6 6838 217-421-2
2-Hydroxy-4-octyloxybenzophenone
Uv absorber for PVC and polyolefins. *Harcros.*

16698 Lankromark® LE296
131-57-7 7088 205-031-5
2-Hydroxy-4-methoxybenzophenone
Uv absorber for PVC and other polymers. *Harcros.*

16699 Lankromark® LE296
131-57-7 7088 205-031-5
Uv absorber for PVC and other polymers. *Harcros.*

16700 Lankromark® LE65
101-02-0 202-908-4
Triphenyl phosphite
Stabilizer for rigid and flexible PVC, PU; epoxy curing agent. *Harcros.*

16701 Lankromark® LZ1034
Organic zinc/epoxy blend; Plasticizer for stabilization of rigid PVC bottles. *Harcros.*

16702 Lankromark® LZ121
Barium/zinc
Nonlubricating stabilizer for PVC emulsion resins. *Harcros.*

16703 Lankromark® LZ187
Barium/zinc
Stabilizer/activator for chemically blown PVC; slow action. *Harcros.*

16704 Lankromark® LZ495
Calcium/zinc
Nonlubricating stabilizer for PVC suspension and emulsion resins. *Harcros.*

16705 Lankromark® LZ616
Barium/zinc
High efficiency selflubricating stabilizer for PVC suspension resins; also for emulsion resins (rotational casting). *Harcros.*

16706 Lankromark® LZ693
Barium/zinc
zinc. High efficiency nonlubricating stabilizers for PVC suspension and emulsion resins. *Harcros.*

16707 Lankromark® OT450
Octyl thiotin
Transparent heat stabilizer for rigid and plasticized PVC applications; suitable for food contact applications. *Harcros.* Name unverified.

16708 Lankromul
Oil spill dispersants. *Harcros.*

16709 Lankroplast®
Lubricants in thermoplastic processing, antifogging agents for PVC film and viscosity modifiers for PVC pastes. *Harcros.*

16710 Lankroplast® L542
Tackifier for use in highly filled calendered flexible PVC formulations. *Harcros.*

16711 Lankroplast® V2012
viscosity modifier for PVC plastisols. *Harcros.*

16712 Lankropol®
Sulfated and sulfonated esters and acids; used as wetting agents, emulsifiers in textiles, paint and emulsion polymers. *Harcros.*

16713 Lankropol® KMA
6001-97-4 1295 227-847-0
$C_{16}H_{29}NaO_7S$
Sulfobutanedioic acid 1,4-bis(1-methylpentyl) ester sodium salt
ethanol; Bis(methylamyl) Sodium Sulfosuccinate; Sodium dihexyl sulfosuccinate. Emulsifier, wetting agent esp. in solutions of electrolytes; solubilizer for soaps, emulsion polymerization aid. *Harcros.*

16714 Lankropol® KO Special
Sodium diisooctyl sulfosuccinate
Clear pale straw liquid containing mineral oil; emulsifier and water carrier in

solvents; used in hydrocarbon solvents, dry cleaning detergent formulations, and as a dewatering aid. *Henkel Chemicals Ltd.* Name unverified.

16715 **Lankropol® KO2**
577-11-7 3460 209-406-4
Sodium dioctyl sulfosuccinate
ethanol. Wetting agent, emulsifier for emulsion polymerization. *Harcros.*

16716 **Lankropol® KPH70**
577-11-7 3460 209-406-4
Sodium dioctyl sulfosuccinate in propylene glycol/water; wetting agent, emulsifier, co-dispersant. *Harcros.*

16717 **Lankropol® KSB 22**
Monoester sulfosuccinate as a clear liquid; primary emulsifier in latex production. *Henkel Chemicals Ltd.* Name unverified.

16718 **Lankropol® KSG 72**
Monoester sulfosuccinate as a clear liquid; low irritancy toiletry intermediate for shampoos, bubble baths, etc. *Henkel Chemicals Ltd.* Name unverified.

16719 **Lankropol® ODS**
Disodium N-octadecyl sulfosuccinamate in liquid or paste form; foaming agent used for latex foam systems in carpet backing. *Henkel Chemicals Ltd.* Name unverified.

16720 **Lankrosol**
Alkylaryl sulfonates
Modified nonionics, phosphate esters; used as hydrotropes, wetters in electrolyte solutions. *Harcros.*

16721 **Lankrosol HS101**
Potassium salt of phosphate ester condensate; low foam hydrotrope for use in highly built liquids.; stable to acids, alkalis, electrolytes. *Harcros.*

16722 **Lankrosol SXS**
Anionic surfactant as a clear, pale straw liquid; solubilizing and viscosity modification agent used in high active liquid detergents and hard surface cleaners; also used for dye leveling in nylon dyeing. *Henkel Chemicals Ltd.* Name unverified.

16723 **Lankrosperse**
Dispersing agents. *Harcros.*

16724 **Lankrostat® 16**
Antistats for plasticized PVC. *Harcros.*

16725 **Lankrostat® LME**
Antistat for polyolefins, PS, crystal PS. *Harcros.*

16726 **Lankrothane**
Polyurethane derivates. *Harcros.*

16727 **Lannate®**
16752-77-5 6062 240-815-0
Methomyl insecticide *DuPont UK.*

16728 **Lanocerin®**
68201-49-0 269-220-4
Lanolin wax
water-oil emulsifier, emollient, conditioner used in cosmetics. *Amerchol Corp.*

16729 **Lanogel® 21**
61790-81-6
PEG-27 lanolin
Emollient, emulsifier, dispersant, wetting agent, solubilizer, foam stabilizer, used in cosmetics, personal care products, pharmaceuticals, facial tissues, antiperspirants, germicidal hand soaps. *Amerchol Corp.*

16730 **Lanogel® 31**
Ethoxylated (40 mol) lanolin
Nonionic surfactant, more water soluble than Lanogel 21. *Amerchol Corp.* Discontinued.

16731 **Lanogel® 41**
Ethoxylated (75 mol) lanolin
50% aqueous gel, water soluble, suitable for aqueous systems; conditioner in soaps and shampoos. *Amerchol Corp.*

16732 **Lanogel® 61**
Ethoxylated (85 mol) lanolin
Nonionic 50% gel soluble in water and alcohol; conditioner in hydroalcoholic systems. *Amerchol Corp.* Discontinued.

16733 **Lanogene®**
Lanolin oil; emollient, moisturizer, and emulsifier which imparts oil sol. and spreading properties. *Amerchol Corp.*

16734 **Lanoiac**
Lanolin paint. *Croda Chem. Ltd.* Discontinued.

16735 **Lanol 14 M**
59686-68-9
Myreth-3 myristate
Cetiol® 1414E; Schercemol MEM-3. Cosmetic emulsifier. *Seppic; Scher; Henkel/Cospha; Henkel canada.*

16736 **Lanol 1688**
261-619-1
Cetearyl octanoate
Cosmetic emulsifier. *Seppic.*

16737 **Lanol P**
4219-49-2 224-160-8
Glycol palmitate
Cosmetic emulsifier. *Seppic.*

16738 **Lanolex L-40**
61790-81-6
PEG-50 lanolin
Reforming agent, emollient for shampoos, hair conditioners. *Nihon Emulsion.*

16739 **Lanolic Acid**
Distilled lanolin fatty acids; cosolvent, emollient for skin and hair cosmetics, makeup, shaving foams and creams. *Croda Chem. Ltd.* Discontinued.

16740 **lanolin**
8006-54-0 (anhyd.) 8020-84-6 (hyd.) 5371 232-348-6
Anhydrous lanolin; wool wax; wool fat; lanalin; lanain; laniol; lanesin; lanichol. Derivative. of unctuous fatty sebaceous secretion of sheep consistg. of complex mixt. of esters of high molecular weight aliphatic, steroid, or triterpenoid alcohol and fa *Amerchol; Croda; Henkel/Emery; Lanaetex; RITA; Stevenson Bros.; Westbrook Lanolin.*

16741 **Lanolin alcohol**
8027-33-6 232-430-1
alcohol, lanolin
wool wax alcohol; Agrobase 125; Agrowax Standard; Ceralan® Hartolan; Hartolite; Ivarlan 3310; Ritawax; Sululan; Super Hartolan;. Mixture of organic alcohols obtained from hydrolysis of lanolin; water-oil emulsifier, stabilizer, softener, emollient, gelling agent, thickener, plasticizer, moisturizer for absorption bases, cosmetics, cleansing preparations. Chemmark; Croda; Henkel/Emery; Heterene; Fanning; Amerchol Corp.; Croda Chem. Ltd.; Brooks industries; RITA.

16742 **Lanoline**
A formulation of lanolin; an emollient to soothe and soften the skin. *The Wellcome Foundation Ltd.*

16743 **Lanoplast**
A proprietary cellulose acetate. No manufacturer.

16744 **Lanoquat® 1756**
Quatermium-33
ethyl hexanediol. Conditioner; emulsifier for skin moisturizers; substantive to hair; provides lubricating, conditioning, antistatic properties. *Henkel/Cospha; Henkel Canada.*

16745 **Lanosol**
Lanolized mineral oil. *Croda Chem. Ltd.* Discontinued.

16746 **Lanosterol**
Lanosterol. *Croda Chem. Ltd.*

16747 **Lanotein AWS 30**
Propylene glycol, hydrolyzed animal protein, PPG-12-PEG-65 lanolin oil; conditioner, film former, lubricant, humectant, and emollient used in hair products. *Fanning.*

16748 **Lanoxicaps**
20830-75-5 3210 244-068-1
A proprietary formulation of digoxin; for treatment of heart failure, atrial fibrillation, atrial flutter, and paroxysmal atrial tachycardia. *The Wellcome Foundation Ltd.*

16749 **Lanoxin**
20830-75-5 3210 244-068-1
Proprietary formulation of digoxin; tablets, injections, oral solution; for management of chronic cardiac failure and certain supraventricular arrhythmias, particularly atrial fibrillation and atrial flutter. *The Wellcome Foundation Ltd.*

16750 **Lanoxine-PG**
20830-75-5 3210 244-068-1
A proprietary preparation of digoxin. *The Wellcome Foundation Ltd.* Discontinued.

16751 **Lanpharsol**
Mild emollient and emulsifier; used for fine cosmetics; nonirritative barrier creams. *Solaver SA.* Name unverified.

16752 **Lanpol**
Polyxyethylated lanolin fatty acids
Croda Chem. Ltd.

16753 **Lanpolamide 5**
PEG-5 lanolinamide
PEG-5 lanolate. water-oil emulsifier forming stable emulsions; emulsion stabilizer, corrosion inhibitor for aerosols. *Croda Inc.* Discontinued.

16754 **Lanstar**
Antistatic agents, emulsifiers; used for polystyrene; synthetic fibers; textile auxiliaries. *Lanstar Ltd.* Discontinued.

16755 Lanstar NP100/50
Nonylphenol ethoxylate in aqueous solution; nonionic surfactant effective at high temperatures and in concentrated electrolyte solutions; used in emulsion polymerization; latex stabilization; waxes and polishes. *Lanstar Ltd.* Name unverified.

16756 Lanstar NP2 and NP4
Nonylphenol ethoxylate nonionic surfactant in the form of an oil-soluble liquid; foam depressors; emulsifiers. *Lanstar Ltd.* Name unverified.

16757 Lanstar NP40, NP50 and NP100
Nonylphenol ethoxylate nonionic surfactant in wax form; water-soluble surfactants effective at high temperatures and in concentrated electrolyte solutions; used in emulsion polymerization; latex stabilization; waxes and polishes. *Lanstar Ltd.* Name unverified.

16758 Lanstar PCH, PC2 and PCO
Calcium petroleum sulfonate in liquid form; emulsifier, dispersing and wetting agent used in cutting oil, lube oil and fuel oil additives; rust preventatives; leather and textile industry. *Lanstar Ltd.* Name unverified.

16759 Lanstar PS
Sodium petroleum sulfonate in liquid form; emulsifier, dispersing and wetting agent for cutting oil, lube oil and fuel oil additives; rust preventatives; leather and textile industry. *Lanstar Ltd.* Name unverified.

16760 Lanstar PSW
Sodium petroleum sulfonate in liquid form; water-soluble anionic surfactant used in ore flotation; building industry; demulsification. *Lanstar Ltd.* Name unverified.

16761 Lantecsol
Mild emollient and emulsifier for technical uses; rust inhibitor; bind material; for printing ink; rust inhibitors; petroleum; rubber adjuvants; lubricants; soaps; leather protection; painting industry. *Solaver SA.* Name unverified.

16762 Lanthanol LAL
Sodium alkyl sulfoacetate in powder or flake form; anionic surfactant used in shampoos and bubble baths. *KWR Chemicals Ltd.* Name unverified.

16763 lanthanum
7439-91-0 5374 231-099-0
La
Metallic element
Lanthanum salts, electronic devices, pyrophoric alloys, rocket propellants, reducing agent catalyst for conversion of nitrogen oxides to nitrogen in exhaust gases, phosphors in x-ray screens. mp = 920°; bp = 3464°; d = 6.162; dec slowly in cold water, more readily on heating. *Aldrich; Atomergic Chemetals; Cerac; Rhône-Poulenc Basic.*

16764 lanthanum chloride
10099-58-8 233-237-5
$LaCl_3$
Lanthanum chloride anhydrous
Anhydrous trichloride used to prepare the rare-earth metal. mp = 860°; bp = 1000°. *Atomergic Chemetals; Cerac; Spectrum Chem. Mfg.*

16765 lanthanum nitrate
10099-59-9 233-238-0
$La(NO)_3{\cdot}6H_2O$
Lanthanum nitrate hexahydrate
Antiseptic, gas mantles. *Atomergic Chemetals; Cerac; Rhône-Poulenc Basic.*

16766 lanthanum oxide
1312-81-8 5374 215-200-5
La_2O_3
Lanthana
Lanthanum trioxide; Lanthanum sesquioxide. Calcium lights, optical glass, technical ceramics, cores for carbon-arc electrodes, fluorescent phosphors, refractories. mp = 2315°; bp = 4200°; d = 6.5100. *Atomergic Chemetals; Cerac; Rhône-Poulenc Basic; Seimi Chem.*

16767 Lantrol® 1673
5371 232-348-6
Lanolin oil
Emollient and moisturizer for makeup, creams, lotions, hair care products, bath oils, medicinal preparations; pigment dispersant. *Henkel/Cospha; Henkel Canada.*

16768 Lantrol® AWS 1692
68458-58-8
PPG-12-PEG-65 lanolin oil; emollient, plasticizer, solubilizer, and conditioner for hair products, shaving lotions, antiperspirants, body colognes. *Henkel/Cospha; Henkel Canada.*

16769 Lantrol® HP-2073
8006-54-0 5371 232-348-6
Lanolin oil; emulsifier, emollient, conditioner, moisturizer, pigment dispersant for personal care products, pharmaceuticals. *Henkel/Cospha; Henkel Canada.*

16770 Lantrol® PLN
8039-09-6
Ethoxylated lanolin
Cleaning and wetting agent; solubilizer for perfumes and germicides; conditioner for shampoos; superfatting agent for soaps; oil-water emulsifier for nonfatty preparations; plasticizer for aerosols and hair sprays. *Pulcra SA.*

16771 Lanvis
154-42-7 9473 205-827-2
A proprietary preparation of 6-thioguanine; used in the treatment of acute myelogenous leukemia and acute lymphoblastic leukemia. *The Wellcome Foundation Ltd.*

16772 Lanxide
Ceramic matrix composites. *DuPont UK.*

16773 Lapis Smiridis
Emery. No manufacturer.

16774 Lapix
A proprietary trade name for a flux used in steel molding; contains carbon and clay. No manufacturer.

16775 Lapofloc
Water treatment chemicals. *Laporte Industries Ltd.*

16776 Laponite®
Synthetic clay products resembling hectorite with thixotropic gelling properties; used as gels in toothpastes and cosmetics; in surface coatings and sprays, conferring antistatic properties; used as retentive medium due to absorbancy. *Laporte Industries Ltd.*

16777 Laponite® B
Sodium magnesium fluorosilicate
Imparts shear sensitive structure to waterborne formulations. *Laporte/Southem Clay.*

16778 Laponite® D
53320-86-8 258-476-2
Sodium magnesium silicate
Used in conjunction with other thickeners for imparting a shear sensitive structure to clear gel and conventional toothpastes. *Laporte/Southem Clay.*

16779 Laponite® S
Sodium magnesium fluorosilicate
tetrasodium pyrophosphate. Imparts a shear sensitive structure to waterborne formulations including paints and cleansers; may be coated onto paper or other surfaces to give electrical conductive films. *Laporte/Southem Clay.*

16780 Laponite® XLG
53320-86-8 258-476-2
Sodium magnesium silicate
Inert base/carrier for act. ingreds.; suspending agent; promotes thixotropy giving stable suspensions; thickens cosmetic, toiletry creams, lotions, toothpaste products. *Laporte/Southem Clay.*

16781 Lapotan
10043-01-3 381 233-135-0
Basic aluminum sulfate. *Laporte Industries Ltd.*

16782 Lapyrium Chloride
6272-74-8 5379 228-464-1
$C_{21}H_{35}ClN_2O_3$
1-[[2-Oxo-2-[(1-oxododecyl)oxy]ethyl]amino]ethyl]pyridinium chloride
Emcol® E-607L; 1-[[(2-hydroxyethyl)carbamoyl]methyl]pyridinium chloride laurate (ester;)N-(lauroylcolaminoformylmethyl)pyridinium chloride; N-(acylcolaminoformylmethyl)pyridinium chloride; emulsepr (obsolete); E-607; Emcol E-607;. Emollient, emulsifier, conditioner, foamer, cleanser, substantive agent, deodorant for cosmetics, toiletries, industrial applications; hair conditioner; antistat; detergent-germicide; *Witco/Organics; Witco SA.* Discontinued.

16783 Laquanol
Water thinnable resins. *Croda Resins Ltd.* Discontinued.

16784 Laractone®
52-01-7 8917 200-133-6
Tablets containing 25 mg and 100 mg spironolactone BP; used in congestive heart failure, hepatic cirrhosis with ascites and edema, malignant ascites, nephrotic syndrome, diagnosis and treatment of primary hyperaldosteronism. *Lagap Pharmaceuticals Ltd.*

16785 Laraflex®
22204-53-1 6504 244-838-7
Laraflex®. Tablets containing 250 mg and 500 mg naproxen BP; for treatment of rheumatoid arthritis, osteoarthritis, ankylosing spondylitis, juvenile rheumatoid arthritis, acute gout and acute musculoskeletal disorders (such as sprains and strains, direct trauma, lum *Lagap Pharmaceuticals Ltd.*

16786 Largactil
69-09-0 2238 200-701-3
Chlorpromazine hydrochloride
A sedative. *Rhône-Poulenc Rorer Ltd.*

16787 Larocin
26787-78-0 617 248-003-8

Proprietary preparation of amoxycillin; antibacterial. *Roche Products Ltd.* Name unverified.

16788 Larodur®
Self-crosslinking, heat-curable polyacrylate resins; for single-coat baking finishes with good resistance to light, heat and chemicals ; for painting domestic appliances. *BASF AG.*

16789 Laroflex®
Vinyl chloride copolymers; for alkali- and acid resistance, lightfast, weather-resistance finishes on metal, concrete; binders for special gravure inks. *BASF AG.*

16790 Laromer®
Dipropylene glycol diacrylate
Reactive diluent for radiation-curable systems *BASF AG.* Name unverified.

16791 Laromer® POEA
48165-04-6 256-360-6
Phenoxyethyl acrylate
Reactive diluent for radiation-hardenable systems *BASF AG.* Name unverified.

16792 Laromer® TPGDA
68901-05-3 272-647-9
Tripropylene glycol diacrylate
Reactive thinner for radiation-curing systems *BASF AG.* Name unverified.

16793 Laromid®
Polymines
Hardeners for epoxy resins. *BASF AG.*

16794 Laropal®
Lightfast aldehyde and ketone resins; for surface coatings, production of pigment pastes, for flexographic and gravure inks. *BASF AG.*

16795 Larostat® 143
Oleyldimethylethyl ammonium ethosulfate
Surface active antistat. *PPG/Specialty Chem.* Discontinued.

16796 Larostat® 377 DPG
Lauric myristic dimethylethyl ammonium ethosulfate
dipropylene glycol. Noncorrosive mold release agent; antistat; useful in polyurethanes; surface active. *PPG/Specialty Chem.* Discontinued.

16797 Larostat® 451
Stearyldimethylethyl ammonium ethosulfate
Noncorrosive release agent forming a hard film; imparts gloss and antistatic properties; post treatment in polystyrenes and fiberglass; surface active. *PPG/Specialty Chem.* Discontinued.

16798 Larostat® 88
68308-67-8 269-663-3
Modified soyadimethylethyl ammonium ethosulfate; noncorrosive mold release agent; antistat; surface active. *PPG/Specialty Chem.* Discontinued.

16799 Larostat® JMR
Surface active antistat, lubricant. *PPG/Specialty Chem.* Discontinued.

16800 Larotid
26787-78-0 617 248-003-8
Amxicillin
Antibacterial. *SmithKline Beecham.* Name unverified.

16801 Laroxyl
549-18-8 208-964-6
A proprietary preparation of amitriptyline hydrochloride; an antidepressive agent. *Roche Products Ltd.* Name unverified.

16802 Larvacide
76-06-2 2208 200-930-9
A proprietary trade name for chloropicrin; used as an insecticide. No manufacturer.

16803 Larvex
A solution of sodium fluosilicate; a proprietary clothes-moth remedy. No manufacturer.

16804 Laser
Emulsifiable concentrate containing 200 g cycloxydim per liter; used to control weeds in grass. *BASF plc.*

16805 Lasilium C
131081-39-5
Lactoyl methylsilanol elastinate
Provides hydrating, anti-inflammatory action, tissue regeneration, slimming action for hand creams, slimming products, toothpastes, cosmetic and health products. *Exsymol.*

16806 Lasilso
6834-92-0 8788 229-912-9
Sodium metasilicate
Laporte Industries Ltd.

16807 Lasix
54-31-9 4331 200-203-6
A proprietary preparation of furosemide; a diuretic. *Hoechst UK.*

16808 Lasso
2-Chloro-2', 6'-diethyl-N-(methoxymethyl) acetanilide
Pre-emergence herbicide. *Monsanto Co.*

16809 Lastex
Proprietary rubber latex threads spun with fiber. No manufacturer.

16810 Lastil
Fungicide for bonded cork. *BDH Chemicals Ltd.*

16811 Lastilac
A proprietary molding compound. No manufacturer.

16812 Latekoll®
Polyacrylic derivatives; thickener for polymer dispersions and latexes; for paints, adhesives, sealants, production of fiber webs. *BASF AG.*

16813 Latene
A solution of trimene base in rubber latex; a proprietary rubber vulcanization accelerator. No manufacturer.

16814 Latensol AP8
A nonionic surfactant. *BASF plc.* Name unverified.

16815 Latex Foam
A proprietary trade name for a type of cellular sponge rubber made from latex by a special method. No manufacturer.

16816 Lathanol® LAL
Sodium lauryl sulfoacetate
Emulsifier, wetting agent, detergent, foaming agent, thickener used in cosmetics and personal care products. *Stepan; Stepan Canada; Stepan Europe.*

16817 Latibon®
544-17-2 1711 208-863-7
Calcium formate; as feed additive and as silage additive. *Bayer AG.*

16818 Latkem
Two-component adhesive incorporating natural rubber latex for use in applications requiring high resilience and flexural strengths; excellent adhesion steel and glass. *Feb Ltd.*

16819 Latol 4
Low-rosin grade fatty acid derived from tall oil; raw material for manufacture of emulsifiers used in disinfectants, cleaners and detergents; air-drying and baking alkyds; gloss oils and varnishes; toy enamels; metallic driers; core oils; masonry cements. *Actrachem.*

16820 Laur 101B
63394-02-5
Laur 101A. Silicone rubber dispersion; used to render glass objects shatter resistance; fully compounded., ready to use; requires heat cure; long pot life. *Laur Silicone Rubber Compding.*

16821 Laur 676U
63394-02-5
Silicone rubber; used for o-rings; oil-resistance, high tear strength, low compression set; peroxide-curable. *Laur Silicone Rubber Compding.*

16822 Latol 1550
Monocyclic C_{21} dicarboxylic acid
Biodegradable synthetic surfactant, coupling agent for surfactants; forms high solids fluid soaps; hydrotrope; for all-purpose household/industrial cleaners, disinfectant cleaners, laundry products; lubricity additive in metalworking formulations; curing agent. *Actrachem.*

16823 Latol MOD
Blown tall oil; primary emulsifier for oil muds. *Actrachem.*

16824 Laur Q-1331
63394-02-5
Silicone dispersion base; used for coating of cloth and for electrical sleeving. *Laur Silicone Rubber Compding.*

16825 Laur Red 10 Silicone Pigment
Iron oxide pigment in silicone rubber base; features excellent handling, improved heat stability, good dispersion. *Laur Silicone Rubber Compding.* Discontinued.

16826 Laural
Fatty alcohol sulfates and ether sulfates; used for detergents, toiletries. *Elf Atochem UK/Ceca.*

16827 lauramide 11
Cocamide DEA; foam booster, thickener, superfatting agent. *Zohar Detergent Factory.*

16828 lauramide DEA
120-40-1 204-393-1
$C_{16}H_{33}O_3N$
lauric diethanolamide
N,N-bis(2-hydroxyethyl)dodecanamide; Diethanolamine lauric acid amide. Mixture of ethanolamides of lauric acid; foam booster/stabilizer, detergency and viscosity builder, emulsifier, wetting agent for personal care products, household and institutional detergents. *Chemron; Karishamns; Mona; Norman, Fox; Sandoz; Scher; Sherex; Stepan; Witco/Humko.*

16829 lauramide EG
Ethylene glycol stearate
Opacifying agent. *Zohar Detergent Factory.*

16830 lauramidopropyl betaine
4292-10-8 224-292-6
Lauramidopropyl betaine
Lexaine® LM; N-(Carboxymethyl)N,N-dimethyl-3-[(1-oxododecyl)amino]-1-propanaminium hydroxide, inner salt; Afaine™ LMB; Amido Betaine-L; Chembetaine L; Chemoxide L; Chimin LX; Mackam™ LMB-LS, LMB; Mafo® LMAB; Mirataine® BB; Monateric LMAB; Tego®-Betaine L-10S,. Mild surfactant, foam booster for bath products, shampoos, liq. soaps, dishwash liqs. Inolex; McIntyre; Zohar Detergent Factory; Chemron; Auchem SpA; PPG/Specialty Chem.; Rhône-Poulenc Surf. & Spec.; Mona Industries; Goldschmidt; Goldschmidt AG.

16831 lauramidopropyl dimethylamine
3179-80-4 221-661-3
Lauramidopropyl dimethylamine
Lexamine L-13; Chemidex L; Mackine 801; Schercodine L; N-[3-(Dimethylamino)propyl]dodecanamide; Dimethylaminopropyl lauramide. Conditioner, emulsifier for hair and skin care products. *Inolex; Scher.*

16832 lauramine oxide
1643-20-5 216-700-6
Empigen® OB; Lauryl dimethylamine oxide; N,N-Dimethyl-1-dodecanamine-N-oxide; Ablumox LO; Ammonyx® DMCD-40; Ammonyx® LO; Amyx LO 3594; Aromox® DMMC-W; Chemoxide LM-30; Emcol® DMCD-40; Emcol® L; Emcol® LO; Empigen® OB; Incromine Oxide L; Lilaminox M24; Mackamine™ LO; Mazox® LDA; Oxamin LO; Radiamox® 6804; Rewominox L 408; Rhodamox®; Schercamox DML; Unimox LO; Varox® 270; Varox® 365; Varox® 375; Zoramox LO;,. Coactive, detergent, antistat, foam booster/stabilizer and viscosity modifier for personal care products, surgical scrubs, fire fighting foam concs., foamed rubbers, bleach additive; solubilizer. d = 0.98; visc^{20} = 25 cps; pH = 7.5. *Albright & Wilson; Albright & Wilson UK;.*

16833 laurel oil
Bayberry oil, obtained from the berries of the laurel tree, *Laurus nobilis.*

16834 Laurel PEG 400 DT
61791-01-3
PEG-8 ditallate
Emulsifier and solubilizer for mineral oils, fats, solvents; for latex paints, metalworking fluids, industrial lubricants, textile specialties. *Reilly-Whiteman.*

16835 Laurel PEG 400 MO
9004-96-0
PEG-8 oleate
Emulsifier for metalworking, paints, solvents, textile chem. specialties. *Reilly-Whiteman.*

16836 Laurel PEG 400 MT
61791-00-2
PEG-8 tallate
Emulsifier for sol. oils, industrial lubricants; softener component for textile industry. *Reilly-Whiteman.*

16837 Laurel R-50
8002-33-3 232-306-7
Sulfated castor oil; penetrant, lubricant, emulsifier used in detergent cleaners, metalworking lubricants, paint, textile lubricants, low-irritation and ethnic hair preparations and dyes, skin cleaners and lotions. *Reilly-Whiteman.*

16838 Laurel SBT
42808-36-6 255-950-0
Sulfated butyl tallate
Industrial lubricant, emulsifier in textile formulations, rewetting agent for corrugated medium. *Reilly-Whiteman.*

16839 Laurel SD-101
Cocamide DEA (2:1)
Detergent, viscosity builder, foamer for hard surface cleaners, laundry products, textile scouring, metal cleaners. *Reilly-Whiteman.*

16840 Laurel SD-400
93-83-4 202-281-7
Oleamide DEA (2:1)
Emulsifier, viscosity builder in metalworking fluids. *Reilly-Whiteman.*

16841 Laurel SD-520T
Erucamide TEA (2:1)
Surfactant, EP lubricant for metalworking formulations. *Reilly-Whiteman.*

16842 Laurel SMR
Sulfated methyl rapeseed ester; fiber-to-metal lubricant for textile fibers; improved wetting, high smoke pt. *Reilly-Whiteman.*

16843 Laurel SRO
617788-68-9
Sulfated rapeseed oil
Used in metalworking lubricants and extreme pressure lubricants. *Reilly-Whiteman.*

16844 Laurel wax
Myrtle wax, bayberry tallow. Myrtle berry wax, obtained from *Myrica cerifera.*

16845 Laurelphos 39
Phosphate ester, free acid; surfactant for oil and water-based metalworking lubricants and synthetic cutting fluids; EP lubricant; neutralized with caustic as wetting agent. *Reilly-Whiteman.*

16846 Laurelphos 400
39464-69-2
Aromatic phosphate ester, free acid; lubricant additive, emulsifier, rust inhibitor for metalworking lubricants. *Reilly-Whiteman.*

16847 Laurelphos RH-44
52623-95-7
Aliphatic phosphate ester, free acid; detergent, hydrotrope, viscosity builder for strong alkali formulations; deduster for powd. alkalis. *Reilly-Whiteman.*

16848 Laurelterge 837, 1390
Detergent for textile scouring. *Reilly-Whiteman.*

16849 Laureltex 308, 308 LF
Detergent for textile scouring. *Reilly-Whiteman.*

16850 Laurent's acid
$C_{10}H_8SO_2$
L-acid
1-Naphthyl-amine-5-sulfonic acid. Used as an azo dye.

16851 Laurent's aluminum solder
The hard solder contains 63-74% zinc and 19-30% tin. The soft solder contains 60-70% zinc, 16-27% tin, and 12% lead.

16852 Laureth-2
Genapol® 24-L-3, 42-L-3; Marlipal® 24/20; Surfonic® L24-2; Tergitol® 24-L-45; C12-14 pareth-3;. C12-14 pareth-2.9; biodegradable detergent intermediate for sulfation for use in cosmetics, shampoos, light duty detergents; emulsifier, prewash spotter, agricultural adjuvant, hydrocarbon-based cleaning systems. *Hoechst Celanese/Colorants & Surf; Hüls AG; Texaco; Union Carbide.*

16853 Laureth-2
68002-97-1
Empilan® KB 2; Diethylene glycol dodecyl ether; PEG-2 lauryl ether; 2-[2-(Dodecyloxy)ethoxy]ethanol; AkyporoxRLM Nikko Chem;. Emulsifier, foam booster, superfatting agent, used in detergents and emulsifying systems; mortar plasticizer. Albright & Wilson UK; Chem-Y GmbH; Henkel KGA/Cospha; Stepan Canada; Auchem SpA; Lonza; Pulcra SA; Heterene; Witco/H-I-P; Zschimmer & Schwarz; Protex; Universal Preserv-A-Chem.

16854 Laurex®
A proprietary trade name for a series of primary fatty alcohols. *Marchon France SA.* Name unverified.

16855 Laurex® CS
Cetearyl alcohol BP
manufacture of surfactants; raw material for ethoxylation, sulfation, etc.; stabilizer in emulsion polymerization; lubricant in rigid PVC, also for pharmaceutical creams, hand lotions, bath oils, shaving creams. *Albright & Wilson Am.; Albright & Wilson UK.*

16856 lauric acid
143-07-7 5396 205-582-1
$C_{12}H_{24}O_2$
n-dodecanoic acid
dodecanoic acid; dodecoic acid; Emery® 650; Hystrene® 9512; Philacid 1200; Prifrac 2920. Fatty acid; alkyd resins, wetting agents, soaps, detergents, cosmetics, insecticides, food additives. d = 0.869; mp = 44°; bp = 225°; n_D^{20} = 1.4183; insoluble in water; soluble in alcohol; freely soluble in benzene, ether; LD_{50} (mus iv) = 131 mg/kg. *Akzo; Henkel/Emery; Mirachem Srl; Unichema; Witco/Humko.*

16857 Lauridit
Foam stabilizers and skin protecting additives; for liquid and powder formulations. *Akzo Chemie Nederland BV.* Name unverified.

16858 Laurodin
Bactericidal preparations of laurolinium acetate. *Allen & Hanburys Ltd.* Name unverified.

16859 Laurox®
105-74-8 203-326-3
Dilauroyl peroxide
Initiator for elevated-temperature polyester cures, for production of PVC resins, and acrylates. *Akzo.* Name unverified.

16860 lauroyl chloride
112-16-3 203-941-7
$C_{12}H_{22}OCl$
dodecanoyl chloride
Surfactant, polymerization initiator, antienzyme agent, foamer; synthesis of lauroyl peroxide, sodium lauroyl sarcosinate, other sarcosinates. *Atochem N. Am.; Hüls AG; PPG Industries.*

16861 lauroyl sarcosine
97-78-9 202-608-3
$C_{15}H_{29}O_3N$
Hamposyl® L; Crodasinic L; Vanseal® LS; Nikkol Sarcosinate LH; Oramix L; N-Methyl-N-(1-oxododecyl) glycine; Sarcosyl® L,. Detergent, wetting and foaming agent, foam stabilizer, emulsifier, anticorrosive agent, conditioner for hair and rug shampoos, cosmetics, skin cleansers; biodegradable W R Grace/Hampshire; Chemplex Chems.; R. T. Vanderbilt Co Inc; Croda Chem Ltd.;Nikko Chem Co. Ltd; Seppic; Ciba-Geigy AG;.

16862 Laurydol
105-74-8 203-326-3
Lauroyl peroxide
Akzo Chemie UK Ltd.

16863 lauryl acrylate
2156-97-0 218-463-4
$C_{14}H_{28}O_2$
dodecyl acrylate
Monomer. Uv-curable reactive diluent in inks and coatings, adhesives, viscosity index improver, finishing aid for leather. *CPS; Sartomer.*

16864 lauryl alcohol
112-53-8 3464 203-982-0
$C_{12}H_{26}O$
1-dodecanol
C_{12} linear primary alcohol; dodecyl alcohol; Fatty alcohol. Manufacture of sulfuric acid esters which are used as wetting agents; synthetic detergents, lube additives, pharmaceuticals, rubber, textiles, perfumes, flavoring agents. *Ethyl; M. Michel; Procter & Gamble; Vista.*

16865 lauryl hydroxyethyl imidazoline
136-99-2 205-271-0
1-Hydroxyethyl-2-lauric imidazoline
Hodag C-100-L; Mackazoline L; Schercozoline L. Intermediate for quaternary ammonium compounds; strongly absorbed on textiles, paper and many metal surfactants; for agricultural, asphalt, cleaners, corrosion inhibitors, demulsifiers, flotation, metalworking, paints, pigment grinding, inks, textiles, wax emulsions *Calgene; McIntyre; Scher.*

16866 lauryl PCA
30657-38-6 250-275-8
Lauryl PCA
Laurydone®; Lauryl pyrrolidonecarboxylate; Pyrrolidone carboxylic acid, lauryl ester. Emulsifier in oily or aqueous continuous phase; lipophilic moisturizer for skin care products, hair care products, toiletries, makeup; exceptionally soft touch. *UCIB.*

16867 laurylpyridinium chloride
104-74-5 203-232-2
$C_5H_5NClC_{12}H_{25}$
1-dodecylpyridinium chloride
Ledmin LPC; Dehyquart C Crystals; Ledmin LPC. Quaternary ammonium compound; Cationic detergent, dispersing and wetting agent; ingredient of fungicides and bactericides. *Schweizerhall; Hoechst UK.*

16868 Lausofan
A hexamethylene ketone; used for destroying insects of the vermin type.

16869 Lautal
A proprietary trade name for an aluminum alloy containing 4% copper and 2% silicon. No manufacturer.

16870 Lauth's violet
581-64-6 9483 209-470-3
$C_{12}H_9N_3S\cdot HCl$
thionine
Aminoiminoiminodiphenyl sulfide hydrochloride. Used as a microscopic stain.

16871 Lavalloy
A proprietary trade name for a ceramic product made from mullite and alumina. No manufacturer.

16872 Lavarock
A heat-treated steatite.

16873 Lavasul
A mixture of 40% coke dust with sulfur; used for tank linings.

16874 Lavender drops
Compound tincture of lavender.

16875 Laventin® CW
Alkanol polyglycol ether
Water-free wetting agent and detergent used in textile industry. *BASF AG.*

16876 Lavite
A heated steatite product.

16877 Lavrex
Fatty alcohols. *Surfachem Ltd.*

16878 Lawinit 100
Emulsion explosives (slurries), in which none of the ingredients is classified as an explosive; cap sensitivity is obtained by the mixing process; characterized by greatest handling safety; can be supplied in cartridges or can be mixed in a pump truck *Dynamit Nobel Wien GmbH.* Name unverified.

16879 Lawn Food
Lawn fertilizer. *Fisons plc, Horticultural Div.* Name unverified.

16880 Lawn Plus
Fertilizer and selective weedkiller. *ICI Chem & Polymers Ltd.*

16881 Lawn Spot Weeder
Selective weedkiller aerosol. *Fisons plc, Horticultural Div.* Name unverified.

16882 Lawn Weed Gun1
Contains 2,4-D and dicamba; ready-for-use herbicide spray for control of common lawn weeds. *ICI Garden Products.*

16883 Lawn Weedkiller
Selective weedkiller. *Murphy Chemical Co Ltd.* Discontinued.

16884 Lawn Weeds Killer
Selective weedkiller. *Fisons plc, Horticultural Div.* Name unverified.

16885 Lawnsman
Range of lawn aids for garden use such as fertilizers, weedkillers, and a spreader. *ICI Garden Products.*

16886 Lawnsman Mosskiller
Mosskiller for lawns. *ICI Chem & Polymers Ltd.*

16887 Lawnsman Spring Feed
Spring/summer lawn food. *ICI Chem & Polymers Ltd.*

16888 Lawnsman Weed and Feed
Fertilizer combined with selective weedkiller. *ICI Chem & Polymers Ltd.*

16889 Lawnsman Winterizer
Autumn lawn feed. *ICI Chem & Polymers Ltd.* Name unverified.

16890 Lawnturf®
Artificial turf made from polyvinylidene chloride filament, etc. *Asahi Chem. Industry.*

16891 Lawsone
83-72-7 5406 209-496-3
$C_{10}H_6O_3$
2-Hydroxy-1,4-naphthalenedione
2-hydroxy-1,4-naphthoquinone; henna. Egyptian privet; flower of paradise, derived from the leaves and roots of *Lawsonia inermis* or *L. Alba*; used as a dye, and for staining the hair, in medicine as an antifungal agent. dec 195-196°.

16892 Laysa
Acoustic absorbers. *Carl Freudenberg.* Name unverified.

16893 Laysa Plan
Acoustic absorbers. *Carl Freudenberg.* Name unverified.

16894 Laytex
A proprietary insulating material derived from rubber latex. No manufacturer.

16895 LCA-1
25928-94-3
Filled gyro-grade epoxy resin adhesive; adhesive with low coefficient of linear expansion; for bonding and sealing aluminum and other metals; mix ratio: 1.07 phr with Activator BA-4. *Bacon.*

16896 LCA-20
25928-94-3
Filled gyro-grade epoxy resin adhesive; semiflexible adhesive for sealing joints between dissimilar metals difficult to seal with regular gyrograde adhesives; mix ratio: 27.4 phr with Activator BA-40. *Bacon.*

16897 LCA-127
25928-94-3
Epoxy adhesive; thermally conductive, electrical insulating. *Bacon.*

16898 LCP-20CF/000
Liq. crystal polymer, 20% carbon fiber. *Compounding Tech.* Name unverified.

16899 Le Sage cement
(Plaster cement). A natural cement obtained from nodules found at Boulogne.

16900 lead
7439-92-1 5410 213-100-4
Pb
Haro® Mix CE-701, Mix CK-711, Mix MH-204. Lead one-pack system; one-pack lead heat stabilizer systems for rigid PVC applications, UPVC pressure and nonpressure pipes; construction material; radiation protection; alloys; manufacture of pigments. mp = 327°; bp = 1740°. *Harcros.*

16901 lead
7439-92-1 5410 231-100-4
Metallic element
Pb. Storage batteries, gasoline additive, radiation shielding, cable covering, ammunition, chemical reaction equip., solder and fusible alloys. *Asarco; Cerac; Noah Chem.*

16902 lead acetate
301-04-2 5411 206-104-4

$C_4H_6O_4Pb$
(Acetic acid, lead salt; sugar of lead; salt of Saturn) Inorganic salt; dyeing of textiles, waterproofing varnishes, insecticides, lead driers, chrome pigments, hair dye, weighting silks. *Am. Biorganics; Cerac; Chemetall GmbH; Hoechst Celanese; Mallinckrodt.*

16903 Lead ashes
The skimmings formed during the melting of lead. It consists mainly of oxide.

16904 lead bronze
An alloy of from 70-90% copper, 6-16% lead, and 4-13% tin.

16905 lead chloride
7758-95-4 5421 231-845-5
Cl_2Pb
Leclo. Pigment; solder; flux. d = 5.85; mp = 501°; bp = 950°; soluble inH_2O, NH_4Cl, NH_4NO_3; MLD (guinea pig orl) = 1.5-2.0 g/kg. *Mechema Chemicals Ltd.* Name unverified.

16906 lead diacetate
6080-56-4 5411
$C_4H_6O_4Pb{\cdot}3H_2O$
lead(II) acetate trihydrate
acetic acid, lead salt; sal Saturni; salt of Saturn; neutral lead acetate, normal lead acetate; sugar of lead. Anhydrous form [301-04-2]. mp = 75°, d = 2.55; soluble in H_2O (630 mg/ml), poorly soluble in organic solvents; LD_{50} (rat ip) = 150 mg/kg.

16907 lead fluoborate
13814-96-5 237-486-0
$B_2F_8{\cdot}Pb$
Salt for electroplating lead. *Elf Atochem N. Am.; Atomergic Chemetals; Hoechst Celanese; M&T Harshaw.*

16908 lead formate
811-54-1 5426 212-371-8
$C_2H_2O_4Pb$
Lead formate
Ledfo. d = 4.63; dec. at 190°; soluble in water; insoluble in alcohol. *Mechema Chemicals Ltd.* Name unverified.

16909 lead molybdate
10190-55-3 5432 233-459-2
$PbMoO_4$
Analytical chemistry, pigments. Insoluble inH_2O; soluble in HNO_3, NaOH. *AAA Molybdenum Prods.; Atomergic Chemetals; Cerac.*

16910 lead monoxide
1317-36-8 5433 215-267-0
PbO
Litharge; Massicot; Lead Oxide yellow; plumbous oxide; Lead protoxide,. Pigments consisting of lead monoxide, PbO. Litharge is obtained in silver refining, and has a more reddish color than Massicot, which is made by roasting lead. d = 9.53; mp = 888°; insoluble in water and alcohol, soluble in acetic acid, diluted HNO_3; LD_{50} (rat ip) = 430 mg/kg.

16911 lead nitrate
10099-74-8 5434 233-245-9
N_2O_2Pb
Ledni. Lead salts, mordant for dyeing and printing textiles and staining mother of pearl, matches, oxidizer in dye industry, sensitizer for photography, explosives, tanning, process engraving, lithography. mp = 470°; d = 4.5300; poisonous; soluble in water, alcohol, methanol; insoluble in concentrated HNO_3. *Blythe, William Ltd; Cerac; Mallinckrodt; Noah Chem.; Spectrum Chem. Mfg.*

16912 lead ocher
PbO
Lead monoxide
Found naturally.

16913 lead peroxide
1309-60-0 5424 215-174-5
PbO_2
Lead(IV) oxide
lead dioxide. d = 9.38; insoluble in H_2O, LD_{50} (gpg ip) = 200 mg/kg.

16914 lead phthalate basic
1344-40-7
EPlthal 120. Heat stabilizer in 90 and 105 C PVC wire and cable compounds *Eagle-Picher.* Name unverified.

16915 lead silicate
10099-76-0 233-246-4
O_3PbSi
lead metasilicate
BSWL 202. Ceramics, fireproofing fabrics, paints, electrode position process in the automatic industry. *Eagle-Picher; Hammond Lead Prods.*

16916 lead solder
An alloy of 50% lead and 50% tin used for soldering lead.

16917 lead stearate
1072-35-1 5442 214-005-2
$C_{36}H_{70}O_4Pb$
lead(II) n-octadecanoate
lead distearate; lead (II) stearate. Used in extreme pressure lubricants and as a dryer in varnishes. mp = 125°; insoluble in H_2O, slightly soluble in EtOH. *Ore & Chem. Corp.; Syn. Prods.; R. T. Vanderbilt Co Inc.*

16918 lead subcarbonate
598-63-0 5442 209-943-4
$2PbCO_3{\cdot}Pb(OH)_2$
White lead
Dibasic lead carbonate. Exterior paint pigment, ceramic glazes. mp = 125°; insoluble in H_2O, slightly soluble in EtOH. *Halstab.*

16919 lead tungate
A preparation obtained from lead acetate and tung oil; used as a drier in the preparation of paints.

16920 Lead vinegar
$Pb(C_2H_3O_2)_2{\cdot}PbO{\cdot}2H_2O$
A basic lead acetate.

16921 Lead water
A 1% solution of basic lead acetate.

16922 leaded bronze
An alloy of 88.5% copper, 10% zinc, and 1.5% lead. Another source gives the following proportions: 80% copper, 10% tin, and 10% lead. It has a melting point of 945 C.

16923 leaded gun metal
An alloy of 85.5% copper, 2% zinc, 9.5% tin, and 3% lead.

16924 Leaded zinc oxides
Pigments containing 20-35% lead sulfate and 60-80% zinc oxide.

16925 Leadoxe
Lead dioxide
Polysulfide sealant. *Mechema Chemicals Ltd.* Name unverified.

16926 Leak Detector
Aqueous special soap solutions; bubble forming fluids that detect and indicate leaks in any pressurized gas piping. *Dylon Industries Inc.* Name unverified.

16927 Lean coal
Coal of the poorest quality.

16928 Leantin
A proprietary trade name for a bearing alloy of lead and tin. No manufacturer.

16929 Leatherlubric
A proprietary trade name for a sulfonated sperm oil. No manufacturer.

16930 Lebanon No. 34
A proprietary trade name for an alloy of 20% chromium, 30% nickel, 3.25% molybdenum, 5% copper, and 3.25% silicon; stated to be resistant to hydrochloric and sulfuric acid. No manufacturer.

16931 Lebanon No. 48
A proprietary trade name for an alloy of 30% chromium, 30% nickel, 0.4% carbon, and the remainder iron. No manufacturer.

16932 Lebaycid®
55-38-9 4044 200-231-9
Fenthion
Insecticidal spray used for control of biting and sucking pests and particularly effective against fruit flies, leafhoppers, cereal bugs and rice stem borers. *Bayer AG.*

16933 Lebbin Salt
A proprietary mixture containing nitrite; used for curing meat. No manufacturer.

16934 Lecin
An iron albuminate, stated to be an easily assimilable form of iron. It contains 20% albumin and 0.6% iron.

16935 Lecipon
A 10% lecithin in a soluble form.

16936 Lecitase
A commercial preparation of phospholipase A-2 (phosphatide-2-acyl-hydrolase, E.C.3.1.1.4), manufactured from porcine pancreatic glands; used for the hydrolysis of lecithins of both animal and vegetable origin for improvement of emulsifying power. *Novo Nordisk.*

16937 Lecithan
A trade name for lecithin. No manufacturer.

16938 Lecithcerebrin
Lecithin prepared from brain substance.

16939 lecithin
8002-43-5 5452 232-307-2
$C_8H_{17}O_5NRR'$
Yolk powder. Mixture of the diglycerides of stearic, palmitic and oleic acids

linked to the choline ester of phosphoric acid; found in plants and animals; R and R' are fatty acid groups; edible surfactant and emulsifier for food use, pharmaceuticals, cosmetics, leather treatment, textiles. *Am. Lecithin; Central Soya; W.A. Cleary; Duphar BV; Lucas Meyer GmbH; Soice King; U.S. Biochemical.*

16940 Lecithin L-Range
8002-43-5 5452 232-307-2
Lecithin
Food emulsifier and coemulsifier for baking additives. *Grünau.*

16941 Lecithin Water Dispersible CLR
8002-43-5 5452 232-307-2
Hydrolyzed soya lecithin
Mild refatting agent, emollient for aqueous skin and hair care preparations, face cleansers, shampoos. *Dr. Kurt Richter; Henkel/Cospha.*

16942 lecithmedullan
A lecithin preparation from bone marrow.

16943 lecithol
An emulsion of brain lecithin containing 1.5% lecithin.

16944 Leclo
7758-95-4 5421 231-845-5
Lead chloride
Mechema Chemicals Ltd. Name unverified.

16945 Lectricon
A proprietary trade name for alkyl ammonium compounds dispersed in mixed aromatic/aliphatic oils with film strength agents; for electrostatically spraying onto form work for the casting of concrete. *British Solvent Oils.* Name unverified.

16946 Lectro Cast
A proprietary trade name for an alloy of iron with 2.75% nickel, and 0.7% chromium. No manufacturer.

16947 Lecutyl
A combination of lecithin and copper cinnamate, containing 1.5% copper.

16948 Ledac
301-04-2 5411 206-104-4
Lead acetate
Mechema Chemicals Ltd. Name unverified.

16949 Ledate®
19010-66-3 242-748-2
$C_6H_{12}N_2PbS_4$
Lead dimethyldithiocarbamate
Ledate; Lead, bis(dimethylcarbamodithioato-S,S')-, (T-4)-; bis(dimethyldithiocarbamato)lead; dimethyldithiocarbamic acid, lead salt; methyl ledate. Lead dimethyldithiocarbamate 50% in oil; liquid; rubber accelerator recommended for improved dynamic properties in natural and polyisoprene rubbers. mp = 310°; d = 2.43; insoluble in H_2O. *RT Vanderbilt Co Inc.*

16950 Ledca
598-63-0 5442 209-943-4
Lead carbonate.
Mechema Chemicals Ltd. Name unverified.

16951 Leddel alloy
An alloy of 86% zinc, 9.5% antimony, and 4.5% copper. Another alloy used for bearings consists of 87.5% zinc, 6.25% copper, and 6.25% aluminum.

16952 Ledeburite
Austenite-cementite eutectic.

16953 Ledebur's metal
Bearing metals. One contains 85% zinc, 10% antimony, and 5% copper. Another consists of 77% zinc, 17.5% tin, and 5.5% copper.

16954 Ledercort
124-94-7 9727 204-718-7
A proprietary preparation of triamcinolone; a steroid. *Lederle Laboratories.* Name unverified.

16955 Lederfen
36330-85-5 4003 252-979-0
Proprietary preparation containing fenbufen; nonsteroidal anti-inflammatory. *Lederle Laboratories.* Name unverified.

16956 Ledermix
Combination dental kit consisting of a paste and cement with democlocyline and triamcinolone acetonide and a hardener; dental treatment combining antibiotic and anti-inflammatory agents. *Lederle Laboratories.* Name unverified.

16957 Lederplex
A proprietary preparation of vitamin B complex. *Lederle Laboratories.* Name unverified.

16958 Lederspan
5611-51-8 227-031-4
A proprietary preparation of triamcinolone hexacetonide; a steroid injection. *Lederle Laboratories.* Name unverified.

16959 Ledfo
811-54-1 5426 212-371-8
Lead formate. *Mechema Chemicals Ltd.* Name unverified.

16960 Ledmin LPC
104-74-5 203-232-2
A proprietary trade name for lauryl pyridinium chloride; an emulsifier for waxes giving bacteriostatic polishes. *Hoechst UK.* Name unverified.

16961 Ledni
10099-74-8 5434 233-245-9
Lead nitrate
Mechema Chemicals Ltd. Name unverified.

16962 Ledrite Brass
A proprietary trade name for a leaded brass containing 60-63% copper and 2.5-3.7% lead. No manufacturer.

16963 Leecure B Series
7637-07-2 1379 231-569-5
Boron trifluoride-based hardeners, epoxy curing agents providing water, heat and chem. resistance compounds with high physical strength and electrical props.; used in electronic potting compounds, electrical varnishes, adhesives, filament winding, fiberglass composit *Leepoxy Plastics.*

16964 Leefex
A hop defoliant. *Plant Protection.* Name unverified.

16965 Leegen®
A blend of a sulfonated petroleum product and selected mineral oil on an inert carrier; an effective dry foam plasticizer and processing aid for NR, SBR, IR, and IIR rubbers. *R. T. Vanderbilt Co Inc.* Discontinued.

16966 Lees
Yeast and various suspended matters of the must produced during the fermentation of grape juice.

16967 Leffmann and Beam's glycerol reagent
For Reichert-Meissl number. It is a mixture of 180 ml pure glycerol and 20 ml of 50% sodium hydroxide solution.

16968 Legion
A suspension concentrate containing 50 g carbendazim [10605-21-7] and 320 g maneb [12427-38-2] per liter; systemic fungicide for cereals. *Tripart Farm Chemicals Ltd.*

16969 Legumex Extra
A solution concentrate containing 27 g benazolin, 237 g 2,4-DB and 42.3 g MCPA per liter; a post-emergence herbicide. *Schering Agrochemicals Ltd.* Discontinued.

16970 Legupren®
A foam system based on Legural (unsaturated polyester resins) and a blowing agent for the system; for combination with light fillers to produce Legupren-based lightweight concrete for use in building construction. *Bayer AG.*

16971 Leguval®
Unsaturated polyester resins for the production of SMC, BMC and DMC, etc.; used for the manufacture of glass reinforced corrugated sheet, light domes and light dome supports, cable distribution boxes, covers, light housings, bulk containers, boats, swimmi *Bayer AG.*

16972 Leinsaat oils
Linseed oil obtained by extracting the residues from the presses.

16973 Lekutherm®
A range of epoxy resins and hardeners; for the production of adhesives, for use in the construction of models, form plates and core boxes used in foundries, for jigs and tool construction. *Bayer AG.*

16974 Lemarquand's alloy
An alloy made from 75% copper, 14% nickel, 2.0% cobalt oxide, 1.8% tin, and 7.2% zinc. It is stated to be very resistant to oxidation.

16975 Lembergite
An artificial hydrous aluminum-sodium silicate.

16976 Lemberg's solution
A solution of aluminum chloride and extract of log wood. It colors calcite mineral surfaces violet.

16977 Lemco
A proprietary meat extract. No manufacturer.

16978 Lemnian earth
Lemnos earth. A red, yellow, or grey, earthy substance consisting of a hydrated silicate of aluminum, found at Lemnos. It is an ocher, and is used as a pigment.

16979 Lem-O-Fos®
7758-29-4 8846 231-838-7
Sodium tripolyphosphate
Food additive for meat, poultry, and seafood industries *Rhône-Poulenc Food Ingreds.*

16980 Lemol
Lemon oil substitute for flavoring. *Bush Boake Allen Ltd.*

16981 Lemolac
A proprietary preparation; a very light mercurous chloride. No manufacturer.

16982 Lemon Delph
Cleansing milk and skin freshener, for daily care of the skin. *Richardson-Vicks Inc.* Name unverified.

16983 Lemon Lac
Orangelac
garnet lac. Terms referring to the color of lac, determined to some extent upon the tree from which it is obtained.

16984 lenacil
2164-08-1 5459 218-499-0
$C_{13}H_{18}N_2O_2$
3-cyclohexyl-6,7-dihydro-1H-cyclopentapyrimidine-2,4(3H,5H)-dione
lenacile; Adol; Du Pont 634; Elbatan; Venzar; Vizor. Selective herbicide, inhibits photosynthesis. Used for control of annual grass and broad-leaved weeds in a variety of crops. mp = 315-317°; sg = 1.32; soluble in H_2O (6 mg/l), more soluble in organic solvents; LD_{50} (rat orl) > 11000 mg/kg.

16985 Leneta
Paint testing panels and opacity charts; for testing of paints and inks. *Cornelius Chemical Co Ltd.* Name unverified.

16986 Lenetol
Textile auxiliary chemicals. *ICI Chem & Polymers Ltd.*

16987 Lenium
A proprietary preparation of selenium sulfide and bithionol; a skin cleanser. *Bayer AG.* Discontinued.

16988 Lenka
Strongly alkaline detergent powder. *Shell UK.* Name unverified.

16989 Lensine
Alkali detergent powders. *Shell UK.* Name unverified.

16990 Lentagran
35512-33-9
Wettable powder containing 45% w/w pyridate; for annual weed control for cereals, oilseed rape and maize. *Ciba-Geigy Agrochemicals.*

16991 Lentana
A red earthy ironstone of the type commonly called Laterite.

16992 Lentizol
549-18-8 208-964-6
A proprietary preparation of amitryptyline hydrochloride; an antidepressant. *Warner.* Name unverified.

16993 Lentopen
6130-64-9 7226
Penicillin G procaine
Antibacterial. *Wyeth Laboratories.* Name unverified.

16994 Lenzing P84
Aromatic polyimide; high performance nonflammable, thermally stable resin with excellent chemical resistance, used for production of needle felts for high temperature filtration, as reinforcement for PTFE, as compounding ingredient. for injection moldable *Lenzing AG.*

16995 Lenzol
A pale cedar-wood oil of known viscosity and refractive index; used for oil immersion in microscopy.

16996 Leo K
7447-40-7 7783 231-211-8
A proprietary preparation of potassium chloride; used in potassium replacement therapy. *Leo Laboratories.* Name unverified.

16997 Leomin AN
Alkyl phosphonate
Surfactant for textile processing; antistat for fiber manufacture and processing, plastics processing. *Hoechst Celanese/Colorants & Surf; Hoechst AG.*

16998 Leomin FANF
Fatty alkyl quaternary ammonium salt; surfactant for textile processing. *Hoechst Celanese/Colorants & Surf.*

16999 Leona®
32131-17-2
Nylon 66 resin. *Asahi Chem. Industry.*

17000 Leonil DB Powd
Sodium diisobutyl naphthalene sulfonate
Wetting agent and dyeing auxiliaries for textiles, agricultural.

17001 Leonil EBL
Oxidizing agents in detergents; fast desizing and wetting agent for cellulosic fibers and cellulosic/synthetic blends. *Hoechst AG.*

17002 Leonil OS
577-11-7 3460 209-406-4
Sodium dioctyl sulfosuccinate
Surfactant for agricultural formulations. *Hoechst Celanese/Colorants & Surf.*

17003 Leophen® BN
Alcohol sulfonate
Foam-free wetting agent in mercerizing textiles. *BASF AG; BASF plc.*

17004 Lepandin
Highly dispersed pyrogenic silica; used for nonslip finishing. *Degussa AG.* Name unverified.

17005 Lepro
1309-60-0 5424 215-174-5
O_2Pb
lead dioxide
Lead peroxide. *Mechema Chemicals Ltd.* Name unverified.

17006 Lepton®
Polymer dispersions; binders for pigment and other finishes in leather and fur industry. *BASF AG.*

17007 Lerbek
Coccidiostat for poultry, turkeys and rabbits. *RMB Animal Health Ltd.*

17008 Lerifond
Phenolic resin. *Winchem Ltd.*

17009 Leriphen
Phenolic resin. *Winchem Ltd.*

17010 Lerite
Phenolic molding resin. *Winchem Ltd.*

17011 Leromoll
Plasticizer with minimum volatility and good resistance to alkalis and acids; for use in combination with unsaponifiable binders in the formulation of chemical-resistant coating. *Bayer AG.*

17012 Lesan
Fungicide; used for control of soil borne diseases on ornamentals. *Bayer AG.*

17013 Lethalbine
A lecithin albuminate, containing 20% lecithin.

17014 Lethane
Insecticide concentrates supplied in petroleum distillate; used in industrial insecticide sprays and mosquito larvicides. *Rohm & Haas UK.*

17015 Lethi
Lead thiosulfate
Mechema Chemicals Ltd. Name unverified.

17016 Leucaniline
$C_{20}H_{21}N_3$
Triamino-diphenyl-tolyl methane

17017 Leucarsone
4-Carbaminophenyl arsonic acid
May & Baker Ltd. Name unverified.

17018 Leucaurin
$C_{19}H_{16}O_3$
Trihydroxytriphenylmethane

17019 Leuchtol
A proprietary synthetic resin. No manufacturer.

17020 Leucine
61-90-5 5475 200-522-0
$C_6H_{13}NO_2$
L-Leucine
α-aminoisocaproic acid; α-amino-γ-methylvaleric acid; 1-leucine; Leu; L; 2-amino-4-methyl-valeric acid; α-aminoisocaproic acid. Nutrient and dietary supplement, biochemical research. Sublimes at 145-148°; mp 293-295° (dec); d = 1.293; $[\alpha]_D^{25}$ = -10.8°; pKa (25°) = 9.6; soluble in water, alcohol, acetic acid, insoluble in ether. *Degussa; R.W. Greeff; Nippon Rikagakuyakuhin; Tanabe USA; U.S. Biochemical.*

17021 l-leucine
61-90-5 5475 200-522-0
$C_6H_{13}NO_2$
α-amino-γ-methylvaleric acid
α-aminoisocaproic acid; leucine. Nutrient and dietary supplement, biochemical research. *Degussa; R.W. Greeff; Nippon Rikagakuyakuhin; Tanabe USA; U.S. Biochemical.*

17022 Leucobenzaurin
$C_{19}H_{16}O_2$
Di-p-hydroxy-triphenylmethane

17023 Leucogen
7631-90-5 8731 231-548-0
$NaHSO_3$
Sodium hydrogen sulfite
Used in bleaching and paper making.

17024 Leucol
91-22-5 8253 202-051-6
An impure form of quinoline.

17025 Leucomycin®
Macrolide antibiotic
Asahi Chem. Industry.

17026 Leuconine
leukonin
An antimony preparation containing 98% of sodium metantimoniate. Recommended as a substitute for tin oxide for enameling.

17027 Leucophor KNR
Fluorescent whitener for commercial/industrial laundry detergents, rug/upholstery cleaners, fabric softeners, laundry bleach, whitening soap, brightening polymers and plastics; esp. for cellulosics. *Sandoz.*

17028 Leucopure EGM Powd
Fluorescent whitener for commercial/industrial laundry detergents, rug/upholstery cleaners, fabric softeners, laundry bleach, whitening soap, brightening polymers and plastics; esp. for synthetics and wool. *Sandoz.*

17029 Leukeran Tablets
305-03-3 206-162-0
A proprietary formulation of chlorambucil; for treatment of chronic lymphocytic leukemia, Hodgkins disease, certain forms of non-Hodgkins lymphoma, Walderstroms macroglobuliremia and advanced ovarian adrenocarcinoma. *The Wellcome Foundation Ltd.*

17030 Leukomycin®
56-75-7 2120 200-287-4
Chloramphenicol
Broadspectrum antibiotic; veterinary preparation. *Bayer AG.*

17031 Leukon
A proprietary synthetic resin molding powder. No manufacturer.

17032 Leukonöl LBA-2
Emulsifier for emulsion polymers; wetting agent for alkaline systems to pH 13. *Münzing Chemie GmbH.*

17033 Leukotan® 974
Acrylic syntan
Retanning agent for leather. *Rohm & Haas.* Name unverified.

17034 Leukotrop® W
Textile assistant for producing white discharge prints on cellulose fibers. *BASF AG; BASF plc.*

17035 Leuna Gas
A proprietary trade name for compressed propane. No manufacturer.

17036 Leuna saltpeter
A double salt of ammonium sulfate and nitrate; a fertilizer similar to Chilean nitrate in its action.

17037 Leunaphos
A mixture of phosphate, nitrate, and sulfate of ammonia. fertilizer. No manufacturer.

17038 Leutalux
Inorganic and organic fluorescent substances; fluorescent screens for electron tubes and fluorescent lamps. *Leuchstoffwerk GmbH.* Name unverified.

17039 Levacast®
Reactive PUR systems for coating leather and other substrates. *Bayer AG.*

17040 Levacell®
Acid and direct powder and liquid dyestuffs for coloring paper; anionic direct liquid dyestuffs for the production of printing inks for tissue paper. *Bayer AG.*

17041 Levaderm®
Anionic dyestuffs for finishing and for drum dyeing and dyeing on the Multima machine. *Bayer AG.*

17042 Levafil
Dope dyes for polyamide fibers. *Bayer plc.*

17043 Levafix®
Reactive dyes; for dyeing cotton as well as textiles made from regenerated cellulose. *Bayer AG; Bayer plc.*

17044 Levaflex®
Thermoplastic rubber. *Bayer plc.* Discontinued.

17045 Levaform®
A silicone release agent; used for the rubber, plastics, man-made fibers, metal, food and pharmaceutical industries. *Bayer AG; Bayer plc.*

17046 Levagard®
Flame retardants for rigid PU foams. *Bayer AG.*

17047 Levagel®
Polyurethane gels and their raw materials; ready-to-use gels covered with film or fabric; used in medical sector, e.g., for wheelchair cushions, mattress overlays to prevent pressure sores. *Bayer AG.*

17048 Levair
Lithium salts mixed with inhibitors compounded in aqueous solution; for dehumidifiers and chiller. *Leverton-Clarke Ltd.*

17049 Levair®
Sodium aluminum phosphate
Acidic, leavening agent for baking, cereals. *Rhône-Poulenc Food Ingreds.*

17050 Levalaine
Scouring agent. *Stephenson Thompson Textile Chemicals.*

17051 Levalan N
1:2 Metal complex dyestuffs used for dyeing wool and polyamide with great coloring strength, good solubility and outstanding fastness properties. *Bayer AG.*

17052 Levalin®
A range of auxilliaries for economical, continuous processes for dyeing wool, cellulosics and synthetic fibers. *Bayer AG; Bayer plc.*

17053 Levanox®
Aqueous preparations of organic pigments. *Bayer AG.*

17054 Levanyl®
Aqueous preparations of organic pigments. *Bayer AG; Bayer plc.*

17055 Levapon®
Degreasing agents, emulsifiers, cleaning and wetting agents used in the textile industry. *Bayer AG; Bayer plc.*

17056 Levapren®
EVA copolymer; Includes electric wire and cable insulations that must withstand high temperatures or present good flame-retardant behavior. *Bayer AG; Bayer plc.*

17057 Levapren® 400
EVA copolymer; synthetic rubber for tech. moldings and extrudates, lamp seals, cable sheathings and insulations, cellular rubber goods, footwear soles, waterproof sheeting, hot-melt and pressure-sensitive hot-melt adhesives etc,; suited for cables. *Bayer; miles.*

17058 Levapren® K
EVA copolymer; grades for adhesives applications. *Bayer AG.*

17059 Levasil®
Mixing liquid for dental investment compounds; nonslip finishing for textiles; spinning lubricant additive; ceramic coating of magnetic steel plates; special ceramic mortars; sprayable compositions for formation of a protective layer on ingot mold basepla *Bayer AG.*

17060 Levasint®
A fluidized-bed coating powder comprising saponified ethylene/vinylacetate copolymer; uses include maintenance coatings on metal parts, especially for fencing, cable rigs, traffic furniture and applications in water area. *Bayer AG; Bayer plc.*

17061 Levasol®
Solubilizers and fixation auxiliaries for phthalogen dyes. *Bayer AG; Bayer plc.*

17062 Levcarb
Noncyanide heat treatment salts and other compounded dry salts; for metals heat treatment. *Leverton-Clarke Ltd.*

17063 Levegal®
Leveling auxiliaries for dyeing synthetic and cellulosic fibers; dye carriers for polyesters. *Bayer AG; Bayer plc.*

17064 Levelan P14B
Nonionic surfactant of the nonylphenol ethoxylate type in liquid form; stabilizer for latex. *Henkel Chemicals Ltd.* Name unverified.

17065 Levelan P208
Nonionic surfactant of the nonylphenol ethoxylate type in liquid form; wetting agent for use at high temperatures; stabilizer for latex; emulsion polymerization aid. *Henkel Chemicals Ltd.* Name unverified.

17066 Levelan P307
Nonionic surfactant of the nonylphenol ethoxylate type in liquid form; wetting agent at high temperatures. *Henkel Chemicals Ltd.* Name unverified.

17067 Levelan P357
Nonionic surfactant of the nonylphenol type in liquid form; primary emulsifiers for emulsion polymerization. *Henkel Chemicals Ltd.* Name unverified.

17068 Levelan® P208
Nonyl phenol ethoxylate aqueous dilution; wetting agent, scouring agent, antistat, dye leveling agent at high temperature and high electrolyte conc. in textile industry; latex stabilizer and emulsifier in emulsion polymer industry. *Harcros.*

17069 Levepox®
Liquid epoxy resins for solvent free coatings. *Bayer AG.*

17070 Levesol
Solubilizers and fixation auxiliaries for phthalogen dyes. *Bayer plc.*

17071 Levn-Lite
Sodium aluminum phosphate
Leavening agent. *Monsanto Co.* Name unverified.

17072 Levochrom®
Cobalt-chrome-molybdenum alloy for dental model cast prosthetics. *Bayer AG.*

17073 Levodip®
Liquid cold-dip hardener for dental industry. *Bayer AG.*

17074 Levofin®
Fine investing compound for model cast techniques; used in dentistry. *Bayer AG.*

17075 Levogel®
Duplicating material for model cast techniques; used in dentistry. *Bayer AG.*

17076 Levogen®
Aftertreating agents; used for improving the wet fastness properties of dyeings. *Bayer AG; Bayer plc.*

17077 Levogen® LF
Used in the leather industry to enhance the surface dyeing. Amphoteric dyeing auxiliary *Bayer AG.*

17078 Levopress
Dental preparation; color-stable autopolymerizate for dentures, orthodontic appliances, repairs, etc. *Bayer AG.*

17079 Levotan®
Anionic or weakly cationic polyurethane retanning materials for soft leather with a tight grain and good drumming or staining properties. *Bayer AG.*

17080 Levotherm®
Rapid-setting precision investing compound for model cast constructions with an exact fit; for dentistry. *Bayer AG.*

17081 Levothroid
55-03-8 5497 200-221-4
Levothyroxine sodium
Thyroid hormone. *USV Pharmaceutical Corp.* Unverified.

17082 Levoxin®
Corrosion inhibition in closed water-steam boilers. *Bayer AG.*

17083 Levuline
A proprietary preparation used in the textile industry for finishing. No manufacturer.

17084 levulinic acid
123-76-2 5498 204-649-2
$C_5H_8O_3$
γ-ketovaleric acid
acetylpropionic acid; 4-oxopentanoic acid; levulic acid. Intermediate for plasticizers, solvents, reins, flavors, pharmaceuticals, acidulant, and preservative; chrome plating; solder flux; stabilizer for calcium greases; control of lime deposits. mp = 33-35°; bp = 245-246°; d = 1.1447; n_D^{18}= 1.442; soluble in H_2O, EtOH, insoluble in hydrocarbon solvents. *Chemie Linz UK Ltd; Otsuka Chem.; Penta Mfg.; QO; Schweizerhall.*

17085 levulose
57-48-7 4295 200-333-3
$C_6H_{12}O_6$
levulose. Fruit sugar, diabetin, levo glucose, sucro-levulose, fructose. Naturally occurring sugar; used in food stuffs, medicine, and as a preservative.

17086 Lewasorb®
Powdered ion exchange resins regenerated for immediate use; used for water purification. *Bayer AG; Bayer plc.*

17087 Lewatit®
Cation and anion exchange resins based on polymerization type synthetic resins of differing mesh size and/or macroporous structure and with active groups of various acidities and basicities or selective properties; used for water treatment, waste water tr *Bayer AG; Bayer plc.*

17088 Lewis metal
An alloy of 1 part tin and 1 part bismuth having the property of expanding when cooling. It has a meltingpoint of 138 C and is used for sealing and holding die parts.

17089 Lewisite
3112 541-253-3
$C_2H_2AsCl_3$
β-Chlorovinyldichloroarsine
A military poison gas; the name Lewisite is also used for a mineral, $5CaO{\cdot}2TiO_2{\cdot}3SbO_3$. No manufacturer.

17090 Lewisol 28
Maleic-modified glycerol ester of rosin; used for Type C gravure inks and for nitrocellulose-based paper and wood coatings. *Hercules.*

17091 Lexaine® C
Cocamidopropyl betaine
viscosity builder, foam booster, thickener in conditioners, specialty shampoos, personal care products, dishwash, sanitizers. *Inolex.*

17092 Lexaine® CSB-50
68139-30-0 268-761-3
Cocamidopropythydroxysultaine
Surfactant, foaming and wetting agent used in personal care products (shampoos, conditioners, bath products), industrial (heavy-duty alkaline cleaners, paint strippers, metal cleaners); stable in systems containing acids, alkali, and electrolytes; lime so *Inolex.*

17093 Lexaine® IS
Isostearamidopropyl betaine
Mild surfactant, foam booster, and thickener for shampoos, bubble baths, foaming conditioners; stable in acid and alkaline systems. *Inolex.*

17094 Lexaine® LM
4292-10-8 224-292-6
Lauramidopropyl betaine
Mild surfactant, foam booster for bath products, shampoos, liq. soaps, dishwash liqs. *Inolex.*

17095 Lexaine® O
25054-76-6 246-584-2
Oleamidopropyl betaine
Surfactant used in personal care products. *Inolex.*

17096 Lexamine 22
16889-14-8 240-924-3
Stearamidoethyl diethylamine
Conditioner, emulsifier for hair and skin care products; conditioning shampoo. *Inolex.*

17097 Lexamine C-13
68140-01-2 268-771-8
Cocamidopropyl dimethylamine
Emulsifier for creams and lotions; when neutralized as conditioners for hair care products. *Inolex.*

17098 Lexamine L-13
3179-80-4 221-661-3
Lauramidopropyl dimethylamine
Conditioner, emulsifier for hair and skin care products. *Inolex.*

17099 Lexamine O-13
109-28-4 203-661-5
Oleamidopropyl dimethylamine
Emulsifier for creams and lotions; when neutralized as conditioners for hair care products. *Inolex.*

17100 Lexamine S-13
7651-02-7 231-609-1
Stearamidopropyl dimethylamine
Emulsifier for creams and lotions; when neutralized as conditioners for hair care products. *Inolex.*

17101 Lexamine S-13 Lactate
55819-53-9 259-837-7
Stearamidopropyl dimethylamine lactate
Conditioner for hair care products. *Inolex.*

17102 Lexan®
BPA polycarbonates
Used for plastic components for automotive, electrical, electronics, lighting, medical, packaging, audio, etc. *GE Plastics Ltd.*

17103 Lexan® 121
PC resin; low viscosity, general purpose grade for molding intricate, hard-to-fill parts. *GE Plastics.*

17104 Lexan® 141L
PC resin; lower viscosity, general purpose grade. *GE Plastics.*

17105 Lexan® 144LR
PC resin; injection blow molding grade; FDA compliance; applications including baby bottles and other liq. pkg. *GE Plastics.*

17106 Lexan® 151, 153, 154
PC resin; extrusion blow molding grade; for water bottle applications requiring high impact strength, glass-like transparency, high gloss, and high heat distort. temperature; avail. as general purpose grade (151), UV-stabilized(153), food additive grade(154). *GE Plastics.*

17107 Lexan® 241
PC resin; General purpose injection molding grades. *GE Plastics.*

17108 Lexan® 500, 503
PC resin, High modulus grade; for applications requiring high ridigidity along with toughness and impact strength; offers dimensional stability, low mold shrinkage for production of precise parts; 503 grade is uv-stabilized for outdoor applications. *GE Plastics.*

17109 Lexan® 920
PC resin; low viscosity flame retardant resin used for electrical, electronic, appliance, aircraft, lighting, computer, communications, business equipment, and hardware applications. *GE Plastics.*

17110 Lexan® 3412
PC resin; 20% glass fiber-reinforced; provides higher mech. props., reduced mold shrinkage over unreinforced grades; able to produce very precise parts. *GE Plastics.*

17111 Lexan® 8040
PC film; High-clarity, heat-resistant film for appliances, business machines, and transportation applications; FDA-approved grade, smooth both sides, polish texture. *GE Plastics.*

17112 Lexan® FL400
PC; engineering structural foam for applications where load bearing capability at elevated temps. is required; for appliance, automotive, telecommunications, material handling, and business machine industries. *GE Plastics.*

17113 Lexan® GR1310
PC resin; medical grade featuring superior gamma resistance. *GE Plastics.*

17114 Lexan® HF1110, HF1130, HF1140
PC resin; high flow resins providing enhanced processability and longer flow lengths; inherent mold release; ideal for disposables or where life cycles are not demanding, and in applications where clarity, heat resistance, and dimensional stability are impor *GE Plastics.*

17115 Lexan® HP1, HPS1
PC resin; resin formulated for the health care industry and meeting FDA and USP standards; inherent mold release; HPS series are gamma radiation stabilized; HP avail. in natural, transparent tints, and custom opaque colors. *GE Plastics.*

17116 Lexan® OQ1020
PC resin, optical quality grade; offers optical clarity and resin purity for the most demanding optical applications; OQ1020 for compact disc market. *GE Plastics.*

17117 Lexan® PK2040
PC resin; packaging grade *GE Plastics.*

17118 Lexan® PPC4501
Polyphthalate carbonate copolymer resin; engineered for high heat and moisture resistance; FDA grades. *GE Plastics.*

17119 Lexan® SP1010
Polycarbonate resin; superior flow and high impact grade for thin wall and intricate parts. *GE Plastics.* Name unverified.

17120 Lexan® WR1210, WR1240
PC resin, PTFE-filled; wear-resistant grade providing greater lubricity and reduced friction and wear between moving parts; for business equip., electrical/electronics applications including keyboard frames, swivel bases, paper drives, printers, camera internal pa *GE Plastics.*

17121 Lexate BPQ
Lauramidopropyl betaine
TEA-coco-hydrolyzed collagen; Oleamidopropyl dihydroxypropyl dimonium chloride. Blended detergent, conditioner, and protein used as economical base for shampoo, bath gel, liq. soaps, dishwash, bubble baths, cleansers. *Inolex.*

17122 Lexate CRC
Stearamidopropyl dimethylamine
Glycol stearate, and ceteth-2; cream rinse conc. and conditioner, emulsifier. *Inolex.*

17123 Lexate PX
Petrolatum, lanolin, and ozokerite; oil-water and auxiliary emulsifier, lanolin cream base, conditioner in soaps and shaving cream, emollient. *Inolex.*

17124 Lexate TA
Glyceryl stearate, IPM, stearyl stearate; auxiliary lipophilic emulsifier in water-oil emulsions, emollient in cosmetic creams and lotions; provides barrier properties and slip. *Inolex.*

17125 Lexate TL
Glyceryl stearate, butyl stearate, stearyl stearate; superfatting agent in milled bar soap, auxiliary lipophilic emulsifier in water-oil systems, emollient; enhances barrier properties in creams and lotions. *Inolex.*

17126 Lexe®
A registered trademark for a cellulose acetatebutyrate insulating tape; it is flame retardant, has low moisture absorption and has a high dielectric strength. No manufacturer.

17127 Lexein® A200
72319-06-3
Myristoyl hydrolyzed collagen; film-forming collagen protein derivative; resin modifier for hair sprays, makeups, protection skin lotions and creams. *Inolex.*

17128 Lexein® A520
68918-77-4
TEA-abietoyl hydrolyzed collagen; sebum control additive; causes delay in refatting of the scalp when used in shampoos. *Inolex.*

17129 Lexein® CP-125
Oleamidopropyl dimethylamine hydrolyzed collagen
Protein for shampoos, cream rinses, hair conditioners. *Inolex.*

17130 Lexein® S620S/Superpro 5A
TEA-coco-hydrolyzed animal protein and sorbitol; mild, high foaming cleansing agent, humectant, moisturizer for skin care products. *Inolex.*

17131 Lexein® X-250HP
9015-54-7 310-296-6
Hydrolyzed collagen
Low odor and color version; cosmetic industry protein exhibiting substantivity; for hair and skin care products. *Inolex.*

17132 Lexell
Hydrocarbon or oxyhydrocarbon and aqueousbased solvents and equipment for industrial cleaning processes. *ICI Chem & Polymers Ltd.*

17133 Lexemul® 503
Glyceryl stearate
Emulsifier, stabilizer, thickener, opacifier in emulsions or surfactant systems, cosmetics, toiletries, pharmaceuticals. *Inolex.*

17134 Lexemul® 561
Glyceryl stearate and PEG-100 stearate; Cosmetic grade emulsifier for oil-water cream or lotion systems. *Inolex.*

17135 Lexemul® CS-20
Cetearyl alcohol
ceteareth-20. Emulsifier used in personal care products, topical preparations *Inolex.*

17136 Lexemul® EGDS
627-83-8 211-014-3
Glycol distearate
Lubricant, opacifier and pearling agent for cosmetic surfactant systems, liq. hand soaps, light duty liqs. *Inolex.*

17137 Lexemul® EGMS
Glycol stearate
Opacifier and pearling agent for personal care products, light duty liqs.; emulsifier for creams, lotions, topicals; secondary suspending agent in oil-water system. *Inolex.*

17138 Lexemul® GDL
27638-00-2 248-586-9
Glyceryl dilaurate
Lipid for creams and lotions formulated to produce a dry skin feel; melts just below body temperature; emollient; coupling agent for more lipophilic materials; emulsion stabilizer. *Inolex.*

17139 Lexemul® P
Propylene glycol stearate SE
For low viscosity emulsions for personal care products. *Inolex.*

17140 Lexemul® PEG-200 DL
9005-02-1
PEG-4 dilaurate
Emulsifier, emollient, lubricant for cosmetics, pharmaceuticals, metalworking fluids, paints, polishes and misc. industrial formulations. *Inolex.*

17141 Lexemul® PEG-400ML
9004-81-3
PEG-8 laurate
Emulsifier, emollient, lubricant dispersant for creams, lotions, spreading bath oils, cosmetic, pharmaceutical and industrial applications. *Inolex.*

17142 Lexemul® T
Glyceryl stearate SE
For use as emulsifier, opacifier, stabilizer, and emollient in alkaline anionic systems, cosmetics. *Inolex.*

17143 Lexgard B
94-26-8 1619 202-318-7
Butylparaben
Cosmetics preservative. *Inolex.*

17144 Lexgard M
99-76-3 6182 202-785-7
Methylparaben USP
Cosmetics preservative. *Inolex.*

17145 Lexgard P
94-13-3 8051 202-307-7
Propylparaben USP
Cosmetics preservative. *Inolex.*

17146 Lexite Granular Carpet
All-solids epoxy fortified with quartz granules; used to provide a tough wearing surface over industrial and commercial floors to make them both skid-proof and reasonably easy to clean. *Metalcrete Mfg Co.*

17147 Lexol 3975
isopropyl palmitate
isopropyl myristate; isopropyl stearate. Emollient; replacement for IPM. *Inolex.*

17148 Lexol EHP
29806-73-3 249-862-1
Octyl palmitate
Emollient for nonocclusive creams and lotions, bath oils, antiperspirants, other cosmetic and topical formulations. *Inolex.*

17149 Lexol GT 855, GT 865
65381-09-1 265-724-3
Caprylic/capric triglyceride
Emollient with nonoily skin feel; moisturizer; for creams, lotions, bath oils, lipstick, makeup; solvent for perfume and flavor ingreds.; vehicle for medicinals, antibiotics, vitamins; solubilizer; oxidative stability. *Inolex.*

17150 Lexol IPM-NF
110-27-0 5234 203-751-4
Isopropyl myristate
NF emollient oil for cosmetic applications; colorless and essentially odorless, low viscosity , low f.p., outstanding spreading, good sol. *Inolex.*

17151 Lexol IPP
142-91-6 205-571-1
IPP; emollient in conditioning cosmetics; solubilizer for cosmetic and topical pharmaceuticals. *Inolex.*

17152 Lexol PG 800
56519-71-2
Propylene glycol dioctanoate
Emollient with nonoily feel, oxidation stability; for creams, lotions, topicals, lipsticks, glossers, makeup bases, bath oils, aftershaves; carrier/vehicle for fragrance. *Inolex.*

17153 Lexol PG 855
Propylene glycol dicaprylate/dicaprate
Emollient, solubilizer for cosmetic creams and lotions; carrier/vehicle for flavors, fragrances, pigmented cosmetics, antibiotics. *Inolex.*

17154 Lexol PG 900
41395-83-9 255-350-9
Propylene glycol dipelargonate
Emollient for bath oils, preshave lotions, aerosol systems, lipsticks, glosses, makeup bases; carrier for frangrances. *Inolex.*

17155 Lexol SS
2778-96-3 220-476-5
Stearyl stearate
Emollient for bath oils, creams, lotions. *Inolex.*

17156 Lexolube® 2J-237
Tetraethylene glycol dicocoate
Lubricant for polyester filament yarns. *Inolex.*

17157 Lexolube® 2T-237
PEG-4 di 2-ethylhexanoate
Lubricant for polyester and nylon textile and industrial yarns; softener for synthetic rubber and other elastomers. *Inolex.*

17158 Lexolube® 2X-109
Ditridecyl adipate
Base stock for crank-case and compressor oils; fiber lubricant in yarns. *Inolex.*

17159 Lexolube® 3G-310
Trimethylolpropane trioleate
Lubricant component for hydraulic and metalworking oils. *Inolex.*

17160 Lexolube® 4N-415
Pentaerythrityl tetra C_8-C_{10} ester
High temperature lubricant for filament yarn production and processing, tire cord lubricant. *Inolex.*

17161 Lexolube® B-108
Isodecyl stearate
Lubricant for synthetic fibers, metalworking, and industrial applications. *Inolex.*

17162 Lexolube® B-109
31556-45-3 250-696-7
Tridecyl stearate
Drawing and heat setting lubricant for textile/industrial filament yarns, plastic extrusion, magnetic tapes. *Inolex.*

17163 Lexolube® BS-Tech
123-95-5 1625 204-666-5
Butyl stearate
General purpose lubricant for textile synthetic fibers, metalworking, coatings, inks, plastics, rubber industries. *Inolex.*

17164 Lexolube® T-110
22047-49-0 244-754-0
2-Ethylhexyl stearate
Lubricant for textile, metalworking, plastics industries. *Inolex.*

17165 Lexone®
21087-64-9 6239 244-209-7
Metribuzin
Triazinone herbicide for weed control in potatoes. *DuPont UK.*

17166 Lexorez 1100-25
Linear poly (diethylene glycol adipate)
Saturated, linear, aliphatic polyester polyol with all primary hydroxyls; used in adhesives, flexible coatings, and castable elastomers. *Inolex.*

17167 Lexorez 1101-50A
Crosslinked poly (diethylene glycol adipate); saturated, crosslinked primary hydroxyl polyester producing high-quality flexible polyester slab foams for use in textile laminates, pkg., filter media, and in noncellular applications such as coatings, adhesives, *Inolex.*

17168 Lexorez 1130-30
Linear poly (1,6-hexanediol adipate)
saturated aliphatic polyester polyols with all primary hydroxyls; raw material for formulation of urethane prepolymers, coatings, and thermoplastic urethanes which exhibit outstanding low temperature flexibility, high strength, and softness. *Inolex.*

17169 Lexorez 1131-190
Crosslinked poly (1,6-hexanediol adipate); saturated primary hydroxyl functional polyester polyol; used in the formulation of high-performance urethane elastomers. *Inolex.*

17170 Lexorez 1400-35
Linear poly (1,6-hexanediol neopentyl glycol adipate); saturated, linear, aliphatic, slightly branched polyester with primary hydroxyl functionality; used in semiflexible coatings, solution adhesives, prepolymer systems; produces urethanes with excellent solvent *Inolex.*

17171 Lexorez 1721-65P
Crosslinked poly (diethylene glycol, neopentyl glycol, 1,6-hexanediol adipate); branched saturated aliphatic polyester polyol with all primary hydroxyl functionality; used to produce high-quality coatings for fabrics and flexible foams which are resistance to HC. *Inolex.*

17172 Lexorez 1821-50
Crosslinked poly (diethylene glycol adipate); slightly crosslinked, saturated polyester polyol with primary hydroxyl functionality; used for flexible foams, cast elastomers, solution coatings, and adhesives; produces urethanes with excellent flexibility. *Inolex.*

17173 Lexorez 5901-55
Crosslinked poly (dipropylene glycol phthalate adipate); saturated mixed aromatic/aliphatic polyester polyol; used for cast systems and solution coatings where extreme service is required; ideal for structural RIM applications. *Inolex.*

17174 Lexquat® 2240
68039-13-4
Polymethacrylamidopropyl trimonium chloride
Film former, hair fixative, skin protectant, humectant, conditioner, antistat; pH stable; exceptional low use levels. *Inolex.*

17175 Lexquat® AMG-BEO
Behenamidopropyl PG-dimonium chloride
Mild conditioning surfactant, emulsifier for shampoos, bath gels, conditioners, shave products. *Inolex.*

17176 Lexquat® AMG-IS
Isostearamidopropyl PG-dimonium chloride
Mild conditioning surfactant, emulsifier for shampoos, bath gels, conditioners, shave products. *Inolex.*

17177 Lexquat® AMG-M
Lauramidopropyl PG-dimonium chloride
Conditioner, emulsifier for hair and skin products; emollient in bath products, liq. soaps. *Inolex.*

17178 Lexquat® AMG-O
Oleamidopropyl PG-dimonium chloride
Conditioner, emulsifier, emollient for hair and skin products, bath gels; forms clear dilutable gels with other fatty quats. *Inolex.*

17179 Lexquat® AMG-WC
Cocamidopropyl PG-dimonium chloride
Foaming conditioner, emulsifier for hair and skin products, bath gels. *Inolex.*

17180 Lexquat® CH
Polyquaternium-29
Film former, hair fixative, skin protectant, humectant, conditioner, antistat; pH stable; exceptional low use levels. *Inolex.*

17181 Ley Cornox
Selective weedkillers. *The Boots Co plc.* Discontinued.

17182 LG Wax
A montan wax derivative used in the production of nonionic and ionic dry-bright emulsions for floor polishes and similar applications. *Bush Beach Ltd.* Name unverified.

17183 LHS 40% Coconut Oil Soap
Highly refined coconut oil, caustic potash; soap. *Emulsion Systems.*

17184 Liancare
Janitorial and hygiene products. *MTM plc.*

17185 Libavius' fuming spirit
7646-78-8 8929 231-588-9
$SnCl_4$
A solution of stannic chloride.

17186 Libfer®
High purity iron EDDHA chelate in soluble powder and granule form for correcting iron deficiency in soils of adversely high pH. *Allied Colloids Ltd.*

17187 Libfer® SP
High purity iron EDDHA chelate. *Atlas Interlates Ltd.*

17188 Libollite
A variety of asphaltum.

17189 Libral Range
Chelated micronutrients. *Atlas Interlates Ltd.*

17190 Librebor®
An organo-boron complex for correcting boron deficiency in most crops. *Allied Colloids Ltd.*

17191 Librel®
Chelated metallic micronutrients designed for soil or foliar application to correct specific trace element deficiencies in all crops. *Allied Colloids Ltd.*

17192 Libreleaf®
An economical range of copper, iron and manganese lignosulfonate chelates for correcting specific trace element deficiencies in all crops for foliar application only. *Allied Colloids Ltd.*

17193 Libsorb®
A nonionic wetting and spreading agent which increases the effectiveness of foliar applied pesticides and fertilizers. *Allied Colloids Ltd.*

17194 Libspray®
A range of complete foliar fertilizers containing chelated trace elements in powder and liquid form to supplement soil applied fertilizers. *Allied Colloids Ltd.*

17195 LICA 01
Neoalkoxy
trineodecanoyl titanate. Coupling agents which also act as adhesion promoters, antioxidants, antistats, antifoaming agents, accelerators, blowing agent activators, catalysts, curatives, corrosion inhibitors, dispersion aids, emulsifiers, flame retardants, foaming agents, hardeners. *Kenrich Petrochemicals.*

17196 LICA 09
Neoalkoxy
dodecylbenzenesulfonyl titanate. *Kenrich Petrochemicals.*

17197 LICA 12
Neoalkoxy
tri (dioctylphosphato) titanate. *Kenrich Petrochemicals.*

17198 LICA 38
Neoalkoxy
tri (dioctylpyrophosphato) titanate. *Kenrich Petrochemicals.*

17199 LICA 44
Neoalkoxy
tri (N ethylaminoethylamino) titanate. *Kenrich Petrochemicals.*

17200 LICA 97
Neoalkoxy
tri (m-amino) phenyl titanate; phenyl glycol ether. *Kenrich Petrochemicals.*

17201 LICA 99
Neopentyl (diallyl)oxy
trihydroxy caproyl titanate. Coupling agent. *Kenrich Petrochemicals.*

17202 Licella yarn
A product made from narrow strips of wood-pulp paper; used as a jute substitute.

17203 lichen starch
$C_6H_{10}O_5$
Lichenin
A carbohytrate derived from Iceland moss.

17204 lichen sugar
149-32-6 3714 205-737-3
$C_4H_{10}O_4$
Erythritol
mp = 121-122°; bp = 329-331°; very soluble in H_2O, less soluble in organic solvents; LD_{50} (dog iv) = 5.0 g/kg. .

17205 Lichner's blue
A variety of smalt. It is a silicate of cobalt and potassium.

17206 Lichtenbergs's metal
An alloy of 50% bismuth, 30% lead, and 20% tin, melting at 91.5 C.

17207 Licomer
Polyethylene primary dispersions. *Hoechst UK.*

17208 Liconite
A rubber substitute made from bitumen and oils. No manufacturer.

17209 Licowet
Fluorinated surfactants. *Hoechst UK.*

17210 Licryl-55
Licryfilcon A
2-hydroxyethylmethacrylate polymer with ethylene dimethacrylate; contains 55% water as 0.9% saline; contact lens material. *Liquid Crystal Lens Co.* Name unverified.

17211 Licryl-70
Licryfilcon A
2-hydroxyethylmethacrylate polymer with ethylene dimethacrylate; contains 70% water as 0.9% saline; contact lens material. *Liquid Crystal Lens Co.* Name unverified.

17212 Licuado Instante
Nutritional additive to milk. *Richardson-Vicks Ltd.* Name unverified.

17213 Lida-Mantle
137-58-6 5505 205-302-8
Lidocaine
Anesthetic. *Bayer.* Name unverified.

17214 Lidarral
65009-35-0 265-307-6
Lidamidine hydrochloride
Antiperistaltic. *William H Rorer Inc.* Name unverified.

17215 Lidex
356-12-7 4186 206-597-6
Fluocinonide
Glucocorticoid *Syntex Laboratories Inc.*

17216 lidocaine
137-58-6 5505 205-302-8
$C_{14}H_{22}ON_2$
α-Diethylaminoaceto-2,6-xylidide
Medicine (local anesthetic). *Atomergic Chemetals; R.W. Greeff; Wyckoff.*

17217 Lieben solution
A solution of iodine in potassium iodide.

17218 Lieberkuhn's jelly
The jelly formed by mixing egg serum with about one third of its volume of twice normal sodium hydroxide.

17219 Liebmann and Studer's acid
1-Naphthol-7-sulfonic acid (1:7)

17220 Life
Oil analysis services. *Monsanto (Solaris).* Name unverified.

17221 Ligantraal
Methyl-N-(2,4-dimethyl cyclohexene-3-ylidenemethyl)-anthranilate
Quest Int'l. UK Ltd.

17222 Ligdynite
A coal mine explosive consisting of 25-27% nitroglycerin, 27-29% sodium nitrate, 10-12% sodium chloride, and 30-33% wood meal.

17223 Light Water Foam
Fluorochemical surfactants in solution; premium grade fire fighting foam proportioned in water; used for flammable liquid fire fighting and protection. *3M UK plc.*

17224 Light-Weld Multi-Cure Structural Adhesives
Acrylic structural adhesives, ultraviolet curing structural adhesives; used for magnet, metal, glass and fiber optics bonding, assembly adhesives and coatings for electronics. *Dymax Corp.*

17225 Lightfast
Mixed metal oxide pigments. *Miles.* Discontinued.

17226 Lightguard
A line of extruded, expanded polystyrene foam insulation products having a protective coating on one side. *Dow UK.* Discontinued.

17227 Lightning powder
A double base smokeless rifle powder.

17228 Lignasan®
For the agriculture industry. *DuPont UK.*

17229 Lignin dynamites
Nitroglycerin absorbed by a mixture of wood pulp, and a nitrate, usually sodium nitrate; used as explosives.

17230 Lignin sulfonate
lignosulfonate
A metallic sulfonate salt; dispersant for concrete, carbon black-rubber mixes, extender for tanning agents, oil-well drilling mud additives, ore flotation agents, production of vanillin, industrial cleaners, gypsum slurries, dyestuffs, and pesticides. *Borregaard Lignotech; Holmen GmbH; R. T. Vanderbilt Co Inc; Westvaco.*

17231 Lignite
brown coal
It is a low rank of coal and is intermediate between sub-bituminous and peat.

17232 Lignorit
8061-52-7
Unfermented calcium lignosulfonate; extender for U/F resins in particle board manufacture *Borregaard Ligno Tech.*

17233 Lignorosin
Calcium lignosulfonate

A by-product in the manufacture of paper pulp. It is a dark-brown syrup, and is used as an assistant in mordanting wool with chrome.

17234 Lignosite®
8061-52-7
Calcium lignosulfonate
Dispersant, emulsifier and emulsion stabilizer. *Georgia-Pacific.*

17235 Lignosol
Calcium and sodium lignosulfonates; specialty chemicals for use in a variety of binding, dispersing, complexing, stabilizing or copolymerizing applications. *Borregaard LignoTech.*

17236 Lignosol B
8061-52-7
Calcium lignosulfonate
Wetting agent, emulsifier, dispersant, binder; used in refractories, bricks, construction, insecticides, herbicides, fungicides; soil stabilizer. *Borregaard LignoTech.*

17237 Lignosol DXD
8061-51-6
Sodium lignosulfonate
Wetting agent and dispersant in industrial cleaners, insecticides, herbicides and fungicides, emulsifier, emulsion stabilizer for asphalt emulsions. *Borregaard Ligno Tech.*

17238 Lignosol SD-60
Kraft lignin; primary dye dispersant providing excellent heat stability, low staining and good milling and paste chars. *Borregaard Ligno Tech.*

17239 Lignosol SF
8061-52-7
Calcium lignosulfonate
Dispersant for Portland cement and concrete, retanning agent. *Borregaard Ligno Tech.*

17240 Lignosol TS
8061-53-8
Ammonium lignosulfonate
Tembind A 002; Wanin AM. Wetting agent, emulsifier, dispersant, tanning extract, slurry water reducer and grinding aid in cement manufacture *Borregaard Ligno Tech.*

17241 Lignovest
Modified polybutadiene. *Hüls AG.*

17242 Lignum vitae
Guaiacum wood.

17243 Ligro
A proprietary trade name for a crude pine fatty acid mixture. No manufacturer.

17244 Ligroin
8032-32-4 5513 232-354-7
V.M.&P.; Naphta; Benzoline; Canadol; Solvent naphta; refined solvent naphta. A term rather loosely applied. It usually denotes a refined distillate of petroleum oil having a boiling-point of 120-135°., of specific gravity 0.73; used as a polishing oil, and as a turpentine substitute in varnishes. d = 0.850 - 0.870; bp = 90-120°.

17245 Ligulin
The dyestuff of privet berries.

17246 Ligurite
A mineral. It is a variety of sphene.

17247 Ligustral
2,3-Dimethyl-3-cyclohexene-1-carboxy-aldehyde
Quest Int'l. UK Ltd.

17248 Lilaflot
Formulated cationic flotation agents based on primary fatty amines, primary ether amines, the corresponding diamines and quaternary ammonium compounds; mineral flotation for potash; feldspar; quartz; iron, sulfide and phosphate ores; scheelite. *Keno Gard (UK) Ltd.* Unverified.

17249 Lilamin AC-59P
Long-chain fatty nitrogen derivative with C14-22 alkyl chain length; mold release agent, processing aid for manufacture of natural and synthetic rubber products. *Berol Nobel AB.*

17250 Lilamin LSP 33
85632-63-9 288-048-0
Tallow dipropylene triamine
Wetting and adhesion agent, flushing agent, pigment grinding and dispersion agent for pigment industry; curing agent for epoxy paints and coatings; corrosion inhibitor. *Berol Nobel AB.*

17251 Lilaminox M24
85408-49-7 287-011-6
C_{12}-C_{14} alkyl dimethylamine oxide
Foam booster/stabilizer, antistat, softener for hair shampoos; thickener for household bleaches based on sodium hypochlorite, hard surface cleaners. *Berol Nobel AB.*

17252 Lilaminox M4
3332-27-2 222-059-3
Tetradecyl dimethylamine oxide
Detergency booster, thickener for household bleaches based on sodium hypochlorite; foaming agent for hair shampoos. *Berol Nobel AB.*

17253 Lilestralis
p-t-butyl-α-methyldihydrocinnamic aldehyde
Bush Boake Allen Ltd.

17254 Lilion 6
Nylon 6 monofilament; used for sewing threads, belting, filtration fabrics. *SNIA (UK) Ltd.*

17255 Lilion 66
Nylon 6 monofilament; used for sewing threads, belting, filtration fabrics. *SNIA (UK) Ltd.*

17256 Lillite
An earthy mineral resembling glauconite, of Bohemia.

17257 Lilvert
1-Ethoxy-1-hexoxyethane
Bush Boake Allen Ltd.

17258 Lily of valley, artificial
98-55-5 9316 202-680-6
$C_{10}H_{18}O$
Terpineol

17259 LIM6045
63394-02-5
Liq. two-component silicone rubber; for use in liq. injection molding to produce elastomeric parts with high clarity; excellent tear strength, very high tensile strength; for mech. parts, sporting goods, health care equip., camera parts, coating metal rol *GE Silicones.* Name unverified.

17260 Limao
Shellac substitute. *British Traders & Shippers Ltd.*

17261 Limbachite
A Serpentine mineral.

17262 Limbaki
Stretch velour fabric. *ICI Chem & Polymers Ltd.* Discontinued.

17263 Limbitrol
A proprietary preparation of amitriptyline [50-48-6] and chlordiazepoxide [58-25-3]; an antidepressant. *Roche Laboratories; Roche Products Ltd.*

17264 Limbux®
A form of mechanically slaked lime; used in agriculture, building and construction, metallurgical and chemical industries. *ICI Chem & Polymers Ltd.* Discontinued.

17265 lime
1305-78-8 1733 215-138-9
Calcium oxide
Burnt lime; quicklime; caustic lime. (Burnt lime, quicklime, caustic lime). CaO, produced by calcining calcium carbonate. Sometimes a mixture of calcium and magnesium oxides is sold under this name.

17266 lime mortar
Mixtures of slaked lime and sand.

17267 lime nitrate
10124-37-5 1729 233-332-1
$Ca(NO_3)_2 \cdot 4H_2O$
Calcium nitrate
(Lime saltpeter). used an a fertilizer.

17268 lime nitrogen
156-62-7 1702 205-861-8
$CaCN_2$
Calcium cyanamide
Used as a fertilizer, nitrogen products, pesticide, hardening iron and steel.

17269 lime saltpeter

17270 lime water
1305-62-0 1716 215-137-3
$Ca(OH)_2$
A saturated solution of calcium hydroxide, in water.

17271 Limeite
A cement containing rubber, tallow, and lime. The addition of vermilion causes the mixture to harden.

17272 Limeolivine
Ca_2SiO_4
Calcium orthosilicate

17273 limestone
1317-65-3 5515 215-279-6
$CaCO_3$
Calcium carbonate, natural
agricultural limestone; lithographic stone. Building stone, metallurgy (flux),

manufacture of lime; source of CO_2; Portland and natural cement; removal of sulfur dioxide from stack gases and sulfur from coal. ECC International Ltd.; EM Industries; Genstar Stone Prods.; Georgia Marble; J.M. Huber; Mallinckrodt; Nichia Kagaku Kogyo; Pfizer; Tilcon Ltd; Whittaker, Clark & Daniels. Discontinued.

17274 Lime-sulfur dips
Sheep dips for the treatment of scab. They contain flowers of sulfur, lime, and water.

17275 Limex
Sodium phosphate and metal preservatives; used in water using equipment where lime scale and corrosion are a problem. *Delaware Chemical Corp.* Name unverified.

17276 Limex G
10124-56-8 8814 233-343-1
Sodium hexametaphosphates
Used in water treatment to sequester calcium minerals. *Delaware Chemical Corp.* Name unverified.

17277 Limit
N-(Acetylamino) methyl-2-chlor-N-2,6-diethyl-phenylacetamide
Turf grass regulator. *Monsanto Co.* Name unverified.

17278 Limit 33
Defoamer for paper industry; used at the size press to defoam starch and mixtures of starch. *Monsanto Co.* Name unverified.

17279 Limo
(Sablon, tartrate of lime). The raw materials obtained by the precipitation of tartaric acid in tartar works or wine distilleries.

17280 Limonene
5989-27-5 5518 227-813-5
$C_{10}H_{16}$
p-mentha-1,8-diene
R(+)-limonene; (+)carvene; (R)-4-isopropenyl-1-methyl-1-cyclohexene. Terpene; flavoring, fragrance, and perfume materials; solvents, wetting agent, resin manufacture *Allchem Industries; Int'l. Flavors & Fragrances; Langley Smith Ltd; Penta Mfg.; SCM Glidco Organics.*

17281 Limpetite
Air-cured liquid-applied synthetic rubber; for protection of items subject to extreme conditions of corrosion, erosion and electrolytic action. *Protective Rubber Coatings (Limpetite) Ltd.*

17282 Linaqua®
Water-soluble drying oil-based chemical vehicle for paint. *Reichhold.*

17283 Linalux
Gloss paint finishes for marine use. *ICI Chem & Polymers Ltd.*

17284 Lincocin
7179-49-9 5525
Lincomycin hydrochloride
Antibacterial. *Upjohn.*

17285 lincomycin hydrochloride hemihydrate
7179-49-9 5525
$C_{18}H_{34}N_2O_6S{\cdot}HCl{\cdot}0.5H_2O$
Lincomycin hydrochloride; LIncocin; Frademicina; Mycivin; Waynecomycin. Antibacterial. mp = 145-147°; $[\alpha]_D^{25}$ = 137°; soluble in water, ethanol, methanol; sparingly soluble in most organic solvents, insoluble in hydrocarbons; LD_{50} (rat orl) = 4 g/kg. *Upjohn.*

17286 Lindenol
Pure α-terpineol. *Bush Boake Allen Ltd.*

17287 Lindex
Insecticide and fungicide seed dressing. *DowElanco Ltd.*

17288 Lindex-Plus
Flowable concentrate of 43 g fenpropimorph, 545 g γ-HCH and 73 g thiram per liter; combined insecticide and fungicide seed treatment. *DowElanco Ltd.*

17289 Linearil
A linear alkylate. *Montedison UK Ltd.* Name unverified.

17290 Linesman
Line marking paint. *Kalon Chemicals Ltd.*

17291 Linevol 79
Blend of primary alcohols, the notable characteristic being a high straight chain alcohol content; readily esterified by conventional means to give a high quality plasticizer. *Shell.*

17292 Linevol 911
Blend of primary alcohols, the notable characteristic being a high straight chain alcohol content; readily esterified by conventional means to give a high quality plasticizer. *Shell.*

17293 Linevol Phthalates
Plasticizers based on straight chain Linevol alcohols; they are characterized by exceptionally low volatility and excellent low temperature properties; a range of three phthalates allows a versatile plasticizing performance. *Shell.*

17294 Linex® 4L
330-55-2 5534 206-356-5
Linuron suspension; flowable herbicide for control of grasses and broadleaf weeds in certain crops and noncrop areas. *Griffin.*

17295 Linfos
Lindane [58-88-9] and chlorpyrifos [2921-88-2]; insecticide. *Makhteshim Chemical Works Ltd.*

17296 Lingraine
379-79-3 3703 206-835-9
A proprietary preparation of ergotamine tartrate; used for migraine. No manufacturer.

17297 lining metal
Alloys of 70-90% lead, 2-20% tin, and 5-20% antimony; used for bearings.

17298 Linklon
Crosslinkable polyolefin compounds. *Mitsubishi Petrochem.* Name unverified.

17299 Linklon-X
Crosslinkable polyolefin compounds. *Mitsubishi Petrochem.* Name unverified.

17300 Linnalite
Co_3S_4
Linnaeite. Native cobalt sulfide.

17301 Linnet
Emulsifiable concentrate containing 106 g linuron [330-55-2] and 192 g trifluralin [1582-09-8] per liter; herbicide for winter cereals. *Pan Britannica Industries Ltd.*

17302 Linocin
7179-49-9 5525
A proprietary preparation containing lincomycin hydrochloride; an antibiotic. *Upjohn Ltd.* Name unverified.

17303 LinoCure
An alkyd oil urethane no-bake binder system; utilized in the production of foundry cores and molds as a resin binder system. *Ashland Chemical Company.* Name unverified.

17304 Linol
8001-26-1 232-278-6
Linseed oil; Oils, linseed. Conjugated linseed oil; drying oil as a component of finished resin, migrating from food package. *Unilever.* Name unverified.

17305 linoleic acid
60-33-3 5529 200-470-9
$CH_3(CH_2)_4{\cdot}cCHCH_2CH{\cdot}cCH(CH_2)_7COOH$
9,12-octadecadienoic acid
linoleic acid. Unsaturated fatty acid; manufacture of paints, coatings, emulsifiers, vitamins. bp_1 = 230-232°; d_4^{18}= 0.914; insoluble in H_2O, soluble in organic solvents. *Arizona; CasChem; Henkel/Emery; Hercules; Langley Smith Ltd.* Discontinued.

17306 linolenic acid
463-40-1 5530 207-334-8
$C_{18}H_{30}O_2$
(Z,Z,Z)-9,12,15-Octadecatrienoic acid
Industrene® 120; α-linolenic acid. Chemical intermediate. d = 0.914; bp_1 230-232°; insoluble in water; soluble in organic solvents. *Witco/Humko.* Discontinued.

17307 linotype metal
An alloy of 13.5% antimony, 2% tin, and 84.5% lead.

17308 linseed meal
Ground oil-cake.

17309 Linseed oil
8001-26-1 232-278-6
Flaxseed oil; Oils, linseed; Llinol. Conjugated linseed oil; drying oil as a component of finished resin, migrating from food package. d = 0.921-0.936; mp = -19°; bp = 343°; fp = 222°. *Arista; Ferro/Bedford; Penta Mfg.; John L Seaton Ltd; Unilever; Penta Mfg.* Name unverified.

17310 Linseed oil soap
A potash soap.

17311 Linseed oil, artist's
Raw linseed oil which has been allowed to stand for weeks, then treated with litharge, and finally bleached by exposure.

17312 Linseed oil, refined
Raw linseed oil which has been treated with a 1% solution of sulfuric acid.

17313 Linsol
Dyestuff solutions in oleic acid. *Morton Int'l. Ltd.* Discontinued.

17314 Linter's starch
A soluble starch prepared by mixing raw starch with 7.5% hydrochloric acid in water, allowing it to stand for several days, with stirring. The solution is decanted, and the starch washed with water and dried.

17315 Lintex A10
A proprietary aqueous emulsion of a styrene copolymer internally plasticized with a copolymerized ester; contains crosslinkable groups and is used as a binder for emulsion paints. *Hüls AG.*

17316 Linurex
330-55-2 5534 206-356-5
3-(3,4-dichlorophenyl)-1-methoxy-1-methylurea
Active ingredient: linuron; selective herbicide for both pre- and post-emergence application. *Agan Chemical Manufacturers Ltd.*

17317 Linuron
330-55-2 5534 206-356-5
Linuron; a residual herbicide for the control of weeds in field crops. *Pan Britannica Industries Ltd.*

17318 Linuron 15
330-55-2 5534 206-356-5
Linuron; a residual herbicide for the control of weeds in field crops. *Farm Protection Ltd.*

17319 Linuron 450 FL
330-55-2 5534 206-356-5
Linuron; a residual urea herbicide for the control of weeds in field crops. *Farmers Crop Chemicals Ltd.* Discontinued.

17320 Lipacide CCO
Capryloyl hydrolyzed collagen mild surfactant with excellent lathering and wetting props., substantivity to hair and skin; for frequent-use shampoos, bath gels, soaps, shaving creams. *Rhône-Poulenc; R. T. Vanderbilt Co Inc.*

17321 Lipacide DPHP
41672-81-5 255-490-0
Dipalmitoyl hydroxyproline
Mild surfactant with excellent lathering and wetting props., substantivity to hair and skin; for frequent-use shampoos, bath gels, soaps, shaving creams; used in cosmetic preparations for maintenance of skins physiological balance. *Rhône-Poulenc; R.T. Vandervilt.*

17322 Lipacide PCO
Palmitoyl animal collagen amino acids; mild surfactant with excellent lathering and wetting props., substantivity to hair and skin; for frequent-use shampoos, bath gels, soaps, shaving creams; for first aid creams, sunburn lotions. *Rhône-Poulenc; R. T. Vanderbilt Co Inc.*

17323 Lipacide PK
Palmitoyl animal keratin amino acids. mild surfactant with excellent lathering and wetting props., substantivity to hair and skin; for frequent-use shampoos, bath gels, soaps, shaving creams. *Rhône-Poulenc; R. T. Vanderbilt Co Inc.*

17324 Lipacide UCO
Undecylenoyl animal collagen amino acids. mild surfactant with excellent lathering and wetting props., substantivity to hair and skin; for frequent-use shampoos, bath gels, soaps, shaving creams. *Rhône-Poulenc; R. T. Vanderbilt Co Inc.*

17325 Lipal
A proprietary name for a range of polyoxyethylene esters and ethers. *PVO International Inc.* Name unverified.

17326 Lipamide MEAA
142-26-7 205-530-8
Acetamide MEA
Lubricating humectant for used in personal care products; hair conditioner for shampoos, rinses, conditioners; antistat, foam modifier. *Lipo.*

17327 Lipamin®
Cationic emulsified oils; fat liquoring agents for leather. *BASF AG.*

17328 Lipamine SPA
7651-02-7 231-609-1
Stearamidopropyl dimethylamine
Raw material for cosmetics, toiletries, pharmaceuticals. *Lipo.*

17329 Lipase
9001-62-1 5536 232-619-9
Triacylglycerol lipase
Glycerol ester hydrolase; Pancreatic Lipase 250. Enzyme; hydrolyzes fat to glycerol and fatty acids; manufacture of cheese; removal of fat spots in drycleaning; in analytical chemistry of fats. The optimum temperature for enzyme action is between 35° and 37° at pH 5-6. *Atomergic Chemetals; Gist-Brocades Food Ingreds.; U.S. Biochemical.*

17330 Lipaton
Styrene acrylic, and styrene butadiene copolymer dispersions. *Hüls AG.*

17331 Lipex 102
68424-60-2 270-311-6
Shea butter. *Karlshamns.*

17332 Lipinol O
An octyl fatty acid ester; viscosity depressant especially for rotational molding plastisols. *Hüls AG.*

17333 Lipinol T
Dibenzyltoluene
Secondary plasticizer, lowers the viscosity of PVC plastisols to a lesser degree than Lipinol O. *Hüls AG.*

17334 Lipiodul Ultra-Fluid
A proprietary preparation of iodized oil fluid injection; x-ray contrast medium. *May & Baker Ltd.* Name unverified.

17335 Lipo AMS
Almond meal; natural abrasive for facial scrubs, abrasive body scrubs and foot products. *Lipo.*

17336 Lipo APS 40/60
Apricot seed powd. natural abrasive for facial scrubs, body scrubs, foot products. *Lipo.*

17337 Lipo DGLS
9004-81-3
PEG-2 laurate SE
Spreading agent, emulsifier, dispersant, lubricant, opacifier, emulsion stabilizer, emollient, viscosity builder used in bath oils, creams, lotions; defoamer for process applications. *Lipo.*

17338 Lipo EGDS
627-83-8 211-014-3
Glycol distearate
Spreading agent, emulsifier, dispersant, lubricant, opacifier, emulsion stabilizer, emollient, viscosity builder used in bath oils, creams, lotions; defoamer for process applications. *Lipo.*

17339 Lipo EGMS
Glycol stearate
Opacifier, pearlizer for shampoos, detergents; water-oil emulsifier; stabilizer for oil-water systems. *Lipo.*

17340 Lipo Gantrisin
9125
Sulfisoxazole acetyl
Antibacterial. *Hoffmann-LaRoche Inc.* Name unverified.

17341 Lipo GMS 450
Glyceryl stearate
General purpose emulsifier, emollient, opacifier and viscosity builder in creams and lotions; food emulsifier. *Lipo.*

17342 Lipo LUFFA 30/100
Luffa; natural abrasive for facial scrubs, abrasive body scrubs and foot products. *Lipo.*

17343 Lipo PGMS
1323-39-3 215-354-3
Propylene glycol stearate
Emulsifier, stabilizer for oil-water lotions, soft creams. *Lipo.*

17344 Lipo Polyol NC
Hydrogenated starch hydrolysate
Humectant for creams, lotions, antiperspirants, depilatories, wavesets. *Lipo.*

17345 Lipo SS
68334-28-1 269-820-6
Hydrogenated vegetable oil; emollient for skin, sunscreens, lipsticks, balms; enhances gloss, reduces blooming. *Lipo.*

17346 Lipo WSF 35/60, 60/100
Walnut shell flour; natural abrasive for facial scrubs, abrasive body scrubs and foot products. *Lipo.*

17347 Lipobee 102
Synthetic beeswax; raw material for cosmetics, pharmaceuticals, and toiletries. *Lipo.*

17348 Lipocire A, CM, DM
Hydrogenated palm glycerides, hydrogenated palm kernel glycerides. lipstick bases. *Gattefosse.*

17349 Lipoclor
Chlorinated natural fats; additives for lubricating and cutting oils and raw material for the preparation of greasings for leather. *Caffaro SpA.* Name unverified.

17350 Lipoclor S
Chlorinated animal fats. *SNIA (UK) Ltd.*

17351 Lipocol C
36653-82-4 2070 253-149-0
Cetyl alcohol
Emollient, consistency builder for creams, lotions, molded stick products. *Lipo.*

17352 Lipocol C-2
9004-95-9
Ceteth-2
Emulsifier, defoamer, wetting agent, solubilizer, conditioning agent for personal care products and pigment dispersion *Lipo.*

17353 Lipocol L
112-53-8 3464 203-982-0
Lauryl alcohol
Nonoily afterfeel emollient for creams, lotions, makeup. *Lipo.*

17354 Lipocol L-4
Laureth-4
Surfactant for pigment dispersions, antiperspirants, depilatories, creams, lotions; antistat, emulsifier. *Lipo.*

17355 Lipocol O
143-28-2 6968 205-597-3
Oleyl alcohol
Nontacky afterfeel emollient for bath oils, creams, lotions, makeup. *Lipo.*

17356 Lipocol O-2
9004-98-2
Oleth-2
Surfactant for pigment dispersions, antiperspirants, depilatories, creams, lotions; antistat, emulsifier. *Lipo.*

17357 Lipocol S
112-92-5 8960 204-017-6
Stearyl alcohol
Waxy afterfeel emollient, consistency builder for creams, lotions, molded sticks. *Lipo.*

17358 Lipocol S-2
9005-00-9
Steareth-2
Surfactant for pigment dispersions, antiperspirants, depilatories, creams, lotions; antistat, emulsifier. *Lipo.*

17359 Lipocol SC-4
68439-49-6 7737
Ceteareth-4
Surfactant for pigment dispersions, antiperspirants, depilatories, creams, lotions; antistat, emulsifier. *Lipo.*

17360 Lipocol TD-3
24938-91-8
Trideceth-3
Emulsifier, welting and scouring agent, dispersant for essential oils; raw material for sulfation and phosphation. *Lipo.* Name unverified.

17361 Lipocutin®
Water, lecithin, cholesterol, and dicetyl phosphate. *Henkel/Cospha; Henkel Canada.* Name unverified.

17362 Lipodan CDS Kosher
Hydrogenated cottonseed oil/dist, monoglycerides blend; peanut butter stabilizer. *Grindsted Prods.; Grindsted Prods. Denmark.*

17363 Lipodan CRE Kosher
Hydrogenated cottonseed/rapeseed oils and dist. monoglycerides blend; stabilizer for peanut butter and other oil-based systems. *Grindsted Prods.*

17364 Lipodan SET Kosher
68334-28-1 269-820-6
Hydrogenated vegetable oil (rapeseed, cottonseed, soybean); stabilizer for peanut butter and other oil-based systems. *Grindsted Prods.*

17365 Lipoderm®
Fat liquoring agent for leather. *BASF AG.*

17366 Lipolan
8031-44-5 232-452-1
Hydrogenated lanolin; auxiliary water-oil emulsifier; emollient; conditioner; lubricant. *Lipo.*

17367 Lipolan 31
68648-27-1
PEG-24 hydrogenated lanolin
oil-water emulsifiers, solubilizer, emollient, conditioner used in cosmetics toiletries and topical pharmaceuticals. *Lipo.*

17368 Lipolan R
Lanolin oil; emollient, spreading agent, conditioner, cosolvent, plasticizer, and lubricant for personal care products; dispersant for pigments. *Lipo.*

17369 Lipolase
Lipolase hydrolyzed triglycerides into mono-and diglycerides, glycerol and free fatty acids, all of which are more soluble than the original fats; used in detergent formulations to remove fat-containing stains as those resulting from frying fats, salad oi *Novo Nordisk.*

17370 Lipo-Lutin
57-83-0 7956 200-350-6
Progesterone
Progestin. *Parke-Davis.* Name unverified.

17371 Lipomulse 165
Glyceryl stearate and PEG-100 stearate; general purpose emulsifier, emollient, opacifier, and viscosity builder in creams and lotions. *Lipo.*

17372 Liponate 2-DH
PEG-4 diheptanoate
Emollient. *Lipo.*

17373 Liponate CL
35274-05-6 . 252-478-7
Cetyl lactate
Emollient for desired feel and penetration in personal care products; thickener and viscosity controller. *Lipo.*

17374 Liponate CRM
10401-55-5 233-864-4
Cetyl ricinoleate
Glosser, emollient with dry after-feel. *Lipo.* Name unverified.

17375 Liponate DPC-6
68130-24-5 268-581-5
Dipentaerythrityl hexacaprylate/hexacaprate
Nontacky emollient for treatment products. *Lipo.*

17376 Liponate EM
Ethyl morrhuate
Lipo. Name unverified.

17377 Liponate GC
65381-09-1 265-724-3
Caprylic/capric triglyceride
Emollient ester for adjusting rub-in and afterfeel of personal care products; thickener and viscosity controller. *Lipo.*

17378 Liponate IPM
110-27-0 5234 203-751-4
isopropyl myristate
IPM; emollient ester for adjusting rub-in and afterfeel of personal care products; thickener and viscosity controller. *Lipo.*

17379 Liponate IPP
142-91-6 205-571-1
isopropyl palmitate
IPP; emollient ester for adjusting rub-in and afterfeel of personal care products; thickener and viscosity controller. *Lipo.*

17380 Liponate MM
3234-85-3 221-787-9
Myristyl myristate; emollient ester for adjusting rub-in and afterfeel of personal care products; thickener and viscosity controller. *Lipo.*

17381 Liponate NPGC-2
70693-32-2 274-764-0
Neopentyl glycol dicaprylate/dicaprate
Dry feel emollient for creams, lotions, cleansers, antiperspirants. *Lipo.*

17382 Liponate PB-4
61682-73-3 262-895-6
Pentaerythrityl tetrabehenate
Emollient ester for adjusting rub-in and afterfeel of personal care products; thickener and viscosity controller. *Lipo.*

17383 Liponate PC
Propylene glycol dicaprylate/dicaprate; emollient ester for adjusting rub-in and afterfeel of personal care products; thickener and viscosity controller. *Lipo.*

17384 Liponate PE-810
Pentaerythrityl tetracaprylate
tetracaprate. Emollient. *Lipo.*

17385 Liponate PO-4
19321-40-5 242-960-5
Pentaerythrityl tetraoleate
Emollient ester for adjusting rub-in and afterfeel of personal care products; thickener and viscosity controller. *Lipo.*

17386 Liponate PS-4
115-83-3 204-110-1
Pentaerythrityl tetrastearate
Emollient ester for adjusting rub-in and afterfeel of personal care products; thickener and viscosity controller. *Lipo.*

17387 Liponate SPS
8002-23-1 232-302-5
Cetyl esters (synthetic Spermaceti). Emollient ester for adjusting rub-in and afterfeel of personal care products; thickener and viscosity controller. *Lipo.*

17388 Liponate SS
2778-96-3 220-476-5
Stearyl stearate
Emollient ester for adjusting rub-in and afterfeel of personal care products; thickener and viscosity controller. *Lipo.*

17389 Liponate TDS
31556-45-3 250-696-7
Tridecyl stearate
Emollient for creams and lotions. *Lipo.*

17390 Liponate TDTM
70225-05-7
Tridecyl trimellitate
Nontacky emollient for treatment products, hair. *Lipo.*

17391 Liponic 70-NC
50-70-4 8873 200-061-5
Sorbitol
Humectant, plasticizer, softener, and lubricant; adds sweet taste and pleasant mouthfeel to oral hygiene products such as dentifrices and mouthwashes; oral dosage pharmaceutical, also for adhesives, leather, and paper coatings. *Lipo.*

17392 Liponic EG-1
31694-55-0
Glycereth-26
Humectant in creams and lotions, lubricant, plasticizer for hair resins, foam stabilizer, pigment dispersant, hair conditioner, foam modifier, antistat; used in personal care products. *Lipo.*

17393 Liponic SO-20
Sorbeth-20
Humectant and plasticizer for cosmetics and toiletries; nongreasy rich afterfeel with moderate lubricity. *Lipo.*

17394 Lipopeg 2-DL
9005-02-1
PEG-4 dilaurate
Dispersant, emulsifier, spreading agent and lubricant in personal care products, bath oils. *Lipo.*

17395 Lipopeg 4-DO
9005-07-6
PEG-8 dioleate
Dispersant, emulsifier, spreading agent and lubricant in personal care products, bath oils. *Lipo.*

17396 Lipopeg 4-DS
9005-08-7
PEG-8 distearate
Dispersant, emulsifier, spreading agent and lubricant in personal care products, bath oils. *Lipo.*

17397 Lipopeg 4-L
9004-81-3
PEG-8 laurate
Spreading agent, emulsifier, dispersant, lubricant for bath oils, creams and lotions. *Lipo.*

17398 Lipopeg 6000-DS
9005-08-7
PEG-150 distearate
Dispersant, emulsifier, spreading agent, and lubricant in personal care products, bath oils. *Lipo.*

17399 Lipo-Peptide AME 30
Acetamide MEA
lauroyl hydrolyzed collagen; glycerin. Moisturizer, humectant, emollient, softener, auxiliary emulsifier for skin products; conditioner, bodying agent, antistat for hair products; foam booster in shampoos, mousses, liq. hand soaps. *May brook.*

17400 Lipoproteol LCO
Sodium/TEA-luroyl hydrolyzed collagen amino acid ; Mild additive with good foaming props. for rich lather shampoos, facial cleaners, infant shampoos, detergents. *Rhône-Poulenc; R. T. Vanderbilt Co Inc.*

17401 Lipoproteol LK
Sodium
TEA-lauroyl hydrolyzed keratin amino acids. Shampoo base. *Rhône-Poulenc; R. T. Vanderbilt Co Inc.*

17402 Lipoquat R
112324-16-0
Ricinoleamidopropyl ethyldimonium ethosulfate
Conditioner, antistat, emollient, glosser, softener for personal care products and anhyd. systems. *Lipo.*

17403 Liposiliol C
130986-04-8
Dioleyl tocopheryl methylsilanol
Aids tissue regeneration; for anti-aging formulations, oily cosmetics and sticks. *Exsymol.*

17404 Liposorb L
1338-39-2 8872 215-663-3
Sorbitan laurate
Emulsifier, thickener, lubricant, antistat, all-purpose lipophilic surfactant used with POE Liposorb series; also used in defoamers, aerosol water-oil emulsions, corrosion inhibition. *Lipo.*

17405 Liposorb L-10
9005-64-5 8872
PEG-10 sorbitan laurate
Oil-water emulsifier, lubricant, antistat, all-purpose hydrophilic surfactant; used for solubilizing oils and in conjunction with Liposorb esters. *Lipo.*

17406 Liposorb L-20
9005-64-5 8872
Polysorbate 20
Oil-water emulsifier, lubricant, antistat, all-purpose hydrophilic surfactant; used for solubilizing oils and in conjunction with Liposorb esters. *Lipo.*

17407 Liposorb O
1338-43-8 8872 215-665-4
Sorbitan oleate
Emulsifier, thickener, lubricant, antistat, all-purpose lipophilic surfactant used with POE Liposorb series; also used in defoamers, aerosol water-oil emulsions, corrosion inhibition. *Lipo.*

17408 Liposorb O-20
9005-65-6 7742
Polysorbate 80
Surfactant for food processing; flavor and color dispersant for pickles; defoamer for beet sugar, yeast processing; wetting agent for poultry defeathering; crystal control agent for salt. *Lipo.*

17409 Liposorb O-5
9005-65-6 7742
Polysorbate 81
Hydrophilic surfactant used for solubilizing oils; emulsifier, lubricant, antistat. *Lipo.*

17410 Liposorb P
26266-57-9 8872 247-568-8
Sorbitan palmitate
Lipophilic surfactant, emulsifier, thickener, lubricant, antistat; also in defoamers, aerosol water-oil emulsions, corrosion inhibition. *Lipo.*

17411 Liposorb P-20
9005-66-7 8872
Polysorbate 40
Hydrophilic surfactant, solubilizer for oils, emulsifier, lubricant, antistat. *Lipo.*

17412 Liposorb S
1338-41-6 8872 215-664-9
Sorbitan stearate
Emulsifier, thickener, lubricant, antistat, all purpose lipophilic surfactant used with POE Liposorb series; also for defoamers, aerosol water-oil emulsions, corrosion inhibition. *Lipo.*

17413 Liposorb S-20
9005-67-8 8872
Polysorbate 60
Food emulsifier, defoamer. *Lipo.*

17414 Liposorb S-4
9005-67-8 8872
Polysorbate 61
Lipo. Name unverified.

17415 Liposorb SQO
8007-43-0 8872 232-360-1
Sorbitan sesquioleate
Emulsifier, thickener, lubricant, antistat, all-purpose lipophilic surfactant used with POE Liposorb series. *Lipo.*

17416 Liposorb TO
26266-58-0 8872 247-569-3
Sorbitan trioleate
Emulsifier, thickener, lubricant, antistat, all purpose lipophilic surfactant used with POE Liposorb series. *Lipo.*

17417 Liposorb TO-20
9005-70-3 8872
Polysorbate 85
oil-water emulsifier, lubricant, antistat, all-purpose hydrophilic surfactant; used for solubilizing oils and in conjunction with Liposorb esters. *Lipo.*

17418 Liposorb TS
26658-19-5 8872 247-891-4
Sorbitan tristearate
Emulsifier, thickener, lubricant, antistat, all-purpose lipophilic surfactant used with POE Liposorb series. *Lipo.*

17419 Liposorb TS-20
9005-71-4 8872
Polysorbate 65
Food emulsifier, defoamer. *Lipo.*

17420 Liposorb™
Absorbent. *Calbiochem Corp.*

17421 Lipostat Tablets
81131-70-6 7894
Pravastatin sodium
Bristol-Myers Squibb Pharmaceuticals Ltd.

17422 Lipotin 100, 100J, SB
8002-43-5 5452 232-307-2
Refined soy lecithin, filtrated, deodorized; wetting and dispersing agents preventing sedimentation in paints; raw material for textile and leather industry compounds. *Lucas Meyer.*

17423 Lipotin A
Phosphoamino compound.; emulsifier, wetting and dispersing agent for aqueous systems, stabilizer for latex and emulsion paints, leather finishes, water-dispersion and reducible air-dry and stoving alkyds, offset printing inks, textile auxiliaries. *Lucas Meyer.*

17424 Lipotin H
8029-76-3 232-440-6
Hydroxylated soy lecithin
Wetting and dispersing agent for air drying and stoving paints; raw material for textile and leather compounds; emulsifier. *Lucas Meyer.*

17425 Lipovol A
8024-32-6 232-428-0
Avocado oil; conditioner, glosser, emollient imparting a light, nongreasy, silky afterfeel to skin and hair products; high film gloss and rapid spread; used in personal care products. *Lipo.*

17426 Lipovol ALM
8007-69-0
Sweet almond oil; conditioner, glosser, emollient imparting a light, nongreasy, silky afterfeel to skin and hair products; high film gloss and rapid spread; used in personal care products. *Lipo.*

17427 Lipovol G
8024-22-4
Grape seed oil; conditioner, glosser, emollient imparting a light, nongreasy, silky afterfeel to skin and hair products; high film gloss and rapid spread; used in personal care products. *Lipo.*

17428 Lipovol HS
8016-70-4 232-410-2
Hydrogenated soybean oil. *Lipo.*

17429 Lipovol J
61789-91-1
Jojoba oil, refined; emollient, conditioner, and lubricant for cosmetics and toiletries; rapid spread and soft, nontacky afterfeel; used in skin and personal care products and oils, anhyd. and emulsified makeups. *Lipo.*

17430 Lipovol MOS-130
Tridecyl stearate
tridecyl trimellitate; dipentaerythrityl hexacaprylate/hexacaprate. Specialty esters exhibiting tactile props. of mineral oil. *Lipo.*

17431 Lipovol MOS-350
Dipentaerythrityl hexacaprylate/hexacaprate
tridecyl trimellitate; tridecyl stearate; neopentyl glycol dicaprylate/dicaprate. Specialty esters exhibiting tactile props. of mineral oil. *Lipo.*

17432 Lipovol MOS-70
Tridecyl stearate
neopentyl glycol dicaprylate/dicaprate; tridecyl trimellitate. Specialty esters exhibiting tactile props. of mineral oil. *Lipo.*

17433 Lipovol P
Apricot kernel oil; emollient used in cosmetics and pharmaceuticals; soft, nontacky afterfeel and high film gloss. *Lipo.*

17434 Lipovol PAL
8002-75-3 232-316-1
Palm oil. *Lipo.*

17435 Lipovol SAF
8001-23-8 8465 232-276-5
Safflower oil; conditioner, glosser, emollient imparting a light, nongreasy, silky afterfeel to skin and hair products; high film gloss and rapid spread; used in personal care products. *Lipo.*

17436 Lipovol SES
8008-74-0 8614 232-370-6
Sesame oil; emollient, solv, and vehicle used in cosmetics, toiletries, and pharmaceuticals; offers light, nontacky feel and enhances gloss and spread of pigmented sticks and pot products. *Lipo.*

17437 Lipovol SO
Hybrid safflower oil; conditioner, glosser, emollient imparting a light, nongreasy, silky afterfeel to skin and hair products; high film gloss and rapid spread; used in personal care products. *Lipo.*

17438 Lipovol SOY
8001-22-7 232-274-4
Soybean oil; natural emollient oil, lubricant, conditioner for luxury skin products, hair care products, makeup, fine soaps, bath oils, anhyd. systems. *Lipo.*

17439 Lipovol SUN
8001-21-6 232-273-9
Sunflower seed oil; emollient imparting a pleasant, nongreasy feel to skin and hair care products and makeups; adds conditioning, spread, and sheen to personal care products; used in preparation of margarine. *Lipo.*

17440 Lipovol WGO
8006-95-9
Wheat germ oil; emollient imparting perceptible afterfeel to skin and hair care products; gloss to anhyd. and emulsified makeups. *Lipo.*

17441 Lipowax D
Cetearyl alcohol
ceteareth-20. oil-water self-emulsifying wax, used in skin and hair creams and lotions, personal care products. *Lipo.*

17442 Lipowax G
Stearyl alcohol and ceteareth-20; emulsifier for personal care products. *Lipo.*

17443 Lipowax NI
Cetearyl alcohol and ceteth-20. *Lipo.*

17444 Lipowax P
Cetearyl alcohol
Polysorbate 60. oil-water selfemulsifying wax for neutral and mildly acidic and alkaline pH systems. *Lipo.*

17445 Lipowax PR
Cetearyl alcohol
polysorbate 60; PEG-150 stearate; steareth-20. Emulsifying wax for formulation of creams, lotions, and ointments. *Lipo.*

17446 Lipowax P-SPEC
Cetearyl alcohol
polysorbate 60. Emulsifying wax for formulation of creams, lotions, and ointments. *Lipo.*

17447 Lipowitz's alloy
An alloy of 50% bismuth, 27% lead, 13% tin, and 10% cadmium. It has a specific heat of 0.0345 calories per gram at from 5-50 C., and melts at 65 C; used for automatic sprinklers and other purposes.

17448 Lipoxol® 12000
25322-68-3 7729
PEG-240; cosmetic ingred. for lipsticks, deodorant sticks, soap bars, powd. bases, creams, and pastes. *Hüls Am; Hüls AG.*

17449 Lipozyme
An experimental preparation of an immobilized lipase; the fungal lipase is produced by submerged fermentation of a selected strain of *Mucor miehei* for interesterification, alcoholysis of oils and fats and synthesis of esters. *Novo Nordisk.*

17450 Liquamar
435-97-2 7413 207-108-9
Phenprocoumon
Anticoagulant. *Organon Inc.* Name unverified.

17451 LiquaPar® Oil
Isopropylparaben
isobutylparaben; butylparaben. Preservative for cosmetics and topical pharmaceuticals. *Sutton Labs.*

17452 Liquapen
113-98-4 7225 204-038-0
Penicillin G potassium
Antibacterial. *Pfizer Inc.* Discontinued.

17453 Liquemin
9005-49-6 4685 232-681-7
Proprietary preparation of heparin; anticoagulant. *Roche Products Ltd.* Name unverified.

17454 Liqueur de van Swieten
Consists of 1 part mercuric chloride, 100 parts alcohol, and 900 parts water.

17455 Liquibor
Organic borate. *Manchem Ltd.* Name unverified.

17456 Liquibor 169
Potassium glycol borate
Corrosion inhibitor for glycol-based brake fluids. *Rhône-Poulenc UK.*

17457 Liquibor 524
A borax-glycol condensation product which is used at a concentration of 1-2% as a corrosion inhibitor in synthetic hydraulic fluids. *Rhône-Poulenc UK.*

17458 Liquibrom
Cooling water biocide. *Great Lakes Europe.*

17459 Liquical
10043-52-4 1699 233-140-8
Anhydrous calcium chloride
Dow UK. Discontinued.

17460 Liqui-Cee
134-03-2 8723 205-126-1
Sodium ascorbate
Vitamin. *Baxter Health Care.* Name unverified.

17461 Liquid 99
Liquid cleanser. *British Nova Works Ltd.*

17462 Liquid Absorption Base Type T
Mineral oil, lanolin alcohol; emollient for liq. make-up to improve dispersion and application properties of pigments; primary oil phase ingred. in oil-water emulsions; type T is better solvent for oil-sol. dyes. *Croda Inc.*

17463 Liquid Bases
Compounded lanolin-derived sterols. *Croda Chem. Ltd.*

17464 Liquid bronzes
Varnishes in which bronze colors are suspended.

17465 Liquid Code XLR
10090-54-7 7832
Potassium pyroantimonate solution; crosslinker for guar and derivatized guar at pH 3-5. *Rhône-Poulenc/Water Soluble Polymers.*

17466 Liquid Copper Fungicide
Fungicide. *Murphy Chemical Co Ltd.* Discontinued.

17467 Liquid Crystal CN/9
57-88-5 2256 200-353-2
Cholesterol esters; carriers for nutrients in skin care products; provide decorative, functional and aesthetic effects to cosmetics. *Presperse.*

17468 Liquid drier
A name given to a concentrated solution of calcium chloride.

17469 Liquid Feed for Hanging Baskets
Liquid concentrate containing NPK 4:2:6 plus trace elements; liquid fertilizer for hanging baskets, tubs, planters and window boxes. *Vitax Ltd.*

17470 Liquid gold
Contains about 10% of gold (as chloride), resin, lavender oil, and bismuth; used for painting china.

17471 Liquid Growmore
Liquid concentrate containing NPK 7:7:7: general purpose liquid fertilizer. *Vitax Ltd.*

17472 Liquid Latex
9003-20-7
Polyvinyl acetate
Used in oil well cementing applications for fluid loss control, rheology modification, and improved bonding. *Rhône-Poulenc/Water Soluble Polymers.*

17473 Liquid Lightning
Inhibited sulfamic acid; liquid; removes water-formed deposits. *Garvey Chemical Corp.* Name unverified.

17474 Liquid Q4 Borders and Beds Fertiliser
Liquid concentrate containing NPK 5.3:5.3:10 plus trace elements; border and bedding plant fertilizer. *Vitax Ltd.*

17475 Liquid resins
Polyterpene
Sulfate Resin; Talleol. Semi-resinous compounds obtained as by-products in the manufacture of wood pulp by the sulfite and sulfate processes of paper making.

17476 Liquid Tomato Feed
Liquid concentrate containing NPK 4.5:4.5:9 plus Mg; fertilizer. *Vitax Ltd.*

17477 Liquidambar
Copalm balsam. A balsam obtained from a large Mexican tree. It contains cinnamyl cinnamate and styrene. It is also erroneously called liquid storax.

17478 Liquidow
10043-52-4 1699 233-140-8
Liquid calcium chloride
For applications on unpaved roads to control dust. *Dow UK.* Discontinued.

17479 Liquifilm Forte®
Polyvinyl alcohol
Ocular lubricant. *Allergan Inc.*

17480 Liquifilm Tears®
Polyvinyl alcohol
Ocular lubricant. *Allergan Inc.*

17481 Liquigel®
21645-51-2, 1309-42-8
Aluminum hydroxide [21645-51-2] and magnesium hydroxide [1309-42-8] fluid gel; suspension antacids. *Reheis Inc.*

17482 Liquigel-AM
Aluminum magnesium fluid gel. *Reheis Inc.*

17483 Liquimeth
Liquid methionine salt. *Degussa Ltd.*

17484 Liqui-Nox®
Blend; phosphate-free detergent for critical manual and ultrasonic cleaning applications. *Alconox.*

17485 Liquinure
Liquid fertilizer. *Fisons plc, Horticultural Div.* Name unverified.

17486 Liquiwax DC-EFA/SS
Dicetyl dilinoleate
Emollient to soften the skin without greasiness and oiliness. *Brooks Industries.*

17487 Liquiwax DIADD
Diisoarachidyl dodecanedioate
Emollient to soften the skin without greasiness and oiliness. *Brooks Industries.*

17488 Liquiwax DICDD
Diisocetyl dodecanedioate
Emollient to soften the skin without greasiness and oiliness. *Brooks Industries.*

17489 Liquiwax DIEFA
Diisoarachidyl dilinoleate
Emollient to soften the skin without greasiness and oiliness. *Brooks Industries.*

17490 liquorice
58-55-9 9421 200-385-7
The dried root of *Glycyrrhiza glandulifera.*

17491 Lisat
Stearyl lactyl-2-lactylates
Thomas Goldschmidt Ltd.

17492 Liskonum Tablets
554-13-2 5552 209-062-5
Lithium carbonate
Controlled release tablet for treatment of acute episodes of mania or hypomania and for the prophylaxis of recurrent manic-depressive illness. *SmithKline Beecham.* Name unverified.

17493 Lissamine
Acid and direct dyestuffs. *ICI Chem & Polymers Ltd.*

17494 Lissanol
Leather finishes. *ICI Chem & Polymers Ltd.*

17495 Lissapol
Synthetic detergents. *ICI Chem & Polymers Ltd.*

17496 Listab
Metal stearates. *Chemson Ltd.*

17497 Litalbin
Lecithin albuminate

17498 Litefax
Low d insulating sideliner tiles for small to medium steel ingots. *Foseco (F.S.) Ltd.* Discontinued.

17499 Litex
A range of styrene-butadiene copolymers, emulsions and dispersions; used for paints and anticorrosion coatings. *Hüls AG.*

17500 Lithane
554-13-2 5552 209-062-5
A proprietary preparation of lithium carbonate; an antidepressant. *Pfizer International.*

17501 Litharge
1317-36-8 5433 215-267-0
Massicot
Pigments consisting of lead monoxide, PbO. Litharge is obtained in silver refining, and has a more reddish color than Massicot, which is made by roasting lead.

17502 Litharge 33
High-purity lead oxide; acid acceptor, activator, and vulcanizing agent in rubber compounding. *Eagle-Picher.* Name unverified.

17503 Lithargrite
A mixture of oxide of lead and calcined magnesia; used as a rubber filler.

17504 Lithene®
Liquid polybutadiene telemers and compounds; components in chlorinated paint, electrical encapsulants, specialty adhesives, rubber coagent, surface coatings. *Reichhold.* Discontinued.

17505 Lithex
Lead dithiobenzoate precipitated on an inert base such as clay; a proprietary rubber vulcanization accelerator. No manufacturer.

17506 Lithic acid
69-93-2 10014 200-720-7
$C_5H_4N_4O_3$
Uric acid
Used in organic synthesis.

17507 Lithiopiperazine
A preparation of piperazine and lithium salts; used as a solvent for uric acid.

17508 lithium
7439-93-2 5543 231-102-5
Li
Metallic element. Used as a scavenger and degasifier for stainless and mild steels in molten state; deoxidizer in copper and alloys; rocket propellants; pharmaceuticals. mp = 180.54°; bp = 1347°; reacts violently with inorganic

acids; soluble in liquid ammonia; does not react with oxygen at room temperature. *Atomergic Chemetals; Cerac; FMC; Leverton-Clarke Ltd.*

17509 lithium hypochlorite
13840-33-0 237-558-1
LiOCl
HyPure L. Laundry bleach, swimming pool chlorination. *FMC;Olin.* Discontinued.

17510 lithium molybdate
13568-40-6 236-977-7
Li_2MoO_4
AAA Molybdenum Prods.; Atomergic Chemetals; Cerac. Steel coating, petroleum cracking catalyst. mp = 705°.

17511 lithium nitrate
7790-69-4 5562 232-218-9
$LiNO_3$
Ceramics, pyrotechnics, salt baths, heat exchange media, refrigeration systems, rocket propellant. mp = 251°; d = 2.3800. *Atomergic Chemetals; Cerac; Mallinckrodt.*

17512 lithium stearate
4485-12-5 224-772-5
$C_{18}H_{35}LiO_2$
lithium octadecanoate
octadecanoic acid; lithium salt. Lithium salt of stearic acid; cosmetics, plastics, waxes, greases, lubricant in powd. metallurgy, corrosive inhibitor, flatting agent, high-temperature lubricant. *Chemetall GmbH; Schweizerhall; Syn. Prods.; Witco.*

17513 Litho-carbon
A material resembling asphalt, found in Texas.

17514 lithoclastite
An explosive. It is a dynamite.

17515 Lithoform
A pretreatment for zinc. *ICI Chem & Polymers Ltd.* Discontinued.

17516 Lithol
Ammonium ichtho-sulfonate

17517 Lithol® Pigments
Azo dye lakes for letterpress, offset, flexographic and gravure inks, for paints, for coloring plastics. *BASF AG.*

17518 Lithomarge
Lithocolla
leucoargilla. A compact clay found in rock fissures.

17519 Litho-oil
Stand oil, polymerized oil. Raw linseed oil heated in such a manner that practically no oxidation occurs, thus thickening the oil through polymerization alone.

17520 Lithopone
1345-05-7 5572 215-715-5
$ZnS \cdot BaSO_4$
Zinc sulfide/barium sulfate
C.I. Pigment White 5; Griffith's zinc white; Beckton white; Duresco; enamel white; Fulton white; Jersey lily white; Knights patent zinc white; Marbon white; Nevin; Orrs white; oleum white; pinolith; porcelain white; Rosss white; zinc baryta white; zincolith. White pigment for paints and coatings. Used in water and oil paints to provide thixotropy, improve gloss and flow; a mixture of zinc sulfide, barium sulfate, and some zinc oxide. *Ore & Chem. Corp.; Sachtleben GmbH.*

17521 Lithostar
Reprographic chemical. *Octel Chemicals Ltd.* Discontinued.

17522 Lithostat
546-88-3 208-913-8
Acetohydroxamic acid
Enzyme inhibitor. *Mission Pharmacal Co.* Name unverified.

17523 Litmus
1393-92-6 5574 215-739-6
(Lichen blue, lacmus, tournesol, tumsole). The coloring matter derived from different species of lichens; used as an indicator used in analytical chemistry where precision is not required, soil testing.

17524 litnum bronze
An alloy of from 80-85% copper, 10-15% aluminum, and 4% iron.

17525 Littorl
A decolorizing agent for sugar juices.

17526 Liver of antimony
The name applied to the impure double sulfides of antimony, obtained by heating antimony sulfide, S_2S_3, with various metallic sulfides, more especially with the alkali and alkaline earth sulfides.

17527 Liver sugar
9005-79-2 4506 232-683-8
Glycogen
Animal starch; liver starch. Reserve carbohydrate of the animal organism. $[\alpha]_D^{25}$= 196°; soluble in H_2O, insoluble in EtOH. .

17528 Liveroid
A proprietary liver extract. No manufacturer.

17529 Lizetan Spray
Insecticidal spray; used for control of biting and sucking pests. *Bayer AG.*

17530 Lloyd's reagent
A hydrous aluminum silicate prepared from Fuller's earth. It absorbs alkaloids.

17531 LMT
Meat tenderizer. *Atomergic Chemetals Corp.*

17532 LO/MIT™-1
Silver colored. nonthickness dependent, low emissivity, radiant barrier coating for energy conservation and light and heat reflection; high temperature tolerant coating. *Solec.*

17533 Loalin
A proprietary trade name for a polystyrene molding compound. No manufacturer.

17534 Loams
Natural mixtures of clay and sand; used for the manufacture of bricks and tiles. No manufacturer.

17535 Lobak
A proprietary preparation of chlormezanone and paracetamol. No manufacturer.

17536 Lobosol
Oxygenated solvents. *BP Chemicals Ltd.* Discontinued.

17537 Lobra
Partially hydrogenated canola oil. *Karishamns.*

17538 Lockite
Polyolefin
Continuous extruded strip high slip fendering for docks and similar applications. *Stanley Smith & Co Plastics Ltd.*

17539 Locobase Cream and Ointment
A proprietary base for creams and ointments. *Brocades Pharma.*

17540 Locoid
13609-67-1 4828 237-093-4
A proprietary preparation of hydrocortisone 17-butyrate; used as a topical corticosteroid. *Brocades Pharma.*

17541 Locorten
2002-29-1 217-901-1
A proprietary preparation of flumethasone pivalate for dermatological use. *Ciba plc.* Name unverified.

17542 Locoum
A gum-like mass prepared from starch paste and sugar.

17543 Locron
1327-41-9 356 215-477-2
Aluminum chlorohydrate
Anhidrotic. *Hoechst-Celanese.*

17544 Loctite
Single component structural adhesives and thread locking materials possessing the unique property of setting when air is excluded; some are based on oxygenated methacrylic molecules of patented formulation. No manufacturer.

17545 Locust bean gum
9000-40-2 232-541-5
Wakal® J Range. (Carob flour; carob seed gum; algaroba) Polysaccharide plant mucilage; ground seed of the ripe fruit of St. John's Bread (Ceratonia siliqua); in foods as stabilizer, thickener, emulsifier; cosmetics; sizing and finishes for textiles; pharmaceuticals; pain *Grindsted; Hercules; Grünau.*

17546 Lodine
41340-25-4 3920
Etodolac capsules or tablets; for acute or long-term use in rheumatoid arthritis and osteoarthritis. *Wyeth Laboratories.*

17547 Lodol
Dephosphorizing agent for ladle treatment of rimming steels. *Foseco (F.S.) Ltd.* Discontinued.

17548 Lo-Dose
61791-44-4 263-177-5
Cationic surfactant containing 800 g/l tallow amine ethoxylate; wetting agent for phosphonoglycine herbicide sprays. *Quadrangle Agro-chemicals.*

17549 Lodosin
28860-95-9 1843 249-271-9
Carbidopa
A decarboxylase inhibitor for concurrent use with levodopa in the treatment of Parkinson's disease. *Merck & Co Inc.* Discontinued.

17550 Lo-ex
An aluminum alloy containing 14% silicon, 2% nickel, 1% copper, and 1% magnesium.

17551 Lofenalac
Proprietary name for an infant milk feed low in phenylalanine; used in the dietary treatment of phenylketonuria. *Bristol-Myers Squibb Co Inc.*

17552 Lofepramin Hydrochloride
23047-25-8 5587 245-396-8
$C_{26}H_{27}ClN_2O.HCl$
1-(4-Chlorophenyl)-2-[[3-(10,11-dihydro-5H-dibenz[b,f]azepin-5-yl)propyl]methylamino]ethanone
Gamonil®; Leo 640; Amplit; Gamanil; Timelit; Tymelyt; N-methyl-N(4-chlorobenzoylmethyl)-3-(10,11-dihydro-5H-dibenzo[*b,f*]azepin-5-yl)propylamine hydrochloride. mp=152-154°; soluble in methanol, ethanol, chloroform; practically insoluble in water; LD_{50} (rat orl) > 2500 mg/kg. *E Merck.*

17553 Iofetamine ^{123}I hydrochloride
85068-76-4 5066
$C_{12}H_{18}{}^{123}IN$
(±)-4-(iodo-^{123}I)-α-methyl-N-(1-methylethyl) benzene ethanamine hydrochloride
Perfusamine; Spectamine. Lipid-soluble radioactive brain imaging agent. mp = 156-158°.

17554 Loffler's methylene blue
A stain for bacteria. It consists of 100 parts of a solution of sodium hydroxide (1 in 10,000) and 30 parts of a saturated alcoholic solution of methylene blue.

17555 Logas
For degassing copper base alloys. *Foseco (F.S.) Ltd.*

17556 LoGel
637-12-7 379 211-279-5
Aluminum stearate; for nonaqueous coating compositions. *Hüls Am.* Discontinued.

17557 logwood
Campeachy wood, Jamaica wood, Hematine paste and powder, steam black. A natural dyestuff from the wood of *Haematoxylon campechianum*. It is sold as chips or extract. The wood contains haematoxylin, $C_{16}H_{14}O_6{\cdot}3H_2O$; used for dying black with a chrome mordant, for wool; with an iron mordant for silk; and with a chrome or iron or aluminum mordant for cotton. .

17558 Lo-Hetaplas®
For the medical industry. *DuPont UK.*

17559 Lohys steel
A mild steel having a high magnetic permeability.

17560 Lokset
Resin anchor systems. *Foseco (F.S.) Ltd.*

17561 Loloss®
Natural polymeric viscosifier stabilized to prevent thermal degradation; fluid loss control agent/flocculant for drilled solids; encapsulates cuttings, stabilizes the borehole. *Rhône-Poulenc/Water Soluble Polymers.*

17562 Lolotint 97
1317-39-1 2734 215-270-7
Cuprous oxide
CP International Chemicals Ltd.

17563 Lomag
Magnesium removing flux *Foseco (F.S.) Ltd.* Discontinued.

17564 loman steel
An abbreviation of low-manganese steel. It contains from 7-10% manganese.

17565 Lomar® D
9084-06-4
Sodium naphthalene sulfonate
Dispersant for disperse and vat dyes. *Henkel/Textiles.*

17566 Lomar® HP
Condensed potassium naphthalene sulfonate
Dispersant; secondary emulsifier for emulsion polymerization. *Henkel/Functional Prods.*

17567 Lomar® LS
9084-06-4
Condensed sodium naphthalene sulfonate
Dispersant, emulsifier for emulsion polymerization, dyestuff manufacture, agricultural formulations; leveling agent for dyeing fibers; low salt. *Henkel/Functional Prods.; Henkel/Textile.*

17568 Lomar® PWA
Condensed ammonium naphthalene sulfonate
viscosity depressant; for molding and extruding operations in ceramics; dispersant for emulsion paints, agricultural formulations; emulsifier for emulsion polymerization of synthetic elastomers; viscosity reducer for pigment slurries; stabilizer. *Henkel/Emery; Henkel/Functional Prods.*

17569 Lomod® AE2020a, AE2040a, AE2060a
Engineering elastomer; automotive exterior grade *GE Plastics.*

17570 Lomod® FR30125a
Engineering elastomer; flame retardant grade offering outstanding electrical props., chem. resistance, toughness; for mating seals, complex parts. *GE Plastics.*

17571 Lomod® FR5020a
Engineering elastomer; flame retardant grade offering outstanding electrical props., chem. resistance, toughness; for mating seals, complex parts. *GE Plastics.*

17572 Lomod® HG3015a
Engineering elastomer; high gravity grade providing structural integrity, sound-damping capabilities and the ability to be molded into complex parts. *GE Plastics.*

17573 Lomod® ST3090a
Engineering elastomer; provides soft, tactile feel and flexibility for sporting goods, appliances, power tools, and personal care articles; adheres to other resins permitting multi-stiffness or multi-color parts. *GE Plastics.*

17574 Lomod® TE3040a
Thermoplastic elastomer; offers flexibility, toughness, resistance to chems., ozone, hydrocarbons and temperature extremes; for hoses, tubing, boots, bellows, diaphragms, seals, safety equip., sporting goods, athletic footwear components. *GE Plastics.*

17575 Lonacol
Copper-free fungicide; used for tomatoes, beans, potatoes, maize, onions and fruits. *Bayer AG.*

17576 Lonarit
An acetyl cellulose product; can be molded and colored. No manufacturer.

17577 Londal
A proprietary trade name for certain aluminum alloys. No manufacturer.

17578 Londax®
For the agriculture industry. *DuPont UK.*

17579 London paste
A paste made by adding a third of the weight of water to a mixture of equal parts of sodium hydroxide and powdered lime. A caustic.

17580 Long oil varnishes
A classification of varnishes. They contain about 1 part of the solid constituents to 1 1/2 parts of drying oil. Short oil varnishes contain 1 part of solid to 1/2 part oil.

17581 Longifene
569-65-3 209-323-3
Buclizine hydrochloride.

17582 Longlife Plus
Mixture of 2,4-D and dicamba; used to control weeds in turf. *ICI Agrochemicals Professional Products.*

17583 Longlife Turf Foods
Fertilizers containing long-lasting nitrogen for sports grounds and parks. *Scottish Agricultural Industries plc.*

17584 Loniten
38304-91-5 6290 253-874-2
Minoxidil
Antihypertensive. *Upjohn.*

17585 Lonsicar
409-21-2 8636 206-991-8
Silicon carbide
Used for antislip and wear resistant concrete or resin surfaces. *Lonza AG.*

17586 Lontrel
1702-17-6 2462 216-935-4
Herbicides based primarily on clopyralid. *Dow UK.* Discontinued.

17587 Lontrel Plus
Soluble concentrate containing 15 g clopyralid, 420 g dichlorprop and 175 g MCPA per liter; a translocated herbicide for use on cereals. (Sold in UK for DowElanco) *ICI Chem & Polymers Ltd.*

17588 Lonza Insta-Pearl®
Proprietary blend; high intensity pearlescent, opacifier for hair and skin care products. *Lonza Inc.*

17589 Lonza KS
7782-42-5 4560 231-955-3
Graphite
Relatively round particle shape, high electrical and thermal conductivity, good compressibility; for brake linings, plastics, carbon brushes, batteries, electrochemistry, pencils, hard metals, lubricants, ceramics, catalysts. *Lonza G+T.*

17590 Lonzaine®
Betaine amphoterics
Used in shampoos and industrial cleaners. *Lonza AG.*

17591 Lonzaine® 12C
68424-94-2 270-329-4
Coco betaine
Conditioner, foaming agent, viscosity modifier, irritation mitigant used in personal care products and industrial applications; biodegradable *Lonza Inc.*

17592 Lonzaine® C
Cocamidopropyl betaine
Foaming agent, conditioner, viscosity booster, wetting agent, used in cosmetics, toiletries, detergents, metal finishing, textile finishing, etc.; biodegradable *Lonza Inc.*

17593 Lonzaine® CS
68139-30-0 268-761-3
Cocoamidopropylhydroxysultaine
Conditioner used in personal care products and industrial applications. *Lonza Inc.*

17594 Lonzest®
A range of sorbitan esters, polyoxyethylene sorbitan esters and polyethylene glycol esters; used for emulsion formation and stabilization in foods, pharmaceuticals and cosmetics. *Lonza AG.*

17595 Lonzest® 143-S
6221-95-0 226-300-9
Myristyl propionate
Emollient, penetrant, and spreading agent in cosmetics; perfume solvent in personal care products; humectant. *Lonza Inc.*

17596 Lonzold
A proprietary cellulose acetate. No manufacturer.

17597 Loobwax 0651
9002-88-4 7728
Polyethylene wax
Oxidized lubricant for high-shear applications, pipes, profiles, clear films. *Astor Wax.* Name unverified.

17598 Loobwax 0761
Esterified
Oxidized polylethylene wax. Lubricant. *Astor Wax.* Name unverified.

17599 Lopid
25812-30-0 4394 247-280-2
Gemfibrozil
Antihyperlipoproteinemic. *Parke-Davis.* Name unverified.

17600 Lopox®
A registered trademark for epoxy resins. No manufacturer.

17601 Lopresoretic
37350-58-6 6235 253-483-7
Beta-blocker. *Ciba plc.* Name unverified.

17602 Lopressor
56392-17-7 6235 260-148-9
Metoprolol tartrate
Anti-adrenergic. *Ciba-Geigy Corp.* Name unverified.

17603 Lopressor SR
56392-17-7 6235 260-148-9
Beta-blocker. *Ciba plc.* Name unverified.

17604 Loprox
41621-49-2 2325 255-464-9
Ciclopiroxolamine ethanolamine salt (1:1)
Antifungal. *Hoechst-Roussel Pharmaceuticals Inc.* Name unverified.

17605 Lorate®
For the agriculture industry. *DuPont UK.*

17606 Loretine
547-91-1 4872 208-938-4
$C_9H_4NI(OH)SO_3H$
Sulfiolinic acid
Iodohydroxyquinolinesulfonic acid; 8-hydroxy-7-iodoquinoline-5-carboxylic acid. Used as a germicide.

17607 Lorexane
58-89-9 5526 200-401-2
Preparations of γ benzene hexachloride (lindane); an insecticide; antiparasitic hair lotion. *ICI Chem & Polymers Ltd.*

17608 Lorival
A proprietary electrical insulation made from a synthetic resin of the phenol-formaldehyde type. No manufacturer.

17609 Lorol
A mixture of alcohols produced by the reduction of coconut oil; sulfated fatty alcohols. *Hickson & Welch Ltd.*

17610 Lorol C10
112-30-1 2911 203-956-9
Decyl alcohol
Intermediate for surfactant manufacture; component in rolling oils, lubricants. *Henkel/Emery.*

17611 Lorol C12
112-53-8 3464 203-982-0
Lauryl alcohol
Intermediate for surfactant manufacture; component in rolling oils, lubricants. *Henkel/Emery.*

17612 Lorol C12-C14
112-53-8 3464 203-982-0
Lauryl alcohol (65-69% C12,24-28% C14, 4-8% C16)
Intermediate for surfactant manufacture; component in rolling oils, lubricants. *Henkel/Emery.*

17613 Lorol C14
112-72-1 6418 204-000-3
Myristyl alcohol
Intermediate for surfactant manufacture; component in rolling oils, lubricants. *Henkel/Emery.*

17614 Lorol C16
36653-82-4 2070 253-149-0
Cetyl alcohol
Intermediate for surfactant manufacture; component in rolling oils, lubricants. *Henkel/Emery.*

17615 Lorol C18
112-92-5 8960 204-017-6
Stearyl alcohol
Intermediate for surfactant manufacture; component in rolling oils, lubricants. *Henkel/Emery.*

17616 Lorol C8
111-87-5 6849 203-917-6
Octyl alcohol
Intermediate for surfactant manufacture; in rolling oils and other lubricants. *Henkel/Emery.*

17617 Lorol C8-C10 Special
112-53-8 3464 203-982-0
Lauryl alcohol(40-48% c8, 51-59% C10)
Chemical intermediate for detergent manufacturing. *Henkel/Emery.*

17618 Lorol DA
Diethanolamine lauryl sulfate in liquid form; anionic surfactant used in shampoos, milder in action than other alkyl sulfates in the Lorol series. No manufacturer.

17619 Lorol MA and MR
Monoethanolamine lauryl sulfate in liquid form; Lorol MR is a built product; anionic surfactant for shampoos. No manufacturer.

17620 Lorol NH
2235-54-3 218-793-9
Ammonium lauryl sulfate in liquid form; anionic surfactant having a mild action; used in good quality shampoos. No manufacturer.

17621 Lorol Special
112-53-8 3464 203-982-0
Lauryl alcohol (70-75% C12, 24-30% C14)
Chemical intermediate for detergent manufacturing. *Henkel/Emery.*

17622 Lorol TA and TAR
139-96-8 205-388-7
Triethanolamine lauryl sulfate in liquid form; Lorol TAR is a built product containing alkylolamide foam booster; anionic surfactant used in shampoos. No manufacturer.

17623 Lorsban
2921-88-2 2242 220-864-4
Insecticides containing chlorpyrifos; used to control ticks on cattle, mosquitoes, and other insects. *Dow UK.* Discontinued.

17624 Losilphos
Ferophosphorus briquettes
Used in steelmaking for their phosphoros content. *Monsanto Co.* Name unverified.

17625 Losophane
$C_7H_7OI_3$
Triiodo-m-cresol
Used externally in skin diseases, as an antiseptic and astringent.

17626 LoSOx
High activity catalyst for the abatement of carbon monoxide and unburned hydrocarbons in the presence of high levels of sulfur compounds; most suitable for operations where the oxidation of sulfur dioxide must be limited. *Engelhard.*

17627 Lotader® 3210
Acrylic terpolymer; for aluminum coating, coextrusion applications. *Elf Atochem SA.*

17628 Lotader® 3700
Acrylic terpolymer; for modification of thermoplastic polymers. *Elf Atochem SA.*

17629 Lotader® AX 8660
Ethylene-acrylic ester-glycidyl methacrylate terpolymer. *Elf Atochem N. Am.*

17630 Lotader® P3-3200
Acrylic terpolymer; for heat adhesive film and powd., pipe coating applications. *Elf Atochem SA.*

17631 Lotrimin
23593-75-1 2478 245-764-8
Clotrimazole
Antifungal. *Schering Corp.* Discontinued.

17632 Lotryl® 17 BG 04
Ethylene-butyl acrylate copolymer; for films (gloves), coextrusion. *Elf Atochem SA.*

17633 Lotryl® 20 MA 08
Ethylene-acrylic ester (EMA) copolymer; for films, coextrusion, coating, hot-melt adhesives, compounds. *Elf Atochem SA.*

17634 Lotryl® 28 BA 175
Ethylene-butyl acrylate copolymer; for hot-melt adhesives. *Atochem SA.*

17635 Lotusate
115-44-6 9206 204-090-4
Talbutal; sedative. *Sterling Drug Inc.* Name unverified.

17636 Louse Powder
Insecticide. *Schering Agrochemicals Ltd.* Discontinued.

17637 Lo-Vel
Amorphous silica flatting agents. *PPG Industries.*

17638 lovol
A mixture of aliphatic alcohols formed by the high pressure hydrogenation of coconut oil.

17639 Low Crock 18-G-3
Modified organic resin emulsion; excellent binder properties for aqueous pigment printing systems. *Eastern Color & Chem.*

17640 Low Crock FB-I
Finishing auxiliary for reduced bleeding/crocking of dyed and printed fabrics; minimal effect on fabric hand, minimum buildup on equip.; resin and catalyst stable. No manufacturer.

17641 Lowerite
Strippable coating compositions. *Croda Chem. Ltd.* Discontinued.

17642 Lowilite® 22
1843-05-6 6838 217-421-2
Benzophenone-12
Uv absorber for PP, PE, PVC, and other polymers; excellent heat resistance at extrusion temps. *Lowi.*

17643 Lowilite® 55
2440-22-4 3503 219-470-5
2-(2'-Hydroxy-5'-methylphenyl) benzotriazole
Uv absorber for polymers including polyester, PVC, PS, HIPS, rubber, PC, PMMA. *Lowi.*

17644 Lowilite® 62
α-Methylstyrene/N-(2,2,6,6-tetramethylpiperidinyl-4)maleimide/N-stearyl maleimide terpolymer
HALS protecting LDPE, HPDE, LLDPE, and PP against degradation by oxidation and photo-oxidation; very high thermal resistance to 320°. *Lowi.*

17645 Lowilite® 77
52829-07-9 258-207-9
$C_{36}H_{60}N_2O_6$
Bis (2,2,6,6-tetramethyl-4-piperidinyl) sebacate
Decanedioic acid, bis(2,2,6,6-tetramethyl-4-piperidinyl) ester; Bis (2,2,6,6-tetramethyl-4-piperidinyl) decanedioate; Uvaseb 770;Tinuvin® 770; BLS™ 1770. Light stabilizer (HALS) for polyolefins, PS, HIPS, ABS, SAN, ASA; also suitable for PU, polyamides, acetal. *Lowi; Ciba-Geigy/Additives; Enichem Synthesis SpA.*

17646 Lowinox® 001
128-39-2 204-884-0
2,6-di-*t*-butylphenol
Antioxidant for oils. *Lowi.*

17647 Lowinox® 002
4,4-Methylenebis(2,6-di-*t*-butylphenol)
Antioxidant for oils. *Lowi.*

17648 Lowinox® 44B25
4,4'-Butylidene-bis(2-*t*-butyl-5-methylphenol
Antioxidant for rubber, latex, adhesives, ABS, polyamide. *Lowi.*

17649 Lowinox® 44S36
96-69-5 202-525-2
4,4'-Thiobis (2-*t*-butyl-5-methylphenol)
Antioxidant for latex, adhesives, plastics, cross-linked polymers. mp = 161-164°. *Lowi.*

17650 Lowinox® 070
98-54-4 1620 202-679-0
t-Butylphenol
Antioxidant for oils and fuels. *Lowi.*

17651 Lowinox® 22IB46
2,2'-Isobutylidene-bis(4,6-dimethylphenol)
Antioxidant for rubber, latex. *Lowi.*

17652 Lowinox® 22M46
119-47-1 204-327-1
2,2'-Methylenebis (4-methyl-6-*t*-butylphenol. Antioxidant for rubber, latex, adhesives, ABS, polyacetate. *Lowi.*

17653 Lowinox® 22M46
119-47-1 204-327-1
Antioxidant for rubber, latex, adhesives, ABS, polyacetate. *Lowi.*

17654 Lowinox® 241
2,2'-Methylenebis-(4-methyl-6-t-butylphenol)
tris (2,4-di-t-butylphenyl)phosphite. Antioxidant for adhesives. *Lowi.*

17655 Lowinox® 242
31570-04-4 250-709-6
Sterically hindered polynuclear phenol and tris-(2,4-di-t-butylphenyl) phosphite; antioxidant for hot-melts. *Lowi.*

17656 Lowinox® 243
Pentaerythrityl tetrakis-3(3',5'-di-*t*-butyl-4'-hydroxyphenyl)propionate
bis (2,4-di-*t*-butylphenyl)pentaerythritol diphosphite. Antioxidant for adhesives, hot-melts, plastic, polyolefins, PS, PC, SAN. *Lowi.*

17657 Lowinox® 244
4,4·g-Butylidene-bis(2-*t*-butyl-5-methylphenol) and tris-(2,4-di-*t*-butylphenyl) phosphite; antioxidant for plastics, polyolefins, PS. *Lowi.*

17658 Lowinox® 245
4,4'-Butylidene-bis(2-*t*-butyl-5-methylphenol and distearyl-3,3'-thio dipropionate; antioxidant for plastics, polyolefins. *Lowi.*

17659 Lowinox® 246
2,2'-Methylenebis(4-methyl-6-*t*-butylphenol) and distearyl-3,3'-thio dipropionate; antioxidant for APP. *Lowi.*

17660 Lowinox® 247
2,6-Di-*t*-butyl-4-methylphenol and ditridecyl thiodipropionate; Antioxidant for polyurethane. *Lowi.*

17661 Lowinox® 624
2,4-Dimethyl-6-*t*-butylphenol
Antioxidant for fuels. *Lowi.*

17662 Lowinox® ACP
Polymeric 2,2,4-Trimethyl-1,2-dihydroquinoline
Antioxidant for rubber, latex, crosslinked polymers. *Lowi.*

17663 Lowinox® AH25
79-74-3 3359 201-222-2
2,5-Di-*t*-amylhydroquinone
Antioxidant for adhesives. *Lowi.*

17664 Lowinox® BHT
128-37-0 1583 204-881-4
BHT
Antioxidant for rubber, adhesives, plastics, polyolefins, polystyrene. *Lowi.*

17665 Lowinox® CA 22
1843-03-4 217-420-7
1,1,3-Tris-(2-methyl-4-hydroxy-5-*t*-butylphenyl) butane
Phenolic antioxidant. *Lowi.*

17666 Lowinox® DLTDP
123-28-4 204-614-1
Dilauryl-3-3'-thiodipropionate
Antioxidant for plastics, polyolefins. *Lowi.*

17667 Lowinox® DSTDP
693-36-7 211-750-5
distearyl-3,3'-thiodipropionate
Antioxidant for plastics, polyolefins. *Lowi.*

17668 Lowinox® MBMC
2-t-Butyl-*m*-cresol
Intermediate for antioxidants, synthetic musk. *Lowi.*

17669 Lowinox® MBPC
2409-55-4 219-314-6
2-t-Butyl-*p*-cresol
Intermediate for antioxidants, uv absorbers. mp = 51-52°; bp = 244°; fp = 100°. *Lowi.*

17670 Lowinox® ODA
Octylated diphenylamine
Antioxidant for rubber. *Lowi.*

17671 Lowinox® PO35
2082-79-3 218-216-0
Octadecyl-3(3',5'-di-*t*-butyl-4'-hydroxyphenyl)propionate
Antioxidant for adhesives, hot-melts, plastics, polyolefins, polystyrene, PVC, PC, SAN. *Lowi.*

17672 Lowinox® PP35
6683-19-8 229-722-6
Pentaerythrityl tetrakis-3-(3',5'-di-*t*-butyl-4'-hydroxyphenyl)propionate
Antioxidant for adhesives, hot-melts, plastics, polyolefins, PS, PC, SAN. *Lowi.*

17673 Lowinox® PTBT
98-51-1 202-675-9
p-*t*-Butyl toluene
Antioxidant. mp = -52°; bp = 189-192°; d = 0.8530; n_D^{20} = 1.4922; fp = 54° *Lowi.* Name unverified.

17674 Lowinox® SDA
Styrenated diphenylamine
Antioxidant. *Lowi. Name unverified.*

17675 Lowinox® TBMX
98-19-1 202-647-6
t-Butyl-*m*-xylene
Intermediate for perfumes and fragrances. *Lowi.* Name unverified.

17676 Lowinox® TNPP
26523-78-4 247-759-6
Tris-nonylphenyl phosphite
Antioxidant for rubber, adhesives, ABS. *Lowi.*

17677 Lowitite® 26
3896-11-5 223-445-4
2-(2'-Hydroxy-3'-t-butyl-5'-methylphenyl)-5-chlorobenzotriazole
Uv stabilizer for polyolefins, polyester resins and coatings. *Lowi.*

17678 Lowroff phosphor bronze
Alloys of 70-90% copper, 4-13% tin, 5-16% lead, and 0.5-1% phosphorus.

17679 Lox
Liquid oxygen; used in mining explosives and in rocket propellants.

17680 Loxa bark
Pale cinchona bark.

17681 Loxiol G 52
C_{16}-C_{18} fatty alcohol; surfactant for polymerization. *Henkel/Functional Prods.*

17682 Loxiol G-70, G-71, G-72, and G-73
Proprietary names for a range of multifunctional polyesters of high molecular weight; used as additives to PVC compounds to reduce surface tackiness. *Henkel Chemicals Ltd.* Name unverified.

17683 Loxiol P 1420
C_{16}-C_{18} fatty alcohol; surfactant for polymerization. *Henkel/Functional Prods.*

17684 Loxiol VPG 1354
112-92-5 8960 204-017-6
Stearyl alcohol; surfactant for polymerization. *Henkel/Functional Prods.*

17685 Loxiol VPG 1451
661-19-8 8135 211-546-6
Behenyl alcohol; surfactant for polymerization. *Henkel/Functional Prods.*

17686 Loxiol VPG 1743
36653-82-4 2070 253-149-0
Cetyl alcohol; surfactant for polymerization. *Henkel/Functional Prods.*

17687 Loxon
321-55-1 4636 206-289-1
A proprietary preparation of haloxon; a veterinary anthelmintic. No manufacturer.

17688 Lozol
26807-65-8 4969 248-012-7
Indapamide
Antihypertensive; diuretic. *USV Pharmaceutical Corp.* Unverified.

17689 LP® -12
Polysulfide, mercaptan-terminated; used for building, aviation, automotive, insulating glass, and marine sealants; excellent resistance to oils, organic solvents, weathering, low and high temperature, gas permeability, oxidation, ozone, moisture. *Morton Int'l./Polymer Systems.*

17690 LP® -977
Polysulfide polymer; for sealants (aircraft, automotive, construction, marine, insulating glass), fluid membranes, electrical potting, leather impregnation. *Morton Int'l./Polymer Systems.*

17691 LP-100
1309-60-0 5424 215-174-5
O_2Pb
Lead dioxide
Catalyst/curing agent for polysulfide, low molecular weight butyl and polyisoprene rubber; oxidizer in manufacture of dyes and to control burning rate of incendiary fuses or pyrotechnics. *Eagle-Picher.* Name unverified.

17692 LPS 1 Greaseless Lubricant
A blend of solvents, lubricants, moisture displacers and corrosion inhibitors to provide greaseless lubrication of delicate mechanisms to dry out sensitive electrical and electronic circuits; lubricant, penetrant, water displacer, cleans electrical equipm *Holt Lloyd Corporation.* Name unverified.

17693 LPS 2 General Purpose Lubricant
A blend of lubricants and corrosion inhibitors that provide general purpose lubrication, corrosion inhibition and penetration of seized mechanisms; lubricant, penetrant and quick release, corrosion inhibitor. *Holt Lloyd Corporation.* Name unverified.

17694 LPS 3 Heavy Duty Rust Inhibitor
A blend of waxes, solvents, corrosion inhibitors and oils to provide long term protection of metal parts; corrosion inhibitors, chain lubricant and antiseize compound. *Holt Lloyd Corporation.* Name unverified.

17695 LPS 500 Plus
A blend of waxes, oils and corrosion inhibitors used to lubricate and prevent corrosion in severe conditions; as a lubricant, corrosion inhibitor or antiseize compound. *Holt Lloyd Corporation.* Name unverified.

17696 LPS Brake Cleaner
A solvent-based cleaner for brakes and brake components. *Holt Lloyd Corporation.* Name unverified.

17697 LPS Electro Contact Cleaner
A premium blend of solvents which is safe to use on rubber paint, plastics and to clean and degrease sensitive electronic equipment; cleans electrical/electronic components and delicate mechanisms. *Holt Lloyd Corporation.* Name unverified.

17698 LPS Engine Degreaser
A petroleum-base cleaner which will emulsify with water; engine degreaser. *Holt Lloyd Corporation.* Name unverified.

17699 LPS Heavy-Duty Silicone Lubricant
A water-based dry film silicone lubricant; lubricant, safe for use on almost any surface, approved by USDA for use in federally inspected meat and poultry plants, approved for use in areas with incidental food contact. *Holt Lloyd Corporation.* Name unverified.

17700 LPS Instant Cold Galvanize
A combination of solvents, resin binders, and 95% pure zinc metal which provides a galvanic coating similar to hot dipped galvanize; cold galvanizing compound that inhibits corrosion for up to three years, acts as a primer or finish coat and repairs damag *Holt Lloyd Corporation.* Name unverified.

17701 LPS Instant Super Cleaner/Degreaser
A combination of chlorinated solvents to clean and degrease metal parts and electrical equipment; cleaner/degreaser for equipment. *Holt Lloyd Corporation.* Name unverified.

17702 LPS Paint Remover
A blend of solvents, waxes, and stripping aids for removing all types of paint from metal surfaces; for removal of paint and varnishes. *Holt Lloyd Corporation.* Name unverified.

17703 LPS Tap-All
An engineered blend of lubricant aids and corrosion inhibitors that can be used for the machining of all metals; does not contain sulfur or chlorine; as a lubricant when tapping, drilling, grinding, milling, threading, reaming, turning, boring, sawing or *Holt Lloyd Corporation.* Name unverified.

17704 LSD
50-37-3 5665 200-033-2
$C_{20}H_{25}N_3O$
Lysergic acid diethylamide
Hallucinogen.

17705 L-serine
56-45-1 8604 200-274-3
$C_3H_7NO_3$
α-amino-β-hydroxy-propionic acid
L-serine; (S)-(-)-serine; Ser; S; L-2-amino-3-hydroxypropionic acid; L-(-)-serine; 2-amino-3-hydroxypropionic acid; 3-hydroxyalanine. Naturally occurring enantiomer of serine. Used in biochemical research, as a dietary supplement, in culture media, microbiological tests, feed additive. mp = 228° (dec); $[\alpha]_D^{20}$ = -7° (c = 10.41 H_2O); soluble in H_2O, insoluble in organic solvents. *Degussa; Janssen Chimica; Mitsui Toatsu Chem.; Nippon Rikagakuyakuhin; U.S. Biochemical.*

17706 Luaktin®
Antiskinning agents for oil-based paints and baking finishes; improved gloss and leveling. *BASF AG.*

17707 Lubafax
A proprietary formulation of propylene glycol and hydroxypropyl methylcellulose; a surgical lubricant for selected medical implements. *The Wellcome Foundation Ltd.*

17708 Lubasin®
Adhesives for textile printing. *BASF AG; BASF plc.*

17709 Lube-Booster® 1320 II
Lubricity additive for sol. oils; designed not to upset existing HLB balance of sol. oils. *Ferro/Keil.*

17710 Lube-Lok®
Family of oven-cured dry film lubricant coatings containing solid lubricants

and suitable binder system; used for fasteners, slides, pins, clips, and numerous applications requiring dry lubrication and corrosion protection particularly under high loads. *E/M Corporation.*

17711 Lubestat

Water soluble, biodegradable lubricants; fiber lubricants, textile softeners, yarn processing aids. *Milliken.*

17712 Lubestine

14807-96-6 9207 238-877-9

Talc, asbestine substitute; extender for paints and as a general purpose filler. *Bromhead & Denison Ltd.*

17713 Lubit® 64

All-purpose textile lubricant for all fibers; promotes better fabric penetration. *Sybron.*

17714 Lubix

Die lubricant for semicontinuous casting. *Foseco (F.S.) Ltd.*

17715 Lubolid

Dry powder lubricants. *Dow Corning Ltd.* Name unverified.

17716 Lubrajel® CG, DV, MS, TW

Polyglycerylmethacrylate and propylene glycol; autoclavable nondrying water-sol. lubricant for medical and surgical use. *Presperse.* Discontinued.

17717 Lubran 145

Alkyl methacrylate polymer; pour pt. depressant, viscosity index improver for lubricating oils. *Toho Chem. Industry.*

17718 Lubran AD

Hydrocarbon wax/naphthalene condensate; pour pt. depressant, dewaxing aid for lubricating oils. *Toho Chem. Industry.*

17719 Lubrhophos® HR-719

70321-78-7

Aliphatic phosphate ester of ethoxylated butanediol, acid form; biodegradable low foaming emulsifier, EP agent, corrosion inhibitor, cleaning and lubricity aids for metalworking fluids. *Rhône-Poulenc Surf.*

17720 Lubrhophos® LB-400

39464-69-2

Organic phosphate ester of ethoxylated oleyl alcohol, free acid; lubricity and EP additive, rust inhibitor, wetting agent, emulsifier, detergent, dispersant; for lubricating and rolling oils, hydraulic and water-based cutting fluids, glass cutting and polish *Rhône-Poulenc Surf.; Rhône-Poulenc France.*

17721 Lubrhophos® LE-500

51811-79-1

Aromatic phosphate ester of ethoxylated nonyl phenol; EP agent, emulsifier, corrosion inhibitor, detergent, lubricant for metalworking fluids. *Rhône-Poulenc Surf.; Rhône-Poulenc France.*

17722 Lubrhophos® LE-700

51811-79-1

Aromatic phosphate ester of ethoxylated nonylphenol, free acid; high foaming lubricant, emulsifier, EP agent, corrosion inhibitor for water-based cutting fluids. *Rhône-Poulenc Surf.*

17723 Lubrhophos® LF-200

39464-67-0

Organic Phosphate ester of ethoxylated dodecylphenol, free acid; lubricant, emulsifier for oil-based cutting fluids, hydraulic fluids, rolling oils, slushing compounds. *Rhône-Poulenc Surf.; Rhône-Poulenc France.*

17724 Lubrhophos® LM-400

39464-64-7

Aromatic phosphate ester of ethoxylated dinonylphenol, free acid; lubricant, emulsifier, EP agent, corrosion inhibitor, detergent for cutting fluids. hydraulic fluids, rolling oils. *Rhône-Poulenc Surf.*

17725 Lubrhophos® LP-700

39464-70-5

Complex organic phosphate ester of ethoxylated phenol, free acid; low foaming emulsifier, EP agent, corrosion inhibitor, detergent, lubricant for water-based metalworking fluids. *Rhône-Poulenc Surf.; Rhône-Poulenc France.*

17726 Lubrhophos® LS-500

9046-01-9

Phosphate acid ester of ethoxylated tridecanol; aliphatic hydrophobic base; lubricant, emulsifier for oil and water-based cutting fluids, hydraulic fluids, rolling oils. *Rhône-Poulenc Surf.; Rhône-Poulenc France.*

17727 Lubri-Bond®

Family of air curing dry film lubricant coatings containing solid lubricants and suitable binder system; for frictional surfaces requiring dry lubrication particularly on substrates which cannot tolerate oven cure. *E/M Corporation.*

17728 Lubricant EHS

Fatty acid ester; heat-stable boundary lubricant for microemulsions for heavy-duty machining of aluminum. *Hoechst Celanese/Colorants & Surf.*

17729 Lubricin 25

Processing aid for rigid PVC during calendering, injection molding of pipe, fittings, conduit, sheet and profiles and for rubber improving appearance of molded goods. *CasChem.*

17730 Lubricin N-1

Additive for use in motor fuels, penetrating oils, cutting, machining and metalworking fluids providing lubricity and detergent action and reducing rust, corrosion and metal wear. *CasChem.*

17731 Lubrico

A proprietary trade name for an alloy of 75% copper, 20% lead, and 5% tin. No manufacturer.

17732 Lubricomp®

Lubricated thermoplastic compounds. *LNP; ICI Chemicals & Polymers Ltd.*

17733 Lubricomp® 189

9016-75-5

PPS composite, PTFE lubricated; for injection molding applications including hard-to-mold thin-walled parts. *LNP.*

17734 Lubricomp® DFL-4036

PC, 30% glass fiber-reinforced, 15% TFE-lubricated, flame-retardant; thermoplastic compound. for gear and bearing applications requiring low frictional properties, high PV values, and good wear resistance. *LNP.*

17735 Lubricomp® DL-4030

PC resin with 15% PTFE lubrication; internally lubricated composite offering excellent wear and frictional char. for gear, cam, and sliding applications. *LNP.*

17736 Lubrigel

1318-93-0 6341 215-288-5

Organically modified refined montmorillonite clay; for use in the conversion of lubricant base stocks to clay-based greases. *AKZO Chemie UK Ltd.*

17737 Lubril CAT-XIVC

Fatty imidazoline derivative; softener for treatment of glass; improves adhesion of hydrocarbon coating materials. *Rhône-Poulenc Surf.*

17738 Lubril PF-570

aqueous solution of cationic polymer; lubricant for fiberglass manufacture *Rhône-Poulenc Surf.*

17739 Lubril QC

High molecular weight hydrophilic polyester dispersion; imparts soil release and moisture transport props. to polyester filament fabrics; oil scavenger keeping oils and greases in suspension during scourin/dyeing; lubricant for fiber-to-fiber lubrication. *Rhône-Poulenc Surf.*

17740 Lubrimet® P 600, P 900

PPG; lubricant; solubilizer for dyestuffs and surfactants. *BASF AG.*

17741 Lubrisol

Natural oil; textile lubricant. *Scher.*

17742 Lubritab®

68334-00-9 269-804-9

Hydrogenated vegetable oil NF; lubricant for pharmaceutical tablets. *Penwest Pharmaceuticals Co.*

17743 Lubrite B33

Bleaching of mineral oils, greases and waxes, including re-refined/recycled lubricating oil. *Minas de Gádor SA.*

17744 Lubrizol® 2152

61789-86-4 263-093-9

Calcium sulfonate

TLA-256. Pigment dispersant and wetting agent for color concs., paints, coatings, inks; pigment flushing aid for organic and inorganic pigments; viscosity stabilizer/reducer in plastisols. *Lubrizol.*

17745 Lubrizol® 2153

123-56-8 9040 204-635-6

$C_4H_5NO_2$

Succinimide

Pigment dispersant and wetting agent for color concs., inks, plastisols and organosols. *Lubrizol.*

17746 Lubrol

Fatty alcohol ethoxylates with a neutral reaction; used as surfactants. *ICI Chem & Polymers Ltd.*

17747 Lubrol 90

Ester; nonoily press release agent for rubber roll release problems in pulp/paper industry; surface lubricant for tabulator card stock and paper board. *CNC Int'l.L.P.*

17748 Lubrol N13

Nonylphenol ethoxylate nonionic surfactant in semisolid form; wetting and emulsifying agent used in pest control, oleines, metal treatment, and degreasing. *ICI Chem & Polymers Ltd.* Name unverified.

17749 Lubrol N5

Nonylphenol ethoxylate nonionic surfactant in viscous liquid form; used for conversion of premium paraffin to self-emulsifiable solvent; in industrial cleaning and degreasing including metal surfaces. *ICI Chem & Polymers Ltd.* Name unverified.

17750 **Lucalen®**
Ethylene copolymers with polar groups. *BASF AG; BASF plc.*

17751 **Lucantin® CX**
Citranaxanthine
For pigmentation of egg yolks. *BASF AG; BASF plc.*

17752 **Lucaphos® 40, 48**
7757-93-9 1739 231-826-1
Dicalcium phosphate
For the feed industry. *BASF AG.*

17753 **Lucarotin® 10%**
7235-40-7 1902 230-636-6
β-Carotene
For the feed industry. *BASF AG.*

17754 **Lucca oil**
8001-25-0 6973 232-277-0
Olive oil.

17755 **Lucerno**
An alloy containing 67.9% nickel, 27.5% copper, 2.4% iron, and 2.2% manganese.

17756 **Luchem AS-946**
Methyl methacrylate/allyl methacrylate copolymer; antishrink additive. *Atochem.* Discontinued.

17757 **Lucidene**
Range of aqueous acrylic and styrene-acrylic emulsions in printing inks and overprint lacquers. *Morton Int'l. Ltd.*

17758 **Lucidol**
94-36-0 1149 202-327-6
Benzoyl peroxide
Akzo Chemie UK Ltd.

17759 **Lucidol 75FP**
94-36-0 1149 202-327-6
Benzoyl peroxide with 25% water; active constituent in anti-acne creams and soaps and as a polymerization initiator for plastics production. *Atochem.* Discontinued.

17760 **Lucidol GS**
94-36-0 1149 202-327-6
A range of various concentration benzoyl peroxide powders. *Akzo Chemie UK Ltd.*

17761 **Lucidol RM**
94-36-0 1149 202-327-6
Benzoyl peroxide
Used as free radical initiator in polymerization. *Akzo Chemie UK Ltd.*

17762 **Lucidol-78**
94-36-0 1149 202-327-6
Benzoyl peroxide
Initiator for bulk, solution., and suspension polymerization, and high-temperature and room temperature cure of polyester resins. *Atochem.* Discontinued.

17763 **Lucilite**
94-36-0 1149 202-327-6
Silica hydrogel
Rhône-Poulenc UK.

17764 **Lucipal**
94-36-0 1149 202-327-6
Benzoyl peroxide compositions. *Akzo Chemie UK Ltd.*

17765 **Lucirin® BDK**
Uv initiator for radiation-curing finishes and putties. *BASF AG.*

17766 **Luclte®**
Proprietary trade names for a methyl methacrylate and acrylate synthetic resins; the material is more transparent than glass and has a very high refractive index. *DuPont UK.*

17767 **Lucobit®**
Ethylene copolymer/bitumen; for the production of damp-proof courses in buildings above ground and underground construction, for elastic foundations, as protection against corrosion for steel tanks, for improving quality of road bitumen, for production of *BASF AG; BASF plc.*

17768 **Lucofen S A**
151-06-4 2235 205-782-9
A proprietary preparation of chlophentermine hydrochloride; antiobesity agent. *Warner.* Name unverified.

17769 **Luconyl®**
Concs. of organic and inorganic pigments using nonionic dispersants; for coloring polymer emulsion paints, inks, water-based wood glazes. *BASF AG.*

17770 **Lucovyl**
A proprietary polyvinyl chloride. No manufacturer.

17771 **Lucryl®**
9011-14-7
Polymethylmethacrylate
Thermoplastic molding compounds; std. grades for injection molding or extrusion offer high transparency and brilliance, high surface hardness and scratch resistance, excellent weather resistance, mech. strength and rigidity; impactmodified grades; for injection *BASF AG; BASF plc.*

17772 **Ludigol F**
127-68-4 204-857-3
Sodium *m*-nitrobenzene sulfonate
Used to accelerate the stripping of nickel from steel, brass or copper; to control the reaction rate in pickling nickel-chromium-iron alloys and for dissolving nickel and copper; additive in phosphate coating of metals. mp = 350°; fp > 100°. *Rhône-Poulenc Surf.*

17773 **Ludigol®**
Mild oxidizing agent for textile finishing. *BASF AG; BASF plc.*

17774 **Ludiomil**
10347-81-6 5792 233-758-8
maprotiline hydrochloride
Antidepressant. *Ciba plc.*

17775 **Ludlum alloy**
A heat-resisting alloy of iron with from 13-17% chromium, 1% silicon, 1% molybdenum, and 0.4% carbon.

17776 **Ludlum No. 602 Steel**
A proprietary trade name for a steel containing 1.7% silicon, 0.7% manganese, 0.4% molybdenum, 0.12% vanadium, and 0.48% carbon. No manufacturer.

17777 **Ludopal®**
Unsatyrated polyester resins; P grades for paraffin-containing polyester finishes, dissolved in styrene; P 6 M is styrene-free resin for peroxide-containing finishes; E grades for elastifying P grades; U 150 for paraffin-containing and paraffin-free uv finishes, filling compounds. *BASF AG.*

17778 **Ludorum**
15545-48-9 239-592-2
Chlorotoluron; Dicurane 500 FW; Toro; Tripart® Ludorum 700. Suspension concentrate containing 500 g chlorotoluron per liter; a contact urea herbicide. *Tripart Farm Chemicals Ltd; Sipcam UK Ltd.; Tripart Farm Chemicals Ltd.; Ciba-Geigy Agrochemicals.*

17779 **Ludox®**
Colloidal silica. *DuPont UK.*

17780 **Lufibrol® E**
Extraction agent for textile desizing and scouring. *BASF.*

17781 **Lufibrol® FW**
Bleaching agent, wool fiber protection agent for prevention of yellowing during dyeing. *BASF.*

17782 **Lufibrol® NB-7**
Extracting and dispersing agent for detaching and washing impurities in desizing, scouring, and caustic boil-off operations. *BASF.*

17783 **Lufilen®**
Organic and inorganic pigments in polyethylene; for polyolefin spin dyeing. *BASF AG.*

17784 **Luflbrol®**
Extraction assistants in pretreatment of textile goods. *BASF AG; BASF plc.*

17785 **Luftseide**
Celta, soie nouvelle, tubulated silk. Artificial silks of the rayon type made with hollow central spaces. They are formed by adding gas-evolving materials to the viscous solution.

17786 **Lugalvan**
Electroplating additive. *BASF plc.*

17787 **Luganil® Dyes**
Anionic dyes for leather and fur dyeing, for coloring cleaners, etc. *BASF AG; BASF plc.*

17788 **Lugol's solution**
Iodine-potassium iodide solution; 5 g of iodine are titrated with 5 g of potassium iodide, and 100 ml of water, and diluted to 1 liter.

17789 **Lugo's powder**
Powdered cinchona bark.

17790 **Luhydran®**
Water-dilutable binders for electrocoating finishes, baking finishes, wood and plastic coatings, anticorrosion and roadmarking paints, printing inks, overprinting varnishes. *BASF AG.*

17791 **Lukens Bone Wax**
Made from natural bees wax, salicylic acid, and a natural oil; applied by surgeon in order to produce hemostasis in bone during surgery in which bone is cut. *Lukens International Corp.* Unverified.

17792 **Lulea tar**
A variety of Stockholm tar.

17793 Lumarith® EC
A registered trademark for an ethyl cellulose thermoplastic resistant to oils. No manufacturer.

17794 Lumarith® ER
A registered trademark for ethyl rubber. No manufacturer.

17795 Lumattin®
Wax modified silica; matting agents in plant and varnish industry. *BASF AG.*

17796 Lumbang oil
A drying oil obtained by pressing the seeds of *Aleurities moluccana* used in the manufacture of paints and soap.

17797 Lumen Alloy 11-C
A proprietary trade name for an alloy containing copper with 10% aluminum and 1% iron. No manufacturer.

17798 Lumen bronze
An alloy of 86% zinc, 10% copper, 4% aluminum, and 0.1% magnesium. A softer variety contains 88% zinc, 8% aluminum, and 4% copper.

17799 Lumicon 68 Silver Amalgam/Powder
Minimum corroding silver alloy of standardized grain size distribution, mixable in any commercial mixer. *Bayer AG.*

17800 Lumicon Non Gamma 2
Silver alloy, practically corrosion free; dental specialty. *Bayer AG.* Discontinued.

17801 Lumifor® Gluma
Hybrid-filled light-curing plastic filling/dentin bonding system for dental industry. *Bayer AG.*

17802 Lumilux
Luminous pigments. *Hoechst UK.*

17803 Luminal Sodium
57-30-7 7386 200-322-3
Phenobarbital sodium
Anticonvulsant; hypnotic; sedative. *Sterling Drug Inc.*

17804 Luminous®
PMMA optical fiber. *Asahi Chem. Industry.*

17805 Lumite
An aluminum alloy containing 5.6% nickel and 1% iron.

17806 Lumiten® E
Foam suppressants for aqueous emulsion paints, cement systems, dispersion adhesives. *BASF AG.*

17807 Lumiten® I, N
Wetting agent for emulsion paints, for the modification of cement systems, and for adhesives. *BASF AG.*

17808 Lumitol®
Hydroxylic polyacrylate resins; used in combination with polyisocyanates to produce chem.-resistance and solvent-resistance polyurethane finishes having excellent mech. props. *BASF AG.*

17809 Lummer's solution
H_2PtCl_6
Hydrogen platinochloride
(3 g in 100 ml water), with 0.02 gram lead acetate; used for coating platinum electrodes with finely divided platinum.

17810 Lumnite cement
An alumina cement consisting of about 40% alumina, 40% lime, 15% iron oxide, 5% silica, and magnesia. The material is made from bauxite, and is stated to be stronger than Portland cement.

17811 Lumo Stabil S 80
Sodium dodecylbenzene sulfonate
Washing and cleansing agents, industrial cleaners. *Zschimmer & Schwarz.* Discontinued.

17812 Lumo WW 75
Sodium dodecylbenzene sulfonate
Heavy and light duty detergent. *Zschimmer & Schwarz.* Discontinued.

17813 Lumorol 4153
Nonionic/anionic surfactants with phosphate and dissolving agents; cleansing agents. *Zschimmer & Schwarz.* Discontinued.

17814 Lumorol K 28
Disodium laurethsulfosuccinate
cocamidopropyl betaine; magnesium lauryl sulfate. Detergent for cosmetics, shampoos, bath preparations *Zschimmer & Schwarz.* Discontinued.

17815 Lumosäure A
Dodecylbenzene sulfonic acid
For low pH cleansing agents. *Zschimmer & Schwarz.* Discontinued.

17816 lunar caustic
7761-88-8 8661 231-853-9
$AgNO_3$
Silver nitrate
Fused and cast into sticks or rods. An energetic caustic for wounds and sores.

17817 Lunosol
A proprietary preparation; it is a colloidal silver chloride. No manufacturer.

17818 Luo-calcite
A name given to the calcium bicarbonate occurring in solution in natural waters.+ys.

17819 Luo-chalybite
A name given to the ferrous bicarbonate occurring in solution in natural waters.

17820 Luo-diallogite
A name given to the manganese bicarbonate occurring in solution in natural waters.

17821 Luo-magnesite
A name given to the magnesium bicarbonate occurring in solution in natural waters.

17822 Lupeose
$C_{24}H_{12}O_{21}$
Mannotetrose. A tetrasaccharide. It occurs in the tubers of Stachy tubifera.

17823 Luperco 101-P20
78-63-7 201-128-1
20% dispersion of 2,5-dimethyl-2,5-di (t-butylperoxy) hexane on a PP powd. carrier; crosslinking agent for elastomers and thermoplastic resins. *Elf Atochem.*

17824 Luperco 230-XL
995-33-5 213-626-6
N-Butyl-4,4-bis (*t*-butyl-peroxy) valerate on an inert filler; initiator for curing elastomers, and for high-temperature cure of polyester resins. *Elf Atochem.*

17825 Luperco 231-XL
6731-36-8 229-782-3
1,1-Di (*t*-butylperoxy)3,3,5-trimethyl cyclohexane on an inert filler; initiator for curing elastomers, crosslinking agent for thermoplastic modification. *Elf Atochem.*

17826 Luperco 233-XL
Ethyl-3,3-di (*t*-butylperoxy) butyrate on an inert filler; initiator for curing elastomers and for polymer modification thermoplastic cross-linking. *Elf Atochem.*

17827 Luperco 331-XL
3006-86-8 221-111-2
1,1-Di (*t*-butylperoxy) cyclohexane on inert filler; initiator for high-temperature cure of polyester resins and for curing elastomers. *Elf Atochem.*

17828 Luperco 500-40KE
80-43-3 201-279-3
Dicumyl peroxide on Burgess KE clay *Elf Atochem.*

17829 Luperco 801-XL
3457-61-2 222-389-8
Trigonox® T. *t*-butyl cumyl peroxide on inert filler; initiator for high-temperature cure of polyester resins, curing elastomers, and polymer modification thermoplastic cross-linking. *Elf Atochem.*

17830 Luperco 802-40KE
α-α-bis (*t*-butylperoxy) diisopropylbenzene on Burgess clay; crosslinking agent for polymer modification, curing elastomers, high-temperature cure of polyester resins. *Elf Atochem.*

17831 Luperco A
94-36-0 1149 202-327-6
Benzoyl peroxide with an inorganic filler; used for polymerization catalysis, drying accelerator, oxidizing agent and bleaching applications. *Elf Atochem North America Inc/Organic Peroxides Div.*

17832 Luperco AC
94-36-0 1149 202-327-6
Benzoyl peroxide with an organic filler; used for polymerization catalysis, drying accelerator, oxidizing agent and bleaching applications. *Elf Atochem North America Inc/Organic Peroxides Div.*

17833 Luperco AFR
94-36-0 1149 202-327-6
A trademark for 50% benzoyl peroxide in plasticizer. *Wallace & Tieman.* Unverified.

17834 Luperco AFR-250
94-36-0 1149 202-327-6
Benzoyl peroxide
Fire retardant initiator for high-temperature and room temperature cure of polyester resins. *Elf Atochem.*

17835 Luperco AST
94-36-0 1149 202-327-6
Benzoyl peroxide with silicone oil; initiator for curing elastomers. *Elf Atochem.*

17836 Luperfoam 40
Aqueous mixture of *t*-butylhydrazinium chloride and cupric chloride; chemical blowing agent for unsaturated polyester resins. *Atochem.* Discontinued.

17837 Luperox
Solid organic peroxides *Elf Atochem North America Inc/Organic Peroxides Div.*

17838 Luperox 118
2,5-Dimethyl-2,5-bis-(benzoylperoxy) hexane
Initiator for vinyl polymerization. *Elf Atochem.*

17839 Luperox 2,5-2,5
2,5-Dihydroperoxy-2,5-dimethylhexane
Initiator for bulk, solution, emulsion, and suspension polymerization. *Atochem.* Discontinued.

17840 Luperox 500R
80-43-3 201-279-3
Dicumyl peroxide
Initiator for bulk, solution., and suspension polymerization, polymer modification thermoplastic crosslinking, curing elastomers, high-temperature cure of polyester resin. *Elf Atochem.*

17841 Luperox 802
25155-25-3 246-678-3
α-α-Bis (t-butylperoxy) diisopropylbenzene
Vul-Cup; Perkadox® 14; Retilox® F 40 MG; Vul-Cup 40KE and R. Crosslinking agent for thermoplastic modification, curing elastomers, and high temperature cure of polyesters; initiator for vinyl polymerization. *Elf Atochem; Akzo; Hercules; Hercules Ltd.*

17842 Lupersol 10
26748-41-4 247-955-1
t-Butyl peroxyneodecanoate
Initiator for vinyl polymerizations. *Elf Atochem.*

17843 Lupersol 11
927-07-1 213-147-2
t-butyl peroxypivalate in odorless mineral spirits. *Elf Atochem.*

17844 Lupersol 70
107-71-1 203-514-5
Aztec® t-Butyl Peracetate-50 OMS, t-Butyl Peracetate-60 OMS, t-Butyl Peracetate-75 OMS; Esperox® 12MD; Trigonox® F-C50. *t*-Butyl peroxyacetate in odorless mineral spirits *Elf Atochem Organic Peroxides; Catalyst Resources; Witco/Argus; Akzo.*

17845 Lupersol 80
109-13-7 203-650-5
t-Butyl peroxyisobutyrate in odorless mineral spirits *Elf Atochem.*

17846 Lupersol 101
78-63-7 201-128-1
2,5-Dimethyl-2,5-di(*t*-butyl-peroxy) hexane
Elf Atochem.

17847 Lupersol 130
1068-27-5 213-944-5
2,5-Dimethyl-2 5-di(t-butylperoxy) hexyne-3
Initiator for bulk, solution, and suspension polymerization, high-temperature curing of polyester resins, and cure of acrylic syrup. *Elf Atochem.*

17848 Lupersol 188-M75
26748-47-0 247-956-7
α-Cumylperoxy neodecanoate in odorless mineral spirits *Elf Atochem.*

17849 Lupersol 219-M60
Diisononanoyl peroxide in odorless mineral spirits; crosslinking agent for bulk, solution, and suspension polymerization. *Elf Atochem.*

17850 Lupersol 220-D50
2,2-Di(t-butylperoxy) butane in DOP; initiator for bulk, solution, and suspension polymerization, high-temperature curing of polyester resins, and cure of acrylic syrup. *Elf Atochem.*

17851 Lupersol 221
16066-38-9 240-211-7
Di(n-propyl) peroxydicarbonate
Elf Atochem.

17852 Lupersol 223
16111-62-9 240-282-4
Di(2-ethylhexyl) peroxydicarbonate
Elf Atochem.

17853 Lupersol 224
2,4-Pentanedione peroxide
Curing agent for unsaturated polyester thermoset resins. *Elf Atochem.*

17854 Lupersol 225
19910-65-7 243-424-3
Di (sec-butyl) peroxydicarbonate
Initiator for bulk, solution., and suspension polymerization and cure of acrylic syrup. *Elf Atochem.*

17855 Lupersol 230
995-33-5 213-626-6
n-Butyl-4,4-bis (t-butylperoxy) valerate
Luperco 230-XL. Initiator for bulk, solution, and suspension polymerization, for high-temperature cure of polyester resins, for curing elastomers, and for cure of acrylic syrup. *Atochem.* Discontinued.

17856 Lupersol 231
6731-36-8 229-782-3
1,1-Di (t-butylperoxy) 3,3,5-trimethyl cyclohexane
Initiator for bulk, solution., and suspension polymerization, curing elastomers, high-temperature cure of polyester resins, and cure of acrylic syrup. *Elf Atochem.*

17857 Lupersol 233-M75
Ethyl-3,3-di (t-butylperoxy) butyrate in odorless mineral spirits; initiator for bulk, solution, and suspension polymerization, high-temperature cure of polyesters and acrylics. *Elf Atochem.*

17858 Lupersol 256
2,5-Dimethyl-2,5-bis(2-ethylhexanoyl-peroxy) hexane
Elf Atochem.

17859 Lupersol 288-M75
26748-47-0 247-956-7
α-Cumyl peroxyneoheptanoate in odorless mineral spirits. *Elf Atochem.*

17860 Lupersol 331-80B
3006-86-8 221-111-2
1,1-Di (*t*-butylperoxy) cyclohexane in butylbenzyl phthalate; initiator for bulk, solution, and suspension polymerization, high-temperature curing of polyester resins, and cure of acrylic syrup. *Elf Atochem.*

17861 Lupersol 531-80B
1,1-Di-(*t*-amylperoxy) cyclohexane in butyl benzyl phthalate; initiator for bulk, solution, and suspension polymerization, for high-temperature cure of polyester resins, and for cure of acrylic syrup. *Elf Atochem.*

17862 Lupersol 533-M75
67567-23-1 403-320-2
Ethyl 3,3-di(*t*-amylperoxy)butyrate in odorless mineral spirits; initiator for curing of acrylic syrup; crosslinking agent for bulk, solution, and suspension polymerization, thermoplastic modification. *Elf Atochem.*

17863 Lupersol 546-M75
68299-16-1 269-597-5
t-amylperoxy neodecanoate solution in OMS; initiator for bulk, solution, and suspension polymerization, and cure of acrylic syrup. *Elf Atochem.*

17864 Lupersol 553-M75
3052-70-8
2,2-Di(*t*-amylperoxy) propane in odorless mineral spirits; initiator for high temperature cure of polyesters, cure of acrylic syrup; crosslinking agent for bulk, solution, and suspension polymerization, thermoplastic modification. *Atochem.* Discontinued.

17865 Lupersol 554-M50, 554-M75
29240-17-3 249-530-6
t-Amylperoxy pivalate solutions. in OMS; initiators for bulk, solution, and suspension polymerization and for cure of acrylic syrup. *Elf Atochem.*

17866 Lupersol 555-M60
t-Amyl peroxyacetate in odorless mineral spirits; crosslinking agent for bulk, solution, and suspension polymerization, high temperature cure of polyester, and cure of acrylic syrup. *Elf Atochem.*

17867 Lupersol 575
686-31-7 211-687-3
t-Amylperoxy-2-ethyl-hexanoate
Initiator for bulk, solution, and suspension polymerization, for high-temperature cure of polyester resins, and for cure of acrylic syrup. *Elf Atochem.*

17868 Lupersol 665-M50
1,1-Dimethyl-3-hydroxybutyl peroxy-2-ethylhexanoate
Elf Atochem.

17869 Lupersol 688-T50
1,1-Dimethyl-3-hydroxybutylperoxy-neoheptanoate in toluene; crosslinking agent for bulk, solution, emulsion, and suspension polymerization, and cure of acrylic syrup. *Atochem.* Discontinued.

17870 Lupersol DDM
A trademark for 60% methyl ethyl ketone peroxide in dimethyl phthalate. *Wallace & Tiernan.* Unverified.

17871 Lupersol DDM-9
1338-23-4 215-661-2
MEK peroxide; polymerization initiator for cure of promoted unsaturated polyester resins and vinyl ester resins at ambient temperatures.; promoter, transition metal salt, activates decomposition of peroxide initiator. *Elf Atochem.*

17872 Lupersol DEL
A trademark for 60% methyl ethyl ketone peroxide in dimethyl phthalate. *Wallace & Tiernan.* Unverified.

17873 Lupersol DSW
1338-23-4 215-661-2

A proprietary trade name for methyl ethyl ketone peroxide; a liquid fire resistant peroxide containing 11.5% active oxygen. *Kingsley & Keith Chemical Corp.* Name unverified.

17874 Lupersol KDB
Di-*t*-butyl diperoxyphthalate in DBP. *Elf Atochem.*

17875 Lupersol P-31, P-33
t-Butyl peroctoate [3006-82-4] and 1,1-di(*t*-butylperoxy) cyclohexane [3006-86-8] blend in mineral oil; initiator for high-temperature cure of polyester resins. *Elf Atochem.*

17876 Lupersol PDO
t-Butyl peroxy-2-ethylhexanoate in DOP. *Elf Atochem.*

17877 Lupersol TAEC
OO-*t*-amyl O-(2-ethylhexyl) monoperoxycarbonate
Crosslinking agent for emulsion, bulk, solution, and suspension polymerization, high temperature cure of polyester, and cure of acrylic syrup. *Elf Atochem.*

17878 Lupersol TBEC
OO-*t*-butyl O-(2-ethylhexyl) monoperoxycarbonate
Elf Atochem.

17879 Lupersol TBIC-M75
t-Butyl O-isopropyl monoperoxycarbonate in odorless mineral spirits. *Elf Atochem.*

17880 Lupersol® 227
A 50% solution of diisobutyryl peroxide in mineral spirits; an organic peroxide for polymerizing. *Wallace & Tiernan.* Unverified.

17881 Lupetazin
$C_6H_{14}N_2$
Dimethylpiperazine
Used as a uric acid solvent.

17882 Luphen® D
Polyurethane disp.
For production of adhesives. *BASF AG.*

17883 Lupinit
A proprietary casein product. No manufacturer.

17884 Lupolen 1800 H/M/S
A proprietary LDPE used in injection molding. *BASF plc.* Name unverified.

17885 Lupolen 1810E
A proprietary LDPE used in the manufacture of milk containers. *BASF plc.* Name unverified.

17886 Lupolen 1810H
A proprietary LDPE used in blow molding. *BASF plc.* Name unverified.

17887 Lupolen 1812D and 1812EH
Proprietary polyethylenes used in the manufacture of wires and cables. *BASF plc.* Name unverified.

17888 Lupolen 1814E
A proprietary LDPE used in the manufacture of milk containers. *BASF plc.* Name unverified.

17889 Lupolen 1852E/H
A proprietary LDPE used in the manufacture of pipes. *BASF plc.* Name unverified.

17890 Lupolen 2040EX and 2410DX
Proprietary LDPEs used in the manufacture of bags. *BASF plc.* Name unverified.

17891 Lupolen 2410S
A proprietary low d polyethylene used in injection molding. *BASF plc.* Name unverified.

17892 Lupolen 2424H and 2425K
Proprietary LDPEs used in the manufacture of transparent packaging materials. *BASF plc.* Name unverified.

17893 Lupolen 2430H
A proprietary low d polyethylene used in blow molding. *BASF plc.* Name unverified.

17894 Lupolen 2452 E
A proprietary low d polyethylene used in the extrusion of pipes. *BASF plc.* Name unverified.

17895 Lupolen 3010 S
A proprietary low d polyethylene used in injection molding. *BASF plc.* Name unverified.

17896 Lupolen 3020 D
A proprietary low d polyethylene used in blow molding. *BASF plc.* Name unverified.

17897 Lupolen 3020 KX and 3025 KX
Proprietary LDPEs used as over-wrappings. *BASF plc.* Name unverified.

17898 Lupolen 4261 AX
A proprietary HDPE of high molecular weight. *BASF plc.* Name unverified.

17899 Lupolen 5011 K
A proprietary HDPE used in injection molding. *BASF plc.* Name unverified.

17900 Lupolen 5052 C
A proprietary HDPE used in the extrusion of pipes. *BASF plc.* Name unverified.

17901 Lupolen 6011 K
A proprietary HDPE used in injection molding. *BASF plc.* Name unverified.

17902 Lupolen 804H and 1814H
Proprietary LDPEs used in the manufacture of liners and barriers. *BASF plc.* Name unverified.

17903 Lupolen V-2524EX and V-3510K
Proprietary LDPE copolymers. *BASF plc.* Name unverified.

17904 Lupolen®
9002-88-4 7728
HDPE, LDPE, LLDPE, LDPE/EVA copolymer; various grades for injection molding, extrusion, blow molding, powd. grades, compression molding grades, semifinished products. *BASF AG; BASF plc.*

17905 Lupolex
Linear low d polyethylene. *BASF plc.*

17906 Lupranat®
MDI/TDI; for production of PU flexible foam, semirigid and rigid foams, thermoplastic elastomers for the furniture, automotive, construction, pkg., electrical, and shoe industry. *BASF AG.*

17907 Lupranol®
Polyether polyols
For the production of PU flexible foam, semirigid and rigid foam for the furniture, automobile, construction, pkg., electrical, and shoe industry. *BASF AG.*

17908 Lupraphen
Polyester polyols
For the production of PU, ready-to-use PU systems, thermoplastic granules, textile coating systems, finished parts made of PU and casting elastomers; for the automotive, machine and equip. construction industries. *BASF AG.*

17909 Luprenal®
Polyacrylate resins; heat-curable resins employing added crosslinkers; used in combination with melamine or urea resins to give hard, tough, weatherresistance and chem, resist, coatings. *BASF AG.*

17910 Luprimol®
Improves hand of pigment prints on textiles. *BASF AG; BASF plc.*

17911 Luprintan®
Fixing assistants for printing on synthetic fibers. *BASF AG.*

17912 Luprintan® ATP
Printing auxiliary for synthetics. *BASF plc.*

17913 Luprintan® DCA
Fixation accelerator for printing acetate, triacetate and polyester with disperse dyes. *BASF.*

17914 Luprintol®
Auxiliaries for emulsion printing. *BASF plc.*

17915 Luprintol® PE
Aryl polyglycol ether
Emulsifier for prep. of oil-water emulsions, solvent-containing or solvent-free thickenings for textile pigment printing. *BASF AG.*

17916 Luprofil®
Organic and inorganic pigments/PP masterbatches; for PP spin dyeing. *BASF AG.*

17917 Lupromag®
Magnesium propionate
Mineralized single feedstuff for animal feeding. *BASF AG.*

17918 Lupron
74381-53-6 5484
Leuprolide acetate
Antineoplastic. *TAP Pharmaceuticals.*

17919 Luprosil®
79-09-4 8010 201-176-3
Propionic acid
Preservative for mixed feedstuffs. *BASF AG.*

17920 Luprosil® NC
Ammonium propionate
Preservative. *BASF AG.*

17921 Luprosil® Salt
4075-81-4 1745 223-795-8
Calcium propionate
Preservative for feed industry. *BASF AG.*

17922 Luprosil® Sodium Salt
137-40-6 8816 205-290-4
Sodium propionate
Preservative for feed industry. *BASF AG.*

17923 Lurafix® Dyes
Wetting agent-free dyes for transfer printing inks. *BASF AG.*

17924 Luramid®
Polyacrylonitrile-soluble metal complex dyes and acid dyes for dyeing PAN chips in aqueous liquors. *BASF AG.*

17925 Luran®
Styrene acrylonitrile copolymer in granule form; used for moldings for domestic appliances and sanitaryware. *BASF plc.*

17926 Luran® 358 N
9003-54-7
SAN copolymer; very easy flow grade for injection moldings of containers, record player covers, TV filter panels, cassettes. *BASF AG.*

17927 Luran® 378 P G7
9003-54-7
SAN copolymer, 35% glass fiber-reinforced; high rigidity reinforced grade for injection molding and extrusion; used for chassis for assembling electrical and other parts, headlamp casings, parts for record players, disc drives, and tape reels. *BASF AG.* Name unverified.

17928 Luran® 757R
A proprietary acrylonitrile styrene-acrylonitrile copolymer used in injection molding. *BASF plc.* Name unverified.

17929 Luran® 776S
A proprietary acrylonitrile styrene-acrylonitrile used in extrusion and injection molding. *BASF plc.* Name unverified.

17930 Luran® KR 2517
A proprietary acrylo-nitrile-styrene copolymer containing 35% glass fiber. *BASF plc.* Name unverified.

17931 Luran® KR 2556
9003-54-7
SAN copolymer; high heat resistance grade for instrument covers and cassettes; used for injection molding and extrusion. *BASF AG.* Name unverified.

17932 Luran® S
Acrylonitrile/styrene/acrylate polymer; for injection molding or extrusion; esp. suitable for structural parts subject to outdoor exposure, e.g., letterboxes, parts for garden appliances, road signs, boat hulls, windsurfers, vehicle external parts. *BASF AG; BASF plc.*

17933 Lurantin®
Lightfast direct dyes for dyeing and printing cellulose fibers, nylon fibers and wool. *BASF AG; BASF plc.*

17934 Luranyl®
Polyphenylene ether and HIPS blend; std. grades for high-precision components which are tough and resistance to heat distort.; glass fiber-reinforced grades for pumps and fittings, hot water and drinking water; impact-resistance grades for car interior parts. *BASF AG.*

17935 Lurapret®
Antislip agents for textile finishing. *BASF AG.*

17936 Lurazol® Dyes
Anionic dyes for leather and fur dyeing, coloring cleaners. *BASF AG.*

17937 Luredox® BP, PO
7775-14-6 8771 231-890-0
Sodium dithionite
For manufacture of cleaners. *BASF AG.*

17938 Luredur®
Acrylate copolymers and modified polyacrylamides; dry strength agents for strengthening the structure of paper and board. *BASF AG; BASF plc.*

17939 Luresin®
Based on aqueous solutions. of formaldehyde-free cationic polyamido-amine-epichlorohydrin resins; for increasing wet and dry strength of papers, esp. in the neutral pH range. *BASF AG.*

17940 Lurgi metal
An alloy of lead with 2% barium.

17941 Luron® Binder
Albumen-type condensate in aqueous colloidal solution; assistant for laminate printing ink, leather and fur processing. *BASF AG.*

17942 Lurotex®
Textile dyeing and finishing auxiliary. *BASF plc.*

17943 Lurotex® A-25
Hydrophilic agent, crease inhibitor for synthetic fiber piecegoods and hank yarns. *BASF.*

17944 Luscin
Coal dust replacement additives for iron foundry sands. *Foseco (F.S.) Ltd.* Discontinued.

17945 Lusol
General purpose maintenance spray; for lubrication, penetrating seized parts, freeing rusted mechanisms, driving moisture out of electrical parts, corrosion prevention on metal surfaces. *Hermetite Products Ltd.* Name unverified.

17946 Lusol®
Wire drawing compounds for drawing of intermediate and fine copper wire, as well as rod breakdown. *Witco/Allied-Kelite.* Discontinued.

17947 Lusolvan® FBH
Diisobutyl ester of a dicarboxylic acid mixt.; coalescent for aqueous polymer dispersions; for paints and textured finishes. *BASF AG.*

17948 Lustilac
A proprietary molding compound. No manufacturer.

17949 Lustra-Cellulose
Artificial silk. No manufacturer.

17950 Lustral
Pfizer International.

17951 Lustralite®
Synthetic resin for industrial use. *Reichhold.*

17952 Lustralite® 44-444
Gloss enhancer. *Reichhold.*

17953 Lustran
Pfizer International.

17954 Lustran ABS
Acrylonitrile-butadiene-styrene thermoplastics; molding and extrusion grades, refrigerator inner liners, business machine and appliance housings, automotive parts, and industrial components. *Monsanto Co.*

17955 Lustran SAN
Styrene-acrylonitrile molding resins; clear rigid thermoplastics for molding high quality articles requiring transparency, brilliant surface, chemical resistance and stiffness. *Monsanto Co.*

17956 Lustran Ultra ABS
ABS molding grades, superior molded part appearance and gloss; used for telephones, appliances, video cassettes, power tool housings, office equipment, toys, sporting goods, and lawn and garden equipment. *Monsanto Co.*

17957 Lustranyl
Solvent soluble dyes. *Morton Int'l. Ltd.* Discontinued.

17958 Lustra-Pearl®
Mica, titanium dioxide; pearlescent pigments for cosmetic eye, face, lip, and body makeup. *Van Dyk.*

17959 Lustrasol®
Acrylic or styrene based resin solutions. *Reichhold.*

17960 Lustre
Protein binders based on modified casein dispersions; used for leather. *Colour-Chem Ltd.* Name unverified.

17961 Lustrex
Polystyrene molding and extrusion resins; used for packaging, toys, appliance, photographic, furniture, audio cassette, medical and houseware markets, thermoformed containers, trays, cups and lids. *Monsanto Co.* Name unverified.

17962 Lustron
A proprietary trade name for polystyrene molding compounds. No manufacturer.

17963 Lustrose
A proprietary compound used in the textile industry for sizing. No manufacturer.

17964 Lusynton® A
Mixt. of organic and inorganic substances; corrosion inhibitor; chlorite stabilizer with resistance to hard water and acid with wetting and detergent effects. *BASF AG; BASF plc.*

17965 Lutan®
Basic aluminum salts; for tanning leather and fur. *BASF AG.*

17966 Lutate
630-56-8 4886 211-138-8
A proprietary preparation of hydroxyprogesterone caproate. *Savage Laboratories.* Discontinued.

17967 Lutavit
Animal feed vitamins. *BASF plc.*

17968 Lutensit®
Ionic surfactants. *BASF plc.*

17969 Lutensit® A-BO
Sodium dioctylsulfosuccinate in water neopentyl glycol; surfactant for chemical industry, detergents, cleaners; wetting agent, dispersant. *BASF AG.*

17970 Lutensit® A-ES
Sodium alkyl-phenol ether sulfate as a yellow liquid; wetting and dispersing agent with many applications, sometimes in combination with nonionics; emulsifying agent for methylene chloride. *BASF plc.* Name unverified.

17971 Lutensit® A-LBA
Anionic surfactant in yellow liquid form; wetting, dispersing, cleaning and

emulsifying agent used in household rinsing and cleaning formulations, all-purpose cleaners, and various industrial emulsions. *BASF plc.* Name unverified.

17972 Lutensit® AN 10
Mixtures of various nonionic alkoxylates and an anionic, acidic, readily neutralizable surfactant; degreasing agent, cold cleaners. *BASF AG.*

17973 Lutensit® A-PS
Sodium alkane sulfonate as a slightly yellow liquid; alkaline industrial, household and metal cleaners, steam jet cleaners and pickling baths. *BASF plc.* Name unverified.

17974 Lutensit® AS 2230, 2270
Sulfated natural alcohol polyglycol ether, foaming surfactant for chemical and cosmetic industry. *BASF AG.* Name unverified.

17975 Lutensit® K-LC
Dimethyl C_{12}-C_{14} fatty alkylbenzyl ammonium chloride in almost colorless aqueous solution; dispersing and wetting agent with bactericidal and fungicidal effects; used in conjunction with nonionics to produce disinfectant cleaners for the food and beverage industries and for miscellaneous industrial and household uses. *BASF AG.* Name unverified.

17976 Lutensit® K-OC
Benzalkonium chloride
Biocidal, wetting and dispersing agents for production of disinfectant cleaners for beverages and foodstuffs, trading concerns and household. *BASF AG.* Name unverified.

17977 Lutensol®
Low foaming nonionic surfactants; used for domestic machine dishwashing detergents and rinse aids, low foam industrial cleaners. *BASF plc.*

17978 Lutensol® A 7
PEG-7 C_{16}-C_{18} fatty alcohol; water-soluble detergent for production of detergents and in chemical processing industry. *BASF AG.*

17979 Lutensol® AO 10
PEG-10 straight chain synthetic C_{13}-C_{15} fatty alcohol; detergent, emulsifying and dispersing agent used in household and industrial detergents, chemical industry; > 80% biodegradable. *BASF AG.*

17980 Lutensol® AO 3
PEG-3 straight chain synthetic C_{13}-C_{15} fatty alcohol; detergent, emulsifying and dispersing agent used in household and industrial detergents, chemical industry; > 80% biodegradable. *BASF AG.*

17981 Lutensol® AP 10
9016-45-9 6772
Nonoxynol-10
Detergent, wetting, emulsifying and dispersing agent, used in cleaners, detergents, leather, fur, paper, paint and dye industries. *BASF AG.*

17982 Lutensol® AP 20
9016-45-9 6772
Nonoxynol-20
Detergent, wetting, emulsifying and dispersing agent, used in cleaners, detergents, leather, fur, paper, paint and dye industries. *BASF AG.*

17983 Lutensol® AP 6
PEG-6 alkylphenol
Detergent, wetting, emulsifying and dispersing agent, used in cleaners, detergents, leather, fur, paper, paint and dye industries. *BASF AG.*

17984 Lutensol® AT 11
PEG-11 saturated C_{16}-C_{18} alcohol; detergent, wetting, dispersing and emulsifying agent used in chemical industry, for household and industrial detergents. *BASF AG.*

17985 Lutensol® ED 140
EO-PO ethylene diamine compound; antistat, detergent, dispersant, defoamer, wetting agent, gelling agent, solubilizer, emulsifier, demulsifier, lubricant, foam suppressor for household and industrial uses. *BASF AG.* Name unverified.

17986 Lutensol® FA 12
Oleyl amine ethoxylate
Emulsifier, dispersant, wetting and degreasing agent used in heavy-duty and other detergents, shampoos; > 80% biodegradable *BASF AG.*

17987 Lutensol® FSA 10
31799-71-0
Oleic acid amide ethoxylate
Emulsifier, dispersant, detergent, wetting agent, used in fine detergents, hand cleaners, fur dressing; > 80% biodegradable *BASF AG.*

17988 Lutensol® LF 220, 221, 223, 224
Alkoxylated straight chain alcohol; surfactant for dishwashing powders and rinse aids. *BASF AG.* Name unverified.

17989 Lutensol® ON 30
PEG-3 C10 oxo-alcohol
Surfactant for chemical industry and production of detergents. *BASF AG.*

17990 Lutensol® TO 3
PEG-3 C13 oxo-alcohol
Surfactant for chemical industry and production of detergents. *BASF AG.*

17991 luteol
57-83-0 7956 200-350-6
Progesterone

17992 Lutetia
Pigments, for paints and inks. *ICI Chem & Polymers Ltd.*

17993 Lutexal® TX-401
Synthetic thickening agent for textile applications. *BASF; BASF plc.*

17994 Lutexan
Natural yellow color. *Atomergic Chemetals Corp.*

17995 2,6-Lutidine
108-48-5 5643 203-587-3
C_7H_9N
2,6-dimethylpyridine
Lutidine; α,α'-lutidin. Pharmaceuticals, resins, dyestuffs, rubber accelerators, insecticides. d = 0.9252; mp = -5.8°; bp_{760} = 144°; n_D^{20} = 1.49797; soluble in water, alcohol, ether; miscible with dimethylformamide and tetrahydrofuran. *Aldrich; Janssen Chimica; Raschig; Reilly Industries.*

17996 Lutofan®
9002-86-2 7746 206-625-7
Vinyl chloride disp.
For production of laminating and heat-sealable adhesives and packaging materials, binders for bonding fiber webs and for textile coating. *BASF AG; BASF plc.*

17997 Lutonal®
Polyvinyl ether.
BASF plc.

17998 Lutonal® A
Polyvinyl ethyl ether
Nonyellowing, very resilient, soft resin for odorless and taste-free finishes in the paint industry, for gravure and flexographic inks, for producing pressure-sensitive, building adhesives, and as tackifying soft resins. *BASF AG.*

17999 Lutonal® D
Polyvinyl ether dispersion. *BASF plc.* Name unverified.

18000 Lutrabond
Polyester nonwovens for roof covers. *Carl Freudenberg.* Name unverified.

18001 Lutradur
Polyester nonwoven. *Carl Freudenberg.* Name unverified.

18002 Lutranal® LC
Polyvinyl isobutyl ether
BASF plc. Name unverified.

18003 Lutrizol®
Dimetridazole, dimetridazole HCl; active ingred. for control of black head disease in turkeys and bloody diarrhea complaints in swine. *BASF AG.*

18004 Lutrol®
Pharmaceutical grades of polyethylene glycol. *BASF plc.*

18005 Lutrol® E 300
25322-68-3 7729
PEG-6; emulsifier, emollient, lubricant, and solvent for liq. preparations *BASF AG.* Name unverified.

18006 Lutrol® E 400
25322-68-3 7729
PEG-8; emulsifier, emollient, lubricant, and solvent for liq. preparations *BASF AG.* Name unverified.

18007 Lutrol® E 1500
25322-68-3 7729
PEG-32; emulsifier, binder, solubilizer, resorption promoter for substances insol. or sparingly sol. in water. *BASF AG.*

18008 Lutrol® E 4000
25322-68-3 7729
PEG-75; emulsifier, binder, solubilizer, resorption promoter for substances insol. or sparingly sol. in water. *BASF AG.* Name unverified.

18009 Lutrol® E 6000
25322-68-3 7729
PEG-150; emulsifier, binder, solubilizer, resorption promoter for substances insol. or sparingly sol. in water. *BASF AG.* Name unverified.

18010 Lutrol® OP-2000
31394-71-5
PPG-26 oleate
BASF. Name unverified.

18011 Lutrol® W-3520
PPG-28-buteth-35
BASF. Name unverified.

18012 Lutron
Waterless foundry molding material. *Foseco (F.S.) Ltd.*

18013 Lutron
Electroplating additive. *BASF plc.*
18014 Lutropur
Electroplating additive. *BASF plc.*
18015 Luvican M170
Polyvinyl carbazole
BASF plc. Name unverified.
18016 Luviflex® VBM 35
26589-26-4
Film former with excellent hydrocarbon compatibility for weatherproof hairstyles. *BASF AG; BASF plc.* Name unverified.
18017 Luviform® ES 22
Ethyl ester of PVM/MA copolymer; film-forming binders for hair sprays and setting lotions. *BASF AG.* Name unverified.
18018 Luviform® ES 42
Butyl ester of PVM/MA copolymer; film-forming binders for hair sprays and setting lotions. *BASF AG.* Name unverified.
18019 Luviform® FA 119
9011-16-9
PVM/MA copolymer; stabilizing and binding agent for toothpastes, denture retaining agents, shampoos, etc. *BASF AG.*
18020 Luviquat®
Quaternary compounds. *BASF plc.*
18021 Luviquat® FC 370
29297-22-0
Polyquaternium-16(methylvinylimidazolium chloride/vinylpyrrolidone copolymer (30:70 ratio) aqueous solution); substantive cationic polymer used as conditioner in products for hair and skin care; film former; foam stabilizing and lubricating effects. *BASF AG.* Name unverified.
18022 Luviquat® FC 550
29297-22-0
Polyquaternium-16 (methylvinylimidazolium chloride/vinylpyrrolidone copolymer (50:50 ratio) aqueous solution; substantive cationic polymer used as conditioner in products for hair and skin care; film former; foam stabilizing and lubricating effects. *BASF AG.* Name unverified.
18023 Luviquat® FC 905
29297-22-0
Polyquaternium-16
Hair conditioners, rinses, shampoos, bleaches, dyes, liquid soaps, bath preparations, skin care products. *BASF AG.* Name unverified.
18024 Luviquat® HM 552
29297-22-0
Polyquaternium-16
Conditioner and setting resin for hair care. *BASF AG.* Name unverified.
18025 Luviquat® Mono CP
Hydroxyethyl cetyldimonium phosphate
Conditioner for hair cosmetics, esp. cold waves. *BASF.* Name unverified.
18026 Luviset
Resins for hairsprays. *BASF plc.*
18027 Luviset CA 66
25609-89-6
Vinyl acetate/crotonic acid copolymer; hair fixative for aerosols, hair sprays, setting lotions, conditioners. *BASF.* Name unverified.
18028 Luviset CAP
Vinyl acetate/crotonic acid/vinyl propionate copolymer; film-forming agents for hair-sprays and fixing lotions; compatible in aerosols with propane/butane. *BASF.* Name unverified.
18029 Luviset CAP X
Crotonic acid/vinyl acetate/vinyl propionate terpolymer; hair fixative for aerosols, hair sprays, setting lotions, conditioners. *BASF.* Name unverified.
18030 Luviskol®
Cosmetic and technical grades of PVP, vinyl acetate copolymers. *BASF plc.*
18031 Luviskol® K12, K17, K30, K60
9003-39-8 7879
Povidine
PVP; film-forming agent, hair fixative, thickener, protective colloid, suspending agent, and dispersant for cosmetics, adhesives, paints, coatings, paper, detergents, glass fibers, inks, ceramics, nonpharmaceutical tableting, photographic films, crop prot *BASF AG.*
18032 Luviskol® K80, K90
9003-39-8 7879
Povidine
PVP; film-forming agent, hair fixative, thickener, protective colloid, suspending agent, and dispersant for cosmetics, adhesives, paints, coatings, paper, detergents, glass fibers, inks, ceramics, nonpharmaceutical tableting, photographic films. *BASF AG.*
18033 Luviskol® VA 28 E
25086-89-9
PVP/VA copolymer (20:80 ratio), ethanol; film-forming agents for hair-sprays and fixing lotions. *BASF AG.* Name unverified.
18034 Luviskol® VAP 343 E
PVP/VA/vinyl propionate copolymer (30:40:30 ratio), ethanol; film-forming agents for hair-sprays and fixing lotions; compatible in aerosols with propane/butane. *BASF.* Name unverified.
18035 Luvisoft
Fabric softeners. *BASF plc.*
18036 Luvitol EHO
261-619-1
Cetearyl octanoate
Emollient oil component for cosmetics and pharmaceuticals. *BASF.*
18037 Luvitol HP
61693-08-1
Hydrogenated polyisobutene
Emollient oil component for cosmetics and pharmaceuticals. *BASF.* Name unverified.
18038 Luwax A
9002-88-4 7728
LDPE homopolymer; additive for printing inks; flatting and antisettling agent in paints; dispersant and color enhancer; in floor plishes; lubricant for plastics processing. *BASF.*
18039 Luwax AF 30
9002-88-4 7728
Micronized polyethylene wax; additive for printing inks, as flattening and antisettling agent in paints, dispersant in color concs.; improves hardness in wax compounds, black heel mark resistance in floor polishes; lubricant in plastics processing. *BASF.*
18040 Luwax AH 3
9002-88-4 7728
HDPE wax; additive for printing inks, as flattening and antisettling agent in paints, dispersant in color concs.; improves hardness in wax compounds, black heel mark resistance in floor polishes; lubricant in plastics processing. *BASF.*
18041 Luwax EAS 1
9002-88-4 7728
Polyethylene copolymer wax; additive for printing inks, as flattening and antisettling agent in paints, dispersant in color concs.; improves hardness in wax compounds, black heel mark resistance in floor polishes; lubricant in plastics processing. *BASF.*
18042 Luwax EVA 1
9002-88-4 7728
Polyethylene copolymer wax; additive for printing inks, as flattening and antisettling agent in paints, dispersant in color concs.; improves hardness in wax compounds, black heel mark resistance in floor polishes; lubricant in plastics processing. *BASF.*
18043 Luwax OA 2
Oxidized polyethylene wax; additive for printing inks, as flattening and antisettling agent in paints, dispersant in color concs.; improves hardness in wax compounds, black heel mark resistance in floor polishes; lubricant in plastics processing. *BASF.*
18044 Luwipal®
9003-08-1
Melamine formaldehyde resins, butanol or methanol etherified; combined with alkyd, polyacrylate, and epoxy resins to give light-resist, and weather-resistance baking finishes, e.g., automotive finishes. *BASF AG.*
18045 Luxalloy
Dental silver alloys and mercury for mixing amalgams; for dentistry and dental engineering. *Degussa Ltd.*
18046 Luxate
Polyurethane intermediates for coatings, elastomers, adhesives, sealants, thermoplastics. *Olin.* Discontinued.
18047 Luxene
A proprietary synthetic resin of the phenol-formal-dehyde class. No manufacturer.
18048 Luxene 44
A proprietary denture compound made from a vinyl copolymer. No manufacturer.
18049 Luxer®
Nonwoven fabric of synthetic filaments. *Asahi Chem. Industry.*
18050 Luxol
Solvent soluble dyes. *Morton Int'l. Ltd.* Discontinued.
18051 Luxor
A combination of hydrolyzed vegetable proteins and other ingredients; used

as a natural flavor in vegetables, meat, chicken and seafood dishes. *Hercules.* Discontinued.

18052 Luzenac B170
14807-96-6 9207 238-877-9
Talc; ultrabright appearance grade for polyolefins, PVC, elastomers, most other plastics. *R. T. Vanderbilt Co Inc.* Discontinued.

18053 Luzerne
A proprietary trade name for a hard rubber. No manufacturer.

18054 Luzidol
94-36-0 1149 202-327-6
$C_{14}H_{10}O_4$
Benzoyl peroxide.

18055 LX
Heat-reactive petroleum resins. *Neville Chemical Co.*

18056 LX-685® 125
Petrol, hydrocarbon resin; used in adhesives (hot melt pressure sensitive, mastic, pressure sensitive), coatings (alum., can and drum, emulsion, industrial, marine, traffic), inks (gelled varnishes, gravure, heat set, letterpress, lithographic), rubber (cements, mechanical goods, molded goods, tires), concrete-curing compounds andcaulking compounds. *Neville.*

18057 LX-782®
Petrol. hydrocarbon resin; used for inks, adhesives, coatings, rubbers, and concrete cures. *Neville.*

18058 lyargol
Silver proteinate

18059 Lycadex®
9050-36-6 232-940-4
Maltodextrin
Fat replacers in foods. *Roquette (UK) Ltd.*

18060 Lycal®
Lightly calcined sea water magnesia. *Steetley Magnesia Products Ltd.*

18061 Lycanol
339-44-6 4157 206-426-5
A proprietary trade name for glymidine. No manufacturer.

18062 Lycasin
Hydrogenated glucose syrup; a sugar substitute, in foods (confectionery) and pharmaceuticals (syrups and lozenges). *Roquette (UK) Ltd.*

18063 lycetol
Dimethylpiperazine tartrate
Tetradine. Used as a solvent for uric acid.

18064 lycine
The base of *Lycium barbarum.* It is identical to betaine.

18065 Lycopon
7772-98-7 8844 231-867-5
Sodium thiosulfate. A proprietary trade name for sodium hyposulfite. No manufacturer.

18066 Lycra®
Fibers. *DuPont UK.*

18067 Lycresse®
Fibers. *DuPont UK.*

18068 Lyddite
88-89-1 7562 201-865-9
Picric acid
Melinite; Pertite; Shimose. An explosive.

18069 Lydian stone
Lydite, touchstone. A siliceous slate containing about 84% silica, 5% alumina, and 1% ferric oxide; used for testing gold by rubbing it upon the stone and testing the streak of metal produced with acid.

18070 Lye glycerin
56-81-5 4493 200-289-5
Glycerin obtained from soap liquor.

18071 Lynex®
Polyphenylene ether-polyamide alloy *Asahi Chem. Industry.*

18072 Lynite
A trademark for an alloy of 88% aluminum, 10% copper, 1.5% iron, and 0.25% magnesium; specific gravity 2.95. No manufacturer.

18073 Lynx
A cotton fabric grade of Tufnol industrial laminates. *Tufnol Ltd.*

18074 Lyocol
Low foaming scouring, leveling and dispersing agent; combined scouring and dyeing of polyester fibers; beam dyeing and jet application. *Sandoz Products Ltd.*

18075 Lyofix
Finishing agent. *Ciba plc.* Name unverified.

18076 Lyofix 363
A proprietary trade name for a modified urea formaldehyde resin precondensate used in the crease-resistance finishing of cellulosic textiles. *Ciba plc.* Name unverified.

18077 Lyofix F
A proprietary trade name for a melamine-formal-dehyde resin precondensate used in finishing paper makers' felts. *Ciba plc.* Name unverified.

18078 Lyogen
Textile leveling agent; used for wool, nylon, polyurethane and polyamide fibers; vat dyestuffs; acid, milling, chrome and direct dyestuffs. *Sandoz Products Ltd.*

18079 Lyonore
A proprietary trade name for a steel containing 0.2% copper with some chromium and nickel. No manufacturer.

18080 Lyons sugar
8463
Sucramine, the ammonium salt of saccharine.

18081 Lypsyl
Lipsalve. *Ciba plc.* Name unverified.

18082 Lyracamine
Acrylic dyes. *ICI Chem & Polymers Ltd.* Discontinued.

18083 Lyric®, Lyril®
For the agriculture industry. *DuPont UK.*

18084 Lysargine
A colloidal silver, containing 60% of silver; used as an antiseptic.

18085 Lysase
Enzyme preparations; used for conversion of starch or starch hydrolysis products. *Roquette (UK) Ltd.* Discontinued.

18086 Lyse S, S III, S III Diff
Reagents which destroy red blood cells to leave white blood cells in suspension for analysis; used for white blood cell and hemoglobin determination. *Coulter Electronics Ltd.*

18087 lysidine
534-26-9 5666 208-596-6
$C_4H_8N_2$
4,5-dihydro-2-methyl-1H-imidazole
methyl glyoxaidine. mp = 105°; bp = 198-200°; soluble in H_2O, EtOH, less soluble in non-polar solvents.

18088 lysine
56-87-1 5667 200-294-2
$C_6H_{14}N_2O_2$
L-lysine
α, ε-diaminocaproic acid; l-lysine; Lys; K; (S)-2,6-diaminohexanoic acid; α,ε-diaminocaproic acid. Biochemical and nutritional research pharmaceuticals, culture media, fortification of foods and feeds, nutrient supplement, animal feed additive. dec 224.5°; $[\alpha]_D^{20}$= 14.6°; pK_1 = 2.20; soluble in water, insoluble in common neutral solvents. *Degussa; R.W. Greeff; U.S. Biochemical; Walton Pharmaceuticals Ltd.*

18089 L-(+)-lysine hydrochloride
657-27-2 5667 211-519-9
$C_6H_{14}N_2O_2 \cdot HCl$
L-Lysine monohydrochloride
Enisyl; Lysine hydrochloride; K; (S)-2,6-diaminohexanoic acid. Amino acid; enrichment of cereals and feeds Dec at 210°; $[\alpha]_D^{20}$= +14.6°; soluble in water; insoluble in non-polar solvents. *Person & Covey Inc.* Discontinued.

18090 dl-lysine monohydrochloride
70-53-1 5667 200-739-0
$NH_2(CH_2)_4CH(NH_2)COOH \cdot HCl$
Degussa; R.W. Greeff; Spectrum Chem. Mfg.; Tanabe USA.

18091 Lysivane
1094-08-2 3793 214-134-4
Ethopropazine hydrochloride
May & Baker Ltd. Name unverified.

18092 Lysmeral®
2-Methyl-3-(4-t-butylphenyl) propanol
Fragrance (floral, powdery, fresh, lily-of-the-valley-like) *BASF AG.*

18093 Lysochlor
59-50-7 2184 200-431-6
C_7H_7ClO
4-chloro-*m*-cresol
A disinfectant. It is a similar preparation to Eusapyl.

18094 Lysodren
53-19-0 6302 200-166-6
Mitotane
Antineoplastic. *Bristol Laboratories.* Name unverified.

18095 Lysoform
50-00-0 4262 200-001-8
formaldehyde

A solution of formaldehyde in alcoholic potash soap solution; a disinfectant much like Lysol. No manufacturer.

18096 Lysol
High activity, wide spectrum disinfectant. *Coventry Chemicals Ltd.*

18097 Lythol Oil
A commercial phenolated oil used as a disinfectant. No manufacturer.

18098 Lytor & RM
Tall oil resin. *Georgia-Pacific.*

18099 Lytron
Range of aqueous styrene copolymer emulsions with controlled particle size; for use in paper coatings and opacifying liquid detergents. *Morton Int'l. Ltd.*

18100 M B T S
120-78-5 3435 204-424-9
Benzothiazyl disulfide
A general purpose accelerator similar to MBT but with a higher activation temperature for greater processing safety; very active above 287°F (142°) used in both natural and synthetic rubbers. *Uniroyal.* Name unverified.

18101 M Violet 112
Manganese violet and bismuthoxychloride. Inorganic colorant. A mixture of manganese violet and bismuth oxychloride. *Presperse.* Discontinued.

18102 M.B
General chemicals and pharmaceuticals. *May & Baker Ltd.* Name unverified.

18103 M.B.A
110-26-9 203-750-9
$C_7H_{10}N_2O_2$
N,N'-Methylenebisacrylamide
MBA; NAPP. Cross-linking agent for preparation of polyacrylamides. N,N-Methylene-bis-acrylamide is an acrylimide compound cross-reacting with unidentified primary sensitizers in NAPP and Nyloprint UV-cured printing plates. mp > 300°; d = 1.235; soluble in H_2O (0.1 - 1.0 mg/ml), more soluble in organic solvents; LD_{50} (rat orl) = 390 mg/kg. *Cyanamid BV.* Name unverified.

18104 M.F.C
Materials for chromatography. *British Drug Houses.* Name unverified.

18105 M.N.T
1321-12-6 215-311-9
$C_7H_7NO_2$
Mononitrotoluenes
Mixture of three isomeric mononitrotoluenes. Used as an intermediate in the preparation of trinitrotoluene.

18106 M.O.D
Octyldodecyl myristate
Emollient; rancidless additive for cosmetics and pharmaceuticals. *Gattefosse.*

18107 M100
1308-38-9 2283 215-160-9
Cr_2O_3
chromic oxide
Chromic oxide; Chrome oxide green; Chromium (III) oxide; Chromium Sesquioxide; Ultramarine Green; Chrome Green; Chrome oxide; Chromic Oxide Sesquioxide; Chromic oxide pigment; Chromium oxide green pigments; Chromium oxide (Cr2O3); dichromium trioxide; Chromium (III) oxide hydrate; Anadonis Green; Chrome Ocher; Chromia; Green Cinnabar; Green Rouge; leaf green; oil green. Chromic oxide pigment. Also used in abrasives, refractories and semiconductors. d^{25} = 5.2200; mp = 2435°; bp = 4000° . *British Chrome & Chemicals Ltd.*

18108 M33, MN3
Polyimide resins. *Rhône-Poulenc NV.* Name unverified.

18109 M50
11113-70-5 234-347-6
Basic lead silico chromate; anticorrosive pigments for coatings. *Rheox Inc.* Discontinued.

18110 M66
Magnesia alumina spinel
Used with magnesia to produce bricks and monoliths for steel, nonferrous, glass and cement industries where generation of toxic hexavalent chrome from chrome-bearing refractories is a problem. *Alcoa Industrial Chemicals.*

18111 M9030
18769-78-3
$C_{31}H_{66}Si$
Methyltridecylsilane
Thermally stable fluid with superior metal-on-metal lubrication and wear characteristics. *Hüls Am.*

18112 MA 20
Soap powder for launderettes. *Unilever.* Name unverified.

18113 Maali resin
A yellowish-white resin resembling elemi of Samoa.

18114 Macadamite
An alloy of 72% aluminum, 24% zinc, and 4% copper.

18115 Macaja butter
Mocaya oil.

18116 Macaloid
Rheological additives; used for paints, sealants, agricultural products, cosmetics and ceramics. *Rheox Inc.*

18117 Macassar nutmeg butter
Papua nutmeg butter. The fat from *Myristica argentae.*

18118 mace butter
Nutmeg butter, obtained from the seeds of *Myristica officinalis,* of the East.

18119 Maceal
7-formyl-5-isopropyl-2-methyl bicyclo (2.2.2)-oct-2-ene
Quest Int'l. UK Ltd.

18120 Macerase®
Pectinase, an enzyme. *Calbiochem Corp.*

18121 MacFarland's alloy
Heat-resisting alloys. a) Contains 59% nickel, 30% chromium, and 11% copper. b) Contains 46% nickel, 43% chromium, and 11% copper.

18122 Macgill metal
An alloy containing 88% copper, 7% nickel, 4.5% iron, and traces of tin and lead. It resists corrosion.

18123 Machacon juice
An alkaline decoction of the juice of the root of a plant having the same name; used to coagulate rubber latex.

18124 Machete
23184-66-9 1533 245-477-8
Butachlor
Herbicide. *Monsanto Co.* Name unverified.

18125 machine bronze
Variable alloys containing 50-90% copper, 25% nickel, 0-30% tin, and 0.8% lead. Some alloys contain no nickel, and also contain zinc. Another alloy of this type consists of 83% copper, 16% zinc, and 1% tin.

18126 Mach's metal
An alloy of aluminum with 2-10% magnesium.

18127 Macht's metal
An alloy of 57% copper and 43% zinc; used for castings.

18128 M-acid
$C_{10}H_9NO_4S$
1-Amino-5-naphthol-7-sulfonic acid
Used as an azo dye intermediate.

18129 Mackam 1C
Sodium cocoamphoacetate
Surfactant for high alkaline cleansers, shampoos, baby shampoos. *McIntyre.*

18130 Mackam 1L
Sodium lauramphoacetate
Surfactant for nonirritant shampoos, cleaners. *McIntyre.*

18131 Mackam 2C
Disodium cocoamphodiacetate
Surfactant for baby shampoo, high alkaline cleaners. *McIntyre.*

18132 Mackam 2CT
Mixture of disodium caproamphodiacetate, sodium trideceth sulfate, and hexylene glycol; nonirritating shampoo. *McIntyre.*

18133 Mackam 2CY
Disodium capryloamphodiacetate
Low-foaming alkaline cleaner. *McIntyre.*

18134 Mackam 2L
Disodium lauroamphodiacetate
Surfactant for nonirritating shampoos, cleaners. *McIntyre.*

18135 Mackam 2W
Disodium wheat germamphodiacetate
Mild surfactant for baby shampoos; wetting agent in caustic cleaners. *McIntyre.*

18136 Mackam 35
Cocamidopropyl betaine
Foamer for shampoos, bubble baths, dishwash. *McIntyre.*

18137 Mackam 151C
84812-94-2 284-219-9
Cocaminopropionic acid
Mild surfactant, conditioner for shampoos. *McIntyre.*

18138 Mackam 151L
Lauraminopropionic acid
Mild surfactant, conditioner for shampoos. *McIntyre.*

18139 Mackam 160C
Sodium lauriminodipropionate
Surfactants for personal care and industrial applications. *McIntyre.*

18140 Mackam CB-35
68424-94-2 270-329-4
Cocobetaine
Surfactant, conditioner, viscosity builder, foam booster for shampoos, skin cleansers, bath products, heavy duty cleaners. *McIntyre.*

18141 Mackam CET
693-33-4 211-748-4
Cetyl betaine
Surfactant, conditioner, viscosity builder, foam booster for shampoos, skin cleansers, bath products, heavy duty cleaners. *McIntyre.*

18142 Mackam CSF
68919-41-5
Sodium cocoamphopropionate
Surfactant for heavy duty cleanser, metal and all-purpose cleaner, nonirritating shampoos. *McIntyre.*

18143 Mackam HV
25054-76-6 246-584-2
Oleamidopropyl betaine
Hair conditioner, foamer, emollient, viscosity builder. *McIntyre.*

18144 Mackam ISA
Isostearamidopropyl betaine
Surfactant, conditioner, viscosity builder, foam booster for shampoos, skin cleansers, bath products, heavy duty cleaners. *McIntyre.*

18145 Mackam J
Cocamidopropyl betaine
Surfactant, conditioner, viscosity builder, foam booster for shampoos, skin cleansers, bath products, and heavy duty cleaners. *McIntyre.*

18146 Mackam LMB
Lauramidopropyl betaine
Surfactant for shampoos, bubble baths, dishwash formulations. *McIntyre.*

18147 Mackam MLT
Sodium lauroamphoacetate and sodium trideceth sulfate; surfactant for shampoos, cleaners, detergents. *McIntyre.*

18148 Mackam OB-30
871-37-4 212-806-1
Oleyl betaine
Viscosity builder for alkaline cleanser. *McIntyre.*

18149 Mackam RA
86089-12-5 289-181-7
Ricinoleamidopropyl betaine
Surfactant, conditioner, viscosity builder, foam booster for shampoos, skin cleansers, bath products, heavy duty cleaners. *McIntyre.*

18150 Mackam TM
Dihydroxyethyl tallow glycinate
Surfactant, conditioner for shampoos; thickener for alkaline oven cleaners, acid bowl cleaners. *McIntyre.*

18151 Mackam WGB
133934-09-5
Wheat germamidopropyl betaine
Surfactant, conditioner, viscosity builder, foam booster for shampoos, skin cleansers, bath products, heavy duty cleaners. *McIntyre.*

18152 Mackamide AME-75, AME-100
142-26-7 205-530-8
$C_4H_9NO_2$
Acetamide MEA
N-(2-Hydroxyethyl)acetamide; N-acetyl ethanolamine. Humectant; surfactant, thickener, foam booster/stabilizer for personal care and industrial applications. *McIntyre.*

18153 Mackamide C
Cocamide DEA (1:1)
Foam stabilizer and thickener for shampoos, industrial applications. *McIntyre.*

18154 Mackamide CMA
68140-00-1 268-770-2
Cocamide MEA
Surfactant, thickener, foam booster/stabilizer for personal care and industrial applications. *McIntyre.*

18155 Mackamide ISA
52794-79-3 258-193-4
Isostearamide DEA
Lubricant, surfactant, thickener, foam booster/stabilizer for personal care and industrial applications. *McIntyre.*

18156 Mackamide L10
120-40-1 204-393-1
Lauramide DEA
Surfactant, thickener, foam booster/stabilizer for personal care and industrial applications. *McIntyre.*

18157 Mackamide LME
5422-34-4 226-546-1
Lactamide MEA
Conditioner for shampoos; surfactant, thickener, foam booster/stabilizer for personal care and industrial applications. *McIntyre.*

18158 Mackamide LMM
142-78-9 205-560-1
Lauramide MEA
Surfactant, thickener, foam booster/stabilizer for personal care and industrial applications. *McIntyre.*

18159 Mackamide LOL
Linoleamide DEA
Surfactant, thickener, foam booster/stabilizer for personal care and industrial applications. *McIntyre.*

18160 Mackamide MO
93-83-4 202-281-7
Oleamide DEA (1:1)
Surfactant, thickener, foam booster/stabilizer for personal care and industrial applications. *McIntyre.*

18161 Mackamide ODM
Oleamide DEA
DEA oleate. Surfactant, thickener, foam booster/stabilizer for personal care and industrial applications. *McIntyre.*

18162 Mackamide OP
111-05-7 203-828-2
Oleamide MIPA
Surfactant, thickener, foam booster/stabilizer for personal care and industrial applications. *McIntyre.*

18163 Mackamide PK
Palm kernelamide DEA
Surfactant, thickener, foam booster/stabilizer for personal care and industrial applications. *McIntyre.*

18164 Mackamide PKM
Palm kernelamide MEA
Surfactant, thickener, foam booster/stabilizer for personal care and industrial applications. *McIntyre.*

18165 Mackamide R
40716-42-5 255-051-3
Ricinoleamide DEA
Emulsifier; softener; surfactant, thickener, foam booster/stabilizer for personal care and industrial applications. *McIntyre.*

18166 Mackamide S
68425-47-8 270-355-6
Soyamide DEA (1:1)
Surfactant, thickener, foam booster/stabilizer for personal care and industrial applications; hair conditioner. *McIntyre.*

18167 Mackamide SMA
111-57-9 203-883-2
Stearamide MEA
Surfactant, thickener, foam booster/stabilizer for personal care and industrial applications. *McIntyre.*

18168 Mackamine CAO
68155-09-9 268-938-5
Cocamidopropylamine oxide
Detergent, wetting agent for heavy-duty cleaners; hair conditioner, viscosity builder, foam booster for personal care products. *McIntyre.*

18169 Mackamine CO
61788-90-7 263-016-9
Cocamine oxide
Detergent, wetting agent for heavy-duty cleaners; hair conditioner, viscosity builder, foam booster/stabilizer for personal care products. *McIntyre.*

18170 Mackamine IAO
Isostearamidopropylamine oxide
Detergent, wetting agent for heavy-duty cleaners; hair conditioner, viscosity builder, foam booster for personal care products. *McIntyre.*

18171 Mackamine ISMO
Isostearamidopropyl morpholine oxide
Detergent, wetting agent for heavy-duty cleaners; hair conditioner, viscosity builder, foam booster for personal care products. *McIntyre.*

18172 Mackamine LAO
61792-31-2 263-218-7
Lauramidopropylamine oxide
Detergent, wetting agent for heavy-duty cleaners; hair conditioner, viscosity builder, foam booster for personal care products. *McIntyre.*

18173 Mackamine LO
Lauramine oxide
Detergent, wetting agent for heavy-duty cleaners; hair conditioner, viscosity builder, foam booster for personal care products. *McIntyre.*

18174 Mackamine O2
14351-50-9 238-311-0
Oleamine oxide
Detergent, wetting agent for heavy-duty cleaners; hair conditioner, viscosity builder, foam booster for personal care products. *McIntyre.*

18175 Mackamine OAO
25159-40-4 246-684-6
Oleamidopropylamine oxide
Detergent, wetting agent for heavy-duty cleaners; hair conditioner, viscosity builder, foam booster for personal care products. *McIntyre.*

18176 Mackamine SAO
25066-20-0 246-598-9
Stearamidopropylamine oxide
Detergent, wetting agent for heavy-duty cleaners; hair conditioner, viscosity builder, foam booster for personal care products. *McIntyre.*

18177 Mackamine SO
2571-88-2 219-919-5
Stearamine oxide
Detergent, wetting agent for heavy-duty cleaners; hair conditioner, viscosity builder, foam booster for personal care products. *McIntyre.*

18178 Mackamine WGO
Wheat germamidopropylamine oxide
Detergent, wetting agent for heavy-duty cleaners; hair conditioner, visc. builder, foam booster for personal care products. *McIntyre.*

18179 Mackanate A-102
Disodium deceth-6 sulfosuccinate
Mild surfactant for personal care products. *McIntyre.*

18180 Mackanate A-103
Disodium nonoxynol-10 sulfosuccinate
Mild surfactant for personal care products. *McIntyre.*

18181 Mackanate AY-65TD
tridecyl alcohol and diamyl sodium sulfosuccinate. Surfactant, wetting agent for industrial applications. *McIntyre.*

18182 Mackanate CM
Disodium cocamido MEA-sulfosuccinate
Base for rug cleaners, shampoos, bubble baths. *McIntyre.*

18183 Mackanate CP
68515-65-1 271-102-2
Disodium cocamido MIPA-sulfosuccinate
Mild surfactant for personal care products. *McIntyre.*

18184 Mackanate DC-30
Disodium dimethicone copolyol sulfosuccinate
Mild surfactant for personal care products. *McIntyre.*

18185 Mackanate DC-30A
Diammonium dimethicone copolyol sulfosuccinate
Mild surfactant for personal care products. *McIntyre.*

18186 Mackanate DOS-40
577-11-7 3460 209-406-4
Dioctyl sodium sulfosuccinate
Surfactant, wetting agent for industrial applications. *McIntyre.*

18187 Mackanate EL
Disodium laureth sulfosuccinate
Surfactant for shampoos and bubble baths. *McIntyre.*

18188 Mackanate LA
Diammonium lauryl sulfosuccinate
Mild surfactant for personal care products, hand cleaners. *McIntyre.*

18189 Mackanate LM-40
25882-44-4 247-310-4
Disodium lauramido MEA-sulfosuccinate
High foaming surfactant for personal care products. *McIntyre.*

18190 Mackanate LO
Disodium lauryl sulfosuccinate
Mild surfactant for hand and skin cleaners, shampoos, bubble baths. *McIntyre.*

18191 Mackanate O-3
Disodium oleth sulfosuccinate
Mild surfactant for personal care products. *McIntyre.*

18192 Mackanate OD-35
56388-43-3 260-143-1
Disodium oleamido PEG-2 sulfosuccinate
Surfactant for mild conditioning shampoo. *McIntyre.*

18193 Mackanate OM
Disodium oleamido MEA-sulfosuccinate
Nonirritating high foaming surfactant for shampoos. *McIntyre.*

18194 Mackanate OP
43154-85-4 256-120-0
Disodium oleamido MIPA-sulfosuccinate
Mild, skin-protective surfactant for personal care products. *McIntyre.*

18195 Mackanate RM
Disodium ricinoleamido MEA-sulfosuccinate
Mild, skin-protective surfactant for personal care products. *McIntyre.*

18196 Mackanate UM
Disodium undecylenamido MEA-sulfosuccinate
Mild surfactant for personal care products. *McIntyre.*

18197 Mackanate WGD
Disodium wheat germamido PEG-2 sulfosuccinate
Nonirritating emollient surfactant for personal care products. *McIntyre.*

18198 Mackazoline C
61791-38-6 263-170-7
Cocoyl hydroxyethyl imidazoline
Emulsifier; corrosion inhibitor for acid bowl cleaners, pickling systems; salts as antistats, water displacer. *McIntyre.*

18199 Mackazoline CY
37478-68-5 253-521-2
Capryl hydroxyethyl imidazoline
Emulsifier; corrosion inhibitor for acid bowl cleaners, pickling systems; salts as antistats, water displacer. *McIntyre.*

18200 Mackazoline L
136-99-2 205-271-0
Lauryl hydroxyethyl imidazoline
Emulsifier; corrosion inhibitor for acid bowl cleaners, pickling systems; salts as antistats, water displacer. *McIntyre.*

18201 Mackazoline O
Oley hydroxyethyl imidazoline
Emulsifier; corrosion inhibitor for acid bowl cleaners, pickling systems; salts as antistats, water displacer. *McIntyre.*

18202 Mackechnie
7758-98-7 2722 231-847-6
Copper sulfate, pentahydrate, monohydrate and anhydrous. Used as a fungicide. *McKechnie Chemicals Ltd.* Name unverified.

18203 Mackechnie's bronze
Usually an alloy of 57% copper, 41% tin, 1% zinc, 1% iron, and 0.5% lead.

18204 Mackenzie's amalgam
Bismuth (2 parts) and lead (4 parts) are melted separately in crucibles, and each poured into 1 part mercury. These amalgams are then rubbed together.

18205 Mackenzie's metal
An alloy of 70% lead, 17% antimony, and 13% tin; also 68% lead, 16% antimony, and 16% bismuth. Electrotype metals.

18206 Mackernium 006
26062-79-3
Polyquaternium-6
Slip agent for liquid hand soaps, conditioner for shampoos without buildup. *McIntyre.*

18207 Mackernium 007
26590-05-6
Polyquaternium-7
Slip agent for liquid hand soaps; conditioner for shampoos without buildup. *McIntyre.*

18208 Mackernium KP
37139-99-4 253-363-4
Olealkonium chloride
Conditioner, lubricant, antistat for personal care products. *McIntyre.*

18209 Mackernium NLE
Quaternium-84
Conditioner, lubricant, antistat for personal care products. *McIntyre.*

18210 Mackernium SDC-25
122-19-0 204-527-9
Stearalkonium chloride
Conditioner, lubricant, antistat for personal care products. *McIntyre.*

18211 Mackester EGMS
Glycol stearate
Emulsifier, lubricant, antistat, defoamer for metalworking, textile lubricants, plastics, paper; emulsifier, pearlescent, emollient for cosmetics. *McIntyre.*

18212 Mackester IDO
59231-34-4 261-673-6
Isodecyl oleate
Emulsifier, lubricant, antistat, defoamer for metalworking, textile lubricants, plastics, paper; emulsifier, pearlescent, emollient for cosmetics. *McIntyre.*

18213 Mackester IP
Glycol stearate
Other ingredients; emulsifier, lubricant, antistat, defoamer for metalworking, textile lubricants, plastics, paper; emulsifier, pearlescent, emollient for cosmetics. *McIntyre.*

18214 Mackester TD-88
Triethylene glycol dioctoate
Emulsifier, lubricant, antistat, defoamer for metalworking, textile lubricants, plastics, paper; emulsifier, pearlescent, emollient for cosmetics. *McIntyre.*

18215 Mackine 101
68140-01-2 268-771-8
Cocamidopropyl dimethylamine
Intermediate for cationic surfactants, chemical specialties, hair conditioners, mild cleansers; corrosion inhibitor; salts as emulsifier for acid systems. *McIntyre.*

18216 Mackine 201
Ricinoleamidopropyl dimethylamine
Intermediate for cationic surfactants, chemical specialties, hair conditioners, mild cleansers; corrosion inhibitor; salts as emulsifier for acid systems. *McIntyre.*

18217 Mackine 301
7651-02-7 231-609-1
Stearamidopropyl dimethylamine
Intermediate for cationic surfactants, chemical specialties, hair conditioners, mild cleansers; corrosion inhibitor; salts as emulsifier for acid systems. *McIntyre.*

18218 Mackine 321
55852-13-6
Stearamidopropyl morpholine
Intermediate for cationic surfactants, chemical specialties, hair conditioners, mild cleansers; corrosion inhibitor; salts as emulsifier for acid systems. *McIntyre.*

18219 Mackine 401
67799-04-6 267-101-1
Isostearamidopropyl dimethylamine
Intermediate for cationic surfactants, chemical specialties, hair conditioners, mild cleansers; corrosion inhibitor; salts as emulsifier for acid systems. *McIntyre.*

18220 Mackine 421
Isostearamidopropyl morpholine
Intermediate for cationic surfactants, chemical specialties, hair conditioners, mild cleansers; corrosion inhibitor; salts as emulsifier for acid systems. *McIntyre.*

18221 Mackine 501
109-28-4 203-661-5
Oleamidopropyl dimethylamine
Intermediate for cationic surfactants, chemical specialties, hair conditioners, mild cleansers; corrosion inhibitor; salts as emulsifier for acid systems. *McIntyre.*

18222 Mackine 601
60270-33-9 262-134-8
Behenamidopropyl dimethylamine
Intermediate for cationic surfactants, chemical specialties, hair conditioners, mild cleansers; corrosion inhibitor; salts as emulsifier for acid systems. *McIntyre.*

18223 Mackine 701
Wheat germamidopropyl dimethylamine
Intermediate for cationic surfactants, chemical specialties, hair conditioners, mild cleansers; corrosion inhibitor; salts as emulsifier for acid systems. *McIntyre.*

18224 Mackine 801
3179-80-4 221-661-3
Lauramidopropyl dimethylamine
Intermediate for cationic surfactants, chemical specialties, hair conditioners, mild cleansers; corrosion inhibitor; salts as emulsifier for acid systems. *McIntyre.*

18225 Mackine 901
68188-30-7
Soyamidopropyl dimethylamine
Intermediate for cationic surfactants, chemical specialties, hair conditioners, mild cleansers; corrosion inhibitor; salts as emulsifier for acid systems. *McIntyre.*

18226 Mackpro KLP
Quaternium-79 hydrolyzed keratin; conditioner for hair and skin care products. *McIntyre.*

18227 Mackpro MLP
Quaternium-79 hydrolyzed milk protein; conditioner for hair and skin care products. *McIntyre.*

18228 Mackpro NLP, NLP-Special
Quaternium-79 hydrolyzed animal protein; hair conditioner. *McIntyre.*

18229 Mackpro NLW
Quaternium-79 hydrolyzed wheat protein; conditioner for hair and skin care products. *McIntyre.*

18230 Mackpro NSP
Quaternium-79 hydrolyzed silk; conditioner for hair and skin care products. *McIntyre.*

18231 Mackpro SLP
Quaternium-79 hydrolyzed soy protein; conditioner for hair and skin care products. *McIntyre.*

18232 Mackpro WWP
Wheatgermamidopropyl dimethylamine hydrolyzed wheat protein; conditioner for hair and skin care products. *McIntyre.*

18233 Mack's cement
Prepared by adding calcined sodium sulfate or potassium sulfate to dehydrated gypsum.

18234 Mackstat® DM
6440-58-0 229-222-8
$C_7H_{12}N_2O_4$
1,3-bis (hydroxymethyl)-5,5-dimethyl-2,4-imidazolidinedione
Dantoin DMDMH; Dimethyloldimethyl hydantoin; DMDMH; Glydant;Mackgard DM; Nipaguard DMDMH; Dantion DMDMH 55; Dantoguard. A broad spectrum cosmetic preservative for shampoos, skin cleansers, bath products, lotions and creams. A preservative and formaldehyde donor. *McIntyre.*

18235 maclurin
519-34-6 5676 208-268-2
$C_{13}H_{10}O_6$
(3,4-dihydroxyphenyl)(2,4,6-trihydroxyphenyl)-methanone
C.I. Natural Yellow 11; Kino-yellow; Laguncurin; Maclurin; Maklurin; Morintannic acid; Moritannic acid; Patent Fustin; 2,3',4,4',6-Pentahydroxybenzophenone. Used to dye fabrics. mp = 222-223°; soluble in H_2O (0.48 g/100 ml), more soluble in organic solvents;.

18236 Macol® 1
9003-11-6 7721
Poloxamer 181
Defoamer, deduster, emulsifier, detergent, dispersant, dye leveler, gellant, antistat, solubilizer, wetting agent, lubricant base for metalworking and textile lubricants, cosmetics, medical, paper, pharmaceutical, chemical intermediates. *PPG/Specialty Chem.* Discontinued.

18237 Macol® 2
9003-11-6 7721
Poloxamer 182
Detergent, emulsifier, wetting agent, dispersant, antistat, defoamer, gellant, solubilizer, lubricant base for cosmetic, medical, paper, metalworking, pharmaceutical and textile industries. *PPG/Specialty Chem.* Discontinued.

18238 Macol® 4
9003-11-6 7721
Poloxamer 184
Detergent, foaming agent, emulsifier, wetting agent, dispersant, antistat, defoamer, gellant, solubilizer, lubricant, base for cosmetic, medical, paper pharmaceutical and textile industries. *PPG/Specialty Chem.* Discontinued.

18239 Macol® 8
9003-11-6 7721
Poloxamer 188
Detergent, foaming agent, wetting agent, dispersant, antistat, gellant, solubilizer, lubricant base for cosmetic, medical, paper, pharmaceutical and textile industries. *PPG/Specialty Chem.* Discontinued.

18240 Macol® 15
9003-11-6 7721
Meroxapol 105
Defoamer, detergent, emulsifier, pulp and paper additive, dispersant, lubricant, leveling aid, wetting agent. *PPG/Specialty Chem.* Discontinued.

18241 Macol® 16
9003-11-6 7721
Meroxapol 108
Defoamer, deduster, detergent, emulsifier, dispersant, dye leveler, gellant, antistat. *PPG/Specialty Chem.* Discontinued.

18242 Macol® 18
9003-11-6 7721
Meroxapol 171
Defoamer, wetting agent, deduster, demulsifier, detergent, dispersant, dye leveler, gellant, antistat; synthetic lubricant base fluid for metalworking and textile lubricants, chemical intermediates; pulp and paper additive. *PPG/Specialty Chem.* Discontinued.

18243 Macol® 19
9003-11-6 7721
Meroxapol 172
Defoamer, wetting agent, deduster, demulsifier, detergent, dispersant, dye leveler, gellant, antistat; synthetic lubricant base fluid for metalworking and textile lubricants, chemical intermediates; pulp and paper additive. *PPG/Specialty Chem.* Discontinued.

18244 Macol® 23
9003-11-6 7721
Poloxamer 403
Foaming agent, emulsifier, wetting agent, dispersant, antistat, gellant, solubilizer, lubricant base for cosmetic, medical, paper, pharmaceutical and textile industries. *PPG/Specialty Chem.* Discontinued.

18245 Macol® 27
9003-11-6 7721
Poloxamer 407
Emulsifier, wetting agent, dispersant, antistat, defoamer, gellant, solubilizer, lubricant base for cosmetic, medical, paper, pharmaceutical and textile industries. *PPG/Specialty Chem.* Discontinued.

18246 Macol® 33
9003-11-6 7721
Meroxapol 311
Wetting agent, dispersant, antistat, defoamer, gellant, solubilizer, lubricant base for cosmetic, medical, paper, metalworking, pharmaceutical and textile industries. *PPG/Specialty Chem.* Discontinued.

18247 Macol® 34
9003-11-6 7721
Meroxapol 254
Wetting agent, dispersant, antistat, defoamer, gellant, solubilizer, lubricant base for cosmetic, medical, paper, pharmaceutical and textile industries. *PPG/Specialty Chem.* Discontinued.

18248 Macol® 35
9003-11-6 7721
Poloxamer 105
Wetting agent, dispersant, antistat, defoamer, gellant, solubilizer, lubricant base for cosmetic, medical, paper, pharmaceutical and textile industries. *PPG/Specialty Chem.* Discontinued.

18249 Macol® 40
9003-11-6 7721
Meroxapol 252
Wetting agent, dispersant, antistat, defoamer, gellant, solubilizer, lubricant base for cosmetic, medical, paper, metalworking, pharmaceutical and textile industries. *PPG/Specialty Chem.* Discontinued.

18250 Macol® 42
9003-11-6 7721
Poloxamer 122
Wetting agent, dispersant, antistat, defoamer, gellant, solubilizer, lubricant base for cosmetic, medical, paper, pharmaceutical and textile industries. *PPG/Specialty Chem.* Discontinued.

18251 Macol® 44
9003-11-6 7721
Poloxamer 124
Wetting agent, dispersant, antistat, defoamer, gellant, solubilizer, lubricant base for cosmetic, medical, paper, pharmaceutical and textile industries. *PPG/Specialty Chem.* Discontinued.

18252 Macol® 46
9003-11-6 7721
Poloxamer 101
Wetting agent, dispersant, antistat, defoamer, gellant, solubilizer, lubricant base for cosmetic, medical, paper, pharmaceutical and textile Industries. *PPG/Specialty Chem.* Discontinued.

18253 Macol® 72
9003-11-6 7721
Poloxamer 212
Wetting agent, dispersant, antistat, defoamer, gellant, solubilizer, lubricant base for cosmetic, medical, paper, pharmaceutical and textile industries. *PPG/Specialty Chem.* Discontinued.

18254 Macol® 77
9003-11-6 7721
Poloxamer 217
Wetting agent, dispersant, antistat, defoamer, gellant, solubilizer, lubricant base for cosmetic, medical, paper, pharmaceutical and textile industries. *PPG/Specialty Chem.* Discontinued.

18255 Macol® 85
9003-11-6 7721
Poloxamer 235
Wetting agent, dispersant, antistat, defoamer, gellant, solubilizer, lubricant base for cosmetic, medical, paper, pharmaceutical and textile industries. *PPG/Specialty Chem.* Discontinued.

18256 Macol® 101
9003-11-6 7721
Poloxamer 331
Wetting agent, dispersant, antistat, defoamer, gellant, solubilizer, lubricant base for cosmetic, medical, paper, pharmaceutical and textile industries. *PPG/Specialty Chem.* Discontinued.

18257 Macol® 108
9003-11-6 7721
Poloxamer 338
Wetting agent, dispersant, antistat, defoamer, gellant, solubilizer, lubricant base for cosmetic, medical, paper, pharmaceutical and textile industries. *PPG/Specialty Chem.* Discontinued.

18258 Macol® 123
Cetearyl alcohol, ceteareth-20 and ceteareth-10. Emulsifier for cosmetic and pharmaceutical applications. *PPG/Specialty Chem.* Discontinued.

18259 Macol® 125
Stearyl alcohol and ceteareth-20. Emulsifier for cosmetic and pharmaceutical applications. *PPG/Specialty Chem.* Discontinued.

18260 Macol® 300
PPG-7 buteth-10
Detergent for toilet cleaners, laundry detergents, emulsion polymerization, defoamers, metalworking fluids, hydraulic fluids. *PPG/Specialty Chem.* Discontinued.

18261 Macol® 660
PPG-12 buteth-16
Detergent for toilet bowl cleaners, laundry; defoamer, rubber lubricant, intermediate; hydraulic, heat transfer, and metal working fluids; mold release agent; emulsion polymerization. *PPG/Specialty Chem.* Discontinued.

18262 Macol® CA-2
9004-95-9
Ceteth-2
Emulsifier, detergent, wetting agent, dispersant, solubilizer, coupling agent for cosmetics, textile, metalworking, household, industrial and other applications. *PPG/Specialty Chem.* Discontinued.

18263 Macol® CPS
$C_{56}H_{114}O_{21}$
Cetearyl alcohol, polysorbate 60, PEG-150 stearate and steareth-20. Emulsifier for cosmetic and pharmaceutical applications. mp = 35-40°; fp > 149°. *PPG/Specialty Chem.* Discontinued.

18264 Macol® CSA-2
68439-49-6 7737
Ceteareth-2
Detergent, wetting agent, emulsifier, dispersant, solubilizer, and coupling agent for cosmetics, textiles, metalworking lubricants, household products, and industrial applications. *PPG/Specialty Chem.* Discontinued.

18265 Macol® DNP-5
9014-93-1
Nonyl nonoxynol-5
Emulsifier, detergent, wetting agent, dispersant, solubilizer, coupling agent for cosmetics, textile, metalworking, household, industrial and other applications. *PPG/Specialty Chem.* Discontinued.

18266 Macol® E-200
25322-68-3 7729
PEG-4
Chemical intermediate for fatty acid esters, lubricant bases in textile and metalworking, as components in pharmaceutical and cosmetic preparations. *PPG/Specialty Chem.* Discontinued.

18267 Macol® LA-4
Laureth-4
Detergent, wetting agent, emulsifier, dispersant, solubilizer, stabilizer, coupling agent for cosmetics, textiles, metalworking lubricants, household products, industrial uses. *PPG/Specialty Chem.* Discontinued.

18268 Macol® LF-110
Polyalkoxylated aliphatic ether
Biodegradable low foam wetting aid and low surface tension surfactant; synthetic lubricant base fluid for metalworking, hardsurface cleaning, and metal cleaning and degreasing; used in cleaners and rinse aids. *PPG/Specialty Chem.* Discontinued.

18269 Macol® NP-4
9016-45-9 6772
Nonoxynol-4
Emulsifier, detergent, wetting agent, dispersant, solubilizer, coupling agent for cosmetics, textile, metalworking, household, industrial and other applications. *PPG/Specialty Chem.* Discontinued.

18270 Macol® OA-2
9004-98-2
$C_{56}H_{116}O_{21}$
Oleth-2
Detergent, wetting agent, emulsifier, dispersant, solubilizer, stabilizer, coupling agent for cosmetics, textiles, metalworking lubricants, household products, industrial uses. mp = 25-30°; fp > 149° *PPG/Specialty Chem.* Discontinued.

18271 Macol® OP-3
9002-93-1 6858

Octoxynol-3
Polyethylene glycol mono[4-(1,1,3,3-tetramethylbutyl)phenyl] ether. Emulsifier, detergent, wetting agent, dispersant, solubilizer, coupling agent for cosmetics, textile, metalworking, household, industrial and other applications. d = 1.0820; n_D^{20}= 1.4902; fp = 274°. *PPG/Specialty Chem.* Discontinued.

18272 Macol® P-500
PPG-9
Defoamer for aqueous systems, in mold release applications, lubricant bases for textile, paper, metalworking formulations, chemical intermediates for fatty acid esters, components for urethane resins. *PPG/Specialty Chem.* Discontinued.

18273 Macol® SA-2
9005-00-9
Steareth-2
Polyoxyethylene(20) stearyl ether. Detergent, wetting agent, emulsifier, dispersant, solubilizer, stabilizer, coupling agent for cosmetics, textiles, metalworking lubricants, household products, industrial uses. mp = 35-40°; fp > 149°. *PPG/Specialty Chem.* Discontinued.

18274 Macol® TD-3
24938-91-8
Trideceth-3
Detergent, wetting agent, emulsifier, dispersant, solubilizer, stabilizer, coupling agent for cosmetics, textiles, metalworking lubricants, household products, industrial uses. *PPG/Specialty Chem.* Discontinued.

18275 Macol® WSL-2000
Non-butanol functional fluid; lubricant in metalworking, textile, and hydraulic fluids. *PPG/Specialty Chem.* Discontinued.

18276 Macor
Photographic chemicals. *Makhteshim Chemical Works Ltd.*

18277 Macquer's salt
7784-41-0 7767 232-065-8
KH_2AsO_4
potassium arsenate

18278 macrogol stearate
Polyoxyl 40 stearate or polyoxyl 8 stearate.

18279 macrogols
Polyethylene glycols.

18280 Macrolex®
High quality dyestuffs; used for dyeing polystyrene, styrene copolymers, polymethacrylate rigid PVC and polycarbonate. *Bayer AG; Bayer plc.*

18281 Macromite
471-34-1 1697 207-439-9
Calcium carbonate
Very fine wet-ground product for paint and polymer industries. *ECC International Ltd.* Discontinued.

18282 Macrosorb
Range of inorganic matrices for separation and support. *Microporous Materials Ltd.*

18283 Macrospherical® 95
Aluminum chlorohydrate powder with a spherical shell; antiperspirant active. *Reheis Inc.*

18284 Macrynal
Hydroxy acrylic resins for use with polyisocyanates; used for automotive paints and industrial finishes. *Resinous Chemicals Ltd.*

18285 Maculanin
Potassium amylate

18286 Macuprax
Bordeaux/cufraneb. Fungicide composed of basic cupric sulfate and cufraneb *McKechnie Chemicals Ltd.* Name unverified.

18287 Madanite
A product used as a binding material made from 2 parts petroleum jelly and 1 part rubber.

18288 Madar fiber
A bast fiber, known in India by this name. obtained from *Calotropis procera and C. gigantea.*

18289 madder
152-84-1 8437 205-808-9
The powdered root of the plant, *Rubia tinctorum.* The chief constituent is ruberythric acid. This acid is a glucoside, and is split up into alizarin and a sugar by the action of acids.

18290 Maddrell salts
Alkaline meta-phosphates which are insoluble in water. They are made by heating monosodium phosphate at 245-250°C. Used in dentrifices, polishes and abrasive detergents.

18291 Madecassol
16830-15-2 869 240-851-7
$C_{48}H_{78}O_{19}$
(4α)-2α,3β,23-trihydroxyurs-12-en-28-oic acid O-6-deoxy-α-L-monnopyranosyl-(1→4)-O-β-D-glucopyranosyl-(1→6)-O-β-D-glucopyranosyl ester
Asiaticoside; Centelase Dermatologico. A proprietary preparation extract of *Centella asiatica* in an ointment; used for skin protection. mp = 230-233°; $[\alpha]_D^{20}$= -14° (EtOH); soluble in EtOH, C_5H_5N. *Rona Laboratories.* Unverified.

18292 madol oil
An oil obtained from the seeds of *Garcinia echinocarpa* .

18293 Madquat Q-6
5039-78-1 225-733-5
$C_9H_{18}ClNO_2$
Methacrylatoethyl trimethylammonium chloride
Methylchloride quaternary salt of dimethylaminoethyl methacrylate. *Rhône Poulenc Surf.*

18294 Madurit®
Melamine resin products; used for laminate and chipboard production, improvement of paper wet strength, bonding of glass fleeces. *Cassella AG.*

18295 Mafloc® 700
Powdered polymer; dewatering aid in vacuum filters and centrifuge systems; especially effective for municipal and industrial wastewater sludge and potable water clarification. *PPG/Specialty Chem.* Discontinued.

18296 Mafloc® 718
Emulsion polymer; used as sludge thickener, vacuum filter, centrifuge, plate and frame filter press, belt filter press in industrial and municipal waste dewatering applications. *PPG/Specialty Chem.* Discontinued.

18297 Mafloc® 764
Polyacrylate polymer; flocculating power towards suspension of solids. *PPG/Specialty Chem.* Discontinued.

18298 Mafo® 13
Potassium salt of complex N-stearyl amino acid; biodegradable emulsifier, wetting agent, corrosion inhibitor, suspending agent, solubilizer for difficult materials, dairy chain and metal-to-metal lubricant, emollient; used for burnishing and polishing compounds. *PPG/Specialty Chem.* Discontinued.

18299 Mafo® C
Cocamidopropyl betaine
Biodegradable solubilizer, emollient, coupling agent, emulsifier, foam booster for shampoos. *PPG/Specialty Chem.* Discontinued.

18300 Mafo® CB 40
68424-94-2 270-329-4
Coco betaine
Biodegradable foam booster, viscosity builder, solubilizer, emollient, coupling agent, emulsifier for shampoos. *PPG/Specialty Chem.* Discontinued.

18301 Mafo® CSB
68139-30-0 268-761-3
Cocamidopropyl hydroxysultaine
Biodegradable solubilizer, emollient, coupling agent, emulsifier, foam booster for shampoos. *PPG/Specialty Chem.* Discontinued.

18302 Mafo® LMAB
Lauramidopropyl betaine
Biodegradable solubilizer, emollient, coupling agent, emulsifier, foam booster for shampoos. *PPG/Specialty Chem.* Discontinued.

18303 Mafo® OB
871-37-4 212-806-1
Oleyl betaine
Biodegradable solubilizer, emollient, coupling agent, emulsifier, foam booster for shampoos. *PPG/Specialty Chem.* Discontinued.

18304 Maftec®
1344-28-1 369 215-691-6
Alumina fiber; reinforcing agent for high temperature usage and composite materials, e.g., furnace lining. *Mitsubishi Kasei.* Name unverified.

18305 Mafu®
62-73-7 3129 200-547-7
dichlorvos. DDVP; Insecticidal aerosol used for control of stored product pests in filled and empty silos, store rooms, etc.; for flying insects, weevils, mites and bugs. *Bayer AG.*

18306 Mafura fat
A fat obtained from the seeds of *Mafureira oleifera.* It contains from 92-95% fatty acids; used in the manufacture of soaps.

18307 Mag-40
Liquid chemical defoliant of cotton and a desiccant in silverskin onion. *Makhteshim Chemical Works Ltd.* Discontinued.

18308 Magadi soda
An East African soda. It contains sodium carbonate and sodium bicarbonate.

18309 Magala® 0.5E
61632-57-3
Di-n-butyl magnesium/triethyl aluminum complex (0.5:1); used in production

of catalysts for polymerization of olefins or dienes and as an alkylating agent. *Akzo.*

18310 MagChem® 10B
1309-48-4 5713 215-171-9
MgO
Magnesium oxide
Hardburned screened grade specially processed to provide low boron level in fused elec. heating element insulation. *Martin Marietta Magnesia Spec.* Name unverified.

18311 MagChem® 20
1309-48-4 5713 215-171-9
Magnesium oxide
Lightburned reactive grade for manufacture of magnesium chemicals, construction products, lubrication oil additives, fuel additives, oil drilling chemicals, in neoprene and chlorinated polymers, sugar refining, uranium ore processing, acid neutralization, leather tanning and water treatment. *Martin Marietta Magnesia Spec.* Name unverified.

18312 MagChem® 1060
1309-48-4 5713 215-171-9
MgO
Magnesium oxide
Hardburned screened grade for refractories, water treatment, acid neutralization, desilication, heavy metals removal. *Martin Marietta Magnesia Spec.* Name unverified.

18313 Magcoke
Magnesium impregnated coke for the desulfurization and nodularization of cast iron. *Foseco (F.S.) Ltd.*

18314 Magecol
A proprietary lampblack suitable for rubber goods. No manufacturer.

18315 Magicote Masonry Paint
Vinyl copolymer based, water thinned; used for brickwork, cement, stone, concrete, pebbledash, internal and external. *Berger Jenson & Nicholson Ltd.* Name unverified.

18316 Magicote Non Drip and Liquid Gloss
Alkyd resin based, white spirit thinned; for internal and external application in the DIY market; non drip; needs no undercoat, liquid requires Magicote undercoat. *Berger Jenson & Nicholson Ltd.* Name unverified.

18317 Magicote solid Emulsion
Vinyl copolymer based, water thinned; available in Vinyl Matt and Vinyl Silk to apply by roller to interior walls and ceilings; supplied ready-for-use in roller tray. *Berger Jenson & Nicholson Ltd.* Name unverified.

18318 Magicote Vinyl Matt
Vinyl copolymer based, water thinned; washable finish for interior walls and ceilings. *Berger Jenson & Nicholson Ltd.* Name unverified.

18319 Magicote Vinyl Silk
Vinyl copolymer based, water thinned; washable finish for interior walls and ceilings. *Berger Jenson & Nicholson Ltd.* Name unverified.

18320 magister of bismuth
1304-85-4 1326 215-136-8
$BiO \cdot NO_3 \cdot H_2O$
bismuth subnitrate. Basic bismuth nitrate. An antacid.

18321 magister of sulfur
Sulfur precipitated in an amorphous condition from solutions of hyposulfites or polysulfides, by acids.

18322 magistral
An impure copper sulfate containing ferric oxide, sodium sulfate, and sodium chloride.

18323 Maglite Y
1309-48-4 5713 215-171-9
A trademark for magnesium oxide. *E Merck.* Discontinued.

18324 Maglite® D
1309-48-4 5713 215-171-9
Magnesium oxide
Used in cure of polychloroprene compounds to aid aging stability, absorbing HCl. *Marine Magnesium.*

18325 Magmet
Magnetic metal particles; used in coatings for magnetic tapes and discs. *Hercules.* Discontinued.

18326 Magna A
68334-28-1 269-820-6
Partially hydrogenated vegetable oil (cottonseed, soybean); kosher; high melting, high performance fat for no-tempering coatings, centers, vegetable dairy systems; shortening for bakery applications. *Van Den Bergh Foods.*

18327 Magna Flow
Flow improvers and pour point depressants. *Baker Performance Chemicals Ltd.*

18328 Magna Tac
Specialty adhesives for commercial and industrial applications; for pressure sensitives, two-part epoxies, solvent-and water-based systems, hot melts and hat sizings. *Beacon Chemical Company Inc.* Name unverified.

18329 Magnabrite® F
Magnesium aluminum silicate NF
Stabilizing and suspending agent for cosmetics and pharmaceuticals. *Am. Colloid.*

18330 Magnacell
Nitrocellulose paste dispersions. *Tennant-KVK Ltd.*

18331 Magnacide
Biocides and algicides. *Baker Performance Chemicals Ltd.*

18332 Magnaclean
Cleaning chemicals and antifoulants. *Baker Performance Chemicals Ltd.*

18333 Magnaclear
Reverse emulsion breakers and water clarifiers. *Baker Performance Chemicals Ltd.*

18334 Magnacryl
Second generation acrylic adhesives and radiation curable (uv/eb) adhesives and coatings; for pressure sensitive and permanent bonding, specialty overcoating application. *Beacon Chemical Company Inc.* Name unverified.

18335 Magnaflo
Replaces magnesium sulfate in stainblocking treatment of carpets with Stain-Free. *Sybron.*

18336 Magnafloc®
Synthetic flocculants and coagulants designed to improve the separation and handling of finely divided solids in aqueous suspension; a very wide range of products is available in terms of chemical type, molecular weight and ionic character; used in sedimentation, filtration and centrifugation. *Allied Colloids Ltd.*

18337 Magnafloc® LT
High molecular weight polyacrylamides for use as coagulant aids in treatment of drinking water and waterworks sludges; synthetic organic coagulants as alum replacements. *Allied Colloids Ltd.*

18338 Magnakyd
Alkyd paste dispersions. *Tennant-KVK Ltd.*

18339 magnalium
Alloys of magnesium and aluminum. One contains from 1-2% of magnesium. It is lighter than aluminum, and is as hard as brass. Another alloy contains 10% magnesium; is used for parts of machinery, for cooking utensils, and for opt ical mirrors.

18340 Magnamite
Carbonized or graphitized polyacrylonitrile fiber; used to reinforce thermoset and thermoplastic resins, and as a replacement for steel, aluminum, titanium and other metals. *Hercules.* Discontinued.

18341 Magnaphoscal®
Multiple phosphate (sodium, calcium, magnesium phosphate) granulated; for mineral feeds and mixed feeds. *Bayer AG.*

18342 Magnaplas PA213-BF83
Nylon 6, barium ferrite-filled; permanent magnetic material *Bay Resins.*

18343 Magnaset
High pigmented binderless paste dispersions. *Tennant-KVK Ltd.*

18344 Magnasoft
Silicone textile softener. *Union Carbide.*

18345 Magnasol
Pigment solution. *Tennant-KVK Ltd.*

18346 Magnasorb
Glycols/alkanolamines. *Baker Performance Chemicals Ltd.*

18347 Magnasperse
Paste dispersions. *Tennant-KVK Ltd.*

18348 Magnaspheres®
1309-42-8 5706 215-170-3
Magnesium hydroxide for water treatment. *Steetley Magnesia Products Ltd.*

18349 Magnate
35554-44-0 3622 252-615-0
Imazalil
Fungicide *Makhteshim Chemical Works Ltd.*

18350 Magnatreat
Oxygen and sulfide scavengers. *Baker Performance Chemicals Ltd.*

18351 magnesia
1309-48-4 5713 215-171-9
MgO
Magnesium oxide
Used in refractories, for steel furnace linings, polycrystalline ceramic for aircraft windshields, electrical insulation, pharmaceuticals and cosmetics.

18352 magnesia alba
39409-82-0 5696

$C_4H_2Mg_5O_{14}\cdot 5H_2O$
magnesium carbonate hydroxide
Marinco C. A basic magnesium carbonate of variable composition. Used in fireproofing, as an antacid and laxative and in manufacture of magnesium compounds.

18353 magnesia bleaching liquid
O_2MgCl_2
Magnesium oxychloride
Used for bleaching.

18354 magnesia white
Both magnesium oxide and magnesium carbonate are known by this name.

18355 magnesia-citrate mixture
Citric acid (20g), is dissolved in 20% ammonium hydrate, and mixed with 1 liter of magnesia mixture (*qv*); used in the determination of phosphorus magnesia mixture.

18356 magnesium
7439-95-4 5687 231-104-6
Mg
Metallic element; aluminum alloys for structural parts, etc.; pyrotechnics; photography; production of iron, nickel, zinc, titanium, zirconium, steel; gasoline additive; magnesium compounds; cathodic protection; reducing agent; precision instruments optical mirrors. mp = 651°; bp = 1100°; d^{20} = 1.738. *Aldrich; Norsk Hydro A/S; Pechiney Electrométallurgie.*

18357 magnesium acetate
142-72-3 5688 205-554-9
$C_4H_6MgO_4\cdot 4H_2O$
acetic acid magnesium salt tetrahydrate
Cromosan. Dye fixative in textile printing, deodorant, disinfectant, antiseptic. mp = 80°; d = 1.45; LD_{50} (mus iv) = 18 mg/kg. *EM Industries; Hoechst Celanese; Verdugt BV.*

18358 magnesium base alloys
The magnesium in these alloys usually varies from 90-95%.

18359 magnesium carbonate
546-93-0 208-915-9
$CO_3.Mg$
Carbonic acid magnesium salt
Magnesium(II) carbonate(1:1); CI77713; Elastocarb Tech Light, Tech Heavy; Magocarb-33. Inorg. filler providing flame retardancy and smoke suppression to elastomers, plastics, and thermosets incl. EPDM, PP, PE, PVC; used in wire and cable compounds, conduit/tubing, film and sheet Soluble in acids; insoluble in alcohol, H_2O; incompatible with formaldehyde; heated to decomposition emits acrid smoke and irritating fumes. Advance Research Chems; Allchem Ind.; Am. Intl.; Baymag; Chemisphere; EnerChem; Fluka; Giulini intl.; Lohmann; Lonza; Magnesia GmbH; Mallinckrodt Baker; Marie Magnesium; Martin Marietta; Magnesia Spec.; Morton Int'l.

18360 magnesium chloride
7786-30-3 5698 232-094-6
Cl_2Mg
magnesium chloride anhydrous
Magnogene. Source of magnesium; disinfectants, fire extinguishers, fireproofing wood, cement, refrigerating brines, ceramics, cooling drilling tools, textile sizes and lubricants, paper manufacture, dust control on roads, flocculating agent, catalyst. mp = 712°; d = 2.41; soluble in H_2O; LD_{50} (rat orl) = 8.1 g/kg. *Aldrich; Magnesia GmbH; Mallinckrodt; Schaefer Salt & Chem.*

18361 magnesium fluoride
7783-40-6 5701 231-995-1
F_2Mg
Afluon; Sellaite. Used in manufacture of ceramics and glass. mp = 1248°; bp = 2260°; d = 3.148; soluble in H_2O (87 mg/l), insoluble in organic solvents; LD_{50} (gpg orl) = 1 g/kg.

18362 magnesium gluconate
3632-91-5 4464 222-848-2
$C_{12}H_{22}MgO_{14}\cdot 2H_2O$
magnesium D-gluconate
D-gluconic acid magnesium salt; Almora; Ultra-Mg. Mineral source for pharmaceutical and food products *Akzo; Atomergic Chemetals; Spectrum Chem. Mfg.*

18363 magnesium laureth sulfate
67702-21-4 221-450-6
$[CH_3(CH_2)_{10}CH_2(OCH_2CH_2)_nOSO_3]_2\cdot Mg$; avg n=1-4
Magnesium lauryl ether sulfate
Elfan® NS 243 S Mg; Empicol® EGB, EGC, EGC70; Sulfochem® MgLES; Zoharpon MgES. Detergent and toiletry raw material. *Albright & Wilson UK;.*

18364 magnesium lauryl sulfate
3097-08-3 221-450-6
Magnesium monododecyl sulfate
Akyposal MGLS; Carsonol® MLS; Elfan® 2240 Mg;Empicol® ML 26/F; Empicol ML30; Rhodapon® LM; Standapol® MG; Stepanol® MG; Sulfetal MG30; Sulfochem MG; Surfax MG; Texapon® MGLS; Unipol MGLS; Witcolate™ MGLS; Zoharpon MgS. Detergent used in the mfg. of shampoos and toothpaste. d = 8.57 lb/gal; pH = 6.5-7.5. *Albright & Wilson UK; Chem-Y-GmbH; Lonza.*

18365 magnesium nitrate
10377-60-3 5710 233-826-7
MgN_2O_6
Magnesium(II) nitrate
Nitric acid,magnesium salt; nitromagnesite. Used in pyrotechnics and in the manufacture of nitric acid. mp = 95°; soluble in H_2O (1.25 g/ml), more soluble in EtOH. *Blythe, William Ltd; EM Industries; Hoechst Celanese; Mallinckrodt.*

18366 magnesium silicate
1343-88-0 215-681-1
Silicic acid, magnesium salt (1:1).
Florisil. Inorganic salt of variable composition; $3MgSiO_3\cdot 5HOH$ (variable); +zw $MgO\cdot SiO_2\cdot xH_2O$; Rubber filler, ceramics, glass, refractories; absorbent for crude oil spills; animal and vegetable oils (bleaching agent). Used in column chromatography. *Cyprus Industrial Min.; PQ Corp.; R. T. Vanderbilt Co Inc.*

18367 magnesium stearate
557-04-0 5730 209-150-3
$C_{36}H_{70}MgO_4$
magnesium octadecanoate
octadecanoic acid, magnesium salt. Magnesium salt of stearic acid baby dusting powder; lubricants in making tablets; drier in paints and varnishes; stabilizer and lubricant for plastics; emulsifying agent for cosmetics. mp = 130-140°; insoluble in H_2O. *EM Industries; Ferro/Grant; Magnesia GmbH; Mallinckrodt; Norac; Syn. Products; Witco.*

18368 magnesium sulfate
7487-88-9 5731 231-298-2
MgO_4S
magnesium sulfate
sal Catharticum; sal Anglicum; sal Seidlitense; (monohydrate) kieserite; Salts, salts of England, hair salt, bitter salt. Used in fireproofing, textiles (warp sizing, dyeing, etc.), mineral waters, catalyst carrier, paper (sizing), cosmetic lotions. Heptahydrate: d = 1.67; soluble in H_2O. *Blythe, William Ltd; Heico; Mallinckrodt; PQ Corp.*

18369 magnesium sulfonate
71786-47-5
Benzenesulfonic acid, mono- and dialkyl derivatives, magnesium salts. Lube and fuel oil additive, rust preventive. *King Industries; Lubrizol; Witco/Sonnebom.*

18370 Magnesium-Monel
A proprietary trade name for an alloy of 50% magnesium and 50% monel metal. No manufacturer.

18371 Magnesol
Hydrated, amorphous, synthetic magnesium silicate; used for adsorption, catalyst support, edible oil and fat reclamation, anticaking, deodorization. *Reagent Chemical & Research Inc.* Name unverified.

18372 magnet steel
Alloys of iron with from 5-50% cobalt, 5-18% nickel, 0-7% manganese, 1-12% tungsten, and 0-12% chromium.

18373 Magnetic Black
A proprietary trade name for a finely ground magnetic iron oxide for use as a pigment. No manufacturer.

18374 magnetic iron ore
$FeO\cdot Fe_2O_3$
Loadstone
lodestone; magnetite; black oxide of iron; ferroferrite. A black ore of iron. It is a ferroso-ferric oxide, and contains over 72% iron.

18375 magnetic pyrites
Pyrrhotine
pyrrhotite; magnetic sulfide of iron. A mineral. In composition it approximates to ferrous sulfide, FeS, or ferroso-ferric sulfide, Fe_3S_4, but often contains nickel.

18376 Magnevist
86050-77-3 4347
$C_{28}H_{54}GdN_5O_{20}$
Dimeglumine gadopentetate
SHL-451A; ZK-93035; Resovist. MRI agent. Soluble in H_2O; LD_{50} (rat iv) = 9.37 g/kg. *Schering Health Care Ltd.* Discontinued.

18377 Magnifin® H7A
1309-42-8 5706 215-170-3
Magnesium hydroxide
Halogen-free flame retardant/filler for plastics and rubber, esp. EPDM, EPM, XL-PE, XL-EVA; environmentally friendly. *Martinswerk GmbH.*

18378 Magnifin® H10
1309-42-8 5706 215-170-3

Magnesium hydroxide
Halogen-free, general purpose flame retardant/filler for plastics and rubber; environmentally friendly. *Martinswerk GmbH.*

18379 Magnifin® H10C
1309-42-8 5706 215-170-3
Magnesium hydroxide
Halogen-free flame retardant/filler for plastics and rubber, especially PVC, EVA, EPDM, PE copolymer; environmentally friendly. *Martinswerk GmbH.*

18380 Magnisal
10377-60-3 5710 233-826-7
$MgN_2O_6{>}H_2O$
Magnesium (II) nitrate hexahydrate
Magnesium nitrate hexahydrate; Nitric acid, magnesium salt; Magnesium nitrate. Used for agricultural (for curing magnesium deficiency via irrigation system, direct soil application or foliar spray) and technical (in metal, textile, ceramic and other industries) applications. *Haifa Chemicals Ltd.* Name unverified.

18381 Magno
An electrical resistance alloy containing 95% nickel and 5% manganese.

18382 Magnobond 3
Construction adhesive; bonds old to new concrete; cures underwater. *Magnolia Plastics.*

18383 Magnobond 6504
Construction adhesive; airport lighting sealer. *Magnolia Plastics.*

18384 Magno-Ceram
Three-part casting system consisting of resin, curing agent and ceramic grain; for injection and compression dies having low shrinkage and good dimensional stability; low thermal expansion enables embedment of reinforcement rods and cooling or heating coils. *Magnolia Plastics.*

18385 Magnocid
$ClHMgO_2$
Magnesium oxychloride
It has bleaching properties.

18386 magnolia metal
A bearing metal consisting mainly of antimony and lead, and sometimes tin, with small quantities of iron and bismuth.

18387 Magnolium
An alloy of 90% lead and 10% antimony.

18388 Magnoloop I
Construction adhesive; traffic loop sealer. *Magnolia Plastics.*

18389 Magnox
1309-37-1 4072 215-168-2
Magnetic iron oxides; used in coatings for magnetic inks, discs and tapes. *Hercules.* Discontinued.

18390 Magnum
60 in. Membrane, spiral wound configuration; both TFC and Roga available; used for reverse osmosis water treatment. *AlliedSignal Fluid Systems.*

18391 Magnum
A family of acrylonitrile-butadiene-styrene resins (ABS) used in appliances, automotive, recreational, medicine and electronics applications. *Dow Chem Co Ltd, UK & Ireland.*

18392 Magnum 240
9003-56-9 262-706-6
ABS resin; automotive low gloss grade. *Dow Plastics.*

18393 Magnum 275
9003-56-9 262-706-6
ABS resin; sheet extrusion grade. *Dow Plastics.*

18394 Magnum 445 HQ
9003-56-9 262-706-6
ABS resin; for automotive trim applications. *Dow Plastics.*

18395 Magnum 788HP
9003-56-9 262-706-6
ABS resin; high heat, low gloss grade. *Dow Plastics.*

18396 Magnum 2610
9003-56-9 262-706-6
ABS resin; for health care industry. *Dow Plastics.*

18397 Magnum 3661
9003-56-9 262-706-6
ABS resin; ignition-resistant grade. *Dow Plastics.*

18398 Magnum 4420
9003-56-9 262-706-6
ABS resin; ignition-resistant resin with excellent processability and toughness for computer and business equipment. *Dow Plastics.*

18399 Magnum 9450P
9003-56-9 262-706-6
ABS resin; plateable grade. *Dow Plastics.*

18400 Magnum FG960
9003-56-9 262-706-6
ABS resin; pipe fitting grade. *Dow Plastics.*

18401 Magnum® F
Mixture of chloridazon and ethofumesate; selective systemic herbicide. *BASF AG; BASF plc.*

18402 Magnuminium
A proprietary trade name for magnesium-aluminum alloys. No manufacturer.

18403 Magnum-White
Blend of magnesium hydroxide and calcium carbonate; fire retardant/smoke suppressant filler for PVC compounds and SBR latex formulations. *RMc Minerals.*

18404 Magocarb-33
546-93-0 208-915-9
$CMgO_3$
Magnesium carbonate
Flame retardant in plastics. *Kaopolite.*

18405 Magoh-S
1309-42-8 5706 215-170-3
H_2MgO_2
Magnesium hydroxide
Fire retardant filler for plastics; extender pigment for flame retardant coatings. *Kaopolite.*

18406 Magotex
1309-48-4 5713 215-171-9
MgO
Fused magnesium oxide
High thermal conduction coefficient, low oil absorption, good electrical resistivity, low moisture; for thermal conductive molding compounds, brake lining, electrical potting compounds, special refractories. *Kaopolite.*

18407 Magox® 98HR
1309-48-4 5713 215-171-9
MgO
Magnesium oxide
Reactive technical grade for chem. process areas where a rapid reaction rate is necessary, as in detergents and chemical neutralization. *Kaopolite.*

18408 Magox® Super Premium
1309-48-4 5713 215-171-9
MgO
Magnesium oxide
Adsorption, absorption agent and scavenger with high surface area for plastics, rubber industries, chemical neutralization, etc. *Kaopolite.*

18409 Magrex
Fluxes for magnesium alloys. *Foseco (F.S.) Ltd.*

18410 Magrods
1309-48-4 5713 215-171-9
MgO
Magnesium oxide sticks
The Chemical & Insulating Co Ltd.

18411 Magspa
Fertilizer additive. *ICI Chem & Polymers Ltd.*

18412 Magtran
7783-40-6 5701 231-995-1
F_2Mg
magnesium fluoride
Magnesium fluoride of optical quality. *BDH Chemicals Ltd.*

18413 mahogany acid
Crude mixtures of sulfonic acids for the refining of petroleum sludge.

18414 mahogany brown
A sienna which has been ignited, ground wet, and made up in the form of pieces, and dried.

18415 Maillechort
A nickel silver containing copper, zinc, nickel and iron.

18416 Maincote
Acrylic resin emulsions. *Rohm & Haas UK.*

18417 Mainstay®
8001-69-2 11, 2464 232-289-6
Cod liver oil BP; dietary supplement; for relief of joint aches, plains, and stiffness. *Seven Seas Ltd.*

18418 Maizena
A proprietary trademark for corn starch. No manufacturer.

18419 Maizolith
A material resembling hard rubber obtained by the alkaline treatment of corn-stalk or corn-cob. No manufacturer.

18420 Majamin
Sodium-β-tetralin sulfonate
Majamin-kalium, the potassium salt, and majammonium, the ammonium salt,

are similar products. They are added to soap and soap powders to increase lathering power.

18421 **Majammonium**

Ammonium-β-tetralin-sulfonate. Majamin-kalium, the potassium salt, and majammonium, the ammonium salt, are similar products. They are added to soap and soap powders to increase lathering power.

18422 **majolica**

A pottery enamelled with a tin oxide enamel.

18423 **majunga noir**

A rubber yielded by *Landolphia perrieri.*

18424 **Makalot**

A proprietary trade name for a phenol-formaldehyde synthetic resin; used for molding. No manufacturer.

18425 **Makon NP6, NP10 and 4, 8, 12, 14 and 30**

Nonylphenol ethoxylate nonionic surfactant in liquid form; detergents, wetting agents and emulsifiers. *KWR Chemicals Ltd.* Name unverified.

18426 **Makon OP6 and OP9**

Octylphenol ethoxylate nonionic surfactant in liquid form; detergents, wetting agents and emulsifiers. *KWR Chemicals Ltd.* Name unverified.

18427 **Makon® 4**

9016-45-9 6772

Nonoxynol-4

Detergent, emulsifier used in chemical specialties, cosmetic, agricultural, industrial and metal cleaners, textile, paper and petroleum industries. *Stepan; Stepan Canada; Stepan Europe.*

18428 **Makon® 8240**

61791-12-6

PEG-36 castor oil; wetting agent, lubricant, coupling agent, defoamer for metalworking fluids, corrosion inhibitors. *Stepan; Stepan Canada.*

18429 **Makon® NI10, NI20, NI30**

Alkylphenol alkoxylate

Emulsifier, dispersant for agricultural microemulsions and emulsifiable concentrates. *Stepan Europe.*

18430 **Makon® OP-6**

9002-93-1 6858

Octoxynol-6

Detergent, wetting agent, emulsifier for household/industrial cleaners. *Stepan Europe.*

18431 **Makroblend®**

PC/polyester blends, characterized by their chemical resistance and low temperature impact strength. *Miles.* Discontinued.

18432 **Makroblend® DP4-1368**

PC/PET blend, impact-modified, flame-retardant; good chemical resistance and low temperature toughness; for appliances, personal care products, electrical connectors, meter housings and instrument enclosures. *Miles.* Discontinued.

18433 **Makroblend® UT 400**

PC/PET polyester blend; unfilled, general purpose resin for injection molding; offers high melt flow for easy processability, dimensional stability and good surface appearance, high impact strength and toughness, good heat resistance; used for vacuum cleaner housings and nozzles, utility locks, protectice face guards, tractor parts, instrument housings, radio housings and speaker grilles. *Miles.* Discontinued.

18434 **Makrofol®**

Polycarbonate-based electrical insulating film with high dielectric strength and long-term thermal stability up to 130°C; used for dial plates and signalling and warning indicators on motor vehicles. *Bayer AG; Bayer plc.*

18435 **makrofol® BL 2-2**

PC film; technical film; light-diffusing, dazzlefree, greater scratch resistance; natural translucent; both sides very fine matte. *Bayer AG.*

18436 **Makrofol® DE 1-1**

PC film; technical film with high clarity; natural color, both sides gloss. *Bayer AG.* Name unverified.

18437 **Makrofol® FR 6-2**

PC film; technical film, nonoptical grade; dazzle-free, greater scratch resistant; natural translucent, one side fine velvet, one side very fine matte. *Bayer AG.* Name unverified.

18438 **Makrolon®**

Polycarbonate for the manufacture of injection moldings primarily for use in the electrical industry and for industrial mechanical applications. *Bayer AG; Bayer plc.*

18439 **Makrolon® 1006 Tint**

PC; optical grade. *Bayer; Miles.*

18440 **Makrolon® 1143**

PC; uv-stabilized extrusion grade with structural viscosity. *Bayer; Miles.* Name unverified.

18441 **Makrolon® 2400**

PC; injection molding grade used in electronics, electrical engineering, lighting, photographic, optical equipment, mechanical process and precision engineering, office supplies, domestic ware, sports and safety applications; available in all major colors transparent, translucent, opaque cylindrical granules; soluble in a number of commercial solvents. Benzene, acetone and CCl_4 cause a superficial swelling effect. *Bayer; Miles.* Name unverified.

18442 **Makrolon® 2600, 2800**

PC; injection molding grade used in electronics, electrical engineering, lighting, photographic, optical equipment, mechanical process and precision engineering, office supplies, domestic ware, sports and safety applications. *Bayer; Miles.*

18443 **Makrolon® 3100, 3200**

PC; for injection molding and extrusion *Bayer; Miles.*

18444 **Makrolon® 3208**

PC; FDA grade for extrusion applications; more resistant to hydrolysis than 3200 *Bayer AG.* Name unverified.

18445 **Makrolon® 6355, 6455**

PC; flame-retardant grade with release agent; amorphous engineering thermoplastic for automotive, transportation, building/construction, business machine, consumer, electrical/electronic, medical, and optical applications. *Bayer; Miles.*

18446 **Makrolon® 8325**

PC, 20% glass-reinforced; high modulus grade. *Bayer; Miles.*

18447 **Makrolon® AL-2647 1068 Tint**

PC; auto lens grade. *Bayer; Miles.*

18448 **Makrolon® FCR-2405, FCR-2407, FCR-2458**

PC resin; amorphous engineering thermoplastic for automotive, transportation, building/construction, business machine, consumer, electrical/electronic, medical, and optical applications; general purpose grades processable by injection molding and extrusion; FCR grades are general purpose, high productivity with additivies for early releaseand uv stability. *Bayer; Miles.*

18449 **Makrolon® GV**

A trademark for glass filled polycarbonate resin. *Bayer; Miles.* Name unverified.

18450 **Makrolon® HMS-3118**

PC; extrusion grade with FDA compliance. *Bayer; Miles.*

18451 **Makrolon® LQ-2847, LQ-3147, LQ-3187**

PC; optical grade. *Bayer; Miles.*

18452 **Makrolon® LTG-3123**

PC; lighting grade for HID lamps. *Bayer; Miles.*

18453 **Makrolon® Rx-2530**

PC; radiation-stabilized, high-flow resin offering transparency and reduced yellowing after radiation exposure; for injection molding applications requiring minimum color change after sterilizing doses of gamma radiation. *Bayer; Miles.*

18454 **Makrolon® SF-600**

PC structural foam, 5% glass-reinforced; flame-retardant. *Bayer; Miles.*

18455 **Maktion**

Methidathion [950-37-8] and dimethoate [60-51-5]; insecticide. *Makhteshim Chemical Works Ltd.*

18456 **malabar tallow**

A fat obtained from the seeds of *Vateria indica.* It melts at 37.5°C., has a saponification value of 188.7-189.3, and an iodine value of 37.8-39.6.

18457 **malacca primers**

Primers for bonding materials such as rubber, wood, fabric, and metals. They usually have a rubber base.

18458 **malachite**

12069-69-1 2697 235-113-6

$CH_2Cu_2O_5$

basic cupric carbonate

Mountain green; green verditer; copper rust; green spar. A mineral used as a pigment. It is a hydrated basic copper carbonate, $CuCO_3 \cdot Cu(OH)_2$, and when ground is sold as mountain green and mineral green.

18459 **Malapaho**

Oil of Panao, collected from *Dipterocarpus vernicifluus.*

18460 **malathion**

121-75-5 5740 204-497-7

$C_{10}H_{19}O_6PS_2$

diethyl [(dimethoxyphosphinothoiyl)thio]butanedioate

Maldison; Insecticide No. 4049; Carbofos; Mercaptothion; Phosphothion; Cythion; Chemathion; Carbophos; Emmatos; Fosfothion; Fyfanon; Karbofos; Kop-Thion; Malacide; Malagran; Malamar; MLT; Sadofos; Calmathion; Carbetox; Carbethoxy Malathion; Carbetovur; Celthion; Cimexan; Compound 4049; Detmol MA; Malaspray; Ethiolacar; Etiol; Cleensheen; Lice Rid. Cholinesterase inhibitor, behaves as a non-systemic acaricide and insecticide with contact, stomach and respiratory action. Used for broad spectrum control of sucking and chewing insects and mites in a wide variety of crops. Particularly useful on Mediterranean fruit fly. mp = 2°; $bp_{0.7}$ = 156-157°; sg_{25} = 1.23; n_D^{25} = 1.4985; soluble in H_2O (145 mg/l), more soluble in organic solvents; LD_{50} (rat orl) = 1375-2800 mg/kg. *Allchem Industries; Am. Cyanamid; Sariaf SpA.*

18461 Malathion 60
121-75-5 5740 204-497-7
Diethyl 2-(dimethoxyphosphinothioyl-thio)succinate
Broad spectrum organophosphorus insecticide. *Farm Protection Ltd.*

18462 Malathion Dust
Insecticide. *Murphy Chemical Co Ltd.* Discontinued.

18463 Malathion Liquid
Insecticide. *Murphy Chemical Co Ltd.* Discontinued.

18464 Malazide
123-33-1 5745 204-619-9
maleic hydrazide
Maleic hydrazide, a plant growth inhibitor. *Fisons plc, Horticultural Div.* Name unverified.

18465 Maldene
Bonding agent. *GE Plastics ABS Ltd.* Name unverified.

18466 Maldene 285
A proprietary copolymer of butadiene and maleic anhydride supplied as a 25% solution in acetone for use as an intermediate. *Unichem.* Name unverified.

18467 Maldene 286
A copolymer similar to Maldene 285 except that it is a 25% solution of the partial ammonium salt in water. *Unichem.* Name unverified.

18468 Maldene 288
A copolymer similar to Maldene 285 except that the solids content in the solution is 35%. *Unichem.* Name unverified.

18469 Maldene 289
A proprietary copolymer of butadiene and maleic anhydride partially ethyl-esterified and dissolved to 25% in ethyl alcohol. *Unichem.* Name unverified.

18470 Maldene 292
A proprietary copolymer of butadiene and maleic anhydride supplied as a partially-butyl-esterified 25% solution in butanol. *Unichem.* Name unverified.

18471 Maldene 293
A copolymer similar to Maldene 285 except that it is a 25% solution of the partial amide-ammonium salt in water. *Unichem.* Name unverified.

18472 Maldene 300
A copolymer similar to Maldene 285 except that it is a 25% solution of the partial octyl ester in toluene. *Unichem.* Name unverified.

18473 Maldene 631
A copolymer similar to Maldene 285 except that it is an 18% solution of a zinc-ammonium-complexed form of Maldene 286 in water. *Unichem.* Name unverified.

18474 male fern
The rhizome of *Aspidium felix-mas.*

18475 maleic acid
110-16-7 5743 203-742-5
$C_4H_4O_4$
(Z)-Butenedioic acid
toxilic acid; *cis*-1,2-ethylenedicarboxylic acid; *cis*-butenedioic acid; maleinic acid. Used to retard rancidity in fats and oils; dyeing and finishing textiles; as intermediate in synthesis. mp = 138-139°; d = 1.59; freely soluble in H_2O, polar organic solvents. *General Chemical; Penta Mfg.; Thor.*

18476 maleic anhydride
108-31-6 5744 203-571-6
$C_4H_2O_3$
cis-butenedioic anhydride
2,5-Furandione; toxilic anhydride; *cis*-butenedioic anhydride; 2,5-furanedione; maleic acid anhydride; Lytron 810; Lytron 820; dihydro-2,5-dioxofuran; MA. Used as preservative and as intermediate in Diels-Alder synthesis. mp = 52.8°; bp = 202°; d = 1.48; soluble in, reacts with H_2O, soluble in organic solvents. *Amoco; Aristech; Ashland; Elf Atochem SA; Hüls AG; Mitsui Toatsu chemical; Monsanto; Oxychem.*

18477 maleic hydrazide
123-33-1 5745 204-619-9
$C_4H_4N_2O_2$
1,2-Dihydro-3,6-pyridazinedione
maleic acid hydrazide; MH; Fazor; Malazide; Regulox; 6-Hydroxy-2H-pyridazin-3-one; 1,2-dihydropyridazine-3,6-dione; 6-hydroxy-3(2H)-pyridazinone; Burtolin; De-Cut; Fair-2; Fair-Plus; Malzid; Mazide; MH-30; Slo-Gro; Super-De-Sprout; Vondalhyd. Plant growth regulator, absorbed by leaves and roots. Used for suppression of growth of grass, sprouts on potatoes and other vegetables. Used with 2,4-D as a herbicide. mp = 292-298°; d^{25} = 1.60; soluble in H_2O (6 g/kg), more soluble in polar organic solvents; LD_{50} (rat orl) > 5000 mg/kg *Rhône-Poulenc; Fair Products; Uniroyal; Synchemicals; Pennwalt Holland.*

18478 Malenite
A material containing an antimony double salt, $SbF_3 \cdot Na_2SO_4$, in addition to sodium fluoride, and the sodium compound of dinitro-phenol or dinitro-o-cresol; used for impregnating wood.

18479 malethamer
Maleic anhydride-ethylene polymer.

18480 Malezafin 55 Plus
Emulsifiable concentrate of ethyl ester of 2,4-D acid and butoxyethanol ester of dichlorprop; broadleaf herbicide for pasture land. *Invequimica & CIA SCA.* Name unverified.

18481 Malezafin 57 LV
Emulsifiable concentrate of a mixture of butoxyethanol esters of 2,4-D and dichlorprop; low volatile, broad leaf herbicide, brush killer for pasture land. *Invequimica & CIA SCA.* Name unverified.

18482 Malezafin LV-4
Emulsifiable concentrate of butoxyethanol ester of 2,4-D acid; low volatile, broadleaf herbicide for corn crops and pasture land. *Invequimica & CIA SCA.* Name unverified.

18483 Malic Acid
6915-15-7 5747 230-022-8
$C_4H_6O_5$
Hydroxybutanedioic acid
apple acid; N-hydroxysuccinic acid; hydroxysuccinic acid. Manufacture of esters and salts, wines; chelating agent, food acidulant, flavoring. mp = 131-132°; soluble in methanol, diethyl ether, ethanol, acetone, water, dioxane, insoluble in benzene. *Allchem Industries; Haarmann & Reimer; Janssen Chimica; Schweizerhall.*

18484 malladrite
16893-85-9 8769 240-934-8
Na_2SiF_2
Sodium fluosilicate
Used in enamels, as a moth repellent, insecticide and rodenticide.

18485 malleable iron
7439-89-6 5106 231-096-4
Fe
wrought iron
Practically pure iron, through which are scattered particles of slag or oxide.

18486 malleable nickel
7440-02-0 6582 231-111-4
Ni
nickel
Nickel commercially refined, and treated with a deoxidizing agent such as magnesium, and cast into ingots. It is suitable for hot or cold working.

18487 Mallebrein
15477-33-5 347 239-499-7
Aluminum chlorate
Used as an antiseptic and astringent.

18488 Mallet alloy
A brass containing 74.6% zinc and 25.4% copper.

18489 Mallet bark
The bark of *Eucalyptus occidentalis.* It contains from 35-52% tannin.

18490 Malloydium
An alloy of 61% copper, 23% nickel, 14% zinc, and 1% iron. It is stated to be acid-resisting.

18491 malonic acid
141-82-2 5749 205-503-0
$C_3H_4O_4$
methanedicarboxylic acid
methanedicarbonic acid; propanedioic acid; carboxyacetic acid; dicarboxymethane. Intermediate for barbiturates and pharmaceuticals. mp = 135° (dec); d = 1.63; soluble in H_2O (1.53 g/ml), less soluble in organic solvents. *Aldrich; R.W. Greeff; Lonza; Penta Mfg.*

18492 Malonoben
2-(3,5 di-tert-butyl-4-hydroxybenzylidene) malononitrile
Used as a pesticide.

18493 Maloran
13360-45-7 236-411-9
chlorbromuron
Substituted urea herbicide. *Ciba plc.* Name unverified.

18494 Malotte's alloy
An alloy of 46% bismuth, 20% lead, 34% tin. Melting-point is 203°F.

18495 Malros
Saponified rosin size. *Tenneco Malros Ltd.* Name unverified.

18496 maltha
A variety of mineral tallow or wax found in Finland. It is also the name applied to certain types of soft bitumen.

18497 malthactite
A clay of the fuller's earth type.

18498 malthite
A name for viscous bitumens.

18499 Maltisorb
Polyhydric alcohol
A sugar substitute in powder or liquid form for foods (confectionery) and pharmaceuticals. *Roquette (UK) Ltd.*

18500 Maltodextrin
9050-36-6 232-940-4
Maltodextrin
Lycadex®; Maltrin® M040, M510; Microduct®; Mor-Rex® I-920; Wickenol® 550. Fat replacers in foods. *Roquette (UK) Ltd; Grain Processing; CasChem.*

18501 maltose
69-79-4 5753 200-716-5
$C_{12}H_{22}O_{11}{\cdot}H_2O$
4-O-α-D-glucopyranosyl-D-glucose
malt sugar; maltobiose; Maltos; Martos-10. Malt sugar, an isomer of cellobiose; nutrient, sweetener, culture media, stabilizer for polysulfides, brewing. $[\alpha]_D^{20}$ = 112° (c = 4); soluble in H_2O, EtOH, insoluble in organic solvents. *Am. Biorganics; Avebe BV; Penta Mfg.; Pfanstiehl Labs.*

18502 Maltrin® M040
9050-36-6 232-940-4
Maltodextrin
Nonsweet, nutritive polymer useful for wet binding and anticaking; adds solution viscosity, good mouthfeel; for pharmaceuticals, foods; DE 4-7. *Grain Processing.*

18503 Maltrin® M200
68131-37-3 268-616-4
Corn syrup solids; dried glucose syrup with very good coating and binding properties; directly compressible; for pharmaceutical use; DE 20-23. *Grain Processing.*

18504 Maltrin® M510
9050-36-6 232-940-4
Agglomerated maltodextrin
Flowable form; exhibits excellent dispersibility and dissolution; directly compressible binder and diluent; for pharmaceutical use; free-flowing granules; DE 9-12. *Grain Processing.*

18505 Maltrin® QD M600
68131-37-3 268-616-4
Agglomerated corn syrup solids; agglomerated form of Maltrin M200; directly compressible binder and good carrier; for pharmaceutical use; free-flowing granules; DE 20-23. *Grain Processing.*

18506 maltyl
Dry malt extract containing about 90% soluble carbohydrates.

18507 maltzyme
Diastase, an enzyme used in the manufacture of maltose from cellobiose.

18508 maluminum
An alloy of 87% aluminum, 6.4% copper, 4.8% zinc, 1.4% iron, and traces of manganese, silicon, and lead.

18509 Manal
638-38-0 5763 211-334-3
$C_4H_6MnO_4$
manganese acetate
Used as a dye mordant and as a drier for paints and varnishes. d = 1.59; soluble in H_2O, EtOH; LD_{50} (rat orl) = 3.73 g/kg. *Mechema Chemicals Ltd.* Name unverified.

18510 Manalox
Aluminum organic compounds; rheology modifiers, ink industry, water repellent for masonry and damp proofing and water repellent for timber. *Rhône-Poulenc UK.*

18511 Manalox AG
A proprietary trade name for glycalox. No manufacturer.

18512 Manalox AS
A proprietary trade name for sucralox. No manufacturer.

18513 Mancarb Plus
A liquid formulation containing 80 g carbendazim, 150 g chlorothalonil and 200 g maneb per liter as a suspension concentrate; eradicant fungicide for use on cereals. *Ashlade Formulations Ltd.*

18514 Manchem
Metal carboxylates
Drier for paint or printing ink, additives for fuel oil or diesel oil. *Rhône-Poulenc UK.*

18515 Mancobride Mancanese
7789-43-7 2496 232-166-7
Br_2Co
Cobalt(II)Bromide;
cobalt dibromide; cobalt bromide. Cobalt bromide solution. mp = 678°; d_4^{25}= 4.909; soluble in H_2O, organic solvents. *Mechema Chemicals Ltd.* Name unverified.

18516 Mancopper
An ethylene bisdithiocarbamate-mixed metal complex containing about 13.7% manganese and about 4% copper; used as a fungicide and pesticide.

18517 mancozeb
8018-01-7 5756
[[1,2-Ethanediylbis[carbamodithioato]](2-)manganese mxture with [1,2-ethanediylbis[carbamodithioato]](2-)]zinc
manganese-zinc ethylenebis(dithiocarbamate); mancozebe; manzeb; Aimcozeb; Crittox MZ; Dithane 945; Dithane M-45; Fore; Karamate; Mancozin; Manzate 200; Manzin; Nemispor; Penncozeb; Phytox MZ; Riozeb; Vondozeb Plus. dec 192-194°; insoluble in H_2O and organic solvents; LD_{50} > 5000 mg/kg. Agrimont; Akzo; All-India Medical; Crystal; Desarrollo Quimica; Diachem; DuPont; Ercros; Pennwalt Holland; Rohm & Haas, Sanachem.

18518 mandelic acid
90-64-2 5757 210-277-1
$C_8H_8O_3$
phenylglycolic acid
α-phenylhydroxyacetic acid; benzoglycolic acid; amygdalic acid. Used in organic synthesis and in medicine as a urinary antiseptic. Alternate CAS RN [611-72-3]. mp = 119°; d = 1.30; soluble in H_2O (159 mg/ml), more soluble in organic solvents. *W.R. Grace Ltd; R.W. Greeff.*

18519 Mandelin's reagent
An alkaloid-detecting reagent, consisting of 1 g of ammonium vanadate dissolved in 200 ml of concentrated sulfuric acid.

18520 Mandops Barleyquat B
999-81-5 2153 213-666-4
Soluble concentrate containing 620 g/l chlormequat; used to increase barley yields. *Mandops (UK) Ltd.*

18521 Mandops Bettaquat B
999-81-5 2153 213-666-4
Soluble concentrate containing 620 g/l chlormequat; used to increase oats yields. *Mandops (UK) Ltd.*

18522 Mandops Halloween, Hele Stone
Mixture of chlormequat and di-1-*p*-menthene. Plant growth regulator. *Mandops (UK) Ltd.*

18523 Mandops Narsty
7784-25-0 335 232-055-3
Aluminum ammonium sulfate
Bird and animal repellent. *Mandops (UK) Ltd.*

18524 Mandops Podquaternary
Chlormequat and di-1-*p*-menthene. plant growth regulator. *Mandops (UK) Ltd.*

18525 Mandops Spring Poquaternary
999-81-5 2153 213-666-4
Soluble concentrate containing 590 g/l chlormequat; used to increase yields of oilseed rape, pea and bean crops. *Mandops (UK) Ltd.*

18526 Maneb
12427-38-2 5761 235-654-8
$C_4H_6MnN_2S_4$
Manganese, [[1,2-ethanediylbis[carbamodithioato]](2-)]-
Manganese, [ethylenebis[dithiocarbamato]]-; m-Diphar; Amangan; Carbamic acid, ethylenebis[dithio-, manganese salt; Chem neb; Chloroble B; CR 3029; Dithane M 22; Dithane M 22 special; Dithane M-45; Dithane S-31; Ethylene bis[dithiocarbamic acid], manganese salt; EBDC, manganese salt; F 10; Griffin Manex; Kypman 80; Labilite; Lonocol M; M-Diphar; Manam; 249; Maneb 80; Maneb-R; Maneba; manebe; MEB; ENT 14875;Kypman; Man-Zox; Manex; Manox; Manzi; Polyram M; Trimangol. Fungicide with protective action. Used in control of many fungal diseases, e.g. blight, leaf spot, rust, downy mildew, scab etc. mp = 192-204°; insoluble in H_2O and common solvents; LD_{50} (rat orl) = 3000 mg/kg. Rohm & Haas; Cumberland; Crystal; Drexel; BASF; Chiltern Farm Chemicals Ltd; Dupont; Pennwalt Holland.

18527 Manex
Suspension concentrate containing maneb and zinc; a protectant fungicide against potato blight, mildew and rust. *Chiltern Farm Chemicals Ltd; Griffin; Quadrangle Agrochemicals; L W Vass (Agricultural) Ltd.*

18528 Manfloc
1302-42-7 8715 215-100-1
sodium aluminate
Sodium aluminate composition; retention aid for paper water treatment. *Manchem Ltd.* Name unverified.

18529 **Mangabeira rubber**
A rubber from the small tree, *Hancornia speciosa* cultivated in Paraguay and Venezuela.

18530 **Mangal**
Chemical complex of manganese and aluminum; drier for cobalt-free paint or printing ink. *Manchem Ltd.* Name unverified.

18531 **mangaloy**
An alloy of nickel containing iron and manganese.

18532 **manganaluminum bronze**
An alloy containing from 9-10% manganese, 85.5-86% copper, and 4 1/2-5% aluminum.

18533 **Manganar**
manganese arsenate

18534 **manganated linseed oil**
Linseed oil which has been boiled with manganese dioxide to increase its drying properties.

18535 **manganese**
7439-96-5 5762 231-105-1
Mn
Used in ferroalloys, i.e., steel manufacture; improves corrosion resistance and hardness in nonferrous alloys; purifying and scavenging agent in metal production; manufacture of aluminum. mp = 1244°; bp = 2095°. *Atomergic Chemetals; Cerac; Chemetals; Kerr-McGee.*

18536 **manganese acetate**
638-38-0 5763 211-334-3
$C_4H_6MnO_4{\cdot}4H_2O$
acetic acid manganese (2+) salt
Used in textile dyeing, oxidation catalyst, paint and varnish drier, fertilizer, food packaging, feed additive. d = 1.59; soluble in H_2O, EtOH; LD_{50} (rat orl) = 3.73 g/kg. *Atomergic Chemetals; Hoechst Celanese; Nihon Kagaku Sangyo; Spectrum Chem. Mfg.; Verdugt BV.*

18537 **manganese boron**
An alloy of manganese and boron; used for making other alloys.

18538 **manganese brass**
Variable alloys, usually containing 51-69% copper, 1-4% manganese, 0-3% iron, 29-40% zinc, 0-2% tin, 0-2% nickel, and sometimes aluminum.

18539 **manganese bronze**
An alloy made by adding ferro-manganese or manganese to bronze. It usually contains from 82-83.5% copper, 8% tin, 5% zinc, 3% lead, and 0.5-2% manganese; an alloy containing 59% copper, 39% zinc, 1.5% manganese, and 0.5% iron.

18540 **manganese cupro nickel**
This is usually an alloy containing from 65-83% copper, 15-30% manganese, and 2-8% nickel.

18541 **manganese dioxide**
1313-13-9 5770 215-202-6
MnO_2
manganese oxide
manganese binoxide; manganese black; manganese peroxide. Oxidizing agent, depolarizer in dry cell batteries, pyrotechnics, matches, catalyst, laboratory reagent, scavenger and decolorizer, textile dyeing, source of metallic manganese (as pyrolusite). mp = 535° (dec); d = 5.0260; insoluble in cold water, nitric or cold sulfuric acid; LD_{50} (iv rabbits) = 45 mg/kg. *Aldrich; Atomergic Chemetals; Eagle-Picher; Hoechst Celanese; Kerr-McGee; Nichia Kagaku Kogyo.*

18542 **manganese German silver**
An alloy of 80% copper, 15% manganese, and 5% zinc.

18543 **manganese gluconate**
$C_{12}H_{22}MnO_{14}{\cdot}2H_2O$
Mineral source for pharmaceutical and food products, vitamin tablets, feed additive. *Akzo; Spectrum Chem. Mfg.*

18544 **manganese nickel**
An alloy of from 51-82% copper, 14-31% manganese, and 3-16% nickel. An alloy containing 95% nickel and 5% manganese, and melting at 1420 C. is also known as manganese nickel.

18545 **manganese nickel brass**
Alloys containing 51-65% copper, 5-40% zinc, 0-2.78% iron, 1.5-3.24% manganese, and 2-18% nickel.

18546 **manganese nickel silver**
Alloys containing 50-70% copper, 9-40% zinc, 1-20% manganese, and 2-20% nickel.

18547 **manganese ore A**
A standard manganese ore. It contains 51.3% manganese, 14.3% available oxygen, 6.5% silica, 1.3% iron, and 0.22% phosphorus.

18548 **manganese steel**
An alloy of manganese and steel containing up to 20% manganese. Commercial manganese steel contains 11-14% manganese, 1-1.3% carbon, 0.3% silicon. 0.05-0.08% sulfur, and 0.05-0.08% phosphorus.

18549 **manganese sulfate**
7785-87-7 5782 231-960-0
$MnO_4S{\cdot}4H_2O$
manganese(II) sulfate
Used in fertilizers, feed additive, paints and varnishes, ceramics, textile dyes, medicine, fungicide, ore flotation, catalyst in viscose process, synthetic manganese dioxide. mp = 700°; bp = 850°; d = 2.9500. *Aldrich; Chemetals; Mallinckrodt; Nihon Kagaku Sangyo.*

18550 **Manganese Violet**
7782-76-5 5777
$HMnO_4P$
Manganese(II) phosphate
Nuremberg violet; mineral violet; permanent violet. Mineral used as a pigment. *Reckitts Colours Ltd.*

18551 **manganese white**
598-62-9 5766 209-942-9
$CMnO_3$
Manganese(II) carbonate
Manganous carbonate, $MnCO_3$, a pigment. Also used as a drier for varnishes and in animal feeds. d = 3.1; insoluble in H_2O, EtOH.

18552 **manganese-aluminum brass**
An alloy of 56% copper, 40% zinc, 3% manganese, and 1% aluminum.

18553 **manganic**
Nickel containing a small percentage of manganese.

18554 **Manganin**
Alloys usually containing 70-86% copper, 4-25% manganese, and 2-12% nickel. One of the best varieties contains 83.6% copper, 13.6% manganese, 2.5% nickel, and 0.3% iron; used in electrical instruments where constant resistivity is important.

18555 **manganite**
1317-34-6 5780 215-264-4
$Mn_2O_3{\cdot}H_2O$
manganese sesquioxide
Brown manganese ore. A mineral. It is a hydrated oxide of manganese, It is also name for a war gas which consisted of a mixture of hydrocyanic acid and arsenic trichloride.

18556 **Mangan-Neusilber**
A nickel silver containing manganese. It contains from 59-72% copper, 5-20% zinc, 10-18% nickel, and 2-20% manganese.

18557 **Mangano Steel**
A proprietary trade name for a non-shrinking steel containing 1.6% manganese, 0.2% chromium and 0.95% carbon. No manufacturer.

18558 **mangano-titanium**
This is usually an alloy of manganese with 30% titanium, and is used as a deoxidizer in making bronze and brass castings.

18559 **Mangatrace**
7773-01-5 5768 231-869-6
$Cl_2Mn{\cdot}4H_2O$
Manganese chloride tetrahydrate
manganous chloride, manganese dichloride; manganese bister. Mineral supplement. mp = 58°; d = 2.01; soluble in H_2O (1.43 g/ml), EtOH, insoluble in organic solvents; LD_{50} (mus sc) = 180-250 mg/kg. *Armour Pharmaceutical Co.* Name unverified.

18560 **Mangnamite**
Graphite fiber
For structural strength applications. *Hercules.* Discontinued.

18561 **Mangol**
ClMnO
Basic magnesium hypochlorite
Used for testing for alkaloids.

18562 **Mangonic**
A manganese-nickel alloy containing about 3% manganese.

18563 **Mangoxe**
1313-13-9 5770 215-202-6
Manganese dioxide
Polysulfide sealant. *Mechema Chemicals Ltd.* Name unverified.

18564 **Manguard®**
12427-38-2 5761 235-654-8
Water-dispersible granule containing 75% w/w maneb; for use as an agricultural fungicide. *Universal Crop Protection Ltd.*

18565 **Manhardt's Aluminum Bronze**
An alloy of 83.3% aluminum, 6.25% copper, 10.13% tin, 0.16% antimony, 0.05% magnesium, and 0.08% phosphorus. No manufacturer.

18566 **Manjak**
Glance pitch. A bitumen found in Mexico, South America, and West Indies; used as a paint, as a roofing material, and in connection with drilling for oil.

18567 Mannheim gold
A brass containing 80% copper and 20% zinc. A jeweler's alloy. The term is also applied to a bronze consisting of 83.7% copper, 9.3% zinc, 7% tin, with a little phosphorus.

18568 Manoblend
Mixtures of rubber chemicals and fillers *Manchem Ltd.* Name unverified.

18569 Manobond
Cobalt boroacylate
Rubber-to-metal bonding agents. *Rhône-Poulenc UK.*

18570 Manocat
Metal carboxylates
Catalysts used in manufacture of unsaturated polyester and polyurethane foams. *Rhône-Poulenc UK.*

18571 Manofast
1758-73-2 217-157-8
$CH_4N_2O_2S$
aminoiminomethanesulfinic acid
Thiourea dioxide. Reducing agent for dyes. Used in organic synthesis to reduce ketones to secondary alcohols. mp = 126° (dec). *Rhône-Poulenc UK.*

18572 Manofil
Chemically treated fillers for rubber and plastic manufacture. *Manchem Ltd.* Name unverified.

18573 Manomet
Metal carboxylates
Stabilizers for PVC, pigment dispersing aids and kicker for PVC foams. *Rhône-Poulenc UK.*

18574 Manosec
Range of metal carboxylates based on synthetic organic acids; paint driers. *Rhône-Poulenc UK.*

18575 Manosil
Combinations of silica or silicates with speciality chemicals; fillers for use with rubber, plastics. *Manchem Ltd.* Name unverified.

18576 Manosperse
Dispersed rubber chemicals; used in rubber industry. *Rhône-Poulenc UK.*

18577 Manox
7704-34-9 9142 231-722-6
Insoluble sulfur
Used in rubber vulcanizing. *Rhône-Poulenc UK.*

18578 Manox
Iron blue pigments; for printing inks, paints and fungicides. *Manox Ltd.* Name unverified.

18579 Manoxol
Sulfosuccinate surface active agents; wetting agent, dispersing agents. *Manchem Ltd.* Name unverified.

18580 Manoxol MA
Sodium di(methylamyl) sulfosuccinate in 60% water/alcohol solution; powerful wetting agents which can act as aids to dispersion, detergency, and emulsification; used for adhesives, asbestos, agriculture and horticulture, bleaching, clays, determination of anionic detergents, dry cleaning and laundry, dust control, electroplating, emulsion polymerization, etching, germicides, leather, metal technology, oil additives, paint, ink, paper, pharmaceuticals, photography, resins, rubber, textiles, polishes. *Manchem Ltd.* Name unverified.

18581 Manoxol OT, OT/P and OT/B
577-11-7 3460 209-406-4
Sodium dioctyl sulfosuccinate
Available as solid, water/alcohol solution, powder (85% with 15% sodium benzoate), or specially pure grade for pharmaceutical use; powerful wetting agents which can act as aids to dispersion, detergency, and emulsification; used for adhesives, asbestos, agriculture, bleaching, clays, dry cleaning, laundry, electroplating, emulsion polymerization, etching, germicides, leather, metal technology, oil additives, paint, ink, paper, pharmaceuticals, photography, resins, rubber, textiles, polishes. *Manchem Ltd.* Name unverified.

18582 Manoxol OT60
577-11-7 3460 209-406-4
Sodium diisooctyl sulfosuccinate
Used as a wetting and dispersing agent. *Harcros Australia.*

18583 Manoxolot
577-11-7 3460 209-406-4
A proprietary trade name for dioctyl sodium sulfosuccinate, a dispersant and wetting agent. No manufacturer.

18584 Manplex
8018-01-7 5756
Manganese zinc ethylenebisdithiocarbamate complex; for control of potato blight and rust, blight and mildew in winter wheat. *Kommer-Brookwick Ltd.*

18585 Manqueta
Manquta
African names for a fossil gum resin, resembling copal.

18586 Manro
Foam boosters and stabilizer; powder and liquid detergents; hand cleaning jellies and cleaners. *Manro Products Ltd.* Name unverified.

18587 Manro ALS
Ammonium primary alcohol sulfate as a clear, pale yellow liquid; foaming and wetting agent for hair, carpet and upholstery shampoos. *Manro Products Ltd.* Name unverified.

18588 Manro BA and NA
Dodecylbenzene sulfonic acid
Anionic surfactant supplied as a brown viscous liquid, low in free oil and inorganic content; raw material for the production of detergents and emulsifiers such as powder and liquid detergents, hand cleaning gels, machine degreasers and tank cleaners. *Manro Products Ltd.* Name unverified.

18589 Manro BES
Range of anionic surfactants in liquid form; anion: sodium; cation: synthetic alcohol ethoxy sulfate; high foaming agents used in liquid detergent formulations, bubble baths and shampoos; industrial applications include use as a drilling aid. *Manro Products Ltd.* Name unverified.

18590 Manro D Paste
Sodium cetyl/oleyl sulfate as a white/pale yellow stiff paste; wetting, dispersing and emulsifying agent used in textile scouring and kier boiling. *Manro Products Ltd.* Name unverified.

18591 Manro DL28
Sodium primary alcohol sulfate as a water white liquid; foaming and wetting agent with good handling characteristics used in latex processing and emulsion polymerization. *Manro Products Ltd.* Name unverified.

18592 Manro DS 35
Sodium primary alcohol sulfate as a pale yellow liquid; wetting and emulsification agent used in metal cleaner formulations etc. *Manro Products Ltd.* Name unverified.

18593 Manro HA
Dodecyl benzene sulfonic acid
Derived from propylene tetramer; anionic surfactant supplied as a brown viscous liquid, low in free oil and inorganic content; raw material for production of detergents and emulsifiers such as powder and liquid detergents, hand cleaning gels, machine degreasers and tank cleaners. *Manro Products Ltd.* Name unverified.

18594 Manro HCS
Anionic surfactant
Supplied as a pale amber sparkling viscous liquid; used for emulsification of a wide range of chemicals, especially in emulsifiable solvent degreasing, e.g., for hand cleaning jellies or engine cleaners. *Manro Products Ltd.* Name unverified.

18595 Manro KXS
Potassium xylene sulfonate
Anionic surfactant in liquid form; cloud point depressant and solubilization agent, especially for heavy duty detergent formulations. *Manro Products Ltd.* Name unverified.

18596 Manro MA 35
Disodium octadecyl sulfosuccinamate as a pale cream liquid; foaming agent in the manufacture of latex foams. *Manro Products Ltd.* Name unverified.

18597 Manro ML 33
Monoethanolamine primary alcohol sulfate in pale yellow liquid form; foaming and wetting agent for high quality hair shampoos and other toiletry products. *Manro Products Ltd.* Name unverified.

18598 Manro NEC
Anionic surfactant as a colorless liquid; cation: sodium; anion: natural or Ziegler alcohol ethoxy sulfate; foaming and cleaning agent for liquid and lotion hair shampoos. *Manro Products Ltd.* Name unverified.

18599 Manro NP
Range of nonionic surfactants of the nonylphenol ethoxylate type in liquid, paste or wax form; general purpose nonionic detergent bases; emulsifiers for agro-chemicals and pesticides; emulsion polymerization. *Manro Products Ltd.* Name unverified.

18600 Manro PTSA
104-15-4 9671 203-180-0
Toluene sulfonic acid
Anionic surfactant in liquid form; intermediate in detergent manufacture, also has descaling and catalyst properties; used in metal industries for resin bound sand castings; manufacture of esters, acetylation and resin production; hydrotrope production. for detergents. *Manro Products Ltd.* Name unverified.

18601 Manro SBS
Anionic surfactant; supplied as a clear pale yellow liquid with low inorganic

content and controlled levels of minor constituents; detergent and emulsifier, e.g., light duty liquid detergents, hard surface cleaners; scouring and wetting in textile industries, emulsifiable insecticides, plastics, rubber. *Manro Products Ltd.* Name unverified.

18602 Manro SDBS
Anionic surfactant; supplied as a white/pale yellow paste with low inorganic content and controlled levels of minor constituents; used as a detergent and emulsifier e.g. light duty detergents, hard surface cleaners; scouring and wetting in textile industries, emulsifiable insecticides, plastics, rubber. *Manro Products Ltd.* Name unverified.

18603 Manro SIOS
Sodium isooctyl sulfate as a pale yellow, clear mobile liquid; wetting agent with good alkali stability used in metal cleaner formulations and similar products. *Manro Products Ltd.* Name unverified.

18604 Manro SLS28
Sodium primary alcohol sulfate as a very pale yellow liquid; foaming and wetting agent for carpet and upholstery shampoos; foam rubber; emulsion polymerization in plastics and rubber industries. *Manro Products Ltd.* Name unverified.

18605 Manro SLS45
Sodium primary alcohol sulfate as a white paste; foaming and wetting agent for hair shampoos. *Manro Products Ltd.* Name unverified.

18606 Manro STS
757-84-1 9671 211-522-5
Sodium toluene sulfonate
Anionic surfactant in liquid form; used for reduction of slurry viscosity before spray drying in heavy duty detergent powder manufacture. *Manro Products Ltd.* Name unverified.

18607 Manro SXS
1300-72-7 215-090-9
Sodium xylene sulfonate
Anionic surfactant in liquid form; cloud point depressant, coupling and stabilization agent for light and heavy duty detergents. *Manro Products Ltd.* Name unverified.

18608 Manro TDBS
Anionic surfactant, pale amber viscous liquid; a detergent, mild to the skin; used in bubble baths, car shampoos and other detergents. Also used as an emulsifier in emulsification polymerization. *Manro Products Ltd.* Name unverified.

18609 Manro TL40
Triethanolamine primary alcohol sulfate in the form of a clear, pale amber liquid; foaming and wetting agent for high quality hair shampoos and other toiletry products. *Manro Products Ltd.* Name unverified.

18610 Manro XSA
25321-41-9 246-839-8
Xylene sulfonic acid
Anionic surfactant in liquid form; intermediate in detergent manufacture, also has descaling and catalyst properties; used in metal industries for resin bound sand castings; manufacture of esters, acetylation and resin production; hydrotrope production for detergents. *Manro Products Ltd.* Name unverified.

18611 Mansonil®
50-65-7 6602 200-056-8
$C_{13}H_8Cl_2N_2O_4$
5-chloro-N-(2-chloro-4-nitrophenyl)-2-hydroxybenzamide
Niclosamide; Cestocid; Devermine; Dichlosale; Fedal-telmin; Fenasal; Helmiantin; Iomesan; Iomezan; Lintex; Mansonil; MATO; Nasemo; Phenasal; Radeverm; Sagimid; Sulqui; Tredemine; Vermitid; Bayluscid. Veterinary preparation; anthelmintic (Cestodes); used against tapeworm infestation in ruminants, dogs, and cats. The ethanolamine salt is used as a molluscicide. *Bayer AG.*

18612 Mansu
7785-87-7 5782 231-960-0
Manganese sulfate
Used as a nutritional factor. *Mechema Chemicals Ltd.* Name unverified.

18613 Mantin
Organotin/dithiocarbamate. *Ciba plc.* Name unverified.

18614 Mantrilon®
Liquid manganese fertilizer; foliar fertilizer to prevent and cure deficiency of manganese in all agricultural crops, vines and fruit. *BASF AG.*

18615 Mantrilon® FL
Foliar feed. *BASF plc.*

18616 Manucol DM
Sodium alginate
Used for ice creams, chilled desserts. *Kelco Int'l. Ltd.*

18617 Manucol Ester E/RK
9005-37-2
Propylene glycol alginate
Used for fermented milks, fruit drinks, citrus concentrates. *Kelco Int'l. Ltd.*

18618 Manucol Ester EX/LL
9005-37-2
A proprietary trade name for propylene glycol alginate; a food grade emulsifying agent. *Alginate Industries Ltd.* Name unverified.

18619 Manucol, Manucol DH
Sodium alginate
Used for bakery glazes, filling creams, cheesecake toppings, cheese spreads, processed cheeses, instant puddings, and aspics. *Kelco Int'l. Ltd.*

18620 Manucreme
Hand cream, to soothe dry and rough hands. *Richardson-Vicks Inc.* Name unverified.

18621 Manugel PTJ
Specialty algin blend; gelling powder for structured glace fruit, high solid fillings, bakery fillings, bakery jellies. *Kelco Int'l. Ltd.*

18622 Manutex
Sodium alginate
Technical grade. *Kelco Int'l. Ltd.*

18623 Manzanate
39255-32-8 254-384-1
$C_8H_{16}O_2$
Ethyl-2-methyl pentanoate
Oil with a natural, fruity, pineapple odor. Used in specialty perfumes and fruit flavorings. $bp_{15} = 60°$; $n_D^{20} = 1.399\text{-}1.406$; $sg^{25} = 0.861\text{-}0.865$. *Quest Int'l. UK Ltd.*

18624 Manzate®
12427-38-2 5761 235-654-8
Suspension concentrate or wettable powder containing maneb; a dithiocarbamate fungicide to control blight, rusts, and mildew. *DuPont UK.*

18625 Manzate® 200 DF
8018-01-7 5756
Wettable powder containing mancozeb; protective fungicide for fruit, field crops and roses. *DuPont UK.*

18626 Map₄
A coarse starch obtained from the fruit of *Inocarous edulis.*

18627 Mapeg® 200 DL
9005-02-1
PEG-4 dilaurate
Emulsifier, dispersant used in cosmetics, pharmaceuticals, metalworking and fiber lubricants, etc. *PPG/Specialty Chem.* Discontinued.

18628 Mapeg® 200 DO
9005-07-6
PEG-4 dioleate
Emulsifier, dispersant used in cosmetics, pharmaceuticals, metalworking and fiber lubricants, etc. *PPG/Specialty Chem.* Discontinued.

18629 Mapeg® 200 DOT
61791-01-3
PEG-4 ditallate
Surfactant, emulsifier for metalworking lubricants; emollient for hair prepratíons, creams and lotions; solubilizer for bath oils and fragrances. *PPG/Specialty Chem.* Discontinued.

18630 Mapeg® 200 DS
9005-08-7
PEG-4 distearate
Emulsifier, dispersant used in cosmetics, pharmaceuticals, metalworking and fiber lubricants, etc. *PPG/Specialty Chem.* Discontinued.

18631 Mapeg® 200 ML
9004-81-3
PEG-4 laurate
Emulsifier, dispersant used in cosmetics, pharmaceuticals, metalworking and fiber lubricants, etc. *PPG/Specialty Chem.* Discontinued.

18632 Mapeg® 200 MO
9004-96-0
PEG-4 oleate
Emulsifier, dispersant used in cosmetics, pharmaceuticals, metalworking and fiber lubricants, etc. *PPG/Specialty Chem.* Discontinued.

18633 Mapeg® 200 MOT
61791-00-2
PEG-4 tallate
Emulsifier, dispersant used in cosmetics, pharmaceuticals, metalworking and fiber lubricants, etc. *PPG/Specialty Chem.* Discontinued.

18634 Mapeg® 200 MS
9004-99-3
PEG-4 stearate
Emulsifier, dispersant used in cosmetics, pharmaceuticals, metalworking and fiber lubricants, etc. *PPG/Specialty Chem.* Discontinued.

18635 Mapeg® 1500 MS
9004-99-3
PEG-6-32 stearate
Emulsifier, dispersant used in cosmetics, pharmaceuticals, metalworking and fiber lubricants, etc. *PPG/Specialty Chem.* Discontinued.

18636 Mapeg® 1540 DS
9005-08-7
PEG-32 distearate
Emulsifier, dispersant used in cosmetics, pharmaceuticals, metalworking and fiber lubricants, etc. *PPG/Specialty Chem.* Discontinued.

18637 Mapeg® 6000 DS
9005-08-7
PEG-150 distearate
Emulsifier, dispersant used in cosmetics, pharmaceuticals, metalworking and fiber lubricants, etc. *PPG/Specialty Chem.* Discontinued.

18638 Mapeg® CO-16H
61788-85-0
PEG-16 hydrogenated castor oil; surfactant, emulsifier, dispersant, wetting agent, emollient for hair preps., creams and lotions; solubilizer for bath oils and fragrances. *PPG/Specialty Chem.* Discontinued.

18639 Mapeg® CO-5
61791-12-6
PEG-5 castor oil. *PPG/Specialty Chem.* Discontinued.

18640 Mapeg® DGLD
9004-81-3
Diethylene glycol laurate
Surfactant for formation of gels. *PPG/Specialty Chem.* Discontinued.

18641 Mapeg® EGDS
627-83-8 211-014-3
Glycol distearate
Emulsifier, dispersant used in cosmetics, pharmaceuticals, metalworking and fiber lubricants, etc.; thickener, opacifier, pearling additive. *PPG/Specialty Chem.* Discontinued.

18642 Mapeg® EGMS
Glycol stearate
Emulsifier, dispersant used in cosmetics, pharmaceuticals, metalworking and fiber lubricants, etc.; thickener, opacifier, pearling additive. *PPG/Specialty Chem.* Discontinued.

18643 Mapeg® S-40
9004-99-3
PEG-40 stearate
Emulsifier, dispersant used in cosmetics, pharmaceuticals, metalworking and fiber lubricants, etc. *PPG/Specialty Chem.* Discontinued.

18644 Mapeg® TAO-15
61791-00-2
PEG-660 tallate
Emulsifier, dispersant used in cosmetics, pharmaceuticals, metalworking and fiber lubricants, etc. *PPG/Specialty Chem.* Discontinued.

18645 Maphos® 17
Aromatic phosphate ester; emulsifier for emulsion polymerization; solubilizer. *PPG/Specialty Chem.* Discontinued.

18646 Maphos® 33
Aliphatic phosphate ester; emulsifier, lubricant with anticorrosive/antifrictional props. for oil and watersol. lubricant systems, e.g., greases, synthetic cutting oils, drawing compounds, chain-belt lubricants, gear oils, and rust preventatives. *PPG/Specialty Chem.* Discontinued.

18647 Maphos® 60A
Complex organic phosphate acid ester; textile wetting, hard surface detergent; lubricant, anticorrosive, dispersant, hydrotrope, solubilizer, emulsifier; metalworking. *PPG/Specialty Chem.* Discontinued.

18648 Maphos® 66H
Phosphate acid ester, neutralized; hard surface cleaning hydrotrope; antistat for solubilization of low foam and conventional surfactants in alkaline liquids; metalworking. *PPG/Specialty Chem.* Discontinued.

18649 Maphos® 78
Complex organic phosphate acid ester; hydrotrope, detergent. *PPG/Specialty Chem.* Discontinued.

18650 Maphos® 8135
Aromatic phosphate ester; dispersant, hydrotrope, emulsifier, EP lubricant additive for greases, synthetic cutting oils, drawing compounds, and hard surface cleaners. *PPG/Specialty Chem.* Discontinued.

18651 Maphos® FDEO
Phosphate ester; surfactant, hydrotrope, detergency aid, antistat for drycleaning and lubricant systems, emulsion polymerization. *PPG/Specialty Chem.* Discontinued.

18652 Maphos® JA 60
Aliphatic phosphate ester; detergent, coupling agent; compatible with builders; for hard surface cleaners, built detergents, metalworking fluids. *PPG/Specialty Chem.* Discontinued.

18653 Maphos® L 13
Aliphatic phosphate ester; surfactant, lubricant, anticorrosive, coupling agent for metalworking, dry cleaning, hard surface cleaning, dedusting, lubrication, emulsion polymerization. *PPG/Specialty Chem.* Discontinued.

18654 Mapico
1309-37-1 4072 215-168-2
Synthetic iron oxide *Sevalco Ltd.*

18655 maple sugar sand
A by-product in the manufacture of maple sugar. The sap from the maple is evaporated in pans, and a precipitate forms when the water content is about 35%. This precipitate is maple sugar sand. The chief constituent is calcium malate (60-80%) from which malic acid is easily prepared.

18656 Mapo®
57-39-6 5998 200-326-5
$C_9H_{18}N_3OP$
tris[1-(2-methyl-aziridinyl) phosphine oxide
metepa; methyl aphoxide; methapoxide. Used for creaseproofing and flameproofing textiles; resin raw material, crosslinker, adhesion promoter and chemosterilant. $bp_{0.15}$ = 90-92°; LD_{50} (rat orl) = 136 mg/kg. *Aceto.*

18657 Maprenal
Etherified melamine/formaldehyde resins; used for stoving finishes and acid curing lacquers. *Resinous Chemicals Ltd.*

18658 Maprenal®
Melamine and benzoguanamine resin products; for production of stoving (baking) lacquers. *Cassella AG.*

18659 Maprofix 563 and LK.USP
Sodium lauryl sulfate in powder or granular form; detergency, foaming and emulsification agent, food and dentifrice grade; also used in emulsion polymerization. *Millmaster-Onyx UK.* Name unverified.

18660 Maprofix 60S and 60N
Sodium or ammonium lauryl 3EO sulfate in liquid form; foaming surfactant for light duty detergents. *Millmaster-Onyx UK.* Name unverified.

18661 Maprofix ES-2
Sodium lauryl 2EO sulfate in liquid form; foaming and dispersing agent for shampoos, bubble baths and general cosmetics. *Millmaater-Onyx UK.* Name unverified.

18662 Maprofix ESY
Sodium lauryl 1EO sulfate in liquid form; used for emulsification, foaming and wetting agent for emulsion polymerization, shampoos and gels. *Millmaster-Onyx UK.* Name unverified.

18663 Maprofix MG
Magnesium lauryl sulfate in liquid form; wetting, emulsifying, and dispersing agent with low cloud point; used in nonalkaline shampoos and rug shampoos. *Millmaster-Onyx UK.* Name unverified.

18664 Maprofix NH and NHL
2235-54-3 218-793-9
Ammonium lauryl sulfate in liquid form; detergency and foaming agent with high buffering capacity pH 6-7; used in nonalkaline shampoos, for general cosmetic use and in industrial foams. *Millmaster-Onyx UK.* Name unverified.

18665 Maprofix TAS
Sodium tallow alcohol sulfate in paste form; detergent and dispersing agent used in high temperature detergents and ore flotation. *Millmaster-Onyx UK.* Name unverified.

18666 Maprofix TLS
139-96-8 205-388-7
Triethanolamine lauryl sulfate in liquid form; detergent with good foam stability and low cloud point; used in clear shampoos, bubble baths, fine fabric detergent, industrial and household cleaners and rug and upholstery shampoos. *Millmaster-Onyx UK.* Name unverified.

18667 Maprofix WA, WAC and WAQ
Sodium lauryl sulfate in liquid or paste form; detergency, foaming and dispersing agent for liquid and cream shampoos; general cosmetic uses; rug and upholstery shampoos; latices and paint pigments. *Millmaster-Onyx UK.* Name unverified.

18668 Maprofix WAC-LA and LCP
Sodium lauryl sulfate in liquid form; high purity, low salt content and viscosity, light in color. wetting, emulsification and dispersing agent, polymerization grade. *Millmaster-Onyx UK.* Name unverified.

18669 Mapromin
A proprietary trade name for a sulfated fatty alcohol used as a wetting agent. No manufacturer.

18670 Mapron®
Soybean milk; food additive. *Mitsubishi Kasei.* Name unverified.

18671 Maprosyl® 30
137-16-6 4379 205-281-5

Sodium n-lauroyl sarcosinate
Gardol®. Detergent, wetting and foaming agent used in personal care and household detergent products; anticorrosive properties. *Stepan; Stepan Canada.*

18672 Marabond 21
Lignosulfonate
Low temperature, low cost oil well cement retarder. *Borregaard Ligno Tech.*

18673 Marabout silk
A white silk which still contains its gum. It is dyed and used for the manufacture of imitation feathers.

18674 Maracarb
Modified lignosulfonate; dispersant, humectant, chelating agent for manufacture of alkaline metal cleaners. *Borregaard Ligno Tech.*

18675 Maracarb N-1
8061-51-6
Sodium lignosulfonate
Modifier, dispersant, and humectant in dyestuff pastes; industrial cleaners; plant foliar spray; chelates metal ions; slime control agent for paper mill systems. *Borregaard Ligno Tech.*

18676 Maracell XE
8061-51-6
Sodium lignosulfonate, partially desulfonated; sludge conditioner, chelating agent for prevention of scale formation in treatment of boiler water, industrial cleaners. *Borregaard Ligno Tech.*

18677 Maracon
Calcium and sodium lignosulfonates; concrete admixtures. *Borregaard Ligno Tech.*

18678 Maramul SS
Lignosulfonate
Emulsifier for asphalt emulsions. *Borregaard Ligno Tech.*

18679 Maranil DBS
Dodecylbenzene sulfonic acid
Intermediate for manufacture of liquid, powdered or paste detergents, emulsifiers, textile auxiliaries; acid catalyst. *Henkel/Functional Prods.; Pulcra SA.*

18680 Maranil Powd. A
Sodium dodecylbenzene sulfonate
Base for manufacture of detergents, dishwashes, cleaning agents; wetting agent; emulsifier for PVC copolymers, carboxylated S/B latexes. *Henkel/Functional Prods.*

18681 maranta
Arrowroot starch.

18682 Maranyl®
Nylon molding and extrusion compounds. *ICI Chem & Polymers Ltd.*

18683 Maranyl® A125
32131-17-2
Nylon 6/6, unfilled, lubricated; engineering material with high mech. props., abrasion, thermal, chem. and creep resist.; for injection molding or extrusion applications. *ICI Advanced Materials.*

18684 Maranyl® A175S
32131-17-2
Nylon 6/6, 30% glass fiber-reinforced, heat stabilized, lubricated; engineering material with high mechanical properties, abrasion, thermal, chemical and creep resistance; for injection molding or extrusion applications. *ICI Advanced Materials.*

18685 Maranyl® A360
32131-17-2
Toughened nylon 6/6, 24% glass fiber-reinforced, flame retardant; engineering material with high mechanical properties, abrasion, thermal, chemical and creep resistance; for injection molding or extrusion applications. *ICI Advanced Materials.*

18686 Maranyl® TA505HS
32131-17-2
Toughened nylon 6/6, heat stabilized; engineering material with high mechanical properties, abrasion, thermal, chemical and creep resistance; for injection molding or extrusion applications. *ICI Advanced Materials.*

18687 Marasperse 52 CP
8061-51-6
Polymerized sodium lignosulfonate; dispersant for dyestuffs; low stain, high heat stability; pitch dispersant in paper mills. *Borregaard Ligno Tech.*

18688 Marasperse GFC
8061-52-7
Calcium lignosulfonate
Sugar-free; additive to high strength concrete; especially effective with silica fume concretes. *Borregaard Ligno Tech.*

18689 Marasperse N-22
8061-51-6
Sodium lignosulfonate
Purified, oxidized; dispersant, oil-water emulsion stabilizer, emulsifier; manufacture of disperse dyes for dyeing acetate and polyesters; dispersant and sequestering agent in cooling water treatments; agricultural chemical formulations gypsum board additive; industrial cleaners. *Borregaard Ligno Tech.*

18690 Marbalettes®
Socket pellets for extraction wounds in dental industry. *Bayer AG.*

18691 Marbledust
471-34-1 1697 207-439-9
Calcium carbonate
Coarse ground filler for putties, glazes and mild abrasive compounds. *ECC International Ltd.*

18692 Marbleloid
Flooring materials. *Weatherguard/Marblelod Products Inc.* Name unverified.

18693 Marblemite
471-34-1 1697 207-439-9
Calcium carbonate
High brightness filler for high loading and min. black specks in cultured marble application. *ECC International Ltd.* Discontinued.

18694 Marblette
A proprietary trade name for a phenol-formaldehyde cast resin. No manufacturer.

18695 Marbo
A proprietary trade name for a chlorinated rubber. No manufacturer.

18696 Marbon B
A proprietary trade name for a cyclo-rubber. No manufacturer.

18697 Marbon Latex
Marmix reactive SBR and ABS latexes. *GE Plastics ABS Ltd.* Name unverified.

18698 Marbon Resins
ABS polymers; for injection molding, sheet extrusions. *GE Plastics ABS Ltd.* Name unverified.

18699 Marbon White
A proprietary brand of Lithopone (zinc sulfide-barium sulfate). No manufacturer.

18700 marcasite
12068-85-8 235-106-8
FeS_2
white iron pyrites
coxcomb pyrites; radiated pyrites. Iron disulfide, a mineral. The term is also occasionally applied to bismuth.

18701 Marchon® DC 1102
Formulated product; shale and cuttings wash cleaner for oilfield industry. *Albright & Wilson UK.*

18702 marcs
The name given to the residue from wine factories, consisting of the stems and skins of grapes.

18703 Mareepa
Kernels of the fruits of the cokerite palm of British Guiana.

18704 Maretin
1491-41-4 6441 216-078-6
Naftalofos
Anthelmintic. *Bayer AG.*

18705 Marexine-CA
FC-126 strain
Frozen turkey herpes virus Marek's vaccine; for immunization of poultry. *Intervet Inc.* Discontinued.

18706 Marezzo marble
An artificial marble from oxychloride cement; used for building.

18707 Marfanil-Prontalbin®
Combination of Marfanil (mafenide) and Prontalbin (sulfanilamide); used against wound infections in veterinary medicine. *Bayer AG.*

18708 Margalite
A phenol-formaldehyde resin product; used in the manufacture of varnishes and insulators. No manufacturer.

18709 margarodite
A mica having an appearance similar to talc.

18710 margines
Marchies. The residues obtained from the manufacture of olive oil.

18711 margol
A mixture of volatile fatty acids; used as a flavoring material for margarine, to give it the taste of butter.

18712 margosa bark
Indian azadirach.

18713 Margosan-O
11141-17-6 926
Active ingredient is azadirachtin, a tetranortriterpionoid; growth regulator on

many insects in various life stages, due to hormonal disruption preventing normal metamorphesis; repellancy and/or antifeedency through olfactory or gustatory rejection by various flying and crawling pests; currently restricted to non-food crop use. *Vikwood Botanicals Inc.* Name unverified.

18714 Margraff alloy
An alloy consisting of 58% copper, 28% tin, and 14% zinc.

18715 Mariazin® KC 21/50
61789-71-7 263-080-8
Cocoalkonium chloride
Bactericide used in production of disinfectant cleaning agents. *Hüls Am; Hüls AG.*

18716 Maricol
$C_{36}H_{66}MgO_6$
Magnesium ricinoleate

18717 Marignac's salt
27790-37-0 7843 248-659-5
$K_2O_8S_2Sn$
Potassium-stanno-sulfate
Decomposed by H_2O.

18718 marine acid
7647-01-0 4821 231-595-7
HCl
Hydrochloric acid

18719 marine fiber
A fiber obtained by dredging the shallow water of a gulf in South Australia. It is a hydrated lignocellulose.

18720 marine oil
A mixture of blown rape oil and a mineral oil; used for marine engines.

18721 marine plasma extract
Hydrolyzed marine protein, brown algae extract, marine plasma fluid, iodine, silicon, iron, sodium, potassium, magnesium, calcium, ascorbic acid, niacinamide; sea plasma extract containing essential elements for development of ocean life. *Brooks Industries.*

18722 marine salt
7647-14-5 8742 231-598-3
NaCl
sodium chloride
Used as a chemical intermediate and reagent.

18723 marine soap
A soap made from coconut oil, which is soluble in fresh and sea water.

18724 Mark 80
Organic additive system (non coumarin); semibright nickel electroplating (Duplex). *Engelhard Technologies Ltd.*

18725 Mark® 1330
A proprietary sulfur-containing organotin stabilized for use in PVC for injection molding and pipe extrusion. *Argus Chemical Corporation.*

18726 Mark® 1414
A proprietary organotin mercaptide stabilizer for PVC used in the extrusion of rigid pipe. *Argus Chemical Corporation.*

18727 Mark® 2140
95823-35-1
Pentaerythrityl hexylthiopropionate
Stabilizer for use in polyolefins and other polymeric systems; synergistic with primary antioxidants; color which reduces or eliminates the need for phosphite; used at 0.1-0.3 phr in PP, 0.05-0.10 phr in HDPE, 0.25-0.5 phr in elastomers, and 0.5-1.0 phr in S.B. latex. *Witco Corporation.*

18728 Mark® 4700
Ba-Zn salt
Non-Cd heat stabilizer for PVC filled and clear compounds. *Witco Corporation.*

18729 Mark® 5089
96328-09-1
Pentaerythritol alkyl thiodipropionate
Antioxidant/stabilizer for plastics. *Witco/Argus.* Discontinued.

18730 Mark® 5095
Lauryl/stearyl thiodipropionate
Antioxidant for polyolefins and other polymeric systems including synthetic rubber; also for pharmaceuticals, cosmetics, industrial oils, greases, lubricants. *Witco/Argus.* Discontinued.

18731 Mark-A-Leak AW
High foaming leak detector for use under all weather conditions, including freezing temperatures. *Actrachem.*

18732 markasol
marcasol
Bismuth boro-phenate. Used as a substitute for iodoform.

18733 Marksman
Mixture of linuron and trifluralin; herbicide for winter cereals. *Farmers Crop Chemicals Ltd.* Discontinued.

18734 Markstat® AL-12
Quaternary ammonium chloride derivative of polyalkoxy tertiary amines; antistat additive for PU films and thermoplastics. *Witco/Argus.* Discontinued.

18735 Markwet NR-25
Ethoxylated alcohol blend; textile penetrant and detergent imparting nonrewetting characteristics; wetting agent and detergent in alkaline, acid, and neutral media; does not promote color transfer of dispersed dyes. *Ivax Industries.*

18736 Markwet WL-12
Surfactant; wetting agent, dispersant for dyestuffs, pigments and dirt; solubilizer; stable in alkaline and mild acid media. *Ivax Industries.*

18737 Marlamid®
A range of fatty acid alkanolamides; stabilizes foam, increases soil-suspending power and has a superfatting effect; intermediates for textile dressing agents and softeners and starting materials for fabric softeners. *Hüls AG.*

18738 Marlamid® A 18
141-21-9 205-469-7
Stearamidoethyl ethanolamine
Base for textile auxiliary agents and softeners. *Hüls Am; Hüls AG.*

18739 Marlamid® D 1885
93-83-4 202-281-7
Oleamide DEA
Foam stabilizer, thickener, superfatting agent for liquid detergents, shampoos. *Hüls Am; Hüls AG.*

18740 Marlamid® DF 1218
61790-63-4 263-153-4
Cocamide DEA
Foam stabilizer, thickener, superfatting agent for liquid detergents, shampoos. *Hüls Am; Hüls AG.*

18741 Marlamid® DF 1818
68425-47-8 270-355-6
Soyamide DEA
Foam stabilizer, thickener, superfatting agent for liq. detergents, shampoos. *Hüls Am; Hüls AG.*

18742 Marlamid® KL
Cocamidopropyl lauryl ether
Pearlescent surfactant, foam stabilizer, thickener, opacifier for liq. and paste detergents, shampoos. *Hüls Am; Hüls AG.*

18743 Marlamid® KLP
Cocamidopropyl lauryl ether
sodium laureth sulfate. Pearlescent base for shampoos, bubble baths, liq. soaps. *Hüls Am; Hüls AG.*

18744 Marlamid® M 1218
68140-00-1 268-770-2
Cocamide MEA
Foam stabilizer, thickener in household, personal, industrial detergents, superfatting agent. *Hüls Am; Hüls AG.*

18745 Marlamid® M 1618
68153-63-9 268-891-0
Tallowamide MEA
Foam stabilizer, thickener for household, personal, and industrial detergents. *Hüls Am; Hüls AG.*

18746 Marlamid® PG 20
Cocamide MEA
Glycol ditallowate. Pumpable pearlescent base for shampoos, bubble baths, liquid soaps. *Hüls Am; Hüls AG.*

18747 Marlate 2-MR Emulsifiable Insecticide
72-43-5 6070 200-779-9
methoxychlor
24% Methoxychlor in an emulsifiable solvent; for control of insects on livestock, agricultural premises, forest and shade trees, agricultural crops, ornamentals and flowers and for mosquito control. *Kincaid Enterprises Inc.* Name unverified.

18748 Marlate 50 WP
72-43-5 6070 200-779-9
50% Methoxychlor wettable powder; for control of insects in stored grain and for livestock, vegetables and fruits. *Kincaid Enterprises Inc.* Name unverified.

18749 Marlate 300 Flowable
72-43-5 6070 200-779-9
30% Methoxychlor (3.0 lbs/gal); seed treatment for insect infestations. *Kincaid Enterprises Inc.* Name unverified.

18750 Marlate 400 Flowable Concentrate
72-43-5 6070 200-779-9

40.5% Methoxychlor (4.0 lbs/gal); used for elms, forage and field crops, vegetables and seed treatment. *Kincaid Enterprises Inc.* Name unverified.

18751 Marlate Methoxychlor Insecticide
72-43-5 6070 200-779-9
50% Methoxychlor wettable powder; primarily for home use for flowers, gardens, trees and ornamentals. *Kincaid Enterprises Inc.* Name unverified.

18752 Marlazin®
A range of fatty amine polyglycol ethers; used in alkali-resistant and acid-resistant industrial cleaners and dyeing auxiliaries. *Hüls AG.*

18753 Marlazin® KC 30/50
61789-18-2 263-038-9
Cocotrimonium chloride
Quaternary for production of hair conditioning agents. *Hüls Am; Hüls AG.*

18754 Marlazin® L 10
PEG-10 lauramine
Surfactant for production of low-foaming acidic cleaners, textile auxiliaries; very high acid resistance. *Hüls Am; Hüls AG.*

18755 Marlazin® OL 2
26635-93-2
PEG-2 oleamine
Detergent for industrial cleaners, acid cleaners, textile auxiliaries; resistant to acids and alkalies. *Hüls Am; Hüls AG.*

18756 Marlazin® S 10
26635-92-7
PEG-10 stearamine
Detergent for industrial cleaners, acid cleaners, textile auxiliaries; resistant to acids and alkalies. *Hüls Am; Hüls AG.*

18757 Marlazin® T 10
61791-44-4 263-177-5
PEG-10 tallowamine
Detergent for industrial cleaners, textile and dyeing auxiliaries; resistant to acids and alkalies. *Hüls Am; Hüls AG.*

18758 Marlex 1708
A proprietary trade name for a LDPE suitable for the extrusion of heavy duty film; Type 1 Class A Grade 4 resin; d 0.917; melt index 0.8. *Pacific Petroleums (Quebec).* Unverified.

18759 Marlex® BMN 55500
A HDPE; used for injection molding of thin wall containers, toys and overcaps. *Phillips.* Name unverified.

18760 Marlex® BMN TR-880
A HDPE; used for injection molding of milk cases, tote boxes, automotive and industrial components and high quality housewares. *Phillips.* Name unverified.

18761 Marlex® CL-100
A crosslinkable HDPE; used for rotational molding of trash containers, industrial and agricultural chemical storage tanks, small engine, snowmobile and automotive fuel tanks. *Phillips.* Name unverified.

18762 Marlex® CL-200
9002-88-4 7728
Polyethylene resin; rotational molding resin used in agricultural, chemical, and sewage tanks, automotive fuel tanks, military packaging, trash containers; crosslinks during molding; excellent environmental stress cracking resistance and impact strength at low temperatures. *Phillips.*

18763 Marlex® CL-50
A crosslinkable HDPE; used for rotational molding of trash containers, chemicals and sewage tanks, seals, boats, camper tops. *Phillips.* Name unverified.

18764 Marlex® EHM 6003
A HDPE; used for blow molding of large chemical tanks and parts such as trash cans requiring high stiffness; for sheet extrusion and thermoforming of sheet, tote boxes, deep draw thermoformed parts. *Phillips.* Name unverified.

18765 Marlex® EHM 6006
A HDPE; used for blow molding of containers for bottling products such as milk and distilled water, fruit juices, etc. *Phillips.* Name unverified.

18766 Marlex® EHM 6007
A HDPE; used for blow molding of lightweight containers for bottling products such as milk and distilled water. *Phillips.* Name unverified.

18767 Marlex® EMN TR-885
A HDPE; used for injection molding of thin walled containers where higher production rate and rigidity is required. *Phillips.* Name unverified.

18768 Marlex® ER9-0002
A reinforced HDPE; used for sheet extrusion and thermoforming for structural automotive applications, housings, tote boxes. *Phillips.* Name unverified.

18769 Marlex® ER9-0020
A reinforced HDPE; used for sheet extrusion and thermoforming for structural automotive applications, seating, shrouds, housings. *Phillips.* Name unverified.

18770 Marlex® HGH-050
A polypropylene homopolymer; used for injection molding of appliance parts and chemical equipment. *Phillips.* Name unverified.

18771 Marlex® HGL-050-01
A polypropylene homopolymer; used for extrusion of soda straws. *Phillips.* Name unverified.

18772 Marlex® HGL-050-01 (Antistatic)
A polypropylene homopolymer; used for injection molding of food containers, housewares and toys. *Phillips.* Name unverified.

18773 Marlex® HGL-120-01 (Antistatic)
A polypropylene homopolymer; used for injection molding of thin wall containers, medical supplies and housewares. *Phillips.* Name unverified.

18774 Marlex® HGL-200 (Antistatic)
A polypropylene homopolymer; used for injection molding of thin wall containers, medical supplies and housewares. *Phillips.* Name unverified.

18775 Marlex® HGL-350 (Antistatic)
Controlled rheology polypropylenes; used for injection molding of thin wall containers, housewares and closures. *Phillips.* Name unverified.

18776 Marlex® HGN-020-01
A nucleated polypropylene; used for extrusion blow molding for drugs and toiletries; for extrusion of profiles, sheet and solid phase pressure forming. *Phillips.* Name unverified.

18777 Marlex® HGN-120-01 (Nucleated)
A polypropylene homopolymer; used for injection molding of closures and food containers. *Phillips.* Name unverified.

18778 Marlex® HGN-200 (Nucleated)
A polypropylene homopolymer; used for injection molding of thin wall containers, medical supplies and housewares. *Phillips.* Name unverified.

18779 Marlex® HGN-200A
A controlled rheology polypropylene homopolymer; used for injection molding of pill vials and medical jars. *Phillips.* Name unverified.

18780 Marlex® HGN-350 (Nucleated)
Controlled rheology polypropylenes; used for injection molding of thin wall containers, housewares and closures. *Phillips.* Name unverified.

18781 Marlex® HGX-010
A polypropylene homopolymer; used for extrusion, blow molding and fiber extrusion for drugs, toiletries and strappings. *Phillips.* Name unverified.

18782 Marlex® HGX-030
9003-07-0 7741
PP homopolymer resin; for monofilament, slit film filament, and staple applications; low water, carry over, processing stability. *Phillips.*

18783 Marlex® HGX-040
A polypropylene homopolymer used for general extrusion and slit film and monofilament extrusion for woven carpet backing and bags, rope and cordage; Woven carpet backing and bags, rope and cordage. *Phillips.* Name unverified.

18784 Marlex® HGX-330 (Controlled Rheology)
A polypropylene; used for multifilament extrusion for multifilament staple. *Phillips.* Name unverified.

18785 Marlex® HGZ-050-02
A polypropylene homopolymer; used for injection molding of food containers, housewares and toys. *Phillips.* Name unverified.

18786 Marlex® HGZ-120-02
9003-07-0 7741
PP homopolymer resin; for injection molding of thin-walled containers, syringes, medical supplies, and closures; hardness and abrasion resistance, high rigidity, and processing stability. *Phillips.*

18787 Marlex® HGZ-120-04
A polypropylene used for multifilament extrusion; used for multiultifilament staple. *Phillips.* Name unverified.

18788 Marlex® HGZ-200
A polypropylene homopolymer; used for injection molding of thin wall containers, medical supplies and housewares. *Phillips.* Name unverified.

18789 Marlex® HGZ-350
Controlled rheology polypropylenes; used for injection molding of thin wall containers, housewares and closures. *Phillips.* Name unverified.

18790 Marlex® HHM 4903
9002-88-4 7728
HDPE resin; for blow molding, sheet, and thermoforming of large industrial containers, fuel tanks, automotive parts, housings, trays, etc.; ease of processing, stress cracking resistance, surface appearance, and melt strength. *Phillips.*

18791 Marlex® HHM 5202
A HDPE; used for blow molding of bleach and detergent containers and chemical packaging; for sheet extrusion and thermoforming of tote boxes, trays, industrial housings, shrouds. *Phillips.* Name unverified.

18792 Marlex® HHM TR-130
A medium d polyethylene; used for film extrusion of merchant bags, produce bags, trash bags and multiwall bag liners. *Phillips.* Name unverified.

18793 Marlex® HHM TR-140
9002-88-4 7728
Polyethylene resin; film grade resin for produce, merchant, and trash bags, multi-wall bag liners; excellent impact and tear strength, good moisture barrier properties, nonblocking characteristics. *Phillips.*

18794 Marlex® HHM TR-144
A HDPE; used for film extrusion of merchant bags, produce bags, multiwall bag liners and trash bags. *Phillips.* Name unverified.

18795 Marlex® HHM TR-210
A HDPE; used for wire and cable coating for primary insulation for telephone conductors. *Phillips.* Name unverified.

18796 Marlex® HHM TR-226
A HDPE used for wire and cable coating; foam skin insulation on telephone singles. *Phillips.* Name unverified.

18797 Marlex® HHM TR-230 Black
A HDPE used for wire and cable coating; telephone cable jacketing. *Phillips.* Name unverified.

18798 Marlex® HHM TR-232 Black
A HDPE; used for wire and cable coating for power cable jacketing. *Phillips.* Name unverified.

18799 Marlex® HHM TR-250 Black
A HDPE; used for wire and cable coating for aerial cable jacketing of drop wire, line wire and tree wire. *Phillips.* Name unverified.

18800 Marlex® HHM TR-400
9002-88-4 7728
Medium d polyethylene resin; base resin for oil field and industrial pipe and fittings. *Phillips.*

18801 Marlex® HHM TR-418 (Black, Orange)
A polyethylene; used for pipe extrusion and injection molding of pipe fittings for gas distribution pipe, potable water pipe, engineered pipe. *Phillips.* Name unverified.

18802 Marlex® HHM-4515
A HDPE; used for injection molding of 5-gallon shipping containers, institutional seating, fuel tanks and closures. *Phillips.* Name unverified.

18803 Marlex® HLM-020
A polypropylene homopolymer; used for injection blow molding for drugs, toiletries, cosmetics and spices. *Phillips.* Name unverified.

18804 Marlex® HLN-120-01
A polypropylene homopolymer; used for injection molding for closures and food containers. *Phillips.* Name unverified.

18805 Marlex® HLN-200 (Antistatic, Nucleated)
A polypropylene homopolymer used for injection molding; thin wall containers, medical supplies and housewares. *Phillips.* Name unverified.

18806 Marlex® HLN-350 (Antistatic, Nucleated)
Controlled rheology polypropylenes used for injection molding; thin wall containers, housewares and closures. *Philips.* Name unverified.

18807 Marlex® HMN-938
A HDPE used for rotational molding; agricultural and industrial containers, FDA approved drums and tanks, trash containers. *Phillips.* Name unverified.

18808 Marlex® HMN 4550
9002-88-4 7728
HDPE; injection molding resin with high stress crack resistance, impact strength, good warpage resistance; for agricultural/industrial containers, food handling containers, seating, fuel tanks. *Phillips.*

18809 Marlex® HMN 5060
A HDPE used for injection molding; industrial containers, fuel tanks, closures and feeder tubs. *Phillips.* Name unverified.

18810 Marlex® HMN 5580
A HDPE; used for injection molding of pails, housewares, closures and crates requiring good toughness. *Phillips.* Name unverified.

18811 Marlex® HMN 6060
A HDPE; used for injection molding of trays, industrial parts, beverage crates and safety helmets. *Phillips.* Name unverified.

18812 Marlex® HMN 54140
A HDPE; used for injection molding of industrial containers, crates and boxes, large frozen food containers. *Phillips.* Name unverified.

18813 Marlex® HMN TR-942
A HDPE; used for rotational molding for agricultural and industrial containers and food handling containers. *Phillips.* Name unverified.

18814 Marlex® HMX-020-01 (Lubricant)
A polypropylene homopolymer used for injection blow molding; drugs, toiletries, cosmetics and spices. *Phillips.* Name unverified.

18815 Marlex® HNS-080
9003-07-0 7741
PP homopolymer resin; for high clarity blown, quenched, or cast film applications, including soft goods, stationery, bakery goods, candy wrapping; high slip and med. anti-block chars., cleanliness, sparkling clarity, low haze, and high gloss. *Phillips.*

18816 Marlex® HXM 50100
9002-88-4 7728
HDPE resin
For sheet and thermoforming including large formed parts, cattle feeders, pallets, and boats; excellent stress cracking resistance, melt strength, impact strength even at low temperatures. *Phillips.*

18817 Marlex® RGX-020
A polypropylene random copolymer used for extrusion blow molding; syrup bottles, food containers and toiletries. *Phillips.* Name unverified.

18818 Marlex® RGX-020 (Antistat)
A polypropylene random copolymer; used for extrusion blow molding for syrup bottles, food containers and toiletries. *Phillips.* Name unverified.

18819 Marlex® RMX-020
9003-07-0 7741
PP random copolymer; clearer than homopolymer grades. *Phillips.*

18820 Marlex® RMN-020C
A polypropylene random copolymer; used for injection blow molding for drugs, toiletries, cosmetics, and spices. *Phillips.* Name unverified.

18821 Marlex® TR.610
A proprietary trade name for high-d polyethylene; for use as a wire and cable insulation. *Pacific Petroleums (Quebec).* Unverified.

18822 Marlex® TR.885
A proprietary trade name for HDPE; for injection molding thin walled containers; d 0.965 g/ml; melt index 30. *Pacific Petroleums (Quebec).* Unverified.

18823 Marley Bitumen Paint Primer
Bitumen/solvent solution; metal and concrete paint. *Marley Floors Ltd.*

18824 Marley Carpet Cleaner
Neutral detergent solution; for carpet cleaning. *Marley Floors Ltd.*

18825 Marley Cement Accelerator
10043-52-4 1699 233-140-8
Calcium chloride solution; cement admixture. *Marley Floors Ltd.*

18826 Marley Cement Colorant
Pigment dispersion; cement admixture. *Marley Floors Ltd.*

18827 Marley Cement Dustproofer
1344-09-8 8825 215-687-4
Sodium silicate solution; cement/concrete surface treatment. *Marley Floors Ltd.*

18828 Marley Cement Plasticiser
Natural resin solution; cement admixture. *Marley Floors Ltd.*

18829 Marley Cement Waterproofer
143-18-0 7818 205-590-5
Potassium oleate solution; cement admixture. *Marley Floors Ltd.*

18830 Marley Cork Tile Adhesive
Tackified acrylic emulsion; cork tile adhesive. *Marley Floors Ltd.*

18831 Marley Exterior Water Repellent
Silicone, solvent solution; wall treatment. *Marley Floors Ltd.*

18832 Marley Floor Cleaner
Detergents and alkalies; floor cleaner. *Marley Floors Ltd.*

18833 Marley Floor Gloss
Metal cross linked resin and wax emulsion; floor polish. *Marley Floors Ltd.*

18834 Marley Floor Primer
Modified synthetic polymer emulsion; subfloor treatment. *Marley Floors Ltd.*

18835 Marley Homelay Adhesive
Modified bitumen emulsion; flooring adhesive. *Marley Floors Ltd.*

18836 Marley Interior Waterproofer
Polyurethane, solvent solution; wall, floor and ceiling treatment. *Marley Floors Ltd.*

18837 Marley Mastic
Bitumen solvent solution; filler for roofs. *Marley Floors Ltd.*

18838 Marley Patchit®
Mineral filled synthetic latex; filler/leveller for floors. *Marley Floors Ltd.*

18839 Marley Roofbond®
Bitumen solvent solution; roofing felt adhesive. *Marley Floors Ltd.*

18840 Marley Roofseal®
Modified bitumen emulsion; roof treatment. *Marley Floors Ltd.*

18841 Marley Stick and Lift
Modified acrylic emulsion; flooring adhesive. *Marley Floors Ltd.*

18842 Marley Superwax
Acrylic and soft wax emulsion; floor polish. *Marley Floors Ltd.*

18843 Marley Universal Flooring Adhesive
Tackified acrylic emulsion; flooring adhesive. *Marley Floors Ltd.*

18844 Marleybond PVA
Polyvinyl acetate emulsion; for multi use PVC. *Marley Floors Ltd.*

18845 Marlican®
67774-74-7 267-051-0
Straight-chain dodecylbenzene
Detergent intermediate, solubilizer; secondary plasticizer; reduces viscosity of PVC pastes; solvent for carbonless copy papers; biodegradable. *Hüls AG.*

18846 Marlie's alloy
An alloy containing 10% iron, 35% nickel, 25% brass, 20% tin, and 10% zinc, which has been quenched in a mixture of acids.

18847 Marlinat®
A range of sodium salts of sulfosuccinic acid esters; highly effective wetting agents for the textile, paint and paper industries; used for the preparation of cleaners for sensitive textiles. *Hüls AG.*

18848 Marlinat® 242/28
9004-82-4
Sodium laureth (2) sulfate
Strongly foaming base surfactant for detergents, shampoos, liquid soaps. *Hüls Am; Hüls AG.*

18849 Marlinat® CM 40
Laureth-5 carboxylic acid
Surfactant for mild detergents, shampoos, foam baths, cleaners. *Hüls Am; Hüls AG.*

18850 Marlinat® DF 8
577-11-7 3460 209-406-4
Dioctyl sodium sulfosuccinate
Highly active wetting agent for textile, paint and paper industries used in cleaners, cosmetic preparations. *Hüls Am; Hüls AG.*

18851 Marlinat® DFK 30
151-21-3 8782 205-788-1
Sodium lauryl sulfate
Base surfactant for hair shampoos, foam baths, shower foams, liquid soaps. *Hüls Am; Hüls AG.*

18852 Marlinat® DFL 40
139-96-8 205-388-7
TEA lauryl sulfate
Finely porous foaming surfactant. *Hüls AG.*

18853 Marlinat® DFN 30
2235-54-3 218-793-9
Ammonium lauryl sulfate
Base surfactant for hair shampoos, foam baths, shower foams, liq. soaps. *Hüls Am; Hüls AG.*

18854 Marlinat® KT 50
Sodium tallow sulfate
sodium cocosulfate. Surfactant for production of hand-washing pastes. *Hüls Am; Hüls AG.*

18855 Marlinat® SL 3/40
Disodium laureth sulfosuccinate; base surfactant for hair shampoos, foam baths, shower foams, liq. soaps. *Hüls Am; Hüls AG.*

18856 Marlinat® SRN 30
Disodium lauramido MEA-sulfosuccinate
sodium C12-14 olefin sulfonate. Base surfactant for carpet and upholstery cleaners. *Hüls Am; Hüls AG.*

18857 Marlipa® MG
9002-92-0 7717
Laureth-7
Solubilizer for active ingredients and oils in cosmetics. *Hüls Am; Hüls AG.*

18858 Marlipal®
A range of fatty alcohol polyglycol ethers; nonionic surfactants used as bases for detergents and dish-washing preparations, having dispersing, wetting, detergent, cleaning, soil-suspending and homogenizing properties. *Hüls AG.*

18859 Marlipal® 011/30
C_{11}-oxo alcohol ethoxylate (3 EO)
Wetting agent for hard surface cleaners, textile pretreatment. *Hüls AG.*

18860 Marlipal® 013/20
C_{13}-oxo alcohol ethoxylate (2 EO)
Dispersant, wetting agent, detergent, cleaning, soil suspending and homogenizing agent, emulsifier for textiles. *Hüls AG.*

18861 Marlipal® 1/12
PEG methyl ether
Surfactant for production of methyl-terminated fatty acid esters. *Hüls AG.*

18862 Marlipal® 24/20
68439-50-9
Laureth-2
Dispersant, wetting agent, emulsifier, detergent for washing, cleaning, soil suspension, textile pretreating and dyeing. *Hüls Am; Hüls AG.*

18863 Marlipal® 24/939
C12-14 alcohol ethoxylate blend; dispersant, wetting agent, emulsifier, detergent for washing, cleaning, soil suspension, textiles. *Hüls AG.*

18864 Marlipal® 124
9002-92-0 7717
Laureth-4
Surfactant for cosmetics, textile auxiliary agents; solubilizer for oils and perfumes. *Hüls Am; Hüls AG.*

18865 Marlipal® 1012/4
26183-52-8
Deceth-4
Polyethyleneglycol 300 monodecyl ether. Wetting surfactant, especially for hard surface cleaning. bp = 165-210°; d = 0.9990; n_D^{20} = 1.4550; fp = 224°. *Hüls Am.*

18866 Marlipal® 1618/6
68439-49-6 7737
Ceteareth-6
Dispersant; production of powder detergents; binding agent and base material for solid cleaning agents (toilet sticks); coating material for foam suppressant, enzymes, etc,; dyeing auxiliaries. *Hüls AG.*

18867 Marlipal® 1850/5
9004-98-2
Oleth-5
Surfactant for manufacture of washing powds., textile auxiliaries. *Hüls Am; Hüls AG.*

18868 Marlipal® KF
66455-15-0
Deceth-6
Surfactant, detergent raw material, wetting agent, and additive used in dishwashing agents, glass and hard surface cleaners, in manufacture of tablet soaps; degreaser for textiles and leather. *Hüls Am; Hüls AG.*

18869 Marlipal® SU
54045-08-8
Cetoleth-25
Surfactant for manufacture of washing powders, dyeing assistants; plasticizer for soap production; yields finely porous foam. *Hüls Am; Hüls AG.*

18870 Marloid® CAS
Specialty algin blend; milk-soluble product for use in milk-based systems. *Kelco.*

18871 Marlon®
A range of alkylbenzenesulfonates; used for the manufacture of detergents and cleaners. *Hüls AG.*

18872 Marlon® A350
Anionic surfactant in liquid form; high quality products with low salt content used in liquid detergents and cleaning materials for domestic and industrial use. *Hüls UK Ltd.*

18873 Marlon® A360, A365, A375
Anionic surfactant in paste form; high quality products with low salt contents used for paste detergents and cleaning materials for domestic and industrial use. *Hüls UK Ltd.*

18874 Marlon® A365
38411-30-3
Sodium dodecylbenzene sulfonate
Surfactant for production of detergents, cleaning agents, textile auxiliaries; biodegradable. *Hüls Am; Hüls AG.*

18875 Marlon® A390, A396, ARL
Anionic surfactant in solid form; high quality products with low salt contents used for powder detergents and cleaning materials for domestic and industrial use. *Hüls UK Ltd.*

18876 Marlon® AMX
Amine dodecylbenzene sulfonate
Base material for production of drycleaning detergents, degreasing agents for metal industry, floor cleaners; biodeg. *Hüls Am.; Hüls AG; Hüls UK Ltd.*

18877 Marlon® ARL
Sodium dodecylbenzene sulfonate and sodium toluene sulfonate; detergent component, foaming, wetting agent; for powdered detergents and scouring powders. *Hüls Am; Hüls AG.*

18878 Marlon® AS3
85536-14-7 287-494-3
Dodecylbenzene sulfonic acid
Intermediate for manufacture of anionic surfactants, detergents, sulfonates, textile auxiliaries; biodegradable. *Hüls Am.; Hüls AG; Hüls UK Ltd.*

18879 Marlon® PF 40
Sodium C_{13}-C_{17} alkane sulfonate
sodium laureth sulfate. Mild, high foaming detergent for domestic and industrial applications; biodegradable. *Hüls Am; Hüls AG.*

18880 Marlon® PS 30
Sodium C_{13}-C_{17} alkane sulfonate

Surfactant for production of liquid concentrated mild cleaning agents, hair shampoos, foam baths, textile auxiliaries. *Hüls Am; Hüls AG.*

18881 Marlophen®
A range of nonylphenol polyglycol ethers with wetting, detergent, dispersing and homogenizing properties; used as textile and paper auxiliaries, as wetting agents for coal, rock dusts and pigments and as an additive for concrete manufacture. *Hüls AG.*

18882 Marlophen® 81N
85005-55-6 284-987-5
Nonoxynol-1
Wetting agent, detergent, dispersant; homogenizing capacity; polymer remover in floor care. *Hüls AG.*

18883 Marlophen® 85
9002-93-1 6858
Octoxynol-5
Detergent, wetting agent, dispersant with washing and homogenizing capacity; for solvent cleaners. *Hüls Am; Hüls AG.*

18884 Marlophen® 810
9002-93-1 6858
Octoxynol-10
Wetting agent for acidic, neutral, and alkaline cleaning agents; production of textile auxiliaries. *Hüls Am; Hüls AG.*

18885 Marlophen® 810N
9016-45-9 6772
Nonoxynol-10
Wetting agent, detergent, dispersant; homogenizing capacity; for acidic, neutral and alkaline cleaners, textile auxiliaries. *Hüls AG.*

18886 Marlophen® 830N
9016-45-9 6772
Nonoxynol-30
Wetting agent, detergent, dispersant; homogenizing capacity; binding agent and base material for solid cleaning agents such as toilet sticks. *Hüls AG.*

18887 Marlophen® DNP 16
9014-93-1
Nonylnonoxynol-16
Raw material for textile and paper auxiliaries, dispersant. *Hüls Am.; Hüls AG; Hüls UK Ltd.*

18888 Marlophen® P 1
Phenol ethoxylate (1 EO)
Solubilizer, solvent. *Hüls Am; Hüls AG.*

18889 Marlophen® X
Alkylphenol polyglycol ether
Wetting agent for use in binders for coal dust, pigments, and in concrete manufacture. *Hüls AG; Hüls UK Ltd.*

18890 Marlophor
A range of partial phosphate esters in the form of acids and salts; used as low foam wetting agents for alkaline solutions, starting materials for the manufacture of acid cleaners, water and oil-soluble wetting agents for the textile and paper industries, *Hüls AG.*

18891 Marlophor® CS-Acid
76483-21-1 278-477-1
Isopropyl phosphate ester
Base compound for formulating acid cleaners for glass, metal, and ceramics, flame retardants, mercerizing wetting agents and antistats; anticorrosive and rust-removing props. *Hüls Am; Hüls AG.*

18892 Marlophor® DS-Acid
68439-39-4
n-Butyl phosphate ester
Surfactant used as base material for acid cleaners for glass, metal, and ceramics, flame retardants, mercerizing wetting agents and antistats; anticorrosive and rustremoving props. *Hüls Am; Hüls AG.*

18893 Marlophor® FC
Sodium alkylpolyglycol ether phosphate in liquid form; low foam wetting agent for detergents for dishwashers and industrial cleaning. *Hüls UK Ltd.*

18894 Marlophor® HS-Acid
39407-03-9 254-445-2
n-Octyl phosphate ester
Emulsifier for silicone oils. *Hüls Am; Hüls AG.*

18895 Marlophor® IH-Acid
12645-31-7 235-741-0
2-Ethylhexyl phosphate ester
Low foam wetting agent for weakly to medium strong alkaline range in textile pretreatment and finishing. *Hüls Am; Hüls AG.*

18896 Marlophor® LN-Acid
Trilauryl phosphate
Wetting agent for textile, paper, and personal care industries; antistat, drycleaning detergent. *Hüls Am.*

18897 Marlophor® MD
Organic phosphate ester surfactant; wetting agent (oil and water soluble) and antistat; used in paper; textiles; natural and synthetic fibers; dry cleaning detergents. *Hüls UK Ltd.*

18898 Marlophor® MO 3-Acid
39464-66-9
Laureth-3 phosphate
Surfactant for production of textile and dyeing auxiliaries, antistats, drycleaning detergents. *Hüls Am; Hüls AG.*

18899 Marlophor® ND-Acid
Isoalkylphosphate ester/alkylphenol PEG ether blend; detergent; wetting agent; antistat; component for textile auxiliary agents, drycleaning formulations, paper industry. *Hüls Am; Hüls AG.*

18900 Marlophor® T10-Acid
Diceteareth-10 phospate
Special liquid formulation for production of drycleaning detergents with antistatic properties, textile and dyeing auxiliaries. *Hüls Am; Hüls AG.*

18901 Marlopon
A range of alkylbenzenesulfonates; used for the manufacture of cosmetic detergents and dishwashing agents. *Hüls AG.*

18902 Marlopon® ADS 50
26545-53-9 247-784-2
DEA dodecylbenzene sulfonate
Detergent raw material for liquid phosphate-containing detergents and cleaners; biodegradable. *Hüls Am.; Hüls UK Ltd.*

18903 Marlopon® AMS 60
Amine dodecylbenzene sulfonate/nonionic blend; surfactant used for detergents, dishwashing agents, industrial cleaners, car shampoos, hand cleaners. *Hüls Am; Hüls AG.*

18904 Marlopon® AT
29381-93-9
TEA-dodecylbenzene sulfonate
Surfactant for manufacture of cosmetic detergents, dishwashing agents. *Hüls Am.; Hüls UK Ltd.*

18905 Marlosoft® A 18 M
Blend of fatty acid alkanolamide acetate and fatty acid polyglycol ester; for production of cold water-soluble textile softeners. *Hüls AG.*

18906 Marlosoft® IQ 75
Imidazolinium methosulfate
Base for fabric softeners. *Hüls Am.*

18907 Marlosoft® IQ 90
Quaternized tallow fatty imidazolinium methosulfate; surfactant for production of fabric softeners. *Hüls AG.*

18908 Marlosol®
A range of polyglycol esters of fatty acids; starting materials for preparing agents used in the synthetic fiber industry. *Hüls AG.*

18909 Marlosol® 183
9004-99-3
PEG-3 stearate
Raw material for finishing agents in the synthetic fiber industry; emulsifier. *Hüls Am; Hüls AG.*

18910 Marlosol® BS
9005-08-7
PEG-12 distearate
Superfatting agent, viscosity enhancer for hair shampoos, cosmetic preparations, fabric softeners, synthetic fiber finishing. *Hüls Am; Hüls AG.*

18911 Marlosol® FS
9005-07-6
PEG-12 dioleate
Superfatting agent, viscosity enhancer for hair shampoos, cosmetic preparations, fabric softeners, synthetic fiber finishing. *Hüls Am; Hüls AG.*

18912 Marlosol® OL2
9004-96-0
PEG-2 oleate
Raw material for preparation of agents for synthetic fiber industry. *Hüls AG.*

18913 Marlosol® R70
61791-12-6
PEG-70 castor oil; thickener, conditioner for toilet sticks. *Hüls Am; Hüls AG.*

18914 Marlosol® RF3
PEG-3 rapeseed fatty acid ester; preparation agent. *Hüls AG.*

18915 Marlosol® TF3
61791-00-2
PEG-3 tallate
Raw material for preparation agent for synthetic fiber industry. *Hüls AG.*

18916 Marlotherm
A range of benzyltoluenes; suitable for use as heat transfer mediums. *Hüls AG.*

18917 Marlowet®
A wide range of emulsifiers; emulsifiers for mineral oils, hydrocarbons, waxes, oleic acids, solvents, pesticides, cold cleaners, spindle oils, textile lubricants, furniture polishes and leather dressings, etc. *Hüls AG.*

18918 Marlowet® 1072
C_{12}-C_{14} alcohol polyglycol ether carboxylic acid
Emulsifier for metalworking, water-miscible cooling lubricants, drilling oils, textile auxiliaries. *Hüls Am; Hüls AG.*

18919 Marlowet® 4536
Nonylphenol polyglycol ether carboxylic acid
Emulsifier for water-miscible cooling lubricants and drilling oils, production of textile auxiliaries. *Hüls AG.*

18920 · Marlowet® 4538
C_{13} oxo-alcohol polyglycol ether carboxylic acid
Emulsifier for metalworking, water-miscible cooling lubricants, drilling oils, textile auxiliaries. *Hüls Am; Hüls AG.*

18921 Marlowet® 4539
C_9 oxo-alcohol polyglycol ether carboxylic acid
Surfactant for metalworking, water-miscible cooling lubricants, drilling oils. *Hüls Am; Hüls AG.*

18922 Marlowet® 4702
52668-97-0
C_{18} fatty acid polyglycol ester; emulsifier for mineral oils, spindle oils, metalworking, textile lubricants. *Hüls Am; Hüls AG.*

18923 Marlowet® 4800
68439-49-6 7737
C_{16}-C_{18} alcohol polyglycol ether; emulsifier for waxes, car and furniture polishes, textile lubricants, leather care. *Hüls Am; Hüls AG.*

18924 Marlowet® 4900
Nonylphenol PEG ether; emulsifier for mineral oils used in metalworking and in cold cleaners. *Hüls Am; Hüls AG.*

18925 Marlowet® 5311
97999-44-5 308-441-3
Isononanol phosphate ester
Emulsifier used in paint removers; wetting agent for chlorinated hydrocarbons; production of textile auxiliaries. *Hüls Am; Hüls AG.*

18926 Marlowet® 5324
39464-70-5
Phenol polyglycol ether phosphate ester
Emulsifier for mineral oils; lubricant auxiliaries; metal processing. *Hüls Am; Hüls AG.*

18927 Marlowet® 5400
26635-93-2
Alkylamine polyglycol ether
Emulsifier for mineral oils, paraffin oils, car wash rinses, furniture polishes, textile auxiliaries. *Hüls Am; Hüls AG.*

18928 Marlowet® 5440
95-38-5 428 202-414-9
$C_{22}H_{42}N_2O$
2-(8-heptadecenyl)-4,5-dihydro-1H-imidazole-1-ethanol
Amine 220®;. Substituted imidazoline; emulsifier for mineral oils, corrosion protection, car wash rinses. Also used as a fungicide and soil stabilizer *Hüls Am; Hüls AG.*

18929 Marlowet® 5459
Fatty acid DEA; corrosion inhibitor for metalworking. *Hüls Am; Hüls AG.*

18930 Marlowet® BL
9002-92-0 7717
C_{12} alcohol polyglycol ether
Emulsifier for mineral oils, spindle oils, textile lubricants, bitumen. *Hüls Am; Hüls AG.*

18931 Marlowet® FOX
68439-49-6 7737
Ceteareth-28
Emulsifier for oleic acid and waxes, textile lubricants, car and furniture polishes, leather care. *Hüls Am; Hüls AG.*

18932 Marlowet® ISM
37205-87-1
Nonylphenol polyglycol ether
Emulsifier for solvents, pesticides and cold cleaners. *Hüls Am; Hüls AG.*

18933 Marlowet® LVS
C_{18} fatty acid ester of ethoxylated castor oil; emulsifier for vegetable oils; used in metalworking, leather auxiliaries, release agents. *Hüls Am; Hüls AG.*

18934 Marlowet® NF
Mixture of carboxylic acid polyglycol esters; emulsifier for leather, auxiliaries, textile lubricants. *Hüls Am.*

18935 Marlowet® OCM
Fatty acid alkanolamide polyglycol ether
Anticorrosive agent for water-miscible cooling lubricants. *Hüls AG.*

18936 Marlowet® PW
68439-49-6 7737
C_{16}-C_{18} alcohol polyglycol ether
Surfactant, emulsifier for paraffin, lanolin waxes, textile and paper impregnation, mold release agents, furniture polishes. *Hüls Am; Hüls AG.*

18937 Marlowet® R 11
61791-12-6
Ethoxylated castor oil; emulsifier for animal and vegetable oils and neutral fats, leather auxiliaries. *Hüls Am; Hüls AG.*

18938 Marlowet® R 40
61791-12-6
PEG-40 castor oil; emulsifier for fatty acids, solvents, cosmetic oils; textile lubricants, dyeing auxiliaries, pesticides, creams; biodegradable. *Hüls Am; Hüls AG.*

18939 Marlowet® WOE
9004-98-2
Oleth-5
Emulsifier for paraffinic mineral oils, textile lubricants. *Hüls Am; Hüls AG.*

18940 Marlox®
A range of alkylene oxide addition products; components of low-foaming detergents and cleaners particularly for low-foaming dishwashing agents and for low-foaming textile finishing agents and processing aids. *Hüls AG.*

18941 Marlox® 3000
Butyl glycol/alkylene oxide addition product; surfactant for manufacture of textile auxiliaries and finishing agents. *Hüls Am; Hüls AG.*

18942 Marlox® FK 14
Propylene glycol capreth-4
Detergent, antistat, foam controller; for low-foaming detergents and cleaners, dishwash, industrial cleaners, textile auxiliaries. *Hüls Am; Hüls AG.*

18943 Marlox® FK 64
68154-97-2
PPG-6 deceth-4
Detergent, antistat, foam controller; for low-foaming detergents and cleaners, dishwash, industrial cleaners, textile auxiliaries. *Hüls Am; Hüls AG.*

18944 Marlox® L 6
9064-14-6
PPG-7 lauryl ether
Detergent, antistat, foam controller, textile auxiliary agent. *Hüls Am; Hüls AG.*

18945 Marlox® MO 124
68439-51-0
PPG-4 laureth-2
Component for low foaming detergents and cleaners, automatic dishwasher formulations, industrial cleaners, textile auxiliaries. *Hüls Am; Hüls AG.*

18946 Marlox® MS 48
69227-21-0
C12-18 fatty alcohol alkylene oxide addition product; surfactant for manufacture of textile auxiliaries and finishing agents. *Hüls Am; Hüls AG.*

18947 marls
Natural mixtures of clay and chalk (aluminum silicate and calcium carbonate); used in the manufacture of cements. The term is also applied to friable earths which are devoid of chalk, such as those of Staffordshire.

18948 Marme's reagent
Consists of 10 parts cadmium iodide, CdI_2, and 20 parts potassium iodide, KI, dissolved in 80 parts water; used for testing for alkaloids.

18949 Marmite
A yeast extract; it is a food preparation resembling meat extract. No manufacturer.

18950 Marmo Bardiglio de Bergamo
Vulpinite, a variety of anhydrite, mixed with silica; used for ornamental purposes.

18951 Marphos
7664-38-2 7500 231-633-2
Phosphoric acid
GE Plastics ABS Ltd. Name unverified.

18952 Marquat Pigments
Cadmium, cobalt and titanium pigments; for coatings industry. *Degussa.* Discontinued.

18953 Marseilles soap
Olive oil soap.

18954 marsh gas
74-82-8 6019 200-812-7
CH_4

Methane
Light carburetted hydrogen. Used as a fuel and in chemical manufacturing.

18955 Marshal 10G
55285-14-8 259-565-9
Carbosulfan
A systemic carbamate insecticide. *Rhône-Poulenc Crop Protection Ltd.*

18956 Marshal/suSCon
55285-14-8 259-565-9
100 g/kg carbosulfan; for termite control in reforestation programs with Eucalyptus species; control of Hylobius weevils in pine plantations. *Incitec Ltd.*

18957 Marshal/suXon
Incitec Ltd.

18958 Marsipol
Leather finishes. *ICI Chem & Polymers Ltd.* Discontinued.

18959 martensite
A solid solution of carbon in iron, and is a characteristic constituent of steel which has been tempered at a temperature a little above the transformation point.

18960 Martifin
21645-51-2 355 244-492-7
Ground and finely precipitated aluminum hydroxide; used as a filler and coating pigment for the paper and cardboard industry. *Lonza AG.*

18961 Martin steel
Open Hearth Steel. Steel obtained in the Martin process by melting from 75% of cast iron in a reverberatory furnace with the necessary quantity of wrought iron to obtain the required amount of carbon.

18962 Martinal
21645-51-2 355 244-492-7
Ground and finely precipitated aluminum hydroxide; used as a flame retardant for plastics and rubber. *Lonza AG.*

18963 Martinal® OL-111 LE
21645-51-2 355 244-492-7
Aluminum hydroxide
Finely precipitated filler, flame retardant; suitable for automatic bulk handling. *Martinswerk GmbH.*

18964 Martinal® ON-4608
21645-51-2 355 244-492-7
Aluminum hydroxide
Medium particle size filler/flame retardant for thermoset plastics, carpetbacking latexes. *Martinswerk GmbH.*

18965 Martinal® OS
21645-51-2 355 244-492-7
Aluminum hydroxide
Filler/flame retardant for use where coarse fillers are needed, e.g., artificial marble, polymer concrete, resin floorin, chipboard. *Martinswerk GmbH.*

18966 martinite
$H_2Ca_5O_{16}P_4$
A mineral.

18967 Martino's alloys
Alloys containing 17.25% pig iron, 3-4.5% ferro-manganese, 1.5-2% chromium, 5.25-7.5% tungsten, 1.25-2% aluminum, 0.5-0.75% nickel, 0.75-1% copper, and 65-70% wrought iron; used for drilling and cutting tools.

18968 Martin's cement
A similar cement to Keene's, except that potassium carbonate solution is used instead of alumina.

18969 Martipol
1344-28-1 369 215-691-6
aluminum oxide
Speciality aluminum oxide product; used as polishing aluminas for the ceramic industry. *Lonza AG.*

18970 Martisorb
1344-28-1 369 215-691-6
aluminum oxide
Speciality aluminum oxide product; used for the purification of water. *Lonza AG.*

18971 Martoxin
1344-28-1 369 215-691-6
aluminum oxide
Speciality aluminum oxide product; used as a coating pigment for carbonfree selfcopying paper. *Lonza AG.*

18972 Marvaloy 750
Acrylic-modified styrene; offers high clarity, favorable economics for injection molding and extrusion applications; used for specialty medical, pharmaceutical, food, and cosmetic packaging, advertising displays, high-strength toys. *Marval Industries.*

18973 Marvanbrite CF
Heterocyclic stilbene derivative; nonyellowing optical brightener for cellulosic, wool, silk, nylon and surgical whites. *Marlowe-Van Loan.*

18974 Marvanfix® ATA
Nonformaldehyde fixing agent for improved wetfastness of direct dyes on cellulosic fibers. *Marlowe-Van Loan.*

18975 Marvanfix® C
Quaternary ammonium resin complex; textile fixative; improves wetfastness on direct and reactive dyes. *Marlowe-Van Loan.*

18976 Marvanlube® 92
Silicone emulsion; textile lubricant for preboarding, finish boarding, hosiery. *Marlowe-Van Loan.*

18977 Marvanlube® BHC
Complex polymeric; low foaming dyebath lubricant for hosiery and piece good dyeings (nylon hosiery, nylon/spandex body suits, cotton tights). *Marlowe-Van Loan.*

18978 Marvanol® Aftertreat 2AF
Emulsion polymer; dyeing assistant; exhaust pigment aftertreat for garments, hosiery. *Marlowe-Van Loan.*

18979 Marvanol® BAN
Sulfated vegetable oil; dye dispersant; minimizes barré. *Marlowe-Van Loan.*

18980 Marvanol® Carrier BB
136-60-7 1587 205-252-7
butyl benzoate
Self-emulsifiable butyl benzoate; carrier for atmospheric and pressure dyeing. *Marlowe-Van Loan.*

18981 Marvanol® Defoamer AM-2
Concentrated silicone emulsion; foam inhibitor and depressant for textile applications. *Marlowe-Van Loan.*

18982 Marvanol® GAW
Ethoxylated derivative
Textile leveling agent; compatibilizer for basic and acid dyes. *Marlowe-Van Loan.*

18983 Marvanol® LSL
Ethoxylated surfactant blend; compatibilizer for basic and acid dyes on synthetics. *Marlowe-Van Loan.*

18984 Marvanol® Penetrant 35
Amine neutralized sulfonic acid
Detergent, wetting agent, textile dyeing, scouring agent, finishing, leveling and retarding agents for acid dyes. *Marlowe-Van Loan.*

18985 Marvanol® Pretreat GD-P
Polymeric resin; dyeing assistant; exhaust pigment pretreat for garments, hosiery. *Marlowe-Van Loan.*

18986 Marvanol® RD2-1852
Sulfated ester blend; dyeing assistant; dyebath lubricant for direct dyes on cotton. *Marlowe-Van Loan.*

18987 Marvanol® REAC A-213
Polymer salt; low foaming dispersing agent, textile scouring agent for fiber reactives; improves dye fixation. *Marlowe-Van Loan.*

18988 Marvanol® SBO (60%)
Sulfated butyl oleate, sodium salt; detergent, wetting and leveling agent, emulsifier, dyeing assistant, lubricant. *Marlowe-Van Loan.*

18989 Marvanol® SCO (50%)
8002-33-3 232-306-7
Sulfated castor oil; detergent, wetting agent, dyeing assistant and lubricant used in finishing operations; leveling agent for cotton. *Marlowe-Van Loan.*

18990 Marvanol® Scour 2 Base
Alcohol ether condensate; multifunctional biodegradable textile scour. *Marlowe-Van Loan.*

18991 Marvanol® SPO (60%)
Sulfated propyl oleate, sodium salt; detergent, wetting and leveling agent, dyeing assistant, lubricant for textiles. *Marlowe-Van Loan.*

18992 Marvanquest 1022
Organic blend; nonsilicate/hydrogen peroxide bleach stabilizer; prevents calcium deposits. *Marlowe-Van Loan.*

18993 Marvanscour® KW
Solvents and detergents; detergent for prescour and afterscour. *Marlowe-Van Loan.*

18994 Marvanscour® LF
Phosphate ester; low-foaming wetting agent and dyeing assistant for jet dyeing cotton and blends. *Marlowe-Van Loan.*

18995 Marvansoft 1771
Silicone emulsion; softener for textile finishing, printing cellulosics and synthetics. *Marlowe-Van Loan.*

18996 Marvantex RBDS
Inorganic salts; one-bath scour, bleaching agent, and dye. *Marlowe-Van Loan.*

18997 Marvylan
Polyvinylchloride
Used in the plastics processing industry, applications in building construction (tubes and pipes, profiles and cables); in the packaging industry (bottles); as synthetic leather (bags, wall covering, clothing); in hoses etc. *DSM NV.* Discontinued.

18998 Marweld M-17
Epoxy resin, liquid sealant, two component, flexible; pipe and flange thread sealant for sealing threads in flanged ductile iron and cast iron water pipes, EPA approved. *RJ Manufacturing Inc.* Name unverified.

18999 Mascot Clearing
15310-01-7 239-352-7
Benodanil
A systemic fungicide. *Rigby Taylor Ltd.*

19000 Mascot Cloverkiller
7085-19-0 5826 230-386-8
Soluble concentrate containing 300 g/l mecoprop; for control of weeds in cereals and grassland. *Rigby Taylor Ltd.*

19001 Mascot Contact Turf Fungicide
50471-44-8 10122 256-599-6
Vinclozolin
A protectant fungicide for turf. *Rigby Taylor Ltd.*

19002 Mascot Gauntlet
1912-24-9 902 217-617-8
Atrazine
A residual herbicide. *Rigby Taylor Ltd.*

19003 Mascot Highway
Aminotriazole [61-82-5] and simazine [122-34-9]; used for total weed control in non crop areas. *Rigby Taylor Ltd.*

19004 Mascot Moss Killer
97-23-4 3120 202-567-1
Dichlorophen
Fungicide, bactericide and algicide. *Rigby Taylor Ltd.*

19005 Mascot Selective
2,4-D and mecoprop. Soluble concentrate containing 60 g 2,4-D and 200 g mecoprop per liter; used to control weeds in grassland. *Rigby Taylor Ltd.*

19006 Mascot Super Selective
Soluble concentrate of 15 g dicamba, 100 g MCPA and 200 g mecoprop per liter; used for weed control in cereals and grassland. *Rigby Taylor Ltd.*

19007 Mascot Systemic Turf Fungicide
10605-21-7 1836 234-232-0
A liquid formulation containing 500 g carbendazim per liter as a suspension concentrate; systemic fungicide. *Rigby Taylor Ltd.*

19008 Mascot Ultrasonic
Glyphosate + simazine. Glyphosate + simazine; a translocated and residual herbicide for the control of grasses and weeds. *Rigby Taylor Ltd.*

19009 Masil® 173
Silicone fluid dispersed in halogenated organic solvent; defoamer for chemical processing, refiners, solvent cleaning; especially for defoaming light hydrocarbon solvents. *PPG/Specialty Chem.* Discontinued.

19010 Masil® 264
Alkyl methyl polysiloxane
Processing aid, lubricant, mold release; for aerosol packaging, ink/printing industry. *PPG/Specialty Chem.* Discontinued.

19011 Masil® 280
Dimethicone copolyol; antistat, wetting agent for personal care products, etc. *PPG/Specialty Chem.* Discontinued.

19012 Masil® 756
Tetrabutoxypropyl methicone
Cosmetic ingredient providing high gloss to hair; nonoily emollient for skin care. *PPG/Specialty Chem.* Discontinued.

19013 Masil® 1066C
Dimethicone copolyol; lubricant and antistat for plastics, textiles, metal processing; wetting and leveling characteristics; antifog for glass cleaners. *PPG/Specialty Chem.* Discontinued.

19014 Masil® EM 100
Dimethylpolysiloxane fluids aqueous emulsion; emulsifier, release aid in molding, extrusion, laminating, and casting for rubber, plastics, and metals; for leather, glass, and vinyl cleaners, polishes, textile softeners, textile/fiber lubricants; recommended for glass hard surface cleaner applications; imparts non-smearing, low gloss and ease-of-wipe properties to such formulations. *PPG/Specialty Chem.* Discontinued.

19015 Masil® EM 350
Dimethylpolysiloxane fluids aqueous emulsion; food grade release emulsion used in manufacture of articles in contact with food. *PPG/Specialty Chem.* Discontinued.

19016 Masil® EM 100,000
Dimethylpolysiloxane fluids aqueous emulsion; emulsifier, release aid in molding, extrusion, laminating, and casting for rubber, plastics, and metals; for leather, glass and vinyl cleaners, polishes, textile softeners, textile/fiber lubricants. *PPG/Specialty Chem.* Discontinued.

19017 Masil® SF 1,000,000
9016-00-6 3264 215-648-1
Dimethicone
Release aid, defoamer for nonaqueous processes, especially in the petroleum, foods, and printing inks industries; internal lubricant for plastics, rubber, and metal; also in furniture and auto-wax polishes, household and personal care products. *PPG/Specialty Chem.* Discontinued.

19018 Masil® SF 5
9016-00-6 3264 215-648-1
Dimethicone
Release aid, defoamer for nonaqueous processes, especially in the petrol., foods, and printing inks industries; internal lubricant for plastics, rubber, and metal also in furniture and auto-wax polishes, household and personal care products; textile lubricant; lowerviscosity fluids recommended for cosmetic applications, higher viscosity fluids (>10,000 cSt) for formulation of lubricants and mold releases in manufacture of plastics and rubber parts. *PPG/Specialty Chem.* Discontinued.

19019 Masil® SF 201
Vinyl-terminated silicone polymer; functional polymer for compounding silicone elastomers for use as encapsulants in electrical/electronic industry. *PPG/Specialty Chem.* Discontinued.

19020 Masil® SF 500
9016-00-6 3264 215-648-1
Dimethicone
Release aid, defoamer for nonaqueous processes, especially in the petroleum, foods, and printing inks industries; internal lubricant for plastics, rubber, and metal; also in furniture and auto-wax polishes, household and personal care products. *PPG/Specialty Chem.* Discontinued.

19021 Masil® SF 500,000
9016-00-6 3264 215-648-1
Dimethicone
Release aid, defoamer for nonaqueous processes, especially in the petroleum, foods, and printing inks industries; internal lubricant for plastics, rubber, and metal; also in furniture and auto-wax polishes, household and personal care products. *PPG/Specialty Chem.* Discontinued.

19022 Masil® SF-MH
9004-73-3
Methicone
Polymethylhydrosiloxane. Reactive fluid for modification of polyester and methacrylic resins, as waterproofing and impregnating agents for textiles, paper, leather and hydrophobizing agents for powders, silicas, and other fillers. d = 0.9950; n_D^{20} = 1.3950-1.3970; fp = 121°. *PPG/Specialty Chem.* Discontinued.

19023 Masil® SFR 70
31692-79-2
Dimethiconol
Reactive fluid; raw material in compounding silicone room temperature vulcanizing systems, textile and paper coatings, plasticizer/processing aid for silicone elastomers, hydrophobizing silica, in water repellent formulations. *PPG/Specialty Chem.* Discontinued.

19024 Masil® SF-V
69430-24-6
Cyclomethicone
Volatile silicone fluid imparting silky light feel and spreadability to cosmetics (hair care products, skin creams and lotions, antiperspirants, deodorants, suntan preparations). *PPG/Specialty Chem.* Discontinued.

19025 Maslip® 500
Proprietary formula containing no chlorinated or sulfonated compounds; lubricant base for metalworking fluids which require extreme-pressure qualities for heavy-duty work. *PPG/Specialty Chem.* Discontinued.

19026 Masocare
5058
Iodophors. Used as germicides, antiseptics and disinfectants. *Evans Vanodine International Ltd.*

19027 Masodine
5058
Iodophors Used as germicides, antiseptics and disinfectants. *Evans Vanodine International Ltd.*

19028 Masol
Cement mixture for pump packing in mines roadways. *Foseco (F.S.) Ltd.*

19029 Masonry Stain and Seal
Methylmethacrylate acrylic, pigments and polysiloxane resins in an aliphatic

and aromatic solvent vehicle system; for semi-transparent stain for masonry. *Nova Chemical Inc.* Name unverified.

19030 Masoten®

52-68-6 9753 200-149-3
trichlorfon
Veterinary preparation; for the control of ectoparasites in fish. Acts as an anthelmintic (nematodes). *Bayer AG.*

19031 Massa Estarinum.

Neutral hard fats based on mixtures of triglycerides; used for preparation of suppositories. *Hüls AG.*

19032 Massa Estarinum® CM

Hydrogenated palm glycerides, hydrogenated palm kernel glycerides; consistency regulator for decorative cosmetics, sticks, pencils, powders, glosses, and eye shadows. *Hüls Am.*

19033 Massaranduba

Balata rans; brittle balata. A pseudo gutta-percha derived from the sap of the Brazilian cow tree, *Mimusops elata.*

19034 massecuite

The boiled mass of beet sugar syrup. It is a semi-solid mass formed during the evaporation of the sugar juice, and consists of sugar crystals and a thick syrup. It contains from 3.5-7% water.

19035 Master Bond AC82

One-part nickel conductive adhesive coating; for EMI/RFI shielding applications. *Master Bond.*

19036 Master Bond EP11HT

25928-94-3
One-part epoxy; heat-resistant version of EP11; service temperature -60 to 350°F. *Master Bond.*

19037 Master Bond EP21HT

25928-94-3
Two-part epoxy; high temperature resistant version of EP21; excellent chemical resistance; service temp. -60 to 400°F. *Master Bond.*

19038 Master Bond EP30HT

25928-94-3
Two-part epoxy; high temperature resistant version of EP30; exceptional adhesion to glass, ceramics, wood; excellent long-term durability; service temperature -60 to 400°F. *Master Bond.*

19039 Master Bond EP34CA

25928-94-3
Two-part epoxy; heat-resistant laminating resin for filament winding of structural FRP components; service temperature -60 to 450°F. *Master Bond.*

19040 Master Bond EP75

25928-94-3
Two-part epoxy, graphite-filled; electrically conductive adhesive/sealant; service temperature -60 to 300°F. *Master Bond.*

19041 Master Bond Supreme 11HT

25928-94-3
Two-part epoxy; heat-resistant. version of Supreme 11; service temperature. -60 to 400°F. *Master Bond.*

19042 Master Bond Supreme 3HT

25928-94-3
One-part epoxy; high peel strength, heat-resistant version of EP3; service temperature -60 to 350°F. *Master Bond.*

19043 Master Sil 701

Silicone; general purose bonding/sealing agent featuring high temperature resistance; used especially with formed-in-place gaskets. *Master Bond.*

19044 Masterblok

Solid phenolic ester in panels; used for master patterns, N/C trials, models, fixtures and vac forming tools. *J R Technology Ltd.*

19045 Masterbond

Cellulose acetate laminating adhesives. *The Scottish Adhesives Co Ltd.*

19046 Mastercarb

Mineral filled concentrates in polyolefins; used for plastics moldings, sheet and film. *Collinda Ltd.*

19047 Mastercolor

Masterbatches of thermoplastics; color and additive concentrates used in plastics. *Ampacet Corporation.*

19048 Masterflam

Halogenic and nonhalogenic masterbatches for making self-extinguishing thermoplastics *VAMP srl.* Name unverified.

19049 Masterwood

Wood-filled thermoplastic polymers; used for plastics moldings and extrusions. *Collinda Ltd.* Discontinued.

19050 mastic

The name for an important resin obtained from *Pistachia lentiscus,* from various parts of the Mediterranean coast; used in the manufacture of spirit varnishes. The term Mastic is also applied to a mixture of asphalt rock and Trinidad pitch.

19051 Masticillin® C, M

Preparations containing penicillin, streptomycin or penicillin sulfonamide; used for the treatment of mastitis in veterinary medicine. *Bayer AG.*

19052 Masuron

A proprietary trade name for a cellulose acetate plastic. No manufacturer.

19053 Matalex

X-ray developer system. *May & Baker Ltd.* Name unverified.

19054 matali

Rubber obtained from the roots of various species of *Apocynaceae.*

19055 Match

21725-46-2 2755 244-544-9
cyanazine. Suspension concentrate containing 500 g cyanazine per liter; a triazine herbicide. *Shell UK.*

19056 Mateflex

Polyethylene resin; used for tennis courts. *Sumitomo Bakelite Co Ltd.* Name unverified.

19057 Mater-Bi®

Chemically modified com starch alloys; thermoplastic of vegetable origin with synthetic substances; biodegradable plastic for packaging films, bags, 6-pack yokes, disposable diapers, hospital and sanitary products. *Novamont N. Am.*

19058 Matexil

Textile auxiliary chemicals. *ICI Chem & Polymers Ltd.*

19059 Mathesius Metal

A proprietary trade name for an alloy of lead with calcium and strontium in small amounts. No manufacturer.

19060 matico-camphor

A camphor from the Peruvian matico.

19061 Matikus

56073-10-0 1400 259-980-5
Rodenticide containing brodifacoum. *ICI Chem & Polymers Ltd.*

19062 Matrikerb

Mixture of clopyralid and propyzamide; a soil and leaf herbicide for winter oilseed rape. *Pan Britannica Industries Ltd.*

19063 Matrikerb

Clopyralid + propyzamide
A post-emergence herbicide for use on winter oilseed rape. *Rohm & Haas UK.*

19064 Matrimid® 5292 System

Bismaleimide resin with O,O'-diallyl bisphenol A hardener; used for advanced composites, high temperature adhesives, laminating, casting, filament winding. *Ciba-Geigy.* Name unverified.

19065 Matrix alloy

An alloy of 48% bismuth. 28.5% lead. 14.5% tin, and 9% antimony. It expands on cooling and is used to hold tools in position.

19066 Matrix®

For the agriculture industry. *DuPont UK.*

19067 Matt salt

1341-49-7 523 215-676-4
F_2H_5N
Acid ammonium fluoride
ammonium bifluoride.

19068 matteucinol

6,8-dimethyl-5,7-dihydroxy-4'-methoxy-flavanone

19069 Mattheylec

Conductive adhesives and coatings; metallising preparations for the electrical and electronic industries. *Johnson Matthey plc.* Discontinued.

19070 Mattina®

Matte pigments displaying intense color with subtle pearlescent glow for soft luster or matte effects' recommended for pressed powd. makeups. *Mearl.*

19071 maturex

A purified α-acetolactate decarboxylase (ALDC) produced by a strain of *Bacillus subtilis*; added at the beginning of an alcohol fermentation (e.g. for beer), it prevents the formation of diacetyl by catalyzing the decarboxylation of α-acetolactate to acetoin.

19072 maucherite

Ni_3As_2
A mineral.

19073 mawele

A millet of East Africa.

19074 Maxahibit TT-50

64665-57-2 265-004-9
Sodium tolyltriazole
Corrosion inhibitor for copper, brass, nonferrous metals in water treatment, cooling waters, engine coolants, cleaners, metalworking fluids. *Actrachem.*

19075 Maxepa®
Fish oil concentrate; used for triglyceride lowering. *Seven Seas Ltd.*

19076 Maxhete
A steel containing nickel, chromium, tungsten, copper, and silicon.

19077 maxi braun®
Self-tanning product. *Bayer AG.*

19078 Maxicrop Moss Killer & Conditioner
10028-22-5 4079 233-072-9
$Fe_2O_{12}S_3$
Ferric sulfate
Used for moss control in turf. *Maxicrop International Ltd.*

19079 Maxigard
A multicomponent nitrite-borate cooling water treatment; diesel engine water treatment used to prevent corrosion and mineral deposits as well as corrosion due to cavitation in medium and high speed diesel engines. *Ashland Chemical Company.* Name unverified.

19080 Maxilon
Modified basic dyes. *Ciba plc.* Name unverified.

19081 Maxilvry Steel
A proprietary high nickel-chromium steel containing copper stated to be corrosion resisting, and particularly resistant to attack by cider. No manufacturer.

19082 Maxim
10605-21-7 1836 234-232-0
carbendazim
A liquid formulation containing 500 g carbendazim per liter as a suspension concentrate; systemic fungicide. *Farmers Crop Chemicals Ltd.* Discontinued.

19083 MaxiMate
A suspension concentrate containing 62 g carbendazim and 400 g maneb per liter; systemic fungicide for cereals. *Farmers Crop Chemicals Ltd.* Discontinued.

19084 Maximite
An explosive. It is similar to cordite.

19085 maxium metal
Castings of magnesium metal.

19086 Maxon
Polyglyconate
Surgical aid. *Davis & Geck.* Name unverified.

19087 Mayari iron
An iron made from Cuban ores. Small amounts of vanadium and titanium are present which give strength to the metal obtained from these ores.

19088 Mayari steel
A Cuban low nickel-chromium steel.

19089 Mayclene
Selective herbicide
Engelhard Technologies Ltd.

19090 Mayco Base 1351
Sulfurized lard oil; heavy-duty EP lubricant additive for cutting oils, heavy-duty gear oils, waxy lubricants for machine tools; for use with highly paraffinic and re-refined oils. *Mayco Oil & Chem.*

19091 Mayer's albumen
A fixing agent used in microscopy. It consists of 50 ml white of egg, 50 ml glycerin, and 1 g sodium salicylate.

19092 Mayer's solution
Mercuric iodide dissolved in aqueous potassium iodide. An reagent for detecting alkaloids.

19093 Mayphos 45
Organic phosphate ester, free acid; EP lubricant additive, metal wetting agent, surfactant; noncorrosive to ferrous and nonferrous metals; prevents water-induced plugging of diatomaceous earth filter. *Mayco Oil & Chem.*

19094 Maypon
Salts of collagen polypeptide fatty acid condensate; used as hair care additives. *Harcros Australia.*

19095 Maypon 4C
68920-65-0
Potassium coco-hydrolyzed collagen; detergent used in personal care products, general purpose cleansers. *Inolex.*

19096 Maypon 4CT
68952-16-9
TEA-coco-hydrolyzed collagen; viscosity builder, foam modifier. *Inolex.*

19097 Maypon UD
68951-92-8
Potassium undecylenoyl hydrolyzed collagen; detergent used in personal care products, antifungal properties. *Inolex.*

19098 maytee
Fenugreek seeds.

19099 May-Tein C
68920-65-0
Potassium cocoyl hydrolyzed collagen; mild surfactant, conditioner, softener, moisturizer, lubricant, antistat, anti-irritant for hair relaxers, shampoos, conditioners, liq. soaps, depilatories; biodeg. *Maybrook.*

19100 May-Tein CT
68952-16-9
TEA-cocoyl hydrolyzed collagen; mild surfactant, conditioner, anti-irritant for shampoos, conditioners, baby products, mousses, makeup removers, shave creams, liq. soaps; biodeg. *Maybrook.*

19101 May-Tein KK
Potassium cocoyl hydrolyzed keratin; mild biodegradable surfactant providing protection and gentle cleaning to skin and hair products, hand, body and face soaps, bath products; substantive, foaming protein, softener, conditioner, antistat; anti-irritant in depilatories and hair relaxers. *Maybrook.*

19102 May-Tein KT
Sodium cocoyl hydrolyzed keratin; mild biodegradable surfactant providing gentle cleansing to skin and hair; source of cystine for damaged hair; substantive, foaming protein, lubricant, antistat; anti-irritant for harsh ingreds. *Maybrook.*

19103 May-Tein R
Potassium cocoyl rice protein; biodegradable protein surfactant for hair and skin care products; provides gentle cleaning, conditioning, foaming, substantivity, lubricity, antistatic properties; for shampoos, hair relaxers, makeup removers, creams and bath products. *Maybrook.*

19104 Mazak
Die casting alloy. *Pasminco Europe Ltd/ISC Alloys Div.*

19105 Mazam
A dextrin of high molecular weight. No manufacturer.

19106 Mazamide® 68
Cocamide DEA (2:1)
Biodegradable thickener, emulsifier, foam builder; for hard surface cleaners, dishwash, shampoos, metalworking fluids, waterless hand cleaners, automotive specialties. *PPG/Specialty Chem.* Discontinued.

19107 Mazamide® 1214
120-40-1 204-393-1
Lauramide DEA (2:1)
Entprol; Quadrol. Biodegradable thickener, emulsifier, foam builder; for hard surface cleaners, dishwash, shampoos, metalworking fluids, waterless hand cleaners, automotive specialties. bp_1 = 190°; soluble in H_2O, organic solvents. *PPG/Specialty Chem.* Discontinued.

19108 Mazamide® C-2
PEG-3 cocamide MEA
Biodegradable foam builder/stabilizer for cosmetic and pharmaceutical shampoos. *PPG/Specialty Chem.* Discontinued.

19109 Mazamide® CFAM
68140-00-1 268-770-2
Cocamide MEA (1:1)
Biodegradable foam stabilizer, viscosity builder for cosmetic and pharmaceutical shampoos. *PPG/Specialty Chem.* Discontinued.

19110 Mazamide® L-298
120-40-1 204-393-1
2:1 Lauramide DEA
Foam builder and stabilizer, emulsifier, dispersant, viscosity builder, solubilizer for hard surface cleaners, dishwashing, shampoos, metalworking fluids, automotive specialties, fiber and hair conditioners, dry cleaning, agric, sprays, leather/fur preparations, emulsifiable waxes, rust inhibitors, polishes, paint removers, rug shampoos; fuel oil additives and textiledetergents. Biodegradable. *PPG/Specialty Chem.* Discontinued.

19111 Mazamide® L-5
PEG-6 lauramide DEA
Emulsifier, lubricant rust inhibitor, buffing compound, detergent, foam builder, stabilizer for cosmetic and pharmaceutical creams, lotions, bath oils, shampoos; biodegradable. *PPG/Specialty Chem.* Discontinued.

19112 Mazamide® LLD
Linoleamide DEA (2:1)
Biodegradable thickener, emulsifier, foam builder; for hard surface cleaners, dishwash, shampoos, metalworking fluids, waterless hand cleaners, automotive specialties. *PPG/Specialty Chem.* Discontinued.

19113 Mazamide® O 20
93-83-4 202-281-7
2:1 Oleamide DEA
Biodegradable thickener, emulsifier, foam builder, corrosion inhibitor, dispersant; for hard surface cleaners, dishwash, shampoos, metalworking fluids, waterless hand cleaners, automotive specialties. *PPG/Specialty Chem.* Discontinued.

19114 Mazamide® SMEA
111-57-9 203-883-2

Stearamide MEA (1:1)
Biodegradable emulsifier, pearlescent for syndet soap bars. *PPG/Specialty Chem.* Discontinued.

19115 Mazamide® SS 20
2:1 Linoleamide DEA
Biodegradable thickener, emulsifier, foam builder, corrosion inhibitor; for hard surface cleaners, dishwash, shampoos, metalworking fluids, waterless hand cleaners, automotive specialties. *PPG/Specialty Chem.* Discontinued.

19116 Mazawax® 163R
Cetearyl alcohol and polysorbate 60; emulsifier for pharmaceutical and cosmetic applications; base; emolliency and thickening properties. *PPG/Specialty Chem.* Discontinued.

19117 Mazawet® 36
Surfactant; low foaming wetting agent, detergent, and dispersant; for machine dishwashing rinse additives, penetrant for textile finishing and scouring operations. *PPG/Specialty Chem.* Discontinued.

19118 Mazawet® DOSS 70
577-11-7 3460 209-406-4
Dioctyl sodium sulfosuccinate
Fast wetting surfactant, emulsifier, dispersant for drycleaning, paper and textile processing, control agent in coal dedusting, antifog for glass cleaners. *PPG/Specialty Chem.* Discontinued.

19119 Mazclean EP
Proprietary emulsifier; microemulsifier package for use in preparing terpene-based cleaning formulations. *PPG/Specialty Chem.* Discontinued.

19120 Mazeen® 173
102-60-3 3635 203-041-4
Tetrahydroxypropyl ethylenediamine
Insecticide and herbicide emulsifier, antistat and rewetting agent, grease additive, textile lubricant; emulsifier for lubricants, inks, and cosmetics; chelant in electroless deposition formulations for electronics industry. crosslinker for rigid polyurethane. *PPG/Specialty Chem.* Discontinued.

19121 Mazeen® C-2
61791-14-8
PEG-2 cocamine
Emulsifier, rewetting agent, lubricant, coupler used in insecticides and herbicides, grease additives, textile lubricants, water-based inks, cosmetics; plastics antistat; viscosity modifier and rust inhibitor in acid media for metalworking. *PPG/Specialty Chem.* Discontinued.

19122 Mazeen® DAPI
67799-04-6 267-101-1
Isostearylpropyl dimethylamine
Emulsifier. *PPG/Specialty Chem.* Discontinued.

19123 Mazeen® DAPL
68140-01-2 268-771-8
Cocamidopropyl dimethylamine
Emulsifier. *PPG/Specialty Chem.* Discontinued.

19124 Mazeen® S-13
7651-02-7 231-609-1
Stearamidopropyl dimethylamine
Emulsifier. *PPG/Specialty Chem.* Discontinued.

19125 Mazeen® S-15
61791-24-0
PEG-15 soyamine
Emulsifier, rewetting agent, lubricant, coupler used in insecticides and herbicides, grease additives, textile lubricants, water-based inks, cosmetics; plastics antistat. *PPG/Specialty Chem.* Discontinued.

19126 Mazeen® S-2
61791-24-0
PEG-2 soyamine
Emulsifier, rewetting agent, lubricant, coupler used in insecticides and herbicides, grease additives, textile lubricants, water-based inks, cosmetics; plastics antistat. *PPG/Specialty Chem.* Discontinued.

19127 Mazeen® SHCFA
68140-01-2 268-771-8
Cocamidopropyl dimethylamine
Emulsifier. *PPG/Specialty Chem.* Discontinued.

19128 Mazeen® T-2
61791-44-4 263-177-5
PEG-2 tallow amine
Emulsifier, rewetting agent, lubricant, coupler used in insecticides and herbicides, grease additives, textile lubricants, water-based inks, cosmetics, metalworking; plastics antistat; visc. modifier, rust inhibitor in acid media. *PPG/Specialty Chem.* Discontinued.

19129 Mazide 25
123-33-1 5745 204-619-9
Maleic hydrazide, 250 g/liter; a tree growth regulator to control shoots on the trunk and suckers around the base of street trees. *Synchemicals Ltd.*

19130 Mazide Selective
Soluble concentrate of 6 g dicamba, 200 g maleic hydrazide and 75 g MCPA per liter. Plant growth regulator for grass verges. *Synchemicals Ltd.*

19131 Mazin®
Wettable powder containing maneb 80% w/w and zinc oxide; protective fungicide against potato blight. *Universal Crop Protection Ltd.*

19132 Mazol® 80 MG K
51158-08-8
PEG-20 glyceryl stearate
Emulsifier, dough conditioner for cakes, icings, whipped toppings. *PPG/Specialty Chem.* Discontinued.

19133 Mazol® 165C
Glyceryl stearate and PEG-100 stearate; emulsifier blend for cosmetic and pharmaceutical oil-water emulsions; acid-stable. *PPG/Specialty Chem.* Discontinued.

19134 Mazol® 300 K
111-03-5 203-827-7
Glyceryl oleate
GRAS dispersant for oil or solvent systems; antifoam for sugar and protein processing; coemulsifier with T-Maz 60 or 80. Kosher. *PPG/Specialty Chem.* Discontinued.

19135 Mazol® 1400
65381-09-1 265-724-3
Caprylic/capric triglyceride
Carrier for flavors, fragrances, vitamins, antibiotics, pigmented cosmetics, medicinals. *PPG/Specialty Chem.* Discontinued.

19136 Mazol® GMO
111-03-5 203-827-7
Glyceryl oleate
GRAS dispersant for oil or solvent systems; antifoam for food processing; base for cosmetic creams, lotions, ointments; water-oil emulsifier with emolliency, thickening properties; plasticizer; lubricant; antifog for PVC; emulsifier, coupling agent for metalworking applications. *PPG/Specialty Chem.* Discontinued.

19137 Mazol® GMR
141-08-2 205-455-0
Glyceryl ricinoleate
Base for cosmetic creams, lotions, ointments; water-oil emulsifier with emolliency, thickening properties; plasticizer. *PPG/Specialty Chem.* Discontinued.

19138 Mazol® GMS
Glyceryl stearate
Lubricant, emulsifier, plasticizer, and thickener for foods, drugs, and cosmetics. *PPG/Specialty Chem.* Discontinued.

19139 Mazol® PETO
19321-40-5 242-960-5
Pentaerythritol tetraoleate
Emulsifier, coupling agent for metalworking applications. *PPG/Specialty Chem.* Discontinued.

19140 Mazol® PGMS
1323-39-3 215-354-3
Propylene glycol stearate
Coemulsifier for edible oil and shortenings, dispersing aid for nondairy creamers. *PPG/Specialty Chem.* Discontinued.

19141 Mazol® PGO-104
34424-98-1 252-011-7
Decaglyceryl tetraoleate
Emulsifier for food products, cosmetics, toiletries, pharmaceuticals, lubricants, mold release compounds, in plasticizers for synthetic fabrics and plastics. *PPG/Specialty Chem.* Discontinued.

19142 Mazol® PGO-31 K
9007-48-1
Triglyceryl oleate
Emulsifier for food products, in cosmetics, toiletries, pharmaceuticals, lubricants, mold release compounds, plasticizers; solubilizer for essential oils and flavors. Kosher. *PPG/Specialty Chem.* Discontinued.

19143 Mazon® 18A
Proprietary surfactant; viscosity booster for aqueous systems. *PPG/Specialty Chem.* Discontinued.

19144 Mazon® 41
Ammonium salt of alkylphenol ethoxylate; surfactant for cleaning formulations. *PPG/Specialty Chem.* Discontinued.

19145 Mazon® 60T
TEA dodecylbenzene sulfonate
High foaming formulated detergent, emulsifier for shampoos, dishwashing, car wash, textile, hard surface cleaners, metal cleaners and other formulations. *PPG/Specialty Chem.* Discontinued.

19146 Mazon® 85
Modified dodecylbenzene sulfonic acid; self-emulsifying degreaser, detergent, emulsifier for cleaning applications; used with hydrocarbon solvs. *PPG/Specialty Chem.* Discontinued.

19147 Mazon® 1045A
POE sorbitol fatty acid ester; emulsifier for pesticide, herbicide, metalworking, die-cast lubricant formulations, and emulsion polymerization; humectant, emollient. *PPG/Specialty Chem.* Discontinued.

19148 Mazon® RI 4A
Amine-based; ferrous corrosion inhibitor developed to replace sodium nitrite in metalworking fluids. *PPG/Specialty Chem.* Discontinued.

19149 Mazox® CAPA
68155-09-9 268-938-5
Cocamidopropylamine oxide
Surfactant, conditioner, emollient, emulsifier, foam booster, viscosity builder, lime soap dispersant for cosmetics, toiletries, household and industrial uses. *PPG/Specialty Chem.* Discontinued.

19150 Mazox® CDA
7128-91-81 230-429-0
Palmitamine oxide
Surfactant, conditioner, emollient, emulsifier, foam booster, viscosity builder for cosmetics, toiletries, household and industrial uses; textile softener. *PPG/Specialty Chem.* Discontinued.

19151 Mazox® KCAO
Potassium dihydroxyethyl cocamine oxide phosphate
Foam booster/stabilizer, detergent, emollient, lime soap dispersant for dishwash, heavy-duty detergents, caustic solutions; stable in caustic soda; compatible with most surfactants, builders, many solvents. *PPG/Specialty Chem.* Discontinued.

19152 Mazox® LDA
Lauramine oxide
Surfactant, conditioner, emollient, emulsifier, foam booster, viscosity builder for cosmetics, toiletries, household and industrial uses. *PPG/Specialty Chem.* Discontinued.

19153 Mazox® MDA
3332-27-2 222-059-3
Myristamine oxide
Surfactant, conditioner, emollient, emulsifier, foam booster, viscosity builder for cosmetics, toiletries, household and industrial uses. *PPG/Specialty Chem.* Discontinued.

19154 Mazox® ODA
14351-50-9 238-311-0
Oleamine oxide
Surfactant, conditioner, emollient, emulsifier, foam booster, viscosity builder for cosmetics, toiletries, household and industrial uses; textile softener. *PPG/Specialty Chem.* Discontinued.

19155 Mazox® SDA
2571-88-2 219-919-5
Stearamine oxide
Surfactant, conditioner, emollient, emulsifier, foam booster, viscosity builder for cosmetics, toiletries, household and industrial uses; textile softener. *PPG/Specialty Chem.* Discontinued.

19156 Maztreat® 246
Organic defoamer; defoamer for metalworking formulations. *PPG/Specialty Chem.* Discontinued.

19157 Maztreat® BOM
Additive for boiler water treatment. *PPG/Specialty Chem.* Discontinued.

19158 Maztreat® CA Powd
Additive for cooling water treatment. *PPG/Specialty Chem.* Discontinued.

19159 Mazu® 10 P Mod 11
Organic defoamer; defoamer for yeast fermentation. *PPG/Specialty Chem.* Discontinued.

19160 Mazu® 43 C
Organic defoamer; defoamer for enzyme, effluent wastewater applications. *PPG/Specialty Chem.* Discontinued.

19161 Mazu® 142
Organic defoamer; defoamer for whitewater, screen room, bleach plant. *PPG/Specialty Chem.* Discontinued.

19162 Mazu® 252
Organic defoamer; defoamer for brownstock, adhesives, latex paints. *PPG/Specialty Chem.* Discontinued.

19163 Mazu® 5118
Organic defoamer; defoamer for delayed cooking. *PPG/Specialty Chem.* Discontinued.

19164 Mazu® DF 100S
Silicon compound; food-grade defoamer for fermentation, vegetable oils. *PPG/Specialty Chem.* Discontinued.

19165 Mazu® DF 200SP
8050-81-5 3264
Simethicone
Food grade defoamer for fermentation, vegetable oils; also is formulated to meet the specific needs of pharmaceutical industry. *PPG/Specialty Chem.* Discontinued.

19166 Mazu® DF 205SX
Silicone emulsion; industrial defoamer formulated with low viscosity for ease of handling and pumping; for leather finishing, metalworking, carpet cleaning, waste treatment. *PPG/Specialty Chem.* Discontinued.

19167 Mazu® DF 243
Silicone emulsion; emulsifier; defoamer for adhesive, water-based paints; soap manufacture, antifreeze, hot aqueous systems, insecticides, textile, paper, petroleum, vinyl latex binders and emulsions, waste treatment; formulated for rigid dilution requirements. *PPG/Specialty Chem.* Discontinued.

19168 MB 450
68586-07-2 271-606-2
Monoethanolamine borate
Corrosion inhibitor. *Werner G. Smith.* Discontinued.

19169 MBS
A terpolymer of methylmethacrylate, butadiene and styrene. Its properties include rigidity, hardness, high impact strength, heat resistance and good clarity.

19170 MBT
149-30-4 5916 205-736-8
2-Mercaptobenzothiazole
Very active, nondiscoloring organic accelerator. *Akrochem.*

19171 MBT®
149-30-4 5916 205-736-8
2-Mercaptobenzothiazole
Intermediate; medium temperature general-purpose accelerator for natural and synthetic rubbers; very active above 240°F (116°C); moderately low activation temperature. *Uniroyal.*

19172 MBTS
120-78-5 3435 204-424-9
Benzothiazyl disulfide
Nonstaining organic accelerator; very active at temperatures above 280°F. *Akrochem.*

19173 MC 2508
General-purpose mold cleaner for urethane and epoxy residues. *George Mann.*

19174 MC®
25038-54-4 6832
Type 6 cast nylon; used for seals, wear applications. *Polymer Corp.*

19175 MC580
Silicone resin; Semirigid thermosetting transfer molding compound with high temperature and thermal shock resistance; for encapsulating high power electronic devices. *GE Silicones.*

19176 McAlester EGDS
627-83-8 211-014-3
Glycol distearate
Emulsifier, lubricant, antistat, defoamer for metalworking, textile lubricants, plastics, paper; emulsifier, pearlescent, emollient for cosmetics. *McIntyre.*

19177 McGill Metal
A proprietary trade name for a group of copper-aluminum-iron alloys, one of which contains 89% copper, 9% aluminum, and 2% iron. No manufacturer.

19178 M-Clean D
75-09-2 6140 200-838-9
Methylene chloride vapor. cleaner and degreaser. *Occidental Chemical Corp.*

19179 McLube 1700
Fluorocarbon
Solvent-based release coating for epoxies and thermosets, and natural, nitrile, SBR and silicone rubber. *McLube.*

19180 McLube 1777
9002-84-0 7743 204-126-9
TFE polymer dispersion; release coating for hot molds used to form rubber and plastic parts; antistick coating for cure of hose and other extrusions and on tools and process equipment. *McLube.*

19181 McLube 1829
Fluorochemical/resin mixture; release coating for rubber molding processes; antitack coating for uncured rubber. *McLube.*

19182 McNamee Clay®
1332-58-7 5294 296-473-8
Kaolin clay; low cost reinforcer and inert filler for paint, paper, rubber, ceramics, plastics and Specialities. *R. T. Vanderbilt Co Inc.* Discontinued.

19183 MCPA
94-74-6 5803 202-360-6
$C_9H_9ClO_3$
(4-Chloro-2-methylphenoxy)acetic acid
2,4-MCPA; MCP; 4-Chloro-o-cresoxyacetic acid; Methyl chlorophenoxy acetic acid; MCPA Ester; Bordermaster; Metaxon; MCP ester; Weedar MCPA; Weedone MCPA Ester; Mephanac; Chiptox; Agroxon; Netazol; Rhonox; Anicon kombi; Anicon m; BH MCPA; Cekherbex; Chloro-o-cresoxy)acetic acid; Chloro-o-tolyloxy)acetic acid; Chwastox; CMP acetate; Cornox-m; Dedweed; Dicopur-M; Dikotex; Emcepan; Empal; FLUID 4; Hedapur M 52; Hedarex M; Hedonal; Hedonal M; Herbicide M; Hormotuho; Kilsem; Krezone; Leuna M; Linormone; Phenoxylene plus; Rhomene; Selektonon M; Shamrox; Vacate; Weed-rhap; Weedar; Weedone; Zelan,. Selective systemic hormone-like herbicide used for pre- and post-emergence control of annual and perennial broad-leaved weeds. mp = 118-119°; soluble in H_2O (825 mg/l), more soluble in organic solvents; LD_{50} (rat orl) = 700 mg/kg.

19184 MCPB
94-81-5 202-365-3
$C_{11}H_{13}ClO_3$
4-(4-Chloro-2-methylphenoxy) butyric acid
Tropotox; Can-Trol; PDQ; 2,4-MCPB; 4-(4-chloro-2-methylphenoxy)butanoate; Bexane; MCP-butyric; Trifolex; 4-(4-chloro-*o*-tolyloxy)butyric acid. Selective systemic hormone-like herbicide. Used for post-emergence control of annual and perennial broad-leaved weeds in cereal and grassland. mp = 99-100°; soluble in H_2O (44 mg/l), more soluble in organic solvents; LD_{50} (rat orl) = 680 mg/kg.

19185 MCPB
94-81-5 202-365-3
3-phenoxybutyric acid
Used as a systemic fungicide active against chocolate spot disease in broad beans.

19186 MD 50
737-31-5 3040 212-004-1
Diatrizoate sodium
Diagnostic aid. *Mallinckrodt Inc.* Name unverified.

19187 MD 60
131-49-7 5851 205-024-7
meglumine diatrizoate
Diagnostic aid. *Mallinckrodt Inc.* Name unverified.

19188 MDI
101-68-8 202-966-0
$C_{15}H_{10}N_2O_2$
methylene di-*p*-phenylene isocyanate
methylene bisphenyl isocyanate; diphenylmethane-4', 4-diisocyanate; MBI; PMDI; Caradate 30; Desmodur 44; Hylene M50; Isonate 125M; Isonate 125MF; Nocconate 300. MDI is a raw material in polyurethane resins. The polyurethanes are of the thermosetting type of plastic and the polymerization is a polyaddition reaction.Monomer used in the preparation of polyurethane resin; bonding rubber to rayon and nylon. mp = 37-39°; bp_5 = 194°; d = 1.1900. *Allchem Industries; BASF; ICI Polyurethanes; Miles.*

19189 Mearl Film
Irridescent film; for packaging, laminating, visual effect. *Cornelius Chemical Co Ltd.* Name unverified.

19190 Mearlin
Titanium dioxide [13463-67-7] coated mica [12001-26-2] pearl pigments; for plastics, paints, automotive finishes. *Cornelius Chemical Co Ltd.* Name unverified.

19191 Mearlin®
13463-67-7 9612 236-675-5
Titanium dioxide/mica
For pearlescent, metallic, antique, and iridescent effects in powdered coating systems. *Mearl.* Name unverified.

19192 Mearlite® GBU
7787-59-9 1303 232-122-7
Bismuth oxychloride
Pearl pigments and pastes used for high opacitiy and smooth luster providing frosted effects in makeups, lipsticks, blushers, etc. *Mearl.*

19193 Mearlmaid® AA
Water, guanine, isopropyl alcohol, methylcellulose; pearl colors for nail enamels, makeup, lotions, hair care products, natural cosmetics. *Mearl.*

19194 Mearlmica® SVA
12001-26-2
Organic surface-treated mica; imparts smooth texture, soft feel, soft luster or matte effect to cosmetics (pressed powder makeup, eye shadows). *Mearl.*

19195 Measac
Ethanolamine sesquisulfite
Albright & Wilson Ltd. Name unverified.

19196 meat-sugar
87-89-8 5008 201-781-2
$C_6H_{12}O_6$
Inositol
Used as a nutrient.

19197 Mebatreat
31431-39-7 5807 250-635-4
A proprietary preparation containing mebandazole; Veterinary antihelmintic (cats and dogs). *Janssen Pharmaceutical Ltd.* Name unverified.

19198 Mebendazole
31431-39-7 5807 250-635-4
$C_{16}H_{13}N_3O_3$
(5-Benzoyl-1H-benzimidazol-2-yl)-carbamic acid methyl ester
Equivurm Plus; Methyl 5-benzoyl-2-benzimidazole-carbamate; 5-benzoyl-2-benzimidazolecarbamic acid methyl ester; R-17635; Bantenol; Lomper , Noverme, Pantelmin; Vermicidin; Vetrmirax; Mebenvet;Ovitelmin;Telmin;Vermox; Mebatreat;,. A proprietary preparation containing mebendazole; veterinary antihelmintic (horses). mp = 288.5°; insoluble in water, ethanol, ether, chloroform; LD_{50} (rat,orl) >40 mg/kg. *Janssen Pharmaceutical Ltd;.* Name unverified.

19199 Mebenvet
31431-39-7 5807 250-635-4
A proprietary preparation containing mebendazole; veterinary antihelmintic (game birds, poultry). *Janssen Pharmaceutical Ltd.* Name unverified.

19200 Mecadox
6804-07-5 1825 229-879-0
Carbadox
An antibacterial used in swine. *Pfizer International.*

19201 Mecca balsam
Balm of Gilead; Duhnul-balasan; Balsan-katel. An oleoresin obtained from *Balsamodendron gileadense* , of Arabia. It is known in India as Balsan-katel, and is imported there under the name of Duhnul-balasan.

19202 Meccarb
409-21-2 8636 206-991-8
SiC
Silicone carbide
Carborundum. An abrasive. *Winchem Ltd.*

19203 Meco
A cupro-nickel alloy.

19204 Mecoprop
7085-19-0 5826 230-386-8
$C_{10}H_{11}ClO_3$
(±)-2-(4-chloro-2-methylphenoxy)propanoic acid
(RS)-2-(4-Chloro-*o*-tolyloxypropionic acid; Clonotox; Compitox; Iso-cornox; Kilprop; Mecomec; Mecopex; Mepro; Propal; RD 4593; U 46 KV Fluid. Selective, systemic hormone-type herbicide, absorbed by leaves, translocated to roots. Post emergence control of broad-leaf weeds such as clovers, chickweeds, plantains and cleavers. mp = 94-95°; soluble in H_2O (620 mg/l), soluble in organic solvents; LD_{50} (rat orl) = 930-1166 mg/kg. Rhône-Poulenc; Schering; BASF; Akzo; BASF; Fermenta; Marks; Rhône Poulenc; Universal Crop Protection.

19205 Mecpa
94-74-6 5803 202-360-6
Selective weedkillers. *Murphy Chemical Co Ltd.* Discontinued.

19206 Mecufix
An adhesive for polyester film. *May & Baker Ltd.* Name unverified.

19207 Meculon
Metallized polyester film. *May & Baker Ltd.* Name unverified.

19208 Med Gel
637-12-7 379 211-279-5
Aluminum stearate
For nonaqueous coating compositions. *Hüls Am.* Discontinued.

19209 medal bronze
An alloy of from 92-97% copper, 1-8% tin, and 0-2% zinc.

19210 medal metal
An alloy of 84% copper and 16% zinc.

19211 Medang Losoh oil
An oil from the wood of *Cinnamonum parthenoxylon.* It consists mainly of safrole.

19212 Medialan KA
61791-59-1 263-193-2
Sodium cocoyl sarcosinate
Detergent used in cream shampoos, foaming agent, mild toiletries, hair shampoos. *Hoechst Celanese/Colorants & Surf.*

19213 Medialan KF
TEA-palm kernel sarcosinate; detergent used in preparation of clear, liquid shampoos and body lotions. *Hoechst Celanese/Colorants & Surf.*

19214 Medialan LD
137-16-6 4379 205-281-5
Sodium lauroyl sarcosinate
Gardol®. Surfactant for dental care products. *Hoechst Celanese/Colorants & Surf.; Hoechst AG.*

19215 Mediker
Shampoo, to wash and condition hair. *Richardson-Vicks Inc.* Name unverified.

19216 Meditar
A proprietary preparation of coal tar; a dermatological product. *Brocades Pharma.* Discontinued.

19217 Medium 7
A preparation consisting of dry sweet whey, sodium caseinate, disodium phosphate and soluble growth factors derived from *Saccharomyces cerevisiae*; used to prepare culture medium for the growth of thermophilic lactic acid bacteria. *Pfizer International.* Discontinued.

19218 Medium 10
A preparation consisting of dry sweet whey, nonfat dry milk, disodium phosphate and soluble growth factors derived from *Saccharomyces cerevisiae*; used to prepare culture medium for the growth of thermophilic lactic acid bacteria. *Pfizer International.* Discontinued.

19219 Medium VS
A preparation consisting of dry sweet whey, sodium caseinate, sodium citrate and soluble growth factors derived from *Saccharomyces cerevisiae*; used to prepare culture medium for the growth of thermophilic lactic acid bacteria. *Pfizer International.* Discontinued.

19220 Medley®
For the agriculture industry. *DuPont UK.*

19221 Medo
215-293-2
Oil-based soap containing cresylic acid; pruning compound and canker cure for garden trees. *Vitax Ltd.*

19222 medol
A combination of cresols and iodine; recommended as an antiparasitic in skin troubles.

19223 Medolit
A phenol-formaldehyde condensation product. It is a resinous material, and is recommended as a shellac varnish substitute.

19224 Medusa
A waterproofing compound mixed with cement.

19225 Meehanite
A proprietary trade name for a close-grained, pearlitic, sorbitic iron with properties superior to cast iron; has good casting and machining properties. No manufacturer.

19226 Meena Harma
A name given to an opaque variety of bdellium gum-resin.

19227 MEF® -LD
Moldable polyethylene foam. *Asahi Chem. Industry.*

19228 Mefarol®
Disinfectant based on a quaternary ammonium compound; veterinary medicine. *Bayer AG.*

19229 mefenacet
73250-68-7 277-328-8
$C_{16}H_{14}N_2O_2S$
2-(2-benzothiazolyloxy) -N-methyl-N-phenylacetamide
FOE 1976; Hinochloa; NTN 801; Rancho. Selective herbicide, inhibits cell division. Used for control of grass weeds, especially *Echinochloa cur-galli* and cyperaceous weeds, pre- and early post-emergence in rice. mp = 135°; soluble in H_2O (4 mg/l), more soluble in organic solvents; LD_{50} (rat orl) > 5000 mg/kg.

19230 mefluidide
53780-34-0 5846 258-767-4
$C_{11}H_{13}F_3N_2O_3S$
N-[2,4-dimethyl-5-[[(trifluoromethyl)sulfonyl]amino]phenyl]acetamide
Echo; Embark; MBR12325; Mowchem; Trimcut. Plant growth regulator and herbicide which inhibits growth and development of grasses. Used in lieu of grass cutting, e.g. on road verges and embankments. mp = 183-185°; soluble in H_2O (180 mg/l), more soluble in organic solvents; LD_{50} (rat orl) >4000 mg/kg.

19231 Mefranal
3-Methyl-5-phenyl-1-pentanal
Quest Int'l. UK Ltd.

19232 Mefrasol
3-Methyl-5-phenyl-1-pentanol
Quest Int'l. UK Ltd.

19233 Meganite
An explosive containing nitroglycerin, and dinitrocellulose, to which has been added a nitro mixture to ensure complete combustion.

19234 Meganox Plus
An emulsifiable concentrate containing 100 g aminotriazole, 250 g atrazine and 100 g MCPA per liter; used for total weed control in non crop areas. *Agri-Technics Ltd.*

19235 Megaperm 4510
A magnetic alloy containing 45% nickel, 45% iron, and 10% manganese.

19236 Megaperm 6510
A magnetic alloy containing 65% nickel, 25% iron, and 10% manganese.

19237 Megapoly
Formerly Heaveaplus MG. Graft polymer of natural rubber latex with methyl methacrylate; adhesive and bonding agent, in water-based adhesives, latex gloves, pressure-sensitive tapes, surgical tapes, foot-wear, flooring, composite hoses. *H.A. Astlett.*

19238 Megapren C 150
A proprietary chloroprene rubber used in the manufacture of moldings and extrusions resistant to sunlight, weathering and ozone. *Rhein-Chemie Rheinau.* Name unverified.

19239 Megapren Si 10, 20, 30, and 60
A proprietary range of materials based on silicone rubber. *Rhein-Chemie Rheinau.* Name unverified.

19240 Megapren U225
A proprietary polyurethane rubber. *Rhein-Chemie Rheinau.* Name unverified.

19241 Megasil
Silicone dielectric materials. *Midland Silicones.*

19242 Megilp
A mixture of linseed oil and mastic varnish; used in artist's oil paints.

19243 Meglumine
6284-40-8 6154 228-506-9
$C_7H_{17}NO_5$
1-deoxy-1-(methylamino)-D-glucitol
N-Methylglucamine. Used in the synthesis of surfactants, pharmaceuticals and dyestuffs. mp = 129-132°; $[\alpha]_D^{20}$ = -23°; soluble in H_2O (1 g/ml), less soluble in organic solvents.

19244 Megomit
A mica product used as an electrical insulator.

19245 Megum
Bonding agent for the bonding of rubber to metals and other materials under vulcanizing conditions. *Chemetall GmbH.* Name unverified.

19246 Mekad
Antiskinning agent for paints. *Octel Chemicals Ltd.* Discontinued.

19247 Meketone
78-93-3 6149 201-159-0
C_4H_8O
Methyl ethyl ketone
MEK. Solvent in paint and lacquer thinners, natural and synthetic resins, gums and rubbers, printing inks, PVC cloth manufacture, cleaning agent for metal surfaces, adhesives and cements; refining and dewaxing of mineral and lubricating oils. *Sasolchem.* Name unverified.

19248 Mekon® White
63231-60-7 264-038-1
Microcrystalline paraffin waxes and hydrocarbon waxes. Microcrystalline wax; release agent; used in hot-melt coatings and adhesives, paper coatings, printing inks, plastic modification (as lubricant and processing aid), lacquers, paints, varnishes, as binder in ceramics, for potting/filling in electrical/electronic components, in investment casting, rubber and elastomers (plasticizer, antisunchecking, antiozonant), as emulsion wax size in paper making; fabric softener ingredient, in emulsion and latex coatings. *Petrolite.*

19249 Mekor®
Volatile oxygen scavenger/metal passivator, corrosion inhibitor for steam generating systems. *Drew Ind. Div.*

19250 Mekor® 70
Corrosion inhibitor which chemically removes oxygen from feedwater, boiler water and condensate in steam generation systems; passivates iron and copper surfaces. *Drew Ind. Div.*

19251 Mekure T1, T2
Copper chrome arsenate; wood preservative. *Mechema Chemicals Ltd.* Name unverified.

19252 MEL 80-P
9003-08-1
Melamine-formaldehyde resin; produces excellent wash fastness on dacron and chintz; has embossing and dimensional stability. *CNC Int'l L.P.*

19253 Melacos
Bottle washing detergents. *ICI Chem & Polymers Ltd.* Discontinued.

19254 Melafix DM
A proprietary trade name for a melamine formaldehyde resin; used as a shrink-resistant in wool. *Ciba plc.* Name unverified.

19255 Melalith
A steatite-porcelain product.

19256 Melamine formaldehyde resin
9003-08-1
Luwipal®; MEL 80-P; Ufomite® 27-802,. High solids melamine formaldehyde resin; produces stiffness and excellent durability to fabrics in resin finishing baths; requires catalyst. Eastern Color & Chem; Akzo; Am. Cyanamid; Astro Industries; Bakelite GmbH; BASF; Monsanto; Rhône-Poulenc Water Treatment; Sybron.

19257 Melamit 200
A proprietary melamine formaldehyde cellulose molding powder. *Bush Beach Ltd.* Name unverified.

19258 Melampyrite
608-66-2 4350 210-165-2
$CH_2(OH)(CH{\cdot}OH)_4{\cdot}CH_2OH$
galactitol
Dulcitol. Dulcitol.

19259 melaniline
102-06-7 3383 203-002-1
$C_{13}H_{13}N_3$
diphenylguanidine
Used as a basic accelerator for rubber.

19260 melanoid
A colloidal bituminous paint material; used as a preservative paint for metal which is in contact with corrosive gases.

19261 Melatix
A proprietary melamine formaldehyde molding compound. *Nisshin Boseki.* Name unverified.

19262 Melax
Proprietary, polymeric based conditioning and sealing agents for fiberglass molds and other porous surfaces; used for fiberglass tools. *Axel Plastics Research Laboratories Inc.*

19263 Melclif®
Zirconium, magnesium, rare-earth and other chemicals, powders and alloys; for use in industry, including metalworking. *Magnesium Elektron Ltd.*

19264 melco
A synthetic milk made from the peanut.

19265 Melcril 4079
A trademark for tetrahydrofurfuryl acrylate. *Danbert Chemical Co.* Unverified.

19266 Melcril 4083
A trademark for an ethylene glycol acrylate phthalate. *Danbert Chemical Co.* Unverified.

19267 Melcril 4085
2495-35-4 219-673-9
$C_{10}H_{10}O_2$
acrylic acid benzyl ester
A trademark for a benzyl acrylate. bp_8 = 110-111°; n_D^{20} = 1.5200. *Danbert Chemical Co.* Unverified.

19268 Melcril 4087
48145-04-6 256-360-6
$C_{11}H_{12}O_3$
2-Propenoic acid, 2-phenoxyethyl ester
2-Phenoxyethyl acrylate; Phenoxyethyl-acrylate. A trademark for a phenoxy ethyl acrylate. *Danbert Chemical Co.* Unverified.

19269 Melcril 5919
A trademark for a melamine acrylate. *Danbert Chemical Co.* Unverified.

19270 Meld
Broad spectrum systemic fungicide for use in cereals. *BASF plc.*

19271 Meldin® 3000A
Polyimide resin; melt processable resin featuring wear resistance; for bearings, thrust washers, piston rings. *Furon.*

19272 Meldin® 3000D
Polyimide resin; melt processable resin featuring wear resistance, nonhardened matting surface; for bearings, thrust washers. *Furon.*

19273 Meldin® 3000F
Polyimide resin; melt processable resin featuring high modulus, high dimensional stability; for picker finger copier, electrical/electronic parts. *Furon.*

19274 Meldin® 3000G
Polyimide resin; melt processable resin featuring high modulus and strength; for gears, bearing retainers. *Furon.*

19275 Meldin® 3000H
Polyimide resin; melt processable resin featuring high modulus and elec. conductivity; for gears, piston rings. *Furon.*

19276 Melhi N, NS and NLM
Acidic, medium-hard, dark amber-colored resins obtained as coproduct during processing of rosin to modified forms; used in low-cost hot-melt adhesives and construction mastic adhesives. *Hercules.*

19277 Melibiase
An enzyme which splits melibiose into glucose and galactose.

19278 Meligrin
A condensation product of dimethyl-hydroxyquinoline and methyl-phenylacetamide.

19279 Melilot
122-00-9 5862 204-514-8
$C_9H_{10}O$
p-methylacetophenone
4-acetyltoluene; sweet clover; yellow melilot; yellow sweet clover; Esberiven. Dried leaves and flowering tops of *Melilotus officinalis.* It imparts the honey-like fragrance of sweet clover, and is used for perfuming soap. mp = 22-24°; bp = 226°; d = 1.0050; n_D^{20}= 1.5328.

19280 Melimax
For acetonaemia in cattle. *The Wellcome Foundation Ltd.* Name unverified.

19281 Melinar
PET polymer. *ICI Chem & Polymers Ltd.*

19282 Melinex®
Polyethylene terephthalate film; an extremely tough material used for cable lapping, motor insulation and capacitors, valve diaphragms, and conveyor belting. *ICI Chem & Polymers Ltd.*

19283 Melinex® 393
Polyester film; super clear, adhesion pretreated one side for printing; good handelability; for metallizing, facestock, overlaminate applications. *ICI Films.*

19284 Melinex® 505
Polyester film; super clear, adhesion pretreated two sides for printing; good handleability; for facestock and overlaminate labels. *ICI Films.*

19285 Melinex® 994
Polyester film; opaque white, high gloss, adhesion pretreated one side for printing; excellent handleability; for facestock labels. *ICI Films.*

19286 Melinite
The French name for Lyddite. It consists of 70% picric acid and 30% collodion cotton.

19287 melinose
Wulfenite, a mineral.

19288 Meliodent®
Heat-curing denture resin based on methyl methacrylate; used in dentistry. *Bayer AG.*

19289 melioform
A ruby-red liquid containing 25% formaldehyde and 15% aluminum acetate; used as a disinfectant.

19290 Melioran
Sodium alkyl sulfate
Used for textiles, flotation. *Elf Atochem UK/Ceca.*

19291 Melioran F6
Alcohol sulfate in paste form; powerful detergent with wetting-out, dispersing, level dyeing and fiber protection properties used in the processing of wool, cotton, linen, silk, artificial and synthetic fibers. *Rewo Chemicals Ltd.* Name unverified.

19292 Meliose
Fructose containing com syrup; used for foods, e.g., soft drinks and confectionery. *Roquette (UK) Ltd.*

19293 Melit
Melamine resins.

19294 Melite®
Zirconium, magnesium, rare-earth and other chemicals, powders and alloys; for use in industry, including metalworking. *Magnesium Elektron Ltd.*

19295 Mellavax
Calf salmonellosis vaccine. *The Wellcome Foundation Ltd.* Name unverified.

19296 Mellite
Stabilizers for PVC. *Harcros.*

19297 Mellol
60-12-8 7370 200-456-2
Extra pure 2-phenylethyl alcohol. Used in perfumery. *Bush Boake Allen Ltd.*

19298 Melmac
A proprietary trade name for a melamine-formaldehyde synthetic resin and adhesive. No manufacturer.

19299 Melmag®
Zirconium, magnesium, rare-earth and other chemicals, powders and alloys; for use in industry, including metalworking. *Magnesium Elektron Ltd.*

19300 Melmex
Melamine formaldehyde compounds; used for molding. *BIP Chemicals Ltd.*

19301 Melnox®
Zirconium, magnesium, rare-earth and other chemicals, powders and alloys; for use in industry, including metalworking. *Magnesium Elektron Ltd.*

19302 Melocol
A proprietary trade name for a polyamide-formaldehyde product. No manufacturer.

19303 Melolam
Melamine/formaldehyde resins. *Ciba plc.* Name unverified.

19304 Melolanel
Melancil and urea melamine resins; used for laminates and surfacing of wood-based panels. *Dynochem UK Ltd.*

19305 Melopas
Melamine/formaldehyde resins/molding compounds. *Ciba plc.*

19306 Meloprufe®
Zirconium, magnesium, rare-earth and other chemicals, powders and alloys; for use in industry, including metalworking. *Magnesium Elektron Ltd.*

19307 Melox®
Zirconium, magnesium, rare-earth and other chemicals, powders and alloys; for use in industry, including metalworking. *Magnesium Elektron Ltd.*

19308 Meloxide®
Zirconium, magnesium, rare-earth and other chemicals, powders and alloys; for use in industry, including metalworking. *Magnesium Elektron Ltd.*

19309 Melpure®
Zirconium, magnesium, rare-earth and other chemicals, powders and alloys; for use in industry, including metalworking. *Magnesium Elektron Ltd.*

19310 Melrasal®
Zirconium, magnesium, rare-earth and other chemicals, powders and alloys; for use in industry, including metalworking. *Magnesium Elektron Ltd.*

19311 Melsprea®
A range of melamine molding compounds; used for tableware and ashtrays. *AMC SPREA S.p.A.*

19312 Meltan®
Zirconium, magnesium, rare-earth and other chemicals, powders and alloys; for use in industry, including metalworking. *Magnesium Elektron Ltd.*

19313 Meltatox®
31717-87-0 250-778-2
Dodemorph-acetate
For control of powdery mildew in roses and other ornamentals. *BASF AG.*

19314 Meltron®
Zirconium, magnesium, rare-earth and other chemicals, powders and alloys; for use in industry, including metalworking. *Magnesium Elektron Ltd.*

19315 Melurac
A proprietary trade name for a melamine-ureaformaldehyde laminating synthetic resin. No manufacturer.

19316 Melweld®
Zirconium, magnesium, rare-earth and other chemicals, powders and alloys; for use in industry, including metalworking. *Magnesium Elektron Ltd.*

19317 Melwhite®
Zirconium, magnesium, rare-earth and other chemicals, powders and alloys; for use in industry, including metalworking. *Magnesium Elektron Ltd.*

19318 Memosil
Addition-cured silicone based transparent bite registration material; used in dentistry. *Bayer AG.*

19319 menadiol sodium diphosphate
131-13-5 5873 205-012-1
$C_{11}H_8Na_4O_8P_2$
2-methyl-1,4-naphthalenediol diphosphoric acid ester tetrasodium salt
Kappadione; Synkavit; Synkayvit. Vitamin (prothrombogenic). Very soluble in H_2O, insoluble in organic solvents.

19320 menispermin
An extract of Canadian moon-seed.

19321 menthol terpine hydrate
89-78-1 5882 201-939-0
$C_{10}H_{20}O$
5α-Methyl-2β-(1α-methylethyl)cyclohexanol
5-methyl-2-(1-methylethyl) cyclohexanol; racemic menthol; hexahydrothymol; 3-*p*-menthanol; l-menthol; hexahydrothymol; peppermint camphor; Diterpene. Perfumery, cigarettes, liqueurs, flavoring agent, chewing gum, chest rubs, cough drops, nasal inhalers. mp = 41-43°; bp_{760} = 212°; d = 0.890; $[\alpha]_D^{18}$ = -50°, (10% EtOH solution); slightly soluble in H_2O, soluble in organic solvents; LD_{50} (rat orl) = 3180 mg/kg. *Janssen Chimica; Penta Mfg.; Quest Int'l.; Robeco.*

19322 Meothrin
39515-41-8 4033 254-485-0
Fenpropathrin
A contact pyrethroid based acaricide. *Shell UK.*

19323 Mephaneine
101-41-7 202-940-9
$C_9H_{10}O_2$
methyl phenylacetate
Methyl α-toluate. Pure methyl phenylacetate. Used in perfumery: Sweet Floral Fruity Honey Spice. bp = 218°, d = 1.07; n_D^{20} = 1.5075; LD_{50} (rat orl) = 2550 mg/kg. *Bush Boake Allen Ltd.* Discontinued.

19324 Mephetol
Herbicides. *ICI Chem & Polymers Ltd.*

19325 Mephetol Extra
Soluble concentrate of 20.8 g dicamba, 333 g dichloprop and 200 g MCPA per liter; used for weed control in cereals and newly sown grass. *BritAg Industries Ltd.*

19326 mephosfolan
950-10-7 213-447-3
$C_8H_{16}NO_3PS_2$
(4-Methyl-1,3-dithiolan-2-ylidene)phosphoramidic acid diethyl ester
phosphonodithioimidocarbonic acid cyclic propylene P,P-diethyl ester; 2-diethoxyphosphinylimino-4-methyl-1,3-dithiolane; cyclic propylene(diethoxyphosphinyl)dithioimidocarbonate; EI 47470; ENT 25,991; Cytrolane; Cytro-Lane;. Emulsifiable concentrate containing 250 g/l mephosfolan; used for control of damsonhop aphid in hops. $bp_{0.001}$ = 120°; soluble in H_2O (57 mg/l), soluble in organic solvents; LD_{50} (rat orl) = 8.9 mg/kg. *Cyanamid of Great Britain Ltd.*

19327 mepiquat chloride
24307-26-4 5903 246-147-6
$C_7H_{16}ClN$
1,1-dimethylpiperidinium chloride
BAS-083; BAS-85559X; Pix. Plant growth regulator, used in combination with ethephon to control growth of cereal crops. mp > 350°; LD_{50} (rat orl) = 1420 mg/kg. *BASF plc; BASF AG; Clifton Chemicals Ltd.*

19328 Mepron
63-68-3 6053 200-562-9
Protected methionine *Degussa.*

19329 meralluride
8069-64-5 5913
$C_{16}H_{23}HgN_6NaO_8$
[3-[[[(3-carboxylato-1-oxopropyl)amino]carbonyl]amino]-2-methoxypropyl]hydroxymercurate(1-) sodium compound with 3,7-dihydro-1,3-dimethyl-1H-purine-2,6-dione
Dilurgen; Mercardan; Mercuhydrin; Mercuretin. Used as a diuretic. Slightly soluble in H_2O, insoluble in organic solvents; LD_{50} (rat sc) = 28 mg/kg.

19330 Meraneine
105-87-3 203-341-5
$C_{12}H_{20}O_2$
(E)-3,7-dimethyl-2,6-octadien-1-yl acetate
acetic acid, geraniol ester; geraniol acetate. Rosy, green, slightly lavendaceous odor. Used in perfume blends & fruit flavors. d = 0.911; bp = 242°; n_D^{20}= 1.4628; insoluble in H_2O, soluble in organic solvents; LD_{50} (rat orl) = 6330 mg/kg. *Bush Boake Allen Ltd.*

19331 Meranol
106-24-1 4411 203-377-1
$C_{10}H_{18}O$
(E)-3,7-Dimethyl-2,6-octadien-1-ol
trans-3,7-dimethyl octa-2,6-dien-1-ol; lemonol. High-quality rose petal odor used in fragrances. d = 0.8750; bp = 229-230°; n_D^{20}= 1.4766; λ_m = 190-195 nm (ε 18000); insoluble in H_2O, soluble in organic solvents. *Bush Boake Allen Ltd.*

19332 Merantine
Level dyeing wool dyestuffs. *Holliday Dyes & Chemicals Ltd.*

19333 Merbron R
Machine cleaner for aftercleaning of dyed polyester, removal of trimers fibers or equipment. *Sybron.*

19334 2-mercaptobenzimidazole
583-39-1 1112 209-502-6
$C_7H_6N_2S$
2-benzimidazolethiol
2-mercaptobenzimidazole; 2-benzimidazolethiol; *o*-phenylenethiourea; Antioxidant MB; Antiegene MB; AOMB; ASM MB. Antioxidant. mp = 303-304°; slightly soluble in H_2O, more soluble in organic solvents; LD_{50} (rat orl) = 1230 mg/kg. *Aceto.*

19335 mercaptobenzothiazole
149-30-4 5916 205-736-8
$C_7H_5NS_2$
2-mercaptobenzothiazole

2-benzothiazolethiol. Curing accelerator for rubber. Used with 1,3-diphenylguanidine. mp = 179°; d = 1.4200.

19336 2-mercaptoethanol
60-24-2 5917 200-464-6
C_2H_6OS
1-ethanol-2-thiol
thioglycol; 2-hydroxyethyl mercaptan; 2-thioethanol; mercaptoethanol; Emery 5791; monothioethylene glycol; 2-hydroxy-1-ethanethiol; β-mercaptoethanol; thiomonoglycol; Sipomer® 2ME. Solvent for dyestuffs, intermediate for producing dyestuffs, pharmaceuticals, rubber chemicals, flotation agents, insecticides, plasticizers, reducing agent, biochemical reagent, PVC stabilizers, agricultural chemicals, textile auxiliary. mp = -40°; bp = 157°; SG = 1.114; n_D^{20} = 1.5006; soluble in H_2O (>100 mg/ml), organic solvents; LD_{50} (mus orl) = 345 mg/kg. *Rhône-Poulenc Surf.; BASF; Morton Int'l.*

19337 Merce Assist ADB
Low foaming mercerization penetrant. *Sybron.*

19338 Mercerisin OR
Sodium salt of sulfonation product; anionic; yellow-brown liquid; wetting agent for mercerizing. *Thor Chemicals (UK) Ltd.* Discontinued.

19339 mercerized cotton
Cotton which has been immersed in a solution of sodium hydroxide. It has a lustrous appearance.

19340 Mercers
A proprietary trade name for a solution of 0.1% *o*-hydroxy-phenyl-mercuric chloride and 0.1% *sec*-amyltricresol in 50% alcohol, 10% acetone, and water; a germicide. No manufacturer.

19341 Mercer's liquor
A solution containing potassium ferricyanide; used for etching.

19342 Merclor D
7681-52-9 8773 231-668-3
A proprietary trade name for a solution of sodium hypochlorite NaOCl; a bleaching agent. No manufacturer.

19343 mercolloid
A colloidal mercury sulfide.

19344 Mercoloy
A proprietary trade name for a nickel bronze containing 60% copper, 25% nickel, 10% zinc, 1% tin, 2% lead, and 2% iron. No manufacturer.

19345 mercufenol chloride
90-03-9 5920 201-962-6
C_6H_5ClHgO
o-(chloromercuri)phenol
o-hydroxyphenylmercuric chloride. Used as a disinfectant mp = 150-152°; soluble in H_2O, EtOH or C_6H_6.

19346 mercuric acetate
1600-27-7 5922 216-491-1
$C_4H_6HgO_4$
mercury acetate
Mercury (II) acetate; Mercuric diacetate; Mercury Acetate; Diacetoxymercury; Acetic Acid, Mercury(2+) Salt. Catalyst for organic synthesis, pharmaceuticals. Used for mercuration of organic compounds. mp = 178-180°; soluble in H_2O (0.4 g/ml), EtOH; LD_{50} (rat orl)= 4 mg/kg. *Atomergic Chemetals; Cerac; Noah Chem.; Thor.*

19347 mercuric chloride
7487-94-7 5926 231-299-8
$HgCl_2$
mercury(II) chloride
mercury bichloride; mercury perchloride; corrosive mercury chloride; Mercuric bichloride; Bichloride of Mercury; Corrosive Sublimate; Mercury Chloride; dichloromercury; perchloride of mercury; sublimate; TL 898; Mercury chloride (2). Manufacture of calomel (mercurous chloride), other mercury compounds; disinfectant, organic synthesis, analytical reagent, metallurgy, tanning, catalyst, sterilant, fungicide, insecticide, wood preservative (kyanizing), embalming fluids, textile printing, photography and dry batteries.Used as a topical antiseptic and disinfectant. Highly toxic. mp = 277°; bp = 302°; d = 5.4; soluble in H_2O (74 mg/ml), less soluble in organic solvents; highly toxic, LD_{50} (rat orl) = 1 mg/kg. *Aldrich; Atomergic Chemetals; Spectrum Chem. Mfg.; Thor.*

19348 mercuric oxide, red and yellow
21908-53-2 5936 244-654-7
HgO
mercuric oxide
red precipitate; yellow precipitate. Red and yellow forms represebt different physical states of an orthorhombic crystal form. Yellow is more finely divided and more reactive. The two forms can be interconverted. Red: chemicals, paint pigment, perfumery, cosmetics, pharmaceuticals, ceramics, dry batteries, polishes, analytical reagent, antifouling paints, fungicide, antiseptic; Yellow: antiseptic, mercury compounds. Both forms used as topical anti-infectives. d = 11.14; LD_{50} (rat orl) = 18 mg/kg. *Cerac; Noah Chem.; Spectrum Chem. Mfg.; Thor UK.*

19349 mercuric sulfate
7783-35-9 5943 231-992-5
HgO_4S
mercury(II) sulfate
mercury persulfate; Sulfuric acid, mercury(2+)salt(1:1); mercury bisulfate; Mercury (II) Sulfate (1:1); Sulfuric Acid, Mercury Salt. Catalyst in the conversion of acetylene to acetaldehyde, extracting gold and silver from roasted pyrites, battery electrolyte. dec 450°; d = 6.47; reacts with H_2O; LD_{50} (rat orl) = 57 mg/kg. *Atomergic Chemetals; Noah Chem.; Spectrum Chem. Mfg.*

19350 mercuric sulfide, red and black
1344-48-5 5944 215-696-3
HgS
mercury(II) sulfide
Mercury Sulfide; Mercury(II) sulfide, red; Mercury (II) sulfide, black; vermilion; quicksilver vermilion; artificial cinnabar. Red pigment. mp = 583°; d = 8.1. *Atomergic Chemetals; Cerac; Noah Chem.*

19351 mercuricide
$3LiI·HgI_2$
Lithio-mercuric-iodide
Used as a germicide.

19352 mercuriocoleols
A double stearate of cholesterol and mercury.

19353 mercuriol
A mercury amalgam with aluminum.

19354 mercurochrome
129-16-8 5914 204-933-6
(2',7'-dibromo-3',6'-dihydroxy-3-oxospiro[isobenzofuran-1(3H),9'-[9H]xanthen]-4'-yl)hydroxymercury sodium salt
dibromohydroxymercurifluorescein disodium salt; Mercurochrome 220 soluble; Chromargyre; Planochrome; Flavurol; D.O.M.F.; Mercurophage; Mercurocol; Gallochrome; Gynochrome; Mercurome; Asceptichrome; Mercuranine. A fluorescein derivative of mercury; used as a bactericide and antiseptic. bpss4.0 = 72°; n_D^{25} = 1.4881; Soluble in H_2O, EtOH, insoluble in organic solvents.

19355 mercury
7439-97-6 5957 231-106-7
Hg
Quicksilver; hydrargyrum; liquid silver. Metallic element; Hg; Amalgam, catalyst, electrical apparatus, cathodes for production of chlorine and caustic soda, thermometers, barometers. mp = -39°; bp = 357°; insoluble in H_2O, organic solvents. *Aldrich; Atomergic Chemetals; Cerac; Cox Chem. Ltd; Spectrum Chem. Mfg.*

19356 Merecol® FA
Water-soluble gum blend; stabilizer for salad dressing, bakery, confectionery, flavors, emulsions, flavored beverages with pulp. *Meer.*

19357 Merecol® I
Water-soluble gum blend; stabilizer, gelling agent for fillings and dips; substitute for gelatin in kosher marshmallows. *Meer.*

19358 Merfusan
Horticultural product. *May & Baker Ltd.* Name unverified.

19359 Mergal
Fungicides and bactericide. *Hoechst UK.*

19360 Mergamma
Mixture of γ-HCH (lindane) and phenylmercury acetate; fungicide seed dressing for cereals. *ICI Chem & Polymers Ltd.*

19361 Mergital OC 30E
Ethoxylated oleo-cetyl alcohol (30 EO)
Surfactant. *Pulcra SA.*

19362 Mergital ST 30/E
9004-99-3
PEG-30 stearate
Surfactant. *Pulcra SA.*

19363 Merigraph
A range of 18 viscous, uv light-sensitive liquid photopolymers; used to make printing plates for letterpress, flexo and dry offset printing processes. *Hercules.*

19364 Meritol
A photographic developer. *Johnsons of Hendon.* Name unverified.

19365 Merix
A line of aerospace, automotive, industrial, paper, photography, rubber, safety and sporting goods chemicals. *Merix.*

19366 Merlon
Polycarbonate resins characterized by their toughness, heat resistance, dimensional stability and clarity. *Miles.* Discontinued.

19367 Merolan
Fungicide containing dithiocarbonate. *ICI Chem & Polymers Ltd.*

19368 Merpafol
2425-06-1 1814 219-363-3
captafol
Active ingredient: captafol; nonsystemic agricultural fungicide. Also see [2939-80-2]. *Makhteshim Chemical Works Ltd.* Discontinued.

19369 Merpan
133-06-2 1815 205-087-0
captan
Active ingredient: captan; broadspectrum agricultural fungicide. *Makhteshim Chemical Works Ltd.*

19370 Merpectogel
62-38-4 7453 200-532-5
Phenylmercuric acetate
Used as a herbicide and fungicide. *Poythress Laboratories Inc.* Unverified.

19371 Merpelan AZ
Wettable powder formulation for pre-emergent control of weeds. *Bayer AG.*

19372 Merpentine
A proprietary trade name for a sodium alkyl naphthalene sulfonate product used as a wetting agent. No manufacturer.

19373 Merpol®
Surface active agent. *DuPont UK.*

19374 Merpol® 100
Octylphenol ethoxylate
Surfactant, emulsifier, wetting agent, detergent, solubilizer, emulsion stabilizer for metal cleaning, industrial and household detergents, agricultural; stable to acids, bases, heat, freezing. *DuPont.*

19375 Merpol® A
Ethoxylated phosphate
Low foaming wetting agent, surface tensile reducer for chemical manufacture, cosmetics, metal processing, paper, petrol., inks, plastics, soaps, synthetic fibers, textiles; stable to acids, bases, heat to 100°C, freezing. *DuPont.*

19376 Merpol® DSR
A nonionic softener for use in conjunction with acrylic type soil release agents in durable press finishes on textiles. *DuPont UK.*

19377 Merpol® SE
Long-chain fatty alcohol ethoxylate; low foaming wetting and rewetting agent, emulsifier, dispersant for textiles, paper, oil-water emulsions, asbestos; stable to acids, bases, heat, freezing. *DuPont.*

19378 Merpol® SH
24938-91-8
Long-chain alcohol ethoxylate; detergent, wetting agent for textiles, hard surface cleaning, paper, metal processing; biodegradable; stable to acids, bases, heat, freezing to 5°C. *DuPont.*

19379 Merquinox
Fungicide for industrial applications. *Octel Chemicals Ltd.* Discontinued.

19380 Merse® 7F
Catalyst for polyester weight reduction work, for removal of insolubilized WD size; used with caustic. *Sybron.*

19381 Mersil
Turf fungicide. *May & Baker Ltd.* Name unverified.

19382 Mersilene
Nonabsorbable surgical suture. *Ethicon Inc.* Name unverified.

19383 Mersize
Paper sizing agent; used in paper mill to reduce sizing costs, improve resistance to ink, water and other aqueous solutions. *Monsanto Co.* Name unverified.

19384 Mersol
A proprietary trade name for a solvent. No manufacturer.

19385 Mersolat H 30
Sodium alkane sulfonates based on n-paraffin; biodegradable detergent base, wetting agent for manufacture electrolyte-resistant textile and leather auxiliaries, alkaline and acid detergents, floor cleaners, bubble baths, disinfectants and car shampoos. *Miles/Organic Products.*

19386 Mersolat® Cl
Chloroalkane sulfonate
Raw material used in the manufacture of special products, especially fat-liquoring agents. *Bayer AG.*

19387 Mersolat® H
Sodium alkane sulfonate
Detergent for use in the production of special purpose materials, especially electrolyte-resistant wetting agents for textiles, leather auxiliaries, acid and alkaline cleansers. *Bayer AG; Bayer plc.*

19388 Mersolat® W 40
Sodium alkane sulfonates based on n-paraffin; biodegradable detergent base, wetting agent for manufacture electrolyte-resistant textile and leather auxiliaries, alkaline and acid detergents, floor cleaners, bubble baths, disinfectants and car shampoos. *Bayer AG; Miles/Organic Prods.*

19389 Mertan
Polyurethane water emulsions, polyurethane solutions, granule form; thermoplastic polyurethane (heel tops, soles etc.). *Merquinsa.* Name unverified.

19390 Mertec
A proprietary trade name for a chlorinated rubber-base paint. No manufacturer.

19391 Merthiolate
54-64-8 9451 200-210-4
A proprietary preparation of thiomerosal (sodium ethyl-mercurithio-salicylate); a skin antiseptic. *Eli Lilly & Co.*

19392 Meruvax
Rubella virus vaccine live; immunizing agent. *Merck & Co Inc.* Name unverified.

19393 Meruvax II
Rubella vaccine; immunization against German measles. *Merck & Co Inc.*

19394 Mesamoll®
Alkyl sulfonic ester of phenol; universal plasticizer used in PVC calendering, extrusion, injection molding, dip coating, high-pressure foam, rotational and compression moldings, film for linings and food packaging, shower curtains, floorcoverings, imitation leather, tarpaulins, protective clothing, cable insulation, tubing, shoes, toys. Also used with PS, joint sealants, S/B, nitrile-butadiene, chlorinated and butyl rubber, cleansing aent for PU-processing equipment. *Bayer AG; Bayer plc; Miles; Polysar.*

19395 Mesamoll® -Verdingrin
Paraffin sulfonated ester; a vinyl plasticizer. *Bayer AG.*

19396 Mesgamma
Combined insecticide fungicide and seed dressing. *Plant Protection.* Name unverified.

19397 Mesicerin
$C_9H_{12}O_3$
1,3,5-tri(hydroxymethyl)benzene

19398 Mesidine
88-05-1 201-794-3
$C_9H_{13}N$
2,4,6-trimethylaniline
Used as an intermediate in the manufacture of dyestuffs. mp = -5°; bp = 233°; d = 0.9630; n_D^{20}= 1.5510; LD_{50} (rat orl) = 743 mg/kg.

19399 Mesitol®
Range of aftertreating and resist agents; used for polyamide, polyamide/cellulosic blends and half wool. *Bayer AG; Bayer plc.*

19400 mesotartaric acid
147-73-9 9238 205-696-1
$C_4H_6O_6 \cdot H_2O$
meso-tartaric acid monohydrate
internally compensate tartaric acid; unresolvable tartaric acid; Antiweinsäure. Prepared by racemization of L-tartaric acid. mp = 140°; d^{20}= 1.66; soluble in H_2O (1.25 g/ml).

19401 Mesurol®
2032-65-7 6050 217-991-2
Methiocarb
Versatile product formulated for different uses; especially as a moluscicide against slugs and snails as well as a seed dressing for repelling depredating birds; also as insecticide/acaricide against foliar-feeding caterpillars and sucking pests on various crops. *Bayer AG.*

19402 Met
Coated titanium anodes and cathodes. *ICI Chem & Polymers Ltd.*

19403 Meta
108-62-3 5983 202-945-6
Metaldehyde produced by the polymerization of acetaldehyde.

19404 metabisulfite
7681-57-4 8784 231-673-0
$Na_2S_2O_5$
Sodium metabisulfite
Antioxidant used in pharmaceuticals.

19405 Metablen
Acrylic modifiers and processing aids for PVC. *British Traders & Shippers Ltd.*

19406 Metablen® C-301
Impact modifier for PVC blow molding compounds; improves oil resistance of bottles; used for olive oil bottles. *Metablen BV; Metco N. Am.*

19407 Metachloron® A4 Liq
Organo metallic compound; color bleed assistant for direct dyes on wet cotton and rayon goods. *Crompton & Knowles.*

19408 Metachrome
Chrome dyestuffs. *J C Bottomley*. Name unverified.

19409 Metacide
298-00-0 6183 206-050-1
Parathion-methyl
Insecticide and acaricide, used in sprays for control of insect pests especially on cotton. *Bayer AG.*

19410 metacinnabarite
A mineral having the same composition as cinnabar, but black in color.

19411 Metacon
Rust remover. *Croda Chem. Ltd.* Discontinued.

19412 Metacrylene
A proprietary trade name for a styrene-methyl methacrylate copolymer. No manufacturer.

19413 Metacrylene BS
A proprietary trade name for an MBS terpolymer. No manufacturer.

19414 Metacure® T-1
1067-33-0 213-928-8
$C_{12}H_{24}O_4Sn$
Dibutyl tin diacetate
dibutyldiacetoxystannane; bis(acetyloxy)dibutyl stannane; BA 2726; diacetoxybutyltin; diacetoxydibutylstannane; diacetoxydibutyltin; dibutyl tin diacetate; ti(catalyst); tin,dibutyl-,diacetate; Di-n-butyltin diacetate; Di-n-Butyldiacetoxytin. Catalyst for use in production of PU coatings, adhesives, and sealants. mp < 12°; bp_5 = 139°; d = 1.3100; insoluble in H_2O; LD_{50} (rat orl) = 32 mg/kg. *Air Prods & Chems. Inc.*

19415 Metacure® T-5
Tin-based; catalyst for production of PU coatings, adhesives, and sealants; increased pot life over Metacure T-12. *Air Prods & Chems. Inc.*

19416 Metacure® T-9
301-10-0 206-108-6
$C_{16}H_{30}O_4Sn$
Stannous octoate
Stannous-2-ethyl hexanoate; Hexanoic acid, 2-ethyl-, tin(2+) salt; Stannous octoate; Bis(2-ethylhexanoate)tin. Catalyst for use in production of PU coatings, adhesives, and sealants; uniform activity and excellent stability. *Air Prods & Chems. Inc.*

19417 Metacure® T-12
77-58-7 3089 201-039-8
Dibutyl tin dilaurate
Catalyst for production of PU coatings, adhesives, and sealants; formulated to remain liq. > 18 C for easier handling. *Air Prods & Chems. Inc.*

19418 Metacure® T-45
Potassium octoate
Catalyst for PU; catalyzes the trimerization of isocyanates and polyol-isocyanate reaction. *Air Prods & Chems. Inc.*

19419 Metacure® T-120
Organotin catalyst for one-part moisture cure and two-part isocyanate coatings; compatible with amine co-catalysts. *Air Prods & Chems. Inc.*

19420 Metacure® T-125
Organotin catalyst for PU coatings, adhesives, and sealants; suitable for one-part moisture cure and two-part reactions. *Air Prods & Chems. Inc.*

19421 Metacure® T-131
Organotin catalyst for PU coatings, adhesives, and sealants; provides delayed action catalysis of isocyanate/polyol reaction. *Air Prods & Chems. Inc.*

19422 Metaform
A packing material for packing stuffing-boxes, and consisting of powdered white metal, graphite, cylinder oil, and asbestos fiber. No manufacturer.

19423 Metafos
Glassy phosphates. *Peridot Chemicals Inc.*

19424 Metaglo Super K
Metallic print binder for screen and machine printing on all types of fabrics; durable to laundering and drycleaning. *Eastern Color & Chem.*

19425 Metagon
50813-16-6 8814 256-779-4
Sodium hexametaphosphate
Albright & Wilson Ltd., Phosphates & Speciality Business.

19426 Metaiclad
High performance cladding overpaint for a wide range of cladding panels. *Feb Ltd.*

19427 metal argentum
An alloy consisting of 85 1/2% tin and 14 1/2% antimony.

19428 Metal Deactivator S
94-91-7 202-374-2
$C_{17}H_{18}N_2O_2$
N,N'-Disalicylidene-1,2-propane diamine
2,2'-[(1-Methyl-1,2-ethanediyl)bis(nitrilomethylidyne)]bis-phenol; N,N'-(2-Hydroxybenzylidene)-1,2-propandiamine. Copper chelating agent for refinery industry. *Hart Chem. Ltd.*

19429 metal soaps
Salts of the heavy metals with fatty acids.

19430 metalaxyl
57837-19-1 5982 260-979-7
$C_{15}H_{21}NO_4$
N-(2,6-dimethylphenyl)-N-(methoxyacetyl)-DL-alanine methyl ester
metaxanin; CGA-48988; Apron; Ridomil; Subdue. Fungicide. mp = 71-72°; soluble in H_2O (7.1 g/l), organic solvents; LD_{50} (rat orl) = 669 mg/kg.

19431 metaldehyde
108-62-3 5983 202-945-6
$C_8H_{16}O_4$
2,4,6,8-tetramethyl-1,3,5,7-tetraoxacyclooctane
metacetaldehyde; Ariotox; Antimilace; Acetaldehyde; Corry's slug death; Farmon Mini Slug Pellets; Gastratox 6G Slug Pellets; Halizan; Helarion; Metacetaldehyde; Metason; META; MifaSlug; Namekil; Slug-tox; tetramer; Ortho Metaldehyde 4% Bait; Slug-Tox; UN1332; R-2,C-4,C-6,C-8-tetramethyl-1,3,5,7-tetroxocane; 2,4,6,8-tetramethyl-1,3,5,7-tetroxocane; Tetramethyl-1,3,5,7-tetroxocane; Totroxocane, 2,4,6,8-tetramethyl-;. Molluscicide with contact and stomach action. Contact with the foot makes the mollusk torpid and induces an increased secretion of mucus, leading to dehydration. Used for control of slugs and snails. The polymer has CAS RN 9002-91-9 mp = 246°; sublimes 112-115°; soluble in H_2O (200 mg/l), C_6H_6, $CHCl_3$, insoluble in EtOH, Et_2O; LD_{50} (rat orl) = 630 mg/kg. *Truchem Ltd; Farm Protection Ltd; Fisons plc.*

19432 metal-furnace slag
A slag formed and used in the preparation of copper metal. It is essentially a silicate of iron, and contains about 4% copper.

19433 Metalite
Proprietary trade names for aluminum oxide abrasives. No manufacturer.

19434 Metall
General purpose primer for all metal surfaces; suitable for wide range of substrates prior to the application of other polyurethane based systems. *Feb Ltd.*

19435 Metallic 9500
Conductive additive for coatings or composites; used in thermoplastics for EMI shielding and ESD protection for electrical products. *DJ Enterprises, Inc.*

19436 Metallic Sodium
7440-23-5 8710 231-132-9
Used in synthetic organic chemistry. *Foseco (F.S.) Ltd.*

19437 Metallichrome
A finish for motor bodies. *ICI Chem & Polymers Ltd.* Discontinued.

19438 metalline
An alloy consisting of 35% cobalt, 30% copper, 10% iron, and 25% aluminum; used in jeweller's work.

19439 Metallyte 70-2, 70-4, 70-U, 80-2, 80-4, 80-U, 140-2, 140-4, 140-U
9003-07-0 7741
1 side coextruded oriented PP film, 1 side metal, heat sealable; high opacity film with adhesion promoting layer on one side and broad seal range surface on the other. *Mobil.* Name unverified.

19440 Metalset
Epoxy resin base adhesives and cements. *Smooth-On Inc.* Name unverified.

19441 Metalyn 582
Ester of pentaerythritol and tall oil fatty acid; synthetic lubricant additive for formulations requiring high thermal and hydrolytic properties at elevated temperatures; for rolling oils. *Actrachem.*

19442 Metalyn 582
An ester of fatty acids designed for metal-rolling and working lubricants that require particular thermal and hydrolytic properties at elevated temperatures; used as an ingredient in wood preserving agents, as a secondary plasticizer for vinyl copolymer resin, as a plasticizer, softener and tackifier for nitrile rubber compounding and in various low cost adhesives and metal working lubricants. *Hercules.*

19443 Metamitron
41394-05-2 5985 255-349-3
$C_{10}H_{10}N_4O$
4-Amino-3-methyl-6-phenyl-1,2,4-triazin-5(4H)-one
Goltix®; BAY DRW 1139; Gountdown. A water dispersible granular formulation containing 70% w/w metamitron; used to control annual weeds in sugar beet grown on mineral and organic soils and red beet, fodder beet and mangolds grown on mineral soils. mp = 169°; LD_{50}(rat, orl) = 3343 mg/kg. *Bayer AG; Bayer plc.*

19444 metam-sodium
137-42-8 6016 205-293-0

$C_2H_4NNaS_2$
sodium methyldithiocarbamodithioate
Vapam; metham; A7 Vapam; Busan 1020; Karbation; Maposol; Metam-Fluid BASF; Nemasol; Solasan 500; Sometam; Trimaton; VPM; Metam sodium; carbam; SMDC; Metam S.A.U.; Roo-Pru; Gatorooter; Vaporooter; Basamid-fluid; Carbam, sodium salt; N-869; Sistan; Woodfume vapam. Soil-fumigant which acts by decomposition to methyl isothiocyanate. Used as a soil sterilant, controlling soil fungi, nematodes, weed seeds and soil insects. Soluble in H_2O (722 g/l), less soluble in organic solvents; LD_{50} (rat orl) = 1800 mg/kg.

19445 metanilic acid
121-47-1 5988 204-473-6
$C_6H_7NO_3S$
3-aminobenzenesulfonic acid
1-aminobenzene-3-sulfonic acid; m-aniline sulfonic acid; m-sulfanilic acid; aniline-m-sulfonic acid. Used as a chemcial intermediate. d = 1.69; soluble in H_2O (<1 g/100ml), less soluble in organic solvents.

19446 Metanium Ointment
Titanium dioxide 20%, titanium peroxide 5%, titanium salicylate 3%, titanium tannate 0.1% in a siliconized excipient; used for the prevention and treatment of nappy rash and other macerated skin conditions. *Bengue & Co Ltd.* Name unverified.

19447 Metaquest
Sequestering agents. *Fisons plc.* Name unverified.

19448 Metasal
Multipurpose fire extinguishing powders for extinguishing fires of incandescent solid, liquid and gaseous materials. *Degussa AG.* Name unverified.

19449 Metasap® 537
637-12-7 379 211-279-5
Aluminum stearate
Process aid, lubricant for PVC, polyolefins, PS, and ABS. *Syn. Prods.*

19450 Metasil A, A+, Extra A+, B, C, E
A pure diatomaceous silica, each powder having different retentive properties; used for the coarse filtration of suspensions down to very fine filtrations giving brilliant filtrates on most liquids e.g. syrups, oils, chemicals, water, beer, vinegar, etc. *Stella-Meta Filters.*

19451 Metasil AL
Metasil A reinforced with hydrate of aluminum; used for the production of sparkling water, e.g., mineral water manufacturers. *Stelia-Meta Filters.*

19452 Metasil ALAG
Metasil AL carrying a surface coating of silver; used for the production of sparkling water for drinking purposes from low quality water supplies. *Stella-Meta Filters.*

19453 Metasil D
Polystyrene micro beads
Special barrier type precoat material. *Stelia-Meta Filters.*

19454 Metasil DA
7440-44-0 1855 231-153-3
Carbon
Used for decolorizing, dechlorinating and taste removal. *Stella-Meta Filters.*

19455 Metasil MQC
Hydrated silicate
Used for the adsorption of oxides in rolling oil coolants. *Stella-Meta Filters.*

19456 Metasil Purasil
Perlite (volcanic rock) type; used for general clarification of liquids not too strongly acid or alkaline, e.g. swimming pool. *Stella-Meta Filters.*

19457 Metasil R
Chopped rayon fiber; special barrier type precoat material. *Stella-Meta Filters.*

19458 Metasil SA, SB
Pure cellulose
Used for the filtration of products where silica is unsuitable e.g. miscelia oils, removal of oil from condensate water. *Stella-Meta Filters.*

19459 Metasil W/2
7440-44-0 1855 231-153-3
Carbon
Used for the filtration of strong alkali liquids. *Stella-Meta Filters.*

19460 Metasystox 55
919-86-8 6129 213-052-6
An emulsifiable concentrate containing demeton-S-methyl 580 g per liter; used to control aphids, red spider mites and certain other pests on arable crops, fruit, market garden and greenhouse crops and ornamentals. *Farm Protection Ltd.*

19461 Metasystox R
301-12-2 206-110-7
A liquid concentrate of oxydemeton-methyl 570 g per liter; used to control aphids, red spider mites and certain other pests on arable crops, fruit, market garden and greenhouse crops and ornamentals; the choice of the grower who prefers a product with a less penetrating odor; for all crops on which Metasystox is used. *Farm Protection Ltd.*

19462 Metasystox® I
919-86-8 6129 213-052-6
Demeton-S-methyl
Systemic insecticide for control of sucking insects on a wide range of crops. *Bayer AG.*

19463 Metasystox® R
301-12-2 206-110-7
Emulsifiable concentrate containing 570 g/l oxydemeton-methyl; a systemic organophosphorus insecticide used to control aphids, red spider mites and certain other pests. *Bayer plc.*

19464 Metatone
A proprietary preparation containing thiamine hydrochloride, and calcium, potassium, sodium, manganese, and strychnine glycerophosphates; a tonic. *Parke-Davis.* Name unverified.

19465 metazachlor
67129-08-2 266-583-0
$C_{14}H_{16}ClN_3O$
2-chloro-N-(2,6-dimethylphenyl)-N-(1H-pyrazol-1-ylmethyl)acetamide
Butisan S; metazachlore; BAS 47900H; Track. Selective herbicide, inhibits germination. Used for control of annual grasses and broad-leaved weeds in fruit and vegetable crops. mp= 85°; soluble in H_2O (17 mg/l), more soluble in organic solvents; LD_{50} (rat orl) = 2150 mg/kg.

19466 Metazene®
Mixture of fatty alcohol esters of methyl methacrylic acid; odor counteractant. *Pestco.*

19467 Metco 450
A nickel/aluminum composite material for building up metal surfaces by spraying. Similar materials are Metco 451 (nickel-chromium), Metco 44 (nickel base/chromium) and Metco 51 (aluminum bronze). No manufacturer.

19468 Meteor
Mixed metal oxide color pigments used in plastics, coatings, and other applications. *Engelhard.*

19469 Meteor Plus
Fine particle size mixed metal oxide color pigments used in plastics, coatings and other applications. *Engelhard.*

19470 meteorite
An alloy of aluminum with from 1-2% zinc and 1-4% phosphorus.

19471 Metglas®
Amorphous metal alloys. *AlliedSignal.*

19472 methabenzthiazuron
18691-97-9 6002 242-505-0
$C_{10}H_{11}N_3OS$
(benzothiazol-2-yl)-1,3-dimethylurea
Tribunil; N-2-benzothiazolyl-N,N'-dimethylurea;Benzothiazolyl)-1,3-dimethylurea; Benzothiazolyl)-N,N'-dimethylurea; Dimethyl-3-(2-benzothiazolyl)urea; Dimethyl-3-(2-benzthiazolyl)-harnstoff; Methyl-N'-methyl-N'-(2-benzothiazolyl)urea; Preparation 5633; Urea, 1-(2-benzothiazolyl)-1,3-dimethyl-; Bay 72483; Methibenzuron. Selective herbicide, acts as a photosynthesis inhibitor and used for pre- and post-emergence control of annual grasses and broad-leaved weeds in garlic, onions, chives, leeks, peas, field beans, cereals, grass seed crops, lucerne, maize, potatoes, artichokes, stonefruit and tree nurseries. mp = 119-121°; soluble in H_2O (59 mg/l), more soluble in organic solvents; LD_{50} (rat orl) > 2500 mg/kg.

19473 Methacrol
Fabric and yarn lubricants. *DuPont UK.*

19474 methamidophos
10265-92-6 6014 233-606-0
$C_2H_8NO_2PS$
O,S-dimethyl phosphoramidothioate
acephate-met; ENT-27396; Bay 71628; Filitox; Monitor; Patrole; Pillaron; SRA 5172; Tam; Tamanox; Tamaron. Systemic insecticide and acaricide with contact and stomach action. Absorbed by roots and leaves. Cholinesterase inhibitor. Used for control of chewing and sucking insects and spider mites in vegetable and fruit crops. mp = 46°; sg_{20} = 1.31; n_D^{40} = 1.5092; soluble in H_2O (>2 kg/l), less soluble in organic solvents; LD_{50} (rat orl) = 20 mg/kg.

19475 methane
74-82-8 6019 200-812-7
CH_4
methyl hydride
marsh gas. Major constituent of American natural gas. Used as a fuel and in chemical manufacturing. mp = -183°; bp = -161°; soluble in H_2O (3.5 ml/100 ml), more soluble in organic solvents.

19476 Methaplex
Methylmethacrylate acrylic in an aromatic and aliphatic solvent vehicle

system; water repellant and sealer for masonry. *Nova Chemical Inc.* Name unverified.

19477 Methar 30
144-21-8 6020 205-620-7
Selective herbicide, disodium methylarsonate (DSMA), for crabgrass control on grasses. *W A Cleary.*

19478 Methasan
137-30-4 10305 205-288-3
Zinc dimethyldithiocarbamate
Ziram, a vulcanization accelerator. *Monsanto Co.* Name unverified.

19479 Methasol
Dyes for printing inks. *ICI Chem & Polymers Ltd.*

19480 Methazate
137-30-4 10305 205-288-3
Ziram. A proprietary preparation of zinc dimethyl dithiocarbamate. *Naugatuck (US Rubber).* Name unverified.

19481 methazole
20354-26-1 6032 243-761-6
$C_9H_6Cl_2N_2O_3$
2-(3,4-dichlorophenyl)-4-methyl-1,2,4-oxadiazolidine-3,5-dione
oxydiazol; VCS-438; Paxilon; Probe; Probe 75 WP; Bioxone; Tunic; Chlormethazole; Mezopur. Broad spectrum herbicide. mp = 123-124°; slightly soluble in H_2O (1.5 mg/l), more soluble in organic solvents; LD_{50} (rat orl) = 777 mg/kg.

19482 Methic
Basic dyestuffs. *ICI Chem & Polymers Ltd.*

19483 methidathion
950-37-8 6048 213-449-4
$C_6H_{11}N_2O_4PS_3$
phosphorodithioic acidS-[(5-methoxy-2-oxo-1,3,4-thiadiazol-3(2H)-yl)methyl O,O-dimethyl ester
GS-13005; Supracide; Ultracid. Insecticide, acaricide. mp = 39-40°; soluble in H_2O (<1%), soluble in organic solvents; LD_{50} (rat orl) = 31 mg/kg.

19484 methiocarb
2032-65-7 6050 217-991-2
$C_{11}H_{15}NO_2S$
Phenyl-3,5-dimethyl-4-(methylthio)-, methylcarbamate
Carbamic acid, methyl-, 4-(methylthio)-3,5-xylyl ester; B 37344; Bay 9026; Bayer 37344; BAY 37344; BAY 5024; BAY 9026; Draza; DCR 736; Esurol; ENT 25,726; mercaptodimethur; Mesurol; Mesurol phenol; methyl carbamic acid, 4-(methylthio)-3,5-xylyl ester; Metmercapturan; Metmercapturon; Methiocarbe. Molluscicide with neurotoxic action. Used for control of slugs and snails in a wide variety of agricultural situations. mp = 121.5°; soluble in H_2O (27 mg/l), more soluble in organic solvents; LD_{50} (rat orl) = 20 mg/kg. *Bayer; Mobay.*

19485 Methiodal sodium
126-31-8 6051 204-782-6
CH_2INaO_3S
iodomethanesulfonic acid sodium salt
sodium iodomethanesulfonate; Skiodan; Abrodil; Radiographol; Segosin; Diagnorenol. Radiopaque medium used as a diagnostic aid, primarily in urology. Soluble in H_2O (70 g/100 ml), less soluble in organic solvents.

19486 methionine hydroxy analog
583-91-5 6054 209-523-0
$C_5H_{10}O_3S$
2-hydroxy-4-(methylthio)-butanoic acid
Alimet; calcium salt: MHA. Feed additive for livestock, especially poultry.

19487 DL-methionine
59-51-8 6053 200-432-1
$C_5H_{11}NO_2S$
DL-2-amino-4-(methylthio)butyric acid
2-amino-4-(methylmercapto)butyric acid; Methilonin; Acimetion; Amurex; Banthionine; Dyprin; Lobamine; Metion; Pedameth; Urimeth. Used in manufacture of pharmaceuticals, as a feed additive, a vegetable oil enrichment, a single-cell protein. Used medically as a hepatoprotectant, a uinary acidifier and an antidote for acetaminophen poisoning. mp = 270-273°; d = 1.34; soluble in H_2O (18.2 g/l), EtOH, insoluble in organic solvents. *Degussa; Penta Mfg.; U.S. Biochemical.*

19488 Methocel
9004-67-5 6120
Methylcellulose
Used in adhesives, cosmetics, foods, paints, pharmaceuticals and textiles. *Dow UK.*

19489 Methocel® 40-202
9004-65-3 4889
Hydroxypropyl methyl cellulose
Cold water-dispersion grade for personal care applications. *Dow Chemical.*

19490 Methocel® A15-LV
9004-67-5 6120
Methylcellulose
Food gums used as thickener, stabilizer, emulsifier, adhesive, and gellant; also for tablet coating applications. *Dow Chemical.*

19491 Methocel® A4C, A4M
9004-67-5 6120
Methylcellulose
Food gums used as thickener, stabilizer, emulsifier, adhesive, and gellant. *Dow Chemical.*

19492 Methocel® A4MP
9004-67-5 6120
Methylcellulose
For personal care products. *Dow Chemical.*

19493 Methocel® E3 Premium
9004-65-3 4889
Hydroxypropyl methylcellulose
For tablet coating applications. *Dow Chemical.*

19494 Methocel® E4M
9004-65-3 4889
Hydroxypropyl methylcellulose
Food gums used as thickener, stabilizer, emulsifier, adhesive, and gellant. *Dow Chemical.*

19495 Methocel® E5
9004-65-3 4889
Hydroxypropyl methylcellulose
Food gums used as thickener, stabilizer, emulsifier, adhesive, and gellant. *Dow Chemical.*

19496 Methocel® F4M
9004-65-3 4889
Hydroxypropyl methylcellulose
Food gums used as thickener, stabilizer, emulsifier, adhesive, and gellant. *Dow Chemical.*

19497 Methocel® K100MP
9004-65-3 4889
Hydroxypropyl methylcellulose
For personal care products. *Dow Chemical.*

19498 Methocel® K35
9004-65-3 4889
Hydroxypropyl methylcellulose
Food gums used as thickener, stabilizer, emulsifier, adhesive, and gellant. *Dow Chemical.*

19499 Meth-O-Gas
74-83-9 6108 200-813-2
Methyl bromide
Great Lakes Europe.

19500 Methokill
Insecticidal formulation. *Mitchell Cotts Chemicals Ltd.*

19501 Methoklone
75-09-2 6140 200-838-9
Stabilized methylene chloride. *ICI Chem & Polymers Ltd.*

19502 methomyl
16752-77-5 6062 240-815-0
$C_5H_{10}N_2O_2S$
S-methyl-N[(methylcarbamoyl) oxy]thioacetimidate
Nudrin; Lannate; Lannate(R); Lanox; Methomyl 5G; Lannabait; Lannate LB; Insecticide 1179; Lanox 216; LANOX 90; Mesomile; Nu-bait II; Thiobutan-2-one, O-(methylcarbamoyl)oxime; Flytek; Kipsin; Dupont 1179; Memilene; Methavin; Methomex; Nudrin. A cholinesterase inhibitor used as a systemic insecticide and acaricide. Used for control of a wide range of insects and spider mites in many fruits and vegetables. Also for control of flies in animal and poultry houses and in dairies. mp = 78-79°; d_4^{24}= 1.2946; soluble in H_2O (5.8 g/ml), more soluble in organic solvents; LD_{50} (rat orl) = 17 mg/kg. *Makhteshim Chemical Works Ltd.*

19503 Methoxone
24017-47-8 9736 245-986-5
$C_{12}H_{16}N_3O_3PS$
Emulsifiable concentrate containing 420 g/l triazophos; an organophosphorus insecticide. *ICI Chem & Polymers Ltd.*

19504 methoxychlor
72-43-5 6070 200-779-9
$C_{16}H_{15}Cl_3O_2$
1,1'-(2,2,2-trichloroethylidene)bis[4-methoxybenzene]
methoxy DDT; 2,2-bis (*p*-methoxyphenol)-1,1,1-trichloroethane; DMDT;OMS 466; ENT 1716; Higalmetox; Marlate; Prentox. Insecticide effective against mosquito larvae and houseflies; recommended for use in dairy barns. mp = 89°; d = 1.41; soluble in H_2O (0.1 mg/ml), more soluble in organic solvents; LD_{50} (rat orl) = 6000 mg/kg. *Kincaid Enterprises.*

19505 *p*-methoxyphenylacetic acid
104-01-8 203-166-4
$C_9H_{10}O_3$
4-methoxyphenylacetic acid
homoanisic acid; *p*-methoxy-α-toluic acid. Used in manufacture of pharmaceuticals and other organic compounds. mp = 86-89°; bp_3 = 140°; soluble in H_2O (6 g/l), more soluble in organic solvents; LD_{50} (rat orl) = 1550 mg/kg. *Penta Mfg.; Schweizerhall.*

19506 methyl acetate
79-20-9 6089 201-185-2
$C_3H_6O_2$
Acetic acid, methyl ester
Methyl ethanoate; methyl acetic ester; Devoton; Tereton. Ester of methyl alcohol and acetic acid paint remover compounds, lacquer solvent, intermediate, synthetic flavoring; solvent for nitrocellulose, acetylcellulose; manufacture of artificial leather. mp = -98°; bp = 57°; d = 0.9320; n_D^{20} = 1.3610; insoluble in H_2O, soluble in organic solvents; LD_{50} (rbt orl) = 3705 mg/kg. *Akzo; Hoechst Celanese; Penta Mfg.*

19507 methyl acetone
78-93-3 6149 201-159-0
C_4H_8O
methyl ethyl ketone
A crude fraction of wood distillation. Its principal constituents are acetone, methyl alcohol, and methyl acetate; used as a solvent for nitrocellulose, cellulose acetate, rubber, gum, resins; paint and varnish removers; extracting perfumes.

19508 Methyl acetyl ricinoleate
140-03-4 205-392-9
$C_{21}H_{38}O_4$
Methyl 12-acetoxyoleate
Flexricin® P-4; Ricinoleic acid, methyl ester, acetate; Methyl 12-acetoxy-9-octadecenoate. All purpose plasticizer, lubricant for vinyls and lacquers. mp = -15°; fp = 196°; ref index = 1.4545; soluble in most organic solvents; insoluble in water; sp gr = 0.938; LD_{50} (mus orl) = 34,900 mg/kg. *CasChem.* Name unverified.

19509 methyl acrylate
96-33-3 6092 202-500-6
$C_4H_6O_2$
2-propenoic acid methyl ester
acrylic acid methyl ester; methyl propenoate; Methoxycarbonylethylene; Acrylic Acid Methyl Ester; methacrylate; methyl ester acrylic acid; Curithane 103. Ester of methyl alcohol and acrylic acid; acrylic polymers; amphoteric surfactants; chemical intermediate; leather finish resins; textile and paper coatings; plastic films. mp = -75°; bp = 80°; d = 0.9560; n_D^{20}= 1.4021; soluble in H_2O (6 g/100 ml), more soluble in organic solvents; LD_{50} (rat orl) = 277 mg/kg. *BASF; Hoechst Celanese.*

19510 methyl acrylate polymer
BASF; Hoechst Celanese.

19511 methyl alcohol
67-56-1 6024 200-659-6
CH_4O
methanol
carbinol; wood spirit; wood alcohol. Industrial solvent; feedstock in organic synthesis. Additive in antifreeze, gasoline. mp = -98°; bp = 65°; d_4^{20}= 0.7915; n_D^{20} = 1.3292; miscible with H_2O, organic solvents; toxic, usual fatal dose in humans = 100-250 ml.

19512 methyl amyl alcohol
108-11-2 203-551-7
$C_6H_{14}O$
methylisobutyl carbinol
4-methylpentanol-2; MIBC; methylamyl alcohol; 4-methyl-2-pentanol; methylpentanol; 4-Methylpentan-2-ol; *sec*-Hexyl Alcohol; 2-Methyl-4-pentanol; Isobutylmethylcarbinol; Methyl-2-pentanol; Pentanol, 4-methyl-. Solvent for dyestuffs, oils, gums, resins, waxes, nitrocellulose, ethylcellulose; organic synthesis; froth flotation; brake fluids. mp = -90°; bp = 132°; d = 0.8020; n_D^{20} = 1.4100; soluble in H_2O (2 g/100 ml), more soluble in organic solvents; LD_{50} (rat orl) = 2590 mg/kg. *Allchem Industries; Ashland; Shell; Union Carbide.*

19513 methyl anthranilate
134-20-3 6099 205-132-4
$C_8H_9NO_2$
methyl-*o*-aminobenzoate
Neroli oil, artificial; 2-aminobenzoic acid, methyl ester; *o*-aminobenzoic acid methyl ester; *o*-carbomethoxyaniline; 2-(methoxycarbonyl)aniline; methyl *o*-aminobenzoate; Carbomethoxyaniline; Methoxycarbonyl)aniline; *o*-Amino methyl benzoate. Flavoring, fragrance, perfume, cosmetics, pomades; intermediate for pharmaceuticals and dyes. mp = 24°; bp = 256°; d = 1.1680; n_D^{20} = 1.5820; slightly soluble in H_2O, more soluble in organic solvents; LD_{50} (rat orl) = 2910 mg/kg. *Bell Flavors & Fragrances; Haarmann & Reimer; PMC Specialties.*

19514 methyl benzene
108-88-3 9667 203-625-9
C_7H_8
toluene
phenyl methane; retinaphtha; Toluol; phenyl methane; Methylbenzol; methyl-Benzene; Monomethyl benzene;
Methacide; tolu-sol; antisal 1a; Tol. Used as aviation gasoline and high octane blending stock; solvent for paints and coatings. mp = -93°; bp = 110°; d = 0.8650; n_D^{20} = 1.4965; soluble in H_2O (0.05 g/100 ml), more soluble in rganc solvents; LD_{50} (rat orl) = 636 mg/kg.

19515 methyl benzoate
93-58-3 6104 202-259-7
$C_8H_8O_2$
benzoic acid methyl ester
Niobe oil; benzoic acid methyl ester; methyl benzenecarboxylate; clorius; oniobe oil; oxidate le; essence of niobe; Oil of niobe. Ester; Perfumery, solvent for cellulose esters and ethers, resins, rubber; flavoring. mp = -12°; bp = 198°; d = 1.0940; n_D^{20} = 1.5165; insoluble in H_2O, soluble in organic solvents; LD_{50} (rat orl) = 1177 mg/kg. *Morflex; Pentagon Chemicals Ltd; Penta Mfg.; Schweizerhall; Sybron.*

19516 methyl benzoquate
13997-19-8 6558 237-796-6
$C_{22}H_{23}NO_4$
6-butyl-1,4-dihydro-4-oxo-7-(phenylmethoxy)-3-quinolinecarboxylic acid methyl ester
Nequinate; ICI-55052; Statyl. Coccidiostat. mp = 287-288°.

19517 methyl bromide
74-83-9 6108 200-813-2
CH_3Br
bromomethane
Brom-O-Sol; Brom-O-Gas; Haltox; Merth-O-Gas; Terr-O-Gas 100. Multi-purpose fumigant, often used in combination with chloropicrin. Used for insecticidal, acaricidal and rodenticidal control in storage facilities and for soil fumigation to control insects, nematodes, soil-borne diseases and weed seeds. mp = -93°; bp = 4°; d_0 = 1.732; soluble in H_2O 13.4 g/l, more soluble in organic solvents; LC_{100}, (rat inh, 6 hrs.) = 0.63 mg/l air. *Akzo; Ethyl; Great Lakes.*

19518 methyl butynol
115-19-5 6113 204-070-5
C_5H_8O
2-methyl-3-butyn-2-ol
dimethyl ethynyl carbinol; Tertiary acetylenic alcohol. Corrosion inhibitor; reactive intermediate in manufacture of pharmaceuticals, plastics, rubbers, fragrances, agriculture.; as solvent, acid inhibitor, viscosity reducer, and stabilizer in vinyl plastisols, platinum catalyst blocker for silicones. mp = 3°; bp = 104°; d = 0.8680, n_D^{20} = 1.4209; insoluble in H_2O, soluble in organic solvents; LD_{50} (rat orl) = 1950 mg/kg. *Air Prods & Chem; BASF.*

19519 methyl caprate
110-42-9 203-766-6
$C_{11}H_{22}O_2$
methyl decanoate
methyl caprinate. Intermediate for detergents, emulsifiers, wetting agents, stabilizer, resins, lubricants, plasticizer. *Penta Mfg. Procter & Gamble.*

19520 methyl caprylate-caprate
67762-39-4 267-017-5
Emery® 2209. Detergent intermediate; solv. for pesticides and herbicides. *Henkel/Emery.*

19521 methyl cellosolve®
109-86-4 6118 203-713-7
$C_3H_8O_2$
2-Methoxyethanol
ethylene glycol methyl ether; ethylene glycol monomethyl ether; Methyl Icinol; Poly-Solv® EM. A proprietary name for the monomethyl ether of ethylene glycol; a colorless and nearly odorless liquid boiling at 124.5C; has the lowest boiling-point and greatest rate of evaporation of all available glycol ethers; it is a solvent for cellulose acetate. d = 0.9663; bp.6+0 = 34-41°; n_D^{20} = 1.4028; fp = 115° F; miscible with water, alcohol, ether, acetone, dimethylformamide; LD_{50} (rat orl) = 2.46 g/kg. *Arco; Ashland; OxyChem; Union Carbide; ICI Australia; Olin.*

19522 methyl chloride
74-87-3 6121 200-817-4
CH_3Cl
chloromethane
monochloromethane; Artic; R-40; Refrigerant R40. Catalyst carrier in low-temperature polymerization, tetramethyl lead, silicones, refrigerant, fluid for thermometric/thermostatic equipment, methylating agent in organic synthesis, extractant and low-temperature solvent, herbicide, topical anesthetic. mp = -

97°; bp = -24°; d = 0.92; insoluble in H_2O, soluble in organic solvents; LC_{100} (mus ihl) = 3146 ppm. *Air Prods & Chem; Mitsui Toatsu Chem.; OxyChem.*

19523 methyl coconate
61788-59-8 262-988-1
Emery® 2253; Radia® 7117. Detergent intermediate; solv. for pesticides and herbicides. *Henkel/Emery; Fina Chemicals.*

19524 methyl cyanoacetate
105-34-0 203-288-8
$C_4H_5NO_2$
malonic methyl ester nitrile
Organic synthesis, pharmaceuticals, dyes. mp = -13°; bp = 204-207°; d = 1.1230; n_D^{20} = 1.4174; insoluble in H_2O, soluble in organic solvents. *R.W. Greeff; Hüls Am.; Lonza.*

19525 methyl cyclohexane
108-87-2 203-624-3
C_7H_{14}
hexahydrotoluene
Cyclohexylmethane; Toluene Hexahydride. Solvent for cellulose ethers, organic synthesis. mp = -126°; bp = 101°; d = 0.7700; n_D^{20}= 1.4222; insoluble in H_2O, soluble in organic solvents; LD_{50} (mus orl) = 2250 mg/kg. *Janssen Chimica; Penta Mfg.; Phillips 66.*

19526 methyl di-Icinol
111-77-3 6116 203-906-6
2-(2-Methoxyethoxy) ethanol
Solvent for use in protective coatings, inks, cleaning products, agricultural chemicals; aids wetting, penetration, and soil removal; coupling solvent. *ICI Australia.*

19527 methyl ester L
Lard methyl ester
Lubricant and wetting agent for industrial lubricants, e.g., soluble cutting and drawing compounds, motor oils, rolling oils; also for textiles and cosmetics. *Actrachem.* Discontinued.

19528 methyl ester S
Soya methyl ester
Lubricant additive; substitute for triglycerides. *Actrachem.*

19529 methyl ether
115-10-6 6148 204-065-8
C_2H_6O
dimethyl ether
wood ether; methyl ether; oxybismethane. Used in refrigerants, as a solvent and extraction agent, a catalyst and stabilizer in polymerization. 1 mp = -141.5°; bp = -24.8°; d^{25} = 1.91855; fp = -41°.

19530 methyl ethyl ketone
78-93-3 6149 201-159-0
C_4H_8O
2-butanone
2-Butanone; MEK; Ethyl methyl ketone; Butanone; methyl acetone; butan-2-one; meetco; Oxobutane. Solvent and chemical intermediate. mp = -87°; bp = 80°; d = 0.8050; n_D^{20} = 1.3780; soluble in H_2O (27 g/100 ml), organic solvents; LD_{50} (rat orl) = 5.5 g/kg. Elf Atochem N. Am.; BP Chem. Ltd; Exxon; Hoechst Celanese; Mallinckrodt; Shell; Texaco; Union Carbide.

19531 methyl ethyl ketone peroxide
1338-23-4 215-661-2
$C_8H_{18}O_6$
MEK peroxide
ethyl methyl ketone peroxide; 2-butanone peroxide. Initiator/catalyst for cure of unsaturated polyester resins. *Akzo; Elf Atochem; Cook Composites & Polymers; Norac; Witco/Argus.*

19532 methyl ethyl ketoxime
96-29-7 202-496-6
C_4H_9NO
methyl ethyl ketone oxime
2-butanone oxime; butanone oxime; ethyl methyl ketoxime; 2-butoxime; MEK-oxime; SKINO 2; Troykyd anti-skin b; Aron M 1. Antiskinning agent for paint industry. mp = -30°; bp = 152°; d = 0.923; n_D^{20}= 1.4410; soluble in H_2O, organic solvents; LD_{50} (mus ip) = 200 mg/kg. *Akzo; AlliedSignal; KMZ Chem. Ltd.*

19533 methyl ethyl sulfide
624-89-5 210-868-4
C_3H_8S
methylthioethane
ethyl methyl sulfide. Unpleasant stench oder. Used as a leak detecting additive to natural gas. bp = 65-67°; d = 0.8420.

19534 methyl eugenol
93-15-2 202-223-0
$C_{11}H_{14}O_2$
4-allyl-1,2-dimethyoxybenzene
4-allyl veratrole; eugenyl methyl ether; eugenol methyl ether; veratrole methyl ether; Allyl veratrole; Allyl-1,2-dimethoxybenzene; Dimethoxy-4-(2-propenyl)benzene; 3,4-Dimethoxyallyl benzene; o-methyl eugenol ether; Methyl eugenyl ether. Insect attractant, flavoring. bp = 248°; d = 1.0330; n_D^{20} = 1.5338; insoluble in H_2O, soluble in organic solvents; LD_{50} (rat orl) = 1179 mg/kg. *Firmenich; Penta Mfg.; Schweizerhall.*

19535 methyl gluceth-10
68239-42-9
Glucam® E-10. Humectant for personal care products; freezing pt. depressant; emollient in aq. and hydroalcoholic products; moisturizer; foam modifier in detergent and shampoo systems; solv. and solubilizer for topical pharmaceuticals; used in emulsions, toilet articles *Amerchol Corp.*

19536 methyl gluceth-20
68239-42-9
Glucam® E-20. Humectant for personal care products; freezing pt. depressant; emollient in aq. and hydroalcoholic products; moisturizer; foam modifier in detergent and shampoo systems; solv. and solubilizer for topical pharmaceuticals. *Amerchol Corp.*

19537 methyl gluceth-20 distearate
98073-10-0
Glucam® E-20 Distearate. Auxiliary o/w emulsifier, moisturizer, emollient and lubricant for cosmetics and pharmaceuticals; conditioner. *Amerchol Corp.*

19538 methyl glucose dioleate
83933-91-3
Glucate® DO. W/o emulsifier, auxiliary emulsifier for o/w systems; conditioner, emollient, lubricant, plasticizer, and pigment dispersant. *Amerchol; Amerchol Europe.* Discontinued.

19539 methyl glucose sesquistearate
68936-95-8 273-049-0
Glucate® SS. W/o emulsifier used with Glucamate SSE-20 to provide visc. stability, mildness. *Amerchol; Amerchol Europe.* Discontinued.

19540 methyl green
14855-76-6 6157 238-920-1
$C_{27}H_{35}BrClN_3$
4-[[4-(dimethylamino)phenyl][4-(dimethylimino)-2,5-cyclohexadien-1-ylidene]methyl]-N-ethyl-N,N-dimethylbenzeneaminium bromide chloride
C.I. 42590; ethyl green; Iodin green (Griesbach). Green powder, soluble in H_2O. Used as a biological stain and in dyeing and printing of textiles.

19541 methyl hydroxycellulose
Binder in plasters, adhesive, and troweling compounds.

19542 methyl iodide
74-88-4 6161 200-819-5
CH_3I
iodomethane
Halon 10001; Monoiodomethane; Methyl iodine. Organic synthesis, microscopy, testing for pyridine. mp = -66°; bp = 43°; d = 2.2800; n_D^{20} = 1.5304; insoluble in H_2O ,soluble in organic solvents; LC_{50} (rat orl) = 1300 mg/l/4H. *Akzo; Andeno BV; Burlington Bio-Medical; Fiarmount; R.W. Greeff.*

19543 methyl isothiocyanate
556-61-6 6165 209-132-5
C_2H_3NS
methyl mustard oil
Isothiocyanatomethane; Methyl mustard oil; Trapex; Degussa methyl isothiocyanate; MTC; Biomet 33; MENCS. Used as a soil fumigant for control of nematodes, fungi, insects and weed seeds. mp = 35-37°; bp = 119°; d = 1.0690; LD_{50} (rat orl) = 72 mg/kg.

19544 methyl lardate
Solvent, carrier for agricultural. spray products; defoaming component in metalworking, paper deinking, pharmaceutical fermentation; wetting agent, lubricant ingredient for metalworking and lubricating oils. *Anar.*

19545 methyl laurate
111-82-0 203-911-3
$C_{13}H_{26}O_2$
methyl dodecanoate
dodecanoic acid methyl ester; methyl dodecanoate; Methyl dodecylate. Ester of methyl alcohol and lauric acid; intermediate for detergents, emulsifiers, wetting agent, stabilizers, lubricants, plasticizers, textiles, flavoring. d = 0.8700; bp_{766} = 262°; n_D^{20} = 1.4292; insoluble in H_2O, soluble in organic solvents. *Henkel/Emery; Procter & Gamble; Stepan.*

19546 methyl icinol
109-86-4 6118 203-713-7
2-Methoxyethanol
Solvent for use in protective coatings, inks, cleaning products, agricultural, chemicals; aids wetting, penetration, and soil removal; coupling solvent. *ICI Australia.*

19547 methyl ledate
19010-66-3 242-748-2
$C_6H_{12}N_2PbS_4$
Lead dimethyldithiocarbamate

Ledate; Lead, bis(dimethylcarbamodithioato-S,S')-, (T-4)-; bis(dimethyldithiocarbamato)lead; dimethyldithiocarbamic acid, lead salt. Ultra accelerator recommended for NR, SBR, IIR, IR, and BR rubbers; used for ultra acceleration, high speed, high temperature vulcanization. mp = 310°; d = 3.43; insoluble in H_2O, soluble in organic solvents. *R. T. Vanderbilt Co Inc.*

19548 methyl methacrylate
80-62-6 6005 201-297-1
$C_5H_8O_2$
methacrylic acid, methyl ester
Methacrylic Acid Methyl Ester; Methyl-2-Methyl-2-Propenoate; MME; MMA; 2-methylacrylic acid methyl ester; methyl methylacrylate; methyl alpha-methylacrylate; diakon; Methyl-2-methylpropenoate. Monomer for polymethacrylate resins, impregnation of concrete. mp = -48°; bp = 100°; d = 0.9360; n_D^{20}= 1.4140; soluble in H_2O (15.9 g/l), organic solvents; LD_{50} (rat orl) = 7872 mg/kg. Aldrich; Allchem Industries; Cyro Industries; Degussa; Mitsubishi Gas; Rohm & Haas; Transol Chem. UK Ltd.

19549 methyl methacrylate polymer
9011-14-7
$(C_5H_8O_2)_n$
Acrylite; methyl methacrylate resin. Thermoplastic acrylic resin used in coatings, barrier coatings for PS, vinyl topcoats, product finishes, printing inks. *Aristech; Cyro Industries; Sybron.*

19550 methyl myristate
124-10-7 204-680-1
$C_{15}H_{30}O_2$
methyl tetradecanoate
tetradecanoic acid methyl ester; Emery® 2214. Ester of methyl alcohol and myristic acid; intermediate for myristic acid detergents; emulsifiers, wetting agents, stabilizers, resins, lubricants, plasticizers, textiles, animal feeds; standard for gas chromatography; flavoring. mp = 18°; bp = 323°; d = 0.8550; n_D^{20}= 1.4362; insoluble in H_2O, soluble in organic solvents. *Henkel/Emery; Stepan.*

19551 methyl namate®
128-04-1 204-876-7
$C_3H_6NNaS_2$
Sodium dimethyldithiocarbamate
Aceto SDD 40; Alcobam NM; Brogdex 555; carbon s; Dibam; Dibam A; DMDK; methyl namate; Sharstop 204; sodium N,N-dimethyldithiocarbamate; Stafresh 615; Steriseal 40; Thiostop N; Vnstop; Vulnopol NM; Wing Stop B; SDMDTC; sodium dimethylcarbamodithioate; Dimethyldithiocarbamic acid sodium salt; Freshgard 40. Sodium dimethyldithiocarbamate aq. solution; water treatment chemical; readily forms water-insoluble salts with heavy metals such as cadmium, copper, chromium, and nickel; clarification agent for wastewater from plating, photo finishing, ore beneficiation processes. LD_{50} (rat orl) = 1000 mg/kg. *R. T. Vanderbilt Co Inc.*

19552 methyl n-amyl ketone
110-43-0 4698 203-767-1
$C_7H_{14}O$
2-heptanone
MAK; amyl methyl ketone; methyl amyl ketone; 2-heptanone; n-amyl methyl ketone; methyl pentyl ketone; butyl acetone; heptan-2-one; Ketone C-7. Industrial solvent for nitrocellulose lacquers, synthetic flavoring and perfumery. mp = -35°; bp = 149-150°; d = 0.8200; n_D^{20}= 1.4085; soluble in H_2O (4.3 g/l), more soluble in organic solvents; LD_{50} (rat orl) = 1670 mg/kg. *Ashland; Eastman; Union Carbide.*

19553 methyl niclate®
15521-65-0 239-560-8
$C_6H_{12}N_2NiS_4$
Nickel dimethyldithiocarbamate
Bis(dimethyldithiocarbamato) nickel complex. Antioxidant for epichlorohydrin and peroxide vulcanized elastomers. *R. T. Vanderbilt Co Inc.* Discontinued.

19554 methyl oleate
112-62-9 6965 203-992-5
Emery® 2301, 2219; Emerest® 2301; Kemester® 104; Priolube 1400; Witconol 2301,. Solv. for pesticides and herbicides. *Henkel/Emery; Witco/Humko; Unichema; Witco/Organics.*

19555 methyl palmitate
112-39-0 203-966-3
$C_{17}H_{34}O_2$
methyl hexadecanoate
hexadecanoic acid methyl ester; Emery® 2216; Radia® 7120. Ester of methyl alcohol and palmitic acid; chemical intermediate, chemical synthesis; lubricant in mineral, cutting, lamination, textile oils, and rust inhibitors; textile and leather application. mp = 25°; bp_{30} = 211°; d = 0.8520; insoluble in H_2O, soluble in organic solvents. *Henkel/Emery; Penta Mfg.; Stepan.*

19556 methyl parathion
298-00-0 6183 206-050-1
O,O-Dimethyl-O-(4-nitrophenyl) phosphorothioate
Insecticide and acaricide used extensively in cotton producing areas. *A/S Cheminova.* Name unverified.

19557 methyl propyl ketone
107-87-9 6193 203-528-1
$C_5H_{10}O$
Pentan-2-one
2-pentanone; Methyl propyl ketone; Ethyl acetone; MPK; Methyl n-Propyl Ketone; propyl methyl ketone. Solvent, substitute for diethyl ketone, flavoring. mp = -78°; bp = 100-101°; d = 0.8120; n_D^{20}= 1.3897; soluble in H_2O (4 g/100 ml), organic solvents; LD_{50} (rat orl) = 1600 mg/kg. *Ashland; Janssen Chimica; Penta Mfg.; CasChem; Freedom textile; Penta Mfg.; Sigma.*

19558 methyl ricinoleate
141-24-2 205-472-3
$C_{19}H_{36}O_3$
12-hydroxy-9-octadecenoic acid methyl ester
Plasticizer, lubricant, cutting oil additive, wetting agent. *Penta Mfg.; Reilly-Whiteman.*

19559 methyl selenac
144-34-3 205-624-9
$C_{12}H_{24}N_4S_4Se$
Selenium dimethyldithiocarbamate
dimethyl-Carbamodithioic acid, tetra-anhydrosulfide with orthothioselenious acid; Selenium, tetrakis(dimethyldithiocarbamate). Rubber accelerator for NR, SBR, and IIR, vulcanizing agents; effective in low sulfur and sulfurless heat resistant compounds. *R. T. Vanderbilt Co Inc.*

19560 methyl stearate
112-61-8 8959 203-990-4
$C_{19}H_{38}O_2$
methyl octadecanoate
octadecanoic acid methyl ester; kemester 9718; methyl (Z)-9-octadecenoate; Emery® 2218; Kemester® 4516. Intermediate for stearic acid detergents, emulsifiers, wetting agents, stabilizers, resins, lubricants, plasticizers. mp = 37-39°; bp_{15} = 215°; insoluble in H_2O, soluble in organic solvents. *Ferro/Keil; Henkel/Emery; Penta Mfg.; Witco/Humko.*

19561 methyl sulfide
75-18-3 6204 200-846-2
C_2H_6S
Thiobismethane
Exact-S®; Dimethyl sulfide. Coking suppressor for ethylene production and for steel mill furnace walls; odorant for natural gas; presulfiding agent for catalysts in refinery processes. mp = -83°; bp = 36.2°; insoluble in water; soluble in alcohol, ether. *Gaylord Chem.*

19562 methyl tuads®
137-26-8 9510 205-286-2
Tetramethylthiuram disulfide
For NR and synthetic rubbers (esp. IIR, CR); accelerator and vulcanizing agent; cure modifier for Neoprene (retards G types; accelerates vulcanization of W types); nondiscoloring in light stocks. *R. T. Vanderbilt Co Inc.* Discontinued.

19563 methyl zimate®
137-30-4 10305 205-288-3
Zinc dimethyldithiocarbamate
Ultra accelerator for NR and synthetic rubbers; latex accelerator; act. over wide temp. range; generally requires thiazole modifier for safe processing and wide curing range; nondiscoloring in light stocks. *R. T. Vanderbilt Co Inc.* Discontinued.

19564 methylal
109-87-5 6093 203-714-2
$C_3H_8O_2$
dimethoxymethane
Formal; formaldehyde dimethyl acetal; Methylene dimethyl ether; methyl formal; dimethylacetal formaldehyde. Used as a solvent; in organic synthesis; perfumes, adhesives, protective coatings; special fuel. mp = -105°; d_4^{20}= 0.8593; bp = 41°; n_D^{20}= 1.3589; soluble in H_2O (30 g/100 ml), more soluble in organic solvents; LD_{50} (rbt orl) = 5708 mg/kg.

19565 2-methylbenzophenone
131-58-8 7470 205-032-0
$C_{14}H_{12}O$
phenyl tolyl ketone
Phenyl 2-tolyl ketone; (2-methylphenyl)phenyl-methanone. Perfume additive (fixative). $bp_{0.3}$ = 125-127°; d = 1.0830; n_D^{20}= 1.5958; insoluble in H_2O, soluble in organic solvents. *Janssen Chimica; Spectrum Chem. Mfg.*

19566 methylcellulose
9004-67-5 6120
cellulose methyl ether
Methyl ether of cellulose; protective colloid in waterbased paints to prevent flocculation of pigment; film and sheeting; binder in ceramic glazes; leather tanning; dispersing, thickening, and sizing agent; food additive; adhesive;

paper greaseproofing; pharmaceuticals. *Allchem Industries; Aqualon; Courtaulds Water Soluble Polymers; Shin-Etsu Chem.*

19567 methylene bisacrylamide
110-26-9 203-750-9
$C_7H_{10}N_2O_2$
N,N'-Methylenebis(2-propenamide)
MBA; N,N-Methylene-bis-acrylamide; NAPP; N',N'-methylene bis acrylamide; Bis-acrylamide; methylenebisacrylamide; N,N'-methylidenebisacrylamide; N,N'-Methylene-bis acrylamide. Organic intermediate, crosslinking agent. mp > 300°; soluble in H_2O (3 g/l); LD_{50} (rat orl) = 390 mg/kg. *Am. Cyanamid; Bio-Rad Labs; Fluka; Schweizerhall.*

19568 methylene chloride
75-09-2 6140 200-838-9
CH_2Cl_2
dichloromethane
methylene bichloride. Used as a solvent, propellant, degreaser and cleaner, and insecticide. mp = -97°; bp = 39-40°; d = 1.3250; n_D^{20}= 1.4235; soluble in H_2O (2 g/100 ml), soluble in organic solvents; LD_{50} (rat orl) = 2.18 g/kg. Ashland; Elf Atochem N. Am.; Farleyway Chem. Ltd; ICI Specialties; Mallinckrodt; Mitsui Toatsu Chem.; OxyChem.

19569 methylene iodide
75-11-6 6143 200-841-5
CH_2I_2
diiodomethane
Used to separate mixtures of minerals, in determination of specific gravity and in the manufacture of x-ray contrast materials. mp = 6°; bp = 181°; d = 3.3250; n_D^{20}= 1.7370; soluble in H_2O (14 g/l), organic solvents.

19570 methyl-glycocoll
107-97-1 8519 203-538-6
$C_3H_7NO_2$
sarcosine
methyl aminoacetic acid. Used in the synthesis of foaming antienzyme compounds for toothpaste, cosmetics, and pharmaceuticals.

19571 methylhexalin®
A trade name for a mixture of three isomeric methylcyclohexanols; a solvent for fats, resins, oils, and waxes. No manufacturer.

19572 2-methyl imidazole
693-98-1 211-765-7
$C_4H_6N_2$
2-methyl-1H-imidazole
2-methyl glyoxaline; 2MZ. Dyeing auxiliary for acrylic fibers, plastic foams. mp = 142-143°; bp = 267-268°; soluble in H_2O, organic solvents; LD_{50} (mus orl) = 1400 mg/kg. *Allchem Industries; BASF; Janssen Chimica.*

19573 *p*-methyl morpholine
109-02-4 203-640-0
$C_5H_{11}NO$
4-methyl morpholine
methyl morpholine; N-methyl morpholine; 4-Methyl-1-oxa-4-azacyclohexane;. Catalyst in polyurethane foams; extraction solvent; stabilizer for chlorinated hydrocarbons; self-polishing waxes; corrosion inhibitor; pharmaceuticals. Amine-based catalyst for flexible slabstock PU foam. Also used in peptide synthesis protocols. mp = -66°; bp_{750} = 115-116°; d = 0.9200; n_D^{20}= 1.4349; insoluble in H_2O, soluble in organic solvents; LD_{50} (rat orl) = 1960 mg/kg.

19574 α-methylnaphthalene
90-12-0 210-966-8
$C_{11}H_{10}$
1-methylnaphthalene
Carrier for polyester/wool blended fabrics. mp = -22°; bp = 240-243°; d = 1.0010; n_D^{20}= 1.6159; insoluble in H_2O, soluble in organic solvents; LD_{50} (rat orl) = 1840 mg/kg. *Allchem Industries; Crowley Tar Prods; Koch.*

19575 β-methylnaphthalene
91-57-6 202-078-3
$C_{11}H_{10}$
2-methylnaphthalene
Solvent; has been used as an insecticide. mp = 33°; bp = 241-242°; d = 1.0000; insoluble in H_2O, soluble in organic solvents; LD_{50} (rat orl).= 1630 mg/kg. *Allchem Industries; Crowley Tar Prods; Koch.*

19576 Methylon
A range of substituted phenolic condensates; adhesion promoters for polysulfide sealants, chemical resistant drum and can linings. *Comelius Chemical Co Ltd.* Name unverified.

19577 methylparaben
99-76-3 6182 202-785-7
$C_8H_8O_3$
methyl 4-hydroxybenzoate
methyl *p*-hydroxybenzoate; Methylparaben; Methyl Chemosept; Methyl Parasept; Nipagin M; Tegosept M; Aseptoform; Nipagin;. Used as a preservative in foods, beverages and cosmetics. mp = 125-128°; bp = 270=280°; soluble in H_2O (2.5 g/l), more soluble in organic solvents.

19578 N-methyl-2-pyrrolidone
872-50-4 6197 212-828-1
C_5H_9NO
NMP. Solvent for resins, acetylene, etc., pigment dispersant; petroleum processing; spinning agent for PVC; intermediate. *Aldrich; Ashland; BASF; Chemoxy Int'l. plc; ISP; Janssen Chimica.*

19579 α-methylstyrene
98-83-9 202-705-0
C_9H_{10}
isopropenylbenzene
2-phenylpropene; 1-methyl-1-phenyl ethylene; AMS; Isopropenylbenzene; 2-Phenylpropene; Ortho Brush Killer A; 2-Phenyl-1-propene; as-methylphenylethylene; β-phenylpropene; 2-phenylpropylene; β-phenylpropylene; 2-phenylpropylene; α-methylstyrol; 1-phenyl-1-methylethylene; 2-phenyl-2-propene; (1-methylethenyl)benzene. Polymerization monomer, especially for polyesters. mp = -24°; bp = 165-169°; d = 0.9090; n_D^{20}= 1.5375; insoluble in H_2O, soluble in organic solvents; LD_{50} (rat orl) = 4900 mg/kg. *AlliedSignal; Ashland; Honeywill & stein Ltd; Mitsui Petrochem. Ind.; Mitsui Toatsu Chem.*

19580 Metillure
An acid-resistant alloy consisting of 17% silicon, 81% iron, 0.9% manganese, 0.25% aluminum, 0.6% carbon, and 0.17% phosphorus.

19581 Metiloil®
Methyl esters of ricinoleic acid and other fatty acids in castor oil; used for tanning and in textiles and for the manufacture of sulfonated emulsifying agents. *Aquitaine-Organico.* Unverified.

19582 metiram
9006-42-2
Tris(amine)(ethylenebis(dithiocarbamato; zinc(2+)(tetrahydro-1,2,4,7-dithiadiazocene-3,8-dithione), polymer
Polyram; ammoniated EBDCs; Polyran; Carbamodithioic acid, Carbatene; Ethylenebis(dithiocarbamic acid), polymer with ammonia complex of zinc ebdc; Polyram-combi; Zinc metiram. Used to prevent crop damage in the field, during storage, or transport. Effective against a broad spectrum of fungi and is used to protect fruits, vegetables, field crops, and ornamentals from foliar diseases and damping off. Practically non-toxic. mp = 140°; insoluble in H_2O.

19583 Metobromuron
3060-89-7 6224 221-301-5
$C_9H_{11}BrN_2O_2$
N'-(4-bromophenyl)-N-methoxy-N-methylurea
Ciba 3126; C 3126; Patoran; Pattonex. Metobromuron, a substituted urea which inhibits photosynthesis and is used for pre-emergence control of annual broad-leaved weeds and grasses in vegetable crops. mp = 95-96°; soluble in H_2O (330 mg/l), alcohols, more soluble in non-polar organic solvents; LD_{50} (rat orl) = 3875 mg/kg. *Ciba plc.* Name unverified.

19584 Metol
55-55-0 6097 200-237-1
$C_{14}H_{20}N_2O_6S$
methyl-*p*-amino-*m*-cresol sulfate
Photol; Verol; Rhodol; Armol; Elon; Genol; Graphol; Photo-Rex; Pictol; Planetol. Used as a developing agent used in photography. Also used for dyeing furs. mp = 260° (dec); soluble in H_2O (5-15 g/100 ml); less soluble in organic solvents.

19585 metolachlor
51218-45-2 6230 257-060-8
$C_{15}H_{22}ClNO_2$
2-chloro-N-(2-ethyl-6-methylphenyl)- N-(2-methoxy-1-methylethyl)acetamide
Metolaclor; Dual; metelilachlor; CGA 24705; Bicep; Turbo; Bicep 6L; Dual 25G; Dual 8E; Pace 6L; Pennant; Primagram; Primextra; Codal; Ontrack 8E. Selective herbicide used for control of annual grasses and some broad-leaved weeds. $bp_{0.001}$ = 100°; d = 1.12; soluble in H_2O (530 mg/l), more soluble in organic solvents; LD_{50} (rat orl) = 2780 mg/kg.

19586 Metolat FC 355
Low foaming wetting agent for color paints, tinting pastes, organic/inorganic pigments; improves color acceptance in emulsion paints. *Münzing Chemie GmbH.*

19587 Metolat FC 515
Dispersant for extenders and pigments, emulsion paints. *Münzing Chemie GmbH.*

19588 Metolat LA 524
Grinding and wetting agent for lacquers and varnishes; wetting agent for organically modified bentonite. *Münzing Chemie GmbH.*

19589 Metolat TH 75
Emulsifier for mineral oil, especially for cutting oils. *Münzing Chemie GmbH.*

19590 metolhydroquinone
metol-quinone
It contains Metol, hydroquinone, sodium phosphate, sodium sulfite, and potassium carbonate; used as photographic developer.

19591 Metopirone
54-36-4 6246 200-206-2
A proprietary preparation of metyrapone; used to test pituitary gland function. *Ciba plc.* Name unverified.

19592 Metprep
Alkaline and solvent paint strippers. *Brent Chemicals International plc.*

19593 Metral®
For the electrical industry. *DuPont UK.*

19594 Metrax
Industrial detergent. *Crosfield Chemicals Ltd.* Discontinued.

19595 metribuzin
21087-64-9 6239 244-209-7
$C_8H_{14}N_4OS$
4-amino-6-(1,1-dimethylethyl)-3-(methylthio)-1,2,4-triazin-5(4H)-one
Sencor; Bayer 6159H; Lexone; 4-amino-6-tert-butyl-3-(methylthio)-as-triazin-5(4H)-one; Bay 94337; Sencorl;Preview; Salute; metribuzin + chlorimuron; Lexone 4L; Lexone 75DF; Lexone DF; Sencor 4L; Sencor 75DF; Sencor DF; Sencorex; Sencor or metribuzin; Amino-6-(1,1-dimethylethyl)-3-(methylthio)-1,2,4-triazin-5(4H)-one; Amino-6-tert-butyl-3-(methylthio)-s-triazin-5(4H)-one; Triazin-5(4H)-one, 4-amino-6-tert-butyl-3-(methylthio)-. Systemic herbicide used for control of many important grasses and broad-leaved weeds in soybeans, potatoes, tomatoes, sugarcane, alfalfa and asparagus; suitable for pre-and in some cases post emergence application. mp = 125°; SG = 1.28; d_4^{20}= 1.28; slightly soluble in H_2O (1.2 g/l), soluble in MeOH, EtOH; LD_{50} (rat orl) = 2200 mg/kg.

19596 metribuzin
21087-64-9 6239 244-209-7
$C_8H_{14}N_4OS$
4-amino-6-(1,1-dimethylethyl)-3-(methylthio)-1,2,4-triazin-5(4H)-one
Sencor; Bayer 6159H; Lexone; 4-amino-6-tert-butyl-3-(methylthio)-as-triazin-5(4H)-one; Bay 94337; Sencorl;Preview; Salute; metribuzin + chlorimuron; Lexone 4L; Lexone 75DF; Lexone DF; Sencor 4L; Sencor 75DF; Sencor DF; Sencorex; Sencor or metribuzin; Amino-6-(1,1-dimethylethyl)-3-(methylthio)-1,2,4-triazin-5(4H)-one; Amino-6-tert-butyl-3-(methylthio)-s-triazin-5(4H)-one; Triazin-5(4H)-one, 4-amino-6-tert-butyl-3-(methylthio)-. Systemic herbicide used for control of many important grasses and broad-leaved weeds in soybeans, potatoes, tomatoes, sugarcane, alfalfa and asparagus; suitable for pre-and in some cases post emergence application. mp = 125°; SG = 1.28; d_4^{20}= 1.28; slightly soluble in H_2O (1.2 g/l), soluble in MeOH, EtOH; LD_{50} (rat orl) = 2200 mg/kg.

19597 Metro Tiles
Industrial floor tiles for heavy abrasion and impact. *Prodorite Ltd.* Discontinued.

19598 Metro-nite
A refined natural mineral composed of calcium carbonate, and carbonates and silicates of magnesium; used in the paint industry as a pigment.

19599 Metropad
Aqueous pigment dispersions; nonresinated systems for pad dyeing applications; excellent resin/catalyst stability. No manufacturer.

19600 Metrosol AZ
Detergent for textile scouring. *Reilly-Whiteman.*

19601 Metrotect
A proprietary bitumen paint for protective treatment of iron-work, etc. No manufacturer.

19602 Metrotex
Aqueous pigment dispersions; for textile printing applications; contains no formaldehyde derivatives. No manufacturer.

19603 Metrotex Colors
Aqueous pigment dispersions containing no formaldehyde derivatives; for textile printing. No manufacturer.

19604 Metso
6834-92-0 8788 229-912-9
Sodium metasilicate. *Crosfield Chemicals Ltd.*

19605 Metso
Sodium metasilicate (anhydrous or pentahydrate) or sodium orthosilicate; alkaline component of heavy duty household, institutional and industrial cleaning compounds, metal cleaning. *PQ Corp.* Name unverified.

19606 metsulfuron-methyl
74223-64-6 6244
$C_{14}H_{15}N_5O_6S$
2-[[[[(4-methoxy-6-methyl-1,3,5-triazin-2-yl)amino]carbonyl]amino]sulfonyl] benzoic acid methylester
Allie, Brushoff; Ally; DPX-6376; DPX-T6376; Escort; Granstar; Gropper. Selective systemic herbicide, used for pre- or post-emergence control of annual and perennial broad-leaved weeds in wheat, barley and oats. mp = 158°; soluble in H_2O (1.1 mg/l), more soluble in organic solvents; LD_{50} (rat orl) > 5000 mg/kg.

19607 Meturon® 4L
2164-17-2 4189 218-500-4
fluometuron
Fluometuron suspension; flowable herbicide controlling annual grasses and broadleaf weeds in cotton and sugarcane. *Griffin.*

19608 metyrapone
54-36-4 6246 200-206-2
$C_{14}H_{14}N_2O$
2-methyl-1,3-di-3-pyridiyl-1-propanone
methopyrapone; mepyrapone; metopyrone; methbipyranone, SU-4885; Metopirone; Metroprione. Diagnostic aid used to measure pituitary function. mp = 50-51°.

19609 metyridine
114-91-0 6247 204-060-0
$C_8H_{11}NO$
2-(β-methoxyethyl)pyridine
methyridine; Dekelmin; Promintic. Anthelmintic, used in veterinary medicine. d^{20} = 0.988; bp_{17} = 94-96°; n_D^{20}= 1.4975; soluble in H_2O, organic solvents.

19610 Mevantraal
Methyl-2-methyl pentylidene anthranilate
Quest Int'l. UK Ltd.

19611 Mewlon
Polyvinyl acetate fiber filaments. *British Traders & Shippers Ltd.*

19612 Mexapol
Floor polishes. *Evans VAnodine International Ltd.*

19613 Mexican onyx
A variety of calcite.

19614 Mexico seeds
Castor oil seeds.

19615 Mexitil
5370-01-4 6257 226-362-1
A proprietary preparation of mexiletine hydrochloride; an anti-arrhythmic agent. *Boehringer Ingelheim Pharmaceuticals Inc; Boehringer Ingelheim Ltd.*

19616 Mexphalte
Trademark for varieties of bitumen; used for road dressing and other purposes. No manufacturer.

19617 Meyerhofferite
12007-56-6 1691 234-511-7
CaB_4O_7
Calcium borate
calcium pyroborate; calcium tetraborate. An artificially prepared mineral. Used as a flux, in antifreeze preparations and in fire retardant paint. Slightly soluble in H_2O. No manufacturer.

19618 Meyer's solution
Mercury-potassium iodide solution, obtained by dissolving 13.35 g mercuric chloride and 49.8 g potassium iodide separately in water, mixing the solutions and diluting to 1 liter.

19619 Meyprofix® 509 (redesignated Polycare® 509)
Vinyl acetate, isobutyl maleate, vinyl neodecanoate copolymer; hairstyling fixative for superior holding and curl retention properties. *Rhône-Poulenc UK.*

19620 MG2/MG4
Range of granular fertilizers. *Fisons plc, Horticultural Div.* Name unverified.

19621 M-Gard
Wood preservative. *Mooney Chemicals Inc.*

19622 MGH-93
1309-42-8 5706 215-170-3
Magnesium hydroxide
Plastics additive permitting higher processing temperatures for olefins, PP, nylons; absorbs more heat than hydrated alumina; does not generate poisonous gases during combustion; dilutes smoke produced by decomposing polymers. *RMc Minerals.*

19623 Mgoa rubber
Commercial name for the rubber from *Muscarenhasia elastica* of East Africa.

19624 MHA
583-91-5 6054 209-523-0
Methionine hydroxy analog
Calcium feed supplement for poultry and other animal feeds. *Monsanto Co.* Name unverified.

19625 miazine
289-95-2 8170 206-026-0
pyrimidine
Metadiazine. Used in medicine and biochemical research.

19626 mica
12001-26-2

laminated talc; glimmer glist;Muscovite mica; Zinnwaldite; Fluorophlogopite; Margarite; Silicates; Soapstone. Filler/extender for plastics, rubber, coatings, and pearlescent pigment applications; for making fireproof window-panes and lampchimneys, electrical equipment; oil well drilling muds; binder and reinforcement in lipsticks. Feldspar; Franklin Industrial Min.; Mearl; Mykroy/Macalex Ceramics; Norwegian Talc UK; Nyco Min.; Van Dyk.

19627 mica silk
An electrical insulating tape made from mica splittings on a silk cloth.

19628 Micabond
A proprietary trade name for a material consisting of mica, shellac, and resin. No manufacturer.

19629 Micacoat®
12001-26-2
SiO_2, Al_2O_3, K_2O, Fe_2O_3, Na_2O, CaO, TiO_2, MnO_2, P, S
Chemically coupled muscovite mica, coarse and fine grinds: used for polymer composites and high performance coatings. *NYCO® Minerals Inc.* Discontinued.

19630 Micafil B
A proprietary synthetic resin-varnish-paper product used in electrical insulation. No manufacturer.

19631 Micafil G
A proprietary electrical insulator made in the form of tubes from shellac, coated paper, and mica. No manufacturer.

19632 Micafil S
A proprietary trade name for a shellac varnish-paper product used as an electrical insulation. No manufacturer.

19633 Micafolium
A general name for electrical insulators made from mica splittings and paper.

19634 Mica-Kote
A proprietary trade name for a roofing felt made from asphalt impregnated felt. No manufacturer.

19635 micanite
A mica material built up of small plates of mica with an insulating material such as shellac, or on a foundation of paper or cloth; used as an electrical insulating material.

19636 micanite cloth
mica Cambric
toile micanite. Products used for electrical insulation, made from mica splittings on a cotton-cambric backing.

19637 Micarta
A trade name for a range of varnished paper and fabric products using natural and synthetic resin varnishes; used for electrical insulation. No manufacturer.

19638 micarta folium
A similar product to micafolium.

19639 Michel XO-24
Modified amine
Antistat additive, integral release agent for linear polyethylene and PVC. *M. Michel.*

19640 Michel XO-150-12
3913-02-8 223-470-0
$C_{12}H_{26}O$
Isododecyl alcohol
Used in organic synthesis. *M. Michel.*

19641 Michel XO-150-16
36311-34-9 252-964-9
$C_{16}H_{34}O$
Isocetyl alcohol
Isohexadecyl alcohol; isohexadecanol. Used in organic synthesis. *M. Michel.*

19642 Michel XO-150-1620
70693-04-8 248-470-8
Isostearyl alcohol
Used in organic synthesis. *M. Michel.*

19643 Michel XO-150-20
5333-42-6 226-242-9
$C_{20}H_{42}O$
Octyldodecanol
2-Octyl-1-dodecanol; 2-octyldodecan-1-ol. Used in organic synthesis. *M. Michel.*

19644 Michler's Base
101-61-1 6264 202-959-2
$C_{17}H_{22}N_2$
4,4'-Methylene bis(N,N'-dimethylaniline)
Arnold's Base; Methane Base Michler's Hydride; Michler's Methane; 4,4'-Methylenebis(N,N-Dimethyl)-Aniline; Tetra-Base; 4,4'-Tetramethyl diaminodiphenylmethane; *p,p'*-bis(dimethylamino)diphenylmethane; tetramethyldiaminodiphenylmethane; bis(*p*-(N,N-dimethylamino)phenyl) methane; *p,p'*-bis(N,N-dimethylaminophenyl) methane; methylene base; Michler's hydride; reduced Michler's ketone; N,N,N',N'-tetramethyl-p,p'-diaminodiphenylmethane; bis[*p*-(dimethylamino)phenyl] methane; *p,p'*-Tetramethyldiamindiphenylmethane; 4,4'-Methylene bis(N,N-dimethyl)benzenamine. An acetic acid solution of tetramethyldiamino diphenyl-methane. Used in manufacture of dyes and as a reagent for lead. mp = 90-91°; bp = 390°; insoluble in H_2O, soluble in organic solvents.

19645 Michler's hydrol
119-58-4 204-335-5
$C_{17}H_{22}N_2O$
4,4'-Bis(dimethylamino)benzhydrol
4,4'-Bis(dimethylamino)diphenyl carbinol. Used as a dye intermediate and in organic synthesis. mp = 100-102°.

19646 Michler's ketone
90-94-8 6265 202-027-5
$C_{17}H_{20}N_2O$
Tetramethyldiaminobenzophenone
bis[4-(Dimethylamino)phenyl]-methanone. Used for making dyestuffs, especially auramine derivatives. mp = 174-176°; bp = 360°; slightly soluble in H_2O, more soluble in organic solvents.

19647 Mi-Col
A mildew fungicide. *Plant Protection.* Name unverified.

19648 Micracet
Pigment preparations. *Ciba plc.*

19649 Micral® 1000
Alumina trihydrate
Economical smoke suppressor/flame retardant with high brightness; for wire and cable insulation, injection molded polyolefins, coatings, adhesives, rubber goods, PVC, EPDM, EPR, XLPE, EVA, and compression molded thermosets. *J.M. Huber/Solem.*

19650 Micral® 855
Alumina trihydrate
Flame retardant/smoke suppressant for wire and cable jacketing and insulation, injection molded and extruded polyolefins. *J.M. Huber/Solem.*

19651 Micral® 932
Alumina trihydrate
Halogen-free smoke suppressor/flame retardant for wire and cable insulation, injection-molded polyolefins, coatings, adhesives, rubber goods, paper filler and coating, PVC, EPDM, EPR, ABS, XLPE, and compression-molded thermosets. *J.M. Huber/Solem.*

19652 Micro P Extender
Amorphous mineral silicate
Lightweight resin extender, filler for aircraft, military, appliances, business equipment, construction, consumer products, electrical/electronic, land transport, and marine applications. *DJ Enterprises, Inc.*

19653 micro-asbestos
An Austrian asbestos of short fiber unsuitable for the ordinary uses of asbestos.

19654 Microbator PC-78
A tamed and stabilized nontoxic chlorine dioxide complex concentrate formulated for use as an additive deodorant without free chlorine release; effectively arrests malodors caused by viruses, fungi, bacteria, and coliform densities when added to coolants, *Punati Chemical Corp.* Unverified.

19655 Microbiotone
A peptic digest of beef tissue that is a water soluble granular product; used in various fermentations, veterinary biologicals and diagnostics as a nutrient for faster growth of various organisms. *Am. Labs.*

19656 Microbloc®
14807-96-6 9207 238-877-9
Surface-treated talc with proprietary coating; antiblock for plastic film industry; enhanced compat. with polyolefins for improved film clarity and low COF. *Pfizer.*

19657 Microcal
10101-39-0 233-250-6
Ca_2O_4Si
Silicic acid, calcium salt (1:1)
Calcium metasilicate; Silicic acid (H2SIO3), calcium salt (1:1). Calcium silicate, used as a filler and in building materials. *Crosfield Chemicals Ltd.*

19658 Microcal ET
10101-39-0 233-250-6
Series of synthetic calcium silicates of controlled particle size used as extenders in emulsion paint. *Crosfield Chemicals Ltd.*

19659 Microcarb
63-25-2 1831 200-555-0
Carbaryl
Contact insecticide and worm killer. *Micro-Biologicals Ltd.*

19660 Microcarb T
Carbaryl + pyrethins
Insecticide spray for the control of fleas and flies in poultry houses. *Micro-Biologicals Ltd.*

19661 Microcatalase®
9001-05-2 1948 232-577-1
Bacterial catalase derived from *Micrococcus lysodeikticus* enzyme which removes residual hydrogen peroxide after antimicrobial treatment of milk. *Solvay Enzymes.*

19662 Micro-Cel® A
10101-39-0 233-250-6
Synthetic calcium silicate; functional filler used as carriers, grinding aids, anticaking agent, and conditioner in agricultural chemicals; toxicants; carriers for liquid seed inoculants; inert carriers to convert sticky viscose liquids to dry liquids. *Celite.*

19663 Micro-Cel® T-38
10101-39-0 233-250-6
Calcium silicate
Extender pigment, opacifier contributing good hiding power to emulsion paints; flatting properties for flat wall paints. *Celite.*

19664 Micro-Chek® 11
26530-20-1 6853 247-761-7
$C_{11}H_{19}NOS$
2-n-Octyl-4-isothiazolin-3-one
octhilinone. Antimicrobial, mildewcide for PVC, polyurethane, other polymers for use in roofing membranes, exterior automotive trims, awnings, tarpaulins, pond liners, marine upholstery, shower curtains, outdoor furniture. *Ferro.*

19665 microcosmic salt
$Na(NH_4)H \cdot PO_4 \cdot 4H_2O$
Sodium-ammonium-hydrogen Phosphate
Phosphorus salt, fusible salt of urine, fusible salt, essential salt of urine.

19666 Microcrystalline wax
63231-60-7 264-038-1
Forbest MW 23; Be Square® 185; Fortex®; Mekon® White; Multiwax® 180-M, HS; Paracol® 404C; Polymekon®; Starwax® 100; Ultraflex®; Victory®. Microcryst. wax finely dispersed in C17 fatty acid; increases slip and abrasion resistance in clear and pigmented stoving enamels, e.g., industrial paints, coil-coating finishes, printing inks (flexo, letterpress). *Lucas Meyer; Petrolite; Witco/Sonneborn; ICI Chem. & Polymers Ltd.;.*

19667 Microcult® GG
Reagent for detecting pathogenic urinary germs, *Candida species* and *Neisseria in relevant media. Bayer AG.*

19668 Microdol (Extra)
A trade name for ground dolomite. No manufacturer.

19669 Micro-Dry®
1327-41-9 356 215-477-2
Aluminum chlorohydrate powder; antiperspirant active. *Reheis Inc.*

19670 Microduct®
9050-36-6 232-940-4
Maltodextrin
Carrier for fragrances; emollient oils, bath products; food additive. *CasChem.*

19671 Micro-fine®
Fibers. *DuPont UK.*

19672 MicroForm B
1302-78-9 1082 215-108-5
Dry, modified bentonite clay; microfloc coagulant *Hercules.*

19673 MicroForm BCS
1302-78-9 1082 215-108-5
Slurried, modified bentonite clay; microfloc coagulant *Hercules.* Discontinued.

19674 Microgen Plus
Mixture of formaldehyde and lindane
Used for control of pests in poultry houses. *Micro-Biologicals Ltd.*

19675 Microhoba
Jojoba oil derivatives. *A & E Connock (Perfumery & Cosmetics) Ltd.*

19676 Micro-K
7447-40-7 7783 231-211-8
Potassium chloride
Replenisher. *Wyeth Laboratories.* Name unverified.

19677 Microlan
Powdered lanolin. *Croda Chem. Ltd.* Discontinued.

19678 Microlith-A
Pigment preparations. *Ciba plc.* Name unverified.

19679 Microlube A
8002-74-2 7155 232-315-6
Paraffin wax emulsion; for surface application to paper and other materials; improves water resistance. *Hercules.*

19680 micromeritol
An alcoholic solution of yerba buena.

19681 Micromet
A slowly soluble sodium metaphosphate. *Albright & Wilson Ltd.* Name unverified.

19682 Micromid 1022
Polyamide resin dispersion. For use in waterbased adhesives and release coating formulations. *Union Camp.*

19683 Micromite
122-14-5 4017 204-524-2
Emulsifiable concentrate of 500 g fenitrothion per liter; an organophosphorus insecticide. *ICI Agrochemicals.*

19684 Micromulse WIO
Designed to form stable water-oil or oil-water emulsions. *Norman, Fox & Co.*

19685 Micronal®
Aqueous microcapsule dispersions; for production of the donor face of non-carbon copying papers. *BASF AG.*

19686 Micropaque
7727-43-7 1023 231-784-4
X-ray grade barium sulfate. *Nipa Laboratories Ltd.*

19687 MicroPflex 1200
14807-96-6 9207 238-877-9
Surface-modified microtalc. *Pfizer.*

19688 Micropil
Range of spherical inorganic oxide particles for containment and adsorption. *Microporous Materials Ltd.*

19689 Micropil
7631-86-9 8637 231-545-4
Silica for industrial use. *Crosfield Chemicals Ltd.*

19690 Micropoly 520
9002-88-4 7728
Polyethylene
Improves body, texture, visual attributes (luster, coverage, opacity) for wide variety of product formulations. *Presperse.*

19691 Micropore
A proprietary surgical tape of rayon with a hypoallergenic adhesive. *3M.* Name unverified.

19692 microporite
An insulating concrete made from ground silica and lime, hardened by treatment with steam.

19693 Micropur
Tablets containing 0.1 mg silver for water sterilization (domestic use). *The Boots Co plc.* Discontinued.

19694 Micropure® Ultra
872-50-4 6197 212-828-1
N-Methyl-2-pyrrolidone
ISP.

19695 Microsan
A copper fungicide. *Mechema Chemicals Ltd.* Name unverified.

19696 Microseal
Process of deposition of thin dry solid film lubricant coating by impingement; mold release for plastic molds, bearings, sliding surfaces and vacuum lubrication. *E/M Corporation.*

19697 Microsil
Rubber reinforcing agent. *Crosfield Chemicals Ltd.*

19698 Microsized
Micron sized salt. *Akzo Salt.*

19699 Microsperse
7440-44-0 1855 231-153-3
carbon
Range of carbon black dispersions in water; used in paper, concrete, ink, paint etc industries. *Breamhurst Ltd.* Unverified.

19700 Micro-Step® H-301
Sulfonate/nonionic blend; microemulsifier for broad range of pesticides. *Stepan; Stepan Canada; Stepan Europe.*

19701 Microstix®
Reagent for detecting pathogenic urinary germs, *Candida* and *Neisseria* species in relevant media. *Bayer AG.*

19702 Micro-supplex®
Fibers. *DuPont UK.*

19703 Microtex GTZ
Guthrie Latex.

19704 Microthene®
Polyolefin powders. *Quantum Chemical Corp.*

19705 Microthene® FA 150-00
Polyolefin
Powdered coating resin. *Quantum/USI.* Name unverified.

19706 Microthene® MA 530-060
9002-88-4 7728
Polyethylene
For rotational molding. *Quantum/USI.* Name unverified.

19707 Microthene® MN 701-00
9002-88-4 7728
LDPE; powdered coating resin. *Quantum/USI.* Name unverified.

19708 Microthene® MP 625U
9002-88-4 7728
Polyethylene
For rotational molding. *Quantum/USI.* Name unverified.

19709 Microthene® MU 760-00
9002-88-4 7728
Polyethylene
19% vinyl acetate incorporated; for rotational molding. *Quantum/USI. Name unverified.*

19710 Micro-triever®
DuPont UK.

19711 Microtuff 1000
14807-96-6 9207 238-877-9
Surface-treated talc; filler for polyolefins. *Pfizer.*

19712 Micro-White® 07 Slurry
471-34-1 1697 207-439-9
Calcium carbonate
Ultrafine wet-ground product for applications requiring high gloss and opacity in coatings and inks. *ECC International Ltd.*

19713 Micro-White® 10 Codex
471-34-1 1697 207-439-9
Calcium carbonate
Food-grade meeting purity requirements for food and food contact applications (flour, cake mixes, cereals, chewing gum, crackers). *ECC International Ltd.*

19714 Micro-White® 15
471-34-1 1697 207-439-9
Calcium carbonate
Ultrafine wet-ground product for paint and polymer industries. *ECC International Ltd.*

19715 Micro-White® 25
471-34-1 1697 207-439-9
Calcium carbonate
Easy dispersing pigment for paint, rubber, plastics, floor covering and other industries. *ECC International Ltd.*

19716 Micro-White® 40
471-34-1 1697 207-439-9
Calcium carbonate
Fine ground filler for economy applications. *ECC International Ltd.*

19717 Micro-White® 100
471-34-1 1697 207-439-9
Calcium carbonate
White general-purpose filler for paints, rubber goods, putties, and joint compounds. *ECC International Ltd.*

19718 Microx
1314-13-2 10279 215-222-5
Zinc oxide
Pugment with high surface area. *Harcros.*

19719 Micryston
A proprietary name for a group of hormone preparations used for replacement therapy. *LAB Ltd.* Unverified.

19720 Midas Gold®
Colorants in solid form and dispersions. *Reichhold.*

19721 Midas®
For agriculture. *DuPont UK.*

19722 middle oils
Carbolic oils. That fraction of coal tar distilling at 170-230°C.

19723 Midstream
85-00-7 3415 201-579-4
Diquat dibromide
A granular contact herbicide and pre-harvest crop desiccant and defoliant. *ICI Chem & Polymers Ltd.*

19724 Midvale Alloys
Heat-resisting alloys; ATV alloy contains from 33-39% nickel, 10-12% chromium, 1.1-1.8% manganese, and the balance iron; BTG alloy contains 60-62% nickel, 10-11% chromium, 1.2-1.5% manganese, and the balance iron. No manufacturer.

19725 Midvaloy H.R
A proprietary nickel-tungsten-chromium alloy; resists corrosion. No manufacturer.

19726 Mifaslug
108-62-3 5983 202-945-6
metaldehyde
Pellets containing 6% w/w metaldehyde; moluscicide with many crop uses; snail and slug bait. *Farmers Crop Chemicals Ltd.*

19727 Mifatox
919-86-8 6129 213-052-6
An emulsifiable concentrate containing 580 g demeton-S-methyl per liter; a systemic organophosphorus insecticide. *Farmers Crop Chemicals Ltd.* Discontinued.

19728 Migafar
Proofing agent. *Ciba plc.* Name unverified.

19729 Migafar AL
A proprietary trade name for an aqueous emulsion of fatty products suitable for the removal of chafe marks from dyed fabrics. *Ciba plc.* Name unverified.

19730 Migassist® NYL
Leveler for acid dyes. *Sybron.*

19731 Migatex
Finishing agents. *Ciba plc.* Name unverified.

19732 Mi-Gee Brand
75-11-6 6143 200-841-5
Methylene iodide
Used for gem and mineral testing. *Geoliquids.* Name unverified.

19733 Migen
House dust mite vaccine. *SmithKline Beecham.* Discontinued.

19734 Mighty Soft
Dough softener. *Eastman.*

19735 Miglyol®
Special oils/neutral oils, triglycerides; used for production of oily suspensions, suppositories, ointments and creams. *Hüls AG.*

19736 Miglyol® 808
538-23-8 208-686-5
$C_{27}H_{50}O_6$
Tricaprylin
Octanoic acid, 1,2,3-propanetriyl ester; Caprylin; Glycerol trioctanoate; caprylic acid triglyceride; glycerin tricaprylate; glyceryl trioctanoate; MCT; panacete 800; RATO; tricaprylyl glycerin; tricapryloylglycerol; tricaprylic glyceride; trioctanoylglycerol. Good skin spreading/penetrating properties. mp = 6°; bp = 233°; d = 0.9530; insoluble in H_2O, soluble in organic solvents; LD_{50} (rat orl) = 33.3 g/kg *Hüls Am.*

19737 Miglyol® 812
65381-09-1 265-724-3
Caprylic/capric triglyceride
Dispersant, lubricant, anticaking agent, carrier, solvent, solubilizer, suspending agent for cosmetics, dietetic products. *Hüls Am; Hüls AG.*

19738 Miglyol® 818
Caprylic/capric/linoleic triglyceride
Emollient for topicals. *Hüls Am.*

19739 Miglyol® 829
Caprylic/capric/diglyceryl succinate
Emollient, suspending agent for cosmetic and pharmaceutical topicals with densities above 1.0. *Hüls Am.*

19740 Miglyol® 840
Propylene glycol dicaprylate/dicaprate
Emollient, dispersant, lubricant, suspending agent, solubilizer; active ingredient for cosmetics and pharmaceuticals; carrier/vehicle and solvent for injection products, topical ointments, creams, lotions, suppositories; dietetic products. *Hüls Am.*

19741 Miglyol® Gel
Caprylic/capric triglyceride and Stearalkonium hectorite; Stabilizer; improves consistency and thermostability in cosmetic and pharmaceutical creams; emulsion stabilizer; as a base for w/o and o/w creams when combined with suitable emulsifiers; does not m *Hüls Am.*

19742 mignonette-geranium oil
An oil obtained by distilling geraniol over mignonette flowers.

19743 migra iron
A special pig iron for high quality castings obtained by a special heat treatment before casting, which results in a remarkably fine grain.

19744 Migraine Dolviran®
379-79-3 3703 206-835-9
Analgesic preparation with ergotamine tartrate. *Bayer AG.*

19745 Migranil 858
Metallic salt; antimigrant for pigment dyeing. *Sybron.*

19746 Migrassist® D
Benzyl trimethyl ammonium chloride
Migrator/dye leveling agent for cationic dyes; nonretarding. *Sybron.*

19747 Mikacion Dye®
Reactive cold-type dyes. *Mitsubishi Kasei.* Name unverified.

19748 Mikawhite®
Fluorescent whitening agent for acetate, polyester, polyacrylic fiber. *Mitsubishi Kasei.* Name unverified.

19749 Mikolite
1318-00-9 10095 406-060-8
A proprietary material; it is a vermiculite (hydrated magnesium-aluminum-iron silicate) which has been expanded by calcination giving a very fine product; used in paints. No manufacturer.

19750 Mikrobin
2177
$C_7H_4ClNaO_2$
Sodium-*p*-chlorobenzoate
Used as a preservative for wines. Freely soluble in H_2O.

19751 Mikron
Organic and inorganic pigments. *Tennant-KVK Ltd.*

19752 Mila
Skin cream, milk cleanser and lotion, for daily care of the skin. *Richardson-Vicks Inc.* Name unverified.

19753 Milanol
Basic bismuth trichloro-butyl-malonate.

19754 Milcap
Fungicide containing ethirimol and captafol for use on wheat. *ICI Chem & Polymers Ltd.*

19755 Mil-Col
5707-69-7 3499 227-197-8
Drazoxolon
Mildew fungicide and seed dressing. *ICI Agrochemicals; ICI Chem. & Polymers Ltd.*

19756 Milcurb
5221-53-4 3266 226-021-7
Fungicide containing dimethirimol. *ICI Chem & Polymers Ltd.*

19757 mild alkali
497-19-8 8739 207-838-8
Na_2CO_3
Sodium carbonate

19758 mild lime
471-34-1 1697 207-439-9
$CaCO_3$
Calcium carbonate (chalk)
Is known in agriculture as mild lime.

19759 Mildothane
23564-05-8 9489 245-740-7
Suspension concentrate containing 500 g/l thiophanate-methyl; a systemic insecticide. *DowElanco Ltd; Hortichem Ltd.*

19760 Mildothane Turf Liquid
23564-05-8 9489 245-740-7
Suspension concentrate containing 500 g/l thiophanate-methyl; a systemic insecticide. *Rhône-Poulenc Crop Protection Ltd.*

19761 Mildvac Ma5 (Mild Mass. Type)
For immunization of poultry. *Intervet Inc.*

19762 Mildvac-C
Connecticut type bronchitis vaccine; for immunization of poultry. *Intervet Inc.* Discontinued.

19763 Mildvac-M
Mild Massachusetts type bronchitis vaccine; for immunization of poultry. *Intervet Inc.*

19764 Milgard
Cleansing lotion, for babies and sensitive skin. *Richardson-Vicks Inc.* Name unverified.

19765 Milgo
23947-60-6 3786 245-949-3
Fungicide containing ethirimol. *ICI Chem & Polymers Ltd.*

19766 milk glass
A soda or flint glass rendered opaque by the addition of a mineral phosphate.

19767 milk of lime
Slaked lime and water in a thin cream.

19768 milk sugar
63-42-3 5356 200-559-2
$C_{12}H_{22}O_{11}$
Lactose
Used to supplement animal feeds.

19769 milk tree wax
Cow tree wax.

19770 Milkamino 20
Milk amino acids
Substantive moisturizer for cosmetics field; low salt. *Brooks Industries.*

19771 milkstone
A mixture of milk salts and protein obtained from milk.

19772 Mill Creek
Natural products, a line of hair and skin care products containing natural ingredients including keratin, aloe vera, jojoba, henna, elastin, apricot and chamomile. *Richardson-Vicks Inc.* Name unverified.

19773 Millad®
Trade name for family of sorbitol-based clarifying agent additives to improve the transparency of polyolefins, especially polypropylenes; for polypropylene random copolymers and homopolymers used for housewares, protective packaging, cases, medical devices, blow molded bottles and sheet. *Milliken.*

19774 Millad® 3905
Additive to enhance clarity and esthetics of polyolefins, esp. PP, LLDPE, some HDPE. *Milliken.*

19775 Millad® 5L71-10
10% Conc. of Millad 3905 in LDPE; clarifying agent for use in LLDPE. *Milliken.*

19776 Millad® HBPA
80-05-7 1338 201-245-8
Hydrogenated bisphenol A
Used in preparation of alkyd, polyester, and epoxy resins where good color stability and improved weatherability are important; for casting, laminating, coatings and fiber production. *Milliken.*

19777 Millaloy
A proprietary trade name for a nickel-chromium steel containing 4% nickel, 1.5% chromium, and 0.4% carbon. No manufacturer.

19778 Millamine® 5260
93% 1,2-diaminocyclohexane, 6% methylpentamethylenediamine.
Cycloaliphatic diamines used as an epoxy curing agent with outstanding heat distortion temperature, chemical resistance and physical properties in cured system; also as corrosion inhibitor, PU crosslinker, intermediate for polyamides and other chemicals. *Milliken.* Discontinued.

19779 Millathane® 66
Polyester urethane rubber; millable urethane which can be processed and cured by conventional rubber equipment and techniques; excellent resistance to abrasion, ozone, fuels, and oils, very good hot tear strength; peroxide curable. *TSE Industries.*

19780 Millathane® 88
Polyether urethane rubber; millable urethane can produce transparent or brightly colored parts with excellent abrasion resistance, low temperature flexibility; very good uv stability; peroxide curable. *TSE Industries.*

19781 Milldride®
Alkenyl succinic anhydrides, dicarboxylic anhydrides used as epoxy curing agents, starch modifiers, additives for motor oil and transmission fluid. *Milliken.*

19782 Milldride® DDSA
Dodecenyl succinic anhydride
Curing agent for epoxy resins, corrosion inhibitor for nonaqueous lubricating oils, intermediate for preparation of alkyd or unsaturated polyester resins, intermediate in chemical reactions. *Milliken.*

19783 Milldride® HDSA
Hexadecenyl succinic anhydride
Curing agent for epoxy resins, corrosion inhibitor for nonaqueous lubricating oils, intermediate for preparation of alkyd or unsaturated polyester resins, intermediate in chemical reactions. *Milliken.* Discontinued.

19784 Milldride® HHPA
Hexahydrophthalic anhydride
Epoxy curing agent; also for preparation of alkyd and polyester resins where good color stability is important; for casting, laminating, embedding, coating, and impregnating electrical components. *Milliken.*

19785 Milldride® MHHPA
Methyl hexahydrophthalic anhydride
Epoxy curing agent; for casting and impregnation applications where low viscosity, light color and good heat resistance are required. *Milliken.*

19786 Milldride® nDDSA
n-Dodecenyl succinic anhydride (C12-linear)
Curing agent for epoxy resins, corrosion inhibitor for nonaqueous lubricating oils, intermediate for preparation of alkyd or unsaturated polyester resins, intermediate in chemical reactions. *Milliken.* Discontinued.

19787 Milldride® nDSA
n-Decenyl succinic anhydride
Curing agent for epoxy resins, corrosion inhibitor for nonaqueous lubricating

oils, intermediate for preparation of alkyd or unsaturated polyester resins, intermediate in chemical reactions. *Milliken.* Discontinued.

19788 Milldride® ODSA

Octadecenyl succinic anhydride

Curing agent for epoxy resins, corrosion inhibitor for nonaqueous lubricating oils, intermediate for preparation of alkyd or unsaturated polyester resins, intermediate in chemical reactions. *Milliken.*

19789 Milldride® OSA

26680-54-6 247-899-8

Octenyl succinic anhydride

Curing agent for epoxy resins, corrosion inhibitor for nonaqueous lubricating oils, intermediate for preparation of alkyd or unsaturated polyester resins, intermediate in chemical reactions. *Milliken.*

19790 Milldride® TDSA

Tetradecenyl succinic anhydride

Curing agent for epoxy resins, corrosion inhibitor for nonaqueous lubricating oils, intermediate for preparation of alkyd or unsaturated polyester resins, intermediate in chemical reactions. *Milliken.* Discontinued.

19791 Millidet

An electric blasting cap with a bronze or aluminum shell; used in quarries, open-pit mines, underground mining operations, coal stripping, shafts, tunnels and heavy construction projects. *Hercules.* Discontinued.

19792 Millifoam

Foaming agent for plasterboard. *Millmaster Onyx UK.*

19793 milling silver

A nickel silver alloy made up of 56% copper, 27.5-31% zinc, 12-16% nickel, and 0.5-1% lead.

19794 Millithix® 925

32647-67-9 251-136-4

Dibenzylidene sorbitol

Thixotrope and gellant for use in unsaturated polyester and vinyl ester resins; also as clear antiperspirant. *Milliken.*

19795 Millon's mase

$OH{\cdot}Hg_2NH_2O$

Hydroxy-dimercuro ammonium hydroxide

Used for coloring porcelain.

19796 Millon's reagent

Mercury dissolved in an equal weight of nitric acid (specific gravity 1.41), and the solution diluted to twice its volume. After standing, the liquid is decanted from the precipitate; used as a test for proteins.

19797 Millophyline

A proprietary preparation of etamiphylline and camphorsulfonate; used as a cardiac and respiratory stimulant; also as a bronchial antispasmodic; for veterinary use. *Dales Pharmaceuticals Ltd.* Name unverified.

19798 Mills Plastic

A proprietary trade name for a vinylidene chloride synthetic resin. No manufacturer.

19799 Milowite

7631-86-9 8637 231-545-4

A proprietary amorphous silica for paint, polishing, and chemical trades; very white in color, and 90% is below 0.01 mm particle size. No manufacturer.

19800 Milstem

23947-60-6 3786 245-949-3

Liquid ethirimol seed dressing. *ICI Chem & Polymers Ltd.*

19801 Milton

Sterilizing fluid and crystals, to sterilise baby bottles. *Richardson-Vicks Inc.* Name unverified.

19802 Miltopan

Blend of alkylarylsulfonates and solvents, anionic; all purpose detergent. *Henkel Chemicals Ltd.* Name unverified.

19803 Milvan Steel

A proprietary trade name for a high speed tool steel containing 19% tungsten, 4% chromium, and 2% vanadium. No manufacturer.

19804 Milwaloy

A proprietary trade name for a chromium-vanadium steel. No manufacturer.

19805 Minadex

Orange flavored, green syrup containing in each 5 ml the following vitamins and minerals; vitamin A 650 iu, vitamin D 65 iu, iron (as green ferric ammonium citrate) 12 mg, calcium glycerophosphate 11.25 mg, potassium glycerophosphate 1.125 mg, manganese sulfate 0.5 mg, copper sulfate 0.5 mg; a vitamin and mineral supplement and appetite restorative for children and adults, particularly during and after illness. *Evans Medical.* Name unverified.

19806 Minamino

A proprietary preparation of vitamins, minerals, and aminoacids; used as a dietary supplement. *Consolidated Chemicals Ltd.* Name unverified.

19807 minargent

An alloy of copper, nickel, and aluminum; used as a silver substitute.

19808 minargentatum

An alloy of 56.82% copper, 39.77% nickel, 2.84% tungsten, and 0.57% aluminum.

19809 Mindel® A-670

Polysulfone ABS-modified; platable, hot water-resistant, tough, dimensionally stable resin; used for automotive applications where decorative parts pass through bake cycles or service near heat sources; also for potable water use, food service applications, plumbing parts. FDA and NSF compliant. *Amoco Chemical Co.*

19810 Mindel® B-310

25135-51-3

Amorphous polysulfone/crystalline polymer blend, glass-reinforced; offers dimensional stability, low shrinkage and warpage; designed for molded electrical connectors; flame-retardant. *Amoco Chemical Co.*

19811 Mindel® B-322

25135-51-3

Amorphous polysulfone/crystalline polymer blend; Offers dimensional stability, low shrinkage and warpage, high insulation resistance, superior hydrolytic stability; used for connectors, items for service in hot-humid atmospheres. *Amoco Chemical Co.*

19812 Mindel® M-800

25135-51-3

Polysulfone, mineral-filled; improved ESCR; for applications requiring good dimensional stability, excellent creep resistance, and toughness; electrical insulation applications. *Amoco Chemical Co.*

19813 Mindel® S-1000

25135-51-3

Polysulfone-based proprietary resin; offers good hydrolytic stability, high heat distort. temp., chem. resist., steam and hot water resist., autoclavability, excellent elec. props., dimensional stability; used for food processing, filtration, food service, medical and hospital applications where sterilization is necessary. FDA and NSF compliant, available in flame-retardant and glass-filled grades. *Amoco Chemical Co.*

19814 Minder.

Biocide. *Rhône-Poulenc UK.*

19815 Minder

Adjuvant containing 94% rape oil; wetting agent for phosphonoglycine herbicide sprays. *Stoller Chemicals Ltd.*

19816 Mindust Series

Blend; dedusting and wetting agent for use in coal preparation. *Hart Chem. Ltd.*

19817 mineral acid

An inorganic acid. Usually refers to hydrochloric acid.

19818 mineral black

A pigment. It is a shale found naturally, and contains 70% silica and 30% carbonaceous matter.

19819 mineral butters

A term formerly used for several of the metallic chlorides, such as those of antimony, arsenic, bismuth, and zinc.

19820 mineral carbon

7782-42-5 4560 231-955-3

Graphite.

19821 mineral flour

A Florida clay used in rubber mixings.

19822 mineral grey

The ash from lapislazuli after the extraction of ultramarine.

19823 Mineral Jelly No. 5

Petrolatum

Preblended base for cosmetic manufacture; odorless; tasteless; miscible with cosmetic ingredients used in oil-base formulations. *Penreco.*

19824 mineral khaki

A mineral color produced on the fiber by impregnating cotton with a mixture of ferrous and chromic acetates, drying, and then steaming. Mixtures of basic ferric and chromate acetates are formed on the material, which are fixed by passing the fiber through solutions of sodium carbonate and sodium hydroxide.

19825 mineral lake

A basic chromate of tin, prepared by adding potassium chromate solution to stannous chloride solution; used for coloring paper, and in oil painting.

19826 mineral rubber

Bitumens of the gilsonite type.

19827 mineral yeast

Torula, a yeast-like organism; used for fodder production.

19828 Minex 2
Anhydrous sodium potassium aluminosilicate
Filler for paint, plastics, adhesives and sealants, rubber, abrasives. *Unimin.*

19829 Minflo
Mining flotation reagents; depressants which selectively separate gangue materials in mineral flotation, increase the purity of the concentrate, increase efficiency of the collector, and can act as filter aids to agglomerate fines and prevent filter plugging. *Hercules.* Discontinued.

19830 Mini Slugit Pellets
9002-91-9 5983 202-945-6
Metaldehyde
Metaldehyde slug killer. *Murphy Chemical Co Ltd.* Discontinued.

19831 Minite
A similar explosive to Kohlen carbonite, without barium nitrate.

19832 minium tego
1314-41-6 5449 215-235-6
lead tetroxide
A high dispersed red lead marketed in Germany.

19833 mink oil
Oil of mink. Oil obtained from the sub-dermal fatty tissue of the mink; used as an emollient for skin and hair care formulations and makeup removers.

19834 Minlon®
Nylon 66 reinforced with 40% mineral filler; used as an engineering thermoplastic resin. *DuPont UK.*

19835 Minofor
Alloys used by jewelers. They contain from 9-64% antimony, 20-84% tin, 2-10% copper, and 1-10% zinc.

19836 Minolite Antigrisouteuse
A Belgian explosive containing 72% ammonium nitrate, 23% sodium nitrate, 3% trinitrotoluene, and 2% trinitronaphthalene.

19837 Mintite
A patent finish for rubber surfaces; consists of powdered mica. No manufacturer.

19838 Minugel
1337-76-4
Colloidal attapulgite clay; for suspension systems. *Whitecourt Ltd.*

19839 Min-U-Gel® 100
1337-76-4
Attapulgite clay; absorbent and adsorbent; gelling and suspending agent. *Floridin.* Name unverified.

19840 Min-U-Gel® 200
1337-76-4
Colloidal attapulgite clay; suspending agent in suspension fertilizers and other agricultural suspensions; absorbent, adsorbent. *Floridin.* Name unverified.

19841 Min-U-Gel® AR
1337-76-4
Attapulgite clay; thixotropic thickener for asphalt cutbacks *Floridin.* Name unverified.

19842 Min-U-Gel® CW
1337-76-4
Attapulgite clay; stabilizer for coal water slurries *Floridin.* Name unverified.

19843 Min-U-Gel® LF
1337-76-4
Colloidal attapulgite clay; thickener, stabilizer for liquid animal feeds containing immiscible materials (fat emulsions, vitamins, limestone powder); suspending agent. *Floridin.* Name unverified.

19844 Min-U-Sil
Micronized natural silica; used for silicon rubber, surface coatings. *Cornelius Chemical Co Ltd.* Name unverified.

19845 Minyak Kerung
An oleo-resin obtained from *Dipterocarpus* species of Malay. It is obtained as a viscous liquid.

19846 Mipcin®
2631-40-5 220-114-6
2-Isopropylphenyl methylcarbamate
isoprocarb. Insecticide, esp. for leafhoppers and plant hoppers on paddy rice. *Mitsubishi Kasei.* Name unverified.

19847 Mipolam
A proprietary trade name for polyvinyl chloride which, when plasticized, has properties resembling soft rubber; used in molding and for sheathing cables. No manufacturer.

19848 Mira metal
Acid-resisting alloys. One contains 74.7% copper, 16-3% lead, 6.8% antimony, 0.91% tin, 0.62% zinc, 0.43% iron, and 0.24% nickel. Another consists of 75% copper, 16% lead, 8% tin, and 1% nickel.

19849 Miracare® 2MCA
Disodium cocoamphodiacetate, sodium lauryl sulfate and hexylene glycol. Used in formulating nonirritating shampoos. *Rhône-Poulenc Surf.; Rhône-Poulenc France.*

19850 Miracare® 2MCA-SF
Disodium cocoamphodipropionate, sodium lauryl sulfate and hexylene glycol. Surfactant for nonirritating and non-eye-sting shampoos. *Rhône-Poulenc Surf.*

19851 Miracare® ANL
Sodium C14-16 olefin sulfonate, sodium laureth sulfate and lauramide DEA. High foaming surfactant concentrate for high-performance shampoos and skin cleaners. *Rhône-Poulenc Surf.; Rhône-Poulenc France.*

19852 Miracare® BC-10
PEG-80 sorbitan laurate, cocamidopropyl betaine, sodium trideceth sulfate, sodium lauroamphoacetate, PEG-150 distearate and sodium laureth-13 carboxylate. Concentrate for preparation of baby shampoo, bubble bath, bath gel and liquid hand soap products requiring mildness. *Rhône-Poulenc Surf.; Rhône-Poulenc France.* Name unverified.

19853 Miracare® BT
Disodium lauroamphodiacetate and sodium trideceth sulfate. Surfactant for nonirritating and non-eye-sting shampoos. *Rhône-Poulenc Surf.*

19854 Miracare® CT 100
Stearyl alcohol and cetrimonium bromide. Emulsifier for creme rinse/conditioner; base for permanent wave foam neutralizers. *Rhône-Poulenc Surf.*

19855 Miracare® M1
Sodium lauryl sulfate, stearamide MEA, glycol stearate and cocamide MEA. Pearlized base for cream shampoos having high flash foam. *Rhône-Poulenc Surf.*

19856 Miracare® MHT
Sodium lauroamphoacetate and sodium trideceth sulfate. Surfactant for nonirritating and non-eye-sting shampoos, foaming skin cleansing products. *Rhône-Poulenc Surf.*

19857 Miracare® MS-1
PEG-80 sorbitan laurate, sodium trideceth sulfate, PEG-150 distearate, disodium lauramino-propionate, cocamidopropyl hydroxysultaine and sodium laureth-13 carboxylate;. Concentrate for preparation of baby shampoo, bubble bath, bath gel and liquid hand soap products requiring mildness. *Rhône-Poulenc Surf.*

19858 Miracare® NWC
Sodium laureth sulfate, cocamide DEA, TEA-lauryl sulfate. Biodegradable base for shampoos and bubble baths. *Rhône-Poulenc Surf.*

19859 Miracare® SCS
122-19-0 204-527-9
Stearalkonium chloride in compatible emulsifier base; formulated base for simplified production of cream rinse conditioners. *Rhône-Poulenc Surf.*

19860 Miracare® XL
DEA-lauryl sulfate, DEA-lauramino-propionate, sodium lauraminopropionate and propylene glycol. High foaming base for preparation of hair and body shampoos. *Rhône-Poulenc Surf.*

19861 Miracle Man
Waterproofing cleaner and penetrant. *Rhône-Poulenc UK.*

19862 Miraculoy
A proprietary trade name for a steel containing 1.25% nickel, 0.65% chromium. 0.4% molybdenum. and 1.55% manganese. No manufacturer.

19863 Miraflon®
Fluorocarbon rubber. *Asahi Chem. Industry.*

19864 Mirage
67747-09-5 7941 266-994-5
Prochloraz
Fungicide. *Makhteshim Chemical Works Ltd.*

19865 Miralite
A light aluminum alloy which can be cast or rolled, and drawn into wire. It contains 12% copper and 2% tin.

19866 Miramant
A cutting alloy of heat resisting metals with a definite fraction of stable and hard carbides, especially molybdenum and tungsten carbides in eutectic proportions.

19867 Miramine® C
67784-90-1 267-058-9
Coco hydroxyethyl imidazoline
Emulsifier, corrosion inhibitor, softener, antistat, for textiles, asphalt, plastics, petrol, industry, cutting oils; water repellent treatment of cement, concrete, and plaster; antifungal agent for wood; tar emulsion breaker; slime control additive in paperboard. *Rhône-Poulenc Surf.*

19868 Miramine® CODI
68140-01-2 268-771-8

Cocamidopropyl dimethylamine
Emulsifier; base for emulsions for creams, lotions, and hair rinses; conditioner, antistat, viscosity builder. *Rhône-Poulenc Surf.*

19869 Miramine® GS
95-19-2 202-397-8
Stearyl hydroxyethyl imidazoline
Wetting agent for foam wax stripper, degreaser, cleaning formulations; stable to acid and alkali. *Rhône-Poulenc France.*

19870 Miramine® HPS-B
61791-39-7 263-171-2
Tall oil hydroxyethyl imidazoline, a surfactant. *Rhône-Poulenc Surf.*

19871 Miramine® O
21652-27-7 244-501-4
Oleyl hydroxyethyl imidazoline
Emulsifier, corrosion inhibitor, softener, antistat, wetting and flocculating agent, lubricant for textiles, asphalt, car wax emulsions, cleaners, paint manufacture, agricultural applications, synthetic coolants; dispersant for clay and pigments in solvent systems; tar emulsion breaker. *Rhône-Poulenc Surf.*

19872 Miramine® SODI
7651-02-7 231-609-1
Stearamidopropyl dimethylamine
Emulsifier, conditioner; produces cationic emulsions. *Rhône-Poulenc Surf.; Rhône-Poulenc France.*

19873 Miramine® TO
61791-39-7 263-171-2
Tall oil hydroxyethyl imidazoline
Emulsifier, corrosion inhibitor in oil burning systems, pickling bath operations, asphalt emulsions; dispersant for clay and pigments in solvent systems. *Rhône-Poulenc Surf.*

19874 Miranate® B
67990-17-4 268-040-3
Sodium butoxyethoxy acetate
Low foaming wetting agent, emulsifier, lubricant for wax strippers, degreasers, metalworking fluids, other cleaner formulations; compatible with high concentrations of electrolytes; stable to acid and alkali media. *Rhône-Poulenc Surf.; Rhône-Poulenc France.*

19875 Miranate® LEC
70632-06-3
Sodium laureth-13 carboxylate
Auxiliary detergent for shampoo systems, lime soap dispersant; emulsifier, lubricant for metalworking fluids. *Rhône-Poulenc Surf.*

19876 Miranol
Liquid foaming agent with outstanding emulsifying properties. *Foseco (F.S.) Ltd.* Discontinued.

19877 Miranol C2M
Imidazoline based amphoteric with whole coconut as fatty radical; clear aqueous solution; nonirritating surfactants, emulsifiers, solubilizers, stabilizers; used in shampoos, cleaners, pharmaceutical applications. Name unverified.

19878 Miranol C2M-SF
Imidazoline based amphoteric with whole coconut as fatty radical; clear amber liquid; for industrial cleaner formulations. Name unverified.

19879 Miranol CM
Imidazoline based amphoteric with whole coconut as fatty radical; light amber liquid; emulsifies grease, suspends particulate soil. Name unverified.

19880 Miranol DM
Imidazoline based amphoteric with stearic fatty radical, in the form of a creamy white paste; fiber softener, wool lubricant. Name unverified.

19881 Miranol JEM
Imidazoline based amphoteric with caprylic and ethylhexoic fatty radicals; clear aqueous solution; used in bottle washing, wax stripping, degreasing, etc. Name unverified.

19882 Miranol L2M-SF
Imidazoline based amphoteric with linoleic fatty radical, in aqueous solution form; used for wax-type polishes and floor finishes, etc. Name unverified.

19883 Miranol SM
Imidazoline based amphoteric with capric fatty radical; clear aqueous solution; used for medicated, germicidal, rug and upholstery shampoos, hand soaps and surgical soaps. Name unverified.

19884 Miranol® 2CIB
68647-53-0 271-957-1
Disodium cocoamphodiacetate
Detergent for high foaming, nonirritating shampoos, skin cleansers, cosmetics, industrial detergents; emulsifier, solubilizer, coupling agent for heavy-duty liq. cleaners. *Rhône-Poulenc Surf.; Rhône-Poulenc France.*

19885 Miranol® BM Conc
6868-66-2
Disodium lauroamphodiacetate
Surfactant for nonirritating shampoos, skin cleansers. *Rhône-Poulenc Surf.; Rhône-Poulenc France.*

19886 Miranol® C2M Anhyd. Acid
68919-40-4 272-897-9
Cocoamphodipropionic acid
Detergent, wetting agent, high foaming surfactant for dispersion on caustic soda and on powder mixes, etc., leveling agent in tin plating from acid baths; metal cleaning; industrial cleaning. *Rhône-Poulenc Surf.; Rhône-Poulenc France.*

19887 Miranol® C2M Conc. NP-PG
68650-39-5 272-043-5
Disodium cocoamphodiacetate, propylene glycol. Surfactant. *Rhône-Poulenc Surf.*

19888 Miranol® C2M-SF 70%
68604-71-7 271-704-5
Disodium cocoamphodipropionate
Detergent used for heavy-duty liquid cleaning compounds, steam cleaners, nonirritating shampoos, medicated cosmetics. *Rhône-Poulenc Surf.; Rhône-Poulenc France.*

19889 Miranol® CM Conc. NP
68608-65-1 271-793-0
Sodium cocoamphoacetate
Detergent, wetting and foaming agent, sequestrant, emulsifier, dispersant, germicidal, viscosity builder; for extra heavy duty cleaners, steam, pressure, metal, all-purpose cleaners; biodegradable. *Rhône-Poulenc Surf.; Rhône-Poulenc France.*

19890 Miranol® CM-SF Conc
68919-41-5
Sodium cocoamphopropionate
Coemulsifier for emulsion polymerization; emulsifier, wetting agent for industrial, institutional and household cleaners; biodegradable. *Rhône-Poulenc Surf.; Rhône-Poulenc France.*

19891 Miranol® CS Conc
68604-73-9 271-705-0
Sodium cocoamphohydroxypropylsulfonate
Detergent, wetting agent, corrosion inhibitor, emulsifier, sequestrant, foaming agent for shampoos, cold water fabrics, household and industrial cleaners; biodegradable. *Rhône-Poulenc Surf.; Rhône-Poulenc France.*

19892 Miranol® DM Conc. 45%
68608-63-9 271-790-4
Sodium stearoamphoacetate
Viscosifier, lubricant, softener, conditioner for cosmetics, textiles, industrial, institutional and household cleaners. *Rhône-Poulenc Surf.; Rhône-Poulenc France.*

19893 Miranol® Ester PO-LM4
96726-23-9
Pentaerythrityl tetralaurate
Emulsifier, emollient, conditioner, antistat for personal care products. *Rhône-Poulenc Surf.*

19894 Miranol® FA-NP
68608-65-1 271-793-0
Sodium cocoamphoacetate
Surfactant for extra-heavy duty liquid cleaning compounds, steam, pressure, metal, and all-purpose cleaners. *Rhône-Poulenc Surf.; Rhône-Poulenc France.*

19895 Miranol® FB-NP
68650-39-5 272-043-5
Disodium cocoamphodiacetate
Surfactant for nonirritating shampoos, skin cleansers, medicated cosmetics, medium duty liquid cleaners. *Rhône-Poulenc Surf.; Rhône-Poulenc France.*

19896 Miranol® FBS
68604-71-7 271-704-5
Disodium cocoamphodipropionate
Surfactant for extra heavy duty liquid cleaning compounds, e.g., steam, pressure, metal, and all-purpose cleaners. *Rhône-Poulenc Surf.; Rhône-Poulenc France.*

19897 Miranol® H2C-HA
3655-00-3 222-899-0
Disodium N-lauryl iminodipropionate
Foamer and wetting agent used in alkaline cleaners, fire fighting compounds. *Rhône-Poulenc Surf.*

19898 Miranol® H2M Conc
68608-66-2 271-794-6
Disodium lauroamphodiacetate
Surfactant for nonirritating shampoos, skin cleaners; biodegradable. *Rhône-Poulenc Surf.; Rhône Poulenc France.*

19899 Miranol® H2M-SF Conc
68610-43-5 271-864-6

Disodium lauroamphodipropionate
Surfactant for nonirritating shampoos and skin cleaners, especially aerosols; biodegradable. *Rhône-Poulenc Surf.; Rhône-Poulenc France.*

19900 Miranol® HM Conc
68608-66-2 271-794-6
Sodium lauroamphoacetate
Detergent, wetting and foaming agent, sequestrant, emulsifier, dispersant, germicidal for shampoos, dishwashing, paints; biodegradable. *Rhône-Poulenc Surf.; Rhône-Poulenc France.*

19901 Miranol® HM-SF Conc
61901-02-8 263-312-8
Sodium lauroamphopropionate
Emulsion polymerization. *Rhône-Poulenc Surf.* Name unverified.

19902 Miranol® J2M Conc
68608-64-0 271-792-5
Disodium capryloamphodiacetate
Emulsifier, caustic soda wetting agent, food washing and peeling, industrial, institutional and household cleaners, wax stripper, emulsion polymerization of synthetic rubber and resins; biodegradable. *Rhône-Poulenc Surf.; Rhône-Poulenc France.*

19903 Miranol® J2M-SF Conc
68815-55-4 272-383-4
Disodium capryloamphodipropionate
Salt free version of Miranol J2M Conc.; emulsifier, wetting agent, coupling agent, solubilizer; for dispersion on caustic soda and on powdered mixes, metal cleaning, industrial cleaning; higher tolerance for alkalies and/or electrolytes; biodegradable. *Rhône-Poulenc Surf.*

19904 Miranol® JAS-50
68877-55-4 272-383-4
Capryloamphopropionate
Emulsifier, wetting agent for industrial, institutional and household cleaners. *Rhône-Poulenc Surf.*

19905 Miranol® JB
68608-64-0 271-792-5
Disodium capryloamphodiacetate
Wetting agent used in caustic soda based cleaners used for food washing and peeling; also in wax stripper formulations. *Rhône-Poulenc Surf.; Rhône-Poulenc France.*

19906 Miranol® JBS
68815-55-4 272-383-4
Disodium capryloamphodipropionate
Low foaming surfactant for medicated shampoos and skin cleansers; emulsifier, wetting agent for industrial cleaners. *Rhône-Poulenc Surf.; Rhône-Poulenc France.*

19907 Miranol® JS Conc
68610-39-9 271-863-0
Sodium capryloamphohydroxypropylsulfonate
Emulsifier, wetting agent, corrosion inhibitor used in pickling acids, low foam, acid and alkali stable industrial cleaners; biodegradable. *Rhône-Poulenc Surf.; Rhône-Poulenc France.*

19908 Miranol® L2M-SF Conc
68991-88-8
Disodiumtallamphodipropionate
High foaming detergent, emulsifier for oil, wash and wax products, heavy-duty detergents, metalworking fluids. *Rhône-Poulenc Surf.*

19909 Miranol® OS-D
68610-38-8 271-862-5
Sodium oleoamphohydroxypropylsulfonate
Detergent, shampoos. *Rhône-Poulenc Surf.*

19910 Miranol® S2M-SF Conc
68815-45-2 272-374-5
Disodium caproamphodipropionate
Low wetting surfactant for medicated shampoos, cleansers; biodegradable. *Rhône-Poulenc Surf.; Rhône-Poulenc France.*

19911 Miranol® SM Conc
68608-61-7 271-789-9
Sodium caproamphoacetate
Wetting agent, foaming agent, detergent used in medicated and germicidal shampoos and hand soaps, rug and upholstery shampoos, in emulsion polymerization; biodegrdable. *Rhône-Poulenc Surf.; Rhône-Poulenc France.*

19912 Miranol® TBS
68991-88-8
Sodium tallamphodipropionate
Surfactant. *Rhône-Poulenc Surf.*

19913 Mirapol® 9, 95, 175
Polyquaternium-27
Conditioner, antistat, emollient for hair and skin products; improves thickening of Bentonite clays in aqueous systems. *Rhône-Poulenc Surf.*

19914 Mirapol® 550
26590-05-6
Polyquaternium-7
Polymer used in hair and skin care products to impart lubricity, slip, detangling, and luster to hair and a smooth, soft feel to skin *Rhône-Poulenc Surf.; Rhône-Poulenc France.*

19915 Mirapol® 1941
Acrylates/steareth-20 methylacrylate copolymer; pH sensitive thickening agent, emulsifier for liquid detergents, shampoos, and cosmetics; optimum thickening at slightly alkaline pH. *Rhône-Poulenc Surf.*

19916 Mirapol® A-15
68555-36-2
Polyquaternium-2
Softening, conditioning, lubricant, antistat, surface modifying agent used in cream rinses, conditioning-type shampoos, textile processing. *Rhône-Poulenc Surf.; Rhône-Poulenc France.*

19917 Mirapol® AD-1
90624-75-2
Polyquaternium-17
Conditioner and antistat for personal care products. *Rhône-Poulenc Surf.*

19918 Mirapol® AZ-1
90624-76-3
Polyquaternium-18
Conditioner and antistat for personal care products. *Rhône-Poulenc Surf.* Name unverified.

19919 Mirasheen® 202
Glycol stearate, lauramide DEA, cocamidopropyl betaine, glycerin; pearl concentrate for cold blend cosmetic formulations, liquid soaps; contains viscosity building, foam boosting, and mild conditioning agents. *Rhône-Poulenc Surf.; Rhône-Poulenc France.*

19920 Mirasol
A proprietary trade name for alkyd varnish and lacquer resins. No manufacturer.

19921 Mirataine® A2P-TS-30
Disodium N-tallow aminodipropionate
Foamer, wetting agent with tolerance to alkali; for use in alkaline cleaners. *Rhône-Poulenc Surf.*

19922 Mirataine® BB
86438-78-0
Lauramidopropyl betaine
Mild substantive surfactant, viscosity builder and foam booster, wetting agent for shampoos and dishwashing liquids; conditioner, antistat, emollient; as solubilizer, viscosity builder, foam booster with lauryl sulfates; stable to acid and alkali media. *Rhône-Poulenc Surf.; Rhône-Poulenc France.*

19923 Mirataine® BET-C-30
70851-07-9 274-923-4
Cocamidopropyl betaine
Mild foaming agent, conditioner, detergent, emulsifier, foam booster/stabilizer, viscosity builder for shampoos, liquid soaps, facial cleansers, bath gels, bubble baths. *Rhône-Poulenc Surf.*

19924 Mirataine® BET-O-30
25054-76-6 246-584-2
Oleamidopropyl betaine
Conditioner, detergent, emulsifier, foam booster/stabilizer, viscosity builder for shampoos, liquid soaps, facial cleansers, bath gels, bubble baths. *Rhône-Poulenc Surf.; Rhône-Poulenc France.*

19925 Mirataine® BET-P-30
693-33-4 211-748-4
Cetyl betaine
Foaming agent, conditioner, detergent, emulsifier, foam booster/stabilizer, viscosity builder for shampoos, liquid soaps, facial cleansers, bath gels, bubble baths. *Rhon-Poulenc Surf.* Name unverified.

19926 Mirataine® BSC
70851-08-0 274-925-5
Cocamidopropyl hydroxysultaine
Surfactant. *Rhône-Poulenc Surf.*

19927 Mirataine® CBS, CBS Mod
70851-08-0 274-925-5
Cocamidopropyl hydroxysultaine
Mild high foaming surfactant, viscosity builder, foam booster, emulsifier, wetting agent for shampoo formulations, liquid bubble baths, industrial cleaners. *Rhône-Poulenc Surf.; Rhône-Poulenc France.*

19928 Mirataine® H2C-HA
14960-06-6 239-032-7
Sodium lauriminodipropionate
High foaming surfactant, foam booster, wetting agent, dispersant for shampoos and skin cleansers, and hard surface cleaners. *Rhône-Poulenc Surf.; Rhône-Poulenc France.*

19929 Mirataine® HC-Acid
1462-54-0 215-968-1
Lauraminopropionic acid
Surfactant. *Rhône-Poulenc Surf.*

19930 Mirataine® JC-HA
Alkyl iminopropionate
Low foam hydrotrope for spray cleaners, rinse additives, mechanical scrubbing systems, highly built alkaline systems, hard surface cleaners, circulating systems. *Rhône-Poulenc Surf.*

19931 Mirataine® T2C-30
61791-56-8 263-190-6
Disodium tallowiminodipropionate
Detergent, solubilizer, moderate foaming surfactant used in textile, leather, metalworking, industrial and personal care products. *Rhône-Poulenc Surf.*

19932 Mirataine® TM
61791-25-1
Dihydroxyethyl tallow glycinate
Wetting agent, viscosifier for industrial and household cleaners; conditioner for shampoos; HCl thickener. *Rhône-Poulenc Surf.; Rhône-Poulenc France.*

19933 Miravon
Flow modifier. *Rhône-Poulenc UK.*

19934 Miravon B12DF
Ethylene/propylene oxide adducts
Biodegradable surfactant, defoamer for dishwash rinse aids, dairy cleaning, metal degreasing, hard surface cleaners, biocides. *Rhône-Poulenc France; Rhône-Poulenc UK.*

19935 mirbane oil
98-95-3 6685 202-716-0
$C_6H_5NO_2$
Nitrobenzene
oil of mirbane. Used in the manufacture of aniline, solvent for cellulose ether, modifying esterfication of cellulose acetate, ingredient of metal polishes; manufacture of benzoline, quinoline, and azobenzene.

19936 Mirion
Iodo-hexamine, a proprietary preparation;. No manufacturer.

19937 Mirrolac
Calendering varnishes. *The Scottish Adhesives Co Ltd.*

19938 mirror bronze
A copper and tin alloy, containing 28-35% tin. It sometimes contains a little nickel.

19939 Mirvale
101-21-3 2240 202-925-7
Chlorpropham
Potato sprout depressant. *Ciba-Geigy Agrochemicals.*

19940 Mischzinn
A tin alloy. Theoretically it is a eutectic containing 63% tin and 37% lead, but in practice the tin is at least 55% antimony and copper must be more than 3.5 and 0.5% respectively, and zinc is present in traces.

19941 miscible carbon disulfide
A mixture of carbon disulfide with castor oil, caustic potash, denatured alcohol, and water. An insecticide for destroying the Japanese beetle in the soil without serious damage to the plant.

19942 Misco
An alloy of 57.5% iron, 15% chromium, 25% nickel, 1.5% silicon, 0.5% manganese, and 0.5% carbon.

19943 Miscrome
A proprietary trade name for a corrosion-resisting alloy of iron with 28% chromium. No manufacturer.

19944 Missile
13457-18-6 8146 236-656-1
Pyrazophos
Systemic organophosphorus fungicide. *Hoechst UK.*

19945 Mission Prenatal
59-30-3 4253 200-419-0
Folic acid
Vitamin. *Mission Pharmacal Co.*

19946 mistletoe rubber
A rubber obtained from the fruit of certain *Loranthaceae* as parasites on the coffee tree.

19947 Mist-O-Matic
A liquid fungicide seed dressing. *DowElanco Ltd.*

19948 Mist-o-matic Ferrax
Fungicide seed treatment. *Murphy Chemical Co Ltd.* Discontinued.

19949 Mistral
67306-03-0 4035 266-639-4
Emulsifiable concentrate of 750 g fenpropimorph per liter; used for control of mildew in cereals. *Rhône-Poulenc Crop Protection Ltd.*

19950 Mistral CT
A suspension concentrate containing 250 g chlorothalonil and 187 g fenpropimorph per liter; a systemic fungicide for winter wheat. *Rhône-Poulenc Crop Protection Ltd.*

19951 Mistron CB
14807-96-6 9207 238-877-9
Talc; surface treated ultrafine platy talc; offers excellent impact properties to polyolefins, good film properties to films, improved tear and aging properties in rubber. *Luzenac Am.*

19952 Mistron Vapor-RE
14807-96-6 9207 238-877-9
Talc; ultrafine talc as reinforcing filler for elastomers; processing aid for elastomers, plastics; rheological control agent for resins, plastisols, adhesives; nucleating agent for plastic foams. *Luzenac Am.*

19953 Mistron ZSC
14807-96-6 9207 238-877-9
Talc, zinc stearate-coated; flatting agent in solv. -based coatings, rheological control in resins, plastisols and adhesives, nucleating agent in plastic foams, mold release agent, rubber dusting aid. *Luzenac Am.* Discontinued.

19954 Mitaban
33089-61-1 510 251-375-4
Amitraz; insecticide and acaricide. *Upjohn.* Name unverified.

19955 Mitac 20
33089-61-1 510 251-375-4
Amitraz; acaricide and insecticide for use on fruit trees and hops. *Schering Agrochemicals Ltd.* Discontinued.

19956 Mitas®
Compound seasonings; for food industry. *Asahi Chem. Industry.*

19957 Mitchalloy A
A proprietary trade name for an alloy of iron with 2.5% nickel and 0.9% chromium. No manufacturer.

19958 Mitec® GP105A
Isocyanate adduct; yellowing general-purpose hardener for chemically resistant, anticorrosion, concrete, wood finishes, PU finishes. *Mitsubishi Kasei.* Name unverified.

19959 Mitigan
115-32-2 3136 204-082-0
Active ingredient: dicofol; specific acaricide. *Agan Chemical Manufacturers Ltd.*

19960 mitigated caustic
A fused mixture of 1 part silver nitrate with 2 parts potassium nitrate.

19961 Mitin
Moth proofing agents. *Ciba plc.* Name unverified.

19962 Mitine
A base for ointments prepared from an emulsion which is superfatted with a non-emulsifying fat. Wool fat is used as the fat, and milk as the serum-like liquid to the extent of 50%.

19963 Mitrelle
Polyester yarn. *ICI Chem & Polymers Ltd.*

19964 Mitschlich's ammoniacal salt
A double compound of mercuroxy-ammonium nitrate, and mercuriammonium nitrate $(NH_2 \cdot Hg_2O)NO_3 \cdot (NH_2Hg)NO_3 \cdot H_2O$.

19965 Mitsubishi 4300J
9003-07-0 7741
Polypropylene resin
For injection molding applications. *Mitsubishi Kasei.* Name unverified.

19966 Mitsubishi BT002
9002-88-4 7728
HDPE; for blow molding of large bottles. *Mitsubishi Kasei.* Name unverified.

19967 Mitsubishi ET008
9002-88-4 7728
HDPE; for monofilament (rope, net, screen), inflation film (wrapper for foods). *Mitsubishi Kasei.* Name unverified.

19968 Mitsubishi F101A
9002-88-4 7728
LDPE; for film (heavy duty bags), blow molding (large articles). *Mitsubishi Kasei.* Name unverified.

19969 Mitsubishi JS050
9002-88-4 7728
HDPE; for injection molding of pails, bottles, caps. *Mitsubishi Kasei.* Name unverified.

19970 Mitsubishi Kasei GF-PET 6010G15
25038-59-9 7730
PET, 15% glass-reinforced; UL94HB grade. *Mitsubishi Kasei.* Name unverified.

19971 Mitsubishi Kasei PBT 5008
26062-94-2

PBT; high flow grade for injection molding, esp. thin-walled products. *Mitsubishi Kasei.* Name unverified.

19972 **Mitsubishi Kasei PBT 5010F1**
26062-94-2
PBT inorganic filled; reinforced grade for applications requiring warp and heat resistance. *Mitsubishi Kasei.* Name unverified.

19973 **Mitsubishi Kasei PPS 704G40**
9016-75-5
PPS, 40% glass reinforced; UL94V-O grade. *Mitsubishi Kasei.* Name unverified.

19974 **Mitsubishi L300**
9002-88-4 7728
LDPE; for lamination applications. *Mitsubishi Kasei.* Name unverified.

19975 **Mitsubishi UF421**
9002-88-4 7728
LLDPE; for high-clarity film applications. *Mitsubishi Kasei.* Name unverified.

19976 **Mitsubishi Yuka-ECX**
Electroconductive polymer. *Mitsubishi Petrochem.* Name unverified.

19977 **Mitsubishi Yuka-SPX**
Soft polyolefin. *Mitsubishi Petrochem.* Name unverified.

19978 **Mittel AEP**
80-40-0 3903 201-276-7
Ethyl-*p*-toluenesulfonate
A proprietary softening agent for cellulose esters. No manufacturer.

19979 **Mittel KP**
Cresyl-p-toluenesulfonate
A softening agent for cellulose esters.

19980 **Mittel L**
A solvent resembling turpentine.

19981 **Mix**
Stabilizing agent for liquid fertilizer-pesticide application. *Draxel Chemical Company.* Unverified.

19982 **Mixad**
Sand conditioner. *Foseco (F.S.) Ltd.*

19983 **mixed ether**
$C_2H_5 \cdot O \cdot CH_3$
An ether containing two different alkyl radicals, as in ethyl-methyl ether.

19984 **Mixed isopropanolamines myristate**
10525-14-1 234-077-9
Lanamine®. High foaming mild detergent for shampoos, shaving soaps. *Amerchol Corp.*

19985 **mixed metal**
A term used for alloys of cerium, lanthanum, and praseodymium.

19986 **mixed vitriol**
$CuSO_4 \cdot 3FeSO_4 \cdot 28H_2O$
Cupric-ferrous sulfate
Salzburg Vitriol. Salzburg vitriol also is used to describe cupric sulfate.

19987 **Mix-Kit**
Package device for two-part compounds. *Hardman.*

19988 **Mixol**
A timber insecticide. *ICI Chem & Polymers Ltd.* Discontinued.

19989 **Mixxim® BB/100**
103597-45-1
Bis[2-hydroxy-5-t-octyl-3-(benzotriazol-2-yl)phenyl
methane. UV light absorber for processing engineering resins (nylon, PC, PET, PBT, PPO) and PVC, styrene, and acrylics; highly effective where long term permanent uv light stability is required. *Fairmount.*

19990 **Mixxim® BB/50**
Blend; uv light absorber for processing engineering resins (nylon, PC, PET, PBT, PPO) and PVC, styrene, and acrylics; highly effective where long term permanent uv light stability is required. *Fairmount.*

19991 **Mixxim® HALS 57**
Tetrakis (2,2,6,6-tetramethyl-4-piperidyl)-1,2,3,4-butane tetracarboxylate
Hindered amine light and heat stabilizer for polyolefins such as PP, polyethylene, PS, ABS, PVC, PU, and engineering plastics. *Fairmount.* Name unverified.

19992 **Mixxim® HALS 63**
1,2,2,6,6-Pentamethyl-4-piperidyl/β,β,β',β'-tetramethyl-3,9-(2,4,8,10-tertaoxaspiro (5,5) undecane) diethyl
1,2,3,4-butane tetracarboxylate. High molecular weight hindered amine light and heat stabilizer for PP, polyethylene, PS, ABS, engineering plastics, and elastomers, especially in PP monofilament, tapes, molded and extruded products, polyethylene blown film. *Fairmount.* Name unverified.

19993 **Mixxim® HALS 68**
2,2,6,6-Tetramethyl-4-piperidyl/β,β,β',β' -tetramethyl-3,9-(2,4,8,10-tetraoxaspiro(5,5) undecane) diethyl -1,2,3,4-butane tetracarboxylate
High molecular weight hindered amine light and heat stabilizer for PP, polyethylene, PS, ABS, engineering plastics, and elastomers, especially in PP monofilament, tapes, molded and extruded products, polyethylene blown film. *Fairmount.* Name unverified.

19994 **MN powder**
Maxim-Nordenfelt powder. An American guncotton powder gelatinized with ethyl acetate.

19995 **moac**
12001-26-2
Very finely divided mica.

19996 **Mobil 1240**
9003-53-6 9028
Crystal PS
High heat grade polystyrene for injection molding and extrusion. *Mobil/Polystyrene.* Name unverified.

19997 **Mobil 2120**
9003-53-6 9028
Crystal PS
General purpose extrusion grade polystyrene. *Mobil/Polystyrene.* Name unverified.

19998 **Mobil 5350**
9003-53-6 9028
HIPS; super high impact extrusion grade. *Mobil/Polystyrene.* Name unverified.

19999 **Mobil 5600**
9003-53-6 9028
HIPS; environmental stress crack resistant extrusion grade. *Mobil/Polystyrene.* Name unverified.

20000 **Mobil 8020**
9003-53-6 9028
PS; ignition-resistant structural foam for injection molding. *Mobil/Polystyrene.* Name unverified.

20001 **Mobil MX 4354**
9003-53-6 9028
HIPS; injection molding grade for durable goods, appliance parts, and any large part requiring good flow and gloss characteristics. *Mobil/Polystyrene.*

20002 **Mobilrap**
Pallet wrap stretch films including one-side cling films, in thicknesses varying from 17 to 35 microns; for wrapping around pallets for increased product protection during transport and warehousing. *Mobil Plastics Europe.* Name unverified.

20003 **Mobilrapper**
Pallet wrap with stretch film system, which can be handled by one person, offering a variety of stretch films; for wrapping pallets for increased product protection during transport and warehousing. *Mobil Plastics Europe.* Name unverified.

20004 **Mobilsol®**
Modifying diluents for epoxy resins; they give flexibility. *Mobil. Name unverified.*

20005 **Moca®**
Polymer. *DuPont UK.*

20006 **Mocap 10G**
13194-48-4 3792 236-152-1
Granules containing ethoprophos; a nematicide and insecticide used to control cyst nematodes and wireworms in potatoes. *Rhône-Poulenc Crop Protection Ltd.*

20007 **Mocasco Iron**
A proprietary trade name for a nickel-chromium-molybdenum cast iron containing 1-1.35% nickel, 0.25-0.3% chromium, and 0.75% molybdenum. No manufacturer.

20008 **mocaya oil**
Macaja butter. Paraguay palm oil obtained from the kernels of *Acrocomia scelerocarpa.*

20009 **mocha-stone**
Agates of white or brown chalcedony from India, with markings due to oxides of iron and manganese.

20010 **mock lead**
Both tungsten ore found in Cornwall and zinc blend are known by this name.

20011 **mock silver**
An alloy of 84% aluminum, 10% tin, 5.5% copper, and 0.1% phosphorus.

20012 **mock vermilion**
7758-97-6 5422 231-846-0
Lead chromate, a yellow pigment.

20013 **Mod Acid**
104-15-4 9671 203-180-0
Modified toluene sulfonic acid production; hydrotrope, solvent, intermediate, catalyst. *Ruetgers-Nease.*

20014 Modaflow
Resin modifier; additive for improving flow, leveling and adhesion properties of surface coatings. *Monsanto Co.* Name unverified.

20015 Modane Soft
577-11-7 3460 209-406-4
Docusate sodium
Stool softener; pharmaceutic aid; surfactant. *Adria Laboratories Inc.*

20016 Modar
Thermosetting acrylic resins. *ICI Chem & Polymers Ltd.*

20017 Modar 814
Modified acrylic resin; features low viscosity, rapid cure, fast cycle times, optimal fire retardancy, low smoke, low combustion toxicity; used in RTM, hand lay-up, spray-up, and filament winding applications *ICI Acrylics.*

20018 Modar 826HT
Modified acrylic resin; features low profile, nonshrink, optimal fire retardancy/low smoke, low combustion toxicity, optimum surface finish; for pultrusion. *ICI Acrylics.*

20019 Modar 865
Modified acrylic resin; features low visc., rapid cure, fast cycle times, highest strength, chem. resist., fire retardant/low smoke; used primarily in RTM, pultrusion, and filament winding applications *ICI Acrylics.*

20020 Modarez APVC 8
Acrylic polymer; modifier for rigid PVC for manufacture of bottles and hollow casings; improves surface and gloss. *Protex.*

20021 Modarez MFP Powd
Acrylic polymer on silica support; leveling agent for powder paints, varnishes and coatings; improves spreading properties, increases wettability of surfaces. *Protex.*

20022 Moddite
An explosive. It is a variety of Cordite, but is made with a nitro-cellulose partially soluble in ether alcohol.

20023 Moderator
A soluble concentrate containing 300 g atrazine and 12.5 g imazapyr per liter; used for total weed control in non crop areas. *Chipman Ltd.*

20024 Modic
Adhesive polyolefin. *Mitsubishi Petrochem.* Name unverified.

20025 Modicol L
PEG fatty ester; chemical and mechanical stabilizer, coagulant, wetting, viscosity control, and dispersing agent, used in latex and resin emulsions. *Henkel/Functional Products.*

20026 Modicol S
Sulfated fatty acid; stabilizer; in rubber industry to stabilize natural and synthetic latexes during compounding, storage, and application; ensures mechanical and chemical stability; prevents premature coagulation during high-speed agitation, acidification, or pigmentation. *Henkel/Functional Prods.*

20027 Modified Butacite
A proprietary trade name for thermosetting polyvinyl butyral synthetic resin. No manufacturer.

20028 modified soda
A mixture of sodium carbonate and bicarbonate used as a cleaning agent in laundries.

20029 Modified Vinylite X
A proprietary trade name for thermosetting polyvinyl butyral synthetic resin. No manufacturer.

20030 Modinal T
A proprietary trade name for a wetting agent consisting of a long chain alcohol sulfate. No manufacturer.

20031 Modulan®
61788-48-5 262-979-2
Acetylated lanolin; conditioner, emollient, softener, lubricant for cosmetic and pharmaceutical products. *Amerchol Corp.*

20032 Modulex
1333-86-4 1856 231-153-3
A trademark for a carbon black for use as a pigment. No manufacturer.

20033 Moellon R
Synthetic anhydrous moellon from pure refined herring oil; leather additive, fiber lubricant, softener; non-yellowing for use on pastel colored leathers. *Atlas Refinery.*

20034 Mofix
Oxime/triazine herbicide. *Ciba plc.* Name unverified.

20035 Mo-Flo
Liquid acid drain cleaner. *Momar Industrial Services Ltd.*

20036 Mogul® L
1333-86-4 1856 231-153-3
Carbon black
For coloring resistive plastics. *Cabot; Carbot Carbon Ltd.*

20037 mohair
A material made from the hair of the Angora goat; used in fabrics for clothing, draperies and upholstery.

20038 Mohawk Steel
A proprietary trade name for a hot die steel containing about 14% tungsten, 3.5% chromium, 0.7% vanadium, and 0.45% carbon. No manufacturer.

20039 Mohr's salt
10045-89-3 552 233-151-8
$FeH_8N_2O_8S_2$
Ferrous ammonium sulfate
ammonium ferrous sulfate; Sulfuric Acid, Ammonium Iron (2+) Salt (2:2:1); Ammonium Iron Sulfate; Sulfuric acid, ammonium iron(2+) salt. Used in analytical chemistry, photography, as a polymerization catalyst and in metallurgy. d_4^{20} = 1.86; LD_{50} (rat orl) = 3.25 g/kg.

20040 Molaschar
A decolorizing carbon used for sugar juices.

20041 molascuit
A cattle food. It is the fine fiber of the sugar cane or begasse, with cane molasses absorbed by it.

20042 molasocarb
A decolorizing black adsorbent made from molasses.

20043 molasses
68476-78-8 270-698-1
The non-crystallizable residue from sugar. Cane molasses contain 55% sugar, 20% water, and 9% ash, whist beet molasses consists of 50% sugar, 10% salts, 20% water, 10% nitrogenous matter, and 10% non-nitrogenous matter; used as a cattle food; in food, raw materials for various alcohols, acetone, citric acid and yeast propagation.

20044 molassine meal
A mixture of molasses and peat moss; used as a cattle food.

20045 Molco
Spirit based mold and core coatings. *Foseco (F.S.) Ltd.*

20046 Mold Wiz Ext
A series of nonsilicone, nonwax, nonstearate polymeric based mold release agents, solvent or water based; used for reinforced plastics composites, injection molding, polyurethane foam and natural and synthetic rubber. *Axel Plastics Research Laboratories Inc.*

20047 Mold Wiz Int
A series of polymeric based additive lubricants/release agents used as processing aids; used for reinforced plastic composites, melamine/phenolic/urea laminates and overlays, thermoplastic injection molding, natural and synthetic rubbers and urethane elastomers. *Axel Plastics Research Laboratories Inc.*

20048 Moldabaste® Moldabaster S
Plaster of Paris adjusted to work with Moldano, normal and quick setting; dental speciality. *Bayer AG.*

20049 Moldag 200
1317-33-5 6318 215-263-9
molybdenum disulfide
Colloidal MoS_2 in 150 solvent refined paraffinic petrol. oil; lubricant additive for aerosols, machine oils, assembly and thread lubes. *Acheson Colloids.*

20050 Moldano®
Blue hard plaster of high Brinell hardness; dental specialty. *Bayer AG.*

20051 Moldaroc®
Yellow, extra hard plaster, for especially resistant models; dental specialty. *Bayer AG.*

20052 Moldasil
Condensation-cured silicone putty material for use in the dental laboratory. *Bayer AG.*

20053 Moldastone
Thixotropic class IV extra-hard dental stone for especially resistant models; used in dentistry. *Bayer AG.*

20054 Moldasynt
Synthetic class IV extra-hard dental stone for especially resistant models; used in dentistry. *Bayer AG.*

20055 Moldcote
Spirit based mold and core coatings. *Foseco (F.S.) Ltd.*

20056 Moldensite
A proprietary synthetic resin product used for electrical insulation. No manufacturer.

20057 Moldesite®
A range of phenolic molding compounds; used for tool machines, motor industry, pottery, electrotechnical industry, sanitary field and electrical field. *AMC SPREA S.p.A.*

20058 Moldex
Crosslinkable polyethylene resin; molding compounds. *Sumitomo Bakelite Co Ltd.* Name unverified.

20059 MoldPro 613
Mold release for epichlorohydrin, EPDM, fluoro elastomers, TPE. *Witco Corporation.*

20060 MoldPro 759
Internal mold release agent for styrene butadiene block copolymer. *Witco Corporation.*

20061 MoldPro 830
Internal mold release agent for styrene butadiene block copolymer. *Witco Corporation.*

20062 moler
A Danish diatomaceous earth containing 82.6% silica 5.33% aluminum oxide, a small proportion of ferric oxide, and organic matter. It is very light in weight, and is used in the manufacture of heat-insulating materials.

20063 Molera
A heat insulator obtained by mixing fine clay with cork dust, and firing.

20064 Mollescal® C Conc
Bactericidal assistant for soaking hides in leather and fur industry. *BASF AG; BASF plc.*

20065 Mollifex
Embedding aid in microscopy. *BDH Chemicals Ltd.*

20066 Mollin
A base for ointments. It is a soft soap containing 17 percent of uncombined fat.

20067 Mollit
A proprietary trade name for a polystyrene synthetic resin. No manufacturer.

20068 Mollit B
614-33-5 210-379-6
A proprietary trade name for glyceryl tribenzoate. No manufacturer.

20069 mollphorus
A glycerin substitute consisting of raw and invert sugar.

20070 Molochite®
A mixture of mullitic aluminum silicate and amorphous siliceous glass produced by the calcination of kaolin clay; refractory aggregate for producing kiln furniture, investment casting molds and refractory bricks and shapes. *ECC International Ltd.* Discontinued.

20071 Moloie
A proprietary trade name for a manganese-molybdenum steel. No manufacturer.

20072 Molsidolat
Antianginal *Cassella AG.*

20073 Moltopren®
Color pastes; used for coloring ester and ether based polyurethane foams. *Bayer AG.*

20074 Molybdate Red
12656-85-8 235-759-9
molybdenum orange, C.I. Pigment Red 104. A lead chromate pigment consisting of mixed crystals of lead chromate, lead sulfate, and a small proportion of lead molybdate. Its color varies from reddish-orange to scarlet.

20075 molybdenum
7439-98-7 6317 231-107-2
Mo
Metallic element; alloying agent in steels and cast iron; pigments for printing inks, paints, ceramics; catalyst; solid lubricants; missile and aircraft parts; reactor vessels; cermets; die-casting copper-base alloys; special batteries. mp = 2622°; bp = 4825°; d = 10.28. *AAA Molybdenum Prods.; Atomergic Chemetals; Cerac; Climax Molybdenum.*

20076 molybdenum dioxide
18868-43-4 242-637-9
MoO_2
molybdenum(IV) oxide
AAA Molybdenum Prods.; Atomergic Chemetals; Climax Molybdenum.

20077 molybdenum disulfide
1317-33-5 6318 215-263-9
MoS_2
molybdenum(IV) sulfide
molybdenite. Dry lubricant and hydrogenation catalyst. mp = 2375°; sublimes at 450°; d_{15}^{15} = 5.06; insoluble in H_2O. *AAA Molybdenum Prods.; Climax Molybdenum; Dow Corning/E/M Corp.*

20078 molybdenum nickel
An alloy of 75% molybdenum and 25% nickel; used in the manufacture of saws.

20079 Molybdenum Permalloy
A proprietary trade name for an alloy of 81% nickel, 17% iron, and 2% molybdenum; has a higher permeability than Standard Permalloy *(qv)*. No manufacturer.

20080 molybdenum steel
A variable alloy. It usually contains from 0.06-1.73% carbon and 0.23-15% molybdenum.

20081 molybdenum trioxide
1313-27-5 6321 215-204-7
MoO_3
molybdenum(VI) oxide
molybdic oxide; molybdic acid hydride; Molybdenum (VI) oxide; Molybdic anhydride; Molybdic trioxide; Molybdenum oxide; natural molybdite; molybdenum anhydride; molybdenum peroxide; Mo 1202T; molybdic acid anhydride; molybdena. Source of Mo; reagent for analytical chemistry; agriculture; manufacture of metallic Mo; corrosive inhibitor; ceramic glazes; enamels; pigments; catalyst. mp = 795°; bp = 1155°; d_4^{26}= 4.696; soluble in H_2O (0.49 g/l), insoluble in organic solvents; LD_{50} (rat orl) = 2689 mg/kg. *AAA Molybdenum Prods.; Atomergic Chemetals; Cerac; Climax Molybdenum.*

20082 molybdosodalite
A variety of sodalite containing nearly 3% MoO_3. It is green in color.

20083 Molydag
1317-33-5 6318 215-263-9
Dispersions of molybdenum disulfide in resin/solvent systems; used for dry film lubrication, or additives for motor oils and special machine greases. *Acheson Colloids.*

20084 Molydag 204
1317-33-5 6318 215-263-9
MoS_2 in 500 solvent refined paraffinic petroleum oil; lubricant additive for assembly and thread lubes. *Acheson Colloids.*

20085 Molydag 206
1317-33-5 6318 215-263-9
Colloidal MoS_2 in water; lubricant additive in dry gear and dry chain lubes, aerosols, penetrating lubes, machine oils, assembly lubes, thread lubes. *Acheson Colloids.*

20086 Molydag 208
1317-33-5 6318 215-263-9
MoS_2 in 2500 solvent refined bright stock petroleum oil; lubricant additive for greases. *Acheson Colloids.*

20087 Molydag 210
1317-33-5 6318 215-263-9
Colloidal MoS_2 in anhydrous isopropanol; lubricant additive for dry gear and chain lubes, aerosols, assembly and thread lubes. *Acheson Colloids.*

20088 Molydag 211
1317-33-5 6318 215-263-9
Colloidal MoS_2 in trichloroethane; lubricant additive for dry gear and chain lubes, aerosols, assembly and thread lubes. *Acheson Colloids.*

20089 Molydag 214
1317-33-5 6318 215-263-9
Colloids MoS_2 in mineral spirits; lubricant additive for dry chain lubes, aerosols, penetrating lubes, assembly lubes. *Acheson Colloids.*

20090 Molykote®
1317-33-5 6318 215-263-9
Silicone, MoS_2 lubricant base; environmental lubricants. *Dow Corning.*

20091 Molyte
A trade name for a patented mixture of calcium and molybdenum oxides with a flux. No manufacturer.

20092 Molyvan
A series of proprietary molybdenum compounds; friction reducers used as antiwear and extreme pressure agents in lubricants; can be used as antioxidants. *R. T. Vanderbilt Co Inc.*

20093 momea
Mimea. A hemp preparation made in Tibet.

20094 Mona NF-10
Low-foaming detergent, wetting agent, solubilizer for spray, soak tank, in-place pipeline cleaners, floor scrubbing formulations. *Mona Industries.*

20095 monacetin
26446-35-5 6325 247-704-6
$C_5H_{10}O_4$
1,2,3-propanetriol monoacetate
Glyceryl monoacetate; monoacetin; glyceryl monoacetate; acetin; acetic acid, monoglyceraldehyde; glycerol α-monoacetate; glycerol, 1-acetate; 1-monoacetin; monacetin; alpha-monoacetin. Mixture of the 1- and 2-monoacetates. Used for tanning; solvent for dyes; food additive; gelatinizing agent in explosives. bp_{17} = 158°; d_4^{20}= 1.206; soluble in H_2O, EtOH, insoluble in othr organic solvents; LD_{50} (rat sc)= 6.6 g/kg.

20096 Monachit
An explosive containing 12% trinitro-xylene, 1% charcoal, and 1% collodion cotton.

20097 Monacrin
134-50-9 427 205-145-5
$C_{13}H_{11}ClN_2$

Aminacrine hydrochloride
9-aminoacridine hydrochloride; acramine yellow; aminacrine hydrochloride; aminoacridine hydrochloride; 5-aminoacridine hydrochloride; 9-aminoacridine monohydrochloride; monacrin hydrochloride; Acridinamine, monohydrochloride; Aminoacridine monohydrochloride; Aminoacridinium chloride; NSC-7571. Anti-infective, topical. Highly fluorescent. mp = 241°; insoluble in H_2O, soluble in organic solvents; LD_{50} (mus orl) = 78 mg/kg. *Sterling Drug Inc.* Name unverified.

20098 Monafax
Range of surfactants, each of which is a mixture of mono-and di-phosphate esters derived from ethylene oxide based surfactants; mainly viscous liquids which are the acid form of the compound; can be converted to salts of alkali, metal, amine or ammonia by mixing with the desired base; used fro emulsion polymerization; industrial cleaners, e.g. soak tank cleaners, all purpose and steam cleaners, herbicide and insecticide emulsifiers, metalworking lubricants and dry lceaning detergents. *D F Anstead Ltd.* Name unverified.

20099 Monafax 785
Nonoxynol-9 phosphate
Emulsifier, lubricant, antistat, detergent, corrosion inhibitor for emulsion polymerization, agricultural, metalworking lubricants, alkaline cleaners, industrial use; antisoil redeposition for dry cleaning. *Mona Industries.*

20100 Monafax 1214
Deceth-4 phosphate
Hypochlorite-stable surfactant, surface tensile reducer, wetting agent for mildew removers, tile cleaners, bowl cleaner, tire cleaners, bleaching of paper pulp and textiles, dairy cleaners, hard surface cleaners. *Mona Industries.*

20101 Monalube 29-78
Alkanolamide
Corrosion inhibitor, lubricant for metalworking lubricants, fiber lubricants, textile specialties, wire and deep metal drawing. *Mona Industries.*

20102 Monalube 780
Modified alkanolamide
Lubricant for copper drawing; corrosion inhibitor. *Mona Industries.*

20103 Monalux CAO
68155-09-9 268-938-5
Cocamidopropylamine oxide
Detergent, surfactant. *Mona Industries.*

20104 Monamate C-1142
68515-65-1 271-102-2
Disodium cocamido MIPA sulfosuccinate
Foaming/cleaning surfactants for personal care and household products. *Mona Industries.*

20105 Monamate CPA
Sulfosuccinate half ester of an alkanolamide in liquid form; high foaming agent which produces a dense rich lather and imparts a soft and silky feel to the skin and hair and has low irritancy; used in mild shampoos, bubble baths, skin cleaners; rug shampoos, detergent formulations. *D F Anstead Ltd.* Name unverified.

20106 Monamate LA-100
Disodium lauryl sulfosuccinate
High foaming, low irritation surfactant for personal care and household products. *Mona Industries.*

20107 Monamate LNT-40
Ammonium lauryl sulfosuccinate
High foaming, low irritation surfactant used in personal care products; biodegradable. *Mona Industries.*

20108 Monamate OPA
Sulfosuccinate half ester of an alkylolamide as a light yellow liquid; produces flash foam and a rich dense lather with good rinsing and viscosity control; used in shampoos, e.g., baby and family types, gel face cleaners. *D F Anstead Ltd.* Name unverified.

20109 Monamate OPA-100
56388-43-3 260-143-1
Disodium oleamide PEG-2 sulfosuccinate
High foaming, nonirritating surfactant for shampoos, bubble baths, soap bars. *Mona Industries.*

20110 Monamate RMEA-40
Disodium ricinoleamido MEA-sulfosuccinate
Surfactant. *Mona Industries.*

20111 Monamid® 7-100
Cocamide DEA (1:1)
Foam booster/stabilizer for industrial and household detergents; biodegradable. *Mona Industries.*

20112 Monamid® 15-70W
1:1 Linoleamide DEA
Thickener, viscosity builder, hair and fiber conditioner; biodegradable. *Mona Industries.*

20113 Monamid® 150-ADY
1:1 Linoleamide DEA
Thickener for aqueous systems; water-oil emulsifier, corrosion inhibitor for soluble oils; biodegradable. *Mona Industries.*

20114 Monamid® 150-CW
136-26-5 205-234-9
1:1 Capramide DEA
Flash foamer, foam stabilizer for cosmetic preparations; also as emulsifier, solubilizer, viscosity control agent; biodegradable. *Mona Industries.*

20115 Monamid® 150-LMWC
120-40-1 204-393-1
Lauramide DEA
Foam booster/stabilizer for industrial and household detergents; biodegradable. *Mona Industries.*

20116 Monamid® 150-MW
7545-23-5 231-426-7
1:1 Myristamide DEA
Nonirritating foam stabilizer, emulsifier for aqueous or nonaqueous cosmetics and toiletries; thickener for systems containing sodium ions; biodegradable. *Mona Industries.*

20117 Monamid® 718
93-82-3 202-280-1
Stearamide DEA
Emulsifier, thickener, opacifier for creams and lotions; biodegradable. *Mona Industries.*

20118 Monamid® CMA
68140-00-1 268-770-2
1:1 Cocamide MEA
Foamer, thickener for liquid and powder detergents; biodegradable. *Mona Industries.*

20119 Monamid® LIPA
142-54-1 205-541-8
1:1 Lauramide MIPA
Foamer, thickener for liquid and powder detergents; biodegradable. *Mona Industries.*

20120 Monamid® LMA
142-78-9 205-560-1
1:1 Lauramide MEA
Foamer, thickener for liquid and powder detergents; biodegradable. *Mona Industries.*

20121 Monamid® R31-42
Lauramide DEA and propylene glycol; foam stabilizer for cosmetic preparations; biodegradable. *Mona Industries.*

20122 Monamid® S
111-57-9 203-883-2
1:1 Stearamide MEA
Emulsifier, thickener for kerosene, mineral oils; biodegradable. *Mona Industries.*

20123 Monamide
68140-00-1 268-770-2
Cocamide MEA
Foam booster, thickener, superfatting agent. *Zohar Detergent Factory.*

20124 Monamine
Foam booster and stabilizers, emulsifiers, detergents, wetting agents, corrosion inhibitors, viscosity builders, lubricants, dispersants; used for cosmetics; shampoos, dry cleaning detergents; metal cleaners, toiletries; rust inhibitors, metal cutting fluids, emulsifiable waxes, agricultural sprays, fuel oil additives, leather and fur preparations. *D F Anstead Ltd.* Name unverified.

20125 Monamine 779
Cocamide DEA and DEA-laureth sulfate; foaming agent, viscosity builder, detergent, soil suspending agent, solubilizer, wetting and penetrating agent used in shampoos, bubble baths, household, industrial cleaners, germicides, uv absorbers; biodegradable. *Mona Industries.*

20126 Monamine AA-100
1:2 Distilled cocamide DEA and diethanolamine; detergent, emulsifier, thickener; biodegradable. *Mona Industries.*

20127 Monamine ACO-100
Lauramide DEA and diethanolamine; detergent, emulsifier, thickener; biodegradable. *Mona Industries.*

20128 Monamine ADY-100
Linoleamide DEA and diethanolmine; detergent, emulsifier, thickener; biodegradable. *Mona Industries.*

20129 Monamine LM-100
Luramide DEA and diethanolamine; detergent, wetting agent, emulsifier, thickener, corrosion inhibitor; biodegradable. *Mona Industries.*

20130 Monamine T-100
68155-20-4 268-949-5

Tallamide DEA and diethanolamine; detergent, wetting agent, foam booster, thickener for household detergents; biodegradable. *Mona Industries.*

20131 Monamulse 653-C
Alkanolamide, modified; emulsifier, solubilizer, degreaser for solvs. such as mineral spirits, kerosene, pine oil; used for cleaners for engine blocks, garage floor, truck bodies, silk screens, mechanics hand cleaners. *Mona Industries.*

20132 Monamulse CI
Imidazoline, modified; corrosion inhibitor improving water resistance in greases, oil-based lubricant systems, and on cast iron; improves emulsion stability; penetrant. *Mona Industries.*

20133 Monaprin
Pigments. *Imperial Chemical Industries plc.*

20134 Monaquat ISIES
67633-57-2 266-778-0
Isostearyl ethylimidonium ethosulfate
Antistat, lubricant, softener, corrosion inhibitor used in cosmetic industry, in industrial and textile applications; biodegradable. *Mona Industries.*

20135 Monaquat TG
Bishydroxyethyl dihydroxypropyl stearaminium chloride
Surfactant used in personal care products; conditioner for hair rinses; antistat and foaming used in fabric laundering and softening products; thickener for acid bowl cleaners, naval gels. *Mona Industries.*

20136 Monarch® 1100
1333-86-4 1856 231-153-3
Carbon black; for coloring plastics. *Cabot; Cabot Carbon Ltd.*

20137 Monastat 1195
Formulated concentrate; antistat/cleaner for glass and plastic, e.g., TV screens, computers, medical diagnostic equipment, safety goggles. *Mona Industries.*

20138 Monastral
Insoluble phthalocyamine pigments. *ICI Chem & Polymers Ltd.*

20139 Monaterge
Excellent detergency and wetting; low to moderate foaming; alkaline stability; wide range of alkaline cleaners, e.g., floor, wall, drain; liquid laundry detergents. *D F Anstead Ltd.* Name unverified.

20140 Monaterge 85 HF
Cocamide DEA, DEA-acrylinoleate and DEA-dodecylbenzene sulfonate. High foaming detergent and wetting agent, hydrotrope; stable in high electrolyte systems. *Mona Industries.*

20141 Monateric 805
Cocoamphodiacetate
disodium cocamido MIPA-sulfosuccinate. High foaming, mild surfactant for hair conditioners, skin care products. *Mona Industries.*

20142 Monateric 810-A-50
Caprylic/capric carboxylic propionate
Imidazoline-derived, salt-free; surfactant for industrial detergents, cleaners, cosmetics; hydrotrope, coupling agent, and/or solubilizer; corrosion inhibitor in metalworking systems, oil well flooding, and aerosol packagings. *Mona Industries.*

20143 Monateric 811
68815-55-4 272-383-4
Disodium capryloamphodipropionate
Corrosion inhibitor, detergent, wetting agent in noncorrosive cleaners and industrial detergents; biodegradable. *Mona Industries.*

20144 Monateric 811
Amphoteric surfactant based on alkyl imidazolines, in amber liquid form; has corrosion inhibiting properties with detergent and surface active properties; used in a broad range of noncorrosive cleaners and industrial detergents. *D F Anstead Ltd.* Name unverified.

20145 Monateric 951A
Disodium lauroamphodiacetate
Flash foamer for shampoos, hand and body cleansers; biodegradable. *Mona Industries.*

20146 Monateric 985A
Sodium lauroamphoacetate
sodium trideceth sulfate. High foaming, mild shampoo base for adult and baby products; biodegradable. *Mona Industries.*

20147 Monateric 1000
68815-55-4 272-383-4
Disodium capryloamphodipropionate
Corrosion inhibitor, detergent, wetting agent for metal cleaning, cutting fluids, synthetic lubricants; biodegradable. *Mona Industries.*

20148 Monateric 1000
Amphoteric surfactant based on alkyl imidazolines; light amber liquid; detergent and wetting agent with corrosion inhibiting properties, suggested primary surfactant for metal cleaning, soak tank cleaners, etc. *D F Anstead Ltd.* Name unverified.

20149 Monateric 1188M
3655-00-3 222-899-0
Disodium lauryl beta-iminodipropionate
High foaming surfactant, hydrotrope for household and industrial hard surface cleaners, shampoos, bubble bath, mild skin cleansers, down hole foamers, air drilling; textile wetter; biodegradable; stable to acid and alkali. *Mona Industries.*

20150 Monateric 1202
Dihydroxyethyl tallow glycinate
Detergent, substantive conditioner for hair care products; coupling agent, viscosity control agent. *Mona Industries.*

20151 Monateric 1203
Sodium hydrogenated tallow dimethyl glycinate
Surfactant, substantive conditioner for hair care products. *Mona Industries.*

20152 Monateric ADA
Cocamidopropyl betaine
High foaming surfactant used in air drilling, foam drilling, foam blanketing, air entraining agent for cement, gypsum, wallboard; for use in presence of brine and oil; biodegradable. *Mona Industries.*

20153 Monateric CA-35
68919-41-5
Sodium cocoamphopropionate
Detergent, wetting agent, emulsifier, dispersant, foaming agent used in cosmetic, household, and industrial products; coupling agent, solubilizer; biodegradable; stable to acid, alkali, electrolytes. *Mona Industries.*

20154 Monateric CA-35%
Amphoteric surfactant based on coconut imidazoline, in amber liquid form; detergent, wetting, emulsifying and dispersing agent with good foam and lather; used in cosmetics, e.g., shampoos, bubble baths, skin cleaners, hair dye formulations, and in high foam floor and metal cleaners. *D F Anstead Ltd.* Name unverified.

20155 Monateric CAB
A cocamido betaine as a clear yellow liquid containing 4.8% sodium chloride; used in bubble bath and shampoo formulations. *D F Anstead Ltd.* Name unverified.

20156 Monateric CAM-40
68919-41-5
Sodium cocoamphopropionate
Surfactant. *Mona Industries.*

20157 Monateric CDL
Cocoamphodiacetate
sodium laureth sulfate. Mild, high foaming shampoo base. *Mona Industries.*

20158 Monateric CDTD
Disodium cocoamphodiacetate
sodium trideceth sulfate. Surfactant for industrial detergents, cleaners, cosmetics; hydrotrope, coupling agent, and/or solubilizer; corrosion inhibitor in metalworking systems, oil well flooding, and aerosol packagings. *Mona Industries.*

20159 Monateric CDX38
An imidazoline derived dicarboxylic acid amphoteric in the form of a light amber viscous liquid; good flash foam and lathering properties; used in baby shampoos, daily use shampoos and skin cleansers. *D F Anstead Ltd.* Name unverified.

20160 Monateric CDX-38
Disodium cocoamphodiacetate
Detergent, foaming agent, mild base surfactant for shampoos and skin cleansers; biodegradable. *Mona Industries.*

20161 Monateric CEM-38
Disodium cocoamphodipropionate
Detergent, wetting agent, emulsifier, dispersant, solubilizer, hydrotrope used in liquid detergent systems, heavy duty detergents, acid bowl cleaners, cosmetics; biodegradable. *Mona Industries.*

20162 Monateric CEM-38%
Sodium salt of a dicarboxyethyl fatty acid derived from imidazoline, in liquid form, clear to hazy amber in color; detergent, wetting, emulsifying and solubilizing agent used in liquid medium duty all-purpose detergents, toilet bowl cleaners and aluminum brighteners. *D F Anstead Ltd.* Name unverified.

20163 Monateric CM-36S
Sodium cocoamphoacetate
Foaming agent, emulsifier, high foaming detergent, wetting agent, solubilizer, conditioner, coupling agent, fulling agent used in cosmetic and textile industries; biodegradable. *Mona Industries.*

20164 Monateric CSH 32
Sodium salt of a dicarboxymethyl fatty acid derived from imidazoline, supplied as a clear yellow liquid; mild detergent properties used in shampoo formulations, particularly baby and daily use types. *D F Anstead Ltd.* Name unverified.

20165 Monateric CSH-32
Disodium cocoamphodiacetate
High foaming, nonirritating surfactant for baby shampoos; biodegradable. *Mona Industries.*

20166 Monateric CyNa-50
Sodium capryloamphopropionate
Detergent, emulsifier, coupling agent, solubilizer, wetting agent, low to moderate foaming surfactant used in conc. electrolyte systems, bottle washing, wax strippers, steam cleaners, industrial cleaning, textile processing, acid pickling baths; biodegradable; stable to acid, alkalis and electrolytes. *Mona Industries.*

20167 Monateric CyNa-50%
An amber liquid based on a capryl imidazoline; used in bottle washing, steam cleaners, wax strippers, all-purpose cleaners, degreasers, Kier boiling, mercerization and acid-pickling. *D F Anstead Ltd.* Name unverified.

20168 Monateric ISA-35
68630-96-6 271-929-9
Sodium isostearoamphopropionate
Surfactant used in cosmetic and industrial products; conditioner, lubricant, thickener; biodegradable. *Mona Industries.*

20169 Monateric ISA-35%
Amphoteric surfactant based on isostearic imidazoline; supplied as an amber flowable gel; used in shampoos such as protein, low pH and daily use types. *D F Anstead Ltd.* Name unverified.

20170 Monateric L30
Sodium lauroamphoacetate
Surfactant. *Mona Industries.*

20171 Monateric LF
An amber liquid, low foaming detergent with high acid and alkaline stability; low or high temperature alkaline and acid cleaners, automatic car wash detergents, steam cleaners and truck body and aircraft cleaners. *D F Anstead Ltd.* Name unverified.

20172 Monateric LMAB
Lauramidopropyl betaine
High foaming shampoo base. *Mona Industries.*

20173 Monateric LMM-30
Sodium lauroamphoacetate
High foaming detergent for nonirritating shampoos, skin cleansers, other cosmetics; biodegradable. *Mona Industries.*

20174 Monateric TA-35
Sodium tallamphopropionate
Surfactant for industrial detergents, cleaners, cosmetics; hydrotrope, coupling agent, and/or solubilizer; corrosion inhibitor in metalworking systems, oil well flooding, and aerosol pkgs. *Mona Industries.*

20175 Monateric TDB-35
61791-56-8 263-190-6
Disodium tallow betaiminodipropionate
Hydrotrope and detergent for high electrolyte systems such as heavy-duty liquid cleaners and wax strippers. *Mona Industries.*

20176 Monatrope 1250
Sodium alkanoate
Surfactant hydrotrope for formulating alkaline built liq. detergent concs.; coupling agent for nonionic and other surfactants in high concs. of electrolytes; for household and industrial detergents, spray washes, textiles, hypochlorite detergents/sanitize *Mona Industries.*

20177 Monawet 1240
Disodium nonoxynol-10 sulfosuccinate
Wetting agent for emulsion polymerization, adhesives, paints. *Mona Industries.*

20178 Monawet MB-45
127-39-9 3238 204-839-5
Diisobutyl sodium sulfosuccinate
Wetting, dispersing, emulsifying, penetrating and solubilizing agent used in emulsion polymerization of S/B for rug backing, paper coating, water treatment. *Mona Industries.*

20179 Monawet MB-45
127-39-9 3238 204-839-5
Sodium diisobutyl sulfosuccinate as a clear colorless liquid; anionic surfactant for styrene/butadiene emulsion systems for rug backing. *D F Anstead Ltd.* Name unverified.

20180 Monawet MM-80
Dihexyl sodium sulfosuccinate
15% water, 5% IPA; wetting agent, detergent for emulsion and suspension polymerization, rug backing, paper coating, textiles, paint, agric., cosmetic, detergent, mining, water treatment, electroplating baths, and food industries; electrolyte tolerant. *Mona Industries.*

20181 Monawet MM-80
Sodium dihexyl sulfosuccinate as a clear colorless liquid; anionic surfactant for pesticidal sprays, shampoos, detergents, latex paints, coatings and textile fibers. *D F Anstead Ltd.* Name unverified.

20182 Monawet MO
577-11-7 3460 209-406-4
Series of anionic surfactants composed of sodium dioctyl sulfosuccinate in liquid form; used for textiles, e.g., cotton cloth desizing, printing and dyeing processes, wool carbonizing; agriculture, e.g., liquid fertilizers, insecticides and fungicides; cosmetics, e.g. creams and lotions, bath oils and shampoos. *D F Anstead Ltd.* Name unverified.

20183 Monawet MO-65-150
577-11-7 3460 209-406-4
Dioctyl sodium sulfosuccinate
Anhydrous; wetting, penetrating and spreading agent, emulsifier used in oil well cleaning, drycleaning detergents, solvent cleaners and strippers, lubricants, agricultural, paints, mining, water treatment, cosmetic applications. *Mona Industries.*

20184 Monawet MO-70
577-11-7 3460 209-406-4
Dioctyl sodium sulfosuccinate
20% water, 10% diethylene glycol butyl ether; wetting, dispersing, emulsifying, penetrating and solubilizing agent used in emulsion and suspension polymerization, adhesives, paints, textile, fertilizer, mining, water treatment, fire fighting, cosmetic, food industries. *Mona Industries.*

20185 Monawet MO-84R2W
577-11-7 3460 209-406-4
Dioctyl Sodium sulfosuccinate
16% solution in propylene glycol [57-55-6]; wetting agent for general use, agricultural, paints, mining, water treatment, cosmetics applications. *Mona Industries.*

20186 Monawet MT Series
2673-22-5 220-219-7
Anionic surfactants containing ditridecyl sodium sulfosuccinate in liquid form; used for vinyl chloride suspensions and styrene emulsions for coatings, paints. MT-80H2W is used in nonaqueous pigment dispersions for printing inks. *D F Anstead Ltd.* Name unverified.

20187 Monawet MT-70
Ditridecyl sodium sulfosuccinate
12% water, 18% hexylene glycol; wetting, dispersing, emulsifying, penetrating and solubilizing agent used in emulsion and suspension polymerization, paints, coatings, indirect food additives. *Mona Industries.*

20188 Monawet SNO-35
Tetrasodium dicarboxyethyl stearyl sulfosuccinamate
Wetting agent, solubilizer, emulsifier, dispersant, viscosity depressant, mild detergent used in polymerization, paints, coatings, textile, cosmetic, agricultural products; biodegradable. *Mona Industries.*

20189 Monawet SNO-35
Tetrasodium N-(1,2 dicarboxyethyl) N-alkyl (C18) sulfosuccinamate as a clear light amber liquid; mild detergent with good wetting and calcium tolerance; emulsifier; used in industrial detergents; cosmetic and textile products, specialty emulsions or dispersions for polymerization systems. *D F Anstead Ltd.* Name unverified.

20190 Monawet TD-30
Disodium deceth-6 sulfosuccinate
Surfactant, emulsifier, foaming agent used in emulsion polymerization, cosmetic and textile industries. *Mona Industries.* Name unverified.

20191 Monawet TD-30
Half ester of sulfosuccinic acid based on an ethoxylated fatty alcohol. Light yellow liquid; wetting, foaming and emulsifying agent with low inorganic electrolyte content; used in emulsion polymerization, textile wet processing and a broad range of cosmetic and fine fabric formulations. *D F Anstead Ltd.* Name unverified.

20192 Monazoline C
61791-38-6 263-170-7
Cocoyl hydroxyethyl imidazoline
Wetting agent, emulsifier for nonpolar liquids, detergent, thickener, corrosion inhibitor, antistat, softener, bactericide used in paint and textile industries; also dispersant for clay and pigments, in acid dairy cleaners, in oil well acidifying and secondary recovery operaiotns; biodegradable. *Mona Industries.*

20193 Monazoline C, CY, O and T
1-Hydroxyethyl-2 alkylimidazoline where the alkyl portion is caprylic, coconut, oleic and tall oil respectively; amber liquids; cationic surfactants with wetting, emulsifying, detergency, thickening, corrosion inhibiting, antistatic, softening and bactericidal properties; used in agricultural sprays, gravel-asphalt bonding, dairy cleaners, rinse aids, paints, sealants, inks, oil well recovery, sludge dispersion, corrosion control, plastics, metal treatment, textiles, softening, ore flotation. *D F Anstead Ltd.* Name unverified.

20194 Monazoline CY
37478-68-5 253-521-2
Capryl hydroxyethyl imidazoline
Wetting agent, emulsifier for nonpolar liquids, detergent, thickener, corrosion inhibitor, antistat, softener, bactericide used in paint and textile industries; biodegradable. *Mona Industries.*

20195 Monazoline IS
68966-38-1 273-429-6
Isostearyl hydroxyethyl imidazoline
Corrosion inhibitor and lubricant. *Mona Industries.*

20196 Monazoline O
Oleyl hydroxyethyl imidazoline
Wetting agent, emulsifier for nonpolar liq., detergent, thickener, corrosion inhibitor, antistat, softener, bactericide used in paint and textile industries; dispersant for clays and pigments, in agric. preparations, acid dairy cleaners; biodegradable. *Mona Industries.*

20197 Monazoline T
61791-39-7 263-171-2
Tall oil hydroxyethyl imidazoline
Wetting agent, emulsifier for nonpolar liquids, detergent, thickener, corrosion inhibitor, antistat, softener, bactericide used in paint and textile industries; also dispersant for clays and pigments, aids gravel-to-asphalt bonding, rinse aid for automatic car washes, printing ink additive, protective metal coatings, biodegradable. *Mona Industries.*

20198 Monceren®
66063-05-6 266-096-3
Dustable powder containing 12.5% w/w pencycuron; a phenylurea fungicide to control black scurf in potatoes. *Bayer plc.*

20199 Moncler Derma
Facial cream, gel, lotion and stick, to improve complexion. *Richardson-Vicks Inc.* Name unverified.

20200 Mond 70 alloy
An alloy of 70% nickel, 26% copper, and 4% manganese.

20201 Mond gas
A combustible gas produced by passing air and steam over heated coal or peat. It consists of a mixture of carbon monoxide, hydrogen, and nitrogen.

20202 Mondur
Toluene diisocyanate and polymethyl diisocyanate. Monomers. *Miles. Discontinued.*

20203 Monece®
Fibers. *DuPont UK.*

20204 Monel Alloy 400
An alloy of 30% copper, 1% manganese, 2.5% (max) iron and the balance nickel; a general engineering alloy with good resistance to corrosion by sea water, sulfuric, hydrochloric and phosphoric acids. *Wiggin Alloys Ltd.* Unverified.

20205 Monel Alloy 414
Monel 400 with a high carbon contain; improved machining properties. *Wiggin Alloys Ltd.* Unverified.

20206 Monel Alloy K-500
An alloy of 30% copper, 3% aluminum, 0.5% titanium and the balance nickel. *Wiggin Alloys Ltd.* Unverified.

20207 Monel metal
An alloy. The cast metal usually contains from 68-70% nickel, 28% copper, 2% iron, 1% silicon, and 0.25% manganese; and the forged alloy consists of 68% nickel, 28% copper, 2% iron, 1.5 *Huntington Alloys, Inco Alloys Int.* Unverified.

20208 monensin
17090-79-8 6329 241-154-0
$C_{36}H_{62}O_{11}$
2-[5-ethyltetrahydro-5-[tetrahydro-3-methyl-5-[tetrahydro-6-hydroxy-6-(hydroxymethyl)-3,5-dimethyl-2H-pyran-2-yl]-2-furyl]-2-furyl]-9-hydroxy-β-methoxy-α,σ,2,8-tetramethyl-1,6-dioxaspiro[4.5]decane-7-butyric acid
monensic acid; Coban; A-3823A. Antiprotozoal; antibacterial, antifungal. Coccidiostat in chickens. Feed additive. mp = 103-105°; $[\alpha]_D$ = 48°; slightly soluble in H_2O, more soluble in organic solvents; LD_{50} (mus orl) = 46 mg/kg.

20209 monetite
7757-93-9 1739 231-826-1
$CaHPO_4$
A calcium phosphate, found in guano.

20210 Monex
97-74-5 202-605-7
$C_6H_{12}N_2S_3$
Tetramethylthiuram monosulfide
Tetramethylthiuram monosulfide; UNADS; TMTM; Thiodicarbonic diamide, tetramethyl-; Bis(dimethylthiocarbamyl) sulfide. A nonstaining and nondiscoloring delayed action accelerator; has a short sharp curing range with normal to high sulfur in natural rubber; used in natural, SBR, butyl, nitrile and neoprene rubbers for wire insulation, druggist sundries, mechanicals, sponge mp = 106-108°. *Uniroyal.* Name unverified.

20211 Monitan
Polysorbate 80
Pharmaceutic aid. *Ives Laboratories Inc.* Name unverified.

20212 Monite
A proprietary plastic. No manufacturer.

20213 Monitor
10265-92-6 6014 233-606-0
Methamidophos, an insecticide. *Monsanto (Solaris).* Name unverified.

20214 Monnex
A nontoxic compound obtained by reacting urea with potassium carbonate; used as a fire extinguisher. *ICI Chem & Polymers Ltd.*

20215 Mono Ammonium Phosphate (Agricultural Grade)
7722-76-1 577 231-764-5
Monoammonium phosphate
Used for fertilizers, other chemicals. *Rhône-Poulenc NV.* Name unverified.

20216 Mono Thiurad
97-74-5 202-605-7
Tetramethylthiuram monosulfide
Vulcanization accelerator. *Monsanto Co.* Name unverified.

20217 Mono-Baycuten®
23593-75-1 2478 245-764-8
clotrimazole
Broad-spectrum antifungal. *Bayer AG.*

20218 Monobed
Ion exchange resins. *Rohm & Haas.*

20219 Monobel
A2 Monobel. A trademark for a smokeless powder. It is a mixture of 9-11 parts nitroglycerin, 56-61 parts ammonium nitrate, 8-10 parts wood meal, 0.5-1.55 parts magnesium carbonate, and 18.5-21.5 parts potassium chloride; used in mines. No manufacturer.

20220 Monocast® MC 901
Heat-stabilized nylon; produced by direct polymerization of monomer within a mold at atmospheric pressure; used in paper and textile, construction, mining, metalworking, and material handling industries for bearings, valve seats, seals, gears, wheels, guides, tooling fixtures, insulators, wear parts, etc; good machinability. *Polymer Corp.*

20221 monochloroacetic acid
79-11-8 2162 201-178-4
$ClC_2H_3O_2$
chloroacetic acid
MCA; Chloroethanoic acid; Monochloroacetic Acid; Monochloroethanoic acid; α-chloroacetic acid. Herbicide, preservative, bacteriostat; intermediate in production of carboxymethylcellulose, ethyl chloroacetate, glycine, synthetic caffeine, sarcosine, thioglycolic acid, EDTA, 2,4-D, 2,4,5-T. Has 3 forms, α, β and γ. mp = (α) 63°; (β) 55-56°; (Ig) 50°; bp = 189°; d = 1.580; soluble in H_2O, organic solvents; LD_{50} (rat orl) = 76 mg/kg. *Allchem Industries; Elf Atochem; Eka Nobel AB; Hoechst Celanese.*

20222 Mono-Coat®
A nontransferring, semipermanent mold release that gives multiple releases with no transfer to the molded part *Chem-Trend.*

20223 Monocron
6923-22-4 6334 230-042-7
Active ingredient; monocrotophos; contact and systemic agricultural insecticide belonging to the enolphosphates. *Makhteshim Chemical Works Ltd.*

20224 monoformin
$C_4H_8O_4$
The formyl derivative of glycerin.

20225 monogermane
7782-65-2 4414 231-961-6
GeH_4
Germanium tetrahydride
germane. A flammable, toxic, colorless gas used for the deposition of epitaxial and amorphous silicon - germanium alloys. mp = -165°; bp = -88°; d = 2.600; LD_{50} (mus orl) = 1250 mg/kg.

20226 Monoglyceride citric ester
91744-38-6 294-600-1
Imwitor® 369, 370. Food emulsifier. *Hüls Am; Hüls AG.*

20227 monoglyme
110-71-4 3274 203-794-9
$C_4H_{10}O_2$
Ethylene glycol dimethyl ether
2,5-Dioxahexane; Dimethoxyethane; Glycol dimethyl ether; Dimethyl Cellosolve; DME; α,β-dimethoxyethane; ethylene dimethyl ether; monoethylene glycol dimethyl ether; 1,2-ethanediol, dimethyl ether; ansul ether 121; EGDME; GDME; Glyme. Used as a solvent. d = 0.86285; mp = -

58°; bp = 83-85°; d = 0.8683; n_D^{20} = 1.3813; flash point = 4.5°; miscible with H_2O, soluble in organic solvents.

20228 Mono-Kay
84-80-0 7536 201-564-2
Phytonadione
phylloquinone. Vitamin. *Abbott Laboratories.* Name unverified.

20229 Monol
10118-76-0 1735 233-322-7
$CaMn_2O_8$
Calcium permanganate
Antiseptic, deodorizer, disinfectant. Used in the textile industry. Soluble in H_2O.

20230 Monolan®
Ethylene and/or propylene oxide condensates; used in low foam nonionics, lubricants, dishwashing, and cosolvency. *Harcros.*

20231 Monolan® 8000 E/80
9003-11-6 7721
EO/PO block polymer; low foam wetting agent and detergent, dispersant; primary emulsifier in emulsion polymerization. *Harcros.*

20232 Monolan® P222
EO-PO block polymer; low foaming detergent, antifoam for detergents, wetting agents, synthetic lubricants. *Harcros.*

20233 Monolan® PT
Complex EO-PO copolymer; low foam detergent and rinse aid, pigment dispersant; antifoam; dye leveling agent; car shampoos, window cleaners, yarn lubricants. *Harcros.*

20234 Monolastex Smooth
Fast drying water-based external or Internal elastomeric coating; brush or roller applied; used for outside/inside of housing, offices, prefabricated units. *Liquid Plastics Ltd.*

20235 Mono-Line
A monolithic refractory material used for lining steelmaking vessels. *Pfizer International.* Discontinued.

20236 Monolite®
Insoluble lake colors and pigments for paints, inks and plastics. *ICI Chem & Polymers Ltd.*

20237 Mono-Lube®
Custom formulations for use in tire industry; processing aids for tire manufacture. *Chem-Trend.*

20238 Monomax AH90 B
Distilled monoglycerides; emulsifier/stabilizer for ice cream, margarine, shortening, peanut butter, confectionery, other foods; chewing gum plasticizer; starch complexing agent in pastas. No manufacturer.

20239 Monomuls® 90-L12
142-18-7 205-526-6
Glyceryl laurate
Co-emulsifier, refatting agent and thickener for personal care products, bath additives, shampoos. *Henkel/Cospha; Henkel Canada.*

20240 Monomuls® 90-O18
111-03-5 203-827-7
Glyceryl oleate
Emulsifier, stabilizer, refatting agent, thickener for cosmetics, food and drugs. *Henkel/Cospha; Henkel KGaA.*

20241 Monomuls® Range
Fatty acid mono and diglycerides; general purpose food emulsifiers. *Grünau.*

20242 Monoplas 279
A proprietary dialkyl (C_7 - C_9) phthalate plasticizer. No manufacturer.

20243 Monoplex®
Specialty plasticizers; for vinyl and synthetic lubricants. *C P Hall.*

20244 Monoplex® 5
A proprietary trade name for dibenzyl sebacate; a vinyl plasticizer. *C.P. Hall.* Name unverified.

20245 Monoplex® DDA
27178-16-1 248-299-9
Diisodecyl adipate
Lubricant; plasticizer. *C.P. Hall.*

20246 Monoplex® DIOA
1330-86-5 215-553-5
$C_{22}H_{42}O_4$
Diisooctyl adipate
Plasticizer. *C.P. Hall.*

20247 Monoplex® DOA
103-23-1 203-090-1
Di-2-ethylhexyl adipate
Plasticizer for PVC food packaging film. *C.P. Hall.*

20248 Monoplex® DOS
122-62-3 1292 204-558-8
Di-2-ethylhexyl sebacate
Plasticizer for electrical PVC compounds, low temperature sheet, film. *C.P. Hall.*

20249 Monoplex® NODA
n-octyl, n-decyl adipate mixture. Plasticizer. *C.P. Hall.*

20250 Monoplex® S-73
Epoxidized octyl tallate
PVC stabilizer, plasticizer. *C.P. Hall.*

20251 Monoplex® S-75
Epoxidized glycol dioleate
PVC stabilizer, plasticizer. *C.P. Hall.*

20252 monopol soap
avirol
Sulfonated oils similar to Turkey red oil; used as wetting-out agents.

20253 Monopoxy
Single-component epoxy resin adhesive. *Hardman.*

20254 Monoset
Pre-packed ultra rapid hardening cementitious mortars and concretes; used for raising of manhole covers and frames, repairs to motorways, repairs in tidal conditions and wherever a minimum downtime is needed. *Ronacrete Ltd.*

20255 Monosiliol C
Methylsilanol tri PEG-8 glyceryl cocoate
Aids tissue regeneration; for anti-aging formulations, cosmetic and health emulsions, oils, milks, and soaps. *Exsymol.*

20256 Monosorb
7558-80-7 8806 231-449-2
Absorbed sodium phosphate. *FMC.* Discontinued.

20257 Monosulfiram
95-05-6 9123 202-387-3
Tetra-ethylthiuram monosulfide
Used as a parasiticide.

20258 Monosulph
Anionic surfactants. *Henkel Chemicals Ltd.* Name unverified.

20259 Monox
10097-28-6 8640 233-232-8
SiO
Silicon(II) oxide
Silylene, oxo-; silicon monoxide. A product containing mainly silicon monoxide, with some silicon, silicon dioxide, and small quantities of silicon carbide. It is obtained by heating sand with silicon, carborundum, or coke, in the electric furnace; good thermal and electrical insulator.

20260 Monsanto Salt
$Cl{\cdot}C_7H_6{\cdot}SO_3{\cdot}Na$
A proprietary trade name for *o*-chloro-*p*-toluene sodium sulfonate. No manufacturer.

20261 Montago
A German light alloy with a specific gravity lower than aluminum.

20262 Montan acid wax
68476-03-9 270-664-6
Hoechst Wax S. Lubricant and release agent for PVC and other polymers. *Hoechst AG; Hoechst Celanese.*

20263 montan pitch
The residue from the production of montan wax. The crude material gives an ash of 1.7%, has an acid value of 3, and a saponification value of 6.

20264 Montana gold
An alloy of 89% copper, 10.5% zinc, and 0.5% aluminum.

20265 Montane 20
1338-39-2 8872 215-663-3
Sorbitan laurate
Emulsifier. *Seppic.* Discontinued.

20266 Montane 40
26266-57-9 8872 247-568-8
Sorbitan palmitate
Emulsifier. *Seppic.* Discontinued.

20267 Montane 60
1338-41-6 8872 215-664-9
Sorbitan stearate
Emulsifier. *Seppic.* Discontinued.

20268 Montane 65
26658-19-5 8872 247-891-4
Sorbitan tristearate
Emulsifier. *Seppic.* Discontinued.

20269 Montane 80
1338-43-8 8872 215-665-4
Sorbitan oleate
Emulsifier. *Seppic.* Discontinued.

20270 Montane 481
1338-43-8 8872 215-665-4
Sorbitan oleate
Beeswax, and stearic acid; cosmetics ingredient. *Seppic.* Discontinued.

20271 montanine
16961-83-4 4220 241-034-8
fluosilicic acid
A liquid containing 31% hydrofluosilicic acid; Recommended as a disinfectant for the walls of breweries and distilleries. It is obtained from by-products in the pottery industry.

20272 Montanox 20 DF
9005-64-5 8872
Polysorbate 20
Emulsifier. *Seppic.* Discontinued.

20273 Montanox 40 DF
9005-66-7 8872
Polysorbate 40
Emulsifier. *Seppic.* Discontinued.

20274 Montanox 60 DF
9005-67-8 8872
Polysorbate 60
Emulsifier. *Seppic.* Discontinued.

20275 Montanox 61
9005-67-8 8872
PEG-4 sorbitan stearate
Emulsifier. *Seppic.* Discontinued.

20276 Montanox 65
9005-71-4 8872
Polysorbate 65
Emulsifier. *Seppic.* Discontinued.

20277 Montanox 70
66794-58-9
PEG-20 sorbitan isostearate
Emulsifier. *Seppic.* Discontinued.

20278 Montanox 80 DF
9005-65-6 7742
Polysorbate 80
Emulsifier. *Seppic.* Discontinued.

20279 Montanox 81
9005-65-6 7742
Polysorbate 81
Emulsifier. *Seppic.* Discontinued.

20280 Montanox 85
9005-70-3 8872
Polysorbate 85
Emulsifier. *Seppic.* Discontinued.

20281 Montax
A proprietary trade name for a filler for rubber, etc.; a mixture of hydrated magnesium carbonate and silica. No manufacturer.

20282 Monteban
55134-13-9 6506
Narasin; a veterinary medicine used as a coccidiostat and growth stimulant. *Eli Lilly & Co.*

20283 Monteine LCK-32
68920-65-0
Potassium cocohydrolyzed animal protein; surfactant for cosmetics. *Seppic.* Discontinued.

20284 Monteine LCQ
Cocamidopropyldimethylaminohydroxypropyl hydrolyzed animal protein; raw material for shampoos and hair conditioners. *Seppic.* Discontinued.

20285 Monteine LCT
68952-16-9
TEA-coco-hydrolyzed animal protein; surfactant for shampoo and general cosmetic use. *Seppic.* Discontinued.

20286 Monterey 30% Iron
Iron sulfate
30% Iron; ferric sulfate. A granular material used to correct iron deficiency; used on turf, flower beds, vegetables, etc. *Lawn & Garden Products Inc.*

20287 Monterey Bayleton
43121-43-3 9723 256-103-8
Triadimefon
A fungicide for the control of diseases in turf and the control of powdery mildew, rusts and blight of ornamental plants. *Lawn & Garden Products Inc.*

20288 Monterey Bloom Popper
A 8-32-7 liquid fertilizer formulated with humic acid for better foliar uptake; high phosphate levels help in the blooming and fruit products of a plant and the humic acid helps in the uptake of that nutrient. *Lawn & Garden Products Inc.*

20289 Monterey Foliar Nutrient 11-4-6
Used as a foliar nutrient, providing nitrogen, phosphoric acid, potash and chelated zinc, iron and manganese. For use on turf, ornamentals, trees, vegetables and house plants. *Lawn & Garden Products Inc.*

20290 Monterey Herbicide Helper
Petroleum distillate and alkylphenoxy polyethoxy ethanols; a spreader/penetrant that makes weed killers work faster and better. *Lawn & Garden Products Inc.*

20291 Monterey Insulate
Aids in the prevention of frost damage; used on ornamentals, vegetables, citrus, fruit trees, etc.; applied to frost conditions and repeated at 7-21 day intervals. *Lawn & Garden Products Inc.*

20292 Monterey Iron Chelate 10%
Used to correct Iron deficiencies, either as a soil drench or as a foliar spray; may be used on ornamentals, vegetables, fruit trees, etc. *Lawn & Garden Products Inc.*

20293 Monterey Liqui-Cop
Copper-Count-N
A liquid copper formulation used as a dormant copper spray fungicide on stone fruits and citrus; used as a replacement for Bordeaux mixtures. *Lawn & Garden Products Inc.*

20294 Monterey Perc-O-Late Plus
Surfactants plus nitrogen, zinc, iron and manganese; used for water penetration and fertilization of turf, flower beds, potted plants, etc. *Lawn & Garden Products Inc.*

20295 Monterey Signal
A blue colored dye to put into the spray tank to let you know where you are spraying; avoid skips and overlaps; breaks down in sunlight. *Lawn & Garden Products Inc.*

20296 Monterey Stimulator 12
Humic acid derived from completely organic sources; aids the plant in the uptake of nutrients from the soil or through foliar application; helps plants utilize the fertilizer you give them. *Lawn & Garden Products Inc.*

20297 Monthier's blue
$(Fe_2)_2(Fe(CN)_6)_3 \cdot 6NH_3 \cdot 9H_2O$
A colored compound obtained by the oxidation of the precipitate formed by the action of ammoniacal ferrous chloride upon potassium ferrocyanide.

20298 Monthyle
111-60-4 203-886-9
Glycol stearate
Emulsifier, stabilizer for ointments, cream lotions. *Gattefosse; Gattefosse SA.*

20299 Montigel
1302-78-9 1082 215-108-5
bentonite
Bentonites with specific swelling properties for various technical purposes, such as for iron ore pelletisations and for sealing of sanitary land fill. *Süd-Chemie AG.* Name unverified.

20300 Montmorillonite
1318-93-0 6341 215-288-5
Fulcat Catalysts; Fulmont Activated Bleaching Earths; K 129; Lubrigel. Acid treated montmorillonit clays; catalysts for a range of organic reactions notably the alkylation of phenols.

20301 Montothene G50
A proprietary trade name for an ethylene vinyl acetate copolymer; translucent, nontoxic; good mechanical properties; used for film, injection and blow molded articles and extrusions. *Armour Pharmaceutical Co.* Name unverified.

20302 Montreal potash
584-08-7 7781 209-529-3
Commercial potassium carbonate.

20303 Moore Floc
Organic and inorganic flocculants, cationic, anionic, alum replacement products; used for flocculating fine dissolved or suspended solids in potable water plants, color and collodial and turbidin solids, algae precipitating aid. *Benzsay & Harrison Inc.*

20304 Moorland
A proprietary preparation for use as an antacid. *The Boots Co plc.* Discontinued.

20305 Moplefan
A proprietary range of polypropylene films. *Montedison UK Ltd.* Name unverified.

20306 Moplen
Polypropylene
A flexible hard, tough hydrocarbon thermoplastic used for molding domestic ware and for electrical purposes. *Montedison UK Ltd.* Name unverified.

20307 morantel tartrate
26155-31-7 6348 247-481-5
$C_{12}H_{16}N_2S \cdot C_4H_6O_6$
(E)-1,4,5,6-tetrahydro-1-methyl-2-[2-(3-methyl-2-thienyl)ethenyl]pyrimidine tartrate
CP-12009-18; Paratect; Suiminth. Anthelmintic.

20308 Morat white
Moudan white
A white pigment. It is a clay found in Switzerland.

20309 Moreau marble
A marble prepared by immersing soft amorphous limestone in a bath of zinc sulfate, and drying.

20310 Morell's solution
A disinfecting solution containing arsenious acid, caustic soda, and a small quantity of phenol, dissolved in water.

20311 Morestan
2439-01-2 7113 219-455-3
Wettable powder containing 25% w/w quinomethionate; for control of red spider mites, including organophosphorus strains, on apples, gooseberries, strawberries, and marrows and American gooseberry mildew (partial control of leafspot) on gooseberries, powdery mildew on marrows and willow anthracnose. *Hortichem Ltd.*

20312 Morestan®
2439-01-2 7113 219-455-3
A wettable powder containing 25% w/w quinomethionate; fungicide with protective and eradicative action against powdery mildews on pome, stone and small fruits, cucurbits and ornamentals; as acaricide effective against eggs and mobile stages of mites and spider mites on pome and stone fruit, citrus, vegetables, ornamentals and coffee. *Bayer AG.*

20313 Morfast
Liquid dyes for surface coatings. *Morton Int'l. Ltd.*

20314 Morfax
95-32-9 202-410-7
$C_{11}H_{12}N_2OS_3$
4-Morpholinyl-2-benzothiazole disulfide
Benzothiazole, 2-(4-morpholinyldithio)-; Morpholinylmercaptobenzothiazole; 2-(4-Morpholinylmercapto)benzothiazol; 4-Morpholinyl-2-benzothiazyl disulfide; MOR. An accelerator for natural rubber, isoprene butadiene, styrene-butadiene, and nitrile-butadiene rubber products; provides good curing activity; suggested for tires and mechanical goods requiring maximum strength and quality. *Goodyear.*

20315 Morflex 100
27554-26-3 1291 248-523-5
A proprietary trade name for diisooctyl phthalate; a vinyl plasticizer. No manufacturer.

20316 Morflex 125
119-07-3 204-295-9
n-octyl n-decyl phthalate
A proprietary plasticizer. *Pfizer International.* Discontinued.

20317 Morflex 150
84-61-7 201-545-9
Dicyclohexyl phthalate
Plasticizer for adhesives, nitrocellulose lacquers. *Morflex.*

20318 Morflex 175
119-07-3 204-295-9
A proprietary trade name for octyl decyl phthalate; a vinyl plasticizer. No manufacturer.

20319 Morflex 190
85-70-1 201-624-8
n-Butyl phthalyl-n-butyl glycolate
Vinyl plasticizer. *Morflex.*

20320 Morflex 210
Diethyl hexyl phthalate
A proprietary plasticizer. *Pfizer International.* Discontinued.

20321 Morflex 240
84-74-2 1622 201-557-4
Dibutyl phthalate
A proprietary plasticizer. *Pfizer International.* Discontinued.

20322 Morflex 310
117-81-7 1291 204-211-0
Di-2 ethyl hexyl adipate
A proprietary plasticizer. *Pfizer International.* Discontinued.

20323 Morflex 325
n-octyl n-decyl adipate
A proprietary plasticizer. *Pfizer International.* Discontinued.

20324 Morflex 330
Didecyl adipate
A vinyl plasticizer. *Pfizer International.* Discontinued.

20325 Morflex 410
117-81-7 1291 204-211-0
Di-2-ethyl hexyl azelate
A proprietary plasticizer. *Pfizer International.* Discontinued.

20326 Morflex 510
Tri-2-ethyl hexyl trimellitate
A proprietary plasticizer. *Pfizer International.* Discontinued.

20327 Morflex 525
Tri(n-octyl n-decyl) trimellitate
A proprietary plasticizer. *Pfizer International.* Discontinued.

20328 Morflex 530
36631-30-8 253-138-0
Triisodecyl trimellitate
A proprietary plasticizer. *Pfizer International.* Discontinued.

20329 Morflex 560
1528-49-0 216-208-1
Tri-n-hexyl trimellitate
Plasticizer with high efficiency, good permanence. *Morflex.*

20330 Morflex 1129
1459-93-4 215-951-9
$C_{10}H_{10}O_4$
Dimethyl isophthalate
Dimethyl 1,3-benzenedicarboxylate. Chemical intermediate in the synthesis of polyesters. mp= 66-67°; bp_{12} = 124°. *Morflex.*

20331 Morflex MSC
1337-33-3 215-654-4
Monostearyl citrate
stearyl citrate. Chelating agent, surface lubricant. Monostearyl Citrate (MSC) is a mixture of mono- and distearyl citrate esters which show utility as an oil-soluble chelating agent. mp = 47°; d = 0.92. *Morflex.*

20332 Morflex P50
A proprietary n-alkyl phthalate plasticizer. *Pfizer International.* Discontinued.

20333 Morhal resin
A resin obtained from *Vatica lanceoefolia*, of India.

20334 Morhulin
Cod liver oil and zinc oxide; for treatment of wounds, scalds and dermatitis. *Napp Laboratories Ltd.* Name unverified.

20335 Morillol®
3391-86-4 222-226-0
$C_8H_{16}O$
1-Octene-3-ol
Vinyl pentyl carbinol; Flowtron mosquito attractant; Morillol; Octen-3-ol; Pentyl vinyl carbinol; Vinyl hexanol; Matsuka alcohol; 3-Octenol; Mushroom alcohol; Pentyl vintyl carbinol. Fragrance and flavoring. Herbaceous and mushroom odor. bp = 174°; d = 0.8300; n_D^{20} = 1.4361; insoluble in H_2O, soluble in organic solvents. *BASF AG.*

20336 Morkit®
84-65-1 726 201-549-0
Anthraquinone
Seed treatment to reduce rook feeding damage until plant emergence in arable and vegetable crops. *Bayer AG.*

20337 Morland's salt
$(Cr(NH_3)_2(SCN)_4)HNH$
The guanidinium salt of the same complex as Reinecke's salt, formed as a by-product in the preparation of the latter salt.

20338 Moroccan olive oil
Argan oil from *Arganum sideroxylon*.

20339 Morocco gum
Mogador gum; brown barberry gum. A variety of gum acacia in the form of tears.

20340 Morocide
485-31-4 1265 207-612-9
binapacryl
Fungicide for mildew. *Hoechst UK.* Name unverified.

20341 Moroline
Petrolatum, white; pharmaceutic aid; protectant. *Schering Corp.* Discontinued.

20342 Moronal
Aluminum-formaldehyde-sulfite
Used as an antiseptic and astringent.

20343 Morpan
Cationic surfactant, biocide *Rhône-Poulenc UK.*

20344 Morpan BC
8001-54-5 1086 231-635-3
Aqueous solution of benzalkonium chloride; Bactericide, algicide and antistatic used in fields such as food; brewing; hospitals; farming; veterinary; general sterilizing. *ABM Chemicals Ltd.*

20345 Morpan NBB
A proprietary trade name for a 50% active composition of octyldimethylbenzylammonium bromide; a low forming flocculating agent used as a filtering aid. No manufacturer.

20346 morpholine
110-91-8 6362 203-815-1
C_4H_9ON
tetrahydro-1,4-oxazine
tetrahydro-2H-1,4-oxazine; diethylene oximide; diethylenimide oxide; 1-Oxa-4-azacyclohexane. Heterocyclic organic compound; rubber accelerator; solvent; additive to boiler water; optical brightener for detergents; corrosion inhibitor; organic intermediate. mp = -5°; bp = 126-130°; d = 0.9940; n_D^{20} = 1.4540; LD_{50} (rat orl) = 1.05 g/kg. *Air Prods & Chem; BASF; Nippon Nyukazai; PMC Specialties; Texaco.*

20347 Morplas
Heat resistant dyes for plastics. *Morton Int'l. Ltd.* Discontinued.

20348 Mor-Rex® I-920
9050-36-6 232-940-4
Maltodextrin
Nondispersive thinner for water-based lime muds. *Grain Processing.*

20349 Morsep
Cetrimide, vitamin A and vitamin D_2; used for diaper rash. *Napp Laboratories Ltd.* Name unverified.

20350 Morstrip
Paint stripper. *Kalon Chemicals Ltd.*

20351 Mortegg Emulsion
Emulsifiable concentrate containing 600 g/l tar oils; fungicidal winter wash for fruit. *DowElanco Ltd.*

20352 Morto
Weedkiller and destroyer of potato haulm (stems and stalks) . *Murphy Chemical Co Ltd.* Discontinued.

20353 Morton's fluid
A solution containing iodine, potassium iodide, and glycerin.

20354 Morwet EFW
Wetting agents and dispersants; used in pesticide formulations. *Witco Corporation.*

20355 mosaic gold
Mock gold; cat's gold; bronzing powder; tin bronze. Flaky yellow form of disulfide of tin, SnS_2; a substance also called mosaic gold is made from an amalgam of tin, mercury, with ammonium chloride and sulfur. formerly used for gilding and imitating bro.

20356 mosaic silver
An alloy of tin and bismuth.

20357 Moss Gunl
Ready-to-use mosskiller in spray form. *ICI Garden Products.*

20358 Mosskil
Lawn fertilizer with iron sulfate. *Fisons plc, Horticultural Div.* Name unverified.

20359 Mos-Tox
Moss eradicant. *May & Baker Ltd.* Name unverified.

20360 Mota
9002-91-9 5983 202-945-6
Tablets of metaldehyde, a molluscicide.

20361 mother of pearl sulfur
7704-34-9 9142 231-722-6
Nacreous sulfur. A form of monoclinic sulfur obtained by heating sulfur with benzene at 140°C. It is unstable.

20362 Motung Steel
A proprietary trade name for a high speed steel containing 7.5-8.5% molybdenum, 1.25-2% tungsten, 3.5-4.5% chromium, 0.9-1.5% vanadium, 0.8 carbon, 0.2-0.4% manganese, and 0.25-0.5% silicon. No manufacturer.

20363 mou-iéou
Pi-yu. Chinese vegetable tallow.

20364 Mould Release Agent N 32
Brushable mold coating on a plastic base. *Chemetall GmbH.* Name unverified.

20365 Mouldrite
Thermosetting molding powders. *BIP Chemicals Ltd.* Discontinued.

20366 mountain blue
12069-69-1 2697 235-113-6
cupric carbonate basic
Azure blue; mineral blue; mopper blue; Hamburg blue; English blue. A basic copper carbonate; a pigment used by painters.

20367 mountain butter
10043-01-3 381 233-135-0
aluminum sulfate
A hydrated aluminum sulfate found in fibrous masses. Used in leather and paper industries and as an anti-infective.

20368 mountain cork
Elastic asbestos. An asbestos which floats on water.

20369 mountain green
12002-03-8 2692
Cupric acetoarsenite
Mineral green; (Acetato)trimetaarsenitodicopper; Paris green; emerald green; French green; Schweinfurt green; imperial green; Mitis green; Parrott green; C.I. Pigment green 21; C.I. 77410. A pigment prepared by precipitating a boiling solution of alum and copper sulfate with a hot solution of sodium or potassium sulfate.

20370 mountain leather
Mountain paper. Thin, tough types of asbestos.

20371 mountain milk
An earth similar to infusorial earth; used as a rubber filler.

20372 mountain wood
A variety of asbestos.

20373 Mountford's paint
A waterproof paint. it consists of asbestos, ground in water, potassium or sodium aluminate, and potassium or sodium silicate sometimes with oil, and zinc white.

20374 Mourey's aluminum solder
An alloy of 82% zinc and 18% aluminum.

20375 Mouse Killer
Rodenticide. *Murphy Chemical Co Ltd.* Discontinued.

20376 Mouser
56073-10-0 1400 259-980-5
Contains brodifacoum; ready-for-use bait box, mouse killer. *ICI Garden Products.*

20377 Moussett's alloy
An alloy of 60% copper, 27.5% silver, 9.5% zinc, and 3% nickel.

20378 Movin B
Agents for the antimicrobial finishing of textiles. *Bayer AG.*

20379 Mowchem
53780-34-0 5846 258-767-4
A grass growth regulator containing 240 g/l mefluidide; suppresses most grasses for up to 8 weeks; for grassed areas not subject to heavy wear. *Rhône-Poulenc Environmental Prods. Ltd.*

20380 Mowilith
Polyvinyl acetate dispersion and solids. *Hoechst AG.*

20381 Mowilith®
Vinyl acetate homo and copolymer, acrylic and styrene, acrylic copolymer dispersions; used for adhesives and surface coatings. *Harlow Chemical Co Ltd.*

20382 Mowiol
Polyvinyl alcohol
Hoechst UK. Name unverified.

20383 Mowiol®
Polyvinyl alcohol
Partially/fully hydrolyzed polyvinyl acetate. Used for dispersion manufacture, paper manufacture, adhesives soluble film. *Harlow Chemical Co Ltd.*

20384 Mowital
Polyvinyl butyl resin. *Hoechst UK.*

20385 Mow-It-Less
Grass seed mixture. *ICI Garden Products.*

20386 Mozanon®
9005-38-3 240
Sodium alginate
Algin. Fungicide for prevention of TMV infection on tobacco. *Mitsubishi Kasel.* Name unverified.

20387 MP-1
O,O-Dimethyl phosphorodithioic acid
Used as an intermediate for organophosphorus insecticides. *A/S Cheminova.* Name unverified.

20388 MP-2
2524-03-0 219-754-9
$C_2H_6ClO_2PS$
O,O-Dimethyl phosphorochloridothioate
Dimethyl thiophosphoryl chloride; O,O-dimethylphosphorochloridothioate; Dimethyl chlorothiophosphate; Phosphorochloridothioic acid O,O-dimethyl ester; Dimethyl phosphochloridothioate; Methyl PCT; O,O-Dimethyl chlorophosphorothioate; Dimethyl phosphorochlorodithioate. Mainly used in the production of organophosphorus insecticides. bp_{16} = 66-67°; d = 1.3220; n_D^{20}= 1.4819. *A/S Cheminova.* Name unverified.

20389 MP-10CF-4CC/15T
25134-01-4

Modified PPO, 10% carbon fiber, 4% conductive carbon, 15% PTFE. *Compounding Tech.* Name unverified.

20390 M-P-A

Thixotropic agent; used as component of paint for protective coating of a substance. *Rheox Inc.*

20391 MP Diol Glycol

2163-42-0
$C_4H_{10}O_2$
2-Methyl-1,3-propanediol
Propane-1,3-diol, 2-methyl-. Used in unsaturated polyesters, gel coats, saturated polyester and alkyd coatings, plasticizers. *Arco Chemical Co.*

20392 MPEM

O,O-Dimethyl-S-methoxycarbonylmethyl phosphorodithioate
Used as an intermediate for phosphorus pesticides. *A/S Cheminova.* Name unverified.

20393 MPI DMSA Kidney Reagent

304-55-2 9034 206-155-2
(RName unverified.,SName unverified.)-2,3-dimercaptobutanedioic acid
Succimer; DMS; DMSA; Ro-1-7977; Chemet. Succimer; chelating aid, diagnostic aid. Used as an antidote in heavy metal poisoning. The ^{99m}Tc complex is used as a radioactive imaging agent. mp= 192-194°; LD_{50} (mus ip) >3000 mg/kg. *Medi-Physics Inc.* Name unverified.

20394 MPI Indium DTPa ^{111}In

Pentetate indium disodium ^{111}In; diagnostic aid; radioactive agent. *Medi-Physics Inc.* Name unverified.

20395 MPI Indium Oxine ^{111}In

Indium ^{111}In oxyquinoline; radioactive agent; diagnostic aid. *Medi-Physics Inc.* Name unverified.

20396 MPI Krypton ^{81m}Kr Gas Generator

Krypton, isotope of mass 81; radioactive agent. *Medi-Physics Inc.* Name unverified.

20397 MPS 500

A chlorinated fatty acid ester; a vinyl plasticizer. *Occidental Chemical Corp.* Name unverified.

20398 M-Quat® 32

Octadecyl diethanol methyl ammonium chloride
Emulsifier, defoamer, coagulant. *PPG/Specialty Chem.* Discontinued.

20399 M-Quat® 40

26062-79-3
Polyquaternium 6
Homopolymer quaternary for hair and skin products; emollient. *PPG/Specialty Chem.* Discontinued.

20400 M-Quat® 257

61789-80-8 263-090-2
Quaternium-18 - isopropyl alcohol. Quaternary for use as textile softener. *PPG/Specialty Chem.* Discontinued.

20401 M-Quat® 522

67633-63-0 266-778-0
Isostearamidopropyl ethyldimonium ethosulfate
Conditioner for hair conditoners and shampoos. *PPG/Specialty Chem.* Discontinued.

20402 M-Quat® 1033

68308-67-8 269-663-3
Soya ethyidimonium ethosulfate
Conditioner for hair conditioners and shampoos; antistat for cleaners, rug shampoos; low foaming. *PPG/Specialty Chem.* Discontinued.

20403 M-Quat® 2475

61789-77-3 263-087-6
Dicocodimonium chloride and isopropanol; emulsifier, defoamer, coagulant; for auto spray wax. *PPG/Specialty Chem.* Discontinued.

20404 M-Quat® B-25

122-19-0 204-527-9
Stearalkonium chloride
Quaternary for hair conditioners. *PPG/Specialty Chem.* Discontinued.

20405 M-Quat® Dimer 18

Hydroxypropyl bisstearyldimonium chloride
Conditioner for hair and skin, perms, mousses; emulsifier. *PPG/Specialty Chem.* Discontinued.

20406 M-Quat® JN

112324-16-0
Ricinolamidopropyl ethyldimonium ethosulfate
Quatenary for shampoo, hair conditioners, lotions; substantive to hair. *PPG/Specialty Chem.* Discontinued.

20407 M-Quat® JO-50

37139-99-4 253-363-4
Olealkonium chloride
Conditioner, antistat for clear hair rinses. *PPG/Specialty Chem.* Discontinued.

20408 MR-1, MR-1A, MR-17A, MR-17B

Proprietary nomenclature for allyl resins. No manufacturer.

20409 MR-502K, MR-502P, MR-502Y

Dimethyl polysiloxane combination; magnetic rubber inspection material for nondestructive inspection of ferromagnetic parts used for inspection of thread, gear roots, inaccessible areas, coated areas; provides permanent record of the inspection. *Dynamold Inc.*

20410 MRV 1000

Polytetrafluorethylene polymer; release agent for epoxy resins and any molded plastics; lubricants for drive belts, gears and most machinery. *Loes Enterprises Inc.*

20411 MS-4

A silicone electrical insulating compound. *Midland Silicones.* Unverified.

20412 MS-26

Epoxy resin based mastic; moldable liquid shim materials used as a spacer between engines or skin of aircraft or ships; surface conforming structural epoxies used to fill gaps between metal parts and between structural members and composites such as graphite. *Dynamold Inc.*

20413 MS-122

116-14-3 204-126-9
C_2F_4
Tetrafluorethylene telomer
Tetrafluorethene; perfluoroethene; perfluoroethylene; TFE; fluoroplast 4. Release agent, dry lubricant for use on cold molds, esp. for epoxy potting/encapsulating, PU, nylon, acrylics, PP, PC phenolics, PS, foams, rubber molding. *Miller-Stephenson.* Name unverified.

20414 MS-170 1,1,1 Trichloroethane Solvent

71-55-6 9766 200-756-3
1,1,1-Trichloroethane
Miller-Stephenson. Name unverified.

20415 MS-180 Freon® TF Solv

76-13-1 200-936-1
$C_2Cl_3F_3$
Trichlorotrifluoroethane
Freon 113; Fluorocarbon 113; 1,1,2-Trichlorotrifluoroethane; halocarbon 113; refrigerant 113; FC 133; Trichlorotrifluoroethane; CFC-113; Diflon S-3; Freon TF (113); Freon TF; Trichloro 1,2,2-trifluoroethane; 1,1,2-Trifluorotrichloroethane; 1,2,2-Trichlorotrifluoroethane; 1,1,2-Trifluoro-1,2,2-trichloroethane; Frigen 113 TR-T; Chlorinated fluorocarbon; 1,2,2-Trichlorotrifluoroethane. Refrigerant; now proscribed. Also used as a solvent. *Miller-Stephenson.* Name unverified.

20416 MSG

142-47-2 6337 205-538-1
$C_5H_8NNaO_4 \cdot H_2O$
glutamic acid monosodium salt
monosodium glutamate; monosodium L-glutamate monohydrate; mono. Flavor enhancer for foods. mp = 232° (dec); $[\alpha]_D^{20}$ = 25° (c = 5 1N HCl). *Ajinomoto Co Inc; Allchem Industries; Asahi Chem Industry Co Ltd; Penta Mfg.; Schweizerhall.*

20417 MSMA

2163-80-6 6020 218-495-9
CH_4AsNaO_3
monosodium methylarsonate
Ansar; Arsonate Liquid; Bueno; Daconate; Dal-E-Rad; Drexar; Super Arsonate; Versar; Weed-E-Rad; Weed-Hoe; monosodium methylarsonate; monosodium methanearsonate. Used in post-emergence control of grass weeds in cotton, sugar cane and under trees, mp = 113-116°; soluble in H_2O (1.4 kg/kg), most organic solvents; LD_{50} (rat orl) = 900 mg/kg. *Lawn & Garden Products Inc; Drexel; Fermenta; Inter-Ag; Pamol; Shinung; Vertac; Vineland.*

20418 M-Soft-1

Silicone-enriched cationic surfactant; softener for all fibers, yarns, fabrics; especially for brushed or raised fabrics. *Zohar Detergent Factory.*

20419 M-Soft-10

Amino functional silicone plus cationic surfactant; softener producing an elastomeric silicone finish with excellent soft handle on all kinds of fibers, yarns, and fabrics. *Zohar Detergent Factory.*

20420 MSP

7558-80-7 8806 231-449-2
Used as a urinary acidifier.

20421 MSS 2,4-D Amine

94-75-7 2865 202-361-1
2,4-D; translocated herbicide for cereals and established grassland. *Mirfield Sales Services Ltd.*

20422 MSS 2,4-D Ester

94-75-7 2865 202-361-1
2,4-D; translocated herbicide for cereals and grass. *Mirfield Sales Services Ltd.*

20423 MSS 2,4-DB + MCPA
2,4-DB [94-75-7] and MCPA [94-74-6]; translocated herbicide applied to cereals and undersown clovers. *Mirfield Sales Services Ltd.*

20424 MSS 2,4-DP
120-36-5 3128 204-390-5
Soluble concentrate of 500 g dichlorprop per liter; used for control of weeds in barley, wheat and oats. *Mirfield Sales Services Ltd.*

20425 MSS 2,4-DP + MCPA
Mixture of dichlorprop and MCPA; used for weed control in cereals and turf. *Mirfield Sales Services Ltd.*

20426 MSS Aminotriazole
61-82-5 513 200-521-5
Amitrole
Translocated herbicide. *Mirfield Sales Services Ltd.*

20427 MSS Chlormequat 40, 46, 60, 70
999-81-5 2153 213-666-4
Chlormequat chloride
Plant growth regulator. *Mirfield Sales Services Ltd.*

20428 MSS CIPC
101-21-3 2240 202-925-7
Chlorpropham
A carbamate herbicide and sprout depressant in stored potatoes. *Mirfield Sales Services Ltd.*

20429 MSS CMPP
24017-47-8 9736 245-986-5
trazophos
Emulsifiable concentrate containing 600 g/l trazophos; an organophosphorus insecticide. *Mirfield Sales Services Ltd.*

20430 MSS CMPP/DP
Soluble concentrate of 520 g dichlorprop and 130 g mecoprop per liter; used for control of weeds in barley, wheat and oats. *Mirfield Sales Services Ltd.*

20431 MSS IPC 50
122-42-9 8001 204-542-0
propham
Wettable powder containing 50% w/w propham; used for weed control for beet crops and peas. *Mirfield Sales Services Ltd.*

20432 MSS MCPA 50
94-74-6 5803 202-360-6
MCPA; herbicide for cereals and grassland. *Mirfield Sales Services Ltd.*

20433 MSS MH18
123-33-1 5745 204-619-9
Maleic hydrazide
A plant growth regulator for grass and to reduce bud growth in trees, hedges and vegetables. *Mirfield Sales Services Ltd.*

20434 MSS Mircam
Soluble concentrate of 18.7 g dicamba and 300 g mecoprop per liter; used for weed control in cereals and grassland. *Mirfield Sales Services Ltd.*

20435 MSS Mircam Plus
Soluble concentrate of 19.5 g dicamba, 245 g MCPA and 86.5 g mecoprop per liter; used for weed control in cereals and grassland. *Mirfield Sales Services Ltd.*

20436 MSS Simazine/Aminotriazole 43FL
A suspension concentrate containing 155 g aminotriazole [61-82-5] and 275 g simazine [122-34-9] per liter; used for total weed control in non crop areas and fruit orchards. *Mirfield Sales Services Ltd.*

20437 MSS Sugar Beet Herbicide
A suspension concentrate containing 37.5 g chlorpropham, 25 g fenuron and 150 g propham per liter; an herbicide for use on beet crops. *Mirfield Sales Services Ltd.*

20438 Mucicarmin
A solution of 2 parts carmine, 1 part aluminum chloride, and 4 parts water; a staining solution.

20439 Mucogel®
Aluminum hydroxide [21645-51-2] and magnesium hydroxide [1309-42-8]; antacid suspension for antacid therapy in gastric and duodenal ulcer gastritis, heartburn, gastric hyperacidity, indigestion. *Pharmax Ltd.*

20440 mudar gum
A material obtained from *Calotropis giganteas.* It resembles gutta-percha, and contains about 20% of a rubbery material.

20441 Mudge's speculum metal
An alloy of 69% copper and 31% tin.

20442 muga silk
The product of the caterpillar, *Antheraca assama* of Assam.

20443 Mulch-Magic
A brown colored dye used on flower bed mulch to restore the faded color to the new freshly applied color. *Lawn & Garden Products Inc.*

20444 muldan
An orthoclase mineral.

20445 mule gum
A name sometimes applied to Ceara rubber.

20446 Mulgofen
Tallow amine
Water-soluble emulsifier for mineral oils; agricultural chemicals. *GAF Great Britain.* Name unverified.

20447 Muller's fluid
A solution of phosphoric acid in alcohol; dichromate; and sodium sulfate dissolved in distilled water. A soldering fluid for brass and copper. The term is also used for a hardening agent used in microscopy. It consists of potassium.

20448 Mullex
A proprietary refractory material made from mixtures containing various proportions of clay and mullite. No manufacturer.

20449 Mullfrax 301
A mullite-alumina product. *Carborundum.* Name unverified.

20450 mullicite
A mineral. It is blue iron earth.

20451 mullite
A refractory material formed by heating sillimanite to a temperature of 1550ºC. It is also made from the minerals andalusite, dumortierite, and Indian cyanite, and by the electric fusion of alumina and silica.

20452 Mulsifan CB
69227-20-9
Beheneth-10
Emulsifier for creams and lotions. *Zschimmer & Schwarz.*

20453 Mulsifan CPA
5274-68-0 226-097-1
Laureth-4
Oil-water emulsifier for cosmetic creams and lotions. *Zschimmer & Schwarz.*

20454 Mulsifan RT 1
9004-96-0
Fatty acid polyglycol ester
Emulsifier for mineral oils, textile processing, metal processing, drilling and cutting oils, chemo-technical products. *Zschimmer & Schwarz.*

20455 Mulsifan RT 141
9005-64-5 8872
Polysorbate 20
Solubilizer for perfumes and volatile oils. *Zschimmer & Schwarz.* Discontinued.

20456 Mulsifan RT 146
9005-65-6 7742
Polysorbate 80
Solubilizer for perfumes and volatile oils. *Zschimmer & Schwarz.* Discontinued.

20457 Mulsifan RT 203/80
68131-39-5
C_{12}-C_{15} pareth-12
Solubilizer for perfumes and volatile oils. *Zschimmer & Schwarz.*

20458 Mulsifan RT 23
3055-95-6 221-281-8
Laureth-5
Emulsifier for paraffin oils, white oils for formulation of lubricating agents, spin finishes, coning oils, emulsions for cosmetics. *Zschimmer & Schwarz.*

20459 Mulsifan RT 72
Coco fatty acid MEA ethoxylate
Washing agent. *Zschimmer & Schwarz.* Discontinued.

20460 Multaglut
A mixture of persulfate and calcium phosphate; used to improve flour.

20461 Multex
12141-46-7 377 235-253-8
Aluminum silicate
Large particle size hard silicate for heavy-duty abrasives; water filtration aid. *Kaopolite.*

20462 Multi Base
Powdered compost additive containing NPK 2.6:2.2:23 plus 3% Mg and trace elements; compost additive. *Vitax Ltd.*

20463 Multibase ABS 3075
9003-56-9 262-706-6
ABS; high-impact grade. *Multibase.*

20464 Multibase ABS 3525 CL
9003-56-9 262-706-6
ABS; clear grade. *Multibase.*

20465 Multibase ABS 3959
9003-56-9 262-706-6
ABS; plating grade. *Multibase.*

20466 multibrol
An organic combination of bromine consisting primarily of the sodium derivative of brom-oleate with a bromine content of 16%.

20467 Multicel
Dry powder dyes; used in the paper industry. *Multicrom SA.*

20468 Multicet
Dry powder disperse dyes; used in the textile industry for dyeing nylon and acetate. *Multicrom SA.*

20469 Multicoild
Film cement. *May & Baker Ltd.* Name unverified.

20470 Multicrack
Fluid cracking catalyst. *Akzo.*

20471 Multicrom
Dry powder inorganic and organic pigments; used in the manufacture of inks, paints and as dispersions for textile printing. *Multicrom SA.*

20472 Multicuer
Dry powder acid dyes; dyes developed for the leather industry. *Multicrom SA.*

20473 Multiflow
Resin modifier; additive for nonaqueous industrial coatings; improves flow, reduces pinholes and craters and improves substrate wetting. *Monsanto Co.* Name unverified.

20474 Multigreen® II
Chelated micronutrients; water-soluble organic blend of metal chelates of iron, zinc, copper and manganese, to improve root growth, color and stress tolerance. *Regal Chemical Company.*

20475 MultiGuard
Non-nitrite, molybdate formula; for closed system treatment. *Garvey Chemical Corp.* Name unverified.

20476 Multilind
Nystatin and zinc oxide; used for the treatment of fungal skin infections. *Bristol-Myers Squibb Pharmaceuticals Ltd.*

20477 Multilind Ointment
Nystatin and zinc oxide in ointment base. *Bristol-Myers Squibb Pharmaceuticals Ltd.*

20478 Multiluz
Dry powder direct dyes; light fast dyes used in the textile industry. *Multicrom SA.*

20479 Multimet
Mixed metal carboxylates; drier for paint or printing ink, PVC stabilizers. *Manchem Ltd.* Name unverified.

20480 Multionic
Cooling water treatment; for corrosion and scale control. *Garvey Chemical Corp.* Name unverified.

20481 Multisil
Precipitated silica
Used in the rubber industry as a filler. *Multicrom SA.*

20482 MultiSperse
Boiler treatment; for scale and sludge control. *Garvey Chemical Corp.* Name unverified.

20483 Multisperse CP
Highly effective silicate dispersant for prevention of silicate scale build-up in continuous bleaching equipment. *BASF.*

20484 Multispray
Public health insecticide. *The Wellcome Foundation Ltd.* Name unverified.

20485 Multistix®
A proprietary test strip for the detection of pH, protein, glucose, ketones, bilirubin, blood and urobilinogen in urine. *Bayer AG.*

20486 Multiter
Dry powder disperse dyes; used in the textile industry for dyeing polyester fibers. *Multicrom SA.*

20487 MultiTherm IG-2 Heat Transfer Fuild
Paraffinic hydrocarbon
Heat transfer fluid. *MultiTherm Corporation.*

20488 MultiTherm PG-1Heat Transfer Fluid
Naphthenic hydrocarbon
Light white mineral oil USP; heat transfer fluid. *MultiTherm Corporation.*

20489 Multivac 4
Dental vacuum investor. *Degussa Ltd.*

20490 Multivite
A proprietary preparation of vitamins A, D, C and thiamine hydrochloride. Vitamin supplement. *British Drug Houses.* Name unverified.

20491 Multi-W FL
A suspension concentrate containing 50 g carbendazim [83601-81-4] and 320 g maneb [12427-38-2] per liter; systemic fungicide for cereals. *Pan Britannica Industries Ltd.*

20492 Multiwax® 180-M
63231-60-7 264-038-1
Microcrystalline wax NF; plasticizer or modifier for polymeric coatings and adhesives; hot melt adhesives and coatings, chewing gum base, protective coatings; FDA approved. *Witco/Sonneborn.* Name unverified.

20493 Multiwax® HS
63231-60-7 264-038-1
Microcrystalline wax; wax used in foil/tissue laminations, heat-seal laminations, glassine laminations; FDA approved. *Witco/Sonneborn.* Name unverified.

20494 Multronol
Polyether polyol
Miles. Discontinued.

20495 Mulukilivary
A gum-resin obtained from *Balsamodedron berryr,* used as a myrrh substitute.

20496 Mumetal
A patented nickel-iron-copper alloy having the highest permeability of all known commercial materials; its exceptional magnetic properties and low losses make it invaluable for cable loading, instrument transformers, relays, magnetic shields, etc. No manufacturer.

20497 Municol
9005-38-3 240
Sodium alginate
Food/pharmaceutical grade. *Kelco Int'l. Ltd.*

20498 munjeet
The root of *Rubia munjista.* It contains purpurin, and is an important Indian dyestuff.

20499 Munjistin
$C_{15}H_8O_6$
purpuroxanthic acid
Dihydroxyanthraquinonecarboxylic acid.

20500 Muntz metal
A brass containing 60% copper and 40% zinc.

20501 Murac
An insulating material made from the latex of *Sapotaceae* species.

20502 Murald
309-00-2 227 206-215-8
Aldrin insecticides. *Murphy Chemical Co Ltd.* Discontinued.

20503 Murcurite
A mercury-containing fungicide. *Murphy Chemical Co Ltd.* Discontinued.

20504 Murdiel
60-57-1 3152 200-484-5
Dieldrin insecticides. *Murphy Chemical Co Ltd.* Discontinued.

20505 Murex
A proprietary trade name for a manganese steel containing 3% manganese, 1% carbon, and 0.85% nickel. No manufacturer.

20506 murexide
3051-09-0 6386 221-266-6
$C_8H_8N_6O_6$
5-[(hexahydro-2,4,6-trioxo-5-pyrimidinyl)imino]-2,4,6(1H,3H,5H)-pyrimidinetrione monoammonium salt
Naples red; acid ammonium purpurate; ammonium purpurate;. An red basic dyestuff, now obsolete, obtained by the action of nitric acid upon guano, and subsequently treating the product with ammonia. Used as an indicator in complexometric titrations. λ_m = 520 nm (H_2O); slightly soluble in H_2O, insoluble in organic solvents.

20507 Murfite
Acaricide. *Murphy Chemical Co Ltd.* Discontinued.

20508 Murfixtan
A mercury fungicide. *Murphy Chemical Co Ltd.* Discontinued.

20509 Murfly
An insecticide. *Murphy Chemical Co Ltd.* Discontinued.

20510 Murfotox
Organo-phosphorus insecticides. *Murphy Chemical Co Ltd.* Discontinued.

20511 Murfume
Pesticidal smoke generators. *Murphy Chemical Co Ltd.* Discontinued.

20512 Murfume Grain Store Smoke
58-89-9 5526 200-401-2
$C_6H_6Cl_6$
lindane
γ-HCH. An organochlorine insecticide. *DowElanco Ltd.*

20513 muriate of potash
7447-40-7 7783 231-211-8
KCl
Potassium chloride

20514 muriate of soda
7647-14-5 8742 231-598-3
NaCl
Sodium chloride

20515 Muritan
Compound used to control mice. *Bayer AG.*

20516 Murlin Premium Ladle Wash
A blend of finely divided, inorganic solids dispersed in an aqueous medium; used to coat and protect surfaces, tools and equipment coming in contact with molten aluminum. *Murlin Chemical Inc.*

20517 Murman's alloy
One contains 92% aluminum, 4.4% zinc, and 3.6% magnesium. Another consists of 72% aluminum, 14.5% zinc, and 13.5% magnesium.

20518 Murnil®
Veterinary preparation; for activation of the skin metabolism of dogs and cats. *Bayer AG.*

20519 Murphex
Disinfestation products. *Murphy Chemical Co Ltd.* Discontinued.

20520 Murphicol
Pesticide suspensions. *Murphy Chemical Co Ltd.* Discontinued.

20521 Murphos
56-38-2 7167 200-271-7
parathion
Parathion insecticides. *Murphy Chemical Co Ltd.* Discontinued.

20522 Murvin 85
63-25-2 1831 200-555-0
Carbaryl
Contact insecticide and worm killer. *DowElanco Ltd.*

20523 muscarine
$C_{18}H_{15}N_2Cl{\cdot}O_2$
Campanuline. A dyestuff. It is the dihydroxy derivative of Meldolás blue; dyes cotton mordanted with tannin and tartar emetic blue; used for calico printing.

20524 muscarine
300-54-9 6389 206-094-1
$[C_{19}H_{20}NO_2]^+$
[2S-(2α,4β,5α)]-tetrahydro-4-hydroxy-N,N,N,5-tetramethyl-2-furanmethanaminium
Toxic principle of the mushroom *Amanita muscaria*. .

20525 Muscatox
56-72-4 2626 200-285-3
coumaphos
Paint-on-bait insecticide for fly control, e.g., in farms. *Bayer AG.*

20526 muscle sugar
87-89-8 5008 201-781-2
$C_6H_{12}O_6$
Inositol.

20527 muscone
541-91-3 6392 208-795-8
$C_{16}H_{30}O$
3-methylcyclopentadecanone
muskone, methylexaltone. Odiferous secretion of the musk. No longer used in perfumery. bp = 328°; $bp_{0.5}$ = 130°; n_D^{17} = 1.4802; $[\alpha]_D^{17}$ = -13°; insoluble in H_2O, soluble in organic solvents.

20528 Mushet steel
Self-hardening steel. Steels containing from 0.7-1.2% carbon and 2-3% tungsten. They require no quenching or tempering.

20529 mushroom sugar
69-65-8 5788 200-711-8
$C_6H_{14}OH_6$
Mannitol
Used as base for dietetic foods, diluent, determination of boron, pharmaceutical products, medicine, thickener, and stabilizer in food products.

20530 Musiv gold
An alloy of from 66-70% copper and 30-34% zinc.

20531 musk
541-91-3 6392 208-795-8
muscone
The dried animal secretion of the musk deer. It has been practically superseded by synthetic compounds but was once prized for its role in cosmetics and perfumery, fragrances, and mothproofing agent.

20532 musk ambrette
83-66-9 201-493-7
$C_{12}H_{16}N_2O_5$
2,6-dinitro-3-methoxy-4-*tert*-butyltoluene
6-*t*-butyl-3-methyl-2,4-dinitroanisole; 5-*t*-butyl-1,3-dinitro-4-methoxy-2-methylbenzene; 1-(1,1-dimethylethyl)-2-methoxy-4-methyl-3,5-dinitrobenzene; 1-Methyl-4-*t*-butyl-3-methoxy-2,6-dinitrobenzene; Musk amberette; Ambrottolide. Used in perfumery. mp = 84-86°; bp = 135°.

20533 musk baur
$C_{11}H_{13}N_2O_4$
2,4,6-trinitro-1-methyl-3-*tert*-butyl-toluene
Musk B; tonquinol. An artificial musk; used for soap and toilet purposes.

20534 Musk R-1
11-Oxahexadecanolide
Quest Int'l. UK Ltd.

20535 Musketeer
Suspension concentrate containing 50 g ioxynil, 250 g isoproturon and 180 g mecoprop per liter; used for control of weeds in wheat and barley. *Hoechst UK.*

20536 Musol 20
Soluble yeast mucins; cosmetic ingredient for moisturizing. *Brooks Industries.*

20537 mussanin
An extract from *Albizzin anthelmintica*; used as a vermifuge.

20538 mustard gas
505-60-2 6395
$C_4H_8Cl_2S$
1,1'-thiobis[2-chloroethane]
Yperite; yellow cross gas; β,β-dichloroethyl sulfide; Kampstoff Lost; Senfgas; Sulfur mustard; S-lost; HS; Iprit; S-yperite; Lost; HD. Used as a military poison gas. Inactivated by sodium or calcium hypochlorite. mp = 13-14°; bp_{10} = 98°; d_4^{20}= 1.2741; n_D^{20}= 1.53125; LD_{50} (rat iv) = 3.3 mg/kg.

20539 mustard oil
542-85-8 3861 208-831-2
C_2H_5NCS
Ethyl isothiocyanate
Military poison gas.

20540 Mustone
A Japanese chloroprene polymer.

20541 Muthmann's liquid
79-27-6 9328 201-191-5
$C_2H_2Br_4$
1,1,2,2-tetrabromoethane
Acetylene-tetrabromide; Tetrabromoacetylene; tetrabromoethane; *sym*-Tetrabromoethane; TBE; *s*-Tetrabromoethane; 1,1,2,2-tetrabromoethylene. Used as a solvent and in flotation of minerals for separation. mp = 0°; bp_{54} = 151°; d = 2.964; n_D^{20} = 1.638; insoluble in H_2O, soluble in organic solvents; LD_{50} (mus ip) = 443 mg/kg.

20542 MVP
Methacrylate
Permanent structural adhesive for bonding metals, plastics and ceramics. *ITW Devcon.*

20543 MXM-7500
A proprietary epoxy putty used for filling in difficult radii and depressions in epoxy-fiber glass components. *Fiberite West Coast Corp.* Unverified.

20544 Myacide® AS Plus
52-51-7 1470 200-143-0
2-Bromo-2-nitropropane-1,3-diol
bronopol. Water-soluble antimicrobial, preservative for papermaking, adhesives, coatings, cooling towers, process waters, oilfield water-flooding operations, deodorizing, other aqueous applications. *Angus; Boots Co, PLC.*

20545 Myacide® S-1, S-2
52-51-7 1470 200-143-0
2-Bromo-2-nitropropane-1,3-diol
bronopol. 2-Bromo-2-nitropropane-1,3-diol in dipropylene glycol methyl ether/water; water treatment antimicrobial for recirculating cooling towers and process water; preservative for oilfield drilling-fluid and water-flooding operations. *Angus; Boots Co, PLC.*

20546 Myacide® SP
1777-82-8 3110 217-210-5
2,4-Dichlorobenzyl alcohol
Antifungal agent, preservative. *Inolex; Boots Co. PLC.*

20547 My-B-Den
61-19-8 157 200-500-0
Adenosine monophosphate
5'-adenylic acid. Used in biochemical research.

20548 Mybond®
Formulated natural and synthetic adhesives in aqueous, hot melt, solventless and solvent-borne forms for use in industry; for bonding a wide range of substrates for the paper and board, packaging, woodworking, building and decorating, textile, footwear, automotive, bookbinding, electrical and product assembly markets. *Mydrin Ltd.*

20549 myclobutanil
88671-89-0 6402

$C_{15}H_{17}ClN_4$
α-butyl-α-(4-chlorophenyl)-1H-1,2,4-triazole-1-propanenitrile
RH-3866; Nova; Rally; Systhane. Systemic fungicide with protective and curative action. Used in control of Ascomycetes, Fungi Imperfecti and Basidomycetes in a wide variety of crops. mp = 63-68°; bp_1 = 202-208°; soluble in H_2O (142 mg/ml), polar organic solvents; LD_{50} (rat orl) = 1600 mg/kg. *Rohm & Haas.*

20550 Mycocide
A slime control agent. *Great Lakes Europe.* Discontinued.

20551 Mycose
99-20-7 9713 202-739-6
$C_{12}H_{22}O_{11}$
Trehalose
α-D-Glucopyranosyl-α-D-glucopyranoside; α,α'-Trehalose. Used in biochemical research. mp = 96-98°; d = 1.54; $[\alpha]_D^{20}$= 178° (c = 7); soluble in H_2O, EtOH, less soluble in organic solvents.

20552 Mycotal
A fungal parasite from *Verticillium lecanil.* Used for control of aphids and whitefly. *Koppert (UK) Ltd.*

20553 Mycovac-L
Mycoplasma gallisepticum. For immunization of poultry. *Intervet Inc.*

20554 Mydochrome
Color reversal processing system. *May & Baker Ltd.* Name unverified.

20555 Mydoneg
Color film processing system. *May & Baker Ltd.* Name unverified.

20556 Mydoprint
Color paper processing system. *May & Baker Ltd.* Name unverified.

20557 myelin
A white, fatty substance obtained from various animal and vegetable tissues.

20558 Myer's naphthol green
Standard. Naphthol green B (50 mg) dissolved in 1,000 ml water; used for cholesterol determination.

20559 Myflam®
Formulated aqueous emulsion polymer coatings for use in the textile industry; coating and impregnating compounds imparting fire retardant and other technical performance requirements; for upholstery, drapes, blinds, carpet, bedding, automotive, aerospace, industrial, wallcoverings, apparel and flooring industries. *Mydrin Ltd.*

20560 Mykon
Wetting and detergent, dispersing and emulsifying properties; used for textiles; general purpose detergent and wetting agent. *Warwick International Ltd.* Name unverified.

20561 Mykon 817
Phosphated alcohol anionic surfactant as a clear liquid; wetting and dispersing agent, scouring and bleach assistant; used primarily in the preparation and dyeing of cotton and synthetic fibers. *Warwick Chemical Ltd.* Name unverified.

20562 Mykonaid
Dispersing and leveling agent; used for textiles especially in high temperature dyeing. *Warwick International Ltd.* Name unverified.

20563 Mykroy/Mycalex
Machinable moldable ceramic material made from glass and mica powder. The material is completely inorganic and is supplied in sheet, rods or parts molded to specification; machinable with standard shop tools and no after-firing is required; for use in areas where dimensional stability is required over a wide temperature range; e.g. high voltage switch gears, arc barriers, asbestos-filled material replacement, thermal barrier for molding presses and thermal switch covers. *Mykroy/Mycalex.* Name unverified.

20564 Mylanta
1309-42-8 5706 215-170-3
H_2MgO_2
magnesium hydroxide
Antacid preparations. *ICI Chem & Polymers Ltd.* Discontinued.

20565 Mylar
A polyester film used for electrical insulation, cable lapping, magnetic tape; features very high tensile strength. *ICI Chem & Polymers Ltd.* Discontinued.

20566 Mylar®
Polymer. *DuPont UK.*

20567 Mylocon
A proprietary preparation containing methyl polysiloxane in a flavored vehicle. *Parke-Davis.* Name unverified.

20568 Mylol
An insect repellant. *The Boots Co plc.* Discontinued.

20569 Myodil
99-79-6 5074 202-787-8
Ethyl iodophenylundecylate
Radiopaque material used as a diagnostic reagent. *Glaxo Laboratories.* Name unverified.

20570 Mypolex®
For the chemical industry. *DuPont UK.*

20571 myrabola oil
A German soap-making material. It is an oil stated to be a mixture of different fatty acids and their glycerides. It is obtained from fat waste.

20572 Myras
Internal gear pump; for light oil and viscous liquid pumping. *Sihi Pumps (UK) Ltd.*

20573 Myrickite
A trade name for a chalcedony, a form of silicon dioxide. No manufacturer.

20574 Myristic acid
544-63-8 6416 208-875-2
$C_{14}H_{28}O_2$
Tetradecanoic acid
Emery® 654; Hystrene ® 9014; Hystrene® 9514; Philacid 1400; Prifac 2940; Univol U320. Detergent surfactant. Organic acid used in the manufacture of soaps, cosmetics, synthesis of esters for flavors and perfumes; component of food-grade additives. mp = 58.5°; bp = 250.5°; d = 0.8622; soluble in abs alcohol, methanol, ether, petr ether, benzene, chloroform; neutralization value = 245.68; LD_{50} (mus iv) = 432.6 mg/kg. *Akzo; Aldrich; Henkel/Emery; Mirachem Srl; Unichema; Witco/Humko.*

20575 myristica
Nutmeg.

20576 Myristoyl hydrolyzed collagen
72319-06-3
Lexein® A200; Pro-Tein SM-20. Myristoyl hydrolyzed collagen; film-forming collagen protein deriv.; resin modifier for hair sprays, makeups, protection skin lotions and creams. *Inolex.*

20577 myristyl alcohol
112-72-1 6148 204-000-3
$C_{14}H_{30}O$
1-tetradecanol
tetradecyl alcohol; Fatty alcohol. C14 linear primary alcohol; surfactant intermediate; organic synthesis; antifoam agent; perfume fixative for soaps and cosmetics; specialty cleaning preparations; emollient for cold creams. *Ethyl; W.R. Greeff; M. Michel; Vista.*

20578 myristyl lactate
1323-03-1 215-350-1
$C_{17}H_{34}O_3$
2-hydroxypropanoic acid, tetradecyl ester
tetradecyl 2-hydroxypropanoate. Ester of myristyl alcohol and lactic acid; Imparts lubricity, sheen, etc. to cosmetic and pharmaceutical applications. *Dinoval Chem. Ltd.*

20579 Myritol
Skin compatible oil component and fattening agent; used for cosmetic preparations. *Henkel Chemicals Ltd.* Name unverified.

20580 Myritol 318
65381-09-1 265-724-3
Caprylic/capric triglyceride
Emollient for pharmaceutical and cosmetic preparations in emulsion form; oily component with solvent capacity; solubilizer. *Henkel/Cospha; Henkel Canada.*

20581 Myrj®
Polyoxyethylene alkylesters
ICI Am.

20582 Myrj® 45
9004-99-3
PEG-8 stearate
General purpose oil-water emulsifiers for cosmetics, pharmaceuticals, etc. *ICI Spec. Chem.; ICI Surf. Belgium.*

20583 Myrj® 52
9004-99-3
PEG-40 stearate
Pharmaceutical aid. *ICI Am.*

20584 Myrj® 53
9004-99-3
PEG-50 stearate
Pharmaceutical aid. *ICI Am.*

20585 Myrj® 59
9004-99-3
PEG-100 stearate
Oil-water emulsifier for cosmetic and pharmaceutical applications. *ICI Spec. Chem.; ICI Surf. Belgium.*

20586 myrmekite
A quartz mineral.

20587 myrobalans
The fruit of *Terminalia chebula* of India. This is the chief variety of this product, but there are at least five varieties of the commercial article which are named after the district where they are marketed.

20588 myrrh
6421
Known as Herabol myrrh, a gum resin obtained from various species of *Balsamodedron* and *Cammiphora*; used in perfumery, incense, and toiletries.

20589 myrtrimonium bromide
1119-97-7 6419 214-291-9
$C_{17}H_{38}BrN$
N,N,N-trimethyl-1-tetradecanaminium bromide
myristryltrimethylammonium bromide; tetradecyltrimethylammonium bromide; trimethyltetradecylammonium bromide; tetradonium bromide. A cationic germicidal detergent, disinfectant and deodorant. mp = 245-250°; soluble in H_2O (20 g/100 ml); LD_{50} (rat iv) = 15 mg/kg.

20590 Mystery gold
An alloy of 1 part platinum and 2 parts copper, with a little silver.

20591 Mystic metal
An alloy of 88.7% lead, 10.85 antimony, 0.4% iron, and 0.1% bismuth.

20592 mystin
A mixture of formaldehyde and sodium nitrite; used as a preservative.

20593 Mystolene
Water repellents and proofers. *Catomance Group.*

20594 Mystox
Preservatives for textiles. *Catomance Group.*

20595 Mytab®
1119-97-7 6419 214-291-9
Myrtrimonium bromide
Surfactant used as emulsifier and antistat in hair rinses; antimicrobial for cosmetics, topicals. *Zeeland.*

20596 Mytex
Formulated aqueous emulsion polymer coatings for use in the textile industry; coating and impregnating compound to impart technical properties such as fabric stabilization, seam slippage resistance, abrasion resistance, tuft lock, stain resistance, shower repellency and waterproofing, air and moisture permeability, thermal and electrical conductivity, blackout; for upholstery, soft furnishings, carpet, bedding, wallcovering, automotive, aerospace, footwear, apparel, label, motif, industrial markets. *Mydrin Ltd.*

20597 Mytolac
Acne lotion and cream, to clear acne blemishes. *Richardson-Vicks Inc.* Name unverified.

20598 Myvacet®
Acetylated monoglycerides
Chemical intermediates. *Eastman.*

20599 Myvacet® 5-07K
Acetylated hydrogenated vegetable glyceride, distilled; emulsifier, emollient; forms films with good moisture vapor barrier properties for nuts, dried fruits, sausages. *Eastman.*

20600 Myvacet® 7-07K
Acetylated hydrogenated vegetable glyceride; forms thin transparent film to serve as oxygen moisture barrier for food pkg. *Eastman.*

20601 Myvacet® 9-08K
Acetylated hydrogenated coconut oil glyceride, distilled; emulsifier, emollient for food processing. *Eastman.*

20602 Myvacet® 9-40
8029-92-3
Acetylated lard glyceride, distilled; emulsifier; food-grade lubricant and emollient; deaerator in some systems. *Eastman.*

20603 Myvacet® 9-45K
Acetylated hydrogenated soybean oil glyceride; emulsifier, lubricant, solvent for icings and shortenings. *Eastman.*

20604 Myvaplex®
Glyceryl monostearate
Surfactant. *Eastman.*

20605 Myvaplex® 600K
Glyceryl stearate, food grade; emulsifier in macaroni and cereal products; starch complexing agent, lubricant, processing aid for foods. *Eastman.*

20606 Myvatem® 30
Diacetyl tartaric acid ester of distilled monoglycerides (from edible tallow); emulsifier, dispersant for food, pharmaceutical, and cosmetic applications. *Eastman.*

20607 Myvatex®
Food emulsifiers. *Eastman.*

20608 Myvatex® 3-50K
Gyceryl stearate (distilled) and propylene glycol stearate (distilled); emulsifier in food applications. *Eastman.*

20609 Myvatex® 8-06K
Hydrogenated soybean oil monoglyceride; food emulsifier. *Eastman.*

20610 Myvatex® 40-06S K
Distilled propylene glycol esters, distilled monoglycerides, lactylic esters of stearic acid, and water; food emulsifier for cakes. *Eastman.*

20611 Myvatex® 90-10K
Hydrogenated rapeseed oil, hydrogenated cottonseed oil; food emulsifier. *Eastman.*

20612 Myvatex® 600PK
Glyceryl monostearate (from hydrogenated soybean oil); starch complexing agent, lubricant, processing aid for foods. *Eastman.*

20613 Myvatex® Do Control K
Succinylated palm oil monoglycerides, palm oil monoglycerides, distilled; dough strengthener for yeast-raised bakery goods. *Eastman.*

20614 Myvatex® Mighty Soft®
Distilled monoglyceride prepared from edible vegetable oil; softener, bodying agent, extension aid for yeast-raised bakery goods, ice cream, frozen desserts, pasta. *Eastman.*

20615 Myvatex® Monoset® K
Rapeseed oil monoglycerides, cottonseed oil monoglycerides, distilled; food emulsifier. *Eastman.*

20616 Myvatex® Super DO
Succinylated monoglycerides and distilled monoglycerides; emulsifier for baked goods. *Eastman.* Name unverified.

20617 Myvatex® Texture Lite® K
Soybean oil monoglycerides, propylene glycol monoesters, sodium stearoyl lactylate, silicon dioxide; emulsifier for cakes, icings, cream fillings, whipped toppings, sauces. *Eastman.*

20618 Myverol®
Propylene glycol emulsifier. *Eastman.*

20619 Myverol® 18-00
Distilled hydrogenated animal glyceride
Emulsifier for baked goods, confectionery products, cosmetics, dehydrated potatoes, etc. *Eastman.* Name unverified.

20620 Myverol® 18-04K
67784-87-6 267-057-3
Hydrogenated palm oil glyceride; food emulsifier for candy, infant formula, margarine, peanut butter. *Eastman.*

20621 Myverol® 18-06K
Hydrogenated soy glyceride; food emulsifier for candy, chewing gum base, cake mixes, infant formula, whipped toppings. *Eastman.*

20622 Myverol® 18-07K
Hydrogenated cottonseed glyceride; food emulsifier for candy, chewing gum base, cake mixes, confectionery coatings, infant formula, whipped toppings. *Eastman.*

20623 Myverol® 18-30
61789-13-7 263-035-2
Tallow glyceride, distilled; emulsifier for baked goods, confectionery products, cosmetics, etc. *Eastman.* Name unverified.

20624 Myverol® 18-35
Palm oil glyceride, distilled; emulsifier for baked goods, confectionery products, cosmetics, etc. *Eastman.* Name unverified.

20625 Myverol® 18-40
61789-10-4 263-032-6
Lard glyceride, distilled; emulsifier for baked goods, confectionery products, cosmetics, etc. *Eastman.*

20626 Myverol® 18-50K
61789-08-0 263-030-5
Hydrogenated vegetable glyceride, distilled; food emulsifier used in icings, cream fillings, cake mixes, shortenings. *Eastman.*

20627 Myverol® 18-85K
8029-44-5 232-438-5
Cottonseed glyceride; food emulsifier for icings, cream fillings, shortenings. *Eastman.*

20628 Myverol® 18-92, 18-92K
Sunflower seed oil glyceride; food emulsifier for icings, cream fillings, shortenings. *Eastman.*

20629 Myverol® 18-98
Safflower glyceride, distilled; emulsifier for diet margarines, icing and cream-fillings shortenings. *Eastman.* Name unverified.

20630 Myverol® 18-99K
Low-erucic rapeseed oil monoglyceride, distilled; food emulsifier for icings, cream fillings, shortenings. *Eastman.*

20631 Myverol® P-06K
Distilled monoester from hydrogenated soybean oil and propylene glycol; food emulsifier, stabilizer, aerating agent. *Eastman.*

20632 Myverol® SMG VK
Hydrogenated palm oil or palm stearine succinylated monoglyceride; emulsifier for shortenings, dough strengthener, softener for bread baking. *Eastman.*

20633 N.
1344-09-8 8824 215-687-4
Na_4O_4Si
sodium orthosilicate
Water glass; Soluble glass; Silicate of soda; Silicic Acid Sodium Salt; Sodium silicate glass. Sodium silicate solution (3.22 SiO_2:Na_2O ratio; 41 Be density); builder for laundry detergents and cleaners, adhesive for corrugated, spiral wound and laminated paper products, binder for foundry cores and molds, manufacture of silica gels, zeolites and catalysts. *PQ Corp.* Name unverified.

20634 N.C.T
Nitrocellulose tutular, a pyro-collodion powder, made from a gelatinized nitrocellulose, pressed in the form of rods. No manufacturer.

20635 N.E. powder
A 36-grain powder, containing metallic nitrates, and nitrohydrocarbons.

20636 N.P.L. alloy
An alloy of 94.5% aluminum, 2.5% nickel, and 1.5% magnesium.

20637 N.S. fluid
A mixture of sodium chloride, aluminum chloride, and iron chloride.

20638 N.S.B
A proprietary preparation of noxytiolin, and vinyl pyrrolidone/vinyl acetate copolymer in an aerosol; wound dressing. *Geistlich Sohne AG.* Name unverified.

20639 N33
An alloy of cast iron with additions of nickel, copper, and chromium.

20640 N-521® Biocide
533-74-4 2892 208-576-7
$C_5H_{10}N_2S_2$
Tetrahydro-3,5-dimethyl-2H-1,3,5-thiadiazine-2-thione
Basimid G; Basamid; Basamid P; basamid-puder; carbothialdine; CRAG; Crag 974; Crag 85W; Dazomet; dimethylformocarbothialdine; DMTT; Fennosan B 100; Micofume; Mylone; Mylone 85; N 521; Nalcon 243; Nefusan; Prezervit; Stauffer N 521; thiazone; Tiazon; Troysan 142; UCC 974. Fungicide, bactericide and herbicide for use in leather, paint, glue, casein, starch, and paper manufacturing and processing. mp = 104-105; SG = 1.3; insoluble in H_2O (< 1mg/ml), EtOH; λ_m = 242, 289 nm (ε7150, 9900 cyclohexane); LD_{50} (rat orl) = 363 mg/kg. *Akzo.*

20641 N-948® Biocide
6317-18-6 228-652-3
$C_3H_2N_2S_2$
Methylene bis (thiocyanate)
thiocyanic acid, methylene ester; dithiocyanatomethane; Antiblu 3737; Busan 110; Cytox; Nalco D-1994; Slimicide MC. Industrial biocide controlling algae, bacteria, yeast and fungi. mp = 102-106°. *Akzo.*

20642 Na Ta
76-03-9 9756 200-927-2
$C_2HCl_3O_2$
trichloroacetic acid
TCA; for control of weeds in field crops. mp = 54-58°; d = 1.629 *Hoechst UK.*

20643 Nacap®
2492-26-4 219-660-8
$C_7H_6NNaS_2$
2(3H)-benzothiazolethione, sodium salt
sodium 2-mercaptobenzothiazole; 2-benzothiazolethiol, sodium salt; duodex; 2-mercapto-benzothiazole, sodium; Nacap; vancide 51; sodium 2-benzothiazolethiolate; sodium 2-mercaptobenzothiazolate; benzothiazolethiol, sodium salt; benzothiazolethione, sodium salt; Nuodex; sodium 2-mercaptobenzothiolate; sodium benzothiazolethiolate; sodium, (2-benzothiazolylthio)-; sodium benzothiazol-2-yl sulfide; sodium benzothiazole-2-thiolate. Metal deactivator; corrosion inhibitor for water, alcohol, and glycol systems; used in antifreeze; chemical intermediate. mp = -6°; bp = 103°; SG = 1.25; soluble in H_2O (>100 mg/ml). *R. T. Vanderbilt Co Inc.* Discontinued.

20644 Nacconal
Sodium alkylbenzene sulfonate
Detergent, wetting agent. *Stepan.* Name unverified.

20645 Nacconol 35SL
Anionic surfactant in liquid form; used for liquid detergents. *KWR Chemicals Ltd.* Name unverified.

20646 Nacconol 90F and 40F
Anionic surfactant in cream flake form; used for powdered detergents. *KWR Chemicals Ltd.* Name unverified.

20647 Nacconol® 40G
25155-30-0 8757 246-680-4
$C_{18}H_{29}NaO_3S$
Sodium dodecylbenzene sulfonate
AA-9; AA-10; Abeson Nam; Bio-soft D-35x; Calsoft L-60; Conco AAS-90; Conoco C-50; Detergent HD-90; Mercol 30; Naccanol SW; Nacconol 40F; Neccanol SW; pilot SP-60; Richonate 60b; Santomerse no. 85; Solar 90; Sulfapol; Sulframin 85; Sulframin 1238 slurry; Sulframin 1250 slurry; Ultrawet K; Stepan DS 60; Ultrawet 1t; Marlon A 350; Marlon A; Maranil; Marlon a 375; Siponate DS 10; Trepolate F 40; Nansa SL; Santomerse ME; Merpisap AP 90P; Nansa SS; Trepolate F 95; Nansa HS 80; Deterlon; Ultrawet 99lS; Sulfuril 50; F 90; Elfan WA; Sandet 60; Steinaryl NKS 50; Sinnozon; Nansa HS 85S; C 550; KB; HS 85S; Nansa HF 80; Arylan SBC; Marlon 375a; X 2073; Conco AAS 35H; Neopelex 05; Richonate 40b; DS 60; Pelopon A; Sulframin 1240; 35SL; SDBS; Nacconol; Santomerse; Sulframin 1238; Ultrawet XK,. Foamer, dispersant, wetting agent, detergent for agriculture, cement, dyeing, emulsion polymerization, textile, metal cleaning, metalworking, mining, paper industries; biodegradable. LD_{50} (rat orl) = 2.0 g/kg. *Stepan; Stepan Canada.*

20648 Nadavin®
Agents to increase the wet strength of papers. *Bayer AG; Bayer plc.*

20649 Nafion®
Perfluorosulfonic acid membrane. *DuPont UK.*

20650 Naftifine
65473-14-5 6442
$C_{21}H_{21}N$
(E)-N-Methyl-N-(3-phenyl-2-propenyl)-1-naphtalenemethanamine
Exoderil; Naftin; Naftifine hydrochloride; N-methyl-N-(1-naphthylmethyl)-3-phenylpropen-1-amine; naftifungin(E)-N-cinnamyl-N-methyl-1-naphthalenemethylamine,. Squalene epoxidase inhibitor. Blocks ergosterol biosynthesis and is used as a topical antifungal. Colorless oil; bp_{11} = 162-167°. *Sandoz Pharmaceuticals.* Name unverified.

20651 Naftocit
Vulcanizing accelerators for latex and rubber processing. *Chemetall GmbH.* Name unverified.

20652 Naftogran®
Polymer bounded vulcanizing accelerators for improved and nonpolluting processing in rubber chemicals. *Chemetall GmbH.*

20653 Naftolen
Plasticizers and extender oils for natural and synthetic rubber. *Chemetall GmbH.* Name unverified.

20654 Naftolen R 100, 510, 530, 550, 570, X413, X414 X10, 134
Vinyl plasticizers. *Wilmington Chemical Co.* Name unverified.

20655 Naftomix
PVC stabilizer and lubricant. *Chemson Ltd.*

20656 Naftonox
Antioxidants for the production and processing of natural and synthetic rubber and their latexes, thermoplastics, and adhesives. *Chemetall GmbH.* Name unverified.

20657 Naftopast
Solid dispersions for improved and nonpolluting processing in rubber compounds. *Chemetall GmbH.* Name unverified.

20658 Naftovin
PVC stabilizer. *Chemson Ltd.*

20659 Naftozin
Stearic acids as processing auxiliaries in the production of compounds and as lubricants in the processing of thermoplastics. *Chemetall GmbH.* Name unverified.

20660 Naganol
129-46-4 9181 204-949-3
$C_{51}H_{34}N_6Na_6O_{23}S_6$
8,8'-[carbonylbis[imino-3,1-phenylenecarbonylimino(4-methyl-3,1-phenylene)carbonylimino]]bis-1,3,5-naphthalenetrisulfonic acid hexasodium salt
suramin sodium; Bayer 205; Fourneau 309; Antrypol; Germanin; Moranyl; Naphuride. Veterinary preparation; for the control of trypanosomiasis of domestic animals. Soluble in H_2O, insoluble in organic solvents; LD_{50} (mus iv) = 620 mg/kg. *Bayer AG.*

20661 Nageli's solution
A solution containing a mixture of zinc chloride and iodide; used as a disinfectant.

20662 Naglusol
527-07-1 8766 208-407-7
Sodium gluconate solution. *Jungbunzlauer Inc.*

20663 nahcolite
A native sodium bicarbonate.

20664 Nalan
Water repellent. *DuPont UK.* Name unverified.

20665 Nalcite
A proprietary trade name for a water softener containing an organic zeolite type exchanger material. No manufacturer.

20666 Nalco® 131
Glycol-type surfactants; food grade antifoam for beet sugar and enzyme aqueous operations. *Nalco.*

20667 Nalco® 2300
Silica/silicone
Defoamer for coatings, latex and high solids systems. *Nalco.*

20668 Nalco® 2340
Silica-organic; nonsilicone antifoam for aqueous resin and polymer processing; short term persistency. *Nalco.*

20669 Nalco® 8669
Silica-organic; nonsilicone antifoam for aqueous fiberglass mat resin processing; effective over wide temperature range. *Nalco.*

20670 Nalfleet
Chemical treatments for the marine industry. *ICI Chem & Polymers Ltd.* Name unverified.

20671 Nalfloc
Water and effluent treatment chemicals. *ICI Chem & Polymers Ltd.* Name unverified.

20672 Nalidone®
28874-51-3 249-277-1
Sodium PCA; moisturizer for dermatological soap, shampoo, after-sun gel, nutritive and regenerative creams and lotions, hair comb-out balm. *UCIB.*

20673 Nalzin
55799-16-1
Zinc hydroxy phosphite
Nontoxic anticorrosive pigment for paint. *Rheox Inc.*

20674 Nametal
7440-23-5 8710 231-132-9
Foil wrapped metallic sodium. *Foseco (F.S.) Ltd.*

20675 Nandel®
Fibers. *DuPont UK.*

20676 Nangawhite
A nonstaining antioxidant for rubber. *Rubber Regenerating Co.* Unverified.

20677 Nankor
299-84-3 8415 206-082-6
A proprietary preparation of fenchlorphos; an insecticide. No manufacturer.

20678 Nansa
Alkylaryl sulfonates
Albright & Wilson Ltd, Detergents Div.

20679 Nansa UC
A range of detergent powders containing a sodium alkyl benzene sulfonate as the active ingredient. No manufacturer.

20680 Nansa UCA/S and UCP/S
Spray-dried built powders based on straight chain alkylbenzene sulfonate; blue, free flowing powder, dust free; anionic for built detergents. *Albright & Wilson Ltd, Detergents Div, Marchon.* Name unverified.

20681 Nansa® 1042
68584-22-5 271-528-9
Dodecylbenzene sulfonic acid, calcium salt
C10-16-Alkylbenzenesulfonic acid, calcium salt; benzenesulfonic acid, C10-16-alkyl derivs. Intermediate used in manufacturing of detergents, laundry products, emulsifiers for emulsion polymerization. *Albright & Wilson UK.* Name unverified.

20682 Nansa® 1106/P
Sodium dodecylbenzene sulfonate
Surfactant for emulsion polymerization. *Albright & Wilson UK.* Name unverified.

20683 Nansa® 1169/P
Alkylaryl sulfonate anionic
Supplied as a golden liquid; surfactant for emulsion polymerization. *Albright & Wilson Ltd, Detergents Div, Marchon.* Name unverified.

20684 Nansa® AS 40
1331-61-9 215-559-8
$C_{18}H_{33}NO_3S$
Ammonium dodecylbenzene sulfonate
Formulation of domestic and industrial liquid detergents. *Albright & Wilson UK.*

20685 Nansa® BMC
Blended anionic nonionic foaming agent, in yellow liquid form; air entraining agent for mortar and concrete. *Albright & Wilson Ltd, Detergents Div, Marchon.* Name unverified.

20686 Nansa® EVM50
68953-96-8 273-234-6
calcium dodecylbenzene sulfonate
Calcium dodecylbenzene sulfonate in aromatic solvent; emulsifier, dispersant for agrochemicals, textiles, surface coatings, polymerization, leather industries. *Albright & Wilson UK.*

20687 Nansa® HS80S
25155-30-0 8757 246-680-4
Sodium dodecylbenzene sulfonate
Formulation of detergents, hard surface and bottle cleaners; metal treatment and paper processing; scouring and wetting agent for textile industry; foamer and mortar plasticizer in building industry. *Albright & Wilson UK.*

20688 Nansa® LES42
Predominantly straight chain sodium dodecylbenzene sulfonate, lauryl ether sulfonate and magnesium xylene sulfonate blend giving an anionic in the form of a golden liquid; used for dishwashing detergents; hard surface cleaners. *Albright & Wilson Ltd, Detergents Div, Marchon.* Name unverified.

20689 Nansa® LSS38/A
68439-57-6 270-407-8
Sodium C_{14}-C_{16} olefin sulfonate
General purpose detergent base; wetting agent for agricultural wettable powders, textile processing; emulsifier for enhanced oil recovery. *Albright & Wilson UK.*

20690 Nansa® MA30
Sodium dodecylbenzene sulfonate and ethoxylated nonionic blend; detergent base for dishwashing and hard surface cleansers; scouring agent for textiles; mortar plasticizer in manufacturing of masonry cement. *Albright & Wilson UK.*

20691 Nansa® SB30
Branched sodium alkylbenzene sulfonate
Surfactant. *Albright & Wilson UK.*

20692 Nansa® SBA
68608-88-8 271-807-5
Branched dodecylbenzene sulfonic acid
Detergent intermediate. *Albright & Wilson UK.*

20693 Nansa® SL 30
Sodium dodecylbenzene sulfonate
Surfactant used in detergent formulations. *Albright & Wilson UK.* Name unverified.

20694 Nansa® SS 30
Sodium dodecylbenzene sulfonate
Surfactant for detergent formulations; emulsifier for agricultural emulsifiable concentrates; curing/foaming agent for thermosets. *Albright & Wilson UK.*

20695 Nansa® SSA
68584-22-5 271-528-9
Dodecylbenzene sulfonic acid
Detergent intermediate; in preparation of emulsifiers for emulsion polymerization. *Albright & Wilson UK.*

20696 Nansa® TDB
25496-01-9 247-036-5
$C_{19}H_{32}O_3S$
Tridecylbenzene sulfonic acid
Detergent ingredient. *Albright & Wilson UK.* Name unverified.

20697 Nansa® TS 50
68584-25-8 271-532-0
TEA alkylbenzene sulfonate
Detergent raw material. *Albright & Wilson UK.*

20698 Nansa® YS94
68584-24-7 271-531-5
Isopropylamine dodecylbenzene sulfonate
Emulsifier for solvent-based hand cleaners, agricultural emulsifiable concentrates; coupling agent for water in charge detergent systems; biodegradable. *Albright & Wilson UK.*

20699 Nansen
A high-speed tungsten steel containing 18% tungsten.

20700 napalm
An aluminum soap consisting of a mixture of oleic naphthenic and coconut fatty acids. Makes petrol thicken and gel; used in flame throwers and fire bombs.

20701 napalite
A dark-red wax which melts at 42°C. It is a hydrocarbon and occurs naturally.

20702 Napelec
Nondraining impregnant (obtained by mixing Napvis with water). *BP Chemicals Ltd.*

20703 Napgel
Antifreeze. *BP Chemicals Ltd.*

20704 naphalane
A crude naphtha product containing soap. It is similar to naphthalene (qv).

20705 naphtha
8030-30-6 7329 232-443-2
coal tar naphtha
petroleum naphtha; benzin. Petroleum distillate; the less volatile portion obtained in redistilling benzine, boiling from about 95-100°C; as a source of

gasoline, special naphthas, petroleum chemicals, especially ethylene; thinners in paint, drycleaning fluid; blending with natural gas. *Ashland; Mobil; Monsanto; Norsk Hydro A/S; Texaco.*

20706 naphthalene
91-20-3 6457 202-049-5
$C_{10}H_8$
naphthalin; naphthene; tar camphor. Intermediate in manufacture of phthalic anhydride, naphthol, chlorinated naphthalenes, dyes, etc.; moth repellent, fungicide, smokeless powder, lubricant, synthetic resin, tanning, preservative, textile auxiliary, emulsion breaker, scintillation counter, antiseptic. mp = 80°; bp = 218°; d_4^{20}= 1.162; n_D^{100}= 1.58212; insoluble in H_2O, soluble in organic solvents. *AlliedSignal; Aristech; Crowley Tar Prods.; Koch; Stan Chem Int'l. Ltd; Texaco.*

20707 β-naphthalene sulfonic acid
120-18-3 6464 204-375-3
$C_{10}H_8O_3S{\cdot}H_2O$
Naphthalene-2-sulfonic acid
2-Naphthalenesulfonic acid monohydrate; Naphthalene-2-sulfonic acid hydrate. Used in manufacture of α-naphthol and to solubilize phenols in water. mp = 24°; d = 1.4400; LD_{50} (rat orl) = 4440 mg/kg.

20708 2,6-naphthalenedisulfonic acid
581-75-9 6461 209-471-9
$C_{10}H_8O_6S_2$
Ebert and Merz's β-acid. Crystals; very soluble in H_2O, EtOH; practically insoluble in Et_2O.

20709 1,6-naphthalenediol
575-44-0 209-386-7
$C_{10}H_8O_2$
1,6-dihydroxynaphthalene
C.I. 76630; naphthalene, 1,6-dihydroxy-; 1.6-dihydroxynaphthalene; 2,5-naphthalenediol; 6-hydroxy-1-naphthol. Polycyclic phenol. Used in chemical synthesis. *Aceto.*

20710 2,3-naphthalenediol
92-44-4 202-156-7
$C_{10}H_8O_2$
2,3-dihydroxynaphthalene
Complexing reagent. mp = 162-164°.

20711 2,7-naphthalenediol
582-17-2 209-478-7
$C_{10}H_8O_2$
2,7-dihydroxynaphthalene
Reagent. Used in synthesis. mp = 187°. *Aceto.*

20712 naphthalol
613-78-5 6502 210-355-5
$C_{17}H_{12}O_3$
2-naphthyl salicylate
betol; 2-hydroxybenzoic acid 2-naphthalenyl ester; naphthosalol; Salinaphthol. mp = 95°; insoluble in H_2O, soluble in organic solvents.

20713 naphthazarin
475-38-7 207-495-4
$C_{10}H_6O_4$
5,8-dihydroxy-1,4 naphthoquinone
5,8-dihydroxy-1,4-naphthalenedione. Used in manufacture of dyestuffs. mp = 237°;.

20714 naphthenic acid
1338-24-5 215-662-8
Paint dryers, fungicides, metal catalysts, corrosion inhibitors, lubricants, fracturing fluids, cellulose preservatives, solvents, detergents, rubber reclaiming agent. Cobalt, manganese, calcium and other salts of naphthenic acids are used for oil-based paint driers, catalysts and lube additives, 45%; copper and zinc naphthenate wood preservatives, 20%; tire adhesion promoter, 20%; oil field corrosion inhibitors and other uses, including surfactants and cutting oils, 15%. bp_6 = 160-198°; d = 1.0340; *Crowley Chem.; Orange Chem. Ltd.*

20715 naphthionic acid
84-86-6 6490 201-567-9
1-Naphthylamine 4-sulfonic acid
Used as an intermediate for azo dyes.

20716 naphthite
Trinitro-naphthalene
Used in explosives.

20717 Naphthochrome
Wool dyestuffs. *Clayton Aniline Co Ltd.* Name unverified.

20718 Naphthocyanine
Fast acid wool dyestuffs. *J C Bottomley.* Name unverified.

20719 naphthocyanole
A homologue of pinacyanol (a red sensitizer for silver bromide plates), prepared by the condensation of β-naphtho-quinaldine ethiodide with formaldehyde in the presence of alcoholic potash; used as a red sensitizer for photographic plates.

20720 naphthoformol
A product of α-naphthol and formaldehyde; used as a dusting powder.

20721 Naphthol
Azoic dyestuffs. *Hoechst AG.*

20722 Naphthol Aristol
$C_{10}H_6I_2O_2$
Iodo-naphthol, used as an antiseptic.

20723 Naphtholite
A proprietary trade name for a light petroleum distillate used as a solvent, etc. No manufacturer.

20724 naphtholith
A bituminous shale.

20725 Naphthopone E
Dispersing agent; used in naphthol AS developing baths for aftersoaping fast dyeings. *Bayer AG.*

20726 naphthoresorcin
132-86-5 6483 205-079-7
$C_{10}H_8O_2$
1,3-dihydroxy-naphthalene
1,3-dihydroxynaphthalene; naphthoresorcinol; naphthalene-1,3-diol. Inhibitor of prostaglandin synthase. Used to assay sugars, oils and glucuronic acid in urine. mp = 123-125°; soluble in H_2O, EtOH, Et_2O.

20727 Naphthoride
A proprietary preparation of naphthalene tetrachloride. No manufacturer.

20728 naphthosultone
The anhydride of 1-naphthol-8-sulfonic acid.

20729 α-naphthylamine
134-32-7 6485 205-138-7
$C_{10}H_9N$
1-naphthalenamine
naphthalidine; naphthalidam; 1-aminonaphthalene;. Used primarily in manufacture of dyes and as a chemical feedstock. mp = 50°; bp = 301°; d = 1.13; soluble in H_2O (1.7 mg/l), more soluble in organic solvents;.

20730 Naphtopon E
Dispersing agent in naphthol AS developing baths. *Bayer plc.*

20731 Naphtopone® E
Dispersing agent in naphthol AS developing baths for aftersoaping fast dyeings. *Bayer AG.*

20732 Napisan
Nappy sterilizer, to sanitize baby diapers. *Richardson-Vicks Inc.* Name unverified.

20733 Naples yellow
13510-89-9 5412 236-845-9
$Pb_3Sb_2O_8$
lead antimonate(V)
Paris yellow, lead antimonate. An orange-yellow pigment. A mixture of this body with carbonate and chromate of lead is also sold under this name cadmium sulfide, CPS, and a pale yellow ocher have been identified by this term; used to stain glass, crockery and porcelain. Insoluble in H_2O.

20734 Naplithin
Lithium-β-hydroxynaphthalene-α-monosulfonate.

20735 Napliwi
A sand cemented together by the rubber latex from the roots of the Chondrilla plant in Russia. The roots are attacked by the larvae of certain insects, when the latex exudes and runs into the sandy soil where it coagulates. The sand usually contains about 2-2.5% rubber and 10% resin.

20736 Napryl
Polypropylene. *BP Chemicals Ltd.* Discontinued.

20737 Napsoft FL
Synthetic softener for cellulose fibers to be napped; produces a lush, rich suede finish. *CNC Int'l L.P.*

20738 Naptel
Polybutenes. *BP Chemicals Ltd.*

20739 Naptol
133-90-4 2115 205-123-5
Chloramben; residual herbicide for use in ornamentals. *Synchemicals Ltd.*

20740 Napvis
Polybutenes. *BP Chemicals Ltd.*

20741 narasin
55134-13-9 6506
$C_{43}H_{72}O_{11}$
4S-methylsalinomycin
Compd. 79891; Antibiotic A-28086 factor A; C-7819B; Monteban. Main component of an antibiotic complex produced by *Streptomyces aureofaciens*

NRRL 5758 and NRRL 8092. A veterinary medicine used as a coccidiostat and growth stimulant. mp = 98-100°; λ_m = 285 nm (ε 58); $[\alpha]_D^{25}$= -54° (c = 0.2 MeOH); insoluble in H_2O, soluble in organic solvents; LD_{50} (mus ip) = 7.15 mg/kg.

20742 Narceol
140-39-6 205-413-1
$CH_3{\cdot}C_6H_4{\cdot}O{\cdot}COCH_3$
p-tolylacetate;
p-cresyl acetate; *p*-methylphenyl acetate; *p*-cresylic acetate;. *p*-tolyl acetate, a synthetic perfume.

20743 Nargentol
A protein compound of silver, containing 24% silver; an antiseptic.

20744 Narki
An acid-resisting silicon-iron alloy.

20745 Narlex EP-2
Dispersant; fluid loss control additive effective even at saturation level of calcium ions; for high pressure (500 psi), high temperature (175°C) requirements; bentonite extender. *Hart Chem. Ltd.*

20746 Narlex LD 42
Nonfoaming non-surface acting polymer; dispersant, plasticizer for pigments, clays, particulates; optimum performance > pH 6.5. *Hart Chem. Ltd.*

20747 NAS® 10
Styrene methyl methacrylate copolymer; featuring sparkling clarity, good scratch resistance; for cosmetics packaging, housewares, displays and office accessories. *Novacor.*

20748 NAS® 30
Styrene methyl methacrylate copolymer; offers sparkling clarity, good scratch resist., easy processing; exhibits less yellowing after gamma sterilization; for medical applications (catheter tubes, closures, IV flow regulators, thermoformed medical packaging *Novacor.*

20749 NAS® 50
Methyl methacrylate styrene copolymer; offers sparkling clarity, excellent scratch resist., and resist. to household detergents; used for salad bowls, tumblers, mugs, pitchers, cutlery handles; FDA compliance. *Novacor.*

20750 Natac®
Resin acids-amine resin soaps blend; tackifier for IIR, NR, SBR; NBR, SBR molding aid; slightly activates thiazole and thiuram accelerators. *Whitney & Oettler.* Name unverified.

20751 Natene
A trademark for Ziegler-type polyethylenes. No manufacturer.

20752 Nathin
Bread improvers. *Rhône-Poulenc UK.*

20753 Natopherol
59-02-9 10159 200-412-2
Vitamin E supplement. *Abbott Laboratories.* Name unverified.

20754 natrena®
Calorie-free sweetener for hot and cold drinks, yogurt, low-fat curd cheese, desserts. *Bayer AG.*

20755 Natritope Chloride
7647-14-5 8742 231-598-3
Sodium chloride ^{22}Na
Radioactive agent. *Bristol-Myers Squibb Co Inc.* Name unverified.

20756 natrium
7440-23-5 8710 231-132-9
Na
sodium
Latin name for sodium, Na.

20757 Natrolith
A water-softening material said to consist of granulated clay which removes lime and magnesium salts from hard water when used as a filter. No manufacturer.

20758 Natrosol® 250
9004-62-0 4707
Hydroxyethylcellulose
Thickener, protective colloid, binder, stabilizer, suspending agent for coatings, cosmetics, toiletries, textile printing pastes, inks, adhesives, electroplating, ceramics, textile and paper sizing, acid thickening for acidizing oil wells. *Aqualon.*

20759 Natrosol® FPS
Fluidized polymer suspension; latex paint thickener. *Aqualon.*

20760 Natrosol® Hydroxyethylcellulose
9004-62-0 4707
Hydroxyethylcellulose
Thickener for hair care products, creams and lotions, latex paints; protective colloid in emulsion polymerization; suspending aid in joint and tile cements; binder for welding rods; soluble in hot and cold water, DMSO; tolerates up to 70% polar organic solvents in water. *Aqualon.*

20761 Natrosol® Plus CS
Cetyl hydroxyethylcellulose
Thickener and viscosity stabilizer for aqueous and surfactant systems in personal care field; binder, stabilizer, film-former, suspending agent. *Aqualon.*

20762 Natrovis® Water-Soluble Polymer
Hydroxypropyl hydroxyethylcellulose
Water-soluble polymer; additive to hydraulic binder and latex-modified building materials, e.g., gypsum plasters, cement stuccos, mortars, masonry cements; controls water balance, workability, adhesion, consistency, tackiness. *Aqualon.* Discontinued.

20763 Natsyn® 2200
9003-31-0
$(C_5H_8)_n$
Rubber; 1,3-butadiene, 2-methyl-, homopolymer; trans-polyisoprene; poly(isoprene), *cis;* Rubber (all-*cis*). Polyisoprene, solution polymerized; nonstaining synthetic elastomer for light colored goods, adhesives, footwear, sponge products, tires, pharmaceutical goods, rubber bands, molded and mechanical goods. *Goodyear.* Name unverified.

20764 Natural Bone Ash, BCP 400
Calcium hydroxyapatite
Used to coat and protect all surfaces contacted by molten nonferrous metals; used extensively in both aluminum and copper industries. *Murlin Chemical Inc.*

20765 Natural Extract AP
trimethylglycine
Emollient, humectant, pigment, conditioner, lubricant for personal care formulations. *RITA.*

20766 natural rubber
9003-31-0 8435
Used for adhesive, dipping, coating, foam, and molded materials, latex applications., foamed rubber, textile, medical, cement, asphalt; processing aid for blending with other rubber. *Hardman; A Schulman.*

20767 Naturchem® CR
10401-55-5 233-864-4
Cetyl ricinoleate
Mild, noncomedogenic, nonoily emollient for cosmetics. *CasChem.*

20768 Naturechem® CAR
Cetyl acetyl ricinoleate
Mild, noncomedogenic, nonoily emollient for skin care products. *CasChem.*

20769 Naturechem® EGHS
33907-46-9 251-732-4
Glycol hydroxystearate
Auxiliary emulsifier, emollient, thickener, opacifier for cosmetics, household products. *CasChem.*

20770 Naturechem® GMHS
1323-42-8 215-355-9
Glyceryl hydroxystearate
Auxiliary emulsifier, emollient, opacifier, bodying and thickening agent for cosmetics, household products. *CasChem.*

20771 Naturechem® GTH
27233-00-7 248-351-0
Glyceryl triacetyl hydroxystearate
Mild, noncomedogenic emollient, wetting agent, stabilizer for pigmented products; imparts gloss. *CasChem.*

20772 Naturechem® GTR
101-34-8 202-935-1
Glyceryl triacetyl ricinoleate
Mild emollient, pigment wetter, cosolvent; softener for waxes and resins. *CasChem.*

20773 Naturechem® MAR
140-03-4 205-388-7
Methyl acetyl ricinoleate
Light emollient; reduces greasiness of emollients such as mineral oil; cosolvent properties; solubilizer for benzophenone-3; superior freeze/thaw properties. *CasChem.*

20774 Naturechem® MHS
141-23-1 205-471-8
Methyl hydroxystearate
Opacifier, pearlescent, emulsifier, visc. builder for surfactant systems. *CasChem.*

20775 Naturechem® OHS
Octyl hydroxystearate
Emollient, softener for cosmetics; refatting additive for soaps, cleansers. *CasChem.*

20776 Naturechem® PGHS
33907-47-0 251-734-5
Propylene glycol hydroxystearate

Auxiliary emulsifier, dispersant, opacifier, thickener, emollient, stabilizer for cosmetics, household products. *CasChem.*

20777 Naturechem® PGR
26402-31-3 247-669-7
Propylene glycol ricinoleate
Wetting agent, stabilizer, pigment/dye dispersant providing emolliency, gloss, plasticization to cosmetics, household products. *CasChem.*

20778 Naturechem® THS-200
PEG-200 trihydroxystearin
Emulsifier, emollient, thickener, stabilizer for cosmetics, household products; stable over broad pH range. *CasChem.*

20779 Nature's Own Spray Helper
A spreader sticker made from cottonseed oil; used in fungicide sprays and others to prolong their life on the plant. *Lawn & Garden Products Inc.*

20780 Naturvue
Hefilcon B
2-hydroxyethyl methacrylate polymer with 1-vinyl-2-pyrrolidinone and ethylene dimethacrylate; contact lens material. *Milton Roy Co.* Name unverified.

20781 Naubuc
A nickel-silver. It contains 58% copper, 25% nickel, 16.25% zinc, and 0.75% iron.

20782 Naugalube® 403
101-96-2 202-992-2
$C_{14}H_{24}N_2$
N,N'-di-*sec*-butyl-*p*-phenylenediamine
N,N-g-di-*s*-butyl-*p*-phenylenediamine; Tenamene 2; N,N'-bis(1-methylpropyl)-1,4-benzenediamine. Acrylic polymerization inhibitor; gasoline antioxidant/sweetener. mp= 18°; insoluble in H_2O (<1 mg/ml), soluble in organic solvents; LDLo (rat orl)= 200 mg/kg. *Uniroyal.*

20783 Naugalube® 438
101-67-7 202-965-5
$C_{28}H_{43}N$
Dioctyl diphenylamine
4-octyl-N-(4-octylphenyl)benzenamine; 4,4'-dioctyldiphenylamine. Antioxidant in automatic transmission fluids, turbine oil, and synthetic lubricants used in jet turbine engines; thermal stabilizer for automatic transmission fluid at high temperatures. mp= 96-97°; insoluble in H_2O (<1 mg/ml); LD_{50} (rat orl) = 7.58 g/kg. *Uniroyal.*

20784 Naugalube® 470
N,N'-Di-isopropyl-*p*-phenylenediamine in methanol; acrylic polymerization inhibitor. *Uniroyal.*

20785 Naugalube® PDA
N-phenyl-*p*-phenylendiamine
Intermediate for lubricants. *Uniroyal.*

20786 Nauganlite
Alkylated phenol; a proprietary antioxidant. *Rubber Regenerating Co.* Unverified.

20787 Nauganlite Powder
A proprietary name for alkylated bisphenol. *Rubber Regenerating Co.* Unverified.

20788 Naugard® 10
6683-19-8 229-722-6
$C_{73}H_{108}O_{12}$
Tetrakis[methylene(3,5-di-*tert*-butyl-4-hydroxyhydrocinnamate)methane]
Methane; antioxidant effective against thermal oxidative degradation during long term heat aging; for polyolefins, styrenics, elastomers, adhesives, lubricants, and oils. mp = 110-125°; *Uniroyal.*

20789 Naugard® 76
2082-79-3 218-216-0
$C_{35}H_{62}O_3$
octadecyl-3,5-di-*tert*-butyl-4-hydroxyhydrocinnamate
Octadecyl 3-(3,5-di-*tert*-butyl-4-hydroxyphenyl)propionate; 3,5-bis(1,1-dimethylethyl)-4-hydroxybenzenepropanoic acid octadecyl ester; 3,5-di-*tert*-butyl-4-hydroxyhydrocinnamate; Irganox 1076. Antioxidant for stabilizing polymeric substances such as polyolefins, styrenics, EPDM, and PVC; provides good thermal and color stability. *Uniroyal.*

20790 Naugard® 445
10081-67-1 233-215-5
4,4'-Bis (α,α-dimethyl-benzyl) diphenylamine
Nondiscoloring antioxidant for polyolefins, styrenics, polyether polyols, hot-melt adhesives, lubricant additives, nylon and other polymers. *Uniroyal.*

20791 Naugard® 524
31570-04-4 250-709-6
C42H63O3P
Tris (2,4-di-*t*-butyl phenyl) phosphite
Phenol, 2,4-bis(1,1-dimethylethyl)-, phosphite (3:1); tris(2,4-di-*tert*-butylphenyl) phosphite. Antioxidant used in thermoplastic and thermoset polymers where color and processing stability are critical. *Uniroyal.*

20792 Naugard® I-2
$C_{20}H_{36}N_2$
N,N'-Bis(1,4-dimethylpentyl)-p-phenylenediamine
N-Cyclohexylbenzothiazyl sulfenamide; CBTS, CBS; Santocure; Eastozone 33; Eastozone;
Tenamene; Santoflex 77; Vulkanox 4030. Styrene polymerization inhibitor; column antifoulant. *Uniroyal.*

20793 Naugard® I-3
N-(1,4-Dimethylpentyl)-N'-phenyl-p-phenylenediamine
Styrene polymerization inhibitor; column antifoulant. *Uniroyal.*

20794 Naugard® I-4
N-phenyl, N'-isopropyl-*p*-phenylenediamine
Styrene polymerization inhibitor; column antifoulant. *Uniroyal.*

20795 Naugard® I-6
N-phenyl, N'-cyclohexyl-*p*-phenylenediamine
Column antifoulant. *Uniroyal.*

20796 Naugard® J
74-31-7 3388 200-806-4
$C_{18}H_{16}N_2$
N,N'-Diphenyl-*p*-phenylenediamine
Agerite DPPD; 1,4-Dianilinobenzene; Diphenyl PPD; DPPD; Flexamine g; JZF; Nonox DPPD; Diafen; Diafen FF; Altofane DIP; Nonflex H; Permanax 18; Stabilizer DPPD; Nocrac DP; Permanax DPPD; DFFD; Antage DP; Ekaland DPPD; Naugard J; N,N'-Diphenyl-*p*-phenylene diamine. Acrylic polymerization inhibitor. mp= 145-152°; bp = 220-225°; SG= 1.2; insoluble in H_2O (<1 mg/ml). *Uniroyal.*

20797 Naugard® NBC
13927-77-0 237-696-2
$C_{18}H_{36}N_2NiS_4$
Nickel dibutyldithiocarbamate
Vanguard N; Nickel, bis(dibutylcarbamodithioato-S,S')-, (SP-4-1)-; Bis(dibutyldithiocarbamate)nickel complex. Nickel chelating uv stabilizer for polyolefins. *Uniroyal.*

20798 Naugard® PANA
90-30-2 201-983-0
$C_{16}H_{13}N$
N-phenyl-α-naphthylamine
N-(1-Naphthyl)aniline; Aceto Pan; Additin 30; C.I. 44050; Neozone A; phenylnaphthylamine; α-phenylnaphthylamine; Algerite;. Antioxidant for lubricants. mp = 60-62°; bp_{15} = 226°; *Uniroyal.*

20799 Naugard® XL-1
70331-94-1 274-572-7
2,2'-oxamido bis-[ethyl 3-(3,5-di-tert-butyl-4-hydroxyphenyl) propionate
Antioxidant and metal deactivator; used in polymerization, processing, and in end use applications, wire and cable insulation, pipe and injection parts for automobiles and appliances; processing stabilizer for polyolefins, film, sheet, and blow molded bottles. *Uniroyal.*

20800 Naugawhite
Phenolic/bisphenolic
Antioxidant for thermoplastics; hot melt adhesives. *Uniroyal.*

20801 Naugex SD-1
103-34-4 3437 203-103-0
$C_8H_{16}N_2O_2S_2$
4,4'-dithiodimorpholine
morpholine, N,N'-disulfide; dimorpholine N,N'-disulfide; Sulfasan R. A sulfur donor used as a partial or total replacement of sulfur for resistance to heat and ageing in NR, SBR, NBR, and EPDM. mp= 124-125°. *Uniroyal.* Name unverified.

20802 Nauli gum
An oleoresin from a tree found in the Solomon Islands. It contains 10% volatile oil, 8% resin, and 3% water-soluble matter containing anisic acid.

20803 Navac
Vacuum processed metallic sodium. *Foseco (F.S.) Ltd.*

20804 naval bronze
An alloy of 88.1% copper, 9.74% tin, and 2.04% zinc.

20805 navy green paint
A mixture of barium sulfate, lead chromate, and an organic blue.

20806 Naxchem CD-6M
Surfactant blend; detergent, foam booster/stabilizer for cosmetics, carpet shampoos, diswash, laundry detergents, textile processing, pigment dispersions. *Ruetgers-Nease.*

20807 Naxchem Detergent CNB
Surfactant blend; detergent, foam booster/stabilizer for lotions and creams, carpet and upholstery shampoos, dishwash, laundry, textile processing, pigment disps. *Ruetgers-Nease.*

20808 Naxchem Dispersant K.
Blend of alkanolamides and synthetic detergents; biodegradable dispersant, emulsifier for oil slicks, naphthas, kerosene, other solvents; general cleaning of crude oil, marine and land transport, and manufacturing plants. *Ruetgers-Nease.*

20809 Naxchem Emulsifier 700
Blend of esters and surfactants; antistatic base for emulsifying low viscosity pale mineral oils; for metal cutting, textile fiber lubricants. *Ruetgers-Nease.*

20810 Naxchem N-Foam 802
Blend of crosslinking and emulsifying agents; curing and foaming agent in mixtures with urea-formaldehyde resins to mask odors from landfills and for foam insulation; as cleaning compound when neutralized. *Ruetgers-Nease.*

20811 Naxel AAS-40S
Sodium dodecylbenzene sulfonate
Biodegradable surfactant, foamer, wetting agent, detergent for household and industrial detergents, rug shampoos, textile wet processing, metal cleaners, dairy cleaners, cosmetics. *Ruetgers-Nease.*

20812 Naxel AAS-60S
TEA dodecylbenzene sulfonate
Foaming agent for bubble baths, shampoos, household and industrial detergents, textile dyeing compounds, etc. *Ruetgers-Nease.*

20813 Naxel AAS-98S
Dodecylbenzenesulfonic acid
Biodegradable detergent intermediate, wetting agent, emulsifier. *Ruetgers-Nease.*

20814 Naxel AAS-Special 3
Isopropylamine dodecylbenzene sulfonate
Emulsifier for drycleaning, metal cleaning, emulsifiable solvent cleaners, fuel oil additives, mop treatments. *Ruetgers-Nease.*

20815 Naxel DDB 500
Dodecylbenzene
Biodegradable detergent intermediate for production of dishwash, laundry, all-purpose and industrial cleaners, in specialty coatings and other industrial applications. *Ruetgers-Nease.*

20816 Naxell
Alkylaryl sulfonates. *Ruetgers-Nease Chemical Co.*

20817 Naxide 1230
61788-90-7 263-016-9
Cocamine oxide
Detergent, foam booster/stabilizer, viscosity builder, conditioner for shampoos, hand cleaners, dishwash, light duty detergents, textiles, lubricants, paper coating. *Ruetgers-Nease.*

20818 Naxolate WA-97
151-21-3 8782 205-788-1
Sodium lauryl sulfate USP, BP
Biodegradable detergent, wetting agent, foamer, emulsifier for cosmetic and household products including shampoos, bubble baths, rug shampoos, toothpaste, dishwash, laundry detergents. *Ruetgers-Nease.*

20819 Naxonac
Phosphate esters. *Ruetgers-Nease Chemical Co.*

20820 Naxonac 510
Nonyl phenol ether phosphate; detergent, wetting agent, emulsifier, lubricant, hydrotrope for heavy-duty and household detergents, waterless hand cleaners, solvent degreasers, emulsion polymerization, paint and wax strippers, electrolytic cleaners. *Ruetgers-Nease.*

20821 Naxonat® 4ST
657-84-1 9671 211-522-5
$C_7H_7NaO_3S$
Sodium toluene sulfonate
Benzenesulfonic acid, methyl-, sodium salt. Hydrotrope, stabilizer, solubilizer used in formulating detergent, inks, electroplating baths, dyestuffs, polymers. *Ruetgers-Nease.*

20822 Naxonate® 4AX
26447-10-9 247-710-9
$C_8H_{13}NO_3S$
Ammonium xylene sulfonate
Hydrotrope, stabilizer, solubilizer used in formulating detergents, inks, electroplating baths, dyestuffs, polymers. *Ruetgers-Nease.*

20823 Naxonate® 4KT
30526-22-8 206-067-4
$C_7H_7KO_3S$
Potassium toluene sulfonate
Benzenesulfonic acid, methyl-, potassium salt. Hydrotrope, stabilizer, solubilizer used in formulating detergents, inks, electroplating baths, dyestuffs, polymers. *Ruetgers-Nease.*

20824 Naxonate® 4L
1300-72-7 215-090-9
$C_8H_9NaO_3S$
Sodium xylene sulfonate
xylenesulfonic acid, sodium salt; dimethylbenzenesulfonic acid, sodium salt; sodium dimethylbenzenesulfonate; Conco SXS; Cyclophil SXS30; Eltesol SX 30; Hydrotrope; Naxonate; Naxonate G; Stepanate X; Surco SXS; Ultrawet 40SX; Calsoft SXS 96; Alkatrope SX-40; Carsosulf SXS; Eltesol SX93; Reworyl NXS40; Rchonate SXS; Witconate SXS; sodium m-xylene sulfonate. Hydrotrope, stabilizer, solubilizer used in formulating detergents, inks, electroplating baths, dyestuffs, polymers. mp= 27°; bp= 157°, d = 1.23; soluble in H_2O (>100 mg/ml), less soluble in organic solvents; LD_{50} (rat orl)= 1000 mg/kg. *Ruetgers-Nease.*

20825 Naxonate® 5KT
Potassium toluene sulfonate
Hydrotrope, stabilizer, solubilizer used in formulating detergents, inks, electroplating baths, dyestuffs, polymers. *Ruetgers-Nease.*

20826 Naxonate® 45SC
32073-22-6 250-913-5
Sodium cumene sulfonate
Hydrotrope, stabilizer, solubilizer used in formulating detergents, inks, electroplating baths, dyestuffs, polymers. *Ruetgers-Nease.*

20827 Naxonate® SC
32073-22-6 250-913-5
Sodium cumene sulfonate
Hydrotrope, stabilizer, solubilizer used in formulating detergents, inks, electroplating baths, dyestuffs, polymers. *Ruetgers-Nease.*

20828 Naxonate® SX
1300-72-7 215-090-9
Sodium xylene sulfonate; hydrotrope, stabilizer, solubilizer used in formulating detergents, inks, electroplating baths, dyestuffs, polymers. *Ruetgers-Nease.*

20829 Naxonic NI-40
9016-45-9 6772
Nonoxynol-4
Wetting agent, dispersant, penetrant, emulsifier, detergent, solubilizer for textile wet processing, agricultural, cosmetic, industrial and household detergents, latex and polymers, wax/polishes; demulsifier for petroleum *Ruetgers-Nease.*

20830 Naxonol CO
Cocamide DEA (2:1); detergent, wetting agent, thickener for household and industrial detergents, shampoos, textile wet processing, leather and fur processing, metal, cleaning, solvent emulsification. *Ruetgers-Nease.*

20831 NBC
13927-77-0 237-696-2
$C_{18}H_{36}N_2NiS_4$
nickel dibutyl dithiocarbamate
Vanguard N; Nickel, bis(dibutylcarbamodithioato-S,S')-, (SP-4-1)-; Bis(dibutyldithiocarbamate)nickel complex. A proprietary name for nickel dibutyl dithiocarbamate. *DuPont UK.* Name unverified.

20832 NC-4
Bis (p-ethylbenzylidene) sorbitol
Nucleating agent improving optical and physical properties of PP for container, packaging film, injection syringe and other applications. *Mitsui Toatsu.*

20833 NE
79-24-3 6694 201-188-9
$C_2H_5NO_2$
Nitroethane
Nitroetan. Intermediate, stabilizer for halogenated solvents, as fuels, explosives, and solvents for coatings or industrial processes. mp = -90°; bp = 114°; SG = 1.045; soluble in H_2O (4.5 g/100 ml). *Angus.*

20834 Neacid
A pickling agent for gold-and silversmiths and in the jewelry industry. *Degussa.*

20835 Neantina
Seed dressing for control of fungal diseases on cereals, rice, cotton and vegetables. *Bayer AG.*

20836 neatsfoot oil
8002-64-0 6518 232-314-0
oil babuluam. Oil from feet of cattle. It is also the name for a mixture of 1 part lard, 3 parts Colza oil; used in the leather industry for fat liquoring and softening, lubricant, oiling wool. d = 0.915; n_D^{20}= 1.4695-1.4708;.

20837 neatsfoot oil
8002-64-0 6518 232-314-0
A fixed oil obtained by boiling ox or cow's feet in water. It is also the name for a mixture of 1 part lard and 3 parts colza oil; used in the leather industry for fat liquoring and softening, lubricant, oiling wool.

20838 Nebony® 100, L-55
Aromatic petroleum hydrocarbon resin; used in wire/cable coatings, rubber

(cements, mechanical goods, molded goods, tires), and caulking compounds. *Neville.*

20839 Nebulin

117-18-0 204-178-2

Tecnazene in liquid fogging solution; used for controlling sprouting and dry rot in stored potatoes. *Wheatley Chemical Co Ltd.; Dean Agrochemicals Ltd.* Name unverified.

20840 Neburex

555-37-3 6523 209-096-0

Active ingredient: neburon; selective herbicide for both pre-and post-emergence application. *Agan Chemical Manufacturers Ltd.*

20841 neburon

555-37-3 6523 209-096-0

$C_{12}H_{16}Cl_2N_2O$

N-butyl-N'-(3,4-dichlorophenyl)-N-methylurea

Granurex; Neburex; Herbalt; 1-Butyl-3-(3,4-dichlorophenyl)-1-methylurea; Kloben; Butyl-3-(3,4-dichlorophenyl)-1-methylurea; Butyl-N'-(3,4-dichlorophenyl)-N-methylurea. Selective herbicide, absorbed through the roots, inhibits photosynthesis. Used for pre-emergence control of annual broad-leaved weeds and grasses in beans, peas, lucerne, garlic, cereals, beets, strawberries and ornamentals, and in forestry. mp = 102-103°; soluble in H_2O (5 mg/l), sparingly soluble in hydrocarbon solvents; LD_{50} (rat orl) >11000 mg/kg.

20842 Necol

Cellulose lacquers, organic enamels, adhesives and plastic wood. *ICI Chem & Polymers Ltd.*

20843 Nectandra bark

Bebeera bark.

20844 Nedi

Polyvinyl acetate water soluble film and contract packing service. *Production Chemicals Ltd.*

20845 needle bronze

An alloy of 84.5% copper, 8% tin, 5.5% zinc, and 2% lead.

20846 needle tin ore

Acute pyramidal crystals of cassiterite (tin oxide).

20847 Neem bark

Indian azadirach.

20848 Nefomolit

A proprietary trade name for a plastic made from mineral oil, formaldehyde, etc. No manufacturer.

20849 Neganol®

For the control of trypanosomiasis in domestic animals. *Bayer AG.*

20850 Negasunt®

Dusting powder against bacterial infection and fly-larvae (myiasis); veterinary preparation controlling fly maggots in wounds. *Bayer AG.*

20851 Negex

Chemical substances used in industry; negative expanders for lead acid batteries. *Associated Lead Manufacturers Ltd.* Name unverified.

20852 Neguvon®

52-68-6 9753 200-149-3

Insecticide against cattle grubs, mange and worm infestation, in particular for control of ectoparasites in the poultry house; veterinary medicine. *Bayer AG.*

20853 Neillite

A proprietary phenol-formaldehyde synthetic resin molding compound. No manufacturer.

20854 Neisser's stain

A microscopic stain. a) Solution contains 0.1 g methylene blue, 2 ml alcohol, 5 ml glacial acetic acid, and 95 ml water. b) Solution contains 0.2 g bismarck brown in 100 ml boiling water.

20855 Nekal

Anionic surfactants. *BASF plc.*

20856 Nekal® BX

Alkylaryl sulfonate anionic surfactant; wetting, dispersing, emulsifying and foaming agents with widespread industrial applications e.g. in the pigment, paint and paper industries; in leakage control in pipes, fittings and valves. *BASF plc.* Name unverified.

20857 Nekal® SBS

Anionic surfactant in free acid form; wetting, dispersing, emulsifying and foaming agent with widespread industrial uses e.g. in the paper, paint and pigment industries; in leakage control in pipes, fittings and valves. *BASF plc.* Name unverified.

20858 Nekanil® 907

Low-ethoxylated alkyl phenol

Detergent, wetting agent used in textile industry. *BASF AG; BASF plc.*

20859 Nelco

A proprietary trade name for whiting. No manufacturer.

20860 Nelio Resin

A proprietary trade name for a purified wood resin. No manufacturer.

20861 Nema

127-18-4 9332 204-825-9

Tetrachloroethylene

Anthelmintic. *Parke-Davis.* Name unverified.

20862 Nemacin

148-79-8 9426 205-725-8

Thiabendazole and an iodophor (a germicidal agent) in dry granular form; used for controlling soil and seed borne diseases in onions. *Wheatley Chemical Co Ltd.* Name unverified.

20863 Nemacur®

22224-92-6 244-848-1

phenamiphos. Nematicide.

20864 Nemafax

23564-06-9 9489 245-741-2

Thiophanate-methyl

A veterinary anthelmintic. *RMB Animal Health Ltd.*

20865 Neo Heliopan® 303

6197-30-4 228-250-8

$C_{24}H_{27}NO_2$

2-ethylhexyl-2-cyano-3,3-diphenylacrylate

Octocrylene; Octocrilene. UV-B absorber for cosmetics, waterproof sunscreens. Insoluble in H_2O, soluble in organic solvents. *Haarmann & Reimer GmbH.*

20866 Neo Heliopan® AV

5466-77-3 6864 226-775-7

$C_{18}H_{26}O_3$

2-Ethylhexyl-4-methoxycinnamate

2-Ethylhexyl p-methoxycinnamate; Escalol 557; Ethylhexylcinnamate; Neo Heliopan AV.; Octyl methoxycinnamate; Parsol MCX; Parsol MOX. UV-B absorber for cosmetic applications, waterproof sunscreens. mp , -25°; bp = 198-200°; SG = 1.007; insoluble in H_2O (<1 mg/ml). *Haarmann & Reimer GmbH.*

20867 Neo Heliopan® BB

131-57-7 7088 205-031-5

$C_{14}H_{12}O_3$

2-Hydroxy-4-methoxybenzophenone

Benzophenone 3; Escalol 567; Eusolex 4360; MOB; Oxybenzone; Uvinul M-40; Spectra-Sorb UV-9; 2-benzoyl-5-methoxyphenol; cyasorb uv 9; syntase 62; UF 3; advastab 45; anuvex; chimassorb 90; cyasorb uv 9 light absorber; MOD; ongrostab hmb; sunscreen uv 15; uvinul 9; uvistat 24; HMB. Uv-A and uv-B broad spectrum absorber for sunscreen formulations. mp= 66°; LD_{50} (rat orl) >13 g/kg. *Haarmann & Reimer GmbH.*

20868 Neo Heliopan® Hydro

27503-81-7 248-502-0

$C_{13}H_{10}N_2O_3S$

2-Phenylbenzimidazole-5-sulfonic acid

phenylbenzimidazol-5-sulfonic acid; Eusolex 232; Novantisol; 2-phenyl-1H-benzimidazole-5-sulfonic acid. UV-B filter for sunscreen formulations. Used in creams, lotions and subscreens. mp >300°. *Haarmann & Reimer GmbH.*

20869 Neo Heliopan® MA

134-09-8 205-129-8

Menthyl anthranilate

UV-A absorber for waterproof sunscreen formulations. *Haarmann & Reimer GmbH.*

20870 Neo Heliopan® OS

118-60-5 204-171-4

$C_{15}H_{22}O_3$

2-Ethylhexyl salicylate

Sunarome O; Sunarome WMO; 2-Ethylhexyl salicylate. Uv-B absorber for cosmetic applications, waterproof sunscreens; solubilizer for oxybenzone. *Haarmann & Reimer GmbH.*

20871 Neoarsycodyl

2163-80-6 6020 218-495-9

CH_4AsNaO_3

sodium methanearsonate

Methanearsonic acid sodium salt; Arsonate; Bueno 6; Ansar; Ansar 6.6; Clean Crop MSMA 6 Plus; Clean Crop MSMA 6.6; Daconate 6; Drexel MSMA 6 Plus; Drexel MSMA 6.6; MSMA; Ansar 170; Dal-E-Rad; Drexar 530; Herb-All; Merge 823; Mesamate; Target MSMA; Trans-Vert; Weed-Hoe; Vertac MSMA 400; Vertac MSMA 600; Vertac MSMA 660; Bueno; Merge; phyban; silvisar 550; sodium acid methanearsonate; weed 108; Ansar 529; Arsanote; Deconate; Monate merge 823; Phyban H.C.; Weed-s-rad. Selective contact herbicide with some systemic properties. Used for pre-emergent control of grass weeds in cotton, sugar cane, under trees, and on non-crop areas. mp = 113-116°; bp (dec); SG = 1.5; soluble in H_2O (>100 mg/ml); LD_{50} (rat orl) = 1105 mg/kg.

20872 Neobee® 18
8001-23-8 8465 232-276-5
Safflower oil
Hi-oleic safflower oil; carthamus tinctorius oil; lipovol saf(lipo); tri-ol sa(tri-k); obesitol; safflor. Hybrid safflower oil; emollient oil for cosmetics and pharmaceuticals, solubilizer, solvent. SG = 0.9211; insoluble in H_2O (<1 mg/ml), soluble in organic solvents. *Stepan/PVO; Stepan Europe.*

20873 Neobee® 20
Propylene glycol dicaprylate/dicaprate
Emollient oil for cosmetics and pharmaceuticals, carrier for flavors and colors in foods; solubilizer, cosolv. *Stepan/PVO.*

20874 Neobee® 62
555-43-1 9885 209-097-6
$C_{57}H_{110}O_6$
Tristearin
glycerol tristearate; octadecanoic acid, 1,2,3-propanetriyl ester; 1,2,3-propanetriyl trioctadecanoate. Solubilizer, stabilizer used in food applications. Used to make candles. mp= 71-73°; d_4^{80}= 0.862; n_D^{80}= 1.4385; insoluble in H_2O, soluble in organic solvents. *Stepan/PVO.*

20875 Neobee® M-5
65381-09-1 265-724-3
Caprylic/capric triglyceride
Diluent vehicle/carrier, solubilizer, cosolv. for flavoring, medicinals, colorings, used in foods, beverages, cosmetic and pharmaceutical product, emollient. *Stepan/PVO; Stepan Europe.*

20876 Neobor
Disodium tetraborate pentahydrate
Pentahydrate Borax. Disodium tetraborate decahydrate, partially dehydrated to effect economy of transport and handling. Used as a preservative and antiseptic, alkalizer and in soldering metals. *Borax Europe Ltd.*

20877 Neo-Calglucon
299-28-5 1712 206-075-8
$C_{12}H_{22}CaO_{14}$
Calcium gluconate
Calcium glubionate; Calciofon; Calglucon; Ebucin; Glucal; Glucobiogen. Replenisher. Used in sewage purification and in coffee powders as an anticaking agent. $[\alpha]_D^{20}$= 10° (c = 1 H_2O). *Dorsey Pharmaceuticals.*

20878 Neo-Cebitate®
6381-77-7 5141 228-973-9
$C_6H_7NaO_6$
Sodium erythorbate
isoascorbate; D-erythro-hex-2-enonic acid, σ-lactone, monosodium salt; isoascorbic acid, sodium salt; neo-cebitate; mercate 20; Isona; sodium D-isoascorbate; monosodium erythorbate; araboascorbic acid, monosodium salt, D-; hex-2-enonic acid, σ-lactone, monosodium salt; sodium D-araboascorbate. Food additive (antioxidant and antimicrobial agent) for meat, poultry, and seafood industries. mp = 168-170°; $[\alpha]_D^{?}$ = 95° (c = 10 H_2O); soluble in H_2O 916 g/100 ml) *Rhône-Poulenc Food Ingreds.*

20879 Neocid
50-29-3 2898 200-024-3
A proprietary preparation containing 5% DDT; an insecticide. No manufacturer.

20880 Neocidol Veterinary Powder
333-41-5 3043 206-373-8
Organophosphorus wound dressing. Active principle is diazinon. *Ciba plc.* Name unverified.

20881 Neocosal
An oil binder; used for cleaning bodies of water and ground surfaces that have been contaminated with oil. *Degussa AG.* Name unverified.

20882 Neocrest
Saturated polyester resin products. *ICI Resins BV.*

20883 NeoCryl
Thermoplastic acrylic resins in solid form, homo and copolymers of acryl-methacrylic and styrene; used in various paints, e.g., marine paints, road paints and paints on plastic; flexo, gravure and screen painting inks; dry toner resins, aerosols and other applications. *Polyvinyl Chemie Holland BV.*

20884 Neocryl
Acrylic polymer resin products. *ICI Resins BV.*

20885 Neo-Cytamen
13422-51-0 6125 236-533-2
A proprietary preparation of hydroxycobalamin. *Duncan Flockhard Ltd.*

20886 Neodene® 6
592-41-6 209-753-1
C_6H_{12}
1-hexene
hexene; hexylene; butylethylene; 1-n-hexene; hexene-1; hex-1-ene. Linear C_6 α olefin; intermediate for biodegradable surfactants and specialty industrial chemicals. mp = -140°; bp = 62-63°; d = 0.673; n_D^{20} = 1.3880; insoluble in H_2O, soluble in organic solvents; *Shell.*

20887 Neodene® 6/12
C_6-C_{12} alpha olefin blend; intermediate for biodegradable surfactants and specialty industrial chemicals. *Shell.*

20888 Neodene® 6/8
C_6-C_8 alpha olefin
Mixture of 1-hexene, 1-heptene and 1-octene. Intermediate for biodegradable. Surfactants and specialty industrial chemicals. *Shell.*

20889 Neodene® 6/8/10
C_6-C_{10} alpha olefin blend; intermediate for biodegradable surfactants and specialty industrial chemicals. *Shell.*

20890 Neodene® 8
111-66-0 1807 203-893-7
1-octene
Linear C_8 α olefin; intermediate for biodeg. surfactants and specialty industrial chemicals. *Shell.*

20891 Neodene® 10
872-05-9 212-819-2
$C_{10}H_{20}$
1-decene
decene-1, dec-1-ene. Linear C_{10} α olefin; intermediate for biodegradable surfactants and specialty industrial chemicals. bp = 170-171°; d = 0.741; n_D^{20}= 1.4210. *Shell.*

20892 Neodene® 10/12/1314
Alpha olefin/internal olefin blend; intermediate for biodegradable surfactants and specialty industrial chemicals. *Shell.*

20893 Neodene® 10-11/12-13
Blend of decene-1, undecene, dodecene and tridecene; intermediate for biodegradable surfactants and specialty industrial chemicals. *Shell.*

20894 Neodene® 10-11/12-13/14
Blend of decene-1, undecene, dodecene, tridecene and tetradecene; intermediate for biodegradable surfactants and specialty industrial chemicals. *Shell.*

20895 Neodene® 12
112-41-4 203-968-4
$C_{12}H_{24}$
1-dodecene
dodec-1-ene; dodecene-1. Linear C_{12} α olefin; intermediate for biodegradable surfactants and specialty industrial chemicals. bp = 212-213°; d = 0.758; n_D^{20}= 1.4290. *Shell.*

20896 Neodene® 12/1314
Blend of dodecene-1, tridecene and tetradecene; intermediate for biodegradable surfactants and specialty industrial chemicals. *Shell.*

20897 Neodene® 14
1120-36-1 214-306-9
$C_{14}H_{28}$
1-tetradecene
tetradec-1-ene; tetradecene-1. Linear C_{14} α olefin; intermediate for biodegradable surfactants and specialty industrial chemicals. bp = 247-250°; d = 0.773; n_D^{20}= 1.4360. *Shell.*

20898 Neodene® 16
629-73-2 211-105-8
$C_{16}H_{32}$
1-hexadecene
hexadec-1-ene; hexadecene-1. Linear C_{16} α olefin; intermediate for biodegradable surfactants and specialty industrial chemicals. bp = 273-274°; d = 0.781; n_D^{20}= 1.4420. *Shell.*

20899 Neodene® 18
112-88-9 204-012-9
$C_{18}H_{36}$
1-octadecene
octadec-1-ene; octadecene-1. Linear C_{18} α olefin; intermediate for biodegradable surfactants and specialty industrial chemicals. mp = 15-17°; bp_{15} = 178-179°; d = 0.789; n_D^{20}= 1.4440. *Shell.*

20900 Neodene® 20
3452-07-1 222-374-6
$C_{20}H_{40}$
1-icosene
icos-1-ene; eicos-1-ene. Linear C_{20} α olefin; intermediate for biodegradable surfactants and specialty industrial chemicals. mp = 29°; bp = 342°; SG = 0.795. *Shell.*

20901 Neodene® 810
Blend of octene-1 [111-66-0] and decene-1 [872-05-9]; intermediates for biodegradable surfactants and specialty industrial chemicals. *Shell.*

20902 Neodene® 1012
Blend of decene-1 [872-05-9] and dodecene-1 [112-41-4]; intermediates for biodegradable surfactants and specialty industrial chemicals. *Shell.*

20903 Neodene® 1014
C_{10-14} α olefin blend; intermediates for biodegradable surfactants and specialty industrial chemicals. *Shell.*

20904 Neodene® 1112
68411-00-7 270-095-3
C_{11}-C_{12} internal olefins; intermediates for biodegradable surfactant and specialty industrial chemicals. *Shell.*

20905 Neodene® 1420
64743-02-8 265-207-2
C_{14-20} α olefin blend; intermediates for biodegradable surfactants and specialty industrial chemicals. *Shell.*

20906 Neodene® 1624
64743-02-8 265-207-2
C_{16}-C_{24} α olefin blend; intermediates for biodegradable surfactants and specialty industrial chemicals. *Shell.*

20907 Neodene® 2024
64743-02-8 265-207-2
C_{20}-C_{24} α olefin blend; intermediates for biodegradable surfactants and specialty industrial chemicals. *Shell.*

20908 Neodol® 1
112-42-5 203-970-5
$C_{11}H_{24}O$
undecan-1-ol
undecyl alcohol; 1-undecanol; 1-hendecanol; decyl carbinol; alcohol C_{11} undecylic. Detergent intermediate. mp = 11°; bp_{30} = 146°; d = 0.8300; n_D^{20} = 1.4400. *Shell.*

20909 Neodol® 1-3
Undeceth-3
Detergent intermediate, emulsifier for general industrial usage, hard surface cleaning; biodegradable. *Shell.*

20910 Neodol® 5
629-76-5 211-107-9
$C_{15}H_{32}O$
Pentadecyl alcohol
Detergent intermediate. mp = 45-46°; bp = 269-271°. *Shell.*

20911 Neodol® 23
75782-86-4 278-306-0
C_{12}-C_{13} alcohols
Detergent intermediate. *Shell.*

20912 Neodol® 23-6.5
66455-14-9
C_{12}-C_{13} alcohols, ethoxylated
C12-13 pareth-7; detergent intermediate used in preparation of sulfates for high foaming liquid detergents, household and industrial use; biodeg. *Shell.*

20913 Neodol® 25
63393-82-8 264-118-6
C_{12}-C_{15} alcohols
Detergent, emulsifier intermediate. *Shell.*

20914 Neodol® 25-3
68131-39-5
C_{12}-C_{15} pareth-3; detergent intermediate used in preparation of sulfates for high foaming liquid detergents; emulsifier; for cosmetic, industrial, dishwashing and liquid detergents; biodegradable. *Shell.*

20915 Neodol® 45
75782-87-5
C_{14}-C_{15} alcohol
C_{14}-C_{15} alcohol; detergent intermediate. *Shell.*

20916 Neodol® 45-7
C_{14}-C+&1s pareth-7
C_{14}-C_{15} pareth-7; detergent, wetting agent, emulsifier for general industrial and household detergent products; biodegradable. *Shell.*

20917 Neodol® 91
66455-17-2 266-367-6
C_9-C_{11} alcohols
C_9-C_{11} alcohols; detergent intermediate. *Shell.*

20918 Neodol® 91-2.5
68439-46-3
C_9-C_{11} pareth-3 (2.5 EO)
C_9-C_{11} pareth-3 (2.5 EO); oil-soluble emulsifier and wetting agent, detergent intermediate, dispersant, surfactant; for general industrial usage, hard surface cleaning; biodegradable. *Shell.*

20919 Neo-Duroterm® 3, 5 and 7
Investment compound system for precious metal casts; dental specialty. *Bayer AG.*

20920 Neo-Duroterm® L
Soldering investment material for the precious metal technique; dental specialty. *Bayer AG.*

20921 Neoflex® 11
112-42-5 203-970-5
$C_{11}H_{24}O$
1-undecanol
undecan-1-ol; 1-hendecanol; decyl carbinol; alcohol c-11 undecylic. C_{11} primary alcohol; intermediate. *Shell.*

20922 Neoflex® 9
68527-05-9 271-250-8
$C_9H_{20}O$
Isononyl alcohol.
Shell.

20923 Neogen
An alloy of 58% copper, 12% nickel, 27% zinc, 2% tin, and 0.5% aluminum. It is a nickel silver *German silver.*

20924 Neolan
Metal complex dyes. *Ciba plc.* Name unverified.

20925 Neoleukorit
A proprietary synthetic resin of the phenolformaldehyde class. No manufacturer.

20926 Neolyn
Elastomeric alkyd type rosin-based resins with good grease resistance; used to impart flexibility and grease resistance to adhesives, is used for vinyl floor tiles and vinyl inks, in vinyl based and type T-gravure inks. *Hercules.* Discontinued.

20927 Neolysol
Lysol made with chlorocresol; used as an antiseptic.

20928 Neomodelon 100
Blend; neutral sizing agent for paper manufacturing. *Toho Chem. Industry.*

20929 neomycin palmitate
55298-68-5 259-582-1
Neomycin hexadecanoate

20930 Neonalium
An alloy of aluminum with 6-14% copper, 1% nickel, and small amounts of other metals.

20931 Neonite
A 33-grain sporting rifle powder. It contains 10% barium or potassium nitrate, 6% petroleum jelly, and insoluble nitrocellulose.

20932 Neonite® EG60/6mm, EG61/12mm
25928-94-3
Epoxy molding compound; for electrical engineering, encapsulation of solenoids, mech. engineering (pump and valve parts, flanges). *Ciba-Geigy GmbH.* Name unverified.

20933 Neopac
Acrylic dispersions. *ICI Chem & Polymers Ltd.*

20934 Neopelline
$C_{32}H_{45}NO_8$
An alkaloid obtained from *Aconitum napellus* .

20935 Neopen SS
A proprietary trade name for a wetting agent containing sodium abietene sulfonate. No manufacturer.

20936 Neopen® Dyes
Metal complex dyes; for inks, ball-pen pastes, copying toners. *BASF AG.*

20937 neopentyl glycol
126-30-7 6547 204-781-0
$C_5H_{12}O_2$
2,2-dimethyl-1,3-propanediol
dimethyltrimethylene glycol; 2,2-dimethylpropane-1,3-diol. Resin intermediate; insect repellent. Used in manufacture of plasticizers and polyesters. mp = 127°; bp= 208°; soluble in H_2O (65 g/ml), organic solvents. *BASF; Eastman; Hüls AG; Mitsubishi Gas.*

20938 Neophax FA
A proprietary brown factice added to polychloroprene, nitrile rubber, Hypalon and polyurethane when maximum resistance to oil is required. *Hubron Rubber Chemicals.* Name unverified.

20939 neopine
467-14-1 6549 207-387-7
$C_{18}H_{21}NO_3$
8,14-didehydro-4,5α-epoxy-3-methoxy-17-methylmorphinan-6α-ol
β-codeine. Alkaloid from opium, isomeric with codeine. mp = 127°; $[\alpha]_D^{23}$ = -28° (c = 7.5 $CHCl_3$); soluble in H_2O, polar organic solvents;.

20940 Neoplen
Foamed polyethylene slabstock and moldable beads; for cushion packaging

of fragile items, antirattle stowage of tools in automotives. *BASF plc.* Name unverified.

20941 Neopolen® E
9002-88-4 7728
Polyethylene foam; for foamed moldings, packaging, building trade, maritime sector, automotive industry, recreation and leisure equipment. *BASF AG; BASF plc.*

20942 Neopolen® P
9003-07-0 7741
Polypropylene foam; for foamed moldings, packaging, building trade, maritime sector, automotive industry, recreation and leisure equipment. *BASF AG; BASF plc.*

20943 Neopon
Surfactant for cosmetics, toiletries, pharmaceutical, processing, agricultural and other industries. *Baxenden Chemicals Ltd.*

20944 Neopon 33
Sodium olefin sulfonate and sodium ether sulfate as a clear amber liquid; foaming agent and detergent for cosmetic and household uses. *Witco Chemical Ltd.* Discontinued.

20945 Neopon LAM
2235-54-3 218-793-9
Ammonium lauryl sulfate, which may be based on natural or synthetic alcohols; liquid form; for hair and carpet shampoos. *Witco Chemical Ltd.* Discontinued.

20946 Neopon LOA/F
Ammonium alcohol 2EO sulfate in liquid form; for hair shampoos. *Witco Chemical Ltd.* Discontinued.

20947 Neopon LOS, LOS/F and LOS/NF
Sodium alcohol 2EO sulfate in liquid form; used for hair shampoos, foam baths, emulsion polymerization. *Witco Chemical Ltd.* Discontinued.

20948 Neopon LOT/F
Triethanolamine alcohol 2EO sulfate in liquid form; for hair shampoos, foam baths, liquid detergents. *Witco Chemical Ltd.* Discontinued.

20949 Neopon LS
Sodium lauryl sulfate, which may be based on natural or synthetic alcohols; liquid form; for hair and carpet shampoos; emulsion polymerization. *Witco Chemical Ltd.* Discontinued.

20950 Neopon LT
139-96-8 205-388-7
TEA lauryl sulfate
Triethanolamine lauryl sulfate which may be based on natural or synthetic alcohols; liquid form; for hair shampoos. *Witco Chemical Ltd.* Discontinued.

20951 Neopralac
Pigments for textile printing. *ICI Chem & Polymers Ltd.* Discontinued.

20952 Neoprene Latex 115
126-99-8 204-818-0
Polychloroprene latex
Stable dispersion, very resistant to deterioration by prolonged intensive mixing; for products requiring high modulus properties; used for adhesives, bonded batts, coatings, saturants; exhibits high mechanical strength, mechanical and chemical stability; vulcanized with ZnO and thiocarbanilide or Thiuram E and Tepidone. *DuPont.*

20953 Neoprene Latex 400
Anionic colloidal system containing a copolymer of chloroprene and 2,3 dichloro-1,3-butadiene; general purpose dispersion for products needing high modulus, abrasion and heat resistance, improved tear strength, and exceptional ozone and weathering resistance. *DuPont.*

20954 Neoprene Latex 750
Anionic colloidal system containing a medium-modulus highly crystallization-resistant copolymer of chloroprene and 2,3 dichloro-1,2-butadiene; general purpose latex for dipped goods requiring low modulus, wet/dry flexibility, and excellent crystallization resistance; adhesives. *DuPont.*

20955 Neoprene NPG 6856
126-99-8 204-818-0
Polychloroprene homopolymer; adhesive grade with improved rheology. *DuPont.* Name unverified.

20956 Neo-protosli
A proprietary colloidal silver preparation; contains about 20% silver iodide, AgI, combined with a protein base, and is a germicide. No manufacturer.

20957 Neopybuthrin
Synergized synthetic pyrethroids. *The Wellcome Foundation Ltd.* Name unverified.

20958 Neorad
Radiation-curable oligomers and formulations. *ICI Resins BV.*

20959 Neoresit
A phenol-formaldehyde resin. No manufacturer.

20960 NeoRez
Water and solvent based polyurethane resins; used in paints, inks and special coatings on all substrates, e.g. wood, metal, plastic, concrete, etc. *Polyvinyl Chemie Holland BV.*

20961 Neorez
Polyurethane resin products. *ICI Resins BV.*

20962 Neorode
Anti-fouling resin. *ICI Chem & Polymers Ltd.*

20963 Neoscan
41183-64-6 4364 255-248-4
$C_6H_5{}^{67}GaO_7$
Gallium citrate-^{67}Ga
Diagnostic aid; radioactive agent. *Medi-Physics Inc.* Name unverified.

20964 Neoscoa 203C
POE alkyl ether
Detergent, scouring agent for cotton, wool, and synthetic fibers; deinking agent for paper. *Toho Chem. Industry.*

20965 Neosolve® AD-1
Wetting agent for asbestos. *M.S. Paisner.*

20966 Neosorb
Sorbitol powders or syrups *Roquette Corp.*

20967 Neosorexa
56073-07-5 259-978-4
$C_{31}H_{24}O_3$
Biphenyl-4-yl-1,2,3,4-tetrahydro-1-naphthyl)-4-hydroxy-1(2H)-benzopyran-2-one
Difenacoum; Ratak; Neosorexa PP580; WBA 8107. A ready-to-use anticoagulant rodenticide. *Sorex Ltd.*

20968 Neosote
The phenoloids of blast-furnace tar. It contains a small quantity of phenol, and a large amount of cresols.

20969 Neospectra
7440-44-0 1855 231-153-3
C
A proprietary trade name for carbon black. No manufacturer.

20970 Neospinol 264
Blend of special anionics, nonionics, and mineral oil; dispersant for sulfur in rayon production; spinning oil for nylon. *Toho Chem. Industry.*

20971 Neostar
Organic brightener system; for acid zinc electroplating. *Harshaw Chemicals Ltd.* Name unverified.

20972 Neosyl®
7631-86-9 8634 231-545-4
Silica for industrial use. *Crosfield Chemicals Ltd.*

20973 Neosyn
Synthetic tannins and auxiliaries. *Hodgson Chemicals Ltd.*

20974 Neotac
Acrylic emulsion. *ICI Chem & Polymers Ltd.*

20975 Neotex
7440-44-0 1855 231-153-3
C
Carbon black. *Sevalco Ltd.*

20976 Neothane
A proprietary polyester based polyurethane elastomer cross linked with diamine. *Goodyear.* Name unverified.

20977 Neothene
Solvent; for gem and mineral testing. *Geoliquids.* Name unverified.

20978 NeoVac
Miscellaneous chemical compositions; for coatings and printing inks. *Polyvinyl Chemie Holland BV.* Name unverified.

20979 Neovadine
Dyeing and printing assistant. *Ciba plc.* Name unverified.

20980 Neo-Voronit®
Fuberidazole, sodium-N-dimethyldithiocarbamate; Nonmercurial liquid seed dressing for treatment of cereal seed against smuts and snow mold. *Bayer AG.*

20981 Neozapon® Dyes
Metal complex dyes with good solubility in polar solvents; for production of flexographic and gravure inks. *BASF AG.*

20982 Neozone A
90-30-2 201-983-0
phenyl-α-naphthylamine
A proprietary antioxidant for rubber. No manufacturer.

20983 Neozone B
A proprietary trade name for meta-toluylene diamine; an antioxidant. *Imperial Chemical Industries plc; Du Pont (UK) Ltd.* Name unverified.

20984 Neozone C
A proprietary antioxidant for rubber; resembles Neozone standard, but contains less m-toluylene diamine. No manufacturer.

20985 Neozone D
90-30-2 201-983-0
phenyl-α-naphthylamine
A proprietary antioxidant for rubber. No manufacturer.

20986 Neozone E
A proprietary trade name for an antioxidant for rubber containing 75 parts of phenyl-β-naphthylamine and 25 parts of meta-toluylene diamine. *Imperial Chemical Industries plc; Du Pont (UK) Ltd.* Name unverified.

20987 NEP
2687-91-4 220-250-6
N-ethyl-2-pyrrolidone
Solvent, reaction intermediate, textile auxilliaries, cosmetic ingredient. *ISP.* Name unverified.

20988 NEPD
597-09-1 209-893-3
$C_5H_{11}NO_4$
2-Nitro-2-ethyl-1,3-propanediol
Chemical and pharmaceutical intermediate, in tire cord adhesives, as formaldehyde release agents, deodorants, antimicrobials. *Angus.*

20989 Nephroflow
133-17-5 5055 205-097-5
$C_9H_7{}^{123}INNaO_3$
o-Iodohippurate sodium ^{123}I
Hippodin; Jodairol; Nephroflow. Diagnostic aid; radioactive agent. Soluble in H_2O. *Medi-Physics Inc.* Name unverified.

20990 Nepol® PP40
9003-07-0 7741
PP, 40% Long glass fiber-reinforced; for injection molded parts requiring high stiffness and good impact resist. *Neste Composite Materials.* Name unverified.

20991 NEPS
Polystyrene expandable beads. *Asahi Chem. Industry.*

20992 Nepton EXT
Durable water repellent and fluorocarbon extender; approved for Quarpel type water repellent treatments; nonsmoking; contains no solvents; excellent resin/catalyst stability. No manufacturer.

20993 Neptun® Bases
Dye bases for production of ball-pen pastes and typewriter ribbon inks. *BASF AG.*

20994 neradol
Synthetic tannins generally prepared by the condensation of phenolsulfonic acids with formaldehyde, under conditions that allow the formation only of water-soluble products. No manufacturer.

20995 Neradol
Synthetic tannins generally prepared by the condensation of phenol-sulfonic acids with formaldehyde, under conditions that only water-soluble products are formed. No manufacturer.

20996 Neral
5392-40-5 2383 226-394-6
$C_{10}H_{16}O$
Z-3,7-dimethyl-2,6-octadiene-1-al
3,7-Dimethyl-2,6-octadienal; Citral A; *cis*-3,7-dimethyl-2,6-octadienal; *cis*-citral; Lemarome n;. Fragrance and flavoring; fresh, lemon-like, green, slightly lime-like. Natural citral is a mixture of the *cis* isomer (neral) and the *trans* isomer (geranial). bp = 228-229°; d = 0.888; n_D^{20} = 1.4876; insoluble in H_2O, soluble in organic solvents. *BASF AG.*

20997 Neramine
Wood stain dyes and pigments for plastics. *ICI Chem & Polymers Ltd.*

20998 Nercol
Soluble cutting oils. *ABM Chemicals Ltd.* Name unverified.

20999 Nercolan
Solvents and cutting oils. *ABM Chemicals Ltd.* Name unverified.

21000 Nercosol
Viscosity depressant, antifoaming; used for vinyl plastisols. *ABM Chemicals Ltd.* Name unverified.

21001 Nerfinol
Textile finishing agent. *ABM Chemicals Ltd.* Name unverified.

21002 Nergandin
An alloy containing 70% copper, 28% zinc, and 2% lead; used for condenser tubes.

21003 Nericur Gel 5
94-36-0 1149 202-327-6
5% Benzoyl peroxide in aqueous gel. *Schering Health Care Ltd.* Discontinued.

21004 Nerloate
Paint dryers. *ABM Chemicals Ltd.* Name unverified.

21005 neroli oil
Oil obtained from the fresh flowers of the bitter orange; used as a perfume and flavoring substance.

21006 Nerolidol
7212-44-4 6561 230-597-5
$C_{15}H_{26}O$
3,7,11-trimethyl-1,6,10-dodecatrien-3-ol
Fragrance and flavoring; sweetly floral, green, woody, lilly-like. Occurs in *cis* and *trans* forms. (*trans*): $bp_{0.15}$ = 78°; n_D^{25} = 1.4792. (cis): $bp_{0.10}$ = 70°; n_D^{25} = 1.4775. (*cis/trans*): bp_3 = 122°; n_D^{25} = 1.4769; d_4^{25} = 0.8720. *BASF AG.*

21007 nerolin
93-04-9 6076 202-213-6
$C_{12}H_{12}O$
β-naphthol-methyl ether
yara-yara; β-naphthyl ethyl ether; Nerolin bromelia; ethyl β-naphthyl ether; ethyl 2-naphthyl ether; 2-ethoxynaphthalene; nerolin old. β-naphthol-ethyl ether is also known under this name. Used in perfumery. mp = 37-38°; bp = 282°; d_{20}^{28} = 1.0640; $n^{47.3}_D$ = 1.5932; insoluble in H_2O, soluble in organic solvents.

21008 Nervan
Wetting agents and finishing oils. *Rhône-Poulenc UK.*

21009 Nervan CP
Anionic surfactant. *Rhône-Poulenc UK.*

21010 Nervanaid
Sequestering agent, metal chelate, filter aid powder. *Rhône-Poulenc UK.*

21011 Nervanzse
9001-19-8 9203 232-588-1
Bacterial α-amylase. *Rhône-Poulenc UK.*

21012 nervin
An extract of meat.

21013 Nesfield's triple tablets
A water sterilizer. It consists of a) a tablet containing an iodide and an iodate; b) one containing citric or tartaric acid; and c) one containing sodium sulfite.

21014 Neste Polyethylene
Polyethylene of low, medium and high density as well as linear low pressure polyethylene; for packaging, distribution, household and construction films, insulation and jacketing of wire and cable, production of pipes, molded materials and paper coating. *Neste OY Chemicals & Unifos Kemi AB.* Name unverified.

21015 netilmicin sulfate

21016 Nettolin
Humus complex-fertilizer on peat basis for the culture of wine, hops, fruit and vegetables, for flowers and lawns. *Süd-Chemie AG.* Name unverified.

21017 Neudorfite
A resinous hydrocarbon found in Bavarian coal pits.

21018 Neuphor®
Neuphor® 100. An anionic rosin emulsion with 35% solids used as a sizing agent; used in paper and paperboard to impart resistance to water and aqueous solutions. *Hercules.*

21019 Neuphor® 635
Emulsion size for use in manufacturing of paper/paperboard; improved stabilizing system. *Hercules.* Discontinued.

21020 Neuro-Phosphates
A proprietary preparation of calcium glycerophosphate, sodium glycerophosphate, and strychnine; a tonic. *SmithKline Beecham.* Name unverified.

21021 Neustrene® 045
68002-72-2 268-085-9
Hydrogenated menhaden oil; textile lubricant, pharmaceutical intermediate, emulsifier, mold release agent, buffing compound. *Witco/Humko.* Discontinued.

21022 Neustrene® 059
67701-27-3 266-945-8
Hydrogenated tallow glycerides
Used in manufacturing of alkali metal soaps, monoglycerides, textile auxiliaries, greases. *Witco Corporation.*

21023 Neustrene® 064
68002-71-1 268-084-3
Hydrogenated soybean oil; textile lubricant, pharmaceutical intermediate, emulsifier, mold release agent, buffing compound. *Witco Corporation.*

21024 neutral alum
A neutral basic alum obtained by the addition of sodium hydroxide to a solution of alum, until the precipitate produced is just redissolved.

21025 Neutral Degras
68815-23-6 272-363-5
Fatty esters of wool grease; lubricant with EP and slip chars.; wire drawing compounds, slushing and cutting oils, lubricants; rust preventative; plasticizer and lubricant for adhesives; textile lubricant; inhibits crystallization of wax components; *Fanning.*

21026 Neutral oils
The name given to the lightest lubricating oils from American petroleum. The term is also applied to refined coal-tar oils.

21027 neutral phosphate
A fertilizer prepared by digesting mineral phosphate, bonemeal, or a mixture of both, with small amounts of sulfuric acid. This renders the P_2O_5 more available. The product contains 20-25% P_2O_5, and is neutral.

21028 neutral red
$C_{15}H_{17}N_4Cl$
toluylene red
A dyestuff; the hydrochloride of dimethyldiaminotoluphenazine; dyes cotton mordanted with tannin and tartar emetic, bluishred; used as an acid-base indicator in the range of pH 6.8-8.0; red in acid, yellow-brown in alkali; also as a biological.

21029 Neutralaleisen
A Swedish silicon-iron alloy, which is stated to resist the action of acids.

21030 Neutralite
An asphaltic material made in Germany.

21031 Neutramag®
1309-42-8 5706 215-170-3
A suspension of magnesium hydroxide for neutralising acidic effluents and reducing sludge volumes. *Steetley Magnesia Products Ltd.*

21032 Neutrase
A bacterial proteinase made by submerged fermentation of a selected strain of Bacillus subtilis; can be used in any case where the aim is breakdown of proteinaceous matter for production of peptides and amino acids. *Novo Nordisk.*

21033 Neutrichrome
Premetallized dyes for wool and wool blends. *ICI Chem & Polymers Ltd.* Discontinued.

21034 Neutrigan®
Alkaline salts and salt mixtures; basifying, deacidifying, and masking agents for chrome, leather, and furs. *BASF AG.*

21035 Neutrogene
Azoic dyes. *ICI Chem & Polymers Ltd.* Discontinued.

21036 Neutrol® TE
102-60-3 3635 203-041-4
Tetrahydroxypropyl ethylenediamine
Neutralizing agent for cosmetics industry. *BASF; BASF AG.*

21037 Neutrolactis
A proprietary preparation containing aluminum hydroxide gel, magnesium trisilicate, calcium carbonate and milk solids; an antacid. *Sandoz.* Name unverified.

21038 Neutronyx®
Nonyl phenol polyglycol ethers. *Millmaster-Onyx UK.*

21039 Neutronyx® 656
9016-45-9 6772
Nonoxynol-11
Detergent, dispersant, wetting agent, emulsifier for household detergents, dishwashing, fine fabrics, metal cleaning and degreasing, industrial cleaning, sanitizers, insecticides, herbicides; silicone emulsifier; lime dispersant; stable in acid, alkali and hard water. *Stepan; Stepan Canada.*

21040 Nevastain
An alloy of 86% iron, 9.5% chrondum, 4% silicon, and 0.43% carbon; it is non-corrosive.

21041 Nevastain R.A
A stainless steel alloy containing iron with approximately 16% chromium, 1% copper, 1% silicon, 0.4% manganese, 0.03% phosphorus and sulfur, and 0.1% carbon (maximum).

21042 Nevastain® A
Hindered phenolic compound; antioxidant used in mastic adhesives, rubber goods such as meach. and molded goods, caulking compounds, cement, and antiskinning agents. *Neville.*

21043 Nevchem® 70
Alkylated petroleum hydrocarbon resin; resin used in adhesives (hot melt pressure sensitive, mastic); coatings (aluminum, emulsion, industrial, marine, paper, traffic, wire/cable), inks, rubber (cements, mechanical and molded goods, tires), concrete-curing and caulking compounds. *Neville.*

21044 Nevex® 100
Modified hydrocarbon resin; enhances compatibility of wax and ethylene-vinyl acetate systems; in adhesives (hot melt, hot melt pressure sensitive, mastic, pressure sensitive), coatings (can & drum, industrial, paper), rubber (cements, mechanical and molded goods, tires. *Neville.*

21045 Nevidene
A proprietary trade name for a coumarone resin. No manufacturer.

21046 Nevile and Winther's acid
84-87-7 6478 201-568-4
$C_{10}H_8O_4S$
α-naphthol-4-sulfonic acid
4-hydroxy-1-naphthalenesulfonic acid; Nevile and Winther's acid. Used in manufacture of azo dyes. dec 170°; very soluble in H_2O.

21047 Nevillac® 10° XL
Hydroxy modified resin; plasticizer and tackifier in nitrile rubber; used in adhesives (hot melt, hot melt pressure sensitive, pressure sensitive), epoxy coatings, rubber cements, antiskinning agents. *Neville.*

21048 Neville
A proprietary trade name for coumarone-indene resins. *Neville Chemical Co.* Name unverified.

21049 Nevillite
A proprietary trade name for a hydrocarbon. No manufacturer.

21050 Nevin
A proprietary brand of lithopone. No manufacturer.

21051 Nevindene
A proprietary trade name for coumarone-indene resins. No manufacturer.

21052 Nevinol
A proprietary coumarone plasticizing oil; a viscous liquid polymer practically nondrying at room temperature. No manufacturer.

21053 Nevolin®
75-09-2 6140 200-838-9
Methylene chloride
Carrier and anti-evaporation agent for agrochemicals being applied by thermal fogging machine. *Fargro Ltd.*

21054 Nevoxy® EPX-L
Hydroxy modified resin; excellent color stability; especially suited for use in white and pastel coatings; produces formulations with good properties retention on aging. *Neville.*

21055 Nevpene® 9500
Modified hydrocarbon resin; aromatic producing an exceptional tackifier with modification; used in adhesives, coatings, rubber, and caulking compounds. *Neville.*

21056 nevraltein
A name for sodium-p-phenetidine-methane-sulfonate.

21057 nevrosthénine
An alkaline solution of the glycero-phosphates of sodium, potassium, and magnesium.

21058 Nevroz® 1420
Modified hydrocarbon resin; reactive, unsaturated character; replacement resins for rosin derivatives, and in inks. *Neville.*

21059 Nevtac® 80
Synthetic polyterpene resin; synthetic tackifier used in adhesives, coatings, rubber products, concrete-curing compounds, and caulking compounds. *Neville.*

21060 New 5C Cycocel
A mixture of chlormequat chloride and choline chloride; a plant growth regulator for cereals, linseed, winter oilseed rape, ornamentals. *BASF plc; BASF AG.*

21061 New 5C Cycocel
Soluble concentrate containing 645 g chlormequat and 32 g choline chloride per liter; plant growth regulator for use in cereals and ornamentals. *Cyanamid of Great Britain Ltd.*

21062 New Brick
Blended organic and inorganic acids in combination with surfactants and wetting agents; cleaner for brick in new construction (mortar smears, dirt etc). *Nova Chemical Inc.* Name unverified.

21063 New Brick (Heavy Duty)
Blended organic and inorganic acids in combination with surfactants and wetting agents; cleaner for brick in new construction (mortar smears, dirt etc). *Nova Chemical Inc.* Name unverified.

21064 New Formulation SBK Brushwood Killer
Liquid concentrate containing 2,4-D, mecoprop and dicamba; selective herbicide for use on coarse and woody weeds. *Vitax Ltd.*

21065 New Hickstor 6
117-18-0 204-178-2
Dustable powder containing 6% w/w tecnazene; protectant fungicide and potato sprout suppressant. *Hickson & Welch Ltd.*

21066 New Hystor
117-18-0 204-178-2

Granules containing 5% w/w tecnazene; protectant fungicide and potato sprout suppressant. *Agrichem (International) Ltd.*

21067 New Kotol
58-89-9 5526 200-401-2
γ-HCH
An organochlorine insecticide. *Embetec crop Protection Ltd.*

21068 New Legumex
Selective weedkillers. *Fisons plc, Horticultural Div.* Name unverified.

21069 New Murbetex
Pre-emergence herbicide. *Murphy Chemical Co Ltd.* Discontinued.

21070 New Verdone
Selective weedkiller. *Plant Protection.* Name unverified.

21071 New Zealand dammar
A name given to Kauri copal resin, obtained from *Dammara Australis*.

21072 Newagit
A trademark for abrasive and refractory materials consisting essentially of alumina. No manufacturer.

21073 Newaloy
A proprietary trade name for a steel containing copper. No manufacturer.

21074 Newcavac, -T
Newcastle disease, inactivated vaccine; for immunization of poultry. *Intervel Inc.*

21075 Newdamp Balancing Fluids
Blends of halogenated alkyl aryl hydrocarbons; controlled high density, low viscosity damping fluid when balancing gyroscope floats; stability, low volatility, noncorrosiveness. *Bacon.*

21076 Newloy
An alloy of 64% copper, 35% nickel, and 1% tin.

21077 Newton's alloy
An alloy of 50% bismuth, 31.2% lead, and 18.7% tin. It is a fusible alloy, and melts at 94.5 C.

21078 New-wrap
A proprietary viscose packing material. No manufacturer.

21079 Nex
1563-66-2 1851 216-353-0
Carbofuran
A systemic insecticide in the form of 5% w/w granules for the soil treatment for beetles and nematodes. *Tripart Farm Chemicals Ltd.*

21080 NFB
7664-38-2 7500 231-633-2
Phosphoric acids 80% and 85%; used in chemical polishing of aluminum. *Monsanto Co.* Name unverified.

21081 NFT Fertilizer
Soluble fertilizer. *Fisons plc, Horticultural Div.* Name unverified.

21082 Ngai camphor
A camphor obtained from *Blumea balsamifera*. It is closely related to borneol.

21083 N-Hance® 3000
65497-29-2
Guar hydroxypropyltrimonium chloride
Conditioner, viscosifier, substantive polymer for hair and skin care products, shampoos, lotions, liquid soaps. *Aqualon.*

21084 N'hangellite
An elastic bitumen.

21085 NI-20GF
9008-66-6
Nylon 6/10 resin, 20% glass fiber-reinforced. *Compounding Tech.*

21086 Niac
59-67-6 6612 200-441-0
Niacin; vitamin. *O'Neal, Jones & Feldman Pharmaceuticals.* Name unverified.

21087 Niacet
Metal acetates and propionates. *Niacet Corp.*

21088 Niacet Calcium Acetate Tech
62-54-4 1683 200-540-9
Calcium acetate
Sequestrant, thickener, pH control agent for petroleum and textile industries. *Niacet.*

21089 Niacet Sodium Acetate Anhyd. Tech
127-09-3 8711 204-823-8
Sodium acetate
Used in chemistry, photography, dyeing and as an acidulant and pharmaceutical aid. *Niacet.*

21090 niacin
59-67-6 6612 200-441-0
nicotinic acid
pyridine-3-carboxylic acid. Used in medicine as a cholesterol lowering agent, nutrition, feeds, enriched flours, dietary supplement.

21091 Niagara blue
72-57-1 9923 200-786-7
$C_{34}H_{24}N_6Na_4O_{14}S_4$
3,3'-[(3,3'-dimethyl[1,1'-biphenyl]-4,4'-diyl)bis(azo)]bis[5-maino-4-hydroxy-2,7-naphthalenedisulfonic acid] tetrasodium salt
C.I. Direct Blue 14; C.I. 23850; Benzamine Blue; Diamine Blue; Benzo Blue; Congo Blue; Dianil Blue; Naphthylamine Blue; Niagara Blue. Trypan blue. Soluble in H_2O, insoluble in organic solvents; LD_{100} (rat iv) = 300 mg/kg.

21092 Niaproof® Anionic Surfactant 08
126-92-1 204-812-8
$C_8H_{18}O_4S$
mono(2-ethylhexyl)sulfate sodium salt
sodium 2-ethylhexyl sulfate; sodium ethasulfate; Tergemist; Tergitol 08; Tergitol anionic 08; 08-Union Carbide; Emersal 6465; Nia proof 08; Propaste 6708; Sipex bos; Emcol d 5-10; Sole Tege TS-25; Sulfirol 8; Pentrone on; sodium etasulfate. Detergent, wetting agent, penetrant, emulsifier used in textile mercerizing, metal cleaning, electroplating, photo chemicals, adhesives, emulsion polymerization, household and industrial cleaners, agricultural, pharmaceuticals; stable to high concentrations of electrolytes. bp = 96-104°; soluble in H_2O (>100 mg/ml); $d_{21.7}$ = 1.114; insoluble in organic solvents; LD_{50} (rat orl) = 4 g/kg. *Niacet.*

21093 Niaproof® Anionic Surfactant 4
139-88-8 8838 205-380-3
$C_{14}H_{29}NaSO_4$
7-ethyl-2-methyl-4-undecanol hydrogen sulfate sodium salt
Sodium tetradecyl sulfate; Sotradecol; Tergitol 4; Trombavar; Trombovar. Detergent, wetting agent, penetrant, emulsifier used in adhesives and sealants, coatings, photo chemicals, emulsion polymerization, metal processing, electrolytic cleaning, pickling baths, plating, pharmaceuticals, leather, textiles. Soluble in H_2O and organic solvents; LD_{50} (rat orl) = 4.95 g/kg. *Niacet.*

21094 Niax
Urethane polymerization catalyst. *Union Carbide.*

21095 Nibiol
4008-48-4 6753 223-662-4
A proprietary preparation of nitroxoline; for veterinary use. *Dales Pharmaceuticals Ltd.* Name unverified.

21096 Nibren
Flame-retardant impregnating, encapsulating and dipping waxes with high dielectric constants; for use in the manufacture of paper capacitors. *Bayer AG.*

21097 Nibren wax
1321-65-9 215-321-3
$C_{10}H_5Cl_3$
1,2,8-trichloronaphthalene
Halowax; Seekay wax. Chlorinated naphthalene. mp = 93°; bp = 304-354°; insoluble in H_2O.

21098 Nibrite
Barrel bright nickel plating process. *Hanshaw Chemicals.* Unverified.

21099 Nicar
Nickel carbonate. *Mechema Chemicals Ltd.* Name unverified.

21100 Nicat
Nickel catalysts and Raney-type nickel catalysts; hydrogenation catalysts. *Crosfield Chemicals Ltd.*

21101 Nicfo
3349-06-2 6592 222-101-0
$C_2H_2NiO_4$
nickel formate
Used in manufacture of nickel compounds, including catalysts. dec 180-200°; $d^{20.2}$ = 2.154; soluble in H_2O, insoluble in organic solvents. *Mechema Chemicals Ltd.* Name unverified.

21102 Ni-chillite
A proprietary trade name for a nickel-chromium-molybdenum cast iron. No manufacturer.

21103 Nichroloy
Electrical resistance alloys containing 23-75% nickel, 7-20% chromium, 7-50% iron, and 1-3% manganese. by steam and dilute acids; They are used for electrical resistance wires, heat resistant products and chemical equipment.

21104 Nichrome®
Alloys of 54-80% nickel, 10-20% chromium, 7-27% iron, 0-11% copper, 0-5% manganese, 0.3-4.6% silicon, and sometimes 1% molybdenum, and 0.25% titanium; used as electrical resistance metals; resists acids. No manufacturer.

21105 Nichrosi
Name used for two different nickel alloys. 1) Contains 25-30% chromium, 16-

18% sillicon, balance nickel. 2) Contains 15-25% chromium, 16-18% silicon, balance nickel.

21106 nickel
7440-02-0 6582 231-111-4
Ni
Metallic element, used in electroplating, as a hydrogenation catalyst and in iron- and copper-based alloys. mp = 1453°; bp (calc) = 2732°; d= 8.908. *Aldrich; Atomergic Chemetals; Cerac; Inco Europe; Spectrum Chem. Mfg.*

21107 Nickel 200
7440-02-0 6582 231-111-4
A grade of nickel containing 99.0% nickel. *Wiggin Alloys Ltd.* Unverified.

21108 Nickel 201
7440-02-0 6582 231-111-4
A grade of nickel containing 99.0% nickel (min) and 0.02% carbon (max). *Wiggin Alloys Ltd.* Unverified.

21109 Nickel 204
7440-02-0 6582 231-111-4
Nickel containing 4% cobalt. *Wiggin Alloys Ltd.* Unverified.

21110 Nickel 205
7440-02-0 6582 231-111-4
Nickel containing 99.0% min. nickel and low carbon. *Wiggin Alloys Ltd.* Unverified.

21111 Nickel 211
7440-02-0 6582 231-111-4
Nickel containing 5% manganese. *Wiggin Alloys Ltd.* Unverified.

21112 Nickel 212
7440-02-0 6582 231-111-4
Nickel containing 2% manganese. *Wiggin Alloys Ltd.* Unverified.

21113 Nickel 213
7440-02-0 6582 231-111-4
Nickel with improved machining properties; nickel 96% min., manganese 2%; high carbon and silicon content. *Wiggin Alloys Ltd.* Unverified.

21114 Nickel 222
7440-02-0 6582 231-111-4
Nickel containing 99.5% min. nickel, 0.06-0.09% magnesium and very low impurity levels. *Wiggin Alloys Ltd.* Unverified.

21115 Nickel 223
7440-02-0 6582 231-111-4
Nickel containing 99.5% min. nickel, 0.035-0.065% magnesium and very low impurity levels. *Wiggin Alloys Ltd.* Unverified.

21116 Nickel 229
7440-02-0 6582 231-111-4
Nickel containing 97.5% min. nickel, 1.8-2.2% tungsten, 0.35-0.65% magnesium and 0.02-0.04% aluminum. *Wiggin Alloys Ltd.* Unverified.

21117 Nickel 270
7440-02-0 6582 231-111-4
Nickel 99.9% pure. *Wiggin Alloys Ltd.* Unverified.

21118 nickel acetate
373-02-4 6583 206-761-7
$C_4H_6NiO_4 \cdot 4H_2O$
nickel acetate tetrahydrate
Textile mordant, catalyst. d = 1.744; soluble in H_2O (167 mg/ml), EtOH. *Ashland; Atomergic Chemetals; Mallinckrodt; Nihon Kagaku Sangyo.*

21119 nickel aluminum bronze
An alloy of 10-40% nickel, 10-88% copper, and 2-30% aluminum. One alloy contains 20% tin.

21120 Nickel Babbitt
A proprietary trade name for a tin-copper-nickel alloy used as a bearing metal for high speeds. No manufacturer.

21121 nickel brass
A nickel silver. One alloy contains 55% copper, 43% zinc, and 2% nickel, and another 50% copper, 34% zinc, 15% nickel, and 0.1% aluminum.

21122 nickel bronze
A nickel silver. It usually contains from 20-30% nickel, 50-86% copper, and 8-25% tin, but other alloys contain 11-18% zinc, and 0-18% lead.

21123 nickel carbonate
3333-67-3 222-068-2
$CNiO_3$
Nickel (II) carbonate
carbonic acid, nickel salt; nickel(II)carbonate basic; nickel carbonate; carbonic acid, nickel(2+) salt (1:1). Chemical intermediate for nickel oxide, nickel powder, and nickel catalysts. Used in vacuum tubes and transistor cans, as a catalyst to remove organic contaminants from wastewater or potable water in the preparation of colored glass of nickel pigments, as a neutralizing
compound in nickel electroplating solution, and in the preparation of many specialty nickel
compounds. mp= 57°.

21124 nickel carbonate, basic
12607-70-4 6586 235-715-9
$NiCO_3 \cdot 2Ni(OH)_2 \cdot 4H_2O$
Electroplating, preparation of nickel catalysts, ceramic colors and glazes. *Ashland; Elf Atochem N. Am.; Farleyway Chem. Ltd; M&T Harshaw; Nihon Kagaku Sangyo.*

21125 nickel chloride
7718-54-9 6588 231-743-0
$NiCl_2$
nickel(II) chloride
nickelous chloride; Hexahydrate [7791-20-0]. Electroplated nickel coatings, chemical reagent. bp = 987°; soluble in H_2O (1 g/ml), EtOH; LD (dog iv) = 40-80 mg/kg. *Ashland; Elf Atochem N. Am.; Atomergic Chemetals; Mallinckrodt; Nihon Kagaku Sangyo.*

21126 nickel glance
$Ni(AsS)_2$
A mineral.

21127 nickel manganese bronze
An alloy containing 2.5% nickel, 53.4% copper, 39% zinc, 1.7% manganese, and 2.6% tin, with small quantities of aluminum and lead.

21128 nickel nitrate
13478-00-7 6596 236-068-5
$N_2NiO_6 \cdot 6H_2O$
nickel(II) nitrate
nickelous nitrate. Nickel plating, preparation of nickel catalysts, manufacture of brown ceramic colors. mp = 57°; bp = 137°; d = 2.05; soluble in H_2O (2.5 g/ml), EtOH; LD_{50} (rat orl) = 1.62 g/kg. *Mallinckrodt; Nihon Kagaku Sangyo; Noah Chem.*

21129 nickel oreide
A nickel silver. It is an alloy of from 63-87% copper, 6-33% zinc, and 2-7% nickel.

21130 nickel oxide
1313-99-1 6595 215-215-7
NiO
nickel (II) oxide
nickelous oxide; nickel monoxide; nickel protoxide; green nickel oxide; bunsenite. Used in preparation of nickel salts, in porcelain painting and fuel cell electrodes. mp = 1960°; d = 6.6700; insoluble in H_2O; soluble in acids. *Atomergic Chemetals; Cerac; Nihon Kagaku Sangyo; Noah Chem.*

21131 nickel silver solder
An alloy of 35% copper, 57% tin, and 8% nickel.

21132 nickel silvers
Albatra; Alfenide; Alpacca; Amberoid; Ambrac; American silver; Aphit; Argentan; Argentin; Argiroide; Argentan solder; Argyroide; Argyrolith; Ateri; Benedict plate; Bismuth bronze; brazing solder; China silver; Carbondale silver; Craig gold; electropate; electrum; Elner's German silver; Nevada silver; nickel brass; nickel bronze; nickelin; silverite; Silveroid; Victoria silver; white button alloy; white solder. Ternary alloys of copper, nickel, and zinc, the standard of which is determined by the nickel content.

21133 nickel steel
An alloy of nickel with steel, usually containing from 3-5% of nickel, but sometimes a larger amount. One alloy contains 30% of nickel, 1% manganese, and 1% chromium. Nickel steels are used for armor plates ships screws, boiler plates, cable wires and gun barrels.

21134 nickel sulfamate
13770-89-3 237-396-1
$H_4N_2NiO_6S_2 \cdot 2H_2O$
nickel (II) sulfamate dihydrate
Sulfamic acid, nickel(2+) salt (2:1). For nickel electroforming and plating; aerospace and electronics application. *Elf Atochem N. Am.; Atomergic Chemetals; M&T harshaw; Witco/Allied-Kelite.*

21135 nickel sulfate
7786-81-4 6600 232-104-9
NiO_4S
nickel(II) sulfate
single nickel salt; nickelous sulfate; nickel(2+) salt, sulfurinc acid; nickel monosulfate. Also occurs as the hexahydrate [10101-97-0; 232-104-9] and the heptahydrate [10101-98-1; 205-788-1]. Used in the electroplating trade. Hexahydrate loses $5H_2O$ at 100° and becomes anhydrous at 280°. Soluble in H_2O (0.7 g/ml); LD_{50} (gpg sc) =62 mg/kg.

21136 nickel zirconium
An alloy of 86.4% nickel, 6% aluminum, 6% silicon, and 1.5% zirconium; used for cutting tools.

21137 Nickeladium
A proprietary trade name for a nickel-vanadium cast steel. No manufacturer.

21138 nickelene
A name suggested for nickel silver (German silver) alloys.

21139 Nickelin
Electrical resistance alloys of nickel and copper, usually with zinc. One alloy contains 55.3% copper, 31%, 13%, 0.4% iron, and 0.2% lead; another 68% copper and 32% nickel; and a third 74.5% copper, 25% nickel, and 0.5% iron.

21140 nickel-manganese-copper
An alloy of 73% copper, 24% manganese, and 3% nickel; used for electrical resistances.

21141 nickel-molybdenum steels
Alloys containing 0.13-0.54% carbon, 0.12-4.4% molybdenum, and 1.8% nickel.

21142 Nickeloid
A proprietary trade name for a dual metal; it is zinc faced with nickel. No manufacturer.

21143 Nickeloy
An alloy of 1.5% nickel, 4% copper, and 94% aluminum.

21144 Niclad
A proprietary trade name for a duplex metal in which nickel or nickel alloy is deposited on steel or iron. No manufacturer.

21145 Niclate®
Nickel dialkyldithiocarbamate
Rubber antioxidants offering antiozonant and antioxidant protection in epichlorohydrin; used for optimum aging properties. *R. T. Vanderbilt Co Inc.* Discontinued.

21146 niclosamide
50-65-7 6602 200-056-8
$C_{13}H_8Cl_2N_2O_4$
5-chloro-N-(2-chloro-4-nitrophenyl)-2-hydroxybenzamide
Bayer 2353; Cestocide; Niclocide; Ruby; Yomesan; Trédémine. Used in medicine and veterinary medicine as an anthelmintic. mp = 225-230°; insoluble in H_2O, sparingly soluble in organic solvents.

21147 Nico
1313-99-1 6595 215-215-7
Nickel monoxide
Used for painting on porcelain. *Mechema Chemicals Ltd.* Name unverified.

21148 Nico-400
59-67-6 6612 200-441-0
Niacin; vitamin. *Marion Merrell Dow Inc.* Name unverified.

21149 Nicobid
59-67-6 6612 200-441-0
Niacin; vitamin. *USV Pharmaceutical Corp.* Unverified.

21150 Nicocap
59-67-6 6612 200-441-0
Niacin; vitamin. *ICN Nutritional Biochemicals Corp.* Name unverified.

21151 Nicolar
59-67-6 6612 200-441-0
Niacin; vitamin. *USV Pharmaceutical Corp.* Unverified.

21152 Nicolle's Carbol-thionin blue
A microscopic stain. It consists of a mixture of 10 ml of a saturated solution of thionin blue in 50% alcohol, and 100 ml of a 2% phenol solution.

21153 Nicolmelt®
Hot-melt adhesives. *Reichhold.*

21154 Nicon
An alloy of 70% iron and 30% nickel.

21155 Nicor
Mixture of nickel oxide and cobalt oxide. *Mechema Chemicals Ltd.* Name unverified.

21156 Nicotine 40% Shreds
Insecticide smoke. *Murphy Chemical Co Ltd.* Discontinued.

21157 Nicral alloys
Aluminum alloys containing varying percentages of nickel, chromium, and copper.

21158 nicro-copper
An alloy of 98% copper and 2% nickel.

21159 Nicrolan
Wool grease-based compositions; anticorrosive coating materials for metal protection. *Westbrook Lanolin.* Name unverified.

21160 Nicroman
A proprietary trade name for a tool steel containing 1% chromium, 1.65% nickel, 0.35% copper, and 0.7% carbon; an oil-hardening hob steel. No manufacturer.

21161 Nicron 325
14807-96-6 9207 238-877-9
Platy talc; general purpose economical filler for industrial coatings and plastics that require maximum filler loadings. *Montana Talc.*

21162 Nicron 665
14807-96-6 9207 238-877-9
Platy talc; ultrafine, high purity talc with excellent reinforcing and dielec. properties; for extrusion wire and cable applications, additive for gloss control in paints and coatings. *Montana Talc.*

21163 Nicron JS 422
14807-96-6 9207 238-877-9
Platy talc; filler providing mineral oil absorp. for low VOC and high solids paints/coatings applications. *Montana Talc.*

21164 Nicrosil
A proprietary trade name for an alloy of iron with 18% nickel and 4-6% silicon. No manufacturer.

21165 Nicrosilal
An alloy of 71.2% iron, 18% nickel, 6% silicon, 2% chromium, 1.8% carbon, and 1% manganese. It is a nonmagnetic grey cast iron which is resistant to staining and has great ductility.

21166 Nicu steel
A nickel steel containing 2.13% nickel, 0.2% copper, 0.51% manganese, 0.03% sulfur, 0.03% silicon, and 0.006% phosphorus.

21167 Niello silver
Russian tula; blue silver. An alloy of silver, copper, lead, and bismuth with a bluish color.

21168 Niflor
Precision nickel/PTFE composite; precision thickness coating for hard, self-lubricating and corrosion resistant surfaces on engineered components. *Fothergill & Harvey plc.* Unverified.

21169 Nifuraldezone
3270-71-1 6622 221-890-9
$C_7H_6N_4O_5$
Aminooxoacetic acid [(5-nitro-2-furanyl)methylene]hydrazide
Furamazone; 5-nitro-2-furaldehyde semioxamazone;. Antibacterial. MW=226.15; crystal powder; mp=270°; *Norwich Eaton Pharmaceuticals Inc.* Name unverified.

21170 Nifurpirinol
13411-16-0 6628 236-503-9
$C_{12}H_{10}N_2O_4$
6-[2-(5-Nitro-2-furanyl)ethenyl]-2-pyridinemethanol
Furanace; 6-[2-(5-nitro-2-furyl)vinyl]-2-pyridine-methanol; furpirinol; Furpyrinol; P-7138. Used as an antibacterial for treatment of diseases in fish. MW=246.22; yellow needles from acetone; mp=170-171°; LD_{50}(eels,orl)=1780 mg/kg *Dainippon Pharmaceutical Co.* Name unverified.

21171 Nigagin
99-76-3 6182 202-785-7
Methyl-*p*-hydroxybenzoate
methyl paraben. A preservative. mp = 125-128°.

21172 Night of Olay Nightcare
Cream for nightly care of the skin. *Richardson-Vicks Inc.* Name unverified.

21173 nigraniline
$C_{30}H_{25}N_5$
Aniline black, a dyestuff.

21174 nigre
The impure soap remaining after the good soap has been removed by running out. It contains iron soaps, caustic soda, and sodium chloride.

21175 nigrol
The residue obtained after the removal of kerosene, gasoline, and light oil from petroleum naphtha.

21176 Nigrosin
Water-soluble acid dyestuffs for a variety of special applications. *Bayer AG.*

21177 Nigrosin Bases
For shoe polish, printing ink, office supplies and plastics industries. *Bayer plc.*

21178 Nigroth Metal
A proprietary trade name for a heat-resisting nickel-chromium cast iron. No manufacturer.

21179 nigrotic acid
Dihydroxysulfonaphthoic acid
Used in the manufacture of dyestuffs.

21180 Ni-hard
Alloys of iron with from 4-5% nickel, 1.5% chromium, and varying amounts of silicon and carbon.

21181 Nikal®
Nickel compound; antiseize compound for extreme temperature applications; acid and chemical resistance. *Jet-Lube.*

21182 Niklad
Electrodeless nickel systems for plating onto prepared substrates including stainless steel, steel alloys, cast iron, copper and alloys, aluminum, titanium, beryllium, nickel alloys, nonconductors such as glass and ceramics. *Witco/Allied-Kelite.* Discontinued.

21183 Nikrome
Proprietary trade name for nickel-chromium steels. Nikrome M contains

2.25% nickel, 1% chromium, 0.45% molybdenum, and 0.4% carbon. No manufacturer.

21184 Nikro-trimmer Steel

A proprietary trade name for a nickel-chromium steel containing 0.3% nickel, 0.55% chromium, and 0.85% carbon. No manufacturer.

21185 Nilex

A proprietary 36% nickel steel used for pendulums, etc., on account of its low coefficient of expansion. No manufacturer.

21186 Nilfom 2X

Nonsilicone defoamer; defoamer for atmospheric and pressure dyeing; compatible with all systems. *Am. Emulsions.*

21187 Nilfom DF-155

Nonsilicone defoamer; defoamer which does not contribute to flammability. *Am. Emulsions.*

21188 Nilo Alloy 36

A trademark for controlled expansion alloys; 36% nickel, balance iron. *Wiggin Alloys Ltd.* Unverified.

21189 Nilo Alloy 42

A trademark for controlled expansion alloys; 42% nickel, balance iron. *Wiggin Alloys Ltd.* Unverified.

21190 Nilo Alloy 48

A trademark for controlled expansion alloys; 48% nickel, balance iron. *Wiggin Alloys Ltd.* Unverified.

21191 Nilo Alloy 51

A trademark for controlled expansion alloys; 51% nickel, balance iron. *Wiggin Alloys Ltd.* Unverified.

21192 Nilo Alloy 475

A trademark for controlled expansion alloys; 47% nickel, 5% chromium, balance iron. *Wiggin Alloys Ltd.* Unverified.

21193 Nilo Alloy K

A trademark for controlled expansion alloys; 29% nickel, 17% cobalt, balance iron. *Wiggin Alloys Ltd.* Unverified.

21194 Nilo Alloy K45

A trademark for controlled expansion alloys; 32% nickel, 13% cobalt, 54% iron. *Wiggin Alloys Ltd.* Unverified.

21195 Nilo Alloy P50

A trademark for controlled expansion alloys; 50% nickel, 50% iron. *Wiggin Alloys Ltd.* Unverified.

21196 Nilomag Alloy 48

A trademark for magnetic alloys made by a powder metallurgy process. It is 48% nickel and the balance is iron. *Wiggin Alloys Ltd.* Unverified.

21197 Nilomag Alloy 51

A trademark for magnetic alloys made by a powder metallurgy process. It is 51% nickel and the balance is iron. *Wiggin Alloys Ltd.* Unverified.

21198 Nilomag Alloy 471

A trademark for magnetic alloys made by a powder metallurgy process. It is 47% nickel, 50% iron, and 3% molybdenum. *Wiggin Alloys Ltd.* Unverified.

21199 Nilomag Alloy 475

A trademark for magnetic alloys made by a powder metallurgy process. It is 47% nickel, 5% chromium, and the balance is iron. *Wiggin Alloys Ltd.* Unverified.

21200 Nilomag Alloy K

A trademark for magnetic alloys made by a powder metallurgy process. It is 29% nickel, 17% cobalt, and the balance is iron. *Wiggin Alloys Ltd.* Unverified.

21201 Nimate

13770-89-3 237-396-1

$H_4N_2NiO_6S_2$

Nickel sulfamate

Sulfamic acid, nickel(2+) salt (2:1). For nickel electroforming of foils, perforated parts, moldings, and record stampers; also resizing of worn parts and nickel plating for aerospace and electronics applications. *Albright & Wilson UK.* Name unverified.

21202 Nimocast Alloy 242

A patented alloy of 21% chromium, 10% cobalt, 10.5% molybdenum and the balance nickel. *Wiggin Alloys Ltd.* Unverified.

21203 Nimocast Alloy 713

13.4% chromium, 4.5% molybdenum, 1% titanium, 6.2% aluminum, 2.3% niobium, balance nickel. *Wiggin Alloys Ltd.* Unverified.

21204 Nimocast Alloy 771

77% nickel, 14% iron, 5% copper, 4% molybdenum. *Wiggin Alloys Ltd.* Unverified.

21205 Nimocast Alloy PE10

20% chromium, 6% molybdenum, 6.5% niobium, 2.5% tungsten, balance nickel. *Wiggin Alloys Ltd.* Unverified.

21206 Nimocast Alloy PK24

An alloy consisting of 10% chromium, 15.2% cobalt, 3% molybdenum, 5.2% titanium, 5.5% aluminum, and the balance is nickel. *Wiggin Alloys Ltd.* Unverified.

21207 Nimol

An alloy of cast iron with 20% monel metal and 2-4% chromium. It is non-magnetic and has high resistance to corrosion by acid and sea water.

21208 Nimonic Alloy 75

A trademark for an alloy of 20% chromium, 0.4% titanium, and the balance is nickel. *Wiggin Alloys Ltd.* Unverified.

21209 Nimonic Alloy 80A

A trademark for an alloy of 20% chromium, 2.3% titanium, 1.3% aluminum, and the balance is nickel. *Wiggin Alloys Ltd.* Unverified.

21210 Nimonic Alloy 90

A trademark for an alloy of 20% chromium, 17% cobalt, 2.5% titanium, 1.5% aluminum, and the balance is nickel. *Wiggin Alloys Ltd.* Unverified.

21211 Nimonic Alloy 93

A trademark for an alloy of 20% chromium, 17% cobalt, 2.5% titanium, 1.5% aluminum, and the balance is nickel; closer control maintained than Alloy 90. *Wiggin Alloys Ltd.* Unverified.

21212 Nimonic Alloy 105

A trademark for an alloy of 15% chromium, 20% cobalt, 5% molybdenum, 1.2% titanium, 4.7% aluminum, and the balance is nickel *Wiggin Alloys Ltd.* Unverified.

21213 Nimonic Alloy 115

A trademark for an alloy of 15% chromium, 15% cobalt, 3.5% molybdenum, 4% titanium, 5% aluminum, and the balance is nickel *Wiggin Alloys Ltd.* Unverified.

21214 Nimonic Alloy 118

A trademark for an alloy of 15% chromium, 15% cobalt, 3.5% molybdenum, 4% titanium, 5% aluminum, balance nickel; a fully vacuum-melted and cast version of Alloy 115. *Wiggin Alloys Ltd.* Unverified.

21215 Nimonic Alloy PE 11

A trademark for an alloy of 18% chromium, 5.2% molybdenum, 2.3% titanium, 0.8% aluminum, 38% nickel, and the balance is iron. *Wiggin Alloys Ltd.* Unverified.

21216 Nimonic Alloy PE 13

A trademark for an alloy of 22% chromium, 1.5% cobalt, 9% molybdenum, 18.5% iron, 0.6% tungsten, balance nickel. *Wiggin Alloys Ltd.* Unverified.

21217 Nimonic Alloy PK 31

A trademark for an alloy of 20% chromium, 14% cobalt, 4.5% molybdenum, 2.3% titanium, 0.4% aluminum, 5% niobium, and the balance is nickel. *Wiggin Alloys Ltd.* Unverified.

21218 Nimonic Alloy PK 33

A trademark for an alloy of 19% chromium, 14% cobalt, 7% molybdenum, 2% titanium, 2% aluminum, and the balance is nickel. *Wiggin Alloys Ltd.* Unverified.

21219 Nimox

Nickel molybdenum oxides on alumina. *Laporte Industries Ltd.*

21220 Nimrod

41483-43-6 1519 255-391-2

An emulsifiable concentrate containing 250 g bupirimate per liter; a systemic fungicide to control powdery mildew. *ICI Agrochemicals; ICI Garden Products.*

21221 Nimrod T

An emulsifiable concentrate containing 62.5 g bupirimate [41483-43-6] and 62.5 g triforine [26644-46-2] per liter; a systemic fungicide to control powdery mildew in ornamental plants. *ICI Agrochemicals; ICI Garden Products.*

21222 Ninate 401

Anionic surfactant as a dark viscous liquid; emulsifier used in oil additives. *KWR Chemicals Ltd.* Name unverified.

21223 Ninate 411

Anionic surfactant as a light viscous liquid; emulsifier for solvent degreasers and dry cleaning detergents. *KWR Chemicals Ltd.* Name unverified.

21224 Ninate 415

Anionic surfactant in light amber liquid form; emulsifier for solvent degreasers and dry cleaning detergents. *KWR Chemicals Ltd.* Name unverified.

21225 Ninate®

Alkyl benzene sulfonate; emulsifier. *Stepan.* Name unverified.

21226 Ninate® 401

Calcium alkylbenzene sulfonate

Emulsifier, dispersant used in pesticide formulations and in self-dispersing liquid; foaming agent. *Stepan; Stepan Canada; Stepan Europe.*

21227 Ninate® 401-A

Calcium alkylbenzene sulfonate. Emulsifier for agricultural formulations. *Stepan; Stepan Canada; Stepan Europe.*

21228 Ninate® 411

Amine dodecylbenzene sulfonate

Emulsifier, solvent degreaser, drycleaning detergent, surface tension reducer, defoamer; emulsifier used in emulsifiable kerosene formulations, agricultural formulations, textiles, metalworking. *Stepan; Stepan Canada.*

21229 Ninate® DS 70
577-11-7 3460 209-406-4
Sodium dioctyl sulfosuccinate
Dispersant, wetting agent, emulsifier for agricultural flowables, suspension and emulsifiable concentrates, granular and powder formulations. *Stepan Europe.*

21230 ninhydrin
485-47-2 6646 207-618-1
$C_9H_6O_4$
2,2-dihydroxy-1H-indene-1,3(2H)-dione
Triketohydrindene hydrate; 1,2,3-indantrione monohydrate;. A colorimetric reagent for aminoacids. dec 241°; soluble in H_2O; LD_{50} (mus ip) = 78 mg/kg.

21231 Ninol®
Alkylolamides
Thickeners, foam stabilizer, detergent. *Stepan.* Name unverified.

21232 Ninol® 11-CM
Cocamide DEA, modified; detergent base for synthetic cleaners; emulsifier, lubricant, antistat for textile applications; emulsifier, corrosion inhibitor in cutting fluids, drawing compounds, metal cleaning. *Stepan; Stepan Canada.*

21233 Ninol® 201
93-83-4 202-281-7
Oleamide DEA; emulsifier, corrosion inhibitor in industrial lubricant systems, cutting fluids, drawing compounds, metal cleaners; thickener for personal care and liquid detergent products; emulsifier, lubricant, antistat for textiles. *Stepan; Stepan Canada.*

21234 Ninol® 30-LL
120-40-1 204-393-1
Lauramide DEA
Foam booster/stabilizer, viscosity builder/modifier for liquid detergents, shampoos, hand soaps, bath products. *Stepan; Stepan Canada.*

21235 Ninol® CMP
68140-00-1 268-770-2
Cocamide MEA
Foam booster, viscosity builder for liquid detergents, detergent blocks or bars. *Stepan; Stepan Canada.*

21236 Ninol® CNR
68140-00-1 268-770-2
Cocamide MEA
Emollient, thickener, foam booster/stabilizer for shampoos, bubble baths, liquid soaps, shower gels, toilet soaps. *Stepan Europe.*

21237 Ninol® LMP
142-78-9 205-560-1
Lauramide MEA
Foam booster/stabilizer, thickener, emollient, detergent for dishwash, liquid detergents, detergent blocks or bars, shampoos, hand soaps, bath products *Stepan; Stepan Canada; Stepan Europe.*

21238 Ninol® M10
Monoisopropanolamide
Thickener, emollient, anticorrosive agent, foam booster/stabilizer for shampoos, bubble baths, liquid soaps, shower gels, toilet soaps. *Stepan Europe.*

21239 Ninox
Foam stabilizer; shampoos; foam baths; detergents. *KWR Chemicals Ltd.* Name unverified.

21240 Ninox®
Amine oxide
Foam enhancers and thickeners. *Stepan.* Name unverified.

21241 Ninox® FCA
68155-09-9 268-938-5
Cocamidopropylamine oxide
Thickener, foam booster/stabilizer, detergent, antistat for scale-removing liquids, liquid soaps, cleaning foams, personal care products. *Stepan Europe.*

21242 Ninox® L
Lauramine oxide
Thickener, foam booster/stabilizer, detergent, antistat for scale-removing liquids, liquid soaps, cleaning foams. *Stepan Europe.*

21243 Ninox® M
3332-27-2 222-059-3
Myristamine oxide
Thickener, foam booster/stabilizer, detergent, antistat for scale-removing liquids, liquid soaps, cleaning foams. *Stepan Europe.*

21244 Ninox® SO
2571-88-2 219-919-5
Stearamine oxide
Thickener, emollient, mild detergent for scouring pastes or creams; additive for hypochlorite. *Stepan Europe.*

21245 niobium
7440-03-1 6648 231-113-5
Nb
Columbium. Metallic element; used in ferrous metallurgy; for superconducting and magnetic alloys, cermets, missiles and rockets, cryogenic equipment, ferroniobium for alloy steels. mp = 2468°; bp = 4927°; d = 8.57. *Atomergic Chemetals; Cabot; Cerac; Noah Chem.*

21246 niobium oxide
1313-96-8 6651 215-213-6
Nb_2O_5
niobium pentoxide
niobium (V) oxide. Used as a chemical intermediate, in electronics fabrication. mp = 1520°; d= 4.6; insoluble in H_2O, soluble in HF or H_2SO_4. *Atomergic Chemetals; Cabot; Cerac; Noah Chem.*

21247 Nipa salt
A material obtained by ignition of the plant *Nipa fructicans* ; used to coagulate rubber latex.

21248 Nipabenzyl
94-18-8 202-311-9
$C_{14}H_{12}O_3$
benzoic acid, 4-hydroxy-, phenylmethyl ester
Benzylparaben. Among the most widely used preservatives in food, drugs, and cosmetics. Preservative, bactericide, fungicide for pharmaceuticals, cosmetics, foods, medicinal preparations, industrial applications. mp = 110-112°. *Nipa Laboratories Ltd.*

21249 Nipabutyl
94-26-8 1619 202-318-7
$C_{11}H_{14}O_3$
butyl-4-hydroxybenzoate
n-butyl-paraben; Butoben; Butyl Chemosept; Butyl Parasept; 4-hydroxybenzoic acid butyl ester; Tegosept B; aseptoform butyl; Butyl Tegosept; Nipabutyl; Solbrol b; Preserval b; SPF; Tegosept butyl. Antifungal agent used as a preservative in foods. mp = 67-69°; slightly soluble in H_2O, soluble in organic solvents. *Nipa Laboratories Ltd.*

21250 Nipabutyl Potassium
38566-94-8 254-009-1
Potassium butylparaben
Preservative. *Nipa Laboratories Ltd.*

21251 Nipabutyl Sodium
36457-20-2 253-049-7
Sodium butylparaben
Preservative, bactericide, fungicide for pharmaceuticals, cosmetics, foods, medicinal preps., industrial applications. *Nipa Laboratories Ltd.*

21252 Nipacide® BCP
120-32-1 2469 204-385-8
o-benzyl-*p*-chlorophenol
chlorophene; Santophen; Septiphene; benzylchlorophenol; 5-chloro-2-hydroxydiphenylmethane; clorofene; ketolin-h; santophen 1; Bio-clave;Neosobenil; Preventol B; Sentiphene. Germicide in disinfectant cleaners for hospitals, schools, homes, etc.; readily degradable. mp = 48°; $bp_{3.5}$ = 160-162°; $d^{55}_{15.5}$ = 1.186-1.190. *Nipa Laboratories Ltd.*

21253 Nipacide® BIT
2634-33-5 220-120-9
C_7H_5NOS
1,2-Benzisothiazolin-3-one
Proxel; Proxel XL; Proxil; Benzisothiazol-3(2H)-one; IPX; Proxan. Used as a preservative in 156 products, including cleaning agents, polishes,and paints. The concentration is less than 0.01% in 46% of products and greater than 0.1% in 24%. The concentration is lowest in shampoos, skin care agents, cleaning agents, printing inks, and polishes. Sensitized patients have reacted to 0.0099% concentration in aqueous and alcohol. *Nipa Laboratories Ltd.*

21254 Nipacide® BK
4719-04-4 225-208-0
$C_9H_{21}N_3O_3$
1,3,5-Triazine-1,3,5(2H,4H,6H)-triethanol
tris-hydroxyethyl-*s*-triazine; Grotan BK; Grotan B; Kalpur TE; Onyxide 200; Grotan; Miliden X-2. Preservative, bacteriostat, fungistat for sol. cutting fluids and coolants and other products. *Nipa Laboratories Ltd.*

21255 Nipacide® DX
133-53-9 3126 205-109-9
$C_8H_8Cl_2O$
2,4-dichloro-3,5-dimethylphenol
2,4-Dichloro-*m*-xylenol; DCMX; Ottacide. Bacteriostat and preservative. mp = 95-96°; soluble in H_2O (200 mg/l), more soluble in organic solvents. *Nipa Laboratories Ltd.*

21256 Nipacide® F
2,2-Dihydroxy-5,5-dichloro diphenyl monosulfide. *Nipa Laboratories Ltd.* Discontinued.

21257 Nipacide® MX
88-04-0 2228 201-793-8
C_8H_9ClO
4-chloro-3,5-dimethylphenol
p-chloro-*m*-xylenol; Benzytol; Dettol. Antimicrobial, preservative for disinfectant, algicide, slimicide, and water treatment pesticide products, polymer emulsions, adhesives, latex paints, metalworking cutting fluids. mp = 115°; bp = 246°; soluble in H_2O (0.3 g/l), more soluble in organic solvents. *Nipa Laboratories Ltd.*

21258 Nipacide® OPP
90-43-7 7458 201-993-5
$C_{12}H_{10}O$
o-phenyl phenol
2-biphenylol; 2-hydroxybiphenyl; Dowicide 1. Disinfectant, preservative for detergents, cooling lubricants, textile/leather finishing, adhesives, paper, citrus fruit, polishes, wax emulsions, ceramic glazes, soap solutions mp = 57-59°; bp = 282°; d = 1.2130; insoluble in H_2O, soluble in organic solvents; LD_{50} (rat orl) = 2.48 g/kg. *Nipa Laboratories Ltd.*

21259 Nipacide® PTAP
80-46-6 7284 201-280-9
p-t-amyl phenol
Disinfectant. *Nipa Laboratories Ltd.*

21260 Nipacide® PX
133-53-9 3126 205-109-9
2,4-di-4-chloro-3,5-xylenol.
Nipa Laboratories Ltd.

21261 Nipacombin SK
$C_7H_5NaO_3$
p-hydroxybenzoic acid sodium salt
Compounded sodium-4-hydroxybenzoic acid; preservative for liquid antacid suspensions and other alkaline solutions *Nipa Laboratories Ltd.*

21262 Nipafax
Chemicals for thermal paper. *Nipa Laboratories Ltd.*

21263 Nipagin A
120-47-8 3883 204-399-4
Ethylparaben
Preservative, bactericide, fungicide for pharmaceuticals, cosmetics, foods, medicinal preps., industrial applications. *Nipa Laboratories Ltd.*

21264 Nipagin A Potassium
36457-19-9 253-048-1
Potassium ethylparaben
A preservative. *Nipa Laboratories Ltd.*

21265 Nipagin A Sodium
35285-68-8 252-487-6
Sodium ethylparaben
A preservative, bactericide, fungicide for pharmaceuticals, cosmetics, foods, medicinal preperations and industrial applications. *Nipa Laboratories Ltd.*

21266 Nipagin M
99-76-3 6182 202-785-7
Methylparaben
Preservative, bactericide, fungicide for pharmaceuticals, cosmetics, foods, medicinal preps., industrial applications. *Nipa Laboratories Ltd.*

21267 Nipagin M Potassium
26112-07-2 247-464-2
Potassium methylparaben
A preservative *Nipa Laboratories Ltd.*

21268 Nipagin M Sodium
5026-62-0 225-714-1
Sodium methylparaben
Preservative, bactericide, fungicide for pharmaceuticals, cosmetics, foods, medicinal preparations and industrial applications. *Nipa Laboratories Ltd.*

21269 Nipaguard® DMDMH
6440-58-0 229-222-8
$C_7H_{12}N_2O_4$
1,3-bis (hydroxymethyl)-5,5-dimethyl-2,4-imidazolidinedione
DMDM hydantoin; Dantoin DMDMH; DMDMH; Glydant; Glydant Plus; Mackgard DM; Nipaguard DMDMH; Dantoin DMDMH 55; Dantoguard. Antimicrobial for cosmetics and personal care products. *Nipa Laboratories Ltd.*

21270 Nipaguard® TCC
101-20-2 9786 202-924-1
Nipaguard® DME. A blend of esters of 4-hydroxybenzoic acid and DMDM hydantoin; used for preservation of cosmetics and toiletries. *Nipa Laboratories Ltd.* Discontinued.

21271 Nipaguard® MPS
A blend of esters of 4-hydroxybenzoic acid; used for preservation of cosmetics and toiletries. *Nipa Laboratories Ltd.* Discontinued.

21272 Nipaheptyl
1085-12-7 214-115-0
$C_{14}H_{20}O_3$
n-heptyl *p*-hydroxybenzoate
Preservative, bactericide, fungicide for pharmaceuticals, cosmetics, foods, medicinal preparations and industrial applications. *Nipa Laboratories Ltd.*

21273 Nipanox® BHT
128-37-0 1583 204-881-4
butylated hydroxytoluene; BHT. Antioxidant. *Nipa Laboratories Ltd.*

21274 Nipanox® S-1
Propyl gallate (20%), citric acid (10%) in propylene glycol; antioxidant for vegetable oil industry. *Nipa Laboratories Ltd.*

21275 Nipanox® Special
BHA, propyl gallate, citric acid in propylene glycol; antioxidant for foods; fat and oil stabilizer. *Nipa Laboratories Ltd.*

21276 Nipantiox
25013-16-5 1582 246-563-8
$C_{11}H_{16}O_2$
2-*tert*-butyl-4-hydroxyanisole
butylated hydroxyanisole; BHA; Butyl Hydroxyanisole; Vertac; Antioxyne B; BOA; Antrancine 12; Embanox; Nipantiox 1-F; Sustane 1-F; Tenox. Antioxidant and preservative for foods. mp = 48-55°; bp_{733} = 264-270°; insoluble in H_2O, soluble in organic solvents; LD_{50} (rat orl) = 2200 mg/kg. *Nipa Laboratories Ltd.* Discontinued.

21277 Nipantiox 1-F
25013-16-5 1582 246-563-8
BHA; antioxidant. *Nipa Laboratories Ltd.* Discontinued.

21278 NiPar 640
Blend of nitroethane and 1-nitropropane; wetting additive for solvent-based coatings including inks, adhesives, etc.; provides improved film integrity, solvent release, and drying times. *Angus.*

21279 NiPar S-10
108-03-2 6724 203-544-9
1-Nitropropane
Intermediate, solvents for coatings. *Angus.*

21280 NiPar S-20
79-46-9 6725 201-209-1
2-nitropropane
Used in solvent blends for inks and coatings; especially for NC, chlorinated rubber, vinyl, epoxy, acrylic, PU, polyamide systems; automotive finishes. *Angus.*

21281 Nipasept
Compounded esters of 4-hydroxybenzoic acid; a preservative. *Nipa Laboratories Ltd.*

21282 Nipasol M
94-13-3 8051 202-307-7
$C_{10}H_{12}O_3$
Propylparaben
Nipasol; Chemocide PK; Propyl Chemosept; Solbrol P; Propyl Parasept; Chemoside PK; Parabens; Aseptoform P; Betacide P; Bonomold Op; Nipasol M; Paraben; Paseptol; Preserval P; Propyl Aseptoform; Protaben P; Tegosept P; Pulvis Conservans; Propagin; Propyl Butex; Nipazol; Nipagin P; Nipasol P; Chemacide Pk; Propyl Chemsept; N-propylparaben. Preservative, bactericide, fungicide for pharmaceuticals, cosmetics, foods, medicinal preps., industrial applications. mp = 95-98°; bp = 133°; soluble in H_2O (0.5 g/l), organic solvents. *Nipa Laboratories Ltd.*

21283 Nipasol M Potassium
84930-16-5 212-819-2
Potassium propylparaben
Preservative. *Nipa Laboratories Ltd.*

21284 Nipasol M Sodium
35285-69-9 252-488-1
$C_{10}H_{11}NaO_3$
Sodium propylparaben
Paradept; sodium propyl paraben. Preservative, bactericide, fungicide for pharmaceuticals, cosmetics, foods, medicinal preps., industrial applications. *Nipa Laboratories Ltd.*

21285 Nipastat
Methylparaben, butylparaben, ethylparaben, and propylparaben. preservative, bactericide, fungicide for pharmaceuticals, cosmetics, foods, medicinal preparations and industrial applications. *Nipa Laboratories Ltd.*

21286 Nipol® 1000X132
9003-18-3
Butadiene-acrylonitrile copolymer, slightly staining antioxidant; used for hose, sheet packing, molded and extruded goods, oil seals, sundries, adhesives;

excellent oil and fuel resist.; vulcanized with sulfur-accelerator, sulfur donor, or peroxide systems. *Zeon.*

21287 Nipol® 1001 CG
9003-18-3
Butadiene-acrylonitrile copolymer, slightly staining antioxidant; cement grade used for coated materials, sheet packing, molded and extruded goods, o-rings, oil seals, adhesives, plastic modification; excellent oil and fuel resist.; vulcanized with sulfur-accelerator, sulfur-donor or peroxide systems. *Zeon.*

21288 Nipol® 1022X59
9003-18-3
Pre-crosslinked butadiene-acrylonitrile copolymer; used in plastics compounding to impart rubbery properties to thermoplastics. *Zeon.*

21289 Nipol® 1072
9003-18-3
Carboxy-modified butadieneacrylonitrile copolymer, nonstaining antioxidant; used for heels and soles, rubber-covered rolls, molded and extruded goods, oil seals, adhesives; improved low temp. brittleness, hottear and abrasion resist.; vulcanized with bivalent metal compounds such as zinc oxide or with sulfur. *Zeon.*

21290 Nipol® 1203F60
NBR/PVC prefluxed blend (70:30), nonstaining antioxidant; used for coated materials, hose, flat belts, molded and extruded goods, shells and soles, rubber-covered rolls, wire and cable jacketing, matting, foam products, oil seals, adhesives; excellent resistance to ozone, weathering, fuels, oils, solvents, abrasion and heat; excellent colorability. *Zeon.*

21291 Nipol® 1204X22
NBR/PVC/DOP prefluxed blend (100/60/120); used for soft rolls, low durometer applications. *Zeon.*

21292 Nipol® 1411
9003-18-3
Butadiene-acrylonitrile copolymer; slightly staining antioxidant; used for friction products, plastic modification; vulcanized with sulfuraccelerator systems; finely divided powder; nonsoluble. *Zeon.*

21293 Nipol® 2782
Block polymer; emulsifier for agricriculttural formulations. *Stepan; Stepan Canada.*

21294 Nipol® AR-31
Acrylic rubber; offers resistance to high and low temperatures; used in automobile gaskets, etc. *Zeon.*

21295 Nipol® AR-42
Acrylic rubber; low temperature-resistant rubber with shortened post-cure time; better compression set, low corrosion and improved extrusion moldability. *Zeon.*

21296 Nipol® AR-74
Acrylic rubber; elastomer with ultra low temperature resistance; used with soap-sulfur cure system; gives excellent low temperature properties for o-rings, seals and gaskets. *Zeon.*

21297 Nipol® DP5123P
Nitrile rubber; nonmigratory, nonextractable, nonvolatile modifier for plastics for improved oil and chemical resistance, low temperature and heat aging properties, superior physical properties, increased abrasion resistance and enhanced rubber-like feel. *Zeon.*

21298 Nippon Ant and Crawling Insect Killer
Aerosol containing permethrin [52645-53-1] and tetramethrin [7696-12-0]; residual contact insecticidal surface spray. *Vitax Ltd.*

21299 Nippon Ant Killer Liquid
Liquid bait material containing borax; used for control of black ants. *Vitax Ltd.*

21300 Nippon Ant Killer Powder
52645-53-1 7321 258-067-9
Permethrin on talc dust carrier; contact residual insecticide. *Vitax Ltd.*

21301 Nippon Fly Killer Spray
Aerosol containing permethrin [52645-53-1] and tetramethrin [7696-12-0]; domestic insecticide. *Vitax Ltd.*

21302 Nippon Ready For Use Ant and Crawling Insect Killer
52645-53-1 7321 258-067-9
Trigger spray pack containing permethrin as a ready-for-use liquid. residual contact insecticidal surface spray. *Vitax Ltd.*

21303 Nipro (i)
105-60-2 1805 203-313-2
Caprolactam
Used for nylon manufacturing. *Columbia Nitrogen Corporation.* Name unverified.

21304 Nipro (ii)
7783-20-2 590 231-984-1
Ammonium sulfate
Fertilizer and chemical intermediate. *Columbia Nitrogen Corporation.* Name unverified.

21305 Ni-resist
A corrosion and heat-resisting castiron containing 12-15% nickel, 5-7% copper, and 1.5-4% chromium. No manufacturer.

21306 Nirex
A proprietary trade name for an alloy of 80% nickel, 14% chromium, and 6% iron. No manufacturer.

21307 Nirez® 7002
Terpene phenolic resin; resin for paste ink applications. *Arizona.* Discontinued.

21308 Nirez® 9007
Ester of selected tall oil fatty acids; specialty ink solvent for use in high solids, low VOC ink systems and in soybean oil systems to reduce viscosity and adjust tack. *Arizona.* Discontinued.

21309 Nirez® 9011
Ester of selected tall oil fatty acids; specialty ink solvent for use in high solids, low VOC ink systems and in soybean oil systems to reduce viscosity and adjust tack. *Arizona.* Discontinued.

21310 Nirostaguss
A non-rusting and heat-resisting 34% chromium cast iron.

21311 Nispan Alloy C-902
A trademark for an alloy of 42% nickel, 5.2% chromium, 2.4% titanium, 0.6% aluminum, and the balance iron. *Wiggin Alloys Ltd.* Unverified.

21312 Nisso PB
Hydroxyl terminated polybutadiene.
British Traders & Shippers Ltd.

21313 niter cake
A residue from the manufacture of nitric acid consisting of a mixture of normal and acid, sodium sulfate.

21314 Nitofol®
10265-92-6 6014 233-606-0
Methamidophos
Broad spectrum systemic insecticide used for treatment of sucking and biting insects and spider mites. *Bayer AG.*

21315 Nitolac
A proprietary polyurethane flooring material. *Tercol, Shifnal, Shropshire.* Name unverified.

21316 Nitoman
58-46-8 9325 200-383-6
A proprietary preparation of tetrabenazine. *Roche Products Ltd.* Name unverified.

21317 Nitracc
Fertilizers. *ICI Chem & Polymers Ltd.*

21318 Nitral
10024-97-2 6751 233-032-0
N_2O
nitrous oxide
dinitrogen monoxide; Laughing gas; hyponitrous acid anhydride; nitrogen Oxide; nitrogen oxide; factitious air;. A trade name for moist nitrous oxide used as a bactericide. mp= -91°; bp = -88°; soluble in H_2O, organic solvents. No manufacturer.

21319 nitralin
4726-14-1 6662 225-219-0
$C_{13}H_{19}N_3O_6S$
4-(methylsulfonyl)-2,6-dinitro-N,N-dipropylbenzeneamine
SD 11831; Planavin. Herbicide. mp = 150-151°; slighty soluble in H_2O (0.6 mg/l), more soluble in polar organic solvents; LD_{50} (rat orl) >2 g/kg.

21320 Nitraline
7697-37-2 6671 231-714-2
Nitric acid used as a dairy pipeline cleaner. *Ciba plc.*

21321 Nitralloy
A nitrided aluminum-chromium-molybdenum steel.

21322 Nitram
6484-52-2 567 229-347-8
Ammonium nitrate fertilizer. *ICI Chem & Polymers Ltd.*

21323 Nitrammite
6484-52-2 567 229-347-8
A native ammonium nitrate.

21324 Nitrammomkalk
A mixture of ammonium and calcium nitrates in a granular form of Norwegian manufacture; used as a fertilizer.

21325 Nitraniline N
Nitraniline mixed with sufficient sodium nitrite necessary for its diazotization. Used in manufacture of dyestuffs.

21326 Nitrapo
A product obtained from crude caliche by crystallization. It contains about 66% sodium nitrate, 29% potassium nitrate, and a little sodium chloride; used as a fertilizer.

21327 Nitraprill
34.5% Nitrogen prilled fertilizer. *Kemira Ince Ltd.* Discontinued.

21328 nitrated oils
Thick syrupy liquids obtained by treating castor oil or linseed oil with a mixture of concentrated sulfuric and nitric acids; they form homogenous mixtures with nitrocellulose; dissolved in acetone, these oils form varnishes, which are used for enameling leather or similar material, and mixing paints.

21329 nitrating acid
Mixed acid. A mixture of sulfuric acid and nitric acid; consists of 36% nitric acid and 61% sulfuric acid; used for nitrating in the manufacture of explosives and plastics.

21330 Nitrene 11230
Modified cocamide DEA (2:1); detergent, wetting agent, emulsifier, thickener, foam stabilizer for industrial and specialty cleaners, floor strippers and degreasers, household hard surface cleaners. *Henkel/Cospha; Henkel Canada.*

21331 Nitrene C Extra
Cocamide DEA (1:1); emulsifier, dispersant, wetting agent, foam booster/stabilizer, detergent, and visc. builder for industrial/household detergents, metalworking; intermediate for liquid detergents; clarifier for liquid soaps; solv. cleaners, drycleanin *Henkel/Cospha; Henkel Canada.*

21332 nitric acid
7697-37-2 6671 231-714-2
HNO_3
Engravers acid; aqua fortis; azotic acid; salpetersäure; Hydrogen nitrate; Azotic acid; Rfna; aqua fortis; fuming nitric acid; red fuming nitric acid; Nital; Nitryl Hydroxide; nitric acid red fuming. Manufacture of ammonium nitrate for fertilizer and explosives, organic synthesis (dyes, drugs, explosives, cellulose nitrate, nitrate salts), metallurgy, photo-engraving, etching steel, ore flotation, urethanes, rubber chemicals, nuclear fuel. mp = -42°; bp = 83°; d_4^{25}= 1.50269. Aceto; Air Prods & Chem; Am. Cyanamid; Angus; Asahi Chem industry Co Ltd; Du Pont; Miles; Monsanto; Nissan Chem. Ind.; Norsk Hydro A/S.

21333 nitric ether
625-58-1 210-903-3
$C_2H_5NO_3$
ethyl nitrate
bp = 17°; reacts with H_2O.

21334 nitrided steel
Steel which has been treated with ammonia gas at a temperature of 950 C, whereby nitrogen is absorbed on the surface giving a hard, nonbrittle surface; steels containing form 0.5-2.0% aluminum and 0.5-4.0% of other elements are used.

21335 Nitrile 10 D
1975-78-6 217-830-6
$C_{10}H_{19}N$
n-decanenitrile
nonyl cyanide; decanonitrile. Chemical intermediate. *Berol Nobel AB.*

21336 Nitrile 12
2437-25-4 219-440-1
$C_{12}H_{23}N$
dodecanenitrile
n-undecyl cyanide; Undecyl cyanide; Dodecanonitrile. Chemical intermediate. *Berol Nobel AB.*

21337 Nitrile BG
Tallow nitrile
Chemical intermediate. *Berol Nobel AB.*

21338 nitrile C₄
109-74-0 1633 203-700-6
C_4H_7N
n-butanenitrile
1-cyanopropane, butyronitrile. mp = -112°; bp = 115-117°; d = 0.7940; n_D^{20}= 1.3840fp = 16°.

21339 nitrilotriacetic acid
139-13-9 6675 205-355-7
$N(CH_2COOH)_3$
N,N-bis(carboxymethyl)glycine
triglycollamic acid; triglycine. NTA; chelating and sequestering agent; builder in synthetic detergents. *W.R. Grace/Hampshire; R.W. Greeff.*

21340 nitrite rubber
A product obtained by treating latex with a nitrite and coagulating with an acid.

21341 nitro base
138-89-6 6736 205-343-1
C8H10N2O
p-nitrosodimethylaniline
p-Nitrosodimethylaniline; *p*-nitroso-N,N-dimethylaniline; accelerene; paranitrosodimethylanilide. An intermediate for dyes. Also used as an accelerator in vulcanizing. mp = 85-87°; sg = 1.145; insoluble in H_2O, soluble in organic solvents;.

21342 Nitro BT
298-83-9 206-067-4
$C_{40}H_{30}Cl_2N_{10}O_6$
2H-tetrazolium, 3,3'-(3,3'-dimethoxy[1,1'-biphenyl]-4,4'-diyl)bis[2-(4-nitrophenyl)-5-phenyl-, dichloride
Nitro BT; Nitro BT monohydrate; NBT. Nitro blue tetrazolium. Activity stain for electrophoresis. Activity stain for histochemistry. Substrate for dehydrogenase. mp = 200°. *Monomer-Polymer & Dajac Laboratories, Inc.*

21343 Nitro Fast
Specialty dye for automotive, shoe, and furniture polishes, oils, waxes, and solvs. *Sandoz.*

21344 *Nitro-26*
A nitrogeneous fertilizer. *Fisons plc.* Name unverified.

21345 *p*-nitroaniline
100-01-6 6681 202-810-1
$C_6H_6N_2O_2$
4-nitroaniline
p-nitraniline; PNA; C.I. 37035; Azofix Red GG Salt; Azoic Diazo Component 37; C.I. Developer 17; Developer P; Devol Red GG; Diazo Fast Red GG; Fast Red Base 2J; Fast Red Base GG; Fast Red 2G Base; Fast Red 2G Salt; Shinnippon Fast Red GG Base; Fast Red Salt 2J; Fast Red Salt GG; Nitrazol CF Extra; Red 2g Base; Fast Red GG Base; Fast Red MP Base; Fast Red P Base; Fast Red P Salt; Naphtoelan Red GG Base; Azoamine Red 2H; C.I. Azoic Diazo Component 37. Intermediate for dyes, antioxidants; gasoline gum inhibitors; corrosion inhibitor. mp = 46°; bp = 332° (calc); soluble in H_2O (0.8-20 mg/ml), more soluble in organic solvents. *Enichem Am.; Hoechst Celanese; Monsanto.*

21346 *o*-nitrobenzaldehyde
552-89-6 6684 209-025-3
$C_7H_5NO_3$
2-nitrobenzaldehyde
Used in the synthesis of dyes, pharmaceuticals, surface active agents. mp = 46°; bp_{23} = 153°; slightly soluble in H_2O, soluble in organic solvents. *Penta Mfg.; Schweizerhall.*

21347 *p*-nitrobenzaldehyde
555-16-8 6684 209-084-5
$C_7H_5NO_3$
4-nitrobenzaldehyde
Used in the synthesis of dyes, pharmaceuticals, surface active agents. mp = 108°. *Penta Mfg.; Schweizerhall.*

21348 nitrobenzene
98-95-3 6685 202-716-0
$C_6H_5NO_2$
Nitrobenzol; Mirbane oil; Essence of mirbane; oil of mirbane; Essence of Myrbane; oil of myrbane. Used as a chemical intermediate and as a solvent. mp = 5-6°; bp = 210-211°; d= 1.1960; LD_{50} (rat orl) = 780 mg/kg.

21349 3,5-dinitrobenzoic acid
99-34-3 3328 202-751-1
$C_7H_4N_2O_6$
Used in chemical analysis. *Exchem Organics; Nobel; Schweizerhall.*

21350 *p*-nitrobenzoic acid
62-23-7 6686 200-526-2
$C_7H_5NO_4$
4-nitrobenzoic acid
p-nitrodracylic acid. Organic synthesis. mp = 239-241°; d = 1.58; soluble in H_2O (42 mg/ml), more soluble in organic solvents; *Du Pont; Hüls Am.; Nobel Chem. Ltd; Schweizerhall.*

21351 Nitrobutylmorpholine
2224-44-4 218-748-3
$C_8H_{16}N_2O_3$
4-(2-Nitrobutyl)morpholine
Bioban P 1487. Bioban P 1487 is a preservative composed of 4,4-(2-Ethyl-2-nitro-trimethylene) dimorpholine (30%) and 4-(2-Nitrobutyl)morpholine (70%).

21352 nitrocalcite
10124-37-5 1729 233-332-1
CaN_2O_6
Calcium nitrate
Wall saltpeter.

21353 nitrocellulose
9004-70-0 8195
$C_{12}H_{16}(ONO_2)_4O_6$
cellulose, nitrate
nitrocotton, Pyroxylin; guncotton. Cellulose derivative; in lacquers, high explosives, rocket propellant; printing ink base; leather finishing, etc. *Allchem Industries; Aqualon; Asahi Chem Industry Co Ltd; Hercules; SNPE Chimie.*

21354 nitrocellulose varnishes
Varnishes used in the manufacture of artificial leather; they consist of nitrocelluloses dissolved in amyl acetate, and colored; used also for painting iron work.

21355 Nitro-chalk
A proprietary fertilizer consisting of an intimate mixture of chalk and ammonium nitrate in the form of a fine powder. *ICI Chem & Polymers Ltd.*

21356 nitro-dextrin
A similar product to nitro-starch; used in explosives.

21357 nitroethane
79-24-3 6694 201-188-9
$C_2H_5NO_2$
Nitroparaffin
Nitroetan. Intermediate for synthesis; stabilizer for chlorinated solvs.; fuel additive; solvent. mp = -90°; bp = 114°; soluble in H_2O (4.6 ml/100 ml), more soluble in organic solvents; LD_{50} (rat orl)= 1100 mg/kg. *Angus; W.R. Grace/Nitroparaffins; Spectrum Chem. Mfg.*

21358 Nitroferrite
An explosive containing 93% ammonium nitrate, 2% trinitronaphthalene, 2% potassium ferricyanide, and 3% sugar.

21359 Nitroform
517-25-9 9859 208-236-8
Tetranitromethane
Used in the manufacture of explosives.

21360 Nitrofuel®
75-52-5 6708 200-876-6
Brand of racing nitromethane containing Nitroguard®; fuel for automotive racing and model engines. *Angus.*

21361 5-nitroisophthalic acid
618-88-2 210-568-3
$C_8H_5NO_6$
5-nitrobenzene-1,3-dicarboxylic acid
5-nitroisophthalic acid; 4-nitroisophthalic acid. mp = 260-261°; *Pfister; Schweizerhall.*

21362 Nitrolac
A German nitrocellulose lacquer.

21363 nitrolignin
Wood which has been nitrated.

21364 nitrolim
Commercial nitrolim contains 57-63% calcium cyanamide, 20% lime, 14% graphite, and 7-8% silica, iron oxide, and alumina; a fertilizer. No manufacturer.

21365 Nitrolite
An explosive containing nitroglycerin.

21366 nitromethane
75-52-5 6708 200-876-6
CH_3NO_2
Nitroparaffin
nitrocarbol; NM. Stabilizer for chlorinated solvents; chemical intermediate; solvent for cellulosic compounds, polymers, waxes; rocket fuel, gasoline additive; in coatings industry. mp = -29°; bp = 101°; SG = 1.1371; soluble in H_2O (100 mg/ml), soluble in organic solvents; LD_{50} (rat orl) = 1.44 g/kg. *Angus; W.R. Grace/Nitroparaffins.*

21367 Nitron
2218-94-2 6710 218-724-2
$C_{20}H_{16}N_4$
1,4-diphenyl-3-(phenylamino)-1H-1,2,4-triazolium inner salt
A base which forms a nitrate almost insoluble in water; used for the determination of nitric acid; also used as a rubber vulcanization accelerator; also a proprietary trade name for a cellulose nitrate plastic. mp = 189°, insoluble in H_2O, soluble in organic solvents;. No manufacturer.

21368 *p*-nitrophenol
100-02-7 6718 202-811-7
$C_6H_5NO_3$
4-Nitrophenol
4-Hydroxynitrobenzene; Niphen; PNP; mononitrophenol. Important chemical intermediate; also used as an indicator, colorless below pH 5.6, yellow above pH 7.6. mp = 113-115°; bp = 279°; d_4^{120}= 1.270; moderately soluble in H_2O, freely soluble in organic solvents; LD_{50} (rat orl) = 616 mg/kg.

21369 Nitrophos® 20-20-0
Complex fertilizer with 20% nitrogen, 20% phosphate; for all agricultural and horticultural crops. *BASF AG.*

21370 Nitrophoska® 10-15-20
P_2O_4, 20% K_2O
Complex fertilizer with 10% N, 15% for agricultural and horticultural crops with low basal demand for nitrogen. *BASF AG.*

21371 nitrophosphate
A fertilizer sometimes wrongly called ammonium superphosphate. It is prepared by mixing calcium superphosphate with ammonium sulfate giving mixtures containing ammonium phosphate and calcium sulfate.

21372 Nitropore® ATA
Blend of 1,1'-Azobisformamide [123-77-3] and dinitrosopentamethylenetetramine [101-25-7]; a chemical blowing agent. *Uniroyal.*

21373 1-nitropropane
108-03-2 6724 203-544-9
$C_3H_7NO_2$
1-nitropan; n-nitropropane. Used as a solvent for cellulose acetate and vinyl resins; also as an intermediate and propellant. A solvent for coatings and inks, cellulose acetate, vinyl resins, lacquers, synthetic rubbers, fats, oils, dyes; chemical intermediate; a rocket propellant and gasoline additive. mp = -108°; bp = 129-131°; d = 1.002; n_D^{20} = 1.4020; slightly soluble in H_2O (1.3 g/100 ml), soluble in organic solvents. *Angus; Ashland; W.R. Grace/Nitroparaffins.*

21374 2-nitropropane
79-46-9 6725 201-209-1
$CH_3CH(NO_2)CH_3$
sec-nitropropane. Nitroparaffin; solvent for coatings and inks, cellulose acetate, vinyl resins, lacquers, synthetic rubbers, fats, oils, dyes; chemical intermediate; rocket propellant; gasoline additive. bp = 118-120°; d = 0.990; n_D^{20} = 1.3940; soluble in H_2O (1.7 g/100 ml), soluble in organic solvents. *Angus; Ashland; W.R. Grace/Nitroparaffins.*

21375 nitropropiol
6722
$C_9H_4NNaO_4$
3-(2-nitrophenyl)-2-propynoic acid sodium salt
Sodium-o-nitro-phenyl-propiolate. An analytical reagent for alkaloids. mp = 157°; moderately soluble in H_2O, less soluble in organic solvents.

21376 Nitrosin Saltpetre
A proprietary preparation containing nitrite; used for curing meat. No manufacturer.

21377 nitro-starch
9056-38-6 8195
$C_{12}H_{12}N_8O_{26}$
xyloidin
pyroxylin. A nitric ester of starch, probably the octonitrate used for blasting explosives, either alone, or by mixing 10% of it with a mixture of sodium nitrate and carbonaceous material.

21378 nitrosyl silver
$Ag_2N_2O_2$
Silver hyponitrite;.

21379 nitrosylsulfuric acid
7782-78-7 6746 231-964-2
HNO_5S
nitrosyl sulfate
Lead chamber crystals; Nitro-sulfonic acid; nitrososulfuric acid; nitroxylsulfuric acid; nitrosulfonic acid; nitro acid sulfite; Nitrose; Weber's acid. Used to bleach flour. dec 73°.

21380 nitrothal-isopropyl
10552-74-6 234-139-5
$C_{15}H_{19}NO_6$
bis(1-methylethyl) 5-nitro-1,3-benzenedicarboxylate
nitrothale-isopropyl; BAS 30000F. Non-systemic contact fungicide with curative action. Used in combination with other fungicides to control powdery mildews on apples, vines, hops, vegetables and ornamentals. mp = 65°; poorly soluble in H_2O (0.39 mg/l), more soluble in organic solvents; LD_{50} (rat orl) >6400 mg/kg.

21381 nitrous gas or air
10102-44-0 6700 233-272-6
N_2O_4
Nitrogen dioxide
dinitrogen tetroxide. Used as a chemical intermediate, oxidizing agent and propellant. mp = -9°; bp = 21°; d_4^{20}= 1.448.

21382 nitrovin
804-36-4 6752 212-358-7
$C_{14}H_{12}N_6O_6$
2-[3-(5-nitro-2-furanyl)-1-[2-(5-nitro-2-furanyl)ethenyl]-2-propenylidene]hydrazinecarboximidamide
Panazon; Payzone. Growth promoter and antibacterial. mp = 217° (dec).

21383 Nitroxan
A compound of barium metaplumbate and barium manganate; a catalyst used for the conversion of ammonia into nitric acid. The ammonia is oxidized to nitric acid, which is retained as barium nitrate.

21384 nitroxynil
1689-89-0 6754 216-884-8
$C_7H_3IN_2O_3$

4-hydroxy-3-iodo-5-nitrobenzonitrile
Dovenix. Anthelmintic. mp = 137-138°; sparingly soluble in H_2O, soluble in organic solvents.

21385 Nitto Nitoflon
Fluorocarbon plastic products, tapes, films, tubes and moldings; used to modify nonadhesive fluoroplastic surfaces into bondable surfaces; used for electric applications and electronics, food production, paper and pulp, etc. *Nitto Electric Industrial Co Ltd.* Name unverified.

21386 Nitto SPV
Surface protection adhesive films and sheets; protects the surface of stainless steel, aluminum, decorated laminates, etc., from damage in transportation, storage and fabrication; used for architectural structures, rolling stock, household articles. *Nitto Electric Industrial Co Ltd.* Name unverified.

21387 Nix Creme Rinse
52645-53-1 7321 258-067-9
A proprietary formulation of permethrin; for treatment of infestation with *pediculus humanus* variety capitis (the head louse) and its nits (eggs). *The Wellcome Foundation Ltd.*

21388 Nix Dermal Cream
52645-53-1 7321 258-067-9
A proprietary formulation of permethrin; for treatment of infestation with *Sarcoptes scabiei. The Wellcome Foundation Ltd.*

21389 Nix Stix L-515
Silicone dispersion; mold release for rubber and urethane products. *Dwight Prods.*

21390 Nix Stix X-9021
Mold release spray for urethane and rubber products. *Dwight Prods.*

21391 Njatuo tallow
A fat obtained from *Palaquium oblongifolium* .

21392 Njave butter
Nari oil; Noumgou oil; Adjab fat. A fat obtained from the seeds of *Mimusops njave* or *djave*, also from *Bassia toxisperma* and *Bassia djave*. The nuts are known as Abeku, Bako, or Mahogany nuts.

21393 NL Chem
Industrial chemicals, specialty chemicals, namely organic and inorganic chemicals generally including polymers and argochemicals. *Rheox Inc.*

21394 NL-10GF
Nylon 6/12 resin, 10% glass fiber-reinforced; outstanding mechanical properties, chemical resistance, dimensional stability, low moisture absorption. *Compounding Tech.*

21395 NLA-10
84-74-2 1622 201-557-4
Dibutylphthalate
A vinyl plasticizer. *National Lead Co.* Name unverified.

21396 NLA-20
117-81-7 1291 204-211-0
di-2-ethylhexylphthalate
A vinyl plasticizer. *National Lead Co.* Name unverified.

21397 NLA-30
68515-49-1 271-091-4
Diisodecylphthalate
A vinyl plasticizer. *National Lead Co.* Name unverified.

21398 NLA-40
84-77-5 201-561-6
$C_{28}H_{46}O_4$
Didecylphthalate.
Plasticizer. mp = 4°; d= 0.9700. *National Lead Co.* Name unverified.

21399 N-Labstix
A proprietary test-strip used to detect pH, protein, glucose, ketones, blood and nitrite in urine. *B. C. Ames.* Name unverified.

21400 NLF-32
A mixed adipate; a vinyl plasticizer. *National Lead Co.* Name unverified.

21401 NLF-33
A modified adipate; a vinyl plasticizer. *National Lead Co.* Name unverified.

21402 NM
75-52-5 6708 200-876-6
Nitromethane
Intermediate, stabilizer for halogenated solvents, as fuels, explosives, and solvents for coatings or industrial processes. *Angus.*

21403 NM-55®
75-52-5 6708 200-876-6
Blend of nitromethane; fuel for automotive racing and model engines; solvent. *Angus.*

21404 NM-AMD
924-42-5 213-103-2
$C_4H_7NO_2$
N-Methylolacrylamide
N-(Hydroxymethyl)acrylamide; N-(Hydroxymethyl)-2-propenamide; N-methanolacrylamide; monomethylolacrylamide; uramine t 80; N-methyloacrylamide. Across-linking agent used in adhesives, binders for paper, crease-resistant textiles, resins, latex film, and sizing agents. mp= 74-75°; insoluble in H_2O, soluble in organic solvents; *Cyanamid BV.* Name unverified.

21405 NMP, NMP Conc
76-39-1 232-306-7
$C_4H_9NO_3$
2-Nitro-2-methyl-1-propanol
Chemical and pharmaceutical intermediate, in tire cord adhesives, as formaldehyde release agents, deodorants, antimicrobials. Chemical intermediate, formaldehyde donor, textile reactant; reduces formaldehyde on finished cloth. *Angus.*

21406 NMP
872-50-4 6197 212-828-1
C_5H_9NO
2-Methyl-2-pyrrolidone
Methylpyrrolidone; 1-Methyl-5-Pyrrolidinone; N-Methylpyrrolidone; N-Methyl-2-Pyrrolidone; 1-Methyl-2-pyrrolidone; Deuterated N-Methyl-2-Pyrrolidone; m-pyrrole; 1-Methylpyrrolidinone; N-methylpyrrolidinone; NMP; N-Methyl-2-pyrrolidinone. Used as coatings solvent, in stripping and cleaning of paints and varnishes, industrial cleaning, mold cleaning, petrochemical processing, agricultural solvent, polymer solvent. mp = -24°; bp = 202°; d = 1.033; n_D^{20} = 1.4700; soluble in H_2O, organic solvents; LD_{50} (rat orl) = 3.9 g/kg. *Arco Chemical Co.*

21407 NN-10GF
32131-17-2
Nylon 6/6, 10% glass fiber-reinforced; tough, strong, abrasion-resistant resins resistant to most industrial chemicals and solvents. *Compounding Tech.*

21408 NN-20CF
32131-17-2
Nylon 6/6 resin, 20% carbon fiber-reinforced; used for applications requiring ultra high strength, fatigue endurance, thermal conductivity, wear resistance and/or electrical conductivity; suitable for aerospace industry applications. *Compounding Tech.*

21409 No Vein Compound
1309-37-1 4072 215-168-2
A proprietary blend of red iron oxides; for use in the foundry industry to eliminate veining and other casting expansion defects. *DCS Color & Supply Co Inc.* Name unverified.

21410 No. 1 White
471-34-1 1697 207-439-9
Calcium carbonate; general-purpose pigment where fineness is of secondary importance. *ECC International Ltd.* Discontinued.

21411 No. 3 White
471-34-1 1697 207-439-9
Calcium carbonate
White general-purpose filler for low-cost paints, rubber goods, mild abrasive compounds, and putties. *ECC International Ltd.* Discontinued.

21412 Noatac
Tackifier resins. *Georgia-Pacific.* Name unverified.

21413 Nobble
Aluminum sulfate + copper sulfate + sodium tetraborate; used for slug and snail control in field and glasshouse crops. *Fieldspray.*

21414 Nobecutane
A proprietary preparation of acrylic resin dissolved in acetic esters used in aerosol form as a plastic wound dressing. *Astra Chemicals Ltd.* Unverified.

21415 Nobel Ardeer powder
A dynamite containing 33% nitroglycerin, 49% magnesium sulfate, 13% kieselguhr, and 5% potassium nitrate.

21416 Nobel Polarite
A mixture of potassium perchlorate and ammonium nitrate, with trinitrotoluene, a little starch, and wood meal; an explosive.

21417 Noblen
Polypropylene. *Montedison UK Ltd.* Name unverified.

21418 Nodulant
Magnesium/iron tablets for production of S.G. cast irons. *Foseco (F.S.) Ltd.*

21419 Nofome 2510
Nonsilicone water-based; textile defoamer for atmospheric and pressure applications. *Sybron.*

21420 Nogos
62-73-7 3129 200-547-7
Dichlorvos. Organophosphorus insecticide. *Ciba plc.* Name unverified.

21421 Noheet metal
Tempered lead. An alloy of 98.4% lead, 1.4% sodium, 0.11% antimony, and 0.08% tin.

21422 Noil
An alloy of 80% copper and 20% tin.
21423 Noiret's aluminum solder
An alloy of 80% zinc and 20% tin.
21424 Nokol
Coal dust replacement additives for iron foundry green sands. *Foseco (F.S.) Ltd.* Discontinued.
21425 Nolibond® 1
Polyamide-imide resin; high-temperature adhesive for aeronautic, aerospace, electrotech. and mechanical industries to produce metal or composite assemblies that require high tensile and shear strength. *Rhône-Poulenc.* Name unverified.
21426 Nolicoat FE 71008
Polyamide-imide resin; used as protective varnish or formulated to give pigmented coatings; provides self-lubricating, high heat resistance, and decorative properties. *Shell.* Name unverified.
21427 no-mag
A nonmagnetic alloy of 77% iron, 12% nickel, 6% manganese, 3% carbon, and 2% silicon. It has a high specific resistance, is close-grained, and has good mechanical properties.
21428 No-max
A proprietary trade name for a high speed molybdenum-tungsten steel for cutting tools. No manufacturer.
21429 Nomel®
Zirconium, magnesium, rare-earth and other chemicals, powders and alloys; for use in industry, including metalworking. *Magnesium Elektron Ltd.*
21430 Nomelle®
Fibers. *DuPont UK.*
21431 Nomelt®
High-temperature greases with excellent water resistance, shear stability, minimal oil separation, outstanding resistance to shrinkage; suitable for high-temperature applications such as coke oven door hinges and roll-out table bearings in steel mills. *Witco/Allied-Kelite.* Discontinued.
21432 Nomex®
Nylon fiber specially fabricated to withstand temperatures up to 500°F; does not melt or drip. *DuPont UK.*
21433 Nonad Tulle
Paraffin gauze dressing. *Allen & Handurys Ltd.* Name unverified.
21434 Nonaid
Nonionic surfactant. *Rhône-Poulenc UK.*
21435 Nonal 206
9016-45-9 6772
Nonylphenol ethoxylate
Detergent, penetrant, emulsifier, scouring agent, wetting agent, dispersant; for textiles. *Toho Chem. Industry.*
21436 Nonanol
3452-97-9 222-376-7
$C_9H_{20}O$
3,5,5-trimethylhexan-1-ol
Nonylol; trimethyl-1-hexanol; trimethylhexyl alcohol; 3,5,5-trimethyl-1-hexanol. Plasticizer, primarily 3,5,5-trimethylhexanol. *ICI Chem & Polymers Ltd.* Discontinued.
21437 Nonanol N
Mixture of isomeric primary nonyl alcohols; starting material for production of plasticizers for PVC and vinyl chloride copolymers; low volatile leveling agent for baking finishes; antifoam. *BASF AG.*
21438 Nonasol 3922
Blend of sulfates, sulfonates, and amide; detergent base for car wash products, etc. *Hart Chem. Ltd.*
21439 Nonasol N4AS
67762-19-0
Ammonium laureth sulfate
Detergent base for dishwash, hard surface cleaners, shampoos; wetting agent; biodegradable. *Hart Chem. Ltd.*
21440 Nonasol N4SS
9004-82-4
Sodium laureth sulfate
Detergent base for high foaming liquid detergents, dishwash, hard surface cleaners, shampoos; wetting agent; biodegradable. *Hart Chem. Ltd.*
21441 Noncorrodite
A proprietary trade name for a chromium steel. No manufacturer.
21442 Nonex
Surfactants. *BP Chemicals Ltd.* Discontinued.
21443 Nonex 04E
9004-96-0
PEG 400 oleate
Emulsifier for mineral and animal oils. *Hart Chem. Ltd.*
21444 Nonex C5E
61791-29-5
PEG 500 cocoate
Lubricant for textile applications. *Hart Chem. Ltd.*
21445 Nonex DL-2
9005-02-1
PEG 200 dilaurate
Coemulsifier and lubricant in industrial and textile oils. *Hart Chem. Ltd.*
21446 Nonex DO-4
9005-07-6
PEG 400 dioleate
Emulsifier, solubilizer for oils, fats, solvs. *Hart Chem. Ltd.*
21447 Nonex S3E
9004-99-3
PEG 300 stearate
Lubricant and softener for textiles. *Hart Chem. Ltd.*
21448 Nonflammable Decobest DA
131-17-9 205-016-3
$C_{14}H_{14}O_4$
Diallyl phthalate
Allyl Phthalate; 1,2-Benzenedicarboxylic acid, di-2-propenyl ester; diallylester phthalic acid; diallyl ester o-phthalic acid; dapon r; dapon 35; DAP. For decorative laminates. mp = -70°; bp_5 = 165-167°; d = 1.1200; n_D^{20} = 1.5194 *Sumitomo Bakelite Co Ltd.* Name unverified.
21449 nongo
A gum resembling tragacanth, obtained from *Albizzia brownei* of Uganda.
21450 non-gran metal
A bronze containing 87% copper, 11% tin, and 2% zinc.
21451 Nonidet
Ethylene oxide condensate. *Shell UK.* Name unverified.
21452 Nonipol 20
9016-45-9 6772
POE nonyl phenyl ether
Penetrant, wetting agent, spreader-sticker; base material for detergents; emulsifier for agricultural pesticides and emulsion polymerization, organic solvents, machine oils and liquid paraffins. *Sanyo Chem. Industries.*
21453 Nonlonic Emulsifier T-9
Proprietary; low foam emulsifier for oils and emulsion systems; 100% biodegradable. *Werner G. Smith.* Discontinued.
21454 Nonox
Group of antioxidants. *ICI Chem & Polymers Ltd.*
21455 No-Nox
Motor fuel. *Monsanto (Solaris).* Name unverified.
21456 Nonox ZA
4-Isopropylamine diphenylamine
A proprietary antioxidant. *ICI Chem & Polymers Ltd.* Discontinued.
21457 non-pareil metal
An alloy of 78% lead, 17% antimony, and 5% tin.
21458 Nonsepara®
Waterproof greases for high-moisture applications where temperatures do not exceed 77°C; used in spring loaded reservoirs, grease cups, guns, pressurized automatic lubrication systems. *Witco/Allied-Kelite.* Discontinued.
21459 Nontoxol
Froth flotation reagents containing hydrocarbons. *Coal Products Ltd.*
21460 nonyl phenol
25154-52-3 246-672-0
$C_{15}H_{24}O$
A mixture of isomeric monoalkyl phenols; nonionic surfactant (nonbiodegradable), lube oil additives, stabilizers, petroleum demulsifiers, fungicides, antioxidants for plastics and rubber. bp = 293-297° *Allchem Industries; Ashland; Berol Nobel AB; GE Specialty; Hüls AG; Mitsui Toatsu Chem.; Texaco.*
21461 Nopalcol
Nonionic surfactants. *Henkel Chemicals Ltd.* Name unverified.
21462 Nopalcol 10-COH
61788-85-0
PEG-20 hydrogenated castor oil; general purpose emulsifier, plasticizer, lubricant, wetting agent, dispersant, binding and thickening agent for emulsion polymerization, dry cleaning, leather, mineral oil emulsions, paper industry, wall-tile mastics, *Henkel/Functional Prods.*
21463 Nopalcol 1-L
9004-81-3
PEG-2 laurate
General purpose emulsifier, plasticizer, lubricant, wetting agent, dispersant, binding and thickening agent for emulsion polymerization, dry cleaning, leather, mineral oil emulsions, paper industry, wall-tile mastics, solv. emulsions. *Henkel/Functional Prods.*

21464 Nopalcol 1-TW
68153-64-0
PEG-2 tallowate
Emulsifier, plasticizer, lubricant, wetting agent, defoamer, binding and thickening agent, used in cosmetics, dry cleaning, leather, textile industries. *Henkel/Functional Prods.*

21465 Nopalcol 2-DL
9005-02-1
PEG-4 dilaurate
General purpose emulsifier, plasticizer, lubricant, wetting agent, dispersant, binding and thickening agent for emulsion polymerization, dry cleaning, leather, mineral oil emulsions, paper industry, wall-tile mastics, solv. emulsions. *Henkel/Functional Prods.*

21466 Nopalcol 4-0
9004-96-0
PEG-8 oleate
Emulsifier; dispersant for leather pigments; paper coating defoamer, plasticizer and leveling agent. *Henkel/Functional Prods.*

21467 Nopalcol 4-C
61791-29-5
PEG-8 cocoate
General purpose emulsifier, plasticizer, lubricant, wetting agent, dispersant, binding and thickening agent for emulsion polymerization, dry cleaning, leather, mineral oil emulsions, paper industry, wall-tile mastics, solv. emulsions. *Henkel/Functional Prods.*

21468 Nopalcol 4-DTW
PEG-8 ditallowate
Emulsifier. *Henkel/Functional Prods.*

21469 Nopalcol 4-L
9004-81-3
PEG-8 laurate
General purpose emulsifier, plasticizer, lubricant, wetting agent, dispersant, binding and thickening agent for emulsion polymerization, dry cleaning, leather, mineral oil emulsions, paper industry, wall-tile mastics, solv. emulsions. *Henkel/Functional Prods.*

21470 Nopalcol 4-S
9004-99-3
PEG-8 stearate
General purpose emulsifier, plasticizer, lubricant, wetting agent, dispersant, binding and thickening agent for emulsion polymerization, dry cleaning, leather, mineral oil emulsions, paper industry, wall-tile mastics, solv. emulsions. *Henkel/Functional Prods.*

21471 Nopalcol 6-DO
9005-07-6
PEG-12 dioleate
General purpose emulsifier, plasticizer, lubricant, wetting agent, dispersant, binding and thickening agent for emulsion polymerization, dry cleaning, leather, mineral oil emulsions, paper industry, wall-tile mastics, solv. emulsions. *Henkel/Functional Prods.*

21472 Nopalcol 6-R
9004-97-1
PEG-12 ricinoleate
General purpose emulsifier, plasticizer, lubricant, wetting agent, dispersant, binding and thickening agent for emulsion polymerization, dry cleaning, leather, mineral oil emulsions, paper industry, wall-tile mastics, solv. emulsions. *Henkel/Functional Prods.*

21473 Nopalcol 30-TWH
PEG-60 hydrogenated tallowate; general purpose emulsifier plasticizer, lubricant, wetting agent, dispersant, binding and thickening agent for emulsion polymerization, dry cleaning, leather, mineral oil emulsions, paper industry, wall-tile mastics, so *Henkel/Functional Prods.*

21474 Nopalcol 200
25322-68-3 7729
PEG 200
General purpose emulsifier, plasticizer, lubricant, wetting agent, dispersant, binding and thickening agent for emulsion polymerization, dry cleaning, leather, mineral oil emulsions, paper industry, wall-tile mastics, solv. emulsions. *Henkel/Functional Prods.*

21475 Nopalcol 400
25322-68-3 7729
PEG 400
General purpose emulsifier, plasticizer, lubricant, wetting agent, dispersant, binding and thickening agent for emulsion polymerization, dry cleaning, leather, mineral oil emulsions, paper industry, wall-tile mastics, solv. emulsions. *Henkel/Functional Prods.*

21476 Nopalcol 600
25322-68-3 7729
PEG 600
General purpose emulsifier, plasticizer, lubricant, wetting agent, dispersant, binding and thickening agent for emulsion polymerization, dry cleaning, leather, mineral oil emulsions, paper industry, wall-tile mastics, solv. emulsions. *Henkel/Functional Prods.*

21477 Nopco Worsted Oil 12
Textile processing aid. *Henkel Chemicals Ltd.* Name unverified.

21478 Nopco® 1179
Fatty amido condensate; wool and worsted fulling and scouring; general purpose detergent. *Henkel/Functional Prods.*

21479 Nopco® 2031
Sulfonated fatty product, sodium neutralized; emulsifier for emulsion polymerization. *Henkel/Functional Prods.: Henkel-Nopco.*

21480 Nopco® 2272-R
Highly sulfated fatty ester; wetting agent in paper towels, latex impregnation and dipping, in metal treating and textile dyeing. *Henkel/Functional Prods.; Henkel-Nopco.*

21481 Nopco® Colorsperse 188-A
Ethoxylated fatty acid; pigment wetting and dispersing. *Henkel-Nopco.*

21482 Nopco® Foamaster
Antifoams/defoamers. *Henkel Chemicals Ltd.* Name unverified.

21483 Nopco® NXZ
Defoamer for synthetic latex emulsions, paint and adhesives from SBR, PVAc, acrylic, water-soluble resins. *Henkel/Cospha.*

21484 Nopco® PD1-D
Defoamer for adhesives, paints, joint compounds, plaster; off-white powder; water-wettable. *Henkel.* Name unverified.

21485 Nopcocastor
8002-33-3 232-306-7
Sulfated castor oil; emulsifier; superfatting agent for cosmetics. *Henkel/Functional Prods.*

21486 Nopcochex RA
Fatty amide condensate; oil-soluble corrosion preventer. *Henkel/Cospha.*

21487 Nopcocide® N-40-D
1897-45-6 2219 217-588-1
Chlorothalonil
Mildewcide, fungicide for trade sales paints. *Henkel.* Name unverified.

21488 Nopcocide® N-96
1897-45-6 2219 217-588-1
Tetrachloroisophthalonitrile
Broad spectrum microbicide for control of fungi in latex exterior and interior emulsion paints, solv.-based paints. *Henkel.* Name unverified.

21489 Nopcocide® N-96-S
1897-45-6 2219 217-588-1
2,4,5,6-Tetrachloroisophthalonitrile
Antimicrobial for marine antifouling coatings. *Henkel.* Name unverified.

21490 Nopcofloc
Water treatment chemical. *Henkel Chemicals Ltd.* Name unverified.

21491 Nopcogen
Speciality surfactant. *Henkel Chemicals Ltd.* Name unverified.

21492 Nopcogen 14-S
93-82-3 202-280-1
Stearamide DEA; textile softener. *Henkel-Nopco.*

21493 Nopcogen 16-L
Lauric polyamine condensate; emulsifier and textile finishing agent. *Henkel Nopco.*

21494 Nopcogen 22-0
Oleylimidazoline
Emulsifier for mineral oil, kerosene, wetting agent, corrosion inhibitor, used in textile, asphalt, paper, agricultural industries, car wax formulations, acid detergents. *Henkel/Functional Prods.*

21495 Nopcolan SHR3
Thiosulfate, organic; shrink resister for wool. *Henkel.*

21496 Nopcolene
Leather processing aid. *Henkel Chemicals Ltd.* Name unverified.

21497 Nopcolube
Textile processing aid. *Henkel Chemicals Ltd.* Name unverified.

21498 Nopcone
Textile processing aid. *Henkel Chemicals Ltd.* Name unverified.

21499 Nopcosant
9084-06-4
Sodium salt of condensed naphthalene sulfonate; dispersant for NR, SR latexes, pigments; used for paints, cements, sealants. *Henkel.* Name unverified.

21500 Nopcosant L
Sulfated naphthalene; NR, SR latex dispersant for use in aqueous systems, paints, paper coatings, textiles, and leather. *Henkel.* Name unverified.

21501 Nopcosize
Speciality sizes. *Henkel Chemicals Ltd.* Name unverified.

21502 Nopcosperse 2B-B
Textile dispersant for pigments providing high grinding efficiency and superior color yield. *Henkel/Textiles.*

21503 Nopcostat 237
Flurochemical-compatible polypropylene finish for carpet fiber. *Henkel/Textiles.*

21504 Nopcosulf CA-60, -70
8002-33-3 232-306-7
Sulfated castor oil; softener in finishing starches, gums; plasticizer for starches; furniture base polish. *Henkel-Nopco.*

21505 Nopcosulf TA-30
Sulfated tallow; softener for cotton goods. *Henkel/Functional Prods.*

21506 Nopcosulph
Anionic surfactant. *Henkel Chemicals Ltd.* Name unverified.

21507 Nopcosurf CA
8002-33-3 232-306-7
Sulfated castor oil; softener in finishing starches, gums; plasticizer for starches; furniture polish base. *Henkel/Functioal Prods.*

21508 Nopcotan
Leather processing aid. *Henkel Chemicals Ltd.* Name unverified.

21509 Nopcote
Paper processing aid. *Henkel Chemicals Ltd.* Name unverified.

21510 Nopcote C-104
1592-23-0 1750 216-472-8
$C_{36}H_{70}CaO_4$
Octadecanoic acid, calcium salt
Calcium stearate dispersion; lubricant for paper coatings. mp= 147-149°; insoluble in H_2O, organic solvents. *Henkel.* Name unverified.

21511 Nopcotex
Textile softener lubricants. *Henkel Chemicals Ltd.* Name unverified.

21512 Nopcowax
Synthetic waxes. *Henkel Chemicals Ltd.* Name unverified.

21513 Nopcowax 22-DS
110-30-5 203-755-6
$C_{38}H_{76}N_2O_2$
Octadecanamide, N,N'-1,2-ethanediylbis-
ethylene bis(stearamide); N,N'-ethylenedi(stearamide); ethylene bisstearamide. High melting synthetic wax; binder, thickener for latex formulation, coatings, adhesives; used in powd. metallurgy as internal lubricant. *Henkel.* Name unverified.

21514 Nora
Rubber flooring and soling. *Carl Freudenberg.* Name unverified.

21515 Noraflor
Spun-bonded flooring for carpets. *Carl Freudenberg.* Name unverified.

21516 Noralastic
Acoustic flooring. *Carl Freudenberg.* Name unverified.

21517 Noralen
Elastic roof cover material. *Carl Freudenberg.* Name unverified.

21518 Noram
Fatty primary, secondary and tertiary monoamines derived from coco, tallow and oleyl alkyl chains; amine salts are cationic emulsifiers, also intermediate in chemical synthesis, mineral flotation, anticaking of fertilizers, corrosion inhibitors. *Elf Atochem UK/Ceca.*

21519 Noramac
Acetate salts of fatty mono amines; bactericides, flotation, flushing agents for pigments, polishes, hydrophobic treatment of textiles, wood, metals, etc. *Elf Atochem UK/Ceca.*

21520 Norament
Special designed floors. *Carl Freudenberg.* Name unverified.

21521 Noramid
Flooring material. *Carl Freudenberg.* Name unverified.

21522 Noramium
Quaternary ammonium compounds derived from fatty amines with coco, tallow or oleyl alkyl chains; fabric softeners, organophilic clays, biocides, antistatics. *Elf Atochem UK/Ceca.*

21523 Noramox
Ethoxylated fatty monoamines; emulsifiers, corrosion inhibitors, dispersing and wetting agents, textile auxiliaries, fuel additives. *Elf Atochem UK/Ceca.*

21524 Noraplan
Flooring material. *Carl Freudenberg.* Name unverified.

21525 Noratex
A proprietary preparation of talc, kaolin, zinc oxide, cod liver oil and wool fat used as a protective skin cream. *H N Norton & Co Ltd.* Unverified.

21526 Norbide
A proprietary trade name for an amorphous boron carbide; an abrasive. No manufacturer.

21527 Norbo
A proprietary trade name for phenolic resin. No manufacturer.

21528 Norcast 142 Systems
25928-94-3
Unfilled 100% solids epoxy resin; casting and impregnating materials *R.H. Carlson.* Name unverified.

21529 Norcast 154FR
25928-94-3
Epoxy resin; flame retardant casting resin *R.H. Carlson.* Name unverified.

21530 Norcast 1460-1
25928-94-3
One-component epoxy resin system; dipping compound *R.H. Carlson.* Name unverified.

21531 Norcast 3220G-1
25928-94-3
One-component epoxy system; dip and brush coating *R.H. Carlson.* Name unverified.

21532 Norcast 3258
25928-94-3
Epoxy
Potting/encapsulating system *R.H. Carlson.* Name unverified.

21533 Norcast 3705
25928-94-3
Two-part epoxy system; variable flexibility system for use as casting resin or high peel strength adhesive; high moisture resistance and ability to withstand physical and thermal shock; mix ratio range from 40-200 parts B/100 parts A; viscosity 10,000 cps (mixed). *R.H. Carlson.* Name unverified.

21534 Norcast 4914-1
25928-94-3
Epoxy, silver-filled; adhesive *R.H. Carlson.* Name unverified.

21535 Norcure 131
Aliphatic amine
Curing agent (used with DER 741) for compatibility and long pot life at room temperature; fast cure; mix ratio; 15 phr with DER 741. *R.H. Carlson.* Name unverified.

21536 Norcure 3298
Epoxy hardener; used as primary hardener for Resin 220-0320 for soft, flexible epoxy system; easily repaired when silica filled; mix ratio: 2/1 resin/hardener. *R.H. Carlson.* Name unverified.

21537 Nordel®
Ethylene-propylene synthetic rubber. *DuPont UK.*

21538 Nordel® 2744
A trademark for a fast curing EPDM hydrocarbon rubber possessing high green strength. *DuPont UK.* Name unverified.

21539 Nordot 101F
Urethane adhesive; solvent-free/water-free adhesive for installing flooring and sport surfaces; for wood, vinyl, rubber, concrete and asphalt. *Syn. Surfaces.*

21540 Nordox
1317-39-1 2734 215-270-7
Cuprous oxide, paint grade, red, micro milled; active ingredient in antifouling paints. *Nordox Industrier AS.* Name unverified.

21541 Noreplast
A proprietary trade name for laminated thermosetting, plastic thermosetting and thermoplastic synthetic resins. No manufacturer.

21542 Norepol
A proprietary trade name for vulcanizable vegetable polymers. No manufacturer.

21543 Noreseal
A proprietary trade name for a cork substitute made from low cost domestic raw materials; stated to be equal in strength to cork. No manufacturer.

21544 Norfox® 1101
61789-30-8 263-049-9
Potassium cocoate
Flash foamer, emulsifier for shampoo bases, liquid hand soaps; lubricant for conveyors; coupling agent for heavily built liquid alkali systems such as steam cleaners and whitewall tire cleaners. *Norman, Fox & Co.*

21545 Norfox® Agent 2A-2S
Air entraining agent, wetting agent for production of mortar, stucco, and plastic cement. *Norman, Fox & Co.*

21546 Norfox® ALPHA XL
68439-57-6 270-407-8
Sodium C14-16 alpha olefin sulfonate
Base for shampoos, hand soaps, bath products, home and janitorial cleaners, dishwash, and light duty liqs. *Norman, Fox & Co.*

21547 Norfox® Coco Betaine
Cocamidopropyl betaine
Mild substantive surfactant, flash foamer; base for mild shampoos; electrolyte tolerance over wide pH range. *Norman, Fox & Co.*

21548 Norfox® DC
61791-31-9 263-163-9
Cocamide DEA and diethanolamine; intermediate for detergent manufacturing, liquid dishwash, cosmetics; suds stabilizer and dedusting agent for dry products. *Norman, Fox & Co.*

21549 Norfox® F-221
93-83-4 202-281-7
Oleamide DEA
Invert emulsifier with hydraulic properties; wetting agent, penetrant; salt tolerant when dissolved in oil phase over a wide temperature range; for herbicides, drilling fluids for petroleum industry, penetrating oils, specialty lubricants. *Norman, Fox & Co.*

21550 Norfox® IM-38
62449-33-6
N-Oleyl imidazolinium hydrochloride
High-foaming wetting agent and emulsifier; base for automotive rinse-wax formulations, corrosion inhibition compounds. *Norman, Fox & Co.*

21551 Norfox® SLES-60
9004-82-4
Sodium laureth sulfate, 14% denatured alcohol; base for shampoos and light duty liquid formulators; flash foam enhancer for automotive, household, personal care, and industrial cleaners; in manufacturing of gypsum wallboard and gas drilling of deep wells. *Norman, Fox & Co.*

21552 Norfox® SLS
151-21-3 8782 205-788-1
Sodium lauryl sulfate
Biodegradable foamer and wetting agent for household, industrial and personal care products, shampoos, hand and body soaps, fabric care products. *Norman, Fox & Co.*

21553 Norfox® Sorbo T-60
9005-67-8 8872
PEG-20 sorbitan stearate
Emulsifier for frozen desserts, salad dressings, cake mixes, icings; dough conditioner. *Norman, Fox & Co.*

21554 Norfox® T-60
TEA dodecylbenzene sulfonate
Detergent, wetting agent and foamer for agricultural and industrial/household use, light duty detergents, hard surface cleaners, shampoos, wool and fine fabric washing. *Norman, Fox & Co.*

21555 Norfox® TLS
139-96-8 205-388-7
TEA lauryl sulfate
Biodegradable mild ingredient for shampoo, light duty liquids, fine fabric detergents. *Norman, Fox & Co.*

21556 Norfox® Unimulse OW
Polymer/surfactant; emulsifier for oily liquids into oil and water emulsions. *Norman, Fox & Co.* Discontinued.

21557 Norfox® X
Cocamide DEA
Modified; used with acidic builders; wash-wax formulations, glass and appliance cleaners. *Norman, Fox & Co.*

21558 Norfroth
Mixed propylene glycol isobutyl ethers and surfactants; ore flotation chemicals. *Chemoxy International Ltd.*

21559 Norgine
The sodium-ammonium salt of laminaric acid from seaweed, *Laminaria digitata* and *Saccharinus digitatus*; used in the treatment of textiles.

21560 Norit, C, PK, R, RO
7440-44-0 1855 231-153-3
C
Activated carbon; used for purification and decolorization of food products, pharmaceutical products, water (potable, process and waste), recovery of gold (CIP and CIL) and air purification (removal of mercury, H_2S, SO_2, etc) *R W Greeff & Co Inc.*

21561 Norithene
7440-44-0 1855 231-153-3
C
Absorptive carbons for use in air conditioning systems. *NORIT UK Ltd.*

21562 Norlig
Calcium and sodium lignosulfonates; specialty chemicals for use in a variety of binding, dispersing, complexing, stabilizing, or copolymerizing applications. *Borregaard Ligno Tech.*

21563 Norlig 11 DA
8061-52-7
Calcium lignosulfonate
Binder for pesticide/herbicide carrier substitutes; aids in subsequent dispersion. *Borregaard Ligno Tech.*

21564 Norlig 415
Modified sodium-calcium lignosulfonate; dispersant and water reducer for gypsum wallboard manufacturing; stucco dispersant giving low retardation. *Borregaard Ligno Tech.*

21565 Norlig A
8061-52-7
Calcium lignosulfonate
Dispersant, binder, soil and dust stabilizer, carbon black, briquetting and pelletizing of coal and charcoal, ceramic additive. *Borregaard Ligno Tech.*

21566 Normacol
5296
Karaya gum; kadaya, katilo; kullo; kuteera; Indian tragacanth; mucara. A proprietary preparation containing sterculia; a laxative. Also used as a dental adhesive. *Norgine Ltd.*

21567 normal powder
A gelatinized gun-cotton powder.

21568 Nor-Mer 020
25928-94-3
Epoxy system; uv-curable adhesive *R.H. Carlson.*

21569 Normet
Metallized biaxially oriented polypropylene film for use in manufacture. *Quantum Chemical Corp.*

21570 No-Roma
A proprietary preparation of paraformaldehyde, sodium carboxy-methyl cellulose, sodium hydroxide and methylene blue; a medical deodorant. *Salt.* Name unverified.

21571 Norox® MCP
MEK peroxides and cumyl hydroperoxide in plasticizers; polymerization initiator for room temperature cure of unsat. polyester and vinyl ester resins; low peak exotherm, longer gel time. *Norac.*

21572 Norpar®
High purity normal paraffin solvent. *Exxon Intl.*

21573 Norpar® 12
64771-72-8 265-233-4
Normal paraffin; solvent. *Exxon.* Name unverified.

21574 Norplex laminates
Woven fiberglass fabric impregnated with thermoset resin, with thin copper foil on one or two sides, pressed under high temperature and heat to form plastic laminated sheets; substrate material for printed circuit boards. *Norplex.* Name unverified.

21575 Norprop
Biaxially oriented polypropylene film for use in manufacturing. *Quantum Chemical Corp.*

21576 Norsil 1000
Optical coupling compound; reduces internal reflections and refractures in optical equipment; optically clear gel coupling in fiber optics. *R.H. Carlson.* Name unverified.

21577 Norsil RTV 811
63394-02-5
Silicone rubber; for broad range of mechanical and electrical/electronic applications; potting and encapsulation, cast-in-place gasket and molds. *R.H. Carlson.* Name unverified.

21578 Norsil SG 131
Dimethyl silicone compound; semitransparent compound used as release agent at -40 to 400°F; lubricant for rubber, plastic and leather parts; insulates and repels water from electrical connections. *R.H. Carlson.* Name unverified.

21579 Norsil SG 169
Dimethyl silicone compound polyethylene-filled; release agent, lubricant for plastic and rubber surfaces; very low dielectric constant. *R.H. Carlson.* Name unverified.

21580 Norsolene®
Coumarone and polyindene resins. *CdF Chimie.* Name unverified.

21581 Norsomix®
A trademark for polyester compounds. No manufacturer.

21582 Norsophen®
Phenolic resins and compounds; composites offering strength, durability, excellent fire performance, high heat resistance; used for aerospace, construction, mass transit, mining, and automotive applications. *Norold Composites.*

21583 Norsorex
Polynorbomene. *Sartomer.* Discontinued.

21584 Nortech
Pre-colored plastic resins, polyolefin foam concentrates specialty filled plastic resins and polyolefin resins. *Quantum Chemical Corp.*

21585 Northovan
13718-26-8 8849 237-272-7
NaO_3V
sodium metavanadate
Sodium vanadate; Sodium vanadate (meta); vanadic acid, monosodium salt. Used in manufacture of inks and in photography. mp = 630°; insoluble in H_2O; LD_{75} (rat ip) = 4-5 mg V/kg. No manufacturer.

21586 Nortron
26225-79-6 3788 247-525-3
Emulsifiable concentrate of 200 g ethofumesate per liter; used for weed control in field crops. *Schering Agrochemicals Ltd.* Discontinued.

21587 Nortron Leyclene
Bromoxynil, ethofumesate, ioxynil; herbicide mixture for new grass lays. *Schering Agrochemicals Ltd.* Discontinued.

21588 Nortuff RA 1700-MO
9003-07-0 7741
PP resin; extrusion grade *Quantum/US.* Name unverified.

21589 Nortuff RA 7020-KO
9002-88-4 7728
Polyethylene
Extrusion and injection molding grade *Quantum/USI.* Name unverified.

21590 Nortuff RC 1700-MO
9003-07-0 7741
PP resin; injection molding grade *Quantum/USI.* Name unverified.

21591 Noruben
555-37-3 6523 209-096-0
Neburon
Herbicide. *Agan Chemical Manufacturers Ltd.*

21592 Norunil
330-55-2 5534 206-356-5
Linuron
Herbicide. *Agan Chemical Manufacturers Ltd.*

21593 Norust
Specialty blends of fatty amines and derivatives; corrosion inhibitors, fuel oil additives. *Elf Atochem UK/Ceca.*

21594 Norval
577-11-7 3460 209-406-4
A proprietary preparation o-dioctyl-sodium sulfosuccinate in gelatine; a fecal softener. *Horlicks.* Name unverified.

21595 Norvinyl
Polyvinyl chloride homo and copolymers (S-PVC); used as a main product or as an extender for production of plastic articles. *Norsk Hydro AS.*

21596 Norvinyl DX 550
9002-86-2 7746 206-625-7
Crosslinkable plasticized PVC compound; for manufacturing high temperature-resistant cables and wires. *Hydro Plast.*

21597 Noryl®
Modified polyphenylene oxide; used for plastic components for automotive, electrical, electronics, lighting, medical, packaging, audio etc. *GE Plastics Ltd.*

21598 Noryl® 731
25134-01-4
Modified PPO resin; thermoplastic general purpose injection molding grade resin, excellent dimensional stability; heat resistance to 265°F, good impact strength, low moisture resistance; used in computers, business equipment, automotive, electrical, electronics, construction, telecommunications, applianes and other industries. *GE Plastics.*

21599 Noryl® 1402B
Engineered blow molding grade for demanding automotive exterior body parts, fluid handling equipment, medical products, motor casings. *GE Plastics.*

21600 Noryl® BN25
25134-01-4
Modified PPO resin; engineered blow molding grade with high heat resistance, impact strength, dimensional stability, moisture resistance, electrical insulating properties; used for computers, appliances, business equipment, automotive, fluid handling, telecommunications equipment. *GE Plastics.*

21601 Noryl® EM5101
25134-01-4
Modified PPO resin; energy management grade for automotive interior applications; superior processability, improved flow characteristics. *GE Plastics.*

21602 Noryl® EN185
25134-01-4
PPO resin; extrusion grade; combines excellent impact and heat resistance, UL recognition, formability and processability; used for electronics housings, beverage cases, transportation components. *GE Plastics.*

21603 Noryl® FN150
25134-01-4
Modified PPO foam resin; engineering structural foam resin for business machine housings and their structural bases, weather-resistant electrical enclosures, and lightweight structural components for the transportation industry which resist the greenhouse effect. *GE Plastics.*

21604 Noryl® FN215X
25134-01-4
Modified PPO structural foam resin; high mechanical strength, heat-and impact-resistance; used in business machine housings and structural bases, weather-resistant electrical enclosures, and lightweight structural components for transportation. *GE Plastics.*

21605 Noryl® GTX810
25134-01-4
PPO resin; unreinforced engineering thermoplastic resin; offers excellent chemical resistance, mechanical performance, high heat resistance, dimensional stability; for automotive body panels, exterior hardware, and other applications requiring high heat resistance, paintability. *GE Plastics.*

21606 Noryl® HM3020
25134-01-4
Modified PPO resin; high modulus grade. *GE Plastics.*

21607 Noryl® HS1000X
25134-01-4
Modified PPO resin; injection molding grade thermoplastic resin for high-strength, thin-wall applications. *GE Plastics.*

21608 Noryl® N190X
25134-01-4
Modified PPO resin, flame retardant; thermoplastic injection molding grade for business machines, small appliance housings, current carrying parts, and telecommunication internal componentry; highly stable under load at elevated temperatures; UL listed. *GE Plastics.*

21609 Noryl® PC180X
25134-01-4
Modified PPO resin; injection molding grade; computer and business equipment grade thermoplastic resin for portable computer applications. *GE Plastics.*

21610 Noryl® PN235
25134-01-4
Modified PPO resin; thermoplastic, platable, injection molding grade that offers improved dimensional stability; lighter weight, more economical than metals; peel strength, low creep, and low temp, impact resistance; in automobile grilles, wheel covers and headlamp bezels, appliance range knobs and decorative trim areas. *GE Plastics.*

21611 Noryl® PX0722
25134-01-4
Modified PPO resin; automotive grade thermoplastic resin with electrical insulating capabilities for instrument panels, speaker grilles, glove box doors, connectors, fuse blocks, instrument panel components, mirrors, exterior trim, grilles, and wheel covers; PX0722 for interior trim. *GE Plastics.*

21612 Noryl® SE100
25134-01-4
Modified PPO resin, flame retardant; thermoplastic injection molding grade for applications requiring moderate heat deflection and excellent dimensional stability. *GE Plastics.*

21613 Noryl® SE1GFN2
25134-01-4
Modified PPO resin, 20% glass reinforced; thermoplastic injection molding grade, broad UL recognition, excellent mechanical and dimensional characteristics; used in electrical applications, connectors, structural parts, television components, liquid handling, business machines. *GE Plastics.*

21614 Nosifeed 40
Feed additive (growth promoter) for swine and poultry. *Mitsubishi Kasei.* Name unverified.

21615 Nosiheptide
A peptide obtained from cultures of *Streptomyces actuosus* 40037 or the same substance obtained by other means; a veterinary antibiotic.

21616 Nosiheptide
Feed additive (growth promoter) for swine and poultry. *Mitsubishi Kasei.* Name unverified.

21617 Nosil 711
Methyl phenol-based silicone; rubber for use at -115 to 204 C; allows for greater flexibility in high temperature applications. *R.H. Carlson.* Name unverified.

21618 No-Swab
Mold coating containing solid lubricants and suitable binder system; molds for glass containers in I.S. machines for extension of mold swabbing cycles. *E/M Corporation.*

21619 Notak
Products for nondestructive detection of surface flaws in iron and steel components. *Foseco (F.S.) Ltd.*

21620 Nouraid
Polyunsaturated fatty acids. *Akzo Chemie UK Ltd.*

21621 Nourycryl
Range of methacrylate monomers for use in adhesive and coating bases. *Akzo Chemie UK Ltd.*

21622 Nourydrier
Range of metal-based drying agents. *Akzo Chemie UK Ltd.*

21623 Nourymix
Additive concentrate for polyolefins. *Akzo Chemie UK Ltd.*

21624 Nouryset®
Allyl-based monomers; for use in optical and ophthalmic resins. *Akzo Chemie UK Ltd.*

21625 Nouryset® 156
Styrene proprietary product; used in the manufacturing of high-quality optical materials, e.g., high minus lenses. *Akzo.* Name unverified.

21626 Nouryset® 200 HV 250
Prepolymerized diethylene glycol bis(allyl carbonate); polymer offers lower shrinkage on curing; used for optical applications. *Akzo.* Name unverified.

21627 Nova 2001
Emulsion stripper. *British Nova Works Ltd.*

21628 Nova Furnipol
Furniture polish. *British Nova Works Ltd.*

21629 Nova Highgloss
Paste wax polish. *British Nova Works Ltd.*

21630 Nova Lime-Lite
Floor maintainer. *British Nova Works Ltd.*

21631 Nova Long-Life
Floor dressing. *British Nova Works Ltd.*

21632 Nova One
Liquid cleanser. *British Nova Works Ltd.*

21633 Nova One Plus
Bactericidal liquid cleanser. *British Nova Works Ltd.*

21634 Nova PC-1000BK
PC resin. *Nova Polymers.*

21635 Nova Pine Fluid
Disinfectant. *British Nova Works Ltd.*

21636 Nova Starbight
Emulsion dressing remover. *British Nova Works Ltd.*

21637 Nova Stripper Super
Seal remover. *British Nova Works Ltd.*

21638 Nova Supercote
Floor dressing. *British Nova Works Ltd.*

21639 Nova Superphaite
Buffable emulsion dressing. *British Nova Works Ltd.*

21640 Nova Supratreet
Carpet shampoo. *British Nova Works Ltd.*

21641 Nova Tri-Power
Degreasant and cleanser. *British Nova Works Ltd.*

21642 Novablend® 501
9002-86-2 7746 206-625-7
PVC compound; general-purpose rigid injection grade; provides an economical alternative in opaque injection molded parts; also for extruded profiles. *Novatec Plastics & Chem.* Name unverified.

21643 Novablend® 5555
9002-86-2 7746 206-625-7
PVC compound, impact-modified; rigid extrusion blow molding compound providing excellent processing and thermal stability; for water bottle applications. *Novatec Plastics & Chem.*

21644 Novabloo
Toilet bowl cleanser. *British Nova Works Ltd.*

21645 Novabold
Toilet bowl cleanser. *British Nova Works Ltd.*

21646 Novacarb
7440-44-0 1855 231-153-3
C
Carbon black; black concentrate and impact modifier for ABS and PVC. *Nova Polymers.*

21647 Novacare
Floor maintainer. *British Nova Works Ltd.*

21648 novacetoform
8006-13-1 332
Burow's solution; Domeboro. Aluminum acetate (5% aqueous solution). An astringent and antiseptic. .

21649 Novacleer
Window cleanser. *British Nova Works Ltd.*

21650 Novacorn
An emulsifiable concentrate containing 240 g bromoxynil and 160 g ioxynil per liter; a post-emergence contact herbicide for cereal crops. *Farmers Crop Chemicals Ltd.* Discontinued.

21651 Novacote
Intermediates for plastics, for use in wrapping and packaging. *Imperial Chemical Industries plc.*

21652 Novacrete
Floor sealer. *British Nova Works Ltd.*

21653 Novacross
Sanitizing fluid. *British Nova Works Ltd.*

21654 Novacryl
Acrylic sealer and primer. *British Nova Works Ltd.*

21655 novaculite
A quartz rock used as an abrasive.

21656 Novacut
Traffic lane cleaner. *British Nova Works Ltd.*

21657 Novadelox
94-36-0 1149 202-327-6
Benzoyl peroxide based; bleaching agent for baking flour. *Akzo Chemie UK Ltd.*

21658 Novadine
Iodophor sanitizer for brewery plant; for sanitizing fermenting vessels, lager tanks, maturation tanks, bright beer tanks, fillers and mains. *Harshaw Chemicals Ltd.* Name unverified.

21659 Novafil
Polybutester; surgical aid. *Davis & Geck.* Name unverified.

21660 NovaFlo
Liquid rosin size; internal sizing agent to give water resistance to paper and paperboard. *Georgia-Pacific.* Name unverified.

21661 Novafrost
Chewing gum remover. *British Nova Works Ltd.*

21662 Novagem
Furniture polish. *British Nova Works Ltd.*

21663 Novagrip
Penetrating seal. *British Nova Works Ltd.*

21664 novalak resins
Synthetic resins of the formaldehydephenol type.

21665 Novalar
Impact modifier for ABS, PVC, vinyl chloride copolymers, PC, PU, epoxies, PBT, acrylics; improves impact strength, ductility, low temperature properties, eliminates brittleness, upgrades scrap. *Nova Polymers.* Name unverified.

21666 Novalast 5000, 9000
Thermoplastic elastomer; low-cost material with good compression set and heat aging properties. *Nova Polymers.*

21667 Novalene
Impact modifier for homopolymer PP and HDPE; upgrades scrap PP homopolymer to copolymer; improves ductility and low temp. properties; eliminates brittleness. *Nova Polymers.*

21668 Novalift
Kitchen grease remover. *British Nova Works Ltd.*

21669 Novalite
Carpet shampoo. *British Nova Works Ltd.*

21670 Novaloy 6521
A proprietary one component epoxy resin supplied in powder form for the coating of electronic components. *Rogers Corp.*

21671 Novaloy® 9000
ABS/PVC alloy; engineering alloy for injection molding, profile and sheet extrusion; provides excellent processing, high heat distortion temperatures, excellent retention of properties on aging; used for appliance housings, electrical boxes, communications, lawn and garden applications, recreational products. *Novatec Plastics & Chem.*

21672 Novamid® 1010C
25038-54-4 6832
Nylon 6; general purpose injection molding grade suitable for fast cycling injection molding grade suitable for fast cycling injection molding; low viscosity. *Mitsubishi Kasei.* Name unverified.

21673 Novamid® 1020VA2
25038-54-4 6832
Nylon 6

Extrusion molding grade for co-extrusion film applications, suitable for both air and water-cooled blown film; medium viscosity. *Mitsubishi Kasei.* Name unverified.

21674 Novamid® 2020A, 2420A
Nylon copolymer; extrusion molding grade offering excellent transparency, high flexibility; for monofilament applications; medium viscosity. *Mitsubishi Kasei.* Name unverified.

21675 Novamyl
A maltogenic amylase produced by a strain of *Bacillus subtilis*; added to flour it modifies the starch in the breadmaking process so that retrogradation is less likely to occur and staling is retarded 2 days or more. *Novo Nordisk.*

21676 Novanyl
A range of acid dyes; for dyeing of polyamide fibers. No manufacturer.

21677 Novapel
Leather finishes. *ICI Chem & Polymers Ltd.*

21678 Novaphalte
Asphalt sealer. *British Nova Works Ltd.*

21679 Novapint
Universal pigment dispersions. *Tennant-KVK Ltd.*

21680 Novaplaste
Leather dyes. *ICI Chem & Polymers Ltd.* Discontinued.

21681 Novapol
Liquid wax polish. *British Nova Works Ltd.*

21682 Novapol® GF-0218-F
9002-88-4 7728
LLDPE with process stabilizer; features toughness, strength, ease of extrusion; for blending. *Novacor Ltd.*

21683 Novapol® GI-2024-A
9002-88-4 7728
LLDPE; features medium flow, easy processability, improved dispersion, good ESCR; for additive and color concentrates, masterbatches, compounding., houseware, trash cans, industrial containers, toys. *Novacor Ltd.*

21684 Novapol® HB-L455-A
9002-88-4 7728
HDPE with process stabilizer; features excellent processability on high speed, multiple head blow molding lines; for household and industrial chemical bottles, automotive fluid bottles, toys. *Novacor Ltd.*

21685 Novapol® LC-0517-A
9002-88-4 7728
LDPE; features excellent adhesion, high draw down, excellent sealability; for dry soup pouches, pharmaceutical packaging, potato chip bags, snack food packaging. *Novacor Ltd.*

21686 Novapol® LE-0220-A
9002-88-4 7728
LDPE; features easy processing, high melt strength; for profile extrusions, tape and sheet extrusion, foam, concentrates. *Novacor Ltd.*

21687 Novapol® LF-0223-B
9002-88-4 7728
LDPE, medium slip and antiblock; features good clarity, stiffness and strength; for consumer packaging, food wrap, bakery bags, liners, blends with LLDPE. *Novacor Ltd.*

21688 Novapol® LF-Y819-D
9002-88-4 7728
LDPE, high slip and antiblock; features high strength and toughness, excellent processability and shrink film characteristics; for industrial packaging, liners, shrink film, blends with LLDPE. *Novacor Ltd.*

21689 Novapol® PF-0118-B
9002-88-4 7728
LLDPE, medium slip and antiblock; features toughness and strength; for trash bags, liners, general packaging applications. *Novacor Ltd.*

21690 Novapol® PR-0636-UG
9002-88-4 7728
Linear MDPE with process and uv stabilizer; rotational molding grade featuring high impact strength, very good ESCR; for toys, custom molding, carts. *Novacor Ltd.*

21691 Novaquik
Barrier seal. *British Nova Works Ltd.*

21692 Novaract
Defoamer. *British Nova Works Ltd.*

21693 Novarex® 7022A
PC; injection molding grade, especially for thinwalled parts; good moldability; low viscosity. *Mitsubishi Kasei.* Name unverified.

21694 Novarex® 7025G10
PC, 10% short glass fiber-reinforced; suitable for injection molding where good dimensional stability is required; medium viscosity. *Mitsubishi Kasei.* Name unverified.

21695 Novarex® 7025NB
PC, nonflammable; self-extinguishing grade. *Mitsubishi Kasei.* Name unverified.

21696 Novasan
Deodorant blocks. *British Nova Works Ltd.*

21697 Novasheen
Heavy duty seal. *British Nova Works Ltd.*

21698 Novashield
Floor dressing. *British Nova Works Ltd.*

21699 NovaSize, Dark Fortified
Tall oil rosin; internal sizing agent to give water resistance to paper and paperboard. *Georgia-Pacific.* Name unverified.

21700 Novasolve
Wax remover. *British Nova Works Ltd.*

21701 NovaSperse
Rosin aqueous dispersion; internal sizing agent to give water resistance to paper/paperboard. *Georgia-Pacific.* Name unverified.

21702 Novastet
Antistatic treatment. *British Nova Works Ltd.*

21703 Nova-T
Polyethylene IUCD made radio-opaque by barium sulfate, with silver-cored copper wire. *Schering Health Care Ltd.* Discontinued.

21704 Novata 299, A, AB, B, BBC, BC, BCF, BD, C, D, E
Cocoglycerides
Suppository bases; consistency giving agent for ointments, creams, and stick preps. *Henkel; Henkel KGaA.*

21705 Novatec
Polyolefin resins. *British Traders & Shippers Ltd.*

21706 Novatec 240H
Extrudable adhesive polyolefin resin; for barrier and HDPE bonding. *Mitsubishi Kasei.* Name unverified.

21707 Novatec-AP
Extrudable adhesive polyolefin resin; for coextrusion of film, bottle, tube, and sheet. *Mitsubishi.* Name unverified.

21708 Novathion
122-14-5 4017 204-524-2
C9H12NO5PS
O,O-dimethyl O-4-nitro-*m*-tolyl phosphorothioate
All-round low toxic insecticide for forest protection, agriculture and public health; used especially where long term effect is desired. *A/S Cheminova.* Name unverified.

21709 Novatreet
Carpet shampoo. *British Nova Works Ltd.*

21710 Novaways
All purpose cleanser. *British Nova Works Ltd.*

21711 Novazole
26097-80-3 1774 247-459-5
Cambendazole
Anthelmintic. *Merck & Co Inc.* Name unverified.

21712 Noveloid
Proprietary products of cellulose esters and ethers. No manufacturer.

21713 Novemol
A trade name for a wetting agent containing sulfonated terpene alcohols. No manufacturer.

21714 Novester
Disperse dyes. *ICI Chem & Polymers Ltd.* Discontinued.

21715 Novex
Low density polyethylene. *BP Chemicals Ltd.*

21716 Novex
Benzalbisdimethyldithiocarbamate. It has a specific gravity of 1.365, is a white crystalline powder, and melts at 175 C; used as a rubber vulcanization accelerator.

21717 Novidium
4767
$C_{21}H_{20}ClN_3$
3,8-diamino-5-ethyl-6-phenylphenthridinium chloride
Homidium chloride. Antiprotozoal. Soluble in H_2O (200 mg/ml). *May & Baker Ltd.* Name unverified.

21718 Novitane
A proprietary polyurethane elastomer. *BF Goodrich.* Name unverified.

21719 Novite
A proprietary trade name for an alloy of iron with about 1.5% nickel and 0.5% chromium. No manufacturer.

21720 Novocoll® NC222 US
Isocyanate-terminated polyester urethane adhesive; laminating adhesive

giving high adhesion on aluminum and transparent packaging films. *Pierce & Stevens.*

21721 Novodur®
Acrylonitrile/butadiene/styrene copolymer; engineering thermoplastic for the manufacture of injection moldings; good toughness, strength, stiffness and resistance to chemicals; excellent surface quality; problem free processing; range of standard grades, increased heat resistance, glass-fiber reinforced, flame retardant, electroplating, extrusion grades, used for housings, covers, automobile components, radio, TV and phono equipment, office machinery, photographic and film sectors and toys. *Bayer AG; Bayer plc.*

21722 Novodur® L3FR
9003-56-9
ABS; flame-retardant grade. *Bayer AG.*

21723 Novodur® P2H-AT
9003-56-9
ABS; standard injection molding grade. *Bayer AG.* Name unverified.

21724 Novodur® P2HE
9003-56-9
ABS; extrusion grade. *Bayer AG.* Name unverified.

21725 Novodur® P2HGV
9003-56-9
ABS; glass fiber reinforced. *Bayer AG.* Name unverified.

21726 Novodur® P2T, P2T-AT
9003-56-9
ABS graft polymer; injection molding grade polymers with higher heat deflection temps.; *Bayer AG.* Name unverified.

21727 Novodur® PMTM
9003-56-9
ABS; increased heat distortion grade. *Bayer AG.*

21728 Novofix
Photographic fixer. *May & Baker Ltd.* Name unverified.

21729 Novogel® ST
637-12-7 379 211-279-5
$C_{54}H_{105}AlO_6$
octadecanoic acid aluminum salt
aluminum tristearate. Aluminum stearate, mineral oil. mp = 117-120°; insoluble in H_2O, soluble in organic solvents. *RhônePoulenc.*

21730 Novol
143-28-2 6968 205-597-3
$C_{18}H_{36}O$
oleyl alcohol
Super refined oleyl alcohollient, emulsion stabilizer, superfatting agent, pigment suspending aid, used in cosmetics, personal care products; lipsticks, sunscreens, antiperspirants, bath oils. *Croda Inc.*

21731 novolac resin
Novolak resin; 2-stage phenolic resin, Phenol-formaldehyde type resin; molding materials; bonding agent in brake linings, abrasive grinding wheels, electrical insulation; reinforcing agent modifier for nitrile rubber; air-drying varnishes; bonding materials.

21732 Novolen®
A range of polypropylenes and polypropylene copolymers supplied in granular form; used for injection molding of technical parts and packaging, extrusion into special soft film, extrusion into fibers and tapes, blow molded bottles. *BASF AG; BASF plc.*

21733 Novoline, Novolith, Novomatic
Photographic developer. *May & Baker Ltd.* Name unverified.

21734 Novon® 2020-6001
Biodegradable polymer based on corn/potato starch; water-dispersion free-foam extrusion polymer; for biodegradable interior cushioning materials. *Novon Prods.*

21735 Novon® 3001
Biodegradable polymer based on corn/potato starch; water-disintegrating injection molding grade for biodegradable golf tees. *Novon Prods.*

21736 Novon® M0121
Biodegradable polymer based on corn/potato starch; injection molding grade for ultrasound needle guides and other single-use medical devices which deform in steam autoclave, solution sterilizing, or exposure to water. *Novon Prods.*

21737 Novon® M0289
Biodegradable polymer based on corn/potato starch; injection molding grade for biodegradable stakes for golf course resodding. *Novon Prods.*

21738 Novonasco
A proprietary trade name for a wetting agent containing modified sodium alkylanaphthalenesulfonate. No manufacturer.

21739 Novoperm
Pigments powders; used for paints and inks. *Hoechst AG.*

21740 Novoplas
A proprietary trade name for an organic polysulfide synthetic elastic material. No manufacturer.

21741 Novoprotin
A proprietary preparation of crystalline vegetable albumin. No manufacturer.

21742 Novor
Crosslinker for rubber. *Durham Chemicals Ltd.* Name unverified.

21743 Novor 950
Methylene bis(4-phenylisocyanate)-based; vulcanizing agent for natural and synthetic rubbers giving exceptional reversion and high temperature aging resistance, reduced emission levels. *Akrochem; Rubber Consultants.*

21744 Novotak
Photographic developer. *May & Baker Ltd.* Name unverified.

21745 Novozone®
1335-26-8 215-559-8
MgO_2
magnesium dioxide
magnesium peroxide. A German registered name for a magnesium peroxide, MgO_2, prepared for medical purposes; an antiseptic used internally, and externally as an ointment, for wounds and gatherings. No manufacturer.

21746 Novozym
A proteolytic enzyme prepared by submerged fermentation of a selected strain of *Bacillus licheniformis* ; used in the manufacture of effective denture cleaners. *Novo Industri A/S.* Discontinued.

21747 Noxamine
Ethoxylated fatty amine oxides
Used for detergents, toiletries. *Elf Atochem UK/Ceca.*

21748 Noxamium
Ethoxylated quaternary ammonium compounds; used for detergents, paints, crude oil demulsifier. *Elf Atochem UK/Ceca.*

21749 Nozolex
Free-flowing refractory fillers used in sliding gate nozzles. *Foseco (F.S.) Ltd.*

21750 NP-10
Aromatic plasticizer; chemically inert, non-saponifiable grade, with low reactivity; used in adhesives (mastic, pressure sensitive), rubber (cements, mechanical and molded goods, tires), and caulking compounds. *Neville.*

21751 NPG® Glycol
126-30-7 6547 204-781-0
$C_5H_{12}O_2$
Neopentyl glycol
2,2-dimethyl-1,3-Propanediol; Dimethyltrimethylene glycol; 2,2-dimethylpropane-1,3-diol. Resin intermediate. *Eastman.*

21752 NPP-1
O,O-Di-n-propyl phosphorodithioic acid
Used as a high grade intermediate for organophosphorus insecticides and other products. *A/S Cheminova.* Name unverified.

21753 NPR 3911
Neoprene latex
Low viscosity polymer which can accept large quantities of fillers while retaining good workability; for formulation of mastics and low-cost facing adhesives. *DuPont.*

21754 NPR 5587
Neoprene latex
Used where rapid bond development is required, as a quick-break adhesive. *DuPont.*

21755 NS®
Used for higher PV applications. *Polymer corp.*

21756 NSAE Powder
Sodium alkyl-naphthalene sulfonate
Anionic surfactant in powder form; wetting and dispersing agent with high electrolyte tolerance, e.g., agricultural dispersing agent for sulfur. *Millmaster-Onyx UK.* Name unverified.

21757 N-Serve
A line of nitrogen stabilizers based primarily on nitropyrin. *Dow UK.* Discontinued.

21758 NT-15GF/000
32131-17-2
Toughened nylon 6/6, 15% glass fiber-reinforced. *Compounding Tech.* Name unverified.

21759 NTA
5064-31-3 225-768-6
Sodium nitrilotriacetate powder and 40% solution; for domestic and industrial laundry detergents, hard surface cleaning formulations, boiler treatment, textile auxiliary. *Monsanto Co.* Name unverified.

21760 Nuact
Loss-of-dryness inhibitors; prevents loss of drying of nonaqueous coating compositions. *Hüls Am.*

21761 Nuade
Slip and mar agents; for water-borne baking finishes. *Hüls Am.*

21762 nuarimol
63284-71-9 264-071-1
$C_{17}H_{12}ClFN_2O$
α-(2-chlorophenyl)-α-(4-fluorophenyl)-5-pyrimidinemethanol
Cidorel; EL-228; Gandural; Gauntlet; Murox; Tridal; Trimidal; Triminol. Systemic foliar fungicide wit curative and protective action. Ergosterol biosynthesis inhibitor. Used for control of a wide range of pathogenic fungi. Used as a foliar spray or seed treatment. mp = 126-127°; soluble in H_2O (26 mg/l), more solble in organic solvents; LD_{50} (rat orl) = 1250 mg/kg.

21763 Nuba
A proprietary trade name for a thermoplastic coal-tar pitch and a cumarone-indene resin for paints. No manufacturer.

21764 Nubex
Remedial treatments for buildings. *Tenneco Organics Ltd.* Name unverified.

21765 Nubrite
A bright nickel plating process. *Hanshaw Chemicals.* Unverified.

21766 Nubun
A proprietary trade name for a synthetic rubber latex insulation for power and communication cables; it is made from a special modification of buna S synthetic rubber. No manufacturer.

21767 Nucleant
An aluminum alloy grain refiner. *Foseco (F.S.) Ltd.*

21768 Nucol
Wide range of high build, conventional and modified chlorinated rubber and vinyl coatings. *Sigma Coatings.* Name unverified.

21769 Nucrel®
Ethylene-methacrylic acid copolymer. *DuPont UK.*

21770 Nufol
Nitrogenous fertilizers. *ICI Chem & Polymers Ltd.* Discontinued.

21771 Nuglas
Tubes molded with glass reinforcement; for electrical, structural, and thermal applications, jigs, fixtures and tooling, lightweight support structures for antennas, lighting, and surveillance. *Fothergill Tygaflor Ltd.* Name unverified.

21772 Nulomoline
A solution of partly inverted sugar; used for some purposes as a substitute for glycerine.

21773 NuoCide®
1897-45-6 2219 217-588-1
Tetrachloroisophthalonitrile
Fungicide and mildewcide for both water and solvent-based coatings, and as a marine antifouling agent in marine coatings. *Hüls Am.*

21774 Nuocure
Metal salts of organic acids; driers for water-borne coating compositions, catalyst for polyurethane foam. *Hüls Am.*

21775 Nuocure 28
Catalyst for polyurethane foams. *Harcros.*

21776 Nuodex
Metal salts of organic acids, fungicides and bactericides, bodying agents, dispersing agents; driers for coating compositions, curing catalysts, biocides for coating compositions, industrial biocides, thickeners, pigment dispersants. *Hüls Am.*

21777 Nuodex 100
Compositions containing the dodecyldimethylbenzylammonium salt of naphthenic acid; antimicrobial preservatives for textiles, shoe linings and plasticized PVC compositions. *Hüls Am.*

21778 Nuodex 321 Extra
Mercurial biocide. *Harcros.*

21779 Nuodex 84
2492-26-4 219-660-8
$C_7H_6NNaS_2$
2(3H)-benzothiazolethione, sodium salt
duodex; Nacap; vancide 51; Nuodex. 50% aqueous solution of the sodium salt of 2-mercaptobenzothiazole; antimicrobial preservative for fibrous substrates, adhesives and aqueous emulsions. mp = -6°; bp = 103°; SG = 1.25; soluble in H_2O 9>100 mg/ml). *Hüls Am.*

21780 Nuodex 87
Biocidal wall washing compound. *Harcros.*

21781 Nuodex NA
A pigment dispersant; used in vinyl and acrylic lacquer formulations. *Hüls Am.*

21782 Nuolate
Metal salts of tall oil fatty acids; driers for nonaqueous coating compositions. *Hüls Am.*

21783 Nuophene
97-23-4 3120 202-567-1
$C_{13}H_{10}Cl_2O_2$
bis(5-chloro-2-hydroxyphenyl)methane
dichlorophen; Antiphen; Cuniphen; Dicestal; Dichlorophen; G-4 (Compound G4); Hyosan; Parabis; Teniathane; Teniatol; Westpuril; Diphenthane 70; Fungicide Fx; Korium; Panacide; Plath-lyse; Prevental; Preventol; Preventol Gd; Preventol GDC; Taeniatol; Didroxan; Dichlorophen B; Wespuril; Super Mosstox; Gefir; G 4; Fungicide GM; Fungicide M; Dichlorofen; Antifen; Difentan; Teniotol; Gingivit; Cordocel; Halenol; Vermithana; Embephen; Palacel; DDDM; GH;. Fungicide for use on textiles, cordage, and hair felt. mp = 177-178°; insoluble in H_2O, soluble in organic solvents; LD_{50} (rat orl)= 1506 mg/kg. *Hüls Am.*

21784 Nuoplaz
Plasticizers; used for plasticizing PVC and other polymers. *Hüls Am.* Discontinued.

21785 Nuosept
Antimicrobial preservatives; used for preservation of aqueous coating compositions, pigment dispersions, inks, adhesives, caulks, metalworking fluids, paper coatings, drilling muds. *Hüls Am.*

21786 Nuosperse
Dispersing agents; used for coating compositions. *Hüls Am.*

21787 Nuostabe
Stabilizers; for heat and light stabilization of PVC compositions. *Hüls Am.* Discontinued.

21788 Nuostabe 1317
A proprietary trade name for a Ba/Cd/Zn stabilizer for PVC. *Durham Raw Materials.* Name unverified.

21789 Nuostabe 1374
A proprietary trade name for a non sulfide staining stabilizer for PVC used in blown foams. *Durham Raw Materials.* Name unverified.

21790 Nuostabe 1605
A stabilizer for PVC based on a barium/cadmium/zinc complex. *Durham Raw Materials.* Name unverified.

21791 Nuosyn
Metal salts of organic acids; driers for nonaqueous coating compositions. *Hüls Am.*

21792 Nupel®
Rosin emulsion sizes. *Reichhold.* Discontinued.

21793 Nurac
Stated to be diphenyl-guanidine; also been said to be thiocarbanilide; a proprietary rubber vulcanization accelerator. No manufacturer.

21794 Nuram
Nitrogenous fertilizers. *ICI Chem & Polymers Ltd.* Discontinued.

21795 Nuremberg gold
An alloy of 90% copper, 7.5% aluminum, and 2.5% gold. A jeweler's alloy.

21796 Nurolon
Nonabsorbable surgical suture. *Ethicon Inc.* Name unverified.

21797 Nusat
A satin finish nickel plating process. *Engelhard Technologies Ltd.*

21798 Nu-Seals Aspirin
50-78-2 886 200-064-1
Aspirin in an enteric coated form; an analgesic. *Eli Lilly & Co.* Discontinued.

21799 Nu-Seals Sodium Salicylate
54-21-7 8819 200-198-0
Sodium salicylate used as an analgesic and antipyretic. *Eli Lilly & Co.* Discontinued.

21800 Nu-Set
Bristle-setting cement. *Hardman.*

21801 Nusolv ABP-103
69009-90-1 273-683-8
$C_{18}H_{22}$
Bis(methylethyl)-1,1'-biphenyl
bis(1-methylethyl)-1,1'-Biphenyl; Diisopropylbiphenyl. Solvent possessing excellent solvency, chemical and thermal stability, nonvolatility; for aerosols, adhesives, electronic parts manufacturing and cleaning, industrial cleaning, inks and paper coatings, metal cleaning, paints, plastics, sealants, tile manufacturing. *Ridge Tech.*

21802 Nusolv ABP-62
25640-78-2 247-036-5
$C_{15}H_{16}$
(1-methylethyl)-1,1'-biphenyl
Isopropylbiphenyl; 2-Isopropylbiphenyl. Alkyl biphenyl mixture; solvent possessing excellent solvency, chemical and thermal stability, nonvolatility; for aerosols, adhesives, electronic parts manufacturing and cleaning, industrial cleaning, inks and paper coatings, metal cleaning, paints, plastics, sealants, tile manufacturing. *Ridge Tech.*

21803 **Nustar®**
For agriculture. *DuPont UK.*

21804 **nut oil**
A term used in China for tung oil (Chinese wood oil). The same name is also used for walnut and arachis oils.

21805 **Nutimalt® Range**
8002-48-0 232-310-9
Malt extract; taste and volume improver in baked goods. *Grünau.*

21806 **nutmeg butter**
8007-12-3
Expressed oil of nutmeg.

21807 **Nutralys**
Animal feedstuff. *Roquette (UK) Ltd.*

21808 **Nutramigen**
A proprietary artificial lactose-free infant milk food. *Bristol-Myers Squibb Co Inc.*

21809 **Nutramin**
Flour improver; vitamin enrichment additive. *Akzo Chemie UK Ltd.*

21810 **Nutramon**
Aqueous ammonia. *ICI Chem & Polymers Ltd.* Discontinued.

21811 **Nutranel**
Protein, fat, carbohydrate, vitamins, minerals, and trace elements; liquid feed. *Roussel Laboratories Ltd.* Discontinued.

21812 **Nutraphos**
Nutrients for effluent treatment. *Scottish Agricultural Industries plc.*

21813 **Nutrapon AL 1**
67762-19-0
Ammonium lauryl ether (1) sulfate; surfactant for shampoos, bubble baths, and other cosmetic products below pH 7. *Clough.*

21814 **Nutrapon AL 60**
67762-19-0
Ammonium laureth-3 sulfate
Surfactant for shampoo concentrates, bubble baths, dishwashing, and light duty detergents. *Clough.*

21815 **Nutrapon B 1365**
Sodium lauryl sulfate and ethylene glycol stearate; shampoo and detergent blend with pearlizing agent for pearlescent formulations; for shampoos, bubble baths, liquid hand soaps. *Clough.*

21816 **Nutrapon BM 3960**
Sodium laureth-3 sulfate
Surfactant for clear liquid shampoos, bubble baths, other cosmetics. *Clough.*

21817 **Nutrapon DL 3891**
151-21-3 8782 205-788-1
Sodium lauryl sulfate
High foaming surfactant for personal care and industrial formulations. *Clough.*

21818 **Nutrapon ES-60 3568**
9004-82-4
Sodium laureth (3) sulfate
Mild surfactant for personal care and industrial products; flash foam in hard water. *Clough.*

21819 **Nutrapon ESY 2299**
Sodium laureth-1 sulfate
Surfactant for shampoos, bubble baths, cosmetics; high tolerance to hard water. *Clough.*

21820 **Nutrapon FA-50 0066**
Ammonium deceth sulfate. High foaming surfactant for specialty applications, e.g., secondary oil recovery. *Clough.*

21821 **Nutrapon HA 3841**
2235-54-3 218-793-9
Ammonium lauryl sulfate; surfactant for personal care products and detergent cleaners. *Clough.*

21822 **Nutrapon KF 3846**
Sodium laureth-3 sulfate
Buffered sulfate for use in mild shampoos, bubble baths, shower gels; flash foam in hard water. *Clough.*

21823 **Nutrapon KPC 0156**
9004-82-4
Sodium laureth-3 sulfate
Surfactant for shampoos, bubble bath, dishwashing detergents. *Clough.*

21824 **Nutrapon PP 3563**
2235-54-3 218-793-9
Ammonium lauryl sulfate
Used in nonalkaline shampoos, bubble baths, mild detergents and cleaners below pH 7.0. *Clough.*

21825 **Nutrapon RS 1147**
Sodium lauryl sulfate
sodium cocoyl sarcosinate. Surfactant for rug and upholstery shampoo; high foaming, min. wetting, low cloud pt.; leaves dry residue for easy removal. *Clough.*

21826 **Nutrapon TK 3603**
Sodium lauryl sulfate
glycol stearate. Pearlized base for shampoos and liquid hand soaps. *Clough.*

21827 **Nutrapon TLS-500**
139-96-8 205-388-7
TEA-lauryl sulfate
Mild ingredient in cosmetic and industrial formulations. *Clough.*

21828 **Nutrapon TW 3987**
TEA-lauryl sulfate
sodium lauryl sulfate. Surfactant for shampoos, bubble baths, liquid hand soaps. *Clough.*

21829 **Nutrapon W 1367**
151-21-3 8782 205-788-1
Sodium lauryl sulfate
High foaming surfactant for personal care and industrial formulations. *Clough.*

21830 **Nutrapon WAQE 2364**
151-21-3 8782 205-788-1
Sodium lauryl sulfate
High foaming surfactant for personal care and industrial formulations. *Clough.*

21831 **NutraponDE3796**
DEA-lauryl sulfate
Mild bubble bath and shampoo concentrate. *Clough.*

21832 **NutraSweet**
22839-47-0 874 245-261-3
Aspartame; sweetener. *G D Searle & Co.* Name unverified.

21833 **Nutrifos**
7722-88-5 9377 231-767-1
Food grade tetrasodium pyrophosphate; used in meat curing. *Monsanto Co.* Name unverified.

21834 **Nutrilan® FPK, H, M**
9015-54-7 310-296-6
Hydrolyzed animal protein; substantive to skin and hair. *Henkel KGaA/Cospha.* Discontinued.

21835 **Nutrilan® I-50**
9015-54-7 310-296-6
Hydrolyzed animal protein; protein for shampoos, conditioning rinses, hair care products. *Henkel/Cospha; Henkel Canada.*

21836 **Nutrilan® Keratin W**
69430-36-0 274-001-1
Hydrolyzed animal keratin; protective protein for hair care products. *Henkel/Cospha; Henkel Canada.* Name unverified.

21837 **Nutrilan® L**
9015-54-7 310-296-6
Hydrolyzed animal protein; protein for shampoos, conditioning rinses, hair care products, cold waves, bath and shower preps. *Henkel/Cospha; Henkel Canada.*

21838 **Nutrilife® Range**
Enzyme-protein compounds; synergistic effects with emulsifiers for baking processes. *Grünau.*

21839 **Nutrisoft® 55, 100**
Distilled monoglyceride; food emulsifier for baking additives, confectionery; antistaling effect. *Grünau.*

21840 **Nutrol 100**
9002-93-1 6858
Octoxynol-9
Wetting agent, dispersant for metal and acid cleaners, pesticides. *Clough.*

21841 **Nutrol 600**
9016-45-9 6772
Nonoxynol-9
Detergent, emulsifier, wetting agent, dispersant. *Clough.*

21842 **Nutrol 611**
9016-45-9 6772
Nonoxynol-8
Detergent, emulsifier, wetting agent, dispersant. *Clough.*

21843 **Nutrol 622**
9016-45-9 6772
Nonoxynol-4
Emulsifier, light duty detergent, moderate foaming agent. *Clough.*

21844 **Nutrol 640**
9016-45-9 6772

Nonoxynol-15
Emulsifier, light duty detergent, moderate foaming agent. *Clough.*

21845 Nutrol 656
9016-45-9 6772
Nonoxynol-11
Emulsifier, light duty detergent, moderate foaming agent. *Clough.*

21846 Nutrol Betaine MD 3863
Cocamidopropyl betaine
Surfactant, foaming agent, foam stabilizer, wetting agent for shampoos, bubble baths, liquid hand soaps. *Clough.*

21847 Nutrol Betaine OL 3798
Cocamidopropyl betaine
Foamer, foam stabilizer, wetting agent for industrial and household cleaners, dishwashing, liquid hand soaps. *Clough.*

21848 Nutrol S-60 5350
Ammonium nonyl phenoxy polyethoxy sulfate
Emulsifier for use in polymers; high foaming detergent, wetting agent for textiles. *Clough.*

21849 Nutrol SXS 5418
1300-72-7 215-090-9
Sodium xylene sulfonate
Hydrotrope, coupling agent, solubilizer. *Clough.*

21850 Nuvan
62-73-7 3129 200-547-7
Emulsifiable concentrate of 500 g dichlorvos per liter; a fumigant organophosphorus insecticide. *Ciba-Geigy Agrochemicals.*

21851 Nuvan Fly Spray
Dichlorvos/pyrethrin insecticide. *Ciba plc.* Name unverified.

21852 Nuvan Top Aerosol
62-73-7 3129 200-547-7
Dichlorvos, an organophosphorus insecticide. *Ciba plc.* Name unverified.

21853 Nuvanol
62-73-7 3129 200-547-7
Dichlorvos, an organophosphorus insecticide. *Ciba plc.*

21854 Nuvis
Thixotropic bodying agents; for control of sag and flow of nonaqueous coating compositions. *Hüls Am.*

21855 Nuxtra
Metal salts of synthetic organic acids; driers for nonaqueous coating compositions. *Hüls Am.*

21856 NY-10GF
25038-54-4 6832
Nylon 6, 10% glass fiber-reinforced; versatile thermoplastic. *Compounding Tech.*

21857 NY-30CF
25038-54-4 6832
Nylon 6 resin, 30% carbon fiber-reinforced; offers balance of strength to weight for use in aerospace and automotive industries. *Compounding Tech.*

21858 Nyacol®
Colloidal dispersions of silica, antimony pentoxide, alumina and other metal oxides; binders for investment casting molds and fibrous refractories, polishing agent for semiconductor wafers, rigid disks, etc., antislip coating for paper, textile fibers, etc. Flame retardant for plastics (antimony oxide dispersion). *PQ Corp.*

21859 Nyacol® A-1530
1314-60-9 739 215-237-7
Colloidal dispersion of antimony pentoxide in water; flame retardant additive to latex emulsions; durable treatment for fabrics, nonwovens, fiberfill, paper, fiberglass, vinyls; suitable for FR adhesives. *PQ Corp.*

21860 Nyad®
13983-17-0 1749 237-772-5
calcium silicate [1344-95-2]
Wollastonite; Wollastonite Calcium Silicates; Cab-o-lite; Cab-o-lite 100; Cab-o-lite 130; Cab-o-lite 160; Cab-o-lite F 1; Dab-o-lite P 4; F 1; Fw 50; Fw 200 (Mineral); Nyad 10; Nyad 325; Nya G; Nycor 200; Nycor 300; Vansil W 10; Vansil W 20; Vansil W 30; Wollastokup. Reinforcer for polyvinylsiloxane used for dental impressions. *NYCO® Minerals Inc.*

21861 Nyad® Wollastonite
13983-17-0 1749 237-772-5
High aspect ratio and fine particle size calcium metasilicate minerals: CaO, SiO_2, Fe_2O_3, Al_2O_3, MnO, MgO, TiO_2; used for plastics, coatings, refractories, fire resistant board, adhesives, rubber and elastomers and polymer concrete. *NYCO® Minerals Inc.*

21862 Nyala
Acidulant; used for chemically leavened bread. *Albright & Wilson Ltd.*

21863 Nybex 12034 BKFR
25038-54-4 6832
Nylon 6, 10% glass fiber-reinforced; flame retardant thermoplastic *Ferro/Engineering Thermoplastics.*

21864 Nybex 12056 BKFR
25038-54-4 6832
Nylon 6, 45% glass fiber-reinforced; flame retardant thermoplastic *Ferro/Engineering Thermoplastics.*

21865 Nybex 13001 BKC
25038-54-4 6832
Nylon 6, 30% carbon fiber-reinforced; conductive grade. *Ferro/Engineering Thermoplastics.*

21866 Nybex 15011 NA
25038-54-4 6832
Nylon 6, 30% mineral/glass fiber-reinforced; low warpage grade. *Ferro/Engineering Thermoplastics.*

21867 Nybex 17000 NAX
25038-54-4 6832
Nylon 6, 30% glassreinforced; EMI shielding grade. *Ferro/Engineering Thermoplastics.*

21868 Nybex 22008 BKUT
32131-17-2
Nylon 66, 33% glass-reinforced, plasticized; automotive grade. *Ferro/Engineering Thermoplastics.*

21869 Nybex 42002 BKHS
25038-54-4 6832
Nylon 6, 13% glass fiber-reinforced, heat-stabilized; blow molding grade. *Ferro/Engineering Thermoplastics.*

21870 Nybex 52000 NA
32131-17-2
Nylon 66 alloy; moisture-resistant grade. *Ferro/Engineering Thermoplastics.*

21871 Nycoa® 438
25038-54-4 6832
Nylon 6; high flow, fast-setting, injection molding grade resin for very fast cycle operation in difficult-to-fill molds; NSF-approved for use in molded pipe fittings (potable water). *Nylon Corp. of Am.*

21872 Nycoa® 446
25038-54-4 6832
Nucleated nylon 6; high viscosity nucleated nylon for extrusion of tubing, rods, sheet, and film with fast set-up time. *Nylon Corp. of Am.*

21873 Nycoa® 500
32131-17-2
Nylon 6/6; standard grade with good moldability, toughness, and chemical resistance; for molding and extrusion. *Nylon Corp. of Am.*

21874 Nycoa® 528
32131-17-2
Nylon 6/6, nucleated; nucleated grade with controlled crystallinity, excellent moldability and cycle, increased stiffness and abrasion resistance. *Nylon Corp. of Am.*

21875 Nycoa® 567
25038-54-4 6832
Plasticized nylon 6; high impact strength extrusion and molding resin, flexibility and toughness; used in injection molded shoe heel lifts, extruded softhand fishing line and monofilament. *Nylon Corp. of Am.*

21876 Nycoa® 714
25038-54-4 6832
Plasticized nylon 6 copolymer; softness and flexibility for extrusion and molding for specialty products, e.g., molded athletic shoe soles, fishing line, etc. *Nylon Corp. of Am.*

21877 Nycoa® 870
25038-54-4 6832
Heat-stabilized nylon 6; extrusion resin for jacketing (wire, cable) including thermoplastic-insulated building wires and gasoline-resistant types. *Nylon Corp. of Am.*

21878 Nycoa® 1417
25038-54-4 6832
Impact-modified nylon 6 resin; medium high impact resin for applications where notch sensitivity of standard nylons produces inconsistent part performance. *Nylon Corp. of Am.*

21879 Nycoa® 4015
25038-54-4 6832
Nylon 6, 15% glass-rein-forced; reinforced grade with excellent heat distortion, high rigidity and strength. *Nylon Corp. of Am.*

21880 Nycoa® 5015
32131-17-2
Nylon 6/6, 15% glassreinforced; reinforced grade with excellent heat distortion, high rigidity and strength. *Nylon Corp. of Am.*

21881 Nycoat®
Family of chemically coupled minerals; e.g. alumina trihydrate, barytes and celestite (varies depending on substrate); for specialty applications: plastics,

high performance coatings, adhesives, rubber and elastomers. *NYCO® Minerals Inc.*

21882 Nycor® Barytes
200 + 235 mesh barium sulfate
Barium sulfate, silica, ferric oxide, manganese and lead; for coatings, rubber and elastomers, friction products, refractories and sound deadening compounds. *NYCO® Minerals Inc.* Discontinued.

21883 Nycor® Celestite
200 + 325 mesh barium sulfate
Strontium sulfate, barium sulfate, calcium carbonate and alumina; for coatings, rubber and elastomers, friction products, refractories and sound deadening compounds. *NYCO® Minerals Inc.* Discontinued.

21884 Nycor® R
13983-17-0 1749 237-772-5
High aspect ratio wollastonite, untreated; for used in coatings, refractory, friction, construction board and plastics. *NYCO® Minerals Inc.*

21885 Nydur
Polyamide resins (Nylon 6). impact modified with low temperature impact, as well as abrasion, heat and chemical resistance. *Miles.* Discontinued.

21886 Nye Tact 520
6-Ring polyphenylether in trichlorotrifluoroethane; electrical connector lubricant. *Wm F Nye.*

21887 Nyebar
A solution of a low surface energy fluorocarbon polymer in a fluorinated solvent; provides a nonwettable surface which controls or prevents the migration or creep of lubricants or other fluids. *Wm F Nye.* Name unverified.

21888 Nyflake®
12001-26-2
Muscovite mica; coarse and fine grinds: SiO_2, Al_2O_3, K_2O, Fe_2O_3, Na_2O_3, CaO, TiO_2, MnO_2, P, S; for plastics, coatings, oil well drilling mud, adhesives. *NYCO® Minerals Inc.* Discontinued.

21889 Nyglas®
Chemically coupled ground glass; soda lime glass, platey, various particle sizes: SiO_2, Na_2O, CaO, MgO, Al_2O_3, K_2O, Fe_2O_3; for plastics and coatings. *NYCO® Minerals Inc.* Discontinued.

21890 Nykon
Corrosion resistant rheological additive; used for greases. *Rheox Inc.*

21891 Ny-Kon® I
Nylon 6/12, <5% MoS_2 lubricant. *LNP.*

21892 Ny-Kon® P
25038-54-4 6832
Nylon 6, <5% MoS_2 lubricant. *LNP.*

21893 Ny-Kon® Q
9008-66-6
Nylon 6/10, <5% MoS_2. *LNP.*

21894 Ny-Kon® R
32131-17-2
Nylon 6/6, < 5% MoS_2. *LNP.*

21895 Ny-Kon® V
High-impact nylon, <5% MoS_2. *LNP.*

21896 Nylander's reagent
An alkaline solution of bismuth subnitrate and Rochelle salt obtained by dissolving 40 g Rochelle salt and 20 g bismuth subnitrate in 1,000 ml of 8% caustic soda; used for the detection of glucose in urine by boiling 5 parts of the glucose solution with 1 part of the reagent when reduction occurs and a black precipitate is produced.

21897 Nylatron®
Nylon and molybdenum disulfide; molding compounds for bearing and wear applications (bearings, bushings, wear pads, etc.) *Polymer Corp.*

21898 Nylatron® 1018 HS
32131-17-2
Nylon 6/6 resin, 33% fiberglass-reinforced, heat-stabilized; injection molding resin. *DSM.* Discontinued.

21899 Nylatron® 1024 HS
32131-17-2
Nylon 6/6 resin, heat-stabilized; general purpose, heat-stabilized injection molding resin with internal lubricant. *DSM.* Discontinued.

21900 Nylatron® GS-63
25038-54-4 6832
Nylon 6 resin with molybdenum disulfide lubricant, uv-stabilized; injection molding resin containing carbon black for uv stabilization. *DSM.* Discontinued.

21901 Nylatron® NSB-90
32131-17-2
Nylon 6/6 resin; injection molding resin specially formulated for use in bearing and wear applications. *DSM.* Discontinued.

21902 Nylatrrn® GS
Nylon, MoS_2-lubricated; wear resistance, low surface friction, and high strength and rigidity; used in bearings, valve seats, thrust washers, wear surfaces, rollers, gears, forming dies, tooling fixtures, etc.; available in rod, disc, strip, etc. *DSM.* Discontinued.

21903 Nylocrom
Dry powder acid dyes; specially developed acid dyes used in the textile industry for dyeing nylon. *Multicrom SA.*

21904 Nylofixan
Aromatic sulfonate, black-tanning agent; for polyamides. *Sandoz Products Ltd.* Name unverified.

21905 Nyloflex
Flexographic printing plates. *BASF plc.*

21906 Nylok® 170
Amino-functional calcined clay; filler for reinforcement of polyamide resin systems. *J.M. Huber/Clay Div.*

21907 Nylomine
Dyestuffs for nylon and polyamide fibers. *ICI Chem & Polymers Ltd.*

21908 nylon
63428-83-1 6831
A family of polyamide polymers characterized by presence of amide group CONH; thermoplastic resin for tire cord, hosiery, wearing apparel, brush bristles, cordage, fish lines, tennis rackets, rugs, artificial turf, parachutes, composites, sails, film, gears/bearings, insulation, surgical sutures, metal coating, fuel tanks. Ashley Polymers; Bamberger Polymers; BASF; DSM; EMS-Am. Grilon; Hoechst Celanese; Hüls Am.; ICI GmbH; LNP; Miles; Monsanto.

21909 nylon 6
25038-54-4 6832
$(C_6H_{13}NO_2)_n$
poly[imino(1-oxo-1,6-hexanediyl)]
poly(iminocarbonylpentamethylene) Polyamide; Nylon 6; Perlon; Poly[imino(1-oxo-1,6-hexanediyl)]; Poly(caprolactam); Poly-epsilon-caprolactam; Policapram. Used in tire cord; fishing lines; tow ropes; hose manufacture; woven fabrics. *Snia UK.*

21910 nylon 6/6
32131-17-2
$[NH(CH_2)_6NHCO(CH_2)_4CO]_n$
poly[imino(1,6-dioxo-1,6-hexanediyl) imino-1,6-hexanediyl
poly(hexamethyleneadipamide) Polyamide. Polymeric amide formed by the reaction of adipic acid with hexylenediamine. Thermoplastic resin for injection molding, extrusion. *Asahi Chem Industry Co Ltd; Snia Uk.*

21911 nylon 6/10
9008-66-6
Electrafil® J-2/CF/30; NI-20GF; Ny-Kon® Q; RTP 201B; Texalon 1600A Nat; Thermocomp® QF-1006FR; Ultramid® S3,. Nylon 6/10, 30% PAN carbon fiber-reinforced; static dissipative and conductive thermoplastic *Akzo Engineering Plastics.* Name unverified.

21912 nylon 12
25038-74-8
azacyclotridecane-2-one polyamide
Poly(laurolactam; Polyamide derived from 12-aminododecanolc acid; thermoplastic resin for injection molding and extrusion applications. *Elf Atochem SA; Daicel-Hüls AG.*

21913 Nylon N-012
32131-17-2
Nylon 6/6 and methicone; imparts soft, lubricious feel to cosmetic formulations, anhydrous systems, emulsions, and powders. *Presperse.* Discontinued.

21914 Nylon Resist NCO
Nylon dye resist concentrate for level and even dyeing; retards dyeing rate. *Eastern Color & Chem.*

21915 Nyloprint
Flexible printing plates. *BASF plc.*

21916 Nylosan®
Synthetic organic acid dyestuffs; specialty dye for aqueous mediums, textiles, fertilizers. *Sandoz; Sandoz Products Ltd.*

21917 Nyloset Finish
Amide resin; nylon builder and softener. *Scher Chemicals Inc.* Discontinued.

21918 Nylosolv
Environmentally friendly solvent for printing plates. *BASF plc.*

21919 Nylox
Adhesive composition for nylon, etc. *Hardman.*

21920 Nyogel
An inorganically gelled series of greases based upon synthetic lubricating oils; for specialty lubrication of delicate machinery and engineered components. *Wm F Nye.* Name unverified.

21921 NyoGel® 744
Fluorocarbon-filled low friction grease for cams and sliding parts. *Wm F Nye.*

21922 NyoSil
Halogenated silicone oil; potentiometer lubricant for wide variety of wiper/substrate combinations. *Wm F Nye.*

21923 Nypel
Reinforced and unreinforced recycled nylon 6 resins utilizing select feedstocks to combine quality, performance and processability at a lower price. *AlliedSignal Inc.*

21924 Nypene
A proprietary trade name for a polyterpene hydrocarbon resin used in adhesives, paints, and varnishes. No manufacturer.

21925 Nyrim
Reaction injection molding nylon, prepared by in-mold polymerization of caprolactam (feedstock for nylon) to a nylon-6 block polymer with utilization of a catalyst 2.0 agents; used for industrial, agriculture, and automotive applications. *DSM NV.* Discontinued.

21926 Nyspheres®
Fine particle hollow glass spheres, light weight: silica, alumina, iron oxides, calcium, magnesium, and alkalis; for plastics, coatings, adhesives, cement products, refractories. *NYCO® Minerals Inc.* Discontinued.

21927 NYsyn® 30-5
Acrylonitrile-butadiene copolymer, cold polymerized; for applications requiring easy processing and excellent low temp. flexibility, e.g., industrial and automotive hose and seals, cable jackets, molded and extruded mechanical goods. *Copolymer Rubber.*

21928 NYsyn® 33-5HM
Acrylonitrile-butadiene copolymer, cold polymerized; nonblooming; general purpose polymer with good oil resist; used for molded products; compounds exhibit high modulus, tensile strength and tear; fast cure. *Copolymer Rubber.*

21929 NYsyn® 305V
NBR/PVC (70:30); recommended for hose, shoe soles, wire and cable compounds, close cell sponge, colored products. *Copolymer Rubber.*

21930 NYsynblak® 9010
Acrylonitrile-butadiene copolymer black masterbatch, cold polymerized (NYsyn 35-8 with 50 parts N-550 carbon black); used for extruded and molded goods requiring medium-high solvent and oil resistant. *Copolymer Rubber.*

21931 NYsynblak® DN 120
Acrylonitrile-butadiene copolymer black masterbatch, cold polymerized (NYsyn 33-3 with 50 parts N-234 carbon black); used for extruded and molded goods requiring medium-high solvent and oil resistance; also as black concentrate impact modifier for ABS. *Copolymer Rubber.*

21932 Nytal® 100
14807-96-6 9207 238-877-9
Hydrous magnesium silicate (talc); reinforcing filler for PVC, vinyl asbestos tile, polyester in match molded articles, body patching compounds, PP, nylon, phenol formaldehyde, polyethylene, ceramic wall tile and artware; improves stiffness; auxiliary flux in vitreous ceramic bodies. *R. T. Vanderbilt Co Inc.* Discontinued.

21933 NZ 01
Neoalkoxy trisneodecanoyl zirconate, IPA; coupling agents which sometimes also act as adhesion promoters, antioxidants, antistats, antifoaming agents, accelerators, blowing agent activators, catalysts, curatives, corrosion inhibitors, dispersion aids, emulsifiers, flame retardants, foamers, grinding aids, hardeners, internals lubes, metal primers, pigment intensifiers, release agents, peroxide activators, retarders, stabilizers, surfactants, suspension aids, and thixotropes. *Kenrich Petrochemicals.*

21934 NZ 09
Neoalkoxy tris (dodecyl) benzene sulfonyl zirconate, IPA
Kenrich Petrochemicals.

21935 NZ 12
Neoalkoxy tris (dioctyl) phosphato zirconate, IPA
Kenrich Petrochemicals.

21936 NZ 33
Neopentyl diallyloxy, trimethacryl zirconate, IPA
Coupling agent. *Kenrich Petrochemicals.*

21937 NZ 38
Neoalkoxy tris (dioctyl) pyrophosphato zirconate, IPA
Kenrich Petrochemicals.

21938 NZ 39
Neopentyl (diallyl) oxy, tri(9,10 epoxy stearoyl) zirconate
Coupling agent. *Kenrich Petrochemicals.*

21939 NZ 44
Neoalkoxy tris (ethylene diamino) ethyl zirconate
Kenrich Petrochemicals.

21940 NZ 89
Neopentyl (diallyl) oxy, trimercapto-phenyl zirconate
Coupling agent. *Kenrich Petrochemicals.*

21941 NZ 90
Neopentyl (diallyl) oxy, tri(dodecyl)benzene-sulfonyl zirconate, IPA
Coupling agent. *Kenrich Petrochemicals.* Name unverified.

21942 NZ 97
Neoalkoxy tris (m-amino) phenyl zirconate, phenyl glycol ether *Kenrich Petrochemicals.*

21943 O.N.V
Diphenylcarbamyl dimethyldithiocarbamate
A proprietary trade name for a rubber vulcanization accelerator. No manufacturer.

21944 O.O.D
22801-45-2 245-228-3
2-Octyldodecyl oleate
Neutral oil improving adhesion, gloss, conditioning in lipsticks, foundation, eye shadow stick, hair creams. *U.S. Cosmetics.*

21945 O9810
556-67-2 6843 209-136-7
$C_8H_{24}O_4Si_4$
Octamethylcyclotetrasiloxane
octamethyltetrasiloxane. As cleaning, polishing and damping media; offers low toxicity, inertness. Used to deactivate glass chromatography columns. mp = 17°; bp = 175-176°; d = 0.9558; n_D^{20}= 1.3960; *Hüls Am.*

21946 O9816
107-51-7 6844 203-497-4
$C_8H_{24}O_2Si_3$
Octamethyltrisiloxane
As cleaning, polishing and damping media; offers low toxicity, inertness. Also used as a foam suppressant. mp= -82°; bp = 153°; d = 0.8200; n_D^{20}= 1.3848. *Hüls Am.*

21947 OA 40-30
Chlorinated paraffin
For industrial applications. *Witco/Argus.* Discontinued.

21948 OA-100A
Hydrocarbon-derived; pour point depressant exhibiting exceptional shear stability; for formulating motor oils, hydraulic oils, paraffinic or paraffinic/naphthenic-based electrical insulating oils. *Witco/Argus.* Discontinued.

21949 OA-154
Alkanolamide
Rust inhibitor, emulsifier for coolant or water-based products. *Witco/Argus.* Discontinued.

21950 OA-252
Sulfurized fatty acid
Additive for industrial oils and greases; used for formulating coolant or water-based systems. *Witco/Argus.* Discontinued.

21951 OA-270
Sulfurized methyl ester
Additive for industrial oils and greases. *Witco/Argus.* Discontinued.

21952 OA-300
Sulfurized sperm oil replacement; additive for industrial oils and greases. *Witco/Argus.* Discontinued.

21953 OA-502
Nonylated diphenylamine
Antioxidant for industrial and automotive oil, grease, and fluid; stabilizer in hydraulic brake fluids. *Witco/Argus.* Discontinued.

21954 OA-505
122-39-4 3375 204-539-4
Diphenylamine
High-temperature oxidant for formulating high-temperature greases, automotive, diesel, and turbine lubricants; synergistic with other antioxidants and additives, enhancing thermal and oxidation stability. *Witco/Argus.* Discontinued.

21955 OA-700
Chlorinated ester; lubricant imparting exceptional metal wetting characteristics to industrial oils. *Witco/Argus.* Discontinued.

21956 OA-770
Sulfur-chlorinated additive; additive for industrial oils and greases. *Witco/Argus.* Discontinued.

21957 OA-951
Chlorinated fatty acid
For formulating coolant or water-based systems. *Witco/Argus.* Discontinued.

21958 Oak Draw 720, 728, 830A
A line of soluble oil, semisynthetic and synthetic metal forming lubricants; for drawing, stamping, bending, forming, piercing. *Oak International Inc.*

21959 Oak Kool 625, 632A, 648
A line of soluble oil, semi-synthetic and synthetic machining and grinding coolants; for CNC machining, screw machining, turning. *Oak International Inc.*

21960 oak moss resin
A resin obtained from lichens; used as a fixative in perfumery.

21961 Oak Oils
Petroleum oils; for drawing, stamping, grinding, and machining. *Oak International Inc.* Discontinued.

21962 oak red
$C_{28}H_{22}O_{11}$
A coloring matter, phlobaphene, obtained by the hydrolysis of quercitannic acid.

21963 Oak Syncrolube
Synthetic (oil free) lubricant; for stamping and machining. *Oak International Inc.* Discontinued.

21964 Oakite Defoamant
Oil-water emulsion containing organic esters, alcohols, silicone, hydrocarbons, and stabilizers; foam control agent for industrial applications, e.g., paper mill stock systems, gas dehydration units, amine scrubbing units, propane deasphalting units. *Oakite Prods.*

21965 Oakite Ladd
Blend of organic defoamer, detergent, solvents; low temperature cleaning/foam controlling additive for acidic or alkaline detergents, spray washing. *Oakite Prods.*

21966 Obanol 516
Polyoctyl polyamino ethyl glycine and POE alkylphenol ether; germicide, disinfectant, deodorant, fungicidal cleaning aid. *Toho Chem. Industry.*

21967 Obazoline 662Y
Imidazoline derivative; antistat and softener for synthetic fibers; base material for shampoos and hair rinse. *Toho Chem. Industry.*

21968 Obermayer's reagent
A solution of ferric chloride in concentrated hydrochloric acid (4 g ferric chloride dissolved in 1 liter of concentrated hydrochloric acid); used for the detection of indoxyl in urine, indigo being formed if this substance is present.

21969 obsidene
A plastic residue from the distillation of petroleum.

21970 OBTS
102-77-2 203-052-4
$C_{11}H_{12}N_2OS_2$
N-Oxydiethylene-2-benzothiazole sulfenamide
N-oxydiethylenebenzothiazole-2-sulfenamide; Amax; 2-Benzothiazolyl-N-morpholinosulfide; Morpholine, 4-(2-benzothiazolylthio)-; 2-(morpholinothio)benzothiazole; 2-(Morpholinthio)-benzothiazole; N-oxydiethylene-benzothiazole sulfenamide. Primary accelerator for natural, SBR, nitrite, and other general-purpose rubbers. *Akrochem.*

21971 Obturin
Soluble fluoresceine.

21972 Occidine
Contains copper sulfate, iron sulfate, sulfur, naphthalene, and calcium carbonate; used as a fungicide.

21973 Occlusin
Light cured dental filling composite. *ICI Chem & Polymers Ltd.* Discontinued.

21974 Occultest
A proprietary test tablet of o-toluidine, strontium peroxide, calcium acetate, tartaric acid, and sodium bicarbonate; used for the detection of blood in urine. *BC Ames.* Name unverified.

21975 Ocenol
The mixture of fatty alcohols derived from sperm oil; also a proprietary trade name for technical oleic acid. No manufacturer.

21976 ocher
1309-37-1 4072 215-168-2
Fe_2O_3
Yellow ocher
oxide yellow; Chinese yellow. A natural pigment consisting of hydrated oxides of iron and manganese, mixed with clay and sand. The term is frequently restricted to a pale, yellowish-brown variety.

21977 ochermatite
$3(3Pb(AsO_4)_2 \cdot PbCl_2)4Pb_2MoO_5$
A mineral.

21978 ochran
A yellow bole (or clay-earth).

21979 ocota cocota gum
Muccocota gum. Names applied in West Africa to varieties of copal.

21980 OctaBoost® 620
Alumina controlled zeolite; catalyst providing superior octane performance and higher gasoline yields in refining processes. *Akzo.*

21981 Octabromodiphenyl oxide
32536-52-0 251-087-9
$C_{12}H_2Br_8O$
OCTA; OBDPO; FR-1208; Great Lakes DE-79™; Saytex®111. Flame retardant for thermoplastics, e.g., ABS, HIPS, LDPE, PP random copolymer; recommended for injection moldings. White powder to off-white powder; sg = 2.9; mp = 125-165; LD_{50} (rat orl) > 5000 mg/kg. *AmeriHaas; Lbemarle; Allchem Ind.; Dead Sea Bromine; Great Lakes.*

21982 octadecenylN-9-octadecenyl hexadecanamide
oleyl palmitamide

21983 octadecyl mercaptopropionate
31778-15-1 250-801-6
Evangard® 18MP; Octadecyl 3-mercaptopropionate. *Evans Chemetics.*

21984 octadecyl methacrylate
32360-05-7 251-013-5
$C_{22}H_{42}O_2$
2-Methyl-2-propenoic acid, octadecyl ester
Octadecyl methacrylate; Stearyl methacrylate. Lube oil additive, pour point depressant. Used for paper caotings, textile finishes, paints, varnishes and pressure-sensitive adhesives. bp_5 = 181°. *CPS; Rohm & Haas, Sartomer.*

21985 octadecylbenzenemethanaminium chloride
122-19-0 204-527-9
$C_{27}H_{50}ClN$
stearyl dimethylbenzylammonium chloride
stearalkonium chloride; benzyldimethyloctadecylammonium chloride; benzyldimethylstearylammonium
chloride; benzylstearyldimethylammonium chloride; dimethylbenzyloctadecylammonium chloride; tallow benzyl dimethyl ammonium chloride; ammonyx 4; ammonyx 485; ammonyx 490; ammonyx 4002; ammonyx ca special; 2B; barquat sb-25; carsoquat sdq-25; carsoquat sdq-85; intexan sb-85; intexsan sb-85; j soft c 4; katamine ab; orthosan mb; quaternol 1; stebac; stedbac; triton x-40; triton x-400; varisoft sdc; arquad dm18b-90; dehyquart stc-25; nissan cation s2-100; Dimethyl-n-octadecylbenzenemethanaminium chloride. Surfactant usd in cosmetics and antimicrobials. Insoluble in H_2O (< 1 mg/ml); LD_{50} (rat orl)= 1250 mg/kg. *Ferrosan Fine Chem.; A/S; Lonza; Mason; McIntyre; Sherex.*

21986 N-octadecyl-13-docosenamide
10094-45-8 233-226-5
$C_{40}H_{79}NO$
stearyl erucamide
Surfactant. *Croda Universal; Witco/Humko; Zeeland.*

21987 Octamine® Flake, Powd
Octylated diphenylamine
A solid amine-type antioxidant, gives minimum discoloration with maximum protection; effective in natural, SBR, BR, neoprene, and nitrile rubbers. *Uniroyal.*

21988 1-octanol
111-87-5 6849 203-917-6
$C_8H_{18}O$
n-octyl alcohol
capryl alcohol; caprylic alcohol. Used as a basis for silicone oils, also as a foam suppressant. bp = 193-195°; d = 0.825; n_D^{20}= 1.4290; insoluble in H_2O, soluble in organic solvents;.

21989 octaphen
78-05-7 7389 201-078-0
$C_{27}H_{42}ClNO$
N,N-diethyl-N-[2-[4-(1,1,3,3-tetramethylbutyl)phenoxy]ethyl]benzenemethanaminium chloride
phenoctide. Used as a topical anti-infective and as a source of its orthophosphate, a lubricant. mp = 112-114°.

21990 Octave
67747-09-5 7941 266-994-5
$C_{15}H_{16}Cl_3N_3O_2$
N-propyl-N-[2-(2,4,6-trichlorophenoxy)ethyl]-1H-imidazole-1-carboxamide
Wettable powder containing 50% w/w prochloraz; a broad-spectrum fungicide for cereal crops. *Fisons plc.*

21991 octene-1
111-66-0 1807 203-893-7
C_8H_{16}
oct-1-ene
Caprylene; octylene. Intermediate for surfactants and specialty industrial chemicals. mp = -102°; bp = 121°. d_4^{20}= 0.7149; n_D^{20}= 1.4087; insoluble in H_2O, soluble in organic solvents. *Air Prods & Chem; Aldrich; Chevron; Ethyl; Shell; Texaco.*

21992 Octoate Z
136-53-8 205-251-1
$C_{16}H_{30}O_4Zn$

Zinc di-2-ethylhexoate
Ethylhexanoic acid zinc salt; Zinc 2-ethylhexanoate. Rubber activator used in soluble cure systems in place of stearic acid and partial replacement of zinc oxide for natural and synthetic rubbers. *R. T. Vanderbilt Co Inc.*

21993 Octocure 456
7704-34-9 9142 231-722-6
S
An aqueous dispersion of sulfur; rubber accelerator; also for vulcanization processes in aqueous latex compounds. *Tiarco.*

21994 Octocure 553
1314-13-2 10279 215-222-5
zinc oxide
Aqueous dispersion of French process zinc oxide; pigment and accelerator for vulcanization of rubber; well-suited for use in gelled latex compounds. *Tiarco.*

21995 Octocure ZDB-50
136-23-2 205-232-8
$C_{18}H_{36}N_2S_4Zn$
Zinc dibutyldithiocarbamate
Zinc N,N-dibutyldithiocarbamate; Butyl zimate; Di-n-butyldithiocarbamic Acid Zinc Salt; Butasan; Butazate; Butazin; Nocceler BZ; Soxinol BZ; ZBC; Zinc, bis(dibutylcarbamodithioato-S,S')-, (T-4)-; Zinc bis(dibutyldithiocarbamate;)Butasan; Butazate; Butazin; Butyl Zimate. Latex and rubber accelerator. An activator, antidegradant, and accelerator for natural rubber, butadiene, styrene-butadiene, nitrile-butadiene, butyl rubber, and ethylene-propylenediene terpolymers. *Tiarco.*

21996 Octocure ZDE-50
14324-55-1 238-270-9
$C_{10}H_{20}N_2S_4Zn$
Zinc diethyldithiocarbamate
Ethyl Ziram; bis(diethylcarbamodithioato-S,S') Zinc; Zinc diethyldithiocarbamate; Ethyl zimate; Zinc, bis(diethylcarbamodithioato-S,S')-, (T-4)-; Bis(diethyldithiocarbamate)zinc complex; Ethazate; Ethyl cymate; Ethyl Ziram; Ethylzimate; Hermat ZDK; Vulcacure; Vulcacure ZE; Vulkacit LDA; Vulkacit ZDK; Zimate, ethyl; Zinc bis(diethyldithiocarbamate); Zinc diethylcarbamodithioate; Zinc diethyldithiocarbamate; Zinc-N,N-diethyldithiocarbamate; Zinc, bis(diethyldithiocarbamato)-; Zinc, tetrakis(diethylcarbamodithioato)di. Latex and rubber accelerator. An accelerator and activator for natural rubber, styrene-butadiene, nitrile-butadiene and butyl rubber *Tiarco.*

21997 Octocure ZDM-50
137-30-4 10305 205-288-3
Zinc dimethyldithiocarbamate
Latex and rubber accelerator. *Tiarco.*

21998 Octocure ZMBT-50
155-04-4 205-840-3
$C_{14}H_8N_2S_4Zn$
Zinc mercaptobenzothiazole
Zetax; 2(3H)-Benzothiazolethione, zinc salt; Mercaptobenzothiazole, zinc salt; Benzothiazolethiol, zinc salt; Benzothiazolethione, zinc salt; Bis(mercaptobenzothiazolato)zinc; Hermat ZN-MBT; OXAF; Pennac ZT; Tisperse MB-58; Vulkacit ZM; Zenite; Zinc 2-mercaptobenzothiazolate; Zinc benzothiazolethiolate; Zinc benzothiazolylmercaptide; Zinc bis(mercaptobenzothiazole;)Zinc mercaptobenzothiazole; ZMBT; ZNMB. Latex and rubber accelerator. *Tiarco.*

21999 Octoguard FR-01
1163-19-5 214-604-9
$C_{12}Br_{10}O$
decabromodiphenyl ether
dentabromodiphenyl ether;1,1'-oxybis(2,3,4,5,6-pentabromobenzene); bis(pentabromophenyl) ether; DPBPO; FR 300BA; FRP 53; Berkflam B 10E; bromkal 82-ODE; BR 55N; FR 300; DE 83R; bromkal 83-10DE; pentabromophenyl ether; saytex 102; saytex 102E; tardex 100; DBDPO. Decabromodiphenyloxide aq. disp.; flame retardant for water-based polymer compounds such as latex adhesives, binders, coatings, and foams. mp > 300°; bp = 425°, insoluble in H_2O (< 1 mg/l). *Tiarco.*

22000 Octoguard FR-10
1309-64-4 752 215-175-0
O_6Sb_4
Antimony (III) oxide
Antimony trioxide; Antimony white; biantimony trioxide; flowers of antimony; Antimonius Oxide; Antimony Peroxide; Antimony Sesquioxide; Antimony oxide; diantimony trioxide; senarmontite; exitelite; weisspiessglanz; A1530; A1582; a1588 lp; AP 50; chemetron fire shield; ci 77052; ci pigment white 11; dechlorane a-o; nyacol a 1530; thermoguard b; thermoguard s; timonox. Flame retardant for water-based polymer compounds such as latex adhesives, binders, coatings, and foams. mp = 655°; bp = 1425°; d = 5.2000; slightly soluble in H_2O, soluble in acids; LD_{50} (rat orl) > 20 g/kg. *Tiarco.*

22001 Octoguard FR-15
Antimony trioxide [1309-64-4] and decabromodiphenyloxide [1163-19-5] (1:5 ratio) in an aqueous dispersion; flame retardant for water-based polymer compounds such as latex adhesives, binders, coatings, and foams. *Tiarco.*

22002 Octoil
117-81-7 1291 204-211-0
Dioctyl phthalate
A proprietary trade name for a plasticizer. No manufacturer.

22003 Octoil S
122-62-3 1292 204-558-8
Dioctyl sebacate
A proprietary trade name for a plasticizer. No manufacturer.

22004 Octojet 104
7440-44-0 1855 233-153-3
C
carbon black
Carbon Lampblack; Acetylene black; Animal bone charcoal; Lampblack; Fullerene tubes; Carbon soot. A dispersion of carbon black; medium jetness carbon black. *Tiarco.*

22005 Octolite 544
95-54-5 7438 202-430-6
Phenylenediamine
Antioxidant and metal inhibitor. *Tiarco.*

22006 Octolite 561
$C_{12}H_{12}O_2$
Aqueous dispersion of bisphenol; antioxidant suitable for all white or light-colored latex compounds where color is critical; also for rubber intended for repeated or continuous contact with food. *Tiarco.*

22007 Octolite AO-28
Polymeric hindered phenol/thioester; antioxidant. *Tiarco.*

22008 Octomer DBM
105-76-0 203-328-4
$C_{12}H_{20}O_4$
Dibutyl maleate
Dibutyl Maleate; 2-Butenedioic acid (Z)-, dibutyl ester. Plasticizer, intermediate. bp_4 = 129°; sg = 0.993; *Tiarco.*

22009 Octomer DIBM
14234-82-3 238-102-4
Diisobutyl maleate
Plasticizer, intermediate. *Tiarco.*

22010 Octomer DIOM
1330-76-3 215-547-2
$C_{20}H_{36}O_4$
Diisooctyl maleate
Plasticizer, intermediate. *Tiarco.*

22011 Octomer DOM
2915-53-9 220-835-6
$C_{20}H_{36}O_4$
2-Butenedioic acid (Z)-, dioctyl ester
Dioctyl maleate. Plasticizer, intermediate. *Tiarco.*

22012 Octopol NB-47
136-30-1 205-238-0
$C_9H_{18}NNaS_2$
Sodium dibutyldithiocarbamate
dibutylcarbamodithioic acid, sodium salt. Ultra accelerator for SBR and natural rubber latex compounds; in polymerization of chloroprene rubber, especially for latex compounds where copper staining of zinc salt dithiocarbamates is a problem. *Tiarco.*

22013 Octopol SDE-25
148-18-5 3443 205-710-6
$C_5H_{10}NNaS_2 \cdot 2H_2O$
Sodium diethyldithiocarbamate
ditiocarb sodium; sodium diethyldithiocarbamate; SDDC; Sodium N,N-Diethyldithiocarbamate; DTC; Imuthiol; Diethylcarbamodithioic acid, sodium salt; DEDC; DDTC; DeDTC; Dithiocarb; Diethyldithiocarbamic acid sodium salt; sodium salt of N,N-diethyldithiocarbamic acid; dithiocarbamate sodium; DEDK; sodium dedt; cupral; Sodium N,N-diethyldithiocarbamate trihydrate. Natural rubber latex preservative; precipitant for heavy metals in waste water treatment. mp = 94-102°; soluble in H_2O, polar organic solvents; λ_m = 257, 290 nm (ε 1200, 13000 EtOH); LD_{50} (rat orl) = 2830 mg/kg. *Tiarco.*

22014 Octopol SDM-40
128-04-1 204-876-7
$C_3H_6NNaS_2$
Sodium dimethyldithiocarbamate
Sodium dimethyldithiocarbamate; Aceto SDD 40; Alcobam NM; brogdex 555; carbon s; Dibam; Dibam A; DMDK; methyl namate; Sharstop 204; sodium N,N-dimethyldithiocarbamate; Stafresh 615; Steriseal 40; Thiostop n; Vinstop; Vulnopol nm; Wing Stop B; SDMDTC; sodium

dimethylcarbamodithioate; Dimethyldithiocarbamic acid sodium salt; Freshgard 40. Polymerization shortstop in SBR rubber; precipitant for heavy metals in waste water treatment. LD_{50} (rat orl) = 1000 mg/kg. *Tiarco.*

22015 Octoran
Low foaming nonionic surfactant of the alkylphenol ethoxylate type in liquid form; wetting, detergency, emulsification and antistatic agent. *Roehm Ltd.* Name unverified.

22016 Octorez
Chemical modified rosins. *Tenneco Malros Ltd.* Name unverified.

22017 Octosol
Modified rosin soap. *Tenneco Malros Ltd.* Name unverified.

22018 Octosol 449
143-18-0 7818 205-590-5
$C_{18}H_{33}KO_2$
potassium oleate
oleic acid potassium salt; trenamine d 200; trenamine d 201; FR 14; nonsoul ok 1; (Z)-9-octadecanoic acid, potassium salt; Octadecenoic acid (Z)-, potassium salt; potassium salt of oleic fatty acid. Foaming agent, stabilizer, emulsifier, dispersant; primary frothing aid in gelled latex foam compounds. Soluble in H_2O (> 100 mg/ml), EtOH. *Tiarco.*

22019 Octosol 474
112-03-8 203-929-1
$C_{21}H_{46}ClN$
Octadecyl trimethyl ammonium chloride
Emulsifier for cationic or cationic/anionic emulsion systems. *Tiarco.*

22020 Octosol 562
112-00-5 203-927-0
$C_{15}H_{34}ClN$
Lauryl trimethyl ammonium chloride
N,N,N-trimethyl-1-dodecanaminium chloride; trimethyl-1-dodecanaminium chloride; n-dodecyl trimethylammonium chloride. Gel sensitizer for latex foam rubber. mp = 235°. *Tiarco.*

22021 Octosol A-1
58353-68-7 261-222-3
Disodium N-[3-(dodecyloxy)propyl]sulfosuccinamate
Emulsifier, dispersant, wetting agent, foaming agent for frothed latex compounds and adhesives; suspending agent in emulsion polymerization; textile softener. *Tiarco.*

22022 Octosol A-18
14481-60-8 238-479-5
Disodium N-octadecyl sulfosuccinamate
Emulsifier, dispersant, foaming agent for latex compounds, cleaners; textile softener; suspending agent in emulsion polymerization; stable in acid and alkaline aq. systems. *Tiarco.*

22023 Octosol A-18-A
68128-59-6 268-577-3
Diammonium N-octadecyl sulfosuccinamate
Emulsifier, stabilizer, foaming agent for acrylic latex frothed compounds, formulations where reduced sodium ion content is desirable; suspending agent in emulsion polymerization. *Tiarco.*

22024 Octosol ALS-28
2235-54-3 218-793-9
$C_{12}H_{29}NO_4S$
Ammonium lauryl sulfate
ammonium lauryl sulfate; dodecyl ester of sulfuric acid, ammonium salt. Stabilizer, emulsifier, dispersant, foaming agent for industrial aq. systems; generates stable, high foam. *Tiarco.*

22025 Octosol HA-80
3006-15-3 221-109-1
$C_{16}H_{29}NaO_7S$
Sodium dihexyl sulfosuccinate
Butanedioic acid, sulfo-, 1,4-dihexyl ester, sodium salt; Sodium dihexyl sulfosuccinate. Emulsifier for latex emulsion polymerization. *Tiarco.*

22026 Octosol IB-45
127-39-9 3238 204-839-5
Sodium diisobutyl sulfosuccinate
Emulsifier for latex emulsion polymerization. *Tiarco.*

22027 Octosol SLS
151-21-3 8782 205-788-1
Sodium lauryl sulfate
Stabilizer, frothing aid, emulsifier, dispersant for latex compounds; low cloud pt., low salt content. *Tiarco.*

22028 Octosol TH-40
23386-52-9 245-629-3
Sodium dicyclohexyl sulfosuccinate
Surfactant for emulsion polymerization, used in manufacture of carboxylated latexes. *Tiarco.*

22029 Octosperse TS-10
Silicone emulsion; defoamer. *Tiarco.*

22030 Octotint 103
1309-37-1 4072 215-168-2
Yellow iron oxide; color dispersion. *Tiarco.*

22031 Octotint 138
13463-67-7 9612 236-675-5
Anatase titanium dioxide. *Tiarco.*

22032 Octovit Tablets
Vitamin A, thiamine mononitrate, riboflavine, nicotinamide, pyridoxine, cyanocobalamin, ascorbic acid, cholecalciferol, tocopherol, calcium, ferrous sulfate, magnesium and zinc; a multivitamin/mineral product indicated where supplementation with vitamins and minerals may be of benefit. *SmithKline Beecham.* Name unverified.

22033 Octowax 321
8002-74-2 7155 232-315-6
Refined paraffin wax emulsion; heat-stable, nondiscoloring wax emulsion with 125-130°F melting range; for use in paper and wood coating, sizing, textile lubrication, as processing aid in latex foams; antiozonant characteristics. *Tiarco.*

22034 Octowax 518
9002-88-4 7728
HDPE emulsion; wax emulsion for use in paper and wood coating, sizing, textile lubrication, as processing aid in latex foams; antiozonant characteristics. *Tiarco.*

22035 Octowet 40
577-11-7 3460 209-406-4
$C_{20}H_{37}NaO_7S$
Di(2-ethylhexyl)sulfosuccinic acid, sodium salt
Sodium dioctyl sulfosuccinate; Aerosol OT; Sulfo-butanedioic acid 1,4-bis(2-ethylhexyl)ester sodium salt; Dioctyl sodium sulfosuccinate; docusate sodium; Bis(2-ethylhexyl) sulfosuccinate, sodium salt; Colace; Bis(2-ethylhexyl)sodium sulfosuccinate. Wetting agent, emulsifier, penetrant for textile washing and dyeing operations, agriculture, mining, paper, printing. mp = 153-157°. *Tiarco.*

22036 Octowet 70A
Ammonium dioctyl sulfosuccinate
High speed wetting agent, solubilizer, penetrant for textile processing, agriculture, mining, paper, printing. *Tiarco.*

22037 octyl acrylate
103-11-7 203-080-7
$C_{11}H_{20}O_2$
acrylic acid, 2-ethylhexyl ester
2-ethylhexyl acrylate; 1-Hexanol, 2-ethylacrylate; 2-Ethylhexylpropenoate; Octyl Acrylate; EHA; acrylic acid
2-ethylhexyl ester; 2-Propenoic acid 2-ethylhexyl ester; Ethyl hexyl acrylate; 2-ethylhexyl 2-propenoate; 2-Propenoic acid octyl ester. mp = -90°; bp = 213-215°; d = 0.885; n_D^{20}= 1.4360; insoluble in H_2O (< 1 mg/ml). *BASF; Hoechst Celanese; Sartomer; Union Carbide.*

22038 octyl dodecanol
5333-42-6 226-242-9
Octyldodecanol
Eutanol G; Exxal® 20; Michel XO-150-20. Lubricant, emollient for cosmetics and pharmaceuticals; for oil-sol. active ingreds.; pigment dispersant. *Henkel/Cospha; Henkel Canada; M. Michel; Exxon.*

22039 octyl methoxycinnamate
5466-77-3 6864 226-775-7
$C_{18}H_{26}O_3$
3-(4-Methoxyphenyl)-2-propenoic acid 2-ethylhexyl ester
Escalol® 557; 2-ethylhexyl p-methoxy-cinnamate; Octyl methoxy cinnamate; Parsol MCX; Parsol MOX; Neo Heliopan, Type AV; Beclovent Inhaler,. Sunscreen. MW=290.40; *Van Dyk; Haarmann & Reimer GmbH; Glaxo Inc.*

22040 octyl palmitate
29806-73-3 249-862-1
$CH_3(CH_2)_{14}COOCH_2(CH_2CH_3)CH(CH_2)_3CH_3$
2-Ethylhexyl palmitate
2-Ethylhexyl hexadecanoate. Ester of 2-ethylhexyl alcohol and palmitic acid; ester for sunscreens, antispersipirants, bath oils, liquid make-up; imparts gloss; binder for pressed powders; solubilizer for benzophenone-3. *Inolex; Van Dyk.*

22041 octyl stearate
22047-49-0 244-754-0
2-Ethylhexyl stearate
Lexolube® T-110. Lubricant for textile, metalworking, plastics industries. *Inolex.*

22042 octyldecyl phthalate
119-07-3 204-295-9

Good-rite® GP-265; Morflex 125, 175. A vinyl plasticizer. *BF Goodrich; Morflex.* Name unverified.

22043 2-octyldodecyl erucate
88103-59-7
E.O.D. Neutral oil improving adhesion, gloss, conditioning in lipsticks, foundtion, eye shadow stick, hair creams. *U.S. Cosmetics.*

22044 ocuba wax
Ocuba fat, from *Myristica ocuba*.

22045 Odoron
Disinfectants. *Evans Vanodine International Ltd.*

22046 Odylen®
135-58-0 5977 205-202-4
$C_{14}H_{12}S_2$
2,7-dimethylthianthrene
mesulfen; mesulphen; Mitigal; Sudermo; Peligal; Neosulfine. Scabicide and anti-pruritic veterinary preparation; for external use against ectoparasites. mp = 123°; bp_{14}= 228-231°; insoluble in H_2O, soluble in organic solvents. *Bayer AG.*

22047 OFA
Fuel additive. *Monsanto (Solaris).* Name unverified.

22048 OFHC Copper
Oxygen-free, high conductivity copper; nickel 0.0006%, bismuth 0.001%, cadmium 0.0001%, lead 0.001%, mercury 0.0001%, oxygen 0.001% (max), phosphorus 0.003%, selinium 0.001%, sulfur 0.0018%, tellurium 0.001%, zinc 0.0001%, iron 0.0005%; 0.00 to 70 total max (tin, antimony, Arsenic, bismuth, manganese, selenium, tellurium). *Amax Inc.* Name unverified.

22049 Oftanol®
25311-71-1 5187 246-814-1
Isofenphos
Soil-applied insecticide used for control of insects in rice crops and pear sucker. *Bayer AG.*

22050 Oftentral
A proprietary preparation containing miconazole nitrate, polymyxin B sulfate; for eye infections in cats and dogs. *Janssen Pharmaceutical Ltd.* Name unverified.

22051 OGA
Gasoline additive. *Monsanto (Solaris).* Name unverified.

22052 Ogtac 85 V
3033-77-0 221-221-0
C6H14ClNO
Glycidyl trimethyl ammonium chloride
(2,3-Epoxypropyl)trimethyl ammonium chloride; glycidyl trimethyl ammonium chloride; N,N,N-trimethyl oxirane methaminium chloride. Intermediate for cationic surfactants; modifier for synthetic polymers such as starch, cellulose, gelatins, polyacrylic acids, epoxy resins; used for production of emulsion layers on photographic plates. *Chem-Y GmbH.*

22053 Ogwin
A mixture of lime and starch used to increase the rate of sedimentation of solids in water.

22054 O-hi-o
A proprietary trade name for a die steel containing 12% chromium, 1.55% carbon, 0.85% vanadium, 0.4% cobalt, and 0.8% manganese. No manufacturer.

22055 OHlan®
68424-66-8 270-315-8
Hydroxylated lanolin
Primary water-oil emulsifier, auxiliary emulsifier and stabilizer, pigment wetting and dispersing agent, emollient and conditioner in personal care products, absorption bases, pharmaceuticals. *Amerchol; Amerchol Europe.*

22056 ohm oil
A mineral oil which has been treated or contains in solution an antioxidant, thereby stabilizing the oil and increasing its electrical resistivity.

22057 Ohmal
A resistance alloy similar in composition to manganin. It usually contains 87.5% copper, 9% manganese, and 3.5% nickel.

22058 ohmlac kapak
A refined elaterite.

22059 Ohmoid
A proprietary trade name for a phenol-formaldehyde synthetic resin laminated product used for electrical insulation. No manufacturer.

22060 Oil asphalt
A thick fluid remaining after distilling crude petroleum; used for roofing materials, and for paving when mixed with natural asphalt.

22061 Oil blue
A pigment prepared by introducing copper filings into boiling sulfur, and after cooling, boiling the mass with sodium hydroxide to remove excess of sulfur. It is applicable as an oil color only.

22062 Oil Die
A proprietary trade name for tool steel containing 1.6% chromium, 0.45% tungsten, and 0.9% carbon. No manufacturer.

22063 Oil Gard
Blend of petroleum oil, viscosity index improvers; reduces oil consumption in automotive vehicles, increases oils' viscosity index. *Gard Corporation.* Name unverified.

22064 Oil of mink
Mink oil.

22065 Oil of Palma Christi
Castor oil.

22066 Oil of Pennyroyal
8007-44-1 7236
$C_{10}H_{16}O$
Oil of Poley. Oil of pulegium. It consists chiefly of pulegone.

22067 Oil of Peter
Oil of Petre. Rock oil, or a mixture of 1 part of oil rosemary, 4 parts turpentine, and 4 parts of Barbados tar.

22068 Oil of Petitgrain
The oil obtained from the leaves of the bitter orange tree.

22069 Oil of Pompillon
An ointment of poplar seeds, also green elder ointment.

22070 Oil of Portugal
Oil of sweet orange peel.

22071 Oil of Ptychotis
Oil of ajowan.

22072 Oil of Rhodium
Oil of duty. The oil obtained from the root of *Genista canariensis*. Also a mixture of sandalwood oil, and otto of rose, or oil of rose geranium.

22073 Oil of Spike
The volatile oil from *Lavandula spica*. It is also the name for a mixture of lavender oil and oil of turpentine, colored with alkanet.

22074 oil of sweet birch
Oil of Betula from *Betula lenta*.

22075 oil of tar
Oil of liquid pitch. Creosote.

22076 oil of tartar
K_2CO_3
Deliquescent potassium carbonate.

22077 oil of tea
The oil obtained from the seeds of *Camellia species*.

22078 oil of turpentine
Spirits of turpentine, essence of turpentine, turps. Derived from the pine, *Pinus palustris* and *P. Taeda*, and from the Scotch fir, *Pinus sylvestris*.

22079 oil of verbena
The oil obtained from *Verbena triphylla*. Also the name for oil of lemongrass.

22080 oil of wheat
The oil obtained from bruised wheat.

22081 Oil Paalsgaard
Schou oil.

22082 oil pulp
Thickener; fluid gelatin; viscom. A gelatinous material made by heating aluminum oleate with mineral oil. It is sold to give increased viscosity to mineral oils.

22083 oil shale
A sedimentary rock which yields from 12-60 gallons of shale oil per ton on distillation.

22084 oil skin
A waterproof material made by impregnating cotton or other fabric with hardening oils.

22085 oil varnishes
Solutions of resins in linseed oil.

22086 oil white
Light white; leukarion; albanol; diamond white; Edelweiss; snow white; anti-white lead; blenda; condor; fixopone; nivan. White lead substitutes. They consist chiefly of lithopone mixed with white lead or zinc white, also with whiting, gypsum, magnesia or silica.

22087 Oildag
Colloidal graphite in 750 hydrotreated refined naphthenic petrol. oil; lubricant additive for chain lubes, aerosols, penetrating lubes, assembly lubes, thread lubes. *Acheson Colloids.*

22088 oiled silk
Thin silk fabric which has been impregnated with oil, usually linseed.

22089 Oilfos
7558-80-7 8806 231-449-2

Glassy sodium phosphate
Controls viscosity of oil well drilling muds. *Monsanto Co.* Name unverified.

22090 Oilsol
Dyestuffs soluble in oils, waxes and plastics. *Morton Int'l. Ltd.*

22091 oil-soluble resins
Synthetic resins of the phenol-formaldehyde type obtained by a fixing process, using rosin in the mixing. They dissolve in hydrocarbon solution and are no longer soluble in alcohol.

22092 Oil-Treet
Fuel oil conditioner and combustion catalyst. *Schaefer Technologies Inc.*

22093 Ointment Base No. 3, 4, 6
White petrolatum USP; ointment base for eye and skin medications; carriers for medical materials. *Penreco.*

22094 oiticia
8016-35-1 232-406-0
Oil derived from the seeds of the Brazilian oiticica tree; used as a drying oil in plants and varnishes.

22095 Okerin
Antiozonant waxes, blends of petroleum waxes; for use in tire sidewall compounds and general rubber products. *Astor Chemical Ltd.*

22096 Oko®
Household insecticide for combating domestic flying insects. *Bayer AG.*

22097 Okol
A disinfectant consisting of an emulsion containing phenols. It is miscible with water.

22098 Okstan M 62
Butyltin carboxylate
Mercaptide-boosted; stabilizer for plastisol processing; provides heat and light stability, superior transparency, excellent color; very suitable for coating of canvas material and metal surfs. *Bärlocher GmbH.*

22099 Okstan M 69 S
Butyltin mercaptide
Stabilizer for plastisol processing; provides superior stability and color properties; resistant to water staining; especially low-odor; suitable for automotive dashboard panels. *Bärlocher GmbH.*

22100 Okstan X3
A trade name for a butyl tin mercaptide; a stabilizer for PVC with an exceptionally high heat stability. *Otto Bärlocher GmbH, Chemische Werke Munchen.* Name unverified.

22101 Okstan XO
58229-88-2
A di-n-octyl tin mercaptide for stabilization of nontoxic PVC compounds. *Croda Resins Ltd.* Discontinued.

22102 Olapon ND-9, SW
Detergent for textile scouring. *Reilly-Whiteman.*

22103 Olay Beauty Bar
A cleansing bar to clean, soften and smooth the skin. *Richardson-Vicks Inc.* Name unverified.

22104 Olay Beauty Cleanser
Gentle, greaseless facial cleanser to leave skin soft and smooth. *Richardson-Vicks Inc.* Name unverified.

22105 olcotrop leather.
A leather made from a species of shark skin.

22106 Old Plantation
6484-52-2 567 229-347-8
Ammonium nitrate
Fertilizer. *Columbia Nitrogen Corporation.* Name unverified.

22107 oleamide
301-02-0 206-103-9
$C_{18}H_{35}NO$
(Z)-9-octadecenamide
oleyl amide; 9-octadecenamide, (Z)-; oleylamide; cis-9,10-octadecenoamide; Aliphatic amide. Slip/antiblock agent for extrusion of polyethylene; wax additive; ink additive. A brain lipid that induces physiological sleep when injected into rats. This lipid may represent a new class of biological signaling molecules. *Akzo; Chemax; Chemron; Croda Universal; Henkel/Emery; Mona; Syn. Prods.; Witco/Humko.*

22108 Oleamidopropyl betaine
25054-76-6 246-584-2
N-(Carboxymethyl)-N,N,-dimethyl-3-[(1-oxooctadecenyl)amino)-1-propanaminium hydroxide, inner salt
Incronam OP-30; Lexaine® O; Mackam™ HV; Mirataine® BET-O-30; Oleamidopropyl dimethyl glycine. Viscosity builder, high foaming surfactant for shampoos, bubble baths, cleansing lotions, hand cleaners, skin care. Yellow viscous liquid; stable in acid and alkaline systems; pH=6.5. *Croda Inc.; Croda Surf. Ltd.; McIntyre; Inolex.* Discontinued.

22109 oleamidopropyl dimethylamine
109-28-4 203-661-5
Lexamine O-13; Dimethylaminopropyl oleamide; N-[3-Dimethylaminopropyl]-9-octadecenamide; Chemidex O; Incromine OPM, OPB; Mackine™ 501; Schercodine O; Unizeen OA,. Emulsifier for creams and lotions; when neutralized as conditioners for hair care products. *Inolex; Chemron; Croda Inc.; McIntyre; Scher; Universal Preserv-A-Chem.*

22110 oleamidopropylamine oxide
25159-40-4 246-684-6
9-Octadecenamide, N-[3-(dimethylamino)propyl]-,N-oxide
Incromine Oxide O; Mackamine™ OAO; Oleamidopropyl dimethylamine oxide; N-[3-(Dimethylamino)propyl]-9-octadecenamide-N-oxide. Foam booster/stabilizer, thickener, lubricant, and visc. builder used in cosmetic, household, and janitorial prods.; wetting agent for conc. electrolyte solutions Amber gel Ph=7.0. *Croda Inc.; Croda Surf. Ltd.; McIntyre.* Discontinued.

22111 oleandocyn
7060-74-4 230-351-7
A proprietary preparation of oleandomycin phosphate; an antibiotic. *Pfizer International.* Discontinued.

22112 oleic Acid
112-80-1 6965 204-007-1
$C_{18}H_{34}O_2$
(Z)-9-Octadecanoic acid
Emersol® 210, 6333 NF, 7021; Industrene® 104; Pamolyn 100, 100 FG, 100, FGK, 125; Pamolyn® 125; Priolene 6900. Detergent intermediate for personal care, emollient, household and industrial applications. MW=282.47; liq.; odorless, colorless; crystals, mp=4°; sol.in alcohol, benzene, chloroform, ether; insol.in water; LD_{50}(I.v., mise)=230+-18mg/kg *Akzo; Arizona; Henkel/Emery; Hercules; Unichema; Union Derivan SA; Witco/Humko.*

22113 olein of saponification
Commercial oleic acid prepared by the saponification of pure fats, and the separation from stearine, by pressing.

22114 oleite
68187-76-8 269-123-7
Sodium-sulfo-ricinoleate
Castor oil, sulfated, sodium salt; Sodium salt of turkey red oil; Sodium sulforicinoleate; Turkey red oil, sodium salt.

22115 oleo
Premier jus. The oil expressed from beef-fat; used in margarine manufacture.

22116 Oleo Keratin ISO
AMP isostearoyl hydrolyzed keratin
Used with isostearic acid, myristyl myristate, isopropyl palmitate as a cosmetics ingredient. *Brooks Industries.*

22117 oleobismuth
An oil suspension of bismuth oleate.

22118 Oleo-Coll LP
123-95-5 1625 204-666-5
butyl stearate
Lecithin; oleoyl sarcosine; lanolin alcohol. Coco-hydrolyzed animal protein, sesame oil; cosmetics ingredient. *Brooks Industries.*

22119 oleoguaiacol
$C_{25}H_{30}O_3$
Guaiacol oleate
Used as an antiseptic.

22120 oleoresins
Resins mixed with a volatile oil; they are semisolid and tacky at room temperature and soft and sticky at high temperatures.

22121 Oleosol
Acheson Colloids. Name unverified.

22122 Oleo-Soy C
Coco-hydrolyzed soy protein
Cosmetic ingredient for hair and skin care products. *Brooks Industries.*

22123 Oleoyl sarcosine
110-25-8 203-749-3
Hamposyl® O; Crodasinic O; Vanseal® OS; N-Methyl-N-(1-oxo-9-octadecenyl)glycine; Oleyl methylaminoethanoic acid; Hamposyl® O; Nikkol Sarcosinate OH; Oramix O;,. Detergent, wetting and foaming agent, foam stabilizer, emulsifier, corrosion inhibitor, mold release agent, conditioner for hair and rug shampoos, cosmetics, skin cleansers; ceramic dispersant; biodeg. W R Grace/Hampshire; Chemplex Chems.; Hampshire; Chemplex Chems.; Nikko Chem. Co. Ltd; Seppic; Croda Chem Ltd.; R. T. Vanderbilt Co Inc.

22124 Olepal ISO
56002-14-3
PEG-6 isostearate
Gattefosse SA.

22125 oleth-2
9004-92-2
Hetoxol OL-2. Intermediate, emulsifier, wetting agent, solubilizer, coupling agent. *Heterene.*

22126 oleth-8
9004-98-2
Emalex 508; Ablunol OA-6; Akyporox RO 90; Akyporox RTO 70; Ameroxol® OE-2, OE-5; Brij® 93; Chemal OA-4, OA-5, OA-20/70CWS; Emalex 508; Ethal OA-10; Eumulgin M8, O5, PWM2, WM5; Genapol® O-020; Hetoxol OA-3 Special; Hostacerin O-20; Lipocol O-2; Macol® OA-2; Marlipal® 1850/5; Marlowet® WOE; Prox-onic OA-1/04; Rhodasurf® ON-870; Ritoleth 2; Simulsol 98; Trycol® 5971; Varonic® 32-E20; Volpo 3; Volpo 5, 10, 20, N3, O3.,. Emulsifier, dispersant, solubilizer for cosmetics; suitable for hair tonics and hair care products as solubilizer for perfumes. Nihon Emulsion; Taiwan Surf; Chem-Y GmbH; Amerchol; ICI Spec. Chem.; Chemax; Ethox; Henkel HGaA; Hoechst; Heterene; Lipo; Croda Inc.; Protex; Huls AG;.

22127 oleth-phosphate Series (2-20)
39464-69-2
Oleyl ether phosphate
Brophos OL-3; Chemfac PB-184; Chemphos TR-505, TR-515, TR-541; Crodafos N10 Acid, N3-Acid, N5 Acid, O2 Acid;Empicol® 0216; Hetphos OA-3; Laurelphos 400; Rhodafac® RB-400; Hodag PE-1803. Fatty alcohol ethoxy phosphate ester; foam stabilizer, conditioner, and antistatic agent for toiletries; detergent used in textile processing. Liquid *Albright & Wilson UK; Brook Industries; Chemron; Rhône--Poulec Surf & Spec; Calgene.*

22128 oleum spirits
A petroleum distillate.

22129 oleum white
Zinc sulfide mixed with a small proportion of barium sulfate (a lithophone); used as a pigment and rubber filler.

22130 Olex
Fat soluble powdered flavorants. *Bush Boake Allen Ltd.*

22131 oleyl alcohol
143-28-2 6968 205-597-3
$C_{18}H_{36}O$
(Z)-9-octadecen-1-ol
9-octadecen-1-ol. Used to manufacture detergents and wetting agents; as an antifoam agent, a cutting oil and a plasticizer. mp = 13-19°; bp_{13} = 207°; d = 0.8490; n_D^{20} = 1.4610; insoluble in H_2O, soluble in organic solvents; *Croda; R.W. Greeff; Lanaetex; M. Michel; Ronsheim & Moore; Sherex.*

22132 oleyl betaine
871-37-4 212-806-1
N-(Carboxymethyl)-N,N-dimethyl-9-octadecen-1-aminium hydroxide, inner salt
Incronam OD-50; Chembetaine OL; Chembetaine OL-30; Mackam™ OB-30; OLB-50; Velvetex® OLB-30; Velvetex® OLB-50; Oleyl dimethyl glycine. High foaming surfactant, visc. builder for conditioning shampoos, clear rinses, cleansing creams and cosmetic lotions. Amber translucent gel; soluble in water; sp gr=0.953; pH=6.0-8.0 *Croda Surf. Ltd.; Ikeda; Chemron.*

22133 oleyl palmitamide
16260-09-6 240-367-6
$C_{34}H_{67}NO$
hexadecanamide
N-9-octadecenyl hexadecanamide. Substituted aliphatic amide; release agent providing slip, antiblocking to thermoplastics incl. PP film, nylon. *Croda Universal; Witco/Humko.*

22134 olibanum
Indian frankincense; Salaigugl. A gum-resin obtained from *Boswellia* species. It contains from 8-18% of essential oil, 55-57% of resin, and 20-23% gum (carbohydrates).

22135 Olicat® C
Polyisobutylene in mineral oil; tackiness agent, lubricant in lubricating oils and greases. *Alox.* Name unverified.

22136 Oligan
Shrink-resist additive. *Ciba plc.* Name unverified.

22137 Olinor
Blend of hydrocarbons and sulfated fatty alcohols; softening agent for raising cotton goods. *Henkel Chemicals Ltd.* Name unverified.

22138 olive oil
8001-25-0 6973 232-277-0
Oils; olive; olea europaea oil; mixed glycerides of oleic (83%), palmitic (9%), linoleic (4%), stearic (2%) and arachidic (1%) acids. Fixed oil obtained from the ripe fruit of *Olea europaea* ; pale yellow to greenish liquid, nondrying with slight odor and taste; salad dressing and other foods; ointments, liniments, soaps, as lubricant and emol emollient, in cosmetics and pharmaceuticals. mp = 6°; d_{25}^{25} = 0.909 - 0.915; n_D^{25} = 1.467; insoluble in H_2O, slightly soluble in EtOH, soluble in organic solvents. *Arista Industries; Croda; Penta Mfg.; Reilly-Whiteman.*

22139 olive oil PEG-6 esters
103819-46-1
Labrafil M 1980 CS. Hydrophilic oil for pharmaceutical and cosmetic formulations. *Gattefosse; Gattefosse SA.*

22140 olivenite
$4CuO{\cdot}As_2O_5$
A mineral.

22141 olives of Java
Kaloempang beans; Beligno seeds; Sterculia kernels. Seeds of *Sterculi foetida* , the source of Sterculia oil.

22142 olivite
A substance having a rubber base; used as an acidproofing material in pumps.

22143 Olminat
688-37-9 367 211-702-3
$C_{54}H_{99}AlO_6$
aluminum oleate
9-octadecenoic acid, aluminum salt, oleic acid aluminum salt. A trade name for a commercial aluminum oleate; contains 1.5% aluminum. Used as lacquer for metals. Insoluble in H_2O, soluble in organic solvents. No manufacturer.

22144 OLOA
Lubricating oil additive. *Monsanto (Solaris).* Name unverified.

22145 Olobintin
A German preparation; a 10% solution of a mixture of various rectified turpentine oils.

22146 Olympic
14807-96-6 9207 238-877-9
talc; soapstone; stearite;. Talc; for cosmetic applications including baby powders, creams and lotions, foot powders, dusting powders. *Cyprus Industrial Minerals.*

22147 Olympic Bronze
A proprietary trade name for an alloy of copper with 3% silicon and 1% zinc. No manufacturer.

22148 Olympic Bronze G
A proprietary trade name for an alloy of copper with 22% zinc and 1% silicon. No manufacturer.

22149 Omacide® P-BBP-5
5% zinc pyrithione, 80% butyl benzyl phthalate plasticizer; antimicrobial dispersion for protection of PVC systems. *Olin.*

22150 Omacide® P-DIDP-5
5% zinc pyrithione, 65% diisodecyl phthalate plasticizer, antimicrobial dispersion for protection of PVC systems. *Olin.*

22151 Omacide® P-ESO-5
5% zinc pyrithione, 75% epoxidized soybean oil plasticizer; antimicrobial dispersion for protection of PVC systems. *Olin.*

22152 Omadine® MDS
3696-28-4 223-024-5
$C_{10}H_8N_2O_2S_2$
Bispyrithione
2,2-Dithiodipyridine-1,1'-dioxide; 2,2'-dithiobis(pyridine) 1,1'-dioxide; 2,2'-Dithiobis(pyridine-N-oxide); ithiobis(pyridine N-oxide); Omadine disulfide; Omadine DS. Bispyrithione and magensium sulfate. An antidandruff agent for nonalkaline hair care products; antimicrobial agent for Gram-negative and Gram-positive bacteria; also inhibits the growth of fungi. mp = 205° (dec). *Olin.*

22153 Omarsan
A detergent and sterilizer. *PPF International Ltd.* Name unverified.

22154 Ombrelub FC 533
Waterproofing agent for aqueous systems, e.g., printing inks, cement, and concrete mixtures. *Münzing Chemie GmbH.*

22155 Omethoate
1113-02-6 214-197-8
$C_5H_{12}NO_4PS$
O,O-dimethyl S-[2-(methylamino)-2-oxoethyl] phosphorothioate
dimethoate-met; Bay 45432; Folimat; S-6876. Systemic insecticide and acaricide with contact and stomach action. Used for control of spider mites, aphids, beetles, caterpillars, scale insects, thrips, suckers, fruit flies etc. in fruit and vegetable crops and in forestry. dec 135°; d^{20} = 1.32; n_D^{20} = 1.4987; soluble in H_2O, organic solvents; LD_{50} (rat orl) = 50 mg/kg. *Bayer.*

22156 Omnicryl
Aqueous color dispersions for use in water based ink systems and rubber latex applications. *Engelhard.*

22157 Omnilac
Non-nitrocellulose overlacquers. *The Scottish Adhesives Co Ltd.*

22158 Omnilube
A series of synthetic oils and greases based on polyalphaolefins and various organic compounds; a variety of oils used for OEM applications in appliances,

power tools, computers, automotive electrical motors, and industrial maintenance applications. *Ultrachem Inc.* Name unverified.

22159 Omnipaque
66108-95-0 266-164-2
Iohexol
Diagnostic aid. *Sterling Drug Inc.* Name unverified.

22160 Omnisorb®
Cells for use in biological research. *Calbiochem Corp.*

22161 OMTS
102-77-2 203-052-4
$C_{11}H_{12}N_2OS_2$
N-oxydiethylenebenzothiazole-2-sulfenamide
Amax; 2-Benzothiazolyl-N-morpholinosulfide; Morpholine, 4-(2-benzothiazolylthio)-; 2-(morpholinothio)benzothiazole; 2-(Morpholinthio)-benzothiazole; N-oxydiethylene-benzothiazole sulfenamide. Delayed-action accelerator for SBR, NR, and nitrile rubbers. *Akrochem.*

22162 Omyastab
A secondary stabilizer for chlorinated polyester resins. It is a colorless cyclic organic compound containing ether linkages which are compatible with the polyester resin.

22163 On The Ball
Superwash for golfballs, reducing skin friction and adding substantial distance to a golfball drive. *Merix.*

22164 Onamine 12
112-18-5 203-943-8
$C_{14}H_{31}N$
N,N-dimethyldodecylamine
ADMA-2; armeen dm-12d; barlene 125; N,N-dimethyllaurylamine; i-dodecanamine, N,N-dimethyl; dodecyldimethylamine; n-dodecyldimethylamine; lauryldimethylamine; n-lauryldimethylamine; monolauryl dimethylamine; Dimethyldodecylamine; N,N-dimethyl-1-aminododecane; N,N-dimethyl-1-dodecanamine. Dodecyl dimethylamine in liquid form; intermediate in the synthesis of surfactants, antioxidants, oil and grease additives. mp = -20°; d = 0.7870. *Millmaster-Onyx UK.* Name unverified.

22165 Onamine 14
112-75-4 204-002-4
$C_{16}H_{35}N$
N,N-dimethyltetradecylamine
dimethyltetradecylamine in liquid form; intermediate in the synthesis of surfactants, antioxidants, oil and grease additives. *Millmaster-Onyx UK.* Name unverified.

22166 Onamine 16
N,N-dimethylhexadecylamine
dimethylhexadecylamine in liquid form; intermediate in the synthesis of surfactants, antioxidants, oil and grease additives. *Millmaster-Onyx UK.* Name unverified.

22167 Onamine 18
124-28-7 3525 204-694-8
$C_{20}H_{43}N$
N,N-dimethyl stearylamine
Dymanthine, N,N-dimethyl octadecanamine; (hydrochloride salt) Dimantine; Dimethyloctadecylamine hydrochloride; Dymanthine hydrochloride; Dymanthine Hydrochloride; DODA-hydrochloride; GS-1339; N,N-Dimethyloctadecylamine hydrochloride; Octadecylamine, N,N-dimethyl-, hydrochloride; Stearyldimethylammonium chloride; Thelmesan. dimethyl stearylamine in liquid form; intermediate in the synthesis of surfactants, antioxidants, oil and grease additives. Both Onamine 18 and its hydrochloride salt ($C_{20}H_{43}N{\cdot}HCl$; GS-1339, NSC-5547; Thelmesan) are used as anthelmintics. mp = 23°; SG = 0.8. *Millmaster-Onyx UK.* Name unverified.

22168 Onamine 65, 835 and 1214
Alkyl dimethylamines in liquid form; intermediate in the synthesis of surfactants, antioxidants, oil and grease additives. *Millmaster-Onyx UK.* Name unverified.

22169 Onamine RO
1-(2-hydroxyethyl) 2-n-heptadecenyl-2-imidazoline in liquid form; acid-stable emulsifier used in corrosion inhibition, solvent degreasing, defoaming, and demulsifying. *Millmaster-Onyx UK.* Name unverified.

22170 Once
Hand cleanser. *Hardman.*

22171 Once
Decorative waterborne high opacity paint. *ICI Chem & Polymers Ltd.*

22172 Oncol
82560-54-1
Insecticide containing benfuracarb. *ICI Chem & Polymers Ltd.*

22173 Oncol 10G
82560-54-1
Benfuracarb
Soil applied insecticide and nematicide for use on sugar beet. *Farm Protection Ltd.*

22174 Oncomouse®
For the medical industry. *DuPont UK.*

22175 Oncor
Basic lead silico pigments; anti-corrosive pigment for paint. No manufacturer.

22176 Oncor 75
Flame retardant compositions. *Anzon Ltd.* Name unverified.

22177 Ondene®
Insecticide with fast killing activity and good penetrating action particularly for the control of sucking insects and bugs, cocoa flat bugs, leafhoppers and other pests; for use on cereals, pome and stone fruit, citrus fruit, rice, sugar cane. *Bayer AG.*

22178 Ondoita
A proprietary synthetic resin molding powder. No manufacturer.

22179 Ongard
Smoke suppressants. *Anzon Ltd.* Name unverified.

22180 Oniachlor
Chlorinated isocyanates. *Sartomer.* Discontinued.

22181 Onion's alloy
A fusible alloy containing 50% bismuth, 30% lead and 20% tin.

22182 Onslaught
Suspension concentrate containing 160 g linuron [330-55-2] and 320 g trifluralin [1582-09-8] per liter; herbicide for winter cereals. *Quadrangle Agrochemicals.*

22183 Ontario Steel
A proprietary trade name for a non-shrinking steel containing 11% chromium, 0.75% molybdenum, 0.25% vanadium, 0.35% silicon, 0.30% manganese, and 1.45% carbon. No manufacturer.

22184 onyx
Consists chiefly of silica.

22185 Onyx Classica
21645-51-2 355 244-492-7
Hydrated alumina
Developed as a synthetic onyx for synthetic marble industry; features high whiteness, purity, consistency, flame retardancy, compatibility. *Alcoa Industrial Chemicals.*

22186 onyx marble
A marble containing fossil shells.

22187 onyx of Tecali
A variety of alabaster. The color varies from milk-white to pale yellow and pale green.

22188 Onyx Premier WP-31
Alumina trihydrate
Filler for cultured onyx; provides highest whiteness, excellent translucency; flame retardant, smoke suppressant. *Alcan.*

22189 Onyxide® 75
Alkenyl (90% C18, 10% C16) dimethyl ethylammonium bromide in paste form; cationic surfactant algicide used in recirculating water systems; swimming pools; humidifiers. *Millmaster-Onyx UK.* Name unverified.

22190 Onyxide® 172
Alkyl dimethyl ethyl benzyl ammonium cyclohexyl sulfamate in liquid form; antifungal agent and preservative used in latex emulsions; paints; adhesives; coated fabrics; cutting oils. *Millmaster-Onyx UK.* Name unverified.

22191 Onyxide® 200
Hexahydro-1,3,5-tris(2-hydroxethyl)-*s*-triazine; preservative for sol. cutting fluids and coolants; bactericide for oilfield drilling fluids, enhanced oil recovery operations. *Stepan; Stepan Canada.*

22192 Onyxide® 3300
Alkyl dimethylbenzyl ammonium saccharinate in powder form; germicide, conditioner and disinfectant with low skin and eye irritation; used in cosmetics and pharmaceuticals; hair preparations; detergent-sanitizers; disinfectants. *Millmaster-Onyx UK.* Name unverified.

22193 Onyxol
Alkanolamides
Millmaster-Onyx UK.

22194 Oolitic limestone
A massive variety of calcium carbonate; used for building purposes.

22195 Opacicoat
471-34-1 1697 207-439-9
Calcium carbonate
Ultrafine ground coated pigment with unique particle size distribution; permits higher loadings in polymer systems while still maintaining properties. *ECC International Ltd.* Discontinued.

22196 Opacimite
471-34-1 1697 207-439-9
Calcium carbonate

Finely ground product with an engineered particle size distribution for use in paints and polymers. *ECC International Ltd.*

22197 opacite
10026-06-9 8929 231-588-9
$SnCl_4$
Stannic chloride.

22198 Opacode
An edible ink for pharmaceutical or food use. *Colorcon Ltd.* Name unverified.

22199 Opacolor
Coloring system for sugar coated confectionery pieces. *Colorcon Ltd.* Name unverified.

22200 Opadry
Complete film coating system in a dry form for reconstitution; Aqueous film coating, organic film coating and enteric coating. *Colorcon Ltd.* Name unverified.

22201 opal Jasper
It is silica, resembling jasper.

22202 Opalite
A proprietary trade name for an amorphous silica. No manufacturer.

22203 Opalon
A proprietary trade name for phenol-formaldehyde cast resins. No manufacturer.

22204 Opalon 740
A trade name for a graft copolymer used as a semi rigid wire insulation. No manufacturer.

22205 Opalux
Coloring system for sugar coated tablets; sugar coating for pharmaceutical products. *Colorcon Ltd.* Name unverified.

22206 opalwax
Hydrogenated castor oil.

22207 Opaspray
Coloring system for film coated tablets; aqueous film coating colorant, organic film coating colorant. *Colorcon Ltd.* Name unverified.

22208 Opatint
A multipurpose liquid color dispersion formulated for the coloring of all types of food and confectionery. *Colorcon Ltd.* Name unverified.

22209 Opazil
Acid and alkaline activated bentonites for the adsorption of detrimental substances in the papermaking process. *Süd-Chemie AG.* Name unverified.

22210 Opazil®
Activated bentonite-based; for absorption of interfering substances in papermaking. *BASF AG.*

22211 Opbipol
Polymeric optical fiber cable. *BASF plc.*

22212 Opera
Suspension concentrate containing 150 g terbutryn and 200 g trifluralin per liter; for weed control in winter cereals. *Tripart Farm Chemicals Ltd.*

22213 Opex® 80
Dinitro; chemical blowing agent. *Uniroyal.*

22214 Ophorite
An ignition powder for projectiles. It consists of magnesium powder and potassium chlorate.

22215 Ophthalgan
Glycerin
A pharmaceutic aid. *Wyeth Laboratories.* Name unverified.

22216 Opogard 500
Suspension concentrate containing 150 g terbuthylazine and 350 g terbutryn per liter; weed germination inhibitor. *Ciba-Geigy Agrochemicals.*

22217 Opol
Chemical polishers for plastics. *Laporte Industries Ltd.*

22218 OPPalyte®
Range of polypropylene films; used in the packaging of food and nonfood products and special industrial applications. *Mobil Plastics Europe.* Name unverified.

22219 OPPalyte® 233 TW, 278 TW, 350 TW
9003-07-0 7741
Oriented PP film; nonheat sealable; high opacity expanded core with modified OPP skin layers both sides for durability and WVTR; treated both sides for ink and adhesive anchorage. *Mobil.* Name unverified.

22220 OPPalyte® 250 ASW, 350 ASW
Oriented PP film, 1 side acrylic coated, 1 side PVDC coated, heat sealable; high opacity pearlescent core; acrylic coating provides excellent machinability, gloss, aroma barrier, and superior surface for solv. and water-based inks; PVDC coating provides strong seals, hot tack, moderate oxygene barrier and excellent adhesion to cold seal; useful in single wall or lamination, and overwrap applications. *Mobil.* Name unverified.

22221 Oppanol®
Polyisobutylene in the form of viscose liquids or rubbery solid crumb, depending on molecular weight; used in sheet or film form for waterproofing or corrosion resistant coatings, to modify properties of polyolefin polymers and waxes used in film or impregnation; for sealants and adhesives. *BASF plc.*

22222 Oppanol® B
Polyisobutylene
For the adhesive and sealant industry, elec. insulating oils, bases for chewing gums, for production of damp-proof courses containing fillers in construction industry. *BASF AG.*

22223 Oppanol® D
Polyisobutylene dispersion. *BASF plc.* Name unverified.

22224 Oppasin®
Inorganic and organic. pigment concentrates; for coloring rubber compounds and solutions; optical brighteners for textile fibers. *BASF AG.*

22225 OPS®
Biaxially oriented polystyrene sheet and film. *Asahi Chem. Industry.*

22226 Opsan
Antibacterial eye drops. *The Boots Co plc.* Discontinued.

22227 optannin
Basic calcium tannate.

22228 Optemet
Prefabricated disposable pouring trumpets for uphill teemed steel ingots. *Foseco (F.S.) Ltd.* Discontinued.

22229 Op-Thal-Zin
7446-20-0 10293
$O_4SZn{\cdot}7H_2O$
Zinc sulfate
Astringent (ophthalmic). *Alcon Laboratories Inc.* Name unverified.

22230 Optibor Boric Acid
Borates
Borateem. *Borax Consolidated Ltd.*

22231 optical bronze
An alloy of 89% copper, 6.5% zinc, and 4.5% tin.

22232 Opticite
Polystyrene-based packaging labelling material. *Dow Chem Co Ltd, UK & Ireland.*

22233 Opticlean® L-1000
Bacterial protease; enzyme for hydrolysis of protein; excellent stability in presence of oxygen bleaches; for use in highly alkaline laundry detergents. *Solvay Enzymes.*

22234 Opticorten
Steroid veterinary ethicals. *Ciba plc.* Name unverified.

22235 Optigard®
A water blocking agent and lubricant used as an additive in the manufacture of optical and fiber cables. *Dow Corning.*

22236 Optigel
Thixotropic clay gellant. *Production Chemicals Ltd.*

22237 Optigel WM
Bentonite, cellulose gum; thixotrope for aq. cosmetic/toiletry systems, putty, caulk, cleaning compounds, waxes, polishes. *United Catalysts.*

22238 Optima 23B
68334-28-1 269-820-6
Partially hydrogenated cottonseed or soybean oil; center fat for confectionery and bakery applications. *Van Den Bergh Foods.*

22239 Optima C
Activated bleaching earth; used for refining of vegetable, animal and mineral oils, fats and waxes. *Minas de Gador SA.*

22240 Optimase® APL, M-440, PAG
Alkaline protease; enzyme for hydrolysis of proteins; used in moderately alkaline laundry detergents, also for recovery of silver from x-ray and photographic film. *Solvay Enzymes.*

22241 Optimask®
For the electrical industry. *DuPont UK.*

22242 Optimem
Plastic (or other polymeric material) membranes used for separation and related liquid treatments. *ICI Chem & Polymers Ltd.*

22243 Optinol
Carriers used for dyeing polyester. No manufacturer.

22244 Option®
For the agriculture industry. *DuPont UK.*

22245 Optipol®
Polymeric optical fiber with PC core and special cladding; high temp. resist. and flexibility suited for use in harsh conditions with relatively short transmission lengths; for machine and appliance automation, traffic signals,

auto instrument illumination, process control engineering. *Bayer AG.* Name unverified.

22246 **Optivest**
Dental chrome investment. *Degussa Ltd.*

22247 **Optiwhite®**
12141-46-7 377 235-253-8
Aluminum silicate
Thermooptic; pigment retaining whiteness and hiding when embedded in binders; for use in paper coating, paints, rubber, and plastics to extend TiO_2 or other costly pigments; excellent dispersion, good electricals, high hiding, improved film propert *Burgess Pigment.*

22248 **Optiwite Series**
Optical whites for use on lycra, nylon, acetate, and wool or blends. *CNC Int'l L.P.*

22249 **Optodent®, Optognath®**
Artificial teeth. *Bayer AG.*

22250 **Optosil® Liquid**
Elastomeric isolating film on silicone basis; used for dental techology. *Bayer AG.*

22251 **Optosil® Optosil Hard**
Elastomeric impression material and ancilliaries; dental speciality for practice and laboratory. *Bayer AG.* Discontinued.

22252 **Optosil® Plus**
Organic silicone polymer; used for dental impressions. *Bayer AG.*

22253 **Optox**
Melted lead oxide granules of high purity; basic raw material for optical glass. *Chemson Polymer Additive GmbH.* Name unverified.

22254 **Optran**
High grade optical chemicals. *BDH Chemicals Ltd.*

22255 **ORA**
Refinery process stream additive. *Monsanto (Solaris).* Name unverified.

22256 **Orabase**
A proprietary preparation containing sodium carboxy-methyl cellulose, pectin, and gelatin in a liquid paraffin polyethylene base. *Bristol-Myers Squibb Pharmaceuticals Ltd.* Name unverified.

22257 **Oracet**
Solvent soluble dyes. *Ciba plc.*

22258 **Oracle®**
Isoproturon and metsulfuron-methyl; used for annual weed control in wheat and barley. *DuPont UK.*

22259 **Ora-Cop**
Poultry nutrition supplement. *Mineral Research & Development Corp.*

22260 **Oragrafin Calcium**
1151-11-7 5087 214-565-8
Ipodate calcium
Diagnostic aid. *Bristol-Myers Squibb Co Inc.* Name unverified.

22261 **Oragrafin Sodium**
1221-56-3 5087 214-945-3
Ipodate sodium
Diagonostic aid. *Bristol-Myers Squibb Co Inc.* Name unverified.

22262 **Orahesive**
A proprietary preparation containing sodium carboxymethyl cellulose, pectin, and gelatin. *Bristol-Myers Squibb Pharmaceuticals Ltd.* Name unverified.

22263 **Oraldene**
141-94-6 4741 205-513-5
A proprietary preparation of hexetidine; an antiseptic mouth wash. *Warner.* Name unverified.

22264 **oralith**
Organic pigments.

22265 **Oramide DL 200 AF**
Cocamide DEA
Cosmetic emulsifier. *Seppic.* Discontinued.

22266 **Orange III**
A coloring for all types of food and confectionery. *Colorcon Ltd.* Name unverified.

22267 **orange lead**
Orange mineral; orange red; sandix; Saturn red. A red lead obtained by calcining powdered white lead. It is a better red lead than crystal minium.

22268 **orange tungsten**
$Na_2WO_4 \cdot W_2O_5$.
Saffron bronze, tungsten-sodium tungstate.

22269 **oranium bronze**
Dirigold
An aluminum bronze. It contains from 87-97% copper and 3-11% aluminum.

22270 **Orasol**
Solvent soluble dyes. *Ciba plc.*

22271 **Oratrast**
7727-43-7 1023 231-784-4
Barium sulfate
Diagnostic aid. *Armour Pharmaceutical Co.* Name unverified.

22272 **Oravue**
41473-08-9 5077
$C_{15}H_{18}I_3NO_5$
Iopronic acid
Diagnostic aid. *Bristol-Myers Squibb Co Inc.* Name unverified.

22273 **Orbinamon**
A proprietary preparation of thiothixene; a psychotherapeutic agent. *Pfizer International.*

22274 **Orbitol**
Photographic developer. *May & Baker Ltd.* Name unverified.

22275 **Orca**
A French synthetic resin prepared from acrolein; used for electrical insulation. No manufacturer.

22276 **Orchard Herbide**
Amitrole [61-82-5] and diuron [330-54-1]. Used for total weed control in fruit orchards. *Hoechst UK.*

22277 **Orchidee**
Gold bearing on glaze colors for porcelain, bone china, and earthenware. *Degussa Ltd.*

22278 **Orchidee**
87-20-7 5139 201-730-4
$C_{12}H_{16}O_3$
Sanfoin
The isoamyl ester of salicylic acid (*o*-hydroxy-benzoic acid), used in perfumery.

22279 **orchindone**
87-19-4 201-729-9
$C_{11}H_{14}O_3$
Isobutyl salicylate
isobutyl *o*-hydroxybenzoate; 2-methylpropyl *o*-hydroxybenzoate. Used in perfumery. mp = 22°; bp = 260°; SG_{25} = 1.063.

22280 **orcinol**
504-15-4 6995 207-984-2
$C_7H_8O_2$
1,3-dihydroxy-5-methylbenzene
methylresorcinol; dihydroxytoluene; orcinol; 3,5-dihydroxytoluene; 5-methylresorcinol; 5-methyl-1,3-benzenediol; 3,5-toluenediol. Reagent for beet sugar, lignin, and pentoses. mp = 109-111°; bp = 290°; Soluble in H_2O, organic solvents; LD_{50} (rat orl) = 844 mg/kg.

22281 **ordeal bark**
8529
Sassy bark. The bark of *Erythrophloeum guincense* .

22282 **ordeal bean**
7539
Split nut; Calabar bean; Physostigma. A source of eserine (physostigmine).

22283 **Ordnance 204, 500**
Oil-based, water-soluble cutting oil for use under extreme pressure. Multipurpose oil-based soluble cutting oil for all metal removing operations where physical and chemical extreme pressure assistance is required to assist functional cooling properties on many metals where a soluble oil is preferred. *Summer Oil Industries.* Unverified.

22284 **ordonezite**
A mineral; $2(ZnSb_2O_6)$.

22285 **Ordoval**
A synthetic tannin made from sulfonated anthracene. No manufacturer.

22286 **ore-furnace slag**
A slag obtained when roasted copper sulfide ore is mixed with oxidized ores and slag and fused for the production of coarse metal. It often contains unfused quartz and less than 1% copper, and is mainly a silicate of iron.

22287 **Oregon balsam**
The true oleoresin is obtained from *Pseudotsuga mucronata* , but another product sold under the same name consists of a mixture of rosin and turpentine.

22288 **Oreide**
A yellow alloy resembling gold. It usually contains from 80-90% copper, 10-14.5% zinc, and 0-4.5% tin.

22289 **orellin**
A yellow coloring matter found in the vegetable dye, annatto. It is probably an oxidation product of bixin, another coloring matter of annatto.

22290 **Oresmasin**
Pigments for textiles. *Ciba plc.* Name unverified.

22291 Orevac® 9309
EVA terpolymer; for cast film coextrusion with PE/PA, ionomers/PA. *Elf Atochem.*

22292 Orevac® 18211
EVA terpolymer; for sheet coextrusion with PS, EVOH, PE, adhesive emulsions (aluminum). *Elf Atochem.*

22293 Orevac® 18302
9002-88-4 7728
PE; for blown film, and bottle blow molding coextrusion; for PE/PA, PS/EVOH/PE adhesives for barrier materials, thermo adhesive film. *Elf Atochem.*

22294 Orevac® PP-C
9003-07-0 7741
PP, grafted; for blown film coextrusion, blow molding coextrusion, steel pipe coating, underfloor heating pipes. *Elf Atochem.*

22295 Orgamide R
A proprietary trade name for nylon 6. No manufacturer.

22296 Organelle
Dyes for the petroleum industry. *ICI Chem & Polymers Ltd.*

22297 Organopol®
A range of retention aids used in papermaking specific to DIP/TMP/CTMP and highly contaminated stocks where normal retention aids have no effect. *Allied Colloids Ltd.*

22298 Organosilane Si203, Si208
For penetrating sealers that protect concrete and wood from water and chloride ion intrusion. *Degussa.*

22299 Orglas
Glass-reinforced plastic linings and coatings. *Prodorite Ltd.*

22300 Orgozon CC 1118
Fatty phosphate ester, potassium salt; detergent and emulsifier for use in mild to strong alkaline conditions. *Clough.*

22301 Orgozon Conc. 0680
Phosphate ester, free acid form; detergent, emulsifier for use in mild to strong alkaline conditions; may be neutralized with sodium hydroxide, potassium hydroxide, etc. *Clough.*

22302 Oriental powder
An explosive used in fireworks. It is a mixture of the gum resin gamboge and potassium nitrate.

22303 Oriodide-131
7790-26-3 8778
Sodium iodide I 131
Antineoplastic; diagnostic aid; radioactive agent. *Abbott Laboratories.* Name unverified.

22304 Orisan
Seed dressing for control of fungal diseases on cereals, rice, cotton and vegetables. *Bayer AG.*

22305 Orkan
Floor cleaning degreaser. *Gansow UK Ltd.*

22306 Orlon®
Fiber. *DuPont UK.* Discontinued.

22307 Ormolu
One alloy contains 58% copper, 25.3% zinc, and 16% tin, and another consists of 90.5% copper, 3% zinc, and 6.5% tin.

22308 Ornamental Weeder
1582-09-8 9815 216-428-8
Trifluralin
A preemergence herbicide; granular material for use on ornamentals, shrubs, trees, roses, and flower beds; control grasses and many broadleaf weeds. *Lawn & Garden Products Inc.*

22309 Ornithite
$Ca_3P_2O_8{\cdot}2H_2O$
A tricalcium phosphate.

22310 Oroglas DR
An impact modified polymethacrylate; used for injection molding, extension. *Rohm & Haas.*

22311 Oromid®
Nylon 6 and 66 granules; engineering plastic. *SNIA (UK) Ltd.*

22312 Oronal LCG
Sodium coceth sulfate
PEG-40 glyceryl cocate. Mild surfactant for cosmetics. *Seppic.* Discontinued.

22313 Oronite
Lubricating oil additive. *Monsanto (Solaris).* Name unverified.

22314 Oropon
Composition of the enzymes of pancreas absorbed in sawdust or kieselguhr, and intimately mixed with ammonium chloride or boric acid; a bate for leather. *Rohm & Haas.*

22315 Orotan
Pigment dispersing agent; used for coatings. *Rohm & Haas.*

22316 Orovite
Oral vitamins B and C. *SmithKline Beecham.*

22317 orpiment
1303-33-9 846 215-117-4
As_2S_3.
arsenic trisulfide
Yellow sulfide of arsenic. A mineral.

22318 orris camphor
Essential oil of orris.

22319 orris root
The rhizome of *Irin florentina* .

22320 Ortegol®
Polyurethane crosslinking agents. *Thomas Goldschmidt Ltd.*

22321 Ortegol® 204
Delayed reaction crosslinking additive for production of high resilience PU slabstock foams. *Goldschmidt AG.* Discontinued.

22322 Orthene
30560-19-1 31 250-241-2
Acephate insecticide. *Monsanto (Solaris).* Name unverified.

22323 Orthenex
Contact and systemic insecticide. *Monsanto (Solaris).* Name unverified.

22324 Ortho
Contact and systemic insecticide. *Monsanto (Solaris).* Name unverified.

22325 Orthochrom
Leather finishes. *Rohm & Haas.*

22326 Orthochrome T
p-toluquinaldine p-toluquinoline-ethylcyanine bromide
A red sensitizer for silver bromide plates.

22327 ortho-chrysotile
$2[Mg_3Si_2O_5(OH)_4]$
A mineral.

22328 Orthocide
133-06-2 1815 205-087-0
Captan fungicide. *Monsanto (Solaris).* Name unverified.

22329 Ortho-Clear
Leather finishes. *Rohm & Haas.*

22330 Ortho-Gro
Liquid plant food. *Monsanto (Solaris).* Name unverified.

22331 Ortho-Klor
57-74-9 2129 200-349-0
Insecticide. *Monsanto (Solaris).* Name unverified.

22332 Ortholate
88-41-5 201-828-7
$C_{12}H_{22}O_2$
2-*t*-Butyl cyclohexyl acetate
Verdox; Green acetate; Ylanat ortho. Used in perfumery. Clear yellow oil with a fresh fruity woody green apple odor. $sg^{25} = 0.937$; $n_D^{20} = 1.4500$. *Quest Int'l. UK Ltd.*

22333 Ortholeum
Grease stabilizer and lubricant assistant. *DuPont UK.*

22334 Ortholite
Leather finishes. *Rohm & Haas.*

22335 Orthomatic
Lawn sprayer. *Monsanto (Solaris).* Name unverified.

22336 Orthophen® 278
80-46-6 7284 201-280-9
$C_{11}H_{16}O$
p-tert-amylphenol
Pentaphen; *p-tert*-amyl phenol; *p*-(α,α-dimethylpropyl)phenol; *p*-(1,1-dimethylpropyl)phenol; PTAP; 1-hydroxy-4-(2-methyl-2-butyl)benzene; dimethylpropyl)phenol; para-tertiary amylphenol;. Intermediate for chemical specialties; also in manufacture of photographic chemicals, oil demulsifiers, phenolic resins, agricultural surfactants and antiskinning agents. mp = 91-94°; bp = 255°; insoluble in H_2O, soluble in organic solvents. *Elf Atochem.*

22337 Orthorix
1344-81-6 215-709-2
CaS_x
calcium polysulfide
lime-sulfur; Eau grison; Neviken; Security Lime Sulphur. Lime-sulfur fungicide. Acts directly and alo by decomposition to sulfur which is also a fungicide. Used for control of powdery mildews, anthracnose, scab and other diseases in benas, clover and fruits. Control of insects and spider mite eggs in fruit trees. SG15.6:kls > 1.28; soluble in H_2O. *Monsanto (Solaris).* Name unverified.

22338 Orthosil
A proprietary trade name for an anhydrous sodium orthosilicate; a detergent. 18. No manufacturer.

22339 Orthotrol
Drift retardant. *Monsanto (Solaris).* Name unverified.

22340 Orth's stain
A microscopic stain. It contains 1 g lithium carbonate and 2.5 g carmine in 100 ml water.

22341 Ortol
Dyestuffs for wool and polyamide. *BASF plc. Name unverified.*

22342 *Ortol*
methyl-*o*-aminophenol and hydroquinone (2:1). A photographic developer. Methyl-*o*-amino-phenol, (2 mols), combined with hydroquinone (1 mol), forms the basis of this developer.

22343 Ortolan
A proprietary thermal insulation. No manufacturer.

22344 Ortolan® Dyes
1:2 Metal complex dyes for dyeing and printing of wool and nylon fibers. *BASF AG; BASF plc.*

22345 Ortosol
It consists of 10% chlorobenzene, 88% *o-* and *m*-dichlorobenzenes, and 2% *p*-dichlorobenzene. A mixture for dry cleaning.

22346 Orvus WA
151-21-3 8782 205-788-1
$C_{12}H_{25}NaO_4S$
Sodium lauryl sulfate
A proprietary trade name for a wetting agent. No manufacturer.

22347 oryzalin
19044-88-3 7015 242-777-0
$C_{12}H_{8}N_4O_6S$
4-(dipropylamino)-3,5-dinitrobenzenesulfonamide
3,5-dinitro-N^4, N^4-dipropyl
sulfanilamide; Dirimal; EL-119; Ryzelan; Surflan; Weed-Stopper. Selective herbicide; inhibits cell division and germination. Used for pre-emergence control of annual grasses. mp = 141-142°; slightly soluble in H_2O (2.5 mg/l), more soluble in organic solvents; LD_{50} (rat orl) > 10000 mg/kg *Lawn & Garden Products Inc.*

22348 Orzan
A range of ammonium and sodium lignin sulfonates; used as cement grinding aids and as thinners for drilling muds. *Harcros Australia.*

22349 Orzol
White mineral oil. *Witco Corporation.*

22350 *os sepiae*
White fish-bone. The calcareous shell lying within the back of the cuttle fish. It consists mainly of calcium carbonate.

22351 OS-2
Emulsifiable ester; forms stable emulsions for buffing, lapping, drawing, grinding compounds; rust preventative. *Werner G. Smith.* Discontinued.

22352 Oscodal
A proprietary preparation of vitamin cod-liver oil product. No manufacturer.

22353 Oscrete
Range of concrete admixtures. *Oils & Soaps Ltd.*

22354 Osimol
Dispersing and leveling agent; used for dyeing. *Degussa AG.* Name unverified.

22355 osmic acid
20816-12-0 7024 244-058-7
O_4Os
osmium tetroxide
perosmic oxide; perosmic acid anhydride. Used as a microscopic stain, in photography, as a oxidation catalyst in organic synthesis. mp = 40°; bp = 130°; soluble in H_2O (7.24 g/100 g), more soluble in organic solvents; highly toxic.

22356 osmium
7440-04-2 7021 231-114-0
Metallic element; Os; extremely dense and hard. Used as a hardener for iridium and platinum, inpen points, instrument pivots, catalyst. mp = 2700°; bp = 5500°; d_4^{20} = 22.61. *Aldrich; Atomergic Chemetals; Degussa; Noah Chem.*

22357 osmium tetroxide
20816-12-0 7024 244-058-7
OsO_4
osmic acid
perosmic acid anhydride; perosmic oxide. Microscopic staining, photography, oxidation catalyst in organic synthesis. *Aldrich; Atomergic Chemetals; Degussa; Janssen Chimica; Spectrum Chem. Mfg.*

22358 Osmoglyn
56-81-5 4493 200-289-5
Glycerin; pharmaceutic aid. *Alcon Laboratories Inc.*

22359 osmo-kaolin
A preparation of kaolin obtained by a patented electro-osmosis process. It has a high covering power and clinging properties, and is used in toilet powders.

22360 osmondite
The stage in the transformation of austentite, at which the solution in dilute sulfuric acid reaches its maximum rapidity.

22361 osmo-sil
A dye absorbent. It is a very pure form of silica, SiO_2.

22362 Osnol
A proprietary preparation of liquid paraffin with vitamin D. No manufacturer.

22363 OSO® 440
1309-37-1 4072 215-168-2
Fe_2O_3
iron(III) oxide
Synthetic yellow pigment with good suspension and stabilization properties for coatings, automotive finishes, appliance enamels, paints, plastics, building products, rubbers, inks. *Hitox.*

22364 OSO® 1905
1309-37-1 4072 215-168-2
Fe_2O_3
iron(III) oxide
Synthetic red iron oxide; pigment with good suspension and stabilization properties for coatings, automotive finishes, appliance enamels, paints, plastics, building products, rubbers, inks. *Hitox.*

22365 osram
An alloy of osmium and tungsten. Used in light bulb filaments made by, amongst others, the Osram Company.

22366 ossein
A variety of gelatin prepared from bones.

22367 Ossivite
A proprietary preparation of bonemeal and vitamins A and D. *Wyeth Laboratories.* Name unverified.

22368 Ostamer
Polyurethane foam; prosthetic aid. *Merrell Dow Pharmaceuticals Inc.* Name unverified.

22369 Ostan
A trade name for pure sodium and potassium hydroxides in disk form (5 mm diameter), suitable for analytical work and convenience in weighing. No manufacturer.

22370 Ostelin
A proprietary preparation; It is a vitamin cod-liver oil product. No manufacturer.

22371 Ostrilan
Sulfonamide quinoline hormone veterinary ethical. *Ciba plc.* Name unverified.

22372 Oswego
A proprietary trademark for cornflour. No manufacturer.

22373 Osyrol
41890-92-0 255-574-7
$C_{11}H_{24}O_2$
3,7-Dimethyl-7-methoxyoctan-2-ol
Used in perfumery. *Bush Boake Allen Ltd.*

22374 ote seeds
Acoomo seeds. Seeds of *Myristica angolensis,* of Nigeria. The seeds yield Kombe fat.

22375 otoba butter
Otoba wax; American mace butter. Otoba fat, obtained from the fruit of *Myristica otoba .*

22376 Otoryl
Compound ear drops, veterinary. *May & Baker Ltd.* Name unverified.

22377 Ottasept
For chemical compositions having antiseptic, germicidal and fungicidal properties such as *para*-chloro-, *meta*-xylenol, in Class 18. *Ferro.* Discontinued.

22378 Oulu 102
61790-12-3 263-107-3
Tall oil fatty acids. Tall oil fatty acid; raw material. *Veitsiluoto Oy.*

22379 Oulu 331
Tall oil rosin; raw material. *Veitsiluoto Oy.*

22380 Oulumer 70
Tall oil rosin; tackifier for pressure-sensitive and hot-melt adhesives; raw material for high softening point rosin esters; also in printing inks. *Veitsiluoto Oy.*

22381 Oulupale XB 100
Pale rosin; tackifier with excellent thermal stability for disposable hot melts in hygiene sector, pressure-sensitive adhesives, bookbinding adhesives; compatible with EVA, SB, SIS, and NR elastomers and waxes. *Veitsiluoto Oy.*

22382 Oulures
Printing ink resins. *Veitsiluoto Oy.*

22383 Oulutac 20 D
Modified tall oil rosin aqueous dispersion; plasticizer for other resin dispersions; gives very soft, water-resistant tacky film; good storage and mechanical stability and resistance against slight freezing. *Veitsiluoto Oy.*

22384 Oulutac 20 EP
Triethylene glycol ester of tall oil rosin; adhesive tackifier and plasticizer. *Veitsiluoto Oy.* Name unverified.

22385 Oulutac 30
Ethylene glycol ester of tall oil rosin; adhesive tackifier. *Veitsiluoto Oy.* Name unverified.

22386 Oulutac 30 D
Aqueous emulsion of tall oil rosin based ethylene glycol ester; adhesive tackifier. *Veitsiluoto Oy.* Name unverified.

22387 Oulutac 80
Glycerol ester of tall oil rosin; adhesive tackifier. *Veitsiluoto Oy.* Name unverified.

22388 Oulutac 80 D
Aqueous dispersion of tall oil rosin based glycerol ester; adhesive tackifier. *Veitsiluoto Oy.* Name unverified.

22389 Oulutac 80 D/HS
Glyceryl rosinate aqueous dispersion; tackifier for natural and synthetic rubber-based latex adhesives; good mechanical stability and resistance to slight freezing. *Veitsiluoto Oy.*

22390 Oulutac 90
Pentaerythritol ester of tall oil rosin; adhesive tackifier. *Veitsiluoto Oy.* Name unverified.

22391 Oulutac 90 D
8050-26-8 232-479-9
Pentaerythrityl tall oil rosinate; tackifier for natural and synthetic rubber-based latex adhesives; good mechanical and storage stability and resistance to slight freezing. *Veitsiluoto Oy.*

22392 Oulutac 90 D
Aqueous dispersion of tall oil rosin based pentaerythritol ester; adhesive tackifier. *Veitsiluoto Oy.* Name unverified.

22393 Oulutac 105
Pentaerythritol ester of tall oil rosin; adhesive tackifier. *Veitsiluoto Oy.* Name unverified.

22394 ounce metal
A bronze consisting of 85% copper, 5% tin, 5% zinc, and 5% lead.

22395 Oust
Descaler products. *Dylon International Ltd.*

22396 ouvarovite
A mineral. It is a lime-chrome-garnet.

22397 Ovac
A blend of the thiazole derivatives of two specially selected aldehyde-amines; a proprietary trade name for a rubber vulcanizing accelerator. No manufacturer.

22398 Ovacryl
Floor polishes. *Evans Vanodine International Ltd.*

22399 Overnite
A slug killer. *Murphy Chemical Co Ltd.* Discontinued.

22400 Ovicide
Miscible tar oil winter wash. *ICI Chem & Polymers Ltd.*

22401 Ovigest
For the weakly lamb. *The Wellcome Foundation Ltd.* Name unverified.

22402 Ovitelmin
31431-39-7 5807 250-635-4
A proprietary preparation containing mebendazole; veterinary antihelmintic (sheep and goats). *Janssen Pharmaceutical Ltd.* Name unverified.

22403 Ovucire WL 2944
Semi-synthetic glycerides. *Gattefosse SA.*

22404 Oxaban®-A
51200-87-4 257-048-2
$C_5H_{11}NO$
4,4-dimethyl oxazolidine
dimethyl-1-oxa-3-aza-cyclopentane; dimethyloxazolidine; Bioban CS-1135; Cosan 101; Nuosept 101; Troysan 192,. Cosmetic preservative, antimicrobial. *Angus.*

22405 Oxaban®-E
7747-35-5 231-810-4
7-Ethyl bicyclooxazolidine
Ethyldihydro-1H,3H,5H-oxazolo(3,4-c)oxazole; Oxazolidine E; Oxazolo(3,4-c)oxazole, 7α-ethyldihydro-; Zoldine ZE; Chemtan A-60. Antibacterial for cosmetics and toiletries. *Angus.*

22406 oxadiazon
19666-30-9 7038 243-215-7
$C_{15}H_{18}Cl_2N_2O_3$
3-[2,4-Dichloro-5-(1-methylethoxy)phenyl]-5-(1,1-dimethylethyl)-1,3,4-oxadiazol-2(3H)-one
Oxydiazon; Ronstar; Ronstar 2G; Ronstar 50W; Scotts OH I; RP-17623. Selective systemic herbicide. mp = 88-89°; insoluble in H_2O (<1 mg/l), soluble in organic solvents; LD_{50} (rat orl) = 3500 mg/kg.

22407 oxadixyl
77732-09-3
$C_{14}H_{18}N_2O_4$
N-(2,6-dimethylphenyl)-2-methoxy-N-(2-oxo-3-oxazolidinyl)acetamide
Anchor; SAN 371F; Sandofan. Systemic fungicide with protective and curative action. Used in combination with contact fungicides (e.g. mancozeb; captofol etc.) for control of downy mildews, late blights and rusts in vines, potatoes, maize, tobacco, hops, sunflowers, citrus, fruits and vegetables. mp = 104-105°; soluble in H_2O (3.4 g/kg), more soluble in organic solvents; LD_{50} (rat orl) = 3480 mg/kg.

22408 Oxaf
155-04-4 205-840-3
$C_{14}H_8N_2S_4Zn$
Zinc-2-mercaptobenzothiazole
Zetax; 2(3H)-Benzothiazolethione, zinc salt; Mercaptobenzothiazole, zinc salt; Benzothiazolethiol, zinc salt; Benzothiazolethione, zinc salt;Bis(mercaptobenzothiazolato)zinc; Hermat ZN-MBT; OXAF; Pennac ZT; Tisperse MB-58; Vulkacit ZM; Zenite; Zinc 2-mercaptobenzothiazolate; Zinc benzothiazolethiolate; Zinc benzothiazolylmercaptide; Zinc bis(mercaptobenzothiazole); Zinc mercaptobenzothiazole; ZMBT; ZNMB. A medium temperature accelerator widely used in latex compounding; also used in proofing, wire and druggist sundries where fast curing and a minimum of odor are required. *Uniroyal.* Name unverified.

22409 Oxaf 50D
A 50% water active dispersion; ready-to-use form for latex compounding. *Uniroyal.* Name unverified.

22410 oxalan
$C_3H_5N_3O_3$
Oxaluramide.

22411 oxalantin
$C_6H_6N_4O_6$
Leucoturic acid.

22412 oxalate blasting powder
A safety explosive powder containing 71% niter, 14% charcoal, and 15% ammonium oxalate.

22413 oxalic acid
144-62-7 7043 205-634-3
$C_2H_2O_4$
oxalic acid
ethane dioic acid; ethanedionic acid; ethane-1,2-dioic acid; ethanedioic acid. Automobile radiator cleanser, metal and equipment cleaning, purifying agent, intermediate, leather tanning, catalyst, laboratory reagent, stripping agent, textile bleaching, rare-earth processing, printing and dyeing auxiliary. mp = 189-190°; d_4^{17}= 1.90; soluble in H_2O (6.71 g/100 ml); LD_{50} (rat orl) = 4.5 mg/ml. *Ashland; General Chem.; Hoechst Celanese; KMZ Chem. Ltd; Mallinckrodt; Mitsubishi Gas.*

22414 oxalic acid dihydrate
6153-56-6 7043 205-634-3
$C_2H_6O_6$
ethanedioic acid, dihydrate
ethandionic acid, dihydrate; ethanedionic acid dihydrate. mp = 101°.

22415 oxalic acid tin (II) salt
tin(II) oxalate
tin oxalate.

22416 oxalic ether
95-92-1 3174 202-464-1
$C_6H_{10}O_4$
Ethyl oxalate
ethyl oxalate; diethyl ester, oxalic acid; ethanedioic acid, diethyl ester. Solventand intermediate in manufacture of chemicals and pharmaceuticals. mp = -41°; bp = 186°; SG = 1.076; n_D^{20}= 1.4101; poorly soluble in H_2O, soluble in organic solvents.

22417 oxalumina
An abrasive consisting of small crystals of alumina.

22418 oxalyl bis (benzylidenehydrazide)
$C_{16}H_{14}O_2N_4$

Eastman® Inhibitor OABH; Aquastab PH 502. Stabilizer used in polyolefins in contact w/copper or copper-containing alloys; copper deactivator. White powder; sp gr=0.216 *Eastman*. Name unverified.

22419 oxazolidine
497-25-6 207-840-9
$C_3H_5NO_2$
2-oxazolidine
oxazolidin-2-one; 2-oxazolidinone. Preservative, antibacterial agent for water-based paints, latexes, emulsions, metalworking fluids; for oilfield water-flooding operations; corrosion inhibitor; crosslinking agent, catalyst for resin systems. mp = 86-89°; bp_{48} = 220°; *Angus*.

22420 Oxetal 500/85
68439-46-3
Fatty alcohol polyglycol ether (5 EO); detergent, dispersant, emulsifier, wetting agent for household and industrial use, textile, paper, and leather industries. *Zschimmer & Schwarz*.

22421 Oxetal D 104
26183-52-8
Deceth-4
Detergent, dispersant, emulsifier, wetting agent for household and industrial use, textile, paper, and leather industries. *Zschimmer & Schwarz*.

22422 Oxetal ID 104
Isodeceth-4
Detergent, dispersant, emulsifier, wetting agent for household and industrial use, textile, paper, and leather industries. *Zschimmer & Schwarz*. Discontinued.

22423 Oxetal O 108
Cetoleth-8
Detergent, dispersant, emulsifier, wetting agent for household and industrial use, textile, paper, and leather industries. *Zschimmer & Schwarz*. Discontinued.

22424 Oxetal TG 111
61791-28-4
Talloweth-11
Detergent, dispersant, emulsifier, wetting agent for household and industrial use, textile, paper, and leather industries. *Zschimmer & Schwarz*.

22425 Oxetal VD 20
3055-95-6 221-281-8
Laureth-2
Washing and cleansing agent. *Zschimmer & Schwarz*.

22426 oxfendazole
53716-50-0 7069 258-714-5
$C_{15}H_{13}N_3O_3S$
[5-(phenylsulfinyl)-1H-benzimidazol-2-yl]carbamic acid methyl ester
RS-8858; Autoworm; Benzelmin; Repidose; Synanthic; Systamex. Anthelmintic. mp = 253° (dec); LD_{50} (rat) >6400 mg/kg.

22427 Oxi-Chek
Antioxidants for plastics and elastomers. *Ferro*. Discontinued.

22428 Oxilube
Polyoxyalkylene diols and derivatives; base components for compressor oils, greases, gear oils, and aviation turbine lubricants. *Shell UK*.

22429 Oximony
A proprietary red oxide of iron used as a rubber pigment. No manufacturer.

22430 Oxi-tan
A proprietary tanning compound. No manufacturer.

22431 Oxitex
Textile fiber lubricant. *Shell UK*.

22432 Oxitol
2-ethoxyethanol. A colorless, slightly hygroscopic liquid with mild odor; used as a solvent in paints, varnishes, inks and stains; also effective as an extraction agent for antibiotics and as a degreasing solvent. *Shell UK*.

22433 Oxivent
30286-75-0 7077 250-113-6
Oxitropium bromide
An antimuscarinic bronchodilator. *Boehringer Ingelheim Ltd*.

22434 Oxolin
A patented material made from oxidized oil, jute fiber, and sulfur; it is a rubber substitute. No manufacturer.

22435 Oxone
1313-60-6 8800 215-209-4
O_2Na_2
sodium peroxide
Oxolin; sodium dioxide; sodium superoxide; solozone. Compressed sodium peroxide, used in washing powders, scouring powders, metal cleaners, hair wave neutralizers, general oxidizing reactions. mp = 440°. *DuPont*. Unverified.

22436 Oxonite
An explosive. It is made from 54% of nitric acid (specific gravity 1.5), and 46% of picric acid.

22437 Oxsoralen-Ultra
298-81-7 6068 206-066-9
$C_{12}H_8O_4$
8-methoxy-2',3',6,7-furocoumarin
Oxsoralen; 8-Methoxypsoralen; Xanthotoxin; 8-MP; Methoxa-Dome; Oxsoralen Ultra; 8-Methoxy-4',5':6,7-furocoumarin; 8-Methoxy(furano-3', 2':6-7-coumarin; Ammoidin; Meloxine; Uvadex; 6-hydroxy-7-methoxy-5-benzofuranacrylic acid δ-lactone; meladinin; methoxalen; oxypsoralen; psoralone-mop; 9-methoxy-7H-furo(3,2-g)benzopyran-7-one; xanthoxin; 7-furocoumarin; zanthotoxin; psoralen-mop; 8-Methoxyfuranocoumarin. Pigmentation agent. mp = 148-150°; λ_m = 219, 249, 300 nm (log ε 4.32, 4.35, 4.06); poorly soluble in H_2O, soluble in organic solvents; LD_{50} (rat ip) = 470 mg/kg. *ICN Pharmaceuticals Inc*.

22438 oxyalizarin
81-54-9 8132 201-359-8
$C_{14}H_8O_5$
1,2,4-trihydroxy-9,10-anthracenedione
C.I. 58205; C.I. 75410; C.I. Natural Red 8; C.I. Natural Rd 16; Hydroxylizaric Acid; Purpurin; Purpurine; Smoke Brown G; Verantin; 1,2,4-Trihydroxyanthraquinone; Purpurin, CI 58205; 1,2,4-Trihydroxy-9,10-anthraquinone; 1,2,4-Trihydroxyanthrachinon; C.I. 1037. Purpurin, an anthraquinone-based dyestuff. mp = 253-256°; insoluble in H_2O, soluble in organic solvents.

22439 oxyammonia
7803-49-8 4874 232-259-2
NH_2OH
Hydroxylamine.

22440 oxyanthracene
529-86-2 725
$C_{14}H_{10}O$
9-anthranol
9-anthrol; 9-anthracenol;. Used in the manufacture of dyestuffs. mp = 120°, 152°.

22441 oxycarboxin
5259-88-1 226-066-2
$C_{12}H_{13}NO_4S$
5,6-dihydro-2-methyl-N-phenyl-1,4-Oxathiin-3-carboxamide 4,4-dioxide
Plantvax; Carboxin sulfone; DCMOO; Oxykisvax; Ringmaster. A systemic fungicide with curative action. Used for the control of rust diseases in ornamentals, nursery trees and wheat. mp = 127-130°; soluble in H_2O (1 g/l), more soluble in organic solvents; LD_{50} (rat orl) = 2000 mg/kg.

22442 oxychloride
7681-52-9 8773 231-668-3
It is a solution of sodium hypochlorite, containing 10-12% available chlorine; a disinfectant.

22443 oxychloride of tin
6389
Tin salts.

22444 Oxyclozanide
2277-92-1 7092 218-904-0
$C_{13}H_6Cl_5NO_3$
2,3,5-trichloro-N-(3,5-dichloro-2-hydroxyphenyl)-6-hydroxybenzamide
Zanil. Anthelmintic)Trematodes). Used in combination with Haloxon. mp = 209-211°.

22445 oxycymol
499-75-2 1923 207-889-6
$C_{10}H_{14}O$
2-methyl-5-(1-methylethyl)phenol
Carvacrol; carvacrol; phenol, 2-methyl-5-(1-methylethyl)-; cymenol; hydroxy-p-cymene; isopropyl-o-cresol; isothymol; methyl-5-(1-methylethyl)phenol. Used in organic synthesis and as a disinfectant. mp = 0°; bp = 237-238°; d_4^{20}= 0.976; n_D^{20}= 1.5229; insoluble in H_2O, soluble in organic solvents; LD (rbt orl) = 100 mg/kg.

22446 oxydase
An oxidizing enzyme.

22447 Oxydasine
Consists mainly of a 0.05% solution of vanadic acid; an antiseptic recommended for wounds.

22448 oxydislin
A compound of silicon having the formula Si_2H_2O, prepared by treating calcium silicide with cold diute alcoholic hydrochloric acid in the dark. It is a white solid spontaneously inflammable in air.

22449 Oxydpech
Oxide pitch obtained as a residue from the distillation of the fatty acids obtained from the oxidation of paraffin.

22450 Oxydurit®
Anionic product (oxidation agents/dispersants); auxiliary in the dyeing and printing of textiles. *Cassella AG.*

22451 oxygen
7782-44-7 7098 231-956-9
O_2
oxygen
Nonmetallic gaseous element; O; copper smelting, steel production; in manufacture of ammonia, methyl alcohol, acetylene; oxidizer for rocket propellants; resuscitation, heart stimulant; decompression chambers; spacecraft; chemical intermediate; in oxidation of municipal and industrial wastes. mp = -218°; bp = -183°; d^0 (gas) = 1.429 g/l. *Air Prods & Chem; Norsk Hydro A/S; Showa Denko.*

22452 oxygen cubes
Made by mixing sodium peroxide and bleaching powder together and compressing into tablets. They contain 100 parts of bleaching powder (33-35% available chlorine), and 39 parts of sodium peroxide. On contact with water, oxygen is evolved.

22453 oxygen powder
1313-60-6 8800 215-209-4
Na_2O_2
Sodium peroxide
mp = 460° (dec); d = 2.8050.

22454 oxygenated oil
Olive oil through which chlorine has been passed for several days.

22455 oxygenite
A mixture of perchlorates or nitrates with a combustible substance. When ignited the mixture produces oxygen, and the material is used for this purpose.

22456 Oxy-Gro
1305-79-9 1736 215-139-4
CaO_2
Calcium peroxide
calcium dioxide. For agricultural and horticultural uses. Use as a rubber stabilizer and as an antiseptic. Slightly soluble in H_2O. *Solvay Interox Ltd.*

22457 Oxyguard
128-37-0 1583 204-881-4
$C_{15}H_{24}O$
2,6-di-*t*-butyl-*p*-cresol
2,6-di-*tert*-butyl-1-hydroxy-4-methylbenzene; BHT; butylhydroxytoluene; dibutylated hydroxytoluene; butylated hydroxytoluene; di-n-butyl hydroxytoluene; 4-methyl-2,6-di-t-butyl-phenol; 2,6-bis(1,1-dimethylethyl)-4-methylphenol; Annulex BHT; Antracine 8; Catalin CAO-3; Dalpac; DBPC; Embanox BHT; Hydagen DEO; Impruvol; Ionol CP; Sustane; Tenox BHT; Topanol OC and 0; Vianol; di-tert-butyl-p-cresol; bis(1,1-dimethylethyl)-4-methylphenol; Topanol. A proprietary antioxidant and antiskinning agent. mp = 69-70°; bp = 264-265°; d_4^{20}= 1.048; insoluble in H_2O, soluble in organic solvents; LD_{50} (mus orl) = 1040 mg/kg. *Naugatuck (US Rubber).* Name unverified.

22458 Oxyliquit
A blasting explosive. It is formed by rapidly mixing liquid air, rich in oxygen, with powdered charcoal, petroleum residues, or cotton wool.

22459 oxylith
A compressed powder. It is a mixture of sodium peroxide and bleaching powder, which evolves oxygen on treatment with water.

22460 Oxymaster
79-21-0 7293 201-186-8
$C_2H_4O_3$
peroxyacetic acid
acetic peroxide; acetyl hydroperoxide; PAA; ethaneperoxoic acid; 4,4'-di(methoxy)-azobenzene; peracetic acid; Desoxon 1; Osbon AC. Peracetic acid for sewage treatment. Strong oxidizing agent. Soluble in H_2O, EtOH. *Solvay Interox Ltd.*

22461 oxymel
Clarified honey (80%) mixed with acetic acid (10%) and water (10%).

22462 oxymuth saca
A colloidal suspension of bismuth oxyhydrate.

22463 Oxyneurine
107-43-7 1225 203-490-6
$C_5H_{11}NO_2$
(Carboxymethyl)trimethylammonium hydroxide inner salt
Betaine; Methanaminium, 1-carboxy-N,N,N-trimethyl-, inner salt; 2-(Trimethylammonio)ethanoic acid, hydroxide, inner salt; (Carboxymethyl)trimethylammonium hydroxide inner salt; Trimethylaminoacetate; Glycine betaine. Used as a soldering flux, in chemical synthesis and as a hepatoprotectant. mp >300°; soluble in H_2O (160 g/100 g), less soluble in organic solvents;.

22464 Oxynone
537-65-5 3022 208-673-4
$C_{12}H_{13}N_3$
1,4-Benzenediamine, N-(4-aminophenyl)-
2,4-Diamino-diphenylamine; 4,4'-diaminodiphenylamine. A proprietary trade name for a rubber antioxidant. Also used to dye fur and to detect hydrogen cyanide. mp = 158°. No manufacturer.

22465 Oxyper
497-19-8 8739 207-838-8
Sodium carbonate peroxyhydrate
Solvay Interox Ltd.

22466 Oxyphenine
Direct cotton dyestuffs. *Clayton Aniline Co Ltd.* Name unverified.

22467 Oxypon 288
70914-02-2
Olive oil PEG-10 esters; solubilizer and refatting for cosmetics. *Zschimmer & Schwarz.*

22468 Oxypon 306
Mink oil PEG-13 esters; solubilizer and refatting agent for cosmetics. *Zschimmer & Schwarz.* Discontinued.

22469 Oxypon 328
PEG-26 jojoba acid
PEG-26 jojoba alcohol. Refatting agent for cosmetics. *Zschimmer & Schwarz.* Discontinued.

22470 Oxypon 365
70914-02-2
Avocado oil PEG-11 esters; refatting agent for cosmetics. *Zschimmer & Schwarz.*

22471 Oxypon 2145
PEG-15 glyceryl isostearate
Emollient, superfatting agent for cosmetics. *Zschimmer & Schwarz.* Discontinued.

22472 oxytoluol
C_7H_8O
Cresol.

22473 Oxytracyl
Antibiotic veterinary ethical. *Ciba plc.* Name unverified.

22474 oxytri
Polymerized β-hydroxy-trimethylene sulfide.

22475 Oxytril
Selective weedkiller. *May & Baker Ltd.* Name unverified.

22476 Oxytril CM
An emulsifiable concentrate containing 200 g bromoxynil and 200 g ioxynil per liter; a post-emergence contact herbicide for cereal crops. *Rhône-Poulenc Crop Protection Ltd.*

22477 Oxytril P
Selective herbicide. *Murphy Chemical Co Ltd.* Discontinued.

22478 Oxzone
7722-84-1 4839 231-765-0
Hydrogen peroxide
Oxogen. *ABM Chemicals Ltd.* Name unverified.

22479 Ozasol
Offset printing plates. *Hoechst AG.*

22480 Ozatec
Photoresists for acid plating and etching. *Hoechst AG.*

22481 ozogen
H_2O_2
Hydrogen peroxide.

22482 Ozokerine
Trade name applied to varieties of soft paraffin. No manufacturer.

22483 ozokerite
8021-55-4
ozocerite
Mineral wax; fossil wax. Hydrocarbon wax derived from mineral or petroleum sources; electrical insulation, rubber products, paints, leather products, printing inks, floor and furniture polishes, cosmetics, ointments; substitute for carnauba and beeswax. *ISP.*

22484 ozole
A term applied to volatile aromatic odors contained in certain dextrins and plant extracts. Dextrinozole is the name applied to the body giving the scent of commercial dextrin.

22485 P 210-D
Acrylic copolymer; processing aid for PVC; improved material handling, faster fusion, increased production output. *Novacor.*

22486 P 506
Styrene methyl methacrylate copolymer; features sparkling clarity, scratch

resistance, easy processing; for appliances, displays, medical devices, toys, office accessories. *Novacor.*

22487 P.A.C
A proprietary manufacture of formaldehyde. No manufacturer.

22488 P.B.N
135-88-6 205-223-9
Phenyl β-naphthylamine
A rubber antioxidant. No manufacturer.

22489 P.H.D
A proprietary trade name for a plasticising oil. No manufacturer.

22490 P.I.B
An abbreviation for polyisobutylene.

22491 P.M.G.Metal
A proprietary trade name for an alloy of copper with 3-4% silicon, 2% iron, and 2% zinc. No manufacturer.

22492 P.M.T. Alloy
A proprietary alloy made as a substitute for Admiralty gun metal; contains 88% copper, 2% zinc, and 10% silicon, manganese, and iron. No manufacturer.

22493 P.N.P
100-02-7 6718 202-811-7
p-nitrophenol
Used as fungicide in the rubber industry.

22494 P.O.T.G
Phenyl-o-tolyl-guanidine
Used as a rubber vulcanization accelerator.

22495 P.P.D
Piperidine-pentamethylene-dithio-carbamate
A rubber vulcanization accelerator.

22496 P.P.S
9016-75-5
A proprietary polyphenylene sulfide; a cross-linkable aromatic thermoplastic with a high modulus used as a coating material capable of withstanding temperatures in the range 200-260°C. *LNP.* Name unverified.

22497 P.S.E. No. 15 powder
An explosive. It is a mixture of ammonium perchlorate and rosin.

22498 P.U.R.E.-CMC
PU rubber elastomer (Part A); isocyanate (Part B); cold molding compound for making rubber molds, flexible blocks and cases, and for applications requiring unusual resistance to abrasion, impact, aging, and severe wear; the cured material has outstanding tear and tensile strength to withstand extensive flexing and use; for use with plaster, gypsum and cements; mix ratio, 1:1 by weight. *Perma-Flex Mold.*

22499 P® -10 Acid
141-22-0 8378 205-470-2
C18H34O3
Ricinoleic acid
12-hydroxy-(cis)-9-octadecenoic acid; 9-Octadecenoic acid, 12-hydroxy-, [R-(Z)]-. Chemical intermediate; imparts lubricity and rust-proofing to soluble cutting oils; basis for grease, soaps, resin plasticizers, and ethoxylated derivatives. *CasChem.*

22500 P0820
Propyltris(trimethylsiloxy)silane
Low viscosity silicone fluid with high control over volatility, viscosity, density and refractive index, and unique solvent properties; used as lubricant. *Hüls Am.*

22501 P-11
25928-94-3
Epoxy resin compound; highly-filled gyro grade potting compound; absence of volatiles, usable in BFC [poly(bromotrifluoroethylene)] damping fluids, unaffected by poly(chlorotrifluoroethylene) oil; excellent machinability for general purpose applications; mix ration: 3.20 phc with Activator BA-1 *Bacon.*

22502 P13 N
A trademark for a range of polyimide varnishes and coatings. *TRW Inc.* Name unverified.

22503 P-2003-K and P-2020-T
Low-density polyethylenes, of Soviet origin. No manufacturer.

22504 P-289
1072-35-1 5442 214-005-2
A proprietary name for a PVC plasticizer of the lead stearate type. *Haagen Chemie BV.* Name unverified.

22505 P3
A proprietary degreasing material; a mixture of waterglass and trisodium phosphate in solid form; used for cleaning metals, glass, and textiles; stated to have no corrosive action on aluminum, aluminum alloys, tin, zinc, and brass. No manufacturer.

22506 P3-Almeco
Chemicals for aluminum anodizing process. *Nickerson Chemicals Ltd.*

22507 P3-Armourbond
Heat resistant anticorrosive coatings. *Nickerson Chemicals Ltd.*

22508 P3-Carclin
Detergent-based vehicle wash. *Nickerson Chemicals Ltd.*

22509 P3-Croni
Spray booth paint denaturants. *Nickerson Chemicals Ltd.*

22510 P3-Ferroclene
Derusting chemical system. *Nickerson Chemicals Ltd.*

22511 P3-Ferromede
Metal protection system. *Nickerson Chemicals Ltd.*

22512 P3-Maxan
Paint strippers. *Nickerson Chemicals Ltd.*

22513 P3-Rodine
Acid inhibitors. *Nickerson Chemicals Ltd.*

22514 P3-Stripalene
Paint strippers. *Nickerson Chemicals Ltd.*

22515 P3-Suncorrite
Anti-corrosive coatings. *Nickerson Chemicals Ltd.*

22516 P-400 Series
High bake phenolic coatings; for immersion service in acids, ammonias, petrol, products, dairy, food products, industrial chemicals; for tank linings, heat transfer equipment, piping, fans, blowers, ductwork, exhaust hoods, tank cars, process equipment. *Heresite Protective Coatings.*

22517 P-51
25928-94-3
Unfilled epoxy compound; general-purpose potting compound, room temperature cure to clear, slightly flexible material; for encapsulation of electronic components for use up to 250 F; mix ratio: 19.5 phc with Activator BA-21. *Bacon.*

22518 P-51
1072-35-1 5442 214-005-2
$C_{36}H_{70}O_4Pb$
lead stearate
A proprietary name for dibasic lead stearate used as a heat stabilizer for PVC, a dier in varnishes and an extreme pressure lubricant. *BF Goodrich UK.* Name unverified.

22519 P-56
25928-94-3
Epoxy resin compound; thermally conductive potting compound; P-56 settles but offers maximum thermal conductivity for settled portion; mix ratio: 71 phc with Activator BA-22. *Bacon.*

22520 P-60-10 and P-60-20 Cold Molding Compounds
Two-component polysulfide system; molding material, excellent transfer of pattern detail for flexible molds (fine detail with low dimensional tolerance); for ceramic applications (reproduces high quality plaster molds from flexible blocks or cases to increase production and reduce breakage); mix ratio: 100 parts P-60 or P-60-S Base a/3 parts Activator C; bulk liquid; Shore hardness A20 and 10 respectively. *Perm-Flex Mold.*

22521 P-80F
25928-94-3
Highly filled epoxy compound; instrument grade potting compound; contains; different ceramic filler than P-80C; preferable for new applications due to marginally better properties; mix ratio: 6.0 Fr BA-42 activator. *Bacon.*

22522 P-85
25928-94-3
Highly filled epoxy compound; low cost, heat-curing potting compound for casting applications requiring ease of handling, long work life, short cure time; offers low coefficient of thermal expansion, high heat distilled temperature, good electrical properties; in electronic applications involving high voltages; mix ratio: 8.0 phc with Activator BA-60. *Bacon.*

22523 PA-111
32131-17-2
Nylon 6/6
General purpose molding compound for mechanical parts, gears, bearings, etc.; UL recognized. *Bay Resins.*

22524 PA-111CF30
32131-17-2
Nylon 6/6
30% carbon fiber-reinforced. *Bay Resins.*

22525 PA-111G13
32131-17-2
Nylon 6/6, 13% glass fiberreinforced; molding compound of exceptional strength and stiffness; recommended for replacement of alloy castings, gears, brackets, etc. *Bay Resins.*

22526 PA-121
32131-17-2
Nylon 6/6, impact-modified; molding compound with increased impact resistance and toughness; used for clips, automotive components, power tool housings, recreational parts, etc. *Bay Resins.*

22527 PA-211
25038-54-4 6832
Nylon 6 homopolymer; molding compound for mechanical components, fittings, gears, handles, and UL electrical applications such as switches, receptacles, and plugs. *Bay Resins.*

22528 PA-211G13
25038-54-4 6832
Nylon 6, 13% glass fiberreinforced; molding compound for maximum strength and stiffness; used for automotive under-hood components such as timing chain covers, vacuum reservoirs, painted exterior body parts, door and window hardware, gears, and connectors. *Bay Resins.*

22529 PA-211N40
25038-54-4 6832
Nylon 6, 25% mineral-and 15% glass fiber-reinforced; molding compound offering balance of engineering properties, low warpage, resistance to sink-mark formation; for painted exterior body parts, door and window hardware. *Bay Resins.*

22530 PA-221
25038-54-4 6832
Nylon 6 copolymer, impact-modified molding compound recommended for applications requiring improved toughness over homopolymer; used for trim clips, fasteners, elec. connectors, highimpact fact masks, safety helmet headbands, medical clamps, power tool co *Bay Resins.*

22531 PA-57
9006-04-6 232-689-0
Lightly cross-linked natural rubber extended with a light-colored, nonstaining oil; processing aid for extrusions, calendering, and open steam curing. *Akrochem.*

22532 PA-80
9006-04-6 232-689-0
Lightly cross-linked natural rubber extended with a light-colored, nonstaining oil; processing aid when blended with natural rubber, SBR, neoprene, or nitrile rubber; for extrusions, calendering, and open steam cure. *Akrochem.*

22533 Pabagel
150-13-0 443 205-753-0
p-aminobenzoic acid
UV screen. *Galderma Laboratories.* Name unverified.

22534 Pabanol
150-13-0 443 205-753-0
p-aminobenzoic acid
UV screen. *ICN Pharmaceuticals Inc.* Discontinued.

22535 Pacer
Soluble concentrate containing 360 g chlormequat and 180 g 2-chloroethylphosphonic (ethephon) acid per liter; plant growth regulator for use in winter wheat. *Farm Protection Ltd.*

22536 pacherite
$BiVO_4$
bismuth vanadate
A mineral, bismuth vanadate.

22537 Pacific Sea Kelp Glycolic Extract B-1063
Source of minerals, vitamins, and amino acids; used in hair care cosmetics adding luster to hair and helping to keep scalp healthy. *Bell Flavors & Fragrances.*

22538 Packfong
A nickel silver. It contains from 26-44% copper, 16-37% zinc, and 32-41% nickel. One alloy contains 40.4% copper, 25.4% zinc, 31.6% nickel, and 2.6% iron.

22539 Packman
Closed fill system pack opener. *Schering Agrochemicals Ltd.* Discontinued.

22540 paclobutrazol
76738-62-0 7118
$C_{15}H_{20}ClN_3O$
(RName unverified.,RName unverified.)-(±)-β-[(4-Chlorophenyl)methyl]-α=(1,1-dimethylethyl-1H-1,2,4-triazole-1-ethanol
Bonzi; Bounty; Clipper; Club; Cultar; Drize; Holdfast; Inevitan; Molass; Oryze; Parlay; PP 333; Predict; Smarect; Trimmid. Suspension concentrate containing 250 g/l paclobutrazol; plant growth regulator. Used to inhibit vegetative growth in fruit trees. mp = 165°; d = 1.22; soluble in H_2O (35 mg/l), more soluble in organic solvents; LD_{50} (rat orl) = 2000 mg/kg *ICI Chem & Polymers Ltd.*

22541 paclobutrazol
76738-62-0 7118
$C_{15}H_{20}ClN_3O$
(RName unverified.,RName unverified.)-(±)-β-[(4-chlorophenyl)methyl]-α-(1,1dimethylethyl)-1H-1,2,4-triazole-1-ethanol
Bonsai; Bonzi; Bounty; Club; Clipper; Cultar; Holdfast; Inevitan; Drize; Molass; Oryze; Parlay; PP 333; Predict; Smarect; Trimmid. Plant growth regulator, gibberellin biosynthesis inhibitor. Used on fruit trees, flowers, turf and rice. mp = 165°; soluble in H_2O (35 mg/l), more soluble in organic solvents; LD_{50} (rat orl)= 2000 mg/kg.

22542 paclobutrazol
76738-62-0 7118
$C_{15}H_{20}ClN_3O$
β-((4-chlorophenyl)methyl)-α-(1,1-dimethylethyl)-1H-1,2,4-triazole-1-ethanol, -(+-)-; Bonsai; Cultar; ICI-PP-333; PP-333; Clipper, Parlay. Plant growth regulator. mp = 165-166°; d = 1.22; soluble in H_2O (35 mg/l), more soluble in organic solvents.

22543 paco
Pacos. A Peruvian term for ferruginous earth containing small quantities of metallic silver.

22544 Pacvac
Liquid ring vaccuum pump, baseplate mounted complete with liquid/air separator; used for vacuum applications where the customer requires a pump set where only electrical pipe connections have to be made. *Sihi Pumps (UK) Ltd.*

22545 Pacwet
Dust suppressant (especially in coal mines both above and underground). *Pacific Chemical Industries Pty Ltd.* Name unverified.

22546 Padac®
β-Lactamase substrate
Calbiochem Corp.

22547 Paddox
Herbicide containing the sodium/potassium salts of MCPA, mecoprop, and dicamba. *ICI Chem & Polymers Ltd.*

22548 Paddox
Soluble concentrate of 18 g dicamba, 237 g MCPA, and 80 g mecoprop per liter; used for weed control in cereals and grassland. *Farm Protection Ltd.*

22549 Padimate O
21245-02-3 3282 244-289-3
C17H27NO2
2-ethylhexyl-4-dimethylaminobenzoate
octyl dimethyl PABA; Escalol 507; Eusolex 6007; octyl Dimethyl-PABA; octyl Dimethylaminobenzoate; Padimate 0; Sundown. A PABA derivative that is used as a sun-screening agent. Of the PABA group it is the most commonly used agent in sunscreens marketed in the US. It is also used in cosmetics for skin, hair, and nails such as moisturizers and lipsticks and lip balms. *Lipo.*

22550 Pafra
Synthetic emulsion adhesives. *Pafra Ltd.* Name unverified.

22551 Pagid
Asbestos and nonasbestos friction material; used for drum brake linings for trucks and passenger cars, disc brake pads for passenger cars, clutch facings for trucks and passenger cars, rolled material for industrial application, slip bearings for tooling machines; repair, servicing and maintenance of vehicle brakes. *Caramba Chemie GmbH.* Name unverified.

22552 painter's naphtha
A petroleum distillate. It has a boiling point of 105-200 C.

22553 pala gum
Indian gutta-percha. A product from a Ceylon tree. The coagulated juice resembles gutta-percha.

22554 palaite
7782-76-5 5777
$HMnO_4P$
Hydrated manganese phosphate, a mineral.

22555 Palamid®
Masterbatches of organic and inorganic dyes in polyamide; for polyamide spin dyeing (fibers and filaments). *BASF AG.*

22556 Palamoll®
Plasticizers for plasticized PVC products resistance to oil, gasoline and bitumen with little migration tendency. *BASF AG; BASF plc.*

22557 Palamoll® 632
A polyester of adipic acid and propanediol; a plasticizer for PVC. *BASF plc.* Name unverified.

22558 Palamoll® 644 and 646
A polyester of adipic acid and butanediol; a plasticizer for PVC. *BASF plc.* Name unverified.

22559 Palamoll® 645 and 647
Propriety polyadipates having viscosities of 60 mPa·s and 10,000 mPa·s at 20°C respectively; used as polymeric plasticizers. *BASF plc.* Name unverified.

22560 Palamoll® 855
A proprietary polymeric plasticizer having a viscosity of 5000 mPa·s. *BASF plc.* Name unverified.

22561 Palanil®
Dyestuffs for polyester fibers. *BASF plc.*

22562 Palanil® Carrier
Auxiliaries for dyeing polyester fibers. *BASF plc.* Name unverified.

22563 Palanil® Dyes
Disperse dyes for dyeing polyester fibers and for printing acetate, triacetate, polyester, polyamide, and acrylic fibers. *BASF.*

22564 Palanil® P Dyes
Disperse dyes with special finish for printing polyester, acetate, triacetate, and acrylic fibers, esp. in combination with synthetic thickening agents of the Lutexal type. *BASF.*

22565 Palanil® T Dyes
Disperse dyes especially for the thermosol dyeing of polyester fibers and polyester/cellulosic fiber blends. *BASF.*

22566 Palanthrene®
Vat dyes for dyeing cellulosic fibers, textile printing. *BASF.*

22567 Palanthrene® T
Specialty vat dyes meeting demands imposed in the thermosol, pad-steam dyeing of polyester/cellulosic fiber blends. *BASF.*

22568 palao amarillo
A rubber obtained from the Mexican *Euphorbia fulva* . It has a high resin content.

22569 Palapreg®
Unsaturated polyester resins; SMC/BMC resins and resin systems for applications in electrical, automotive, building, and sanitary industries; light-hardening, glass fiber-reinforced, semifinished grades available. *BASF AG; BASF plc.*

22570 Palatal®
Unsaturated polyester resins and vinyl ester resins; for glass fiber-reinforced and unreinforced applications, e.g., building sector, containers, pipes, silos, boat and ship building, motor vehicle sector, electrical industry, chemical industry and mechanical engineering. *BASF AG; BASF plc.*

22571 Palatal® KR 1397
A proprietary unsaturated polyester based on chlorendic acid dissolved in monostyrene; used in the manufacture of articles made of glass-reinforced plastics. *Dexine Rubber Co Ltd.* Name unverified.

22572 Palatal® P5
A proprietary unsaturated polyester resin dissolved in monostyrene; it has low viscosity and is used in the manufacture of articles made of glass-reinforced plastics. *BASF plc.* Name unverified.

22573 Palatal® P8, P50T and P52TL
Proprietary polyester resins used in the manufacture of articles made of glass reinforced plastics. *BASF plc.* Name unverified.

22574 Palatal® S333
A proprietary polyester resin used in the manufacture of articles made of glass-reinforced polyesters. *BASF plc.* Name unverified.

22575 Palatase
A fungal lipase produced by fermentation of a selected strain of *Aspergillus niger*; used for production of certain Italian cheese types and other specialty cheeses in which a modest lipolysis is desired. *Novo Nordisk.*

22576 Palatex® NB-2
Crease inhibitor during dyeing of cotton fabrics in becks and jets. *BASF.*

22577 Palatin® Fast Dyes
Sulfo-group containing 1:1 metal complex dyes for dyeing and printing wool, polyamide, and silk fibers. *BASF; BASF AG; BASF plc.*

22578 palatinit
A mixture of sodium hydrosulfite (blankit) and zinc dust; used as a bleaching agent.

22579 Palatinol
Phthalate plasticizers. *BASF plc.*

22580 Palatinol 1C
84-69-5 201-553-2
Diisobutyl phthalate
A nonvolatile softening agent for cellulose esters. *BASF AG.* No manufacturer.

22581 Palatinol® 11
Undecyl phthalate
Plasticizer for plasticized PVC *BASF AG.*

22582 Palatinol® 711
Phthalic acid ester consisting of C7-11 alcohols; plasticizer with low volatility for plasticized PVC products with low brittle temps. *BASF AG.*

22583 Palatinol® A
84-66-2 201-550-6
Diethyl phthalate
Plasticizer for cellulose lacquers; dissolves cellulose nitrate, ester gum, coumarone, etc. *BASF.* Name unverified.

22584 Palatinol® AH
Di-2-ethyl benzylphthalate and dioctylphthalate. PVC plasticisers for general application. *BASF plc.* Name unverified.

22585 Palatinol® C
84-74-2 1622 201-557-4
Dibutyl phthalate
Plasticizer for nitrocellulose dyes, surface coatings, for leather/fur industry *BASF AG.*

22586 Palatinol® D10
27554-26-3 248-523-5
Diisooctyl phthalate
Plasticizer for PVC. *BASF plc.* Name unverified.

22587 Palatinol® DBP
84-74-2 1622 201-557-4
Di-n-butylphthalate
Plasticizer for PVAc-based coatings and adhesives; used to bond paper and in lamination of vinyl wall coverings to gypsum board; NC lacquers and cellophane coatings; manufacturing of smokeless powders for military use, processing aid for nitrile rubbers, carrtier solvent for pigmetns and dyes, solvent for printing ink vehicles, fixative for perfumes, modifier of phenolic laminates, component of floor waxes. *BASF.* Name unverified.

22588 Palatinol® DIDP
68515-49-1 271-091-4
Diisodecylphthalate
Primary monomeric plasticizer for vinyls; used in hightemp. processing in application such as wire and cable formulations, automotive formulations. *BASF.* Name unverified.

22589 Palatinol® DN
68515-48-0 271-090-9
Diisononyl phthalate
Plasticizer for PVC. *BASF plc.* Name unverified.

22590 Palatinol® DOA
103-23-1 203-090-1
Di (2-ethylhexyl) adipate
Plasticizer used in PVC, NC, and rubber; food contact applications; plastisols; dip coating formulations. *BASF.* Name unverified.

22591 Palatinol® DOP
117-81-7 1291 204-211-0
Di (2-ethylhexyl) phthalate
General-purpose plasticizer for vinyl compositions; solvating power for PVC. *BASF.* Name unverified.

22592 Palatinol® K
Dibutylglycol phthalate
Plasticizer for PVC. *BASF plc.* Name unverified.

22593 Palatinol® M
131-11-3 3304 205-011-6
Dimethyl phthalate
Plasticizer. *BASF.* Name unverified.

22594 Palatinol® N
68515-48-0 271-090-9
Diisononyl phthalate
Primary plasticizer for PVC, plastisols; used for film for the building trade and engineering. *BASF.* Name unverified.

22595 Palatinol® TOTM
3319-31-1 222-020-0
Tri (2-ethylhexyl) trimellitate
Primary, monomeric plasticizer for PVC homopolymer and copolymer resins; used where extreme low volatility of a plasticizer is required; used in vinyl compositions, in high temperature wire insulation and critical automotive applications. *BASF.* Name unverified.

22596 Palatinol® Z
68515-49-1 271-091-4
Diisodecyl phthalate
Plasticizer for PVC. *BASF plc.* Name unverified.

22597 Palatone
A proprietary trade name for an acrylic denture material. No manufacturer.

22598 Palau
A platinum substitute. It is an alloy of gold and palladium, usually 80% gold and 20% palladium. Another alloy termed Palau contains 60% nickel, 20% platinum, 10% palladium, and 10% vanadium.

22599 Pale 4
Polymerized castor oil; pigment wetting/dispersing agent; plasticizer for resins, gums, polymers; lubricant, penetrant; coupling solv.; adhesion promoter; for cellulose lacquers, inks, adhesives, industrial lubricants, polishes, caulks, leather dressing, hydraulic fluids, rubber compounding and gasket cement. *CasChem.*

22600 pale acid
Nitric acid containing less than 0.1% nitrogen oxides.

22601 pale oils
A name applied to a distillate from the residue of petroleum which has been treated with acid and soda, washed or filtered to a certain degree of refining or color. They have a light and medium viscosity; used lubricants for rapid motion machines.

22602 Palegal® A
Dispersant and leveling agent for polyester dyeing. *BASF.*

22603 Palegal® NB-SF
Nonfoaming leveling agent, auxiliary for high temperature dyeing of polyester fibers. *BASF.*

22604 Palenine
Leveling and stripping agent for textiles. *Catawba-Charlab.*

22605 Palette 70
On glaze colors for porcelain. *Degussa Ltd.*

22606 Paliocrom
Special effect copper phthalocyanine pigment. *BASF plc.*

22607 Paliogen®
Organic pigments; for high-quality paints, coloring plastics, tin plate inks, artists' paints and crayons. *BASF AG.*

22608 Paliotan®
Co-finish pigments; for high-quality industrial finishes and polymer dispersions. *BASF AG.*

22609 Paliotol®
Organic pigments; for special surface coatings, for coloring plastics, for production of artists' paints and crayons, for tin plate inks. *BASF AG.*

22610 Palite
22128-62-7 244-793-3
$C_2H_2Cl_2O_2$
Chloromethyl chloroformate
A military poison gas.

22611 palladium
7440-05-3 7121 231-115-6
Metallic element; Pd; alloys for electrical relays and switching systems in telecommunications equip.; air craft spark plugs; protective coatings. mp = 1555°; bp = 3167°; d_4^{20} = 12.02. *Aldrich; Atomergic Chemetals; Degussa; Noah Chem.*

22612 palladium asbestos
An asbestos coated with palladium used in gas analysis for the absorption of hydrogen.

22613 palladium black
7440-05-3 7121 231-115-6
A finely divided palladium used as a catalyst in the hydrogenation of oils.

22614 palladium gold
White gold. An alloy of 90% gold and 10% palladium. An alloy of 40% copper, 31% gold, 19% silver, and 10% palladium, is also known by this name.

22615 palladium oxide
1314-08-5 7125 215-218-3
Opd
palladium monoxide
Reduction catalyst in organic synthesis. d = 8.3; insoluble in H_2O. *Atomergic Chemetals; Degussa.*

22616 palladium red
Ammonio-chloride of palladium; a red pigment.

22617 palladium(II) chloride
7647-10-1 7122 231-596-2
Cl_2Pd
palladium chloride (ous)
palladous chloride; palladium dichloride; palladium chloride anhydrous. Analytical chemistry, electrodeless coatings for metals, photography, leak detection in gas lines, indelible inks, catalyst. mp = 678-680°; soluble in H_2O, EtOH, Me_2CO; MLD (rbt iv) = 18.6 mg/kg. *Aldrich; Atomergic Chemetals; Dajac Labs; Degussa; Spectrum Chem. Mfg.*

22618 palladium(II) nitrate
10102-05-3 7124 233-265-8
N_2O_6Pd
palladous nitrate
palladium nitrate (ous). Analytical reagent, catalyst. Poorly soluble in H_2O. *Atomergic Chemetals; Degussa.*

22619 Pallgrip
Adhesives for use as an antislip agent on bags to be stacked. *Norsk Hydro AS.* Name unverified.

22620 Palliag
A range of precious metal alloys; for dentistry and dental engineering. *Degussa Ltd.*

22621 Pallinal®
Metiram, nitrothal-isopropyl; for control of powdery mildew in fruit, vegetables, hops, ornamentals *BASF AG.*

22622 Pallitop®
Nitrothal-isopropyl, metiram; for control of powdery mildew in apples, pears *BASF AG; BASF plc.*

22623 Pallitop® S
Nitrothal-isopropyl-sulfur mixture. Contact fungicide against powdery mildew in apples *BASF AG.*

22624 palm kernel oil
8023-79-8 232-425-4
Oils, palm kernel. Oil obtained from seeds of *Elaeis guineensis. Alba Int'l.; Karlshamns; Penta Mfg.; Stevenson Bros.*

22625 palm oil
8002-75-3 232-316-1
Oils, palm; palm butter; palm grease; palm acidulated soapstock. Natural oil obtained from *Elaeis guineensis*; yellow brown buttery edible solid; soap manufacture, pharmacy, food shortening; cutting tool lubricant; cosmetics; softener in rubber processing; cotton goods fini *Alba Int'l.; Karlshamns; Penta Mfg.; Stevenson Bros.*

22626 Palm oil
8002-75-3 232-316-1
Lipovol PAL; Oils, palm; Palm butter; Palm grease. Fat replacer; coating agent; emulsifier; lubricant; texturizer, food shortening, margarine. Sol.in alcohol, chloroform, carbon disulfide; dens. 0.952; mp=26-30°. *Lipo.*

22627 palm pitch
A pitch obtained by the treatment of palm oil with sulfuric acid.

22628 palm wax
A yellow wax from *Ceroxylon andicola*; used as a beeswax substitute.

22629 palmerite
Aluminum-potassium phosphate, a mineral.

22630 Palmitamidopropyl dimethylamine
39669-97-1 254-585-4
N-[3-(Dimethylamino)propyl]hexadecanamide
Incromine PB; Chemidex P; Schercodine P; Dimethylaminopropyl palmitamide. Intermediate, substantive conditioner, thickener, and emulsifier used in hair care prods. Cationic; waxy solid; mw=340; soluble in organic solvents; mp = 55-60°; alkali value 160-170. *Chemron.*

22631 palmitic acid
57-10-3 7128 200-312-9
$C_{16}H_{32}O_2$
hexadecanoic acid
hexadecylic acid; cetylic acid; Univol U332. A proprietary trade name for 90% palmitic acid. Used in manufacture of soaps. mp = 63-64°; bp_{15} = 215°; d_4^{62} = 0.853; n80:s$_D$= 1.4273; insoluble in H_2O, soluble in organic solvents; LD_{50} (mus iv) = 57 mg/kg. *UOP Inc.* Name unverified.

22632 palmitic acid
57-10-3 7128 200-312-9
$C_{16}H_{32}O_2$
hexadecanoic acid
hexadecylic acid; cetylic acid; Univol U332. A proprietary trade name for 90% palmitic acid. Used in manufacture of soaps. mp = 63-64°; bp_{15} = 215°; d_4^{62} = 0.853; n80:s$_D$= 1.4273; insoluble in H_2O, soluble in organic solvents; LD_{50} (mus iv) = 57 mg/kg. *UOP Inc.* Name unverified.

22633 palmitic acid
57-10-3 7128 200-312-9
$C_{16}H_{32}O_2$
hexadecanoic acid
hexadecylic acid; cetylic acid; Univol U332. A proprietary trade name for 90% palmitic acid. Used in manufacture of soaps. mp = 63-64°; bp_{15} = 215°; d_4^{62} = 0.853; n80:s$_D$= 1.4273; insoluble in H_2O, soluble in organic solvents; LD_{50} (mus iv) = 57 mg/kg. *UOP Inc.* Name unverified.

22634 palmitic acid
57-10-3 7128 200-312-9
$C_{16}H_{32}O_2$
hexadecanoic acid
cetylic acid; hexadecylic acid. A saturated fatty acid, used in the manufacture of metallic palmitates, soaps, lube oils, waterproofing, food-grade additives. mp = 63-64°; bp_{15} = 215°; d_4^{62} = 0.853; n_D^{80} = 1.4273; insoluble in H_2O, soluble in organic solvents; LD_{50} (mus iv) = 57 mg/kg. *Akzo; Ashland; Henkel/Emery; Unichema; Witco/Humko.*

22635 Palmitic Acid
57-10-3 7128 200-312-9
$C_{16}H_{32}O_2$
Hexadecanoic acid
Emersol® 143; Hexadecylic acid; cetylic acid; Glycon® P-45; Hystrene® 8016; Industrene® 4516; Prifac 2960; Univol U332. Detergent intermediate; opacifier in cosmetics, soaps, emulsifiers, chemical specialties. White

crystalline scales; mp=63-64°; insol.in water; sol.in alcohol, ether, chloroform; LD_{50}(I.v., mise)=57+-3.4 mg/kg *Henkel/Emery; Lonza; UOP Inc.; Witco/Humko; Unichema.*

22636 palmitin
Commercial palmitic acid is incorrectly called by this name.

22637 Palomar®
Phthalocyanine pigments for the surface coatings industry, especially for automotive coatings. *Bayer AG.*

22638 Palorium
A platinum substitute. It is a white alloy of gold and platinum only distinguished from platinum with difficulty. It is a ductile, homogeneous alloy with a melting-point of 1310 C, and it remains stronger than platinum on heating.

22639 Palormone
Aqueous solution containing 50% w/v 2,4-D as the amine salt; for use as an agricultural herbicide. *Universal Crop Protection Ltd.*

22640 Palusol
Intumescent fireboard based on sodium silicate; used for fire and smoke seals at fire resistant doors and service penetrations of fire walls. *BASF plc.*

22641 Pamak
A range of tall oil fatty acids; used in chemicals and chemical processing, construction and building, floor coverings, paints and coatings, rubber, soaps, detergents and household products. *Hercules.*

22642 Pamolyn
61790-12-3 263-107-3
Tall oil fatty acids. *Hercules Ltd.*

22643 Pamolyn
60-33-3 5529 200-470-9
Conjugated linoleic acid derived from vegetable sources; used as epoxy ester resin intermediates for adhesives and sealants, as chemical intermediates for conjugated double-bond reactions, as modifiers of stirenated, vinylated and methacrylated alkyds. *Hercules.*

22644 Pamolyn 100 FGK
112-80-1 6965 204-007-1
Pamolyn 100, 100 FG. Oleic acids from vegetable sources; used as plasticizers, emulsifiers and textile and wet-processing aids, in the food and drink industry, in the manufacture of personal care products and cosmetics. *Hercules.*

22645 Pamolyn 125
60-33-3 5529 200-470-9
Pamolyn 200, 240. Technical grade linoleic acid derived from vegetable sources; used in caulking and sealant compositions; used to produce oleoresinous printing ink vehicles, for making epoxy resin ester coatings and pale color-retentive fast drying alkyds. *Hercules.*

22646 Pamolyn 327B
A partially bodied, conjugated fatty acid; used as a replacement for G-H-viscosity dehydrated castor oil in short to medium oil alkyds and copolymer alkyd resins. *Hercules.*

22647 Pamolyn 380
60-33-3 5529 200-470-9
Conjugated linoleic acid derived from vegetable sources; used as epoxy ester resin intermediates for adhesives and sealants, as chemical intermediates for conjugated double-bond reactions, as modifiers of styrenated, vinylated, and methacrylated alkyds. *Hercules.* Discontinued.

22648 Pamolyn® 125
112-80-1 6965 204-007-1
Oleic acid
For conversion to soaps and sulfonates for use as textile processing aids, automotive additives, mold lubricants, surfactants, agents for production of synthetic rubber. *Actrachem.* Discontinued.

22649 pan scale
The calcium sulfate, containing some sodium chloride, which settles out during the crystallization of salt from brine. It is sold as salt lick for cattle, also for manuring purposes.

22650 Panabath
Panacide preparation for water baths. *BDH Chemicals Ltd.*

22651 Panablock
Biocide. *BDH Chemicals Ltd.*

22652 Panacide
97-23-4 3120 202-567-1
Dichlorophen
Fungicide, bactericide and algicide used as a moss-killer. *BDH Chemicals Ltd.*

22653 Panaclean
Biocidal cleaner. *BDH Chemicals Ltd.*

22654 Panacryl
Cationic dyestuffs
For dyeing acrylic fibers, modacrylic fibers and basic dyeable polyesters and nylons. *Holliday Dyes & Chemicals Ltd.*

22655 Panacur
43210-67-9 4000 256-145-7
Fenbendazole
An anthelmintic. *Hoechst-Roussel Pharmaceuticals Inc.* Name unverified.

22656 Panagran
Biocide *BDH Chemicals Ltd.*

22657 Panalane® L-14E
61693-08-1
Hydrogenated polyisobutene
Specifically designed for cosmetics applications; outstanding feel and moisturizing ability *Amoco Chemical Co.*

22658 Panama bark
Quillaia bark.

22659 Panama crimson
A coloring matter obtained from the leaves of a vine called china.

22660 Panasand
Biocide *BDH Chemicals Ltd.*

22661 Panasol
Aromatic solvents. *Amoco Chemical Co.*

22662 Panaspray
Biocide. *BDH Chemicals Ltd.*

22663 Panastat
Biocide. *BDH Chemicals Ltd.*

22664 Panatest
Bacterial test kit. *BDH Chemicals Ltd.*

22665 Panazyme
Fungal protease. *Rhône-Poulenc UK.*

22666 Pancil T
26530-20-1 6853 247-761-7
Paste containing 1% w/w octhilinone; fruit tree canker paint. *Rohm & Haas UK.*

22667 Pancoxin
59-40-5 9109 200-423-2
A proprietary preparation containing sulfaquinoxaline; a veterinary coccidiostat. No manufacturer.

22668 Pancoxin
121-25-5 631 204-458-4
A proprietary preparation of amprolium; an antiprotozoan for veterinary use. No manufacturer.

22669 Pancoxin
59-06-3 3791 200-414-3
A proprietary preparation of ethopabate; a veterinary anti-protozoal. No manufacturer.

22670 pancreas diastase
Amylopsin
An enzyme.

22671 Pancrease
8049-47-6 7137 232-468-9
Pancrelipase
A concentrate of pancreatic enzymes standardized for lipase content; enzyme; used as a digestive adjunct. *McNeil Pharmaceuticals.* Name unverified.

22672 Pancreatic Lipase 250
9001-62-1 5536 232-619-9
Lipase; enzyme for hydrolysis of triglycerides to glycerol and fatty acids; used for development of flavors. *Solvay Enzymes.*

22673 pancreatokinase
A mixture of Eukinase and pancreatin.

22674 Pancreol
A trypsin preparation; used for bating skins. *Hodgson Chemicals Ltd.*

22675 Pancrex V
8049-47-6 7137 232-468-9
A proprietary preparation of concentrated pancreatin; used in the treatment of cystic fibrosis. *Paines & Byme Ltd.* Name unverified.

22676 Pancrolin
Enzyme preparation; used for leather processing. *Hodgson Chemicals Ltd.*

22677 Pandex
A selenium preparation for rubber vulcanization.

22678 Panelyte
A proprietary trade name for phenol-formaldehyde laminated products and paper, fabric, wood veneer, fiber glass, and asbestos base thermosetting plastics for structural work. No manufacturer.

22679 Panex
Diacetyltartaric ester of edible mono-diglycerides; used as

emulsifier/antistaling additive for bread fats, mayonnaise and sauces. *Harcros Australia.*

22680 Panilax®
A trade name for materials made from anilineformaldehyde synthetic resin; they are thermoplastic but have a softening point about 100°C. No manufacturer.

22681 Panodan 235, FDP-K, SD, SD-K
Monoglyceride diacetyl tartaric acid ester; food emulsifier. *Grindsted Products.*

22682 Panogen M
151-38-2 205-790-2
$C_5H_{10}HgO_3$
2-Methoxyethylmercury acetate
(acetato-O)(2-methoxyethyl)mercury; methoxyethyl mercury acetate; MEMA; Panogen; acetoxy(2-methoxyethyl)mercury. Cereal seed treatment. Formulated as aqueous solutions or dusts and used chiefly as a seed protectant. Use of alkyl mercury fungicides in the United States has been virtually prohibited for several years. Phenyl mercuric acetate is still used to control diseases of turf, but other applications have been sharply restricted. *Embetec Crop Protection Ltd.*

22683 Pan-O-Lite
7785-88-8 232-090-4
Sodium aluminum phosphate
Phosphoric acid, aluminum sodium salt. Food leavening agent. *Monsanto Co.* Name unverified.

22684 Pansorbin®
Cells. *Calbiochem Corp.*

22685 Pantal
An aluminum alloy containing 0.8-2.0% magnesium, 0.4-1.4% manganese, 0.5-1.0% silicon, and 0.3% titanium. It resists corrosion, and has a tensile strength of from 18-33 $kgmm^2$.

22686 Pantalast® 1120
VC-EVA copolymer; engineered material; low smoke, low flammability, good abrasion and high temperature resistance, low HCl generation, good weathering properties; for demanding wire/cable applications; nonmigratory; available in standard colors. *Pantasote Polymers.* Name unverified.

22687 Pantarol
A proprietary product for the protection of metal; it is applied by brushing, spraying, and dipping; resists the action of light, sea air, steam, acid fumes, but is destroyed by concentrated acids and alkalis. No manufacturer.

22688 Pantene
Hair tonic, shampoo and conditioner to keep hair healthy and attractive. *Richardson-Vicks Inc.* Name unverified.

22689 Pantene Grooming Lotion
Grooming lotion. *Roche Products Ltd.* Name unverified.

22690 d-Panthenol
81-13-0 2988 201-327-3
$C_9H_{19}NO_4$
2,4-dihydroxy-N-(3-hydroxypropyl)-3,3-dimethylbutanamide
pantothenylol; N-pantoyl-3-propanolamine; pantothenol; pantothenyl alcohol; Alcopan-250; Intrapan; Pantenyl; Panthoderm; Motilyn; Bepanthen; Cozyme; Ilopan; Urupan. Nutrient, humectant for hair and skin care formulations; hair repair agent; soothing to skin. Nutritional factor, source of pantothenic acid. $bp_{0.02}$ = 118-120°; d_{20}^{20} = 1.2; $[\alpha]_D^{20}$ = 30° (c = 5); n_D^{20} = 1.497; soluble in H_2O, EtOH.

22691 Panthenylethyl ether
667-83-4 211-569-1
Humectant SD-35; Panthenyl ethyl ether. Humectant. *Presperse.* Discontinued.

22692 Panther
Suspension concentrate containing 50 g diflufenican and 500 g isoproturon per liter; used for control of weeds in winter cereals. *Rhône-Poulenc Crop Protection Ltd.*

22693 Pantholin
137-08-6 7147 205-278-9
Calcium pantothenate
Vitamin, nutritional factor. *Eli Lilly & Co.* Discontinued.

22694 Pantolit
A proprietary synthetic resin of the phenol-formaldehyde type. No manufacturer.

22695 Pantopaque
99-79-6 5074 202-787-8
$C_{19}H_{29}IO_2$
ethyl 10-(*p*-iodophenyl)undecylate
iophendylate; iofendylate. Diagnostic aid. d_{20}^{20} = 1.240-1.263; n_D^{20} =1.5230-1.5280; insoluble in H_2O, soluble in organic solvetns; LD_{50} (rat ip) = 19 g/kg. *Alcon Laboratories Inc.* Name unverified.

22696 Pantosept
A German antiseptic in which the active agent is hypochlorous acid.

22697 pantothenyl alcohol
81-13-0 2988 201-327-3
N-pantoyl-3-propanolamine
pantothenylol; dexpanthenol. The alcohol corresponding to pantothenic acid with vitamin activity; used in biochemical research, as a food additive and dietary supplement.

22698 papain
9001-73-4 7148 232-627-2
vegetable pepsin; papayotin; papoid; Arbuz; Nematolyt; Summetrin; Tromasin; Velardon; Vermizym. Pepsin-like proteolytic enzyme; a vegetable digestive ferment obtained from the unripe fruit of the papaw tree; meat tenderizer, tobacco, pharmaceutical; antihazing agent for beer; cosmetics, leather, textiles. *EM Industries; Dr. Madis Labs; Meer; Spice King; Stan Chem Int'l. Ltd.*

22699 Papain Conc
Protease; enzyme for hydrolysis of proteins; for brewing (stabilizes and chillproofs beer); meat tenderizers; pharmaceutical (digestion aids); protein modification. *Solvay Enzymes.*

22700 Paperad
Finely divided aluminum trihydrate paper pigment; used in papers. *Reynolds Metal Co.*

22701 paperhanger's alum
10043-01-3 381 233-135-0
Aluminum sulfate
Used for sizing paste.

22702 Paperine
A starch product used in paper manufacture.

22703 Paper-Pac®
Paper manufacturing additive for neutral sizing. *Sachtleben Chemie GmbH.*

22704 paper-spar
A variety of calcite, a mineral.

22705 Papi 27
Polymeric MDI
For polyurethane industry for molded elastomers; for use in structural foam molding formulations, appliances. *Dow Chemical.*

22706 Papi 4901
Polymeric MDI
For polyurethane industry for molded flexible foam. *Dow Chemical.*

22707 Papite
Acrolein with stannic chloride; a tear gas.

22708 paposite
A mineral that contains ferric sulfate.

22709 Pappenheim's stain
Pyronin stain. A microscopic stain. It consists of 1 part of concentrated solution of pyronine B and 3 parts of a concentrated solution of methyl green.

22710 paprika
Cayenne pepper.

22711 Papyrus
A proprietary casein product. No manufacturer.

22712 Par
No manufacturer.

22713 Par Clay®
1332-58-7 5294 296-473-8
Hydrated aluminum silicate (Kaolin clay); mineral filler used as filler, extender or reinforcing agent for paint, paper, rubber, ceramics, plastics and specialities. *R. T. Vanderbilt Co Inc.* Discontinued.

22714 para palm oil
Para butter. Pinot oil, a semi-drying oil from the seeds of *Euterpe oleracea.*

22715 para toner
Paranitraniline red
Used as a toner for lakes.

22716 Parabar
Oxidation inhibitors. *Exxon UK.* Name unverified.

22717 Parabis
80-05-7 1338 201-245-8
bisphenol A
A pure form of bisphenol A; used as an intermediate in the production of polycarbonate and polysulfone resins. *Dow Cheml Co Ltd, UK & Ireland.*

22718 Parable
Soluble concentrate of 100 g diquat and 100 g paraquat per liter; used for weed control in field crops. *ICI Agrochemicals.*

22719 Parabolix® 100
Degreaser, destaticizer for electronics, parabolic light fixtures. *Merix.*

22720 Paracol
Herbicide containing diuron and paraquat. *ICI Chem & Polymers Ltd.*

22721 Paracol
Wax and wax-rosin emulsions with excellent shelf, shear and chemical stability; used in paper and paperboard and to impart resistance to water and aqueous solutions; also in many types of wet or dry-formed building products. *Hercules.*

22722 Paracol® 1886
8002-74-2 7155 232-315-6
Paraffin-based aqueous emulsion; acid-breaking-type emulsion used in specialized internal sizing applications. *Hercules.* Discontinued.

22723 Paracol® 403A6
Paraffin wax-pale rosin emulsion; sizing agent for surface sizing of paper and paperboard where all-wax emulsions might cause excessive slipperiness; improves sizing and finish of writing and printing papers; used in food-pkg. grades of paper/paperboard. *Hercules.* Discontinued.

22724 Paracol® 404C
63231-60-7 264-038-1
Microcrystalline paraffin waxes and hydrocarbon waxes. Refined microcrystalline wax emulsion; sizing agent for improving surface properties of paper and paperboard; detackifier; antiblocking agent. *Hercules.* Discontinued.

22725 Paracol® 800N
8002-74-2 7155 232-315-6
Paraffin wax emulsion; sizing particleboard; superior pumping, mechanical, and spraying stability. *Hercules.* Discontinued.

22726 Paracol® 810N
8002-74-2 7155 232-315-6
Paraffin wax emulsion; sizing agent for particleboard, hardboard, and insulation board; mechanical, pumping, and spraying stability for trouble-free operation in metering pumps and spray nozzles. *Hercules.* Discontinued.

22727 Paracol® M161
8002-74-2 7155 232-315-6
Paraffin wax emulsion; for surface sizing to improve sheet water resistance and finish where sheet brightness is of concern. *Hercules.* Discontinued.

22728 paracon
A generic name for polyester elastomers. No manufacturer.

22729 paracoumarone resin
Cumar resin; cumar gum; coumarone resin; benzo-furane resin; cumar. A synthetic resin produced from coal-tar distillates; used in the production of varnishes, polishes, artificial leathers.

22730 Paracril® 1880, 1880LM
Nitrile-butadiene rubber, hot polymerized, slightly staining antioxidant; improved low temperature performance, good heat resistance and resilience; used for many military applications. *Uniroyal.*

22731 Paracril® 2813
Nitrile-butadiene rubber, cold polymerized, nonstaining antioxidant; used for continuously cured extruded goods including sponge and PVC blends; FDA compliance; excellent processing; fast cure rate. *Uniroyal.*

22732 Paracril® BJLT M-30
Nitrile-butadiene rubber, low temperature polymerized, nonstaining stabilizer; easy and safe processing, low mold fouling, low water absorption; ideal balance of low temperature flexibility and oil resistance for most automotive applications; used in seals, o-rings, shoe soles, molded and extruded goods. *Uniroyal.*

22733 Paracril® OZO
Nitrile-butadiene rubber/PVC fluxed blend (70/30), nonstaining stabilizer; used for shoe soles, rolls, wire and cable jacketing, hose tubes and covers, sponge, belting, and gasketing; excellent ozone resistance, good fuel and oil resistance, high abrasion resist, excellent color stability, easy to mix. *Uniroyal.*

22734 Paracure
Dispersions and emulsions of various chemical additives, primarily rubber latex antioxidants, accelerators and curatives; chemical additives for rubber latex compounding. *Testworth Laboratories Inc.*

22735 Paradene®
A proprietary trade name for coumaroneindene resins. *Neville.* Name unverified.

22736 Paradene® No.2
Petrol. hydrocarbon resin; used for inks, adhesives, coatings, rubbers, and concrete cures. *Neville.*

22737 Paradol
A range of fat liquors; for modification of leather handle. No manufacturer.

22738 Paradow
106-46-7 3107 203-400-5
p-dichlorobenzene
For commercial and industrial solvents, deodorants and sanitary products. *Dow UK.* Discontinued.

22739 Paradura
A proprietary trade name for a phenolic synthetic resin for varnish and lacquers. No manufacturer.

22740 Paradyne
Fuel oil additives; flow improvers. *Exxon Int'l.; Exxon UK.* Name unverified.

22741 Paraffagar
A proprietary preparation of liquid paraffin and agar-agar. No manufacturer.

22742 paraffin
8002-74-2 7155 232-315-6
C_nH_{2n-22}
Paraffin wax
hard paraffin; petroleum wax; crystalline; Hydrocarbon,. Solid mixture of hydrocarbons obtained from petroleum; characterized by relatively large crystals; Candles, paper coating, protective sealant for food products; lubricants; hot-melt carpet backing, floor polishes, cosmetics, chewing gum base; raising melting points of ointments. *EM Industries; Exxon; Humphrey; Jonk BV; Koster Keunen; Phillips 66; Texaco; Vista.*

22743 Paraffin
8002-74-2 7155 232-315-6
Glycolube® VL; Petroleum wax; Microlube A; Octowax 321; Paracol® 800N, 810N, 1886, M161; Paratulle®; Press-Aid; Ross Wax 100, 145, 165; Sasolwaks; Sasolwaks M3; Synwax; Vybar® 103,. Synthetic wax; internal and external lubricant, release agent for thermoplastics; also used for textiles. Lonza Inc; Hercules; Tiarco; Seton Healthcare Group plc; Presperse; Frank B. Ross; Sasolchem; Petrolite.

22744 paraffin, liquid
Paroline; Paroleine. A mixture of liquid hydrocarbons obtained by the distillation of the liquid remaining after the lighter hydrocarbons have been removed from petroleum. It is decolorized and purified.

22745 Parafil
Fibre reinforced thermoplastic rope/cable. *ICI Chem & Polymers Ltd.* Discontinued.

22746 Parafilm
A proprietary trade name for a rubber composition. No manufacturer.

22747 Parafix
Photographic fixer. *May & Baker Ltd.* Name unverified.

22748 Paraflow
Dewaxing oils and pour point depressants for lubricants and power transmission fluids. *Exxon Int'l.* Name unverified.

22749 Para-Flux® 4156
Aromatic process oil; plasticizer for CR compounds *C.P. Hall.*

22750 Paraform
Molded thermal insulation shapes. *The Chemical & Insulating Co Ltd.*

22751 paraformaldehyde
30525-89-4 7158 200-001-8
$HO(CH_2O)_nH$
para-formaldehyde
polyoxymethylene; paraform. A polymer of formaldehyde in which n = 8-100; used in fungicides, bactericides, disinfectants, adhesives, hardener and waterproofing agent for gelatin contraceptive creams. Soluble in H_2O, insoluble in organic solvents. *Andrulex Trading Ltd; Degussa; Hoechst Celanese; Mitsubishi Gas; Mitsui Toatsu Chem.*

22752 Paraglas
Cast acrylic sheet; used for production of light domes, illuminated signs, wash basins and bathtubs, safety coverings for machines, showcases, graduated dials, models etc. *Degussa Ltd.*

22753 Paragon II
Mixture of propylene glycol, DMDM hydantoin, methylparaben and propylparaben. Patented dual action cosmetics preservative with bactericidal and fungicidal properties. *McIntyre.*

22754 Paragon Steel
A proprietary trade name for a non-shrinking steel contains 1.55% manganese, about 0.6% chromium, and 0.25% vanadium. No manufacturer.

22755 Paragon-15
Polysiloxane resin in an aliphatic vehicle system; water repellant for masonry. *Nova Chemical Inc.*

22756 Paragrid
Synthetic reinforcement and support materials for coil engineering applications. *ICI Chem & Polymers Ltd.* Discontinued.

22757 Paragutta
A patented insulating compound for use in the manufacture of submarine telegraph and telephone cables; made from deproteinized rubber obtained by the heat treatment of rubber latex to hydrolyze the protein and washing. This rubber has reduced water-absorbing properties and is mixed with gutta percha or balata and suitable waxes to produce paragutta. No manufacturer.

22758 Paralac
Solvated solutions of synthetic rubber and resins; used for contact adhesives and coatings. *Testworth Laboratories Inc.*

22759 paraldehyde
123-63-7 7160 204-639-8
$C_6H_{12}O_3$
2,4,6-trimethyl-1,3,5-trioxane;
Paracetaldehyde; trimethyl trioxane; acetaldehyde, trimer; elaldehyde; paraacetaldehyde; PARAL; PCHO; 2,4,6-trimethyl-s-trioxane; s-trimethyltrioxymethylene. Polymer of acetaldehyde. Use in the manufacture of organic chemicals and also as a sedative and hypnotic. mp = 13°; bp = 124°; d_{25}^{25} = 0.994; n_D^{20}= 1.4049; LD_{50} (rat orl) = 1.65 g/kg.

22760 Paralene
A range of synthetic tanning agents; for leather industry. No manufacturer.

22761 Paralink
Synthetic reinforcement and support materials for coil engineering applications. *ICI Chem & Polymers Ltd.* Discontinued.

22762 Paraloid® BTA-702
methacrylate/butadiene styrene. Impact modifier. *Rohm & Haas.* Name unverified.

22763 Paraloid® EXL 2607
Acrylic and MBS additives for engineering plastics; toughening polymers particularly for automotive and leisure industries. *Rohm & Haas.*

22764 Paraloid® EXL-3330
Butyl acrylate-based; toughener for PC, PBT and PET. *Rohm & Haas.* Name unverified.

22765 Paraloid® HT-510
Polyglutarimide acrylic copolymer; heat distortion temperature modifier for PVC *Rohm & Haas.* Name unverified.

22766 Paraloid® K-120N
9011-14-7
Methyl methacrylate polymer; hard acrylic resin with excellent exterior durability and abrasion resistance; processing aid for vinyl topcoats. *Rohm & Haas.*

22767 Paraloid® KM-318F
Acrylic; impact modifier for vinyl foam applications such as sheet, profile and pipe; controls cell structure. *Rohm & Haas.* Name unverified.

22768 Paraloop
Synthetic reinforcement and support materials for coil engineering applications. *ICI Chem & Polymers Ltd.* Discontinued.

22769 Paralux
Lube oils. *Monsanto (Solaris).* Name unverified.

22770 Paramel
A range of synthetic resins; used for tanning of leather. No manufacturer.

22771 Paramel
Modified melamine formaldehyde resin; used in the paper industry. *Cyanamid BV.*

22772 Paramet Ester Gum
A proprietary trade name for rosinglycerol synthetic resin for lacquer and varnish manufacture. No manufacturer.

22773 Paramid
Lube oils. *Monsanto (Solaris).* Name unverified.

22774 Paramin DF
1698-60-8 216-920-2
Chloridazon
A pyridazinone herbicide for beet crops. *BASF plc.*

22775 Paramins
Additives for the petroleum industry. *Exxon UK.*

22776 Paramos
Benzalkonium chloride
Algicide and moss killer for paths and flower pots. *Chemsearch (UK) Ltd.*

22777 Paramount
Lube oils. *Monsanto (Solaris).* Name unverified.

22778 Paramount B
68990-82-9 273-627-2
Partially hydrogenated palm kernel oil; kosher; center fat for confectioner's coatings, vegetable dairy systems, candy centers, icings, cosmetic/pharmaceutical applications. *Van Den Bergh Foods.*

22779 Paramul® SAS
Stearamide di-isobutyl adipate-stearate; emulsifier, pearlizing agent, opacifier. *Bernel.*

22780 paranitraniline red
$C_{16}H_{11}N_3O_3$
Azophor red; para red; nitrazol; discharge lake R and RR; p-Nitro-benzeneazo beta-naphthol. *p*-Nitro-benzeneazo β-naphthol, used for dyeing cotton, and in the preparation of lakes for paper-staining.

22781 Paranol
A proprietary trade name for a phenol-formaldehyde synthetic resin. No manufacturer.

22782 Paranox
Detergents and dispersants. *Exxon Int'l.* Name unverified.

22783 paraoxon
311-45-5 7164 206-221-0
$C_{10}H_{14}NO_6P$
phosphoric acid diethyl 4-nitrophenyl ester
phosphacol; E-600; Ester 25; Eticol; Fosfakol; Mintacol; Miotisal A; Soluglaucit. Cholinesterase inhibitor. Used as an insecticide. $bp_{1.0}$ = 169-170°; d_4^{25}= 1.2683; n_D^{20}= 1.50959; λ_m = 264 nm (ε 8900); soluble in H_2O (2.4 mg/ml), more soluble in organic solvents; LD_5_70 (rat orl) = 1.8 mg/kg.

22784 Parapak
Industrial oil additives. *Exxon UK.* Name unverified.

22785 Parapel® HC-85
Linoleamidopropyl ethyldimonium ethosulfate
dimethyl lauramine isostearate. Emulsifier compatible with most anionic surfactants; hair conditioner. *Bernel.*

22786 Parapel® LAM-100
5422-34-4 226-546-1
Lactamide MEA
Hair conditioner, humectant; compatible with thioglycolate. *Bernel.*

22787 Parapel® LIS
70729-87-2 274-834-0
Dimethyl lauramine isostearate
Shampoo and conditioner additive; silky, velvety emollient. *Bernel.*

22788 Paraplast 8100
Inorganic powder; used to cast mandrels or cores used in production of hollow plastic articles (reinforced plastic ducts, filament-wound pressure vessels, etc.); offers heat resistance to 275°F. *Hexcel.*

22789 Paraplex®
Series of polymeric plasticizers; used for flexible vinyl and rubber. *C P Hall.*

22790 Paraplex® G.62
8013-07-8 232-391-0
Epoxidized soya bean oil; vinyl plasticizer and vinyl stabilizer. *Rohm & Haas.*

22791 Paraplex® G-25 100%
High molecular weight polyester sebacate; plasticizer for electrical tapes, high temperature insulation, coaxial cable, upholstery and coated fabric. *C.P. Hall.*

22792 Paraplex® G-50
Polyester adipate; pigment grinding medium; PVC plasticizer; for insulation, upholstery, window channels, liners, gaskets and coated fabrics. *C.P. Hall.*

22793 Paraplex® G-60
8013-07-8 232-391-0
Epoxidized soybean oil; polymeric type plasticizer for coating formulations based on PVC and copolymers, NC, chlorinated rubber and paraffin; permanence in surface coating films under severe exposure, high plasticizing efficiency; good flexibility at low temperatures; stabilizes materials with acid-producing components. *C.P. Hall.*

22794 Parapoid
Extreme pressure additives for gear lubricants. *Exxon Int'l.* Name unverified.

22795 Parapol
Liquid polyisobutylene
Exxon.

22796 paraquat
4685-14-7 7165 225-141-7
$[C_{12}H_{14}N_2]^{2+}$
1,1'-dimethyl-4,4'-bipyridinium
methyl viologen(2+). Non-selective contact herbicide.

22797 Paraquat + Plus
4685-14-7 7165 225-141-7
Herbicide
Monsanto (Solaris). Name unverified.

22798 pararosaniline
569-61-9 209-321-2
C19H18ClN3
4-((4-aminophenyl)(4-imino-2,5-cyclohexadien-1-ylidene)methyl)benzenamine monohydrochloride
Triamino-triphenyl-carbinol hydrochloride; p-fuchsin; C.I. 42500; Magenta; Basic Fuchsin; C. I. Basic Red 9.HCl; Paramagenta; Fuchsine, Acid; Pararosaniline Hydrochloride; C.I. Basic Red 9; C.I. Basic Red 9 monohydrochloride; Parafuchsine; Calcozine Magenta N; Pararosaniline Chloride; Basic Red 9; Basic Red 9, monohydrochloride. Dyes wool and silk, purple-red, and cotton with mordants. mp = 268-270; insoluble in H_2O.

22799 Paraset
Printing ink distillates. *Carless Refining & Marketing Ltd.*

22800 Parasiticine
A fungicide containing 57% copper sulfate, sodium carbonate and sodium bicarbonate.

22801 Parasol 17
A wide distillation range white spirit substitute with 17% aromatic content. Used as a solvent. *Sasolchem.*

22802 Paratac
Additive for lubricating oils and greases. *Exxon Int'l.* Name unverified.

22803 paratartaric acid
133-37-9 9236 205-105-7
$C_4H_6O_6$
Racemic tartaric acid

22804 Paratect Bolus
26155-31-7 6348 247-481-5
Morantel tartrate
A cattle antihelminitic. *Pfizer International.*

22805 Paratemp
10101-39-0 233-250-6
Asbestos-free calcium silicate; thermal insulator. *The Chemical & Insulating Co Ltd.*

22806 Paratherm NF
8042-47-5 232-455-8
White mineral oil; nonfouling heat transfer fluid. *Paratherm.*

22807 parathion
56-38-2 7167 200-271-7
$C_{10}H_{14}NO_5PS$
O,O-diethyl O-*p*-nitrophenyl phosphorothioate
diethoxy, nitro-phenoxy phosphorothioate; DNTP; S.N.P.; E-605; AC-3422; ENT-15108; Alkron; Alleron; Aphamite; Etilon; Folidol; Fosferno; Niran; Paraphos; Rodiatox; Thiophos. A powerful insecticide and acaricide. mp = 6°; bp = 375°; d_4^{25}= 1.26; n_D^{25}= 1.5370; insoluble in H_2O, soluble in organic solvents; LD_{50} (rat orl) = 13 mg/kg.

22808 Parathion-methyl
298-00-0 6183 206-050-1
$C_8H_{10}NO_5PS$
O,O-dimethyl O-(4-nitrophenyl) phosphorothioate
methyl parathion; metaphos; OMS 213; ENT 17292; Bladan M; Cekumethion; Devithion; Folidol-M; Fulkil; Metacide; Methyl-bladan; Nitrox; Parataf; Paratox; Partron M; Penncap-M; Tekwaisa; Wofatox. Cholinesterase inhibitor which serves as a non-systemic insecticide and acaricide with contact and stomach action. Used for control of sucking and chewing insects and mites in a wide range of crops. mp = 35-36°; bp_1 = 154°; soluble in H_2O (55 mg/l), more soluble in organic solvents; LD_{50} (rat orl)= 14-24 mg/kg.

22809 Paratie
Synthetic reinforcement and support materials for coil engineering applications. *ICI Chem & Polymers Ltd.* Discontinued.

22810 Paratol
Natural and synthetic rubber latex based adhesives and coatings; cohesive, pressure-sensitive and synthetic resin emulsions; used for numerous adhesive bonding applications. *Testworth Laboratories Inc.*

22811 Paratone
Viscosity index improver. *Exxon UK.* Name unverified.

22812 Paratulle®
8002-74-2 7155 232-315-6
White soft paraffin; tulle primary wound dressing in the treatment of burns, cuts, wounds and abrasions. *Seton Healthcare Group plc.*

22813 Paraweb
Fiber-reinforced thermoplastic webbing for civil engineering applications and used as cargo slings, snow fencing, windbreaks. *ICI Chem & Polymers Ltd.* Discontinued.

22814 Parazol
Crude dinitrodichlorobenzene
It contains *m*-dinitro-*p*-dichlorobenzene, *o*-dinitro-*p*-dichlorobenzene, and *p*-dinitro-*p*-dichlorobenzene; used as a high explosive.

22815 Parco
Wear reducing processes. *Brent Chemicals International plc.*

22816 Parco® 58-C-55
Vinyl acrylic copolymer emulsion; used in paints (interior wall, gloss and semigloss exterior, masonry); clarity, high gloss. *Thibaut & Walker.* Name unverified.

22817 Parcolene
Chemical treatment for use after bonderizing and parkerizing treatments. *Brent Chemicals International plc.*

22818 Parcryl® 250
100% Acrylic emulsion
Used in caulking compounds and masonry paints; wet adhesion, high gloss. *Thibaut & Walker.* Name unverified.

22819 Parcryl® 311
100% Acrylic emulsion; extremely flexible polymer for high pigment loading; for roof and tennis-court coatings; wet adhesion, high gloss. *Thibaut & Walker.* Name unverified.

22820 pareira
The dried root of *Chondrodendron tomentosum.*

22821 Parel 58
Sulfur vulcanizable, elastomeric copolymer of polypropylene oxide and alkyl glycidyl ether; used for flexible wing seals in supersonic aeroplanes, for motor mounts and other noise suppression applications requiring long term durability and as a blending rubber in tubes and tyres. *Hercules.* Discontinued.

22822 Parel®
Propylene oxide - allyl glycidyl ether copolymer
Elastomer with outstanding balance of low and high temperature resistance; used for dynamic mounts and high flex applications. *Zeon.*

22823 Parenamine
9015-54-7 310-296-6
Protein hydrolysate
Replenisher. *Sterling Drug Inc.* Name unverified.

22824 Parenol
Consists of 65% soft paraffin, 15% wool fat, and 20% distilled water.

22825 Parenol liquid
Consists of 70% liquid paraffin, 5% white bees wax, and 25% distilled water.

22826 Parentrovite
Parenteral vitamins B and C. *SmithKline Beecham.*

22827 Parenzyme
9002-07-7 9926 232-650-8
Trypsin, crystallized; enzyme (proteolytic). *Merrell Dow Pharmaceuticals Inc.* Name unverified.

22828 Parez
Modified resins of melamine and formaldehyde; used in the paper industry. *Cyanamid BV.*

22829 Parez 631NC
Cationic modified polyacrylamide; used in the paper industry. *Cyanamid BV.*

22830 Parfenac
2438-72-4 1497 219-451-1
bufexamac
A proprietary preparation of bufexamac used as a skin cream. *Lederle Laboratories.* Name unverified.

22831 Parian cement
A cement which is similar to Keene's cement, except that a solution of borax is used instead of alum.

22832 parianite
An asphaltum from the pitch lake at Trinidad.

22833 Paricin® 1
141-23-1 205-471-8
Methyl hydroxystearate
Lubricant/processing aid for butyl rubber; wax firming agent in cosmetics and specialty inks; source of hydroxystearic acid for glycerin-free lithium greases. *CasChem.* Name unverified.

22834 Paricin® 13
1323-42-8 215-355-9
Glyceryl hydroxystearate
Wax modifier, emollient; physical properties similar to those of beeswax. *CasChem.* Name unverified.

22835 Paricin® 15
33907-46-9 251-732-4
Ethylene glycol hydroxystearate
Wax modifier, emollient; physical properties similar to those of candelilla wax. *CasChem.* Name unverified.

22836 Paricin® 18
Stearyl 12-hydroxystearate
Saturated, low melting point wax. *CasChem.* Name unverified.

22837 Paricin® 210
N-Stearyl 12-hydroxystearamide
Lubricant/antistat for plastics, metals; mold release, antiblocking agent for textile coatings; slip agent for varnishes and lacquers; also for electrical potting compounds, crayons, wax blends, high-temperature greases; release agent in contact with food. *CasChem.*

22838 Paricin® 220
N (2-Hydroxyethyl) 12-hydroxystearamide
Internal mold release agent, lubricant for polyolefins, PVC, styrenics. *CasChem.*

22839 Paricin® 285
N,N·g-Ethylene bis 12-hydroxystearamide
Internal lubricant, mold release, slip additive for PVC. *CasChem.*

22840 Paricin® 6
Butyl acetoxystearate
Oxidation-stable plasticizer for vinyls. *CasChem.* Name unverified.

22841 Paricin® 8
Glyceryl tri(acetoxystearate)
Lubricant plasticizer for vinyls, esp. high-temp. wire jacketing; heat and oxidation-stable; grinding medium for pigment dispersions; plasticizer for nitrocellulose. *CasChem.* Name unverified.

22842 Paricin® 9
33907-47-0 251-734-5
Propylene glycol hydroxystearate
Wax modifier, emollient; physical properties similar to those of spermaceti wax. *CasChem.* Name unverified.

22843 Parilene
A proprietary trade name for polyparaxylyene; a plastics material used for film manufacture for electrical purposes. No manufacturer.

22844 Paris salts
A disinfectant containing 50 parts zinc sulfate, 50 parts ammonia alum, 1 part potassium permanganate, and 1 part lime.

22845 Parkerised Steel
A patented process for the treatment of steel with iron and manganese phosphates to give the surface resistance to corrosion. No manufacturer.

22846 Parlay
76738-62-0 7118
paclobutrazol
Growth regulator for growers of grass seed. Contains paclobutrazol, a plant growth regulator. *ICI Chem & Polymers Ltd.*

22847 Parlodion
A trademark for a shredded form of pure collodion, a nitrocellulose used as a cement. No manufacturer.

22848 Parlon
Chlorinated rubber available in six viscosity grades; used as film-formers in adhesives, in corrosion resistant coatings for wood, metal, and concrete, for wood floor finishes and sealers, in inks, etc. *Hercules.* Discontinued.

22849 Parlon P
A range of chlorinated polypropylenes; forms clear, hard, protective films that are resistant to chemicals, salt solutions and water; used as film formers in adhesives and sealants, the construction and building industry, for lumber and wood products, for floor coverings etc. *Hercules.* Discontinued.

22850 Parmentine
A mixture of glycerin, gelatin, dextrine, sodium sulfite, and zinc sulfate; used for sizing and finishing cotton, wool, and silk.

22851 Parmetol
Range of preservatives; used in the paint, ink, and adhesives industries. *Sterling-Winthrop Group Ltd.*

22852 Parmr
The trade name for a blown bitumen residue; it is a mineral rubber for use in the rubber industry; Grade 1 melts from 190-310°F and Grade II at above 300°F. No manufacturer.

22853 Paroa-caxy Oil
The seed-oil of *Penta-clethra flamentosa.* No manufacturer.

22854 Parogen
oleogen
oxygenated paraffin; vasogen. Consists of 2 parts liquid paraffin, 2 parts oleic acid, and 1 part ammoniated alcohol (5%).

22855 Paroil®
Chlorinated paraffin, liquid; used in cutting oils, industrial lubricant, flame retardant and plasticizer in plastics. *Dover.*

22856 Paroil® 10, 145, 1061
Chlorinated paraffin; lubricant additive. *Dover.*

22857 Paroil® 152
Doverguard® 152. Chlorinated paraffin; flame retardant for plastics, rubbers, adhesives, paints, fabric coatings. d = 1.270; viscosity = 15 poise. *Dover.*

22858 Paroil® 5761
Doverguard® 5761. Chlorinated paraffin; flame retardant for plastics, rubbers, adhesives, paints, fabric coatings. d = 1.355; viscosity = 20 poise. *Dover.*

22859 Parol 70
8042-47-5 232-455-8
White mineral oil, petroleum
Mineral oil technology; used for animal feed dedusting, food packaging materials, meat packaging, household cleaners and polishes. *Penreco.*

22860 Par-o-lac
An impregnating compound.

22861 Parolite
24887-06-7 246-515-6
Zinc formaldehyde sulfoxylate
Stripping agent for removing dyes from wool and nylon. *Henkel/Textiles.*

22862 Parraynite
A trade name for a rubber compound which is used by X-ray operators to protect them from injury by exposure to the rays. No manufacturer.

22863 parrot coal
Cannel coal.

22864 Parr's alloys
Anti-corrosion alloys. One alloy contains 80% nickel. 15% chromium, and 5%, copper; another contains 66.6% nickel, 18% chromium, 8.5% copper, 3.3% tungsten, 2% aluminum, and 1% manganese.

22865 parsley camphor
523-80-8 777 208-349-2
$C_{12}H_{14}O_4$
4,7-dimethoxy-5-(2-propenyl)-1,3-benzodioxole
Camphre de Persil; parsley apiole; apiol; apioline;. Crystallized apiole. Synergistic activity with insecticides. mp = 29°; bp = 294°; n_D^{20} = 1.536-1.538; insoluble in H_2O, soluble in organic solvents.

22866 Parsol® 1789
70356-09-1 1616 274-581-6
$C_{20}H_{22}O_3$
4-*tert*-butyl-4'-methoxy-dibenzoylmethane
butyl methoxydibenzoylmethane; Avobenzone; Parsol A. UV-A absorber for production photostability; used in sunscreens. An ultraviolet A absorbing agent, used in cosmetics, lipsticks and lip balms, moisturizers, nail polishes, shampoos and other hair care products and sunscreens. mp = 83°. *Bernel; Givaudan.*

22867 Parsol® MCX
5466-77-3 6864 226-775-7
$C_{18}H_{26}O_3$
2-ethylhexyl methoxycinnamate
Escalol 557; Ethylhexylcinnamate; Neo Heliopan AV.; Octyl methoxycinnamate; Parsol MCX; Parsol MOX. UV absorber, sunscreening agent in the wavelength range of 2900-3200 A which causes sunburn and skin damage, stimulates tanning. A UVB absorbing agent that belongs to the cinnamates. Cross-reactions to other cinnamates that are used as fragrances or flavoring agents is possible. $bp_{0.76}$ = 185-195°. *Bernel; Givaudan.*

22868 Parsolin
Organophosphorus insecticide. *Ciba plc.* Name unverified.

22869 Parson's Alloy
A proprietary trade name for an alloy of 56% copper, 41.5% zinc, 1.2% iron, 0.7% tin, 0.1% manganese, and 0.46% aluminum; specific gravity 8.4. No manufacturer.

22870 Partagon
A German preparation. It consists of rods containing silver chloride and - sodium-silver chloride.

22871 Partinium
An aluminum alloy; varies in composition, and often contains tungsten, copper, tin, zinc, and magnesium; one alloy contains 96% aluminum, 2.4% antimony, 0.8% tungsten, 0.64% copper, and 0.16% tin; another alloy consists of 88.5% aluminum, 7.4% copper, 1.7% zinc, 1.3% iron and .1% silicon. No manufacturer.

22872 PartsPrep™ Degreaser
872-50-4 6197 212-828-1
N-Methyl-2-pyrrolidone and surfactants; degreaser formulations. *ISP.*

22873 Parvol
A range of auxiliary products; for use as assistants in tanning processes. No manufacturer.

22874 parvoline
$C_9H_{13}N$
Dimethylethylpyridine

22875 Parylene N
A plastic material used to make thin film membranes, 2-1000 Angstroms thick. *Union Carbide (UK) Ltd.* Name unverified.

22876 Pasilex
12141-46-7 377 235-253-8
Precipitated aluminum silicate; used as filler for the paper industry. *Degussa AG.* Name unverified.

22877 Passini's solution
An aqueous solution of mercury and sodium chlorides and glycerine; used to preserve animal tissue.

22878 Passow's slag cement
Prepared by blowing into liquid slag as it issues from the blast furnace, when it becomes granulated. It is then finely ground.

22879 pastaccio
A residue from the manufacture of calcium citrate. It consists of vegetable cellulose with some hydrocarbons.

22880 Pasturol Plus
Soluble concentrate of 25 g dicamba, 200 g MCPA, and 400 g mecoprop per liter; broad-spectrum herbicide used for weed control in cereals and grassland. *Farmers Crop Chemicals Ltd.*

22881 Patafol
58810-48-3 261-451-9
$C_{14}H_{16}ClNO_3$
2-chloro-N-(2,6-dimethylphenyl)-N-(tetrahydro-2-oxo-3-furanyl)acetamide
milfuram. Fungicide containing ofurace for the control of potato blight. (Sold in UK for Chevron Chemical Co) *ICI Chem & Polymers Ltd.*

22882 Patafol Plus
Manganese zinc ethylenebisdithiocarbamate (maneb, zineb) and ofurace (Patafol) mixture; for control of potato blight. (Sold in UK for Chevron Chemical Co.) *ICI Chem & Polymers Ltd.*

22883 patava oil
Batana oil; Coumou oil. A semi-drying oil obtained from the kernels of the Brazilian palm tree, *Oenocarpus batava.*

22884 patchouli
An Indian herb, *Pogostemon patchouly*, used in perfumery.

22885 Patco® 3
Blend of Emplex sodium stearoyl lactylate and Verv® calcium stearoyl-2-lactylate; conditioner/softener; starch and protein complexing agent for use in yeast-leavened bakery products. *Am. Ingredients/Patco.*

22886 Patcote® 305
Silicone emulsion, 10% filled; defoamer for general food and industrial applications. *Am. Ingredients/Patco.*

22887 Patcote® 306
20% Filled silicone emulsion; defoamer for general food and industrial applications; FDA compliant. *Am. Ingredients/Patco.*

22888 Patcote® 309
Nonsilicone aqueous emulsion; defoamer for food and industrial use; FDA compliant. *Am. Ingredients/Patco.*

22889 Patcote® 315
10% Silicone emulsion; defoamer for food and industrial applications; FDA compliant. *Am. Ingredients/Patco.*

22890 Patcote® 337
Nonsilicone; defoamer for use in alcohol production via grain fermentation; FDA compliant. *Am. Ingredients/Patco.*

22891 Patcote® 500
Silicone-containing; process defoamer, dispersing aid for latex trade sales. *Am. Ingredients/Patco.*

22892 Patcote® 512
Silicone-containing; defoamer for use in urethane-modified resins. *Am. Ingredients/Patco.*

22893 Patcote® 525
Silicone-containing; defoamer for use in water-reducible alkyd and industrial acrylic systems. *Am. Ingredients/Patco.*

22894 Patcote® 555K
100% filled silicone; kosher grade; defoamer for food and nonfood processing; FDA compliant. *Am. Ingredients/Patco.*

22895 Patcote® 803
Nonsilicone; defoamer for acrylic and terpolymer emulsions for trade sales; FDA compliant. *Am. Ingredients/Patco.*

22896 Patcote® 811
Nonsilicone; defoamer for graphic arts waterbased acrylic systems. *Am. Ingredients/Patco.*

22897 patent black
An acid dyestuff. It is a substitute for logwood.

22898 patent zinc white
A pigment made by adding a soluble sulfide to a zinc chloride or zinc sulfate solution, filtering off the precipitate, drying it, and then calcining it. It has the composition, 5ZnS ·g. ZnO.

22899 patents
The small portion of very white flour obtained from wheat. It is poor in proteins, and is used for fancy breads.

22900 Pat-Fix RBD
Fixative for direct dyes and prints. No manufacturer.

22901 Path Gun+se
Ready-for-use herbicide spray. *ICI Chem & Polymers Ltd.*

22902 Pathclear
Contains aminotriazole, diquat, paraquat and simazine; long-acting weedkiller for paths, drives, and patios. *ICI Garden Products.*

22903 patina
The green film which forms on copper and bronze moldings. It consists of basic copper carbonate or other basic copper salts.

22904 Pationic 122A
29051-57-8
Sodium capryl lactylate
Emulsifier/foam booster for facial cleansers and personal care products; microbial inhibitor properties. *RITA.*

22905 Pationic 138
13557-75-0 236-942-6
Sodium lauroyl lactylate
Detergent, conditioner, foam booster, viscosity builder, lipophilic emulsifier, cleansing agent used in shampoos, facial cleansers. *RITA.*

22906 Pationic CSL
5793-94-2 11, 1711 227-335-7
Calcium stearoyl-2-lactylate
Water-oil emulsifier and protein complexer for cosmetics. *RITA.*

22907 Pationic ISL
66988-04-3 266-533-8
Sodium isostearoyl-2-lactylate
Surfactant, emulsifier for cosmetics; perfume solubilizer; substantive conditioner for hair and skin. *RITA.*

22908 Pationic SSL
25383-99-7 246-929-7
Sodium stearoyl 2-lactylate
Emulsifier, viscosity builder, protein complexer for cosmetics products. *RITA.*

22909 Patlac® IL
42131-28-2 255-674-0
Isostearyl lactate
Surfactant, emollient for cosmetics. *Am. Ingredients/Patco; RITA.*

22910 Patlac® LA
50-21-5 5349 200-018-0
Lactic acid
Moisture binder, humectant. *Am. Ingredients/Patco; RITA.*

22911 Pat-Lube Series
Lubricants for jet and beck dyeing offering fiber-to-metal and fiber-to-fiber lubricity. No manufacturer.

22912 Patogen 311
Low foam wetting agent/scour for bleach baths. No manufacturer.

22913 Patogen 353
Wetting and scouring agent for bleach baths, dye baths. No manufacturer.

22914 Patogen AO-30
68155-09-9 268-938-5
Cocamidopropyl dimethylamine oxide
Fugitive wetter, foam booster, detergent, emulsifier, foaming agent for textile applications. No manufacturer.

22915 Patogen P-10 Acid
Alkyl phosphate
Scouring agent for textiles. No manufacturer.

22916 Patoran
3060-89-7 6224 221-301-5
Metobromuron
Metobromuron, a substituted urea herbicide. *Ciba plc.*

22917 Patoran® FL
3060-89-7 6224 221-301-5
Metobromuron
For preemergence weed control in potatoes, soybeans, tobacco, tomatoes *BASF AG.*

22918 Pat-Quest CS
Chelating/sequestering agent for textile bleaching and scouring operations; excellent caustic stability. No manufacturer.

22919 Patrol
67306-00-7 4034
Emulsifiable concentrate of 750 g fenpropidin per liter; a systemic fungicide. *ICI Agrochemicals.*

22920 Pat-Soft 1442
Softener and lubricant for textile fabrics; excellent heat stability; recommended for finishing prior to printing. No manufacturer.

22921 pattern metal
An alloy of 83% copper, 10% zinc, 4% tin, and 3% lead.

22922 Pattinson's white lead
$PbCl_2 \cdot 2Pb(OH)_2$
A pigment. It is basic lead chloride,.

22923 Pattonex
3060-89-7 6224 221-301-5
3-(4-bromophenyl)-1-methoxy-1-methyl-urea

Active ingredient: metobromuron; residual herbicide for use as a selective weedkiller. *Agan Chemical Manufacturers Ltd.*

22924 Pattrex
Pattern plaster. *Foseco (F.S.) Ltd.*

22925 Pattrit
Casting plaster. *Foseco (F.S.) Ltd.* Discontinued.

22926 Pat-Wet LF-55
Low foaming wetting agent with non-rewetting properties for textile processing. No manufacturer.

22927 Pat-Wet SP
Alkyl phosphate
Low foaming wetting agent, bath stabilizer for textile dyeing; excellent for sulfur and indigo dye baths. No manufacturer.

22928 pavlin
524-30-1 4290 208-355-5
$C_{16}H_{18}O_{10}$
8-(-D-glucopyranosyloxy)-7-hydroxy-6-methoxy-2H-1-benzopyran-2-one
Fraxin; fraxetin-8-glucoside; fraxoside. A substance which occurs in the bark of the common ash. mp = 205°; sparingly soluble in H_2O, soluble in EtOH, insoluble in Et_2O.

22929 Pavy's solution
A modified Fehling's solution used for the determination of sugar.

22930 Paxbestos
Asbestos products bonded with hydraulic cement and impregnated with bitumen; used as insulating materials.

22931 Paxolin
A synthetic resin bonded paper product used for insulating purposes.

22932 Payne's grey
An oil and water color prepared from black alizarin, madder, and indigo.

22933 Payne's solution
A solution of sodium hypobromite.

22934 Payzone
A proprietary preparation of nitrovin hydrochloride; a veterinary growth promoter. No manufacturer.

22935 Pazo
An insecticide. *Murphy Chemical Co Ltd.* Discontinued.

22936 PBI Slug Pellets
108-62-3 5983 202-945-6
Pellets containing 6% w/w metaldehyde; snail and slug bait. *Pan Britannica Industries Ltd.*

22937 PBI Spreader
Nonionic wetting agent for use in agrochemical sprays. *Pan Britannica Industries Ltd.*

22938 PBT-1100
26062-94-2
Polybutylene terephthalate; molding compound for engineering parts requiring heat resistance, dimensional stability, rigidity, str., chem. resistance; used for elec., automotive, appliance, and telecommunications parts. *Bay Resins.*

22939 PBT-1100G15
26062-94-2
Polybutylene terephthalate, 15% glass-reinforced. *Bay Resins.*

22940 PBT-1300
26062-94-2
Polybutylene terephthalate, highly impact-modified. *Bay Resins.*

22941 PBT-1700
26062-94-2
Polybutylene terephthalate, flame retardant. *Bay Resins.*

22942 PC-000/5T
Polycarbonate, 5% PTFE-lubricated; improved surface wear characteristics, reduced static and dynamic coefficients of friction. *Compounding Tech.*

22943 PC-1100
Polycarbonate; general purpose molding compound offering high precision, dimensional stability, and toughness. *Bay Resins.*

22944 PC-1100G10
Polycarbonate, 10% glass-reinforced; molding compound offering additional rigidity for frames, cases, appliance housings. *Bay Resins.*

22945 PC-1100H30
Polycarbonate, 30% carbon fiber-reinforced; electrically dissipative molding compound with excellent mech. properties; for business machine components. *Bay Resins.*

22946 PC-1244
Defoamer; used to suppress foaming tendencies in lubricating oil formulations. *Monsanto Co.* Name unverified.

22947 PC-1344
Defoamer in nonaqueous hydrocarbon and solvent systems. *Monsanto Co.* Name unverified.

22948 PC-1700G10FR
Polycarbonate, 10% glass-reinforced, flame-retardant; molding compound for elec./electronic, business machine applications. *Bay Resins.*

22949 PC-20CF
Polycarbonate resin, 20% carbon fiber-reinforced; offers exceptional combination of dimensional stability and mech. strength and stiffness; used for wide range of precision electro-mech. parts and compounds. *Compounding Tech.*

22950 PC-20GF/15T
Polycarbonate, 20% glass fiber-reinforced, 15% PTFE-lubricated; improved mechanical strength and stiffness, greater chemical resistance, thermal resistance, superior accuracy in molding, dramatically increased wear resistance; used for precision electro-mechanical components and gears. *Compounding Tech.*

22951 PCA 301
Isocyanate terminated polyester based PU prepolymer; yields cured elastomers with abrasion and tear resistance, high load bearing capacity; for solid industrial tires, rollers, sheet goods applications; ideal for pouring large parts. *Polyurethane Corp. of Am.* Name unverified.

22952 PCA 4-1
Polyether (PTMEG) based urethane liquid polymer; prepolymer producing elastomers with good aging properties when subjected to heat, humidity, fungus, and other environments, good abrasion and tear resistance, low-temperature flexibility, toughness, ease of processing. *Polyurethane Corp. of Am.* Name unverified.

22953 PCI
Corrosion inhibitors used in deicers. *Georgia-Pacific.*

22954 PCMX
For chemical compositions having antiseptic and germicidal properties, in Class 18. *Ferro Corporation.* Discontinued.

22955 PDS
Polydioxanone
Surgical aid. *Ethicon Inc.* Name unverified.

22956 PDX-82427
PEEK (polyetheretherketone) with 30% nickel; conductive attenuating composite effectively shielding electromagnetic and/or radio frequency interference; used in avionics housings, business machine enclosures, and other electronic devices. *LNP.*

22957 PDX-84367
61128-46-9
PEI
Antistatic grade composite for protection of moderately sensitive electronic components from low voltages. *LNP.*

22958 PDX-84368
Polycarbonate; antistatic grade composite for protection of moderately sensitive electronic components from low voltages. *LNP.*

22959 PDX-84369
26062-94-2
Polybutylene terephthalate; antistatic grade composite for protection of moderately sensitive electronic components from low voltages. *LNP.*

22960 PE 100 228 FH-VP
Halogen-free flame-retardant compound; extrusion compound for processing at +zw 185°C. *Zipperling Kessler.*

22961 PE 1017
9002-88-4 7728
LDPE homopolymer resin; extrusion resin; high-speed coating and laminating resin. *Monsanto (Solaris).*

22962 PE 4517
9002-88-4 7728
LDPE homopolymer resin; coating and laminating resin for carton stock coating and photographic paper. *Monsanto (Solaris).*

22963 PE 5222
LDPE copolymer resin with 6% EVA and high slip and antiblock additives; film resin offering good heat seal; used for frozen foods and ice bags. *Monsanto (Solaris).*

22964 PE 5554-H
9002-88-4 7728
LDPE homopolymer resin with medium slip and antiblock additive; high-clarity film resin. *Monsanto (Solaris).*

22965 PE 5861
9002-88-4 7728
LDPE homopolymer resin with high slip and medium antiblock additive; film resin for garment applications. *Monsanto (Solaris).*

22966 Pétrole Hahn
Hair tonic, shampoo and conditioner to keep hair healthy and attractive. *Richardson-Vicks Inc.* Name unverified.

22967 Pea Pro-Tein BK
9008-99-8
Hydrolyzed pea protein; substantive protein, film-former, moisturizer for skin and hair car products (shampoos, conditioners, creams, lotions, liquid hand soaps); anti-irritant in anionic formulations. *Maybrook.*

22968 peach black
A variety of carbon black similar to lampblack.

22969 peanut oil
8002-03-7 7191 232-296-4
Arachis oil; groundnut oil; Katchung oil. Refined fixed oil obtained from seed kernels of one or more cultivated varieties of *Arachis hypogaea*; edible oil; substitute for olive oil; vehicle for medicine; in manufacture of margarine, soaps, paints *Croda; Karlshamns; Penta Mfg.*

22970 peanut ore
A mineral. It is a variety of wolframite. Contains Fe, Mn, W.

22971 pear oil
123-92-2 5125 204-662-3
$C_7H_{14}O_2$
Isoamyl acetate
3-Methylbutyl acetate; Isoamyl ethanoate; Isoamyl Acetate; amyl acetate ester; 3-methyl-1-butanol acetate; Banana Oil; 3-Methyl-1-Butyl Acetate; 3-methyl butyl ester acetic acid; isopentyl ester acetic acid; isopentyl alcohol, acetate; pear oil; β-methyl butyl acetate; Amyl acetate, common; mixed isomers: [628-63-7}. Used in the manufacture of fruit essences for flavoring confectionery. Has a pear-like odor. bp = 142°; d_4^{15} = 0.875; n_D^{21} = 1.400; soluble in H_2O (2.5 mg/ml), more soluble in organic solvents;.

22972 Pearex-L®
9032-75-1 232-885-6
Fungal pectinase
Enzyme used to prevent haze formation in pear and apple processing. *Solvay Enzymes.*

22973 Pearistick 46-10/06
9009-54-5
Linear polyurethane
Features high rate of crystallization; for two-component adhesives used in footwear, wood, pkg. *Merquinsa.*

22974 Pearistick 65-05
Linear polycaprolactone-polyurethane
Features very high rate of crystallization; additive for two-component adhesives used in wood and automotive industries; plasticizer in EVA, PVC, ASA hot melts for impregnation and textile coating. *Merquinsa.*

22975 Pearl
8008-20-6 232-366-4
Kerosene. *Monsanto (Solaris).* Name unverified.

22976 pearl alum
10043-01-3 381 233-135-0
aluminum sulfate
A specially prepared aluminum sulfate used in the paper industry.

22977 pearl ash
584-08-7 7781 209-529-3
CK_2O_3
potassium carbonate
salt of tartar. A variety of potassium carbonate. Used in chemical manufacturing and in medicine as an alklaizer and diuretic.

22978 Pearl Dust®
584-08-7 7781 209-529-3
A registered trade name for a form of potassium carbonate, K_2CO_3; used as a filler. No manufacturer.

22979 Pearl I, II, III
7787-59-9 1303 232-122-7
Bismuth oxychloride
Presperse.

22980 pearl spar
$C2CaMgO_6$
A double carbonate of magnesium and calcium.

22981 Pearl Supreme UVS
7787-59-9 1303 232-122-7
Bismuth oxychloride
Pearlescent pigment. *Presperse.*

22982 pearl white
7787-59-9 1303 232-122-7
BiOCl
Bismuth oxychloride
flake white; Blanc de perle. A basic bismuth nitrate, $Bi(OH)_2NO_3$ is also known as pearl white, a pigment. The term is sometimes used in connection with a white lead which has been tinted with Paris blue or indigo.

22983 Pearlex GC 0311
Proprietary blend; pearlizing and opacifying agent for shampoos, bubble baths, liquid hand soaps at low addition levels. *Clough.*

22984 Pearl-Glo®
7787-59-9 1303 232-122-7
Bismuth oxychloride
Pearlescent pigment powders and dispersions for cosmetic eye, face, lip, and body makeup. *Van Dyk.*

22985 pearl-hardening
7778-18-9 1753 231-900-3
$CaSO_4$
calcium sulfate
Used as a loading for paper.

22986 pearlite
Iron carbide eutectoid, consisting of alternate masses of ferrite and cementite.

22987 Pearlstick
Polyurethane pellets
Raw materials for adhesives. *Merquinsa.*

22988 Pearlstick 45-05/40
Linear polycaprolactone-polyurethane
Features very high rate of crystallization; additive for adhesives for wood and automotive industry and plastics; plasticizer for EVA, PVC, ASA hot-melts for impregnation and textile coating. *Merquinsa.*

22989 Pearsall
Flame retardant. *Witco.* Discontinued.

22990 Pearsol
Synthetic phenolic germicides in a terpeneol vegetable oil soap base; for general disinfection/antisepsis; a substitute for chloroxylenol solution. *William Pearson Ltd.* Name unverified.

22991 Pearson's cerate
Consists of 4 parts lead plaster, 1 part beeswax, and 3 parts almond oil.

22992 Pearson's solution
A solution of dried sodium arsenate 0.1% - 1.0%.

22993 Peat
The partially decayed remains of plants; used as fuel.

22994 peat coal
An intermediate between peat and lignite.

22995 Peaweed
Suspension concentrate containing 152 g prometryn and 304 g terbutryn per liter; for weed control in peas, beans and potatoes. *Pan Britannica Industries Ltd.*

22996 Pebax®
Polyether block amides; for sparking applications, gears. *Elf Atochem.*

22997 Pebax® 2533 SA 00, 2533 SD 00, 2533 SN 00
Polyether block amide polymers; SA indicates additive-free; SD is uv stabilized with mold release additive; SN is uv stabilized; extrusion/molding grade polymer used in sport, automobile, elec./electronics, medical, agriculture, and mechanical handling I industries; also as a basis for hot-melt adhesive formulations. *Elf Atochem N. Am./Plastics.*

22998 Pebax® 4011 MA 00
Polyether block amide polymer, antistatic, hydrophilic grade; molding grade polymer used in sport, automobile, electrical and electronics, medical, agriculture, and mechanical handling industries; also as a basis for hot-melt adhesive formulations. *Elf Atochem N. Am./Plastics.*

22999 Pebax® 4033 SN 70, 5533 SN 70
Polyether block amide polymers, uv stabilized with antistatic fillers; extrusion and molding grade polymer. *Elf Atochem N. Am./Plastics.*

23000 Pebax® 6333 SA 00, 6333 SD 00, 6333 SN 00
Polyether block amide polymers; SA indicates additive-free; SD is uv stabilized with mold release additive; SN is uv stabilized; *Elf Atochem N. Am./Plastics.*

23001 pecan oil
Oil obtained from the seed of the North American walnut, *Juglans niger.*

23002 Peceol Isostearique
32057-14-0
Glyceryl isostearate
Water-oil coemulsifier; pigment dispersant; additive for lipsticks; superfatting agent for emulsified preparations *Gattefosse SA.*

23003 Pecosil CAS-36
Octyldodecyl/dimethicone copolyol citrate
Emollient. *Phoenix.*

23004 Pecosil DAS 36
Dilinoyl dimethicone copolyol citrate
Emollient. *Phoenix.*

23005 Pecosil GSA 36
Octyldodecyl dimethicone copolyol adipate
Emollient. *Phoenix.*

23006 Pecosil OS-100B
Dimethicone propylethylenediamine behenate
Emollient. *Phoenix.*

23007 Pecosil OS-100DA
Dimethicone propylethylenediamine dilinoleate
Emollient. *Phoenix.*

23008 Pecosil OS-100HS
Dimethicone propylethylenediamine hydroxystearate
Emollient. *Phoenix.*

23009 Pecosil OS-100L
Dimethicone propylethylenediamine laurate
Emollient. *Phoenix.*

23010 Pecosil OS-100M
Dimethicone propylethylenediamine myristate
Emollient. *Phoenix.*

23011 Pecosil OS-100U
Dimethicone propylethylenediamine undecylenate
Emollient. *Phoenix.*

23012 Pecosil PS-100
Dimethicone copolyol phosphate
Emollient. Imparts silky feel to aq. formulations, skin creams/lotions, mousses, hair conditioners, shaving products, bath products and gels. *Phoenix.*

23013 Pecosil PS-100K
Potassium dimethicone copolyol phosphate
Emollient. *Phoenix.*

23014 Pecosil SG-20
Octyldodecoxy dimethicone
Emollient. *Phoenix.*

23015 Pecosil SSP
Hydrolyzed soy protein/dimethicone copolyol phosphate copolymer. Emollient. *Phoenix.*

23016 pectase
A clotting enzyme, which produces vegetable and fruit jellies.

23017 pectin
9000-69-5 7194 232-553-0
Citrus pectin. Purified polysaccharide obtained from the dilute acid extract of the inner portion of the rind of citrus fruits or from apple pomace; jellies, foods, cosmetics, drugs, protective colloids, emulsifying agents, dehydrating agents. *Hercules; Penta Mfg.; Pomosin GmbH; Spice King; U.S. Biochemical.*

23018 pectinase
9032-75-1 232-885-6
Enzyme from *Rhizopus* spp.; used in biochemical research and in the juice and jelly industries. *Gist-Brocades Food Ingreds.; U.S. Biochemical.*

23019 Pectinase AT
9032-75-1 232-885-6
Fungal pectinase
Enzyme used for cranberry juice depectinization. *Solvay Enzymes.*

23020 Pectinex
A purified enzyme preparation produced from a selected strain of *Aspergillus niger*; used where the aim is breaking down of soluble and insoluble pectins with varying degrees of esterification, for reduction of viscosity, clarification, maceration of plant tissue and depectinization. *Novo Nordisk.*

23021 Pecutrin®
Vitaminized mineral salt mixture; for individual dosing and as a feed additive for all animals kept for use. *Bayer AG.*

23022 Peerless alloy
A heat-resisting alloy containing 78.5% nickel, 16.5% chromium, 3% iron, and 2% manganese.

23023 Peerless®
1332-58-7 5294 296-473-8
kaolin
Kaolinite; China clay; Bolus alba; Porcelain clay; Aluminum silicate hydroxide; Kaopectate; Aluminum silicate (hydrated); Aluminum silicate dihydrate. Kaolin clay; filler and extender. *R. T. Vanderbilt Co Inc.* Discontinued.

23024 Peerless® No. 1
1332-58-7 5294 296-473-8
kaolin
Secondary kaolin clay; filler used in adhesives, wallboard, paint, paper, fertilizer, roofing granules, crayons, powdered soaps, pharmaceuticals, ceramics; sanitaryware, artware, generalware, floor tile, electrical and chemical porcelain, and special refractories; Imparts more plasticity to cast piece. *R. T. Vanderbilt Co Inc.* Discontinued.

23025 PEG lanolin series
61790-81-6
Ethoxylan® 1685. Emollient, emulsifier, dispersant, foam stabilizer, resin plasticizer for cosmetic and pharmaceutical preparations, textile processing. *Henkel/Cospha; Henkel/Textile; Henkel Canada.*

23026 PEG stearamine series
10213-78-2 233-520-3
Chemeen 18-2; Chemstat® 273-E; Ethomeen® 18/12; PEG-2 stearamine. Emulsifier, dispersant used in textile processing. *Akzo; Chemax;.*

23027 PEG -12 PEG-50 lanolin
68458-88-8
PPG-12-PEG-50 lanolin
Laneto AWS. Auxiliary emulsifier, moisturizer, emollient for personal care products; plasticizer for hair spray, resins. *RITA.*

23028 PEG ditallate series
61791-01-3
Ethox DTO-9A; PEG-8 ditallate; Laurel PEG 400 DT; Mapeg® 200 DOT; Pegosperse® 400 DOT. Emulsifier for oils and solvs; used for industrial lubricants. *Ethox; Reilly-Whiteman; PPG/Specialty Chem.; Lonza Inc.*

23029 PEG hydrogenated tallow series
68155-24-8
PEG-13 hydrogenated tallow amide
Ethomid® HT/23, HT/60; Ethomid® HT/23. Emulsifier, dispersant, detergent, dye leveling agent; for silicone finishing agents, sizing lubricants. *Akzo;.*

23030 PEG isostearate series
56002-14-3
Ethox MI-9. Emulsifier and lubricant. *Ethox.*

23031 PEG ricinoleate series
9004-97-1
PEG-8 ricinoleate
Hodag 40-R; Nopalcol 6-R. Emulsifier, wetting agent, plasticizer for general cosmetic, pharmaceutical, and other uses. *Calgene; Henkel/Functional Prods.*

23032 PEG sorbitan hexaoleate series
57171-56-9
Ethox HO-50; PEG-50 sorbitol hexaoleate. Emulsifier and lubricant for heat-stable systems. *Ethox.*

23033 PEG soyamine series
61791-24-0
PEG-10 soyamine
Accomeen S2, S10, S15; Chemeen S-2; Ethomeen® S/12, S/15,S/25; Hetoxamine S-2; Mazeen® S-2, S-15; Teric 16M2; Tomah E-S-2; Ethomeen® S/20,. Emulsifier, dispersant used in textile processing. *Azko; Karlshamns; Chemax;Heterene; PPG/Specialty Chem.; ICI Australia; Exxon/Tomah.*

23034 PEG stearate series
9004-99-3
Ethox MS-8. Lubricant, wax and oil emulsifier, detergent, softener; for aq. processing. *Ethox.*

23035 PEG tallate series
61791-00-2
EM-600; PEG 600 monotallate; Aconol X6; Actrol 6M25P; Chemax TO-8; Chemax TO-10; Chemax TO-16; EM-600; Ethofat® 242/25; Ethox TO-8; Genaden TA-080; Hetoxamate FA-5; Industrol® TO-16; Laurel PEG 400 MT; Mapeg® 200 MOT; Mapeg® TAO-15; Marlosol® TF3; Prox-onic TA-1/08; Sellig T 3 100; Sellig T 14 100; Sellig T 1790; Trydet 2682,. Surfactant, emulsifier, wetting agent, detergent for industrial applications, sol. cutting oils and drawing compounds Ferro/Keil; Hart Chem. Ltd.; Climax Fluids Additives; Chemax; Heterene; BASF; Reilly - Whiteman; PPG/Specialty Chem.; Huls Am.;Ceca SA; Henkel/Emery; Henkel/Textile.

23036 PEG-10 sorbitan laurate
9005-64-5 8872
PEG-10 sorbitan laurate
Liposorb L-10. O/w emulsifier, lubricant, antistat, all-purpose hydrophilic surfactant; used for solubilizing oils and in conjunction with Liposorb esters. *Lipo.*

23037 PEG-150 distearate series
9005-08-7
Emalex 6300 Di-ST; PEG-150 distearate; Marlosol® BS; Rewopal® PEG 6000 DS; Ritapeg 400 DS; Witconol 2642; Genapol® TS Powd; Hetoxamate 400 DS; Hodag 602-s; Kessco® PEG 200 DS; Lipopeg 4-DS, 6000-DS; Mapeg® 200 DS, 1540 DS, 6000 DS;,. Thickener, stabilizer for shampoos, hair conditioners, cleansing foams. *Nihon Emulsion; Hoechst; Heterene; Calgene; Stepan; Lipo; PPG/Spec.Chem.; Hüls AM; Witco.*

23038 PEG-2 cocobenzonium chloride
61789-68-2 263-078-7
Ethoquad® CB/12. IPA; industrial surfactant for agriculture, textiles,

protective coatings, inks, pigment dispersions, acid pickling baths, metalworking, electroplating, plastics mfg. *Akzo.*

23039 PEG-2 cocomonium nitrate
71487-00-8
Ethoquad® C/12 Nitrate; PEG-2 cocomethyl ammonium nitrate. IPA; industrial surfactant for agricultural, textiles, protective coatings, inks, pigment dispersions, acid pickling baths, metalworking, electroplating, plastics mfg. *Azko.*

23040 PEG-2 dioleate series
9005-07-6
Emalex DEG-di-O; PEG-2 dioleate; Alkamuls® 400-DO, 600-DO; Chemax PEG 200 DO, PEG 400 DO, PEG 600 DO; CPH-211-N; Dyafac PEG 6DO; Emalex 200 di-O; Ethox DO-2; Hodag 42-O; Kessco® PEG 200 DO; Lipopeg 4-DO; Mapeg® 200 DO; Marlosol® FS; Nonex DO-4; Nopalcol 6-DO; Secoster® DO 600; Witconol 2648, H33,. Oilphase ingred., emulsifier, dispersant with good spreadability for cosmetics, creams, milky lotions, foundations. Nihon Emulsion; Rhone-Poulec Surf & Spec; Chemax; Ethox; Calgene; Stepan; Lipo; PPG/Spec. Chem.; Huls AM; Huntsman; Henkel/Chems. Group; Stepan Europe; Witco/Oleo-Surf.

23041 PEG-2 laurate
141-20-8 3169 205-468-1
$CH_3(CH_2)_{10}CO(OCH_2CH_2)_nOH$, avg. n ·c 2
dodecanoic acid 2-(2-hydroxyethoxy)ethyl ester
diglycol laurate; PEG 100 monolaurate; diethylene glycol laurate. Water-oil emulsifier, dispersant, antistat, defoamer and plasticizer for textile, paper processing, cutting oils, polishes, emulsion cleaners, rubber latex, wool lubricants, paints. mp = 17-18°; bp = 270° (dec); d^{20}_{20} = 0.963-0.968; insoluble in H_2O, soluble in organic solvents. *Henkel/Emery; Inolex; Karlshamns; Lonza; Mona; Stepan; Witco/Humko.*

23042 PEG-2 oleate
106-12-7 203-364-0
$C_{22}H_{42}O_4$
diethylene glycol monooleate
diglycol oleate. Emulsifier, dispersant, antistat for cosmetic, textile, paper processing, cutting oils, polishes, emulsion cleaners, rubber latex, wool lubricants; leather softener. *Henkel/Emery; Inolex; Karlshamns; Lipo; Lonza; Mona; Witco/Humko.*

23043 PEG-2 stearate
106-11-6 203-363-5
$C_{26}H_{30}O_4$
diethylene glycol stearate
diglycol stearate. Emulsifier, plasticizer, lubricant, wetting agent, binding and thickening agent, dispersant, antistat, opacifier, pearlescent, stabilizer used in cosmetics, dry cleaning, leather, textile industries, paper processing, rubber. *Henkel/Emery; Inolex; Karlshamns; Lipo; Lonza; Stepan; Witco/Humko.*

23044 PEG-2 tallowmonium chloride
67784-77-4 267-052-6
ethanol; Ethoquad® T/12; PEG-2 tallowalkyl methyl ammonium chloride. Industrial surfactant for agriculture, textiles, protective coatings, inks, pigment dispersions, acid pickling baths, metalworking, electroplating, plastics mfg. *Azko.*

23045 PEG-20 hydrogenated lanolin
68648-27-1
Fancol HL-20; Ivarlan 3450; Lipolan 31; Satexlan 20; Super-sat AWS-4. Solubilizer, superfatting agent, gelling agent for cosmetics, pharmaceuticals, makeup, nail polish, night creams, microemulsions. *Fanning; Lipo; Brooks industries; Croda Inc.; RITA.*

23046 PEG-20 methyl glucose sesquistearate
68389-70-8
Glucamate® SSE-20. O/w emulsifier, solubilizer used with Glucate SS; effective at low concs. *Amerchol; Amerchol Europe.* Discontinued.

23047 PEG-3 lauramine oxide
59355-61-2
Empigen® OY; POE (3) lauryl dimethyl amine oxide; PEG (3) lauryl dimethyl amine oxide. Detergent, antistat, foam booster/stabilizer for foamed rubbers, fire fighting, bleach additive. *Albright & Wilson; Albright & Wilson UK.*

23048 PEG-3 oleamide
31799-71-0
Oleic acid amide ethoxylate
Lutensol® FSA 10; Dionil® OC. Emulsifier, dispersant, detergent, wetting agent, used in fine detergents, hand cleaners, fur dressing; > 80% biodeg. *BASF AG; Hüls AG.*

23049 PEG-4
112-60-7 203-989-9
$C_8H_{18}O_5$
(2,2·g-[oxybis(2,1-ethanediyloxy)bisethanol)
3,6,9-Trioxaundecan-1,11-diol. Lubricant for rubber molds, textile fibers, metalworking; in food and food pkg.; in cosmetics and hair preparations; pharmaceutic aid; in gas chromatography; in paints, paper coatings, polishes, ceramics. mp = -4°; bp = 324-330°; SG = 1.125; n_D^{20}= 1.4590; LD_{50} (rat orl) = 29 g/kg.

23050 PEG-4 dilaurate series
9005-02-1
Emalex 200 di-L; Emalex DEG-di-L; Emalex TEG-di-L; PEG-4 dilaurate; Emerest® 2622; Emerest® 2704; Ethox DL-5; Hetoxamate 200 DL; Hodag 22-L; Kessco® PEG-200DL; Lexemul® PEG-200 DL; Lipopeg 2-DL; Mapeg® 200DL; Nonex DL-2; Nopalcol 2-DL; Pegosperse® 200 DL; Witconol 2622. Oil-phase cosmetic ingred., emulsifier for creams, milky lotions, hair conditioners; cleaner, superfattening agent, thickener, reforming agent. Nihon Emulsion; Henkel/Emery; Henkel/Functional Products; Henkel/Cospha; Henkel Canada; Ethox; Heterene;Calgene; Stepan; Inolex; Lipo; PPG/Spec. Chem.; Hart Chem Ltd.; Lonza; Witco/Organics.

23051 PEG-7 palmitate
9004-94-8
Genagen P-070. Surfactant for textile processing. *Hoechst Celanese/Colorants & Surf.*

23052 PEG-75 lanolin
8039-09-6
PEG-75 lanolin
Lan-Aqua-Sol 100; Lantrol® PLN. Emulsifier for cosmetic and pharmaceutical emulsions; emollient, superfatting agent, conditioner for skin and hair care products, household detergents; solubilizer, wetting agent, dispersing aid. *Fanning; Pulcra SA.*

23053 PEG-8 caprylic/capric glyceride
85536-07-8; 57307-99-0
Labrasol; PEG 400 caprylate/caprate glycerides; L.A.S. Hydrophilic oil; excipient, solubilizer for pharmaceutical and cosmetic formulations; surfactant for microemulsions; wetting agent; penetration enhancer. *Gattefosse; Gattefosse SA.*

23054 PEG-9 pelargonate
31621-91-7
Ethox 1122; Genagen PL-090. High cohesion lubricant for synthetic fiber production and processing. *Ethox; Hoechst Celanese/Colorants & Surf.*

23055 pegmatite
A felspathic rock, similar to Cornish stone.

23056 pegnin
A preparation of lactose and rennet, which yields a finely divided curd from cows' milk; used in infant food.

23057 Pegol® L-10
EO/PO block copolymer; emulsifier for agricultural formulations, cutting and grinding fluids, asphalt emulsions. *Rhône-Poulenc Surf.*

23058 Pegosperse®
Polygycol esters and glycol esters. Emulsifiers. *Lonza Inc.*

23059 Pegosperse® 100 L
9004-81-3
PEG-2 laurate
Emulsifier for oil-water emulsions; dispersant; for industrial use. *Lonza Inc.*

23060 Pegosperse® 100 O
9004-96-0
PEG-2 oleate
Emulsifier for oil-water emulsions; dispersant; for industrial use. *Lonza Inc.*

23061 Pegosperse® 100 S
9004-99-3
PEG-2 stearate
Emulsifier for oil-water emulsions; dispersant; for industrial use. *Lonza Inc.*

23062 Pegosperse® 1750 MS
9004-99-3
PEG-40 stearate
Emulsifier for oil-water emulsions; dispersant; for industrial use. *Lonza Inc.*

23063 Pegosperse® 200 DL
9005-02-1
PEG-4 dilaurate
Emulsifier, dispersant, opacifier, viscosity control agent, defoamer for cosmetics, household products, textiles, plastics, water treatment. *Lonza Inc.*

23064 Pegosperse® 400 DOT
61791-01-3
PEG-8 ditallate
Emulsifier, dispersant, opacifier, viscosity control agent, defoamer for cosmetic, household products, textiles, paper, water treatment. *Lonza Inc.*

23065 Pegosperse® 50 DS
627-83-8 211-014-3
Glycol distearate
Emulsifier, opacifier, stabilizer for suspensions and dispersions; emollient, lubricant and pigment dispersant in pharmaceuticals and cosmetics; thickener, wetting agent and plasticizer in hair products. *Lonza Inc.*

23066 Pegosperse® 50 MS
9004-99-3

Glycol stearate
Dispersant, emulsifier for oil-water emulsions for industrial use. *Lonza Inc.*

23067 Pegosperse® PMS CG
1323-39-3 215-354-3
Propylene glycol stearate
Emulsifier, dispersant, opacifier, visc. control agent, defoamer for cosmetics, household products, textiles, plastics, water treatment. *Lonza Inc.*

23068 Peka Glas
A proprietary safety glass. No manufacturer.

23069 Pekafill
Resin based light-curing small particle size hybrid composite filling material; used in dentistry. *Bayer AG.*

23070 Pekafix
Cold curing denture resin on the basis of methyl methacrylate; dental preparation. *Bayer AG.* Discontinued.

23071 Pekalux
Resin based micro-filled light-curing composite filling material; used in dentistry. *Bayer AG.*

23072 Pekatop
Heat-curing denture resin on the basis of methyl methacrylate; dental preparation. *Bayer AG.* Discontinued.

23073 Pekatray®
Cold curing plastic for the preparation of individual impression trays; dental specialty. *Bayer AG.*

23074 Pelamag
Salt coated magnesium granules for desulfurizing blast furnace iron; (sold under license from Dow Chemical Company). *Foseco (F.S.) Ltd.*

23075 Pelamagsalt
Coated magnesium granules; used for desulfurization of blast furnace iron. *Dow Chemical.*

23076 Pelargene
2,4-Dimethyl-6-phenyldihydropyrane
Quest Int'l, UK Ltd.

23077 Pelargone
A trade name for nylon 9. No manufacturer.

23078 Pelargonic acid
112-05-0 7198 203-931-2
$C_9H_{18}O_2$
Nonanoic acid
Emery® 1202; nonylic acid; nonoic acid. Detergent intermediate. MW=158.24; oily liq.; colorless; mp=12.5°; bp=252°; insol.in water; sol.in alcohol, chloroform,ether; LD_{50}(I.v., mise)= 224+-4.6 mg/kg *Henkel/Emery.*

23079 Pelaspan
Expandable polystyrene resin used in manufacture of loose-fill packing. *Dow Cheml Co Ltd, UK & Ireland.*

23080 Pelaspan 333FR
A flame retardant expandable polystyrene. *Dow Cheml Co Ltd, UK & Ireland.*

23081 Pelaspan GP
General purpose expandable polystyrene. *Dow Chemical.*

23082 Pelaspan Mold-a-Pac
For packaging products using Pelaspan PAC loose fill coated with an adhesive to form a resillient molded cushion. *Dow Cheml Co Ltd, UK & Ireland.*

23083 Pelaspan PAC
Expanded polystyrene; loose-fill packaging material. *Dow Cheml Co Ltd, UK & Ireland.*

23084 pelionite
A coal of the cannel type.

23085 Pellethane®
A wide range of polyurethane elastomers, 'rubber-plastics' used cured or uncured fabricated into various shapes and forms by conventional methods. *Dow Cheml Co Ltd, UK & Ireland.*

23086 Pellethane® 2102-55D
PU elastomer (polyester polycaprolactone resin); can be used uncured or cured; tough, high-performance, wear-resistance material providing clarity, chemical resistance, flexibility for automotive (fascia, brake cable jacketing, body side trim, etc.) health care (tubing, catheter components, transdermal patches), in film or sheets as bladders, lining or belting, in fabricated products for casters; athletic shoes, bushings, wheels or extruded profiles. *Dow Plastics.*

23087 Pellethane® 2354-45DGA
PU elastomer; automotive grade. *Dow Plastics.*

23088 Pellethane® 2363-55D
PU elastomer; health care resin. *Dow Plastics.*

23089 Pelonit D
Contains aluminum, has a strong shoving effect and is used where coarse fragmentation is required. *Dynamit Nobel Wien GmbH.* Name unverified.

23090 Pels
1310-73-2 8772 215-185-5
Sodium hydroxide
PPG Industries.

23091 Peltex
Ferro-chromium lignosulfate, modified; wetting agent, emulsifier, dispersant; oil well drilling mud thinner. *Borregaard Ligno Tech.*

23092 Penacolite® B-18-S
65876-95-1
Resorcinol-formaldehyde resin; dry bonding agent formulated for rubber to-wire adhesion; for tires, industrial belts, and retreads. *Indspec.*

23093 Penacolite® R-2170
24969-11-7
Resorcinol-formaldehyde resin; bonding agent for dipping formulas for bonding synthetic industrial fabrics such as aramid, polyester and glass, to rubber. *Indspec.*

23094 Penaryl A
Amyldiphenyl
A proprietary trade name for a plasticizer. No manufacturer.

23095 Penaryl B
Diamyldiphenyl
A proprietary trade name for a plasticizer. No manufacturer.

23096 Penchlor
A proprietary acid-proof cement made from cement powder and sodium silicate solution; used for lining tanks. No manufacturer.

23097 pencycuron
66063-05-6 266-096-3
$C_{19}H_{21}ClN_2O$
N-((4-chlorophenyl)methyl)-N-cyclopentyl-N'-phenylurea
Bay NTN 19701; Monceren; [(chlorophenyl)-methyl]-N-cyclopentyl-N'-phenylurea; Trotis. Non-systemic fungicide with protective action. Used for control of *Rhizoctonia and Pellicularia* species. mp = 129°; insoluble in H_2O (0.4 mg/l), soluble in organic solvents; LD_{50} (rat orl) > 5000 mg/kg. *Bayer.*

23098 pencycuron
66063-05-6 266-096-3
$C_{19}H_{21}ClN_2O$
N-[(4-chlorophenyl)methyl]-N-cyclopentyl-N'-phenylurea
Bay NTN 19701; Monceren; Trotis. Non-systemic fungicide used fro control of *Rhizoctonia and Pellicularia* spp. in potatoes, rice, cotton and vegetables. In particular, control of black scurf of potatoes, sheath blight of rice, and damping-off of ornamentals. mp = 129°; slightly soluble in H_2O (0.4 mg/l), more soluble in organic solvents; LD_{50} (rat orl) >5000 mg/kg.

23099 pencycuron
66063-05-6 266-096-3
$C_{19}H_{21}ClN_2O$
N-((4-chlorophenyl)methyl)-N-cyclopentyl-N'-phenylurea
Bay NTN 19701; Monceren; [(chlorophenyl)-methyl]-N-cyclopentyl-N'-phenylurea; Trotis. Non-systemic fungicide with protective action. Used for control of *Rhizoctonia and Pellicularia* species. mp = 129°; insoluble in H_2O (0.4 mg/l), soluble in organic solvents; LD_{50} (rat orl) > 5000 mg/kg. *Bayer.*

23100 pendare
A name for Venezuelan chicle.

23101 pendecamaine
N,N-dimethyl-(3-palmitamidopropyl)-glycine betaine
A surface-active agent present in tegobetaines.

23102 pendimethalin
40487-42-1 7211 254-938-2
$C_{13}H_{19}N_3O_4$
N-(1-Ethylpropyl)-3,4-dimethyl-2,6-dinitrobenzenamine
Penoxalin; Herbadox; Prowl; Pentagon; Stomp; Pre-M 60DG; Prowl 3.3E; Prowl 4E; Accotab; Herbodox; Go-Go-San; Way Up; Pay-off; Sipaxol;. Selective herbicide, absorbed by roots and leaves. Used for control of most annual grasses and many broad-leaved weeds in cereal, vegetable and fruit crops. mp = 56-57°; SG = 1.19; soluble in H_2O (0.3 mg/l), more soluble in organic solvents; LD_{50} (rat orl) = 1250 mg/kg.

23103 Peneteck
8042-47-5 232-455-8
Mineral oil technical; emollient. *Penreco.*

23104 Penetral NA 20
Low-foam wetting agent for textile mercerizing; stable in strongly alkaline solution *Ceca SA.*

23105 Penetrol
A sulfonated oxidation product of petroleum; used as an insecticide against aphids.

23106 Penetrol
A compound used as a textile detergent.

23107 **Penetrol 2-EHS**
Ammonium 2-ethylhexyl sulfate
Solubilizer and wetting agent especially at high pH and temperature. *Clark.*

23108 **Penetron OT-30**
2-Ethylhexyl sulfosuccinate
Penetrant, wetting agent, surface tension depressant for textiles, paper, paint, plastic, rubber and metal industries. *Hart Products Corp.*

23109 **Penicillin G potassium**
113-98-4 7225 204-038-0
$C_{16}H_{17}KN_2O_4S$
[2S-(2α,5α,6β)]-3,3-Dimethyl-7-oxo-6-[(phenylacetyl)amino]-4-thia-1-azabicyclo[3.2.0]heptane-2-carboxylic acid monopotassium salt
benzylpenicillin potassium; potassium penicillin G; potassium benzylpenicillinate; benzylpenicillinic acid potassium salt; Notaral; Crystapen; Hipercilina; Pentid; Tabilin; Eskacillin; Forpen; Hylenta; Cosmopen; Falapen; Hyasorb; Cristapen; M-Cillin; Monopen; Megacillin Tablets; Scotcil. Antibacterial mp = 214-217° (dec); $[\alpha]^{22}_D$ = 285-310° (c = 0.7); soluble in H_2O; *Bristol-Myers Squibb Co. Inc.*

23110 **Penicillin G potassium**
113-98-4 7225 204-038-0
$C_{16}H_{17}KN_2O_4S$
[2S-(2α,5α,6β)]-3,3-Dimethyl-7-oxo-6-[(phenylacetyl)amino]-4-thia-1-azabicyclo[3.2.0]heptane-2-carboxylic acid monopotassium salt
benzylpenicillin potassium; potassium penicillin G; potassium benzylpenicillinate; benzylpenicillinic acid potassium salt; Notaral; Crystapen; Hipercilina; Pentid; Tabilin; Eskacillin; Forpen; Hylenta; Cosmopen; Falapen; Hyasorb; Cristapen; M-Cillin; Monopen; Megacillin Tablets; Scotcil. Antibacterial mp = 214-217° (dec); $[\alpha]^{22}_D$ = 285-310° (c = 0.7); soluble in H_2O; *Bristol-Myers Squibb Co. Inc.*

23111 **Penicillin V**
87-08-1 7230 201-722-0
$C_{29}H_{38}N_4O_6S\text{-}H_2O$
[2S-(2α,5α,6β)]-3,3-Dimethyl-7-oxo-6-[(phenylacetyl)amino]-4-thia-1-azabicyclo-[3.2.0]heptane-2-carboxylic acid compound with 2-(diethylamino)ethyl 4-aminobenzoate(1:1) monohydrate
G.P.V; Benzylpenicillin procaine; procaine benzylpenicillinate; Procaine penicillin G; Abbocillin-DC; Afsillin; Ampinnate; Aquacillin; Aquasuspen; Avloprocil; Cilicaine; Crysticillin; Despacilina; Depocillin; Distaquaine; Dorallin; A.R; Duracillin; Flo-Cillin Aqueous; Hydracillin; Ilcocillin P; Kabipenin; Ledercillin; Lenticillin; Mammacillin; Megapen; Mylipen; Neoproc; Penaquacaine G; Pen-Fifty; Premocillin; Procanodia; Pro-Pen; Wycillin. Antibacterial Monoclinic hemimorphic crystals from methanol-water; mw=588.73; mp=106-110°; d=1.255-1.256; pH of saturated aq solution=5-7.5; 1 gram dissolves in 250 ml water; 30 ml alcohol, 60 ml chloroform; LD_{50}=sc in mice 2.3 g/kg. *Galen Ltd.* Discontinued.

23112 **Penicillinase**
9001-74-5 7219 232-628-8
β-lactamase; Neutrapen. An enzyme obtained from cultures of *Bacillus cereus* which hydrolyzes the amide bond in the lactam ring of benzylpenicillin to penicilloic acid; it antagonizes the antibacterial action of penicillin; used in biomedial research.

23113 **Pennad 150**
100-37-8 3161 202-845-2
Diethylaminoethanol
Intermediate, emulsifier, catalyst in urethane foams, curing agent, corrosion inhibitor. *Elf Atochem.*

23114 **Pennchem® Mortar**
Two-component, silica-filled, vinyl ester resin-based mortar; for field mixing for setting brick and tile or for masonry construction. *Atochem.* Discontinued.

23115 **Penncozeb**
8018-01-7 5756
Wettable powder containing mancozeb; protectant fungicide for fruit, field crops, and roses. *Shell UK.*

23116 **Pennfloat® 3-2277**
Mercaptan; intermediate. *Atochem.* Discontinued.

23117 **Pennodorant® 1013**
110-01-0 9357 203-728-9
Tetrahydrothiophene
Odorant for natural gas to permit detection of leaks. *Atochem.* Discontinued.

23118 **Pennstop® 1866**
3710-84-7 223-055-4
$C_4H_{11}NO$
N,N-Diethylhydroxylamine
Free radical scavenger used by the rubber industry as an emulsion polymerization inhibitor; vapor phase inhibitor for olefin or styrene monomer recovery systems; in-process inhibitor for production of styrene, divinyl benzene, butadiene, isoprene; interme mp = -25- -26°; bp = 125-130°; d = 0.8670; n_D^{20}= 1.4195. *Elf Atochem.*

23119 **pennyroyal oil**
8007-44-1 7236
Pulegium oil
hedeoma oil; Oils, pennyroyal; hedeoma pulegioides; Pennyroyal oil. Yellow to reddish essential oil; used in the manufacture of pulegone, flavoring alcoholic beverages.

23120 **Pennzone B 0685**
109-46-6 203-674-6
$C_9H_{20}N_2S$
1,3-dibutylthiourea
Thiourea, N,N'-dibutyl-; N,N-dibutylthiourea; N,N'-di-N-butylthiourea; 1,3-dibutyl-2-thiourea. Accelerator for mercaptan-modified chloroprene rubber, an activator for ethylenepropylenediene terpolymers and natural rubber, an antidegradant for natural rubber-latex and thermoplastic styrene-butadiene rubber. mp = 63-65°. *Elf Atochem N. Am.*

23121 **Pennzone E 0686**
105-55-5 203-308-5
$C_5H_{12}N_2S$
1,3-Diethylthiourea
1,3-diethyl-2-thiourea; diethylthiourea; thiourea, N,N-diethyl-, N,N-diethylthiourea; thiate H; N,N'-diethylthiocarbamide; Pennzone E; diethyl-2-thiourea. Accelerator for mercaptan-modified chloroprene rubber. Antidegradant for natural, nitrile-butadiene, styrene-butadiene, and chloroprene rubbers. mp = 76-78°; soluble in H_2O (1-5 mg/ml), organic solvents; LD_{50} (rat orl) = 316 mg/kg. *Elf Atochem N. Am.*

23122 **Penreco 1520, 3070**
Petrolatum, technical; used in rubber processing aids, carbon papers, buffing and polishing compounds, corrosion preventatives, general purpose lubricants, printing inks, solder pastes. *Penreco.*

23123 **Penreco 2251 Oil**
64742-14-9 265-114-7
Petroleum distillates; high purity hydrocarbon processing solvent, foam control agent, in waterless hand cleaners, agricultural sprays, polishes, fruit and vegetable processing, cleaning oils. *Penreco.*

23124 **Penreco Amber**
Petrolatum USP; emollient, base for cosmetic and pharmaceutical preparations; waterproofing agent for butcher paper; lubricant, water repellent, moisture barrier for textile and paper; carrier for modeling clays, soldering paste and flux; pigment carrier for carbon paper, binder and conditioner for crayons. *Penreco.*

23125 **Penreco Red**
Technical petrolatum; technical-grade base; base and binder for polishes. *Penreco.*

23126 **Pensa's rubber**
A rubber substitute made from coal tar, petroleum tar, oil of turpentine, and boric or phosphoric acids.

23127 **Pensil® 100**
63394-02-5
One-component silicone rubber sealant; sealant offering high strength and excellent primerless adhesion to building substrates; seals against spread of smoke and fire; autobondable. *GE Silicones.*

23128 **Penta G.P. 79**
Rust preventatives. *Croda Chem. Ltd.* Discontinued.

23129 **pentabromodiphenyl oxide**
32534-81-9 251-084-2
$C_{12}H_5Br_5O$
2,3,4,6,2'-pentabromodiphenyl ether
Flame retardant. Insoluble in H_2O, soluble in organic solvents. *Ethyl; Great Lakes.*

23130 **Pentabromodiphenyl oxide**
32534-81-9 251-084-2
$C_{12}H_5Br_5O$
FR-1205; FR-1215; Great Lakes DE-71™; Saytex® 115; PBDPO. Flame retardant for use in laminates (both epoxy and phenolic), unsat. polyesters, synthetic fibers, and flexible PU foams; suitable for textiles. Liquid 50-60°; soluble in CCl_4; methylene chloride, benzene, acetone; insoluble in water; slightly soluble in methanol; mw=565; sp gr=2.25; LD_{50}=rat oral 5200 mg/kg; rabbit dermal>2000 mg/kg *AmeriHaas; Albemarle; Dead Sea Bromine; Great Lakes.*

23131 **pentabromotoluene**
87-83-2 201-774-4
$C_7Br_5H_3$
2,3,4,5,6-pentabromotoluene
Flame retardant for unsaturated polyesters, polyethylene, PP, PS, SBR latex, textiles, rubbers. *Great Lakes.*

23132 **Pentabromotoluene**
87-83-2 201-774-4
$C_7H_3Br_5$

2,3,4,5,6-Pentabromotoluene
FR-705. Flame retardant Mw=486.65; mp=285-286°; LD_{50}=rat oral>5000 mg/kg; irritant *Aldrich; AllmeriBrom; Chem one; Dead Sea Bromine; Great Lakes.*

23133 Pentac Aquaflow
2227-17-0 3154 218-763-5
Emulsifiable concentrate of 480 g dienochlor per liter; an acaricide. *DowElanco Ltd.*

23134 pentachlorethane
76-01-7 7241 200-925-1
C_2HCl_5
ethane pentachloride
pentalin; NCI-C53894; UN 1669. Solvent, degreaser, cleaning agent. bp = 159-162°; d = 1.680; n_D^{20} = 1.5040; insoluble in H_2O, miscible with organic solvents; LC_{50} (rat ihl) = 4238 ppm/2 hr.

23135 pentachlorophenol
87-86-5 7242 201-778-6
C_6HCl_5O
pentachlorophenol
PCP, Penta; penchlorol;Santophen 20. Insecticide used in termite control, as a defoliant and general herbicide. Also used for wood preservation. mp = 190-191°; bp = 310°; d^{22}_4 = 1.9780; poorly soluble in H_2O (80 mg/l), soluble in organic solvents; LD_{50} (rat orl) = 146 mg/kg.

23136 pentachlorophenol
87-86-5 7242 201-778-6
C_6HCl_5O
PCP; penta; penchlorol; Santophen 20. Obtained from chlorination of phenol; fungicide, bactericide, algicide, herbicide; preservation of wood, starches, dextrins, glues. *Penta Mfg.; Vulcan.*

23137 Pentacizers
Proprietary trade names for plasticisers. No manufacturer.

23138 pentaerythritol
115-77-5 7245 204-104-9
$C_5H_{12}O_4$
2,2-bis(hydroxymethyl)-1,3-propanediol
monopentaerythritol; Hercules P 6; PE 200; tetramethylolmethane; THME; PETP; pentaertyhritol; tetrakis(hydroxymethyl)methane. Used in alkyds; synthetic resins, paints, varnishes and as a chemical intermediate. mp = 260°; bp_{30} = 276°; SG = 1.396; soluble in H_2O (55 mg/ml), alcohols, insoluble in non-polar organic solvents. Aqualon; Degussa; Hoechst Celanese; Mitsubishi Gas; Mitsui Toatsu Chem.; Penta Mfg.; Perstorp Polyols.

23139 Pentaerythritol
115-77-5 7245 204-104-9
$C_5H_{12}O_4$
2,2-Bis(hydroxymethyl)-1,3-propanediol
Hercules® Mono-PE; tetrakis(hydroxymethyl)methane; tetramethylolmethane; Metab-Auxil; Penetek; Pentek. Used in production of alkyd resins, rosin esters, urethane resins, drying oils, synthetic lubricants, plasticizer, intumescent paints, plastics, stabilizers for plastics, explosives. MW=136.15; crystals from HCl; mp=260°; sol.in water, ethanol, glycerol, formamide; insol.in acetone, benzene, paraffin, ether, carbon tetrachloride *Hercules.*

23140 Penta-erythritol Tetrastearate (PET)
A proprietary release agent used in injection-molding processes. *DuPont UK.* Name unverified.

23141 Pentaerythrityl tetraoleate
19321-40-5 242-960-5
Pentaerythrityl tetraoleate
Liponate PO-4; Edenor PTO; Mazol® PETO. Emollient ester for adjusting rub-in and afterfeel of personal care products; thickener and visc. controller. *Lipo; Henkel/GosphaPPG/Specialty Chem.*

23142 pentaerythrityl tetrastearate
115-83-3 204-110-1
$C_{77}H_{148}O_8$
2,2-Bis(octadecanoyloxymethyl)-1,3-propanediyl dioctadecanoate
Pentaerythritol tetrastearate. Tetraester of pentaerythritol and stearic acid; used in polishes, coatings, textile finishes. mp = 71°; d = 0.940; *Hercules; Lipo; Lonza.*

23143 Pentaerythrityl tetrastearate
115-83-3; 91050-82-7 204-110-1; 293-029-5
Liponate PS-4; Pentaerythritol tetrastearate; Alkamuls® PETS; Crodamol PETS; Kessco PTS; Radia® 7176. Emollient ester for adjusting rub-in and afterfeel of personal care products; thickener and visc. controller. *Lipo; Rhône-Poulenc Surf. & Spec.; Croda Surf. Ltd.; Akzo BV; Fina Chemicals.*

23144 pentaerythrityl triacrylate
3524-68-3 222-540-8
$C_{14}H_{18}O_7$
(2-propenoic acid-2-(hydroxymethyl)-2-(((1-oxo-2-propenyl) oxy) methyl)-1,3-propanediyl ester)
Crosslinking agent used in adhesives, coatings, inks, textile products, photoresists, castings, modifiers for polyester, fiberglass, or polymers. A trifunctional cross-linking acrylic monomer cured by UV light used in the production of polyfunctional aziridine, added to paint primer and floor top coatings as a self-curing cross-linker or hardener. bp = 205-215° (detonates); insoluble in H_2O *Sartomer.*

23145 Pentagan
Plant growth regulator containing chlormequat and choline chloride. *Agan Chemical Manufacturers Ltd.*

23146 pentagastrin
5534-95-2 7250 226-889-7
$C_{37}H_{49}N_7O_9S$
N-[(1,1-dimethylethoxy)carbonyl]-β-alanyl-L-tryptophanyl-L-methionyl-L-α-aspartyl--L-phenylalaninamide
ICI-50123; AY-6608; Gastrodiagnost; Peptavlon. Gastric secretion stimulasnt used as a diagnostic aid. mp = 229-230°; $[\alpha]_D^{22}$ = -29° (DMF); λ_m = 280, 289 nm (=le 5340, 4590, 2N NH_4OH); almost insoluble in H_2O, organic solvents.

23147 pental
513-35-9 649 208-156-3
C_5H_{10}
2-methyl-2-butene
trimethylethylene; β-isoamylene;. Colorless, volatile liquid; used in organic synthesis, high-octane fuel manufacture. mp = -134°; bp = 35-38°; d = 0.6620; n_D^{20}=1.3870; insoluble in H_2O, soluble in organic solvents.

23148 Pental 28
A range of rosin esters; used to modify nitrocellulose based coatings and lacquers. *Hercules.*

23149 Pental 802A
A range of rosin esters; used in heat set lithographic inks. *Hercules.*

23150 Pental 8D
A range of rosin esters; used to produce chewing gum base. *Hercules.*

23151 Pental A
A range of rosin esters; a tackifying agent for natural rubber and SBR based pressure sensitive adhesives. *Hercules.*

23152 Pental G, X
A range of rosin esters; used in heat set lithographic inks. *Hercules.*

23153 Pentalan
Pentaerythritol ester of woolwax fatty acids. *Croda Chem. Ltd.* Discontinued.

23154 pentaline
76-01-7 7241 200-925-1
$CHCl_2 \cdot CCl_3$
Pentachlorethane
Used as a solvent for oil and grease in oil cleaning; for separation of coal from impurities by density difference.

23155 Pentalyn
Pentaerythritol esters of rosin, modified rosins, dibasic acid- and phenolic-modified rosins; synthetic resins used in heat-set offset inks, letterpress printing inks, flexographic inks. *Hercules.*

23156 Pentalyn 255, 261, 856
Alkali-soluble resins used in emulsion floor polishes. *Hercules.*

23157 Pentalyn 344, C, K
Tackifying and reinforcing resins for adhesives and sealants. *Hercules.*

23158 Pentalyn 802A, 802A Pale, 833, G, X
Used in overprint varnishes. *Hercules.*

23159 Pentalyn A
Used in the production of chewing gum base. *Hercules.*

23160 Pentalyn H
Tackifying and reinforcing resin for adhesives and sealants; also in the production of chewing gum base. *Hercules.*

23161 Pentamid C12
Fatty acid polydiethanolamide
Metal-working corrosion inhibitor. *Pentagon Urethanes Ltd.*

23162 Pentamid KH
Methyl N-octadecyl terephthalamate
Gelling agent for high-temperature greases. *Pentagon Urethanes Ltd.*

23163 Pentamin BDMA etc
103-83-3 203-149-1
$C_9H_{13}N$
N,N-Dimethyl benzylamine
N-benzyldimethylamine; BDMA; benzyl dimethylamine; benzenemethanamine, N,N-dimethyl-; N-benzyl-N,N-dimethylamine; N-benzyldimethylamine; dimethyl benzylamine. Polyurethane catalyst, epoxy curing agent. mp = -75°; bp = 183-184°; d = 0.900; n_D^{20};ks = 1.5010; LD_{50} (rat orl) = 265 mg/kg. *Pentagon Chemicals Ltd.*

23164 n-pentane
109-66-0 7255 203-692-4

C_5H_{12}
pentane
C_5 alkane; amyl hydride. Solvent; artificial ice manufacture, low-temperature thermometers, solvent extract processes, blowing agent in plastics (expandable polystyrene), pesticide. mp = -130°; bp = 36°; d = 0.626; insoluble in H_2O, soluble in organic solvents; LC_{100} (mus ihl) = 128200 ppm. *Ashland; Phillips 66.*

23165 pentanochlor
2307-68-8 8851 218-988-9
$C_{13}H_{18}ClNO$
N-(3-Chloro-4-methylphenyl)-2-methylpentanamide
3'-Chloro-2-methyl--toluidide; pentanochlore; solan; CMMP; CMA; Croptex Bronze; FMC 4512. Emulsifiable concentrate containing 400 g/1 pentanochlor; used to control weeds in horticultural crops. mp = 85-86°; d_{20} = 1.106; soluble in H_2O (8-9 mg/l), freely soluble in organic solvents; LD_{50} (rat orl) > 10000 mg/kg *Hortichem Ltd; Atlas-Interlates.*

23166 Pentanox 4X, 24X
Alkyl dimethyl amine oxides
Thickening and foaming agents. *Pentagon Chemicals Ltd.*

23167 Pentaphane
A proprietary trade name for a film made from a chlorinated polyether (polymerized 3,3-bis(chloromethyl)oxetane). *British Cellophane.* Unverified.

23168 Pentaphen® 67
80-46-6 7284 201-280-9
$C_{11}H_{16}O$
1-hydroxy-4(2-methyl-2-butyl)benzene
Pentaphen; *p*-tert-amyl phenol; *p*-(α,α-dimethylpropyl)phenol; *p*-(1,1-dimethylpropyl)phenol; PTAP; ucar amyl phenol 4t; *p-t*-amylphenol; dimethylpropyl)phenol; *para-tertiary* amylphenol. Intermediate for chemical specialties; in germicidal formulations; also in manufacturing of photographic chemicals, oil demulsifiers, phenolic resins, agricricultural surfactants, antiskinning agents. mp = 91-94°; bp = 255°; d_4^{20}= 0.962; insoluble in H_2O, soluble in organic solvents; LD_{50} (rat orl) = 3.08 g/kg. *Elf Atochem.*

23169 Pentaquest Extra 0685
140-01-2 205-391-3
$C_{14}H_{18}N_3Na_5O_{10}$
Bis(2-(bis(carboxymethyl)amino)ethyl)glycine pentasodium salt
Pentasodium pentetate; CHEL 330; Detarex PY; HAMP-EX 80; Kiresuto P; Pentasodium DTPA; Pentasodium pentetate; Penthanil; Perma Kleer 140; Plexene D; Syntron C; Versenex 80. Chelating agent for iron chelation up to pH 11.5. mp = -40°; bp = 106°; SG = 1.299; soluble in H_2O. *Clough.*

23170 Pentaquest OPAC 0201
67-43-6 7266 200-652-8
$C_{14}H_{23}N_3O_{10}$
[[(Carboxymethyl)imino]bis(ethylenenitrilo)]-tetra-acetic acid
DTPA; Diethylenetriaminepentaacetic acid; Diethylenetriamine-N,N,N',N'',N''-pentaacetic acid; Pentetic acid; N,N-Bis(2-(bis-(carboxymethyl)amino)ethyl)-glycine. DTPPA; chelating agnt with affinity for iron; scale and corrosion inhibitor for aqueous systems; stable to hydrogen peroxide. mp = 220° (dec); soluble in H_2O (5 g/l). *Clough.*

23171 Pentaquest OPNA 0256
140-01-2 205-391-3
Pentasodium DTPPA
Scale and corrosion inhibitor for aq. systems; stable to peroxide. *Clough.*

23172 Pentasodium pentetate
140-01-2 205-391-3
Hamp-Ex® 80; Cheelox® 80; Chel DTPA-41; Kalex Penta; Pentaquest Extra 0685, OPNA 0256; Polyquest 80; Trilon® C Liq,. Chelating agent for alkaline earth and heavy metal ions; peroxide bleaching. *W R Grace/Hampshire; Rhône-Poulenc Surf.; Ciba-Geigy; Hart Chem Ltd.; Clough; BASF.*

23173 Pentasol
n-amyl alcohol [71-41-0] (75%) and *sec*-amyl alcohol [6032-29-7] (25%). A mixture of pure amyl alcohols containing 75% primary alcohol and 25% secondary alcohol, and is obtained from the pentane fraction of gasoline; and is used as a varnish and lacquer solvent.

23174 Pentasol
577-11-7 3460 209-406-4
Sodium dioctyl sulfosuccinate
Wetting agent. *Pentagon Chemicals Ltd.*

23175 Pentateric 24B, B, BLG
Alkyl betaines; used for toiletries. *Pentagon Chemicals Ltd.*

23176 Pentek
115-77-5 7245 204-104-9
pentaerythritol
A proprietary trade name for a technical grade of pentaerithritol used in synthetic resins and in the paint and varnish industry. No manufacturer.

23177 Pentelex
Photographic developer. *May & Baker Ltd.* Name unverified.

23178 Pentex® 40
Alkylaryl sulfonate
Wetting agent, penetrant, emulsifier; textile and industrial processing; improves dye uniformity; aids rewetting of leather, paper, and paper-mill felts. *Rhône-Poulenc/Latex & Spec. Polymer.*

23179 Penthrinit
It is a plastic mixture of 80% pentaerythritol-tetranitrate and 20% nitroglycerine; an explosive.

23180 Penthrit
78-11-5 7249 201-084-3
Pentaerythritol tetranitrate

23181 Pentine 1185 5432
Isopropylamine dodecylbenzene sulfonate
Emulsifier for emulsion degreasers and dry cleaning soaps; agricultural emulsifier. *Clough.*

23182 Pentine Acid 5431
27176-87-0 248-289-4
$C_{18}H_{30}O_3S$
Dodecylbenzene sulfonic acid
laurylbenzenesulfonic acid. Base for dishwashing and laundry detergents, industrial and institutional cleaners. *Clough.*

23183 Pentol
Timber fungicides. *Plant Protection.* Name unverified.

23184 Pentonate DB
Fatty alcohol benzoate
Substitute for isopropyl myristate; emollient. *Pentagon Chemicals Ltd.* Name unverified.

23185 Pentonite
Sodium benzoate/sodium nitrite mixture; corrosion inhibitor for aqueous systems. *Pentagon Urethanes Ltd.* Name unverified.

23186 Pentonium 50, 80
alkyl dimethyl benzalkonium chlorides and alkyl trimethyl ammonium chlorides. Biocides. *Pentagon Chemicals Ltd.* Name unverified.

23187 Pentopan
A pentosanase; added to the flour to improve dough handling properties during baking. *Novo Nordisk.*

23188 Pentoxyl M
Perfumery specialty. *Bush Boake Allen Ltd.*

23189 Pentrex
Phenolic-modified and maleic modified esters of rosin; Pentrex G and X rosin esters are used in heat set lithographic inks; Pentrex 28 is used in nitrocellulose-based coatings and lacquers. *Hercules.*

23190 Pentrone
Anionic surfactant. *Rhône-Poulenc UK.*

23191 Pentrone ON
Sodium 2-ethylhexyl sulfate in liquid form; anionic surfactant for dispersing and alkaline wetting agent used in electroplating and lye peeling. *Rhône-Poulenc UK.*

23192 Pentrone S
Range of anionic surfactants of the disodium mono-alkyl polyalkylene sulfosuccinate type; supplied as liquids; dispersing, foaming, and emulsification agents with low toxicity; used in emulsion polymerization, surgical scrubs, and shampoos. *Rhône-Poulenc UK.*

23193 Pentrosan
Dispersing and solubilizing agents. *Rhône-Poulenc UK.*

23194 pentyl
C_5H_{11}
Synonym for the amyl group.

23195 Pentylol
Mixture of amyl alcohol isomers; solvent. *Sasolchem.*

23196 Penzold's reagent
A solution of diazo-benzosulfonic acid and potassium hydroxide; a reagent for sugar in urine.

23197 Pep
Polyester promoters. *Air Prods & Chems. Inc.*

23198 Pep Set
Resins and catalysts associated with the production of foundry cores and molds; when applied to foundry sand, provide a binder system useful in the production of foundry cores and molds; these binder systems fall into the large category of no-bake binders used within the foundry industry and cure at room temperature. *Ashland Chemical Company.* Name unverified.

23199 Pepper Dust
Powdered pepper; animal deterrent. *Vitax Ltd.*

23200 pepsin
9001-75-6 7289 232-629-3
Pepsinum

A digestive enzyme of gastric juice; it decomposes albuminous bodies into peptone; medicine (digestive ferment); substitute for rennet in cheesemaking. *Am. Biorganics; EM Industries; G Fiske & Co Ltd; R.W. Greeff; Worthington Biochemical.*

23201 Peptavlon
5534-95-2 7250 226-889-7
Pentagastrin
For diagnostic testing of gastric secretions. *Imperial Chemical Industries plc.*

23202 Peptein® 2000®
9015-54-7 310-296-6
Hydrolyzed animal protein; conditioner for cosmetic applications; especially substantive to hair. *Hormel.* Name unverified.

23203 Peptein® VgW
Hydrolyzed wheat protein; for cosmetic hair and skin care products *Hormel.*

23204 Peptizer 566
Naphthenic oil/sulfonate ester blend. *C.P. Hall.*

23205 Peptizer 7010
Mineral oil/sulfonate blend. *C.P. Hall.*

23206 Peptoil
Petroleum oil; spray adjuvant for enhancing herbicide activity and defoliation performance with foam eliminator. *Draxel Chemical Company.* Unverified.

23207 Peptorub
Pre-plasticized comminuted rubber.

23208 Peptrex
Rubber peptizer and rubber. *Miles.* Discontinued.

23209 Peradinol
Textile dyeing auxiliaries. *Fine Dyestuffs & Chemicals Ltd.* Name unverified.

23210 Peralfan T Concentrate
Combination of volatile solvents of specific activity with nonionic emulsifiers and anionic substances; yellowish liquid; spotting, dispersing and scouring agent to remove stubborn soilings. *Thor Chemicals (UK) Ltd.* Discontinued.

23211 Perapret®
Additives for textile resin finishing. *BASF plc.*

23212 Perapret® PE-2
Softener and lubricant improving abrasion and sewability in durable press finishing. *BASF.*

23213 perborax
7632-04-4 8797 231-556-4
$NaBO_3 \cdot 4H_2O$
Sodium perborate
A washing and bleaching agent.

23214 perborin
7632-04-4 8797 231-556-4
$NaBO_3$
Sodium perborate
A constituent of washing powders.

23215 Perbunan®
Acrylonitrile-butadiene rubber. *Bayer plc; Miles Inc.*

23216 Perbunan® N
Acrylonitrile butadiene rubber; used for rubber goods that must withstand oils, greases, and petrol and must have good resistance to heat, ageing and abrasion, e.g., seals, roller covers, hoses, cable sheathings, plant linings, conveyor belting. *Bayer AG.*

23217 Perbunan® N 1807 NS
NBR; low permeability to gases and good physiological properties; used in technical moldings e.g., seals, sleeves, diaphragms, bellows, valves, vibration dampers, footwear soles; roll covers, printing blocks and blankets, belting, fabric proofings (e.g., for flexible silos); technical hoses, cable sheathings, brake & clutch linings, open and closed cell sponge rubber, punching blocks, gloves, adhesives; nonstaining. *Bayer; Miles.*

23218 Perbunan® N/VC70
NBR/PVC fluxed blend (70% Perbunan N 2807 NS/30% PVC); nonstaining stabilizer; elastomer used in rubber goods with excellent resistance to mineral oil, grease, fuels, superior resistance to ozone and weathering; applications including hose, cable jackets, diaphragms, roll covers, thermal blown insulation, spinning aprons. *Bayer; Miles.*

23219 Percarbamid
7722-84-1 4839 231-765-0
A hydrogen peroxide preparation; used in the cosmetics and pharmaceuticals industry. *Degussa AG.*

23220 perchlorethylene
127-18-4 9332 204-825-9
C_2Cl_4
Tetrachloroethylene
Solvent, degreaser.

23221 perchloric acid
7601-90-3 7296 231-512-4
$ClHO_4$
Analytical chemistry, catalyst, manufacture of esters, ingredient of electrolytic bath in deposition of lead, electropolishing, explosives. mp = -112°; bp_{11} = 19°; d^{22} = 1.768. *Spectrum Chem. Mfg.*

23222 perchloroethylene
127-18-4 9332 204-825-9
$Cl_2C{\cdot}cCCl_2$
tetrachloroethylene
tetrachloroethene; ethylene tetrachloride. Dry-cleaning solvent; vermifuge; drying agent; degreasing metals. *Asahi-Penn; Ashland; Elf Atochem; General Chem.; ICI Spec.; OxyChem; PPG Industries.*

23223 Perchloron
7778-54-3 1717 231-908-7
A technical calcium hypochlorite containing 68.1% available chlorine.

23224 Percist
Insecticidal formulation. *Mitchell Cotts Chemicals Ltd.*

23225 Perclene
127-18-4 9332 204-825-9
Perchloroethylene
Cleaner and degreasant. *Occidental Chemical Corp.*

23226 Perclene TG
127-18-4 9332 204-825-9
Perchloroethylene
Transformer grade. *Occidental Chemical Corp.*

23227 Percol®
Polyacrylamide-based high efficiency retention aids for paper manufacture. *Allied Colloids Ltd.*

23228 Percolaye
Attapulgite or sepiolite clay; used for refining of minerals and chemicals. *Bromhead & Denison Ltd.*

23229 Percresan
A mixture of cresols, soap, and water; used as a disinfectant in 1-2% solution.

23230 Percumyl D
80-43-3 201-279-3
$C_{18}H_{22}O_2$
Dicumyl peroxide
Bis(1-methyl-1-phenylethyl) peroxide; bis(α,α-dimethylbenzyl) peroxide; cumyl peroxide; di-α-cumyl peroxide; Di-cup; diisopropylbenzene peroxide; isopropyl benzene peroxide; Cumene peroxide. Oxidizer, catalyst. bp = 130°; insoluble in H_2O (<1 mg/ml), soluble in organic solvents; LD_{50} (rat orl) = 4100 mg/kg. *British Traders & Shippers Ltd.*

23231 Perduren
An organic polysulfide synthetic rubber.

23232 Perdynamine
A compound of albumen and hemoglobin.

23233 Perecot
A copper containing fungicide. *ICI Chem. & Polymers Ltd.*

23234 Peregal®
Leveling agent when dyeing cellulose fibers with vat dyes. *BASF AG.*

23235 Peregal® O
Polyoxyethylated fatty alcohol (nonionic)
Dyeing assistant for use with basic and direct colors; assistant for the dyeing, leveling and stripping of vat dyes; leveling agent in acetate printing. *ISP.*

23236 Peregal® OK
Methylpolyethanol quaternary amine (cationic); used as a vat dye retarder. *ISP.*

23237 Peregal® ST
9003-39-8 7879
Polyvinylpyrrolidone
Stripping assistant for cotton and rayon yarns or fabrics that have been dyed or printed with vat, sulfur or direct colors; rag stripping assistant in high grade paper. *ISP.*

23238 Pereman
A copper fungicide. *Plant Protection.* Name unverified.

23239 Perenol EI
Polyvinyl isobutyl ether
Defoamer for petroleum solvent-based paints. *Henkel KgaA.* Discontinued.

23240 Perenox
1317-39-1 2734 215-270-7
A cuprous oxide-based fungicide. *ICI Chem. & Polymers Ltd.*

23241 Perenyl's fluid
chromo-nitric acid
It contains 3 parts of 92% alcohol, 4 parts of 10% nitric acid, and 3 parts of 0.5% chromic acid. A fixing agent used in microscopy; the objects are treated with alcohol after fixing.

23242 Perfecta
Petrolatum USP. *Witco.*

23243 **Perfection®**
7758-16-9 8713 231-835-0
Sodium acid pyrophosphate
Food grade leavening acid for baking, cereals. *Rhône-Poulenc Food Ingreds.*

23244 **Perfekthion®**
60-51-5 3269 200-480-3
Dimethoate
Systemic insecticide for control of sucking and biting insects *BASF AG.*

23245 **Perfix**
High speed photographic fixer. *May & Baker Ltd.* Name unverified.

23246 **Perflex**
Partially hydrogenated vegetable oil (soybean, cottonseed), propylene glycol mono and diesters of fats and fatty acids, mono and diglcyerides, lecithin, BHA; kosher; multipurpose shortening for cakes, nonstandard yeast-leavened products. *Van Den Bergh Foods.*

23247 **Perflex**
A proprietary trade name for unstretched polyvinylidene chloride. No manufacturer.

23248 **Perfluidone**
37924-13-3 7300 253-718-3
$C_{14}H_{12}F_3NO_4S_2$
1,1,1-Trifluoro-N-[2-methyl-4-(phenylsulfonyl)phenyl]methanesulfonamide
Lancer; Flamprop-methyl; 1,1,1-trifluoro-4'-(phenylsulfonyl)methanesulfono-*o*-toluidide; MBR-8251; Destun; 2-methyl-4-phenylsulfonyltrifluoromethanesulfonanilide,. Herbicide for wild oat control. MW=379.38; solid,crystals; mp=142-144°; sol.in water, acetone, benzene, dichloromethane, methanol. *ICI Chem. & Polymers Ltd.*

23249 **PerforMax® 403**
Polymeric; deposit inhibitor for open recirculating cooling water systems. *Drew Ind. Div.*

23250 **perfumery oil**
Refined petroleum, of specific gravity 0.880-0.885; used in perfumery.

23251 **Perfusamine**
85068-76-4 5066
Iofetamine ^{123}I hydrochloride
Diagnostic aid; radioactive agent. *Medi-Physics Inc.* Name unverified.

23252 **Pergacid**
Dyes for paper. *Ciba plc.* Name unverified.

23253 **Pergamin**
Pergamyn
A grease-proof paper made from cellulose pulp.

23254 **Pergantine**
Pigment dispersions for paper. *Ciba plc.* Name unverified.

23255 **Pergaprint**
Crosslinking agents for starch. *Ciba plc.* Name unverified.

23256 **Pergascript**
Chemicals for carbonless paper. *Ciba plc.* Name unverified.

23257 **Pergasol**
Dyes for paper. *Ciba plc.* Name unverified.

23258 **pergenol**
A mixture of sodium perborate and bitartrate. It releases hydrogen peroxide upon addition of water.

23259 **Perglanz-Konzen-Trat B48 and B30**
Blend of amphoterics and anionics, in the form of a white liquid; sheen additive for shampoos, bath, and shower preparations *Th Goldschmidt Ltd.* Unverified.

23260 **Perglazmittel GM 4006**
Nonionics and fatty alcohol ether sulfate; pearlescent for hair shampoos and bath additives. *Zschimmer & Schwarz.* Discontinued.

23261 **Perglow**
A bright nickel plating process. *Hanshaw Chemicals.* Unverified.

23262 **perglycerol**
72-17-3 8781 200-772-0
An aqueous solution of sodium lactate; Glycerol substitute in antifreeze, calico printing and anti-corrosion agents. Also used medically as an electrolyte replenisher.

23263 **Pergopak**
Organic fillers. *Ciba plc.* Name unverified.

23264 **Pergopak**
Urea formaldehyde resins. *Lonza Inc.*

23265 **Pergut®**
Chlorinated rubber; used for the formulation of coatings with high water, chemical, and low-temperature resistance as well as for use in printing inks. *Bayer AG; Bayer plc.*

23266 **Perhydrate**
No manufacturer.

23267 **Perhydrit**
No manufacturer.

23268 **Perhydrol**
7722-84-1 4839 231-765-0
H_2O_2
Hydrogen peroxide
One volume of 30% hydrogen peroxide giving 100 volumes of oxygen; used for bleaching, also as a disinfectant.

23269 **Peri acid**
82-75-7 6492 201-437-1
$C_{10}H_9NO_3S$
1-Naphthylamine-8-sulfonic acid
8-amino-1-naphthalenesulfonic acid. Used as an azo dye intermediate.

23270 **Perichthol**
8029-68-3 4929 232-439-0
ichthammol
ammonium bituminosulfonate; ammonium ichthosulfonate; ammonium sulfobituminate; bitumol; bituminol; ichthammonium; ichthosulfol; Ichthyol; Hirathiol; Ichden; Ichtammon; Ichthadone; Ichthymall; Ichthysalle; Ichthium; Ichthalum; Ichthopur; Ichthosan; Ichthynat; Ichthyopon; Lithol; Petrosulpho; Perichthol; Piscarol; Pisciol; Saurol; Subitol; Sulfogenol; Thilaven; Thiolin; Thiozin; Trasulphane; Tumenol; Leukochthol; Ichthosauran; Amsubit; Bitulan. A proprietary preparation of ammonium ichthosulfonate made by sulfonation and ammoniation of a distillate from mineral deposits. Used as a demulcent, emollient, antiseptic and topical anti-infective. Miscible with H_2O, organic solvents;. No manufacturer.

23271 **Perikol**
A sensitizer for silver bromide plates, prepared by treating the addition product of toluquinaldine and the ethyl ester of toluene-sulfonic acid with alcoholic potassium hydroxide.

23272 **perilla oil**
An oil obtained from the seed of the Asiatic mint, *Perilla ocymoides*; used as a drying oil in substitution for linseed oil.

23273 **periodic acid**
10450-60-9 7312 233-937-0
H_5IO_6
Periodic Acid Dihydrate
Orthoperiodic Acid. Oxidizing agent, increases wet strength of paper, photographic paper. Used in organic synthesis. mp = 122°; soluble in H_2O (30 g/ml), less soluble in organic solvents. *Atomergic Chemetals; Blythe, William Ltd; EM Industries; Janssen Chimica; Spectrum Chem. Mfg.*

23274 **Periograf**
1306-06-5 3519 215-145-7
Durapatite
Calcium phosphate derivative. Used as a prosthetic aid in artifical bones and teeth. *Sterling Drug Inc.* Name unverified.

23275 **Periygel®**
94-36-0 1149 202-327-6
A trademark for benzoyl peroxide. No manufacturer.

23276 **Perizin®**
56-72-4 2626 200-285-3
C14H16ClO5PS
O-(3-chloro-4-methyl-2-oxo-2H-1-benzopyran-7-yl) O,O-diethyl phosphorothioate
coumaphos; OMS 495; ENT 17957; Asuntol; Bayer 21/199; Co-Ral. Non-systemic insecticide used for the control of ectoparasites (Varroa jacobsoni) in bees. mp = 95°; d = 1.474; soluble in H_2O (1.5 mg/l), poorly soluble in organic solvents; LD_{50} (rat orl) = 41 mg/kg. *Bayer AG.*

23277 **Perkacit® CBS**
95-33-0 202-411-2
$C_{13}H_{16}N_2S_2$
N-Cyclohexyl-2-benzothiazole sulfenamide
N-Cyclohexyl-2-benzothiazolesulfenamide; Sufenax CB; Rhodifax 16; Accelerator CZ; Durax; CBTS; CBS; Cyclohexylbenzothiazyl sulphenamide; Durax; Santocure. An accelerator used in natural rubber and styrene-butadienethiazyl sulfenamide rubber. mp = 93-100°; insoluble in H_2O, soluble in organic solvents. *Akzo.* Name unverified.

23278 **Perkacit® CDMC**
137-29-1 205-287-8
$C_6H_{12}CuN_2S_4$
Copper dimethyldithiocarbamate
Copper dimethyldithiocarbamate; Cumate; Copper, bis(dimethylcarbamodithioato-S,S')-, (SP-4-1)-. Accelerator for rubber *Akzo.* Name unverified.

23279 **Perkacit® DCBS**
N,N·g-Dicyclohexyl-2-benzothiazole sulfenamide
Accelerator for rubber *Akzo.* Name unverified.

23280 Perkacit® DOTG
97-39-2 202-577-6
$C_{15}H_{17}N_3$
N,N-Di-*o*-tolylguanidine
Accelerator for rubber *Akzo.* Name unverified.

23281 Perkacit® DPG
102-06-7 3383 203-002-1
$C_{13}H_{13}N_3$
N,N-Di-phenylguanidine
1,3-Diphenylguanidine; DPG; Melaniline; Nocceler D; Sanceler D; Soxinol D; *sym*-diphenylguanidine; Vulkazit; DPG accelerator; Vulcacid D; Vulkacit d/c. Accelerator for rubber. A medium accelerator for use with thiazoles and sulfenamides in various rubber products. mp = 150°; dec 170°; d = 1.13; soluble in H_2O, more soluble in organic solvents; LD_{50} (rat orl) = 375 mg/kg. *Akzo.* Name unverified.

23282 Perkacit® DPTT
120-54-7 204-406-0
$C_{12}H_{20}N_2S_6$
Dipentamethylenethiuram tetrasulfide
1,1'-(tetrathiodicarbonothioyl)-bis-Piperidine. Accelerator for rubber *Akzo.* Name unverified.

23283 Perkacit® ETU
96-45-7 3849 202-506-9
$C_3H_6N_2S$
Ethylene thiourea
2-mercapto-imidazoline; ETU; Mercozen; Mercaptoimidazoline; 2-Imidazolidimethione; 2-Imidazolidinethione; Imidazoline-2-thiol; 2-Mercaptoimidazoline; Akrochem ETU-22; NA-22; Robac 22; Sanceller 22; Vulkacit NPV/C; Imidazolidinethione; 4,5-dihydroimidazole-2(3H)-thione; N,N'-ethylenethiourea; 1,3-ethylene-2-thiourea; sodium-22 neoprene accelerator; 2-thiol-dihydroglyoxaline; pennac cra; Warecure C; NA-22-D; Vulkacit NPV/C2; Rhodanin S 62. An accelerator in synthetic rubber productions and a degradation product of ethylenebisdithiocarbamate fungicides such as mancozeb, maneb, and zineb. mp = 203°; soluble in H_2O (1-5 mg/ml), more soluble in organic solvents; LD_{50} (rat orl) = 1832 mg/kg. *Akzo.* Name unverified.

23284 Perkacit® MBS
102-77-2 203-052-4
$C_{11}H_{12}N_2OS_2$
2-(Morpholinothio) benzothiazole
N-oxydiethylenebenzothiazole-2-sulfenamide; Amax. Accelerator for rubber. *Akzo.* Name unverified.

23285 Perkacit® MBT
149-30-4 5916 205-736-8
2-Mercaptobenzothiazole
Accelerator for rubber *Akzo.* Name unverified.

23286 Perkacit® MBTS
120-78-5 3435 204-424-9
$C_{14}H_8N_2S_4$
Dibenzothiazole disulfide
Altax; MBTS; Naugex MBT; Thiofide; Vulkacit DM; MBTS rubber accelerator; Royal MBTS; Accel TM; Ekagom GS; Pneumax DM; Vulcafor MBTS; Vulkacit DM/C;. Accelerator for rubber. Added to many types of rubber before vulcanization. It is a component of mercapto mix. mp = 168°; SG = 1.5; insoluble in H_2O (<0.1 mg/ml); LD_{50} (rat orl) > 12 g/kg. *Akzo.* Name unverified.

23287 Perkacit® NDBC
13927-77-0 237-696-2
$C_{18}H_{36}N_2NiS_4$
Nickel dibutyldithiocarbamate
Vanguard N; Nickel, bis(dibutylcarbamodithioato-S,S')-, (SP-4-1)-; Bis(dibutyldithiocarbamate)nickel complex. Nickel dibutyl dithiocarbamate; accelerator for rubber *Akzo.* Name unverified.

23288 Perkacit® SDMC
128-04-1 204-876-7
$C_3H_6NNaS_2$
Sodium dimethyldithiocarbamate
Dimethyl-carbamodithioic Acid, Sodium Salt; Sodium Dimethyldithiocarbamate; Aceto SDD 40; Alcobam NM; Brogdex 555; Carbon S; Dibam; Dibam A; DMDK; Methyl Namate; Sharstop 204; Sodium N,N-dimethyldithiocarbamate; Stafresh 615; Steriseal 40; Thiostop N; Vinstop; Vulnopol NM; Wing Stop B; SDMDTC; Freshgard 40. Accelerator for rubber industry; shortstopper for polymer production LD_{50} (rat orl) = 1000 mg/kg. *Akzo.* Name unverified.

23289 Perkacit® TBBS
95-31-8 202-409-1
$C_{11}H_{14}N_2S_2$
Butyl 2-benzothiazole sulfenamide
NTBBTS; Nocceler NS; Santocure NS; Vulkacit NZ. Accelerator for rubber. mp = 105°. *Akzo.* Name unverified.

23290 Perkacit® TDEC
20941-65-5 244-121-9
$C_{20}H_{40}N_4S_8Te$
Tellurium diethyl dithiocarbamate
tellurium diethyldithiocarbamate; tellurac; Tellurium (IV) diethyldithiocarbamate. Accelerator for rubber mp = 108-118°; SG = 1.44; insoluble in H_2O (<0.1 mg/ml), soluble in organic solvents. *Akzo.* Name unverified.

23291 Perkacit® TETD
97-77-8 3428 202-607-8
$C_{10}H_{20}N_2S_4$
Tetraethylthiuram disulfide
Accelerator for rubber. *Akzo.* Name unverified.

23292 Perkacit® TMTD
137-26-8 9510 205-286-2
$C_6H_{12}N_2S_4$
Tetramethylthiuram disulfide
Accelerator for rubber *Akzo.* Name unverified.

23293 Perkacit® TMTM
97-74-5 202-605-7
$C_6H_{12}N_2S_4$
Tetramethylthiuram monosulfide
Accelerator for rubber *Akzo.* Name unverified.

23294 Perkacit® ZBEC
14726-36-4 238-778-0
$C_{30}H_{28}N_2S_4Zn$
Zinc dibenzyl dithiocarbamate
Zinc dibenzyldithiocarbamate. Accelerator for rubber. mp = 186°. *Akzo.* Name unverified.

23295 Perkacit® ZDBC
136-23-2 205-232-8
$C_{18}H_{36}N_2S_4Zn$
Zinc di-n-butyl dithiocarbamate
Butyl zimate; Di-n-butyldithiocarbamic Acid Zinc Salt; Butasan; Butazate; Butazin; Nocceler BZ; Soxinol BZ; ZBC; Zinc, bis(dibutylcarbamodithioato-S,S')-, (T-4)-; Zinc bis(dibutyldithiocarbamate). Accelerator for rubber *Akzo.* Name unverified.

23296 Perkacit® ZDEC
14324-55-1 238-270-9
$C_{10}H_{20}N_2S_4Zn$
Zinc diethyl dithiocarbamate
Ethyl Ziram; bis(diethylcarbamodithioato-S,S') Zinc; Zinc diethyldithiocarbamate; Ethyl zimate; Zinc, bis(diethylcarbamodithioato-S,S')-, (T-4)-; Bis(diethyldithiocarbamate)zinc complex. Accelerator for rubber *Akzo.* Name unverified.

23297 Perkacit® ZDMC
137-30-4 10305 205-288-3
Zinc dimethyl dithiocarbamate
Ziram. Accelerator for rubber *Akzo.* Name unverified.

23298 Perkacit® ZMBT
155-04-4 205-840-3
$C_{14}H_8N_2S_4Zn$
Zinc-2-mercaptobenzothiazole
Zetax; Hermat ZN-MBT; OXAF; Pennac ZT; Tisperse MB-58; Vulkacit ZM; Zenite; Zinc 2-mercaptobenzothiazolate; ZMBT; ZNMB. Accelerator for rubber *Akzo.* Name unverified.

23299 Perkadox®
Organic peroxides; crosslinking agents *Akzo Chemie UK Ltd.*

23300 Perkadox® 14
25155-25-3 246-678-3
$C_{20}H_{34}O_4$
Di-(2-t-butylperoxyisopropyl)benzene
bis(*tert*-butyldioxyisopropyl)benzene. Low reactivity peroxide useful as a finishing initiator at high temperatures for styrenics; synergist for some halogen-containing flame retardants; also for cross-linking of olefin copolymers, EPDM, SBR, Neoprene, Hypalon. *Akzo.* Name unverified.

23301 Perkadox® 16
15520-11-3 239-557-1
$C_{22}H_{38}O_6$
Bis(4-*t*-butylcyclohexyl) peroxydicarbonate
Ultrafast initiator for polyester cure above 180°F; pultrusion, matched die molding; short ambient temperature compound shelf life. *Akzo.* Name unverified.

23302 Perkadox® 16-W40-GB5
15520-11-3 239-557-1
Di-(4-t-butylcyclohexyl) peroxydicarbonate

Initiator for polymerization of vinyl chloride, acrylates, methacrylates, etc.; recommended for manufacturing of PVC for electrical applications. *Akzo.*

23303 Perkadox® 20
3034-79-5 221-231-5
$C_{16}H_{14}O_4$
Di(2-methylbenzoyl) peroxide
Initiator. *Akzo.* Name unverified.

23304 Perkadox® 26-fl
53220-22-7 258-436-4
$C_{30}H_{58}O_6$
Dimyristyl peroxydicarbonate
Peroxydicarbonic acid, ditetradecyl ester. Catalyst, initiator. *Akzo.* Name unverified.

23305 Perkadox® 30
1889-67-4 217-568-2
$C_{18}H_{22}$
2,3-Dimethyl-2,3-diphenylbutane
Benzene, 1,1'-(1,1,2,2-tetramethyl-1,2-ethanediyl)bis-. *Akzo.* Name unverified.

23306 Perkadox® 58
10192-93-5 233-474-4
3,4-Dimethyl-3,4-diphenylhexane
Akzo. Name unverified.

23307 Perkadox® BC
80-43-3 201-279-3
Dicumyl peroxide
High-temperature initiator used as a flame retardant synergist; also as cross-linking agent for a variety of natural and synthetic rubbers and olefins. *Akzo.* Name unverified.

23308 Perkadox® GS
Initiator for the polymerization of vinyl chloride, vinyl acetate, acrylates, etc. *Akzo Chemie UK Ltd.*

23309 Perkadox® RM
Solid organic peroxides
Used as free radical initiators in polymerization. *Akzo Chemie UK Ltd.*

23310 Perkadox® SE-8
762-16-3 212-094-2
$C_{16}H_{30}O_4$
Dioctanoyl peroxide
Di-n-octanoyl peroxide; Peroxide, bis(1-oxooctyl). Initiator. *Akzo.* Name unverified.

23311 Perkaglycerol
An aqueous solution of potassium lactate; used as a substitute for glycerol for medical and cosmetic purposes.

23312 Perkalink® 300
101-37-1 202-936-7
$C_{12}H_{15}N_3O_3$
Triallyl cyanurate
1,3,5-Triazine, 2,4,6-tris(2-propenyloxy)-; 2,4,6-Triallyloxy-1,3,5-triazine. Co-agent to improve efficiency of peroxide-induced crosslinking of rubber; sensitizer for radiation-cured compounds. *Akzo.* Name unverified.

23313 Perkalink® 301
101-37-1 202-936-7
Triallyl isocyanurate
Co-agent to improve efficiency of peroxide-induced cross-linking of rubber; sensitizer for radiation-cured compounds. *Akzo.* Name unverified.

23314 Perkalink® 400
3290-92-4 221-950-4
$C_{18}H_{26}O_6$
2-methyl-2-propenoic acid 2-ethyl-2-[[(2-methyl-1-oxo-2-propenyl)oxy]methyl]-1,3-propanediyl ester
Trimethylolpropane trimethacrylate
trimethylolpropane trimethacrylate; 2,2-bis(methacryloxymethyl)butyl methacrylate; 1,1,1-Trimethylol propane trimethacrylate. Co-agent to improve efficiency of peroxide-induced cross-linking of rubber; sensitizer for radiation-cured compounds. *Akzo.* Name unverified.

23315 Perkalink® 401
97-90-5 202-617-2
$C_{10}H_{14}O_4$
Ethylene glycol dimethacrylate
Glycol Dimethacrylate; Ethylene Glycol Dimethacrylate; Ethylene Methacrylate; EGDMA. Co-agent to improve efficiency of peroxideinduced cross-linking of rubber; sensitizer for radiationcured compounds. bp = 85°; d = 1.0510; LD_{50} (rat orl) = 3300 mg/kg. *Akzo.* Name unverified.

23316 Perkasil® KS 207
Aluminum-magnesium-sodium silicate
Reinforcing filler for rubber industry; medium reinforcing properties. *Akzo.* Name unverified.

23317 Perkasil® KS 300
silica
Precipitated silica; highly dispersible reinforcing filler with excellent processing; for rubber industry *Akzo.* Name unverified.

23318 Perkasil® KS 404
silica
Precipitated silica; highly dispersible reinforcing filler with excellent transparency; for rubber industry *Akzo.* Name unverified.

23319 Perkasil® VP 406
silica
Precipitated silica; reinforcing filler for silicone rubber applications. *Akzo.* Name unverified.

23320 Perkin's Base
p-Tolylaminoditolyl-p-toluquinone diimine

23321 Perklone
127-18-4 9332 204-825-9
Perchloroethylene
A solvent for dry cleaning. *ICI Chem. & Polymers Ltd.*

23322 Perlankrol®
Anionic alcohol-, alcohol ether-, amide ether-and phenol ether-sulfates; used in toiletries, as foam boosters, emulsifiers, cement and gypsum foamers. *Harcros.*

23323 Perlankrol® ADP3
Sodium laureth (3) sulfate
Biodegradable surfactant; base for preparation of high foaming shampoos and toiletries. *Harcros.*

23324 Perlankrol® ATL40
139-96-8 205-388-7
TEA lauryl sulfate
Akyposal TLS; Cycloryl tawf; Dodecyl sulfate, triethanolamine salt; Drene; Elfan 4240 T; EMAL T; Maprofix TLS; Melanol LP 20 T; Propaste T; Rewopol TLS-40; Richonol T; Sipon lt; Standapol TLS 40; Sterling wat; Texapon T-35; Tylorol LT-50. Biodegradable surfactant for preparation of high foaming shampoos and toiletries; emulsifier for emulsion polymerization. *Harcros.*

23325 Perlankrol® DAF25
2235-54-3 218-793-9
Ammonium lauryl sulfate
Foaming agent, base; formulation of toiletries and carpet shampoos. *Harcros.*

23326 Perlankrol® DSA
151-21-3 8782 205-788-1
Sodium lauryl sulfate
Foaming agent for synthetic latexes, emulsion polymerization aid; base for preparation of high foaming shampoos and toiletries; wetting agent; industrial detergent additive. *Harcros.*

23327 Perlankrol® ESD
Synthetic primary alcohol ether sulfate, sodium salt; biodegradable high foam additive for liquid household and industrial detergents; emulsifier for emulsion polymerization. *Harcros.*

23328 Perlankrol® ESK32
Sodium primary alcohol ethoxylate sulfate; foam booster/stabilizer for detergent formulations, foam cleaning, fire fighting foams, plasterboard production *Harcros.*

23329 Perlankrol® PA Conc
30416-77-4 250-188-5
Alkyl phenol ether sulfate, ammonium salt; foam booster/stabilizer, detergent base, frothing agent used in liquid detergent formulations; emulsifier for cresylic acid, emulsion polymerization. *Harcros.*

23330 Perlatum 400, 410, 410CG, 420, 510
White petrolatum USP; basic ingredient in cosmetics, ointments, lubricants; dust control agents, and lubricants in baking industry, animal feed, pharmaceuticals, food processing. *IGI Petroleum Spec.* Name unverified.

23331 Perlatum 415, 415 CG, 425
White petrolatum USP; basic ingredient in cosmetics, ointments, lubricants; dust control agents, and lubricants in baking industry, animal feed, pharmaceuticals, food processing. *IGI Petroleum Spec.* Name unverified.

23332 Perlex
7787-59-9 1303 232-122-7
Bismuth oxychloride pearls
Used for decorative cosmetic products. *Morton Int'l. Ltd.* Discontinued.

23333 Perlextra
Morton Int'l. Ltd. Discontinued.

23334 Perlit®
Water repellents based on silicone, paraffin or fatty acid condensation products; used for textile finishing. *Bayer AG; Bayer plc.*

23335 perlite
A eutectic product resulting from an alloy of ferrite and cementite in steel.

23336 Perlon
Monofil; used for fishing lines, zip fasteners, woven fabric and industrial uses. *Bayer AG.*

23337 Perma Shield
An elastomeric coating; suitable for application to concrete, brick, concrete block, stucco, metal, wood, sheetrock, masonite and plywood as a protective and decorative finish. *Secure Inc.* Discontinued.

23338 Perma Sta
Ethylene glycol based antifreeze. *Chemcentral Corp.*

23339 Permabond
Oriented polypropylene laminating adhesives. *The Scottish Adhesives Co Ltd.*

23340 Permador
A precious metal alloy; for dentistry and dental engineering. *Degussa AG.*

23341 Perma-Flex Blak-Stretchy®
Polysulfide system; two-component system for making room temperature setting rubbery, flexible, extremely elastic molds and patterns, for precision casting where max. stretch and good dimension control are important; in plaster casting and foundry pattern material; mix ratio: 100A:15B by weight. The addition of about 0.25% pink or yellow curative C will speed up final set and give rubbery bounce to Blak Stretchy. *Perma-Flex Mold.*

23342 Perma-Flex Blak-Tufy®
Polysulfide system with carbon black pigment; three-component room temperature vulcanizing molding synthetic rubber reinforced with blk. reinforcing pigment with good toughness, low visc. and pouring properties in uncured form; in plaster shops for waste molds, intermediate working molds for prototype reproductions and prior to heavy duty thermoplastic Koroseal mold fabrication; mix ratio: 100A:20B:2C by weight. *Perma-Flex Mold.*

23343 Perma-Flex Blu-Sil
Silicone-type cold molding compound; two-component system, sets to flexible rubbery solid; for molds to cast Perma-Flex CMCU-2PU, hot melt polyethylenes, epoxy resins, and acrylic compounds formed cold, and later cured at elevated temps.; mix ratio: 10A:25B by weight. *Perma-Flex Mold.*

23344 Perma-Flex Green-Sil
Silicone cold molding compound; two-component, high strength, low visc., room temperature curing elastomer for general mold making applications, e.g., wheels, gaskets, liners; for limited reprods. when casting polyester, epoxy, and PU rigid foam resins, and for electronic applications such asd potting and encapsulating. Mix ratio: 100A:5B by weight. *Perma-Flex Mold.*

23345 Permafuse
Modified phenolic adhesives; for bonding friction materials to their backing, e.g., brake shoes, clutch facings, transmission linings etc. *The Permafuse Corp.* Name unverified.

23346 permalba
A composite pigment consisting mainly of barium sulfate. An artist's color.

23347 Perma-Leaf
Aluminium pigment paste; used for protective coatings. *Reynolds Metal Co.*

23348 Permalens
Perfilcon
Contact lens material. *Coopervision Inc.* Unverified.

23349 Permali
A proprietary trade name for laminated products containing wood or paper impregnated with synthetic resin; some are made from thin wood coated with synthetic resin solution and compressed under heat, others are impregnated under pressure, solvent removed and then compressed. No manufacturer.

23350 Permalloy®
A trademark for alloys of nickel and iron containing more than 30% nickel; they are prepared by certain heat treatment and show unusual magnetic properties, giving a high initial permeability; one of the best alloys contains 78.5% nickel and 21.5% iron; another contains 78.5% Ni18% Fe, 3% Mo and 0.5% Mn; a typical analysis gives 78.23% Ni, 21.35% Fe, 0.04% C, 0.03% Si, 0.035% S, 0.22% Mn, 0.37% Co, 0.1% Cu and traces of P. No manufacturer.

23351 Permalon
A proprietary trade name for stretched vinylidene chloride. No manufacturer.

23352 Permalose
Textile auxiliary chemicals. *ICI Chem. & Polymers Ltd.*

23353 Permalux
Neoprene accelerator. *DuPont UK.*

23354 Permalyn® 7085
Polyhydroxy ester of rosin; thermoplastic resin for use in hot-melt adhesives where light color and resistance to thermal degradation are required; improved oxidation resistance. *Hercules.*

23355 Perma-Mold®
Release agents especially formulated for rigid urethane insulation. *Chem-Trend.*

23356 Permanax 6PPD
N-1,3-Dimethylbutyl-N·g-phenyl-p-phenylene diamine
Staining antiozonant for rubber. *Akzo.* Name unverified.

23357 Permanax BL, BLN
Acetone/diphenylamine condensate; staining antioxidant for rubber. *Akzo.* Name unverified.

23358 Permanax BLW
Acetone/diphenylamine condensate on inert carrier; Staining antioxidant for rubber. *Akzo.* Name unverified.

23359 Permanax CNS
Nonstaining antioxidant for rubber. *Akzo.* Name unverified.

23360 Permanax CR
Staining antiozonant and antioxidant for chloroprene rubber. *Akzo.* Name unverified.

23361 Permanax DPPD
74-31-7 3388 200-806-4
$C_{18}H_{16}N_2$
N,N'-Diphenyl-p-phenylenediamine
Agerite DPPD; Diphenyl PPD; DPPD; JZF; Nonox DPPD; Diafen; Diafen FF; Altofane DIP; Nonflex H; Permanax 18; Stabilizer DPPD; Nocrac DP; Permanax DPPD; DFFD; Antage DP; Ekaland DPPD; Naugard J;. Staining antiozonant for rubber. mp = 145-152°; $bp_{0.5}$ = 220-225°; d = 1.200; insoluble in H_2O, soluble in organic solvents; LD_{55} (rat orl) = 2370 mg/kg. *Akzo.* Name unverified.

23362 Permanax HD
Heptylated diphenylamine
Staining antioxidant for rubber. *Akzo.* Name unverified.

23363 Permanax HD (SE)
Heptylated diphenylamine
Self-emulsifying grade; staining antioxidant for rubber. *Akzo.* Name unverified.

23364 Permanax IPPD
101-72-4 202-969-7
$C_{15}H_{18}N_2$
N-Isopropyl-N·g-phenyl-p-phenylene diamine
Akrochem Antioxidant PD1; Anto H; Flexone 3c; IPPD; Permanex IPPD; Santoflex IP; Cyzone; Elastozone 34; Nonox ZA; Santoflex 36; Cyzone IP; Vulkanox 4010 NA. Staining antiozonant for rubber. N-Isopropyl-N'-phenyl-4-phenylenediamine is an antidegradant in natural rubber, styrene-butadiene, nitrile-butadiene, butadiene, and chloroprene rubber. mp = 72-76°; bp_1 = 161°; insoluble in H_2O, soluble in organic solvents; LD_{50} (rat orl) = 555 mg/kg. *Akzo.* Name unverified.

23365 Permanax OD
Octylated diphenylamine
Staining antioxidant for rubber. *Akzo.* Name unverified.

23366 Permanax OZNS
Nonstaining antiozonant and antioxidant for rubber. *Akzo.* Name unverified.

23367 Permanax TQ
Polymerized 2,2,4-trimethyl-1,2-dihydroquinoline
Staining antioxidant for rubber. *Akzo.* Name unverified.

23368 Permanax WSL
2,4-Dimethyl-6-(1-methyl cyclohexyl) phenol
Nonstaining antioxidant for rubber. *Akzo.* Name unverified.

23369 Permanax WSL Pdr
2,4-Dimethyl-6-(1-methyl cyclohexyl) phenol on inert carrier; nonstaining antioxidant for rubber. *Akzo.* Name unverified.

23370 Permanax WSO
High molecular weight phenolic compound nonstaining antioxidant for rubber. *Akzo.* Name unverified.

23371 Permanax WSP
2,2·g-Methylene-bis(6-(1-methyl-cyclohexyl)-p-cresol
Nonstaining antioxidant for rubber. *Akzo.* Name unverified.

23372 Permanax WSP (PQ)
2,2·g-Methylene-bis(6-(1-methyl-cyclohexyl)-p-cresol
Nonstaining antioxidant for use in polyethylene. *Akzo.* Name unverified.

23373 Permanent
Pigment powders; used in paints and inks. *Hoechst AG.*

23374 Permanent Encapsulant 185N
Urethane compound; fastsetting, low exotherm castable elastomer for encapsulating telecommunications cable to provide moisture and mechanical protection and elec. insulation to cable splices; nonexpanding. *Hexcel.*

23375 Permanite
A cobalt steel which has very high magnetic properties.

23376 Permaplex
An ion exchange membrane. *The Permutit Co.* Name unverified.

23377 Permasect
52645-53-1 7321 258-067-9

Permethrin
A pyrethroid insecticide. *Mitchell Cotts Chemicals Ltd.*

23378 Permasep®
Polymer. *DuPont UK.*

23379 Perma-Slik
Family of air curing dry lubricant coatings containing solid lubricants and suitable binder system; for frictional surfaces requiring dry lubrication particularly on substrates which cannot tolerate oven cure. *E/M Corporation.*

23380 Permatag
Insecticidal eartag for cattle . *Mitchell Cotts Chemicals Ltd.*

23381 Permathin
Tetrafilcon A
Contact lens material. *UCO Optics Inc.* Name unverified.

23382 Permatol A
A proprietary trade name for a preservative for wood; it contains pentachlorphenol in oil. No manufacturer.

23383 Permax
A nickel steel containing 76% nickel, of French manufacture. It has magnetic properties.

23384 permethrin
52645-53-1 7321 258-067-9
$C_{21}H_{20}Cl_2O_3$
3-(2,2-dichloroethenyl)-2,2-dimethylcyclopropanecarboxylic acid(3-ohenoxyphenyl)methyl ester
FMC-33297; NIA-33297; NRDC-143; PP-557; SBP-1513; S-3151; Ambush; Corsair; Adion; Ambushfog; Atroban; Bio Flydown; Coopex; Dragnet; Dragon; Ectiban; Eksmin; Epigon; Expar; Flee; Imperator; Jureong; Kafil; Kestrel; LE 79-519; Outflank; Perigen; Permanone; Permasect; Permetrina; Permit; Perthrine; Picket; Pounce; Pramex; Pynosect; Qamlin; Stockade; Stomoxin; Talcord; Tornade; Torpedo. Non-systemic insecticide with contact and stomach action. Used for control of the larvae of lepidopterous and coleopterous insect pests in fruit and vegetable crops. mp = 34-35°; $bp_{0.01}$= 200°; SG_{20} = 1.19-1.27; insoluble in H_2O (0.2 mg/l), soluble in organic solvents; LD_{50} (rat orl) = 4000 mg/kg.

23385 Permethrin
52645-53-1 7321 258-067-9
Fumite Permethrin; Ambush; Cooper Coopex; Darmycel Agarifume Smoke; Elimite®; Kafil; Nippon Ant and Crawling Insect Killer; Nippon Ant Killer Powder; Nyppon Fly Killer Spray; Nippon Ready for Use Ant and Crawling Insect Killer; Nix Creme Rinse; Nix Delmal Cream; Perigen; Permasect; Picket; Quamilin; Turbair Permethrin,. Smoke insecticide (active ingredient permethrin); for use in enclosed areas against whitefly and other pests of protected crops, cockroaches on stored produce, domestic insect pests. Octavius Hunt Ltd; ICI Chem. & Polymers Ltd; The Welcome Foundation Ltd; darmycel UK; Allergan Inc; Vitax Ltd; Mitchell Cotts Chemicals Ltd; ICI Garden Products; Pan Britannica Industries Ltd.

23386 permethrin
52645-53-1 7321 258-067-9
$C_{21}H_{20}Cl_2O_3$
3-(2,2-dichloroethenyl)-2,2-dimethylcyclopropanecarboxylic acid(3-ohenoxyphenyl)methyl ester
FMC-33297; NIA-33297; NRDC-143; PP-557; SBP-1513; S-3151; Ambush; Corsair; Adion; Ambushfog; Atroban; Bio Flydown; Coopex; Dragnet; Dragon; Ectiban; Eksmin; Epigon; Expar; Flee; Imperator; Jureong; Kafil; Kestrel; LE 79-519; Outflank; Perigen; Permanone; Permasect; Permetrina; Permit; Perthrine; Picket; Pounce; Pramex; Pynosect; Qamlin; Stockade; Stomoxin; Talcord; Tornade; Torpedo. Non-systemic insecticide with contact and stomach action. Used for control of the larvae of lepidopterous and coleopterous insect pests in fruit and vegetable crops. mp = 34-35°; $bp_{0.01}$= 200°; SG_{20} = 1.19-1.27; insoluble in H_2O (0.2 mg/l), soluble in organic solvents; LD_{50} (rat orl) = 4000 mg/kg.

23387 Permethyl 101A
4390-04-9 224-506-8
$C_{16}H_{34}$
2,2,4,4,6,8,8-Heptamethylnonane
HMN. Isohexadecane, cosolubilizer for nonhydrocarbon materials; in eyeliners, mascaras, sun care and skin products; cleanser for eye and face makeup. bp = 240°; d = 0.7930; n_D^{20}= 1.4391. *Presperse.*

23388 Permethyl 102A
93685-79-1 297-627-7
$C_{20}H_{42}$
Isoicosane, cosolubilizer for nonhydrocarbon materials; for skin care and sun care products; plasticizer for mascara. *Presperse.*

23389 Permethyl 104A
9003-29-6
Polyisobutene
isooctahexacontane. Cosolubilizer for nonhydrocarbon materials; for lipsticks and glossers to improve wear and impart sheen; in skin care and sun care products *Presperse.*

23390 Permethyl 99A
13475-82-6 236-757-0
$C_{12}H_{26}$
2,2,4,6,6-Pentamethylheptane
Isododecane, cosolubilizer for nonhydrocarbon materials; for mascara, eyeliner, antiperspirant and where residual film is not desirable; solvent for debris on skin. *Presperse.*

23391 Permidan
$C_5H_9N_3O$
Dimethylaminopyrazolone

23392 Perminal
Textile auxiliary chemicals. *ICI Chem. & Polymers Ltd.*

23393 Perminvars
Proprietary alloys having exceptional magnetic properties; particularly suited for use in electrical communication circuits; one alloy contains 45% nickel, 25% cobalt, and 30% iron, and has a high initial permeability. No manufacturer.

23394 Permobel
Alkyd-based top coat for automobiles. *ICI Chem. & Polymers Ltd.*

23395 Permonite
An explosive used in mines. It is a mixture of potassium perchlorate and ammonium nitrate, with trinitro-toluene, a little starch, and wood meal.

23396 Permutite
An artificially made zeolite, prepared by igniting together china clay (aluminum silicates), and (sometimes) quartz or sand, with alkali carbonates; used for removing calcium and magnesium salts, sodium and potassium salts, and manganese and iron, from water. No manufacturer.

23397 Permyl B-100
For stabilizers for halogenated hydrocarbon resins in US Class 6. *Ferro.*

23398 Pernax
An artificial gutta-percha made from rubber, wax, and rosin. No manufacturer.

23399 Perone
7722-84-1 4839 231-765-0
A proprietary trade name for pure hydrogen peroxide. No manufacturer.

23400 Peronoid
A trade name for a mixture of copper sulfate and lime; a fungicide. No manufacturer.

23401 Peropal®
41083-11-8 255-209-1
$C_{20}H_{35}N_3Sn$
1-(tricyclohexylstannyl)-1H-1,2,4-triazole
Azocyclotin; Bay BUE 1452. Long-lasting acaricide with contact action, used for control of motile stages of spider mites on pome and stone fruit, grapes, citrus, vegetables, etc. mp = 219°; insoluble in H_2O (<1 mg/l), more soluble in organic solvents; LD_{50} (rat orl) = 99 mg/kg. *Bayer AG.*

23402 Perox
Dyes for plastics. *Morton Int'l. Ltd.* Discontinued.

23403 Peroxal
7722-84-1 4839 231-765-0
H_2O_2
hydrogen peroxide
A trade name for hydrogen peroxide. No manufacturer.

23404 peroxidase
9003-99-0 232-668-6
An enzyme found in most plant cells and some animal cells; promotes the oxidation of various substrates such as phenols, aromatic amines, etc. by means of hydrogen peroxide. *Am. Int'l. Chem.; Sigma; Spectrum Chem. Mfg.*

23405 Peroxide RH-2
A proprietary trade name for a high melting, stable, aromatic organic peroxide; used as a polymerization catalyst. No manufacturer.

23406 Peroxidol®
Epoxidized oils. *Reichhold.*

23407 Peroximon
A proprietary range of organic peroxides. *Montedison UK Ltd.* Name unverified.

23408 Peroximon® DC 40 MG
80-43-3 201-279-3
Dicumyl peroxide/EPM masterbatch; vulcanizing and crosslinking agent for ethylene-propylene elastomers, polyethylene, EVA copolymers, nitrile and PU elastomers, SBR, PVC, neoprene; for rubber car parts, injection molded articles, electrical cable insulation, conveyor belts, shoe soles. *AkroChem.*

23409 Peroximon® DC-40
80-43-3 201-279-3

$C_{18}H_{22}O_2$
Dicumyl peroxide
Bis(1-methyl-1-phenylethyl) peroxide; bis(alpha,alpha-dimethylbenzyl) peroxide; cumyl peroxide; di-alpha-cumyl peroxide; Di-cup; diisopropylbenzene peroxide; isopropyl benzene peroxide; Cumene peroxide. Initiator and catalyst for polymerization reactions. bp= 130° (dec); insoluble in H_2O, more soluble in organic solvents; LD_{50} (rat orl) = 4100 mg/kg. *Akrochem.*

23410 Peroximon® S-164/40P
6731-36-8 229-782-3
$C_{17}H_{34}O_4$
1,1-Di(t-butylperoxy)-3,3,5-trimethylcyclohexane
Peroxide, (3,3,5-trimethylcyclohexylidene)bis[(1,1-dimethylethyl); 1,1-Di-(tert-butylperoxy)-3,3,5-trimethyl cyclohexane. Initiator and catalyst for polymerization reactions. *Akrochem.*

23411 Peroxol
7722-84-1 4839 231-765-0
H_2O_2
hydrogen peroxide
Disinfectant solutions containing hydrogen peroxide, sometimes mixed with other disinfectants. No manufacturer.

23412 peroxydol
7632-04-4 8797 231-556-4
$BNaO_3 \cdot 4H_2O$
Sodium perborate
An antiseptic, deodorant, and bleaching agent.

23413 Peroxyl
7722-84-1 4839 231-765-0
Hydrogen peroxide
Bleach and oxidizer. *May & Baker Ltd.* Name unverified.

23414 Perpentol
A tetralin preparation used for cleaning wool.

23415 Perrindo®
High quality organic pigments for the surface coatings industry, especially for automotive coatings. *Bayer AG.*

23416 Persellig T
Leveling agent for dispersed dyes on polyamide and polyester. *Ceca SA.*

23417 Persian balsam
119-53-9 1124 209-441-5
Compound tincture of benzoin. Used in organic synthesis.

23418 Persian berry carmine
Dutch Yellow. A pigment consisting of the aluminum and calcium lakes of the Persian berry coloring matters.

23419 Persian yellow
$C_{14}H_{10}N_4O_7$
Nitro-tolueneazonitrosalicylic acid
Dyes chromed wool yellow; also used in cotton printing giving yellow shades with chromium acetate.

23420 Persiderm®
Preparation for fashionable writing and luster effects on suede; used in leather industry. *Bayer AG.*

23421 Persiderm® Black
Aniline pigment for the lustering of black suedes. *Bayer plc.*

23422 persionin
An acetone extract of cud-bear, a lichen also known as crottle. A source of salts of orcein, including litmus.

23423 Persistol®
Hydrophobic and oleophobic agents for textile finishing. *BASF AG; BASF plc.*

23424 Persistol® E
Zircon paraffin emulsion; for rendering suede furs water-repellent. *BASF AG.*

23425 persodine
A mixture of ammonium and potassium persulfates. Oxidizer and bleach.

23426 Persoftal®
Softeners for textiles; selected brands simultaneously impart an antistatic effect. *Bayer AG.*

23427 Persoftal® PE Special
Oligomer binder; used in polyester dyeing. *Bayer AG; Bayer plc.*

23428 Persoz's reagent
Zinc oxide (2g), is added to a solution of zinc chloride (10g), in 10 ml water. It dissolves silk, and detects silk in the presence of wool.

23429 Perspex
Acrylic methyl methacrylate resins in sheet form. *ICI Chem. & Polymers Ltd.*

23430 Perstoff
503-38-8 3395 207-965-9
$C_2Cl_4O_2$
Trichloromethyl-chloroformate
diphosgene. A poison gas. No manufacturer.

23431 Perstorp Phenolic Moulding Compound
Toilet seats, pan handles, meter cases, automotive, domestic accessories, law bowls, electrical accessories. *Perstorp Ferguson Ltd.* Name unverified.

23432 Perstorp Urea Moulding Compound
Electrical accessories, toilet seats, closures. *Perstorp Ferguson Ltd.* Name unverified.

23433 Persulon
Fungicide; used for control of powdery mildew on cereals, fruit, vegetables and ornamentals. *Bayer AG.*

23434 Persyst
8022-00-2 6129
Emulsifiable concentrate containing 580 g demeton-S-methyl per liter; a systemic organophosphorus insecticide. *Ashlade Formulations Ltd.*

23435 Perthane
72-56-0 200-785-1
$C_{18}H_{20}Cl_2$
diethyldiphenyldichloroethane
DDD. Trademark for an agricultural insecticide based on diethyldiphenyldichloroethane, supplied as a wettable powder or emulsifiable concentrate; controls insects on plants and livestockp; also used as a moth protection for textiles. No manufacturer.

23436 Pertinit
A proprietary synthetic resin of the urea-formaldehyde type. No manufacturer.

23437 Pertite
88-89-1 7562 201-865-9
An Italian explosive; the main constituent is picric acid.

23438 Pertscan-99m
23288-60-0 8802
$NaO_4{}^{99m}Tc$
sodium pertechnitate ^{99m}Tc
Ultratechnekow. Sodium pertechnetate ^{99m}Tc; radioactive agent. Used in brain scans, thyroid function tests. *Abbott Laboratories.* Name unverified.

23439 Peruol®
A registered trademark currently awaiting reallocation by its proprietors to cover a range of pharmaceuticals. *Cassella AG.* Name unverified.

23440 Peruscabin
120-51-4 1162 204-402-9
$C_6H_5 \cdot CO_2 \cdot CH_2 \cdot C_6H_5$
Benzyl benzoate
It is the active constituent of Peru balsam, and is used in the same manner, and for the same purpose as Peruol.

23441 Peruvian balsam
The oleoresin of *Myroxylon Pereirae* , of Central America.

23442 Peruvlan bark
Cinchona bark.

23443 Peruvin
104-54-1 2362 203-212-3
$C_9H_{10}O$
Cinnamyl alcohol.

23444 Pervon
Fully reacted polyurethane/pitch coating. *Sigma Coatings.* Name unverified.

23445 Pescola oil
An oil used in the tanning industry.

23446 Pest-B-Gon
Roach bait. *Monsanto (Solaris).* Name unverified.

23447 Pestex
An insecticide. *Fisons plc, Horticulture Div.* Name unverified.

23448 Pestilizer®
Phosphate esters
Compatibilizers for liquid fertilizers. *Stepan.* Name unverified.

23449 Pestilizer® B Series
Amphiphilic alkyl sulfosuccinate
Emulsifiers for liquid fertilizers; high stability in electrolytic systems. *Stepan Europe.* Name unverified.

23450 Petalin
Distilled tall oils and fatty acids. *Surfachem Ltd.*

23451 Petameth
59-51-8 6053 200-432-1
Racemethionine
Acidifier. *O'Neal, Jones & Feldman Pharmaceuticals.* Name unverified.

23452 Petcat R-9
1309-64-4 752 215-175-0
Antimony trioxide
Catalyst in PET polyester production. *Laurel Industries.*

23453 Petiole
Phenyl ethyl isopropyl ether
Quest Int'l. UK Ltd.

23454 Petlon®
Thermoplastic PET polyesters; modified polyethylene terephthalate grades which offer good dimensional stability, high rigidity, high heat resistance, good electrical properties and chemical resistance. *BASF AG; Miles Inc.*

23455 Petra
Range of polyethylene terephthalate polymers, including glass reinforced compounds; used for automotive ignition and carburetor components; electrical connectors, bobbins, relays and switches; intricate mechanical parts that maintain dimensional stability under load, or at temperatures above 150°C, or in the presence of moisture. *AlliedSignal Inc.*

23456 Petra® 130
25038-59-9 7730
PET polyester, 30% glass-reinforced; for automotive parts, e.g., mirror housing assembly, chair shells; features strength, rigidity. *AlliedSignal Engineered Plastics.*

23457 Petra® 130FR
25038-59-9 7730
PET polyster, 30% glass-reinforced, flame-retarded; for electrical connectors; features processability, dimensional stability, rigidity. *AlliedSignal Engineered Plastics.*

23458 Petra® 230
25038-59-9 7730
PET polyester, 35% mineral/glass-reinforced; for automotive grill reinforcement, band saw pulleys; features strength and stiffness, minimum warpage. *AlliedSignal Engineered Plastics.*

23459 Petra® 242
25038-59-9 7730
PET polyester, 40% mineral/glass-reinforced, impact-modified; for automotive roof rack; impact modified; features rigidity, toughness, paintability, strength. *AlliedSignal Engineered Plastics.*

23460 Petrac® 165
Petroleum wax; lubricant used in rigid PVC compounds to control external lubrication and fusion. *Syntetics Products Co.*

23461 Petrac® 215
Oxidized polyethylene wax; lubricant used to control external lubrication of rigid PVC compounds and provide high gloss surface. *Syntetics Products Co.*

23462 Petrac® 270
57-11-4 8959 200-313-4
stearic acid
External lubricant in flexible PVC processing; chemical intermediate for metallic stearates, esters, etc.; dispersant, plasticizer, activator, lubricant in rubber compounding; thicknener for greases. *Syntetics Products Co.*

23463 Petrac® CP-11
1592-23-0 1750 216-472-8
Calcium stearate
Mold release agent, lubricant, pigment suspension aid, and flow control agent for plastics, paints, inks, waterproofing of cement and clay tile, lubrication and glossing in paper coatings; plastics application. *Syntetics Products Co.*

23464 Petrac® Eramide®
112-84-5 204-009-2
$C_{22}H_{43}NO$
(Z)-13-Docosenamide
Erucylamide. Erucamide; slip, release, antitack, and/or internal mold release agent; used in PP for extrusion of sheets, in injection molding; antistat; in polyvinyls for films and sheeting; in polyethylene it imparts slip and antiblock characteristics in film application; internal mold release agent in molded products, lamination of PE to cellophane and in PE extrusion coatings; withstands high processing temperatures, food contact applications. mp = 79°. *Syntetics Products Co.*

23465 Petrac® GMS
Glyceryl stearate
Lubricant and mold release agent for plastics industry, pigment suspension in paints and inks, waterproofing agent for cement and clay tile, lubricant and glossing aid in paper coatings; used in rigid PVC pipe, foundry resins. *Syntetics Products Co.*

23466 Petrac® MG-20 NF
557-04-0 5730 209-150-3
Magnesium stearate NF
Dry lubricant and anticaking agent used in food products, and filling of pharmaceutical capsules; antistick properties in tabletting; improves stability, smoothness, and texture of cosmetic emulsions, creams, ointments; improves texture and water-repellenc *Syntetics Products Co.*

23467 Petrac® PHTA
61790-38-3 263-130-9
Partially hydrogenated tallow fatty acids; used in industrial products including synthetic lubricants, bar soaps, cosmetics, rubber tires. *Syntetics Products Co.*

23468 Petrac® Slip-Eze
301-02-0 206-103-9
Oleamide
Slip, release, antitack, and/or internal mold release agent; polyethylene film and sheeting; injection or extruded molded products; polyvinyl film; PVC plastisol systems where quick slip characteristics are required; food packaging materials. *Syntetics Products Co.*

23469 Petrac® Vyn-Eze®
124-26-5 204-693-2
Stearamide
Slip, antitack, antiblock, and/or internal mold release agent; polyvinyl molded products and PVC plastisol systems; polyethylene film; food pkg. materials. *Syntetics Products Co.*

23470 Petrac® ZN-41
557-05-1 10292 209-151-9
Zinc stearate USP
Lubricant and mold release agent in PS, melamine, U-F, phenolformaldehyde, and polyester molding resins; dusting agent for uncured rubber slabs; suspending and flattening agent in solvent- and water-based paints; anticaking agent used in extinguishers; lubricant in powders and metallurgy. *Syntetics Products Co.*

23471 Petralon
A German name for a preparation of wood tar; an antiseptic.

23472 Petramin
Special disperse dyes for dyeing nickel-modified polypropylene fibers. *Bayer plc.*

23473 Petrasul
A synthetic stone material made from asbestos and Portland cement, which is sulfur-impregnated; the sulfur content varies from 15-35% and the material can be colored; suitable for counter-tops, or similar purposes. No manufacturer.

23474 Petre
7757-79-1 7815 231-818-8
KNO_3
Potassium nitrate
Used in pyrotechnics. No manufacturer.

23475 Petrex
A proprietary trade name for a polybasic acid used in synthetic resin manufacture, the essential constituent of which is 3-isopropyl-6-methyl-3:6 endo-ethylene-+sx$_4$-tetrahydrophthalic anhydride. No manufacturer.

23476 Petrex 7-75T
A solution of an alkyd-type resin derived from a terpene polybasic acid; used in coatings for cellophane. *Hercules.*

23477 Petrinex
Propylene glycol mixtures. *ICI Chem. & Polymers Ltd.*

23478 Petro 11
Alkyl naphthalene sodium sulfonate; hydrotrope and surfactant. *Witco Corporation.*

23479 Petro 22
Modified alkyl napthalene sodium sulfonate; low foam surfactant for various cleaning formulations. *Witco Corporation.*

23480 Petro AG Special
Alkyl naphthalene sulfonate; water soluble anticaking agent for detergents, fertilizers and various salts. *Witco Corporation.*

23481 Petro BAF
Linear alkyl naphthalene sodium sulfonate; hydrotrope and surfactant for detergent and cleaning formulations. *Witco Corporation.*

23482 Petro P
Alkyl napthalene sodium sulfonate; wetting agent for pesticide and fertilizer formulations. *Witco Corporation.*

23483 Petro S
Alkyl napthalene sodium sulfonate; soil conditioner. *Witco Corporation.*

23484 Petro ULF
Modified alkyl naphthalene sodium sulfonate; low foam surfactant for various cleaning applications. *Witco Corporation.*

23485 Petro WP
Modified alkyl naphthalene sodium sulfonate; wetting agent for industrial cleaning formulations and pesticide formulations. *Witco Corporation.*

23486 Petroacid
A proprietary trade name for a mixture of fatty acids obtained from petroleum distillates. No manufacturer.

23487 petrobenzol
A petroleum distillate solvent; it has a boiling point of 61-96 C.

23488 Petroclastite
Petroklastite
An explosive. It contains potassium nitrate, sulfur, coal tar pitch, and potassium dichromate.

23489 Petrofibe 201, 210, 215, 235
Petrolatum
Industrial grades for applications requiring good moisture resistance, corrosion protection and lubricity. *IGI Petroleum Spec.* Name unverified.

23490 Petrofracteur
An explosive consisting of 10% nitrobenzene, 67% potassium chlorate, 20% potassium nitrate and 3% antimony pentasulfide.

23491 Petrogils
Drilling muds. *Rhône-Poulenc UK.*

23492 petrol
A product of the distillation of petroleum. The term is synonymous with gasoline and petroleum spirit. Other names for the same product are naphtha, petroleum naphtha or mineral naphtha, benzoline, benzine, and carburine.

23493 Petrolagar
A proprietary emulsion of liquid paraffin and agar-agar. No manufacturer.

23494 Petrolane
Liquefied gas for general purposes. *Quantum Chemical Corp.*

23495 petrolatum
8009-03-8 (NF); 8027-32-5 (USP)
Petroleum hydro carbons
Petroleum jelly; petrolatum amber; petrolatum white; mineral jelly; paraffinum molle. Semisolid mixture of hydrocarbons obtained from petroleum; consists of the yellow, semi-solid, purified residue left when petroleum is distilled; specific gravity 0.87 *Exxon; Harcros; Magie Bros. Oil; Mobil; Penreco; Stevenson Bros.; Witco/Sonneborn.*

23496 petrolenes
Malthenes. Constituents of bitumens which are soluble in hexane.

23497 petroleum ether
Canadol; light ligrin; gasoline, solene. A distillate from petroleum oil. It consists essentially of pentane and hexane, and is a solvent for resins.

23498 petroleum naphtha
8030-30-6 7329 232-443-2
A term very loosely applied. It often denotes the first fraction of boiling-point up to 150°C., obtained from the distillation of crude petroleum oil, but is sometimes applied to any low boiling petroleum product.

23499 petroleum pitch
8052-42-4 232-490-9
Asphalt.

23500 petroleum spirit
8032-32-4 5513 232-354-7
(Light petroleum; benzine; naphtha) Both benzoline and naphtha are sold under this name. They are used as motor spirits, and for drycleaning cloths.

23501 petroleum wax
8002-74-2 7155 232-315-6
Petroleum hydrocarbon derived from petroleum; lubricant for formulating PVC and electrical wire and cable compounds; protectant for elastomers. *Astor Wax; Mobil.*

23502 Petrolig
Dispersant for oil well drilling muds. *Borregaard Ligno Tech.*

23503 petroline
A fraction of petroleum distillation of boiling-point 120-150 C., of specific gravity 0.722-0.737; used for defatting, or cleaning. The term is also used for a volatile oil yielded by asphalt, when it is distilled with water.

23504 Petrolit
A German explosive containing potassium chlorate and mineral oil. No manufacturer.

23505 Petrolite® C-400
Salt of an oxidized polyethylene; wax used as a gelling agent for solvents in the formulation of paste products, e.g., shoe or floor polishes. *Petrolite.*

23506 Petrolite® C-7500
Oxidized synthetic wax; used in the formulation of emulsions having hard films with good gloss and dry-bright polish properties; oil-binding properties make it useful in solvent-based polishes and release agents. *Petrolite.*

23507 Petromix 9
Balanced blend of sodium sulfonates, auxiliary fatty acid soaps and coupling agents with antifoaming agents; strong emulsification agent used in textile processing oils, cutting oils, metal degreasers. *Witco Corporation.*

23508 Petromor
Petroleum sulfonates
Burmah-Castrol Ltd. Name unverified.

23509 Petronate® L, HL, K, CR and S
Series of anionic surfactants in which the cation is sodium and the anion is petroleum sulfonate; emulsifiers, dispersing and wetting agents; CR is rust preventative; used in printing inks; dry cleaning soaps; leather oils; lubricating grease; metal worki *Witco Corporation/Sonneborn; Witco Chemical Ltd.*

23510 Petronate® RP
Blend of sodium and calcium petroleum sulfonates; anionic surfactant used in rust preventative formulations. *Witco Corporation.*

23511 Petronauba® C
Oxidized microcrystalline wax; used in the formulation of polishes and emulsions; camauba substitute. *Witco Corporation.*

23512 Petrone A4 and A6C
Amine salt of alkylaryl sulfonic acid; brown viscous liquid; solubilization, antistat, dewatering, formulation, pigment dispersion, emulsification; corrosion inhibition properties. *ABM Chemicals Ltd.* Name unverified.

23513 Petropul
A proprietary trade name for a synthetic resin. No manufacturer.

23514 Petro-Rez 801
Cycloaliphatic hydrocarbon resin
Tackifier for rubber compounding applications including molded mechanical goods, extrusions, shoe soling, electrical insulation, flooring; processing aid for mineral-filled elastomers; extender in black loaded compounds. *Akrochem.*

23515 Petrostep
Petroleum sulfonates
232-373-2 For enhanced oil recovery. *Stepan.* Name unverified.

23516 Petrostep A-70
Branched chain alkylate sulfonic acid, anionic surfactant properties, viscous amber liquid; an emulsifier intermediate for speciality products. *KWR Chemicals Ltd.* Name unverified.

23517 Petrosul® H-50
Sodium petrol. sulfonate
Surfactant and corrosion inhibitor; used as motor and fuel oil additives, rustproofing formulations; also in dry cleaning solvents, leather processing, printing inks, oil well drilling fluids. *Penreco.*

23518 Petrosul® M-50, M-60, M-70
Sodium petroleum sulfonate
Surfactant, emulsifier, and corrosion inhibitor for metalworking fluids; also for dry cleaning solvents, leather processing, textile oils, printing inks, oil well drilling fluids. *Penreco.*

23519 Petrothene®
Polyethylene. *Quantum Chemical Corp.*

23520 Petrothene® GA 501
9002-88-4 7728
LLDPE, butene co-monomer; film extrusion resin for heavy duty shipping sacks, liners, consumer packaging; excellent puncture resistance, elongation, and heat-seal strength *Quantum/USI.*

23521 Petrothene® GA 564
9002-88-4 7728
LLDPE; resin for injection molding of industrial containers, trash cans, large parts; features high rigidity, good low temperature impact strength, high ESCR. *Quantum/USI.*

23522 Petrothene® GA 808-090
9002-88-4 7728
High molecular weight LLDPE resin; wire and cable compound. *Quantum/USI.* Name unverified.

23523 Petrothene® HD 5903B
9002-88-4 7728
High molecular weight HDPE; wire and cable compound. *Quantum/USI.* Name unverified.

23524 Petrothene® LB 5003-00
9002-88-4 7728
HDPE; blow molding resin. *Quantum/USI.* Name unverified.

23525 Petrothene® LB 6001-00
9002-88-4 7728
HDPE; sheet and profile extrusion grade *Quantum/USI.* Name unverified.

23526 Petrothene® LF 6030-00
9002-88-4 7728
HDPE copolymer; injection molding resin with good toughness and warp resistance for spray pump parts and overcaps; FDA compliance. *Quantum/USI.*

23527 Petrothene® LP 5102-00
9002-88-4 7728
MMW-HDPE; film resin with high stiffness and strength, excellent appearance, good draw down for merchandise bags; FDA compliance. *Quantum/USI.*

23528 Petrothene® LS 3150-00
9002-88-4 7728
HDPE copolymer; injection molding resin with excellent impact resistance,

good flow for containers, closures, housewares; FDA compliance. *Quantum/USI.*

23529 Petrothene® LT 5704-00
9002-88-4 7728
HDPE, high load melt index; blow molding resin *Quantum/USI.* Name unverified.

23530 Petrothene® NA 155-000
9002-88-4 7728
LDPE; film resin with good stiffness, optics for industrial films with improved clarity. *Quantum/USI.*

23531 Petrothene® NA 204-000
9002-88-4 7728
LDPE resin; extrusion coating and injection molding resin *Quantum/USI.*

23532 Petrothene® NA 341-000
9002-88-4 7728
LDPE homopolymer; high clarity film resin with good melt strength for clarity bundling. *Quantum/USI.*

23533 Petrothene® PA 436
9002-88-4 7728
MDPE butene co-monomer; resin for rotational molding of agric. and chem. storage containers; excellent ESCR, low temperature impact strength, warp resistance. *Quantum/USI.*

23534 Petrothene® PP 1510-HC
9003-07-0 7741
PP impact copolymer; heat-stabilized resin with chemical resistance for film, profile, tubing, coating applications. *Quantum/USI.*

23535 Petrothene® PP 2004-MR
9003-07-0 7741
PP homopolymer; extrusion grade resin with excellent elongation, high melt flow for film and filament. *Quantum/USI.*

23536 Petrothene® PP 7300-KF
9003-07-0 7741
PP random copolymer; resin with excellent clarity, impact strength, stiffness for general purpose blow molding, sheet extrusion. *Quantum/USI.* Name unverified.

23537 Petrothene® PP 8000-GK
9003-07-0 7741
PP homopolymer; injection molding resin with toughness, excellent surface appearance for general purpose containers, housewares. *Quantum/USI.*

23538 Petrothene® PP 8770-HU
9003-07-0 7741
PP copolymer; heat-stabilized super-impact injection molding resin for tool boxes, pet carriers, toys. *Quantum/USI.*

23539 Petrothene® XL
Crosslinkable polyethylene resins. *Quantum Chemical Corp.*

23540 Petrowet® R
Sodium alkyl sulfonate
Wetting agent, detergent, dispersant, penetrant, foamer for petrol., metalworking, textile, and paper industries, chemical manufacturing, oil well servicing; industrial formulations and cleaning. *DuPont.*

23541 peucedanin
482-44-0 4960 207-581-1
Imperatorin
Imperatorin. It occurs in the root of masterwort.

23542 Pevafix
An adhesive for polyvinyl alcohol film. *May & Baker Ltd.* Name unverified.

23543 Pevalon
Polyvinyl alcohol film. *May & Baker Ltd.* Name unverified.

23544 Pevikon
Polyvinyl chloride homo and copolymers (E-PVC); used in organosols and plastisols for production of plastic articles. *Norsk Hydro AS.*

23545 pewter
A variable alloy of from 73-89% tin, 1.6-6.7 antimony, 1-6.8% copper, and 0-20.5% lead, and sometimes zinc.

23546 Pexalyn
Polar, acidic hydrocarbon-based synthetic resins; Pexalyn A500 and A600 are used as tackifier resins for solvent and emulsion adhesives and as modifier resins for ethylene/vinyl acetate copolymer and wax-based hotmelt coatings and adhesives. *Hercules.*

23547 Pexate
Metal resinates. *Hercules Ltd.* Discontinued.

23548 Pexite
Wood rosins; used in wax modification, as a chemical intermediate, in solder fluxes and in wax based coatings. *Hercules.*

23549 Pexol® 245
Fortified dark paste rosin size; sizing agent for paper/paperboard producing high level of water resistance; used in unbleached pulps and secondary fiber pulps used in multi-ply board. *Hercules.*

23550 Pexol® 50, Dark Fluid
Fortified rosin size; sizing agent for paper/paperboard producing resistance to water and aqueous solutions; for use on unbleached kraft grades. *Hercules.*

23551 Peyton powder
It is a nitrocellulose and nitroglycerin powder, containing 20% ammonium picrate; an explosive.

23552 PF-10GF/15T
25135-51-3
Polysulfone resin, 10% glass fiber-reinforced, 15% PTFE-lubricated; improved mechanical properties, reduced surface wear characteristics, wide temperature service range; used for applications. with moving parts requiring retention of properties, especially creep resistance. *Compounding Tech.*

23553 PF-20GF
25135-51-3
Polysulfone resin, 20% glass fiber-reinforced; features low creep and shrinkage; used for electrical connectors, business machine components, coffee maker bodies, microwave ovenware, sterilizer trays. *Compounding Tech.*

23554 Pfeilringspalter
A catalyst used in the decomposition of fats. It is prepared by treating a mixture of hydrogenated ricinoleic acid and naphthalene with sulfuric acid.

23555 Pferrico
1309-37-1 4072 215-168-2
A cobalt treated ferric oxide for magnetic media use. *Pfizer International.* Discontinued.

23556 Pferrisperse
1309-37-1 4072 215-168-2
A high-solids iron oxide pigment slurry. *Pfizer International.* Discontinued.

23557 Pferritan
A zinc ferrite compound for high temperature color pigment use. *Pfizer International.* Discontinued.

23558 Pferrocal
A steel-clad calcium wire effective in the production of high quality specialty steels. *Pfizer International.* Discontinued.

23559 Pferromet
A metallic iron particle for magnetic tapes and disks. *Pfizer International.* Discontinued.

23560 Pferrox
1309-37-1 4072 215-168-2
A gamma ferric oxide for magnetic tape and disk applications. *Pfizer International.* Discontinued.

23561 Pfico$_2$-Hop
A nonisomerized carbon dioxide extract of hops; used by adding to the brew kettle during boiling in the manufacture of malt beverages, to add bitterness to this product. *Pfizer International.*

23562 Pfico$_2$-Isohop
A modified aqueous hop extract produced from a liquid carbon dioxide base concentrate and standardized to 35% isomerized α-acids; for addition to malt beverage after fermentation to standardize bitterness. *Pfizer International.*

23563 Pfico$_2$-Redihop
A modified aqueous hop extract produced from a liquid carbon dioxide base concentrate and standardized to 35% reduced α-acids; for addition to malt beverage after fermentation to standardize bitterness. *Pfizer International.*

23564 Pfiklor
7447-40-7 7783 231-211-8
ClK
Potassium chloride
Replenisher. *Pfizer Inc.* Discontinued.

23565 Pfinodal
A copper-nickel-tin alloy used in the fabrication of electronic connectors. *Pfizer International.* Discontinued.

23566 P-Flakes
Hydrogenated palm oil. *Karlshamns.*

23567 Phaltan
133-07-3 4255 205-088-6
Folpet, an agricultural fungicide. *Monsanto (Solaris).* Name unverified.

23568 Phamosan
Skin protection ointments, lotions and soaps; for protection and care for the skin under environmental stress. *Dynamit Nobel Wien GmbH.* Name unverified.

23569 Phanteine
115-95-7 5521 204-116-4
$C_{12}H_{20}O_2$
3,7-dimethyl-1,6-octadien-3-yl acetate
linalyl acetate; Bergamol. Pure linalyl acetate. Found in volatile oils such as

bergamot and lavendar oils. Used in perfumery. bp = 220°; d_4^{20}= 0.895; n_D^{20}= 1.4460, insoluble in H_2O, soluble in organic solvents. *Bush Boake Allen Ltd.* Discontinued.

23570 Phantol
78-70-6 5520 201-134-4
$C_{10}H_{18}O$
3,7-dimethyl-1,6-octadien-3-ol
linalool; Linalool ex orange oil; Linalool ex bois de rose oil; Linalool ex ho oil; linalol. Pure linalool. Used in perfumery. *Bush Boake Allen Ltd.* Discontinued.

23571 Phantom
23103-98-2 7651 245-430-1
pirimicarb
Granules containing 50% w/w pirimicarb; used for control of aphids. *Bayer plc.*

23572 Pharmagel
9000-70-8 4388 232-554-6
gelatin
Gelfoam; Puragel. A proprietary trade name for pure gelatin. Mixturee of partially hydrolyzed proteins, used as a thickener in foods. No manufacturer.

23573 Pharmasorb
1337-76-4
Pharmaceutical grade attapulgite (hydrous magnesium aluminum silicate); used for tablet or liquid antidiarrheals, as tableting aid, pharmaceutical carrier, inert, in cosmetics. *Engelhard.*

23574 Pharmaton
A proprietary preparation containing multivitamins and minerals; also contains ginseng. *Windsor Healthcare Ltd.*

23575 Pharmatone
A hydrolyzed pork tissue that is spray dried; over 90% protein; water soluble; used in veterinary biologicals and food supplements, for its nutrient and high nitrogen content. *Am. Labs.*

23576 Pharmolin
1332-58-7 5294 296-473-8
Fine particle sized pharmaceutical grade hydrous kaolin; used as an extender in pharmaceuticals, tableting, cosmetics. *Engelhard.*

23577 Phase Alpha®
Resins for automotive fabrication. *Ashland.* Name unverified.

23578 Phase II®
SMC resin for automotive fabrication. *Ashland.* Name unverified.

23579 phaseomannite
87-89-8 5008 201-781-2
$C_6H_{12}O_6$
inositol
Inositol, used as a nutrient.

23580 Phathalogen
Phthalocyanine dyes for blue and green shades for cotton and regenerated cellulose. *Bayer AG.*

23581 Phemerol Chloride
121-54-0 1103 204-479-9
Benzethonium chloride
Antiinfective, topical; pharmaceutic aid, used primarily in cosmetics for its antimicrobial and cationic surfactant
properties. *Parke-Davis.* Name unverified.

23582 Phemox
Mercury containing fungicide. *Murphy Chemical Co Ltd.* Discontinued.

23583 Phenac
A proprietary trade name for a phenolic synthetic resin for varnish and lacquer. No manufacturer.

23584 phenald resins
A general term for phenol-formaldehyde resins.

23585 Phenaldine
102-06-7 3383 203-002-1
Diphenylguanidine
A proprietary trade name for a rubber vulcanization accelerator. No manufacturer.

23586 Phenegol
The mercury-potassium salt of nitro-*p*-phenolsulfonic acid. A bactericide.

23587 Phenester
A proprietary trade name for a synthetic resin of the coumarone-indene type. No manufacturer.

23588 phenethyl mustard oil
C_9H_9NS
p-Ethylphenyl-thiocarbimide.

23589 Phenetidine
156-43-4 7373 205-855-5
$C_8H_{11}NO$
4-ethoxybenzeneamine
Amino-phenyl ethyl ether; *p*-phenetidine; 4-ethoxybenzenamine; 4-aminophenetole; 4-ethoxyaniline; *para*-aminoethoxybenzene. An antipyretic. Used in the manufacture of acetophenetidin. mp = 3°; bp = 253-255°; d_4^{16}= 1.0652; insoluble in H_2O, soluble in organic solvents; LD_{50} (rat orl) = 540 mg/kg.

23590 phenetole
103-73-1 7374 203-139-7
$C_8H_{10}O$
ethoxy benzene
phenyl ethyl ether. mp = -30°; bp = 169-170°; d = 0.9660; n_D^{20} = 1.5076; insoluble in H_2O, soluble in organic solvents; LD_{50} (mus orl) = 2200 mg/kg.

23591 Phenex
An aldehyde-amine; a proprietary trade name for a rubber vulcanization accelerator. No manufacturer.

23592 phenic acid
108-95-2 7390 203-632-7
C_6H_5OH
phenic alcohol
Phenol. Used in rubber manufacture, solvents, in phenolic resins.

23593 Phenistix
A proprietary preparation of ferric ammonium sulfate, magnesium sulfate and cyclohexylsulfamic acid impregnated on a test strip; used for the detection of phenylketonuria and ingestion of salicylates. *BC Ames.* Name unverified.

23594 Phenitol
62-38-4 7453 200-532-5
Phenylmercuric nitrate
Pharmaceutic aid. Bactericide, germicide. *Alcon Laboratories Inc.* Name unverified.

23595 phenmedipham
13684-63-4 7384 237-199-0
$C_{16}H_{16}N_2O_4$
3-[(methoxycarbonyl)amino]phenyl (3-methylphenyl)carbamate
Alegro; Beetomax; Beetup; Beta; Betaflow; Betalion; Betanal; Betanal E; Betosip; Fender; Goliath; Gusto; Kemifam; Medipham; Pistol; Protrum K SN 38584; Spn-aid; Suplex; Vangard. Selective systemic herbicide, absorbed through the leaves, inhibits photosynthesis. Used for post-emergence control of annual broad-leaved weeds in sugar beet, fodder beet, beetroot, mangels, spinach and strawberries. bp > 290°; d^{25} = 1.06; soluble in H_2O (2 mg/l), more soluble in organic slvents; LD_{50} (rat orl)= >10000 mg/kg.

23596 Phenmedipham
13684-63-4 7384 237-199-0
Headland Dephend; Beetomax; Beetup; Betalion; Betanal E; Betanal Tandem; Goliath; Gusto; Pistol; Pistol 400; Protrum K; Suplex; Tripart® Beta, Beta 2; Vangard; Vanguard,. Emulsifiable concentrate containing 114 g/l phenmedipham; for weed control for beet crops. SBC Technology Ltd; Fine Agrochemicals Ltd; MTM Agrochemicals Ltd; Schering Agrochemicals Ltd; ABM Chemicals Ltd; Farm Protection Ltd; Rhone-Poulenc UK; Atlas Interlates Ltd; Universal Crop Protection Ltd; Tripart Farm Chemicals Ltd;Farmers Crop Chem. Ltd. Unverified.

23597 phenmedipham
13684-63-4 7384 237-199-0
$C_{16}H_{16}N_2O_4$
3-(methylphenyl)carbamic acid 3-[(methoxycarbonyl)amino]phenyl ester
Spin-Aid; Betamix; Betanal; EP-452; Schering-38584; SN 4075; SN-38584; Fenmedifam; Schering 4072; Alegro; Beetomax; Beetup; Beta; Betaflow; Betalion; Betosip; Fender, Goliath; Gusto; Kemifam; Medipham; Pistol; Protrum K; Suplex; Vangard. Herbicide. mp = 139-142°; SG_{20} = 0.25-0.30; insoluble in H_2O (<10 ppm); LD_{50} (rat orl) >8 g/kg.

23598 phenmedipham
13684-63-4 7384 237-199-0
$C_{16}H_{16}N_2O_4$
3-[(methoxycarbonyl)amino]phenyl (3-methylphenyl)carbamate
Alegro; Beetomax; Beetup; Beta; Betaflow; Betalion; Betanal; Betanal E; Betosip; Fender; Goliath; Gusto; Kemifam; Medipham; Pistol; Protrum K SN 38584; Spn-aid; Suplex; Vangard. Selective systemic herbicide, absorbed through the leaves, inhibits photosynthesis. Used for post-emergence control of annual broad-leaved weeds in sugar beet, fodder beet, beetroot, mangels, spinach and strawberries. bp > 290°; d^{25} = 1.06; soluble in H_2O (2 mg/l), more soluble in organic slvents; LD_{50} (rat orl)= >10000 mg/kg.

23599 Phenmerzyl Nitrate
55-68-5 7455 200-242-9
Phenylmercuric nitrate
Pharmaceutic aid. Bactericide, germicide. *Merrell Dow Pharmaceuticals Inc.* Name unverified.

23600 Phenobarbitone sodium
57-30-7 7386 200-322-3
$C_{12}H_{11}N_2NaO_3$
Gardenal Sodium. A proprietary preparation of phenobarbitone sodium; a

hypnotic, may be habit forming; a contoled substance. *Rhône-Poulenc Rorer Ltd.*

23601 Phenodip

Liquid, compounded esters of 4-hydroxybenzoic acid; preservative. *Nipa Laboratories Ltd.* Discontinued.

23602 Phenodur

Heat hardening phenol/formaldehyde resins; used in industrial paints, brake and clutch linings and abrasives. *Hoechst AG.*

23603 Phenodur

Heat hardening phenol/formaldehyde resins; used for industrial paints, brake and clutch linings and abrasives. *Resinous Chemicals Ltd.*

23604 Phenol Carbonate

102-09-0 7433 203-005-8

$C_{13}H_{10}O_3$

Carbonic acid diphenyl ester

2-ethylhexyl acetate; Diphenyl carbonate. Solvent for nitrocellulose, resins, lacquers, baking finishes. Lustrous needles; mw=214.22; mp=80-81°; bp=302-306°; bp_{15}=168°; practically insoluble in water; soluble in hot alcohol, benzene, ether, glacial acetic acid. *Eastman; Hüls AG; MTM Speciality Chem. Ltd; Penta Mfg.*

23605 phenol red

143-74-8 7397 205-609-7

$C_{19}H_{14}O_5S$

Phenol-sulfonphthalein

Used as an acid-base indicator; diagnostic reagent medicine, laboratory reagent.

23606 Phenolene, Phenolene Supra

Tin plating bath brighteners. *Ciba plc.* Name unverified.

23607 phenolic resin

Phenol-formaldehyde; one step: A-stage resin, one-stage resin, resole; two step: novolac resin, novolak resin, two-stage resin. Thermosetting resin from condensation of phenol or substituted phenol with aldehydes such as formaldehyde, acetaldehyde, and furfural; thermosetting resin for applications requiring heat resistance, exhaust duct systems, wiring devices, switch gears, ovens, toasters, pot handles, electrical devices, coil bobbins, coatings, laminates. 3M; Akzo; Arakawa; Arizona; Asahi Yukizai Kogyo; Bakelite GmbH; BP Chem. Ltd; Georgia-Pacific; Hüls AG; PMC Specialties; QO; Raschig.

23608 Phenoline

A disinfectant identical with Lysol. It is a cresol made soluble in water by saponification.

23609 Phenolite

A proprietary trade name for a phenol-formaldehyde synthetic resin laminated product. No manufacturer.

23610 Phenolite

A proprietary trade name for a phenol-formaldehyde synthetic resin laminated product. No manufacturer.

23611 phenolsulfonphthalein

143-74-8 7397 205-609-7

$C_{19}H_{14}O_5S$

4,4'-(3H,2,1-benzoxathiol-3-ylidene)bisphenol S,S-dioxide

phenol red, P.S.P.; Sulfonphthal. The lactone of dioxy-triphenyl-carbinol-carboxylic acid; used as an indicator in alkalimetry, red at pH >8.4, yellow at pH <6.8. Also used in diagnosis of renal function. Soluble in H_2O (0.8 mg/ml), EtOH, Me_2CO, insoluble in other organic solvents; LD_{50} (rat orl) >600 mg/kg.

23612 Phenonip

Phenoxyethanol and methyl, ethyl, propyl, butyl esters of *p*-hydroxybenzoic acid. Fully active liquid multi-component preservative system with low toxicity and wide spectrum activity, especially against pseudomonads; for cosmetics and pharmaceuticals. *Nipa Labs.*

23613 Phenopreg

A proprietary trade name for phenolic impregnated fabrics and papers. No manufacturer.

23614 Phenoro

Proprietary preparation of canthaxanthin and β-carotene. photoprotective agent. *Roche Products Ltd.* Name unverified.

23615 phenosalyl

A mixture of phenol, salicylic acid, menthol, and lactic acid; an antiseptic.

23616 Phenosept

Mixture of propylene phenoxetol and *p*-chloro-*m*-xylenol; an antiseptic. *Nipa Laboratories Ltd.*

23617 Phenosulfonic acid

1333-39-7 215-587-0

Eltesol® PSA 65; Phenol sulfonic acid. Catalyst for foundry resins; descaling agent for metal cleaning; anti-stress additive and plating aid in electroplating bath; curing aid in the plastics industry; raw material in the mfg. of dyes and pigments; detergents industry; pharmaceutical chemicals *Albright & Wilson UK.*

23618 phenothiazine

92-84-2 7404 202-196-5

$C_{12}H_9NS$

Dibenzothiazine

Agrazine; Antiverm; Biverm; Dibenzothiazine; Orimon; Thiodiphenylamine; Lethelmin; Souframine; Nemazene; Padophene; Fenoverm; Fentiazine; Contaverm; Dibenzo-p-thiazine; 10H-Phenothiazine; afi-tiazin; contraverm; dibenzo-1,4-thiazine; Feeno; helmetina; nemazine; phenegic; phenosan; phenoverm; phenovis; phenoxur; phenthiazine; reconox; wurm-thional; XL-50; phenothiazone; Contavern; Helmetine; Phenovarm; Vermitin. Insecticide, anthelmintic and intermediate in chemical, pharmaceutical synthesis. mp = 180-185°; bp = 371°; $bp_{26.6}$ = 235°; insoluble in H_2O, soluble in organic solvents; LD_{50} (mus orl) = 5 g/kg.

23619 phenothrin

26002-80-2 7405 247-404-5

$C_{23}H_{26}O_3$

2,2-dimethyl-3-(2-methyl-1-propenyl)cyclopropanecarboxylic acid (3-phenoxyphenyl)methyl ester

phonothrin; Sumithrin; S-2539; Fenothrin; Forte; phenoxybenzyl (1R)-*cis/trans* chrysanthemate; phenoxybenzyl chrysanthemate; phenoxybenzyl-(±)-*cis,trans*-chrysanthemate; phenoxybenzyl-D-Z/E-chrysanthemate; Phenoxythrin. Non-systemic insecticide with rapid knockdown action. Used for control of insect pests, including mosquitoes in public health and in stored grain. Mixture of isomers. bp > 290°; d_{25}^{25} = 1.06; n_D^{25}= 1.5483; insoluble in H_2O (<2 mg/ml), soluble in organic solvents; LD_{50} (rat orl) > 10000 mg/kg.

23620 Phenoweld

Phenolic adhesives for nylon, etc. *Hardman.*

23621 Phenox

phenylmercury hydroxide

A proprietary trade name for phenylmercury hydroxide. No manufacturer.

23622 Phenoxetol

122-99-6 7410 204-589-7

Phenoxyethanol

Antimicrobial preservative for cosmetics and pharmaceuticals. *Nipa Labs.*

23623 phenoxin

56-23-5 1864 200-262-8

CCl_4

Carbon tetrachloride

Solvent, dry-cleaning agent.

23624 phenoxy resin

$[OC_6H_4C(CH_3)_2C_6H_4OCH_2CH(OH)CH_2]_n$

High, molecular weight plastics copolymer of Bisphenol A and epichlorohydrin; coatings and adhesives; blow-molded containers, pipe, ventilating ducts, and other molded parts. *Aldrich; Union Carbide.*

23625 1-Phenoxy-2-propanol

770-35-4 212-222-7

$C_9H_{12}O_2$

1-phenoxypropan-2-ol

propylenephenoxythol. d = 1.0510; n_D^{20}= 1.5235.

23626 phenoxyacetic acid

122-59-8 7407 204-556-7

$C_8H_8O_3$

phenoxyethanoic acid

glycolic acid phenyl ether; *o*-phenylglycolic acid. Intermediate for dyes, pharmaceuticals, pesticides, other organics, fungicides, flavoring, laboratory reagent, precursor in antibiotic fermentations especially penicillin V. mp = 98°; bp = 285°; soluble in H_2O (13 mg/ml), more soluble in organic solvents; LD_{50} (rat orl) = 1500 mg/kg. *Chemie Linz UK; Great Lakes; Penta Mfg.; Schweizerhall.*

23627 2-Phenoxyethanol

122-99-6 7410 204-589-7

$C_8H_{10}O_2$

Emeressence® 1160 Rose Ether; Phenoxyethanol; 1-Hydroxy-2-phenoxyethane; ethylene glycol monophenyl ether; β-hydroxyethylphenyl ether; Phenoxethol; Phenoxetol; Phenyl Cellosolve,. Cosmetic preservative; effective against gram negative microorganisms. MW=138.17; crystals from ethanol, mp=71-73°; sl.sol.in water; sol.in ethanol, methanol; LD_{50}(mise, orl)=960 mg/kg *Henkel/Emery.* Name unverified.

23628 Phenoxylene 50

94-74-6 5803 202-360-6

MCPA; herbicide for cereals and grassland. *Schering Agrochemicals Ltd.* Discontinued.

23629 Phenoxylene Plus

94-74-6 5803 202-360-6

MCPA

A selective weed killer using MCPA as the active principle. *Fisons plc, Horticulture Div.* Name unverified.

23630 Phenprocoumon
435-97-2 7413 207-108-9
$C_{18}H_{16}O_3$
4-Hydroxy-3-(1-phenylpropyl)-2H-1-benzopyran-2-one
Liquamar; 3-(1-phenylpropyl)-4-hydroxycoumarin; Falithrom; Marcoumar. Anticoagulant. MW=280.32; mp=179-180°. *Organon Inc; Roshe Products Ltd.* Name unverified.

23631 Phenyform
$[C_6H_4(OH)CH_2OH]_xCH_2O$
An antiseptic powder prepared from phenol and formaldehyde; used as an indicator and for denaturing purposes.

23632 phenyl dimethicone
2116-84-9 218-320-6
$C_{15}H_{32}O_3Si_4$
methyl phenyl polysiloxane
polyphenylmethyl siloxane; Phenyltris(trimethylsiloxy)silane. Silicone polymer.

23633 phenyl ether
101-84-8 7442 202-981-2
$C_{12}H_{10}O$
diphenyl ether
1,1'-oxybisbenzene; Diphenyl Oxide; phenoxybenzene. Used as a heat transfer medium, in perfumery and in organic synthesis. mp = 27-28°; bp = 259°; d = 1.0730; insoluble in H_2O, soluble in organic solvents; LD_{50} (rat orl) = 3370 mg/kg.

23634 phenyl salicylate
118-55-8 7464 204-259-2
$C_{13}H_{10}O_3$
2-hydroxybenzoic acid phenyl ester
salol. Used as a UV absorber in plastics and as an analgesic, antipyretic and anti-inflammatory. mp = 41-43°; bp_{12} = 173°; d = 1.25; soluble in H_2O (150 mg/l); more soluble in organic solvents.

23635 Phenyl trimethicone
2116-84-9 218-320-6
Emalex MTS-30E; Phenyldimethicone; Abil® AV 20-1000, AV 8853;. Oil-phase cosmetic ingred. for alcoholic milky-wh. lotions. *Nihon Emulsion; Goldschmidt; Goldschmidt AG.*

23636 phenyl-γ acid
2-Phenylamino 8-naphthol-6-sulfonic acid.

23637 phenylacetic acid
103-82-2 7422 203-148-6
$C_6H_5CH_2COOH$
α-toluic acid
Perfume, precursor in manufacture of penicillin G, fungicide, flavoring, laboratory reagent. *Calaire Chimie SNC; Penta Mfg.; Schweizerhall.*

23638 phenylamine
62-53-3 696 200-539-3
$C_6H_5NH_2$
benzeneamine
aniline; aniline oil; aminobenzene; aminophen; Kyanol. Isolated from coal tar. Used as an intermediate in the manufacture of dyestuffs and pharmaceuticals. mp = -6°; bp = 184°; d = 1.0220; n_D^{20} = 1.5855; soluble in H_2O (3 g/100 ml), more soluble in organic solvents; LD_{50} (rat orl) = 0.44 g/kg.

23639 *m*-phenylenediamine
108-45-2 7437 203-584-7
$C_6H_8N_2$
1,3-benzenediamine
1,3-phenylenediamine; *m*-diaminobenzene; 3-aminoaniline. Aromatic amine; manufacture of dyes, hair dyes; rubber curing agent; in photography; as reagent for gold and bromine. mp = 62-63°; bp = 284-287°; d = 1.139; soluble in H_2O, organic solvents; LD_{50} (rat orl) = 650 mg/kg. *Du Pont.*

23640 *o*-phenylenediamine
95-54-5 7438 202-430-6
$C_6H_8N_2$
1,2-phenylenediamine
1,2-benzenediamine; *o*-diaminobenzene; *o*-aminoaniline. Used in dye manufacture mp = 103-104°; bp = 256-258°; soluble in H_2O (1%), more soluble in organic solvents; LD_{50} (rat orl) = 1070 mg/kg. Name unverified.

23641 *p*-phenylenediamine
106-50-3 7439 203-404-7
$C_6H_8N_2$
1,4-benzenediamine
1,4-phenylenediamine. Aromatic amine; azo dye intermediate; photographic developing agent; intermediate in manufacture of antioxidants, accelerators for rubber, synthetic fibers; dyeing hair and fur. mp = 145-147°; bp = 267°; soluble in H_2O (1 g/100 ml), more soluble in organic solvents; LD_{50} (rat orl) = 80 mg/kg. *Du Pont; Hoechst Celanese; Janssen Chimica.*

23642 2-phenylethyl alcohol.
60-12-8 7370 200-456-2
$C_8H_{10}O$
2-Phenylethanol
Phenyl ethyl alcohol; Benzyl carbinol; β-Phenyl ethyl alcohol; Benzeneethanol; Hydroxyethylbenzene; β-p.e.a.; Mellol; 1-Phenyl-2-ethanol. Has the smell of roses. Used in perfumery and as an antimicrobial agent. mp= -27°; bp_{750} = 219-221°; d_{25}^{25} = 1.017; n_D^{20}= 1.530; soluble in H_2O (2 g/100 ml), more soluble in organic solvents; LD_{50} (rat orl)= 1790 mg/kg.

23643 2-phenyl imidazole
670-96-2 211-581-7
$C_9H_8N_2$
2-phenylimidazole
Epoxy curing agent for printed circuit boards, molding compounds, potting; accelerator for dicyandiamide and anhydrides. mp = 146-148°; bp = 335-337°; insoluble in H_2O, soluble in organic solvents. *BASF; Janssen Chimica; Schweizerhall.*

23644 phenylmercuric acetate
62-38-4 7453 200-532-5
$C_6H_5HgOCOCH_3$
phenylmercury acetate
(acetato)phenylmercury; acetoxyphenylmercury. PMAC; Metallo-organic compound; fungicide, herbicide, mildewcide for paints; slimicide in paper mills. *Allchem Industries; Atomergic Chemetals; W.A. Cleary; EM Industries.*

23645 phenylmercuric oleate
104-60-9 203-218-6
$C_{24}H_{38}HgO_2$
phenylmercury oleate
Mercury, (9-octadecenoato-O)phenyl-, (Z)-; Octadecenoato-O)phenylmercury; Phenylmercuric oleate. Mildewproofing agent for paints; fungicide, germicide. *Noah Chem.; Thor.*

23646 phenylmercury acetate
62-38-4 7453 200-532-5
$C_8H_8HgO_2$
(acetato)phenylmercury
PMA; PMAC; PMAS; Ceresan slaked lime; Gallotox; Liquiphene; Mersolite; Nylmerate; Phix; Riogen; Scutl; Tag Fungicide; Tag HL-331; Single Purpose. Organomercury fungicide seed dressing for cereals and fodder beet. mp = 149°; soluble in H_2O (1.6 mg/ml), more soluble in organic solvents; LD_{50} (rat orl) = 22 mg/kg. *DowElanco Ltd.*

23647 phenyl-peri acid
82-76-8 699 201-438-7
$C_{16}H_{13}NO_3S$
1-Phenyl-naphthylamine-8-sulfonic acid
8-Anilino-1-naphthalenesulfonic acid; 1,8-ANS; Phenylperi acid; 8-(phenylamino)-1-naphthalenesulfonic acid. Fluorescent probe used in protein conformation studies. mp = 215-217°.

23648 *o*-phenylphenol
90-43-7 7458 201-993-5
$C_{12}H_{10}O$
1,1'-biphenyl-2-ol
2-phenyl phenol; 2-Hydroxybiphenyl; *o*-Hydroxydiphenyl; 2-Hydroxydiphenyl; 2-Phenylphenol; Torsite; Orthoxenol; 2-Biphenylol; *o*-Xonal; 1,1'-Biphenyl-2-ol; (1,1-Biphenyl)-2-ol; orthohydroxydipbenyl; oxenol; biphenyl-2-ol; Biphenylol; Dowicide 1; Hydroxdiphenyl; Hydroxy-2-phenylbenzene; Hydroxybiphenyl; OPP; Phenylphenol; Preventol O extra; Remol TRF; Tumescal OPE; Xenol. Substituted aromatic compound.; Intermediate; dyes, germicides, fungicides, rubber chemicals; laboratory reagents; food pkg.; disinfectant. mp = 57-59°; bp = 282°; d = 1.2130; soluble in H_2O (0.7 g/l), more soluble in organic solvents; LD_{50} (rat orl) = 2 g/kg. *Coalite Chem. Div.; Nipa Labs.*

23649 5-phenyltetrazole
18039-42-4 241-950-8
$C_7H_6N_4$
Blowing agent for foaming plastics and elastomers at elevated temps. *Hüls Am.*

23650 phenyltrichlorosilane
98-13-5 202-640-8
$C_6H_5Cl_3Si$
trichlorophenylsilane
Intermediate for silicones; laboratory reagent. bp = 201°; d = 1.324; n_D^{20} = 1.5230; LD_{50} (rat orl) = 2390 mg/kg. *Aldrich; Hüls Am.; PCR.*

23651 phenyltrimethoxysilane
2996-92-1 221-066-9
$C_9H_{14}O_3Si$
(trimethoxysilyl)benzene
Trimethoxyphenylsilane; Phenylmethoxysilane. Coupling agent, release

agent, lubricant, blocking agent, chemical intermediate. *Hüls Am.; Janssen Chimica; PCR.*

23652 Phenylurethane
101-99-5 7473 202-995-9
$C_9H_{11}NO_2$
Phenylcarbamic acid ethyl ester
Ethyl phenylcarbamate; ethyl carbanilate; phenyl urethan; N-phenylurethane; ethyl N-phenylcarbanilate. Made by reaction of ethyl chloroformate and aniline. mp = 52-53°; bp_{760} = 238° (dec); 1.106; n^{20}_D = 1.5376; slightly soluble in H_2O, soluble in EtOH, Et_2O.

23653 pheophytin
The brownish derivative obtained by treating chlorophyll with acid.

23654 pherocon Adox-O Adoxamone
Pheromones; for control of tortrix moth in tree fruit. *DowElanco Ltd.*

23655 Pherocon Archemone
Pheromones; for control of tortrix moth in tree fruit. *DowElanco Ltd.*

23656 Pherocon CM Codelemone
Pheromones; for control of codling moth in apples. *DowElanco Ltd.*

23657 Pherocon GFun Funemone
Pheromones; for control of plum tree moth in plums. *DowElanco Ltd.*

23658 Philacid 0810
67762-36-1 267-013-3
Caprylic/capric acid (C_{8-10})
Intermediate for manufacturing of synthetic specialty detergents, esters for use as lubricants and emollients. *United Coconut Chem.*

23659 Philacid 0818
Whole distilled coconut fatty acid (C_{8-18}); intermediate for manufacturing of soap, synthetic detergents, fatty amines. *United Coconut Chem.*

23660 Philacid 1200
143-07-7 5396 205-582-1
$C_{12}H_{24}O_2$
Lauric acid
Dodecanoic acid; n-dodecanoate; laurostearic acid; C-1297; duodecyclic acid; hydrofol acid 1255; hydrofol acid 1295; neo-fat 12; neo-fat 12-43; ninol aa62 extra; 1-undecanecarboxylic acid; wecoline 1295; undecane-1-carboxylic acid; dodecoic acid; hystrene 9512. Intermediate for manufacturing of toilet soaps, synthetic detergents, cosmetics and pharmaceuticals. mp = 44-46°; bp_{100} = 225°; d = 0.8830; insoluble in H_2O, soluble in organic solvents; LD_{50} (rat orl) = 12 g/kg. *United Coconut Chem.*

23661 Philacid 1214
Lauric/myristic acid (C_{12-14})
Intermediate for manufacturing of soaps, detergents. *United Coconut Chem.*

23662 Philacid 1400
544-63-8 6416 208-875-2
$C_{14}H_{28}O_2$
Myristic acid
n-Tetradecanoic Acid; Myristic Acid; Tetradecanoic acid; Crodacid; Emery 655; hydrofol acid 1495; n-tetradecoic acid; neo-fat 14; 1-tridecanecarboxylic acid; Univol U 3165; Hystrene 9014. Intermediate for manufacturing of toilet soaps, cosmetics, esters for use as flavors and perfumes. mp = 54-55°; bp_{100} = 250°; insoluble in H_2O, soluble in organic solvents. *United Coconut Chem.*

23663 philanized cotton
Cotton material which has been treated with concentrated nitric acid to convert it into a wool-like fabric. *United Coconut Chem.*

23664 Philcohol 1200
112-53-8 3464 203-982-0
Lauryl alcohol
Intermediate for manufacturing of synthetic detergents (e.g., lauryl sulfates, ethoxylated lauryl alcohol, lauryl ether sulfates, lauric alkanolamides, etc.). *United Coconut Chem.*

23665 Philcohol 1214
68425-37-6 270-351-4
C_{12-14} coconut fatty alcohol
Intermediate for surfactant manufacturing. *United Coconut Chem.*

23666 Philcohol 1400
112-72-1 6418 204-000-3
C_{14} alcohols
Intermediate for surfactant manufacturing. *United Coconut Chem.*

23667 Philcohol 1600
36653-82-4 2070 253-149-0
C_{16} alcohols
Intermediate for surfactant manufacturing. *United Coconut Chem.*

23668 Philcohol 1618
C_{16-18} alcohols
Intermediate for surfactant manufacturing. *United Coconut Chem.*

23669 Philcohol 1800
112-92-5 8960 204-017-6
C_{18} alcohol
Intermediate for surfactant manufacturing. *United Coconut Chem.*

23670 philosopher's wool
1314-13-2 10279 215-222-5
Flowers of zinc. The zinc oxide produced in a flocculent condition by burning zinc. A pigment.

23671 Phisomed
18472-51-0 2140 242-354-0
chlorhexidine gluconate
A proprietary preparation of chlorhexidine gluconate, an antimicrobial, in a detergent base used as a skin cleanser. *Winthrop Laboratories.* Name unverified.

23672 phlobaphenes
Red or brown coloring matter from barks, usually oak bark.

23673 phloba-tannins
Tannins which give the phlobaphene reaction.

23674 phloroglucinol
108-73-6 7482 203-611-2
$C_6H_6O_3$
1,3,5-Benzenetriol
1,3,5-trihydroxybenzene; 1,3,5-THB; phloroglucine. Analytical chemistry, decalcifying agent for bones, preparation of pharmaceuticals and dyes, resins, preservative for cut flowers, textile dyeing and printing, available in anhydrous and dihydrate forms. mp = 203°; soluble in H_2O (10 g/l), more soluble in organic solvents; LD_{50} (rat orl) = 4 g/kg. *ISK Europe SA; Schweizerhall; Spectrum Chem. Mfg.; Schering Berlin Polymers.*

23675 phlorol
90-00-6 7483 201-958-4
$C_8H_{10}O$
o-ethylphenol
1-ethyl-2-hydroxybenzene; Ethylphenol. Organic intermediate. mp = -18°; bp = 195-197°; d = 1.0370; n_D^{20} = 1.5372; insoluble in H_2O, soluble in organic solvents; LD_{50} (mus orl) = 600 mg/kg.

23676 phlorone
137-18-8 205-283-6
$C_8H_8O_2$
p-xyloquinone
metaphlorone; 2,5-Dimethyl-p-benzoquinone; 2,5-dimethyl-2,5-Cyclohexadiene-1,4-dione. Organic intermediate. mp = 124-126°; LD_{50} (mus orl) = 290 mg/kg.

23677 Phobol
Finishing agent. *Ciba plc.* Name unverified.

23678 Phobotex FTN
A proprietary trade name for a fat modified melamine resin with outstanding water-repellancy; fast to repeated washing in soap and soda; provides a durable water repellant finish for textiles. *Ciba plc.* Name unverified.

23679 Phobotone
Proofing agent. *Ciba plc.* Name unverified.

23680 Phoenix alloy
An electrical resistance alloy containing 25% nickel and 75% iron.

23681 Phoenix powder
An explosive containing 28-31% nitroglycerin, 0-1% nitro-cotton, 30-34% potassium nitrate, and 33-37% wood meal.

23682 Phoenixite
A proprietary pyroxylin product. No manufacturer.

23683 Pholin's alloy
An alloy of 77% tin, 19% bismuth, and 4% copper.

23684 Phono Bronze
A proprietary trade name for certain copper alloys containing about 1.25% tin and small amounts of silicon and cadmium. No manufacturer.

23685 Phos Flex
Mixed triaryl phosphate ester fire retardant plasticizers containing halogen and possessing better flame retardant properties in PVC than tricresyl phosphate. *Akzo Chemie UK Ltd.*

23686 Phos-Ad 100
126-73-8 9749 204-800-2
Tributyl phosphate
Defoamer concentrate; for drilling fluids. *Chemron.*

23687 Phosal
7722-76-1 577 231-764-5
Monoammonium phosphate fertilizer. *Scottish Agricultural Industries plc.*

23688 Phosal
8002-43-5 5452 232-307-2
Lecithin fractions. *Rhône-Poulenc UK.*

23689 Phosal 25 SB
5452
Phosphatidylcholine

Phosphatidyl-N-trimethylethanolamine. Natural emulsifying aid, hair conditioner, oil restorer, source of essential fatty acids for cosmetics (shampoos, creams, oil baths, soap additive, hair rinses); so-called vitamin F. *Seppic.* Discontinued.

23690 Phosal 53 MCT
5452
Phosphatidylcholine
Phosphatidyl-N-trimethylethanolamine. Production of liposomes for dermatology and cosmetics (improves skin moisturization and penetration); solubilizer/dissolving intermediary for lipophilic substances; source for dietetics. *Seppic.* Discontinued.

23691 phosalone
2310-17-0 7489 218-996-2
$C_{12}H_{15}ClNO_4PS_2$
S-[(6-chloro-2-oxo-3(2H)-benzoxazolyl)methyl] O,O-diethylphosphorodithioate
ENT 27163; 11974 RP; Azofene; Rubitox; Zolone. Non-systemic insecticide and acaricide with contact and stomach action. Used for control of sucking and chewing insects, spider mites, Colorado beetles, aphids, bollwormsand stem borers. mp = 45-48°; soluble in H_2O (10 mg/l), more soluble in organic solvents; LD_{50} (rat orl) = 120-175 mg/kg.

23692 Phosbrite®
Chemical brightening solutions. *Albright & Wilson Ltd, Phosphates & Speciality Business.*

23693 Phosbrite® 172
Patented product for chemically polishing aluminum, producing a reflective surface on a wide range of alloys. *Albright & Wilson Am.* Name unverified.

23694 Phos-chek
68333-79-9 269-789-9
ammonium polyphosphate
Fire retardant, ammonium polyphosphate; catalyst used in intumescent coatings for paints, used to help meet flammability requirements for rigid and flexible polyurethane foams. *Monsanto Co.* Name unverified.

23695 Phosclene
Metal finishing cleaners. *Albright & Wilson Ltd, Phosphates & Speciality Business.* Name unverified.

23696 Phosclere
Stabilizers and antioxidants. *Harcros.*

23697 Phos-copper
A proprietary welding alloy composed essentially of copper with from 5-10% phosphorus; melts at 700°C, becoming extremely fluid at 750°C. No manufacturer.

23698 Phosfetal 201 K
Calcium alkyl polyglycol ether phosphate
Cosmetics surfactant. *Zschimmer & Schwarz.*

23699 Phosfetal 205
Alkyl polyglycol ether phosphate, acid
Cleansing agent. *Zschimmer & Schwarz.* Discontinued.

23700 Phosfetal 603
Alkylaryl polyglycol ether phosphate, acid
Cleansing agent. *Zschimmer & Schwarz.* Discontinued.

23701 phosgene
75-44-5 7491 200-870-3
$COCl_2$
carbonyl chloride
carbon oxychloride; chloroformyl chloride; carbonic dichloride; chloroformyl chloride; Carbonyl dichloride. Used in the synthesis of isocyanates, polyurethane, and polycarbonate resins; organic carbonates and chloroformates; pesticides and herbicides; dye manufacture. Used as a chemical warfare agent. mp = -118°; bp = 8°; SG = 1.37; slightly soluble in H_2O.

23702 phospham
PN_2H
Phosphorus imidonitride

23703 phosphammite
Native ammonium phosphate. Used as a fireproofing agent.

23704 Phosphanol series
Phosphate ester
Emulsifier for textile spinning oils, polymerization; antistat; anticorrosive agent. *Toho Chem.*

23705 Phosphazote
It is an intimate mixture of superphosphate and urea containing 4-11% nitrogen and 10-14% P_2O_5; a fertilizer.

23706 PhosPho 642
8029-76-3 232-440-6
Hydroxylated lecithin
Surfactant, suspending agent for nail polish, hair/skin conditioner. *Fanning.*

23707 PhosPho E-100
8002-43-5 5452 232-307-2
Lecithin
Emulsifier, suspending agent, solubilizer, superfatting agent for face and hand creams. *Fanning.*

23708 PhosPho F-97
8002-43-5 5452 232-307-2
Lecithin
For makeup, cleansers, body creams and lotions, shampoos, moisturizing creams. *Fanning.*

23709 PhosPho H-00
Hydrogenated soya lecithin
For liposome applications. *Fanning.*

23710 PhosPho H-150
Hydrogenated lecithin
For liposome applications. *Fanning.*

23711 PhosPho LCN-TS
8002-43-5 5452 232-307-2
Lecithin
Phospholipid used as surfactant/emulsifier and skin conditioning agent for makeup, cleansers, body creams, shampoos, moisturizing creams, face creams, cosmetics, pharmaceuticals. *Fanning.*

23712 PhosPho PL-50
Phospholipids
Emulsifier, refatting agent, source of vitamin F (essential fatty acids); for soft gelatin capsules, lotions, creams, oil baths, natural cosmetics. *Fanning.*

23713 PhosPho S-85
8002-43-5 5452 232-307-2
Phospholipids
Wetting agent for makeup, cleansers, body creams and lotions, shampoos, moisturizing creams. *Fanning.*

23714 PhosPho T-20
8002-43-5 5452 232-307-2
Phospholipids
Hydrophillic cosmetic ingredient for makeup, cleansers, body creams and lotions, shampoos, moisturizing creams. *Fanning.*

23715 Phospho-gélose
A German clarifier for sugar juice. It consists of 70% phosphate of lime, and 30% kieselguhr.

23716 Phosphola® ALF5
Tallow fatty alcohol phosphate ester on sodium carbonate carrier; foam limiter for use with anionic surfactants. *Harcros.*

23717 Phospholan®
Alkylaryl sulfonates, modified nonionics, phosphate esters; used as hydrotropes and wetters in electrolyte solutions. *Harcros.*

23718 Phospholan® KPE4
Arylethoxy phosphate potassium salt
Hydrotrope for solubilization of low foaming surfactants into highly built liquids; industrial and domestic detergents and cleaners; fertilizer and toxicant formulations. *Henkel/Cospha; Harcros UK.*

23719 Phospholan® PHB 14
Phosphate ester, free acid; hydrotrope and compatibility agent for agrochemical formulations; solubilizers anionic and nonionic surfactants into high electrolyte concentrates. *Harcros.*

23720 Phospholan® PTP7
Phosphate ester
Surfactant offering excellent stability in high concentrations of acid, alkali and electrolyte. *Harcros.*

23721 Phospholipid EFA
Linoleamidopropyl PG-dimonium chloride phosphate
Patented component of skin and hair formulations; emulsifier, antimicrobial, moisturizer; solubilizer and fixative for fragrances; suitable for hypoallergenic products *Mona Industries.*

23722 Phospholipid PTC
Cocamidopropyl PG-dimonium chloride phosphate
Patented biodegradable substantive, mild foamer, hydrotrope, viscosity builder, wetter, surface tension reducer, bactericide for personal care products, baby products, feminine washes, ophthalmic preparations, disinfectant cleansers. *Mona Industries.*

23723 Phospholipid PTD
83682-78-4 280-518-3
Lauramidopropyl PEG-dimonium chloride phosphate
Bactericidal, conditioner, antistat, detergent, foamer, emulsifier, solubilizer, dispersant, thickener and wetting agent for personal care, household, pharmaceutical, veterinary products, fire fighting foams, petroleum production, photographic processes, agriculture, mining and textiles. *Mona Industries.*

23724 Phospholipid PTS
Stearamidopropyl PG-dimonium chloride phosphate
Patented mild substantivity agent, skin conditioner, thickener for personal care products; forms stable, low pH, smooth and elegant cosmetic emulsions. *Mona Industries.*

23725 Phospholipid SV
Stearamidopropyl PG-dimonium chloride phosphate and cetyl alcohol; patented substantive emulsifier, skin conditioner for skin and personal care products; moisturizer. *Mona Industries.*

23726 Phospholipon® 90/90G
7677
Phosphatidylcholine
Emulsifier for dermatology and cosmetics, solubilizer for parenteralia, raw material for liposomes; source for drugs and dietetics. *Seppic.* Discontinued.

23727 Phospholipon® CC
3436-44-0
1,2-Dicaproyl-sn-glycero(3) phosphatidylcholine
Seppic. Discontinued.

23728 Phospholipon® MC
18194-24-6 242-085-9
1,2-Dimyristoyl-sn-glycero(3) phosphatidylcholine.
Seppic. Discontinued.

23729 Phospholipon® PC
2644-64-6 220-153-9
1,2-Dipalmitoyl-sn-glycero(3) phosphatidylcholine.
Seppic. Discontinued.

23730 Phospholipon® SC
816-94-4 212-440-2
1,2-Distearoyl-sn-glycero(3) phosphatidylcholine
Seppic. Discontinued.

23731 phosphomolybdic acid
11104-88-4 7498 234-336-6
$24MoO_3 \cdot P_2O_5 \cdot xH_2O$
phospho-12-molybdic acid
molybdophosphoric acid; dodecamolybdophosphoric acid; PMA. Reagent for alkaloids; pigments; catalyst; fixing agent in photography; additive in plating processes; imparts water resistance to plastics, adhesives, and cement. *AAA Molybdenum Prods.; Atomergic Chemetals; Noah Chem.*

23732 Phosphomort
Organo-phosphorous insecticides. *Murphy Chemical Co Ltd.* Discontinued.

23733 phosphor bronze
A bearing metal. It is an alloy of from 70-97% copper, 3-13% tin, 0-16% lead, and 0.1-1.0% phosphorus and sometimes a little zinc. The alloys for casting usually contain 85-92% copper, 7-13% tin 0.3-1.0 phosphoros, and traces of lead and zinc.

23734 phosphor copper
An alloy of copper with from 5-15% of phosphorus; used as an addition to other metals, and in the manufacture of phosphor bronze.

23735 phosphor resin
A resin prepared from triphenyl phosphamide by heating and passing carbon dioxide through the compound.

23736 phosphor steel
An alloy of steel with phosphorus.

23737 phosphor tin
An alloy of tin and phosphorus, containing up to 10% phosphorus.

23738 phosphoric acid
7664-38-2 7500 231-633-2
H_3O_4P
orthophosphoric acid
white phosphoric acid; Sonac. Inorganic acid; in manufacture of inorganic phosphates, fertilizers, detergents, chemical polishing, priming metals, petroleum refining; acid catalyst; as acidulant and flavor, antioxidant and sequestrant in food; pharmaceutic acid; in dental cements; as an analytical reagent. mp = 42°; bp = 260°; SG = 1.685; soluble in H_2O. Albright & Wilson; Ashland; Farleyway Chem. Ltd; FMC; Mallinckrodt; Mitsui Toatsu Chem.; Monsanto; Rasa Ind.; Rhone-Poulenc Basic.

23739 phosphoric acid
7664-38-2 7500 231-633-2
H_3O_4P
orthophosphoric acid
white phosphoric acid; orthophosphoric acid; Sonac. Used in manufacture of superphosphates and detergents. Used as an acid catalyst and an acidulant, antioxidant and sequestrant. hemihydrate: mp = 29°; bp = 260°; SG = 1.685, soluble in H_2O, less soluble in organic solvents.

23740 phosphoric ether
78-40-0 9806 201-114-5
$C_6H_{15}O_4P$
Triethyl phosphate

23741 phosphorus acid
13598-36-2 7502 237-066-7
H_3O_3P
orthophosphorous acid
phosphonic acid. Intermediate for manufacture of diphosphonic acids and phosphite salts used as pesticides, chelates, and plastic additives; restricts color formation in esterification and condensation reactions (in small quantities); chemical reducing agent. Usually marketed as a 20% aqueous solution. mp = 73°; d_4^{21} = 1.65; soluble in H_2O, EtOH. *Albright & Wilson; Janssen Chimica; Lonza; Rasa Ind.; Rhône-Poulenc Basic; Witco/Argus.*

23742 phosphorus ether
122-52-1 204-552-5
$C_6H_{15}O_3P$
Triethyl phosphite
Phosphorous acid, triethyl ester; Triethoxyphosphine. Used as a plasticizer and a reducing agent. mp = -112°; bp = 155°; d = 0.97; insoluble in H_2O; LD_{50} (rat orl) = 1840 mg/kg.

23743 phosphorus oxychloride
10025-87-3 7506 233-046-7
Cl_3OP
phosphoryl chloride
phosphorus chloride. Chlorinating agent, used in the manufacture of insecticides, phosphate esters, pharmaceuticals; gasoline additives; dopant for semiconductor grade silicon, tricresyl phosphate, and fire-retarding agents. mp = 1°; bp = 105°; d = 1.6450; n_D^{20} = 1.4610; *Albright & Wilson; Aldrich; Cerac; FMC; Hoechst Celanese; Rhône-Poulenc Basic.*

23744 phosphorus pentachloride
10026-13-8 7508 233-060-3
PCl_5
phosphoric chloride
phosphoric perchloride; phosphorus perchloride. Chlorinating and dehydrating agent, catalyst. mp = 148°; bp = 160°; d = 1.6000. *Aldrich; Atomergic Chemetals; Hoechst Celanese; Noah Chem.*

23745 phosphorus pentoxide
1314-56-3 7512 215-236-1
O_5P_2
phosphoric anhydride
phosphoric oxide. Exists as P_4O_{10}. Used in manufacture of chemicals, dessicant (drying agent), surfactants, condensing agent in organic synthesis, sugar refining, lab reagent, fire extinguishing, special glasses. mp = 340°; d = 2,30; very soluble in H_2O. Albright & Wilson; Aldrich; Elf Atochem N. Am.; Cerac; Hoechst Celanese; Rasa Ind.; Rhone-Poulenc Basic.

23746 phosphorus pentoxide
1314-56-3 7512 215-236-1
O_5P_2
phosphorus(V) oxide
phosphoric anhydride; diphosphorus pentoxide. Used as a drying and dehydrating agent. mp = 340°; d = 2.390; reacts with H_2O to form phosphoric acid.

23747 phosphorus trichloride
7719-12-2 7515 231-749-3
Cl_3P
phosphorus chloride
phosphorous chloride; Phosphorus (III) chloride; Chloride of phosphorus; PICl; Phosphorus chloride. Source of phosphorus in manufacture of phosphite and phosphonate esters; chlorinating agent in manufacture of organic acid chlorides; analysis; catalyst. mp = -112°; bp = 76°; d_4^{21}= 1.574; reacts with H_2O, EtOH; soluble in organic solvents; LD_{50} (rat orl) = 18 mg/kg. *Albright & Wilson; Aldrich; Atomergic Chemetals; FMC; Rhône-Poulenc Basic.*

23748 Phosphoteric® T-C6
Sodium dicarboxyethylcoco phosphoethyl imidazoline
Patented surfactant, hydrotrope; synergizes detergency with ethoxylated nonionics; improves wetting, penetrating, and detergency; for high electrolyte industrial cleaners. *Mona Industries.*

23749 phosphotungstic acid
12067-99-1 235-087-6
$24(WO_3) \cdot 2H_3PO_4 \cdot 48H_2O$
phosphotungstic acid
tungstophosphoric acid. Used asa reagent to detect alkaloids and other nitrogenous bases.

23750 phosphotungstic acid
12067-99-1 7521 235-087-6
$H_3P(W_3O_{10})_4 \cdot H_2O$
phosphotungstic acid hydrate
tungstophosporic acid; phosphowolframic acid. Reagent in analytical chemistry and biology; manufacture of organic pigments; additive in plating

industry; imparts water resistance to plastics, adhesives, cement; catalyst for organic reactions; photographic fixative; textile antistat. *Atomergic Chemetals; Noach Chem.; Spectrum Chem. Mfg.*

23751 phosphotungstic acid
12067-99-1 7521 235-087-6
$24(WO_3)\cdot 2H_3PO_4\cdot 48H_2O$
phosphotungstic acid
tungstophosphoric acid. Used as a reagent to detect alkaloids and other nitrogenous bases.

23752 Phosteem
Metal treating compositions. *ICI Chem. & Polymers Ltd.* Discontinued.

23753 Phostin
Tin-plating process. *Albright & Wilson Ltd, Phosphates & Speciality Business.* Name unverified.

23754 Phostoxin
20859-73-8 372 244-088-0
Aluminum phosphide
Used for gasing of rabbits and moles. *Rentokil Ltd.*

23755 Photal
A photographic developer.

23756 Photine
2606-93-1 1349 220-021-0
$C_{28}H_{22}N_4Na_2O_8S_2$
2,2'-(1,2-ethenediyl)bis[5-[[(phenylamino)carbonyl]amino]benzenesulfonic acid] disodium salt
C.I. Fluorescent Brightener 30; C.I. 40600; Blancol C; Blankophor R; Leucophor R; Lumisol RV; Phorwite RN; Photine R; Pontamine White BR; Tintophen X. Fluorescent whitening agents for paper textiles and detergents. *Hickson & Welch Ltd.*

23757 Photocure 51
24650-42-8 3453 246-386-6
$C_{16}H_{16}O_3$
2,2-Dimethoxy-2-phenylacetophenone
DMPA; Benzil dimethyl ketal. UV photoinitiator. mp = 67-70°. *Aceto.*

23758 Photoglaze
UV and eb cure coatings of varying viscosities and chemical composition; used to protect and preserve the original appearance of a wide range of materials, e.g., vinyls, plastics, paper, wood, even metallized surfaces, giving good resistance to abrasion, *Lord Corporation (UK) Ltd.*

23759 Photomer®
Radiation-initiated curing chemicals. *Harcros.*

23760 Photomer® 3005
Acrylated epoxidized oil; radiation-cured chemicals for eb coatings, uv inks and varnishes. *Harcros.*

23761 Photomer® 5029
Polyester acrylate
Radiation-curing chemicals for eb coatings, uv varnishes. *Harcros.*

23762 Photomer® 6360
Polyurethane acrylate
Radiation-curing chemicals for eb coatings, uv inks and varnishes. *Harcros.*

23763 Photophor®
1305-99-3 1742 215-142-0
Ca_3P_2
calcium phosphide
Photophor. A trade name for a calcium phosphide, used for signal fires and as a rodenticide. mp = 1600°; d = 2.51; decomposed by H_2O. No manufacturer.

23764 Photozinc
1314-13-2 10279 215-222-5
Photoconductive zinc oxide *Pigment & Chemical Inc.*

23765 phoxim-methyl
see 7523
O-α-cyanobenzylideneamino O,O-di-methyl phosphorothioate
Used as an insecticide.

23766 Phtalofix® FN
Premordant; used for Phtalogen K dyestuffs. *Bayer AG.*

23767 Phtalogen®
Phthalocyanine dyes; used to produce blue and green shades of outstanding fastness on cotton and regenerated cellulose. *Bayer AG.*

23768 Phtalogen® K, N1
Heavy metal donor for phthalogen dyestuffs; used in dyeing and textile printing. *Bayer AG.*

23769 Phtalotrop® B
Resist agent; for phtalogen resist printing. *Bayer AG.*

23770 phthalic anhydride
85-44-9 7528 201-607-5
$C_8H_4O_3$
1,3-isobenzofurandione
Alkyd resins, plasticizers, hardener for resins, polyester, insecticides, laboratory reagent; manufacture of phthaleins, phthalates, benzoic acid. Aldrich; Aristech; Elf Atochem SA; BASF; Exxon; Mitsubishi Gas; Mitsui Toatsu Chem.; OxyChem; Stepan; UCB SA.

23771 Phthalopal®
Oil-free phthalate resins; for spirit-based and NC finishes; in hydrolyzed form, binder for aqueous flexographic inks and ballpen inks. *BASF AG.*

23772 Phycon 15, 18, 25
Modified polybasic buffered acids; acid buffers for dyeing and printing operations. *Am. Emulsions.*

23773 Phycon LPH
Proprietary blend; pH control agent for stain-resistant finished nylon carpet. *Am. Emulsions.*

23774 Phylatol
Biocidal preparation. *BDH Chemicals Ltd.*

23775 Phyomone
86-87-3 6458 201-705-8
1-naphthaleneacetic acid
Growth promoting hormone. *ICI Chem. & Polymers Ltd.*

23776 physiological salt solution
Normal salt solution; isotonic salt solution; surgical solution. Consists of 8.5 g of sodium chloride in 1,000 ml of distilled water. It is sterilised and used for intravenous injection.

23777 phytic acid
83-86-3 7542 201-506-6
$C_6H_{18}O_{24}P_6$
phytinic acid
Inositolhexaphosphoric acid; Inositol-hexaphosphoric acid; Myo-inositol hexaphosphate; myo-Inositol, hexakis(dihydrogen phosphate); Inositol Hexaphosphate; myo-Inositol hexakisphosphate; Fytic acid. Used for chelation of heavy metals in processing of animal fats and vegetable oils; corrosion inhibitor, metal treating, in the treatment of hard water. d= 1.2850; n_D^{20}= 1.3910; miscible with H_2O, insoluble in organic solvents.

23778 Phyt'iod
Ethylic esters of the iodized fatty acid of poppy seed oil (ethiodized oil); slendering product for cosmetics. *Alban Muller.*

23779 Phytoforol
59-02-9 10159 200-412-2
Vitamin E capsules. *British Drug Houses.* Name unverified.

23780 phytol
150-86-7 7545 205-776-6
$C_{20}H_{40}O$
3,7,11,15-Tetramethyl-2-hexadecen-1-ol
A primary alcohol, $C_{20}H_{39}OH$, obtained by the decomposition of chlorophyll, the coloring matter of plants. Used in preparation of vitamins E and K. bp_{10} = 202-204°; d = 0.8500; n_D^{20}= 1,4701; LD_{50} (rat orl) >5 g/kg.

23781 PI-20GF/000
61128-46-9
Polyetherimide
20% glass fiber-reinforced. *Compounding Tech.* Name unverified.

23782 Pibiter
A proprietary range of molding products based on saturated thermoplastic polyesters. *Montedison UK Ltd.* Name unverified.

23783 Picaltal®
Mixture of aromatic sulfonic acids; nonswelling, complex-forming acid in chrome tanning. *BASF AG.*

23784 Picamar
$C_{11}H_{16}O_3$
Propylpyrogallol dimethyl ether
Used in perfumery.

23785 Picco® 5000, 6000
Aromatic hydrocarbon resin derived from petroleum; used in higher-vinyl acetate content ethylene/vinyl acetate copolymer-based hot melt adhesives; used in caulking compounds, sealants, and concrete curing resins, and in news ink and oil-based lithographic ink. *Hercules.*

23786 Piccodiene®
Aliphatic hydrocarbon resins, composed mainly of polydicyclopentadienes; used in concrete curing compounds, in metallic paints and varnishes and as a compounding ingredient and process aid in many rubber goods. *Hercules.*

23787 Piccolastic®
Proprietary trade name for vinyl plasticizers based upon polymerized styrene and homologues. *Hercules.*

23788 Piccolastic® D125, D150
A range of styrene monomer hydrocarbon resins; used for pressure sensitive, hot melt and solvent adhesives and as epoxy resin extenders. *Hercules.*

23789 Piccolyte C
A series of polyterpene hydrocarbon resins derived from d-limonene; used as tackifiers for natural rubber based pressure sensitive adhesives and can sealants used in the production of chewing gum base. No manufacturer.

23790 Piccolyte®
A proprietary trade name for thermoplastic terpene resins. *Hercules.*

23791 Piccolyte® A, C
Polyterpene hydrocarbon resins derived from α-pinene; used in adhesives for tapes and labels, in construction adhesives, in packaging adhesives, as tackifiers for styrene-butadiene rubber and styrene block-copolymer rubber. *Hercules.*

23792 Piccolyte® HM110
A hydrocarbon-modified terpene resin; a tackifier resin of particular value in adhesives requiring high shear and high tack. *Hercules.*

23793 Piccolyte® S
A range of polyterpene hydrocarbon resins derived from β-pinene; used in natural rubber-based pressure sensitive adhesives and styrene-butadiene rubber can sealants and in textile dry sizes. *Hercules.*

23794 Piccomer® XX
A range of low molecular weight resins produced from petroleum-derived monomers; binder and plasticizer resins in putty and sealants, as felt saturants in felt-based sheet goods, in waterproofing treatments for paperboard, as softeners for elastomers in rubber compounding. *Hercules.* Discontinued.

23795 Picconol®
A range of aliphatic hydrocarbon resin emulsions; tackifiers in adhesives and in the production of waterproof finishes. *Hercules.*

23796 Picconol® A200
Anionic terpene resin emulsions based largely on low molecular weight thermoplastic terpene resins produced from β-pinene or δ-limonene monomers; used in water-based laminating, case-sealing adhesives, in laminating paper, and as tackifiers for natural rubber latex. *Hercules.* Discontinued.

23797 Picconol® A300
Anionic terpene resin emulsions based largely on low molecular weight thermoplastic terpene resins produced from β-pinene or d-limonene monomers; used in water-based laminating, case-sealing adhesives, in laminating paper and as tackifiers for natural rubber latex. *Hercules.* Discontinued.

23798 Picconol® A400
A low molecular weight anionic pure monomer resin emulsion; used with other aqueous thermoplastic and/or elastomeric systems to produce excellent adhesives and coatings. *Hercules.* Discontinued.

23799 Picconol® A500 A600
A range of aromatic hydrocarbon resin emulsions; used as adhesives, laminants and tackifiers. *Hercules.* Discontinued.

23800 Piccopale®
Range of aliphatic hydrocarbon resins manufactured from petroleum-derived monomers; used in high ethylene/vinyl acetate hot-melt and natural rubber-based adhesives, in can coatings for packaging, in varnishes to impart gloss and flowability, in paper saturation, and as waterproofing agents for paper and textiles. *Hercules.*

23801 Piccopyn
A range of phenolic-modified terpene resins; used as laminating agents and modifiers in ethylene/vinyl acetate copolymer-based packaging and tray forming adhesives, in specialty coatings, to modify specific polyurethanes, as tackifiers and process aids in specialty rubber compounding. *Hercules.* Discontinued.

23802 Piccotac®
A range of aliphatic hydrocarbon resins produced from mixed monomers of petroleum origin; developed especially for adhesives, particularly pressure sensitive and hot-melt types. *Hercules.*

23803 Piccotex®
Vinyltoluene copolymer hydrocarbon resins; used in hot-melt product assembly adhesives, for wax based coatings, in transparentizing paper, as dry sizing agents for textiles. *Hercules.*

23804 Piccotoner
Styrene acrylic copolymer hydrocarbon resins; used as toner resins for dry reproduction inks. *Hercules.*

23805 Piccoumaron
A proprietary trade name of terpene varnish and lacquer resins. No manufacturer.

23806 Piccoumarone Resins
Proprietary trade name for coumarone-indene resins. No manufacturer.

23807 Piccovar® AB
Aliphatic hydrocarbon resin; used as reinforcing agents in adhesives with high-temperature requirements, in heat set printing ink applications, in specialty coatings, in rubber compounding. *Hercules.*

23808 Piccovar® AP
Aromatic hydrocarbon resins used as plasticizers, softeners and tackifiers; suitable for use in various adhesive systems based on natural rubber, styrene-butadiene rubber, and poly-chloroprene. *Hercules.*

23809 Piccovar® L
Alkylated aromatic hydrocarbon resins; nonpolar, low molecular weight resins used as saturants, plasticizers, and tackifiers in adhesives and hot melts. *Hercules.*

23810 pichurim camphor
A substance resembling laurel camphor obtained from pichurim beans.

23811 picked turkey gum
9000-01-5 232-519-5
White sennaar gum
The best variety of gum acacia.

23812 Picket
52645-53-1 7321 258-067-9
Garden insecticide based on permethrin. *ICI Garden Products.*

23813 pickle alum
10043-01-3 381 233-135-0
$Al_2O_{12}S_3$
Aluminum sulfate
Used for packing and preserving.

23814 pickle green
10290-12-7 2693 233-644-8
A commercial variety of Scheelés green (cupric arsenite).

23815 α-picoline
109-06-8 7554 203-643-7
C_6H_7N
2-methylpyridine
2-picoline; α-methylpyridine. Organic intermediate for pharmaceuticals, dyes, rubber chemicals, solvent, source for vinyl pyridine, laboratory reagent. mp = -70°; bp = 128-129°; d = 0.9430; n_D^{20}= 1.5000; insoluble in H_2O, soluble in organic solvents; LD_{50} (rat orl) = 790 mg/kg. *Lonza; Nepera; Schweizerhall.*

23816 β-picoline
108-99-6 7555 203-636-9
C+66H7N
3-picoline
3-methylpyridine; β-methylpyridine. Solvent in synthesis of pharmaceuticals, resins, dyestuffs, rubber accelerators, insecticides; preparation of nicotinic acid, nicotinic acid amide, waterproofing agents; laboratory reagent. mp = -19°; bp = 143-144°; d = 0.9750; n_D^{20}= 1.5060; insoluble in H_2O, soluble in organic solvents; LD_{50} (rat orl) = 400 mg/kg. *Nepera; Schweizerhall.*

23817 γ-picoline
108-89-4 7556 203-626-4
C_6H_7N
4-methylpyridine
4-picoline; σ-Picoline; G-Picoline. Solvent in synthesis of pharmaceuticals, resins, dyestuffs, rubber accelerators, pesticides, waterproofing agents; laboratory reagent; manufacture of isoniazid; catalyst; curing agent. mp = 2°; bp = 145°; d = 0.9570; n_D^{20}= 1.5045; insoluble in H_2O, soluble in organic solvents; LD_{50} (rat orl) = 1290 mg/kg. *Lonza; Schweizerhall.*

23818 Pi-cone
A proprietary lithopone containing 15% titanium oxide, 25% zinc sulfide, and 60% precipitated barium sulfate; stated to have a much higher covering power than ordinary lithopone. No manufacturer.

23819 picrasmin
21293-20-9 5239
$C_{22}H_{28}O_6$
Isoquassin
Bitter principle from Jamaica quassia; isomer of quassin. mp = 222-225°; λ_m = 258 nm (ε 12500); {=la].

23820 picrate powder
The name given to explosive powders, in which the main constituent is the potassium or ammonium salt of picric acid.

23821 picric acid
88-89-1 7562 201-865-9
$C_6H_3N_3O_7$
2,4,6-Trinitrophenol
picronitric acid; nitroxanthic acid; carbazotic acid;. Explosive, used in matches, for etching copper and production of colored glass. mp = 121-123°; soluble in H_2O (1 g/78 ml), more soluble in organic solvents;.

23822 picric powder
An explosive consisting of ammonium picrate and potassium nitrate.

23823 picro-aniline blue
A stain used in microscopy. It is prepared by adding aniline blue to a saturated solution of picric acid in 92% alcohol until the liquid becomes deep blue-green in color.

23824 picrocarmine
A microscopic stain obtained by mixing 1 gram carmine in 10 ml water and 3 ml strong ammonia solution, and adding the mixture to 200 ml of a saturated solution of picric acid.

23825 picrocrocin
138-55-6 7563
$C_{16}H_{26}O_7$
(R)-4-(β-D-glucopyranosyloxy)-2,6,6-trimethyl-1-cyclohexene-1-carboxaldehyde
From the stigmas of *Crocus Sativus L. Iridaceae*. Important in growth control of the plant. mp = 154-156°; $[\alpha]_D^{20}$= -58° (c = 0.6); soluble in H_2O, EtOH, less soluble in organic solvents.

23826 picronigrosine
An alcoholic solution of picric acid and nigrosin; a microscopic stain.

23827 picro-sulfuric acid
A liquid made by adding to 100 volumes water, 2 volumes sulfuric acid, and about 0.25% picric acid; used in microscopy as a fixing agent.

23828 pictet crystals
$SO_2 \cdot xH_2O$
White crystals, formed when liquid sulfur dioxide evaporates.

23829 Pictet's fluid
Liquid carbon dioxide; used for freezing machines.

23830 Pictet's liquid
A mixture of liquid carbon dioxide and sulfur dioxide; used for producing low temperature.

23831 pictolin
A mixture of liquid carbon dioxide and sulfur dioxide.

23832 Pidolidone®
98-79-3 8185 202-700-3
L-pyroglutamic acid
PCA; cellular penetration vector for amino acids or mineral salts; used in aqueous phases of skin and hair care formulations. *UCIB.*

23833 Pielanase
Enzyme. *ABM Chemicals Ltd.* Name unverified.

23834 Pierrot metal
An alloy consisting mainly of zinc, with smaller amounts of copper, tin, antimony, and lead.

23835 Pif-Paf
Insecticide preparations. *The Wellcome Foundation Ltd.* Name unverified.

23836 pig iron
Crude form of cast iron; product of blast furnace ore reduction cast into ingots called pigs.

23837 pig lead
Lead is obtained from galena by heating in a reverberatory furnace with a silica flux, then heated with coke, and sometimes lime in a cupola furnace. The lead drawn off is called pig lead.

23838 Pigmentar
A proprietary trade name for standardized pine tar prepared for use as a rubber softener in rubber compounding. No manufacturer.

23839 pilasonite
Bi_3Te_2
A mineral.

23840 Piliogrip Adhesive System for Styructural Bonding
100% Reactive urethane structural adhesives designed for bonding thermosets, thermoplastics and metals; SMC to SMC for automotive and truck body assemblies; SMC to metal for automotive body assemblies. *Ashland Chemical Company.* Name unverified.

23841 Pilot
76578-14-8 8269
Quizalofop-ethyl
Selective herbicide for control of grasses in mustard, rape and beet crops. *Schering Agrochemicals Ltd.* Discontinued.

23842 Pilot
Rolling oils for aluminum. *Schering Agrochemicals Ltd.* Discontinued.

23843 Pilot SXS-40
1300-72-7 215-090-9
$C_8H_9NaO_3S$
Sodium xylene sulfonate
Xylenesulfonic Acid, Sodium, Salt; Dimethylbenzenesulfonic Acid, Sodium Salt; Sodium Dimethylbenzenesulfonate; Conco SXS; Cyclophil SXS30; Eltesol SX 30; Hydrotrope; Naxonate; Naxonate G; Stepanate X; Surco SXS; Ultrawet 40SX; Calsoft SXS 96; Alkatrope SX-40;Carsosulf SXS; Eltesol SX93; Reworyl NXS40; Richonate SXS; Witconate SXS; Sodium *m*-xylene Sulfonate. Coupling agent, hydrotrope, solubilizer, solvent, stabilizer; used in liquid cleaners, organic polymers and dyestuffs, petroleum industry, pulping, animal glues. mp = 27°; bp = 157°; soluble in H_2O (> 100 mg/ml). *Pilot.*

23844 Pimel®
Photosensitive polyimide. *Asahi Chem. Industry.*

23845 pimelite
A mineral. It is meerschaum (magensium trisilicate) containing nickel.

23846 pimple metal
A term used for a type of copper metal produced from the coarse metal obtained from sulfide ores which have been fused with an excess of copper oxide in their purification.

23847 pinachrom
p-Ethoxy-quinaldine-*p*-methoxy-quinoline-ethyl-cyanin-bromide
A red sensitizer for silver bromide plates.

23848 pinacolone
75-97-8 7594 200-920-4
$C_6H_{12}O$
3,3-Dimethyl-2-butanone
methyl-*t*-butyl-ketone; pinacoline; 3,3-dimethylbutan-2-one; *tert*-butyl methyl ketone; Pinacolin. Used in chemical synthesis, manufacture. mp = -53°; bp = 106°; d = 0.8010; n_D^{20}= 1.3964; soluble in H_2O (2.5 g/100 ml), more soluble in organic solvents; LD_{50} (rat orl) = 610 mg/kg.

23849 pinacyanol
A red sensitizer for silver bromide plates obtained by treating quinaldinium salts with formaldehyde followed by alkali.

23850 pinaflavol
A basic dye used as a green sensitizer in photography.

23851 pinakol
A pyrogallol photographic developer in which part of the alkali usually employed is replaced by sodium amino-acetate.

23852 Pinaverdol
A green sensitizer for silver bromide plates.

23853 pinchbeck
An alloy of from 83-93% copper and 6-17% zinc. A brass.

23854 Pincoffin
72-48-0 247 200-782-5
Commercial alizarin. No manufacturer.

23855 pine gum
Pine resin; white pine resin; Cypress pine resin. Names applied to Australian sandarac resin, obtained from *Callitris quadrivalis* and *C. calcarata.*

23856 pine oil
8002-09-3
Oils, pine; yarmor. Volatile oil obtained from distillation of the *Pinus* Spp; the name was originally applied to turpentine oils obtained from pine trees; the term is used in U.S. to designate turpentine obtained by distilling pine wood; it is also used for the lighter oils of pine tar, a refined rosin oil, and as an odorant, disinfectant; penetrant; wetting agent, presevative, reagent and fragrance. *Allchem Industries; Arizona; Hercules; Langley smith Ltd; Penta Mfg.*

23857 α-pinene
80-56-8 7599 232-077-3
$C_{10}H_{16}$
2-pinene
2,6,6-trimethylbicyclo(3.1.1)-2-hept-2-ene;)Terpene hydrocarbon. Solvent for protective coating, synthesis of camphene, pine oil, odorant, lube oil additives, flavoring, insecticides. *Aldrich; Arizona; Hercules; SCM Glidco Organics; Veitsiluoto Oy.*

23858 β-l-pinene
127-91-3 7600 204-872-5
$C_{10}H_{16}$
6,6-dimethyl-2-methylenebicyclo[3.1.1]heptane
Nopinene; Pseudopinene; Bicyclo[3.1.1]heptane, 6,6-dimethyl-2-methylene-; 2(10)-Pinene. Found in most essential oils; used as a chemical intermediate. mp = -61°; bp = 162-163°; d^{15} = 0.874; n_D^{15}= 1.4872; $[\alpha]_D$ = -22°.

23859 β-pinene
127-91-3 7600 204-872-5
$C_{10}H_{16}$
6,6-dimethyl-2-methylenebicyclo [3.1.1] heptane
nopinene; Terpene hydrocarbon. Polyterpene resins; intermediate for perfumes and flavorings. *Arizona; Penta Mfg.; SCM Glidco Organics.*

23860 Pineotrene K
Detergent for textile scouring. *Reilly-Whiteman.*

23861 piney tallow
Malabar tallow. An edible fat obtained from the seeds of *Vateria indica*, of East Indies.

23862 pinguin
$C_{10}H_{16}O$

Alantol
Obtained from the roots of *Inula elecampane.*

23863 pink salt
$SnCl_4 \cdot 2NH_4Cl$
Ammonium-stannic-chloride
Formerly used as a mordant for dyes.

23864 Pinnacle
Mixture of imazamethabenz-methyl and isoproturon; used for control of weed grasses in winter cereals. *Cyanamid of Great Britain Ltd.*

23865 pinolin
A name for rosin oil; the first distillate from rosin, boiling at from 78-250°C.

23866 Pinwire brass
An alloy of from 66-73% copper and 27-34% zinc.

23867 Pioloform
Polyvinyl butyral resins. *Wacker Chemicals Ltd.*

23868 Pioneer
Single-stage centrifugal glandless circulator; for domestic and industrial central heating circulator. *Sihi Pumps (UK) Ltd.*

23869 Pioneer
12-20-20 Compound fertilizer. *Kemira Ince Ltd.* Discontinued.

23870 Pioneer alloy
An alloy of 20% copper, 38% nickel, 4% silicon, 3% molybdenum, 2% tungsten, and the remainder iron.

23871 pip pip
An abbreviated name for piperidinium-pentamethylene-dithiocarbamate. An accelerator for rubber vulcanization.

23872 pipeclay
12141-46-7 377 235-253-8
Aluminum silicate
An abrasive.

23873 piperazine
110-85-0 7617 203-808-3
$C_4H_{10}N_2$
piperazine anhydrous
diethylene diamine; pyrazine hexahydride; piperazidine. Corrosion inhibitor, anthelmintic, insecticide, accelerator for curing polychloroprene. *Allchem Industries; BASF; Janssen Chimica.*

23874 piperidine
110-89-4 7621 203-813-0
$C_5H_{11}N$
hexahydropyridine
pentamethyleneamine; azacyclohexane; cyclopentimine; cypentil; hexazane. Solvent and intermediate, curing agent for rubber and epoxy resins, catalyst for condensation reactions, ingredient in oils and fuels, complexing agent. mp = -13°; bp = 106°; d = 0.861; n_D^{20} = 1.4534; LD_{50} (rat orl) = 400 mg/kg. *Aldrich; Janssen Chimica; Nepera; Schweizerhall.*

23875 piperonyl butoxide
51-03-6 7629 200-076-7
$C_{19}H_{30}O_5$
4,5-methylenedioxy-2-propylbenzyldiethylene glycol butyl ether
Butacide; 6-Propylpiperonyl Butyl Diethylene Glycol Ether; Pyrenone 606; NIA 5273; butocide; PB; Pybuthrin; Synpren-Fish; Vex; Obilique; Scourge; Alleviate; ENT 14250;. Synergist in insecticides, especially pyethrins and rotenone; used in combinations with pyrethrins (Derringer, Duracide, Grovex, Prentox; Scourge). $bp_{1.0}$ = 180°; d = 1.04-1.07; n_D^{20} = 1.50; insoluble in H_2O, soluble in organic solvents; LD_{50} (rat orl) = 7500 mg/kg. *Burlington Bio-Medical; Wellcome Foundation Ltd.*

23876 Pipricide
Veterinary anthelmintic *The Wellcome Foundation Ltd.* Name unverified.

23877 Piral
87-66-1 8184 201-762-9
$C_6H_3(OH)_3$
Pyrogallol
Chemical intermediate.

23878 Piria's acid
84-86-6 6490 201-567-9
$C_{10}H_9NO_3S$
α-Naphthylamine-4-sulfonic acid
4-aminonaphthalene-1-sulfonic acid; naphthionic acid. Intermediate in manufacture of dyestuffs. The sodium salt is used as a hemostatic. d_4^{25} = 1.673; decomposes without melting; soluble in H_2O (0.28 g/l), sparingly soluble in organic solvents;.

23879 pirimicarb
23103-98-2 7651 245-430-1
$C_{11}H_{18}N_4O_2$
dimethylcarbamic acid 2-(dimethylamino)-5,6-dimethyl-4-pyrimidinyl ester
Pirimor; Abol; Aficida; Aphox; Fernos; Rapid; 2-(dimethylamino)-5,6-dimethyl-4-pyrimidinyl dimethylcarbamate; PP-062; ENT-27766. Insecticide.

23880 Pirimor
23103-98-2 7651 245-430-1
Wettable powder containing 50% w/w pirimicarb; for control of aphids. *ICI Chem. & Polymers Ltd.*

23881 Piror
Biocide. *Union Carbide.*

23882 Pirsch-Baudoin's alloy
An alloy of 71% copper, 16.5% nickel, 1.75% cobalt oxide, 2.5% tin, and 7% zinc.

23883 Pistol
13684-63-4 7384 237-199-0
Formulated phenmedipham, a herbicide. *Rhône-Poulenc UK.*

23884 Pistol 400
13684-63-4 7384 237-199-0
Emulsifiable concentrate containing 118 g/l phenmedipham; for weed control for beet crops. *Quadrangle Agrochemicals.*

23885 pitch barm
A cement made from casein, water-glass, and caustic lime.

23886 Pit-ite No. 2
An explosive consisting of 23-25% nitroglycerin, 28-31% potassium nitrate, 33-36% wood meal, and 7-9% ammonium oxalate.

23887 Pitralon
A wood tar derivative; used in Germany as an antiseptic.

23888 Pittabs
7778-54-3 1717 231-908-7
$CaCl_2O$
Calcium hypochlorite
Oxidizer and bleach. *PPG Industries.*

23889 pittaccal
$C_{25}H_{26}O_9$
Eupittonic acid.

23890 Pittclor
7778-54-3 1717 231-908-7
Calcium hypochlorite
Oxidizer and bleach. *PPG Industries.*

23891 Pitteliene
A mixture of coal tar and oil; used as an insecticide.

23892 pittinite
A mineral. It is a variety of gummite.

23893 pittylen
A mixture of pine tar with formaldehyde; used in skin diseases.

23894 Pivaloxycyclene
Tricyclodecenyl dimethyl propanoate
Quest Int'l. UK Ltd.

23895 Pivarose
2-Phenylethyl pivalate
Quest Int'l. UK Ltd.

23896 Pivofax
A dried yeast.

23897 pix solubilis
soluble pitch
A soluble modification of the tar obtained by sulfonating the tar obtained from peat.

23898 Pix® ULV
24307-26-4 5903 246-147-6
Mepiquat chloride
Plant growth regulator for reduction of undesired vegetative growth of cotton, better boil retention, earlier maturity; improves yield and market quality of garlic and onions *BASF AG.*

23899 Pixol
A form of wood tar soluble in water, made from tar and soap; used as a disinfectant.

23900 Pixtonet
A trade name for a slate substitute; used as an electrical insulator. No manufacturer.

23901 PJ1 Chain Lube, Blue Label
Keeps o-rings soft and pliable to maintain chain's seal against wear-causing dirt and moisture. *PJ1 Corporation.* Unverified.

23902 PJ1 Octane Plus
Together with racing fuel additives, it boosts the power of lower octane gasolines with complete safety; gas stabilizers keep the gas in tank from going stale and prevent gum and varnish build-up in the carburetors and fuel system. *PJ1 Corporation.* Unverified.

23903 PJ1 Super Cleaner
Cleans and degreases all metal parts, also disperses water and works on all electrical parts that need to be cleaned or dried. *PJ1 Corporation.* Unverified.

23904 PK-10GF/000
PEEK, 10% glass fiber-reinforced. *Compounding Tech.* Name unverified.
23905 PK-20CF/000
PEEK, 20% carbon fiber-reinforced. *Compounding Tech.* Name unverified.
23906 PKWF
Printing ink distillates and oils. *Haltermann Ltd.*
23907 Placentaliquid Oil-Soluble
Extract of unborn bovine placentas; products for aging skin. *Dr. Kurt Richter; Henkel/Cospha.*
23908 Placet alloy
An alloy of 60% nickel, 20% iron, 15% chromium, and 5% manganese; an electrically resistant alloy.
23909 Planavin
4726-14-1 6662 225-219-0
A trademark for a nitrogenous weed-killer. *Shell Chemie GmbH.* Name unverified.
23910 Planell Oil
Squalene, squalane, glycolipids, phytosterol, tocopherol. Natural plant lipid extract for use as cosmetic emollient; maintains normal skin function by augmenting skin's own naturally occurring lipid membrane. *Brooks Industries.*
23911 Planet
Nonionic wetting agent containing 85% alkyl polyglycol ether and fatty acid; for use in agrochemical sprays. *Ideal Manufacturing Ltd.*
23912 Planetol
Photographic preparations. *May & Baker Ltd.* Name unverified.
23913 Planofix
86-87-3 6458 201-705-8
Plant growth regulator, use as a pre-harvest fruit drop inhibitor. *May & Baker Ltd.* Name unverified.
23914 Planotox
1929-73-3 217-680-1
$C_{14}H_{18}Cl_2O_4$
(2,4-dichlorophenoxy)acetic acid 2-butoxyethyl ester
Planotox; 2,4-D, butoxyethanol ester; 2,4-D, 2-butoxyethyl ester; Aqua-kleen; BEE; Bladex-B; Brush killer 64; Butoxyethanol ester of 2,4-D; Weedone LV 4. Selective weedkiller. *May & Baker Ltd.* Name unverified.
23915 plant indican
487-60-5 4976
$C_{14}H_{17}NO_6$
1H-indol-3-yl-β-D-glucopyranoside
indican. Found in leaves of *Indigofera tinctoria.* mp = 57-58°; $[\alpha]^{18}_{546}$ = -66°; soluble in H_2O, organic solvents;.
23916 Plantaren 1200
C12-16 alkyl polyglycoside
Mild surfactant, foamer, cleanser for low-irritation personal care products, hair shampoos, bath and shower gels, foam baths, facial cleansers; biodegradable. *Henkel/Cospha; Henkel Canada.*
23917 Plantaren 2000
Decyl polyglucose
Mild surfactant, foamer, cleanser for low-irritation personal care products, hair shampoos, bath and shower gels, foam baths, facial cleansers; biodegradable. *Henkel/Cospha; Henkel Canada.*
23918 Plantaren 600 CS UP
C12-14 fatty alcohol glycoside
Surfactant for dishwashing agents, laundry detergents, cleaners. *Henkel KGaA.* Discontinued.
23919 Plantaren CG 60
C8-10 fatty alcohol glycoside
Solubilizer, wetting agent for alkali bottle washing, neutral and acidic cleaners. *Henkel KGaA. Discontinued.*
23920 Plantex®
Fiber. *DuPont UK.*
23921 Plantvax
5259-88-1 226-066-2
Oxycarboxin
A systemic fungicide. *ICI Agrochemicals.*
23922 Plantvax 20
5259-88-1 226-066-2
Oxycarboxin
A systemic fungicide for the control of rust in wheat. *Uniroyal Chemical Ltd.*
23923 Plantvax 75
5259-88-1 226-066-2
Oxycarboxin
A systemic fungicide for the control of rust in ornamental plants. *Fargro Ltd.*
23924 Plasadd™
Granulated concentrate designed as an additive to improve the physical properties of polyolefins. *Cabot Plastics Ltd.*
23925 Plasblak® EV 1755
EVA masterbatch with 40% SRF carbon black; black masterbatch offering adequate weathering properties, good opacity, easy processing for molding and extrusion applications; compatible with LDPE, HDPE, PP, PS. *Cabot Plastics Ltd.*
23926 Plasblak® EV 3524
50% Carbon black (average particle size 27 nm) in LDPE copolymer carrier; black masterbatch for molding and extrusion of articles for packaging (boxes, lids), electronic equipment cases (cassettes, VCRs), radios, TVS, domestic appliances; designed for coloring PS, ABS, Nylon 6 and PET. *Cabot Plastics Ltd.* Name unverified.
23927 Plasblak® masterbatches
Granulated concentrates of carbon black in plastics; used for coloring of thermoplastic resins and for imparting resistance to weathering degradation. *Cabot Plastics Ltd.*
23928 Plasblak® PE 1851
50% SRF carbon black (average particle size 60 nm) in LDPE carrier with calcium carbonate extender as antiblocking agent; black masterbatch for film, molding, and extrusion applications; compatible with LDPE, HDPE, PP, and ethylene copolymers. *Cabot Plastics Ltd.*
23929 Plasbumin
Human albumin; blood volume supporter. *Cutter Laboratories, Miles Laboratories Inc.* Name unverified.
23930 Plas-Chek
For plasticizers for vinyl halide resins, in Int Class 1. *Ferro.* Discontinued.
23931 Plascoat Plasinter
Polyethylene coating powders; used for coating domestic wirework, display stands, tools, clips, WRC approved, insulation of electrical components. *Plascoat Systems Ltd.*
23932 Plascoat PPA
Polyolefin alloy; thermoplastic coating powder; used for coating wire dishwasher baskets, process valves, pipework and units, electrical switchgear, seat frames and the lining of fire extinguishers. *Plascoat Systems Ltd.*
23933 Plascoat PPA 31 series
Resistant to aqueous chemicals at temperatures up to 100°C. *Plastic Coatings Ltd.* Name unverified.
23934 Plasdeg™
Granulated concentrate designed to impart controlled degradability to polyolefin-based products. *Cabot Plastics Ltd.*
23935 Plasdone®
9003-39-8 7879
Povidone. Pharmaceutic aid (dispersing and suspending agent). *ISP.*
23936 Plasdone® C
9003-39-8 7879
A range of pyrogen-free polyvinylpyrrolidones; solubilizer, stabilizer, protective colloid for veterinary pharmaceuticals which minimizes toxic side effects and reduces irritation at site of infection. *ISP.*
23937 Plasdone® K
9003-39-8 7879
A range of polyvinylpyrrolidones; tablet binder and coating agent, cohesive agent, stabilizer and protective colloid, detoxicant for many poisons and irritants, drug vehicle and retardant, film forming agent in medicinal aerosols. *ISP.*
23938 Plasgon
A proprietary trade name for plastic gasket and joint cement. No manufacturer.
23939 Plasgrey®
Granulated concentrates of carbon black in plastics; used in small amounts as colorants for naturally uncolored plastics. *Cabot Plastics Ltd.*
23940 Plaslube® AC-80/TF/10
Acetal copolymer, 10% PTFE; internally lubricated engineering thermoplastic. *Akzo Engineering Plastics.* Name unverified.
23941 Plaslube® G-1/30/SI/2
32131-17-2
Nylon 6/6, 2% silicone; internally lubricated engineering thermoplastic. *Akzo Engineering Plastics.* Name unverified.
23942 Plaslube® G-3/40/MS/5
25038-54-4 6832
Nylon 6, 5% molybdenum disulfide; internally lubricated engineering thermoplastic. *Akzo Engineering Plastics.* Name unverified.
23943 Plaslube® G-50/20/TF/10
PC, 10% PTFE; internally lubricated engineering thermoplastic. *Akzo Engineering Plastics.* Name unverified.

23944 Plaslube® J-1/30/MS/5
32131-17-2
Nylon 6/6, 5% molybdenum disulfide; internally lubricated engineering thermoplastic. *Akzo Engineering Plastics.* Name unverified.

23945 Plaslube® J-1/33/TF/13/SI/2
32131-17-2
Nylon 6/6, 13% PTFE, 2% silicone; internally lubricated engineering thermoplastic. *Akzo Engineering Plastics.* Name unverified.

23946 Plaslube® J-1/CF/15/TF/20
32131-17-2
Nylon 6/6, 15% PAN carbon fiber, 20% PTFE; internally lubricated engineering thermoplastic. *Akzo Engineering Plastics.* Name unverified.

23947 Plaslube® J-1300/30/TF/15
9016-75-5
PPS, 15% PTFE; internally lubricated engineering thermoplastic. *Akzo Engineering Plastics.* Name unverified.

23948 Plaslube® J-1300/CF/20/MS/10/TF/15
9016-75-5
PPS, 20% PAN carbon fiber, 10% molybdenum disulfide, 15% PTFE; internally lubricated engineering thermoplastic. *Akzo Engineering Plastics.* Name unverified.

23949 Plaslube® J-3/30/MS/5
25038-54-4 6832
Nylon 6, 5% molybdenum disulfide; internally lubricated engineering thermoplastic. *Akzo Engineering Plastics.* Name unverified.

23950 Plaslube® J-4/30/TF/15
Nylon 6/12, 15% PTFE; internally lubricated engineering thermoplastic. *Akzo Engineering Plastics.* Name unverified.

23951 Plaslube® J-4/CF/30/TF/10
Nylon 6/12, 30% PAN carbon fiber, 10% PTFE; internally lubricated engineering thermoplastic. *Akzo Engineering Plastics.* Name unverified.

23952 Plaslube® J-50/20/TF/10
PC, 10% PTFE; internally lubricated engineering thermoplastic. *Akzo Engineering Plastics.* Name unverified.

23953 Plaslube® J-50/30/SI/2
PC, 2% silicone; internally lubricated engineering thermoplastic. *Akzo Engineering Plastics.* Name unverified.

23954 Plaslube® J-50/30/TF/10
PC, 10% PTFE; internally lubricated engineering thermoplastic. *Akzo Engineering Plastics.* Name unverified.

23955 Plaslube® J-50/CF/10/TF/15/FR
PC, 10% PAN carbon fiber, 15% PTFE; flame retarded, internally lubricated engineering thermoplastic. *Akzo Engineering Plastics.* Name unverified.

23956 Plaslube® J-77/30/TF/15
32131-17-2
Nylon 6/6 resin, 30% chopped fiber reinforcement, 15% PTFE lubricant; *Akzo.* Name unverified.

23957 Plaslube® J-80/20/TF/15
Acetal copolymer, 15% PTFE; internally lubricated engineering thermoplastic. *Akzo Engineering Plastics.* Name unverified.

23958 Plaslube® J-80/CF/10/TF/10
Acetal copolymer, 10% PAN carbon fiber, 10% PTFE; internally lubricated engineering thermoplastic. *Akzo Engineering Plastics.* Name unverified.

23959 Plaslube® NY-1/MS/5/TF/30
32131-17-2
Nylon 6/6, 5% molybdenum sulfide, 30% PTFE; internally lubricated engineering thermoplastic. *Akzo Engineering Plastics.* Name unverified.

23960 Plaslube® NY-1/SI/5
32131-17-2
Nylon 6/6, 5% silicone-lubricated; internally lubricated engineering thermoplastic. *Akzo Engineering Plastics.* Name unverified.

23961 Plaslube® NY-1/TF/10
32131-17-2
Nylon 6/6, 10% PTFE; internally lubricated engineering thermoplastic. *Akzo engineering Plastics.* Name unverified.

23962 Plaslube® NY-4/TF/10
Nylon 6/12, 10% PTFE; internally lubricated engineering thermoplastic. *Akzo Engineering Plastics.* Name unverified.

23963 Plaslube® PC-50/TF/10
PC, 10% PTFE; internally lubricated engineering thermoplastic. *Akzo Engineering Plastics.* Name unverified.

23964 Plasmosan
9003-39-8 7879
A proprietary trade name for povidone. No manufacturer.

23965 Plastacele
A proprietary trade name for a plasticized cellulose acetate compound. No manufacturer.

23966 Plastamid
Polyamides. *Croda Resins Ltd.* Discontinued.

23967 Plastammone
An explosive containing ammonium nitrate, glycerin, mononitro-toluene, and nitro-semicellulose.

23968 Plastamol
Plasticizers for PVC. *BASF plc.* Name unverified.

23969 Plastazote
Crosslinked foamed polyethylene. *BXL Plastics Ltd.* Name unverified.

23970 Plastech™
A compound based on a modified thermoplastic resin; designed for technical applications. *Cabot Plastics Ltd.*

23971 Plastech™ EP 8126
9003-07-0 7741
PP copolymer, rubber-modified; rubber masterbatch modifying impact resistance. *Cabot Plastics Ltd.*

23972 Plastech™ PP 3344
9003-07-0 7741
40% mineral-filled PP homopolymer compound; designed for manufacturing of injection molded parts for automotive industry, e.g., dashboards, door moldings, lamp housings, under-the-hood applications such as fans and fan housings. *Cabot Plastics Ltd.*

23973 plaster cement
Cements made from gypsum.

23974 plaster of Paris
$2(CaSO_4)\cdot H_2O$
A partially dehydrated gypsum, It is made from gypsum by heating the latter from 212-400°F, when 3 parts of the water of crystallization is released.

23975 Plasteryl
An explosive. It is a mixture of 99.5% trinitrotoluene and 0.5% resin.

23976 Plast-E-Tint
Pigment dispersions; for coloration of plastisols and organisols. *Pacific Dispersions Inc.* Name unverified.

23977 Plasthall®
Series of monomeric and specialty plasticizers; used for rubber and flexible vinyl. *C P Hall.*

23978 Plasthall® 100
Isooctyl tallate
Plasticizer. *C.P. Hall.*

23979 Plasthall® 200
117-83-9 204-213-1
$C_{20}H_{30}O_6$
bis(2-butoxyethyl) phthalate;
1,2-Benzenedicarboxylic acid, bis(2-butoxyethyl) ester; Dibutoxy ethyl phthalate; Dibutoxyethyl phthalate. Plasticizer. *C.P. Hall.*

23980 Plasthall® 201
Dibutoxyethyl glutarate
Plasticizer. *C.P. Hall.*

23981 Plasthall® 203
Dibutoxyethyl adipate
Plasticizer. *C.P. Hall.*

23982 Plasthall® 205
Dibutoxyethyl azelate
Plasticizer. *C.P. Hall.*

23983 Plasthall® 207
Dibutoxyethyl sebacate
Plasticizer. *C.P. Hall.*

23984 Plasthall® 220
Dibutoxyethoxyethyl phthalate
Plasticizer. *C.P. Hall.*

23985 Plasthall® 224
Dibutoxyethoxyethyl glutarate
Plasticizer. *C.P. Hall.*

23986 Plasthall® 226
Dibutoxyethoxyethyl adipate
Plasticizer. *C.P. Hall.*

23987 Plasthall® 325
Butoxyethyl oleate
Plasticizer. *C.P. Hall.*

23988 Plasthall® 4141
PEG-3 caprate-caprylate
Plasticizer for rubber goods; lubricant for aluminum can industry. *C.P. Hall.*

23989 Plasthall® 503
142-77-8 205-559-6
$C_{22}H_{42}O_2$
Butyl oleate
9-Octadecenoic acid (Z)-, butyl ester; Butyl oleate; (Z)-9-Octadecenoic acid

butyl ester. Textile surface finisher, softener, thread lubricant and antistat; plasticizer. *C.P. Hall.*

23990 Plasthall® 6-10P
Plasticizer for PVC. *C.P. Hall.*

23991 Plasthall® 8-10 TM-E
n-Octyl
n-decyl trimellitate. Plasticizer for polychloroprene compounds. *C.P. Hall.*

23992 Plasthall® 83SS
Dibutoxyethoxyethyl sebacate substitute; plasticizer for rubber industry. *C.P. Hall.*

23993 Plasthall® BSA
3622-84-2 222-823-6
N,N-butyl benzene sulfonamide
Plasticizer for emulsion adhesives, packaging, caulk, printing ink, surface coatings. *C.P. Hall.*

23994 Plasthall® DBS
109-43-3 203-672-5
$C_{18}H_{34}O_4$
Dibutyl sebacate
n-Butyl Sebacate; decanedioic acid dibutyl ester; Dibutyl sebacate; Butyl sebacate; Dibutyl decanedioate. Plasticizer. Used as an excipient in various pharmaceutical coating formulations. mp = -12°; bp_3 = 180°; d = 0.9400; n_D^{25}= 1.429 - 1.441. *C.P. Hall.*

23995 Plasthall® DBZZ
Dibenzyl azelate
Plasticizer. *C.P. Hall.*

23996 Plasthall® DIBA
141-04-8 205-450-3
$C_{14}H_{26}O_4$
Diisobutyl adipate
Hexanedioic acid, bis(2-methylpropyl) ester. Plasticizer. *C.P. Hall.*

23997 Plasthall® DIBZ
Diisobutyl azelate
Plasticizer. *C.P. Hall.*

23998 Plasthall® DIDA
27178-16-1 248-299-9
Diisodecyl adipate
Plasticizer. *C.P. Hall.*

23999 Plasthall® DIDG
Diisodecyl glutarate
Plasticizer; lubricant additive. *C.P. Hall.*

24000 Plasthall® DIOA
1330-86-5 215-553-5
$C_{22}H_{42}O_4$
Diisooctyl adipate
Plasticizer. *C.P. Hall.*

24001 Plasthall® DIODD
Diisooctyl dodecanedioate
Lubricant, additive; plasticizer. *C.P. Hall.*

24002 Plasthall® DOA
103-23-1 203-090-1
$C_{22}H_{42}O_4$
Dioctyl adipate
BEHA; DEHA; DOA; Dioctyl Adipate; Octyl Adipate; Hexanedioic Acid Bis(2-ethylhexyl) Ester; Adipic Acid Bis (2-ethylhexyl) Ester; Adipol 2EH; Bisoflex DOA; Effemoll DOA; Effomoll DOA; Ergoplast Addo; Flexol A 26; Flexol Plasticizer 10-A; Flexol Plasticizer A-26; Kemester 5652; Kodaflex DOA; Mollan S; Monoplex DOA; Plastomoll DOA; PX-238; Reomol DOA; Rucoflex Plasticizer DOA; Sicol 250; Staflex DOA; Truflex DOA; Uniflex DOA; Vestinol OA; Wickenol 158; Witamol 320; Bis(2-ethylhexyl)hexanedioate. Plasticizer. mp = -67°; bp = 417°; d = 0.9240. *C.P. Hall.*

24003 Plasthall® DODD
Dioctyl dodecanedioate dioate
Lubricant additive; useful as textile surface finishes, softeners, thread lubricants and/or antistats; plasticizer. *C.P. Hall.*

24004 Plasthall® DOP
117-81-7 1291 204-211-0
di-2-ethylhexyl phthalate
C.P. Hall.

24005 Plasthall® DOS
122-62-3 1292 204-558-8
$C_{26}H_{50}O_4$
Dioctyl sebacate
Bis(2-Ethylhexyl) Ester Decanedionic Acid; Octyl Sebacate; Bis(2-Ethylhexyl) Ester Sebacic Acid; bis(2-ethylhexyl) sebacate; bisoflex DOS; DOS; 2-ethylhexyl sebacate; 1-hexanol, 2-ethyl-, sebacate; Monoplex DOS; Octoil S; PX 438; Staflex DOS; Plexol 201; bis(2-ethylhexyl) decanedioate; Edenol 888; Ergoplast SNO; Reolube DOS; DEHS. Plasticizer. mp = -67°; bp = 248°; d = 0.914; insoluble in H_2O, soluble in organic solvents. *C.P. Hall.*

24006 Plasthall® DOSS
Dioctyl sebacate substitute; plasticizer for rubber industry. *C.P. Hall.*

24007 Plasthall® DOZ
103-24-2 203-091-7
$C_{25}H_{48}O_4$
Dioctyl azelate
dioctyl azelate; Nonanedioic acid, bis(2-ethylhexyl) ester; Di(ethylhexyl)azelate. Plasticizer. *C.P. Hall.*

24008 Plasthall® ESO
8013-07-8 232-391-0
Epoxidized soybean oil; plasticizer. *C.P. Hall.*

24009 Plasthall® HA7A
Polyester adipate; plasticizer. *C.P. Hall.*

24010 Plasthall® NODA
n-octyl, n-decyl adipate. Plasticizer. *C.P. Hall.*

24011 Plasthall® P-550
Polyester glutarate; plasticizer for PVC applications; flexibilizing, permanent, nonmigrating; adhesive for film backing and varieties of tape. *C.P. Hall.*

24012 Plasthall® P-7035
Polyester glutarate; plasticizer for flexible PVC. *C.P. Hall.*

24013 Plasthall® P-7068
Polyester phthalate; plasticizer. *C.P. Hall.*

24014 Plasthall® R-9
Octyl tallate
Transfer aid on correctable ribbon; penetration and tack agent for computer ribbons, carbon paper; plasticizer. *C.P. Hall.*

24015 Plasthall® TIOTM
53894-23-8 258-847-9
Triisooctyl trimellitate
Plasticizer. *C.P. Hall.*

24016 Plasthall® TOTM
3319-31-1 222-020-0
Tricotyl trimellitate
Plasticizer. *C.P. Hall.*

24017 Plastibase
Ointment base containing polyethylene and mineral oil. *Bristol-Myers Squibb Pharmaceuticals Ltd.*

24018 Plastic
Solvent dyestuffs; for coloration of plastics. *Holliday Dyes & Chemicals Ltd.* Discontinued.

24019 Plastic A
614-33-5 210-379-6
A proprietary trade name for glyceryl tribenzoate. No manufacturer.

24020 plastic bronze
An alloy of 64% copper, 30% lead, 5% tin, and 1% nickel.

24021 Plastic Magnet
Thermosetting material; for molded products. *Sumitomo Bakelite Co Ltd.* Name unverified.

24022 plastic metal
An alloy of 80.5% tin, 9.5% copper, 8.6% antimony, and 1.4% iron.

24023 Plastic Steel
Paste containing 80% powdered steel and 20% epoxy resin capable of being hardened. *ITW Devcon.*

24024 plastic sulfur
Prepared by heating sulfur to 225°C; used to a limited extent as a material for preparing molds for electrotyping.

24025 plastic wood
Plastic plant product. A material prepared by cooking vegetable matter with neutral salt solution followed by mechanical treatment to break down intercellular binding material. The name is also used to describe wood cellulose in solution with certain added solvents such as ether or acetone; used for filling holes in wood or other materials.

24026 Plastic X
78-30-8 9893 201-103-5
$C_{21}H_{21}O_4P$
Phosphoric acid
tris(2-methylphenyl) ester
Tricresyl phosphate; tricresylphosphate; *o*-trioyl phosphate; TCP; TOCP; Phosphoric acid tris(2-methylphenyl) ester; *o*-cresyl phosphate; Phosflex 179-C; phosphoric acid, tri-*o*-cresyl ether; TOFK; o-tolyl phosphate; TOTP; tris(*o*-methylphenyl) phosphate; tris(*o*-tolyl) phosphate; tri-*o*-tolyl ester phosphoric acid; Triorthocresyl phosphate. A proprietary plasticizer with specific gravity from 1.177-1.18. Generally used in lacquers and varnishes. mp = 25°; bp_{20} = 265°; d = 1.1955; insoluble in H_2O, slightly soluble in organic solvents;. No manufacturer.

24027 Plasticalk
A proprietary trade name for a plastic resin used as an adhesive and filler. No manufacturer.

24028 Plasticede
A proprietary trade name for a plasticizer for clay; it contains tannins and lignin. No manufacturer.

24029 Plasticizer 13
The ester of *p*-hydroxybenzoic acid and 2-ethyl hexanol; a plasticizer for polyamides. *BASF plc.*

24030 Plasticizer 28P
117-81-7 1291 204-211-0
Dioctyl phthalate
A vinyl plasticizer. No manufacturer.

24031 Plasticizer 9
Glycerol ether alcohol; plasticizer for naphtharesistance and benzene-resistance nitro lacquers. *BASF AG.*

24032 Plasticizer CEL
Phthalic acid glycol ester; plasticizer for cellulose acetate coatings, nitrocellulose lacquers; low volatility, good light stability; also as a desensitizer for peroxides. *Bayer AG.*

24033 Plasticizer E
A proprietary trade name for a chlorinated paraffin plasticizer. No manufacturer.

24034 Plasticizer REO
Mixture of paraffinic/naphthenic process oils and petroleum sulfonates; plasticizer, process aid for synthetic and natural rubber compounds; peptizer; reclaiming agent; improves filler dispersion in dry compounds. *Akrochem.*

24035 Plasticote
Vinyl/resin lacquer
Used for decoration of vinyl articles. *W J Ruscoe Co.* Discontinued.

24036 Plastifix
Polychloroprene
Bayer AG.

24037 Plastigel®
Chemical thickeners for polyester SMC, BMC, TMC. *Plasticolors.*

24038 Plastigen® G
Carbamide resin; very lightfast, nonhydrolyzable resin for nitrocellulose and chlorinated rubber finishes. *BASF AG.*

24039 Plastikon
A proprietary trade name for a rubber putty for filling and adhesive purposes. No manufacturer.

24040 Plastilit® 3060
Polypropylene glycol alkylphenol ether
Plasticizer for polymer dispersions in the paint, construction, chemical, adhesive and sealant industries. *BASF AG.*

24041 Plastisorb
UV and weather-resistant colorants and additives for polymer systems. *Plasticolors.*

24042 Plastisperse®
Low dusting, high loaded, dry concentrates for thermoplastic elastomers, bulk molding compounds. *Plasticolors.*

24043 Plastite
A name applied to a vulcanite, a hardened rubber.

24044 Plastitoy
Vinyl/resin lacquer; used for decoration of vinyl toys. *W J Ruscoe Co.* Discontinued.

24045 Plastitube
A proprietary trade name for cellulose acetate tubing. No manufacturer.

24046 Plastodent
Dental wax. *Degussa Ltd.*

24047 Plastogen
A mixture of an oil-soluble sulfonic acid of high molecular weight with a paraffin oil; an effective plasticizer in all elastomers; used in sponge and low durometer stocks. *King Industries.*

24048 Plastokyd
A range of alkyds for industrial paints and inks. *Croda Resins Ltd.*

24049 Plastokyd SC
A range of alkyds and polyesters modified with silicone for heat resistance and improved exterior exposure. *Croda Resins Ltd.*

24050 Plastol®
Low-viscosity polymer plasticizers for PVC; for products which must have moderate resistance to chemicals. *BASF AG.*

24051 Plastolin I
140-11-4 1158 205-399-7
$C_9H_{10}O_2$
Acetic acid phenylmethyl ester
phenyl methyl acetate; α-acetoxytoluene; (acetoxymethyl)benzene. A proprietary solvent; benzyl acetate. A natural constituent of several essential oils and flower absolutes extracted from jasmine, hyacinth, gardenia, tuberose, ylang-ylang,cananga, and neroli. Commercial benzyl acetate, a liquid prepared synthetically from benzyl chloride, acetic acid, and triethylamine is used primarily as a component of perfumes for soaps and as a flavoring ingredient. mp = -51°; bp = 206°; d = 1.0400; n_D^{20} = 1.5006; insoluble in H_2O, soluble in organic solvents;. No manufacturer.

24052 Plastomag® 170
1309-48-4 5713 215-171-9
Oil-dispersed magnesium oxide; chemical thickener for polyester resins; anticaking agent; used in synthetic rubber compounding, adhesives, fuel oil additives, and as acid acceptor for specialty plastics. *Morton Int'l./Plastics Additives.*

24053 Plastomenite
A German explosive powder made by incorporating 1 part nitro-lignin with 5 parts fused dinitrotoluene, and granulating. it may contain barium nitrate.

24054 Plastomoll®
Adipic acid esters; for low-temperature resistance and light-stable nitro lacquers; plasticizers for plasticized PVC products with low brittle temperatures. *BASF AG; BASF plc.*

24055 Plastomoll® 34
An aliphatic acid ester mixture; vinyl plasticizer. *BASF plc.* Name unverified.

24056 Plastomoll® BMB
3622-84-2 222-823-6
$C_{10}H_{15}NO_2S$
N-butylbenzene sulfonamide
N-n-butylbenzenesulfonamide; 4-n-Butylbenzenesulfonamide. A plasticizer for polymides. *BASF plc.* Name unverified.

24057 Plastomoll® DMA
Adipic ester of hydrogenated higher cyclic alcohols; vinyl plasticizer. *BASF plc.* Name unverified.

24058 Plastomoll® DOA
103-23-1 203-090-1
Di-2-ethylhexyl adipate
PVC plasticizer. *BASF plc.* Name unverified.

24059 Plastomoll® NA
33703-08-1 251-646-7
Diisonoyl adipate
PVC plasticizer. *BASF plc.* Name unverified.

24060 Plastomoll® TAH
Thiodibutyric acid ester of a synthetic octyl alcohol; a vinyl plasticizer. *BASF plc.* Name unverified.

24061 Plastomoll® WH
Ester of an aliphatic dicarboxylic acid; vinyl plasticizer. *BASF plc.* Name unverified.

24062 Plastone
Methylene diphenyldiamine mixed with a small amount of stearic acid or with naphthalene or naphthalene oil; a proprietary rubber vulcanization accelerator. No manufacturer.

24063 Plastone A
A proprietary trade name for a phenolic molding compound containing cotton seed hull. No manufacturer.

24064 Plastone B
A proprietary trade name for an inorganic and phenolic molding compound. No manufacturer.

24065 Plastopal®
Urea-formaldehyde resins, butanol or methanol etherified; binders for coatings and paints. *BASF AG.*

24066 Plastopal® 11
A modified urea-formaldehyde condensation product in the form of a 65% butanol-white spirit solution. *BASF plc.*

24067 Plastoprene
A proprietary trade name for an isomerized rubber resin used for printing inks and chemical resistant coatings. *Croda Resins Ltd.*

24068 Plastorit
Natural coalescance of quartz, chlorite and mica; used for surface coatings including anticorrosive paints, textured and powder. *Mercian Minerals & Colours Ltd.* Unverified.

24069 Plastose
Proprietary synthetic resin molding powders. No manufacturer.

24070 Plastosperse
A range of pigments dispersed in plasticisers for use in plastics. *Collinda Ltd.* Discontinued.

24071 Plastosperse 40
40% Carbon black dispersed in a plasticizer. *Collinda Ltd.* Discontinued.

24072 Plastpak
Priming preparations and plasticizing additive. *ICI Chem. & Polymers Ltd.*

24073 Plastplate
A proprietary trade name for molded plastics plated with chromium, copper, gold, or nickel. No manufacturer.

24074 Plastrotyl
An explosive. It is a plastic product prepared from trinitrotoluene, resin, collodion cotton, and crude dinitrotoluene. Sometimes larch turpentine is used.

24075 Plastules with Liver
A proprietary preparation of ferrous sulfate, yeast and liver extract; a hematinic. *Wyeth Laboratories.* Name unverified.

24076 Plastyrol
A proprietary trade name for acrylated, styrenated and vinylated alkyd resins for fast air drying paints. *Croda Resins Ltd.*

24077 Plastyrol E6X
A styrenated epoxide ester resin; a rapid air drying resin. *Croda Resins Ltd.*

24078 Plasvita
Formaldehyde casein; tablet disintegration agent. *Dynamit Nobel Wien GmbH.* Name unverified.

24079 Plasvita® TSM
Methylene casein; tablet disintegration agent; high capillary activity, low swelling effect. *Hüls AG.*

24080 Plaswite® LL 7014
70% Rutile (titanium dioxide) in LLDPE carrier; white masterbatch for high quality LLDPE film pigmentation for applications requiring good dispersion, thermal and light stability, and which need to run at fast LLDPE extrusion rates; compatible with LDPE, LLDPE, HDPE, PP and EVA. *Cabot Plastics Ltd.*

24081 Plaswite® LL 7105
70% Titanium dioxide/lithopone blend in LLDPE carrier; white masterbatch for high quality LLDPE film pigmentation for applications requiring good dispersion, thermal stability, and which need to run at fast LLDPE extrusion rates; compatible with LDPE, LLDPE, HDPE, PP and EVA. *Cabot Plastics Ltd.*

24082 Plaswite® masterbatches
Granulated concentrates of titanium dioxide in plastics; used for coloring of thermoplastic resins. *Cabot Plastics Ltd.*

24083 Plaswite® PS 7174
50% Rutile (titanium dioxide) in crystal PS carrier, with additional toner; white masterbatch for manufacturing of various articles including thin-walled containers such as yogurt pots, disposable drinking cups, domestic appliances, toys, packaging materials; compatible with PS. *Cabot Plastics Ltd.*

24084 Platalargan
An alloy of aluminum and silver with some platinum. It is similar to Alargan, except that it contains platinum; used as a platinum substitute.

24085 Platamid
Polymides for extrusion and injection molding.

24086 Platamid® Series
Hot melt adhesive produced from a copolyamide base; for extrusion of fusible film, monofilament, netting, web, and multifilament; for thin film fusible coatings; for application by dispersion techniques used in textile industry; for fusion bonding of textiles, leather, wood, glass and metals and miscellaneous applications, including automotive interiors, belts, color concentrates, hats, hose construction, rainwear, shoes, television tubes, wire coating, etc. *Elf Atochem.*

24087 Plataril
Monofilament
For grass trimmer and fishing line. *Atochem.* Discontinued.

24088 plate pewter
An alloy of 90% tin, 6% antimony, and 2% each bismuth and copper.

24089 plate powder
$Ca_{32}O_8P$
Bone ash of which calcium phosphate, forms 80% is sold as a non-mercurial plate powder under the name of white rouge.

24090 plate sulfate
$K_2SO_4{\cdot}Na_2SO_4$
The double sulfate, is called plate sulfate. It crystallizes from hot water, a flash of light accompanying the separation of each crystal.

24091 platina
An alloy of 53.5% tin, and 46.5% copper.

24092 platine
A brass containing 43% copper and 57% zinc.

24093 Platine-autitre
A proprietary trade name for a platinum substitute containing 65-83% silver and platinum. No manufacturer.

24094 Platinite
A proprietary trade name for a nickel steel containing 46% nickel, and 0.15% carbon; has a low coefficient of expansion, and can be sealed in glass. No manufacturer.

24095 platinized asbestos
Loosely fibered asbestos moistened with a concentrated solution of platinum chloride, dried, dipped into ammonium chloride solution, again dried, then brought to a red heat. It usually contains 8-8.5% platinum; and is used in the manufacture of sulfuric anhydride in the contact process for sulfuric acid.

24096 Platino
An alloy of 11% platinum and 80% gold. It is resistant to fused potassium nitrate, and to alkalis.

24097 platinoid
An alloy of 60% copper, 24% zinc, 2% tungsten, and 14% nickel; used as the material which connects filaments with outside wires of electric lamps. It has the same coefficient of expansion as glass.

24098 Platinor
An alloy of 2 parts platinum, 5 parts copper, 1 part silver, and 1 part nickel.

24099 Platinum
7440-06-4 7689 231-116-1
Pt
Pt; Metallic element; catalyst, laboratory ware, rayon and glass fiber manufacture, jewelry, dentistry, electrical contacts, thermocouples, surgical wire, bushings, electroplating, electric furnace windings, chemical reaction vessels, permanent magnets. mp = 1773°; bp = 3827°; d = 21.447. *Aldrich; Degussa; Handy & Harman; Noah Chem.*

24100 platinum black
7440-06-4 7689 231-116-1
Pt
Finely divided platinum metal; used as a catalyst; to absorb hydrogen, oxygen.

24101 platinum bronze
An alloy consisting of 90% nickel, 9% tin, and 1% platinum.

24102 platinum gold
Variable alloy containing from 12-81% copper, 9-58% platinum, 0-70% gold, 0-37% silver, and 0-4% zinc.

24103 platinum iridium
An alloy usually containing 90% platinum and 10% iridium; used in jewelry, electrical contacts, fuse wire, and hypodermic needles, and in other applications where high corrosion resistance is required.

24104 platinum silver
An alloy containing 66.6% silver and 33.3% platinum.

24105 platinum solder
An alloy usually consisting of 73% silver and 27% platinum.

24106 platinum substitute
An alloy of 72% nickel 23.6% aluminum, 3.7% bismuth, and 0.7% gold.

24107 platinum yellow
A barium or other alkaline chloroplatinate; used as a coating for fluorescent screens in X-ray work.

24108 platnam
An alloy consisting of 56% nickel, 31% copper, 12% lead. 0.48% iron, and 0.32% aluminum.

24109 platnik
An alloy of nickel and platinum. A platinum substitute.

24110 Platol II
Spray-on coating for arc furnace electrodes. *Foseco (F.S.) Ltd.* Discontinued.

24111 Platone
A peptone powder that is water soluble; used in the electroplating industry. *Am. Labs.*

24112 Plessite
A German explosive containing potassium chlorate and mineral oil. It is also the name for the mineral Gersdorffite. *NiAsS.*

24113 Plessy's Green
7789-04-0 2284 232-141-0
CrO_9P_3
Chromic phosphate, used as a pigment.

24114 Plexar
Adhesive; for multilayer film and plate between polyolefins and polar plastics (polyamide, polyester) and metals. *DSM NV.*

24115 Plexar®
For polyolefin based extrudable adhesives; for use in multilayer barrier packaging, reactive polyolefin adhesives for use in boding polar plastics to nonpolar polyolefins, and modified polyolefin adhesive that form reaction bonds with polar plastics and metals. *Quantum Chemical Corp.*

24116 Plexar® PX 108
9% EVA-based resin adhesive; tie-layer resin for bonding various substrates for barrier packaging in coextruded bottles, films, and sheet. *Quantum/USI.*

24117 Plexiglas® DR
Impact-modified acrylic resin; molding pellets for injection molding and extrusion; good long-term outdoor performance. *Rohm & Haas.*

24118 Plexiglas® HFI-10
Impact-modified acrylic resin; high impact molding pellets with improved injection molding characteristics; good long-term outdoor performance. *Rohm & Haas.*

24119 Plexiglas® V045
Acrylic resin; injection molding and extrusion pellets with high heat resistance, environmental stability; transparent, translucent, and opaque pellets. *Rohm & Haas.*

24120 Plexiglas® VS
Acrylic resin; maximum flow, lowest heat resistance for applications where ease of injection molding is the major factor. *Rohm & Haas.*

24121 Plexiglo
A proprietary trade name for a polish and cleaner for transparent plastics. No manufacturer.

24122 Plexigum
Pure acrylic resins; for surface coatings, inks, firing lacquers, heat seal lacquers, road marking paints, cementitious coatings. *Cornelius Chemical Co Ltd.* Name unverified.

24123 Pleximon
Polymeric/oligomeric methacrylate based resins; thickeners, dental etc. *Cornelius Chemical Co Ltd.* Name unverified.

24124 Plexisol
Acrylic resin solution; for paints, inks etc. *Cornelius Chemical Co Ltd.* Name unverified.

24125 Plexol
Oil additives. *Rohm & Haas UK.*

24126 Plexophor
Textile sequestering agents. *Sandoz Products Ltd.*

24127 Plextol
Pure acrylic resin emulsions; for paints, concrete add mixtures, textile finishes, leather dressings, wood coatings, etc. *Cornelius Chemical Co Ltd.* Name unverified.

24128 Plialite
A proprietary product that is rubber resin. No manufacturer.

24129 Plictran
13121-70-5 2829 236-049-1
A line of miticides based primarily on cyhexatin. *Dow UK.* Discontinued.

24130 Plictran
13121-70-5 2829 236-049-1
Acaracide containing cyhexatin; (Sold in UK on behalf of Dow Chemical Co). *ICI Chem. & Polymers Ltd.*

24131 Plimmer and Paine's stain
A microscopic stain. It contains 10 g tannic acid, 18 g aluminum chloride, 18 g zinc chloride, 1.5 g rosaniline hydrochloride, and 40 ml 60% alcohol.

24132 plinthite
A red clay from Ireland.

24133 Pliobond Adhesives
Rubber/solvent base adhesives; adhesive for bonding and sealing a variety of substrates (metal, rubber, wood, glass, ceramic, leather, cork, canvas, fiberglass and aluminum). *W J Ruscoe Co.* Name unverified.

24134 Pliobond®
A broad line of solvent and water borne adhesives based on various polymer systems. Includes contact adhesives, heat reactive systems, pressure sensitives and elastomeric sealants; used for bonding forms, metals, plastics, wood products, fiberglass, rubber in various construction, consumer, industrial and automotive applications. *Ashland Chemical Company.* Name unverified.

24135 Plio-Caulk
Butyl and acrylic caulking compounds; supplied 1/10 gallon cartridges; used to caulk and seal windows, doors, siding, stone and masonry. For interior and exterior use. *Ashland Chemical Company.* Name unverified.

24136 Pliocord® LVP-4668
Vinyl pyridine-styrene-butadiene latex; used for cord adhesion in tires, conveyor belts, hose, etc.; exhibits excellent adhesion between fabric and rubber; used alone or with resorcinol-formaldehyde dip formulations. *Goodyear.* Name unverified.

24137 Plioflex
A proprietary trade name for polyvinyl chloride. No manufacturer.

24138 Plioflex® 1006
SBR, hot polymer, FA emulsifier, nonstaining stabilizer; general purpose grade. *Goodyear.* Name unverified.

24139 Plioflex® 1028
SBR, FA emulsifier; food grade. *Goodyear.* Name unverified.

24140 Plioflex® 1905
SBR resin masterbatch (80:20 SBR/high styrene resin), FA emulsifier, nonstaining stabilizer; used in sponge products, rug underlay, shoe soles, household goods; gives high styrene reinforcement; vulcanized with standard curing systems including sulfur and peroxide. *Goodyear.* Name unverified.

24141 Plioform
A proprietary type of rubber plastic obtained by the action of halogenated acids on rubber. No manufacturer.

24142 Pliogrip®
Industrial and structural adhesives. *Ashland.*

24143 Pliolite
Modified isomerized rubber, rubber derivatives and rubber-like resins. *Rhône-Poulenc UK.*

24144 Pliolite® 7103, 7104
Carboxylated styrene-acrylic aqueous dispersion; designed to replace solvent-borne resins in demanding specialty and OEM coating applications. *Goodyear.* Name unverified.

24145 Pliolite® AC-80
Styrene/acrylate emulsion copolymer; high uv resistance; for high build maintenance coatings and architectural paints. *Goodyear.* Name unverified.

24146 Pliolite® LPF-2108
SBR latex; for spread foam on carpet or fabric, molded foam for automotive and home furnishings, binders for mastic and adhesive formulations, asphalt modifier for construction. *Goodyear.* Name unverified.

24147 Pliolite® S-6B, S-6F
9003-55-8 8534
S/B polymer; reinforces to increase hardness, stiffness, and abrasion resistance. *Goodyear.* Name unverified.

24148 Pliolite® VT
Vinyl-toluene/butadiene emulsion copolymer; produces films with clarity, strength, hardness and chem. resistance; for traffic paints, block fillers, paper coatings, adhesives, laminates, printing inks, hot melts, abrasion-resistance coatings. *Goodyear.* Name unverified.

24149 Pliolite® VTAC
Vinyl-toluene/acrylate emulsion copolymer; for texture coatings, interior and exterior masonry paints, intumescent fire-retardant paints. *Goodyear.* Name unverified.

24150 Plio-Nail
High performance synthetic rubber-based adhesives; certified by PFS Corporation to meet the American Plywood Associations AFG-ol specification; used for joints, sub floors, siding, decorative panels, gypsum board, fixtures and a wide variety of materials. *Ashland Chemical Company.* Name unverified.

24151 Plio-Seam
Elastomeric rubber sealant; provides a strong and durable seal for aluminum, steel, glass, wood, masonry, ceramics, fiberglass joints; remains flexible and tough; may be painted; available in clear, white, aluminum, black and architectural bronze; used to install or repair downspouts, rain troughs, roof flasjings, shower stalls and metal or wood structures; weather seals glass in metal or wood storm windows or doors. *Ashland Chemical Company.* Name unverified.

24152 Plio-Tac
Neoprene contact adhesive supplied in aerosol spray cans; will not cavitate polystyrene foams; also used to bond wood, carpet, vinyl fabric, metal, rubber and many other materials; used for bonding insulation to various surfaces, craft assembly projects, rebonding carpet or vinyl flooring, counter or table top laminates, mounting signs and photos and attaching labels to various surfaces. *Ashland Chemical Company.* Name unverified.

24153 Plio-Tac 38
Contact cement designed for DIY; meets CPSA guidelines; for bonding decorative laminates to wood and metal surfaces. Also bonds leather, fabrics, unglazed ceramics, wall boards, cove base and carpets to themselves and each other. *Ashland Chemical Company.* Name unverified.

24154 Pliovic® DR-450, DR-453, DR-454, DR-600, DR-602, DR-652
Vinyl dispersion resin; developed for plastisol and organosol applications. *Goodyear.* Name unverified.

24155 Pliovic® M-50, M-70, M-70SC, M-90
Vinyl homopolymer resin; blending resin for modifying plastisol compounds resulting in lower viscosity, improved viscosity stability, increased hardness, minimized surface gloss. *Goodyear.* Name unverified.

24156 Pliovic® WO-1, WO-2, WO-3, WO-S
Vinyl homopolymer dispersed resin; used in plastisol and organosol compounds for automotive air and oil filters, dipped goods including gloves, coatings for awnings, rainwear, wall coverings, laminating plastisols for fabric-to-fabric adhesion, foams for garments, upholstery and capret backings, weatherstripping, rootcast and slushmolded toys, boots and novelties. *Goodyear.* Name unverified.

24157 Plioway® EC1
Acrylic resin; film-forming resin for odorless interior paints, exterior masonry and concrete coatings. *Goodyear*. Name unverified.

24158 Plitex
A proprietary trade name for a wood and phenolic resin. No manufacturer.

24159 Plombit
A German acid-resisting material made from hard rubber, oleic acid, sulfuric acid, and sulfur.

24160 Plondrel
5131-24-8 225-875-8
$C_{12}H_{14}NO_4PS$
O,O-diethyl (1,3-dihydro-1,3-dioxo-2H-isoindol-2-yl)phosphonothioate
Diethyl phthalimidophosphonothioate; Dowco 199; Laptran; Plondrel;. Fungicide containing ditalimfos; for the control of powdery mildew and scab. (Sold in UK on behalf of Dow Chemical Co). *ICI Chem. & Polymers Ltd.*

24161 plumbago grease
A mixture of plumbago and tallow; used for lubricating.

24162 plumber's solder
Usually a mixture of lead and tin, sometimes with a little antimony. Coarse solder contains 75% lead and 25% tin, and melts at 250°C. Ordinary solder (slicker solder) usually consists of 67% lead and 33% tin, and melts at 227°C. Plumber's fine or soft solder contains 50% lead and 50% tin and melts at 188°C.

24163 plumber's white alloy
An alloy of from 54-58% copper, 25-27% zinc, 13-17% nickel, 1-7% lead, and sometimes 1% tin and 1% iron.

24164 plumboxan
$Na_2MnO_4{\cdot}Na_2PbO_3$
A compound or solid solution of sodium manganate and sodium metaplumbate. It gives up oxygen when treated with steam.

24165 Plumbral
Copper/lead alloys flux. *Foseco (F.S.) Ltd.*

24166 Plumbrex
Flux for lead and alloys. *Foseco (F.S.) Ltd.*

24167 Plumbrit
Cleansing flux for lead alloys. *Foseco (F.S.) Ltd.* Discontinued.

24168 plumose mica
A variety of muscovite.

24169 Pluracol® 220
Polyether polyol
Specialty polyol for low modulus foams and blown elastomers. *BASF*. Name unverified.

24170 Pluracol® 355
Amine-based polyol; cross-linking agent for semiflexible urethane foams, coatings, adhesives, and polymers. *BASF*. Name unverified.

24171 Pluracol® 450
POP derivative of pentaerythritol; crosslinking agent for rigid urethane foams. *BASF*. Name unverified.

24172 Pluracol® 581
Polymer reinforced (graft polyol) polyether; high-resilience molding polyol for high firmness and strength in flexible foam applications; used in automotive molded, carpet underlay, semiflexible applications. *BASF*. Name unverified.

24173 Pluracol® E1000
25322-68-3 7729
PEG-20
Dispersant. *BASF*.

24174 Pluracol® E1500
25322-68-3 7729
PEG-6-32
Dispersant. *BASF*.

24175 Pluracol® E200
25322-68-3 7729
PEG-4
Intermediate for preparation of nonionic surfactants; binder, base, coating, stabilizer, solvent, vehicle, extender, and coupling agent for pharmaceutical, cosmetic, and toiletries; lubricant for metal applications, rubber industry; wood treatment; textile conditioning, antistat and sizing agent. *BASF*.

24176 Pluracol® E2000
25322-68-3 7729
PEG-40
Dispersant. *BASF*.

24177 Pluracol® E300
25322-68-3 7729
PEG-6
Also dispersant in food tablets and preparations; plasticizer. *BASF*.

24178 Pluracol® E400
25322-68-3 7729
PEG-8
Dispersant. *BASF*.

24179 Pluracol® E400 NF
25322-68-3 7729
PEG-8
Chemical intermediate, base, coupler, thickener, lubricant, mold release agent, defoamer, softener, conditioner, antistat, sizing agent, dispersant for pharmaceutical, cosmetic, and oral care preparations, in metal polishing and cleaning formulations, rubber products, paper and wood products, textile processing and ink formulations. *BASF*.

24180 Pluracol® E4000
25322-68-3 7729
PEG-75
Dispersant. *BASF*.

24181 Pluracol® E4500
PEG
Dispersant. *BASF*.

24182 Pluracol® E6000
25322-68-3 7729
PEG-150
Dispersant. *BASF*. Name unverified.

24183 Pluracol® E600NF
25322-68-3 7729
PEG-12
Dispersant. *BASF*.

24184 Pluracol® E8000
25322-68-3 7729
PEG
Dispersant. *BASF*.

24185 Pluracol® V-10
Polyoxyalkylene glycol polyol
Thickening agent to control the viscosity of water-glycol type, fire-resistant hydraulic fluids; resistance to shearing stresses and will not hydrolyze or degrade under use conditions; noncarbonizing and nongumming at high temperatures. *BASF*.

24186 Pluracol® W170
74623-31-7
PPG-5-buteth-7
Component in demulsifying and wetting formulations; brake and metalworking fluids; rubber and fiber lubricant; textile applications; defoamer for hot and cold applications, food and chemical processing; cosmetic formulations. *BASF*. Name unverified.

24187 Pluracol® W2000
74623-31-7
PPG-20-buteth-30
Wetting agent, demulsifying agent. *BASF*. Name unverified.

24188 Pluracol® W3520N
9038-95-3
PPG
Wetting agent, demulsifying agent. *BASF*.

24189 Pluracol® W5100N
74623-31-7
PPG-33-buteth-45
Wetting agent, demulsifying agent. *BASF*.

24190 Pluracol® W660
74623-31-7
PPG-12-buteth-16
Wetting agent, demulsifying agent. *BASF*. Name unverified.

24191 Plurafac®
Low foam nonionic surfactants. *BASF plc.*

24192 Plurafac® A-38
68439-49-6 7737
Ceteareth-27
Detergent, dispersant, wetting agent, emulsifier for heavy-duty detergents, metal cleaners, detergent tablets, electrolytic cleaning; biodegradable. *BASF*.

24193 Plurafac® A-39
68439-49-6 7737
Ceteareth-55
Detergent, dispersant, wetting agent, emulsifier for heavy-duty detergents, metal cleaners, detergent tablets, electrolytic cleaning; biodegradable. *BASF*.

24194 Plurafac® B-25-5
Straight chain primary aliphatic oxyalkylated alcohol; detergent, dispersant, wetting agent, emulsifier for heavy-duty detergents, all-purpose liquids, detergent tablets, sanitizers, metal cleaners; biodegradable. *BASF*.

24195 Plurafac® D-25
PPG-6 C12-18 pareth-11

Detergent, dispersant, wetting agent, emulsifier, deduster for heavy duty detergents, all-purpose liquids, detergent tablets, sanitizers, metal cleaners, hard surface cleaners; biodegradable. *BASF.*

24196 Plurafac® LF 120
Alkoxylated fatty alcohol
Surfactant for acid and alkaline low foaming cleaners. *BASF AG.*

24197 Plurafac® LF 220
Alkoxylated straight chain alcohol; low foaming surfactant for dishwashing powders, rinse aids. *BASF AG.*

24198 Plurafac® RA-20
Straight chain primary aliphatic oxyalkylated alcohol; detergent, dispersant, wetting agent, emulsifier, defoamer, deduster used in rinse aids and dishwashing products. *BASF.*

24199 Pluraflo® E4A E5G, N5G
Formulated product; dispersant, wetting agent for pesticides. *BASF.*

24200 Plurasafe
Glycol hydraulic fluid. *BASF.*

24201 Pluriol®
Polyglycols. *BASF plc.*

24202 Pluriol® E 200
PEG; solubilizer, impregnating agent, humectant, mold release agent; flow improver, thermal and hydraulic fluid, organic intermediate; detergent and cleaner; dye and pigment dispersant; inks; textile and coatings industry; coloring ceramics; softener in paper industry; plasticizer in adhesives industry and in production of cellulose film; ceramics and metalworking lubricant. *BASF AG.* Name unverified.

24203 Pluriol® E 4000
PEG; solubilizer, humectant; binder and hardener in personal care products; dispersing dyes and pigments; inks; textile and coating industry; coloring ceramics; paper industry softener; plasticizer in adhesives industry and production of cellulose film. *BASF AG.* Name unverified.

24204 Pluriol® P 2000
PPG; mold release agent, additive for oils and fluids; lubricant and antifoam for rubber; consistency improver and solubilizer; intermediate in industrial applications. *BASF AG.* Name unverified.

24205 Pluriol® P 600
PPG; mold release agent, additive for oils and fluids; lubricant and antifoam for rubber; consistency improver and solubilizer; intermediate in industrial applications *BASF AG.* Name unverified.

24206 Plurivite
Vitamin preparations. *The Boots Co plc.* Name unverified.

24207 Plurol Isostearique
Polyglyceryl-6 isostearate
Emulsifier; cosurfactant for microemulsions. *Gattefosse SA.*

24208 Plurol Oleique WL 1173
Polyglyceryl-6 dioleate
Emulsifier; cosurfactant for microemulsions. *Gattefosse SA.*

24209 Plurol Stearique WL 1009
Polyglyceryl-6 distearate
Consistency agent and stabilizer for heated oil-water emulsions; food emulsifier. *Gattefosse SA.*

24210 Pluronic® 10R5
9003-11-6 7721
Meroxapol 105
Emulsifier, wetting agent, binder, stabilizer, plasticizer, lubricant, solubilizer, dispersant, viscosity control agent, defoamer, intermediate for hard surface detergents, rinse aids, automatic dishwashing, textile processing; cosmetics; pharmaceuticals, pulp, paper and petroleum industries, agricultural products, in iodophores, water treatment systems, fermentation and cutting and grinding fluids. *BASF.*

24211 Pluronic® 10R8
9003-11-6 7721
Meroxapol 108
BASF.

24212 Pluronic® 17R1
9003-11-6 7721
Meroxapol 171
Also for foam control in paper sizing operations. *BASF.*

24213 Pluronic® 17R2
9003-11-6 7721
Meroxapol 172
BASF.

24214 Pluronic® 17R4
9003-11-6 7721
Meroxapol 174
BASF.

24215 Pluronic® 17R8
9003-11-6 7721
Meroxapol 178
Also dry toilet bowl cleaners, dye levelers, solubilizer of drugs, stick type cosmetics; soap bars, dispersant in de-inking operations. *BASF.*

24216 Pluronic® 25R1
9003-11-6 7721
Meroxapol 251
Also foam control in paper sizing operations and antifreeze. *BASF.*

24217 Pluronic® 25R2
9003-11-6 7721
Meroxapol 252
Also wetting and rinse aid; lubricant and leveling agent for paper coating. *BASF.*

24218 Pluronic® 25R4
9003-11-6 7721
Meroxapol 254
BASF.

24219 Pluronic® 25R5
9003-11-6 7721
Meroxapol 255
Also thickener for cosmetic pastes and creams. *BASF.*

24220 Pluronic® 25R8
9003-11-6 7721
Meroxapol 258
Also dry toilet bowl cleaners, dye levelers for fabrics; solubilizer for drugs; thickener for cosmetics; de-inking operations; felt washing operations. *BASF.*

24221 Pluronic® 31R1
9003-11-6 7721
Meroxapol 311
Also floating bath oils; foam control in antifreeze. *BASF.*

24222 Pluronic® 31R2
9003-11-6 7721
Meroxapol 312
Paper coating color additive; deinking and felt washing operations. *BASF.*

24223 Pluronic® 31R4
9003-11-6 7721
Meroxapol 314
BASF.

24224 Pluronic® F108
9003-11-6 7721
Poloxamer 338
BASF.

24225 Pluronic® F127
9003-11-6 7721
Poloxamer 407
BASF.

24226 Pluronic® F38
9003-11-6 7721
Poloxamer 108
Wetting agent, emulsifier, demulsifier, foam and viscosity control agent, dispersant, antistat, gelling agent, dyeing assistant, leveler, lubricant for agricultural, cosmetics, pharmaceuticals, metal cleaning, pulp/paper, textile scouring and water treatment. *BASF.* Name unverified.

24227 Pluronic® F68
9003-11-6 7721
Poloxamer 188
BASF.

24228 Pluronic® F68LF
9003-11-6 7721
Poloxamer 108
BASF.

24229 Pluronic® F77
9003-11-6 7721
Poloxamer 217
BASF.

24230 Pluronic® F87
9003-11-6 7721
Poloxamer 237
BASF. Name unverified.

24231 Pluronic® F88
9003-11-6 7721
Poloxamer 238
BASF.

24232 Pluronic® F98
9003-11-6 7721
Poloxamer 288
BASF.

24233 Pluronic® L101
9003-11-6 7721
Poloxamer 331
BASF.

24234 Pluronic® L121
9003-11-6 7721
Poloxamer 401
BASF.

24235 Pluronic® L122
9003-11-6 7721
Poloxamer 402
BASF.

24236 Pluronic® L31
9003-11-6 7721
Poloxamer 101
BASF.

24237 Pluronic® L35
9003-11-6 7721
Poloxamer 105
BASF.

24238 Pluronic® L42
9003-11-6 7721
Poloxamer 122
BASF.

24239 Pluronic® L43
9003-11-6 7721
Poloxamer 123
BASF.

24240 Pluronic® L44
9003-11-6 7721
Poloxamer 124
BASF.

24241 Pluronic® L61
9003-11-6 7721
Poloxamer 181
BASF.

24242 Pluronic® L62
9003-11-6 7721
Poloxamer 182
BASF.

24243 Pluronic® L62D
9003-11-6 7721
Poloxamer 108
BASF.

24244 Pluronic® L62LF
9003-11-6 7721
Poloxamer 108
BASF.

24245 Pluronic® L63
9003-11-6 7721
Poloxamer 183
BASF.

24246 Pluronic® L64
9003-11-6 7721
Poloxamer 184
BASF.

24247 Pluronic® L72
9003-11-6 7721
Poloxamer 212
BASF.

24248 Pluronic® L81
9003-11-6 7721
Poloxamer 231
BASF.

24249 Pluronic® L92
9003-11-6 7721
Poloxamer 282
BASF.

24250 Pluronic® P103
9003-11-6 7721
Poloxamer 333
BASF.

24251 Pluronic® P104
9003-11-6 7721
Poloxamer 334
BASF.

24252 Pluronic® P105
9003-11-6 7721
Poloxamer 335
BASF.

24253 Pluronic® P123
9003-11-6 7721
Poloxamer 403
BASF.

24254 Pluronic® P65
9003-11-6 7721
Poloxamer 185
BASF.

24255 Pluronic® P75
9003-11-6 7721
Poloxamer 215
BASF.

24256 Pluronic® P84
9003-11-6 7721
Poloxamer 234
BASF.

24257 Pluronic® P85
9003-11-6 7721
Poloxamer 235
BASF.

24258 Pluronic® P94
9003-11-6 7721
Poloxamer 284
BASF.

24259 Pluronic® PE 3100
PO/EO block polymer; for defoaming and controlled foam detergents. *BASF AG.*

24260 Plus
Range of fertilizers for garden use. *ICI Chem. & Polymers Ltd.*

24261 Plusbrite
A chemical for bright nickel plating. *Albright & Wilson Ltd.* Name unverified.

24262 Plus-Gas C
Noncorrosive cutting fluids. *Foseco (F.S.) Ltd.* Discontinued.

24263 Plus-Pac®
Palladium-based product; for recovery of platinum lost during the catalytic oxidation of ammonia to make nitric acid. *Johnson Matthey plc.*

24264 Pluviusin
A proprietary trade name for a synthetic resin of the urea type. No manufacturer.

24265 Plyamul®
Polyvinyl acetate emulsions, adhesives. *Reichhold.*

24266 Plyamul® 40305-00
108-05-4 10130 203-545-4
Vinyl acetate emulsion copolymer; adhesive base for high-speed applications. *Reichhold/Emulsion Polymers.*

24267 Plymul 98-759
A range of polyvinyl acetate thermosetting emulsions used for bonding cellulose to cellulose. *Reichhold.*

24268 Plyophen
A proprietary trade name for a phenolic laminating resin and varnish. No manufacturer.

24269 Plyothene
A proprietary phenolic molding material. *Reichhold.* Discontinued.

24270 Ply-Pro 25
Tissue-converting adhesive. *Aqualon.*

24271 Plyron
Continuous fiber-reinforced thermoplastics. *ICI Chem. & Polymers Ltd.*

24272 Plysolene
Modified polyisobutylene sheet; waterproof membrane for insulation. *Plysolene Ltd.*

24273 Plytron
Semi-finished plastic. *ICI Chem. & Polymers Ltd.*

24274 PMA 18, 60
62-38-4 7453 200-532-5
Solubilized form and powdered form, respectively, of phenylmercuric acetate; preservative and fungicide for aqueous paints. *Hüls Am.* Discontinued.

24275 PMAS
Mercurial fungicide for prevention and control of pink and gray snow mold. *W A Cleary.*

24276 Pneulec Core Gum
A proprietary product; it is a linseed oil and wood extract material, and is used as a binder for the sand for cores in metal casting. No manufacturer.

24277 pneumatogen
A mixture of the peroxides of potassium and sodium.

24278 Poast®
74051-80-2 8620 277-682-3
Sethoxydim
Post-emergence graminicide against annual and perennial grasses. *BASF AG.*

24279 Pocan®
Thermoplastic polyester based on polybutylene terephthalate; for the manufacture of injection moldings with brilliant surface finish; uses include household appliances, office machines and electrical components. *Bayer AG; Bayer plc; Miles.*

24280 Pocan® B 1300
26062-94-2
PBT polyester; FDA-grade with medium viscosity. *Bayer AG.*

24281 Pocan® B 1305
26062-94-2
PBT polyester; heat-stabilized engineering plastic, easy mold release, medium viscosity. *Bayer AG.* Name unverified.

24282 Pocan® B 1505
26062-94-2
PBT polyester; engineering thermoplastic, high heat deflection temps., stiffness and hardness, min. surface friction and cold flow, abrasion resistance, low moisture absorp., good dynamic fatigue strength, resistance to chemicals; and performance in fire tests; flow and fast cycling. *Bayer AG.* Name unverified.

24283 Pocan® KU1-7033
26062-94-2
PBT polyester, 30% glass-reinforced, impact-modified; general-purpose grade. *Bayer AG.* Name unverified.

24284 Pocan® S 1506
26062-94-2
PBT polyester, elastomer modified; high impact even at low temperatures. *Bayer AG.* Name unverified.

24285 Poilite
A trade name for an asbestos cement product used for building work. No manufacturer.

24286 poison flour
1327-53-3 844 215-481-4
As_4O_6
Arsenious oxide
Primary material for all arsenic compounds. Used in chemical manufacturing and also as a parasiticide and rodenticide.

24287 Poivrette
The ground stones of olive fruit; used as an adulterant and toning agent in spices.

24288 Pokalon
Polycarbonate film. *Lonza AG.*

24289 Polacure® 740M
Diamine curative; curative for high-performance elastomers. *Air Prods & Chem/Polyurethanes.* Name unverified.

24290 Polamine® 250
Polytetramethylene ether glycol (PTMEG)-diamine; curative for high-performance room temperature cast and cured elastomers for cast prototypes, coatings, adhesives, sealants, and spray systems; epoxy flexibilizer. *Air Prods & Chem/Polyurethanes.* Name unverified.

24291 Polamine® 650, 1000, 2000
54667-43-5
Polytetramethyleneoxide-di-*p*-aminobenzoate
Curative for high performance room temperature cast and cured elastomers for cast prototypes, coatings, adhesives, sealants, and spray systems; epoxy flexibilizer. *Air Prods & Chem/Polyurethanes.* Name unverified.

24292 Polaqua
Water based primer/adhesive with low solids and containing no organic solvents; for extrusion and lamination of film to film, paper and foil adhesion promoter to films, paper and foil. *ADM Tronics Unlimited Inc.*

24293 Polar
Acid dyes for wool. *Ciba plc.* Name unverified.

24294 Polar Dynobel
Polar Monobel No. 2; Polar Rex; Polar Saxonite; Polar Stomonal; Polar Super Clifite ; Polar Thames Powder; Polar Viking. Proprietary low freezing explosives containing a mixture of nitrated glycerin and polyglycerin or glycerin and ethylene glycol, ammonium nitrate, sodium chloride, wood meal, etc. No manufacturer.

24295 Polargel® HV
1302-78-9 1082 215-108-5
Bentonite USP/NF
High viscosity white montmorilonite used as disintegrant, binder, suspension agent and thickener. *Am. Colloid.*

24296 Polargel® NF
1302-78-9 1082 215-108-5
Purified white bentonite USP/NF; thickener and suspending agent for cosmetics and pharmaceuticals. *Am. Colloid.*

24297 Polarin® Range
Glycerol and its derivatives; solvents, humectants for food industry. *Grünau.*

24298 Polaris
Boiler and cooling water treatment. *Laporte Industries Ltd.*

24299 Polarite®
Surface-treated metakaolinitic aluminum silicate produced by the calcination of kaolin clay; surface-treating agents include substituted silanes and elastomers; functional fillers for rubbers and plastics. *ECC International Ltd.*

24300 Polarite® 420E(W)
Low-profile additive for DMC and BMC applications; for low shrink compounds. *ECC.*

24301 Polarite® 420G(W)
Low-profile additive for DMC and BMC applications; for zero shrink compounds. *ECC.*

24302 Polarite® 880E(W)
21645-51-2 355 244-492-7
Aluminum hydroxide
Aluminum trihydrate. Low profile flame retardant/smoke suppressant additive for BMC moldings. *ECC.*

24303 Polarwhite
A headless white paint. *J C Bottomley.* Name unverified.

24304 Polathane
Polyester and polyether toluene diisothiocyanate prepolymers. *Air Prods & Chems. Inc.*

24305 Polathane STE-73D, STE-83A, STE-90A, STE-95A
TDI-PTMEG; polyurethane prepolymer *Air Prods & Chem/Polyurethanes.* Name unverified.

24306 Polathane STS-55
TDI-ester; polyurethane prepolymer *Air Prods & Chem/Polyurethanes.* Name unverified.

24307 Polathane XPE-10, XPE-20, XPE-30
TDI-PTMEG; polyurethane prepolymer *Air Prods & Chem/Polyurethanes.* Name unverified.

24308 Polawax®
Emulsifying wax NF; emulsifier, thickener, opacifier, suspending agent; stabilizer for oil-water emulsions; for cosmetics, pharmaceuticals, hair straighteners, moisturizers, nail preparations, sunscreens, antibiotic creams and lotions, acne preparations, analgesic rubs. *Croda Inc.; Croda Surface Ltd.*

24309 Polawax® A31
Emulsifying wax NF; emulsifier used in quick-breaking foams and mousses, cosmetics, pharmaceuticals. *Croda Inc.; Croda Surf. Ltd.*

24310 Polectron 430
Vinylpyrrolidone/styrene copolymer; a stable opacifier especially used in high pH systems, e.g., detergents and cold wave lotions; in surface coatings, provides adhesion, film toughness, color receptivity and noncorrosiveness. *ISP.*

24311 Polidene 528F
A proprietary polyvinlyidene chloride emulsion. *A E Staley Manufacturing Co.* Name unverified.

24312 Polidene®
Vinylidene chloride copolymer emulsions; used for paints, binders and membrane coatings. *Scott Bader.*

24313 Polidene® 33-001
Vinylidene chloride copolymer; paper coating; binder or coating for bonding of glass nonwoven. *Scott Bader.*

24314 Polidene® 33-004
Vinylidene chloride copolymer; fire retardant coatings, adhesives, and fiber impregnants; pigment binder. *Scott Bader.*

24315 Polidene® 33-055
Carboxylated vinylidene chloride copolymer emulsion; highly resistance rustproof aqueous primers for steel. *Scott Bader.*

24316 Polifil® C-10
9003-07-0 7741
PP homopolymer, 10% calcium carbonate-reinforced; features high impact, stiffness, heat agent, good colorability, ESCR, low mold shrinkage; for appliances, electrical components, housewares, toys, automotive and utility products. *Ralco Industries.*

24317 Polifil® CAS-40
9003-07-0 7741
PP homopolymer, 40% calcium sulfate-reinforced; features higher whiteness

level, reduced mold shrinkage, easier colorability and processability. *Ralco Industries.*

24318 Polifil® GFPP-10
9003-07-0 7741
PP homopolymer, 10% glass fiber-reinforced; provides high impact, stiffness, hardness, higher continuous use temperature; for appliances, electrical components, automotive and utility products *Ralco Industries.*

24319 Polifil® GFPPCC-10
9003-07-0 7741
PP homopolymer, 10% chemically coupled glass fiber-reinforced; offers superior strength, stiffness, improved high temp. performance, higher impact strength; for chemical resistance applications, appliances electrical components, automotive, ignition and utility products. *Ralco Industries.*

24320 Polifil® M-20
9003-07-0 7741
PP homopolymer, 20% mica-reinforced; used where high modulus values are required; improved high stiffness values and low mold shrinkage. *Ralco Industries.*

24321 Polifil® RMC-10
9003-07-0 7741
High impact PP, 10% calcium carbonate-reinforced; provides highest impact, and good stiffness, heat aging resistance, solvent resistance, ESCR, and surface quality; for automotive, appliances, electrical components, housewares, utility products. *Ralco Industries.*

24322 Polifil® RMT-10
9003-07-0 7741
High impact PP, 10% talc-reinforced; provides high impact strength, flexibility modulus, stiffness and heat deflection resistance; used in automotive applications. *Ralco Industries.*

24323 Polifil® T-10
9003-07-0 7741
PP homopolymer, 10% talc-reinforced; features high flexibility modulus, stiffness and deflection temperatures, maximum stiffness, low shrinkage, good colorability; used in underhood automotive, major appliances, electrical goods, housewares and utility products. *Ralco Industries.*

24324 Poligen
Polymer emulsion dispersions. *BASF plc.*

24325 Poligen MMV
A proprietary aqueous dispersion of styrene/acrylic copolymers. *BASF plc.* Name unverified.

24326 Poligen PE
9002-88-4 7728
Polyethylene aqueous dispersion; dry-bright floor polish emulsions, release coats. *BASF AG; BASF plc.*

24327 Poligen WE 1
High molecular weight polyethylene wax aqueous secondary emulsion; dry-bright floor polish emulsions. *BASF AG.* Name unverified.

24328 Polimex TR
High performance expanded polymer foam core material; dimensionally stable at elevated temperatures; for aircraft interior panels, antennas and radome structures, insulated freight containers, fuel tank insulation, sporting goods, marine structures and tooling. *Polimex.*

24329 Polisax
Polyethylene sacks. *ICI Chem. & Polymers Ltd.* Discontinued.

24330 polishing oil
A fraction of petroleum oil, having a boiling point of 130-160°C., and a specific gravity of 0.74-0.77; used as a turpentine substitute.

24331 Politarp
Polyethylene sheets. *ICI Chem. & Polymers Ltd.*

24332 Politec
Cold box resin sand process. *Foseco (F.S.) Ltd.*

24333 Politint 1, 2
Proprietary dyes for plastics. No. 1 for methylmethacrylate. No. 2 for cellulose acetate, cellulose acetate butyrate, and ethyl cellulose. No manufacturer.

24334 Politol®
Sodium lignate
Protein coagulant in purification of fats and oils. *Westvaco.*

24335 Pollack's cement
A cement composed of glycerin, litharge, and red lead.

24336 Pollopas
Urea resin molding compounds; used for electrical engineering, sealing caps and sanitary equipment. *Dynamit Nobel Wien GmbH.* Name unverified.

24337 Polnac
A range of polyester resins; for GFRP building coverings, buttons, electrical industry, various applications in GFRP, silos for fodder, varnishes for wood, pieces for industrial coachwork, prefab for the building industry, tanktrucks for raw material stockage, marble aggregates for floorings and decoration for furniture, etc. *AMC SPREA S.p.A.*

24338 Polomyx
Modified acrylate based coarse particle size suspension; architectural coating that is spray applied to gypsum board and other substrates and produces a textured, tone-on-tone or multi toned, seamless wallcovering. *Polomyx Industries Inc.* Name unverified.

24339 Poloxalkol
A polymer of ethylene oxide, propylene oxide, and propylene glycol. No manufacturer.

24340 Poloxyl Lanolin
A polyoxyethylene condensation-product of anhydrous lanolin. No manufacturer.

24341 poly (ethylene glycol adipate)
Functional polyol for formulating polyurethane for solution; coatings, adhesives, and castable elastomers. *Werner G. Smith.*

24342 Poly (pentabromobenzyl) acrylate
594477-57-3
$(C_{10}H_5Br_5O_2)_x$
FR-1025; PBB-PA. Polymeric flame retardant for engineering thermoplastics, PET, PBT, nylon, PP, and PS. White to off-white powder; insoluble in common organic solvents; density=2.05; mp=190-220°; LD_{50}=rat oral >5000 mg/kg; mildly irritating to eyes and skin *AmeriHaas; Dead Sea Bromine.*

24343 Poly bd® R-45HT
9003-17-2
Butadiene polymer, hydroxyl-terminated; for electrical potting and encapsulating applications including cable connectors, high voltage capacitors, shock absorbers, transformers, voltage regulators. *Elf Atochem.*

24344 Poly C4M
Oil and grease additive. *Crowley Chem.*

24345 Poly Check
Boiler water treatment. *Dearbom Chemicals Ltd.* Name unverified.

24346 Poly/Bed 812
Replacement for Shell's discontinued Epon 812; embedding kit for light microscopy. *Polysciences Inc.*

24347 Poly/Sep 47
Mixture of 47 buffers; electrofocusing buffer for electrophoresis; generates a stable linear pH in polyacrylamide gels in the pH range of 2.5-10; avoids inconsistencies of conventional SCAM's. *Polysciences Inc.*

24348 polyacrylamide
9003-05-8
$[CH_2CHCONH_2]_x$
2-propenamide Homopolymer
Polyamide of acrylic monomers. thickener, dispersant, antiprecipitant, solubilizer, binder, sizing, flocculating, suspending, crosslinking agent, filtering aid, lubricant; used in adhesives, agriculture, cement, coatings, cosmetics, detergents, latex manufacture, printing ink. *Aldrich; Allied Colloids; Cyanamid BV; Calgon; Rhône-Poulenc Water Treatment.*

24349 polyacrylate
Acrylic resin. Solutions of polyacrylic acid; dispersant for pesticides; anticaking agent; coatings additive; flocculant. *Polysar BV; Scott Bader Ltd.*

24350 Polyaldo® 2010 KFG
Polyglyceryl-10 dioleate
Food emulsifier, kosher grade; surfactant; replacement for polysorbate 80. *Lonza Inc.*

24351 Polyaldo® 2P10 KFG
Polyglyceryl-10 dipalmitate
Food emulsifier, surfactant, gelling agent; replacement for polysorbate 60. *Lonza Inc.*

24352 Polyaldo® 2S6 KFG
Polyglyceryl-6 distearate
Kosher food grade emulsifier; aerating and whipping agent. *Lonza Inc.*

24353 Polyaldo® DGDO KFG
11094-60-3 234-316-7
Polyglyceryl-10 decaoleate
Kosher grade food dispersing agent. *Lonza Inc.*

24354 Polyaldo® HGDS KFG
Polyglyceryl-6 distearate
Kosher food grade cake emulsifier. *Lonza Inc.*

24355 Polyaldo® TGMS KFG
37349-34-1
Polyglyceryl-3 stearate
Kosher food grade emulsifier; aerating and whipping agent. *Lonza Inc.*

24356 Polyalk
A proprietary preparation of dimethicone and aluminum hydroxide gel; an antacid. *Galen Ltd.* Discontinued.

24357 polyamide
A high molecular weight polymer in which amide linkages (CONH) occur along the molecular chain; may be either natural or synthetic; Natural polyamides include casein, soybean, and peanut proteins, zein; synthetic polyamides typified by various nylons; used for plastics, textile fibers and adhesives. Aldrich; Arizona; Ashley Polymers; BASF; EMS-Am. Grilon; Georgia-Pacific; Hoechst Celanese; Hüls AG; Miles; Monsanto; SNIA UK; Union Camp.

24358 Polyanthrene KS Liq. New
Polymeric aliphatic compound; formaldehyde-free fixing agent for improving washfastness of direct dyes on cellulose. *Crompton & Knowles.*

24359 Polybead
Monodisperse latex polymer beads; an identification tag and a size reference for agglutination tests, flow cytometry, instrument calibration, gel filtration, light scattering, and phagocytosis. *Polysciences Inc.*

24360 Polybilt
Asphalt modifiers. *Exxon UK.*

24361 Polyblack
Nitriles. *BP Chemicals Ltd.* Discontinued.

24362 Polyblends
Nitrile/PVC blends. *BP Chemicals Ltd.* Discontinued.

24363 Polybond®
Polyolefins
BP Chemicals Ltd. Discontinued.

24364 Polybond® 1000
PP homopolymer, acrylic acid modified; thermoplastic for use as a chemical coupling agent, compatibilizing agent, metal adhesive and nucleating agent. *BP Performance Polymers.* Discontinued.

24365 Polybond® 1009
9002-88-4 7728
HDPE homopolymer, acrylic acid modified; thermoplastic with good adhesion to a wide variety of substrates, excellent compatible with fillers and reinforcers; as chemical coupling agent for glass and mica-reinforced polyethylene; reduces moisture sensitivity in polyamides. *BP Performance Polymers.* Discontinued.

24366 Polybond® 1011
PP copolymer, high impact, acrylic acid modified; impact modifier in reinforced PP. *BP Performance Polymers.* Discontinued.

24367 Polybond® 2005
PP, acrylic acid-modified; additive for adhesion to metals and polar polymers; extrusion coating; FDA applications. *BP Performance Polymers.* Discontinued.

24368 Polybond® 2021
LLDPE, maleic anhydride; additive for adhesion to metals and polar polymers; film and sheet extrusion, injection molding *BP Performance Polymers.* Discontinued.

24369 Polybor®
12280-03-4
$H_2B_8O_{13}$
Disodium octaborate tetrahydrate
Boric acid, disodium salt, tetrahydrate; Disodium octaborate; Tim-Bor; Bora-Care; Disodium octaborate tetrahydrate. Fire retardant for treatment of lumber. *U.S. Borax & Chem.; Borax Consolidated Ltd.*

24370 Polybut
Polybutenes. *British Traders & Shippers Ltd.*

24371 polybutad1,2-polybutadiene
29406-96-0
Syndiotactic form. Thermosetting resin for wire coating, EPDM peroxide-cured modifier, coatings, processing aid.

24372 polybutadiene
9003-17-2
$(C_4H_6)_n$
butadiene rubber
cis-polybutadiene. BR; elastomer for tire industry, footwear, molded goods; blending ingredient in SBR; additive for plastics; coating resin in liquid form. Asahi Chem Industry co Ltd; BASF; Fireston Syn. Rubber & Latex; Goodyear; Phillips 66; Reichhold/Emulsion Polymers; Revertex Ltd.

24373 Polybutene
9003-28-5
$[C_4H_8]_n$
polybutylene
PIB; 1-butene, Homopolymer. Polymer formed by polymerization of a mixture of iso and normal butenes. Thermoplastic resin; used as tackifier, strengthener, and extender in adhesives, as plasticizer for rubber, as vehicle and fugitive binder for coatings, *Amoco; Ashland; BP Chem. Ltd; Harcros.*

24374 Polycarbafil®
A trade name for flame retardant polycarbonate materials. No manufacturer.

24375 polycarbonate resin
PC resin. Thermoplastic resin for molded products, solutions in cast or extruded film, structural parts, tube and piping. prosthetic devices, optical parts, windows, computer and business equipment, household appliances, compact disks, food contact and medical applications. *Aldrich; Ashland; LNP; Miles; Westlake Plastics.*

24376 Polycarboxylate AMC 60
Acrylic acid-maleic acid copolymer; high calcium dispersion power; co-builder in phosphate-free washing powders. *Hüls AG.*

24377 Polycare® 509
Vinyl acetate
isobutyl maleate. Vinylneodecanoate copolymer; hair styling fixative resin providing superior hold and curl retention even under extremely humid conditions. *Rhône-Poulenc Surf.*

24378 Polycat®
Amine catalysts for polyurethane foam. *Air Prods & Chems. Inc.*

24379 Polycat® 12
Catalyst for improved processing of flexible slabstock PU foam; for methylene chloride systems. *Air Prods & Chem/Polyurethanes.*

24380 Polycat® 5
Amine-based catalyst for improved processing of flexible slabstock PU foam; promotes blow reaction. *Air Prods & Chem/Polyurethanes.*

24381 Polycat® 58
Amine-based catalyst for PU flexible molded foam; for improved surface cure. *Air Prods & Chem/Polyurethanes.*

24382 Polycat® 77
Amine-based catalyst for improved processing of flexible slabstock PU foam; promotes gel reaction. *Air Prods & Chem/Polyurethanes.*

24383 Polycat® 77
Amine-based catalyst for PU flexible molded foam. *Air Prods & Chem/Polyurethanes.*

24384 Polycat® 91
Amine-based catalyst for improved processing of flexible slabstock PU foam; gives delayed action. *Air Prods & Chem/Polyurethanes.*

24385 Poly-Chek
For additives for polymers, such as heat and light stabilizers, lubricants and the like, in Int Class 1. *Ferro.* Discontinued.

24386 polychloroprene
126-99-8 204-818-0
neoprene
2-chlorobutadiene 1,3; chloroprene rubber. CR; elastomer for molding, extrusion, and calendering for adhesive compounding, construction, automotive, hose and cable jackets, conveyor belts, closed cell sponge, etc.

24387 Polychol 5
61791-20-6
Laneth-5
O/w emulsifier, dispersant, gellant, emollient, solubilizer for personal care products, hair straighteners, pharmaceuticals; bromo dye solvent. *Croda Inc.; Croda Chem. Ltd.*

24388 polychrome blue of unna
A dyestuff prepared by the action of potassium carbonate on methylene blue; used in microscopy.

24389 Polycin 12
Urethane polyol
CasChem. Name unverified.

24390 Polycizer DOP
117-81-7 1291 204-211-0
Dioctyl phthalate
Polycizer 162; Merrol DOP. Plasticizer, used with polyvinyls. *Harwick Distribution Corporation.*

24391 Polycizer DOA
103-23-1 203-090-1
Dioctyl adipate
Polycizer 332; Merrol DOA. Plasticizer, used with polyvinyls. *Harwick Distribution Corporation.*

24392 Polyclar® 10
Polyvinylpolypyrrolidone homopolymer; insoluble cross-linked polymer; stabilizer for beverage clarification and stabilization; adsorbent in thin-layer and column chromatography. *ISP.*

24393 Polyclear
Textile reduction clearing additive. *Hoechst AG.*

24394 Polyclear 32-F
Clearing agent for use on polyester and polyester/cotton blends; removes loose dye from disperse dyes on polyester and nylon fabrics. *Henkel/Textiles.*

24395 Polycomp® 139
9016-75-5

PPS-filled PTFE composite; excellent wear properties; not for moderate to highly loaded applications due to low resistance to deformation. *ICI Fluoropolymers.*

24396 Polycomp® 185
9016-75-5
PPS-filled PTFE composite; good deformation resistance with excellent wear properties, very low abrasiveness. *ICI Fluoropolymers.*

24397 Polycon
Photographic developer. *May & Baker Ltd.* Name unverified.

24398 Polycon II
Silafocon A
Contact lens material. *Syntex Ophthalmics Inc.* Name unverified.

24399 Polycon S-60 K
1338-41-6 8872 215-664-9
Sorbitan stearate
Lipophilic food emulsifier used in emulsions where weaker water-binding properties and enhanced aeration are desired; crystallization promoter, surface film former. *Witco Corporation.*

24400 Polycon S-80 K
1338-43-8 8872 215-665-4
Sorbitan oleate
Lipophilic food emulsifier used in emulsions where weaker water-binding properties and enhanced aeration are desired; crystallization promoter, surface film former. *Witco Corporation.*

24401 Polycon T-60 K
9005-67-8 8872
Polysorbate 60
Hydrophilic food emulsifier, surfactant for formulating oil-water emulsions. *Witco Corporation.*

24402 Polycon T-80 K
9005-65-6 7742
Polysorbate 80
Hydrophilic food emulsifier, surfactant; used in pickle industry due to very low melting point. *Witco Corporation.*

24403 Polycote Pedigree
Benomyl + iodofenphos + metalaxyl; a fungicide and insecticide seed coating for seeds. *Seedcote Systems Ltd.*

24404 Polycote Prime
Powder mixture of iprodione, metalaxyl and thiabendazole; a polymer seed coating for carrots. *Seedcote Systems Ltd.*

24405 Polycoupler IMP RFB X-353
Textile resin copolymer utilizing multifunctional groups of a hydantoin-acetylenic derivative to achieve resin fixation. *CNC Int'l L.P.*

24406 polycroit
8466
The coloring matter of saffron.

24407 Polycrol
A proprietary preparation containing methyl-polysiloxane, aluminum hydroxide gel and magnesium hydroxide; an antacid. *Nicholas Laboratories Ltd.* Name unverified.

24408 Polycron
Disperse dyestuffs; for coloring of polyester fibers. *Holliday Dyes & Chemicals Ltd.*

24409 Polycryl
Range of acrylic polymer emulsions; for textile finishing. *Morton Int'l. Ltd.*

24410 Polycure
Polyester catalysts. *Mooney Chemicals Inc.*

24411 Polydis® TR 121
301-02-0 206-103-9
$C_{18}H_{35}NO$
Oleamide
9-Octadecenamide, (Z)-; Oleylamide; cis-9,10-Octadecenoamide. Slip agent for polyethylene films; lubricant and mold release for injection molding applications, processing thermoplastic resins, thermoplastic elastomers, and thermoset rubber systems; dispersant. *Struktol.*

24412 Polydis® TR 131
112-84-5 204-009-2
Erucamide
Slip agent for polyolefin films; release agent for polymer systems; dispersant in color concs. and printing inks; process aid/lubricant for thermoplastic elastomers, thermoplastic resins, thermoset rubber systems. *Struktol.*

24413 Polydur
Polyester resin molding compounds; for electronics and electrical engineering. *Dynamit Nobel Wien GmbH.* Name unverified.

24414 polydymite
$(Ni{\cdot}Co)_4S_5$
nickel-linnaeite
A mineral.

24415 Polydyol 30-G
Dye carrier giving good color value, level and spot-free dyeings. *Eastern Color & Chem.*

24416 Polyeite
A proprietary polyester laminating resin. *Reichhold.* Discontinued.

24417 Poly-Em
Polyethylene wax emulsions. *Rohm & Haas UK.*

24418 Polyester 1606
Complex polyester; emulsion preventer/breaker base for oil treatment; base for oil-emulsion prevention. *Chemron.*

24419 Polyester N-95
Surfactant; emulsion preventer/breaker base for oil treatment; well stimulation additive. *Chemron.*

24420 Polyestren®
Blended dyestuffs (dispersion/vat or sulfur/vat dyes); for dyeing and printing of blended textiles. *Cassella AG.*

24421 Poly-Eth
Plastic resins. *Monsanto (Solaris).* Name unverified.

24422 Poly-Eth Hi-D
Plastic resins. *Monsanto (Solaris).* Name unverified.

24423 Polyetherimide
61128-46-9
Electrafil® J-1106/CF/30; PDX-84347; PI-20GF/000; Thermocomp® EC-1006; Ultem® 1000; Ultem® 2100; Ultem® 2212; Ultem® 4000; Ultem® 6000; Ultem® CRS5001; Ultem® CRS5011; Ultem® FXU100. 30% PAN carbon fiber-reinforced; static dissipative and conductive thermoplastic *Akzo Engineering Plastics.* Name unverified.

24424 polyethylene
9002-88-4 7728
$[C_2H_4]_x$
ethene homopolymer
Polymer of ethylene monomers. laboratory tubing; prostheses; electrical insulation; packaging materials; kitchenware; tank and pipe linings; paper coatings; textile stiffeners. Asahi Chem Industry Co Ltd; Ashland; Elf Atochem SA; Eastman; EniChem UK; Exxon; LNP; Mitsubishi Petrochem.; Quantum/USI.

24425 polyethylene glycol
25322-68-3 7729
polyglycol; polyether glycol. PEG; Condensation polymers of ethylene glycol; chemical intermediates, plasticizers, softeners, humectants, ointments, polishes, paper coating, mold lubricants, bases for cosmetics and pharmaceuticals, solvents, binders, metal and rubber processing, food additives, laboratory reagent. BASF; BP Chem. Ltd; Calgene; Dow; DuPont; Harcros; Henkel; Hüls; Inolex; Olin; Rhone-Poulenc Surf.; Texaco; Union Carbide.

24426 Polyethylene terephthalate
25038-59-9 7730
Electrafil® J-1800/CF/30; Celstran® PETG30-01-4; Cleartuf Series; Crastine® XMB 1068; Electrafil® J-1800/CF/30; Ertalyte®; Grilpet® EV-30; Impet® 330; Kodapak® 5214A; Kodapak® PET; Copolyester 13339; Kodar® PETG Copolyester 6763; Mitsubishi Kasei GF-PET 6010G15; Petra® 130; Petra® 130FR; Petra® 230; Petra® 242; RTP 1105FR; Tenite° PET 9902; Traytuf Ultra-Clear; Ultralen® SP 3700 S; Ultralen® SP 3705; Valox® 9215,. PET, 30% PAN carbon fiber-reinforced; static dissipative and conductive thermoplastic *Akzo Engineering Plastics.* Name unverified.

24427 polyethylene wax
Wax for polishes, plastics and rubber processing, printing inks, pigment masterbatches, hot melts; PVC lubricant. *Hoechst Celanese; Hüls AG; IGI Boler; Sartomer; Stevenson Bros.; Syn. Prods.*

24428 polyethylene, high-density
HDPE
Blow-molded products, injection-molded items, film and sheet, piping, fibers, gasoline and oil containers. AlliedSignal; BP Chem. Ltd; Chevron; Chisso; DuPont; Exxon; Hüls Am.; OxyChem; Quantum/USI; Solvay Polymers; Westlake plastics.

24429 polyethylene, linear low density
LLDPE
Multi-purpose polymer. *BP Chem. Ltd; Neste Polyeten AB.*

24430 polyethylene, low-density
LDPE
Packaging film, food packaging, paper coating, liners for drums, wire and cable coating, toys, cordage, waste bags, chewing gum base, squeeze bottles, electrical insulation. AlliedSignal; Chevron; Eastman; Exxon; Hüls Am.; Neste Polyeten AB; Quantum/USI; Westlake Plastics.

24431 Polyfeed
Water soluble, chlorine free N-P-K fertilizer, with chelated micro-nutrients; for

direct soil application, via irrigation system or foliar spray. *Haifa Chemicals Ltd.* Name unverified.

24432 Polyfil® WC
Anhydrous organofunctional pigment; pigment reducing water vapor transmission, yielding excellent wet and dry electrical properties and good long-term stability in EPR and crosslinked polyethylene. *J.M. Huber/Clay Div.*

24433 Polyfilm
Polyethylene film; used in packaging and industrial applications. *Dow UK.* Discontinued.

24434 Polyfine MF15C
Thermoplastic resin; formulated as economical substitute for polyacetal resin in applications where friction and wear properties are required; recommended for VTR tape reel, noiseless gear, bearing surface, etc.; for injection molding; MF15C offers high rigidity and high gloss for audio tape reels. *Advanced Web Prods.* Discontinued.

24435 Polyflex
91-53-2 3800 202-075-7
$C_{14}H_{19}NO$
6-Ethoxy-2,2,4-trimethyl-1,2-dihydroquinoline
EMQ; Santoflex; Santoquin;EQ; Santoflex A; Santoflex Aw; Stop-scald; Nix-scald; Niflex; Antioxidant EC; Permanax 103; Amea 100; Dawe's Nutrigard; Quinol Ed; Nocrac AW; Niflex D; Antox; Aries Antox; Antage AW;. A proprietary antioxidant. *Naugatuck (US Rubber).* Name unverified.

24436 Polyflex®
A registered trademark for a flexible polystyrene sheet and fiber. No manufacturer.

24437 Polyflo
Oil additives. *Hoechst AG.*

24438 Polyflon
A proprietary brand of polytetrafluoroethylene (PTFE). *Daikin Kogyo Co.* Unverified.

24439 Polyfon
8061-51-6
A range of sodium lignin sulfonates; used as concrete retard additives and solids dispersion additives. *Harcros Australia.*

24440 Polyfon® F
8061-51-6
Sodium lignosulfonate
Dispersant for industrial applications from agriculture to ceramics. *Westvaco.*

24441 Polyfusor Solutions
A range of sterile pyrogen-free solutions. *The Boots Co plc.* Discontinued.

24442 Poly-G Fluids
Polyalkylene glycols
Olin UK. Discontinued.

24443 Poly-G Polyols
Polyether polyols for urethane foam. *Olin UK.* Discontinued.

24444 Poly-G® 200
25322-68-3 7729
PEG 200; chemical intermediate for production of surfactants for cleaners, textiles, paper, cosmetics; carrier for pharmaceuticals; also in cosmetics and personal care products, textiles, rubber mold releases, printing inks and dyes, metalworking fluids, foods, paints, paper, wood products, adhesives, agricultural products, ceramics, electrical equipment, petroleum products, photographics products and resins. *Olin.* Discontinued.

24445 Poly-G® 20-28
Polyether polyol (diol)
For caulks, sealants, coatings, tire fill, clay pipe seal, high modulus reaction injection molding applications. *Olin.* Discontinued.

24446 Poly-G® 20-56
Polyether polyol (diol)
For adhesives, caulks, sealants, coatings, castable elastomers, tire fill, clay pipe seal. *Olin.* Discontinued.

24447 Poly-G® 30-56
Polyether polyol (triol)
For adhesives, coatings. *Olin.* Discontinued.

24448 Poly-G® 55-28
Polyether polyol (VHP diol)
Suitable for applications where reactivity is important; for adhesives, caulks, sealants, thermoplastic urethanes and hot melts, one-shot elastomers, reaction injection molding applications *Olin.* Discontinued.

24449 Polygard
Tris (mixed mono and dinonyl phenyl) phosphite
Antioxidant, processing and color stabilizer for PP, LDPE, LLDPE, HDPE, HIPS, ABS, PVC, PC, and EVA and polyamide hot-melt adhesives. *Uniroyal.*

24450 polygeline
A polymer of urea and polypeptides derived from denatured gelatin.

24451 Polygeline™
9007-34-5 2543 232-697-4
Modified collagen. *Calbiochem Corp.*

24452 polyglactin
A synthetic suture capable of being absorbed by the patient's body. It is a mixture of lactic acid polyester with glycolic acid.

24453 Polygliceryl-10 decaoleate
11094-60-3 234-316-7
Hodag PGO-1010 (formerly Hodag SVO-10107;)Caprol® 10G10O; Polyaldo® DGDO KFG; Decaglyceryl decaoleate; Decaglycerol decaoleate; Caprol® 10G10O; Drewmulse® 10-10-O; Drewpol® 10-10-O; Nikkol Decaglyn 10-O; Santone_so 10-10-O; Unitolate PGO-1010,. Emulsifier; solubilizer; emulsion stabilizer; dispersant aid in flavors; cosmetic ingradient Calgene; Karlshamns; Stepan/PVO; Calgene; Universal Preserv-A-Chem; Nikko Chem Co. Ltd; Van der Bergh Foods;.

24454 Polygloss
Flexographic printing inks; for printing flexible film. *AlliedSignal Inc/Sinclair and Valentine Division.* Name unverified.

24455 Polyglucadyne
1-3 and 1-6 β-glucans copolymer; patented material for protection of skin in cosmetics; macrophage stimulating factor. *Brooks Industries.*

24456 Polyglyceryl isostearostearate
Lafil WL 3254. *Gattefosse SA.*

24457 Polyglyceryl stearate series
37349-34-1
Emalex MSG-2; Diglyceryl monostearate; Caprol® 3GS; Hodag PGS-101, PGS-61; Polyaldo® TGMS KFG; Santone® SDD,. Emulsifier for cosmetics and foods. *Nihon Emulsion; Calgene; Karlshamns; Lonza Inc.; Van Den Bergh Foods.*

24458 Polyglycol B-11-50
EO/PO random copolymer based on butanol; surfactant for textile processing. *Hoechst Celanese/Colorants & Surf.*

24459 polyglycolic acid
Poly(oxycarbonylmethylene).
A synthetic suture capable of being absorbed by the patient's body.

24460 Polygon
7758-29-4 8846 231-838-7
Sodium tripolyphosphate
Albright & Wilson Ltd., Phosphates & Speciality Business. Name unverified.

24461 Polygrade
Degradable plastic materials; additive concentrates to cause degradation (after useful lifetime). *Ampacet Corporation.*

24462 Polyguide®
DuPont UK.

24463 polyhalite
isobelite
A mineral which occurs in the Strassfurt deposits. It is a crystalline mixture of the sulfates of calcium, magnesium, and potassium, and is found with rock salt. It has the formula, $K_2SO_4{\cdot}MgSO_4{\cdot}2CaSO_4{\cdot}2H_2O$.

24464 Polyhall® 21J
9003-05-8
Very high molecular weight polyacrylamide; flocculant for setting coal floation residue and slime tailings; settling aid for phosphate slimes, gold and copper tailings; clarification and thickening aid. *Rhône-Poulenc/Water Soluble Polymers.*

24465 Polyhipe
Range of organic microporous matrices for immobilization. *Microporous Materials Ltd.*

24466 polyhydrite
Silicate of iron, a mineral.

24467 polyimide, thermoplastic
Thermoplastic resin for compression molding, injection molding and film casting; structural composites, adhesives, film, insulation, coatings.

24468 polyimide, thermoset
Thermoset resin for structural and nonstructural aerospace applications., bearings, automotive components, films for electrical equipment.

24469 polyisobutene
9003-27-4; 9003-29-6
$[C_4H_8]_x$
polyisobutylene homopolymer
Homopolymer of isobutylene. synthetic rubber; polymeric additive; thickener for lubricating oils; for the adhesive and sealant industry, electrical insulating oils, bases for chewing gums, for production of damp-proof courses containing fillers in construction industry. *BASF; Rit-Chem.*

24470 polyisoprene
9003-31-0
$[C_5H_8]_x$

2-methyl-1,3-butadiene
isoprene rubber; IR; cis-1,4-polyisoprene rubber. Thermoplastic polymer of isoprene. Elastomer for light colored goods, adhesives, footwear, sponge products, tires, pharmaceutical goods, rubber bands, molded and mechanical goods. *Goodyear; A. Schulman.*

24471 Polyisoprene
9006-04-6 232-689-0
DPNR; Dynatex GTZ; Hyflo NS, S; PA-57; PA-80; Unitex; ENR 25; Natural Rubber. Natural rubber; modified by epoxidation reaction to achieve greater oil resist., gas permeability, and damping chars.; used for specialty tires, milking inflations, hoses, footwear. *Guthrie Latex;H.A. Astlett; Akrochem; Cray Valley Ltd.*

24472 Polylite
Alkylated diphenylamine
Closely related to octamine in chemical structure and may be considered a liquid form of octamine; has the same broad protective action with the same minimum of discoloration and staining; for use as an antioxidant and stabilizer in the manufacture of SBR and nitrile polymers and as an antioxidant in latex compounding. *Uniroyal.* Name unverified.

24473 Polylite
A mineral; it is a variety of pyroxene. No manufacturer.

24474 Polylite®
Polyester resins or polyurethane foam; also chemicals used in industry, science, photography, agriculture, horticulture, forestry, unprocessed artificial resins, unprocessed plastics, manures, fire extinguishing compositions, tempering and soldering prepa preparations, etc. *Reichhold.*

24475 Polylite® 32-162
Isophthalic polyester resin, acrylic-modified; produces Corian®-like product with good stain resistance, high heat distortion temperature; used for vanities, etc. *Reichhold.*

24476 polylithionite
A mineral. It is a variety of zinnwaldite containing lithium.

24477 Polyloy® 6
Polyamide 6
Granular form; used in automotive, machinery, electrical/electronics, building industries. *EMS-Chemie AG.* Discontinued.

24478 Polyloy® A
Polyamide 6/6
Granular form; used in automotive, machinery, electrical/electronics, building industries. *EMS-Chemie AG.* Discontinued.

24479 Polylube 1105
Blend of fatty acid esters, polyoxyalkylene derivs.; lubricant, antistat for nylon. *Hart Chem. Ltd.*

24480 Polylube ASTL
Polyoxyalkylene alcohol phosphate
Antistat component for lubricant formulations for all fibers. *Hart Chem. Ltd.*

24481 Polylube DDL
Blend of polyoxyalkylene alcohol derivatives and polyalkylene glycol ethers; heat-stable lubricant for glass extrusion applications. *Hart Chem. Ltd.*

24482 Polylube Wax
Blend of polyoxyalkylene derivatives; wax for sizing operations; cohesive agent for short staple fiber in woolen systems. *Hart Chem. Ltd.*

24483 Polymate
Cooling water treatment. *Dearborn Chemicals Ltd.*

24484 Polymekon
Silicone antifoam emulsions. *Thomas Goldschmidt Ltd.*

24485 Polymekon®
63231-60-7 264-038-1
Modified microcrystalline wax; used in the formulation of inks and coatings and as binder, antislip and antimar agent. *Petrolite.*

24486 Polymel 7
Modified polyethylene wax; low molecular weight wax incorporated into rubber batches giving excellent mold release. *Frank B. Ross.*

24487 Polymene AZ
61791-12-6
Blend; for textile use. *Rhône-Poulenc Surf.*

24488 Polymer C
Microcrystalline wax replacement. *Crowley Chem.*

24489 Polymeric Sealant Gun
Thermal extruder for rubber-like sealants. *Hardman. Name unverified.*

24490 Polymet®
Polymer Corp.

24491 polymethylmethacrylate
9011-14-7
$[C_5H_8O_2]_n$
2-propenoic acid methyl ester, homopolymer
Polymer of methyl methacrylate; thermoplastic used as main constituent of acrylic sheet, molding and extrusion compounds. *Elf Atochem SA; Shuman Plastics.*

24492 Polymethylmethacrylate
9011-14-7
Polymethylmethacrylate
Lucryl®; Polymethyl methacrylate; Methyl methacrylate copolymer; Acryloid® A-30, B-44; Paraloid® K-120N,. Thermoplastic molding compounds; std. grades for injection molding or extrusion offer high transparency and brilliance, high surface hardness and scratch resist., excellent weather resist., mech. strength and rigidity; impactmodified grades. *BASF AG; BASF plc; Rohm & Haas.*

24493 Polymica 200, 325, 400
12001-26-2
Wet-processed musscovite mica; filler for coating and performance polymer applications where high brightness, color, particle size, and consistency are important. *Franklin Industrial Minerals.*

24494 polymignite
$4(Ca \cdot Ce \cdot Fe)O \cdot (Ti \cdot Zr)O_2 \cdot CaO \cdot Nb_2O_5$
A mineral.

24495 Polymin®
Polyacrylamide-based, modified polyethylene imines; retention, dewatering, and flocculating agents. *BASF AG.*

24496 Polymist® F-5
9002-84-0 7743 204-126-9
Polytetrafluoroethylene
Additive in elastomers or plastics where improved lubricity and/or wear resistance are required. *Ausimont.*

24497 Polymoist Mask
9007-34-5 2543 232-697-4
Collagen fiber material; moisturizer and vehicle for cosmetic active agents. *Henkel/Cospha; Henkel Canada.*

24498 Polymon
Soluble dyes for plastics. *ICI Chem & Polymers Ltd.*

24499 Polymone
94-75-7; 120-36-5
Aqueous solution containing 10% w/v 2,4-D and 40% w/v Dichlorprop; for use as an agricultural herbicide. *Universal Crop Protection Ltd.*

24500 Polymul
Range of polyethylene emulsions; variety of protective and decorative coatings. *Henkel Chemicals Ltd.* Name unverified.

24501 Polyoil Hüls 110
9003-17-2
Polybutadiene resin; improves water resistance and dry characteristics of alkyds; used in anticorrosion and electrodeposition coatings. *Hüls Am.*

24502 Polyox
A range of water soluble, high molecular weight polymers of ethylene oxide; used in adhesives, binders, pharmaceuticals, and lubricants. *Union Carbide (UK) Ltd.*

24503 Polyox® WSR 3333
25322-68-3 7729
PEG-9M; a water-soluble polymer. *Amerchol Corp.* Discontinued.

24504 Polyox® WSR N-10
25322-68-3 7729
PEG-2M; thickener. *Amerchol Corp.*

24505 Poly-Pale®
Polymerized rosin and rosin esters; used for hot melt coating and adhesives. *Hercules; Hercules Ltd.*

24506 Poly-Pale®
Poly-Pale® Ester 10. Polymerized rosin esters; tackifying resin for solvent and emulsion pressure-sensitive adhesives and for hot-melt packaging adhesives. *Hercules.*

24507 Polypeg-E
Polyethylene glycol fatty ester. *Olin UK.* Discontinued.

24508 Polypel
Fertilizer. *Monsanto (Solaris).* Name unverified.

24509 Polypenco® Cast Acrylic Rod
Cast acrylic
Optical clarity, high tensile strength, good electrical properties; used in displays, signs, furniture components, lenses, electrical/electronic parts. *Polymer Corp.*

24510 Polypenco® Nylon 101
32131-17-2
Type 6/6 nylon; strong, stiff nylon for use in food processing, machinery, electronics, military and other industries for bearings, bushings, valve seats, seals, rollers, gears, insulators, fasteners, liners, tooling fixtures, forming dies, etc. *Polymer Corp.*

24511 Polypenco® PEEK
PEEK; thermoplastic offering high continuous service temperature (480°F), low emission of smoke and toxic fumes on exposure to flame; for electrical/electronic components, aircraft components, microwave, automotive applications, pump and valve parts and chemical processing. *Polymer Corp.*

24512 Polypenco® Polycarbonate
PC; thermoplastic high impact. strength, good electrical properties; used in stand-off insulators, coil forms, optical and transparent structural components. *Polymer Corp.*

24513 Polypenco® Polysulfone
25135-51-3
Polysulfone
Semi-transparent, heat, chemical and hydrolysis resistance, high-performance engineering thermoplastic; for medical tubing, food and beverage contact parts, dairy equipment, aerospace components, circuit boards and connectors. *Polymer Corp.*

24514 Polypenco® Q200.5
9003-53-6 9028
Crosslinked PS; rigid insulating material; outstanding dielectric properties, good impact strength; used in UHF, VHF, and microwave insulators, communication and electronic equipment. *Polymer Corp.*

24515 Polypentek
A proprietary trade name for polypentaerythritol. No manufacturer.

24516 Polypeptide 10
9015-54-7 310-296-6
Hydrolyzed collagen
Protective colloid, anti-irritant, substantivity agent, moisturizer for hair and skin care products, dish detergents; dye leveler and fiber protectant in textile industry. *May-brook.*

24517 Polypeptide 37
9015-54-7 310-296-6
Hydrolyzed collagen
Surfactant for personal care products *Inolex.*

24518 Polyphenylene oxide
25134-01-4
Electrafil® G-1704/SS/5; Electrafil® F-1700/CF/10/A; Electrafil® J-1700/CF/10; MP-10-CF-4CC/15T; Noryl® BN25; Noryl® EM5101; Noryl®EN185; Noryl®FN150; Noryl®FN215X; Noryl® GTX810; Noryl® HM3020; Noryl® HS1000X; Noryl® N190X; Noryl® PC180X; Noryl® PN235; Noryl® PX0722; Noryl® SE100; Noryl®SE1GFN2; Stat-Kon® ZC-1003; Thermocomp® ZF-1004; Stylex 72001 NA;,. PPO, modified, 5% stainless steel fiber filled; static dissipative and conductive thermoplastic *Akzo Engineering Plastics.* Name unverified.

24519 Polyphos
50813-16-6 8814 256-779-4
Sodium hexametaphosphate
sodium polymetaphosphate. *Olin UK.* Discontinued.

24520 polyphosphoric acid
8017-16-1 7740 232-417-0
$H_{n+2}P_nO_{3n+1}$
superphosphoric acid
phospholeum; tetraphosphoric acid. Acid used in the manufacture of phosphates, phosphate esters, catalysts, fuel cell electrolytes, metal cleaning and brightening, organic reactions. d = 2.1000. *Albright & Wilson.*

24521 Polyplasdone®
Crospovidone; pharmaceutic aid. *ISP.* Name unverified.

24522 Polyplasdone® XL
Polyvinylpolypyrrolidones
Insoluble crosslinked polymer of N-vinyl-2-pyrrolidone used as a tablet disintegrant, complexing agent, detoxifier and antidiarrhea agent; adsorbent in thin-layer chromatography. *ISP.*

24523 Polyplate 90
1332-58-7 5294 296-473-8
Kaolin clay; delaminated functional filler with high brightness, finer particle size; for adhesives, paints, plastics, and inks. *J.M. Huber.*

24524 Polypor
Silica gel. *Quantum Chemical Corp.*

24525 Polyprene
Weber's name for rubber. No manufacturer.

24526 Poly-Pro
Plastic resins. *Monsanto (Solaris).* Name unverified.

24527 polypropylene
9003-07-0 7741
$[C_3H_6]_x$
1-propene, homopolymer
PP; propylene polymer; polypropene. Polymer of propylene monomers; three forms; isotactic (fiber-forming), syndiotactic, atactic (amorphous). (nucleated); Isotactic; fishing gear, ropes, filter cloths, laundry bags, protective clothing, blankets, fabrics, carpets, yarns. Amoco; Aristech; Ashland; Chisso; Eastman; Exxon; Fina; Hüls; LNP; Mitsubishi Petrochem.; Mitsui Toatsu Chem.; Neste UK; Quantum/USI; Shell; Solvay Polymers.

24528 polypropylene glycol
$HO(C_3H_6O)_nH$
Hydraulic fluids, rubber lubricants, antifoam agents, intermediates for urethane foams, adhesives, coatings, elastomers, plasticizers, paint formulations, lab reagent. Aldrich; Arco; Ashland; BASF; BP Chem. Ltd; Calgene; Dow; Harcros; Hüls AG; Miles; Olin; PPG Industries; Rhone-Poulenc Surf.; Texaco; Witco.

24529 Polyquart H
PEG-15 tallow polyamine
Surfactant used in personal care products; hair conditioner, antistat. *Henkel/Cospha; Henkel Canada.*

24530 Polyquaternium-11
53633-54-8
Vinylpyrrolidone/dimethylaminoethyl methacrylate copolymer/diethyl sulfate reaction product
Gafquat® 734; Gafquat® 755; Gafquat® 755N; Quaterium 23. Film-forming substantive polymer, conditioner for formulation of hair conditioners, rinses, sprays, shampoos, dyes, semipermanents, deodorants, antiperspirants, shaving preparations, antiseptics, toilet soaps, skin creams, sunburn remedies. Liquid; 20% aqueous solution *ISP.*

24531 Polyquaternium-16
29297-22-0
Polyquaternium-16
Luviquat® HM 552; Luviquat® FC 370; Luviquat® FC 905; Luviquat® FC 550. Conditioner and setting resin for hair care. *BASF AG.* Name unverified.

24532 Polyquest
Boiler water treatment. *Grace Dearbom Ltd.*

24533 Polyquest 80
140-01-2 205-391-3
Pentasodium diethylenetriamine pentaacetate
Sequestering agent for iron. *CNC Int'l L.P.*

24534 Polyrad
5-and 11-mole ethylene oxide adducts of Amine D dehydroabietylamine; Polyrad rosin is used as a corrosion inhibitor and detergent in petroleum-processing agents and for inhibiting hydrochloric acid used in industrial and household cleaners. *Hercules.*

24535 Polyram
Alkyl propylene polyamines based on coco and tallow alkyl chains; bitumen adhesion agents, intermediate for chemical synthesis. *Elf Atochem UK/Ceca.*

24536 Polyram® -Combi, DF
9006-42-2
Metiram; for control of fungus diseases in fruit, hops, vines, vegetables, ornamentals. *BASF AG.*

24537 Polysalt
Based on salts of polycarboxylic acids; dispersants for extenders, papercoating pigments; for stabilizing coating mixtures and slurries; grinding assistants for chalk. *BASF AG.*

24538 Polysar Bromobutyl 2030
Brominated isobutylene/isoprene copolymer; nonstaining stabilizer; low Mooney viscosity version of Polysar Bromobutyl X2 for easier processing. *Bayer Corp; Polysar.*

24539 Polysar Butyl 100
9010-85-9
Isobutylene/isoprene copolymer rubber; nonstaining stabilizer; butyl rubber for med. and high voltage insulation, membranes for roofing and reservoir linings; vulcanizable with sulfur, quinoid, resin; slow curing. *Bayer Corp; Polysar.*

24540 Polysar Chlorobutyl 1240
Chlorinated isobutylene/isoprene copolymer, nonstaining stabilizer; low Mooney viscosity version of Polysar Chlorobutyl 1255 for easier processing; vulcanized with zinc oxide, sulfur, quinoid, resin, peroxide. *Bayer Corp; Polysar.*

24541 Polysar EPDM 227
EPDM terpolymer; used alone to promote heat resistance or in blends to increase hardness while decreasing compound viscosity; for injection molding and peroxide-cured heat-resistance compounds; slow cure rate. *Bayer Corp; Polysar.*

24542 Polysar EPDM 6463
EPDM, with 50 parts per hundred nonstaining paraffinic oil; oil-extended rubber for optimum heat stability and weather resistance; standard cure rate; used for heat-resistant applications, hose, electrical insulation, roofing membranes, and molded, extruded, and calendered goods. *Bayer Corp; Polysar.*

24543 Polysar EPDM XG 006
EPDM terpolymer extended with paraffinic oil; high viscosity product designed to improve uv and color stability in low flexibility modulus, high flow polyolefin blends; for impact modification of polyolefins and for use in thermoplastic elastomers. *Bayer Corp; Polysar.*

24544 Polysar EPM XF 004
Ethylene-propylene copolymer; low viscosity product designed to improve flow and gloss in polyolefin blends, for use in impact modification in polyolefins, for use in thermoplastic elastomers. *Bayer Corp; Polysar.*

24545 Polysar XL 30102
9010-85-9
Lightly crosslinked butyl terpolymer; used for preformed tape in automotive and architectural applications, solvent release sealants, pressure-sensitive adhesives, hot-applied sealants. *Polysar.*

24546 Polyseal
Polyurethane concrete sealer; suitable for use on any concrete floor that is subject to heavy abrasive traffic or is exposed to chemicals or oils. *Secure Inc.* Discontinued.

24547 Polyset 100
Acrylic triazine resin solution; stiffening agent for polyester. *CNC Int'l L.P.*

24548 Polysil
Silicone rubber gums and compounds. *Midland Silicones.* Unverified.

24549 Polysoft 35
Emulsion synthetic softener/lubricant for precure, post-cure, and crease resistance finishes with excellent resistance to yellowing. *CNC Int'l L.P.*

24550 Polysoft B
9002-88-4 7728
Polyethylene emulsion; textile softener improving physical properties and sewability in durable press applications. *Sybron.*

24551 Polysoft CA
Polyethylene
Antistatic agent, lubricant and softener compatible with resin finishes. *Scher Chemicals Inc.* Discontinued.

24552 Poly-Solv®
Glycol ether solvents. *Olin UK.* Discontinued.

24553 Poly-Solv® DE (High Gravity)
Diethylene glycol monoethyl ether (75%) and ethylene glycol (25%); solvent for brake fluids, hard-surface cleaners, leather dyeing, paints, coatings, printing inks, textile vat dyeing and printing, adhesives, antifreeze, floor waxes/polishes, insect repel *Olin.* Discontinued.

24554 Poly-Solv® DM
111-77-3 6116 203-906-6
$C_5H_{12}O_3$
Diethylene glycol monomethyl ether
methyl carbitol. Solvent for brake fluids, hard-surface cleaners, leather dyeing, paints, coatings, printing inks, textile vat dyeing and printing, adhesives, antifreeze, floor waxes/polishes, insect repellents; solubilizer for dyes; plasticizer. *Olin.* Discontinued.

24555 Poly-Solv® DPM
34590-94-8 3407 252-104-2
$C_7H_{16}O_3$
Dipropylene glycol monomethyl ether
dipropylene glycol monomethyl ether; Dowanol 50b; Oxybispropanol, Methyl Ether; Dowanol DPM; bis-(2-Methoxypropyl) ether; Arcosolve DPM; Methoxymethylethoxy)propanol; PPG-2 methyl ether; DPGME; DPM;. Solvent for brake fluids, hard-surface cleaners, leather dyeing, paints, coatings, printing inks, textile vat dyeing and printing, adhesives, antifreeze, floor waxes/polishes, insect repellents; solubilizer for dyes; plasticizer. mp = -83°; bp = 190°; d_4^{25}= 0.948; n_D^{25}= 1.419; soluble in H_2O, organic solvents; LD_{50} (rbt orl) = 5.4 mg/kg. *Olin.* Discontinued.

24556 Poly-Solv® EE
110-80-5 3797 203-804-1
$C_4H_{10}O_2$
Ethylene glycol monoethyl ether
2-ethoxyethanol. Solvent for brake fluids, hard-surface cleaners, leather dyeing, paints, coatings, printing inks, textile vat dyeing and printing, adhesives, antifreeze, floor waxes/polishes, insect repellents; solubilizer for dyes; plasticizer. *Olin.* Discontinued.

24557 Poly-Solv® EM
109-86-4 6118 203-713-7
$C_3H_8O_2$
Ethylene glycol monomethyl ether
methyl cellosolve. Solvent for brake fluids, hard-surface cleaners, leather dyeing, paints, coatings, printing inks, textile vat dyeing and printing, adhesives antifreeze, floor waxes/polishes, insect repellents; solubilizer for dyes; plasticizer. *Olin.* Discontinued.

24558 Poly-Solv® MPM
107-98-2 203-539-1
$C_4H_{10}O_2$
Monopropylene glycol monomethyl ether
1-methoxy-2-propanol. Solvent for brake fluids, hard-surface cleaners, leather dyeing, paints, coatings, printing inks, textile vat dyeing and printing, adhesives, antifreeze, floor waxes/polishes, insect repellents; solubilizer for dyes; plasticizer. *Olin.* Discontinued.

24559 Poly-Solv® TM
112-35-6 203-962-1
Triethylene glycol monomethyl ether
Solvent for brake fluids, hard-surface cleaners, leather dyeing, paints, coatings, printing inks, textile vat dyeing and printing, adhesives, antifreeze, floor waxes/polishes, insect repellents; solubilizer for dyes; plasticizer. *Olin.* Discontinued.

24560 Poly-Solv® TPM
PPG-3 methyl ether
Solvent for brake fluids, hard-surface cleaners, leather dyeing, paints, coatings, printing inks, textile vat dyeing and printing, adhesives, antifreeze, floor waxes/polishes, insect repellents; solubilizer for dyes; plasticizer. *Olin.* Discontinued.

24561 Polysolvan E
A proprietary trade name for a solvent mixture comprising the acetates of propyl, isobutyl, and amyl alcohols. No manufacturer.

24562 Polysolvan O
A solvent composed of the ester of isobutyl alcohol with glycollic and butyl glycollic acids.

24563 Polysolvan O
Coalescing solvent for adhesives. *Hoechst UK.* Name unverified.

24564 Polysolvan SHS
A proprietary trade name for acetic acid esters of alcohols up to C_{11}. No manufacturer.

24565 polysorbate 20
9005-64-5 8872
POE (20) sorbitan monolaurate
PEG-20 sorbitan laurate; Sorbimacrogol laurate 300. Mixture of laurate esters of sorbitol and sorbitol anhydrides, with +zw 20 moles ethylene oxide; (generic); oil-water emulsifier, solubilizer; used in agriculture, cosmetics, leather, metalworking, and textile industries.

24566 polysorbate 21
9005-64-5 8872
POE (4) sorbitan monolaurate
PEG-4 sorbitan laurate. Mixture of laurate esters of sorbitol and sorbitol anhydrides, with +zw 4 moles ethylene oxide; emulsifier for PVC polymerization, solubilizer for colorants, dye leveling agent.

24567 polysorbate 40
9005-66-7 8872
POE (20) sorbitan monopalmitate
sorbimacrogol palmitate 300; sorbitan, monohexadecanoate, poly(oxy-1,2-ethaneidyl) derivatives. Mixture of palmitate esters of sorbitol and sorbitol anhydrides, with +zw 20 moles of ethylene oxide; o/w emulsifier, solubilizer; used in agriculture, cosmetics, leather, metalworking, and textile industries.

24568 polysorbate 60
9005-67-8 8872
$C_6+64H_{126}O_{26}$
POE (20) sorbitan monostearate
PEG-20 sorbitan stearate; sorbimacrogol stearate 300. Mixture of stearate esters of sorbitol and sorbitol anhydrides, with +zw 20 moles ethylene oxide; (generic); industrial chemicals, solvent, emulsifier, pharmaceuticals, veterinary drug.

24569 polysorbate 61
9005-67-8 8872
POE (4) sorbitan monostearate
PEG-4 sorbitan stearate. Mixture of stearate esters of sorbitol and sorbitol anhydrides, with +zw 4 moles ethylene oxide; (generic); emulsifier, solubilizer, lubricant for textile use, household formulations, suppositories in pharmaceuticals.

24570 polysorbate 65
9005-71-4 8872
POE (20) sorbitan tristearate
PEG-20 sorbitan tristearate; sorbimacrogol tristearate 300. Mixture of stearate esters of sorbitol and sorbitol anhydrides, with +zw 20 moles ethylene oxide; oil-water emulsifier, solubilizer; used in agriculture, cosmetics, leather, metalworking, and textile industries.

24571 Polysorbate 65
9005-71-4 8872
Ethsorbox TS-20; PEG-20 sorbitan tristearate;Sorbax PTS-20; T-Maz® 65; Tween® 65, 65K; Crillet 35; Disponil STS 120 F1; Drewpone® 65K; Durfax® 65; Glycosperse® TS-20 FG; Hetsorb TS-20; Hodag PSTS-20; Ice No. 12K;

Liposorb TS-20; Montanox 65,. Lubricant, softener for textile goods; emulsifier for fats and oils. *Ethox; Lipo;.*

24572 polysorbate 80
9005-65-6 7742
POE (20) sorbitan monooleate
PEG-20 sorbitan oleate; Sorbimacrogol oleate 300. Mixture of oleate esters of sorbitol and sorbitol anhydrides, with +zw 20 moles ethylene oxide; pharmaceutic aid (surfactant); as emulsifier and dispersant in medicinal products; as defoamer and emulsifier in foods.

24573 polysorbate 81
9005-65-6 7742
POE (5) sorbitan monooleate
PEG-5 sorbitan oleate. Mixture of oleate esters of sorbitol and sorbitol anhydrides, with +zw 5 moles ethylene oxide; oil-water emulsifier, solubilizer, used in agriculture., cosmetics, leather, metalworking, and textile industries.

24574 polysorbate 85
9005-70-3 8872
POE (20) sorbitan trioleate
PEG-20 sorbitan trioleate; sorbimacrogol trioleate 300. Mixture of oleate esters of sorbitol and sorbitol anhydrides, with +zw 20 moles ethylene oxide; used as a surfactant.

24575 polysphaerite
$(Pb\cdot Ca)_3\cdot(PO_4)_2(Pb\cdot Ca)_2Cl(PO_4)$
A mineral.

24576 Polysphere 3000 SP
9003-53-6, 111-01-3
Polystyrene, squalane; binder for pressed powders; lubricious, lusterous, high-grade filler for liquid and powdered formulations. *Presperse.* Discontinued.

24577 Polyspin MP-7-29
Polyoxyalkylene derivative; heat-stable lubricant for polypropylene filament. *Hart Chem. Ltd.*

24578 Polyspin PA
Blend of synthetic oils and polyoxyethylene derivatives; spin finish for nylon 6 and polypropylene. *Hart Chem. Ltd.*

24579 Polystab
Polymer additives. *Harcros.*

24580 Polystal®
Composites (profiles, tape, sheet, etc.) of thermosets and thermoplastics with glass or carbon fibers; for aeronautic and aerospace engineering, traffic engineering, elec. engineering/electronics, machine, plant and appliance engineering. *Bayer AG.*

24581 Polystat
Antistatic agent. *Crosfield Chemicals Ltd.* Discontinued.

24582 Polystate C
9004-99-3
PEG-6 stearate
Base for cosmetic lotions. *Gattefosse; Gattefosse SA.*

24583 Polystay AA-1
Anilino-phenyl methacrylamide
Antioxidant in emulsion polymers, NBR, SBR, BR, ABS, and CR. *Goodyear.* Name unverified.

24584 Polystep®
Various surfactants; for emulsion polymerization. *Stepan.* Name unverified.

24585 Polystep® A-11
Isopropylamine dodecylbenzene sulfonate (branched)
Emulsifier, pigment dispersant; emulsion polymerization (S/B, vinyl chloride, vinylidene chloride latexes). *Stepan; Stepan Canada; Stepan Europe.*

24586 Polystep® A-13
dodecylbenzene sulfonic acid (linear)
Emulsion polymerization surfactant; catalyst in acid catalyzed reactions. *Stepan; Stepan Canada.*

24587 Polystep® A-15-30K
27177-77-1 248-296-2
Potassium dodecylbenzene sulfonate (linear)
Surfactant for styrene-butadiene, vinyl chloride, and vinylidene chloride latexes; thermal and hydrolytic stability. *Stepan; Stepan Canada.*

24588 Polystep® A-18
68439-57-6 270-407-8
Sodium alpha olefin (C_{14}, C_{16}) sulfonate
Surfactant for vinyl and vinylidene chloride, acrylic, styrene-acrylaic, SBR polymerization. *Stepan; Stepan Canada; Stepan Europe.*

24589 Polystep® A-7
Sodium dodecylbenzene sulfonate (linear)
Emulsifier for emulsion polymerization, S/B, vinyl chloride, vinylidene chloride latexes. *Stepan; Stepan Canada.*

24590 Polystep® B-1
9051-57-4
Ammonium nonoxynol-4 sulfate
Emulsifier for emulsion polymerization. *Stepan; Stepan Canada.*

24591 Polystep® B-11
67762-19-0
Ammonium laureth sulfate (4 EO)
Emulsifier for emulsion polymerization (acrylics, styrene-acrylic, vinyl acrylics). *Stepan; Stepan Canada; Stepan Europe.*

24592 Polystep® B-12
9004-82-4
Sodium laureth sulfate (4 EO)
Emulsifier for polymerization (acrylics, styrene-acrylics, vinyl acrylics). *Stepan; Stepan Canada; Stepan Europe.*

24593 Polystep® B-25
Sodium decyl sulfate
Emulsifier for emulsion polymerization (S/B, vinyl chloride, acrylic), high surface tension latex; hydrophilic. *Stepan; Stepan Canada; Stepan Europe.*

24594 Polystep® B-27
9014-90-8
Sodium nonoxynol-4 sulfate
Emulsifier for acrylics, SBR, vinyl chloride, and butyl rubber. *Stepan; Stepan Canada; Stepan Europe.*

24595 Polystep® B-29
Sodium octyl sulfate
Low-foaming emulsifier for vinyl chloride systems. *Stepan; Stepan Canada; Stepan Europe.*

24596 Polystep® B-3
151-21-3 8782 205-788-1
Sodium lauryl sulfate
Emulsifier for emulsion polymerization, vinyl chloride, soluble acrylics. *Stepan; Stepan Canada; Stepan Europe.*

24597 Polystep® B-7
2235-54-3 218-793-9
Ammonium lauryl sulfate
Emulsion polymerization surfactant; latex foaming agent; water resistance in coatings. *Stepan; Stepan Canada.*

24598 Polystep® C-OP3S
Sodium octoxynol-3 sulfate
Emulsifier for controlling particle size in vinyl acetate specialty copolymers. *Stepan; Stepan Canada; Stepan Europe.*

24599 Polystep® F-1
9016-45-9 6772
Nonoxynol-4
Nonfoaming pigment dispersant; emulsifier for emulsion polymerization (styrene acrylic, acrylic, vinyl acrylic, S/B). *Stepan; Stepan Canada; Stepan Europe.*

24600 Polystep® F-95B
9016-45-9 6772
Nonoxynol-34
Emulsifier for acrylics and vinyl acetate; blended with other surfactants to increase the latex particle size. *Stepan; Stepan Canada; Stepan Europe.*

24601 polystyrene
9003-53-6 9028
$(C_8H_8)_x$
styrene polymer
PS; ethenylbenzene, homopolymer; benzene, ethenyl-, homopolymer. Polymer; grades: crystal, impact, expandable. thermoplastic resin for injection molding, extrusion of egg carton foam, pill bottles, packaging, appliances, electronics, toys, recreation and construction, expendable polystyrene for insulation and protective packaging. Asahi Chem Industry Co Ltd; Ashland; Elf Atochem SA; BASF; Chevron; Fina; Hüls AG; LNP; Mitsubishi Petrochem.; Mitsui Toatsu Chem.; Westlake Plastics.

24602 Polystyrene 101
9003-53-6 9028
Crystal polystyrene
Heat-resistance extrusion and injection molding grade with high clarity, good dimensional stability; for medical molding, thick-wall housewares, sheet glazing, coextrusion. *Novacor Ltd.*

24603 Polystyrene 220
9003-53-6 9028
Crystal polystyrene
Soft flow injection molding grade for housewares, drinkware, cutlery, toys, molded packaging, cosmetic and medical molding, as blending resin for transparent impact and molded packaging. *Novacor Ltd.*

24604 Polystyrene 410
9003-53-6 9028
Impact polystyrene; medium impact injection molding grade with excellent

gloss, high flow; for cassettes, housewares, closures, hangers, toys, compact disc case inserts. *Novacor Ltd.*

24605 Polystyrene P 2122
9003-53-6 9028
HIPS; extrusion and injection molding grade with superior gloss. *Novacor Ltd.*

24606 Polystyrol
Polystyrene granules; used for extrusion (packaging) and injection molding (household appliances). *BASF plc.*

24607 Polystyrol 143E
A proprietary polystyrene having easy melt-flow and mechanical properties. *BASF plc.* Name unverified.

24608 Polystyrol 165H
A proprietary polystyrene similar to Polystyrol 143E but possessing greater mechanical strength. *BASF plc.* Name unverified.

24609 Polystyrol 168N
A proprietary polystyrene stabilized against ultraviolet light. *BASF plc.* Name unverified.

24610 Polystyrol 427M
An impact-resistant polystyrene with good resistance to deformation at high temperatures. *BASF plc.* Name unverified.

24611 Polystyrol 432F
A proprietary impact resistant styrene-butadiene copolymer. *BASF plc.* Name unverified.

24612 Polystyrol 466 I, 472 D
A proprietary styrene/butadiene copolymer with high impact resistance. *BASF plc.* Name unverified.

24613 Polystyrol 473 E
A proprietary styrene-butadiene copolymer offering high impact resistance and easy flow properties. *BASF plc.* Name unverified.

24614 Polystyrol 475 K
A proprietary styrene butadiene copolymer offering high resistance to impact. *BASF plc.* Name unverified.

24615 Polystyrol KR 253 and KR 2538
Proprietary styrene-butadiene copolymers offering very high resistance to impact at low temperatures. *BASF plc.* Name unverified.

24616 Polystyrol KR 2536
A proprietary styrene/butadiene copolymer offering good resistance to impact and to deformation at high temperatures. *BASF plc.* Name unverified.

24617 polysulfide
Synthetic polymer; elastomer for use in sealants, adhesives, potting and encapsulating compounds, gasoline and oil-loading hose, casting of molds, barrier coatings, binder in solid rocket fuel. *Morton Int'l.*

24618 polysulfone
25135-51-3
Engineering thermoplastic; amorphous thermoplastic with low flammability and smoke emission, good electrical properties; for injection molding and extrusion of food and chemical processing equipment, electrical/electronic components, medical and hospital parts requiring sterilization. *Aldrich; BASF; LNP.*

24619 Polysulfone
25135-51-3
Electrafil® J-1500/CF/20; Polysulfon; Mindel® B-310; Mindel® B-322; Mindel® M-800; Mindel® S-1000; PF-10GF/15T; PF-20GF; Polypenco® Polysulfone; RTP 901; Thermocomp® GF-1004; Udel® GF-110; Udel® P-1700; Ultrason® S 1010,. 20% PAN carbon fiber-reinforced; static dissipative and conductive thermoplastic *Akzo Engineering Plastics.* Name unverified.

24620 PolySurf
Cetyl hydroxyethyl cellulose
Hydrophobically modified hydroxyethylcellulose; thickener and viscosity stabilizer for aqueous and surfactant systems in personal care applications. *Aqualon.*

24621 Polysystems
Urethane foam chemical systems. *Olin UK.* Discontinued.

24622 Polytac
Sealant, adhesive. *Crowley Chem.*

24623 Polytac® 100
Rosin-based resinate; tackifier resin for adhesives. *Arizona.* Discontinued.

24624 PolyTalc 445
14807-96-6 9207 238-877-9
Surface-modified platy talc; filler for polymer applications where color is critical; provides enhanced long term heat stability. *Pfizer.*

24625 Polytar
A proprietary preparation of liquid paraffin, tar crude oil, coal tar, and arachid oil extract of coal tar used in the treatment of skin diseases. *Stiefel Laboratories Inc; Stiefel Laboratories (UK) Ltd.*

24626 polytelite
A mineral, $(Pb{\cdot}Ag_2)_4Sb_2S_7$, with $(Zn{\cdot}Fe)_4Sb_2S_7$.

24627 Polyterge PAT
Surfactant for removal and suspension of residual disperse dyes on polyester. *CNC Int'l L.P.*

24628 Poly-Tergent® 2A1 Acid
Dodecyl diphenyl ether disulfonic acid
Industrial cleaners for textile industry. *Olin.* Discontinued.

24629 Poly-Tergent® 2A1-L
Sodium dodecyl diphenyl ether disulfonate
Industrial cleaners for textile industry. *Olin.* Discontinued.

24630 Poly-Tergent® 3B2
36445-71-3 253-040-8
Sodium decyl diphenyl ether disulfonate
Decyl(sulfophenoxy)benzenesulfonic acid, disodium salt. Surfactant for industrial and institutional detergents and textiles. *Olin.* Discontinued.

24631 Poly-Tergent® 3B2 Acid
Decyl diphenyl ether disulfonic acid
Surfactant for industrial and institutional detergents and textiles. *Olin.* Discontinued.

24632 Poly-Tergent® 4C3
70191-76-3
Sodium alkyl diphenyl ether disulfonate
Surfactant for paper, textiles, metalworking, industrial and household cleaners. *Olin.* Discontinued.

24633 Poly-Tergent® B-150
9016-45-9 6772
Nonoxynol-4.5
Surfactant for a wide variety of applications. *Olin.* Discontinued.

24634 Poly-Tergent® CS-1
Carboxylated linear alcohol alkoxylate, sodium salt; builder surfactant, emulsifier, sequestrant for laundry detergents, hard surface cleaners, bottle washing, dairy and food service, alkaline, metal, transportation cleaners; high temperature electrolyte and alkaline stability. *Olin.* Discontinued.

24635 Poly-Tergent® E-17A
EO/PO block polymer; foam control agent, solubilizer, dispersant, wetting agent, spreading agent for automatic dishwashing, rinse aids, industrial laundry, metal cleaning, water treatment, textiles, agricultural, paper processing. *Olin.* Discontinued.

24636 Poly-Tergent® P-17A
EO/PO block polymer; emulsifier, detergent, wetting agent, foam control agent, deduster, binder for rinse aids, automatic dishwashing, coatings, sizes, water treatment, egg washing, textiles, foods. *Olin.* Discontinued.

24637 Poly-Tergent® S-305LF
Alkoxylated linear alcohol; defoamer, wetting agent, dispersant for rinse aids, dairy cleaners, textiles, hard surface cleaners, automatic dishwashing, metal cleaning; biodegradable. *Olin.* Discontinued.

24638 Poly-Tergent® SL-42
Alkoxylated linear alcohol; wetting agent, emulsifier, for prespots, laundry detergents, transportation cleaners, toilet bowl cleaners, glass cleaners, textiles, metal cleaners, degreasers, abrasive cleaners; biodegradable. *Olin.* Discontinued.

24639 Poly-Tergent® SLF-18
Alkoxylated linear alcohol; lowfoaming biodegradable detergent, dispersant, wetting agent, emulsifier, deduster; for commercial detergents, rinsing aid in machine dishwashing; used in removing proteintype soils. *Olin.* Discontinued.

24640 polytetrafluoroethylene
9002-84-0 7743 204-126-9
$9C_2F_4)_n$ (n > 20,000)
Ethene, tetrafluoro-, homopolymer
PTFE; Teflon; Politef; Tetrafluoroethylene Resin; Fluon; Tetran. Versatile, chemically inert polymer. White solid, usable between -270° and 265°;.

24641 polytetrafluoroethylene
9002-84-0 7743 204-126-9
$[CF_2CF_2]_x$, x +zw 20,000
PTFET; TFE; tetrafluoroethene homopolymer; tetrafluoroethylene polymer; polytetrafluoroethylene resin. Thermoplastic homopolymer. as tubing or sheeting for chemical laboratory and process work; gaskets and pump packings; as electrical insulators especially in high frequency applications, filtration fabrics, protective clothing, prosthetic aids. *Janssens NV.*

24642 Polytetrafluoroethylene
9002-84-0 7743 204-126-9
Tetrafluoroethene homopolymer
Emralon 8301-01, 304; Electrafil® TR-1900/EC; Fluorocomp® FC-101, FC-144, FC-174, FC-182; Fluoromelt®; Fluorosint® 500; Polytetrafluoroethylene resin; Hostaflon® TF 1101, 1620, 2071, 5032, 5537; McLube 1777; Polymist® F-5; polytef; PTFE; PTFE-19; RT/Duroid® M; SLA 1611, 1612;

Fluon; Fluon® AD1, AD1L, AD1H, CDI, G170; Teflon; Tetran;. Colloidal PTFE in water; lubricant additive for dry gear and chain lubes, aerosols, assembly and thread lubes. High polymer; nonflamable; solid; *Acheson; Akzo Engineering Plastics; ICI Fluoropolymers; Hoechst AG; McLube; Ausimont; Presperse;.*

24643 polytetramethylene ether glycol
24979-97-3; 25190-06-1
$HO[(CH_2)_4O]_nH$
poly (oxy-1,4-butanediyl)-α-hydro-ω-hydroxy)
Polyether glycol. PTMEG; PTMG; for polyurethane formulation for automotive hose and gaskets, tires, industrial belts, tank and pipe liners, floor and roof coatings, medical devices. *BASF; QO.*

24644 polythene
The general term for a range of solid polymers of ethylene. Name unverified.

24645 Polytrap®
A polymer used for entrapping solid and/or liquid materials. *Dow Corning.*

24646 Polytrap® Q5-6035
Cyclomethicone and acrylates copolymer; delivers volatile silicone fluid in powder form; provides lubricious skin feel for skin care and cosmetics formulations. *Dow Corning.*

24647 Polytrap® Q5-6603
Acrylates copolymer; adsorbs high levels of lipopholic and certain hydrophilic liquids; for skin care, sun care, and cosmetic formulations. *Dow Corning.*

24648 Polytrend
Colorant dispersions, polyester extender and pigment vehicle; for coloring plastic compositions, extender for unsaturated polyester compositions. *Hüls Am.*

24649 Polytrix
25322-68-3 7729
Solution of hyaluronic acid in PEG-115M. Moisturizer used in skin care. *Amerchol Corp.* Discontinued.

24650 Polytron SMV 9081
Static dissipative alloy featuring durability, flame retardance, colorability, impact, chemical resistance and processability; used for circuit cards, tote bins, paper handling equipment; rapid decay series can dissipate 5000 V to O V in <2 seconds; also available as standard decay series and low dusting series. *BF Goodrich/Geon Vinyl.* Name unverified.

24651 Polytron SMV 9804
Alloy; static dissipative alloy for injection molding; standard decay. *BF Goodrich/Geon Vinyl.*

24652 Polytrope
Rheological additives; used for unsaturated polyester resins. *Rheox Inc.*

24653 Polyurax
Urethane intermediates polyols. *BP Chemicals Ltd.* Discontinued.

24654 Polyurethane elastomer or rubber
Elastomer for extrusion, injection molding, and calendering; for sealants, caulks, adhesives, film and sheet, shoes, encapsulation of electronic components, automotive parts, flexible and rigid casting shapes. Air Products; Crowley; DSM UK; Ferro/Bedfore; BF Goodrich; Hardman; Miles; Morton Int'l.; Polyurethane Corp. of Am.; Soluol; UCB SA.

24655 polyurethane foam
9009-54-5
Foam insulation for thermal insulation, fabricated shapes, pipe covering, void filling, and cold storage applications. *Grace NV.*

24656 Polyval B
Vat dye stripper. *Ceca SA.*

24657 Polyvel CR-5F
78-63-7 201-128-1
$C_{16}H_{34}O_4$
2,5-Dimethyl-2,5-di(*t*-butylperoxy) hexane
1,1,4,4-tetramethyl-1,4-butanediyl)bis[(1,1-dimethylethyl) peroxide; 2,5-Dimethyl-2,5-di(*tertiary*-butylperoxy)-hexane. Flow modifier and processing aid for PP. mp = 8°; bp = 50-52°; insoluble in H_2O, soluble in organic solvents. *Polyvel.*

24658 Polyvel CR-5P
Bis(*t*-butylperoxyisopropyl) benzene
Flow modifier and processing aid for PP. *Polyvel.*

24659 Polyvel CR-5T
614-45-9 210-382-2
$C_{11}H_{14}O_3$
t-butyl perbenzoate
tert-butyl perbenzoate; Benzenecarboperoxoic acid, 1,1-dimethylethyl ester; *tertiary*-Butyl perbenzoate. Additive for grafting and extrusion reactions. mp = 8°; bp = 113°; d = 1.021; n_D^{20}= 1.4990; insoluble in H_2O, soluble in organic solvents; *Polyvel.*

24660 Polyvel CR-L10
1068-27-5 213-944-5
$C_{16}H_{30}O_4$
2,5-Dimethyl-2,5-di(*t*-butylperoxy)hexyne-3
2,5-Dimethyl-2,5-di-(*tert*-butylperoxy)hexyne-3; 3-hexyne, 2,5-dimethyl-2, 5-di(*t*-butylperoxy)-; 2,5-Dimethyl-2,5-di(*tertiary*-butylperoxy)-3-hexyne. Additive for roto molding and crosslinking. *Polyvel.*

24661 Polyvel PCL-20
80-43-3 201-279-3
$C_{18}H_{22}O_2$
Dicumyl peroxide
Bis(1-methyl-1-phenylethyl) peroxide; bis(α,α-dimethylbenzyl) peroxide; cumyl peroxide; di-α-cumyl peroxide; Di-cup; diisopropylbenzene peroxide; isopropyl benzene peroxide; Cumene peroxide. Additive for crosslinking polyethylene. bp = 130°; insoluble in H_2O, soluble in organic solvents. *Polyvel.*

24662 Polyvest
Polybutadienes used for treatment of minerals. *Hüls AG.*

24663 Polyvest 25
A polymeric filler activator; used for activating light-colored silicate fillers in EPDM compounds. *Hüls AG.*

24664 Polyvest C70
A polymeric chalk filler activator; used for activating carbonate fillers in EPDM compounds. *Hüls AG.*

24665 polyvinyl acetate homopolymer
9003-20-7
$[C_4H_6O_2]_x$
PVAc; acetic acid, ethenyl ester, homopolymer; acetic acid vinyl ester polymers; ethenyl acetate, homopolymer. Homopolymer of vinyl acetate. Resin with weathering resistance; used for paints; adhesives for food packaging, paper, wood, glass, and metals; primer sealers, dry wall cement; intermediate for conversion to polyvinyl alcohol and acetals; paper coating, component of lacquers and inks. *Aldrich; H.B. Fuller; Monsanto; Nat'l. Starch & Chem.; Wacker-Chemie GmbH.*

24666 polyvinyl alcohol
9002-89-5 7745
$[C_2H_4O]_x$
ethenol, homopolymer
PVA; PVAL. Used in plastics industry in molding compounds, surface coatings, films resistant to gasoline, textile sizes; for elastomers (artificial sponges, fuel hoses); printing inks; pharmaceutical finishing; cosmetics; film and sheeting; ophthalmic lubricant. *British Traders & Shippers; Dajac Labs; Honeywill & Stein Ltd.*

24667 polyvinyl butyral
63148-65-2; 9003-62-7
$(-C_4H_6O_2-)_n$
PVB; vinyl acetal polymers, butyrals; vinyl acetyl polymers, butyrates; polyvinyl butyral resin. Polymer produced by condensation of polyvinyl alcohol and butyraldehyde; thermoplastic for extrusion, molding, coating, and casting processes; for adhesives, paints, lacquers, films, as sheet interlayer in safety glass and shatter-resistant protection in aircraft. *Caim Chem Ltd; Monsanto; Wacker.*

24668 polyvinyl chloride
9002-86-2 7746 206-625-7
$[C_4H_6Cl_2]_n$
chloroethene homopolymer
PVC; chloroethylene polymer. Rubber substitutes; electrical wire and cable coverings; pliable thin sheeting; film finishes for textiles; nonflammable upholstery; raincoats; tubing; belting; gaskets; shoe soles. Aldrich; ashland; Atochem SA; Chisso; Georgia Gulf; BF Goodrich; Goodyear; Hüls Am.; Mitsui Toatsu Chem.; Norsk Hydro AS; OxyChem; Vista; Wacker-Chemie GmbH.

24669 Polyvinyl methyl ether
9003-09-2
Gantrez® M-154. Polymer functioning as tackifier, binder, and plasticizer; used in printing inks, textile sizes and finishes, latex modification. *ISP.*

24670 Polyviol
Polyvinyl alcohol in a range of viscosities and degrees of saponification; used as protective colloid for emulsions, dispersions and suspensions, textile auxiliaries (finishes, impregnating agents, sizes); thickening agent for glues and adhesives; release agent in the processing of polyester resins. *Wacker-Chemie GmbH.*

24671 Polywax® 500
9002-88-4 7728
Polyethylene homopolymer. Release agent in carbon ink formulations; modifier for paraffin waxes; component in hot-melt coatings, adhesives, and chewing gum base; lubricant in plastics and rubber processing, electrical insulation, powdered coatings, and printing inks, antiblocking agent and binder. *Petrolite.*

24672 Polywet® ND-2
Functionalized oligomer, sodium salt; dispersant for pigments, minerals,

extenders, fillers in aqueous systems; latex paints and enamels; coatings, adhesives, paper and paperboard; boiler water. *Uniroyal.*

24673 Polywet® Z1766

Sodium salt of a polyfunctional oligomer; dispersant for titanium dioxide, other pigments. *Uniroyal.*

24674 Poly-zole AZDN

78-67-1 949 201-132-3

$C_8H_{12}N_4$

Azo diisobutyronitrile

Azobisisobutyronitrile; Azodiisobutyrodinitrile; AIBN; α,α'-Azodiisobutyronitrile; 2,2'-Dicyano-2,2'-azopropane; Porofor-57; 2,2'-azo-bis(isobutyronitrile). A vinyl polymerization catalyst; gives freedom from side reactions and is not readily poisoned. Used as a blowing agent and as an initiator for free radical reactions. mp = 102-103° (dec); λ_m = 345 nm (EtOH); insoluble in H_2O, soluble in EtOH, MeOH; LD_{50} (mus orl) = 700 mg/kg. *National Polychemicals.* Unverified.

24675 Polyzote

A proprietary trade name for nitrogen-expanded synthetic resin plastics. No manufacturer.

24676 pomace

The residue from the extraction of apple juice in cider manufacture; a cattle food. No manufacturer.

24677 Pomarsol®

137-26-8 9510 205-286-2

Thiram

Fungicidal spray used for control of scab, storage rots, skin and other diseases on pome and stone fruit, grapes and vegetables; also as a seed dressing. *Bayer AG.*

24678 Pombe

A beer made from Sorghum millet.

24679 Pommetrol M

Mixture of chlorpropham and propham; a plant growth regulator for potato sprout growth suppression. *Sam Fletcher Agricultural Specialists.*

24680 Pomoloy

A proprietary trade name for a cast iron made by a special process. No manufacturer.

24681 Pondermite

$Ca_2B_6O_{11}\cdot 4H_2O$

A mineral, source of boric acid.

24682 Ponder's stain

A microscopic stain. It consists of 0.02 g toluidine blue, 1 ml glacial acetic acid, and 2 ml absolute alcohol in 100 ml distilled water.

24683 Pondicherry oil

Nut oil. Arachis oil.

24684 ponite

A variety of rhodocrosite containing iron, a mineral.

24685 ponolith

Sunolith; superlith. Lithopone pigments.

24686 Pontallor

A range of precious metal alloys; for dentistry and dental engineering. *Degussa Ltd.*

24687 Pool-Chem

Consumer product line for control of water chemistry in swimming pools. *Puma Chemical Co Inc.* Name unverified.

24688 poonac

Coconut cake. A cattle food. The term is also used for the residue from castor oil seeds after cold and hot pressing and solvent extraction; used for caulking timber.

24689 poonahlite

Hydrated aluminum-calcium silicate, a mineral.

24690 Pope's solution

A solution of 1 part in 10,000 of a mixture of 10 parts 2 : 7-dimethyl-3 : 6-diaminoacridinium-methylchloride hydrochloride and 1 part crystal violet; an antiseptic for wounds.

24691 poppy capsules or heads

The dried, immature fruit of *Papaver somniferum* . Source of opium alkaloids.

24692 porcelain

A mixture of clay, quartz, and felspar. A normal mix consists of 50% clay, 25% quartz, and 25% felspar.

24693 porcelain clay

Synonym for Kaolinite.

24694 porcelanite

A fused clay and shale found in burned coal seams.

24695 Porcelave

A proprietary trade name for a ceramic material. No manufacturer.

24696 Porocel

A proprietary trade name for a carefully prepared and screened bauxite. No manufacturer.

24697 Porofor®

A range of chemical blowing agents for the production of cellular plastics primarily based on PVC, polyethylene, ABS and polystyrene. *Bayer AG; Bayer plc.*

24698 Porofor® ADC/E

123-77-3 950 204-650-8

Azodicarbonamide

Chemical blowing agent for production of plastic foams; used in expanded UPVC pipe and sections; extrusion of foamed polyethylene. *Bayer; Miles; Polysar.*

24699 Poron® 4701-01

High-density microcellular PU; used for gaskets, seals, product protection, cushion pads, vibration mounts, motor mounts, RF shielding, PCB cushions, spacers, foam-backed tapes, athletic padding in automotive, electrical/electronic industries, etc. *Rogers.* Name unverified.

24700 Poron® S2000-80-24031

Cellular silicone material; for use at temp. and performace extremes; excellent flame resistance, low smoke generation, low toxicity of combustion by-products; resistant to uv and ozone. *Rogers.*

24701 Porosil-Clarcel

Diatomite fillers; for painting, fertilizers coating and defluoration. *Elf Atochem UK/Ceca.*

24702 porous alum

10102-71-3 378 233-277-3

$Al_2(SO_4)_3\cdot Na_2SO_4\cdot 24H_2O$

Sodium aluminum sulfate

soda alum. Astringent.

24703 porpezite

A native alloy of gold and palladium; commonly contains silver as well.

24704 porphyry

A building stone having the same composition as felspar.

24705 porporino

An alloy of mercury, tin, and sulfur; used for decorating purposes.

24706 Portagen

A proprietary artificial infant food containing medium chain triglycerides and nonlactose carbohydrates, for use in cases of intolerance of fat and lactose. *Bristol-Myers Squibb Co Inc.*

24707 Portland arrowroot

The starch from *Arum maculatum.*

24708 Portland cement

65997-15-1 266-043-4

Made by heating an intimate mixture of argillaceous and calcareous substances, such as lime and clay, and pounding the product. The material does not slake with water, and has energetic hydraulic properties.

24709 Portland stone

1317-65-3 5515 215-279-6

A limestone.

24710 Portman 5C Chormequat

Chlormequat + choline chloride; plant growth regulator for use in cereals and ornamentals. *Portman Agrochemicals Ltd.*

24711 Portman Chlormequat 400, 460, 600, 700

999-81-5 2153 213-666-4

Chlormequat; plant growth regulator. *Portman Agrochemicals Ltd.*

24712 Portman Isotop

34123-59-6 5237 251-835-4

Suspension concentrate containing 500 g isoproturon per liter; used for annual weed control in cereals. *Portman Agrochemicals Ltd.*

24713 Portman propachlor 50FL

1918-16-7 7977 217-638-2

Propachlor; a pre-emergence herbicide for various horticultural crops. *Portman Agrochemicals Ltd.*

24714 Portman Supaquat

Chlormequat + choline chloride; plant growth regulator for use in cereals and ornamentals. *Portman Agrochemicals Ltd.*

24715 Portsmouth Accelerator No. 3

Phenyl-o-tolyl-guanidine

A proprietary rubber vulcanization accelerator. No manufacturer.

24716 Portugallo oil

Essential oil of orange peel.

24717 Pos O Print

Ammonium hydroxide - 26 Baumé-29.4% concentrate, 24 Baumé-25.5% concentrate, 23 Baumé-23.5% concentrate, 20 Baumé-17.7% concentrate. Developing solution for blue prints in the engineering/drafting industries; developing solution for microfilm. *W D Service Company Inc.*

24718 Posistac
53003-10-4 8488 258-290-1
Salinomycin
An ionophorous antibiotic used as a growth stimulating nutritional aid in cattle and swine. *Pfizer International.*

24719 Poskydal
Unsaturated polyester resins; used for the formulation of furniture finishes with and without paraffin wax as well as for use in fillers. *Bayer AG.*

24720 Post-4
Anti-settling additives; used for paints. *Rheox Inc.*

24721 Post-Kite
Suspension concentrate containing 50 g ioxynil, 250 g isoproturon and 180 g mecoprop per liter; used for control of weeds in wheat and barley. *Schering Agro-chemicals Ltd.* Discontinued.

24722 pot metal
An alloy of lead and copper.

24723 potarite
A mineral (Pd·Hg).

24724 potash alum
10043-67-1 373 233-141-3
$AlKO_8S_2$
aluminum potassium sulfate
A double sulfate of potassium and aluminum, used in paper, matches, paints, tanning agents, waterproofing agents, purification of water, astringent, cement hardener.

24725 potash bordeaux mixture
Contains 6 lb copper sulfate, 2 lb potassium hydroxide, and 50 gallons water. A fungicide.

24726 potash glass
A glass containing silicate of potassium.

24727 potash water-glass
A mixture of potassium silicates.

24728 pot-ashes
Impure potassium carbonate.

24729 potash-lead glass
A glass usually containing from 40-50% SiO_2, 28-53% PbO, 8-11% K_2O, and 1% of both Al_2O_3 and Fe_2O_3.

24730 potassalumite
A mineral that is a potash alum.

24731 potassic superphosphate
A fertilizer made by combining calcium superphosphate with potash salts.

24732 potassium
7440-09-7 7763 231-119-8
K
Kalium. Metallic element; K; intermediate for potassium peroxide, heatexchange alloys; laboratory reagent; component of fertilizers (as potassium chloride). Used, with sodium, as a heat transfer agent in nuclear reactors. mp = 63°; bp = 765°; d^{20} = 0.856; reacts with H_2O, soluble in liquid NH_3, ethylenediamine, aniline. *Aldrich; Atomergic Chemetals.*

24733 potassium acetate
127-08-2 7764 204-822-2
$C_2H_3KO_2$
acetic acid potassium salt
Dehydrating agent, textile conditioner, analytical reagent, medicine, cacodylic derivatives, crystal glass, synthetic flavors. Am. Int'l. Chem.; EM Industries; General Chem.; Heico; Honeywill & Stein Ltd; Niacet; Schaefer Salt & Chem.

24734 potassium acetate
127-08-2 7764 204-822-2
$C_2H_3KO_2$
potassium acetate
An alkalizer; has been used to treat veterinary cardiac arrhythmias. mp = 292°; d = 1.8000; soluble in H_2O (2 g/ml), poorly soluble in organic solvents; LD_{50} (rat orl) = 3.25 g/kg.

24735 potassium alginate
9005-36-1 241
potassium polymannuronate
alginic acid, potassium salt; Stercofuge. Potassium salt of alginic acid; gellant, emulsifier, and stabilizer in food and indust. application; gum, bodying agent for creams and lotions, dental impression materials; used for water holding in foods and industry. *Atomergic Chemetals; Kelco Int'l.*

24736 potassium amalgam
An alloy of potassium and mercury, formed by the combination of the elements.

24737 potassium arsenate
7784-41-0 7767 232-065-8
AsH_2KO_4
potassium dihydrogen arsenate
potassium acid arsenate; KDA; Arsenic acid, monopotassium salt. Used in textile, tanning and paper industries. Also in insecticide formulations. mp = 288°; d = 2.8; slightly soluble in H_2O (0.2 g/ml), insoluble in EtOH.

24738 potassium benzoate
582-25-2 1122 209-481-3
$C_7H_5KO_2 \cdot 3H_2O$
benzoic acid potassium salt trihydrate
Anti-corrosive, preservative, fermentation-inhibitor, anti-fungal agent for tobacco production, pyrotechnical additive. Soluble in H=72O, EtOH. *Am. Biorganics; Mallinckrodt; Pentagon Chemicals Ltd; Schweizerhall; Verdugt BV.*

24739 potassium bichromate
7778-50-9 7790 231-906-6
$Cr_2K_2O_7$
potassium dichromate(VI)
Oxidizing agent. Used in tanning leather, dyeing, painting, printing, photolithography, staining wood, pyrotechnics, in safety matches, as a bleach and corrosion inhibitor and in electric batteries. mp = 398°; d = 2.6760; soluble in H_2O (4-50%).

24740 potassium bitartrate
868-14-4 7776 212-769-1
$KHC_4H_4O_6$
Cream of tartar; potassium acid tartrate. Baking powder, preparation of other tartrates, galvanic tinning of metals, food additive. *Penta Mfg.; Spectrum Chem. Mfg.*

24741 potassium bromate
7758-01-2 7779 231-829-8
$BrKO_3$
bromic acid, potassium salt
Laboratory reagent, oxidizing agent, permanent wave compounds, dough conditioner, food additive. mp = 350°; d = 3.2700; soluble in H_2O (80 mg/ml), insoluble in EtOH; *Allchem Industries; Gist-Brocades Food Ingreds.*

24742 potassium bromide
7758-02-3 7780 231-830-3
KBr
Photography, process engraving and lithography, special soaps, spectroscopy, infrared transmission, lab reagent. Also used as a sedative and anticonvulsant. mp = 730°; d = 2.75; soluble in H_2O (0.67 g/ml), less soluble in organic solvents; *Aldrich; Great Lakes; Mallinckrodt; Morton Int'l.*

24743 potassium carbonate
584-08-7 7781 209-529-3
CK_2O_3
salt of tartar; pearl ash. Used in chemical manufacturing and in medicine as an alkalizer and diuretic. mp = 891°; d = 2.9; soluble in H_2O (1 g/ml), insoluble in organic solvents; LD_{50} (rat orl) = 1.87 g/kg.

24744 potassium carbonate
584-08-7 7781 209-529-3
K_2CO_3
Inorganic salt
Carbonic acid, dipotassium salt; dipotassium carbonate; potash. Special glasses (optical, TV tubes), potassium silicate, fertilizer manufacture, dehydrating agent, pigments, printing inks, lab reagent, general purpose food additive, textile, dyeing. *Hüls; Mallinckrodt; OxyChem.*

24745 potassium carbonate
584-08-7 7781 209-529-3
$C_2K_2O_3$
Carbonic acid, potassium salt..

24746 potassium chlorate
3811-04-9 7782 223-289-7
$ClKO_3$
chloric acid, potassium salt
Berthollet salt; chlorate of potash; potash chlorate; potcrate. Used in explosives, fireworks and matches. Also in printing and dyeing of cotton and wool. mp = 368°; bp = 400°; SG = 2.32; soluble in water (7 g/100 ml), insoluble in organic solvents;.

24747 potassium chloride
7447-40-7 7783 231-211-8
KCl
Fertilizer; foods, pharmaceuticals; in photography; in buffer solutions, electrode cells. *Aldrich; Heico; Mallinckrodt; Morton Int'l.; Reheis.*

24748 Potassium chloride
7447-40-7 7783 231-211-8
KCl
Emplets Potassium Chloride; Chloropotassuril; Duffi-K; Enseal; Kaleorod; Kalitabs; Kalium-Duriles; Kaon-Cl; Kaskay; Kayback; Kay-Cee-L; K-Contin; Klor-Con; K-Norm; K-Tab; Lento-Kalium; Micro K; Nu-K; Peter-Kal; PfiKlor; Rekawan; Repone K; Slow-K; Span-K; ClK; Leo K; K-Lyte/C1; Kay Ciel; Selora; Trona Potassium Chloride.,. Replenisher. MW=74.55; white crystals or powder; d=1.98; mp=773°; sol.in water, glycerol, alcohol; insol.in ether,

acetone; Parke-Davis;Abbott Laboratories; Berlex Laboratories Inc.; Adria Laboratories Inc.; Kerr-McGee Chemical Corp.; Leo Laboratories;Wyeth Laboratories;. Name unverified.

24749 potassium chloride
7447-40-7 7783 231-211-8
ClK
potassium chloride
Chloropotassuril ;Diffu-K; Enseal; Kaleorid; Kalitabs; Kalium-Durlles; Kaon-Cl; Kaskay; Kayback; Kay-Cee-L; K-Contin; Klor-Con; K-Norm; K-Tab; Lento-Kalium; Micro K; Nu-K; Peter-Kal; PfiKlor; Rekawan; Repone K; Slow-K; Span-K; sal digestnum Sylvii. Used in buffers and electrolytes; used medicially as an electrolyte replenisher and as a potassium supplement. mp = 773°; d = 1.9800; soluble in H_2O (0.36 g/ml), poorly soluble in organic solvents;.

24750 potassium citrate
866-84-2 7785 212-755-5
$C_6H_5K_3O_7$
1,2,3-Propanetricarboxylic acid, 2-hydroxy-, tripotassium salt
Tripotassium citrate; Urocit-K. Food additive; antacid. Soluble in H_2O (1.5 g/ml), less soluble in alcohols, insoluble in organic solvents. *Mallinckrodt; Pfizer SA; Schweizerhall.*

24751 potassium cyanide
151-50-8 7789 205-792-3
CKN
Gold and silver extraction, analytical reagent, insecticide, fumigant, electroplating. mp = 634°; d = 1.52; soluble in H_2O (0.5 g/ml), less soluble in organic solvents; LD_{50} (rat orl) = 10 mg/kg. *Elf Atochem N. Am.; Degussa; Du Pont; W.R. Grace Ltd; Mallincrodt.*

24752 potassium d-gluconate
299-27-4 7796 206-074-2
$C_6H_{11}KO_7$
D-gluconic acid potassium salt
Gluconsan K; Kalimozan; Kaon; Katorin; Potasoral; Potassuril; K-IAO; Tumil-K. User as a replenisher, a mineral supplement and a mineral source for pharmaceutical and food products. dec 180°; freely soluble in H_2O, insoluble in organic solvents. *Akzo; R.W. Greeff.*

24753 potassium dichloroisocyanurate
2244-21-5 9896 218-828-8
$C_3HCl_2N_3O_3{\cdot}K$
dichloroisocyanuric acid potassium salt
potassium dichloro-s-triazinetrione. Bleaching compound., sanitizer/disinfectant, oxidizer in dishwashing compositions.

24754 potassium dichromate
7778-50-9 7790 231-906-6
$Cr_2K_2O_7$
potassium dichromate(VI)
red potassium chromate. Oxidizing agent, analytical reagent, brass pickling, electroplating, pyrotechnics, explosives, matches, textile dyeing and printing, adhesives, tanning leather, wood stains, lithography, synthetic perfumes, pigments, alloys, ceramics, batteries. mp = 398°; d = 2.6760; soluble in H_2O (4-50%). *Hoechst Celanese; Mallinckrodt.*

24755 Potassium gluconate
299-27-4 7796 206-074-2
$C_6H_{11}KO_7$
Gluconic acid potassium salt
Gluconal® K; Glucosan K; Kalimozan; Kaon; Potasoral; Potassuril; K-IAO; Tumil-K; Jaon; Katorin;,. Pharmaceutical/food grade mineral source for human and veterinary pharmaceutical preps., dietary supplements, fortified foods and animal feed. MW=234.25; crystals; dec 180°; sol.in water; insol.in abs alcohol, ether, benzne, chloroform *Akzo Chemie; Adria Laboratories Inc; The Boots Co plc.*

24756 potassium hydroxide
1310-58-3 7806 215-181-3
HKO
Caustic potash
potassium hydrate; lye; potassa. Inorganic base; Soap manufacture, bleaching, manufacture of chemicals, electrolyte, absorbent, liquid fertilizers, food additive, herbicides, electroplating, mercerizing, paint removers, reagent. mp= 360°; soluble in H_2O (1.1 g/ml), less soluble in EtOH; LD_{50} (rat orl) = 1.23 g/kg. *Hüls Am.; ICI Specialties; Olin; OxyChem.*

24757 Potassium hypochlorite
7778-66-7 231-909-2
HyPure K. For liq. bleach, water treatment, hard surface cleaners. *Olin.*

24758 potassium iodide
7681-11-0 7809 231-659-4
IK
Jodid, Thyroblock; Thyrojod. Reagent in analytical chemistry; photographic emulsions; animal feed additive; dietary supplement; in table salt; nylon stabilizer. mp = 680°; d = 3.12; soluble in H_2O (1.4 g/ml), less soluble in organic solvents; LD_{50} (rat iv) = 285 mg/kg. *Aldrich; Atomergic Chemetals; R.W. Greeff; Mitsui Toatsu.*

24759 Potassium Iodide
7681-11-0 7809 231-659-4
IK
Embamix; Jodid; Thyroblock; Thyrojod,. Potassium iodide mixtures; antifungal; expectorant; iodine supplement; vet. Medicine MW=166.00; crystals; mp=680°; sol.in water alcohol, methanol, acetone, glycerol, glycol; *May & Baker Ltd.*

24760 potassium lignosulfonate
Used as a dispersant. *Wesco Tech. Ltd.*

24761 potassium metabisulfite
16731-55-8 7811 240-795-3
$K_2O_5S_2$
disulfurous acid, dipotassium salt
potassium pyrosulfite; dipotassium disulfite. Inorganic salt; antiseptic; reagent; source of sulfurous acid; photographic developing agent; brewing, wine making; food preservative; bleaching agent. Freely soluble in H_2O, insoluble in organic solvents. *Allchem Industries; Farleyway Chem. Ltd; Mallinckrodt.*

24762 potassium nitrate
7757-79-1 7815 231-818-8
KNO_3
niter
nitre; saltpeter. Pyrotechnics, explosives, matches, fertilizer, reagent, glass manufacture, tempering steel, curing foods, oxidizer in solid rocket propellants. *Am. Biorganics; EM Industries; Mallinckrodt; SanYuan Chem. Co. Ltd.; Whiting, Peter Ltd.*

24763 potassium oleate
143-18-0 7818 205-590-5
$CH_3(CH_2)_7CH{\cdot}cCH(CH_2)_7COOK$
potassium 9-octadecenoate
oleic acid; potassium salt. Potassium salt of oleic acid; Liq. soap for hand cleaners, tire mounting lubricant; emulsifier and corrosion control in paint strippers. *Emkay; Norman, Fox.*

24764 Potassium oleate
143-18-0 7818 205-590-5
$C_{18}H_{33}KO_2$
Oleic acid pottasium salt
Emkapol PO-18; Marley Cement Waterproofer; Octosol 449. Detergent, soap, emulsifier; stabilizer for natural latex; biodeg. Yellowish or brownish soft mass; soluble in water and alcohol. *Emkay.*

24765 potassium oxalate
6487-48-5 7820
$C_2K_2O_4{\cdot}2H_2O$
potassium oxalate monohydrate
Oxalic acid, dipotassium salt monohydrate. Used as an anti-coagulant, in photography and for cleaning and bleaching straw. d = 2.1270; soluble in H_2O (30 g/100 ml).

24766 potassium oxalate
6487-48-5 7820
$C_2K_2O_4{\cdot}H_2O$
potassium oxalate monohydrate
oxalic acid potassium salt; Neutral oxalate of potash; oxalic acid, dipotassium salt. Analytical reagent, source of oxalic acid, bleaching and cleaning, removing stains from textiles, photography, anticoagulant. d = 2.1270; soluble in H_2O (0.3 g/ml) *Am. Int'l. Chem.; General Chem.; Heico; Verdugt BV.*

24767 potassium perchlorate
7778-74-7 7822 231-912-9
$ClKO_4$
Perchloric acid, potassium salt
Explosives, oxidizing agent, photography, pyrotechnics and flares, reagent, oxidizer in solid rocket propellants. mp = 400°; d = 2.52; soluble in H_2O (0.75 g/100 ml), insoluble in organic solvents. *Am. Int'l. Chem.; Eka Nobel AB; Mallincrodt; San Yuan Chem. Co. Ltd.*

24768 potassium periodate
7790-21-8 7823 232-196-0
IKO_4
potassium metaperiodate
periodic acid potassium salt; Potassium Periodate Meta. Analysis, oxidizing agent. mp = 582°; d = 3.6180. *Atomergic Chemetals; Cerac; Spectrum Chem. Mfg.*

24769 potassium permanganate
7722-64-7 7824 231-760-3
$KMnO_4$
Oxidizer, disinfectant, deodorizer, bleach, dye, tanning, radioactive decontamination of skin, analytical reagent, medicine (antiseptic),

manufacture of organic chemicals, air and water purification. *Am. Biorganics; Am. Int'l. Chem.; Blythe, William Ltd; Mallinckrodt; Schweizerhall.*

24770 potassium permanganate
7722-64-7 7824 231-760-3
$KMnO_4$
permanganc acid potassium salt
chameleon mineral. Powerful oxidizing agent. Used in bleaches, dyeing wood brown, priniting fabrics, washing CO_2, in photography, tanning, purifying water and as a chemical reagent. dec 240°; d = 2.7; soluble in H_2O (70-280 mg/ml); LD_{50} (rat orl) = 1.09 g/kg.

24771 potassium persulfate
7727-21-1 7825 231-781-8
$K_2S_2O_8$
peroxydisulfuric acid dipotassium salt
potassium peroxydisulfate. Bleaching, oxidizing agent, reducing agent in photography, antiseptic, soap manufacture, analytical reagent, polymerization promoter, pharmaceuticals, starch modifier, flour-maturing agent, textile desizing. *Allchem Industries; Du Pont; FMC; Mallinckrodt; San Yuan Chem. Co. Ltd.; Transol Chem. UK Ltd.*

24772 potassium phosphate, monobasic
7778-77-0 7829 231-913-4
KH_2PO_4
potassium phosphate, monobasic
potassium dihydrogen orthophosphate; monopotassium orthophosphate. MKP; Inorganic salt; Food products, baking powder, nutrient solutions, water treatment, buffer and sequestrant, laboratory reagent. *Albright & Wilson; Aldrich; FMC; Heico; Monsanto.*

24773 potassium phosphate, dibasic
7758-11-4 7828 231-834-5
K_2HPO_4
dipotassium phosphate
dipotassium hydrogen orthophosphate; dipotassium orthophosphate. DKP; food and automotive industry; buffer in antifreezes; nutrient; humectant; pharmaceuticals. Very soluble in H_2O (1.5 g/ml). *Albright & Wilson; Aldrich; FMC; Heico; Monsanto; U.S. Biochemical.*

24774 potassium phosphate, tribasic
7778-53-2 7830 231-907-1
K_3O_4P
tripotassium phosphate
TKP. Detergent, water treatment, automotive products, fertilizer, foods as emulsifier. mp = 1340°; d_4^{17}= 2.564; soluble in H_2O (40 - 60%). *Albright & Wilson; Ashland; FMC; Monsanto.*

24775 potassium polyacrylate
25608-12-2
$(C_3H_4O_2)_x{\cdot}xK$
Polyacrylic acid, potassium salt. Potassium salt of polyacrylic acid; dispersant for latex paints and coatings, pigments.

24776 Potassium pyroantimonate, Acid
10090-54-7 7832
$K_2H_2Sb_2O_7.4H_2O$
Potassium antimonate(V)
Liquid Code XLR. Potassium pyroantimonate solution; crosslinker for guar and derivatized guar at pH 3-5. Crystalline powder; insol.in alcohol; sol.in cold water. *Rhône-Poulenc/Water Soluble Polymers.*

24777 potassium pyrophosphate
7320-34-5 7833 230-785-7
$K_4P_2O_7{\cdot}3H_2O$
tetrapotassium pyrophosphate
potassium pyrophosphate, normal. TKPP; soap and detergent builder, sequestering agent, peptizing and dispersing agent. *Elf Atochem N. Am.; FMC; Monsanto.*

24778 Potassium ricinoleate
7492-30-0 231-314-8
Emulgeen P; Kricinol 35; Solricin® 135. Mold release; detergent; emulsifier; mild germicide; rubber lubricant; foam stabilizer *S & D Chemicals Ltd; Climax Fluids Additives;CasChem.* Name unverified.

24779 potassium silicate
1312-76-1 7838 215-199-1
$K_2Si_2O_5$ to $K_2Si_2O_7$
soluble potash glass; soluble potash water glass;. Used as a binder in carbon electrodes, lead pencils etc., also in protective coatings, detergents and glass. Poorly soluble in H_2O, insoluble in organic solvents.

24780 potassium sorbate
24634-61-5 7841 246-376-1
$CH_3CH{\cdot}cCHCH{\cdot}cCHCOOK$
2,4-hexadienoic acid, potassium salt
sorbic acid, potassium salt; potassium 2,4-hexadienoate. As mold and yeast inhibitor. dec 270°; d_{20}^{25} = 1.363; soluble in H_2O (58 g/100 ml), less soluble in organic solvents. *Chisso Am.; Gist-Brocades Food Ingreds.; Hoechst Celanese; Pfizer Spec. Chem.; Protameen.*

24781 potassium sorbate
24634-61-5 7841 219-258-2
$C_6H_7KO_2$
2,4-hexadienoic acid potassium salt
sorbic acid potassium salt; (E,E)-hexadienoic acid, potassium salt; potassium (E,E)-hexa-2,4-dienoate. Used as a mold and yeast inhibitor. mp = 270° (dec); d_{20}^{25} = 1.363; soluble in H_2O (58 g/100 ml), less soluble in organic solvents.

24782 potassium stannate
12142-33-5 235-255-9
$K_2O_3Sn{\cdot}3H_2O$
potassium stannate(IV)
Textile dyeing and printing, alkaline tinplating bath. d = 3.197; soluble in H_2O (1 g/ml), insoluble in EtOH. *Allchem industries; Elf Atochem N. Am.; Blythe, William Ltd; M&T Harsaw; Nihon Kagaku Sangyo.*

24783 potassium stearate
593-29-3 7844 209-786-1
$C_{18}H_{35}KO_2$
stearic acid potassium salt
Used in manufacture of textile softeners. Slightly soluble in H_2O.

24784 potassium stearate
593-29-3 7844 209-786-1
$C_{18}H_{35}KO_2$
octadecanoic acid, potassium salt
stearic acid, potassium salt. Potassium salt of stearic acid; used in manufacture of textile softeners. *Original Bradford Soap Works; Witco.*

24785 potassium sulfate
7778-80-5 7845 231-915-5
K_2SO_4
Analytical reagent, medicine (cathartic), gypsum cements, fertilizer, manufacture of alum and glass, food additive. *Chisso; General Chem.; Heico; Mallinckrodt.*

24786 potassium tetroborate
$2ZnO{\cdot}3B_2O_3{\cdot}3.5H_2O$
Zinc Borate 2335. A specialty flame retardant additive for plasicized PVC and other polymers. Reduces smoke and afterglow. *Borax Europe Ltd.*

24787 potassium tripolyphosphate
13845-36-8 237-574-9
$K_5O_{10}P_3$
pentapotassium tripolyphosphate
Potassium triphosphate; KTPP. Used in detergents, paints, cleaners, specialty fertilizers, sequestrant. *Albright & Wilson; FMC.*

24788 Potassium undecylenoyl hydrolyzed collagen
68951-92-8
Lamepon UD; Maypon UD. Potassium undecylenoyl hydrolyzed animal protein; detergent for hair preparations, shampoos, skin cleansers for damaged skin, antidandruff shampoos. *Henkel/Cospha; Henkel Canada; Henkel KGaA.*

24789 potato gum
Almadina; euphorbia gum. Almeidina gum, stated to be derived from *Euphorbia rhipsaloides*, of West Africa. The latex contains about 10% rubber, 32% water, 51% resin, 1% protein, 6% insoluble matter, and gives an ash of 2.5%. The dry material contains 14.3 % rubber and 75.8% resin.

24790 potato oil or spirit
The alcohol obtained from potato starch.

24791 Potato-Pro EN-15
Hydrolyzed potato protein; cosmetic ingredient for skin and hair care products. *Brooks Industries.*

24792 Potazote
A French fertilizer containing 14% nitrogen, as ammonium chloride, and 20% potassium oxide as potassium chloride.

24793 Potenzol V
Alkyl aryl polyether alcohol as an emulsifiable compound; wetting agent for herbicides *Invequimica & CIA SCA.* Unverified.

24794 Potin
An alloy of 72% copper, 25% zinc, 2% lead, and 1% tin.

24795 Potkem
1312-76-1 7838 215-199-1
potassium silicate
Two-component products based on potassium silicate; as mortars for use with acid resistant bricks and paviors in heavy-duty chemical applications, e.g., brick chimney linings. *Feb Ltd.*

24796 potstone
An impure steatite (talc).

24797 Potting Base
Powdered compost additive containing NPK 3.6:2.2:2.9 plus 3% Mg and trace elements; compost additive. *Vitax Ltd.*

24798 Pouckpong gum
Touchpong gum. A rubber gum of British Guiana.

24799 Poudre B
Vieille powder. A French explosive. It is a smokeless powder made from a mixture of soluble and insoluble nitrocellulose, thoroughly gelatinized with a mixture of ether and alcohol, rolled into sheets, and cut into strips.

24800 Poudre EF
A French explosive made from nitro-cellulose and binding material.

24801 Poudre J
A French explosive containing 83% guncotton and 17% potassium bichromate.

24802 Poudre Pyroxulée
A French sporting powder; consists of insoluble nitro cellulose, with 35% barium and potassium nitrates.

24803 Poulenc 309
sym-disodium-*m*-amino-benzoyl-*m*-amino-*p*-methyl-benzoyl-1-naphthyl-amino-4,6,8-trisulfonate-urea
No manufacturer.

24804 Poutet's reagent
Consists of 1 ml mercury dissolved in 12 ml nitric acid, specific gravity 1.42; used for testing oils.

24805 Povidone
9003-39-8 7879
Poly(vinylpyrrolidone)Plasmosan

24806 povidone-iodine
25655-41-8 7880
A complex produced by reacting iodine with poly(vinylpyrrolidone); betadine.

24807 Powaspray Glymark
An inert white carrier marker for use with completel;y denatured alcohol; an amenity and industrial weed control system that allows approved pesticides to be added to Glymark and applied at lower rates, e.g., glyphosate/Kerts flowable. *Polymer Corp.*

24808 Powax
Additive to hot rinse tanks used after acid pickling of steel sheet and rod. *Foseco (F.S.) Ltd.* Discontinued.

24809 Powder 19/04/15H Black 904
A proprietary polyethylene used in rotational molding and carpet-backing applications; can be used in contact with foodstuffs. No manufacturer.

24810 Powder 215 Natural
A proprietary 400-micron powder used in flame-retardant rotational moldings. No manufacturer.

24811 Powder 22/04/00A 400
A proprietary polyethylene powder of micron size having a low melting point; used for making interliners for fabrics and as carpet backing. No manufacturer.

24812 Powder 26/04/00
A proprietary polyethylene powder possessing good rigidity; used in rotational molding. No manufacturer.

24813 powder of Algaroth
Basic chloride; English powder; powder of Algarotti. A mixture of antimony oxychloride, SbOCl, and antimony oxide, Sb_2O_3; used in the preparation of tartar emetic.

24814 Powdered Aloe Vera (1:200) Food Grade
Aloe vera gel; for personal care products, suntan, sun treatment, burn gels, first aid creams, soaps, hair care, cosmetics, weight control, oral hygiene. *Tri-K Industries.*

24815 powdered hydrocyanic acid
592-01-8 1704 209-740-0
C_2CaN_2
calcium cyanide
The name applied to a calcium cyanide prepared from calcium carbide and hydrocyanic acid. It evolves hydrocyanic acid with moisture, hence the name. A fumigator.

24816 powellite
A mineral, calcium molybdate, $CaMoO_4$.

24817 Power 64, 640, 700
999-81-5 2153 213-666-4
Chlormequat chloride
Plant growth regulator. *Kommer-Brookwick Ltd.*

24818 Power Chlorothalonil 50
1897-45-6 2219 217-588-1
Chlorothalonil
A fungicide for a wide range of agricultural crops. *Kommer-Brookwick Ltd.*

24819 Power Demo
23103-98-2 7651 245-430-1
pirimicarb. Granules containing 50% w/w pirimicarb; for control of aphids. *Kommer-Brookwick Ltd.*

24820 Power Diquat
85-00-7 3415 201-579-4
Diquat dibromide
A contact herbicide and pre-harvest crop desiccant. *Kommer-Brookwick Ltd.*

24821 Power Drive
50471-44-8 10122 256-599-6
Vinclozolin
A protectant fungicide for oilseed rape, peas and beans. *Kommer-Brookwick Ltd.*

24822 Power DSM
919-86-8 6129 213-052-6
demeton-S-methyl
Emulsifiable concentrate containing 580 g demeton-S-methyl per liter; a systemic organophosphorus insecticide and acaricide. *Kommer-Brookwick Ltd.*

24823 Power Ethephon 48
16672-87-0 3777 240-718-3
Soluble concentrate containing 480 g 2-chlorethylphosphonic acid per liter; plant growth regulator for cereal crops. *Kommer-Brookwick Ltd.*

24824 Power Flame
52756-22-6 258-154-1
$C_{19}H_{19}ClFNO_3$
Isopropyl-N-benzoyl-N-(3-chloro-4-fluorophenyl)alanine
flamprop-isopropyl; Suffix BW; WL 29762; Barnon; Flufenprop-isopropyl;. Emulsifiable concentrate of 200 g flamprop-isopropyl per liter; used for control of wild oats in cereal crops. *Kommer-Brookwick Ltd.*

24825 Power Flamprop
52756-22-6 258-154-1
Emulsifiable concentrate of 200 g flamprop-M-isopropyl per liter; used for control of wild oats in cereal crops. *Kommer-Brookwick Ltd.*

24826 Power Gard
Proprietary blend of fuel additives; used for improved fuel consumption, reduced carbon deposits, cleaner carburator, reduced valve deposits. *Gard Corporation.* Name unverified.

24827 Power Gro-Stop
Mixture of chlorpropham and propham; a plant growth regulator for potato sprout growth suppression. *Kommer-Brookwick Ltd.*

24828 Power MCPA
94-74-6 5803 202-360-6
Herbicide for cereals and grassland. *Kommer-Brookwick Ltd.*

24829 Power Non-ionic Wetter
Nonionic wetting agent containing 900 g/l alkylphenol ethylene oxide concentrate; for use in herbicides and pesticides. *Kommer-Brookwick Ltd.*

24830 Power Phosphine Pellets
20859-73-8 372 244-088-0
Aluminum phosphide
Used for gasing of rabbits and moles. *Kommer-Brookwick Ltd.*

24831 Power Platoon
Soluble concentrate containing 155 g 2-chlorethylphosphonic acid (ethephon) and 305 g mepiquat chloride per liter; plant growth regulator for cereal crops. *Kommer-Brookwick Ltd.*

24832 Power Propiconazole
60207-90-1 8003 262-104-4
Propiconazole
A systemic triazole fungicide for control of powdery mildew and rust in cereals. *Kommer-Brookwick Ltd.*

24833 Power Spray Save
61791-44-4 263-177-5
Cationic surfactant containing 800 g/l tallow amine ethoxylate; wetting agent for phosphonoglycine herbicide sprays. *Kommer-Brookwick Ltd.*

24834 Power Swing
34123-59-6 5237 251-835-4
Suspension concentrate containing 500 g isoproturon per liter; used for annual weed control in cereals. *Kommer-Brookwick Ltd.*

24835 Power Task
67306-03-0 4035 266-639-4
Emulsifiable concentrate of 750 g fenpropimorph per liter; used for control of mildew in cereals. *Kommer-Brookwick Ltd.*

24836 Powers Terebine
Drier solution; for addition to certain paints and varnishes to speed drying. *Llewellyn Ryland Ltd.*

24837 Powerspire
60207-90-1 8003 262-104-4

Propiconazole
A systemic triazole fungicide for control of powdery mildew and rust in cereals. *Kommer-Brookwick Ltd.*

24838 Powmet
Metal powders and premixes. *McKechnie Chemicals Ltd.* Name unverified.

24839 PP Captan 83
133-06-2 1815 205-087-0
Captan; a dicarboximide fungicide. *ICI Agrochemicals.*

24840 PP-10GF/000
9003-07-0 7741
PP, 10% glass fiber-reinforced. *Compounding Tech.* Name unverified.

24841 PPG butyl ether series
9003-13-8
butyl ether; Hodag PB-285; Ucon® Fluid AP; PPG-15. *Calgene; Amerchol Corp.*

24842 PPG methyl glucose ether series
61849-72-7
PPG-20 methyl glucose ether
Glucam® P-20, P-10. Humectant for personal care products; freezing pt. depressant; emollient in aq. and hydroalcoholic products; moisturizer; foam modifier in detergent and shampoo systems; solv. and solubilizer for topical pharmaceuticals. *Amerchol Corp.*

24843 PPG-12-PEG-65 lanolin oil;
68458-58-8
Fluilan AWS; Ivarlan AWS; Lantrol® AWS 1692; Ritalan® AWS; Vigilan AWS; Laneto AWS; Lanexol AWS. Emollient, solubilizer; plasticizer and film modifier for hair sprays. *Croda Inc.; Brooks Industries; Henkel/Cospha; Henkel Canada; RITA; Croda Inc.; Fanning.*

24844 PPG-3 myristyl ether
63793-60-2
Hetoxol MP-3; Carsonon® 144-P; Promyristyl PM-3; Witconol APM. Lubricant; emollent; solubilizer for cosmetics; deodorant; wetting agent; penetrant; dye dispersant; antistat; scouring agent; textile surfactant *Heterene; Lonza Inc; Croda Inc; Witco Organics; Witco SA.*

24845 PPG-30 cetyl ether series
9035-85-2
Fancol 707; Carsonon® 169-P; Procetyl 10; Wickenol® 707. Skin conditioner, emollient for permanent waves, shampoos, blushers, indoor tanning preparations. *Fanning; Lonza Inc.; Croda Inc.; CasChem.*

24846 PPG-40 diethylmonium chloride
9076-43-1
Emcol® CC-42. Pigment dispersant, particle suspension aid, emulsifier, solv., conditioner, antistat, lubricant, corrosion inhibitor for toiletries, cosmetics, germicides, synthetic fibers and plastics, textiles, industrial processes; ore flotation additive. *Witco/Organics; Witco Chemical Ltd.* Discontinued.

24847 PPO
Polyphenylene oxide.

24848 PP-Vac
Pigeon pox; for immunization of poultry. *Intervet Inc.*

24849 PPX-30GF/000 HC
9003-07-0 7741
PP, highly chemically coupled, 30% glass fiber-reinforced. *Compounding Tech.* Name unverified.

24850 Pradone Plus
34205-21-5 251-879-4
$C_{15}H_{19}ClN_4O_3$
N'-(3-chloro-4-(5-(1,1-dimethylethyl)-2-oxo-1,3,4-oxadiazol-3(2H)-yl)phenyl)-N,N-dimethylurea
dimefuron; Pradone. Herbicide. *May & Baker Ltd.* Name unverified.

24851 Praepagen
Textile softening materials. *Hoechst AG.*

24852 Praestol® 186K
Polydimethyl-dialkyl ammonium chloride
Polymer for use as wastewater coagulant in mining/mineral processing industries, water plants, biological wastewater facilities; filtration aid. *Stockhausen.*

24853 Praestol® A3010L
Acrylamide emulsion polymer; flocculating agent for effluent treatment (papermill, textile mill, food processing waste, petrochemical wastewater, steel mill effluent), mineral processing, brine clarification, sand/gravel washing and sugar juice clarification. *Stockhausen.*

24854 Praestol® K2001
7446-70-0 348 231-208-1
$AlCl_3$
aluminum chloride
Basic aluminum chloride; coagulant for municipal and industrial water treatment; effective for turbidity reduction, phosphorus removal, water clarification, flotation, oil/water demulsification. *Stockhausen.*

24855 Pragmatar
A proprietary preparation of cetyl alcohol and coal tar distillate with sulfur and salicylic acid; used in the treatment of dandruff. *SmithKline Beecham.* Name unverified.

24856 Pragmatar Ointment
Combination of cetyl alcohol-coal tar distillate, precipitated sulfur and salicylic acid; for treatment of dandruff, seborrheic conditions and common scaly skin disorders where skin is unbroken. *SmithKline Beecham.* Name unverified.

24857 Prapagen WK, WKL, and WKT
Cationic surfactants of the quaternary ammonium chloride type in liquid or past form; antistatic agents, fabric conditioner and softener, fiber finishers, water-repellant agents and dewatering agents; wetting agents for oils, dispersants for pigments, flushing, foaming and wetting agents, spinning bath and viscous additives; flotation chemicals and anti-caking agents for rendering salts free-flowing, corrosion inhibitors, anchoring agents for tar, bitumen, coatings, lacquers, adhesives, disinfectants. *Hoechst UK.* Name unverified.

24858 prase
A mineral, SiO_2.

24859 PRC
Motor fuel additive. *Monsanto (Solaris).* Name unverified.

24860 Preact
Galvanizing fluxes. *Mineral Research & Development Corp.*

24861 Prebane 500
886-50-0 212-950-5
Suspension concentrate containing 500 g/l terbutryn; for weed control in cereals. *Ciba-Geigy Agrochemicals.*

24862 Prebond
Pre-phosphate refining additive. *Brent Chemicals International plc.*

24863 Prechem 90
Phosphate detergent; detergent, wetting agent, dispersant, soil suspending agent for textile preparation and dyeing; leveling agent; penetrant. *Ivax Industries.*

24864 Prechem NPX
Nonylphenoxy polyethoxy alcohol
Detergent concentrate, wetting agent, emulsifier for textile wet processing of cotton, wool and synthetics; aids dyebath stability. *Ivax Industries.*

24865 Precifac ATO
540-10-3 2073 208-736-6
Cetyl palmitate
Gattefosse SA.

24866 precipitated phosphate
7758-87-4 1741 231-840-8
Insoluble calcium phosphate. Used in the manufacture of fertilizers and as a calcium replenisher.

24867 Precirol ATO
8067-32-1 232-514-8
Glyceryl di/tripalmito stearate
Additive for tablets, binder, lubricant, sustained release. *Gattefosse SA.*

24868 Precirol WL 2155
8067-32-1 232-514-8
Glyceryl ditristearate
Additive for tablets manufacturing. *Gattefosse; Gattefosse SA.*

24869 predazzite
A mineral. It is a mixture of calcite (calcium carbonate) and brucite (magnesium hydroxide).

24870 Predef
338-98-7 5190 206-423-9
Isoflupredone acetate
Veterinary anti-inflammatory. *Upjohn.* Name unverified.

24871 Preditec®
For agriculture industry. *DuPont UK.*

24872 prednisolone
50-24-8 7901 200-021-7

24873 Pree®
67129-08-2 266-583-0
Metazachlor
Selective herbicide, used for control of grasses and broadleaf weeds in maize *BASF AG.*

24874 Pre-Empt
Emulsifiable concentrate containing 46 g linuron, 54 g trietazine and 200 g trifluralin per liter; herbicide for winter cereals. *Schering Agrochemicals Ltd.* Discontinued.

24875 Prefera® SSL/CSL
Sodium or calcium stearyl-2-lactylates; food emulsifier for improvement of

fermentation tolerance, volume and texture of yeast-raised baked goods; antistaling effect. *Grünau.*

24876 Prefix D
1194-65-6 3093 214-787-5
A granular formulation containing 6.75% dichlobenil; provides season-long weed control of both annual and perennial grasses and broad-leaved weeds. *Burts & Harvey; Hoechst UK; Shell UK.* Name unverified.

24877 Pregaday
A proprietary preparation of ferrous fumarate and folic acid; hematinic for use in pregnancy. *Duncan Flockhard Ltd.*

24878 Pregeflo®
Pre-cooked starches; thickener for foods; binding agent in pharmaceuticals. *Roquette (UK) Ltd.*

24879 Pregl's solution
A solution of potassium iodide and sodium iodate with a small amount of sodium chloride and bicarbonate.

24880 pregolan
A chlorine compound giving 65-72% available chlorine. Used in bleaches.

24881 pregrattite
A variety of paragonite, a mineral.

24882 Pregwood
A proprietary synthetic resin-impregnated wood, made by impregnating and then subjecting the wood to heat and pressure. No manufacturer.

24883 prehnite
$2CaO{\cdot}Al_2O_3{\cdot}3SiO_2$
Jacksonite
Calcium aluminum silicate, a mineral.

24884 Pre-Kite
Selective herbicide. *Schering Agrochemicals Ltd.* Discontinued.

24885 Prelude
67747-09-5 7941 266-994-5
Prochloraz
Fungicide used to protect against seed-borne diseases. *Agrichem (International) Ltd.*

24886 Premaline
Herbicide based upon linuron. *May & Baker Ltd.* Name unverified.

24887 Premalox
Herbicide and growth regulator based upon propham. *May & Baker Ltd.* Name unverified.

24888 Premerge
88-85-7 3341 201-861-7
dinoseb
Herbicide used for the control of broadleaf weeds in peas, soybeans, potatoes, and orchards. Active ingredient is dinoseb. *Dow UK.* Discontinued.

24889 Premier alloy
A heat-resisting alloy containing 61% nickel, 11% chromium, 25% iron, and 3% manganese.

24890 Premix
Masterbatches in thermoplastics; for pigmentation of thermoplastics. *Cornelius Chemical Co Ltd.* Name unverified.

24891 Prenite
A proprietary trade name for an asbestos sheet bonded with neoprene; a packing material. No manufacturer.

24892 Prenol
556-82-1 209-141-4
$C_5H_{10}O$
3-Methyl-2-butene-1-ol
Fragrance and flavoring (fresh, herbal, green, fruity, slightly lavender-like). mp = 0°; bp = 140°; d = 0.8480; n_D^{20}= 1.4412; soluble in H_2O (170 g/l), organic solvents; LD_{50} (rat orl) = 810 mg/kg. *BASF AG.*

24893 Prepagen
Textile softening agents. *Hoechst UK.* Name unverified.

24894 Prepagen WK
107-64-2 203-508-2
$C_{38}H_{80}ClN$
distearyldimethylammonium chloride;
Di(hardened tallow)dimethylammonium chloride; Aerosurf TA-100; Dimethyl distearyl ammonium chloride; Dimethyl-n-octadecyl-1-octadecanaminium chloride; dimethyldioctadecylammonium chloride; Distearyldimonium Chloride. A proprietary trade name for 75% distearyl dimethyl ammonium chloride in isopropanol; used as a conditioner, anti-stat and softener in the laundry trade. *Hoechst UK.* Name unverified.

24895 prepared calamine
1675
Obtained by calcining and powdering negative zinc carbonate or calamine, and freeing the product from gritty particles. It consists of zinc oxide with some oxide of iron. Used as an astringent and a topical protectant.

24896 prepared chalk
Creta proeparata B.P.. Washed chalk or whiting.

24897 prepared cobalt oxide
1307-96-6 2507 215-154-6
CoO
Cobaltous oxide
Obtained by heating the black oxide, Co_2O_3; used as a pigment in the ceramic industry.

24898 preparing salt
12058-66-1 8826 235-030-5
$Na_2SnO_3{\cdot}3H_2O$
sodium stannate(IV)
Used as a mordant in dyeing and calico printing.

24899 Preperite
Compound used to remove rust from iron and steel and corrosion products from aluminum. *Chemicals International plc.* Unverified.

24900 Prepon®
Casting and modelling wax, blue, green and ivory; dental preparation. *Bayer AG.*

24901 PrepRite™ Coating Remover
Mixture of N-methyl-2-pyrrolidone, butyrolactone, and other solvents. *ISP.*

24902 Preservaline
Iceline; Freezine. Names for formaldehyde used as a preservative for milk.

24903 Preservals®
Parabens
Preservatives for cosmetics and pharmaceuticals. *Laserson & Sabetay.*

24904 Preservative
Mixture of antimicrobials; preservative; for cosmetic preparations. *Rewo Chemicals Ltd.* Name unverified.

24905 Preservol
Creosote, partially emulsified with pyroligneous acid; used for preserving wood. No manufacturer.

24906 Preservo-O-Sote
Wood preservatives. *Crowley Chem.*

24907 Preservotabs
Water bath sanitizing tablets. *The Boots Co plc.* Discontinued.

24908 Presite®
For agriculture industry. *DuPont UK.*

24909 Presol
Damp-proofer and wood preserver solvents. *Carless Solvents Ltd.* Name unverified.

24910 Presol W
Mercury fungicide solution. *Great Lakes Europe.* Discontinued.

24911 Presomet
Black bituminous paint. *Thomas Ness Ltd.* Name unverified.

24912 Press N Seal
Caulking tape in roll form; used to caulk windows, doors and other openings in construction buildings and homes. *Chemseco.* Name unverified.

24913 Press-Aid
8002-74-2, 9010-66-6
Synthetic wax, corngluten protein; binder for pressed powders. *Presperse.*

24914 press-cake
The mill-cake formed by mixing the ingredients of gunpowder in the incorporating mill, is subjected to a high pressure to make press-cake.

24915 Pressolith
Sillimanith
Earthenware porcelain products.

24916 Pressphan
A German name for press-boards made from wood pulp; used as insulating materials.

24917 Presszell
A German synthetic resin varnish-paper product used as an electrical insulator. No manufacturer.

24918 Presto Steel
A proprietary trade name for a steel containing 1.4% chromium. No manufacturer.

24919 Prestochlor
7778-54-3 1717 231-908-7
$CaCl_2O$
Calcium hypochlorite
Bleach and oxidizer. *P B & S Chemical Co Inc.* Unverified.

24920 Prestogen® K
Stabilizer for low silicate peroxide bleaching of cellulosic fibers and blends. *BASF; BASF plc.*

24921 Prestogen® TX-155
Multifunctional stabilizer for peroxide bleaching; high alkali stability. *BASF.*

24922 Prest-o-lite
A proprietary brand of acetylene gas compressed in cylinders. No manufacturer.

24923 Prestone
Ethylene glycol antifreeze. *National Carbon Company Inc.* Name unverified.

24924 Pretolone
Dyeing and printing assistant. *Ciba plc.* Name unverified.

24925 Prevail 3050
Thermoplastic PU/ABS blend; engineering thermoplastic providing unique combination of flexibility and toughness; suggested for bumpers and fairings on commercial trucks, on snowmobiles, ATVs and campers, and in the automotive market for paintable soft fascia. *Dow Plastics.*

24926 Preventol®
97-23-4 3120 202-567-1
Wide range of microbicides/preservatives and corrosion inhibitors. Active principal is dichlorophen. *Bayer AG; Bayer plc.*

24927 Preventol® A2
An organic inhibitor; for use in the formulation of fungicidal interior paints (emulsion and solvent based paints) excluding air-drying paints. *Bayer AG.*

24928 Preventol® Cl 5
621-08-9 210-668-7
$C_{14}H_{14}OS$
Dibenzylsulfoxide
Benzene, 1,1'-[sulfinylbis(methylene)]bis-; Benzyl sulfoxide. An organic inhibitor for use in cleansing acids and in the surface treatment of metals. *Bayer AG.*

24929 Preventol® Cl7-100
29385-43-1 249-596-6
$C_7H_7N_3$
Tolyl triazole
tolyltriazole; tolutriazole; methyl-1H-benzotriazole; 5-methylbenzotriazole; 5-methyl-1,2,3-benzotriazole; Cobratec TT-100. Corrosion inhibitor for copper, copper alloys and other metals; particularly suitable for antifreezes, coolants, cutting fluids and hydraulic fluids. mp = 76-87°; bp = 160°; d = 1.24; poorly soluble in H_2O (<100 mg/l), more soluble in organic solvents. *Bayer AG.*

24930 Preventol® Cl 8
95-14-7 1140 202-394-1
$C_6H_5N_3$
Benzotriazole
Benzotriazole; Cobratec 99; 1H-benzotriazole; 1,2,3-Benzotriazole; Azimidobenzene; benzene azimide; 1,2,3-triaza-1H-indene; U-6233; aziminobenzene; 1,2-aminoazophenylene; cobratec 99; 2,3-diazaindole; 1,2,3-triazaindene. Corrosion inhibitor for copper, copper alloys and other metals; particularly suitable for antifreezes, coolants, cutting fluids and hydraulic fluids. mp = 96-97°; bp_{15} = 201-204°; soluble in H_2O (20 g/l), organic solvents; LD_{50} (rat orl) = 600 mg/kg. *Bayer AG.*

24931 Preventol® O Extra
90-43-7 7458 201-993-5
o-phenyl phenol
Preservative for chemical and industrial products; raw material for disinfectants. *Bayer AG.*

24932 Prevex® BJA
Phenylene ether copolymer alloy; flameretardant structural foam molding grade polymer for both thin- and thick-wall sections; used for CRT and printer housings, copier panels, instrument bases, card cages, and workstations. *GE Plastics.*

24933 Prevex® PMA
Polyphenylene ether resin; good heat resistance and dimensional stability; for small appliances, lawn care tools, power tools, industrial devices. *GE Plastics.*

24934 Prevex® PQA
Phenylene ether copolymer alloy; general purpose injection molding grade polymer; used for power tools, portable appliances, transportation components, hot water appliances, and liquid handling devices. *GE Plastics.*

24935 Prevex® VFA
Polyphenylene ether resin; good impact strength, dimensional stability for use in business machine applications, internal parts, other enclosures. *GE Plastics.*

24936 Prevex® VGA
Polyphenylene ether resin, flame-retardant; business machine applications, internal parts, other enclosures. *GE Plastics.*

24937 Prevex® W20
Phenylene ether polymer alloy; automotive polymer with high flow, high impact, and high heat resistance used for instrument panels and interior trim components. *GE Plastics.*

24938 Prevex® W30
Phenylene ether copolymer alloy; heatresistant automotive polymer used for instrument panels and interior trim components. *GE Plastics.*

24939 Priamid
Lather booster and stabilizer, emulsifier; antistatic agents; used for powdered detergents; toilet and shaving soap; shampoos; bubble baths; thermoplastics and synthetic fibers; liquid paste cleaners. *Unichema International.* Name unverified.

24940 pribramite
A mineral. It is a variety of sphalerite.

24941 Pricat 9900
Catalyst for hydrogenation of vegetable and animal fats and oils. *Hart Chem. Ltd.*

24942 Pricerine 9071
56-81-5 4493 200-289-5
Glycerol; polyol solvent for low color alkyd resins. *Unichema.*

24943 pricite
$3CaO{\cdot}4BO_3{\cdot}6H_2O$
Bechilite
A mineral. It is a calcium borate.

24944 priderite
$(K,Ba)_{1.3}(Ti,Fe)_8O_{16}$
A mineral.

24945 Prifac 7912
Groundnut fatty acid; used in manufacturing of alkyd resins for stoving enamels, acid curing lacquers and NC lacquers; in wood and metal varnishes. *Unichema.*

24946 Prifac 7920
Tallow acid
Raw material for surfactants, soaps, nitrogen derivatives, buffing formulations. *Unichema.*

24947 Prifac 7951
Soybean fatty acid; used as raw material for high quality long oil alkyds for consumer paints, for medium and short oil alkyd resins for air and oven-drying lacquers with high flexibility. *Unichema.*

24948 Prifac 7960
Sunflower fatty acid; for high quality medium to fast drying alkyd resins, suitable for nonyellowing lacquer systems with excellent heat stability. *Unichema.*

24949 Prifac 9428
61790-38-3 263-130-9
Hydrogenated tallow acid
Chemical intermediate for surfactants, stabilizers, detergents, fabric softeners. *Unichema.*

24950 Prifrac 2901
124-07-2 1808 204-677-5
Caprylic acid
Intermediate for ester production, synthetic lubes, latex stabilizers, substituted glycerides used as skin protectors. *Unichema.*

24951 Prifrac 2906
334-48-5 1802 206-376-4
Capric acid
Detergent raw material for soaps, industrial cleaners. *Unichema.*

24952 Prifrac 2920
143-07-7 5396 205-582-1
Lauric acid
Emulsifier for hot emulsion polymerization of NBR, and NR latex stabilization. *Unichema.*

24953 Prifrac 2940
544-63-8 6416 208-875-2
Myristic acid
Intermediate for ester, detergent, and surfactant products *Unichema.*

24954 Prifrac 2960
57-10-3 7128 200-312-9
Palmitic acid
Intermediate for surfactants, soap and cosmetic formulations. *Unichema.*

24955 Prifrac 2980
57-11-4 8959 200-313-4
Stearic acid
Intermediate for surfactants, soap and cosmetic formulations. *Unichema.*

24956 Prifrac 2989
112-85-6 1051 204-010-8
Behenic acid
Intermediate for surfactants, soap and cosmetic formulations. *Unichema.*

24957 Prifrac 2990
112-86-7 3713 204-011-3
Erucic acid
Chemical intermediate. *Unichema.*

24958 Primacor
A family of adhesive polymers used for extrusion coating and layers in flexible packaging. *Dow Cheml Co Ltd, UK & Ireland.*

24959 Primacor 1320
Adhesive polymer; for blown film; used as adhesive layer or sealant layer in flexible packaging structures; FDA compliant. *Dow Chemical.*

24960 Primacor 2912
Adhesive polymer; extrusion coating resin used for aseptic, medical, pharmaceutical and condiment packaging, dry mix, and moist and dry food packagings, laminated tubes, lidding stock. *Dow Chemical.*

24961 Primacor 4990 Dispersion
9010-77-9
Ethylene acrylic acid copolymer dispersion; dispersion polymer for use as binder for nonwoven fibers including PP, polyester, glass, and nylon; provides soft fabrics having excellent antisoil redeposition properties. *Dow Chemical.*

24962 Primafloc
Polyelectrolyte coagulants, flocculants. *Rohm & Haas UK.*

24963 Primal
Acrylic emulsions; used for decorative and industrial coatings, binders for textile and nonwoven applications, floor polishes, leather, adhesives, cement modifiers. *Rohm & Haas.*

24964 Primallor
Gold casting alloy for dental applications. *Degussa.*

24965 Primapel
Soil retardant and leather chemicals. *Rohm & Haas UK.*

24966 Primasol® AMK
Padding auxiliary to prevent dye migration. *BASF; BASF plc.*

24967 Primasol® FP
Aliphatic sulfonic acids, organic phosphate aqueous solution; detergent, wetting agent, textile auxiliary, pigment padding, mercerizing and bleaching processes; migration inhibitor; low foaming; stable to electrolytes, alkalies. *BASF/Fibers; BASF AG.*

24968 Primasol® NB-NF
Phosphate ester; low foaming wetting and deaerating agent for dyeing, desizing, bleaching, pad bath application; modified surface tension to minimize foam formation. *BASF/Fibers.*

24969 Primatol AA
1912-24-9 902 217-617-8
atrazine
Herbicide. *Ciba plc.* Name unverified.

24970 Primatol AD 85WP
Aminotriazole + atrazine + 2,4-D; used for total weed control in non crop areas. *Ciba-Geigy Agrochemicals.*

24971 Primatol AP
Triazine, picloram total herbicide. *Ciba plc.* Name unverified.

24972 Primatol SE 500FW
61-82-5, 122-34-9
A suspension concentrate containing 180 g aminotriazole and 300 g simazine per liter; used for total weed control in non crop areas and fruit orchards. *Ciba-Geigy Agrochemicals.*

24973 Primax
Textile and leather chemicals. *Rohm & Haas UK.*

24974 Primax
Surface modified, ultrahigh molecular weight polyethylene. *Air Prods & Chems. Inc.*

24975 Primax UH-1060
9002-88-4 7728
Ultrahigh molecular weight polyethylene, modified; surface modified particles producing cast elastomer composites with high abrasion resistance and low coeff. of friction; imparts improved abrasion, corrosion and chem. resistance in coatings. *Air Prods & Chem/Polyurethanes.* Name unverified.

24976 Primax UH-1080
9002-88-4 7728
Ultrahigh molecular weight polyethylene, modified; surface modified particles producing cast elastomer composites with high abrasion resistance and low coeff, of friction; imparts improved abrasion, corrosion and chem. resistance in coatings. *Air Prods & Chem/Polyurethanes.* Name unverified.

24977 Primax UH-1250
9002-88-4 7728
Ultrahigh molecular weight polyethylene, modified; surface modified particles producing cast elastomer composites with high abrasion resistance and low coeff. of friction; imparts improved abrasion, corrosion and chem. resistance in coatings. *Air Prods & Chem/Polyurethanes.* Name unverified.

24978 Primazin Fixing Agent RP
Fixing agent for resist printing with fiber reactive dyes. *BASF.*

24979 Primazin®
Reactive dyes for dyeing and printing cellulosic fibers. *BASF; BASF plc.*

24980 Prime Flavours
Meat and savory flavors; for food products (soups, sauces etc) and processed meats. *Fries & Fries.* Name unverified.

24981 Primecoat®
For the electrical industry. *DuPont UK.*

24982 Primene
Tert-alkylamines; stabilizers. *Rohm & Haas UK.*

24983 Primicid
23505-41-1 7652 245-704-0
Soil applied insecticide containing pirimiphos-ethyl. *ICI Chem & Polymers Ltd.*

24984 Primor
Lubricating and industrial oils. *Burmah-Castrol Ltd.* Name unverified.

24985 Primorol 1511
Branched chain C_{24}, C_{26} and C_{28} alcohols. *Henkel/Cospha.*

24986 Primotec
23505-41-1 7652 245-704-0
Insecticide and seed dressing containing pirimiphos-ethyl. *ICI Chem & Polymers Ltd.*

24987 primrose smokeless
A smokeless 42 grain powder.

24988 primuline base
p-toluidine heated with sulfur.

24989 Primus
Fiber-rich bar to help regularity. *Richardson-Vicks Inc.* Name unverified.

24990 Primus®
For the automotive industry. *DuPont UK.*

24991 Prince's blue
A mineral. It is a blue variety of Sodalite. The slabs are polished for ornamental purposes.

24992 Prince's metal
Prince Rupert's Metal. An alloy. It is a variety of brass containing from 61-83% copper and 17-39% zinc. Another alloy, also called Prince's metal, consists of 84.75% tin and 15.25% antimony.

24993 Prince's mineral
Prince's metallic. A clay containing about 40% iron oxides.

24994 Prinsyl
Antiseptic. *Coventry Chemicals Ltd.*

24995 Printel
Aqueous dispersions of pigments for textile printing. *European Colour (Pigments) Ltd.*

24996 Printer's acetate
8006-13-1 332
$AlC_6H_9O_6$
Aluminum acetate.

24997 printer's iron liquor
A deep black solution of ferrous acetate, containing some ferric acetate. It contains about 10% iron.

24998 Printex
1333-86-4 1856 231-153-3
Carbon black; used for printing inks. *Degussa.*

24999 Printex 25 Beads and Powd
1333-86-4 1856 231-153-3
Furnace black; especially developed for uv inks, low uv absorption; glossy; improves shelf-life of inks. *Degussa.*

25000 Printex P
1333-86-4 1856 231-153-3
Furnace black; for plastics formulations requiring highest uv absorption. *Degussa.*

25001 Printex U Beads and Powd
1333-86-4 1856 231-153-3
Channel-type carbon black; general purpose industrial black coating systems *Degussa.*

25002 Printing Black for Wool
A dyestuff produced by the reduction of a mixture of 1:5-and 1:8-dinitronaphthalene by means of glucose in alkaline solution in the presence of sodium sulfite. Dyes wool violet black from an acid bath. Employed in printing.

25003 printing inks
Inks consisting of pigments incorporated with varnish made by heating linseed oil.

25004 Printlok 1046
Vinyl acrylic
Self-crosslinking pigment printing binder. *Catawba-Charlab.*

25005 Printogen
Printing oils and fixation accelerators used for textile printing. *Degussa AG.* Name unverified.

25006 Printol®
Vegetable soaps, xylenols, pine oils; soluble phenolic disinfectant for hospital use. *Coventry Chemicals Ltd.*

25007 Printosol
Prinking ink distillates and oils. *Haltermann Ltd.*

25008 Printsolve™ Ink Remover
N-methyl-2-pyrrolidone and dipropylene glycol methyl ether
Solvents which serve as an ink remover for the pulp and paper industry, flexographic/rotogravure industries. *ISP.*

25009 Printwash
Mixture of esters, alcohols and toluene; for cleaning of print rollers. *Solrec Ltd.* Name unverified.

25010 Prinza® Range
9000-30-0 4601 232-536-8
Guar gum (*cyamopsis tetragonolobus*); gum guar; Burtonite V-7-E; cyamopsis gum; guar flour; gum cyamopsis; Jaguar no.124; Jaguar gum A-20-D; Jaguar plus; Regonal; Supercol U; Decorpa; Guarem; Guaran. Guar gum; thickener and stabilizer for food industry. Insoluble in H_2O, organic solvents; LD_{50} (rat orl) = 7060 mg/kg. *Grünau.*

25011 Prioderm
121-75-5 5740 204-497-7
A proprietary preparation of malathion; used in the treatment of infestation by lice. *Napp Laboratories Ltd.* Name unverified.

25012 Priolene 6900
112-80-1 6965 204-007-1
Oleic acid
Intermediate for ethoxylates, esters, nitrogen derivatives, surfactants; used in soaps, personal care products, lubricant and metalworking fluids, for NR latex stabilization. *Unichema.*

25013 Priolube 1400
112-62-9 6965 203-992-5
Methyl oleate
Synthetic lubricant-based fluids, metalworking, industrial applications. *Unichema.* Name unverified.

25014 Priolube 1405
142-77-8 205-559-6
n-butyl oleate
Plasticizer for chloroprene rubber. *Unichema.*

25015 Priolube 1407
Glyceryl oleate
Lubricant for PVC processing; antifog agent for PVC film. *Unichema.*

25016 Priolube 1409
25637-84-7 247-144-2
Glyceryl dioleate
Lubricant for PVC processing; antifog for PVC film. Lipophilic emulsifier; main component is a diglyceride, an oil component with water-holding capacity. *Unichema.*

25017 Priolube 1414
84988-79-4 284-868-8
Isobutyl oleate
Additive to improve ozone resistance of chloroprene rubber. *Unichema.*

25018 Priolube 1429
85049-34-9 285-203-4
Propylene glycol dioleate
Synthetic lubricant-based fluids, metalworking, industrial applications. *Unichema.* Name unverified.

25019 Priolube 1435
122-32-7 9861 204-534-7
Glyceryl trioleate
Synthetic lubricant-based fluids, metalworking, industrial applications. *Unichema.* Name unverified.

25020 Priolube 1451
123-95-5 1625 204-666-5
n-butyl stearate
Plasticizer for butyl rubber; lubricant for PVC processing. *Unichema.*

25021 Priolube 1458
91031-48-0 292-951-5
Isooctyl stearate
Lubricant for PVC processing. *Unichema.*

25022 Priormatt
Synthetic emulsion paint; for decorative purposes on walls; for coating the interior of GRP boats. *Llewellyn Ryland Ltd.*

25023 Priplast 1431
Epoxidized oleate
Plasticizer for PVC. *Unichema.*

25024 Priplast 1562
Plasticizer for nitrile rubber to be used for petroleum hose connections. *Unichema.*

25025 Priplast 3013
109-31-9 203-664-1
Di-n-hexyl azelate
Plasticizer for PVC. *Unichema.*

25026 Priplast 3018
103-24-2 203-091-7
Di-2-ethylhexyl azelate
Plasticizer for chloroprene and nitrile rubbers. *Unichema.*

25027 Priplast 3157
Polymeric plasticizer; plasticizer for rubber with excellent heat stability and good solvent extraction resistance. *Unichema.*

25028 Priplast 3191
Dimer-based polyester; used for flexible PU foams with outstanding hydrolytic stability, low fogging properties, suitable for automotive applications. *Unichema.*

25029 Pripol 1004
C44 dimer acid
Modifier for nylon; polyester fibers; polyamide for hot melts; urethane elastomer. *Unichema.*

25030 Pripol 1009
Dimer acid
Building block/modifier permitting production of higher molecular weight condensation polymers with flexibility, toughness, improved impact resistance, low moisture absorption, hydrolytic resistance; used in polyester fibers, polyamides and urethane elastomers. *Unichema.*

25031 Pripol 1025
Hydrogenated dimer acid
For hot melt adhesives; polyamide for thermographic inks. *Unichema.*

25032 Pripol 1040
Trimer acid
Used in polyamino-amides for PVC plastisols to be used as car underbody coating and in water-soluble alkyd resins. *Unichema.*

25033 Prism
Gas separators; separates industrial gases through use of gas permeable hollow fiber membranes. *Monsanto Co.*

25034 Prism
Polyurethane reaction injection molding system. *Miles.* Discontinued.

25035 prismatic niter
7757-79-1 7815 231-818-8
KNO_3
potassium nitrate
The name derives from the form of the crystals.

25036 Prisorine 3508
2724-58-5 220-336-3
Isostearic acid
Polysulfide rubber additive; results in use of less solvent or plasticizer, easier preparation, better cure retarding, better heat stability in final products *Unichema.*

25037 Pristane
1921-70-6 7932 217-650-8
$C_{19}H_{40}$
2,6,10,14-tetramethylpentadecane
norphytane, Robuoy. Isolated from Shark liver oil. A chemically inert oil used as a lubricant and heat transfer oil in transformers. bp = 296°; d_4^{20}= 0.78267; n_D^{20}= 1.43848; insoluble in H_2O, soluble in organic solvents.

25038 Pristene 180
Mixed tocopherols; food-grade antioxidant; natural source of vitamin E. *UOP.*

25039 Pristene R20
84604-14-8 283-291-9
Extract of *Rosmarinus officinalis.* Foodgrade natural flavoring and stabilizer. Toner for skin, scalp uses, dandruff control. *UOP.*

25040 Pristerene 4904
57-11-4 8959 200-313-4
Stearic acid
Intermediate for ethoxylates, esters, nitrogen derivs., personal product formulations; heat-stable. *Unichema.*

25041 Pristine
Natural mixed tocophenols; antioxidants. *UOP Inc.*

25042 Pro Seal
Silicone sealant, gasketing compounds; for making formed-in-place gaskets and for use within cut gaskets. *Novest Inc.* Unverified.

25043 Pro Weld
Two-part cold welding compound; for bonding aluminum, brass and steel castings. *Novest Inc.* Unverified.

25044 Proaid 9802
Nonstaining, nondiscoloring processing and dispersing aid for polychloroprene and EPDM. *Akrochem.*

25045 Proaid 9814
Homogenizing agent and softening resin for use in most elastomers; improves processing. *Akrochem.*

25046 Proaid 9904
Specialty processing aid for use in CM (CPE) and CSM (Hypalon) polymers; gives some mold release properties; stabilizes hot air aging properties. *Akrochem.*

25047 Proban
Textile flame retardant. *Albright & Wilson Ltd., Phosphates & Speciality Business.*

25048 Probe
20354-26-1 6032 243-761-6
methazole
Wettable powder containing 75% methazole; for post-emergence weed control. *ICI Chem & Polymers Ltd.*

25049 proberite
$2(Na,Ca,B_8O_9 \cdot 5H_2O)$
A mineral.

25050 Probimer
Photo-crosslinkable synthetic resin. *Ciba plc.*

25051 Probimide
Polyimides. *Ciba plc.*

25052 Procal
Vacuum-formed ceramic fiber insulation. *Foseco (F.S.) Ltd.*

25053 Procetyl 10
9035-85-2
PPG-10 cetyl ether
Emollient, coupler, cosolvent, plasticizer, superfatting, wetting and spreading agent, penetrant; lubricant in cosmetics and personal care products; alcoholic and aqueous alcoholic compositions. *Croda Inc.*

25054 Procetyl AWS
9087-53-0
PPG-5 ceteth-20
Emulsifier, plasticizer, coupler, humectant, dispersant, emollient, and fragrance solubilizer in aqueous and aqueous alcoholic systems, personal care products, antiperspirants, bath oils. *Croda Inc.; Croda Chem. Ltd.*

25055 Procetyl AWS
PPG-8-ceteth-2
Emollient and emulsifier. *Croda Inc.*

25056 Prochinor
Specialty blends of fatty amines and derivatives; demulsifiers for crude oil production and industrial oils. *Elf Atochem UK/Ceca.*

25057 prochloraz
67747-09-5 7941 266-994-5
$C_{15}H_{16}Cl_3N_3O_2$
N-propyl-N-[2-(2,4,6-trichlorophenoxy)ethyl]-1H-imidazole-1-carboxamide
Ascurit; BTS 40542; Octave; Omega; Prelude; Sporgon; Sportak. Fungicide wth protective and eradicative action. Applied as a foliar spray to control *Rhynchosporium, Helminthosporium, Septoria, Fusarium, Pseudocercosporella, Erysiphe and Pyrenophora* spp. in crops. mp = 38-41°; $bp_{0.2}$ = 208-218°; soluble in H_2O (55 mg/l), more soluble in organic solvents; LD_{50} (rat orl) = 1600 mg/kg.

25058 prochlorite
A mineral similar to chlorite.

25059 Procilene
Polyester/cotton dyes. *ICI Chem & Polymers Ltd.*

25060 Procinyl
Reactive disperse dyes. *ICI Chem & Polymers Ltd.* Discontinued.

25061 Procion
Reactive dyestuffs. *ICI Chem & Polymers Ltd.*

25062 Procol®
An extensive range of flotation reagents for both sulfide and nonsulfide minerals; includes collectors, frothers, depressants and modifiers. *Allied Colloids Ltd.*

25063 Procom
Polypropylene compound. *ICI Chem & Polymers Ltd.*

25064 Procond-101
9003-07-0 7741
Electrically conductive PP copolymer; extrusion grade *United Composites.*

25065 Procor 75 AB-X
Two-side acrylic coated biaxially oriented PP film; for horizontal, vertical and general overwrap applications; excellent seal strength, moisture/flavor/odor barrier properties, wide sealing range. *Mobil/Films.* Name unverified.

25066 procythol
A liver extract. Used as a nutrient.

25067 Prodag
7782-42-5 4560 231-955-3
Graphite in water; lubricant additive for machine oils, assembly lubes. *Acheson Colloids.*

25068 Prodew 100
Preparation containing sodium lactate, sodium PCA, sorbitol, hydrolyzed animal protein and proline. Formulated moisturizer for cosmetics, soaps, hair care products; humectant. *Ajinomoto.* Name unverified.

25069 Prodex
10265-92-6 6014 233-606-0
Methamidophos
Insecticide. *Makhteshim Chemical Works Ltd.*

25070 Prodoraqua
Liquid waterproofer and hardener. *Prodorite Ltd.*

25071 Prodorbond
A range of polymer modified heavy duty industrial floor finishes. *Prodorite Ltd.*

25072 Prodorcrete GT
A range of urethane resin based floor toppings for acid, alkali and chemical resistance. *Prodorite Ltd.*

25073 Prodorfilm
Series of special light stable wall coatings and potable linings for vessels and tanks. *Prodorite Ltd.*

25074 Prodorflor
Acid and chemical resisting epoxy resin based floor finish. *Prodorite Ltd.*

25075 Prodorglas
A range of stoved coatings for vessels and tanks to resist chemical corrosion. *Prodorite Ltd.*

25076 Prodorglaze
A range of textured decorative multi-colored wall coatings. *Prodorite Ltd.*

25077 Prodorguard
Epoxy resin based floor coatings for application to concrete floors and walls. *Prodorite Ltd.*

25078 Prodorite
An acid-resisting material; it is a concrete with a hardened pitch binder; the mineral part is carefully graded and mixed with the pitch; stated to be suitable for plants containing corrosive gases. No manufacturer.

25079 Prodorlac
Heavy bituminous paint. *Prodorite Ltd.*

25080 Prodorshield
Self-leveling epoxy resin based floor topping. *Prodorite Ltd.*

25081 Prodox
Alkyl phenols; intermediate chemicals for antidegradants, agricultural chemicals and resins. *PMC Specialities Group Inc.* Name unverified.

25082 Product AAS 90
Anionic surfactant in the form of a free flowing powder; for powder type detergents and cleaning formulations where free flowing product is required. *Henkel Chemicals Ltd.* Name unverified.

25083 Product MB320
Cyclohexylamine lauryl sulfate in the form of a solid block which is water and oil soluble; emulsifier for insecticides; printing ink manufacture. No manufacturer.

25084 Productol
215-293-2
Cresylic acid
Generic name for the three isomers of cresol. Used as a disinfectant. *PMC Inc.*

25085 Produkt 2058
Fatty alcohol/fatty acid ester; industrial surfactant. *Zschimmer & Schwarz.*

25086 Produkt GM 4210
MIPA ammonium fatty alcohol ether sulfate in propylene glycol. Industrial surfactant. *Zschimmer & Schwarz.* Discontinued.

25087 Produkt GS 5001
Fatty acid DEA, modified; for hair shampoos, bath additives, dishwash, cleansing agents. *Zschimmer & Schwarz.* Discontinued.

25088 Pro-Etch
Patented glass etching system; for etching vehicle registration number onto glass car windows. *Hermetite Products Ltd.* Name unverified.

25089 Profalon
Emulsifiable concentrate containing 200 g chlorpropham and 100 g linuron per liter; used to control weeds in bulb crops. *Hoechst UK.*

25090 Profax
Insulating refractory sideliner tiles for killed steel ingots. *Foseco (F.S.) Ltd.*

25091 Pro-fax® 6323
9003-07-0 7741
PP homopolymer resin, heat-stabilized; high flow, maximum stiffness, general-purpose resin used in automotive, hospital and institutional ware, housewares, closures by injection and fiber extrusion processes; UL rated at 110°C. *Himont.* Name unverified.

25092 Pro-fax® 65F4-4
9003-07-0 7741
Polypropylene; 40% talc-filled; maximum stiffness and high temperature performance; UL (115°C continuous use); in automotive, appliances, industrial components. *Hercules.* Discontinued.

25093 Pro-fax® 65F5-4
9003-07-0 7741
Polypropylene, 40% calcium carbonate-filled; best high flex modulus/impact balance, good colorability, surface finish; in housewares, small appliances. *Hercules.* Discontinued.

25094 Pro-fax® 7523
9003-07-0 7741
PP copolymer resin, heat-stabilized; general-purpose resin for injection, fiber extrusion; improved low temperature impact, greater toughness allowing wider design latitude; UL rated at 110°C; used in totes, trays, automotive, furniture. *Himont.* Name unverified.

25095 Pro-fax® 8523
9003-07-0 7741
Polypropylene copolymer resin, heat-stabilized; high impact grade, toughness, low temperature impact resistance, long high tempatrue life; UL rated at 110°C; used in automotive parts, luggage, totes, bins, appliances. *Himont.* Name unverified.

25096 Pro-fax® HB-301
9003-07-0 7741
Polypropylene resin; ultra-high molecular weight polymer with consistent particle size; for manufacturing of porous products. *Himont.* Name unverified.

25097 Pro-fax® PC-072PM
9003-07-0 7741
Polypropylene homopolymer modified for chemical coupling with glass; high heat performance; used in appliances, automotive, chemical processing equipment. *Himont.* Name unverified.

25098 Pro-fax® PD-064
9003-07-0 7741
Polypropylene homopolymer resin; film resin for biaxially oriented film. *Himont.* Name unverified.

25099 Pro-fax® PF-101
9003-07-0 7741
Polypropylene homopolymer resin; for thermoforming sheet; features clarity, low odor and taste, good processability; used for cookie trays, packaging. cups; available with antistat; FDA compliant. *Himont.* Name unverified.

25100 Pro-fax® SA-747M
9003-07-0 7741
Polypropylene random copolymer; features outstanding clarity, fast cycling, FDA compliance for housewares, food containers. *Himont.* Name unverified.

25101 Pro-fax® SB-242
9003-07-0 7741
Polypropylene resin; injection molding grade for housewares; very high flow; good impact/stiffness balance. *Himont.* Name unverified.

25102 Pro-fax® SD-062
9003-07-0 7741
Polypropylene copolymer resin; high-speed coating/laminating resin for films; used for fabric and paper coating; FDA compliant. *Himont.* Name unverified.

25103 Pro-fax® SE-191
9003-07-0 7741
Polypropylene copolymer resin; for wire and cable primary insulation. *Himont.* Name unverified.

25104 Pro-fax® SV-256M
9003-07-0 7741
Polypropylene random copolymer, clarifier; for blow molding applications requiring optimum clarity. *Himont.* Name unverified.

25105 Pro-fax® Z-39S
9003-07-0 7741
Polypropylene fiber resin; low gas fading for staple. *Himont.* Name unverified.

25106 Proferdex
Iron dextran; hematinic. *Fisons plc, Pharmaceuticals Div.* Name unverified.

25107 Progacyl® ADG
High viscosity thickener for continuous dyeing, random dyeing, and printing of tufted carpet using acid, premetallized acid, dispersed, direct, basic and nickel chelating dispersed colors; also for other industries where alkaline stability required. *Rhône-Poulenc Surf.*

25108 Progacyl® CP-7
9000-30-0 4601 232-536-8
Guar; polymeric thickener for flatbed screen printing of carpets, rotary screen printing. *Rhône-Poulenc Surf.*

25109 Progacyl® CP-82
Synthetic hydrocolloid; print thickener for use on carpets. *Rhône-Poulenc Surf.*

25110 Progallin LA
1166-52-5 214-620-6
$C_{19}H_{30}O_5$
Dodecyl gallate
Lauryl gallate; Dodecyl 3,4,5-trihydroxybenzoate. Antioxidant for cosmetics. *Nipa Laboratories Ltd.* Discontinued.

25111 Progallin P
121-79-9 8044 204-498-2
propyl gallate
Propyl gallate; antioxidant for cosmetics. *Nipa Laboratories Ltd.*

25112 Pro-Gas (Gas Disclaimed)
71-23-8 8027 200-746-9
Propanol. *Monsanto (Solaris).* Name unverified.

25113 Progasol® COG
Low-foaming detergent for scouring cotton, synthetics, and blends; for use in jet dyeing machines. *Rhône-Poulenc Surf.*

25114 Progene
Liquid detergents. *Unilever.* Name unverified.

25115 Progilite
A proprietary trade name for phenol-formaldehyde resins. No manufacturer.

25116 proglumide
6620-60-6 7958 229-567-4
$C_{18}H_{26}N_2O_4$
(±)-4-(benzoylamino)-5-(dipropylamino)-5-oxopentanoic acid
xylamide; CR-242; W-5219; Milid; Milide; Promid. Anticholinergic. mp = 142-145°; LD_{50} (mus orl)= 7350-8861 mg/kg.

25117 Progressite
An explosive containing 89% ammonium nitrate, 4.7% aniline hydrochloride, 6% ammonium sulfate, and 0.2% coloring matter.

25118 proidonite
A mineral, SiF_4.

25119 Proil
Rust preventatives. *Croda Chem. Ltd.* Name unverified.

25120 proiodin
A combination of iodine with protein, containing 4.4% iodine.

25121 Project® 70 Stainless Type 316
Molybdenum bearing austenitic steel with increased percentages of nickel; higher tensile and creep strength at elevated temperatures; for paper/pulp handling equipment, process equipment for producing photographic chemicals, inks, rayon, rubber, textile bleaches and dyestuffs, high temperature equipment. *Carpenter Tech.*

25122 Prokayvit Oral
Tablets of acetome-naphthone (vitamin K). *British Drug Houses.* Name unverified.

25123 Prolan®
Preparations of gonadotropins; for the treatment of reproductive disorders in animals. *Bayer AG.*

25124 Prolaurin
A proprietary trade name for propylene glycol monolaurate, a surfactant. No manufacturer.

25125 Proleaf
Horticultural foliar feed fertilizer. *Schering Agrochemicals Ltd.* Discontinued.

25126 Prolein
1330-80-9 215-549-3
A proprietary trade name for propylene glycol monooleate, a surfactant. No manufacturer.

25127 Prolene
Nonabsorbable surgical suture. *Ethicon Inc.* Name unverified.

25128 Prolex
1918-16-7 7977 217-638-2
2-chloro-N-isopropyl-acetanilide
Active ingredient; propachlor; pre-emergence weed control of annual weeds. *Agan Chemical Manufacturers Ltd.*

25129 proline
147-85-3 7963 205-702-2
$C_5H_9NO_2$
L-proline
2-pyrrolidine carboxylic acid; Pro; P;. A nonessential amino acid; in moisturizers, biochemical and nutritional research, microbiological tests, culture media, dietary supplements, lab reagent. dec 220-222°; $[\alpha]_D^{20}$= -52.6°

(c = 0.57 in 0.5N HCl); soluble in H_2O (127 g/100 ml), less soluble in organic solvents. *Am. Biorganics; Degussa; R.W. Greeff; Nippon Rikagakuyakuhin; Penta Mfg.*

25130 Pro-Line
Universal color dispersions for use in architectural coatings. *Engelhard.*

25131 Prolit
Bleaching earths (activated clays); for refining vegetable and mineral oils, vegetable and animal fats and in purifying solvents; as color developing agent for carbonless copying papers. *Caffaro SpA.* Name unverified.

25132 Prolit
Activated bleaching earths; for the decolorizing of vegetable and mineral oils, animal and vegetable fats and solvents. *SNIA (UK) Ltd.*

25133 Prolith
Photographic developer. *May & Baker Ltd.* Name unverified.

25134 Prolongal®
Iron preparations; for prophylaxis and treatment of piglet anemia. *Bayer AG.*

25135 Prolugen
57-55-6 8040 200-338-0
Propylene glycol
Solvent. *Octel Chemicals Ltd.* Discontinued.

25136 Promax® HV
68153-28-6
Soy protein; concentrate with superior fat emulsification, emulsion stabilization and water binding capability; highly digestible protein-rich food ingredient for portion meats, poultry, seafood products, protein beverages. *Central Soya.*

25137 prometal
A variety of cast iron; used in the construction of furnace parts.

25138 Promethus
A blasting powder. It contains potassium chlorate, manganese dioxide, iron oxide, nitrobenzene, turpentine oil, and naphtha.

25139 Prometrex
7287-19-6 7973 230-711-3
2,4-bis-(isopropylamino)-6-methylthio-1,3,5-triazine
prometryne. Active ingredient: prometryne; selective pre- and post-emergence herbicide for the control of broadleaf and grass weeds in a variety of crops. *Agan Chemical Manufacturers Ltd.*

25140 prometryn
7287-19-6 7973 230-711-3
$C_{10}H_{19}N_5S$
1,3,5-Triazine-2,4-diamine, N,N'-bis(1-methylethyl)-6-(methylthio)-
Cotton-Pro®; *s*-triazine, 2,4-bis(isopropylamino)-6-methylthio-; *s*-triazine, 2,4-bis(isopropylamino)-6-methylmercapto-; A 1114; Caparol; G 34161; Gesagard; Gesagard 50; Gesagarde 50 Wp; Mercasin; Mercazin; Merkazin; Polisin; Primatol Q; Prometrex; Prometrin; Prometryne; Selectin; Selectin 50; Selektin; Sesagard; Uvon; 2-(methylmercapto)-4,6-bis(isopropylamino)-*s*-triazine; 2-(methylthio)-4,6-bis(isopropylamino)-*s*-triazine;. Prometryn suspension; flowable herbicide for selective weed control in cotton and celery crops. mp = 118-120°; d_{20} = 1.157; soluble in H_2O (48 mg/l), more soluble in organic solvents; LD_{50} (rat orl) = 1800 mg/kg *Griffin.*

25141 prometryn
7287-19-6 7973 230-711-3
$C_{10}H_{19}N_5S$
N,N'-bis(1-methylethyl)-6-methylthio-1,3,5-triazine-2,4-diamine
Caparol; Gesagard; Caparol 4L; Caparol 80W; Cotton-Pro; Primatol Q; Prometrex; G-34161; Selectin; Uvon;. Selective systemic herbicide, absorbed by roots and foliage. Used for pre- and post-emergence control of most annual grasses and broad-leaved weeds. mp = 118-120°; soluble in H_2O (48 mg/l), more soluble in organic solvents; L770 (rat orl) = 3.75 g/kg.

25142 Promicrol
Ultra fine grain photographic developer. *May & Baker Ltd.* Name unverified.

25143 Promintic
114-91-0 6247 204-060-0
A proprietary preparation of methyridine; a veterinary anthelmintic. No manufacturer.

25144 Promodan SP
1323-39-3 215-354-3
$C_{21}H_{42}O_3$
Propylene glycol stearate
Food emulsifier. *Grindsted Prods.*

25145 Promol
Release agent for rubber and plastics. *Rhône-Poulenc UK.*

25146 Promoloid
5727
A fertilizer containing collodial magnesium silicate; a Japanese product.

25147 Promosoy® 20/60
68153-28-6
Soy protein; high digestible protein-rich food ingredient. *Central Soya.*

25148 Promotor 301
A metal-containing compound dissolved in a hydrocarbon solvent; accelerator used in combination with peresters to achieve an optimal cure in hot press molding and a fast cure at room temperature and elevated temperatures. *Akzo.* Name unverified.

25149 Promozyme
A heat stable debranching enzyme obtained from a novel species of *Bacillus* by submerged fermentation; belongs to the group of debranching enzymes known as pullulanases; used in the production of dextrose and maltose. *Novo Nordisk.*

25150 Promulgen® D
Cetearyl alcohol and ceteareth-20; gelling agent; oil-water emulsifier, emollient, and stabilizer for cosmetics and pharmaceuticals; highly resistant to acidic and alkaline conditions. *Amerchol Corp.*

25151 Promulgen® G
Stearyl alcohol and ceteareth-20; gelling agent; oil-water emulsifier, emollient, and stabilizer for cosmetics and pharmaceuticals; highly resistant to acidic and alkaline conditions. *Amerchol Corp.*

25152 Promyr
110-27-0 5234 203-751-4
Isopropyl myristate
Emollient and solvent for cosmetics, toiletries, makeups; nongreasing rub in. *Amerchol Corp.*

25153 Promyristyl PM-3
63793-60-2
PPG-3 myristyl ether
Low-viscosity emollient for clear analgesic, deodorant, and fragrance sticks. *Croda Inc.*

25154 Pronal 502, 502A
Polyalkylene glycol ester
Defoaming agent for paper, latex, and water paint manufacturing. *Toho Chem. Industry.*

25155 Pronal EX-100
Silicone emulsion; antifoaming agent for petrochemical industry. *Toho Chem. Industry.*

25156 Pronalys
Analytical grade reagents. *May & Baker Ltd.* Name unverified.

25157 Pronase®
Protease. *Calbiochem Corp.*

25158 Pronel Capsules
9000-70-8 4388 232-554-6
A proprietary preparation of gelatin; used in the treatment of flaking fingernails. *Bioglan Laboratories.* Name unverified.

25159 Pronova
9067-32-7 4793
Sodium hyaluronate
Used for medical, pharmaceutical, veterinary, botanical, microbiological, and cosmetic uses, as well as for bioreactor processes. *Protan.*

25160 proof spirit
A term originally intended to denote alcohol that was just strong enough to ignite gunpowder, when burnt upon it. It is alcohol containing 49.24 parts of alcohol to 50.76 parts of water by weight, or 100 volumes of alcohol to 81.82 volumes of water.

25161 proof vinegar
A vinegar containing 66% acetic acid.

25162 Proofite
A proprietary trade name for a product similar to Aquatec. No manufacturer.

25163 Propachlor
1918-16-7 7977 217-638-2
$C_{11}H_{14}ClNO$
2-Chloro-N-(1-methylethyl)-N-phenylacetamide
Bexton 4L; Niticid; Satecid; Propachlor; 2-chloro-N-isopropylacetanilide;CP 31393; Bexton; Prolex;Propaclor; Ramrod-atrazine; Ramrod 20G; Ramrod Flowable; propachlor + atrazine; Chloro-N-isopropylacetanilide; Isopropyl-2-chloroacetanilide; Albrass; Cp 31393; Croptex; Amber; Niticid; Orange; Prolex; Ramrod; Satecid; Sentinel. Selective herbicide, absorbed by seedling shoots and roots. Used for control of annual grasses and some broad-leaved weeds in vegetable crops. mp = 77°; $bp_{0.03}$ = 110°; soluble in H_2O (613 mg/l), more soluble in organic slvents; LD_{50} (rat orl)= 1800 mg/kg.

25164 Propachlor
1918-16-7 7977 217-638-2
$C_{11}H_{14}ClNO$
2-Chloro-N-(1-methylethyl)-N-phenylacetamide
Bexton 4L; Niticid; Satecid; Propachlor; 2-chloro-N-isopropylacetanilide;CP 31393; Bexton; Prolex;Propaclor; Ramrod-atrazine; Ramrod 20G; Ramrod Flowable; propachlor + atrazine; Chloro-N-isopropylacetanilide; Isopropyl-2-chloroacetanilide; Albrass; Cp 31393; Croptex; Amber; Niticid; Orange; Prolex; Ramrod; Satecid; Sentinel. Selective herbicide, absorbed by seedling

shoots and roots. Used for control of annual grasses and some broad-leaved weeds in vegetable crops. mp = 77°; $bp_{0.03}$ = 110°; soluble in H_2O (613 mg/l), more soluble in organic slvents; LD_{50} (rat orl)= 1800 mg/kg.

25165 Propafilm
Balanced biaxially oriented polypropylene film. *ICI Chem & Polymers Ltd.*

25166 Propafoil
Metallized oriented polypropylene film. *ICI Chem & Polymers Ltd.*

25167 Propaklone
Industrial solvent. *ICI Chem & Polymers Ltd.*

25168 Propal
142-91-6 205-571-1
$C_{19}H_{38}O_2$
Isopropyl palmitate
Emollient and solvent for cosmetics, toiletries, makeups; nongreasing rub in. Colorless liquid. Low viscosity; good penetration and spreading properties; non-oily feel, low viscosity, good penetration and spreading properties. *Amerchol Corp.*

25169 Propamine
Catalysts for polyurethane foams. *Harcros.*

25170 Propamine D
110-18-9 203-744-6
$C_6H_{16}N_2$
Tetramethylethylenediamine
N,N,N',N'-tetramethyl-1,2-diaminoethane; Tetramethylethylenediamine; Temed; 1,2-Di-(dimethylamino)-ethane; N,N,N',N'-Tetramethyl-1,2-ethanediamine; TMEDA; N,N,N',N'-tetramethylethenediamine. A liquid catalyst miscible with both water and organic liquids mp = -55°; bp = 120-122°; d = 0.7700; n_D^{20}= 1.4179; soluble in H_2O, organic solvents; LD_{50} (rat orl) = 1020 mg/kg. *Harcros.*

25171 Propamocarb hydrochloride
25606-41-1 247-125-9
Filex. An aqueous concentrate containing propamocarb hydrochloride; a protective fungicide for use on all ornamentals and some edible crops against *Pythium, Peronospora* and *Phytophthora. Fisons plc.*

25172 propanal
123-38-6 8008 204-623-0
C_3H_6O
propionaldehyde
propyl aldehyde; propanal; Propionic aldehyde; Methylacetaldehyde; Propyl aldehyde; propaldehyde; propylic aldehyde; Proprionaldehyde. Used in the manufacture of propionic acid, polyvinyl and other plastics, synthesis of rubber chemicals, and preservatives. mp = -81°; bp = 46-50°; n_D^{20}= 1.3650; d = 0.8050; soluble in H_2O (350 g/l), organic solvents; LD_{50} (rat orl) = 1410 mg/kg.

25173 propane
74-98-6 7982 200-827-9
C_3H_8
dimethylmethane
propyl hydride; Hydrocarbon propellant. Used as a propellant and in organic synthesis, household and industrial fuel, manufacture of ethylene, extractant, solvent, refrigerant, gas enricher. mp = -188°; bp = -42°; insoluble in H_2O, soluble in organic solvents. *Air Prods & Chem; Fina; Phillips 66.*

25174 Propanil
709-98-8 7987 211-914-6
$C_9H_9Cl_2NO$
N-(3,4-dichlorophenyl)propanamide
Propanil; DPA; FW-734; Stam; Stampede; Rogue; Chem Rice; Surcopur; Bay 30130; Cekupropanil; Erbanil; Herbax; Prop Job; Propa; Propal; Propanex; Prostar; Riselect; Strel; Supernox; Surpur; Wham. Post-emergence applied herbicide with no residual effect for control of numerous grasses and broad-leaved weeds in rice crops. Also used as a nematocide. mp = 91-93°; soluble in H_2O (225 mg/l); LD_{50} (rat orl) = 1384 mg/kg. *Bayer AG.*

25175 propanol
71-23-8 8027 200-746-9
C_3H_8O
n-propyl alcohol
propyl Alcohol; 1-propanol; ethyl carbinol; Propan-1-ol. Normal propyl alcohol. Used as a chemical intermediate and as a solvent, especially for resins, cellulose esters etc. mp = -126°; bp = 97°; d = 0.8040; n_D^{20}- 1.3837; soluble in H_2O, organic solvents; LD_{50} (rat orl) = 1870 mg/kg.

25176 propantheline bromide
50-34-0 7989 200-030-6
$C_{23}H_{30}BrNO_3$
N-methyl-N-(1-methylethyl)-N-[2-[(9H-xanthen-9-ylcarbonyl)oxy]ethyl]-2-propanaminium bromide
Corrigast; Ercotina; Pro-Banthine; Neo-Metantyl; Pantheline. Anticholinergic agent. mp = 159-161°; very soluble in H_2O, EtOH, $CHCl_3$, insoluble in Et_2O, C_6H_6.

25177 Propasol
Propyl oxide glycol ether, a solvent. *Union Carbide.*

25178 Propathene
Polypropylene, the lightest of the thermoplastics; features good rigidity and tensile strength which are retained at elevated temperatures, excellent resistance to chemicals with no tendency to environmental stress cracking; used in plastics manufacture and in molding and extrusion compounds. *ICI Chem & Polymers Ltd.*

25179 propazine
139-40-2 7996 205-359-9
$C_9H_{16}ClN_5$
6-chloro-N,N'-bis(1-methylethyl)-1,3,5-triazine-2,4-diamine
G-30028; Gesamil; Milogard; Prozinex. Pre-emergent selective systemic herbicide used for control of annual grasses and broad-leaved weeds in sorghum and crops such as carrots, chervil and parsley. mp = 213°; soluble in H_2O (8.6 mg/ml), poorly soluble in organic solvents; LD_{50} (rat orl) >7000 mg/kg.

25180 Propcorn
Chemical products for the treatment of corn. *BP Chemicals Ltd.*

25181 propenol
107-18-6 294 203-470-7
C_3H_6O
allyl alcohol
Allylic alcohol; Propenol; 2-propen-1-ol; 2-propenol; vinyl carbinol; 3-Hydroxypropene; 2-Propenyl Alcohol; Propenyl alcohol; propen-1-ol-3; weed drench; 1-Propenol-3-ol; Propene-1-ol; Propenol-3. Intermediate for pharmaceuticals and other organic chemicals, herbicide. mp = -129°; bp = 96-98°; d = 0.8540; n_D^{20}= 1.4119; soluble in H=72O. Organic solvents; LD_{50} (rat orl) = 64 mg/kg.

25182 Propetal 241
68154-97-2
Fatty alcohol EO/PO adduct; biodegradable detergent, wetting agent, emulsifier, antifoam; intermediate for preparation of low foaming detergents for household and industry, dishwash, sanitary and floor tile cleaners, industrial spray cleaners, metal pickling, paper, textile, leather. *Zschimmer & Schwarz.*

25183 Propetamphos
31218-83-4 8000 250-517-2
$C_{10}H_{20}NO_4PS$
(E)-3-[[(ethylamino)methoxyphosphinothioyl]oxy]-2-butenoic acid 1-methylethyl ester
SAN-322I; Blotic; Safrotin; Seraphos; Zoecon; TSAR. Used as an ectoparasiticide. $bp_{0.005}$ = 87-89°; d_4^{20}= 1.1294; n_D^{20}= 1.495; soluble in H_2O (110 mg/l), more soluble in organic solvents; LD_{50} (rat orl) = 82 mg/kg.

25184 propezite
A natural alloy of palladium and gold, containing 7% gold.

25185 Propham
122-42-9 8001 204-542-0
$C_{10}H_{13}NO_2$
1-Methylethyl phenylcarbamate
Agermin; Ban-Hoe; Beet-Kleen; Birgin; Collavin; IFC; IFK; INPC; IPC. IPPC; ISO.PPC; Isopropyl phenylcarbamate; Isopropyl carbanilate. Selective sytemic herbicide and growth regulator. Used to control annual grasses and some broad-leaf weeds. mp = 87-88°; d_{20} = 1.09; soluble in H_2O (250 mg/l), soluble in organic solvents; LD_{60} (rat orl) = 1000 mg/kg.

25186 Propham
122-42-9 8001 204-542-0
$C_{10}H_{13}NO_2$
phenylcarbamic acid 1-methylethyl ester
INPC; IPC; IsoPPC; Chem-Hoe; Chem Hoe FL4; Isopropyl carbanilate; Agermin; Birgin; Collavin; IFC; IFK; INPC; IPPC; Isoppc; Isopropyl phenyl urethane; Ortho grass killer; Phenyl isopropyl carbamate; Premalox; Profam; Tixit; Triherbide. Propham; plant growth regulator for control of sprouting in stroed potatoes and in some cases as herbicide against weeds in vegetables. mp = 90°; insoluble in H_2O, soluble in organic solvents; LD_{50} (rat orl) = 3724 mg/kg. *Bayer AG.*

25187 Propiofan®
Polymer dispersions based on vinyl propionate; binder for exterior and interior paints, for composite thermal insulation systems; modifier for cement mortar; raw material for production of building adhesives, laminate adhesives and packaging adhesives; binders for textile coatings. *BASF AG; BASF plc.*

25188 Propiofan® D
Polyvinyl propionate dispersion. *BASF plc.* Name unverified.

25189 propiolic acid
471-25-0 8006 207-437-8
$C_3H_2O_2$
propargylic acid
2-propynoic acid; acetylenecarboxylic acid;. Used in chemical synthesis. mp

= 9°; bp_{50} = 70-75; bp = 144° (dec); d_4^{20}= 1.1380; $n^{20.3}_D$ = 1.4302; LD_{50} (rat orl) = 100 mg/kg.

25190 propionic acid

79-09-4 8010 201-176-3

C_2H_5COOH

methylacetic acid

propanoic acid; ethylformic acid. Esterifying agent; in production of cellulose propionates, etc.; as mold inhibitors and preservatives; in manufacture of ester solvents, fruit flavors, perfume bases; antifungal. *BASF; Eastman; Hoechst Celanese; Penta Mfg.; Union Carbide.*

25191 Proplatinum

An alloy of 72% nickel, 23.6% silver, 3.7% bismuth, and 0.7% gold.

25192 Propocon

Formulated polyurethane systems for rigid, semi-rigid, microcellular foams; used in insulation, construction, automotive and shoe soling. *Harcros.*

25193 Propomeen 2HT-11

71060-61-2

Bis-hydrogenated tallowalkyl-2-hydroxypropyl amines

Industrial surfactant. *Akzo.*

25194 Propomeen C/12

68516-06-3 271-130-5

Dipropylene glycol cocamine

Industrial surfactant. *Akzo.*

25195 Propomeen T/12

68951-72-4 273-160-4

Dipropylene glycol tallowamine

Industrial surfactant. *Akzo.*

25196 Propoquad® 2HT/11

68554-09-6 271-401-8

Di(hydrogenated tallowalkyl) (2-hydroxy-2-methylethyl) quaternary ammonium chloride

Solution in aqueous isopropanol used as an industrial surfactant. *Akzo.*

25197 Propoquad® T/12

79770-97-1

Tallowalkylmethyl-bis(2-hydroxy-2-methylethyl) quaternary ammonium methylsulfates

Solution in diethylene glycol used as an industrial surfactant. *Akzo.*

25198 Propoxur

114-26-1 8022 204-043-8

$C_{11}H_{15}NO_3$

2-(1-methylethoxy)phenyl methyl carbamate

2-isopropoxyphenyl methyl carbamate; *o*-isopropoxyphenyl methyl carbamate; 58 12 315; arporcarb; OMS 33; ENT 25671; Bay 39007; Baygon; Blattanex; Brifur; Invisi-Gard; Pillargon; Propogon; Rhoden; Sendran; Suncide; Tendex; Tugen; Unden; Undene; Undeen. Non-systemic insecticide used for treatment of sucking and biting insects, e.g., aphids, mealybugs, scales, leafhoppers, caterpillars on vegetables, pome and stone fruit, cocoa, rice, oil palms and other crops. Used for protection of flowers, fruits and mp = 84-87°; d = 1.12; soluble in H_2O (2 g/l), more soluble in organic solvents; LD_{50} (rat orl) = 8350 mg/kg. *Bayer AG.*

25199 Propoxur

114-26-1 8022 204-043-8

$C_{11}H_{15}NO_3$

2-(1-methylethoxy)phenyl methyl carbamate

2-isopropoxyphenyl methyl carbamate; *o*-isopropoxyphenyl methyl carbamate; 58 12 315; arporcarb; OMS 33; ENT 25671; Bay 39007; Baygon; Blattanex; Brifur; Invisi-Gard; Pillargon; Propogon; Rhoden; Sendran; Suncide; Tendex; Tugen; Unden; Undene; Undeen. Non-systemic insecticide used for treatment of sucking and biting insects, e.g., aphids, mealybugs, scales, leafhoppers, caterpillars on vegetables, pome and stone fruit, cocoa, rice, oil palms and other crops. Used for protection of flowers, fruits and mp = 84-87°; d = 1.12; soluble in H_2O (2 g/l), more soluble in organic solvents; LD_{50} (rat orl) = 8350 mg/kg. *Bayer AG.*

25200 Propoxur

114-26-1 8022 204-043-8

$C_{11}H_{15}NO_3$

2-(1-methylethoxy)phenyl methyl carbamate

2-isopropoxyphenyl methyl carbamate; *o*-isopropoxyphenyl methyl carbamate; 58 12 315; arporcarb; OMS 33; ENT 25671; Bay 39007; Baygon; Blattanex; Brifur; Invisi-Gard; Pillargon; Propogon; Rhoden; Sendran; Suncide; Tendex; Tugen; Unden; Undene; Undeen. Non-systemic insecticide used for treatment of sucking and biting insects, e.g., aphids, mealybugs, scales, leafhoppers, caterpillars on vegetables, pome and stone fruit, cocoa, rice, oil palms and other crops. Used for protection of flowers, fruits and mp = 84-87°; d = 1.12; soluble in H_2O (2 g/l), more soluble in organic solvents; LD_{50} (rat orl) = 8350 mg/kg. *Bayer AG.*

25201 propoxur

114-26-1 8022 204-043-8

$C_{11}H_{15}NO_3$

2-(1-methylethoxy)phenol methylcarbamate

aprocarb; Bay 39007; Bay 9010; Baygon; Blattanex; Propyon; Suncide; Unden. Non-systemic insecticide with contact and stomach action. Used for control of cockroaches, flies, fleas, mosquitoes, bugs, ants, millepedes and other insect pests in houses and food storage areas. mp = 91°; soluble in H_2O (0.2 g/100 ml), more soluble in organic solvents; LD_{50} (rat orl) = 83 mg/kg.

25202 Propoxyol® 1695

PPG-5 lanolin wax glyceride

Emollient, stabilizer, and pigment dispersant for anhydrous makeups; cosmetic additive. *Henkel/Cospha; Henkel Canada.*

25203 propyl acetate

109-60-4 8026 203-686-1

$C_5H_{10}O_2$

acetic acid propyl ester

propylacetate; 1-Acetoxypropane; 1-Propyl Acetate. Flavoring agent, perfumery, solvent for nitrocellulose and other cellulose derivatives, natural and synthetic resins, lacquers, plastics, organic synthesis, lab reagent. mp = -92°; bp = 101°; d_4^{20}= 0.836; n_D^{20}= 1.3862; soluble in H_2O (1.6 g/100 ml), more soluble in organic solvents; LD_{50} (rat orl) = 9370 mg/kg. *BASF; BP Chem. Ltd; Eastman; Hoechst Celanese; Union Carbide.*

25204 propyl gallate

121-79-9 8044 204-498-2

$(HO)_3C_6H_2COOCH_2CH_2CH_3$

3,4,5-trihydroxybenzoic acid propyl ester

gallic acid propyl ester; n-propyl 3,4,5-trihydroxybenzoate. Ester of propyl alcohol and gallic acid; food and feed antioxidant, flavor and packaging material. *Aceto; Eastman; Nipa Labs; UOP.*

25205 propyl oleate

Ester of propyl alcohol and oleic acid; base for industrial lubricants; mold release agent, defoamer, flotation agent, plasticizer for cellulosic plastics, needle lubricants; when sulfated is useful as wetting, rewetting, and dye leveling agent in textile and leather industries. *Witco/Humko.* Discontinued.

25206 Propyl Zithate®

1000-90-4 213-680-0

Zinc isopropyl xanthate

Rubber accelerator used in natural and synthetic cements and doughs; nondiscoloring in presence of copper or iron. *R. T. Vanderbilt Co Inc.* Discontinued.

25207 Propylan

Polyether polyols for flexible, semirigid and rigid polyurethane foams; used in upholstery, bedding, insulation, automotive, elastomers, coatings, and as pigment dispersing media. *Harcros.*

25208 Propylan A350

A proprietary amine initiated polyether used in the manufacture of polyurethane foam. *Harcros.*

25209 Propylan G600

A proprietary polyoxypropylene trial of low molecular weight used in the production of rigid urethane foams and other urethane compositions, including elastomers. *Harcros.*

25210 Propylan RF55

A proprietary modified sorbitol-based polyether for making rigid, flame-proof urethane foams. *Harcros.*

25211 propylan-propyl alcohol

71-23-8 8027 200-746-9

$CH_3CH_2CH_2OH$

1-propanol

ethyl carbinol. Organic synthesis; chemical intermediate; solvent for waxes, vegetable oil, natural and synthetic resins. *Eastman; Hoechst Celanese; Mallinckrodt; Union Carbide.*

25212 propylene

115-07-1 8034 204-062-1

C_3H_6

propene

1-Propylene; Methylethylene; 1-Propene; methylethene. Chemical intermediate for manufacture of isopropyl alcohol, polypropylene, synthetic glycerol, acrylonitrile, propylene oxide, heptene, cumene, polymer gasoline, acrylic acid, vinyl resins, oxo chemicals. mp = -185°; bp = -47°; d_4^{20}= 0.5193; soluble in H_2O (23 g/100 ml), more soluble in organic solvents. Air Prods & Chem; Amoco; BP Chem. Ltd; Chevron; Exxon; Fina; Mobil; OxyChem; Phillips 66; Shell; Texaco; Vista.

25213 propylene dichloride

78-87-5 8039 201-152-2

$C_3H_6Cl_2$

1,2-dichloropropane

propylene chloride; 1,2-dichloropropane; dichloropropane; α,β-dichloropropane; α,β-propylene dichloride; dichloropropanes. Intermediate for perchloroethylene, CCl_4; lead scavenger for antiknock fluids; solvents for fats, oils, waxes, gums, resins, cellulose esters and ethers; scouring compounds; metal degreasers; soil fumigant for nematodes. bp = 95-96°; d_{25}^{25} = 1.159; n_D^{20}= 1.4388; slightly soluble in H_2O, soluble in organic solvents; LD_{50} (rat orl) = 1.38 g/kg. *Naphtachimie SA.*

25214 propylene glycol
57-55-6 8040 200-338-0
$CH_3CHOHCH_2OH$
1,2-propanediol
1,2-dihydroxypropane; methyl glycol; Aliphatic alcohol. Solvent, emulsifier, production paints, resins, foods, drugs, antifreeze in breweries and dairies; substitute for ethylene glycol and glycerol; mold growth and fermentation inhibitor. *Aldrich; Arco; Asahi Denka Kogyo; Ashland; BP Chem. Ltd; Hüls AG; Olin; Texaco.*

25215 propylene glycol
57-55-6 8040 200-338-0
$C_3H_8O_2$
1,2-propanediol
1,2-dihydroxypropane; methylethylene glycol; trimethyl glycol; 1,2-propylene glycol; monopropylene glycol; propane-1,2-diol; α-propyleneglycol; Dowfrost; PG 12; sirlene; solar winter ban; Propanediol. Metabolized to pyruvic & acetic aicds, ths not toxic like ethylene glycol. Used as an antifreeze and solvent and a non-toxic substitute for ethylene glycol or glycerol. Used in veterinary medicine as an oral glucogen in ruminants. dl form: mp = -59°; bp = 188°; SG = 1.036; n_D^{20} = 1.4319; soluble in H_2O (>100 mg/ml), organic solvents; LD_{50} (rat orl) = 26 g/kg.

25216 propylene glycol alginate
9005-37-2
$(C_9H_{14}O_7)_8$
hydroxypropyl alginate
alginic acid, ester with 1,2-propanediol. Mixture of propylene glycol esters of alginic acid. food additive (human). *Kelco Int'l.; Meer.*

25217 Propylene glycol dioleate
85049-34-9 285-203-4
Emalex PG-di-O; Priolube 1429; Radia® 7204. Cosmetic ingred. *Unichema; Fina Chemicals.*

25218 Propylene glycol isostearate
68171-38-0 269-027-5
Emerest® 2384; Hydrophilol ISO. Solubilizer for fragrances in low alcohol or oil preparations; emollient for personal care products *Henkel/Cospha; Henkel Canada; Gattefosse SA.*

25219 propylene glycol monolaurate
10108-22-2 233-292-5
$C_{15}H_{30}O_3$
Surfactant.

25220 propylene glycol monomethyl ether
107-98-2 203-539-1
$C_4H_{10}O_2$
1-Methoxypropan-2-ol
Dowanol 33B; Dowanol PM; Dowtherm 209; Glycol Ether PM; PGME; Polysolve MPM; Propasol Solvent M; UCAR Solvent LM. Used as a solvent. mp = -97°; bp = 120°; SG = 0.924; soluble in H_2O (>100 mg/ml), organic solvents; n_D^{20}= 1.4030; LD_{50} (rat orl) = 5660 mg/kg.

25221 Propylene glycol oleate
1330-80-9 215-549-3
Emalex PGO; Propylene glycol monooleate; Prolein; Radiamuls® PG 2206,. Surfactant for food and cosmetics. *Nihon Emulsion; Fina Chemicals.*

25222 Propylene glycol ricinoleate
26402-31-3 247-669-7
$C_{21}H_{40}O_4$
12-Hydroxy-9-octadecenoic acid, monoester with 1,2propanediol
Flexricin® 9; Cithrol PGMR N/E; Naturechem® PGR. Wetting agent, dye solv., wax plasticizer, stabilizer for textile, household, and cosmetic applications, rewetting dried skins. Sp gr=0.960; mp<-16°; flash pt=221°; ref index *CasChem.*

25223 propylene oxide
75-56-9 8041 200-879-2
C_3H_6O
1,2-epoxypropane
S(-)-methyloxirane; Propene oxide; methyloxirane; methyl ethylene oxide; 1,2-propylene oxide; epoxypropane; propylene epoxide; 2,3-epoxypropane; AD 6. Intermediate; polyols for urethane foams, propylene glycols, surfactants, detergents, isopropanolamines, fumigant, synthetic lubricants, synthetic elastomers, solvent. mp = -112°; bp = 34°; d = 0.8300; n_D^{20}= 1.3660; soluble in H_2O (40 g/100 ml), organic solvents; LD_{50} (rat orl) = 380 mg/kg. *Arco; Ashland; Hüls AG; Texaco.*

25224 Propylex
Polypropylene plastics in the form of sheets, bands, strips, films, plates, slabs, blocks, tubes, etc. *Courtaulds Fibres Ltd.*

25225 Propyliodone
587-61-1 8048 209-603-5
$C_{10}H_{11}I_2NO_3$
N-Propyl 3,5-di-iodo-4-pyridone-N-acetate
3,5-diodo-1-propoxycarbonyl-methylpyrid-4-one; Bronchodiagnostin; Bronkhodiagnostin; Brosombra; Dionosil; Diostril; Propiodone;. Radiopaque medium. Used in medicine as a diagnostic aid. dec 186-187°; soluble in H_2O (14 mg/100 ml), more soluble in organic solvents; LD_{50} (mis iv) = 300 mg/kg.

25226 Propyliodone
587-61-1 8048 209-603-5
$C_{10}H_{11}I_2NO_3$
3,5-diiodo-4-oxo-1(4H)-pyridineacetic acid propyl ester
Bronchodiagnostin; Bronkhodiagnostin; Brosombra; Dionosil; Diostril; Propiodone. Radiopaque medium used as a diagnostic aid. mp = 186-187° (dec); soluble in H_2O (14 mg/100 ml); LD_{50} (mus iv) = 300 mg/kg.

25227 n-propyl mercaptan
107-03-9 203-455-5
C_3H_8S
1-mercaptopropane
1-propanethiol. Unpleasant stench oder. Used as a leak detecting additive to natural gas. mp = -113°; bp = 67-68°; d = 0.8410; n_D^{20}= 1.4380.

25228 n-propyl mercaptan
107-03-9 203-455-5
C_3H_8S
1-propanethiol
1-propanethiol; 1-Mercaptopropane; n-propyl mercaptan; propanethiol. Chemical intermediate, herbicide. mp = -113°; bp = 67-68°; d = 0.8410; n_D^{20}= 1.4380; insoluble in H_2O, soluble in organic solvents. *Elf Atochem N. Am.; Phillips 66.*

25229 n-propyl mercaptan
107-03-9 203-455-5
C_3H_8S
1-mercaptopropane
1-propanethiol. Unpleasant stench oder. Used as a leak detecting additive to natural gas. mp = -113°; bp = 67-68°; d = 0.8410; n_D^{20}= 1.4380.

25230 Propylol
Normal propyl alcohol containing approximately 12% secondary butyl alcohol; solvent in paints and lacquers, foundries, deoiling waxes, grinding media, for glass forming, manufacture of hair lacquers, floor polishes, latex rubber production, disinfectants, hand cleaners, degreasers, rust removers, printing inks, production of xanthates and other mining chemicals. *Sasolchem.*

25231 propylparaben
94-13-3 8051 202-307-7
$C_{10}H_{12}O_3$
propyl *p*-hydroxybenzoate
4-hydroxybenzoic acid propyl ester; propyl parahydroxybenzoate. Organic ester of n-propyl alcohol and *p*-hydroxybenzoic acid; food preservatives, fungicide, mold control in sausage casings, pharmaceutic aid. *Allchem Industries; R.W. Greef; Mipa Labs; Penta Mfg.*

25232 n-propyltrichlorosilane
141-57-1 205-489-6
$C_3H_7Cl_3Si$
trichloropropylsilane
Intermediate for silicones. *Hüls Am.; PCR.*

25233 Prosan
Biocide. *Union Carbide.*

25234 Prosil®
Organosilinanes. *PCR.* Discontinued.

25235 Prosil® 178
18395-30-7 242-272-5
$C_6H_{16}O_3Si$
Isobutyltrimethoxysilane
isobutyltrimethoxysilane; Dynasylan IBTMO. Coupling agent for hydrophobic treatment. *PCR.*

25236 Prosil® 196
4420-74-0 224-588-5
$C_6H_{16}O_3SSi$
γ-mercaptopropyl trimethoxy silane
(3-mercaptopropyl)trimethoxysilane; trimethoxysilylpropanethiol; 1-Propanethiol,3-(trimethoxysilyl)-; σ-mercaptopropyl trimethoxy silane; Dynasylan MTMO. Coupling agent having both organic and inorganic reactivity; for acrylic, epichlorohydrin, nitrile, polysulfone, PS, PVC, urethane thermoplastics; thermoset acrylic, epoxy, nitrile/phenolic, phenolic, polybutadiene; and elastomerics. bp = 215°; d = 1.039; n_D^{20}= 1.4436; LD_{50} (rat orl) = 2940 mg/kg. *PCR.* Discontinued.

25237 Prosil® 220
919-30-2 213-048-4
$C_9H_{23}NO_3Si$

γ-aminopropyl triethoxy silane
1-propanamine, 3-(triethoxysilyl)-; 3-aminopropyl-triethoxysilane; 3-triethoxysilylpropylamine; AMEO; Dynasylan AMEO. Technical grade coupling agent enhancing and promoting chemical bonding between inorganic and organic molecules; for acetal, acrylic, epichlorohydrin, nitrile, NC, polyamide, PC, polyethylene, polyimide, polymethacrylate, PP, polysulfone, PS, PVC, urethane and vinyl thermoplastics; thermoset acrylic (thermoset, latex), alkyd, epoxy, furan, melamine, nitrile/phenolic, phenolic, polyester, vinyl butyral/phenolic and elastomers. Reacts with glass of columns. mp < -70°; bp = 217°; d = 0.942; n_D^{20}= 1.4210; LD_{50} (rat orl) = 1780 mg/kg. *PCR.* Discontinued.

25238 Prosil® 248
2530-85-0 219-785-8
$C_{10}H_{20}O_5Si$
γ-Methacryloxypropyl trimethoxy silane
Silane A-174; 3-(trimethoxysilyl)propyl methacrylate; 3-methacryloxypropyltrimethoxysilane; MEMO; Dynasylan MEMO; σ-methacryloxypropyl trimethoxysilane; methacryloxypropyltrimethoxy silane. Coupling agent having reactive methacrylate and trimethoxysilyl groups; improves adhesion of organic thermoset resins to inorganic materials such as fiberglass, clay, quartz, and other siliceous surfaces; for ABS, acrylic, polyethylene, polyimide, polymethacrylate, PP, PS, silicone, SAN, urethane thermoplastics, alkyd, DAP, epoxy, polybutadiene, polyester, cross-linked polyethylene thermosets and elastomers. bp = 190°; d = 1.0450; n_D^{20} = 1.4313. *PCR.*

25239 Prosil® 3128
1760-24-3 217-164-6
$C_8H_{22}N_2O_3Si$
N-(2-Aminoethyl)-3-aminopropyl trimethoxysilane
N-[3-(trimethoxysilyl)propyl]-1,2-ethanediamine; N-β-(aminoethyl)-σ-aminopropyl trimethoxy-silane; N-[3-(Trimethoxysilyl)propyl]ethylenediamine; [3-(2-Aminoethyl)aminopropyl]trimethoxysilane; DAMO; N-(2-Aminoethyl-3-aminopropyl)trimethoxysilane; Dynasylan DAMO. Coupling agent for epoxies, phenolics, melamines, nylons, PVC, urethanes, acrylics. bp_{15} = 146°; d = 1.0100; n_D^{20}= 1.4450. *PCR.* Discontinued.

25240 Prosil® 5136
2530-83-8 219-784-2
$C_9H_{20}O_5Si$
3-Glycidoxypropyl trimethoxysilane
Glycidoxypropyltrimethoxysilane; Glymo; σ-Glycidoxypropyltrimethoxysilane; Dynasylan GLYMO. Coupling agent for epoxies, urethanes, acrylics, polysulfides. bp_2 = 120°; d = 1.0700; d = 1.1170; n_D^{20}= 1.4315. *PCR.*

25241 Prosil® 9202
2943-75-1 220-941-2
$C_{14}H_{32}O_3Si$
n-Octyltriethoxysilane
triethoxy-n-octyltriethoxysilane; Octyltriethoxysilane; Dynasylan OCTEO. Coupling agent for hydrophobic treatment. *PCR.*

25242 Prosil® HMDS
999-97-3 4725 213-668-5
$C_6H_{19}NSi_2$
Hexamethyldisilazane
HMDS; 1,1,1,3,3,3-Hexamethyldisilazane; hexamethyl disilizane; 1,1,1-Trimethyl-N-(trimethylsilyl)silanamine; Dynasylan HMDS. Coupling agent; silica treatment for silicone elastomers, novolac photoresist adhesion promoter. bp = 125°; d = 0.7650; n_D^{20}= 1.4071. *PCR.* Discontinued.

25243 Prosobee
A proprietary artificial baby milk derived from soya; used in cases of intolerance to cows' milk. *Bristol Myers Squibb Co Inc.*

25244 Prosol 4692
Emulsified mineral oil; lubricant for garnetting operations; suitable for oil fibers. *Hart Chem. Ltd.*

25245 Prosol 525
Emulsified mineral oil with selected antistatic agents; antistatic oil for woolen and worsted processes. *Hart Chem. Ltd.*

25246 prosopite
$Ca(F{\cdot}OH)_2{\cdot}Al_2(F{\cdot}OH)_3$
A mineral.

25247 Prosparol
A proprietary preparation of arachis oil and water emulsion; a high calorie food. *Duncan Flockhard Ltd.* Name unverified.

25248 Prospect
Processing oils. *Carless Refining & Marketing Ltd.*

25249 Prospect®
For the agriculture industry. *DuPont UK.*

25250 Prostearyl 15
25231-21-4
PPG-15 stearyl ether
Emollient, lubricant for cosmetics, bath oils, sunscreens, hair products, aerosol antiperspirants, hand and body lotions; coupler for fragrances. *Croda Inc.* Discontinued.

25251 protargentum
9015-51-4 8671
silver protein
proganol; protargin; Argentoproteinum; silver proteinate; silver nucleate; silver nucleinate. A compound of gelatin and silver. It contains 8% silver, and is used in aqueous solution in medicine as an antiseptic and also as a microscopic stain.

25252 Protargolgranulat
A German product. It consists of 1 part protargol and 2 parts urea. It has the advantage of easy solubility. Used as an antiseptic.

25253 Protars
A dry arsenical fungicide prepared from talc, lime, and arsenic oxide.

25254 Protasan
Denture adhesive (powder, cream and liquid), for securing dentures. *Richardson-Vicks Inc.* Name unverified.

25255 Protavic
Conductive resins. *Protex.*

25256 protease
9014-01-1 232-752-2
Enzyme which breaks down protein; used in detergent compositions, food processing, tanning, protein hydrolysis, desizing textiles. *Am. Biorganics; PMP Fernentation Prods.; Schweizerhall; Solvay Enzymes GmbH; U.S. Biochemical.*

25257 Protectoid
A proprietary trade name for a cellulose acetate plastic in the form of a nonflammable film. No manufacturer.

25258 Protectol
A brown, syrupy liquid used for the protection of animal products such as hair, wool, silk, skin, and leather from the action of alkaline liquids. It contains sodium lignin sulfonate.

25259 Protectol®
Biocides
BASF plc.

25260 Protectol® DMT
696-59-3 211-797-1
$C_6H_{12}O_3$
Tetrahydro-2,5-dimethoxyfuran
Dimethoxytetrahydrofuran. Biocide for disinfectants mp = -45°; bp = 145-147°; d = 1.0200; n_D^{20}= 1.4154-1.4157. *BASF AG.*

25261 Protectol® GDA, GT 50
111-30-8 4480 203-856-5
Glutaraldehyde
Biocide for aqueous systems and cleaners. *BASF AG.*

25262 Protectol® GL 40
107-22-2 4519 203-474-9
Glyoxal
Biocide for aqueous systems and cleaners. *BASF AG.*

25263 Protectol® KLC 50, 80
1086
Benzalkonium chloride
Biocidal surfactant for chemical industry, detergent manufacturing. *BASF AG.*

25264 Protectol® TOE
Thiadiazine derivative; biocide for use in aq. systems. *BASF AG.*

25265 Protectyl
A solution containing 0.2% mercury, 1% salicylic acid, 3% glycerin, and 95.8% water; used as a disinfectant.

25266 Protegin®
Mineral oil, petrolatum, ozokerite, glyceryl oleate, lanolin alcohol, emollient, emulsifier. *Goldschmidt; Goldschmidt AG.*

25267 Protegin® W, WX
Petrolatum, ozokerite, hydrogenated castor oil, glyceryl isostearate, polyglyceryl-3 oleate; SE water-oil emulsifier, emollient, absorptive base for cosmetics and pharmaceuticals. *Goldschmidt; Goldschmidt AG.*

25268 Pro-Tein ES-20
68951-89-3
Hydrolyzed collagen; ethyl ester; substantive to hair; plasticizer for styling resins; improves film properties and gloss in cream rinses, hair conditioners, mousses; anti-irritant and protectant in facial toners, antiperspirants. *Maybrook.*

25269 Protein Grade®
Detergents. *Calbiochem Corp.*

25270 Pro-Tein SA-20
68952-15-8

Lauroyl hydrolyzed collagen in ethanol; adds body, mitigates drying effects of alcohol in skin and hair care products (hair sprays, aftershaves, facial toners, antiperspirants); plasticizer for hair styling resins. *Maybrook.*

25271 **Pro-Tein SM-20**
72319-06-3
Myristoyl hydrolyzed collagen
Cosmetic ingredient. *Maybrook.*

25272 **proteinase**
Bacterial protease. *Rhône-Poulenc UK.*

25273 **protelds**
The same as albuminoids.

25274 **Proteodermin**
Soluble proteoglycans; for skin care preparations for aging or sun damaged skin. *Dr. Kurt Richter; Henkel/Cospha.*

25275 **Proteol**
A combination of casein with formaldehyde; an antiseptic dusting powder.

25276 **Proteosilane C**
133101-79-8
Methylsilanol elastinate
Aids tissue regeneration; for regenerative creams, anti-aging formulations, stretch mark preventives, other cosmetic and health emulsions, creams, alcoholic lotions. *Exsymol.*

25277 **Prote-pon P 2 EHA-02-Z**
Alkyl ether phosphoric acid
Wetting agent, detergent, hydrotrope, emulsifier, rust inhibitor, EP lubricant for alkaline detergents, metal cleaners/lubricants, hard surface cleaners, textile scours/lubricants, emulsion polymerization, agricultural, and drycleaning formulations. *Protex.*

25278 **Prote-sorb SML**
1338-39-2 8872 215-663-3
Sorbitan laurate
Emulsifier for food, cosmetic, household and industrial, agricultural, leather, metalworking, and textile industries. *Protex.*

25279 **Prote-sorb SMO**
1338-43-8 8872 215-665-4
Sorbitan oleate
Emulsifier for food, cosmetic, household and industrial, agricultural, leather, metalworking, and textile industries. *Protex.*

25280 **Prote-sorb SMP**
26266-57-9 8872 247-568-8
Sorbitan palmitate
Emulsifier for food, cosmetic, household and industrial, agricultural, leather, metalworking, and textile industries. *Protex.*

25281 **Prote-sorb SMS**
1338-41-6 8872 215-664-9
Sorbitan stearate
Emulsifier for food, cosmetic, household and industrial, agricultural, leather, metalworking, and textile industries. *Protex.*

25282 **Prote-sorb STO**
26266-58-0 8872 247-569-3
Sorbitan trioleate
Emulsifier for food, cosmetic, household and industrial, agricultural, leather, metalworking, and textile industries. *Protex.*

25283 **Prote-sorb STS**
26658-19-5 8872 247-891-4
Sorbitan tristearate
Emulsifier for food, cosmetic, household and industrial, agricultural, leather, metalworking, and textile industries. *Protex.*

25284 **Protex**
A proprietary safety glass. No manufacturer.

25285 **Pro-Tex**
Maneb (32.63%) and triphenyltin hydroxide (4.72%) solution; flowable fungicide for potatoes and sugar beets; restricted use. *Griffin.*

25286 **Protexulate**
Loose-fill mineral powder; for underground pipework for heat insulation and as a water barrier to stop corrosion. *Croxton & Garry Ltd.* Name unverified.

25287 **protheite**
A mineral. It is a variety of pyroxene.

25288 **Prothera™**
White petrolatum; skin cleanser and protectant. *ICN Pharmaceuticals Inc.*

25289 **protocatechuic acid**
99-50-3 8080 202-760-0
$C_7H_6O_4$
3,4-dihydroxybenzoic acid
Used in chemical synthesis. mp = 200-202°; d = 1.54; soluble in H_2O (2 g/100 ml), organic solvents. *Dinoval Chem. Ltd.*

25290 **protogest**
9015-54-7 310-296-6
Protein hydrolysate veterinary. *The Wellcome Foundation Ltd.* Name unverified.

25291 **Proto-Lan 20**
PEG-75 lanolin, propylene glycol, ceteth-16, hydrolyzed animal protein and lanolin oil. Substantive emollient, softener, conditioner, moisturizer, lubricant, base for skin and hair products (shampoos, conditioners, ethnic products, creams, lotions, face masks, and bath products). *Maybrook.*

25292 **Proto-Lan 30**
Propylene glycol, PPG-12-PEG-65 lanolin oil and hydrolyzed animal protein. Substantive emollient, moisturizer, adjunct emulsifier for skin and hair care products (shampoos, conditioners, creams, lotions, bath products, antiperspirants, facial toners). *Maybrook.*

25293 **proto-Lan 4R**
Mineral oil, coco-hydrolyzed collagen, cetyl alcohol, myristyl myristate, ceteth-16 and hydrogenated lanolin. Emollient, moisturizer, auxiliary emulsifier, antistat, conditioner, base for skin and hair care products (hair conditioners, creams, lotions, lipsticks); provides lubricious film, water emulsifying and binding properties. *Maybrook.*

25294 **Proto-Lan 8**
Lecithin, butyl stearate, coco-hydrolyzed animal protein, oleoyl sarcosine, sesame oil and lanolin alcohol. Nongreasy emollient, moisturizer, emulsifier, conditioner, antistat, lubricant for creams, lotions, ethnic formulations, conditioners, baby oils, bath oils, suntan products; shave creams, makeup and face soaps. *Maybrook.*

25295 **Proto-Lan IP**
2724-58-5 220-336-3
Isostearic acid and isostearoyl hydrolyzed collagen. Nongreasy emollient, moisturizer for ethnic formulations, conditioners, hair treatments, lip care, skin care, nail polish, soap bars, and antiperspirants. *Maybrook.*

25296 **Proto-Lan KT**
Isostearic acid, sorbitan oleate and cocoyl hydrolyzed keratin. Ingredient in cosmetics. *Maybrook.*

25297 **Protopet**
Petrolatum USP
Hydrocarbon oil. *Witco Corporation.*

25298 **Protovit**
Proprietary preparation containing the B complex vitamins, vitamins A, C, D and E, biotin and nicotinamide. *Roche Products Ltd.* Name unverified.

25299 **Protrum K**
13684-63-4 7384 237-199-0
phenmedipham
Emulsifiable concentrate containing 114 g/l phenmedipham; for weed control for beet crops and strawberries. *Atlas Interlates Ltd.*

25300 **Protugan**
34123-59-6 5237 251-835-4
Isoproturon
Herbicide *Agan Chemical Manufacturers Ltd.*

25301 **Proventin**
Protective agent for bleaching polyamide fibers. *Degussa.*

25302 **Proventin 7**
Antioxidant for use with nylon to prevent attack of peroxide compounds during bleaching and washing. *Henkel/Textiles.*

25303 **Proventin 7**
An active oxygen compound; used in the textile industry as specific protection agent for polyamide fibers during peroxide bleaching and in the dyeing process. *Degussa AG.* Name unverified.

25304 **Provil®**
Addition-cured silicone based precision impression material in four different viscosities; for dentistry. *Bayer AG.*

25305 **Provol**
A compound used for electrical insulation. It is a mixture of pitch, bitumen or similar materials, and mineral matter.

25306 **Provol 50**
52581-71-2
PPG-50 oleyl ether
Emollient, superfatting agent, lubricant, pigment dispersant, coupler for personal care products, ethnic hair products; aids spreading and pigment dispersion in make-up systems. *Croda Inc.* Discontinued.

25307 **Proxel**
Industrial microbiocides. *ICI Chem & Polymers Ltd.*

25308 **Proxitane**
79-21-0 7293 201-186-8
Peracetic acid
Oxidizing agent. *Solvay Interox Ltd.*

25309 Prox-onic 2EHA-1/02
PEG-2 2-ethylhexyl ether
Detergent, wetting agent, emulsifier, dispersant, solubilizer, defoamer for textiles, metal cleaners, industrial, institutional and household cleaners, hand cleaners. *Protex.*

25310 Prox-onic BP-02 P
POP (2) bisphenol
Aromatic dihydric alcohol for manufacturing unsaturated polyester and alkyd resins. *Protex.*

25311 Prox-onic CC-05
61791-29-5
PEG-5 cocoate
Emulsifier, lubricant additive for metalworking, textiles, cosmetics, defoamers; viscosity control agent in plastisols. *Protex.*

25312 Prox-onic CSA-1/04
68439-49-6 7737
Ceteareth-4
Emulsifier, emulsion stabilizer, detergent, wetting agent, dispersant, solubilizer, defoamer, dye assistant, leveling agent for cosmetics, textiles, metal cleaners, industrial, institutional and household cleaners, emulsion polymerization. *Protex.*

25313 Prox-onic DA-1/04
26183-52-8
Deceth-4
Wetting agent, emulsifier, detergent, dispersant, solubilizer, defoamer for textiles, metal cleaners, industrial, institutional and household cleaners. *Protex.*

25314 Prox-onic DDP-09
9014-92-0
Dodoxynol-9
Detergent, wetting agent for industrial and heavy-duty detergents. *Protex.*

25315 Prox-onic DNP-08
9014-93-1
Nonyl nonoxynol-8
Emulsifier, detergent, wetting agent, dispersant, solubilizer, coupling agent for textiles, metalworking, household, industrial, agricultural, paper, paint and other industries. *Protex.*

25316 Prox-onic DT-03
PEG-3 tallow diamine
Surfactant, emulsifier, lubricant additive, antistat, detergent for textile, metal, plastics, dyeing assistants, degreasers, corrosion inhibitor, agricultural; intermediate for quaternaries. *Protex.*

25317 Prox-onic EP 1090-1
Difunctional block polymer ending in primary hydroxyl groups; defoamer for metalworking, cosmetic, paper, textiles; base for low foaming surfactants, antifoams, dishwash, dispersing and wetting agents for paints, drilling muds, emulsifiers, petroleum demulsifiers, emulsion polymerization. *Protex.*

25318 Prox-onic HR-05
61791-12-6
PEG-5 castor oil; emulsifier, pigment dispersant, leveling agent, softener, rewetting agent, degreaser, antistat, emulsion stabilizer, lubricant for leather, paint, paper, plastics, textile, and cosmetics industries; solubilizer for perfumes. *Protex.*

25319 Prox-onic HRH-05
61788-85-0
PEG-5 hydrogenated castor oil; emulsifier, pigment dispersant, leveling agent, softener, rewetting agent, degreaser, antistat, emulsion stabilizer, lubricant for leather, paint, paper, plastics, textile, and cosmetics industries; solubilizer for perfumes. *Protex.*

25320 Prox-onic L 081-05
POE (5) linear alcohol ether; biodegradable low foam detergent, wetting agent, emulsifier for household, agricultural and industrial cleaners; coupling agent and solubilizer for perfumes and organic additives. *Protex.*

25321 Prox-onic LA-1/02
Laureth-2
Coupling agent, solubilizer, emulsion stabilizer for cosmetic and hair care products; with anionic surfactants for emulsion polymerization; in coning and textile spin finishes. *Protex.*

25322 Prox-onic MC-02
61791-14-8
PEG-2 cocamine
Surfactant, emulsifier, lubricant additive, antistat, detergent for textile, metal, plastics, dyeing assistants, degreasers, corrosion inhibitor, agricultural; intermediate for quaternaries. *Protex.*

25323 Prox-onic MG-020 p
61849-72-7
POP (20) methyl glucoside
Cosolvent, perfume fixative. *Protex.*

25324 Prox-onic MHT-015
61791-26-2
PEG-15 hydrogenated tallow amine
Surfactant, emulsifier, lubricant additive, antistat, detergent for textile, metal, plastics, dyeing assistants, degreasers, corrosion inhibitor, agricultural; intermediate for quats. *Protex.*

25325 Prox-onic MO-02
PEG-2 oleamine
Surfactant, emulsifier, lubricant additive, antistat, detergent for textile, metal, plastics, dyeing assistants, degreasers, corrosion inhibitor, agricultural; intermediate for quaternaries. *Protex.*

25326 Prox-onic MS-05
26635-92-7
PEG-5 stearamine
Surfactant, emulsifier, lubricant additive, antistat, detergent for textile, metal, plastics, dyeing assistants, degreasers, corrosion inhibitor, agricultural; intermediate for quaternaries. *Protex.*

25327 Prox-onic MT-02
61791-44-4 263-177-5
PEG-2 tallow amine
Surfactant, emulsifier, lubricant additive, antistat, detergent for textile, metal, plastics, dyeing assistants, degreasers, corrosion inhibitor, agricultural; intermediate for quaternaries. *Protex.*

25328 Prox-onic NP-04
9016-45-9 6772
Nonoxynol-4
Surfactant, detergent, defoamer. *Protex.*

25329 Prox-onic OA-1/04
9004-98-2
Oleth-4
Coupling agent, solubilizer, emulsion stabilizer for cosmetic and hair care products; with anionic surfactants for emulsion polymerization; in coning and textile spin finishes. *Protex.*

25330 Prox-onic OCA-1/06
Cetoleth-6
Detergent, wetting agent, emulsifier, dispersant, solubilizer, defoamer for textiles, metal cleaners, industrial, institutional and household cleaners, hand cleaners. *Protex.*

25331 Prox-onic OL-1/05
9004-96-0
PEG-5 oleate
Surfactant for cuttting oils, degreasing solvents, metal cleaners, textiles, leather, cosmetics; dyeing assistant; emulsifier for mineral oils, fatty oils. *Protex.*

25332 Prox-onic OP-09
9002-93-1 6858
Octoxynol-9
Emulsifier for metal, textile processing, household and industrial cleaners, vinyl and acrylic polymerization. *Protex.*

25333 Prox-onic PEG-2000
25322-68-3 7729
PEG-2M
Low foam wetting in paper pulping; emulsifier for metal degreasing; bottle cleaner defoamer; binder for tobacco; polyurethane manufacturing; mold release agent; agriculture. *Protex.*

25334 Prox-onic PH-01
122-99-6 7410 204-589-7
Phenoxyethanol
Protex.

25335 Prox-onic PPG-900
PPG (900); agricultural emulsifier concentrate. *Protex.*

25336 Prox-onic SA-1/02
9005-00-9
Steareth-2
Detergent, wetting agent, emulsifier, dispersant, solubilizer, defoamer for textiles, metal cleaners, industrial, institutional and household cleaners, hand cleaners. *Protex.*

25337 Prox-onic SA1-015/P
25231-21-4
POP (15) stearyl alcohol
Surfactant. *Protex.*

25338 Prox-onic SML-020
9005-64-5 8872
Polysorbate 20
Emulsifier, solubilizer for petroleum oils, solvents, vegetable oils, waxes, silicones, etc.; for agriculture, cosmetic, leather, metalworking and textile industries. *Protex.*

25339 Prox-onic SMO-020
9005-65-6 7742
Polysorbate 80
Emulsifier, solubilizer for petroleum oils, solvents, vegetable oils, waxes, silicones, etc.; for agriculture, cosmetic, leather, metalworking and textile industries. *Protex.*

25340 Prox-onic SMO-05
9005-65-6 7742
Polysorbate 81
Emulsifier, solubilizer for petroleum oils, solvents, vegetable oils, waxes, silicones, etc.; for agriculture, cosmetic, leather, metalworking and textile industries. *Protex.*

25341 Prox-onic SMP-020
9005-66-7 8872
Polysorbate 40
Emulsifier, solubilizer for petroleum oils, solvents, vegetable oils, waxes, silicones, etc.; for agriculture, cosmetic, leather, metalworking and textile industries. *Protex.*

25342 Prox-onic SMS-020
9005-67-8 8872
Polysorbate 60
Emulsifier, solubilizer for petroleum oils, solvents, vegetable oils, waxes, silicones, etc.; for agriculture, cosmetic, leather, metalworking and textile industries. *Protex.*

25343 Prox-onic ST-05
9004-99-3
PEG-5 stearate
Emulsifier, lubricant additive for metalworking, textiles, cosmetics, defoamers; viscosity control agent in plastisols. *Protex.*

25344 Prox-onic TA-1/08
61791-00-2
PEG-8 tallate
Emulsifier, detergent for degreasers and neutral or mildly alkaline detergents. *Protex.*

25345 Prox-onic TD-1/03
24938-91-8
Trideceth-3
Intermediate for surfactants; emulsifier, detergent, foam builder, dispersant, wetting agent, solubilizer for mech. dishwash, alkaline cleaners, pulp and paper, textiles, corrosion inhibition. *Protex.*

25346 Prox-onic TM-06
POE (6) isolauryl mercaptan
Wetting agent, surfactant, detergent for metal cleaning, household and industrial cleaners/scours, degreasers; emulsifier for herbicides/insecticides, greases, soils. *Protex.*

25347 Prox-onic UA-03
Isoundeceth-3
Detergent, wetting agent, emulsifier, dispersant, solubilizer, defoamer for textiles, metal cleaners, industrial, institutional and household cleaners, hand cleaners. *Protex.*

25348 Proxy
7722-84-1 4839 231-765-0
H_2O_2
hydrogen peroxide
A hydrogen peroxide solution. Bleach and oxidizer.

25349 Proxyl
A proprietary trade name for a pyroxylin-based denture material. No manufacturer.

25350 Prozinex
139-40-2 7996 205-359-9
propazine
Active ingredient; propazine; 2-chloro-4, 6-bis-(isopropylamino)-1,3,5-triazine; selective pre-emergent herbicide. *Agan Chemical Manufactures Ltd.*

25351 Prozone
Solvent for cleaning for electronic applications. *BP Chemicals Ltd.*

25352 prussic acid
74-90-8 4836 200-821-6
HCN
hydrocyanic acid
formonitrile. Used in the manufacture of acrylonitrile, acrylates, adiponitrile; chelates; pesticides.

25353 Pruteen
Single-cell protein used as a feed additive. *ICI Chem & Polymers Ltd.* Discontinued.

25354 Pruv
4070-80-8 223-781-1
Sodium stearyl fumarate
Lubricant for pharmaceutical tablets. *Penwest Pharmaceuticals Co.*

25355 Pryfon
25311-71-1 5187 246-814-1
Isofenphos
Insecticide for treatment of soil against subterranean termites. *Bayer AG.*

25356 Prystal
A formaldehyde-urea condensation product, a French proprietary rubber vulcanization accelerator; also a proprietary trade name for a cast clear phenolic molding. No manufacturer.

25357 Prystaline
A proprietary molding compound of the urea formaldehyde type of synthetic resin. No manufacturer.

25358 PS021
Tetrachlorophenyl-siloxane dimethyl siloxane copolymer, branched; thermal silicone fluid for metal-to-metal lubrication; base for high temperature grease lubricants; lubricant for instruments. *Hüls Am.*

25359 PS034
9006-65-9 3264
Dimethicone
Antifoam, leveling and flow control agent for coatings; heat exchangers, baths and thermostats; dielec. media; for cooling in transduction applications; ultrasonic applications. *Hüls Am.*

25360 PS071
Ethylene oxide-modified polydimethylsiloxane; surfactant, wetting agent for photographic plates; antifog for glass and plastic optics; slip agent for flexographic and gravure inks. *Hüls Am.*

25361 PS072
Ethylene oxide/propylene oxide-modified polydimethylsiloxane; surfactant, lubricant for fibers and plastics, metal-to-plastic wear interfaces; antitack and mar resistance aid for urethane coatings; slip agent in flexographic and gravure inks. *Hüls Am.*

25362 PS073
Ethylene oxide-modified polydimethylsiloxane; surfactant, antifoam in water-based coatings. *Hüls Am.*

25363 PS-10GF/000
Thermoplastic polyester, 10% glass fiber-reinforced. *Compounding Tech.* Name unverified.

25364 PS130
68607-75-0
Polyoctadecylmethylsiloxane
Component in thread and fiber lubricants; process aid in melt spinning. *Hüls Am.*

25365 PS140
68440-90-4
Polyoctylmethylsiloxane
Lubricant for soft metals; rubber and plastic lubricant especially when mated against steel or aluminum; for aluminum machining operations; process aid and plasticizer in polyolefin rubbers. *Hüls Am.*

25366 PS181
63148-56-1
Polymethyl-3,3,3-trifluoropropyl siloxane
Lubricant for EP applications, automotive and aerospace lubes, electrical contacts, precision timers; flotation medium for inertial guidance systems; in sonar lens, mechanical vacuum pumps. *Hüls Am.*

25367 PS187
Polymethyl-3,3,3-trifluoropropylsiloxane- (50%) dimethylsiloxane copolymer; lubricant for EP applications, automotive and aerospace lubes, electrical contacts, precision timers; flotation medium for inertial guidance systems; in sonar lens, mechanical vaccum pumps. *Hüls Am.*

25368 PS3/PS4/PS5
Range of granular fertilizers. *Fisons plc, Horticultural Div.* Name unverified.

25369 PS-30GM/000
26062-94-2
PBT polyester, 30% milled glass-reinforced. *Compounding Tech.* Name unverified.

25370 PSA
Phenol sulfonic acid
Electrolyte for tin plating; catalyst for resins. *Hart Chem. Ltd.*

25371 PSDF04
1,1,5,5-Tetraphenyl-1,3,3,5-tetramethyltrisiloxane
Diffusion pump fluid with excellent resistance to heat oxidation and chemicals. *Hüls Am.*

25372 PSDF05
1,1,3,5,5-Pentaphenyl-1,3,5-trimethyltrisiloxane
Diffusion pump fluid with excellent resistance to heat, oxidation and chemicals. *Hüls Am.*

25373 pseudo-alums
Double sulfates of aluminum and another metal containing a bivalent metal

sulfate instead of a monovalent one. A type is $MnSO_4 \cdot Al_2(SO_4)_3 \cdot 24H_2O$. They are not isomorphous with the alums.

25374 Pseudocollagen
High molecular weight matrix oligosaccharides and soluble proteins; cosmetic ingredient forming moisture-retentive films on skin, leaving skinsoft and supple. *Brooks Industries.*

25375 PTAL
104-87-0 203-246-9
C_8H_8O
p-Tolualdehyde
4-methylbenzaldehyde; *p*-formyltoluene. Additive for resins; intermediate for pharmaceuticals; fragrance. mp = -6°; bp = 204-205°; d = 1.0190; n_D^{20} = 1.5447; slightly soluble in H_2O, more soluble in organic solvents. *Mitsubishi Gas.*

25376 PTFE-20
Mixture of powdered PTFE and titanium dioxide. Binder for pressed powders; excellent thermal and chemical resistance, good skin adhesion; imparts luxurious, lubricious feel. *Presperse.*

25377 PTSA 70
104-15-4 9671 203-180-0
p-toluenesulfonic acid
Catalyst for resins. *Hart Chem. Ltd.*

25378 PTZ® Phenothiazine Purified
92-84-2 7404 202-196-5
Phenothiazine
Antioxidant, monomer stabilizer, chemical and pharmaceutical intermediate. *ICI Polymer Additives.*

25379 pulegone
89-82-7 8124 201-943-2
$C_{10}H_{16}O$
1-methyl-4-isopropylidene-3-cyclohexanone
$\Delta^{4,8}$-*p*-menthen-3-one; 1-isopropylidene-4-methyl-2-cyclohexanone; (+)-4(8)-*para*-menthen-3-one; (R)-(+)-Pulegone; D-Pulegone. Found in natural oils such as pennyroyal oil. bp_{12} = 97°; $[\alpha]_D^{20}$= 21° (neat); d = 0.9300; insoluble in H_2O, soluble in organic solvents.

25380 Pulluzyme
Pullulanase, an enzyme. *Rhône-Poulenc UK.*

25381 Pulmolite
Technetium, ^{99m}Tc in aggregated albumin; used as a diagnostic aid and a radioactive agent. *DuPont NEN Medical Products.* Name unverified.

25382 Pulpex E and P
Polypropylene pulps
Special fibrous additives that can be mixed in any proportion with natural pulps and can be handled on conventional equipment; very wide range of applications. *Hercules.* Discontinued.

25383 Pulpzyme
xylanase
A liquid xylanase preparation for reducing the need of bleaching chemicals in kraft pulp bleaching. *Novo Nordisk.*

25384 Pulsar
7778-54-3 1717 231-908-7
Calcium hypochlorite
Dry chlorinating agent. *Olin.*

25385 Pulsar
A solution concentrate containing 200 g bentazone and 200 g MCPA per liter; a post-emergence herbicide. *BASF plc.*

25386 Pulse
A family of polycarbonate resin blends used for interior automotive applications. *Dow Cheml Co Ltd, UK & Ireland.*

25387 Pulse 1310
Engineering thermoplastic; engineering thermoplastic providing easy processing, toughness, resistance to temperature extremes; suitable for snowmobile components that must withstand freezing conditions as adjacent parts become red-hot, cellular phones exposed to heat inside closed automobiles; for durables. *Dow Chemical.*

25388 Pulse 1735
PC/ABS resin; uv-stable resin with high heat distortion, easy processing, practical toughness for computer and business equipment. *Dow Plastics.*

25389 Pulse 600
Engineering thermoplastic; engineering thermoplastic providing easy processing, toughness, resistance to temperature extremes; suitable for snowmobile components that must withstand freezing conditions as adjacent parts become red-hot, cellular phones; exposed to heat inside closed automobiles, automotive grade. *Dow Chemical.*

25390 Pulse®
Fish oil concentrate; dietary supplement of ω 3 fatty acids. *Seven Seas Ltd.*

25391 Pultac
Strippable protective for paint spray walls, windows and floor grills. *Brent Chemicals International plc.*

25392 Pulvatex
A trademark for rubber (raw or partly prepared) for manufacture. No manufacturer.

25393 Pumice Plus
Abrasive solvent base hand cleaner gel. *Momar Industrial Services Ltd.*

25394 pumice stone
(Obsidian). A volcanic mineral, lava froth, consisting mainly of aluminum silicate; used as an abrasive.

25395 Pumiline
A proprietary preparation of pine oil. No manufacturer.

25396 Pump Repair Putty
Ceramic filled epoxy resin; used to repair, rebuild and protect pumps. *ITW Devcon.*

25397 Punch® C
carbendazim and flusilazole. Eradicant fungicide for use on cereal crops. *DuPont UK.*

25398 Punctilious® Ethyl Alcohol
64-17-5 3806 200-578-6
Ethanol
Solvent and extraction medium. *Quantum/USI.*

25399 punicin
A coloring matter obtained from *Purpura capillus* and other shell fish.

25400 Purac
7440-44-0 1855 233-153-3
Decolorizing and absorptive activated carbon. *Lancashire Chemical Works Ltd.*

25401 Puraspec
Range of catalysts and absorbents used for purification. *ICI Chem & Polymers Ltd.*

25402 Puratronic
High purity chemicals suitable for electronic materials; used in electronic device materials, crystal growing, epitaxy. *Johnson Matthey plc.*

25403 Purbeck stone
A limestone used in building.

25404 Purdox
1344-28-1 369 215-691-6
A proprietary trade name for high purity recrystallized alumina. *Morgan Refractories Ltd.* Name unverified.

25405 Pure-Dent® B700
9005-25-8 232-679-6
Corn starch USP, NF; binder and diluent for granulations and tablets when used wet or dry; disintegrant; for pharmaceutical use. *Grain Processing.*

25406 Purez
Polyester polyols based on adipic acid. *ICI Polyurethanes.* Name unverified.

25407 Purgatol
Consists mainly of the anhydrides and lactones of fatty acids; used for dressing hides.

25408 purging nut oil
Physic nut oil; Curcas oil. Oil from the seeds *Jatropha Curcas*; used in soap-making, and for lubricating.

25409 Purifloc
Polyacrylamides used as flocculants. *Dow Cheml Co Ltd, UK & Ireland.*

25410 Purisol
Chemicals and equipment for treating water, effluent and sewage. *ICI Chem & Polymers Ltd.*

25411 Purity® 21
9005-25-8 232-679-6
Corn starch; binder, filler and disintegrant for cosmetic and pharmaceutical formulations, body powders, footpowders, dry shampoos, makeup, eye liner, mascara. *Nat'l. Starch.*

25412 Pur-Oba®
Natural wax ester; water-white jojoba oil for high quality cosmetics. *Goldschmidt AG.* Discontinued.

25413 Purochem
Organotin bactericides. *Akzo Chemie.* Name unverified.

25414 Puromix
Food phosphate mixtures. *Albright & Wilson Ltd., Phosphates & Speciality Business.* Name unverified.

25415 Puron
Food grade phosphates. *Albright & Wilson Ltd., Phosphates & Speciality Business.*

25416 Purozone
A proprietary preparation consisting of an alcoholic solution of the sodium

salts and acids of wood tar or similar materials; a disinfectant. No manufacturer.

25417 Purple Copp
1317-39-1 2734 215-270-7
Purple cuprous oxide
Fungicide and antiseptic, used in antifouling paints. *Am. Chemet.*

25418 Purplecopp 97N Premium
1317-39-1 2734 215-270-7
Cuprous oxide
Fungicide and antiseptic, used in antifouling paints. *CP International Chemicals Ltd.*

25419 Purton CFD
61791-31-9 263-163-9
Cocamide DEA
Foam stabilizer, thickener and superfatting agent for cosmetics, cleaners. *Zschimmer & Schwarz.*

25420 Purton SFD
68425-47-8 270-355-6
Linoleamide DEA
Foam stabilizer, thickener and superfatting agent for cosmetics, cleaners. *Zschimmer & Schwarz.*

25421 Purzaust® Catalysts
Catalysts for broad range of air purification applications *AlliedSignal.*

25422 Pusher
Secondary oil recovery polymer; an additive to water injected into petroleum reservoirs; decreases the mobility of the water flood in relation to the mobility of the crude. *Dow UK.* Discontinued.

25423 putrescine
110-60-1 8134 203-782-3
$C_4H_{12}N_2$
1,4-butanediamine
Tetramethylenediamine. Ubiquitous biological chemical formed by decarboxylation in tissue of ornithine or arginine. Used in biochemical and biological research. mp = 27-28°; bp = 158-160°; d = 0.8770; n_D^{20} = 1.4569; soluble in H_2O (40 g/l), organic solvents.

25424 putty powder
SnO_2
stannic oxide. An impure stannic oxide, used for polishing glass. It is also used sometimes in rubber mixing specific gravity of 6.6.

25425 puzzuolana
A volcanic material found in various parts of Italy, especially Puzzuoli. It is employed for the conversion of pure lime into a hydraulic lime.

25426 PV, PV Fast
Pigment powders for plastics. *Hoechst AG.*

25427 Pvacote
Compounded polymeric emulsion; for spray coating of asbestos stripped rooms to bind and localize asbestos fibers. *Howlett Adhesives Ltd.* Name unverified.

25428 PVC Deodorant 5417, OS
Mixture of fragrance materials; masks odors during processing and in finished product *Andrea Aromatics.* Name unverified.

25429 PVP
9003-39-8 7879 201-800-4
$C_6H_9NO_x$
polyvinylpyrrolidone
povidone; 1-ethenyl-2-pyrrolidinone, homopolymer. Polymer of 1-vinyl-2-pyrrolidone monomers. film-forming agent, hair fixative, thickener, protective colloid, suspending agent, and dispersant for cosmetics industry, technical applications.; drug vehicle and retardant; tablet binder, pharmaceutical excipient; used in adhesives and detergents. *Allchem Industries; BASF; ISP.*

25430 PVP/eicosene copolymer
28211-18-9
1-Ethenyl-2-pyrrolidinone polymer with 1 eicosene
Ganex® V-220; Anatron®V-220; Unimer U-15. PVP/eicosene copolymer; used in cosmetics and toiletries as moisture barrier, adhesive, protective colloid, and microencapsulating resin; as dispersant for pigments; as solubilizer for dyes; in petroleum industry as sludge and detergent dispersant. Off white waxy soild; soluble in mineral oil; kerosene; organic solvents an other polymers; mw=8600; solid pt=35-40°; HLB=8.0; LD_{50} rat oral >17.1 g/kg. *ISP.*

25431 PVP/hexadecene copolymer
32440-50-9
Polyvinylpyrrolidone/hexadecene copolymer
Ganex® V-216; Anatron® V-216; Unimer U-151. PVP/hexadecene copolymer; used in cosmetics and toiletries as moisture barrier, adhesive, protective colloid, and microencapsulating resin; as dispersant for pigments; as solubilizer for dyes; in petroleum industry as sludge and detergent dispersant. Pale yellow visc liquid; soulbe in mineral oil, kerosene, organic solvents and other polymers; partially soluble in ethanol; mw=7300; HLB=10.0; LD_{50} rat oral>17.1 g/kg. *ISP.*

25432 PX-104
84-74-2 1622 201-557-4
Dibutyl phthalate
Reagent grade plasticizer *Aristech.* Name unverified.

25433 PX-109
68515-48-0 271-090-9
Diisononyl phthalate
Reagent grade plasticizer *Aristech.* Name unverified.

25434 PX-10GF/000
Polyphenylene ether
10% glass fiber-reinforced. *Compounding Tech.*

25435 PX-111
3648-20-2 222-884-9
$C_{30}H_{50}O_4$
1,2-Benzenedicarboxylic acid diundecyl ester
Diundecyl phthalate. Reagent grade plasticizer Insoluble in H_2O, slightly soluble in organic solvents. *Aristech.* Name unverified.

25436 PX-120
68515-49-1 271-091-4
Diisodecyl phthalate
Reagent grade plasticizer *Aristech.* Name unverified.

25437 PX-126
119-06-2 204-294-3
$C_{34}H_{58}O_4$
Ditridecyl phthalate
1,2-Benzenedicarboxylic acid ditridecyl ester; Polycizer 962-BPA; Staflex DTDP; 1-tridecanol, phthalate; Truflex DTDP. Reagent grade plasticizer bp_5 >285°; d = 0.9500. *Aristech.* Name unverified.

25438 PX-138
117-81-7 1291 204-211-0
Dioctyl phthalate
Reagent grade plasticizer *Aristech.* Name unverified.

25439 PX-209
33703-08-1 251-646-7
$C_{24}H_{46}O_4$
Hexanedioic acid, diisononyl esteri
diisononyl adipate. Reagent grade plasticizer *Aristech.* Name unverified.

25440 PX-238
103-23-1 203-090-1
$C_{22}H_{42}O_4$
Bis(2-ethylhexyl)hexanedioate
BEHA; DEHA; DOA; dioctyl adipate; octyl adipate; adipol 2EH; Bisoflex DOA; Effemoll DOA; Effomoll DOA; Ergoplast ADDO; Flexol A 26; Flexol Plasticizer 10-A; Flexol Plasticizer A-26; Kemester 5652; Kodaflex DOA; Mollan S; Monoplex DOA; Plastomoll DOA; PX-238; Reomol Doa; Rucoflex Plasticizer DOA; Sicol 250; Staflex DOA; Truflex DOA; Uniflex DOA; Vestinol OA; Wickenol 158; Witamol 320. Reagent grade plasticizer mp = -67°; bp = 417°; d = 0.9240. *Aristech.* Name unverified.

25441 PX-338
3319-31-1 222-020-0
$C_{33}H_{54}O_6$
Trioctyl trimellitate
tris(2-ethylhexyl) trimellitate; tris(2-ethylhexyl) ester 1,2,4-benzenetricarboxylic acid; Trioctyl; Trimellitate tris(2-ethylhexyl) ester; Kodaflex TOTM; Tri(2-ethylhexyl)trimellitate ester; 2-Ethylhexyl Trimellitate. Reagent grade plasticizer d = 0.9890; LD_{50} (mus orl) >60 g/kg. *Aristech.* Name unverified.

25442 PX-339
53894-23-8 258-847-9
Triisononyl trimellitate
Reagent grade plasticizer *Aristech.* Name unverified.

25443 PX-504
105-76-0 203-328-4
$C_{12}H_{20}O_4$
Dibutyl maleate
Dibutyl Maleate; 2-Butenedioic acid (Z)-, dibutyl ester. Reagent grade plasticizer bp_4 = 129°; d = 0.9930; LD_{50} (rat orl) = 3730 mg/kg. *Aristech.* Name unverified.

25444 PX-538
2915-53-9 220-835-6
$C_{20}H_{36}O_4$
Dioctyl maleate
2-Butenedioic acid (Z)-, dioctyl ester. Reagent grade plasticizer and chemical intermediate, e.g. for the production of sulfosuccinates. *Aristech.* Name unverified.

25445 PX-800
8013-07-8 232-391-0
Epoxidized soya oil; reagent grade plasticizer *Aristech.* Name unverified.

25446 PX-914
84-78-6 201-562-1
$C_{20}H_{30}O_4$
Butyl octyl phthalate
Reagent grade plasticizer *Aristech.* Name unverified.

25447 PY Garden insect Killer
Liquid concentrate containing pyrethrum and piperonyl butoxide (a synergist); contact insecticide for use on crop and noncrop plants. *Vitax Ltd.*

25448 PY Garden Insecticide
Aerosol containing pyrethrins and piperonyl butoxide; contact insecticide for use on crop and noncrop areas. *Vitax Ltd.*

25449 PY Powder Garden & Household Insect Killer
Dust containing pyrethrins and piperonyl butoxide; contact domestic and garden insecticide. *Vitax Ltd.*

25450 Pybuthrin
Pyrethrins. Used as insecticides. *The Wellcome Foundation Ltd.* Name unverified.

25451 Pylkrome
A composite alloy steel. The base is mild steel, and it has a corrosion-resisting surface of high chrome or high chromium-nickel-iron alloys.

25452 Pylumin
Process for bonding paint to aluminum and its alloys. *Brent Chemicals International plc.*

25453 Pynol
Disinfectants. *Evans Vanodine International Ltd.*

25454 Pynosect 30
Mixture of pyrethrins and resmethrin; an insecticide for greenhouse and horticultural crops. *Mitchell Cotts Chemicals Ltd.*

25455 Pyoctanin
548-62-9 4401 208-953-6
Pyoktanin
Gentian Violet. The name given to different coal-tar colors: Yellow pyoctanin (auramine) and Blue pyoctanin (methyl violet); used in surgery as bactericides.

25456 Pyracur® FL
chloridazon and metolachlor. Pre-emergence herbicide for control of grasses and broadleaf weeds in sugar beet and fodder beet *BASF AG.*

25457 Pyradex® T
Chloridazon and triallate. Pre-plant incorporated herbicide for control of broadleaf weeds and grasses in sugar and fodder beet *BASF AG.*

25458 Pyradiolin
A proprietary synthetic plastic used as a dielectric material in wireless telegraphy, and for other purposes; it is a modified pyroxylin plastic. No manufacturer.

25459 Pyradur®
Chloridazon + metolachlor; pre-emergence herbicide for control of grasses and broad-leaved weeds in sugar and fodder beet crops. *BASF AG.*

25460 Pyra-Fog 100
8148
Pyrethrin
Contact insecticide. *Chemsearch (UK) Ltd.*

25461 Pyralin
9004-70-0 8195
A celluloid product available in transparent, translucent, opaque colored and colorless forms; resistant to hydrocarbons and oils.

25462 Pyralin®
Polyimide high temperature-resistant materials. *DuPont UK.*

25463 pyraloxin
Oxidized pyrogallic acid
A dark brown powder obtained by oxidation of pyrogallic acid with air and ammonia.

25464 Pyralux®
Flexible materials. *DuPont UK.*

25465 Pyramid
Potassium silicates
Granular potassium silicates and sodium disilicates. *Crosfield Chemicals Ltd.*

25466 Pyramin®
1698-60-8 216-920-2
Chloridazon
For pre-and post-emergence control of weeds in sugar beet, fodder beet, Swiss chard, some ornamentals. *BASF AG; BASF plc.*

25467 Pyramol
Silica sols for industry. *Crosfield Chemicals Ltd.*

25468 Py-Ran
7758-23-8 1740 231-837-1
Anhydrous monocalcium phosphate
Used as acid component in leavening agents for self-raising flours, baking powders, cake mixes and corn meal. *Monsanto Co.* Name unverified.

25469 Pyranet
Detergent sanitizer. *Crosfield Chemicals Ltd.* Discontinued.

25470 Pyrasteel
A proprietary trade name for a heat-resisting alloy of iron with 25% nickel, 14% chromium, and 2.5-3.0% silicon. No manufacturer.

25471 Pyratex
2-Vinylpyridine-styrene-butadiene terpolymer; used to treat tire cord and other textiles to improve their adhesion to rubber and enhance the compatability of the textile-rubber system. *Bayer AG; Bayer plc.*

25472 Pyraton
123-42-2 3008 204-626-7
A proprietary trade name for diacetone alcohol. No manufacturer.

25473 Pyrax talcs A and B
Qualities of pure white talc mineral from deposits in America; used as a filler and extender in the rubber, textile, and ceramic industries.

25474 Pyrax® A
12269-78-2 377
Pyrophyllite
hydrated aluminum silicate. Inert filler/extender for NR and synthetic rubbers and latexes. *R. T. Vanderbilt Co Inc.*

25475 Pyre-ML®
Wire enamel and insulating varnish; used in automotive industry. *DuPont UK.*

25476 Pyrene
Range of cleaners, phosphating process and paint shop care products. *Brent Chemicals International plc.*

25477 pyrene oil
Begasses oil
An inferior olive oil.

25478 pyrethrum powder
83-79-4 8427 201-501-9
An insecticide made from the powdered flowers of some species of pyrethrum plants. The active principle is rotenone.

25479 Pyricit
13755-29-8 8761 237-340-6
$NaBF_4$
Sodium fluoroborate
Used as a fluorinating agent.

25480 Pyridate
35512-33-9
Lentagran. Wettable powder containing 45% w/w pyridate; for annual weed control for cereals, oilseed rape and maize. *Ciba-Geigy Agrochemicals.*

25481 pyridine
110-86-1 8153 203-809-9
C_5H_5N
azabenzene
Azine; Pyr. Organic compound; used as a solvent and in synthesis of agrochemicals, pharmaceuticals, photographic materials, coatings, curing agents, rubber chemicals, plastics, antidandruff shampoos, textiles, dyestuffs. mp = -42°; bp = 114-116°; d_4^{20}= 0.98272; n:20_D = 1.50920; soluble in H_2O, organic solvents; LD_{50} (rat orl) = 1.58 g/kg. *Nepera; Penta Mfg.; Schweizerhall; Whitecourt Ltd.*

25482 Pyrido rubber
A polymerized acrolein-methyl-amine. A rubber-like material.

25483 pyridoxine hydrochloride
58-56-0 8166 200-386-2
$C_8H_{11}NO_3 \cdot HCl$
5-hydroxy-6-methyl-3,4-pyridinedimethanol
Vitamin B_6; Adermin hydrochloride; Bonasanit; Hexabione hydrochloride; Hexabetalin; Hexavibex; Pyridipca; Pyridox; Benadon; Hexermin; Campoviton 6; Hexobion. Vitamin, nutritional factor. mp = 214° (dec); λ_m = 290 nm (ε 8400 0.1N HCl); soluble in H_2O (0.22 g/ml), less soluble in organic solvents.

25484 pyrimidine
289-95-2 8170 206-026-0
$C_4H_4N_2$
1,3-diazine
metadiazine; miazine; Pyr; Py. Used in biochemical research. mp = 20-22°; bp = 123-124°; d = 1.0160; n_D^{20}= 1.5035; soluble in H_2O, organic solvents.

25485 pyrimithate
5221-49-8 8172 226-020-1
$C_{11}H_{20}N_3O_3PS$
phosphorothioic acid O-[2-(dimethylamino)-6-methyl--4-pyrimidinyl] O,O-diethyl ester

ICI-29661; Diothyl; pyrimitate. Used as an acaricide and an insecticide. $bp_{0.04}$ = 128-132°; d= 1.165; insoluble in H_2O, soluble in organic solvents.

25486 Pyrinex
2921-88-2 2242 220-864-4
Active ingredient: chlorpyrifos; organophosphorous agricultural insecticide effective against a broad range of insects. *Makhteshim Chemical Works Ltd.*

25487 Pyro
64-17-5 3806 200-578-6
Denatured alcohol, chemical intermediate and solvent. *Quantum Chemical Corp.*

25488 pyro alcohol
67-56-1 6024 200-659-6
CH_3OH
Methanol
Chemical intermediate and solvent.

25489 pyro cotton
A nitrated cellulose, not so fully nitrated as guncotton.

25490 Pyroban G
Durable fire retardant for 100% polyester industrial products *CNC Int'l L.P.*

25491 pyro-bitumen
Bitumen which is insoluble in carbon tetrachloride.

25492 Pyroblak
Oxide blackening process for iron and steel. *Brent Chemicals International plc.*

25493 Pyrobond
Wet blast anticorrosion additive. *Brent Chemicals International plc.*

25494 Pyrobor
Anhydrous sodium dimetaborate
Kerr-McGee Chemical Corp. Discontinued.

25495 Pyrobrite
Copper plating processes. *Albright & Wilson Ltd., Phosphates & Speciality Business.* Name unverified.

25496 Pyrocast
A proprietary trade name for a nickel-chromium cast iron. No manufacturer.

25497 pyrocatechin
120-80-9 8183 204-427-5
$C_6H_4(OH)_2$
catechol
pyrocatechol. Used as an antiseptic, in photography, dyestuffs, electroplating; specialty inks, light stabilizers.

25498 pyrocatechol
120-80-9 8183 204-427-5
$C_6H_4(OH)_2$
1,2-benzenediol
pyrocatechin; 1,2-dihydroxybenzene; Phenol. In photography; dyeing furs; as reagent. *Aldrich; Coalite Chem. Div.; James River; Spectrum Chem. Mfg.*

25499 pyrocatechol arsenic acid
o-hydroxyphenyl arsenate
A reagent for alkaloids.

25500 Pyro-Chek® 68PB
Brominated polystyrene
Flame retardant for plastics (especially engineering plastics); suitable for thermosetting resins and polyolefins; synergist with antimony oxide. *Ferro.*

25501 Pyro-Chek® LM
Brominated polystyrene
Flame retardant for styrenics. *Ferro.*

25502 Pyrochlor
Fire-resistant and wood preservative. *Hickson & Welch Ltd.*

25503 Pyrocide
Pyrethrin based chemical intermediates. *McLaughlin Gormley King Co.*

25504 Pyroclean
Acid, alkaline, and emulsion cleaners. *Brent Chemicals International plc.*

25505 Pyroclense
Chassis and body cleaners for public transport and commercial haulage operators. *Brent Chemicals International plc.*

25506 pyrocollodion
A soluble nitrocellulose containing the highest practicable percentage of nitrogen, about 12.5%.

25507 Pyrodialite
An explosive containing 80-88% potassium chlorate, 5-6% charcoal, 10-18% gas tar, and 3-4% sodium and ammonium bicarbonates.

25508 Pyroforane
Halon fire extinguishing agents. *Elf Atochem SA.* Discontinued.

25509 pyrofulmin
A yellow substance obtained by heating mercuric fulminate. It is probably a mixture of mercuric oxycyanide and oxide.

25510 pyrogallic acid
87-66-1 8184 201-762-9
$C_6H_6O_3$
Pyrogallol
1,2,3-Trihydroxybenzene; 1,2,3-Benzenetriol; Pyrogallic acid; C.I. 76515; C.I. Oxidation Base 32; Fouramine Brown AP; Fourrine 85; Fourrine PG; Pyro; Piral; Fouramine base AP; Benzenetriol. Absorbs oxygen, used in gas analysis. Chemical intermediate. mp = 131-133°; bp = 309°; d = 1.4600; soluble in H_2O (440 g/l), organic solvents; LD_{50} (mus orl) = 300 mg/kg.

25511 pyrogallol
87-66-1 8184 201-762-9
$C_6H_6O_3$
1,2,3-Trihydroxybenzene
1,2,3-Benzenetriol; pyrogallic acid. Protective colloid in preparation of metallic colloidal solutions, used in photography, in preparation of dyes, intermediates, medicine, as a reducing agent and antoxidant. mp = 131-133°; bp_{760} = 309°; d = 1.45; soluble in H_2O (1 g/1.7 ml), soluble in organic solvents; L770 (rbt orl) = 1.6 g/kg *Burlington Biomedical; Fuji chemical Ind.; Hoechst Celanese; Mallinckrodt; Schering Berlin Polymers.*

25512 pyrogallol
87-66-1 8184 201-762-9
$C_6H_6O_3$
1,2,3-benzenetriol
1,2,3-trihydroxybenzene, pyrogallic acid; C.I. 76515; C.I. Oxidation Base 32; Fouramine Brown AP; Fourrine 85; Fourrine PG; Pyro; Piral; Fouramine Base AP. Used as a photographic developer, dyeing wool, staining leather, in metallurgy and analytical chemistry. mp = 131-133°; bp = 309°; d = 1.45; soluble in H_2O (0.59 g/ml), slightly more soluble in organic solvents; LD_{50} (rbt orl) = 1.76 g/kg.

25513 pyrogallol
87-66-1 8184 201-762-9
$C_6H_6O_3$
1,2,3-benzenetriol
pyrogallic acid; 1,2,3-trihydroxybenzene; Phenol. Protective colloid in preparation of metallic colloidal solution, photography, dyes, intermediate, synthetic drugs, medicine, lab reagent, reducing agent, antioxidant in lubricating oils. *Burlington Bio Medical; Fuji Chem. Ind.; Hoechest Celanese; Malincekrodt; Schering Berlin Polymers.*

25514 Pyrogallol
87-66-1 8184 201-762-9
1,2,3-Trihydroxybenzene
Chemical intermediate for electronics; photographic chemicals. *Schering Berlin Polymers.* Discontinued.

25515 pyrolignite of iron
3094-87-9 221-441-7
$C_4H_6FeO_4$
Ferrous acetate
iron liquor; black mordant; black liquor; liqueur de ferraile. Prepared by the action of pyroligneous acid upon iron turnings. The solution also contains ferric acetate; used in calico printing, and in dyeing, for the preparation of blue, violet, black, and brown colors. mp= 190°.

25516 pyrolignite of lime
62-54-4 1683 200-540-9
Calcium diacetate
Used as a chemical intermediate and in dyeing and tanning.

25517 Pyrolith
Fire retardant treatment for timber; for internal use only. *Hickson & Welch Ltd.*

25518 Pyromet® Alloy 625
Nonmagnetic nickel-base alloy; corrosion and oxidation-resistance alloy with high strength and toughness; for heat shields, furnace hardware, gas turbine engine ducting, combustion liners, chemical plant hardware, seawater applications. *Carpenter Tech.*

25519 pyromic
A particularly pure form of induction melted nickel-chromium used for electric furnaces, heaters, ovens, etc.

25520 pyromorphic phosphorus
Prepared by heating red phosphorus with a trace of iodine at 280 C, or *in vacuo*. Used as a chemical intermediate.

25521 Pyronate
Alkylaryl petroleum sulfonate, completely hydrophilic, water soluble; dark in color; for demulsification; froth flotation of nonmetallic ores. *Witco Chemical Ltd.* Discontinued.

25522 pyronine B
2150-48-3 8189 218-429-9
$C_{21}H_{27}ClN_2O$
3,6-bis(diethylamino)xanthylium chloride

C.I. 45010. With ferric chloride, forms a green complex which is used as a stain for bacteria, molds and RNA.

25523 pyronine Y
92-32-0 8190 202-147-8
$C_{17}H_{19}ClN_2O$
3,6-bis(dimethylamino)xanthylium chloride
C.I. 450095; pyronine G. With ferric chloride, gives a green complex which is used as a bacteriological and biological stain. Soluble in H_2O (9 g/100 ml), EtOH; λ_m = 576 nm.

25524 Pyronium
A proprietary substitute for tin oxide in enamels; used in conjunction with tin oxide. No manufacturer.

25525 Pyrophan
A combination of pyrogallol and dimethylamine.

25526 pyroretin
A brown resin found in lignite.

25527 Pyros
A paramagnetic alloy consisting of nickel with 7% chromium, 5% tungsten, 3% manganese, and 3% iron. It is suitable for expansion pyrometers.

25528 Pyroset
Organic phosphonium salt in water; flame retardants for textile industry. *Cyanamid BV.*

25529 Pyroter CPI-40
PEG-40 hydrogenated castor oil PCA isostearate; emulsifier, solubilizer and thickener used in personal care products and detergents; low irritation, nontoxic. *Ajinomoto; Nihon Emulsion.*

25530 Pyroter GPI-25
Glycereth-25 PCA isostearate
Moisturizer, emulsifier, solubilizer, dispersant and emollient. *Ajinomoto; Nihon Emulsion.*

25531 Pyrovatex
Proofing agent. *Ciba plc.* Name unverified.

25532 Pyrox®
Mineral iron sulfide (pyrite); 48-50% S, 45-46% Fe; filler for resin-bonded grinding wheels, brake linings, etc. *Metallgesellschaft AG.*

25533 pyrrhol
109-97-7 8198 203-724-7
C_4H_5N
Pyrrole
1H-Pyrrole; Pyrroline. Used in the manufacture of pharmaceuticals.

25534 pyrrodiazole
288-88-0 9735 206-022-9
$C_2H_3N_3$
1H-1,2,4-triazole
CGA-71019; 1,2,4-1H-Triazole. Chemical intermediate. mp= 119-121°; bp = 260°; soluble in H_2O (1250 g/l), organic solvents; LD_{50} (rat orl) = 1750 mg/kg.

25535 Pyrrole
109-97-7 8198 203-724-7
C_4H_5N
1H-Pyrrole
Azole; Imidole; Divinylenimine. Chemical intermediate. mp = -23°; bp = 129-131°; d = 0.9670; n_D^{20}= 1.5082; soluble in H_2O (60 g/l), organic solvents.

25536 pyrrolidone2-pyrrolidone
616-45-5 8200 210-483-1
$CH_2CH_2CH_2C(O)NH$
butyrolactam
pyrrolidone-2. Plasticizer and coalescing agent; solvent for veterinary medicine. *Allchem Industries; BASF; ISP; UCB SA.*

25537 pyruvic acid
127-17-3 8205 204-824-3
$C_3H_4O_3$
Chemical intermediate. mp = 12°; bp= 165°; d = 1.2650; n_D^{20}= 1.4315; soluble in H_2O, organic solvents.

25538 pyruvic acid
127-17-3 8205 204-824-3
$C_3H_4O_3$
2-oxopropionic acid
α-ketopropionic acid; acetylformic acid; pyroracemic acid. Biochemical research. *Penta Mfg.; Schweizerhall; U.S. Biochemical.*

25539 Python
An emulsifiable concentrate containing 105 g aminotriazole, 207 g atrazine and 100 g 2,4-D per liter; used for total weed control in non crop areas. *CDA Chemicals Ltd.*

25540 pyxol
An emulsion of coal-tar acids with soap; used as a disinfectant.

25541 Q-1300
N-Nitrosophenylhydroxylamine ammonium salt; polymerization inhibitor; chelating agent; antioxidant; germicides, fungicides, agricultural chems.; lustering agent for plating; corrosion inhibitor for metals; heat stabilizer for chlorosulfonated polyethylene; raw material for dye synthesis; analytical reagent. *Wako Pure Chem. Ind.; Wako Chem. USA.* Discontinued.

25542 Q-1301
15305-07-4 239-341-7
N-Nitrosophenylhydroxylamine aluminum salt; polymerization inhibitor; ideal for uv ink stabilizer. *Wako Pure Chem. Ind.; Wako Chem. USA.* Discontinued.

25543 QA-555
1337-76-4
Attapulgite clay; absorbent, adsorbent. *Floridin.* Name unverified.

25544 Qaulineg
Color film processing system. *May & Baker Ltd.* Name unverified.

25545 Qazul
Polyamide fibers for luggage. *ICI Chem & Polymers Ltd.* Discontinued.

25546 Q-Broxin
Drilling mud additive. *Georgia-Pacific.*

25547 QC-8800
Vinyl ester-based sheet molding compound; for compression molding of components requiring high structural strength and fatigue resist. *Quantum Composites.*

25548 Q-Cast
A castable refractory monolithic material for high temperature applications. *Pfizer International.* Discontinued.

25549 Q-Cel® 300
Inorganic silicate hollow microspheres; extender/filler for plastics, fiberglass-reinforced plastics, cultured marble, cast polyester furniture and decorative parts, bowling ball cores, cast urethane and epoxy systems, autobody repair fillers, marine putties, PVC plastisol compounds, slurry explosives. *PQ Corp.*

25550 Q-Crete
A monolithic refractory concrete for high temperature applications. *Pfizer International.* Discontinued.

25551 Q-Gum
A gunnable refractory monolithic material for high temperature applications. *Pfizer International.* Discontinued.

25552 QO® Furan
110-00-9 4316 203-327-3
Furan
A chemical intermediate in the manufacturing of herbicides, pharmaceuticals, plastics, and fine chemicals. *QO Chem.*

25553 QO® Furcarb®
Modified furan-phenolic resins; offers storage stability, low volatility; can be cured with acidic or basic catalysts; suitable for molding or extrusion; used in bonding carbon, graphite, basic refractory grains, silicon carbide, sand, and other aggregates to form shaped articles; applications include preformed carbon and graphite shapes, extruded electrodes, refractory shapes, tap hole mixes, grinding wheels, ramming mixes, crucibles and heat exchangers. *QO Chem.*

25554 QO® Furfural
98-01-1 4324 202-627-7
2-furaldehyde
furfural. Chemical intermediate for manufacture of derivatives (furan and THF); solvent for separating saturated from unsaturated compounds in petroleum lubricating oil, gas oil, and diesel fuel; extractive distillation of C4 and C5 hydrocarbons for the manufacture of synthetic rubber, decolorizing agent for wood rosin, solvent, processing aid for anthracene, ingredient in resins, reactive solvent and wetting agent in abrasive wheels and brake linings. *QO Chem.*

25555 QO® Furfuryl Alcohol (FA®)
98-00-0 4325 202-626-1
Furfuryl alcohol
Used in the production of foundry sand binders and corrosion-resistant resins; intermediate for esterification and etherification; impregnating solution and carbon binder; wood adhesive component; solvent and temporary plasticizer for phenolic resins in the nabufacture of cold-molded abrasive wheels; viscosity reducer, cure promoter and carrier in amine-cured epoxy resins. *QO Chem.*

25556 QO® Polymeg® 650
Polytetramethylene ether glycol polyol
Used in urethane elastomers, fibers, coatings, and adhesives, in the production of high-performance thermoset and thermoplastic elastomers, elastomeric polyesters, as polyester modifiers. *QO Chem.*

25557 QO® Quacorr® Resin/Catalyst Systems
Modified furfuryl alcohol-based (furan) resins and liquid catalyst; produces laminates exhibiting excellent chemical resistance; inherently fire resistant; used to fabricate fiberglass-reinforced plastic equipment with outstanding corrosion resistance, retention of strength at elevated temperatures, low

flame spread, smoke emission; applications include chemical process equipment, such as tanks, process vessels, ducting, stacks, scrubbers, hoods and pipe. *QO Chem.*

25558 QO® Tetrahydrofuran (THF)
109-99-9 9351 203-726-8
Tetrahydrofuran
Industrial solvent with high solvency for broad range of materials and low boiling point; also used in the production of heterocyclic and open-chain compounds; chemical intermediate. *QO Chem.*

25559 QO® Tetrahydrofurfuryl Alcohol (THFA®)
97-99-4 9353 202-625-6
Tetrahydrofurfuryl alcohol
High boiling solvent and carrier for pesticides; FDA approved for use in paper processing; chemical intermediate; also in industrial and consumer cleaners, leather and textile dyeing, epoxies, coatings, inks, paints, and adhesives; plasticizer and vinyl stabilizer carrier. *QO Chem.*

25560 Q-Therm
High temperature fluids with excellent thermal conductivity, temperature range and dielectric properties; used for the design of new functional fluids and lubricants with low toxicity values, e.g., heat transfer fluids for process equipment. *Anderson Development Company.* Name unverified.

25561 Quab 151
2,3-epoxypropyltrimethylammonium chloride. Used in manufacture of cationic biopolymers (e.g. starch, cellulose, guar, gelatine, protein) and synthetic polymers (e.g., polyacrylic acid, acrylamide-acrylic acid copolymers, polyaminoamides, polyethylene imide. *Degussa.*

25562 Quab 188
3327-22-8 222-048-3
$C_6H_{15}Cl_2NO$
(3-Chloro-2-hydroxypropyl)trimethylammonium chloride
For quaternization of compounds with hydroxyl, amino, and other functional groups, especially corresponding polymers, production of cationic polyelectrolytes. mp = 191-193°; *Degussa.*

25563 Quabond® 210
9003-20-7
High molecular weight polyvinyl acetate homopolymer aqueous emulsion; provides good adhesion to metal, glass, wood, cork, leather, textile, paper, and plastic surfaces; stiffener and hand builder on cotton and cotton/polyester fabrics. *Rhône-Poulenc Surf.*

25564 Quabond® 230
Acrylic copolymer aqeous emulsion; self-crosslinking emulsion for use on synthetics and cotton fabrics. *Rhône-Poulenc Surf.*

25565 Quad DSM
919-86-8 6129 213-052-6
Emulsifiable concentrate containing 500 g demeton-S-methyl per liter; a systemic organophosphorus insecticide. *Quadrangle Agrochemicals.*

25566 Quad MCPA 50%
94-74-6 5803 202-360-6
MCPA; herbicide for cereals and grassland. *Quadrangle Agrochemicals.*

25567 Quad Mini Slug Pellets
9002-91-9 5983 202-945-6
Pellets containing 6% w/w metaldehyde; snail and slug bait. *Quadrangle Agrochemicals.*

25568 Quad Store
117-18-0 204-178-2
Granules containing 5% w/w tecnazene; protectant fungicide and potato sprout suppressant. *Quadrangle Agrochemicals.*

25569 Quadban
Soluble concentrate of 19.5 g dicamba, 245 g MCPA and 86.5 g mecoprop per liter; used for weed control in cereals and grassland. *Quadrangle Agrochemicals.*

25570 Quadefome® MAB
Redesignated Foamex MAB. *Rhône-Poulenc Surf.*

25571 Quad-Fast
Surfactant containing dl-*p*-menthene; coating agent for contact herbicides, pesticides, growth regulators, and foliar feeds. *Quadrangle Agrochemicals.*

25572 Quad-Keep
117-18-0 204-178-2
Dustable powder containing 3% w/w tecnazene; protectant fungicide and potato sprout suppressant. *Quadrangle Agrochemicals.*

25573 Quadrafos
Polyphosphate
Water softener. *Marlowe-Van Loan.*

25574 Quadrafos
Glassy phosphates. *Peridot Chemicals Inc.*

25575 Quadrangle Chlormequat 700
999-81-5 2153 213-666-4
Soluble concentrate containing 700 g/l chlormequat; plant growth regulator. *Quadrangle Agrochemicals.*

25576 Quadrangle Cropspray 11E
Adjuvant containing 99% highly refined mineral oil; wetting agent for herbicide and fungicide sprays. *Quadrangle Agrochemicals.*

25577 Quadrangle Cyper
66841-24-5 266-492-6
Emulsifiable concentrate containing 100 g cypermethrin per liter; a pyrethroid insecticide. *Quadrangle Agrochemicals.*

25578 Quadrangle Super-Tin 4L
76-87-9 9875 200-990-6
Suspension concentrate containing 480 g fentin hydroxide per liter; used for control of potato blight. *Quadrangle Agrochemicals.*

25579 Quadrilan® AT
Quaternized fatty amine ethoxylate; a clear amber liquid; antistatic agent for PVC and other polymers. *Harcros.*

25580 Quadrilan® BC
8001-54-5 1086 231-635-3
Benzalkonium chloride BP grade; bactericide, fungicide, germicide in disinfectants and detergent sanitizers, emulsifier in the dyeing industry; biodegradable. *Harcros.*

25581 Quadrilan® MY 211
Specially developed cationic surfactant as a hazy amber liquid; used in alkaline spray cleaning concentrates for vehicle and chassis cleaning and crate washing. *Harcros.*

25582 Quadrol®
102-60-3 3635 203-041-4
Tetra (2-hydroxypropyl) ethylenediamine
Polyol; chelating agent; intermediate used in resins, emulsifiers, surfactants, pharmaceuticals, herbicides, fungicides, insecticides, adhesives, and plasticizers. *BASF; BASF plc.*

25583 Qualamox
26787-78-0 617 248-003-8
Amoxycillin injection *RMB Animal Health Ltd.*

25584 Qualidot
Machine lith developer system. *May & Baker Ltd.* Name unverified.

25585 Qualifix
Photographic fixer. *May & Baker Ltd.* Name unverified.

25586 Qualitol
Photographic developer. *May & Baker Ltd.* Name unverified.

25587 Qualloflex®, Quallofil®, Quallofirm®
Fibers. *DuPont UK.*

25588 Quamectant AM-50
6-(N-acetylamino)-4-oxahexyltrimonium chloride
Substantive humectant with emollient feel; nonoily and nongreasy; skin moisturizer; hair care products. *Brooks Industries.*

25589 Quamilin
52645-53-1 7321 258-067-9
Formulations for permethrin. *The Wellcome Foundation Ltd.* Name unverified.

25590 Quantacure®
A range of photoinitiators and coinitiators for photo-curing formulations. *Octel Chemicals Ltd.*

25591 Quantum
Protective coatings for concrete which provide mechanical strength, abrasion resistance, chemical corrosion resistance and weather resistance for floors, bridge and parking decks, garages, livestock pens, swimming pools, etc. *Quantum Chemical Corp.*

25592 Quantum
Refining catalysts. *Crosfield Chemicals Ltd.*

25593 Quantum
FCC catalysts. *Crosfield Chemicals Ltd.*

25594 Quantum (LOGO)
A full line of petrochemicals and oleochemical preparations for industrial use. *Quantum Chemical Corp.*

25595 Quantum®
For the agriculture industry. *DuPont UK.*

25596 quartz
14808-60-7 8637 238-878-4
SiO_2
Silicon dioxide; silica; crystallized silicon dioxide; there are three types. a) crystalline, such as tridymite and cristobalite. b) crypto-crystalline, such as chalcedony, and c) hydrated silica or opal; electronic components; TV components. *Unimin; U.S. Silica; Westo Industrial Prods. Ltd.*

25597 quartz glass
60676-86-0 262-373-8
Fused silica glass.

25598 quartzilite
A metallic carbide formed by action of silica and carbon at 2000-3000 C. It is suitable for electrical resistances.

25599 quarzal
An alloy of aluminum with 15% copper, 6% manganese, and 0.5% silicon; used for the cylinders of internal combustion engines.

25600 Quasar
FCC catalysts. *Crosfield Chemicals Ltd.*

25601 Quasilan
Pure MDI prepolymers; curing components for use with Propocon polyether systems. *Harcros.*

25602 Quat Keratin WKP
68915-25-3
Cocodimonium hydroxypropyl hydrolyzed animal keratin; cosmetics ingredient. *Brooks Industries.*

25603 Quat-Coll CDMA 40
Cocodimonium hydroxypropyl hydrolyzed animal protein; cosmetics ingredient. *Brooks Industries.*

25604 Quat-Coll IP10-30
Hydroxypropyltrimonium gelatin; cosmetics ingredient. *Brooks Industries.*

25605 Quat-Coll QS
11174-62-0
Steartrimonium hydrolyzed animal protein; cosmetics ingredient. *Brooks Industries.*

25606 quaternary steels
Steels containing two special elements in addition to the iron and carbon.

25607 Quaternium-27
86088-85-9 289-151-3
$R_2C_8H_{15}N_3O_5S$, Rrepresents tallow alkyl
Methyl-1-tallow amido ethyl-2-tallow imidazolinium methyl sulfate
ACCOSOFT® 808-90; Empigen® FRC90S; Incrosoft S-75; Incrosoft S-90; Incrosoft S-90M; Varisoft® TIMS; Ditallow imidazoline methyl sulfate; Tallow imidazolinium methosulfate. Softener base, lubricant, antistat and rewetting agent for fabrics and syns. Pale amber soft paste; mw=723; density=0.95 g/cc; set point=20°; pH=4.0-6.5 *Albright & Wilson; Stepan.*

25608 Quat-Pro E
111174-64-2
Triethonium hydrolyzed collagen ethosulfate
Substantive protein for hair and skin care products (shampoos, leave-on conditioners, mousses, creams, lotions, face tonics, liquid soaps). *Maybrook.*

25609 Quat-Pro S, S-30
11174-62-0
Steartrimonium hydrolyzed animal protein; substantive film-former, moisturizer, conditioner, protectant for hair and skin care products (shampoos, conditioners, creams, lotions, liquid hand soaps, bath products). *Maybrook.*

25610 Quatramine
A range of quaternary ammonium salts; used in antiseptics, disinfectants, bactericides and algicides. *Harcros Australia.*

25611 Quatrene 7670
Fatty amidoamine quaternary compound; wetting agent, demulsifier, and corrosion inhibitor for petrol. production. *Henkel/Cospha.*

25612 Quatrene C-5-6
Coco benzyl imidazolinium chloride
Corrosion inhibitor for continuous treatment in oil and gas production, pipelines. *Henkel/Cospha.*

25613 Quatrene CB-50
Benzalkonium chloride
Corrosion inhibitor for continuous treatment in oil and gas production, pipelines. *Henkel.* Name unverified.

25614 Quatrex
Epoxy resins used for the electronics industry. *Dow Cheml Co Ltd, UK & Ireland.*

25615 Quatrex 152
Quaternized imidazoline
Emulsifier, stabilizer for oilfield applications. *Chemron.*

25616 Quatrex 162
Surfactant
Wetting agent, additive for reverse emulsion breakers, oil treating; emulsion preventer in acids; clay stabilizer. *Chemron.*

25617 Quatrex 182
Quaternized imidazoline
Corrosion inhibitor concentrate; surfactant, solubilizer for other corrosion inhibitor components; oilfield applications; emulsion preventer; clay stabilizer. *Chemron.*

25618 Quatrex 1010
25928-94-3
Epoxy resin; electronic grade resin for laminating applications. *Dow Chemical.*

25619 Quatrex 2410
Pheno-epoxy novolac
Electronic grade characterized by extremely low hydrolyzable chlorine content. *Dow Chemical.*

25620 Quatrex 5010
One-component brominated epoxy resin; ignition-resist. grade with high glass transition temperature. (180°C). *Dow Chemical.*

25621 Quatrex 6410
Brominated epoxy resin; electronic grade designed for use where maximum bromine is needed to offset large ratios of hardener or nonbrominated resin. *Dow Chemical.*

25622 Quatrex CRC
Cetearyl alcohol, stearalkonium chloride, PEG-40 castor oil. concentrated surfactant for cream rinses; fully compatible with esters, fatty alcohols and proteins. *Chemron.*

25623 Quatrex CT-100
Stearyl alcohol, cetrimonium chloride
Cream rinse concentrate, conditioner for quality salon and professional hair care products. *Chemron.*

25624 Quatrex CTAC
112-02-7 203-928-6
Cetrimonium chloride
Surfactant, conditioner for hair treatment applications. *Chemron.*

25625 Quatrex S
Soyamidopropalkonium chloride
Substantivity and conditioning agent for conditioning shampoos, sprays, mousses, setting gels, conditioners. *Chemron.*

25626 Quatrex STC-25
122-19-0 204-527-9
Stearalkonium chloride
Personal care surfactant for after-shampoo cream rinses. *Chemron.*

25627 Quatrisoft Polymer LM-200
107987-23-5
Polyquaternium-24
Stabilizer for emulsions; thickener for surfactant systems; substantive conditioner for hair and skin care products. *Amerchol Corp.*

25628 Quat-Silk QTM-10
Hydroxypropyltrimonium hydrolyzed silk. Cosmetics ingredient. *Brooks Industries.*

25629 Quat-Soy CDMA-25
977039-11-4
Cocodimonium hydroxypropyl hydrolyzed soy protein; cosmetic ingredient for hair and skin care products. *Brooks Industries.*

25630 Quat-Soy LDMA-30
Lauryldimonium hydroxypropyl hydrolyzed soy protein; cosmetic ingredient for skin and hair care products. *Brooks Industries.*

25631 Quat-Veg Q-30
Hydroxypropyltrimonium hydrolyzed vegetable protein; cosmetic ingredient for skin and hair care products. *Brooks Industries.*

25632 Quat-Wheat CDMA-30
Cocodimonium hydroxypropyl hydrolyzed wheat protein; cosmetic ingred. for skin and hair care products. *Brooks Industries.*

25633 Quat-Wheat QTM-20
Hydroxypropyltrimonium hydrolyzed wheat protein; cosmetic ingred. for hair and skin care products. *Brooks Industries.*

25634 Quat-Wheat SDMA-30
Soya dimonium hydrolyzed wheat protein; substantivity agent for hair care products; enhances compability, softness, conditioning; especially as prewrap before chemical treatment or as post-perm treatment. *Brooks Industries.*

25635 quebrachite
$C_6H_{14}O_6$
quebrachitol
levo-inositol monomethyl ether. The monomethyl ether of levo-inositol, It is found in the latex of *Hevea brasiliensis* (the rubber tree) to the extent of 1-2%.

25636 quebracho
1401-55-4 9221 215-753-2
The wood of this tree (*Loxopteryngium lorenzii*) contains about 20% tannin, and is used in the form of an extract for tanning.

25637 Quecodur AE
Dimethylol ethylene urea
Reactant for wash and wear finishes. *Thor Chemicals (UK) Ltd.* Discontinued.

25638 Quecodur B Granular
Urea-formaldehyde condensation product; used for textile crease-proofing and resilient finishing. *Thor Chemicals (UK) Ltd.* Discontinued.

25639 Quecodur CW Conc
Modified dimethylol dihydroxy ethylene urea; clear colorless liquid; used for low formaldehyde and chlorine resistant textile wash-and-wear finishes. *Thor Chemicals (UK) Ltd.* Discontinued.

25640 Quecophob HPA
Fluorocarbon resin emulsion; cationic; oleophobic and water-repelling agent; may be used in combination with synthetic resin precondensates, reactants and nonionic finishing agents. *Thor Chemicals (UK) Ltd.* Discontinued.

25641 Queen's metal
A jeweler's alloy, very variable in composition. It contains from 50-85% tin, 7-16% antimony, 0-16% lead, 0-3.5% copper, and 1-12% zinc.

25642 Quelicin
71-27-2 9044 200-747-4
Succinylcholine chloride
Blocking agent. *Abbott Laboratories.* Name unverified.

25643 Quell Oil
Oil spill dispersants. *Harcros.*

25644 Quellada
A proprietary preparation of γ-benzene hexachloride; antiparasitic for dermatological use. *Stafford-Miller.* Unverified.

25645 Quenty®
Functional cosmetics containing collagen. *Bayer AG.*

25646 Quenty® forty
Functional cosmetics containing Elastin-Bayer and collagen. *Bayer AG.*

25647 quercetin
117-39-5 8216 204-187-1
$C_{15}H_{10}O_7$
2-(3,4-dihydroxyphenyl)-3,5,7-trihydroxy-4H-1-benzopyran-4-one
meletin; sophoretin; cyanidenolon 1522; 3,3',4',5,7-pentahydroxyflavone; 2-(3,4-dihydroxyphenyl)-4H-1-benzopyran-4-one; cyanidelonon 1522; C.I. natural yellow 10; C.I. natural yellow 10 & 13; C.I. natural red 1; C.I. 75670; meletin; quercetol; quertine; sophoretin; t-gelb bzw. grun 1; xanthaurine. The aglycone of quercitrin. Used in medicine as a capillary protectant; possible formation of epoxy resins. mp = 314° (dec); λ_m = 258, 375 nm (log ε 2.75, 2.75 EtOH); insoluble in H_2O, slightly soluble in EtOH, AcOH, insoluble in non-polar organic solvents; LD_{50} (mus orl) = 160 mg/kg. *EM Industries; Schweizerhall; U.S. Biochemical.*

25648 quercitron
$C_{15}H_{10}O_7$
A coloring matter, sold as chips, or as a coarse powder, obtained by grinding the bark of *Quercus tinctoria Q. nigra.* The dyeing principle is quercitrin or flavin, which forms yellow lakes with aluminum and tin salts; used for calico printing and wool dyeing.

25649 Querton 14Br-40
1119-97-7 6419 214-291-9
Myrtrimonium bromide
Germicidal disinfectant, detergent for hospital and industrial cleaning and disinfection. *Berol Nobel AB.*

25650 Querton 16Cl-29
112-02-7 203-928-6
Cetrimonium chloride
Emulsifier, dispersant, antistat for hair conditioners, shampoos, detergent sanitizers. *Berol Nobel AB.*

25651 Querton 210Cl-50
7173-51-5 3149 270-331-5
$C_{22}H_{48}ClN$
Didecyldimethylammonium chloride
Arquad 10; Bardac 22; N-decyl-N,N-dimethyldecanaminium chloride; didecyl dimethylammonium chloride; Q uaternium 12. Bactericide, fungicide for food processing industry, breweries, catering and hospitals; in detergent sanitizers. *Berol Nobel AB.*

25652 Querton 280
68391-03-7 269-922-0
Trimethyl alkyl ammonium chloride
General purpose quaternary surfactant. *Berol Nobel AB.*

25653 Querton 441-BC
61789-72-8 263-081-3
Hydrogenated tallow dimethylbenzyl ammonium chloride
Agent for organophilic bentonites production. *Berol Nobel AB.*

25654 Querton 442
61789-80-8 263-090-2
Dihydrogenated tallow dimethyl ammonium chloride
Imparts soft feel and antistatic properties to textile softeners for commercial laundry and textile manufacturing applications; also for hair conditioners, paper chemicals, manufacture of organoclays. *Berol Nobel AB.*

25655 Querton GCl-50
61789-71-7 274-022-6
1-Cocoalkylguanidinium chlorides
Bactericide, fungicide for detergent sanitizers. *Berol Nobel AB.*

25656 Querton KKBCl-50
61789-71-7 263-080-8
Cocoalkyldimethylbenzyl ammonium chloride
Corrosion inhibitor for oil drilling; emulsifier and dispersant for sludge in oil drilling and waste water treatment. *Berol Nobel AB.*

25657 Quesbrom
126-06-7 204-766-9
$C_5H_6BrClN_2O_2$
1-Bromo-3-chloro-5,5-dimethylhydantoin
BCDMH; Bromo-1-chloro-5,5-dimethyl-2,4-imidazolinedione; Bromo-1-chloro-5,5-dimethylhydantoin; Hydantoin, 3-bromo-1-chloro-5,5-dimethyl-; Imidazolidinedione, 3-bromo-1-chloro-5,5-dimethyl-. 95% active tablet; an extremely effective microbiocial bactericide, fungicide and algicide used in cooling towers, once-thru and closed loop water system. *Ques Industries.*

25658 Queschlor
87-90-1 9188 201-782-8
$C_3Cl_3N_3O_3$
1,3,5-trichloro-1,3,5-triazine-2,4,6(1H,3H,5H)-trione,
1,3,5-Trichloro-S-triazine-2,4,6-trione; Trichloro-s-triazinetrione; 1-Trichloro-s-triazinetrione; 1,3,5-trichloro-1,3,5-triazine-2,4,6(1H,3H,5H)-trione; 1,3,5-Trichloroisocyanuric acid; Symclosene; 1,3,5-Trichloro-1-triazine-2,4,6(1H,3H,5H)-trione; Isocyanuric chloride; Triazine-2,4,6(1H,3H,5H)-trione, 1,3,5-trichloro-. A chlorinated isocyanurate containing 90% available chlorine; a high performance chlorine microbiocide used to control algae, bacteria, and fungi growth in recirculating water and cooling towers. mp = 246-247° (dec); soluble in H_2O (2 g/100 ml), insoluble in non-polar organic solvents. *Ques Industries.*

25659 Quesfloc
A series of organic polymers suitable for waste water clarification; available in both dry and liquid compositions; used for industrial and municipal waste water (effluent) clarification. *Ques Industries.*

25660 Quesfloc F11283-1
A high molecular weight anionic polymer which performs as a coagulant aid and sludge conditioning agent in a variety of solid-liquid separation processes; in its dry form, it is a white, dustless free-flowing granular powder; used *Ques Industries.*

25661 Questal DI 0770
139-33-3 3556 205-358-3
$C_{10}H_{16}N_2O_8$
Disodium EDTA
Edetate disodium. Chelating agent for use in mildly acidic dry formulations. *Clough.*

25662 Questal Extra Powd. Conc. 0780
64-02-8 3557 200-573-9
Tetrasodium EDTA
Chelating agent for dry formulations. *Clough.*

25663 Questal FEC 0800
139-89-9 10102 205-381-9
Trisodium HEDTA
Chelating agent effective for iron to pH 12. *Clough.*

25664 Questal Special 0860
64-02-8 3557 200-573-9
Tetrasodium EDTA
Chelating agent over wide pH range. *Clough.*

25665 Questric Acid 5286
60-00-4 3559 200-449-4
EDTA; chelating agent for use in dry formulations where sodium salt is undesirable. *Clough.*

25666 Quevenne's iron
Reduced iron.

25667 Quiacryl
Acrylic water emulsions; for leather finishes, textile, coatings, adhesives. *Merquinsa.*

25668 Quiana®
Fibers. *DuPont UK.*

25669 Quick Cure®
For the electrical industry. *DuPont UK.*

25670 quickening liquid
A solution of mercuric nitrate or cyanide; used in electro-plating.

25671 Quick-pach
A plastic fire-clay for making monolithic linings and quick repairs. No manufacturer.

25672 Quickset® Extra
1338-23-4 215-661-2

MEK peroxide
High purity initiator for room temperature curing of polyester resins; increased reactivity with lower peroxide conc. *Witco Corporation.*

25673 Quickstir
Ceramic glaze and body stains offering easy dispersability; used for ceramic tile, sanitaryware, hobby ware, and porcelain enamel. *Engelhard.*

25674 Quidur
Urethane prepolymers; binder for cork-compositions, cross linkers for PU adhesives. *Merquinsa.*

25675 Quikote
Release agent. *Morton Int'l.*

25676 Quikset
Blended metallic complex accelerator for one-bath soil-release finishes on textiles. *CNC Int'l L.P.*

25677 Quilastic
Polyurethyane water emulsions, polyurethane solutions, granule form; raw materials for adhesives, leather finishes, synthetic leather, textile. *Merquinsa.*

25678 quinaldine
91-63-4 8228 202-085-1
$C_{10}H_9N$
2-methylquinoline
α-methylquinoline; chinaldine. Manufacture of dyes, pharmaceuticals, fine organic chemicals, acid-base indicators. bp = 246-247°, d = 1.06; insoluble in H_2O, soluble in organic solvents; LD_{50} (rat orl) = 1.23 g/kg. *Allchem Industries.*

25679 quinalphos
13593-03-8 237-031-6
$C_{12}H_{15}N_2O_3PS$
O,O-diethyl O-2-quinoxalinyl phosphorothioate
Bayrusil; Ekalux; Diethquinalphion; Diethyl O-(2-quinoxalyl) phosphorothioate; Diethyl O-(quinoxalin-2-yl) thiophosphate; Diethyl O-2-quinoxalinyl phosphorothioate; Diethyl O-quinoxalin-2-yl thionophosphate; SRA 7312; Wie oben; Savall. Insecticide and caricide with contact and stomach action. Used for control of insect pests of the orders Lepidotera, Coleoptera, Diptera, Hemiptera etc. mp = 31-32°; $bp_{0.0003}$ = 142°; SG_{20} = 1.235; n_D^{25} = 1.5624; soluble in H_2O (22 mg/l), more soluble in organic solvents; LD_{50} (rat orl) = 71 mg/kg.

25680 Quindex
Fungicides containing copper 8-quinolinolate; for preservation of textiles, cordage, paper, adhesives and caulking compounds. *Hüls Am.* Discontinued.

25681 Quindo®
High-quality quinacridone pigments for the surface coatings and plastics industries. *Bayer AG.*

25682 quindoxin
2423-66-7 219-352-3
$C_8CuH_6N_2O_2$
quinoxaline-1,4-dioxide copper(II) salt
A growth promoter.

25683 quinizarin
81-64-1 8250 201-368-7
$C_{14}H_8O_4$
1,4-dihydroxy-9,10-anthracenedione
1,4-dihydroxyanthraquinone; C.I. 58050. Antioxidant in synthetic lubricants, dyes. mp = 200-203°; insoluble in H_2O, soluble in organic solvents. *BASF; Sandoz.*

25684 Quinn's Rubber
A patented rubber substitute made from rapeseed oil, petroleum, and chloride of sulfur. No manufacturer.

25685 quinol
123-31-9 4853 204-617-8
$C_6H_6O_2$
Hydroquinone
p-dihydroxybenzene; *p*-hydroxyphenol; 1,4-benzenediol; hydroquinol; quinol; Aida; Eldoquin; Eldopaque; Tecquinol. Used as a photographic developer. mp = 170-171°; bp = 285-287°; d_{15} = 1.332; soluble in H_2O (70 mg/ml), more soluble in organic solvents;LD_{50} (rat orl) = 320 mg/kg.

25686 quinomethionate
2439-01-2 7113 219-455-3
$C_{10}H_6N_2OS_2$
6-methyl-1,3-dithiolo[4,5-b]quinoxalin-2-one
chinomethionat; oxythioquinox; quinoxalines; ENT 25606; Morestan; Bay 36205; Chinomethionate; Forstan; SS 2074. Selective non-systemic contact fungicide and acaricide. Used for control of powdery mildews and spider mites on fruits, ornamentals, curcubits, coffee and tea and various other crops. mp = 169-170°; soluble in H_2O (1 mg/l), more soluble in organic solvents; LD_{50} (rat orl) = 2500-3000 mg/kg.

25687 Quinosol
134-31-6 4890 205-137-1
$C_{18}H_{14}N_2O_2{\cdot}H_2SO_4{\cdot}H_2O$
8-quinolinol sulfate monohydrate
Sunoxol; Chinosol. The potassium salt of oxyquinoline sulfate; used as an antiseptic. mp = 175-178°; freely soluble in H_2O, poorly soluble in organic solvents.

25688 Quinovasugar
$C_6H_{12}O_4$
Quinovitol.

25689 Quinta-Pro Conc
Hydrolyzed collagen, triethonium hydrolyzed collagen ethosulfate, cationic collagen polypeptides, hydrolyzed keratin, collagen amino acids; protein blend for hair and skin care products (shampoos, conditioners, hair treatment products, creams, lotions, bath products, liquid hand soaps, shaving products. *Maybrook.*

25690 Quintesse
Polyamide fibers for upholstery. *ICI Chem & Polymers Ltd.*

25691 Quintiofos
O-Ethyl O-8-quinolyl phenylphosphonothioate
Used as an insecticide.

25692 quintozene
82-68-8 8264 201-435-0
$C_6Cl_5NO_2$
pentachloronitrobenzene
Avicol; Botrilex; PCNB; PKhNB; Brassicol; Earthcide; Folosan; Kobu; Kobutol; Pentagen; Saniclor; Terraclor; Terrazan; Tilcarex; Tritisan; Tubergran; Turfcide. Seed and soil-fungicide used for control of fungal diseases in fruits and vegetables. mp = 143-144°; bp = 328° (dec); insoluble in H_2O (0.44 mg/l), soluble in organic solvents; LD_{50} (rat orl) > 12000 mg/kg. *Rhône-Poulenc Environmental Prods. Ltd.*

25693 3-quinuclidinone hydrochloride
1193-65-3 214-776-5
$C_7H_{12}ClNO$
1-azabicyclo [2.2.2] octan-3-one hydrochloride
Janssen Chimica; Schweizerhall.

25694 quisqueite
An asphaltum-like compound containing much sulfur. It is found in Peru.

25695 Quiver
Suspension concentrate containing 200 g cyanazine and 350 g isoproturon per liter; a residual herbicide for winter cereals. *Shell UK.*

25696 Quixalud
8067-69-4 4637
Halquinol, an anti-infective, in a feed additive. *Ciba plc.* Name unverified.

25697 quizalofop-ethyl
76578-14-8 8269
$C_{19}H_{17}ClN_2O_4$
2-[4-[(6-chloro-2-quinoxalinyl)oxy]phenoxy]propanoic acid ethyl ester
quinofop-ethyl; DPX-Y6202; NCI-96683; NC-302; Assure; Targa; Pilot; EXP-3864; FBC-32197; INY-6202. Post-emergence herbicide used for control of grasses in broad-leaved crops. mp = 92-93°; $bp_{0.2}$ = 220°; insoluble in H_2O, soluble in organic solvents; LD_{50} (rat orl) = 1670 mg/kg.

25698 quorn
Mycoprotein. *Marlow Foods Ltd.*

25699 Quso® G27, G29, G35, G38, WR55, WR55-FG, WR83
Precipitated silica; thickener for pastes, creams, lotions in cosmetics and toiletries; suspending agent; improves free-flowing chars. of fine powds.; for pulp and paper defoaming. *Degussa.*

25700 Quso® WR55-FG
Precipitated silica; thickener for pastes, creams, lotions in cosmetics and toiletries; suspending agent; improves free-flowing chars. of fine powds.; for food grade defoamers. *Degussa.*

25701 Q-Vibe
A vibratable refractory monolithic material for high temperature applications. *Pfizer International.* Discontinued.

25702 R Type Solvent®
Adipate polyester plasticizer; suitable for softening hot melt adhesive of packaging grade. *Nordson (UK) Ltd.*

25703 R.A.E. 57 alloy
An alloy of aluminum with 4% copper, 2% iron, and 0.5% magnesium; its specific gravity is 2.8.

25704 R.O.D
2-octyldodecyl ricinoleate
Neutral oil improving adhesion, gloss, conditioning in lipsticks, foundation, eye shadow stick, hair creams. *U.S. Cosmetics.*

25705 R-1007
One-part silicone; RTV dispersion coating, ink. *McGhan NuSil.*

25706 R-2 Crystals
A proprietary trade name for an ultra accelerator for latex, etc.; it is the reaction product of carbon disulfide with methylenedipiperidine. No manufacturer.

25707 R-502, MR-502K, MR-502Y, MR-502P
Dimethyl polysiloxane combination; magnetic rubber inspection material for nondestructive inspection of ferromagnetic parts used for inspection of thread, gear roots, inaccessible areas, coated areas; provides permanent record of the inspection. *Dynamold Inc.*

25708 Rabalon
Thermoplastic elastomer. *Mitsubishi Petrochem.* Name unverified.

25709 Rachromate-51
10039-53-9 8745
Sodium chromate(VI) ^{51}Cr
Diagnostic aid; radioactive agent. *Abbott Laboratories.* Name unverified.

25710 R-Acid
148-75-4 6474 205-724-2
$C_{10}H_8O_7S_2$
2-Naphthol-3,6-disulfonic acid
3-hydroxy-2,7-naphthalenedisulfonic acid; R acid. Used as an azo dye intermediate; the disodium salt is used as a reagent in detecting nitrogen dioxide in the air. Soluble in H_2O, EtOH, insoluble in Et_2O.

25711 Rackarock
A blasting explosive, consisting of 79% potassium chlorate and 21% nitrobenzene, mixed sometimes with picric acid or sulfur.

25712 Rackarock Special
An explosive similar to Rackarock, but containing 12-16% picric acid.

25713 Racumin®
5836-29-3 227-424-0
$C_{19}H_{16}O_3$
2H-1-Benzopyran-2-one, 4-hydroxy-3-(1,2,3,4-tetrahydro-1-naphthalenyl)-
Coumetralyl; 2H-1-Benzopyran-2-one, 4-hydroxy-3-(1,2,3,4-tetrahydro-1-naphthalenyl)-; Coumarin, 4-hydroxy-3-(1,2,3,4-tetrahydro-1-naphthyl)-; BAY 25634; Bay ene 11183 B; Bayer 25 634; Coumatetralyl; Cumatetralyl; Endox; Endrocid; Endrocide; Ene 11183 B; 4-Hydroxy-3-(1,2,3,4-tetrahydro-1-naftyl)-cumarine; 4-Hydroxy-3-(1,2,3,4-tetrahydro-1-naphthalenyl)-2H-1-benzopyran-2-one; 4-Hydroxy-3-(1,2,3,4-tetrahydro-1-naphthyl)coumarin; 4-Hydroxy-3-(1,2,3,4-tetrahydro-1-naphthyl)cumarin; 3-(1,2,3,4-Tetrahydro-1-naphtyl)-4-hydroxycoumarine; 3-(α-Tetralinyl)-4-hydroxycoumarin; 3-(α-Tetral)-4-oxycoumarin; 3-(D-Tetralyl)-4-hydroxycoumarin; 3-(α-Tetralyl)-4-hydroxycoumarin; Rodentin; 3-(1,2,3,4-Tetrahydro-1-naphthyl)-4-hydroxycumarin,. A ready-to-use anticoagulant rodenticide. *Bayer AG; Bayer plc.*

25714 Radar, Radar Propiconazole
60207-90-1 8003 262-104-4
$C_{15}H_{17}Cl_2N_3O_2$
1-[[2-(2,4-dichlorophenyl)-4-propyl-1,3-dioxolan-2-yl]methyl]-1H-1,2,4-triazole
Propiconazole; CGA-64250; proconazole; Banner; Desmel; Orbit; Tilt. A systemic triazole fungicide for control of powdery mildew and rust in wheat and barley. $bp_{0.1}$ = 180°; soluble in H_2O (110 mg/l), more soluble in most organic solvents; LD_{50} (rat orl) = 1517 mg/kg. *Farm Protection Ltd; Ciba-Geigy Agrochemicals; ICI Chem. & Polymers Ltd.*

25715 Radel® A-100
Polyarylsulfone resin; excellent toughness, good chemical resistance, hydrolytic stability; for medical devices, chemical processing, electrical/electronic, food packaging, aviation applications; *Amoco Chemical Co.*

25716 Radex
Insulating powders for ladles and continuous casting tundishes. *Foseco (F.S.) Ltd.*

25717 Radia® 7040
84988-74-9 284-863-0
Butyl oleate
Chemical intermediate, lubricant; chemical synthesis; carbon source in antibiotic culture broths; lubricity improvers in mineral oils; formulation of cutting, lamination, and textile oils, rust inhibitors; textile and leather industry. *Fina Chemicals.*

25718 Radia® 7051
123-95-5 1625 204-666-5
$C_{22}H_{44}O_2$
Butyl stearate
octadecanoic acid butyl ester. Chemical intermediate, lubricant, plasticizer, chemical synthesis; lubricant in mineral, cutting, lamination, and textile oils, rust inhibitors; textile and leather application. mp = 27°; bp = 343°; d_{25}^{25} = 0.855-0.875 *Fina Chemicals.*

25719 Radia® 7060
67762-38-3 6956 267-015-4
$C_{18}H_{34}O_2$
(Z)-9-octadecenoic acid methyl ester
methyl oleate; oleic acid methyl ester. Chemical intermediate, lubricant; chemical synthesis; lubricity improvers in mineral oils; formulation of cutting, lamination, and textile oils; rust inhibitors; textile and leather industry. bp_2 = 168-170°; d_4^{18}= 0.879; n_D^{26}= 1.4510; soluble in EtOH, Et_2O. *Fina Chemicals.*

25720 Radia® 7108
85409-09-2 287-075-5
Glyceryl C_8-C_{10} triester
Chemical intermediate, chemical synthesis; lubricant in mineral, cutting, lamination, textile oils, and rust inhibitors; textile and leather application; *Fina Chemicals.*

25721 Radia® 7110
85586-21-6 8959 287-824-6
$C_{19}H_{38}O_2$
Methyl stearate
Chemical intermediate, chemical synthesis; lubricant in mineral, cutting, lamination, textile oils, and rust inhibitors; textile and leather application. mp = 38-39°; bp_{15} = 215°; insoluble in H_2O, soluble in organic solvents. *Fina Chemicals.*

25722 Radia® 7117
61788-59-8 262-988-1
Methyl cocoate
Emollient, plasticizer, lubricant for cosmetics, pharmaceuticals, plastics, lubricating oils, textile and leather additives, cutting oils for metallurgy; chemical intermediate. *Fina Chemicals.*

25723 Radia® 7120
112-39-0 203-966-3
$C_{17}H_{34}O_2$
Methyl palmitate
Chemical intermediate, chemical synthesis; lubricant in mineral, cutting, lamination, textile oils, and rust inhibitors; textile and leather application. mp = 28°; d = 0.8520. *Fina Chemicals.*

25724 Radia® 7131
91031-48-0 292-951-5
Isooctyl stearate
Cosmetics emollient, solvent; plasticizer for PVC; lubricant for PS. *Fina Chemicals.*

25725 Radia® 7171
68604-44-4 271-694-2
Pentaerythritol tetraoleate
Lubricant, chemical intermediate; formulation of cutting, lamination, and textile oils; corrosion inhibitors; chemical synthesis. *Fina Chemicals.*

25726 Radia® 7176, Radiasurf® 7175
115-83-3 204-110-1
$C_{77}H_{148}O_8$
Pentaerythritol tetrastearate
Lubricant, chemical intermediate; formulation of cutting, lamination, and textile oils; corrosion inhibitors; chemical synthesis. mp = 71°; d = 0.9400 *Fina Chemicals.*

25727 Radia® 7185
111-61-5 203-887-4
$C_{20}H_{40}O_2$
Ethyl stearate
ethyl octadecanoate. Chemical intermediate, chemical synthesis; lubricant in mineral, cutting, lamination, textile oils, and rust inhibitors; textile and leather application; also as emollient, plasticizer, solubilizer of active components in cosmetics and pharmaceuticals. mp = 33-35°; bp_{15} = 213-215°; insoluble in H_2O, soluble in organic solvents. *Fina Chemicals.*

25728 Radia® 7187
85049-36-1 6965 285-206-0
$C_{20}H_{38}O_2$
Z)-9-octadecenoic acid ethyl ester
ethyl oleate. Chemical intermediate, chemical synthesis; lubricant in mineral, cutting, lamination, textile oils, and rust inhibitors; textile and leather application; also as emollient, plasticizer, solubilizer of active components in cosmetics and pharmaceuticals. bp = 205-208°; d = 0.87; insoluble in H_2O, soluble in organic solvents. *Fina Chemicals.*

25729 Radia® 7190
110-27-0 5234 203-751-4
$C_{17}H_{34}O_2$
Isopropyl myristate
tetradecanoic acid 1-methylethyl ester; Estergel. Chemical intermediate, chemical synthesis; lubricant in mineral, cutting, lamination, textile oils, and rust inhibitors; textile and leather application; also as emollient, plasticizer, solubilizer of active components in cosmetics and pharmaceuticals. mp = 3°; bp_9 = 167°; d = 0.8530; n_D^{25}= 1.433 *Fina Chemicals.*

25730 Radia® 7200
142-91-6 205-571-1
$C_{19}H_{38}O_2$

Isopropyl palmitate
hexadecanoic acid isopropyl ester; palmitic acid isopropyl ester. Chemical intermediate, chemical synthesis; lubricant in mineral, cutting, lamination, textile oils, and rust inhibitors; textile and leather application; mp = 11-13°; d = 0.852; n_D^{20}= 1.4380 *Fina Chemicals.*

25731 Radia® 7204
85049-34-9 285-203-4
Propylene glycol dioleate
Chemical intermediate, chemical synthesis; lubricant in mineral, cutting, lamination, textile oils, and rust inhibitors; textile and leather application; also as emollient, plasticizer, solubilizer of active components in cosmetics and pharmaceuticals. *Fina Chemicals.*

25732 Radia® 7230
84988-79-4 284-868-8
Isobutyl oleate
Fatty acids, C_{16-18} and C_{18}-unsatyrated, iso-butyl esters. Chemical intermediate, lubricant; chemical synthesis; carbon source in antibiotic culture broths; lubricity improvers in mineral oils; formulation of cutting, lamination, and textile oils, rust inhibitors; textile and leather industry. *Fina Chemicals.*

25733 Radia® 7231
85116-87-6 285-540-7
Isopropyl oleate
Chemical intermediate, lubricant; chemical synthesis; lubricity improvers in mineral oils; formulation of cutting, lamination, and textile oils; rust inhibitors; textile and leather industry. *Fina Chemicals.*

25734 Radia® 7241
85865-69-6 288-668-1
Isobutyl stearate
Cosmetics emollient, solvent; plasticizer for PVC; lubricant for PS. *Fina Chemicals.*

25735 Radia® 7266
91031-31-1 292-932-1
Ethylene glycol distearate
Chemical intermediate, chemical synthesis; lubricant in mineral, cutting, lamination, textile oils, and rust inhibitors; textile and leather application; also as emollient, plasticizer, solubilizer of active components in cosmetics and pharmaceuticals. *Fina Chemicals.*

25736 Radia® 7331
85049-37-2 285-207-6
Ethylhexyl oleate
Chemical intermediate, lubricant; chemical synthesis; lubricity improvers in mineral oils; formulation of cutting, lamination, and textile oils; rust inhibitors; textile and leather industry. *Fina Chemicals.*

25737 Radia® 7345
26658-19-5 8872 247-891-4
Sorbitan tristearate
Chemical intermediate, chemical synthesis; lubricant in mineral, cutting, lamination, textile oils, and rust inhibitors; textile and leather application; also as emollient, plasticizer, solubilizer of active components in cosmetics and pharmaceuticals. *Fina Chemicals.*

25738 Radia® 7355
26266-58-0 8872 247-569-3
Sorbitan trioleate
Chemical intermediate, chemical synthesis; lubricant in mineral, cutting, lamination, textile oils, and rust inhibitors; textile and leather application; also as emollient, plasticizer, solubilizer of active components in cosmetics and pharmaceuticals. d_4^{20}= 0.952. *Fina Chemicals.*

25739 Radia® 7363
122-32-7 9861 204-534-7
$C_{57}H_{104}O_6$
9-octadecenoic acid 1,2,3-propanetriyl ester
glyceryl trioleate; olein; triolein. Lubricant, chemical intermediate; formulation of cutting, lamination, and textile oils; corrosion inhibitors; chemical synthesis; as carbon source in antibiotic culture broths. mp = -4°; bp_{15} = 235-240°; d_4^{15} = 0.915; n_D^{20} = 1.4561; insoluble in H_2O, soluble in organic solvents. *Fina Chemicals.*

25740 Radia® 7370
68002-79-9 268-093-2
Trimethylpropane trioleate
Fatty acids, C_{14-18} and C_{16-18} unsaturated, triesters with trimethylolpropane; trimethylolpropane tallowate; (C_{14}-C_{18}) and (C_{16}-C_{18}) Unsatured trialkylcarboxylic acid, trimethylolpropane ester. Lubricant, chemical intermediate; formulation of cutting, lamination, and textile oils; corrosion inhibitors; chemical synthesis. *Fina Chemicals.*

25741 Radia® 7371
68002-79-9 268-093-2
Trimethylol propane triester, unsaturated
Fatty acids, C_{14-18} and C_{16-18} unsaturated, triesters with trimethylolpropane; trimethylolpropane tallowate; (C_{14}-C_{18}) and (C_{16}-C_{18}) Unsatured trialkylcarboxylic acid, trimethylolpropane ester. Chemical intermediate, chemical synthesis; lubricant in mineral, cutting, lamination, textile oils, and rust inhibitors; textile and leather application; also as emollient, plasticizer, solubilizer of active components in cosmetics and pharmaceuticals. *Fina Chemicals.*

25742 Radia® 7500
540-10-3 2073 208-736-6
$C_{32}H_{64}O_2$
Cetyl palmitate
hexadecanoic acid hexadecyl ester; palmitic acid hexadecyl ester; hexadecyl palmitate. Chemical intermediate, chemical synthesis; lubricant in mineral, cutting, lamination, textile oils, and rust inhibitors; textile and leather application; also used for wax formulation due to its high melting point mp = 54°; d^{20} = 0.989; n_D^{20}= 1.4398; insoluble in H_2O, soluble in organic solvents. *Fina Chemicals.*

25743 Radia® 7501
2778-96-3 220-476-5
$C_{36}H_{72}O_2$
Stearyl stearate
Chemical intermediate, chemical synthesis; lubricant in mineral, cutting, lamination, textile oils, and rust inhibitors; textile and leather application; also used for wax formulation due to its high melting point *Fina Chemicals.*

25744 Radia® 7505
90193-76-3 290-580-3
Distearyl phthalate
Chemical intermediate, chemical synthesis; lubricant in mineral, cutting, lamination, textile oils, and rust inhibitors; textile and leather application; also as emollient, plasticizer, solubilizer of active components in cosmetics and pharmaceuticals. *Fina Chemicals.*

25745 Radia® 7506
68955-45-3 203-755-6
Ethylene bis-stearamide
Chemical intermediate, chemical synthesis; lubricant in mineral, cutting, lamination, textile oils, and rust inhibitors; textile and leather application; also as emollient, plasticizer, solubilizer of active components in cosmetics and pharmaceuticals. *Fina Chemicals.*

25746 Radia® 7510
91031-57-1 292-960-4
Isononyl stearate
Chemical intermediate, chemical synthesis; lubricant in mineral, cutting, lamination, textile oils, and rust inhibitors; textile and leather application *Fina Chemicals.*

25747 Radia® 7514
84539-90-2 283-078-0
Pentaerythritol tetrabehenate
Lubricant, chemical intermediate; formulation of cutting, lamination, and textile oils; corrosion inhibitors; chemical synthesis. *Fina Chemicals.*

25748 Radiamac 6149
61790-59-8 263-149-2
Hydrogenated tallow amine acetate
Flotation reagent, anticaking aid for fertilizer, corrosion inhibitor, emulsifier. *Fina Chemicals.*

25749 Radiamac 6169
61790-57-6 263-147-1
Coconut oil amine acetate; flotation reagent, anticaking aid for fertilizer, corrosion inhibitor, emulsifier. *Fina Chemicals.*

25750 Radiamine 6140
61788-45-2 262-976-6
Hydrogenated tallow amine; mineral flotation, corrosion inhibitor, pigment dispersant; cosmetics; lubricant and mold release for hard rubber, textile chemical, chemical synthesis; antistat and antifog additive for plastic foils. *Fina Chemicals.*

25751 Radiamine 6141
61788-45-2 262-976-6
Hydrogenated tallow amine, distilled; mineral flotation, corrosion inhibitor, pigment dispersant; cosmetics; lubricant and mold release for hard rubber, textile chemical, chemical synthesis; antistat and antifog additive for plastic foils. *Fina Chemicals.*

25752 Radiamine 6160
61788-46-3 262-977-1
Coconut oil amine; mineral flotation, corrosion inhibitor, pigment dispersant; cosmetics; lubricant and mold release for hard rubber, textile chemical, chemical synthesis; antistat and antifog additive for plastic foils. *Fina Chemicals.*

25753 Radiamine 6161
61788-46-3 262-977-1
Coconut oil amine, distilled; mineral flotation, corrosion inhibitor, pigment dispersant; cosmetics; lubricant and mold release for hard rubber, textile

chemical, chemical synthesis; antistat and antifog additive for plastic foils. *Fina Chemicals.*

25754 Radiamine 6164
124-22-1 204-690-6
$C_{12}H_{27}N$
n-dodecylamine
1-aminododecane. Lauramine, distilled; mineral flotation, corrosion inhibitor, pigment dispersant; cosmetics; lubricant and mold release for hard rubber, textile chemical, chemical synthesis; antistat and antifog additive for plastic foils. mp = 28-30°; bp = 247-249°. *Fina Chemicals.*

25755 Radiamine 6170
61790-33-8 263-125-1
Tallow amine
Mineral flotation, corrosion inhibitor, pigment dispersant; cosmetics; lubricant and mold release for hard rubber, textile chemical, chemical synthesis; antistat and antifog additive for plastic foils. *Fina Chemicals.*

25756 Radiamine 6171
61790-33-8 263-125-1
Tallow amine, distilled; mineral flotation, corrosion inhibitor, pigment dispersant; cosmetics; lubricant and mold release for hard rubber, textile chemical, chemical synthesis; antistat and antifog additive for plastic foils. *Fina Chemicals.*

25757 Radiamine 6172
112-90-3 204-015-5
$C_{18}H_{37}N$
cis-1-amino-9-octadecene
Oleylamine. Mineral flotation, corrosion inhibitor, pigment dispersant; cosmetics; lubricant and mold release for hard rubber, textile chemical, chemical synthesis; antistat and antifog additive for plastic foils. mp = 348-350°; bp = 18-26°; d_4^{20}= 0.83. *Fina Chemicals.*

25758 Radiamine 6240
16789-79-5
Dihydrogenated tallow amine
Intermediate for chemicals used in production of fabric softeners, household products, disinfectants. *Fina Chemicals.*

25759 Radiamine 6260
61789-76-2 263-086-0
Dicocamine
Intermediate for wide range of chemicals used as fabric softeners, household products, disinfectants. *Fina Chemicals.*

25760 Radiamine 6270
68783-24-4 272-191-0
Ditallowamine
Intermediate for wide range of chems. used as fabric softeners, household products, disinfectants. *Fina Chemicals.*

25761 Radiamine 6310
1120-49-6 214-312-1
$C_{20}H_{43}N$
Di-n-decylamine
Intermediate for wide range of chems. used as fabric softeners, household products, disinfectants. mp = 42-45° *Fina Chemicals.*

25762 Radiamine 6343
61788-63-4 262-991-8
Dihydrogenated tallow methylamine
Intermediate for wide range of chemicals used as fabric softeners, household products, disinfectants. *Fina Chemicals.*

25763 Radiamine 6346
4088-22-6 223-819-7
Distearyl methylamine
Intermediate for wide range of chemicals used as fabric softeners, household products, disinfectants. *Fina Chemicals.*

25764 Radiamine 6360
61788-62-3 262-990-2
Dicocomethylamine
Intermediate for wide range of chemicals used as fabric softeners, household products, disinfectants. *Fina Chemicals.*

25765 Radiamine 6540
68603-64-5 271-696-6
Hydrogenated tallow propanediamine
Corrosion inhibitor, dispersant, emulsifier, intermediate for chemical synthesis. *Fina Chemicals.*

25766 Radiamine 6560
61791-63-7 263-195-3
Coconut oil propane diamine
Corrosion inhibitor, dispersant, emulsifier, intermediate for chemical synthesis. *Fina Chemicals.*

25767 Radiamine 6570
68439-73-6 270-416-7
Tallow propanediamine
Corrosion inhibitor, dispersant, emulsifier, intermediate for chemical synthesis. *Fina Chemicals.*

25768 Radiamine 6572
68037-97-8 268-221-7
Oleyl propanediamine
Corrosion inhibitor, dispersant, emulsifier, intermediate for chemical synthesis. *Fina Chemicals.*

25769 Radiamuls® 2602
Glyceryl soyate
Food emulsifier, plasticizer, homogenizer, foam inhibitor, rehydrator, anticaking agent. *Fina Chemicals.*

25770 Radiamuls® Acetem 2021, 2134
Acetylated mono-diglycerides
Alpha-tending emulsifier for food industry. *Fina Chemicals.*

25771 Radiamuls® Citrem 2931, 2932
Citrylated mono-diglycerides
Anti-spattering aids for cooking margarines; emulsifier for meat industry; 2932 is self-emulsifying. *Fina Chemicals.*

25772 Radiamuls® CSL 2980
5793-94-2 11, 1711 227-335-7
Calcium stearoyl lactylate
Food emulsifier; dough structure improver for baked goods. *Fina Chemicals.*

25773 Radiamuls® Datem 2001, 2008
Diacetyl tartaric monoglycerides
Food emulsifier; 2008 is also a dough structure improver for bread and other baked goods. *Fina Chemicals.*

25774 Radiamuls® GTH 2375, GTH 2376
Glyceryl triheptanoate
Butter tracers in food industry. *Fina Chemicals.*

25775 Radiamuls® Lactem 2950
lactylated mono-diglycerides
Alpha-tending emulsifier for food industry. *Fina Chemicals.*

25776 Radiamuls® MCT 2108
538-23-8 208-686-5
$C_{27}H_{50}O_6$
Glyceryl tricaprylate-caprate
trioctanoin. Edible oil; forms very thin films for coating confectionery and dried fruits; mold release aid for bakery, confectionery; lubricant for food processing equipment; viscosity depressant, carrier of actives in oleoresins; lipid for dietetic foods. mp = 9-10°; bp_1 = 23°; d_4^{20}= 0.954; n_D^{20}= 1.447. *Fina Chemicals.*

25777 Radiamuls® MG 2141, MG 2142, MG 2600, MG 2900
31566-31-1 4498 250-705-4
Glyceryl stearate SE
monostearin. Food emulsifier, plasticizer, foam inhibitor, homogenizer, rehydrator, anticaking agent, lubricant for extruded foods and for food manufacturing equipment *Fina Chemicals.*

25778 Radiamuls® MG 2152
Glyceryl oleate
Food emulsifier, plasticizer, homogenizer, foam inhibitor, rehydrator, anticaking agent. *Fina Chemicals.*

25779 Radiamuls® PG 2201
1323-39-3 215-354-3
Propylene glycol stearate; food emulsifier. *Fina Chemicals.*

25780 Radiamuls® PG 2206
1330-80-9 215-549-3
Propylene glycol oleate
Food emulsifier. *Fina Chemicals.*

25781 Radiamuls® Poly 2248
Polyglyceryl monostearate
Aerator and stabilizer for food products. *Fina Chemicals.*

25782 Radiamuls® Poly 2253
Polyglycerol polyricinoleate
Water-oil emulsifier for production of mold release aids for bakery and other foodstuffs, lubricants for food processing equipment; improves chocolate fluidity. *Fina Chemicals.*

25783 Radiamuls® Sorb 2145, Sorb 2161, Sorb 2166
1338-41-6 8872 215-664-9
$C_{24}H_{46}O_6$
Sorbitan stearate
sorbitan monostearate; Alkamuls SMS; Arlacel 60; Glycomul S; Span 60. Food emulsifier, emulsion stabilizer; solubilizer and dispersant for flavors and other ingreds. Sorb 2161 is self-emulsifying. mp = 49-65°; insoluble in H_2O, Me_2CO, soluble in other organic solvents. *Fina Chemicals.*

25784 Radiamuls® Sorb 2147
9005-67-8 8872
PEG-20 sorbitan stearate

Food emulsifier, emulsion stabilizer; solubilizer and dispersant for flavors and other ingreds. *Fina Chemicals.*

25785 Radiamuls® Sorb 2157
9005-65-6 7742
PEG-20 sorbitan oleate
Food emulsifier, emulsion stabilizer; solubilizer and dispersant for flavors and other ingreds.; emulsifier for calf milk replacers. *Fina Chemicals.*

25786 Radiamuls® Sorb 2344, Sorb 2345
26658-19-5 8872 247-891-4
Sorbitan tristearate
Food emulsifiers. Emulsion stabilizers; solubilizers and dispersants for flavors and other ingredients. *Fina Chemicals.*

25787 Radiamuls® Sorb 2345
26658-19-5 8872 247-891-4
$C_{60}H_{116}O_8$
sorbitan tristearate
Non-ionic surfactant. *Fina Chemicals.*

25788 Radiamuls® SSL 2990
25383-99-7 246-929-7
Sodium stearoyl lactylate
Food emulsifier; dough structure improver for bread and other baked goods. *Fina Chemicals.*

25789 Radiaquat 6410, 6412
7173-51-5 3149 230-525-2
$C_{22}H_{48}ClN$
Didecyldimethylammonium chloride
N-decyl-N,N-dimethyl-1-decanaminium chloride; dimethyldidecylammonium chloride; Bardac 2250/2280; BTC-1010; Dodigen 1881; Querton 210CL. Bactericidal quaternary for hard surface cleaners, mildew preventers for commercial and industrial laundries. Soluble in organic solvents. *Fina Chemicals.*

25790 Radiaquat 6412
7173-51-5 3149 230-525-2
C22H48ClN
didecyldimethylammonium chloride
Arquad 10; Bardac 22; Bardac 2250/2280; Decanaminium, N-decyl-N,N-dimethyl-, chloride; didecyl dimethyl ammonium chloride; Quaternium 12; BTC 1010; Dodigen 1881; Querton 210CL. Fungicide and bactericide, general purpose disinfectant, sanitizer and mildew preventative. Used in commercial laundries. Soluble in Me_2CO, C_6H_6, insoluble in hexane.

25791 Radiaquat 6442
61789-80-8 263-090-2
Dihydrogenated tallow dimethyl ammonium chloride; surfactant, softener for laundry applications. *Fina Chemicals.*

25792 Radiaquat 6444
112-02-7 203-928-6
$C_{19}H_{42}ClN$
Palmityl trimethyl ammonium chloride
Softener, antistat, detergent for textile, leather, detergent, and cosmetic industries; clay modifier. *Fina Chemicals.*

25793 Radiaquat 6462
61789-77-3 263-087-6
Dicocodimethylammonium chloride
Detergent, antistat, softener, bactericide for detergents, textiles, fabric softeners, cosmetics. *Fina Chemicals.*

25794 Radiaquat 6470
68153-32-2 268-877-4
Ditallowdimethylammonium chloride
Surfactant, softener for laundry applications. *Fina Chemicals.*

25795 Radiaquat 6471
68002-61-9 268-074-9
Tallowtrimethylammonium chloride
Detergent, antistat, softener for textile, leather, detergent and cosmetic products. *Fina Chemicals.*

25796 Radiaquat 6475, 6480
61789-80-8 263-090-2
Dihydrogenated tallow dimethyl ammonium chloride
Surfactant, softener for laundry products. *Fina Chemicals.*

25797 Radiasurf® 7125
68154-36-9 8872 268-910-2
$C_{18}H_{34}O_6$
Sorbitan laurate
sorbitan monolaurate; Alkamuls SML; Arlacel 20; Emsorb 2515; Glycomul L; Span 20. Emulsifier, descouring aid, antistat; anticorrosive agent for pipelines; cleaner for metallic surfaces; superfatting, bodying and antifog aid; pigment dispersant; detergent; emulsion of solvents; Insoluble in H_2O, soluble in organic solvents. *Fina Chemicals.*

25798 Radiasurf® 7135
26266-57-9 8872 247-568-8
Sorbitan palmitate
Emulsifier, descouring aid, antistat; anticorrosive agent for pipelines; cleaner for metallic surfaces; superfatting, bodying and antifog aid; pigment dispersant; detergent; emulsion of solvent; cutting oils; textile lubricant additive. *Fina Chemicals.*

25799 Radiasurf® 7137
9005-64-5 8872
Polysorbate 20
Emulsifier, descouring aid, antistat; anticorrosive agent for pipelines; cleaner for metallic surfaces; superfatting, bodying and antifog aid; pigment dispersant; detergent; emulsion of solvent; cutting oils; textile lubricant additive. *Fina Chemicals.*

25800 Radiasurf® 7147
9005-67-8 8872
Polysorbate 60
Emulsifier, descouring aid, antistat; anticorrosive agent for pipelines; cleaner for metallic surfaces; superfatting, bodying and antifog aid; pigment dispersant; detergent; emulsion of solvent; cutting oils; textile lubricant additive, *Fina Chemicals.*

25801 Radiasurf® 7150
111-03-5 203-827-7
$C_{21}H_{40}O_4$
Glyceryl oleate
α-monoolein; Aldo HMO; Aldo MO; glycerin-1-monooleate; glycerol-α-*cis*-9-octadecenate; glycerol-α-monooleate; glycerol-1-monooleate; Monomuls 90O18; Monoolein; glycerol-1-oleate; glyceryl monooleate; 1-glyceryl oleate; 1-monoolein; 1-monooleoylglycerol; 1-oleoylglycerol; 1-oleylglycerol. Internal lubricant for PVC; biodeg. surfactant, wetting agent, emulsifier for cosmetics, pharmaceuticals, agriculture, chemical synthesis, explosives, polymers, glass fibers, surface coatings, textiles and leather. *Fina Chemicals.*

25802 Radiasurf® 7155
1338-43-8 8872 215-665-4
$C_{24}H_{44}O_6$
Sorbitan oleate
sorbitan monooleate; Alkamuls SMO; Arlacel 80; Capmul O; Emsorb 2500; Glycomul O; Span 80. Emulsifier, descouring aid, antistat; anticorrosive agent for pipelines; cleaner for metallic surfaces; superfatting, bodying and antifog aid; pigment dispersant; detergent; emulsion of solvent; cutting oils; textile lubricant additive. Insoluble in H_2O, soluble in organic solvents *Fina Chemicals.*

25803 Radiasurf® 7156
12772-47-3 288-305-7
Pentaerythritol oleate
9-octadecenoic acid ster with 2,2-bis(hydroxymethyl)-1,3-propanediol; Pentol; Pentol(emulsifier). Corrosion inhibitor for lubricating oils and greases; biodeg. surfactant, wetting agent, emulsifier for cosmetics, pharmaceuticals, agriculture, chemical synthesis, explosives, polymers, glass fibers, surface coatings, textiles and leather. *Fina Chemicals.*

25804 Radiasurf® 7157
9005-65-6 7742
Polysorbate 80
polyoxyethylene (20) sorbitan monooleate; POE (20) sorbitanmonooleate; Emsorb 6900; Liposorb O-20; Monitan; Sorlate; T-Maz 80; Tween 80. Emulsifier, descouring aid, antistat; anticorrosive agent for pipelines; cleaner for metallic surfaces; superfatting, bodying and antifog aid; pigment dispersant; detergent; emulsion of solvent; cutting oils; textile lubricant additive. d = 1.06-1.09; soluble in H_2O, polar organic solvents; LD_{50} (rat ip) = 6.3 mg/kg. *Fina Chemicals.*

25805 Radiasurf® 7175
85116-93-4 285-547-5
pentaerythritol stearate
Chemical intermediate, emulsifier, detergent, wetting agent, lubricant, used in manufacture of chemicals and pharmaceuticals, pearlescent shampoos, detergency and cleaning products, cutting, lamination and textile.

25806 Radiasurf® 7270
97281-23-7 306-522-8
Ethylene glycol stearate
Wetting aid, lubricant, opacifier, antistat, dispersant, w/o emulgent, scouring and detergent aid, defoamer, plasticizer, rust inhibitor; cosmetics and pharmaceuticals, lubricating and cutting oils, textile and leather aids, pigment grinding *Fina Chemicals.*

25807 Radiasurf® 7400
93455-78-8 297-364-8
Diethylene glycol oleate
Emulsifier, wetting agent, defoamer, rust inhibitor, pigment grinder, antistat. *Fina Chemicals.*

25808 Radiasurf® 7402
85736-49-8 288-459-5
PEG-4 oleate
Wetting aid, lubricant, opacifier, antistat, dispersant, o/w emulcent, scouring and detergent aid, defoamer, plasticizer, rust inhibitor, visc. modifier, antifog aid; cosmetics and pharmaceuticals; lubricating and cutting oils; textile and leather aids; *Fina Chemicals.*

25809 Radiasurf® 7403
85736-49-8 288-459-5
PEG-8 oleate
Wetting aid, lubricant, opacifier, antistat, dispersant, o/w emulgent, scouring and detergent aid, defoamer, plasticizer, rust inhibitor, visc. modifier, antifog aid; cosmetics and pharmaceuticals; lubricating and cutting oils. *Fina Chemicals.*

25810 Radiasurf® 7404
85736-49-8 288-459-5
PEG-12 oleate
Wetting aid, lubricant, opacifier, antistat, dispersant, o/w emulgent, scouring and detergent aid, defoamer, plasticizer, rust inhibitor, visc. modifier, antifog aid; cosmetics and pharmaceuticals; lubricating and cutting oils. *Fina Chemicals.*

25811 Radiasurf® 7410
85116-97-8 285-550-1
Diethylene glycol stearate
Wetter, lubricant, opacifier, antistat, dispersant, detergent, defoamer, plasticizer, rust inhibitor; for cosmetics, pharmaceuticals, textiles, leather, paints, inks, plastic, waxes, maintenance products, insecticides. *Fina Chemicals.*

25812 Radiasurf® 7414
97281-23-7 306-522-8
PEG-12 stearate
Wetting aid, lubricant, opacifier, antistat, dispersant, o/w emulgent, scouring and detergent aid, defoamer, plasticizer, rust inhibitor, visc. modifier, antifog aid; cosmetics and pharmaceuticals; lubricating and cutting oils. *Fina Chemicals.*

25813 Radiasurf® 7417
97281-23-7 306-522-8
PEG-1500 stearate; wetting aid, lubricant, opacifier, antistat, dispersant, o/w emulgent, scouring and detergent aid, defoamer, plasticizer, rust inhibitor, visc. modifier, antifog aid; cosmetics and pharmaceuticals; lubricating and cutting oils. *Fina Chemicals.*

25814 Radiasurf® 7423
37318-14-2 253-458-0
PEG-8 laurate; wetting aid, lubricant, opacifier, antistat, dispersant, o/w emulgent, scouring and detergent aid, defoamer, plasticizer, rust inhibitor, visc. modifier, antifog aid; cosmetics and pharmaceuticals; lubricating and cutting oils. *Fina Chemicals.*

25815 Radiasurf® 7443
85736-49-8 288-459-5
PEG-8 dioleate
Wetting aid, lubricant, opacifier, antistat, dispersant, o/w emulgent, scouring and detergent aid, defoamer, plasticizer, rust inhibitor, visc. modifier, antifog aid; cosmetics and pharmaceuticals; lubricating and cutting oils. *Fina Chemicals.*

25816 Radiasurf® 7444
85736-49-8 288-459-5
PEG-12 dioleate; biodeg. surfactant, wetting agent, emulsifier for cosmetics, pharmaceuticals, agriculture, chemical synthesis, explosives, polymers, glass fibers, surface coatings, textiles and leather. *Fina Chemicals.*

25817 Radiasurf® 7453
97281-23-7 306-522-8
PEG-8 distearate; wetting aid, lubricant, opacifier, antistat, dispersant, o/w emulcent, scouring and detergent aid, defoamer, plasticizer, rust inhibitor, visc. modifier, antifog aid; cosmetics and pharmaceuticals; lubricating and cutting oils. *Fina Chemicals.*

25818 Radiasurf® 7454
97281-23-7 306-522-8
PEG-12 distearate; wetting aid, lubricant, opacifier, antistat, dispersant, o/w emulgent, scouring and detergent aid, defoamer, plasticizer, rust inhibitor, visc. modifier, antifog aid; cosmetics and pharmaceuticals; lubricating and cutting oils. *Fina Chemicals.*

25819 Radiasurf® 7473
97281-23-7 306-522-8
PEG-8 stearate; biodeg. surfactant, wetting agent, emulsifier for cosmetics, pharmaceuticals, agriculture, chemical synthesis, explosives, polymers, glass fibers, surface coatings, textiles and leather. *Fina Chemicals.*

25820 Radiasurf® 7600
85251-77-0 286-490-9
Glyceryl stearate; wetter, defoamer, rust inhibitor, pigment grinder, emulsifier, antistat. *Fina Chemicals.*

25821 Radiasurf® 7900
91052-47-0 293-208-8
Glyceryl stearate
Biodegradable surfactant, wetting agent, emulsifier for cosmetics, pharmaceuticals, agriculture, chemical synthesis, explosives, polymers, glass fibers, surface coatings, textiles and leather. *Fina Chemicals.*

25822 Radiflam A AE
Nylon; self-extinguishing; molding compound. *Radilon.* Name unverified.

25823 Radilon A CP300
Nylon, reinforced and filled; molding compound. *Radilon.* Name unverified.

25824 Radilon A, A 32E
Nylon; molding and extrusion compound. *Radilon.* Name unverified.

25825 Radilon S BHS200/201
Nylon; for molding. *Radilon.* Name unverified.

25826 Radilon S, S 35FL/FLC
Nylon; for extrusion. *Radilon.* Name unverified.

25827 Radiocaps-131
7790-26-3 8778
NaI
Sodium iodide-^{131}I
sidum radioiodide; Iodotope; Oriodide; Radiocaps-131; Theriodide-131. Antineoplastic; diagnostic aid; radioactive agent. ^{131}I has a half-life of 8 days. *Abbott Laboratories.* Name unverified.

25828 Radio-malt
A proprietary preparation consisting of malt extract with irradiated ergosterol (radiostol); contains vitamins A, B, and D. No manufacturer.

25829 Radiometal
A nickel-iron-copper alloy having high incremental permeability and low losses. It is largely used for radio transformers, relays, etc.

25830 Radiopaque
7727-43-7 1023 231-784-4
BaO_4S
barium sulfate
Blanc fixe; Actybaryte; Bakontal; Baridol; Baritop; Barosperse; Citobaryum; Esophotrast; E-Z-Paque; Intestibar; Microbar; Micropaque; Microtrast; Mixobar; Neobar; Oratrast; Polybar; Prontobario; Radiopaque; Telebar; Unibaryt. A special barium sulfate. Used in radiology. mp > 1600° (dec); d = 4.2-4.5.

25831 Radiose
A French nitrocellulose lacquer.

25832 Radmolite
An electrical insulating material for supporting heating coils; largely consists of diatomaceous earth; has a slower rate of heat absorption; replaces fireclay. No manufacturer.

25833 Radspor FT, 65WP
2439-10-3 3468 219-459-5
$C_{15}H_{33}N_3O_2$
dodecylguanidine monoacetate
Dodine; AC-5223; Carpene; Cyprex; Melprex. A fungicide for the control of scab in apples and pears. mp = 136°; soluble in hot H_2O, EtOH, less soluble in other solvents; LD_{50} (rat orl) = 566 mg/kg. *Truchem Ltd.*

25834 raffinate
Material from which a soluble substance has been extracted, e.g. petroleum oil from which higher hydrocarbons have been removed.

25835 raffinose
512-69-6 8279 208-146-9
$C_{18}H_{32}O_{16}\cdot 5H_2O$
β-D-fructofuranosyl-O-α-D-galactopyranosyl-(1→6)-α-D-glucopyranoside
gossypose; melitose; mellitriose. A sugar; used in bacteriology, in the preparation of other saccharides. The pentahydrate [17629-30-0] has EINECS Number 208-146-9. mp = 80°; d = 1.465; $[\alpha]_D^{20}$ = 105° (c = 4); soluble in H_2O (0.14 g/ml), less soluble in organic solvents.

25836 Rainbow Custom Colored Mortars
1309-37-1 4072 215-168-2
ferric oxide
Blends of iron oxide colors and mortars; preblended colored mortar to produce more uniform color in the mortar joints of a building. *DCS Color & Supply Co Inc.* Name unverified.

25837 Rainbow Ware
A proprietary synthetic resin of the ureaformaldehyde type. No manufacturer.

25838 RAK®
Pheromones
BASF AG.

25839 Rakel's alloy
An aluminum, bronze. It contains 87.5% copper, 10.5% aluminum, 1% manganese, and 1% lead or zinc.

25840 Rakusol®
Organic and inorganic pigments in paraffin oil and glycerol ester; used for coloring plastics; less suitable for PVC. *BASF AG.*

25841 Ralox® 02
4,4'-Methylene-bis-(2,6-di-t-butylphenol)
Non-staining antioxidant. *Raschig.*

25842 Ralox® 46
119-47-1 204-327-1
2,2'-Methylene-bis-(6-t-butyl-4-methylphenol)
Nonstaining antioxidant for plastics, rubber, latex, adhesives, hot melts, cables. *Raschig.*

25843 Ralox® 530
2082-79-3 218-216-0
Octadecyl-3-(3,5-di-*t*-butyl-4-hydroxyphenyl) propionate
Nonstaining antioxidant for plastics, hot melts, cables, mineral oil processing (lubricants). *Raschig.*

25844 Ralox® 630
6683-19-8 229-722-6
Tetrakis [methylene (3,5-di-*t*-butyl-4-hydroxyhydrocinnamate)] methane
Nonstaining antioxidant for plastics, hot melts, cables. *Raschig.*

25845 Ralox® BHT food grade
128-37-0 1583 204-881-4
$C_{15}H_{24}O$
2,6-bis(1,1-dimethylethyl)-4-methylphenol
2,6-di-*tert*-butyl-*p*-cresol; 2,6-di-*tert*-butyl-4-methylphenol; BHT; antrancine 8; Tenox BHT; Ionol CP; Sustane; Dalpac; Impruvol; Vianol. BHT; nonstaining antioxidant for plastics in contact with food, rubber, latex, adhesives, hot melts, mineral oil processing (lubricants), petroleum, feedstuffs. mp = 70°; bp = 265°; d_{4}^{20}= 1.048; insoluble in H_2O, soluble in organic solvents; LD_{50} (mus orl) = 1040 mg.kg. *Raschig.*

25846 Ralox® TMQ-R
2,2,4-Trimethyl-1,2-dihydroquinoline polymer; staining antioxidant for rubber processing. *Raschig.*

25847 Ralufon® 414
Cocamidopropyl betaine
Surfactant. *Raschig.*

25848 Ralufon® DL
N,N-Dimethyl-N-lauryl-N-(3-sulfopropyl)-ammonium betaine
Surfactant. *Raschig.*

25849 Ralufon® DT
N,N-Dimethyl-N-tallow-N-(3-sulfopropyl)ammonium betaine
Surfactant. *Raschig.*

25850 Ralufon® N
Nonylphenol polyethylene oxide sulfopropyl ether, potassium salt; surfactant. *Raschig.*

25851 Ralufon® TA
N,N-Dimethyl-N-stearic acid-amidopropyl-N-(3-sulfopropyl)-ammonium betaine
Surfactant. *Raschig.*

25852 Raluquin®
6-Ethoxy-2,2,4-trimethyl-1,2-dihydroquinoline
Staining antioxidant. *Raschig.*

25853 Ramasit® KGT
Water repelling agent for textiles. *BASF; BASF plc.*

25854 Ramenti ferri
Iron filings.

25855 Ramet
A proprietary cutting material for steel alloys, cast iron, etc.; consists of tantalum carbide with nickel; melts at 4100°. No manufacturer.

25856 Rametin
1491-41-4 6441 216-078-6
$C_{16}H_{16}NO_6P$
2-[(Diethoxyphosphinyl)oxy]-1H-benz[de]isoquinoline-1,3(2H)-dione
N-hydroxynaphthalimide diethyl phosphate; Naftalofos; O,O-diethyl O-naphthaloximide phosphate; Bay 9002; Bayer 25820; ENT-25567; S-940; Maretin. A proprietary preparation of naphthalophos; a veterinary anthelmintic. No manufacturer.

25857 ramie
Chinese grass; rhea; rhea ramine; green ramie. A fiber obtained from *Boehmeria tenacissima.*

25858 Ramix
A proprietary magnesite refractory.

25859 Ramos
A proprietary phenol-formaldehyde resin.

25860 Rampart
1563-66-2 1851 216-353-0
$C_{12}H_{15}NO_3$
2,3-dihydro-2,2-dimethyl-7-benzofuranol methylcarbamate
7-Benzofuranol, 2,3-dihydro-2,2-dimethyl-, methylcarbamate; Carbamic acid, methyl-, 2,3-dihydro-2,2-dimethyl-7-benzofuranyl ester; BAY 70143; BAY 78537; C2292-59a; Carbamic acid, methyl-, 2,2-dimethyl-2,3-dihydrobenzofuran-7-yl ester; Carbamic acid, methyl-, 2,2-dimethyl-2,3-dihydro-7-benzofuranyl ester; Carbofuran; Carbofuran; Carbofuran; Carbofurane; Chinufur; Crisfuran; Curaterr; D 1221; 2,3-Dihydro-2,2-dimethylbenzofuran-7-yl methylcarbamate; 2,3-Dihydro-2,2-dimethyl-7-benzofuranyl methylcarbamate; 2,3-Dihydro-2,2-dimethylbenzofuranyl-7-N-methylcarbamate; 2,2-Dimethyl-7-coumaranyl N-methylcarbamate; 2,2-Dimethyl-2,2-dihydrobenzofuranyl-7 N-methylcarbamate; ENT 27,164; FMC 10242; Furacarb; Furadan; Furadan 3g; Furadan 3G; Furadan 4f; Furadane; Furadan; Furadan 75 wp; Furodan; Karbofuranu; Me f248; Methyl carbamic acid 2,3-dihydro-2,2-dimethyl-7-benzofuranyl ester; NIA 10242; Niagara 10242;OMS 864;Yaltox,. Cholinesterase inhibitor. A systemic carbamate insecticide and nematode in the form of granules for soil treatment. mp = 150-153°; soluble in H_2O (700 mg/l); LD_{50} (rat orl) = 2 mg/kg. *Sipcam UK Ltd; Universal Crop Protection Ltd.*

25861 Ramrod
1918-16-7 7977 217-638-2
$C_{11}H_{14}ClNO$
2-chloro-N-(1-methylethyl)-N-phenylacetamide
Acetamide, 2-chloro-N-(1-methylethyl)-N-phenyl-; Acetanilide, 2-chloro-N-isopropyl-; Bexton; Bexton 4L; Chloressigsaeure-N-isopropylanilid; 2-Chloro-N-isopropylacetanilide; α-Chloro-N-isopropylacetanilide; 2-Chloro-N-isopropyl-N-phenylacetamide; 2-Chloro-N-(1-methylethyl)-N-phenylacetamide; CIPA; CP 31393; N-Isopropyl-2-chloroacetanilide; N-Isopropyl-α-chloroacetanilide; Nitacid; Niticid; Propachlor; Propachlore; Ramrod; Ramrod 65; Satecid. A pre-emergence herbicide for various horticultural crops. mp = 67-76°; soluble in H_2O (700 mg/l), organic solvents; LD_{50} (rat orl) = 710 mg/kg. *Monsanto plc.*

25862 Ramtap
Fibrous tap hole plugs for electric arc and basic oxygen steel-making furnaces. *Foseco (F.S.) Ltd.*

25863 Rancho®
Herbicide for control of grass weeds and some broad-leaved weeds in irrigated rice crops. *Bayer AG.*

25864 Randanite
Ceyssatite, Memilite
Varieties of hydrated silica or opal.

25865 Raneoff® S
Silicone resin emulsion; durable water repellent. *Eastern Color & Chem.*

25866 Ranestol
5714-82-9 9787
Triclofenol piperazine
Anthelmintic. *Parke-Davis.* Name unverified.

25867 Raney nickel
7440-02-0 6582 231-111-4
Ni
Raney Nickel
A form of finely divided nickel containing traces of aluminum, used in hydrogenating certain organic compounds; used as a catalyst for hydrogenation.

25868 Ranide
22662-39-1 8280 245-148-9
$C_{19}H_{11}Cl_2I_2NO_3$
N-[3-chloro-4-(4-chlorophenoxy)phenyl]-2-hydroxy-3,5-diiodobenzamide
Rafoxanide; MK-990; Bovanide; Duofas; Flukanide. Fasciolicide, anthelmintic. mp = 168-170°; insoluble in H_2O, soluble in Me_2CO, CH_3CN. *Merck & Co Inc.* Name unverified.

25869 Ranotex
Textile softening agents. *Rhône-Poulenc UK.*

25870 Ransome's stone
An artificial stone made by mixing sand with sodium silicate and a little chalk, or other similar material. The product is molded to shape, and immersed in a solution of calcium chloride.

25871 Raolein 131
Triolein-^{131}I
Radioactive agent. *Abbott Laboratories.* Name unverified.

25872 RAP
Polypropylene film, for use by industry in overwrap packaging of consumer items. *Quantum Chemical Corp.*

25873 Rapadex
Photographic developer. *May & Baker Ltd.* Name unverified.

25874 rapeseed oil
8002-13-9 8289 232-299-0
Brassica campestris oil; oil of rapeseed; Colza oil. Vegetable oil expressed from seeds of *Brassica campestris* ; metal lubricant additive; edible oil for salad dressings, margarine; soft soaps. mp = -2 - -10°; d = 0.915; n_D^{20} = 1.4752, soluble in organic solvents. *Arista Industries; Climax Performance; Penta Mfg.; Reilly-Whiteman; Werner G. Smith; Witco/Humko.*

25875 Rapiblend
Predispersed rubber chemicals. *Anchor Chemical Group plc.*

25876 rapic acid
A name which has been applied to the fatty acids of rape oil; identical to oleic acid.

25877 Rapi-Cure BHC
Hydroxybutyl vinyl ether carbonate
ISP.

25878 Rapi-Cure CHMVE
Cyclohexane dimethanol, monovinyl ether
ISP.

25879 Rapi-Cure CHVE
1,4-Cyclohexane dimethanol divinyl ether
ISP.

25880 Rapi-Cure CVE
Cyclohexyl vinyl ether
ISP.

25881 Rapi-Cure DVE-3
Triethylene glycol divinyl ether
Reactive diluent *ISP.*

25882 Rapi-Cure EHVE
103-44-6 203-111-4
2-Ethylhexyl vinyl ether
ISP.

25883 Rapi-Cure HBVE
Hydroxybutyl vinyl ether.
ISP.

25884 Rapid
23103-98-2 7651 245-430-1
$C_{11}H_{18}N_4O_2$
dimethylcarbamic acid 2-(dimethylamino)-5,6-dimethyl-4-pyrimidinyl ester
Pirimicarb; 5,6-dimethyl-2-dimethylamino-4-dimethylcarbamoyloxypyrimidine; PP-062; EN-27766; Aphox; Fernos; Pirimor. Garden insecticide. mp = 91°; soluble in H_2O (2.7 g/l), organic solvents; LD_{50} (rat orl) = 147 mg/kg. *ICI Chem & Polymers Ltd.*

25885 Rapid Purge 2
Nonabrasive purging compound effective on all thermoplastics; for cleaning injection molding, hot runner tools, extrusion, multi-layer dies, blow molding equipment. *Rapid Purge.*

25886 Ra-Pid-Gro
Plant food. *Monsanto (Solaris).* Name unverified.

25887 Rapidogen
Dyestuffs; used for dyeing cotton by the warp sizing process. *Bayer AG.*

25888 Rapidosept®
A hand disinfectant; gentle to skin. *Bayer AG.*

25889 Rapier
23950-58-5 8058 245-951-4
$C_{12}H_{11}Cl_2NO$
3,5-dichloro-N-(1,1-dimethyl-2-propynyl)benzamide
Propyzamide; pronamid; RH-315; Kerb. A residual herbicide for oil seed rape. mp = 155-156°; soluble in H_2O (15 mg/l), organic solvents; LD_{50} (rat orl) = 8350 mg/kg. *Farmers Crop Chemicals Ltd.* Discontinued.

25890 Rappor
13516-27-2
Guazatine
A fungicide seed dressing for wheat. *DowElanco Ltd.*

25891 Rappor Plus
Guazatine [13516-27-2] combined with imazalil [73790-28-0]. A fungicide seed dressing for barley and oats. *DowElanco Ltd.*

25892 Raschit
59-50-7 2184 200-431-6
$C_9H_9ClO_3$
p-Chloro-*m*-cresol
4-chloro-*m*-cresol; 4-chloro-3-methylphenol; 3-methyl-4-chlorophenol; *para*-chlorometacresol; 6-chloro-*m*-cresol; 6-chloro-3-hydroxytoluene; 2-chloro-5-hydroxytoluene. Used as a preservative for latex. mp = 62-65°; bp = 234-235°; soluble in H_2O (3.85 mg/ml), organic solvents.

25893 Rassamix CDA
Combination of amitrole + atrazine + diuron. A mixture of herbicides for weed control. *Denoon CDS.*

25894 Rassapron
Combination of amitrole + atrazine + diuron. A mixture of herbicides for weed control. *BP Oil Ltd.*

25895 Rat Flip
Trade name for a line of rat and mouse bait products for both indoor and outdoor use; anticoagulant active ingredients are safer around domestic animals and pets; vitamin K is an antidote. *Colonial Products Inc.*

25896 Ratak
56073-07-5 259-978-4
$C_{31}H_{24}O_3$
Difenacoum
3-[3-(1,1'-Biphenyl)-4-yl-1,2,3,4-tetrahydro-1-naphthalenyl]-4-hy droxy-2H-1-benzopyran-2-one; 3-(3-Biphenyl-4-yl-1,2,3,4-tetrahydro-1-naphthyl)-4-hydroxycoumarin; 3-(3-(4-Biphenylyl)-1,2,3,4-tetrahydro-1-naphthyl)-4-hydroxycoumarin; Coumarin, 3-(3-(4-biphenylyl)-1,2,3,4-tetrahydro-1-naphthyl)-4-hydroxy-; Difenacoum; Difenakum; Diphenacoum; Neosorex; Neosorexa; Neosorexa pp580; PP 580; Rastop; Ratak; WBA 8107. A ready-to-use anticoagulant rodenticide. *ICI Garden Products.*

25897 Ratio
82558-50-7 5256
$C_{18}H_{24}N_2O_4$
N-[3-(1-ethyl-1-methylpropyl)-5-isoxazolyl]-2,6-dimethoxybenzamide
benzamizole; EL-107; NA-8318; Flexidor; Gallery. Suspension concentrate containing 125 g isoxaben per liter; used for control of annual dicotyledons in cereals, grass and fruit. mp = 176-179°; poorly soluble in H_2O (1 mg/l), more soluble in organic solvents; LD_{50} (rat orl) > 10000 mg/kg. *Tripart Farm Chemicals Ltd.*

25898 Ratox
81-81-2 10174 201-377-6
Warfarin-based rodenticide. *The Wellcome Foundation Ltd.* Name unverified.

25899 Rauxite
Proprietary trade name for urea-formaldehyde varnish and lacquer resins. No manufacturer.

25900 Rauxone
A proprietary trade name for alkyd varnish and lacquer resins. No manufacturer.

25901 Rauzene
A proprietary trade name for a phenolic varnish and lacquer resin. No manufacturer.

25902 Rauzene Ester
A proprietary trade name for an ester gum. No manufacturer.

25903 Raven
7440-44-0 1855 231-153-3
Carbon black. *Sevalco Ltd.*

25904 Ravinil
PVC suspension and mass polymers. *European Vinyls Corporation Ltd.*

25905 Ravolen
Plasticizer and rubber extender. *Burmah-Castrol Ltd.* Name unverified.

25906 Ravolen 11(T)
A proprietary trade name for a decolorized petroleum aromatic extractive *Burmah-Castrol Ltd.* Name unverified.

25907 Rawstol
Radarsan. Vermicides consisting of solutions of fluosilicic acid.

25908 Raxil
107534-96-3 9253 403-640-2
$C_{16}H_{22}ClN_3O$
(±)-α-[2-(4-chlorophenyl)ethyl]-α-(1,1-dimethylethyl)-1H-1,2,4-triazole-1-ethanol
Tebuconazole; (RS)-1-(4-chlorophenyl)-4,4-dimethyl-3-(1H-1,2,4-triazol-1-ylmethyl)pentan-3-ol; ethyltrianol; fenetrazole; terbuconazole; terbutrazole; BAY HWG 1608; HWG 1608; Corail; Elite; Folicur; Horizon; Lynx; Sivacur. Cereal seed dressing with systemic properties for control of seed-borne diseases such as stinking smut, loose smuts and covered smut; highly effective at low dosage rates. mp = 104-107°; soluble in H_2O (32 mg/l), more soluble in organic solvents. *Bayer AG.*

25909 Raybar
7727-43-7 1023 231-784-4
A proprietary preparation of barium sulfate. *Fleet Co Inc.* Name unverified.

25910 Rayo
An electrical resistance alloy consisting of 85% nickel and 15% chromium.

25911 Rayon
Semisynthetic fiber composed of regenerated cellulose which has been cogulated or solidified from a solution of cellulose xanthate, cellulose nitrate or from a solution of cellulose in ammoniacal copper oxide; used in nonwoven fabrics, surgical dressings,.

25912 Rayox
13463-67-7 9612 236-675-5
O_2Ti
A proprietary trade name for titanium dioxide. No manufacturer.

25913 RC 7
Fluorocarbon; mold release agent and lubricant for silicone rubber molding operations, thermoset plastic molding. *Releasomers.*

25914 RCR Grey Squirrel Killer Concentrate
81-81-2 10174 201-377-6
Warfarin; bait for grey squirrels. *Leo Fay Ltd.*

25915 RD10
Roadway dust suppression agent. *Foseco (F.S.) Ltd.*

25916 RDPE
Dimethyl ester solvent. *Orange Chemicals Ltd.*

25917 Réamur's alloy
An alloy of 70% antimony and 30% iron.

25918 Réboulet's solution
An aqueous solution of calcium chloride, potassium nitrate, and alum; used to preserve anatomical specimens.

25919 Reach® 101, 201, 501
1327-41-9 356 215-477-2
$H_5O_5Al_2Cl{\cdot}2H_2O$
Aluminum chlorohydrate
basic aluminum chloride; aluminum chlorohydroxide; aluminum chlorohydrate; Astringen; Chlorhydrol; Hyperdrol; Locron; Phosphonorm. Antiperspirant for increased wetness protection, especially for aerosols. *Reheis.*

25920 Reach® AZP-701, AZP-703
Aluminum zirconium tetrachlorohydrex glycine
Enhanced efficacy antiperspirant. *Reheis.*

25921 Reacrone
Acrylic resins; for paints, inks and metal decorating. *Resinas Sinteticas SA.* Name unverified.

25922 Reactal
Alkyds; for paints, lacquers and varnishes, inks and metal decorating. *Resinas Sinteticas SA.* Name unverified.

25923 Reactint®
Polymeric colorants for flexible urethane foams. *Milliken.*

25924 Reactobond
Chemically reactive oils for tube drawing. *Brent Chemicals International plc.*

25925 Reacton
Rare earth metals and compounds; for electronic materials, phosphors, magnetic materials, sputtering targets. *Johnson Matthey plc.*

25926 Reafor
Etherified amino resins; for paints, lacquers and varnishes, inks and metal decorating. *Resinas Sinteticas SA.* Name unverified.

25927 Reafree
Saturated polyester; for paints, inks and metal decorating. *Resinas Sinteticas SA.* Name unverified.

25928 Reagens-CF2
3327-22-8 222-048-3
$C_6H_{14}ClNO{\cdot}HCl$
3-Chloro-2-hydroxypropyltrimethyl ammonium chloride
1-Propanaminium, 3-chloro-2-hydroxy-N,N,N-trimethyl-, chloride; Ammonium, (3-chloro-2-hydroxypropyl)trimethyl-, chloride; (3-Chloro-2-hydroxypropyl)trimethylammonium chloride; 2-Hydroxy-3-chloropropyltrimethylammonium chloride; Trimethyl(3-chloro-2-hydroxypropyl)ammonium chloride; Trimethyl(2-hydroxy-3-chloropropyl)ammonium chloride;. *Synthetic Chemicals Ltd.*

25929 Reakt
Colorless dye intermediates for non-carbon copying paper. *BASF AG.*

25930 realgar
1303-32-8 834
As_2S_2
Arsenic disulfide
Ruby sulfur; red orpiment; red arsenic; red arsenic glass; ruby arsenic; arsenic orange. Red pigment used in the leather industry, paint, pyrotechnics, and taxidermy. mp = 320°; bp = 565°; insoluble in H_2O.

25931 Realox®
1344-28-1 369 215-691-6
aluminum oxide
Reactive calcined alumina; thermally reactive alumina for ceramic applications requiring high density to be attained at lower firing temps. with a minimum of fluxing additives. *Alcoa Industrial Chemicals.*

25932 Reamul
Aqueous dispersions (polyvinyl acetate, styrene. acrylic etc.); used for paints and inks. *Resinas Sinteticas SA.* Name unverified.

25933 Rearguard
Suspension concentrate containing 64% w/v sulfur, 16% w/v maneb, 1% w/v copper oxychloride; for use as an agricultural fungicide. *Universal Crop Protection Ltd.*

25934 Reatane
Polyisocyanates
Used for lacquers and varnishes. *Resinas Sinteticas SA.* Name unverified.

25935 Reater
Modified polyester resins; used for lacquers and varnishes. *Resinas Sinteticas SA.* Name unverified.

25936 Reatint
Trade name for family of polymeric, liquid, reactive colorants used to color polyurethane polymers; colorants for flexible polyether/polyester urethane foam used for carpet underlay, packaging, sponges, etc., and RIM polyurethane systems for automotive an *Milliken.* Discontinued.

25937 reaumerite
$CaNa_2O{\cdot}3SiO_2$
A compound obtained by heating glass at its softening temperature.

25938 Reax® 45A
8061-51-6
Sodium lignosulfonate with anionic wetting agent; wetting agent and dispersant for pesticides; recommended for use with chlorinated hydrocarbons and organophosphates. *Westvaco.*

25939 Reax® 80C
8061-51-6
Sodium lignosulfonate, modified; dispersant, suspending agent for dyestuffs, lead acid storage batteries. *Westvaco.*

25940 Recoil
A wettable powder containing 10% w/w oxadixyl and 56% w/w mancozeb; fungicide used to control foliar and tuber blight in potatoes. *Schering Agrochemicals Ltd.* Discontinued.

25941 Recoil
A wettable powder containing 10% w/w oxadixyl and 56% w/w mancozeb; fungicide to control foliar and tuber blight in potatoes. *Schering Agrochemicals Ltd; Bayer.* Discontinued.

25942 Recoura's sulfate
$Cr_2S_3O_{12}{\cdot}6H_2O$
A chromium hexahydrated sulfate.

25943 recovered grease
The oil used to lubricate wool during spinning is recovered from the washwater; used to manufacture a low-grade stearin.

25944 Recresal
7558-80-7 8806 231-449-2
H_2NaO_4P
sodium phosphate, monobasic
A proprietary preparation of acid sodium phosphate in tablets. No manufacturer.

25945 Recrete NRC
Flowable replacement concrete used to replace damaged, defective or deteriorated surfaces. *Feb Ltd.*

25946 Rectified Spirit S. V. R
A specially rectified ethyl alcohol 68-69 over proof, containing 96-97% ethyl alcohol by volume; used in perfumery and in pharmaceutical extracts and tinctures.

25947 Recupex
Flux for recovering nonferrous scrap. *Foseco (F.S.) Ltd.*

25948 Red 139
1309-37-1 4072 215-168-2
Iron oxides, bismuthoxychloride
Inorganic colorant. *Presperse.* Discontinued.

25949 red acid
Nitric acid of 40 Bé, or stronger. It contains dissolved nitrogen oxides.

25950 red algar
1303-32-8 834
As_2S_2
Arsenic disulfide.

25951 red antimony
$Sb_2O_3{\cdot}2Sb_2S$
antimony blende
pyrantimonite; pyrostibnite; antimony cinnabar. A mineral; an oxysulfide of antimony; is also obtained by treating antimony chloride with sodium thiosulfate in aqueous solution; used as a pigment to replace ordinary cinnabar.

25952 red brass
A brass containing 90% copper and 10% zinc. Also Tombac, which has been pickled in acid.

25953 red charcoal
A wood charcoal made at low temperature. It contains hydrogen and oxygen.

25954 red cobalt
Erythrite, a mineral that contains cobaltous arsenate, a pigment.

25955 Red Dot
A smokeless powder; designed for light and standard shotshell loads of all gauges; can also be used in specific handgun cartridges. *Hercules.* Discontinued.

25956 red drops
Red lavender. Compound tincture of lavender.

25957 red gold
A jeweler's alloy containing 75% gold and 25% copper.

25958 Red Hermetite
Red paste, semi hardening; used as gasket jointing compound to supplement all gaskets flanged or threaded applications; ensures leak-free joints in most environments. *Hermetite Products Ltd.* Name unverified.

25959 Red Hot Pellets
1305-62-0 1716 215-137-3
Pelleted calcium hydroxide. *Schaefer Salt & Chemical Co.*

25960 red lead
1314-41-6 5449 215-235-6
O_4Pb_3
lead tetroxide
Red lead oxide, minium, Paris red, Satum red; lead tetroxide; lead orthoplumbate; lead oxide red; mineral orange; mineral red; Saturn red; C.I. Pigment Red 105; C.I. 77578. (). a pigment, It is oxide of lead, made by heating litharge, PbO. There are several kinds on the market distinguished by their color and amount of lead dioxide they contain; used in storage batt mp = 500° (dec); d = 9.1; insoluble in H_2O, EtOH; LD_{50} (gpg ip) = 220 mg/kg.

25961 red liquor
$Al_2(C_2H_3O_2)_6$
Mordant rouge. A solution which appears to consist of a diacetate of aluminum, and acetic acid. Red liquor is; largely used in dyeing and calico printing, especially for the production of red colors, for the manufacture of.

25962 red metal
A term usually applied to an alloy of 90% copper, and 10% zinc.

25963 red nickel ore
A mineral that is niccolite (NiAs) or nickeline.

25964 red oil
112-80-1 6595 204-007-1
oleic acid (70%), 15% of linolic acid (15%), stearic acid (15%). Commerical grade of oleic acid; these oils are used for general lubrication.

25965 red oxide of chromium
1333-82-0 2293 215-607-8
CrO_3
chromic oxide
Chromium trioxide.

25966 red oxide of mercury
21908-53-2 5936 244-654-7
HgO
Mercuric oxide, red
Red precipitate; yellow precipitate. Used in marine bottom paints, ceramic paints; in dry batteries and as a reagent. mp = 500° (dec); d = 11.14; insoluble in H_20.

25967 red salts
Both crude sodium acetate and crude sodium carbonate, colored red by ferric oxide, are known as red salts.

25968 red soda
A solution of red ink containing a little gum arabic and sodium carbonate; used as a marking ink for blueprints.

25969 Red Star Powder
It is a 33-grain smokeless powder containing metallic nitrates, nitro-hydrocarbons, and petroleum jelly; used as an explosive.

25970 red storax
Solid storax. An artificial product obtained by mixing poor storax with sawdust, and pressing the mixture; used for fumigating candles and powders.

25971 red vitriol
Botryogen. A native ferroso-ferric sulfate from Sweden.

25972 red wash
A zinc sulfate solution containing red coloring matter.

25973 red water bark
8529
Sassy bark; Saucy bark; Mancona bark; ordeal bark; red-water tree bark; casca bark; Saxon bark; doom bark;teli; bondou. Bark of *Erythrophleum guineense* G. Don *Leguminosae*. Contains erythrophleine, tannin and resins.

25974 Redalloy
A proprietary trade name for brass containing 85% copper, 14% zinc, and 1% tin. No manufacturer.

25975 Redd Citrus Specialties
A range of essential citrus oils, natural citrus aroma and natural citrus specialty products; used for a wide range of applications in the food and beverage industries, personal care products, cosmetics, soaps, detergents and household products. *Hercules.* Discontinued.

25976 reddingite
$Fe_3Mn_3{\cdot}P_2O_5$
A mineral.

25977 Rediclear
Hydrosulfite/dispersing agent blend. *RV Chemicals Ltd.* Discontinued.

25978 Redicote
Specialized cationic bitumen emulsifiers. *Akzo.*

25979 Redray
An electrical resistance alloy containing 85% nickel and 15% chromium.

25980 Reduce® -150
Blend of sodium stearoyl lactylate, calcium sulfate, and sodium sulfite; dough conditioner for use in flour tortillas, pie crusts, and pizza shells. *Am. ingredients/Patco.*

25981 reduced turpentine
A mixture of turpentine oil and petroleum.

25982 reducin
$C_6H(NH_2)_3{\cdot}(OH)_2$
Triaminoresorcinol
Used in photography as a developer.

25983 Reductone
7775-14-6 8771 231-890-0
$Na_2O_4S_2$
sodium hydrosulfite
sodium dithionite; sodium sulfoxylate. A reducing agent, sodium hydrosulfite solution is used in continuous vat dyeing and afterscouring. mp = 52° (dec) *Olin.*

25984 Redurit
Electrically fused standard grade corundum; used for production of abrasives, abrasive paper, discs and cloth. *Hüls UK Ltd.*

25985 Redux® 501
25928-94-3
Two-component epoxy adhesive; ambient or warm curing epoxy resin matrix system; mix ratio 100: 15 pbw *Ciba-Geigy Plastics UK.* Name unverified.

25986 Reduxol Z
24887-06-7 246-515-6
Soluble zinc formaldehyde sulfoxylate. *RV Chemicals Ltd.* Discontinued.

25987 redwoods
Red dye woods. These dye woods are divided into two classes: a) Soluble, which comprise Brazil, Pernambuco or Fernambuco wood, Peach wood, Lima wood, Sapan wood, Bimas redwood, and Nicaragua wood. All of them contain the coloring principle brazilin, $C_{16}H_{14}O_5$.

25988 Reed C-ABS-17415
ABS-based black concentrate. *Reed Plastics.*

25989 Reed C-NY-261
Nylon-based color concentrate; for use in nylon 6 and 6/6. *Reed Plastics.*

25990 Reed C-NY-4892
Nylon-based color concentrate with 20% carbon black; for use in nylon 6 and 6/6 especially for filled and toughened products; superior hiding power. *Reed Plastics.*

25991 Reed C-PBT-1338
PBT polyester-based color concentrate, 20% carbon black; concentrate for use in PBT polyester. *Reed Plastics.*

25992 ReedLite C-NY
Heavy metal-free nylon color concentrates; for automotive, electronic and mechanical applications. *Reed Plastics.*

25993 ReedLite CPC
Heavy metal-free PC color concentrates; for electrical, mechanical, automotive, appliance and business machine applications. *Reed Plastics.*

25994 Rees' thionin stain
A microscopic stain. It consists of 1.5 g thionin and 10 ml alcohol in 100 ml of 5% solution of carbolic acid; used at the rate of 5 ml in 20 ml water.

25995 Reese's alloy
An alloy used in dental work. It contains 87% tin, 8.6% silver, and 4.4% gold.

25996 refikite
A resin found in lignite.

25997 Refine®
For the agriculture industry. *DuPont UK.*

25998 refined silver
A silver usually containing from 99.7-99.9% metal.

25999 Refinex
1337-76-4
Powdered attapulgite clay; used for oil refining. *Whitecourt Ltd.*

26000 Refkon
A range of water-based coatings having high solids content. *Foseco (F.S.) Ltd.*

26001 Reflectafoam
Closed cell polyethylene backing with highly reflective aluminized polyester surface; used for preventing heat loss behind radiators and insulating airing cupboards. *Piccadilly Products Ltd.* Name unverified.

26002 Reflite
A proprietary synthetic resin molding powder. No manufacturer.

26003 reflorit
88-89-1 7562 201-865-9
$C_6H_3N_3O_7$
2,4,6-Trinitrophenol
Picronitric acid; Melinite; pertite; Carbazotic Acid; 2,4,6-Trinitrophenyl; trinitrophenol; 1,3,5-Trinitrophenol; lyddite; shimose; phenol trinitrate; 2-hydroxy-1,3,5-trinitrobenzene; nitroxanthic acid; C.I. 10305. Picric acid; used in the disinfection of seed-corn. mp = 122°; bp > 300°; d = 1.763; soluble in H_2O (1-10 mg/ml).

26004 reform phosphate
Rock phosphate which has been treated with small quantities of dilute acid to render it more porous, converting calcium carbonate into calcium hydrogen carbonate.

26005 Refrax
Bricks made from recrystallized silicon carbide. A refractory material.

26006 Refuse Trol
Liquid waste dispersant and deodorant. *Momar Industrial Services Ltd.*

26007 Regal Crown
Combination of growth stimulators which enhances plant growth by accelerating root growth. *Regal Chemical Company.*

26008 Regal® 400R
7440-44-0 1855 231-153-3
Carbon black; for coloring higher dielec. PVC cable compounds. *Cabot; Cabot Carbon Ltd.*

26009 Regalite
Hydrogenated hydrocarbon resins. *Hercules Ltd.*

26010 Regalox
A sintered material comprising 88% alumina.

26011 RegalStar
Herbicide to control crabgrass, goosegrass and other annual weeds; applied in early spring to turfgrass and cultivated nursery fields prior to weed seed germination. *Regal Chemical Company.*

26012 regenerated turpentine
A product of synthetic camphor manufacture. bp = 170°.

26013 Regenex
Copper and nickel alloy flux. *Foseco (F.S.) Ltd.* Discontinued.

26014 Regent® 12XX
7758-23-8 1740 231-837-1
Monocalcium phosphate, monohydrate
Leavening agent for baking, cereal, beverages. *Rhône-Poulenc Food Ingreds.*

26015 Reginal
Proofing agent. *Ciba plc.* Name unverified.

26016 Reginol 2701
Surfactant blend; scouring agent for dyeing; biodegradable. *Hart Prods. Corp.*

26017 Reglone, Reglox
85-00-7 3415 201-579-4
Diquat dibromide. A contact herbicide and preharvest crop desiccant. *ICI Chem & Polymers Ltd.*

26018 Regnis
Machine lubricant for glass forming machinery. *Specialty Products Co.* Name unverified.

26019 Regulex
77-06-5 4426 201-001-0
$C_{19}H_{22}O_6$
Gibb-3-ene-1,10-dicarboxylic acid, 2,4a,7-trihydroxy-1-methyl-8-methylene-, 1,4a-lactone, (1α,2β,4aα,4bβ,10β)-
Gibberellic acid; Gibberellin A3; Activol; Berelex; Cekugib; Gibberellin; Gibrel; Pro-Gibb; Pro-Gibb Plus; GA3; brellin; gibberellin x; gibbrel; gib-sol; gib-tabs; 2,4a,7-Trihydroxy-1-methyl-8-methylenegibb-3-ene-1,10-dicarboxylic acid 1,4a-lactone; Gibberellin 1; 2beta-Hydroxygibberellin 1; Gibb-tabs; Grocel; Regulex; Trihydroxy-1-methyl-8-methylenegibb-3-ene-1,10-dicarboxylic acid, 1,4a-lactone. A plant growth regulator; increases cropping in apples and pears. mp = 233-235°; $[\alpha]_D^{19}$ = 86° (c = 2.12); soluble in H_2O, Et_2O, more soluble in EtOH, Me_2CO. *ICI Chem & Polymers Ltd.*

26020 Regulox K
123-33-1 5745 204-619-9
Maleic hydrazide
A plant growth regulator for grass and to reduce bud growth in trees, hedges, and vegetables. *Rhône-Poulenc Environmental Prods. Ltd.*

26021 Regulus
Thermoplastic polyimide film; super heat resistant film for wire and cable, thermoplastic composites, pressure-sensitive adhesive tapes, primary insulation film. *Advanced Web Prods.* Discontinued.

26022 Regulus metal
This is usually a 5-12% antimony with lead.

26023 regulus of antimony
Produced by heating antimony ore, Sb_2S_3. It contains about 10% of iron.

26024 regulus of Venus
An alloy of copper and antimony, $SbCu_2$.

26025 Rehydragel® Compressed Gel
21645-51-2 355 244-492-7
Aluminum hydroxide
Adsorbent gel for pharmaceuticals (enhances suspensions, builds viscosity); carrier for toxins in veterinary vaccines. *Reheis.*

26026 Rehydrol®
Aluminum chlorohydrate and propylene glycol; antiperspirant active. *Reheis Inc.*

26027 Reich's bronze
An aluminum bronze containing 85.2% copper, 7.52% iron, 6.6% aluminum, 0.5% manganese, and 0.15% lead.

26028 Reicolit
A proprietary insulation. No manufacturer.

26029 Reillex 202
Cross-linked poly-2-vinylpyridine. *Reilly Industries Inc.*

26030 Reillex 402 and 425
Cross-linked poly-4-vinylpyridine. *Reilly Industries Inc.*

26031 Reilline 2200 and 240
Linear poly-2-vinylpyridene. *Reilly Industries Inc.*

26032 Reilline 4200 and 450
Linear poly-4-vinylpyridene. *Reilly Industries Inc.*

26033 Reinecke's salt
13573-16-5 8298 237-003-3
$C_4CrH_{10}N_7S_4 \cdot H_2O$
chromate(1-), diaminetetrakis(thiocyanato-n)-, ammonium, (OC-6-11)-ammonium reineckate; ammonium tetrathiocyanodiaminochromate; ammonium reineckate trihydrate. Produced when ammonium cyanate is melted and ammonium bichromate added; used as a precipitating agent for organic bases and amino acids and as a reagent for mercury. mp = 270° (dec).

26034 Reiset's first base
$Pt(NH_3 \cdot NH_3 \cdot OH)_2$
Plato-diamine hydroxide.

26035 Reiset's first chloride
$Pt(NH_3 \cdot NH_3 \cdot Cl)_2 \cdot H_2O$.
Plato-diamine-chloride,.

26036 Reith alloy
An alloy of 75% copper, 10% tin, 10% lead, and 5% antimony.

26037 Reldan 50
5598-13-0 2242 227-011-5
Chlorpyrifos-methyl
An organophosphate insecticide for the treatment of pests in stored grain and oilseed rape. *DowElanco Ltd.*

26038 Release Agent NL-1
Wax mixture, hydrocarbon solvent release agent for molded products; used as a first thin layer on the mold surface *Akzo.* Name unverified.

26039 Release Agent NL-2
9002-89-5 7745
Polyvinyl alcohol solution, alcohol water solvent film-forming release agent. *Akzo.* Name unverified.

26040 Release Agent NL-10
Wax mixture, paraffin oil solvent release agent for molded products; used as a first thin layer on the mold surface *Akzo.* Name unverified.

26041 Releasil
Silicone release agents. *Dow Corning Ltd.* Name unverified.

26042 Releez
Methyl oleate, methyl stearate, methyl palmitate, methyl laurate, methyl myristate; asphalt release agent *Alzo.* Discontinued.

26043 Relief®
polyvinyl alcohol and phenylephrine hydrochloride. Redness reliever, ocular lubricant. *Allergan Inc.*

26044 Relimate®, Relipress®
For the electrical industry. *DuPont UK.*

26045 Reloder 7
A smokeless powder; designed for rifle loads and 'benchrest' type reloads. *Hercules.* Discontinued.

26046 Relugan GT
111-30-8 4480 203-856-5
An aqueous solution of glutaraldehyde; used as a tanning agent.

26047 Relugan®
Resin or aldehyde tanning agents for leather. *BASF AG; BASF plc.*

26048 Remafin
Pigment masterbatches for plastics. *Hoechst UK.*

26049 Remazol
Reactive dyestuffs. *Hoechst UK.*

26050 Remcoil
Leather oils. Hodgson Chemicals Ltd.

26051 Remcopal
Nonionic surfactants (ethylene oxide derivatives); emulsifiers, wetting agents, antifoams. *Elf Atochem UK/Ceca.*

26052 Remcopal 4
Laureth-4
Emulsifier for beeswax; intermediate for polyethoxy ether sulfate. *Ceca SA.*

26053 Remcopal 6
9004-96-0
PEG-6 oleate
Emulsifier for mineral oil and solvents; stabilizer for polyurethane foams. *Ceca SA.*

26054 Remcopal 10
Cetoleth-10
Base for emulsifiers and viscous spindle oils. *Ceca SA.*

26055 Remcopal 18
Cetoleth-18
Retarder for colors; emulsifier for fatty alcohol and olein. *Ceca SA.*

26056 Remcopal 20
Laureth-20
Emulsifier for fatty alcohol and olein. *Ceca SA.*

26057 Remcopal 25
Cetoleth-25
Raw material for degreaser compounds; emulsifier for fatty alcohol, olein, and waxes. *Ceca SA.*

26058 Remcopal 29
9016-45-9 6772
Nonoxynol-8.5
Emulsifier. *Ceca SA.*

26059 Remcopal 40
61791-12-6
PEG-31 castor oil; emulsifier for olein. *Ceca SA.*

26060 Remcopal 40 S3
61791-12-6
PEG-40 castor oil; emulsifier for vitamins and essential oils. *Ceca SA.*

26061 Remcopal 121
Laureth-3
Intermediate for sulfation. *Ceca SA.*

26062 Remcopal 207
9004-96-0
PEG-4.5 oleate
Emulsifier for vitamins. *Ceca SA.*

26063 Remcopal 21411
Laureth-11
Emulsifier for light hydrocarbons, raw material shampoos. *Ceca SA.*

26064 Remcopal 21912 AL
Laureth-12
Surfactant. *Ceca SA.*

26065 Remcopal 220
Cetoleth-25
Emulsifier. *Ceca SA.*

26066 Remcopal 229
68439-49-6 7737
Ceteareth-25
Emulsifier for emulsion polymerization, washing detergents. *Ceca SA.*

26067 Remcopal 234
Cetoleth-4
Surfactant. *Ceca SA.*

26068 Remcopal 238
68439-49-6 7737
Ceteareth-20
Surfactant for degreaser compounds. *Ceca SA.*

26069 Remcopal 258
Laureth-9
Wetting agent. *Ceca SA.*

26070 Remcopal 273
Isodeceth-3
Wetting agent for pigments and fillers. *Ceca SA.*

26071 Remcopal 306
9002-93-1 6858
Octoxynol-5.5
Wetting agent for carbon black, emulsifier for aromatic hydrocarbons, turpentine oil, and tallow. *Ceca SA.*

26072 Remcopal 334
9016-45-9 6772
Nonoxynol-4
Emulsifier, dispersant for pigments. *Ceca SA.*

26073 Remcopal 349
9016-45-9 6772
Nonoxynol-8
Emulsifier, dispersant for pigments. *Ceca SA.*

26074 Remcopal 666
9016-45-9 6772
Nonoxynol-6
Emulsifier for turpentine oil, heavy aromatic solvents; wetting agent and dispersant for pigments. *Ceca SA.*

26075 Remcopal 3112
9016-45-9 6772
Nonoxynol-2
Antifoam, emulsifier. *Ceca SA.*

26076 Remcopal 3712
9016-45-9 6772
Nonoxynol-12
Emulsifier, wetting agent. *Ceca SA.*

26077 Remcopal 3820
9016-45-9 6772
Nonoxynol-20
Surfactant. *Ceca SA.*

26078 Remcopal 4000
61791-12-6
PEG-31 castor oil; surfactant. *Ceca SA.*

26079 Remcopal 4018
61791-12-6
PEG-23 castor oil; surfactant. *Ceca SA.*

26080 Remcopal 6110
9016-45-9 6772
Nonoxynol-9
Emulsifier, wetting agent. *Ceca SA.*

26081 Remcopal 31250
9016-45-9 6772
Nonoxynol-50
Emulsifier for epoxy resin. *Ceca SA.*

26082 Remcopal 33820
9016-45-9 6772
Nonoxynol-20
Emulsifier for emulsion polymerization. *Ceca SA.*

26083 Remcopal D
Cetoleth-23
Surfactant. *Ceca SA.*

26084 Remcopal HC 7
61788-85-0
PEG-7 hydrogenated castor oil; surfactant. *Ceca SA.*

26085 Remcopal HC 20
61788-85-0
PEG-20 hydrogenated castor oil; surfactant. *Ceca SA.*

26086 Remcopal HC 33
61788-85-0
PEG-33 hydrogenated castor oil; surfactant. *Ceca SA.*

26087 Remcopal HC 40
61788-85-0
PEG-40 hydrogenated castor oil; surfactant; solubilizer for essential oils and vitamins. *Ceca SA.*

26088 Remcopal HC 60
61788-85-0
PEG-60 hydrogenated castor oil; surfactant. *Ceca SA.*

26089 Remcopal L9
Laureth-9
Emulsifier for heavy hydrocarbons, degreaser, antistat for synthetic fibers. *Ceca SA.*

26090 Remcopal L12
Laureth-10.5
Surfactant. *Ceca SA.*

26091 Remcopal L30
9016-45-9 6772
Nonoxynol-27
Surfactant. *Ceca SA.*

26092 Remcopal LO 2B
Isodeceth-3
Surfactant. *Ceca SA.*

26093 Remcopal LP
Laureth-9
Surfactant. *Ceca SA.*

26094 Remcopal NP 30
9016-45-9 6772
Nonoxynol-27
Emulsifier. *Ceca SA.*

26095 Remcopal O9
9002-93-1 6858
Octoxynol-9
Solubilizer for essential oils; emulsifier. *Ceca SA.*

26096 Remcopal O11
9002-93-1 6858
Octoxynol-11
Solubilizer for essential oils; emulsifier. *Ceca SA.*

26097 Remcopal O12
9002-93-1 6858
Octoxynol-12
Solubilizer for essential oils; emulsifier. *Ceca SA.*

26098 Remcopal PONF
9016-45-9 6772
Nonoxynol-11
Emulsifier. *Ceca SA.*

26099 Remex
Desulfurizing flux for injection into steel melted in electric arc furnaces. *Foseco (F.S.) Ltd.*

26100 Remsynol
Leather oils. *Hodgson Chemicals Ltd.*

26101 Remtal
Trietazine [1912-26-1] and simazine [122-34-9]; selective herbicide for peas and beans. *Schering Agrochemicals Ltd.* Discontinued.

26102 Renacit® 4
Nonstaining antioxidant. *Bayer AG; Bayer plc.*

26103 Renacit® 7
133-49-3 205-107-8
C_6HCl_5S
pentachlorothiophenol
pentachlorobenzenethiol. Pentachlorothiophenol absorbed on clay; peptizing agent facilitating open mill and internal mixer mastication in rubber industry. *Bayer Corp; Polysar.*

26104 Renault alloy
An aluminum alloy containing 88% aluminum, 10% zinc, and 2% copper.

26105 Rencal
7205-52-9 7542
$C_6H_9Na_9O_{24}P_6$
myo-inositol hexakis(dihydrogen phosphate) sodium salt
Phytate sodium; sodium phytate; Phytat D.B.; inositol hexaphosphoric acid hexasodium salt. Chelating agent. *Bristol-Myers Squibb Co Inc.* Name unverified.

26106 Rendells
A proprietary preparation of nonoxynol used in the form of contraceptive pessaries. *W J Rendell Ltd.* Name unverified.

26107 Rendrock
An explosive. It is a modification of Lithofracteur, consisting of 40% potassium nitrate, 40% nitroglycerin, 13% wood pulp, and 7.0% paraffin or pitch.

26108 Renektan
Leather dyes. *ICI Chem & Polymers Ltd.*

26109 Renex
Polyexyethylene alkyl or alkyl acryl esters. *ICI Chem & Polymers Ltd.*

26110 rennet
9001-98-3 8303 232-796-2
rennin
chymosin; rennase; lab; abomasal enzyme;. Rennet is a dried extract containing rennin.

26111 Rennilase
A milk clotting enzyme produced by a selected nonpathogenic strain of the fungus *Mucor miehel*; used in cheese-making for coagulation as an alternative to calf rennet. *Novo Nordisk.*

26112 Renol
Pigment and dyestuff materials for plastics. *Hoechst UK.*

26113 Reno-M
131-49-7 5851 205-024-7
$C_{18}H_{26}I_3N_3O_9$
1-deoxy-1-(methylamino)-D-glucitol 3,5-bis(acetylamino)-2,4,6-triiodo benzoate salt
Diatrizoate meglumine; urografic acid methylglucamine salt; diatrizoate methylglucamine; meglumine amidotrizoate; Angiografin; Cardiografin; Cystografin; Hypaque Cysto; Hypaque Meglumine; Renografin; Urovist. Diagnostic aid. mp = 189-193°; soluble in H_2O (89 g/100 ml). *Bristol-Myers Squibb Co Inc.*

26114 Renova
Solvent cleaner and degreaser. *British Nova Works Ltd.*

26115 Renovue-65, Renovue-DIP
18656-21-8 5031 242-480-6
$C_{19}H_{28}I_3N_3O_9$
3-(acetylamino)-5-[(acetylamino)methyl]-2,4,6-triiodobenzoic acid N-methyl-D-glucamine salt
Iodamide meglumine; Isteropac E.R.; Jodomiron; Opacist E.R.; Uromiro. Diagnostic aid. Sparingly soluble in H_2O, alcohol, insoluble in other organic solvents; LD_{50} (rat iv) = 11.4 g/kg *Bristol-Myers Squibb Co Inc.*

26116 Rentokil Deadline
28772-56-7 1403 249-205-9
$C_{30}H_{23}BrO_4$
3-[3-(4'-bromo[1,1'-biphenyl]-4-yl)-3-hydroxy-1-phenylpropyl]-4-hydroxy-2H-1-benzopyran-2-one
Bromadiolone; Bromone; Canadien 2000; Contrac; Maki; Ratimus; Tamogam; Boldo; Bromo-4-biphenylyl)ethyl)benzyl)-4-hydroxycoumarin; LM-637; Super-caid; 3-(3-(4'-Bromo-(1,1'-biphenyl)-4-yl)-3-hydroxy-1-phenylpropyl)-4-hydroxycoumarin. An anticoagulant rodenticide as a concentrated bait. mp = 200-210°; λ_m = 260 nm ($E^{1\%}_{1cm}$ 560 EtOH); slightly soluble in H_2O (19 mg/l), more soluble in organic solvents; LD_{50} (rat orl) = 1.125 mg/kg. *Rentokil Ltd.*

26117 Renyx
A proprietary trade name for an alloy of aluminum with nickel, copper, and silicon. No manufacturer.

26118 Reochlor (LF and 54)
Chlorinated paraffins; an extender and plasticizer for PVC *Ciba plc.* Name unverified.

26119 Reoflam
A range of proprietary plasticizers used with PVC. *Ciba plc.* Name unverified.

26120 Reofos
A range of synthetic organic phosphates. *Ciba plc.* Name unverified.

26121 Reogen
A mixture of an oil soluble sulfonic acid of high molecular weight with a paraffin oil; plasticizer effective in all elastomers. *King Industries.*

26122 Reolube
A trade name for a range of synthetic organic phosphates. *Ciba plc.* Name unverified.

26123 Reolube FAD
A proprietary trade name for a long chain fatty acid mixture, the principal components being C_{14}, C_{16}, and C_{18} acids. *Ciba plc.* Name unverified.

26124 Reomet®
Metal treatment additives. *Ciba plc.* Name unverified.

26125 Reomol
Plasticizers. *Ciba plc.* Name unverified.

26126 Reomol 4PG
85-70-1 201-624-8
$C_{18}H_{24}O_6$
Butyl phthalyl butyl glycolate
Butyl phthalyl butyl glycollate; 1,2-benzenedicarboxylic acid 2-butoxy-2-oxoethyl butyl ester; phtahlic acid, butyl ester, ester with butyl glycolate; butoxycarbonyl methyl butyl phthalate. A proprietary plasticizer. *Ciba plc.* Name unverified.

26127 Reomol BCF
Butyl carbinol formal; a proprietary plasticizer. *Ciba plc.* Name unverified.

26128 Reomol D79S
A mixture of heptyl and nonyl sebacates; a vinyl plasticizer. No manufacturer.

26129 Reomol DBS
109-43-3 203-672-5
$C_{18}H_{34}O_4$
n-butyl sebacate
Dibutyl sebacate; decanedioic acid dibutyl ester; butyl sebacate; dibutyl decanedioate. Vinyl plasticizer. mp = -12°; d = 0.9400. No manufacturer.

26130 Reomol DCP
117-81-7 1291 204-211-0
$C_{24}H_{38}O_4$
bis(2-ethylhexyl)phthalate
Dicapryl phthalate; DEHP; DOP; Diethylhexyl phthalate; Dioctyl Phthalate;1,2-Benzenedicarboxylic acid bis(2-ethylhexyl) ester; Octoil; Ethyl hexyl phthalate; 2-Ethylhexyl phthalate; bis(2-ethylhexyl)ester phthalic acid; bis-(2-ethylhexyl)-1,2-benzenedicarboxylate; di(2-ethylhexyl)orthophthalate; octyl phthalate; phthalic acid dioctyl ester; BEHP; bisoflex 81; bisoflex DOP; compound 889; DAF 68; ergoplast FDO; eviplast 80; eviplast 81; fleximel; flexol DOP; flexol plasticizer DOP; good-rite gp 264; hatcol DOP; hercoflex 260; kodaflex DP; mollan o; nuoplaz DOP; palatinol AH; Pittsburgh PX-138; platinol AH; platinol DOP; rc plasticizer DOP; reomol DOP; reomol D 79P; sicol 150; staflex dop; truflex dop; vestinol ah; vinicizer 80; witcizer 312; Bi(2-ethylhexyl)trimellitate ester; Benzenedicarboxylic acid, bis(2-ethylhexyl) ester; Union carbide flexol 380,. A vinyl plasticizer mp = -50°; bp = 384°; d = 0.9810; n_D^{20}= 1.4853. No manufacturer.

26131 Reomol DOS
122-62-3 1292 204-558-8
$C_{26}H_{50}O_4$
bis(2-ethylhexyl) sebacate
Di-2-ethylhexyl sebacate; Octyl Sebacate; Bis(2-Ethylhexyl) Ester Sebacic Acid; bis(2-ethylhexyl) sebacate; bisoflex dos; DOS; 2-ethylhexyl sebacate; 1-hexanol, 2-ethyl-, sebacate; monoplex dos; octoil s; PX 438; staflex dos; plexol 201; bis(2-ethylhexyl) decanedioate; edenol 888; ergoplast sno; reolube dos; DEHS. A vinyl plasticizer mp = -67°; bp = 248°; d_{25}^{25} = 0.914; n_D^{25} = 1.4496; insoluble in H_2O. No manufacturer.

26132 Reomol P
117-82-8 204-212-6
$C_{14}H_{18}O_6$
1,2-benzenedicarboxylic acid, bis(2-methoxyethyl) ester
methoxy ethyl phthalate; 2-methoxy ethyl phthalate; phthalic acid bis(2-methoxyethyl) ester. Reomol P is a proprietary trade name for dimethoxy ethyl phthalate; a chemical bonding agent for cellulose acetate staple fiber. *Ciba plc.* Name unverified.

26133 Reomol PBPS
A sebacic acid polyester with a small proportion of nonpolymeric ester. *Ciba plc.* Name unverified.

26134 Reomol TC9
A proprietary trade name for a chemical bonding agent for terylene and cellulose triacetate fibers. No manufacturer.

26135 Reoplast
Epoxy plasticizers/stabilizers. *Ciba plc.* Name unverified.

26136 Reoplex
Polyester plasticizers. *Ciba plc.* Name unverified.

26137 Reoplex 200, 220, 300
Vinyl plasticizers of the polyester type. *Ciba plc.* Name unverified.

26138 Reoplex 901
A trademark for a plasticizer for PVC sheeting intended for manufacture of surgical and electrical tapes. *Ciba plc.* Name unverified.

26139 Reoplex 902
A plasticizer with good resistance to extraction by petroleum. *Ciba plc.* Name unverified.

26140 Reostene
A nickel-iron alloy.

26141 Repak
7778-54-3 1717 231-908-7
$CaCl_2O_2$
Calcium hypochloride
Losantin; Calcium hypochloride; Hypochlorous Acid, Calcium Salt; BK Powder; Hy-Chlor; Lo-Bax; Chlorinated lime; Lime chloride; Chloride of lime; Calcium oxychloride; HTH; Mildew remover X-14; Perchloron; Pittchlor. Oxidizing agent, used as an algicide, bactericide, fungicide, deodorant, disinfectant and bleaching agent. mp = 100° (dec); d = 2.3500; *PPG Industries.*

26142 Repelit
A proprietary synthetic resin varnish-paper product used for electrical insulation. No manufacturer.

26143 Repello DC
Resin-wax blend; fabric water repellent. *Scher Chemicals Inc.* Discontinued.

26144 Repel-O-Tex® QCJ
High molecular weight hydrophilic polyester emulsion; imparts soil release and moisture transport properties to polyester filament fabrics; oil scavenger keeping oils and greases in suspension during scouring/dyeing processes; fiber-to-fiber lubricant. *Rhône-Poulenc Surf.*

26145 Replay RP 2177
9003-53-6 9028 .203-066-0
Impact polystyrene; contains controlled amounts of post-consumer recycled polystyrene. *Huntsman.*

26146 Replay RP 2236
9003-53-6 9028 203-066-0
Polystyrene
General purpose grade containing controlled amounts of post-consumer recycled PS. *Huntsman.*

26147 Replicast CS
Patented process for making casting in which an expanded polystyrene pattern is coated with ceramic material then the pattern is burnt out. *Foseco (F.S.) Ltd.* Discontinued.

26148 Replicast FM
Patented process for making castings using expanded polystyrene patterns in unbonded sand compacted under vacuum. *Foseco (F.S.) Ltd.*

26149 Reprodin®
73523-00-9 5638
$C_{21}H_{29}ClO_6S$
[1S-[1α(Z),2β(RName unverified.), 3α,5α]]-7-[2-[[3-(3-chlorophenoxy)-2-hydroxypropyl]thio]-3,5-dihydroxycyclopentyl]-5-heptenoic acid
EMD-34946; Pronilin; Prosolvin;. A luteolytic prostaglandin for use with cattle, horses, pigs and sheep. Affects regulation of estrus cycle. *Bayer AG.*

26150 Reproxal
Trimellitate plasticizers. *RWE-DEA Chemicals UK Ltd.*

26151 Repulse
1897-45-6 2219 217-588-1
$C_8Cl_4N_2$
2,4,5,6-Tetrachloro-1,3-benzenedicarbonitrile
Forturf; Bravo; Exotherm; m-TCPN; Sweep; TCIN; Termil; TPN; Daconil; ; tetrachloroisophthalonitrile;m-tetrachlorophthalodinitrile; 2,4,5,6-tetrachloro-1,3-dicyanobenzene; 1,3-dicyano-2,4,5,6-tetrachlorobenzene; DAC-2787; Daconil 2787; Exotherm Termil; Chlorothanonil; Bombardier; Farber; Jupital; Ole; Pillarich; Repulse; Taloberg; Tuffcide; Black Leaf Lawn & Garden Fungicide; Bonide; ClortoCaffaro; Clortosip; Dexol Fungicide Containing Daconil; Dragon Daconil 2787; Ferti-lome; Green Charm Multi-Purpose Fungicide; Green Thumb Lawn & Garden Fungicide; Ortho Multi-Purpose Fungicide Daconil 2787; Pennington's Pride Multi-Purpose Fungicide; Pro-Care Multi-Purpose Fungicide; Rigo's Best Lawn & Garden Fungicide; Security Fungi-Gard; 2,4,5,6-tetrachloroisophthalonitrile; m-tetrachlorophthalonitrile; bravo-w-75; chloroalonil; Dacobre; Echo 75; Vanox; Benzenedicarbonitrile, 2,4,5,6-tetrachloro-; Tetrachloro-1,3-dicyanobenzene,. A fungicide for a wide range of agricultural crops. mp = 250-241°; bp = 350°; d^{25}_4 = 1.7; insoluble in H_2O, soluble in organic solvents; LD_{50} (rat orl) > 10.o g/kg. *ICI Agrochemicals.*

26152 resacetophenone
89-84-9 8308 201-945-3
$C_8H_8O_3$
ethanone, 1-(2,4-dihydroxyphenyl)-
2,4-Dihydroxyacetophenone; 4-acetylresorcinol; 2',4'-dihydroxyacetophenone. Reagent specific for iron. mp = 143-145°; d = 1.1800.

26153 Resad
Polymer emulsions, polyvinyl acetate homopolymers and copolymers, acrylic and styrene acrylic polymers; used in adhesives, textile treatments and surface coatings. *Resadhesion Ltd.*

26154 Resamine
A proprietary trade name for formaldehyde resins. *Vianova Resins.* Discontinued.

26155 Resarit
Acrylic molding compound; for double and triple walled sheets, rear lights, automotive parts, instrument covers, lampshades, condenser lenses, casings, covers for measuring instruments, etc. *Resart-IHM AG.* Name unverified.

26156 Resarix SF
Scratch resistant coating system for surface treatment; for optical industry (lenses, magnifying glasses, lenses for sunglasses, scales) and for head protection (visors for crash helmets and astronaut helmets). *Resart-IHM Ag.* Name unverified.

26157 Resart
A range of melamine molding compounds; used for molded parts with tracking resistance, molded parts with high-grade dimensional stability and electrical components such as switches and relays. *Resart-IHM AG.* Name unverified.

26158 Resartglas GS
Cast acrylic sheets LDII; used for roof windows, light domes, windscreens, caravan windows, displays, advertising gifts, furniture, for solarium equipment and solar beds - transparent to uv a light. *Resart IHM AG.* Name unverified.

26159 Resartherm
Glass fiber-reinforced polyester; used for electrical engineering, household appliances, electric tools, car ignition systems. *Resart-IHM AG.* Name unverified.

26160 Resart-PMMA XT
Standard extruded acrylic sheets 500 high impact; used for advertising aids (letters, displays, advertising transparencies), engineering components (housings, machine covers etc.), lighting fittings (cover for long-field light fittings, *Resart IHM AG.* Name unverified.

26161 Resazurin
550-82-3 8309 263-718-5
$C_{12}H_7NO_4$
7-hydroxy-3H-Phenoxazin-3-one 10-oxide
diazoresorcin; diazoresorcinol; resazoin. An acid-base indicator; pH 3.8 orange, pH 6.5 dark violet. Also used in detection of hyposulfite, and also in food research.

26162 Rescon
Disposable devices used for obtaining samples of molten steel. *Foseco (F.S.) Ltd.* Discontinued.

26163 Rescue
Fuel additives. *UOP Inc.*

26164 Resibon
Phenolic resins; used for metal decorating, adhesives, abrasives, thermal insulation refractories, interior can coatings. *Resinas Sinteticas SA.* Name unverified.

26165 Resicart
Wet strength resins. *Ciba plc.* Name unverified.

26166 Residuren
101-21-3 2240 202-925-7
Chlorpropham
Herbicide. *ICI Chem & Polymers Ltd.*

26167 Residuren Extra
Chlorpropham [101-21-3] and diuron [330-54-1]; an herbicide for treatment of grass in bulbs, peas, and beans. *Farm Protection Ltd.*

26168 Resigum
Hydrogenated rosin esters. *Barron Chemicals Ltd.* Unverified.

26169 Resilia
A proprietary trade name for a silico-manganese spring steel. No manufacturer.

26170 Resilita
Polyamide resins: used for inks, paints, and adhesives. *Resinas Sinteticas SA.* Name unverified.

26171 Resilla
A proprietary trade name for a special silicon-manganese spring steel. No manufacturer.

26172 Resilon
409-21-2 8636 206-991-8
Silicon carbide
Used in the electrical industry. *Lonza AG.*

26173 Resimene
Melamine-formaldehyde resin; used as binders in adhesives, molded products, and foundry cores, paint coating ingredients and printing ink products. *Monsanto Co.* Name unverified.

26174 Resin 18
Poly-α-methylstyene.
Amoco Chemical Co. Discontinued.

26175 Resin 164
9009-54-5
Polyether PU compound; nonexpanding compound for making low-cost moisture blocks and pressure dams in paper, pulp, and plastic insulated telecommunications cable (direct sheath injection method); service proven in all climates and environments; *Hexcel.*

26176 Resin 731D
Modified dehydrogenated (disproportionated) rosin; pale, oxidation-resistant, thermoplastic resin used in hot-melt-applied adhesives and coatings for paper and paperboard substrates, as tackifier and processing aid for rubber-based adhesives *Hercules.* Discontinued.

26177 Resin 885, 3072
Mixed emulsifiers containing both disproportionated fatty acid and rosin acid; used in emulsion polymerization of elastomers calling for mixed emulsifier system. *Hercules.* Discontinued.

26178 resin blende
Zinc blende, ZnS, of a yellow color, is sometimes called by this name.

26179 Resin Ether L
A proprietary synthetic resin for use as a cellulose-lacquer plasticizer; nondrying, not susceptible to atmospheric oxidation, and has a low acid value. No manufacturer.

26180 Resin EX
Mixed glycol ester of rosin plus emulsifiers; self-emulsifying liquid rosin ester; plasticizer and tackifier in water-based adhesives. *Blueminster Ltd.*

26181 Resin H
Hydrogenated rosin. *Barron Chemicals Ltd.* Unverified.

26182 resin lutea
Acaroid balsam. A name applied to yellow acaroid balsam, a yellow resin obtained from *Xanthorrhoea Hastile*.

26183 Resin M.S.2.
Cyclohexanone condensation products. *Laporte Industries Ltd.*

26184 Resin NC-11
Acidic resin derived from rosin; pale, noncrystalline, oxidation-resistant, thermoplastic resin used in formulating adhesive masses for surgical and industrial tapes, as resin component of various hot melt coating and adhesive compositions; *Hercules.* Discontinued.

26185 resin oil
That fraction of the distillation of resin (colophony), which distills over from 300-400°. It consists principally of terpineol, $C_{10}H_{18}O$; used as a lubricant.

26186 Resin Release N
Mold release agent for fiberglass reinforced hand layup molding or casting. *Specialty Products Co.* Name unverified.

26187 Resin WP
A proprietary trade name for a melamine based thermosetting resin used for crease-resisting finishes. *Ciba plc.* Name unverified.

26188 Resinall 153
Zincated modified rosin; high melting point resin with excellent sol. in low KB aliphatic solvs.; used in adhesives and rubber compounding; also as modifier for phenolic resins, printing inks and gloss oil. *Resinall Corp.*

26189 Resinase
A liquid lipase containing preparation used to eliminate pitch/resin related problems in the paper industry. *Novo Nordisk.*

26190 Resinette
A synthetic resin obtained from phenol and formaldehyde. No manufacturer.

26191 Resinoid 1324
Phenolic resin, glass-reinforced; thermoset for injection and transfer molding; excellent creep resistance and retention of properties on prolonged exposure to elevated temperatures; *Resinoid Engineering.*

26192 Resinoid 2002-4
Phenolic resin, fabric-reinforced; thermoset for transfer and compression molding; features good strength with excellent mech. shock resist.; used for automotive timing gears, aircraft and automotive pulleys, butterfly valve housings, *Resinoid Engineering.*

26193 resinol
A varnish substitute obtained by the dehydrogenation of petroleum, distillation, and polymerization.

26194 resinous silica
A variety of hydrated silica or opal.

26195 Resinox
A proprietary synthetic resin molding powder of the phenol-formaldehyde type. No manufacturer.

26196 resins, acrolein
Resins obtained by the polymerization of acrolein by means of inorganic and organic bases or salts of iron and lead. Orea is a trade name for a resin of this type. Acrolein also condenses with phenols to form resins. Name unverified.

26197 Resipol
Unsaturated polyester; used for paints, lacquers and varnishes, reinforced plastics. *Resinas Sinteticas SA.* Name unverified.

26198 Resipol DL
Glyceryl dilactate
A proprietary trade name for a plasticizer. No manufacturer.

26199 Resipol ML
Glyceryl monolactate
A proprietary trade name for a plasticizer. No manufacturer.

26200 Resiren®
Sublimable disperse dyestuffs; used for heat transfer printing preferably on PES and other synthetics. *Bayer AG.*

26201 Resisco
A proprietary trade name for an alloy of 91% copper, 7% aluminum, and 2% nickel. No manufacturer.

26202 Resista
A glass similar in composition to Pyrex glass. It contains 70% silica and 13.5% boric oxide.

26203 Resista steel
An alloy of iron, nickel, and manganese, which is ductile at low temperatures.

26204 Resistac
A proprietary trade name for an alloy of copper with 9% aluminum and 1% iron. No manufacturer.

26205 Resistal
A heat-resisting alloy containing 63.5% iron, 16.6% nickel, 15% chromium, 4.5% silicon, and 0.3% carbon.

26206 resistance bronze
A term for an alloy of from 84-86.5% copper, 11.5-13.5% manganese, and 2% iron.

26207 Resistherm®
Raw materials used for the formulation of wire enamels with high thermal stability. *Bayer AG.*

26208 Resistin
An electrical resistance alloy containing 84-86% copper, 2% iron, and 11-13% manganese.

26209 Resistoflex
A proprietary trade name for polyvinyl alcohol synthetic resins. No manufacturer.

26210 Resistolac
Heat resistant cigarette carton lacquers. *The Scottish Adhesives Co Ltd.*

26211 Resistone
Cationic surfactant, biocide. *Rhône-Poulenc UK.*

26212 Resistone QD, Resitone QD
Alkylaryl quaternary ammonium salt in pale yellow aqueous solution; bactericide and algicide, useful in alkaline media; used in static suppression. *Rhône-Poulenc UK.*

26213 Resistox
An aldehyde-amine condensation product; a proprietary antioxidant for rubber. No manufacturer.

26214 Resithren
Combinations of disperse and vat dyes; used for the one-bath dyeing of polyester/cellulosic blends. *Bayer AG.*

26215 Resitone QD
Alkylaryl quaternary ammonium salt in pale yellow aqueous solution. Bactericide and algicide, useful in alklaine media and used in static suppression. *Rhône-Poulenc UK.*

26216 Reslin
28434-01-7 1271 249-014-0
$C_{22}H_{26}O_3$
(1R trans)-2,2-dimethyl-3-(2-methyl-1-propenyl)cyclopropanecarboxylic acid [5-(phenylmethyl)-3-furanyl] methyl ester
5-benzyl-3-furylmethyl-(+)-*trans*-chrysanthemate; (+)-*trans*-resmethrin; NRDC-107; NIA-18739; SBP-1390; Resbuthrin; Biobenzylfuroline. Synergized bioresmethrin. An insecticide. $bp_{0.0008}$ = 174°; n_D^{20} = 1.5346; $[\alpha]_D^{20}$ = -8° (c = 5 Me_2CO); LD_{50} (rat orl) = 1244 mg/kg. *The Wellcome Foundation Ltd.* Name unverified.

26217 Reslin S
Synergized pyrethroid/S-bioallethrin *The Wellcome Foundation Ltd.* Name unverified.

26218 resmethrin
10453-86-8 233-940-7
$C_{22}H_{26}O_3$
5-(Phenylmethyl)-3-furanyl]methyl 2,2-dimethyl-3-(2-methyl-1-propenyl)cyclopropanecarboxylate
Benzofuroline; Chryson; Chrysron; Pynosect; Synthrin; Crossfire; 5-Benzylfurfuryl chrysanthemate; Bioresmethrin D-*trans* isomer); For-Syn; NIA-17370; Premgard; Pyretherm; Raid Flying Insect Killer; Scourge; Sun-Bugger 4; SPB-1382; Syntox; Vectrin; Whitmire PT-110. Non-systemic insecticide with contact action. Used as a household and garden insecticide and in agricultural premises. mp = 56°; SG = 1.044; insouble in H_2O (<1 mg/ml), very soluble in organic solvents; LD_{50} (rat orl) >2500 mg/kg.

26219 Resocoton
Mixtures of disperse and reactive dyestuffs; used for printing polyester/cotton blends. *Bayer AG.*

26220 Resogen® 35 Conc
Amino-aldehyde condensate; after-treatment fixing agent for improving wetfastness of direct dyes on cellulose; good penetration, excellent lightfastness; low formaldehyde (·5 1.5%). *Crompton & Knowles.*

26221 Resoglaz
A proprietary trade name for a polymerized styrene. No manufacturer.

26222 Resolamin
Dyestuffs; for the one-bath dyeing of wool and polyester blends. *Bayer AG.*

26223 Resolin®
Disperse dyestuffs; used for polyester fibers. *Bayer AG; Bayer plc.*

26224 Resolin® P
A range of disperse dyestuffs; used for the dyeing of polyamide fibers. *Bayer AG.*

26225 Resoltex
Phenolic and melaminic resins; used for bonding of wood, foundry sands, and abrasives. *RWE-DEA Chemicals UK Ltd.*

26226 Resopol
A proprietary polyester laminating resin. *DSM Resines France.* Discontinued.

26227 Resorband
Phenolic resin adhesives. *Georgia-Pacific.*

26228 resorcin
108-46-3 8323 203-585-2
$C_6H_6O_2$
resorcinol
m-dihydroxybenzene; 1,3-dihydroxybenzene; 3-hydroxyphenol;. Used in the manufacture of resorcinol-formaldehyde resins, dyes, cosmetics, pharmaceuticals, as a crosslinking agent in the manufacture of neoprene. mp = 109-110°; bp_{16} = 178°; soluble in H_2O (1.1 g/ml), organic solvents. *Fairmount; Indspec; Janssen Chimica; Penta Mfg.; Whitecourt Ltd.*

26229 resorcinal
A mixture of equal parts of resorcinol and iodoform; used as an antiseptic dusting-powder.

26230 resorcinol
108-46-3 8323 203-585-2
$C_6H_6O_2$
1,3-dihydroxybenzene
Resorcin; *m*-dihydroxybenzene; 1,3-benzenediol; 3-hydroxyphenol; C.I. 76505; *m*-hydroquinone; Eskamel. Used in tanning, manufacture of resins and resin adhesives, explosives, dyes and cosmetics, in dyeing and porinting of textiles, as a reagent for zinc and medically, as a keratolytic and antisbeorrheic. mp = 109-110°; bp = 280°; solubl in H_2O (1.1 g/ml), more soluble in organic solvents; *Fairmount; Indspec; Janssen Chimica; Penta Mfg.; Whitecourt Ltd.*

26231 Resorcinol benzoate
136-36-7 205-241-7
$C_{13}H_{10}O_3$
Resorcinol monobenzoate
Eastman® Inhibitor RMB. Industrial grade uv absorber/stabilizer for cellulosic plastics and PVC formulations; uv absorber. White crystalline solid; insoluble in water, benzene; soluble in acetone, ethanol; mw=214.23; sp gr>1.0; mp=132-135°; bp=140°; poison by intraperitoneal route; moderatle toxic by ingestion. *Eastman; Monomer-Polymer & Dajac.*

26232 Resorufin
635-78-9 211-241-8
$C_{12}H_7NO_3$
Hydroxyphenazone.

26233 Resovin
A proprietary synthetic resin of the vinyl type. No manufacturer.

26234 Resovyl
Textile auxiliary chemicals. *ICI Chem & Polymers Ltd.* Discontinued.

26235 Responsar
Insecticide for control of all insect pests, e.g., cockroaches, beetles, flies, mosquitoes. *Bayer AG.*

26236 Respumit®
Antifoam; dyeing and printing auxiliary. *Bayer AG; Bayer plc.*

26237 Restoration Cleaner
Blended organic and inorganic acids in combination with surfactants and wetting agents; for heavier duty concentrated cleaner for building exteriors. *Nova Chemical Inc.* Name unverified.

26238 Restoration Cleaner (Heavy Duty)
Blended organic and inorganic acids in combination with surfactants and wetting agents; used for heavier duty concentrated cleaner for building exteriors. *Nova Chemical Inc.* Name unverified.

26239 Restoration Cleaner (Super Heavy Duty)
Blended organic and inorganic acids in combination with surfactants and wetting agents; for heavier duty concentrated cleaner for building exteriors. *Nova Chemical Inc.* Name unverified.

26240 Restoration Rinse
Blended organic and inorganic acids in combination with surfactants and wetting agents; for light duty cleaning for historical building exteriors. *Nova Chemical Inc.* Name unverified.

26241 Restor-E (Restoration Chemical Products)
Commercial line of products for professional and do-it-yourself restoration of a variety of surfaces (wood, floors, soft goods, odor control, corrosion control) following fire, floods, etc. *Puma Chemical Co Inc.* Name unverified.

26242 Restore-X Exterior Paint Remover
Green color, heavybodied liquid, nonflammable, water soluble, sodium hydroxide remover; surface preparation tool for removal of deteriorated, exterior paints and heavy-bodied stains. *Restech Industries Inc.* Name unverified.

26243 Restore-x Weathered Wood Renewer
Blue color, heavy Bodied liquid, nonflammable, water soluble, sodium hydroxide remover; removes semi-transparent stain and the grey, weathered look from wood. *Restech Industries Inc.* Name unverified.

26244 Resydrol
Water soluble synthetic resins; used for paints and printing inks. *Resinous Chemicals Ltd.*

26245 Resyn® 28-1310
25609-89-6
Vinyl acetate/crotonic acid copolymer. Used as a hair fixative; for hair sprays, setting lotions, conditioners. *Nat'l. Starch.*

26246 Resyn® 28-2913
55353-21-4
Vinyl acetate/crotonic acid/vinyl neodecanoate copolymer. Used as a hair fixative providing excellent holding power, manageability, gloss; for aerosol and pump hairsprays, setting lotions, spritzes, cosmetics. *Nat'l. Starch.*

26247 Retain PE-1001
9002-88-4 7728
Recycle content polyethylene; for injection molding applications. *Dow Plastics.*

26248 Retain PE-5009
9002-88-4 7728
Recycle content polyethylene; for overwrap film applications and blown film extrusion. *Dow Plastics.*

26249 Retain PS-4000
9003-53-6 9028 203-066-0
Recycled content polystyrene resin; rubber modified; for injection molding applications *Dow Plastics.*

26250 Retain Rp-120
9002-88-4 7728
Pigmented recycle content polyethylene; for film, injection molding and blow molding applications. *Dow Plastics.*

26251 Retaminol®
Paper auxiliaries; used to increase the filler and pigment yields in paper manufacture, to improve drainage speed and for backwater clarification. *Bayer AG.*

26252 Retard
Potassium salt of maleic hydrazide; growth retardant for trees, shrubs, ivy and grass. *Draxel Chemical Company.* Unverified.

26253 Retarder AK
85-44-9 7528 201-607-5
$C_8H_4O_3$
1,3-isobenzofurandione
phthalic anhydride. Modified phthalic anhydride; nondiscoloring retarding agent to reduce scorching of rubber compounds at processing temps; also acts as an activator for certain blowing agents. mp = 131°; bp = 295°; d = 1.53; soluble in H_2O (6 mg/ml), soluble in organic solvents. *Akrochem.*

26254 Retarder BA, BAX
65-85-0 1122 200-618-2
Predominantly benzoic acid
Retarding agent for natural and synthetic rubbers and latexes; nonstaining; acts as an activator for certain blowing agents; processing aid with certain cis-polybutadiene rubbers; Bax is oil treated. *Akrochem.*

26255 Retarder ESEN
85-44-9 7528 201-607-5
Surface treated phthalic anhydride; a nondiscoloring retarder of vulcanization in all stock with all accelerators at processing temperature with a minimum retarding action at curing temperatures. *Uniroyal.* Name unverified.

26256 Retarder N
139-07-1 203-351-5
Dimethyl lauryl benzyl ammonium chloride
Retarder in dyeing of acrylics and polyesters; antistatic agent. *Hart Prods.Corp.*

26257 Retarder OC
Quaternary base; retarding agent for dyeing of acrylic fibers; substantive. *Eastern Color & Chem.*

26258 Retarder PX
85-44-9 7528 201-607-5
Phthalic anhydride, oil treated; nondiscoloring retarding agent to reduce scorching of rubber compounds at processing temps. *Akrochem.*

26259 Retarder SAFE
Treated aromatic sulfonamide; nonstaining, nondiscoloring retarder for natural and synthetic rubber compounds; also useful for replasticizing of slightly scorched stocks by cold mill mixing. *Akrochem.*

26260 Retarder SAX
69-72-7 8484 200-712-3
Tech. salicylic acid (90%) and light process oil treatment (10%); retarder; vulcanization inhibitor for SBR and natural rubber compounds; also as accelerator for W types of Neoprene; blowing agent activator in sponge rubber compounds *Akrochem.*

26261 Retarder V-48
Textile retarder; promotes leveling on fast striking cationic dyes. *Sybron.*

26262 Retardine
Retarder/leveling agent for dyeing cellulosics with vat, direct and sulfur dyes. *Henkel/Textiles.*

26263 Retardit A
Organic condensate; retarder and leveling agent offering good performance with improved economy. *Eastern Color & Chem.*

26264 Retardol
Flame retardants. *Albright & Wilson Ltd.*

26265 Retargal
Leveling agent; for dyeing acrylic fibers with basic dyes. *Sandoz Products Ltd.* Name unverified.

26266 Retariox
A proprietary trade name for an aldehyde-amine condensation product. No manufacturer.

26267 Reten®
High molecular weight synthetic water-soluble polymers; used as thickeners, flocculants, antistatic agents, film formers, adhesives, slip agents, solids-suspending agents and crosslinking agents. *Hercules.* Discontinued.

26268 Reten® 157
Acrylamide-based copolymer; retention aid for paper industry; provides maximum retention of fiber fines, paper chemicals, and fillers. *Hercules.* Discontinued.

26269 Reten® 763
Polymer solution; retention aid for fiber, fines, and fillers and provides drainage and flocculation of wh-water solids in paper machine and flotation save-all operations; functions over a wide pH range in alum and alum-free systems; *Hercules.* Discontinued.

26270 Reten® 1232
Acrylamide-based copolymer; retention aid and flocculant. *Hercules.* Discontinued.

26271 retene
483-65-8 8328 207-597-9
$C_{18}H_{18}$
1-methyl-7-(1-methylethyl)phenanthrene
methylisopropylphenanthrene; 1-methyl-7-isopropylphenanthrene. mp = 99°; bp = 390-394°; poorly soluble in H_2O (> 1 mg/ml), soluble in organic solvents.

26272 Reticusol
Hydrolyzed reticulin
Moisturizer; conditioner for skin care products. *Croda Inc.*

26273 Retilox
A proprietary range of organic peroxides suitable for polymer crosslinking. *Montedison UK Ltd.* Name unverified.

26274 Retilox® F 40 MG
25155-25-3 246-678-3
$C_{20}H_{34}O_4$
α,α'-Bis(*t*-butylperoxy)-*p*-diisopropylbenzene
peroxide, [phenylenebis(1-methylethylidine)bis(1,1-dimethylethyl; (phenylene diisopropylidine)bis(*tert*-butylperoxide). α,α'-Bis(t-butylperoxy)-m/p-diisopropylbenzene/EPM masterbatch; curing and crosslinking agents for EPM, EPDM, polyethylene, silicone rubbers, NBR, EVA copolymers, SBR, chlorosulfonated polyethylene, PVC, PU rubbers, polybutadiene rubbers, neoprene, n mp = 44-48°; insoluble in H_2O (< 1 mg/ml). *Akrochem.*

26275 Retingan® R6, R7, R48
Resin tanning materials; used for filling aftertreatment of chrome upper leather, particularly from cattle hides and sheepskins. *Bayer AG; Bayer plc.*

26276 retinite
retin asphalt; walchowite. A fossil resin found in brown coal. It occurs in Derbyshire and in Walchow. The material found near Walchow is a polymeric resin made up chiefly of sesquiterpenes.

26277 retinol
68-26-8 10150 200-683-7
$C_{20}H_{29}OH$
Codoil, rosinol, rosin oil. a) A product obtained by the distillation of rosin; b) vitamin A alcohol; used as an antiseptic and a vehicle for ointments.

26278 Retnolite
A proprietary trade name for a phenol-formaldehyde synthetic resin. No manufacturer.

26279 Retrocure® G
136-85-6 205-265-8
$C_7H_7N_3$
5-methyl-1H-benzo-1,2,3-triazole
5-methyl-1H-benzotriazole;. Blend of 4- and 5- methylbenzo triazole(tolyltrizole); prevulcanization retarder for sulfur modified (G type) polychloroprene rubbers; also for NBR systems where MBTS/sulfur cure systems are used. mp = 80-82°; bp_{12} = 210-212° *Akrochem.*

26280 Retz alloy
An alloy of 75% copper, 10% lead, 10% tin, and 5% antimony.

26281 reussinite
A reddish-brown resin found in certain coal deposits.

26282 Re-Vac II
Reovirus disease, inactivated vaccine; for immunization of poultry. *Intervet Inc.*

26283 Revacryl®
Acrylic and styrene acrylic copolymer dispersions; used for adhesives, surface coatings and textiles. *Harlow Chemical Co Ltd.*

26284 Revalon
A proprietary trade name for an alloy of 76% copper, 22% zinc, and 2% aluminum. No manufacturer.

26285 Revatol
Reserving agent; for printing. *Sandoz Products Ltd.* Name unverified.

26286 Revatol S
A proprietary trade name for sodium *m*-nitrobenzenesulfonate. No manufacturer.

26287 Revertex®
Range of evaporated natural rubber latex concentrates; binders for reconstituted leatherboard, modifiers for asphalt, bitumen, etc. *Revertex Ltd.*

26288 Revlen
Esterifilcon A
Butyl methacrylate polymer with butyl acrylate and ethylene dimethacrylate; contact lens material. *BioContacts Inc.* Unverified.

26289 Revolex
Antistatic formulations. *Rhône-Poulenc UK.*

26290 Revolite
A proprietary phenol-formaldehyde synthetic resin impregnated cloth. No manufacturer.

26291 Revona
A proprietary trade name for a water-soluble aminoplast; a very effective pitch dispersant in paper making. *Ciba plc.* Name unverified.

26292 Revuitex®
Proprietary range of prevulcanized natural rubber latex concentrates; used for dipped goods, e.g., medical, balloons, etc. *Revertex Ltd.*

26293 Rewagit
1344-28-1 369 215-691-6
Crystalline aluminum oxide as blasting corundum; especially suited for descaling, derusting, roughening of work piece surfaces and blasting of austenitic steels. *Hüls UK Ltd.*

26294 Rewo-amid
Deodorant additive, perfume extender additive, versatile additive; foam booster - pearlizing agent, thickening agent, superfatting agent; for synthetic detergents, soaps, shampoos, bubble baths, cosmetics, toiletries. *Rewo Chemicals Ltd.* Name unverified.

26295 Rewocid®
Undecylenic acid derivatives; surfactants. *Rewo Chemicals Ltd.*

26296 Rewocid® DU 185
Undecylenamide DEA and diethanolamine; detergent, emulsifier used as bacteriocide, thickener, foam stabilizer in shampoos. *Rewo GmbH.*

26297 Rewocid® SBU 185 P
Disodium undecylenamido MEA-sulfosuccinate
Antidandruff shampoo surfactant, fungicide. *Rewo GmbH.*

26298 Rewocid® U 185
Undecylenamide MEA; fungicide, antimycotic agent; detergent; improves foam quality, stability, superfatting, increases viscosity. *Rewo GmbH.*

26299 Rewocid® UTM 185
Undecylenamidopropyltrimonium methosulfate
Bactericide, fungicide for toiletries; conditioner for shampoos, antistat. *Rewo GmbH.*

26300 Rewocor
Anti-corrosive additive; for soluble cutting oils, etc. *Rewo Chemicals Ltd.* Name unverified.

26301 Rewocoros
Corrosion inhibitors. *Rewo Chemicals Ltd.*

26302 Rewocoros B 2045
Alkenyl sulfosuccinic acid anhydride
Corrosion inhibitor, tar adhesive agent. *Rewo GmbH.*

26303 Rewocoros B 3032
Alkenyl succinic acid, TEA salt; rust preventive additive in aqueous media. *Rewo GmbH.*

26304 Rewocoros BAC
Butyl ammonium caprylate
Corrosion inhibitor. *Rewo GmbH.*

26305 Rewocoros RA 280
Oleic acid dibutylamide
Corrosion inhibitor. *Rewo GmbH.*

26306 Rewocoros RAB 90
Modified boric DEA; low foaming corrosion inhibitor for soluble aqueous metalworking oils, synthetic cold lubricants, water glycol hydraulic fluids, grinding lubricants. *Rewo GmbH.*

26307 Rewocoros TPAC 100
Tallow propylene diammonium caprylate; corrosion inhibitor for oils and boiler feed water systems. *Rewo GmbH.*

26308 Rewocors B 3010
Alkenyl succinic acid, disodium salt; rust preventive additive in aqueous media. *Rewo Gmbh.*

26309 Rewoderm® ES 90
68201-46-7
PEG-7 glyceryl cocoate
Emulsifier for cosmetics, superfatting agent, solubilizer. *Rewo GmbH.*

26310 Rewoderm® LI 48
PEG-80 glyceryl tallowate
Mild surfactant, thickener and superfatting agent for cosmetics. *Rewo GmbH.*

26311 Rewoderm® S 1333
Disodium ricinoleamido MEA-sulfosuccinate
Detergent; used for skin protection, washing up liquids, personal care products; decreases irritancy of alkylbenzene sulfonate and other surfactants; emulsifier for emulsion polymerization. *Rewo GmbH.*

26312 Rewoderm® SPS
Disodium sitosteareth-14 sulfosuccinate. Surfactant for shampoos, shower and foam baths, mild skin cleaners, baby baths, skin care products. *Rewo GmbH.*

26313 Rewolan®
Lanolin derivatives; surfactants. *Rewo Chemicals Ltd.*

26314 Rewolan® 5
68890-92-6
Disodium laneth-5 sulfosuccinate
Detergent, moisturizer, superfatting agent for personal care products; skin protective agent. *Rewo GmbH.*

26315 Rewolan® AWS
PEG-75 lanolin oil; superfatting agent for personal care products. *Rewo GmbH.*

26316 Rewolan® LP
Isopropyl lanolate and lanolin; emollient for skin care, toiletries. *Rewo GmbH.*

26317 Rewolub KSM 80
Dicarboxylic acid diamide modified; surfactant for synthetic cooling oils, metalworking lubricants. *Rewo GmbH.*

26318 Rewolub TMP 275
Trimethylolpropane oleic acid ester
Lubricant in metalworking fluids and textile auxiliaries. *Rewo GmbH.*

26319 Rewomid®
Fatty acid alkylolamides. *Rewo Chemicals Ltd.*

26320 Rewomid® C 212
68140-00-1 268-770-2
Cocamide MEA
Detergent, thickener, foam booster/stabilizer, superfatting agent used in detergent products; stabilizer of emulsions. *Rewo GmbH.*

26321 Rewomid® DL 203 S
Lauramide DEA and diethanolamine; foam booster/stabilizer, superfatting agent, and thickener for personal care products, floor cleaners, general purpose cleaners, textile lubricants. *Rewo GmbH.*

26322 Rewomid® DO 280
93-83-4 202-281-7
$C_{22}H_{43}NO$
oleic acid diethanolamide
9-octadecenamide N,N-bis(2-hydroxyethyl)- (Z); oleic acid diethanolamine. Oleamide DEA and diethanolamine; detergent; products in the cosmetic, cleaning, and detergent industries. *Rewo GmbH.*

26323 Rewomid® DO 280 SE
93-83-4 202-281-7
Oleamide DEA (1:1)
Detergent, thickener, foam booster/stabilizer, superfatting agent for cosmetic products; conditioner for shampoos, bath oils. *Rewo GmbH.*

26324 Rewomid® F
Linoleamide DEA and diethanolamine; foam booster, superfatting agent and thickener for shampoos. *Rewo GmbH.*

26325 Rewomid® IPE 280
111-05-7 203-828-2
Oleamide MIPA
Detergent, emulsifier, foam stabilizer, thickener, superfatting agent; additive for skin protecting products. *Rewo GmbH.*

26326 Rewomid® IPL 203
142-54-1 205-541-8
Lauramide MIPA
Foam stabilizer, detergent for shampoos, shaving foams, and dishwashing liqs.; additive for solid and paste end products; improved washing power; stabilizer of emulsions. *Rewo GmbH.*

26327 Rewomid® IPP 240
Cocamide MIPA
Detergent, foam stabilizer, thickener, additive for solid and paste end products; improved washing power; emulsion stabilizer. *Rewo GmbH.*

26328 Rewomid® L 203
142-78-9 205-560-1
$C_{14}H_{29}NO_2$
Lauramide MEA
copramyl; crillon lme; 2-dodecanamidoethanol; N-(2-hydroxyethyl)lauramide; lauric acid ethanolamide; lauric ethylolamide; laurylamidoethanol; lauroylethanolamide; stabilor cmh; N-(2-hydroxyethyl)dodecaneamide; Lauryl monoethanolamide. Detergent, thickener, foam booster/stabilizer, superfatting agent for detergent preparations; fixation of perfumes; stabilizer of emulsions. *Rewo GmbH.*

26329 Rewomid® R 280
106-16-1 203-368-2
Ricinoleamide MEA
Surfactant for synthetic soap bars; foam stabilizer, thickener. *Rewo GmbH.*

26330 Rewomid® S 280
111-57-9 203-883-2
Stearamide MEA
Detergent, thickener, foam booster, superfatting agent; stabilizer of emulsions; synthetic soap bars; anti-inflammatory agent. *Rewo GmbH.*

26331 Rewomin
Amine oxides
Surfactants. *Rewo Chemicals Ltd.*

26332 Rewominox
Amine oxides
Surfactants. *Rewo Chemicals Ltd.*

26333 Rewominox B 204
68155-09-9 268-938-5
Cocamidopropylamine oxide
Foam booster, thickener for personal care products; antistat; hair conditioner; skin compatible. *Rewo GmbH.*

26334 Rewominox L 408
Lauramine oxide
Foam booster, thickener for personal care products; hair conditioner, antistat. *Rewo GmbH.*

26335 Rewominox S 300
2571-88-2 219-919-5
Stearamine oxide
Foam booster, antistat, hair conditioning agent. *Rewo GmbH.*

26336 Rewominoxid
Foam booster; conditioner; fabric softener; for shampoos and foam baths. *Rewo Chemicals Ltd.* Name unverified.

26337 Rewominoxid L 408
Lauramine oxide
Foam booster for personal care products. *Rewo GmbH.* Name unverified.

26338 Rewominoxid S 300
2571-88-2 219-919-5
Stearamine oxide
Foam booster, antistat, conditioner. *Rewo GmbH.* Name unverified.

26339 Rewomul MG SE
977053-96-5
Glyceryl stearate SE
Emulsifier for cosmetics. *Rewo GmbH.*

26340 Rewopal® BN 13
PEG-13 naphthole
Wetting agent for electroplating baths. *Rewo GmbH.*

26341 Rewopal® C 6
61791-08-0
PEG-6 cocamide
Dispersant, emulsifier, foam booster, wetting agent for calcium soap, personal care products; solubilizer for perfume oils. *Rewo GmbH.*

26342 Rewopal® HV 4
9016-45-9 6772
Nonoxynol-4
Emulsifier for mineral oils, petroleum, aliphatic hydrocarbons. *Rewo GmbH.*

26343 Rewopal® LA 3
Laureth-3
Emulsifier for solvents (cold water detergents), cosmetic oils (bath oils), mineral oils (textile auxiliary, metalworking), coupler, raw material for the production of ether sulfates; shampoos. *Rewo GmbH.*

26344 Rewopal® MPG 10
122-99-6 7410 204-589-7
$C_8H_{10}O_2$
2-Phenoxyethanol
Phenoxetol; Phenoxyethyl alcohol; Arosol; Ethylene glycol phenyl ether; 1-Hydroxy-2-phenoxyethane; β-Hydroxyethyl phenyl ether; Ethylene glycol mono phenyl ether; Euxyl K 400; Phenyl cellosolve; Phenoxethol; Phenoxyl ethanol; glycol monophenyl ether; phenoxytol; phenylmonoglycol ether; 2-hydroxyethyl phenyl ether; β-phenoxyethyl alcohol; dowanol ep; dowanol eph; emeressence 1160; emery 6705; rose ether; Ethanol-2-phenoxy. Solvent, solubilizer for preservatives. mp = 14°; bp = 245°; d = 1.102; n_D^{20} = 1.534; soluble in H_2O (10-50 mg/ml), soluble in organic solvents; LD_{50} (rat orl) = 1.26 g/kg. *Rewo GmbH.*

26345 Rewopal® MT 65
Fatty alcohol/PEG methyl ether
Low foaming detergent for strong acidic cleaners, textile auxiliaries; acid-stable. *Rewo GmbH.*

26346 Rewopal® O 8
PEG-9 oleamide
Detergent; wetting agent, oil-water emulsifier, and dispersant for calcium soap; suitable for machine washing formulations. *Rewo GmbH.*

26347 Rewopal® PEG 6000 DS
9005-08-7
PEG-150 distearate
Thickener for toiletries. *Rewo GmbH.*

26348 Rewopal® PG 280
627-83-8 211-014-3
$C_{38}H_{74}O_4$
Glycol distearate
ethylene glycol distearate. Pearlizing agent for cosmetics. *Rewo GmbH.*

26349 Rewopal® RO 40
Castor oil ethoxylate; emulsifier for metalworking fluids, textile auxiliaries, agricultural chemicals. *Rewo GmbH.*

26350 Rewopal® TA 11
61791-28-4
Talloweth-11
Detergent, wash-active base, wetting agent, dispersant, emulsifier for waxes; biodegradable. *Rewo GmbH.*

26351 Rewophat
Phosphate esters
Surfactants. *Rewo Chemicals Ltd.*

26352 Rewophat EAK 8190
39464-66-9
Laureth-3 phosphate
Corrosion inhibitor, emulsifier, dispersant, wetting agent; for metalworking fluids, textile auxiliaries and cleaners; antistat; high pressure additive; biodegradable. *Rewo GmbH.*

26353 Rewophat NP 90
Nonylphenol polyglycol ether phosphate
Emulsifier for emulsion polymerization; textile auxiliaries; antistat, raw material for industrial cleaners. *Rewo GmbH.*

26354 Rewophos EAK 8190
Phosphate ester alkyl polyglycol ether in liquid form; corrosion inhibitor, emulsifier, antistat for metal treatment agents. *Rewo Chemicals Ltd.* Name unverified.

26355 Rewophos TD40
Phosphate ester alkyl polyglycol ether in liquid form; low foaming surfactant for metal treatment and antistatic applications. *Rewo Chemicals Ltd.* Name unverified.

26356 Rewophos TD70 and OP80
Phosphate ester alkykl polyglycol ether in liquid form; surfactant/hydrotrope used in metal treatment, antistatic applications and textile auxillary applications. *Rewo Chemicals Ltd.* Name unverified.

26357 Rewopol
Anionic surfactants. *Rewo Chemicals Ltd.*

26358 Rewopol® B 1003
90268-48-7 290-850-0
Disodium tallow sulfosuccinamate

Foaming and antigelling agent for latex foam backings and coatings; emulsifier for emulsion polymerization; flotation agent. *Rewo GmbH.*

26359 Rewopol® B 2003
Tetrasodium dicarboxyethyl stearyl sulfosuccinamate
Flotation reagent; emulsifier for emulsion polymerization; foaming agent for latex emulsion (carpet backing); antigelling agent, cleaning agent for paper mill felts. *Rewo GmbH.*

26360 Rewopol® CHT 12
Coco-EDTA-amide
Sequestering agent, complexing surfactant for detergents. *Rewo GmbH.*

26361 Rewopol® CL 30
Laureth-3 carboxylic acid
Surfactant for household cleaners and toiletries. *Rewo GmbH.*

26362 Rewopol® CLN 100
33939-64-9
Sodium laureth-11 carboxylate
Surfactant for toiletries. *Rewo GmbH.*

26363 Rewopol® CT 65
Trideceth-7 carboxylic acid
Acid-stable cleaner for household and industrial use, textile auxiliaries, mineral oil emulsions, tertiary oil recovery; emulsifier, wetting agent for personal care products. *Rewo GmbH.*

26364 Rewopol® DLS
DEA-lauryl sulfate in liquid form; raw material for shampoos, detergents, etc. *Rewo Chemicals Ltd.*

26365 Rewopol® MLS 30
MEA-lauryl sulfate
Surfactant for personal care products, foam baths, shampoos, liquid detergents; detergent raw material. *Rewo GmbH.*

26366 Rewopol® NEHS 40
Sodium octyl sulfate
Low foaming wetting agent for alkaline cleaners, mercerizing, electroplating; hydrotrope. *Rewo GmbH.*

26367 Rewopol® NL 2-28
9004-82-4
Sodium laureth sulfate
Surfactant for shampoos, shower gels, foam baths, liquid soaps, dishwashing liquids, emulsion polymerization, air entrainment agent, textile auxiliaries. *Rewo GmbH.*

26368 Rewopol® NLS 15 L
151-21-3 8782 205-788-1
$C_{12}H_{25}NaO_4S$
Sodium lauryl sulfate
Emulsifier for emulsion polymerization. *Rewo GmbH.*

26369 Rewopol® NOOSE 5
Nonyl phenol polyglycol ether sulfate
Emulsion polymerization surfactant especially for styrene. *Rewo GmbH.*

26370 Rewopol® SBC 212
Disodium cocamido MEA-sulfosuccinate
Surfactant for foam cleaners, light duty detergents, and personal care products. *Rewo GmbH.*

26371 Rewopol® SBDB 45
127-39-9 3238 204-839-5
$C_{12}H_{21}NaO_7S$
sulfobutanedioic acid 1,4-bis(2-methylpropyl) ester sodium salt
Diisobutyl sodium sulfosuccinate; sodium dibutyl sulfosuccinate; Aerosol IB; Alphasol IB. Emulsifier for emulsion polymerization; stabilizer for dispersions; pigment dispersant. *Rewo GmbH.*

26372 Rewopol® SBDC 40
23386-52-9 245-629-3
Dicyclohexyl sodium sulfosuccinate
Emulsifier for emulsion polymerization; stabilizer for disps.; pigment dispersant. *Rewo GmbH.*

26373 Rewopol® SBDD 65
Diisodecyl sodium sulfosuccinate
Emulsifier for emulsion polymerization; stabilizer for disps.; pigment dispersant. *Rewo GmbH.*

26374 Rewopol® SBDO 75
577-11-7 3460 209-406-4
Dioctyl sodium sulfosuccinate
Wetting agent, solubilizer; emulsion polymerization. *Rewo GmbH.*

26375 Rewopol® SBF 12
Disodium lauryl sulfosuccinate
Detergent raw material for personal care products; carpet and upholstery shampoos; toilet and syndet soaps. *Rewo GmbH.*

26376 Rewopol® SBFA 30
Disodium laureth sulfosuccinate
Detergent raw material for personal care products, cleansing agents. *Rewo GmbH.*

26377 Rewopol® SBL 203
25882-44-4 247-310-4
Disodium lauramido MEA-sulfosuccinate
Detergent for cosmetics; aids spray-drying; carpet and upholstery shampoos; soaps. *Rewo GmbH.*

26378 Rewopol® SBMB 80
Diisohexyl sulfosuccinate
Emulsion polymerization surfactant; stabilizer for dispersions; pigment dispersant. *Rewo GmbH.*

26379 Rewopol® SBZ
Disodium PEG-4 cocamido MIPA-sulfosuccinate
Surfactant for mild foam baths, shampoos, light duty detergents. *Rewo GmbH.*

26380 Rewopol® SK 275
Oleyl sarcosinic acid
Emulsifier and corrosion inhibitor for mineral oils. *Rewo GmbH.*

26381 Rewopol® SLS
Sodium lauryl sulfate. Sodium lauryl sulfate in paste form; raw material for shampoos, detergents, etc. *Rewo Chemicals Ltd.*

26382 Rewopol® SMS
Alkyl disodium sulfosuccinamate as a clear liquid; spreading and penetrating agent for latex emulsions. *Rewo Chemicals Ltd.*

26383 Rewopol® TLS 40
139-96-8 205-388-7
TEA-lauryl sulfate
Surfactant raw material for shampoos, foam baths, liquid detergents. *Rewo GmbH.*

26384 Rewopol® TLS 90 L
661-61-6
TIPA-Lauryl sulfate
Raw material for high active ingredient cosmetic products, foaming bath oils. *Rewo GmbH.*

26385 Rewopol® TMS and ODS
Alkyl disodium sulfosuccinamate in liquid or paste form; foaming agent for latex emulsions, emulsifier for emulsion polymerization. *Rewo Chemicals Ltd.*

26386 Rewopon®
Imidazolines/amines
Surfactants. *Rewo Chemicals Ltd.*

26387 Rewopon® IM OA
1-Hydroxyethyl-2-alkyl-imidazoline
Corrosion inhibitor, emulsifier, antistat. *Rewo GmbH.*

26388 Rewopon® IM-BT
Fatty acid quaternary imidazoline in liquid/paste form; cationic surfactant which improves adhesion of bitumens and other binding and coating agents. *Rewo Chemicals Ltd.*

26389 Rewoquat B 10
7173-51-5 3149 230-525-2
$C_{22}H_{48}ClN$
didecyldimethylammonium chloride
Didecyldimonium chloride; Bardac 2250-2280; BTC-1010; Dodigen 1881; Querton 210CL. Disinfectant for cleaners, e.g., for dairies and food industry. Soluble in Me_2CO, C_6H_6, insoluble in hexane. *Rewo GmbH.*

26390 Rewoquat B 50
8001-54-5 1086 231-635-3
Benzalkonium chloride
Disinfectant for cleaners, dairy and food industries; algicide; textile dyeing auxiliary *Rewo GmbH.*

26391 Rewoquat CPEM
68989-03-7
PEG-5 cocomonium methosulfate
Hair conditioner for shampoos, emulsifier in emulsion polymerization, antistat. *Rewo GmbH.*

26392 Rewoquat CR 3099
Difatty acid ester dimethyl ammonium methosulfate as a viscous liquid; softener, dry-cleaning agent, textile auxiliary, leather auxiliary. *Rewo Chemicals Ltd.*

26393 Rewoquat DQ 35
93572-63-5 297-495-0
PEG-3 tallow propylenedimonium dimethosulfate
Antistat and wetting agent. *Rewo GmbH.*

26394 Rewoquat RTM 50
Ricinoleamidopropyltrimonium methosulfate
Conditioner for personal care products, liquid soaps; antistat. *Rewo GmbH.*

26395 Rewoquat W 75 H
91723-55-6 294-563-1
Quatermium-83
Fabric softener. *Rewo GmbH.*

26396 Rewoquat W 75 PG
Quaternium-27, free of isopropanol; fabric softener. *Rewo GmbH.*

26397 Rewoquat W 222 LM
Ditallow amidoammonium methosulfate
Fabric softener. *Rewo GmbH.*

26398 Rewoquat W 3690, W 3690 PG
Dioleyl imidazoline methosulfate
Fabric softener. *Rewo GmbH.*

26399 Rewoquat W 7500
Quaternary imidazoline in liquid form; cationic surfactant used as a fabric softener. *Rewo Chemicals Ltd.*

26400 Rewoquaternary
Cationic fabric softeners. *Rewo Chemicals Ltd.*

26401 Reworyl®
Alkylaryl sulfonates
Surfactants. *Rewo Chemicals Ltd.*

26402 Reworyl® ACS
37475-88-0 253-519-1
Ammonium cumene sulfonate
Anionic surfactant in liquid form; hydrotrope. *Rewo Chemicals Ltd.*

26403 Reworyl® C
28631-63-2 249-112-3
Cumene sulfonic acid
Anionic surfactant in liquid form; catalyst for synthetic resins; hydrotropes. *Rewo Chemicals Ltd.*

26404 Reworyl® K
Dodecylbenzene sulfonic acid
Biodegradable raw material for anionic detergent systems. *Rewo GmbH.*

26405 Reworyl® KXS
Potassium xylene sulfonate
Anionic surfactant in liquid form; hydrotrope. *Rewo Chemicals Ltd.*

26406 Reworyl® NCS
32073-22-6 250-913-5
Sodium cumene sulfonate
Anionic surfactant in liquid form; hydrotrope. *Rewo Chemicals Ltd.*

26407 Reworyl® NKS 100
Sodium dodecylbenzene sulfonate
Detergent and cleaner, textile auxiliaries. *Rewo GmbH.*

26408 Reworyl® NTS
Sodium toluene sulfonate
Anionic surfactant in liquid form; hydrotrope. *Rewo Chemicals Ltd.*

26409 Reworyl® NXS 40
1300-72-7 215-090-9
Sodium xylene sulfonate
Hydrotrope for detergent systems. *Rewo GmbH.*

26410 Reworyl® T
657-84-1 9671 211-522-5
p-toluene sulfonic acid
Anionic surfactant in liquid form; catalyst for foundry resins. *Rewo Chemicals Ltd.*

26411 Reworyl® TKS 90/L
TEA dodecylbenzene sulfonate
Biodegradable raw material for liquid detergents. *Rewo GmbH.*

26412 Reworyl® X
25321-41-9 246-839-8
Xylene sulfonic acid
Anionic surfactant in liquid form; catalyst for synthetic resins. *Rewo Chemicals Ltd.*

26413 Rewoteric®
Amphoteric baby shampoos; surfactants. *Rewo Chemicals Ltd.*

26414 Rewoteric® AM 2C NM
61791-32-0 263-164-4
Disodium cocoamphodiacetate
Mild nonirritating raw material for personal care products. *Rewo GmbH.*

26415 Rewoteric® AM 2C SF
Disodium cocoamphodipropionate
High-foaming surfactant for baby and intimate hygiene products; excellent for high pH systems; coupler and detergent in high electrolyte formulations. *Rewo GmbH.*

26416 Rewoteric® AM 2L-40
14350-97-1 238-306-3
Disodium lauroamphodiacetate
Mild nonirritating raw material for baby shampoos, mild shampoos, foam baths, intimate hygiene products. *Sherex/Div. of Witco.*

26417 Rewoteric® AM B-13
61789-40-0 263-058-8
Cocamidopropyl betaine
Mild raw material, foam booster, viscosity builder for personal care products (shampoos, skin cleansers), liquid soaps, all-purpose cleaners; lime soap dispersant; coemulsifier. *Sherex/Div. of Witco.*

26418 Rewoteric® AM B-14LS
Cocamidopropyl betaine
Low-salt surfactant for personal care products. *Sherex/Div. of Witco.*

26419 Rewoteric® AM B-15
Cocamidopropyl betaine
Foam booster for shampoos, viscosity builder, low-irritation skin cleanser, lime soap dispersant; aids in deposition of protein and cationic polymers on hair; coemulsifier. *Sherex/Div. of Witco.*

26420 Rewoteric® AM CA
Disodium lauroamphodiacetate and sodium laureth sulfate; raw material for mild foam baths, hair shampoos, baby preps. *Rewo GmbH.*

26421 Rewoteric® AM CAS
68139-30-0 268-761-3
Cocamidopropyl hydroxysultaine
High-foaming mild detergent for shampoos, bubble bath; coemulsifier. *Sherex/Div. of Witco.*

26422 Rewoteric® AM CAS-15
68139-30-0 268-761-3
Cocamidopropyl hydroxysultaine
Surfactant used in personal care products and acid and alkaline cleaners. *Sherex/Div. of Witco.*

26423 Rewoteric® AM DM-35L
11140-78-6 234-401-9
Lauryl betaine
High-foaming mild surfactant for baby shampoos, hard surface cleaners, steam jet cleaners, and pickling baths. *Sherex/Div. of Witco.*

26424 Rewoteric® AM G30
Sodium lauroamphoacetate, sodium lauryl sulfate and hexylene glycol; surfactant for baby and child care cosmetic formulations. *Rewo GmbH.*

26425 Rewoteric® AM HC
13197-76-7 236-164-7
Lauryl hydroxysultaine
Mild detergent foam booster, skin conditioning agent; hydrotrope; excellent for high pH systems. *Sherex/Div. of Witco.*

26426 Rewoteric® AM R40
86089-12-5 289-181-7
Ricinoleamidopropyl betaine
Mild cosmetic and medicated cleaning agents; baby and child care cosmetics. *Rewo GmbH.*

26427 Rewoteric® AM TEG
61791-25-1
Dihydroxyethyl tallow glycinate
High-foaming mild surfactant for acidic and alkaline cleaning products, conditioning shampoos; thickener. *Rewo GmbH.*

26428 Rewoteric® AM V
13039-35-5 235-907-2
Sodium capryloamphoacetate
Raw material, wetting agent, rust inhibitor for pickling baths, acid and alkaline cleaners. *Sherex/Div. of Witco.*

26429 Rewoteric® AM VSF
Sodium capryloamphopropionate, salt-free; wetting agent for alkaline and acidic cleaners. *Rewo GmbH.*

26430 Rewoteric® AMKSF 40
93820-52-1 298-632-7
Sodium cocoamphopropionate, salt-free; raw material, high-foaming surfactant, wetting agent used in personal care products (baby and intimate hygiene products) and industrial cleaners. *Rewo GmbH.*

26431 Rewoteric® AMLP
Sodium lauryliminodipropionate
High foaming surfactant for baby and intimate hygiene products; excellent for high pH systems. *Sherex/Div. of Witco.*

26432 Rewoteric® QAM 50
100085-64-1 309-206-8
Cocobetainamido amphopropionate
Surfactant for hard surface disinfectants, deodorants, cleaning and bacteriocidal products for use in industrial catering, hotels, hospitals, and households; biodegradable. *Rewo GmbH.*

26433 Rewowax
Synthetic waxes. *Rewo Chemicals Ltd.*

26434 Rewowax CG
540-10-3 2073 208-736-6
$C_{32}H_{64}O_2$
Cetyl palmitate
hexanoic acid hexadecyl ester; hexadecyl palmitate. Spermaceti wax replacement; emulsifier for cosmetic creams and lotions. mp = 54°; d^{20} = 0.989; n_D^{70}= 1.4398; insoluble in H_2O, soluble in EtOH, Et_2O. *Rewo GmbH.*

26435 Rex
1344-28-1 369 215-691-6
A trademark for abrasive goods consisting essentially of alumina. No manufacturer.

26436 Rex95
A proprietary trade name for a cobalt steel containing 5% cobalt, 14% tungsten, and 4% chromium, 2% vanadium, and 0.5% molybdenum. No manufacturer.

26437 Rex-blak
A special preparation of carbon black, containing carbon, rubber, and glue, in varying proportions; used in rubber mixings.

26438 Rexene® 11S12
9003-07-0 7741
PP homopolymer; superclean grade for electronics, medical and food packaging, material handling equipment for cleanroom environments, tooling for healthcare devices. *Rexene Prods.*

26439 Rexene® 13S10A
9003-07-0 7741
Copolymer; radiation-resistance improved clarity grade for injection molded parts in medical market, food packaging, see-through reusable household storage containers, electronics, consumer products *Rexene Prods.*

26440 Rexene® 14S4A
9003-07-0 7741
PP med.-impact copolymer; easy processing for injection molding. *Rexene Prods.*

26441 Rexene® 18C3A
9003-07-0 7741
PP impact copolymer; very high impact grade for underhood automotive applications, consumer products. *Rexene Prods.*

26442 Rexene® PE 1903
Ethylene-vinyl acetate (9%) copolymer; excellent toughness and flexibility at temperatures, good heat seal response; for laminating film, controlled atmosphere and heavy-duty packaging, medical tubing. *Rexene Prods.*

26443 Rexene® PE 6010
9002-88-4 7728
LDPE; good stress crack resistance; for blow molding of small bottles, containers, vials. *Rexene Prods.*

26444 Rexene® PE6076
9002-88-4 7728
LDPE homopolymer; for blow molding of bottles with good clarity and excellent stiffness. *Rexene Prods.*

26445 Rexene® PP 12R10A
9003-07-0 7741
PP random copolymer; good contact clarity, impact resistance, stiffness, high deflection temperature; for rigid containers, single-use food packaging, laboratoryware, medicine vials, hypodermic syringes, radiation sterilization resistance applications. *Rexene Prods.*

26446 Rexene® PP 23S2
9003-07-0 7741
PP copolymer; blow molding grade for bottles. *Rexene Prods.*

26447 Rexene® PP 9234
9003-07-0 7741
PP copolymer; radiation-resistance grade for thermoformed trays, basins, and utensils for medical field; gamma sterilizable with little change in color, strength, or flexibility; FDA compliance. *Rexene Prods.*

26448 Rexene® PP41E2
9003-07-0 7741
PP homopolymer resin; for extrusion into highest quality biaxially oriented capacitor film with stiffness and excellent optical and barrier properties; high dielectric strength, low electrical loss. *Rexene Prods.*

26449 Rexenite
A proprietary trade name for a cellulose acetate butyrate plastic. No manufacturer.

26450 Rexhide
A rubber-glue stock material for use in rubber mixings.

26451 Rexin X
Liquid mixed glycol ester of rosin; plasticizer and tackifier in adhesive compositions. *Blueminster Ltd.*

26452 Rexite
A blasting explosive, containing 6.5-8.5% nitroglycerin, 64-68% ammonium nitrate, 13-16% sodium nitrate, 6.5-8.5%, trinitro-toluene, and 3-5% wood meal.

26453 Rexobase BAT
Alkylaryl detergent; economical detergent and emulsifier which can be cut 50-50 with water for use as detergent and 50-50 with varsol to make a solvent scour. *Emkay.*

26454 Rexobase EN
Emulsifier for naphtha or toluene. *Emkay.*

26455 Rexobase HD
Degreaser for hide tanning; stable in brine solution *Emkay.*

26456 Rexobase XX
Amide condensate; emulsifier for varsol or Stod.; forms bright soluble oil used in water solutions for the removal of stubborn soil and grease. *Emkay.*

26457 Rexobond
Resin gum; prepared weighter and body builder finish compatible with softeners and glues; excellent lining finish (textiles). *Emkay.*

26458 Rexoclean
Synthetic detergents, terpenes, and solvents; complete scouring agent for stubborn oil and grease stains and for removal of nylon warp size. *Emkay.*

26459 Rexoclean 200N
Emulsified blend of solvents and detergents; multipurpose cleaner for heavy duty cleaning including rubber marks and stains, in soap dispensers. *Emkay.*

26460 Rexodull CT
Multipurpose substantive delustrant with a titanium dioxide base. *Emkay.*

26461 Rexofos WS
Mixture of water softeners and mild alkalies for softening and conditioning water. *Emkay.*

26462 Rexogel
Purified gelatin base product; economical stiff finish for textiles requiring no curing. *Emkay.*

26463 Rexogum GL
Blend of selected glues and gums; finishing agent for textile fabrics; gives a clear film without discoloration. *Emkay.*

26464 Rexol
A mixture of ammonium perchlorate, potassium chlorate, rosin, zinc or aluminum, and mineral oil or wax; an explosive.

26465 Rexol 2000 HWM
Amide ethoxylate, modified; antistatic emulsifier for mineral oils, manufacturing of textile lubricants; anticorrosive and cohesive properties *Hart Chem. Ltd.*

26466 Rexol 25/10
9016-45-9 6772
Nonoxynol-10
Detergent, dispersant, emulsifier, wetting agent; for paint, textiles, pulp/paper industries, emulsion polymerization; degreaser for leather; scouring agent for raw wool. *Hart Chem. Ltd.*

26467 Rexol 25/4
9016-45-9 6772
Nonoxynol-4
Detergent, dispersant, stabilizer; low foaming emulsifier for oils and petroleum solvents, emulsion polymerization; intermediate for anionic sulfonates; pigment dispersant. *Hart Chem. Ltd.*

26468 Rexol 35/3
Alcohol ethoxylate
Emulsifier for oil-water mixtures; intermediate for sulfation and phosphation. *Hart Chem. Ltd.*

26469 Rexol 45/1
9002-93-1 6858
Octoxynol-1
Emulsifier, detergent, dispersant; coemulsifier for surfactant and solventpreparation blends. *Hart Chem. Ltd.*

26470 Rexol 65/4
9014-92-0
Dodoxynol-4
Ingredient for solvent cleaner and drycleaning formulations; emulsifier for agricultural oils; degreaser. *Hart Chem. Ltd.*

26471 Rexol AE-1
Linear alcohol ethoxylate
Intermediate for shampoo and detergent manufacturing. *Hart Chem. Ltd.*

26472 Rexole 612
Plasticizer and softener for use in textile sizing formulations. *Emkay.*

26473 Rexolene
Sulfonated base dispersing agent for acetate dyes; dyeing assistant. *Emkay.*

26474 Rexoloid
Alkyd resin; clear resin requiring no curing; used as a ribbon finish on acetates; imparts stiff hand with high luster. *Emkay.*

26475 Rexolube
Lubricant and chafe eliminator for use in the diazotizing bath to eliminate chafes on developed black and navy colors. *Emkay.*

26476 Rexonic 1006
Linear alcohol ethoxylate
Wetting and scouring agent for natural and synthetic fibers; leveling and dispersing aid for acid and disperse dyes; foaming agent for kuster dyeing carpets; compatible with dyestuffs over wide pH range; biodegradable. *Hart Chem. Ltd.*

26477 Rexonic N23-3
Linear alcohol ethoxylate
Detergent, emulsifier, dispersant, and chemical intermediate used in solvent emulsion cleaners and dry cleaning detergents; biodegradable. *Hart Chem. Ltd.*

26478 Rexonic P-1
Propoxylated ethoxylated linear alcohol; lowfoaming detergent, wetter, emulsifier, dispersant for natural and synthetic fibers, industrial and commercial low foam detergents; base for rinse aids and machine dishwashing formulations. *Hart Chem. Ltd.*

26479 Rexonic RL
Alcohol ethoxylate
Detergent, post-scouring agent. *Hart Chem. Ltd.*

26480 Rexonit D
Protein-resin mixture; heavy finish for rayon and acetate fabrics; compatible with salt weighters for finishing knit goods. *Emkay.*

26481 Rexopal 3928
Modified alcohol ethoxylate; low foaming wetting and scouring agent for natural and synthetic fibers; biodegradable. *Hart Chem. Ltd.*

26482 Rexopal SM-5
Nonfoaming surfactant, wetting agent, detergent for prescouring. *Hart Chem. Ltd.*

26483 Rexopene
Sodium alkylaryl sulfonate
Wetting, dye leveling agent, penetrant, scouring assistant. *Emkay.*

26484 Rexophos 25/67
Alkylaryl phosphate ester
Emulsifier for solvents used over wide temperature range; surfactant for hard surface cleaners, alkaline metal cleaners. *Hart Chem. Ltd.*

26485 Rexophos 4668
Linear alcohol phosphate ester
Detergent; stable to highly alkaline cleaning solutions. *Hart Chem. Ltd.*

26486 Rexophos BP-2
Linear alcohol phosphate ester
Scouring agent and cleaner for textile industry; high alkaline stable. *Hart Chem. Ltd.*

26487 Rexopon E
Amide amine condensate, modified; synthetic detergent; after wash for prints. *Emkay.*

26488 Rexopon SK
Silk degumming and scouring agent; used with soap or detergent to accelerate degumming and boil-off of silk goods. *Emkay.*

26489 Rexoscour SF
Scour and fulling agent for wool. *Emkay.*

26490 Rexoslip AS
Nonyellowing nonslip agent and weighter used primarily as satin finish. *Emkay.*

26491 Rexosolve 150
Sulfonated oils/emulsifier blend; detergent, scouring assistant; oil and grease remover. *Emkay.*

26492 Rexoteric XCE
Carboxyethylated coco amphoteric in liquid form; for alkaline industrial cleaners. No manufacturer.

26493 Rexoteric XCG
Specially designed amphoteric in liquid form; used in nonirritating conditioning shampoos, bath and toiletry products when high viscosity is required. No manufacturer.

26494 Rexoteric XCO
Carboxymethylated coco imidazoline derivative in liquid form; for high foaming, nonirritating shampoos, bath and toiletry products. No manufacturer.

26495 Rexoteric XJO
Carboxymethylated caprylic imidazoline derivative in liquid form; detergent and wetting agent for low foaming alkaline and acid cleaners. No manufacturer.

26496 Rexoteric XOO
Carboxymethylated oleic imidazoline derivative as a viscous liquid; imparts softening properties and is used especially in high viscosity hand cleaners and detergent formulations. No manufacturer.

26497 Rexoteric YCB
Carboxyethylated coco imidazoline derivative; high foaming detergent for use in all-purpose cleaners. Viscosity modulator, gel-builder and conditioner for low pH nonirritating shampoos, bath and toiletry products. No manufacturer.

26498 Rexoteric YCE
Carboxyethylated coco amine derivative; high foaming light to heavy duty cleaners in acid to alkaline pH range. No manufacturer.

26499 Rexoteric YJE
Carboxyethylated short chain amine derivative; moderately foaming light to heavy duty cleaners, giving good surface tension reduction over the whole pH range. No manufacturer.

26500 Rexoteric YOB
Carboxyethylated alkyl imidazoline derivative; specially designed for use in corrosion inhibitors and as a softener. No manufacturer.

26501 Rexoteric ZXCO
A blend of Rexoteric XCO and an anionic surfactant, in liquid form; ready-made base for nonirritating shampoos, bath and toiletry products. No manufacturer.

26502 Rexowax CNN
Water-dispersible wax and resin; recommended as satin finish. *Emkay.*

26503 Rexowet 500
Sulfonated dioctyl succinate
Surface tension depressant; stable in acid, alkaline, and neutral baths. *Emkay.*

26504 Rexowet 77
Heptadecyl sodium sulfate
Fast wetting and penetrating agent for textiles; resistant to acid and alkaline media. *Emkay.*

26505 Rexowet ASG-81
577-11-7 3460 209-406-4
Sodium dioctyl sulfosuccinate
Wetting agent for textile use. *Emkay.*

26506 Rexowet CR
Sulfonated isopropyl oleate and cresylic acid; wetting agent and penetrant for cotton fabrics; scouring agent for removing mineral oil stains. *Emkay.*

26507 Rextox
A proprietary trade name for a material consisting of copper with a layer of cuprous oxide formed on the surface of the metal at high temperatures. No manufacturer.

26508 Rextrude
A proprietary trade name for a cellulose-acetate-butyrate plastic. No manufacturer.

26509 Reyalite, Reyalith
Aluminum alloys containing lithium. *Reynolds Metal Co.* Discontinued.

26510 Reyalith
Aluminum alloys containing lithium. *Reynolds Metal Co.* Discontinued.

26511 Reycomp
Laminated sheets of aluminum having plastic inner cores. *Reynolds Metal Co.* Discontinued.

26512 Reydox
Aluminum and aluminum alloy castings. *Reynolds Metal Co.*

26513 Reynolds Wrap
Aluminum foil sheets and rolls; for consumer use in cooking, wrapping, etc. *Reynolds Metal Co.*

26514 Reynolon
Plastic film, laminated and unsupported; for consumer and food use. *Reynolds Metal Co.*

26515 Reynolon
PVC with various plasticizers and stabilizers; shrink film for presentation applications. *S Kempner Ltd.*

26516 REZ 300
Urea-formaldehyde resin; theromsetting resin paste producing superior crease resistance and shrinkage control on cottons and synthetics with a smooth, round hand. *CNC Int'l L.P.*

26517 Rezal®
Aluminum-zirconium tetrachlorohydrex glycine or pentachlorohydrox glycine; antiperspirant actives. *Reheis Inc.*

26518 Rezal® 36GP
Aluminum zirconium tetrachlorohydrexglycine
Antiperspirant active. *Reheis Inc.*

26519 Rezcat 2, 3
Highly reactive polyfunctional aziridine products; crosslinkers in polymeric systems containing carboxyl or hydroxyl functionalities; improves water and solvent resistance and increases adhesion to many substrates. *CNC Int'l L.P.*

26520 Rezex
Construction auxiliaries. *Crosfield Chemicals Ltd.*

26521 Rezifilm
137-26-8 9510 205-286-2
$C_6H_{12}N_2S_4$
tetramethlthiuram disulfide
TMTD; Arasan; sq 1489; Fernasan; Nomersan; Pomarsal; Puralin; Rezifilm; Tersan; Tetramethylthioperoxydicarbonic diamide; Thiosan; Thiurad; Thiuramyl; Thylate; Tiuramyl; Tuads; Tulisan; Pomarsol; AAtack; Aules; Chipco Thiram 75; Fermide 850; Hexathir; Mercuram; Nomersam; Polyram-Ultra; Pomarsol forte; Spotrete-F; Spotrete WP 75; Tetrapom; Thimer; Thioknock; Thiotex; Thiramad; Thirasan; Thiuramin; Tirampa; TMTDS; Trametan; Tripomol; Vancide TM; tetramethylthiuram bisulfide; accelerator thiuram; arasan 70; arasan-m; cyuram ds; ekagom tb; falitiram; fermide; fernacol; fernasan a; fernide; hermal; hermat tmt; Heryl; kregasan; nobecutan; normersan; panoram 75; pomasol; royal tmtd; sadoplon; spotrete; tersan 75; tetrasipton; thillate; thiotox; thiram 75; thiram b; thiulix; tridipam; TUEX; vulcafor tmtd; vulkacit mtic; vulkacit thiuram;Vancide TM-95,. Vulcanizer, bacteriostat, antifungal and animal repellent. mp = 155-156°; bp_{20} = 129°; d = 1.43; insoluble in H_2O (30 mg/l), soluble in organic solvents *Bristol-Myers Squibb Co Inc.* Name unverified.

26522 Rezistal
Corrosion and heat-resisting steels consisting of iron with up to 0.4% carbon, 1-5.5% silicon, 8-26% chromium, and 7-35% nickel.

26523 Rezolin 164
Polyether PU compound; nonexpanding compound for making low-cost moisture blocks and pressure dams in paper, pulp, and plastic insulated telecommunications cable (direct sheath injection method); service proven in all climates and environments; *Hexcel.*

26524 Rez-O-Sperse® 3
Chlorinated paraffin(Paroil 170-HV) aqueous emulsion; flame retardant for plastics, rubbers, adhesives, paints, fabric coatings, inks, carpet backings, paper coatings; plasticizer, tackifier. *Dover.*

26525 Rez-O-Sperse® A-1
Resinous chlorinated paraffin aqueous dispersion; flame retardant for plastics, rubbers, adhesives, paints, fabric coatings, inks, carpet backings, paper coatings; used where increased hardness is desired. *Dover.*

26526 Rezthane Series
Polyurethane emulsion coatings; topcoat, basecoat, water-repellent, low crock, abrasion-resistance, oily and waxy textile finishes. *CNC Int'l.I.P.*

26527 Rezyl
A proprietary trade name for an alkyd synthetic resin. No manufacturer.

26528 RH Maneb 80
Protectant fungicide. *Rohm & Haas UK.*

26529 rhamnose
3615-41-6 8338 222-793-4
$C_6H_{12}O_5 \cdot H_2O$
L-(+)-rhamnose monohydrate
6-deoxy-L-mannose; L-mannomethylose; isodulcit. Sweetener. mp = 93-95°; d_4^{20}= 1.4708; $[\alpha]_D^{20}$ = 8° (c = 10 H_2O). *Penta Mfg.; Pfanstiehl Labs; U.S. Biochemical.*

26530 rhapontin
155-58-8 8340 205-845-0
$C_{21}H_{24}O_9$
4'-methoxy-3,3',5-stilbenetriol-3-glucoside
ponticin; rhaponticum. The crystalline substance from the common English rhubarb. mp = 236° (dec); $[\alpha]_D^{22}$ = -59° (Me_2CO); soluble in EtOH, H_2O, Me_2CO, less soluble in other organic solvents;.

26531 Rhapsodie®
Fibers. *DuPont UK.*

26532 Rhapsody®
For the agriculture industry. *DuPont UK.*

26533 rhatany
The dried root of *Krameria triandra* and *K. argentea.*

26534 rhein
478-43-3 8342 207-521-4
$C_{15}H_8O_6$
9,10-dihydro-4,5-dihydroxy-9,10-dioxo-2-anthracenecarboxylic acid;
9,10-dihydro-4,5-dihydroxy-9,10-dioxo-2-anthroic acid; cassic acid; chrysazin-3-carboxylic acid; 1,8-dihydroxyanthraquinone-3-carboxylic acid; monorhein; rheic acid; rhubarb yellow; 4,5-dihydroxy-2-anthraquinonecarboxylic acid; 1,8-dihydroxy-3-carboxyanthraquinone. Found in rhubarb root. mp = 321°; λ_m = 229, 258, 435 nm (ε 36800, 20100, 11100, MeOH); insoluble in H_2O (< 1 m/ml), soluble in organic solvents.

26535 Rhenalkote
Alkyd resins and polyesters; used in surface coating industry. *RWE-DEA Chemicals UK Ltd.*

26536 Rhenalyd
Alkyd resins and polyesters; used in surface coating industry. *RWE-DEA Chemicals UK Ltd.*

26537 Rhenania phosphate
Vesta phosphate
Prepared by sintering together in a furnace at 1200-1300 C, a mixture of raw phosphate, limestone, and alkali silicate. The resulting product approximates to the formula $Ca_2KNa(PO_4)_2$.

26538 Rhenanit V
A German explosive containing nitroglycerin, ammonium nitrate, vegetable meal, nitro-compounds, and potassium perchlorate.

26539 Rhenappret B
Synthetic polymer solution; handle improvement agent for rayon and viscose staple fibers and synthetics. *Thor Chemicals (UK) Ltd.* Discontinued.

26540 Rheniforming
Catalyst. *Monsanto (Solaris).* Name unverified.

26541 Rhenish dynamite
A solution of 75% nitroglycerin in naphthalene, 2% chalk or barium sulfate, and 23% kieselguhr.

26542 Rhenital
Phenolic and melaminic resins; used for the bonding of wood, foundry sands and abrasives. *RWE-DEA Chemicals UK Ltd.*

26543 Rhenoblend
Polymer blends. *Bayer plc.*

26544 Rhenocure
Accelerators for polymers. *Bayer plc.*

26545 Rhenodiv
Separating agents. *Bayer plc.*

26546 Rhenodiv
A combination of surface active agents with film forming substances, partly enriched with corrosion inhibitors; prevents the sticking together of raw rubber, uncured rubber compounds, blanks and extrudates. *Rhein-Chemie Rheinau.*

26547 Rhenofit
Activators. *Bayer plc.*

26548 Rhenoflex
Chlorinated PVC; used for lacquer industry, adhesives industry, pyrotechnics. *Dynamit Nobel Wien GmbH.* Name unverified.

26549 Rhenogran
A range of polymer-bound rubber chemicals with an activity content of 80%; for technical molded and extruded articles, tires and cable coverings. *Rhein-Chemie Rheinau.*

26550 Rhenogran
Predispersed polymer-bound rubber chemicals. *Bayer plc.*

26551 Rhenomag
1309-48-4 5713 215-171-9
A range of magnesium oxide prepartions of several qualities; acid acceptor and vulcanization activator used in the rubber industry. *Rhein-Chemie Rheinau.*

26552 Rhenopor
Blowing agents for the rubber and plastic industries. *Bayer plc.*

26553 Rhenopren
A polymeric processing promoter for the rubber industry; for technical molded and extruded articles. *Rhein-Chemie Rheinau.* Name unverified.

26554 Rhenosin
A range of thermoplastic hydrocarbon resins; used as softening resins and homogenizers in the rubber industry; used for light and dark colored compounds, e.g., tires and conveyor belts. *Rhein-Chemie Rheinau.* Name unverified.

26555 Rhenosin
Range of hydrocarbons for use as processing promoters in the rubber industry. *Bayer plc.*

26556 Rhenosorb
Dessiccants. *Bayer plc.*

26557 Rheocin
139-44-6 205-364-6
Trihydroxystearin
Thixotrope for aliphatic solvent systems, trade sales alkyds, stains, industrial alkyds, other coatings. *United Catalysts.*

26558 Rheocin
Antisettling and thickening agents for paints, varnishes, lubricants, adhesives, coatings, putties, and cosmetics. *Süd-Chemie AG.* Name unverified.

26559 Rheolate
Rheological additives; used for water borne coatings, adhesives and paper coatings. *Rheox Inc.*

26560 Rheolube 350SBG-2
Mixed-soap petroleum-based; multipurpose bearing grease excellent for high speeds. *Wm F Nye.*

26561 Rheostene
A nickel-iron alloy. It has a specific resistance of 77 microhms per cm³ at 0 C.

26562 Rheotan I
An electrical resistance alloy containing 84% copper, 12% manganese, and 4% zinc.

26563 Rheotan II
An alloy for electrical resistances. It contains 25% nickel, 52% copper, 5% iron, and 18% zinc. This alloy or a similar one has been called Rheostan.

26564 Rheotemp 500
Polyester-based; bearing grease for high speed and wide temperature range (-65 to 350°F); for computer cooling fans, aircraft applications. *Wm F Nye.*

26565 Rheothik Polymer 80-11
Polysulfonic acid
Thickener, lubricant, suspending agent, slip agent for aqeous lubricants, synthetic cutting oils; thickens HF and other acids, alkaline systems; stable in strong acids. *Henkel/Cospha.*

26566 Rheotix AS
Antisetting and antisag agent for high solids/low VOC formulations; does not appreciably increase apparent visc. *United Catalysts.*

26567 Rheotol®
Polymerized alkyl phosphate in solvents; rheology modifier, wetting and dispersing agent. *R. T. Vanderbilt Co Inc.*

26568 Rheovis®
Associative thickening agents designed primarily for emulsion paints. *Allied Colloids Ltd.*

26569 Rheox
Rheological additives; used for paints, coatings, paper, plastic, ink, grease, cosmetics, adhesives, sealants and caulks. *Rheox Inc.*

26570 Rhesonativ
Human immunoglobulin for prophylaxis of rhesus immunization. *KabiVitrum AB.* Name unverified.

26571 Rhevois
Associate rheology modifiers. *Allied Colloids Ltd.*

26572 Rhine metal
An alloy of 97% tin and 3% copper. It has a specific gravity of 7.35 and melts at 300 C.

26573 Rhizopon A, AA
87-51-4 4994 201-748-2
$C_{10}H_9NO_2$
1H-Indole-3-acetic acid
Indol-3-yl acetic acid; Heteroauxin; 3-Indoleacetic Acid; β-Indoleacetic Acid; IAA; Indoleacetate; IAA; Indol-3-yl)acetic acid; Indolyl-3-acetic acid; Indolylacetic acid; Rhizopin; Skatole carboxylic acid. A root growth promoter in either powder or tablet form. mp = 165-169°; poorly soluble in H_2O, $CHCl_3$, soluble in EtOH, Me_2CO, Et_2O. *Fargro Ltd.*

26574 Rhizopon B
86-87-3 6458 201-705-8
$C_{12}H_{10}O_2$
1-Naphthylacetic acid
1-naphthaleneacetic acid; naphthylacetic acid; Planofix; α-naphthylacetic acid; Fruitone; NAA; naphthalene-1-acetic acid; Tre-Hold; Phyomone. A plant growth regulator used to control suckering in fruit trees. mp = 132°; soluble in H_2O (380 mg/l), similarly soluble in organic solvents; LD_{50} (rat orl) = 1000 mg/kg. *Fargro Ltd.*

26575 Rhodacal® 70/B
Calcium dodecylbenzene sulfonate
Emulsifier, dispersant for herbicides and pesticides. *Rhône-Poulenc Surf.; Rhône-Poulenc Geronazzo.*

26576 Rhodacal® 301-10
68439-57-6 270-407-8
Sodium C_{14}-C_{16} olefin sulfonate
Emulsion polymerization surfactant. *Rhône-Poulenc Surf.*

26577 Rhodacal® 330
26264-05-1 247-556-2
Isopropylamine dodecylbenzene sulfonate
Emulsifier, wetting agent, grease/pigment dispersant, lubricant, solubilizer, solvent, penetrant, high foaming base for shampoos and cleaners, metalworking fluids, *Rhône-Poulenc Surf.*

26578 Rhodacal® 2283
Amine dodecylbenzene sulfonate
Emulsifier for agricultural formulations. *Rhône-Poulenc Surf.; Rhône-Poulenc Geronazzo.*

26579 Rhodacal® ABSA
27176-87-0 248-289-4
Dodecylbenzenesulfonic acid
Emulsifier. *Rhône-Poulenc Surf.*

26580 Rhodacal® CA, 70%
Calcium dodecylbenzene sulfonate
Biodegradable oil-water emulsifier for agriculture, industrial applications; dispersant for polyester yarn dyeing. *Rhône-Poulenc Surf.; Rhône-Poulenc Surf. Canada.*

26581 Rhodacal® DDB 60T
TEA-dodecylbenzenesulfonate
Surfactant. *Rhône-Poulenc Surf.*

26582 Rhodacal® DDB-40
25155-30-0 8757 246-680-4
Sodium dodecylbenzene sulfonate
High foaming emulsifier, dispersant, wetting agent for industrial, institutional and household cleaners, agricultural formulations. *Rhône-Poulenc Surf.*

26583 Rhodacal® DOV
TEA dodecylbenzene sulfonate
Emulsion polymerization surfactant. *Rhône-Poulenc Surf.*

26584 Rhodacal® DSB
28519-02-0 249-063-8
Sodium dodecyl diphenyloxide disulfonate
Detergent, foamer, emulsifier, textile dye leveling agent, coupling agent, solubilizer; dispersant in metal and other industrial cleaners; coemulsifier in emulsion polymerization; wetting agent, dispersant for agricultural formulations; *Rhône-Poulenc Surf.; Rhône-Poulenc France.*

26585 Rhodacal® IN
1322-93-6 215-343-3
Sodium isopropyl naphthalene sulfonate
Aerosol OS; Alkanol B; Nekal A; Novonacco; NSAE; Petroll; Seoolgen W; sodium diisopropylnaphthalenesulfonate; Vatsol OS. Dispersant and suspending agent for pesticide formulations. *Rhône-Poulenc France.*

26586 Rhodacal® IPAM
26264-05-1 247-556-2
Isopropylamine dodecylbenzene sulfonate
dodecylbenzene sulfonate, isopropylamine salt. Emulsifier for drycleaning charge soaps, solvent degreasers; solubilizer in fuel oil; forms clear blends of water and kerosene. *Rhône-Poulenc Surf.*

26587 Rhodacal® LA Acid
Dodecylbenzene sulfonic acid
Biodegradable detergent, wetter, emulsifier, penetrant, foamer, intermediate used in liquid dishwashing detergents, cleaners, personal care products, and degreasers. *Rhône-Poulenc Surf.*

26588 Rhodacal® Liquid, Rhodacal® N
9084-06-4
Sodium polynaphthalene sulfonate
Process aid in paper, leather; dispersant for pulp/paper, metal cleaning; emulsifier, wetting agent for industrial cleaners, pesticides; moisture reducer in concrete; textile dye leveling agent. *Rhône-Poulenc Surf.*

26589 Rhodacal® N
9084-06-4
Naphthalenesulfonic acid, polymer with formaldehyde, sodium salt
Sodium polynaphthalenesulfonate. Processing aid in paper and leather industries. Dispersant for pulp and paper, used in metal cleaning, as an emulsifier and wetting agentfor industrial cleaners and pesticides; as a moisture reducer in concrete and as a *Rhône-Poulenc Surf.*

26590 Rhodacal® RM/210
26264-58-4 247-561-6
Polynaphthalene-methane sulfonate, sodium salt; fluidizing and plasticizer agent for concrete and mortar. *Rhône-Poulenc Geronazzo.*

26591 Rhodacal® RM/77-D
Dinaphthalene-methane sulfonate sodium salt. Dispersant, protective colloid, stabilizer of natural and synthetic elastomers; pigment grinding aid for paints. *Rhône-Poulenc Geronazzo.*

26592 Rhodacal® T
TEA dodecylbenzene sulfonate
Base for clear liquid detergents and bubble baths. *Rhône-Poulenc Surf.*

26593 Rhodafac® BG-510
108818-88-8
Aliphatic phosphate ester of ethoxylated isodecyl alcohols, free acid; detergent, emulsifier, wetting agent, dispersant for liquid industrial alkaline cleaners, hard surface detergents, soak-tank metal cleaning, steam cleaning, household cleaning; compati *Rhône-Poulenc Surf.; Rhône-Poulenc France.*

26594 Rhodafac® BP-769
39464-70-5
Aromatic phosphate ester of ethoxylated phenol. Low foaming surfactant, hydrotrope for industrial, institutional and household cleaners with electrolyte content; good alkali stability. *Rhône-Poulenc Surf.*

26595 Rhodafac® GB-520
68186-34-5
Sodium oleth-7 phosphate
Emulsifier, lubricant, softener, textile finishing aid for wool and synthetic fibers, metalworking fluids. *Rhône-Poulenc Surf.; Rhône-Poulenc France.*

26596 Rhodafac® LO-529
37340-60-6
Sodium nonoxynol-6 phosphate
Detergent, foamer, corrosion inhibitor, emulsifier for detergent concentrates, floor cleaners, agricultural concentrates, metalworking fluids. *Rhône-Poulenc Surf.; Rhône-Poulenc France.*

26597 Rhodafac® MC-470
42612-52-2
Sodium laureth-4 phosphate
Detergent, emulsifier, viscosity builder for creams and lotions, polymerization and stabilization of latexes; fatliquoring of leathers; metalworking fluids; textile antistat, lubricant, softener; emulsifier, for mineral oils. *Rhône-Poulenc Surf.*

26598 Rhodafac® MD-12-116
Aliphatic phosphate ester, free acid; low foaming emulsifier, dispersant, wetting agent with good rinsability; for industrial cleaners, pesticides, oil well cleanout; hydrotrope for metal cleaners, rinse aids. *Rhône-Poulenc Surf.*

26599 Rhodafac® PA-15
Complex organic phosphate ester, free acid; hydrophilic detergent for liquid industrial cleaners; low foaming; stable to alkalies. *Rhône-Poulenc France.*

26600 Rhodafac® PE-510
68412-53-3
Nonoxynol-6 phosphate
Detergent for drycleaning, waterless hand cleaners; intermediate for textile lubricants; emulsifier for emulsion polymerization (PVAc and acrylic films); emulsifier, dispersant for agricultural formulations; corrosion inhibitor; good electrolyte tolerance *Rhône-Poulenc Surf.; Rhône-Poulenc France.*

26601 Rhodafac® PEH
12645-31-7 235-741-0
$C_8H_{19}O_4P$
2-Ethylhexyl phosphate
mono(2-ethylhexyl)phosphate. Detergent, dispersant, and wetting agent in textile wet processing. *Rhône-Poulenc Surf.*

26602 Rhodafac® PL-620
68649-29-6
C_{10}-C_{16} ethoxy, propoxy, phosphate; surfactant for well cleanout (asphaltene, paraffin and scale removal). *Rhône-Poulenc Surf.*

26603 Rhodafac® RA-600
68130-47-2
Deceth-4 phosphate
Detergent, emulsifier, wetting agent, foamer, dispersant for hard surface cleaners, industrial alkaline detergents, textile wet processing; coupler used in liquid alkali detergents. *Rhône-Poulenc Surf.; Rhône-Poulenc France.*

26604 Rhodafac® RB-400
39464-69-2
Oleth-4 phosphate
Lubricant, antistat, emulsifier for fibers and metal. *Rhône-Poulenc Surf.*

26605 Rhodafac® RD-510
39464-66-9
Laureth-4 phosphate
Emulsifier, antistat, lubricant, solubilizer for fibers, metals, cosmetics, agricultural formulations. *Rhône-Poulenc Surf.*

26606 Rhodafac® RE-610
68412-53-3
Nonoxynol-9 phosphate
Detergent, emulsifier, wetting agent, dispersant, antistat, lubricant, dedusting agent for drycleaning, pesticides, emulsion polymerization, textile wet processing, metals, household and industrial detergents. *Rhône-Poulenc Surf.; Rhône-Poulenc France.*

26607 Rhodafac® RM-410
39464-64-7
Nonyl nonoxynol-7 phosphate
Detergent, emulsifier, antistat for drycleaning, pesticides; rust inhibitor; textile wetting agent. *Rhône-Poulenc Surf.*

26608 Rhodafac® RS-610
9046-01-9
Trideceth-6 phosphate
Emulsifier for emulsion polymerization, waterless hand cleaners, pesticides; detergent for drycleaning formulations, textile wetting agent, lubricant for fiber and metal treatment, paper-mill felt washing; antistat for aerosols. *Rhône-Poulenc Surf.; Rhône-Poulenc France.*

26609 Rhodameen® 0-12
PEG-2 oleamine
Coemulsifier in emulsifiable concentrates (herbicides), water-oil emulsions of solvents, waxes, and oils, catiaonic asphalt emulsions. *Rhône-Poulenc France.*

26610 Rhodameen® HT-50
68783-22-2
PET-50 hydrogenated tallow amine; wetting agent, penetrant, emulsifier, stabilizer, dispersant, antistat, lubricant. *Rhône-Poulenc Surf.*

26611 Rhodameen® OA-860
PEG-30 oleamine
Hydrophilic emulsifier, leveling agent; textile dyeing assistant; antiprecipitant for dyeing processes. *Rhône-Poulenc Surf.*

26612 Rhodameen® OS-12
PEG-2 oleyl/stearyl amine
Coemulsifier and antistat for cosmetics, textile and plastic processing. *Rhône-Poulenc France.*

26613 Rhodameen® PN-430
61791-26-2
PEG-5 hydrogenated tallowamine
Emulsifier, corrosion inhibitor, lubricant for metalworking fluids, agricultural formulations; acid corrosion inhibitor for ferrous alloys. *Rhône-Poulenc Surf.; Rhône-Poulenc France.*

26614 Rhodameen® T-5
61791-26-2
PEG-5 hydrogenated tallow amine
Emulsifier, dispersant; textile scouring and desizing assistant; softener and antistatic agent. *Rhône-Poulenc Surf.*

26615 Rhodameen® VP-532/SPB
61791-26-2
PEG-8 hydrogenated tallowamine
Wetting agent, penetrant, emulsifier, stabilizer, antistat, and lubricant; retarder, antiprecipitant, leveling agent in dyeing; dispersant for fiberglass mat manufacturing. *Rhône-Poulenc Surf.*

26616 Rhodamine B
81-88-9 8349 201-383-9
$C_{28}H_{31}ClN_2O_3$
9-(2-carboxyphenyl)-3,6-bis(diethylamino)xanthylium chloride
Basic violet 10; C.I. 45170; tetraethylrhodamine hydrochloride; D&C Red No. 19; C.I. Food Red15; Rhodamine B Extra. The hydrochloride of diethyl-m-aminopheno-phthalein; a dyestuff; dyes wool and silk bluish-red with fluorescence also tannined cotton, violet-red. mp = 210-211°; soluble in H_2O, EtOH; LD_{50} (rat iv) = 89.5 mg/kg. No manufacturer.

26617 Rhodamine G and G Extra
989-38-8 213-584-9
C28H31ClN2O3
Xanthylium, 9-[2-(ethoxycarbonyl)phenyl]-3,6-bis(ethylamino)-2,7-dimethyl-, chloride;
ethyl-o-(6-(ethylamino)-3-(ethylimino)-2,7-dimethyl-3H-xanthen-9-yl)benzoate monohydrochloride; basic red 1; calcozine red 6g; C.I. basic red 1, monohydrochloride; rhodamine 6gex ethyl ester; silosuper pink b; fanal pink gfk; rhodamine f4g; rhodamine 6gb; rhodamine gdn; rhodamine 6 gdn; rhodamine 4gh; rhodamine 6gx; rhodamine lake red 6g; rhodamine y 20-7425; rhodamine zh; rhodamine 6zh. Dyestuffs, consisting chiefly of triethyl-rhodamine; dyes wool, silk, and tannined cotton, red. Insoluble in H_2O (<1 mg/ml). No manufacturer.

26618 Rhodamine S
The hydrochloride of dimethyl-m-aminophenol-succineine; a dyestuff; dyes cotton red, and is used for dyeing half silk goods, and for coloring paper pulp and wool. No manufacturer.

26619 Rhodamox® CAPO
68155-09-9 268-938-5
Cocamidopropyl dimethylamine oxide
Foaming agent/stabilizer, thickener, emollient for shampoos, bath products, dishwash, rug shampoos, fine fabric detergents, shaving creams, lotions, foam rubber, electroplating, paper coatings; used in toiletries for mildness. *Rhône-Poulenc Surf.*

26620 Rhodamox® LO
1643-20-5 216-700-6
$C_{14}H_{31}NO$
N,N-dimethyldodecylamine-N-oxide
1-Dodecanamine, N,N-dimethyl-, N-oxide; Dimethyldodecylamine oxide; dimethyldodecylamine-N-oxide; ammonyx lo; ammonyx ao; aromox dmmc-w; conco xal; DDNO; N,N-dimethyldodecylamine oxide; dodecycldimethylamine oxide; n-dodecyldimethylamine oxide; lauryldimethylamine oxide; Lauramine oxide. Foaming agent/stabilizer, thickener, emollient for shampoos, bath products, dishwash, fine fabric detergents, shaving creams, lotions, textile softeners, foam rubber, in electroplating, paper coatings; used in toiletries for mildness. mp = 130-131°. *Rhône-Poulenc Surf.; Rhône-Poulenc France.*

26621 Rhodapex® 674/C
25446-78-0 246-985-2

$C_{19}H_{39}NaO_7S$
2-[2-[2-(tridecyloxy)ethoxy]ethoxy]ethanol hydrogen sulfate sodium salt
Sodium trideceth sulfate. Surfactant. *Rhône-Poulenc Surf.*

26622 Rhodapex® AB-20
67762-19-0
Ammonium laureth sulfate
Emulsifier for emulsion polymerization and mineral oils; detergent. *Rhône-Poulenc France.*

26623 Rhodapex® CO-415
9051-57-4
Ammonium nonoxynol-4 sulfate
Emulsifier. *Rhône-Poulenc Surf.*

26624 Rhodapex® CO-433
68891-39-4
Sodium nonoxynol-4 sulfate
High foaming detergent, wetting agent, dispersant for dishwashing, scrub soaps, car washes, rug and hair shampoos; emulsifier for emulsion polymerization, petrol. waxes; antistat for plastics and synthetic fibers; lime soap dispersant. *Rhône-Poulenc Surf.; Rhône-Poulenc France.*

26625 Rhodapex® CO-436
9051-57-4
Ammonium nonoxynol-4 sulfate
High foaming detergent, wetting agent, dispersant for dishwashing, scrub soaps, car washes, rug and hair shampoos; emulsifier for emulsion polymerization, petrol. waxes; antistat for plastics and synthetic fibers; lime soap dispersant. *Rhône-Poulenc Surf.; Rhône-Poulenc France.*

26626 Rhodapex® EA
67762-19-0
Ammonium laureth (3) sulfate
High foaming emulsifier for industrial, institutional and household cleaners. *Rhône-Poulenc Surf.*

26627 Rhodapex® EAY
67762-19-0
Ammonium laureth sulfate
Detergent for shampoos, bubble baths. *Rhône-Poulenc Surf.*

26628 Rhodapex® ES
9004-82-4
Sodium laureth (3) sulfate
Emulsifier, high foaming base for shampoos, light duty detergents, bubble baths; polymerization surfactant. *Rhône-Poulenc Surf.*

26629 Rhodapex® EST-30
25446-78-0 246-985-2
Sodium trideth sulfate
Emulsifier, wetting agent, dispersant for baby shampoo, other personal care products, household, industrial, institutional and industrial formulations, emulsion polymerization of styrene systems, textile scouring, dishwash. *Rhône-Poulenc Surf.; Rhône-Poulenc France.*

26630 Rhodapex® F-85/SD
Sodium octylphenol ethoxy sulfate
Surfactant. *Rhône-Poulenc France.*

26631 Rhodapon® 101-10
151-21-3 8782 205-788-1
$C_{12}H_{25}NaO_4S$
Sodium lauryl sulfate
avirol 101; avirol 118; berol 452; carsonol sls; carsonol sls paste b; carsonol sls special; conco sulfate wa; conco sulfate wa-1200; conco sulfate wa-1245; conco sulfate wag; conco sulfate wan; conco sulfate was; conco sulfate wn; cycloryl 21; cycloryl 31; cycloryl 580; cycloryl 585n; detergent 66; Dreft; duponal waqe; dupanal; duponol; Emal o; emal 10; emersal 6400; empicol lpz; empicol ls 30; empicol lx 28; emulsifier no. 104; gardinol; hexamol sls; lrium; lanette wax-s; maprofix 563; melanol cl; monogen y 100; montopol la paste; neutrazyme; nikkol sls; odoripon al 95; orvus wa paste; p and g emulsifier 104; perlandrol l; quolac ex-ub; rewopol nls 30; richonol a; sinnopon ls 95; sintapon l; sipex op; sipon ls; SLS; sodium monolauryl sulfate; solsol needles; standapol 112; steinapol nls 90; stepanol me; sterling wa paste; sulfetal l 95; sulfopon wa 1; sulfotex wa; swascol 3l; syntapon; tarapon k 12; texapon DL; trepenol wa; tvm 474; ultra sulfate sl-1; WAQE,. Surfactant. mp = 204-207°; insoluble in H_2O (<1 mg/ml). *Rhône-Poulenc Surf.*

26632 Rhodapon® BOS
126-92-1 204-812-8
$C_8H_{18}O_4S$
Sodium 2-ethylhexyl-sulfate
Sodium 2-Ethylhexyl Sulfate; Sulfuric acid, mono(2-ethylhexyl) ester, sodium salt; 2-ethyl-1-hexanol hydrogen sulfate sodium salt; mono(2-ethylhexyl)sulfate sodium salt; 2-ethyl-1-hexanol sulfate sodium salt; sodium ethasulfate; tergemist; tergimist; tergitol 08; tergitol anionic 08; 08-union carbide; emersal 6465; nia proof 08; propaste 6708; sipex bos; emcol d 5-10; sole tege ts-25; sulfirol 8; pentrone on; Sodium etasulfate. Biodegradable wetting agent, emulsifier, detergent, foamer with high electrolyte tolerance for industrial, institutional, and household cleaners; latex stabilizer, nickel brightener; metal treatment; bp = 96-104°; soluble in H_2O (>100 mg/ml).g. *Rhône-Poulenc Surf.; Rhône-Poulenc France.*

26633 Rhodapon® CAV
68299-17-2 269-598-0
Sodium isodecyl sulfate
Wetting agent, emulsifier, detergent, foamer, rinse aid, visc. control agent; post-stabilizer in latex paints; metal treatment; textile and plywood manufacturing; fruit/vegetable washing; hard surface cleaners; emulsion polymerization of vinyl, acrylic, S *Rhône-Poulenc Surf.*

26634 Rhodapon® EC111
59186-41-3
Sodium cetyl/stearyl sulfate
Emulsifier, detergent, flotation agent; collector in ore flotation; softener/lubricant for textiles; cosmetics, and toiletries. *Rhône-Poulenc Surf.; Rhône-Poulenc France.*

26635 Rhodapon® L-22, L-22/C
2235-54-3 218-793-9
$C_{12}H_{29}NO_4S$
Ammonium lauryl sulfate
dodecyl ester of sulfuric acid, ammonium salt. High foaming detergent, emulsifier for shampoo, bubble bath, pet shampoos, industrial and institutional cleaners, wool scouring, fire fighting foams, assistant for pigment dispersion; emulsion polymerization aid. *Rhône-Poulenc Surf.; Rhône-Poulenc France.*

26636 Rhodapon® LCP
151-21-3 8782 205-788-1
Sodium lauryl sulfate
Low cloud point emulsion polymerization surfactant. *Rhône-Poulenc Surf.*

26637 Rhodapon® LM
3097-08-3 8782 221-450-6
Magnesium lauryl sulfate
High-foaming emulsifier, wetting agent, dispersant, detergent for rug and upholstery shampoos, bubble baths, shampoos; food packaging applications. *Rhône-Poulenc Surf.*

26638 Rhodapon® LT-6
139-96-8 205-388-7
TEA-lauryl sulfate
Akyposal TLS; Cycloryl tawf; Dodecyl sulfate, triethanolamine salt; Drene; Elfan 4240 T; EMAL T; Lauryl sulfate ester, triethanolamine salt; Maprofix TLS; Melanol LP 20 T; Propaste T; Rewopol TLS-40; Richonol T; Sipon lt; Standapol TLS 40; Sterling wat; Sulfuric acid, monododecyl ester, compd. with 2,2',2''-nitrilotris(ethanol) (1:1;)Texapon T-35; Triethanolamine salt of lauryl sulfate; Tylorol LT-50; tris(2-hydroxyethyl)ammonium decyl sulfate. Emulsifier, high foaming base for industrial and household detergents, shampoos, bubble baths; obtains creamy, mild lather. *Rhône-Poulenc Surf.*

26639 Rhodapon® OLS
142-31-4 205-535-5
$C_8H_{17}NaO_4S$
Sodium octyl sulfate
octyl sodium sulfate. Wetting agent; rinse aid for industrial, institutional and household cleaners; mercerizing agent for cotton goods; surfactant in electrolyte baths for metal cleaning; hard surface cleaning; neoprene dispersant; emulsifier for emulsion polymerization. *Rhône-Poulenc Surf.*

26640 Rhodapon® OS
1847-55-8 217-430-1
Sodium oleyl sulfate
Specialty emulsifier for emulsion polymerization. *Rhône-Poulenc Surf.*

26641 Rhodapon® TDS
3026-63-9 221-188-2
Sodium tridecyl sulfate
Emulsifier, wetting agent, emulsion polymerization of vinyl chloride, styrene, and styrene/acrylic monomers; detergent formulations; base for shampoos, foaming bath oils, cosmetic emulsions. *Rhône-Poulenc Surf.*

26642 Rhodaquat® M214B/99
1119-97-7 6419 214-291-9
$C_{17}H_{38}BrN$
tetradecanaminium, N,N,N-trimethyl-, bromide
myrystyltrimethylammonium bromide; tetradecyltrimethylammonium bromide; Myrtrimonium bromide; tetradonium bromide. Conditioner with superior antistatic properties and light feel. mp = 245-250°. *Rhône-Poulenc Surf.*

26643 Rhodaquat® M242B/99
57-09-0 2068 200-311-3
$C_{19}H_{42}BrN$
Hexadecanaminium, N,N,N-trimethyl-, bromide
Cetab; Centimide; Cetyltrimethylammonium Bromide; N-Hexadecyltrimethylammonium Bromide; N,N,N-trimethyl-1-Hexadecanaminium bromide; HTAB; CTAB; Cetrimonium bromide; CTABr;

Cetrimide. Quaternary ammonium compound used in conditioners to increase antistatic and comb out effects. mp = 218°. *Rhône-Poulenc Surf.*

26644 Rhodaquat® M242C/29
112-02-7 203-928-6
$C_{19}H_{42}ClN$
Hexadecanaminium, N,N,N-trimethyl-, chloride
Cetyltrimethylammonium Chloride; 1-Hexadecanaminium, N,N,N-trimethyl-, chloride; Cetrimonium chloride; C16-alkyltrimethylammonium chloride. Surfactant with conditioning and emolliency effect on hair. *Rhône-Poulenc Surf.*

26645 Rhodaquat® M270C/18
122-19-0 204-527-9
Stearalkonium chloride
Surfactant. *Rhône-Poulenc Surf.*

26646 Rhodasurf® 25-7
68131-39-5
C_{12}-C_{15} pareth-7
Biodegradable detergent, wetting agent, emulsifier for household and industrial cleaners; coupler and solubilizer for perfumes and organic additives. *Rhône-Poulenc Surf.*

26647 Rhodasurf® 91-6
68439-46-3
C_9-C_{11} pareth-6
Biodegradable detergent, wetting agent, emulsifier for household and industrial cleaners; coupler and solubilizer for perfumes and organic additives. *Rhône-Poulenc Surf.*

26648 Rhodasurf® 860/P
Isodeceth-6
Wetting agent, foamer, emulsifier, detergent for textile processing, metal degreasing, herbicides, waxes, resins; stable to strong acidic and alkaline media. *Rhône-Poulenc France.*

26649 Rhodasurf® A-1P
68439-49-6 7737
Ceteareth-23
Surfactant. *Rhône-Poulenc France.*

26650 Rhodasurf® B-1
Laureth-16
Surfactant. *Rhône-Poulenc France.*

26651 Rhodasurf® BC-420
24938-91-8
Trideceth-3
Emulsifier, detergent, dispersant for petrol. oils, agricultural formulations; intermediate for manufacturing of high-foaming anionic surfactants. *Rhône-Poulenc Surf.*

26652 Rhodasurf® C-2
9004-95-9
Ceteth-2
Surfactant, emollient for creams and lotions. *Rhône-Poulenc Surf.*

26653 Rhodasurf® DA-4
26183-52-8
Isodeceth-4
Wetting agent, penetrant for pressure dyeing of fabrics, textile and industrial applications. *Rhône-Poulenc Surf. Canada.*

26654 Rhodasurf® DA-530
61827-42-7
Isodeceth-4
Low foaming rapid wetting agent for industrial, institutional and household cleaners, textile compounding; scouring agent, emulsifier for defoamers. *Rhône-Poulenc Surf.*

26655 Rhodasurf® E 400
25322-68-3 7729
PEG-8
Surfactant intermediate; binder and lubricant in compressed tablets; softener for paper, plasticizer for starch pastes and polyethylene films; coupling agent for skin care lotions. *Rhône-Poulenc Surf.*

26656 Rhodasurf® L-4
68002-97-1
Laureth-4
Emulsifier, thickener, wetting agent, pigment dispersant, lubricant, solubilizer for cosmetic and industrial emulsions; textile scouring agent, emulsion polymerization, metal cleaning, monomer systems, floor waxes, paper finishes, rubber; emollient for ph *Rhône-Poulenc Surf.*

26657 Rhodasurf® LA-3
68131-39-5
C_{12}-C_{15} pareth-3
Detergent base and emulsifier for dishwash, personal care products, industrial applications; textile lubricant. *Rhône-Poulenc Surf.*

26658 Rhodasurf® ON-870
9004-98-2
Oleth-20
High foaming emulsifier, stabilizer, dispersant, wetting agent, solubilizer for mineral oils, fatty acids, waxes; for industrial cleaners, metal cleaners, agriculture, paints, adhesives, textile, leather, cosmetic, pharmaceutical industries. *Rhône-Poulenc Surf.; Rhône-Poulenc France.*

26659 Rhodasurf® PEG 3350
25322-68-3 7729
PEG-75
Mold release and antistat for rubber products; binder. *Rhône-Poulenc Surf.*

26660 Rhodasurf® PEG 400
25322-68-3 7729
PEG-9
Intermediate, plasticizer, solvent, lubricant, coupler in cosmetic lotions. *Rhône-Poulenc France.*

26661 Rhodasurf® S-2
9005-00-9
Steareth-2
Emollient, detergent, emulsifier for creams and lotions, facial cleansers, bath and toiletries. *Rhône-Poulenc Surf.*

26662 Rhodasurf® T-95
24938-91-8
Trideceth-10
Surfactant. *Rhône-Poulenc France.*

26663 Rhodasurf® TB-970
9005-00-9
Steareth-200
Detergent, viscosifier, foam booster, dispersant, surfactant for most dry blending operations in detergent formulations, sanitizers; controls dissolution rate of solid or block type hard surface cleaners; textile wet processing aid; biodeg. *Rhône-Poulenc Surf.; Rhône-Poulenc France.*

26664 Rhodasurf® TDA-6
24938-91-8
Trideceth-6
Coemulsifier, wetting agent, dispersant for cutting oils, low temperature wool scouring. *Rhône-Poulenc Surf. Canada.*

26665 Rhodasurf® TMP-3
Trimethylolpropane triethoxylate
Monomer for coatings; reactive diluent; useful in alkyds and polyurethanes. *Rhône-Poulenc Surf.*

26666 Rhodaterge® 206C
Proprietary; high foaming upholstery cleaner concentrate. *Rhône-Poulenc Surf.*

26667 Rhodaterge® LD-50Q
Blend; detergent concentrate for dishwashing and general purpose cleaning; relatively mild to skin. *Rhône-Poulenc Surf. Canada.*

26668 Rhodaterge® SSB
Sodium lauryl sulfate
disodium lauryl sulfosuccinate - propylene glycol. Surfactant. *Rhône-Poulenc Surf.*

26669 Rhodaterge® WHC-347
Formulated blend; high foaming surfactant for formulating waterless hand cleaners; wetting agent. *Rhône-Poulenc Surf.*

26670 Rhodeftal
Polyamide-imide wire enamel. *Rhône-Poulenc UK.*

26671 rhodeoretin
Jalapin, the chief constituent of Jalap resin.

26672 Rhodester
Polyester resins. *Rhône-Poulenc NV.* Name unverified.

26673 Rhodialite
A proprietary cellulose acetate. No manufacturer.

26674 Rhodialux
Ultraviolet stabilizers for polymers. *Rhône-Poulenc UK.*

26675 Rhodialux A
131-57-7 7088 205-031-5
$C_{14}H_{12}O_3$
2-Hydroxy-4-methoxyphenyl)phenylmethanone
Benzophenone-3; 4-Methoxy-2-hydroxybenzophenone; Escalol 567; Eusolex 4360; MOB; Oxybenzone; Uvinul M-40; Spectra-Sorb UV-9; 2-benzoyl-5-methoxyphenol; cyasorb uv 9; syntase 62; UF 3; advastab 45; anuvex; chimassorb 90; cyasorb uv 9 light absorber; MOD; ongrostab hmb; sunscreen uv 15; uvinul 9; uvistat 24; HMB; oxybenzone. Ultraviolet stabilizer for polymers. mp = 66°; bp_5 = 150-160°; Insoluble in H_2O (<1 mg/ml), soluble in most organic solvents; LD_{50} (rat orl) > 12.8 g/kg. *Rhône-Poulenc.*

26676 Rhodialux S
4065-45-6 9157 223-772-2

$C_{14}H_{12}O_6S$
2-Benzoyl-5-methoxy-1-phenol-4-sulfonic acid
Sulisobenzone; Benzophenone-4; 3-Benzoyl-4-hydroxy-6-methoxybenzenesulfonic acid; 5-Benzoyl-4-hydroxy-2-methoxybenzenesulfonic acid; Cyasorb UV 284; Spectra-Sorb UV 284;Sungard; Uval; Uvinul MS-40; Uvistat 1121; 2-Hydroxy-4-methoxybenzophenone-5-sulfonic acid; Sulisobenzone. UV absorber. *Rhône-Poulenc France.*

26677 Rhodia-Phos
7758-29-4 8846 231-838-7
Sodium tripolyphosphate
Used for detergents, water treatment, paper industry, food industry, textile industry, animal food. *Rhône-Poulenc NV.* Name unverified.

26678 Rhodiarome
121-32-4 3904 204-464-7
$C_9H_{10}O_3$
Ethyl vanillin
3-ethoxy-4-hydroxybenzaldehyde; Bourbonal; ethyl protocatechualdehyde; ethyl protal; Ethavan; Ethovan. Vanilla flavor for baking, cereal, diary, cheese, processed foods, beverages, confections. mp = 77-78°; poorly soluble in H+2O, soluble in organic solvents; LD_{50} (rat orl) >2000 mg/kg. *Rhône-Poulenc Food Ingreds.; Rhône-Poulenc UK.*

26679 Rhodiasolv RPDE
Dibasic ester solvent. *Rhône-Poulenc UK.*

26680 Rhodiastab 50, 83
Food grade PVC heat stabilizers. *Rhône-Poulenc UK.*

26681 Rhodiastab 83
Technical grade PVC heat stabilizer. *Rhône-Poulenc UK.*

26682 Rhodicare XC
11138-66-2 10191 234-394-2
Xanthan gum; thickening and stabilizing hydrocholloid; emulsion/foam stabilizer; imparts viscosity and suspends insoluble additives in cosmetics and toiletry formulations. *Rhône-Poulenc Surf.; Rhône-Poulenc France.*

26683 Rhodigel® EZ
11138-66-2 10191 234-394-2
xanthan
xanthan gum; polysaccharide B-1459; Keltrol F; Kelzan. Used in cosmetic applications to suspend insoluble additives. Possesses excellent pseudo-plastic properties. Used in baking, cereals, meat, poultry, dairy, cheese, processed foods, beverages and confections. *Rhône-Poulenc Food Ingreds.*

26684 Rhodigel®; Rhodigel® EZ
11138-66-2 10191 234-394-2
Xanthan gums, food grade; emulsion stabilizer, suspending agent, thickener for cosmetic and pharmaceutical applications. *Rhône-Poulenc Food Ingreds.; Rhône-Poulenc UK.* Name unverified.

26685 rhodinol
6812-78-8 8352 229-887-4
$C_{10}H_{20}O$
S-(-)-3,7-dimethyl-7-octen-1-ol
A terpene alcohol prepared from the oils of rose, geranium, and citronella. It is practically pure geraniol; used in perfume manufacture and as a flavoring agent. bp_{12} = 114-115°; d_4^{20}= 0.8549; n_D^{20}= 1.4556; $[\alpha]_D^{20}$= -2.9°; λ_m = 186-189 nm (ε 9000); slightly soluble in H_2O, soluble in organic solvents.

26686 rhodium
7440-16-6 8353 231-125-0
Rh
Rhodium
Metallic element in the platinum group; alloy with platinum for high temperature thermocouples, furnace windings, laborabory crucibles, spinerets for rayon, electrical contacts, jewelry, catalyst, mp = 1966°; d^{20} = 12.41. *Aldrich; Atomergic Chemetals; Degussa; Noah Chem.*

26687 rhodium chloride
10049-07-7 8355 233-165-4
Cl_3Rh
rhodium trichloride
Manufacture of rhodium trifluoride. Insoluble in H_2O; LD_{50} (rat iv) = 198 mg/kg. *Aldrich; Atomergic Chemetals; Degussa; Noah Chem.*

26688 rhodium gold
rhodite
A native alloy of from 57-66% gold and 34-43% rhodium.

26689 Rhodizite
A borate of lime, $3CaO{\cdot}4B_2O_3$, imported from the West Coast of Africa.

26690 Rhodocap
Cyclo dextrins. *Rhône-Poulenc UK.*

26691 Rhodold
A proprietary plastic material made from cellulose acetate. No manufacturer.

26692 rhodole
Products intermediate between fluorescein-phthaleine and the rhodamines.

26693 Rhodopas
Polyvinyl acetates. *Rhône-Poulenc NV.* Name unverified.

26694 Rhodopas AX
Polyvinyl acetate/polyvinyl chloride copolymers. *Rhône-Poulenc NV.*

26695 Rhodopas X
Polyvinyl chlorides. *Rhône-Poulenc NV.*

26696 Rhodopol® 23, XGD
11138-66-2 10191 234-394-2
polysaccharide B-1459; Keltrol F; Kelzan. Xanthan gum; thickener for food and agricultural formulations. XGD is used as a viscosifier in drilling. *Rhône-Poulenc Surf.; Rhône-Poulenc Chemicals Ltd; Rhône-Poulenc France.*

26697 Rhodopol® XGD
11138-66-2 10191 234-394-2
xanthan
xanthan gum; polysaccharide B-1459; Keltrol F; Kelzan. Viscosifier used in drilling. *Rhône-Poulence/Water Soluble Polymers.*

26698 Rhodorsil®
Silicones. *Rhône-Poulenc UK.*

26699 Rhodorsil® 10336
Methyl silicone resin; used for high temperature paints; for high air-drying speed. *Rhône-Poulenc Silicones.*

26700 Rhodorsil® 1505
Methylphenyl silicone resin; used for hightemperature paints; features excellent adherence to metal. *Rhône-Poulenc Silicones.*

26701 Rhodorsil® AF 422, AF 426R
Dimethylpolysiloxane emulsions; antifoams for aqueous systems, agriculture, textile, chemical, rubber, metallurgy industries. *Rhône-Poulenc Silicones.*

26702 Rhodorsil® AF 426R
Silicone emulsion. Antifoam for agricultural formulation. *Rhône-Poulenc Silicones.*

26703 Rhodorsil® RS 44, RS 48
63394-02-5
Silicone rubber; for molding, calendering, or extrusion applications; recommended for general purpose applications including sheet goods, gasketing, oil seals, miscellaneous parts; peroxide-cured; no post-cure required. *Rhône-Poulenc Silicones.* Name unverified.

26704 Rhodorsil® SC 5020
Silicone emulsion; antifoam for agricultural formulation. *Rhône-Poulenc Silicones.*

26705 Rhodoviol
Polyvinyl alcohol. *Rhône-Poulenc NV.*

26706 Rhometal
A complex nickel-iron alloy having high electrical resistivity and retaining its permeability up to very high frequencies; used for television transformers, special radio transformers, HF alternators, etc.

26707 Rhon'Sec
Laboratory absorbent. *Rhône-Poulenc UK.*

26708 Rhoplex®
An acrylic resin for textile finishes. *Rohm & Haas UK.*

26709 Rhoplex® 2133
Acrylic polymer emulsion; for interior concrete and resilient tile sealers. *Rohm & Haas.*

26710 Rhoplex® AC-64
Acrylates copolymer; vehicle for exterior and interior latex paints. *Rohm & Haas.*

26711 Rhoplex® B-15
Acrylic latex
Semireactive emulsion used for coated materials, adhesives; excellent cohesive and tensile strength, durability; will crosslink with melamine or urea/formaldehyde resins. *Rohm & Haas.*

26712 Rhoplex® K-3
Acrylates copolymer; provides softness, resilience and low temperature flexibility to treated fabrics; hand modifier, glass fabric adhesive, fabric finishing, pigment printing and dyeing, fabric backing. *Rohm & Haas.*

26713 Rhoplex® Multilobe 100
Acrylates copolymer; vehicle for exterior flat, sheen and semigloss paints. *Rohm & Haas.*

26714 Rhoplex® NT-2624
Acrylic polymer emulsion; provides excellent gloss and durability for floor polishes. *Rohm & Haas.*

26715 Rhotanium
A series of alloys, consisting mainly of gold (60-90%) and palladium, and in some cases with a small proportion of rhodium. They are said to be more resistant to hot concentrated sulfuric acid and fused caustic soda than lead.

26716 Rhovinal B
Polyvinyl butyrals. *Rhône-Poulenc NV.* Name unverified.

26717 Rhovinal F
Polyvinyl formals. *Rhône-Poulenc NV.* Name unverified.

26718 Rhubafuran
2,4-Dimethyl-4-phenyltetrahydrofuran
Quest Int'l. UK Ltd.

26719 rhyolite
A volcanic rock. It usually contains lime and iron.

26720 Riamat
51-48-9 9555 200-101-1
Thyroxine ^{125}I
Radioactive agent. *Mallinckrodt Inc.* Name unverified.

26721 Ribiene
9002-88-4 7728
A trade name for polyethylene. *ABCD Petrochimica.* Unverified.

26722 riboflavin
83-88-5 8367 201-507-1
$C_{17}H_{20}N_4O_6$
7,8-dimethyl-10-(1' d-ribityl)isoalloxazine
vitamin B_2; flavaxin; riboflavine; Ovoflavin; flavin; lactoflavine, Zinvit-G. Organic compound; crystalline pigment; dietary supplement; principal growth-promoting factor of vitamin B_2 complex (functions as flavor protein in tissue respiration). mp = 278-282° (dec); $[\alpha]_D^{25}$ = -112- -122° (dil. NaOH); λ_m = 220-225, 266, 371, 444, 475 nm; soluble in H_2O (0.3 mg/ml), less soluble in organic solvents; LD_{50} (rat orl) >10 g/kg. Am. Biorganics; BASF; Bio-Rad Labs; EM Industries; Hoffmann-La Roche; Honeywill & Stein Ltd; Takeda USA.

26723 Ricaccel
Cure rate accelerator for chloroprene elastomers; imparts good storage life and gives outstanding cured properties *Ricon Resins.*

26724 rice paper
A paper made from plant pith, particularly in China and Japan.

26725 rice rubber
An elastic cellulose product made from Japanese rice.

26726 Rice-Pro EN-20
Hydrolyzed rice protein; cosmetic ingredient for skin and hair care products. *Brooks Industries.*

26727 Rice's bromide solution
A solution containing 125 parts bromine, 125 parts sodium bromide, and 1,000 parts water; used in the determination of urea.

26728 Richard's aluminum solder
An alloy of 71.5% tin, 25% zinc, and 3.5% aluminum.

26729 Richardson's speculum metal
An alloy of 65.3% copper, 30% tin, 2% silicon, 2% arsenic, and 0.7% zinc.

26730 Riché gas
A gas obtained during the dry distillation of wood. It contains, on an average, about 60% carbon dioxide, 25% carbon monoxide, 15% methane, and a very small quantity of hydrogen.

26731 ricin
9009-86-3 8376
Toxic constituent of castor beans; an N-glucosidase used as a reagent for pepsin and trypsin.

26732 Ricinion
61791-12-6
polyoxyethylene (20) castor oil (ether, ester)
ethoxylated castor oil; POE (20) castor oil (ether, ester). PEG-33 castor oil. solvent, emulsifier for cosmetic/pharmaceutical emulsions. *Gattefosse.*

26733 Ricinoleamidopropyl ethyldimonium ethosulfate
112324-16-0
Lipoquat R; M-Quat® JN; Surfactol® Q4. Conditioner, antistat, emollient, glosser, softener for personal care products and anhyd. systems. *Lipo; PPG/Specialty Chem.; Cas Chem.*

26734 ricinoleic acid
141-22-0 8378 205-470-2
$C_{18}H_{34}O_3$
9-Octadecenoic acid, 12-hydroxy-, [R-(Z)]-
12-hydroxy-(cis)-9-octadecenoic acid; P® -10 Acid; d-12-hydroxyoleic acid. Chemical intermediate; imparts lubricity and rust-proofing to sol. cutting oils; basis for grease, soaps, resin plasticizers, and ethoxylated derivs. mp = 5°; bp_{120} = 245°; $d^{27.44}$ = 0.940; n_D^{20} = 1.4716; $[\alpha]_D^{22}$ = 6.7° (c = 5 Me_2CO); insoluble in H_2O, soluble in organic solvents. *S & D Chemicals Ltd; CasChem;.*

26735 Ricobond 1031, 1731, 1756
Polymeric
Reactive adhesive promoter for compounding with elastomers to give in creased adhesion to metal, elastomer, plastic, mineral, fabric, fiber, etc. *Ricon Resins.*

26736 Ricon 100
SBR random copolymer; thermosetting liquid resin system, outstanding electrical properties, thermal stability, moisture and age resistance, adhesion, and chemical resistance; *Ricon Resins.*

26737 Ricon 130MA8
Polybutadiene maleic anhydride adducted resin; resin for electrical applications (low dielectric and moisture absorption flexible epoxy formulations), room temperature cured elastomers for sealant and coating applications. *Ricon Resins.*

26738 Ricon 159
9003-17-2
1,2-Polybutadiene homopolymer resin; thermosetting resin used in laminates, radomes, mica sheets, bonding flexible mica parts; active homopolymer used for continuous production of laminate; forms prepregs which block easily; *Ricon Resins.*

26739 Ricon P30/Dispersion
1,2-Vinylpolybutadiene dispersion in Micro-Cel E; additive for carbon-reinforced natural rubber; nonextractable plasticizer and processing aid resulting in improved carbon black loading; crosslinkable peptizer. *Ricon Resins.*

26740 Ricoroof
Two-part room temperature curing elastomer; formulated for outdoor applications such as roofing, flashing repair, parking garage concrete sealants. *Ricon Resins.*

26741 Ricoseal
Two-part elastomer; room temperature curable soft sealant and potting compound; resistance to penetration by moisture, flexible at low temperatures, very impact energy absorbent. *Ricon Resins.*

26742 Ricotuff, Ricotuff L.V.
Two-part anhydride/epoxy system built on the polybutadiene chain; strong adhesive for structural applications; excellent for electrical applications; for manufacturing of reinforced laminates; *Ricon Resins.*

26743 Ric-Syn Wax
Hydrogenated castor oil. *United Catalysts.*

26744 Ridacto®
Activator. *Kenrich Petrochemicals.*

26745 Rid-A-Roach
A cockroach trap which contains an attractant to bring the roach to the sticky trap; a nonchemical method of control that is safe and easy-to-use. *Lawn & Garden Products Inc.*

26746 Ridene
3691-35-8 2204 223-003-0
2-[(4-chlorophenyl)phenylacetyl]-1H-Indene-1,3(2H)-dione
chlorophacinone; Topitox; Chlorophenyl)phenylacetyl)-1,3-indandione; Indandione, 2-((p-chlorophenyl)phenylacetyl)-; Indene-1,3(2H)-dione, 2-((4-chlorophenyl)phenylacetyl)-; Partox; 2-((p-Chlorophenyl)phenylacetyl)-1,3-indandione; LM-91; Caid; Drat; Liphadione; Quick; Raviac; Rozol. An oil formulation containing 2.5% of chlorophacinone; an anticoagulant rodenticide; bait to control black rats, brown rats, house mice and voles. mp = 140°; λ_m = 325 nm (Me_2CO); poorly soluble in H_2O, soluble in organic solvents; LD_{50} (rat orl) = 20 mg/kg. *Ace Chemicals Ltd.*

26747 Ridoline
Metal treating compositions. *ICI Chem & Polymers Ltd.* Name unverified.

26748 Ridomil MBC 60WP
Carbendazin + metalaxyl; a fungicide mixture used to prevent the spread of fruit and vegetable storage diseases. *Ciba-Geigy Agrochemicals.*

26749 Ridomil Plus 50WP
Copper oxychloride + metalaxyl; a systemic fungicide mixture used to prevent mildew and root rot in a range of fruit and vegetables. *Ciba-Geigy Agrochemicals.*

26750 Ridosol
Metal treating compositions. *ICI Chem & Polymers Ltd.* Discontinued.

26751 Rifleite
Nitrocellulose gelatinzed by acetone; an explosive.

26752 Rigidex
9002-88-4 7728
Polyolefins and derivatives; used for manufacture of plastics and plastic articles. *BP Chemicals Ltd.*

26753 Rigidex 3
9002-88-4 7728
High density polyethylene copolymer having a density of 0.946, melt index 0.3. *BP Chemicals Ltd.*

26754 Rigidex 9
9002-88-4 7728
High density polyethylene with a density of 0.960 and a melt index of 0.9. *BP Chemicals Ltd.*

26755 Rigidex X4RR
9002-88-4 7728
High density polyethylene having a density of 0.946. *BP Chemicals Ltd.*

26756 Rigidite®
Advanced composites based on reactive resins and glass, carbon, and aramid fibers. *BASF AG.*

26757 Rigidoll
Oil spill clean up products. *BP Chemicals Ltd.* Discontinued.

26758 Rigillene
9002-88-4 7728
Polyethylene; sheet and block for dockside fendering. *Stanley Smith & Co Plastics Ltd.*

26759 Rigipore
9003-53-6 9028 203-066-0
Expandable polystyrene. *BP Chemicals Ltd.*

26760 rilan wax
Waxy acids and vegetable oils hardened by hydrogenation. It is obtained from the higher alcohols of the fat series.

26761 Rilata
Vulcanized bitumen linseed oil mixture containing 16% sulfur.

26762 Rilsan®
Nylon 11 and nylon 12, molding goods, extrusion goods and fine powder coating for metal; used for medical and sporting goods. *Elf Atochem.*

26763 Rilsan® BESHVO, BESVO
25035-04-5
Nylon 11; extrusion grade resin for tubular structures for applications (hoses, coiled air tubing, fuel lines); in blow molding of automotive gas tanks, fuel containers, and hollow products; low temperature toughness, resistance to hydraulic fluids and chemical *Elf Atochem N. Am./Plastics.*

26764 Rimflex A/A
Thermoplastic elastomer
Provides high mechanical properties within a given hardness range; suitable to replace thermoplastic PU, partially cured thermoplastic vulcanizates, thermoset rubber, and high performance PVC compounds; *Syn. Rubber Tech.*

26765 RIMline® 8711B/8700A
Two-component modified polyurethane system; for open or closed pour molding of lowdens. structural RIM composites; for automotive interior trim, e.g., door panels, package trays, sunshades, seat backs. *ICI Polyurethanes.*

26766 RIMline® GMR-5000
Two-component modified polyurethane system; for structural reaction injection molding of fiber-reinforced composites; for radiator or battery supports or anywhere load-bearing thermal, environmental, and structural requirements are important. *ICI Polyurethanes.*

26767 RIMline® GMR-8711
Two-component PU system; low dens. structural reaction injection molding system providing design flexibility and tailored thickness and glass content. *ICI Polyurethanes.*

26768 Rimplast®
Silicone-containing thermoplastic compositions; for injection molding and extrusion of polymers having interpenetrating networks. *Hüls Am.*

26769 Rimthane
Multi-component systems used in the manufacture of polyurethane products via reaction injection molding. *Dow UK.* Discontinued.

26770 Ringer solution
An isotonic solution containing 0.7% sodium chloride, 0.03% potassium chloride, and 0.025% calcium chloride in water; used in physiological experiments.

26771 Ringmaster
5259-88-1 226-066-2
$C_{12}H_{13}NO_4S$
5,6-dihydro-2-methyl-N-phenyl-1,4-Oxathiin-3-carboxamide 4,4-dioxide
Plantvax; Carboxin sulfone; DCMOO; Dihydro-2-methyl-1,4-oxathiin-3-carboxanilide 4,4-dioxide; Oxathiin-3-carboxamide, 5,6-dihydro-2-methyl-N-phenyl-, 4,4-dioxide; Oxathiin-3-carboxanilide, 5,6-dihydro-2-methyl-, 4,4-dioxide. A systemic fungicide for the control of fairy rings in grass. mp = 119-121°. *Rhône-Poulenc Environmental Prods. Ltd.*

26772 Rinoxin
A rodenticide. *Gerhardt Pharmaceuticals.* Name unverified.

26773 Rinsan
Dairy hygiene circulation cleaner. *The Wellcome Foundation Ltd.* Name unverified.

26774 Rintal®
58306-30-2 3982 261-205-0
$C_{20}H_{22}N_4O_6S$
[[2-[(methoxyacetyl)amino]-4-(phenylthio)phenyl]carbonimidoyl]biscarbamic acid dimethyl ester
febantel; Bay Vh 5757; Bay h 5757. Broad-spectrum anthelmintic for use in sheep, goat, camels, cattle, horses; veterinary medicine. mp = 129-130°. *Bayer AG.*

26775 Rintal® Plus
Broad-spectrum anthelmintic plus boticide; used for horses. *Bayer AG.*

26776 Rio Resin
A proprietary amorphous resin. *R. T. Vanderbilt Co Inc.* Discontinued.

26777 Ripercol
A proprietary preparation containing levamisole hydrochloride; veterinary antihelmintic (cattle). *Janssen Pharmaceutical Ltd.* Name unverified.

26778 Ripping ammonal
It contains 84-87% ammonium nitrate, 7-9% aluminum, 2-3% charcoal, and 3-4% potassium dichromate; used as an explosive.

26779 Rippite
An English explosive containing 56-63% nitroglycerin gelatinized with a small quantity of collodion cotton, potassium nitrate, wood meal, castor oil, ammonium oxalate, with the addition of calcium or magnesium carbonate, and a petroleum jelly.

26780 Rismavac-CR6
CV1 988, clone, CR-6 strain, frozen chicken herpes virus; for immunization of poultry. *Intervet Inc.*

26781 Riso
A material consisting of ammonium carbonate ground in a mixture of mineral and vegetable oils (cycline oil); used in the manufacture of sponge rubber.

26782 Risolex
57018-04-9 260-515-3
$C_9H_{11}Cl_2O_2PS$
O-(2,6-dichloro-4-methylphenyl) O,O-dimethyl phosphorothioate
Tolclofos-methyl; Rizolex; S-3349; Dichloro-4-methylphenyl-O,O-dimethyl phosphorothioate; Dimethyl O-(2,6-dichloro-4-methylphenyl) phosphorothioate. An organophosphorus fungicide which gives protection against soil-borne diseases. mp = 78-80°. *Schering Agriculture.* Discontinued.

26783 Rissicol
A castor oil powder, containing 49% castor oil and 36% of inorganic matter, mainly magnesia.

26784 Ristin
A 25% solution of ethylene-glycol monobenzol ester.

26785 Riston®
For the electrical industry. *DuPont UK.*

26786 Risunal
A proprietary preparation of β-diethylaminobutyric acid, aniline hydrochloride, isopropylphenazone, and ethyl and benzyl nicotinate; a rubefacient skin ointment. *Geistlich Sohne AG.* Name unverified.

26787 Rita AZ
Guay-azulene sodium sulfonate
Emollient, humectant, pigment, conditioner, lubricant for personal care formulations. *RITA.*

26788 Rita CA
36653-82-4 2070 253-149-0
$C_{16}H_{34}O$
1-hexadecanol
Cetyl alcohol; ethal; ethol; palmityl alcohol. Thickener, opacifier, emollient for cosmetic formulations. mp = 49°, bp = 344°, d = 0.811; insoluble in H_2O, soluble in organic solvents. *RITA.*

26789 Rita EDGS
627-83-8 211-014-3
$C_{38}H_{74}O_4$
ethylene glycol distearate
Emulsifier, emulsion stabilizer, emollient, pearlescent, opacifier, visc. builder for cosmetic systems. *RITA.*

26790 Rita EGMS, GMS
Glycol stearate
Emulsifier, emulsion stabilizer, emollient, pearlescent, opacifier, visc. builder for cosmetic systems. *RITA.*

26791 Rita HA C-1
9067-32-7 4793
$(C_{14}H_{20}NNaO_{11})_x$
Sodium hyaluronate
Protein for use in skin and hair care preps. *RITA.*

26792 Rita KA
Emollient, humectant, pigment, conditioner, lubricant for personal care formulations. Also used as a flavorant and food additive. *RITA.*

26793 Rita SA
112-92-5 8960 204-017-6
Stearyl alcohol
Thickener, opacifier, emollient for cosmetic formulations. *RITA.*

26794 Ritabate 20
Polysorbate 20
Emulsifier for personal care products. *RITA.*

26795 Ritabate 60
9005-67-8 8872
Polysorbate 60
Emulsifier for personal care products. *RITA.*

26796 Ritabate 80
9005-65-6 7742
Polysorbate 80
Emulsifier for personal care products. *RITA.*

26797 Ritacet-20
68439-49-6 7737
Ceteareth-20
Emulsifier for personal care products. *RITA.*

26798 Ritaceti
8002-23-1 232-302-5
Cetyl esters
Raw material for personal care products. *RITA.*

26799 Ritacetin
117-39-5 8216 204-187-1
Quercetin
Nutrient, humectant for hair and skin care formulations; hair repair agent; soothing to skin. *RITA.*

26800 Ritacetyl®
61788-48-5 262-979-2
Acetylated lanolin
Superfatting agent for soaps, shampoos; film-former for creams and lotions, water resistant films. *RITA.*

26801 Ritachlor 50%
1327-41-9 356 215-477-2
$Al_2H_5O_5Cl \cdot 2H_2O$
Aluminum chlorohydrate
Raw material for personal care products. *RITA.*

26802 Ritachol®
Mineral oil and lanolin alcohol; used in cosmetics and toiletries; stabilizer for emulsions, dispersions, and suspensions; primary or auxiliary emulsifier; epidermal moisturizer, lubricant, and emollient. *RITA.*

26803 Ritachol® 1000
Cetearyl alcohol, polysorbate 60, PEG-150 stearate, and steareth-20. Emulsifier for personal care products, pharmaceuticals, and household specialties. *RITA.*

26804 Ritachol® SS
2778-96-3 220-476-5
Stearyl stearate
Raw material for personal care products. *RITA.*

26805 Ritacollagen BA-1
9007-34-5 2543 232-697-4
Broad-spectrum anthelmintic plus boticide; protein for use in skin and hair care preps. *RITA.*

26806 Ritaderm®
Petrolatum, lanolin, sodium PCA, and polysorbate 85. Emollient, moisturizer, and lubricant used in cosmetics. *RITA.*

26807 Ritahydrox
68424-66-8 270-315-8
Hydroxylated lanolin
Water-oil emulsifier, hypoallergenic emollient. *RITA.*

26808 Ritalafa®
68424-43-1 270-302-7
Lanolin acid
Film-former, emollient; rewetting of makeup preparations. *RITA.*

26809 Ritalan®
Lanolin oil USP; moisturizer, plasticizer, penetrant, emollient; hypoallergenic, nonsensitizing skin lubricant. *RITA.*

26810 Ritalan® AWS
68458-58-8
PPG-12-PEG-65 lanolin oil; auxiliary emulsifier, moisturizer, emollient. *RITA.*

26811 Ritalan® C
IPP and lanolin oil; blending agent, emollient, epidermal penetrant, rewetting agent, and solubilizer for waxes and other oil-sol. or disp. materials; personal care products. *RITA.*

26812 Ritalanine
56-41-7 205 200-273-8
L-Alanine
Aminoacid for use in skin and hair care preps. *RITA.*

26813 Ritalastin EL-10
Hydrolyzed elastin
Protein for use in skin and hair care preps. *RITA.*

26814 Ritaloe 1X
Aloe vera; emollient, humectant, pigment, conditioner, lubricant for personal care formulations. *RITA.*

26815 Ritamectant K2
Dipotassium glycyrrhizinate
Humectant for personal care products. *RITA.*

26816 Ritamectant PCA
28874-51-3 249-277-1
Sodium PCA
Humectant for personal care products. *RITA.*

26817 Ritapan CAP
137-08-6 7147 205-278-9
Calcium pantothenate
Nutrient, humectant for hair and skin care formulations; hair repair agent; soothing to skin. *RITA.*

26818 Ritapan D, DL
81-13-0 2988 201-327-3
$C_9H_{19}NO_4$
d-Panthenol
Nutrient, humectant for hair and skin care formulations; hair repair agent; soothing to skin. *RITA.*

26819 Ritapan NAP
Sodium pantothenate
Nutrient, humectant for hair and skin care formulations; hair repair agent; soothing to skin. *RITA.*

26820 Ritapan TA
Panthenyl triacetate
Nutrient, humectant for hair and skin care formulations; hair repair agent; soothing to skin. *RITA.*

26821 Ritapeg 150 DS
9004-99-3
PEG-150 stearate
Emulsifier for personal care products. *RITA.*

26822 Ritapeg 400 DS
9005-08-7
PEG-8 distearate
Emulsifier for personal care products. *RITA.*

26823 Ritaphenone 3
131-57-7 7088 205-031-5
Benzophenone-3
Raw material for personal care products. *RITA.*

26824 Ritaplast
Mineral oil and polyethylene; raw material for personal care products. *RITA.*

26825 Ritaplast R
Lanolin oil, polyethylene; raw material for personal care products. *RITA.*

26826 Ritaplast TN
C_{12}-C_{15} alcohols benzoate and polyethylene; raw material for personal care products. *RITA.*

26827 Ritapro 100
Cetearyl alcohol, steareth-20, and steareth-10; oil-water emulsifier. *RITA.*

26828 Ritapro 165
Glyceryl stearate and PEG-2 stearate; oil-water emulsifier for creams, lotions. *RITA.*

26829 Ritaquat Q
Steartrimonium hydrolyzed animal protein; protein for use in skin and hair care preparations. *RITA.*

26830 Ritasilk
96690-41-4 306-235-8
Hydrolyzed silk protein; protein for use in skin and hair care preps. *RITA.*

26831 Ritasilk Powd
9009-99-8
Silk powder; protein for use in skin and hair care preps. *RITA.*

26832 Ritasol
63393-93-1 264-119-1
Isopropyl lanolate
Emollient, spreading agent, water-resistant film former for lip preparations. *RITA.*

26833 Ritasynt IP
Glycol stearate and other ingredients; foam booster, thickener, opacifier for cosmetics. *RITA.*

26834 Ritatin
58-85-5 1272 200-399-3
biotin
Biotin; nutrient, humectant for hair and skin care formulations; hair repair agent; soothing to skin. *RITA.*

26835 Ritawax
8027-33-6 232-430-1

Lanolin alcohol
Emulsifier, emollient for skin preparations. *RITA.*

26836 Ritawax 5
61791-20-6
Laneth-5
Emulsifier, solubilizer, emollient for creams and lotions. *RITA.*

26837 Ritawax AEO
Polysorbate 80, acetylated lanolin alcohol, cetyl acetate; emollient, lubricant, moisturizer, penetrant, solubilizer, dispersant, plasticizer for personal care products. *RITA.*

26838 Ritawax ALA
Cetyl acetate and acetylated lanolin alcohol; emollient, lubricant, moisturizer, penetrant. *RITA.*

26839 Rit-Cizer 8
N-Ethyl o/p toluenesulfonamide
Plasticizer for adhesives, paints, printing inks, epoxy resins, polyamide resins, phenolics, melamine resins. *Rit-Chem.*

26840 Rit-Cizer 9
o/p Toluene sulfonamide
Plasticizer for thermosetting resins (melamine, urea, phenolics), nylon, casein, PVAc; imparts good gloss, improves wetting action. *Rit-Chem.*

26841 Riteflex®
Thermoplastic polyester elastomer. *Hoechst AG.*

26842 Riteflex® 347ZS
Thermoplastic polyester elastomer
Processable by blow molding or extrusion; improved thermal stability for improved long-term prop. retention. *Hoechst Celanese.* Name unverified.

26843 Riteflex® 540
Thermoplastic polyester elastomer
Softest, most flexible grade; processable on conventional injection molding and extrusion equipment; recommended for colors and polymer modification. *Hoechst Celanese.* Name unverified.

26844 Riteflex® BP 8929
Thermoplastic polyester elastomer alloy; high impact grade with excellent processability, especially good mold filling characteristics in injection molding applications. *Hoechst Celanese.* Name unverified.

26845 Riteflex® BP 9057
Thermoplastic polyester elastomer alloy; most flexible grade with excellent low temperature impact properties; excellent processability and mold filling capabilities. *Hoechst Celanese.* Name unverified.

26846 Rit-Ester B-100
Functional polyester; basic polymer or properties enhancer for adhesives and coatings; contains functional hydroxyl groups which can be reacted with melamines, isocyanates, and other resin systems. *Rit-Chem.*

26847 Ritha
An alkaline deposit found on the land in India; used as a soap substitute.

26848 Ritoleth 2
9004-98-2
Oleth-2
Oil-water emulsifier, solubilizer; stable over wide pH range. *RITA.*

26849 Rit-O-Lite MHP-S
Formaldehyde/toluenesulfonamide polymer; modifier, adhesion promoter for coatings, lacquers, printing inks, adhesives; compatible with alkyds, acrylics, urethanes, nitrocellulose and vinyls. *Rit-Chem.*

26850 Rit-O-Lite MS-80
Promotes, modifies, and improves adhesion, gloss retention, solvent release, and lowers visc. of other resin systems. *Rit-Chem.*

26851 Ritox 35
Laureth-23
Emulsifier for personal care products. *RITA.*

26852 Ritox 52
9004-99-3
PEG-40 stearate
Emulsifier for personal care products. *RITA.*

26853 Ritox 721
9005-00-9
Steareth-21
Emulsifier for personal care products. *RITA.*

26854 Rival
Glyphosate + simazine; a translocated and residual herbicide for the control of grasses and weeds. *Chipman Ltd.*

26855 Rivalit P
A German explosive containing nitro-glycerin, ammonium nitrate, vegetable meal, nitro-compounds, and potassium perchlorate.

26856 rivet metal
An alloy of copper and tin, to which zinc is sometimes added.

26857 RIX 90149
Modified cycloaliphatic amine adduct; curing agent for room temperature cure of epoxy resins; suggested end uses including industrial floor toppings, high build glaze, sealer or gel coatings, general purpose castings and encapsulations. *Shell.* Name unverified.

26858 Rizolex
Tolclofos-methyl
Fungicide. *Schering Agrochemicals Ltd.* Discontinued.

26859 RJ-100
Styrene-allyl alcohol resins; paint coating ingredient. *Monsanto Co.* Name unverified.

26860 RMD
High molecular weight polymers based on sodium acrylate; used in the mining industry. *Cyanamid BV.*

26861 RMR
High molecular weight polymers based on sodium acrylate; used in the mining industry. *Cyanamid BV.*

26862 Roach Stoppers
Blatticide. *Atomergic Chemetals Corp.*

26863 Roachban
A cockroach insecticide. *Murphy Chemical Co Ltd.* Discontinued.

26864 Roanoid
A proprietary urea-formaldehyde molding compound. No manufacturer.

26865 roaster slag
A slag produced in the purification of copper metal. It contains from 17-40% copper as silicate and metal.

26866 Ro-A-Vit
68-26-8 10150 200-683-7
Proprietary preparation of vitamin A (retinol). *Roche Products Ltd.* Name unverified.

26867 Robac
Principally rubber accelerators and vulcanizing agents; some activators and antioxidants; for vulcanization and protection of natural and synthetic rubbers. *Robinson Brothers Ltd.*

26868 Robac T.B.Z
Zinc thiobenzoate
A low temperature peptizing agent for natural rubber.

26869 Robacure
Principally rubber accelerators and vulcanizing agents; some activators and antioxidants; for vulcanization and protection of natural and synthetic rubbers. *Robinson Brothers Ltd.*

26870 Robane®
111-01-3 8923 203-825-6
$C_{30}H_{62}$
Squalane NF
Moisturizer, emollient, lubricant, humectant; aids spread of topical agents over the skin, increases skin respiration, prevents insensible water loss, imparts suppleness to skin without greasy feel; cosmetics and pharmaceuticals. *Robeco.*

26871 Robengatope I-131
50291-21-9 8421
$C_{20}H_2Cl_4I_4Na_2O_5$
4,5,6,7-tetrachloro-3',6'-dihydroxy-2',4',5',7'-tetraiodospiro[isobenzofuran-1(3H),9'-[9H]xanthen]-3-one-disodium salt
Rose Bengal Extra; Rose bengal sodium ^{125}I;. Radioactive agent. Used to detect corneal trauma and in hepatic function tests. *Mallinckrodt Inc.* Name unverified.

26872 Robert's reagent
For proteins. It consists of 1 volume of pure nitric acid with 5 volumes of a 40% (saturated) solution of magnesium sulfate.

26873 Robertson alloy
Consists of 1 part gold, 3 parts silver, and 2 parts tin. It is mixed with mercury for use as a dental filler.

26874 Ro-Bile
Pancrelipase; a concentrate of pancreatic enzymes standardized for lipase content; enzyme. *Rowell Laboratories Inc.* Name unverified.

26875 Robond
Acrylic polymer based adhesives. *Rohm & Haas.*

26876 Roburite
An explosive used in mines. It consists of 86% ammonium nitrate and 14% chlorodinitro-benzene.

26877 Roburite I
An explosive containing 87.5% ammonium nitrate, 7% dinitro-benzene, 0.5% potassium permanganate, and 5% ammonium sulfate.

26878 Roburite III
An explosive containing 87% ammonium nitrate, 11% dinitro-benzene, and 2% chloro-naphthalene.

26879 Rocagel
Soil stabilization formulations. *Rhône-Poulenc UK.*

26880 Roccal
8001-54-5 1086 231-635-3
A proprietary preparation of benzalkonium chloride; a preoperative skin cleanser. *Bayer AG.*

26881 Rocel
Plastics in the form of sheets, bands, strips, films, plates, slabs, blocks, tubes, etc. *Courtaulds plc.* Discontinued.

26882 Rochelle salt
304-59-6 7840 206-156-8
Rochdale salt.

26883 rock asphalt
Limestone or other material and found naturally, impregnated with bitumen.

26884 rock cork
A variety of asbestos.

26885 rock crystal
7631-86-9 8637 231-545-4
silica. Transparent and colorless quartz.

26886 rock dammar
A variety of dammar resin derived from *Hopea odorata* of Burma.

26887 rock salt
7647-14-5 9742 231-598-3
NaCl
Sodium chloride
halite. From sea water.

26888 rock wool
Rock cork; Rocktex. A furnace product made from self-fluxing siliceous and argillaceous dolomite in which the basic and acidic constituents are present in proportions that their fluxing action is nearly balanced.

26889 Rocket® Ultra
Tridemorph, fenpropiomorph; systemic fungicide for control of cereal diseases *BASF AG.*

26890 Rocksil
A proprietary trade name for rockwool insulating materials; withstands 760°. *Cape Insulation Cape Asbestos Co.* Name unverified.

26891 Rocktex
A proprietary trade name for rock wool. No manufacturer.

26892 Roclys®
Glucose syrup; sweetening agent for general food use. *Roquette (UK) Ltd.*

26893 Rocol P.R
A proprietary silicone-based spray used for mold release. *Rocol Ltd.* Name unverified.

26894 Rocol R.S.7
A proprietary nonsilicone-based wet film spray used for mold release. *Rocol Ltd.* Name unverified.

26895 Rocryl
Specialty acrylic and methacrylic monomers. *Rohm & Haas UK.*

26896 Rocsol
Oxidized hydrocarbon waxes.
Croda Chem. Ltd.

26897 rod wax
A wax-like mass deposited on the drill rods in many petroleum oil wells.

26898 Rodea
40716-42-5 255-051-3
Ricinoleic diethanolamide
Emulsifier, thickener, solvating agent, corrosion inhibitor. *Rhône-Poulenc Surf.* Discontinued.

26899 Rodeo
38641-94-0 254-056-8
$C_6H_{17}N_2O_5P$
Phosphonomethyl)glycine, isopropylamine salt
Roundup; Quick; Revoke; Glyphosate isopropylamine salt; Drat; Liphadione; Glyphosate Amine; Glifonox; Glycel; Glycine, N-(phosphonomethyl)-, compd. with 2-propanamine (1:1;)Isopropylamine glyphosate; Phosphonomethyl)glycine, isopropylamine salt; Rodeo; Rondo. Aquatic herbicide. *Monsanto Co.* Name unverified.

26900 Rodinal
123-30-8 482 204-616-2
A photographic developer. The active constituent is *p*-aminophenol hydrochloride.

26901 Rodo
Blend of essential oils (deodorants); neutralizes typical dry rubber odors and in emulsions acts as a deodorant in finished latex products. *R. T. Vanderbilt Co Inc.*

26902 Roebaryt
7727-43-7 1023 231-784-4
A barium sulfate prepared for use in X-ray work.

26903 Roesch's aluminum solder
An alloy of 50.2% zinc, 50% tin, 0.7% antimony, and 0.2% copper.

26904 Roferose
50-99-7 4467 200-075-1
Dextrose monohydrate
Used for food, pharmaceutical, and industrial applications. *Roquette (UK) Ltd.*

26905 Roga
Cellulose acetate spiral wound membrane; reverse osmosis water treatment. *AlliedSignal Fluid Systems.* Name unverified.

26906 Rogé Cavaills
Soap, bath additive, and shampoo; for cleansing care of skin and hair. *Richardson-Vicks Inc.* Name unverified.

26907 Roghan
Afridi wax. Obtained by boiling safflower oil for 2 hours, then putting it into vessels partly filled with water.

26908 Rogor E
dimethoate. Insecticide. *Schering Agrochemicals Ltd.* Discontinued.

26909 Rohafloc
Methacrylate based ionic/nonionic flocculants; for sludge dewatering, industrial waste treatment, paper sizing, etc. *Cornelius Chemical Co Ltd.* Name unverified.

26910 Rohagit® SM V
Acrylic polymer; thickener. *Rohm Tech.*

26911 Rohamere® 8662
Thermoplastic acrylic polymer emulsion; forms a medium film; as finishing agent in textile, leather, paper applications. *Rohm Tech.*

26912 Rohatol® BV 382
Self-crosslinking acrylic polymer emulsion; for nonwoven bonding for household, hospital, hygiene applications, flocking adhesives for textiles. *Rohm Tech.*

26913 Rohatol® D 362
Thermoplastic acrylic polymer emulsion; for textile coating, protective foils, paper finishes. *Rohm Tech.*

26914 Rohn alloys
Heat-resisting alloys containing nickel, chromium, and iron, sometimes with the addition of manganese and molybdenum.

26915 Rohrbach's solution
10048-99-4 1011 233-160-7
$BaHgI_4$
barium mercuric iodide
A solution of barium and mercuric iodides (100 g barium iodide and 130 g mercuric iodide heated with 20 ml water to 150-200 C.). The solution is allowed to cool, when a double salt is deposited. The liquid is decanted. Very soluble in H_2O.

26916 Roica®
Elastic polyurethane filament. *Asahi Chem. Industry.*

26917 RokLok® B-3
Polyurethane
Binders to control water and broken strata in mines, tunnels, shafts and stream sealing applications; upon injection they permeate the strata and form expanded binder with excellent mechanical properties. *Flexible Prods.*

26918 Rokon
149-30-4 5916 205-736-8
2-Mercaptobenzothiazole
Corrosion inhibitor for coolants, greases and fuel; metal deactivator and extreme pressure agent for petroleum lubricants; chemical intermediate. *R. T. Vanderbilt Co Inc.*

26919 Rolafix
Photographic fixer. *May & Baker Ltd.* Name unverified.

26920 roll sulfur
7704-34-9 9142 231-722-6
S
sulfur
sulfur, sulphur. Sulfur which has been melted and poured into molds.

26921 Rollit
7782-42-5 4560 231-955-3
C
graphite. Graphite powder; used as mandrel bar lubricant. *Lonza AG.*

26922 Rollofix X100
One-component expanding polyurethane foam comprising polyol, methyl di-isocyanate, freon and niax; used for filling cavities (domestic and industrial), bonding (sealing against noxious vapors and moisture). *Piccadilly Products Ltd.* Name unverified.

26923 Rol-man Steel
A proprietary trade name for a high-manganese steel containing 11-14% manganese and 1-1.4% carbon. No manufacturer.

26924 Rolox
Two-part epoxy compounds. *Hardman.*

26925 Roman alum
$Al_2(SO_4)_3 \cdot K_2SO_4 \cdot 24H_2O$
Potash alum.

26926 Roman bronze
An alloy of 90% copper and 10% tin.

26927 Roman cement
Parker's cement. A natural cement made by calcining the modules of argillaceous limestone mixed with calcareous spar, which occurs in London and other clays.

26928 Romane
111-01-3 8923 203-825-6
Squalane. *A & E Connock (Perfumery & Cosmetics) Ltd.*

26929 Romanium
An alloy of 97.43% aluminum, 1.75% nickel, 0.25% copper, 0.25% antimony, 0.17% tungsten, and 0.15% tin.

26930 Romensin
17090-79-8 6329 241-154-0
Monensin
Antiprotozoal agent. *Dista Products Ltd.*

26931 Romite
Ammonium nitrate mixed with a solid, melted hydrocarbon (paraffin or naphthalene), gelatinized with a liquid hydrocarbon (paraffin oil), and contains gelatinized potassium chlorate; an explosive.

26932 Rompel's alloy
An antifriction metal, containing 62% copper, 10% zinc, 10% tin, and 18% lead.

26933 romperit G
A German explosive containing nitroglycerin, ammonium nitrate, vegetable meal, nitro-compounds, and potassium perchlorate.

26934 Rompun®
10213
xylazine. Sedative, analgesic, anesthetic, and muscle relaxant; for use in cattle. *Bayer AG.*

26935 Ronabond®
Pre-packed ready-to-use cementitious mortar; used in concrete repair, the laying of thin screeds and floors and the fixing of building components. *Ronacrete Ltd.*

26936 Ronafix
Styrene butadiene waterproof bonding additive; used in concrete repair, the laying of thin screeds and floors and the fixing of building components. *Ronacrete Ltd.*

26937 Ronaset®
Pre-packed high-strength mortars, concretes, and transition strip; used for bedding preformed bridge expansion joints, transition strips, motorway repairs, floors, and screeds. *Ronacrete Ltd.*

26938 Rondis
Disproportionated resin esters and salts; emulsifier in synthetic latex. *Akzo Chemie UK Ltd.*

26939 Ronfalin
Acrylonitrile butadiene styrene; plastic for general purposes and for special applications in the automotive and electrical industries; for toys, domestic appliances, telematic equipment and extruded products; uses include dashboard components and radiato *DSM NV.* Discontinued.

26940 Ronfaloy E
Blends/compounds with ABS; for use in the open air. *DSM NV.* Discontinued.

26941 Ronfaloy V
Blends/compounds with ABS; for the telematic industry. *DSM NV.* Discontinued.

26942 Ronfusil Steel
A proprietary trade name for a manganese-steel containing 12% manganese. No manufacturer.

26943 Rongal®
Reducing agents for dyeing cellulose fibers with vat dyes. *BASF AG.*

26944 Rongalit®
Textile discharging and reducing agents. *BASF plc.*

26945 Rongalit® C
870-72-4 4263 212-800-9
CH_3NaO_4S
Sodium hydroxymethane sulfonate
hydroxymethanesulfonic acid monosodium salt; sodium formaldehyde bisulfite; methylolsulfonic acid sodium salt. Reducing and discharge agents for textile printing. mp = 200°. *BASF AG.*

26946 Rongalite C
A combination of the sodium salt of the unstable sulfoxylic acid and formaldehyde; a reducing agent used in the dye industry, and as a photographic developer.

26947 Ronia metal
A brass containing small quantities of cobalt, manganese, and phosphorus.

26948 Ronicol
6164-87-0 6614 202-864-6
A proprietary preparation of nicotinyl tartrate; used in the treatment of circulatory disorders. *Roche Products Ltd.* Name unverified.

26949 Ronilan® DF, FL
50471-44-8 10122 256-599-6
Vinclozolin
Contact fungicide for use in vines, fruit, strawberries, vegetables, ornamentals, hops, etc. *BASF AG.*

26950 Ronilon
50471-44-8 10122 256-599-6
Vinclozolin
A protectant fungicide for a wide range of vegetables and fruit. *BASF plc.*

26951 ronnel
299-84-3 8415 206-082-6
$C_8H_8Cl_3O_3PS$
O,O-dimethyl O-(2,4,5-trichlorophenyl) phosphorothioate
Fenchlorophos; dimethyltrichlorophenylthiophosphate; Korlan; Ectoral; Etrolene;Fenchlorfos; Trichlormetaphos; Blitex; Dermafos; Dimethyl (2,4,5-trichlorophenyl) phosphorothionate; Dimethyl O-(2,4,5-trichlorophenyl) thiophosphate; Gesektin K; Moorman's medicated RID-EZY; Nankor; Phenol, 2,4,5-trichloro-, O-ester with O,O-dimethyl phosphorothioate; Remelt; Ronnel; Rovan; Trichlorometafos; Trolene; Viozene; Fenclofos. Insecticide with systemic action. mp = 41°; insoluble in H_2O, soluble in organic solvents; LD_{50} (rat orl) = 2630 mg/kg.

26952 Ronoxan
Antioxidant pastes for food. *Roche Products Ltd.* Name unverified.

26953 Ronseal
Varnishes for general wood protection. *Sterling Ronseal Ltd.*

26954 Ronstar 2G
19666-30-9 7038 243-215-7
Oxadiazon
A contact herbicide to control weeds and grasses in fruit and ornamental crops. *Embetec Crop Protection Ltd.*

26955 Ronstar TX
Carbetamide [16118-49-3] and oxadiazon [19666-30-9]; a residual herbicide for pre-weed emergence control for container grown nursery plants. *Hortichem Ltd.*

26956 Roofcover LM
One-pack moisture-curing paint providing protection to roofing materials. *Feb Ltd.*

26957 Root Guard
333-41-5 3043 206-373-8
Diazinon insecticide. *Murphy Chemical Co Ltd.* Discontinued.

26958 Rooting Powder
86-87-3 6458 201-705-8
Powder containing 1-naphthylacetic acid; rooting powder. *Vitax Ltd.*

26959 Root-Out
7773-06-0 589 231-871-7
Ammonium sulfamate 98%; herbicide to control weeds and grasses in vegetables and ornamentals prior to planting; tree, weed and brushwood killer. *Dax Products Ltd.*

26960 Ropaque
Opaque polymer emulsion; opacifying agent for decorative coatings, paper coatings, etc. *Rohm & Haas.*

26961 Ro-Pel
Animal repellent. *Atomergic Chemetals Corp.*

26962 rose B
erythrosin.

26963 rose bengal
632-69-9 8241 211-183-3
$C_{20}H_2Cl_4I_4Na_2O_5$
4,5,6,7-Tetrachloro-3',6'-dihydroxy-2',4',5',7'-tetraiodospiro[isobenzofuran-1(3H),9'[9H]xanthen]-3-one dipotassium salt
4,5,6,7-Tetrachloro2',4',',7'-tetraiodofluorescein potassium derivative potassium salt; C.I. Acid Red 94; Rose Bengale B; C.I. 45440.

26964 Rose Ester
90-17-5 201-972-0
$C_{10}H_9Cl_3O_2$

trichloromethyl phenyl carbinyl acetate
α-trichloromethylbenzyl acetate; benzenemethanol, α-(trichloromethyl)-, acetate. Rose ester is a trade name for trichloromethyl phenyl carbinyl acetate; a rose scent used in the perfumery industry. mp = 85-87° *Nipa Laboratories Ltd.* Name unverified.

26965 Rose Food
Granular fertilizer. *Fisons plc, Horticultural Div.* Name unverified.

26966 Rose Plus
Granular fertilizer containing magnesium. *ICI Garden Products.*

26967 rose quartz
14808-60-7 238-878-4
A mineral, it is a variety of quartz, SiO_2, and is stated to owe its color to manganese.

26968 Rosé oil
Roshé oil; oil of geranium; oil of rose-geranium; oil of pelargonium; ginger grass oil; Turkish geranium oil; oil of palmarosa. (Andropogon oils, obtained from a grass, *Andropogon nardus*. They contain geraniol, $C_{10}H_{18}O$.

26969 Roseclear
Contains bupirimate, pirimicarb and triforine; combined insecticide and fungicide for garden use. *ICI Garden Products.*

26970 rosein
An alloy of 44.4% nickel, 33.3% aluminum, 11.1% silver, and 11.1% tin. Another alloy contains 40% nickel, 30% aluminum, 20% tin, and 10% silver; used by jewelers.

26971 Roseline
Finishing agent. *Ciba plc.* Name unverified.

26972 Roselle fiber
A Malay fiber similar to jute.

26973 roseol
reuniol
Names applied to citronellol, $C_9H_{17}\cdot CH_2OH$, or to mixtures of citronellol and geraniol; used in the manufacture of perfumes.

26974 Rose's metal
Fusible bismuth alloys. a) Consists of 42% bismuth, 42% lead, and 16% tin. It melts at 79 C. b) Consists of 33.3% bismuth, 33.3% lead, and 33.3% tin. c) Consists of 48.9% bismuth, 27.5% lead, and 23.6% tin.

26975 Rosette copper
This is obtained in thin films by throwing water on to the surface of molten copper and removing the crusts formed.

26976 rosin
08050-09-7 8424 232-475-7
colophony
gum rosin; rosin gum. Residue from distilling off the volatile oil from the oleoresin obtained from *Pinus palustris* and other species of *Pinaceae*; manufacture of varnishes, paint driers, printing inks, cements, soap, sealing wax, wood polishes, paper, plastics, *Am. Cyanamid Arakawa; Arizona; Georgia-Pacific; Hercules BV; Meer; Natrochem; Veitsiluoto Oy.*

26977 rosin grease
A combination of rosin oil with lime.

26978 rosin oil
8002-16-2 232-300-4
An oil obtained by the distillation of rosin from 300-400°; used as an adulterant for olive oil, also as a lubricant for iron bearings.

26979 rosin spirit
Essence of resin, resin spirit; the first distillate from rosin, 78-250°. It is a complex mixture of hydrocarbons somewhat resembling turpentine; used as a substitute for turpentine.

26980 rosin tin
A yellow variety of the mineral cassiterite or tinstone.

26981 Rosinal
Rubber softener and tackifier. *Crowley Chem.*

26982 Rosintene
Terpene-rosin polymer. *Crowley Chem.*

26983 Rosite® 4030FS
Thermoset polyester SMC, 30% glass fiber-reinforced; zero shrink material providing surface smoothness for demanding applications, e.g., auto headlamp retainers. *Rostone.*

26984 Rosite® ESD Cond. C
Thermoset polyester BMC, 15% glass fiber-reinforced; electrostatic dissipative compound for molded paper guides in printers and plotters. *Rostone.*

26985 Roskens
Hand conditioner. *Fisons plc, Pharmaceuticals Div.* Name unverified.

26986 Roskydal®
Unsaturated polyester resins; used for formulation of furniture waxes with and without paraffin wax. *Bayer AG; Bayer plc.*

26987 RosIte® 3250A
Thermoset polyester BMC, 5% glass fiberreinforced; outstanding electrical characteristics; for applications where molds designed for high shrinkage are employed. *Rostone.*

26988 RosIte® 4010ES
Thermoset polyester SMC, 22-25% glass fiber-reinforced; SMC for business equipment applications; high strength, dimensional control, and surface smoothness. *Rostone.*

26989 rosolene
A rosin oil obtained by the distillation of rosin.

26990 Ross alloy
A bronze containing 68% copper and 32% tin.

26991 Ross Japan Wax Substitute 525
Fish glycerides, tallow glycerides, oleostearine, microcrystalline wax; for coatings, textile finishes, cosmetics, lubricants, crayons, wax modeling, buffing compounds, pharmaceuticals. *Frank B. Ross.*

26992 Ross Spermaceti Wax Substitute 573
8002-23-1 232-302-5
Cetyl esters
For adhesives, cosmetic creams, lotions, and soaps, medicinals (ointments, salves), candles, textiles, coatings. *Frank B. Ross.*

26993 Ross Wax 100
8002-74-2 7155 232-315-6
Synthetic wax
Mold release in rubber and plastics; external lubricant for PVC; increases melting point, opacity, gloss in wax blends, floor waxes, textiles, coatings, candles, hot melts, paints, inks, asphalt. *Frank B. Ross.*

26994 Ross Wax 145, 165
8002-74-2 7155 232-315-6
High melting point fully refined petroleum wax; lubricant for formulating PVC and electrical wire and cable compounds. *Frank B. Ross.*

26995 Rota
pH indicator papers. *Rhône-Poulenc Laboratory Prods. Ltd.*

26996 Rotal
Electrolytic coloring of anodized aluminum. *Albright & Wilson Ltd.*

26997 Rotalin
330-55-2 5534 206-356-5
Herbicide containing linuron. *ICI Chem & Polymers Ltd.*

26998 Rotalin
330-55-2 5534 206-356-5
Linuron
A residual urea herbicide for the control of weeds in field crops. *Farm Protection Ltd.*

26999 Rotax®
149-30-4 5916 205-736-8
2-Mercaptobenzothiazole
Accelerator; primary accelerator for natural and synthetic rubbers; nonstaining and nondiscoloring; used in proofing compounds where lowest odor is desired; corrosion inhibitor in automotive chems., industrial cleaners; protects silverware from sulfur bla *R. T. Vanderbilt Co Inc.*

27000 rotenone
83-79-4 8427 201-501-9
$C_{23}H_{22}O_6$
(2R,2α,6aα,12aα)-1,2,12,12a-tetrahydro-8,9-dimethoxy-2-(1-methylethenyl)[1]benzopyrano[3,4-b]furo[2,3-H][1]benzopyran-6(6aH)-one
Tubatoxin; Derris; Rotacide; Noxfish; Noxfire; Foliafume; Nusyn-Noxfish; PB-Nox; Chem-Fish; (-)-rotenone; barbasco; chem-mite; cube root; Cubor; dactinol; Deril; derrin; dri-kil; extrax; fish-tox; green cross warble; haiari; mexide; nicouline; paraderil; pro-nox fish; RO-KO; ronone; rotefive; rotefour; rotessenol; rotocide; tubotoxin; Synpren; Prenfish; Nusyn; Hydrogenated rotenone; Rotenone, hydrogena. A crystalline material found in the roots of derris, a plant grown in the rubber plantations of the Malay Peninsula; also found in the South American cube plant. It occurs up to 5.5% in derris and up to 7% in cube. mp = 165-166°; $[\alpha]_D^{20}$= -228° (c = 2.22 C_6H_6); insoluble in H_2O, soluble in organic solvents; LD_{50} (mus ip) = 2.8 mg/kg.

27001 Roter
A proprietary preparation containing bismuth subnitrate, magnesium carbonate, sodium bicarbonate and frangula; an antacid used in the treatment of peptic ulcers. *FAIR Laboratories.* Unverified.

27002 Rotersept
18472-51-0 2140 242-354-0
$C_{34}H_{54}Cl_2N_{10}O_{14}$
chlorhexidine digluconate
Hibidil; Hibicare; Hibisol; Chlorhexidine digluconate; Bacticlens; Hibiclens; Hibiscrub; Hibitane; Corsodyl; Plac Out; Peridex; pHisoMed; Plurexid; Rotersept; Unisept; Chlorhexidine gluconate; Hibistat; Orahexal; Hexamethylene bis(5-(4-chlorophenyl)biguanide) digluconate; Hibitane chlorhexidine gluconate; Tetraazatetradecanediimidamide, N,N''-bis(4-

chlorophenyl)-3,12-diimino, di-D-gluconate. A proprietary aerosol spray containing chlorhexidine digluconate; used in the prophylaxis of puerperal mastitis. soluble in H_2O (>50% w/v); LD_{50} (mus iv) = 1800 mg/kg. *FAIR Laboratories*. Unverified.

27003 Rotoval
Gravure printing inks; used for package printing. *AlliedSignal Inc/Sinclair and Valentine Division*. Name unverified.

27004 Rotoxit
Resistant high-silicon-copper alloy; resistant to uric, fluosilicic, fatty acids, dilute hydrochloric, sulfuric, and acetic acids, 30% phosphoric acid, hydrogen peroxide, ammonia, lyes, and sulfates, but not to chromic, nitric, and lactic acids.

27005 Rotra bark
The bark of *Rotra fotsy and R.* meno. The bark contains 12.6% tannin.

27006 Rotten stone
A soft, friable aluminum silicate, containing a little organic matter; used as a polishing material. The term is also sometimes applied to Tripoli.

27007 Rotuba H
9004-35-7 2013
Cellulose acetate
Molding material. *Rotuba*. Name unverified.

27008 Rouen white
A pigment. It is a clay found near Rouen.

27009 rouge
1309-37-1 4072 215-168-2
Good qualities of rouge consist of very fine iron oxide, Fe_2O_3, and are used as abrasives. The finest rouge is prepared from safflower.

27010 Roumanite
Romanite
The amber of Roumania, which much resembles the Prussian variety.

27011 Roundup
1071-83-6 4522 213-997-4
Glyphosate
Herbicide. *Schering Agrochemicals Ltd.* Discontinued.

27012 Rousselot Gelatine
9000-70-8 4388 232-554-6
Gelatin derived from animal tissue available in several mesh sizes; used for photographic emulsions, pharmaceutical hard, soft and micro capsules, edible gelatins for confectionery, meat, dairy and dessert industries. *SKW Biosystems Ltd.*

27013 Roussel's solution
A solution of sodium phosphate.

27014 Roussin's black salt
A compound, $Fe_3H_2N_4O_4S_5$, obtained by adding a solution of ferrous or ferric chloride slowly to the mixed solution of potassium nitrite and ammonium sulfide, and then boiling.

27015 Roussin's red salt
A salt, $Fe_2S_4N_2O_2Na_2 . H_2O$, Obtained by treating the sodium salt of Roussin's black salt with excess of acid after boiling, and then evaporating.

27016 Roussin's salts
Salts of the type $KFe_4(NO)_7S_3$, Obtained when nitric oxide is passed through a suspension of ferrous sulfide in a sulfide solution.

27017 Roux's stain
A microscopic stain that contains 0.5 g gentian violet, 1.5 g methyl green, and 200 ml distilled water.

27018 Rovace 571
9003-20-7
High molecular weight, highly branched vinyl acetate homopolymer, PVAL stabilized; emulsion for wood and packaging adhesives, textile hand stiffeners, masonry coatings; high wet tack, ease of compounding, machinability. *Rohm & Haas.*

27019 Rovace 2113
9003-20-7
High molecular weight vinyl acetate homopolymer; emulsion used in joint cement, carpet back size, acoustical coatings, textile sizing, corrugated adhesives; high filler load capacity, fast setting, water resistance, starch compatibility. *Rohm & Haas.*

27020 Rovace 9100
Vinyl/acrylic copolymer emulsion; binder for interior flat and semigloss paints. *Rohm & Haas.*

27021 Roval Dust
36734-19-7 5093 253-178-9
Iprodione
A fungicide with protectant activity for lettuce and glass house crops. *Hortichem Ltd.*

27022 Roval Flo
36734-19-7 5093 253-178-9
Iprodione
A fungicide with protectant activity for use in field crops, cereals, fruit trees and bulbs. *Rhône-Poulenc Crop Protection Ltd.*

27023 Roval Green
36734-19-7 5093 253-178-9
Iprodione
A fungicide with protectant activity for use in turf and amenity grasses. *Rhône-Poulenc Environmental Prods. Ltd.*

27024 Roval WP
36734-19-7 5093 253-178-9
Iprodione
A fungicide with protectant activity for use in nursery stock and fruit trees. *Embetec Crop Protection Ltd.*

27025 Rovel
Weatherable polymers used to manufacture spas, truck toppers, automotive trim. *Dow UK.* Discontinued.

27026 Rovigon
Proprietary vitamin preparation containing vitamins A and E. *Roche Products Ltd.* Name unverified.

27027 Rovimix
Vitamin supplements for animal feeds. *Roche Products Ltd.* Name unverified.

27028 Rovisol
Water miscible vitamins for animals. *Roche Products Ltd.* Name unverified.

27029 Rovral
Fungicide. *Rhône-Poulenc Rorer Ltd.*

27030 Rovral WP
Horticultural fungicide. *Embetec Crop Protection Ltd.*

27031 Roxite
A synthetic resin product used for electrical insulation. No manufacturer.

27032 Roxon
A proprietary synthetic resin of the phenol-formaldehyde type. No manufacturer.

27033 Roxotit
A copper-silicon acid-resisting alloy.

27034 Royal Slo-Gro
123-33-1 5745 204-619-9
Maleic hydrazide
A plant growth regulator for grass and to reduce bud growth in trees, hedges and vegetables. *Uniroyal Chemical Ltd.*

27035 Royalac 133
A modified dithiocarbamate. produces excellent vulcanizates in EPDM polymers when used with normal sulfur levels; develops high degree of tensile strength at shore cure times while retaining equivalent compression set properties; the processing safety is *Uniroyal.* Name unverified.

27036 Royalac 136
A proprietary ultra-accelerator developed for use in EPDM polymers. *Uniroyal.* Name unverified.

27037 Royalac 140
Zinc (hexadecyl-octadecyl) isopropyl dithiocarbamate
Improves physical properties of EPDM diene blends; EPDM-NBR blends vulcanized with sulfur curing systems employing Royalac 139 and 140 compare favorably with CR compound in heat ageing, have essentially the same oil resistance, *Uniroyal.* Name unverified.

27038 Royalcast® 3105
Castable thermoset plastic; provides rigidity of a structural plastic with advantage of liquid castability such as low tooling costs, production of large parts or parts with thick or varied wall cross-sections. *Uniroyal.*

27039 Royalene 301-T
EPDM terpolymer; used as an antiozonant in tire sidewalls and coverstrip and in blends with butyl rubber to improve heat and ozone resistance in inner tubes; vulcanizable with sulfur, peroxide, or radiation. *Uniroyal.*

27040 Royalene 306
EPDM terpolymer, nonstaining stabilizer; very good low temperature impact, excellent weathering properties; blends readily with polyolefins; also for wire and cable applications. *Uniroyal.*

27041 Royalene 521
EPDM terpolymer, nonstaining stabilizer; used for high quality molded mechanical goods, press-cured sponge, wire and cable, coated fabric; suitable for mill mixing, calendering; exceptional flexibility. *Uniroyal.*

27042 Royalene 622
EPDM terpolymer, 40 phr nonstaining naphthenic oil, nonstaining stabilizer; accepts high filler loadings; high gm. strength, good shape retention in extrusions; used in extruded hose and molded mech. goods. *Uniroyal.*

27043 Royaltherm® 1411
Silicone-modified EPDM elastomer; general-purpose elastomer used for automotive applications such as belts, diaphragms, protective boots, CV

window gaskets, and for solar tubing, architectural gaskets, expansion joints; sulfur- or peroxide-curable. *Uniroyal.*

27044 Roydalox
A proprietary trade name for alumina porcelain with good resistance to corrosion and thermal shock; used in ball mills and grinding balls. *Doulton.* Unverified.

27045 Roydazide
12033-89-5 8641 234-796-8
N_4Si_3
silicon nitride
trisilicon tetranitride. A proprietary trade name for silicon nitride; used for the manufacture of turbine blades and generally where a temperature of up to 1650 C is present. *Doulton.* Unverified.

27046 RPDE
Dibasic ester solvent. *Rhône-Poulenc UK.*

27047 RPM
Motor oil. *Monsanto (Solaris).* Name unverified.

27048 RR 5
Semipermanent mold release agent for most thermosetting rubber and plastic materials; available in cold and hot formulations. *Releasomers.*

27049 RR 53 alloy
A aluminum alloy containing 91.85% aluminum, 2.25% copper, 1.3% nickel, 1.5% magnesium, 1.5% iron, 1.5% silicon, 0.1% titanium; used for die casting pistons.

27050 RRV
Resorcylindene
A proprietary antioxidant for rubber. No manufacturer.

27051 RS Nitrocellulose
9004-70-0 8195
Nitrocellulose
Used as a clear, tough, fast drying film former. *Hercules.* Discontinued.

27052 R-salt
The sodium salt of R-acid.

27053 RT/Duroid® M
9002-84-0 7743 204-126-9
PTFE, glass microfiberreinforced; offers high heat resistance, dimensional stability creep resistance, low coeff. of friction, chemical, thermal and wear resistance; for seals, bushings, wear strips, gaskets, bearings, and piston rings. *Rogers.*

27054 RTF 762
63394-02-5
Two-component silicone rubber foam; cures at room temperature to medium density foam with the addition of a curing agent; applications including potting compound, cast-in-place thermal or electrical insulation, *GE Silicones.* Name unverified.

27055 RTP 100 GB 10
9003-07-0 7741
PP, 10% glass bead-reinforced, heat-stabilized; thermoplastic resin to exhibit optimum flow and warpage control; used in larger flat parts and thin-walled parts; bonds at knit lines are improved. *RTP.*

27056 RTP 200FR
32131-17-2
Nylon 6/6, flame-retardant nonreinforced; thermoplastic material exhibiting maximum moldability, thermal stability, and flexibility for unfilled flame-retardant nylon 6/6; non-corrosive. *RTP.*

27057 RTP 201A
25038-54-4 6832
Nylon 6/10, glass-reinforced; thermoplastic resin, improved strength, moduli, and deflective tempe over the base resin; improved moldability with less shrinkage, plus superior surface finish over nylon 6/6 materials; in automotive parts, *RTP.*

27058 RTP 201B
9008-66-6
Nylon 6/10, glass-reinforced; thermoplastic resin, improvement in strength, moduli, and deflective temperature over base resin; very low water aosorption compared to other nylons; dimensional stability and high physical strengths. *RTP.* Name unverified.

27059 RTP 201C
25035-04-5
Nylon 11, 10% glass fiberreinforced; thermoplastic resin, lower water absorption, lower specific gravity, better chemical resistance, and better low temperature resistance than most glass-filled nylons; physical strengths are reduced. *RTP.*

27060 RTP 201D
Nylon 6/12, 10% glass fiber-reinforced; thermoplastic; better dimensional stability, toughness, and lower water absorption than most glass-filled nylons; good moldability and ease of flow; in pumps, electronic applications, etc. *RTP.*

27061 RTP 301
PC, 10% glass fiber-reinforced; thermoplastic material; dimensional stability and improved rigidity, heat resistance, and strengths over the base resin; self-extinguishing characteristics maintained; in business machine parts, power tool components, gears *RTP.*

27062 RTP 401
9003-53-6 9028
PS, 10% glass fiber-reinforced; thermoplastic, improvements in strength, moduli, and dimensional stability over base resin; very rigid with low shrink; used in housings, automotive applications, etc. *RTP.*

27063 RTP 501
9003-54-7
SAN resin, 10% glass fiber-reinforced; thermoplastic; improved strengths over base resin and RTP 400 Series; balance of properties with strengths, low shrinkage, dimensional stability, and cost effectiveness; in tape hubs, housings, pumps, fan blades, etc *RTP.* Name unverified.

27064 RTP 601
9003-56-9
ABS resin, 10% glass fiberreinforced; thermoplastic; improved properties over base resin; in automotive applications. *RTP.*

27065 RTP 701
9002-88-4 7728
HDPE, 10% glass fiber-rein-forced; thermoplastic; improved strengths and dimensional stability over base resin; flows very well and has ·5 1/10 shrinkage of base resin; used when molding large semi-irregularly shaped parts of up to 25 lb in wt.; *RTP.*

27066 RTP 801
Acetal copolymer resin, 10% glass fiber-rein-forced; thermoplastic material with improved dimensional control, strength, and stiffness over base resin; shrinkage reduced and electrical properties maintained; creep and wear resistance; *RTP.* Name unverified.

27067 RTP 901
25135-51-3
Polysulfone, 10% glass fiber-reinforced; strong, rigid, thermoplastic material used in applications requiring good heat resistance (up to 300°F); self-extinguishing; good electrical properties; *RTP.*

27068 RTP 1001
26062-94-2
PBT, 10% glass fiber-rein-forced; thermoplastic polyester having balance of properties (good strengths, electrical and thermal properties, low water absorption); used in connectors and automotive applications requiring 300°F heat resistance. *RTP.*

27069 RTP 1105FR
25038-59-9 7730
PET, 30% glass fiber-reinforced, flame-retardant; thermoplastic material; strengths, electrical properties, and smooth, molded surface. *RTP.*

27070 RTP 1201-80D
Thermoplastic PU elastomer, 10% glass fiber-reinforced; 80-D durometer thermoplastic material; abrasion and chemical resistance, controlled flexibility, and inherent toughness *RTP.* Name unverified.

27071 RTP 1301
9016-75-5
PPS, 10% glass fiber-rein-forced; thermoplastic material; chemical and heat resistance, excellent strengths, flame retardant; used in low load applications up to 500°F, (hostile environment pump parts and connectors). *RTP.*

27072 RTP 1378
9016-75-5
Lubricated PPS
Thermoplastic material combining strength, heat and wear resistance, chemical inertness, and flame retardance; can operate at temps. over 400°F. *RTP.*

27073 RTP 1401
PES, 10% glass fiber-reinforced; thermoplastic material; excellent physical properties including impact strength and tensile elongation; flame retardant; operates to temperatures up to 400°F; used in connectors and aerospace applications. *RTP.*

27074 RTP 1501
Thermoplastic polyester elastomer, 10% glass fiber-reinforced; thermoplastic elastomer with strength, controlled resiliency, impact resistance, high and low temperature capabilities, and good chemical and electrical properties; *RTP.*

27075 RTP 2301A
9009-54-5
Rigid thermoplastic polyurethane, 10% glass fiber-reinforced; offers dimensional stability and excellent molding characteristics *RTP.*

27076 RTP 2381A
9009-54-5
Rigid thermoplastic polyurethane, 10% PAN carbon fiber-reinforced; offers

dimensional stability, moldability, good impact strength, and conductive properties *RTP*.

27077 RTP 3403-3

Liquid crystal polymer, 20% glass fiber-rein-forced; offers outstanding heat and chemical resistance, inherent flame retardance; excellent flowability. *RTP*.

27078 RTP 3405-3 TFE 15

Liquid crystal polymer, glass fiber-reinforced, 15% PTFE; outstanding heat resistance, chemical resistance, intumescent flame retardance for wear-resistance applications. *RTP*.

27079 RTP 4001

Polyphthalamide, 10% glass fiber-reinforced; high heat distortion, excellent strength and stiffness at room and elevated temps. *RTP*.

27080 RTP 4081

Polyphthalamide, 10% PAN carbon fiber-reinforced; offers high strength, stiffness, heat distortion temperature and electrical conductivity. *RTP*.

27081 RTP ESD-300 EM FR

PC, glass-reinforced, flame-retardant, static dissipative; flame retardant material with static dissipative characteristics; excellent toughness and flow. *RTP*.

27082 RTV 11

RTV silicone

General purpose encapsulant; with the addition of a curing agent, cures to durable silicone compound at room temperature with service temperature of -60 to 204°; 11 is easily pourable, self-leveling grade for general purpose potting; *GE Silicones*.

27083 RTV 31

RTV silicone

High-temperature encapsulant offering moderate strength for applications requiring -54 to 260° continuous operating performance with intermittent performance to 316°; 31 is a pourable grade for electrical *GE Silicones*.

27084 RTV 133

One-component RTV silicone adhesive sealant; cures to a strong, durable, resilient silicone rubber at room temperature and on exposure to atmospheric moisture; offers high- and low-temperature performance; *GE Silicones*.

27085 RTV 511

63394-02-5

Two-part RTV silicone; low temperature encapsulant offering moderate strength for extended low temperature performance (to -115°); 511 is a pourable, self-leveling product for potting, encapsulating, *GE Silicones*.

27086 RTV 615

63394-02-5

Two-part RTV silicone rubber; encapsulant; addition-cure kits offering reliable deepsection cure even in closed assemblies; room temperature or heat curing; 615 is a general purpose grade providing *GE Silicones*.

27087 RTV 6156

Two-part methyl-phenyl silicone gel; dielectric gel which cures to form soft elastomer to preserve dielectric integrity and provide protection from vibration and shock for delicate electronic assemblies operating in harsh environments; solventless; *GE Silicones*. Name unverified.

27088 Rubalt

A proprietary compound of rubber, bitumen, and benzene; it is a waterproof, rust-proof, and acid-resisting paint, and is stated to be highly resistant to mineral acids, alkalis, chlorine, ammonia, and salt solution. No manufacturer.

27089 rubber

9006-04-6 232-689-0

An elastic material contained in the latex of certain plants. The most important plant is *Hevea brasiliensis* of South America, which yields the para rubber of commerce; used in tires, conveyor belts, foam rubber, electrical insulation.

27090 rubber cements

These are made by dissolving rubber in suitable solvents, such as coal-tar naphtha or carbon disulfide. Sometimes rosin or turpentine is added. Rubber cement is often a mixture of rubber and sulfur dissolved in oil.

27091 rubber formolite

A product obtained by the action of formaldehyde on a petroleum ether solution of a pale crepe rubber to which has been added concentrated sulfuric acid.

27092 rubberite

An artificial rubber made from asphalt, oxidized oil, petroleum jelly, and sulfur.

27093 Rubberlene

A white refined petroleum product; used as a solvent, and can be substituted for carbon disulfide for dissolving rubber. It boils at 145-300°F.

27094 Rubbermakers Sulfur

7704-34-9 9142 231-722-6

Sulfur

For rubber compounding. *Akrochem*.

27095 rubber-sulfur

Amorphous plastic sulfur, obtained from the Kobui sulfur mine, Japan.

27096 Rubbone

A patented composition stated to be rubber resin prepared by oxidizing rubber catalytically; used in paints, varnishes, etc., in electrical insulation, and in the impregnation of coils. No manufacturer.

27097 Rubel metal

An alloy of 55% copper, 40% zinc, and 5% aluminum-iron-manganese-nickel alloy. Another alloy contains 51% copper, 40% zinc, and 5% aluminum-iron-manganese-nickel alloy, and 4% ferro-manganese.

27098 rubellan

A mineral that is a variety of mica.

27099 rubeosine

A nitrochlorofluorescein, obtained by the action of nitric acid upon aureosin.

27100 ruberite

A name for red copper ore, Cu_2O.

27101 Rub-erok

A proprietary trade name for hard rubber for electrical insulation. No manufacturer.

27102 Rub-er-red

A red iron oxide of fine particle size, which is acid and alkali free, and contains no soluble salts; a proprietary pigment for rubber. No manufacturer.

27103 rubianic acid

152-84-1 8437 205-808-9

$C_{25}H_{26}O_{13}$

1-hydroxy-2-[(6-O-β-D-xylopyranosyl-β-D-glucopyranosyl)oxy]-9,10-anthracenedione

ruberythric acid; β-2-akizarinprimeveroside; ruberythrinic acid; rubian. mp = 259-261°; soluble in H_2O, less soluble in organic solvents.

27104 rubidium carbonate

584-09-8 209-530-9

CO_3Rb_2

Rubidium carbonate

Special glass formulations. mp = 837° *Atomergic Chemetals; Cabot; Cerac; Noah Chem*.

27105 rubidium chloride

7791-11-9 8441 232-240-9

RbCl

Analysis (testing for perchloric acid), source of rubidium metal. Also used as a gasoline additive and as an antidepressant. mp = 718°; bp = 1390°; d = 2.8000. *Atomergic Chemetals; Cabot; Cerac; Noah Chem*.

27106 rubidium iodide

7790-29-6 8443 231-198-1

Irb

Rubidium iodide

Used as a source of iodine. mp = 642°, bp = 1300°; d = 3.55; soluble in H_2O (1.5 g/ml), EtOH. *Atomergic Chemetals; Cabot; Cerac; Noah Chem*.

27107 rubidium sulfate

7488-54-2 231-301-7

Rb_2SO_4

Rubidium sulfate

Atomergic Chemetals; Cabot; Cerac; Noah Chem.

27108 Rubinate® LF-168

101-68-8 202-966-0

$C_{15}H_{10}N_2O_2$

4,4'-diphenylmethane diisocyanate

MBI; 4,4'-Methylenediphenyl diisocyanate; PMDI; bis(1,4-isocyanatophenyl)methane; methylenebis(p-phenylene isocyanate); caradate 30; desmodur 44; 4,4'diisocyanatodiphenylmethane; hylene m50; isonate 125M; isonate 125MF; 4,4'-methylenediphenylene isocyanate; nocconate 300; Methylenedi-p-phenyl diisocyanate; Diphenylmethyl Diisocyanate; Methylene Bisphenyl Isocyanate. Modified 4,4' diphenylmethane diisocyanate; for manufacturing of high performance polyurethane elastomeric materials including reaction injection molding processed and cast elastomers, sealants, coatings, and adhesives. mp = 37-39; bp_5 = 194°; d = 1.1900. *ICI Polyurethanes*. Name unverified.

27109 Rubini's essence

A saturated solution of camphor in alcohol.

27110 rubio ore

A brown ore of iron, from Bilbao, in Spain.

27111 Rubmag

A proprietary trade name for a light magnesium carbonate used for rubber reinforcing. No manufacturer.

27112 Rubout

Deodorants for rubber processing and products. *CPL Group Ltd*.

27113 Rubox

1314-13-2 10279 215-222-5

Zinc oxide, rubber grade; used in the rubber industry. *Manchem Ltd.* Name unverified.

27114 Rubramin PC
68-19-9 10152 200-680-0
Cyanocobalamin
Vitamin. *Bristol-Myers Squibb Co Inc.*

27115 Rubratope-57
13115-03-2 10153
Cyanocobalamin ^{57}Co
Rubratope-57. Diagnostic aid; radioactive agent, a γ emitter with half-life of 271 days. *Bristol-Myers Squibb Co Inc.*

27116 rubrax
A mineral rubber containing 98-99% of material soluble in chloroform, with an ash less than 0.5%.

27117 rubrescin
An indicator prepared from resorcinol and chloral hydrate.

27118 rubrica
A red pigment. It is a natural burnt ocher, containing varying quantities of iron.

27119 Rub-tex
A proprietary trade name for a hard rubber. No manufacturer.

27120 ruby ore
1317-39-1 2734 215-270-7
Cu_2O
Cuprous oxide

27121 ruby powder
A sporting 42-grain powder containing 50% nitro-cellulose, metallic nitrate, 8% nitro-hydrocarbon, and 6% starch.

27122 ruby tin
A red variety of the mineral Cassiterite, or tinstone.

27123 Rucoflex® F-2014
Slightly branched, hydroxyl-terminated polyester polyol; specialty polyester designed for reaction with diisocyanates to produce crosslinked elastomeric PUs; used in solution laminating adhesives, solution coatings, prepolymers, and one-shot castables; lo *Ruco Polymer.*

27124 Rucoflex® S-101 Series
Saturated, aliphatic, linear, hydroxyl-terminated polyester diol based on short chain glycols and polyester; polyester used for solution coatings and adhesives, thermoplastic elastomers, and castable prepolymers; PUs have high strength, solvent, *Ruco Polymer.*

27125 Rucote 102
Polyester powdered coating resin; yields coatings with superior flow and flexibility; cured coatings feature outstanding weatherability, excellent flexibility and impact resistance. *Ruco Polymer.*

27126 Rucothane 2010L
PU resin latex; solvent-free, water dispersion of high molecular weight, aliphatic-based, thermoplastic elastomer; polymer is extremely supple, exhibits good adhesion to a variety of substrates, and can be readily frothed by mechanical means; *Ruco Polymer.*

27127 Rudol
White mineral oil. *Witco Corporation.*

27128 rufiopin
$C_{14}H_8O_6$
Tetrahydroxyanthraquinone.

27129 Ruge's solution
A solution containing 1 ml glacial acetic acid, 2 ml formalin, and 100 ml distilled water.

27130 Rulan®
Polymer. *DuPont UK.*

27131 Runa
13463-67-7 9612 236-675-5
Rutile type titanium dioxide. *Laporte Industries Ltd.* Discontinued.

27132 Runaway
Color remover. *Dylon International Ltd.*

27133 Ruolz alloys
Nickel silvers.

27134 Ruselite
A proprietary trade name for an alloy of 94% aluminum, 4% copper, 2% chromium, and 2% molybdenum; it is stated to be resistant to corrosion. No manufacturer.

27135 rusma
A mixture of arsenic sulfide (orpiment), As_2S_3, with lime. It is made into a paste with water, and used for unhairing skins prior to tanning them.

27136 Ruspini's solution
A styptic containing tannic acid, rose water, alcohol, and water.

27137 Russian tallow
A mixture of beef and mutton fat.

27138 Russian turpentine
The oleo-resin from *Pinus sylvestris* and *P. Ledebouril.* .

27139 Russian white lead
A white lead having the composition, $5PbCO_3 \cdot 2Pb(OH)_2 \cdot PbO$.

27140 Rustban
Protective coatings. *Exxon UK.* Name unverified.

27141 Rustlan Oil
Yellow to amber colored liquid used as rust preventive and stamping oil; used in wire mills and steel mills to protect rust while shipping and inside storage; for metal stamping, *Rustlan Chemical Co.* Unverified.

27142 rustlessIron
A rustless steel containing about 0.1% carbon. It is made in the electric furnace by means of practically carbon-free ferro-chrome.

27143 Rust-Tap
A light yellow nonflammable liquid used as tapping oil; residual coating acts as rust preventive; used in machine shops for tapping operation of all metals except aluminum; used as drilling lubricant. *Rustlan Chemical Co.* Unverified.

27144 Rutaform
Phenolic, melamine and granular polyester molding powders. *Bakelite Polymers (UK) Ltd.* Name unverified.

27145 Rutamod
Black extenders for epoxy and polyurethane resins. *Collinda Ltd.*

27146 Rutasolv DI
Diisopropyl naphthalene high boiling solvent. *Collinda Ltd.*

27147 Rutenol
96-69-5 202-525-2
$C_{22}H_{30}O_2S$
4,4-Thiobis-(6-*t*-butyl-3-methyl-phenol)
bis(4-hydroxy-5-*tert*-butyl-2-methylphenyl) sulfide; Santonox; disperse mb-61; santowhite; 4,4'-thiobis(2-*tert*-butyl-5-methylphenol); santonox bm; santonox r; sumilizer wx; sumilizer wx-r; thioalkofen bm4; thioalkofen mbch; thioalkofen bmch; thioalkophene bm-4. mp = 161-164°; insoluble in H_2O. *Collinda Ltd.*

27148 ruthenium red
1307-52-4 8453
$Cl_6H_{42}N_{14}O_2Ru_3$
Ammoniated ruthenium oxychloride
C.I. 77800;. Used as a microscopic stain, and as a reagent for pectin, plant mucin, and gum. Soluble in H_2O, NH_3.

27149 Ruthmol
A proprietary preparation of potassium chloride, lactose and gluten-free starch; a sodium-free table salt. *Larkhall Laboratories plc.* Name unverified.

27150 rutin trihydrate
153-18-4 8456 205-814-1
$C_{27}H_{30}O_{16} \cdot 3H_2O$
3-[[6-O-(6-deoxy-α-L-mannopyranosyl)-β-D-glucopyranosyl]oxy]-2-(3,4-dihydroxyphnyl)-5,7-dihydroxy-4H-1-benzopyran-4-one
quercetin 3-rutinoside; rutoside; 3,3',4',5,7-pentahydroxyflavone-3-rutinoside; melin; phytomelin; eldrin; ilixathin; sophorin; globularicitrin; paliuroside; osyritrin; myrticolorin; violaquercitrin; Birutan. Protects capillary structure. mp = 214-215° (dec); $[\alpha]_D^{23}$ = 14° (EtOH); insoluble in H_2O, more soluble in polar organic solvents; LD_{50} (mus iv) = 950 mg/kg.

27151 Rutiox
13463-67-7 9612 236-675-5
A rutile titanium dioxide pigment. *Laporte Industries Ltd.* Discontinued.

27152 Ruvea®
Fibers. *DuPont UK.*

27153 RVPaba Lipstick
150-13-0 443 205-753-0
p-aminobenzoic acid
Ultraviolet screen. *ICN Pharmaceuticals Inc.*

27154 RX® 1-501N
131-17-9 205-016-3
$C_{14}H_{14}O_4$
1,2-benzenedicarboxylic acid, di-2-propenyl ester
allyl phthalate; diallyl ester of phthalic acid; Dapon R; Dapon 35; DAP. Diallyl phthalate, mineral reinforced; thermoset molding material mp = -70°; bp = 290°; SG - 1.121; n_D^{20} = 1.5190; insoluble in H_2O, soluble in organic solvents. *Rogers.* Name unverified.

27155 RX® 1906
25928-94-3
Epoxy, mineral/glass-reinforced; thermoset molding material *Rogers.* Name unverified.

27156 RX® 3-1-530
131-17-9 205-016-3
Diallyl phthalate, long glass-reinforced; thermoset molding material. *Rogers.* Name unverified.

27157 RX-56
Porofocon A
Contact lens material. *Rynco Scientific Corp.* Unverified.

27158 RXXL
High boiling tar acids. *Coalite Fuels & Chemicals Ltd.* Name unverified.

27159 Ryax C
Center-line suspended process pump; for process pumping applications. *Sihi Pumps (UK) Ltd.*

27160 Ryax F, O
Back pull out process pump; for process pumping application. *Sihi Pumps (UK) Ltd.*

27161 Ryflex
Belt driven single-stage centrifugal pump capable of being stacked one unit on top of another to a maximum of two; for heating and cooling water applications where floor space is small and a stacked pump unit is required. *Sihi Pumps (UK) Ltd.*

27162 Rylex
UV light absorber. *DuPont UK.*

27163 Rynite®
Polyester resin. *DuPont UK.*

27164 Ryoto Ester KA
Sugar ester, calcium carbonate, potassium carbonate carbohydrate; quality improver for fish paste products. *Mitsubishi Kasei.* Name unverified.

27165 Ryoto Ester SP
Sugar ester, monoglyceride, sorbitol, propylene glycol; batter aerating agent for sponge and pound cake. *Mitsubishi Kasei.* Name unverified.

27166 Ryoto Sugar Ester LWA-1570
25339-99-5 246-873-3
β-D-fructopyranosyl-α-D-glucopyranoside monodecanoate
Sucrose laurate; Sucrose monolaurate. Oil-water and water-oil emulsifier, softener, conditioner, and aerating agent in foods. *Mitsubishi Kasei.* Name unverified.

27167 Ryoto Sugar Ester OWA-1570
Sucrose oleate
Emulsifier, conditioner, softener, detergent for foods, drugs, cosmetics; tablet lubricant. *Mitsubishi Kasei.* Name unverified.

27168 Ryoto Sugar Ester P-1570, P-1670
26446-38-8 247-706-7
Sucrose palmitate
Emulsifier, conditioner, softener, detergent for foods, drugs, cosmetics; tablet lubricant. *Mitsubishi Kasei.* Name unverified.

27169 Ryoto Sugar Ester S-1170
25168-73-4 246-705-9
Sucrose stearate
Emulsifier, conditioner, softener, detergent for foods, drugs, cosmetics; tablet lubricant. *Mitsubishi Kasei.* Name unverified.

27170 Ryoto Sugar Ester S-170
Sucrose di, tristearate
Emulsifier, conditioner, softener, detergent for foods, drugs, cosmetics; tablet lubricant. *Mitsubishi Kasei.* Name unverified.

27171 Ryoto Sugar Ester S-570, S-770
27195-16-0 248-317-5
Sucrose distearate
Emulsifier, conditioner, softener, detergent for foods, drugs, cosmetics; tablet lubricant. *Mitsubishi Kasei.* Name unverified.

27172 Rytherm
Condensate extraction pump set with low mounted tank; several sizes of set suit various sizes of industrial central heating systems, pumps extract condensate at or near 100 C, thus reducing heat loss. *Sihi Pumps (UK) Ltd.*

27173 Ryton® A-200
9016-75-5
PPS; High performance engineering thermoplastic with inherent chemical and flame resistance; for injection molding. *Phillips.*

27174 Ryton® R-4
9016-75-5
Glass fiber-reinforced PPS resin; injection molding engineering thermoplastic possessing chemical resistance and mechanical properties even at elevated temperature; dimensional stability, good moldability, and elec. properties unaffected by moisture; for inject *Philips.*

27175 Ryton® V-1
9016-75-5
PPS
High melt flow grade for thin coatings on metal substrates or to improve flow of highly filled PPS; for cookware, industrial release applications, pumps, valves where chemical resistance is required. *Phillips.*

27176 Ryvin
Single-stage centrifugal pump vertical in-line; for in-line circulating duties. *Sihi Pumps (UK) Ltd.*

27177 S Monel
A Monel metal with 3.75% silicon used in valves, etc., which are subject to corrosion.

27178 S.21
Nonflammable solvent. *Stowlin Ltd.*

27179 S.O.S.®
Range of household cleaning products. *Bayer AG.*

27180 S.S.T® Sump Saver Tablets
126-11-4 9883 204-769-5
$C_4H_9NO_5$
Tris (hydroxymethyl) nitromethane
2-nitro-2-(hydroxymethyl)-1,3-propanediol; tris(hydroxymethyl)nitromethane; Trimethylolnitromethane; Tris-Nitro; Cimcool wafers; Nitroisobutylglycerol; Nitrotris(hydroxymethyl)methane; 2-hydroxymethyl-2-nitropropanediol. Antibacterial agent, preservative for metalworking fluids. mp = 214°; soluble in H_2O (220 g/100 ml), alcohols, less soluble in organic solvents. *Angus.*

27181 S160 Beads and Powder
1333-86-4 1856 231-153-3
Carbon black; for medium quality PVC and calendered systems. *Degussa.*

27182 S-201
1344-28-1 369 215-691-6
Alumina
Claus catalyst for use where H_2S is the primary compound to be removed from feed gas; offers lower pressure drop. *La Roche Chem.*

27183 S-60 RVM
1337-76-4
Attapulgite clay; absorbent, adsorbent. *Floridin.* Name unverified.

27184 S945, S975, S987, S992, S994
1314-23-4 10313 215-227-2
O_2Zr
Monoclinic zirconium dioxide with a zirconia content (including hafnia) of 94.5%, 97.5% or 98.7%, 99.2% or 99.4%; for manufacture of ceramic pigments, welding fluxes and insulating material. *Ferro.*

27185 S975
Monoclinic zirconium dioxide with a ZrO_2 content (including hafnia) of 97.5%. Used in the manufacture of pigments for ceramics and enamels and in welding fluxes. *Ferro.*

27186 S987
Monoclinic zirconium dioxide with a ZrO_2 content (including hafnia) of 98.7%. Used in the manufacture of pigments for ceramics. *Ferro.*

27187 S992
Monoclinic zirconium dioxide with a ZrO_2 content (including hafnia) of 99.2%. Used in the manufacture of lead-zirconate titanate piezo electric ceramics, zirconates, zirconia technical ceramics, oxygen sensors, milling media and ceramic pigments. *Ferro.*

27188 S994
Monoclinic zirconium dioxide with a ZrO_2 content (including hafnia) of 99.4%. Used in the manufacture of lead-zirconate titanate piezo electric ceramics, zirconates, zirconia technical ceramics and oxygen sensors. *Ferro.*

27189 Sabalith®
Thermal salt mixture of inorganic nitrates ($LiNO_3 + KNO_3$); nitrite-free thermal salt for the salt bath curing in open or closed circulations, e.g., in LCM installations for the vulcanization of rubber. *Chemetall GmbH.*

27190 Sabeco Metal
A proprietary trade name for copper with 21% lead and 9% tin. No manufacturer.

27191 Sabre
34123-59-6 5237 251-835-4
Suspension concentrate containing 553 g isoproturon per liter; used for annual weed control in cereals. *Schering Agrochemicals Ltd.* Discontinued.

27192 Sabre 1628
PC/polyester resin; engineering resin offering extra margin of heat resistance for improved processing and paintability as well as excellent chemical resist., lowtemperature toughness, easy processing; applications including automotive wheelcovers, lawn and garden equipment and telecommunications devices. *Dow Plastics.*

27193 Sabulite
An explosive containing ammonium nitrate, charcoal, and calcium silicide.

27194 Sabutol
Mixture of normal butyl alcohol, isobutyl alcohol, and secondary pentyl alcohol; solvent in paints, printing inks, dyes, foundries, manufacture of butyl acetate and xanthate. *Sasolchem.*

27195 saccharin
81-07-2 8463 201-321-0

$C_7H_5NO_3S$
1,1-dioxide-1,2-benzisothiazol-3(2H)-one
1,2-benzothiazol-3(2H)-one 1,1-dioxide; *o*-benzoic sulfimide; benzoic sulfimide; benzo-2-sulfimide; benzosulfinide; *o*-benzoyl sulfimide; 3-hydroxybenzisothiazole-S,S-dioxide; 1,2-benzisothiazolin-3-one, 1,1-dioxide; glycophenol; anhydro-*o*-sulfaminebenzoic acid; *o*-benzosulfimide; natreen; sacarina; sucre edulcor; sucrette; 2-sulfobenzoicimide; syncal; zaharina; 1,2-dihydro-2-ketobenzisosulfonazole; glycosin; 2,3-dihydro-3-oxobenziso sulfonazole; kandiset; garantose; glucid; gluside; hermesetas; saccharin acid; 550 saccharine; saccharin insoluble; saccharinol; saccharinose; neosaccharin; saccharol; *o*-sulfobenzimide; *o*-sulfobenzoic acid imide; benzosulfimide; insoluble saccharin; saccharimide; 1,1-Diox-1,2-benzisothiazol-3-one,. Saccharin, insoluble; saccharinol; saccharinose; saccharol, saccharose; sycorin; sykose. Organic compound; noncaloric, non-nutritive sweetener (500 times sweeter than cane sugar), pharmaceutic aid; in formulations for electro mp = 229°; d = 0.828; slightly soluble in H_2O (100 mg/ml), soluble in organic solvents. Aisan chem Co Ltd; Dinoval Chem. Ltd; R.W. Greeff; Maruzen Fine Chem.; PMC Specialties; Rit-Chem; Spice King.

27196 saccharin sodium
128-44-9 8463 204-886-1
$C_7H_8NNaO_5S$
1,2-Benzisothiazol-3(2H)-one 1,1-dioxide, sodium salt dihydrate
Saccharin soluble; Sodium saccharin; artificial sweetening substance Gendorf 450; crystallose; dagutan; madhurin; Saxin; sodium o-benzosulfimide; sodium benzosulfimide; sodium saccharide; sodium saccharinate; soluble saccharin; soluble gluside; succaril; Sucra; o-sulfonbenzoic acid imide sodium salt; sulfobenzoic imide, sodium salt; Sweeta; sykose; Willosetten. Non-caloric sweetener, pharmaceutic aid (flavorant). Soluble in H_2O (0.83 g/ml), less soluble in organic solvents; LD_{50} (rat orl)= 17.0 g/kg.

27197 Sachsse's solution
A solution containing 18 g mercuric iodide, 25 g potassium iodide, and 80 g potassium hydroxide in a liter; used for the determination of reducing sugars.

27198 Sachtocup
Cupreous pyrites
25-26% Cu, 32-34% S, 28-30% Fe; filler for resin-bonded grinding wheels, brake linings, etc. *Metallgesellschaft AG.*

27199 Sachtoklar®
Primary flocculant agent. *Sachtleben Chemie GmbH.*

27200 Sachtolen®
Masterbatches and compounds. *Sachtleben Chemie GmbH.*

27201 Sachtolith®
1314-98-3 10294 215-251-3
ZnS
Zinc sulfide
White inorganic pigment for paints and coatings. 1023 *Sachtleben Chemie GmbH.*

27202 Sachtoperse® HU
7727-43-7 1023 231-784-4
$BaSO_4$
Barium sulfate
Ultra fine solid state additive. *Sachtleben Chemie GmbH.*

27203 Sachtopyr®
Polycrystalline iron sulfide (pyrite); 47-48% S, 40-41% Fe; filler for resin-bonded grinding wheels, brake linings, etc. *Metallgesellschaft AG.*

27204 Sachtosil®
Chemical products for industrial purposes, especially pigments for paints and colors, fibers, cosmetics and catalysts. *Sachtleben Chemie GmbH.*

27205 SACI
Rust preventative. *Witco Corporation.*

27206 Sacon®
7727-43-7 1023 231-784-4
$BaSO_4$
Transparent conductive pigment. *Sachtleben Chemie GmbH.*

27207 sacred bark
The bark of *Rhamnus purshianus* .

27208 Saduren®
Synthetic resin solutions; precondensates based on melamine-formaldehyde; binders for bonding fiber webs, especially glass fibers. *BASF AG.*

27209 Safari®
For the automotive industry. *DuPont UK.*

27210 Safebond 3
Modified bitumen emulsion; flooring adhesive. *Marley Floors Ltd.*

27211 Safe-Break
Glass bottles with plastic safety coating. *BDH Chemicals Ltd.*

27212 Safe-FR
Nonhalogenated flame retardant polyolefins. High loading of magnesium hydroxide provides flame retardancy. Low smoke, low toxicity, low emission of acid gas upon exposure to fire. *Uvtec.*

27213 Safepak
Plastic safety containers. *BDH Chemicals Ltd.*

27214 safety dynamite
An explosive consisting of 24% nitroglycerin, 1% guncotton, and 75% ammonium nitrate.

27215 safety nitro-powder
An explosive similar in composition to Giant powder.

27216 Safety-Cool
Heavy-duty cutting and grinding fluids specifically designed to cool, lubricate and protect metal surfaces in a variety of machining applications; metalworking fluid. *Chem-Trend.* Discontinued.

27217 Safety-Lube®
Water-based lubricant formulations for highpressure aluminum and zinc die casting. *Chem-Trend.*

27218 Safex
Stated to be dinitro-phenyl-dimethyl-dithio-carbamate; used with zinc oxide; a proprietary super-accelerator for rubber vulcanization. No manufacturer.

27219 Safezone® Cleaning Solvent & Flux Remover
Solvent cleaner. *Miller-Stephenson.* Name unverified.

27220 Saffil
High-temperature inorganic fibers. *ICI Chem & Polymers Ltd.*

27221 Saffil®
1344-28-1 369 215-691-6
High purity alumina fibers; refractory bulk fibers for high performance at elevated temps. *Thermal Ceramics.*

27222 safflower oil
8001-23-8 8465 232-276-5
Hi-Oleic safflower oil
Carthamus tinctorious oil; safflower oils; lipovol saff(lipo); tri-ol sa(tri-k); obesitol; safflor. Oily liquid obtained from seeds of *Carthamus tinctorius* consisting principally of triglycerides of linoleic acid; used in alkyd resins, paints, varnishes, medicine; salad oil blend sg = 0.9211; insoluble in H_2O. *Arista Industries; Clagene; Croda; Lipo.*

27223 saffron
A coloring matter obtained from the dried and powdered flowers of the saffron plant, *Crocus sativus*; used for coloring confectionery.

27224 saffron bronze
$Na_2W_3O_9$
gold bronze
Tungsten-sodium bronze. Used as a pigment. The corresponding potassium salt is known as violet bronze or magenta bronze.

27225 saffron oil
8001-23-8 8465 232-276-5
Safflower oil from the seeds of Carthamus tinctorius . It has an acid value of 9.8 and a saponification value of 197.3; used in alkyd resins, paints, varnishes, medicine, dietetic foods, margarine, hydrogenated shorten.

27226 saffron sugar
$C_6H_{12}O_6$
Crocase.

27227 Saflex
Polyvinyl butyral film; interlayer for safety glass. *Monsanto plc.*

27228 Safoam
Endothermic chemical nucleating and blowing agent for thermoplastics which process above 340 F. *Reedy Int'l.*

27229 safrol
94-59-7 8468 202-345-4
$C_{10}H_{10}O_2$
5-(2-propenyl)-1,3-benzodioxole
4-allyl-1,2-methylenedioxybenzene; allyl catechol methylene ether; allyldioxybenzene methylene ether; *m*-allylpyrocatechin methylene ether. The methylene ether of allyl-pyrocatechol, It is found in oil of sassafras, and is obtained from red oil of camphor; used in the place of oil of sassafras, in perfumes, soaps, and medicine. mp = 11°; bp = 232-234°; d = 1.0950; n_D^{20} = 1.5370; insoluble in H_2O, soluble in organic solvents; LD_{50} (rat orl) = 1750 mg/kg.

27230 Saf-T-Side
A petroleum oil insecticide that kills by contact; used as both a dormant spray and as a summer spray; controls whiteflies, aphids, mites, mealybugs, etc.; used on vegetables, ornamentals, citrus and fruit trees. *Lawn & Garden Products Inc.*

27231 Saf-T-Sol
Aromatic solvent for paints and varnishes. *Crowley Chem.*

27232 Sag
Silicone antifoam. *Union Carbide.*

27233 Sahli's reagent
A mixture of equal parts of a 48% solution of potassium iodide and an 8% solution of potassium iodate; used to test for free hydrochloric acid in stomach contents.

27234 Sahli's stain
A solution of borax and methylene blue in water; used to stain nervous tissues and cell nuclei.

27235 Saisan
5707-69-7 3499 227-197-8
Liquid drazoxolan seed dressing. *ICI Chem & Polymers Ltd.* Discontinued.

27236 Sajji
An alkaline deposit found on the land in India; used as a soap substitute.

27237 sakaloid
A synthetic resin. It is a polymerized sugar product obtained from sugar, dextrose, levulose, etc. It can be used for varnishes and lacquers and, when extruded, as an artificial silk.

27238 Sakarat
81-81-2 10174 201-377-6
Warfarin
A rodenticide. *Killgerm Chemicals Ltd.*

27239 Sakarat Special
3691-35-8 2204 223-003-0
An oil formulation containing 2.5% of chlorophacinone; an anticoagulant rodenticide; a bait to control black rats, brown rats, house mice and voles. *Killgerm Chemicals Ltd.*

27240 sakoa oil
An oil obtained from the seeds of *Sclerocarpa caffra* . It is a nondrying oil, and has a saponification value of 193.5.

27241 Sakresote
Creosote. *Sasolchem.*

27242 sal absinthii
584-08-7 7781 209-529-3
K_2CO_3
Salt of wormwood, potassium carbonate.

27243 sal amarum
7487-88-9 5731 231-298-2
$MgSO_4$
Magnesium sulfate
sal Catharticum; sal Anglicum; sal Seidlitense.

27244 sal commune
7647-14-5 8742 231-598-3
NaCl
Sodium chloride.

27245 sal Culinaris
7647-14-5 8742 231-598-3
NaCl
Sodium chloride.

27246 sal diureticum
127-08-2 7764 204-822-2
$C_2H_3KO_2$
Potassium acetate.

27247 sal Martis
7782-63-0 4105 233-072-9
$FeO_4S{\cdot}7H_2O$
iron(II) sulfate heptahydrate
Ferrous sulfate.

27248 sal mineral
1309-37-1 4072 215-168-2
Fe_2O_3
iron(III) oxide
Ferric oxide.

27249 sal Mirabil
7757-82-6 8829 231-820-9
Na_2SO_4
Sodium sulfate.

27250 sal niter
7757-79-1 7815 231-818-8
KNO_3
Potassium nitrate.

27251 sal Prunella
7757-79-1 7815 231-818-8
KNO_3
Potassium nitrate
In balls.

27252 sal Rupellensis
304-59-6 7840 206-156-8
Sodium-potassium tartrate.

27253 sal Saturni
301-04-2 5411 206-104-4
$C_4H_6O_4Pb$
Lead acetate.

27254 sal sedativus
10043-35-3 1364 233-139-2
H_3BO_3
Boric acid.

27255 sal soda
497-19-8 8739 207-838-8
CNa_2O_3
Sodium carbonate.

27256 sal succini
110-15-6 9038 203-740-4
$HOOC{\cdot}CH_2{\cdot}CH_2{\cdot}COOH$
Succinic acid.

27257 sal tartar
133-37-9 9236 205-105-7

27258 sal volatile
506-87-6 534 208-058-0
Commercial ammonium carbonate, $(NH_4)_2CO_3$.

27259 Salargyl
A protein-silver preparation.

27260 Salcare®
Body-friendly polymers for hair conditioning and skin care. *Allied Colloids Ltd.*

27261 salenixon
Crude potassium sulfate, obtained in the manufacture of nitric acid.

27262 saleratus
144-55-8 8726 205-633-8
$NaHCO_3$
Sodium hydrogen carbonate.

27263 salesthin
75-09-2 6140 200-838-9
CH_2Cl_2
Methylene chloride.

27264 Sal-ethyl
118-61-6 3894 204-265-5
Ethyl salicylate.

27265 Salfax 77
Vegetable and animal fats, aliphatic alcohols, fatty acids and cationics; hair conditioner; base for cream rinses. *Chem-Y GmbH.*

27266 Salfuride
A proprietary preparation of nifursol; a veterinary anti-protozoan. No manufacturer.

27267 Salge metal
An alloy of 4% copper, 9.9% tin, 1.1% lead, and 85% zinc.

27268 Salhar gum
A gum-resin obtained from *Boswellia serrata* .

27269 salicylaldehyde
90-02-8 8478 201-961-0
$C_7H_6O_2$
salicylal
salicylic aldehyde; 2-hydroxybenzaldehyde; *o*-hydroxybenzaldehyde. Analytical chemistry, perfumery, synthesis of coumarin, auxiliary fumigant, flavoring. mp = -7°; bp = 196-197°; d_4^{20}= 1.167; n_D^{20}= 1.5735; slightly soluble in H_2O, soluble in organic solvents; MLD (rat sc) = 1 g/kg. *Janssen Chimica; Penta Mfg.; Seimi Chem.*

27270 salicylic acid
69-72-7 8484 200-712-3
HOC_6H_4COOH
2-hydroxybenzoic acid
o-hydroxybenzoic acid; benzoic acid, 2-hydroxy-. Aromatic acid; preservative for foods; manufacture of methyl salicylate, acetylsalicylic acid, etc., dyes; reagent in analytical chemistry; topical keratolytic agent. Used as a synthetic intermediate and as a topical keratolytic. mp = 158-160°; bp_{20} = 211°; d = 1.44; soluble in H_2O (2 mg/ml), more soluble in organic solvents; LD_{50} (mus iv) = 500 mg/kg. Allchem Industries; EM Industries; Hilton Davis; Jansse Chimica; Mitsui Toatsu Chem.; PMC Specialties; Rhone-Poulenc Santé SA.

27271 Salinomycin
53003-10-4 8488 258-290-1
$C_{42}H_{70}O_{11}$
Coxistac. An ionophorous antibiotic used as an anticoccidial agent in poultry. Used in veterinary medicine as an anticoccidial agent. mp = 112.5-113.5°; $[\alpha]^{25}_D$ = -63° (c = 1, EtOH); λ_m = 284 nm (ε = 126, EtOH-H_2O, 2:1); LD_{50} (mus orl) = 50 mg/kg *Pfizer International.*

27272 Saliretins
Saliretin resins. Resins obtained from saligenin by either heating or treating it with formaldehyde. They are similar to phenol-formaldehyde resins.
27273 Saliter
7631-99-4 8792 231-554-3
$NaNO_3$
Sodium nitrate.
27274 Salkowski's solution
A solution of phospho-tungstic acid; used for the detection in urine.
27275 Sallit's speculum metal
An alloy of 64.6% copper, 31.3% tin, and 4.1% nickel.
27276 Sally Nixon
Fused niter cake (acid sodium sulfate).
27277 Salmocid
9011-05-6 7735
poly[methylenedi(hydroxymethyl)urea]
oxymethyleneurea; polynoxyline; polynoxylin; polyoxymethyleneurea; Anaflex; Larex; Ponoxylan. A proprietary preparation of polynoxylin, a polymer of urea and formaldehyde. Used as a topical antibacterial. Amorphous powder; soluble in H_2O (0.28-0.31%). *Geistlich Sohne AG.* Name unverified.
27278 Salodine
A proprietary preparation of an iodized salt. No manufacturer.
27279 Salsorb®
Superabsorbent polymers for body fluids. *Allied Colloids Ltd.*
27280 salt
Sodium chloride.
27281 salt cake
7757-82-6 8829 231-820-9
Na_2SO_4
Crude sodium sulfate
Produced in the Leblanc soda process; used in paper, pulp, detergents, sodium salts, ceramic glazes, plate and window glass.
27282 salt of Alembroth
$2NH_4Cl{\cdot}HgCl_2{\cdot}H_2O$
Salt of wisdom;, sal-Alembroth. A compound of mercuric chloride and ammonium chloride,.
27283 salt of amber
110-15-6 9038 203-740-4
$C_2H_4(COOH)_2$
succinic acid.
27284 salt of Hartshorn
506-87-6 534 208-058-0
$(NH_4)_2CO_3$
Ammonium carbonate.
27285 salt of Lemery
13454-96-1 236-645-1
K_2SO_4
Potassium sulfate.
27286 salt of Norton
$PtCl_4{\cdot}5H_2O$
Platinum tetrachloride.
27287 salt of Saturn
301-04-2 5411 206-104-4
$Pb(C_2H_3O_2)_2$
Normal lead acetate
Sugar of Saturn.
27288 salt of soda
497-19-8 8739 207-838-8
Na_2CO_3
Sodium carbonate.
27289 salt of Sorrel
potassium hydrogen oxalate
Salts of Sonel; salts of Lemon. The two acid salts of potassium oxalate, C_2O_4HK and $C_2O_4KH{\cdot}C_2H_4O_2{\cdot}2H_2O$, are both sold under these names.
27290 salt of steel
7782-63-0 4105 233-072-9
$FeSO_4$
Ferrous sulfate.
27291 salt of tin
10025-69-1 8939 231-588-9
$SnCl_4$
Stannous chloride.
27292 salt of wormwood
584-08-7 7781 209-529-3
Sal Absinthii. Impure potassium carbonate, K_2CO_3, made from plant ash.
27293 salt perlate
7558-79-4 8805 231-448-7
HNa_2PO_4
Sodium phosphate.
27294 Salt, Amido-G
86-65-7 421 201-689-2
$C_{10}H_9NO_6S_2$
7-amino-1,3-naphthalenedisulfonic acid
2-naphthylamine-6,8-disulfonic acid; Amido-G Acid. Used as an intermediates in manufacture of chemicals, particularly dyestuffs. mp > 300°; soluble in H_2O (92 g/l), less soluble in organic solvents. No manufacturer.
27295 Salt, Amido-R
92-28-4 423 202-143-6
$C_{10}H_9NO_6S_2$
3-amino-2,7-naphthalenedisulfonic acid
2-naphthylamine-3,6-disulfonic acid. Used in manufacture of dyes. Soluble in H_2O. No manufacturer.
27296 Saltex
7647-14-5 8742 231-598-3
Sodium chloride solution
Omex Agriculture Ltd.
27297 Saltialgine H8
9005-32-7 241 232-680-1
Alginic acid
Tablet disintegrant for pharmaceutical compressed tablets. *Mendell.* Discontinued.
27298 saltpeter, saltpeter flour
7757-79-1 7815 231-818-8
KNO_3
niter
Potassium nitrate. Used in explosives, matches, pyrotechnics, fertilizers, reagent for tobacco, tempering steel, curing foods.
27299 saltpeter rot
13477-33-4 233-332-1
$Ca(NO_3)_2{\cdot}4H_2O$
Calcium nitrate
Causes the rapid disintegration of mortar.
27300 saltpeter superphosphate
A fertilizer made by mixing niter with calcium superphosphate.
27301 salufer
16893-85-9 8769 240-934-8
F_6Na_2Si
sodium hexafluorosilicate
sodium fluosilicate; sodium silicofluoride. The sodium salt of hydrofluosilicic acid, used as an antiseptic and as a rodenticide and insecticide. Also used to make enamels. d = 2.68; soluble in H_2O (7-25 mg/ml).
27302 salunol
7681-52-9 8576 231-668-3
An aqueous solution of sodium hypochlorite; used as a disinfectant.
27303 Salut®
Chlorpyrifos - dimethoate
Insecticide with systemic, contact, stomach, and vapor action; used against hidden insects such as leaf miners, leaf rollers, wooly aphids. *BASF AG.*
27304 Salvarom
The diethyl ester of phthalic acid; used as a solvent.
27305 Salvex
Compounded plastic material. *Mitsubishi Petrochem.* Name unverified.
27306 Salysal
552-94-3 8491 209-027-4
Salicyl-salicylate.
27307 samarium
7440-19-9 8498 231-128-7
Sm
Samarium
A rare-earth metallic element; neutron absorber, dopant for laser crystals, metallurgical research, permanent magnets. mp = 1074°; bp = 1794°; d = 7.536. *Aldrich; Atomergic Chemetals; Cerac; Rhône-Poulenc Basic.*
27308 samarium oxide
12060-58-1 8498 235-043-6
O_3Sm_2
Catalyst in the dehydrogenation of ethanol, infrared-absorbing glass, neutron absorber, preparation of samarium salts. d = 8.347. *Noah Chem.; Rhône-Poulenc Basic.*
27309 Samaron
Disperse dyestuffs for synthetic surfaces. *Hoechst UK.* Name unverified.
27310 Sambarin
Suspension concentrate containing 500 g chlorothalonil and 250 g

propiconazole per liter; a systemic fungicide for winter wheat. *Ciba-Geigy Agrochemicals.*

27311 Samite
A trade name for a carborundum product; an abrasive. No manufacturer.

27312 samli
A clarified butter from East Africa.

27313 Samson Steel
A proprietary trade name for nickel-chromium steel containing 1.25% nickel and 0.6% chromium. No manufacturer.

27314 Samsonite
An explosive containing nitroglycerin, collodion cotton, potassium nitrate, wood meal, and ammonium oxalate; used in coal mines.

27315 Sanachlor
Hypochlorite for sterilizing dairy equipment. *Ciba plc.*

27316 Sanaklenz
Agricultural disinfectant. *The Wellcome Foundation Ltd.* Name unverified.

27317 Sanatank
Bulk milk tank sanitizer. *The Wellcome Foundation Ltd.* Name unverified.

27318 Sanatogen
Casein sodium glycerophosphate
Fisons plc, Pharmaceuticals Div. Name unverified.

27319 Sanction®
For the agriculture industry. *DuPont UK.*

27320 Sancure® 776
Waterborne aliphatic urethane polymer; ink grinding vehicle forming tough, flexible films; excellent adhesion to various substrates including nylon and polyester; high modulus, medium hardness and elongation. *Sanncor Industries.*

27321 sand acid
16961-83-4 4220 241-034-8
H_2SiF_6
Hydrofluosilicic acid.

27322 Sandacid
Textile dying buffers. *Sandoz Products Ltd.*

27323 sandalwood
Redwoods.

27324 sandarac resin
A resin obtained from the North West African tree *Callitris quadrivalis* ; used in the manufacture of spirit varnishes. Pine gum or Australian sandarac is obtained from *Callitris* species in Australia, and resembles the African variety.

27325 sandaracha
1303-32-8 834
realgar, arsenic disulfide.

27326 sandiver
Glass gall. The scum formed on the surface of molten glass. It consists of calcium and sodium sulfates, with about one-tenth of its weight of glass.

27327 Sandobet SC
68139-30-0 268-761-3
Cocoamidopropylhydroxy sultaine
High foaming, mild surfactant for cosmetic and toiletry applications; acid and alkali stable. *Sandoz.*

27328 sandoce
$C_8H_7NO_3$
Methyl-saccharin
A sweetening substance.

27329 Sandocryl®
Specialty dye for aqueous mediums. *Sandoz.*

27330 Sandofix
Fixing agents; for direct and reactive dyes. *Sandoz Products Ltd.* Name unverified.

27331 Sandofluor
General water and oil repellents. *Sandoz Products Ltd.*

27332 Sandogen
Aromatic sulfonate
Leveling agent. *Sandoz Products Ltd.* Name unverified.

27333 Sando-K
7447-40-7 7783 231-211-8
A proprietary preparation of potassium chloride. *Sandoz.* Name unverified.

27334 Sandol
Synthetic organic dyestuffs used on leather. *Sandoz Products Ltd.*

27335 Sandolan®
Synthetic organic acid dyestuffs; for aqueous media, textiles, fertilizers. *Sandoz; Sandoz Products Ltd.*

27336 Sandolube
Softening/lubricating agent; improves handling of natural and synthetic fibers. *Sandoz Products Ltd.*

27337 Sandopac
Oil drilling auxiliaries. *Sandoz Products Ltd.*

27338 Sandopan®
Very low foam detergent; used for removal of winding preparations from polyester fibers. *Sandoz Products Ltd.*

27339 Sandopan® DTC Linear P
70632-06-3
Sodium C_{12}-C_{15} pareth-6 carboxylate
Detergent, emulsifier, wetting agent, solubilizer, viscosity booster for industrial, personal care, and household products; stable in alkali high temperatures. *Sandoz.*

27340 Sandopan® DTC Linear P Acid
88497-58-9
C_{12}-C_{15} pareth-7 carboxylic acid
Detergent, emulsifier, wetting agent for industrial, personal care, and household use; oil solubilizer. *Sandoz.*

27341 Sandopan® DTC-100
68891-17-8
Sodium trideceth-7 carboxylate
Detergent, emulsifier, wetting agent for liquid detergents, solvent cleaners, all-purpose cleaners, germicidal cleaners, shampoos, bubble baths; oil solubilizer for aqeous systems, acid bowl cleaners. *Sandoz.*

27342 Sandopan® DTC-Acid
Trideceth-7 carboxylic acid
Detergent, emulsifier, wetting agent for industrial and personal care, conditioning shampoos, liquid soaps, household and industrial cleaners. *Sandoz.*

27343 Sandopan® JA-36
Trideceth-19 carboxylic acid
Moderate foaming mild surfactant, oil solubilizer, wetting agent cosmetics/toiletries, laundry products, cleaners, industrial specialties. *Sandoz.*

27344 Sandopan® KST
33939-65-0
Sodium ceteth-13 carboxylate
Mild emulsifier, detergent, lime soap dispersant for use in sticks, bar soaps, antiperspirants, other personal care products, industrial specialties, laundry products; inhibits sodium stearate crystal formation. *Sandoz.*

27345 Sandopan® LA-8
Laureth-5 carboxylic acid
Surfactant for cosmetics/toiletries, laundry products, cleaners, industrial specialties. *Sandoz.*

27346 Sandopan® LS-24
33939-64-9
Sodium laureth-13 carboxylate
Mild detergent, emulsifier, solubilizer for baby shampoos and personal care products. *Sandoz.*

27347 Sandopan® MA-18
28212-44-4
Nonoxynol-10 carboxylic acid
Detergent, wetting agent, solubilizer for cosmetics/toiletries, laundry products, cleaners, industrial specialties. *Sandoz.*

27348 Sandopan® TA-10
Isosteareth-6 carboxylic acid
Surfactant for cosmetics/toiletries. *Sandoz.*

27349 Sandopan® TS-10
Sodium isosteareth-6 carboxylate
Surfactant for cosmetics/toiletries. *Sandoz.*

27350 Sandopur
Washing off assistant; dyestuff complexing and fixing agent; for printed polyamide fabrics; improvement of wet fastness of metal complex and acid milling dyes on wool. *Sandoz Products Ltd.* Name unverified.

27351 Sandorin
Synthetic organic pigment colors used in printing inks and paints. *Sandoz Products Ltd.*

27352 Sandosperse
Synthetic organic pigment colors used in dispersions. *Sandoz Products Ltd.*

27353 Sandostab P-EPQ
38613-77-3 254-037-4
tetrakis (2,4-di-tertbutylphenyl) 4,4-biphenylenediphosphonite
Effective processing stabilizer, secondary antioxidant for polymers including polyolefins, ABS, PS, polybutylene, PBT, and nitrile barrier resins; peroxide decomposer for plastics manufacturing.; prevents polymer yellowing; synergizes uv light stability; reduces equipment corrosion in flame retardant applications. *Sandoz.*

27354 Sandoteric CFL
68604-73-9 271-705-0
Cocamphohydroxypropyl sulfonate
Extremely mild surfactant producing synergistic viscosity increase with alkyl sulfates; weak ampholyte. *Sandoz.*

27355 Sandoteric TFL Conc
68610-38-8 271-862-5
Sodium oleoamphohydroxypropyl sulfonate
Extremely mild surfactant producing synergistic viscosity increase with alkyl sulfates; weak ampholyte. *Sandoz.*

27356 Sandotex
Low soiling antistatic agent; for finishing synthetic fiber carpets. *Sandoz Products Ltd.* Name unverified.

27357 Sandoxylate® AC-46, AD-4
Alcohol ethoxylate
Nonfoaming wetting agent, dispersant for cleaning products and processes; stable in alkaline and acid media. *Sandoz.*

27358 Sandoxylate® C-32
Castor oil ethoxylate
Nonfoaming wetting agent, dispersant for cleaning products and processes; stable in alkaline and acid media. *Sandoz.*

27359 Sandoxylate® NT-15
Amine ethoxylate
Nonfoaming wetting agent, dispersant for cleaning products and processes; stable in alkaline and acid media. *Sandoz.*

27360 Sandoxylate® SX-408
PPG-2-isodeceth-4
Wetting agent for household and industrial applications, textile wet processing; intermediate for anionic surfactants. *Sandoz.*

27361 Sandozin
Powerful and economic wetting agent; for neutral and weakly acid media; many uses in textile processing. *Sandoz Products Ltd.* Name unverified.

27362 sandscale
471-34-1 1697 207-439-9
$CaCO_3$
Impurities formed in the pan during the concentration of brine; consists mainly of calcium carbonate,.

27363 sandstone
A stone consisting of grains of sand cemented together by a cementing material, e.g., silica, carbonate of iron, and iron oxide.

27364 sanfoin
87-20-7 5139 201-730-4
$C_{12}H_{16}O_3$
isoamyl salicylate
orchidee. Used in perfumes and soaps. bp = 274-278°; d_{15}^{18} = 1.048; n_D^{20} = 1.506; insoluble in H_2O, soluble in organic solvents.

27365 Sangajol
A trademark for a fraction of Bormeo petroleum distillate boiling at 160-170°C; it contains cyclic hydrocarbons, and is used as a turpentine substitute and resin solvent. No manufacturer.

27366 Saniblanket, Sanifoam
Urea resin-based foams; cover for refuse and hazardous waste. *Sanifoam Inc.* Unverified.

27367 Sani-Soil-Set
Dust laying composition. *Monsanto (Solaris).* Name unverified.

27368 Sanitant
Dairy hygiene detergent sterilizer. *The Wellcome Foundation Ltd.* Name unverified.

27369 Sanitary 1700
Silicone sealant; mildew-resistant sealant for ceramic tile, tubs, spas, showers, and plumbing fixtures. *GE Silicones.*

27370 sanitas
An aqueous liquid prepared by blowing air through warm oil of turpentine, in contact with water. It contains hydrogen peroxide and thymol, and is used as a disinfectant.

27371 Sanoform
A disinfectant consisting of a mixture of the disinfecting constituents of various tar oils, with calcium chloride and magnesium chloride in a saponified form.

27372 Sanogran®
Pigment for coloring gels, powders, solids. *Sandoz.*

27373 Sanoleum
A mixture of crude cresols with hydrocarbons; used in disinfecting urinals.

27374 Sanoscent
Camphortar. Preparations containing camphor as the main ingredient; used as disinfectants.

27375 Sansalid
Uredofos
Veterinary anthelmintic. *SmithKline Beecham.* Name unverified.

27376 sanse
Sanse oil. The residual cakes obtained from pressed Italian olives. When dried, and the oil extracted with carbon disulfide, it gives the so-called sulfocarbon oil, or sulfur olive oils; used in the manufacture of green soap for use in the textile industry.

27377 Sansilic 11
Silicone-free; antifoaming agent for textile processing, all fibers. *Ceca SA.*

27378 Sansorbin®
Cells. *Calbiochem Corp.*

27379 Sanspor
2425-06-1 1814 219-363-3
Captafol
2939-80-2. Fungicide for use on potatoes. *ICI Chem & Polymers Ltd.*

27380 Santac 52
9003-28-5
High molecular weight polybutene tackifier in LLDPE; cling additive for stretch cling film suitable as pallet wrap, silage wrap, food wrap, etc. *Santech.*

27381 santalwood
$C_{15}H_{14}O_5$
It contains 16% of santalin, The extract is used to color confectionery and liqueurs.

27382 Santechem 21-21
123-77-3 950 204-650-8
$C_2H_4N_4O_2$
1,1'-azobisformamide
Azodicarbonamide; diazenedicarboxamide; azodicarboxamide; azobiscarbonamide; azobiscarboxamide; 1,1'-azobiscarbamide. Blowing agent for polyolefins, extruded and injection molded structural foam; elminates die sink marks in injection molding. mp = 225° (dec); soluble in hot H_2O, EtOH; *Santech.*

27383 Santechem Grey F.R. P.E. Conc
Halogen compound/antimony oxide in LLDPE; flame retardant concentrate for polyethylene products. *Santech.*

27384 Santel
Water-based acrylic resins and coatings; for surface coatings for paper, board, foil and films, gloss enhancers, surface protection, barrier coatings, grease and oil resistance for paper. *ADM Tronics Unlimited Inc.*

27385 Santiciser
A proprietary trade name for vinyl plasticizers coded as follows: M-17: Methyl phthalyl ethyl glycollate. 1-H: N-cyclohexyl paratoluenesulfonamide. 107: Di-2-ethyl hexyl phthalate. 140: Cresyl diphenyl phosphate. No manufacturer.

27386 Santiciser SC
The triglycol ester of a vegetable oil fatty acid; a vinyl plasticizer. *Harwick Standard Chemical Co.* Discontinued.

27387 Santicizer 10
A proprietary trade name for a plasticizer; *o*-Cresyl-*p*-toluene sulfonate. No manufacturer.

27388 Santicizer 97
Dialkyl phthalate
Plasticizer for PVC film, sheet and coatings; gives low temperature flexibility. *Monsanto plc.* Name unverified.

27389 Santicizer 141
Alkylaryl phosphate
Used in PVC films, sheets, extrusions, moldings, organosols and plastisols, flame retardant plasticizer. *Monsanto plc.*

27390 Santicizer 143
Modified triaryl phosphate
Plasticizer compatible with PVC, vinyl nitrile elastomers, latex emulsions and cellulosic materials. *Monsanto plc.*

27391 Santicizer 148
Alkyl diaryl phosphate
Flame-retardant plasticizer for PVC resins. *Monsanto plc.*

27392 Santicizer 154
Triaryl phosphate
Flame-retardant plasticizer, compatible with PVC resins and vinyl nitrile rubber, PVA emulsions and cellulosics. *Monsanto plc.*

27393 Santicizer 160
85-68-7 201-622-7
Butyl benzyl phthalate
General purpose plasticizer for PVC, used in flooring industry. *Monsanto plc.*

27394 Santicizer 711
Dialkyl adipate
General purpose plasticizer for PVC resins, outperforms DOP. *Monsanto plc.*

27395 Santicizer DUP
Plasticizer. *Monsanto plc.* Name unverified.

27396 Santion
85009-19-9
flusilazole. Emulsifiable concentrate of 400 g flusilazole per liter; systemic insecticide for cereals, apples , vines and sugar beet. *DuPont UK.*

27397 Santobrite
7242
A proprietary trade name for sodium pentachlorphenate; a preservative used in paints, adhesives, etc. No manufacturer.

27398 Santocel
7631-86-9 8637 231-545-4
A proprietary trade name for silica gel, a porous form of silica; used as a heat insulator, drying agent, etc. No manufacturer.

27399 Santochlor
106-46-7 3107 203-400-5
A proprietary trade name for *p*-dichlorobenzene; used as a deodorizer, moth preventative, etc. No manufacturer.

27400 Santocure
95-33-0 202-411-2
$C_{13}H_{16}N_2S_2$
N-Cyclohexyl-2-benzothiazole-sulfenamide
Sufenax CB; Rhodifax 16; Accelerator CZ; Durax; N-cyclohexylbenzothiazole-2-sulfenamide; CBTS; CBS. N-Cyclohexylbenzothiazyl sulphenamide is an vulcanization accelerator used in natural rubber and styrene-butadienethiazyl sulfenamide rubber. mp = 93-100°; insoluble in H_2O, soluble in benzene *Monsanto Co.* Name unverified.

27401 Santocure MOR/MOR90
4-Benzothiazole-2-sulfenyl morpholine
Rubber accelerators. *Monsanto plc.*

27402 Santocure NS
95-31-8 202-409-1
$C_{11}H_{14}N_2S_2$
N-t-Butylbenzothiazole-2-sulfenamide rubber. mp = 105-110°; *Monsanto plc.*

27403 Santoflex
Antidegradant, antioxidant, antiozonant; rubber processing chemicals. *Monsanto Co.*

27404 Santoflex 13, 77
Antiozonant. *Monsanto plc.* Name unverified.

27405 Santoflex 1P
Antiozonant/antioxidant. *Monsanto plc.* Name unverified.

27406 Santoflex A
A proprietary trade name for a mixed ketoneamine and diphenyl-*p*-phenylene-diamine; a rubber antioxidant. No manufacturer.

27407 Santoflex AW
Rubber antiozonant. *Monsanto plc.* Name unverified.

27408 Santoflex B
A proprietary trade name for the condensation product of acetone and *p*-amino-diphenyl; a rubber antioxidant. No manufacturer.

27409 Santoflex BX
A proprietary trade name for a constant composition blend of Santoflex B and diphenylparapheny lenediamine; a rubber antioxidant. No manufacturer.

27410 Santoflex DD. DPA
Rubber antioxidant. *Monsanto plc.* Name unverified.

27411 Santogard PVI
17796-82-6 241-774-1
N-(Cyclohexylthio)phthalimide
Prevulcanization inhibitor for natural and synthetic rubber. *Monsanto Co.* Name unverified.

27412 Santolite
A proprietary trade name for synthetic resins of the sulfonamide aldehyde type, e.g., toluene sulfonamide-formaldehyde, for use in lacquers, etc. No manufacturer.

27413 Santomerse
A proprietary trade name for an alkylated aryl sulfonate; used as a wetting agent. No manufacturer.

27414 Santone® 3-1-S
37349-34-1
Polyglyceryl-3 stearate
Emulsifier; aeration of lipid systems; solubilizer; color/flavor dispersant in water. *Van Den Bergh Foods.*

27415 Santone® 3-1-SH
9007-48-1
Polyglyceryl-3 oleate
Emulsifier and aerating agent used in food industry, textile and plastic lubricant; color dispersant. *Van Den Bergh Foods.*

27416 Santonox
4,4'-Thiobis(6-*t*-butyl-*m*-cresol)
Antioxidants for polyethylene and other plastic resins. *Monsanto Co.* Name unverified.

27417 Santophen 20
87-86-5 7242 201-778-6
A proprietary trade name for pentachlorphenol; a preservative for paints, wood, adhesives, etc. No manufacturer.

27418 Santoprene® 181-55, 181-64, 281-55, 281-64
Thermoplastic rubbers; medical grade for use in syringe sals and caps, medical tubing, sterilizable applications. *Advanced Elastomer Systems.*

27419 Santoprene® 281-87, 283-40
Thermoplastic rubber; medical grade for use in syringe sals and caps, medical tubing, sterilizable applications. *Advanced Elastomer Systems.*

27420 Santoquin
91-53-2 3800 202-075-7
Ethoxyquin
Feed preservative. *Monsanto Co.*

27421 Santo-Res
Wet strength paper resins; cationic retention aid for paper. *Monsanto Co.* Name unverified.

27422 Santoresin
A proprietary trade name for a synthetic resin. No manufacturer.

27423 santorin earth
A volcanic ash found in the island of Santorin; used to convert lime into hydraulic lime.

27424 Santosite
7757-83-7 8831 231-821-4
A proprietary trade name for anhydrous sodium sulfite; a reducing agent. No manufacturer.

27425 Santotan KR
15244-38-9 2287
$Cr_2O_{12}S_3$
A proprietary trade name for a basic chromium sulfate, a tanning agent. No manufacturer.

27426 Santotrac
Synthetic hydrocarbons
High temperature lubricant. *Monsanto Co.* Name unverified.

27427 Santovac
Polyphenyl ether
Vacuum diffusion pump fluid. *Monsanto Co.* Name unverified.

27428 Santovar
79-74-3 3359 201-222-2
$C_{16}H_{26}O_2$
2,5-di-*t*-pentylhydroquinone
2,5-di(*t*-amyl) hydroquinone; Santovar A. Antioxidant for unvulcanized rubber. mp = 179-180°. *Monsanto Co.*

27429 Santowax
Mixed isomeric terphenyls; high melting hydrocarbons. *Monsanto Co.* Name unverified.

27430 Santoweb
Treated cellulosic short fiber; rubber processing material. *Monsanto Co.*

27431 Santowhite
85-60-9 201-618-5
$C_{26}H_{38}O_2$
4,4' -Butylidenebis(6-t-butyl-m-cresol)
4,4-butylidenebis[2-(1,1-dimethylethyl)-5-methylphenol; SWP. Antioxidant for polypropylene, polyethylene, nylon molding powders, and other polymer resins. *Monsanto Co.*

27432 sanyan
A silk from a wild silkworm of Nigeria.

27433 Sanylene
Synthetic organic pigment colors used in plastics. *Sandoz Products Ltd.*

27434 Sapamine
Finishing agent. *Ciba plc.* Name unverified.

27435 Sapamine
A proprietary trade name for diethylaminoethyloleylamino acetate and similar compounds; used in conjunction with dyes. No manufacturer.

27436 Sapecron
470-90-6 2137 207-432-0
Emulsifiable concentrate containing 240 g chlorfenvinphos per liter; a soil-applied organophosphorus insecticide. *Ciba-Geigy Agrochemicals.*

27437 Saphire
Lube oil. *Monsanto (Solaris).* Name unverified.

27438 sapin
A mixture of Japan wax with heavy mineral oil (soft or liquid paraffin); a superfatting agent for soaps.

27439 sapoform
A product containing oleic acid, alcohol, potassium hydroxide, formalin, and distilled water.

27440 Sapogenat
Alkylphenol polyglycol detergent base. *Hoechst UK.*

27441 Sapogenat T
Range of nonionic surfactants of the tributyl phenol ethoxylate type in liquid, paste or wax form; auxiliaries in textile and paper manufacturing, domestic and industrial cleaning agents, emulsifiers and plant protection agents. *Hoechst UK.*

27442 sapogenin
A decomposition product ($C_{14}H_{22}O_2$) of saponin.

27443 saponine
A name usually applied to the active constituent of Panama bark, which is used instead of soap, for washing and producing a lather. The term is also used for a boring and cutting oil.

27444 saponins
11006-75-0
sapogenin glycosides
Water soluble high molecular weight glycosidal substances occurring naturally in plants; forms colloidal solutions. on shaking with water; Foam producer in fire extinguishers, detergent in textile industries, sizing, substitute for soap, emulsification agent for fats and oils. *Penco of Lyndhurst.*

27445 Saporin®
Feed industry additive. *BASF AG.*

27446 Sapp 4
7758-16-9 8713 231-835-0
Sodium acid pyrophosphate
Leavening agent for baking, cereals. *Rhône-Poulenc Food Ingreds.*

27447 Saprol
26644-46-2 9819 247-872-0
Emulsifiable concentrate containing 190 g/l triforine; a systemic insecticide. *Hoechst UK.*

27448 saprol
Disinfection oil. A mixture of crude cresols, hydrocarbons, and pyridine bases; used for disinfecting lavatories.

27449 Sar Gel®
Water/alcohol indicating paste showing presence of water bottoms in storage tanks containing solvent, gasoline, oil, etc. *Sartomer.*

27450 Saran
9002-86-2 7746 206-625-7
A range of polyvinylidene chloride plastics. *Dow Cheml Co Ltd, UK & Ireland.*

27451 Saran 313
9002-86-2 7746 206-625-7
Vinylidene chloride polymer; melt-processable extrusion resins with excellent gas and moisture barrier properties, broad chemical and ignition resist.; used for rigid multilayer coextruded containers and in extrusion/ coextrusion of film and sheet. *Dow Plastics.*

27452 Saran 510
9002-86-2 7746 206-625-7
Vinylidene chloride polymer; monofilament extrusion resin with excellent gas and moisture barrier properties, broad chemical and ignition resistance; used for rigid and flexible barrier packaging of medicines, foods, and cosmetics. *Dow Plastics.*

27453 Saran F-239, F-278, F-310
9002-86-2 7746 206-625-7
Vinylidene chloride polymer; excellent gas and moisture barrier properties, broad chemical and ignition resist.; used for rigid and flexible barrier packaging of medicines, foods, and cosmetics; solvent-soluble polymer for barrier coatings. *Dow Plastics.*

27454 Saran Wrap
9002-86-2 7746 206-625-7
Vinylidene chloride-vinyl chloride film. *Dow Chemical.*

27455 Saran® Filament
9002-86-2 7746 206-625-7
Polyvinylidene chloride fiber. *Asahi Chem. Industry.*

27456 Saranex
Coextruded multiayered films. *Dow Cheml Co Ltd, UK & Ireland.*

27457 Sarapron
An emulsifiable concentrate containing 98.8 g aminotriazole, 197.5 g atrazine and 20 g dicamba per liter; used for total weed control in non crop areas. *BP Oil Ltd.* Discontinued.

27458 Saratoga Steel
A proprietary trade name for a non-shrinking steel containing small quantities of manganese, chromium, tungsten, and carbon. No manufacturer.

27459 Sarbox® SB 400
Carboxylated acid terminated resin; features acrylate, anhydride and carboxyl functionality; for photoimaging, inks, metal coatings, plastic coatings. *Sartomer.*

27460 sarcine
68-94-0 4917 200-697-3
$C_5H_4N_4O$
Hypoxanthine; Sarkine.

27461 sarco
A material made from elaterite (a mineral rubber); used in rubber mixings.

27462 sarcocoll
A gum resin from *Penoea sarcocolla* , of Africa.

27463 sarcosine
107-97-1 8519 203-538-6
$C_3H_7NO_2$
N-methylglycine
methyl glycocoll; methylaminoethanoic acid; Sar; N-methylaminoacetic acid; methylaminoacetic acid. Synthesis of foaming antienzyme compounds for toothpaste, cosmetics, pharmaceuticals. mp = 212° (dec); soluble in H_2O (2.8 g/ml). *BASF; W.R. Grace/Hampshire; Schweizerhall; SWS Oilchemicals BV.*

27464 Saret® 500, 515
Trifunctional crosslinking agents which minimize scorch and offer processing flexibility when used with the peroxide cure of elastomers; suitable for injection, transfer, and compression molding. *Sartomer.*

27465 Sarkosyl NL30
137-16-6 4379 205-281-5
Sodium lauroyl sarcosinate
Gardol. Anionic surfactant for corrosion inhibition, enzyme inhibition, bacteristatic activity; used for cosmetics, toilet goods, pharmaceuticals, particularly useful in dentifrices, hair, carpet and upholstery shampoos; specialty and alkaline detergents; window cleaners, hand dishwash, fine fabric detergents, synthetic toilet soap, emulsion polymerization, letal processing, food products, *Ciba plc.* Name unverified.

27466 Sarkosyl O
Oleoyl sarcosine in acid form; anionic surfactant for corrosion inhibition, enzyme inhibition, bacteristatic activity; used for cosmetics, toilet goods, pharmaceuticals, particularly useful in dentifrices, hair, carpet and upholstery shampoos; specialty and alkaline detergents; window cleaners, hand dishwash, fine fabric detergents, synthetic toilet soap, emulsion polymerization, letal processing, food products, *Rhône-Poulenc UK.*

27467 Sarlink 1000
Thermoplastic elastomer; excellent oil and chemical resistance, bondability and abrasion resistance; for hose and tubing, coated fabrics, weatherstripping, diaphragms, gaskets, seals, extruded sheet and linings, grips, ducting, belts and trays. *DSM Thermoplastic Elastomers.* Discontinued.

27468 Sarlink 2000
Thermoplastic elastomer; low permeability to moisture and gases, excellent damping and compression creep properties; for use in tank linings, waterproofing membranes, sports grips, ball bladders, seals, gaskets, belts, hoses, tubing, rollers, medical stoppers. *DSM Thermoplastic Elastomers.* Discontinued.

27469 Sarlink 3000
Thermoplastic elastomer; high resiliency, low tensile and compression creep, excellent flexibility, weatherability; for weatherstripping, seals, o-rings, boots, bellows, switch covers, furniture parts, flashlight housings, belts, plugs, connectors, hoses, tubing. *DSM Thermoplastic Elastomers.* Discontinued.

27470 Saroul
Hand cleaners. *Brent Chemicals International plc.*

27471 Sarpol
A preparation of crude phenol (carbolic acid); a disinfectant. No manufacturer.

27472 Sartomer
Functional acrylics. *Sartomer.* Discontinued.

27473 Sascol
Collectors for mineral recovery. *Sasolchem.*

27474 Sasetone
67-64-1 64 200-662-2
Acetone
Solvent in paints, varnishes, lacquer thinners, printing inks, nail polish removers, acetylene in filling of cylinders, bituminous paints, polyester resins, PVC cloth manufacture, explosives, adhesives; raw material for manuf *Sasolchem.* Name unverified.

27475 Sasfroth
Frothers for mineral flotation. *Sasolchem.*

27476 Sasolwaks
8002-74-2 7155 232-315-6
Hard, high melting point crystalline paraffin wax, average molecular formula $C_{50}H_{102}$, constituent in polishes, plastics to enhance gloss, color disperiosn etc., insulating components in elelctrical cables, paper conversion, chewing

gums, carbon paper backing, priniting inks, paints, hot-melt adhesives, lubricant in rubber, laundry machines and plastic molding. *Sasolchem.*

27477 Sasolwaks M3
8002-74-2 7155 232-315-6
Semi-refined paraffin wax *Sasolchem.*

27478 Satco Metal
A proprietary trade name for a lead-base bearing alloy modified by the addition of tin, calcium, magnesium mercury, aluminum, potassium lithium, all in very small quantities except tin which may rise to 1%. No manufacturer.

27479 Satessa
A highly dispersed pyrogenic silica preparation; used in the textile industry as a lubricating additive; to increase strength of yarn; to increase fiber friction; to increase nonslip characteristics. *Degussa AG.* Name unverified.

27480 Satexlan 20
68648-27-1
PEG-20 hydrogenated lanolin
Oil-water emulsifier, emollient, thickener; perfume solubilizer; imparts superfatting properties. *Croda Inc.* Discontinued.

27481 Satialgine H8
9005-32-7 241 232-680-1
Alginic acid NF
Tablet disintegrant for compressed tablets. *Penwest Pharmaceuticals Co.*

27482 satin rouge
A variety of lamp-black used for polishing.

27483 satin white
12004-14-7 234-448-5
A pigment consisting of gypsum mixed with alumina. A mixture of calcium sulfate with aluminum sulfate is also known under this name.

27484 Satina 44
Partially hydrogenated palm kernel oil, lecithin; kosher; coating fat for butterscotch and chocolate flavored confectionery coatings. *Van Den Bergh Foods.*

27485 Satina®
Nonirritant skin and body care products for sensitive and problem skin. *Bayer AG.*

27486 Satintone
1332-58-7 5294 296-473-8
Very fine to coarse particle size calcined anhydrous kaolin (aluminum silicate); used as a reinforcer and extender for plastics, PVC, color concentrates, rubber, coatings. *Engelhard.*

27487 Satinwood
Satin finish alkyd paint for wood. *ICI Chem & Polymers Ltd.*

27488 Satisfar
38260-54-7 3936 253-855-9
Etrimfos
Used to control pests in stored grain. *Nickerson Seeds Ltd.*

27489 sativic acid
$C_{18}H_{35}O_5$
Trihydroxy-stearic acid.

27490 Satrapol
A photographic developer containing monomethyl *p*-amino-phenolsulfate.

27491 Satric
443-48-1 6242 207-136-1
Metronidazole
An antiprotozoal. *Savage Laboratories.* Discontinued.

27492 Satulan
8031-44-5 232-452-1
Hydrogenated lanolin
Water-oil emulsifier, emollient used in personal care products. *Croda Inc.* Discontinued.

27493 Saturn Glace
Polymers, fluorocarbons composition; applied to new and used vehicles. *Adasco Inc.*

27494 Saturseal
Joint sealants. *Crowley Chem.*

27495 sauconite
A clay containing zinc.

27496 Sauflon PW
Lidofilcon B
Contains 79% of water; contact lens material. *Visiontech Inc.* Unverified.

27497 Savall
13593-03-8 237-031-6
Insecticide containing quinalphos. *ICI Chem & Polymers Ltd.*

27498 Savall
13593-03-8 237-031-6
Emulsifiable concentrate containing 25% w/w quinalphos per liter; an organophosphorus insecticide. *Farm Protection Ltd.*

27499 Savan
Sodium ammonium metavanadate. *Kerr-McGee Chemical Corp.*

27500 Savinase
A preparation containing a proteolytic enzyme; used in the detergent industry as an additive to powder detergents to improve the detergency towards protein containing stains and as an additive to nonbuilt liquid detergents. *Novo Nordisk.*

27501 Savinyl®
Specialty dye for coloring adhesives, oils, waxes, solvents, wood stains. *Sandoz.*

27502 Savloclens
55-56-1 2140 200-238-7
Chlorhexidine
Broad spectrum antiseptic with added detergent properties *ICI Chem & Polymers Ltd.*

27503 Savlon
A proprietary preparation of chlorhexidine and cetrimide; used for a range of antiseptics. *ICI Chem & Polymers Ltd.* Discontinued.

27504 Savlon Babycare
55-56-1 2140 200-238-7
A range of products containing chlorhexidine; used principally for antiseptics for use by babies and infants. *ICI Chem & Polymers Ltd.* Discontinued.

27505 savol
A medicated soap containing salol (phenyl salicylate) with perfumes.

27506 Savona
A soap concentrate; used to control insects in greenhouses. *Koppert (UK) Ltd.*

27507 savonette oil
A mixture of vegetable fatty acids and resin acids, a by-product of paper manufacture. It is recommended as a substitute for oleic acid in soap manufacture.

27508 Savoselling REAC 4
Soaping agent for fiber reactive dyes on cotton. *Ceca SA.*

27509 Saxifragin
An explosive mixture containing 76% barium nitrate, 2% potassium nitrate, and 22% charcoal.

27510 Saxin
Artificial sweetening agent. *The Wellcome Foundation Ltd.* Name unverified.

27511 Saxon
Lotion for men (woodspice and musk), aftershave skin conditioner. *Richardson-Vicks Inc.* Name unverified.

27512 Saxonite
An explosive similar to Samsonite in composition. The term is also used for a mineral which is a mixture of olivine and enstatite.

27513 Saytex®
Brominated flame retardants. *Ethyl Corp; Ethyl SA.*

27514 Saytex® 102E
1163-19-5 214-604-9
$C_{12}Br_{10}O$
1,1'-oxybis(2,3,4,5,6-pentabromobenzene
Decabromodiphenyl oxide; Decabromodiphenyl ether; Pentabromodiphenyl ether; bis(pentabromophenyl) ether; DPBPO; fr 300ba; FRP 53; berkflam b 10E; bromkal 82-ode; BR 55N; FR 300; DE 83R; bromkal 83-10de; pentabromophenyl ether; saytex 102; saytex 102E; tardex 100; DBDPE.
Flame retardant; high-purity, elec. grade for wire and cable insulation application. mp = 302°; bp = 425°; insoluble in H_2O (<1 mg/ml). *Ethyl Corp.*

27515 Saytex® 111
32536-52-0 251-087-9
$C_{12}H_2Br_8O$
Octabromodiphenyl oxide
octabromom-1,1'-oxybisbenzene; 2,2',3,3',4,4',5,5'-octabromodiphenyl ether. Flame retardant for ABS, HIPS, polyamides, elastomers, adhesives, and coatings; semiplasticizing additive for styrenic polymers and copolymers such as ABS. *Ethyl Corp.*

27516 Saytex® 120
58965-66-5 261-526-6
Tetradecabromodiphenoxy benzene
Flame retardant for nylon, alloys of styrenics and engineering plastics, engineering thermoplastic polyesters, ABS, crosslinked polyethylene, elastomers, and high-impact PS. *Ethyl Corp.*

27517 Saytex® 8010
Flame retardant for HIPS and other styrenics, thermoplastic polyolefins, electronic applications *Ethyl Corp.*

27518 Saytex® BCL-462
3322-93-8 222-036-8
$C_8H_{12}Br_4$
1,2-dibromo-4-(1,2-dibromoethyl)cyclohexane

Dibromoethyldibromocyclohexane. Flame retardant for expandable, crystalline and high-impact PS, SAN resins, adhesives, coatings, textile treatment, PU. mp = 68-90°; insoluble in H_2O (<1 mg/ml). *Ethyl Corp.*

27519 Saytex® BN-451
52907-07-0 258-250-3
Ethylenebis dibromonorbornane dicarboximide
Thermally stable flame retardant for PP, polyamides, PU elastomers and coatings; off-white powd. *Ethyl Corp.*

27520 Saytex® BT-93®
32588-76-4 251-118-6
$C_{18}H_4Br_8N_2O_4$
2,2'-(1,2-ethanediyl)bis[4,5,6,7-tetrabromo-1H-isoindole-1,3(2H)-dione
Ethylene bis-tetrabromophthalimide; 1,2-bis(tetrabromophthalimido)ethane. Flame retardant for high-impact PS, polyethylene, PP, thermoplastic polyesters, nylon, EPDM, rubbers, PC, ethylene copolymers, ionomer resins, textile treatment. mp = 446°; SG = 2.67; insoluble in H_2O (<1 mg/ml). *Ethyl Corp.*

27521 Saytex® FR-1138
Mixture of monobromopentaerythritol, dibromoneopentyl glycol and tribromoneopentyl alcohol. Used as a flame retardant for unsaturated polyesters, rigid and flexible PU foam and PU elastomers. *Ethyl Corp.*

27522 Saytex® HBCD-LM
3194-55-6 221-695-9
$C_{12}H_{18}Br_6$
1,2,5,6,9,10-hexabromocyclododecane
hexabromocyclododecane. Low melting flame retardant mp = 173-177°. *Ethyl Corp.*

27523 Saytex® RB-100
79-94-7 201-236-9
$C_{15}H_{12}Br_4O_2$
2,2',6,6'-tetrabromo-4,4'-isopropylidene phenol
Tetrabromobisphenol A. Reactive or additive source of bromine for flame retardancy; reactive intermediate for preparation of brominated epoxy resins, polycarbonates, and unsaturated polyesters; additive for ABS, PS, and phenolic resins; intermediat mp = 180-184°, insoluble in H_2O. *Ethyl Corp.*

27524 Saytex® RB-49
632-79-1 211-185-4
$C_8Br_4O_3$
4,5,6,7-tetrabromo-1,3-isobenzofurandione
3,4,5,6-tetrabromophthalic anhydride. Flame retardant; monomer for unsaturated polyester; reactive intermediate for preparation of polyols, esters, and imides. mp = 279-281°; insoluble in H_2O. *Ethyl Corp.*

27525 Saytex® VBR
593-60-2 209-800-6
C_2H_3Br
vinyl bromide
bromoethylene. Flame retardant; intermediate in organic synthesis and in the manufacturing. of flame retardants, polymers, copolymers, pharmaceuticals, fumigants, and other chemicals; also used in textiles, adhesives, coating, photographic plates and films. mp = -139°; bp = 16°; d = 1.493; n_D^{20}= 1.4320; *Ethyl Corp.*

27526 SB 70/52P6
Mixture of ethanol and isopropanol; solvent. *Sasolchem.*

27527 SB-136
1344-28-1 369 215-691-6
Alumina trihydrate
Filler featuring flame retarding and smoke suppressing properties; resin extender in polyester, vinyls, PU, latex, neoprene foam systems, wire and cable insulation, vinyl wall and floor coverings, epoxies. *J.M. Huber/Solem.*

27528 SB-336
1344-28-1 369 215-691-6
Alumina trihydrate
Filler with flame and smoke suppression; low viscosity, high filler loading and glass wetout; for spray-up or land lay-up FRP application, filament winding, panel production, resin injection, SMC/BMC/acrylic sheet rigidizing, and cast polyester parts. *J.M. Huber/Solem.*

27529 SB-632
1344-28-1 369 215-691-6
Alumina trihydrate
Filler featuring flame retarding and smoke suppressing properties; also used for plastic and rubber application; suitable only for low viscosity systems or low levels of loading; particle suspension; poor processing characteristi *J.M. Huber/Solem.*

27530 SBP
Special boiling point solvents. *Carless Refining & Marketing Ltd.*

27531 SB-VAC
SB-1 strain, frozen chicken herpes virus marek's vaccine; for immunization of poultry. *Intervet Inc.*

27532 SB-VAC Plus Marexine-CA
For combination use; for immunization of poultry. *Intervet Inc.*

27533 SC-10
Filled silicone-based resins; thermally conductive resin, cures to resilient solid; high thermal conductivity, thermal stability, and elec. properties; used in electronic applications requiring heat conductive encapsulant; mix ratio: 2.38 phc with Activator BA-52. *Bacon.*

27534 SC-17
Filled silicone-based compounds; thermally conductive polymers which cure to resilient solids; used in electronic applications as heat conductive encapsulant. *Bacon.*

27535 SC-53
471-34-1 1697 207-439-9
Calcium carbonate
Coarse ground filler for polyolefins, carpet backing, caulks, sealants, putties, as mild abrasives in cleaners. *ECC International Ltd.* Discontinued.

27536 Scabene Lotion
58-89-9 5526 200-401-2
Lindane
Pediculicide; scabicide *Stiefel Laboratories Inc.* Name unverified.

27537 Scadoplast RA3L, RA350
A proprietary trade name for an adipic acid polyester vinyl plasticizer. *DSM Kunstharze GmbH.* Discontinued.

27538 Scadoplast RS 20, RS 150
A proprietary trade name for a sebacic acid polyester vinyl plasticizer. *DSM Kunstharze GmbH.* Discontinued.

27539 Scagliola
A stone manufactured from Keene's cement mixed with coloring matter, to which is added water containing dissolved glue or isinglass.

27540 S-CAL
Stabilized and treated human blood cells in an artificial plasma medium; calibrator for automatic blood cell analyzers. *Coulter Electronics Ltd.*

27541 scale
Crude Scotch paraffin wax is known in commerce by this name.

27542 Scale Cleen
Dry acid descaler. *Dearborn Chemicals Ltd.* Name unverified.

27543 Scalol
C_7H_9NO
A photographic developer containing methyl-*p*-amino-phenol, as the active constituent.

27544 scandia
12060-08-1 8537 235-042-0
Sc_2O_3
Scandium oxide
Used in the preparation of scandium fluoride.

27545 scandium
7440-20-2 8537 231-129-2
Sc
scandium
Metallic element; no major industrial use; some application in semiconductor field; an artificial radioactive isotope has been used in tracer studies and leak detection. mp = 1541; bp = 2836°; d = 2.9890. *Atomergic Chemetals; Cerac; Rhône-Poulenc.*

27546 scandium oxide
12060-08-1 8537 235-042-0
O_3Sc_2
scandium(III) oxide
Used in preparation of scandium fluoride. Sc_2O *Aldrich; Atomergic Chemetals; Cerac; Noah Chem.*

27547 Scarab
Urea formaldehyde molding compounds. *BIP Chemicals Ltd.*

27548 Scarat
A proprietary trade name for a synthetic resin of the urea type. No manufacturer.

27549 S-Carb
533-96-0 8823 208-580-9
$C_2HNa_3O_6$
carbonic acid, sodium salt (2:3)
trisodium hydrogendicarbonate ($NaHCO_3$:Na_2CO_3); Urao; trona. Feed grade sodium sesquicarbonate. d = 1.2112; soluble in H_2O (13 g/100 ml). *FMC.* Discontinued.

27550 scarlet vermillon
Extract of vermilion; Chinese vermilion; orange vermilion; Field's orange vermilion. Varieties of vermilion,.

27551 Scav-Ex® 235
Synthetic pulp; fibrous processing aid for paper industry (setting of stickies in

tissue and toweling, gypsum paper, boxboard, fine paper manufacturing). *Hercules.* Discontinued.

27552 Scav-Ox® 35%
302-01-2 4809 206-114-9
Hydrazine aqueous solution; corrosion protector in low-, medium-and high-pressure boilers; oxygen scavenger in feedwater. *Olin.*

27553 Scav-Ox® II
302-01-2 4809 206-114-9
Catalyzed hydrazine
Corrosion protector in industrial boilers. *Olin.*

27554 Scent Off
Wax pellet containing naphthalene and essential oils; animal deterrent. *Vitax Ltd.*

27555 Scent Sticks
Flammable, scented, oil saturated, compressed wood pulp on a sandalwood stick (incense); used as a scented air freshener (incense), novelty item. *Ambrosia Scents.* Name unverified.

27556 ScentCap
Microencapsulated fragrance oils for use in soap, clothing, personal care products, cosmetics, emollients in skin creams/powders. *M-CAP Tech. Int'l.*

27557 Scentinel
Gas leak warning odorants. *Phillips.*

27558 Schaeffer's acid
93-01-6 6479 202-209-4
$C_{10}H_8O_4S$
2-naphthol-6-sulfonic acid
Baum's acid; Armstrong acid. α-Naphthol-2-sulfonic acid also β-naphthol-6-sulfonic acid; used as an azo dye intermediate. mp = 125°; soluble in H_2O, EtOH, insoluble in non-polar organic solvents.

27559 Schaeffer's salt
6479
$C_{10}H_7NaO_4S$
The sodium salt of β-naphthol-6-sulfonic acid (Schaeffer's acid). Soluble in H_2O, EtOH.

27560 schallerite
$9MnSiO_3 \cdot Mn_3As_2O_8 \cdot 7H_2O$
An arseno-silicate found in the Franklin furnace. It approximates to the formula.

27561 Scheele's acid
A 4% solution of hydrocyanic acid, HCN.

27562 Scheele's green
10290-12-7 2693 233-644-8
$CuHAsO_3$
cupric arsenite
Mineral green, Swedish green; Scheele's mineral; acid copper arsenite; arsenious acid, copper(2+) salt (1:1); arsonic acid, copper(2+) salt;. A pigment consisting of copper arsenite.

27563 scheeletine
$PbWO_4$
Scheelinite, lead tungstate. A tungstate of lead.

27564 scheelinite

27565 scheerite
A mineral wax resembling ozokerite.

27566 Scheiber oil
The glyceride of dehydrated ricinoleic acid; suitable for varnishes.

27567 Scheibler's reagent
51312-42-6 8811 257-132-9
$Na_4O_2 \cdot P_2O_5 \cdot W_{12}O_{36} \cdot H_{36}O_{18}$
sodium phosphotungstate
sodium tungstophosphate. Obtained by dissolving 100 g sodium tungstate and 70 g sodium phosphate in 500 ml water, and acidifying with nitric acid; used as a testing reagent for alkaloids.

27568 Scheiderite
A mixture of trinitronaphthalene and ammonium nitrate; an explosive.

27569 Schellan solution
A colloidal solution of a synthetic resin made from urea and formaldehyde, and kept from gelatinizing by means of sodium acetate; used as a dressing material for textiles.

27570 Schensand
Resin coated foundry sand. *Schenectady - Midland Ltd.*

27571 Schenvar
Insulating varnish. *Schenectady-Midland Ltd.*

27572 Schercamox C-AA
68155-09-9 268-938-5
Cocamidopropylamine oxide
Conditioner, detergent, wetting agent, antistat used in personal care products and light dishwashing detergents; biodegradable. *Scher.*

27573 Schercamox CMA
61791-47-7 263-180-1
Dihydroxyethyl cocamine oxide
Softener, and wetting agent for cosmetics; builds viscosity and stabilizes foam in personal care products; soft emollient feel on the skin; emolliency, lubricity, and slip to shave creams; conditions and prevents fly-away in hair shampoos. *Scher.*

27574 Schercamox DMC
61788-90-7 263-016-9
Cocamine oxide
Wetting agent, foam stabilizer, viscosity enhancer. *Scher.*

27575 Schercamox DML
1643-20-5 216-700-6
Lauramine oxide
Antistat; emulsifier, emulsion stabilizer for used in cosmetics industry; foam booster and viscosity builder for shampoos. *Scher.*

27576 Schercamox DMM
3332-27-2 222-059-3
Myristamine oxide
Wetting and foaming agent, surfactant for light duty dishwashing compounds, shampoos; emulsifier and emulsion stabilizer for mineral oils. *Scher.*

27577 Schercamox DMS
2571-88-2 219-919-5
Stearamine oxide
Skin emollient, softener, viscosity controller, foam stabilizer, and hair conditioner in personal care products. *Scher.*

27578 Schercassist AC
Quaternary; dye-leveling agent for acrylics. *Scher.* Discontinued.

27579 Schercemol 65
5434-57-1 226-602-5
Isohexyl neopentanoate
Penetrating emollient with good solubility in hydro-alcoholic systems; constituent in aroma chemicals. *Scher.*

27580 Schercemol 145
144610-95-5
Myristyl neopentanoate
Light emollient; reduces tackiness in skin and hair preparations. *Scher.*

27581 Schercemol 185
58958-60-4 261-521-9
Isostearyl neopentanoate
Substantive emollient with low cloud point; aids as cloud and freeze point depressant, emulsion and freeze-thaw stabilizer; cosmetic preparations for skin care especially near eyes; low level of skin and eye irritation; binder for pigmented makeup. *Scher.*

27582 Schercemol 318
68171-33-5 250-651-1
Isopropyl isostearate
Emollient for bath oils, creams, lotions, and lipsticks; lubricity without oiliness. *Scher.*

27583 Schercemol 1688
59130-69-7 261-619-1
Cetearyl octanoate
Emollient; spreads evenly on skin imparting velvety softness; functions as waterproofing agent due to adhesion properties. *Scher.*

27584 Schercemol 1818
41669-30-1 255-485-3
Isostearyl isostearate
Substantive emollient imparting luxurious softness to skin; used in cosmetics imparting slip and lubricity, luster, and sheen; cosolv. and solubilizer in perfumes. *Scher.*

27585 Schercemol BE
18312-32-8 242-201-8
Behenyl erucate
Emollient base for lip care cosmetics, skin creams and lotions; chemically comparable to one of main constituents of jojoba oil; melts close to body temperature; nontoxic. *Scher.*

27586 Schercemol CM
2599-01-1 220-001-1
Cetyl myristate
Solid emollient, lubricant and body builder. *Scher.* Discontinued.

27587 Schercemol CO
59130-69-7 261-619-1
Cetyl octanoate
Solvency properties for use in make-up removers. *Scher.*

27588 Schercemol CP
540-10-3 2073 208-736-6

Cetyl palmitate
Synthetic spermaceti wax. *Scher.* Discontinued.

27589 Schercemol CS
1190-63-2 214-724-1
Cetyl stearate
Waxy emollient for creams and lotions, thickener and body builder. *Scher.* Discontinued.

27590 Schercemol DEGMS
9004-99-3
PEG-2 stearate
Primary emulsifier in creams and lotions. *Scher.* Discontinued.

27591 Schercemol DEIS
Decyl isostearate
Emollient, lubricant and penetrant with unusual pigment-dispersing properties. *Scher.* Discontinued.

27592 Schercemol DIA
6938-94-9 248-299-9
Diisopropyl adipate
Nonoily penetrating emollient, lubricant, and solv. with mild drying effects used in hydroalcoholic cosmetic formulations. *Scher.*

27593 Schercemol DICA
Diisocetyl adipate
Low viscosity emollient, useful in skin and hair preparations. *Scher.* Discontinued.

27594 Schercemol DID
103213-20-3
Diisopropyl dimer dilinoleate
Nonoily, glossy emollient producing a cushiony feel and body to skin and makeup preparations; improves disp. and spreading of pigments; binder for pressed powd.; offers sheen, emolliency in lip preparations; highly substantive; suitable for suntan preparations requiring some water repellency. *Scher.*

27595 Schercemol DIS
7491-02-3 231-306-4
Diisopropyl sebacate
Nonoily emollient, lubricant, solubilizer with mild drying effects used in hydro-alcoholic personal care products; solv. and coupling properties; fast spreading action. *Scher.*

27596 Schercemol DISD
103213-19-0
Diisostearyl dimer dilinoleate
Emollient offering lingering effect retained on skin after washing; used in personal care products. *Scher.*

27597 Schercemol DISF
113431-53-1
Diisostearyl fumarate
Lubricant, conditioner. *Scher.*

27598 Schercemol DO
3687-46-5 222-981-6
Decyl oleate
High molecular weight, low freeze point, nonoily lubricant, emollient, penetrant, and moisturizer for cosmetic and personal care products. *Scher.*

27599 Schercemol EE
27640-89-7 248-587-4
Erucyl erucate
Emollient ester for use in skin, hair and suntanning preparations. *Scher.* Discontinued.

27600 Schercemol EGMS
111-60-4 203-886-9
Glycol stearate
Emulsifier, opacifier, and pearlescent for cosmetic and personal care products,; thickener and viscosity controller for cosmetic preparations. *Scher.* Discontinued.

27601 Schercemol GMIS
66085-00-5 266-124-4
Glyceryl isostearate
Emulsifier and emollient for creams and lotions. *Scher.*

27602 Schercemol GMS
Glycerol stearate
Primary emulsifier for creams and lotions. *Scher.* Discontinued.

27603 Schercemol ICS
25339-09-7 246-868-6
Isocetyl stearate
Nongreasy emollient used in creams; imparts elegant feel to makeup, lotions, bath preparations; remains liquid even at low temperatures. *Scher.*

27604 Schercemol IDO
59231-34-4 261-673-6
Isodecyl oleate
Emollient, lubricant, penetrant with pigment dispersing properties; for makeup and makeup removers. *Scher.*

27605 Schercemol ISE
84605-09-4
Isostearyl erucate
Lubricating emollient for skin and bath preparations. *Scher.*

27606 Schercemol MEL-3
84605-13-0 283-390-7
Myreth-3 laurate
Nonoily, rich, penetrating emollient for cosmetic and personal care products,; dispersibility and spreadability in bath oils; coupler in hydro-alcoholic systems; emulsifier and solubilizer in lotions. *Scher.*

27607 Schercemol MEM-3
59686-68-9
Myreth-3 myristate
Nonoily, rich, penetrating emollient for cosmetic and personal care products; dispersibility and spreadability in bath oils; coupler in hydro-alcoholic systems; emulsifier and solubilizer in lotions. *Scher.*

27608 Schercemol MEP-3
84605-14-1 293-391-2
Myreth-3 palmitate
Nonoily, rich, penetrating emollient for cosmetic and personal care products; dispersibility and spreadability in bath oils; coupler in hydro-alcoholic systems; emulsifier and solubilizer in lotions. *Scher.*

27609 Schercemol MM
3234-85-3 221-787-9
Myristyl myristate
Soft, waxy emollient that melts near body temperature; viscosity builder; imparts substantivity to personal care products; ease of combing of hair preparations; velvety feel on skin. *Scher.*

27610 Schercemol MP
6221-95-0 226-300-9
Myristyl propionate
Emollient for antiperspirants, body oils, creams, and lotions. *Scher.*

27611 Schercemol MS
17661-50-6 241-640-2
Myristyl stearate
Waxy emollient for creams and lotions. *Scher.* Discontinued.

27612 Schercemol NGDC
27841-06-1 248-688-3
Neopentyl glycol dicaprate
Solvency properties for use in make-up removers. *Scher.*

27613 Schercemol NGDL
10525-29-0 234-081-0
Neopentyl glycol dilaurate
Emollient and skin conditioner for creams and lotions. *Scher.*

27614 Schercemol NGDO
28510-23-8 249-060-1
Neopentyl glycol dioctanoate
Low freeze point emollient; solvent for makeup remover. *Scher.*

27615 Schercemol OHS
29710-25-6 249-793-7
Octyl hydroxystearate
Emollient producing slip, lubricity, and tackiness reduction in skin preparations. *Scher.*

27616 Schercemol OLO
3687-45-4 222-980-4
Oleyl oleate
Nonoily emollient for cosmetic formulations contributing luster, softness, and high degree of lubricity in skin and hair preparations; cosolv. and solubilizer in perfumes; lubricant for metal working and wire drawing. *Scher.*

27617 Schercemol OP
29806-73-3 249-862-1
Octyl palmitate
Nonoily emollient ester for cosmetic and personal care products giving sheen without greasiness; anticlogging and suspending agent in antiperspirants; soft velvety feel in skin creams, lotions, and aftershaves. *Scher.*

27618 Schercemol OPG
59587-44-9 261-819-9
Octyl pelargonate
Dry, nonoily rich penetrating emollient for cosmetic and personal care products; anticlogging agent in antiperspirants; soft, luxurious feel in creams and aftershaves. *Scher.*

27619 Schercemol PGDP
41395-83-9 255-350-9
Propylene glycol dipelargonate
Emollient offering low f.p.; cosolv. for perfumed bath oils, creams, and lotions. *Scher.*

27620 Schercemol PGML
27194-74-7 205-542-3
Propylene glycol laurate
Emollient and solvent; stable base for cosmetics; emulsion stabilizer; solubilizes and couples ingredients such as perfumes, coloring and flavoring agents, sunscreen compounds into natural fatty vegetable or mineral oils; solvent for organic pesticides; plasticizer and stabilizer in copolymers made from PVAc and PVC; defoaming agent in vinyl PVAc emulsions. *Scher.*

27621 Schercemol PGMS
1323-39-3 215-354-3
Propylene glycol stearate
Emulsifier for creams and lotions. *Scher.*

27622 Schercemol SE
289-256-4
Stearyl erucate
Emollient wax with the look and feel of real cocoa butter. *Scher.*

27623 Schercemol TISC
113431-54-2
Triisostearyl citrate
High viscosity ester imparting gloss to lipsticks and lip gloss preparations. *Scher.*

27624 Schercemol TIST
103213-22-5
Triisostearyl trilinoleate
Emollient; superior gloss and moisturizing characteristics; emolliency, shine, viscosity, and good binding characteristics. *Scher.*

27625 Schercemol TT
Triisopropyl trimerate
Binder for pigmented products, imparts gloss and sheen in makeup and hair preparations. *Scher.* Discontinued.

27626 Scherco Finish AL
Resin dispersion; lubricant for sewing, cutting, napping, and softening all textiles. *Scher.* Discontinued.

27627 Scherco Softener 1
Quaternary; softener and finishing agent for orlon and acrilan. *Scher.* Discontinued.

27628 Scherco Softener 2
Quaternary; softening agent for acrylics and synthetics. *Scher.* Discontinued.

27629 Schercoat OE-44
Fatty oil, edible grade; coating for liquor and food glass containers. *Scher.*

27630 Schercoat OE-44K
High purity fatty oil; protective coating for glass food containers. *Scher.*

27631 Schercoat P-110
Modified polyethylene emulsion; protective coatings for glass bottles. *Scher.*

27632 Schercoat PC-550
Substantive poly emulsion; lubricant for glass containers. *Scher.* Discontinued.

27633 Schercoat S-220
Modified vinyl resin ionomer emulsion; protective coatings for glass containers. *Scher.* Discontinued.

27634 Schercoat S-330
Stabilized vinyl resin ionomer emulsion; roller coating of glass bottles to impart scuff resistance. *Scher.* Discontinued.

27635 Schercodine B
60270-33-9 262-134-8
Behenamidopropyl dimethylamine
Emulsifier with conditioning properties for hair and skin preparations. *Scher.*

27636 Schercodine C
68140-01-2 268-771-8
Cocamidopropyl dimethylamine
Good foaming surfactant for hair and bath preparations; emulsifier, intermediate for betaine amphoterics. *Scher.*

27637 Schercodine I
67799-04-6 267-101-1
Isostearamidopropyl dimethylamine
Versatile liquid oil-water emulsifier for creams and lotions; lubricant for hair rinses and conditioners. *Scher.*

27638 Schercodine L
3179-80-4 221-661-3
Lauramidopropyl dimethylamine
Emulsifier; surfactant, intermediate for betaine amphoterics. *Scher.*

27639 Schercodine M
45267-19-4 256-214-1
Myristamidopropyl dimethylamine
Oil-water emulsifier, conditioner, viscosity builder. *Scher.*

27640 Schercodine O
109-28-4 203-661-5
Oleamidopropyl dimethylamine
Oil-water emulsifier; emollient conditioner for hair and skin preparations. *Scher.*

27641 Schercodine S
7651-02-7 231-609-1
Stearamidopropyl dimethylamine
Softener, emulsifier, and conditioner in hair and skin preparations. *Scher.*

27642 Schercodine T
68650-79-3 272-047-7
Tallamidopropyl dimethylamine
Conditioner for cationic emulsions; substantivity and thickening properties. *Scher.*

27643 Schercolene SB
Detergent; heavy duty textile scouring agent. *Scher.* Discontinued.

27644 Schercolube 707
Fatty ester ethoxylate; softener for knit goods and lubricant for nylon separator threads in sweater bodies. *Scher.* Discontinued.

27645 Schercomid 304
Modified coco amide; dry-cleaning detergent. *Scher.* Discontinued.

27646 Schercomid 1214
120-40-1, 7545-23-5 204-393-1
Lauramide DEA and diethanolamine; foam booster/stabilizer for detergent compositions; good detergency by itself and works synergistically with other surfactants; thickening agent and viscosity builder; used in personal care items and cleaners for hard surfaces. *Scher.*

27647 Schercomid AC-S
Mixed fatty amide; cold water textile detergent and wool fulling agent. *Scher.*

27648 Schercomid AME
142-26-7 205-530-8
Acetamide MEA
Solubilizer, humectant, skin and hair conditioner, intermediate, coupling agent, pigment dispersant, solubilizer. *Scher.*

27649 Schercomid CDA
68603-42-9 271-657-0
Cocamide DEA and diethanolamine; foam stabilizer, soil suspender, lime soap dispersant, and detergency booster for industrial and household cleaners. *Scher.*

27650 Schercomid EAC
Modified coco amide; wool scouring and fulling agent effective at low temperature. *Scher.*

27651 Schercomid HT-60
68783-22-2
PEG-50 hydrogenated tallow amide
Thickener, detergent, emulsifier, dispersant with foam characteristic. *Scher.*

27652 Schercomid LME
5422-34-4 226-546-1
Lactamide MEA
Humectant, skin and hair conditioner, coupling agent, emollient. *Scher.*

27653 Schercomid ODA
93-83-4 202-281-7
$C_{22}H_{43}NO$
oleic diethanolamide
Oleamide DEA and diethanolamine; water-oil emulsifier, pigment dispersant, conditioner, corrosion inhibitor, and viscosity builder; emulsifier for aromatic and aliphatic hydrocarbon solvent; used in gel-type pine cleaners, shampoo formulations, hair conditioning agent. *Scher.*

27654 Schercomid OME
111-58-0 203-884-8
Oleamide MEA
Water-oil emulsifier, conditioner, and thickener. *Scher.*

27655 Schercomid OMI
111-05-7 203-828-2
Oleamide MIPA
Thickener, foam stabilizer for shampoos; hair conditioning agent; emulsifier for mineral oil, IPP, IPM, butyl stearate, creams and lotions; imparts slip, lubrication, some emolliency, and softening effects upon the skin. *Scher.*

27656 Schercomid SAP
185123-36-8
Apricotamide DEA
Thickener, foam stabilizer for natural and herbal shampoos. *Scher.*

27657 Schercomid SCE
68603-42-9 271-657-0
Cocamide DEA (1:1)
Detergent, viscosity builder and foam stabilizer for cosmetic formulations bubble baths, liquid dish wash detergents, and rug shampoos. *Scher.*

27658 Schercomid SLE
56863-02-6 260-410-2
Linoleamide DEA (1:1)
Solubilizer, thickener, w/o emulsifier, conditioner, and emollient for personal care products; emulsion stabilizer for oil-water emulsions. *Scher.*

27659 Schercomid SL-Extra
120-40-1 204-393-1
Lauramide DEA (1:1)
Thickener, foam booster/stabilizer for hair shampoos, soaps, synthetic detergent for mulations; bubble bath applications; industrial applications including manual dishwashing formulations, liquid heavy-duty laundry detergents all purpose cleaning products; emulsifier for aromatic and aliphatic hydrocarbons and oils in oil-water emulsions. *Scher.*

27660 Schercomid SLS
68425-47-8 270-355-6
Soyamide DEA (1:1)
Conditioner and emollient for personal care products; emulsifier for water-oil systems and hydrocarbons; dispersant for pigments and mineral clays; viscosity builder; emulsion stabilizer. *Scher.*

27661 Schercomid SO-A
93-83-4 202-281-7
Oleamide DEA (1:1)
Water-oil emulsifier, lubricant, conditioner. *Scher.*

27662 Schercomid SO-T
68155-20-4 268-949-5
Tallamide DEA (1:1)
Water-oil emulsifier. *Scher.*

27663 Schercomid SWG
124046-39-5
Wheatgermamide DEA
Thickener, foam stabilizer for natural and herbal shampoos. *Scher.*

27664 Schercomid TO-2
68155-20-4 268-949-5
Tallamide DEA and diethanolamine; water-oil emulsifier, viscosity builder, pigment and mineral clay dispersant, corrosion inhibitor; emulsifier for aromatic and aliphatic hydrocarbon solvents; used in shampoos where it generates a creamy, luxurious foam, stabilizes foam when used with surfactants and detergents; hair conditioning agent. *Scher.*

27665 Schercopol CMS-Na
68784-08-7 272-219-1
Disodium cocamido MEA-sulfosuccinate
Nonirritating surfactant, foam-stabilizer, solubilizer, softener for personal care products; home and industrial detergent cleaning formulations; anti-irritant for other surfactants; biodegradable. *Scher.* Discontinued.

27666 Schercopol DOS-70
577-11-7 3460 209-406-4
Dioctyl sodium sulfosuccinate
Wetting agent, surface tension depressant. *Scher.*

27667 Schercopol DS-120
Modified ethoxylated alkylamine; dispersant for glass fibers in aqueous media; for textile industry. *Scher.*

27668 Schercopol LPS
39354-45-5 255-062-3
Disodium laureth sulfosuccinate
Mild high foaming surfactant; viscosity enhancer. *Scher.*

27669 Schercopol OMS-Na
68479-64-1 270-864-3
Disodium oleamido MEA-sulfosuccinate
Solubilizer; nonirritating surfactant imparting soft, emollient feel on skin and conditioning effect on hair; foamer in tolletries, hand dishwashing detergents, and personal care products. *Scher.* Discontinued.

27670 Schercopon 2WD
Ethoxylated sulfosuccinate
Detergent and wetting agent for high electrolyte cleansers, dry-cleaning detergent. *Scher.*

27671 Schercoquat ALA
90283-04-8
Dilauryl acetyl dimonium chloride
Water-oil and oil-water emulsifier; conditioner for hair and skin products. *Scher.*

27672 Schercoquat APAS
115340-78-8
Apricotamidopropyl ethyldimonium ethosulfate
Natural, mild conditioner; imparts good slip and shine. *Scher.*

27673 Schercoquat BAS
68797-65-9 258-377-8
Behenamidopropyl ethyldimonium ethosulfate
Conditioner for dry and over-processed hair. *Scher.*

27674 Schercoquat CAS
113492-03-8
Cocamidopropyl ethyl dimonium ethosulfate
Water-oil and oli-water emulsifier, conditioner, antistat for hair care products. *Scher.* Discontinued.

27675 Schercoquat COAS
Quaternary based on canola oil and ethyl sulfate; natural, mild hair and skin conditioner; imparts good slip and velvety feel. *Scher.* Discontinued.

27676 Schercoquat DAS
111905-55-6
Quaternium-61; conditioner for personal care products. *Scher.*

27677 Schercoquat FOAS
113492-04-9
Saffloweramidopropyl ethyldimonium ethosulfate
Quaternary effective in hair conditioners; good slip, shine, and compatibility. *Scher.*

27678 Schercoquat IALA
134112-42-8
Isostearamidopropyl laurylacetodimonium chloride
Water-oil and oli-water emulsifier; conditioner for hair and skin care products. *Scher.*

27679 Schercoquat IAS
67633-63-0 266-778-0
Isostearamidopropyl ethyldimonium ethosulfate
Conditioner for personal care products. *Scher.*

27680 Schercoquat IB
Isostearamidopropyl alkonium chloride
Liquid quaternary, possessing some bactericidal activity; used in conditioners, hair rinses and skin lotions. *Scher.* Discontinued.

27681 Schercoquat IEP
84605-15-2
Isostearamidopropyl epoxypropyl dimonium chloride
Conditioning quaternary offering good water-solubility and good compatibility with many anionic surfactants. *Scher.* Discontinued.

27682 Schercoquat IIB
Isostearyl benzyl imidonium chloride
Quaternary possessing bactericidal activity. *Scher.* Discontinued.

27683 Schercoquat IIS
67633-57-2 266-778-0
Isostearyl ethylimidonium ethosulfate
Conditioner for personal care products. *Scher.*

27684 Schercoquat ROAS
94552-41-7 305-488-1
Rapeseedamidopropyl ethyldimonium ethosulfate
Conditioner for personal care products. *Scher.*

27685 Schercoquat ROEP
112324-11-5
Rapeseedamidopropyl epoxypropyl dimonium chloride
Condioner for conditioning shampoos and hair sprays. *Scher.* Discontinued.

27686 Schercoquat SAS
67846-16-6 267-360-0
Stearamidopropyl ethyl dimonium ethosulfate
Conditioner for hair rinses. *Scher.*

27687 Schercoquat SOAB
Soyamidopropyl benzyldimonium chloride
Conditioning agent used in hair preparations. *Scher.* Discontinued.

27688 Schercoquat SOAS
90529-57-0 291-990-5
Soyamidopropyl ethyldimonium ethosulfate
Conditioner for personal care products. *Scher.*

27689 Schercoquat WOAS
115340-80-2
Wheatgermamidopropyl ethyldimonium ethosulfate
Surfactant, mild conditioner imparting body, bounce, antistatic properties, shine to hair. *Scher.*

27690 Schercosol DS
Chlorinated solvent; dry side rapid stain remover. *Scher.* Discontinued.

27691 Schercosol NL
Modified coco amide; wet side spotter and fiber lubricant. *Scher.* Discontinued.

27692 Schercosol P
Sulfated amide; protein stain remover. *Scher.* Discontinued.

27693 Schercosol T
Acid-stable detergent; tannin stain remover. *Scher.* Discontinued.

27694 Schercotaine APAB
133934-08-4

Apricotamidopropyl betaine
Mild detergent, conditioner, emollient, viscosity enhancer. *Scher.*

27695 Schercotaine CAB
61789-40-0 263-058-8
Cocamidopropyl betaine
Detergent, wetting agent, foamer, cloud point depressant, antistat and softener in personal care products. *Scher.*

27696 Schercotaine CAB-A
Cocamidopropyl betaine, ammonium chloride; mild surfactant with higher foam than the sodium counterpart, decreased defatting properties. *Scher.* Discontinued.

27697 Schercotaine IAB
6179-44-8 228-227-2
Isostearamidopropyl betaine
Conditioner and detergent for shampoos and emollient body treatments; viscosity control agent; textile softener. *Scher.* Discontinued.

27698 Schercotaine MAB
59272-84-3 261-684-6
Myristamidopropyl betaine
Detergent, thickener, wetting agent with antistatic properties for cosmetic and toiletry preparations. *Scher.*

27699 Schercotaine PAB
32954-43-1 251-306-8
Palmitamidopropyl betaine
Thickening agent, good hair and skin conditioner for lotions and cream rinses. *Scher.*

27700 Schercotaine SCAB
68139-30-0 268-761-3
Cocamidopropyl hydroxysultaine
Detergent, wetting agent and foamer, cloud point depressant used in personal care products. *Scher.*

27701 Schercotaine UAB
133798-12-6
Undecylenamidopropyl betaine
Surfactant with germicidal/bactericidal activity; for shampoos. *Scher.*

27702 Schercotaine WOAB
133934-09-5
Wheat germamidopropyl betaine; detergent with emulsification properties, conditioner, surfactant with vitamin E; imparts good body to hair. *Scher.*

27703 Schercotarder
Fatty amide
Low temperature wool scouring and fulling agent; post-scouring agent for dyed or printed goods. *Scher.* Discontinued.

27704 Schercoterge 140
Ethoxylated amide
Detergent, wetting and textile scouring agent, emulsifier, wool fulling; dyeing assistant; post scouring agent; dye bath stabilizer. *Scher.*

27705 Schercoteric CY-2
7702-01-4 231-721-0
Disodium capryloamphodiacetate
Low foaming surfactant for household and industrial cleaning products. *Scher.* Discontinued.

27706 Schercoteric I-AA
68630-96-6 271-929-9
Sodium isostearoamphopropionate
Surfactant for cosmetic and industrial cleaners. *Scher.*

27707 Schercoteric MS
68334-21-4 269-819-0
Sodium cocoamphoacetate
Foamer, mild detergent, conditioner used in personal care products and industrial cleaners. *Scher.*

27708 Schercoteric MS-2
68650-39-5 272-043-5
Disodium cocoamphodiacetate
Mild detergent used in personal care products and industrial cleaners. *Scher.*

27709 Schercoteric MS-EP
68604-73-9 271-705-0
Sodium cocoamphohydroxypropylsulfonate
Surfactant for personal care products; low skin irritation, low cloud point. *Scher.* Discontinued.

27710 Schercoteric O-AA
67892-37-9 267-569-7
Sodium oleoamphopropionate
Surfactant for drycleaning industry, other industrial cleaners. *Scher.*

27711 Schercowet DOS-70
577-11-7 3460 209-406-4
Sodium dioctyl sulfosuccinate
Emulsifier for emulsion polymerization; wetting and rewetting agent for textile wet processing. *Scher.*

27712 Schercozoline C
61791-38-6 263-170-7
Cocoyl hydroxyethyl imidazoline
Antistat, dispersant, wetting agent, emulsifier, microbicide, detergent, intermediate for quaternary ammonium compounds, primer paints, emulsion cleaning, cleaners, polishes, surface treatment, textile and leather processing, agriculture and cosmetic. *Scher.*

27713 Schercozoline I
68966-38-1 273-429-6
Isostearyl hydroxyethyl imidazoline
Surfactant, softener, antistat, dye assistant for textiles, paper, cutting oils, metal lubricants, polishes, cosmetics, agriculture, corrosion inhibitors, building materials. *Scher.*

27714 Schercozoline L
136-99-2 205-271-0
Lauryl hydroxyethyl imidazoline
Surfactant, softener, dye assistant, antistat for textiles, paper, cutting oils, metal lubricants, polishes, cosmetics, corrosion inhibitors, building materials; intermediate for quaternaries. *Scher.*

27715 Schercozoline O
95-38-5 428 202-414-9
Oleyl hydroxyethyl imidazoline
Surfactant, softener, dye assistant, antistat, water-oil emulsifier, corrosion inhibitor, intermediate for quatenary ammonium compounds, textiles, paper, cutting oils, metal lubricants, polishes, cosmetics, agriculture, building materials. *Scher.*

27716 Schercozoline S
95-19-2 202-397-8
Stearyl imidazoline
Surfactant, softener, and antistatic agent. *Scher.* Discontinued.

27717 Scheroba Oil
Isostearyl-erucyl erucate
Similar properties to jojoba oil, but has advantages of low price, product consistency and availability. *Scher.* Discontinued.

27718 Scherpol LSB
Ethoxylated alcohol
Nonfoaming, jet-dyeing assistant for polyester; minimizes subsequent smoke formation. *Scher.* Discontinued.

27719 Schersoftoil P
Ester of natural oils; winding lubricant applied in a package dye machine. *Scher.* Discontinued.

27720 Schiff's reagents
a) Consists of a solution of rosaniline hydrochloride, decolorized by sulfur dioxide, and is used to test for aldehydes. b) Furfuraldehyde and hydrochloric acid, employed for testing for urea. c) Concentrated sulfuric acid, followed by ammonia; used as a test for cholesterol.

27721 schlempe
Beet sugar waste. It is the thick brown liquor remaining after the extraction of all possible sugar. It is also called Vinasse.

27722 Schlichte®
Sizes for cellulose fibers and synthetic fibers, fiber blends and filaments. *BASF AG.*

27723 Schlippe's salt
13776-84-6 8840 237-414-8
$Na_3S_4Sb{\cdot}9H_2O$
Sodium thioantimonate(V)
sodium sulfantimonate. Soluble in H_2O (300 mg/ml), insoluble in EtOH. Source of antimony sulfide.

27724 Schneiderite
An explosive. It contains 88% ammonium nitrate, 11% dinitronaphthalene, and 1% resin.

27725 schoenite
$K_2SO_4{\cdot}MgSO_4{\cdot}6H_2O$
Potassium-magnesium sulfate
Schonite.

27726 Scholikopf's acids
82-75-7 6492 201-437-1
$C_{10}H_9NO_3S$
1-Naphthol-4,8-disulfonic acid
1-naphthylamine-8-sulfonic acid; 8-amino-1-naphthalenesulfonic acid; Peri acid. Used in manufacture of dyestuffs. Soluble in H_2O (0.2 - 4 mg/ml).

27727 Schonberg's alloy
A die-casting alloy containing 87% zinc, 10% tin, and 3% copper.

27728 schorl
shorle
A black tourmaline.

27729 **schorl rock**
An aggregate of black tourmaline and quartz.
27730 **Schou oil**
Paalsguard oil
An emulsifier made from soyabean oil.
27731 **Schradan**
152-16-9 8540 205-801-0
$C_8H_{24}N_4O_3P_2$
octamethyldiphosphoramide
OMPA; Pestox III; Sytam. Insecticide. mp = 14-20°; bp_2 = 154°; d_4^{25}= 1.09; soluble in H_2O, polar organic solvents; LD_{50} (rat orl) = 9.1 mg/kg.
27732 **schraufite**
A fossil resin found in Carpathian sandstone.
27733 **schreibersite**
dyslytite
An iron nickel phosphide found in meteorites. A chromium sulfide has also been called Schreibersite.
27734 **Schultze's reagents**
A) Phospho-antimonic acid, made from sodium phosphate and antimony pentachloride, an alkaloidal reagent. b) Consists of 25 parts dry zinc chloride, 8 parts potassium iodide, 8 1/2 parts water, and iodine. It gives a blue colo.
27735 **Schultze's stain**
A microscopic stain. It consists of equal parts of a 2% solution of β-naphthol sodium and a 2% solution of dimethyl-p-phenylenediamine hydrochloride. The solutions are mixed and filtered.
27736 **schungite**
A mineral that is carbon in an amorphous form.
27737 **Schutzenberger's salt**
$NaHSO_2$
sodium hydrosulfite
Distinct from dithionite.
27738 **Schweitzer's reagent**
20427-59-2 2709 243-815-9
CuH_2O_2
copper(II) hydroxide
Kocide; copper hydroxide; Comac parasol; copper dihydroxide; copper hydrate; Criscobre; Cudrox; Cuidrox; Cupravit blue; Kocide 101. A solution of copper hydroxide, in strong ammonia; a solvent for cellulose.
27739 **schwelkohle**
A brown coal of Germany. It is light brown in color.
27740 **Scintillase**
9001-73-4 7148 232-627-2
Papain. *Rhône-Poulenc UK.*
27741 **Scintran**
Reagents used in scintillation counting. *BDH Chemicals Ltd.*
27742 **Sclair®**
Polyethylene resins. *DuPont UK.*
27743 **sclerolac**
A suggested name for hard lac resin.
27744 **Sclomo**
Sulfur chlorinated base; extreme pressure agent for use in threading and tapping operations, cutting and grinding oils. *Ferro/Keil.*
27745 **Scolaban**
1055-55-6 1510 213-890-2
Bunamidine hydrochloride
The Wellcome Foundation Ltd. Name unverified.
27746 **Scopacron**
A proprietary trade name for thermosetting acrylic resins modifiable by means of epoxy resins. *Stirene Co-Polymers Ltd.* Unverified.
27747 **Scopacron 50, 75 and 80**
A proprietary trade name for a thermosetting acrylic resin capable of crosslinking with amino and epoxy compounds; primarily intended for use with melamine formaldehyde resin for motor car top coats. *Stirene Co-Polymers Ltd.* Unverified.
27748 **Scopacryl**
A proprietary trade name for thermoplastic acrylic resin solutions used for wall paints and road marking applications. *Stirene Co-Polymers Ltd.* Unverified.
27749 **Scopasol 550**
A proprietary trade name for a waterdilutable thermosetting acrylic resin; used for high performance white gloss coatings to be applied by electrophoresis techniques. *Stirene Co-Polymers Ltd.* Unverified.
27750 **Scopol 58M, 58SP**
A proprietary trade name for a vinyl toluene modified alkyd resin. *Stirene Co-Polymers Ltd.* Unverified.
27751 **Scopol 85X**
A proprietary trade name for a styrene modified alkyd resin; used for quick drying coatings with exceptional adhesion properties. *Stirene Co-Polymers Ltd.* Unverified.
27752 **Scopolux 221SP**
A proprietary trade name for a medium oil alkyd based on linseed oil. *Stirene Co-Polymers Ltd.* Unverified.
27753 **Scorchex**
1309-48-4 5713 215-171-9
Magnesium oxide products; used in rubber goods. *Croxton & Garry Ltd.* Name unverified.
27754 **Scorchguard O**
1309-48-4 5713 215-171-9
Magnesium oxide
Bayer AG.
27755 **Scorchguard-bound**
1309-48-4 5713 215-171-9
Magnesium oxide
Bayer plc.
27756 **Scotch cement**
A cement prepared from feebly hydraulic limes, by the addition of 5% plaster of Paris, and grinding.
27757 **Scotch foundry pig**
A pig iron made for foundry purposes from Scotch clay-band or black-band ores. It usually contains from 0.7-1% phosphorus, and 2.5% silicon.
27758 **scotch soda**
497-19-8 8739 207-838-8
Na_2CO_3
Impure sodium carbonate.
27759 **Scotch topaz**
Golden topaz, a yellow variety of quartz.
27760 **Scotchkote® 213, 214**
25928-94-3
Fusion-bonded one-part epoxy coatings; spray grade thermosetting coatings providing maximum corrosion protection for wire fabric and reinforcing steel. *3M.*
27761 **Scotchlite**
Glass bubbles; engineered fillers for industry. *3M.*
27762 **Scotphos**
Fertilizers containing phosphate. *Scottish Agricultural Industries plc.*
27763 **Scour 1161**
Low foaming multipurpose surfactant. *Catawba-Charlab.*
27764 **Scour KSV, KSV Special**
Emulsified solvent containing stain remover and wetting and scouring agent. *Catawba-Charlab.*
27765 **scouring slag**
A slag produced in making spiegel. It is black in color and contains up to 8% of oxide of iron.
27766 **Scram**
Dog and cat repellent. *Monsanto (Solaris).* Name unverified.
27767 **scrap rubber**
Formed by the drying of the latex on the bark at the tapping cut. It is variable in quality and color.
27768 **Scratch-Guard**
Abrasion resistant coating; protective coating for various films i.e., polyester, polycarbonate, etc. *Custom Coating & Laminating Corp.* Name unverified.
27769 **Screen**
Seed protectant. *Monsanto Co.* Name unverified.
27770 **Screen Star Photo Emulsion**
Direct screen making emulsion for professional screen printing; for textiles, electronic circuits and paper stock. *Bond Adhesives Co.* Name unverified.
27771 **Screte**
Sulfur concrete. *Monsanto (Solaris).* Name unverified.
27772 **screw bronze**
An alloy of 93.5% copper, 5% zinc, 1% tin, and 0.5% lead.
27773 **Scripset**
Styrene-maleic anhydride copolymers; paper coating and specialty coating resins. *Akzo.*
27774 **Scripset 520**
9011-13-6
Styrene/maleic anhydride copolymer; emulsifier, binder, sizing agent, viscosity modifier, stabilizer; starch modifier; pigment dispersant, protective colloid, sizing, coating, water-paint calsomines, adhesives, printing, preparation of emulsifier paints. *Monsanto Co.*
27775 **Scripset 720**
9011-13-6
Styrene/maleic anhydride copolymer; emulsifier, dispersant, thickener for high solids systems, latex manufacturing. *Monsanto Co.*

27776 Scuranate
584-84-9 9668 209-544-5
Toluene diisocyanate
Rhône-Poulenc UK.

27777 Scurane V
Polyurethane varnishes. *Rhône-Poulenc NV/CdF Chimie AZF.* Name unverified.

27778 Scuttle
Sulfonated cod liver oil; an animal repellent. *Fine Agrochemicals Ltd.*

27779 Scythe
1910-42-5 7165 217-615-7
Soluble concentrate containing 200 g/l paraquat; a pre-emergence bipyridilium herbicide to control weeds in field crops and ornamentals. *Cyanamid of Great Britain Ltd.*

27780 SD-1, -2
Super dispersible rheological additive; for paints. *Rheox Inc.*

27781 SD-376
9003-07-0 7741
Med. impact polypropylene copolymer; for end-use applications requiring excellent stiffness/impact and improved injection moldability; for consumer products, toys, appliances, crates and totes. *Himont.*

27782 SE Wax
A montan wax ester containing an emulsifier; used in the preparation of nonionic, self-polishing emulsions for floors. *Bush Beach Ltd.* Name unverified.

27783 SE-458
A proprietary silicone rubber compound used for bonding to unprimed surfaces during the curing process. *GE Silicones.* Name unverified.

27784 SeaBuffer
A dry, granular mixture of salts for maintenance of normal pH in recirculating seawater systems; marine aquarium water supplement. *Aquarium Systems Inc.*

27785 SeaCure
A solution of copper sulfate and citric acid for treatment of protozoan parasites of marine fishes; used for treatment of marine aquariums. *Aquarium Systems Inc.*

27786 Sea-Gard® Formula FP-91
7758-29-4 8846 231-838-7
Sodium tripolyphosphate
Food additive for meat, poultry, and seafood industries. *Rhône-Poulenc Food Ingreds.*

27787 SeaGarden
Soluble nutrients for algae, particularly in marine aquariums; aquarium water supplement. *Aquarium Systems Inc.*

27788 Seair
Nontoxic, concentrated solution of neutralized resin; an admixture for concrete to increase workability, reduce bleeding of the mixing water, provide a more uniform concrete mix and reduce frost damage and scaling. *Secure Inc.* Discontinued.

27789 Seal and Heal
Fungicide and pruning paint. *May & Baker Ltd.* Name unverified.

27790 Sealac
Sealers. *The Scottish Adhesives Co Ltd.*

27791 sealite
A liquid containing glucose, corn starch, glycerol, calcium chloride, and glue; used to prevent evaporation from oil storage tanks.

27792 Sealum
Mastic rubber tape; tape sealant for metal buildings. *Chemseco.* Name unverified.

27793 SeaTest
A series of colorimetric tests for analyzing seawater; used for aquarium water testing and natural seawater testing. *Aquarium Systems Inc.*

27794 sea-water bronze
Sheathing bronze. An alloy of 32.5% nickel, 45% copper, 5.5% zinc, 16% tin, and 1% bismuth. It resists sea water.

27795 sebacic acid
111-20-6 8558 203-845-5
$C_{10}H_{18}O_4$
decanedioic acid
1,8-octanedicarboxylic acid; 1,10-decanedioic acid; n-decanedioic acid. Stabilizer; raw material in manufacture of alkyd resins, maleic and other polyesters, plasticizers, polyester rubbers, synthetic polyamide fibers. mp = 134°; bp_{10} = 232°; d_4^{20}= 1.207; n_D^{134}= 1.422; soluble in H_2O (1 mg/ml), soluble in polar organic solvents. *Janssen Chimica; Penta Mfg.; Union Camp.*

27796 Sebacil®
14816-18-3 7523 238-887-3
$C_{12}H_{15}N_2O_3PS$
4-Ethoxy-7-phenyl-3,5-dioxa-6-aza-4-phosphaoct-6-ene-8-nitrile 4-sulfide phenylglyoxylonitrile oxime O,O-diethyl phosphorothioate; O,O-diethyl O-(alpha-cyanobenzylideneamino)phosphorothioate; Phoxim; α-[[(diethoxyphosphinothioyl)oxy]imino]benzeneacetonitrile; Bay 5621; Bay 77488;Baythion; Sebacil; Volation; Volaton; Benzoyl cyanideO-(diethoxyphosphinothioyl)oxime; Diethoxyphosphinothioyl)oxy)imino)benzeneacetonitrile; Valexone. For control of all ectoparasites, especially mange mites of domestic animals; veterinary medicine. $bp_{0.01}$ = 102°; d_4^{20}= 1.176; n_D^{20}= 1.5405; soluble in organic solvents; LD_{50} (mus orl) >2000 mg/kg. *Bayer AG.*

27797 Sebase
Ethoxylated lanolin plus fatty alcohols and hydrocarbons; base, emollient, lubricant for oil-water emulsions; viscosity stabilizer for cosmetics. *Westbrook Lanolin.*

27798 Sebastine
A dynamite explosive.

27799 sebkanite
A crude potassium chloride obtained by the evaporation and crystallization of the water of the salt lake in Tunis.

27800 Sebond
Modified acrylic emulsion; used to bond new concrete to either new or old concrete. For patching and resurfacing precast architectural panels, industrial concrete floors, highway and bridge deck repair. *Secure Inc.* Discontinued.

27801 Sebrite
A clear, transparent, penetrating liquid sealer; for protecting and beautifying mechanically textured concrete, exposed aggregate and stone surfaces. *Secure Inc.* Discontinued.

27802 Secol
Fibre lubricant. *Stephenson Thompson Textile Chemicals.*

27803 Secolan S-1, BA-1, BA-1G
9007-34-5 2543 232-697-4
Soluble animal collagen. *RITA.* Name unverified.

27804 Secolat
Disodium alkyl sulfosuccinamate as a clear yellow liquid; anionic surfactant used in latex foams. *KWR Chemicals Ltd.* Name unverified.

27805 Secomine TA 02
Tallow amine ethoxylate
Lubricating, wetting, antistatic agent, emulsifier, anticorrosive agent for detergents, surface treatments. *Stepan Europe.*

27806 Secomix® E40
Surfactant blend; detergent for industrial/household cleaners. *Stepan Europe.*

27807 Secosol AL 959
Disodium monolauryl sulfosuccinate as a white paste; anionic surfactant for shampoos; foam baths; creams. *KWR Chemicals Ltd.* Name unverified.

27808 Secosol AL/MG 50
Anionic surfactant in which the anion is monolauryl sulfosuccinate and the cations are sodium and magnesium; supplied as a white paste; used for shampoos; foam baths; creams. *KWR Chemicals Ltd.* Name unverified.

27809 Secosol ALL/40
Disodium monolauryl ether sulfosuccinate as a water-white liquid; anionic surfactant; used in foam baths, shampoos, and liquid soaps. *KWR Chemicals Ltd.* Name unverified.

27810 Secosol DOS/70
577-11-7 3460 209-406-4
Sodium dioctyl sulfosuccinate in liquid form; wetting and emulsifying agent. *KWR Chemicals Ltd.* Name unverified.

27811 Secosol EA/40
Disodium monoalkyl ethanolamide sulfosuccinate as a clear yellow liquid; anionic surfactant for special shampoos; foam baths; liquid soaps. *KWR Chemicals Ltd.* Name unverified.

27812 Secosol® AL 959
Sodium lauryl sulfosuccinate
Mild foamer for shampoos, bubble baths, liquid soaps, shower gels, bath salts. *Stepan Europe.*

27813 Secosol® ALL40
Sodium laureth sulfosuccinate
Mild foamer for shampoos, bubble baths, liquid soaps, shower gels, bath salts; emulsifier for emulsion polymerization. *Stepan Europe.*

27814 Secosol® DOS 70
577-11-7 3460 209-406-4
Sodium dioctyl sulfosuccinate
Emulsifier, dispersant and wetting/rewetting agent for household/industrial cleaners, emulsion polymerization, paints, inks, oilfield production; textile additive. *Stepan Europe.*

27815 Secosov
Emulsifiers; for mineral oils; solvents; dispersant for phytosanitaires; cosmetic creams and milks; insecticides. *KWR Chemicals Ltd.* Name unverified.

27816 Secoster® A
Fatty acid ethoxylate
Dispersant, solvent and oil emulsifier for household/industrial cleaners, oilfield production. *Stepan Europe.*

27817 Secoster® DMS
627-83-8 211-014-3
Glycol distearate
Emollient, pearlescent, emulsifier, opacifier for creams, cleansing milks, shampoos. *Stepan Europe.*

27818 Secoster® DO 600
9005-07-6
PEG 600 dioleate
Additive for cutting oils; solvent, emulsifier for solvents and oils, creams, cleansing milks, pesticides, textile lubricants, oilfield production: dispersant; for household/industrial cleaners. *Stepan Europe.*

27819 Secoster® EMS
Glycol stearate
Emulsifier for creams and cleansing milks; emollient, pearlescent, and opacifier. *Stepan Europe.*

27820 Secoster® MA 300
PEG 300 abietate
Additive for cutting oils; solvent, emulsifier for solvents and oils, creams, cleansing milks and pesticides; dispersant; for household/industrial cleaners. *Stepan Europe.*

27821 Secoster® MO 400
9004-96-0
PEG 400 oleate
Lubricant, antistat, solvent and oil emulsifier; for household/industrial cleaners, oilfield production. *Stepan Europe.*

27822 Secoster® SDG
Glyceryl stearate
Antistat and lubricant for polyolefins. *Stepan Europe.*

27823 Secosyl
137-16-6 4379 205-281-5
Sodium N-lauroyl sarcosinate
Gardol®. Detergent, foaming agent, base, anticorrosion additive for rug shampoos, mild dishwash, household cleaners, personal care products; stable in hard water. *Stepan Europe; KWR Chemicals Ltd.*

27824 Secretan
An alloy of from 91-95% copper, 5-9% aluminum, 1.5% magnesium, and 0.5% phosphorus.

27825 Secretol
A fat-splitting material similar and equal to Twitchell's reagent in its action.

27826 Securamid®
For the electrical industry. *DuPont UK.*

27827 Secure
Sprayable liquid resins, concrete curing compound; for application to freshly placed concrete following the finishing. For use in preventing cracking and crazing caused by rapid moisture loss due to hot, windy weather. *Secure Inc. Discontinued.*

27828 Securite
It is a mixture of 26% *m*-dinitrobenzene and 74% ammonium nitrate. It sometimes contains dinitronaphthalene and potassium nitrate; a safety explosive for mines.

27829 Securitol
A sodium silicate used to hasten the setting of cements.

27830 Securon 540
Multifunctional bleaching aid for silicate bleaching; complexes metal ions and prevents formation of silicate scale build up on equipment. *Henkel/Textiles.*

27831 Sedanox
470-90-6 2137 207-432-0
Chlorfenvinphos
A soil-applied organophosphorus insecticide. *Bayer plc.*

27832 Sedaplant Richter
Polyvalent herbal extract plus antiirritants (fennel, hops, camomile, balm mint, mistletoe, yarrow, urea, and urea deriv.) in water-alcohol medium; emollient for aq. and hydroalcoholic herbal cosmetics, skin and hair protection products; emulsified preparations. *Dr. Kurt Richter; Henkel/Cospha.*

27833 Sedefos 75®
Glycol stearate - PEG-2 stearate - trilaneth-4 phosphate. Self-emulsifying base for cosmetics and pharmaceuticals. *Gattefosse; Gattefosse SA.*

27834 Sedex
Ceramic foam filters to prevent nonmetallic inclusions in iron castings. *Foseco (F.S.) Ltd.*

27835 Sedifloc Flocculant Aids
Organic polymer, polyacrylamide, water soluble, polymers, cationic, anionic, nonionic; used for dewatering, settling and flotating municipal and industrial solids found in their waste water treatment plant. *Benzsay & Harrison Inc.*

27836 Sedipol®
Higher aliphatic alcohols; antifoams for industrial and communal sewage works. *BASF AG.*

27837 Sedipur®
High molecular weight water-soluble anionic, cationic and nonionic polymers; flocculants for industrial and communal water treatment, sewage sludge treatment, and in the mining industry. *BASF AG; BASF plc.*

27838 Seed Base
Powdered compost additive containing NPK 2.1:2.2:2 plus 3% Mg, 0.2% Fe and trace elements; compost additive. *Vitax Ltd.*

27839 seed-lac
Stick-lac, after washing free from the coloring matter soluble in water.

27840 Seedox SC
22781-23-3 1063 245-216-8
Bendiocarb
A contact, systemic insecticide. *Schering Agrochemicals Ltd.* Discontinued.

27841 Seedtect
Mixture of imazalil and thiabendazole; fungicide treatment of potatoes at planting time. *MSD Agvet.*

27842 Seekay Pitch®
A registered trade name for chlorinated naphthalene products available in various grades. No manufacturer.

27843 Seenox 412S
29598-76-3 249-720-9
Pentaerythrityl tetrakis
β-laurylthiopropionate. Antioxidant for polyolefins, thermoplastic elastomers, engineering thermoplastics; synergistic with primary antioxidants; outstanding long-term, heat aging performance. *Witco Corporation.*

27844 SEF
Modacrylic fibers. *Monsanto Co.* Name unverified.

27845 Segetan
A silver cyanide with a copper complex; used as a seed preservative.

27846 Seifert solder
An alloy of 73% tin, 21% zinc, 5% lead, 0.5% phosphorus and 0.5% tin.

27847 Seignette salt
Rochelle salt. Cathartic.

27848 Sekawrap
Polypropylene with various stabilizers; shrink film for presentation applications. *S Kempner Ltd.*

27849 Sekicel
Cellulose starch dispersion; for food industry. *Asahi Chem. Industry.*

27850 sel d'Angleterre
7487-88-9 5731 231-298-2
$MgSO_4$
Magnesium sulfate.

27851 Selar®
Barrier resin. *DuPont UK.*

27852 Selastin EL-10, EL-30, SE EM 95
Hydrolyzed animal elastin. *RITA.* Name unverified.

27853 Selazate
5456-28-0 226-713-9
$C_{10}H_{20}N_2S_4Se$
Selenium diethyl dithiocarbamate
ethyl selenac. A proprietary accelerator. *Naugatuck (US Rubber).* Name unverified.

27854 Selbana 2001
Synthetic lubricant for wool; provides antistatic and lubricant properties. *Henkel/Textiles.*

27855 Selbax
Synthetic lubricant for textiles. *Crosfield Chemicals Ltd.* Discontinued.

27856 Select-A-Sorb
14807-96-6 9207 238-877-9
Hydrous magnesium silicate (industrial talc); filler, extender, and reinforcing agent for rubber, paper (pitch control), plastics. *R. T. Vanderbilt Co Inc.*

27857 Selectrol
Soluble concentrate of 18 g dicamba, 252 g MCPA and 84 g mecoprop per liter; used for weed control in cereals and grassland. *R P Adams Ltd.*

27858 Select-Trol
Soluble concentrate containing 6.6% w/w 2,4-D and 250 g mecoprop; used to control weeds in grassland. *Chemsearch (UK) Ltd.*

27859 Selek
Sealants to prevent metal penetration between ingot molds and bottom plates. *Foseco (F.S.) Ltd.* Discontinued.

27860 Selenac®
Selenium dialkyldithiocarbamate
Accelerators and vulcanizing agents for rubber. *R. T. Vanderbilt Co Inc.* Discontinued.

27861 selenium
7782-49-2 8572 231-957-4
Se
colloidal selenium
A nonmetallic element; electronics, colorant for glass (ceramics), rectifiers, relays, solar batteries. Has three forms, black, grey and red selenium. *Appleby Group Ltd; Asarco; Atomergic Chemetals; Cerac; Shinko Chem.; R. T. Vanderbilt Co Inc.*

27862 selenium dioxide
7446-08-4 8576 231-194-7
O_2Se
selenous acid anhydride
selenious anhydride. Analysis (testing for alkaloids), oxidizing agent, antioxidant in lubricating oils, catalyst. mp = 340°; d_{15}^{15} = 3.954; soluble in H_2O (38 g/100 ml), less soluble in organic solvents; *Aldrich; Atomergic Chemetals; Cerac; Shinko Chem.*

27863 selenium sulfide
8580
Se_4S_4
Red crystals, mp = 113° (dec); d = 3.20; soluble in CS_2, less soluble in C_6H_6 (0.4 g/l).

27864 selenium trisulfide
8580
Se_2S_6
Orange crystals, mp = 121°; d = 2.44; soluble in CS_2, C+66H_6 (12 g/l).

27865 Seleron
Aeron
A group of aluminum alloys containing 85% aluminum, with copper, nickel, zinc, manganese, silicon, and lithium, as the other ingredients. They are claimed to be useful for electrical apparatus.

27866 Selexsorb® COS
1344-28-1 369 215-691-6
Activated alumina; selective adsorbent for catalytic reforming, polyethylene and polypropylene production, isomerization processes. *Alcoa Industrial Chemicals.*

27867 Selin® O
Trioleate
Basic oil for release agents, bread cutting oil. *Grünau.*

27868 Seliwanoff's reagent
A solution of 0.05 gram resorcinol in 100 ml dilute (1:2) hydrochloric acid. It gives a red color with fructose.

27869 Seljut
Special processing emulsifying agent. *Crosfield Chemicals Ltd.* Discontinued.

27870 Sellacron, Sella Acid, Sella Fast, Sellaflor
Acid dyes for leather. *Ciba plc.* Name unverified.

27871 Sellaflor
Dyes for leather. *Ciba plc.* Name unverified.

27872 Sellasol
Synthetic tanning agents for leather. *Ciba plc.* Name unverified.

27873 Sellifix Helios
Formaldehyde-free fixative for fiber-reactive dyes. *Ceca SA.*

27874 Sellig
Specialty blends of anionic surfactants; used in textile industry. *Elf Atochem UK/Ceca.*

27875 Sellig Antimousse S
Silicone-based; antifoaming agent for textiles. *Ceca SA.*

27876 Sellig AO 15 100
9004-96-0
PEG-15 oleate
Surfactant for emulsions, detergents. *Ceca SA.*

27877 Sellig AO 25 100
9004-96-0
PEG-25 oleate
Surfactant. *Ceca SA.*

27878 Sellig AO 6100
9004-96-0
PEG-6 oleate
Surfactant. *Ceca SA.*

27879 Sellig DN 10 100
9014-93-1
Nonyl nonoxynol-10
Surfactant. *Ceca SA.*

27880 Sellig DN 22 100
9014-93-1
Nonyl nonoxynol-22
Surfactant. *Ceca SA.*

27881 Sellig HR 18 100
61791-12-6
PEG-21 castor oil
Surfactant. *Ceca SA.*

27882 Sellig LA 1150
Laureth-20
Surfactant for shampoos, degreaser, textile applications. *Ceca SA.*

27883 Sellig N 10 100
9016-45-9 6772
Nonoxynol-10
Surfactant. *Ceca SA.*

27884 Sellig N 1050
9016-45-9 6772
Nonoxynol-40
Surfactant. *Ceca SA.*

27885 Sellig N 11 100
9016-45-9 6772
Nonoxynol-11
Surfactant. *Ceca SA.*

27886 Sellig N 12 100
9016-45-9 6772
Nonoxynol-12
Surfactant. *Ceca SA.*

27887 Sellig N 15 100
9016-45-9 6772
Nonoxynol-16
Surfactant for emulsions, oils, greases. *Ceca SA.*

27888 Sellig N 1780
9016-45-9 6772
Nonoxynol-17
Surfactant. *Ceca SA.*

27889 Sellig N 20 80
9016-45-9 6772
Nonoxynol-19
Surfactant. *Ceca SA.*

27890 Sellig N 30 70
9016-45-9 6772
Nonoxynol-25
Surfactant. *Ceca SA.*

27891 Sellig N 4 100
9016-45-9 6772
Nonoxynol-4
Surfactant. *Ceca SA.*

27892 Sellig N 5 100
9016-45-9 6772
Nonoxynol-5
Surfactant. *Ceca SA.*

27893 Sellig N 50 100
9016-45-9 6772
Nonoxynol-52
Surfactant. *Ceca SA.*

27894 Sellig N 6 100
9016-45-9 6772
Nonoxynol-6
Surfactant for emulsions, petroleum products, chlorinated and aromatic solvents, greases, silicones. *Ceca SA.*

27895 Sellig N 8 100
9016-45-9 6772
Nonoxynol-8
Surfactant for emulsions, petroleum products, mineral oils, greases, silicones; base for low foam household detergents. *Ceca SA.*

27896 Sellig N 9 100
9016-45-9 6772
Nonoxynol-9
Surfactant for emulsions, mineral oils, greases, silicones; base for household, industrial and textile detergents. *Ceca SA.*

27897 Sellig O 11 100
9002-93-1 6858
Octoxynol-11
Surfactant. *Ceca SA.*

27898 Sellig O 12 100
9002-93-1 6858

Octoxynol-12
Surfactant. *Ceca SA.*

27899 Sellig O 20 100
9002-93-1 6858
Octoxynol-20
Surfactant. *Ceca SA.*

27900 Sellig O 4 100
9002-93-1 6858
Octoxynol-4
Surfactant. *Ceca SA.*

27901 Sellig O 5 100
9002-93-1 6858
Octoxynol-5
Surfactant. *Ceca SA.*

27902 Sellig O 6 100
9002-93-1 6858
Octoxynol-6
Surfactant. *Ceca SA.*

27903 Sellig O 8 100
9002-93-1 6858
Octoxynol-8
Surfactant. *Ceca SA.*

27904 Sellig O 9 100
9002-93-1 6858
Octoxynol-9
Surfactant. *Ceca SA.*

27905 Sellig R 20 100
61791-12-6
PEG-20 castor oil
Surfactant. *Ceca SA.*

27906 Sellig R 3395
61791-12-6
PEG-33 castor oil
Surfactant. *Ceca SA.*

27907 Sellig R 3395 SP
61791-12-6
PEG-30 castor oil
Surfactant. *Ceca SA.*

27908 Sellig R 3395-C435
61791-12-6
PEG-32 castor oil
Surfactant. *Ceca SA.*

27909 Sellig R 4095
61791-12-6
PEG-40 castor oil
Surfactant. *Ceca SA.*

27910 Sellig R 4495
61791-12-6
PEG-44 castor oil
Surfactant. *Ceca SA.*

27911 Sellig S 30 100
9004-99-3
PEG-30 stearate
Surfactant. *Ceca SA.*

27912 Sellig SP 16 100
Cetoleth-16
Surfactant for emulsions. *Ceca SA.*

27913 Sellig SP 20 100
Cetoleth-18
Surfactant. *Ceca SA.*

27914 Sellig SP 25 50
Cetoleth-27
Surfactant. *Ceca SA.*

27915 Sellig SP 30 100
Cetoleth-30
Surfactant. *Ceca SA.*

27916 Sellig SP 3020
Cetoleth-30
Surfactant. *Ceca SA.*

27917 Sellig SP 8 100
Cetoleth-8
Surfactant for oil-water emulsions, oils, organic solvents, mineral oil, paraffin; stable to electrolytes. *Ceca SA.*

27918 Sellig Stearo 6
9004-99-3
PEG-6 stearate
Surfactant. *Ceca SA.*

27919 Sellig SU 18 100
68439-49-6 7737
Ceteareth-18
Surfactant. *Ceca SA.*

27920 Sellig SU 25 100
68439-49-6 7737
Ceteareth-20
Surfactant. *Ceca SA.*

27921 Sellig SU 30 100
68439-49-6 7737
Ceteareth-32
Surfactant, dispersant, household and industrial detergent base; emulsions. *Ceca SA.*

27922 Sellig SU 4 100
68439-49-6 7737
Ceteareth-4
Surfactant for emulsions, oils. *Ceca SA.*

27923 Sellig SU 50 100
68439-49-6 7737
Ceteareth-46
Surfactant. *Ceca SA.*

27924 Sellig T 14 100
61791-00-2
PEG-14 tallate
Detergent base with controlled foam; solubilizer for essential oils. *Ceca SA.*

27925 Sellig T 1790
61791-00-2
PEG-17 tallate
Surfactant. *Ceca SA.*

27926 Sellig T 3 100
61791-00-2
PEG-3 tallate
Surfactant. *Ceca SA.*

27927 Selligon SP
Softening detergent for wool and cotton. *Ceca SA.*

27928 Selligor 860 SP
For washing out of acidic dyes. *Ceca SA.*

27929 Sellogen DFL
9084-06-4
Sodium alkyl naphthalene sulfonate
Wetting and dispersing agent for pesticide formulations. *Henkel/Functional Prods.*

27930 Sellogen HR-90
9084-06-4
Sodium alkyl naphthalene sulfonate
Wetting and dispersing agent for insecticides, acid and alkaline media. *Henkel/Functional Prods.*

27931 Selora
7447-40-7 7783 231-211-8
A proprietary preparation of potassium chloride; used as a substitute for table salt. *Winthrop Laboratories.* Name unverified.

27932 Seloxone
Soluble concentrate containing 15 g clopyralid and 510 g mecoprop per liter; a translocated herbicide for cereals and grassland. *ICI Chem & Polymers Ltd.*

27933 Selsun
A proprietary preparation of selenium sulfide and a detergent; used as a treatment for dandruff. *Abbott Laboratories.* Name unverified.

27934 seltzers
Usually consist of 25 parts sodium carbonate, 5 parts sodium chloride, 6 parts sodium sulfate, and 1,000 parts water.

27935 Sembonit
Erostabil. Several compositions of natural caoutchouc, synthetic caoutchouc, filling material, synthetic resins, depending on resistance demands; for surface protection against corrosion and erosion of vessels, tanks, tubes, and industrial equipment. *Schaumstoff and Kunststoff GmbH.* Name unverified.

27936 Semeron
1014-69-3 213-800-1
Wettable powder containing 25% w/w desmetryn; a triazine herbicide. *Ciba-Geigy Agrochemicals.*

27937 Semfreeze
Reactive or curable chemical materials for use in industry. *Courtaulds Aerospace Ltd.*

27938 semicoke
A fuel made from coal by low temperature carbonization. It is a smokeless fuel with a low ash.

27939 **seminose**
3458-28-4 5791 222-392-4
$C_6H_{12}O_6$
Mannose.

27940 **Semirit**
Electrically fused corundum, semifriable grade; used in production of abrasives, abrasive paper, discs and cloth. *Hüls AG.*

27941 **semi-steel**
A metal having properties between cast iron and cast steel; used for filter-press plates. The term is applied to grey cast irons of low carbon content.

27942 **Sempatap**
Insulating material for walls, ceilings and floors; glass fiber nonwoven with SBR-latex backing; for interior acoustic and thermal insulation. *Ebnother AG.* Name unverified.

27943 **Sempollan**
Cast polyurethane; for components required to display high strength, lasting resilience, high wear and oil resistance and a maximum useful life. *Schaumstoff und Kunststoff GmbH.* Name unverified.

27944 **Semtol**
White mineral oils, technical grade. *Witco Corporation.*

27945 **Senate**
Lube oil. *Monsanto (Solaris).* Name unverified.

27946 **Senate**
Suspension concentrate containing 250 g terbutryn [886-50-0] and 250 g trietazine [1912-26-1] per liter; herbicide for weed control in potatoes, peas and field beans. *Schering Agrochemicals Ltd.* Discontinued.

27947 **Sencor®**
21087-64-9 6239 244-209-7
Metribuzin
Herbicide for control of many important grasses and broad-leaved weeds in soybeans, potatoes, tomatoes, sugarcane, alfalfa and asparagus; suitable for pre-and in some cases post emergence application. *Bayer AG.*

27948 **Sencorex® WG**
21087-64-9 6239 244-209-7
Water dispersible granular formulation containing 70% w/w metribuzin; used to control annual weeds in early and maincrop potatoes. *Bayer plc.*

27949 **Sendust**
A proprietary trade name for an iron-silicon aluminum alloy. No manufacturer.

27950 **Senegal gum**
West African gum. A gum arabic ranking second to Khordofan gum. It is derived from *Acacia senegal and other species of Acacia*. It gives a good adhesive mucilage.

27951 **Sengite**
An American explosive. It has a guncotton base and is similar to Tonite, except that sodium nitrate replaces barium nitrate.

27952 **Sensitizer**
Diazo photosensitizer. *ABM Chemicals Ltd.* Name unverified.

27953 **Sensolve BEA**
Butoxy ethyl acetate
Solvent. *Sasolchem.*

27954 **Sensolve EEA**
111-15-9 3798 203-839-2
Ethoxy ethyl acetate
Solvent. *Sasolchem.*

27955 **Sensolve EPA**
Ethoxy propyl acetate
Solvent. *Sasolchem.*

27956 **Sensolve MPA**
Methoxy propyl acetate
Solvent. *Sasolchem.*

27957 **Sentinel**
Water treatment used in hot water boiler systems. *Grace Dearborn Ltd.*

27958 **Sentry Cyclomethicone**
69430-24-6
Cyclomethicone
Pharmaceutic aid. *Union Carbide.* Name unverified.

27959 **Sentry Dimethicone**
9016-00-6 3264 215-648-1
Dimethicone
silicone rubber; Latex; Dimethylpolysiloxane; Simethicone; Dimethyl silicone; Dermafilm; Dimethicream; Silbar. Prosthetic aid. SG = 0.98 *Union Carbide.* Name unverified.

27960 **SEP 55**
A general purpose farm disinfectant which can be diluted with water or oil. *Coventry Chemicals Ltd.*

27961 **Sep 6**
Isolating solution for waxes; dental preparation. *Bayer AG.* Discontinued.

27962 **Sepabase**
Mineral demulsifier concentrates. *BASF plc.*

27963 **Sepabase A Grades**
High molecular weight EO/PO adduct; for dehydration of crude oil emulsions and removal of residual salts. *BASF AG.*

27964 **Sepabeads® FP Series**
High porous type hydrophilic polymer; for industrial purification of protein and enzyme by chromatography. *Mitsubishi Kasei.* Name unverified.

27965 **Sepacid® CE 5209**
Quaternary ammonium compound; biocide for oilfield applications. *BASF AG.*

27966 **Sepacid® CE 5265**
Glutaraldehyde derivative; biocide for oilfield applications. *BASF AG.* Name unverified.

27967 **Sepaclear®**
For removal of residual oil from water in refineries and in crude oil production. *BASF AG; BASF plc.*

27968 **Sepacorr®**
Mineral oil corrosion inhibitors. *BASF plc.*

27969 **Sepacorr® HT**
Nitrogen-containing condensation production; corrosion inhibitor. *BASF AG.* Name unverified.

27970 **Sepaflood®**
Water-soluble polymers; for tertiary crude oil production. *BASF AG.*

27971 **Sepaflux®**
Pour point and/or viscosity depressants for crude oils and residual oils. *BASF AG; BASF plc.*

27972 **Sepakoll®**
Protective colloids for drilling fluids and borehole cementing. *BASF AG; BASF plc.*

27973 **Sepapar® P**
Paraffin inhibitor. *BASF AG.*

27974 **Separan**
Polyacrylamides used as flocculants. *Dow Cheml Co Ltd, UK & Ireland.*

27975 **Separit**
Parting powder. *Foseco (F.S.) Ltd.*

27976 **Separol**
Liquid parting medium. *Foseco (F.S.) Ltd.*

27977 **Separol**
Demulsifiers for crude oils. *BASF plc.*

27978 **Separol AF 27**
High molecular weight EO/PO adduct; demulsifier used in crude oil emulsions and dehydrating equipment. *BASF AG.*

27979 **Sepascale®**
Scale inhibitors for petroleum production and processing. *BASF AG; BASF plc.*

27980 **Sepasolv® MPE**
For purification of natural gas. *BASF AG.*

27981 **Sepawet®**
Surfactants for petroleum production and pipeline transport. *BASF AG; BASF plc.*

27982 **sepia**
A brownish-black pigment derived from the ink-bag of the cuttle-fish; used as water color.

27983 **Sepicide HB**
Mixture of phenoxyethanol, methylparaben, ethylparaben, propylparaben and butylparaben. Cosmetic preservative. *Seppic.* Discontinued.

27984 **Sepigel 305**
Polyacrylamide
C_{13}-C_{14} isoparaffin; laureth-7. Cosmetic ingredient. *Seppic.*

27985 **Sepigel A**
1319-41-1 215-289-0
Sepiolite mineral; gelling and suspending agent; drilling muds, foundation drilling, slurry trench cut-off walls, soil admixture for impermeable barriers. *Floridin.* Name unverified.

27986 **Sepramar**
Reagents for amino acids analyzis. *BDH Chemicals Ltd.*

27987 **Seprate K**
Fluid emulsion parting agent. *Foseco (F.S.) Ltd.*

27988 **Septal**
12427-38-2 5761 235-654-8
Carbendazim [10605-21-7] and maneb[12427-38-2]; systemic fungicide for cereals. *Schering Agrochemicals Ltd.* Discontinued.

27989 Seqlene® 270
Reaction mixture forming sodium-α-δ-glucoheptonate, sodium-β-δ-glucoheptonate, aldobionates, and other complex carbohydrates; sequestrant forming nonionic chelates; used in metal applications; scavenger for antioxidants, bactericides, bottle washing, alkaline cleaning and textile applications. *Pfanstiehl Labs.*

27990 Sequalog
Reagents for peptide sequence and synthesis. *Scheweizerhall.*

27991 Sequenase
Range of modified or genetically engineered DNA polymerases. *U.S. Biochemical.*

27992 Sequest-All
A dry nontoxic potable water treatment used to control minerals in water, to prevent red water, scale, build-up and corrosion in the distribution system; used for municipal water systems, irrigation systems, cooling towers, boilers, apartments, hotels. *SPER Chemical Corporation.* Name unverified.

27993 Sequestrene®
Iron chelates-food additives foliar feeds. *Ciba plc.* Name unverified.

27994 Sequestrene® 30A
64-02-8 3557 200-573-9
Tetrasodium EDTA
Chelating agent used in water softening, liquid soaps, detergents, chemical cleaning, scale removal, beerstone removal, processing of textile, paper, and leather, in metal treatment, and for synthetic rubber. *Ciba-Geigy/Dyestuffs.*

27995 Sequestrene® 220
64-02-8 3557 200-573-9
Tetrasodium EDTA dihydrate
Chelating agent used in powder cleaning compounds; scale removal, hair rinses; processing of synthetic fibers and textiles, and industrial cleaning preparations. *Ciba-Geigy/Dyestuffs.*

27996 Sequestrene® AA
60-00-4 3559 200-449-4
EDTA
Chelating agent for photographic developer baths, shampoos, cosmetics, electroplating, rare earth separations, metal determinations, liquid soaps, germicides, herbicide sprays. *Ciba-Geigy/Dyestuffs.*

27997 Sequestrene® NA2
139-33-3 3556 205-358-3
Disodium EDTA dihydrate
Chelating agent for control of trace metal contamination in pharmaceutical and cosmetic products. *Ciba-Geigy/Dyestuffs.*

27998 Sequestrene® NA2 Edetate USP
139-33-3 3556 205-358-3
Disodium edetate USP
Chelating agent for control of trace metal contamination in pharmaceutical manufacturing, ophthalmic solutions, high purity cosmetics; analytical reagent for metals. *Ciba-Geigy/Dyestuffs.*

27999 Sequestrene® NA2Ca
Disodium-calcium EDTA dihydrate
Chelating agent for control of trace metal contamination in pharmaceutical and cosmetic products. *Ciba-Geigy/Dyestuffs.*

28000 Sequestrene® NA3
150-38-9 3558 205-758-8
Trisodium EDTA trihydrate
Chelating agent used in personal care products; processing of synthetic fibers and textiles; stabilizer for resin systems; photographic baths, electrolytic and electroless plating; foam stabilizer, water treatment. *Ciba-Geigy/Dyestuffs.*

28001 Sequestrene® NAFe 13% Fe
Na(FeEDTA)
Chelating agent; micronutrient; animal feeds, photographic uses, polymerization catalyst for synthetic rubber. *Ciba-Geigy/Dyestuffs.*

28002 Sequestrene® NH4Fe
Ferric ammonium EDTA
Chelating agent in photographic baths, blix baths; in fertilizer formulations. *Ciba-Geigy/Dyestuffs.*

28003 Sequestrene® Tetraammonium
Tetraammonium EDTA
Chelating agent used in photographic developer baths, replenishment of blix baths, water treatment, and for boiler cleaning. *Ciba-Geigy/Dyestuffs.*

28004 Seracelle
A proprietary cellulose acetate packing material. No manufacturer.

28005 Seradix
133-32-4 4995 205-101-5
4-indol-3-ylbutyric acid
A root growth promoter. *Embetec Crop Protection Ltd.*

28006 Sera-Pak®
Test combinations for determining uric acid, cholesterol, HDL cholesterol, and urinary proteins in clinical chemistry. *Bayer AG.*

28007 Serdet DCK
Sodium alkyl ether sulfate, based on a natural alcohol (C12-C14) in paste or liquid form; anionic surfactant used in shampoos and bubble bath formulations and dishwashing. *Chemische Fabriek Servo BV.* Name unverified.

28008 Serdet DFK
Sodium alkyl sulfate, based on a natural alcohol (C12-C14), in liquid or paste form; detergent and emulsifier for shampoos and bubble baths; toothpaste; dishwashing; emulsion polymerization. *Chemische Fabriek Servo BV.* Name unverified.

28009 Serdet DFL, DFM, and DFN
Anionic surfactants in liquid form; foaming agents for shampoos and bubble baths. *Chemische Fabriek Servo BV.* Name unverified.

28010 Serdet DM and DMK
Dodecylbenzene sulfonate in acid form or as sodium salt; biodegradable primary emulsifiers used in scouring powders, liquid detergents, and emulsion polymerization. *Chemische Fabriek Servo BV.* Name unverified.

28011 Serdet DML
Triethanolamine dodecylbenzene sulfonate in liquid form; biodegradable anionic surfactant used in shampoos and bubble baths. *Chemische Fabriek Servo BV.* Name unverified.

28012 Serdet DNK
9014-90-8
Sodium nonylphenol 4EO-sulfate in liquid form; detergent base for liquid detergent formulations; emulsifier in emulsion polymerization. *Chemische Fabriek Servo BV.* Name unverified.

28013 Serdet DPK
Sodium alkyl ether sulfate based on a synthetic alcohol (C12-C15) in liquid or paste form; anionic surfactant used in shampoo and bubble bath formulations and dishwashing. *Chemische Fabriek Servo BV.* Name unverified.

28014 Serdet DSK
Sodium 2-ethylhexyl sulfate in liquid form; alkali-stable wetting agent used for latex stabilization. *Chemische Fabriek Servo BV.* Name unverified.

28015 Serdolamide
Foam stabilizer; refatting agent; viscosity modifier and improver; for detergent, shampoo and bubble bath formulations. *Chem-Y, Fabriek van Chemische Producten BV.* Name unverified.

28016 Serdox
Emulsifier; antistatic agent; for crude and vegetable oils; plastics; textiles processing; cosmetic emulsions. *Chem-Y, Fabriek van Chemische Producten BV.* Name unverified.

28017 Serdox NNP4
Nonylphenol ethoxylate nonionic surfactant in liquid form; oil-soluble detergents; emulsifier for insecticides and herbicides; dispersing agent. *Chemische Fabriek Servo BV.* Name unverified.

28018 Serdox NNP5 and NNP6
Nonylphenol ethoxylate nonionic surfactant in liquid form; emulsifier for insecticides and herbicides; oil soluble detergents. *Chemische Fabriek Servo BV.* Name unverified.

28019 Serdox NNP7, NNP8.5 and NNP9
Nonylphenol ethoxylate nonionic surfactant in liquid form; for scouring of textiles; soaking assistant for leather; household and industrial detergents; emulsifier for insecticides and herbicides; plasticizer for mortar and c *Chemische Fabriek Servo BV.* Name unverified.

28020 Serdox NNP10, NNP12
Nonylphenol ethoxylate nonionic surfactant in liquid form; for scouring of textiles; soaking assistant for leather; household and industrial detergents; emulsifier for insecticides and herbicides; plasticizer for mortar and c *Chemische Fabriek Servo BV.* Name unverified.

28021 Serdox NNP15, NNP20, NNP25
Nonylphenol ethoxylate nonionic surfactant in liquid or solid form; detergents and wetting agents for use at high temperatures and electrolyte concentrations; emulsifier for fatty acids and waxes; NNP20 is used as a stabilize *Chemische Fabriek Servo BV.* Name unverified.

28022 Serdox NNP30, NNP30/70
Nonylphenol ethoxylate nonionic surfactant in solid or liquid form; dyeing assistant; lime soap dispersing agent; emulsifier and stabilizer for emulsion polymerization. *Chemische Fabriek Servo BV.* Name unverified.

28023 Serdox NNPQ 7/11
Nonionic surfactant of the alkylphenol ethoxylate type in liquid form; low foaming wetting agent. *Chemische Fabriek Servo BV.* Name unverified.

28024 Serdox NOP 30/70
Octylphenol ethoxylate nonionic surfactant in liquid form; emulsifier and

stabilizer used in emulsion polymerization. *Chemische Fabriek Servo BV.* Name unverified.

28025 Serdox NOP9
Octylphenol ethoxylate nonionic surfactant in liquid form; for scouring of textiles; soaking assistant for leather; household and industrial detergents; emulsifier for insecticides and herbicides; plasticizer for mortar and concrete, paper manufacture. *Chemische Fabriek Servo BV.* Name unverified.

28026 seretin
56-23-5 1864 200-262-8
CCl_4
carbon tetrachloride
A chlorinated hydrocarbon; used as a solvent, in refrigerants, as an agricultural fumigant, metal degreasing.

28027 Serfene
Polyvinylidene chloride dispersion coatings. *Morton Int'l. Ltd.*

28028 sericine
Silk size; silk rubber. The gum surrounding the silk from the silk spinner.

28029 Sericite PHN
12001-26-2
Mica; filler imparting softness, smooth feel, skin adhesion, and spreadability. *Presperse.*

28030 Sericite SL-012
Mica [12001-26-2] mixed with methicone [9004-73-3]. A surface-treated powder providing creamier, more lubricious feel, improved spreadability and skin adhesion; ideal for wet and dry applications. *Presperse.*

28031 Sericose
Cellulose acetate
Used for making artificial silk and dope.

28032 Serilan
A range of disperse and acid dye mixtures; for dyeing of polyester/wool blended fibers. No manufacturer.

28033 Serilene Dyes
Disperse dyes for atmospheric and high temperature exhaust, pad-thermosol and print applications on polyester and polyester blend fabrics. No manufacturer.

28034 Serilube Series
Lubricants for jet dyeing various blended and 100% cotton fabrics. No manufacturer.

28035 L-serine
56-45-1 8604 200-274-3
$C_3H_7NO_3$
α-amino-β-hydroxy-propionic acid
L-serine; (S)-(-)-serine; Ser; S; L-2-amino-3-hydroxypropionic acid; L-(-)-serine; 2-maino-3-hydroxypropionic acid; 3-hydroxyalanine. Naturally occurring enantiomer of serine. Used in biochemical research, as a dietary supplement, in culture media, microbiological tests, feed additive. mp = 228° (dec); $[\alpha]_D^{20}$ = -7° (c = 10.41 H_2O); soluble in H_2O, insoluble in organic solvents. *Degussa; Janssen Chimica; Mitsui Toatsu Chem.; Nippon Rikagakuyakuhin; U.S. Biochemical.*

28036 Serinyl
A range of disperse dyes; for dyeing of polyamide fibers. No manufacturer.

28037 Seriplast Dyes
Selected disperse dyes formulated for heat transfer printing. No manufacturer.

28038 Seripol Series
Dispersants, levelers, and retardants for disperse dyeing. No manufacturer.

28039 Seriprint Dyes
Disperse dyes formulated for use in printing with synthetic printing thickeners. No manufacturer.

28040 Serisol Dyes
Disperse dyes for acetate; also suitable for polyester, triacetate, nylon, and acrylic fibers. No manufacturer.

28041 Seritox
Selective weedkiller. *May & Baker Ltd.* Name unverified.

28042 Seritox
Mixture of dichlorprop and MCPA; used for weed control in cereals and turf. *Rhône-Poulenc Crop Protection Ltd.*

28043 Serizyme
A proprietary trade name for an enzyme used in desizing acetate fabrics and similar materials which contain protein. No manufacturer.

28044 Sermag®
1309-48-4 5713 215-171-9
Liquid magnesium oxide oil fuel additive. *Steetley Magnesia Products Ltd.*

28045 Sermix
Alkaloid animal feed additive. *Ciba plc.* Name unverified.

28046 Sermul
Emulsifier, biodegradable; for mineral oils; pesticide formulations; white spirit and turpentine; vegetable and animal oils. *Chem-Y, Fabriek van Chemische Producten BV.* Name unverified.

28047 Sermul EA 88
Calcium dodecylbenzene sulfonate in liquid form; biodegradable emulsifier for pesticide formulations. *Chemische Fabriek Servo BV.* Name unverified.

28048 Sermul EA129
2235-54-3 218-793-9
Ammonium lauryl sulfate in liquid form; emulsifier in emulsion polymerization. *Chemische Fabriek Servo BV.* Name unverified.

28049 Sermul EA150
Sodium lauryl sulfate in paste form; emulsifier in emulsion polymerization. *Chemische Fabriek Servo BV.* Name unverified.

28050 Sermul EA176
Sodium mono nonylphenol 10-EO sulfosuccinate in liquid from; emulsifier for emulsion polymerization. *Chemische Fabriek Servo BV.* Name unverified.

28051 Sermul EA188, EA136 and EA205
Anionic surfactants of the phosphate ester type; supplied as liquids in acid form; emulsifiers for emulsion polymerization. *Chemische Fabriek Servo BV.* Name unverified.

28052 Sermul EA54, EA151 and EA146
Anionic surfactants of the ether sulfate type in liquid form; emulsifiers for emulsion polymerization. *Chemische Fabriek Servo BV.* Name unverified.

28053 SeroClear®
Reagent. *Calbiochem Corp.*

28054 Serotulle®
White soft paraffin chlorhexidine acetate 0.5% w/w; gauze wound dressing to prevent infection in minor wounds, injuries with minor skin loss and ulcerative lesions. *Seton Healthcare Group plc.* Name unverified.

28055 Serramix CDA
An emulsifiable concentrate containing 100 g aminotriazole, 200 g atrazine and 100 g MCPA per liter; used for total weed control in non crop areas. *Denoon CDS; Powaspray (CDA) Ltd.*

28056 Serseal
Heat conserving compound for use in metal pretreatment. *ICI Chem & Polymers Ltd.* Discontinued.

28057 Serumpro EN-10
Serum protein; cosmetics ingred.; contains all essential amino acids. *Brooks Industries.*

28058 Servamine KAC 422
N-Coco N-N-dimethyl-N-benzalkonium chloride in liquid form; cationic surfactant used as a bactericide, fungicide, sanitizer and germicide. *Chemische Fabriek Servo BV.* Name unverified.

28059 Servamine KEP 4527
N-(palmitylamidopropyl)-N-N-N-trimethylammonium chloride in liquid form; cationic surfactant emulsifier with bactericide properties. *Chemische Fabriek Servo BV.* Name unverified.

28060 Servamine KET 350
N-(Tall oil amidopropyl)-N-N-dimethylamine based on tall oil; supplied as a liquid; cationic surfactant used as an adhesion agent and corrosion inhibitor in bitumen. *Chemische Fabriek Servo BV.* Name unverified.

28061 Servamine KET 4542
N-(Alkylamidopropyl)-N-ethyl N-N-dimethylammonium ethosulfate
Based on tall oil, in liquid form; cationic emulsifier. *Chemische Fabriek Servo BV.* Name unverified.

28062 Servamine KOO 330
Amino ethyl oleyl imidazoline in liquid form; cationic adhesion agent and corrosion inhibitor for bitumen. *Chemische Fabriek Servo BV.* Name unverified.

28063 Servamine KOO 330B
Oleylamido ethyl oleyl imidazoline in liquid form; basic material in the manufacture of quaternary imidazolines. *Chemische Fabriek Servo BV.* Name unverified.

28064 Servamine KOO 360
Hydroxyethyl oleyl imidazoline in liquid form; cationic adhesion agent and corrosion inhibitor for bitumen. *Chemische Fabriek Servo BV.* Name unverified.

28065 Servil® Range
Wax esters; release agents for confectionery and baked goods. *Grünau.*

28066 Servit® Range
Citric acid esters of mono and diglycerides of fatty acids with other emulsifiers; food emulsifier for manufacturing of instant dry yeast. *Grünau.*

28067 Servo Ampholyt (B) JA110
Modified imidazoline in liquid form; non eye-irritating for shampoo formulations. *Chemische Fabriek Servo BV.* Name unverified.

28068 Servo Ampholyt (B) JA140
Modified imidazoline in liquid form; non eye-irritating and non skin-irritating for hair shampoos. *Chemische Fabriek Servo BV.* Name unverified.

28069 Servo Ampholyt (B) JB130
Betaine structure, liquid form; mild shampoo raw material with hair stipulating properties, for baby shampoos. *Chemische Fabriek Servo BV.* Name unverified.

28070 Servo Brilliant Oil B AZ 75
Sodium castor oil sulfonate in liquid form; anionic surfactant used in softener and finishing oils, and pasting oil for dyestuffs. *Chemische Fabriek Servo BV.* Name unverified.

28071 Servoxyl VLA 2170
577-11-7 3460 209-406-4
Sodium di-2-ethylhexyl sulfosuccinate in liquid form; wetting and rewetting agent. *Chemische Fabriek Servo BV.* Name unverified.

28072 Servoxyl VLB 1123
Sodium monoalkyl polyglycol ether sulfosuccinate in liquid form; raw material for high quality baby shampoos and mild hair shampoos; cleaning agent. *Chemische Fabriek Servo BV.* Name unverified.

28073 Servoxyl VLE 1159
Sodium mono nonylphenol 10-EO sulfosuccinate in liquid form; emulsifier for emulsion polymerization. *Chemische Fabriek Servo BV.* Name unverified.

28074 Servoxyl VP
Range of anionic surfactants of the phosphate ester type, supplied mainly as liquids in acid form; those based on alcohol and alkyl polyglycol ethers are biodegradable; used in dry-cleaning; formulation of metal cleaners, emulsion polymerization, pesticide formulation, cosmetic preparations. *Chemische Fabriek Servo BV.* Name unverified.

28075 Sesame oil
8008-74-0 8614 232-370-6
Gingili oil
teal oil; teel oil; gingelly oil; til oil, beni oil; benne oil, beniseed oil; Lipovol SES; Super refined sesame oil,. Sesame oil obtained from the seeds of *Sesamum indicum* and of *S orientate*; used in the manufacture of margarine and soap, and as a burning oil. *Lipo.*

28076 Seseal
A thermoplastic acrylic sealer; seals, hardens and dustproofs concrete. *Secure Inc.* Discontinued.

28077 Seseal 8
Cure, seal, hardener and dustproofer for concrete; prevents mortar and concrete droppings from bonding to floors, reducing clean-up cost, base for mastic adhesives. *Secure Inc.* Discontinued.

28078 Sesolvan® L
Dye solvent for textile dyeing. *BASF AG.*

28079 Sestrip
Form release agent; for application to all types of concrete forms prior to concrete placement to ensure release of the forms and to minimize form clean up. *Secure Inc.* Discontinued.

28080 Setac
Waxes for dental laboratory technology in bead form and block form; casting and modeling waxes blue, green, sticky wax red. *Bayer AG.* Discontinued.

28081 Setacin 103 Spezial
39354-45-5 255-062-3
Disodium laureth sulfosuccinate
Detergent for personal care products; cleaning agent. *Zschimmer & Schwarz.*

28082 Setacin F Spezial Paste
13192-12-6 236-149-5
Disodium lauryl sulfosuccinate
Detergent, cosmetics, washing and cleaning agent. *Zschimmer & Schwarz.*

28083 Setacure
Multifunctional acrylic monomers and prepolymers; for uv, ebc, radiation curing applications. *Synthese BV.* Name unverified.

28084 Setafix
Acrylic and polyester resins; photocopy toners. *Synthese BV.* Name unverified.

28085 Setair
A clear or colored solution of PVC polymer; coating zinc sprayed, shot or sand blasted steel, asbestos, concrete, brick, plywood, softboard, strawboard, chipboard and hardboard. *Llewellyn Ryland Ltd.*

28086 Setal
Saturated polyester and alkyd resins; for decorative and industrial paints. *Synthese BV.* Name unverified.

28087 Setalana
A proprietary name for a natural nest silk produced by worms of the genus *Anaphe*, introduced into Germany from Africa; the fiber resembles tussah silk, but is stated to be not so strong. No manufacturer.

28088 Setalin
Modified phenolic resins, modified hydrocarbon resins, water thinnable acrylic resins, acrylic dispersions, alkyd resins, modified phenolic resins, modified hydrocarbon resins; for varnishes, rotogravure inks, packaging inks and offset inks. *Synthese BV.* Name unverified.

28089 Setalux
Acrylic resins; air-drying, thermosetting, isocyanate curing resins for industrial paint (automotive, refinishing and general industry). *Synthese BV.* Name unverified.

28090 Setamine US
Melamine-formaldehyde resins; for ovendrying industrial paints (automotive and general industry). *Synthese BV.* Name unverified.

28091 Setamol®
Dispersants for dye dispersions used in dyeing synthetic and cellulose fibers with vat dyes. *BASF AG; BASF plc.*

28092 Setarol
Unsaturated polyester resins; for fiber-reinforced polyesters. *Synthese BV.* Name unverified.

28093 Setatack A
Acrylic polyols; for pressure-sensitive adhesives. *Synthese BV.* Name unverified.

28094 Setatack AF
Thermoplastic/thermosetting acrylics; for hotmelt adhesives. *Synthese BV.* Name unverified.

28095 Setatack LP
Linear polyester polyols; PUR elastomers and prepolymers. *Synthese BV.* Name unverified.

28096 Setatack P
Polyesterpolyols; for two-pack PUR adhesives. *Synthese BV.* Name unverified.

28097 Setatack T
Modified rosin esters; for lamination/paint lamination adhesives. *Synthese BV.* Name unverified.

28098 Sethotope
1187-56-0 8585
$C_5H_{11}NO_2{}^{75}Se$
2-amino-4-(methyl-^{75}Se-seleno)butanoic acid
Selenomethionine Se 75; diagnostic aid, a radioactive imaging agent. mp = 265° (dec). *Bristol-Myers Squibb Co Inc.* Name unverified.

28099 Sethoxydim
74051-80-2 8620 277-682-3
$C_{17}H_{29}NO_3S$
(±)-(EZ)-2-(1-ethoxyiminobutyl)-5-[2-(ethylthio)propyl]-3-hydroxycyclohex-2-enone
sethoxydime; BAS 90520H; Checkmate; Expand; Fervinal; Grasidim; Nabu; NP-55; Poast; SN 81742. Selective systemic herbicide absorbed by foliage, used in control of annual and perennial grasses in broad-leaved crops. $bp_{0.00003}$ > 90°; d^{25} = 1.043; soluble in H_2O (25 mg/l), soluble in organic solvents; LD_{50} (rat orl)= 3200 mg/kg *BASF; Rhône-Poulenc; Ewos; Schering; Sipcam; Nippon Soda;.*

28100 Setic
Waxes for dental laboratory technology in bead and block form. *Bayer AG.*

28101 Setilon
Fatty alcohol compound with emulsifiers, nonionic; softener with scrooping effect especially developed for absorbent cotton, silky finish of cotton fabrics. Softening agent for the raising of synthetic goods. *Henkel Chemicals Ltd.* Name unverified.

28102 Setilose
A French cellulose acetate artificial silk.

28103 Setilthe
Cellulose acetate molding granules. *Rhône-Poulenc UK.*

28104 Setirene
Styrenated acrylated alkyds; for quick-drying industrial paints. *Synthese BV.* Name unverified.

28105 Setit®
Activated dithiocarbamate blend; primary and ultra accelerators for latex compounds. *R. T. Vanderbilt Co Inc.* Discontinued.

28106 Setreat
High-solids linseed oil-based penetrating concrete sealer; for treating rough finished, porous concrete to penetrate and seal the surfaces to prevent the absorption of water, salts and other contaminants harmful to concrete. *Secure Inc.* Discontinued.

28107 Sett®
Hydrogenated triglycerides
Viscosity enhancer, hardstock for margarine, coating agent. *Grünau.*

28108 Setter 33
A solution concentrate containing 50 g benazolin, 237 g 2,4-DB and 43 g MCPA per liter; a post-emergence herbicide. *DowElanco Ltd.*

28109 Sevacarb
7440-44-0 1855 231-153-3
Carbon black. *Sevalco Ltd.*

28110 Sevamine KOV 4342B
Cationic surfactant composed of quaternary imidazoline in liquid form; raw material in the preparation of laundry softeners. *Chemische Fabriek Servo BV.* Name unverified.

28111 Sevefilm 20
Surfactant blend; film-forming, anticorrosive, antiredeposition agent for water cooling circuit treatment products. *Stepan Europe.*

28112 Sevelyte K
Potassium lauryl amino propionate, in orange yellow liquid form; pigment dispersant for paints and inks. *KWR Chemicals Ltd.* Name unverified.

28113 Seven Seas®
Cod liver oil, various nutritional supplements, herbal remedies; for relief of joint aches and pains; dietary supplement. *Seven Seas Ltd.*

28114 Sevestat ML 300
Surfactant blend; lubricant, antistat, emulsifier for textile lubricants for mineral and synthetic fibers. *Stepan Europe.*

28115 Sevin
63-25-2 1831 200-555-0
1-naphthyl N-methylcarbamate
A proprietary preparation of carbaryl; used as a veterinary insecticide. No manufacturer.

28116 Sevin® Brand SL
63-25-2 1831 200-555-0
Insecticide for control of insects such as grasshoppers, gypsy moths, mosquitoes, ticks, ants. *Rhône-Poulenc/Ag; W.A. Cleary.*

28117 Sewarin
81-81-2 10174 201-377-6
Warfarin
A rodenticide. *Killgerm Chemicals Ltd.*

28118 Sextone
108-94-1 2795 203-631-1
Cyclohexanone
Laporte UK Trading. Discontinued.

28119 Seymourite
A proprietary trade name for an alloy of 64% copper, 18% nickel, and 18% zinc. No manufacturer.

28120 SF18
3264
Dimethicone fluid; lubricant, antifoam, mold release; rubber and plastic lubricant; base fluid for grease; mold release for rubber, plastic, and food application; antifoam in food application and aq. defoaming formulations. *GE Silicones.*

28121 SF-20CF
9016-75-5
PPS, 20% carbon fiber-reinforced; offer strength, stiffness, lubricity, conductivity, thermal and chemical resistance; used for precision electromechanical components, corrosive chemical and fluid handling equipment. *Compounding Tech.*

28122 SF69
3264
Dimethyl silicone fluid; film modifier, antifloat, flow control, anticrating, and pigment control agent in varnishes, paints, enamels, surface coatings. *GE Silicones.*

28123 SF81
Dimethyl silicone fluid; used as low temperature damping media, heat transfer media, dielectric coolants; base fluid for silicone compounds. *GE Silicones.*

28124 SF96®
3264
Polydimethylsiloxane
Emollient, lubricant for polishes, antifoams, textiles, chemical specialties; plastic and rubber lubrication; dampening or heat transfer fluids; oil defoamer, paint additives; mold release for tires, rubber, plastics; textile softener and modifier. *GE Silicones.*

28125 SF99
Reactive polydimethylsiloxane fluid; forms water-repellent fluids with heat or heat/catalyst; used in textiles, particle treatment, magnesium oxide, and Calrod® units. *GE Silicones.*

28126 SF1023
Phenyl-containing silicone fluid; flow control agent in polyester coatings; pigment dispersant, gloss aid in organic finish systems. *GE Silicones.*

28127 SF1154
Polydimethyldiphenyl siloxane
Lubricant, dielectric coolant, coupler; high temp heat transfer application; base fluid in high temperature greases; high temperature ultrasonic coupler; high temperature bath and oxide protector for solder baths; outstanding heat resistance. *GE Silicones.*

28128 SF1173
69430-24-6
Cyclomethicone
Emollient, lubricant used in antiperspirants, skin care products, sunscreen products, hair conditioners, facial makeup, particle treatment. *GE Silicones.*

28129 SF1188
Dimethicone copolyol
Emollient, lubricant, and release agent for textiles, cosmetics and toiletries, paint, plastic mold release, and rubber lubricants. *GE Silicones.*

28130 SF1250
Chlorophenyl methyl siloxane
Lubricant, mold release agent, hydraulic systems; fluid transmission; servomotors and mechanisms; instruments, clocks, and timers; machine tool components; antifriction, rolling, and sliding mechanisms; shock absorbers and damping devices. *GE Silicones.*

28131 S-Flakes
8016-70-4 232-410-2
Hydrogenated soybean oil. *Karlshamns.*

28132 Shadeacrete
Dry powdered colorants, iron oxides, ochres, umbers and composite pigments; colors for mortars, concrete roofing tiles, floor tiles, sand-lime bricks, concrete blocks, reconstructed stone, split blocks, paving slabs and cement sheets. *W Hawley & Son Ltd.*

28133 Shadow
Clay; sun reflector for protection against sunburn on agronomic and ornamental crops. *Draxel Chemical Company.* Unverified.

28134 Shaku-do
A Japanese alloy. It usually contains 94-96% copper, 3.76-4.16% gold, and 0.08-1.55% silver.

28135 shale
A dark-grey or black mineral containing 73-80% mineral matter and 20-27% organic matter. It is a source of oil for lubricating purposes.

28136 shale oil
68308-34-9 269-646-0
The tarry oil obtained by the distillation of certain bituminous shales. It contains unsaturated hydrocarbons.

28137 Shatah
Surfactant used with agrochemicals. *Makhteshim Chemical Works Ltd.*

28138 Shawinigan Black
Acetylene black. *Monsanto (Solaris).* Name unverified.

28139 Shawplas
Abrasive compounds and polishes for plastic articles; for fabricated components, turned parts, buttons, buckles, spectacle frames in acrylic, cellulose acetate, casein, polyester. *Shawplas Ltd.* Name unverified.

28140 Shcherbokov's solder
An aluminum solder. It contains 49% zinc, 46% tin, and 1.5% aluminum.

28141 Shea Butter
68424-60-2 270-311-6
Fancol Karite Butter; Cetiol® SB45; Lipex 102; Shebu, Refined. Shea butter; ointment base, anti-irritant for skin, skin conditioner, occlusive agent, solv. for suntan preparations, body lotions, winter sports products, wrinkle creams, soaps, shave foams, shampoos, balsams. *Fanning; Henkel/Cospha; Henkel Canada; Karlshamns; RITA.*

28142 shea butter
Bambuk butter. The fat obtained from the seeds of *Butyrospernum parkii* or *Bassio parkii* .

28143 Shea butter extract
68424-59-9 270-310-0
Fancol Karite Extract. Emollient with excellent spreadability for suntan preparations, skin toners, lipsticks, eye liners, ointments, suppositories. *Fanning.*

28144 Shebu WS
PEG-50 shea butter; emollient, humectant, pigment, conditioner, lubricant for personal care formulations. *RITA.*

28145 Shebu, Refined
68424-60-2 270-311-6
Shea butter; emollient, humectant, pigment, conditioner, lubricant for personal care formulations. *RITA.*

28146 Shell 5A18Z
9003-07-0 7741
PP homopolymer, nucleated, antistat; for injection molding; provides fast cycle times, good opticals; especially suited for cutlery. *Shell.*

28147 Shell 5A95
9003-07-0 7741
PP homopolymer; for cast unoriented film. *Shell.*

28148 Shell 5C64
9003-07-0 7741
PP homopolymer; for stretched tape processes which produce woven fabrics for carpet backing, industrial and geotextile fabrics. *Shell.*

28149 Shell 6A01K
9003-07-0 7741
PP random copolymer, antistat; for injection molding of high-quality housewares where high clarity is required. *Shell.*

28150 Shell 6C20S
9003-07-0 7741
PP random copolymer, antistat; for extrusion blow molding; especially suited for multi-layered barrier bottles. *Shell.*

28151 Shell 7C55H
9003-07-0 7741
PP extra high impact copolymer; heat resistant product for extrusion and injection molding, especially applications requiring high impact strength at ambient and sub-freezing temperatures; excellent base stock for compounding filled/reinforced grades. *Shell.*

28152 Shell D 50
94-75-7 2865 202-361-1
2,4-D; translocated herbicide for cereals and grass. *Shell UK.*

28153 Shell DS 6C46L, Shell DS 7C04N
9003-07-0 7741
PP high-ethylene random copolymer; for cast unoriented film. *Shell.*

28154 Shell DS 7C04N
9003-07-0 7741
Polypropylene impact copolymer. For cast, unoriented film. *Shell.*

28155 Shell Elexar
Thermoplastic rubber; used for wire and cable insulations. *Shell.*

28156 Shell JF 6100
9003-07-0 7741
PP homopolymer; for heat-set oriented film. *Shell.*

28157 shell limestone
A variety of calcium carbonate in massive form.

28158 Shell PDC 1120
9003-07-0 7741
PP low-ethylene random copolymer; for heat-set oriented film. *Shell.*

28159 Shell WRS 6-198
9003-07-0 7741
PP high-ethylene random copolymer; for shrink oriented film. *Shell.*

28160 Shell WRS 6-205
9003-07-0 7741
PP low-ethylene random copolymer; for heat-set oriented film. *Shell.*

28161 shellackose
An alcohol-soluble phenol-formaldehyde resin; used in the preparation of lacquers.

28162 Shellflex Process and Extender Oils
Hydrocarbon solvents which generally have lower aromatic content than the Dutrex grades and cover a wide viscosity range; used as extender and process oils for natural and synthetic rubber. *Shell.*

28163 Shellite
It is a mixture of ammonium perchlorate and paraffin wax; an explosive.

28164 Shellsol D40, D60, D70
Highly refined solvents with a very low aromatic content and a very slight sweet odor; principal applications similar to white spirits, in low odor paints and metal cleaning products. *Shell.* Name unverified.

28165 Shellsol E, A, AB, R
High-boiling aromatic hydrocarbons; versatile solvents used in paints, varnishes and in the preparation of agricultural chemical formulations. *Shell.* Name unverified.

28166 shellsol T
A high-boiling, isoparaffinic, aliphatic solvent with high flash point; virtually without odor; useful in odorless paints, household aerosols, fragrant polishes, and cosmetic creams; also as a catalyst carrier in polymerization reactions. *Shell.* Name unverified.

28167 Shellswim 11T
A polymeric additive dissolved in toluene to facilitate its effective distribution in crude oil; enables waxy crude oils to be transported at temperatures below their pour point; also facilitates restarting of a pipeline following a shut-down. *Shell.* Name unverified.

28168 Shellswim 5X
A polymeric additive dissolved in xylene to facilitate its homogeneous distribution in fuel and crude oils; can be added to waxy residual fuels and waxy crude oils to permit storage, handling and pumping below their natural pour point. *Shell.* Name unverified.

28169 Shellvis 50 (SAP 150)
A styrene-based hydrocarbon viscosity index improver for engine lubricants; normally supplied in the form of bales which require shredding before dissolving in lubricating oil, although also available as a concentrate or in a crumb form. *Shell.* Name unverified.

28170 Sherbelizer®
9005-37-2
Propylene glycol alginate blend; used for sherberts, soft-serve mixes, dispensed milk shakes, ice cream, sour cream and dips. *Kelco.*

28171 Sherpa
An insecticide based upon cypermethrin. *May & Baker Ltd.* Name unverified.

28172 Shibu-ichi
A Japanese alloy containing 51-67% copper, 32-49% silver, and traces of gold and iron.

28173 Shield
1702-17-6 2462 216-935-4
Clopyralid. Selective herbicide based upon clopyralid. *Murphy Chemical Co Ltd.* Discontinued.

28174 shikimole
94-59-7 8468 202-345-4
$C_{10}H_{10}O_2$
Safrole, the chief constituent of oil of sassafras. Used in perfumery.

28175 shikon
The dried roots of *Lithospermum erythrorhizon* .

28176 Shilajatu
An Indian mineral gum.

28177 Shimosite
A Japanese explosive, the chief constituent of which is picric acid.

28178 shinnamu
A vegetable dye obtained from a species of maple found in Korea.

28179 shio liao
A Chinese cement for marble, porcelain, etc., made from 54%, slaked lime, 6% alum, and 40% blood.

28180 Ship Shape® Resin Cleaner
N-Methyl-2-pyrrolidone, butyrolactone, deceth-6, fragrance
Resin cleaner for fiberglass fabrication, furniture refinishing, rubber industry. *ISP.*

28181 Shipley's solutions
Solutions of pyro-gallol and caustic soda in water, usually 10 ml of 1:1 caustic soda solution, 1 and 4 ml water, and 2 and 10 g pyrogallol; used to measure the absorption of oxygen.

28182 Shiro Bishi®
Coke; for general grade and malleable cast iron. *Mitsubishi Kasei.* Name unverified.

28183 SHL Lawn Sand Plus
Dichlorophen + ferrous sulfate; moss killer/fertilizer mixture for turf. *Sinclair Horticulture & Leisure Ltd.*

28184 Shoemaker's black
7782-63-0 4105 233-072-9
$FeSO_4 \cdot 7H_2O$
Ferrous sulfate

28185 Shoemaker's paste
A paste made by allowing the gluten from flour to putrefy, rolling it out thin, and making it into a paste; used for securing leather to leather, paper, or other material.

28186 Shokusen SE
Sugar ester, propylene glycol, ethanol, sodium citrate; detergent for foods. *Mitsubishi Kasei.* Name unverified.

28187 Shostakovsky Balsam
111-34-2 203-860-7
$C_6H_{12}O$
1-(ethenyloxy)butane
vinyl butyl ether; vinyl-n-butyl ether; Polyvinox. A proprietary preparation of synthetic vinyl butyl ether. mp = -92°; bp = 94°; SG = 0.774; insoluble in H_2O. *Leopold Charles & Co.* Unverified.

28188 shot metal
Shot lead, bullet metal. An alloy of lead with not more than 3% arsenic. One alloy contains 99.8% lead and 0.2% arsenic.

28189 Shur-Coal® FCA
56-81-5 4493 200-289-5
Glycerin; freeze conditioner for coal industry; pumpable temp.-25°C. *Sherex/Div. of Witco.*

28190 Si 69
40372-72-3 254-896-5
3,16-dioxa-8,9,10,11-tetrathia-4,15-disilaoctacane

Bis(3-triethoxysilylpropyl) tetrasulfane. Crosslinking/coupling agent for rubber industry. *Degussa.*

28191 Si 264
3-Thiocyanatopropyltriethoxy silane
Reinforcing agent improving properties of fillers (silicas, silicates, clays and whitings) in unsaturated polymers with double bonds (NR, IR, SBR, BR, NBR, EPDM). *Degussa.*

28192 sialonite
A mineral, Be_8Se_3.

28193 Siapton
Liquid organic foliar feed. *ICI Chem & Polymers Ltd.*

28194 *Sibercizer C6*
N-ethyl *o-/p*-toluene sulfonamide.
Siba Hegner Ltd.

28195 Siberez
Toluene sulfonamide resins. *Siba Hegner Ltd.*

28196 Sibley alloy
An alloy of 67% aluminum and 33% zinc.

28197 Sibor
A proprietary safety glass. No manufacturer.

28198 Sibutol®
Bitertanol [55179-31-2] and fuberidazole [3878-19-1]; seed dressing for cereals for control of loose smuts, stinking smuts, flag smuts; especially effective against dwarf bunt of wheat and seed-borne snow mold. *Bayer AG.*

28199 Sical
An alloy of from 22-29% aluminum, 50-51% silicon, 2-4% titanium, 1% calcium, 0.2-0.3% carbon, and the remainder iron.

28200 siccative
12228-91-0 5764 235-446-7
$MnB_4O_7 \cdot 8H_2O$
Manganese borate
Used as a drying agent mixed with linseed oil and resin, for impregnating leather. Insoluble in H_2O, organic solvents.

28201 Siccatol
A range of metal-based drying agents. *Akzo Chemie UK Ltd.*

28202 Siccolam
13463-67-7 9612 236-675-5
Compound titanium dioxide paste; desiccant for exudatory dermatoses. *British Drug Houses.* Name unverified.

28203 Sickle
Bromoxynil + fluroypyr; post-emergence contact herbicide for cereals. *DowElanco Ltd.*

28204 Siclor
1897-45-6 2219 217-588-1
Tetrachloroisophthalonitrile
Fungicide for the preparation of antifouling marine paints, aqueous paints, wood primers, adhesives. *Caffaro SpA.* Name unverified.

28205 Sico®
Predominantly azo pigments; for baking and air-drying finishes, for letterpress, offset, flexographic and special gravure inks, for coloring plastics. *BASF AG.*

28206 Sicocab®
Organic and inorganic pigments in a cellulose acetobutyrate vehicle; for wood stains with excellent fastness. *BASF AG.*

28207 Sicodop®
Organic and inorganic pigments in DOP; for coloring plasticized PVC. *BASF AG.*

28208 Sicoflex 80, 85
A proprietary ABS terpolymer possessing very high flow properties. *Mazzucchelli Celluloide Spa.* Name unverified.

28209 Sicoflex 85
A proprietary ABS terpolymer possessing high flow properties. *Mazzuccheli Celluloid Spa.* Name unverified.

28210 Sicoflex 90
A proprietary ABS terpolymer having high impact strength. *Mazzucchelli Celluloide Spa.* Name unverified.

28211 Sicoflex 93
A proprietary general purpose ABS terpolymer. *Mazzucchelli Celluloide Spa.* Name unverified.

28212 sicoflex 95
A proprietary ABS terpolymer possessing high tensile strength. *Mazzucchelli Celluloide Spa.* Name unverified.

28213 Sicoflex 99
A proprietary ABS terpolymer offering high resistance to heat. *Mazzucchelli Celluloide Spa.* Name unverified.

28214 Sicoflex MBS
A thermoplastic material based on methyl methacrylate, butadiene and styrene. *Mazzucchelli Celluloide Spa.* Name unverified.

28215 Sicoflush® A
Organic and inorganic pigments in medium oil soy alkyd resin; for air-drying and stove-drying finishes which are compatible with alkyd resins. *BASF AG.*

28216 Sicolen®
Organic and inorganic pigments in polyethylene; for coloring polyethylene hollow articles, injection molding and extrusion articles. *BASF AG.*

28217 Sicolub®
Waxes. *BASF plc.*

28218 Sicolub® DSP
distearyl phthalate
Lubricant for PVC. *BASF AG.*

28219 Sicolub® E
8002-53-7 6339 232-313-5
Montan wax derivative; lubricant for PVC. *BASF AG.* Name unverified.

28220 Sicolub® EDS
Ethylene diamine distearyl amide;
Lubricant for PVC and PS. *BASF AG.* Name unverified.

28221 Sicolub® OA2, OA4
Oxidized polyethylene wax; lubricant for PVC. *BASF AG.* Name unverified.

28222 Sicolub® TDS
Isotridecyl stearate
Lubricant for PVC. *BASF AG.* Name unverified.

28223 Sicomet®
Anionic dyes, cationic dyes, fat-soluble dyes, organic pigments, inorganic pigments, pigment concentrates; colorants for cosmetics. *BASF AG.*

28224 Sicomin®
Chrome yellow and molybdate orange pigments; for paints and surface coatings, for coloring plastics, for flexographic and gravure inks for packaging materials, laminated paper coloring. *BASF AG.*

28225 Sicomix®
Pigment concentrates for surface coatings. *BASF AG.*

28226 Sicopal®
Inorganic pigments having a spinel structure based on various metal oxides; for industrial finishes with excellent fastness and for coloring plastics. *BASF AG.*

28227 Sicopharm®
Soluble dyes and pigments for coloring pharmaceuticals. *BASF AG.*

28228 Sicoplast®
Predispersed pigment mixtures for mass coloring of thermoplastics. *BASF AG.*

28229 Sicopos®
Organic and inorganic pigment concentrates for coloring PET. *BASF AG.*

28230 Sicopur®
High purity iron oxides for paper refiners. *BASF AG.*

28231 Sicopurol®
Organic and inorganic pigment concentrates in ester polyol; for coloring polyurethane foams based on polyester or polyether polyols. *BASF AG.*

28232 Sicorin®
Anticorrosion pigments. *BASF AG.*

28233 Sicostab®
Heat stabilizers for rigid and plasticized PVC, for internal and external applications, PVC foams. *BASF AG.*

28234 Sicostyren®
Organic and inorganic pigment concentrates in polystyrene; for coloring polystyrene and styrene copolymers. *BASF AG.*

28235 Sicotan®
Inorganic mixed-phase pigments with structure of rutile titanium dioxide and other metal oxides; in combination with organic pigments for luminous hues for surface coatings and plastics with high processing temperatures. *BASF AG.*

28236 Sicotherm®
Cadmium sulfide/zinc sulfide or cadmium sulfide/selenide mixed crystals; for coloring plastics and paints. *BASF AG.*

28237 Sicotrans®
Transparent inorganic pigments of extremely fine particle size; for high-quality paint systems, especially metallics, and for coloring plastics. *BASF AG.*

28238 Sicoversal®
Organic and inorganic pigment concentrates; for coloring thermoplastics. *BASF AG.*

28239 Sicovinyl®
Organic and inorganic pigments in plasticized PVC; for mass coloring of plasticized PVC. *BASF AG.*

28240 Sicovit®
Soluble colorants and pigments for coloring foodstuffs. *BASF AG.*
28241 Sicromo Steel
A proprietary trade name for a chromium-silicon-molybdenum steel containing from 2.25-2.75% chromium, 0.5-1.0% silicon, 0.4-0.6% molybdenum and up to 0.15% carbon. No manufacturer.
28242 Sidanyl
Polyamide film; for incorporation into laminates, well suited for thermoforming. *UCB nv Film Sector.* Discontinued.
28243 Sident 15
Silica
Abrasiveness and thickening agent for toothpastes. *Degussa.*
28244 Sident 22LS, 22S
Silica
Nonabrasive thickening agent for toothpastes. *Degussa.*
28245 Sideraphthite
An alloy resembling silver. It contains 64.5% iron, 22.5% nickel, 4.5% each aluminum and copper, and 4% tungsten; nonoxidizable.
28246 sidero cement
A cement in which iron ores are wholly or partly substituted for the clay.
28247 Sidot's blende
A phosphorescent zinc sulfide.
28248 Siemensite
A refractory material produced by fusing a mixture of chromite, bauxite, magnesite, and a reducing agent in the arc furnace to obtain a slag containing from 20-40% Cr_2O_3, 25-45% Al_2O_3, 18-30% MgO, and 8-14% other constituents. No manufacturer.
28249 sienna
1309-37-1 4072 215-168-2
Hydrated oxide of iron, mixed with a little manganese, and clay. It contains from 50-70% Fe_2O_3, 8-12% SiO_2, 2-8% Al_2O_3, 2-5% $CaSO_4$ or CaO, and water; a pigment in oil paints, stains, pastels.
28250 Sifbronze
A proprietary trade name for a brass containing some ferromanganese and tin. No manufacturer.
28251 Siflox
Highly dispersed precipitated silica; applied in shoe soling materials, hoses, cable sheeting, profiles, etc. *Chemische Fabriken Oker und Braunschweig AG.* Name unverified.
28252 Siflural
A trade name for a solution of aluminum fluosilicate; a disinfectant. No manufacturer.
28253 Sigal
A proprietary alloy of 10% Si and 90% Al; a pigment. No manufacturer.
28254 Sigma
Lube oil. *Monsanto (Solaris).* Name unverified.
28255 Sigmalium
A proprietary trade name for an alloy of aluminum containing 1% silicon, 4% copper, and 0.7% magnesium. No manufacturer.
28256 Sigmathane
A proprietary single-pack moisture-cured polyurethane coating. *Sigma Coatings.* Name unverified.
28257 Signal Red
A dyestuff; the British equivalent of Lithol red B. No manufacturer.
28258 SIL Therm
Thermoset polyester; bulk and sheet molding compounds; excellent corrosion resistance, superior moldability. *Industrial Dielectrics.* Name unverified.
28259 Silacros
Silicates for use in industry. *Crosfield Chemicals Ltd.*
28260 Silacto®
Activator. *Kenrich Petrochemicals.*
28261 Silal
A proprietary trade name for a grey iron with 5% silicon and 2.5% total carbon; stated to resist oxidation, growth, and scaling up to 750°C. No manufacturer.
28262 Silamide DCA-100
Conditioner for personal care products. *Siltech.* Discontinued.
28263 Silamine 65
Gloss and conditioning agent for personal care products. *Siltech.* Discontinued.
28264 Silanca
A stainless silver with a high silver content.
28265 silane
7803-62-5 8633 232-263-4
SiH_4
silicane
silicon tetrahydride. Used as a doping agent, production of amorphous silicon.
28266 Silantox
7631-86-9 8637 231-545-4
A proprietary colloidal silicon dioxide. No manufacturer.
28267 silaonite
A mineral; a mixture of bismuth and bismuth trisulfide.
28268 Silasorb
10101-39-0 233-250-6
$Ca_2O_4Si \cdot nH_2O$
silicic acid calcium salt (1:1)
calcium metasilicate. Synthetic calcium silicate; adsorbent; controls free fatty acids. *Celite.*
28269 Silastic® 21145
63394-02-5
Silicone rubber compound; flame-resistant elastomer for compression, transfer, or injection molding applications. *Dow Corning STI.*
28270 Silastic® GP-30
63394-02-5
Silicone rubber; general purpose elastomer for molding, calendering, and blending; FDA compliance. *Dow Corning STI.*
28271 Silastic® GP-950+
63394-02-5
Silicone rubber with extending filler; economical, extendable, general-purpose elastomer for extrusion, molding, calendering; FDA compliance. *Dow Corning STI.*
28272 Silastic® HE-26
63394-02-5
Silicone rubber; specialty elastomer for extrusion, molding, calendering; low modulus; FDA compliance. *Dow Corning STI.*
28273 Silastic® LCS-740
63394-02-5
Silicone rubber; low compression set elastomer for molding, calendering; reversion resist.; no post cure; FDA compliance. *Dow Corning STI.*
28274 Silastic® LT-40
63394-02-5
Silicone rubber; low temperature elastomer for molding, calendering; accepts filler. *Dow Corning STI.*
28275 Silastic® NPC-40
63394-02-5
Silicone rubber; no post cure elastomer for extrusion, molding, calendering; accepts filler; FDA compliance. *Dow Corning STI.*
28276 Silastic® SPG-30
63394-02-5
Silicone rubber; specialty elastomer for molding; low-temp. sponge base; FDA compliance. *Dow Corning STI.*
28277 Silastic® TR-55
63394-02-5
Silicone rubber; tough elastomer for extrusion, molding, calendering; excellent tear and flex life; no post cure; FDA compliance. *Dow Corning STI.*
28278 Silastic® WC-50
63394-02-5
Silicone rubber; wire and cable elastomer for extrusion, molding, calendering; excellent long-term heat stability; accepts filler; no post cure. *Dow Corning STI.*
28279 Silastic® HS-30
63394-02-5
Silicone rubber; high strength elastomer for molding, calendering; FDA compliance. *Dow Corning STI.* Name unverified.
28280 Silastomer
Silicone rubbers. *Midland Silicones.* Unverified.
28281 Silastoseal
Room temperature curing silicone rubber sealants. *Midland Silicones.* Unverified.
28282 Silberit
A jewelry alloy; contains aluminum, nickel, and silver.
28283 Silbione
Silicones for cosmetics. *Rhône-Poulenc UK.*
28284 Silbond® 40
Ethyl polysilicate
Intermediate for binders used for inorganic zinc coatings, investment casting molds, cores, ceramic shapes and coatings. *Akzo.*
28285 Silbond® Condensed
78-10-4 3895 201-083-8
Tetraethyl orthosilicate 90%; intermediate for binders used for investment castings, ceramic shapes, cores, and coatings; chemical intermediate. *Akzo.*

28286 Silcar
A proprietary trade name for a pigment comprising a mixture of silicon dioxide and silicon carbide. No manufacturer.

28287 Silcasil S
An inorganic filler; uses include carrier material for insecticides, auxiliary for improving the free-flowing and grinding properties of powders, thickener for liquids. *Bayer AG.*

28288 Silcat
Crosslinker. *Union Carbide.*

28289 Sil-Cell® 32
Glass microcellular filler (73% silicon dioxide, 17% aluminum oxide, 5% potassium oxide); inert, inorganic, nontoxic hollow irregular shape filler/reinforcement for manufacturing of adhesives, auto body putty, cultured marble, coatings, wall patching compounds, stucco. *Silbrico.*

28290 Silchrome
A heat-resisting alloy containing 86% iron, 9.5% chromium, 4% silicon, and 0.5% carbon.

28291 Silchrome 46M
A proprietary trade name for a chromium steel containing 4-6% chromium, 0.5% molybdenum, and 0.2% carbon. No manufacturer.

28292 Silchrome R.A.
A proprietary trade name for a steel containing 16% chromium, 1% silicon, 1% copper, and 0.12% carbon. No manufacturer.

28293 Silchrome Wire
An alloy of iron with 18% chromium, 3% silicon, 3% tungsten, and 0.3% carbon. No manufacturer.

28294 Silcolapse
Textile auxiliary chemicals. *ICI Chem & Polymers Ltd.* Discontinued.

28295 Silcolease
Silicone coatings. *ICI Chem & Polymers Ltd.* Discontinued.

28296 Silcon
Silafilcon; contact lens material. *Dow Corning Ophthalmics Inc.* Name unverified.

28297 Silcoset
Rubber curing agents. *ICI Chem & Polymers Ltd.* Discontinued.

28298 Silcron®
A fine-particle silica. *SCM.* Name unverified.

28299 Sildura®
A range of silicone-rubber compositions, curable by the application of heat. *GE Silicones.* Name unverified.

28300 Silene
10101-39-0 233-250-6
A proprietary trade name for a precipitated calcium silicate; used in rubber mixes to give wear-resistance. No manufacturer.

28301 Silesia powder
It is a mixture of 75% potassium chlorate, with pure or nitrated resin, and a little castor oil; an explosive.

28302 Silesite
A tin silicate with 55% tin, found in the Bolivian tin deposits.

28303 Silester
78-10-4 3895 201-083-8
$C_8H_{20}O_4Si$
Ethyl silicate
tetraethyl orthosilicate; tetraethoxysilane; TEOS. Used to weatherproof and harden stone and in the manufacture of weather- and acid-proof mortars and cements. mp = -77°; bp = 168°; d = 0.9340; n_D^{20} = 1.3825; insoluble in H_2O, slightly soluble in EtOH. *Monsanto plc.*

28304 silex
A name applied to silica (SiO_2); used also for tripoli employed as a filler in paints. A ground flint is also known as silex.

28305 Silflex
Cosmetic specialties. *Rhône-Poulenc UK.*

28306 Sil-fos
A proprietary trade name for a phosphor-silver brazing solder containing 80% copper, 15% silver, and 5% phosphorus. No manufacturer.

28307 Silfrax
A product obtained by the action of silicon on carbon and consisting of carbon with a coating of silicon carbide and carbon; stated to be tougher and stronger than carborundum; used as a refractory material in the manufacture of pyrometer tubes for electrical heating elements. No manufacturer.

28308 Sil-Free
Chemical additive to prevent silicate deposit in peroxide/silicate bleach system. *Sybron.*

28309 Silglaze® II 2800
Silicone sealant; sealant offering extended tooling time and shorter cure time; for general sealing and glazing, skylights, window installation, and in-shop window fabrication. *GE Silicones.*

28310 Silgrip® PSA590
Polydimethylsiloxane gum/polysiloxane resin reactive copolymer in toluene solution; pressure sensitive adhesive for splicing and repair tape and platers tape for the printed circuit market. *GE Silicones.*

28311 Silgrip® PSA595
Polydimethylsiloxane gum/polysiloxane resin reactive copolymer in xylene solution; pressure sensitive adhesive for platers tape for the printed circuit market. *GE Silicones.*

28312 Silhydrate C
131044-78-5
Copper PCA methylsilanol
Provides skin moisturization, skin restructuring for anti-aging formulations, skin treatments, cosmetic and toiletry emulsions, creams, lotions. *Exsymol.*

28313 silica
7631-86-9 8637 231-545-4
O_2Si
silicon dioxide, fumed
silicon dioxide; silicic anhydride; quartz. Inorganic oxide; occurs in nature as agate, amethyst, chalcedony, cristobalite, flint, quartz, sand, tridymite; manufacture of glass, water glass, refractories, abrasives, ceramics, enamels, petrol. products; filler in cosmetics, rubber reinforcing agent, anticaking and defoaming agent, abrasive and thickener. d^0 = 2.2 (for quartz, 2.65; melts to a glass; insoluble in all solvents except HF. Akzo; BYK-Chemie; Cabot Carbon Ltd; Catalysts & Chemicals Industries; Chisso Am.; Degussa; Du Pont; Geltech; J. M. Huber; Nippon Silica Ind.; Nissan Chem; Ind.; PPG Industries; PQ; Unimin; U. S. Silica.

28314 Silica FK 160, FK 320 DS
Silica
Active filler for hot-vulcanizing silicone rubber. *Degussa AG.*

28315 Silica FK 320 DS
7631-86-9 8637 231-545-4
silica
Active filler for hot-vulcanizing silicone rubber. *Degussa AG.*

28316 silica gel
The name applied to a colloidal form of silica, prepared by treating sodium silicate with acetic or hydrochloric acid, washing the gelatinous silica, and drying. It is highly absorbent, and is used to absorb water vapors.

28317 silicam
Silicon imido-nitride, Si_2N_3H, formed when silicon diimide is heated to 900°C in an atmosphere of dry nitrogen.

28318 Silicane
7803-62-5 8633 232-263-4
H_4Si
Silicon hydride
silane; silicon tetrahydride; monosilane. Gas, source of ultra-pure silicon for electronics fabrication. mp = -185°; bp = -112°; d^{-185} = 0.68; dissolves slowly in H_2O with decomposition, insoluble in all other solvents.

28319 silicated soap
A soap to which water glass (sodium silicate) has been added; a detergent.

28320 Silicex
Silicone products; antifoams, release agents, molding rubbers and sealants. *Siliconas Hispania SA.*

28321 Silicium
A proprietary trade name for silicon used as a pigment in the Atephen system (qv); a chemically resistant coating. No manufacturer.

28322 Silicoderm® F
A preparation for the protection and the care of the skin. *Bayer AG.*

28323 silicol
Ferro-silicon, usually containing 84% silicon, for use in the preparation of hydrogen by the action of caustic soda.

28324 silicolloid
A natural siliceous material free from iron. Suitable for use in paper manufacture, cleansers, and tooth pastes.

28325 silico-manganese
An alloy of silicon and manganese made in the electric are type of furnace. It contains 60-75% manganese, 20-25% silicon, and the rest iron.

28326 silicon
7440-21-3 8635 231-130-8
Si
silicon
Nonmetallic element; Si; semiconductor in solid-state devices; organosilicon compounds; silicon carbide; alloying agent in steels, aluminum, copper, bronze, iron; cermets, refractories; halogenated silanes; deoxidizer in steel manufacture. mp =1410°; d_4^{25} = 2.33; insoluble in H_2O, organic solvents. *Atomergic Chemetals; Cerac; Dow Corning; Eagle-Picher; Pechiney Electrometallurgie; Shin-Etsu Chem.*

28327 silicon brass
An alloy of 81% copper, 14% zinc, and 3% silicon.

28328 **silicon bronze**
Silicum bronze
An alloy of 97.37% copper, 1.32% tin, 1.24% zinc, and 0.7% silicon.

28329 **silicon carbide**
409-21-2 8636 206-991-8
SiC; Abrasive for cutting and grinding metals, refractory in nonferrous metallurgy, ceramic industry, boiler furnaces, composite tubes for steam reforming operations; fibrous form in filament-wound structures and heat-resistant, high-strength composites. *Atomergic Chemetals; Carborundum; Lonza; Mitsui Toatsu; New Metals & Chems. Ltd; Showa Denko.*

28330 **silicon copper**
Alloys of copper with small amounts of silicon; used for the manufacture of telephone and telegraph wires. An alloy with 10% silicon is also called silicon-copper.

28331 **silicon nickel brass**
An alloy of 81% copper, 14% zinc, 3% silicon, and 2% nickel.

28332 **silicon steel**
A steel made by melting steel and ferro-silicon in crucibles; used for making sheets, springs, and acidresisting plants.

28333 **silicon tetrachloride**
10026-04-7 8644 233-054-0
$SiCl_4$
tetrachlorosilane
silicon chloride. Used in generation of smoke screens; manufacture of ethyl silicate, silicones, highpurity silica, fused silica glass; source of silicon, silica, and hydrogen chloride; laboratory reagent. mp = -70°; bp = 59°; d^0_4 = 1.52; decomposed by H_2O, soluble in organic solvents. *Air Products; Atomergic Chemetals; Chisso Am.; Degussa; Dow Corning; Hüls; PCR; Union Carbide.*

28334 **silicon tetrafluoride**
7783-61-1 8645 232-015-5
F_4Si
tetrafluorosilane
silicon fluoride. Manufacture of fluosilicic acid, intermediate in manufacture of pure silicon, to seal water out of oil wells during drilling. mp = -90°; d^{-80} = 1.590. *Air Prods & Chem; Atomergic Chemetals; Hüls Am.; PCR.*

28335 **silicone**
8639
Organosiloxane. Siloxane polymers; thermosetting siloxane polymers used as mold release for plastics and rubber, defoamers for mining, latex, ink, soaps, agriculture., food processing, as lubricants, conditioners, and emollients in personal care products; molding compounds and encapsulants. Dow Corning; GE Silicones; Genesee Polymers; Goldschmidt; Hüls Am.; Miles; Sandoz; Shin-Etsu Chem.; Union Carbide; Wacker Silicones.

28336 **Silicone Antifoam Emulsion SE 9**
Silicone emulsion; antifoam with good storage stability; recommended for food sector, as processing aid in pharmaceutical and cosmetic industries and in fermentation processes; also for manufacturing of plastics in contact with food; for degassing of polymers. *Wacker-Chemie.*

28337 **silicone elastomer**
63394-02-5
polydimethylsiloxane rubber
Room temperature vulcanizing silicone rubber; encapsulation of electronic parts, elec. insulation, gaskets, surgical membranes and implants; automotive engine components; miscellaneous mechanical parts. *Ambersil Ltd; Dow Corning Europe; GE Silicones Europe.*

28338 **Silicone Emulsion E-130**
Silicone emulsion; used for mold release, furniture polish emulsions, cosmetics, textile lubricants and softeners. *Wacker-Chemie.*

28339 **Silicone Systems 820**
One-component silicone systems; optically clear silicone. *R.H. Carlson.*

28340 **Silicone Systems 976**
One-component silicone system; self-leveling high temperature silicone for electrical/electronic potting applications. *R.H. Carlson.*

28341 **silicon-eisen**
Silicon pig. A pig iron containing from 5-15% silicon.

28342 **Silicosehl**
Room temperature vulcanizing two-part silicone rubber systems; for casting and potting requiring exact surface detail, flexibility, stability and excellent electrical characteristics over wide temperature range; chemical and weather resistant; non-toxic. *Solochart Ltd.* Name unverified.

28343 **silico-spiegel**
An alloy of 20% manganese, 12% silicon, and the rest iron.

28344 **silico-superphosphate**
A preparation made by mixing superphosphate with kieselguhr or precipitated silicic acid. It is stated to give better results on medium and light soils.

28345 **silico-titanium**
A titanium-silicon alloy used in the steel industry.

28346 **Siligen®**
Textile finishing auxiliaries. *BASF plc.*

28347 **Siligen® FA**
Antimigrant and softening agent for pigment padding, textile finishing. *BASF.*

28348 **Siligen® GL**
Nonfoaming softening agent for resin finishing cellulosic fibers and their blends with synthetics. *BASF.*

28349 **Siligen® MA**
Wetting agent for mineral application techniques. *BASF.*

28350 **Siligen® TX-510**
Modified silicone elastomer; textile auxiliary. *BASF.*

28351 **Siligran**
A range of inorganic smelted products and their mixtures; used as antiscale and fluxing compounds in steel production. *Bayer AG.*

28352 **Silikoftal®**
Resin binders for surface coatings. *Thomas Goldschmidt Ltd.*

28353 **Silikoftal® CC 3**
Silicone-modified polyester resin; binder for weather-resistant lacquers applied by coil-coating. *Tego GmbH.* Name unverified.

28354 **Silikoftal® HTL 2**
Silicone-modified polyester resin; binder for decorative high-quality coating of cookware and electrical domestic appliances. *Tego GmbH.* Name unverified.

28355 **Silikophen® Nonstick 50**
Reactive release agent and temperature-resistant silicone resin. *Tego GmbH.* Name unverified.

28356 **Silikophen® P 40/W**
68083-14-7
Phenylmethyl polysiloxane resin emulsion; binder for water-based high temperature anticorrosion paints resist. to 650°C, e.g., coatings applied to industrial equipment, chimneys, and exhaust tubes. *Tego GmbH.* Name unverified.

28357 **Silipact**
Elastic sealants; for the construction industry and formed-in-place gaskets. *Lonza AG.*

28358 **Siliporite**
Molecular sieves; for drying of liquids and gases, desulfuration, separation of gases and isomers, in the chemical and petrochemical industries; drying agent for the double glazing industry and for polyurethane formulations. *Elf Atochem UK/Ceca.*

28359 **siliset**
Foundry chemical hardeners. *Foseco (F.S.) Ltd.*

28360 **Silit**
A material made by exposing mixed silicon, silicon carbide, and carbon, to the action of carbon monoxide at 1500+C. It is made in three qualities. a) A material for resistance subjected to permanent losses, b) for electric heating work up to 1400°C and c) a fireproof material capable of withstanding violent changes in temperature.

28361 **Silk grass**
A term applied to pineapple fiber, obtained from the pineapple plant; used for making cloth in the Phihppine islands.

28362 **Silk Pro-Tein**
96690-41-4 306-235-8
Hydrolyzed silk; moisturizer, substantive protein, protective barrier for elegant skin and hair preparations (lotions, creams, bath gels, shampoos, conditioners, treatment products, shave preparations, soap bars). *Maybrook.*

28363 **silk wadding**
The waste from the spinning of silk.

28364 **Silk, Anaphe**
The silk obtained from a caterpillar in German East Africa; it has a specific gravity of 1.282. No manufacturer.

28365 **Silkin**
A proprietary cellulose nitrate silk. No manufacturer.

28366 **Silkiol**
Additive; adhesive for carded worsted and cotton spinning. *Henkel Chemicals Ltd.* Name unverified.

28367 **Silksoft® Supreme**
Modified amine silicone; softener giving excellent soft and slick hand. *Sybron.*

28368 **Sillikolloid P 87**
A filler composed of quartz [14808-60-7] and kaolinite [1318-74-7] *Hoffmann Min.*

28369 **Sillitin N 82**
Hoffmann Mineral.

28370 **Sillitin-Aktisil**
Quartz-kaolinite; filler for rubber and paint, soft abrasive for polishing agent. *Hoffmann Mineral.* Name unverified.

28371 Sillman bronze
An alloy of 86% copper, 10% aluminum, and 4% iron.
28372 Silm
Photographic fixer stain remover. *May & Baker Ltd.* Name unverified.
28373 Silman Steel
A proprietary trade name for a silicon steel containing 2.1% silicon, 0.85% manganese, 0.3% vanadium, 0.25% chromium, and 0.55% carbon. No manufacturer.
28374 Silmar® 901R
Discontinued.
Thermoset polyester; synthetic marble two-way resin. *BP Chemicals Inc.*
28375 Silmar® S249
Discontinued.
Thermoset polyester; ortho laminating resin. *BP Chemicals Inc.*
28376 Silmar® S585
Discontinued.
Thermoset polyester; synthetic marble resin. *BP Chemicals Inc.*
28377 Silmar® S957
Discontinued.
Thermoset polyester; densified matrix resin. *BP Chemicals Inc.*
28378 Silmar® S958
Discontinued.
Thermoset polyester; ortho laminating resin. *BP Chemicals Inc.*
28379 Silmod 20A
Silicon-modified polyether; for formulating one and two-part sealants; excellent weatherability, good long-term storage stability, good paintability, good adhesive strength. *Union Carbide.*
28380 Silmod SAT-30
Silicon-modified polyether; for the formulation of adhesives for bonding large stone or metal panels to concrete, formation of expansion joints, the construction of interior walls, joining of materials with different coeffici *Union Carbide.*
28381 Sil-o-cel
A brand of kieselguhr, also a heat insulator made from kieselguhr. No manufacturer.
28382 Siloid®
A registered trademark for micron-sized silica gels. No manufacturer.
28383 Silopren®
Hot air and room temperature vulcanizing silicone rubber; used for seals, gaskets, dampening components, hoses, profiled strip, electrical insulating material, production of sealants, casting and coating compounds, fabric pro *Bayer AG.*
28384 Silopren® HV
A basic compound of silicone rubber enriched with crosslinking agents, pigments etc., ready for processing; used for press-cured and extruded articles, e.g., seals, damping components, articles for food contact, hoses, profiled strip, electrical insulating material, production of sealants, casting and coating compounds, fabric proofings and impression coatings. *Bayer AG; Rhein Chemie Rheinau.*
28385 Sil-O-Wet™
Inhibited aluminum pigments; used for aqueous paints and coatings used for decorative metallic effects as well as a wide range of protective coatings applications. *Silberline Mfg Co Inc.*
28386 siloxicon
A fireproof material resistant to the action of acids and alkalis; made by heating powdered silica with a small quantity of carbon in an electric furnace, the composition is Si_2C_2O. It is produced with silicon carbide in a carborundum furnace; used alone or with binding materials for making crucibles or muffles.
28387 siloxide
A mixture of silica with a little titanium, or zirconium oxide.
28388 Silpruf® 2000
Silicone sealant; for high-performance weatherproofing and glazing. *GE Silicones.*
28389 Silquat Q-100
Nonrewetting conditioner for personal care products. *Siltech.* Discontinued.
28390 Silres® KX
Methyl silicone resin; binder which dries to tackfree finish at room temperature, for highly pigmented paints (preferably with aluminum powder). *Wacker Silicones.*
28391 Silres® MP 42 E
Methyl phenyl silicone resin emulsion; dries to tack-free finish at room temperature; binder for highly pigmented and gloss paints, clear varnishes; recommended for blending with water-based organic resins; high shear resistance. *Wacker Silicones.*
28392 Silres® MSE 100
Polymethyl siloxane, methoxy-functional; solvent-free room temperature curing binder requiring addition of catalyst; for highly pigmented paints. *Wacker Silicones.*
28393 Silres® REN 50
Methyl phenyl silicone resin; binder which dries to tack-free finish at room temperature, for highly pigmented and gloss paints, clear varnishes; good adhesion to difficult substrates. *Wacker Silicones.*
28394 Silso®
Inorganic semiconductor materials as base material for solar cells. *Wacker-Chemitronic Gesellschaft fur Elektronik-grundstoff.*
28395 Silsoft
Elastofilcon A; contact lens material. *Dow Corning Ophthalmics Inc.* Name unverified.
28396 Silsoft
Silicone textile softener. *Union Carbide.*
28397 Silteg
A range of highly and medium-active aluminum silicates; used for shoe sole materials and technical rubber articles. *Degussa.*
28398 Siltek
CO_2 process binder with exceptional breakdown and fast gassing properties. *Foseco (F.S.) Ltd.*
28399 Siltek® L Polymer
9010-79-1
Ethylene/propylene copolymer; for use in cosmetic stick formulations; offers compatibility, improved oil retention and better stick structure. *Petrolite.*
28400 Siltem® STM1300
Silicone/polyetherimide block copolymer; excellent flame resistance, low smoke properties for applications where combustion and corrosion must be minimized; fast line and extrusion speeds for wire and cable manufacturing. *GE Plastics.*
28401 Siltex
Fused silica; high performance extender pigment and filler industrial coatings, polymer systems, abrasives, electrical molding and potting compounds, translucent/transparent sealants; outstanding electrical properties, low refractive index and thermal expansion. *Kaopolite.*
28402 Siltouch Cotton Plus
Amino functional silicone; durable softener with excellent shear stabilty. No manufacturer.
28403 Silumin
Proprietary alloys of aluminum and silicon containing 12% silicon; they have a specific gravity of 2.63-2.65. No manufacturer.
28404 Siluminite
An electric insulator consisting of 75% of mineral matter (asbestos, calcium silicate, and aluminum silicate), with pitch as the binding material.
28405 Silumin-Y
A proprietary aluminum-silicon alloy with small additions of manganese and magnesium; it has high corrosion resistance. No manufacturer.
28406 silundum
409-21-2 8636 206-991-8
A product similar to carborundum; articles, such as crucibles and tubes, are made by shaping pieces of graphite, embedding them in carborundum, and subjecting them to the action of silicon vapor at high temperatures in the electric furnace. It has a high electrical resistance and is used for making electrodes.
28407 Silva
A name used mainly in Germany for a type of artificial silk.
28408 Silvacur
107534-96-3 9253 403-640-2
tebuconazole
Fungicide with systemic properties and broad spectrum activity against rusts, leaf spot diseases, e.g., *Septoria* spp., powdery mildew and several *Fusarium* species on cereals; whitemold, *Phoma* and various leafspot diseases on oilseed rape. *Bayer AG.*
28409 silvapron
1702-17-6 2462 216-935-4
Clopyralid. Translocated herbicide for cereals and grass. Used in combination with 2,4-D. *BP Oil Ltd.* Discontinued.
28410 Silvatol
Pretreatment agent. *Ciba plc.* Name unverified.
28411 Silvaz
A proprietary trade name for an alloy used for the manufacture of steel, contains iron with 40-45% silicon, 6.0-6.5% vanadium, 6.0-6.5% aluminum, and 6.0-6.5% zirconium. No manufacturer.
28412 Silvel
A proprietary trade name for an alloy containing 67.9% copper, 16% zinc, 6.5% nickel, 0.5% lead, 2.2% iron, and 6.8% manganese. No manufacturer.
28413 silver
7440-22-4 8647 231-131-3

Ag
silver
Silver, colloidal. Metallic element; used in manufacture of silver nitrate, etc.; sterilant; water purification; for coinage; manufacture of tableware, jewelry, ornaments; for electroplating; as catalyst; in dental alloys. mp = 961°; bp = 2000°; d^{15} = 10.49. *Aldrich; Asarco; Cerac; Degussa; Handy & Harman; Mariovilla SpA.*

28414 silver acetate
563-63-3 8648 209-254-9
$C_2H_3AgO_2$
acetic acid silver salt
Laboratory reagent, oxidizing agent. d = 3.26; soluble in H_2O (1-3%). *Atomergic Chemetals; Spectrum Chem. Mfg.*

28415 silver alum
$Al_2(SO_4)_2{\cdot}Ag_2SO_4{\cdot}24H_2O$
An aluminum-silver sulfate.

28416 silver amalgam
An alloy of mercury and silver. It occurs as a mineral, but is also prepared artificially.

28417 silver bell metal
An alloy of 40-42% copper and 58-60% tin.

28418 Silver Bond 30
7631-86-9 8637 231-545-4
Silica
Filler, flatting agent for traffic paint, exterior block fillers, mastics and adhesives, buffing compounds. *Unimin.*

28419 silver bronze
An alloy of 64% copper, 17% manganese, 13% zinc, 5% silicon, and 1% aluminum. An electrical resistance alloy.

28420 silver carbonate
534-16-7 8650 208-590-3
CAg_2O_3
Laboratory reagent. Dec 220°; d = 6.08; slightly soluble in H_2O (0.3 g/l), readily soluble in acids, ammonia. *Aldrich; Atomergic Chemetals; Spectrum Chem. Mfg.*

28421 silver chloride
7783-90-6 8652 232-033-3
Ag_2CrO_4
Photography, photometry and optics, batteries, photochromic glass, silver plating, production of pure silver, antiseptic. mp = 455°; bp = 155o°; d = 5.56; soluble in H_2O (1.93 mg/l), ammonia, acids; *Atomergic Chemetals; Degussa; Noah Chem.*

28422 silver citrate
126-45-4 8654 204-786-8
$C_6H_5Ag_3O_7$
2-hydroxy-1,2,3-propanetricarboxylic acid silver salt
Itrol. Anti-infective dusting powder. soluble in H_2O (0.29 mg/ml).

28423 silver fluoride
7775-41-9 8657 231-895-8
AgF
silver monofluoride
argentous fluoride. Fluorinating agent, also used as a topical anti-infective. mp = 435°; bp = 1150°; d = 5.8520; soluble in H_2O (182 g/100 ml);.

28424 silver foil
An alloy of from 90-97% tin, 0-2.5% copper, and 0.10% zinc, is known by this name.

28425 silver grain
The cochineal insect killed in an oven at three months old is called silver grain.

28426 silver grey
A Bohme dyestuff. It contains extracts of logwood and redwood, together with a chrome and iron mordant.

28427 silver ink
A mixture of gum arabic and ground white mica; used for inlaying buttons.

28428 silver iodide
7783-96-2 8659 232-038-0
AgI
Photography, cloud seeding for artificial rainmaking, laboratory reagent, antiseptic. mp = 552°; d = 5.67; poorly soluble in H_2O (0.03 mg/l), soluble in ammonia. *Aldrich; Atomergic Chemetals; Cerac; Noah Chem.*

28429 silver leaf
An alloy of 91% tin, 8% zinc, 0.35% lead, and 0.2% iron. Another alloy contains 91% tin, 8.25% zinc, and 0.4% antimony.

28430 silver metal
An alloy of 66.5% zinc and 33.5% silver.

28431 silver nitrate
7761-88-8 8661 231-853-9
$AgNO_3$
silver saltpeter. Photographic film, catalyst for ethylene oxide, indelible inks, silver plating, silver salts, silvering mirrors, germicide, hair dyeing, antiseptic, laboratory reagent. mp = 212°; d = 4.35; soluble in H_2O (2.5 - 10 g/ml), EtOH, Me_2CO. *Accurate Chem. & Scientific; Aldrich; Degussa; Johnson Matthey plc; Spectrum Chem. Mfg.*

28432 silver(I) oxide
20667-12-3 8665 243-957-1
Ag_2O
argentous oxide
Polishing glass, coloring glass yellow, catalyst, purifying drinking water, laboratory reagent. Used in veterinary medicine as a germicide and parasiticide. Dec. 200-300°; d_4^{25}= 7.22; soluble in H_2O (25 mg/l), freely soluble in HNO_3, ammonia; LD_{50} (rat orl) = 2.82 g/kg. *Elf Atochem N. Am.; Atomergic Chemetals; Degussa; Spectrum Chem. Mfg.*

28433 silver solder
Variable alloys. Usually they contain silver, copper, and zinc. A soft silver solder contains 67% silver and 33% brass, and is suitable for sheet. A hard silver solder consists of 80% silver and 20% copper. Some alloys contai.

28434 silver sulfate
10294-26-5 8676 233-653-7
Ag_2O_4S
Silver sulfate normal. Laboratory reagent. mp = 657°; dec 1085°; d = 5.45; soluble in H_2O (8 mg/ml), HNO_3 and ammonia. *Atomergic Chemetals; Noah Chem.*

28435 silver sulfide
21548-73-2 8677 244-438-2
Ag_2S
Inlaying in niello metal work, ceramics. mp = 845°; d_4^{20}= 7.234; insoluble in H_2O. *Atomergic Chemetals; Cerac.*

28436 silverine
An alloy of 77% copper, 17% nickel, 2% iron, 2% zinc, and 2% cobalt.

28437 silvering solutions
Typically consist of solutions of silver cyanide and ammonium cyanide in water; used for the electro-deposition of silver.

28438 Silverline 200
14807-96-6 9207 238-877-9
Platy talc
General purpose, economical filler suited for antiblocking and industrial coatings applications. *Montana Talc.*

28439 Silverline 665
14807-96-6 9207 238-877-9
magnesium silicate, hydrated
talcum; French chalk; Platy talc. Ultrafine, high purity talc with excellent reinforcing and dielectric properties; for extrusion of wire and cable; additive for gloss control in paints and coatings. *Montana Talc.*

28440 Silveroid
An alloy of 45% nickel, 54% copper, and 1% manganese.

28441 silver-salt
Sodium anthraquinone monosulfonate
Obtained in alizarin manufacture.

28442 Silverstone Supra®
Polymer. *DuPont UK.*

28443 Silverstone®
Nonstick finishes. *DuPont UK.*

28444 silvestrite
A mineral; siderazote.

28445 Silvet
Aluminum pigments for plastics; used for all plastics, automobiles, toys, bottles, etc. *Siberline Mfg Co Inc.*

28446 Silvital
Fertilizer for vitalizing damaged forest. *Süd-Chemie AG.* Name unverified.

28447 Silvoline
Aluminum paint; for marine and general purposes. *Llewellyn Ryland Ltd.*

28448 Silwax® S
Silicone wax; patented, highly lubricious wax for personal care applications, polishes, textile lubrication and softening, laundry products, dryer sheet softeners, synthetic lubricants, and plastics lubricants. *Siltech.* Discontinued.

28449 Silwet®
Silicone surfactant. *Union Carbide.*

28450 Silwet® L-77
27306-78-1
Polyalkylene oxide-modified polymethylsiloxane
Surfactant, flow/leveling agent, antistat, antifog, dispersant, wetting agent, flotation agent, spreading agent for coatings, printing inks, adhesives, agriculture, automotive, cleaners, antifogging agent, mining, paper, pharmaceutical applications. *Union Carbide.*

28451 Silwet® L-720
68554-65-4
Dimethicone copolyol
Surfactant; anticaking agent; slip additive for paper; also for pharmaceutical use, printing inks. *Union Carbide.*

28452 Silwet® L-7200
68937-55-3
Silicone glycol copolymer; flow, wetting, slip, dispersion, gloss agent, emulsifier for industrial coatings, household and institutional products, textiles, inks. *Union Carbide.*

28453 Silwet® L-7500
68440-66-4
Dimethicone copolyol
Surfactant, antifoam, dispersant, emulsifier, leveling and flow control agent, lubricant, slip additive for adhesives, automotive, chemical processing, coatings, petrol. extraction, paper, personal care products, plastics and rubber, pharmaceutical and textile applications. *Union Carbide.*

28454 Silwet® L-7600
68938-54-5
Dimethicone copolyol
Surfactant, wetting agent for adhesives, window cleaners, textiles, personal care products; internal lubricant for plastics and rubber. *Union Carbide.*

28455 Silwet® L-7605
68938-54-5
Dimethicone copolyol
Defoamer, slip additive for chemical processing, coatings. *Union Carbide.*

28456 Silwet® L-7607
117272-76-1
Polyalkylene oxide-modified polymethylsiloxane; surfactant, wetting agent, leveling and flow control agent, grease cleaner, flotation and spreading agent for adhesives, agriculture, automotive specialties, chemical processing carpet antistat, mining, metal processing, petroleum extraction; printing inks, textiles. *Union Carbide.*

28457 Silwet® L-7622
68938-54-5
Silicone glycol copolymer; flow, wetting, slip, dispersion, gloss agent, emulsifier for industrial coatings, household and institutional compounds, textiles, and inks. *Union Carbide.*

28458 silzin bronze
An alloy of copper with 10-20% zinc and 4.5-5.5% silicon.

28459 Simadex
122-34-9 8681 204-535-2
Simazine
Selective and total herbicide. *Schering Agrochemicals Ltd.* Discontinued.

28460 Simanex
122-34-9 8681 204-535-2
2-chloro-4,6-bis(ethylamino)-1,3,5-triazine
Active ingredient; simazine; pre-emergence herbicide for control of weeds in a variety of crops as well as a soil sterilant. *Agan Chemical Manufacturers Ltd.*

28461 Simapron
122-34-9 8681 204-535-2
Emulsifiable concentrate containing 150 g/l simazine; a triazine herbicide to control weeds and grasses in cane fruit, roses and some vegetables. *BP Oil Ltd.*

28462 Simask®
For the electrical industry. *DuPont UK.*

28463 Simax
409-21-2 8636 206-991-8
Silicon carbide
Used in metallurgical applications. *Lonza AG.*

28464 simazine
122-34-9 8681 204-535-2
$C_7H_{12}ClN_5$
1,3,5-Triazine-2,4-diamine, 6-chloro-N,N'-diethyl
s-triazine, 2-chloro-4,6-bis(ethylamino)-; *s*-triazine, 2,4-bis(ethylamino)-6-chloro-; A 2079; Aktinit S; Amizine; Aquazine; Batazina; Bitemol; Bitemol S 50; Bitemol S-50; Cat (herbicide); Cekusan; Cekuzina-S; CAT; CAT (herbicide); CDT; CET; DCT; ENT-51142; Framed; G 27692; Geigy 27,692; Gesapun; Gesatop; Gesatop-50; H 1803; Herbazin; Herbazin 50; Premazine; Primatol S; Princep; Printop; Radocon; Radokor. Selective systemic herbicide, absorbed through roots; used to control most germinating grasses and broad-leaved weeds in fruit, vegetables, trees mp = 225-227°; d_{20} = 1.302; soluble in H_2O (5 mg/l), soluble in organic solvents; LD_{50} (rat orl) = 971 mg/kg Ciba-Geigy; Sedagril; Hoechst; Pepro; Protex; Sipcam-Phytoeurop.;Griffin; Schering; Schering Agrochemicals Ltd.; Rhône-Poulenc.

28465 Simazol
Active ingredients: azolan (aminotriazole) plus simanex (simazine); multipurpose herbicidal mixture which eradicates a wide spectrum of established weeds, while preventing further weed germination for extended periods. *Agan Chemical Manufacturers Ltd.*

28466 Simchin WS
PEG-40 jojoba oil; emollient, humectant, pigment, conditioner, lubricant for personal care formulations. *RITA.*

28467 Simchin, Natural
61789-91-1
Jojoba oil; moisturizer, emollient, conditioner for skin and hair care products. *RITA.*

28468 Simetite
Sicilian amber of wine-red to garnet-red color.

28469 Simfix
Granular mixture of diuron and simazine; used for control of weeds in woody crops and noncrop areas. *Rhône-Poulenc Environmental Prods. Ltd.*

28470 Simflex
Flexible circuitry. *Carl Freudenberg.* Name unverified.

28471 Simflow
122-34-9 8681 204-535-2
Simazine formulated as a flowable liquid; a soil residual herbicide suitable for selective weed control in shrubs; also used as a weedkiller to keep paths, bare ground and industrial installations free of weeds and grasses. g *Rhône-Poulenc Environmental Prods. Ltd.*

28472 Simflow Plus
A suspension concentrate containing 100 g aminotriazole [61-82-5] and 300 g simazine [122-34-9] per liter; used for total weed control in non crop areas and fruit orchards. *Rhône-Poulenc Environmental Prods. Ltd.*

28473 Similor
A rich-colored brass. It usually contains from 80-89% copper, 9-20% zinc, and 0-7% tin.

28474 Simmering
Rotating shaft seal. *Carl Freudenberg.* Name unverified.

28475 Simoniz
Polishes and valeting products for cars. *Spectra Brands plc.*

28476 Simplex
Sanitizers. *Thomas Goldschmidt Ltd.*

28477 Simplex Steel
A proprietary trade name for a nickelchromium steel containing 1.25% nickel and 0.6% chromium. No manufacturer.

28478 Simply White
Net curtain whitener. *Dylon International Ltd.*

28479 Simrax
Face seals. *Carl Freudenberg.* Name unverified.

28480 Simrit
Seals and packing rings. *Carl Freudenberg.* Name unverified.

28481 Simulsol 98
9004-98-2
Oleth-20
Cosmetic emulsifier. *Seppic.* Discontinued.

28482 Simulsol 989
61788-85-0
PEG-7 hydrogenated castor oil; cosmetic emulsifier. *Seppic.* Discontinued.

28483 Simulsol 58
9004-95-9
Ceteth-20
Cosmetic emulsifier. *Seppic.* Discontinued.

28484 Simulsol 165
PEG-100 stearate
glyceryl stearate. Cosmetic emulsifier. *Seppic.*

28485 Simulsol 1292
61788-85-0
PEG-25 hydrogenated castor oil; solubilizer, emulsifier for cosmetics. *Seppic.* Discontinued.

28486 Simulsol 5719
Ethoxydiglycol
ceteareth-16; nonoxynol-8. Cosmetic emulsifier. *Seppic.* Discontinued.

28487 Simulsol CS
68439-49-6 7737
Ceteareth-33
Cosmetic emulsifier. *Seppic.* Discontinued.

28488 Simulsol M 45
9004-99-3
PEG-8 stearate
Emulsifier. *Seppic.* Discontinued.

28489 Simulsol P4
Laureth-4
Emulsifier. *Seppic.* Discontinued.

28490 sin red
7722-64-7 7824 231-760-3
$KMnO_4$
Potassium permanganate.

28491 Sinapoline
1801-72-5 217-291-7
$C_7H_{12}N_2O$
Diallyl-urea
1,3-diallylurea. mp = 90-93°.

28492 Sinatron
A range of polyester resins which include orthophthalic, isophthalic, bisphenolic, neopentilic, and self-extinguishing resins. *Lonza AG.*

28493 Sinazine
122-34-9 8681 204-535-2
simazine. Selective systemic herbicide, used for control of most germinating annual grasses and broad-leaved weeds. *Murphy Chemical Co Ltd.* Discontinued.

28494 Sinbar®
5902-51-2 9298 227-595-1
Wettable powder containing 80% w/w terbacil; a weedkiller. *DuPont UK.*

28495 Sin-Chu
Japanese brass. An alloy of 66.5% copper, 33.4% zinc, and 0.1% iron.

28496 Sindanyo
A trade name for proprietary asbestos products. No manufacturer.

28497 single muriate of tin
7772-99-8 8939 231-868-0
An acid solution of stannous chloride; used as a mordant.

28498 Single Purpose
62-38-4 7453 200-532-5
phenylmercury acetate
Organomercury fungicide seed dressing for cereals and fodder beet.

28499 Singlex CIP
Heavy-duty cleaner for metallic surface. *Adeka Fine Chem.*

28500 sinigrin
3952-98-5 8691 223-545-8
$C_{10}H_{16}KNO_9S_2$
1-thio-β-D-glucopyranose 1-[N-(sulfoxy)-3-butenimidate] monopotassium salt potassium myronate; sinigroside; allyl glucosinolate. A constituent of black mustard seed and horseradish root. Substrate for thioglucosidase. The monohydrate is [64550-88-5; 223-545-8]. (monohydrate) mp = 127-129°; $[\alpha]_D^{18}$= -16° (c = 1 H_2O); soluble in H_2O, EtOH, insoluble in organic solvents.

28501 Sinodor
A basic magnesium acetate, containing an excess of magnesium hydrate; used for disinfecting purposes.

28502 Sinter-corundum
A proprietary preparation; it is a ceramic material produced from pure alumina at a temperature of about 1800°; the thermal conductivity at 16° is about twenty times as high as that of porcelain, and it is stated to be not attacked by hydrofluroic acid or hot alkali. No manufacturer.

28503 Sinterit
A proprietary trade name for a form of sponge iron (*qv*) used for coupling packings. No manufacturer.

28504 Sinterloy
A proprietary trade name for a steel powder. No manufacturer.

28505 Sintrex®
Airex AG.

28506 Sinvabond
Flexographic printing inks; for printing laminated film structures. *AlliedSignal Inc/Sinclair and Valentine Division.* Name unverified.

28507 Sinvaset
Web heatset offset printing inks; for publication printing. *AlliedSignal Inc/Sinclair and Valentine Division.* Name unverified.

28508 Siogel
7631-86-9 8637 232-545-4
Approximately 99.7% silicon dioxide; for dehydration of air and other gases. *Chemische Fabriken Oker und Braunschweig AG.* Name unverified.

28509 sionon®
A range of foods suitable for diabetics. *Bayer AG.*

28510 Siopel
A preparation of dimethicone and cetrimide; a barrier skin cream. *ICI Chem & Polymers Ltd.*

28511 Sioplas
Polyethylene cross linking technology. *Dow Corning Ltd.*

28512 Sipalin AOC
849-99-0 212-702-6
$C_{18}H_{30}O_4$
Dicyclohexyl adipate
A proprietary solvent for cellulose nitrate. No manufacturer.

28513 Sipalin AOM
Dimethylcyclohexyl adipate
A proprietary solvent for cellulose nitrate and plasticizer for rubber. bp = 225-232°; SG = 1.011. No manufacturer.

28514 Sipalin MOM
Dimethylcyclohexyl β-methyladipate
A proprietary solvent and plasticizer. bp = 216-224°; sg = 1.009. No manufacturer.

28515 Sipcam UK Carbosip 5G
1563-66-2 1851 216-353-0
Carbofuran
A granular systemic carbomate insecticide and nematode for soil treatment. *Sipcam UK Ltd.*

28516 Sipcam UK Rover 500
1897-45-6 2219 217-588-1
Chlorothalonil
A fungicide for a wide range of agricultural crops. *Sipcam UK Ltd.*

28517 Sipenol IT-50-46
61791-26-2
PEG-50 hydrogenated tallow amine
Surfactant. *Rhône-Poulenc Surf. Canada.*

28518 Sipernat®
Range of spray dried hydrophillic precipitated silicas; free flow/anticaking aids; as carrier substances for production of highly concentrated pulverulent formulations of liquid or paste-like active substances. *Degussa.*

28519 Sipernat® 22
Hydrated silica
Adsorbent, anticaking and free-flow agents; used as aid to convert liqs. into powds.; processing aid; hydrophilic; for adhesives, sealants, detergents, food, cosmetics industries. *Degussa.*

28520 Sipernat® 22S
7631-86-9 8637 232-545-4
silicon dioxide
Hydrated silica; free-flow/anticaking agent for powd. detergents, sealants, foodstuffs, pharmaceuticals, fire extinguishers. *Degussa.*

28521 Sipernat® 44
12141-46-7 377 235-253-8
Aluminum silicate
Antiblocking agent in PE and PP blown films. *Degussa.*

28522 Sipernat® 50
7631-86-9 8637 232-545-4
Hydrated silica
Carrier, free-flow agent, anticaking agent for cosmetics, detergents, food industries. *Degussa.*

28523 Sipernat® 283LS
7631-86-9 8637 232-545-4
Precipitated silica; for polishing of silicon wafers in the electronics industry. *Degussa.* Discontinued.

28524 Sipernat® D17
7631-86-9 8637 232-545-4
Precipitated silica; anticaking and free-flow agent; hydrophobic; for fire extinguishers, pesticides, plastics. *Degussa.*

28525 Sipex 30
Anionic surfactant as a liquid paste; general detergent; emulsifier for emulsion polymerization. *Henkel Chemicals Ltd.* Name unverified.

28526 Sipex DS
Anionic surfactant in acid form, supplied as a viscous liquid; basic material for general detergents, e.g., in dishwashers, and in other formulations requiring a cheap source of anionic active matter. *Henkel Chemicals Ltd.* Name unverified.

28527 Sipomer® β-CEA
β-Carboxyethyl acrylate
Monomer for emulsion polymerization, adhesives and coatings; improves latex properties, adhesion. *Rhône-Poulenc Surf.*

28528 Sipomer® 2M1M
2867-47-2 220-688-8
$C_8H_{15}NO_2$
dimethylaminoethyl methacrylate
N,N-dimethylaminoethyl methacrylate; DMAEMA, 2-methyl-2-propenoic acid 2-(dimethylamino)ethyl ester. Monomer to produces polymers with pendant amino function. mp = -30°; bp = 182°; SG = 0.933; n_D^{20}= 1.4391. *Rhône-Poulenc Surf.*

28529 Sipomer® 2ME
60-24-2 5917 200-464-6
2-mercaptoethanol
Rhône-Poulenc Surf.

28530 Sipomer® AAE
Allyl alcohol ethoxylate
Reactive intermediate for silylation; bound protective colloid; copolymerizable stabilizer. *Rhône-Poulenc Surf.*

28531 Sipomer® AGE
106-92-3 203-442-4
$C_6H_{10}O_2$
1-allyloxy-2,3-epoxy propane
AGE; 1,2-epoxy-3-allyloxypropane; allyl-2,3-epoxypropyl ether; [(2-propenyloxy)methyl]oxirane; 1-allyl-2,3-epoxypropane; allyl glycidyl ether-ethylene glycol prepolymer (18/1). Glutamate decarboxylase inhibitor. mp = -100°; bp = 154°; d = 0.9620; n_D^{20}= 1.4332 *Rhône-Poulenc Surf.*

28532 Sipomer® AM
96-05-9 202-473-0
$C_7H_{10}O_2$
Allyl methacrylate
2-methyl-2-propenyl 2-propenoate. Monomer for coatings, elastomers, adhesives, intermediates; contributes hardness and scratch resisance; crosslinker/hardener. mp = -65°; bp = 144°; SG = 0.938; n_D^{20}= 1.4360 *Rhône-Poulenc Surf.*

28533 Sipomer® DCPA
12542-30-2 235-697-2
Dicyclopentenyl acrylate
dihydrodicyclopentadienyl acrylate. Monomer for air-drying adhesives, coatings, caulks, sealants, elastomers; crosslinks acrylics, unsaturated polyesters and alkyds. *Rhône-Poulenc Surf.*

28534 Sipomer® DCPM
51178-59-7 257-033-0
2-methyl-2-propenoic acid 3a,4,7,7a-tetrahydro-4,7-methano-1H-indenyl ester
dicyclopentenyl methacrylate. Monomer for concrete resurfacing, air-drying adhesives and coatings, elastomers; improves cohesive strength through crosslinking. *Rhône-Poulenc Surf.*

28535 Sipomer® HEM-5
25736-86-1
PEG monomethacrylate
Rhône-Poulenc Surf.

28536 Sipomer® HEM-D
868-77-9 212-782-2
$C_6H_{10}O_3$
2-Hydroxyethyl methacrylate
glycol methacrylate; β-hydroxyethyl methacrylate; Mhoromer; heme-a; GMA; Ethylene glycol methacrylate. mp = -12°; bp_9 = 95°; SG = 1.08; *Rhône-Poulenc Surf.*

28537 Sipomer® IBOA
5888-33-5 227-561-6
isobornyl acrylate
Rhône-Poulenc Surf.

28538 Sipomer® IBOMA
7534-94-3 231-403-1
$C_{14}H_{22}O_2$
Isobornyl methacrylate
(1,7,7-trimethyltricyclo[2.2.1]hepten-2-yl)-2-methyl-2-propenoate. mp = -60°; bp = 245°; d = 0.9830. *Rhône-Poulenc Surf.*

28539 Sipomer® IDA
1330-61-6 215-542-5
$C_{13}H_{24}O_2$
isodecyl 2-propenoate
Isodecyl acrylate. bp_{10} = 121°. *Rhône-Poulenc Surf.*

28540 Sipomer® MEM
3990-03-2 223-635-7
Monoethyl maleate
Rhône-Poulenc Surf.

28541 Sipomer® TATM
2694-54-4 220-264-2
$C_{18}H_{18}O_6$
Triallyl trimellitate
trimellitic acid triallyl ester. mp = -30°; bp_4 = 210°; d = 1.160. *Rhône-Poulenc Surf.*

28542 Sipomer® TMPEO
29860-47-7
Ethoxylated trimethylol propane.
Rhône-Poulenc Surf.

28543 Sipomer® TMPTA
15625-89-5 239-701-3
Trimethylolpropane triacrylate
Rhône-Poulenc Surf.

28544 Sipon®
Redesignated Rhodapon® or Rhodapex® *Rhône-Poulenc Surf.*

28545 Sirdate®
For the agriculture industry. *DuPont UK.*

28546 Sirlene
Feed grade propylene glycol, an emulsifying agent and general purpose food additive; as a conditioner in animal feed, a preservative, humectant, energy source, lubricant, extender, palatability improver and fines reducer; in dairy cattle, used for the prevention and treatment of acetonemia (ketosis). *Dow UK.* Discontinued.

28547 Sirius® Sirius Supra, Sirius Supra LL
A range of direct dyestuffs; used for dyeing cellulosics. *Bayer AG; Bayer plc.*

28548 Sirtan
Floor polishes. *Evans Vanodine International Ltd.*

28549 Sisellig
Antislipping agent for all fibers; antiagglomerant for short fibers. *Ceca SA.*

28550 Sistan®
137-42-8 6016 205-293-0
Metam-sodium 38% w/v aqueous solution; for use as a soil sterilant in horticulture and nurseries. *Universal Crop Protection Ltd.*

28551 Sitilan
The methylcyclohexyl ester of adipic acid; a solvent for cellulose and rubber.

28552 Sitol
A proprietary trade name for the sodium salt of *m*-nitrobenzenesulfonic acid; used in dyeing. No manufacturer.

28553 Sitren
Hydrophobic silicone agents. *Thomas Goldschmidt Ltd.*

28554 Siverslice
Proteinaceous substances for use as food or food ingredients. *Courtaulds Aerospace Ltd.*

28555 Sivex
Ceramic foam filters for removal of nonmetallic inclusions from aluminum alloys. *Foseco (F.S.) Ltd.*

28556 Sivex F
Ceramic foam filters used for the production of aluminum and copper based alloys. *Foseco (F.S.) Ltd.*

28557 Size
Textile sizing agents. *BASF plc.*

28558 size
Usually consists of a starch solution containing small amounts of tallow or oil, and China clay or French chalk; to strengthen and smooth the fibers for weaving; used on yarn or cloth to bind the threads together and to weigh.

28559 Size CB
9003-03-6
Polyacrylate
Sizing agent for cellulosic fibers and their blends. *BASF.*

28560 Sizing Wax PA, PT, SM
PEGs
For production of water-soluble sizing auxiliaries; antistats, dyeing auxiliaries. *Hüls AG.*

28561 SKA
A butadiene polymer of Soviet origin, derived from petroleum.

28562 Skane
Mildewcide; used for coatings. *Rohm & Haas.*

28563 Skane® M-8
26530-20-1 6853 247-761-7
$C_{11}H_{19}NOS$
2-octyl-3(2H)-isothiazolone
2-n-octyl-4-isothiazolin-3-one; 2-Octyl-4-isothiazolin-3-one; Octhilinone; Kathon 893; RH
893; Skane M-8; 2-octyl-3-isothiazolone; kathon lp preservative; kathon sp 70; micro-chek 11; micro-chek 11d; micro-chek skane; pancil; pancil-t; skane hq; Kathon; Microbicide M-8; Isothiazolone, 2-octyl-; Octyl-3(2H)-isothiazolone. Mildewcide for paints. Fungicide, biocide in cooling towers, cutting oils, cosmetics and shampoos. Leather preservative. $bp_{0.01}$ = 120°; λ_m = 280 nm (log ε 3.88 MeOH). *Rohm & Haas.*

28564 Skaterpax
3691-35-8 2204 223-003-0
An oil formulation containing 2.5% of chlorophacinone; an anticoagulant rodenticide; a bait to control black rats, brown rats, house mice and voles. *Lever Industrial Ltd.*

28565 SKB
A butadiene polymer of Soviet origin, derived from alcohol.

28566 Skelleftea
A variety of Stockholm tar.

28567 Skellite®
Solvent used as stove and lamp fuel. *Texaco.*

28568 Skellysolve
A proprietary trade name for a series of petroleum solvents. No manufacturer.

28569 Skin Wiz
Proprietary, polymeric film forming coating to protect molds and molded parts which releases easily when required; used for fiberglass tools, injection molds and molded or fabricated plastic or metal parts. *Axel Plastics Research Laboratories Inc.*

28570 Skiodan Sodium
126-31-8 6051 204-782-6
Methiodal sodium
Diagnostic aid. *Sterling Drug Inc.* Name unverified.

28571 Skleron
A proprietary trade name for an aluminum alloy containing 12% zinc, 3% copper, 0.6% manganese, 0.25% silicon, and a small amount of nickel. No manufacturer.

28572 Skliro Distilled
68424-43-1 270-302-7
Lanolin acid. Emollient, superfatting agent for aerosol shave foams, hand soaps; water repellent films; water-oil emulsifier; stable emulsions in preparation of waterproof makeup. *Croda Inc.* Discontinued.

28573 Skybond
Polyamide resin varnish; designed for structural, electrical and specialty applications where extended exposure to high temperature is required. *Monsanto.* Name unverified.

28574 Skydrol
Fire resistant hydraulic fluid; distributed to commercial airlines. *Monsanto.*

28575 Skyllex
Dialkyl dimethyl ammonium chloride
Supplied as a mixture of water and isopropanol suspension forming a stiff paste; cationic surfactant used in fabric softeners and as a bactericide. *Efkay Chemicals Ltd.* Name unverified.

28576 SLA 1208, 1261
1317-33-5 6318 215-263-9
Colloidal MoS_2 in 500 solvent refined paraffinic petroleum oil; lubricant additive for gear oils. *Acheson Colloids.*

28577 SLA 1261
1317-33-5 6318 215-263-9
molybdenum sulfide
Colloidal molybdenum sulfide in 150 sovent-refined paraffinic petroleum oil. A lubricant additive for gear oils and machine oils. *Acheson Colloids.*

28578 SLA 1262, 1275
7782-42-5 4560 231-955-3
Collodial graphite in 150 solvent refined paraffinic petroleum oil; solid lubricant additive for gear oils, engine oils, chain lubes, aerosols, machine oils, etc. *Acheson Colloids.*

28579 SLA 1275
7782-42-5 4560 231-955-3
Colloidal graphite in 150 solvent-refined paraffinic petroleum oil. Solid lubricant additivie for engine oils, chain lubes, aerosols, machine oils etc. *Acheson Colloids.*

28580 SLA 1286
1317-33-5 6318 215-263-9
Colloidal MoS_2 in 150 solvent refined paraffinic petroleum oil; lubricant additive for engine oils, chain lubes, aerosols, penetrating lubes, machine oils, assembly lubes. *Acheson Colloids.*

28581 SLA 1611, 1612
9002-84-0 7743 204-126-9
Colloidal PTFE in 150 solvent refined paraffinic petroleum oil; lubricant additive for engine oils, gear oils, chain lubes, aerosols, machine oils, penetrating oils, assembly and thread lubes. *Acheson Colloids.*

28582 SLA 1612
9002-84-0 7743 204-126-9
Colloidal teflon in 150 solvent-refined paraffinic petroleum oil. Lubricant additivie for engine oils, chain lubes, aerosols, machine oils, penetrating oils, assembly and thread lubes. *Acheson Colloids.*

28583 SLA 2208
1317-33-5 6318 215-263-9
Colloidal MoS_2 in 2500 solvent refined bright stock petroleum oil; lubricant additive for greases. *Acheson Colloids.*

28584 SLA 2239
7782-42-5 4560 231-955-3
Colloidal graphite in 500 solvent refined paraffinic petroleum oil; solid lubricant additive for greases. *Acheson Colloids.*

28585 Slab Dip AC699
471-34-1 1697 207-439-9
Calcium carbonate pigmented powder with dispersing agents; rubber slab dipping system for rubber industry. *Ayers Cliff.* Name unverified.

28586 slack wax
A soft paraffin wax from the pressing of paraffin distillate.

28587 Slag A
A British chemical standard. It is a basic slag containing 44.5% CaO, 16.15% SiO_2, 12.93% P_2O_5, 8.97% Fe, and 6.9% MgO.

28588 slag sand
Blast furnace slag is run out of the furnace to fall into a running stream of water, when it is broken up into a fine sand.

28589 slag wool
Mineral cotton; mineral wool. Blast furnace slag (essentially a glass composed of silicates of aluminum and calcium), which has had air blown through it. It resembles spun glass, and is used for packing steam pipes.

28590 slagbestos
A similar product to slag wool (blast furnace slag).

28591 slaked lime
1305-62-0 1716 215-137-3
H_2O_2Ca
calcium hydroxide.

28592 slate black
mineral black.

28593 slate fust
Slate filler. A ground slate used as a filler in rubber mixings. It usually has a specific gravity of 2.7-2.8.

28594 slate grey
Stone grey; silver grey; mineral grey. Grey pigments obtained by grinding and levigating special kinds of grey slate, which occur in Germany; used as priming paint, and for the preparation of putty. They are imitated by mixtures of white clay, blacks, ochers and ultramarine.

28595 slate lime
A mixture of 60% lime with 40% of calcined slate powder used in the manufacture of porous concrete.

28596 Slax
Slag coagulants. *Foseco (F.S.) Ltd.*

28597 Slaymor
28772-56-7 1403 249-205-9
Bromadiolone
A ready-to-use bait for the control of rats and house mice. *Ciba-Geigy Agrochemicals.*

28598 SLCC-D
Ammonium hydroxide solution; corrosion inhibitor for preboiler and afterboiler sections of steam generating systems; used in dairies and other facilities where use of conventional neutralizing amines is prohibited. *Drew Ind. Div.*

28599 S-lec
Polyvinyl butyral resin; interlayer film for safety glass. *Sekisu (UK) Ltd.* Name unverified.

28600 Slick
A sulfurized processing aid for glass manufacturing. *Specialty Products Co.* Name unverified.

28601 Slick Slide®
Lubricant. *Graphite Products Corp.*

28602 Slip-Ayd
Dispersed slipping agents; for paints, inks etc. *Cornelius Chemical Co Ltd.* Name unverified.

28603 Slip-Ayd® Surface Conditioners.
Polyethylene and polymeric waxes in predispersed or micronized forms; for increasing slip and film hardness, reducing blocking, marring, metal marking and improving other related surface properties. *Elementis Specialties.*

28604 Slipicone
Silicone lubricants. *Dow Corning.*

28605 Slix
Heat resisting refractory cement. *Foseco (F.S.) Ltd.* Discontinued.

28606 slop wax
The wax present in the heavier wax distillates obtained in the refining of petroleum waxes. It is commonly considered unpressable, and therefore different from the paraffin wax pressed from lighter wax distillates.

28607 Slow-Fe
A proprietary preparation of ferrous sulfate in a slow-release base; iron supplement. *Ciba plc.*

28608 Slow-FE Folic
Iron replacement. *Ciba plc.* Name unverified.

28609 Slow-K
7447-40-7 7783 231-211-8
Potassium chloride
Potassium replenisher. *Ciba-Geigy Corp.* Name unverified.

28610 Slow-Sodium
7647-14-5 8742 231-598-3

A proprietary preparation of sodium chloride in sustained-release form. *Ciba plc.* Name unverified.

28611 SLPE

Polyethylene wire and cable coating. *BP Chemicals Ltd.* Discontinued.

28612 sludge acid

Sulfuric acid which has been used in the refining of petroleum.

28613 Slue

Compressor lubricants. *Dow Chemical.*

28614 Slug

Slug Killers. *Murphy Chemical Co Ltd.* Discontinued.

28615 Slug Destroyer

9002-91-9 5983 202-945-6

Pellets containing 6% w/w metaldehyde; snail and slug bait. *Schering Agrochemicals Ltd.* Discontinued.

28616 Slug Pellets

9002-91-9 5983 202-945-6

Metaldehyde slug killer. *ICI Garden Products.*

28617 Slug Snail Killer

Molluscicide pellets. *Fisons plc, Horticultural Div.* Name unverified.

28618 Slug-Geta

Slug and snail bait. *Monsanto (Solaris).* Name unverified.

28619 Slugit Liquid

9002-91-9 5983 202-945-6

Metaldehyde slug killer. *Murphy Chemical Co Ltd.* Discontinued.

28620 Slugoids

9002-91-9 5983 202-945-6

3% Metaldehyde slug killer pellets (containing animal repellant); used for slug/snail control. *Doff Portland Ltd.*

28621 SM 945

1314-23-4 10313 215-227-2

O_2Zr

Lightly milled monoclinic zirconium dioxide with a zirconia content (including hafnia) of 94.5%; for manufacture of ceramic pigments and welding fluxes. *Ferro.*

28622 SM 975

1314-23-4 10313 215-227-2

O_2Zr

Lightly milled monoclinic zirconium dioxide with a zirconia content (including hafnia) of 97.5%; for manufacture of ceramic pigments and welding fluxes. *Ferro.*

28623 SM 987

1314-23-4 10313 215-227-2

O_2Zr

Lightly milled monoclinic zirconium dioxide with a zirconia content (including hafnia) of 98.7%; for manufacture of ceramic pigments. *Ferro.*

28624 SM 992

1314-23-4 10313 215-227-2

O_2Zr

Lightly milled monoclinic zirconium dioxide with a zirconia content (including hafnia) of 99.2%; for manufacture of lead-zirconate-titanate piezo electric ceramics, zirconates, zirconia technical ceramics and oxygen sensors. *Ferro.*

28625 SM 994

1314-23-4 10313 215-227-2

O_2Zr

Lightly milled monoclinic zirconium dioxide with a zirconia content (including hafnia) of 99.4%; for manufacture of lead-zirconate-titanate piezo electric ceramics, zirconates, zirconia technical ceramics and oxygen sensors. *Ferro.*

28626 SM 2059

Amodimethicone

Cationic emulsion of an aminefunctional silicone polymer in water; silicone emulsion cures to a durable, detergent-resistant film for mold release, particle treatment, textile finishes, and polishes applications. *GE Silicones.*

28627 SM 2061

Silicone dimethyl oil-water emulsion; aerosol spray starch, mold release. *GE Silicones.*

28628 SM 2133,2135

Dimethyl silicone oil-water emulsion; for furniture, vinyl, auto polishes. *GE Silicones.*

28629 SM 2140

Dimethyl polysiloxane resin; mold release agent used in rubber and plastic production operations, in foundry release, aerosol spray starch; good lubricity. *GE Silicones.*

28630 S-M 5731 Process-Type Silicone Sealant

One-component silicone sealant; glazing sealant for fenestration fabrications. *Schnee-Morehead.*

28631 SMA

A proprietary milk feed for babies. *Wyeth Laboratories.*

28632 SMA 17352 A

A proprietary copolymer of styrene and maleic anhydride, of low molecular weight; used as a leveling agent in polishes. *Arco Chemical Europe.* Name unverified.

28633 SMA 3840

A proprietary styrene-maleic anhydride copolymer of low molecular weight used for coating cans and drums. *Arco Chemical Europe Inc.* Name unverified.

28634 SMA 5500

A proprietary copolymer of styrene and maleic anhydride, partially esterified and of low molecular weight; used as a vehicle for thermosetting electrodeposited coatings. *Arco Chemical Europe.* Name unverified.

28635 SMA® 1000

9011-13-6

1:1 styrene/maleic anhydride copolymer; soil release agent; used in ammoniacal water solution in carpet shampoos, paints, inks, paper coatings, commercial laundries; in emulsion polymerization, temporary coatings, oven cleaners, as leveling resin for floor polishes, as dispersant. *Elf Atochem.*

28636 SMA® 1440

9011-13-6

1:1 Styrene/maleic anhydride copolymer partially esterfied with organic alcohol; dispersant for titanium dioxide, other pigments in aq. media; used in paints, and inks. *Elf Atochem.*

28637 SMA® 2625

9011-13-6

2:1 Styrene/maleic anhydride copolymer partially esterified with organic alcohol; dispersant; anti-resoil agent for carpet shampoo and carpet manufacturing. *Elf Atochem.*

28638 smalt

Saxony blue; saxon blue; king's blue; royal blue; zaffer; zaffre; bleu d'azure; bleu de saxe; azure blue. A potash glass containing oxide of cobalt; used in paint pigments, ceramic industries (pigment) coloring glass, bluing paper, starch and textiles; coloring rubber.

28639 Smaragdine

A trade name for a solidified alcohol consisting of alcohol and gun-cotton, colored with malachite green. No manufacturer.

28640 S-Maz® 20

1338-39-2 8782 215-663-3

$C_{18}H_{34}O_6$

Sorbitan laurate

sorbitan monolaurate; sorbitan monododecanoate. Lubricant, antistat, textile softener, process defoamer, opacifier, coemulsifier, solubilizer, dispersant, suspending agent, coupler; prepares excellent w/o emulsions; with T-Maz Series used as oil-water emulsifiers in cosmetics, food formulations, industrial oils and household products; lipophilic. *PPG/Specialty Chem.* Discontinued.

28641 S-Maz® 40

26266-57-9 8782 247-568-8

Sorbitan palmitate

Lubricant, antistat, textile softener, process defoamer, opacifier, coemulsifier, solubilizer, dispersant, suspending agent, coupler; prepares excellent w/o emulsions. *PPG/Specialty Chem.* Discontinued.

28642 S-Maz® 60K

1338-41-6 8872 215-664-9

Sorbitan stearate

Food emulsifier for vegetable and dairy products, cakes; gloss aid in chocolate, nondairy creamers. *PPG/Specialty Chem.* Discontinued.

28643 S-Maz® 65K

26658-19-5 8782 247-891-4

Sorbitan tristearate

Lubricant, antistat, textile softener, process defoamer, opacifier, coemulsifier, solubilizer, dispersant, suspending agent, coupler; prepares excellent w/o emulsions. *PPG/Specialty Chem.* Discontinued.

28644 S-Maz® 80

1338-43-8 8872 215-665-4

Sorbitan oleate

Lubricant, antistat, textile softener, process defoamer, opacifier, coemulsifier, solubilizer, dispersant, suspending agent, coupler; prepares excellent w/o emulsions; also as lubricant, rust inhibitor, and penetrant in metal *PPG/Specialty Chem.* Discontinued.

28645 S-Maz® 83R

8007-43-0 8872 232-360-1

Sorbitan sesquioleate

Solubilizer, emulsifier and dispersant. *PPG/Specialty Chem.* Discontinued.

28646 S-Maz® 85

26266-58-0 8872 247-569-3

Sorbitan trioleate

Lubricant, antistat, textile softener, process defoamer, opacifier, coemulsifier, solubilizer, dispersant, suspending agent, coupler; prepares excellent w/o emulsions; also as lubricant, rust inhibitor, penetrant in metal wor *PPG/Specialty Chem.* Discontinued.

28647 S-Maz® 90
8782
Sorbitan tallate
Lubricant, antistat, textile softener, process defoamer, opacifier, coemulsifier, solubilizer, dispersant, suspending agent, coupler; prepares excellent w/o emulsions; also as lubricant, rust inhibitor, penetrant in metalwork *PPG/Specialty Chem.* Discontinued.

28648 S-Maz® 95
8782
Sorbitan tritallate
Antistat, textile softener, lubricant, process defoamer, opacifier, coemulsifier; prepares w/o emulsions; together with T-Maz series, as o/w emulsifier in metalworking fluids and coolants, semisynthetic and oil-based metalwor *PPG/Specialty Chem.* Discontinued.

28649 smithite
A mineral; silver sulfarsenite.

28650 Smithol 22LD
Ester; metal and fiber wetting agent; demulsifier; forms continuous monomolecular film with rust prevention properties; stabilizer for chlorinated systems; for drawing, stamping, rolling lubricants; waste water stripping oils leather treatment, textile spin finishes. *Werner G. Smith.*

28651 Smithol PEG Adipate
68647-16-5
Lubricant for water-based lubricant systems. *Werner G. Smith.* Discontinued.

28652 Smitter-Lenian
An alloy containing 72% copper, 12.75% nickel, 9.75% zinc, 2.3% iron, 2.25% tin, and 1% bismuth.

28653 Smoke
Uncut disperse and solvent dyestuffs; for coloration of smoke grenades mainly for use by army. *Holliday Dyes & Chemicals Ltd.*

28654 smoke black
A carbon black used as a pigment; contains 99.75% carbon.

28655 smokeless diamond powder
A 33 grain powder consisting of insoluble nitrocellulose, with 15% metallic nitrates, 6% charcoal, and 3% petroleum jelly.

28656 smoking salts
Impure hydrochloric acid.

28657 smoky quartz
A quartz containing organic matter or hydrocarbons. It is usually brown in color.

28658 Smoothex
A range of glycerol esters; used as emulsifiers, stabilizers and antifoam agents in the food and cosmetic industries. *Harcros Australia.*

28659 Smooth-On
Iron and foundry cements, epoxy adhesives and cements. Other formulations, based on epoxy, polysulfide and polyurethane polymers; for maintenance and repair, structural bonding and plastic tooling applications (flexible molds cast and laminated plastic tools). *Smooth-On Inc.* Name unverified.

28660 SN-30GF/000 FR
9003-54-7
Styrene-acrylonitrile, 30% glass fiber-reinforced, flame-retardant. *Compounding Tech.* Name unverified.

28661 Snac-Kote
Partially hydrogenated vegetable oil (cottonseed, soybean), sorbitan stearate, polysorbate 60; kosher; coating fat for bakery coatings. *Van Den Bergh Foods.*

28662 Snapper CDA
An emulsifiable concentrate containing 95 g aminotriazole, 190 g atrazine and 99 g 2,4-D per liter; used for total weed control in non crop areas. *ICI Agrochemicals Professional Products.*

28663 Sniafil
An Italian synthetic wool substitute. It is a product obtained in a similar way to viscose silk, but differs from it in the treatment of the viscose solution.

28664 Sniafoam
Glass fiber-reinforced polyester foam. *Lonza AG.*

28665 Sniamid®
Nylon 6 and 66 granules; engineering plastic. *SNIA (UK) Ltd.*

28666 Sniamid® ADS 40 I
25038-54-4 6832
Nylon 6, flame retardant; high viscosity grade with high flexibility and toughness; for injection molding, film extrusion, and extrusion of sheet, rod and tube. *Nylon Corp. of Am.*

28667 Sniamid® ASN 27T
25038-54-4 6832
Nylon 6; transparent grade for injection molding and extrusion including carburetor filters, brake fluid reservoirs, monofilament. *Snia Technopolimeri; Nylon Corp. of Am.*

28668 Sniamid® SSD 300 EP 021
32131-17-2
Nylon 6/6, 30% mineral-filled; toughened, improved impact grade. *Snia Technopolimeri; Nylon Corp. of Am.*

28669 Sniamid® SSD AF
32131-17-2
Nylon 6/6; reduced coefficient of friction; for applications where friction and wear are likely to occur. *Snia Technopolimeri; Nylon Corp. of Am.*

28670 Sniasan®
Acrylonitrile-butadiene-styrene, styrene-acrylonitrile or polystyrene. *SNIA (UK) Ltd.*

28671 Sniatal®
Acetal copolymers for injection molding or extrusion applications. *SNIA (UK) Ltd.*

28672 Sniater®
Polybutylene terephthalate *SNIA (UK) Ltd.*

28673 Snomelt
10043-52-4 1699 233-140-8
Calcium chloride pellets
Used for snow and ice melting, dust control and tire weighting. *Standard Tar Products Co Inc.*

28674 Snow White 200 Mica
12001-26-2
200 mesh mica; chemically inert filler with excellent whiteness; highly refractory; imparts crack resistance and improves moisture impermeability to coatings; improves flexural modulus in plastics; filler for microwave cookware. *Unimin.*

28675 Snowflake P. E.
471-34-1 1697 207-439-9
Calcium carbonate
Medium ground pigment for maximum loading in highly filled systems, e.g., polyester resins, SMC, BMC, TMC, XMC. *ECC International Ltd.*

28676 Snowflake White
471-34-1 1697 207-439-9
Calcium carbonate
General-purpose easy dispersing pigment for protective coatings, rubber, plastics, caulks, glazing compounds, mastics. *ECC International Ltd.*

28677 Snowtack
Adhesive tackifier emulsion. *Tenneco Malros Ltd.* Name unverified.

28678 Snowtack 342A
Gum rosin acid
General-purpose tackifier for SBR and acrylic polymers. *Eka Nobel.*

28679 Snowtack SE 325 A
Gum rosin ester; modifying tackifying resin for low temperature and improved low energy substrate. *Eka Nobel.*

28680 S-nyl
Polyvinyl acetate homopolymer; chewing gum base. *Sekisu (UK) Ltd.* Name unverified.

28681 So/San 30M
Specialty product containing BTC 2125M as active biocide; fabric softener-sanitizer. *Stepan; Stepan Canada.*

28682 Soa
126-14-7 9052 204-772-1
A proprietary trade name for sucrose octa-acetate; used as a plasticizer. No manufacturer.

28683 soap bark
The bark of *Quillaia saponaria* of Chile. The active principle is saponin; used to clean clothes.

28684 Soapearl®
Pearlescent pigments for incorporation into extruded soap bars where they exhibit luxurious color effects. *Mearl.*

28685 Soarblen
Ethylene-vinyl acetate copolymer. *British Traders & Shippers Ltd.*

28686 Soarnol
Ethylene-vinyl alcohol copolymers. *British Traders & Shippers Ltd.*

28687 Sobalg FD 100 Series
Sodium alginate
Food stabilizer, gellant, film-former for dairy products, desserts, beverages, fruits and vegetables. *Grindsted Prods.*

28688 Sobalg FD 200 Series
9005-36-1 241

Potassium alginate
For food industry applications. *Grindsted Prods.*
28689 Sobalg FD 300 Series
9005-34-9
Ammonium alginate
For food industry applications. *Grindsted Prods.*
28690 Sobalg FD 460
9005-35-0 241
Calcium alginate
For food industry applications. *Grindsted Prods.*
28691 Sobee
A proprietary baby feed based on soya; used in cases when an infant is intolerant of milk. *Bristol-Myers Squibb Inc.*
28692 Sobral®
Modified alkyds, acrylic resins, polyurethane elastomers, or vinyl-modified epoxy esters; used for surface coatings, paints, primers, stains, varnishes, and inks. *Scott Bader.*
28693 Sobral 12-101
Alkyd, chain stopped, linoleic rich oil type, xylene solvent; for quick air dry industrial finishes. *Scott Bader.*
28694 Sobral 72-625D
Polyurethane resin, linoleic rich oil type, white spirit solvent; for high build decorative/marine finishes. *Scott Bader.*
28695 Sobral 1321
Alkyd, rosin/phenolic modified, linseed oil type, naphtha solvent; for industrial air drying/stoving primers/finishes, sanding sealers. *Scott Bader.*
28696 Sobral 2911
Alkyd, D.C.O. oil-modified, xylene solv.; pure drying alkyd for general purpose air drying/stoving finishes/primers; good color retention. *Scott Bader.*
28697 Sobral 9257
Alkyd, styrene-modified, D.C.O. oil type, xylene solvent; for economical quick air dry paints. *Scott Bader.*
28698 Sobral AD-002
Alkyd, lineoleic oil-modified; pure drying alkyd for litho inks, high build decorative paints. *Scott Bader.*
28699 Sobral AN-001
Alkyd, synthetic fatty acid oil-modified, xylene solvent; nondrying alkyd for tin printing inks, nonsetting mastics. *Scott Bader.*
28700 Sobral EE-632
25928-94-3
Epoxy ester, vinyl-modified, semidrying D.C.O. oil type, xylene solvent; for automobile primers/surfacers, chemical plant protection, road line paints. *Scott Bader.*
28701 Sobral L90-20A
Polyurethane resin, elastomeric type, IPA solv.; for color-retentive lacquer finishes for flexible substrates and plastics; based on aliphatic isocyanate. *Scott Bader.*
28702 Sobrom
Methyl bromide with chloropicrin; fumigant for soil-borne diseases and stored products. *Brian Jones & Associates Ltd.*
28703 Socal
471-34-1 1697 207-439-9
Precipitated calcium carbonate
Laporte UK Trading. Discontinued.
28704 Socci 3500, 3500-WP
Antimicrobial for textile applications. *Morton Int'l./Plastics Additives.*
28705 Sochamine A 271
Coconut and lauric carboxy/sulfate; for non-eye-stinging shampoos. *Witco Chemical Ltd.* Discontinued.
28706 Sochamine A 7525
Coconut dicarboxylate in liquid form; for hair and baby shampoo. *Witco Chemical Ltd.* Discontinued.
28707 Sochamine A 7527
Coconut dicarboxylate in liquid form; for hard surface cleaners, hair shampoos, textile treatment, strongly acid or alkaline cleaners. *Witco Chemical Ltd.* Discontinued.
28708 Sochamine A 8955
Alkyl imidazoline dicarboxylate in liquid form; for low foam detergents. *Witco Chemical Ltd.* Discontinued.
28709 soda
Sodium carbonate and bicarbonate are both known by this term.
28710 soda ash
497-19-8 8739 207-838-8
Na_2CO_3
sodium carbonate anhydrous
soda; calcined. Manufacture of glass, pulp and paper, chemicals, sodium compounds, soaps and detergents, water treatment, aluminum production, textile processing, cleaning compounds, petroleum refining, sealing ponds from leakage, catalyst in coal liquefaction; a commercial variety of soda ash used for softeninig boiler feed water is known as 58% sdoa ash and contians 58% Na_2O. *FMC; General Chem.; Mallinckrodt; Norsk Hydro A/S; Solvay SA; Texasgulf.* Discontinued.
28711 Soda Ash Blocks
Supplementary fluxes used in the melting of cast iron. *Foseco (F.S.) Ltd.* Discontinued.
28712 soda blue
Gas Blue. Impure Prussian blues prepared by using sodium ferrocyanide instead of the potassium salt.
28713 soda Bordeaux mixture
Made with 6 lb copper sulfate, 2 lb caustic soda, and 50 gallons water.
28714 soda lye
1310-73-2 8772 215-185-5
Obtained by boiling a solution of sodium carbonate with slaked lime.
28715 soda pulp
Wood pulp obtained by means of caustic soda.
28716 soda tar
The name applied to an alkaline solution which has been used to purify petroleum oils after they have been treated with sulfuric acid.
28717 soda water glass
A mixture of sodium silicates.
28718 Sodagrain
Caustic soda for use in making animal feeds. *Imperial Chemical Industries plc; ICI Chem. & Polymers Ltd.*
28719 soda-lime glass
A glass usually containing from 71-78% SiO_2, 12-17% Na_2O, 5-15% CaO, 1-4% Al_2O_3 and Fe_2O_3, and 0-2% K_2O.
28720 soda-olein
A sulfonated castor oil.
28721 Sodaphos
10124-56-8 8814 233-343-1
$Na_6O_{18}P_6$
Sodium hexametaphosphate
Metaphosphoric acid, hexasodium salt; Glassy sodium; Hexasodium metaphosphate;
Metaphosphoric acid (H6P6O18), hexasodium salt; Sodium metaphosphate; Sodium phosphate. *FMC.* Discontinued.
28722 Sodasorb
Soda lime; carbon dioxide absorbent. *W R Grace & Co.* Name unverified.
28723 Sodastraw
Caustic soda for use in making animal feeds. *Imperial Chemical Industries plc; ICI Chem. & Polymers Ltd.*
28724 Sodatol
A mixture of sodium nitrate and trinitrotoluene; an agricultural explosive.
28725 sodium
7440-23-5 8710 231-132-9
Na
natrium
Metallic element; polymerization catalyst for synthetic rubber, laboratory reagent, coolant in nuclear reactors, heat transfer agent, manufacture of tetraethyl and tetramethyl lead, sodium peroxide, sodium hydride; radioactive isotopes in tracer studies and medicine. mp = 98°; bp = 892°; SG = 0.9; reacts with H_2O. *Associated Octel Co Ltd; Foseco (FS) Ltd.*
28726 sodium acetate
127-09-3 8711 204-823-8
$C_2H_3NaO_2$
sodium acetate anhydrous
acetic acid sodium salt anhydrous. Dye and color intermediate, pharmaceuticals, cinnamic acid, soaps, photography, purification of glucose, meat preservation, medicine, electroplating, tanning, dehydrating agent, buffer, laboratory reagent, food additive. Trihydrate: mp = 58°; d = 1.45; soluble in H_2O (1.25 g/ml), less soluble in organic solvents. *Aldrich; EM Industries; General Chem.; Heico; Honeywill & Stein Ltd; Lonza; Niacet; Verdugt BV.*
28727 sodium acid pyrophosphate
7758-16-9 8713 231-835-0
$H_2Na_2O_7P_2{>}H_2O$
disodium dihydrogen pyrophosphate hexahydrate
disodium pyrophosphate; diphosphoric acid disodium salt. Leavening agent for baking, cereals. mp = 220° (dec); d = 1.86; soluble in H_2O.
28728 sodium acid sulfite
7631-90-5 8731 231-548-0
$NaHSO_3$
sodium bisulfite
hydrogen sulfite sodium; sodium hydrogen sulfite; sulfurous acid, monosodium salt; monosodium sulfite; sodium sulfite. Used as a disinfectant

and bleach and as a reducing agent. d = 1.4800; soluble in H_2O (0.28 g/ml), less soluble in organic solvents; LD_{50} (rat iv) = 115 mg/kg.

28729 Sodium Aerofloat Promoter
Sodium diethyl dithiophosphate
Used in the mining industry. *Cyanamid BV.*

28730 sodium alum
10102-71-3 378 233-277-3
$AlNaO_8S_2{\cdot}12H_2O$
aluminum sodium sulfate
SAS; alum; soda alum. Used in textiles as a mordant and for waterproofing, dry colors, ceramics, tanning, paper size precipitant, sugar refining, water purification. An astringent. dodecahydrate: mp = 60°; d = 1.61; soluble in H_2O (1 g/ml).

28731 sodium aluminate
1302-42-7 8715 215-100-1
$AlNaO_2$
aluminum sodium oxide
sodium aluminum dioxide. Mordant, zeolites, water purification, sizing paper, manufacture of milk glass, soap and cleaning compounds. Soluble in H_2O, insoluble in EtOH. *Asada Chem Industry Co Ltd; Hüls AG; Laporte Absorbents; Nalco.*

28732 sodium aluminum chlorhydroxy lactate
8038-93-5
Sodium salt of a complex of lactic acid and aluminum chlorohydrate; *Reheis.*

28733 sodium aluminum phosphate
7785-88-8 232-090-4
$NaAl_3H_{14}(PO_4)_8{\cdot}4H_2O$ or $Na_3Al_2H_{15}(PO_4)_n$
sodium aluminum phosphate acidic
Food additive for baked products. *Monsanto; Rhône-Poulenc Basic; Whiting, Peter Ltd.*

28734 sodium ascorbate
134-03-2 8723 205-126-1
$C_6H_7NaO_6$
L(·2)-ascorbic acid sodium salt
vitamin C sodium salt; Sodium L-Ascorbate; Sodium L-(+)-Ascorbate; Vitamin C Sodium Salt; Sodium Ascorbate; L-Ascorbic acid, monosodium salt; monosodium ascorbate; 3-oxo-L-gulofuranolactone sodium; ascorbicin; ascorbin; cebitate; cenolate; iskia-c; natrascorb; natri-c; sodascorbate. Antioxidant in food products. Source of vitamin C. Dec 218°; [α]β = 104° ; soluble in H=72O (62 g/100 ml). *BASF; EM Industries; Hoffmann-La Roche; Pfizer Spec. Chem.; Spice King; Takeda USA.*

28735 sodium benzoate
532-32-1 8725 208-534-8
$C_7H_5NaO_2$
benzoic acid sodium salt
Sodium salt of benzoic acid; fungicide; preservative in pharmaceuticals and foods, especially in slightly acidic media; clinical reagent (bilirubin assay). Soluble in H_2O (555 mg/ml); LD_{50} (rat orl) = 4.07 g/kg. *Aceto; Dinoval Chem. Ltd; DSM BV; Haarmann & Reimer; Pentagon Chemicals Ltd; Mallinckrodt.*

28736 sodium bicarbonate
144-55-8 8726 205-633-8
$CHNaO_3$
Sodium hydrogen carbonate
Col-Evac; sodium acid carbonate; baking soda. Used in baking, effervescent salts, beverages, fire extinguishers and cleaning compounds. Also used medically as an antacid or alkalizer. mp = 270° (decomposes to sodium carbonate and CO_2 when heated); d = 2.1590 *O'Neal, Jones & Feldman Pharmaceuticals.* Name unverified.

28737 sodium bisulfide
16721-80-5 8730 240-778-0
$HNaS{\cdot}2H_2O$
sodium hydrosulfide
sodium sulfhydrate; sodium hydrogen sulfide; sodium sulfhydrate; Sodium Bisulfide; Sodium sulfide. Paper pulping, dyestuffs processing, rayon and cellophane desulfurizing, dehairing hides, bleaching reagent. mp = 55°; d = 1.79; soluble in H_2O, organic solvents. *Nissan Chem. Ind.*

28738 sodium bisulfite
7631-90-5 8731 231-548-0
$NaHSO_3$
sodium acid sulfite
sulfurous acid, monosodium salt; sodium hydrogen sulfite. Fusion of minerals to make solutions for analysis; pickling metals; carbonizing wool. *Aldrich; BASF; Du Pont; Hoechst Celanese; Hüls AG; Penreco.*

28739 sodium borate anhydrous
1330-43-4 8733 215-540-4
$B_4Na_2O_7$
sodium tetraborate
sodium biborate; sodium pyroborate; borax. Used in heat-resistant glass; porcelian enamel; detergents, herbicides, fertilizers, rust inhibitors, pharmaceuticals, leather, photography, bleaches, paint, boron compounds, flame retardant fungicide for wood, soldering flux, cleaners and lab. reagents. mp = 741°; d = 2.3670 *Spectrum Chem. Mfg.; U.S. Borax & Chem.*

28740 sodium borohydride
16940-66-2 8735 241-004-4
BH_4Na
sodium tetrahydroborate
sodium tetrahydridoborate. Source of H_2 and other borohydrides; bleaching wood pulp; blowing agent for plastics; decolorizer for plasticizers. mp = 36°; d = 1.074. *Atomergic Chemetals; Morton Int'l.; Tennant Trading Ltd.*

28741 sodium bromide
7647-15-6 8737 231-599-9
BrNa
Sedoneural. Photography, medicine (sedative), preparation of bromides. mp = 755°; d = 3.21; soluble in H_2O (0.9 g/ml), less soluble in organic solvents; crystallizes as a dihydrate; LD_{50} (rat orl) = 3.5 g/kg. *Ethyl; Great Lakes; Hawks Chem. Co Ltd; Morton Int'l.*

28742 sodium butyrate
156-54-7 205-857-6
$C_4H_7NaO_2$
butyric acid sodium salt
sodium butanoate; sodium n-butyrate. Used in chemical laboratory as agent for alteration of chromatin structure; enhances transfection efficiency and improved cell survival. mp = 250-253°. *Penta Mfg.*

28743 sodium caprylate
1984-06-1 217-850-5
$C_8H_{15}NaO_2$
caprylic acid sodium salt
Sodium n-octanoate. Source of caprylic acid, an intermediate in manufacture of dyestuffs, surfactants etc. *Aldrich; Hart Prod. Corp.; Penta Mfg.*

28744 sodium carbonate
497-19-8 8739 207-838-8
Na_2CO_3
soda ash
carbonic acid; disodium salt; Inorganic salt. Detergent and cleaning products; pH control of water; intermediate in thermochemical reactions; food additive; manufacture of glass; analytical reagent. *Albright & Wilson; EM Industries; Rhône-Poulenc Basic; Solvay SA.*

28745 sodium carbonate monohydrate
5968-11-6 8739 207-838-8
$CNa_2O_3{\cdot}H_2O$
sodium carbonate monohydrate
soda ash; carbonic acid, disodium salt. bp_{760} = 851°; d = 2.25; soluble in 3 parts H_2O, insoluble in organic solvents *Church & Dwight Co.*

28746 sodium caseinate
9005-46-3
casein sodium salt
Food additive; binder, and extender in sausage, soups, etc.; emulsifier, stabilizer. *Am. Casein; Spectrum Chem. Mfg.; Spice King.*

28747 sodium chlorate
7775-09-9 8741 231-887-4
$ClNaO_3$
Atlacide; Defol; Dervan; Chlorate of soda. Oxidizing agent, pulp bleaching, defoliant and herbicide; leather tanning and finishing; textile mordant. Also used in pyrotechnics. mp = 248°; d = 2.5; soluble in H_2O, less soluble in organic solvents; LD_{50} (rat orl) = 12000 mg/kg. *Albright & Wilson; Elf Atochem; Eka Nobel AB; Georgia Gulf; Kerr-McGee; OxyChem; PPG Industries.*

28748 sodium chloride
7647-14-5 8742 231-598-3
NaCl
Rock salt; salt; common salt. Inorganic salt; occurs in nature as the mineral halite; source of chlorine and sodium; preservative, seasoning for foods; manufacture of soaps, dyes; in freezing mixtures; dyeing and printing fabrics, galzing pottery. Curing hides, metallurgy, herbicide, electrolyte replenisher, emetic, topical anti-inflammatory. mp = 804°; soluble in H_2O (0.36 g/ml), insoluble in organic solvents; LD_{50} (rat orl) = 3.75 g/kg. *Akzo Salt; EM Industries; Heico; Morton Salt; Stan Chem Int'l. Ltd.*

28749 sodium chlorite
7758-19-2 8743 231-836-6
$ClNaO_3$
chlorous acid sodium salt
Improves taste and odor of potable water; oxidizing agent, bleaching agent for textiles, paper pulp, disinfecting. Dec 180-200°; soluble in H_2O (34 g/100 ml). *Albright & Wilson; Elf Atochem N. Am.; Int'l. Dioxcide; Olin.*

28750 sodium chromate
7775-11-3 8745 231-889-5
Cr_2NaO_4
sodium chromate(VI)

neutral sodium chromate; Chromate of soda; Chromic acid, disodium salt; sodium chromate (VI); chromic acid (H2CrO4), disodium salt; chromium disodium oxide; chromium sodium oxide (CrNa2O4); disodium chromate; neutral sodium chromate. Inks, dyeing, paint pigment, leather tanning, other chromates, protection of iron against corrosion, wood preservative. Has tetrahydrate and decahydrate (mp = 20°); soluble in H_2O (1 g/ml). *Elf Atochem N. Am.; OxyChem.*

28751 sodium citrotartrate
A mixture of sodium citrate and tartrate.

28752 sodium cumene sulfonate
32073-22-6 250-913-5
Eltesol® SC 93; Naxonate® 45SC; Naxonate® SC; Reworyl® NCS; Stepanate® SCS; Witconate SCS 45%. Hydrotrope for hard surface cleaners. *Albright & Wilson UK; Ruetgers-Nease; Stepan; Stepan Canada; Rewo Chemicals Ltd;.*

28753 sodium cyanide
143-33-9 8750 205-599-4
CNNa
hydrocyanic acid sodium salt
Cyanide of Sodium; Cymag; cyanobrik; cyanogran. Extraction of gold and silver from ores, electroplating, heat treatment of metals, making hydrogen cyanide, insecticide, metal cleaning, fumigation, dyes and pigments, nylon intermediates, chelating compounds, ore flotation. mp = 563°; soluble in H_2O, poorly soluble in organic solvents; LD_{50} (rat orl) = 15 mg/kg. *Elf Atochem N. Am.; Degussa; DSM UK; Du Pont; FMC; W.R. Grace Ltd; ICI Spec.; Mitsui Toatsu Chem.*

28754 sodium cyclamate
139-05-9 2770 205-348-9
$C_6H_{12}NNaO_3S$
cyclohexylsulfamic acid, monosodium salt
sodium cyclohexanesulfamate; sodium cyclohexyl amidosulfate; sodium cyclohexyl sulfamate; cyclohexanesulfamic acid, monosodium salt; sodium cyclohexylsulfamidate; sodium sucaryl; Asugryn; Dulzor-Etas; Hachi-sugar; Ibiosuc; Sucrosa; Sucrun 7; Sugarin; Sugaron; Assugrin; Assurgrin Feinsuss; Assurgrin Vollsuss; Suessette; Suestamin. Artificial sweetener; about 30x as sweet as cane sugar. mp = 265°; soluble in H_2O, insoluble in organic solvents; LD_{50} (rat orl) = 15.25 g/kg.

28755 sodium decyl sulfate
84501-49-5 282-968-6
$C_{10}H_{21}O_4S.Na$
Empimin® SDS; Atlasol 103; Avirol® SA 4110; Empicol® 0758; Empimin® SDS; Serdet DFK 30; Sulfotex 110; Texapon® 1030; Texapon® NDS. Emulsifier, dispersant, detergent, and wetting agent for industrial and institutional cleansers, mfg. of pigments, alkaline cleansers; dust suppression. MW=260.36; moderately toxic by ingestion Albright & Wilson Australia; Atlas Refinery; Henkel/Chems. Group;Albright & Wilson Asia; Albright & Wilson UK; Servo Delden BV; Henkel Canada;.

28756 sodium dichromate
7789-12-0 8754 234-190-3
$Cr_2Na_2O_7$
sodium bichromate
sodium dichromate dihydrate; dichromic acid, ($H_2Cr_2O_7$), disodium salt, dihydrate; disodium dichromate dihydrate; sodium dichromate dihydrate ($Na_2Cr_2O_7 \cdot 2H_2O$). Colorimetry (copper determination), complexing agent, oxidation inhibitor in ethyl ether. *British Chrome & Chems.; Chemisphere Ltd; OxyChem; Rit-Chem.*

28757 sodium dihydroxyethyl glycinate
139-41-3 205-360-4
Hampshire® DEG; Sodium dihydroxyethylglycinate. Chelating agent used for control of iron only in alkaline sol's. *W R Grace/Hampshire.*

28758 sodium dimethyldithiocarbamate
128-04-1 204-876-7
$C_3H_6NNaS_2$
sodium N,N-dimethyl dithiocarbamate
SDDC; dimethyldithiocarbamic acid sodium salt; methyl namate; sodium dimethyldithiocarbamate; Aceto SDD 40; Alcobam NM; Brogdex 555; Carbon S; Dibam; Dibam A; DMDK; Sharstop 204; sodium N,N-dimethyldithiocarbamate; Stafresh 615; Steriseal 40; Thiostop N; Vinstop; Vulnopol NM; Wing stop b; SDMDTC; sodium dimethylcarbamodithioate; Dimethyldithiocarbamic acid sodium salt; Freshgard 40; Novate SM-40,. Pesticide, fungicide, corrosion inhibitor, rubber accelerator. trihydrate; mp = 94-102°; soluble in H_2O, polar organic solvents; λ_m = 257, 290 nm (ε 1200, 13000 EtOH); LD_{50} (rat orl) = 2830 mg/kg. *NovaChem; Uniroyal; R. T. Vanderbilt Co Inc.*

28759 sodium dinonyl sulfosuccinate
63217-13-0 264-016-1
Geropon® WS-25, WS-25-I. Rewetting agent for textile finishing, in application of resins, softeners, starches; wetting and dispersing agent for latex paints. *Rhône-Poulenc Surf.; Rhône-Poulenc France.*

28760 sodium dodecylbenzenesulfonate
25155-30-0 8757 246-680-4
$C_{18}H_{29}NaO_3S$
sodium dodecyl benzene sulfonate
sodium lauryl benzene sulfonate; Nacconol 35SL; Neccanol sw; Pilot hd-90; Richonate 1850; Santomerse 3; Solar 90; sulfapol; sulframin 85; Sulframin 1238 slurry; Sulframin 1250 slurry; ultrawet k; ultrawet 60k; ultrawet kx; AA-9; AA-10; Abeson Nam; Bio-soft D-40; Bio-soft D-60; Bio-soft D-62; Bio-soft D-35x; Calsoft f-90; Calsoft L-40; Calsoft L-60; Conco AAS-35; Conoco C-50; Conoco c-60; Conoco sd 40; Detergent HD-90; Mercol 25; Mercol 30; ultrawet sk; stepan ds 60; ultrawet 1t; marlon a 350; marlon a; maranil; marlon a 375; siponate ds 10; trepolate f 40; conoco c 550; kb (surfactant); nansa sl; santomerse me; merpisap ap 90p; nansa ss; trepolate f 95; nansa hs 80; deterlon; ultrawet 99ls; sulfuril 50; F 90; elfan wa; sandet 60; steinaryl nks 50; sinnozon; NANSA HS 85S; C 550; KB; HS 85S; nansa hf 80; arylan sbc; marlon 375a; X 2073; conco aas 35H,. Anionic detergent; cosmetic applications. mp >300°; soluble in H_2O (5-10 mg/ml); LD_{50} (mus orl) = 2 g/kg. *Du Pont; Emkay; Norman, Fox; Pilot; Stepan; Unger Fabrikker AS; Witco.*

28761 sodium erythorbate
6381-77-7 5141 228-973-9
Eribate; Neo-Cebitate®. *PMP Fermentation Prods. Inc; Rhône-Poulenc Food Ingreds.*

28762 sodium erythorbate
6381-77-7 5141 228-973-9
$C_6H_7NaO_6$
D-erythro-hex-2-enonic acid γ-lactone monosodium salt
Isoascorbate; D-erythro-hex-2-enonic acid, σ-lactone, monosodium salt; isoascorbic acid, sodium salt; neo-cebitate; mercate 20; Isona; sodium D-isoascorbate; monosodium erythorbate; araboascorbic acid, monosodium salt, D-; hex-2-enonic acid, σ-lactone, monosodium salt; Sodium D-araboascorbate; sodium isoascorbate. Sodium salt of erythorbic acid; antioxidant, preservative. Soluble in H_2O (16 g/100 ml). *PMP Fermentation Prods.; Schweizerhall.*

28763 sodium ferrocyanide
13601-19-9 8760 237-081-9
$Na_4Fe(CN)_6 \cdot 10H_2O$
sodium ferrocyanide decahydrate
sodium hexacyanoferrate; Yellow prussiate of soda; Ferrate(4-), hexakis(cyano-C)-, tetrasodium, (OC-6-11)-; Sodium ferrocyanide; tetrasodium hexacyanoferrate. Yellow prussiate of soda; Manufacture of sodium ferricyanide, blue pigments, blueprint paper, anticaking agent for salt, ore flotation, pickling metals, polymerization catalyst, photographic fixative. Dec 435°; soluble in H_2O (17.5 g/100 ml), insoluble in organic solvents. *Atomergic Chemetals; Degussa Ltd; Rit-Chem.*

28764 sodium fluoride
7681-49-4 8762 231-667-8
Fna
Disodium difluoride; Floridine; Florocid; Villiaumite; NaF; sodium hydrofluoride; sodium monofluoride; trisodium trifluoride; alcoa sodium fluoride; antibulit; cavi-trol; chemifluor; Credo;duraphat; fda 0101; f1-tabs; flozenges; fluoral; fluorident; fluorigard; fluorineed; fluorinse; fluoritab; flurocid; fluor-o-kote; fluorol; fluoros; Flura; flura-gel; flura-loz; flurcare; flursol; fungol b; Gel II; gelution; Gleem; iradicav; karidium; karigel; kari-rinse; lea-cov; lemoflur; luride; luride lozi-tabs; luride-sf; nafeen; nafpak; na frinse; nufluor; ossalin; Ossin; osteofluor; pediaflor; pedident;pennwhite; pergantene; phos-flur; point two; predent; rafluor; rescue squad; Roach salt; sodium fluoride cyclic dimer; So-flo; stay-flo; studafluor; super-dent; t-fluoride; thera-flur; thera-flur-n; zymafluor; Les-cav,. Fluoridation of municipal water, degassing steel, wood preservative, insecticide, fungicide, rodenticide, chemical cleaning, electroplating, glass manufacture, vitreous enamels, preservative for adhesives, toothpastes, disinf toothpastes, disinfectants, dental prophylaxis. mp = 993°; bp = 1704°; d = 2.78; soluble in H_2O (4 g/100 ml); insoluble in organic solvents; LD_{50} (rat orl) = 0.18 g/kg. *Cerac; EM Industries; General Chem.; Hoechst Celanese; Solvay GmbH; Whiting, Peter Ltd.*

28765 sodium fluoroborate
13755-29-8 8761 237-340-6
BF_4Na
sodium tetrafluoroborate
sodium borofluoride;. Used as a fluorinating agent. mp = 384°; d = 2.4700; soluble in H_2O (108 g/100 ml), less soluble in organic solvents;.

28766 sodium formaldehydesulfoxylate
149-44-0 8764 205-739-4
CH_3NaO_3S
Hydroxymethanesulfinic acid sodium salt
Formusol®; Hydro AWC; Sodium formaldehyde sulfoxylate; formaldehyde sodium sulfoxylate; formaldehydesulfoxylic acid sodium salt; sodium hydroxymethanesulfinate; sodium methanalsulfoxylate; Aldanil; Rongalite C;,. Used for textile printing, emulsion polymerization; treatment of mercury

poisoning MW=118.09; crystals; mp=63-64°; sol.in water; insol.in abs alcohol, ether, benzene; LD_{50}(mise, s.c.)=4.0 g/kg *RV Chemicals Ltd.*

28767 sodium formate
141-53-7 8765 205-488-0
$CHNaO_2$
sodium formate
Formic acid, sodium salt; Salachlor; sodium formate, hydrated; sodium formate, hydrate. Reducing agent, manufacture of formic acid, oxalic acid and sodium dithionite, organic chemicals, mordant, complexing agent, analytical reagent (noble metal precipitant), buffering agent. mp = 253°; bp = 300°; d = 1.92; soluble in H_2O (0.77 g/ml), less soluble in organic solvents. *Aqualon; Heico; Hoechst Celanese; Perstorp Polyols; Spectrum Chem. Mfg.*

28768 sodium glucoheptonate
C7HH13O8
Used in metal cleaning, bottle washing, kier boiling, caustic boil-off, paint stripping. *Belzak; Hickson Danchem; Pfanstiehl Labs.*

28769 sodium gluconate
527-07-1 8766 208-407-7
$C_6H_{11}NaO_7$
D-gluconic acid monosodium salt
gluconic acid sodium salt; gluconic acid sodium salt; Developer, part B; D-gluconic acid, monosodium salt. As sequestering agent; in metal plating, tanning of hides, mordants for fabrics, paints; rust remover; mineral source in foods and pharmaceuticals. Soluble in H_2O (59 g/100 ml), less soluble in organic solvents. *Akzo; Albright & Wilson; Pfizer Spec. Chem.; PMP Fermentation Prods.; Rit-Chem.*

28770 sodium hexametaphosphate
10124-56-8 8814 233-343-1
Na6O18P6
Metaphosphoric acid, hexasodium salt
Glassy sodium; Hexasodium metaphosphate; Metaphosphoric acid ($H_6P_6O_1$+88), hexasodium salt; Sodium metaphosphate ($Na_6(PO_3)_6$); Sodium phosphate ($Na_6(PO_3)_6$); sodium polymetaphosphate; Graham's salt; Hy-Phos. Used in water softeners and detergents such as Calgon, Giltex; Quadrafos, Hagan Phosphate and Micromet. Also used in leather tanning, dyeing, laundries and textile processing. mp = 628°; soluble in H_2O. Hart Chem. Ltd.; Delaware Chemical Corp.; Rhone-Poulenc Food Ingreds.; Flexibulk Ltd; Albright & Wilson; Calgon; Farleyway Chem. Ltd.; FMC; Monsanto; Rhone-Poulenc Basic.

28771 sodium hydrosulfite
7775-14-6 8771 231-890-0
$Na_2O_4S_2$
disodium dithionite
sodium hyposulfite; sodium sulfoxylate; sulfoxylate; Virtex L; Reductone; Hydrolin; D-Ox; Vatrolite; Hydrosulfite; dithionous acid, disodium salt; sodium dithionite hydrate. Inorganic salt; chemical reagent for the reduction of aldehydes and ketones to alcohols; vat dying of fibers and textiles; bleaching sugar, soap, oils; oxygen scavenger for synthetic rubbers. mp = 52°; soluble in H_2O, less soluble in organic solvents. Farleyway Chem. Ltd; Henkel/Organic Prods.; Hoechst Celanese; Mitsubishi Gas; Morton Int'l.; Nissan Chem. Ind.; Olin.

28772 sodium hydroxide
1310-73-2 8772 215-185-5
HNaO
caustic soda
sodium hydrate; Caustic soda; Sodium hydrate; soda lye; Lye; White Caustic; lye, caustic; Augus Hot Rod. Inorganic base; chemical manufacture, rayon, cellophane, neutralizing agent in petrol. refining; textile processing; pulp, paper, soaps; vegetable oil refining; reclaiming rubber; food additive; etching and electroplating. mp = 318°; bp = 1390°; SG = 1.045; soluble in H_2O (0.5 g/ml); LD_{50} (rbt orl) = 500 mg/kg (10% solution). Akzo; Asahi Chem Industry Co Ltd; Asahi Denka Kogyo; Elf Atochem N. Am.; Georgia Gulf; Georgia-Pacific; ICI Spec.; Nissan chem. Ind.; Norsk Hydro A/S; Olin; OxyChem; PPG Industries; Rasa Ind.

28773 sodium hypochlorite
7681-52-9 8773 231-668-3
$ClNaO\cdot 5H_2O$
sodium hypochlorite pentahydrate
Clorox; bleach; Liquid bleach; Sodium oxychloride; Javex; Antiformin; Showchlon; Chlorox; B-K; Carrel-dakin solution; Chloros; Dakin's solution; Hychlorite; Javelle water; Mera industries 2MOM3B; Milton; Modified dakin's solution; Piochlor. Bleaching paper, pulp and textiles; intermediate; organic chemicals; water purification; medicine; fungicide; swimming pool disinfectant; laundering; reagent; germicide. mp = 18°; d = 1.250; soluble in H_2O (29 g/100 ml). *George Mann; Norsk Hydro A/S; Olin; OxyChem; Showa Denko.*

28774 sodium iodate
7681-55-2 8776 231-672-5
$INaO_3$
Iodic acid, sodium salt
Antispetic, disinfectant, feed additive reagent. d = 4.28; soluble in H_2O (9 g/100 ml), insoluble in organic solvents; LD (dog iv) = 200 mg/kg. *Atomergic Chemetals; Mallinckrodt.*

28775 sodium iodide
7681-82-5 8777 231-679-3
Ina
sodium monoiodide
sodium iodine. Photography, solvent for iodine, organic chemicals, reagent, feed additive, cloud seeding, scintillation, expectorant. mp = 651°; bp = 1300°; d = 3.67; soluble in H_2O (2 g/ml), less soluble in organic solvents; MLD (rat iv) = 1.3 g/kg. *Aldrich; EM Industries; Mallinckrodt.*

28776 sodium iron edetate
15708-41-5 4076 239-802-2
$C_{10}H_{12}FeN_2NaO_8$
[[N,N'-ethanediylbis[N-(carboxymethyl)glycinato]](4-)]-N,N',O,O',O^{N},O$^{N'}$-ferrate(1-) sodium
ferric sodium edetate; EDTA, iron (III) derivative, sodium salt; Sodium feredetate; Ferric sodium EDTA; Ferrate(1-), Ferrostrane; Ferrostrene; Sybron. Used as a source of iron.

28777 sodium lactate
72-17-3 8781 200-772-0
$C_3H_5NaO_3$
2-hydroxypropanoic acid monosodium salt
Lacolin; lactic acid, monosodium salt; propanoic acid, 2-hydroxy-, monosodium salt; sodium-dl-lactate. Sodium salt of lactic acid; hygroscopic agent; glycerol substitute; plasticizer for casein; corrosion inhibitor in alcoholic antifreeze; electrolyte replenisher. d = 1.326. *Am. Biorganics; EM Industries; R.W. Greeff; Patco; RITA; Verdugt BV.*

28778 sodium laureth sulfate
9004-82-4 221-416-0
$CH_3(CH_2)_{10}CH_2(OCH_2CH_2)_nOSO_3Na$, avg. n ·c 1-4
sodium lauryl ether sulfate (n·c1-4)
PEG (1-4) lauryl ether sulfate sodium salt; poly(oxy-1,2-ethanediyl), α-sulfo-ω-(dodecyloxy)-, sodium salt; sodium laureth sulfate; dodecyl alcohol, monoether with polyethylene glycol, hydrogen sulfate sodium salt; poly(oxyethylene) lauryl ether sulfate sodium salt; polyethylene glycol, mono(hydrogen sulfate), dodecyl ether, sodium salt. Foam stabilizer, detergent, flash foamer, wetter for detergent systems, personal care products; emulsion polymerization; shampoo base. *Chemron; Lonza; Norman, Fox; Pilot; Sandoz; Stepan; Unger Fabrikker AS; Vista.*

28779 sodium lauroyl sarcosinate
137-16-6 4379 205-281-5
$C_{15}H_{28}NNaO_3$
N-methyl-N-(1-oxododecyl)glycine, sodium salt
N-lauroylsarcosine sodium salt; n-lauroylsarcosine, sodium salt; glycine, N-methyl-N-(1-oxododecyl)-, sodium salt; sodium lauroylsarcosinate; Gardol; Hamposyl L 30; lauroylsarcosine, sodium salt; Maprosyl 30; sodium-N-dodecanoyl-N-methylglycinate; Medialan LL-99; methyl-N-(1-oxododecyl)glycine, sodium salt; Sarkosyl NL; Secosyl; sodium N-lauroylsarcosinate. Sodium salt of lauroyl sarcosine; foaming agent, wetting agent, detergent, lubricant, antistat, corrosion inhibitor, bacteriostat, penetrant used in dental, pharmaceutical, shampoos, depilatories, and shaving preparations, fo *W.R. Grace/Hampshire; R. T. Vanderbilt Co Inc.*

28780 sodium lauroyl taurate
70609-66-4 274-695-6
Hostapon KTW New; Hostapon KTW; Sodium-2-[(1-oxododecyl) amino] ethanesulfonate. Detergent base for cosmetics, toothpastes. *Hoechst Celanese/Colorants & Surf.; Hoechst AG.*

28781 sodium lauryl sulfate
151-21-3 8782 205-788-1
$C_{12}H_{25}NaO_4S$
sodium dodecyl sulfate
sulfuric acid monododecyl ester, sodium salt; SDS; detergent 66; Dreft; duponal waqe; dupanal; duponol; duponol c; Emal o; emal 10; emersal 6400; empicol lpz; empicol ls 30; empicol lx 28; emulsifier no. 104; akyposal sds; aquarex me; aquarex methyl; avirol 101; avirol 118; berol 452; carsonol sls; carsonol sls paste b; carsonol sls special; conco sulfate wa; conco sulfate wa-1200; cycloryl 21; dehag sulfate gl emulsion; finasol osr (sub 2); gardinol; hexamol sls; Irium; lanette wax-s; maprofix 563; melanol cl; melanol cl 30; monododecyl sodium sulfate; monogen y 100; montopol la paste; neutrazyme; nikkol sls; odoripon al 95; orvus wa paste; p and g emulsifier 104; perlandrol l; product no. 75; product no. 161; quolac ex-ub; rewopol nls 30; richonol a; sinnopon ls 100; sintapon l; sipex op; sipex sb; sipon ls; sipon ls 100; sipon lsb,. Sodium salt of lauryl sulfate; anionic detergent; surface tension depressant; emulsifier for fats; wetting agent; in textile industry, in toothpastes. Soluble in H_2O (10 g/100 ml); LD_{50} (rat orl) = 1288 mg/kg. Albright & Wilson Ltd; Chemron; Dajac Labs; Du Pont; Lonza; Norman, Fox; Pilot; Sandoz; Stepan; Unger Fabrikker AS; Witco.

28782 sodium magnesium silicate

53320-86-8 258-476-2

Laponite® XLG, D

Sodium magnesium silicate

Inert base/carrier for act. ingreds.; suspending agent; promotes thixotropy giving stable suspensions; thickens cosmetic, toiletry creams, lotions, toothpaste products. *Laporte/Southem Clay.* Discontinued.

28783 sodium metabisulfite

7681-57-4 8784 231-673-0

$Na_2O_5S_2$

disulfurous acid, disodium salt

sodium pyrosulfite; pyrosulfurous acid, disodium salt; sodium pyrosulfite; disodium salt pyrosulfurous acid; sodium disulfite; disodium disulfite; disulfurous acid, disodium salt. Inorganic salt; in foods, as preservative, antioxidant, laboratory reagent. Soluble in H_2O, slightly soluble in EtOH. *BASF; Blythe, William Ltd; EM Industries; General Chem.; Mallinckrodt.*

28784 sodium metaborate

7775-19-1 8785 231-891-6

$NaBO_2$

monosodium metaborate

boric acid (HBO_2), sodium salt; sodium borate ($NaBO_2$); sodium metaborate ($NaBO_2$). Herbicide. mp = 966°; soluble in H_2O. *Ashland; U.S. Borax & Chem.*

28785 sodium metaborate

7775-19-1 8785 231-891-6

$NaBO_2$

Rasorite. Kernite (hydrated sodium borate), and boron ores and products in general . *Borax Europe Ltd.*

28786 sodium metaperiodate

7790-28-5 8786 232-197-6

$INaO_4$

sodium periodate

Source of periodic acid, analytical reagent, oxidizing agent. mp = 300°; d_4^{16}= 3.865; soluble in H_2O; trihydrate: dec 175°; d_4^{18}= 3.219; soluble in H-=72O (0.12 g/ml). *Atomergic Chemetals; Noah Chem.*

28787 sodium metaphosphate

10361-03-2 5680 233-782-9

$(NaPO_3)_n$, n·c 3-10

sodium metaphosphate

(Cyclic) or larger (polymers); dental polishing agents, detergent builders, water softening, sequestrants, emulsifiers, food additives, textile processing, laundering. *Atomergic Chemetals.*

28788 sodium metasilicate

6834-92-0 8788 229-912-9

Na_2O_3Si

silicic acid, disodium slat

sodium metasilicate, anhydrous; Sodium Metasilicate; Disodium Monosilicate; Orthosil; Metso Beads, Drymet; Silicic acid (H_2SiO_3), disodium salt; Silicic acid disodium salt; Waterglass. Inorganic salt; cosmetics, laundry, dairy, and metal cleaning as soap builder and detergent, bleaching agent; as flocculant and dispersant in metallurgy and mining. mp = 1089°; d = 2.614; n_D^{25}= 1.520; soluble in cold H_2O, insoluble in organic solvents. *Eka Nobel AB; PQ; Rhône-Poulenc Basic.*

28789 sodium metasilicate pentahydrate

10213-79-3 229-912-9

$H_{10}Na_2O_8Si$

Ingredient in detergents for food processing equipment and general cleaning in dairies, bakeries, packing houses, in laundry, textile, paper, oil, metal industries. *Oxychem; PQ.*

28790 sodium methallyl sulfonate

1561-92-8 216-341-5

Geropon® MLS/A. Dye improver reactive comonomer for acrylic fibers polymerization; reactive emulsifier or coemulsifier in latex emulsion polymerization. *Rhône-Poulenc Geronazzo.*

28791 sodium methyl cocoyl taurate

61791-42-2 263-174-9

Sodium N-methyl-N-cocoyl taurate

Geropon® TC-42. Foamer, dispersant, detergent for detergent bars, shampoos, bubble baths, cosmetics; chemically stable. *Rhône-Poulenc Surf.; Rhône-Poulenc France.*

28792 sodium methyl stearoyl taurate

149-39-3 205-738-9

Hostapon STT Paste; Sodium stearoyl methyl taurate; Sodium N-methyl-N-stearoyl taurate; Sodium N-stearoyl-N-methyl taurate; Nikkol SMT,. Detergent for high quality cream shampoos. *Hoechst Celanese/Colorants & Surf.; Hoechst AG; Nikko Chem Co. Ltd.*

28793 sodium methyl tall oil acid taurate

61791-41-1 263-173-3

Fenopon TK32; Fenopon TN-74; Geropon® TK-32. Sodium N-methyl N-tall oil acid taurate in liquid form; dispersing and suspending agent used as a precipitation inhibitor for salts of barium, calcium and strontium, and for scale prevention in oil well tubing and flow lines. Clear liquid; VCS=8max; mw=439; soluble in water; pH=8-10. *ISP; Rhône-Poulenc Surf & Spec.* Name unverified.

28794 sodium methyl tall oil acid taurate

61791-41-1 263-173-3

Sodium N-methyl-N-tallowyl taurate

Geropon® TK-32. Detergent, suspending agent, dispersant; precipitation inhibitor for org. and inorg. salts of Ba, Ca, Sr; for petrol. industry. *Rhône-Poulenc Surf.*

28795 sodium m-nitrobenzene sulfonate

127-68-4 204-857-3

Ludigol F. Used to accelerate the stripping of nickel from steel, brass or copper; to control the reaction rate in pickling nickel-chromium-iron alloys and for dissolving nickel and copper; additive in phosphate coating of metals. *Rhône-Poulenc Surf.*

28796 sodium Molybdate, dihydrate

10102-40-6 8790 231-551-7

$Na_2MoO_4{\cdot}2H_2O$

sodium molybdate(VI) dihydrate

Molyhibit 100; molybdic acid, disodium salt, dihydrate; disodium molybdate dihydrate; sodium molybdate (VI) dihydrate. Reagent in analytical chemistry, used in manufacture of paint pigment, corrosion inhibitor, catalyst in dye and pigment production, additive for fertilizers and feeds, micronutrient. Anhydrous sodium molybdate [77631-95-0]. dec 100°; soluble in H_2O (0.6 g/ml) *AAA Molybdenum Prods.; Cerac; Climax Molybdenum; Mallincrkodt; PMC Specialties.*

28797 sodium monochloracetate

3926-62-3 2162 223-498-3

$C_2H_2ClNaO_2$

chloroacetic acid sodium salt

monoxone; sodium chloroacetate; sodium monochloroacetate; SMA; SMCA. Used as a herbicide. Soluble in H_2O (85 g/1oo ml); LD_{50} (rat orl) = 76 mg/kg.

28798 sodium monofluorophosphate

7631-97-2 231-552-2

FNa_2O_3P

SMFP. Used in manufacture of toothpaste. *Albright & Wilson; Elf Atochem N. Am.*

28799 sodium morrhuate

8031-09-2

The sodium salt of the fatty acids of codliver oil.

28800 sodium nitrate

7631-99-4 8792 231-554-3

$NaNO_3$

soda niter

cubic niter; sodium (I) nitrate; nitric acid, sodium salt; soda niter; nitrate of soda; Chile saltpeter. Oxidizing agent; solid rocket propellants; fertilizer, flux, glass manufacture, pyrotechnics, dynamites; color fixative and preservative in cured meats, fish, enamel for pottery; modifying burning properties of tobacco. mp = 308°; d = 2.26; soluble in H_2O (0.9 g/ml), less soluble in organic solvents; LD_{50} (rbt orl) = 1.96 g/kg. *BASF; Faesy & Besthoff; Farleyway Chem. Ltd; Mallinckrodt; Nissan Chem. Ind.; Spice King.*

28801 sodium nitrite

7632-00-0 8793 231-555-9

$NNaO_2$

nitrous acid, sodium salt; Anti-rust; Filmerine; diazoting salts; erinitrit. Diazotization, rubber accelerators, color fixative and preservative in cured meats, meat products, fish; pharmaceuticals, photographic and analytical reagent, dye manufacture, Vasodilator and antidote for cyanide poisoning. mp = 271°; dec > 320°; d = 2.17; soluble in H_2O (0.7 g/ml), less soluble in organic solvents; LD=75_0 (rat orl) = 180 mg/kg. *BASF; DuPont; EM Industries; Farleyway Chem. Ltd; General Chem.; ICI Spec.; PMC Spec.*

28802 sodium oleate

143-19-1 6965 205-591-0

$C_{18}H_{33}NaO_2$

sodium 9-octadecenoate

9-octadecenoic acid, sodium salt; oleic acid, sodium salt; olate; 9-octadecenoic acid (Z)-, sodium salt; octadecenoic acid sodium salt; Eunatrol. Sodium salt of oleic acid; ore flotation, waterproofing textiles, emulsifier of oil/water systems. mp = 232-235°; soluble in H_2O (10 g/l), less soluble in EtOH, organic solvents. *Hart Prod. Corp.; Norman, Fox; Scheweizerhall; Witco.*

28803 Sodium Omadine® 40% Aq. Sol'n

3811-73-2 8178 223-296-5

$C_5H_4NNaOS{\cdot}H_2O$

2-Mercaptopyridine-N-oxide, sodium salt monohydrate

sodium omadine; sodium pyrithione; N-hydroxy-2-pyridinethione, sodium salt;

2-pyridinethiol-1-oxide, sodium salt; mercaptopyridine-N-oxide sodium salt. Sodium 2-pyridinethione in water; industrial microbiostat; chelating agent; used in aqueous metal coolant and cutting fluids, latex emulsion, inks, fiber lubricants. mp -25 - -30°; bp = 109°; d = 1.2200 *Olin.*

28804 sodium *para*-aminohippurate
94-16-6 462 202-309-8
$C_9H_{11}N_2NaO_4$
p-hippuric acid sodium salt
Nephrotest; 4-aminohippuric acid, sodium salt monohydrate; sodium 4-aminohippurate hydrate. For intravenous use to measure effective renal plasma flow and tubular secretory capacity. mp = 123-125° *Merck & Co Inc.*

28805 sodium pentachlorphenate
C_6Cl_5NaO
pentachlorophenol sodium salt
sodium pentachlorophenate; sodium pentachlorophenoxide; Santobrite; Dowicide G. Insecticide, fungicide and non-selective contact herbicide. Used in control of termites, as a wood preservative, a pre-harvest defoliant and a general pre-emergence herbicide. Soluble in H_2O; free pentachlorophenol: LD_{50} (rat orl) = 210 mg/kg.

28806 sodium perborate
7632-04-4 8797 231-556-4
$BNaO_3$
sodium perborate anhydrous
Dexol. Topical antiseptic, denture cleaner, oxygen source. *Degussa; DuPont; Eka Nobel AB; ICI Spec.; Mitsubishi Gas.*

28807 sodium percarbonate
15630-89-4 239-707-6
$C_2H_6Na_4O_{15}$
Carbonic acid disodium salt, compd. with hydrogen peroxide (H2O2) (2:3)
disodium carbonate, hydrogen peroxide (2:3); disodium carbonate, compound with hydrogen peroxide (2:3); hydrogen peroxide sodium carbonate adduct. Bleaching agent for domestic and industrial use, denture cleaner, mild antiseptic. *Chemoxal SA; Degussa; Interox Am.*

28808 sodium permutite
A sodium zeolite, made artificially.

28809 sodium persulfate
7775-27-1 8801 231-892-1
$Na_2O_8S_2$
sodium peroxydisulfate
sodium peroxydisulfate; peroxydisulfuric acid, disodium salt; peroxydisulfuric acid disodium salt; disodium peroxodisulfate. Bleaching agent (fats, oils, fabrics, soaps), battery depolarizers, emulsion polymerization. Soluble in H_2O (549 g/l; MLD (rbt iv) = 178 mg/kg. *Aldrich; Chemoxal SA; Degussa; FMC.*

28810 sodium phosphate
7558-80-7 8806 231-449-2
$NaH_2PO_4{\cdot}2H_2O$
monosodium dihydrogen phosphate dihydrate
sodium phosphate, monobasic; sodium biphosphate; MSP. Controls pH in mildly acidic solutions; food products, water treatment, metal treatment. mp = 60°; d = 1.915; *Albright & Wilson; Aldrich; BritAg Industries Ltd.*

28811 sodium polyacrylate
9003-04-7
polycrylic acid, sodium salt
poly(podium acrylate); 2-propenoic acid, homopolymer, sodium salt; propenoic acid, sodium carbonate polymer; Calnox; propenoic acid, homopolymer, sodium salt. Thickener, stabilizer, protective colloid for natural and synthetic latexes for paints, films, coatings, and adhesives; dispersant; antiredeposition agents; antiscalant. *Alco; Allchem Industries; Arakawa; Rhône-Poulenc; 3-V.*

28812 sodium propionate
137-40-6 8816 205-290-4
$C_3H_5NaO_2$
propanoic acid, sodium salt
sodium propionate; propanoic acid, sodium salt. Sodium salt of propionic acid; food additive (preservative); fungicide. Has been used in veterinary medicine as a ketosis treatment. Soluble in H_2O (1 g/ml), less soluble in organic solvents. *Gist-Brocades Food Ingreds.; Niacet; Spectrum Chem. Mfg.; Verdugt BV.*

28813 sodium pyrithione
3811-73-2 8178 223-296-5
C_5H_4NNaOS
N-Hydroxy-2-pyridinethione, sodium salt
Sodium omadine; sodium pyrithione; 2-Mercaptopyridine-N-oxide, sodium salt; 2-Pyridinethiol-1-oxide, sodium salt; Mercaptopyridine-N-oxide sodium salt; Fonderma. Antibacterial and antifungal agent mp = -25-30° (dec); bp 109°; d = 1.2200;.

28814 sodium pyrophosphate
7722-88-5 9377 231-767-1
$Na_4O_7P_2$
tetrasodium pyrophosphate
TSPP; sodium pyrophosphate, normal; Pyrophosphoric acid, tetrasodium salt; Tetrasodium Pyrophosphate; Phosphotex; Pyrophosphate; TSPP; Sodium Diphosphate; Diphosphoric acid; tetrasodium salt; Sodium diphosphate (Na4P2O7); Sodium phosphate (Na4P2O7); Sodium pyrophosphate (4:1); Sodium pyrophosphate (Na4P2O7); Tetrasodium diphosphate; Victor TSPP. Water softener, synthetic detergent builder, dispersant, emulsifier, metal cleaner, boiler water treatment, viscosifier for drilling muds, deinking newsprint, synthetic rubber, textile dyeing, wool scouring, buffer, sequestra mp = 988°; d = 2.534; soluble in H_2O (6.23 g/100 ml), insoluble in organic solvents. *Farleyway Chem. Ltd; FMC; Mitsui Toatsu Chem.; Monsanto; Spectrum Chem. Mfg.*

28815 sodium selenite
26970-82-1 8822 233-267-9
$Na_2O_3Se{\cdot}5H_2O$
sodium selenite pentahydrate
Selenious acid, disodium salt, pentahydrate. Glass manufacture (removes green color during manufacture), reagent in bacteriology, testing germination of seeds, decorating porcelain. Soluble in H_2O, insoluble in organic solvents; LD_{50} (rat orl) = 7 mg/kg. *Atomergic Chemetals; Degussa; Noah Chem.*

28816 sodium sesquicarbonate
533-96-0 8823 208-580-9
$Na_2CO_3{\cdot}NaHCO_3{\cdot}2HOH$
carbonic acid, sodium salt (2:3)
carbonic acid, sodium salt (2:3); trisodium hydrogendicarbonate. Inorganic salt; detergent and soap builder; mild alkaline agent for general cleaning and water softening; bath crystal; alkaline agent in leather tanning; food additive. d = 2.112; soluble in H_2O (13-42 g/100ml). *FMC.* Discontinued.

28817 sodium silicate
1344-09-8 8824 215-687-4
Na_4O_4Si
silicic acid, sodium salt
Water glass; Soluble glass; Silicate of soda; Sodium Orthosilicate; Silicic Acid Sodium Salt; Sodium silicate glass. Used for lining Bessemer converters, acid concentrators; manufacture of grindstones, abrasive wheels; as solution; preserving eggs; fireproofing fabrics; detergent in soaps; as adhesive, waterproofing walls, in cements and paints. bp = 102°. *Aichi Silicate Chem Co Ltd; Asahi Denka Kogyo; Crosfield; OxyChem; PQ; Spectrum Chem. Mfg.*

28818 sodium silicofluoride
16893-85-9 8769 240-934-8
F6Na2Si
sodium hexafluorosilicate
sodium fluosilicate; sodium silicofluoride; sodium fluosilicate; Earwig bait; silicate(2-), hexafluoro-, disodium; disodium hexafluorosilicate. Fluoridation, laundry soaps, opalescent glass, vitreous enamel frits, metallurgy (aluminum and beryllium), insecticides, rodenticides, chemical intermediate, glue, leather and wood preservative, moth repellent, manufacture of manufacture of pure silicon. d = 2.68; soluble in H_2O (7 mg/ml); LD_{50} (rat orl) = 125 mg/kg. *Faesy & Besthoff; La Roche Chem.; Mitsui Toatsu Chem.; Whiting, Peter Ltd.*

28819 sodium stannate
12058-66-1 8826 235-030-5
$H_6Na_2O_6Sn$
sodium stannate(IV)
sodium tin oxide. Dyeing mordant, ceramics, glass, source of tin for electroplating, textile fireproofing, stabilizer for hydrogen peroxide, blueprint paper, laboratory reagent. Soluble in H_2O (0.59 g/ml). *Elf Atochem N. Am.; Blythe, William Ltd; M&T Harshaw; Spectrum Chem. Mfg.*

28820 sodium stearate
822-16-2 8827 212-490-5
$C_{18}H_{35}NaO_2$
sodium octadecanoate
octadecanoic acid, sodium salt; stearic acid sodium salt. Waterproofing and gelling agent; toothpaste, cosmetics; stabilizer in plastics; emulsifier and stiffener in pharmaceuticals; in glycerol suppositories. *Elf Atochem N. Am.; Magnesia GmbH; Norman, Fox; Original Bradford Soap Works; Witco.*

28821 sodium sulfate
7757-82-6 8829 231-820-9
Na_2O_4S
sodium sulfate, anhydrous
Disodium sulfate; Sulfuric acid, sodium salt; Sulfuric acid, disodium salt. Manufacture of Kraft paper, etc.; filler in synthetic detergents; processing textile fibers, dyes, tanning, pharmaceuticals, lab reagent, food additive. The decahydrate is known as Glauber's salt. Used as a cathartic. mp = 884°; SG = 2.68; soluble in H_2O (0.28 g/ml), insoluble in organic solvents; *Akzo Salt; Atochem N. Am.; Kemira Kemi AB; Lenzing AG; Occidental.*

28822 sodium sulfide
1313-82-2 8830 215-211-5
Na_2S

sodium monosulfide
sodium sulfuret; sodium sulfide (Na_2S); disodium sulfide
Organic chemicals, sulfur dyes, intermediates, viscose rayon, leather depilatory, paper pulp, hydrometallurgy of gold ores, sulfiding oxidized lead and copper ores, sheep dips, photographic reagent, engraving and lithography, analytical reagent. Nonahydrate [16721-80-5]. mp = 1180°; d_4^{14}= 1.856; soluble in H_2O (18,6 g/100 ml), less soluble in organic solvents; *Aldrich; Cerac; Farleyway Chem. Ltd; PPG Industries.*

28823 sodium sulfite
7757-83-7 8831 231-821-4
Na_2O_3S
sulfurous acid, sodium salt (1:2)
sulfurous acid, disodium salt; disodium sulfite; exsiccated sodium sulfite; sulfurous acid, disodium salt; sulftech;
sulfurous acid, sodium salt (1:2); sodium sulfite (Na_2SO_3). Inorganic salt; paper industry; reducing agent (dyes); food preservative and antioxidant; textile bleaching; in photographic developers. Soluble in H_2O (0.31 g/ml), less soluble in organic solvents. BASF; Blythe, William Ltd; EM Industries; Ferro/Grant; Indspec; Nissan Chem. Ind.; Rhone-Poulenc Basic.

28824 sodium tallowate
8052-48-0 232-491-4
Emery® 2895 Foamaster Soap L; Foamaster Soap L. Defoamer for dry agricultural formulations. *Henkel/Emery.*

28825 sodium tartrate
868-18-8 8833 212-773-3
$C_4H_4Na_2O_6$
disodium tartrate
sal tartar; tartaric acid disodium salt; Butanedioic acid, 2,3-dihydroxy-, [R-(RName unverified.,RName unverified.)]-, disodium salt. Reagent, food additive, sequestrant, stabilizer. Used as a cathartic. d = 1.82; soluble in H_2O (0.3 g/ml), insoluble in organic solvents; *Novarina Sri; Schweizerhall.*

28826 sodium thiosulfate
7772-98-7 8844 231-867-5
$Na_2S_2O_3 \cdot 5H_2O$
sodium thiosulfate pentahydrate
sodium subsulfite; sodium hyposulfite; Hyporice; sodium subsulfite; HYPO; antichlor; thiosulfuric acid, disodium salt; thiosulfuric acid ($H_2S_2O_3$), disodium salt. Photographic fixative, chrome tanning, chlorine removal in bleaching and papermaking, extraction of silver, dechlorination of water, mordant, reagent, bleaching, reducing agent in chrome dyeing, sequestrant in salt, antidote mp = 48°; d = 1.69; soluble in H_2O, insoluble in organic solvents; LD_{50} (rat iv) >2.5 g/kg. *Aldrich; Blythe, William Ltd; Ferro/Grant; General Chem.; Nissan Chem. Ind.*

28827 sodium *p*-toluenesulfonate
657-84-1 9671 211-522-5
$C_7H_7NaO_3S$
methylbenzenesulfonic acid, sodium salt
Hydrotrope, solubilizer, coupling agent, and viscosity modifier in liq. formulations; cloud pt. depressant in detergent formulations. Substituted aromatic compound; hydrotropic solvent. Soluble in H_2O. *Albright & Wilson; Hüls AG; Ruetgers-Nease.*

28828 sodium tripolyphosphate
7758-29-4; 13573-18-7 8846 231-694-5
$Na_5O_{10}P_3$
STPP
sodium triphosphate, tripoly; pentasodium triphosphate; sodium triphosphate; triphosphoric acid, pentasodium salt; pentasodium triphosphate; pentasodium tripolyphosphate. Water softener, sequestrant, deflocculating agent, food additive, texturizer. Soluble in H_2O (20 g/100 ml); LD_{50} (rat orl) = 6.5 g/kg. *Albright & Wilson; FMC; Kemira Kemi AB; Mitsui Toatsu Chem.; Monsanto; Rhône-Poulenc Basic.*

28829 sodium tungstate
13472-45-2 8847 236-743-4
Na_2O_4W
sodium tungstate(VI) dihydrate
sodium wolframate; tungstate, disodium, (T-4)-; sodium tungsten oxide. Intermediate for tungsten compounds (e.g., phosphotungstate), reagent, fireproofing textiles, alkaloid precipitant. Soluble in H_2O (0.9 g/ml), insoluble in organic solvents; LD_{50} (gpg orl) = 990 mg/kg. *Cerac; Noah Chem.; Spectrum Chem. Mfg.*

28830 sodium undecylenate
3398-33-2 222-264-8
$C_{11}H_{19}O_2Na$
10-undecenoic acid, sodium salt
Sodium salt of undecylenic acid; bacteriostat and fungistat in cosmetics and pharmaceuticals. *Elf Atochem N. Am.*

28831 sodium versenate
139-33-3 3556 205-358-3
$C_{10}H_{14}N_2Na_2O_8$
Glycine, N,N'-1,2-ethanediylbis[N-(carboxymethyl)-, disodium salt
edetate disodium; disodium dihydrogen ethylenediaminetetraacetate; disodium ethylenediaminetetraacetate; ethanediylbis(N-(carboxymethyl)glycine) disodium salt; Versene disodium salt. Chelating agent; pharmaceutic aid. mp = 252° (dec); soluble in H_2O; LD_{50} (rat orl) = 2 g/kg. *3M Pharmaceuticals.*

28832 sodium xylene sulfonate
1300-72-7 215-090-9
Eltesol® SX 30; Esi-Terge SXS; Hartotrope SXS 40, Powd; Manro SXS; Naxonate® 4L; Naxonate® SX; Nutrol SXS 5418; Pilot SXS-40; Reworyl® NXS 40; Stepanate® SXS; Sulfotex SXS-40; Witconate SXS 40%,. Hydrotrope, cloud pt. depressant used in the detergent mfg.; solubilizer, coupler. mp = 27°; bp = 157°; SG_{20} = 1.23; soluble in H_2O (>100 mg/ml), insoluble in organic solvents; LD_{50} (rat orl) = 1000 mg/kg. Albright & Wilson UK; Manro Products Ltd.; Ruetgers-Nease; Clough; Pilot; Stepan; Stepan Canada; Rewo Chemicals Ltd;Hart Chem. Ltd;Henkel/Cospha; Henkel Canada; Mitsubishi Gas; Pilot; Rutgers-Nease; Stepan; Witco.

28833 sodusec
Caustic soda. *ICI Chem & Polymers Ltd.*

28834 Sofanate
A fungicide for fruit storage. *Plant Protection.* Name unverified.

28835 Sofibex
10039-56-2 8775 216-948-5
H_4NaO_3P
Sodium hypophosphite monohydrate
Used for surface treatment, electrodeless nickel plating. mp = 90°; soluble in H_2O (1 g/ml), alcohols, glycerol, insoluble in other organic solvents; *Elf Atochem UK/Ceca.* Discontinued.

28836 Soflens
Polymacon; contact lens material. *Bausch & Lomb, Professional Products Div.* Name unverified.

28837 Sofnol Soda-lime G
A proprietary form of soda-lime containing a little manganic acid; it is stated to absorb much more carbon dioxide than ordinary soda-lime, and to change color as the degree of saturation is approached. No manufacturer.

28838 Sofnon 105G
Imidazoline derivative; softener for cotton, wool, and synthetic fibers. *Toho Chem. Industry.*

28839 soft copal
A name applied to varieties of Australian sandarac resin.

28840 soft platinum
Commercially pure platinum, containing about 1% iridium.

28841 Soft Resin P 65
Nonreactive polyester resin for the formulation of pigment pastes. *Bayer AG.*

28842 Soft Touch 1052
Silicone. Softener for prefinishing printed fabrics; excellent napping lubricant and shear stability. No manufacturer.

28843 Soft-Clad®
Polyester polyol/urethane color bead particles; for use in the manufacture of synthetic leather-like coatings. *Reichhold.*

28844 Softcon
Vifilcon
Contact lens material (hydrophilic). *Parke-Davis.* Name unverified.

28845 Softenol®
Fatty acid esters; for technical applications. *Dynamit Nobel Wien GmbH.* Name unverified.

28846 Softenol®
Antistatics and lubricants. *Hüls UK Ltd.*

28847 Softenol® 3100
67701-26-2 266-944-2
C_{12}-C_{18} fatty acid triglyceride
Lubricant for machinery used in manufacturing of foodstuffs; additive to textile and glass fiber finishing, sizing agent. *Hüls Am.*

28848 Softenol® 3114
C_{14} fatty acid triglyceride
Compressing aid in tableting technical substances; hydrophobing agent against moisture; lubricant for machinery, lacquer and varnish formulations; additive to liquid lubricating systems. *Hüls Am.*

28849 Softenol® 3408
C_8-C_{10} fatty acid-1,2-propanediol ester; additive in manufacturing of textile and glass fibers; lubricant for cutting devices; mold release agent in processing of plastics. *Hüls Am.*

28850 Softenol® 3829
C_8-C_{10} fatty acids triglyceride; modified; oil in manufacturing of aluminum foils; lubricant for machinery, medical/technical equipment, instruments; additive to cutting oils and lubricants. *Hüls Am.*

28851 Softenol® 3991
Glyceryl stearate
Lubricant, emulsifier, antistat, pigment dispersant, plasticizer, antiblocking agent for plastics, textile and leather processing aids, drilling and cutting oils, polishes, rubber; biodegradable. *Hüls Am.*

28852 Softex
Proprietary trade name for pure red oxide. No manufacturer.

28853 Softex
Stearoyl lactylates
Used as dough conditioners and antistaling additives in bread and baked goods. *Harcros Australia.*

28854 Softigen® 701
141-08-2 205-455-0
Glyceryl ricinoleate
Emollient; refatting and skin protecting agent; emulsifier; personal care products and pharmaceuticals. *Hüls Am; Hüls AG.*

28855 Softigen® 767
52504-24-2
PEG-6 caprylic
capric glycerides. Refatting and wetting agent, solubilizer, emollient used in cosmetics and pharmaceuticals, liquid soaps. *Hüls Am; Hüls AG.*

28856 Softisan®
Ointment and cream bases; for creams, ointments, emulsions and lipsticks. *Hüls AG.*

28857 Softisan® 100
Hydrogenated cocoglycerides
Consistency regulator, emollient, ointment base for personal care products and pharmaceutical industry. *Hüls Am.*

28858 Softisan® 378
Caprylic/capric/stearic triglyceride; ointment base, emollient, moisturizer, stabilizer; neutral ointment base with good skin compatibility and resorption characteristics, for the preparation of nonaq. ointments and creams. *Hüls Am.*

28859 Softisan® 601
Glyceryl cocoate
Hydrogenated coconut oil, and ceteareth-25; emollient; cosmetic and pharmaceutical self-emulsifying base for oil-water products. *Hüls Am.*

28860 Softisan® 649
Bis-diglyceryl caprylate/caprate/isostearate/stearate/hydroxystearate adipate; emollient, ointment base for creams, lipsticks, other decorative cosmetics; lanolin substitute. *Hüls Am.*

28861 Softisan® Gel
Bis-diglyceryl caprylate/caprate/isostearate/hydroxystearate adipate, propylene glycol dicaprylate/dicaprate, stearalkonium hectorite, propylene carbonate; consistency regulator for creams; high temperature stabilizer for water-oil emulsions and anhydrous skin care and decorative formulations. *Hüls Am.*

28862 SoftMate DW
Hefilcon A
2-hydroxyethyl methacrylate polymer with 1-vinyl-2-pyrrolidinone and ethylene dimethacrylate; contact lens material. *Barnes-Hind Inc.* Name unverified.

28863 Softrite
A proprietary rubber softener; it has a zinc laurate base. No manufacturer.

28864 Softyne H
Amido-fatty quaternary; textile softening agent. *Hart Prods. Corp.*

28865 Soil Pests Killer
Granular insecticide. *Fisons plc, Horticultural Div.* Name unverified.

28866 Soil TRIGGRR
A liquid containing cytokinin, a plant growth regulator; used to increase crop yields and quality; applied to the soil; used for a wide variety of crops including corn, peanuts, sorghum, soybeans, fruits, and vegetables. *Westbridge Research Group.* Unverified.

28867 soilime
A lime residue from cyanamide manufacture. It contains 50% of lime.

28868 S-oils
Sulfur-containing oils obtained by the distillation of crude petroleum oil in the presence of sulfur. They have a strong antiseptic action against wood-destroying fungi.

28869 Soiltex®
Nonhazardous soil conditioners in liquid and granular form for maintaining soil structure and preventing capping. *Allied Colloids Ltd.*

28870 Sokalan® CP 2
Sodium salt of a maleic anhydride/methyl vinylether copolymer; dispersant, anti-incrustation agent. *BASF; BASF AG.*

28871 Sokalan® CP 5
Sodium salt of maleic acid/acrylic acid copolymer; dispersant, anti-incrustation agent. *BASF; BASF AG.*

28872 Sokalan® CP 7
9003-04-7
Sodium salt of a modified polyacrylic acid; auxiliary in phosphate-reduced or phosphate-free detergents to improve primary and secondary (antiredeposition and anti-incrustation) washing effects; dispersant. *BASF; BASF AG.*

28873 Sokalan® HP 22
109464-53-1
Acetic acid ethenyl ester, polymers, polymer with oxirane, graft; antiredeposition agent and soil shield polymer for laundry products. *BASF.*

28874 Sokalan® HP 50
9003-39-8 7879
Polyvinylpyrrolidone
Antiredeposition inhibitor for detergents in laundry applications; dispersant. *BASF; BASF AG.*

28875 Sokalan® HP 53
9003-39-8 7879
Polyvinylpyrrolidone
Antiredeposition inhibitor for detergents; carbon dispersant. *BASF; BASF AG.*

28876 Sokalan® PA 110 S
9003-01-4
Polyacrylic acid
Dispersant for water treatment, laundry detergents, agriculture, paints, coatings, industrial and institutional cleaners. *BASF; BASF AG.*

28877 Sokalan® PA 13 PN
Polyacrylic
Dispersant for water treatment, laundry detergents, agriculture, paints, coatings, industrial and institutional cleaners. *BASF.*

28878 Sokalan® PA 15
9003-04-7
Sodium salt of polyacrylic acid; dispersant for water treatment, laundry detergents, agriculture, paints, coatings, industrial and institutional cleaners. *BASF; BASF AG.*

28879 Sokalan® PM 10
Copolymeric carboxylate
Dispersant. *BASF.*

28880 Sokoff
Industrial grease solvent. *The Wellcome Foundation Ltd.* Name unverified.

28881 Sokolan
Polymeric dispersants. *BASF plc.*

28882 sol rubber
The portion of rubber which enters solution when unmilled raw rubber is treated with a solvent.

28883 Solactol
97-64-3 3863 202-598-0
A proprietary trade name for ethyl lactate. No manufacturer.

28884 Soladox
Chlorosulfonated polyethylene in a solvent base, with dense pigmentation; a corrosion preventative for offshore oil, all marine, roofing, any corrosive environment where complete protection is needed. *Liquid Plastics Ltd.* Discontinued.

28885 Soladox 112
Chlorosulfonated polyethylene in a solvent base, with dense pigmentation; weatherproofing compound; reflects infrared radiation; for use in defense related environments. *Liquid Plastics Ltd.* Discontinued.

28886 Solan
61790-81-6
PEG-75 lanolin
Surfactant, emollient, conditioner, superfatting agent, emulsifier, solubilizer, foam stabilizer, plasticizer, humectant for soaps, detergent bars, shampoos, skin cleansers, hair sprays, deodorants, chemical specialties. *Croda Inc.*

28887 Solan
2307-68-8 8851 218-988-9
$C_{13}H_{18}ClNO$
chloro-2-methyl-*p*-valerotoluidide
chloro-4-(methylphenyl)-2-methylpentamide; CMMP; Dakuron; Solan; pentanochlor. Emulsifiable concentrate containing 400 g/l pentanochlor; used to control weeds in horticultural crops. mp = 79-80°; insoluble in H_2O, soluble in orghanic solvents; LD_{50} (rat orl) = 10 g/kg. *Atias Interlates Ltd.*

28888 Solan 50
61790-81-6
PEG-60 lanolin
Hydrophilic surfactant, emollient, conditioner, thickener, superfatting agent,

foam stabilizer, plasticizer, humectant for personal care products, soaps, pharmaceuticals, chemical specialties; fragrance solubilizer. *Croda Inc.*

28889 Solan E
Polyethoxylated lanolin
Croda Chem. Ltd.

28890 Solane
Sulfur dyes. *ICI Chem & Polymers Ltd.*

28891 Solangel 401
61790-81-6
PEG-75 lanolin
Emulsifier, humectant for soap. *Croda Inc.* Discontinued.

28892 Solanthrene
Vat dyes. *ICI Chem & Polymers Ltd.*

28893 Solar
Synthetic organic direct dyestuffs. *Sandoz Products Ltd.*

28894 Solar
Nonionic spreader containing 75% polypropoxypropanol and 15% alkyl polyglycol ether; adjuvant for plant growth regulator sprays. *Ideal Manufacturing Ltd.*

28895 solar oil
The name given to various hydrocarbons obtained as by-products in the treatment of brown coal tar in paraffin works.

28896 solar salt
Salt (sodium chloride) obtained by the evaporation of sea-water.

28897 Solar Steel
A proprietary trade name for a silicon steel containing 1% silicon, 0.5% molybdenum, 0.4% manganese, and 0.5% carbon. No manufacturer.

28898 Solarchem® O
21245-02-3 3282 244-289-3
$C_{17}H_{27}NO_2$
2-Ethylhexyl-4-dimethylaminobenzoate
octyl dimethyl PABA; Escalol 507; Eusolex 6007; Octyl Dimethyl-PABA; Octyl Dimethylaminobenzoate; Padimate 0. Selective uv absorber for suncare and cosmetic products. 2-Ethylhexyl-4-dimethylaminobenzoate is a PABA derivative that is used as a sun-screening agent. Of the PABA group it is the most commonly used agent in sunscreens marketed in the United States. It may also be found in cosmetics for skin, hair, and nails such as moisturizers and lipsticks and lip balms. *CasChem.*

28899 Solasol
Vay dyes. *ICI Chem & Polymers Ltd.* Discontinued.

28900 Solatene
7235-40-7 1902 230-636-6
β-Carotene
UV screen. *Hoffmann-LaRoche Inc.* Name unverified.

28901 Solatol
A preparation of crude phenol (carbolic acid); a disinfectant.

28902 Solbrol®
p-Hydroxybenzoic acid ester
Used for the preservation of pharmaceuticals, cosmetics, foodstuffs, and technical products. *Bayer AG.*

28903 Solcod
A proprietary trade name for sulfonated cod oil. No manufacturer.

28904 Solcornol
A proprietary trade name for a sulfonated corn oil. No manufacturer.

28905 Soldaflux
Soft solder flux. *Degussa Ltd.*

28906 Soldamoll
Special soft solder. *Degussa Ltd.*

28907 solder
The various alloys or mixtures which constitute solder are usually classified as hard or soft, according to their melting point. Hard solder includes brazing solder, silver solder, and gold solder, while the soft solders usually consist of tin and lead and melt below 300°. Used for joining metals.

28908 Solderel®
Solder paste for the electrical industry. *DuPont UK.*

28909 soldering acid
7647-01-0 4821 231-595-7
HCl
Hydrochloric acid.

28910 soldering salt
Ammonium and zinc chlorides.

28911 soldering solution
A solution of zinc chloride.

28912 Soldis
A disinfectant containing phenolic and cresylic bodies; miscible with water.

28913 soldo
A flux used for tinning metals. It is mixed with powdered tin.

28914 Sole Terge 8
43154-85-4 256-120-0
Disodium oleamido MIPA-sulfosuccinate
High foaming base for bubble baths, shampoos, etc.; frother in specialized ore flotation; emulsifier and stabilizer in emulsion polymerization. *Calgene.*

28915 Soledon
Solubilized vat dyes. *ICI Chem & Polymers Ltd.*

28916 Solef®
$(CH_2CF_2)_n$;
Polyvinylidene fluoride (PVDF) homopolymers and copolymers, used for high purity and corrosion resistant applications in the chemical processing and semiconductor manufacturing industries, including pipe and fittings, valves, pumps and vessels, protective coatings, wire and cable jacketing, fiber optics buffer tubing and optical fiber raceway applications. *Solvay Polymers.*

28917 Solef® 1008
24937-79-9
PVDF homopolymer; for injection molding for general use and complicated shapes or thin walls; for extrusion of tubes; transfer and centrifugal molding, film extrusion. *Solvay Polymers.*

28918 Solef® 1010
24937-79-9
PVDF homopolymer; for general extrusion of tubes, films, sheets, thin panels, injection molding, compression and transfer molding, blow molding of films and hollow objects. *Solvay Polymers.*

28919 Solef® 5008
24937-79-9
PVDF homopolymer; top coat for electrostatic powder spraying. *Solvay Polymers.*

28920 Solef® 6010
24937-79-9
PVDF homopolymer; improved thermal stability resin for thick semi-finished items. *Solvay Polymers.*

28921 Solef® 8808
24937-79-9
PVDF homopolymer, carbon fiber-reinforced; reinforced grade for applications requiring extremely high rigidity. *Solvay Polymers.*

28922 Solef® 11008/0003
24937-79-9
PVDF copolymer; for rotational molding. *Solvay Polymers.*

28923 Solef® 11010
24937-79-9
PVDF copolymer; for use where more flexibility and very high elongation at break are required, e.g., electric and telephone cable sheathing, extrusion of sheets. *Solvay Polymers.*

28924 Sole-Mulse B
Modified ethoxylate
Emulsifier for emulsion degreasers, industrial cleaners. *Calgene.*

28925 Solenhofen stone
A porous limestone containing clay.

28926 Solenite
An explosive. It is an Italian smokeless powder, and contains 30% nitroglycerin, 40% insoluble nitrocellulose, and 30% soluble nitrocellulose.

28927 Solester
Vat dyes. *ICI Chem & Polymers Ltd.* Discontinued.

28928 Solfa
7704-34-9 9142 231-722-6
S
sulfur
Formulation of sulfur, a fungicide for mildew control in a wide range of crops. *ICI Chem & Polymers Ltd.*

28929 Solfa
7704-34-9 9142 231-722-6
Sulfur
Used for mildew control in a wide range of crops. *Farm Protection Ltd.*

28930 Solfac
68359-37-5 2826 269-855-7
cyfluthrin. Insecticide for the control of stored product pests, vector-and nuisance insects. *Bayer AG.*

28931 Solflex® 1216
Solution S/B vinyl, nonstaining stabilizer; developed for tire formulations; also suitable for molded goods. *Goodyear.* Name unverified.

28932 Solgen
Industrial detergent. *Crosfield Chemicals Ltd.* Discontinued.

28933 Solicum
A material made from waste rubber and oil.

28934 solid alcohol
A soapy mass containing about 20% water, 20% sodium stearate, and 60% alcohol.
28935 Solidarol
Reactive dyestuffs. *Hoechst AG.*
28936 Solidegal®
Leveling agents for textile dyeings and printings. *Cassella AG.*
28937 Solidermin®
Sulfur dyestuffs; used for leather dyeing. *Cassella AG.*
28938 Solidex
Photographic developer. *May & Baker Ltd.* Name unverified.
28939 solidified alcohol
A name applied to a solution of nitrocellulose in ethyl alcohol for use in heaters, etc., as a fuel.
28940 solidified linseed oil
oxidized linseed oil; linoxyn. A flexible solid mass obtained when linseed oil is exposed to oxidation.
28941 Solidite
A proprietary trade name for a range of molded products made from shellac, bitumen, and fillers; used for electrical insulation. No manufacturer.
28942 Solidogen LT-13
Cationic resin; improves wet fastness of direct and developed dyeings on cellulosic fibers; on suede leathers for garments or gloves to increase fastness to washing and drycleaning and promote level drying. *ISP.* Name unverified.
28943 Solidogen®
Dyeing auxiliaries and improving agents for textile dyeing and printing. *Cassella AG.*
28944 Solidokoil®
Padding auxiliaries with migration resistant properties. *Cassella AG.*
28945 Soligen
A proprietary trade name for certain metallic naphthenates used as paint driers. No manufacturer.
28946 Solimide® Foam
Flexible polyimide foam; fire-resistant, lightweight, low smoke, flexible insulation foams for thermal and acoustical insulation applications; flexible and resilient from - 184 to 260°C. *Ethyl Corp; Ethyl SA.*
28947 Solintor
Inorganic and organic pigments. *Tennant-KVK Ltd.*
28948 Solinure
Range of soluble fertilizers. *Fisons plc, Horticultural Div.* Name unverified.
28949 Solka-Floc® BW-40, BW-100, BW-200, BW-2030, UF-900-FCC & NF
9004-34-6 2012 232-674-9
Cellulose
Binder, diluent, disintegrant, stabilizer, absorption aid, stabilizer, tablet filler for pharmaceutical formulations. *Mendell.* Discontinued.
28950 Solklean™ 101
7664-38-2 7500 231-633-2
Phosphoric acid
Solvents and surfactants; liquid formulated for cleaning of metallic solar absorber plates prior to painting. *Solec.*
28951 Solkote™ Hi/Sorb™ -II
Selective optical coating specifically formulated for solar applications; high and low temperature air and liquid absorbers; trombe walls; photographic applications; high temperature applications. *Solec.*
28952 Sollacaro's aluminum solder
An alloy of 64% zinc, 30% tin, and 6% lead.
28953 Solmed 100
Gloss finish film for thermoforming, heat contact and impulse welding processes. *Solvay Ind. Films.*
28954 Solo
Emulsifiable concentrate containing 1250 g linuron [330-55-2] and 240 g trifluralin [1582-09-8] per liter; herbicide for winter cereals. *MTM Agrochemicals Ltd.*
28955 Solochrome
After-chrome dyes. *ICI Chem & Polymers Ltd.*
28956 Solok®
For the electrical industry. *DuPont UK.*
28957 Solon Conc
64-02-8 3557 200-573-9
Tetrasodium EDTA
Chelating of all metals, iron specialized. *Eastern Color & Chem.*
28958 Solon Fe Special
Sodium dihydroxy-ethyl glycine type material; chelating agent for high concentrates of ion under alkaline conditions. *Eastern Color & Chem.*
28959 Solophenyl
Direct dyes. *Ciba plc.* Name unverified.
28960 Solosil
Foundry binder for the CO_2 process. *Foseco (F.S.) Ltd.*
28961 Solox
A proprietary trade name for an alcohol-type solvent, a fuel for alcohol lamps, blow torches, portable stoves, etc.; it contains mostly ethyl alcohol, with small quantities of ethyl acetate and petrol. No manufacturer.
28962 Solozone
1313-60-6 8800 215-209-4
A proprietary trade name for a sodium peroxide containing 20.5% available oxygen. No manufacturer.
28963 Solpolac
Chloropolyethylene
Resin for the preparation of paints and varnishes. *Caffaro SpA; SNIA (UK) Ltd.* Name unverified.
28964 Solricin® 135
7492-30-0 231-314-8
$C_{18}H_{33}KO_3$
hydroxy-9-octadecenoic acid, monopotassium salt
potassium ricinoleate. Detergent, emulsifier, mild germicide, glycerized rubber lubricant, foam stabilizer in foamed rubber; making of cutting and sol. oils, household and cosmetic products. *CasChem.* Name unverified.
28965 Solricin® 235
8013-05-6 232-388-4
Potassium castor soap solution in water; emulsifier, dispersant, mild germicide; glycerized rubber lubricant; emulsifier, foam stabilizer for foamed rubber. *CasChem.*
28966 Solricin® 285
Ammonium ricinoleate solution in water; emulsifier, dispersant, rustproofing agent (leaves corrosion resistant film on exposure to air); lubricant; for both oil and water systems. *CasChem.*
28967 Solricin® 435
5323-95-5 8378 226-191-2
Sodium ricinoleate aqueous solution; emulsifier, stabilizer, defoamer for emulsion polymerization of resins (PVC, PVAc). Also used as a mild germicide; glycerized rubber lubricant; emulsifier and foam stabilizer for foamed rubber. *CasChem.*
28968 Solsaf-T-Solv™ 403
For cleaning or precleaning of aluminum. *Solec.*
28969 Solsolv™ 301
For cleaning spray equipment. *Solec.*
28970 Solsperse
Dispersants. *Imperial Chemical Industries plc.*
28971 Solstar
Computerized design programs to stimulate casting solidification prior to making patterns and designing feeding aids. *Foseco (F.S.) Ltd.*
28972 Soltair
Mixture of diquat, paraquat, and simazine; total herbicide. *ICI Chem & Polymers Ltd.*
28973 Soltrol
Isoparaffinic solvents. *Phillips.*
28974 Solu Kera-Tein M
Soluble keratin; cosmetics ingredient. *May-brook.*
28975 Solubilisant Gamma 2420
Octoxynol-11
polysorbate 20. Surfactant blend. *Gattefosse.*
28976 Solubilisant Gamma 2428
PEG-40 hydrogenated castor oil, polysorbate 20, octoxynol-11; surfactant blend. *Gattefosse.*
28977 Solu-Biloptin
1151-11-7 5087 214-565-8
3 g Calcium ipodate in powder form; used as an x-ray contrast media. *Schering Health Care Ltd.* Discontinued.
28978 soluble algin
240
Alginate of soda, obtained from seaweed.
28979 soluble cream of tartar
868-14-4 7776 212-769-1
Cream of tartar (potassium bitartrate, $C_4H_5O_6K$) dissolved in a solution of boric acid or borax.
28980 soluble glass
Soluble soda-glass; water glass; glass liquor. A syrupy solution containing 50% of sodium silicate, Na_4SiO_4, and Na_2SiO_3; used to impregnate articles to render them fire-resistant, as an adhesive for glass and porcelain, as an adulterant in soap, dyeing and egg preserving.

28981 soluble phenyle
A fluid containing coal-tar creosote, rosin oil, potassium oleate, and caustic soda. It gives an emulsion with water, and is used as a sheep dip.

28982 Soluble Plant Feed
Soluble powder containing NPK 19:19:19 plus 0.2% Mg and trace elements; a fertilizer. *Vitax Ltd.*

28983 soluble potash glass
1312-76-1 7838 215-199-1
K_2SiO_3
Potassium silicate.

28984 Soluble Rose Feed
Soluble powder containing NPK 16:8:32 plus 0.15% Mg and trace elements; fertilizer. *Vitax Ltd.*

28985 soluble salumin
$Al_2(C_6H_4(ONH_4)CO_2)_6 \cdot 2H_2O$
Aluminum ammonium salicylate
Used as an astringent.

28986 soluble starch
9004-84-9 8955
amylodextrin
amylogen. Obtained by heating starch with glycerin and adding alcohol; a soluble starch is also made by boiling starch in water and adding a little caustic soda to clear it; an emulsifying agent.

28987 soluble tartar
921-53-9 7849 213-067-8
$C_4H_4K_2O_6$
potassium tartrate

28988 Soluble Tomato Feed
Soluble powder containing NK 18:36 and trace elements; a fertilizer. *Vitax Ltd.*

28989 Solublon
Packaging materials and plastic film of polyvinyl alcohol which dissolves either in cold or hot water; for packaging for toxic, skin-irritating or strongly colored materials used in aqueous solution, to protect the handling handling personnel; water-soluble laundry bags for packing the contaiminated lines in the hospital before laundering in hot water. *Aicello Chemical Co Ltd.* Name unverified.

28990 Solubor
1330-43-4 8733 215-540-4
$Na_2B_8O_{13} \cdot 4H_2O$
sodium borate
A highly soluble form of sodium borate ; used to correct boron deficiency in plants by applying either as a foliar spray, in nutrient feeds or with herbicides. *Borax Europe Ltd.*

28991 Solubor DF
BR Destral. Heavy-duty nonselective herbicide; used for weedkilling. *Borax Europe Ltd.*

28992 Solu-Coll
9007-34-5 2543 232-697-4
Soluble collagen; cosmetic ingredient for moisturizing applications. *Brooks Industries.*

28993 Solu-Coll P
Procollagen; cosmetic ingredient for moisturizing applications. *Brooks Industries.*

28994 Solucryl
Acrylic solutions. *UCB Chemical Sector.*

28995 Soluene 100 and 350
0.5N quaternary ammonium hydroxides in toluene; cationic surfactants with the ability to solubilize a wide variety of biological samples. *Packard Instrument BV.* Name unverified.

28996 Solufeed
Soluble fertilizer. *ICI Chem & Polymers Ltd.*

28997 Soluhoba
Jojoba oil derivatives. *A & E Connock (Perfumery & Cosmetics) Ltd.*

28998 Solulan
8027-33-6 232-430-1
lanolin alcohols
Amerchol L 101; lanolin alcohol. Amerchol L 101 and Solulan are trade names of a product containing lanolin alcohols obtained from hydrolysis of lanolin. It is used as an emulsifier and emollient. Used in cosmetics, toiletries, pharmaceuticals, cutting oils, frniture polish, inks, leather, metal corrosion prevention, papers, textiles, topical medications and waxes. *D F Anstead Ltd.* Name unverified.

28999 Solulan 5
Ethoxylated complex of lanolin alcohols and related fatty alcohols; nonionic water-oil emulsifier, stabilizer for oil-water systems; nontacky lubricant, emollient and moisturizer; wetting and dispersing aid for cosmetic pigments. *Amerchol Corp.* Discontinued.

29000 Solulan® 5
Laneth-5
ceteth-5; oleth-5; steareth-5. Mixture of laneth 5, cteth 5, oleth 5 and steareth 5. Used as an emulsifier, wetting agent, dispersant, lubricant, solvent, conditioner, plasticizer, emollient used in personal care and dermatology products; foam stabilizer for detergent systems. *Amerchol; Amerchol Europe.* Discontinued.

29001 Solulan® 16
Ethoxylated complex of lanolin alcohols and related fatty alcohols; nonionic solubilizer, wetting agent and oil-water emulsifier; excellent foam stabilizer and conditioning agent in shampoos. *Amerchol Corp.*

29002 Solulan® 75
61790-81-6
PEG-75 lanolin
Emulsifier, wetting agent, dispersant, lubricant, solvent, conditioner, plasticizer, emollient used in personal care and dermatology products; foam stabilizer for detergent systems. *Amerchol; Amerchol Europe.* Discontinued.

29003 Solulan® 97
Polysorbate 80 acetate
cetyl acetate; acetylated lanolin alcohol. Dispersant, lubricant, emollient, conditioner for cosmetics and pharmaceuticals. *Amerchol; Amerchol Europe.* Discontinued.

29004 Solulan® 98
Partially acetylated complex of Polysorbate 80 and Acetulan; nonionic oil-water emulsifier, solubilizer and pigment wetting agent; conditioner for shampoos. *Amerchol Corp.*

29005 Solulan® C-24
choleth-24
ceteth-24. Emulsifier, wetting agent, dispersant, lubricant, solvent, conditioner, plasticizer, emollient used in personal care and dermatology products; foam stabilizer for detergent systems. *Amerchol; Amerchol Europe.*

29006 Solulan® PB-10
Propoxylated (10 mol) lanolin alcohols; spreading agent, pigment wetting agent, glosser and hydrophobic emollient. *Amerchol Corp.* Discontinued.

29007 Solulan® PB-2
68439-53-2
PPG-2 lanolin alcohol ether; spreading agent, dispersant, plasticizer; emollient and conditioner for personal care products, detergents, pharmaceuticals, waxes, polishes, leather treatment. *Amerchol; Amerchol Europe.* Discontinued.

29008 Solulan® PB-20
Propoxylated (20 mol) lanolin alcohol; plasticizer, spreading agent, and conditioner. *Amerchol Corp.* Discontinued.

29009 Solulan® PB-5
Propoxylated lanolin alcohols; water-resistant conditioner for skin and hair; pigment wetter and glosser. *Amerchol Corp.*

29010 Solu-Lastin 10
Hydrolyzed animal elastin. Structural protein giving flexibility and elasticity to tissues; for cosmetics use. *Brooks industries.*

29011 Solu-Lastin 30
Hydrolyzed elastin
Structural protein giving flexibility and elasticity to the tissues; for cosmetics use. *Brooks Industries.*

29012 Solulys®
Corn steep liquor or powder; growth medium for fermentation industry. *Roquette (UK) Ltd.*

29013 Solu-Mar EN-30
Hydrolyzed marine protein; film-former and moisturizer for skin and hair cosmetics; substantive protein. *Brooks Industries.*

29014 Solu-Mar Native
Soluble marine protein; film-former and moisturizer for skin cosmetics. *Brooks Industries.*

29015 Solumin
Anionic surfactant. *Rhône-Poulenc UK.*

29016 Solumin F
A range of sodium sulfated alkylphenol ethoxylates supplied as aqueous solutions; wetting, foaming, and detergency, dispersion, emulsification, e.g., in emulsion polymerization; pigment dispersion; and cleaning formulations. *Rhône-Poulenc UK.*

29017 Solumin PFN
A range of anionic surfactants consisting of phosphate esters of ethoxylated alkylphenols, as viscous liquids; emulsion hydrotroping, corrosion inhibition and conductivity improvement agents used in emulsion polymerization and as a conducitivity additive. *Rhône-Poulenc UK.*

29018 Solumin PV27
Phosphate ester of ethoxylated alcohol in yellow liquid form; anionic surfactant with good stability, detergency and corrosion resistance; used in industrial cleaners, hydrotroping and lubricants. *Rhône-Poulenc UK.*

29019 Solumin T45S
Sodium sulfated synthetic alcohol ethoxylate in aqueous solution; foaming, wetting and dispersing agent used in shampoos, bubble baths, and surgical scrubs. *Rhône-Poulenc UK.*

29020 Solumin V27SD
Sodium sulfated synthetic alcohol ethoxylate in aqueous solution; foaming, wetting, dispersing and emulsification agent. *Rhône-Poulenc UK.*

29021 so-luminum
An aluminum solder consisting of 55% tin, 33% zinc, 11% aluminum, and 1% copper.

29022 Solumix
Mixture of toluene, xylene, and aliphatics; an aromatic solvent. *Sasolchem.*

29023 Soluphor® P
616-45-5 8200 210-483-1
C_4H_7NO
Pyrrolidone-2
α-pyrrolidone; σ-butyrolactam; 2-pyrrolidinone. Solvent for polymers, pesticides and sugars. Used in printers inks and as a plasticizer and coalescing agent for floor polishes. mp = 23-25°; bp_{10} = 128-130°; d = 1.107; $n_D^{?}$ = 1.4870; *BASF AG.*

29024 Solu-Silk 25
Silk amino acids
Cosmetics ingredient. *Brooks Industries.*

29025 Solu-Silk Protein
96690-41-4 306-235-8
Hydrolyzed silk; protein. *Brooks Industries.*

29026 Solu-Soy EN-25
Hydrolyzed soy protein; cosmetic ingredient for skin and hair care products; low salt. *Brooks Industries.*

29027 Sol-U-Tein 6861
Hydrolyzed soy protein; skin/hair conditioner for permanent waves, rinses, shampoos, tonics, dressings, cleansers, face, body and hand creams and lotions, moisturizing creams. *Fanning.*

29028 Sol-U-Tein EA
9006-59-1 216 232-692-7
Albumen
Albumin. Binder, coagulant; used in pharmaceuticals and personal care products; dye mordant in textiles, adhesives, veneers, sizing and making papers; gilding leather; book binding; and food application. *Fanning.*

29029 Sol-U-Tein FS-1000
Hydrolyzed soy protein; skin/hair conditioner for permanent waves, rinses, shampoos, tonics, dressings, cleansers, face, body and hand creams/lotions, moisturizing creams. *Fanning.*

29030 Sol-U-Tein PS-1000
Hydrolyzed soy protein; skin/hair conditioner for permanent waves, rinses, shampoos, tonics, dressings, cleansers, face, body, and hand creams/lotions, moisturizing creams. *Fanning.*

29031 Sol-U-Tein VG
Hydrolyzed soy protein; skin/hair conditioner for permanent waves, rinses, shampoos, tonics, dressings, cleansers, face, body and hand creams/lotions, moisturizing creams. *Fanning.*

29032 Solutene
Textile auxiliary chemicals. *ICI Chem & Polymers Ltd.* Discontinued.

29033 Solutol
An alkaline solution of sodium cresylate in an excess of cresol, obtained by treating cresol with caustic soda; used as a disinfectant.

29034 Solutol® HS 15
PEG-660 hydroxystearate
Solvent for injection solutions. *BASF AG.*

29035 Solu-Veg EN-35
Hydrolyzed vegetable protein; cosmetic ingredient for skin and hair care products; low salt. *Brooks Industries.*

29036 Soluvit Richter
Multivitamin herbal complex (vitamins A, B, E, H, essential fatty acids, horse-chestnut extract), hydro-alcohol solubilized; broad spectrum vitamin treatment for skin and hair protection. *Dr. Kurt Richter; Henkel/Cospha.*

29037 Solux
p-hydroxy phenylmorpholine
A proprietary trade name for a rubber antioxidant. No manufacturer.

29038 Solva
Emulsifying agents used for cheese. *Giulini Corp.*

29039 Solvaperm
Dyestuffs for plastics. *Hoechst UK.*

29040 Solvatone
A mixture of approximately 80% acetone, 10% isopropyl alcohol, and 10% toluene; used as a solvent for lacquers.

29041 Solvay® Soda
497-19-8 8739 207-838-8
A registered trade name for a sodium carbonate for water softening. No manufacturer.

29042 Solvene
A heavy grade coal-tar naphtha; a proprietary solvent for ester gum, pitches, etc. No manufacturer.

29043 Solvenol 2, 226
Solvenol 1. Terpene liquids; used as thinners and antiskinning agents in paints; reclaiming agents for natural and synthetic rubbers. *Hercules.*

29044 Solvenon® BB
Volatile solvent for surface coatings and paints; additive in cleaning and pickling agents. *BASF AG.*

29045 Solvenon® DIP
Mixture of isomers of dipropylene glycol monoisopropyl ethers; solvent for resins and dyes, surface coatings industry; leveling agent for water-dilutable baking finishes; film-binding assistant for dispersions; component in cleaning agents and printing inks. *BASF AG.*

29046 Solvenon® DPM
Mixture of isomeric dipropylene glycol monomethyl ethers; solvent for domestic cleaners, acid rust remover, production of printing inks; film-binding assistant for aqueous polymer dispersions in paint and varnish industry. *BASF AG.*

29047 Solvenon® IPP
3944-37-4 223-534-8
$C_6H_{14}O_2$
1-Isopropoxy-2-propanol
2-isopropoxy-1-propanol. Solvent for resins and dyes, surface coatings, cleaning agent for printing plates *BASF AG.*

29048 Solvenon® I
85% Isobutyl formate, 15% isobutanol; solvent and diluent for adhesives, paints, varnishes *BASF AG.*

29049 Solvenon® PC
108-32-7 203-572-1
$C_4H_6O_3$
propylene carbonate
4-methyl-1,3-dioxolan-2-one; 1,2-propanediol cyclic carbonate; PC; 1,2-propylene carbonate. Solvent for pigments and dyes, in screen printing dyes; extracting agent; washing liquid for natural and synthetic gases; intermediate for organic syntheses. mp = -55°; bp = 240°; d = 1.1890; n_D^{20} = 1.4210. *BASF AG.*

29050 Solvenon® PM, Solvent PM
107-98-2 203-539-1
$C_4H_{10}O_2$
1-Methoxypropan-2-ol
methoxypropanol, α isomer; propylene glycol methyl ether; 1-methoxypropan-2-ol; 1-methoxy-2-propanol; methoxy ether of propylene glycol; α-propylene glycol monomethyl ether; polypropylene glycol methyl ether; propylene glycol 1-methyl ether; (+/-)-1-methoxy-2-propanol; Dowanol 33b; Dowanol pm; Dowtherm 209; glycol ether pm; PGME; poly-solve MPM; propasol solvent M; UCAR solvent LM. Solvent for cleaning agents, printing inks, surface coatings. mp = -97°; bp = 118-119°; d = 0.9220; n_D^{20} = 1.4030. *BASF AG.*

29051 Solvenon® PP
1-Phenoxypropanol-2 and 2-phenoxypropanol-1; solvent in surface coatings, e.g., electrodipping varnishes by cathodic deposition; component of cleaning agents. *BASF AG.*

29052 Solvent 78
Mixture of esters, alcohols, ketones, hydrocarbons; paint thinners, wash solvent. *Solrec Ltd.* Name unverified.

29053 Solvent 401
Mixture of esters, alcohols, ketones, hydrocarbons; paint thinner for car refinishing. *Solrec Ltd.* Name unverified.

29054 Solvent GC
Mixture of ethylene glycol acetates; solvent for printing inks and rubber stamp inks. *BASF AG.*

29055 solvent naphtha
A fraction of coal tar distillation of specific gravity 0.875; also the name for the wood naphtha recovered from grey acetate of lime, prepared from the distillation of wood.

29056 Solvent Scour 25/27
Alkanolamides and alkyl phenol ethoxylate; scouring agent and prescour for greasy and oil soiled fabrics. *Hart Chem. Ltd.*

29057 Solveol
Cresols made soluble in water by the addition of sodium cresotinate; a disinfectant and substitute for guaiacol and creosote.

29058 Solvesso
Aromatic solvents of high purity. *Exxon Int'l.*

29059 Solvetek
Pigment dispersions; for coloration of solvent-based coatings. *Pacific Dispersions Inc.* Name unverified.

29060 Solvethane
71-55-6 9766 200-756-3
Trichloroethane
Laporte Industries Ltd. Discontinued.

29061 Solvic
PVC resins. *Laporte Industries Ltd.* Discontinued.

29062 Solvifog (N.R.I.)
Diluent and carrier for thermal fogging of pesticides. *Makhteshim Chemical Works Ltd.* Discontinued.

29063 Solvigran
298-04-4 3429 206-054-3
Insecticide containing disulfoton. *ICI Chem & Polymers Ltd.*

29064 Solvoclarin
Combination of special detergents, finishing agents, antistats, and deodorants; detergent for chlorinated and fluorinated hydrocarbons. *Henkel Chemicals Ltd.* Name unverified.

29065 Solvol
Tetrahydro-naphthol acetate
A proprietary solvent. No manufacturer.

29066 Solvtext
Solvent publications. *Exxon UK.* Name unverified.

29067 Somacount
Stabilized and treated blood cells in a milk-like medium; reference controls for automated milk cell analyzers. *Coulter Electronics Ltd.*

29068 Somafix
Formaldehyde-based cell fixative; allows the automated counting of somatic cells in milk. *Coulter Electronics Ltd.*

29069 Somali gum
An acacia gum from *Acacia glaucophylla* and *Acacia abyssinica.*

29070 Somaton
Alcoholic saline diluent
Allows the counting of somatic cells in milk by automated analyzers. *Coulter Electronics Ltd.*

29071 Somepon T25
Sodium methyl cocoyl taurate
Surfactant for cosmetics. *Seppic.* Discontinued.

29072 Somon
3926-62-3 2162 223-498-3
Powder containing 96% w/w sodium monochloroacetate; annual dicotyledons control in various horticultural crops. *Hortichem Ltd.*

29073 Sonacide
111-30-8 4480 203-856-5
$C_5H_8O_2$
Glutaric dialdehyde
glutaraldehyde; pentanedial; 1,3-diformylpropane; 1,5-Pentanedial; Alhydex; Cidex; Dioxopentane; Glutaral; Glutardialdehyde; Glutarol; Sporicidin; Ucarcide; Veruca-sep; Gluteraldehyde; 1,5-pentanedione; potentiated acid glutaraldehyde; sonacide; Aldesan; Coldcide-25 microbiocide; Glutaralum; Hospex. Disinfectant. mp = -6°; bp = 101°; d = 1.0600; n_D^{25}= 1.4338; soluble in H_2O, LD_{50} (0.6 g/kg, aqueous solution). *Wyeth Laboratories.*

29074 Sonnenschein's reagent
An alkaloid assay reagent prepared by adding phosphoric acid to a warm solution of ammonium molybdate in nitric acid, boiling the precipitate produced in *aqua regia*, evaporating to dryness, and dissolving in 10% nitric acid.

29075 Sonojell
Petroleum jellies. *Witco Corporation.*

29076 Sonora gum
Arizona shellac
A variety of shellac obtained from *Larrea Mexicana.*

29077 Sonostat 1111
Warp size lubricant and plasticizer for synthetic and gelatin type sizes designed to reduce static and buildup during weaving. *Henkel/Textiles.*

29078 Sonostat NTL
Finish for nylon carpet yarn requiring resistance to discoloration. *Henkel/Textiles.*

29079 Sontara®
Fiber. *DuPont UK.*

29080 Sontex
White mineral oils. *Pennzoil Products Co, Penreco Div.*

29081 Sontique®
Fibers. *DuPont UK.*

29082 Sopanox
Soap antioxidant. *Monsanto plc.* Name unverified.

29083 Soprodac
Food preservative. *BP Chemicals Ltd.*

29084 Soprophor® 37
Ethoxylated polystyrylphenol
Emulsifier for styrene copolymer; latex stabilizer; dispersant and antioxidant for phenolic slurries. *Rhône-Poulenc Geronazzo.*

29085 Soprophor® 3D33
Ethoxylated tristyrylphenol phosphate
Acid form; emulsifier, dispersant for agricultural formulations. *Rhône-Poulenc Surf.; Rhône-Poulenc Geronazzo.*

29086 Soprophor® BSU
PEG-16 tristyrylphenol
Emulsifier, dispersant for agricultural formulations. *Rhône-Poulenc Surf.; Rhône-Poulenc Geronazzo.*

29087 Sorane
Two-component polyurethane systems. *ICI Polyurethanes.* Name unverified.

29088 Sorban
50-70-4 8873 200-061-5
$C_6H_{14}O_6$
sorbitol
A strong solution of sorbitol.

29089 Sorbax HO-40
POE sorbitol ester; emulsifier for agricultural pesticide/herbicide, emulsion polymerization, metalworking lubricants, die-cast lubricants. *Chemax.*

29090 Sorbax HO-50
POE sorbitol ester; oil-water emulsifier for solvents, vegetable and petroleum oils used in the textile, agriculture, emulsion polymerization, and metal lubricant industries. *Chemax.*

29091 Sorbax PML-20
9005-64-5 8872
Polysorbate 20; oil-water emulsifier, solubilizer for perfumes, flavors; for agriculture, cosmetic, leather, metalworking, and textile industries. *Chemax.*

29092 Sorbax PMO-20
9005-65-6 7742
Polysorbate 80; oil-water emulsifier, solubilizer for perfumes and flavors; emulsifier for petroleum oils, solvents, vegetable oils, waxes, silicones in agriculture, cosmetic, leather, metalworking, and textile industries. *Chemax.*

29093 Sorbax PMO-5
9005-65-6 7742
Polysorbate 81; oil-water emulsifier, solubilizer for perfumes and flavors; emulsifier for petrol. oils, solvs., vegetable oils, waxes, silicones in agriculture, cosmetic, leather, metalworking, and textile industries. *Chemax.*

29094 Sorbax PMP-20
9005-66-7 8872
Polysorbate 40; oil-water emulsifier, solubilizer for perfumes and flavors; emulsifier for petroleum oil, solvents, vegetable oils, waxes, silicones in agriculture, cosmetic, leather, metalworking, and textile industries. *Chemax.*

29095 Sorbax PMS-20
9005-67-8 8872
Polysorbate 60; oil-wateremulsifier, solubilizer for perfumes and flavors; emulsifier for petroleum oils, solvents, vegetable oils, waxes, silicones in agriculture, cosmetic, leather, metalworking, and textile industries. *Chemax.*

29096 Sorbax PTO-20
9005-70-3 8872
Polysorbate 85; oil-water emulsifier, solubilizer for perfumes and flavors; emulsifier for petroleum oils, solvents, vegetable oils, waxes, silicones in agriculture, cosmetic, leather, metalworking, and textile industries. *Chemax.*

29097 Sorbax PTS-20
9005-71-4 8872
Polysorbate 65; oil-water emulsifier, solubilizer for perfumes and flavors; emulsifier for petroleum oils, solvents, vegetable oils, waxes, silicones in agriculture, cosmetic, leather, metalworking, and textile industries. *Chemax.*

29098 Sorbax SML
1338-39-2 8872 215-663-3
Sorbitan laurate
Emulsifier for petrol. oils, solvents, vegetable oils, waxes, silicones in agriculture, cosmetic, leather, metalworking, and textile industries. *Chemax.*

29099 Sorbax SMO
1338-43-8 8872 215-665-4
Sorbitan oleate

Emulsifier, surfactant for oil-water emulsion stabilizers and thickeners used in cosmetic, agriculture, metalworking, leather, and textile industries. *Chemax.*

29100 Sorbax SMP
26266-57-9 8872 247-568-8
Sorbitan palmitate
Emulsifier, surfactant for o/w emulsion stabilizers and thickeners used in cosmetic, agriculture, leather, metalworking, and textile industries. *Chemax.*

29101 Sorbax SMS
1338-41-6 8872 215-664-9
Sorbitan stearate
Emulsifier, surfactant for o/w emulsion stabilizers and thickeners used in cosmetic, agriculture, leather, metalworking, and textile industries. *Chemax.*

29102 Sorbax STO
26266-58-0 8872 247-569-3
Sorbitan trioleate
Emulsifier, surfactant for oil-water emulsion stabilizers and thickeners used in cosmetic, agriculture, metalworking, leather, and textile industries. *Chemax.*

29103 Sorbax STS
26658-19-5 8872 247-891-4
Sorbitan tristearate
Emulsifier, surfactant for oil-water emulsion stabilizers and thickeners used in cosmetic, agriculture, metalworking, leather, and textile industries. *Chemax.*

29104 Sorbelite C
50-70-4 8873 200-061-5
Crystalline sorbitol NF; provides sweetness and cooling mouthfeel to direct compression tablet manufacturing; improved flowability, excellent cohesion. *Mendell.* Discontinued.

29105 sorbic acid
110-44-1 8869 203-768-7
$C_6H_8O_2$
trans,trans-2,4-hexadienoic acid
2,4-hexadienoic acid; 2-propenyl acrylic acid; Preservastat; Sorbistat; hexadienoic acid;1,3-pentadiene-1-carboxylic acid; *trans,trans*-sorbic acid; α-*trans*-σ-*trans*-sorbic acid; (E,E)-2,4-hexadienoic acid; panosorb; (2-butenylidene)acetic acid; crotylidene acetic acid; hexadienoic acid, (E,E); hexa-2,4-dienoic acid. Organic acid; animal feed preservative. mp = 134°; bp = 228° (dec); soluble in H_2O (0.25 g/100 ml), more soluble in organic solvents; LD_{50} (rat orl) = 7.36 g/kg. *Allchem Industries; Chisso; Hoechst Celanese; Honeywill & Stein Ltd; Penta Mfg.; Spice King.*

29106 Sorbismal
A German preparation of finely divided bismuth in oil.

29107 Sorbistat
110-44-1 8869 203-768-7
Sorbic acid
Pfizer Ltd.

29108 Sorbistat K
590-00-1
Potassium sorbate
Pfizer Ltd.

29109 Sorbitan diisostearate
68238-87-9 269-410-7
$C_{42}H_{80}O_7$
Emsorb® 2518; Anhydrohexitol diisostearate. Auxiliary emulsifier, solubilizer, corrosion inhibitor in lubricants, metal protectants and cleaners, emulsion polymerization. *Henkel/Emery.*

29110 Sorbitan isostearate
71902-01-7 8872 276-171-2
Emalex SPIS-100; Crill 6; Emsorb® 2516. Oil-phase cosmetic ingred., surfactant; for creams, milky lotions, hair conditioners. *Henkel/Emery;Croda Inc; Croda Surf. Ltd.*

29111 sorbitan tristearate
26658-19-5 8872 247-891-4
$C_{60}H_{114}O_8$
anhydrosorbitol tristearate
sorbitan trioctadecanoate; STS; sorbitan trioctadecanoate. Triester of stearic acid and hexitol anhydrides derived from sorbitol; emulsifier for foods, cosmetics, household products, industrial applications. *Henkel/Emery; ICI Spec.; Lonza.*

29112 sorbite
A constituent of iron, which is formed in the transformation of austenite; the stage following trootsite and osmondite, and preceding pearlite.

29113 sorbitol
50-70-4 8873 200-061-5
$C_6H_{14}O_6$
Hexahydric alcohol
D-glucitol; D-sorbitol; D-sorbite; sorbite; D-sorbit; L-gulitol; glucitol; gulitol; D-galactitol. Nutrient and dietary supplement, food additive; bodying agent for paper, textile, liquid pharmaceuticals; in manufacture of sorbose, ascorbic acid, propylene glycol, synthetic plasticizers, resins; as humectant, sequestrant. mp = 110-112°; $[\alpha]_D^{20}$ = -2° (H_2O); soluble in H_2O, polar organic solvents. *Aldrich; Cerestar UK; EM Industries; Fanning; ICI Spec.; Lipo; Lonza.*

29114 Sorbitol (EGIC)
50-70-4 8873 200-061-5
A proprietary preparation of sorbitol used in intravenous nutrition. *Servier Laboratories Ltd.* Name unverified.

29115 Sorbo®
50-70-4 8873 200-061-5
Sorbitol
Cosmetics ingred. *ICI Am.*

29116 Sorbolene®
The trademark for a fat liquor for the leather trade; used in the tanning and dyeing process to give greater elasticity to the leather, and enables fuller shades to be obtained in dyeing. No manufacturer.

29117 Sorbon S-20
1338-39-2 8872 215-663-3
$C_{18}H_{34}O_6$
Sorbitan laurate
Span 20; sorbitan monolaurate; sorbitan, monododecanoate; sorbitan laurate; Alkamuls SML; Arlacel 20; Emsorb 2515; Glycomul L. Emulsifier, dispersant for water-oil emulsion. Insoluble in H_2O, soluble in organic solvents. *Toho Chem. Industry.*

29118 Sorbon S-40
26266-57-9 8872 247-568-8
$C_{22}H_{42}O_6$
Sorbitan palmitate
Span 40; sorbitan monopalmitate; sorbitan, monohexadecanoate; sorbitan palmitate. Used with Sorbon T series. Applied topically, can act as a tumor promoter. *Toho Chem. Industry.*

29119 Sorbon S-60
1338-41-6 8872 215-664-9
$C_{24}H_{46}O_6$
Sorbitan stearate
Span 60; sorbitan monostearate; Alkamuls SMS; Arlacel 60; Glycomul S. Used with Sorbon T series. mp = 49-65°; insoluble in H_2O, Me_2CO, soluble in EtOH, CCl_4, C_7H_8. *Toho Chem. Industry.*

29120 Sorbon S-66
36521-89-8 8872 253-084-8
sorbitan distearate
Toho Chem. Industry. Name unverified.

29121 Sorbon S-80
1338-43-8 8872 215-665-4
$C_{24}H_{44}O_6$
Sorbitan oleate
Span 80; sorbitan monooleate; Alkamuls SMO; Arlacel 80; Capmul O; Emsorb 2500; Glycomul O. Used with Sorbon T series. Insoluble in H_2O, propylene glycol, soluble in organic solvents. *Toho Chem. Industry.*

29122 Sorbon T-20
9005-64-5 8872
PEG-20 sorbitan laurate
Emulsifier, dispersant, solubilizer for oil-water emulsions. *Toho Chem. Industry.*

29123 Sorbon T-40
8872
POE sorbitan palmitate
Used with Sorbon S series. *Toho Chem. Industry.*

29124 Sorbon T-60
8872
POE sorbitan stearate
Used with Sorbon S series. *Toho Chem. Industry.*

29125 Sorbon T-80
8872
POE sorbitan oleate
Used with Sorbon S series. *Toho Chem. Industry.*

29126 Sorbon TR 814
POE sorbitol oleate
Detergent, emulsifier for cosmetics. *Toho Chem. Industry.*

29127 Sorbon TR 843
POE sorbitol oleate
Detergent, emulsifier for cosmetics. *Toho Chem. Industry.*

29128 Sorbonorit
7440-44-0 1855 231-153-3
carbon
Highly activated granular carbon; used for solvent recovery. *NORIT NV.*

29129 sorbose
87-79-6 8874 201-771-8

$C_6H_{12}O_6$
L-xylo-hexulose
L-(-)-sorbose; sorbin; sorbinose. Made from sorbitol (itself made by reduction of glucose) by fermentation and used in the manufacture of ascorbic acid (vitamin C). mp = 158-160°; $[\alpha]_D^{30}$ = -43° (c = 5); soluble in H_2O, insoluble in organic solvents;.

29130 Sorbosil
Thickening and polishing agents for toothpastes. *Crosfield Chemicals Ltd.*

29131 Sorbothane®
Visco-elastic polymer; features quasi-liquid properties enabling high mechanical damping and energy absorption, return to original shape, stable properties over broad temperature range; for applications in shock/impact absorption, vibration isolation and acoustical damping. *Sorbothane.*

29132 Sorbsil
Dessicant silica gels. *Crosfield Chemicals Ltd.*

29133 Soreflon
A proprietary range of PTFE polymers. *Montedison UK Ltd.* Name unverified.

29134 Sorel Cement
A magnesium oxychloride cement, made from magnesite (magnesium carbonate, $MgCO_3$) and magnesium chloride, $MgCl_2$. No manufacturer.

29135 Sorel's gutta-percha substitutes
Substitutes containing rosin, pitch, rosin oil, slaked lime, and gutta-percha. Some are filled with china clay, and in others coal tar is used.

29136 Sorensen's salt
$Na_2HPO_4 \cdot 2H_2O$
Sodium phosphate.

29137 Sorex Golden Fly Bait
16752-77-5 6062 240-815-0
Granules containing 1% w/w methomyl; used for fly control in livestock houses. *Sorex Ltd.* Discontinued.

29138 Sorex Super Fly Spray
Mixture containing phenothrin and tetramethrin; for control of flying insects in agricultural premises. *Sorex Ltd.*

29139 Sorex Wasp Nest Destroyer
Mixture of resmethrin and tetramethrin; wasp destroyer. *Sorex Ltd.*

29140 Sorexa CD
Caciferol + difenacoum; used for the control of mice in farm buildings. *Sorex Ltd.*

29141 Sorexa Plus
81-81-2 10174 201-377-6
Warfarin; a rodenticide. *Sorex Ltd.* Discontinued.

29142 Sorgan
Propachlor + propazine; herbicide. *Agan Chemical Manufacturers Ltd.*

29143 Soricinol 40
5323-95-5 8378 226-191-2
$C_{18}H_{33}NaO_3$
[R-(Z)]-12-hydroxy-9-octadecenoic acid sodium salt
sodium ricinolate; sodium ricinoleate; castor oil, sodium salt; sodium 12-hydroxy-(cis)-9-octadecenoate; Soricin; Colidosan. Mold release. Soluble in H_2O, alcohols. *Actrachem.*

29144 Sorlate
9005-65-6 7742
Polysorbate 80; pharmaceutic aid. *Abbott Laboratories.* Name unverified.

29145 Soromin®
Sizing agents for synthetic fiber industry. *BASF AG; BASF plc.*

29146 Soromine AT
Complex fatty amido compound; imparts an excellent hand, good body and draping qualities, and lubricity to synthetic, cellulosic and animal fibers and leathers. *ISP.* Name unverified.

29147 Sorpol 320
POE alkylaryl ether, POE sorbitan alkylate, and sulfonate; emulsifier for Malathion emulsifiable concentrates and other pesticides. *Toho Chem. Industry.*

29148 Sorpur®
709-98-8 7987 211-914-6
Propanil. Post-emergence applied herbicide with no residual effect for control of numerous grasses and broad-leaved weeds in rice crops. *Bayer AG.*

29149 Sorrel's alloy
An alloy of 98% zinc, 1% iron, and 1% copper. Another alloy contains 80% zinc, 10% iron, and 10% copper.

29150 Sorvall®
For medical applications. *DuPont UK.*

29151 Soubieran's ammonical salt
$(NH_2 \cdot Hg_2O)NO_3$
Mercury ammonium-nitrate

29152 Soucol
Industrial chemical. *Crosfield Chemicals Ltd.* Discontinued.

29153 Souesite
A nickel-iron alloy which occurs naturally.

29154 Soulan's cement
Consists of 7 parts resin, 10 parts ether, 15 parts collodion, and aniline red. A semi-transparent varnish; used for sealing corks into bottles.

29155 Southalite
A phenol-formaldehyde condensation product, with a filler of paper; used for insulating purposes.

29156 Sovatex C1
Alkylaryl sulfonate in liquid form; anionic scouring and milling agent. *Standard Chemical Company.*

29157 Sovatex EP 5288
61791-38-6 263-170-7
Coconut imidazoline amphoteric in liquid form; antistats; lubricants; corrosion inhibitors, detergents. *Standard Chemical Company.*

29158 Sovatex IM12H
61791-38-6 263-170-7
Hydroxyethyl imidazoline of coconut fatty acid in the form of a semi-liquid; corrosion inhibitor; lubricant; antistatic agent; used as a base for cationic surfactants. *Standard Chemical Company.*

29159 Sovatex IM12N
Aminoethyl imidazoline of coconut fatty acid in the form of a semi-liquid; corrosion inhibitor; lubricant; antistatic agent; used as a base for cationic surfactants. *Standard Chemical Company.*

29160 Sovatex IM17H
95-38-5 428 202-414-9
$C_{22}H_{42}N_2O$
2-(8-Heptadecenyl)-4,5-dihydro-1H-imidazole-1-ethanol
Hydroxyethyl imidazoline of oleic acid in liquid form; corrosion inhibitor; lubricant; antistatic agent; used as a base for cationic surface active agents. *Standard Chemical Company.*

29161 Sovatex IM17N
Aminoethyl imidazoline of oleic acid in liquid form. corrosion inhibitor; lubricant; antistatic agent; used as a base for cationic surface active agents. *Standard Chemical Company.*

29162 Sovatex MP/1
Oleyl imidazoline amphoteric in liquid form; antistats; lubricants; corrosion inhibitors; detergents. *Standard Chemical Company.*

29163 Sovatex WA
Sulfosuccinate surfactant in liquid form; concentrated anionic wetting agent. *Standard Chemical Company.*

29164 Sovermol POL 1008
Polyester polyol; for coatings, elastomers, rigid foam formulations, PU dispersions. *Henkel/Functional Prods.*

29165 Sovermol POL 1012
Branched polyalcohol; solvent-free polyol for potting and casting, adhesives, rigid PU foams. *Henkel/Functional Prods.*

29166 Sovprene
9010-98-4
A Russian chloroprene synthetic rubber obtained by the polymerization of acetylene to form divinyl-acetylene and then the formation of chloroprene by treating with hydrogen chloride, followed by polymerization.

29167 Soxhlet's solution
A modified Fehling's solution. It consists of a solution of 34.639 g copper sulfate in 500 ml water and 50 g caustic soda, 173 g potassium sodium tartrate in 500 ml water; used for the determination of sugars.

29168 Soy Flour
68513-95-1
Emcosoy®; Soy polysaccharides; Soyarich® 115 W; Soyafluff® 200 W;,. Tablet disintegrant for direct compression preparation *Mendell; Central Soya;.*

29169 soya bean oil
8001-22-7 232-274-4
Soybean oil; Chinese bean oil; soy oil; Acidulated soybean soapstock; Refined soybean oil; Refined undeodorized soybean oil; Soya oil; Soybean acidulated soapstock; Soybean deodorizer distillate; Soybean oil, bleached; Soybean oil bleaching; Soybean oil, degummed; Soybean oil deodorization; Soybean oil, deodorized; Soybean oil fatty acids, glycerol triester; Soybean vegetable oil, winter fraction. An oil obtained from the seeds of *Soja hispida* by expression or extraction with a solvent; used as an edible oil, and is also used in soap-making, paints and varnishes, and in the linoleum industry.

29170 Soyafluff® 200 W
68513-95-1
Soy flour; functional soy flour for food industry. *Central Soya.*

29171 Soy-Amino Quat L/O
Lauryloleylmethylamine soy amino acids
Cosmetic ingredient for skin and hair care products; conditioning agent; gives excellent gloss in hairsprays; highly compatible with popular resins. *Brooks Industries.*

29172 Soyarich® 115 W
68513-95-1
Soy flour; functional soy flour for food industry. *Central soya.*

29173 Soybean oil
8001-22-7 232-274-4
Lipovol SOY; Super Refined Soybean Oil. Soybean oil; natural emollient oil, lubricant, conditioner for luxury skin products, hair care products, makeup, fine soaps, bath oils, anhyd. systems. *Lipo.*

29174 Soy-che
Zinc, iron, copper, manganese, sulfur; chelated micronutrient for soybeans. *Draxel Chemical Company.* Unverified.

29175 Soy-Quat C
977039-11-4
Cocodimonium hydroxypropyl hydrolyzed soy protein; substantive conditioner, moisturizer for hair and skin care products (shampoos, conditioners, creams, lotions, bath products, face and body cleansers). *Maybrook.*

29176 Soy-Tein NL
70084-94-5 274-308-0
Hydrolyzed soy protein; protein providing moisture retentive, protective and sealing films, substantivity to hair and skin care products (shampoos, conditioners, treatment products, lotions, creams, bath products); mitigates *Maybrook.*

29177 SP-33
1332-58-7 5294 296-473-8
Coarse particle sized metakaolin (aluminum silicate) specially processed for polyvinyl chloride electrical insulation compounds. *Engelhard.*

29178 SP-731
Acetate/amine blend; line flushing and clean-up solvent for adhesives, resins, and urethane foam; contains no methylene chloride. *Specialty Prods.*

29179 SP-2205
Ethylene-methyl acrylate copolymer resin; film resin for specialty coextrusions, blends. *Monsanto (Solaris).*

29180 SP-2207
Ethylene-methyl acrylate copolymer resin; coating and laminating resin with excellent adhesion to OPP, PVDC coated surfaces and SPT, low temp. heat seal. *Monsanto (Solaris).*

29181 SP-6700
Oil-modified phenol formaldehyde resin; resin in NBR, SBR, NR, and CR as plasticizer during processing; acts as hardener with 8% hexa after cure; use where scorching problems preclude use of hexa-containing materials such as materials such as SP-6600. *Schenectady.*

29182 Spac
137-40-6 8816 205-290-4
Sodium propionate
Feed additive (growth promoter) for swine and poultry. *Mitsubishi Kasei.* Name unverified.

29183 SpaceRite S-11
21645-51-2 355 244-492-7
Hydrated alumina; optimum spacer material for titanium dioxide pigment used in surface coatings such as trade and commercial paints, powder coatings, inks. *Alcoa Industrial Chemicals.*

29184 Spalerite
1314-98-3 10294 215-251-3
ZnS
Zinc blende. Used as a pigment.

29185 Spallshield®
For automotive applications. *DuPont UK.*

29186 Span®
Sorbitan fatty acid esters. *ICI Am.*

29187 Span® 20
1338-39-2 8872 215-663-3
Sorbitan laurate NF
Emulsifier, stabilizer, thickener, lubricant, softener, antistatic agent; foods, pharmaceuticals, cosmetics, cleaning compounds, textiles. *ICI Spec. Chem.; ICI Surf. Belgium.*

29188 Span® 40
26266-57-9 8872 247-568-8
Sorbitan palmitate NF
Emulsifier, stabilizer, thickener, lubricant, softener, antistatic agent; foods, pharmaceuticals, cosmetics, cleaning compounds, textiles. *ICI Spec. Chem.; ICI Surf. Belgium.*

29189 Span® 60, 60K
1338-41-6 8872 215-664-9
Sorbitan stearate NF
Emulsifier, stabilizer, thickener, lubricant, softener, antistatic agent; foods, pharmaceuticals, cosmetics, cleaning compounds, textiles; also dispersant for inorganic pigments in thermoplastics. *ICI Spec. Chem.; ICI Surf. Belgium.*

29190 Span® 65
26658-19-5 8872 247-891-4
Sorbitan tristearate
Emulsifier, stabilizer, thickener, lubricant, softener, antistatic agent; foods, pharmaceuticals, cosmetics, cleaning compounds, textiles. *ICI Spec. Chem.; ICI Surf. Belgium.*

29191 Span® 80
1338-43-8 8872 215-665-4
Sorbitan oleate NF
W/o emulsifier, oil additive for corrosion inhibition; fiber lubricant and softener. *ICI Spec. Chem.; ICI Surf. Belgium.*

29192 Span® 85
26266-58-0 8872 247-569-3
Sorbitan trioleate
Emulsifier, stabilizer, thickener, lubricant, softener, antistatic agent; foods, pharmaceuticals, cosmetics, cleaning compounds, textiles. *ICI Spec. Chem.; ICI Surf. Belgium.*

29193 Spandofoam
Rigid polyurethane block foam. *Baxenden Chemicals Ltd.*

29194 Spandra Transparent Dressing
Fabric supported tecoflex polyurethane film; moisture vapor permeable, hypoallergenic dressing; a transparent intravenous dressing. *Thermedics Inc.* Name unverified.

29195 Spangite
This is phillipsite, a zeolite containing potassium.

29196 Spanish oxide
A natural red pigment; a red oxide of iron that contains over 80% Fe_2O_3.

29197 Spanish soap
An olive oil soap.

29198 Spannit
2921-88-2 2242 220-864-4
Chlorpyrifos
An organophosphate insecticide. *Pan Britannica Industries Ltd.*

29199 Spanscour EFS
Surfactant; detergent for Spandex scouring. *CNC Int'l L.P.*

29200 Spanscour GR
Scour and inhibitor combination for nylon/Spandex fabrics with a tendency to yellow due to atmospheric conditions. *CNC Int'l L.P.*

29201 Spanscour N20
Detergent for textile scouring. *Reilly Whiteman.*

29202 Sparkaloy
A proprietary trade name for silicon-manganese-nickel alloy used for spark-plug wire. No manufacturer.

29203 Spark-L® HPG
9032-75-1 232-885-6
Fungal pectinase; enzyme for depectinization and pulp washing; for fruit juices, citrus juice, citrus oil. *Solvay Enzymes.*

29204 Sparkle Silver®
Nonleafing aluminum pigments; used for metallic colors (aesthetics), automobiles, trucks, bicycles, furniture etc. *Silberline Mfg Co Inc.*

29205 Sparkle Silver® Premier (SSP)
Nonleafing aluminum pigments; aluminum pigments for paints and coatings used to achieve metallic effects most often in automotive and other decorative finishes. *Silberline Mfg Co Inc.*

29206 Sparkle Silvet
Aluminum pigments for plastics; used for all plastics, automobiles, toys, bottles etc. *Silberline Mfg Co Inc.*

29207 Sparkle Silvex
Silberline Mfg Co Inc.

29208 Sparkolac
High gloss nitrocellulose lacquers. *The Scottish Adhesives Co Ltd.*

29209 Sparrnite
7727-43-7 1023 231-784-4
A bleached barium sulfate pigment. *Pfizer International.* Discontinued.

29210 Spartase
Potassium aspartate and magnesium aspartate; nutrient. *Wyeth Laboratories.* Name unverified.

29211 Sparticide®
41205-21-4
2,3-dichloro-N-4-fluorophenylmaleimide
fluoroimide. Fungicide for apple fruit spot, melanose, and scab of citrus,

coffee berry disease, and pink disease on rubber. *Mitsubishi Kasei.* Name unverified.

29212 Spartrix
42116-76-7 1897 255-663-0
$C_8H_{12}N_4O_3S$
carnidazole
A proprietary preparation containing carnidazole; veterinary antihelmintic (pigeons). *Janssen Pharmaceutical Ltd.* Name unverified.

29213 Spasor
1071-83-6 4522 213-997-4
glyphosate. A nonresidual herbicide containing 360 g/liter glyphosate for the control of annual and perennial broad-leaved weeds and grasses; used for clearing ground prior to planting, weed control in all hard surfaces; control of floating and emergent aquatic weeds. *Rhône-Poulenc Environmental Prods. Ltd.*

29214 Spauldite
A proprietary trade name for a phenol-formaldehyde synthetic resin laminated product. No manufacturer.

29215 Specflex
Urethane system; microcellular foam system for use in dynamic elastomers for shoe soles, industrial tires, industrial rollers, mechanical good, wheels, power transmission belts, and sports equipment. *Dow Chemical.*

29216 Special Black 100 Powd
7440-44-0 1855 231-153-3
carbon
Carbon black; for paints and coatings industries. *Degussa.*

29217 Special Black 4, 4A Beads and Powd
7440-44-0 1855 231-153-3
Carbon black with oxidized surface; for water-reducible coating systems, packaging inks. *Degussa.*

29218 Special Extender
14807-96-6 9207 238-877-9
Talc, lamellar structure; extender for paints and as a general purpose filler. *Bromhead & Denison Ltd.* Discontinued.

29219 Special Fat 42/44
Hydrogenated coconut oil. *Hüls Am.*

29220 Special Oil 107
Enanthic acid triglyceride
Tracer oil for butter; release agent for conveyor belts. *Hüls AG.*

29221 Special Oil 619
Triisostearin
Lubricant in emulsions, lipsticks, hair care products; adds fine sheen to skin; pigment dispersant in sticks and liners. *Hüls Am.*

29222 Specpure
Spectrographically standardized metals and chemicals. *Johnson Matthey plc.*

29223 Spectra
Paints and maintenance products for cars. *Spectra Brands plc.*

29224 Spectra®
Fibers for diverse applications including electrical transformers, medical implants. *AlliedSignal.*

29225 Spectraban
21245-01-2 3282 244-288-8
$C_{14}H_{21}NO_2$
4-(dimethylamino)benzoic acid isoamyl ester
Padimate A; Escalo 506. A proprietary preparation of isoamyl-*p*-N, N-dimethylaminobenzoate in ethanol; a lotion used to protect skin from uv light. *Stiefel Laboratories (UK) Ltd.* Name unverified.

29226 Spectracote
Solvent-free high-performance coatings for formulated for barrier, corrosion or chemical resistance, concrete coating, textiles/wovens, encapsulation and adhesives uses. *Flexible Prods.*

29227 Spectraguard
Plastics materials for packaging and wrapping. *Courtaulds plc.* Discontinued.

29228 Spectra-Pearl®
13463-67-7 9612 236-675-5
Colored titanium dioxide/mica; pearlescent giving sparkle, luster, and color to cosmetic eye, face, lip, and body makeup. *Van Dyk.*

29229 Spectra-Sorb UV9
131-57-7 7088 205-031-5
Oxybenzone
Ultraviolet screen. *Am. Cyanamid.*

29230 Spectra-Sorb UV 24
131-53-3 3357 205-026-8
$C_{14}H_{12}O_4$
(2-hydroxy-4-methoxyphenyl)(2-hydroxyphenyl)methanone
Dioxybenzone; benzophenone-8; Cyasorb UV 24;. UV screen. mp = 68°; insoluble in H_2O, soluble in organic solvents. *Am. Cyanamid.*

29231 Spectra-Sorb UV 284
4065-45-6 9157 223-772-2
Sulisobenzone
UV screen. *Am. Cyanamid.*

29232 Spectra-Sorb UV 531
1843-05-6 6838 217-421-2
$C_{21}H_{26}O_3$
[2-hydroxy-4-(octyloxy)phenyl]phenylmethanone
Octabenzone; benzophenone-12;. UV screen. mp = 45-46°. *Am. Cyanamid.*

29233 Spectra-Sorb UV 5411
3147-75-9 221-573-5
$C_{20}H_{27}N_3O$
Octrizole
UV screen. *Am. Cyanamid.*

29234 Spectratech® CM 10540
Dual uv inhibitor for extrusion molding. *Quantum/USI.* Name unverified.

29235 Spectratech® CM 10608, 10634, 10778, 10779, 11013, 11056, 11513
Antiblock concentrates for film applications; FDA approved; CM 11513 also slip agent. *Quantum/USI.* Name unverified.

29236 Spectratech® CM 10777, 11045, 11638, 77242
Antistat concentrates for film and sheet; CM 11638 and 77242 for electronics packaging; CM 11045 also for injection and blow molding. *Quantum/USI.* Name unverified.

29237 Spectratech® CM 11014, 11126, 11172, 11174, 11194
Slip concentrates for film applications. *Quantum/USI.* Name unverified.

29238 Spectratech® CM 11053, 11489, 11591
Flame retardant concentrates; CM 11053 for LDPE film, 11489 for HDPE film, 11591 for LDPE, HDPE, PP film and wire and cable. *Quantum/USI.* Name unverified.

29239 Spectratech® CM 11246
UV absorber for packaging, agricultural films, molded items. *Quantum/USI.* Name unverified.

29240 Spectratech® CM 11340, KM 11264
128-37-0 1583 204-881-4
butylated hydroxytoluene
BHT. BHT; antioxidant concentrate for HDPE cereal liners. *Quantum/USI.* Name unverified.

29241 Spectratech® CM 11357, 11367
UV inhibitor for agricultural film, pool coverings, greenhouse film. *Quantum/USI.* Name unverified.

29242 Spectratech® CM 11698
Optical brightener for film applications. *Quantum/USI.* Name unverified.

29243 Spectratech® FM 1035H
EVA foam concentrate, 20% azodicarbonamide blowing agent; for structural foam, extrusion, injection moldings. *Quantum/USI.* Name unverified.

29244 Spectratech® FM 1150H
9002-88-4 7728
LDPE foam concentrate, 20% azodicarbonamide blowing agent; structural foam, extrusion, PP sheet. *Quantum/USI.* Name unverified.

29245 Spectratech® FM 1776H
PS foam concentrate, 10% sodium borohydride blowing agent; for structural foam, business machine and computer housings. *Quantum/USI.* Name unverified.

29246 Spectrathene
Color and additive concentrates for polyethylene and other plastics; colorants, additives; antistat, slip, antiblock, uv inhibitors, processing aids and flame retardants. *Quantum Chemical Corp.*

29247 Spectraveil
Chemical compositions for absorbing uv light. *Tioxide Group plc.*

29248 Spectrim
Multicomponent systems used in the manufacture of polyurethane products via reaction injection molding. *Dow Cheml Co Ltd, UK & Ireland.*

29249 Spectrim 5
Urethane system; RIM polymer for automotive applications, industrial/consumer RIM applications, and dynamic elastomers. *Dow Chemical.*

29250 Spectrim MM 310
PU composite; for structural RIM applications; excellent for load-bearing applications requiring excellent impact strength and high-temp. stability, e.g., bumper beams. *Dow Chemical.*

29251 Spectrim Polyurea HF85
Urethane system; RIM polymer for automotive applications, industrial/consumer RIM applications, and dynamic elastomers. *Dow Chemical.*

29252 **Spectroflux**
Buffer mixtures for spectrographic analysis. *Johnson Matthey plc.*

29253 **Spectromel**
Powder mixtures for spectographic analysis of relatively pure materials. *Johnson Matthey plc.* Discontinued.

29254 **Spectron**
Suspension concentrate containing 211 g chloridazon [1698-60-8] and 200 g ethofumesate [26225-79-6] per liter; a residual herbicide for beet crops. *Schering Agrochemicals Ltd.* Discontinued.

29255 **Spectrosol**
Materials for spectroscopy. *BDH Chemicals Ltd.*

29256 **specularite**
1309-37-1 4072 215-168-2
Fe_2O_3
Specular hematite. Iron oxide, with a bright metallic luster.

29257 **speculum metal**
An alloy of 66% copper and 34% tin, with a little arsenic. An alloy of 64% copper, 32% tin, and 4% nickel. It has a specific gravity of 8.6 and a melting-point of 750 C; used for making mirrors of reflection.

29258 **Speed X Accelerator**
A proprietary rubber vulcanization accelerator containing 60% diphenylguanidine and 40% zinc oxide. No manufacturer.

29259 **Speedcure BEDB**
2-n-Butoxyethyl 4-(dimethylamino) benzoate
Liquid photo activator. *Lambson Ltd.*

29260 **Speedcure BMDS**
4-Benzoyl-4-methyl diphenyl sulfide
Lambson Ltd.

29261 **Speedcure EDB**
10287-53-3 233-634-3
$C_{11}H_{15}NO_2$
ethyl-p-dimethyl aminobenzoate
UV photoinitiator. mp= 63-65°; d = 1.0610; insoluble in H_2O, soluble in organic solvents. *Aceto; Lambson Ltd.*

29262 **Speedcure ITX**
Isopropylthioxanthone
Photo activator. *Lambson Ltd.*

29263 **Speedcure ITX.**
Isopropyl thioxanthone
Uv photoinitiator. *Aceto.*

29264 **Speedway**
1910-42-5 7165 217-615-7
Granules containing 8% w/w paraquat; a pre-emergence bipyridilium herbicide to control weeds in field crops and ornamentals. *ICI Chem & Polymers Ltd.*

29265 **SpeeDye**
Aqueous pigment dispersions; for pigment garment dyeing. No manufacturer.

29266 **Speetan SB60**
Special 58/60 basic chrome tanning powder. *Lancashire Chemical Works Ltd.*

29267 **spelter**
Zinc used in galvanizing. The term is also used for hard solder.

29268 **Spenbond®**
Waterborne adhesives; for film, foil and paper lamination. *Reichhold.*

29269 **Spence metal**
A material obtained by melting ferrous sulfide with sulfur; used as a jointing material.

29270 **Spenkel®**
Modified vegetable oil; used for manufacture of resins, copolymer resins, urethane resins, and waterthinnable paint vehicles. *Reichhold.*

29271 **Spenkel® F18-M-60**
Oil-modified urethane. *Reichhold.*

29272 **Spenlite®**
Aliphatic urethane resins; used for paints, adhesives and textiles. *Reichhold.*

29273 **Spenlite® M22-X-40**
Aliphatic moisture curing urethane. *Reichhold.*

29274 **Spensol®**
Water-reducible urethane resins; used for paints, adhesives and textiles. *Reichhold.*

29275 **Spensol® F74-70**
Water-dispersible oil-modified urethane. *Reichhold.*

29276 **spermaceti**
8002-23-1 232-302-5
$C_{32}H_{64}O_2$
A wax obtained from the head of the sperm whale. The crude product is obtained by chilling the head and blubber oils. It consists principally of cetyl palmitate.

29277 **spermolin**
A linseed oil product. It is a proprietary binder for sands used as cores in metal casting.

29278 **spermoline oil**
A compound spindle oil used for lubrication.

29279 **Spermwax**
Synthetic spermaceti cetyl ester waxes. *Robeco Chemicals Inc.*

29280 **Spersol**
Sodium and ammonium polyacrylates; used as antiredeposition agents in laundry detergents, as detergents, as water reducing agents in slurries and as dispersants in boiler water. *Harcros Australia.*

29281 **SPG Gelatine**
9000-70-8 4388 232-554-6
Gelatin powder; used for granulation of fine powders to improve free flowing and dispersion characteristics. *SKW Biosystems Ltd.*

29282 **Sphagni**
An insulating material prepared from the white moss found on the Swedish peat moors.

29283 **Sphericel 110P8**
Hollow glass spheres; spheres which withstand molding pressures and reduce weight in engineering grade plastic compounds and molded parts. *Potters Industries.*

29284 **Spheriglass® A-Glass**
Solid glass spheres; additive for thermoplastic and thermoset resins; provides improved flow properties, high resin displacement, low shrinkage and warpage, better molded parts, dimensional stability. *Potters Industries.*

29285 **Spheron P-1500**
Silica [7631-86-9] and methicone [9004-73-7]; spherical hollow microbeads as carriers for sunscreens, fragrances, emollients; imparts lubricity. *Presperse.*

29286 **Spidax**
Specialty polyolefin resin; molding compounds. *Sumitomo Bakelite Co Ltd.* Name unverified.

29287 **Spiegeleisen**
A ferro-manganese alloy containing from 10-35% manganese, 60-85% iron, 1% silicon, and 4-5% carbon.

29288 **Spiegler Jolle's reagent**
A solution of 2 g mercuric chloride, 4 g succinic acid, and 4 g sodium chloride in 100 ml water; a reagent for albumin in urine.

29289 **Spiegler's reagent**
This consists of 40 g mercuric chloride and 20 g tartaric acid dissolved in 500 ml water. To this solution is added 100 g glycerol and 50 g, sodium chloride, and the whole made up to 1,000 ml. Used for analysis of proteins.

29290 **Spiller's resin**
The oxidation products of rubber are sometimes called by this name.

29291 **Spinflam**
A proprietary range of flame retardants for polymers. *Montedison UK Ltd.* Name unverified.

29292 **Spinnaker**
55219-65-3 9724 259-537-6
Emulsifiable concentrate containing 250 g/l triadimenol; used to control powdery mildew, rusts and rhychosporium in winter and spring crops of cereals, beet and brassicas. *Shell UK.*

29293 **Spinuvex**
A proprietary range of hindered amine light stabilizers for polymers. *Montedison UK Ltd.* Name unverified.

29294 **Spiragas**
Dimethyl silicone membrane; used for gas separations, primarily oxygen enrichment. *AlliedSignal Fluid Systems.* Name unverified.

29295 **spirit of alum**
7664-93-9 9147 231-639-5
H_2SO_4
Sulfuric acid

29296 **spirit of Hartshorn**
7664-41-7 517 231-635-3
A solution of ammonia.

29297 **spirit of red lavender**
Compound tincture of lavender.

29298 **spirit of salt**
7647-01-0 4821 231-595-7
Strong impure hydrochloric acid.

29299 **spirit of vitriol**
7664-93-9 9147 231-639-5
H_2SO_4
Sulfuric acid

Used in fertilizers, dyes, and pigments, chemicals, electroplating baths, nonferrous metallurgy.

29300 spirit of wood
67-56-1 6024 200-659-6
CH_3OH
Methyl alcohol
Used in the manufacture of formaldehyde, acetic acid, as a solvent, in chemical synthesis of methyl amines, methyl chloride and methyl methacrylate.

29301 spirit oil
The first fraction from the distillation of Yorkshire Grease (*qv*); used for making black varnish.

29302 spirit varnishes
Prepared by mixing resins with such solvents as methylated spirit or turpentine.

29303 spirit vinegar
Made from potato or grain spirit. It contains up to 12% of acetic acid.

29304 spirits of wine
64-17-5 3806 200-578-6
C_2H_5OH
ethanol, silent spirit
ethyl alcohol. Commercial spirits of wine contain 84% by weight of alcohol.

29305 Spirittine
Soft wood tar creosote; a wood preservative.

29306 Spiroflor
3-Ethyl-2,4-dioxaspiro (5.5) undec-8-ene
Complex, natural odor fragrance raw material. *Henkel/Cospha.*

29307 Spirolite
Plastic pipe. *Monsanto (Solaris).* Name unverified.

29308 Spitfire
Mixture of cyanazine and fluroxypyr; a post-emergence herbicide. *DowElanco Ltd.*

29309 sponge iron
A finely porous form of iron obtained by reducing iron oxide at a temperature where no sintering or fusion takes place; a reagent for the precipitation of copper, lead, and other metals from solution.

29310 Spongolit® Range
Esterified glycerides; aerating emulsifier for Madeira and sponge cakes, flavors. *Grünau; Henkel/Cospha; Henkel Canada.*

29311 Sponto® 101
Sulfonate/POE ether blend; emulsifier for phosphate toxicant mixtures. *Witco Corporation.*

29312 Sponto® 150T
Sulfonate/POE ether blend; agricultural emulsifier for highly concentrate organophosphate pesticides. *Witco Corporation.*

29313 Sponto® 168-D
Phosphoric acid mono and diesters/alkylphenoxy polyethoxyethanol blend; emulsifier and compatibility agent for pesticides and fertilizers, high electrolyte systems. *Witco Corporation.*

29314 Sponto® AG3-55T
Sulfonate and POE ether blend; specialty emulsifier for phenoxy-ester herbicides. *Witco Corporation.*

29315 Sponto® N-140B
Sulfonate and POE ether; emulsifier for organic phosphate insecticides. *Witco Corporation.*

29316 Sporak
67747-09-5 7941 266-994-5
Wettable powder containing 50% w/w prochloraz; a broad-spectrum fungicide for cereal crops. *Darmycel UK.*

29317 Sporgon
67747-09-5 7941 266-994-5
Wettable powder containing 50% w/w prochloraz; a broad-spectrum fungicide for cereal crops. No manufacturer.

29318 Sporocide
A wood preservative mainly consisting of potassium-*o*-dinitro-cresylate.

29319 Sportak
67747-09-5 7941 266-994-5
Emulsifiable concentrate containing prochloraz; a broad-spectrum fungicide for cereal crops. *Schering Agrochemicals Ltd.* Discontinued.

29320 Sportak Alpha
A suspension concentrate containing 266 g prochloraz [67747-09-5] and 100 g carbendazim [83601-81-4] per liter; systemic fungicide for cereals. *Schering Agrochemicals Ltd.* Discontinued.

29321 Sportak Delta
67747-09-5 7941 266-994-5
Prochloraz
cyproconazole. Fungicide. *Schering Agrochemicals Ltd.* Discontinued.

29322 sporting ballistite
A smokeless powder consisting of 37.6% nitroglycerin and 62.3% nitrocotton.

29323 Spotleak® 1001
80% *t*-butyl mercaptan, 20% dimethyl sulfide; odorant for natural gas to permit detection of leaks. *Elf Atochem.*

29324 Spotleak® 1003
76.5% t-Butyl mercaptan, 23.5% isopropyl mercaptan *Elf Atochem.*

29325 Spotleak® 1007
80% t-Butyl mercaptan, 20% methyl ethyl sulfide *Elf Atochem.*

29326 Spotleak® 1009
79% *t*-butyl mercaptan, 15% isopropyl mercaptan, 6% normal propyl mercaptan *Elf Atochem.*

29327 Spotleak® 1044
75% tetrahydrothiophene, 25% *t*-butyl mercaptan *Elf Atochem.*

29328 Spotleak® 2323
50% *t*-butyl mercaptan, 50% methyl ethyl sulfide *Elf Atochem.*

29329 Spotret 75 WDG
137-26-8 9510 205-286-2
Thiram fungicide for control of dollar spot, brown patch, snow mold and animal repellent. *W A Cleary.*

29330 Spray Guard
Splash and spray suppressing rain flaps. *Monsanto Co.* Name unverified.

29331 Spray-Add 77
Agricultural spreader-sticker. *Monsanto (Solaris).* Name unverified.

29332 Sprayfast
Surfactant adjuvant containing di-l-p-menthene and nonylphenol ethylene oxide condensate; coating agent for herbicides, pesticides, growth regulators and foliar feed sprays. *Mandops (UK) Ltd.*

29333 Spraymate Activator 90
Nonionic wetting agent containing alkylphenyl hydroxypolyoxyethylene; spreader for use in agricultural sprays. *Newman Agrochemicals Ltd.*

29334 Spraymate Bond
Extender containing 450 g/l synthetic latex; sticking agent for use with contact herbicides, fungicides, and insecticide sprays. *Newman Agrochemicals Ltd.*

29335 Spraymate LI-700
Acidifying surfactant containing 750 g/l soyal phospholids; wetting agent for systemic pesticides and foliar feeds. *Newman Agrochemicals Ltd.*

29336 SprayN Save Christmas Tree Spray
Aerosol containing di-1-p-menthene; antitranspirant spray. *Vitax Ltd.*

29337 Spraypover
Adjuvant containing 800 g/l refined mineral oil; wetting and spreading agent for use with residual herbicide sprays. *Fine Agrochemicals Ltd.*

29338 Sprayset® MEKP
1338-23-4 215-661-2
MEK peroxide in anhydrous ethyl acetate; catalyst for spray gun applications in the curing of polyesters. *Witco/Argus.* Discontinued.

29339 Spreading Agent
Complex alkyl ether. *Croda Chem. Ltd.*

29340 Spreitan
Blend of fatty acid esters and emulsifiers, nonionic; universal coning oil for mono and multifilament and for textures synthetic yarns. *Henkel Chemicals Ltd.* Name unverified.

29341 Sprengel's explosives
Cakes of potassium chlorate which have absorbed combustible liquids.

29342 Sprengsalpeter
An explosive consisting of 75% sodium nitrate, 15% brown coal, and 10% sulfur.

29343 Sprills
Pelleted pesticide. *Monsanto (Solaris).* Name unverified.

29344 Springbok
Bakery phosphates. *Albright & Wilson Ltd, Phosphates Speciality Business.* Name unverified.

29345 Springclene 2
Selective herbicide. *Schering Agrochemicals Ltd.* Discontinued.

29346 Springcorn Extra
Soluble concentrate of 18 g dicamba, 360 g MCPA and 160 g mecoprop per liter; used for weed control in cereals and grassland. *Farmers Crop Chemicals Ltd.* Discontinued.

29347 Sprint
Emulsifiable concentrate of 375 g fenpropimorph [67306-03-0] and 225 g prochloraz [67747-09-5] per liter; used for mildew control in cereals. *Schering Agrochemicals Ltd.* Discontinued.

29348 Sprodco
Textile yarn lubricant processing aid. *Specialty Products Co.* Name unverified.

29349 SPS
1332-58-7 5294 296-473-8
kaolin
Kaolinite; China clay; Bolus alba; Porcelain clay; Aluminum silicate hydroxide; Kaopectate; Aluminum silicate (hydrated); Aluminum silicate dihydrate. Fine kaolin clay sold in powder, bulk or aqueous slurry form; paper coating pigment. *ECC International Ltd.*

29350 Spud-Nic®
101-21-3 2240 202-925-7
Chloropropham
Herbicide. *Aceto.*

29351 Spudweed
Suspension concentrate containing 152 g prometryn and 304 g terbutryn per liter; used for weed control in peas, beans and potatoes. *Pan Britannica Industries Ltd.*

29352 Spuncote
Coatings for centrifugal dies used in the spinning of cast iron pipes. *Foseco (F.S.) Ltd.*

29353 Spurso
Dispersing agent. *Mooney Chemicals Inc.*

29354 Sputamin
A German preparation; a powder containing 80% chloramine; used as an antiseptic.

29355 Sputolysin®
Reagent. *Calbiochem Corp.*

29356 Squadron
A suspension concentrate containing 100 g carbendazim [83601-81-4] and 275 g maneb [12427-38-8] per liter; systemic fungicide for cereals. *Quadrangle Agrochemicals.*

29357 squalane
111-01-3 8923 203-825-6
$C_{30}H_{62}$
dodecahydrosqualene
spinacane; 2,6,10,15,19,23-hexamethyltetracosane; perhydrosqualene. Saturated branched chain hydrocarbon obtained by hydrogenation of shark liver oil or other natural oils; high-grade lubricating oil, perfume fixative, gas chromatographic analysis, transformer oil; in cosmetics and pharmaceut mp = -38°; bp_{10} = 263°; d_4^{15}= 0.8115; n_D^{15}= 1.4530; insoluble in H_2O, soluble in organic solvents. *Arista Industries; Robeco.*

29358 squalene
111-02-4 8924 203-826-1
$C_{30}H_{50}$
(all-E)-2,6,10,15,19,23-hexamethyl-2,6,10,14,18,22-tetracosahexaene
trans-squalene; 2,6,10,15,19,23-hexamethyl-2,6,10,14,18,22-tetracosa hexaene; spinacene; all-*trans-squalene; trans*-spinacene; squalen; Supraene. Biochemical and pharmaceutical research; a precursor of cholesterol in biosynthesis; chemical intermediate for manufacture of pharmaceuticals, organic colorants, rubber chemicals, aromatics, surfactants; bactericide. bp_{25} = 285°; d_4^{20}= 0,8584; n_D^{20} = 1.4965; insoluble in H_2O, soluble in organic solvents. *Arista Industries; Robeco.*

29359 squill
Scilla; sea onion. Bulb of *Urginea maritima.*

29360 SR-201
96-05-9 202-473-0
Allylmethacrylate
With 50-185 ppm hydroquinone inhibitor [123-31-9]; monomeric acrylic ester capable of polymerizing to hard, infusible resin that is water white, clear, and glass-like; as co-monomer with vinyl-type monomers to produce crosslinked polymers; unsymmetrical cross-linking agent where a two-stage polymerizing or drying action is desired. *Sartomer.*

29361 SR-203
Tetrahydrofurfuryl methacrylate
With 100±25 ppm hydroquinone inhibitor; high boiling, low viscosity monomeric ester; more reactive than methacrylates with equivalent molecular weight; polymerization initiated by conventional methods (peroxide catalysts, thermally, ionizing and uv radiation); room temperature cures by promoting peroxide system with aromatic amines. *Sartomer.*

29362 SR-205
Triethylene glycol dimethacrylate
With 80±20 ppm hydroquinone inhibitor; noncorrosive, low visc., high boiling crosslinking monomeric ester; in vinyl plastisols reduces initial visc. and oil extractability, and improves ultimate hardness, heat distort., hot tear strength and stain resistance; in cast acrylic sheet and rod, in the button, watch crystal and contact lens industries, in synthetic rubber and ion-exchange resins and in dental compositions. *Sartomer.*

29363 SR-206
Ethylene glycol dimethacrylate
With 40-150 ppm hydroquinone inhibitor; crosslinking agent used in emulsion polymerization, cast acrylic sheet, fiberglass-reinforced polyesters, ion exchange resins, rubber compound. *Sartomer.*

29364 SR-208
Cyclohexyl methacrylate
With 80 ± 20 ppm hydroquinone inhibitor; monomer; polymerizes into hard polymers of superior optical qualities and high ref. index; one-half the shrinkage on polymerization of methyl methacrylate; can undergo vinyl polymerization and co-polymerization; homopolymers are thermosetting; using azo or peroxide catalyst, can be free-radically polymerized in bulk to give hard, rigid, colorless polymer useful in optical applications. *Sartomer.*

29365 SR-209
Tetraethylene glycol dimethacrylate
With 75 ± 25 ppm hydroquinone inhibitor; crosslinking agent used in castings, plastisols, coatings, fibers, papers, and other fabrications. *Sartomer.*

29366 SR-211
n-Hexyl methacrylate
With 80 ± 20 ppm of hydroquinone inhibitor; polymerizes to soft polymer soluble in aliphatic hydrocarbons, copolymerized with other monomers for hard or soft compositions; used in preparation of homopolymers and copolymers by emulsion, suspension or bulk polymerization; good aging characteristics, internal plasticization, highly soluble in aliphatic hydrocarbons; suitable for adhesives, binders, finishes, sizes, waxes and emulsions. *Sartomer.*

29367 SR-212
1,3-Butylene glycol diacrylate
With 500 ± 20 ppm hydroquinone inhibitor; low viscosity monomer; curing agent; polymerizes to hard, insoluble, infusible, thermoset resin; exothermic polymerization reaction, initiated thermally or by common free radical initiators, i.e. high energy and uv radiation, peroxide compounds. *Sartomer.*

29368 SR-220
Cyclohexyl acrylate (monomer)
Curing agent. *Sartomer.*

29369 SR-231
Diethylene glycol dimethacrylate
With 80 ± 20 ppm hydroquinone inhibitor; high boiling monomeric ester; polymerized by common free radical initiators, i.e., peroxidic compounds, heat, uv, and ionizing radiation. *Sartomer.*

29370 SR-239
1,6-Hexanediol methacrylate
With 100 ± 25 ppm hydroquinone inhibitor; cross-linking agent used in casting compounds, glass fiber-reinforced plastics, adhesives, coatings, ion-exchange resins, textile products, plastisols, dental polymers and rubber compounding. *Sartomer.*

29371 SR-247
Neopentyl glycol diacrylate (monomer)
With 225 ± 25 ppm *p*-methoxyphenol inhibitor; curing agent; diluent for uv irradiation-cured systems (coatings, printing inks, etc.). *Sartomer.*

29372 SR-256
2-(2-Ethoxyethoxy)-ethyl acrylate
With 100 ppm MEHQ inhibitor; monofunctional monomer which can be polymerized by the use of heat, catalysts, and/or irradiation; especially sensitive to uv light, and is a good solv., making it very suitable for radiation-curable ink and coating formulations. *Sartomer.*

29373 SR-259
PEG 200 diacrylate
With 100-150 ppm hydroquinone inhibitor; cross-linking agent used in radiation-cured coatings, inks, adhesives, textile products, photoresists. *Sartomer.*

29374 SR-272
1680-21-3 216-853-9
Triethylene glycol diacrylate (monomer)
Curing agent. *Sartomer.*

29375 SR-285
Tetrahydrofurfuryl acrylate
With 500 ppm hydroquinone inhibitor; monofunctional monomeric ester of acrylic acid, polymerizes to hard, infusible, insoluble thermoset resin; used in uv irradiated coatings, because of low viscosity and sensitivity to uv; high boiling with low potential for cross-linking helpful in providing more flexible coatings; curing agent. *Sartomer.*

29376 SR-295
Pentaerythritol tetraacrylate
With 300-400 ppm MEHQ inhibitor; cross-linking agent in adhesives, coatings, inks, textile products, photoresists, castings, or as modifiers for polyester or polymers. *Sartomer.*

29377 SR-306
Tripropylene glycol diacrylate
With 125 ppm hydroquinone inhibitor; monomer acrylic ester, polymerizes to hard, infusible resin that is clear and glass-like; used as comonomer with

vinyl-type monomers to produce crosslinked polymers; polymerization reaction is midly exothermic and initiated by common free radical initiators, e.g. high energy and uv radiation, heat, peroxides etc; curing agent. *Sartomer.*

29378 SR-335
2156-97-0 218-463-4
N-lauryl acrylate (monomer)
Curing agent. *Sartomer.*

29379 SR-339
48145-04-6 256-360-6
$C_{11}H_{12}O_3$
2-Phenoxyethyl acrylate
2-propenoic acid, 2-phenoxyethyl ester; phenoxyethyl-acrylate. With 100 ppm HQ inhibitor; high boiling monomeric ester of acrylic acid; polymerization initiated by heat, catalysis, and/or radiation; copolymerization with other acrylic-type monomers is easily achieved; curing agent. *Sartomer.*

29380 SR-350
3290-92-4 221-950-4
$C_{18}H_{26}O_6$
Trimethylolpropane trimethacrylate
With 80 ± 20 ppm hydroquinone inhibitor; cross-linking agent. *Sartomer.*

29381 SR-351
15625-89-5 239-701-3
$C_{15}H_{20}O_6$
2-Propenoic acid 2-ethyl-2-[[(1-oxo-2-propenyl)oxy]methyl]-1,3-propanediyl ester
trimethylolpropane triacrylate. With 100-150 ppm hydroquinone inhibitor; high boiling monomeric ester polymerized by common free radical initiators; curing agent. *Sartomer.*

29382 SR-379
106-91-2 203-441-9
$C_7H_{10}O_3$
Glycidyl methacrylate
2,3-epoxypropyl methacrylate; CP-105; glycidyl alpha-methylacrylate; 1-propanol, 2,3-epoxy-, methacrylate. With 50 ppm inhibitor; polyfunctional monomer polymerized by applications of heat, heat and peroxidic catalysts, and irradiation by uv, β, σ, or x-ray; in hydrogels for contact lenses and membranes, molding and casting compounds; impregnating paper, concrete and wood, coatings and printing inks, adhesives and sealents, elastomers etc. *Sartomer.*

29383 SR-440
29590-42-9 249-707-8
$C_{11}H_{20}O_2$
Isooctyl acrylate (monomer)
2-propenoic acid, isooctyl ester; isoctyl acrylate. Curing agent. *Sartomer.*

29384 SR-444
3524-68-3 222-540-8
$C_{14}H_{18}O_7$
Propenoic acid 2-(hydroxymethyl)-2-[[(1-oxo-2-propenyl)oxy]methyl]-1,3-propanediyl ester
PETA; Pentaerythritol triacrylate. With 300-400 ppm MEHQ inhibitor; cross-linking agent used in adhesives, coatings, inks, textile products, photoresists, castins, modifiers for polyester, fiberglass, or polymers. *Sartomer.*

29385 SR-7475
Stereospecific butadiene/styrene copolymer; blendable modifier for thermoplastic resins and asphalt. *Firestone Syn. Rubber.* Discontinued.

29386 SS 24049, 24519
Nickel pigment, urethane binder in water; EMC shielding coating for interior walls; protects sensitive electronic equipment. *Acheson Colloids.*

29387 SS 24656
Silver/nickel pigment, polyester binder in SB-1; EMC shielding coating for plastics; protects sensitive electronic equipment. *Acheson Colloids.*

29388 SS Nitrocellulose
9004-70-0 8195
Nitrocellulose
Used in flexographic inks where an alcohol-rich solvent is desirable and is used in heat sealing coatings. *Hercules.* Discontinued.

29389 SS-30
Copper flake; anti-seize compound protecting stainless steel and high-strength alloys; for electrical connections. *Jet-Lube.*

29390 SSF
16893-85-9 8769 240-934-8
F_6Na_2Si
Sodium silico-fluoride
sodium silicofluoride; sodium hexafluorosilicate; sodium fluosilicate; Earwig bait; silicate(2-), hexafluoro-, disodium; disodium hexafluorosilicate.

29391 St. Helen's powder
An explosive consisting of 92-95% ammonium nitrate, 2-3% aluminum powder, and 3-5% trinitrotoluene.

29392 St. Ignatius bean
The seed of *Strychnos ignatii.*

29393 St. Johns Wort Oil CLR
Fatty oil extract of St. Johns wort blossoms; general skin protection, especially for sensitive skin. *Dr. Kurt Richter; Henkel/Cospha.*

29394 Sta
Methanol-based antifreeze. *Chemcentral Corp.*

29395 Stabaxol
Polycarbodiimide. *Bayer plc.*

29396 Stabgel
Soil consolidation agents. *ICI Chem & Polymers Ltd.* Discontinued.

29397 Stabicol®
Blend of stabilizers and sequestrants for use in hydrogen peroxide bleaching of textiles and as a pitch control aid in papermaking. *Allied Colloids Ltd.*

29398 Stabifix
Dyeing auxiliary; fixing agent for direct dyes to improve water and wash fastness. *Henkel Chemicals Ltd.* Name unverified.

29399 Stabil-9
7785-88-8 232-090-4
Sodium aluminum phosphate
phosphoric acid, aluminum sodium salt. Leavening agent for self-raising flour, corn meal, prepared mixes. *Monsanto Co.* Name unverified.

29400 Stabilator A.R
135-88-6 205-223-9
$C_{16}H_{13}N$
N-phenyl-β-naphthylamine
N-phenyl-2-naphthalenamine; PBNA; Agerite; PBN; Aceto PBN; 2-anilinonaphthalene; antioxidant 116; antioxidant PBN; Neozon D; Neozone; Nilox PBNA; Nonox d; Stabilizator ar; Neosone D; Vulkanox PBN; Nonox DN;Stabilizer AR; Nocrac D; Naftam 2;. Melts at 108°C; recommended for white mixings; a proprietary antioxidant for rubber. An antidegradant for various rubber products such as natural rubber, styrene-butadiene, nitrile, butadiene and chloroprene. mp = 108°; bp = 395°; SG = 1.24; insoluble in H_2O, soluble in organic solvents;. No manufacturer.

29401 Stabileze 06
Gantrez (methylvinyl ether/maleic anhydride copolymers) cross-linked with 1,9-decadiene. *ISP.*

29402 Stabilisal®
Stabilizing agents used in the dyeing and printing of textiles with sulfur dyestuffs. *Cassella AG.*

29403 Stabilite Alba
Di-*o*-tolyl-ethylenediamine
A proprietary rubber vulcanization accelerator. No manufacturer.

29404 Stabilizer
37067-27-9 253-332-5
$C_6H_5KO_5S$
Potassium hydroquinone monosulfate
Potassium 4-hydroxyphenyl sulfate. *Rhône-Poulenc UK.*

29405 Stabilizer 1097
Organic acid chloride in butyl acetate; stabilizer used to extend the pot life of the bonding agent system/plastisol mixture *Bayer AG; Miles; Polysar.*

29406 Stabilizer 2013-P®
622-16-2 210-721-4
$C_{13}H_{10}N_2$
Carbodiimide
Benzenamine, N,N'-methanetetraylbis-. Activator; improves resistance to hydrolysis by acids, bases, and hot water in vulcanizates based on millable PU. *TSE Industries.*

29407 Stabilizer C
102-08-9 3393 203-004-2
Diphenylthiourea
For economical heat stabilization of PVC containing alkalis and emulsifiers. *Bayer AG.*

29408 Stabilizer No. 1
A proprietary trade name for 1:3:5-isopropyl-cresol. No manufacturer.

29409 Stabilizer NS
Organic/inorganic; peroxide stabilizer. *Marlowe-Van Loan.*

29410 Stabiloid
Colorant dispersions; for coloring of paper coating and saturation compositions, textile inks and latex paints. *Hüls Am.*

29411 Stabilor
A range of precious metal alloys; for dentistry and dental engineering. *Degussa Ltd.*

29412 Stabilosol®
A range of Hydron dyestuffs with particular solubility properties; used for textile dyeing and printing. *Cassella AG.*

29413 Stabinol
3567-08-6
$C_{13}H_{17}N_3O_3S_2$
N-(5-Isobutyl-1,3,4-thiadiazol-2-yl)-*p*-methoxybenzenesulfonamide
Glysobuzole; Isobuzol. A proprietary trade name for isobuzole. No manufacturer.

29414 Stabiram
Specialty blends of quaternary ammonium compounds; bitumen emulsifiers. *Elf Atochem UK/Ceca.*

29415 Stabismol
A proprietary solution of α-carbonyl-cyclohexanyl acetate in olive oil. No manufacturer.

29416 Stablex
A stabilized bitumen used for protective coatings to resist acids.

29417 Stabochlor
A proprietary chloride of lime specially prepared. No manufacturer.

29418 Sta-Clad®
Synthetic resins. *Reichhold.*

29419 Stacol
A complex sodium borophosphate, an inorganic water-soluble resin stable to acids and alkalis.

29420 Stadis®
Fuel oil antistatic additive. *DuPont UK.*

29421 staff
A mixture of plaster of Paris and tow; used for moldings.

29422 staffelite
A mineral containing calcium phosphate with calcium chloride or fluoride.

29423 Staffordshire all mine pig
A pig iron made in Staffordshire from ore. It contains about 0.5-0.75% of phosphorus.

29424 Staflex CP
A mixed alkyl phthalate; vinyl plasticizer *Miles.* Discontinued.

29425 Stahl's sulfur salt
10117-38-1 7847 233-321-1
$K_2SO_3{\cdot}2H_2O$
Potassium sulfite
Used as a photographic developer, medicine, in food and wine as a preservative.

29426 Stain Resist
Stain resistant treatment for carpets. *Rohm & Haas UK.*

29427 Stainaway L2B
Phenol sulfonate
Stain-blocker for nylon 6/6 and nylon 6 carpet fibers; used in beck or continuous methods. *Am. Emulsions.*

29428 Stain-Free
Aromatic condensate; stain resist auxiliary for nylon carpet. *Sybron.*

29429 Stainguard GYS
Aromatic condensate; dye fixative for acid dyes on nylon; reserving agent for nylon and nylon/polyester blends. *Am. Emulsions.*

29430 stainless invar
A Japanese alloy of 54% cobalt, 36.5% iron, and 9.5% chromium. It has a low coefficient of expansion and with stands corrosion well.

29431 stainless iron
This is really stainless steel, and usually contains from 0.1-0.2% carbon, 12-27% chromium, and up to 0.5% silicon.

29432 stainless silver
This is usually an alloy of 92.5% silver with copper and antimony, and is used for tableware.

29433 stainless steel
12597-68-1
Rustless steel. A chromium-steel alloy containing 12-15% chromium, and not more than 0.45% carbon; used for cutlery, acid pumps, turbine blades, and exhaust valves for engines. Some alloys contain 12-18% chromium, 8-12% nickel and 74-75% iron.

29434 stalactites
Deposits of calcium carbonate in the form of icicles, formed when water containing calcium carbonate drips from the roofs of caves.

29435 stalagmites
Similar deposits to stalactites, except that they are formed on the floors of caves.

29436 Stalloy
A proprietary trade name for an alloy containing 3.5-4.0% silicon and 0.1-0.2% aluminum; has a specific resistance of about 55 michroms; its magnetic hysteresis is much lower than that of pure iron; used in the construction o. No manufacturer.

29437 Stamere®
9000-07-1 1914 232-524-2
Carrageenan; suspending agent, stabilizer, thickener, gelling agent for food industry; lubricant, emollient for pharmaceutical jellies, laxatives; tablet binder; emulsion stabilizer; pigment suspender in ceramic glazes; also for cosmetics, tothpaste, wire-drawing lubricants, electroplating baths. *Meer.*

29438 Stamford powder
An explosive containing from 68-72% ammonium nitrate, 21-23% sodium nitrate, 3-4% trinitrotoluene, and 3 1/2-4 1/2% ammonium chloride.

29439 Stamglan®
Low density polyethylene. *AKU Holand.* Unverified.

29440 Stamid HT 3901
Cocamide DEA
Foam stabilizer, emulsifier and thickener used in a variety of household, industrial and cosmetic formulations. *Clough.*

29441 Stamid LS 5487
68425-47-8 270-355-6
Soyamide DEA
Stabilizer, emulsifier, thickener for shampoos and bubble bath formulations. *Clough.*

29442 Stamylan HD
High density polyethylene; used in the plastics processing industry for production of crates, household articles, bottles, containers, tubes and pipes, cables, nets, packaging film, toys, etc. *DSM NV.* Discontinued.

29443 Stamylan LD
Low density polyethylene; used in the plastics processing industry for production of packaging film, heavy-duty bags, extrusion-coated cardboard and paper, tubes and pipes, cable sheathing, household articles, toys, agricultural film, foamed board, etc. *DSM NV.* Discontinued.

29444 Stamylan P
Polypropylene
Used in the plastic processing industry for a wide variety of applications: injection molding of car components (bumpers, accumulator cases, boot linings, housings), electronic equipment, furniture and thin-walled containers (dairy products); for extrusion of film, fibers and non-woven fabrics, tape and belts; for the blow molding of bottles and containers. *DSM NV.* Discontinued.

29445 Stamylex PE
Special linear low density and medium density polyethylene; used for very special high-performance applications, e.g., packaging articles such as tanks, containers, covers, special food packaging films, leisuretime products, cables, monofilamernts, fasteners, caps, stoppers and special technical applications. *DSM NV.* Discontinued.

29446 Stancard 5219
Mineral oil-based lubricant for use as yarn lubricant for cotton dust control. *Henkel/Textiles.*

29447 Stanclere®
Organotin heat stabilizers for PVC. *Akzo Chemie UK Ltd.*

29448 Stanclere® T4817
Sulfur-containing butyltin stabilizer. *Akzo.*

29449 Stanclere® T-883
Sulfur-containing octyltin stabilizer. *Akzo.*

29450 stand oil
Standöl varnish; Dicköl varnish; lithographer's varnish. Linseed oil boiled strongly, and allowed to burn until it has the desired thickness.

29451 Standacol
United Kingdom foodstuffs colors. *Morton Int'l. Ltd.* Discontinued.

29452 Standafin FCX
Extender for fluorocarbon finishes for nylon and polyproylene carpet yarn. *Henkel/Textiles.*

29453 Standalloy
Dental silver alloys and mercury for mixing amalgams; for dentistry and dental engineering. *Degussa AG.* Name unverified.

29454 Standamid® CD
136-26-5 205-234-9
$C_{14}H_{29}NO_3$
bis(2-hydroxyethyl)decanamide
Capric diethanolamide. Capramide DEA (2:1) and diethanolamine; detergent, foam enhancer for anionic systems; solubilizes fragrances into hydroalcoholic systems; secondary emulsifier in oil-water systems, perfume stabilizer; personal care products, industrial cleaners. *Henkel/Cospha; Henkel Canada.*

29455 Standamid® KD
Cocamide DEA (1:1)
Foaming and thickening agent for liquid or gel shampoos and bubble baths,

industrial cleaners, etc.; superfatting agent, foam stabilizer, emulsifier, intermediate; detergent and solubilizer for oily components; hair conditioner. *Henkel/Cospha; Henkel Canada.*

29456 Standamid® KDS
120-40-1 204-393-1
Lauramide DEA (1:1)
Detergent, solubilizer for oils; enhances foam density, lubricity, stability for shampoos, skin cleansers, bath and shower products. *Henkel/Cospha; Henkel Canada.*

29457 Standamid® PD
Cocamide DEA (2:1)and diethanolamine more efficient foam builder, less efficient visc. builder than 1:1 superamides; stabilizer; used in shampoos, bubble baths with anionics, nonionics, and amphoterics; industrial cleaning; solubilizer for oily additives. *Henkel/Cospha; Henkel Canada.*

29458 Standamid® SOMD
Linoleamide DEA (1:1)
Viscosity builder with foam enhancement characteristics; produces especially high viscosity when used with ethoxylated anionics; used in gel shampoos, bubble baths, liquid hand soaps, industrial cleaners, and formulation; where the amount of electrolyte must be kept to a minimum; low cost shampoo concentrates. *Henkel/Cospha.*

29459 Standamox 01
14351-50-9 238-311-0
Oleamine oxide
Biodegradable thickener, bacteriostat, dye assistant, lubricant, softener used in industrial applications, hair products, plating compounds, lube oils. *Henkel/Cospha; Henkel Canada.*

29460 Standamox CAW
68155-09-9 268-938-5
Cocamidopropylamine oxide
Wetting agent, foam builder, stabilizer, thickener, lubricant, emollient for low irritation baby shampoos, bubble baths, skin care preparations. *Henkel/Cospha; Henkel Canada.*

29461 Standamox PCAW
68155-09-9 268-938-5
Cocamidopropyl dimethylamine oxide
Detergent, emulsifier, wetting agent, foam stabilizer; stable over wide pH range. *Pulcra SA.*

29462 Standamox PL
Lauramine oxide and myristamine oxide; detergent, emulsifier, wetting agent, foam stabilizer; stable over wide pH range. *Pulcra SA.*

29463 Standamul® Conc. 1002
Cetearyl alcohol, PEG-40 hydrogenated castor oil, stearalkonium chloride; conditioner and softener used in hair care preparations. *Henkel/Emery.*

29464 Standapol® 1610
73138-79-1 277-298-6
Sulfated peanut oil; emulsifier, wetting agent, dispersant, antistat for nylon finishes. *Henkel/Textiles.*

29465 Standapol® 7088
Ammonium myreth sulfate
cocamide MEA. Foaming surfactant base for personal care cleansing and bath products *Henkel/Cospha; Henkel Canada.*

29466 Standapol® 7092
Sodium laureth sulfate
glycol stearate. Pearlizing shampoo base; excellent stability over wide temperature ranges. *Henkel/Cospha; Henkel Canada.*

29467 Standapol® A
2235-54-3 218-793-9
$C_{12}H_{29}NO_4S$
Ammonium lauryl sulfate
Sulfuric acid, monododecyl ester, ammonium salt; Ammonium lauryl sulfate; Dodecyl ester of sulfuric acid, ammonium salt. Detergent, foamer, suspending agent, base for shampoos, cleaning compounds with near neutral pH. *Henkel/Cospha; Henkel Canada.*

29468 Standapol® AP Blend
9004-82-4
Sodium laureth sulfate
cocamide DEA; cocamidopropyl betaine; poly(oxy-1,2-ethanediyl), α-sulfo-ω-(dodecyloxy)-, sodium salt; sodium laureth sulfate; dodecyl alcohol, monoether with polyethylene glycol, hydrogen sulfate sodium salt; poly(oxyethylene) lauryl ether sulfate sodium salt; Polyethylene glycol, mono(hydrogen sulfate), dodecyl ether, sodium salt. Concentrate for shampoo, bath and cleansing products, liquid soaps; excellent foaming and viscosity response. *Henkel/Cospha; Henkel Canada.*

29469 Standapol® CAT
Mixture of glycol stearate, lauramine oxide, propylene glycol and ceteareth-20. Pearlescent surfactant for personal care products. *Henkel/Cospha; Henkel Canada.*

29470 Standapol® CS Paste
Mixture of sodium lauryl sulfate, sodium cetyl sulfate and laureth-3; laureth-3. Detergent, foamer; formulated concentrate for pearlescent cream shampoos, bubble baths. *Henkel/Cospha; Henkel Canada.*

29471 Standapol® DEA
DEA-lauryl sulfate
Base, detergent, foamer used in personal care products *Henkel/Cospha; Henkel Canada.*

29472 Standapol® EA-1
67762-19-0
Ammonium laureth sulfate
Surfactant for clear liquid shampoos, bubble baths. *Henkel/Cospha; Henkel Canada.*

29473 Standapol® ES-1
Sodium laureth sulfate
Detergent, foamer, base for liquid shampoos and bubble baths. *Henkel/Cospha; Henkel Canada.*

29474 Standapol® LF
Sodium octyl sulfate
Wetting agent for metal degreasers, hard surface cleaners, food equipment cleaners, dust control; solubilizer, hydrotrope; resistant to hard water. *Henkel/Cospha; Henkel Canada.*

29475 Standapol® MG
3097-08-3 221-450-6
$C_{24}H_{50}MgO_8S_2$
Magnesium lauryl sulfate
sulfuric acid monododecyl ester magnesium salt; magnesium lauryl sulfate; dodecyl sulfate, magnesium salt. Foamer with good dermatological properties for personal care products. *Henkel/Cospha; Henkel Canada.*

29476 Standapol® Pearl Conc. 7130
Mixture of glycol distearate, sodium laureth sulfate and propylene glycol.cocamide MEA; laureth-9,. Pearlescent surfactant used in personal care products. *Henkel/Cospha; Henkel Canada.*

29477 Standapol® S
Mixture of sodium lauryl sulfate, sodium laureth sulfate, lauramide MIPA and cocamide MEA. Glycol stearate, and coceth-8; pearlescent shampoo and liquid soap base. *Henkel/Cospha; Henkel Canada.*

29478 Standapol® SCO
8002-33-3 232-306-7
sulfonated castor oil
Turkey Red Oil. Sulfated castor oil; emulsifier and dispersant, wetting agent, foam depressant, emollient surfactant and solubilizer used in personal care products. *Henkel/Cospha; Henkel Canada.*

29479 Standapol® SH-100
56388-43-3 260-143-1
Disodium oleamido PEG-2 sulfosuccinate
Detergent used in personal care products; nonirritating shampoo base; anti-irritant for other surfactants. *Henkel/Cospha; Henkel Canada.*

29480 Standapol® SHC-101
Disodium oleamido PEG-2 sulfosuccinate and sodium lauryl sulfate. Base for low-irritation shampoos for babies and adults; anti-irritant for other surfactants. *Henkel/Cospha; Henkel Canada.*

29481 Standapol® T
139-96-8 205-388-7
Triethanolamine lauryl sulfate
TEA-lauryl sulfate; Akyposal TLS; Cycloryl tawf; Dodecyl sulfate, triethanolamine salt; Drene; Elfan 4240 T; EMAL T; Lauryl sulfate ester, triethanolamine salt; Maprofix TLS; Melanol LP 20 T; Propaste T; Rewopol TLS-40; Richonol T; Sipon lt; Standapol TLS 40; Sterling wat; Sulfuric acid, monododecyl ester, compd. with 2,2',2"-nitrilotris(ethanol) (1:1;)Texapon T-35; Triethanolamine salt of lauryl sulfate; Tylorol LT-50; tris(2-hydroxyethyl)ammonium decyl sulfate. Detergent, foamer, base for mild shampoos, aerosols. *Henkel/Cospha; Henkel Canada.*

29482 Standapol® WAQ-LC
151-21-3 8782 205-788-1
Sodium lauryl sulfate
Foaming agent, detergent, suspending agent for personal care products, liquid cleaners; low salt content for improved corrosion resistance. *Henkel/Cospha; Henkel Canada.*

29483 Standapon 4149 Conc
Concentrated scour for soaping-off dyed or printed goods. *Henkel/Textiles.*

29484 standard benzine
Light petroleum spirit of specific gravity 0.695-0.705 at 15°C., of which 95% boils between 65 and 95°C; used for the determination of asphalt in oils.

29485 standard gold
sterling gold
A 22-carat gold containing 91.6% gold with 8.4% other metals, usually

copper, to render it harder, American standard gold contains 90% gold and 10% copper.

29486 standard silver
sterling silver
Silver, 92.5% with another metal, usually copper, to harden it. American standard silver contains 90% silver and 10% copper.

29487 Standup
999-81-5 2153 213-666-4
Preparation containing 40% chlormaquat; growth regulator for cereals. *L W Vass (Agricultural) Ltd.* Discontinued.

29488 Stan-Fast
Chemical additive to a coloring bath for metals such as anodized aluminum; used in anodizing aluminum. *Reynolds Metal Co.*

29489 Stanlev R-276
Leveling agent for use with disperse or neutral acid dyes on nylon. *Henkel/Textiles.*

29490 Stanleys Crow Repellant
Active ingredients; refined coal tar and creosote oil; seed protectant to prevent sprout pulling by birds in newly planted corn. *Borderland Products Inc.* Discontinued.

29491 Stannal
Electrolytic coloring for aluminum. *Albright & Wilson Ltd.*

29492 stannekite
$C_{20}H_{22}O_3$
A resinous hydrocarbon, found in coal deposits in Bohemia.

29493 Stannex
Covering and cleansing fluxes for tin and tin-lead alloys. *Foseco (F.S.) Ltd.* Discontinued.

29494 stannic chloride
7646-78-8 8929 231-588-9
Cl_4Sn
tin tetrachloride
tin perchloride; tin tetrachloride; tin (IV) chloride; tin chloride; stannane, tetrachloro-; tin (IV) chloride anhydrous. Coatings that are electroconductive and electroluminescent, textile dye mordant, perfume stabilizer, manufacture of fuchsin, blueprint paper, color lakes, ceramic coatings, bleaching agent for sugar, stabilizer for resins, tin salts, soap bactericide and fungicide. mp = -33°; bp_{770} = 114°; d = 2.2260; soluble in H_2O, organic solvents. *Aldrich; Elf Atochem N. Am.; Chemisphere Ltd; M & T Harshaw; Nihon Kagaku Sangyo; Witco/Argus.*

29495 stannic chromate
38455-77-5 8930 253-946-3
Cr_2O_8Sn
stannic chromate(VI)
tin(IV) chromate(VI). Rose-violet pigment used to color paper, porcelain and china. Soluble in H_2O.

29496 stannic sulfide
1315-01-1 8936 215-252-9
S_2Sn
Tin(IV) sulfide
tin disulfide; mosaic gold; tin bronze. Used for gilding and bronzing metals and other surfaces. d = 4.5; insoluble in H_2O.

29497 Stannicide
Fungicides, bactericides and algicides. *Akzo Chemie.* Name unverified.

29498 Stannicide
Formulated organotin compounds; fungicides, biocides and algicidal application in water based coatings and adhesives and in the water treatment industry. *Thomas Swan & Co Ltd.* Discontinued.

29499 Stannine
Acid inhibitor. *Rhône-Poulenc UK.*

29500 Stanniol
An alloy of 96.2% tin, 2.4% lead, 1% copper, 0.3% nickel, and 0.1% iron.

29501 stannite
A mineral; tin pyrites.

29502 stannous chloride
7772-99-8 8939 231-868-0
$SnCl_2$
Tin (II) chloride anhydrous
tin crystals; tin salt; tin dichloride; tin protochloride,. Reducing agent for intermediates, dyes, polymers, phosphors; manufacture of lakes; textile dyeing and printing; tin galvanizing; analytical reagent; silvering mirrors; antisludge for lubricants; food preservative; perfume stabilizer, soldering flux. mp = 37-38°, bp = 247°; d = 3.95; soluble in H_2O, polar organic solvents; LD_{50} (mus ip) = 66 mg/kg. *Aldrich; Elf Atochem N. Am.; Blythe, William Ltd; Cerac; M & T Harshaw; Noah Chem.*

29503 Stansoft 626-B
Nonyellowing towel softener imparting very soft hand and excellent absorbency. *Henkel/Textiles.*

29504 Stansperse 506
Dyebath stabilizer improving stability of softeners in textile dyebaths. *Henkel/Textiles.*

29505 Stantex Antistat F
Low melting point antistat and emulsifier for nylon finishes. *Henkel/Textiles.*

29506 Stantex PENE 20
Wetting agent for latex carpet backing. *Henkel/Textiles.*

29507 stantienite
A brown resin found with Prussian amber.

29508 Sta-Nut EE
68334-00-9 269-804-9
Hydrogenated cottonseed oil;
Cottonseed oil, hydrogenated; Cottonseed oil, partially hydrogenated; Partially hydrogenated cottonseed oil; Bath Wax. Partially hydrogenated coattonseed oil, fatty acid mono and diglycerides; kosher; peanut buffer stabilizer. *Van Den Bergh Foods.*

29509 Stanvis
Pyroxylin embedding solution for microscopy. *BDH Chemicals Ltd.*

29510 Stanyl®
Engineering plastics with excellent impact strength and high temperature resistance; used in the electrical and automotive industries and as material for technical yarns. *DSM NV.* Discontinued.

29511 Stanyl® TE200F6
50327-22-5
Nylon 4/6, 30% chopped glass fiber-reinforced; special heat stabilizer for elec. applications; high heat engineering plastic offering outstanding wear, friction, creep, fatigue, chemical resist., toughness, and stiffness properties; used in the automotive field, electrical/electronic and auto under-the-hood applications, high performance mechanical components. *DSM.* Discontinued.

29512 Stanyl® TE300
50327-22-5
Nylon 4/6; special heat stabilizer for electrical applications. *DSM.* Discontinued.

29513 Stanyl® TE350
50327-22-5
Nylon 4/6, flame-retarded; special heat stabilizer for electrical applications. *DSM.* Discontinued.

29514 Stanyl® TQ200F6
50327-22-5
Nylon 4/6, 30% chopped glass fiber-reinforced; oil-resistant heat stabilizer. *DSM.* Discontinued.

29515 Stanyl® TW300
50327-22-5
Nylon 4/6; standard heat stabilizer. *DSM.* Discontinued.

29516 Stanza
Fungicide. *Schering Agrochemicals Ltd.* Discontinued.

29517 stanzaite
A mineral. It is a variety of Andalusite.

29518 staple fiber
Staple artificial silk; artificial wool; artificial chappe. This fiber consists of artificial threads of cellulose or cellulose compounds possessing a definite medium length. It is worked up by ordinary spinning machinery and is suitable for mixing with cotton or wool.

29519 Stapron S SG340
9011-13-6
Styrene/maleic anhydride copolymer, rubber modified, 20% glass fiber-reinforced; amorphous engineering plastic. *DSM.* Discontinued.

29520 Stapron S SM300
9011-13-6
Styrene/maleic anhydride copolymer, rubber modified; amorphous engineering plastic. *DSM.* Discontinued.

29521 Star
56-81-5 4493 200-289-5
Glycerin USP
Humectant; in pharmaceuticals, toiletries, tobacco, alkyds, food products, explosives, cellophane, urethane foam, other industries. *Procter & Gamble.*

29522 star bowls
Antimony metal obtained by refining with iron. The metal containing about 91% antimony with about 7% iron is mixed with crude antimony and salt and heated. The product is known as star bowls. It contains about 99.5% antimony.

29523 Star Chlormequat
999-81-5 2153 213-666-4
Soluble concentrate containing 700 g/l chlormequat; plant growth regulator. *Star Agrochem Ltd.*

29524 Star DSM
8022-00-2 6129
Emulsifiable concentrate containing 580 g demeton-S-methyl per liter; a systemic organophosphorus insecticide. *Star Agrochem Ltd.*

29525 Star MCPA
94-74-6 5803 202-360-6
MCPA
Herbicide for cereals and grassland. *Star Agrochem Ltd.*

29526 Star Stran 748
Continuous filament glass strand; reinforcement for PP, PPS, PEI. *Schuller.*

29527 Staralox
A trademark for abrasive goods made essentially of alumina. No manufacturer.

29528 Starane
81406-37-3 4238 279-752-9
Emulsifiable concentrate of 200 g fluroxypyr meptyl per liter; used for weed control in cereals and grassland. *DowElanco Ltd.*

29529 Starane 2
Selective post-emergence herbicide. *Murphy Chemical Co Ltd.* Discontinued.

29530 starch glazes
Made by adding borax, powdered stearic acid, or paraffin to potato starch.

29531 starch glue
Prepared by adding 3 pints water and 1/2 lb nitric acid to 2 1/2 lb starch, warming, then heating.

29532 starch syrups
Glucose mixed with dextrine used in the place of sugar for various purposes.

29533 Stardrops
Household cleaner. *Thornton & Ross Ltd.*

29534 Starfol® BB
17671-27-1 241-646-5
Behenyl behenate
High temperature ester lubricant with high smoke point for fiber and yarn lubrication applications. *Sherex/Div. of Witco.*

29535 Starfol® CP
540-10-3 2073 208-736-6
Cetyl palmitate
Emollient for cosmetic creams and lotions. *Sherex/Div. of Witco.*

29536 Starfol® IS
41669-30-1 255-485-3
Isostearyl isostearate
High-temperature ester lubricant with high smoke point, low viscosity; replacement for butyl stearate and mineral oil for fiber and yarn lubrication applications. *Sherex/Div. of Witco.*

29537 Starfol® OO
3687-45-4 222-980-4
Oleyl oleate
High-temperature ester lubricant with high smoke point, low viscosity; replacement for butyl stearate and mineral oil in fiber and yarn lubrication applications. *Sherex/Div. of Witco.*

29538 Starfol® OS
22766-82-1 245-204-2
Octyldodecyl stearate
Emollient and moisturizer for creams and lotions; imparts luxurious, conditioned feel to the skin without greasiness; high-temp. lubricant for textiles. *Sherex/Div. of Witco.*

29539 Starfol® Wax CG
8002-23-1 232-302-5
Cetyl esters; synthetic spermaceti wax; emollient for cosmetics, creams, lotions; opacifier and feel modifier. *Sherex/Div. of Witco.*

29540 Starglo
Blend of alcohols, aldehydes and nonionic wetters; tin electroplating additive. *Taskem Inc.* Name unverified.

29541 Starim
Reaction injection molding nylon, prepared by in-mold polymerization of caprolactam (feedstock for nylon) to a nylon-6 block polymer with utilization of a catalyst 2.0 agents; used for general industrial, agriculture, and automotive applications (e.g. body parts). *DSM NV.* Name unverified.

29542 Starlite
A proprietary synthetic resin. No manufacturer.

29543 Starpass
A proprietary urea-formaldehyde synthetic resin. No manufacturer.

29544 Starplex® 90
High-purity, molecularly distilled monoglyceride prepared from edible fats or oils and glycerin with TBHQ and citric acid; provides increased hydration and improved functionality of the monoglyceride; improves softness and extends shelf life in baked goods; emulsifies and stabilizes fat in sauces and gravies; starch complexing agent. *Am. Ingredients/Patco.*

29545 Starter Flowable
1698-60-8 216-920-2
Suspension concentrate containing 430 g chloridazon per liter; a pyridazinone herbicide for beet crops. *Truchem Ltd.*

29546 Starwax® 100
63231-60-7 264-038-1
Hard microcrystalline wax consisting of n-paraffinic, branched paraffinic, and naphthenic hydrocarbons; wax used in hot-melt coatings and adhesives, paper coatings, printing inks, plastic modification (as lubricant and processing aid), lacquers, paints and varnishes, binder in ceramics, in electronic components, rubber, elastomers, as emulsion wax size, fabric softener ingredient, in emulsion and latex coatings, hand creams and lipsticks. *Petrolite.*

29547 Starycide
64628-44-0 9809 264-980-3
Triflumuron. Insect growth regulator for the control of household pest larvae (especially flea and cockroach larvae). *Bayer AG.*

29548 stasite
$8UO_3{\cdot}4PbO{\cdot}3P_2O_5{\cdot}12H_2O$
A mineral, hydrated phosphate of uranium and lead, found in Katanga.

29549 Stasoft J
68585-05-7 271-548-8
Sulfated neatsfoot oil; fatliquor for chrome, chrome-alum, and chrome-bleached side leathers, production of full-grain soft upper leather and baseball glove leather. *Reilly-Whiteman.*

29550 staszicite
A mineral from Meidzianka, containing 39% As_2O_5, 26.5% CuO, 20.8% CaO, and 7.3% ZnO.

29551 Sta-Tac® B, T
Mixed olefin hydrocarbon resin; for pressure-sensitive adhesive, hot-melt, and laminating adhesives. *Arizona.* Discontinued.

29552 Statex
7440-44-0 1855 231-153-3
C
carbon
Carbon black. *Sevalco Ltd.*

29553 Statexan®
Antistatic agents; used for the textile industry. *Bayer AG; Bayer plc.*

29554 Statexan® K1
Sulfonated aliphatic hydrocarbon
Internal antistat additive or external coating for PVC, PS. *Bayer AG; Miles; Polysar.*

29555 Staticide®
Water-based topical antistat for spray, dip, or wipe-on application to any material; for industrial, commercial, and institutional facilities, clean room environments; biodeg. *ACL.*

29556 Statik-Blok® FDA-3
Polyether type; surface-active antistatic solution for food contact surfaces. *Amstat Industries.*

29557 Statil
Aldose reductase inhibitor. *ICI Chem & Polymers Ltd.* Discontinued.

29558 Stat-Kon®
Electrically conducting thermoplastic compounds. *LNP; ICI Chemicals & Polymers Ltd.*

29559 Stat-Kon® AC-1003
9003-56-9
ABS, 15% carbon fiber-reinforced; statically dissipative thermoplastic composite for protection against electrostatic discharge damage; used in electronic packaging systems and functional components. *LNP.*

29560 Stat-Kon® AS
9003-56-9
ABS, SS fiber-reinforced; statically dissipative composite for electronic packaging systems and functional components where protection from ESD damage is required. *LNP.*

29561 Stat-Kon® C
Carbon powd. grade; statically dissipative composite for electronic packaging systems and functional components where protection from ESD damage is required. *LNP.*

29562 Stat-Kon® DC-1002 FR
PC, 10% carbon fiber-reinforced; statically dissipative composite for electronic packaging systems and functional components where protection from ESD damage is required. *LNP.*

29563 Stat-Kon® FE
9002-88-4 7728
HDPE; extrusion grade, statically dissipative thermoplastic composite for

protection against electrostatic discharge damage; used in electronic packaging systems and functional components. *LNP.*

29564 Stat-Kon® M-1 HI
9003-07-0 7741
High impact PP; statically dissipative thermoplastic composite for protection against electrostatic discharge damage; used in electronic packaging systems and functional components. *LNP.*

29565 Stat-Kon® OC-1006
9016-75-5
PPS with 30% carbon fiber reinforcement; statically dissipative thermoplastic composite for protection against electrostatic discharge damage; used in electronic packaging systems and functional components. *LNP.*

29566 Stat-Kon® P
25038-54-4 6832
Nylon 6, carbon powd. grade; statically dissipative composite for electronic packaging systems and functional components where protection from ESD damage is required. *LNP.*

29567 Stat-Kon® PDX-84440
32131-17-2
Nylon 6/6; statically dissipative thermoplastic composite for protection against electrostatic discharge damage; used in electronic packaging systems and functional components installed close to sensitive elec./electronic devices. *LNP.*

29568 Stat-Kon® R
32131-17-2
Carbon powder grade nylon 6/6; statically dissipative thermoplastic composite for protection against electrostatic discharge damage; used in electronic packaging systems and functional components. *LNP.*

29569 Stat-Kon® RC-1002
32131-17-2
Nylon 6/6, 10% carbon fiber-reinforced; statically dissipative thermoplastic composite for protection against electrostatic discharge damage; used in electronic packaging systems and functional components. *LNP.*

29570 Stat-Kon® RF-15
32131-17-2
Carbon powder grade nylon 6/6 with 15% glass fiber reinforcement; statically dissipative thermoplastic composite for protection against electrostatic discharge damage; used in electronic packaging systems and functional components. *LNP.*

29571 Stat-Kon® W
26062-94-2
PBT polyester, carbon powd. grade; statically dissipative composite for electronic packaging systems and functional components where protection from ESD damage is required. *LNP.*

29572 Stat-Kon® ZC-1003
25134-01-4
Modified PPO, 15% carbon fiber-reinforced; statically dissipative thermoplastic composite for protection against electrostatic discharge damage; used in electronic packaging systems and functional components. *LNP.*

29573 Stat-Rite® C-2300
Chain extended low molecular weight polyoxirane; static dissipative polymer for alloying wide variety of thermoplastics. *BF Goodrich/Spec. Polymers.*

29574 statuary bronze
A variable alloy. It usually contains from 75-95% copper, 1-10% tin, 0-5% zinc, 0-6% lead, 0.12-0.34% phosphorus, and 0.19-0.7% nickel.

29575 statuary marble
471-34-1 1697 207-439-9
$CaCO_3$
Marble, with a crystalline or saccharoid structure.

29576 Stature
Static control additive. *Dow Chemical.*

29577 Status®
For agricultural applications. *DuPont UK.*

29578 Statyl
13997-19-8 6558 237-796-6
A proprietary preparation of methyl benzoquate; a veterinary antiprotozoan. No manufacturer.

29579 Staufen
PVC film. *ICI Chem & Polymers Ltd.* Discontinued.

29580 Stauffer N-1386®
3064-70-8 221-310-4
$C_2Cl_6O_2S$
Bis(trichloromethyl) sulfone
Hexachlorodimethyl sulfone; sulfonylbis(trichloromethane). Industrial biocide for control of algae, bacteria and fungi; slimicide for paper/paperboard production; preservative for adhesives, latexes, secondary oil well recovery. *Akzo.*

29581 staurolite
$HFeAl_5Si_2O_{13}$
A mineral; a basic aluminum ferrous iron silicate.

29582 Staybelite®
Hydrogenated rosin; used as a modifier for wax-elastomer ethylene adhesive compositions; used in electrical cable paper saturants, in ceramic ink vehicles, in metal resinates and soldering fluxes. *Hercules; Hercules Ltd.*

29583 Staybrite
A proprietary trade name for stainless steels containing chromium and nickel; they usually contain 18% chromium, 8% nickel, 74% iron, sometimes with molybdenum and occasionally with titanium and tungsten; they possess extreme. No manufacturer.

29584 steadite
$3(CaO{\cdot}P_2O_5){\cdot}2CaO\ (2CaO{\cdot}SiO)72)$
Iron-phosphorus eutectic, consisting of about 61% iron-phosphide, Fe_3P, with iron, a constituent of cast iron. The same name has been applied to a basic calcium-silico phosphate, found in the basic slag of the Thomas- Gilchrist process for the dephosphorization of iron.

29585 Stead's reagent
A reagent: 100 ml methyl alcohol, 18 ml water, 2 ml concentrated hydrochloric acid, 1 g copper chloride ($CuCl_2{\cdot}2H_2O$), 4 g magnesium chloride ($MgCl_2{\cdot}6H_2O$); an etching reagent used in the examination of steels.

29586 steam glue
Russian steam glue. A preparation of glue made by treating glue with nitric acid.

29587 Steamate
Boiler water treatment. *Grace Dearborn Ltd.*

29588 steamed bone meal
A fertilizer consisting of crushed bones, which have been treated with superheated steam and benzene, to remove fat and glue and contains about 1% nitrogen.

29589 Steamfilm FG
124-30-1 204-695-3
$C_{18}H_{39}N$
1-octadecylamine
Stearylamine; 1-Octadecanamine; octadecylamine; Adogenen 142; Alamine 7; Armeen 1180; n-octadecylamine. Aqueous emulsion of octadecylamine; corrosion inhibitor for control of corrosion caused by carbon dioxide and oxygen in the afterboiler section of steam-generating systems by forming a nonwettable, monomolecular film on metal surfaces. mp = 50-52°; bp = 349°; insoluble in H_2O (<1 mg/ml). *Drew Ind. Div.*

29590 steapsin
Pancreatic lipase. Fat-digesting enzyme contained in the pancreatic juice.

29591 Stearal
112-92-5 8960 204-017-6
$C_{18}H_{38}O$
Stearyl alcohol
1-octadecanol; stearyl alcohol; Aldol 62; Alfol 18; Atalco S; Cachalot S 43; CO 1895F; Conol 1675; Conol 30F; Crodacol S; 1-hydroxyoctadecane; Kalcohl 80; Lanol S; Lorol 28; n-octadecanol; octadecyl alcohol; Sipol S; Siponol S; Siponol SC; Stearol; Steraffine; Stenol; octadecan-1-ol. Emollient, auxiliary emulsifier, texturizer; nonoily, velvety feel; higher viscosity in emulsions. mp = 56-60°; bp_{15} = 210°; soluble in organic solvents. *Amerchol Corp.*

29592 stearalkonium chloride
122-19-0 204-527-9
$C_{27}H_{50}ClN$
N,N-dimethyl-N-octadecylbenzenemethanaminium chloride
dimethylbenzyloctadecylammonium chloride; N-octadecyl-N-benzyl-N,N-dimethylammonium chloride; octadecyldimethylbenzylammonium chloride; stearalkonium chloride; stearyldimethylbenzylammonium chloride; tallow benzyl dimethyl ammonium chloride; ammonyx 4; ammonyx 485; ammonyx 490; ammonyx 4002; ammonyx ca special; 2B; Barquat sb-25; Carsoquat sdq-25; Carsoquat sdq-85; Intexan sb-85; Intexsan sb-85; J soft C 4; Katamine ab; Orthosan mb; Quaternol 1; Stebac; Stedbac; Triton x-40; Triton x-400; Varisoft sdc; Arquad DM18b-90; Dehyquart stc-25; Nissan cation S2-100; Benzenemethaminium, N-octadecyl-N,N-dimethyl, chloride; Dimethyl-n-octadecylbenzenemethanaminium chloride,. *Ferrosan Fine Chem. A/S; Lonza; Mason; McIntyre; Sherex.*

29593 stearamide
124-26-5 204-693-2
$C_{18}H_{37}NO$
octadecanamide
stearic acid amide; amide C_{18}; stearoylamide; stearic acid amide; octadecanamide. Slip/antiblock agent for LDPE, HDPE, PP. mp = 102-104°; bp_{12} = 250-251°; *Akzo; Astor Wax; Chemax; Croda Universal; Henkel/Emery; Syn. Prods.; Witco/Humko.*

29594 stearamidoethyl diethylamine
16889-14-8 240-924-3
Lexamine 22. Conditioner, emulsifier for hair and skin care products; conditioning shampoo. *Inolex.*

29595 stearamidopropyl dimethylamine lactate
55819-53-9 259-837-7
Emcol® 3780; Hetamine 5L-25; Incromate SDL; Lexamine S-13 Lactate,. Cosmetics and toiletry surfactant used as antistat, conditioner, emollient, foaming and substantive agent. *Croda Inc.; Inolex; Heterene; Witco Corporation.*

29596 steareth series
9005-00-9
Steareth-40; Ablunol SA-7; Brij® 72; Brij® 700S; Brij® 721; Emalex 640; Hetoxol STA-2; Hodag Nonionic S-2; Lipocol S-2; Macol® SA-2; Prox-onix SA-1/02; Rhodasurf® S-2, TB-970; Ritox 721; Trycol® 5888; Volpo S-2,. Emulsifier, dispersant, thickener for cosmetics, creams, milky lotions. Nihon Emulsion; taiwan Surf; ICI Spec. Chem; Nihon Emulsion; Heterene; calgene; Lipo; PPG/Specialty Chem.; Protex; Rhône-Poulenc Surf; RITA; Henkel/Emery; Croda Inc.

29597 Stearex
57-11-4 8959 200-313-4
A trademark for a standardized stearic acid; it is a commercially pure, free fatty acid prepared for rubber manufacture; two grades; double pressed stearic acid, and single pressed stearic acid; the single pressed grade contains more oleic acid. No manufacturer.

29598 steargillite
A variety of the mineral montmorillonite.

29599 stearic acid
57-11-4 8959 200-313-4
$C_{18}H_{36}O_2$
n-octadecanoic acid
Emersol 132; 1-heptadecanecarboxylic acid; stearophanic acid; n-octadecylic acid; cetylacetic acid; barolub fta; century 1210; century 1220; century 1230; century 1240; dar-chem 14; emersol 120; emersol 132; emersol 150; emersol 153; emersol 6349; formula 300; glycon dp; glycon s-70; glycon s-80; glycon s-90; glycon tp; groco 54; groco 55; groco 55l; groco 58; groco 59; humko industrene r; hydrofol acid 150; hydrofol acid 1655; hydrofol acid 1855; hydrofol 1895; hy-phi 1199; hy-phi 1205; hy-phi 1303; hy-phi 1401; hystrene 80; hystrene 4516; hystrene 5016; hystrene 7018; hystrene 9718; hystrene s 97; hystrene t 70; industrene 5016; industrene 8718; industrene 9018; industrene r; kam 1000; kam 2000; kam 3000; loxiol g 20; lunac s 20; naa 173; neo-fat 18; neo-fat 18-s; neo-fat 18-53; neo-fat 18-54; neo-fat 18-55; neo-fat 18-59; neo-fat 18-61; PD 185; pearl stearic; promulsin; proviscol wax,. Used in cosmetics, chemicals, as a dispersant and softener in rubber compounds, in food packaging; suppositories and onitments. mp = 67-69°; bp = 361°; d = 0.8450; slightly soluble in H_2O, more soluble in organic solvents; LD_{50} (rat iv) = 21 mg/kg. Akrochem.; Akzo; Henkel/Emery; Lonza Inc.; Sherex/Div of Witco; Witco/Humco; Syn. Products; Unichema; UOP Inc.

29600 stearin pitch
Candle pitch; candle tar. A pitch obtained in the sulfuric acid treatment of fats. After distillation in steam of the washed acids (stearic, palmitic, and oleic), stearin pitch remains to the extent of 2%.

29601 Stearite
57-11-4 8959 200-313-4
A proprietary trade name for synthetic stearic acid produced by hydrogenation of certain oils. No manufacturer.

29602 stearopodis
Magnesium stearate
Used in the preparation of soap and face creams.

29603 stearosan
A compound of santalol and stearic acid.

29604 stearyl alcohol
112-92-5 8960 204-017-6
$CH_3(CH_2)_{16}CH_2OH$
n-octadecanol
1-octadecanol; C18 linear alcohol. Fatty alcohol; perfumery, cosmetics, intermediate, surfactants, lubricants, resins, antifoam agent. mp = 56-58°; bp = 340-355°; SG = 0.812; insoluble in H_2O, soluble in organic solvents. Aarhus Oliefabrik A/S; Amerchol; Chemron; Croda; Ethyl; Lipo; Lonza; M. Michel; Procter & Gamble; Sherex; Vista.

29605 stearyl erucamide
10094-45-8 233-226-5
$C_{40}H_{78}NO$
Stearyl erucamide
HTSA 3. Substituted aliphatic amide. Release agent providing slip, antiblocking to thermoplastics incl. PP film, nylon. *Hexcel; Croda Universal; Witco/Humko; Zeeland.*

29606 stearyl Hydroxyethyl imidazoline
95-19-2 202-397-8
$C_{22}H_{44}N_2O$
1-Hydroxyethyl-2-stearic imidazoline; Stearyl Imidazoline; 2-Heptadecyl-4,5-dihydro-1H-imidazole; 2-Heptadecyl-2-imidazoline-1-ethanol; Hodag C-100-S; Calgene C-100-S; Crodazoline S; Unamine® S; Imidazoline SOH; Monazoline S; Schercozoline S,. Intermediate for quaternary ammonium compounds; strongly absorbed on textiles, paper and many metal surfs.; for agric., asphalt, cleaners, corrosion inhibitors, demulsifiers, flotation, metalworking, paints, pigment grinding, inks, textiles, wax emulsions MW=354.70; moderately toxic by ingestion; *Calgene; Croda Universal Ltd.; Lakeland Labs Ltd.; Mona Ind; Scher; Lonza.*

29607 stearyl methacrylate
32360-05-7 251-013-5
$C_{22}H_{42}O_2$
octadecyl methacrylate
Lube oil additive, pour point depressant, paper coatings, textile finishes, paints, varnishes, pressure-sensitive adhesives. *CPS; Rohm & Haas; Sartomer.*

29608 stearyl/lauryl thiodipropionate
13103-52-1 236-025-0
Evans Chemetics.

29609 Stedbac®
122-19-0 204-527-9
Stearalkonium chloride
Hair conditioner, emulsifier imparting softness, antistatic properties to cream rinse formulations. *Zeeland.*

29610 steel
Iron containing combined carbon up to 1.5%. High carbon steels contain from 0.5-1.5% carbon, and mild steels from a trace to 0.5% carbon.

29611 Steel 01
A carbon steel containing 0.333% carbon, 0.162% silicon, 0.032% sulfur, 0.031% phosphorus, 0.617% manganese, 0.024% arsenic, 0.162% nickel, 0.017% chromium, and 0.037% copper.

29612 Steel A2
A British chemical standard carbon steel containing 0.037% carbon, 0.034% silicon, 0.020% sulfur, 0.008% phosphorus, 0.043% manganese, 0.031% arsenic, 0.059% nickel, 0.013% chromium, 0.067% copper, 0.04% oxygen, 99.72% iron.

29613 Steel B4
A British chemical standard carbon steel containing 0.400% carbon, 0.026% silicon, 0.046% sulfur, 0.103% phosphorus, 0.735% manganese, and 0.140% arsenic.

29614 Steel C
A British chemical standard carbon steel containing 0.093% carbon.

29615 Steel E
A standard steel containing 0.115% carbon and 0.491% manganese; used as the colorimetric standard for the determination of carbon in steels containing more than 0.100% carbon.

29616 Steel F
A German steel containing 0.67-1.1% silicon and 0.1-0.14% carbon.

29617 Steel Guard
Petroleum base; rust inhibitors for steel, industrial, commercial and vehicle applications. *Adasco Inc.*

29618 Steel H
A British chemical standard carbon steel containing 0.428% carbon, 0.047% sulfur, and 0.035% phosphorus.

29619 Steel I
A British chemical standard carbon steel containing 0.521% carbon and 0.726% manganese.

29620 Steel M
A British chemical standard. It is a carbon steel containing 0.228% carbon and 0.057% silicon.

29621 Steel N
A British chemical standard carbon steel containing 0.17% carbon, 0.117% silicon, 0.034% sulfur, 0.037% phosphorus, 0.432% manganese, and 0.029% arsenic.

29622 Steel N1
A carbon steel containing 0.153% carbon, 0.176% silicon, 0.050% sulfur, 0.036% phosphorus, 0.527% manganese, 0.030% arsenic, 0.260% nickel, and 0.04% copper. It is a British chemical standard.

29623 Steel O
A British chemical standard; a nickel steel containing 0.325% carbon, 0.590% manganese, and 3.985% nickel.

29624 steel ore
A variety of cinnabar containing 75% mercury.

29625 **Steel P**
A high silicon and phosphorus steel. It is a British chemical standard.
29626 **Steel R**
A British chemical standard. It is a carbon steel containing 0.786% carbon, 0.053% sulfur, and 0.914% manganese.
29627 **Steel S1**
A British chemical standard. It is a carbon steel containing 0.921% carbon and 0.051% phosphorus.
29628 **Steel T**
A British chemical standard nickel steel containing 3.367% nickel.
29629 **Steel U**
A British chemical standard carbon steel containing 1.203% carbon, 0.472% manganese, and 0.608% nickel.
29630 **Steel V**
A British chemical standard alloy steel containing 0.548% carbon, 0.161% silicon, 0.063% sulfur, 0.024% phosphorus, 0.542% manganese, 0.861% chromium, and 0.273% vanadium.
29631 **Steel V2A**
A rustless steel containing iron with 20% chromium, 7% nickel, and 0.2% carbon.
29632 **Steel W**
A British chemical standard. It is an alloy steel containing 0.695% carbon, 0.187% silicon, 0.075% sulfur, 0.028% phosphorus, 0.101% manganese, 0.44% nickel, 3.01% chromium, 0.791% vanadium, 4.76% cobalt, and 16.21% tungsten.
29633 **Steel W2**
A British chemical standard high-speed alloy steel containing 0.17% carbon, 0.14% silicon, 0.051% sulfur, 0.220% manganese, 3.29% chromium, 0.82% vanadium, 16.12% tungsten, 4.35% cobalt, 0.43% nickel, and 0.55% molybdenum.
29634 **steelite**
An explosive consisting of potassium chlorate, mixed with oxidized resin, and a little castor oil.
29635 **Steinazid SBU 185**
Undecylenic acid alkylolamide sulfosuccinate in liquid form; antidandruff agent, fungicidal, and bacteriostatic additive for shampoos, hair lotions, foam baths, etc. *Rewo Chemicals Ltd.* Name unverified.
29636 **Stelex**
Ceramic foam filter for steel and reactive alloy castings. *Foseco (F.S.) Ltd.*
29637 **Stellak**
Chemicals for boiler water treatment. *Steetley Chemicals Ltd.* Name unverified.
29638 **Stellar 500**
14807-96-6 9207 238-877-9
magnesium silicate
Talc; for plastics applications where good color and brightness are required; antiblocking agent in polyolefin films. *Cyprus Industrial Minerals.*
29639 **Stellite®**
Cobalt-base alloys; excellent wear resist. with limited corrosion resist. *Haynes Int'l.*
29640 **Stellited metal**
Metals treated with an alloy consisting chiefly of chromium, tungsten, and cobalt (stellite). The metals treated are usually steel, cast iron, malleable iron, and semi-steel. It is an economical method for these treated metals are rendered suitable for wear-resisting parts of machinery.
29641 **Stellos**
7778-54-3 1717 231-908-7
Calcium hypochlorite
Used for chlorination of water. *Stella-Meta Filters.*
29642 **Stellox 380EC**
An emulsifiable concentrate containing 190 g bromoxynil and 190 g ioxynil per liter; a post-emergence contact herbicide for cereal crops. *Ciba-Geigy Agrochemicals.*
29643 **Stelogen**
Flux for degassing steel. *Foseco (F.S.) Ltd.*
29644 **Stelopack**
Uphill teeming flux for killed steel ingots. *Foseco (F.S.) Ltd.*
29645 **Stelorit**
Covering and cleansing fluxes for steels. *Foseco (F.S.) Ltd.* Discontinued.
29646 **Stelotol**
Powder flux for uphill teeming. *Foseco (F.S.) Ltd.*
29647 **Stelpur**
Insulating feeding sleeve with Stelex ceramic foam filter incorporated within it. *Foseco (F.S.) Ltd.*
29648 **Stempor DG**
10605-21-7 1836 234-232-0
Carbendazim
Systemic fungicide. *ICI Agrochemicals.*
29649 **Stenol 1618**
Saturated fatty alcohols (45-55% C_{16}, 45-55% C_{18}); intermediate for surfactant manufacturing. *Henkel/Emery.*
29650 **Stenorol**
64924-67-0 4627
Halofuginone hydrobromide. Antiprotozoal. *Roussel UCLAF, Fine Chemicals.* Name unverified.
29651 **Stentor Steel**
A proprietary trade name for a non-shrinking steel containing 1.6% manganese, 0.25% silicon, and 0.9% carbon. No manufacturer.
29652 **Steol 3OS**
Sodium lauryl ether sulfate as a clear yellow viscous liquid; anionic surfactant used in shampoos, foam baths, and liquid detergents. *KWR Chemicals Ltd.* Name unverified.
29653 **Steol 4N**
Sodium fatty ether sulfate as a nearly water white liquid; for shampoo, bubble bath, liquid cleaner. *KWR Chemicals Ltd.* Name unverified.
29654 **Steol 7T**
Triethanolamine fatty ether sulfate as a pale yellow liquid; very mild base for shampoos and bubble baths. *KWR Chemicals Ltd.* Name unverified.
29655 **Steol CA-460 and KA-460**
Ammonium fatty ether sulfate in liquid form; for shampoo, bubble baths, dish detergents, and degreasers. *KWR Chemicals Ltd.* Name unverified.
29656 **Steol CS-460 and KS-460**
Sodium fatty ether sulfate in liquid form; for shampoo, bubble baths, dish detergents and degreasers. *KWR Chemicals Ltd.* Name unverified.
29657 **Steol CS-760 and 7N**
Sodium fatty ether sulfate as a pale yellow liquid; very mild base for shampoos and bubble baths. *KWR Chemicals Ltd.* Name unverified.
29658 **Steol FA**
Ammonium fatty ether sulfate as a pale yellow liquid; for general detergent uses and manufacture of gypsum board. *KWR Chemicals Ltd.* Name unverified.
29659 **Steol®**
Alkyl ether sulfates
Detergents, emulsifiers, foaming agents. *Stepan.* Name unverified.
29660 **Steol® 4N**
9004-82-4
Sodium laureth sulfate
Detergent, emulsifier, foamer, wetting agent used in personal care products; car wash, dishwash; textile mill applications; emulsion polymerization. *Stepan; Stepan Canada.*
29661 **Steol® CA-460**
67762-19-0
Ammonium laureth sulfate
Detergent, emulsifier, foamer, dispersant, and wetting agent used in shampoos, dishwashers, car washers, textile mill applications, emulsion polymerization. *Stepan; Stepan Canada.*
29662 **Steol® COS 433**
9014-90-8
Sodium nonoxynol-4 sulfate
Emulsifier for acrylics, SBR, vinyl chloride, butyl rubber. *Stepan Canada.*
29663 **Steol® CS-130**
Sodium laureth sulfate
Surfactant for personal care applications. *Stepan; Stepan Canada.*
29664 **Steol® OS 28**
9004-82-4
Sodium laureth sulfate
Detergent, foaming agent for all-purpose and specialty household/industrial cleaners. *Stepan Europe.*
29665 **Stepan**
Emollient; for bath oils; antiperspirants etc. *KWR Chemicals Ltd.* Name unverified.
29666 **Stepan C-40**
Methyl laurate
Intermediate for manufacturing of detergents, emulsifiers, wetting agents, stabilizers, lubricants, plasticizers, resins, and textile specialties; lubricant for metalworking formulations. *Stepan; Stepan Canada.*
29667 **Stepan C-65**
Methyl palmitate-oleate
Intermediate for manufacturing of detergents, emulsifiers, wetting agents, stabilizers, lubricants, plasticizers, resins, and textile specialties; lubricant for metalworking formulations. *Stepan; Stepan Canada.*
29668 **Stepan C-68**
Methyl oleate/stearate

Intermediate for manufacturing of detergents, emulsifiers, wetting agents, stabilizers, lubricants, plasticizers, resins, and textile specialties; lubricant for metalworking compounds. *Stepan; Stepan Canada.*

29669 Stepan Pearl Series
Surfactant blend; pearlescent, satining or opacifying agent, conditioning agent for shampoos, liquid soaps, bubble baths, shower gels. *Stepan Europe.*

29670 Stepan TAB® -2
Dihydrogenated tallow phthalic acid amide
Emulsifier, suspending agent for silicones, zinc pyrithione, sulfur, selenium sulfide, coal tar, oil extracts; opacifier; especially for use in conditioning and antidandruff shampoos. *Stepan; Stepan Europe.*

29671 Stepanate®
Hydrotropes
Coupling agent, cloud point depressant. *Stepan.* Name unverified.

29672 Stepanate® AXS
26447-10-9 247-710-9
Ammonium xylene sulfonate
Hydrotrope, solubilizer, coupler in detergent field; heavy duty cleaners, wax strippers, dishwashing detergents; solvent or fluidizer; freeze-thaw stabilizer; additive for high electrolyte or brine systems. *Stepan; Stepan Canada.*

29673 Stepanate® SCS
32073-22-6 250-913-5
Sodium cumene sulfonate
Coupler or solubilizer for liquid cleaners; detergent slurries. *Stepan; Stepan Canada.*

29674 Stepanate® SXS
1300-72-7 215-090-9
Sodium xylene sulfonate
Solubilizer, coupling agent, cloud point depressant, visc. reducer for industrial and household detergents, textile applications. *Stepan; Stepan Canada.*

29675 Stepanflo
Surfactant; for enhanced oil recovery. *Stepan.* Name unverified.

29676 Stepanflote® 85L
Sodium alkyl ether sulfate
Flotation reagent for molybdenum ore. *Stepan; Stepan Canada.*

29677 Stepanform® 1440
Blend; foaming agent, air entrainer for cellular concrete. *Stepan; Stepan Canada.*

29678 Stepanform® 1750
Blend; foamer for drilling applications. *Stepan; Stepan Canada.*

29679 Stepanhold® Extra
26589-26-4
PVP/ethyl methacrylate/methacrylic acid terpolymer. Hair care fixative for super-hold formulations. *Stepan; Stepan Canada; Stepan Europe.*

29680 Stepanhold® R-1
26589-26-4
Acrylates/PVP copolymer. Hair fixative for hair sprays, setting lotions, conditioners. *Stepan; Stepan Canada.*

29681 Stepan-Mild® LSB
Sodium lauryl sulfoacetate
disodium laurethsulfosuccinate. Surfactant for shampoos, hand soaps, bubble baths, facial cleansers, baby products, sensitive skin products. *Stepan; Stepan Canada; Stepan Europe.*

29682 Stepan-Mild® SL3
Disodium laurethsulfosuccinate
Surfactant for low-irritation shampoos, bubble baths, dishwashing detergents. *Stepan; Stepan Canada.*

29683 Stepanol AM
Ammonium fatty alcohol sulfate as a pale yellow liquid; for shampoo; bubble bath; liquid detergents. *KWR Chemicals Ltd.* Name unverified.

29684 Stepanol DEA
Diethanolamine fatty alcohol sulfate as a pale yellow liquid; for shampoo. *KWR Chemicals Ltd.* Name unverified.

29685 Stepanol ME
Sodium fatty alcohol sulfate as a white powder; for powdered detergents. *KWR Chemicals Ltd.* Name unverified.

29686 Stepanol Mg
magnesium fatty alcohol sulfate
A pale yellow liquid; for rug and upholstery shampoos. *KWR Chemicals Ltd.* Name unverified.

29687 Stepanol WA, WAC, WAQ
Ionic surfactants from the Stepanol WA range; for shampoos, bubble baths, liquid and paste detergents. *KWR Chemicals Ltd.* Name unverified.

29688 Stepanol WA-100
Sodium fatty alcohol sulfate as a white powder; anionic surfactant used as a dentifrice, and in the pharmaceutical industry. *KWR Chemicals Ltd.* Name unverified.

29689 Stepanol WAT
Triethanolamine fatty alcohol sulfate as a nearly water white liquid; for shampoos. *KWR Chemicals Ltd.* Name unverified.

29690 Stepanol®
Alkyl sulfates
Mild detergent, foaming agent. *Stepan.* Name unverified.

29691 Stepanol® AEG
Mixture of ammonium lauryl sulfate, ammonium laureth sulfate, cocamidopropyl betaine and cocamide DEA. Base for liquid hand and body soaps, shampoos. *Stepan; Stepan Canada.*

29692 Stepanol® AEM
Mixture of ammonium laureth sulfate and cocamide MEA. Concentrate for shampoo and bath products *Stepan; Stepan Canada.*

29693 Stepanol® AM
2235-54-3 218-793-9
Ammonium lauryl sulfate
Detergent, foamer used in personal care products; rug and upholstery shampoos; household, metal, and industrial cleaners; fruit washing; insecticides; textile and leather processing; pharmaceuticals. *Stepan; Stepan Canada.*

29694 Stepanol® LX
Mixture of DEA lauryl sulfate, DEA lauraminopropionate and sodium lauraminopropionate. Mild blended concentrate for shampoos, baby soaps, bubble baths; foaming power; substantive to hair and skin. *Stepan; Stepan Canada.*

29695 Stepanol® ME Dry
151-21-3 8782 205-788-1
Sodium lauryl sulfate
Detergent, foamer used in personal care products; rug and upholstery shampoos; household, metal, and industrial cleaners; fruit washing; insecticides; textile and leather processing; pharmaceuticals. *Stepan; Stepan Canada.*

29696 Stepanol® MG
Magnesium lauryl sulfate
Detergent, foamer used in personal care products; rug and upholstery shampoos; household, metal, and industrial cleaners; fruit washing; insecticides; textile and leather processing; pharmaceuticals. *Stepan; Stepan Canada.*

29697 Stepanol® SPT
139-96-8 205-388-7
TEA lauryl sulfate
Mild detergent, foaming agent for household cleaners, liquid soaps. *Stepan Europe.*

29698 Stepanol® WA Extra
151-21-3 8782 205-788-1
Sodium lauryl sulfate
Detergent, foamer used in personal care products; rug and upholstery shampoos; household, metal, and industrial cleaners; fruit washing; insecticides; textile and leather processing; pharmaceuticals. *Stepan; Stepan Canada.*

29699 Stepanon CG
Surfactant; brine tolerant foamer for acid or alkaline media. *Stepan; Stepan Canada.*

29700 Stepanquat® 6585
Dipalitmoyletyl hydroxyethylmonium methosulfate
Mild surfactant for cream rinses and conditioners. *Stepan; Stepan Canada.*

29701 Stepantan A
Anionic surfactant in powder form; dispersant, tanning agent. *KWR Chemicals Ltd.* Name unverified.

29702 Stepantan NP 80
Anionic surfactant in powder form; dispersant for phyto-sanitary products and wettable powders. *KWR Chemicals Ltd.* Name unverified.

29703 Stepantan®
Alkyl sulfonate
Dispersant. *Stepan.* Name unverified.

29704 Stepantan® AS-12
Sodium alpha olefin sulfonate
Foamer for soft and hard waters, fresh water, and moderate brine conditions. *Stepan; Stepan Canada.*

29705 Stepantan® DS-40
Sodium dodecylbenzene sulfonate
Detergent, wetting agent, foaming agent; used specifically in dust control applications. *Stepan; Stepan Canada.*

29706 Stepantan® DT-60
TEA dodecylbenzenesulfonate

Detergent, wetting agent, foaming agent; used specifically as air entraining agent in concrete applications. *Stepan; Stepan Canada.*

29707 Stepantan® H-100
Dodecylbenzene sulfonic acid
Detergent intermediate; emulsifier for oils, solvs., waxes and oil field applications; air entraining agent and foamer for cellular concrete; wetting agent. *Stepan; Stepan Canada.*

29708 Stepantex Q90B
Dialkyl methoxysulfate as an amber viscous liquid; cationic surfactant used in textile softeners. *KWR Chemicals Ltd.* Name unverified.

29709 Stepantex® B-29
Sodium octyl sulfate
Wetting and mercerizing agent for textiles. *Stepan; Stepan Canada.*

29710 Stepantex® CO-30
61791-12-6
PEG-30 castor oil; lubricant, emulsifier for fiber finish and wet processing. *Stepan; Stepan Canada.*

29711 Stepantex® TD14
Tall oil ester; biodegradable textile auxiliary. *Stepan; Stepan Canada.*

29712 Stepantex® VS 90
Diester quaternary ammonium methyl sulfate
Good hand, excellent rewet and antistatic properties for household and commercial rinse-added fabric softeners; textile processing aid for natural and synthetic fibers. *Stepan; Stepan Canada.*

29713 Stepfac® 8170
Ethoxylated nonylphenol phosphate
Hydrotrope for nonionics; emulsifier for agriculture, emulsion polymerization, oils, metalworking lubricants, corrosion inhibitors, pigment dispersants; heavy-duty industrial/household alkali cleaners; compatibility agent for for liquid fertilizers. *Stepan; Stepan Canada.*

29714 Stepfac® 8171
Acid form; compatibility agent for agricultural formulations. *Stepan; Stepan Canada.*

29715 Stepfac® PN 10
Phosphate ester; emulsifier, wetting and dispersing agent for agricultural wettable powders and flowables. *Stepan Europe.*

29716 Step-Flow 21
Nonionic dispersant; surfactant for aqueous agricultural flowables. *Stepan; Stepan Europe.*

29717 Steposol®
Alkyl ether sulfate
Foaming agent. *Stepan.* Name unverified.

29718 Steposol® CA-60H
Ammonium ether sulfate
High flash point foamer for heavy brine conditions. *Stepan; Stepan Canada.*

29719 Steposol® CA-207
Ammonium ether sulfate
Foaming agent used for oilfield applications; also for gypsum board, cellular concrete, air drilling, foam cleaners; excellent for heavy brine conditions; stable in soft or hard water. *Stepan; Stepan Canada; Stepan Europe.*

29720 Stepsperse® DF-100
Surfactant blend; dispersant for agricultural flowables and dry flowables. *Stepan; Stepan Europe.*

29721 Stepwet® DF-60
Surfactant blend; wetting agent for agricultural flowables and dry flowables. *Stepan; Stepan Europe.*

29722 Stereon® 840A
Stereospecific SBR block copolymer; rubber modifier for thermoplastic resins, especially for HIPS, flame-retardant HIPS, and PP. *Firestone Syn. Rubber.*

29723 Stereon® 881
Butadiene-styrene copolymer; offers low gel, high gloss, toughness, clarity, and processing ease for injection molded items including medical devices, containers, toys, food containers; FDA compliance. *Firestone Syn. Rubber.* Discontinued.

29724 Stereon® 900
Multiblock styrene-butadiene copolymer; designed to blend easily with general purpose PS to yield extruded and thermoformed products with excellent clarity, sparkle, impact resist., and flexibility; FDA compliance. *Firestone Syn. Rubber.* Discontinued.

29725 stereotype plate
An alloy of 85% lead and 14% antimony, sometimes with the addition of small amounts of tin.

29726 steresol
An antiseptic varnish made by dissolving 270 parts purified shellac, 10 parts benzoin, 10 parts balsam of tolu, 100 parts phenol, 6 parts oil of cinnamon, and 6 parts saccharin in alcohol, to make 1000 parts.

29727 Sterethox
75-71-8 3114 200-893-9
Sterilizer containing dichlorodifluoromethane. *ICI Chem & Polymers Ltd.*

29728 Sterets Pre-Injection Swabs®
67-63-0 5227 200-661-7
Isopropyl alcohol 70%
Used for injection site cleansing. *Seton Healthcare Group plc.*

29729 Steribath
An antiseptic solution containing an iodophore. *ICI Chem & Polymers Ltd.* Discontinued.

29730 Steridex
Fungicidal water-based elastomeric protective coating, applied by brush or spray; for totally eradicating mold growth and bacteria in all hygiene sensitive environments - hospitals, food factories, breweries. *Liquid Plastics Ltd.* Discontinued.

29731 Steriflux
A range of sterile nonpyrogenic intravenous infusions. *The Boots Co plc.* Discontinued.

29732 SteriLine 200
14807-96-6 9207 238-877-9
Platy talc USP; general purpose, coarse talc for use in dusting and baby powders, soaps, antiperspirant sticks, color extensions. *Montana Talc.*

29733 SteriLine 665
14807-96-6 9207 238-877-9
Platy talc USP; ultrafine talc for cosmetic applications requiring high oil absorption, gloss control, and smoothness; viscosity control additive in pharmaceutical excipient applications. *Montana Talc.*

29734 Sterilite Hop Defoliant
120-12-7 721 204-371-1
$C_{14}H_{10}$
anthracene
Paranaphthalene; anthracin; green oil; tetra olive n2g. Anthracene oil; used for chemical stripping in hop vines. *Coventry Chemicals Ltd.*

29735 Sterilite®
Emulsifiers, tar oils and high boiling tar acids; fungicidal winter wash for fruit; disinfectant for animal health field. *Coventry Chemicals Ltd.*

29736 Sterillium
Synthetic phenolic germicides in a detergent base; a disinfectant for laundry use. *William Pearson Ltd.* Name unverified.

29737 Sterisafe
Food sterilizer treatments. *Grace Dearborn Ltd.*

29738 Sterisheen
Modified acrylic semi-gloss, tough, flexible, waterbased coating; applied by brush or spray; for totally eradicating mold, micro-organisms, fungi, and bacteria in all hygiene sensitive environments, hospitals, food factories, breweries and kitchens. *Liquid Plastics Ltd.*

29739 Sterisil
141-94-6 4741 205-513-5
hexetidine
A proprietary trade name for hexetidine, an antifungal agent. No manufacturer.

29740 Steritile® Plus
A two-component, water-based, acrylic epoxy coating; forms a tough, durable surface; easy to clean and resists growth of mold and bacteria; applied by brush, roller airless spray; ideal for hospitals, food factories, brewerie *Liquid Plastics Ltd.*

29741 Steriwipe®
Cetrimide 0.15% and chlorhexidine acetate 0.015%; used for topical disinfection, cleansing minor wounds and abrasions. *Seton Healthcare Group plc.*

29742 Sterline
An alloy of 68% copper, 17-18% nickel, 13-14% zinc, 0.75-0.8% iron, and 0-0.8% lead. It is a nickel silver. *German silver.*

29743 Sterling
7440-44-0 1855 231-153-3
Carbon black. *Cabot Carbon Ltd.*

29744 sterling solder
An alloy of 61.6% tin, 15.2% zinc, 11.2% aluminum, 8.3% lead, 2.5% copper, and 1.2% antimony.

29745 Sterlite
A proprietary trade name for a nickel brass containing 25% nickel, 20% zinc, and small amounts of iron, manganese, silicon, and carbon. No manufacturer.

29746 Sterlith
A trademark for materials of the refractory and abrasive type; they consist essentially of crystalline alumina. No manufacturer.

29747 **Sternite**
Phenol formaldehyde resin; molding powders. *Manchem Ltd.* Name unverified.

29748 **Sternite**
Phenolic and polystyrene molding materials. *Sterling Moulding Materials.* Discontinued.

29749 **Stero WW**
Wool grease with low free fatty acids; lubricant, emulsifier, rust inhibitor; base for wire drawing and metalworking compounds, component for metal polishes, cationic fat liquors, leather stuffing compounds; fingerprint inhibitor in rust preventatives. *Actrachem.*

29750 **Sterocoll**
Alkali-soluble dispersion. *BASF plc.*

29751 **Steron 210**
Duradene® 710. Solution-polymerized B/S copolymer, nonstaining; processing and extrusion aid to be used with other elastomers for such applications as shoe soling; also for asphalt modification and adhesives. *Firestone Syn. Rubber.*

29752 **Sterotabs**
Water treatment tablets. *The Boots Co plc.* Discontinued.

29753 **Sterotex®**
68334-28-1 269-820-6
Hydrogenated vegetable oil; binder and internal lubricant for pressed powds. *Karlshamns.*

29754 **Sterotex® HM**
68334-28-1 269-820-6
Hydrogenated vegetable oils; lubricant in pharmaceutical tableting, powd. compression applications. *Karlshamns.*

29755 **Sterox**
alkylaryl polyoxyethylene ether
Dodecylphenol-ethylene oxide condensate nonionic surface active agent. *Monsanto Co.*

29756 **Sterox DF, DJ**
Anionic surface active agent. *Monsanto plc.*

29757 **Steroxin-Hydrocortisone**
A proprietary preparation of chlorquinaldol and hydrocortisone; used in dermatology as an antibacterial agent. *Ciba plc.* Name unverified.

29758 **Steroxol**
Chlorinated detergent. *Rhône-Poulenc UK.*

29759 **Sterpon**
A proprietary polyester laminating resin. *Convert (Ets G).* Name unverified.

29760 **Ster-Zac**
70-30-4 4716 200-733-8
A proprietary preparation of hexachlorophene; a topical antiseptic. *Hough, Hoseason & Co Ltd.* Name unverified.

29761 **stibium**
7440-36-0 733 231-146-5
Sb
Latin name for antimony.

29762 **Stickstoffoxydbaryt**
13465-94-6 1013 236-709-9
$BaN_2O_4 \cdot H_2O$
Barium nitrite

29763 **Stik-It**
Nonionic wetting agent for use in a wide range of fungicides and insecticides. *Quadrangle Agrochemicals.*

29764 **stilbene**
588-59-0 8972 209-621-3
$C_{14}H_{12}$
1,1'-(1,2-Ethenediyl)bis[benzene]
Eccobrite RB; bibenzal; bibenzylidene. Whitening agent for cotton and acetates. *Eastern Color & Chem.*

29765 ***cis*-stilbene**
645-49-8 8972 211-445-7
$C_{14}H_{12}$
1,1'-(1,2-ethanediyl)bis[benzene]
cis-stilbene. Used in chemical synthesis. mp = -5°; bp_{10} = 135°; n_D^{25} = 1.6188; λ_m = 278 nm (ε 10200 95% EtOH); insoluble in H_2O, soluble in organic solvents.

29766 **Stillingia oil**
An oil obtained by crushing the kernel of *Stillingia sebifera*.

29767 **Stimufol**
Soluble fertilizer. *ICI Chem & Polymers Ltd.*

29768 **Sting**
1071-83-6 4522 213-997-4
Glyphosate. Non-selective systemic herbicide. Used for control of a wide variety of annual, biennial and perennial grasses, sedges, broad-leaved weeds and woody shrubs. *Monsanto plc.*

29769 **stink quartz**
Fetid quartz. A quartz which has a bad odor, due to organic matter.

29770 **stink-stone**
Oil-stone. A bituminous schist found in the Tyrol. A source of ichthammol.

29771 **Stirene**
Polystyrene. *Dow Chemical.*

29772 **Stiresol®**
Styrenated glycerol phthalate resin solution. *Reichhold.* Discontinued.

29773 **Stirling's gentian violet**
A microscopic stain. It contains 5 g gentian violet, 10 ml 95% alcohol, 2 ml aniline, and 88 ml water.

29774 **Stockalite**
A proprietary product; it is a very highly refined china clay used as a filler for tires, cables, and high grade mixes. No manufacturer.

29775 **Stockholm**
Pine tar. A tar obtained principally from pinewood distillation. It is obtained from *Pinus sylvestris* and other species of *Pinus*; used as a preservative paint for ships and roofing and as a rubber softener.

29776 **Stockholm pitch**
Pine-wood tar pitch. It is soluble in alkalies, and is used in the preparation of varnishes, in the rubber and gutta-percha trades, and in the preparation of impervious cements.

29777 **Stoco**
A proprietary bituminous plastic. No manufacturer.

29778 **Stoddard Solvent**
A proprietary trade name for a refined petroleum product for dry cleaning. No manufacturer.

29779 **Stoffertite**
$CaHPO_4 \cdot 5H_2O$
A calcium phosphate, it occurs in guano.

29780 **stoic metal**
An alloy similar in composition to invar.

29781 **Stoke's reagent**
A reducing agent prepared by dissolving 30 g ferrous sulfate and 20 g tartaric acid in 1 liter of water. When required for use, strong ammonia is added until the precipitate first formed is redissolved.

29782 **Stoller Flowable Sulphur**
7704-34-9 9150 231-722-6
sulfur
Suspension concentrate containing 720 g/l sulfur; a protectant fungicide. *Stoller Chemicals Ltd.*

29783 **Stomahesive**
A proprietary preparation containing gelatin pectin, carboxy-methyl cellulose and polyisobutylene on a protective film; used for the protection of skin around surgical stomata. *Bristol-Myers Squibb Pharmaceuticals Ltd.* Name unverified.

29784 **Stomp**
40487-42-1 7211 254-938-2
Emulsifiable or suspension concentrate containing pendimethalin; a dinitroaniline herbicide for cereals and bush fruit. *Cyanamid of Great Britain Ltd.*

29785 **Stomp H**
40487-42-1 7211 254-938-2
Emulsifiable or suspension concentrate containing pendimethalin; a dinitroaniline herbicide for cereals and bush fruit. *Hortichem Ltd.*

29786 **stone wax**
A name applied to carnauba wax.

29787 **Stoner A500**
Citrus distillates
Biodegradable cleaner/degreaser for cleaning built-up, baked-on residue from molds, tools, machinery, and other equipment. *Stoner.*

29788 **Stoner E800**
Proprietary nonsilicone release blend; mold release lubricant for molds for natural rubber, nitrile, silicone, neoprene, SBR, and other compounds. *Stoner.*

29789 **Stoner K206**
Proprietary silicone release blend; mold release for release of plastics, rubber, and waxes. *Stoner.*

29790 **stone's bronze**
An alloy of 87% copper, 11% tin, and 2% phosphor-copper.

29791 **Stonite**
An explosive consisting of 68% nitroglycerin, 20% kieselguhr, 8% potassium nitrate, and 4% wood meal.

29792 **Stoodite**
A proprietary trade name for a high manganese steel. No manufacturer.

29793 Stora
A proprietary trade name for a Swedish charcoal iron used for making malleable iron. No manufacturer.

29794 Storaid Dust, Storite SS
Dustable powder containing 6% w/w tecnazene [117-18-0] and 1.8% w/w thiabendazole [148-79-8]; a protectant fungicide and potato sprout suppressant. *MSD Agvet.*

29795 storax calamita
The powdered bark of *Liquidambar styracflua* , most of the resin being first extracted. The product has no connection with Storax.

29796 Storite
148-79-8 9426 205-725-8
Thiabendazole
A systemic insecticide. *MSD Agvet.*

29797 Storm®
Bentazon
acifluorfen. For post-emergence control of broadleaf weeds in soybeans and peanuts. *BASF AG.*

29798 Stortex
Malt extract. *ABM Chemicals Ltd.* Name unverified.

29799 Stowite
An explosive containing 58-61% nitroglycerin, 4.5-5% nitrocotton, 18-20% potassium nitrate, 6-7% wood meal, and 11-15% ammonium oxalate.

29800 Strandex
Chemical compositions for use in industry as additives in polymer and plastics processing, all containing metal compounds. *Associated Lead Manufacturers Ltd.* Name unverified.

29801 Strandol
Mould lubricant for continuous casting of steel billets. *Foseco (F.S.) Ltd.* Discontinued.

29802 strass
Paste. A kind of glass used to imitate precious stones. It is made from 100 parts sand, 40 parts minium, 24 parts potassium carbonate, 20 parts borax, and 12 parts potassium nitrate. This gives a colorless product, and various oxides are added to color it.

29803 Strassburg turpentine
The oleoresin from the silver fir, *Pinus picea* .

29804 Strasser solder
An alloy of 62% tin, 12% zinc, 4% aluminum, 8% lead, 5% copper, 5% bismuth, and 4% cadmium.

29805 Strata-Fire
A fuel additive used to reduce engine wear, improve engine performance, increase fuel economy and reduce emission of air pollutants; used for all types of internal combustion engines, both diesel and gasoline. *SN Corp/Appropriate Technology Ltd.*

29806 Stratos®
101205-02-1
Cycloxydim
Post-emergence graminicide against annual and perennial grasses; selective in broadleaf crops, e.g., sugar beet, cotton, soybean, vegetables, onions. *BASF AG.*

29807 Stratton
Solvent stripper. *Dow Chemical.*

29808 Stratyl
A proprietary polyester laminating compound. No manufacturer.

29809 Strawlink
Mineral/vitamin animal feed supplement for straw. *ICI Chem & Polymers Ltd.*

29810 Strelax
Road nosing compounds. *ICI Chem & Polymers Ltd.*

29811 Strenes Metal
A proprietary trade name for a nickel-chromium-molybdenum cast iron. No manufacturer.

29812 Strepsils
Proprietary preparations containing 2,4-dichlorobenzyl alcohol and amylmetacresol; antibacterial. *Crookes Healthcare.*

29813 Stresnil
1649-18-9 931 216-715-8
A proprietary preparation containing azaperone; veterinary sedative (pigs). *Janssen Pharmaceutical Ltd.* Name unverified.

29814 strewing smalt
The coarsest powdered smalt (*qv*).

29815 Strim
Rimming agent for steel ingots. *Foseco (F.S.) Ltd.*

29816 Stripcote
Liquid parting agents and release agents. *Foseco (F.S.) Ltd.*

29817 Strite
Ion control resins. *Rohm & Haas.*

29818 Strobane
terpene polychlorinates; Dichloricide Aerosol; Dichloricide Mothproofer; Insecticide 3960-X14. A trademark for an insecticide and acaricide; it is based on mixed polychlorinated terpene isomers and contains 66% chlorine. No manufacturer.

29819 Strodex® MO-100
2-Ethylhexyl polyphosphoric ester acid anhydride
Emulsifier; pigment dispersant used in oil-based paints and leather coating specialties. *Dexter.*

29820 Strodex® MOK-70
Phosphated alcohol
Dispersant/stabilizer for carbon black pigments, zinc pigments, lead silicate in exterior latex paints; wetting and rewetting agent; rust inhibitor. *Dexter.*

29821 Strodex® P-100
Phosphated coester of alcohol and aliphatic ethoxylate acid anhydride; emulsifier; dispersant for pigments in polar and nonpolar solvents, coupler for emulsifiers in aqueous and nonaqueous systems; used in metal and tile cleaners; pigment grinding aid. *Dexter.*

29822 Strodex® PK-90
Phosphated coester of alcohol and aliphatic ethoxylate; emulsifier, dispersant, wetting agent for extender pigments in latex paints, barium sulfate, and iron oxides; oxidation-corrosion inhibitor; used in heavy-duty alkaline cleaners. *Dexter.*

29823 Strodex® SE-100
Phosphated aliphatic ethoxylate acid anhydride
Dispersant, emulsifier and coemulsifier in aqueous and nonaqueous systems. *Dexter.*

29824 Strodex® Super V-8
Phosphate ester, potassium salt; patented emulsifier, detergent, dispersant, wetting agent used in textile processing applications. *Dexter.*

29825 Stronscan-85
10476-85-4 9000 233-971-6
$Cl_2{}^{85}Sr$
Strontium chloride-^{85}Sr
strontium chloride hexahydrate;. Radioactive agent. Hexahydrate loses $5H_2O$ at 100°, loses last H_2O at 150°. mp = 61°; d = 1.96; anhydrous: mp = 868°; soluble in H_2O (1.2 g/ml); LD_{50} (mus iv) = 148 mg/kg. *Abbott Laboratories.* Name unverified.

29826 strontia
1314-11-0 9009 215-219-9
Osr
Strontium oxide
strontium monoxide. Used in the manufacture of strontium salts, pyrotechnics, greases, and soaps. mp = 2430°; d = 4.7; reacts with H_2O.

29827 strontium
7440-24-6 8994 231-133-4
Sr
Metallic element; alloys of strontium used in electron tubes as a 'getter' to combine chemically with active gases and to hold inactive gases by adsorption. Used in fireworks and tracer bullets. mp = 757°; bp = 1366°; d = 2.6; reacts with oxygen; *Atomergic Chemetals; Degussa; Noah Chem.*

29828 strontium carbonate
1633-05-2 8998 216-643-7
CO_3Sr
Catalyst, in radiation-resistant glass for color TV tubes, ceramic ferrites, pyrotechnics. Used in sugar refining. dec 1100°; d = 3.5; soluble in H_2O (10 mg/l). *Atomergic Chemetals; Cerac; Mallinckrodt; Solvay GmbH.*

29829 strontium chloride
10476-85-4 9000 233-971-6
Cl_2Sr or $Cl_2Sr{\cdot}6H_2O$
strontium chloride hexahydrate
Strontium salts, pyrotechnics, electron tubes. mp = 61°; d = 1.96; anhydrous: mp = 868°; soluble in H_2O (1.2 g/ml); LD_{50} (mus iv) = 148 mg/kg. *Aldrich; Atomergic Chemetals; Hoechst Celanese.*

29830 strontium nitrate
10042-76-9 9007 233-131-9
N_2O_6Sr
Pyrotechnics, marine signals, railroad flares, matches. mp = 570°; d = 2.99; soluble in H_2O (0.66 g/ml), slightly soluble in EtOH, Me_2CO; LD_{50} (rat ip) = 540 mg/kg. *Bernardy Chimie SA; Hoechst Celanese; Noah Chem.; Solvay GmbH.*

29831 strontium titanate
12060-59-2 235-044-1
O_3SrTi
strontium titanium oxide
Electronics, electrical insulation. *Atomergic Chemetals; Ferro/Transelco; TAM Ceramics.*

29832 Strotope
10042-76-9 9007 233-131-9
$N_2O_6{}^{85}Sr$
Strontium nitrate ^{85}Sr
Radioactive agent. mp = 570°; d = 2.99; soluble in H_2O (0.66 g/ml), less soluble in rganic solvents; LD_{50} (rat ip) = 540 mg/kg. *Bristol-Myers Squibb Co Inc.* Name unverified.

29833 Struktol® 40 MS Flakes
Modified mixture of rubber-compatible nonhardening synthetic resins; resin plasticizer; homogenizing agent for elastomer blend compounds. *Struktol.*

29834 Struktol® Activator 73
Mixture of zinc salts of aliphatic and aromatic carboxylic acids; vulcanization activator for natural rubber. *Struktol.*

29835 Struktol® HP 55
Process aid for the tire industry; improves mixing and milling characteristics without adversely affecting cured properties, abrasion resistance or tear strength. *Struktol.*

29836 Struktol® PE H-100
9002-88-4 7728
Low molecular weight polyethylene homopolymer wax; improves flow and processability in natural and synthetic elastomers; provides release from equip. with improved pigment dispersion and finish to molded articles. *Struktol.*

29837 Struktol® SU 109
7704-34-9 9142 231-722-6
sulfur
Coated insoluble sulfur; used for rubber compounding. *Struktol.*

29838 Struktol® T.M.Q
Polymerized 2,2,4-trimethyl-1,2-dihydroquinoline
Antioxidant in rubber tires, belts, mechanical goods, sponge and retreading; minimally discoloring. *Struktol.*

29839 Struktol® WB 300
Blend of high molecular aliphatic and aromatic polyesters; plasticizer for nitrile rubber and chloroprene; end uses including petrol hoses, printing rollers, oil seals, etc.; processing aid. *Struktol.*

29840 struthiin
$C_{19}H_{30}O_{10}$
Githagin
polygalin; polygallic acid; senegin. Saponin, a glucoside found mainly in the common soapwort.

29841 Stryden Forte Rapid
Streptomycin-penicillin bovine intramammary antibiotic. *RMB Animal Health Ltd.*

29842 ST-Size
Modified rosin emulsion size. *Hercules.* Discontinued.

29843 stucco
A specially hard plaster which can be polished. There are two kinds: a) made from plaster-of-Paris; and b) made from lime. They are usually mixed with size.

29844 Stuk
Adhesives for footwear industry. *ICI Polyurethanes.* Name unverified.

29845 stupp
A mercurial soot condensed in the chambers during the treatment of mercury ores. It contains about 20% mercury as metal, and sulfate.

29846 Sturcal®
471-34-1 1697 207-439-9
$CaCO_3$
calcium carbonate
Precipitated calcium carbonate pharmaceutical, dentifrice. *Rhône-Poulenc Sturge Lifford.*

29847 Sturcarb
Whiting. *Rhône-Poulenc Sturge Lifford.* Discontinued.

29848 Stycast® 1090
25928-94-3
epoxy resin
Epoxy resin filled with glass microballoons; low-density syntactic foam which withstands high hydrostatic pressure; for room temperature cure; converts to thermoset solid by addition of a catalyst. *Emerson & Cuming Polymer Group.* Name unverified.

29849 Stycast® 1210
25928-94-3
epoxy resin
Filled epoxy resin; casting resin adapts to single-cure impregnation and potting, and exceeds MIL-Spec thermal shock requirements; not for room temperature cure; converts to thermoset solid by addition of second compound. *Emerson & Cuming Polymer Group.* Name unverified.

29850 Stycast® 1266
25928-94-3
epoxy resin
Unfilled epoxy resin; room temperature curing, high clarity casting epoxy; converts to thermoset solid by addition of second compound. *Emerson & Cuming Polymer Group.* Name unverified.

29851 Stycast® 1467
25928-94-3
epoxy resin
Filled epoxy resin; fire retardant casting resin offers machinability with high strength; self-extinguishing; for room temperature cure; converts to thermoset solid by addition of catalyst. *Emerson & Cuming Polymer Group.* Name unverified.

29852 Stycast® 2850-FT
25928-94-3
epoxy resin
Filled epoxy resin; low expansion, high thermal conductivity, good thermal shock, and high temperature resistance; castings can be machined by grinding; for room temperature cure; converts to solid thermoset by addition of catalyst *Emerson & Cuming Polymer Group.*

29853 Stycond-109
Conductive/static dissipative thermoplastic composite. *United Composites.*

29854 Stygene Series
Polynuclear aromatic polymers derived from specially prepared petroleum stream; resins in rubber compounding, joint cements, plastic compounds, protective coatings, fiber board, inks, epoxy potting compounds, adhesives, insecticides, briquettes, floor tile etc.; soft grades as rubber plasticizers, hard grades as rubber extenders, carrier for insecticidal toxicants, in calendered and extruded good and mechanical goods. *Chemfax.*

29855 Stylac® ABS
9003-56-9
Styrene-acrylonitrile-butadiene copolymer. *Asahi Chem. Industry.*

29856 Stylac® AS
9003-54-7
Styrene-acrylonitrile copolymer *Asahi Chem. Industry.*

29857 Styphen I
A mixture of stirenated phenols; a proprietary antioxidant. *Corning Glass Works, Zircoa Products.* Name unverified.

29858 styphnic acid
82-71-3 9026 201-436-6
$C_6H_3N_3O_8$
2,4,6-trinitro-1,3-benzenediol
2:4-dihydroxy-1,3,5-trinitrobenzene; trinitroresorcinol;. Used in explosives as a priming agent. mp = 175°; soluble in H_2O (6-12 g/l), soluble in organic solvents.

29859 Styquin
54400-62-3 1540
Butamisole hydrochloride
Anthelmintic. *Am. Cyanamid.* Name unverified.

29860 styracin
122-69-0 2364 204-566-1
$C_{18}H_{16}O_2$
3-phenyl-2-propen-1-yl 3-phenyl propenoate
Cinnamyl cinnamate; phenyl allyl cinnamate; 3-phenyl allyl cinnamate; cinnamyl β-phenyl acrylate. Used in chemical synthesis. (*trans,trans* form): mp = 44°; λ_m = 216, 223 nm (log ε 3.45, 3.25 95% EtOH); insoluble in H_2O, soluble in organic solvents.

29861 Styrafil®
A trade name for flame-retardant polystyrene. No manufacturer.

29862 Styraloy 22, 22A
A proprietary trade name for an elastomeric styrene derivative. No manufacturer.

29863 Styramic H.T. and M.T
Proprietary trade names for polystyrene thermoplastics possessing a higher softening point than usual; they are stated to be polydichlorstyrenes. No manufacturer.

29864 styrene-acrylonitrile copolymer
9003-54-7
(SAN; SAN copolymer; ACS) Injection molding and extrusion resin for cosmetic packaging, fan blades, toys and games, business machines, interior refrigerator parts, medical parts, beverage tumblers, food containers, tableware, dinnerware, containers, automotive parts, cassettes. *BASF; LNP; Monsanto; Reichhold/Emulsion Polymers.*

29865 Styresol 13-031
Styrene monomer-modified alkyd. *Reichhold.*

29866 Styrid
Styrene suppressant for fiberglass molding; reduces emissions and odors without affecting interlaminar bonding. *Specialty Prods.*

29867 Styrocell
Expanded and expandable polystyrene. *Shell UK.*

29868 Styrochrom®
Polystyrene masterbatches; antistatic finishing for consumer goods; improves surface appearance and scratch resistance of housings and technical parts. *BASF AG.*

29869 Styrodur®
9003-53-6 9028
Extruded rigid PS foam; contains flame retardant; used for thermal insulation of roofs, floors, walls and ceilings in domestic, industrial and farm buildings, as frost protection of subsoils below roads, aircraft runways etc. for thermal and low temperature insulation in refrigerated vehicles, containers. *BASF AG; BASF plc.*

29870 Styrofan®
Polymer dispersions based on styrene; gloss binders for paper and board coating; raw material for production of laminating adhesives; binders for bonding fiber webs. *BASF AG; BASF plc.*

29871 Styrofill®
Loose padding and filling material for cushioning and packaging uses. *BASF AG; BASF plc.*

29872 Styroflex
A proprietary trade name that is a flexible polymer of styrene. No manufacturer.

29873 Styrofoam Brand Insulation
9003-53-6 9028
Extruded rigid PS foam insulation board; available in residential sheathing, tongue and groove, score board, and square edge material grades in various thicknesses and sizes; provides high strength, high resistance to moisture and water vapor, long term R value to the building industry. *Dow Chemical.*

29874 styrogallol
$C_{16}H_8O_5$
Dihydroxyanthracoumarin
A yellow dyestuff.

29875 styrol
100-42-5 9028 202-851-5
C_8H_8
styrene
cinnamene, cinnamol; styrolene; styrene monomer; phenylethylene; styrol; ethenylbenzene; Annamene; vinyl benzene; cinnamenol; Diarex HF 77; phenethylene; Styron; Styropol; Styropor; vinylbenzol. Used in the manufacture of polystyrene, SBR, ABS, and SAN resins. mp = -31°; bp = 145-146°; d^{20} = 0.9059; n_D^{25}= 1.60.

29876 Styrolit®
For producing formed foam plastic parts, for fullmold-casting and for thin-walled formed parts. *BASF AG.*

29877 Styrolux®
9003-55-8 8534
S/B block copolymer; used for injection molding, extrusion, thermoforming, and blow molding, primarily for transparent, impact-resistant packaging material, domestic goods, toys, technical parts, medical applications. *BASF AG; BASR plc.*

29878 styrolyl alcohol
93-56-1 9029 202-258-1
$C_8H_{10}O_2$
1-phenyl-1,2-ethanediol
styryl alcohol; phenyl glycol; styrene glycol. Used in perfumery. Its esters used as plasticizers. mp = 67-68°; bp_{755} = 272-274°; soluble in H_2O, organic solvents.

29879 Styromol
Refractory coatings for expanded polystyrene patterns used in Replicast FM process. *Foseco (F.S.) Ltd.*

29880 Styron
General purpose and high impact polystyrene resins used in packaging, housewares, toys, medical and electronics. *Dow Cheml Co Ltd, UK & Ireland.*

29881 Styron 421
9003-53-6 9028
High impact polystyrene; offers excellent extrusion performance, excellent rigidity, good impact, good gloss, excellent heat resist. for extrusion, thermoforming, injection molding, and blow molding; end uses including dinnerware, cups, packaging, etc. FDA compliant. *Dow Plastics.*

29882 Styron 478
9003-53-6 9028
High impact polystyrene; offers enhanced gloss with a good balance of flow and toughness for injection and blow molding of toys, housewares, and small and large appliances; FDA compliant. *Dow Plastics.*

29883 Styron 479
9003-53-6 9028
High impact polystyrene; offers good gloss for injection and blow molding of toys and small battery cases; FDA compliance. *Dow Plastics.*

29884 Styron 697
9003-53-6 9028
Polystyrene; general purpose resin with high heat resist., excellent clarity; designed for highspeed, in-line extrusion thermoforming of clear, strong, lightweight food containers; FDA compliant. *Dow Plastics.*

29885 Styron 6075
9003-53-6 9028
Polystyrene; ignition-resistant grade; offers flame retardance, high heat resistance, high modulus, good processability for injection molding applications. *Dow Plastics.*

29886 Styron 6087 SF
9003-53-6 9028
Polystyrene structural foam resin; ignition-resistant grade; offers flame retardance, excellent moldability for structural foam applications. *Dow Plastics.*

29887 Styronal®
Polymer dispersions based on butadiene and styrene; binders for paper and board coating. *BASF AG.*

29888 styrone
104-54-1 2362 203-212-3
$C_9H_{10}O$
Cinnamyl alcohol
Used in perfumery and in deodorants.

29889 Styroplus®
9003-55-8 8534
S/B copolymer; for sealable packaging film and high impact strength lids for packing containers; highly resistant to thermoforming. *BASF AG.*

29890 Styropor
Expandable polystyrene. *Mitsubishi Petrochem.* Name unverified.

29891 Styropor® F
9003-53-6 9028
Expandable PS; difficultly flammable insulating material for thermal insulation, for structural insulating units and system solutions in building trade; for noise absorption; for blocks and boards as road foundations; for insulation of deep-freeze rooms, refrigerators; profiles, strips, boards for furnishing and decorative purposes. *BASF AG; BASF plc.*

29892 Styropor® FH
9003-53-6 9028
Expandable polystyrene; difficultly flammable foam for the production of packaging, pallets, boards, maritime articles; resistance to oils, fats, and petroleum fractions free from aromatic hydrocarbons. *BASF AG; BASF plc.*

29893 Styropor® P
9003-53-6 9028
Expandable polystyrene; load bearing, shock absorbent, and thermal insulation packaging for machinery, appliances, glass, porcelain, chemical, pharmaceuticals, food, etc.; toys, seating; moldings for full mold casting; aggregates for lightweight concrete; drainage boards for the agricultural and building trades. *BASF AG; BASF plc.*

29894 Styrothane 5329/5330
Aromatic urethane coating; rapid curing, solventless, sprayable coating meeting EPA emission levels for VOC; used for protection of EPS, Styrofoam, plywood, urethane and phenolic foam boardstocks. *Futura Coatings.* Name unverified.

29895 Styvex 22000 NA
9003-53-6 9028
Polystyrene, 20% glass-reinforced. *Ferro/Engineering Thermoplastics.*

29896 Styvex 32000 BK
9003-54-7
Styrene-acrylonitrile polymer, 20% glass-reinforced. *Ferro/Engineering Thermoplastics.*

29897 Styvex 40007 BKL2
9003-56-9
ABS, self-lubricated; high-impact grade. *Ferro/Engineering Thermoplastics.*

29898 Styvex 42023 NAFR
9003-56-9
ABS, 10% glass fiber-reinforced; flame retardant thermoplastic. *Ferro/Engineering Thermoplastics.*

29899 Styvex 72001 NA
25134-01-4
Modified PPO, 20% glass-reinforced. *Ferro/Engineering Thermoplastics.*

29900 Südflock
Inorganic precipitants, flocculants and adsorbents for the purification of industrial and domestic effluents. *Süd-Chemie AG.* Name unverified.

29901 Suakin gum
Talca gum; talka gum; Sennaar gum. A brittle variety of gum acacia from *Acacia fistula.* It gives a ropy mucilage.

29902 subacetate of lead
1335-32-6 5443 215-630-3
$C_4H_{10}O_8Pb_3$
Monobasic lead acetate
lead subacetate; basic lead acetate; bis(acetato)tetrahydroxytrilead; bis(acetato-O)tetrahydroxytrilead; bis(aceto)tetrahydroxytrilead; bis(aceto)dihydroxytrilead; BLA; monobasic lead acetate; subacetate lead. Used as a decolorizing agent. Soluble in H_2O (62-250 mg/ml).

29903 suberic acid
505-48-6 9031 208-010-9
$C_8H_{14}O_4$
octane-1,8-dioic acid
octanedioic acid; 1,6-hexanedicarboxylic acid. Used in plastics manufacture. mp = 142-144°; bp_{100} = 279°; soluble in H_2O (1.6 mg/ml), more soluble in EtOH, Et_2O, insoluble in $CHCl_3$. *BASF; Hüls AG; Penta Mfg.*

29904 Sublaprint
Uncut disperse dyestuffs; for the production of inks for printing heat transfer papers. *Holliday Dyes & Chemicals Ltd.*

29905 sublimed blue lead
A pigment produced by heating mixed ores of zinc and lead in a furnace with an air blast. It usually contains 50% lead sulfate, 20% lead oxide, PbO, 11% lead sulfide, PbS, 8% lead sulfite, and 3% zinc oxide.

29906 Sublimed Blue Lead
Basic blue lead sulfate; lubricating aid and friction aid for the manufacturing of brake linings and clutch facings; lubricant additive in oil and high-pressure greases; rust inhibitive pigment for structural steel. *Eagle-Picher.* Name unverified.

29907 sublimed white lead
A white lead manufactured from mixed ores of galena and zinc blende. They are roasted in the presence of an air blast, and the lead sulfate, lead oxide, and zinc oxide formed is collected in large chambers. The average composition is 75% lead sulfate; 20% lead oxide and 5% zinc oxide. Used as a pigment.

29908 sublimoform
A mercury-formaldehyde preparation; used as an antifungal seed preservative.

29909 subox
A protective coating paint consisting of a suspension of colloidal lead in linseed oil.

29910 Suburban Propane (and Design)
Liquefied petroleum gas. *Quantum Chemical Corp.*

29911 Sub-Vitralen
High molecular weight high density polyethylene sheet; used for orthopedic splints. *Stanley Smith & Co Plastics Ltd.*

29912 Sucaryl
Saccharin sodium
Sweetener. *Abbott Laboratories.* Name unverified.

29913 Sucaryl
A proprietary trade name for sodium cyclamate. No manufacturer.

29914 Succinellite
110-15-6 9038 203-740-4
succinic acid
Obtained from amber.

29915 succinic acid
110-15-6 9038 203-740-4
$C_4H_6O_4$
Dicarboxylic acid
butanedioic acid; amber acid; ethylene succinic acid. Organic synthesis; manufacture of lacquers, dyes, esters for perfumes, photography, in foods as a sequestrant, buffer, neutralizing agent. *Am. Biorganics; Du Pont; General Chem.; Hüls AG; Pentagon Chemicals Ltd; L30012Mallinckrodt.*

29916 succinic anhydride
108-30-5 9039 203-570-0
$C_4H_6O_4$
dihydro-2,5-furandione
succinyloxide; butanedioic anhydride; dihydro-2,5-diketotetrahydrofuran; succinic acid anhydride; tetrahydro-2,5-dioxofuran; 2,5-dioxotetrahydrofuran; succinyl peroxide; SAA. Manufacture of chemicals, pharmaceuticals, esters; hardner for resins, starch modifier in foods. mp = 119°; bp = 261°; d = 1.503; slightly soluble in H_2O, Et_2O, more soluble in organic solvents. *Hüls AG; Humphrey; Penta Mfg.; Schweizerhall.*

29917 succinimide
123-56-8 9040 204-635-6
$C_4H_5NO_2$
2,5-diketopyrrolidine
2,5-pyrrolidinedione; butanimide; Orotric;. Growth stimulants for plants, organic synthesis. mp = 125-127°; bp = 287-289°; d = 1.41; soluble in H_2O (0.3-1.4 g/ml), less soluble in organic solvents; LD_{50} (rat orl) = 14 g/kg. *Chemie Linz UK; Penta Mfg.; Schweizerhall.*

29918 succinite
Baltic amber which contains succinic acid. Also the name for a mineral, a lime-aluminum garnet.

29919 succinol
An oil obtained by the distillation of amber.

29920 Suchar
A decolorizing carbon used for sugar juices; prepared from waste sulfite-cellulose liquors.

29921 Sucker Plucker Concentrate
Fatty alcohol mixture; contact tobacco sucker control. *Draxel Chemical Company.* Unverified.

29922 Sucker Stuff
123-33-1 5745 204-619-9
Potassium salt of maleic hydrazide; systemic control of tobacco suckers. *Draxel chemical Company.* Unverified.

29923 Suconox-4®
101-91-7 4859 202-988-0
$C_{10}H_{13}NO_2$
N-butyryl-*p*-aminophenol
4'-hydroxybutyranilide. Processing aid (antioxidant) for thermoplastics. mp = 139-140°; soluble in H_2O, EtOH. *Zeeland.*

29924 Suconox-9®
N-Pelargonoyl-p-aminophenol
Processing aid for thermoplastics. *Zeeland.*

29925 Suconox-12®
N-Lauroyl-p-aminophenol
Processing aid for thermoplastics. *Zeeland.*

29926 Suconox-18®
N-Stearoyl-p-amino phenol
Processing aid and antioxidant in thermoplastics. *Zeeland.*

29927 Sucostrin
71-27-2 9044 200-747-4
Succinylcholine chloride
Blocking agent. *Bristol-Myers Squibb Co Inc.* Name unverified.

29928 sucramine
8463
$C_7H_8N_2O_3S$
saccharin ammonium
Lyons sugar; Daramin; Sucline. The ammonium salt of saccharin; used in France as a sweetening substance.

29929 sucrase
9001-57-4 5025 232-615-7
Invertase
invertin; Saccharase. Invertase, an enzyme which decomposes saccharose into glucose and fructose; used in the production of invert sugar for candy and syrups, and as an analytical reagent for sucrose.

29930 sucrate of hydrocarbonate of lime
sucro-carbonate of lime.

29931 Sucro Ester 7
25168-73-4 246-705-9
Saccharose distearate
Food emulsifier; oil-water emulsifier wetting agent, crystallization inhibitor for vegetable oils and fats; inhibitor of thermal denaturation of proteins. *Gattefosse SA.*

29932 Sucro Ester 15
25168-73-4 246-705-9
Saccharose palmitate; food emulsifier; oil-water emulsifier and wetting agent, crystallization inhibitor for vegetable oils and fats; inhibitor of thermal denaturation of proteins. *Gattefosse SA.*

29933 sucro-carbonate of lime
Sucrate of hydrocarbonate of lime
A complex compound of lime, calcium sucrate, and calcium carbonate formed in the production of sugar from the beet, when carbon dioxide gas is passed into a solution of sucrate of lime.

29934 sucrose
57-50-1 9051 200-334-9
$C_{12}H_{22}O_{11}$
β-D-fructofuranosyl-α-D-glucopyranoside
saccharose; sugar; cane sugar; beet sugar; table sugar. Disaccharide; sweetening agent in food, pharmaceuticals; in fermentation; as flavor, preservative, antioxidant (in form of invert sugar); granulation agent and excipient for tablets; in plastics and cellulose industry, rigid polyurethane foams and manufacture of ink. dec 160-186°; d_4^{25}= 1.587; $[\alpha]^{25}_D$ = 66.5°; soluble in H_2O (2 g/ml), less soluble in organic solvents. *Am. Biorganics; Mallinckrodt; Mendell; Pfanstiehl Labs; Vista.*

29935 **sucrose octa-acetate**
126-14-7 9052 204-772-1
$C_{28}H_{38}O_{19}$
D-(+)-sucrose octa-acetate
Used as an adhesive and for impregnating and insulating paper. mp = 82-85°; bp_1 = 260°; n_D = 1.4660; $[\alpha]_D^{20}$= 58° (c = 1 $CHCl_3$); SG = 1.28; soluble in H_2O (9 mg/ml), more soluble in organic solvents.

29936 **suction gum**
Suction powder. Powdered gum tragacanth.

29937 **Sudafed Nasal Spray**
2315-02-8 7100 219-015-0
A proprietary preparation of oxymetazoline hydrochloride; for the relief of nasal congestion. *The Wellcome Foundation Ltd.*

29938 **Sudan I**
842-07-9 212-668-2
$C_{16}H_{12}N_2O$
1-phenylazo-2-naphthol
C.I. solvent yellow 14; Scarlet B; Oil Orange;solvent yellow 14; C.I. 12055; Atul orange R; brasilazina oil orange; brilliant oil orange r; Calcogas M, orange NC, oil orange 7078, oil orange, oil orange Z-7078; Carminaph; Ceres orange R; Cerotinorange g; Dispersol yellow PP; dunkelgelb; enial orange I; fast oil orange; fettorange 4a, lg, r; grasal orange; grasan orange r; hidaco oil orange; lacquer orange vg; motiorange r; orange a l'huile; orange 3RA soluble in grease; orange resenole no. 3; orange r fat soluble; organol orange; orange soluble a l'huile; orient oil orange ps; petrol orange y; plastoresin orange f4a; pyronalorange; resoform orange g; resinol orange r; sansel orange g; Scharlach b; Silotras orange tr; Somalia orange i; Soudan i; spirit orange; spirit yellow i; stearix orange; Sudan j; Sudan orange r; Tertrogras orange sv; Toyo oil orange,. A dyestuff, used for coloring oils and varnishes. mp = 131-133°; insoluble in H_2O.

29939 **Sudan II**
3118-97-6 221-490-4
$C_{18}H_{16}N_2O$
1-(2,4-dimethylphenylazo)-2-naphthol
C.I. Solvent Orange 7; 1-[(2,4-dimethylphenyl)azo]-2-Naphthalenol; C.I. 12140; Solvent Orange 7. Used for coloring oils and varnishes. mp = 156-158°.

29940 **Sudan III**
85-86-9 9054 201-638-4
$C_{22}H_{16}N_4O$
1-[4-(phenylazo)phenylazo]-2-naphthol
Benzeneazo-benzeneazo-β-naphthol; Solvent red 23; Sudan Red III; Oil Scarlet; C.I. 26100; Sudan Red; D&C red No. 17; Tony Red. Used for coloring oils and varnishes. Approved by FDA for external use. mp = 199°; insoluble in H_2O, soluble in organic solvents.

29941 **Sudan IV**
85-83-6 8538 201-635-8
$C_{24}H_{20}N_4O$
1-[[2-methyl-4-[(2-methylphenyl)azo]phenyl]azo]-2-naphthalenol
o-tolueneazo-o-toluene-azo-β-naphthol; Oil Red; Scarlet Red; C.I. Solvent Red 24; C.I. 26105; Scarlet Red Scharlach; Solvent Red 24; Lipid crimson; Oil Red BB; Biebrich Scarlet Red; Fat Ponceau R. Used as a dyestuff to stain fats. mp = 199° (dec); insoluble in H_2O, soluble in organic solvents.

29942 **Sudan®**
Oil- and fat-soluble dyes; for marking mineral oil products, e.g., engine fuels, heating oil, lubricating greases; also for coloring shoe polishes, floor polishes and waxes, for producing smoke dyes. *BASF AG.*

29943 **Sudol®**
Vegetable soaps and xylenols; general purpose clear soluble disinfectant for hospital and animal health fields. *Coventry Chemicals Ltd.*

29944 **Sudranol**
Polyethylene wax emulsions. *Suddeutsche Emulsions GmbH.*

29945 **Sufatone SCS/B**
Concentrated cationic surfactant in paste form; softening and antistatic agent used for all fabrics, particularly synthetics including acrylics. *Standard Chemical Company.* Discontinued.

29946 **Sufatone SCS/CL**
Cationic surfactant in liquid form; general purpose mild softening and antistatic agent used for all fibers. *Standard Chemical Company.* Discontinued.

29947 **Sufatone SMC/L**
Cationic surfactant in liquid form; softening agent for most fibers particularly wool and acrylics. *Standard Chemical Company.*

29948 **Sufatone SMC/W**
Concentrated cationic surfactant in the form of a semi-liquid; mild softening agent for most fibers, particularly wool and chlorinated wool. *Standard Chemical Company.* Discontinued.

29949 **Suffa**
7704-34-9 9142 231-722-6
Sulfur
Fungicide for fruit and vegetable crops. *Draxel Chemical Company.* Unverified.

29950 **SufuSorb 8**
Activated carbon; for vapor phase applications. *Calgon Carbon.* Name unverified.

29951 **sugamo**
A Japanese seaweed suggested for use in papermaking.

29952 **sugar cane wax**
A wax obtained from the dried filter press cake from sugar mills by benzine extraction; the African cake contains 14-17% of wax, the Java cake 4%. The wax obtained is not a pure product, but is a mixture of wax and fatty material with 7% glycerol.

29953 **sugar charcoal**
Lampblack. Amorphous carbon.

29954 **sugar house black**
A bone black pigment. It is a by-product of the sugar mills.

29955 **sugar of lead**
301-04-2 5411 206-104-4
$C_4H_6O_4Pb$
Normal lead acetate, used as a mordant in dyeing and printing, and for the preparation of lead salts and paints.

29956 **Sugartab®**
57-50-1 9051 200-334-9
Sucrose (90-93%), invert sugar; inert base for directly compressible pharmaceutical tablets. *Penwest Pharmaceuticals Co.*

29957 **Suhler white copper**
A nickel silver alloy of 40% copper, 32% nickel, 25% zinc, 2.6% tin, and 0.6% cobalt.

29958 **Sulconazole**
61318-91-0 9062
$C_{18}H_{15}Cl_3N_2S$
1-[2-[[(4-Chlorophenyl)methyl]-thio]-2-(2,4-dichlorophenyl)ethyl]-1H-imidazole
Exelderm. Topical antifungal preparation containing sulconazole nitrate. MW=397.75; *ICI Chem & Polymers Ltd.*

29959 **Sulfacide**
Acid dyes. *ICI Chem & Polymers Ltd.* Discontinued.

29960 **Sulfads®**
120-54-7 204-406-0
$C_{12}H_{20}N_2S_6$
bis(pentamethylene)thiuram tetrasulfide
1,1'-(tetrathiodicarbonothioyl)-bis-piperidine. Essentially dipentamethylene thiuram hexasulfide; ultra accelerator for NR and synthetic rubbers; vulcanizing agent. *R. T. Vanderbilt Co Inc.* Discontinued.

29961 **Sulfa-Hitech® 0382**
624-92-0 210-871-0
$C_2H_6S_2$
dimethyldisulfide
2,3-dithiabutane; (methyldithio)methane; sulfa-hitech; DMDS. DMDS-based; solvent used to dissolve sulfur in the production of sour gas wells, in sour-gas pipelines, and in refinery and chemical plant flowlines. mp = -85°; bp = 109°; d = 1.0625; n_D^{20}= 1.5250. *Elf Atochem.*

29962 **sulfamethazine**
57-68-1 9083 200-346-4
$C_{12}H_{14}N_4O_2S$
4-Amino-N-(4,6-dimethyl-2-pyrimidinyl)benzenesulfonamide
N^1-(4,6-dimethyl-2-pyrimidinyl)sulfanilamide; N^1-(4,6-dimethyl-2-pyrimidinyl)sulfanilamide; N^1-(4,6-dimethyl-2-pyrimidinyl)sulfanilamide; N^1-(4,6-dimethyl-2-pyrimidyl)sulfanilamide; sulfamexathine; sulfadimerazine; sulfadimidine; sulfamidine; sulfadimethylpyrimidine; Diazil; Dimezathine; Mefenal; Sulphix; S-Mez; Sulfadine; S-Dimidine; Dimidim-R; Vertolan; Neazina; Pirmazin; Sulmet; Azolmetazin. Bacteriostatic agent. mp = 176°; λ_m = 241 nm ($E^{1\%}_{1cm}$ 670, H_2O, pH 6.6); soluble in H_2O (150 mg/100 ml); LD_{50} (mus ip)= 1.06 g/kg *Merck & Co Ltd.*

29963 **sulfamic acid**
5329-14-6 9090 226-218-8
H_3NO_3S
amidosulfonic acid
Amidosulfuric acid; Sulfamidic acid; Aminosulfonic acid. Metal/ceramic cleaning, nitrite removal in azo-dyeing, gas-liberating compositions, organic synthesis, analytical standard, chlorine stabilizer, bleaching paper pulp and textiles, catalyst for ureaformaldehyde resins, sulfonating agent, used for pH control and as a weedkiller. mp = 205° (dec); d = 2.15; soluble in H_2O (0.15 g/ml), less soluble in organic solvents; MLD (rat orl) = 1.6 g/kg. *General Chemical; PMC Specialties; Schweizerhall; Transol Chem. UK Ltd.*

29964 Sulfamin
Anionic surfactant in liquid form; for liquid cleaners. *Berol Kemi (UK) Ltd.* Name unverified.

29965 sulfammonium
A solution of sulfur in liquid ammonia to form a purple solution.

29966 sulfanilic acid
121-57-3 9096 204-482-5
$C_6H_7NO_3S$
p-aminobenzenesulfonic acid
p-anilinesulfonic acid. Dyestuffs, organic synthesis, medicine, reagent. mp = 288°; SG = 1.485; soluble in H_2O (1 g/100 ml), insoluble in organic solvents. *Am. Cyanamid; Penta Mfg.; 3-V; U.S. Biochemical.*

29967 Sulfanol
Sulfur dyes. *ICI Chem & Polymers Ltd.*

29968 sulfaquinoxaline
59-40-5 9109 200-423-2
$C_{14}H_{12}N_4O_2S$
4-amino-N,2-quinoxalinylbenzenesulfonamide
sulfabenzpyrazine; Compd 3-120; S.Q.; Sulquin. A coccidiostat, used in combination with diaveridine. mp = 247-248°; λ_m = 252, 360 nm ($E^{1\%}_{1cm}$ 1110, 275 H_2O pH 6.6); soluble in H_2O (7.5 mg/l), more soluble in organic solvents.

29969 Sulfarine
A mixture of magnesium sulfate with 15% sulfuric acid; used against potato scab.

29970 Sulfasan
103-34-4 3437 203-103-0
$C_8H_{16}N_2O_2S_2$
4,4·g-Dithiodimorpholine
4,4'-dithiobis[morpholine]; morpholine, N,N'-disulfide; dimorpholine N,N'-disulfide; Sulfasan R. Vulcanizing agent for natural and synthetic rubbers. Also used as a fungicide. mp = 124-125°; *Monsanto Co.*

29971 sulfate pulp
Wood pulp obtained by the treatment of wood with alkali liquors containing sodium sulfate.

29972 sulfated butyl tallate
42808-36-6 255-950-0
Sulfated butyl tallate
Laurel SBT. Industrial lubricant, emulsifier in textile formulations, rewetting agent for corrugated medium. *Reilly-Whiteman.*

29973 Sulfatine
A mixture of 73% sulfur, 20% lime, and 7% copper sulfate; a fungicide used against black rot.

29974 Sulfato de Cobre Valles
7758-98-7 2722 231-847-6
Crystallized copper sulfate
For manufacture of agricultural fungicides and many industrial products. *Industrias Quimicas Del Valles SA.*

29975 Sulfatol CL
Sodium sulfated fatty alcohol in paste form; anionic surfactant which is stable to hard water and disperses lime soaps; used for various industrial applications including textiles and leather. *Standard Chemical Company.*

29976 Sulfatol E3
Sodium lauryl ether sulfate (coconut/palm kernel C12-C14 alcohol) as a clear almost colorless low viscosity liquid; for production of all types of liquid and lotion shampoos and bubble baths; also a raw material for light duty liquid detergents, dishwashing detergents and auto shampoos. *Efkay Chemicals Ltd.* Name unverified.

29977 Sulfatol LS3
Combined sulfated fatty alcohol and nonionic surfactant in liquid form; detergent and washing off liquid for textiles etc. *Standard Chemical Company.*

29978 Sulfatol LX/B
Sodium sulfated fatty alcohol in paste form; detergent, wetting, leveling and softening agent for industrial applications including textiles and leather. *Standard Chemical Company.*

29979 Sulfatol PD/B
Potassium sulfated fatty alcohol in paste form; detergent with good water solubility, for industrial applications including textiles and leather. *Standard Chemical Company.*

29980 Sulfatol TL/B
Anionic surfactant with a blend of cations and a sulfated fatty alcohol as anion; detergent for industrial applications including textiles and leather. *Standard Chemical Company.*

29981 Sulfetal 4105
126-92-1 204-812-8
Sodium isooctyl sulfate
Low-foaming detergent, wetting agent in acid and alkaline media; for industrial and metal cleaners. *Zschimmer & Schwarz.*

29982 Sulfetal C 38
151-21-3 8782 205-788-1
Sodium lauryl sulfate
Light duty detergent, dishwashing agent, cold wash agent. *Zschimmer & Schwarz.* Discontinued.

29983 Sulfetal CJOT 38
21142-28-9 244-238-5
MIPA-lauryl sulfate
Basic material for hair shampoos, bath additives, cosmetics, liquid detergents. *Zschimmer & Schwarz.* Discontinued.

29984 Sulfetal FA 40
Sodium isooctyl sulfate
Modified; low-foaming detergent and wetting agent; used in industrial cleaners, metal and tank cleaning; highly resistant to alkalies. *Zschimmer & Schwarz.* Discontinued.

29985 Sulfetal KT 400
139-96-8 205-388-7
TEA-lauryl sulfate
For detergents, cosmetics, hair shampoos. *Zschimmer & Schwarz.*

29986 Sulfetal MG 30
Magnesium lauryl sulfate
For detergents, cosmetics. *Zschimmer & Schwarz.* Discontinued.

29987 Sulfetal TC 50
Sodium tallow sulfate and sodium lauryl sulfate; detergent; washing and cleaning agents; hand cleaning pastes. *Zschimmer & Schwarz.* Discontinued.

29988 sulfide dyestuffs
A class of dyestuffs prepared by the fusion of organic amines and other substances with sulfur and sodium sulfide. They are used for cotton dyeing, and are usually fixed by oxidizing agents.

29989 Sulfiformin
Formaldehyde-sulfurous acid
An antiseptic; a 1% solution has been used for spraying vines.

29990 Sulfil®
A registered trade name for flame retardant polysulfone. No manufacturer.

29991 sulfite carbon
7440-44-0 1855 231-153-3
A decolorizing carbon used for sugar juices. It is prepared from sulfite-cellulose liquors.

29992 sulfite pulp
Wood pulp obtained by means of calcium bisulfide. It is made by digesting the disintegrated wood under pressure with the calcium bisulfite, which gives a mass of cellulose fibers free from lignocellulose amounting to about 45% of the wood.

29993 sulfite turpentine
A by-product obtained from the pulping of spruce by the sulfite process. The main constituent is *p*-cymene.

29994 sulfite turpentine oil
Cellulose turpentine oil. A by-product obtained in the manufacture of cellulose. When decolorized it resembles turpentine oil.

29995 Sulfochem 436
9051-57-4
Ammonium nonoxynol-4 sulfate
High-foaming detergent for dishwashing, carwash, and carpet shampoo formulations; emulsion polymerization surfactant for SBR, acrylic, vinyl acrylic systems. *Chemron.*

29996 Sulfochem ALS
2235-54-3 218-793-9
Ammonium lauryl sulfate
Surfactant, mild detergent for use in low pH systems; foaming and suspending agent; for personal care and industrial applications. *Chemron.*

29997 Sulfochem DLS
DEA-lauryl sulfate
Surfactant foamer with soap-like characteristics; for shampoos, bubble baths, skin cleaners, syndet bars, shower gels. *Chemron.*

29998 Sulfochem EA-1
67762-19-0
Ammonium laureth sulfate
Surfactant for low pH shampoo systems; mildness, high flash foam, low cloud point, viscous response; for shampoos, cleansers, bath products, gels. *Chemron.*

29999 Sulfochem EA-2, EA-3
67762-19-0
Ammonium laureth sulfate
Surfactant for low pH shampoo systems; mildness, high flash foam, low cloud point, viscous response. *Chemron.*

30000 Sulfochem EA-60
67762-19-0
Ammonium laureth sulfate
Surfactant for shampoo systems, bubble baths, cleansers. *Chemron.*

30001 Sulfochem EA-70
67762-19-0
Ammonium laureth sulfate (2 mole); surfactant for toiletries and cosmetics. *Chemron.*

30002 Sulfochem ES-1
Sodium laureth-1 sulfate
Surfactant for personal care cleansers. *Chemron.*

30003 Sulfochem ES-2
9004-82-4
Sodium laureth-2 sulfate
Flash foamer and detergent for personal cleansing products, specialty cleaning products. *Chemron.*

30004 Sulfochem ES-3
Sodium laureth-3 sulfate
Surfactant for personal cleansing products *Chemron.*

30005 Sulfochem ES-60
Sodium laureth-3 sulfate
Surfactant for shampoo systems, bubble baths, cleansers. *Chemron.*

30006 Sulfochem ES-70
9004-82-4
Sodium laureth-2 sulfate
Surfactant for toiletries, cosmetics, specialty industrial compounds. *Chemron.*

30007 Sulfochem K
4706-78-9 225-190-4
Potassium lauryl sulfate
Detergent, foamer. *Chemron.*

30008 Sulfochem MG
Magnesium lauryl sulfate
Mild detergent, foamer for bubblebath, shampoos, cleansing preparations, carpet shampoos. *Chemron.*

30009 Sulfochem MLS
MEA-lauryl sulfate
Surfactant foamer with soap-like characteristics; for shampoos, bubble baths, skin cleaners, shower gels, syndet bars. *Chemron.*

30010 Sulfochem SAC
Sodium lauryl sulfate
Foamer for lotion and paste shampoos; detergent base for pearlescent shampoos, bubble baths, shower gels, cleansers. *Chemron.*

30011 Sulfochem SLC
Sodium lauryl sulfate
Surfactant for clear shampoos, bath products, cleaners. *Chemron.*

30012 Sulfochem SLN
151-21-3 8872 205-788-1
Sodium lauryl sulfate
Foamer, dispersant, wetting agent, detergent for dry blends used in cleaning compounds, carpet shampoos, shampoo concentrates, bubble baths, cosmetic cleansers. *Chemron.*

30013 Sulfochem SLP-95
151-21-3 8872 205-788-1
Sodium lauryl sulfate
Foamer, dispersant, wetting agent, detergent for high-act., cleaning concentrates, dentifrices, high purity cleansers. *Chemron.*

30014 Sulfochem SLS
151-21-3 8872 205-788-1
Sodium lauryl sulfate
Detergent, foamer, suspending agent for hard surface cleaners, carpet shampoos, upholstery cleaners, spot removers, personal care products. *Chemron.*

30015 Sulfochem SLX
151-21-3 8872 205-788-1
Sodium lauryl sulfate
Detergent, foaming and suspending agent for rug and upholstery shampoos, spot removers. *Chemron.*

30016 Sulfochem TLS
139-96-8 205-388-7
TEA-lauryl sulfate
Surfactant foamer, wetting agent with soap-like characteristics; for liquid soaps and shampoos, industrial applications; good tolerance to hard water. *Chemron.*

30017 sulfoform
$C_{18}H_{15}SbS$
triphenylstibine sulfide

30018 sulfogenol
A crude mineral oil obtained from bituminous shale, is saturated with sulfur, and sulfonated. It is the ammonium salt of the sulfonated product and has similar properties to ichthyol.

30019 Sulfokyl DAS
Alkylbenzene sulfonate
Anionic; yellowish liquid; washing agent for all fibers, particularly for wool; gives excellent washing effects at low temperatures. *Thor Chemicals (UK) Ltd.* Discontinued.

30020 Sul-fon-ate AA-10
25155-30-0 8757 246-680-4
Sodium dodecylbenzene sulfonate
Wetting agent; air pollution control; cement, food, commercial laundry and industrial industries; cosmetics; fertilizers, insecticides; leather, paper, petroleum, and rubber processing; metal cleaning, electroplating, etching, pickling, mining. *Boliden Intertrade.*

30021 Sul-fon-ate OA-5R
67998-94-1 268-062-3
Sulfonated oleic acid
Sodium salt; surfactant. *Boliden Intertrade.*

30022 sulfonic acid
Organic compound containing one or more sulfo radicals; dispersant, wetting agent, leveling agent, protective colloid for dyestuffs. No manufacturer.

30023 Sulfopon 101, 101 Special
151-21-3 8872 205-788-1
Sodium lauryl sulfate
Surfactant for personal care products and light duty detergents. *Pulcra SA.*

30024 Sulfopon 101/POL
151-21-3 8872 205-788-1
Sodium lauryl sulfate
Surfactant for emulsion polymerization. *Pulcra SA.*

30025 Sulfopon LS
Sodium lauryl sulfate (C12-C18) in liquid/paste form; for cream shampoos and bubble baths. *Henkel Chemicals Ltd.* Name unverified.

30026 Sulfopon O 680
1847-55-8 217-430-1
Sodium oleyl sulfate
Low foaming surfactant. *Henkel/Functional Prods.*

30027 Sulfopon P-40
151-21-3 8872 205-788-1
Sodium lauryl sulfate
Dispersant and emulsifier for acrylates, styrene acrylic, vinyl chloride, vinyl acetate copolymers; also for cream shampoos, specialty cleaners, rug shampoos. *Pulcra SA.*

30028 Sulfosept oil
The next higher fraction to thiosept oil (*qv*); used as an insecticide.

30029 Sulfosoft
Anionic surfactant in acid form; for detergent production. *Berol Kemi (UK) Ltd.* Name unverified.

30030 Sulfotex 110
Sodium n-decyl sulfate
Wetting agent and emulsifier used in cleaning formulations, rug shampoos; used in sealing food containers; effective in cold and hard water; biodegradable. *Henkel/Cospha; Henkel Canada.*

30031 Sulfotex LAS-90
Sodium dodecylbenzene sulfonate
Surfactant for all-purpose cleaners, laundry products, hard surface cleaners; penetrates and removes grease. *Henkel/Cospha; Henkel Canada.*

30032 Sulfotex LCX
151-21-3 8872 205-788-1
Sodium lauryl sulfate
Detergent, foaming agent, wetting agent, and emulsifier; primary surfactant for rug and upholstery shampoos, hard surface cleaners; biodegradable. *Henkel/Cospha; Henkel Canada.*

30033 Sulfotex LMS-E
9004-82-4
Sodium lauryl ether sulfate
Detergent, wetting agent, foamer for industrial applications, liquid dishwashing detergents; biodegradable. *Henkel/Cospha; Henkel Canada.*

30034 Sulfotex OA
Sodium octyl sulfate
Foaming agent, wetting agent for animal glue on paper and paperboard, high electrolyte concentrates, and food processing; detergent and dispersant in industrial cleaners; mercerizing agent in dyeing of fibers; biodegradable. *Henkel/Cospha; Henkel Canada.*

30035 Sulfotex OT
67762-19-0
Ammonium lauryl ether sulfate

Detergent, emulsifier, foaming agent used in household and industrial detergents, car shampoos; biodegradable. *Henkel/Cospha; Henkel Canada.*

30036 Sulfotex SXS-40
1300-72-7 215-090-9
Sodium xylene sulfonate
Hydrotrope, solubilizer for liquid detergents, pine oil in water, inks to prevent gumming. *Henkel/Cospha; Henkel Canada.*

30037 Sulfotex T-65
TEA-dodecylbenzene sulfonate
Henkel/Cospha.

30038 sulfourea
62-56-6 9505 200-543-5
CH_4N_2S
Thiourea
Used as a chemical feedstock and also as a photographic fixer.

30039 Sulframin
Surfactant for cosmetics, toiletries, pharmaceutical, processing, agriculture, and other industries. *Baxenden Chemicals Ltd.*

30040 Sulframin 14-16 AOS
Sodium olefin sulfonate (C14-C16) as a clear amber liquid; foaming agent and detergent for cosmetic and household uses. *Witco Chemical Ltd.* Discontinued.

30041 Sulframin 33
Sodium olefin sulfonate and sodium ether sulfate as a clear amber liquid; foaming agent and detergent for industrial and household liquid detergents. *Witco Chemical Ltd.* Discontinued.

30042 Sulframin 1250
Sodium alkylaryl sulfonate in liquid form; anionic surfactant. *Witco Chemical Ltd.* Discontinued.

30043 sulfur
7704-34-9 9150 231-722-6
S
sulfur
Brimstone. Nonmetallic element; S; sulfuric acid manufacture, petroleum refining, dyes and chemicals, fungicide, insecticides, explosives, detergents, rubber vulcanization. *Norsk Hydro A/S; Shell; Texaco BV.*

30044 sulfur dioxide
7446-09-5 9144 231-195-2
SO_2
sulfurous oxide
sulfurous anhydride. Chemicals, sulfite paper pulp, ore and metal refining, soybean protein, intermediates, solvent extraction, bleaching oils and starch, sulfonation of oils, disinfectant, fumigant, food additive, reducing agent, antioxidant. mp = -72°; bp = -10°; soluble in H_2O (10 g/100 ml), more soluble in organic solvents. *Air Prods & Chem; Boliden Intertrade; Hoechst Celanese; Outokumpu Oy; Rhône-Poulenc Basic.*

30045 sulfur gold
1315-04-4 738 215-255-5
$S_{10}Sb_4$
Antimony pentasulfide
antimonic sulfide; antimonial saffron; antimony red. Used for vulcanizing and imparting a red color to rubber. Also used in fireworks and matches. Insoluble in H_2O; LD_{50} (rat ip) = 150 mg/kg.

30046 sulfur hexafluoride
2551-62-4 9146 219-854-2
F_6S
Used in electrical circuit breakers and in electronic ultra-high frequency piping. mp = -51°; soluble in oil.

30047 sulfur hypochlorite
A mixture of sulfur and sulfur chloride; used in rubber vulcanizing.

30048 sulfur olive oils
A name for the oil dissolved out of residual olive oil cake by means of carbon disulfide. It is also called sulfocarbon oil. It is rich in stearin.

30049 sulfur soap
Usually a yellow medicated soap to which has been added about 10% powdered sulfur.

30050 sulfur waste
The residue from the distillation of iron pyrites.

30051 sulfurated antimony
antimony crocus, saffron of antimony. A mixture of antimony pentasulfide, Sb_2S_5, with a little oxide, Sb_4O_6, and some free sulfur; formerly used in making tartar emetic.

30052 sulfurated potash
Liver of sulfur. A mixture of sulfides, mainly $K_2S_2O_3$ and K_4S_3 obtained by heating potassium carbonate with half its weight of sulfur. When fresh and carefully prepared it is the color of liver, and was called liver of sulfur and is used in the leather industry as a depilatory.

30053 sulfuretted hydrogen
7783-06-4 4843 231-977-3
H_2S
Hydrogen monosulfide
hepatic acid; Stink Damp; sulfur hydroxide; hydrosulfuric acid; sulfur hydride; Sewer gas; Sour gas. Used as a chemiucal feedstock and as an analytical reagent. mp = -85°; bp = -60°; soluble in H_2O (4.1 g/l), similarly soluble in organic solvents; LC_{50} (rat inh, 4 hr) = 444 ppm.

30054 Sulfur-F
7704-34-9 9150 231-722-6
S
Sulfur
Micronized flowable dispersion of elemental sulfur. Used in chemical manufacturing and in medicine as a scabicide. *W A Cleary.*

30055 sulfuric acid
7664-93-9 9147 231-639-5
H_2SO_4
hydrogen sulfate
battery acid; electrolyte acid; Inorganic acid. Fertilizers, chemicals, dyes and pigments, laboratory reagents, electroplating baths. Akzo; Amax; Am. Cyanamid; Boliden Intertrade; Du Pont; Metallgesellschaft AG; Nissan Chem. Ind.; OxyChem; Pasminco Europe; Rasa Ind.; Rhône-Poulenc Basic.

30056 sulfuric ether
Phosphoric ether; diethyl ether. Different names for diethyl ether.

30057 sulfurite
A name applied to a sulfur from Java, which contains 29% arsenic.

30058 Sulmet
Industrial detergent. *Crosfield Chemicals Ltd.* Discontinued.

30059 Sul-Perm® 10
Sulfur bases; replacement for sulfurized sperm oil products; lubricant for gear oils, slideway oils, cutting oils. *Ferro/Keil.*

30060 Sul-Perm® 110
Sulfur bases; replacement for sulfurized sperm oil products; used in industrial oils including slideway oils, industrial gear lubricants; compatible with heavy metal salts used as antioxidants and corrosion inhibitors. *Ferro/Keil.*

30061 Sul-Perm® C
Sulfur base; replacement for sulfurized sperm oil products; used in grease applications; imparts lubricity and film strength to EP greases. *Ferro/Keil.*

30062 Sulphatol 33
Sodium lauryl sulfate
Derived from the C12-C14 fraction of coconut/palm kernel fatty alcohol; clear liquid or thin paste; raw material for liquid cream and egg shampoos; emulsifier for cosmetic products. *Efkay Chemicals Ltd.* Name unverified.

30063 Sulphatol 33 MO
Monoethanolamine lauryl sulfate
Derived from the C12-C14 fraction of coconut/palm kernel fatty alcohol; clear pale yellow liquid; raw material for clear, oil and other liquid shampoos. *Efkay Chemicals Ltd.* Name unverified.

30064 Sulphatol B6
Ammonium/triethanolamine lauryl sulfate
Derived from the C12-C14 fraction of coconut/palm kernel fatty alcohol; clear, golden yellow viscous liquid; raw material for all types of liquid shampoos and bubble baths. *Efkay Chemicals Ltd.* Name unverified.

30065 Sulphol
Sulfur dyestuffs. *James Robinson & Co Ltd.*

30066 Sulphol High Fast
Sulfur and sulfurized vat dyes; liquid dispersions; used to dye cellulosic fibers alone and in blends with polyester; compatible with disperse and vat dyes. *James Robinson & Co Ltd.*

30067 Sulphonic Acid LS
Alkylbenzene sulfonic acid
Base for wide variety of detergents; can be neutralized with metallic bases or amines. *Hart Chem. Ltd.*

30068 Sulphonol
A range of acid dyes; for dyeing of wool and similar fibers. No manufacturer.

30069 Sulphophone
A trademark for a mixture of zinc sulfide and calcium sulfate; it is an analogous product to lithopone, a white pigment used in paints. No manufacturer.

30070 Sulphosol
Solubilized sulfur dyestuffs. *James Robinson & Co Ltd.*

30071 Sulphramin B and TPB
Alkylaryl sulfonic acid in liquid form; emulsifier for powder and liquid detergents. *Witco Chemical Ltd.* Discontinued.

30072 Sulsol
A proprietary trade name for a colloidal sulfur preparation for horticultural purposes. No manufacturer.

30073 Sulveol DC
Alkyl polyglycol ether; nonionic; yellowish pourable paste; rapid wetting agent, fully effective with excellent emulsifying and dispersing properties in cold and hot liquors. *Thor Chemicals (UK) Ltd.* Discontinued.

30074 sumacel
A diatomaceous earth containing 80% SiO_2, 5-3% Fe_2O_3 and Al_2O_3, 2.02% CaO, and 8.16% H_2O; used as a filtering medium for sugars.

30075 sumach
Sumac. The dried and finely powdered leaves and shoots of species of *Rhus* ; used for tanning leather, also for dyeing and printing.

30076 sumalban
Alban obtained from sumatra gutta-percha.

30077 sumaphos
A mixture of diatomaceous earth and acid phosphate, containing 36.22% P_2O_5.

30078 Sumet Processed lead
An alloy of 70-80% copper and 15-30% lead.

30079 Sumibond PA
Phenolic resin; for adhesives. *Sumitomo Bakelite Co Ltd.* Name unverified.

30080 Sumicidin
51630-58-1 4051 257-326-3
Emulsifiable concentrate of 100 g fenvalerate per liter; a pyrethroid insecticide. *Shell UK.*

30081 Sumicool
EVA resin; for mats. *Sumitomo Bakelite Co Ltd.* Name unverified.

30082 Sumiflex
Polyvinyl chloride family; molding compounds. *Sumitomo Bakelite Co Ltd.* Name unverified.

30083 Sumikon
A proprietary range of phenolic molding materials. *Sumitomo Bakelite Co Ltd.* Name unverified.

30084 Sumikon AM
Diallyl phthalate resin; molding compounds. *Sumitomo Bakelite Co Ltd.* Name unverified.

30085 Sumikon EM, EME
Epoxy resin; molding compounds. *Sumitomo Bakelite Co Ltd.* Name unverified.

30086 Sumikon IM
Polyimide resin; molding compounds. *Sumitomo Bakelite Co Ltd.* Name unverified.

30087 Sumikon PM
Phenolic resin; molding compounds. *Sumitomo Bakelite Co Ltd.* Name unverified.

30088 Sumikon TM
Polyester resin; molding compounds. *Sumitomo Bakelite Co Ltd.* Name unverified.

30089 Sumikon VM
Polyvinyl chloride resin; molding compounds. *Sumitomo Bakelite Co Ltd.* Name unverified.

30090 Sumilac PC
Phenolic resin; industrial resins. *Sumitomo Bakelite Co Ltd.* Name unverified.

30091 Sumilite CEL
Composite resin; for sheets. *Sumitomo Bakelite Co Ltd.* Name unverified.

30092 Sumilite EI
Epoxy resin; for laminate materials. *Sumitomo Bakelite Co Ltd.* Name unverified.

30093 Sumilite EL
Epoxy resin; for laminated sheets. *Sumitomo Bakelite Co Ltd.* Name unverified.

30094 Sumilite ELC
Epoxy resin; copper clad laminates. *Sumitomo Bakelite Co Ltd.* Name unverified.

30095 Sumilite FS
PES, PEI resins; for sheets. *Sumitomo Bakelite Co Ltd.* Name unverified.

30096 Sumilite IL
Polyimide resin; for laminated sheets. *Sumitomo Bakelite Co Ltd.* Name unverified.

30097 Sumilite ILC
Polyimide resin; for copper clad laminates. *Sumitomo Bakelite Co Ltd.* Name unverified.

30098 Sumilite ILI
Polyimide resin; for laminate materials. *Sumitomo Bakelite Co Ltd.* Name unverified.

30099 Sumilite NS
Polypropylene resin; for sheets. *Sumitomo Bakelite Co Ltd.* Name unverified.

30100 Sumilite PL
Phenolic resin; for laminated sheets. *Sumitomo Bakelite Co Ltd.* Name unverified.

30101 Sumilite PLC
Phenolic resin; for copper clad laminates. *Sumitomo Bakelite Co Ltd.* Name unverified.

30102 Sumilite Resin PR
Resorcinol resin; for adhesives. *Sumitomo Bakelite Co Ltd.* Name unverified.

30103 Sumilite Resin PR
Phenolic resin; industrial resins. *Sumitomo Bakelite Co Ltd.* Name unverified.

30104 Sumilite Resin PR
Epoxy resin; industrial resins. *Sumitomo Bakelite Co Ltd.* Name unverified.

30105 Sumilite STS
Polystyrene resin; for sheets. *Sumitomo Bakelite Co Ltd.* Name unverified.

30106 Sumilite TFC
Polyimide and other resins; for copper clad laminates. *Sumitomo Bakelite Co Ltd.* Name unverified.

30107 Sumilite TFP
Polyimide and other resins; for flexible printed circuit boards. *Sumitomo Bakelite Co Ltd.* Name unverified.

30108 Sumilite VSL
Polyvinyl chloride and metal foil; for sheets. *Sumitomo Bakelite Co Ltd.* Name unverified.

30109 Sumilite VSS
Polyvinyl chloride resin; for sheets. *Sumitomo Bakelite Co Ltd.* Name unverified.

30110 Sumine® 2005
100-46-9 1160 202-854-1
N-Benzylamine
Zeeland.

30111 Sumine® 2015
Tertiary catalyst; epoxy curative. *Zeeland.*

30112 Suminet
Polyethylene resin; materials for land improvements. *Sumitomo Bakelite Co Ltd.* Name unverified.

30113 Sumipipe
Polyethylene resin; materials for land improvement. *Sumitomo Bakelite Co Ltd.* Name unverified.

30114 Sumitac EA
Epoxy resin; for adhesives. *Sumitomo Bakelite Co Ltd.* Name unverified.

30115 Sumitac GA
Polyurethane resin; for adhesives. *Sumitomo Bakelite Co Ltd.* Name unverified.

30116 Sumitac VA
Polyvinyl chloride resin; for adhesives. *Sumitomo Bakelite Co Ltd.* Name unverified.

30117 Sumithrin
26002-80-2 7405 247-404-5
Emulsifiable concentrate containing 103 g/1 phenothrin; a pyrethroid insecticide. *Sumito Chemical (UK) plc.*

30118 Summit
55219-65-3 9724 259-537-6
Triadimenol
Fungicide used to control powdery mildew, rusts, and rhychosporium in winter and spring crops of wheat, barley, oats, and rye. *Bayer AG.*

30119 Summit
Suspension concentrate containing 150 g terbutryn and 200 g trifluralin per liter; for weed control in winter cereals. *Ashlade Formulations Ltd.*

30120 Sumquat® 2355
56-37-1 200-270-1
$C_{13}H_{22}ClN$
Benzyl triethyl ammonium chloride
TEAC; N,N,N'-triethylbenzenemethanaminium chloride. Fibre dyeing auxilliary, phase transfer catalyst mp = 185° (dec). *Hexcel.*

30121 Sumquat® 6020
124-03-8 2071 204-672-8
$C_{20}H_{44}BrN$
cetyldimethylethylammonium bromide
Cetethyldimonium bromide; ethylhexadecyldimethylammonium bromide; ethyl cetab; CTA; Ammonyx DME; Bretol. Used as a disinfectant, laboratory reagent and antiseptic in detergents. mp = 178-186° (dec); soluble in H_2O, EtOH, less soluble in organic solvents; LD_{50} (rat orl) = 500 mg/kg. *Hexcel.*

30122 Sumquat® 6030
57-09-0 2068 200-311-3
$C_{19}H_{42}BrN$
N,N,N-trimethyl-1-hexadecanaminium bromide
Cetrimonium bromide; hexadecyltrimethylammonium bromide; Bromat; Cetab; Cetavlon; Cetylamine; c.T.A.B.; Lissolamine V; Micol; Quamonium. Cationic detergent and antiseptic. Also used as a laboratory reagent. mp = 237-243°; soluble in H_2O (10 g/100 ml), EtOH, less soluble in organic solvents; LD_{50} (rat iv) = 44 mg/kg. *Hexcel.*

30123 Sumquat® 6045
107-64-2 203-509-2
$C_{38}H_{80}ClN$
distearyldimethylammonium chloride
Di(hardened tallow)dimethylammonium chloride; Aerosurf TA-100; dimethyl distearyl ammonium chloride; dimethyl-n-octadecyl-1-octadecanaminium chloride; dimethyldioctadecylammonium chloride; Distearyldimonium chloride. Cationic detergent and antiseptic. Also used as a laboratory reagent. *Hexcel.*

30124 Sumquat® 6050
122-18-9 2059 204-526-3
$C_{25}H_{46}ClN$
N-hexadecyl-N,N-dimethylbenzenemethanaminium chloride
Cetalkonium chloride; Banicol; Acetoquat CDAC; Acquat CDAC; Ammonyx G; Zettyn; Ammonyx T; Cetol. Cationic quaternary ammonium surfactant germicide and fungicide. Topical anti-infective. mp = 59°; soluble in H_2O, organic solvents. *Hexcel.*

30125 Sumquat® 6110
1119-97-7 6419 214-291-9
$C_{17}H_{38}BrN$
N,N,N-trimethyl-1-tetradecanaminium bromide
myrtrimonium bromide; myristyltrimethylammonium bromide; tetradecyltrimethylammonium bromide; trimethyltetradecylammonium bromide; tetradecanaminium, N,N,N-trimethyl-, bromide; tetradonium bromide. Germicidal detergent. mp = 245-250°. *Hexcel.*

30126 Sumquat® 6210
122-19-0 204-527-9
$C_{27}H_{50}ClN$
N,N-dimethyl-N-octadecylbenzenemethanaminium chloride
Stearalkonium chloride; Ammonyx 4; Ammonyx 485; Ammonyx 490; Ammonyx 4002; Ammonyx CA Special; 2b; Barquat Sb-25; Carsoquat SDQ-25; Carsoquat SDQ-85; Intexan SB-85; Intexsan SB-85; J Soft C 4; Katamine AB; Orthosan MB; Quaternol 1; Stebac; Stedbac; Triton X-40; Triton X-400; Varisoft SDC; Arquad DM18b-90; Dehyquart STC-25; Nissan Cation S2-100. Germicidal detergent. Insoluble in H_2O, soluble in organic solvents; LD_{50} (rat orl) = 1250 mg/kg. *Hexcel.*

30127 sun bronze
An alloy of from 40-60% copper, 30-40% tin, and 10% aluminum; used in jeweler's work. The name is also used for an alloy of from 50-60% cobalt, 30-40% copper, and 10% aluminum.

30128 Sun Wrap®
Polyvinylidene chloride wrapping film. *Asahi Chem. Industry.*

30129 Sunaptic Acids
High molecular weight naphthenic acids; corrosion inhibitor, oil well drilling mud formulations, emulsifiers, foundry binders. *Sun Refining & Marketing Co.* Unverified.

30130 Sunaptol
Textile auxiliary chemicals. *ICI Chem & Polymers Ltd.*

30131 Sunaptol NP55
Nonionic surfactant of the nonylphenol ethoxylate type in liquid form; emulsifier and intermediate used in natural waxes, mineral oils; ethoxysulfate manufacture; emulsifiable solvents for metal degreasing; general cleaning applications. No manufacturer.

30132 Sunaptol NP65, NP70
Nonionic surfactants of the nonylphenol ethoxylate type in liquid form; low temperature wool scouring; emulsifier for silicone and mineral oils; emulsifier and stabilizer for kerosine based hand cleaning gels. No manufacturer.

30133 Sunaptol NP80, NP95
Nonylphenol ethoxylate nonionic surfactant in liquid form; detergency, wetting and emulsifying agent used in wool scouring; metal cleansing; mineral oils, usually in combination with a hydrophobic surfactant. No manufacturer.

30134 Sunaptol NP100
Nonylphenol ethoxylate nonionic surfactant in liquid form; for solubilization of essential oils and perfumes. No manufacturer.

30135 Sunaptol NP140
Nonylphenol ethoxylate nonionic surfactant in the form of a white waxy solid; iodine complexing for iodofor sterilizers; production of self emulsifiable oleines. No manufacturer.

30136 Sunaptol NP350
Nonylphenol ethoxylate nonionic surfactant in the form of a white solid; ready molding into solid block for detergent blocks and tablets; emulsifier and stabilizer for vinyl acetate and acrylic polymer emulsions. No manufacturer.

30137 Sunbrite
Coke. *ICI Chem & Polymers Ltd.*

30138 Suncide®
114-26-1 8022 204-043-8
Propoxur
Insecticide used for treatment of sucking and biting insects. *Bayer AG.*

30139 Sundora
A proprietary trade name for cellulose acetate. No manufacturer.

30140 Sunett
33665-90-6 35 251-622-6
$C_4H_5NO_4S$
6-methyl-1,2,3-oxathiazin-4(3H0-one 2,2-dioxide
Acesulfame. Sweetening agent. *Hoechst AG.*

30141 sunflower seed oil
8001-21-6 232-273-9
Oil expressed from seeds of the sunflower, *Helianthus annuus*; emollient for skin and hair care products and makeups; adds conditioning, spread, and sheen to personal care products; used in preparation of margarine. *Arista Industries; Lipo; Penta Mfg.*

30142 Sunfort®
Photosensitive dry film resist. *Asahi Chem. Industry.*

30143 Sungard
4065-45-6 9157 223-772-2
Sulisobenzone
UV screen. *Bayer.* Name unverified.

30144 Sunimac ECR
Epoxy resin; industrial resins. *Sumitomo Bakelite Co Ltd.* Name unverified.

30145 Sunimac GCR
Polyurethane resin; industrial resins. *Sumitomo Bakelite Co Ltd.* Name unverified.

30146 Sunnis
Refrigeration oil, premium quality. *Witco.* Discontinued.

30147 Sunnol
A range of alkyl aryl sulfonates; used as scouring agents, dyeing assistants, and in emulsion polymerization. *Harcros Australia.*

30148 Sunolith
A proprietary trade name for a pigment containing 71% barium sulfate and 29% zinc sulfide. No manufacturer.

30149 Sunproof
Blend of waxes for all types of stock to inhibit static atmospheric cracking and frosting. *Uniroyal.* Name unverified.

30150 Sunproofing Wax 1343
Complex hydrocarbon mixture; antiozonant, anticracking, and sunchecking wax for the rubber industry. *Frank B. Ross.*

30151 SunShade 18-89
UV stabilizer for polyethylene sections greater than 22 mils in thickness. *Santech.*

30152 Sunshine
Resistant glaze decorating colors for porcelain, bone china and earthenware. *Degussa Ltd.*

30153 Suntec® -HD
9002-88-4 7728
High-density polyethylene *Asahi Chem. Industry.*

30154 Suntec® -LD
9002-88-4 7728
Low-density polyethylene *Asahi Chem. Industry.*

30155 suntei tallow
A white sweetish fat expressed from the seeds of *Palaquium oleosum* .

30156 Sunveil®
High-shrink polystyrene film. *Asahi Chem. Industry.*

30157 Sunvex®
Water gel explosive. *Asahi Chem. Industry.*

30158 Supaclean
Universal power cleaner. *Kalon Chemicals Ltd.*

30159 Suparamin 30, 120
Blend; wet strength resin for paper industry. *Toho Chem. Industry.*

30160 Suparen
Fermentation-derived rennet. *Pfizer Ltd.* Name unverified.

30161 Suparex
Superplasticizers and auxiliaries; used for the construction industry. *GE Plastics Ltd.*

30162 Supec® G401
9016-75-5
PPS crystalline polymer; high-strength, high-performance resin with

outstanding heat and chemical resistance; ideal metal replacement material; inherently flame retardant; used for industrial, electrical/electronic, and aircraft applications. *GE Plastics.*

30163 **Super A**
68-26-8 10150 200-683-7
Vitamin A; antixerophthalmic. *Upjohn.* Name unverified.

30164 **Super AD-IT**
Di(phenylmercuric) dodecenyl succinate
Preservative and fungicide for aqueous coating compositions. *Hüls Am.* Discontinued.

30165 **Super Alkyd® 574-75TK**
Alkyd resin, air dry; very fast set to touch air dry short oil used in drum enamels and other fast air dry applications; low temperature bakes. *Thibaut & Walker.* Name unverified.

30166 **Super Beckacite® 2000**
Terpene phenolic
Tackifier resin for adhesives. *Arizona.* Discontinued.

30167 **Super Beckamine®**
Synthetic resins. *Reichhold.*

30168 **Super Beckosol®**
Synthetic resins. *Reichhold.*

30169 **Super Bowl**
Toilet cleaner. *Momar Industrial Services Ltd.*

30170 **super bronze**
An alloy of from 57-69% copper, 1.2-5.1% aluminum, 1.3-2% iron, 21-37% zinc, and 3-3.2% manganese.

30171 **Super Cat**
Metal salts of organic acids; fuel oil additives. *Hüls Am.*

30172 **super cement**
An ordinary Portland cement to which has been added a waterproofing material.

30173 **Super Corona**
Refined anhydrous lanolin USP; superfatting emollient and emulsifier for cosmetics and pharmaceuticals; improves spreading, penetration, and esthetic properties of these products; plasticizer in aerosol hairsprays; stable water-oil emulsions. *Croda Inc.*

30174 **Super D**
8001-69-2 11, 2464 232-289-6
Cod liver oil; source of vitamins A and D. *Upjohn.* Name unverified.

30175 **Super Die**
A proprietary trade name for a tool steel containing 10.5% chromium, 1% tungsten, and 1% silicon. No manufacturer.

30176 **Super Glue**
Cyanoacrylate adhesive; for bonding two similar or dissimilar materials in seconds. *Novest Inc.* Unverified.

30177 **Super Green**
Various granular fertilizer blends; fertilizers for lawns, gardens, and flowers. *Horn's Crop Service Center.* Name unverified.

30178 **Super Hartolan**
8027-33-6 232-430-1
Lanolin alcohol; spreading agent, dispersant, stabilizer, plasticizer, oil-water emulsifier, emulsion stabilizer, and emolient for cosmetics and pharmaceuticals. *Croda Inc.; Croda Chem. Ltd.*

30179 **Super Lacolene**
Aliphatic solvent. *Ashland.*

30180 **Super Lubestine**
14807-96-6 9207 238-877-9
Talc. Paint extender. *Bromhead & Denison Ltd.*

30181 **Super Lubracon**
A water-based food industry lubricant, with cleaning and antimicrobial properties; bottle conveyor lubricant, can conveyor lubricant, keg conveyor lubricant, crate conveyor lubricant. *Harshaw Chemicals Ltd.* Name unverified.

30182 **Super Moss Killer & Lawn Fungicide**
97-23-4 3120 202-567-1
Dichlorophen fungicide/moss killer. *Murphy Chemical Co Ltd.* Discontinued.

30183 **Super Mosstox**
97-23-4 3120 202-567-1
A liquid formulation containing 34% dichlorophen; controls moss in fine turf, footpaths, hard tennis courts, playgrounds, roof and other affected hard surfaces. *Burts & Harvey.*

30184 **Super Mosstox**
97-23-4 3120 202-567-1
Dichlorophen
Fungicide, bactericide, and algicide used as a moss-killer. *Rhône-Poulenc Environmental Prods. Ltd.*

30185 **Super Nevtac® 99**
Synthetic polyterpene resin; synthetic tackifier used in adhesives, coatings, rubber products, concrete-curing compounds, and caulking compounds. *Neville.*

30186 **Super Nickel**
A proprietary trade name for alloys of 20-30% nickel with 70-80% copper; they are corrosion resisting. No manufacturer.

30187 **Super Refined Almond NF**
Almond oil; emollient; provides elegant skin feel and promotes spreading in creams, lotions, bath oils. *Croda Inc.*

30188 **Super Refined Coconut Oil**
8001-31-8 232-282-8
Coconut oil; emollient; improved color and odor for sunscreen products. *Croda Inc.* Discontinued.

30189 **Super Refined Crossential EPO**
Evening primrose oil; emollient; contains essential fatty acids vital to health of skin. *Croda Inc.* Discontinued.

30190 **Super Refined Grapeseed Oil**
8024-22-4
Grapeseed oil; emollient. *Croda Inc.* Discontinued.

30191 **Super Refined Menhaden**
8002-50-4 232-311-4
Menhaden oil; emollient; contains essential fatty acids vital to health of skin. *Croda Inc.*

30192 **Super Refined Mink Oil**
Mink oil; emollient oil for makeup remover systems. *Croda Inc.* Discontinued.

30193 **Super Refined Olive**
8001-25-0 6973 232-277-0
Olive oil; cosmetic emollient; lubricant for hair care products. *Croda Inc.*

30194 **Super Refined Peanut**
8002-03-7 7191 232-296-4
Peanut oil; emollient in skin care products; also for nutritional supplements, pharmaceutical delivery systems. *Croda Inc.*

30195 **Super Refined Safflower USP**
8001-23-8 8465 232-276-5
Safflower oil; emollient. *Croda Inc.*

30196 **Super Refined Sesame**
8008-74-0 8614 232-370-6
Sesame oil; emollient oil used in skin care preparations, nutritional supplements, pharmaceutical delivery systems. *Croda Inc.*

30197 **Super Refined Shark**
68990-63-6 273-616-2
Shark liver oil; emollient, moisture repellent for skin protection. *Croda Inc.*

30198 **Super Refined Soybean**
8001-22-7 232-274-4
Soybean oil; emollient oil used in skin care preparations, nutritional supplements, pharmaceutical delivery systems. *Croda Inc.*

30199 **Super Refined Sunflower Oil**
8001-21-6 232-273-9
Sunflower seed oil. *Croda Inc.* Discontinued.

30200 **Super Solan**
61790-81-6
PEG-75 lanolin; emollient, conditioner, superfatting agent, solubilizer. *Croda Inc.*

30201 **Super Solvitax®**
Cod liver oil B, vet C; dietary supplement for animals; conditioning oil; for relief of stiffness in animals. *Seven Seas Ltd.*

30202 **Super Sta-Tac® 80**
Specialty hydrocarbon based on mixed olefins; for pressure-sensitive adhesive, hot-melt, and sealants; tackifier for S-I-S and S-B-S polymers. *Arizona.* Discontinued.

30203 **Super Sta-Tac® 100**
Specialty hydrocarbon based on mixed olefins; for pressure-sensitive adhesive, hot-melt, and sealants. *Arizona.* Discontinued.

30204 **Super Sterol Ester**
C_{10}-C_{30} cholesterol/lanosterol esters; emollient, lubricant, and moisturizer for dry skin, cosmetics, pharmaceuticals. *Croda Inc.*

30205 **Super Sulfur No. 1**
The oxidized zinc salt of dimethyl dithiocarbamic acid; a proprietary rubber vulcanization accelerator. No manufacturer.

30206 **Super Sulfur No. 2**
19010-66-3 242-748-2
Lead dimethyldithiocarbamate; a proprietary rubber vulcanization accelerator. No manufacturer.

30207 Super Thane® 975-70
Oil-modified urethane; water-reducible urethane for general-purpose varnish applications. *Thibaut & Walker.*

30208 Super Tin® 4L
76-87-9 9875 200-990-6
Triphenyltin hydroxide solution; flowable fungicide for pecans, potatoes, sugar beets; restricted use. *Griffin.*

30209 Super Verdone
Soluble concentrate containing 72 g 2,4-D, 12 g dicamba and 48 g ioxynil per liter; used to control weeds in turf. *ICI Chem & Polymers Ltd.*

30210 Super Vilex
Denatonium saccharide
Bittering aversive agent. *Atomergic Chemetals Corp.*

30211 Super Weedex
Total weedkiller. *Murphy Chemical Co Ltd.* Discontinued.

30212 Super Wet
Wetting agent and emulsifier for pesticides; soil penetrant; enhances pesticide and fertilizer efficiency. *W A Cleary.*

30213 Super White Fonoline®
White petrolatum USP
Low melting point grade with superior snow white color, exhibiting elegant feel and texture; for premium cosmetic formulations and food related applications. *Witco/Sonneborn.*

30214 Super-A
Chemical products for general industrial use; organometallic chemicals, organic chemicals, catalysts, styrene butadiene polymers, urethane and epoxy polymers, urethane and epoxy curatives and activated carbons. *Anderson Development Company.* Name unverified.

30215 Superadoplast
Textile softener used in padding machines. *Ceca SA.*

30216 Superam
A fertilizer obtained by neutralizing the acids of ordinary super phosphate with ammonia gas.

30217 superargol
Tartar cake; substitute of tartar. Preparations containing simply acid sodium sulfate. Some contain oxalates, and others contain tartaric acid and sulfuric acid.

30218 Super-ascoloy
A ferrous alloy containing 8% nickel and 18% chromium.

30219 Superba
7440-44-0 1855 231-153-3
A proprietary trade name for a carbon black. No manufacturer.

30220 superbasique metal
A modification of cast iron. It is resistant to alkalis.

30221 Superbond®
High-performance carrier for bonding high ring crush or recycled medium and liner. *Grain Processing.*

30222 Superbrillantoline®
Hydromethylabietate
Raw material for cosmetics. *Laserson & Sabetay.*

30223 Supercadoplast
Textile softener to facilitate raising; for use in winch and autoclave. *Ceca SA.*

30224 Super-Ceram
Three-part casting system consisting of resin, curing agent, and ceramic grain; for injection and compression dies having low shrinkage and good dimensional stability; low thermal expansion enables embedment of reinforcement rods and cooling or heating coils. *Magnolia Plastics.*

30225 Superchlon
Chlorinated polymers. *British Traders & Shippers Ltd.*

30226 Superclear 80-N
Aqueous colloidal dispersion of a branched polysaccharide macromolecule containing active carbonyl and hydroxyl groups; natural gum for textile dyeing and printing; water modifier controlling mobility of discrete dye particles in aqueous systems; hinders migration of dyestuffsduring drying of fabrics. *Henkel/Textiles.*

30227 Super-cliffite
Explosives. No. 1 contains 10% nitroglycerin, 1% collodion cotton, 60% ammonium nitrate, 16% sodium chloride, 11% ammonium oxalate, and 6% wood meal. No. 2 has the sodium chloride increased to 20% and the sodium oxalate reduced to 6%.

30228 Supercoat®
1317-65-3 5515 215-279-6
Calcium carbonate
Limestone. Coated ultrafine ground pigment offering easy incorporation and dispersion in plastics with improved physical properties. *ECC International Ltd.*

30229 Supercol® Guar Gum
9000-30-0 4601 232-536-8
Guar gum. *Hercules.*

30230 Supercore® S13F
9005-25-8 232-679-6
Modified corn starch; used in stucco slurry to protect the critical core-to-paper bond from breakdown during drying. *Grain Processing.*

30231 Supercut
Engineering cutting fluids. *Lambson Ltd.* Discontinued.

30232 Super-excellite
An explosive containing 73.5-77% ammonium nitrate, 6.5-8% potassium nitrate, 2-4% wood meal, 3.5-5% nitro-glycerin, and 9-11% ammonium oxalate.

30233 Superfast Power Pack
Liquid epoxy resin and hardener; fast setting system; general purpose adhesive. *Wessex Resins & Adhesives Ltd.*

30234 Superfiltchar
A proprietary product; it is an active decolorizing carbon made from sawdust. No manufacturer.

30235 Superfine Lanolin USP
5371
Lanolin; superfatting emollient with some emulsifying properties. *Croda Inc.*

30236 Superfloc
High molecular weight polymers based on acrylamide in powder, solution or emulsion form; nonionic, anionic, cationic. *Cyanamid BV.*

30237 Superfloc C507
Melamine formaldehyde resin. *Cyanamid BV.*

30238 Superfloc C521
Monomethylamine-epichlorohydrin condensation products. *Cyanamid BV.*

30239 Superfloc C567, C573, C577, C581
Demethylamine-epichlorohydrine condensation products. *Cyanamid BV.*

30240 Superforcite
A Belgian gelatin dynamite containing 64% nitroglycerin.

30241 Supergreen
Lawn fertilizer. *May & Baker Ltd.* Name unverified.

30242 Superinone
25301-02-4 9964
$C_8H_{11}NO$
Tyloxapol
p-(1,1,3,3-tetramethylbutyl)phenol, polymer with ethylene oxide and formaldehyde. Detergent. *Sterling Drug Inc.* Name unverified.

30243 Superior alloy
A heat-resisting alloy containing 78% nickel, 19.5% chromium, 2% manganese, and 0.5% iron.

30244 Superior® Granulated
7647-14-5 8737 231-598-3
Sodium chloride
Evaporated granulated salt refined by closed pan (vacuum) evaporation of purified brine. *Akzo Salt.*

30245 Superite
An explosive consisting of 80-84% ammonium nitrate, 9-11% potassium nitrate, 2-5% starch, and 3.5-4.5% nitroglycerin.

30246 Superjet
7440-44-0 1855 231-153-3
A carbon black pigment. *Pfizer International.* Discontinued.

30247 Super-karma
An alloy wire containing 80% nickel and 20% chromium.

30248 Superkleen C
Polymer dispersion; imparts durable hydrophilic and soil release properties to 100% polyester. *CNC Int'l L.P.*

30249 Super-kolax No. 2
An explosive containing nitroglycerin, collodion cotton, potassium nitrate, barium nitrate, wood meal, starch, and ammonium oxalate.

30250 Super-ligdynite
A coal mine explosive containing from 15-17% nitroglycerin, 15-17% ammonium nitrate, 23-25% sodium nitrate, 10-12% flour, 19-21% wood pulp, and 9-11% sodium chloride.

30251 Superlit
A proprietary synthetic resin. No manufacturer.

30252 Superlite
18282-10-5 8933 242-159-0
tin(IV) oxide.
Keeling & Walker Ltd. Name unverified.

30253 Superlock
Anaerobic; use to seal and lock parts in place. *ITW Devcon.*

30254 Superloid®
9005-34-9
Refined ammonium alginate; gum used as gelling agent, thickener, emulsifier, filmforming agent, suspending agent, and stabilizer in food, pharmaceutical, and industrial applications; stabilizer in paper and textile industry. *Kelco.*

30255 Supermagnesia 66
Mineral fiber-free thermal insulator. *The Chemical & Insulating Co Ltd.*

30256 Supermite®
1317-65-3 5515 215-279-6
Calcium carbonate
Ultrafine pigment developing superior physical properties and surface gloss in plastics, elastomers, and coatings. *ECC International Ltd.*

30257 Superneutral metal
A silicon-iron alloy; used for nitric acid plants.

30258 Superol
56-81-5 493 200-289-5
Glycerin USP; humectant; in pharmaceuticals, toiletries, tobacco, alkyds, food products, explosives, cellophane, urethane foam, other industries. *Procter & Gamble.*

30259 Superox®
Oxidation catalysts for synthetic resins. *Reichhold.*

30260 Superpalite
76-02-8 200-926-7
C_2Cl_4O
Trichloromethyl chloroformate
Diphosgene; green cross gas; trichloroacetyl chloride. A military poison gas. mp = -57°; bp = 118°; sg = 1.629, reacts with H_2O.

30261 Super-Pflex® 100
471-34-1 1697 207-439-9
Surface-modified precipitated calcium carbonate; filler; provides excellent extrusion processing, impact strength retention, and enhanced color stability on outdoor exposure in rigid PVC siding and profiles. *Pfizer.*

30262 Superphosphate
1314-56-3 7512 215-236-1
Soluble powder containing 18% phosphorus pentoxide; phosphate fertilizer for use throughout the garden. *Vitax Ltd.*

30263 superphosphate
$CaH_4O_8P_2$
Mineral superphosphate; superphosphate of lime. Consists of mono-calcium phosphate, mixed with calcium sulfate, and contains 25-28% soluble phosphate; used as a fertilizer.

30264 Superpolystate
9004-99-3
PEG-6 stearate SE
Self-emulsifying gelling base for pharmaceutical and cosmetic lotions. *Gattefosse SA.*

30265 Superprill
57-13-6 10005 200-315-5
Urea (prilled); fertilizer. *Columbia Nitrogen Corporation.* Name unverified.

30266 Super-Pro 5A
TEA coco-hydrolyzed collagen, sorbitol; detergent, conditioner, emulsifier, moisturizer, foamer used in personal care products. *Inolex.*

30267 Super-Quench
Metal working oil. *Checron.* Name unverified.

30268 Super-rippite
A smokeless powder containing from 51-53 parts nitroglycerin, 2-4 parts nitrocotton, 13.5-15.5 parts potassium nitrate, 15.5-17.5 parts dried borax, and 7-9 parts potassium chloride.

30269 Super-rippite No.2
An explosive for coal mines containing 51% nitroglycerin, 3% nitrocotton, 11% potassium perchlorate, 24% borax, and 10% potassium chloride.

30270 Super-Sat
8031-44-5 232-452-1
Hydrogenated lanolin; plasticizer, emollient; cosmetics and pharmaceuticals; makeups, night creams, shaving creams. *RITA.*

30271 Super-Sat AWS-4
68648-27-1
PEG-20 hydrogenated lanolin; emollient, emulsifier, plasticizer for emulsion systems. *RITA.*

30272 Superseal
A high performance EPDM rubber-based single layer roofing membrane. *Feb Ltd.*

30273 Supersevtox
Selective herbicide. *Schering Agrochemicals Ltd.* Discontinued.

30274 Supersoft NI
Silicone emulsion; synthetic softener for elasticized fabrics; excellent resistance to yellowing. *CNC Int'l L.P.*

30275 Supersol ADM
Blend; wire drawing lubricant. *Rhône-Poulenc Surf.*

30276 Supersol ICS
Textile solvent scour to remove wax and oil; caustic stable. *Sybron.*

30277 Supersols
Wire drawing lubricants. *Rhône-Poulenc UK.*

30278 Super-Sorb C Water Absorbant
Copolymer acrylamide sodium acrylate; increases water holding capacity of soils and horticultural media; used in greenhouses, nurseries, and landscaping. *Aquatrols Corp of Am.* Name unverified.

30279 Super-Sorb F Water Absorbant
Copolymer acrylamide sodium acrylate; holds water as a gel around plant roots; used for transplanting and transporting of bare root plant material, in reforestation, landscaping, and crop production. *Aquatrols Corp of Am.* Name unverified.

30280 Supersorbon
Molded activated carbon for the recovery of solvents. *Degussa AG.* Name unverified.

30281 supersoy
A high grade soya bean flour.

30282 Superspray
55-56-1 2140 200-238-7
Chlorhexidine teat spray. *Ciba plc.* Name unverified.

30283 Supersurf AFX
Blend; wetting agent for fourth generation nylon. *Am. Emulsions.*

30284 Supersurf FL/4
Penetrant, wetting/rewetting agent, emulsifier for bleaching, dyeing, finishing. *Am. Emulsions.*

30285 Supertac
Sealant adhesive. *Crowley Chem.*

30286 Supertherm 2003
25928-94-3
One-part epoxy adhesive; ultra-high thermally conductive diamond-filled epoxy adhesive for semiconductor die-attach and other micro-electronic applications where very high thermal conductivity and excellent electrical insulation over wide temperature range are required. *Tra-Con.*

30287 Superthin
Tetrafilcon A
Contact lens material. *Am. Optical.* Name unverified.

30288 Supertox
Selective weedkiller. *May & Baker Ltd.* Name unverified.

30289 Supertox
Mixture of 2,4-D and mecoprop; used to control weeds in grassland. *Rhône-Poulenc/Agri.*

30290 superturpentine
Spirits of turpentine specially rectified *in vacuo*. It boils at 155°C, and distills completely below 160°C.

30291 Superwipes
Chlorhexidine/cetrimide udder wipes. *Ciba plc.* Name unverified.

30292 Suplex
13684-63-4 7384 237-199-0
Emulsifiable concentrate containing 114 g/l phenmedipham; used for weed control in beet crops. *Universal Crop Protection Ltd.*

30293 Supplex®
Fibers. *DuPont UK.*

30294 Suppocire AIP
Saturated polyglycolized glycerides. *Gattefosse SA.*

30295 Supra EF
14807-96-6 9207 238-877-9
Platy talc; for cosmetic applications including dusting and pressed powders., creams and lotions, antiperspirants, bath and loose powders. *Luzenac Am.* Discontinued.

30296 Suprac
Decolorizing and absorptive activated carbons; used for dry cleaning and chemical purification. *Lancashire Chemical Works Ltd.*

30297 Supracen®
A range of acid wool dyes with outstanding leveling power and very good lightfastness. *Bayer AG; Bayer plc.*

30298 Supracet
Disperse dyestuffs; for dyeing of acetate, triacetate and nylon fibers and blends. *Holliday Dyes & Chemicals Ltd.*

30299 Supracide
950-37-8 6048 213-449-4
methidathion. Organophosphorus insecticide. *Ciba plc.* Name unverified.

30300 Supradyn
Proprietary multivitamin preparation containing the B complex vitamins, vitamins A, C, D and F, biotin, nicotinamide, folic acid, (in addition to minerals and trace elements Ca, Fe, Mg, Mn, Cu, Zn, Mo and P). *Roche Products Ltd.* Name unverified.

30301 Supraene
111-02-4 8924 203-826-1
Squalene. *A & E Connock (Perfumery & Cosmetics) Ltd.*

30302 Supraene®
111-02-4 8924 203-826-1
Purified squalene; natural emollient. *Robeco.*

30303 Suprafino A
14807-96-6 9207 238-877-9
Talc; for cosmetic applications including antiperspirants, eye shadows, soaps, creams and lotions, nail polish, aerosols. *Cyprus Industrial Minerals.*

30304 Suprafix Paste
Vat dyes. *Bayer plc.*

30305 Suprafrax
A clay with a high percentage of alumina; used as a furnace lining.

30306 Supragil® GN
Sodium phenyl sulfonate
Dispersant, suspending agent for pesticide wettable powders and water dispersible granules; nonfoaming. *Rhône-Poulenc Geronazzo.*

30307 Supragil® MNS/90
26264-58-4 247-561-6
Sodium methyl naphthalene sulfonate condensate
Surfactant, dispersing/suspending agent for pesticide wettable powders and water-dispersion granules. *Rhône-Poulenc Surf.; Rhône-Poulenc France; Rhône-Poulenc Geronazzo.*

30308 Supragil® NK
25417-20-3 246-960-6
Sodium dibutyl naphthalene sulfonate
Wetting agent for pesticides; emulsifier, wetting and dispersing agent, foamer for dyes, pigments, emulsion polymerization, industrial use. *Rhône-Poulenc Surf.; Rhône-Poulenc France; Rhône-Poulenc Geronazzo.*

30309 Supragil® WP
1322-93-6 215-343-3
Sodium diisopropyl naphthalene sulfonate
Wetting agent, dispersant, stabilizer without detergent properties, foaming agent used in textiles, paints, pesticides, latex emulsions. *Rhône-Poulenc Surf.; Rhône-Poulenc Geronazzo.*

30310 Supralated LS
1847-55-8 217-430-1
Duponol LS. A proprietary trade name for sodium oleyl sulfate, a wetting agent. *Witco Corporation.*

30311 Supralated ME
151-21-3 8782 205-788-1
sodium lauryl sulfate
Duponol ME. A proprietary trade name for sodium lauryl sulfate, a wetting agent. *Witco Corporation.*

30312 Supralated WA
sodium lauryl sulfate-lauryl alcohol
Duponol WA. A proprietary trade name for a mixture of sodium salt of sulfated lauryl alcohol and lauryl alcohol. *Witco Corporation.*

30313 Supramin®
A range of acid wool dyestuffs with superior fastness to water, washing and perspiration. *Bayer AG; Bayer plc.*

30314 Supranol®
Acid wool dyes; used in dyeings which are fast to washing, water and seaweed. *Bayer AG; Bayer plc.*

30315 Suprapal®
Styrene copolymer; for roadmarking paints, gravure and flexographic printing inks, paper coatings, zinc dust primers. *BASF AG.*

30316 supraresen
The residue obtained when dammar is prepared for use in lacquers and is soluble in hydrocarbons; used in the varnish industry.

30317 Suprasec
Isocyanates for general application. *ICI Chem & Polymers Ltd.*

30318 Suprathion
950-37-8 6048 213-449-4
Active ingredient: methidathion; organophosphorous insecticide with high degree of insecticidal activity; also efficient in controlling scales. *Makhteshim Chemical Works Ltd.*

30319 Suprex white
A highly purified precipitated calcium carbonate for use as a rubber filler in the place of blanc fixe. It has a specific gravity of 2.7.

30320 Suprexcel
Fast-to-light direct cotton dyestuffs. *Holliday Dyes & Chemicals Ltd.* Discontinued.

30321 Suprofix
High speed photographic fixer. *May & Baker Ltd.* Name unverified.

30322 Suprol
Photographic developer. *May & Baker Ltd.* Name unverified.

30323 Supronal®
Sulfonamide mixture; chemotherapeutic drug for the treatment of bacterial infections in animals. *Bayer AG.*

30324 Supronic
Nonionic surfactant. *Rhône-Poulenc UK.*

30325 Supronic B10, B25, B50, B75 and B100
Polyoxy alkylated polyalkylene glycols
Low-foam surfactants. *Glover (Chemicals) Ltd.* Unverified.

30326 Supro-Tein R
Sodium cocoyl hydrolyzed rice protein, sorbitol; biodegradable, mild, foaming substantive protein for conditioning, cleansing, thickening in hair and skin care products (baby shampoos, ethnic shampoos, hairstying products, lotions, hands/face cleanser, bath products, depilatories. *Maybrook.*

30327 Supro-Tein V
TEA-coco-hydrolyzed animal protein, sorbitol; mild surfactant with foam stability; emulsifier, solubilizer, moisturizer; for ethnic and mild shampoos, conditioners, lotions, hair and skin mousses, liquid soap, shave creams. *Maybrook.*

30328 Surbex T
A proprietary multi-vitamin preparation. *Abbott Laboratories.*

30329 Surco
Alkanolamides. *Millmaster-Onyx UK.*

30330 Surcol®
Acrylic and vinyl polymers in bead form for use in paints, adhesives, polishes, printing inks, lacquers, concrete sealers. *Allied Colloids Ltd.*

30331 Surcopur
709-98-8 7987 211-914-6
Herbicide and nematocide. Post-emergence applied herbicide with no reidual effect. Used for control of numerous grasses and broad-leaved weeds in rice crops. *Bayer AG.*

30332 Sure-Curd
A standardized solution of fermentation derived milk clotting enzyme elaborated by *Endothia parasitica* ; used in the manufacture of cheese, especially Swiss and Italian varieties. *Pfizer International.*

30333 Suresperse® 911
Low-foaming antifoulant for dispersing oil, lint, biological matter, silt, mud, textile fibers, tobacco dust and other debris in industrial air washers; also removes deposits. *Drew Ind. Div.*

30334 Sure-Step
10043-52-4 1699 233-140-8
$CaCl_2$
Chip calcium chloride
Calcosan; Calcium dichloride. *Schaefer Salt & Chemical Co.*

30335 Suretex® 920
Static control agent for air washers for use in textile mills. *Drew Ind. Div.*

30336 Surf Ac 820
Alkylaryl polyethoxyethanol and n-butanol; nonionic biodegradable surfactant. *Draxel Chemical Company.* Unverified.

30337 Surfactant N-42
Nonylphenol ethoxylate nonionic surfactant in the form of a clear oil; general surfactant. *Rohm & Haas UK.* Name unverified.

30338 Surfactant XQS20
Phosphate ester in free acid form; aqueous solution; wetting agent and detergent; textile lubricant; antistatic agent; industrial emulsifiers. *Rohm & Haas UK.* Name unverified.

30339 Surfactol® 13
1323-38-2 11, 1904 215-353-8
Castor oil, modified; wetting agent, emulsifier, solubilizer, dispersant, wax plasticizer, mold release agent, antifoamer for textiles, leather, paints, household, cosmetics, dyeing, tanning, finishing, sizing, making of cutt *CasChem.*

30340 Surfactol® 318
61791-12-6
PEG-5 castor oil
Emulsifier for oils, waxes; solubilizer for fragrances; emollient; used in textiles, paints, household, cosmetics, dyeing, tanning, finishing, sizing, insecticides, herbicides, fungicides, kier boiling, making of cutting and

soluble oils, dispersing waxes, pigments, resins, rewetting dried skins. *CasChem.*

30341 Surfactol® Q1
Ricinoleamidopropyltrimonium chloride
propylene glycol. Surfactant, emollient, refatting agent, antistat, emulsifier for skin and hair care products, clear shampoos. *CasChem.*

30342 Surfactol® Q2
Hydroxystearamidopropyl trimonium chloride
propylene glycol. Surfactant, emollient, emulsifier, refatting agent, antistat for conditioners, liquid hand soaps, shampoos, skin lotions. *CasChem.*

30343 Surfactol® Q3
Hydroxystearamidopropyl trimonium methosulfate, propylene glycol
Surfactant, emulsifier, emollient, antistat for aerosols and mousses, liquid hand soaps, shampoos, skin lotions; compatible in aerosol systems without corroding cans. *CasChem.*

30344 Surfactol® Q4
112324-16-0
Ricinoleamidopropyl ethyldimonium ethosulfate
Emollient, emulsifier for personal care products (shampoos, conditioners, aerosols). *CasChem.*

30345 Surfadone LP-100
Caprylyl pyrrolidone
ISP.

30346 Surfadone LP-300
Lauryl pyrrolidone
ISP.

30347 Surfadone QSP
Dimethyl stearamidopropyl [(2-pyrrolidonyl) methyl] ammonium chloride
ISP.

30348 Surfadone WSP
Stearamidopropyl pyrrolidonylmethyl dimonium chloride
ISP.

30349 Surfageen
Anionic surfactant of the alkyl ether phosphate type in solid form; emulsifier for emulsion polymerization; deinking of waste paper; rust inhibitor. *Chem-Y, Fabriek van Chemische Producten BV.* Name unverified.

30350 Surfageen S30
Fatty alcohol ether sulfosuccinate as a clear liquid; detergent, foaming, wetting and emulsifying agent used as a mild raw material for shampoos and foam baths. *Chem-Y, Fabriek van Chemische Producten BV.* Name unverified.

30351 Surfagene FAD 106
Sodium nonoxynol-6 phosphate
Wetting agent for weakly acidic and alkaline cleaners, metalworking cooling lubricants. *Chem-Y GmbH.*

30352 Surfagene FAZ 109
Nonoxynol-9 phosphate
Emulsifier for emulsion polymerization. *Chem-Y GmbH.*

30353 Surfagene FDD 402
Laureth-2 phosphate
Emulsifier for cosmetic formulations, oil baths; additive for mineral collectors. *Chem-Y GmbH.*

30354 Surfagene FGD 600
69331-39-1 273-968-7
DEA-cetyl phosphate
Chem-Y GmbH. Name unverified.

30355 Surfagene FGZ 608
Ceteth-8 phosphate
Chem-Y GmbH. Name unverified.

30356 Surfagene FHD 704 NV
Sodium steareth-4 phosphate
Chem-Y GmbH. Name unverified.

30357 Surfagene FMD-12-03 KAM
Trideceth-3 phosphate
Chem-Y GmbH. Name unverified.

30358 Surfagene FPG 50
Sodium glycereth-1 polyphosphate
Chem-Y GmbH. Name unverified.

30359 Surfagene FPT
TEA phosphate ester
Scale inhibitor for cooling water systems. *Chem-Y GmbH.*

30360 Surfagene MB 1705
Disodium undecylenamido MEA-sulfosuccinate
Chem-Y GmbH. Name unverified.

30361 Surfagene S 30
39354-45-5 255-062-3
Disodium laureth-3 sulfosuccinate
Mild raw material for personal care products, baby products, shampoo, foam baths; mild to skin and eyes; biodeg. *Chem-Y GmbH.*

30362 Surfak
128-49-4 3459 204-889-8
Docusate calcium
Stool softener. *Hoechst-Roussel Pharmaceuticals Inc.* Name unverified.

30363 Surf-A-Seis
A suface explosive; used for surface energy sources in portable seismic operations. *Hercules. Discontinued.*

30364 Surfonic®
A series of *p*-nonylphenol ethoxylates; surfactants in agricultural chemicals, industrial cleaners, heavy-duty detergents, paper industry. *Texaco.*

30365 Surfonic® HDL
9016-45-9 6772
Nonoxynol-8 (86%) and triethanolamine (14%); biodegradable, emulsifier, wetting agent, dry cleaning detergent, penetrant, solubilizer , lime soap dispersant, antifoamer used in agriculture, cosmetics, industrial cleaners, ceramics, concrete, dust control, wallpaper removal, photographic film developing, fire fighting, emulsion polymerization, indirect food additives, cutting oil emulsifiers, degreaser for leather industry. *Texaco.*

30366 Surfonic® JL-80X
Alkoxypolyalkoxyethanol
Biodegradable surfactant used as emulsifier, wetting agent, detergent, penetrant, solubilizer, dispersant for household detergents, industrial products, agricultural sprays, dry cleaning, metal cleaners, ceramics, concrete, textile processing, paper manufacture. *Texaco.*

30367 Surfonic® L12-3
C_{10}-C_{12} pareth-3
Biodegradable surfactant for detergent, laundry prespotters, hard surface cleaners, emulsifiers, personal care products, agricultural pesticides. *Texaco.*

30368 Surfonic® L24-2
68439-50-9
C_{12}-C_{14} pareth-2
Biodegradable surfactant for detergents, laundry prespotters, hard surface cleaners, personal care products, agricultural pesticides. *Texaco.*

30369 Surfonic® L46-7
C_{14}-C_{16} pareth-7
Biodegradable surfactant, detergent, emulsifier, foamer, penetrant, intermediate for detergent, laundry prespotters, hard surface cleaners, personal care products, agricultural pesticides. *Texaco.* Name unverified.

30370 Surfonic® LF-17
69013-18-9
Primary alcohol-EO adduct, modified; biodegradable low foaming wetting agent, detergent for aqueous systems, metal cleaners, latex paints, textiles, paper, rinse aids, industrial and home mechanical dishwashing compounds; defoamer in some systems. *Texaco.*

30371 Surfonic® N-10
9016-45-9 6772
Nonoxynol-1; biodegradable emulsifier, wetting agent, detergent, penetrant, solubilizer, dispersant for household cleaners, textile, agriculture, metal cleaning, petrol, cosmetic, latex paint, cutting oil, janitorial supply industries. *Texaco.*

30372 Surfynol®
A range of acetylenic glycols; used as wetting agents with very low foam for paints. *Air Prods & Chems. Inc.*

30373 Surfynol® 61
107-54-0 203-500-9
$C_8H_{14}O$
3,5-Dimethyl 1-hexyn-3-ol
Surfactant, wetting agent used for paper coatings, inks, floor polishes, and glass cleaning formulations; cleaner in silicon wafer industry. bp = 150°; d= 0.859; n_D^{20}= 1.4340. *Air Prods & Chems. Inc.*

30374 Surfynol® 82
78-66-0 201-131-8
$C_{10}H_{18}O_2$
3,6-Dimethyl-4-octyne-3,6-diol
Surfactant, defoamer, wetting agent, visocsity reducer used with aqueous systems, pesticide concentrates, shampoos, vinyl plastisols, starch solutions, flexographic inks, electroplating baths. Detergent for radiator cleaners, mp = 53-55°; bp_{680} - 208-214°. *Air Prods & Chems Inc.*

30375 Surfynol® 82, 82S
78-66-0 201-131-8
$C_{10}H_{18}O_2$
3,6-Dimethyl-4-octyne-3,6-diol
3,6-Dimethyl-4-octyne-3,6-diol on amorphous silica carrier; defoamer/wetting agent in pesticide wettable powders, electroplating baths, cement, plastics,

coatings; solubilizer and clarifier in shampoos. mp = 53-55°; bp_{680} = 208-214°. *Air Prods & Chems. Inc.*

30376 Surfynol® 104
126-86-3 204-809-1
$C_{14}H_{26}O_2$
Tetramethyl decynediol
Defoamer and dye dispersant in paints, inks, dyestuffs, pesticides; surfactant in rinse aids; substrate pigment wetting agent for industrial coatings and adhesives; wetting agent for industrial cleaners; viscosity reducer for vinyl dispersions. mp = 40-42°; bp = 254-255° *Air Prods & Chems. Inc.*

30377 Surfynol® 104A
126-86-3 204-809-1
Tetramethyl decynediol and 2-ethyl hexanol; defoamer, wetting agent for pesticides, coatings, dyestuffs, aq. systems; lubricity additive for metalworking formulations. *Air Prods & Chems. Inc.*

30378 Surfynol® 104BC
126-86-3 204-809-1
Tetramethyl decynediol in 2-butoxyethanol; wetting and defoaming agent in water-based systems, e.g., coatings, adhesives, inks, cements, metalworking fluids, latex dipping and paper coatings. *Air Prods & Chems. Inc.*

30379 Surfynol® 104E
126-86-3 204-809-1
$C_{15}H_{28}O_2$
2,4,7,9-Tetramethyl-5-decyn-4,7-diol
Tetramethyl decynediol; used with ethylene glycol as a wetting agent, defoamer, dispersant, viscosity stabilizer. mp = 40-42°; bp = 254-255°. *Air Prods & Chems. Inc.*

30380 Surfynol® 104H
126-86-3 204-809-1
Tetramethyl decynediol in ethylene glycol; wetting and defoaming agent in water based systems, e.g., coatings, adhesives, inks, cements, metalworking fluids, latex dipping and paper coatings. *Air Prods & Chems. Inc.*

30381 Surfynol® 104PA
126-86-3 204-809-1
Tetramethyl decynediol in IPA; wetting and defoaming agent in water-based systems, e.g., coatings, adhesives, inks, cements, metalworking fluids, latex dipping and paper coatings. *Air Prods & Chems. Inc.* Name unverified.

30382 Surfynol® 104PG
126-86-3 204-809-1
Tetramethyl decynediol in propylene glycol; wetting and defoaming agent in water-based systems, e.g., coatings, adhesives, inks, cements, metalworking fluids, latex dipping and paper coatings. *Air Prods & Chems. Inc.* Name unverified.

30383 Surfynol® 104S
126-86-3 204-809-1
Tetramethyl decynediol on amorphous silica; wetting agent, defoamer, dispersant, visc. stabilizer for agricultural formulations. *Air Prods & Chems. Inc.*

30384 Surfynol® 420
9014-85-1
Poly(oxy-1,2-ethanediyl), α,α'-[1,4-dimethyl-1,4-bis(2-methylpropyl)-2-butyne-1,4-diyl]bis[ω-hydroxy-
PEG-30 tetramethyl decynediol; ethoxylated-2,4,7,9-tetramethyl-5-decyne-4,7-diol. Wetting agent, defoamer, dispersant for aqueous coatings, inks, adhesives, agriculture, electroplating, oilfield chemicals and paper coatings. *Air Prods & Chems. Inc.*

30385 Surfynol® 440
9014-85-1
Poly(oxy-1,2-ethanediyl), α,α'-[1,4-dimethyl-1,4-bis(2-methylpropyl)-2-butyne-1,4-diyl]bis[ω-hydroxy-
PEG-30 tetramethyl decynediol; ethoxylated-2,4,7,9-tetramethyl-5-decyne-4,7-diol. Water-based industrial finishes; defoamer, rewetting, and leveling agent for paperboard coatings, agricultural formulations; metal cleaning and plating bath additive. *Air Prods & Chems. Inc.*

30386 Surfynol® 465
9014-85-1
Poly(oxy-1,2-ethanediyl), α,α'-[1,4-dimethyl-1,4-bis(2-methylpropyl)-2-butyne-1,4-diyl]bis[ω-hydroxy-
PEG-30 tetramethyl decynediol; ethoxylated-2,4,7,9-tetramethyl-5-decyne-4,7-diol. Wetting agent, defoamer for aqueous coatings, inks, adhesives; surfactant for emulsion polymerization; electroplating additive. *Air Prods & Chems. Inc.*

30387 Surfynol® 485
9014-85-1
Poly(oxy-1,2-ethanediyl), α,α'-[1,4-dimethyl-1,4-bis(2-methylpropyl)-2-butyne-1,4-diyl]bis[ω-hydroxy-
PEG-30 tetramethyl decynediol; ethoxylated-2,4,7,9-tetramethyl-5-decyne-4,7-diol. Wetting agent, defoamer for aq. coatings, inks, adhesives, agriculture, electroplating, oilfield chems., paper coatings. *Air Prods & Chems. Inc.*

30388 Surfynol® CT-136
Proprietary blend; wetting agent, defoamer, grind aid and dispersant for water and glycol-based inks and pigments. *Air Prods & Chems. Inc.*

30389 Surfynol® D-101, D-201
Nonsilicone; defoamer/antifoamer for aqueous systems, coatings, adhesives; effective for PVAc, ethylene-vinyl acetate systems. *Air Prods & Chem; Air Prods & Chem Nederland BV.*

30390 Surfynol® DF-08
Surfactant; wetting agent, defoamer for waterborne coatings, inks, adhesives, and latex dipping. *Air Prods & Chem; Air Prods & Chem Nederland BV.*

30391 Surfynol® DF-110, DF-110S
Silicone-free acetylenic derivs.; defoamer, de-air entrainment agent for inks, metalworking fluids, coatings, cement, ceramics, adhesives. *Air Prods & Chem; Air Prods & Chem Nederland BV.*

30392 Surfynol® DF-110D, DF-110L
126-86-3 204-809-1
Higher molecular weight acetylenic glycol in dipropylene glycol; defoamer, de-air entrainment aid for water-based coatings, inks, adhesives, and highly pigmented systems (concrete, paper coatings, grouts, ceramics). *Air Prods & Chems. Inc.*

30393 Surfynol® DF-210
Silicone-free organic defoamer, defoamer for aqeous formulations, printing inks, adhesives, and coatings. *Air Prods & Chems. Inc.*

30394 Surfynol® DF-34
Proprietary blend; wetting agent, defoamer for water-based coatings, inks, and adhesives; dewebbing agent and defoamer for latex gloves and other dipped goods. *Air Prods & Chems. Inc.*

30395 Surfynol® DF-37
Silicone-free organic defoamer; defoamer minimizing web formation during latex glove, waterborne coating dipping processes, and other aqueous systems (inks, adhesives, coatings, agriculture, cement, metalworking fluids); wetting agent for low energy substrates. *Air Prods & Chem; Air Prods & Chem Nederland BV.*

30396 Surfynol® DF-574
Organo and organo-modified silicone compound; rapid knockdown defoamer for aqueous coatings and inks. *Air Prods & Chems. Inc.*

30397 Surfynol® DF-58
Organo-modified silicone-based defoamer; self-emulsifying defoamer used in water-based applics, e.g., inks, coatings, adhesives, cements, latex containing formulations. *Air Prods & Chems. Inc.*

30398 Surfynol® DF-60
Silicon-based defoamer; for aqueous coating and ink systems, especially for grinding and dispersion stage. *Air Prods & Chems. Inc.*

30399 Surfynol® DF-695
Silicone emulsion; defoamer for aqueous ink systems. *Air Prods & Chems. Inc.*

30400 Surfynol® DF-70
Organic defoamer; defoamer for aqueous systems, printing inks, coatings, adhesives, especially acrylic and styrene-acrylic formulations. *Air Prods & Chems. Inc.*

30401 Surfynol® DF-75
Silicone-free; defoamer for water-based systems, e.g., inks, adhesives, coatings, overprint varnishes, paper coatings; especially effective in acrylic systems. *Air Prods & Chem; Air Prods & Chem Nederland BV.*

30402 Surfynol® GA
Acetylenic diol
Pigment wetting agent, grinding aid in coatings and other pigmented systems. *Air Prods & Chems. Inc.*

30403 Surfynol® PC
126-86-3 204-809-1
Acetylenic glycol
Defoamer used in paper coating and adhesive latexes; antishock agent for paper coatings. *Air Prods & Chems. Inc.*

30404 Surfynol® PG-50
126-86-3 204-809-1
Tetramethyl decynediol in propylene glycol; wetting and defoaming agent in water based systems, e.g., coatings, adhesives, inks, cements, metalworking fluids, latex dipping and paper coatings. *Air Prods & Chems. Inc.* Name unverified.

30405 Surfynol® PSA-204
Surfactant; low or nonfoaming wetting agent for pressure-sensitive adhesives. *Air Prods & Chem; Air Prods & Chem Nederland BV.*

30406 Surfynol® SE
Acetylenic diol
Wetting agent, foam control agent in pressure-sensitive adhesives, aqueous

lubricants, water-based paints, inks, dye processing, agricultural formulations, and paper coatings. *Air Prods & Chems. Inc.*

30407 Surfynol® TG
126-86-3 204-809-1
Tetramethyl decynediol and ethylene glycol; pigment and substrate wetting agent used in latex and water-reducible paints, adhesives, paper coatings, and pigmented aq. systems. *Air Prods & Chems. Inc.*

30408 Surfynol® TG-E
126-86-3 204-809-1
Acetylenic glycol and propylene glycol; low foaming and wetting agent used in pesticide formulations. *Air Prods & Chems. Inc.*

30409 Surlyn®
A trademark for a range of ionomer resins. *DuPont UK.*

30410 Surmabond Lining
Pigmented solvent-free epoxy systems; for seamless flooring and tank lining composition. *Surmak Products Ltd.*

30411 Surmabond Roadway
Tar modified solvent-free epoxy system with special aggregate; light-weight screed for footbridges, concrete floors. *Surmak Products Ltd.*

30412 Surmabond Screeding
Pigmented solvent-free epoxy system with special aggregates; chemical resistant heavy duty, light-weight screed and concrete repair material. *Surmak Products Ltd.*

30413 Surmafil
Bituminous mixture with liquid resins and aggregates; for instant repair of roads and parking areas, cold applied. *Surmak Products Ltd.*

30414 Surmaglaze U12
Moisture cured polyurethane; clear or colored dust-proofing sealer for concrete, stone, and timber floors. *Surmak Products Ltd.*

30415 Surmaplast
Solvent-free epoxy knifing systems; gap filling of blowholes in castings, repair of petrol, oil, and water tanks; quick setting. *Surmak Products Ltd.*

30416 Surmaseal 101
Chlorinated rubber paint; acid resistant paint for steelwork, concrete floors, and swimming pools. *Surmak Products Ltd.*

30417 Surmaseal 102
Solvent-based pigmented epoxy paint, polyamide cured; chemical and wear-resistant floor coating, also for corrosion protection of steel. *Surmak Products Ltd.*

30418 Surmatar
Solvent-based epoxy pitch coating; heavy duty anticorrosive coating. *Surmak Products Ltd.*

30419 Surmax® CS-504
Surfactant; alkaline-stable surfactant for formulating detergent concentrates; moderate foamer, wetting agent, detergent; for metalworking formulas, paint strippers, tire cleaners, transportation cleaners, dairy and food plant cleaners, paper felt washing, sanitizers and wax strippers. Biodegradable. *Chemax.*

30420 Surolan
A proprietary preparation containing miconazole nitrate, prednisolone; used for veterinary ear and skin infections (cats and dogs). *Janssen Pharmaceutical Ltd.* Name unverified.

30421 Surophosphate
Dasag. A German fertilizer made from sewage, other waste material, and peat.

30422 Surpassol 53
Secondary emulsifier for oil muds. *Actrachem.*

30423 Surpassol NT-57
Secondary emulsifier for oil muds; low toxicity. *Actrachem.*

30424 Survac®
Polymer. *DuPont UK.*

30425 suSCon® Blue
2921-88-2 2242 220-864-4
140 g/kg chlorpyrifos; for control of white grub in sugarcane. *Incitec Ltd.*

30426 suSCon® Green
2921-88-2 2242 220-864-4
100 g/kg chlorpyrifos; for control of black vine weevil in hardy ornamental stock; pasture grub in pasture. *Incitec Ltd.*

30427 Suscovax
Pig samonellosis vaccine. *The Wellcome Foundation Ltd.* Name unverified.

30428 Susini
An alloy of aluminum containing from 1.5-4.5% copper, 0.5-1.5% zinc, and 1-8% manganese.

30429 Suspend-Ayd® Gels
Dispersions of modified Bentones; for pigment suspension and flow control of solvent-thinned coatings. *Elementis Specialties.*

30430 Suspendex
Expanded polystyrene foam loose-fill used for cushioning and packaging applications. *Dow UK.* Discontinued.

30431 Sustain
Floor polishes. *Evans Vanodine International Ltd.*

30432 Sustane® 1-F
25013-16-5 1582 246-563-8
$C_{11}H_{16}O_2$
2(3)-*tert*-butyl-4-hydroxyanisole
BHA; Butyl Hydroxyanisole; *tert*-Butyl *p*-hydroxyanisole; Vertac; *tert*-butyl-4-hydroxyanisole; antioxyne b; *tert*-butyl-4-methoxyphenol; BOA; Antrancine 12; Embanox; Nipantiox 1-F; Tenox BHA. BHA; preservative and antioxidant for foods, flavors, cosmetics, vitamins, oils, waxes, essential oils, tallow, sausage, chewing gum base, shortening, lard, food packaging materials, potatoes, and cereals; inhibits oxidation mp = 48-55°; bp_{733} = 264-270°; insoluble in H_2O, soluble in organic solvents; LD_{50} (rat orl) = 2200 mg/kg. *UOP.*

30433 Sustane® 3
Propylene glycol, BHA, propyl gallate, citric acid, ratio 70:20:6:4; preservative and antioxidant used in snack foods, cosmetics, and spices. *UOP.*

30434 Sustane® 20
TBHQ, citric acid, propylene glycol, ration 20:10:70; preservative and antioxidant used for edible fats and oils. *UOP.*

30435 Sustane® BHA
25013-16-5 1582 246-563-8
BHA antioxidant, stabilizer for fats, oils, and other foods. *UOP.*

30436 Sustane® PG
121-79-9 8044 204-498-2
$C_{10}H_{12}O_5$
3,4,5-trihydroxybenzoic acid propyl ester
propyl gallate; n-propyl gallate; gallic acid propyl ester; Progallin P; Tenox PG; n-propyl 3,4,5-trihydroxybenzoate; Nipagallin P; PG; Nipa 49. Preservative and antioxidant for fats and oils. mp = 148-150°; soluble in H_2O (0.35 g/100 ml), more soluble in organic solvents; LD_{50} (rat orl) = 2.1-7.0 g/kg *UOP.*

30437 Sustane® TBHQ
1948-33-0 217-752-2
$C_{10}H_{14}O_2$
1,4-Benzenediol, 2-(1,1-dimethylethyl)-
mono-*tert*-butyl hydroquinone; TBHQ; 2-*tert*-Butylhydroquinone. Preservative and antioxidant for foodstuffs and meat products; color stable and useful as substitute for reactive antioxidants that tend to form purple complexes with iron or copper. mp = 127-129°; bp = 273°; LD_{50} (rat orl) = 700 mg/kg. *UOP.*

30438 Sustilan® N
Fiber preserving agent; dyeing and printing auxiliary. *Bayer AG.*

30439 Sutermeister's stain
For paper a) Contains 1.3 g iodine and 1.8 g potassium iodide in 100 ml water, and b) consists of a clear saturated solution of calcium chloride.

30440 Suttocide® A
70161-44-3 274-357-8
$C_3H_6NNaO_3$
Glycine, N-(hydroxymethyl)-, monosodium salt
sodium hydroxymethylglycinate; sodium hydroxymethyl glycinate; sodium hydroxymethylamino acetate; Suttocide A. Antimicrobial preservative for cosmetics. *Sutton Labs.*

30441 Suva®
For the chemical industry. *DuPont UK.*

30442 suXon
Insecticide, acaricide. *Incitec Ltd.*

30443 Swale
Coatings and adhesives; used for flexible packaging. *Swale Coatings & Inks Ltd.*

30444 Swale powder
An explosive containing potassium perchlorate, nitro-glycerin, collodion cotton, ammonium oxalate, wood meal, and a little nitroluene.

30445 Swalite
An explosive for coal mines, similar to Swale powder *(qv)*. .

30446 Swan
A paper-based grade of Tufnol industrial laminates. *Tufnol Ltd.*

30447 Swardsman
25-5-5, compound fertilizer. *Kemira Ince Ltd.* Discontinued.

30448 Swedelec
A Swedish charcoal iron with a high magnetic permeability.

30449 Swedex AR58P-15AC
Diacetylated tartaric acid ester of mono and diglycerides; emulsifier. No manufacturer.

30450 Swedex SSL-5AC
25383-99-7 246-939-7
Sodium stearoyl-2-lactylate with 5% anticaking agent; emulsifier. No manufacturer.

30451 Swedish factory tar
A tar obtained from waste wood in charcoal kilns, as a by-product in charcoal burning.

30452 Sweet Almond oil
8007-69-0
Lipovol ALM; Oil of sweet almond; Sweet Almond Oil BP 73; Nikkol Sweet Almond Oil; Tri-Ol ALM,. Sweet almond oil; conditioner, glosser, emollient imparting a light, nongreasy, silky afterfeel to skin and hair products; high film gloss and rapid spread; used in personal care products. Colorless liquid; sl.sol.in alcohol; misc.with benzene, chloroform, ether; insol.in water; dens. 0.910-0.915 *Lipo; Tri-K Industries; Nikko Chem. Co. Ltd.*

30453 sweet bark
Sweet wood bark, Eleuthera bark. Cascarilla; used for extracting cascarilla oil and as an ingredient in insecticides, etc.

30454 Sweeta
81-07-2 8463 201-321-0
Saccharin
Pharmaceutical aid. *Bristol-Myers Squibb Co Inc.* Name unverified.

30455 Sweetex
Low calorie sweetener. *Crookes Healthcare.*

30456 Sweetex Plus
Low calorie sweetener. *Crookes Healthcare.*

30457 Sweetrex®
Dextrose, fructose, maltose, isomaltose, other polysaccharides; directly compressible chewable tablet base with high sweetness, coolness, and mouthfeel. *Mendell.* Discontinued.

30458 sweet-water
Consists of glycerin and water, obtained in the distillation of crude glycerol.

30459 Sweetzyme
An immobilized glucose isomerase produced from a selected strain of *Bacillus coagulans*; Type Q developed specially for long-term use in a continuous fixed-bed column process for production of fructose syrup; Type A used for batch operation characterized by much higher residence times and enzyme cost compared to the fixed bed operation. *Novo Nordisk.*

30460 Swim clear
7778-54-3 1717 231-908-7
Calcium hypochlorite
Used for swimming pools. *Schaefer Salt & Chemical Co.*

30461 Swipe 560 EC
An emulsifiable concentrate containing 56 g bromoxynil, 56 g loxynil and 448 g mecoprop per liter; a post-emergence contact herbicide for cereal crops. *Ciba-Geigy Agrochemicals.*

30462 Swirl
Adjuvant containing 590 g/l refined mineral oil; for use as a wetting agent with arylalanine herbicides. *Shell UK.*

30463 Swiss Polyamid Grilon
Copolyamide
Used for tape weaving, filter fabric, and filter manufacturing process, string, embroidery etc. *EMS-Chemie AG.* Discontinued.

30464 Swiss Polyamid Grilon
Polyamide 6
Used for clothes, nonwovens, technical fabrics; and paper felts. *EMS-Chemie AG.* Discontinued.

30465 Swiss Polyester Grilene
Polyester; used for woven and knitted fabrics, home textiles and sewing threads, technical applications, various nonwovens and fiber-fill. *EMS-Chemie AG.* Discontinued.

30466 SWS-101
Dimethylpolysiloxane terminated with nonreactive trimethylsiloxy groups; features inertness, heat and oxidative stability, excellent elec. properties for dielec. coolants, brake fluids, lubricants, auto care products, release, heat transfer, aerosols, damping media, household products, antifoam agents, cosmetics, shock absorbers. *Wacker Silicones.*

30467 SWS-290
Silica-filled dimethylpolysiloxane compound; features inertness, resistance to most organic salts, diluted acids and alkalies, water, oxidation; used in water repellents, sealants, coatings for electrical equipment, lubricants for rubber/plastic molding, damping media. *Wacker Silicones.*

30468 SWS-725
Silicone rubber base; all-purpose material designed to accept nonreinforcing fillers; features low compression set, good heat resistance, excellent extrusion qualities. *Wacker Silicones.*

30469 SWS-7532u
63394-02-5
Silicone rubber compound; nonmilling specialty compound used as wire and cable insulation; normally catalyzed with Cadox TS-50; high strength, good elec. resistance, heat age retention of properties, high modulus, good shelf life and good extrusion rates; meets requirements of UL Code 62 Class 22, UL code 44, Class SA and MIL-W-8777. *Wacker Silicones.*

30470 SWS-7655u
63394-02-5
Silicone rubber compound; high tear strength material; high resilience, high green strength, good calenderability, and extrusion; when properly cured, meets specifications of ZZR-765, Class 3b, and AMS-3347; recommended catalyst levels including 1.2 phr of TS-50, 0.8 phr of Varox, or 0.4 phr of DiCup R. *Wacker Silicones.*

30471 SWS-7675u
63394-02-5
Silicone rubber compound; uncatalyzed high strength and high tear compound; high resilience and modulus, high grn. strength, low surface tack, and good mold flow and calenderability; when properly cured, meets specifications ZZR-765, Class 3b, Grade 70 amd AMS 3349. *Wacker Silicones.*

30472 SWS-7865u
63394-02-5
Silicone rubber compound; easy processing conductive compound for gaskets for EMI/RFI shielding. *Wacker Silicones.*

30473 SWS-03314
69430-24-6
Cyclomethicone
Slip agent, lubricant, release agent for personal care industry; emollient for skin care products; plasticizer for hair spray resins; replacement for IPM in aerosol antiperspirants. *Wacker Silicones.*

30474 SWS-06545u
63394-02-5
Silicone elastomer; wire and cable compound. *Wacker Silicones.*

30475 Sybol
29232-93-7 7652 249-528-5
Contains pirimiphos-methyl; garden insecticide. *ICI Garden Products.*

30476 Sydex
Soluble concentrate containing 125 g 2,4-D and 250 g mecoprop per liter; used to control weeds in grassland. *Synchemicals Ltd.*

30477 Syford
94-75-7 2865 202-361-1
2,4-D; translocated phenoxy herbicide applied to cereals and grass. *Synchemicals Ltd.*

30478 Syl
A proprietary preparation of dimethicone 350, benzalkonium solution, and nitrocellulose; used in dermatology as an antibacterial agent. *Lloyd, Hamol.* Name unverified.

30479 Sylade
Silage preservative. *ICI Chem & Polymers Ltd.*

30480 Sylfam 2082
Imidazoline
Corrosion inhibitor for oilfield servicing, refinery operations and oil transport pipes, surfactant in metalworking formulations, pigment dispersant, chemical intermediate in the synthesis of amines. *Arizona.* Discontinued.

30481 Sylfan 20
Corrosion inhibitor for oilfield servicing, refinery operations and oil transport pipes, surfactant in metalworking formulations, pigment dispersant, chemical intermediate in the synthesis of amines. *Arizona.* Discontinued.

30482 Sylfat® D-1
Monomeric fatty acid distillate from dimerization of tall oil fatty acid; dispersant for drilling muds, component of cutting oils, grinding compounds, textile drawing lubricants, defoamers, plasticizers, thickeners in greases *Arizona.*

30483 Sylfat® DX, MM
Tall oil fatty acid ester; for plasticizers, extenders, surfactants in grinding and cutting oils, specialty lubricant additives, corrosion inhibitors, specialty solvs. for printing inks, metalworking, and oil well servicing. *Arizona.* Discontinued.

30484 Sylfat® RD-1
Monomeric fatty acid distillate from dimerization of tall oil fatty acid; dispersant for drilling muds, component of cutting oils, grinding compounds, textile drawing lubricants, defoamers, plasticizers, thickeners in greases *Arizona.* Discontinued.

30485 Sylgard® 170
Two-part silicone system; low viscosity, flame retardant electrical/electronic insulating resin with temperature rating to 200°C; cure can be heat accelerated; mix ratio: 1:1 by weight. *Dow Corning.*

30486 Sylgard® 182, 184
Two-part silicone system; clear, repairable electrical/electronic insulating resin with temperature rating to 200°C, Sylgard 184 for quicker cure; mix ratio: 100:10 by weight. *Dow Corning.*

30487 Syl-off®
Silicone release coatings for paper and films. *Dow Corning.*

30488 Syloid 72
A proprietary trade name for a silica gel for addition (2%) to plasticized vinyls to prevent plate out. No manufacturer.

30489 Sylphane S
PVC shrink film; for sales display and bundling. *UCB nv Film Sector.* Discontinued.

30490 Sylphrap
A proprietary trade name for a regenerated cellulose transparent sheet. No manufacturer.

30491 Syltherm®
Silicone heat transfer fluid. *Dow Corning.* Discontinued.

30492 Syltherm® 444
Silicone-based; heat transfer liquid for flat plate and concentrating collectors. *Dow Corning.* Discontinued.

30493 Syltherm® XLT
Dimethylsiloxane polymer; used as low temperature, liquid-phase heat transfer medium for -73 to 260°C service; ideal for single-fluid processing, heating, cooling systems in the pharmaceutical and fine chemical industries. *Dow Corning.* Discontinued.

30494 Sylvacote® K
Corrosion inhibitor for oilfield servicing, refinery operations and oil transport pipes, surfactant in metalworking formulations, pigment dispersant, chemical intermediate in the synthesis of amines. *Arizona.*

30495 Sylvadym® M-35
Tall oil fatty acid-based dimer acid; for amidation and esterification reactions, manufacturing of imidazolines and amidoamine derivatives for use as corrosion inhibitors in oilfield applications, chemical intermediate in production of polyamide resins, alkyds, epoxies, lubricant additives and quaternaries. *Arizona.* Discontinued.

30496 Sylvamid®
Nonreactive polyamide resins. *Arizona.* Discontinued.

30497 Sylvan
534-22-5 208-594-5
C_5H_6O
2-methylfuran
α-methylfuran; 5-methylfuran. A constituent of wood tar. Used in perfumery. mp = -89°; bp = 62-64°; d = 0.915; n_D^{20}= 1.4330; insoluble in H_2O, soluble in organic solvents.

30498 Sylvania Cellophane
A proprietary trade name for regenerated cellulose. No manufacturer.

30499 Sylvaros® 20
Tall oil rosin; printing ink binder as resin or salt, paper sizing agent, emulsifier for SBR polymerization as soap, tackifier resin in adhesives, imidazoline modifier in corrosion inhibitors, elastomer modifier in emulsion polymerization, dust control additive, film former and plasticizer in lacquers and varnishes. *Arizona.*

30500 Sylvaros® 315
Lower acid number tall oil rosins; for manufacturing of paper size, intermediate in rosin derivative production, printing ink binders as resins, tackifier resin in sealants and mastics, starting point rosin for resin esters. *Arizona.*

30501 Sylvaros® R
Polymerized rosin; tackifier for pressuresensitive adhesives, construction adhesives, pick-up gums for labeling, adhesion promoter for difficult-to-bond substrates. *Arizona.*

30502 Sylvatac® AC
Modified tall oil rosin; for manufacturing of paper size, intermediate in rosin derivative production, printing ink binders as resins, tackifier resin in sealants and mastics, as starting point rosin for resin esters. *Arizona.* Discontinued.

30503 sylvic acid
514-10-3 3 208-178-3
$C_{19}H_{29}COOH$
Impure abietic acid; a major active ingredient of rosin; used as varnish driers, in soaps, and in the fermentation industry.

30504 Sylvid®
A range of silica fillers for plastics processing. *W R Grace & Co.* Name unverified.

30505 Symalit GM 20 PP
9003-07-0 7741
20% Glass mat-reinforced polypropylene; composite material featuring high stiffness, good impact and toughness, high energy absorption; for automotive parts. *Symalit.*

30506 Symax
Laminates of nomex, presspaper, leatheroid, paper, melinex, mylar and kapton; slot liner and closure material used in the manufacture and repair of electrical motors, transformers, and other electrical equipment. *Fothergill Tygaflor Ltd.* Name unverified.

30507 Symax
Electrical insulation material. *Courtaulds Advanced Materials (Holdings) Ltd.* Discontinued.

30508 Symel
Extruded silicone elastomer sleeving, fusible silicone rubber tapes and fabrics; for high temperature electrical applications such as lead out wires, marker sleeves, and peristaltic pumps. *Fothergill Tygaflor Ltd.* Name unverified.

30509 Symel
Electrical insulation material. *Courtaulds Advanced Materials (Holdings) Ltd.* Discontinued.

30510 Sympathy®
For the agricultural industry. *DuPont UK.*

30511 Symphony®
For the agricultural industry. *DuPont UK.*

30512 Syn Fac®
Bisphenol-A polyols
Polyester resin intermediates, reactive diluents. *Milliken.*

30513 Syn Fac® 222
Aryl polyoxyether
Low foaming, high surface tension; dispersant for latex paint, textile printing systems; fiber finish component; dispersant for degreasing solvs. used in cleaning textiles and metals; scouring aid for cellulosics; intermediate for anionics. *Milliken.* Discontinued.

30514 Syn Fac® 334
Aryl polyoxyether
Dispersant/emulsifier for pigments, insecticides, solvents, and cleaning compounds. *Milliken.*

30515 Syn Fac® 8009, 8017
Aromatic polyether polyols; reactive diluent, dispersant for waterborne coatings and other applications. *Milliken.*

30516 Syn Fac® TEA-97
Ethoxylated amine
Solvent and dispersant in the dye industry. *Milliken.*

30517 Syn Lube
Water soluble, biodegradable lubricants; fiber lubricants, textile softeners, yarn processing aids. *Milliken.*

30518 Synacril
Acrylic dyes. *ICI Chem & Polymers Ltd.*

30519 Synacto
Emulsifiers and corrosion preventives. *Exxon UK.* Name unverified.

30520 Synanthic
53716-50-0 7069 258-714-5
Oxfendazole
Anthelmintic. *Syntex Laboratories Inc.* Name unverified.

30521 Synaqua
Water soluble resins. *Cray Valley Ltd.*

30522 Synasol
64-17-5 3806 200-578-6
Denatured ethyl alcohol. *Union Carbide.*

30523 Syn-Chek 1203
Chlorinated lubricity and EP additive for metalworking lubricants, drawing and cutting oils. *Ferro/Keil.*

30524 Synchemicals Dalapon
75-99-0 2869 200-923-0
Water soluble powder containing 85% dalapon; a translocated herbicide. *Synchemicals Ltd.*

30525 Synchemicals Total Weed Killer
A suspension concentrate containing 53 g aminotriazole [61-82-5] and 110 g simazine [122-34-9] per liter; used for total weed control in non crop areas and fruit orchards. *Synchemicals Ltd.*

30526 Synclyst
Refining catalysts. *Crosfield Chemicals Ltd.*

30527 Syncol®
A range of sizes based on polyacrylic acid with specific properties and application to all types of nylon. *Allied Colloids Ltd.*

30528 Syncrolube
Fatty acid esters; for use as plastics lubricants. *Croda Surf. Ltd.*

30529 Syncrowax
Synthetic waxes. *Croda Surf. Ltd.*

30530 Syncrowax AW1-C
C_{18}-C_{36} acid
Emulsifier, emollient, opacifier. *Croda Inc.*

30531 Syncrowax BB4
Synthetic beeswax; emulsifier, emollient, opacifier; also suspending agent for anhydrous systems, auxiliary water-oil emulsifier, thickener for oils and waxes; used in creams and sticks. *Croda Inc.*

30532 Syncrowax ERL-C
C_{18}-C_{36} acid glycol ester
Emulsifier, emollient, opacifier; also lubricant, stabilizer, suspending agent for anhydrous systems, thickener, reducer of bleeding and sweating; gloss improver; sticks, creams. *Croda Inc.*

30533 Syncrowax HGL-C
C_{18}-C_{36} acid triglyceride
Emulsifier, emollient, opacifier; also lubricant, suspending agent, strength improver, stabilizer, gloss improver; used in cosmetic makeup. *Croda Inc.*

30534 Syncrowax HR-C
18641-57-1 242-471-7
$C_{69}H_{134}O_6$
Docosanoic acid, 1,2,3-propanetriyl ester
Glyceryl tribehenate. Emulsifier, emollient, opacifier; also suspending agent, thickener, gloss improver used in personal care products. *Croda Inc.*

30535 Syncrowax HRS-C
Glyceryl tribehenate
In mixture with calcium behenate. Emulsifier, emollient, opacifier; also suspending agent for anhyd. systems, auxiliary w/o emulsifier, gellant, thickener for oils and waxes. *Croda Inc.*

30536 Syndane
57-74-9 2129 200-349-0
Chlordane
Used to control earthworms in turf. *Synchemicals Ltd.*

30537 Syndite
An explosive consisting of 10-22% nitroglycerin, 0.1-0.3% collodion cotton, 45-49% ammonium nitrate, 7-9% sodium nitrate, 2-5% glycerin, 2-5% starch, and 26-28% sodium chloride.

30538 Syndraw
Water reducible lubricants containing no mineral oil; additives include surfactants, lubricity additives, fatty acid soaps, synthetic corrosion inhibitors, biocides and polymers. *Franklin Oil Corporation (Ohio).*

30539 Synektan
Leather dyes. *ICI Chem & Polymers Ltd.*

30540 Synergol
Mixture of dichlorophen, 4-indol-3-yl butyric acid, and 1-naphthylacetic acid; used to promote rooting of cuttings. *Hortichem Ltd.*

30541 Syn-Fab DC-1
Emulsifiable hydrocarbon used as a dye carrier. *Crowley Chem.*

30542 Synfluid
Synthetic base stocks for lubes and hydraulic fluids. *Monsanto (Solaris).* Name unverified.

30543 Syngran
122-34-9 8681 204-535-2
Granules containing 2% w/w simazine; a triazine herbicide to control weeds and grasses in cane fruit, roses, and some vegetables. *Synchemicals Ltd.*

30544 Synkad® 100
Borate salt; corrosion inhibitor for metalworking coolants. *Ferro/Keil.*

30545 Synkad® 200
Bromide; corrosion inhibitor for metalworking coolants. *Ferro/Keil.*

30546 Synkad® 303
Carboxylic acid salt; corrosion inhibitor for synthetic grinding fluids. *Ferro/Keil.*

30547 Synkad® 6000
Complex derivative of a carboxylate salt; rust and corrosion inhibitor for synthetic drawing and stamping fluids, synthetic and semi-synthetic cutting and grinding fluids, hydraulic fluids, and for protection of parts during short-term indoor storage. *Ferro/Keil.*

30548 Synkavit
131-13-5 5873 205-012-1
Menadiol sodium diphosphate
Vitamin K analog preparations. *Roche Laboratories; Roche Products Ltd.*

30549 Synkayvite
131-13-5 5873 205-012-1
Menadiol sodium diphosphate
Vitamin. *Hoffmann-LaRoche Inc.* Name unverified.

30550 Synkrolith
13775-53-6 237-410-6
AlF_6Na_3
Aluminate(3-), hexafluoro-, trisodium, (OC-6-11)-
synthetic cryolite; sodium hexafluoroaluminate; trisodium hexafluoroaluminate. 55% F, 13-14% Al, 32-33% Na; powder; filler for synthetic resin-bonded grinding wheels, brake linings, etc. *Metallgesellschaft AG.*

30551 Synmold
Phenol formaldehyde resin; molding powders. *Manchem Ltd.* Name unverified.

30552 Synmold
Phenolic molding powders. *Sterling Moulding Materials.* Discontinued.

30553 Syn-O-Ad® 8412
126-73-8 9749 204-800-2
tri-n-butyl phosphate
Antiwear and extreme pressure agents in non-crankcase lubricants; ferrous metal passivator; fluid base stock where inhibition of flame propagation is desired. *Akzo.* Name unverified.

30554 Syn-O-Ad® 8475M
25155-23-1 246-677-8
$C_{24}H_{27}O_4P$
dimethylphenol phosphate (3:1)
trixylenyl phosphate; tri(dimethylphenyl)phosphate. Trixylenyl (mixed) phosphate; antiwear and extreme pressure agents in non-crankcase lubricants; ferrous metal passivator; fluid base stock where inhibition of flame propagation is desired. bp_{10} = 243-265°; SG = 1.155; insoluble in H_2O, soluble in organic solvents. *Akzo.* Name unverified.

30555 Syn-O-Ad® 8478
56803-37-3 260-391-0
$C_{22}H_{23}O_4P$
Phosphoric acid (1,1-dimethylethyl)phenyl diphenyl ester
Butylated triphenyl phosphate; *tert*-Butylphenyl diphenyl phosphate; diphenyl mono(*p-tert*-butylphenyl)phosphate; (1,1-dimethylethyl)phenyl diphenyl ester phosphoric acid. Antiwear and EP agents in non-crankcase lubricants; ferrous metal passivator; fluid base stock where inhibition of flame propagation is desired. bp_5 = 245-260°; SG_{25} = 1.15; insoluble in H_2O, soluble in organic solvents. *Akzo.* Name unverified.

30556 Syn-O-Ad® 8479
29761-21-5 249-828-6
$C_{22}H_{31}O_4P$
Isodecyl diphenyl phosphate
Phosphoric acid isodecyl diphenyl ester; Isodecyldiphenyl Phosphate; IDPP; Diphenyl isodecyl-phosphate. Antiwear and extreme pressure agents in non-crankcase lubricants; ferrous metal passivator; fluid base stock where inhibition of flame propagation is desired. mp = -35°; insoluble in H_2O, soluble in organic solvents. *Akzo.* Name unverified.

30557 Syn-O-Ad® 8480
28108-99-8 248-848-2
$C_{21}H_{21}O_4P$
Phosphoric acid (1-methylethyl)phenyl diphenyl ester
Propylated triphenyl phosphate; Isopropylphenyl diphenyl phosphate. Antiwear and EP agents in non-crankcase lubricants; ferrous metal passivator; fluid base stock where inhibition of flame propagation is desired. Insoluble in H_2O, soluble in organic solvents. *Akzo.* Name unverified.

30558 Syn-O-Ad® 8484
1330-78-5 9892 215-548-8
$C_{21}H_{21}O_4P$
phosphoric acid tris (methylphenyl) ester
Tricresyl phosphate; TCP; PX-917; Celluflex 179; Kronitex TCP; Lindol. Antiwear and EP agents in non-crankcase lubricants; ferrous metal passivator; fluid base stock where inhibition of flame propagation is desired. bp_{10} = 265°; d_{25}^{25} = 1.16; n_D^{25} = 1.55; insoluble in H_2O, soluble in organic solvents. *Akzo.* Name unverified.

30559 Syn-O-Ad® 8485
56803-37-3 260-391-0
$C_{22}H_{23}O_4P$
Butylated triphenyl phosphate
Phosphoric acid (1,1-dimethylethyl)phenyl diphenyl ester; *tert*-Butylphenyl diphenyl phosphate; diphenyl mono(*p-tert*-butylphenyl)phosphate; (1,1-dimethylethyl)phenyl diphenyl ester phosphoric acid. Antiwear and EP agents in non-crankcase lubricants; ferrous metal passivator; fluid base stock where inhibition of flame propagation is desired. bp_5 = 245-260°; insoluble in H_2O, soluble in organic solvents. *Akzo.* Name unverified.

30560 Syn-O-Ad® P-310
3658-48-8 222-904-6
$C_{16}H_{35}O_3P$
Dioctyl (2-ethylhexyl) phosphite
bis(2-ethylhexyl) hydrogen phosphite. Used as amine salts in mineral and synthetic base stocks; as load-carrying additives with secondary activity as low-temperature stabilizers and metal deactivators *Akzo.* Name unverified.

30561 Syn-O-Ad® P-312
102-85-2 203-061-3

$C_{12}H_{27}O_3P$
Tri-n-butyl phosphite
Tributyl phosphite; Phosphorous acid, tributyl ester. Antioxidant and antiwear agent in gear and transmission oils mp = -80°; bp_7 = 118-125°; SG = 0.925; n_D^{25}= 1.4326. *Akzo.* Name unverified.

30562 Syn-O-Ad® P-316
1809-19-4 217-316-1
$C_8H_{19}O_3P$
Di-n-butyl phosphite
Phosphonic acid, dibutyl ester; Dibutyl phosphite. Used as amine salts in mineral and synthetic base stocks; as load-carrying additives with secondary activity as low-temperature stabilizers and metal deactivators. bp_{11} = 118-119°; d = 0.9950; n_D^{20}= 1.4231. *Akzo.* Name unverified.

30563 Syn-O-Ad® P-374
301-13-3 222-020-0
$C_{24}H_{51}O_3P$
tris(2-ethylhexyl) phosphite
trioctyl phosphite; phosphorous acid, tris(2-ethylhexyl) ester. Antioxidant and antiwear agent in gear and transmission oils $bp_{0.3}$ = 163-164°; SG = 0.902; insoluble in H_2O, soluble in organic solvents. *Akzo.* Name unverified.

30564 Syn-O-Ad® P-399
101-02-0 202-908-4
$C_{18}H_{15}O_3P$
Triphenyl phosphite
Phosphorous acid, triphenyl ester; EFED. Antioxidant and antiwear agent in gear and transmission oils mp = 22-24°; bp = 360°; d = 1.1840; n_D^{20}= 1.5905; LD_{50} (rat orl) = 1600 mg/kg. *Akzo.* Name unverified.

30565 Syn-O-Ad® P-408
Tridecyl acid phosphate
Corrosion inhibitor for iron alloys in lubricants; as extreme pressure additives when neutralized. *Akzo.* Name unverified.

30566 Syn-O-Ad® P-412
Octyl (2-ethylhexyl) acid phosphate
Corrosion inhibitor for iron alloys in lubricants; as EP additives when neutralized. *Akzo.* Name unverified.

30567 Syn-O-Ad® P-415
Isoamyl acid phosphate
Corrosion inhibitor for iron alloys in lubricants; as EP additives when neutralized. *Akzo.* Name unverified.

30568 Syn-O-Ad® P-417
n-butyl acid phosphate
Corrosion inhibitor for iron alloys in lubricants; as EP additives when neutralized. *Akzo.* Name unverified.

30569 Synocryl
Thermoplastic and thermosetting acrylic resins. *Cray Valley Ltd.*

30570 Synocure
Crosslinking acrylic resins. *Cray Valley Ltd.*

30571 Synogist
A proprietary preparation of sodium sulfosuccinate and undecylenic monoalkyl amide; used as a shampoo for dandruff. *Maltown.* Unverified.

30572 Synolac
Alkyds, unsaturated polyester resins, epoxy esters. *Cray Valley Ltd.*

30573 Synolide
Polyamide resins. *Cray Valley Ltd.*

30574 Synotex 800
Cyclized rubber resins for coatings. *Elementis Specialties.*

30575 Synouryn
Dehydrated castor oil products. *Akzo Chemie UK Ltd.*

30576 Synova
Mixed coal tar products with different viscosity with and without fillers; used for protective paint, roof paint, injection paint, etc. *Caramba Chemie GmbH.* Name unverified.

30577 Synox
Soluble concentrate containing 75 g ioxynil and 225 g mecoprop per liter; used for weed control in turf. *Synchemicals Ltd.*

30578 Synperonic
Range of nonionic surfactants. *ICI Chem & Polymers Ltd.*

30579 Synperonic 3S27 and 3S60S
Sodium 3EO sulfate of Synprol in liquid form; emulsifier, detergent, wetting and foaming agent, stable in high electrolyte concentrations and hard water, but tends to hydrolyze in acid solution; used in liquid household detergents; industrial and domestic cleaning formulations, emulsifying systems, shampoos and bubble baths. *ICI Chem & Polymers Ltd.* Name unverified.

30580 Synperonic 3S60A
Ammonium 3EO Sulfate of Synprol in liquid form; emulsifier, detergent, wetting, and foaming agent, stable in high electrolyte concentrations and hard water, but tends to hydrolyze in acid solution; used in liquid household detergents, industrial and domestic cleaning formulations, emulsifying systems, shampoos and bubble baths. *ICI Chem & Polymers Ltd.* Name unverified.

30581 Synperonic N, NX, NXP and NDB
Nonylphenol ethoxylate nonionic surfactants in liquid form; general purpose detergents and wetting agents for textile processing, metal treatment, dust suppression and general cleaning applications; emulsification of medium polarity oils and solvents. *ICI Chem & Polymers Ltd.* Name unverified.

30582 Synperonic NP10 and NP12
Nonylphenol ethoxylate nonionic surfactants in liquid form; water-soluble detergents, detergent additives, solubilizers, dispersants, and stabilizers. *ICI Chem & Polymers Ltd.* Name unverified.

30583 Synperonic NP13 and NP15
Nonylphenol ethoxylate nonionic surfactants in liquid or paste form; used in conjunction with an oil-soluble anionic surfactant, they are good emulsifiers for a range of solvents, agrochemical pesticides and herbicides. *ICI Chem & Polymers Ltd.* Name unverified.

30584 Synperonic NP20 and NP30
Nonylphenol ethoxylate nonionic surfactants in liquid or solid form; solubilizing agents and emulsifiers or coemulsifiers for highly polar substrates. *ICI Chem & Polymers Ltd.* Name unverified.

30585 Synperonic NP4, NP5 and NP6
Nonylphenol ethoxylate nonionic surfactants in liquid form; oil-soluble detergents and emulsifiers; used as intermediates for sulfation and phosphorylation to give anionic detergents, lubricants and antistatic agents; emulsifying agents for wide range of oils, waxes and solids; compatible with all other surfactants. *ICI Chem & Polymers Ltd.* Name unverified.

30586 Synperonic NP8 and NP9
Nonylphenol ethoxylate nonionic surfactants in liquid form; water-soluble, high performance detergents and wetting agents; used in textile scouring; emulsifiers for medium polarity oils and solvents. *ICI Chem & Polymers Ltd.* Name unverified.

30587 Synperonic OP
Range of nonionic surfactants of the octylphenol ethoxylate type in liquid form; water-soluble general purpose detergents, wetting agents and emulsifiers with good solution properties in the presence of alkalis and at higher temperatures. *ICI Chem & Polymers Ltd.* Name unverified.

30588 Synpro®
Metallic soaps of naturally occurring fatty acid; includes metals: calcium, zinc, magnesium, aluminum, barium and cadmium and fatty acids: stearic, palmitic and lauric; plastic lubricants and stabilizers for thermoplastics and thermosets, used for PVC, ABS, polystyrene, polyolefins and phenolics; also as lubricants in powdered metals and cosmetics; as thixotropic agents and bodying materials in greases, oils and oil well drilling muds and paints. *Synthetic Products Company.* Name unverified.

30589 Synpro® 8
557-05-1 10292 209-151-9
Zinc stearate
Process aid, lubricant for PVC, polyolefins, PS, ABS. *Syn. Prods.*

30590 Synpro® 15F
1592-23-0 1750 216-472-8
Calcium stearate
Process aid, lubricant for PVC, polyolefins, PS, ABS. *Syn. Prods.*

30591 Synpro® 90
557-04-0 5730 209-150-3
Magnesium stearate
Process aid, lubricant for PVC, polyolefins, PS, ABS. *Syn. Prods.*

30592 Synpro® 303
300-92-5 206-101-8
$C_{36}H_1AlO_5$
Aluminum distearate
Aluminum Stearate, Di; Aluminum, hydroxybis(octadecanoato-O)-; Hydroxyaluminum distearate. Process aid, lubricant for PVC, polyolefins, PS, ABS. *Syn. Prods.*

30593 Synpro® 404
637-12-7 379 211-279-5
$C_{54}H_{105}AlO_6$
Aluminum tristearate
Aluminum stearate; Aluminum tristearate; Monoaluminum stearate; Stearic acid, aluminum salt; Octadecanoic acid, aluminum salt. Process aid, lubricant for PVC, polyolefins, PS, ABS. mp = 117-120°, SG = 1.01; insoluble in H_2O, soluble in organic solvents; LD_{50} (rat orl) >5 g/kg. *Syn. Prods.*

30594 Synpro® 505 USP
Aluminum monostearate
Process aid, lubricant for PVC, polyolefins, PS, ABS. *Syn. Prods.*

30595 Synpro® Aluminum Octoate
Aluminum octoate
Process aid, lubricant for PVC, polyolefins, PS, ABS. *Syn. Prods.*

30596 Synpro® Barium Stearate
6865-35-6 229-966-3
$C_{36}H_{70}BaO_4$
Barium stearate
Octadecanoic acid, barium salt. Process aid, lubricant for PVC, polyolefins, PS, ABS. *Syn. Prods.*

30597 Synpro® Cadmium Stearate
2223-93-0 218-743-6
Cadmium stearate
Process aid, lubricant for PVC, polyolefins, PS, and ABS. *Syn. Prods.*

30598 Synpro® Calcium Pelargonate
Calcium pelargonate
Process aid, lubricant for PVC, polyolefins, PS, and ABS. *Syn. Prods.*

30599 Synpro® Stannous Stearate
Stannous stearate
Process aid, lubricant for PVC, polyolefins, PS, and ABS. *Syn. Prods.*

30600 Synprol
Detergent alcohols C9-11, C13-15. *ICI Chem & Polymers Ltd.*

30601 Synprol Sulphate
Anionic surfactant as a cream viscous liquid; for shampoos, bubble baths, liquid detergents and emulsifying systems. *ICI Chem & Polymers Ltd.* Discontinued.

30602 Synprolam
Range of synthetic fatty amines and derivative. *ICI Chem & Polymers Ltd.*

30603 Synprolam 35
Synthetic (C13/C15) alkyl primary amine in liquid form; surfactant intermediate; corrosion inhibitor; flotation agent; fertilizer anticaking agent. *ICI Chem & Polymers Ltd.* Name unverified.

30604 Synprolam 35 BQC
Benzyl quaternary ammonium chloride of a synthetic (C13/C15) dimethyl tertiary amine, in liquid form; emulsifier; general sanitizer; biocide; corrosion inhibitor; textile dyeing auxilliary; timber preservative. *ICI Chem & Polymers Ltd.* Name unverified.

30605 Synprolam 35 DM
Synthetic (C13/C15) dimethyl tertiary amine in liquid form; cationic surfactant intermediate. *ICI Chem & Polymers Ltd.* Name unverified.

30606 Synprolam 35 DMA
Acetic acid salt of a synthetic (C13/C15) dimethyl tertiary amine; emulsifier; biocide; timber preservative. *ICI Chem & Polymers Ltd.* Name unverified.

30607 Synprolam 35 N3
n-(C_{13}-C_{15}) alkyl-1,3-propane diamine in liquid form; corrosion inhibitor; bitumen adhesion agent/emulsifier. *ICI Chem & Polymers Ltd.* Name unverified.

30608 Synprolam 35A
Acetic acid salt of a synthetic (C13/C15) alkyl primary amine in solid form; emulsifier; fertilizer anticaking agent; mineral flotation. *ICI Chem & Polymers Ltd.* Name unverified.

30609 Synpron
Proprietary mixtures of various metallic soaps and salts of organic acid, antioxidants, organophosphites and lubricants supplied as solids or liquids; heat and light stabilizers for PVC flexible and rigid compounds. *Synthetic Products Company.* Name unverified.

30610 Synpron 241
Phosphite chelator
Auxiliary PVC heat stabilize providing crisp initial color and long-term stability to rigid and plasticized nontoxic formulations. *Syn. Prods.*

30611 Synpron 1009
77-58-7 3089 201-039-8
Dibutyltin dilaurate
Stabilizer providing long term stability in bottle compouds, PVC pipe, injection molded fittings. *Syn. Prods.*

30612 Synpron 1027
Antimony mercaptide
PVC heat stabilizer especially for pipe and conduit applications; recommended for use with calcium stearate for optimum stabilization; may be used in NSF potable water pipe at 1.0 phr maximum. *Syn. Prods.*

30613 Synpron 1032, 1033
A proprietary range of liquid antimony mercaptides used as heat stabilizers. *Dart Industries Inc.* Name unverified.

30614 Synpron 1538
Zinc soap; PVC stabilizer for low odor and taste applications, e.g., refrigerator gasketing. *Syn. Prods.*

30615 Synpron 1800
Barium/cadmium
Extrusion stabilizer for clear applications. *Syn. Prods.*

30616 Synpro-Ware
Dispersions of rubber or plastic chemicals in elastomers, silicones, pastes, wetted powder or pellet form; chemical dispersion for ease of handling, incorporation and safety for use in rubber, wire and cable and plastic compounding. *Synthetic Products Company.* Name unverified.

30617 Synresin RD 461
A proprietary blocked, one-component polyurethane resin; it is thermosetting and is used as a rubber flock adhesive. *Synres International NV, Holland.* Name unverified.

30618 Synsoft
Polymacon; contact lens material. *Syntex Ophthalmics Inc.* Name unverified.

30619 Synsolve
General name for range of proprietary cleaning and maintenance products; used for many areas of routine and maintenance cleaning. *Synthite Ltd.* Discontinued.

30620 Synstryp
General name for a range of proprietary paint removers and surface coating products; for removal of paint-surface coatings. *Synthite Ltd.* Discontinued.

30621 Syntase® 62
131-57-7 7088 205-031-5
Benzophenone-3
UV absorber for protection of PS, PVC, methacrylate polymers, and polyesters, against uv degradation over prolonged exposure; useful in protecting clear varnishes and lacques, linseed oil-based alkyds, and phenolic coatings intended for use on uv-sensitive surfaces. *Rhône-Poulenc Surf.*

30622 Syntase® 230
4065-45-6 9157 223-772-2
Benzophenone-4
UV absorber used in water-based cosmetics, including sun tan lotions, body creams, shampoos, hair sprays, and hair dyes, and in wool fabrics. *Rhône-Poulenc Surf.*

30623 Syntergent 55-A
Fulling and scouring agent for wool and wool/synthetic blends; provides excellent alkaline stability, effective felting action and detergency. *Henkel/Textiles.*

30624 Syntex
A proprietary trade name for oil modified alkyd resin (*qv*). No manufacturer.

30625 Synteze
Proprietary mixture; penetrating oil. *Synthite Ltd.* Discontinued.

30626 Synthacalk
Polysulfide sealant; used for general caulking and sealing in the construction industry, vertical joints, and perimeters of doors and windows, etc. *Pecora Corporation.*

30627 Synthacryl
Thermoplastic and thermosetting acrylic resins; used for automotive stoving finishes, industrial stoving finishes, and pressure sensitive adhesives. *Resinous Chemicals Ltd.*

30628 Synthamel
Air drying and stoving finishes. *ICI Chem & Polymers Ltd.* Discontinued.

30629 Synthamica
$SiO_2 \cdot Al_2O_3 \cdot MgO \cdot K_2O$ F with 3% impurities; additive to oil, paint, grease, plastics, glass and other inorganic materials which require a high dielectric or high heat resistant material; a coating for welding rods requiring special atmospheric controls; a thermal barrier and high electrical resistivity. *Mykroy/Mycalex.* Name unverified.

30630 Synthane
A proprietary trade name for phenol-formaldehyde synthetic resin laminated products and other plastics. No manufacturer.

30631 Synthappret®
Special products for the nonfelting finishing of wool; used for the textile industry. *Bayer AG.*

30632 Synthaprufe
Pitch rubber emulsion for damp proofing. *Thomas Ness Ltd.* Name unverified.

30633 Syntharesin®
Products for obtaining nonslip effects; used by the textile industry. *Bayer AG.*

30634 Synthasil
Colorless silicone water repellant for masonry brick, etc. *Thomas Ness Ltd.* Name unverified.

30635 Synthawax
Hydrogenated castor oil. *Unilever.* Name unverified.

30636 synthecite
A distillate from vulcanized rubber containing vegetable oils and waxes; used as a rubber softener.

30637 Synthemul®
Emulsifiable synthetic resins and synthetic resin emulsions. *Reichhold.*

30638 Synthemul® 40-422
Styrene-acrylic emulsion polymer; industrial air-dry maintenance coatings; outstanding adhesion, chemical resist, corrosion resistant. *Reichhold/Emulsion Polymers.*

30639 Synthemul® 40-425
Solution acrylic; thermosetting; general industrial grade with high gloss, hardness/flexibility balance and solv. system versatility. *Reichhold/Emulsion Polymers.*

30640 Synthemul® 40850-00
Vinyl-acrylic emulsion polymer; for nonwoven industry; excellent recovery, good filler acceptance; for filtration media, high loft automotive carpet. *Reichhold/Emulsion Polymers.*

30641 synthe-plastic
A reaction product of a terpene base; a rubber plastic containing no pitches or waxes.

30642 Synthetic Bone Ash
Calcium hydroxyapatite
tricalcium phosphate. Used to coat molds when casting molten purified copper and copper aloys. *Murlin Chemical Inc.*

30643 Synthetic Rutile
13463-67-7 9612 236-675-5
94% titanium dioxide. *Kerr-McGee Chemical Corp.*

30644 synthin
A product obtained by heating synthol (*qv*), at 400 C in an autoclave. A liquid results which contains saturated hydrocarbons and sulfuric acid; used as a liquid fuel.

30645 Synthite
A proprietary trade name for formaldehydes. *Synthite Ltd.* Discontinued.

30646 synthocarbone
A specially prepared charcoal for use as a fuel.

30647 synthol
A liquid fuel containing hydrocarbons, acids, alcohols, aldehydes, and esters. It is obtained by reducing carbon monoxide in water gas at high temperatures and under pressure, using iron borings coated with potassium carbonat.

30648 Syntholvar
A proprietary trade name for extruded polyvinyl chloride. No manufacturer.

30649 Syntol K77 and N77
Sodium lauryl ether sulfate (synthetic C12-C15 alcohol) as a pourable gelled paste or a low viscosity liquid; for production of all types of liquid and lotion shampoos and bubble baths; also a raw material for light duty liquid detergents, dishwashing detergents and auto shampoos. *Efkay Chemicals Ltd.* Name unverified.

30650 Synton PAO-100
37309-58-3
Polydecene
Synthetic lubricant. *Uniroyal.*

30651 Syntopon 8
Series of nonionic surfactants of the octylphenol ethoxylate type in liquid form; surfactants with range of properties and uses. *Witco Chemical Ltd.* Discontinued.

30652 Syntopon A, B, C and D
Nonylphenol ethoxylate in liquid form; nonionic surfactant. *Witco Chemical Ltd.* Discontinued.

30653 Syntopon F, G and N
Nonylphenol ethoxylate in solid form; nonionic surfactant. *Witco Chemical Ltd.* Discontinued.

30654 Syntox Total Weed Killer
A suspension concentrate containing 53 g aminotriazole [61-82-5] and 100 g simazine [122-34-9] per liter; used for total weed control in non crop areas and fruit orchards. *Syntex Manufacturing Ltd.*

30655 Syntroil
A series of polyol ester base fluid oils with various organic compounds; for various OEM applications, primarily small electric motors. *Ultrachem Inc.* Name unverified.

30656 Syntron
Phenol formaldehyde resin; molding powders. *Manchem Ltd.* Name unverified.

30657 Synvaren
A proprietary trade name for a phenol formaldehyde resin adhesive. No manufacturer.

30658 Synvarol
A proprietary trade name for an urea formaldehyde resin adhesive. No manufacturer.

30659 Synwax
8002-74-2 7155 232-315-6
Paraffin Wax; Paraffin waxes; Paraffin wax (petroleum); Poly(methylene)wax; Wax extract; Paraffin wax fume; Fischer-tropsch wax; Cream E45; Derma-Oil; Duratears; Granugen; Parachoc; Replens. Synthetic wax from long-chain saturated fatty acids. mp = 53°. *Reilly-Whiteman.*

30660 Syrian asphalt
A natural asphalt containing about 100% bituminous matter. It has a specific gravity of about 1.06, a melting point of about 100°C, and practically no mineral matter.

30661 Sys Tec® 1998
23564-05-8 9489 245-740-7
Thiophanate-methyl
A broad-spectrum systemic fungicide and wound protectant which controls a variety of diseases for use on turf and ornamentals. *Regal Chemical Company.*

30662 Systamex
53716-50-0 7069 258-714-5
Oxfendazole-based veterinary anthelmintic. *The Wellcome Foundation Ltd.* Name unverified.

30663 Systemic Fungicide
Garden fungicide. *Murphy Chemical Co Ltd.* Discontinued.

30664 Systemic Insecticide
60-51-5 3269 200-480-3
dimethoate. Insecticide. *Murphy Chemical Co Ltd.* Discontinued.

30665 Systhane
88671-89-0 6402
Myclobutanil
Light yellow solid; broad spectrum systemic fungicide. *Hoechst UK; Pan Britannica Industries Ltd; Rohm & Haas UK.*

30666 Systol M
Mixture of cymoxanil and mancozeb; used to control potato blight. *Quadrangle Agrochemicals.*

30667 Sytam
152-16-9 8540 205-801-0
Schradan. Systemic organo phosphorous insecticide. *Murphy Chemical Co Ltd.* Discontinued.

30668 Sytobex
68-19-9 10152 200-680-0
Cyanocobalamin
Vitamin. *Parke-Davis.* Name unverified.

30669 Syton
Silica sol. *Monsanto plc.*

30670 Sytron
A proprietary preparation of sodium iron edetate; a hematinic. *Parke-Davis.* Name unverified.

30671 T metal
An alloy of 95% aluminum, 4% magnesium, 0.5% silicon, 0.5% iron, and 0.1% copper.

30672 T.A.M
97692-58-5 307-710-2
Thenoyl methionine
Cosmetic ingredient rich in organic sulfur; contributes to formation of disulfur bridges in skin tissues; for scalp treatments, prevention of hair fallout. *Exsymol.*

30673 T-64
1344-28-1 369 215-691-6
Al_2O_3
aluminum oxide
Tabular alumina; used for high-temperature refractory bricks, shapes, monolithic liners; as catalyst support where high purity, low porosity and high-temperature strength are desired. *Alcoa Industrial Chemicals.*

30674 T-1061
1344-28-1 369 215-691-6
Al_2O_3
aluminum oxide
alumina. Tabular alumina; converted by heat to corundum state; very hard and dense; used in high temperature refractories in steel and aluminum industries, manufacturing of tech. ceramics, investment castings, as bed support in catalytic converters. *La Roche Chem.*

30675 T1750
4766-57-8 225-305-8
$C_{16}H_{36}O_4Si$
Tetra-n-butoxysilane
tetrabutyl orthosilicate. For heat exchange applications, as dielectric fluids; lubricant in airborne radar. bp_3 = 114-116°; d = 0.899; n_D^{20}= 1.4130. *Hüls Am.*

30676 T1920
Tetrakis (2-ethylbutoxy) silane
For heat exchange applications, as dielectric fluids; lubricant in airborne radar. *Hüls Am.*

30677 T4250
Phenyltris(trimethylsiloxy) silane

Low viscosity silicone fluid with high control over volatility, viscosity, density and refractive index, and unique solvent properties; used as a lubricant. *Hüls Am.*

30678 *TABS*

p-Menthadiene biodegradable solvent. British Traders & Shippers Ltd.

30679 Tacamahac resin

West Indian amine resin. A resin obtained from various plants, usually from *Calophyllum* species.

30680 TACC 104

Silicone; thermally conductive, nondripping heat sink grease to reduce impedance of air gap between semiconductor and heat sink. *TACC Int'l.*

30681 TACC 524

Two-component acrylic adhesive; nonflammable, fast-cure, structural adhesive for bonding wide variety of dissimilar substrates; excellent peel and tensile strength with steel, aluminum, ABS, FRP, PVC, PC, glass, wood, and ceramic. *TACC Int'l.*

30682 TACC 700-82

One-component epoxy; self-leveling, heat cured sealing compound; features rapid cure at moderate temperatures to give high heat and chemical resistance; for filter media sealing or bonding ceramics or metals. *TACC Int'l.*

30683 TACC AR-1001

Epoxy adhesive; high strength adhesive; variable curing agent level for rigid to flexible systems. *TACC Int'l.*

30684 TACC CR-3200

One-component polyurethane; conformal coating varnish which forms a resilient high gloss finish; excellent salt spray resistance, moisture barrier for ICs and PCBs. *TACC Int'l.*

30685 Tachigaren 70

10004-44-1 4905 233-000-6

$C_4H_5NO_2$

5-methyl-3(2H)-isoxazolone

Hymexazol; 3-hydroxy-5-methylisoxazole; F-319; RTY-319;. Fungicide for pelleting sugar beet seed. mp = 84-85°; soluble in H_2O (8.5 g/100 ml), very soluble in organic solvents; LD_{50} (rat orl) = 4678 mg/kg. *Sumito Chemical (UK) plc.*

30686 tachiol

7775-41-9 8657 231-895-8

AgF

silver(I) fluoride

Tachyol; silver fluoride; silver monofluoride; argentous fluoride. Anti-infective, antiseptic. mp = 435°; bp = 1150°; d = 5.8520; soluble in H_2O (1.82 g/ml).

30687 Tachryrate

Methyl-1 methyl-3-cyclohexene carboxylate

Quest Int'l. UK Ltd.

30688 tachyiite

A dark volcanic glass.

30689 Tack

7440-44-0 1855 231-153-3

C

carbon

Carbon black pastes; used for simple and dust-free dyeing of paints, lacquers, paper, cardboard, plastics, synthetic fibers, printing inks, and mineral binders. *Degussa AG.* Name unverified.

30690 Tackidex

Dextrinified starch; used for a wide range of applications in the adhesives and food industries according to viscosity requirements. *Roquette (UK) Ltd.*

30691 tackol

A mixture of oils and resins; used as a rubber plasticizer.

30692 Tacolyn

Tackifying resin dispersions. *Hercules Ltd.*

30693 Tactel

Polyamide textile fibers. *ICI Chem & Polymers Ltd.*

30694 Tactesse

Polyamide carpet fibers. *Imperial Chemical Industries plc.*

30695 Tactix

Epoxy resins used for matrix resins for the fabrication of aerospace components. *Dow Cheml Co Ltd, UK & Ireland.*

30696 Tactix 123

25928-94-3

High-purity liquid bisphenol A epoxy; thermoset performance polymer for aerospace applications. *Dow Plastics.*

30697 Tactix 556

25928-94-3

Hydrocarbon epoxy

Thermoset performance polymer for aerospace applications. *Dow Plastics.*

30698 Tactix 742

25928-94-3

Tris (hydroxyphenyl) methane-based epoxy; high thermal oxidative stability; performance polymer for aerospace applications. *Dow Plastics.*

30699 Tactix H31

Thermoset performance polymer; epoxy curing agent, hardener. *Dow Plastics.*

30700 taffy

A residue from the neutralization of the mixed organic acids produced by the fermentation of kelp-seaweed in the production of acetone. It consists chiefly of calcium propionate.

30701 Tafigel PUR 40

Polyurethane thickener for gloss emulsion paints, wood preservative stains, and adhesives. *Münzing Chemie GmbH.*

30702 Taflite 900

A proprietary range of weather-resistant high impact polystyrenes made from an EPDM graft polymerized with styrenes and dispersed as spherical microgels in polystyrene phases. *Mitsui Toatsu.* Name unverified.

30703 Tag

Aqueous emulsion of natural and synthetic waxes; fruit coating wax. *Makhteshim Chemical Works Ltd.* Discontinued.

30704 Tagat®

Polyoxyethylene glycerol fatty acid esters

Solubilizers for water insoluble substances such as flavors, perfumes, vitamin oils; dispersing and antistatic agents for technical purposes. *Goldschmidt Ltd.*

30705 Tagat® 12

69468-44-6

PEG-20 glyceryl isostearate

Preparation of oil-water emulsions; solubilizer for flavors, perfumes, vitamin oils; dispersant and antistat. *Goldschmidt; Goldschmidt AG.* Discontinued.

30706 Tagat® I

69468-44-6

PEG-30 glyceryl isostearate

Preparation of oil-water emulsions; solubilizer for flavors, perfumes, vitamin oils; dispersant and antistat. *Goldschmidt; Goldschmidt AG.* Discontinued.

30707 Tagat® L

51248-32-9

PEG-30 glyceryl laurate

Preparation of o/w emulsions; solubilizer for flavors, perfumes, vitamin oils; dispersant and antistat. *Goldschmidt; Goldschmidt AG.*

30708 Tagat® L2

51248-32-9

PEG-20 glyceryl laurate

Preparation of oil-water emulsions for pharmaceuticals; solubilizer for flavors, perfumes, vitamin oils; dispersant and antistat. *Goldschmidt; Goldschmidt AG.*

30709 Tagat® O

51192-09-7

PEG-30 glyceryl oleate

Preparation of oil-water emulsions; solubilizer for flavors, perfumes, vitamin oils; dispersant and antistat. *Goldschmidt; Goldschmidt AG.*

30710 Tagat® O2

51192-09-7

PEG-20 glyceryl oleate

Preparation of oil-water emulsions for pharmaceuticals; solubilizer for flavors, perfumes, vitamin oils; dispersant and antistat. *Goldschmidt; Goldschmidt AG.*

30711 Tagat® R40

61788-85-0

PEG-40 hydrogenated castor oil; solubilizer for water-insoluble substances, e.g., essential oils, perfumes, vitamins, cosmetic/pharmaceutical active ingredients; coemulsifier for o/w emulsions. *Goldschmidt; Goldschmidt AG.*

30712 Tagat® R60

61788-85-0

PEG-60 hydrogenated castor oil; solubilizer for water-insoluble substances, e.g., essential oils, perfumes, vitamins, cosmetic/pharmaceutical active ingredients; coemulsifier for oil-water emulsions. *Goldschmidt; Goldschmidt AG.*

30713 Tagat® R63

61788-85-0

PEG-60 hydrogenated castor oil and propylene glycol; solubilizer for water-insoluble substances, e.g., essential oils, perfumes, vitamins, cosmetic/pharmaceutical active ingredients; coemulsifier for o/w emulsions. *Goldschmidt; Goldschmidt AG.* Discontinued.

30714 Tagat® S

51158-08-8

PEG-30 glyceryl stearate

Preparation of o/w emulsions; solubilizer for flavors, perfumes, vitamin oils; dispersant and antistat. *Goldschmidt; Goldschmidt AG.*

30715 Tagat® S2
51158-08-8
PEG-20 glyceryl stearate
Preparation of oil-water emulsions for pharmaceuticals; solubilizer for flavors, perfumes, vitamin oils; dispersant, and antistat. *Goldschmidt; Goldschmidt AG.*

30716 Tagat® TO
68958-64-5
PEG-25 glyceryl trioleate
Preparation of oil-water emulsions; solubilizer for flavors, perfumes, vitamin oils; dispersant and antistat; refatting agent for hair/bath preparations. *Goldschmidt; Goldschmidt AG.*

30717 Tagit®
Reagent. *Calbiochem Corp.*

30718 ta-hong
A lead glass containing ferric oxide; used by the Chinese as a red enamel on porcelain.

30719 TAHP-80
3425-61-4 222-321-7
80% *t*-amyl hydroperoxide solution in *t*-amyl alcohol/water; initiator. *Witco Corporation.*

30720 Taifun
Alkaline floor cleaner. *Gansow UK Ltd.*

30721 tailor's chalk
French chalk (magnesium silicate) mixed with a little China clay.

30722 Tak
Mold sealing compound. *Foseco (F.S.) Ltd.*

30723 Taka-Sweet®
Glucose isomerase
Enzyme for production of fructose syrups from glucose; immobilized. *Solvay Enzymes.*

30724 Taka-Therm® L-340
9000-92-4 640 232-567-7
Bacterial α-amylase
Enzyme for starch liquefaction and textile desizing; to produce high fructose corn syrup; in liquefaction of starch fermentation media. *Solvay Enzymes.*

30725 Takatol
123-30-8 482 204-616-2
C_6H_7NO
p-Aminophenol
4-hydroxyaniline; *p*-hydroxyaniline; 4-aminophenol; 4-amino-1-hydroxybnzene; Activol; Azol; Certinal; Citol; Paranol; Rodinal; Unal; Ursol P. mp = 188-190°; bp= 284°; soluble in H_2O (0.39%), more soluble in EtOH, methyl ethyl ketone, insoluble in C_6H_6, $CHCl_3$;.

30726 takizolit
1332-58-7 5294 296-473-8
A red micro-crystalline kaolin found in Japan, and having the composition, $2Al_2O_3{\cdot}7SiO_2{\cdot}7H_2O$. It also contains appreciable amounts of rare earth oxides.

30727 ta-kong
A lead glass containing ferric oxide and used by the Chinese as a red enamel.

30728 Taktene 220
9003-17-2
Solution polybutadiene, nonstaining; used in passenger car and truck tires for improved treadwear in blends with SBR/NR; excellent low temperature properties; also for golf centers, footwear, belting, hose, floor tile, molded and extruded goods; mill processable; FDA compliant; vulcanized with sulfur-containing accelerator systems or with peroxide for maximum resilience. *Bayer Corp; Polysar.*

30729 Taktic
33089-61-1 510 251-375-4
$C_{19}H_{23}N_3$
N-(2,4-dimethylphenyl)-N-[[(2,4-dimethylphenyl)imino]methyl]-N-methylmethaniminamide
N,N'[(methylimino)dimethylidyne]di-2,4-xylidine; N-methylbis(2,4-xylyliminomethyl)amine; N,N-bis(2,4-xylyliminomethyl)methylamine; N-methyl-N'-2,4-xylyl-N-(N,2,4-xylylformimidoyl)formamidine; 1,5-bis(2,4-dimethylphenyl)-3-methyl-1,3,5-triazapenta-1,4-diene; amitraz; amitraze; OMS 1820;ENT 27967; BTS 27419; Acadrex; Acarac; Azadieno; BAAM; Bumetran; Danicut; Ectodex; Edrizar; Istambul; Maitac; Mitac; Ovasyn; Topline; Triatix; Triatox; Tudy. Animal health insecticide. Non-systemic acaricide and insecticide with contact and respiratory action. Thought to interact with octopamine receptors in the nervous system, causing an increase in nervous activity. Used for control at all stages of mites, pear suckers,scale insects, mealy bugs, whitefly, aphids and lepidoptera. mp = 86-87°; d = 1.128; soluble in H_2O (1 mg/l), more soluble in organic solvents; LD_{50} (rat orl) = 800 mg/kg. *Schering Agrochemicals Ltd; Atabay; NOR-AM; Quimica Estrela;.* Discontinued.

30730 Talbor's powder
Cinchona bark in powder form.

30731 talc
14807-96-6 9207 238-877-9
$3MgO{\cdot}4SiO_2{\cdot}H_2O$
hydrated magnesium silicate
Hydrous magnesium silicate; industrial, cosmetic, or platy talc; talcum; soapstone; steatite; talcum. Native, hydrous magnesium silicate somethimes containing small portion of aluminum silicate; ceramics, cosmetics, pharmaceuticals; as filler and pigment in rubber, paints, soaps, etc. Dusting agent; lubricant; electrical insulation. Insoluble in H_2O, organic solvents. *Cyprus Industrial Min.; Pfizer; L.A. Salomon; R. T. Vanderbilt Co Inc; Whittaker, Clark & Daniels.*

30732 Talc MS
14807-96-6 9207 238-877-9
Talc; cosmetic ingredient contributing soft, smooth feel, skin adhesion, and spreadability. *Presperse.* Discontinued.

30733 Talent
Herbicide. *May & Baker Ltd.* Name unverified.

30734 talipot
Raw palmira root flour. A starch obtained from a palm, *Corypha umbraculifera* .

30735 Talisman
15545-58-9 239-592-2
$C_{10}H_{13}ClN_2O$
N'-(3-chloro-4-methylphenyl)-N,N-dimethylurea
3-(3-chloro-*p*-tolyl)-1,1-dimethylurea; 3-(3-chloro-4-methylphenyl)-1,1-dimethylurea; chlorotoluron; chlortoluron; C 2242; Clortokem; Deltarol; Dicuran; Dicurane; Higaluron; Highuron; Ludorum; Tolurane; Tolurex; Toro. Suspension concentrate containing 500 g chlorotoluron per liter; a selective cereal herbicide, absorbed by roots and foliage. Inhibits photosynthesis. Used for pre- and post-emergence control of annual grasses. mp = 47-48°; soluble in H_2O (70 mg/l), more soluble in organic solvents; LD_{50} (rat orl) > 10,000 mg/kg. *Farmers Crop Chemicals Ltd; Agrolinz; Ciba-Geigy; Diachem; Hightex; Kemichrom; Makhteshim-Agan.*

30736 talite
A siliceous earth containing 84% silica with small quantities of oxides of iron and aluminum; used as a rubber filler.

30737 tall oil
8002-26-4 232-304-6
By-product of sulfate pulp manufacture; contains 2.2% material soluble in petroleum ether, 12.4% unsaponifiable matter, 30.4% resin acid, and 54.9% fatty acids. The resin acid consists of abietic acid, and the fatty acids contain oleic, linoleic, and linolenic acids; used as paint vehicles, in oil drilling muds, lubricants and greases, asphalt derivatives; in rubber reclaiming and as chemical intermediates,.

30738 talloel
A Swedish liquid resin obtained as a by-product in the production of cellulose from Swedish fir by the soda process; consists mainly of resin acids, and is closely related to rosin.

30739 tallow
61789-97-7 263-099-1
Beef tallow; mutton tallow. Fat derived from fatty tissue of sheep or cattle; consists primarily of fatty acid glycerides; soap stock, leather dressing, candles, greases, manufacture of stearic and oleic acids, animal feeds; adherent in tire molds. *Atlas Refinery; Norman Fox; Geo. Pfaus Sons; Reilly-Whiteman; Witco/Humko.*

30740 tallow amine
61790-33-8 263-125-1
Tallow primary amine, technical; surfactant, wetting agent, ore flotation agent, asphalt emulsifier, corrosion inhibitor, anticaking agent, pigment modifier; oil production; petrol. production additives. *Exxon/Tomah.*

30741 tallow clays
Clays containing varying proportions of zinc silicate.

30742 tallow diamine
61791-55-7 263-189-0
Tallow diamine
Surfactant, wetting agent, ore flotation agent, asphalt emulsifier, corrosion inhibitor, anticaking agent, pigment modifier; oil production; petrol. product additives. *Exxon/Tomah.*

30743 tallow fatty acid
61790-37-2 263-129-3
Industrene® 143; Fatty acids, tallow; Acids, tallow. Intermediate used in alkyd resins, rubber compounding, water repellents, polishes, soaps, abrasives, cutting oils, candles, crayons, emulsifiers; FG grades as lubricant,

release agent, binder, defoamer in foods, intermediate for food emulsifiers. *Witco Corporation.*

30744 tallow seed oil
Stillingia oil, obtained from the seeds of *Stillingia sebifera* .

30745 tallow tetramine
68911-79-5 272-787-0
Tallow tetramine
Surfactant, wetting agent, ore flotation agent, asphalt emulsifier, corrosion inhibitor, anticaking agent, pigment modifier; oil production; petrol. product additives. *Exxon/Tomah.*

30746 tallow triamine
61791-57-9 263-191-1
Tallow triamine
Surfactant, wetting agent, ore flotation agent, asphalt emulsifier, corrosion inhibitor, anticaking agent, pigment modifier; oil production; petroleum product additives. *Exxon/Tomah.*

30747 tallow trimonium chloride
8030-78-2 232-447-4
Tallow trimethyl ammonium chloride
Jet Quat T-50, T-27W; Arquad® T-27W;. Bactericide, textile softener, asphalt emulsifier, petroleum processing; dispersant; emulsifier; antistat; used in corrosion inhibitor formulations; base for hair conditioners; in agriculture. *Jetco.*

30748 tallowamine
61790-33-8 263-125-1
Tallow fatty acid amine
Genamin TA Grades; Adogen®; Amine BG; Armeen® T, TD; Crodamine 1.T; Genamin TA Grades; Jet Amine PT; Radiamine 6170, 6171; Tallow amine. Surfactant for agric. formulations. *Hoechst Celanese/Colorants & Surf.*

30749 Tally® 100 Plus
Glyceryl stearate, PEG-20 glyceryl stearate, hydrogenated soybean oil; emulsifier, dough strengthener and crumb softener used in breads; also as textile lubricant and fabric softener. *Van Den Bergh Foods.*

30750 Talon
56073-10-0 1400 259-980-5
$C_{21}H_{33}BrO_3$
3-[3-(4'-bromo[1,1'-biphenyl]-4-yl)-1,2,3,4-tetrahydro-1-naphthalenyl]-4-hydroxy-2H-1-benzopyran-2-one
Brodifacoum; PP-581; WBA-8119; Ratak.+ Rodenticide containing brodifacoum. Anticoagulant. mp = 228-230°; insoluble in H_2O, slightly soluble in organic solvents; LD_{50} (rat orl) = 270 μg/kg. *ICI Agrochemicals; ICI Chem. & Polymers Ltd.*

30751 Talon
2921-88-2 2242 220-864-4
$C_9H_{11}Cl_3NO_3PS$
phosphorothioic acid O,O-diethyl O-(3,5,6-trichloro-2-pyridinyl) ester
chlorpyrifos; O,O-diethylO-3,5,6-trichloro-2-pyridyl phosphorothioate; chlorpyrifos-ethyl; Dowco 179; ENT-27311; Dursban; Lorsban; Pyrinex. Emulsifiable concentrate containing 228g chlorpyrifos per liter; broad-spectrum insecticide with many crop uses. mp = 41-42°; insoluble in H_2O (2 ppm), very soluble in organic solvents; λ_m = 208, 230, 290 nm; LD_{50} (rat orl) 145 mg/kg. *Farmers Crop Chemicals Ltd.*

30752 Talotalo gum
Kau drega. A gum somewhat resembling gutta-percha, from Fiji.

30753 ta-lou
The Chinese term for a glass flux used for enameling on porcelain. It is mainly a silicate of lead with a little copper.

30754 Talpex
Titanium-aluminum organic complex; thixotropic agent for latex paints. *Manchem Ltd.* Name unverified.

30755 Talpheno
50-06-6 7386 200-007-0
Phenobarbital
Anticonvulsant. hypnotic, sedative. *Merrell Dow Pharmaceuticals Inc.* Name unverified.

30756 Talstar
82657-04-3 1257
An emulsifiable concentrate containing 100 g bifenthrin per liter; a residual herbicide for the control of weeds in winter cereals, oilseed rape and peas. *DowElanco Ltd.*

30757 Talunex
20859-73-8 372 244-088-0
AIP
Aluminum phosphide
Celphos; Detia; Phostoxin. Reatcs with water to give phosphine (PH_3). Used for gasing of rabbits and moles. d^{15}_4 = 2.85; stable below 1000°. *Kommer-Brookwick Ltd.*

30758 talwaan
A tanning material. It is the root of *Elephantorrhiza burchelli* .

30759 tamarac
The dried bark of *Larix larcina* , an American larch. The extract is used as an astringent and stimulant.

30760 Tamaron®
10265-92-6 6014 233-606-0
$C_2H_8NO_2PS$
O,S-dimethyl phosphoramidothioate
Methamidophos;acephate-met; Bay 71628; Filitox; Monitor; Patrole; Pillaron; SRA 5172; Tam; Tamanox;. Broad spectrum systemic insecticide with stomach and contact action used for treatment of sucking and biting insects and spider mites on a wide range of crops including cotton, tobacco, vegetables, potatoes, sugar beets. mp = 46°; d = 1.31; n_D^{40} = 1.5092; very soluble in H_2O (> 2 kg/l), less soluble in organic solvents; LD_{50} (rat orl) = 20 mg/kg. *Bayer AG.*

30761 tambookie grass
The product of *Hyperrhenice glauca* ; used for paper-making.

30762 Tamclad 7200
A proprietary PVC organosol. *Tamite Industries Inc.* Unverified.

30763 Tamguard 840, 840H and 840S
A proprietary range of PVC plastisols used for coating electroplating racks; Shore hardnesses are A90, D35 and A70 respectively. *Tamite Industries Inc.* Unverified.

30764 Tamol®
Dispersing agents; used as dispersing agents. *Rohm & Haas.*

30765 Tamol® 731-25%
26426-80-2
Isobutylene/maleic anhydride copolymer; dispersant for dyes and pigments. *Rohm & Haas.*

30766 Tamol® 850
Sodium polymethacrylate
Low foaming dispersant for inorganic pigments; for caulks, sealants, paper industry, paints. *Rohm & Haas.*

30767 Tamol® L Conc
9084-06-4
Sodium naphthalene sulfonate
Dispersant for cement, pigments, carbon black; tanning agent for leather; secondary dispersant for polymerization of rubber, in paper/pulp slurries. *Rohm & Haas.*

30768 Tamolan®
Synthetic laking agent for alkaline flexographic inks. *BASF AG.*

30769 tampicin
A resin, obtained from *Ipomoea simulans* .

30770 Tamsil 8
7631-86-9 8637 231-545-4
O_2Si
silicon dioxide
Silica. Filler for finishes, enamels, maintenance paints, plastic film antiblocks, urethane rubber, and polishes. *Unimin.*

30771 Tamsil Gold Bond
7631-86-9 8637 231-545-4
O_2Si
Silicon dioxide
silica. Filler for traffic paint, interior/exterior coatings, maintenance and marine coatings, mastics and adhesives, elec. epoxy compounds, buffing compounds. *Unimin.*

30772 Tamtam
An alloy of 78% copper and 22% tin.

30773 Tanabond STA
Durable, pressure sensitive, aqueous screen printing table adhesive. *Sybron.*

30774 Tanabron W
Wetter, detergent with good oil and wax emulsification; textile auxiliary; caustic stable. *Sybron.*

30775 tanacetone
546-80-5 9533 208-912-2
$C_{10}H_{16}O$
4-methyl-1-(1-methylethyl)bicyclo[3.1.0]hexan-3-one
thujone; 6-ketosabinane; 3-thujanone. A monoterpene ketone; used as a solvent. λ_m = 300 nm (ε 23); insoluble in H_2O, soluble in organic solvents; LD_{50} (mus sc)= 134.2 mg/kg.

30776 Tanafresh HFO
Odor masking agent. *Sybron.*

30777 Tanal
Chromium/aluminum complexes; used for tanning. *Lancashire Chemical Works Ltd.*

30778 Tanalev® 221
Anionic polymer; direct dye leveler with retarding properties. *Sybron.*

30779 Tanalith
Copper/chrome/arsenate waterborne wood preservative to prevent fungal decay and insect attack; for pressure treated timber for construction, fencing, agriculture and any application where timber requires protection. *Hickson & Welch Ltd.*

30780 Tanalon® EFA
Orthochlorotoluene biphenyl
Carrier giving good dye yield; can be used at atmospheric and higher pressures. *Sybron.*

30781 Tanalube® RF
Lubricant for rayon and polyester; prevents crease and rope marks. *Sybron.*

30782 Tanapal® LD-3, LD-3T
Leveling agent for disperse dyes. *Sybron.*

30783 Tanapel® 54
High molecular weight thermosetting resin; produces very durable water repellent textile finishes; works well with fluorocarbons. *Sybron.*

30784 Tanapon NF-200
Phosphate ester
Low foaming wetting agent for alkaline cleaning compounds, pigments, adhesive. *Sybron.*

30785 Tanapure® AC
Leveling agent for acid dyes on nylon fibers. *Sybron.*

30786 Tanasoft® PNL
Quaternary fatty amine; durable softener producing slick full hand. *Sybron.*

30787 Tanassist® JCR
Low foaming dye migrator for pressure equipment. *Sybron.*

30788 Tanastat® PH
Antistat for all fibers at low use levels. *Sybron.*

30789 Tanaterge® SCP
Nonenzymatic desizing agent for cotton and poly/cotton woven goods; stable to caustic and hard water. *Sybron.*

30790 Tanatex® Nostick
Specialty additive preventing polymer buildup on dry cans or equipment in textile industry. *Sybron.*

30791 Tanavol® URC
trichlorobenzene/biphenyl mixture. Carrier for textile industry repair work; effective for atmospheric and pressure equipment. *Sybron.*

30792 Tanawet® AR
Nonrewetting wetting agent for water repellent and fluorocarbon textile finishes. *Sybron.*

30793 Tanawet® RCN
Wetter and scouring agent for textile processing, peroxide bleaching; removes oils and waxes. *Sybron.*

30794 Tanbase®
1309-48-4 5713 215-171-9
Magnesium oxide for leather tanning. *Steetley Magnesia Products Ltd.*

30795 Tandearil
129-20-4 7106 204-936-2
Oxyphenbutazone
Anti-inflammatory; antirheumatic. *Ciba-Geigy Corp.* Name unverified.

30796 Tandem
58138-08-2 9795
$C_{10}H_7Cl_5O$
(±)-2-(3,5-dichlorophenyl)-2-(2,2,2-trichloroethyl)oxirane
(RS)-2-(3,5-dichlorophenyl)-2-(2,2,2-trichloroethyl)oxirane; Dowco 356; Nelpon. A line of herbicides based primarily on tridiphane, a selective non-systemic herbicide. Used for control of annual grass seedlings and broad-leaved weeds in maize. mp = 43°; soluble in H_2O (1.8 mg/l), very soluble in organic solvents; LD_{50} (rat orl) = 1743-1918 mg/kg. *Dow UK.* Discontinued.

30797 Tandem 11H K
Hydrated mono- and diglycerides, polysorbate 60 with sodium propionate and lactic acid; food emulsifier, conditioner/softener for yeast-raised baked goods. *Witco Corporation.*

30798 Tandem 5K, 8
9005-67-8 8872
Mono and diglycerides, polysorbate 60 with BHA, citric acid; food emulsifier, conditioner/softener for yeast-raised baked goods. *Witco Corporation.*

30799 tanekaha
The bark of *Phyllocladus trichomanoides* ; used in tanning leather.

30800 Tanigan®
Range of tanning materials comprising advanced syntans which make the tanning process safe and economical and improve the quality of the leather. *Bayer AG.*

30801 tanked oil
Linseed oil from which the moisture and other matter has settled out. It has a higher iodine value than the ordinary oil.

30802 tannal
$AlH_2O_2 \cdot C_{14}H_9O_9 \cdot 5H_2O$
Tannalum; basic aluminum tannate. An astringent used as a dusting powder.

30803 tannaline films
Gelatin films hardened by formaldehyde; used for photographic purposes.

30804 tanner's wool
Glover's wool. A wool pulled from the carcasses of slaughtered sheep with the assistance of lime. It does not dye well.

30805 Tannesco
Tanning agents for leather. *Ciba plc.* Name unverified.

30806 Tannex® MGP
Organic stabilizer/chelate for peroxide bleaching of textiles. *Sybron.*

30807 tannic acid
1401-55-4 9221 215-753-2
$C_{76}H_{52}O_{46}$
Glycerite; tannin; gallotannic acid; digallic acid; gallotannin. A mixture of organic acids occurring in the bark and fruit of many plants, e.g., oak species, sumac; mordant in dyeing; manufacture of ink, imitation tortoise shell; sizing paper; printing fabrics; tanning; clarifying beer; in photography; as coagulant in rubber; analytical chemistry reagent; astringent. *Burlington Bio-Medical; Crompton & Knowles; Fuji Chem. Ind.; Mallinckrodt; Thiem; Ulrich GmbH.*

30808 Tanolin
A proprietary trade name for a basic chromium chloride for use in chrome tanning baths. No manufacturer.

30809 Tanret's reagent
To a solution of 1.35 g mercuric chloride in 25 ml water is added a solution of 3.32 g potassium iodide in 25 ml water. This is made up to 60 ml with water and 20 mlc glacial acetic acid.

30810 Tansel
A specially prepared salt for curing hides. No manufacturer.

30811 Tansul
Clarifying agent; for beer. *Rheox Inc.*

30812 Tansul-7
A substance for the protection and stabilization against freezing of malt drinks like beer. *Rheox Inc.*

30813 tantalum
7440-25-7 9223 231-135-5
Element; Ta; used in capacitors, chemical equipment, dental and surgical instruments, rectifiers, vacuum tubes, furnace components, high-speed tools, catalyst, 'getter'. Also used in alloys in electron tubes, sutures, body implants, electronic circuitry, and thin-film components. mp = 2996°; bp = 5429°; d = 16.69; very insoluble in H_2O, all other solvents. *Aldrich; Atomergic Chemetals; Cabot; Cerac; NRC.*

30814 tantalum carbide
TaC
Cutting tools and dies, cemented carbide tools. *Atomergic Chemetals; Cerac; Noach Chem.*

30815 tantalum oxide
1314-61-0 9226 215-238-2
O_5Ta_2
tantalum pentoxide
tantalic acid anhydride. Production of tantalum, tantalum carbide; optical glass; piezoelectric and laser applications; dielectric layers in electronic circuits. Insoluble in H_2O, soluble in HF; LD_{50} (rat orl) = 8000 mg/kg. *Aldrich; Atomergic Chemetals; Cabot; Cerac; NRC.*

30816 tantcopper
A copper alloy analogous to tantiron.

30817 tantiron
An alloy of 84% iron, 15% silicon, and 1% carbon. It has a specific gravity of 6.8 and is acid-resisting.

30818 Tap Aid
ASTM S-215 Oil, 1,1,1-trichlorethane, mixed with mask odor No 3; for small hole, drilling, reaming and tapping; excellent for wire drawing and grinding. *Doyle Specialties.* Name unverified.

30819 tap cinder
The basic silicate of iron constituting the slag, and flowing through the tap-hole of the pudding furnace.

30820 Taquence
Taq DNA sequencing kit. *U.S. Biochemical.*

30821 tara
The tannin from the pods of *Coesalpinia tinctoria* .

30822 Tardex
A range of brominated organics; flame retardant additives. *ISC Chemicals Ltd.* Name unverified.

30823 Tardomyocel®, Tardomyocel L
Long-acting broad-spectrum antibiotic for animals. *Bayer AG.*

30824 Tarmac
A proprietary preparation of blast furnace slag, refined tar, and other ingredients; used for road dressing. No manufacturer.

30825 Tarmex
A proprietary name for a combination of prepared tar and mexphalte; used for road dressing. No manufacturer.

30826 tarnowitzite
(Ca·Pb) CO_3
Plumbocalcite; tarnovicite. A mineral.

30827 tarola
A coal-tar product used as a sheep dip.

30828 Taroma
Hydrated vegetable protein flavors. *Giulini Corp.*

30829 Tarot®
For the agricultural industry. *DuPont UK.*

30830 Tarslag
A proprietary preparation of cold blast slag which has been treated with a bituminous compound; used as a road dressing. No manufacturer.

30831 *meso*-tartaric acid
147-73-9 9238 205-696-1
$C_4H_6O_6$
mesotartaric acid
internally compensated tartaric acid; unresolvable tartaric acid; Antiweinsäure. Prepared by racemization of L-tartaric acid. mp = 140°; d_4^{20}= 1.666; soluble in H_2O (125 g/100 ml).

30832 tartaric acid
133-37-9 9236 205-105-7
$C_4H_6O_6$
DL-2,3-dihydroxybutanedioic acid
dihydroxysuccinic acid; dl-tartaric acid anhydrous; racemic tartaric acid; racemic acid; uvic acid; resolvable tartaric acid; paratartaric acid; dl-Weinsäure; Vogesensäure; Traubensäure. Manufacture of cream of tartar, tartar emetic, acetaldehyde; sequestrant, tanning, effervescent beverages, baking powder, fruit esters, ceramics, galvanoplastics, photography, textiles, silvering mirrors, coloring metals, acidulant in foods. mp = 206°; $d^{20}{}_4$ = 1.697; soluble in H_2O; *Bromhead & Denison Ltd; R.W. Greeff; Mallinckrodt; Penta Mfg.; RitChem.*

30833 tartarline
7646-93-7 7774 231-594-1
$KHSO_4$
acid potassium sulfate
potassium hydrogen sulfate; potassium bisulfate; potassium acid sulfate; sal enixum. Used as a substitute for tartaric acid for industrial purposes; conversion of wine lees and tartrates into potassium bitartrate; in the manufacture of fertilizers. mp = 197°; d = 2.24; soluble in H_2O (0.55 g/ml).

30834 tartars
Raw materials which contain more than 40% tartaric acid are termed tartars.

30835 tartrazine
1934-21-0 9239 217-699-5
$C_{16}H_9N_4Na_3O_9S_2$
4,5-dihydro-5-oxo--1-(4-sulfophenyl)-4-[(4-sulfophenyl)azo]-1H-pyrazole-3-carboxylic acid trisodium salt
3-carboxy-5-hydroxy-1-*p*-sulfophenyl-4-*p*-sulfophenylazopyrazole trisodium salt; C.I. Acid yellow 23; hydrazine yellow; C.I. 19140; FD&C Yellow No. 5; C.I. Food Yellow 4. A dyestuff, used to dye silk and wool. mp > 300° (dec); Freely soluble in H_2O.

30836 Tarvia
A proprietary trade name for a specially refined coal tar.

30837 tarwar
The bark of *Cassia auriculate.* A tanning material.

30838 Tasian®
Fibers. *DuPont UK.*

30839 Taski TR101
Highly concentrated dry foam carpet and upholstery shampoo. *Lever Industrial.*

30840 Taslan®
Fibers. *DuPont UK.*

30841 Tasnon®
110-85-0 7617 203-808-3
piperazine
Anthelmintic. *Bayer AG.*

30842 Tasprin
50-78-2 886 200-064-1
Proprietary preparation of soluble aspirin. *Unichem.* Name unverified.

30843 tasteless salts
7558-79-4 8805 231-448-7
HNa_2PO_4.
Sodium phosphate, dibasic
disodium hydrogen phosphate; disodium orthophosphate; disodium phosphate; DSP; phosphate of soda; secondary sodium phosphate. Used as a chelator, emulsifier and buffer. Also as a cathartic or (vet) laxative. soluble in H_2O (0.2 g/ml),.

30844 Taterfos®
7758-16-9 8713 231-835-0
$H_2Na_2O_7P_2$
Sodium acid pyrophosphate
disodium dihydrogen pyrophosphate. Additive for processed foods. mp= 220° (dec); d = 1.86; soluble in H_2O. *Rhône-Poulenc Food Ingreds.*

30845 taurine
107-35-7 9241 203-483-8
$C_2H_7NO_3S$
2-aminoethanesulfonic acid
Biochemical research, pharmaceuticals, wetting agents. mp = 300° (dec); soluble in H_2O (0.065 g/ml), insoluble in organic solvents. *Chemisphere Ltd; Mitsui Toatsu Chem.; Penta Mfg.; Schweizerhall; Tanabe USA.*

30846 Taylor
A proprietary trade name for a phenol-formaldehyde synthetic resin laminated product. No manufacturer.

30847 Taylor Oil
A patented binding material obtained by boiling raw linseed oil with driers (litharge), then forcing air through the oil when heated to 300 F, and finally heating it for some time at 500-600 F. No manufacturer.

30848 Taylor solder
An alloy of 60% tin, 12% lead, 12% silver, 8% zinc, 4% aluminum, and 4% copper.

30849 Taylors Lawn Sand
7782-63-0 4105 233-072-9
Ferrous sulfate heptahydrate
Used for moss control in turf. *Rigby Taylor Ltd.*

30850 Tazoline
A proprietary preparation of antazoline hydrochloride, octaphonium chloride, titanium dioxide and calaroine. *Rybar Laboratories Ltd.* Name unverified.

30851 TBAB
1643-19-2 216-699-2
$C_{16}H_{36}BrN$
tetrabutyl ammonium bromide
mp = 103-104°; *Hexcel; Pentagon Chemicals Ltd.*

30852 TBEP
78-51-3 201-122-9
$C_{18}H_{39}O_7P$
Tributoxyethylphosphate
Nonfoaming emulsifier, wetting agent, dispersant, leveling agent; acid and electrolyte stable. bp_4 = 215-228°; d = 1.0060; n_D^{20}= 1.4359. *Rhône-Poulenc Surf.; Rhône-Poulenc France.*

30853 TBHP-70
75-91-2 1604 200-915-7
$C_4H_{10}O_2$
t-butyl hydroperoxide
1,1-dimethylethylhydroperoxide; TBHP. *t*-butyl hydroperoxide solution with di-*t*-butyl peroxide and small amounts of *t*-butyl alcohol and water as diluents; stable below 75°; an initiator. mp = 5°; bp_{15} = 37°; d = 0.9400; n_D^{20}= 1.3840 *Witco Corporation.*

30854 Tc 99m Lungaggregate
Technetium-^{99m}Tc-aggregated albumin; diagnostic aid; radioactive agent. *Medi-Physics Inc.* Name unverified.

30855 T-Carb
471-34-1 1697 207-439-9
Calcium carbonate. *Harcros Organics.*

30856 TCC
101-20-2 9786 202-924-1
$C_{13}H_9Cl_3N_2O$
3,4,4'-Trichlorocarbanilide
N-(4-chlorophenyl)-N'-(3,4-dichlorophenyl)urea; 3,4,4'-trichlorocarbanilide; 1-(3',4'-dichlorophenyl)-3-(4'-chlorophenyl)urea; Cutisan; Nobacter; Solubacter. Bacteriostatic agents for bar soaps. mp = 255-256°. *Monsanto Co.* Name unverified.

30857 TCP
1330-78-5 9892 215-548-8
$C_{21}H_{21}O_4P$
tritolyl phosphate
phosphoric acid tris(methylphenyl) ester; tricresyl phosphate; TCP; PX-917; Celluflex 179; Kronitex TCP; Lindol. Trade name for tricresyl phosphate, a

plasticizer for cellulose lacquers, and polyvinyl chloride; it has a specific gravity of 1.185-1.189, a boiling range of 430-440 C and a flash point of 215 C; mixture of :ito,m,p+ro isomers (1:65:35). The term TCP also applies to an aqueous solution of trichlorophenyl-iodomethylsalicyl, an antiseptic and germicide. bp_{10} = 265°; d = 1.185-1.189; n_D^{25} = 1.55; soluble in organic solvents. No manufacturer.

30858 TDA-1

Phase transfer catalyst. *Rhône-Poulenc UK.*

30859 T-Det® 25-3A

Ammonium C_{12}-C_{15} alkyl ether sulfate (3EO)

High foaming detergent, general industrial cleaner, dishwashes, laundry products, car wash. *Harcros Organics.*

30860 T-Det® 25-3S

Sodium C_{12}-C_{15} alkyl ether sulfate (3 EO)

High foaming detergent, general industrial cleaner, dishwashes, laundry products, car wash. *Harcros Organics.*

30861 T-Det® C-40

61791-12-6

PEG-40 castor oil; emulsifier, solubilizer, degreaser, lubricant, dispersant, penetrant used in leather, paper, textile and metal processing, rubber, paint; dispersant for pigment slurries; leveling agent, defoamer, stabilizer; wax and polish preparations *Harcros Organics.*

30862 T-Det® D-150

9014-93-1

PEG-150 dinonyl phenyl ether

Surfactant used in hard surface cleaners. *Harcros Organics.*

30863 T-Det® DD-5

9014-92-0

Dodoxynol-5

Emulsifier, surfactant for aromatic and aliphatic hydrocarbon solvs., solvent and emulsion cleaners, agriculture, degreasers; dry-cleaning soap additive. *Harcros Organics.*

30864 T-Det® EPO-61

EO/PO block copolymer; wetting agent, defoamer for paper manufacturing; water treating compounds, metal cleaning, rinse aids, mechanical dishwash, textile; crude oil demulsifier; emulsifier and wetting agent in agricultural toxicant formulation. *Harcros Organics.*

30865 T-Det® N-1007

9016-45-9 6772

Nonoxynol-100

Emulsifier for asphalt, emulsion polymerization, textiles, high temp. and high electrolyte applications. *Harcros Organics.*

30866 T-Det® N-4

9016-45-9 6772

Nonoxynol-4

α-(4-nonylphenyl)-ω-hydroxytetra(oxy-1,2-ethanediyl). Detergent, wetting agent; agricultural toxicant formulation; intermediate for household detergents; dry cleaning soap additive, sludge dispersant additive in fuel oils; gasoline additive; solv. cleaner for metal processing; color development aid stabilize *Harcros Organics.*

30867 T-Det® N-40

9016-45-9 6772

Nonoxynol-40

Emulsifier for emulsion polymerization, asphalt, elevated temperature applications. *Harcros Organics.*

30868 T-Det® O-4

9002-93-1 6858

Octoxynol-4

Surfactant for agriculture, leather and metal processing, paint, drycleaning detergents, degreasing applications. *Harcros Organics.*

30869 T-Det® O-407

9002-93-1 6858

Octoxynol-40

Detergent, emulsifier, stabilizer for various applications. *Harcros Organics.*

30870 T-Det® RQ1

Block copolymer; low foam surfactant for mechanical dishwash, metal cleaners, rinse aids, hard surface cleaners, textile, paper; defoamer. *Harcros.*

30871 TEA-abietoyl hydrolyzed collagen

68918-77-4

Lexein® A520; Lamepon PA-TR. TEA-abietoyl hydrolyzed collagen; sebum control additive; causes delay in refatting of the scalp when used in shampoos. *Inolex.*

30872 Teaberry Oil

119-36-8 6200 204-317-7

$C_8H_8O_3$

Methyl salicylate

2-hydroxybenzoic acid methyl ester; wintergreen oil; betula oil; sweet birch oil;. Used as a flavor in foods and beverages, pharmaceuticals, and in perfumery. mp = -9°; bp = 220-224°; d_{25}^{25} = 1.184; n_D^{20}= 1.536; poorly soluble in H_2O (0.7 mg/l), solble in organic solvents; LD_{50} (rat orl) = 887 mg/kg.

30873 TEA-lauryl sulfate

139-96-8 205-388-7

$C_{18}H_{41}NO_7S$

triethanolammonium lauryl sulfate

triethanolamine lauryl sulfate; Sulfuric acid, monododecyl ester, compound with 2,2',2 -nitrilotris [ethanol] (1:1); Triethanolamine salt of lauryl sulfuric acid. Emulsifier, detergent, wetting agent, dispersant, foaming agent used for household cleaning products, cosmetics, emulsion polymerization. *Chemron; Lonza; Norman, Fox; Sandoz; Stepan.*

30874 tea-lead

An alloy of from 97-99% lead and 1-3% zinc. Also an alloy of lead with 2% tin used for wrapping tea.

30875 Team

61791-44-4 263-177-5

ethoxylated tallow amine. Cationic surfactant containing 800 g/l ethoxylated tallow amine; for use as a wetting agent for phosphonoglycine herbicides. *Monsanto plc.*

30876 Tears Plus®

9003-39-8 7879 294-352-4

1-ethenyl-2-pyrrolidinone polymers

Polyvinylpyrrolidone; povidone; PVP; 1-vinyl-2-pyrrolidinone polymers; polyvidone; P.V.P.; RP-143; Kollidon; Periston; poly[1-(2-oxo-1-pyrrolidinyl)ethylene]; Plasdone; Plasmosan; Protagent; Subtosan; Vinisil; Haemodyn. Ocular lubricant. Soluble in H_2O, EtOH, $CHCl_3$, insoluble in Et_2O; *Allergan Inc.*

30877 Teatcote Plus

A proprietary preparation of polyhexanide; a veterinary antibacterial. No manufacturer.

30878 Tebol 88, 99

71-36-3 1575 200-751-6

$C_4H_{10}O$

n-butyl alcohol

n-butanol; 1-butanol; butyl alcohol; propyl carbinol. Butyl alcohol (99: high purity); solvent, cosolvent, compatibilizer, coupling agent, processing aid for pharmaceuticals, personal care products, aqueous coatings and adhesives, agricultural formulations, polymer processing, cleaners/disinfectants; chlorinated hydrocarbon stabilizer. mp = -90°; bp = 117-118°; d_4^{20}= 0.810; n_D^{20} = 1.3993; soluble in H_2O (0.091 ml/ml), more soluble in organic solvents; LD_{50} (rat orl) = 4.36 g/kg. *Arco Chemical Co; Arco Chemical Europe.*

30879 tebutam

35256-85-0 252-470-3

$C_{15}H_{23}NO$

2,2-Dimethyl-N-(1-methylethyl)-N-(phenylmethyl)propanamide

N-Benzyl-N-isopropylpivalamide; N-Benzyl-N-isopropyltrimethylacetamide; tebutame; butam; Comodor; GPC-5544; Ro 14-9480/000. Selective herbicide, acting by inhibition of weed germination. Used for pre-emergence control of annual grasses and broad-leaved weeds. BP;si0.1 = 95-97°; d_{25} = 0.975; soluble in H_2O (0.79 mg/l. 25°), readily soluble in organic slvents; LD_{50} (rat orl) = 6210 mg/kg *Maag; ICI Farm Protection; La Quinoleine; Roche.*

30880 Tec

A proprietary trade name for cellulose acetate varnish resins. No manufacturer.

30881 Tecagg

Very low density mineral aggregates comprising foamed waste products or clays; used for insulating building products, refractory insulation. d= 0.3 - 0.8 *Filtec Ltd.*

30882 Tecane

Selective herbicide. *Schering Agrochemicals Ltd.* Discontinued.

30883 Teccel

Hollow glass microspheres, expanded minerals; white, mono or multicellular lightweight fillers of density 0.15 - 0.6 gm/cc; used for explosives, deep submergence buoyancy, paints, and cultured marble. *Filtec Ltd.*

30884 Tec-Char

A proprietary trade name for a granular charcoal. No manufacturer.

30885 Tecfil

Name of a range of lightweight mineral fillers in particular cenospheres; hollow ceramic microspheres derived from fly-ash; composed primarily of silica and alumina, size 5-300 μm, density 0.5 to 0.8 gm/cc; frequently shortened to 'T' with a suffix, e.g *Filtec Ltd.*

30886 Tecgran

117-18-0 204-178-2

$C_6HCl_4NO_2$

1,2,4,5-tetrachloro-3-nitrobenzene

Arena; Bygran; Easytec; Fusarex; Hickstor; Hystore; Hytec; Nebulin; Tubodust; Tubostore. Fungicide with protective and curative action. Tecgran is provided as granules or dispersible powder containing tecnazene; protectant fungicide and potato sprout suppressant. mp = 99°; bp = 304° (dec); soluble in H_2O (0.44 mg/l), more soluble in roganic solvents; LD_{50} (rat orl) = 2047 mg/kg. *Atlas Interlates Ltd; ICI.*

30887 Techmate
Resilient molded foam used for cushion packaging applications. *Dow Cheml Co Ltd, UK & Ireland.*

30888 Techne Coll
9256
Technetium Tc 99m sulfur colloid; radioactive agent. *Mallinckrodt Inc.* Name unverified.

30889 TechneScan MAA
Albumin, aggregated; diagnostic aid. *Mallinckrodt Inc.* Name unverified.

30890 TechneScan PYP
15578-26-4 8945 239-635-5
Stannous pyrophosphate
Complex of ^{99m}technetium with stannous pyrophosphate; used as a diagnostic aid. *Mallinckrodt Inc.* Name unverified.

30891 TechneScan SSC
Stannous sulfur colloid; diagnostic aid. *Mallinckrodt Inc.* Name unverified.

30892 Technyl
Nylon injection molding compounds. *Rhône-Poulenc UK.*

30893 Techroline
Gasoline additive. *Monsanto (Solaris).* Name unverified.

30894 Techron
Gasoline additive. *Monsanto (Solaris).* Name unverified.

30895 Techster
Polybutylonterephthalate injection molding compounds. *Rhône-Poulenc UK.*

30896 Techtron™ PPS
9016-75-5
polyphenylene sulfide
PPS; stock shapes for demanding applications requriing resistance to high heat and hostile chemical environments; for high pressure liquid chromatography, chemical processing, automotive, electrical/electronic, industrial, consumer goods, medical devices. *Polymer Corp.*

30897 Teclam®
For the electrical industry. *DuPont UK.*

30898 tecnazene
117-18-0 204-178-2
$C_6HCl_4NO_2$
1,2,4,5-tetrachloro-3-nitrobenzene
Arena; Bygran; Easytec; Fusarex; Hickstor; Hystor; Hytec; Nebulin; Tubodust; Tubostore. Tecnazene (6% w/w); used to control dry rot in both ware and seed potatoes, and sprouting in ware potatoes. mp = 99°; bp = 304° (dec); soluble in H_2O (0.44 mg/l), more soluble in organic solvents; LD_{50} (rat orl) = 2047 mg/kg. *Tripart Farm Chemicals Ltd.*

30899 Tecnocin
A proprietary range of curing agents for fluorocarbon rubbers. *Montedison UK Ltd.* Name unverified.

30900 Tecnoflon®
A proprietary range of fluorocarbon based rubbers. *Montedison UK Ltd.* Name unverified.

30901 Tecnoflon® FOR-45C2/R
Fluoroelastomer copolymer; with curative; low cross-link density polymer providing good hot tear resistance for complicated shapes, for low modulus and high elongation items; used for valve stem seals, bellows, custom molded goods. *Ausimont.* Name unverified.

30902 Tecnoflon® FOR-65BI/R
Fluoroelastomer copolymer; with curative; for compression or transfer molding of general purpose items. *Ausimont.* Name unverified.

30903 Tecnoflon® FOR-LHF
Fluoroelastomer polymer; with curative; for low-hardness articles (Shore A45-50) with good low-temp. resistance; used as processing aid when blended with standard copolymers or terpolymers; excellent processing and extrudability. *Ausimont.* Name unverified.

30904 Tecnoflon® NH, NM, NMB, NML, NMLB
Fluoroelastomer copolymer; without curative; polymers for use alone or together to achieve required visco, mold flow, tear strength for custom formulations; curable with M1 (2-5 phr) and M2 (1-2 phr) for general-purpose compounds; also curable with diamin *Ausimont.* Name unverified.

30905 Tecnoflon® P-1, P-2, P-2HV, P-40
Fluoroelastomer polymer; without curative; peroxide-curable grades with improved chem. and acid resistance, improved resistance to steam and bases, and amine-containing fluids; P-40 developed with excellent flow characteristics for extrusion. *Ausimont.* Name unverified.

30906 Tecnoflon® TN-LATEX
Fluoroelastomer polymer; with curative; 70% FKM solids emulsion used to make hightemp. chem.-resist. coatings; applied by spraying, dipping, or silk screening; excellent substitute for solv.-based fluoroelastomer fabric coatings; resistant to oils and cor *Ausimont.* Name unverified.

30907 Tecnoprene
A proprietary range of filled polymer molding granules. *Montedison UK Ltd.* Name unverified.

30908 Tecoflex Polyurethane
Linear segmented aliphatic polyether polyurethane, medical grade elastomer; for medical products, tubing. *Thermedics Inc.* Name unverified.

30909 Tecpril
High quality foamed clay aggregates, in form of white, regular pellets; density 0.1 - 0.6 gm/cc; for high tech insulation products, fireproof composites, aerospace/hydrospace composites. *Filtec Ltd.*

30910 Tecquinol® Tech. Grade
123-31-9 4853 204-617-8
$C_6H_6O_2$
p-hydroxybenzene
Hydroquinone, 1,4-benzenediol; hydroquinol; quinol; Aida; Black and White Bleaching Cream; Eldoquin; Eldopaque. Antioxidant for synthetic latexes, fats, oils, monomers, polyester resins. mp = 170-171°; bp = 285-287°; d_{15} = 1.332; soluble in H_2O (0.07 g/ml), more soluble in organic solvents; LD_{50} (rat orl) = 320 mg/kg. *Eastman.* Name unverified.

30911 Tecsol®
Denatured alcohol. *Eastman.*

30912 Tectilon
Acid dyes. *Ciba plc.* Name unverified.

30913 Tecto, Tecto 60%
148-79-8 9426 205-725-8
$C_{10}H_7N_3S$
2-(4-thiazolyl)-1H-benzimidazole
2-(thiazol-4-yl)benzimidazole; thiabendazole; 4-(2-benzimidazolyl)thiazole; MK-360; Omnizole; Thiaben; Thibenzole; Bovizole; Eprofil; Equizole; Mintezol; Top Form Wormer; Mertect; Lombristop; Minzolum; Nemapan; Polival; TBZ; 2-(1,3-thiazol-4-yl)benzimidazole; 2-(4-thiazolyl)benzimidazole. A systemic fungicide with protective and curative action. Used for control of various fungal species in fruits and vegetables. Tecto 60% is a wettable powder containing 60% w/w thiabendazole. mp = 304-305°; insoluble in H_2O; soluble in organic solvents; LD_{50} (rat orl) = 3300 mg/kg. *MSD Agvet.*

30914 Tectrode
Cathodic protection. *Imperial Chemical Industries plc.*

30915 TEDA-D007
Stannous octoate; standard catalyst for PU flexible foam; also for slabstock and elastomer shoe soles. *Tosoh.*

30916 TEDA-L33
280-57-9 9801 205-999-9
$C_6H_{12}N_2$
triethylenediamine
Dabco; 1,4-diazabicyclo[2.2.2]octane. TEDA-L33 is triethylenediamine in dipropylene glycol; gelling catalyst for PU slabstock and molded flexible, rigid, and elastomer shoe shoe applications. mp = 158°; bp = 174°; soluble in H_2O (45 g/100g), similarly soluble in organic solvents. *Tosoh.*

30917 TEDA-T411
77-58-7 3089 201-039-8
$C_{32}H_{64}O_4Sn$
Dibutyltin dillaurate
Standard catalyst for PU elastomer applications; also for slabstock, semirigid, and rigid applications. d = 1.066; n_D^{20}= 1.4710. *Tosoh.*

30918 Tedimon
Isocyanates. *Montedison UK Ltd.* Name unverified.

30919 Tedion V-18
116-29-0 9339 204-134-2
$C_{12}H_6Cl_4O_2S$
1,2,4-trichloro-5-[(4-chlorophenyl)sulfonyl]benzene
4-chlorophenyl-2,4,5-trichlorophenyl sulfone; 2,4,4',5-tetrachlorodiphenyl sulfone; *p*-chlorophenyl 2,4,5-trichlorophenyl sulfone; Acaroil TD; Acarvin; Agrex T-7.5; Aracnol K; Mitifon; Tetranol; V-18. Tetradifon; emulsifiable concentrate containing 125 g propiconazole, 350 g tridemorph per liter; non-systemic selective acaricide for use against mite infestation in orchards, citrus fruit plantations, hop fields, groundnut plantations, vegetable plots, cotton fields and on ornamental plants; red spider mite control in horticultural crops. mp = 148-149°; d_{20} = 1.515; soluble in H_2O 0.08 mg/l), more soluble in organic solvents *Duphar BV; Hortichem Ltd.*

30920 Tedlar®
Clear or pigmented polyvinyl fluoride film with high resistance to weathering; generally chemically inert. *DuPont UK.*

30921 Tedur®
Polyphenylene sulfide resins. *Bayer AG; Miles.*

30922 Tedur® KU1-9510-1
9016-75-5
Polyphenylene sulfide, 40% glass reinforced; offers high strength, heat distortion temperature, excellent rigidity, good electrical properties, and outstanding chemical resistance; suggested for parts exposed to high service temperatures and mechanical loads in chemically aggressive environments, e.g., automotive, electrical/electronic (halogen lamp sockets, switch housings, fuse boxes, capacitors), industrial/mechanical (pump components, heavy equipment parts, replacement for metals). *Miles.* Discontinued.

30923 Tedur® KU1-9511
9016-75-5
Polyphenylene sulfide, 45% glass fiber-reinforced; injection molding grade. *Bayer AG.* Name unverified.

30924 Tedur® KU1-9530
9016-75-5
Polyphenylene sulfide, 60% mineral/glass fiber-reinforced; embedding material. *Miles.* Discontinued.

30925 Tedur® KU1-9552
9016-75-5
Polyphenylene sulfide, 45% glass fiber/mineral-reinforced; conductive injection molding grade. *Bayer; Miles.*

30926 Tedur® KU1-9561
9016-75-5
Polyphenylene sulfide, 60% mineral/glass fiber-reinforced; reflector grade. *Miles.* Discontinued.

30927 Teebrix
Preformed inserts for extending bottom plate lives. *Foseco (F.S.) Ltd.*

30928 Teefroth
Polyglycol ethers; froth flotation agent. *ICI Chem & Polymers Ltd.*

30929 Teepol CM44
4378
C_9-C_{13} alkyl benzene sulfonate, sodium salt
Gardinol. An aqueous solution of sodium salt of C_9-C_{13} alkyl benzene sulfonate; mainly used as a constituent for other detergent blends. *Shell.* Name unverified.

30930 Teepol FC5
Gardinol. A high foam detergent for repackers; a solution of a primary alcohol ethoxysulfate. *Shell.* Name unverified.

30931 Teepol GD53
Gardinol. A highly biodegradable, clear, pale amber liquid containing formalin as a preservative; specially formulated for dishwashing and general cleaning. *Shell.* Name unverified.

30932 Teepol HB6
C_9-C_{13} primary alcohol sulfate, sodium salt
Gardinol. A highly biodegradable aqueous solution of the sodium salts of C_9-C_{13} primary alcohol sulfate; contains formalin as a preservative; used as a solubilizer and emulsifier component of hard surface cleaners and germicidal cleaners. *Shell.* Name unverified.

30933 Teepol PB
primary alcohol ethoxy sulfate, sodium salt
Gardinol. Sodium primary alcohol ethoxy sulfate as a clear aqueous solution; foaming agent used in gypsum wall-board manufacture. *Shell UK.* Name unverified.

30934 teerlack
Coal tar pitch.

30935 Tefaire®
Fibers. *DuPont UK.*

30936 Teflon®
7743
Polytetrafluoroethylene
Polytetrafluoroethylene (PTFE) plastic material having good resistance to high temperatures. *DuPont UK.*

30937 Teflon® F.E.P
Hexafluoropropylene copolymers. *DuPont UK.*

30938 Tefose® 63
PEG-6-32 stearate and glycol stearate; selfemulsifying base for cosmetics, oil-water pharmaceutical ointments; excellent skin and mucosal tolerance; especially for anti-mycotic preparations. *Gattefosse; Gattefosse SA.*

30939 Tefose® 1500
9004-99-3
PEG-6-32 stearate; self-emulsifying base for oil-water cosmetic or pharmaceutical emulsions. *Gattefosse; Gattefosse SA.*

30940 Tefose® 2000
PEG-6 stearate, ceteth-20, steareth-20; self-emulsifying base for oil-water cosmetic or pharmaceutical emulsions. *Gattefosse; Gattefosse SA.*

30941 Tefose® 2561
PEG-6 stearate, glyceryl stearate, and ceteth-20; base for oil-water cosmetic or pharmaceutical emulsions. *Gattefosse; Gattefosse SA.*

30942 Tefzel® 200
A proprietary ETFE fluoro-polymer resin extruded for use as wire insulation. *DuPont UK.*

30943 Tegiloxan®
Methylsilicone oils. Antifoams for mineral oils, in rubber and plastics industry; lubricant for tire production; additive for polishes. *Goldschmidt.*

30944 Tegin®
Partial esters of glycerol with various fatty acids. *Thomas Goldschmidt Ltd.*

30945 Tegin® 4011
31566-31-1 4498 250-705-4
Glyceryl stearate; emulsifier for pharmaceuticals. *Goldschmidt; Goldschmidt AG.* Discontinued.

30946 Tegin® O
25496-72-4 247-038-6
Glyceryl mono/dioleate; emulsifier for water-oil emulsions for pharmaceuticals. *Goldschmidt; Goldschmidt AG.*

30947 Tegin® V
Glyceryl stearate SE; self-emulsifying grade forming oil-water emulsion for creams and lotions. *Goldschmidt.*

30948 Teginacid
Glycerol mono-distearates with other nonionics. *Thomas Goldschmidt Ltd.*

30949 Teginacid C
68439-49-6 7737
Emulgator E 2568 SE. Ceteareth-25; emulsifier and stabilizer for cosmetic and pharmaceutical oil-water emulsions. *Goldschmidt; Goldschmidt AG.*

30950 Teglac
A proprietary trade name for an alkyd synthetic varnish and lacquer resin. No manufacturer.

30951 TegMeR® 703
PEG-3 diheptanoate
Lubricant for aluminum can industry. *C.P. Hall.*

30952 TegMeR® 803
Triethylene glycol di-2-ethylhexoate; plasticizer for adhesives. *C.P. Hall.*

30953 TegMeR® 804 Special
PEG-4 di-2-ethylhexoate
Lubricant for aluminum can industry. *C.P. Hall.*

30954 TegMeR® 903
PEG-3 dipelargonate
Lubricant for aluminum can industry. *C.P. Hall.*

30955 Tego® Airex 900
Polysiloxane modified with organic groups; deaerator for solvent-containing and solvent-free systems based on epoxide and unsaturated polyesters. *Tego GmbH.* Name unverified.

30956 Tego® Airex 960
Silicone modified organic polymer; deaerator for medium and high polar coating systems, especially polyurethane paints; also for high solids alkyds, ES spray systems, and urethane-polyesters. *Tego GmbH.* Name unverified.

30957 TEGO Amid S18
7651-02-7 231-609-1
Stearamidopropyl dimethylamine
Tegamine® 18. Surfactant, conditioner for hair care and bath products; auxiliary emulsifier for creams and lotions; provides skin feel. *Goldschmidt.*

30958 Tego® -Antiflamm® N
Flame retardant for PU foams. *Goldschmidt.* Discontinued.

30959 Tego® -Antifoam
Organic antifoam for waste water treatment. *Goldschmidt.*

30960 Tego Betain CK D
Cocamidopropyl betaine
Tegotain D. Mild surfactant. *Goldschmidt AG.*

30961 Tego® -Betaine C
61789-40-0 263-058-8
Cocamidopropyl betaine
Surfactant used as foam stabilizer and viscosity builder in personal care products, dishwash, liquid soap. *Goldschmidt.* Discontinued.

30962 Tego® -Betaine E
61789-40-0 263-058-8
Cocamidopropyl betaine
Tego® -Betaine L-5351. Low-salt surfactant for use in salt-critical formulations, e.g., hair dyes, fixatives, reactive products, aerosols, electrolyte-sensitive formulations. *Goldschmidt; Goldschmidt AG.*

30963 Tego® -Betaine HS
Cocamidopropyl betaine and glyceryl laurate; for nonirritating shampoos, baby care products; good refatting properties. *Goldschmidt; Goldschmidt AG.*

30964 Tego® -Betaine L-7
61789-40-0 263-058-8
Cocamidopropyl betaine
Surfactant, detergent, emulsifier, foam stabilizer, viscosity builder in low-irritation personal care products, pharmaceuticals. *Goldschmidt; Goldschmidt AG.*

30965 Tego® -Betaine L-90
Lauramidopropyl betaine
Surfactant, foam stabilizer, viscosity builder for shampoos, bath, dishwash, liquid soap, and cream and lotion products. *Goldschmidt.* Discontinued.

30966 Tego® -Betaine S
Cocamidopropyl betaine
Surfactant, foam stabilizer, viscosity builder for shampoos, bath, dishwash, liquid soap, cream and lotion products. *Goldschmidt.* Discontinued.

30967 Tego® Care 150, Care 300
31566-31-1 4498 250-705-4
Glyceryl stearate
steareth-25; ceteth-20; stearyl alcohol. Wax-like oil-water balanced emulsifier systems for creams and lotions; optimized for ease of formulation. *Goldschmidt; Goldschmidt AG.*

30968 Tego® Care 450
Stearyl glucoside
Oil-water emulsifier producing stable emulsions with cosmetics oils and waxes; emulsifier for creams, lotions, hair conditioners, sun products; for systems with pH of 6-9. *Goldschmidt.* Discontinued.

30969 Tego® Dispers 610
Higher molecular weight unsaturated polycarboxylic acid; wetting and dispersion additive to counter sedimentation and flooding of pigments; produces selective flocculation of pigments and extenders; stabilizer for pigment dispersions; used in binder systems such as alkyd, acrylate, polyester/melamine, polyisocyanate; nitrocellulose paints; chlorinated polymers. *Tego GmbH.* Name unverified.

30970 Tego® Dispers 630
Salt of a higher molecular weight polycarboxylic acid and amine derivative; wetting and dispersion additive against sediment, sagging, and flooding; produces selective flocculation of pigments and extenders, stabilizes pigment dispersion; for binder systems such as alkyd, acrylate, polyester/melamine, polyisocyanate; nitrocellulose paints; chlorinated polymers. *Tego GmbH.* Name unverified.

30971 Tego® Effect L 104
Polyurethane-based solution; thickener for aqueous polymer dispersions. *Tego GmbH.* Name unverified.

30972 Tego® Emulsion 3454
Silicone emulsion; release agent for production of food pkg. *Goldschmidt AG.*

30973 Tego® Emulsion ASL
Silicone emulsion; release agent for natural and synthetic rubbers and the foundry industry. *Goldschmidt AG.*

30974 Tego® Emulsion PK
Silicone emulsion; release agent for coating of building seals; avoids stress cracking in acrylic glass. *Goldschmidt AG.*

30975 Tego® Flow 425
68937-54-2
Polysiloxane-polyether copolymer; flow and leveling additive for clear water and solv.-based paint systems (alkyd, acrylate, PU, alkyd-melamine, PU-acrylic), building protection paints, wood/furniture varnishes. *Tego GmbH.* Name unverified.

30976 Tego® Foamex 800
68937-54-2
Polysiloxane-polyether copolymer oil-water emulsion; defoamer for water-based emulsion paints and water-thinnable systems, PU paints, PU-acrylics, furniture paint, wood varnish, adhesives, dispersion paints, polymer dispersions. *Tego GmbH.* Name unverified.

30977 Tego® Foamex 3062
67762-96-3
Dimethicone copolyol
Defoamer for water-based emulsion paints and water-thinnable systems, building protection coatings (acrylic, styrene-acrylic, PVA), wood/furniture varnishes, printing inks. *Tego GmbH.* Name unverified.

30978 Tego® Foamex KS 10
Paraffin-based mineral oil, polysiloxane polyether copolymer; defoamer for emulsion paints, building protection coatings (styrene-acrylate, PVAc, vinyl propionate, S/B). wood/furniture varnishes. *Tego GmbH.* Name unverified.

30979 Tego® Foamex N
8050-81-5 3264
Simethicone
Defoamer concentrate for medium, high solid and solvent-free paint systems, building protection coatings, anticorrosive coatings, high viscosity paints. *Tego GmbH.* Name unverified.

30980 Tego® Glide 100
Dimethicone copolyol
Surfactant, mar resistance aid and flow additive for water and solvent-based paints (alkyd, saturated polyester, polyacrylate), building protection coatings, anticorrosive paints, car finishes, printing inks. *Tego GmbH.* Name unverified.

30981 Tego® Glide 410
Dimethicone copolyol
Mar resistant additive for solv.-based paints, primers, extenders, automotive paints, building protection coatings, anticorrosion paints, wood/furniture varnishes, industrial paints, lacquers, aq. and solv.-based printing inks; reduces surface tension, im *Tego GmbH.* Name unverified.

30982 Tego® Hammer 300000
Methyl silicone oil; hammer finish additive for solvent-based paints; low surface tension. *Tego GmbH.* Name unverified.

30983 Tego® Heat Conductive Paste Z
Conductive paste for fitting diodes, transistors and embedding thermocouples. *Goldschmidt AG.*

30984 Tego® IMR 918
Polysiloxane polyoxyalkylene block copolymer; additive in production of reaction injection molding parts and integral skin foams of high density. *Goldschmidt AG.* Discontinued.

30985 Tego® -Pearl B-48
Cocamidopropyl betaine, glycol distearate, cocamide DEA, cocamide MEA
Pearlescent and opacifier for hair care products. *Goldschmidt; Goldschmidt AG.*

30986 Tego® -Pearl S-33
Sodium C_{14}-C_{16} olefin sulfonate, glycol distearate, cocamidopropyl betaine, sorbitan laurate
Tego Pearl N100, Tego Pearl N 300. Pearlescent for shampoos, bath and shower preparations with excellent superfatting properties; cold processable. *Goldschmidt AG.*

30987 Tego® Phobe 1030
Solvent-free silicone resin; additive to regulate the uptake of foundation solution during offset printing. *Tego GmbH.* Name unverified.

30988 Tego® Phobe L 1004
Fluorocarbon resin aqueous emulsion; permanent water and soil repellent for leather finishing. *Tego GmbH.* Name unverified.

30989 Tego® Release Agent M 379
Organic modified polysiloxane; internal release agent for thermosetting resins. *Goldschmidt AG.*

30990 Tego® Silicone Acrylate RC
Silicone acrylate; for release coatings to be cured by electron beam or uv-radiation in the paper and film industries. *Goldschmidt.*

30991 Tego® Silicone Paste A
Lubricant with low load bearing properties for screw threads, rubber and plastic seals and gaskets. *Goldschmidt AG.*

30992 Tego® WetKL 245
Dimethicone copolyol
Substrate wetting and spreading additive for solvent and water-based systems, wood/furniture varnishes, industrial/household appliance paints, heat-resistant coatings. *Tego GmbH.* Name unverified.

30993 Tegoamin®
Amine catalysts. *Thomas Goldschmidt Ltd.*

30994 Tegoamin® 33
280-57-9 9801 205-999-9
Triethylene diamine
Catalyst for the manufacturing of PU foams and elastomers. Solution in water; sp. gr. 1.033±0.005. *Goldschmidt.* Discontinued.

30995 Tegoamin® DMEA
108-01-0 2900 203-542-8
$C_4H_{11}NO$
2-(dimethylamino)ethanol
Deanol; Dimethylethanolamine; N,N-dimethyl-2-hydroxyethylamine; β-dimethylaminoethyl alcohol. Catalyst for the production of flexible PU foams. bp = 134-136°; d = 0.886; n_D^{20}= 1.4294; soluble in H_2O, EtOH, Et_2O. *Goldschmidt.* Discontinued.

30996 Tegoamin® PMD
Tertiary amine; catalyst for the manufacturing of flexible PU foams; for slabstock and molded foams. *Goldschmidt.* Discontinued.

30997 Tegoamin® SMP
Tertiary amine; catalyst for the manufacturing of conventional flexible PU slabstock and molded foams (hot-cured foam). *Goldschmidt.* Discontinued.

30998 Tegochrome® 22
2,2-Ethylene dithiodiethanol

Organic intermediate for producing color developers in the photographic industry. *Goldschmidt.*

30999 Tegocoll®

One and two-component polyurethane adhesives; for bonding of glass, metal, and thermoplastics to other substrates, for use with PU foam, in wood processing industry. *Goldschmidt.*

31000 Tegocolor®

Pigment dispersions in a polyether polyol; heat-stable color pastes for polyether polyurethane foams. *Goldschmidt.* Discontinued.

31001 Tegodont®

Chlorinated agent for water treatment. *Goldschmidt.*

31002 Tegodor

Disinfectants. *Thomas Goldschmidt Ltd.*

31003 Tegoglätte

A litharge having smaller particles than the ordinary type; used in rubber mixings.

31004 Tegoglas

Glass coating agents. *Thomas Goldschmidt Ltd.*

31005 Tegold

Disinfectants. *Thomas Goldschmidt Ltd.*

31006 Tegomag

Permanent magnet alloys. *Thomas Goldschmidt Ltd.*

31007 Tegoman

Sanitizers. *Thomas Goldschmidt Ltd.*

31008 Tegomuls®

Mono-diglycerides of edible fatty acids. *Thomas Goldschmidt Ltd.*

31009 Tegomuls® 19

61789-10-4 263-032-6

Lard mono/diglyceride; food emulsifier; for shortenings. *Goldschmidt AG.* Discontinued.

31010 Tegomuls® B

Glyceryl monostearate

Food emulsifier; for margarine, ice cream, pasta; self-emulsifying. *Goldschmidt AG.* Discontinued.

31011 Tegomuls® P 411

1,2-Propylene glycol monostearate

Food emulsifier; for whipped sponge mixtures, toppings, shortenings. *Goldschmidt AG.* Discontinued.

31012 Tegopren®

Organic modified siloxanes; surfactants used as antistats, wetting and leveling agents, emulsifiers, dispersants, and for the improvement of lubricity; additive for polishes. *Goldschmidt.*

31013 Tegosil®

Silicone-based; aerosol for release of injection molded plastic and rubber moldings. *Goldschmidt.*

31014 Tegosipon®

Silicone. Antifoam for manufacturing of stomach and intestinal preparations. *Goldschmidt.*

31015 Tegosivin® HL 250

Solvent-free low molecular weight silane-modified siloxane; for preparation of impregnation solutions. for treatment of fresh concrete. *Goldschmidt AG.* Discontinued.

31016 Tegosoft® CI

Cetearyl isononanoate

Cosmetic esters for emolliency, skin softening, and moisture retention in hair and skin care products, creams and lotions, hair conditioners/glossing systems, sunscreen preparations, soaps. *Goldschmidt.*

31017 Tegosoft® CO

59130-69-7 261-619-1

Cetyl octanoate

Cosmetic esters for emolliency, skin softening, and moisture retention in hair and skin care products, creams and lotions, hair conditioners/glossing systems, sunscreen preparations, soaps. *Goldschmidt.*

31018 Tegosoft® CT

65381-09-1 265-724-3

Caprylic/capric triglycerides

Cosmetic esters for emolliency, skin softening, and moisture retention in hair and skin care products, creams and lotions, hair conditioners/glossing systems, sunscreen preparations, soaps. *Goldschmidt.*

31019 Tegosoft® DO

3687-46-5 222-981-6

Decyl oleate

Cosmetic esters for emolliency, skin softening, and moisture retention in hair and skin care products, creams and lotions, hair conditioners/glossing systems, sunscreen preparations, soaps. *Goldschmidt.*

31020 Tegosoft® EE

Octyl octanoate

Cosmetic esters for emolliency, skin softening, and moisture retention in hair and skin care products, creams and lotions, hair conditioners/glossing systems, sunscreen preparations, soaps. *Goldschmidt.*

31021 Tegosoft® GC

68201-46-7

PEG-7 glyceryl cocoate

Mild surfactant for shampoos, bath products and personal cleansers; refatting and conditioning agent; foam booster/stabilizer; rinses clean. *Goldschmidt.*

31022 Tegosoft® Liquid

59130-69-7 261-619-1

Cetearyl octanoate

Cosmetic esters for emolliency, skin softening, and moisture retention in hair and skin care products, creams and lotions, hair conditioners/glossing systems, sunscreen preparations, soaps. *Goldschmidt.*

31023 Tegosoft® Liquid M

Cetearyl octanoate and isopropyl myristate. Cosmetic esters for emolliency, skin softening, and moisture retention in hair and skin care products, creams and lotions, hair conditioners/glossing systems, sunscreen preparations, soaps. *Goldschmidt.*

31024 Tegosoft® M

110-27-0 5234 203-751-4

Isopropyl myristate

Cosmetic esters for emolliency, skin softening, and moisture retention in hair and skin care products, creams and lotions, hair conditioners/glossing systems, sunscreen preparations, soaps. *Goldschmidt.*

31025 Tegosoft® OP

29806-73-3 249-862-1

Octyl palmitate

Cosmetic esters for emolliency, skin softening, and moisture retention in hair and skin care products, creams and lotions, hair conditioners/glossing systems, sunscreen preparations, soaps. *Goldschmidt.*

31026 Tegosoft® P

142-91-6 205-571-1

Isopropyl palmitate

Cosmetic esters for emolliency, skin softening, and moisture retention in hair and skin care products, creams and lotions, hair conditioners/glossing systems, sunscreen preparations, soaps. *Goldschmidt.*

31027 Tegosoft® S

112-10-7 203-934-9

Isopropyl stearate

Cosmetic esters for emolliency, skin softening, and moisture retention in hair and skin care products, creams and lotions, hair conditioners/glossing systems, sunscreen preparations, soaps. *Goldschmidt.*

31028 Tegosoft® SH

66009-41-4 266-065-4

Stearyl heptanoate

Cosmetic esters for emolliency, skin softening, and moisture retention in hair and skin care products, creams and lotions, hair conditioners/glossing systems, sunscreen preparations, soaps. *Goldschmidt.*

31029 Tegostab®

Foam stabilizers for polyurethanes. *Thomas Goldschmidt Ltd.*

31030 Tegostab® B 1048

Silicone foam stabilizer, stabilizer for continuously laminated PU boardstock, refrigeration and pipe insulation, rigid block foams. *Goldschmidt AG.* Discontinued.

31031 Tegostab® B 2219

Polysiloxane-polyoxyalkylene copolymer; stabilizer for continuously laminated PU boardstock, rigid block foams. *Goldschmidt AG.* Discontinued.

31032 Tegostab® B 4900

Polysiloxane-polyoxyalkylene block copolymer; foam stabilizer with wide processing latitude for hot-cured PU foam. *Goldschmidt AG.* Discontinued.

31033 Tegostab® B 8406

Polysiloxane-polyether copolymer surfactant; stabilizer for PU foams with integral skins, especially shoe sole systems. *Goldschmidt AG.* Discontinued.

31034 Tegostab® BF 2270

Polysiloxane-polyoxyalkylene block copolymer; foam stabilizer for manufacturing of flexible polyether polyurethane foams (hot cured foams). *Goldschmidt AG.* Discontinued.

31035 Tegotens 4100

Palmitic/stearic acid mono/diglycerides. Surface coating for expanded polystyrene beads containing a propellant; antistat for PE/PP; fabric softener; coemulsifier. *Goldschmidt AG.* Discontinued.

31036 Tegotens I.

Polyoxyethylated glyceryl isostearate

Solubilizer, coemulsifier, wetting agent. *Goldschmidt AG.* Discontinued.

31037 Tegotrenn® LH 157 A, 525

Release agent for demolding of flexible hot-cured polyurethane foam. *Goldschmidt.* Discontinued.

31038 Tegovakon
Sandstone strengtheners. *Thomas Goldschmidt Ltd.*

31039 Tegul
A proprietary sulfur jointing compound for bell and spigot pipes; it contains sulfur and sand. No manufacturer.

31040 Tegula
Bitumen sheet; for underneath stretcher strip. *Vedag GmbH.* Name unverified.

31041 Teka Oil
A proprietary trade name for an extract form stand oil (a linseed oil derivative) from which bases and acids have been removed. No manufacturer.

31042 Tekblend
One-shot high yield grout. *Foseco (F.S.) Ltd.*

31043 Tekemail
Classified ceramic glaze granules. *BASF AG.*

31044 Tekfoam
Lightweight high yield grout for cavity filling. *Foseco (F.S.) Ltd.*

31045 Teknol
Photographic developer. *May & Baker Ltd.* Name unverified.

31046 Tekpak-Tekbent
Reactor part of anhydrite/bentonite mixture. *Foseco (F.S.) Ltd.*

31047 Tekpak-Tekcem
Alumina cement for pump packing. *Foseco (F.S.) Ltd.*

31048 Tekstim 8504
Corrosion inhibitor for use in acid cleaning applications where corrosion inhibition on copper is critical. *Exxon.*

31049 Tekstim 8741
Surfactant, self-demulsifying detergent for truckwash applications. *Exxon/Tomah.*

31050 Telconax
A patented insulating compound made from selected bitumen, waxes, and rubber. No manufacturer.

31051 Telconite
A proprietary insulating material made in various colors. No manufacturer.

31052 Telconstan
A non-magnetic nickel-copper alloy prepared in induction furnaces and having exceptional purity and very low temperature coefficient of resistance; used for resistances where standard of resistance with temperature is important. No manufacturer.

31053 Telcothene®
Polythene powder, tube, and sheet. *Telcon Plastics Ltd.* Name unverified.

31054 Telcovin®
Polyvinyl chloride tube and sheet. *Telcon Plastics Ltd.* Name unverified.

31055 Teleblock
A proprietary range of thermoplastic rubbers. *Phillips.* Name unverified.

31056 telegraph bronze
telegraph metal; electric metal. An alloy of 80% copper, 7.5% lead, 7.5% zinc, and 5% tin.

31057 Telepaque
96-83-3 5072 202-539-9
$C_{11}H_{12}I_3NO_2$
3-amino-α-ethyl-2,4,6-triiodobenzenepropanoic acid
Iopanoic acid; 3-amino-α-ethyl-2,4,6-triiodohydrocinnamic acid; Cistobil; Colepax; Telepaque; Teletrast. Diagnostic aid. (dl) mp = 155-157°; soluble in H_2O, organic solvents; LD_{50} (rat orl)= 3870 mg/kg. *Sterling Drug Inc.* Name unverified.

31058 Telloy®
13494-80-9 9273 236-813-4
Te
tellurium
Aurum paradoxum; metallum problematum;. Tellurium powder; vulcanizing agent for rubber. m = 450°; bp = 990°; d = 6.11-6.27. *R. T. Vanderbilt Co Inc.*

31059 Tellurac®
Tellurium diethyldithiocarbamate
Accelerator for elastomers. *R. T. Vanderbilt Co Inc.* Discontinued.

31060 telluretted hydrogen
7783-09-7 4844 231-981-5
H_2Te.
Hydrogen telluride
mp = -49°; bp = -2°; d^{12}_{liq}= 2.68; soluble in H_2O.

31061 Tellurit
Chill producing mold dressing containing tellurium. *Foseco (F.S.) Ltd.*

31062 tellurium
13494-80-9 9273 236-813-4
Te
tellurium
Aurum paradoxum; metallum problematum;. Nonmetallic element; Te; alloys, secondary rubber vulcanizing agent, manufacture of iron and steel casting; coloring agent for glass and ceramics; thermoelectric devices. m = 450°; bp = 990°; d = 6.11-6.27. *Asarco; Atomergic Chemetals; Cabot; Cerac; R. T. Vanderbilt Co Inc.*

31063 tellurium lead
An alloy containing 0.05% tellurium with lead; resists sulfuric acid.

31064 tellurium oxide
7446-07-3 9276 231-193-1
O_2Te
tellurium dioxide
tellurous acid anhydride. mp= 733°; d = 5.75 or 6.04; slightly soluble in H_2O. *Asarco; Atomergic Chemetals; Cerac; Noah Chem.*

31065 Tellurium Tubes
Copper tubes containing pre-determined quantities of tellurium. *Foseco (F.S.) Ltd.* Discontinued.

31066 Telmin
31431-39-7 5807 250-635-4
$C_{16}H_{13}N_3O_3$
(5-benzoyl-1H-benzimidazol-2-yl) carbamic acid methyl ester
R-17635; Bantenol; Equi-Vurm Plus; Lomper; Mebenvet; Noverme; Ovitelmin; Pantelmin; Telmin; Vermicidin; Vermirax; Vermox. A proprietary preparation containing mebendazole; veterinary antihelmintic (horses). mp = 288.5°; insoluble in H_2O, most organic solvents, soluble in formic acid; LD_{50} (rat orl) > 40 mg/kg. *Janssen Pharmaceutical Ltd.* Name unverified.

31067 Telogen
1:2 Metal complex disperse dyes for fast dyeing of polyamide fibers. *Bayer plc.* Discontinued.

31068 Telon®
Selected acid dyestuffs; for the dyeing of polyamide fibers and wool/polyamide blends. *Bayer AG; Bayer plc.*

31069 Telone
542-75-6 3125 208-826-5
$C_3H_4Cl_2$
1,3-dichloro-1-propene
1,3-dichloropropene; Anema; D-D92; D-D95; Dedisol; Dorlone II; Sepisol; Telone 2000; Telone II. Agricultural products containing 1,3-dichloropropene as the active ingredient; soil fumigants added to soil prior to planting to control soil pests such as nematodes which feed on the roots of plants and reduce yields. mp < -50°; bp = 108°; d^{20} = 1.220; soluble in H_2O (2 g/l), soluble in organic solvents; LD_{50} (rat orl) = 150 mg/kg. *DowElanco Ltd.*

31070 Telopar
68813-55-8 7055 272-332-6
$C_{49}H_{48}N_4O_8$
(E)-3-[2-(1,4,5,6-tetrahydro-1-methyl-2-pyrimidinyl)ethenyl]phenol pamoate
Oxantel pamoate; Oxantel embonate; CP-14445-16. Anthelmintic. *Pfizer Inc.* Discontinued.

31071 Telsit
A gelatin explosive containing from 10-15% dinitrotoluene or liquid trinitrotoluene.

31072 Teluran
Acrylonitrile-butadiene-styrene polymers and related products. *BASF plc.* Name unverified.

31073 Telvar®
For the agriculture industry. *DuPont UK.*

31074 Temadex
Veterinary skin dressing. *The Wellcome Foundation Ltd.* Name unverified.

31075 Temasept I
Dibromo/tribromo salicylanilide
Antimicrobial for resins, latex emulsions, plastics. *Hexcel.*

31076 Temasept IV
87-10-5 9747 201-723-6
$C_{13}H_8Br_3NO_2$
3,5-dibromo-N-(4-bromophenyl)-2-hydroxybenzamide
Tribromosalicylanilide; tribromsalan; 3,4',5-tribromosalicylanilide; TBS; Tuasol 100. Antimicrobial for resins, latex emulsions, plastics. mp = 227-228°; insoluble in H_2O, soluble in Me_2CO, DMF. *Hexcel.*

31077 Tembind A 002
8061-53-8
Ammonium lignosulfonate
Emulsifier and stabilizer for asphalt; road dust suppressant; coat dust suppressant; binder and pelletizing agent. *Temfibre.*

31078 Temlock
A proprietary trade name for a board made from wood fibers impregnated with resin and subjected to pressure. No manufacturer.

31079 Tempaloy
A patented alloy of approximate composition, 95% copper, 4% nickel, and 1% silicon. No manufacturer.

31080 temper
Alloys of arsenic and lead or copper, and tin; used as hardening materials for shot or pewter.

31081 Temperite Alloys
A proprietary trade name for alloys of lead, tin, and cadmium. No manufacturer.

31082 Tempo
Proprietary cellulose esters. No manufacturer.

31083 Tempo
Herbicide containing linuron and terbutryn. *ICI Chem & Polymers Ltd.*

31084 Tempo
Suspension concentrate containing 150 g linuron and 150 g terbutryn per liter; used for control of annual dicotyledons in potatoes. *Farm Protection Ltd.*

31085 Tempro
A water-based acrylic copolymer temporary coating and remover. *ICI Chem & Polymers Ltd.*

31086 Temsperse S 001
8061-51-6
Sodium lignosulfonate
Dyestuff dispersant, water reducer, slurry thinner; binder; concrete admixtures; emulsion stabilizer. *Temfibre.*

31087 Tem-Tuf
Aluminious metal sheet. *Reynolds Metal Co.*

31088 Tenac®
Polyacetal homopolymer and copolymer resin *Asahi Chem, Industry.*

31089 Tenacity
Fluxes for silver alloy brazing. *Johnson Matthey plc.*

31090 Tenamine 1
N-butylated-*p*-aminophenol
A proprietary antioxidant. *Eastman.* Name unverified.

31091 Tenamine 2
101-96-2 202-992-2
N,N'-di-*sec*-butyl-*p*-phenylenediamine
A proprietary antioxidant. *Eastman.* Name unverified.

31092 Tenamine 3
128-37-0 1583 204-881-4
$C_{15}H_{24}O$
2,6-di-*t*-butyl-*p*-cresol
2,6-bis(1,1-dimethylethyl)-4-methylphenol; 2,6-di-*tert*-butyl-4-methylphenol; BHT, Antrancine 8; Tenox BHT; Ionol CP; Sustane; Dalpac; Impruvol; Vianol; butylated hydroxytoluene; DBPC. A proprietary antioxidant used with foods. Antiskinning agent. mp = 70°; bp = 265°; d_4^{20} = 1.048; soluble in organic solvents; LD_{50} (rat orl) = 1040 mg/kg. *Eastman.* Name unverified.

31093 Tenasco
A proprietary trade name for synthetic fiber resembling nylon. No manufacturer.

31094 Tenase® 1200, L-340, L-1200
9000-92-4 640 232-567-7
Bacterial α-amylase
Enzyme for starch liquefaction; for syrups, textile desizing. *Solvay Enzymes.*

31095 Tenatine
Thermoplastic polyester molding compounds. *Ciba plc.* Name unverified.

31096 Tenax metal
A zinc alloy containing from 0.35-2.56% copper, 0.2-4.42% aluminum, 0-0.35% iron, and up to 1.2% lead; used for the manufacture of guide rings.

31097 Tenax Wax
Mineral grease, wax and resin formulated as a grafting wax. *Vitax Ltd.*

31098 Tenaxatex VA 632
A proprietary trade name for a high molecular weight vinyl acetate homopolar water emulsion containing 55% solids; used as an adhesive base. *H A Smith.* Name unverified.

31099 Tenaxatex VA 956
A proprietary trade name for a vinyl acetate/acrylate copolymer emulsion containing 55% solids; a medium viscosity adhesive base. *H A Smith.* Name unverified.

31100 Tenazit
A proprietary trade name for laminated bakelite or similar synthetic resin. No manufacturer.

31101 Tendrelle
Nylon. *ICI Chem & Polymers Ltd.*

31102 Tendril®
For the agriculture industry. *DuPont UK.*

31103 Tenephrol
10377-51-2 5561 233-822-5
LiI
lithium iodide
A proprietary preparation of lithium iodide (31% solution). Used in photography. mp = 446°; bp = 1171°; d = 3.4900; soluble in H_2O (2 g/ml), EtOH, Me_2CO. No manufacturer.

31104 Tenite® 105-MS
9004-35-7 2013
Cellulose acetate
Good processibility and chemical resistance; for tool handles, optical applications, toothbrushes, personal care items, tubing, pipe, medical devices, automotive parts, appliance parts, toys, sporting goods, sheeting, furniture profiles. mp = 240°; n^{20}:kiD = 1.4750. *Eastman.*

31105 Tenite® 154DF
9002-88-4 7728
Polyethylene
For film extrusion. *Eastman.*

31106 Tenite® 264-MH
9004-36-8
Cellulose acetate butyrate
Good processibility and chemical resistance; for tool handles, optical applications, toothbrushes, personal care items, tubing, pipe, medical devices, automotive parts, appliance parts, toys, sporting goods, sheeting, furniture profiles. mp = 110-125°. *Eastman.*

31107 Tenite® 360-H2
9004-39-1
Cellulose acetate propionate
Good processibility and chemical resistance; for tool handles, optical applications, toothbrushes, personal care items, tubing, pipe, medical devices, automotive parts, appliance parts, toys, sporting goods, sheeting, furniture profiles. mp = 188-210°. *Eastman.*

31108 Tenite® Cellulosic Acetate
9004-35-7 2013
Cellulose acetate
Thermoplastic ester offering toughness and clarity in virtually unlimited color range at reasonable price; easy moldability. *Eastman.* Name unverified.

31109 Tenite® Cellulosic Butyrate
Cellulose butyrate
Thermoplastic ester offering toughness and clarity in virtually unlimited color range at reasonable price; easy moldability. *Eastman.* Name unverified.

31110 Tenite® Cellulosic Propionate
9004-39-1
Cellulose propionate
Thermoplastic ester offering toughness and clarity in virtually unlimited color range at reasonable price; easy moldability. *Eastman.* Name unverified.

31111 Tenite® PET 9902
25038-59-9 7730
PET polyester for extrusion into crystallizable sheet that can be thermoformed into food trays for microwave or conventional ovens. *Eastman.*

31112 Tenite® Polyethylene
9002-88-4 7728
Polyethylene for injection molding, film extrusion, extrusion coating, profile and other specialty extrusions. *Eastman.*

31113 Tennafast
Inorganic and organic pigments. *Tennant-KVK Ltd.*

31114 Tennafast, Tennalaks
Inorganic and organic pigments. *Tennant-KVK Ltd.*

31115 Tennal
A proprietary trade name for certain aluminum alloys for casting purposes. No manufacturer.

31116 Tenncol
Saponified rosin size. *Tenneco Malros Ltd.* Name unverified.

31117 Tennessee phosphates
Mineral phosphates containing from 60-70% calcium phosphate; used as a fertilizer.

31118 Tenoban
Arecoline-acetarsol. Anthelmintic-cathartic in veterinary medicine. *The Wellcome Foundation Ltd.* Name unverified.

31119 Tenoran
1982-47-4 217-843-7
$C_{15}H_{15}ClN_2O_2$
N-[4-(4-chlorophenoxy)phenyl]-N,N-dimethylurea
Chloroxuron; 3-[4-(4-chlorophenoxy)phenyl]-1,1-dimethylurea; 3-[*p*(*p*-chlorophenoxy)phenyl]-1,1-dimethylurea; C 1983; Gesamoos. Urea herbicide for on strawberries and ornamentals. mp = 151-152°; soluble in H_2O (3.7 mg/l), more soluble in organic solvents; LD_{50} (rat orl) = 700 mg/kg. *Ciba-Geigy Agrochemicals.*

31120 Tenox® 2
Propylene glycol, BHA, propyl gallate, citric acid; food-grade antioxidant. *Eastman.*

31121 Tenrez
Rosin esters. *Tenneco Malros Ltd.* Name unverified.

31122 Tensabit
Emulsifier; alkaline emulsion; for bitumen. *Tensia SA.* Name unverified.

31123 Tensactol
Self-emulsifying wax; for cosmetics and pharmaceuticals. *Tensia SA.* Name unverified.

31124 Tensadal
Reviving and softening agent; for cotton and cellulose fibers. *Tensia SA.* Name unverified.

31125 Tensagex
Liquid blend; for baby shampoos. *Hickson & Welch Ltd.*

31126 Tensagex BV
Anionic surfactant in the form of a low viscosity liquid; for liquid detergents; liquid shampoos; bubble baths. *Hickson & Welch Ltd.* Name unverified.

31127 Tensagex DMY
Sodium alkyl ether sulfate as a low viscosity liquid; low irritant anionic surfactant used in baby shampoos and foam baths. *Hickson & Welch Ltd.* Name unverified.

31128 Tensagex DP24
Sodium alkylphenol ether sulfate in liquid form; wetting and dispersing agent for pigments and metal degreasing. *Hickson & Welch Ltd.* Name unverified.

31129 Tensagex EOC
Sodium lauryl alcohol/2.5EO sulfate in liquid or gel form; high foaming anionic surfactant used in shampoos, bubble baths, liquid detergents and emulsion polymerization. *Hickson & Welch Ltd.* Name unverified.

31130 Tensagex SPDL
Sodium alcohol ether sulfate (C12-C15) in liquid form; high foaming anionic surfactant used in shampoos, bubble baths, liquid detergents and emulsion polymerization. *Hickson & Welch Ltd.* Name unverified.

31131 Tensami 1/05
Lecithin, xanthan gum; natural emulsifier. *Alban Muller.*

31132 Tensami 3/06
Casein, xanthan gum; natural emulsifier. *Alban Muller.*

31133 Tensami 4/07
Soy protein, xanthan gum; natural emulsifier. *Alban Muller.*

31134 Tensami 8/09
Corn oil, egg yolk extract; natural emulsifier. *Alban Muller.*

31135 Tensami 10/06
Saponins, xanthan gum; natural emulsifier. *Alban Muller.*

31136 Tensamina
Adhesion improver; for bitumen on wet stones. *Tensia SA.* Name unverified.

31137 Tensamine C, O, S, and SH
Cationic surfactants in the form of primary amines, with the alkyl portion being coconut, oleic, tallow, and hydrogenated tallow respectively; liquid or solid form; for emulsification, pigment dispersion, synthesis intermediate. *Tensia SA.* Name unverified.

31138 Tensaminox
Foaming agents; dispersing agents; thickening agents; for shampoos; liquid detergents; pigments, textiles. *Tensia SA.* Name unverified.

31139 Tensarane SBTE
TEA-dodecylbenzene sulfonate, supplied as a liquid; anionic surfactant, wetting agent and dyeing auxiliary for cottons. *Tensia SA.* Name unverified.

31140 Tensaryl
Low foaming detergent base; for light duty detergents. *Tensia SA.* Name unverified.

31141 Tensaryl 40CC, 50B, 80B, and 82F
Sodium tetrapropylene benzene sulfonate as powder, beads or flakes; detergent, anticaking agent, and wetting agent, for emulsion polymerization, wettable powders, and various industrial uses. *Tensia SA.* Name unverified.

31142 Tensaryl DF90
Sodium dodecylbenzene sulfonate
Foaming agent for fine fabrics washing. *Tensia SA.* Name unverified.

31143 Tensaryl DX54Sp. and DX62
Sodium dodecylbenzene sulfonate
DX54Sp also contains perborate; anionic surfactant for heavy duty products. *Tensia SA.* Name unverified.

31144 Tensaryl KD
Tetrapropylene benzene sulfonate. Tetrapropylene benzene sulfonate as a viscous liquid; intermediate for liquid and powder detergents when biodegradability is not requested. *Tensia SA.* Name unverified.

31145 Tensaryl L48
Sodium dodecylbenzene sulfonate
Low foaming, anionic surfactant for fine fabrics washing. *Tensia SA.* Name unverified.

31146 Tensaryl S30P and S70P
Sodium tetrapropylene benzene sulfonate. Sodium tetrapropylene benzene sulfonate in the form of a liquid or paste with low salt content; anionic surfactant used for emulsion polymerization. *Tensia SA.* Name unverified.

31147 Tensaryl SB
dodecylbenzene sulfonate. Straight chain dodecylbenzene sulfonate in acid form; supplied as a viscous liquid; raw material for the manufacture of liquid, pasty or solid surfactants. *Tensia SA.* Name unverified.

31148 Tensaryl SB Ca
Calcium dodecylbenzene sulfonate. Calcium dodecylbenzene sulfonate in powder form; water-insoluble anionic surfactant used as a co-emulsifier in organic systems. *Tensia SA.* Name unverified.

31149 Tensaryl SB85P
Sodium dodecylbenzene sulfonate. Sodium dodecylbenzene sulfonate in paste form; wetting and dispersing agent with low salt content; used in emulsion polymerization. *Tensia SA.* Name unverified.

31150 Tensaryl SBD
Triethanolamine dodecylbenzene sulfonate. Triethanolamine dodecylbenzene sulfonate in liquid form; liquid detergent for dishwashing, textiles and car shampoos. *Tensia SA.* Name unverified.

31151 Tensatil D100
Sodium octylsulfate. Sodium octylsulfate in liquid form; anionic surfactant used as a wetting agent in polymerization. *Tensia SA.* Name unverified.

31152 Tensatil DA120
Ammonium octyl/decyl-sulfate. Ammonium octyl/decyl-sulfate in liquid form; anionic surfactant used as a wetting and foaming agent in froth flotation. *Tensia SA.* Name unverified.

31153 Tensatil DB120
Short chain alcohol sulfate in liquid form; anionic surfactant used as a wetting agent in electrolyte solutions. *Tensia SA.* Name unverified.

31154 Tensatil DEH120
Sodium 2-ethylhexyl sulfate. Sodium 2-ethylhexyl sulfate in liquid form; anionic surfactant used as a wetting agent in metal cleaning preparations and styrene polymerization. *Tensia SA.* Name unverified.

31155 Tensiamix
Range of low foaming detergent base; for heavy duty detergents. *Tensia SA.* Name unverified.

31156 Tensianol
Sensitive skin special washing composition; for cosmetics; synthetic toilet bars. *ICI plc.* Name unverified.

31157 Tensibet 50
N-Alkylbetaine in liquid form; foam booster and detergent with antistatic effect; used in cosmetics and baby shampoos. *Tensia SA.* Name unverified.

31158 Tensibet 55
Alkylamidobetaine in liquid form; foam stabilizer and thickening agent which is mild to the skin; used in lauryl ether sulfate formulations. *Tensia SA.* Name unverified.

31159 Tensidef
Antifoaming agents; for paper industry. *Tensia SA.* Name unverified.

31160 Tensidye
Nonionic liquid; wetting agent for cosmetics. *Tensia SA.* Name unverified.

31161 Tensilac 39
A proprietary rubber vulcanization accelerator; it is a soft form of ethylidene-aniline. No manufacturer.

31162 Tensilac 40
A proprietary rubber vulcanization accelerator; it is a resinous condensation product. No manufacturer.

31163 Tensilac 41
A proprietary rubber vulcanization accelerator; it is a hard form of ethylidene-aniline. No manufacturer.

31164 Tensilite
An aluminum bronze. It contains from 64-67% copper, 3.1-4.4% aluminum, 0-1.2% iron, 2.5-3.8% manganese, and 24-29% zinc.

31165 Tensimul
Emulsifier; for oils, solvents, natural and synthetic waxes. *Tensia SA.* Name unverified.

31166 Tensiofix
Range of emulsifiers; for pesticide formulations. *Tensia SA.* Name unverified.

31167 Tensioquat C50
Benzalkonium chloride in aqueous solution; cationic surfactant used in disinfection and emulsification. *Tensia SA.* Name unverified.

31168 Tensioquat C75
Benzalkonium chloride in liquid IPA form; cationic surfactant used in the paint industry. *Tensia SA.* Name unverified.

31169 Tensiorex
Conditioning agent; concentrated pearling agent; for after shampoo; shampoos. *Tensia SA*. Name unverified.

31170 Tensiostat
Antistatic agent; for cellulosic fibers and polymers. *Tensia SA*. Name unverified.

31171 Tensipar
Antifoaming agent; for industrial uses including food. *Tensia SA*. Name unverified.

31172 Tensitex
Wetting agent; for caustic lye and mercerizing. *Tensia SA*. Name unverified.

31173 Tensloy
A proprietary trade name for an alloy of iron with approximately 1.5% nickel and 0.5% chromium. No manufacturer.

31174 Tensocide
Antiseptic; for paper mills. *Tensia SA*. Name unverified.

31175 Tensol
Cements for vinyl and acrylic sheets. *ICI Chem & Polymers Ltd.*

31176 Tensol
Photographic activator/stabilizer chemicals. *May & Baker Ltd.* Name unverified.

31177 Tensol
A proprietary trade name for a dispersing and emulsifying agent containing a sulfonated ether. No manufacturer.

31178 Tensoleate
Pigment grinding aid; dispersing agent; for paint industry. *Tensia SA*. Name unverified.

31179 Tensoline
Emulsifier; leveling and dispersing agent; for oiling and fulling of wool; dyeing of woollen and acrylic fibers. *Tensia SA*. Name unverified.

31180 Tensomel
Superamide; foam stabilizer; additive; anticorrosion; for shampoos; liquid detergents; bubble baths; cutting oils. *Tensia SA*. Name unverified.

31181 Tensomin
Leveling agent; for dyeing woolen, acrylic, polyamide, and polyester fibers; dyeing with acid and metallizing dyes. *Tensia SA*. Name unverified.

31182 Tensopac
Wetting and dispersing agents; for paper mill. *Tensia SA*. Name unverified.

31183 Tensopane D
Series of nonionic surfactants of the octylphenol ethoxylate type; emulsification; detergent compounding; industrial cleaning; metal pickling; stabilizer in emulsion polymerization. *Tensia SA*. Name unverified.

31184 Tensophene 2D30
Dinonylphenol ethoxylate nonionic surfactant; emulsifier; detergent; metal cleaning; emulsion polymerization. *Tensia SA*. Name unverified.

31185 Tensophene D12, D15, D18
Nonionic surfactants of the nonylphenol ethoxylate type; for foam control; dispersion; oil emulsifier; coupling agent. *Tensia SA*. Name unverified.

31186 Tensophene H10, I10, DT, D36, D42EC, D45, D60, D90
Nonionic surfactants of the nonylphenol ethoxylate type; used for all types of detergent; metal cleaning; emulsion polymerization. *Tensia SA*. Name unverified.

31187 Tensopol 12A, 12P
Sodium dodecyl sulfate in needle or powder form; anionic surfactant used in pharmaceuticals, toothpastes and shampoos. *Hickson & Welch Ltd.* Name unverified.

31188 Tensopol 30E, LDS
Sodium dodecylbenzene sulfonate
Foaming detergents. The LDS variant also contains an optical dye. *Hickson & Welch Ltd.* Name unverified.

31189 Tensopol A, 7, USP
Sodium lauryl sulfate in the form of needles or powder; detergency and foaming agent for toothpastes, shampoos, pharmaceuticals, emulsion polymerization and pigment dispersion. *Hickson & Welch Ltd.* Name unverified.

31190 Tensopol ACL, PCL
Sodium lauryl sulfate in the form of needles or powder; detergency, foaming, and wetting agent for shampoos, emulsion polymerization, and pigment dispersion. *Hickson & Welch Ltd.* Name unverified.

31191 Tensopol AG, MG
Magnesium lauryl sulfate in the form of needles or powder; anionic surfactant used in shampoos and toothpastes. *Hickson & Welch Ltd.* Name unverified.

31192 Tensopol DX85, FL
Sodium alcohol sulfate in liquid or paste form; foaming agent for liquid shampoos, including carpet types; latex rug backing. *Hickson & Welch Ltd.* Name unverified.

31193 Tensopol LT
139-96-8 205-388-7
Triethanolamine lauryl sulfate in liquid form; anionic surfactant used in liquid shampoos, bubble baths and hair lotions. *Hickson & Welch Ltd.* Name unverified.

31194 Tensopol N
2235-54-3 218-793-9
Ammonium lauryl sulfate in liquid form; anionic surfactant used in liquid shampoos and bubble baths. *Hickson & Welch Ltd.* Name unverified.

31195 Tensopol SPK
4706-78-9 225-190-4
Potassium lauryl sulfate in powder form; anionic surfactant used as a base in synthetic toilet bars. *Hickson & Welch Ltd.* Name unverified.

31196 Tensopol VAL
Sodium lauryl sulfate in liquid form; liquid detergents, shampoos, bubble baths, and carpet shampoos. *Hickson & Welch Ltd.* Name unverified.

31197 Tensoprene
Low foaming surfactant; rinse aid, detergent for dishwashing formulations; industrial cleaners. *Tensia SA*. Name unverified.

31198 Tensostat
Bactericide and fungicides; for paper mills. *Tensia SA*. Name unverified.

31199 Tensovax
Emulsifier, solubilizer; wetting agent; for oils, perfumes, vitamins; metal working; textile specialities. *Tensia SA*. Name unverified.

31200 Tensovyl
Dispersing agent; for post-tinctorial washing of polyester fibers. *Tensia SA*. Name unverified.

31201 Tensuccin D8
577-11-7 3460 209-406-4
Sodium dioctyl sulfosuccinate in liquid form; anionic surfactant used in textile desizing; dry cleaning; cosmetics. *Tensia SA*. Name unverified.

31202 Tensuccin H724, H925
Disodium alkyl ether hemiester sulfosuccinate in water solution; anionic surfactant used as a polymerization stabilizer. *Tensia SA*. Name unverified.

31203 Tensuccin HS40
Disodium stearic hemiester sulfosuccinate in the form of a paste; anionic surfactant used as a dispersing agent. *Tensia SA*. Name unverified.

31204 Tensuccin ML, MO, MS
Anionic surfactants of the sulfosuccinamate type, in liquid or soft paste form; foaming agent used in latex for carpet backing and foam insulation. *Tensia SA*. Name unverified.

31205 Tensyl 30
137-16-6 4379 205-281-5
Sodium lauroyl sarcosinate in liquid form; anionic surfactant used in cosmetics, toothpastes, and baby shampoos. *Tensia Ltd.* Unverified.

31206 Tensynvac
1133 Strain viral arthritis; for immunization of poultry. *Intervet Inc.*

31207 Tephal
A wetting agent and detergent; used in the textile industry for pretreatment, desizing, and dyeing. *Degussa AG*. Name unverified.

31208 Tepperite
A proprietary polystyrene. No manufacturer.

31209 Teralan
Dyes for transfer polyester/wool. *Ciba plc*. Name unverified.

31210 Teraprint
Dyes for transfer printing. *Ciba plc*. Name unverified.

31211 Terasil
Disperse dyes. *Ciba plc*. Name unverified.

31212 Terate 101, 131
Balsamic resins derived from petroleum aromatic hydrocarbons; used in adhesives and sealants, in alkyd coatings, in molded goods and in rubber compounding. *Hercules*. Discontinued.

31213 Terate 202, 203, 204
Thermoplastic resins; aromatic polyester polyols derived from polycarbomethoxy-substituted diphenyls, polyphenyls, and benzyl esters of the toluate family; used to extend reactive polyurethane and polyisocyanurate urethane polyols in the manufacture of rigid urethane foam. *Hercules*. Discontinued.

31214 Terate 203
Thermoplastic resins; aromatic polyester polyols derived from polycarbomethoxy-substituted diphenyls, polyphenyls, and benzyl esters of the toluate family; used to extend reactive polyurethane and polyisocyanurate urethane polyols in the manufacture of rigid urethane foam. *Hercules*. Discontinued.

31215 Terate 203
Thermoplastic resins. Aromatic polyester polyols derived from polycarbomethoxy-substituted diphenyls, polyphenyls and benzyl esters of

the toluate family. Used to extend reactive polyurethane and polyisocyanurate urethane polyols in the manufacture of rigid urethane foam. *Hercules.* Discontinued.

31216 Terate 204

Thermoplastic resins; aromatic polyester polyols derived from polycarbomethoxy-substituted diphenyls, polyphenyls, and benzyl esters of the toluate family; used to extend reactive polyurethane and polyisocyanurate urethane polyols in the manufacture of rigid urethan foam. *Hercules.* Discontinued.

31217 Terate 204

Thermoplastic resins. Aromatic polyester polyols derived from polycarbomethoxy-substituted diphenyls, polyphenyls and benzyl esters of the toluate family. Used to extend reactive polyurethane and polyisocyanurate urethane polyols in the manufacture of rigid urethan foam. *Hercules.* Discontinued.

31218 Terathane

Polyether glycol. *DuPont UK.*

31219 Terathane® 650

Polytetramethylene ether glycol

Functions as a soft segment in PU resins. *DuPont.* Name unverified.

31220 terbacil

5902-51-2 9298 227-595-1

$C_9H_{13}ClN_2O_2$

5-Chloro-3-(1,1-dimethylethyl)-6-methyl-2,4-(1H,3H)-pyrimidinedione

Sinbar; Sinbar 80W; DPX-D732; 3-*tert*-butyl-5-chloro-6-methyluracil; Geonter; butyl-5-chloro-6-methyluracil; chloro-3-*tert*-butyl-6-methyluracil; Herbicide 732; 5-chloro-3-(1,1-dimethylethyl)-6-methyl-pyrimidinedione,. Selective herbicide, used for control of annual broad-leaved weeds, most annual grasses and some perennial weeds. mp = 175-177°; SG - 1.34; soluble in H_2O (710 mg/l), more soluble in organic solvents; LD_{50} (rat orl) > 5000 mg/kg.

31221 Terbalin

Active ingredients; terbutryn plus triflurex; selective pre-emergence herbicidal mixture. *Agan Chemical Manufacturers Ltd.*

31222 terbium

7440-27-9 9300 231-137-6

Tb

Terbium

A lanthanide element; phosphor activator, dope for solid-state devices. mp = 1356°; bp= 3230°; d = 8.27. *Aldrich; Atomergic Chemetals; Cerac; Rhône-Poulenc Basic.*

31223 terbium oxide

12037-01-3 9300 234-856-3

O_3Tb_2

Terbia

Atomergic Chemetals; Cerac; Rhône-Poulenc Basic.

31224 Terblend®

Styrene copolymer blends. *BASF plc.*

31225 Terblend® S

ASA/PC blend; thermoplastic polymer for injection molding of car instrument covers, tail light assemblies, etc., housings for small appliances, transformer housings, switchgear for house wiring, meter housings, etc. *BASF AG.*

31226 Terbufos

13071-79-9 9301 235-963-8

$C_9H_{21}O_2PS_3$

S-*t*-butylthio-methyl-O,O-diethyl phosphorodithioate

S-*tert*-butylthiomethyl O,O-diethyl phosphorodithioate; S-[[(1,1-dimethylethyl)thio]methyl] O,O-diethyl phoshporodithioate; ENT-27920; AC 92100; Aragran; Contraven; Counter; Cyanater; Plydax. A soli insecticide and nematicide with stomach and contact action. A cholinesterase inhibitor. Used for control of soil insects in vegetable and fruit crops. mp = -29°; $bp_{0.01\,mm}$ = 69°; d_{24}= 1.105; n_D^{23}= 1.52; soluble in H_2O (4.5 mg/l), more soluble in organic solvents; LD_{50} (rat orl) = 1.6 mg/kg. *American Cyanamid.* No manufacturer.

31227 Terbuthylazine

5915-41-3 227-637-9

$C_9H_{16}ClN_5$

2-*t*-butylamino-4-chloro-6-ethylamino-1,3,5-triazine

Tyllanex; Gardoprim; GS 13529; Primatol M;. Herbicide, absorbed mainly by roots. Used for broad spectrum weed control. . mp = 177-179°; SG_{20}= 1.188; soluble in H_2O (8.5 mg/l), more soluble in organc solvents; LD_{50} (rat orl)= 2160 mg/kg.

31228 Terbutol

98-54-4 1620 202-679-0

$C_{10}H_{14}O$

p-*tert*-butylphenol

p-*t*-butylphenol; 4-(1,1-dimethylethyl)phenol; butylphen. Used as an intermediate in manufacture of varnishes, as a soap antioxidant, in demulsifiers and as a motor oil additive. mp = 96-100°; bp = 236-238°; d = 0.9080; n_D^{20}= 1.4787; LD_{50} (rat orl) = 3.25 l/kg. *ICI Chem & Polymers Ltd.* Discontinued.

31229 Terbutrex

886-50-0 212-950-5

2-*tert*-butylamino-4-ethylamino-6-methylthio-1,3,5-triazine

terbutryne. Active ingredient; pre-emergence and post-emergence weed control. *Agan Chemical Manufacturers Ltd.*

31230 terbutryn

886-50-0 212-950-5

$C_{10}H_{19}N_5S$

1,3,5-Triazine-2,4-diamine, N(1,1-dimethylethyl)-N'-ethyl-6-(methylthio)-

s-Triazine, 2-(*tert*-butylamino)-4-(ethylamino)-6-(methylthio)-; N-(1,1-Dimethylethyl)-N'-ethyl-6-(methylthio)-1,3,5-triazine-2,4-diamine; A 1866; Clarosan; Igran; Igran 50; Igran 500; Prebane; Shortstop; Shortstop E; Terbutrex; Terbutryne; Athado; GS 14260; Plantonit;. Selective herbicide absorbed by roots and foliage and used for control of most grasses in winter cerals, vegetables and citrus fruit. mp = 104-105°; $bp_{0.06}$ = 154-160°; d_{20} = 1.115; soluble in H_2O (25 mg/l 20°), more soluble in organic solvents; LD_{50} (rat orl) = 2045 mg/kg *Probelte; Ciba-Geigy; Chemolimpex; Makhteshim-Agan.*

31231 Terbytex

Softening and antistatic agent used with natural and synthetic fibers. *Tensia SA.* Name unverified.

31232 Tercoton

Dyes for polyester/cotton. *Ciba plc.* Name unverified.

31233 terebene

Acid-isomerized turpentine consisting of a mixture of dipentene and other hydrocarbons.

31234 terebine

Liquid drier, Japan drier. Made by heating oxides of lead and manganese with linseed oil or rosin, or mixtures of the oil and rosin, and thinning with turpentine or turpentine substitute; a drier for paints.

31235 Terephane

A proprietary name for polyethylene terephthalate film. *French Origin.* No manufacturer.

31236 Terephane

A proprietary name for polyethylene terephthalate film. French origin. No manufacturer.

31237 Terethane®

For the chemical industry. *DuPont UK.*

31238 Terfenol

Rosin derivatives; for paints and inks. *Resinas Sinteticas SA.* Name unverified.

31239 Terg-A-Zyme®

Alkylaryl sulfonate, lauryl alcohol sulfate, phosphate, carbonate, and protease enzyme. Biodegradable detergent, wetting agent, sequestering and synergistic agents; used in hospitals, laboratories, dairies; cleaning agent in dairy and pollution processing. *Alconox.*

31240 Tergenol 1122

Blend; heavy-duty detergent for heavily soiled woven and knitted fabrics; stable to acids, alkalies, and oxidizing agents. *Hart Chem. Ltd.*

31241 Tergitex KW

Polyether solvent blend; for oil removal from garnetted wool and knitted goods. *Scher.* Discontinued.

31242 Tergitol®

A series of biodegradable nonionic intermediates comprising ethoxylates and ethoxysulfates of linear secondary alcohols; used in the production of biodegradable detergents. *Union Carbide (UK) Ltd.* Name unverified.

31243 Tergitol® 15-S-3

68131-40-8

C_{11}-C_{15} pareth-3; biodegradable detergent, emulsifier, wetter, defoamer for aqueous systems, intermediate used in textiles, solvent cleaners, drycleaning, metalworking fluids, water treatment, oilfield chemicals, pulp/paper de-inking, latex emulsions, plastics antistat, agriculture. *Union Carbide.*

31244 Tergitol® 24-L-45

68439-50-9

C_{12}-C_{14} pareth-6(6.3 EO); biodegradable surfactant, detergent, wetting/spreading agent, emulsifier, foaming agent, intermediate, dispersant for household and industrial cleaners, textile wet processing, paper processing, and agricultural formulations. *Union Carbide.*

31245 Tergitol® 26-L-3

C_{12}-C_{16} pareth-3; surfactant, emulsifier, intermediate for sulfation; used for prewash spotters, coning oil, hydrocarbon-based cleaners, agriculture; as sulfated product in cosmetics, hand dishwash, light duty detergents. *Union Carbide.*

31246 Tergitol® D-683

37251-69-7

Alkoxylated alkylphenol
Emulsifier for fiber finishing operations; dispersant for pigments in resins, plastics, and for abrasives in hard surface cleaners; improves wetting of oil-based materials in coatings and adhesives. *Union Carbide.*

31247 Tergitol® Min-Foam 1X
68551-14-4
C_{11}-C_{15} alcohols reacted with EO and PO; biodegradable surfactant, foam depressant, wetting agent, detergent for household/industrial cleaners, drycleaning, textile processing, metal cleaning, circuit board cleaners, leather, paper deinking. *Union Carbide.*

31248 Tergitol® TMN-3
60828-78-6
Isolaureth-3
Emulsifier, wetting agent, coupler, penetrant, leveling agent for textile processing, lubricants, water treatment, solvent cleaners/degreasers, metalworking fluids, drycleaning, oilfield chemicals, pulp/paper deinking; defoamer for aqueous systems; intermediate for anionic surfactants used in household, industrial and personal care products. *Union Carbide.*

31249 Tergitol® XD
9038-95-3
PPG-24-buteth-27; emulsifier, dispersant, stabilizer for agricultural insecticides/herbicides, latex polymerization, iodophor manufacturing for germicidal cleaning, latex paints, dye pigments, leather; emulsifier for silicone oils, diacyl peroxides. *Union Carbide.*

31250 Tergitol® XH
9038-95-3
EO/PO copolymer; emulsifier, dispersant for agriculture, latex polymerization, iodophor manufacturing for germicidal cleaning, latex paints, dye pigments, leather finishes, toilet bowl cleaners; emulsifier for silicone oils, diacyl peroxides. *Union Carbide.*

31251 Tergolix
Leather fat liquoring agents. *Sandoz Products Ltd.*

31252 Tergraf
Rosin derivatives; used in paints and inks. *Resinas Sinteticas SA.* Name unverified.

31253 Tergum
Rosin derivatives; for paints, inks, adhesives, chewing gum. *Resinas Sinteticas SA.* Name unverified.

31254 Teric 9A2
68439-46-3
C_9-C_{11} ethoxylate (2 EO)
Surfactant for sulfation feedstock, polishes and waxes, drycleaning, solvent cleaners/degreasers. *ICI Australia.*

31255 Teric 12A2
68131-39-5
C_{12}-C_{15} pareth-2; intermediate for phosphate ester production, sulfation; surfactant for solvent cleaners/degreasers; fiber lubricant/antistat for textile spinning. *ICI Australia.*

31256 Teric 12M2
61791-14-8
PEG-2 cocamine; dispersant, emulsifier, stabilizer; wetting agent of hydrophobic surfaces; used in metal, stone, paper and textile processing; formulation of lubricants and dye bath auxiliaries; emulsion stabilization; fat liquoring compounds. *ICI Australia.*

31257 Teric 13A5
24938-91-8
Trideceth-5
Emulsifier for agriculture; filter cake dewatering; polishes/waxes; solvent cleaners/degreasers. *ICI Australia.*

31258 Teric 15A11
C_{14}-C_{15} pareth-11; in biodegradable detergent formulation, low foam laundry products, bottle washing, metal soaking. *ICI Australia.*

31259 Teric 16A16
68439-49-6 7737
Ceteareth-16; wetting agent, emulsifier, detergent, solubilizer, dispersant in hydrophobic conditions; dye and pigment carriers; textile applications; manufacturing of wax emulsions and polishes for household and industrial use; antistat. *ICI Australia.*

31260 Teric 16M2
61791-24-0
PEG-2 soya amine; wetting agent, dispersant, emulsifier for waxes and fats; formulation of leather dressing and metal cleaning compounds, fiber lubricant applications in the bldg. industry. *ICI Australia.*

31261 Teric 17A2
Cetoleth-2; emulsifier, manufacturing of textile lubricants, solvent and waterless hand cleaners; coemulsifier; detergent additive in petroleum oils; intermediate for anionic surfactants. *ICI Australia.*

31262 Teric 17M2
61791-44-4 263-177-5
PEG-2 tallow amine; wetting agent and dispersant; dewatering agent in electrical components and road aggregates; emulsifier for fats, waxes, mineral oils, and metal working lubricants; corrosion inhibitor. *ICI Australia.*

31263 Teric 18M5
26635-92-7
PEG-5 stearamine; wetting agent, dispersant; emulsion stabilizer; emulsifier used in agricultural toxicants and processing of textiles, paper, leather, and building. board; corrosion inhibitor in lubricants and greases; softener and antistat in solvent cleaning of textiles and plastics. *ICI Australia.*

31264 Teric BL8
Synthetic alcohol polyalkylene oxide deriv
Detergent; wetting agent and dispersant in printing inks, textiles, wool scouring, dust suppression, household and industrial surface cleaning; emulsifier of solvents, greases and oils; biodegradable. *ICI Australia.*

31265 Teric DD5
9014-92-0
Dodoxynol-5; emulsifier, coemulsifier in solvent emulsion cleaners; formulation of concentrates for metal lubricants, cutting, milling and grinding aids, textile processing aid and mineral drilling lubricants. *ICI Australia.*

31266 Teric G12A4
68131-39-5
C_{12}-C_{15} pareth-4(4.5 EO); wetting agent and dispersant, detergent; dust suppression; aqueous pigment dispersion; coemulsifier for water-oil-type emulsion concs.; liquid and powder detergents for industrial and domestic laundry, dishwashing, metal cleaning, sanitizing, textile processing aids, abrasive cleaners, intermediate in manufacturing of sulfate and phosphate surfactants. *ICI Australia.*

31267 Teric G9A5
68439-46-3
C_9-C_{11} pareth-5; emulsifier for solvents; formulation of hard surface cleaners, degreasers and dispersants in industrial and domestic applications *ICI Australia.*

31268 Teric LA4
68131-39-5
C_{12}-C_{15} pareth-4 (4.5 EO); wetting agent and dispersant, detergent; dust suppression; aqueous pigment dispersion; coemulsifier for water-oil type emulsion concentrates; liquid and powder detergents for industrial and domestic laundry, dishwashing, metal cleaning, sanitizing, textile processing aids, abrasive cleaners, intermediate in manufacturing of sulfate and phosphate surfactants. *ICI Australia.*

31269 Teric N2
9016-45-9 6772
Nonoxynol-2; defoamer, wetting agent and dispersant in solvent- and oil-based systems; coemulsifier for oil-wateremulsions; emulsifier for water-oil emulsions; used in agricultural toxicant products, industrial solvent-cleaning systems; surface coating preparations; intermediate for production of anionic detergents. *ICI Australia.*

31270 Teric OF4
9004-96-0
PEG-4 oleate; emulsifier used in oil-water emulsions of vegetable and mineral oils, fats, waxes and solvents; formulation of cutting oils; metal lubricant in metal cleaning formulations and in solvent degreasers; textile and paper finishing applications; antistat in textile processing and manufacturing of synthetic fibers. *ICI Australia.*

31271 Teric PE61
POP + 5.7 EO; demulsifier, intermediate, emulsifier, detergent, wetting agent, and dispersant; pigments in latex paints; detergent sanitizer and alkaline cleaner; dewatering aid for treatment of crude oil emulsions; defoamer in paper pulp liquors, starch *ICI Australia.*

31272 Teric PEG 200
25322-68-3 7729
PEG 200
Biodegradable; binder for glazing; intermediate for PEG esters, methacrylate resins, PU foams; plasticizer/solvent for cork; toiletries; metalworking lubricants; paints/resins; paper/film; printing inks; textile emulsifier. *ICI Australia.*

31273 Teric PEG 300
25322-68-3 7729
PEG 300
Pesticide solubilizer/carrier; viscosity modifier for brake fluids; intermediate for PEG esters, PU foams; plasticizer/solvent; cosmetics; metalworking lubricants; paints/resins; paper/film; pharmaceuticals; printing inks; rubber; and textile auxiliary *ICI Australia.*

31274 Teric PEG 400
25322-68-3 7729
PEG 400

Biodegradable; emulsifier, antistat for textiles; rubber; inks; cosmetics; pharmaceuticals; paper/film; pesticide solubilizer/carrier; intermediate for PEG esters, PU foams; plasticizer/solvent for cork; metalworking lubricants; paints/resins. *ICI Australia.*

31275 Teric PEG 600
25322-68-3 7729
PEG 600
Biodegradable; emulsifier for textiles; pesticide solubilizer/carrier; binder for ceramics; intermediate for PEG esters, PU foams; cosmetics; pharmaceuticals; metalworking lubricants; resins; paper/film; inks; rubber. *ICI Australia.*

31276 Teric PEG 800
25322-68-3 7729
PEG 800
Biodegradable; textile auxiliary; pharmaceutical tableting; pesticide solubilizer/carrier; intermediate for PEG esters; toiletries; metalworking lubricants. *ICI Australia.*

31277 Teric PEG 1000
25322-68-3 7729
PEG 1000
Biodegradable; intermediate for PEG esters; cosmetics/toiletries/soaps; metalworking lubricants; wood processing. *ICI Australia.*

31278 Teric PEG 1500
PEG 1500
Biodegradable; textile finishing and sizing; wood processing; latex production; printing inks; pharmaceuticals; paper/film. *ICI Australia.*

31279 Teric PEG 3350
PEG 3350
Biodegradable; intermediates for PEG esters; spinning aid for textiles; pharmaceuticals. *ICI Australia.*

31280 Teric PEG 4000
25322-68-3 7729
PEG 4000
Biodegradable; binder/plasticizer for ceramics; intermediate for copolymers, PEG esters; cosmetics; pharmaceuticals; metalworking lubricants and electropolishes; resins; paper/film; printing inks; rubber antistat, release, compounding aid; textile auxiliary. *ICI Australia.*

31281 Teric PEG 6000
25322-68-3 7729
PEG 6000
biodegradable; cosmetics; pharmaceuticals; thickener for inks; wood processing; binder/plasticizer for ceramics; intermediate for copolymers, PEG esters; metalworking lubricants; resins. *ICI Australia.*

31282 Teric PEG 8000
25322-68-3 7729
PEG 8000
Biodegradable; intermediate for copolymers, PEG esters; cosmetics; pharmaceuticals; thickener for inks; release for rubber molding. *ICI Australia.*

31283 Teric PEG 12000
25322-68-3 7729
PEG 12000
Biodegradable. *ICI Australia.*

31284 Teric PPG 400
PPG 400
Biodegradable; intermediate for surfactants, ethers, esters; metalworking lubricants; solvent for paints/varnishes, printing inks, vegetable oils; textile lubricants. *ICI Australia.*

31285 Teric PPG 1000
PPG 1000
Biodegradable; intermediate for surfactants, ethers, esters; antifoam for ceramics, rubber; hydraulic brake fluids; dyeing; metalworking lubricants; plasticizer for plastics; latex coagulant; textile lubricant. *ICI Australia.*

31286 Teric PPG 1650
PPG 1650
Biodegradable; intermediate for surfactants, ethers, esters; antifoam for ceramics, rubber; lubricant/softener for leather; metalworking lubricant; demulsifier for petroleum industry; plasticizer for plastics. *ICI Australia.*

31287 Teric PPG 2250
PPG 2250
Biodegradable; intermediate for surfactants, ethers, esters; hydraulic brake fluids; cosmetics; dyeing; lubricant/softener for leather; metalworking lubricant; demulsifier for petroleum; latex coagulant; rubber release agent; solv. for vegetable oils. *ICI Australia.*

31288 Teric PPG 4000
PPG 4000
Biodegradable; intermediate for surfactants, ethers, esters; cosmetics; lubricant/softener for leather; metalworking lubricants; demulsifier for petroleum industry; mold release for rubber. *ICI Australia.*

31289 Teric X5
9002-93-1 6858
Octoxynol-5
Wetting agent and dispersant, detergent used in alkaline and metal cleaners, agricultural powders, emulsions, pigment and wax dispersions; textile auxiliaries; paints. *ICI Australia.*

31290 Terinda
Polyester yarns. *ICI Chem & Polymers Ltd.*

31291 Terlac
Rosin derivatives; for paints and inks. *Resinas Sinteticas SA.* Name unverified.

31292 Terlan
Polyamide resins; for adhesive/molding applications. *The Terrell Corporation.* Name unverified.

31293 Terluran
ABS color compounds. *Norsk Hydro Polymers Ltd.* Unverified.

31294 Terluran 846 L
A proprietary ABS of medium rigidity and toughness used for injection molding, extrusion and thermoforming. *BASF plc.* Name unverified.

31295 Terluran 886
A tough grade of ABS. *BASF plc.* Name unverified.

31296 Terluran 8760 Galvano
A special grade of ABS used for electroplating. *BASF plc.* Name unverified.

31297 Terluran®
9003-56-9
ABS polymers; injection molding, easy-flow injection molding, heat-resistant, high impact, extrusion, transparent, flame-retardant, reduced gloss, glass-reinforced, and electroplating grades available. *BASF AG; BASF plc.*

31298 Terlux®
9003-56-9
Clear ABS. *BASF.*

31299 Termamyl®
9000-92-4 640 232-567-7
A liquid enzyme preparation containing an outstandingly heat-stable α-amylase expressed in and produced by a selected strain of *Bacillus licheniformis*; used in the starch, alcohol, brewing, sugar, and textile industries. *Novo Nordisk.*

31300 Termamyl® 120L
9000-92-4 640 232-567-7
Bacterial α-amylase; enzyme for liquefaction of gelatinized starch in production of dextrose, high fructose, and other syrups; used in alcohol, brewing, and textile industries. *Novo Nordisk.* Name unverified.

31301 Termex
Laminates of nomex, presspaper, leatheroid, paper, melinex, mylar and kapton; slot liner and closure material used in the manufacture and repair of electrical motors, transformers and other electrical equipment. *Fothergill Tygaflor Ltd.* Name unverified.

31302 Termex
Plastics used in soles and heels of footwear, in manufacture of heel and toe stiffening components of footwear. *Courtaulds Advanced Materials (Holdings) Ltd.* Discontinued.

31303 Terminate
Bacillus thuringiensis wettable powder; applied by spray to control larvae of Lepidopteran insects. *Westbridge Research Group.* Unverified.

31304 Term-X
Various Herbicides/Insecticides employed in do-it-yourself products for household and commercial use and application. *Puma Chemical Co Inc.* Name unverified.

31305 ternary steels
Alloy steels containing one special element in addition to the iron and carbon.

31306 Terne metal
An alloy of 80% lead, 18% tin, and 2% antimony.

31307 Terne plate
An alloy of lead and tin, coated on iron plate, and intended for use in roofing.

31308 Terohane
Polyester film. *Foseco (F.S.) Ltd.*

31309 Terpal®
Soluble concentrate containing 155g 2-chlorethylphosphonic acid and 305g mepiquat chloride per liter; plant growth regulator for use in cereals. *BASF plc; BASF AG; Clifton Chemicals Ltd.*

31310 Terpal® CC, M
999-81-5 2153 213-666-4
Chlormequat chloride
cholinechloride. Bioregulator for improved resistance against lodging in barley, rye, wheat, and flax. *BASF AG.*

31311 Terpanol
Dipentene substitute. *Crowley Chem.*

31312 Terpenato
Rosin derivatives; for paints and inks. *Resinas Sinteticas SA.* Name unverified.

31313 terpene resin
Thermoplastic resin for use in hot-melt adhesives and coatings, as masticatory agents in chewing gum. *Cardolite; Hercules; Langely Smith Ltd.*

31314 terpestrol
A powder containing lactose with 5% oil of turpentine.

31315 Terpex D, K-3, S
A proprietary trade name for terpene vinyl plasticizers. *Glidden Co.* Name unverified.

31316 Terphane®
Polyester film; high quality film for demanding applications incl. food pkg., magnetic tape, elec. cable insulation, graphics arts. *Rhône-Poulenc/Film Div.*

31317 Terpigol
A proprietary trade name for terpinyl monoethylene glycol ether. *Nipa Laboratories Ltd.* Name unverified.

31318 terpine
A turpentine substitute that is a product of the distillation of petroleum.

31319 Terpinoxo
Oxygen forming compound for rubber. *Crowley Chem.*

31320 Terposol No. 3
A proprietary trade name for a solvent consisting of terpene methyl ethers. No manufacturer.

31321 Terposol No. 8
A proprietary trade name for a solvent consisting of terpene glycol ethers. No manufacturer.

31322 Terpurile
Wetting-out agents consisting of soaps with organic solvents. No manufacturer.

31323 terra fullonica
Fuller's earth.

31324 Terra Nova
Lead/cadmium-free colors for use on glazes. *Degussa Ltd.*

31325 terra ponderosa
7727-43-7 1023 231-784-4
BaO_4S
Barium sulfate
Foseco (F.S.) Ltd. Discontinued.

31326 terra verte
A green earthy material found in the Mendip Hills. It consists of a species of ocher, is essentially silica with oxide of iron and small quantities of other oxides and is used as a pigment..

31327 Terracote
Coatings for dies, molds, chills etc. *Foseco (F.S.) Ltd.*

31328 terra-cotta
A building material made from clay.

31329 Terracur® P
115-90-2 4042 204-114-3
$C_{11}H_{17}O_4PS_2$
O,O-diethyl O-[4-(methylsulfinyl)phenyl] phosphorothioate
O,O-diethyl O-4-methylsulfinylphenyl phosphorothioate; DMSP; OMS 37; ENT 24945; Bay 25141; Dasanit; S 767;. Granular insecticide; used for treatment of biting insects and nematodes. Nematicide and insecticide with primarily contact action. Cholinesterase inhibitor. Used for control of nematodes and soil-dwelling insects. $bp_{0.01}$ = 138-141°; d_{20} = 1.202; n_D^{25} = 1.540; soluble in H_2O (1.54 g/l), soluble in most organic solvents; LD_{50} (rat orl)= 10.5 mg/kg. *Bayer AG.*

31330 Terradust
Brass foundry 'caster's flour'. *Foseco (F.S.) Ltd.* Discontinued.

31331 Terrafen®
Special fertilizer incorporated into a gel which operates on the ion exchange principle; for long-term fertilization of all types of houseplants. *Bayer AG.*

31332 Terragloss
Ultraviolet curing coatings for printed paper and board. *Swale Coatings & Inks Ltd.*

31333 Terraklene
Herbicide. *ICI Chem & Polymers Ltd.*

31334 Terralacke
Water-based coatings for printed packaging. *Swale Coatings & Inks Ltd.*

31335 Terram
Nonwoven civil engineering fabric. *ICI Chem & Polymers Ltd.* Discontinued.

31336 Terramix CDA
An emulsifiable concentrate containing 99 g aminotriazole, 190 g atrazine and 95 g 2,4-D per liter; used for total weed control in non crop areas. *Denoon CDS; Powaspray (CDA) Ltd.*

31337 Terraneb SP Turf Fungicide
2675-77-6 220-222-3
$C_8H_8Cl_2O+2$
1,4-dichloro-2,5-dimethoxybenzene
Demosan; Soil Fungicide 1823; Teremec; Tersan SP. Systemic soil- and seed fungicide, absorbed by the roots. Terraneb SP Turf Fungicide is 65% Chloroneb wettable powder; used for control seedling diseases like snow mold (*Typhula*) and pythium blight. mp 133-135°; bp = 268°; soluble in H_2O (8 mg/l), more soluble in organic solvents; LD_{50} (rat orl) > 11000 mg/kg. *Kincaid Enterprises Inc.* Name unverified.

31338 Terranox
An emulsifiable concentrate containing 105g aminotriazole, 207g atrazine and 100g 2,4-D per liter; used for total weed control in non crop areas. *Agri-Technics Ltd.*

31339 Terrapaint
Coatings for foundry molds and cores. *Foseco (F.S.) Ltd.*

31340 Terrapowder
Ferrous mold and core dressings. *Foseco (F.S.) Ltd.*

31341 terrar
A preparation from earthy zirconia, in Brazil; used as an opacifying agent in enamels and glazes.

31342 Terra-Systam
Systemic organo-phosphorus insecticides. *Murphy Chemical Co Ltd.* Discontinued.

31343 Terrathion
298-02-2 7486 206-052-2
$C_7H_{17}O_2PS_3$
O,O-diethyl S-[(ethylthio)methyl]phosphorodithioate
AC 3911; Agrimet; Chim; Forate; Geomet; Granutox; Rampart; Thimet; Vegfru Foratox; Volphor. Terrathion consists of granules containing 10% w/w phorate; an organophosphorus insecticide. mp <-15°; $bp_{0.8}$= 118-120°; d_{25} = 1.167; n_D^{25} = 1.5349; soluble in H_2O (50 mg/l), more soluble in organic solvents; LD_{59} (rat orl) = 3.7 mg/kg. *Farmers Crop Chemicals Ltd.* Discontinued.

31344 Terravest 801
A stereospecific, low molecular weight polybutadiene; consolidates soil against erosion and binds dust, e.g., on heaps of all sorts. *Hüls AG.*

31345 Terr-O-Gas
Methylbromide with chloropicrin. Fumigant mixture, used for control of insects, mites and rodents in stored grain. *Great Lakes Europe.*

31346 Terry's stain
A microscopic stain. It contains 20 ml of a 1% aqueous solution of methylene blue, 20 ml of a 1% solution of potassium carbonate, and 60 ml water. This is boiled, cooled, and 10 ml of a 10% solution of acetic acid added, and the whole made up with water to 100 ml.

31347 Terset
Bromoxynil + ioxynil + isoproturon + mecoprop; a contact herbicide for use in cereal crops. *Rhône-Poulenc Crop Protection Ltd.*

31348 Terylene
7730
polyethylene terephthalate
Polyethylene terephthalate produced from dimethyl terephthalate and ethylene glycol; synthetic polyester textile fiber, resistant to most dry-cleaning solvents, possessing good wear resistance. *ICI Chem & Polymers Ltd.*

31349 Tesal
Propylene glycol stearate SE; self-emulsifying base for cosmetic/pharmaceutical ointments, lotions, creams. *Gattefosse.*

31350 Tescol®
Vinyl or acrylic copolymers; continuous filament polyester sizes for textiles *Allied Colloids Ltd.*

31351 testalin
An aluminum soap, made by treating ordinary soap with aluminum sulfate; used for the cementing together of sandstone to form a solid block.

31352 testifas oil
A fraction of petroleum distillation; used as a burning oil.

31353 Tesuloid
Technetium-^{99m}Tc sulfur colloid
A radioactive agent. *Bristol-Myers Squibb Co Inc.* Name unverified.

31354 Tetanol
A proprietary preparation of calcium hevulinate. No manufacturer.

31355 tetjamer
An aluminum bronze. It contains from 86-93% copper, 5-10% aluminum, 1-3% silicon, and 0.72-0.98% iron.

31356 tetra(2-hydroxypropyl) ethylenediamine
102-60-3 3635 203-041-4
$C_{14}H_{32}N_2O_4$
N,N,N',N'-tetrakis(2-hydroxypropyl)ethylenediamine

(ethylenedinitrilo)tetra-2-propanol; EDTP; Entprol; Quadrol. Chelating agent; intermediate used in resins, emulsifiers, surfactants, pharmaceuticals, herbicides, fungicides, insecticides, adhesives, and plasticizers. $bp_{0.8}$ = 175-181°, d = 1.0130; n_D^{20}= 1.4812; soluble in H_2O, organic solvents;.

31357 Tetrabor®
12069-32-8 1374 235-111-5
Boron carbide
Abrasives, lapping media, nozzles, building materials, drills, saws, bearings, shafts, files, rasps, and whetstones, all made wholly or principally of boron carbide. *Elektroschmelzwerk Kempten GmbH.*

31358 tetrabromo phthalatediol
20566-35-2 243-885-0
Great Lakes PHT4-Diol; Tetrabromophthalatediol. Reactive intermediate used to produce flame retardant rigid urethane foam; can replace chlorinated polyols; for PU elastomers, coatings, adhesives, and fibers. *Great Lakes.*

31359 tetrabromobisphenol-A
79-94-7 201-236-9
$C_{15}H_{12}Br_4O_2$
4,4'-Isopropylidenebis (2,6-dibromophenol)
FR-1524. Reactive flame retardant used in the mfg. of epoxy, PC, ABS, phenolic, PS, and polyester resins, rubber; flame retardant intermediate. White powder; soluble in acetone, benzene, alcohol; mw=543.9; sp gr=2.17; mp=180°; dec at 240°; LD_{50}=rat oral>50,000; rabbit dermal>2000 mg/kg; irritant. *AmeriHaas; Dead Sea Bromine.*

31360 tetrabromodipentaerythritol
109678-33-3
$C_{10}H_{18}Br_4O_3$
FR-1034; TBDPE;. Flame retardant for PP extruded fibers; processing aid for ABS and HIPS. Powder; mw=506; mp=75-82°; LD_{50}=rat oral>5000mg/kg. *AmeriHaas; Dead Sea Bromine.*

31361 tetrabromophthalic anhydride
632-79-1 211-185-4
Great Lakes PHT4; Saytex® RB-49. Flame retardant in production of unsat. polyester resins and rigid PU polyols; cohardener for epoxy resins; cost efficient additive for latex emulsions; derivs. used as flame retardants in diverse application (wire coating, and wool, etc.). *Great Lakes; Ethyl Corp.*

31362 tetrabutyl ammonium bromide
1643-19-2 216-699-2
$C_{12}H_{28}BrN$
tetra-N-butylammonium bromide
TBAB. Quaternary ammonium salt. *Aldrich; Hawks Chem Co Ltd; Schweizerhall; Zeeland.*

31363 tetrabutylthiuram disulfide
1634-02-2 216-652-6
$C_{18}H_{36}N_2S_4$
tetrabutylthioperoxydicarbonic diamide
thioperoxydicarbonic diamide, tetrabutyl-;. Rubber accelerator, sulfur donor, accelerator, vulcanizing agent.

31364 Tetracarnit
A mixture of pyridine and its homologues with Turkey red oil or similar substances; used as a wettingout agent to assist the penetration of textiles by liquids.

31365 tetracycline
60-54-8 9337 200-481-9
$C_{22}H_{24}N_2O_8$
4-(Dimethylamino)-1,4,4a,5,5a,6,11,12a-octahydro-3,6,10,12,12a-pentahydroxy-6-methyl-1,11-dioxo-2-naphthacenecarboxamide
deschlorobiomycin; tsiklomitsin; Abricycline; Achromycin; Agromicina; Ambramicina; Ambramycin; Bio-Tetra; Bristaciclina; Cefracycline suspension; Criseociclina; Cyclomycin; Democracin; Hostcyclin; Omegamycin; Panmycin; Polycycline; Purocyclina; Sanclomycine; Steclin; Tetrabon; Tetracyn; Tetradecin. Anti-amebic, antibacterial, antirickettsial. mp = 170-175° (dec); $[\alpha]^{25}_D$ = -239° (MeOH); λ_m = 220, 268, 355 nm (ε = 13000, 18040, 13320, 0.1N HCl); pKa8.3, 10.2 (50% aqueous DMF); soluble in H_2O (1.7 mg/ml), more soluble in MeOH; LD_{50} (rat orl) = 807 mg/kg.

31366 tetradifon
116-29-0 9339 204-134-2
$C_{12}H_6Cl_4O_2S$
Benzene, 1,2,4-trichloro-5-[(4-chlorophenyl)sulfonyl]-
Sulfone; *p*-chlorophenyl 2,4,5-trichlorophenyl; *p*-chlorophenyl 2,4,5-trichlorophenyl sulfone; Akaritox; Aredion; Duphar; ENT 23,737; FMC 5488; Mition; NIA 5488; NIA-5488; Polacaritox; Roztoczol; Roztoczol Extra; Roztozol; Sulfone, 2,4,4',5-tetrachlorodiphenyl; Tedion; Tedion V-18; Tetradiphon; Tetrafidon; Acaroil TD; Acarvin; Agrex T-7.5; Aracnol K; Mitifon; Teranol; V-18. Long-acting, non-systemic acaricide used to control eggs of phytophagous mites on fruit trees. mp = 148-149°; d^{20} = 1.151; slightly soluble in H_2O, soluble in organic solvents; LD_{50} (rat orl) > 14700 mg/kg *Afrasa; Inagra; Sadisa; Diachem; Hellenic Chemical; Duphar; Diana.*

31367 tetraethylenepentamine
112-57-2 203-986-2
$C_8H_{23}N_5$
1,4,7,10,13-pentaazatridecane
TEPA. Solvent for sulfur, acid gases, various resins and dyes; saponifying agent for acidic materials; manufacture of synthetic rubber; dispersant in motor oils; intermediate for oil additives. mp = -40°; bp = 340°; d = 0.9980; n_D^{20}= 1.5055. *Tosoh; Union Carbide.*

31368 Tetraflon
A proprietary polytetrafluoroethylene (PTFE). *Nitto Chemical Industry Co Ltd.* Name unverified.

31369 tetraform
56-23-5 1864 200-262-8
A specially pure carbon tetrachloride.

31370 tetrahydrofuran
109-99-9 9351 203-726-8
C_4H_4O
tetramethylene oxide
THF; diethylene oxide. Solvent, Grignard reactions, reductions, and polymerizations; chemical intermediate and monomer. mp = -109°; bp = 66°; d_4^{20}= 0.8892; n_D^{20} = 1.4070; miscible with H_2O and organic solvents. *Arco Europe; Ashland; BASF; DuPont; Great Lakes; Hüls UK; Janssen Chimica; QO.*

31371 tetrahydrofurfuryl alcohol
97-99-4 9353 202-625-6
$C_5H_{10}O_2$
tetrahydrofuryl carbinol
tetrahydro-2-furanmethanol; tetrahydro-2-furancarbinol; tetrahydro-2-furylmethanol; THFA. Solvent for vinyl resins, dyes for leather, chlorinated rubber, cellulose esters, coupling agent, solvent-softener for nylon. bp = 178°; d_{20}^{20} = 1.0543; n_D^{20}= 1.4520; miscible with H_2O, organic solvents. *Penta Mfg.; QO; Schweizerhall.*

31372 tetrahydrofurfuryl methacrylate
2455-24-5 219-529-5
$C_9H_{14}O_3$
methacrylic acid tetrahydrofurfuryl ester
Anaerobic adhesives and sealants, printed circuit boards, artificial finger nails, modifier for hard rubber rolls, wire and cable coatings, screen printing inks, emulsion polymerization, plastic modifier, EB-curable coatings. Often stabilized with 4-methoxyphenol/hydroquinone bp_4 = 83-84°; d= 1.040; n_D^{20}= 1.4580 *CPS; Sartomer.*

31373 tetrahydronaphthalene
119-64-2 9360 204-340-2
$C_{10}H_{12}$
1,2,3,4-tetrahydronaphthalene
tetralin. Chemical intermediate, solvent for greases, fats, oils, waxes; substitute for turpentine. bp = 206-207°; d = 0.970; n_D^{20}= 1.5410. *Du Pont; Hüls AG.*

31374 tetrahydrothiophene
110-01-0 9357 203-728-9
C_4H_8S
thiophane
tetramethylene sulfide; tetramethylene sulphide; thiolane. Solvent, intermediate, fuel gas odorant. mp = -96°; bp = 117-119°; d = 0.999; n_D^{20}= 1.5030; LC_{50} (mus inh)27 mg/l. *Elf Atochem N. Am.*

31375 Tetrakal
7320-34-5 7833 230-785-7
$K_4O_7P_2$
potassium pyrophosphate
diphosphoric acid tetrapotassium salt. Tetrapotassium pyrophosphate. Soluble in H_2O, insoluble in EtOH, organic solvents. *Albright & Wilson Ltd.*

31376 Tetralex Plus
Soluble concentrate of 18 g dicamba, 252 g MCPA and 84 g mecoprop per liter; used for weed control in cereals and grassland. *Shell UK.*

31377 Tetralide®
7-Acetyl 1,1,3,4,4,6-hexamethyltetraline
Musky odor used in fragrances. *Bush Boake Allen Ltd.*

31378 Tetralin Extra
A mixture of tetralin and decalin.

31379 tetraline
An old name for tetrachloroethane.

31380 tetralitbenzol
A mixed fuel for internal combustion engines. It contains 50% benzene, 25% tetralin, and 25% of 95% alcohol.

31381 Tetralol
530-91-6 9361 208-497-8
$C_{10}H_{12}O$
1,2,3,4-tetrahydro-2-naphthalenol

ac-tetrahydro-β-naphthol; *ac*-β-tetralol. Used as an antiseptic. mp = 16°; bp_{12} = 140°; LD_{50} (rat orl) = 1.0 ml/kg.

31382 Tetralon®

A range of sequestering agents based on nitriloacetic acid, ethylene tetramine diacetic acid and diethylenetriamine pentaacetic acid; mainly used in soaps and detergents. *Allied Colloids Ltd.*

31383 Tetramet-125

9555

Thyroxine-^{125}I

Radioactive agent. *Abbott Laboratories.* Name unverified.

31384 tetramethrin

7696-12-0 9362 231-711-6

$C_{19}H_{25}NO_4$

2,2-Dimethyl-3-(2-methyl-1-propenyl)cyclopropanecarboxylic acid (1,3,4,5,6,7-hexahydro-1,3-dioxo-2H-isoindol-2-yl)methyl ester

Neopynamin; FMC-9260; SP-1103; phthalthrin; Py-Kill. Non-systemic insecticide with rapid knockdown. Used in combination with synergists such as piperonyl butoxide for control of flies, cockroaches, mosquitoes, wasps and other insect pests. mp = 65-80°; d_{20}^{20} = 1.108; $n^{21.5}_D$ = 1.5175; poorly soluble in H_2O (4.6 mg/l), very soluble in organic solvents; LD_{50} (mus orl) = 1000 mg/kg.

31385 tetramethylammonium hydroxide

75-59-2 9363 200-882-9

$C_4H_{13}NO$

N,N,N-trimethylmethanaminium hydroxide

Uusually marketed as a 10% aqueous solution. d^{25}_4 = 1.00. *Aldrich; Fluka; janssen Chimica.*

31386 tetranitromethane

517-25-9 9859 208-236-8

CHN_3O_6

nitroform

Used in the manufacture of explsives and propellants. mp = 15°; d_4^{25}= 1.469.

31387 Tetranol

Sodium butyl oleate

Wetting and leveling agent for dyeing processes; assistant for vat dyeing; emulsifier and detergent in scouring processes. *Sandoz.*

31388 Tetranyl

2,3,4,6-tetranitro-aniline

An explosive.

31389 tetra-paper

Tetra-base-paper. Paper which has been treated with dimethyl or tetramethyl-*p*-phenylene-diamine; used in testing for ozone.

31390 tetraphosphate

Tetra. Produced by mixing natural phosphate rock powder with 6% of a powder containing equal parts of the carbonates of calcium, sodium, and magnesium, with a little sulfate of soda.

31391 tetrasodium EDTA

64-02-8 3557 200-573-9

$C_{10}H_{12}N_2Na_4O_8$

N,N'-1,2-ethanediylbis[N-(carboxmethyl)glycine]tetrasodium salt

edetate sodium; ethylenediamine tetraacetic acid tetrasodium salt; tetrasodium ethylene diaminetetraacetate; EDTA tetrasodium; Endrate Tetrasodium; Questex; Versene; Sequestrine; Tetrine; Kalex; Trilon B; Komplexon; Nullapon; Aquamollin; Complexone; Distol 8; Irgalon; Calsol; Syntes 12a; Tyclarosol; Nervanaid B. Substituted amine; general-purpose chelating agent. mp > 300°; soluble in H_2O; chelates most di- and trivalent metal ions. *W.R. Grace/Hampshire; Rhône-Poulenc Basic.*

31392 Tetraterge D-101, NFF

Detergent for textile scouring. *Reilly-Whiteman.*

31393 Tetrathal

117-08-8 204-171-4

$C_8Cl_4O_3$

4,5,6,7-tetrachloro-1,3-isobenzofurandione

Tetrachlorophthalic anhydride. Flame retardant for polyester resins and polyols. mp = 254-258°; bp = 349-354°; *Monsanto Co.*

31394 Tetrawet DWN

Wetting agent for textile processing of cellulosics and blends. *Reilly-Whiteman.*

31395 tetrazolium chloride

1871-22-3 9380 217-488-8

$C_{40}H_{32}Cl_2N_8O_2$

2,4,5-triphenyltetrazolium chloride

TTC; tetrazolium salt; tetrazolium blue; Dimethoxy neotetrazolium; Ditetrazolium chloride; Tetrazolium Blue; Blue Tetrazolium; 3,3'-(3,3'-Dimethoxy-4,4'-biphenylene)bis(2,5-diphenyl-2H-tetrazolium chloride); BT; Blue Tetrazolium chloride. Used in germination and viability testing of seeds. Also used as a stain for bacteria and molds and to detect redox enzymes in cells. mp = 242-245° (dec); slightly soluble in H_2O, soluble in MeOH, EtOH, $CHCl_3$, insoluble in non-polar organic solvents *Dajac Labs; U.S. Biochemical.*

31396 Tetron

7722-88-5 9377 231-767-1

Tetrasodium pyrophosphate. *Albright & Wilson Ltd., Phosphates & Speciality Business.*

31397 Tetrone

Rubber accelerator. *DuPont UK.*

31398 Tetronic® 150R1

107397-59-1

EO/PO ethylene diamine block copolymer; *BASF.*

31399 Tetronic® 304

11111-34-5

Poloxamine 304

Emulsifier, thickener, wetting agent, dispersant, solubilizer, stabilizer for cosmetics and pharmaceuticals; demulsifier in petrol. industry; detergent ingredient; antistat for polyethylene and resin molding powds.; metal treatment; emulsion polymerization; used in latex-based paints, aqueous-basedsynthetic cutting fluids and vulcanization of rubber. *BASF.*

31400 Tetronic® 504

11111-34-5

Poloxamine 504

BASF.

31401 Tetronic® 50R1

EO/PO ethylene diamine block copolymer; surfactant series functioning as emulsion stabilizers, solubilizers, dispersants, wetting agents, antistats, penetrants, plasticizers, defoaming agents, demulsifiers in the petrol., paint, paper, cement, ink, cosmetic, drug, plastic, detergent and metalworking industries; rubber activator; R series for low-foaming applications. *BASF.*

31402 Tetronic® 701

11111-34-5

Poloxamine 701

BASF.

31403 Tetronic® 702

11111-34-5

Poloxamine 702

BASF.

31404 Tetronic® 704

11111-34-5

Poloxamine 704. BASF.

31405 Tetronic® 707

11111-34-5

Poloxamine 707. *BASF.*

31406 Tetronic® 901

11111-34-5

Poloxamine 901. *BASF.*

31407 Tetronic® 904

11111-34-5

Poloxamine 904. *BASF.*

31408 Tetronic® 908

11111-34-5

Poloxamine 908. *BASF.*

31409 Tetronic® 1101

11111-34-5

Poloxamine 1101. *BASF.*

31410 Tetronic® 1102

11111-34-5

Poloxamine 1102. *BASF.*

31411 Tetronic® 1104

11111-34-5

Poloxamine 1104. *BASF.*

31412 Tetronic® 1107

11111-34-5

Poloxamine 1107. *BASF.*

31413 Tetronic® 1301

11111-34-5

Poloxamine 1301. *BASF.*

31414 Tetronic® 1302

11111-34-5

Poloxamine 1302. *BASF.*

31415 Tetronic® 1304

11111-34-5

Poloxamine 1304. *BASF.*

31416 Tetronic® 1307

11111-34-5

Poloxamine 1307. *BASF.*

31417 Tetronic® 1501
11111-34-5
Poloxamine 1501. *BASF.*

31418 Tetronic® 1502
11111-34-5
Poloxamine 1502 *BASF.*

31419 Tetronic® 1504
11111-34-5
Poloxamine 1504. *BASF.*

31420 Tetronic® 1508
11111-34-5
Poloxamine 1508. *BASF.*

31421 Tetrosan 3,4 D
Alkyl dimethyl 3:4 dichlorobenzyl ammonium chloride in liquid form; disinfectant, deodorant, and germicide with high biocidal activity; veterinary, pharmaceutical and agricultural uses. *Millmaster-Onyx UK.* Name unverified.

31422 Tetroxone
Selective weedkiller containing bromoxynil, dichlorprop, ioxynil and MCPA as potassium salts. *ICI Chem & Polymers Ltd.*

31423 Tetryl
479-45-8 6664 207-531-9
$C_7H_5N_5O_8$
N-methyl-N,2,4,6-tetranitrobenzenamine
Nitramine; N-methyl-N,2,4,6-tetranitroaniline; picrylmethylnitramine; picrylnitromethylamine; Tetralite. A trade name for trinitrophenylmethylnitramine, used in explosives; a detonator known as tetryl contains 0.4 gram tetranitrophenylmethylnitramine, and 0.3 gram of a mixture of 87.5% mercury fulminate and 12.5% potassium chlorate. Also used as an acid-base indicator. mp = 130-132°; d = 1.57; insoluble in H_2O, soluble in organic solvents. No manufacturer.

31424 Texacar® EC
96-49-1 202-510-0
$C_3H_4O_3$
Ethylene carbonate
1,3-dioxolan-2-one. Solvent for organic and inorganic material; Rule 66 exempt; also used as reactant and plasticizer in fibers and textiles, plastics and resins, aromatic hydrocarbon extraction, electrolytes, hydraulic brake fluids. mp = 35-37°; bp = 244-245°; d = 1.320. *Texaco.*

31425 Texacar® PC
108-32-7 203-572-1
$C_4H_6O_3$
Propylene carbonate
4-methyl-1,3-dioxol-2-one; 1,2-propanediol cyclic carbonate. Solvent for organic and inorganic materials; Rule 66 exempt; also used as reactant and plasticizer in fibers and textiles, hydraulic fluids, plastics and resins, gas treating, aromatic hydrocarbon extraction, metal extraction, surface coatings, foundry sand binders, lubricants, electrolytes, personal care products, gellant for clays in greases and cosmetics. mp = -49°; bp = 235-239°; d = 1.205; n_D^{20} = 1.4210. *Texaco.*

31426 Texacat® DD
34745-96-5 252-182-8
Tertiary amine
Catalyst for semiflexible foam urethanes. *Texaco.*

31427 Texacat® DMDEE
2,2'-Dimorpholinodiethylether
Catalyst for polyurethane flexible and rigid foam, one-component foam, slabstock foam applications. *Texaco.*

31428 Texacat® DME
108-01-0 2900 203-542-8
$C_4H_{11}NO$
N,N-Dimethylethanolamine
2-dimethylaminoethanol; deanol; β-dimethylaminoethyl alcohol; N,N-dimethyl-2-hydroxyethylamine. Catalyst for polyurethane rigid foam, flexible ether slabstock. Also a CNS stimulant. bp = 134-136°; d = 0.886, n_D^{20} = 1.4294; soluble in H_2O, organic solvents. *Texaco.*

31429 Texacat® DMP
106-58-1 203-412-0
$C_6H_{14}N_2$
N,N'-dimethyipiperazine
1,4-dimethylpiperazine. Catalyst for polyurethane molded high resilience flexible foam, ester slabstock. bp = 131-132°; d = 0.844; n_D^{20} = 1.4460. *Texaco.*

31430 Texacat® DPA
N,N-(Dimethyl)-N',N'- diisopropanol-1,3-propanediamine
Catalyst for rigid PU foam, pkg. foam. *Texaco.*

31431 Texacat® NEM
100-74-3 202-885-0
$C_6H_{13}NO$
N-Ethylmorpholine
4-ethylmorpholine. Catalyst for flexible and rigid PU foam, polyester slabstock. bp = 136-138°; d = 0.908; n_D^{20}= 1.4413. *Texaco.*

31432 Texacat® NMM
109-02-4 203-640-0
$C_5H_{11}NO$
N-Methylmorpholine
4-methylmorpholine; NMM. Catalyst for polyester urethane flexible foam, high rise rigid foam panels. mp = -65°; bp = 116-118°; d = 0.919; n_D^{20}= 1.4349. *Texaco.*

31433 Texacat® TD-33
280-57-9 9801 205-999-9
Triethylenediamine in propylene glycol; general purpose catalyst for producing polyurethane foam systems (flexible and rigid). *Texaco.*

31434 Texacat® ZF-10
N,N,N' - Trimethyl-N'-hydroxyethyl-bisaminoethylether
Catalyst for polyurethane molded high resilience flexible foam, ether slabstock, pkg. foam. *Texaco.*

31435 Texacat® ZF-22
Bis-(2-dimethylaminoethyl)ether, dipropylene glycol; general purpose catalyst for flexible and rigid PU foam, polyether slabstock. *Texaco.*

31436 Texacat® ZR-50
N,N-Bis-(3-dimethylaminopropyl)-N-isopropanolamine; catalyst for flexible and rigid PU foam, pkg. foam. *Texaco.*

31437 Texacat® ZR-70
1704-62-7 216-940-1
$C_4H_{15}NO_2$
2-(2-Dimethylaminoethoxy)ethanol
Catalyst for polyurethane molded high resilience flexible foam, ether slabstock, pkg. foam. *Texaco.*

31438 Texaco BQ
An insecticide having a petroleum base; used for killing the boll weevil. No manufacturer.

31439 Texacure EA-24
112-24-3 203-950-6
N,N'-bis(2-aminoethyl)-1,2-ethanediamine
High boiling mixture of triethylenetetramine components; epoxy curing agent featuring rapid room temperature cure, good adhesive and mechanical properties, good electrical and insulating properties, good chemical resistance. *Texaco.*

31440 Texadril 2010
EO/PO block copolymer; low foaming wetting agent and coemulsifier. *Henkel/Cospha; Henkel/Functional Prods.*

31441 Texalon 600A NU
25038-54-4 6832
Nylon 6 resin, nucleated; used for very small parts, rigid applications. *Texapol.*

31442 Texalon 1000A
25038-54-4 6832
Nylon 6 resin; high impact, tough grade. *Texapol.*

31443 Texalon 1200A BK-11
32131-17-2
Nylon 6/6 resin with 2% carbon black; weather-resistant resin for use in wire tires and outdoor parts; uv resistant. *Texapol.*

31444 Texalon 1200A HR-2 BK-16
32131-17-2
Nylon 6/6 resin; hydrolysis-resistant grade. *Texapol.*

31445 Texalon 1308 A
32131-17-2
Nylon 6/6 resin, impact modified; used for automotive applications, mechanical housings. *Texapol.*

31446 Texalon 1600A Nat
9008-66-6
Nylon 6/10 resin; low moisture. *Texapol.*

31447 Texalon GF 600A (6-33)
25038-54-4 6832
Nylon 6 resin, glass-filled; general purpose glass-filled resin used for automotive grade mechanical parts. *Texapol.*

31448 Texalon GF 1200A (13-40)
32131-17-2
Nylon 6/6 resin, glass-filled; general purpose resin used for metal replacement in automotive applications. *Texapol.*

31449 Texalys
Modified starch; used in textile industry in precoating compounds and also in laminating double backs. *Roquette (UK) Ltd.*

31450 Texamine 84(L)
Alkanolamide and ethanolamine alkylbenzene sulfonate; raw material for manufacturing of liquid detergents. *Zohar Detergent Factory.*

31451 Texanol® Ester-Alcohol
25265-77-4 246-771-9
$C_{12}H_{24}O_3$
2,24-Trimethyl-1,3-pentanediol monoisobutyrate
2,2,4-trimethyl-1,3-pentanediolmono(2-methylpropanoate). Slow-evaporating solvent used as coalescing agent in latex finishes and water-base inks, PVAc latexes, PVAc-acrylic latices, and acrylic, EVA, and B/S latexes; used as solvent in electrodeposition coatings, lacquer coatings; defoamer in waterborne systems and drilling muds for oil industry; chemical intermediate for production of esters; used as stain-resistant plasticizers, lubricant base stocks, solvents, synthetic detergents and herbicides. bp = 244°; d = 0.9500 *Eastman.*

31452 Texaphor 277
Quaternary ammonium salt; suspending agent for paints with high chemical and water resistance requirements. *Henkel Canada.*

31453 Texapon ALS
2235-54-3 218-793-9
Ammonium lauryl sulfate
Base for shampoos, shower baths, and bubble baths. *Henkel KGaA.*

31454 Texapon ASV
Sodium laureth sulfate, mixed with magnesium laureth sulfate, sodium laureth-8 sulfate, sodium oleth sulfate and magnesium oleth sulfate
Extremely mild surfactant base for baby shampoos, facial cleansers, and foam baths. *Henkel/Cospha; Henkel Canada; Henkel KGaA.*

31455 Texapon CS Paste
Sodium lauryl sulfate mixed with sodium myristyl sulfate, sodium cetyl sulfate, sodium stearyl sulfate and laureth-10
Concentrate for creamy shampoos, foam baths; has pearlescent effects. *Henkel/Cospha; Henkel Canada; Henkel KGaA.*

31456 Texapon DEA
DEA-lauryl sulfate
Detergent, foamer for personal care products. *Henkel/Cospha; Henkel Canada.*

31457 Texapon EA-1
67762-19-0
Ammonium laureth sulfate
Surfactant, viscosity builder, solubilizer in personal care products. *Henkel/Cospha; Henkel Canada.*

31458 Texapon ES-1
Sodium laureth sulfate
Surfactant and solubilizer for personal care products. *Henkel/Cospha; Henkel Canada.*

31459 Texapon EVR
Sodium lauryl sulfate mixed with sodium laureth sulfate; lauramide MIPA; cocamide MEA; glycol stearate
Laureth-10; detergent, foamer; base for pearlescent shampoos. *Henkel Canada; Henkel KGaA.* Discontinued.

31460 Texapon IES
MIPA-laureth sulfate and cocamide DEA. Concentrate for liquid foam bath preparation with high essential oil content. *Henkel KGaA.*

31461 Texapon K-12, K-1296, L-100
151-21-3 8782 205-788-1
$C_{12}H_{25}NaO_4S$
Sodium lauryl sulfate
sulfuric acid monodecyl ester sodium salt; sodium dodecyl sulfate; SDS; Irium. Emulsifier for emulsion polymerization; detergent for highsolids cleansers; also for dispersants and wettable powders. Additive for mechanical latex foaming. mp = 204-207°; soluble in H_2O (100 g/l); LD_{50} (rat orl)= 1288 mg/kg. *Henkel/Cospha; Henkel/Functional Prods.*

31462 Texapon K-1296
151-21-3 8782 205-788-1
sodium lauryl sulfate
Wetting agent and detergent for cleaning formulations. Additive for mechanical latex foaming. *Henkel/Cospha; Henkel Functional Prods.*

31463 Texapon L20C
Ammonium amine lauryl sulfate in liquid form; basic material for liquid shampoos. *Henkel Chemicals Ltd.* Name unverified.

31464 Texapon LS Highly Conc
151-21-3 8782 205-788-1
Sodium lauryl sulfate (C_{12}-C_{14})
Foaming agent for acrylate dispersions, carpet and upholstery cleaners. *Henkel/Functional Prods.*

31465 Texapon MG
67702-21-4 221-450-6
Magnesium laureth sulfate
Mild surfactant for liquid shampoo and bath preparation base, baby shampoos. *Henkel/Cospha; Henkel KGaA.*

31466 Texapon MG 3
Mixture of magnesium lauryl sulfate and disodium laureth sulfosuccinate. Used for shampoos, shower baths, bubble baths. *Henkel KGaA.*

31467 Texapon N 25, Texapon NSE
9004-82-4
Sodium laureth (2) sulfate
Base for personal care products; dishwashing agent; firefighting foam concs.; solubilizer for perfumes. *Pulcra SA.*

31468 Texapon NA
67762-19-0
Ammonium lauryl ether sulfate. Detergent for liquid shampoos and bubble baths. *Henkel KGaA.*

31469 Texapon NSE
9004-82-4
Sodium laureth sulfate
Surfactant for emulsion polymerization. *Pulcra SA.*

31470 Texapon NSF
Sodium lauryl ether sulfate in liquid form, with extremely low salt contents; basic material for cosmetic preparations such as shampoos, bubble bath preparations, etc. *Henkel Chemicals Ltd.* Name unverified.

31471 Texapon NSO
9004-82-4
Sodium lauryl ether (2) sulfate
Detergent for liq. shampoos, bubble baths, shower gels, manufacturing of household detergents; wetting agent and detergent for textile fibers, especially wool and blends. *Pulcra SA.*

31472 Texapon OT Highly Conc. Needles
151-21-3 8782 205-788-1
Sodium lauryl sulfate C12-C18
Detergent, shampoo base; for bubble bath, soaps; emulsifier for emulsion polymerization, additive for mechanical latex foaming, carpet and upholstery cleaners. *Henkel/Functional Prods.; Henkel KGaA.*

31473 Texapon PLT-227
9004-82-4
Sodium laureth-2 sulfate
For shampoos and liquid detergents. *Pulcra SA.*

31474 Texapon PNA-127
32612-48-9
Ammonium lauryl ether (1) sulfate
Wetting agent for acid or gel-like shampoos, light duty detergents, window cleaners; stable foam; low eye and skin irritation. *Pulcra SA.*

31475 Texapon QLV
Sodium laureth sulfate. Low salt surfactant for shampoos, cleansing preparations, light duty detergents. *Henkel Canada.*

31476 Texapon SB-3
Disodium laurethsulfosuccinate
Mild surfactant, foamer for bubble baths, baby shampoos, cleansing preparations. *Henkel Canada; Henkel KGaA.* Discontinued.

31477 Texapon SBN
Sodium laureth sulfate and disodium laureth sulfosuccinate. Detergent base for personal care products. *Henkel KGaA.*

31478 Texapon SG
Sodium laureth sulfate, PEG-8, cocamide MEA, glycol disterate, and glycerin. Detergent for emulsion and pearl shampoos. *Henkel KGaA.*

31479 Texapon SH 100
56388-43-3 260-143-1
Disodium oleamide PEG-2 sulfosuccinate
Nonirritating detergent base for shampoos and skin cleansers. *Henkel/Cospha; Henkel Canada.*

31480 Texapon T 42
139-96-8 205-388-7
TEA lauryl sulfate
Detergent, emulsifier for hair and rug shampoos, bubble baths, shower gels, fire fighting foams. *Pulcra SA.*

31481 Texapon VHC Needles, ZHC Needles
151-21-3 8782 205-788-1
Sodium lauryl sulfate
Wetting agent, foamer, emulsifier, detergent and cosmetic base for personal care products, scouring agents, pigment dispersions, and emulsion polymerization. *Henkel/Cospha; Henkel/Functional Prods.; Henkel Canada; Henkel KGaA.*

31482 Texapon ZHC Needles
151-21-3 8782 205-788-1
Sodium lauryl sulfate
Foaming agent, dispersant, wetting agent for foaming bubble baths, cosmetic

cleansing creams and emulsions; air entraining agent. *Henkel/Cospha; Henkel KGaA.*

31483 Texapon ZHC Needles

151-21-3 8782 205-788-1

sodium lauryl sulfate

Foaming and wetting agent and dispersant for foaming bubble baths. Cosmetic cleansing creams and emulsions. Air-entraining agent. *Henkel/Cospha; Henkel KGaA.*

31484 Texapret®

Fillers and stiffeners for textile finishing. *BASF AG; BASF plc.*

31485 Texgas

Distributorship services in the field of equipment using liquefied petroleum gas. *Quantum Chemical Corp.*

31486 Texgas

Distributorship services in the field of equipment using liquefied natural gas. *Quantum Chemical Corp.*

31487 Texi TD

Replacement for sodium hydrosulfite for reduction clearing, stripping dyes, and equipment cleaning in textile industry. *CNC Int'l L.P.*

31488 Texicote®

Polyesters, vinyl acetate copolymers, or polystyrene emulsions; used for powder coatings, paints, adhesives, fabric stiffeners, and binders. *Scott Bader.*

31489 Texicote® 03-001

PVAc homopolymer emulsion stabilized with PVAL; base emulsion for wet bond adhesives; high viscosity and wet tack; base for ceramic tile adhesives, for wood and general-purpose adhesives in the building industry, and as a cement additive; can be compounded with PVAL, resin ester emulsions and plasticizers, e.g. DIBP, to increase wet tack viscosity, flexibility and adhesion. *Scott Bader.*

31490 Texicote® 03-019

Vinyl acetate/acrylic copolymer emulsion; general purpose flat and silk finish paints. *Scott Bader.*

31491 Texicote® 1000, 1050

Polyester powder coating; medium and fast reactivity respectively, excellent gloss, high flow. *Scott Bader.*

31492 Texicryl®

Emulsion polymers and copolymers based on acrylic acid esters; used for binders, coatings, membranes, thickeners, and dispersion agents; paints, textiles, adhesives, and paper coating. *Scott Bader.*

31493 Texicryl® 13-002

Acrylic/methacrylic copolymer emulsion; emulsion for surface coating applications, e.g., in self-textured finishes providing good color, stain resistance, and good flexibility, in high-build one-coat finishes, in acrylic wood primers, in paints with exterior durability; added to cement/sand to improve intercoat adhesion; chemical/abrasion resistance; binder in pigmented leatherboardand fabric backings with wet abrasion resistance and flexibility. *Scott Bader.*

31494 Texicryl® 13-011

Carboxylated acrylic copolymer emulsion; aqueous emulsion of alkali-soluble acrylic copolymer used in water-based printing inks, in the manufacturing of high gloss and other types of emulsion paint, and as a binder for nonwoven fabrics. *Scott Bader.*

31495 Texicryl® 13-030

Styrene acrylic copolymer emulsion; used as quality binder for water-based paints (varying from gloss finish to high PVC low-cost paints), crack fillers, textured coatings, and ceramic tile adhesives. *Scott Bader.*

31496 Texicryl® Additive 87-1280

Acrylic polymer, water-borne; transfer characteristics controller for water-based flexographic inks. *Scott Bader.*

31497 Texicryl® Ecobinder

Styrene-acrylic copolymer emulsion; media for nonpolluting water-based paints for interior and exterior use. *Scott Bader.*

31498 Texicryl® Hyperbinder

Styrene-acrylic copolymer emulsion; for cost-effective high pigment loaded interior and exterior matte paints. *Scott Bader.*

31499 Texigel®

Polyacrylate; gel thickeners. *Scott Bader.*

31500 Texileather

A proprietary trade name for pyroxylin-coated leather cloth. No manufacturer.

31501 Texin 3203, 3215, 4203, 4206, 4210, 4215

Thermoplastic PU/PC blends; for injection molding; some are paintable formulations. *Miles.* Discontinued.

31502 Texin 480-A

9009-54-5

Thermoplastic polyester PU; for injection molding, extrusion (hose, tubing, profiles, wire and cable, film and sheet), blow molding. *Miles.* Discontinued.

31503 Texin 985-A

9009-54-5

Thermoplastic polyether PU; for injection molding, extrusion (hose, tubing, profiles, wire and cable, film and sheet), blow molding. *Miles.* Discontinued.

31504 Texin 5286

9009-54-5

Thermoplastic polyether polyurethane; medical grade for flexible tubing and film applications. *Miles.* Discontinued.

31505 Texipol®

Inverse polymer emulsions; used for paper, textiles, adhesives, paints, thickening agents and dispersants. *Scott Bader.*

31506 Texipol® 63-002

Acrylamide copolymer emulsion; thickener for acrylic flocking adhesives and carpet backing compositions; for use at pH 5.5-9.0 *Scott Bader.*

31507 Texlin® 300

112-24-3 9796 203-950-6

$C_6H_{18}N_4$

N,N'-bis(2-aminoethyl)-1,2-ethanediamine

trientene; triethylenetetramine; 1,8-diamino-3,6-diazaoctane. Improves yields and reduces processing time for many applications; intermediate in asphalt additives, corrosion inhibitors, epoxy curing agents, surfactants in fabric softener and textile additives; in paper industry, petrol. products. mp = 12°; bp = 266-267°; d = 0.9780; n_D^{20}= 1.4971; LD_{50} (rat orl)= 2.5 g/kg. *Texaco.*

31508 Texlin® 400

112-57-2 203-986-2

$C_8H_{23}N_5$

Tetraethylenepentamine

Lube oil additive; intermediate in asphalt additives, corrosion inhibitors, epoxy curing agents, surfactants, and in the paper industry. mp = -40°; bp = 340°; d = 0.9980; n_D^{20}= 1.5055. *Texaco.*

31509 Texlin® 500

68173-73-7

Polyethylenepolyamine

Used in fuel and lube oil industry, in asphalt chemistry, corrosion inhibitors, and ore flotation applications. *Texaco.*

31510 Texoderm

A cellulose product; an imitation leather.

31511 Texofor

Nonionic surfactant. *ABM Chemicals Ltd.* Name unverified.

31512 Texofor A and B

A proprietary range of higher fatty alcohol-based polyoxy-alkylene condensates used as nonionic surfactants. *Glover (Chemicals) Ltd.* Unverified.

31513 Texofor C

A proprietary range of unsaturated fatty acid-based polyoxy-alkylene condensates used as nonionic surfactants. *Glover (Chemicals) Ltd.* Unverified.

31514 Texofor D

A proprietary range of glyceride oil-based polyoxyalkylene condensates; used as nonionic surfactants. *Glover (Chemicals) Ltd.* Unverified.

31515 Texofor E and ED

A proprietary range of saturated fatty acid-based polyoxy-alkylone condensates; used as nonionic surfactants. *Glover (Chemicals) Ltd.* Unverified.

31516 Texofor FN, FP and FX

Range of nonionic surfactants of the alkylphenol ethoxylate type; used for industrial and household cleaners; emulsion polymerization; agriculture, etc. *Rhône-Poulenc UK.*

31517 Texofor G

A proprietary unsaturated fatty acid-based polyoxyalkylene condensate; used as a nonionic surfactant. *Glover (Chemicals) Ltd.* Unverified.

31518 Texofor J4

A linear fatty alcohol ethoxylate; proprietary biodegradable nonionic emulsifier. *Glover (Chemicals) Ltd.* Unverified.

31519 Texofor M

A proprietary range of unsaturated fatty acid-based polyoxy-alkylene condensates; used as nonionic surfactants. *Glover (Chemicals) Ltd.* Unverified.

31520 Texofor N

A proprietary range of fatty alcohol-based polyoxyalkylene condensates; used as nonionic surfactants. *Glover (Chemicals) Ltd.* Unverified.

31521 Texofor P

A proprietary range of complex amide-based polyoxyalkylene condensates; used as nonionic surfactants. *Glover (Chemicals) Ltd.* Unverified.

31522 Texofor T

A proprietary range of higher fatty alcohol-based polyoxy-alkylene condensates; used as nonionic surfactants. *Glover (Chemicals) Ltd.* Unverified.

31523 Texogent
Blend of surfactants and solvents; used for degreasing, cleaning, antispotting in textile and engineering industries. *ABM Chemicals Ltd.* Name unverified.

31524 Texowax
Water soluble waxy coatings; release lubricant; softening; used for paper coating; plastics; textiles; and adhesives. *ABM Chemicals Ltd.* Name unverified.

31525 Texox® PPG-400
PPG-400
Intermediate yielding esters; useful as lubricants, defoaming agents in rubber and pharmaceuticals, solvents and humectant modifiers for inks, plasticizers, and functional fluids. *Texaco.*

31526 Texox® WL-440
EO/PO derivs.; functional fluids used as lubricants, plasticizers, solvs., coupling agents, frothing agents, heat transfer fluids, intermediates, defoaming agents; used in solder reflow applications, boiler defoaming, ore flotation, inks, dyes. *Texaco.*

31527 Texsolve
Aliphatic hydrocarbon solvents. *Texaco.* Name unverified.

31528 Texsolve B
Highly refined commercial hexane; solvent used for vegetable oil and pharmaceutical extraction, compounding rubber cements and sealants, and polyolefin production. *Texaco.*

31529 Texsolve C
Mixed heptane fraction; solv. used for compounding rubber cements and sealants, extraction of oils and fats, and recrystallization of organic chemicals. *Texaco.*

31530 Texsolve E
Hexane-heptane fraction; solv. used in rubber tire manufacturing, rubber cements, sealants, inks, lacquers, and adhesives. *Texaco.*

31531 Texsolve S-66
Exempt mineral spirit with maximum aromatic content; solvent used in paint and protective coatings, dry-cleaning, degreasing, wood treating, charcoal lighter fluid; meets or exceeds standards for mineral spirits and Stod. *Texaco.*

31532 Texsolve V
8030-30-6 7329 232-443-2
petroleum benzin; naphtha. VM & P naphtha; solvent used in paints, coatings, rubber compounding, sealants, and chemical absorption. *Texaco.*

31533 Textamine Carbon Detergent K
Fatty amino complex; emulsifier for room temperature solv. degreasing formulations; aids in carbon removal from aircraft, diesels, and metal surfaces. *Henkel/Cospha; Henkel Canada.*

31534 Textamine T-1
61791-39-7 263-171-2
1-Hydroxyethyl-2-tall oil imidazoline; corrosion inhibitor, emulsifier, wetting agent for metalworking fluids; dispersant. *Henkel/Cospha; Henkel Canada.*

31535 Textase
A diastase preparation.

31536 Textile Resin 2309 Conc
Highly reactive crosslinking agent for finishing of cellulosics or blends with synthetics or wool. *BASF.*

31537 Textile Resin NF-U
Crosslinking agent for resin finishing cellulosic fibers and their blends with synthetics. *BASF.*

31538 Textile Wax
Textile preparation and finishing agent. *BASF plc.*

31539 Textile Wax W
Wax-like substance for addition to textile sizing and finishing liquors. *BASF.*

31540 Textol 80 (L)
Ethanolamine alkylbenzene sulfonate
Raw material for manufacturing of liquid detergents. *Zohar Detergen Factory.*

31541 Textone® -50
Sodium chlorite/sodium nitrate blend; for bleaching textiles and stripping dyestuffs for natural and synthetic fibers. *Olin.* Discontinued.

31542 Textulite
A proprietary trade name for phenol formaldehyde laminated synthetic resin and molded compounds. No manufacturer.

31543 Texzyme
Protein digesting enzyme liquid. *PMP Fermentation Products Inc.*

31544 TF Solvent
Mild cleaning agent for removal of flux residue, oil, grease and contaminants from magnetic tape heads, thermal print heads, optical equipment, contacts, relays, and PCB assemblies. *Chemtronics.*

31545 TFC
Thin film composite spiral wound membrane; for reverse osmosis water treatment. *AlliedSignal Fluid Systems.* Name unverified.

31546 tfol
An argillaceous earth containing free gelatinous silica; used as a soap.

31547 TG Buffer
139-13-9 6675 205-355-7
$C_6H_9NO_6$
N,N-bis(carboxymethyl)glycine
nitriloacetic acid; Complexone(I); NTA; Triglycollamic acid. Trisglycine solution or powder. Chelator, sequestrant and metal complexing agent. mp = 246° (dec); soluble in H_2O (1.28 g/l). *Am. Research Prods.* Unverified.

31548 TG-8
Triethylene glycol dicaprylate. *Ruco Div.* Name unverified.

31549 T-gas
Ethylene oxide and carbon dioxide. The commercial mixture of ethylene oxide and carbon dioxide; used an insecticide.

31550 TG-SDS Buffer
Trisglycine-SDS solution or powder. *Am. Research Prods.* Unverified.

31551 Thaio Green No. 1
A halogenated copper phthalocyanine green; extremely light and heat-fast. *Reckitts Colours Ltd.* Name unverified.

31552 thallium
7440-28-0 9391 231-138-1
Tl
Thallium
Metallic element; thallium salts, mercury alloys, low-melting glasses, rodenticides, photoelectric applications, electrodes in dissolved oxygen analyzers. mp = 303°; bp = 1457°; d = 11.85. *Aldrich; Atomergic Chemetals; Cerac; Noah Chem.*

31553 thallium alum
$O_{16}Al_2S_4Tl_2 \cdot 24H_2O$
A double sulfate of thallium and aluminum.

31554 thallium chloride
7791-12-0 9395 232-241-4
ClTl
Thallium(I) chloride-^{201}Tl
Thallous chloride Tl 201. Diagnostic aid; radioactive agent. mp = 430°; bp = 720°; d = 7.0; soluble in H_2O (4 mg/ml), insoluble in EtOH. *Amersham Corp.* Name unverified.

31555 Thalo Blue No. 1
An alpha, solvent sensitive, red shade phthalocyanine blue pigment; used for solventless printing inks. *Reckitts Colours Ltd.* Name unverified.

31556 Thalo Blue No. 2
A beta, solvent stable, green shade phthalocyanine blue pigment. *Reckitts Colours Ltd.* Name unverified.

31557 THAM
77-86-1 9902 201-064-4
$C_4H_{11}NO_3$
2-amino-2-hydroxymethyl-1,3-propanediol
Tromethamine; trimethylol aminomethane; tris(hydroxymethyl)aminomethane; trisamine; tris buffer; trometamol; tromethane; TRIS; Talatrol; Tris Amino; Trizma. Alkalizer. mp = 171-172°; bp_{10} = 219-220°; soluble in H_2O and organic solvents. *Abbott Laboratories.* Name unverified.

31558 Thancat
A series of amine-type urethane catalysts; catalysts for production of flexible and rigid urethane foams, elastomers and sealants. *Texaco.*

31559 Thanecure®
Zinc chloride [14239-57-9] and benzothiazyl disulfide. Vulcanization activator for sulfur curable millable urethane elastomer. *TSE Industries.*

31560 Thanol
A series of urethane polyols; intermediates for production of flexible and rigid urethane foams, elastomers and sealants. *Texaco.*

31561 Thanol E-2103
Urethane polyol; adhesives and sealants raw materials. *Eastman.*

31562 thao
A gelatinous preparation made in Cochin China from seaweed. It has frequently appeared in England under the names of Japanese or Chinese isinglass, and is used for the same purposes as isinglass.

31563 Thapsia resin
The resin of *Thapsia garganica* root. It contains caprylic acid and thapsic acid.

31564 Thawpit
56-23-5 1864 200-262-8
A preparation of carbon tetrachloride used for cleaning materials. No manufacturer.

31565 The Chemistry to Compete
Polyolefins specialty resins, colorants, and compounds for use by the plastics industries as additives to polyolefins and other petrochemicals for use in further manufacturing in a wide variety of industries. *Quantum Chemical Corp.*

31566 The Little Chemical Giant
Synthetic olefin polymers or copolymers in particulate form, denatured alcohol and ethyl alcohol in industry or agriculture. *Quantum Chemical Corp.*

31567 Theic
Tris-(2-hydroxyethyl)isocyanurate
Used in the manufacture of heat-resistant wire lacquers. *BASF plc.*

31568 thenium closylate
4304-40-9 9414 224-318-6
$C_{21}H_{24}ClNO_4S_2$
N,N-Dimethyl-N-(2-phenoxyethyl)-2-thiophenmethanaminium salt with 4-chlorobenzenesulfonic acid (1:1)
Dimethyl(2-phenoxyethyl)-2-thenylammonium *p*-chlorobenzenesulfonate; Canopar; 611C55; Bancaris;. Anthelmintic. mp = 159-160°; soluble in H_2O (0.6% w/v) *Burroughs Wellcome Co.* Name unverified.

31569 Theophyllisilane C
128973-73-9
Methylsilanol theophyllinacetate alginate
Ingredient for slimming and anti-aging formulations, cosmetic and health products. *Exsymol.*

31570 Therabloat
Poloxalene
Liquid nonionic surfactant polymer of the polyethylene-polypropylene glycol type; pharmaceutic aid. *Norden Laboratories Inc.* Name unverified.

31571 Therban®
Hydrogenated nitrile-butadiene rubber; high abrasion resistance, good to excellent hot air resistance and resistance to ozone, technical oils, brake and hydraulic fluids, and crude oil containing H_2S and amines; used for technical rubber goods employed in mineral oil explor *Bayer AG; Bayer plc.*

31572 Therban® 1706
Saturated hydrocarbon ACN copolymer; specialty elastomer offering resist. to heat, oils, ozone, abrasion, hydrogen sulfide, amines; improved processing during mixing, extrusion, and injection molding; applics incl. automotive and industrial areas, e.g., seals, gaskets, power transmission belting, membranes, cable jackets, extruded profiles, rubber linings, hoses, bellows, sleeves, oil field parts. *Bayer; Miles.*

31573 Therlo
An electrical resistance alloy containing 85% copper, 13% manganese, and 2% aluminum.

31574 Thermaflo
PVC compounds; a wide range of moldings and extrusions including applications in footwear, building, cable, automotive, etc. *Evode Plastics Ltd.* Name unverified.

31575 Thermalate H320
Fiberglass-reinforced thermoset polyester composite; for high temperature mold and platen thermal applications. *Haysite Reinforced Plastics.*

31576 Thermalene
An mixture of acetylene and vaporized oils; used for production at high temperatures, in cutting and welding metals.

31577 Thermalloy
A patented form of thermit containing 50% iron oxide, 27% aluminum, and 23% sulfur; the name appears to be applied also to an alloy containing 66.5% nickel, 30% copper, and 2% iron; it has a magnetic permeability which decreases at higher temperature; an. No manufacturer.

31578 Thermalloy A
A proprietary trade name for an alloy containing 67.5% nickel, 0.15% silicon and 30% copper. No manufacturer.

31579 Thermalloy B
A proprietary trade name for an alloy containing 57.8% nickel, 0.15% carbon, 0.15% silicon, and 40% copper. No manufacturer.

31580 thermatomic carbon
A fine carbon produced by cracking natural gas into carbon and hydrogen by passing the gas over heated brickwork; used as a rubber filler.

31581 Thermax® Floform N-990
7440-44-0 1855 231-153-3
Medium thermal carbon black; used in rubber products with high loadings of carbon black, in metallurgy, specialty, and refractory applications. *Cancarb; R. T. Vanderbilt Co Inc.* Name unverified.

31582 Thermax® Stainless
7440-44-0 1855 231-153-3
Medium thermal carbon black; reinforcer for all elastomers, and black stocks. *CanCarb; R. T. Vanderbilt Co Inc.*

31583 Thermax® Stainless Floform N-907
7440-44-0 1855 231-153-3
Medium thermal carbon black; nonstaining product used in seals and gaskets. *Cancarb; R. T. Vanderbilt Co Inc.* Name unverified.

31584 Thermazote
A trademark for an expanded thermosetting plastic, manufactured in densities between 7 and 30 lb/ft^3; it is nonflammable and odorless and withstands temperatures as high as 300°C; has a low thermal conductivity; used in construction, etc. No manufacturer.

31585 Therm-Chek
For heat and light stabilizers for vinyl halide compositions, in Int Class 1. *Ferro.*

31586 Thermex
Heat transfer media. *ICI Chem & Polymers Ltd.*

31587 Thermexo
Highly exothermic metal producing compounds. *Foseco (F.S.) Ltd.*

31588 Thermica
$SO_2AL_2O_3MgOK_2OF$ with 3% impurities
Additive to oil, paint, grease, plastics, glass and other inorganic materials which require a high dielectric or high heat resistant material; a coating for welding rods requiring special atmospheric controls; a thermal barrier and high electrical resistivity. *Mykroy/Mycalex.* Name unverified.

31589 Thermid® EL-5010, EL-5512
Polyimide
Interlayer dielectric coating with exceptional adhesion and low moisture absorption for the fabrication of multichip modules. *Ablestik.*

31590 Thermilin
Heat transfer fluid. *Monsanto Co.*

31591 Therminol®
Heat transfer fluids. *Monsanto Co.* Name unverified.

31592 Therminol® 44
Modified ester-based fluid; liquid phase heat transfer fluid featuring low temperature fluidity, high flash point; for use in nonpressurized systems. *Monsanto Co.*

31593 Therminol® 66
Modified polyphenyl; liquid phase heat transfer fluid for use in nonpressurized systems; high temperature, low pressure fluid. *Monsanto Co.*

31594 Therminol® VP-1
73.5% diphenyl oxide, 26.5% biphenyl; liquid/vapor phase heat transfer fluid for use in nonpressurized systems; ultra high temperature. *Monsanto Co.*

31595 Thermisilio
A proprietary trade name for a chemical resisting iron-silicon alloy in which the brittleness has been diminished. No manufacturer.

31596 Thermisilizid
A Swedish iron-silicon alloy of the acid-resisting type.

31597 Thermit
A bearing metal containing lead, antimony (20%) and small quantities of tin, nickel, and copper. No manufacturer.

31598 Thermit manganese
Manganese metal made by the Thermit reduction method. It contains approximately 98% manganese.

31599 Thermit metal
A German bearing metal alloy containing: 14-16% antimony, 5-7% tin, 0.8-1.2% copper, 0.7-1.5% nickel, 0.3-0.8% arsenic, 0.7-1.5% cadmium, and 72-78.5% lead.

31600 Thermit®
A world-wide registered trademark for aluminothermic mixtures consisting essentially of nearly equal parts of powdered aluminum and metal oxides, usually iron or manganese oxides; these mixtures burn with a high temperature; used for welding metals; also used as an ingredient in incendiary bombs. No manufacturer.

31601 Thermlo F
An organic polysulfide; a proprietary rubber vulcanization accelerator. No manufacturer.

31602 Thermocal
Heat transfer medium. *ICI Chem & Polymers Ltd.*

31603 Thermocal B
Monoethylene glycol-based; corrosion inhibiting heat transfer fluid for industrial cooling and refrigeration installations and moder. high temp. heat exchange systems. *ICI Australia.*

31604 Thermocast
A proprietary trade name for an ethyl cellulose composition. No manufacturer.

31605 Thermocomp®
Filled and reinforced thermoplastic compounds. *LNP; ICI Chemicals & Polymers Ltd.*

31606 Thermocomp® AF-1004
9003-56-9
ABS, 20% glass fiber-reinforced; superior toughness, strength, hardness, and high temp. resistance when compared to glass-fortified PS or SAN; used in business machine components, photographic equipment, and automotive and appliance applications where combined toughness and dimensional stability are required. *LNP.*

31607 Thermocomp® BF-1006
9003-54-7
SAN, 30% glass fiber-reinforced; thermoplastic offering low mold shrinkage. *LNP.*

31608 Thermocomp® BF-1006FR
9003-54-7
SAN, 30% glass fiber-reinforced, flame-retardant; thermoplastic characterized by unequalled stiffness and low temperature creep resistance, processability, and lowest mold shrinkage of any thermoplastic; produces extremely high quality, tight tolerance molded parts withexcellent surface finish and metal-like end user performance; used in the appliance, business machine and camera industries as materials for housings, frames and internal components. *LNP.*

31609 Thermocomp® CF-1004
9003-53-6 9028
PS, 20% glass fiber-reinforced; improved strength, stiffness, dimensional stability; natural and pigmented. *LNP.*

31610 Thermocomp® DF-1004
PC, 20% glass fiber-reinforced; used as replacement for metal in mechanical, structural, and elec. applications, (elec. connector block, computer insulator mount, instrument face plate, desk calculator housing, electronic terminal block); glass fibers increase strength and dimensional stability and retain inherent flame retardancy of the base resin; they lower water absorption, dramatically increase deflection temperature, and substantially increase stiffness. *LNP.*

31611 Thermocomp® EC-1006
61128-46-9
Polyetherimide, 30% PAN carbon fiber. *LNP.*

31612 Thermocomp® GF-1004
25135-51-3
Polysulfone, 20% glass fiber-reinforced; high temperature engineering thermoplastic for structural/insulating applications; mechanical and thermal properties, dimensional stability, increased flammability resistance, resistance to environmental stress-cracking, dip solderability, increased fatigue endurance, moldability, etc. *LNP.*

31613 Thermocomp® HF-1006
25035-04-5
Nylon 11, 30% glass fiber-reinforced; low moisture absorption and good dimensional stability, but with slightly reduced mechanical properties compared to the other fortified nylon series. *LNP.*

31614 Thermocomp® IF-1002
Nylon 6/12, 10% glass fiberreinforced; reinfored nylon similar to fortified nylon 6/10 versions (QF Series), exhibits the same moisture, dimensional stability, and impact characteristics. *LNP.*

31615 Thermocomp® JF-1004
PES, 20% glass fiber-reinforced; outstanding high temp. creep and hydrolysis resistance, self-extinguishing properties, good impact strength and low mold shrinkage. *LNP.*

31616 Thermocomp® KB-1008
105-57-7 36 203-310-6
Acetal, 40% glass bead-fortified; outstanding solvent resistance of base polymer with improved dimensional control. *LNP.*

31617 Thermocomp® MF-1002
9003-07-0 7741
PP, 10% glass fiber-reinforced; resistance to heat, water absorption, and chemical attack; used in pipe fittings, pump housings, and heater components; certain formulations in MF series meet the requirements of MIL-P-46109. *LNP.*

31618 Thermocomp® NF-1004
9003-53-6 9028
Styrenic copolymer, 20% glass fiber-reinforced; improved elevated temp. performance, deflection temp. 30-50°F higher than other conventional glass-reinforced styrenic systems, retains dimensional stability, stiffness, and low moisture absorption; used in camera housings, business machine parts, swimming pool pumps, water softener parts and interior automotive parts; NF series compounds are available with internal lubricants for use in bearings and wear applications. *LNP.*

31619 Thermocomp® OC-1006
9016-75-5
PPS, 30% PAN carbon fiber-reinforced; offers stiffness in addition to generally improved properties; offers reduced mold shrinkage and thermal coefficient of expansion, and increased thermal conductivity; also offers high temperature stability and extreme chemical and solvent resistance inherent in the base resin. *LNP.*

31620 Thermocomp® PC-1006
25038-54-4 6832
Nylon 6, 30% PAN carbon fiber-fortified; reinforced thermoplastic. *LNP.*

31621 Thermocomp® QF-1006FR
9008-66-6
Nylon 6/10, 30% glass fiber-reinforced, flame-retardant; for applications requiring increased chemical resistance and lower water absorption. *LNP.*

31622 Thermocomp® RC-1002
32131-17-2
Nylon 6/6, 10% PAN carbon fiber-fortified; offers stiffness, and improved physical properties; compared to glass fiber-reinforced nylon 6/6, RC Series offers reduced mold shrinkage and thermal coefficient of expansion, and greatly increased thermal conductivity; these systems capitalize on the high HDT, toughness, solvent resistance and moldability inherent in the base resin. *LNP.*

31623 Thermocomp® SF-1006
Nylon 12, 30% glass fiber-reinforced; good dimensional stability and lowest moisture absorption of nylons, slightly reduced mechanical properties compared to other fortified nylon series. *LNP.*

31624 Thermocomp® TF-1004
9009-54-5
Thermoplastic PU, 20% glass fiber-reinforced; tough thermoplastic; resists wear, abrasion, creep, and exposure to petrol. products; for gears, gaskets, bushings, bearings, and other applications, where outstanding toughness and wear resistance are needed. *LNP.*

31625 Thermocomp® VF-1002
Super tough nylon (Zytel), 10% glass fiber-reinforced; offers strength, stiffness, toughness, chemical and thermal resistance, and notched and unnotched izod impact values not formerly acheivable in a resin system; used in tool housing and other industrial components; retains moldability and surface finish of glass-reinforced nylons. *LNP.*

31626 Thermocomp® WC-1006
26062-94-2
PBT polyester, 30% PAN carbon fiber-reinforced; offers stiffness, and generally improved properties, reduced mold shrinkage and thermal coefficient of expansion, and increased thermal conductivity in comparison to glass-fortified polyester; capitalizes on low moisture absorption, moldability and balanced frictional and mechanical properties inherent in the base resin. *LNP.*

31627 Thermocomp® XF-1004
Amorphous nylon, 20% glass fiber-reinforced; combines the features of crystalline and amorphous materials the solvent resistance of nylon with the dimensional control possible from PC. *LNP.*

31628 Thermocomp® YF-1002
Polyester elastomer based on Hytrel, 10% glass fiber-reinforced; very tough, reinforced thermoplastic elastomer with resistance to abrasion and superior resistance to deformation at elevated temps.; resistance to common fuels and lubricating oils; used in hostile environments and high and low temperatures; lubricated versions are suitable gear materials, especially when noise reduction is required; for low temperature applications such as body, drive train parts for snowmobiles, chain saw parts. *LNP.*

31629 Thermocomp® ZF-1004
25134-01-4
Modified PPO, 20% glass fiber-reinforced; tough, rigid thermoplastic, high moisture resistance and dimensional stability; for appliance and business machine components, water softener and industrial filters, pump housings and impellers, radio, TV, and electrical/electronic insulator units; meets MIL-P-46131 specification. *LNP.*

31630 Thermoflex
di-*p*-methoxydiphenylamine
A proprietary trade name for di-*p*-methoxydiphenylamine; an antioxidant. *Imperial Chemical Industries plc; Du Pont (UK) Ltd.* Name unverified.

31631 Thermoflex A
Contains 50 parts of phenyl-β-naphthylamine, 25 parts of methoxydiphenylamine and 25 parts of diphenyl-*p*-phenylene diamine; an antioxidant. *Imperial Chemical Industries plc; Du Pont (UK) Ltd.* Name unverified.

31632 Thermoguard® 505
1163-19-5 214-604-9
$C_{12}Br_{10}O$
Decabromodiphenyloxide
bis(pentabromophenyl)ether; decabromodiphenyl ether. Flame retardant mp = 294-296°. *Elf Atochem.*

31633 Thermoguard® 8218
Decabromodiphenyloxide concentrate in polyethylene; flame retardant *Elf Atochem.*

31634 Thermoguard® CPA
7440-36-0 733 231-146-5
Antimony
Flame retardant for use as replacement for antimony oxide in many formulated plastics especially in flexible PVC applications such as wire and cable insulation and jacketing. *Elf Atochem N. Am./Plastics Additives.*

31635 Thermoguard® FR
1314-60-9 739 215-237-7
antimony pentoxide
Antimony oxide; flame retardant for vinyls and other plastics. *Elf Atochem N. Am./Plastics Additives.*

31636 Thermoguard® L
1309-64-4 752 215-175-0
Antimony trioxide
Flame retardant for use with a halogen-containing compound; flame retardant pigment for PVC; also for use with chlorinated organics for producing flame-retardant polyesters and polyethylene compounds. *Elf Atochem N. Am./Plastics Additives.*

31637 Thermoguard® UF
1327-33-9 215-474-6
Antimony oxide
Flame retardant; superfine grade *Elf Atochem.*

31638 Thermolastic®
Extrudable thermoplastic rubber-like materials based upon styrene butadiene copolymers not requiring vulcanization. *Shell.* Name unverified.

31639 Thermolin
Chlorinated phosphate ester additive for flexible foams. *Olin. Discontinued.*

31640 Thermoloft®
Fiber. *DuPont UK.*

31641 Thermonit
A refractory cement made in the electric furnace; it is stated to be used as a paint or mortar, and to resist high temperatures; Keramonit is the cement reinforced with metal mesh. No manufacturer.

31642 Thermoplast®
Special dyes soluble in plastics; for mass dyeing of thermoplastic and thermosetting plastics, e.g., styrene polymers (PS, S/B, SAN, ABS), rigid PVC, polymethacrylate, cellulose derivatives, polycarbonates, unsaturated polyesters, etc. *BASF AG.*

31643 Thermoplaste
Plastics dyes. *ICI Chem & Polymers Ltd.*

31644 Thermoprene
Products obtained by heating rubber with either an organic sulfonyl chloride or an organic sulfonic acid at 125-135 for several hours. p-toluene-sulfonyl chloride and *p*-toluene sulfonic acid are suitable reagents; used in making protective paints resistant to acids and bases.

31645 Thermorun
Thermoplastic elastomer. *Mitsubishi Petrochem.* Name unverified.

31646 Thermoseal
Heat seal lacquers. *The Scottish Adhesives Co Ltd.*

31647 Thermoset 100
25928-94-3
Two-part epoxy; adhesive system for tooling applications. *Thermoset Plastics.*

31648 Thermoset 300/No. 65 Hardener
25928-94-3
Two-part black epoxy; low viscosity, room temperature-curing system with high gloss; moisture insensitive during cure; for electrical/electronic insulation applications. *Thermoset Plastics.*

31649 Thermoset 310
25928-94-3
Two-part epoxy; ovencuring, semirigid, tough system with very good shock resistance; for electrical/electronic insulation applications; black. *Thermoset Plastics.*

31650 Thermoset DC-232
25928-94-3
One-part epoxy; oven-curing impregnating system with long shelf life for a single-component system.; for electrical/electronic insulation applications. *Thermoset Plastics.*

31651 Thermoset ME-101
Electrically conductive resin; general purpose, low temperature curing. *Thermoset Plastics.*

31652 Thermoset ME-177
One-part dielectric resin; chip-on-board encapsulant, long room temperature life. *Thermoset Plastics.*

31653 Thermoset SC-102
Two-part silicone system; clear, dielectric gel electrical/electronic insulating resin; temperature rating to 200°; mix ratio: 1:1 w/w. *Thermoset Plastics.*

31654 Thermoset SC-113
One-part silicone system; protective sealer for brush, dip, or spray application for electrical/electronic insulation; gray. *Thermoset Plastics.*

31655 Thermoset UR-101
Black, two-part PU; room temperature-curing, tough potting system with good abrasion resistance, excellent low temperature flexibility; for electrical/electronic insulation applications. *Thermoset Plastics.*

31656 Thermoset UR-105
Black, two-part PU; room temperature-curing potting system with excellent low temperature flexibility; repairable; softer version of UR-101; for electrical/electronic insulation applications. *Thermoset Plastics.*

31657 Thermotex
Thermit bottom plate patching compound. *Foseco (F.S.) Ltd.*

31658 Thermount®
Fiber. *DuPont UK.*

31659 Thermolite 139, 380
Sulfur-containing organotin; heat stabilizer for weatherable coextruded siding, window profiles, and other rigid PVC applications. *Elf Atochem N. Am.*

31660 THFA
97-99-4 9353 202-625-6
Tetrahydrofurfuryl alcohol
Great Lakes.

31661 thiabendazole
148-79-8 9426 205-725-8
$C_{10}H_7N_3S$
1H-Benzimidazole, 2-(4-thiazolyl)-
Thiabendazole; benzimidazole, 2-(4-thiazolyl)-; Apl-Luster; Arbotect; Bioguard; Bovizole; Eprofil; Equizole; Lombristop; Mertec; Mertect; Mertect 160; Metasol TK 100; Mintesol; Mintezol; Minzolum; Mycozol; MK 360; MK-360; Nemapan; Omnizole; Polival; Tbz; Tebuzate; Tecto; Tecto RPH; Tecto 10P; Tecto 40F; Tecto 60; Testo; Thiaben; Thiabendazol; Thiprazole; Tiabenda. Systemic fungicide with protective and curative action. Absorbed by leaves and roots; used for control of fungus in vegetabls, fruits and cereals. mp = 300° (dec); soluble in H_2O (250 mg/l at pH 2-5), more soluble in organic solvents; LD_{50} (rat orl) = 2080 mg/kg *MSD Agvet; Pennwalt; Duphar; Agrichem; BASF, Ciba-Geigy; Dow Elanco.*

31662 Thial
Hexamethylenetetramine hydroxymethylsulfonate
Used as an antiseptic.

31663 thiambutosine
500-89-0 207-914-0
$C_{19}H_{25}N_3OS$
Thiourea, N-(4-butoxyphenyl)-N'-[4-(dimethylamino)phenyl]-
carbanilide, 4-butoxy-4'-(dimethylamino)thio-; 4-Butoxy-4'-(dimethylamino)thiocarbanilide; N-(4-butoxyphenyl)-N'-[4-(dimethylamino)phenyl]urea; Ciba 1906; Su 1906; Summit 1906; Thiambutosin; Thaimbutosine.

31664 thiamine
59-43-8 9430 200-425-3
$C_{12}H_{17}ClN_4OS$
Vitamin B_1. Medicine, nutrient, enriched flours; available as thiamine hydrochloride and thiamine mononitrate. *Hoffmann-La Roche SA; Honeywill & Stein Ltd; Parke-Davis.*

31665 thiamine hydrochloride
67-03-8 9430 200-641-8
$C_{12}H_{17}N_4OSCl{\cdot}HCl$
thiamine dichloride
vitamin B_1; aneurine hydrochloride. Medicine, nutrient, enriched flours. *Aldrich; BASF; EM Industries; Hoffmann-La Roche; Takeda USA.*

31666 Thiate E
2489-77-2 219-644-0
$C_4H_{10}N_2S$
trimethyl thiourea
A proprietary accelerator. *K & K Greeff Chemicals Ltd.* Name unverified.

31667 Thiate®
Trimethylthiourea, 1,3-diethylthiourea or 1,3-dibutylthiourea; curing agents for elastomers. *R. T. Vanderbilt Co Inc.*

31668 Thibenzole
148-79-8 9426 205-725-8
A proprietary trade name for thiabendazole. No manufacturer.

31669 thickened mineral oils
Mineral oils which have been thickened by dissolving soap, usually aluminum soap.

31670 Thiel-Stoll solution
A saturated solution of lead chlorate, $Pb(ClO_4)_2$. Its has a density of 2.6 and is used for the determination of the specific gravity of minerals.

31671 Thiersch's antiseptic solution
A solution containing salicylic acid and boric acid.

31672 thiet-sie
A resinous substance used as a varnish by the Burmese.

31673 thifensulfuron-methyl
79277-27-3
$C_{12}H_{13}N_5O_6S_2$
3-[[[[(4-methoxy-6-methyl-1,3,5-triazin-2-yl)amino]carbonyl]amino]sulfonyl]-2-thiophenecarboxylic acid methyl ester
Harmony Extra; DPX-M6316; Harmony; INM-6316; Pinnacle; Thiameturon methyl ester; Thiameturon-methyl;Thifensulfuron Me. Controls annual dicotyledons in cereals.

31674 Thinners
Formulated thinners for use with spirit based coatings. *Foseco (F.S.) Ltd.*

31675 Thinoline
Vulcanized oils. Vulcanized linseed oil; used in rubber mixings.

31676 Thinsec
63-25-2 1831 200-555-0
A suspension concentrate containing 450 g carbaryl per liter. contact insecticide and fruit thinner for apples. *ICI Agrochemicals.*

31677 Thinsol
Lacquer thinners of various glosses. *Sasolchem.*

31678 thioacetamide
62-55-5 9453 200-541-4
CH_3CSNH_2
Replacement for gaseous hydrogen sulfide in qualitative analysis. *Aceto; Burlington Bio-Medical; Penta Mfg.*

31679 thiocamf
A liquid formed by exposing camphor to the action of sulfur dioxide; used as a disinfectant as it evolves sulfur dioxide on exposure to air.

31680 thiodan
115-29-7 3614 204-079-4
$C_9H_6Cl_6O_3S$
6,7,8,9,10,10-hexachloro-1,5,5a,6,9,9a-hexahydro-6,9-methano-2,4,3-benzodioxathiepin 3-oxide
Endosulfan; benzoepin; OMS 750; ENT 23979; Beosit; Cyclodan; Chlortiepin; Devisulphan; Endocel; Endosol; FMC 5462; Hilda; Hoe 2671; Insectophene; Malix; Rasayansulfan; Thifor; Thimul; Thionex; Thiosulfan. Non-systemic insecticide and acaricide with contact and stomach action. Used in control of sucking, chewing and boring insects and mites on a wide variety of crops. The commercial product is a mixture of an α-isomer (mp = 108-110°) anda β-isomer (mp = 208-210°). mp = 109°; $bp_{0.7}$ = 106°; d_{20} = 1.745; soluble in H_2O (0.32 mg/l), more soluble in organic solvents; LD_{50} (rat orl)= 70 mg/kg. *Hoechst UK.*

31681 thiodet
Residual thiosulfate test kit. *May & Baker Ltd.* Name unverified.

31682 thiodiglycol
111-48-8 9466 203-874-3
$C_4H_{10}O_2S$
thiodiethylene glycol
2,2'-thiodiethanol; bis(2-hydroxyethyl) sulfide; dihydroxyethyl sulfide. Intermediate for elastomers and antioxidants, solvent for dyes in textile printing. mp = -16°; bp_{14} = 168°; d_4^{20}= 1.1824; n_D^{20}= 1.519; soluble in H_2O, EtOH. *Morton Int'l.*

31683 thiodiglycolic acid
123-93-3 9467 204-663-9
$C_4H_6O_4S$
2,2'-thiodiacetic acid
2,2'-thiobis[acetic acid]; dimethylsulfide-α,α'-dicarboxylic acid; mercaptodiacetic acid. Analytical reagent, used for detection of metals such as copper, lead, mercury and silver. mp = 129°; soluble in H_2O, EtOH. *Witco/Argus.* Discontinued.

31684 3,3'-thiodipropionic acid
111-17-1 9468 203-841-3
$C_6H_{10}O_4S$
3,3'-thiobis[propanoic acid]
β,β-thiodipropionic acid; thiodihydracrylic acid; diethyl sulfide 2,2'-dicarboxylic acid. Antioxidant in food packaging, soaps, plasticizers, lubricants, fats, and oils. mp = 134°; soluble in H_2O (37 g/l), more soluble in hot H_2O, EtOH, Me_2CO. *Evans Chemetics; Janssen Chimica; Witco/Argus.*

31685 Thiofide
Benzthiazyl disulfide
Rubber accelerator. *Monsanto plc.*

31686 Thiofluor™
13242-44-9 236-221-6
$C_4H_{11}NS\cdot HCl$
N,N-Dimethyl-2-mercaptoethylamine hydrochloride
Nucleophile used in preparation of o-phthalaldehyde reagent for fluorescence detection in HPLC. Metal complexing agent. mp = 158-160°. *Pickering Laboratories Inc.*

31687 thiofurfuran
110-02-1 9490 203-729-4
C_4H_4S
thiophene
thiofuran. Organic synthesis (condenses with phenol and formaldehyde, copolymerizes with maleic anhydride), solvent, dye, pharmaceutical manufacture. *Elf Atochem N. Am.; Penta Mfg.*

31688 thioglycolic acid
68-11-1 9472 200-677-4
$C_2H_4O_2S$
2-mercaptoacetic acid
mercaptoacetic acid; thioglycollic acid; Thiovanic® Acid. In chemical analysis for the spectrophotometric determination of palladium; cosmetics (intermediates for hairwaving, depilatories), vinyl stabilizer intermediate, reaction intermediate for radiation-cured plastics; reagent for iron; manufacture of thioglycolates. mp = 10°; bp_{16} = 117°; d^{15}_4 = 1.220; n_D^{16} = 1.4823; soluble in H_2O, organic solvents. Elf Atochem N. Am.; Chemische Fabrik GmbH; EM Industries; Evans Chemetics; Witco/Argus; Evans Chemetics.

31689 Thiokol® 2135
Polysulfide
One-part, fast-cure joint sealant for sealing, caulking, and glazing applications on buildings. *Morton Int'l./Polymer Systems.*

31690 Thiokol® 2153, 2157
Polysulfide latex; caulk useable alone or with MC 2027, a chemically resistant masonry sealer, R-2100 radon barrier coating; excellent adhesion to masonry surfaces; good water resistance; 2153 is self-leveling version; 2157 is gun grade, nonsag version. *Morton Int'l./Polymer Systems.*

31691 Thiokol® FEC-2232
Blend of liquid polysulfide and epoxy resin; flexibilized coating for use in secondary containment, drum storage, and truck unloading areas, and in fuel storage tanks; abrasion and chemically resistant. *Morton Int'l./Polymer Systems.*

31692 Thiokol® FES-2258
25928-94-3
Two-component flexible epoxy sealant; crack injection system for making concrete walls and floors water-tight. *Morton Int'l./Polymer Systems.*

31693 Thiokol® LP
Liquid polysulfide polymers with SH-terminals; used as a basis polymer for the production of sealants for insulating glass, building joints, caulking ship decks, etc. *Thiokol GmbH.*

31694 Thiokol® MC-2027
Single-component polysulfide water dispersant; excellent barrier coating and moisture barrier for concrete/masonry surfaces; provides a film with excellent resistance to common solvents, chemicals., water and salt spray; applications include highway, coastal, construction applications. *Morton Int'l./Polymer Systems.*

31695 Thiokol® R-2100
Polysulfide water dispersion; radon barrier coating for application to porous masonry walls. *Morton Int'l./Polymer Systems.*

31696 Thiokol® RLP-2078
Two-component reinforced polysulfide system; elastomeric coating curing by chemical reaction to a tough, flexible, chemically resistant lining that acts as a leakproof barrier; for concrete surfaces; bridges nonstructural cracks. *Morton Int'l./Polymer Systems.*

31697 Thiolim
Hypo eliminator. *May & Baker Ltd.* Name unverified.

31698 Thiolite
An insulator prepared from formaldehyde, cresol, and sulfur chloride.

31699 Thiolyte®
Reagent. *Calbiochem Corp.*

31700 thiometon
640-15-3 211-362-6
$C_6H_{15}O_2PS_3$
S-[2-(ethylthio)ethyl] O,O-dimethyl phosphorodithioate
S-2-ethylthioethyl O,O-dimethyl phosphorodithioate; dithiometon; M-81; Bay 23129; Ekatin; Medrin; nimeton. Systemic insecticide and acaricide with contact and stomach action. Cholinesterase inhibitor, used for control of sucking insects in fruit and vegetable crops. $bp_{0.1}$ = 110°; d_{20} = 1.209; n_D^{20}= 1.5515; soluble in H_2O (200 mg/l), more soluble in organic solvents; LD_{50} (rat orl)= 125 mg/kg. *Sandoz.*

31701 Thionalide
$C_{12}H_{11}NOS$
A commercial name for thioglycollic acid, β-amino-naphthalide; used as an analytical reagent.

31702 Thionex
115-29-7 3614 204-079-4
Active ingredient: endosulfan; a chlorinated cyclic sulfurous acid ester having broad spectrum insecticidal activity of long-lasting effect. *Makhteshim Chemical Works Ltd.*

31703 Thionex
97-74-5 202-605-7
Tetramethyl-thiuram-monosulfide
An ultra-accelerator for rubber vulcanization.

31704 Thionol®
A registered trade name for certain dyestuffs. No manufacturer.

31705 Thionoline
$C_{12}H_8N_2OS$
hydroxyaminoiminodiphenyl-sulfide

31706 thiophanate methyl
23564-05-8 9489 245-740-7
$C_{12}H_{14}N_4O_4S_2$
Carbamic acid, [1,2-phenylenebis(iminocarbonothioyl)]bis-, dimethyl ester
Allophanic acid, 4,4'-*o*-phenylenebis[3-thio-, dimethyl ester; *o*-bis(3-methoxy carbonyl-2-thioureidobenzene; BAS 32500F; Cercobin methyl; Cercobin M; Dimethyl[(1,2-phenylene)bis(iminocarbonothioyl)]bis[carbamate]; Enovit methyl; Enovit-Supper; Fungitox; Fungo; Fungo 50; Labilite; Methyl thiophamate; methyl thiophanate; methyl topsin; methylthiofanate; methylthiophanate; Mildothane; Neotopsin; NF 44; NF-44; Pelt 14; Sigma; Thiophanate methyl; Thiophanate-methyl dimethyl; Topsin WP methyl; Trevin. mp = 172°; soluble in H_2O (26 mg/l), more soluble in organic solvents; LD_{50} (rat orl) = 7500 mg/kg.

31707 thiophene
110-02-1 9490 203-729-4
C_4H_4S
thiophene
thiofuran; thiole; thiotetrole; divinylene sulfide; thiofurfuran. Used in organic synthesis, pharmaceutical manufacture, as a dye, solvent mp = -38°; bp = 84°; d_4^{25} = 1.0573; n_D^{25}= 1.5268; insoluble in H_2O, soluble in most organic solvents.

31708 thiophenol
108-98-5 9492 203-635-3
C_6H_6S
benzenethiol
Phenyl mercaptan;. Pharmaceutical synthesis. bp = 168°, $d^{25}{}_4$= 1.0728; n_D^{25}= 1.5860; insoluble in H_2O, soluble in organic solvents. *Aldrich; ICI Am.; Janssen Chimica; Schweizerhall.*

31709 Thiophor Bronze 5G
A dyestuff obtained by the fusion of *p*-phenylene-diamine and *p*-aminoacetanilide with sulfur. No manufacturer.

31710 Thiophor Indigo
A dyestuff obtained by heating the indophenol derivative from α-naphthol and *p*-aminodimethyl aniline with sodium sulfide and sulfur. No manufacturer.

31711 thiophosgene
463-71-8 207-341-6
CCl_2S
thiocarbonyl chloride
Used as an intermediate in organic synthesis. bp = 73°; d = 1.5080; n_D^{20} = 1.542. *Aldrich; Fine Organics Ltd; Fluka; Pfaltz & Bauer.*

31712 Thioprene® -48
An elastomeric mercaptan-terminated polymer; used for sealing glass. *Polymeric Systems Inc.* Name unverified.

31713 thiosept oil
A sulfur-containing distillation product of shale oil.

31714 Thioset® M
15535-29-2 239-580-7
Ethanolamine sulfite
Evans Chemetics.

31715 Thiostab
A proprietary pure sodium thiosulfate in ampoules. No manufacturer.

31716 Thiostop E, N
20624-25-3 205-710-6
$C_5H_{10}NNaS_2 \cdot 3H_2O$
diethyldithiocarbamic acid sodium salt trihydrate
Sodium diethyl dithiocarbamate (E: 25-30%, N: 40% aqueous solution); an ultra-accelerator for NR and SBR latexes; an activator for guanidine type accelerators. Also used in analytical chemistry to assay copper. mp = 95-99°. *Uniroyal.* Name unverified.

31717 Thiostop N
128-04-1 204-876-7
$C_3H_6NNaS_2$
sodium dimethyl dithiocarbamate
Aceto Sdd 40; Alcobam NM; Brogdex 555; Carbon S; Dibam; Dibam A; DMDK; Methyl Namate; Sharstop 204; Stafresh 615; Steriseal 40; Thiostop N; Vinstop; Vulnopol NM; Wing Stop B; SDMDTC; Freshgard 40. Used in a 40% aqueous solution as a non-staining and non-colorizing polymerization short-stop for SBR and similar rubbers. LD_{50} (rat orl)= 1000 mg/kg. *Uniroyal.* Name unverified.

31718 Thiotan
Reserving agent; used for polyamide/elastomer fibers. *Sandoz Products Ltd.*

31719 Thiotax
149-30-4 5916 205-736-8
$C_7H_5NS_2$
2(3H)-benzothiazolethione
2-mercaptobenzothiazole; 2-benzothiazolethiol; MBT; Captax; Dermacid; Mertax. Vulcanization accelerator. Zinc and sodium salts used as a fungicide. mp = 179°; d = 1.4200; insoluble in H_2O, soluble in organic solvents. *Monsanto Co.*

31720 thiourea
62-56-6 9505 200-543-5
CH_4N_2S
sulfourea
thiocarbamide. Photography, photocopy papers, organic synthesis (intermediate, dyes, drugs, hair preparations), rubber accelerator, analytical reagent, amino resins, mold inhibitor. mp = 174-177°; d = 1.4100; soluble in H_2O (91 g/l), soluble in EtOH, porrly soluble in other organic solvents; L_{50} (rat orl) = 1830 mg/kg. *Allchem Industries; Bechem Chemie BV; Dajac Labs; Fairmount; R.W. Greeff.*

31721 thiourea dioxide
1758-73-2 217-157-8
$CH_4N_2O_2S$
formamidinesulfinic acid
aminoiminomethanesulfinic acid. Used in chemical synthesis. *Allchem Industries; Arol Chem Prods.; Degussa.*

31722 Thiovanic® Acid
68-11-1 9472 200-677-4
Thioglycolic acid
Evans Chemetics.

31723 Thiovanol®
96-27-5 9471 202-495-0
$C_3H_8O_2S$
Thioglycerin
2,3-dihydroxypropanethiol; 1-thioglycerol; 3-mercaptopropane-1,2-diol; α-monothioglycerol. Stabilizer for acrylonitrile polymers; crosslinking agent for hard highgloss coatings; accelerator for epoxy-amine condensation reactions; reducing agent; used in hair waving and straightening, hair dyes, depilatories, textiles, furs, pharmaceuticals, surfactants, foam stabilizing additives for detergents, shampoos, insecticides, pesticides, fungicides and dessicants. Uses medically as a vulnerary. bp_5 = 118°; d = 1.247; n_D^{20} = 1.5260; slightly soluble in H_2O, EtOH, insoluble in Et_2O. *Evans Chemetics.*

31724 Thiovit
7704-34-9 9142 231-722-6
S
Sulfur
soufre; Alfa, Aquilite; Cosan; Elosal; Golden Dew; Imber; Kolodust; Kolofog; Kolospray; Kumulus; Magnetic 6; Solfa; Suffa; Sulfex; Sulflox; Sulfospor; Sulphotox; Super Six; That; Thiolux; Thion; This; Tiolene; Uniflow; Zolvis. A protectant fungicide. Used for control of scab in fruits and mildews on a variety of crops. Also used as an acaricide. mp = 113°, 114° or 119°; bp = 444°; d = 2.07; insoluble in H_2O, slightly soluble in organic solvents; non-toxic to humans and animals *Pan Britannica Industries Ltd.*

31725 thioxydant lumire
7727-54-0 575 231-786-5
$H_8N_2O_8S_2$
ammonium peroxydisulfate
Ammonium persulfate. An oxidizer and bleach, used in photography, dyeing, etching, decolorizing and electroplating.

31726 Thiram
137-26-8 9510 205-286-2
$C_6H_{12}N_2S74$
tetramethylthiuram disulfide
bis(dimethylthiocarbamyl) disulfide; TMTD; tetramethylthioperoxydicarbonic diamide; ENT 987; SQ 1489; NSC 1771; Unicrop Thianosan; Arasan; Thiurad; Thiosan; Thylate; Tiuramil; Thiuramyl; Puralyn; Frnasan; Nomersan; Rezifilm; Pomarsol; Tersan; Tuads; Tulisan. Fungicide with animal repellent properties. Also used in vulcanization and as a bacteriostat. mp = 146-148°;

insoluble in H_2O, soluble in non-polar organic solvents; LD_{50} (rat orl) = 650 mg/kg. *Universal Crop Protection Ltd.*

31727 Thisol

Bitumen emulsions; for industrial applications. *Vedag GmbH.* Name unverified.

31728 Thissirol

An aqueous solution of about 57% castor oil soap and 29% chloroxylenol mixture; used as a bactericide.

31729 thitsi

Burma black varnish. A natural lacquer that is the sap of the black varnish tree, *Melanorrhoea visitata* .

31730 Thiurad

137-26-8 9510 205-286-2

$C_6H_{12}N_2S_4$

tetramethylthioperoxydicarbonic diamide

tetramethylthiuram disulfide; bis(dimethylthiocarbamyl) disulfide; bis(dimethylthiocarbamoyl) disulfide; TMTD; ENT-987; SQ-1489; NSC-1771; Thylate; Fernasan; Nomersan; Rezifilm; Pomarsol; Tersan; Tuads; Arasan. Vulcanization accelerator. mp = 146-148°; d = 1.29; insoluble in H_2O, soluble in organic solvents; LD_{50} (rat orl)= 640 mg/kg. *Monsanto Co.*

31731 Thixatrol

Organic derivatives of castor oil; rheological additives designed to impart thixotropy, viscosity, and antisettling properties; used in solvent systems (paints, inks, caulks, mastics, plastisols). *Rheox Inc.*

31732 Thixolan

Viscosity modifiers. *Harcros.*

31733 Thixomen

Thixotropic additive. *ICI Chem & Polymers Ltd.*

31734 Thixon® 300

Vulcanizable bonding system; one-coat for bonding fluoroelastomers to metals when equal parts by volume are mixed. *Morton Int'l./Industrial Adhesives.*

31735 Thixon® 511-T

Vulcanizable bonding system; cover coat for bonding NR, SBR, CR, BR, IR, CSM, NBR, EPDM, and IIR rubbers. *Morton Int'l./Industrial Adhesives.*

31736 Thixon® 753

Vulcanizable bonding system; aqueous primer and one-coat for bonding NBR; especially designed for phosphatized metals. *Morton Int'l./Industrial Adhesives.*

31737 Thixon® 957

Water-based overcoat adhesive; for bonding a variety of elastomers during vulcanization. *Morton Int'l./Industrial Adhesives.*

31738 Thixon® 2000

Solvent-based adhesive; used as one-coat adhesive or two-coat system applied over Thixon P-15 adhesive primer; bonds various elastomers including natural rubber, SBR, neoprene, EPDM, butyl, and nitrile. *Morton Int'l./Industrial Adhesives.*

31739 Thixon® OSN-2

Vulcanizable bonding system; one-coat for bonding NR, SBR, CR, BR, IR, CO/ECO, CSM, and ACM to most substrates; excellent for post-vulcanization bonding. *Morton Int'l./Industrial Adhesives.*

31740 Thixon® P-15

Vulcanizable bonding system; primer for all thixon cover coat adhesives and excellent one-coat for bonding NBR; resists severe environmental conditions; excellent sprayability. *Morton Int'l./Industrial Adhesives.*

31741 Thixseal

Rheological additives for imparting special properties to coating compositions; used for solvent-based sealants, caulks and thick film coatings. *Rheox Inc.*

31742 Thomas meal

Ground slag obtained from the Thomas process for iron; used as a fertilizer.

31743 Thomas slag

Thomas phosphate; Belgian slag. Basic slag containing phosphorus; by-product of steel manufacture.

31744 Thomasite

$Ca_6Fe_2O_{15}P_2Si$

A compound, $6CaO{\cdot}P_2O_5Fe_2SiO_4$. It is a constituent of the basic slag of the Thomas-Gilchrist process for the dephosphorization of iron.

31745 Thompsons

Water seal. *Sterling Roncraft.*

31746 Thonzide

553-08-2 9512 209-032-1

$C_{32}H_{55}BrN_4O$

N-[2-[[(4-methoxyphenyl)methyl]-2-pyrimidinylamino]ethyl]-N,N-dimethyl-1-hexadecanaminium bromide

Thonzonium bromide; Thonzide. Detergent. mp= 91-92°. *Parke-Davis.* Name unverified.

31747 Thoran

An alloy of 96% tungsten with 4% carbon.

31748 Thoren

A technical diamond substitute. It is an alloy made from tungsten and tungsten carbide.

31749 thorium dioxide

1314-20-1 9518 215-225-1

O_2Th

thorium dioxide

thoria; thorium anhydride; thorium oxide; umbrathor. Used in ceramics, gas mantles, nuclear fuel, medicine and non-silica optical glass. mp = 3390°; d = 10.0; insoluble in H_2O; carcinogen.

31750 Thornel® Carbon Fiber T600/50C 12K

Carbon fiber; continuous length, high strength, high modulus fiber consisting of 12,000 filaments in a one-ply construction; its treated surface increases interlaminar shear strength in a resin matrix composite. *Amoco Chemical co.*

31751 Thoroclear

A silicone-based water repellent coating for limestone. *Standard Dry Wall Products Inc.* Unverified.

31752 thoron

10043-92-2 233-146-0

Radon; used in cancer treatment and medical research.

31753 Thorosheen®

An acrylic paint for masonry. *Standard Dry Wall Products Inc.* Unverified.

31754 Thorotrast

A proprietary colloidal thorium dioxide preparation. No manufacturer.

31755 Thorowet G-40 3230

577-11-7 3460 209-406-4

Sodium dioctyl sulfosuccinate

Wetting and rewetting agent; stable to dilute acids and alkali; dewatering agent; stable to dilute acids and alkali; dewatering agent for mining ores. *Clough.*

31756 Thorquest 39

64-02-8 3557 200-573-9

Tetrasodium EDTA

Anionic liquid; complexing agent, particularly for bivalent metal ions. *Thor Chemicals (UK) Ltd.* Discontinued.

31757 Thor-stabilizator BF

Organic/inorganic complex; anionic; yellowish clear liquid; silicate-free stabilizer for peroxide bleaching. *Thor Chemicals (UK) Ltd.* Discontinued.

31758 Thorstat ASA

Phosphoric acid ester; anionic; clear liquid; antistatic agent for textile and fiber industry. *Thor Chemicals (UK) Ltd.* Discontinued.

31759 Thoulet's solution

A concentrated solution of potassium and mercury iodides in water; used to determine the density of minerals.

31760 Thovaline

A proprietary preparation of talc, kaolin, zinc oxide, and cod-liver oil; used in dermatology. *Ilon Laboratories.* Unverified.

31761 Thowless solder

An alloy of tin and zinc with small amounts of aluminum and silver.

31762 Three Elephant Boric Acid

99.8% minimum H_3BO_3; two technical granular grades (granular and fine granular), and powdered grade. *Kerr-McGee Chemical Corp.* Discontinued.

31763 Three Elephant Pyrobor Dehydrated Borax

99% Minimum $Na_2B_4O_7$; standard (-20 to 200 US mesh) and fine ((-100 to + 325 US mesh) technical grades. *Kerr-McGee Chemical Corp.* Discontinued.

31764 Three Elephant V-Bor Refined Pentahydrate Borax

99.8% Minimum $Na_2B_4O_7{\cdot}5H_2O$; standard grade. *Kerr-McGee Chemical Corp.* Discontinued.

31765 Thresh's reagent

Potassium bismuth iodide

Used for testing alkaloids.

31766 Thripstick®

52918-63-5 2934 258-256-6

$C_{22}H_{19}Br_2NO_3$

[1R-[1α(SName unverified.),3α]]-cyano(3-phenoxyphenyl)methyl 3-(2,2-dibromomethenyl)-2,2-dimethylcyclopropanedicarboxylate

Deltamethrin; deltamethrine; Butoflin; Butox; Cislin; Crackdown; Decis; Delsekte; K-Otek; K-Othrin; NRDC 161; RU 22974. A fast-acting non-systemic pyrethroid insecticide with contact and stomach action. Used to control many species of insect in many crops. Non-phytotoxic. mp = 98-101°; insoluble in H_2O (< 0.002 mg/l), very soluble in organic solvents; LD_{50} (rat orl) = 128 mg/kg. *Aquaspersions Ltd; Hoechst UK.*

31767 thrombase

A clotting enzyme.

31768 thsing-hoa-liao
The Chinese name for a cobaltiferous aluminic silicate; used in the manufacture of porcelain.

31769 thulium oxide
12036-44-1 9535 234-851-6
O_3Tm_2
thulium(III) oxide
thulia. Source of thulium metal. *Atomergic Chemetals; Cerac; Rhône-Poulenc Basic.*

31770 Thuricide HP
Bacillus thuringiensis ; a bacterial insecticide for control of caterpillars. *Atlas Interlates Ltd.*

31771 Thurmalox
A line of silicon-based, heat and corrosion resistant coatings for protection of metal structures or vessels subjected to high temperatures up to 1600F (870C); for stacks, breechings, furnaces, heat exchangers, exhaust manifolds, kilns, chemical process equipmet, prevention of stress-corrosion cracking of stainless steel, wood stoves, barbecue grills and solar collector panels. *Dampney Company Inc.* Name unverified.

31772 Thurston's alloy
An alloy of 80% zinc, 14% tin, and 6% copper.

31773 Thwaites' solution
A mixture of alcohol, creosote, and chalk in water; used to preserve animal tissues.

31774 T-Hydro
75-91-2 1604 200-915-7
t-butyl hydroperoxide solution. *Arco Chemical Co.*

31775 thyme camphor
89-83-8 9540 201-944-8
$C_{10}H_{14}O$
5-methyl-2-(1-methylethyl)phenol
thymic acid; thymol; isopropyl-*m*-cresol; 3-*p*-cymenol; 3-hydroxy-*p*-cymene;. Used in the prevention of mold and mildew, in flavoring and perfumery, as a preservative and antioxidant and a topical antiseptic. mp= 49-51°; bp = 232°; d = 0.9650; n_D^{20} = 1.5227; slightly soluble in H_2O (1 mg/ml), more soluble in organic solvents;.

31776 thymene
The residual oils obtained from the preparation of thymol; used as a inexpensive perfume for soaps.

31777 thymine
65-71-4 9539 200-616-1
$C_5H_6N_2O_2$
5-methyl-2,4(1H,3H)-pyrimidinedione
5-methyl-2,4-dioxypyrimidine; 5-methyluracil; 2,4-dihydroxy-5-methylpyrimidine. Obtained by the hydrolysis of nucleic acids; used in biochemical research. mp = 335-337°; λ_m = 205, 264.5 nm (ε 9500, 7900 pH 7); slightly soluble in H_2O (4 g/l), poorly soluble in organic solvents.

31778 thymol
89-83-8 9540 201-944-8
$(CH_3)_2CHC_6H_3(CH_3)OH$
5-methyl-2-(1-methylethyl) phenol
6-isopropyl-m-cresol; 2-isopropyl-5-methylphenol. Substituted phenol; antibacterial and antifungal agent, perfumery, microscopy, preservative, antioxidant, flavoring, lab reagent, synthetic menthol. *Haarmann & Reimer; Janssen Chimica; Quest Int'l.*

31779 Thymoxane
3,3-Dimethyl-1,5-dioxaspiro[5,5] undecane
Fresh thyme, leathery odor; used in fragrances. *Bush Boake Allen Ltd.*

31780 Thyodene
Analytical reagent HS 382200-00-0 for iodine and iodometry; as a white-water soluble powder it is superior to starch solution, it is stable and used direct from bottle to solutions to be titrated. *Campbell Williams & Co.*

31781 thyol
A substitute for ichthyol obtained by treating tar oils with sulfur.

31782 Thytropar
9002-71-5 9931 232-664-4
TSH
thyrotropic hormone; thyroid-stimulating hormone; TTH; Dermathycin. A proprietary preparation of thyrotropin. *Armour Pharmaceutical Co.* Name unverified.

31783 Tibond
Anodes. *ICI Chem & Polymers Ltd.*

31784 Ticevite
A proprietary preparation of vitamins A, D, E, and B complex. *Unichem.* Name unverified.

31785 Tico
An electrical resistance alloy containing 67.5% iron, 30.5% nickel, and small quantities of manganese and copper.

31786 Tiers argent
An alloy containing 66.6% aluminum and 33.3% silver.

31787 Tiform
Unit anode systems. *ICI Chem & Polymers Ltd.*

31788 Tiguvon®
55-38-9 4044 200-231-9
$C_{10}H_{15}O_3PS_2$
O,O-dimethyl O-[3-methyl-4-(methylthio)phenyl]phosphorothioate
Fenthion; MPP; mercaptophos; OMS 2; ENT 25540; Bay 29493; Baycid; Baytex; Entex; Lebaycid; Queletox; S 1752. Veterinary preparation; for use on domestic animals against warble infestation and lice. Also used in agriculture as an insecticide with stomach and contact action. mp = 7.5°; $bp_{0.01}$ = 87°; d20 = 1.246; n_D^{20} = 1.5698; soluble in H_2O (2 mg/l), more soluble in organic solvents; LD_{50} (rat orl) = 252 mg/kg. *Bayer AG.*

31789 Til
Compounds of titanium. *Tioxide Group plc.*

31790 Tilcom
Organic compounds of titanium and zirconium; catalysts, cross-linkers, thixatropes for emulsion paints and adhesion promoters; especially for printing inks onto plastic film. *Tioxide Group plc.*

31791 tile ore
An earthy variety of native cuprous oxide.

31792 Tilite
Self-sinking aluminum grain refiner. *Foseco (F.S.) Ltd.*

31793 Tillantina
Seed dressing for control of fungal diseases on cereals, rice, cotton, and vegetables. *Bayer AG.*

31794 Tilt Turbo
Emulsifiable concentrate containing 125 g propiconazole and 350 g tridemorph per liter; for control of mildew and rust in barley and wheat. *Ciba-Geigy Agrochemicals.*

31795 Timail®
A range of transparent and colored enamels; for use in the surface coating of metals, particularly steel and aluminum. *Bayer AG.*

31796 Timang Steel
A proprietary trade name for a high manganese steel. No manufacturer.

31797 Tim-Bor®
12280-03-4
Disodium octaborate tetrahydrate
Insecticide for use by professional pest control operators; effective against decay fungi and wood infesting pests (termites, carpenter ants); for all interior and exterior wood and wood-foam composite structural components, lumber. *U.S. Borax & Chem.; Borax Consolidated Ltd.*

31798 Tim-bor industrial
Borester. Organic boron compounds. *Borax Europe Ltd.*

31799 Tim-bor Professional
Borocil. Nonselective weedkiller. *Borax Europe Ltd.*

31800 Timbrel
55335-06-3 9789 259-597-3
$C_7H_4Cl_3NO_3$
[(3,5,6-trichloro-2-pyridinyl)oxy]acetic acid
3,5,6-trichloro-2-pyridyloxyacetic acid; Ace-Brush; Crossbow; Dowco 233; Exetor; Garlon; Mutan; Redeem; Rely; Remedy; Turflon. Timbrel is an emulsifiable concentrate containing 480 g/l triclopyr; herbicide to control perennial and woody weeds. mp = 148-150°; bp= 290° (dec); soluble in H_2O (440 mg/l), more soluble in organic solvents; LD_{50} (rat orl)= 713 mg/kg. *DowElanco Ltd.*

31801 Timbrelle
Polyamide carpet yarns. *ICI Chem & Polymers Ltd.*

31802 Time Bomb
Trade name for total release fogger insecticide; household insecticide controlling flying insects, roaches, fleas etc. *Colonial Products Inc.*

31803 Timica
Titanium dioxide coated mica to meet cosmetic standards. *Cornelius Chemical Co Ltd.* Name unverified.

31804 Timica®
Blends of titanium dioxide [13463-67-7] and mica [12001-26-2]; for frosted and iridescent effects in lipsticks, cream makeups, nail enamels, pressed powds. *Mearl.*

31805 Timolate
26921-17-5 9585 248-111-5
timolol maleate
Anti-adrenergic. *Merck & Co Inc.* Name unverified.

31806 Timonox
1314-60-9 739 215-237-7
O_5Sb_2
Antimony pentoxide

Antimony oxide; A 1530; Antimonic acid; Antimonic anhydride; Antimonic oxide; Antimony pentaoxide; Antimony pentoxide; Diantimony pentaoxide; Diantimony pentoxide; Stibic anhydride. Used as a flame retardant in clothing. d = 3.78; slightly soluble in H_2O; LD_{50} (rat ip) = 4 g/kg. *Anzon Ltd.* Name unverified.

31807 Timonox
1327-33-9 752 215-474-6
O_3Sb_2
antimony trioxide
Antimony oxide; A 1530; Antimonious oxide; Antimony oxide, Sb2O3; Antimony peroxide; Antimony sesquioxide; Antimony trioxide; Antimony White; Antox; Chemetron Fire Shield; C.I. Pigment White 11; Dechlorane A-O; Diantimony trioxide; Exitelite; Flowers of antimony; NCI-C55152; Thermoguard B; Thermoguard S. Also RN 1309-64-4. Used as a pigment and flame-proofant and in manufacture of tartar emetic. mp = 655°; bp = 1425°; d = 5.2000 *Anzon Ltd.*

31808 Timonox Blue Star
1327-33-9 215-474-6
A proprietary preparation of pure antimony oxide; the arsenic amounts to 0.0018%. No manufacturer.

31809 tin
7440-31-5 9587 231-141-8
Sn
stanum
Element; Used in manufacture of tin plate, anodes, corrosion-resistant coatings, manufacture of chemicals. mp = 232°; bp= 2507°; d = 7.31. *Aldrich; Atomergic Chemetals; Cerac; M & T Harshaw; Noah Chem.*

31810 tin amalgam
A tin-mercury alloy containing 44-51% tin. It is prepared by electrolysis.

31811 tin ash
18282-10-5 8933 242-159-0
O_2Sn
stannic oxide
stannic acid; stannic anhydride; tin peroxide; tin dioxide; white tin oxide; flowers of tin;. Used as a polishing powder. mp = 1127°; d = 6.9500; insoluble in H_2O;.

31812 tin bronze
An alloy of 89% copper and 11% tin.

31813 tin ore
O_2Sn
A mineral. It is tinstone, predominantly SnO_2.

31814 tin salts
7772-99-8 8939 231-868-0
$SnCl_2$
tin crystals
stannous chloride. Used as a wool mordant for dyeing cochineal scarlet, for dyeing blacks on silk, for weighting silk, and for calico printing.

31815 tin white
$Sn(OH)_4$
tin(IV) hydroxide
stannic hydroxide. A pigment used in enamel and glass-making.

31816 tin(II) oxalate
814-94-8 8943 212-414-0
C_2O_4Sn
stannous oxalate
oxalic acid tin (II) salt. Catalyst for stannous esterification; dying and printing textiles. d = 3.56; insoluble in H_2O. *Elf Atochem N. Am.; Atomergic Chemetals.*

31817 tin(IV) oxide
18282-10-5 8933 242-159-0
O_2Sn
tin dioxide
stannic oxide; cassiterite; white tin oxide; stannic anhydride; flowers of tin. Inorganic oxide; polishing glass and metals; tin salts, catalyst, ceramic glazes and colors, putty, perfume, cosmetic preparations (fingernail polish), manufacture of special glasses. d = 6.95; insoluble in H_2O. *Atomergic Chemetals; Cerac; Goldschmidt.*

31818 Tinaderm
2398-96-1 219-266-6
A proprietary preparation of tolnaftate; a skin fungicide. *Glaxo Laboratories.* Name unverified.

31819 Tinamul®
Partial glycerides
For preparation of bakery goods, margarine, ice cream, release agents, sausages, mayonnaise, toffees, snacks. *Hüls AG.*

31820 tincal
Tinkal. An impure borax.

31821 Tinman
Mixture of fentin hydroxide, maneb, and zineb; used for control of potato blight. *Chiltern Farm Chemicals Ltd.*

31822 Tinoclarite
Bleaching stabilizers. *Ciba plc.* Name unverified.

31823 Tinofil
Dispersed pigments. *Ciba plc.* Name unverified.

31824 Tinopal®
Fluorescent whitening agents for paper and detergents. *Ciba plc.* Name unverified.

31825 Tinopal® 5BM-GX, AMS-GX
Stibene type; whitening agent for powder and liquid anionic and nonionic detergents. *Ciba-Geigy.*

31826 Tinopal® AMS-GX
Stilbene type; whitening agent for powd. anionic and nonionic detergents or laundry soaps. *Ciba-Geigy.*

31827 Tinopal® AMS-GX
Stilbene type of whitening agent for powdered anionic and non-ionic detergents or laundry soaps. *Ciba-Geigy.*

31828 tinopal® CBS-X
Disodium distyrylbiphenyl disulfonate
Highly soluble, low dusting, lightfast, chlorine stable whitener for cotton and other cellulosics; used in anionic and nonionic laundry detergents, dry bleaches, fabric softeners, commercial laundry products toilet bar soaps, liquid products, and products for use at low washing temperatures. *Ciba-Geigy.*

31829 Tinorex
Proofing agents. *Ciba plc.* Name unverified.

31830 Tinosol
Dyeing and printing assistants. *Ciba plc.* Name unverified.

31831 Tinovetin
Biodegradable detergent; highly effective wetting and scouring agent; used for all textile processing; scouring, wetting and emulsifying greases. *Ciba plc.* Name unverified.

31832 Tintacrete®
Dry powdered colorants, iron oxides, ochers, umbers and composite pigments sold in small packages for DIY trade; colors for mortars, concrete roofing tiles, floor tiles, sand-lime bricks, concrete blocks, reconstructed stone, split blocks, paving slabs, a *W Hawley & Son Ltd.*

31833 Tint-Ayd
Pigment dispersions; used for paints, inks, etc. *Cornelius Chemical Co Ltd.* Name unverified.

31834 Tinuvin®
A trademark for uv light absorbers for incorporation in plastics materials. *Ciba plc.* Name unverified.

31835 Tinuvin® 622LD
65447-77-0
Polymeric hindered amine; dimethyl succinate polymer with tetramethyl hydroxy-1-hydroxyethyl piperidine; light and heat stabilizer for polyolefins, PP injection molded bars, HDPE, linear LDPE plaques, PP tape, LDPE film, PP multifilament, ABS polymer syst *Ciba-Geigy/Additives.*

31836 Tinuvin® 770
52829-07-9 258-207-9
Bis(2,2,6,6-Tetramethyl-4-piperidinyl)sebacate
Light stabilizer for polyolefins, incl. natural and pigmented PP multifilament slit film, polyethylene, ethylene-propylene copolymer and terpolymer, PU, ABS, impact PS, and other styrenic polymers and copolymers; synergistic with Tinuvin P in ABS, impact PS, and other styrenic polymers and copolymers; synergistic with Tinuvin P in ABS, impact PS, and PU; useful with costabilizers such as high-performance phenolic antioxidants. *Ciba-Geigy/Additives.*

31837 Tinuvin® P
A substituted benzotriazole derivative having a peak absorption at 340 mμ; recommended for PVC, polystyrene, and acrylics. *Ciba plc.* Name unverified.

31838 Tioga Adhesion Promoter 30-6-600
Conductive water-based adhesion promoter for polypropylene, TPO, and TPR. *Tioga Coatings.*

31839 Tiona®
13463-67-7 9612 236-675-5
Range of titanium dioxide pigments, including both anatase and rutile crystal forms, with surface treated grades for enhanced performance; opacifying and whitening of all paint systems, plastics and floorcoverings, paper, textiles, inks, ceramics, rubber *SCM Chemicals Europe.*

31840 Tiona® HSS
13463-67-7 9612 236-675-5

Anatase titanium dioxide, high solids aqueous dispersion; provides opacity and brightness for highest quality fine papers. *SCM.*

31841 Tiona® RCL-4

13463-67-7 9612 236-675-5

Rutile titanium dioxide, alumina surface treatment; plastics grade additive for faster processing, superior dispersion, higher tint strength, outstanding resistance to PE yellowing. *SCM.*

31842 Tiona® RCL-535

13463-67-7 9612 236-675-5

Titanium dioxide

Coatings grade providing high durability, haze-free gloss, opacity, and tinting strength with low oil absorp. *SCM.*

31843 Tiona® RCS-P

13463-67-7 9612 236-675-5

Rutile titanium dioxide aqueous dispersion; for fine paper, paper-board, and specialty paper applications. *SCM.*

31844 Tioveil

UV protection product. *Tioxide Group plc.*

31845 Tiox

61570-90-9 9599 262-854-2

$C_{12}H_{14}N_2O_3S$

(6-propoxy-2-benzothiazolyl)carbamic acid methyl ester

Tioxidazole; methyl-6-propoxy-2-benzothiazolyl carbamate; Sch-21480. Anthelmintic, used in horses. mp = 178-180°; insoluble in H_2O, soluble in organic solvents. *Schering Corp.* Discontinued.

31846 Tioxide

13463-67-7 9612 236-675-5

O_2Ti

titanium dioxide

Titanium dioxide pigment; used for decorative and industrial paints, plastics, paper, printing inks, ceramics, and man-made fibers. *Tioxide Group plc.*

31847 TIP

18265-54-8 7416

$C_{20}H_8I_4Na_2O_4$

4,5,6,7-tetraiodophenolphthalein sodium

phentetiothalien sodium; phenoltetraiodothalein sodiumIso-Iodeikon. A proprietary preparation; tetraiodophenolphthalein sodium for use in cholecystography. Soluble in H_2O, EtOH. No manufacturer.

31848 Tipoff

86-87-3 6458 201-705-8

$C_{12}H_{10}O_2$

1-Naphthylacetic acid

1-naphthaleneacetic acid; naphthylacetic acid; NAA; Fruitone-N; Phyomone; Planofix; Tre-Hold. A plant growth regulator used to control suckering in fruit trees. mp = 134-135°; soluble in H_2O (380 mg/l), more soluble in organic solvents; LD_{50} (rat orl) = 1000 mg/kg. *ICI Agrochemicals.*

31849 Ti-pure®

Titanium pigments. *DuPont UK.*

31850 Ti-Pure® R-103

13463-67-7 9612 236-675-5

Titanium dioxide

Adhesive grade with improved rheology. *DuPont.*

31851 Tirucalli gum

A product of an Indian plant of the *Euphorbia* species. It somewhat resembles gutta-percha.

31852 Tisco Steel

A proprietary trade name for a high manganese steel containing up to 15% manganese. No manufacturer.

31853 Tisept® Solution

Cetrimide 0.15% and chlorhexidine gluconate 0.015%; antiseptic with detergent properties for swabbing in obstetrics, disinfecting and cleansing traumatic and surgical wounds and burns. *Seton Healthcare Group plc.*

31854 Ti-Sphere AB-15155A

Titanium dioxide [13463-67-7]/silica [7631-86-9]

Spherical powder imparting spreadability and smooth creamy feel to cosmetic powders; may be used as nonchemical sunscreen. *Presperse.*

31855 Tissalys

Modified starch; used in textile industry for sizing natural, artificial, and synthetic fibers. *Roquette (UK) Ltd.*

31856 Tissier's metal

An alloy of 97% copper, 2% zinc, and 1% arsenic.

31857 Tisyn®

12141-46-7 377 235-253-8

Al_2O_5Si

aluminum silicate

andalusite; cyanite; sillimanite; anauxite; dickite; kaolinite; kochite; mullite; newtonite; pyrophyllite; takizolite; termiertie; ton;. Aluminum silicate; thermo-optic; high hiding pigment for paint formulations, paper coatings, plastics, rubber, water and solv. systems. *Burgess Pigment.*

31858 Titan

13463-67-7 9612 236-675-5

Design for titanium dioxide pigments; used for paint, paper, ink, plastics, ceramics and glass. *Rheox Inc.* Discontinued.

31859 Titan

999-81-5 2153 213-666-4

Soluble concentrate containing 667 g/l chlormequat chloride; plant growth regulator. *See chlormequat Schering Agrochemicals Ltd.* Discontinued.

31860 Titan cements

Cements obtained by fusing a mixture of titaniferous iron ore, limestone, and coke. It consists essentially of calcium titanate ($CaTiO_3$) with small amounts of ferrites, aluminates, and calcium silicate, together with from 2-10% ferric oxide.

31861 Titan Design

13463-67-7 9612 236-675-5

Design for titanium dioxide pigments; used for paints, plastics, inks, paper, ceramics and glass. *Rheox Inc.* Discontinued.

31862 Titanital

9612

The trade name for a proprietary titanium white; the golden seal brand contains from 95-98% titanium oxide. TiO_2 and the silver seal grade is a mixture of 80% titanium dioxide with 20% zinc oxide.

31863 Titanite

A proprietary aluminum-manganese alloy containing titanium. No manufacturer.

31864 Titanite No. 1

An explosive consisting of 85-88% ammonium nitrate, 6-8% trinitro-toluene, and 4.5-6.5% charcoal.

31865 titanium

7440-32-6 9610 231-142-3

Ti

Metallic element; alloys; as structural material in aircraft, jet engines, marine equipment, chemical equipment, surgical instruments, orthopedic appliances, food-handling equipment; x-ray tube targets; abrasives; cermets; electrodeposited and dipped coat mp = 1677°; bp= 3277°; d^{25} = 4.506; *Cerac; Inco Alloys Int'l.; New Metals & Chems.*

31866 titanium alloy

A ferro-titanium is called by this name.

31867 titanium dioxide

13463-67-7 9612 236-675-5

O_2Ti

Titanium dioxide

Titanic anhydride; titanic acid anhydride; titanium oxide; Unitane; C.I. Pigment White 6; C.I. 77891; Titania; rutile; anatase; brookite. Inorganic oxide; white pigments in paints, paper, rubber, etc.; opacifying agent, cosmetics; radioactive decontamination of skin. mp = 1855°; d = 4.23, 3.90 or 4.13. *Bayer NV; British Traders & Shippers; Degussa; Du Pont; Ferro/Transelco; Kerr-McGee; Miles; SCM.*

31868 Titanium Dioxide 110

Titanium dioxide [13463-67-7] and bismuthoxychloride [7787-59-9]

Inorganic colorant. *Presperse.* Discontinued.

31869 Titanium Dioxide P25

13463-67-7 9612 236-675-5

Titanium dioxide

Catalyst carrier for fixed bed catalyst; heat stabilizer for HCR silicone rubber and flame retardant; uv absorber for sunscreen lotions. *Degussa.*

31870 Titanium Putty

Titanium reinforced epoxy resin; for repairing worn or gouged parts, rebuilding wear surfaces and reseating worn or oversized bearings. *ITW Devcon.*

31871 titanoferrite

FeO_3Ti

A variety of the mineral ilmenite.

31872 Titanox

13463-67-7 9612 236-675-5

Titanium dioxide pigments; used for paint, paper, plastics, ink, ceramics and glass. *Rheox Inc.* Discontinued.

31873 Titanox Design

13463-67-7 9612 236-675-5

Design for titanium dioxide pigments; used for paint, plastics, paper, glass and ceramics. *Rheox Inc.* Discontinued.

31874 Titanox RA-39

13463-67-7 9612 236-675-5

A stearate coated titanium dioxide pigment easily dispersible in polystyrene and polyolefins. *Laporte Industries Ltd.* Discontinued.

31875 Titanweiss C, Extra T, Standard T, Standard A
13463-67-7 9612 236-675-5
Trade names for titanium dioxide pigments extended with calcium or barium sulfates. No manufacturer.

31876 Titite
A proprietary rubber cement, partly made from rubber; it is waterproof, and is used for mending cloth, paper, rubber, leather, and wood. No manufacturer.

31877 Title®
For agriculture. *DuPont UK.*

31878 Ti-tone
A titanium lithopone containing 15% titanium dioxide, 25% zinc oxide, and 60% barium sulfate. Its specific gravity is 4.25, and it has a covering power 60% greater than ordinary lithopone.

31879 Titus®
For agriculture. *DuPont UK.*

31880 Tixogel
Antisettling and thickening agents for paints, varnishes, lubricants, adhesives, coatings, putties and cosmetics. *Süd-Chemie AG.*

31881 Tixogel LAN
Stearalkonium bentonite (10%), lanolin oil (65%), isopropyl palmitate (22%), propylene carbonate (3%); for lipsticks, suntan products, creams and lotions. *United Catalysts.*

31882 Tixogel OMS
Quaternium-18 bentonite (10%), mineral spirits (87%), propylene carbonate (3%); for eye makeup, mascara, eyeshadow. *United Catalysts.*

31883 Tixogel VP
68953-58-2 273-219-4
Quaternium-18 bentonite; for low to medium polarity systems, e.g., antiperspirants; as suspension aid for active ingredients. *United Catalysts.*

31884 Tixogel VSP
Quaternium-18 bentonite (17.5%), cyclomethicone (79.5%), propylene carbonate (2.5%), water (0.5%); for suntan products, creams, lotions. *United Catalysts.*

31885 Tixogel VZ
Stearalkonium bentonite
For high polarity and oxygenated solvent systems such as nail lacquers. *United Catalysts.*

31886 Tixoton
Bentonites with high swelling properties for the drilling and building industry. *Süd-Chemie AG.*

31887 Tizit
An alloy of 40-80% tungsten, 4-15% titanium, 4% chromium, 2-4% carbon, 1-5% cerium, and 3-40% iron.

31888 TL 4190
Water-based polyurethane laminating adhesive; fast wet tack, high adhesion to various metal foils and films incl. polyester, PU, and vinyl, low foam coatability; for paper, film, fabric, or foil constructions. *Mace Adhesives & Coatings.*

31889 TLA-111B
Alkyl zinc dithiophosphate
Antioxidant, antiwear agent for petroleum products. *Texaco.*

31890 TLA-227
Methacrylate-based economical viscosity index improver and pour point depressant with moderate shear stability and high thickening power; for petroleum products. *Texaco.*

31891 TLA-256
61789-86-4 263-093-9
Slightly basic calcium sulfonate; detergent for petroleum products. *Texaco.*

31892 T-lim
Modified rosins in aqueous emulsions; sizing agent for paper and paperboard. *Hercules.* Discontinued.

31893 T-Maz® 20
9005-64-5 8872
Polysorbate 20 emulsifier, solubilizer, wetting agent, antistat, stabilizer, dispersant, viscosity modifier, suspending agent used in the food, cosmetic, drug, textile, and metalworking industries. *PPG/Specialty Chem.* Discontinued.

31894 T-Maz® 28
9005-64-5 8872
PEG-80 sorbitan laurate
Emulsifier and solubilizer of essential oils, wetting agent, viscosity modifier, antistat, stabilizer and dispersant used in food, cosmetic, drug, textile, and metalworking industries. *PPG/Specialty Chem.* Discontinued.

31895 T-Maz® 40
9005-66-7 8872
Polysorbate 40
Emulsifier, solubilizer, wetting agent, antistat, stabilizer, dispersant, viscosity modifier, suspending agent used in the food, cosmetic, drug, textile, and metalworking industries. *PPG/Specialty Chem.* Discontinued.

31896 T-Maz® 60
9005-67-8 8872
Polysorbate 60
Emulsifier, solubilizer, wetting agent, antistat, stabilizer, dispersant, viscosity modifier, suspending agent used in the food, cosmetic, drug, textile, and metalworking industries. *PPG/Specialty Chem.* Discontinued.

31897 T-Maz® 61
9005-67-8 8872
Polysorbate 61
Emulsifier, solubilizer, wetting agent, antistat, stabilizer, dispersant, viscosity modifier, suspending agent used in the food, cosmetic, drug, textile, and metalworking industries. *PPG/Specialty Chem.* Discontinued.

31898 T-Maz® 65
9005-71-4 8872
Polysorbate 65
Emulsifier, solubilizer, wetting agent, antistat, stabilizer, dispersant, viscosity modifier, suspending agent used in the food, cosmetic, drug, textile, and metalworking industries. *PPG/Specialty Chem.* Discontinued.

31899 T-Maz® 80
9005-65-6 7742
Polysorbate 80
Emulsifier, solubilizer, wetting agent, antistat, stabilizer, dispersant, viscosity modifier, suspending agent used in the food, cosmetic, drug, textile, and metalworking industries. *PPG/Specialty Chem.* Discontinued.

31900 T-Maz® 81
9005-65-6 7742
Polysorbate 81
Emulsifier, solubilizer, wetting agent, antistat, stabilizer, dispersant, viscosity modifier, suspending agent used in the food, cosmetic, drug, textile, and metalworking industries. *PPG/Specialty Chem.* Discontinued.

31901 T-Maz® 85
9005-70-3 8872
Polysorbate 85
Emulsifier, solubilizer, wetting agent, antistat, stabilizer, dispersant, viscosity modifier, suspending agent used in the food, cosmetic, drug, textile, and metalworking industries. *PPG/Specialty Chem.* Discontinued.

31902 T-Maz® 90
PEG-20 sorbitan tallate
Emulsifier, solubilizer, wetting agent, antistat, stabilizer, dispersant, viscosity modifier, suspending agent used in the food, cosmetic, drug, textile, and metalworking industries. *PPG/Specially Chem.*

31903 T-Maz® 95
PEG-20 sorbitan tritallate
Emulsifier, solubilizer, wetting agent, viscosity modifier, antistat, stabilizer, dispersant for food, cosmetic, drug, textile, and metalworking industries. *PPG/Specialty Chem.* Discontinued.

31904 TMP
Amine salts of organic acids, aromatic acid, aromatic and aliphatic petroleum distillate; for use with propanil herbicide in spraying rice to control evaporation, prevent crystallization of propanil and control drift. *Stull Chemical Company.* Name unverified.

31905 TMPD® Glycol
144-19-4 205-619-1
$C_8H_{18}O_2$
2,2,4-Trimethyl-1,3-pentanediol
Resin intermediate. mp = 49°; d = 0.9400 *Eastman.*

31906 TMTM
97-74-5 202-605-7
$C_6H_{12}N_2S_3$
Tetramethyl thiuram monosulfide
bis(dimethylthiocarbamyl) sulfide. Nonstaining, nondiscoloring, fast-curing accelerator for use alone or in combination in NR, SBR, NBR, butyl rubber, neoprene, and reclaim rubber. mp = 106-108°. *Akrochem.*

31907 T-Mulz® 66H
Phosphate ester, potassium salt; hydrotrope for heavy-duty liquid alkaline cleaners. *Harcros Organics.*

31908 T-Mulz® 596
Phosphate ester, free acid; emulsion polymerization surfactant. *Harcros Organics.*

31909 T-Mulz® 1158
Phosphate ester, free acid; emulsifier for mineral oil, metal processing, solvent cleaners, textile processing, agricultural formulations, corrosion inhibition; antistat; oilfield applications. *Harcros Organics.*

31910 T-Mulz® AO2
Blend; emulsifier for spray oils used in conjunction with pesticides. *Harcros Organics.*

31911 T-Mulz® Mal 5
Calcium alkylaryl sulfonate/POE ether blend; high flash emulsifier for 5 lb/gal Malathion. *Harcros Organics.*

31912 Tnegal
Dyeing and printing assistant. *Ciba plc.* Name unverified.

31913 TNX
Abbreviation for tetranitroxylene.

31914 Tobias acid
81-16-3 6493 201-331-5
$C_{10}H_9NO_3S$
2-naphthylamine-1-sulfonic acid
2-amino-1-naphthalene-sulfonic acid. Used as an azo dye intermediate and as an optical brightener. Slightly soluble in H_2O, EtOH, Et_2O.

31915 Tobin bronze
Alloys of 59-83% copper, 3-48% zinc, 0.9-12.4% tin, 0.31-2.14% lead, and 0.1-0.8% iron. One alloy contains 58.79% copper, 40.43% zinc, and 0.88% tin.

31916 Tochlorine
127-65-1 2118 204-854-7
A proprietary preparation; it is chloramine-T. No manufacturer.

31917 Tocopherex
59-02-9 10159 200-412-2
Vitamin E supplement. *BristolMyers Squibb Co Inc.* Name unverified.

31918 tocopherol
59-02-9 10159 200-412-2
$C_{29}H_{50}O_2$
3,4-Dihydro-2,5,7,8-tetramethyl-2-(4,8,12-trimethyltridecyl)-2H-1-benzopyran-6-ol
5,7,8-Trimethyltocol; antisterility vitamin; Vitamin E; Epirolin-S; Epsilan; Ephynal; Syntopherol; E-Vimin; Evipherol; Etavit; Phytogermine; Profecundin; Tokopharm; Viprimol; Viteolin; Esorb; Vascuals; Covitol; Evion; Copherol® F 1300; Covi-Ox T-50; Covipherol T-75. Occurs largely in plants. Used as an antioxidant in vegetable oils and shortening. Antioxidant and moisturizer for sun protection and skin care products mp = 2.5-3.5°; $bp_{).1}$= 200-220°; d^{25}_4 = 0.950; n^{25}_D= 1.5045; λ_m = 294 nm ($E^{1\%}_{1cm}$ = 71); insoluble in H_2O, soluble in organic solvents *Henkel/Cospha; Henkel Canada.*

31919 Tocopherol Oil CLR
Vitamin E carrier with natural tocopherols in soya oil medium; general skin care products. *Dr. Kurt Richter; Henkel/Cospha.*

31920 Toffix
Special hard fat; used for manufacture of caramels and chewing sweets. *Dynamit Nobel Wien GmbH.*

31921 Toffix®
Fats containing an emulsifier; for preparation of toffees and chewing sweets. *Hüls AG.*

31922 Togocoll
One and two-component polyurethanes, sealers, glazing, and structural adhesives, primers, underbody coatings, windshield-adhesives and corrosion preventative coatings (waxes) for automotive industries. *EMS-Chemie AG.* Discontinued.

31923 Togoplast
PVC plastisols; used for automotive underbody coatings, sealants and adhesive systems. *EMS-Chemie AG.* Discontinued.

31924 Togotec
Corrosion protective wax coatings for automotive box section and underbody protection. *EMS-Chemie AG.* Discontinued.

31925 Togotherm
Polyurethane foam systems; for filling car body box section. *EMS-Chemie AG.* Discontinued.

31926 Toho Me-PEG Series
Methoxy polyethylene glycol (m.w. 225, 350, 550, 705, 1000); base material for surfactant, synthetic resin, plasticizer, lubricating industries; wetting, softening, penetrating, lubricating and cleaning agent for textile, paper, ink, pigments, etc. *Toho Chem. Industry.*

31927 Toho PEG Series
Polyethylene glycol (m.w. 200, 300, 400, 600, 1000, 1500, 1540, 2000, 4000); base material for surfactant, synthetic resin, plasticizer, lubricating industries; wetting, softening, penetrating, lubricating, and cleaning agent for textile, paper, ink, pigm *Toho Chem. Industry.*

31928 Toho Salt A-5
Anionic complex; dyeing assistant, dispersant for dyestuffs. *Toho Chem. Industry.*

31929 Tohol N-220
Cocamide DEA
Foam stabilizer and thickener for shampoo, detergent, toothpaste. *Toho Chem. Industry.*

31930 Toisin's solution
A microscopic stain used for staining white blood corpuscles; based on methyl violet.

31931 Toku Bishi®
Coke; for ductile and high quality cast iron. *Mitsubishi Kasei.* Name unverified.

31932 Tokuthion®
34643-46-4 252-125-7
$C_{11}H_{15}Cl_2O_2PS_2$
O-(2,4-dichlorophenyl) O-ethyl S-propyl phosphorodithioate
O-2,4-dichlorophenyl O-ethyl S-propyl phosphorodithioate; Prothiofos; Prothiophos; Bay NTN 8629; Bideron;. Non-systemic insecticide with contact and stomach action; cholinesterase inhibitor, especially effective against leaf-eating caterpillars. $bp_{0.1}$ = 125-128°; d_{20} = 1.3; n^{20}_D= 1.5694; soluble in H_2O (1.7 mg/l), freely soluble in organic solvents; LD_{50} (rat orl) = 1500 mg/kg. *Bayer AG.*

31933 Tolan
501-65-5 9643 207-926-6
$C_{14}H_{10}$
1,1'-(1,2-ethanediyl)bisbenzene
diphenylacetylene; diphenylethyne. Used in organic synthesis. mp = 60-61°; bp = 300°; λ_m = 216, 221, 269, 272, 279, 288, 297 nm (ε 20600, 20300, 23450, 25200, 33000, 23250, 29400); insoluble in H_2O, soluble in organic solvents.

31934 Tolcide MBT
6317-18-6 228-652-3
$C_3H_2N_2S_2$
Methylenedithiocyanate
Methylenebisthiocyanate; dithiocyanatomethane. Biocide for use in water treatment, paper, antifoulant paint, leather, timber preservation. mp = 103-105°. *Albright & Wilson UK.*

31935 Tolgard
Flame retardants. *Albright & Wilson Ltd.*

31936 Tolinase
1156-19-0 9644 214-588-3
Tolazamide
Antidiabetic. *Upjohn.*

31937 Tolkan
34123-59-6 5237 251-835-4
Herbicide. *May & Baker Ltd.* Name unverified.

31938 Tolkan
34123-59-6 5237 251-835-4
Suspension concentrate containing 500 g isoproturon per liter; used for annual weed control in cereals. *Rhône-Poulenc Crop Protection Ltd.*

31939 Tolkan 500
34123-59-6 5237 251-835-4
Suspension concentrate containing 500 g isoproturon per liter; used for annual weed control in cereals. *Farmers Crop Chemicals Ltd.* Discontinued.

31940 Tollen's reagent
9654
A solution of ammoniacal silver nitrate containing free caustic soda; used to test for aldehydes.

31941 Tolochrome
Photographic color developer. *May & Baker Ltd.* Name unverified.

31942 Tolonate
Aliphatic diisocyanates. *Rhône-Poulenc UK.*

31943 Toloy 45
An alloy of 45% nickel and 20% chromium with other materials; used where stress corrosion resistance is required. The material conforms, to BS, 1648 Grade H.

31944 Tolplaz
Specialty plasticizers. *Albright & Wilson Ltd.*

31945 tolu balsam
The oleoresin of *Myroxylon toluifera*, of South America.

31946 toluene
108-88-3 9667 203-625-9
$C_6H_5CH_3$
methylbenzene
phenylmethane; toluol; methacide. Aromatic compound; aviation gasoline additive; solvent for paint; diluent and thinner in nitrocellulose lacquers; adhesive solvent in plastic toys; manufacture of benzoic acid, benzaldehyde, explosives, dyes; in extraction of various principles from plant mp = -95°; bp = 111°; d^{20}_4= 0.866; n^{20}_D= 1.4967; insoluble in H_2O, soluble in organic solvents; LD_{50} (rat orl) = 7.53 g/kg. Ashland; Chevron; Exxon; Fina; Mitsubishi Petrochem.; Mitsui Petrochem. Ind.; Mobil; Phillips 66; shell; Texaco; Unocal.

31947 toluene diisocyanate
584-84-9 9668 209-544-5
$C_9H_6N_2O_2$

toluene 2,4-diisocyanate
TDI; 2,4-tolylene diisocyanate; 2,4-diisocyanatotoluene; Nacconate 100. In manufacture of polyurethane foams, elastomers, and coatings. mp = 20-22°; bp = 751°; d_4^{20}= 1.2244; reacts with H_2O, soluble in organic solvents. *Bayer Hispania Industrial SA; ICI Polyurethanes; Nippon Polyurethane Ind.; Olin.*

31948 o-toluenesulfonamide
88-19-7 201-808-8
$C_7H_9NO_2S$
o-Toluene sulfonamide
Uniplex 171. Plasticizer for thermoplastic and thermoset resins; imparts gloss and wetting to melamine, urea and phenolic resins. Component (with p-isomer) of Uniplex 171. mp = 156-158°. *Unitex.*

31949 *p*-toluenesulfonamide
70-55-3 200-741-1
$C_7H_9NO_2S$
p-toluenesulfonamide
toluene-4-sulfonamide; PTSA; *p*-toluenesulphonamide. Organic synthesis; plasticizers, resins; fungicide and mildewcide in paints and coatings. mp = 135-137°. *Allchem Industries; Honeywill & Stein Ltd; ICI Spec.; Rit-Chem.; Unitex.*

31950 toluene sulfonic acid
104-15-4 9671 203-180-0
$C_7H_8O_3S$
4-methylbenzenesulfonic acid
p-toluene sulfonic acid; tosylic acid; PTSA. Dyes, organic synthesis, acid catalyst. mp = 106-107°; bp_{20} = 140°; soluble in H_2O (67 g/100 ml), soluble in organic solvents. Boliden Intertrade; BYK-Chemie; Eastman; Ferro/Grant; Manro Prods. Ltd; Nissan Chem. Ind.; PMC Spec.; Ruetgers-Nease; Witco.

31951 toluhydroquinone
95-71-6 202-443-7
$C_7H_8O_2$
2,5-dihydroxytoluene
methylhydroquinone. Antioxidant, polymerization inhibitor. mp = 125-128°. *Eastman.*

31952 α-toluic acid
103-82-2 7422 203-148-6
$C_8H_8O_2$
benzeneacetic acid
phenylacetic acid. Synthetic intermediate in prefumery industry. mp = 77°; bp = 266°; d_4^{77}= 1.091; soluble in H_2O, more soluble in organic solvents;.

31953 *m*-toluic acid
99-04-7 202-723-9
$C_8H_8O_2$
3-methylbenzoic acid
m-toluylic acid. Organic synthesis, to form N,N-diethyl-*m*-toluamide, a broad-spectrum insect repellent. mp = 108-110°; bp = 263°; d = 1.0540; slightly soluble in H_2O, soluble in organic solvents. *Mitsubishi Gas; Witco/Argus.*

31954 o-toluic acid
118-90-1 9673 204-284-9
$C_8H_8O_2$
2-methylbenzoic acid
o-toluylic acid. Bacteriostat. mp = 103-105°; bp = 258-259°; d = 1.0620; slightly soluble in H_2O, soluble in organic solvents. *Mitsubishi Gas.*

31955 *p*-toluic acid
99-94-5 202-803-3
$C_6H_4CH_3COOH$
4-methylbenzoic acid
p-toluylic acid. Agricultural chemicals, animal feed supplement. mp = 180-=181°; bp = 274-275°; slightly soluble in H_2O, soluble in organic solvents. *Hüls Am.; Nat'l. Starch & Chem.; Penta Mfg.*

31956 toluol
Obsolete name for toluene.

31957 *p*-toluoyl chloride
874-60-2 212-864-8
C_8H_7ClO
4-methylbenzoyl chloride
Used in chemical synthesis. mp = -2°; bp = 225-227°; d = 1.1690; n_D^{20} = 1.5535. *Aldrich; James River.*

31958 Tolurex
15545-58-9 239-592-2
$C_{10}H_{13}ClN_2O$
N'-(3-chloro-4-methylphenyl)-N,N-dimethylurea
chlortoluron; 3-(3-chloro-*p*-tolyl)-1,1-dimethylurea; C 2242; Clortokem; Deltarol; Dicuran; Dicurane; Higaluron; Highuron; Ludorum; Talisman; Tolurane; Tolurgan; Toro. Selective pre-and post-emergence herbicide absorbed by foliageand roots. Used with winter cereals for control of annual grasses and broad leaved weeds. Inhibits photosynthesis. mp = 147-148°; soluble in H_2O (70 mg/l), more soluble in organic solvents; *Agan Chemical Manufacturers Ltd.*

31959 Tolurgan
15545-58-9 239-592-2
Chlortoluron
Herbicide. *Agan Chemical Manufacturers Ltd.*

31960 toluylene
103-30-0 203-098-5
$C_{14}H_{12}$
1,1'-(1,2-ethanediyl)bis[benzene]
trans-stilbene; *trans*-1,2-diphenylethylene. Used in the manufacture of optical bleaches and dyes. mp = 124°; bp = 306-307°; λ_m = 296, 305 nm (ε 28100 26700 95% EtOH); insoluble in H_2O, soluble in organic solvents.

31961 *p*-tolyl aldehyde
104-87-0 203-246-9
C_8H_8O
4-Methylbenzaldehyde
p-formyltoluene; *p*-tolualdehyde. Intermediate in the perfumes, pharmaceuticals and dyestuffs industries. Also used as a flavoring agent. mp = -6°; bp = 205°; SG = 1.019; n_D^{20} = 1.5447. *BASF; Mallinckrodt; Mitsubishi Gas.*

31962 tolyltriazole
29385-43-1 249-596-6
$C_7H_7N_3$
5-methyl-1,2,3-benzotriazole
tolutriazole; methyl-1H-benzotriazole; 5-methylbenzotriazole; cobratec tt-100. Corrosion inhibitor for copper, brass, bronze and ferrous metals. Used in metalworking fluids. mp = 76-87°; bp_2 = 160°; d = 1.24; insoluble in H_2O, soluble in organic solvents; LD_{50} (rat orl) = 1600 mg/kg. *Dinoval Chem. Ltd.; PMC Spec.; Sandoz.*

31963 Tomah AO-14-2
Bishydroxyethylisodecyloxypropylamine oxide
Foam stabilizers/boosters in liq. detergents, shampoos, hard surface cleaners, laundry detergents; grease emulsifier, soil suspension aid; forms synergistic surfactant base for built household, institutional and industrial cleaners with quaternaries and nin-ionics. *Exxon/Tomah.*

31964 Tomah AO-728 Special
Amine oxide
Detergent, foam booster/stabilizer for industrial and household detergents, dishwash, personal care products. *Exxon/Tomah.*

31965 Tomah BExM-1
Modifier for clay stabilized coal tar and asphalt emulsions. *Exxon.*

31966 Tomah DA-14
72162-46-0 276-432-0
N-Isodecyloxypropyl-1,3-diaminopropane
Intermediate for textile foaming agents, surfactants, ethoxylates, agricultural chemicals; corrosion inhibitor for metalworking fluids; additive for fuels, lubricants, petroleum refining; crosslinking agent for epoxy resins; bactericidal properties. *Exxon/Tomah.*

31967 Tomah DA-16
N-isododecyloxypropyl-1,3-diaminopropane
Intermediate for textile foaming agents, surfactants, ethoxylates, agricultural chemicals; corrosion inhibitor for metalworking fluids; additive for fuels, lubricants, petroleum refining; crosslinking agent for epoxy resins. *Exxon/Tomah.*

31968 Tomah DA-17
N-Isotridecyloxypropyl-1,3-diaminopropane
Intermediate for textile foaming agents, surfactants, ethoxylates, agricultural chemicals; corrosion inhibitor for metalworking fluids; additive for fuels, lubricants, petroleum refining; crosslinking agent for epoxy resins. *Exxon/Tomah.*

31969 Tomah E-14-2
34360-00-4
Bis (2-hydroxyethyl) isodecyloxypropylamine
Emulsifier, corrosion inhibitor, lubricant used in mineral acid inhibition, textile processing. *Exxon/Tomah.*

31970 Tomah E-14-5
PEG-5 isodecyloxypropylamine
Emulsifier, corrosion inhibitor, lubricant used in mineral acid inhibition, textile processing. *Exxon/Tomah.*

31971 Tomah E-18-2
Bis (2-hydroxyethyl)isotridecyloxypropylamine
Emulsifier, corrosion inhibitor, lubricant used in mineral acid inhibition, textile processing. *Exxon/Tomah.*

31972 Tomah E-18-2
Bis (2-hydroxyethyl) octadecyloxypropylamine
Emulsifier, corrosion inhibitor, lubricant used in mineral acid inhibition, textile processing. *Exxon/Tomah.*

31973 Tomah E-18-5
PEG-5 stearyloxypropylamine

Emulsifier, corrosion inhibitor, lubricant used in mineral acid inhibition, textile processing. *Exxon/Tomah.*

31974 **Tomah E-19-2**
Bis (2-hydroxyethyl)
Linear alkyloxypropylamine; surfactant to modify emulsification, surface tension, solubility; for acid thickeners, antistats, petrol. production and refining, agricultural adjuvants, textile processing aids, corrosion inhibition, detergent boosters; chem. *Exxon/Tomah.*

31975 **Tomah E-24-2**
PEG-2 Guerbet C20 alcohol amine
Surfactant, corrosion inhibitor. *Exxon/Tomah.*

31976 **Tomah E-DT-3**
PEG-3 1,3-diaminopropane
Surfactant for acid thickeners, antistat, cationic emulsification, petroleum production and refining, agricultural adjuvants, textile processing aids, corrosion inhibition, detergent boosters; chemical intermediate. *Exxon/Tomah.*

31977 **Tomah E-S-2**
61791-24-0
PEG-2 soyamine
Emulsifier, corrosion inhibitor, lubricant used in mineral acid inhibition, textile processing. *Exxon/Tomah.*

31978 **Tomah E-T-2**
61791-44-4 263-177-5
PEG-2 tallowamine
Surfactant for acid thickeners, antistat, cationic emulsification, petrol. production and refining, agricultural adjuvants, textile processing aids, corrosion inhibition, detergent boosters; chem. intermediate. *Exxon/Tomah.*

31979 **Tomah PA-10**
Hexyloxypropylamine
Corrosion inhibitor for metalworking fluids; antistat; flotation collector; additive for fuel, lubricant, petrol. refining; intermediate for surfactants, textile foaming agents, ethoxylate, and agricultural chem.; crosslinking agent for epoxy resins. *Exxon/Tomah.*

31980 **Tomah PA-12EH**
5397-31-9 226-420-6
2-Ethylhexyloxypropylamine
Intermediate used in manufacture of detergents . *Exxon/Tomah.*

31981 **Tomah PA-13i**
29317-52-0 249-554-7
isononyloxypropylamine
Intermediate used in manufacture of detergents . *Exxon/Tomah.*

31982 **Tomah PA-14**
7617-78-9 231-530-2
Isodecyloxypropylamine
Emulsifier; corrosion inhibitor for metalworking fluids; antistat; flotation collector; additive for fuel, lubricant, petroleum refining; intermediate for surfactants, textile foamers, ethoxylates, and agricultural chemicals; crosslinking agent for epoxies. *Exxon/Tomah.*

31983 **Tomah PA-14 Acetate**
Isodecyl oxypropyl amine acetate
Patented emulsifier, gellation/wetting agent for clays, fillers, and fibers in organic coatings and sealants, e.g., roof coatings, tile adhesives, caulks, pipe coatings, automotive undercoatings, alkyd paints, foundry coatings, polymers/elastomers; has bactericidal properties. *Exxon/Tomah.*

31984 **Tomah PA-16**
Isododecyloxypropylamine
Corrosion inhibitor for metalworking fluids; antistat; flotation collector; additive for fuel, lubricant, petrol. refining; intermediate for surfactants, textile foamers, ethoxylates and agricultural chem.; crosslinking agent for epoxies. *Exxon/Tomah.*

31985 **Tomah PA-17**
Isotridecyloxypropylamine
Corrosion inhibitor, cationic emulsification, and replacement for oleyl and soya amines; chemical intermediate. *Exxon/Tomah.*

31986 **Tomah PA-19**
68610-26-4 271-855-7
Linear C_{12}-C_{15} alkyloxypropylamine; corrosion inhibitor for metalworking fluids; antistat; flotation collector; additive for fuel, lubricant, petroleum refining; intermediate for surfactants, textile foamers, ethoxylates, and agricultural chemicals; crosslinking agent for epoxies. *Exxon/Tomah.*

31987 **Tomah PA-24**
Isoarachidyloxypropylamine
Guerbet C_{20} alcohol primary amine; detergent intermediate; experimental product for research and development. *Exxon/Tomah.*

31988 **Tomah PA-1214**
Octyl/decyloxypropylamine
Corrosion inhibitor for metalworking fluids; antistat; flotation collector; additive for fuel, lubricant, petroleum refining; intermediate for surfactants, textile foaming agents, ethoxylates and agricultural chem.; crosslinking agent for epoxy resins. *Exxon/Tomah.*

31989 **Tomah Q-2C**
61789-77-3 263-087-6
Dicoco dimonium chloride - isopropanol
Detergency booster with biocidal activity. *Exxon/Tomah.*

31990 **Tomah Q-14-2**
125740-36-5
Isodecyloxypropyl dihydroxyethyl methyl ammonium chloride
Quaternary used as acid corrosion inhibitor, plastics and textile antistat, and emulsifier; bactericidal properties. *Exxon/Tomah.*

31991 **Tomah Q-17-2**
Isotridecyloxypropyl dihydroxyethyl methyl ammonium chloride
Emulsifier; boosts efficiency of nonionic surfactants; used in hard surface cleaners, laundry, transportation cleaners; bactericidal properties. *Exxon/Tomah.*

31992 **Tomah Q-18-2**
Octadecyl dihydroxyethyl methyl ammonium chloride - isopropanol
Quaternary surfactant for use as emulsifiers, antistats, corrosion inhibitors, nonionic detergency booster in laundry products, in nonbutyl cleaning systems. *Exxon/Tomah.*

31993 **Tomah Q-24-2**
Methyl dihydroxyethyl isoarachidaloxypropyl ammonium chloride - isopropanol
Guerbet C_{20} alcohol dihydroxyethyl ammonium chloride; detergency booster with biocidal activity; experimental material. *Exxon/Tomah.*

31994 **Tomah Q-311**
Monosoya amidoamine quaternary
Detergency booster with biocidal activity. *Exxon/Tomah.*

31995 **Tomah Q-511**
Monococo amidoamine quatermary
Detergency booster with biocidal activity. *Exxon/Tomah.*

31996 **Tomah Q-C-15**
61791-10-4
PEG-15 cocomonium chloride
Quaternary surfactant for use as emulsifiers, antistats, corrosion inhibitors, nonionic detergency booster in laundry products, in nonbutyl cleaning systems. *Exxon/Tomah.*

31997 **Tomah Q-D-T**
Tallow dimethyl trimethyl propylene diammonium chloride, IPA; quaternary used as an acid corrosion inhibitor, plastics and textile antistat, and emulsifier. *Exxon/Tomah.*

31998 **Tomah Q-DT-HG**
Tallow diamine quaternary in hexylene glycol; quaternary surfactant for use as emulsifiers, antistats, corrosion inhibitors, nonionic detergency booster in laundry products, in nonbutyl cleaning systems. *Exxon/Tomah.*

31999 **Tomah Q-S**
61790-41-8 263-134-0
Soya trimethyl ammonium chloride - isopropanol
Quaternary surfactant for use as emulsifiers, antistats, corrosion inhibitors, nonionic detergency booster in laundry products and in nonbutyl cleaning systems. *Exxon/Tomah.*

32000 **Tomah Q-ST-50**
112-03-8 203-929-1
Steartrimonium chloride
Quaternary surfactant for use as emulsifiers, antistats, corrosion inhibitors, nonionic detergency booster in laundry products, in nonbutyl cleaning systems. *Exxon/Tomah.*

32001 **Tomahawk**
70124-77-5 4160 274-322-7
flucythrinate. An emulsifiable concentrate containing flucythrinate; a pyrethroid insecticide for the control of aphids, whitefly, caterpillars, and red spider mite. *Fisons plc, Horticultural Div.* Name unverified.

32002 **Tomaset**
Active ingredient; N-m-tolylphthalamic acid; a flower and fruit plant growth regulator. *Agan Chemical Manufacturers Ltd.*

32003 **Tomato Setting Spray**
Aerosol spray containing 2-naphthyloxyacetic acid; setting spray for tomatoes. *Vitax Ltd.*

32004 **Tombac**
An alloy usually containing 89% copper, 5.5% zinc, and 5.5% tin.

32005 **Tombasil**
A proprietary trade name for an alloy consisting mainly of tombac metal with silicon. No manufacturer.

32006 **Tombel**
Mixture of quinalphos and thiometon; for control of caterpillers and aphids. *Hortichem Ltd.*

32007 Tomophan
A proprietary viscose packing material. No manufacturer.

32008 Tomorite
Liquid fertilizer. *Fisons plc, Horticultural Div.* Name unverified.

32009 Toncan
A corrosion-resisting alloy containing pure iron, copper, and molybdenum.

32010 Toncas metal
An alloy of 29% nickel, 36% copper, 7.1% iron, 7.1% zinc, 7.1% lead, 7.1% tin, and 7.1% antimony; used for ornamental work.

32011 Tone
Caprolactone derivatives. *Union Carbide.*

32012 Tonite
Potentite. An explosive consisting of mixtures of granulated gun cotton and barium nitrate.

32013 Tonophosphan
575-75-7 9649 209-391-4
$C_9H_{13}NNaO_2P$
(4-dimethylamino-*o*-tolyl)phosphonous acid sodium salt
toldimfos sodium; Foston; Tonofosfan. A proprietary preparation of toldimfos sodium; a source of phosphorus used for veterinary purposes. Soluble in H_2O and EtOH. No manufacturer.

32014 Tonox® 22, Tonox® R
101-77-9 3023 202-974-4
$C_{13}H_{14}N_2$
4,4'-methylenebis[benzeneamine]
p,p'-diaminodiphenylmethane. Crude methylene dianiline; curing agent for epoxy resins. mp = 91-92°; bp = 398-399°; slightly soluble in H_2O, very soluble in organic solvents. *Uniroyal.*

32015 Tonsil
Used for the adsorptive decolorization and purification of oils and fats, hydrocarbons, waxes, and other liquid intermediate products. *Süd-Chemie AG.*

32016 Tool Life
Water-based synthetic cutting fluid (extreme pressure); for all metal removing machining operations where physical and chemical extreme pressure assistance is required to assist functional cooling properties on all metals, except magnesium. *Sumner Oil Industries.* Unverified.

32017 Toolife
A broad line of industrial fluids, both oil and synthetic in nature, water soluble and straight. *Specialty Products Co.* Name unverified.

32018 Top 7 Mosaic and Pebble
Expanded polystyrene veneer 7mm thick; decorative veneers for domestic ceilings. *Vencel Resil Ltd.*

32019 Topanex 100BT
2440-22-4 3503 219-470-5
$C_{13}H_{11}N_3O$
2-(2H-benzotriazol-2-yl)-4-methylphenol
Drometrizole; (2'-Hydroxy-5mg-methylphenyl) benzotriazole; Tinuvin P. Uv absorber, light stabilizer, antioxidant protecting plastics (PVC, styrenics, acrylics, unsaturated polyesters), lacquers; effective at 290-380 nm. mp = 131-133°; bp_{10} = 225°; soluble in organic solvents. *ICI Am.*

32020 Topanex 500H
1,5-Dioxaspiro[5.5]undecane 3,3-dicarboxylic acid, bis(2,2,6,6-tetramethyl-4-piperidinyl)ester
UV stabilizer. *ICI Polymer Additives.*

32021 Topanol®
A range of antioxidants. *ICI Chem & Polymers Ltd.*

32022 Topanol® 205
2,2-Bis[4-(2-(3,5-di-t-butyl-4-hydroxyhydrocinnamoyloxy; ethoxyphenyl] propane
High molecular weight hindered phenolic antioxidant for retardation of thermal and oxidative degradation in polyolefins, styrenics, and engineering polymers. *ICI Polymer Additives.*

32023 Topanol® CA
1843-03-4 217-420-7
1,1,3-Tris (2-methyl-4-hydroxy-5-t-butyl phenyl) butane
Antioxidant for polyolefin and styrenic polymers, plasticizers, hot melt adhesives, SBR latexes, polyamides, polyesters; FDA approved. Antioxidant, stabilizer for PVC wire and cable compounds. *ICI Polymer Additives.*

32024 Topanol® LVT 600
Antioxidant dispersion. *ICI Polymer Additives.*

32025 Topanol® M
101-96-2 202-992-2
N,N-Di-sec-butyl-p-phenylenediamine
Commercial grade. *ICI Chem & Polymers Ltd.*

32026 Topas 100
66246-88-6 266-275-6
$C_{13}H_{15}Cl_2N_3$
1-[2-(2,4-dichlorophenyl)pentyl]-1H-1,2,4-triazole
Award; CGA 71818; Topaz; Topaze; penconazole. Systemic fungicide with protective and curative action. Inhibits biosynthesis of ergosterol in the cell membrane. Topas 100 is an emulsifiable concentrate containing 100 g/l penconazole; for control of powdery mildew. mp = 60°; soluble in H_2O (70 mg/l), more soluble in organic solvents; LD_{50} (rat orl) = 2125 mg/kg. *Ciba-Geigy Agrochemicals.*

32027 Topas C 50WP
Captan + penconazole; protectant fungicide for apple and pear trees. *Ciba-Geigy Agrochemicals.*

32028 Topaz
Cooling water treatment. *Grace Dearborn Ltd.*

32029 Topbraun®
Suntan product without a sun protection factor. *Bayer AG.*

32030 Topclip Dridress
Animal health organophosphorus wound dressing. *Ciba plc.* Name unverified.

32031 Topclip Fly and Scab Dip
Organophosphorus/phenols sheep dip. *Ciba plc.* Name unverified.

32032 Topclip Foot Rot Aerosol
Antiseptic sheep treatment. *Ciba plc.* Name unverified.

32033 Topclip Formalin
Formaldehyde for sheep foot rot control. *Ciba plc.*

32034 Topclip Gold Shield
Scab approved organophosphorus sheep dip. *Ciba plc.*

32035 Topclip Marker Aerosols
Sheep marker dyes. *Ciba plc.* Name unverified.

32036 Topclip Marker Fluid
Sheep marker fluid. *Ciba plc.* Name unverified.

32037 Topclip Parasol
66841-24-5 266-492-6
$C_{22}H_{19}Cl_2NO_3$
cyano(3-phenoxyphenyl)methyl 3-(2,2-dichloroethenyl)-2,2-dimethylcyclopropanecarboxylate
Aimcocyper; Ambush C; Ammo; Arrivo; Barricade; Basathrin; CCN52; Cymbush; Cymperator; Cynoff; Cyper; Cypercopal; Cyperguard; Cyperkill; Cyperscet; Cypertox; Cyrux; Demon; Fenom; Flectron; Fligene Cl; Folcord; Halt;Imperator; Kafil Super; LE 79600; NRDC 149; Nurelle, Polytrin; PP 383; Prevail; Ralothrin; Ripcord; Sherpa; Siperin; Stockade; Toppel; Ustaad; WL 43467. Cypermethrin pour-on; lice control for sheep and cattle. mp = 60-80°; d_{20} = 1.25; insoluble in H_2O, soluble in organic solvents; LD_{50} (rat orl) = 250-4150 mg/kg. *Ciba plc.*

32038 Topclip Scab Dip
Organophoshorus/phenols sheel dips. *Ciba plc.* Name unverified.

32039 Topclip Sheep Dip
Organophosphorus sheep dip. *Ciba plc.* Name unverified.

32040 Topclip Vaccines
Various sheep vaccines. *Ciba plc.* Name unverified.

32041 Topclip Wormer
Organophosphorus anthelmintic for sheep and pigs. *Ciba plc.* Name unverified.

32042 Top-Cop
Mixture of copper sulfate and sulfur; a contact fungicide to control mildew. *Stoller Chemicals Ltd.*

32043 Topex
94-36-0 1149 202-327-6
Buffered acne medication, 10% benzol peroxide; to help clear pimples without overdrying the skin. *Richardson-Vicks Inc.* Name unverified.

32044 Topexane
Antibacterial face wash to help clear acne blemishes. *Richardson-Vicks Inc.* Name unverified.

32045 Topfix
A range of epoxy, polyurethane and cyanoacrylate structural adhesives. *Elf Atochem UK/Ceca.*

32046 Tophet
An electrical resistance alloy containing 61% nickel, 10% chromium, 26% iron, and 3% manganese.

32047 Tophet A
A proprietary trade name for 80% nickel, 20% chrome resistance wire. No manufacturer.

32048 Tophet C
A proprietary trade name for nickel chrome iron resistance wire. No manufacturer.

32049 Toppel
66841-24-5 266-492-6
Insecticide containing cypermethrin. *ICI Chem & Polymers Ltd.*

32050 Toppel
66841-24-5 266-492-6
Emulsifiable concentrate containing 100 g cypermethrin per liter; a pyrethroid insecticide. *Farm Protection Ltd.*

32051 Topshot
Bentazone + cyanazine + 2,4-DB; an herbicide. *Shell UK.*

32052 Topsol
Textile combing lubricant. *Crosfield Chemicals Ltd.*

32053 Topup
61791-44-4 263-177-5
Cationic surfactant containing 800 g/l ethoxylated tallow amine; for use as a wetting agent for phosphonoglycine herbicides. *Farmers Crop Chemicals Ltd.*

32054 Torapron
An emulsifiable concentrate containing 95 g aminotriazole, 190 g atrazine and 99 g 2,4-D per liter; used for total weed control in non crop areas. *BP Oil Ltd.* Discontinued.

32055 torbanite
A variety of cannel coal.

32056 Tordon
1918-02-1 7552 217-636-1
$C_6H_3Cl_3N_2O_2$
4-amino-3,5,6-trichloro-2-pyridine carboxylic acid
4-amino3,5,6-trichloropicolinic acid; piclorame; Grazon. Herbicides based primarily on picloram; broadleaf and brush killers for forestry, grain and corn. mp = 215° (dec); soluble in H_2O (430 mg/l), more soluble in organic solvents; LD_{50} (rat orl) = 8200 mg/kg. *Dow UK.* Discontinued.

32057 Tordon 22K
1918-02-1 7552 217-636-1
Soluble concentrate containing 240 g/l picloram; a picolinic herbicide to control woody weeds in noncrop areas. *Chipman Ltd.*

32058 Torelle
5598-52-7 4283
$C_7H_7Cl_3NO_4P$
dimethylphosphoric acid 3,5,6-trichloro-2-pyridinyl ester
Fospirate; Dowco 217. Anthelmintic. mp = 86-88°. *Dow UK.* Discontinued.

32059 Toric Contact Lens
Hefilcon B
2-hydroxyethyl methacrylate with 1-vinyl-2-pyrrolidinone and ethylene dimethacrylate; contact lens material. *Bausch & Lomb, Professional Products Div.* Name unverified.

32060 Torlon® 4203L
Poly(amide-imide) resin; high-strength resin for connectors, switches, relays, valve seats, mechanical linkages, bushings, wear rings, insulators, cams, ball bearings, rollers, and thermal insulators. *Amoco; Polymer Corp.*

32061 Torlon® 4275
Polyamide-imide engineering resins filed with 20% graphite powder; for bearings, thrust washers, wear pads, strips, piston rings, seals, vanes and valve seats. *Amoco Chemical Co.*

32062 Torlon® 4301
Polyamide-imide engineering resins filled with 12% graphite powder, 3% fluorocarbon; for bearings, thrust washers, wear pads, strips, piston rings, seals, vanes, valves, seats. *Amoco Chemical Co.*

32063 Torlon® 4347
Poly(amide-imide) resin, 12% graphite powd.; wear-resistant resin for reciprocating motion or bearings subject to high loads at low speeds, e.g., for bearings, thrust washers, wear pads, strips, piston rings, seals. *Amoco Chemical Co.*

32064 Torlon® 5030
Poly(amide-imide) resin, 30% glass fiberreinforced; high-strength resin used for burn-in sockets, gears, valve plates, fairings, tube clamps, impellors, rotors, housings, back-up rings, terminal strips, insulators, brackets. *Amoco; Polymer Corp.*

32065 Torlon® 7130
Polyamide-imide engineering resins filled with 30% graphite fiber, 1% fluorocarbon; for metal replacements, housings, mechanical linkages, gears, fasteners, spline linears, cargo rollers, brackets, valves, labyrinth seals, fairings, tube clamps, standoffs *Amoco Chemical Co.*

32066 Torlon® 7330
Poly(amide-imide) resin proprietary blend with carbon fibers and fluorocarbons; high-strength resin used for sliding vanes; potential use for EMI shielding. *Amoco Chemical Co.*

32067 tormentil
The dried rhizome of *Potentilla tormentilla* ; used for tanning and as an astringent.

32068 Tormol
A nickel-chromium-molybdenum steel highly resistant to shock and fatigue. No manufacturer.

32069 Tornac
Hydrogenated nitrile-butadiene rubber. *Bayer plc.*

32070 Tornac B 3850
Hydrogenated (98%) nitrile rubber; nonstaining stabilizer; offers superior resistance to heat, ozone, and nonpolar HC fluids including those containing aggressive additives; vulcanizable by sulfur and peroxides. *Bayer Corp; Polysar.*

32071 Tornado
63-25-2 1831 200-555-0
carbaryl. A suspension concentrate containing 450 g carbaryl per liter; contact insecticide and worm killer. *ICI Agrochemicals Professional Products.*

32072 Tornesit
A trade name for a protective coating base prepared by the chlorination of rubber. No manufacturer.

32073 Tornusil
13597-65-4 10291 237-057-8
O_4SiZn_2
zinc orthosilicate
willemite. A proprietary two-pack moisture cured inorganic zinc silicate primer. Insoluble in H_2O. *Sigma Coatings.* Name unverified.

32074 Toro
15545-48-9 239-592-2
chlorotoluron. Suspension concentrate containing 500 g chlorotoluron per liter; a contact urea herbicide for cereal crops. *Sipcam UK Ltd.*

32075 toron
A sulfur-terpene compound prepared by heating turpentine with sulfur. It is a black viscid liquid or semisolid; used for waterproofing cloth, preparing rubberized cloth, and for attaching or coating metal surfaces with rubber.

32076 Torqseal
Anaerobic thread locking fluid; suitable for thread locking and securing studs and bearings; easily undone with normal hand tools. *Hermetite Products Ltd.* Name unverified.

32077 Torque
13356-08-6 4004 236-407-7
$C_{60}H_{78}OSn_2$
hexakis(2-methyl-2-phenylpropyl)distannoxane
bis[tris(2-methyl-2-phenylpropyl)tin] oxide; Fenbutatin oxide; fenbutatin oxyde; hexakis; ENT 27738; Osadan; SD 14114; Vendex. A non-systemic acaricide with contct and stomach action. Used for control of phytophagous mites in fruit crops. mp = 138-139°; poorly soluble in H_2O (0.005 mg/l), more soluble in organic solvents; LD_{50} (rat orl) = 2631 mg/kg. *ICI Agrochemicals.*

32078 Torrax
Malt flour. *Rhône-Poulenc UK.*

32079 Totablan
Hydrogen peroxide bleaching stabilizer and wetting agent. *Ceca SA.*

32080 Totacol
Herbicide based on diuron and paraquatermary *ICI Chem & Polymers Ltd.*

32081 Totril
1689-83-4 216-881-1
Ioxynil
Contact herbicide for use in onion crops. *Embetec Crop Protection Ltd.; May & Baker Ltd.*

32082 Toucas metal
An alloy of 35.75% copper, 28.56% nickel, 7.1% zinc, 7.2% tin, 7.1% lead, 7.2% antimony, and 7.1% iron.

32083 tough copper
Commercial copper, containing impurities such as arsenic.

32084 Tough Gel
Ion exchange resins. *Dow Chemical.*

32085 toughened caustic
Consists of 95% silver nitrate and 5% potassium nitrate, fused together.

32086 Tournant Oil
A commercial brand of olive oil obtained from fermented marc of expressed olives; contains free fatty acids, and is used as a Turkey-red oil. No manufacturer.

32087 Tournay's metal
An alloy of 82.5% copper and 17.5% zinc.

32088 tous-les-mois starch
Queensland arrowroot. The starch from the rhizomes of *Canna edulis* .

32089 Toval®
Fibers. *DuPont UK.*

32090 Tower Brick
Sodium phosphates, wetting agents, and corrosion inhibitors; used in open

recirculating cooling water systems to prevent lime scale and corrosion form fouling up the system. *Delaware Chemical Corp.* Name unverified.

32091 Tower Treat
Liquid algicide/biocide; used for cooling water treatment. *Scheaefer Technologies Inc.*

32092 Toximul®
Sulfonate/nonionic blend; agricultural blender. *Stepan.*

32093 Toximul® 8240
61791-12-6
PEG-36 castor oil; emulsifier for agricultural formulations. *Stepan; Stepan Canada; Stepan Europe.*

32094 Toximul® 8320
Butyl EO/PO block copolymer; emulsifier component, flowable surfactant, intermediate for pesticides. *Stepan; Stepan Canada.*

32095 Toximul® D
Sulfonate/nonionic blend; matched pair emulsifier with Toximul H-HF for pesticide formulations; dispersant, stabilizer, hydrophobic agent. *Stepan; Stepan Canada; Stepan Europe.*

32096 Toximul® H-HF
Sulfonate/nonionic blend; matched pair emulsifier with Toximul D for pesticides; dispersant, stabilizer, hydrophilic agent. *Stepan; Stepan Canada; Stepan Europe.*

32097 Toximul® SEE-340
PEG-20 sorbitan tritallate; emulsifier for agricultural formulations. *Stepan; Stepan Canada; Stepan Europe.*

32098 Toximul® TA-2
61791-44-4 263-177-5
PEG-2 tallowamine; emulsifier for agricultural formulations. *Stepan; Stepan Canada; Stepan Europe.*

32099 Toximul® 600
Sulfonate/nonionic blend; general purpose emulsifier pesticides. *Stepan; Stepan Canada.*

32100 Toyocat®-DMA
108-01-0 2900 203-542-8
N,N'-Dimethylethanolamine
Catalyst for PU slabstock and molded flexible, semirigid, and rigid applications. *Tosoh.*

32101 Toyocat®-DMCH
N,N'-Dimethylcyclohexylamine
Catalyst for PU rigid applications. *Tosoh.*

32102 Toyocat®-DT
N,N,N',N,N- Pentamethyldiethylenetriamine
Blowing catalyst for PU slabstock and molded flexible and rigid applications. *Tosoh.*

32103 Toyocat®-ET
Bis(dimethylaminoethyl) ether, dipropyleneglycol; blowing catalyst for PU slabstock and molded flexible and rigid applications. *Tosoh.*

32104 Toyocat®-HPW
N-Methyl-N·g-hydroxyethylpiperazine, 10% water; reactive catalyst for PU slabstock and molded flexible, semirigid applications. *Tosoh.*

32105 Toyocat®-MR
N,N,N',N' - Tetramethylhexanediamine
Nonthermosensitive catalyst for PU slabstock and molded flexible, semirigid and rigid applications. *Tosoh.*

32106 Toyocat®-NEM
100-74-3 202-885-0
$C_6H_{13}NO$
N-Ethylmorpholine
Catalyst for PU molded flexible, semirigid, and rigid applications. mp = -63°; bp= 139°; d = 0.9050; n_D^{20}= 1.4410. *Tosoh.*

32107 Toyocat®-NP
N,N',N'-Trimethylaminoethyl piperazine
Thermosensitive catalyst for PU slabstock and molded flexbile, rigid, and elastomer shoe sole applications. *Tosoh.*

32108 Toyocat®-TE
N,N,N',N' - Tetramethylenediamine
Balanced blowing/gelling catalyst for PU molded flexible, semirigid, and rigid applications. *Tosoh.*

32109 Toyocerin®
Probiotic feed additive. *Asahi Chem. Industry.*

32110 TP-35 Solvent
Precision cleaner/defluxer; removes residual moisture from electronic equipment; for critical cleaning of PCBs, components, magnetic tape heads, optical lenses, precision instruments; safe for most plastics. *Chemtronics.*

32111 TPG
Abbreviation for triphenylguanidne.

32112 TPL
1314-60-9 739 215-237-7
Colloidal antimony pentoxide. *Laurel Industries Inc.*

32113 TPS 20
Diterdodecyl trisulfide
EP additive for gear oils, metalworking fluids compatible with copper and silver alloys, and rolling greases. *Elf Atochem SA; Elf Atochem UK Ltd.*

32114 TPS 27
Diternonyl trisulfide
EP additive for gear oils, metalworking fluids compatible. with copper and silver alloys, and rolling greases. *Elf Atochem SA; Elf Atochem UK Ltd.*

32115 TPS 32, 37
Diterdodecyl pentasulfide
EP additive for metalworking fluids; recommended for clear premium EP lubricants; not compatible with copper and silver alloys. *Elf Atochem SA; Elf Atochem UK Ltd.*

32116 TPS 327
Ditertio nonyl pentasulfide
1 *Elf Atochem UK Ltd.*

32117 TPX-80CNI
Polymethylolpentane, 80% carbonyl iron powder. *Compounding Tech.* Name unverified.

32118 Tra-Bond 2151
25928-94-3
Two-part epoxy; thermal conductive electrical insulating compound; used for staking transistors, diodes, resistors, integrated circuits, other heat-sensitive components to printed circuit boards; bonds readily to itself and to metals, silica, alumina, other ceramics, glass, plastics, etc.; provides excellent resistance to salt solutions, mild acids and alkalis, petroleum solvents, lubricating oils and alcohol. *Tra-Con.*

32119 Tra-Bond F113
25928-94-3
Epoxy adhesive; highimpact fiber optic adhesive for bonding opto-electronic lens displays, SMA connectors; excellent glass-glass bonds, superior wicking. *Tra-Con.*

32120 Trabuk
An alloy containing 87.5% tin, 5.5% nickel, 5% antimony, and 2% bismuth. It resists vegetable acids.

32121 Tra-Cast 3103
25928-94-3
Epoxy casting compound; general purpose. *Tra-Con.*

32122 Tracey B
Trace element complex containing boron. *Mandops (UK) Ltd.*

32123 Tracey C
Trace element chelate containing copper. *Mandops (UK) Ltd.*

32124 Tracey M Plus
Trace element chelate containing copper and manganese. *Mandops (UK) Ltd.*

32125 Tracey MG
Trace element chelate containing magnesium. *Mandops (UK) Ltd.*

32126 Tracey SSS
Foliar feed containing sulfur and nitrogen. *Mandops (UK) Ltd.*

32127 Trachine
Fowl laryngotracheitis; for immunization of poultry. *Intervet Inc.*

32128 trachyte
A volcanic rock composed of felspar with some hornblende and mica.

32129 Track
67129-08-2 266-583-0
$C_{14}H_{16}ClN_3O$
2-chloro-N-(2,6-dimethylphenyl)-N-(1H-pyrazol-1-ylmethyl)acetamide
2-chloro-N-(pyrazol-1-ylmethyl)acet-2',6'-xylidide; metazachlor; BAS 47900H; Butisan S. Suspension concentrate containing 500 g/l metazachlor; for weed control in brassicas and ornamental crops. Metazachlor is a selective herbicide, absorbed by the hypocityls and roots and inhibiting germination. mp = 85°; soluble in H_2O (17 mg/l), freely soluble in organic solvents; LD_{50} (rat orl) = 2150 mg/kg. *Kommer-Brookwick Ltd.*

32130 Tracker
1918-00-9 3090 217-635-6
Dicamba
Herbicide used to control bracken. *Shell UK.*

32131 Tra-Duct 2902
25928-94-3
Silver paste epoxy; conductive silver paste epoxy adhesive; maximum use temp. to 110°C. *Tra-Con.*

32132 Traffaid 30 B
Barium/calcium lignosulfonate
Antiscumming agent in brick and tile manufacturing. *Borregaard Ligno Tech.*

32133 tragasol
A gum obtained by steeping locust-bean kernels in water; used as a binding material.

32134 Tramisol®
16595-80-5 5486 240-654-6
$C_{11}H_{12}N_2SCl$
(S)-2,3,5,6-tetrahydro-6-phenylimidazo[2,1-b]thiazole hydrochloride
Levamisole hydrochloride; R-12564; Ascaridil; Decaris; Ergamisol; Levacide; Levadin; Levasole; Meglum; Nemicide; Nilverm; Ripercol; Solaskil; Spartakon. Veterinary anthelmintic. mp = 227-229°; $[\alpha]_D^{20}$= -124° (c= 0.9 H_2O); soluble in H_2O. *Am. Cyanamid.*

32135 Trans Gard
Blend of petroleum additives and petroleum distillates; automatic transmission stop leak and fluid conditioner. *Gard Corporation.* Name unverified.

32136 Transclene
Transformer dielectric solution. *Occidental Chemical corp.*

32137 Transcutol
111-90-0 1847 203-919-7
$C_6H_{14}O_3$
2-(2-ethoxyethoxy)ethanol
Ethyldiglycol; carbitol; ethyldigol; diethylene glycol monomethyl ether. Solvent for active ingredients in pharmaceutical preparations; cosurfactant for microemulsions. bp = 196°; d_{20}^{20} = 1.0273; n_D^{20} = 1.4273; miscible with H_2O, organic solvents; LD_{50} (rat orl) = 11 g/kg. *Gattefosse SA.*

32138 Transjojoba
Jojoba oil derivatives. *A & E connock (Perfumery & Cosmetics) Ltd.*

32139 Translink®
1332-58-7 5294 296-473-8
Fine particle size surface treated calcined kaolin (aluminum silicate); used as a reinforcer and extender for rubber, wire and cable, plastics, gel coats. *Engelhard.*

32140 Translink® 37
Calcined aluminum silicate with vinyl functional surface treatment; reinforcing extender; used in crosslinked polyethylene and the ethylene-propylene rubbers used in electrical compounds. *Engelhard.*

32141 Translink® 555
Calcined aluminum silicate with aminosilane surface treatment; reinforcing extender. *Engelhard.*

32142 Transol
Transformer oils. *Carless Refining & Marketing Ltd.*

32143 Transoxide
1309-37-1 4072 215-168-2
Transparent iron oxides. *PMC Specialities International Ltd.*

32144 Transpafill
Precipitated aluminum silicate; nonreactive transparent filler for printing inks. *Degussa; Degussa AG.*

32145 Transpalene®
9003-07-0 7741
Polypropylene; for high clarity and gloss; for thermoforming, injection and blow molding. *Neste.*

32146 transpar
Lactic acid and buffered lactic acid mixtures.

32147 Transpex
Acrylic sheet used for double glazing. *Imperial Chemical Industries plc.*

32148 Traseolide
5-acetyl-3-isopropyl-1,12,6-tetramethylindane
Quest Int'l. UK Ltd.

32149 Tra-Shield 2867
25928-94-3
Silver-filled solvent borne epoxy coating; conductive coating for EMI shielding, EMI gasket flange coating, surface ground, antistatic charge dissipation, and corona shielding applications. *Tra-Con.*

32150 trass
A volcanic material found on the bank of the Rhine; used in Holland as an addition to lime, to convert it into hydraulic lime.

32151 Trastan LS
8061-52-7
Low sugar calcium lignosulfonate. *Actrachem.* Discontinued.

32152 Traton®
For agriculture. *DuPont UK.*

32153 Travase
Sutilains. Enzyme. *The Boots Co plc.* Discontinued.

32154 travertine
Calc sinter, calcareous tufa. A limestone deposited by calcareous springs.

32155 Traytuf Ultra-Clear
25038-59-9 7730
PET polyester; for high purity, high strength packaging with optimum transparency. *Goodyear.*

32156 Tread-Brite
Embossed aluminum plate. *Reynolds Metal Co.*

32157 Treadfast®
Various PVA, SBR, acrylic, epoxy, polyurethane formulations; flooring adhesives for PVC, rubber, carpets, etc. *Tremco Ltd.*

32158 treble superphosphate
Monocalcium phosphate, containing 48-49% P_2O_5 (41-42% water soluble P_2O_5); used as a fertilizer.

32159 Tree Bug-Lok Adhesive
Polyisobutylene
For the control of Gypsy moth caterpillars; nontoxic, ecologically safe, holds forever, traps any crawling insect including ants and cankerworms. *TACC Int'l.* Name unverified.

32160 tree copal
A name applied to white Zanzibar copal.

32161 tree gum
$C_6H_{10}O_5$
Wood gum; xylan.

32162 Treflan
1582-09-8 9815 216-428-8
$C_{13}H_{16}F_3N_3O_4$
2,6-dinitro-N,N-dipropyl-4-(trifluoromethyl)benzeneamine
trifluralin; trifluraline; Brassix, Digermin; Elancolan; Heritage; Ipersan; L-36352; Olitref; Proflan; Prolan; Sinflouran; Tarene; Tri-4; Triflurex; Trigard; Trimaran; Tristar; Zeltoxone. Emulsifiable concentrate containing 480 g/l trifluralin; a dinitroaniline herbicide to control annual weeds and grasses. mp = 48-49°; $bp_{4.2}$ = 139-140°; insoluble in H_2O, soluble in organic solvents; LD_{50} (rat orl) > 10000 mg/kg. *DowElanco Ltd.*

32163 Trefsin®
Thermoplastic elastomers. *Exxon UK.*

32164 Trefsin® 3201-50, 3201-60
Thermoplastic elastomer; lowgas permeability rubber for medical, industrial and consumer goods; general purpose grade. *Advanced Elastomer Systems.* Name unverified.

32165 Tre-Hold
2122-70-5 218-332-1
$C_{14}H_{14}O_2$
ethyl-1-naphthyl acetate
ethyl 2-(1-naphthyl) acetate; ethyl 1-naphthalene acetate; Tipoff. Plant growth regulator. Used to inhibit sprouting at pruning points. bp_3 = 158-160°; d_{25} = 1.106; insoluble in H_2O, soluble in organic solvents; LD_{50} (rat orl) = 3850 mg/kg. *A H Marks & Co Ltd.* Name unverified.

32166 Trelit
A proprietary pyroxylin plastic. No manufacturer.

32167 Trem-LF-40
Sodium alkylallyl sulfosuccinate
Polymerizable surfactant, solubilizer; provides low foaming emulsions with improved water resistance. *Henkel.*

32168 Tremvac
Calnek strain avian encephalomyelitis; for immunization of poultry. *Intervet Inc.*

32169 Tremvac-FP
Calnek strain avian encephalomyelitis-fowl pox; for immunization of poultry. *Intervet Inc.*

32170 Trenbest 500
Trimethyl hydroxyethyl ammonium ester of a carboxyglycerol phosphoric acid; mold release agent for the plastic and caoutchouc industry; used for molding and extrusion powds., laminated plastics, injection molding powds., ABS. *Lucas Meyer.*

32171 Trenbolone acetate
10161-34-9 9716 233-432-5
$C_{20}H_{24}O_3$
(17β)-17-acetyloxyestra-4,9,11-trien-3-one
Finaplix; 17β-Acetoxy-3-oxoestra-4,9,11-triene. Anabolic (veterinary). Crystals; mp=96-97°. *Roussel UCLAF, Fine Chemicals.* Name unverified.

32172 Trenchmaster
Expanded polystyrene foundation formwork system for building sites. *Vencel Resil Ltd.* Discontinued.

32173 Trench's flameless explosive
An explosive containing ammonium nitrate.

32174 Trend®
For agriculture. *DuPont UK.*

32175 Trenyline
Triphenylguanidine

A proprietary trade name for a rubber vulcanization accelerator. No manufacturer.

32176 Tretobond®
Various building, laminating and structural adhesives based on PVA, neoprene, SBR, epoxy, polyurethane, etc; used for building, laminating, and structural bonding. *Tremco Ltd.*

32177 Tret-o-Lite
A patented preparation for the destruction of petroleum emulsions; it consists of 83% sodium oleate, 5.5% sodium resinate, 5.0% sodium silicate, 4.0% phenol, and 1.5% paraffin wax. No manufacturer.

32178 Trevira
Polyester fiber. *Hoechst UK.*

32179 TRH-Roche
Proprietary preparation of protirelin; thyroid function test. *Roche Products Ltd.* Name unverified.

32180 Tri
An abbreviation for trichloroethylene, C_2HCl_3; used as a solvent.

32181 Triabon® 16-8-12-4
Complex fertilizer for substrates regardless of their pH, and greenhouse crops. *BASF AG.*

32182 triacetin
102-76-1 9721 203-051-9
$C_9H_{14}O_6$
glyceryl triacetate
1,2,3-propanetriol triacetate; triacetyl glycerol; triacetyl glycerin; Enzactin; Fungacetin. Triester of glycerin and acetic acid; plasticizer; fixative in perfumery; manufacture of cosmetics; specialty solvents, manufacture of celluloid, photographic films; topical antifungal. mp = -78°; bp = 258-260°; d_4^{25} = 1.1562; n_D^{20} = 1.4307; miscible with organic solvents; LD_{50} (mus iv) = 1600 mg/kg. *Bayer AG; Eastman; MTM Spec. Chem. Ltd; Penta Mfg.; Spectrum Chem. Mfg.; Unichema.*

32183 Tri-Ad
Gasoline additive. *Monsanto (Solaris).* Name unverified.

32184 Triadimefon
43121-43-3 9723 256-103-8
$C_{14}H_{16}ClN_3O_2$
1-(4-chlorophenoxy)-3,3-dimethyl-1-(1H-1,2,4-triazol-1-yl) butan-2-one
tridemifone; Amiral; Bay MEB 6447; Bayleton. Systemic fungicide with protective and curative action. Used for control of powdery mildews in cereals, fruit and vegetables. mp = 82°; d_{20} = 1.22; soluble in H_2O (260 mg/l), more soluble in organic solvents; LD_{50} (rat orl) = 568 mg/kg.

32185 triadimenol
55219-65-3 9724 259-537-6
$C_{14}H_{18}ClN_3O_2$
β-(4-chlorophenoxy)-α-(1,1-dimethylethyl)-1H-1,2,4-Triazole-1-ethanol
Baytan; Bayfidan; Summit; Spinnaker. Agricultural fungicide. Inhibits ergosterol biosynthesis in fungi.

32186 Triadine® 3
Hexahydro-1,3,5-tris (2-hydroxyethyl)-s-triazine
Bactericide for metalworking fluids. *Olin.*

32187 Triadine® 10
Sodium pyrithione (6.4%) and hexahydro 1,3,5-tris (2-hydroxyethyl)-*s*-triazine (63.6%); broad spectrum microbiostat effective against gram-positive and gram-negative bacteria and fungi; used in aq.-based metalworking fluids. *Olin.*

32188 Triafol
9012-09-3
Cellulose triacetate film; particularly suitable for coil insulation. *Bayer AG.*

32189 Triagran®
Bentazon, dichorprop, MCPA; post-emergence herbicide for control of broad-leaved weeds in winter and spring cereals. *BASF AG.*

32190 triallate
2303-17-5 9726 218-962-7
$C_{10}H_{16}Cl_3NOS$
S-(2,3,3-trichloro-2-propenyl) bis(1-methylethyl)carbamothioate
CP 23426; Avadex BW; Avadex BE; Far-Go; Showdown. Selective herbicide used for control of wild oats and some annual grasses in cereal and vegetable crops. mp = 29-30°; bp_{40} = 117°; d = 1.273; slightly soluble in H_2O (4 mg/l), more soluble in organic solvents; LD_{50} (rat orl) = 1100 mg/kg.

32191 triallyl cyanurate
101-37-1 202-936-7
$C_{12}H_{15}N_3O_3$
1,3,5-Triazine, 2,4,6-tris(2-propenyloxy)-
2,4,6-Triallyloxy-1,3,5-triazine. Polymers as monomers and modifier; organic intermediate. *Akzo; Am. Cyanamid; Degussa; Nat'l. Starch & Chem.*

32192 triallyl isocyanurate
1025-15-6 213-834-7
$C_{12}H_{15}N_3O_3$
Triallyl-s-triazine-2,4,6(1H,3H,5H)-trione
1,3,5-triallylisocyanurate; 1,3,5-triallylisocyanuric acid; 1,3,5-tri-2-propenyl-1,3,5-triazine-2,4,6(1H,3H,5H)-trione; isocyanuric acid triallyl ester; 1,3,5-triallyl-s-triazine-2,4,6(1H,3H,5H)-trione; DIAK 7; TAIC. Co-agent improving peroxide-induced crosslinking of rubber. Stabilized with 100 ppm BHT. bp_4 = 149-152°; n_D^{20} = 1.5129. *Akzo.*

32193 Triameen T
61791-57-9 263-191-1
N-Tallowalkyl dipropylene triamine
Industrial surfactant. *Akzo.*

32194 triaryl phosphate
68937-41-7 273-066-3
Triisopropylated Phenyl Phosphate
Durad 110 hydraulic fluid; Phenol, isopropylated, phosphate (3:1); Kronitex 50 triaryl phosphate. Flame retardant plasticizer for PVC; compatibilizing agent; intermediate; catalyst carrier and pigment vehicle for polyurethane; processing aid in rubber belting and mech. goods. *Akzo; Ashland; FMC; Monsanto.*

32195 triaryl phosphate
68937-41-7 273-066-3
Fyrquel® EHC; Kronitex® 25, 50, 100, 200, 1840. Fire-resist. electro-hydraulic control fluid. *Akzo; FMC.*

32196 triazophos
24017-47-8 9736 245-986-5
$C_{12}H_{16}N_3O_3PS$
O,O-diethyl O-(1-phenyl-1H-1,2,4-triazol-3-yl) phosphorothioate
HOE 2960; Hostathion. Broad spectrum insecticide and acaricide. Used for control of aphids and insects in a wide variety of crops. mp = 2-5°; bp dec; d^{20} = 1.247; n_D^{20} = 1.5501; soluble in H_2O (30 mg/l), more soluble in organic solvents; LD_{50} (rat orl) = 57 mg/kg.

32197 Tribase
Intermediate for production of dyes. *BASF AG.*

32198 Tribonol
Powder coating for direct application to sand molds and cores. *Foseco (F.S.) Ltd.*

32199 Tri-Borne
Paint dip. *ICI Chem & Polymers Ltd.* Discontinued.

32200 Tribovax
Combined cattle vaccine. *The Wellcome Foundation Ltd.* Name unverified.

32201 tribromoneopentyl alcohol
36483-57-5 253-057-0
$C_5H_9Br_3O$
2,2-Bis(bromomethyl)-3-bromo-1-propanol
TBNPA; FR-513;. Flame retardant for flexible and rigid PU; flame retardant intermediate. White to off white flakes; soluble in alchols; mw=324.92; density=2.28; mp=62-67°; LD_{50}=rat oral 2823 mg/kg; irritant to eyes, mild irritant to skin. *AmeriHaas; Dead Sea Bromine.*

32202 2,4,6-tribromophenol
118-79-6 9744 204-278-6
$C_6H_3Br_3O$
TBP; Tribromophenol; Bromol; Emery® 9332; FR-613; Great Lakes PH-73™;. Reactive flame retardant used mainly as an intermediate for polymeric flame retardants. Soft white needles, sweet taste, bromine odor; soluble in alcohol, chloroform, ether, caustic alkiline solutions; almost insoluble in water; mw=330.83; density=2.55; mp=96°; bp=244°; LD_{50}=rat oral>5000 mg/kg; hazard by ingestion, inhalation, skin. *AmeriHaas; Aldrich; Ameribrom; Dead Sea Bromine; Fluka; Great Lakes; Morre-Tec Ind.*

32203 tribromophenyl allyl ether
3278-89-5 221-913-2
$C_9H_7Br_3O$
Allyl 2,4,6-Trobromophenyl ether
TBP-AE; FR-913. Aromatic flame retardant for expandable PS; synergist with hexabromocyclododecane. White to off-white crystal powder; mw=370.88; sp gr=2.20; mp=74-76°; LD_{50}=rat oral>5000mg/kg; rabbit dermal>2000 mg/kg. *AmeriHaas; AmeriBrom; Dead Sea Bromine.*

32204 Tribunil®
18691-97-9 6002 242-505-0
Methabenzthiazuron
Broadspectrum herbicide for pre- and post-emergence application. *Bayer AG; Bayer plc.*

32205 Tribute
Mixture of dicamba, MCPA and mecoprop; used for weed control in cereals and grassland. *Chipman Ltd.*

32206 tributoxyethyl phosphate
78-51-3 201-122-9
$C_{18}H_{39}O_7P$
Tris(2-butoxyethyl) phosphate
Tributyl cellosolve phosphate; tributoxyethyl phosphate; KP-140; TBEP;

tris(2-butoxyethyl)ester phosphoric acid; 2-butoxy-ethanol phosphate (3:1;)TBEP; ethanol, 2-butoxyphosphate (3:1). Primary plasticizer for most resins and elastomers; floor finishes and waxes; flame-retarding agent; latex paints; as defoamer. bp_4 = 215-228°; d = 1.0060; n_D^{20} = 1.4359. *Akzo; Albright & Wilson; FMC.*

32207 tributyl phosphate
126-73-8 9749 204-800-2
$C_{12}H_{27}O_4P$
tri-n-butyl phosphate
TBP; phosphoric acid tributyl ester; Phosphoric acid, tri-n-butyl ester; tri-n-butyl phosphate; Butyl phosphate; Phosphoric acid tributyl ester; celluphos 4. Heat-exchange medium; solvent extraction of metal ions; plasticizer for cellulose esters, lacquers, plastics, vinyl resins; antifoam agent; dielectric. mp = -79°; bp_{22} = 180-183°; d = 0.9790; n_D^{20} = 1.4245. *Akzo; Chemron; FMC.*

32208 tributyl phosphite
102-85-2 203-061-3
$C_{12}H_{27}O_3P$
tri-n-butyl phosphite
Phosphorus derivative. Additive for greases and extreme pressure lubricants; stabilizer for fuel oils and polyamides; gasoline additive. *Albright & Wilson; Janssen Chimica.*

32209 tributyltin oxide
56-35-9 200-268-0
$C_{24}H_{54}OSn_2$
hexabutyldistannoxane
HBD; TBTO; bis(tributyltin) oxide. Bactericide, fungicide, chemical intermediate. mp = -45°; bp_2 = 170-180°; d = 1.172; n_D^{20} = 1.4860. *Akzo; Elf Atochem; KMZ Chem. Ltd.*

32210 Tricap
Triethyleneglycol dicaprylate and caprate. *Croda Chem. Ltd.* Name unverified.

32211 tricaprylin
538-23-8 208-686-5
Emalex O.T.G; Glyceryl trioctanoate; Captex® 8000; Miglyol® 808. Oil-phase cosmetic ingred.; emollient. *Nihon Emulsion.*

32212 trichlorfon
52-68-6 9753 200-149-3
$C_4H_8Cl_3O_4P$
dimethyl (2,2,2-trichloro-1-hydroxyethyl)phosphonate
trichlorphon; DEP; metriphonate; chlorofos; dipterex; OMS 800; ENT 19763; Bay 15922; Bay L13/59; Briten; Cekufon; Danex; Denkaphon; Dep; Ditrifon; Dylox; Ertefon; Nevugon; Proxol; Totalene; Tugon; Zeltivar. Non-systemic insecticide with stomach and contact action. Used for insct control in agriculture and horticulture. mp = 75-79°; $bp_{0.1}$ = 100°; SG = 1.73; n_D^{20}= 1.3439; soluble in H_2O (120 g/l), more soluble in organic solvents; LD_{50} (rat orl) = 560 mg/kg.

32213 trichloroacetic acid
76-03-9 9756 200-927-2
$C_2HCl_3O_2$
TCA; NaTa;. Used for control of weeds in field crops. Also used as a decalcifier and fixative in microscopy and a protein precipitant. mp= 54-58°; bp= 196-197°; d = 1.629; soluble in H_2O (10 g/ml), organic solvents; LD_{50} (rat orl) = 5000 mg/kg.

32214 1,2,4-trichlorobenzene
120-82-1 9760 204-428-0
$C_6H_3Cl_3$
1,2,4-trichlorobenzene
unsym-trichlorobenzene. Solvent in chemical manufacture, dyes, intermediates, dielectric fluid, synthetic transformer oils, lubricants, heat-transfer media, insecticides. mp = 15-16°; bp = 213-214°; d = 1.454; n_D^{20} = 1.5710; insoluble in H_2O, soluble in organic solvents. *Ashland; Stan Chem Int'l. Ltd.*

32215 1,1,1-trichloroethane
71-55-6 9766 200-756-3
$C_2H_3Cl_3$
trichloroethane

32216 trichloroethane
71-55-6 9766 200-756-3
$C_2H_3Cl_3$
1,1,1-trichloroethane
ethane, 1,1,1-trichloro; methylchloroform; chlorothene. Halogenated aliphatic hydrocarbon; solvent for cleaning precision instruments, metal degreasing, pesticide, textile processing. mp = -32°; bp= 74°; d_4^{20}= 1.3376; n_D^{20}= 1.4384; insoluble in H_2O, soluble in organic solvents. Asahi Chem Industry Co Ltd; Asahi-Penn; Ashland; Elf Atochem N.Am.; Chemoxy Int'l. plc; PPG Industries; Vulcan.

32217 trichloroethylene
79-01-6 9769 201-167-4
C_2HCl_3
trichloroethene
ethinyl trichloride; Tri-Clene; Trilene; Trichloren; Algylen; Trimar; Triline; Trethylene; Westrosol; Chlorylen; Gemalgene; Germalgene. Metal degreasing; extraction solvent for oils, fats, waxes; solvent for cellulose esters and ethers; drycleaning; solvent dyeing; manufacture of organic chemicals, pharmaceuticals. mp = -85°; bp = 87°; d_4^{25}= 1.4559; n_D^{20}= 1.4775; slightly soluble in H_2O (1.1 g/l), soluble in organic solvents; LD_{50} (rat orl)= 4.92 ml/kg. *Ashland; Asahi-Penn; Elf Atochem N.Am.; General Chem.; Nickerson Chem. Ltd; PPG Industries.*

32218 trichloroisocyanuric acid
87-90-1 9188 201-782-8
$C_3Cl_3N_3O_3$
isocyanuric chloride
trichlorocyanuric acid; trichloroiminocyanuric acid; ACL-85; Chloreal; N,N',N-trichloroisocyanuric acid; symclosene; 1,3,5-trichloro-s-triazine-2,4,6-trione. Bleaching agent, dishwashing compounds, disinfectant in swimming pools, bactericide, algicide, deodorant; active ingredient in household cleaners (dry). mp = 246-247° (dec). *Allchem Industries; Chlor Chem Ltd; Monsanto; Nissan Chem. Ind.; 3-V.*

32219 N-trichloromethylthiotetrahydrophthalimide

32220 2,3,6-trichlorophenol
933-75-5 213-271-7
$C_6H_3Cl_3O$
2,3,6-trichlorophenol
Used as a disinfectant and fungicide. mp = 55-57°; bp = 253°; $bp_{0.4}$ = 88°; LD_{50} (rat ip) = 308 mg.kg.

32221 2,4,5-trichlorophenol
95-95-4 9772 202-467-8
$C_6H_3Cl_3O$
2,4,5-trichlorophenol
Collunosol; Dowicide 2; Preventol I. Used as a fungicide and bactericide. Sodium salt is Dowicide B, soluble in H_2O and organic solvents. mp = 63-65°; bp = 253°; SG = 1.5; insoluble in H_2O (< 1 mg/ml), soluble in organic solvents; LD_{50} (rat orl) = 820 mg/kg.

32222 2,4,6-trichlorophenol
88-06-2 9773 201-795-9
$C_6H_3Cl_3O$
2,4,6-trichlorophenol
Dowcide 2S; TCP; 2,4,6 T; Omal; Phenachlor; Dowicide 2S; Trichloro-2-hydroxybenzene. Used as a fungicide, bactericide and preservative, Commonly used as disinfectant and topical anti-infective. mp = 64-66°; bp = 246°; poorly soluble in H_2O (< 1 mg/ml), soluble in organic solvents.

32223 trichlorosilane
10025-78-2 9776 233-042-5
Cl_3HSi
trichloromonosilane
silicochloroform. Organic synthesis; purification of silicon; intermediate. mp = -127°; bp = 32°; d_4^{20}= 1.3417; n_D^{20}= 1.4020; soluble in organic solvents; LD_{50} (rat orl) = 1.03 g/kg. *Chisso Am.; Hüls Am.; PCR.*

32224 trichlorotrifluoroethane
76-13-1 200-936-1
$ClCF_2CCl_2F$
1,1,2-trichloro-1,2,2-trifluoroethane
Refrigerant; dry-cleaning solvent; fire extinguishers; manufacture of chlorotrifluoroethylene; blowing agent; polymer intermediate; solvent drying; drying electronic parts and precision equipment. *Air Prods & Chem; AlliedSignal; Elf Atochem N. Am.; PCR.*

32225 Trichrome
Trivalent chromium plating process. *Engelhard Technologies Ltd.*

32226 Triclene
79-01-6 9769 201-167-4
A proprietary trade name for trichlorethylene used in dry- cleaning. *Occidental Chemical Corp.*

32227 Triclopyr
69633-04-1
Garlon; Garlon 2; Garlon 4. Herbicides based primarily on triclopyr. *Dow Cheml Co Ltd, UK & Ireland.* Discontinued.

32228 Tricoid
Cement used with cine film. *May & Baker Ltd.* Name unverified.

32229 tricresol
1319-77-3 2645 215-293-2
A purified mixture of the three cresols. It contains about 35% ortho, 20% meta, and 25% para-cresol.

32230 tricresyl phosphate
78-30-8 9893 215-548-8
$(CH_3C_6H_4O)_3PO$
TCP; phosphoric acid, tris(methylphenyl) ester; tritolyl phosphate; Plasticizer,

fire retardant for plastics, air filter medium, waterproofing, additive to extreme pressure lubricants. *Akzo; Chemron; Daihachi Chem. Ind.; FMC.*

32231 Trideceth phosphate series
9046-01-9
Lubrhophos® LS-500; Crodafos T2 Acid; Rhodafac® RS-610. Phosphate acid ester of ethoxylated tridecanol; aliphatic hydrophobic base; lubricant, emulsifier for oil and water-based cutting fluids, hydraulic fluids, rolling oils. *Rhône-Poulenc Surf.; Rhône-Poulenc France; Croda;.*

32232 Trideceth series
24938-91-8
PEG-6 tridecyl alcohol
Ethal TDA-6, TDA-3, TDA-18; Bio-Soft® TD 400; Chemal TDA-3; Genapol® X-040; Hetoxol TD-3; Hodag Nonionic TD-15; Iconol TDA-3, TDA-6, TDA-8, TDA-9, TDA-10; Lipocol TD-3; Macol® TD-3; Prox-onic TD-1/03; Rhodasurf® BC-420, T-95, TDA-6; Teric 13A5; Trycol® 5874, 5940; Volpo T-3,. Wetting agent, detergent, foamer, dispersant, emulsifier. *Ethox; Lipo;.*

32233 tridecyl alcohol
112-70-9 203-998-8
$C_{13}H_{28}O$
Tridecan-1-ol
Tridecanol; C13 linear primary alcohol; tridecyl alcohol;. Synthetic alcohol; commercial mixture of isomers; esters for synthetic lubricants, detergents, antifoam agents, other tridecyl compounds; perfumery. mp = 34°; bp = 250-270°; d = 0.822 *Allchem Industries; Penta Mfg.*

32234 tridecyl phosphite
$C_{30}H_{63}O_3P$
Chemical intermediate, stabilizer for polyvinyl and polyolefin resins. *GE Specialty; Stave.*

32235 tridecyl trimellitate
70225-05-7
Tridecyl trimellitate
Liponate TDTM. Nontacky emollient for treatment products, hair. *Lipo.*

32236 tridecylbenzene sulfonic acid
25496-01-9 247-036-5
Tridecylbenzene sulfonic acid
LABS 100/H.V.; Nansa® TDB. Higher viscosity detergent intermediate. *Zohar Detergent Factory.*

32237 trieline
79-01-6 9769 201-167-4
A term for trichloroethylene, C_2HCl_3; used as an extraction solvent for oils, fats, waxes, drycleaning, diluent in paints and adhesives, chemical intermediate.

32238 trietazine
1912-26-1 9797 217-618-3
$C_9H_{16}ClN_5$
6-chloro-N,N,N'-triethyl-1,3,5-triazine-2,4-diamine
2-chloro-4-diethylamino-6-ethylamino-*s*-triazine; 2-ethylamino; -4-diethyl amino-*s*-triazine; G-27901; NC-1667; Aventox; Gesafloc. Herbicide. mp = 100-102°; soluble in H_2O (20 mg/l), more soluble in organic solvents; LD_{50} (rat orl) = 1750 mg/kg. *Schering Agrochemicals Ltd.* Discontinued.

32239 Tri-Ethane
71-55-6 9766 200-756-3
Stabilized 1,1,1-trichloroethane. *PPG Industries.*

32240 triethanolamine
102-71-6 9798 203-049-8
$C_6H_{15}NO_3$
2,2',2-nitrilotris(ethanol)
trolamine; tris(-2-hydroxyethyl)amine; trihydroxytriethylamine; Alkanolamine; triethylolamine; TEA. Intermediate in manufacture of surfactants, textile specialties, waxes, polishes, toiletries, cutting oils, fatty acid soaps (drycleaning), cosmetics, household detergents, emulsions; solvent for casein, shellac, dyes; dispersion agent, water repellent; also used as an analgesic. mp = 22°; bp= 335°; d_4^{20}=1.1242; n_D^{20} = 1.4852; soluble in H_2O, polar organic solvents. *Aldrich; Hüls AG; OxyChem; Schweizerhall; Texaco; Union Carbide.*

32241 triethanolamine hydrochloride
637-39-8 9798 211-284-2
$C_6H_{15}NO_3 \cdot HCl$
2,2',2-nitrilotris(ethanol) hydrochloride
mp = 171° *Arol Chem. Prods.; Janssen Chimica; Spectrum Chem. Mfg.; U.S. Biochemical.*

32242 triethyl citrate
77-93-0 201-070-7
Hydagen® C.A.T; Citroflex 2; 2-Hydroxy-1,2,3-propanetricarboxylic acid, triethyl ester; Ethyl citrate; Crodamol TC; Uniflex TEC. Nonmicrobiocidal deodorant active agent. *Henkel/Cospha; Henkel Canada; Pfizer International; Croda Surf. Ltd; Universal Preserv-A-Chem.*

32243 triethyl phosphate
78-40-0 9806 201-114-5
$C_6H_{15}O_4P$
triethyl phosphate
o-phosphoric acid triethyl ester; ethyl phosphate; TEP. Intermediate for agricultural insecticides; in floor polishes, unsaturated polyesters, lubricants; as plasticizer, solvent, and catalyst. bp = 215-216°; d^{19} = 1.0725; n_D^{17} = 1.4067; soluble in H_2O, EtOH, Et_2O. *Eastman; Miles.*

32244 triethyl phosphite
122-52-1 204-552-5
$C_6H_{15}O_3P$
phosphorous acid triethyl ester
Synthesis, plasticizer, stabilizers, lube and grease additives. bp = 155-157°; d = 0.958; n_D^{20}= 1.4130. *Akzo; Albright & Wilson; ICI Am.; Janssen Chimical.*

32245 triethylenetetramine
112-24-3 9796 203-950-6
$NH_2(C_2H_4NH)_2C_2H_4NH_2$
N,N'-bis(2-aminoethyl)-1,2-ethanediamine
trien; trientine. TETA; detergents and softening agents, synthesis of dyestuffs, pharmaceuticals, rubber accelerator; as thermosetting resin; epoxy curing agent; lubricating oil additive; analytical reagent for Cu, Ni. *Rit-Chem; Texaco; Tosoh; Union Carbide.*

32246 Trifarmon
Suspension concentrate containing 160 g linuron [330-55-2] and 320 g trifluralin [1582-09-8] per liter; herbicide for winter cereals. *Farm Protection Ltd.*

32247 Tri-Farmon
Herbicide containing linuron [330-55-2] and trifluralin [1582-09-8]. *ICI Chem & Polymers Ltd.*

32248 triflumuron
64628-44-0 9809 264-980-3
$C_{15}H_{10}ClF_3N_2O_3$
2-chloro-N-[[[4-(trifluoromethoxy)phenyl]amino]carbonyl]benzamide
Alsystin; SIR 8514; Mascot; Trifluron. Chitin synthesis inhibitor, used for control of Lepidoptera, Psyllidae, Dipter and Coleoptera on fruit, soy beans, forest trees and cotton. mp = 195°; insoluble in H_2O (0.25 mg/l), soluble in organic solvents; LD_{50} (rat orl) > 5000 mg/kg.

32249 trifluoroacetic acid
76-05-1 9812 200-929-3
$C_2HF_3O_2$
trifluoroacetic acid
perfluoroacetic acid. Strong nonoxidizing acid, laboratory reagent, solvent, catalyst. mp = -15°; bp = 72°; d^{20} = 1.5351; soluble in H_2O, organic solvents; LD_{50} (mus iv) = 1200 mg/kg. *Aldrich Ltd; AlliedSignal; Atomergic Chemetals; Janssen Chimica; PCR; Solvay GmbH.*

32250 2,2,2-trifluoroethanol
75-89-8 200-913-6
$C_2H_3F_3O$
trifluoroethyl alcohol
TFE. mp = -45°; bp = 77-80°; d = 1.3930; n_D^{20} = 1.2907. *Aldrich; Schweizerhall; Solvay GmbH.*

32251 trifluoromethanesulfonic acid
1493-13-6 216-087-5
CHF_3O_3S
trifluoromethanesulfonic acid
triflic acid. Catalyst or reactant; polymerization of epoxies, styrenes, and THF, alkylation and some acylation reactions; pharmaceuticals, explosives, dyes, and intermediates; electrolytes; formation of biaryls; polymerization reactions. mp = -40°; bp = 162°; d = 1.6960; n_D^{20}= 1.3270. *Schweizerhall.*

32252 Trifluralin
1582-09-8 9815 216-428-8
$C_{13}H_{16}F_3N_3O_4$
Benzeneamine, 2,6-dinitro-N,N-dipropyl-4-(trifluoromethyl)-
p-Toluidine, α,α,α-trifluoro2,6-dinitro-N,N-dipropyl-; Agreflan; Agriflan 24; Crisalin; Digermin; Elancolan; ENT 28203; L-36352; N,N-Di-n-propyl-2,6-dinitro-4-trifluoromethylaniline; N,N-Dipropyl-4-trifluoromethyl-2,6-dinitroaniline; Nitran; Nitrofor; NCI C00443; Olitref; Trefanocide; Treficon; Treflam; Treflan; Treflanocide; Triflan; Trifluoralin; Trifluraline; Trifurex; Trikepin; TRIM; Brassix; Heritage; Ipersan; Prolan; Sinfluoran; Tarene; Tri-4; Triflurex; Trigard; Trimaran; Tristar; Zeltoxone,. Selective soil herbicide. Used for pre-emergence control of annual grasses and broad-leaved weeds. mp = 48.5-49°; $bp_{4.2}$ = 139-140°; insoluble in H_2O (, 1 mg/l), soluble in organic solvents; LD_{50} (rat orl) = 1930 mg/kg *Agrimont; DowElanco; Shell, ICI; Am. Cyanamid; Agan Chemical Manufacturers Ltd.*

32253 Trifolex-Tra
Soluble concentrate containing 34 g MCPA and 216 g MCPB per liter; for control of weeds in undersown cereals and grassland. *Shell UK.*

32254 triforine
26644-46-2 9819 247-872-0
$C_{10}H_{14}Cl_6N_4O_2$
N,N'-[1,4-piperazinediylbis(2,2,2-trichloroethylidene)]bisformamide
Funginex; Ortho Rose Pride Funginex Rose and Shrub Disease Control; Saprol; Denarin; Cela W-524; biformylchlorazin; CA 70203; CA 73021; CELA 50; CW 524; Piperazinediylbis(2,2,2-trichloroethylidine)bis(formamide); Piperazinediylbis(α,α,α-trichloroethylidene; bis (formamide). Fungicide. mp = 155°; slightly soluble in H_2O (28 mg/l), insoluble in organic solvents. *Synchemicals Ltd.*

32255 Trigard
Copper corrosion inhibitors. *Ciba plc.* Name unverified.

32256 Trigard
1582-09-8 9815 216-428-8
Emulsifiable concentrate containing 480 g/liter trifluralin; a residual herbicide for soil incorporation with many crop uses. *Farmers Crop Chemicals Ltd.*

32257 Trigger
Herbicide. *May & Baker Ltd.* Name unverified.

32258 Trigonal®
UV-activated catalyst for polyester resins. *Akzo Chemie UK Ltd.*

32259 Trigonal® 12
Aromatic ketone; uv initiated catalyst; for printing ink and paper coating applications. *Akzo.* Name unverified.

32260 Trigonal® 14
Benzoin ether; initiator used in uv light curing of polyester resins; long shelf life in polyester formulation; uv sensitizer for lacquers and FRP laminates. *Akzo.* Name unverified.

32261 Trigonal® 15
Benzoin ether; general uv sensitizer; mainly used for curing of putties; improved stability in presence of fillers. *Akzo.* Name unverified.

32262 Trigonal® 121
Ketone mixture; UV initiator used in combination with an amine accelerator for UV curing of FRP laminates. *Akzo.* Name unverified.

32263 Trigonox®
Organic peroxides. *Akzo Chemie UK Ltd.*

32264 Trigonox® 17
n-Butyl 4,4-di(*t*-butylperoxy) valerate
Initiator for cure of unsaturated polyester resins *Akzo.* Name unverified.

32265 Trigonox® 21
3006-82-4 221-110-7
$C_{12}H_{24}O_3$
t-Butyl peroctoate
Hexaneperoxoic acid, 2-ethyl-, 1,1-dimethylethyl ester; *tert*-Butyl peroxy-2-ethylhexanoate. Initiator for acrylates, styrenics, and LDPE polymerization; used where presence of water is objectionable. *Akzo.* Name unverified.

32266 Trigonox® 22-BB80
3006-86-8 221-111-2
1,1-Di-*t*-butylperoxycyclohexane in butyl benzyl phthalate
Initiator for SMC and BMC formulations. *Akzo.* Name unverified.

32267 Trigonox® 23
26748-41-4 247-955-1
t-Butyl peroxyneodecanoate
Akzo. Name unverified.

32268 Trigonox® 25-C75
927-07-1 213-147-2
$C_9H_{18}O_3$
tert-butyl peroxypivalate
2,2-dimethylpropaneperoxoic acid, 1,1-dimethylethyl ester. *t*-Butyl peroxypivalate in odorless mineral spirits; initiator for LDPE polymerization, styrenics, and PVC. mp = -17°. *Akzo.* Name unverified.

32269 Trigonox® 29
6731-36-8 229-782-3
$C_{17}H_{34}O_4$
1,1-Di(*t*-butylperoxy)-3,5,5-trimethylcyclohexane
Peroxide, (3,3,5-trimethylcyclohexylidene)bis(1,1-dimethylethyl); 1,1-Di-(*tert*-butylperoxy)-3,3,5-trimethyl cyclohexane. Polymerization initiator. *Akzo.* Name unverified.

32270 Trigonox® 36-C75
3851-87-4 223-356-0
Bis(3,5,5-trimethylhexanoyl) peroxide
Initiator for LDPE polymerization, styrenics, acrylates, PVC polymerization. *Akzo.* Name unverified.

32271 Trigonox® 41-C75
109-13-7 203-650-5
t-butyl peroxyisobutyrate
Polymerization initiator. *Akzo.* Name unverified.

32272 Trigonox® 42
13122-18-4 236-050-7
t-butyl peroxy-3,5,5-trimethyl hexanoate
Initiator for cure of unsaturated polyester resins *Akzo.* Name unverified.

32273 Trigonox® 42 PR
13122-18-4 236-050-7
t-butyl peroxy-3,5,5-trimethyl hexanoate
Initiator for cure of unsaturated polyester resins *Akzo.* Name unverified.

32274 Trigonox® 42S
13122-18-4 236-050-7
t-butyl peroxy-3,5,5-trimethylhexanoate
Polymerization initiator. *Akzo.* Name unverified.

32275 Trigonox® 61, 63, 67
Ketone peroxide mixture; initiator for cure of unsaturated polyester resins *Akzo.* Name unverified.

32276 Trigonox® 93
614-45-9 210-382-2
$C_{11}H_{14}O_3$
t-butyl peroxybenzoate
tert-butyl perbenzoate; benzenecarboperoxoic acid, 1,1-dimethylethyl ester; *tertiary*-butyl perbenzoate. Initiator for cure of unsaturated polyester resins mp = 8°; bp = 113°; d = 1.021. *Akzo.* Name unverified.

32277 Trigonox® 97-C75
22313-62-8 244-906-6
t-butyl peroxy-2-methylbenzoate
Odorless mineral spirits; initiator for styrenics. *Akzo.* Name unverified.

32278 Trigonox® 99-B75
26748-47-0 247-956-7
Cumyl peroxyneodecanoate
Polymerization initiator. *Akzo.* Name unverified.

32279 Trigonox® 101
78-63-7 201-128-1
$C_{16}H_{34}O_4$
2,5-Dimethyl-2,5-di (t-butyl peroxy)hexane
2,5-dimethyl-2,5-bis(*t*-butylperoxy)hexane; (1,1,4,4-tetramethyl-1,4-butanediyl)bis[(1,1-dimethylethyl) peroxide; 2,5-Dimethyl-2,5-di(*tertiary*-butylperoxy)-hexane. Initiator for polyester when molding at elevated temps.; also for crosslinking of olefin copolymers, chlorinated polyethylene, EPDM, SBR, and vinyls; initiator for free radical polymerizations of styrene; crosslinking agent for olefin copolymers, chlorina mp = 8°; bp = 152°; insoluble in H_2O (<1 mg/ml). *Akzo.* Name unverified.

32280 Trigonox® 111-B40
2155-71-7 218-454-5
$C_{16}H_{22}O_6$
di-*t*-butyl peroxyphthalate.
1,2-Benzenedicarboperoxoic acid, bis(1,1-dimethylethyl) ester; di-(*tert*-butylperoxy)phthalate. Polymerization initiator. *Akzo.* Name unverified.

32281 Trigonox® 121
686-31-7 211-687-3
$C_{13}H_{26}O_3$
t-amyl peroxy-2-ethylhexanoate.
tert-amyl peroxy-2-ethylhexanoate; Hexaneperoxoic acid, 2-ethyl-, 1,1-dimethylpropyl ester. Polymerization initiator. *Akzo.* Name unverified.

32282 Trigonox® 123-C75
68299-16-1 269-597-5
$C_{15}H_{30}O_3$
t-amyl peroxyneodecanoate
tert-amyl peroxyneodecanoate; neodecaneperoxoic acid, 1,1-dimethylpropyl ester. Odorless mineral spirits; initiator for PVC. *Akzo.* Name unverified.

32283 Trigonox® 125-C75
29240-17-3 249-530-6
t-amyl peroxypivalate
Odorless mineral spirits; initiator used to provide faster curing of BMC and SMC; initiator for LDPE and PVC polymerization. *Akzo.* Name unverified.

32284 Trigonox® 127
4511-39-1 224-831-5
t-amyl peroxybenzoate
Polymerization initiator. *Akzo.* Name unverified.

32285 Trigonox® 141
13052-09-0 235-935-5
2,5-Dimethyl-2,5-(2-ethylhexanoylperoxy) hexane
Polymerization initiator. *Akzo.* Name unverified.

32286 Trigonox® 145
1068-27-5 213-944-5
2,5-Dimethyl-2,5-di(*t*-butylperoxy)-hexyne-3
Polymerization initiator. *Akzo.* Name unverified.

32287 Trigonox® 151-C70
2,4,4-Trimethylpentyl-2-peroxyneodecanoate
Polymerization initiator. *Akzo.* Name unverified.

32288 Trigonox® 161
Mixture; initiator for cure of unsaturated polyester resins. *Akzo.* Name unverified.

32289 Trigonox® 169-OP50
16580-06-6 240-638-9
Di-t-butyl peroxyazelate
Polymerization initiator. *Akzo.* Name unverified.

32290 Trigonox® 239A
80-15-9 201-254-7
Cumene hydroperoxide
Initiator for cure of unsaturated polyester resins. *Akzo.* Name unverified.

32291 Trigonox® A-80
75-91-2 1604 200-915-7
$C_4H_{10}O_2$
t-butyl hydroperoxide
1,1-dimethylethyl hydroperoxide. Polymerization initiator. mp = - 8°; bp_{20} = 35°; d = 0.937; n_D^{20} = 1.3870; soluble in organic solvents. *Akzo.* Name unverified.

32292 Trigonox® ACS-M28
3179-56-4 221-658-7
Acetyl cyclohexane sulfonyl peroxide in DMP; reactive initiator available for use in producing PVC. *Akzo.* Name unverified.

32293 Trigonox® ADC
19910-65-7 243-424-3
Peroxydicarbonate
Initiator for low-and medium-density polyethylene; initiator for PVC polymerization. *Akzo.* Name unverified.

32294 Trigonox® ADC-NS60
78350-78-4 278-901-5
Peroxydicarbonate mixture. Polymerization initiator. *Akzo.* Name unverified.

32295 Trigonox® B
110-05-4 3515 203-733-6
$C_8H_{18}O_2$
di-*t*-butyl peroxide
DTBP; bis(1,1-dimethylethyl) peroxide. Initiator for LDPE polymerization; lower molecular weight at higher temps.; high-temp. peroxide used as finishing initiator for styrenics; crosslinking agent for olefin copolymers; volatile liq. best suited when used in injected extruder application. mp = -40°; bp_{284} = 80°; d_4^{20} = 0.7940; n_D^{20} = 1.3890; poorly soluble in H_2O (100 mg/l), soluble in organic solvents. *Akzo.* Name unverified.

32296 Trigonox® BPIC
2372-21-6 219-143-7
t-butyl peroxy isopropyl carbonate, odorless mineral spirits; initiator for BMC and SMC molding; initiator for free radical polymerizations of styrene and acrylates; cross-linking agent for elastomers including SBR, urethanes, EPR, EPDM, and nitrile rubbers. *Akzo.* Name unverified.

32297 Trigonox® C
614-45-9 210-382-2
$C_{11}H_{14}O_3$
t-butyl perbenzoate
tert-butyl perbenzoate; Benzenecarboperoxoic acid, 1,1-dimethylethyl ester; *tertiary*-butyl perbenzoate. Initiator for elevated-temperature polyester cures, for LDPE polymerization, and high-temperature polymerization of acrylic emulsion polymers; used in BMC and SMC molding in temperature range 275-325°F. *Akzo.* Name unverified.

32298 Trigonox® D-E50
2167-23-9 218-507-2
2,2-di-(*t*-butylperoxy) butane
Polymerization initiator. *Akzo.* Name unverified.

32299 Trigonox® EHP
16111-62-9 240-282-4
Bis(2-ethylhexyl) peroxydicarbonate
Versatile, highly reactive initiator for PVC production; high molecular weight reduces reactor fouling. *Akzo.* Name unverified.

32300 Trigonox® F-C50
107-71-1 203-514-5
t-butyl peracetate
Odorless mineral spirits; initiator for low-and medium-density polyethylene and for styrenics. *Akzo.* Name unverified.

32301 Trigonox® GS
A range of liquid organic peroxides for the high temperature curing of unsaturated polyester resins. *Akzo Chemie UK Ltd.*

32302 Trigonox® HM
MIBK peroxide
methyl isobutyl ketone peroxide. Initiator for cure of unsaturated polyester resins. *Akzo.* Name unverified.

32303 Trigonox® K-95
80-15-9 201-254-7
Cumene hydroperoxide
Polymerization initiator. *Akzo.* Name unverified.

32304 Trigonox® KSM
t-butyl peroctoate [3006-82-4] and 1,1-bis (*t*-butylperoxy)-3,3,5-trimethylcyclohexane [6731-36-8]; SMC/BMC initiator offering good shelf life in compound and molding temp. range of 230-290 F; polyester cures. *Akzo.* Name unverified.

32305 Trigonox® M-50
3736-26-3
Diisopropylbenzene monohydroperoxide
Polymerization initiator. *Akzo.* Name unverified.

32306 Trigonox® RM
A range of liquid organic peroxides for free radical polymerization. *Akzo Chemie UK Ltd.*

32307 Trigonox® SBP
19910-65-7 243-424-3
Di(s-butyl) peroxydicarbonate
Initiator for LDPE polymerization; fastest conversion to polymer; used for production of PVC. *Akzo. Name unverified.*

32308 Trigonox® T
3457-61-2 222-389-8
t-butyl cumyl peroxide
Crosslinking agent for EPDM, SBR, Neoprene, Hypalon; low reactivity peroxide useful as a finishing initiator at high temperatures for styrenics; a synergist for some halogen-containing flame retardants. *Akzo.* Name unverified.

32309 Trigonox® TAHP-W85
3425-61-4 222-321-7
t-amyl hydroperoxide
Polymerization initiator. *Akzo.* Name unverified.

32310 Trigonox® TMPH
5809-08-5 227-369-2
2,4,4-trimethylpentyl-2-hydroperoxide
Polymerization initiator. *Akzo.* Name unverified.

32311 Trihyde
Alumina trihydrate
Fire retardant for PVC, rubber, polyester, epoxy resins. *Croxton & Garry Ltd.* Name unverified.

32312 triisooctyl phosphite
25103-12-2 246-614-4
$C_{24}H_{51}O_3P$
phosphorous acid tri-isooctyl ester
Intermediate; insecticides; lubricant additive; specialty solvents; stabilizer for acrylics, nylon, unsaturated polyester, PVC; improves antiwear and antifriction properties. d = 0.891. *Albright & Wilson; GE Specialty Stave.*

32313 triisopropanolamine
122-20-3 204-528-4
$C_9H_{21}NO_3$
1,1',1-nitrilopropan-2-ol
1,1'1-nitrilotris-2-propanol; tris(2-hydroxypropyl)amine; TIPA. Emulsifying agent. mp = 48-50°; bp_{32} = 190° *Aldrich; Ashland.*

32314 Trik
Aminotriazole + 2,4-D + diuron; used for total weed control in non crop areas. *Smyth-Morris Chemicals Ltd.*

32315 Triklone
79-01-6 9769 201-167-4
Trichloroethylene solvents. *ICI Chem & Polymers Ltd.*

32316 trilactine
A preparation of lactic acid bacilli.

32317 Trilaurin
538-24-9 208-687-0
$C_{39}H_{74}O_6$
Glyceryl tri-laurate
trilauroylglycerol; Dodecanoic acid, 1,2,3-propanetriyl ester. mp = 45-47°; d = 0.90.

32318 Trilene® 65
25034-71-3
Ethylene-propylene-diene terpolymer; produces liquid elastomers; used alone or with solid elastomers in thermosetting and thermoplastic applications; features low viscosity, crosslinking, oxidation resist., good electrical properties, ozone and uv resistance. *Uniroyal.*

32319 Trilin® 10G
1582-09-8 9815 216-428-8

Trifluralin
Herbicide. *Griffin.*

32320 Trilinoleic acid
68937-90-6
Hystrene® 5460; Trimer acid; Fatty acids; C18, unsaturated, trimers; 9,12-Octadecanoic acid, trimer; Empol® 1045,. Corrosion inhibitor, lubricant; intermediate for mfg. of soaps, emulsions, creams, lotions, ethoxylates, buffing compounds, lubricants. *Witco Corporation; Henkel/Emery.*

32321 trillite
Haidinger's name for the ferrous sulfide which occurs in meteorites.

32322 Trilon®
Complexing agents, water softeners. *BASF plc.*

32323 Trilon® A, A-92
5064-31-3 225-768-6
Trisodium NTA
Low molecular weight general purpose chelates; complexes iron in acid pH range; detergent builder for nonphosphate liquids; in soaps, detergents, water treatment, metal finishing and plating, pulp and paper manufacturing, synthesis of polymers, photography, textiles, chemical cleaning. *BASF.*

32324 Trilon® B Powd
64-02-8 3557 200-573-9
$C_{10}H_{16}N_2Na_4O_{10}$
Tetrasodium EDTA
(Ethylenedinitrilo) tetraacetic acid tetrasodium; Glycine, N,N'-1,2-ethanediylbis[N-(carboxymethyl)-, tetrasodium salt; Ethylenediamine tetraacetic acid, tetrasodium salt; Sodium edetate; Tetrasodium (ethylenedinitrilo)tetraacetate; Ethylenediaminetetraacetic tetrasodium salt. General purpose chelate; complexes most common metals over pH range, iron at acidic pH; chelating agents used in soaps, detergents, water treatment, metal finishing and plating, pulp and paper manufacturing, synthesis of polymers, photographic products, textiles, chemical cleaning for scale removal. mp > 300°. *BASF.*

32325 Trilon® BD
139-33-3 3556 205-358-3
$C_{10}H_{14}N_2Na_2O_8 \cdot 2H_2O$
N,N'-1,2-ethanediylbis[N-(carboxymethyl)glycine] disodium salt dihydrate
Edetate disodium; Disodium EDTA; edathamil disodium; tetracemate disodium; Chelaplex III; Endrate disodium; Sequestrene NA 2; sodium versenate; Titriplex III; Versene disodium salt. Chelating agent. mp = 52° (dec); LD_{50} (rat orl) = 2 g/kg. *BASF.* Name unverified.

32326 Trilon® B-FA
Ferric ammonium EDTA
Chelating agent for photographic applications. *BASF.* Name unverified.

32327 Trilon® BS
60-00-4 3559 200-449-4
$C_{10}H_{16}N_2O_8$
ethylenediamine tetraacetic acid
edetic acid; ethylenedinitrilotetraacetic acid; EDTA, free base; 4H; EDTAfree acid; Ethylenediamine-N,N,N',N'-tetraacetic acid; Hampene; Versene; N,N'-1,2-Ethanediylbis-(N-(carboxymethyl)glycine); EDTA. EDTA; chelating agent; used where sodium ion is undesirable; for soaps, detergents, water treatment, metal finishing and plating, pulp and paper manufacturing, synthesis of polymers, photographic products, textiles, chemical cleaning for scale removal and antioxidant in foods. mp = 220° (dec); soluble in H_2O (500 mg/l). *BASF.*

32328 Trilon® C Liq
140-01-2 205-391-3
$C_{14}H_{18}N_3Na_5O_{10}$
diethylenetriaminepentaacetic acid, pentasodium salt
Pentasodium pentetate; CHEL 330; Detarex PY; Diethylenetrinitrilo)pentaacetic acid, pentasodium salt; Glycine, N,N-bis(2-(bis(carboxymethyl)amino)ethyl)-, pentasodium salt; Diethylenetriaminepentaacetic acid, pentasodium salt; Pentasodium diethylenetriaminepentaacetate; Bis(2-(bis(carboxymethyl)amino)ethyl)glycine pentasodium salt; Carboxymethyl)imino) bis(ethylenenitrilo); tetraacetic acid, pentasodium salt; HAMP-EX 80; Kiresuto P;
Pentasodium dtpa; Pentasodium pentetate; Penthanil; Perma Kleer 140; Plexene D; Syntron C;
Versenex 80; pentasodium (carboxylatomethyl)iminobis(ethylenenitrilo)tetraacetate. Chelating agent used when higher metal chelate stability is required; used in peroxide bleach systems. d = 1.2990 *BASF.*

32329 Trilon® D Liq
139-89-9 10102 205-381-9
$C_{10}H_{15}N_2Na_3O_7$
Trisodium HEDTA
Trisodium N-(2-hydroxyethyl)ethylenediaminetriacetate; N-(2-Hydroxyethyl)ethylenediaminetriacetic acid, trisodium salt; N-(2-(biscarboxymethylamino)ethyl)-N-(2-hydroxyethyl)glycine trisodium salt; Versen-Ol; Carboxymethyl)-N'-(2-hydroxyethyl)-N,N'-ethylenediglycine, trisodium salt; Hydroxyethyl)ethylenediaminetriacetic acid, trisodium salt; Trisodium 2-hydroxyethyl)ethylenediaminetriacetate; Trisodium hedta; Versenol 120. General purpose chelate for control of iron in pH range of 6.5-9.5, Ca, Mg; used in soaps, detergents, water treatment, metal finishing and plating, pulp and paper manufacturing, synthesis of polymers, photographic products, textiles, chemical cleaning. *BASF.*

32330 Trim
82-68-8 8264 201-435-0
Quintozene in dry granular form; used for controlling soil borne diseases in tulips and flower crops. *Wheatley Chemical Co Ltd.* Name unverified.

32331 Trimangol
12427-38-2 5761 235-654-8
Wettable powder containing maneb; a dithiocarbamate fungicide to control blight, rusts, and mildew. *Pennwalt Chemicals Ltd.*

32332 Trimanzone
Mixture of ferbam, maneb and zineb; a fungicide. *Pennwalt Chemicals Ltd.*

32333 Trimaran
1582-09-8 9815 216-428-8
Emulsifiable concentrate containing 480 g/l trifluralin; dinitroaniline herbicide to control annual weeds and grasses. *Ashlade Formulations Ltd.*

32334 Trimastan
Mixture of fentin acetate and maneb; used for control of potato blight. *Pennwalt Chemicals Ltd.*

32335 trimellitic acid
528-44-9 9832 208-432-3
$C_9H_6O_6$
1,2,4-benzenetricarboxylic acid
1,2,4-tricarboxybenzene. Used in organic synthesis. mp = 231°; soluble in H_2O (21 g/l), similarly soluble in polar organic solvents.

32336 Trimene Base
Ethyl chloride, ammonia and formaldehyde reaction product; high temperature accelerator with a medium to long curing range; prevents sagging of stock in early stages of air curing; it is a latex foam stabilizer which prevents foam collapse by causing gelling to take place at a higher pH; used in natural and SBR rubbers and latexes. *Uniroyal.* Name unverified.

32337 trimesic acid
554-95-0 209-077-7
$C_9H_6O_6$
Trimesic acid
1,3,5-benzenetricarboxylic acid, trimesitinic acid. Chemical intermediate. mp > 330°;.

32338 trimethylolpropane
77-99-6 201-074-9
$C_6H_{14}O_3$
1,1,1-tris(hydroxy)propane
2-ethyl-2-hydroxymethyl-1,3-propanediol; hexaglycerol. Conditioner, manufacture of varnishes, alkyd resins, synthetic drying oils, urethane foams and coatings, silicone lube oils, lactone plasticizers, textile finishes, surfactants, epoxidation products. mp = 56-58°; bp_2 = 159-161°. *Hoechst Celanese; Mitsubishi Gas; Perstorp Polyols.*

32339 2,2,4-trimethyl-1,3-pentanediol monoisobutyrate
25265-77-4 246-771-9
$C_{12}H_{24}O_3$
2,2,4-trimethyl-1,3-pentanediolmono(2-methylpropanoate)
Ester alcohol; intermediate; manufacture of plasticizers, surfactants, pesticides, and resins. bp = 244°; d = 0.9500 *Chisso Am.; Eastman.*

32340 4,5,8-trimethylpsoralen
3902-71-4 9864 223-459-0
$C_{14}H_{12}O_3$
2,5,9-trimethyl-7H-Furo[3,2-g][1]benzopyran-7-one
Trixsalen; Trioxsalen; Trisoralen; NSC 71047. Photosensitizer, pigmentation agent. mp = 229-230°; λ_m = 250, 295, 335 nm (log ε 4.35, 3.99, 3.80 MeOH); insoluble in H_2O, soluble in organic solvents. *ICN Pharmaceuticals Inc.*

32341 α,α,α'-trimethyltrimethyleneglycol
hexylene glycol.

32342 1,3,7-trimethylxanthine
Caffeine.

32343 Trimetso
Industrial detergent. *Crosfield Chemicals Ltd.* Discontinued.

32344 Triminol
63284-71-9 264-071-1
Emulsifiable concentrate containing 90 g/l nuarimol; pyrimidine fungicide to control mildew. *DowElanco Ltd.*

32345 Trimmit
Grass growth regulator. *ICI Chem & Polymers Ltd.*

32346 Trinidad asphalt
A natural asphalt obtained from the Trinidad pitch lake. The crude pitch contains from 40-46% bitumen, 24-30% mineral matter (clay), and 21-29% water.

32347 trinitrocresol
28905-71-7
$C_7H_5N_3O_7$
4-hydroxy-2,3,5-trinitrobenzene.

32348 Trinoram
Alkyl propylene triamines based on coco and tallow alkyl chains; intermediate for chemical synthesis. *Elf Atochem UK/Ceca.*

32349 trioctyl phosphate
78-42-2 201-116-6
$C_{24}H_{51}O_4P$
Tris(2-ethylhexyl)phosphate
2-ethylhexoic acid; Octyl phosphate; Amgard=soTOF; Disflamol®TOF; Kronitex® TOF; TOF; TOF™;. Paint and varnish driers (metallic salts); esters as plasticizers. Solvent, antifoaming agent; flame retardant. Liquid; soluble in alcohol, acetone, ether; mw=744.04; density=0.924; bp=220-230(8mm) flash pt>230°F; combustible; toxic by ingestion. *Aldrich; Ashland; Azko Nobel; Albright & Wilson; BASF; Eastman; Neste UK; Union Carbide.*

32350 Triodine
Iodophor low foaming sanitizer. *Ciba plc.* Name unverified.

32351 triolein
122-32-7 9861 204-534-7
$C_{57}H_{104}O+66$
glyceryl trioleate
olein; 9-octadecenoic acid, 1,2,3-propanetriyl ester. Triester of glycerin and oleic acid; lubricant, emollient, emulsifier for cosmetics, metals, leather, textiles; carbon source in antibiotic culture broths. mp = -4°; bp_{15} = 235-240°; d_4^{15}= 0.915; n_D^{20}= 1.4676; insoluble in H_2O, soluble in organic solvents.

32352 Triolein PEG-6 esters
Labrafil M 2735 CS; Triolein PEG-6 complex. Hydrophilic hydrogenated oil; excipient for pharmaceutical and cosmetic formulations. *Gattefosse; Gattefosse SA.*

32353 Trioleotope
Triolein-^{131}I
Radioactive agent. *Bristol-Myers Squibb Co Inc.* Name unverified.

32354 Trione
Ready-to-use ninhydrin reagent solution containing organic modifier and buffer; used for amino acid analysis, amine analysis. *Pickering Laboratories Inc.*

32355 Triox
Vegetation killer. *Monsanto (Solaris).* Name unverified.

32356 Trioxitol
Ethyltrigol
Shell UK.

32357 Trioxone
93-76-5 9194 202-273-3
Herbicide. *ICI Chem & Polymers Ltd.*

32358 trip
1309-37-1 4072 215-168-2
Fe_2O_3
Ferric oxide

32359 Tripart® Acer
Soyal phospholipids 750 g/liter; an adjuvant for use with a wide range of crop protection products. *Tripart Farm Chemicals Ltd.*

32360 Tripart® Arena 6
117-18-0 204-278-2
$C_6HCl_4NO_2$
2,3,5,6-tetrachloronitrobenzene
tecnazene.

32361 Tripart® Arena 6 + TBZ
Dustable powder containing 6% (w/w) tecnazene [117-18-0] and 1.8% thiabendazole [148-79-8]; a protectant fungicide and potato sprout suppressant. *Tripart Farm Chemicals Ltd.*

32362 Tripart® Arena Granules
117-18-0 204-178-2
Tecnazene (5 or 10% w/w); used to control dry rot in both ware and seed potatoes, and sprouting in ware potatoes. *Tripart Farm Chemicals Ltd.*

32363 Tripart® Arena Plus
6% (w/w) tecnazene [117-18-0] and 2% carbendazim [83601-81-4]; protectant fungicide and sprout suppressant for stored potatoes. *Tripart Farm Chemicals Ltd.*

32364 Tripart® Beta
13684-63-4 7384 237-199-0
Phenmedipham 118 g/liter (12.1% w/w) in isophorone; a post-emergence herbicide for the control of a wide range of broad-leaved weeds in sugar beet, red beet, fodder beet and mangels. *Tripart Farm Chemicals Ltd.*

32365 Tripart® Beta 2
13684-63-4 7384 237-199-0
Emulsifiable concentrate containing 114 g/l phenmedipham; for weed control for beet crops. *Tripart Farm Chemicals Ltd.*

32366 Tripart® Brevis
999-81-5 2153 213-666-4
Chlormequat chloride 700 g/liter (61.7% w/w) aqeuous solution; plant growth regulator for use in cereals to reduce lodging and in wheat and barley. *Tripart Farm Chemicals Ltd.*

32367 Tripart® Chlormequat 460
999-81-5 2153 213-666-4
Soluble concentrate containing 460 g/l chlormequat; plant growth regulator. *Tripart Farm Chemicals Ltd.*

32368 Tripart® Chlormequat 5C
Mixture of 40.8% (w/w) chlormequat chloride [999-81-5] 460 g/liter and choline chloride [67-48-1] 320 g/liter (28.3% w/w); plant growth regulator for cereals and ornamentals. *Tripart Farm Chemicals Ltd.*

32369 Tripart® Cropspray 11E
Adjuvant containing 99% highly refined mineral oil; wetting agent for translocated herbicide sprays. *Tripart Farm Chemicals Ltd.*

32370 Tripart® Defensor FL
10605-21-7 1836 234-232-0
A liquid formulation containing 500 g carbendazim per liter as a suspension concentrate; a broad-spectrum systemic fungicide for control of several diseases in a variety of crops. *Tripart Farm Chemicals Ltd.*

32371 Tripart® Faber
1897-45-6 2219 217-588-1
Chlorothalonil 500 g/liter (40.4% w/w); a nonsystemic broad-spectrum fungicide for use in a wide variety of crops. *Tripart Farm Chemicals Ltd.*

32372 Tripart® Gladiator
1698-60-8 216-920-2
Chloridazon 430 g/liter (36.1% w/w); used for the control of annual broad-leaved weeds in sugar beet, fodder beet and mangels. *Tripart Farm Chemicals Ltd.*

32373 Tripart® Granular
1918-16-7 7977 217-638-2
Propachlor
A preemergence granular herbicide for various horticultural crops. *Tripart Farm Chemicals Ltd.*

32374 Tripart® Imber
7704-34-9 9142 231-722-6
Suspension concentrate containing 720 g/l (52.4% w/w) sulfur; used for disease prevention and growth enhancement in cereals, oil seed rape, sugar beet, and others. *Tripart Farm Chemicals Ltd.*

32375 Tripart® Legion
Carbendazim [10605-21-7] 50 g and maneb [12427-38-2] 320 g/liter mixture; for use as a fungicide in cereals. *Tripart Farm Chemicals Ltd.*

32376 Tripart® Lentus
A liquid containing 450 g/liter synthetic latex; for use as an agricultural sticking and extending agent. *Tripart Farm Chemicals Ltd.*

32377 Tripart® Liquid Manganese
7439-96-5 5762 231-105-1
Manganese
Manganese 140 g/liter (10.5% w/w) minimum; for use on crops that are suffering from manganese deficiency or grown on manganese-deficient soils. mp = 1245°; bp = 2097°; d^{20} = 7.47. *Tripart Farm Chemicals Ltd.*

32378 Tripart® Ludorum 700
15545-48-9 239-592-2
Chlortoluron 700 g/liter (58.5% w/w); used for the control of black grass, wild oats, and other annual grasses and a range of broad-leaved weeds in a range of named wheats, winter barleys, durum wheats and triticale. *Tripart Farm Chemicals Ltd.*

32379 Tripart® Mensa
A concentrated complex of manganese salts; for use on crops that are suffering from manganese deficiency or grown on manganese-deficient soils. *Tripart Farm Chemicals Ltd.*

32380 Tripart® Minax
Alkyl phenol ethylene oxide 900 g/liter (87.3% w/w); for use with fungicides, insecticides, and herbicides to improve spreading and penetration; also for use in cleaning and washing out spraying machinery. *Tripart Farm Chemicals Ltd.*

32381 Tripart® Mini Slug Pellets
9002-91-9 5983 202-945-6
Metaldehyde (6% w/w); for the control of slugs and snails in all agricultural and horticultural crops. *Tripart Farm Chemicals Ltd.*

32382 Tripart® Nex
1563-66-2 1851 216-353-0
Carbofuran (5% w/w); a specially formulated microgranule insecticide with systemic action for the control of a range of pests in sugar beet, fodder beet, mangels, brassicas, etc. *Tripart Farm Chemicals Ltd.*

32383 Tripart® Ratio
82558-50-7 5256
Isoxaben 125 g/liter; a residual herbicide for the control of broad-leaved weeds in winter and spring cereals, grass leys and herbage seed crops. *Tripart Farm Chemicals Ltd.*

32384 Tripart® Senator Flowable
Copper oxychloride + maneb + sulfur; a fungicide for wheat and barley which also stimulates yield. *Tripart Farm Chemicals Ltd.*

32385 Tripart® Sentinel
1918-16-7 7977 217-638-2
Propachlor 480 g/liter (43.2% w/w); used for the control of germinating weeds in onion, leek, brassica crops and strawberry. *Tripart Farm Chemicals Ltd.*

32386 Tripart® Systemic Insecticide
8022-00-2 6129
Emulsifiable concentrate containing 500 g demeton-S-methyl per liter; a systemic organophosphorus insecticide. *Tripart Farm Chemicals Ltd.*

32387 Tripart® Trifluralin 48 EC
1582-09-8 9815 216-428-8
Trifluralin 480 g/liter (45.5% w/w); used for the control of certain germinating broad-leaved weeds and annual grasses in beans brassicae, cabbage, and a wide variety of other crops. *Tripart Fram Chemicals Ltd.*

32388 Tripart® Ultrafaber
1897-45-6 2219 217-588-1
Chlorothalonil 720 g/liter (54% w/w); a nonsystemic broad-spectrum fungicide for use in a wide variety of crops. *Tripart Farm Chemicals Ltd.*

32389 Tripart® Victor
A liquid formulation containing 80 g carbendazim, 150 g chlorothalonil and 200 g maneb per liter as a suspension concentrate; eradicant fungicide for use on cereals. *Tripart Farm Chemicals Ltd.*

32390 tripentaerythritol
$C_{15}H_{35}O_8$
Hard resins, varnishes, fast-drying tall oil vehicles. *Allchem Industries.*

32391 triphenyl phosphate
115-86-6 9872 204-112-2
$C_{18}H_{15}O_4P$
phosphoric acid, triphenyl ester
TPP; phenyl phosphate; TPP; triphenyl phosphoric acid ester; celluflex tpp. Fire-retarding agent; noncombustible substitute for camphor in celluloid; plasticizer for cellulose acetate and nitrocellulose; in lacquers and varnishes; impregnating roofing paper. mp = 49-50°; bp_{11} = 245°; insoluble in H_2O, soluble in organic solvents. *Monsanto.*

32392 triphenyl phosphine
603-35-0 9873 210-036-0
$C_{18}H_{15}P$
triphenyl phosphorus
phosphorus triphenyl. Organic synthesis, phosphonium salts, other phosphorus compounds, polymerization initiator. mp = 81°; bp > 360°; d_4^{25} = 1.194; insoluble in H_2O, soluble in organic solvents. *Elf Atochem; BASF; Janssen Chimica; Morton Int'l.*

32393 triphenyl phosphite
101-02-0 202-908-4
$C_{18}H_{15}O_63P$
phosphorous acid, triphenyl ester
TPP. TPP; chemical intermediate, stabilizer systems for resins, metal scavenger, diluent for epoxy resins, antioxidant and antiwear agent in gear and transmission oils. mp = 22-24°; bp = 360°; d = 1.1840; n_D^{20} = 1.5903 *Dover; GE Specialty; Stave; Witco/Argus.*

32394 Triplastic
Prepared by mixing trinitro-toluene, together with some lead nitrate and chlorate, with a gelatin made from dinitrotoluene and nitrocellulose; an explosive.

32395 Triplastite
An explosive consisting of a mixture of di- and trinitrotoluene 70 parts, guncotton 1.2 parts, and lead nitrate 28.8 parts.

32396 triple salts
Trisalytes. Used in the electro-deposition of metals. They consist of the cyanide of the metal to be deposited, potassium cyanide, and potassium sulfite.

32397 Triplematic
Metering, proportioning, mixing and dispensing machines for multicomponent reactive resin systems. *Hardman.* Name unverified.

32398 Triplevac
B_1 type, B_1 strain, Newcastle with Massachuttes and Connecticut types; for immunization of poultry. *Intervet Inc.*

32399 Tripoli
1317-95-9 9876
Tripoli powder; rotten stone. A mineral. It consists mainly of silica associated with small quantities of alumina and iron oxide, but the composition varies. The variety of tripoli powder found in Derbyshire, is called rotten stone; used as an abrasive.

32400 tripropylene glycol
24800-44-0 246-466-0
$C_9H_{20}O_4$
[(1-methyl-1,2-ethanediyl)bis(oxy)]bis(propanol)
[(methylethylene)bis(oxy)]dipropanol;. Intermediate in resins, plasticizers, pharmaceuticals, insecticides, dyestuffs, and mold lubricants. *Union Carbide.*

32401 tripsa
Tribasic phosphate of soda; used for the prevention of incrustation on boilers.

32402 tris (2,3-dibromopropyl) phosphate
126-72-7 9881 204-799-9
$C_9H_{15}Br_6O_4P$
1-Propanol, 2,3-dibromo-, phosphate
TDCPP; Tris(2,3-dibromopropyl)phosphoric acid ester; FR-2406;. A proprietary fire-retardant additive used in acrylics, epoxies, latices, phenolics, polyesters, polystyrenes, polyvinyl chloride, rayon celluloses and polyurethanes. Pale yellow viscous liquid; n_D^{20}= 1.5772; LD_{50} (rat orl) = 2830 mg/kg. *Dow UK.* Discontinued.

32403 tris (hydroxymethyl) aminomethane
77-86-1 9902 201-064-4
$(CH_2OH)_3CNH_3$
2-amino-2-(hydroxymethyl)-1,3-propanediol
tromethamine (CTFA); THAM; tris buffer. emulsifying agent for oils, fats, waxes; absorbent for acidic gases, medicine; chemical intermediate. *Aldrich; Angus; Dajac Labs; W.R. Grace/Hampshire; Heico; Janssen Chimica; Sigma.*

32404 tris (hydroxymethyl) aminomethane hydrochloride
1185-53-1 9902 214-684-5
$C_4H_{11}NO_3Cl$
tris hydrochloride
Am. Biorganics; Janssen Chimica; U.S. biochemical.

32405 Tris Amino®
77-86-1 9902 201-064-4
$C_4H_{11}NO_3$
Tri (hydroxymethyl) aminomethane
TRIS; Tris(hydroxymethyl)aminomethane; THAM; Tromethamine; tris base; trizma base; tris (hydroxymethyl)aminoethane; Tris, free base; Tromethane; Tris buffer; amino-2-(hydroxymethyl)-1,3-propanediol; Hydroxymethyl)-2-amino-1,3-propanediol; 2-amino-2-hydroxymethylpropanediol; Trizma; Trometamol; Tris(hydroxymethyl)methylamine. Pigment dispersant, neutralizing amine, corrosion inhibitor, acid-salt catalyst, pH buffer, chemical and pharmaceutical intermediate, solubilizer. mp = 171-172°; bp_{10} = 219-220°; soluble in H_2O and organic solvents. *Angus.*

32406 Tris Amino® Molecular Biology Grade
77-86-1 9902 201-064-4
Tris (hydroxymethyl) aminomethane
Chemical intermediate, resin synthesis, neutralizing amine in cosmetics, buffer for enzyme and diagnostic testing, pharmaceutical buffer and solubilizer. *Angus.*

32407 Tris Amino® Ultra Pure Standard
77-86-1 9902 201-064-4
Tris (hydroxymethyl) aminomethane
Chemical intermediate, resin synthesis, neutralizing amine in cosmetics, buffer for enzyme and diagnostic testing, pharmaceutical buffer and solubilizer. *Angus.*

32408 Tris Nitro®
126-11-4 9883 204-769-5
$C_4H_9NO_5$
2-Hydroxymethyl-2-nitro-1,3-propanediol
Tris (hydroxymethyl) nitromethane; 2-Nitro-2-(hydroxymethyl)-1,3-propanediol; Tris(hydroxymethyl)nitromethane; Trimethylolnitromethane; Tris-Nitro; Cimcool wafers; Hydroxymethyl)-2-nitro-1,3-propanediol; Hydroxymethyl)-2-nitropropane-1,3-diol; Hydroxymethyl)-2-nitropropanediol; Nitro-2-(hydroxymethyl)-1,3-propanediol; Nitroisobutylglycerol; Nitrotris(hydroxymethyl)methane; 2-hydroxymethyl-2-nitropropanediol. Antibacterial agent, preservative for water treatment, metalworking fluids, oil production, deodorizing; formaldehyde releaser. Chemical intermediate, registered pesticide for use as an antimicrobial in metalworking fluids, cooling water, oil production, drilling muds; formaldehyde donor; deodorant for chemical toilets. mp = 214° (dec); soluble in H_2O (220 g/100 ml), EtOH, poorly soluble in organic solvents. *Angus.*

32409 Trisec
Drying additives. *ICI Chem & Polymers Ltd.*

32410 Trisem
Trisulfapyrimidines (oral suspension); mixture of sulfadiazine, sulfamerazine, and sulfamethazine; an antibacterial. *SmithKline Beecham.* Name unverified.

32411 Trisnonylphenyl phosphite
26523-78-4 247-759-6
Tris-nonylphenyl phosphite
Lowinox® TNPP; Doverphos® 4, 4-HR; Lankromark® LE109; Weston® 399, TNPP; Wytox® 312. Antioxidant for rubber, adhesives, ABS. *Lowi; Dover; Harcros UK; GE Specialty; Uniroyal.*

32412 trisodium phosphate, anhydrous
7601-54-9 8808 231-509-8
Na_3O_4P
sodium phosphate, tribasic
Sodium orthophosphate; Sodium phosphate; Trisodium orthophosphate; Phosphoric acid, trisodium salt; TSP-O; sodium phosphate, tribasic; trisodium orthophosphate; phosphoric acid, trisodium salt. Inorganic salt; For pH adjustment in food systems; cleaning compounds, water treatment, textiles. Soluble in H_2O (0.29 g/ml), insoluble in organic solvents; LD_{50} (rat orl)= 7.40 g/kg. *Albright & Wilson; Kemira Kemi UK; Monsanto.*

32413 trisodium phosphate, dodecahydrate
10101-89-0 8808 231-509-8
$H_{24}Na_3O_{16}P$
Sodium phosphate
TSP-12; sodium phosphate, tribasic dodecahydrate; Phosphoric acid, trisodium, 12-hydrate; Sodium Phosphate 12-Water; Sodium Phosphate Tribasic Dodecahydrate; TSP dodecahydrate. Food products, cleaning compounds, water treatment. mp = 75°; d = 1.6. *Albright & Wilson; Rhône-Poulenc Basic.*

32414 Trisophone
Nonionic surfactant. *Rhône-Poulenc UK.*

32415 Tristar
1582-09-8 9815 216-428-8
Emulsifiable concentrate containing 480 g/l trifluralin; dinitroaniline herbicide to control annual weeds and grasses. *Pan Britannica Industries Ltd.*

32416 Tri-Star Antifoam 27
Synergistic blend of organic chemicals, 100% active, water dispersible; for rapid knockout of existing foam and the prevention of foam formation in systems containing detergents, in latex emulsions, in industrial processes, paints, glues, paper coating fo *Tri-Star Chemical Co Inc.* Name unverified.

32417 Tri-Star Padding Compounds
Modified polyvinyl acetate emulsions; fast setting padding compounds. *Tri-Star Chemical Co Inc.* Name unverified.

32418 Tri-Star White Glues
Modified polyvinyl acetate emulsions; used for packaging, bookbinding, labeling, and wood-working, etc. *Tri-Star Chemical Co Inc.* Name unverified.

32419 Trisul
Leather oils. *Hodgson Chemicals Ltd.*

32420 tritane
519-73-3 9871 208-275-0
$C_{19}H_{16}$
1,1',1''-methylidynetris[benzene]
triphenylmethane. Used as a dye. mp = 93°; bp = 360°; d_4^{100}= 1.0134; n_D^{100}= 1.5955; insoluble in H_2O, soluble in organic solvents.

32421 Trithac
Fungicide. *Murphy Chemical Co Ltd.* Discontinued.

32422 Trithion
786-19-6 1869 212-324-1
Carbophenothion
Insecticide seed treatment. *Murphy Chemical Co Ltd.* Discontinued.

32423 Tritiotope
7732-18-5 10175 231-791-2
$H_2O\text{-}^3H$
Tritiated water; radioactive agent. *Bristol-Myers Squibb Co Inc.* Name unverified.

32424 Tritisol/Tritisol Xm
Soluble wheat protein; film-forming conditioning, moisturizing protein improving skin firmness; for hair and skin care preparations, permanent waves, activated conditioners; binder in mascara formulations. *Croda Inc.*

32425 Triton
Range of alkylaryl polyether alcohol surfactants; used for cosmetics, household products, and industrial cleaners. *Union Carbide; Union Carbide Europe.*

32426 Triton® CF-10
Alkylaryl polyether
Low-foaming detergent for mechanical dishwashing, rinse aids, laundering, metal and dairy cleaning, textile wetting; nylon dyeing assistant; defoamer for food soils. *Union Carbide; Union Carbide Europe.*

32427 Triton® DF-12
Polyethoxylated alcohol, modified; biodegradable low-foaming detergent for mechanical dishwashing, rinse additive, automatic laundering, metal cleaning, floor scrubbing, dairy equipment cleaners, textile wetting; stable in acid, caustic solutions. *Union Carbide; Union Carbide Europe.*

32428 Triton® GR-5M
577-11-7 3460 209-406-4
Dioctyl sodium sulfosuccinate
High speed wetting and rewetting agent, emulsifier, dispersant for paints, textiles. *Union Carbide; Union Carbide Europe.*

32429 Triton® H-55
Phosphate ester, potassium salt; hydrotrope/solubilizer for built lliquid concentrations; as surfactant for alkaline builder solutions. *Union Carbide.*

32430 Triton® N-57
9016-45-9 6772
Polyoxyethylene (9) Nonylphenyl Ether; nonyl phenol ethoxylate
Nonoxynol-5; Tergitol TP-9; polyethylene glycol 450 nonyl phenyl ether; Ethoxylated nonylphenol; antarox; antarox bl-344; macrogol nonylphenyl ether; nonoxinol; nonoxynol; conco ni; dowfax 9n; igepal co; Makon; neutronyx 600's; nonipol no; polytergent b; renex 600's; solar np; triton n; tergitol np; T-DET-N; surfionic n; sterox; arkopal N-090; carsonon N-9; conco ni-90; igepal co-630; neutronyx 600; peg-9 nonyl phenyl ether; protachem 630; rewpol HV-9; Tergitol TP-9 (non-ionic); Glycols, polyethylene, mono(nonylphenyl) ether; Nonylphenoxypoly(ethyleneoxy)ethanol; Tergitol NP-14; Tergitol NP-27; Tergitol NP-35; Tergitol NP-40; Tergitol npx; POE (1.5) nonyl phenol; POE (4) nonylphenol; POE (5) nonylphenol; POE (6) nonylphenol; POE (8) nonylphenol; POE (10) nonylphenol; POE (14) nonylphenol; POE (20) nonylphenol; POE (30) nonylphenol,. Detergent, emulsifier for solvent cleaners; intermediate. *Union Carbide; Union Carbide Europe.*

32431 Triton® QS-44
Phosphate surfactant, free acid; hydrotrope, detergent, wetting agent; solubilizer for nonionic surfactants in alkaline cleaning baths, metal cleaning and for nonionic and anionic surfactants in built concs. *Union Carbide; Union Carbide Europe.*

32432 Triton® RW-20
Alkylamine ethoxylate (2 EO); emulsifier, detergent, degreaser for secondary oil recovery, waste treatment, transport cleaners, pipeline/refinery equip./chem. plant cleaning, metal cleaning, metalworking fluids, textiles. *Union Carbide; Union Carbide Europe.*

32433 Triton® X-15
9002-93-1 6858
$[C_{16}H_{26}O_2]_n$
Octoxynol-1
Alkylaryl polyether alcohol; Octyl phenol ethoxylate; Triton X-100 Surfactant; Polyoxyethylated octyl phenol; α-[4-(1,1,3,3-tetramethylbutyl)phenyl]-ω-hydroxypoly(oxy-1,2-ethanediyl); Octoxinol; Triton X 100; Triton X 102; Ethylene glycol octyl phenyl ether; Polyoxyethylene octyl phenyl ether; Octylphenoxypolyethoxyethanol; Polyethylene glycol mono [4-(1,1,3,3-tetramethylbutyl)phenyl] ether; Poly(oxyethylene)-*p-tert*-octylphenyl ether; POE octylphenol; polyoxyethylene (10) octylphenol; POE (10) octylphenol. Surfactant, coupling agent, emulsifier for industrial/household cleaners, emulsion polymerization, agriculture, latex stabilizer. *Union Carbide; Union Carbide Europe.*

32434 Tritox
An aqueous concentrate containing MCPA, mecoprop, and dicamba; a selective herbicide for the control of broad-leaved weeds in turf. *Fisons plc, Horticultural Div.* Name unverified.

32435 Triumphnetzer ZSG
577-11-7 3460 209-406-4
Diisooctyl sulfosuccinate
Wetting agent for textile and chem-tech products, household/industrial cleaners, drycleaning, textile, ceramic, and varnish industries. *Zschimmer & Schwarz.*

32436 Trixene
Moisture-curing urethane prepolymers for surface coatings, adhesives, mastics and sealants; fully reacted urethane polymers in solution; blocked isocyanates for heat-activated systems; moisture scavengers; solvent and water-based acrylic polymers. *Baxenden Chemicals Ltd.*

32437 Trixidin
An emulsion of antimony trioxide, containing 30% Sb_2O_3.

32438 Trocor
Mechanically resistant material (corundum); aluminum oxide; admixtures for the building industry. *Hüls AG.*

32439 Trodax
1689-89-0 6754 216-884-8

Nitroxynil
Anthelmintic. *RMB Animal Health Ltd.*

32440 Trogamid T
Special polyamide, thermoplastic molding compound; injection, blow and extrusion molding compounds used in various branches of industry. *Dynamit Nobel Wien GmbH.* Name unverified.

32441 Trogamid T G35
Special polyamide, glass fiber reinforced; injection molding compounds for electrical engineering, electronics, telecommunication, mechanical and apparatus engineering, precision mechanics. *Dynamit Nobel Wien GmbH.* Name unverified.

32442 Trojan SC
1698-60-8 216-920-2
Suspension concentrate containing 430 g chloridazon per liter; a pyridazinone herbicide for beet crops. *Schering Agrochemicals Ltd.* Discontinued.

32443 Trolene
299-84-3 8415 206-082-6
Ronnel; insecticide. *Dow UK.* Discontinued.

32444 Trolit F
A proprietary pyroxylin product. No manufacturer.

32445 Trolit S and Special
Proprietary phenol-formaldehyde resin molding compounds. No manufacturer.

32446 Trolit W
A proprietary cellulose acetate product. No manufacturer.

32447 Trolite
A synthetic resin of the phenol formaldehyde type. It is a term also applied to trinitro-toluene.

32448 Trolon
Phenolic resins; main fields of application are molded laminated materials, wood working, metal casting, abrasives, friction linings, molded plastics. *Hüls AG.*

32449 Trolovol®
52-67-5 7214 200-148-8
Penicillamine
For the treatment of rheumatoid arthritis. *Bayer AG.*

32450 Troluoil
A proprietary trade name for a petroleum solvent. No manufacturer.

32451 Trona Anhydrous Sodium Sulphate
7757-82-6 8829 231-820-9
Minimum purity 99% Na_2SO_4 in fine, standard, coarse, and special coarse granulations. *Kerr-McGee Chemical Corp.* Discontinued.

32452 Trona Boron Tribromide
10294-33-4 1377 233-657-9
BBr_3
boron tribromide
99.8% minimum BBr_3. Used in manufacture of pure boron. mp = -46°; bp = 91°; d:s0 = 2.6980; *Kerr-McGee Chemical Corp.* Discontinued.

32453 Trona Boron Trichloride
10294-34-5 1378 233-658-4
BCl_3
boron trichloride
99.9% minimum BCl_3. Used in manufacture of boron. mp = -107°; bp = 12°; d⁰ = 1.3728. *Kerr-McGee Chemical Corp.* Discontinued.

32454 Trona Elemental Boron
7440-42-8 1373 231-151-2
B
Amorphous type, standard grade, fine, dark brown powder meeting specifications PA-PD-451, OS11608 and MiL -B-51092 (ORD); boron content is 90-92%. *Kerr-McGee Chemical Corp.*

32455 Trona Muriate of Potash
7447-40-7 7783 231-211-8
White agricultural grade. 60.5% minimum K_2O in coarse, standard, and fine grades. *Kerr-McGee Chemical Corp.*

32456 Trona Potassium Chloride
7447-40-7 7783 231-211-8
High purity white industrial grade; 96.8% KCl(61.8% K_2O equivalent). *Kerr-McGee Chemical Corp.*

32457 Trona Potassium Sulphate
7778-80-5 7845 231-915-5
High purity white industrial grade. 50% K_2O minimum in granular and standard grades. *Kerr-McGee Chemical Corp.*

32458 Trona Salt Cake
7757-82-6 8829 231-820-9
Minimum purity 98% Na_2SO_4. *Kerr-McGee Chemical Corp.* Discontinued.

32459 Trona Soda Ash
497-19-8 8739 207-838-8
Na_2CO_3
sodium carbonate
Grades: dense-99.7% min. Na_2CO_3; Granular-97·7% min. Na_2CO_3; light-98·3% min. Na_2CO_3 (dry basis). *Kerr-McGee Chemical Corp.*

32460 Trona Sulphate of Potash
7778-80-5 7845 231-915-5
potassium sulfate
Standard white agricultural grade with 50% minimum K_2O. *Kerr-McGee Chemical Corp.*

32461 Tronacarb Sodium Bicarbonate
144-55-8 8726 205-633-8
White granular solid, industrial and animal feed grades. *Kerr-McGee Chemical Corp.* Discontinued.

32462 Tronalight Light Soda Ash
497-19-8 8739 207-838-8
sodium carbonate
Grades: Dense-99.7% mineral Na_2CO_3; Granular-97.7% mineral Na_2CO_3; Light-98.3% min. Na_2CO_3 (dry basis). *Kerr-McGee Chemical Corp.*

32463 Tronamang Electrolytic Manganese Metal
7439-96-5 5762 231-105-1
manganese
Chip form; grades: Low-Hy(0.005% H_2), Extra Low-Hy (0.001%H_2), and Nitor-6 nitrided (6% N_2). *Kerr-McGee Chemical Corp.*

32464 Tronamang-75, 85 Manganese Aluminum Briquettes
75% manganese and 25% aluminum in briquette form for aluminum alloying. *Kerr-McGee Chemical Corp.*

32465 Tronox Titanium Dioxide Pigments, Chloride Process
13463-67-7 9612 236-675-5
titanium dioxide
Nine grades for paint, plastics, printing inks and paper applications. *Kerr-McGee Chemical Corp.*

32466 troostite
A mineral, $2(Zn{\cdot}Mn)O{\cdot}SiO_2$. It is also the name for a constituent of steel tempered at a high temperature. It occurs in the transformation of austenite, the stage following marten site, and preceding sorbite.

32467 Trophysan
A proprietary preparation of aminoacids, minerals, and vitamins in sorbitol used for intravenous feeding. *Servier Laboratories Ltd.* Name unverified.

32468 Tropotox
94-81-5 202-365-3
Soluble concentrate containing 400 g/l MCPB; for control of weeds in undersown cereals and grassland. *Rhône-Poulenc Crop Protection Ltd.*

32469 Tropotox Plus
Soluble concentrate containing 37.5 g MCPA and 262.5 g MCPB per liter; for control of weeds in undersown cereals and grassland. *Rhône-Poulenc Crop Protection Ltd.*

32470 Trosiplast
Rigid and plasticized PVC compounds; injection, blow and extrusion molding compounds used in various branches of industry. *Dynamit Nobel Wien GmbH.* Name unverified.

32471 Trosiplast M
Compound PVC; for hollow body blow molding, injection molding, profile extruding, calendering. *Dynamit Nobel Wien GmbH.* Name unverified.

32472 Trosiplast S
Suspension PVC; further processing into injection molding and extrusion molding compounds. *Dynamit Nobel Wien GmbH.* Name unverified.

32473 Trotis
66063-05-6 266-096-3
Pencycuron
A phenylurea fungicide to control black scurf in potatoes. *Bayer AG.*

32474 Trotyl
118-96-7 9860 204-289-6
$C_7H_5N_3O_6$
Trinitrotoluene
(Trolite; Trilite; Tritolo; Trinol; tolite; Triton; Tritole; T.N.T.). an explosive constituent.

32475 Troykyd® D44
Stearate-modified hydrocarbon; defoamer for aqueous. systems, latex paints, pigment dispersions, adhesives, caulks, sealants, cement compounds. *Troy.*

32476 Troykyd® Perma-Dry
Cobalt compound complexed with rare earth and alkaline earth metals; lead-free; feeder drier for paints. *Troy.*

32477 Troysan® 174
[2-(Hydroxymethyl) amino] ethanol
Preservative for protection against bacterial spoilage in aqueous systems,

paints/coatings, resin emulsions, pigment slurries, adhesives, joint cements, and metalworking fluids. *Troy.*

32478 Troysan® 192
2-[(Hydroxymethyl) amino]-2-methylpropanol
Preservative used in aqeous systems. *Troy.*

32479 Troysan® Polyphase® EC17
3-Iodo-2-propynyl butyl carbamate
Fungicide, mildewcide for paints, leather protection, inks, metalworking fluids, paper and textile coatings, adhesives, caulks, sealants, and plastic coatings. *Troy.*

32480 Troysol AFL
Polymeric ester blend; surfactant, defoamer, antifloat agent, air release agent for nonaqeous coatings (alkyd, acrylic, epoxy, polyester, lacquer), printing inks. *Troy.*

32481 Troysol LAC
Alkyl surfactant; substrate wetting agent, flow control additive, leveling agent for aqueous systems, paints, printing inks, adhesives, and coatings for polyethylene and wax coated film and packaging. *Troy.*

32482 Troysol S366
Modified siloxane copolymer; surface active flow/leveling agent, substrate wetting agent for aqeous and nonaqeous systems, paints/coatings, alkyd, acrylic, epoxy, urethane, polyester, lacquer, printing inks. *Troy.*

32483 Troythix A
Polymerized, chemically inert organic ester; bodying agent for trade sales and industrial coatings thinned with hydrocarbons, ketone, and ester solvent; imparts high thixotrophy resistance to sagging, pigment setting, and penetration. *Troy.*

32484 TRS Rubber
A proprietary air dried fast-curing rubber. *Mitsui Co Ltd.* Name unverified.

32485 Trubin
Macrolide antibiotic
Especially useful against CRD in poultry and enzootic pneumonia in pigs; growth promoter. *Bayer AG.*

32486 Trucal
471-34-1 1697 207-439-9
Calcium carbonate
Foodstuffs for animals, additives, included in Class 31 for use in animal foodstuffs but not including cereals or cereal products. *Tilcon Ltd.* Name unverified.

32487 Trucarb
471-34-1 1697 207-439-9
Calcium carbonate
Industrial limestone powders and granules for use in the manufacture of carpets, paints, glues, glass, PVC products, floor covering, mastics, agrochemicals, ceramics, roofing felt, rubber, resins, pigments, and pharmaceuticals. *Tilcon Ltd.* Name unverified.

32488 Truchem Quintex
Emulsifiable concentrate containing 30 g chlorpropham, 30 g fenuron and 130 g propham per liter; an herbicide for beet crops. *MTM Agrochemicals Ltd.*

32489 Tru-Color
Color anodized aluminum. *Reynolds metal Co.* Discontinued.

32490 Tru-Flake® Salt
7647-14-5 8742 231-598-3
Sodium chloride
Flat rectangular crystalline salt prepared by compacting vacuum pan salt. *Akzo Salt.*

32491 Trufree
A range of special flours for dietary use; gluten-free, wheat-free flour for low salt diets. *Larkhall Laboratories plc.* Name unverified.

32492 Trugreen 0-0-2
Micronutrient formulation which promotes and stimulates chlorophyll production; chelating agent with seven elements. *W A Cleary.*

32493 Trulime
1305-62-0 1716 215-137-3
CaH_2O_2
Hydrated lime, used for horticulture/agriculture, alkalinity, building industry, civil engineering (soil stabilization and soil modification), leather processing, organic and inorganic chemicals, petrochemicals, plasterwork, sewage and water treatment. *Tilcon Ltd.* Name unverified.

32494 Trump
Suspension concentrate containing 236 g isoproturon and 236 g pendimethalin per liter; used for annual weed control in winter wheat, rye, and barley. *Cyanamid of Great Britain Ltd.*

32495 Trustan®
Mixture of cymoxanil and mancozeb; systemic fungicide to control potato blight. *DuPont UK.*

32496 Truzone
7722-84-1 4839 231-765-0
hydrogen peroxide
Hydrogen peroxide for the hairdressing trade. *Solvay Interox Ltd.* Discontinued.

32497 Trycite
Polystyrene film used for packaging. *Dow Cheml Co Ltd, UK & Ireland.*

32498 Trycol® 5874
24938-91-8
α-tridecyl-ω-hydroxy-poly(oxy-1,2-ethanediyl)
Trideceth-14; polyoxyethylene (4) tridecyl alcohol; polyoxyethylene (6) tridecyl alcohol; polyoxyethylene (8) tridecyl alcohol; POE (4) tridecyl alcohol; POE (6) tridecyl alcohol; POE (8) tridecyl alcohol. Hydrophilic general purpose emulsifier; for agricultural applications. *Henkel/Emery.*

32499 Trycol® 5882
Laureth-4
Coemulsifier for silicone in polishes, mold release agents; emulsifier in industrial lubricants, agriculture, textile applications; intermediate for shampoo bases; biodegradable. *Henkel/Emery; Henkel/Textile.*

32500 Trycol® 5888
9005-00-9
Steareth-20
Emulsifier, solubilizer, solvent emulsifier in textile dye carriers, agricultural formulations; stabilizer in latexes; used in dyeing assistants, fruit coatings. *Henkel/Emery; Henkel/Textile.*

32501 Trycol® 5940
24938-91-8
Trideceth-6
Detergent, wetting agent, emulsifier, dispersant, foam builder, solubilizer, coupling agent, rewetting agent for institutional, industrial, and household cleaners, degreasers, cutting oils, wool scouring, agricultural applications; intermediate. *Henkel/Emery; Henkel/Textile.*

32502 Trycol® 5950
26183-52-8
Deceth-4
Wetting agent and penetrant for textile and agricultural applications; intermediate. *Henkel/Emery; Henkel/Textile.*

32503 Trycol® 5971
9004-98-2
Oleth-20
Emulsifier, dispersant, solubilizer; textile lubricant; intermediate for shampoo base. *Henkel/Emery; Henkel/Textile.*

32504 Trycol® 6940
9016-45-9 6772
Nonoxynol-5
Emulsifier, detergent, wetting agent; moderate to low foam; for agricultural and textile formulations. *Henkel/Emery; Henkel/Textile.*

32505 Trycol® 6956
9002-93-1 6858
Octoxynol-40
Surfactant for textile use; salt-free. *Henkel/Textiles.*

32506 Trycol® 6985
9014-93-1
Nonyl nonoxynol-8
Emulsifier in textile dye carrier applications, insecticides, wax emulsions; foam control agent; spreading agent in pigment printing; post-stabilizer in emulsion polymerization; intermediate. *Henkel/Emery; Henkel/Textile.*

32507 Trydet 2610
9004-99-3
PEG-23 stearate
Emulsifier, thickener for agricultural formulations. *Henkel/Emery.*

32508 Trydet 2644
56002-14-3
PEG 400 isostearate
Surfactant, lubricant for fiber lubricants, textile processing aids, fabric softeners. *Henkel/Emery; Henkel/Textile.*

32509 Trydet 2676
9004-96-0
PEG-10 oleate
Emulsifier, lubricant for pesticides, metal cleaners, textile detergents and dyeing assistants, leather; rewetting agent for paper. *Henkel/Emery; Henkel/Textile.*

32510 Trydet 2682
61791-00-2
PEG-16 tallate
Detergent, emulsifier, textile leveling agent; coemulsifier. *Henkel/Emery; Henkel/Textile.*

32511 Tryfac® 5552
Phosphate ester, free acid form; surfactant intermediate; salts used as emulsifiers for aliphatic and aromatic solvents, agricultural formulations, textile processing. *Henkel/Emery; henkel/Textile.*

32512 Tryfac® 5553
potassium salt of tryfac 5552. Phosphate ester; emulsifier, wetting agent, detergent, antistat; used in heavy duty cleaners, metalworking compounds, agriculture, textile applications; corrosion inhibitor, dispersant. *Henkel/Emery; Henkel/Textile.*

32513 Tryfac® 5556
51811-79-1
Complex phosphate ester, free acid form; wetting agent, dispersant, antistat in textile processing; solvent emulsifier for textile scours, detergents, pesticides; drycleaning detergent; also used in emulsion polymerization. *Henkel/Emery; Henkel/Textile.*

32514 Tryfac® 5573
12751-23-4 235-799-7
Phosphate ester, free acid form; mold release agent, antistat, dispersant, emulsifier. *Henkel/Emery.*

32515 Trylon® 6735
Alkyl polyether
Low-foaming emulsifier, wetting agent for industrial, institutional, and consumer detergents, textile scouring, bleaching and jet dyeing systems; coemulsifier for solvs. *Henkel/Emery; Henkel/Textile.*

32516 Trylox® 5900
61791-12-6
PEG-5 castor oil; emulsifier, dispersant, carrier, foam control agent, lubricant for paints, paper coatings, dye carriers, agricultural formulations. *Henkel/Emery.*

32517 Trylox® 5921
61788-85-0
PEG-16 hydrogenated castor oil; emulsifier, lubricant, softener; used in fabric softners and aerosol fabric sprays. *Henkel/Emery; Henkel/Textile.*

32518 Trylox® 6746
57171-56-9
PEG-40 sorbitol hexaoleate; oil-water emulsifier, dispersant, wetting agent, lubricant, plasticizer, solubilizer for household/industrial/institutional products, metal lubricants, textile, and cosmetic use. *Henkel/Emery; Henkel/Textile.*

32519 Trylox® 6753
PEG-20 sorbitol
Humectant, plasticizer; intermediate; used in surfactant solution.; emulsifier for textile and cosmetic use. *Henkel/Emery.*

32520 Trylube 7602
Finish for carpet staple providing effective antistatic properties and low fuming. *Henkel/Textiles.*

32521 Trymeen® 6601
61791-14-8
PEG-10 cocamine
Coemulsifier, dispersant, antistat, emulsifier, softener, lubricant for textile applications, industrial lubricants, pesticides; substantive to metals, fibers, clays, etc. *Henkel/Emery; Henkel/Textile.*

32522 Trymeen® 6606
61791-44-4 263-177-5
PEG-15 tallow amine
Emulsifier, antiprecipitant for dye baths, agricultural formulations; leveling agent; intermediate for quaternaries; antistat for synthetic fiber processing. *Henkel/Emery; Henkel/Textile.*

32523 Trymeen® 6617
26635-92-7
PEG-50 stearyl amine
Emulsifier, antistat for metal buffing compounds, latex rubber compounding, agricultural formulations; anticoagulant; lubricant, leveling agent for textile applications. *Henkel/Emery; Henkel/Textile.*

32524 Trymeen® 6620
PEG-30 oleamine
Surfactant for textile use. *Henkel/Textiles.*

32525 Trymeen® 6657
Stearic acid diamide
Lubricant/softener; hydrophobic component with mineral oils and fatty esters in formulating fabric finishes; maintains good whiteness retention during fabric or yarn processing. *Henkel/Emery; Henkel/Textile.*

32526 Trymeen® OAM 30/60
PEG-30 oleamine
Surfactant for textile use. *Henkel/Textiles.*

32527 Trymeen® SAM-50
26635-92-7
PEG-50 stearyl amine
Antistat, emulsifier in metal buffing compounds; lubricant for fiberglass; leveling agent for dye applications. *Henkel/Emery; Henkel/Textile.*

32528 Trymeen® TAM-15
61791-44-4 263-177-5
PEG-15 tallow amine
Emulsifier; antiprecipitant for mixed dye baths; leveling agent for acid dyes; migrating agent for dispersed dyes; intermediate for quaternary ammonium compounds; antistat for processing synthetic fibers. *Henkel/Emery; Henkel/Textile.*

32529 Trymer
Rigid polyisocyanurate bunstock used in the manufacture of insulation. *Dow UK.* Discontinued.

32530 trypsin
9002-07-7 9926 232-650-8
Trypsin
Parenzyme; Parenzymol; Tryptar; Trypure. Proteolytic enzyme; formed in the small intestine by the action of a peptidase, enterokinase, on the pancreatic cell production, trypsinogen; proteolytic enzyme; dehairing of hides. Soluble in H_2O, insoluble in organic solvents *Am. Biorganics; Armour Pharmaceutical Co.; Unibios SpA; U.S. Biochemical; Worthington Biochemical.*

32531 TSE Mold Release®
Glycol surfactant; mold release especially for Viton parts. *TSE Industries.*

32532 TSE-2000
Diphenylmethane diisocyanate in methylene chloride; one-coat adhesive to bond millathane to a variety of substrates during vulcanization. *TSE Industries.*

32533 tse-hong
A mixture of white lead, aluminia, ferric oxide, and silica; used by the Chinese for painting on porcelain.

32534 tse-leou
ting-yu
An oil expressed from Chinese tallow seeds (seeds of *Sapium sebiferum*).

32535 T-siloxide
Product of silica fused with 0.1-2% titania; a silica glass.

32536 tsing-lieu
A red pigment used in porcelain painting consisting of a mixture of stannic and plumbic silicates, with copper oxide or cobalt and gold.

32537 T-Size
Rosin emulsion size. *Hercules.* Discontinued.

32538 TSP-12
10101-89-0 8808 231-509-8
$Na_3O_4P{\cdot}12H_2O$
trisodium phosphate dodecahydrate.

32539 TSP-O
7601-54-9 8808 231-509-8
Na_3O_4P
trisodium phosphate, anhydrous.

32540 TSPP
7722-88-5 9377 231-767-1
$Na_4O_7P_2$
tetrasodium pyrophosphate.

32541 Türkischrotöl 100%
68187-76-8 269-123-7
Sodium sulforicinoleate
Solubilizer for cosmetics. *Zschimmer & Schwarz.*

32542 Tuads®
137-26-8 9510 205-286-2
Tetrabutylthiuram disulfide, tetramethylthiuram disulfide or 60:40 blend methyl and ethyl; rubber accelerators, sulfur donors, accelerators, vulcanizing agents. *R. T. Vanderbilt Co Inc.* Discontinued.

32543 Tuasol 100
87-10-5 9747 201-723-6
$C_{13}H_8Br_3NO_2$
3,5-dibromo-N-(4-bromophenyl)-2-hydroxybenzamide
Tribromsalan; TBS; Temasept IV; Diaphene; 3,4,5-Tribromosalicylanilide; ASC-4; Bromsalans; Lamar L-300; Polybrominated salicylanilide; Salicyanilide, 3,4,5-tribromo-; Temasept; Tuasal. Disinfectant. mp = 227-228°; insoluble in H_2O, soluble in organic solvents such as Me_2CO and DMF. *Merrell Dow Pharmaceuticals Inc.* Name unverified.

32544 Tubania
Jeweler's alloys of varying composition, usually containing copper or brass, antimony, tin, and bismuth. The English alloy contains 12 parts brass, 12 parts tin, 12 parts antimony and 12 parts bismuth. German Tubania consists of 4 parts copper, 3 1/4 parts tin and 42 parts antimony.

32545 Tubazole
148-79-8 9426 205-725-8
Thiabendazole and iodophors such as Dermevan, Idonyx or Wescodyne as a

foggable solution; used for controlling various fungal diseases in stored potatoes. *Wheatley Chemical Co Ltd.* Name unverified.

32546 Tubazole

Mixture containing nonylphenoxypoly (ethyleneoxy)ethanol-iodine complex and thiabendazole; used for controlling various diseases in stored potatoes. *Dean Agrochemicals Ltd.*

32547 Tubazole M

TBZ and iodophor as a sprayable solution; for controlling various diseases in stored potatoes. *Wheatley Chemical Co Ltd.* Name unverified.

32548 Tubergran

82-68-8 8264 201-435-0

Quintozene in dry granular form; for controlling common scab and rhizoctonia in growing potatoes. *Wheatley Chemical Co Ltd.* Name unverified.

32549 Tubodust, Tubostore

117-18-0 204-178-2

Dustable powder containing respectively, 5% or 6% w/w tecnazene; protectant fungicide and potato sprout suppressant. *Farmers Crop Chemicals Ltd.* Discontinued.

32550 Tubotin

Fungicide with fentin hydroxide as the active component. *May & Baker Ltd.* Name unverified.

32551 Tubotox

88-85-7 3341 201-861-7

Fungicide based on dinoseb. *May & Baker Ltd.* Name unverified.

32552 Tuc-tur metal

A nickel silver. It contains from 59-61% copper, 21-28% zinc, 12-18% nickel, and 0.3% iron.

32553 Tuex

137-26-8 9510 205-286-2

Tetramethylthiuram disulfide

Short curing range in natural with normal to high sulfur. fast curing SBR and is flat curing; accelerator with sulfur for nitrile, butyl and EPDM rubbers. *Uniroyal.* Name unverified.

32554 Tuf Stuf

Two-part epoxy putty; for filling and bonding most materials; when set (rock hard) can be drilled, tapped, filed, sanded, contoured, painted and polished; resistant to oil, water, fuel, bleach and dilute acids. *Hermetite Products Ltd.* Name unverified.

32555 Tufcote

Acrylic emulsion finish. *DuPont UK.*

32556 Tufcote E-50SM

Polyether urethane foam, 1-mil aluminized surface; high acoustical performance for absorbers, barriers, and composite products. *E-A-R.*

32557 Tufdene®

BR, SBR thermoplastic elastomers. *Asahi Chem. Industry.*

32558 Tuf-Draw

Lubricant containing mineral oil; may contain additives such as emulsifiers, corrosion inhibitors, biocides, surfactants, and lubricating additives. *Franklin Oil Corporation (Ohio).*

32559 Tuff Stuff

Two-component flexible adhesive. *Hardman.*

32560 Tufflake™

Degradation resistant aluminum pigments; specifically formulated for automotive OEM coatings systems. *Silberline Mfg Co Inc.*

32561 Tufftride

Metal surface treatment for wear and fatigue. *Degussa Ltd.* Discontinued.

32562 Tuflin HS-7066 NT7

9002-88-4 7728

LLDPE; blown film resin for grocery sacks, shipping sacks, heavy-duty film, trash bags, liners. *Union Carbide.*

32563 Tuf-Lube

Fluorocarbon dry surface lubricant, release agents. *Specialty Products Co.* Name unverified.

32564 Tufnol

Laminated plastics materials bonded with synthetic resins incorporating fillers such as cotton fabric, paper and asbestos fabric; the materials are used for electrical and mechanical components in most manufacturing industries. *Tufnol Ltd.*

32565 Tufprene®

BR, SBR thermoplastic elastomers. *Asahi Chem. Industry.*

32566 Tufseal

A trademark for a range of polymerisable mixtures of asphalt, polyols, and isocyanates; used as adhesives. *Robertson Co.* Unverified.

32567 Tufset

A rigid polyurethane used for engineering purposes. *Tufnol Ltd.*

32568 Tuftane

High performance polyester and polyether based urethane films; for high performance applications such as fabric lamination, belting, protective covers and similar uses where the durability of urethane is required. *Lord Corporation (UK) Ltd.*

32569 Tuftec®

Hydrogenated thermoplastic elastomer. *Asahi Chem. Industry.*

32570 Tugon, Tugon OKO, Tugon sp 80

52-68-6 9753 200-149-3

Trichlorfon

Insecticide for control of adult flies and fly larvae. Used for indoor control of crawling and flying insects in public health. *Bayer AG.*

32571 Tula metal

An alloy of silver, copper, and lead.

32572 Tullanox HM-250

Hydrophobic precipitated silica, modified by organic silazane compound; high surface area, high water repellency; provides reinforcement, rheology control, corrosion resistance, anticaking, thickening to silicone sealants, coatings, powders, polyester resins, liquid systems, elastomers; electrical insulators, defoamers; catalyst carrier; filler/additive for plastics, paints, coatings, inks, pharmaceuticals, cosmetics, fertilizers, metals, adhesives, toners. *Tulco.*

32573 Tumbleblite

Systemic fungicide. *Murphy Chemical Co Ltd.* Discontinued.

32574 Tumblebug

Garden insecticide. *Murphy Chemical Co Ltd.* Discontinued.

32575 Tumblemoss

Moss killer preventer. *Murphy Chemical Co Ltd.* Discontinued.

32576 Tumbleslug

Slug and snail killer *Murphy Chemical Co Ltd.* Discontinued.

32577 Tumbleweed Gel

Weedkiller for spot application. *Murphy Chemical Co Ltd.* Discontinued.

32578 Tuncast

Tunning refractory; used for lining tundishes. *Foseco (F.S.) Ltd.*

32579 Tundak

Preformed insulating refractory cones for lining nozzle wells in tundishes. *Morton Int'l. Ltd.*

32580 tung oil

8001-20-5 9944 232-272-3

Chinese wood oil. The oil obtained by pressure from the seeds of *Aleurites cordata and Aleurites fordii* of China and Japan. The seeds contain from 40-53% of oil, main component eleostearic acid.

32581 tunga resin

A neutral glycerol-rosin ester, made with the aid of tung oil as esterifying catalyst.

32582 Tungophen® B

Condensation product of substituted phenols and formaldehyde resins; for improving the hardness and through-drying of alkyd paints and varnishes. *Bayer AG.*

32583 tungsten

7440-33-7 9945 231-143-9

W

Tunsten

Wolfram. Metallic element; high-speed tool steel, alloys, filaments for electric light bulbs, contact points, x-ray and electron tubes, welding electrodes, heating elements, rocket nozzles, sheet steel, chemical apparatus, high-speed rotors, solar energy devices. mp = 3410°; bp = 5660°; d_4^{20}= 18.7-19.3; LD_{50} (rat ip) = 5 g/kg. *Aldrich; Atomergic Chemetals; Cerac; Noah Chem.*

32584 tungsten blue

O_5W_2

ditungsten pentoxide

A colloidal solution of the blue oxide of tungsten. It may be used for dyeing silk.

32585 tungsten brass

Wolfram brass. An alloy of 60% copper, 22% zinc, 14% nickel, and 4% tungsten. An alloy containing 60% copper, 34% zinc, 2.8% aluminum, 2% tungsten, 0.7% manganese and 0.15% tin is also known by this name.

32586 tungsten bronze

Wolfram bronze. An alloy made by fusing potassium tungstate with pure tin; used for decorative purposes. It is also the name for an alloy of 95% copper, 3% tin, and 2% tungsten. An alloy containing 90% copper and 10% tungsten is also known by this name.

32587 tungsten hexafluoride

7783-82-6 9946 232-029-1

F_6W

tungsten (VI) fluoride

Vapor-phase deposition of tungsten, fluorinating agent. mp= 2°; bp = 17°; d^{15} = 3.441. *Air Prods & Chem; Akzo; Elf Atochem N. Am.; Atomergic Chemetals.*

32588 tungsten steel
A very hard alloy of steel and tungsten. It usually contains from 5-8% tungsten, often 4% chromium, and 1.25% carbon; used for armor plates, projectiles, firearms, and high speed tools. Tool steels contain 1-4% tungsten and a rifle barrel steel contains 3-6% tungsten.

32589 tungsten steel, high
These steels usually contain from 80-85% iron with more than 14% tungsten. Some alloys contain from 77-81% iron, 15-18% tungsten, 3-4% chromium, and 0.15-0.35% silicon.

32590 tungsten steel, low
These alloys usually contain about 96% iron with 1.5-2% tungsten, 0.5-1% chromium, and 0.15-0.35% silicon.

32591 tungsten trioxide
1314-35-8 9947 215-231-4
O_3W
tungstic oxide
tungsten trioxide; tungstic acid anhydride; wolframic acid anhydrous. Forms metals by reduction, alloys, preparation of tungstates for xray screens, fireproofing fabrics, yellow pigment in ceramics. Insoluble in H_2O. *Atomergic Chemetals; Cerac; Climax Molybdenum.*

32592 tungstic acid
7783-03-1 9948 231-975-2
H_2O_4W
tungstic(VI) acid
Wolframic acid; orthotungstic acid. Textiles (mordant, color resist), plastics, tungsten metal, wire, etc. Poorly soluble in H_2O. *Am. Int'l. Chem.; Atomergic Chemetals; Noah Chem.*

32593 turacine
A red coloring matter contained in the feathers of the turaco birds of Africa. The coloring matter contains 8% copper.

32594 turbadium bronze
An alloy of 46% copper, 44% zinc, 5% lead, 2% nickel, 1.5% manganese, and small quantities of tin and aluminum; used for propeller castings.

32595 Turbair
60-51-5 3269 200-480-3
Dimethoate
An organophosphorus insecticide and acaricide. *Pan Britannica Industries Ltd.*

32596 Turbair Grain Store Insecticide
Mixture of tenitrothion, permethrin, and resmethrin; used to control insects in grain stores. *Pan Britannica Industries Ltd.*

32597 Turbair Permethrin
52645-53-1 7321 258-067-9
Permethrin; a pyrethroid insecticide. *Pan Britannica Industries Ltd.*

32598 Turbair Roval
36734-19-7 5093 253-178-9
Iprodione; a fungicide with protectant activity for glass house crops. *Pan Britannica Industries Ltd.*

32599 Turbex
Nonionic surfactant. *ABM Chemicals Ltd.* Name unverified.

32600 turbiston bronze
An alloy containing 55% copper, 41% zinc, 2% nickel, 1% aluminum, 0.84% iron, and 0.16% manganese. It resists sea water.

32601 Turboclean
A blend of detergents and surfactants with inhibitors; a cleaning fluid for compressors of gas turbine engines. *The Kent Chemical Company Ltd.*

32602 Turbo-Grass
Mineral and plant extracts in a water base containing cytokinin, B-vitamin, morphogenic and porphyrin activity to aid in increased plant metabolism and yield; for all agricultural, horticultural, and forestry products. *SN Corp/Appropriate Technology Ltd.* Discontinued.

32603 Turbonit
A German synthetic varnish paper product; used for electrical insulation. No manufacturer.

32604 Turfclear
10605-21-7 1836 234-232-0
A liquid formulation containing 500 g carbendazim per liter as a suspension concentrate; systemic fungicide. *Fisons plc.*

32605 turgolds
A name applied to substances such as textile fibers, hide, tissue, leather, and wood fibers, which swell in water but do not dissolve.

32606 Turgum® S
Turpene-resin acid blend; plasticizer and conditioner for SBR; retarder for high, intermediate, and super abrasion furnace black/natural rubber stocks. *Whitney & Oettler.* Name unverified.

32607 Turkey red oils
9953
Sulfonated castor oils, monopol oil, soluble castor oil, sulfated oil, red oil, oleine. Viscid, transparent liquids; used in the preparation of cotton fiber for printing, alizarin dye assistant, textiles, leather, manufacture of soaps.

32608 turmeric
8024-37-1 9955
Indian saffron, terra Merita, curcuma. A natural dyestuff obtained from the underground stems of rhizome of *Curcuma longa and C. rotunda*. The dyeing principle is curcumin, $C_{21}H_{20}O_6$. It dyes cotton greenish-yellow, and is also a coloring matter for wools, silks, oil, butter, cheese, curry powder, wood and wax.

32609 Turnbull's blue
Gmelin's blue. Ferrous ferricyanide, $Fe_3[Fe(CN)_6]2$.

32610 turpentine
8006-64-2 9957 232-350-7
The exudation from from incisions made in certain varieties of pine, fir, and larch. American, French, German, Mexican, Portuguese, and Spanish turpentine refers to the balsam turpentine; German, Finnish, Polish, Russian, and Swedish refers to wood turpentine. Volatile essential oil whose chief constituents are pinene and diterpene; used as a solvent for paints, varnishes, and lacquers.

32611 turpentyne
Turpenteen. A turpentine substitute composed of rosin spirit, shale spirit, petroleum spirit, and coal tarnaphtha.

32612 turpeth mineral
1312-03-4 5941 215-191-8
$HgSO_4 \cdot 2H_2O$
mercuric subsulfate
Turbith mineral; Queen's yellow. Mercuric subsulfate, a yellow basic sulfate of mercury. Insoluble in H_2O.

32613 Turpex
Finishing agents. *Ciba plc.* Name unverified.

32614 Turrisin
Milk product stabilizer. *Giulini Corp.*

32615 Tursione
Biocidal preparation. *BDH Chemicals Ltd.*

32616 Tusadin
A proprietary agent for protection against frost in motor engines. No manufacturer.

32617 tussar silk
Tasar silk. The product of the caterpillar of *Antheraca paphia* of India.

32618 Tutania
Alloys. An English type contains 91% tin, 8% lead, 0.7% copper, and 0.3% zinc; and another 80% tin, 165 antimony, 2.7% copper, and 1.3% zinc; the German alloy contains 92% antimony, 7% tin, and 1% copper; another consists of 62% antimony.

32619 Tutenag
Tutenague; Tutenay. A nickel silver. It consists of from 44-46% copper, 16-40% zinc, and 15-40% nickel.

32620 Tutol No. 2.
An explosive similar to Rexite (*qv*); it contains sodium nitrate instead of potassium nitrate, and 12% of the explosive base is replaced by sodium chloride. No manufacturer.

32621 Tutol®
A registered trade name for certain explosives. No manufacturer.

32622 Tutor
Glass fabric based products having fire resistance. *Rentokil Ltd.*

32623 tutty powder
Tutia. An impure oxide of zinc, formed during the smelting of lead ores containing zinc. Sometimes a mixture of blue clay and copper filings is sold under this name.

32624 Tween®
Polyexyethylene sorbitan fatty acid esters; surfactants. *ICI Am.*

32625 Tween® 20
9005-64-5 8872
Polysorbate 20 NF
Solubilizer; oil-water emulsifier; detergent for shampoos; antistat and fiber lubricant used in textile industry; flavor emulsifier. *ICI Spec. Chem.; ICI Surf. Belgium.*

32626 Tween® 21
9005-64-5 8872
Polysorbate 21
Emulsifier, solubilizer; antistat, fiber lubricant for textiles. *ICI Spec. Chem.; ICI Surf. Belgium.*

32627 Tween® 40
9005-66-7 8872
Polysorbate 40 NF

Emulsifier, solubilizer; textile antistat, fiber lubricant. *ICI Spec. Chem.; ICI Surf. Belgium.*

32628 Tween® 60, 60K
9005-67-8 8872
Polysorbate 60 NF
Food emulsifier, dough strengthener for food industry; flavor emulsifier and dispersant; foaming agent for beverages. *ICI Spec. Chem.; ICI Surf. Belgium.*

32629 Tween® 61
9005-67-8 8872
Polysorbate 61
Emulsifier, solubilizer for perfume, flavor, vitamin oils. *ICI Spec. Chem.; ICI Surf. Belgium.*

32630 Tween® 65, 65K
9005-71-4 8872
Polysorbate 65
Emulsifier for cake mixes, icings, fillings, coffee whiteners, frozen desserts, whipped toppings; antifoam; 65K is kosher grade. *ICI Spec. Chem.; ICI Surf. Belgium.*

32631 Tween® 80, 80K
9005-65-6 7742
Polysorbate 80 NF
Emulsifier, solubilizer for food, vitamins, oils; antifoam; wetting agent, detergent for cleaning contact lenses, skin care products; deflocculant. *ICI Spec. Chem.; ICI Surf. Belgium.*

32632 Tween® 81
9005-65-6 7742
Polysorbate 81
Emulsifier, solubilizer for perfume, flavor, vitamin oils. *ICI Spec. Chem.; ICI Surf. Belgium.*

32633 Tween® 85
9005-70-3 8872
Polysorbate 85
Emulsifier, solubilizer for perfume, flavor, vitamin oils; floating bath oils. *ICI Spec. Chem.; ICI Surf. Belgium.*

32634 Twent®
For skin impurities. *Bayer AG.*

32635 Twinspan
Mixture of chlorpyrifos and disulfoton; a systemic and fumigant insecticide for brassica crops. *Pan Britannica Industries Ltd.*

32636 Twin-Tak
Herbicide. *May & Baker Ltd.* Name unverified.

32637 Twitchells reagent
Benzenestearo-sulfonic acid; used in the decomposition of fats.

32638 Two Cubed Eight
1309-37-1 4072 215-168-2
ferric oxide
A gamma ferric oxide for magnetic media use. *Pfizer International.* Discontinued.

32639 Twosward
Fertilizers. *ICI Chem & Polymers Ltd.* Discontinued.

32640 Tybrite
Coextruded packaging films. *Dow Chemical.*

32641 Tycel®
Urethane laminating adhesives; for lamination of many polymeric materials to rigid and flexible substrates. *Lord Corporation (UK) Ltd.*

32642 Tycel® 7000 Series
Laminating adhesives for bonding applications in decorative and protective building, automotive, and general purpose laminations. *Lord.*

32643 Tychem®
Synthetic polymers; chemicals and curing agents for polymers and elastomers. *Reichhold.*

32644 Tygacell
Composite molding with carbon, glass, aramid and ceramic reinforcement; structural components for racing cars and power boats, components for civil air-craft, space defence, medical X-ray equipment and high performance applications. *Fothergill Tygaflor Ltd.* Name unverified.

32645 Tygadure
High performance insulated wires and cables and optical fibers; used for avionics and electronics, industrial applications, short haul data communications, secure and hazardous environments. *Courtaulds Advanced Materials (Holdings) Ltd.* Discontinued.

32646 Tygafion
Fluorocarbons with fiber, metal and other fillers; molded and machined custom components for all industries; used for high performance electrical applications, expansion joints and shaft covers. *Courtaulds Advanced Materials (Holdings) Ltd.* Discontinued.

32647 Tygaflor
Fluorinated coated glass and aramid fabrics; for process conveying belt systems used in the baking and packaging industries and other industrial uses, lightweight membrane roofs and radomes. *Courtaulds Advanced Materials (Holdings) Ltd.* Discontinued.

32648 Tygafluor
A proprietary trade name for an aqueous dispersion of PTEE (*qv*) with a curing temperature of 90-140°C. No manufacturer.

32649 Tygalam
Composite molding with carbon, glass, aramid, and ceramic reinforcement; structural components for racing cars and power boats, components for civil air-craft, space defense, medical X-ray equipment, and high performance applications. *Fothergill & Harvey plc.*

32650 Tygan
Polyvinylidene chloride coated fabric; filtration, insect screening, glare-reducing blinds. *Courtaulds Advanced Materials (Holdings) Ltd.* Discontinued.

32651 Tygatape
Engineered, high performance PTEE and silicone tapes; masking applications in light/heavy engineering, electrical and electronics, aerospace. *Courtaulds Advanced Materials (Holdings) Ltd.* Discontinued.

32652 Tygavac
Materials for vacuum bag molding of composite components, TFF aerosol release sprays; Aerospace, automotive, medical engineering, industrial applications. *Courtaulds Advanced Materials (Holdings) Ltd.* Discontinued.

32653 Tyglas
Glass fiber woven fabrics; for electrical insulation, filtration, reinforced plastics, thermal insulations, industrial plant applications. *Courtaulds Advanced Materials (Holdings) Ltd.* Discontinued.

32654 Tygon F
A proprietary trade name for a furan resin. No manufacturer.

32655 Tylac®
Synthetic rubber, latex or synthetic resin emulsions. *Reichhold.*

32656 Tylac® 68151-00
9003-18-3
Butadiene acrylonitrile emulsion polymer; for nonwoven industry; excellent oil and other hydrocarbon resistance; excellent pigment acceptance; suggested for synthetic leather, saturant for needle punched fabrics. *Reichhold/Emulsion Polymers.*

32657 Tylac® 68202-00
9003-55-8 8534
Styrene butadiene latex; multipurpose adhesive base for laminating cloth to cloth; firm film. *Reichhold/Emulsion Polymers.*

32658 Tyllanex
5915-41-3 227-637-9
Terbuthylazine
Agan Chemical Manufacturers Ltd.

32659 Tylose® C, C-p, CB, CB-p
9004-32-4 1877
Sodium CMC
Binder in pencil leads; thickener in batteries, rubber industry, cosmetics, foodstuffs, pharmaceuticals, tobacco and textile industry; dispersant, emulsifier for insecticidal, fungicidal and herbicidal products; plasticizer in ceramics; surface sizing in paper industry; press aid and lubricant in welding electrodes. *Hoechst Celanese/Colorants & Surf.; Hoescht AG.*

32660 Tylose® H Series
9002-62-0
Hydroxyethylcellulose
Binder, thickener, plasticizer, viscosity control agent, protective colloid in ceramics, emulsion polymerization, tobacco, and textile industry, agriculture, cosmetics, soaps, and hand cleaning pastes. *Hoechst Celanese/Colorants & Surf.; Hoechst AG.*

32661 Tylose® MHB
9032-42-2
Methyl hydroxyethylcellulose
Binder, thickener, pigment, foam, and filler stabilizer, dispersant, emulsifier, plasticizer, viscosity control and sedimenting aid, and protective colloid used in coatings, paints, resins, mining, batteries, insecticides, fungicides, herbicides; rubber, texttile and leather industry, ceramics, suspension polymerization and pharmaceuticals. *Hoechst Celanese/Colorants & Surf.*

32662 Tylose® P, P-x, PS-x, P-z Series
Methyl hydroxycellulose
Binder in plasters, adhesive, and troweling compounds. *Hoechst Celanese/Colorants & Surf.*

32663 tylosin
1401-69-0 9963 215-754-8
$C_{46}H_{77}NO_{16}$
Tylan; Tylon. An veterinary antibiotic derived from an actinomycete

resembling *Streptomyces fradioe* . mp = 128-132°; $[\alpha]_D^{25}$ = -46° (c = 2 MeOH); λ_m = 282 nm ($E^{1\%}_{1cm}$ 245); soluble in H_2O (5 mg/l), more soluble in organic solvents.

32664 Tynex®
Polymer. *DuPont UK.*

32665 Tyox®
For the electrical industry. *DuPont UK.*

32666 Typar®
Fibers. *DuPont UK.*

32667 Type 798 Roving
Highly filamentized glass bulked roving sized with silane binder; for use in polyester and vinyl ester resin systems for filament winding and pultrusion applications. *PPG/Fiber Glass.*

32668 type metal
A variable alloy of lead and antimony, frequently with the addition of tin, and sometimes copper or bismuth. The lead is present to the extent of from 50-93% the antimony 4-30% the tin 2-40% copper 0-5% and bismuth 0-29%.

32669 typewriter metal
An alloy of 57% copper, 20% nickel, 20% zinc, and 3% aluminum.

32670 Typly
Rubber and metal bonding agents. *Anchor Chemical Group plc.*

32671 Ty-Ply BN
One-coat adhesive for NBR. *Lord.*

32672 Typophor®
Dye base preparations in olein; for brightening of printing inks. *BASF AG.*

32673 Typro®
Fibers. *DuPont UK.*

32674 Tyrenka
A proprietary trade name for a synthetic fiber resembling nylon. No manufacturer.

32675 Tyril 125
9003-54-7
SAN resin; blending resin. *Dow Plastics.*

32676 Tyril 880
9003-54-7
SAN resin; strong, transparent thermoplastic with good chem. and heat resist., toughness, load-bearing strength, easy processing; injection molding resin for cosmetic containers, industrial battery cases and caps, high-pressure filter housings, packaging, appliances, blood aspirators, other medical parts, connectors, tumblers, dinnerware, utensils. *Dow Plastics.*

32677 Tyril 880B
9003-54-7
SAN random copolymer resin; high heat, high strength resin; superior chemical resistance to other grades and high viscosity; offers the lowest melt flow making it superior for injection blow molding; avail. in a wide range of transparent, translucent, and opaque colored cylindrical pellets. *Dow Plastics.*

32678 Tyril 1011
9003-54-7
SAN resin; automotive resin; *see also Tyril 880. Dow Plastics.*

32679 Tyrin
Chlorinated polyethylene resins and elastomers used in a wide variety of applications including roofing membranes, wire and cable, automotive tubing, ignition wire. *Dow Cheml Co Ltd, UK & Ireland.*

32680 Tyrin CM 0136
Chlorinated polyethylene elastomer; used in extruded and calendered goods, e.g., hose, cable jacketing, linings, gasketing, o-rings; excellent heat aging, ozone resistance, good weathering, good oil, flame, and grease resistance; outstanding chem. resist.; vulcanizable with organic peroxides, aminothiadiazole systems or radiation. *Dow Plastics.*

32681 Tyrin CM 0836
Chlorinated polyethylene elastomer; vulcanizable with organic peroxides, aminothiadiazole systems. *Dow Plastics.*

32682 Tyrin CM 566
Chlorinated polyethylene elastomer; used in cable jacketing, extruded goods requiring low moisture, excellent electrical properties; vulcanizable with organic peroxides, aminothiadiazole systems. *Dow Plastics.*

32683 Tyrin CM 3615
Chlorinated polyethylene elastomer; thermoplastic resin which can be fabricated as the primary polymer, as a major alloy component, or as a modifying agent imparting impact resist. to other resins (e.g., SAN, PP, PE, PVC); excellent chemical resistance, easy processing, used for siding, molded goods, extruded profiles, hoses, sheet, lining, tubing, an impact modifier grade. *Dow Plastics.*

32684 Tyrite® 7412
One-component urethane adhesive; structural moisture-cure adhesive; high performance hose adhesive for Hytrel®, nylon, polyester, Kevlar®, thermoplastics and TPU. *Lord.*

32685 Tyrite® 7660
One-component urethane adhesive; structural moisture-cure adhesive with high solids, grab and flexibility initial high grab, crosslinks over time to form a structural bond to foam, plastic, fabric, rubber, and prepared metals. *Lord.*

32686 Tyrol-2, 32B, 6, CEP
Flame retardant materials for plastics. *Stauffer Chemical.* Name unverified.

32687 Tyrolite
$Cu_3As_2O_8{\cdot}2Cu(OH)_2{\cdot}7H_2O$
Cupriferous calamine; copper froth. A basic copper arsenate of green color, found in the Tyrol.

32688 Tyrosilane C
131044-77-4
Copper acetyl tyrosinate methylsilanol
Tanning activator and anti-aging action for cosmetic and health products. *Exsymol.*

32689 tysonite
a) A mineral that is a fluoride of cerium metallic elements with thorium. b) A blend of Gibsonite and vulcanized vegetable oils.

32690 Tytanpol
13463-67-7 9612 236-675-5
Titanium dioxide. *British Traders & Shippers Ltd.*

32691 Tyvek Practik®, Protech®
Fibers. *DuPont UK.*

32692 Tyvek®
Spunbonded polyolefin. *DuPont UK.*

32693 Tyzor®
Organic titanate. *DuPont UK.*

32694 U 46®
Used for weed control in cereals, maize, sugar cane, rice, grassland, forestry, perennial crops, tree crops. *BASF AG.*

32695 U Blue 104
Ultramarine blue. Inorganic colorant. Mixture of bismuthoxychloride and ultramarine. *Presperse.* Discontinued.

32696 U Pink 113, U Violet 109
7787-59-9 1303 232-122-7
BiClO
Ultramarine pink
bismuthoxychloride. inorganic colorant. *Presperse.* Discontinued.

32697 U35
Polyester-based aliphatic urethane aqueous dispersion; lightfast, medium soft polymer producing a dry surface feel; used as skincoat for transfer coating applications; excellent adhesion to plasticized PVC. *BF Goodrich/Spec. Polymers.* Name unverified.

32698 Uba
16893-85-9 8769 240-934-8
Styxol; tanatol. Preparations containing sodium fluorosilicate, Na_2SiF_6, as the main ingredient.

32699 UBOB®
101-54-2 202-951-9
$C_{12}H_{12}N_2$
p-aminodiphenylamine
N-phenyl-*p*-phenylenediamine; 4-aminodiphenylamine. Intermediate. mp = 73-75°. *Uniroyal.*

32700 UBS
Dry cleaning soap and paint remover. *S & D Chemicals Ltd.* Name unverified.

32701 UC-5500 Series
Two-component polyurethane coatings; for high gloss maintenance coatings, e.g., exterior coatings for process equipment, transport vessels, tanks, structural steel in corrosive environments; excellent weathering, flexibility, good abrasion resistance *Heresite Protective Coatings.*

32702 Ucar® Acrylic 503
Modified acrylic emulsion; surfactant for high-quality exterior, interior, and semigloss house paints. *Union Carbide.* Name unverified.

32703 Ucar® Latex 100
Acrylic latex; for caulks, sealants. *Union Carbide.* Name unverified.

32704 Ucar® Latex 130
9003-20-7
Polyvinyl acetate homopolymer emulsion; multipurpose, large particle size emulsion with high pigment-binding capacity; for ready-to-use tape joint compounds, wall patching compounds, low-cost mastic compounds, and tub and tile caulks. *Union Carbide.* Name unverified.

32705 Ucar® Latex 148
Styrene-acrylic latex; for mastics. *Union Carbide.* Name unverified.

32706 Ucar® Latex 351
Vinyl/acrylic emulsion; high-solids latex for interior and exterior paints. *Union Carbide.* Name unverified.

32707 Ucar® Latex 405
Polyvinyl alcohol homopolymer latex; for building industry applications. *Union Carbide.* Name unverified.

32708 Ucar® Phenoxy Resin PKHM-301
Phenoxy resin; used for flexible and rigid packaging; provides films with high chemical resistance without loss of toughness and durability. *Union Carbide.* Name unverified.

32709 Ucar® Vehicle 435
Acrylic polymer used in trade paints, industrial finishes. *Union Carbide.* Name unverified.

32710 Ucar® Vehicle 451
Styrene-acrylic
Polymer used in industrial finishes. *Union Carbide.* Name unverified.

32711 Ucarcide® 225
111-30-8 4480 203-856-5
$C_5H_8O_2$
glutaraldehyde
pentanedial; glutaral; glutaric dialdehyde; 1,3-diformylpropane; Cidex; glutarol; Verucasep. Preservative, antimicrobial for cosmetic, toiletry, and chemical specialty products. bp = 178-179°; d = 1.060; n^{25}_D = 1.4338; soluble in H_2O; LD_{50} (rat orl, 25% solution) = 2.38 ml/kg. *Union Carbide.*

32712 Ucare® Polymer JR-30M, JR-125, JR-400, LR-30M, LR-400, SR-10
Polyquaternium-10
Hair fixative; for cosmetics, toiletries, hair and skin care prods. *Amerchol Corp.*

32713 Ucarsil® FR-1A, FR-1B
Organosilicon chemical; additives allowing the processing of up to 70% alumina trihydrate into polyolefins. *Union Carbide.*

32714 Ucecoat
Polyurethanes for textile coatings. *UCB Chemical Sector.*

32715 Ucecryl
Acrylic emulsions. *UCB Chemical Sector.*

32716 Ucefix
Modified acrylic copolymers or polyurethanes for adhesives. *UCB Chemical Sector.*

32717 Uceflex
Thermoplastic polyurethanes. *UCB (Chem) Ltd.* Discontinued.

32718 Ucelone
Dispersions of silicone resins; acrylic emulsions. *UCB Chemical Sector.*

32719 Uchatius bronze
Steel bronze. An alloy containing 92% copper and 8% tin.

32720 Ucicline, Ucipol
Boiler and cooling water treatments. *Laporte Industries Ltd.* Discontinued.

32721 Ucinite
A proprietary trade name for a phenol-formaldehyde resin laminated product. No manufacturer.

32722 Ucipol
Boiler and cooling water treatment. *Laporte Industries Ltd.* Discontinued.

32723 Ucon®
Synthetic lubricants and fluids. *Union Carbide (UK) Ltd.*

32724 Ucon® 50-HB-400
PPG-9-buteth-12
Emollient. *Amerchol Corp.* Discontinued.

32725 Ucon® Fluid AP
9003-13-8
PPG-14 butyl ether
Amerchol Corp.

32726 Ucon® Hydrolube HP-5046
Water-glycol; hydraulic fluid designed to operate at pressures to 7000 psi; superior fire retardancy. *Union Carbide.*

32727 Ucrete
A proprietary cement-modified polyurethane resin used for flooring. *ICI Chem & Polymers Ltd.*

32728 Ucuhuba fat
A fat obtained from the seeds of *Myristica bicuhyba* and contains 92% fatty acids.

32729 Udel
A proprietary polysulfone; a high-performance, high temperature thermoplastic resin used for injection molding and extrusion. *Union Carbide (UK) Ltd.* Name unverified.

32730 Udel® GF-110
25135-51-3
Polysulfone, glass-reinforced; resin with high tensile strength, excellent electricals, good chemical resistance, water and steam resistance; useful as alternative to metals in engineering plastics; metals in engineering plastics, printed circuit boards, connectors, electrical housings, sterilizable medical devices, process and food service equipment, automotive and plumbing fixtures and microwave cookware. *Amoco Chemical Co.*

32731 Udel® P-1700
25135-51-3
Polysulfone resin; tough, rigid, high-strength thermoplastic polymer for electrical/electronic, automotive/aerospace applications, consumer products, medical, process and sanitary pipe, plumbing components, milking machine parts, pollution control equipme and milking machine compnents, pollution control equipment. The general purpose grade for injection modling or extrusion. FDA compliant. *Amoco Chemical Co.*

32732 Udet 950
Sodium alkylaryl sulfonate
Fast dissolving, high active sulfonate for detergent powders. *Degussa.*

32733 Udikral
Acrylonitrile-butadiene-styrene polymers; for injection molding, sheet extrusions. *GE Plastics ABS Ltd.* Name unverified.

32734 Ufablend DC
Surfactant blend; biodegradable high foaming detergent conc. for manufacturing of liquid detergents, dishwash, hard surface cleaners; effective in hard and soft water. *Unger Fabrikker AS.* Discontinued.

32735 Ufacem
Surfactant blends; used for cement and concrete admixtures. *Unger Fabrikker AS.* Discontinued.

32736 Ufacid K
n-dodecylbenzene sulfonic acid
Intermediate for manufacturing of sulfonates used in detergent powders, liquids, emulsifiers; biodegradable. *Unger Fabrikker AS.*

32737 Ufacid KA
2-Phenyl alkylbenzene sulfonic acid
Intermediate for manufacturing of sulfonates used in liquid detergents; biodegradable *Unger Fabrikker AS.* Discontinued.

32738 Ufanon K-80
Cocamide DEA (2:1); biodegradable detergent, foaming agent, wetting agent, thickener, foam stabilizer for liquid detergents, shampoos, cosmetics, leather industry; confers some corrosion protection. *Unger Fabrikker AS.* Discontinued.

32739 Ufapol
Sulfonated and sulfated surfactants; used for emulsion polymerization. *Unger Fabrikker AS.* Discontinued.

32740 Ufapore GP
Anionic surfactant blends; foaming agent for gypsum board. *Unger Fabrikker AS.* Discontinued.

32741 Ufarol
Fatty alcohol sulfate, neutralized with different amines; active ingredient in shampoos, bath products and liquid detergents.

32742 Ufarol Am 30
2235-54-3 218-793-9
Ammonium lauryl sulfate
Biodegradable detergent, wetting and foaming agent for shampoos, bath products, general-purpose detergents, laundry cleaners, carpet shampoos, furniture cleaning, textiles, leather, paints; stable in hard water and alkali, moderately stable in acids. *Unger Fabrikker AS.*

32743 Ufarol Na-30
151-21-3 8782 205-788-1
Sodium lauryl sulfate
Biodegradable detergent, wetting, and foaming agent for shampoos, bath products, general-purpose detergents, laundry cleaners, carpet shampoos, furniture cleaning, textiles, leather, paints; stable in hard water and alkali, moderately stable in acids. *Unger Fabrikker AS.* Discontinued.

32744 Ufarol TA-40
139-96-8 205-388-7
TEA lauryl sulfate
Biodegradable surfactant for shampoos, bath products, carpet shampoos, furniture cleaning, laundry, textiles; mild to hair and scalp. *Unger Fabrikker AS.*

32745 Ufaryl
Sodium alkylbenzene sulfonate powder; drum dried, active ingredient in detergent powders; used as emulsifier in herbicide and pesticide systems; air entrainment agent in cement. *Unger Fabrikker AS.* Discontinued.

32746 Ufaryl DB80
Sodium dodecylbenzene sulfonate, branched; detergent, wetting agent, foaming agent, emulsifier for light and heavy-duty detergents, dairy, metal, floor, vehicle and bottle cleaners; wetting agent in insecticides, metal pickling, printing inks, paper proce *Unger Fabrikker AS.*

32747 Ufasan 35
Linear sodium dodecylbenzene sulfonate; biodegradable detergent, wetting agent, foaming agent for liquid detergents, dishwash, hair and car shampoos and in plastics, metal, agriculture, polish, textiles, mining, oil and cement industries; stable in hard water and acids. *Unger Fabrikker AS.* Discontinued.

32748 Ufasan IPA
Isopropylamine alkylbenzene sulfonate
Biodegradable raw material for detergents, dishwash, cleaning liqs., car shampoos, industrial cleaners and in plastic, metal, agriculture, polish, textiles, mining, oil and cement industries; emulsifier in solvent-based waterless hand and industrial cleaners; stable in hard water, diluted acids and alkalis. *Unger Fabrikker AS.*

32749 Ufasan TEA
TEA dodecylbenzene sulfonate, linear; biodegradable surfactant for liquid detergents, dishwash, hair and car shampoos and in plastics, metal, agriculture, polish, textiles, mining, oil and cement industries; stable in hard water, diluted acids and alkalis *Unger Fabrikker AS.*

32750 Ufasoft 75
Quaternary ammonium imidazoline methosulfate; surfactant, softener, fabric conditioner, antistat, bactericide for laundry products. *Unger Fabrikker AS.* Discontinued.

32751 Uffelmann's reagent
Prepared by adding a ferric chloride solution to a 2% phenol solution until it is a violet color. The color of the reagent changes to deep yellow in the presence of lactic acid.

32752 Uformite®
Synthetic resins; used for bonding agents. *Reichhold.*

32753 Uformite® 21-806
Urea formaldehyde resin; similar to Uformite 21-805 but supplied in high-flash naphtha to give less odor and extremely high viscosity; produces baking enamels of low solids content, in automotive enamels and primers, roller-coating and metal-decorating finishes, metal bond-coats and flexible tube enamels; andfor any industrial baking enamel in which butanol solvent has proved troublesome; solvent-reducible type. *Reichhold.*

32754 Uformite® 27-802
9003-08-1
Melamine formaldehyde resin; high-quality, general-purpose thermosetting resin offering fast curing speed coupled with intercoat adhesion; used as a crosslinker with hydroxyl-functional thermosetting acrylics to achieve exterior durability; in enamels for automobiles, appliances and metal furniture; solvent-reducible type. *Reichhold.*

32755 Ufoxane 2
8061-51-6
Purified, desulfonated and fractionated high molecular weight sodium lignosulfonate; dispersant for textile dyestuffs. *Borregaard Ligno Tech.*

32756 Ugikral RA, RB and SN
Proprietary trade names for acrylonitrile-butadiene-styrene terpolymers. No manufacturer.

32757 uhligite
A mineral that is a titanate of zirconium, calcium, and aluminum; found in East Africa.

32758 Ulcatite
A proprietary trade name for a rubber vulcanization accelerator containing hexamethylenetetramine, benzthiazyl disulfide, and diphenylguanidine. No manufacturer.

32759 Ulco
A metal used as a substitute for Babbitt metal. It contains 98-99% lead, the rest being barium and calcium. Used for machinery bearings.

32760 Ulcony
An alloy of 65% copper and 35% lead.

32761 ulexine
485-35-8 2858 207-616-0
$C_{10}H_{14}N_2O$
1,2,3,4,5,6-hexahydro-1,5-methano-8H-pyrido[1,2-a][1,5]diazocin-8-one
baptitoxine; cytisine; Cytiton; sophorine. An alkaloid found in laburnum and furze. mp= 152-153°; bp_2 = 218°; $[\alpha]^{17}_D$ = -120°; soluble in H_2O (0.77 g/ml), more soluble in organic solvents; LD_{50} (mus orl) = 101 mg/kg.

32762 Ulmal
A proprietary trade name for an alloy containing aluminum with 10% magnesium, 1% silicon and 0.5% manganese. No manufacturer.

32763 ulmite
A name for the dark-colored fiber covering the grains of sandstone found on the coast of New South Wales. In its properties it resembles those of humus, obtained from brown peat.

32764 Ulon
A proprietary polyurethane elastomer. *Unitex Ltd.*

32765 Ultara®
Fibers. *DuPont UK.*

32766 Ultem®
61128-46-9
Polyetherimide. Used for plastic components for automotive, electrical, electronics, lighting, medical, packaging, audio, etc. *GE Plastics Ltd.*

32767 Ultem® 1000
61128-46-9
Amorphous polyetherimide resin; thermoplastic resin; superior properties with outstanding processability, machinability, and finishing characteristics; for automotive and aircraft applications exposed to gasoline and oils, for laboratory ware and automotive heat transfer systems, electrical and electronic applications, computer circuitry, microwave components. An unreinforced grade of resin. *GE Plastics; Polymer Corp.*

32768 Ultem® 2100, 2112
61128-46-9
Polyetherimide resin, 10% glass-reinforced; the glass reinforced resins provide greater rigidity and improved dimensional stability while retaining processability. 2112 is a low warpage grade with dimensional stability. *GE Plastics; Polymer Corp.*

32769 Ultem® 2212
61128-46-9
Polyetherimide resin, 10% short-glass reinforced. Low waroage grade for dimensional stability. *GE Plastics.*

32770 Ultem® 4000
61128-46-9
Amorphous thermoplastic polyetherimide resin, reinforced, internal lubricant; bearing-grade resin for applications requiring enhanced wear resistance. *GE Plastics.*

32771 Ultem® 6000
61128-46-9
Thermoplastic polyetherimide copolymer resin; high temperature grade; processable on standard injection molding equipment. *GE Plastics.*

32772 Ultem® CRS5001
61128-46-9
Polyetherimide, thermoplastic; highest chemical resistance in series, excellent performance in hot water; also for extrusion. *GE Plastics.*

32773 Ultem® CRS5011
61128-46-9
Polyetherimide, thermoplastic; improved chemical resistance, processability and performance in hot water. *GE Plastics.*

32774 Ultem® FXU100
61128-46-9
Polyetherimide structural foam resin, unreinforced; high-performance engineering structural foam; used in the aircraft and automotive industries and applications requiring resistance to chemical attack. *GE Plastics.*

32775 Ultem® LTX100A
Polyetherimide/polycarbonate blend; offers heat/chemical resistance and impact strength. *GE Plastics.*

32776 Ultimem
Plastic (or other polymeric material) membranes used for separation and related liquid treatments. *ICI Chem & Polymers Ltd.*

32777 Ulto Accelerator
Zinc salt of a complex dithiocarbamate; a proprietary rubber vulcanization accelerator. No manufacturer.

32778 Ultra 206
Organic brightener system; used for bright nickel electroplating (high leveling, fast brightening). *Harshaw Chemicals Ltd.*

32779 Ultra Anhydrous Lanolin HP-2060
8006-54-0 5371 232-348-6
Lanolin
Cosmetic grade for lip and other products where odor, taste and color are critical; moisturizer; self-emulsifying. *Henkel/Cospha; Henkel Canada.* Name unverified.

32780 Ultra Pure
302-01-2 4809 206-114-9
H_4N_2
hydrazine
hydrazine anhydrous. Hydrazine propellant; propellant for mono and bi-propellant systems. mp = 2°; bp = 113.5°; d^{25}_4 = 1.0036; $n^{22.3}_D$= 1.46979; LD_{50} (mus orl) = 59 mg/kg. *Olin.*

32781 Ultra Touch
Water, glycerin, stearic-TEA, butyl stearate, parasepts, fragrance; moisturizing hand and body lotion. *Virkler Chemical Co.* Name unverified.

32782 Ultra Zinc DMC
137-30-4 10305 205-288-3
$C_6H_{12}N_2S_4Zn$
Zinc dimethyldithiocarbamate

Ziram. A proprietary trade name for a rubber vulcanization accelerator. No manufacturer.

32783 Ultra®

Bismuth-based; highly lustrous pearl pigments available in nitrocellulose paste dispersions for use in frosted nail enamels. *Mearl.*

32784 Ultrabase

White soft paraffin (10%), liquid paraffin (10%), and stearyl alcohol (8%). *Schering Health Care Ltd.* Discontinued.

32785 Ultrablend®

PBT/PC blends; used for technical parts in the automotive and electrical industries. *BASF AG.*

32786 Ultrablend® S

PBT/ASA blends; used for injection molding for automotive and electrical/electronics industries; high resistance to heat distortion, high gloss, good outdoor performance, no tendency to environmental stress cracking. *BASF AG.*

32787 Ultracast PE-35, PE-60

89339-41-3

PDI-PTMEG; polyurethane prepolymer producing PU elastomer with high hydrolytic stability and excellent dynamic performance, e.g., for oil field service, rollers, gaskets, pumps, wheels, rollers, bearings, shock absorbers. *Air Prods & Chem/Polyurethanes.* Name unverified.

32788 Ultracene

A guanidine derivative used as a proprietarey rubber vulcanization accelerator. No manufacturer.

32789 Ultrachem Assembly Fluid 1

A tacky polymer; used for assembly of o-rings used in helicopter transmissions, jet turbines, pumps, etc. *Ultrachem Inc.* Name unverified.

32790 Ultracut

Water-reducible lubricants containing mineral oil, but when diluted in water forms a micro-emulsion which is translucent to transparent; additives include emulsifiers, corrosion inhibitors, surfactants, biocides, and lubricating additives. *Franklin Oil Corporation (Ohio).*

32791 Ultra-DMC

598-64-1 209-945-5

C5H14N2S2

dimethylamine dimethyldithiocarbamate

Carbamodithioic acid, dimethyl-, compd. with N-methylmethanamine (1:1); Carbamic acid, dimethyldithio-, compd. with dimethylamine (1:1); Carbamic acid, dimethyldithio-, dimethylamine salt (1:1;)Dimethylamine dimethyldithiocarbamate; Dimethylammonium dimethyldithiocarbamate; Dimethyl dithiocarbamate dimethylammonium salt; Dimethyldithiocarbamic acid compd. with dimethylamine (1:1;)Dimethyldithiocarbamic acid dimethylamine salt; Dimethyldithiocarbamic acid, dimethylammonium salt. A proprietary trade name for a vulcanization accelerator whose main ingredient is dimethylamine dimethyldithiocarbamate. No manufacturer.

32792 Ultradur®

Polybutylene terephthalate granules; used for injection molding (electrical engineering components, key buttons). *BASF plc.*

32793 Ultradur® B 2550

26062-94-2

Thermoplastic polyester resin (PBT); low-viscosity resin used for coating paper and board for packaging oven-ready deep-freeze and convenience foods; primarily for extrusion. *BASF.*

32794 Ultradur® B 4300 G10

26062-94-2

Thermoplastic polyester resin (PBT), 50% glass fiber-reinforced; used for engineering parts with particularly great rigidity; primarily for injection molding. *BASF.* Name unverified.

32795 Ultradur® B 4500

26062-94-2

Thermoplastic polyester resin (PBT); medium-viscosity resin offering high precision for engineering parts, e.g., cams, number barrel mechanisms and poppet valves; primarily for injection molding. *BASF.* Name unverified.

32796 Ultradur® KR 4001

26062-94-2

Thermoplastic polyester resin (PBT), 25% mineral-reinforced; used for housings and front panels for household appliances; primarily for injection molding. *BASF.* Name unverified.

32797 Ultrafast® 830 Liq

UVabsorber for improvement of lightfastness and photo degradation properties of polyester fibers; especially suitable in combination with cationic dyes. *BASF.*

32798 ultraferran

A colloidal iron.

32799 UltraFine® II

1309-64-4 752 215-175-0

O_3Sb_2

Antimony trioxide

diantimony trioxide; flowers of antimony; Senarmontite; Valentinite; Exitelite; Weisspiessglanz. Flame retardant for liquid systems, e.g., unsaturated polyesters, epoxies, PU, phenolics, textile treatments; produces minimum loss of physical properties in ABS, etc. mp = 655°; bp = 1425°; d = 5.2000; LD_{50} (rat orl) > 20 g/kg. *Laurel Industries.*

32800 Ultra-Flat

Aluminum alloy sheet. *Reynolds Metal Co.* Discontinued.

32801 Ultraflex®

63231-60-7 264-038-1

Microcrystalline wax; plastic wax offering high ductility, flexibility at very low temps., provides protective barrier properties against moisture vapor and gases; uses include hot-melt laminating adhesives for papers, films, and foils; hot-melt coatings; in anti-sun checking agents in rubber goods, electrical insulating agents, leather treating agents, water repellents for textiles, rustproof coatings, cosmetic ingredients, plasticizer for waxes; used in crayons, dental compounds, chewing gum and candles. *Petrolite.*

32802 Ultraflo

A heat-stable multi-active betaglucanase preparation prodiuced by a selected strain of *Humicola insolens* ; used in the mashing process of beer to degrade β-glucan for better filtration. *Novo Nordisk.*

32803 *Ultraform*

Polyacetal. Degussa Ltd.

32804 Ultraform®

Polyoxymethylene/polyacetal granules

Used for injection molding, mechanical engineering, cassette hubs. *BASF plc.*

32805 Ultraform® H 2320

Acetal copolymer resin; high molecular weight resin for extrusion of thin tubes and panels and thick stock for machining; also suitable for thick injection moldings. *BASF.*

32806 Ultraform® N 2200 C4X

Acetal copolymer (POM), carbon fiber-reinforced; higher strength and stiffness than glass-reinforced grades, good electrical conductivity, excellent resistance to fuels and solvs., good frictional properties; used for gas pumps, sliding bearings. *BASF.* Name unverified.

32807 Ultraform® N 2200 G5

Acetal copolymer resin, glass-reinforced; high strength and rigidity injection molding compound. *BASF.* Name unverified.

32808 Ultraform® N 2211 PVX

Acetal copolymer resin, lubricated; injection molding compound with lubrication; permits low coeff. of friction and reduced wear with no adverse effect on mech. properties. *BASF.* Name unverified.

32809 Ultraform® N 2320

Acetal copolymer resin; rapidly solidifying, general purpose injection molding production. *BASF.* Name unverified.

32810 Ultraform® N 2320 BK 11005 MO

Acetal copolymer resin, MoS_2 lubricated; injection molding compound for the production of friction bearings. *BASF.* Name unverified.

32811 Ultraform® W 2320

Acetal copolymer resin; very easy flow, rapidly solidifying product for injection molding to meet extreme demands on processing. *BASF.* Name unverified.

32812 Ultraglaze® 4400

Two-part silicone sealant; high-strength, high-modulus structural glazing sealant. *GE Silicones.*

32813 Ultrahold 8

Acrylates/acrylomide copolymer. Resin used in hair spray to provide outstanding hold and humidity resistance. *BASF; BASF AG.*

32814 Ultrahold, Ultrahold 8

Resins for hydrocarbon propellant hairsprays. Ultrahold 8 has outstanding hold and humidity resistance. *BASF plc.*

32815 Ultralen® SP 3700 S, SP 3705

25038-59-9 7730

PET, light-stabilized; spinning polymers for production of bright textile yarns *BASF AG.*

32816 Ultralen® SP 3705

25038-59-9 7730

PET. Light-stabilized; spinning polymerused for the production of semi-dull textile yarns. *BASF AG.*

32817 ultra-light alloys

Alloys having a specific gravity below 2 are known by this name. Magnesium-aluminum-zinc and magnesium-copper are alloys of this type.

32818 Ultralin

A rosin-fatty acid reaction product.

32819 Ultralog

Grade of chemical reagents with 99.8% to 100% purity. *Schweizerhall.*

32820 Ultralumin
A jeweler's alloy containing more than 90% aluminum, with nickel, copper, and some rare earth metals of the thorium group. It is specially resistant to sea water.

32821 ultramarine
1317-97-1 9980
Lapis-lazuli blue, oriental blue, brilliant ultramarine, French blue, new blue, permanent blue, French ultramarine, soda ultramarine; C.I. Pigment Blue 29; C.I. 77007; Ultramarine soda. Blue pigment from lapis lazuli. Mixture of sodium aluminosulfosilicates. Formerly prepared from the rare mineral lapis-lazuli, by powdering and washing. It is prepared artificially by fusing together kaolin and sulfur with soda or with a mixture of sodium sulfate and charcoal.

32822 Ultramarine ash
When preparing ultramarine from lapis-lazuli, a blue product is first yielded, then a pale blue, and finally a pale bluish-grey material, which is called ultramarine ash.

32823 Ultramid®
Polyamides. *BASF; BASF plc.*

32824 Ultramid® A3
32131-17-2
Nylon 6/6 resin; low-viscosity resin for extrusion of monofilaments and bristles. *BASF*.

32825 Ultramid® A3G5
32131-17-2
Nylon 6/6 resin, 25% glass fiber-reinforced; for machine parts and housings of great rigidity and dimensional stability. *BASF*. Name unverified.

32826 Ultramid® A3K
32131-17-2
Nylon 6/6 resin, stabilized; for injection molding of all kinds of engineering parts subjected to high loads, e.g., bearings and gears, and dielectric parts, e.g., terminal strips and cable binders. *BASF*. Name unverified.

32827 Ultramid® A3X1G10
32131-17-2
Nylon 6/6 resin, 50% glass fiber-reinforced, flame-retarded; for parts with utmost rigidity and enhanced fire resistance. *BASF*. Name unverified.

32828 Ultramid® A4
32131-17-2
Nylon 6/6 resin; medium-viscosity resin for extrusion; stock for machining, monofilaments for spiral zip fasteners, bristles, and injection-molded engineering parts. *BASF*. Name unverified.

32829 Ultramid® B3
25038-54-4 6832
Nylon 6 resin; low-viscosity resin for extrusion of monofilaments, bristles, fishing lines, nets, film, stretched tapes and coatings. *BASF*. Name unverified.

32830 Ultramid® B35G3
25038-54-4 6832
Nylon 6 resin, 15% glass fiber-reinforced; for housings of enhanced impact strength, e.g., for car outer rear-view mirrors and wheels for cross-country bicycles. *BASF*. Name unverified.

32831 Ultramid® B3EG10, B3WG10
25038-54-4 6832
Nylon 6 resin, 50% glass fiber-reinforced, stabilized; for machine parts, housings, and sheathed magnets of very great rigidity, dimensional stability, and resistance to hightemp aging; B3EG10 preferred for dielectrics. *BASF*. Name unverified.

32832 Ultramid® B3G10
25038-54-4 6832
Nylon 6 resin, 50% glass fiber-reinforced; for machine parts, housings of very great rigidity, dimensional stability. *BASF*. Name unverified.

32833 Ultramid® B3K
25038-54-4 6832
Nylon 6 resin, stabilized; for injection molding of machine parts, mainly with wall thicknesses > 2 mm. *BASF*. Name unverified.

32834 Ultramid® B5
25038-54-4 6832
Nylon 6 resin; high-viscosity resin for extrusion of tubes, sections, extruded stock of very high impact strength for machining, tubular film, panels, tapes, thick monofilaments, and blow moldings, e.g., tanks for hydraulic fluids. *BASF*. Name unverified.

32835 Ultramid® BS 3300
25038-54-4 6832
Polyamide 6 (polycaprolactam); fiber grade polymer for production of industrial yarns. *BASF AG*.

32836 Ultramid® BS 400 S
25038-54-4 6832
Polyamide 6 (polycaprolactam); fiber grade polymer for production of textile yarn. *BASF AG*.

32837 Ultramid® BS 416
25038-54-4 6832
Polyamide 6 (polycaprolactam), light-stabilized; fiber grade polymer for production of dull textile yarn. *BASF AG*.

32838 Ultramid® C35
Copolyamide 6/66 coextruded with polyethylene; composite film for packing foodstuffs, multi-ply blow moldings, extremely tough and flexible fishing lines. *BASF*. Name unverified.

32839 Ultramid® KR 4205
Nylon resin, stabilized, flame-retarded; for injection molding of dielectric parts with improved fire resistance, e.g., contactor bases and plug boards. *BASF*. Name unverified.

32840 Ultramid® S3
9008-66-6
Nylon 6/10 resin; medium viscosity resin for extruded stock for machining, tubular, and flat film, monofilaments and bristles. *BASF*. Name unverified.

32841 Ultramite
471-34-1 1697 207-439-9
Calcium carbonate
Ultrafine wetground product available in slurry form for water-based systems, primarily coatings. *ECC International Ltd*. Discontinued.

32842 Ultramoll® I, II, III
A range of polyadipates; polymeric plasticizers. *Bayer AG; Bayer plc.*

32843 Ultramoll® M
Polyadipate
A polymeric plasticizer for use in the formulation of flame retardant pastes for polyester based Moltopren. *Bayer AG*.

32844 Ultramoll® PP
Polyphthalate
Polymeric plasticizer for PVC; resistant to oil and bitumen, largely resistant to migration; also for PVC plastisols. *BASF AG*.

32845 Ultramoll® TGN
Polyphthalate
Polymeric plasticizer for PVC; relatively low viscosity; particularly suitable for use in pigment mill bases and as a desensitizer for peroxides. *BASF AG*.

32846 Ultranox® 236
96-69-5 202-525-2
$C_{22}H_{30}O_2S$
4,4'-Thio-bis (2-t-butyl-5-methylphenol)
5-*tert*-butyl-4-hydroxy-2-methylphenyl sulfide. Antioxidant for use in adhesives, rubber articles for repeated use, polymers include polyolefins, PVC, acrylic ethyl cellulose; antioxidant for lubricants, cutting oils, water-sol. oils, hydraulic oils. mp = 161-164°. *GE Specialty*.

32847 Ultranox® 246
119-47-1 204-327-1
$C_{23}H_{38}O_2$
2,2'-Methylene-bis-(4-methyl-6-t-butylphenol)
2,2'-methylenebis(6-*tert*-butyl-4-methylphenol). Stabilizer, antioxidant for PP, polyethylene, Polyoxymethylene copolymer, polyoxymethylene homopolymer, styrenics, and rubber. mp = 123-127°; *GE Specialty*.

32848 Ultranox® 257
68610-51-5 271-867-2
Polymeric sterically hindered phenol; butylated reaction product of *p*-cresol and dicyclopentadiene; antioxidant for white, transparent, and colored natural and synthetic rubber goods; especially for latex applications such as carpet backing, dipping articles; also for EVA-based hot melts, thermoplastic rubber and stryenics. *GE Specialty*.

32849 Ultranox® 276
2082-79-3 218-216-0
$C_{35}H_{62}O_3$
Octadecyl 3,5-di-t-butyl-4-hydroxyhydrocinnamate
octadecyl-3-(3,5-di-*tert*-butyl-4-hydroxyphenyl)propionate. High molecular weight antioxidant/stabilizer for styrenics, polyolefins, PVC, urethane and acrylic coatings, adhesives, and elastomers; effective replacement for BHT in polyolefins. mp = 50-52°. *GE Specialty*.

32850 Ultranox® 626
26741-53-7 247-952-5
Bis(2,4-di-t-butylphenyl) pentaerythritol diphosphite
Mixed with 0.5-1.2% triisopropanolamine; antioxidant for polyolefin, PVC, PET, styrenics, ABS, and PC polymers; stabilizer. *GE Specialty*.

32851 Ultranox® 627A
26741-53-7 247-952-5
Bis (2,4-di-t-butylphenyl) pentaerythritol diphosphite
Antioxidants providing color stability, hydrolytic stability, reduction in polymer degradation during processing of ABS, PVC, PC. *GE Specialty*.

32852 Ultranyl®
Polyamide and polyphenylene ether alloy; used in injection molding for

technical parts, e.g., in automotive engineering and for office machines. *BASF AG; BASF plc.*

32853 **Ultrapas**
Melamine resin molding compounds; for manufacture of tableware, bathroom and publicity items, screw tops. *Dynamit Nobel Wien GmbH.* Name unverified.

32854 **Ultrapek® KR 4176**
Polyaryletherketone
Low viscosity, easy flow for injection molding, blow molding, film extrusion, and spinning fibers. *BASF AG.*

32855 **Ultrapek® KR 4177 G4**
Polyaryletherketone, 20% glass fiber-reinforced; general-purpose injection molding grade; also suitable for extrusion. *BASF AG.* Name unverified.

32856 **Ultra-Pflex®**
471-34-1 1697 207-439-9
Ultrafine surface-treated precipitated calcium carbonate; impact modifier in rigid PVC applications. *Pfizer.*

32857 **Ultraphan**
Acetate film. *Lonza AG.*

32858 **Ultraphor® CW**
Optical brighteners for cotton. *BASF.*

32859 **Ultraphor® SFG liquid, SFR Liq**
Optical brighteners for polyester fibers. *BASF.*

32860 **Ultraprene®**
9002-86-2 7746 206-625-7
PVC elastomer; features conformability, resistance to compression set, abrasion, high tensile strength, outstanding weatherability; for architectural glazing seals, weatherstripping, wire/cable applications, tool handle grips, automotive interior and exte *Teknor Apex.*

32861 **Ultrasil® VN3SP**
High purity silica; catalyst carrier; reinforcement for rubber compounding. *Degussa.*

32862 **Ultrasol**
Solvents doubly distilled in glass. *British Drug Houses.* Name unverified.

32863 **Ultrason® E 1010**
Polyethersulfone
Very easy flow grade for moldings with adverse ratios of runner length to wall thickness. *BASF AG.*

32864 **Ultrason® E 1010 G2**
Polyethersulfone, 10% glass fiber-reinforced; very easy flow for moldings with adverse ratios of runner length to wall thickness. *BASF AG.* Name unverified.

32865 **Ultrason® E 2010**
Polyethersulfone
Easy flow, general-purpose grade for injection molding; also suitable for extrusion, film extrusion, blow molding. *BASF AG.* Name unverified.

32866 **Ultrason® E 3010**
Polyethersulfone
High viscosity grade for injection molding, extrusion, film extrusion, blow molding. *BASF AG.* Name unverified.

32867 **Ultrason® E 6010**
Polyethersulfone
Very high viscosity grade for extrusion, film extrusion, blow molding. *BASF AG.* Name unverified.

32868 **Ultrason® KR 4101**
Polyethersulfone, 30% mineral-reinforced; easy-flow, general-purpose injection molding grade; also suitable for extrusion. *BASF AG.* Name unverified.

32869 **Ultrason® S 1010**
25135-51-3
Polysulfone
Very easy flow grade for moldings with adverse ratios of runner length to wall thickness. *BASF AG.*

32870 **Ultrastyr**
A proprietary range of high impact polystyrene resins. *Montedison UK Ltd.* Name unverified.

32871 **Ultratex**
Finishing agent. *Ciba plc.* Name unverified.

32872 **Ultrathane**
Hot, cast, thermosetting polyurethane from various isocyanates combined with polyesters or polyethers and crosslinked with polyols or diamines to form elastomeric products; used for solid tires, roller coverings, squeegee blades, abrasion resistant lining *Watts Urethane Products Ltd.*

32873 **Ultrathene®**
Polyolefin copolymers. *Quantum Chemical Corp.*

32874 **Ultrathene® UE 630-000**
EVA copolymer; resins with good flow and flexibility, toughness, processability, and low temperature properties; used in flexible and semiflexible parts. *Quantum/US.*

32875 **Ultrathene® UE 632-000**
EVA copolymer; extrusion coating resin; film extrusion *Quantum/US.* Name unverified.

32876 **Ultrathene® UE 659-04**
EVA copolymer; for adhesives and coatings; FDA compliance. *Quantum/US.* Name unverified.

32877 **Ultravist**
73334-07-3 5076 277-385-9
Aqueous solutions of iopromide; x-ray contrast media. *Schering Health Care Ltd.* Discontinued.

32878 **Ultravon**
Pretreatment agent. *Ciba plc.* Name unverified.

32879 **Ultravon An**
A proprietary trade name for a fatty acid amide derivative; it emulsifies oils and fats more effectively and A special detergent primarily for wool scouring. It maintains the handle and color of wool better than conventional detergents. No manufacturer.

32880 **Ultrawet**
High foaming versatile nonionic surfactant; for liquid detergent formulations. *Cornelius Chemical Co Ltd.* Name unverified.

32881 **Ultrawet DS**
Anionic surfactant supplied as cream colored flakes; detergency, wetting, sudsing, dispersing and emulsifying agent for speciality cleaning and industrial processing. *Cornelius Chemical Co Ltd.* Name unverified.

32882 **Ultrawet K and AOK**
Anionic surfactant supplied as cream colored flakes; various industrial and heavy-duty house-hold detergents. *Cornelius Chemical Co Ltd.* Name unverified.

32883 **Ultrax®**
Self-reinforcing liquid crystalline polymers; for extrusion and injection molding; very high dimensional stability of moldings, low water absorption, low thermal expansion coefficient. *BASF AG.*

32884 **Ultrazine CA**
8061-52-7
Calcium lignosulfonate
Used in gypsum board manufacturing *Borregaard Ligno Tech.*

32885 **Ultrazine NA**
8061-51-6
Purified high molecular weight sodium lignosulfonate; high dispersing efficiency for textile dyestuffs, pesticides, gypsum board, and concrete additives. *Borregaard Ligno Tech.*

32886 **Ultrazym**
A purified pectolytic enzyme preparation produced from a selected strain of *Aspergillus niger*; can be used in any case where the aim is breaking down of soluble and insoluble pectins with varying degrees of esterification for reduction of viscosity. *Novo Nordisk.*

32887 **Ultrmoll NR**
Polybutadiene-acrylonitrile
Improves abrasion resistance and dimensional stability of PVC profiles used for shoe soles, automotive profiles, cables, window frames and jointing profiles. *Bayer AG.*

32888 **Ultrmoll PP**
Polyphthalate
Low viscosity polymeric plasticizer for PVC with high resistance to oil and bitumen. *Bayer AG.*

32889 **Ultrmoll PU**
Linear polyurethane; compatible with PVC giving high abrasion resistance and used for shoe soles, cables, film, injection moldings and profiles. *Bayer AG.*

32890 **Ultrmoll TGN**
Polyphthalate
Low viscosity polymeric plasticizer for PVC for use in pigment mill bases and as a desensitizer for peroxides. *Bayer AG.*

32891 **Ultroil**
A sulfonated vegetable oil; a proprietary trade name for a wetting agent for textiles. No manufacturer.

32892 **Ultrol®**
Calbiochem Corp.

32893 **Ultryl 6010**
A proprietary nonstabilized PVC resin of the suspension type. *Phillips Petroleum Int'l.* Name unverified.

32894 **Ultryl 6500**
A proprietary PVC polymer of the suspension type; used in the production of glass-clear film, tube, etc. *Phillips Petroleum Int'l.* Name unverified.

32895 Ultryl 6800
A proprietary plasticizer-free PVC resin used in the manufacture of pipe and profiles. *Phillips Petroleum Int'l.* Name unverified.

32896 Ultryl 7100
A proprietary PVC resin of the suspension type with easy processing properties. *Phillips Petroleum Int'l.* Name unverified.

32897 Ultryl 7150
A proprietary PVC resin of the suspension type containing additives to give high clarity. *Phillips Petroleum Int'l.* Name unverified.

32898 Ulvio Cocoa
A German proprietary food material prepared by exposing cocoa to uv radiation. No manufacturer.

32899 umbelliferone
1317-97-1 9980
$C_9H_6O_3$
7-hydroxy-2H-1-benzopyran-2-one
7-hydroxycoumarin; hydrangin; skimmetin. Aglucon of skimmin, obtained by distillation of *umbelliferae* resins. Formed by metabolism in coumarin in humans. Used in sunscreens and as medical reagent. mp = 225-228°; soluble in H_2O (10 g/l); more soluble in organic solvents.

32900 umber
Umber brown, mineral brown, velvet brown, chestnut brown, manganese velvet brown, burnt umber. Mineral varieties of umber. They are ochers colored brown by oxides of manganese, and containing varying amounts of clay. Burnt umber is umber calcined at low.

32901 umbrenal
A 25% solution of lithium iodide in ampoules.

32902 Umbrite A
An explosive containing 49% nitroguanidine, 38% ammonium nitrate, and 13% silicon.

32903 Umbrite B
An explosive containing 37.5% nitroguanidine, 49.5% ammonium nitrate, and 13% silicon.

32904 umburana seed
The product of *Amburana Claudii* ; used in Brazil for perfuming tobacco.

32905 umea tar
A pale Swedish pine-wood tar. It is a good variety of Stockholm tar produced in the Umea district.

32906 UN-28 and UN-32
Fertilizer solutions containing 28% and 32% nitrogen respectively; designed for direct agricultural use as a three-way source of nitrogen. *Hercules.* Discontinued.

32907 Unads®
97-74-5 202-605-7
$C_6H_{12}N_2S_3$
bis(dimethylthiocarbamyl) sulfide
tetramethylthiuram monosulfide. mp = 106-108°. *R. T. Vanderbilt Co Inc.* Discontinued.

32908 Unal
123-30-8 482 204-616-2
C_6H_7NO
4-amino-1-hydroxybenzene
p-hydroxyaniline; *p*-aminophenol; Activol; Azol; Certinal; Citol; Paranol; Rodinal; Ursol P. A photographic developer. It is Rodinal (*p*-aminophenol) in a solid form, containing, besides *p*-aminophenol, the ingredients necessary for solidification. mp = 188-190°; bp = 284°; soluble in H_2O (0.8%), methyl ethyl ketone, EtOH, insoluble in C_6H_6, $CHCl_3$.

32909 Unamide® C-5
61791-08-0
PEG-6 cocamide
Foam stabilizer, viscosity builder, emulsifier for shampoos, light and heavy-duty detergents; stable over broad pH range. *Lonza Inc.*

32910 Unamide® C-72-3
Cocamide DEA (2:1); viscosity builder, foam stabilizer; used for light duty liqs., industrial, household hard surface cleaners, surfactant, emulsifier, corrosion inhibitor, lubricant and personal care products. *Lonza Inc.*

32911 Unamine® C
61791-38-6 263-170-7
Coco hydroxyethyl imidazoline
Emulsifier, surfactant, fungicide; textile antistat; leather treating; base for quaternaries; acid detergent and wetting agent; corrosion inhibitor; pigment flushing agent. *Lonza Inc.*

32912 Unamine® O
Oleyl hydroxyethyl imidazoline
Emulsifier, demulsifier; antistat for drycleaning fluids; water displacing agent; corrosion inhibitor; base for quaternaries; pigment flushing agent. *Lonza Inc.*

32913 Unasyn® IM/IV
69-53-4 628 200-709-7
Ampicillin
Bactesyn; Becalin; Duocid; Unacid; Unasyna; Unasyne. Antibiotic. Also known as *Pfizer International.*

32914 Unburn
A proprietary antiburn cream containing benzocaine, hexachlorophane, orthophenyl phenol, menthol and lanolin. *Leeming.* Name unverified.

32915 undecyl dodecyl phthalate
68515-47-9 271-089-3
$C_{34}H_{58}O_4$
Jayflex® UDP. Pasticizer d = 0.959; pour point = -40°; flash point = 437°F (TCC) *Exxon.* Name unverified.

32916 undecylenic acid
112-38-9 9983 203-965-8
$C_{11}H_{20}O_2$
undecenoic acid
10-undecenoic acid; 9-undecenoic acid; 10-hendecenoic acid; Declid; Renselin; Sevinon. Perfumery, flavoring, plastics, modifying agent, medicine (antifungal agent); intermediate in chemical synthesis, polyamide plastics and fibers, synthetic floors. mp = 24.5°; bp = 275° (dec); d^{25}_{25} = 0.9102; n^{25}_D = 1.4486; insoluble in H_2O, soluble in EtOH, $CHCl_3$, Et_2O; LD_{50} (mus orl) = 8.15 g/kg. *Elf Atochem; CasChem; Schweizerhall.*

32917 Unden®, Undene®
114-26-1 8022 204-043-8
propoxur
Insecticide used for control of sucking and chewing insects (aphids, mealybugs, scales, leafhoppers or caterpillars on vegetables, pome and stone fruit, cocoa, rice, oil palms and other crops. *Bayer AG.*

32918 Unger A-50
Softener, antistat for rayon and cotton fibers and yarns; stable in hard water; medium stability in acids and alkalis. *Unger Fabrikker AS.* Discontinued.

32919 Ungerol
Sodium fatty alcohol ether sulfate
Active ingredient in shampoos, bath products and liquid detergents, emulsifying agent in polymerization. *Unger Fabrikker AS.* Discontinued.

32920 Ungerol AM3-75
67762-19-0
Ammonium lauryl ether (3 EO) sulfate
Biodegradable detergent, wetting and foaming agent, emulsifier used in liquid detergents, car shampoos, bath foams; stable in hard water, moderately stable in acids. *Unger Fabrikker AS.* Discontinued.

32921 Ungerol N2-28
9004-82-4
Sodium laureth sulfate (2 EO)
Biodegradable detergent, wetting and foaming agent, emulsifier for liquid detergents, shampoos, bath products, wallboard manufacturing, textiles, drilling auxiliary; excellent stability in hard water, alkalis; moderately stable in acids. *Unger Fabrikker AS.* Discontinued.

32922 Unguentum
A proprietary preparation of silicilic acid, liquid paraffin, soft paraffin, cetostearyl alcohol, polysorbate-40, glycerol, oil, sorbic acid and propylene glycol; a protective skin cream. *E Merck.* Discontinued.

32923 Uni G
Motor oil. *Monsanto (Solaris).* Name unverified.

32924 Unibind Series
Polymer; binding agent for coal and mineral transportation or storage to prevent wind blown losses. *Hart Chem. Ltd.*

32925 Uni-Cal 66
Colorant dispersions; for coloring of nonaqueous industrial and maintenance coating compositions. *Hüls Am.*

32926 Unicast
One-component epoxy casting resins include general purpose, thermally conductive and fire retardant. *Emerson & Cuming Polymer Group.*

32927 Unicell
Lightweight waterproof mortars for high build and overhead applications; used for repairs to concrete and stonework. *Ronacrete Ltd.*

32928 Unicoat
One-component conformal coatings for electronics, dip coats for resistors and capacitors, and spray or brush coatings for general purpose applications. *Emerson & Cuming Polymer Group.*

32929 Unicor
Corrosion inhibitors. *Universal-Matthey Products Ltd.*

32930 Unicrop 6% Mini Slug Pellets
108-62-3 5983 202-945-6
Pellets containing 6% w/w metaldehyde; snail and slug bait. *Universal Crop Protection Ltd.*

32931 Unicrop Leatherjacket Pellets
γ-HCH
An organochlorine insecticide, contains lindane. *Universal Crop Protection Ltd.*

32932 Unicrop Zineb
12122-67-7 10300 235-180-1
Zineb
A dithiocarbamate fungicide. *Universal Crop Protection Ltd.*

32933 Unicrylic
Solvent and acrylic sealant; used for glazing of windows, panels and general caulking. *Pecora Corporation.*

32934 Unidem
Demulsifiers. *Universal-Matthey Products Ltd.* Name unverified.

32935 Unidene
Solution butadiene styrene rubbers. *EniChem Elastomers Ltd.*

32936 Unidri A-74
Dewatering agent for vacuum filtration of minerals. *Hart Chem. Ltd.*

32937 Unidri M-75
Low foaming surfactant for dewatering metallic oxides, coal and other minerals. *Hart Chem. Ltd.*

32938 Uniflex® 192, Uniflex® EHT
Octyl tallate
Plasticizer for synthetic rubbers; metalworking. *Union Camp.* Name unverified.

32939 Uniflex® 300
Polymeric plasticizer; plasticizer for PVC; high temperature electrical applications; high grade vinyl upholstery; adhesives for electrical tapes or wall coverings; gasketing and tubing; low migration. *Union Camp.* Name unverified.

32940 Uniflex® BYO
142-77-8 205-559-6
Butyl oleate
Monomeric plasticizer and processing aid for plastics, rubber; textile lubricant, mold lubricant, in water-proofing agents, polishes, and metalworking lubricants. *Union Camp.* Name unverified.

32941 Uniflex® BYS-Tech, Unimate® BYS
123-95-5 1625 204-666-5
$C_{22}H_{44}O_2$
Butyl stearate
octadecanoic acid butyl ester. Ester with low viscosity, good color, low odor for plasticizers, textile fiber lubricants, metalworking oils. mp = 27°; bp = 343°; slightly soluble in H_2O, soluble in organic solvents; d_{25}^{25} = 0.855-0.875. *Union Camp.* Name unverified.

32942 Uniflex® DBS
109-43-3 203-672-5
$C_{18}H_{34}O_4$
Dibutyl sebacate
Monomeric plasticizer for plastics and rubber (cellulosics, PVC food wraps, nitrite and neoprene rubbers). mp = -12°; d = 0.9400 *Union Camp.* Name unverified.

32943 Uniflex® DCA, Unimate® DCA
105-97-5 203-349-9
$C_{18}H_{34}O_4$
Dicapryl adipate
Monomeric plasticizer for plastics and rubber. *Union Camp.* Name unverified.

32944 Uniflex® DCP
Dicapryl phthalate
Monomeric plasticizer for plastics and rubber. *Union Camp.* Name unverified.

32945 Uniflex® DOA
103-23-1 203-090-1
$C_{22}H_{42}O_4$
Dioctyl adipate
bis(2-ethylhexyl)adipate; di-(2-ethylhexyl)adipate; adipic acid bis(2-ethylhexyl) ester. Monomeric plasticizer for plastics and rubber. bp = 166-168°; d = 0.990; n^{20}_D = 1.4470. *Union Camp.* Name unverified.

32946 Uniflex® DOS
122-62-3 1292 204-558-8
$C_{26}H_{50}O_4$
Dioctyl sebacate
di-(2-ethylhexyl)sebacate; sebacic acid bis(2-ethylhexyl) ester. Monomeric plasticizer for plastics, rubber, and lubricants. *Union Camp.* Name unverified.

32947 Uniflex® EHT
octyl tallate
Plasticizer used in synthetic rubbers and in metalworking. *Union Camp.* Name unverified.

32948 Uniflex® IBYS
646-13-9 5165 211-466-1
$C_{22}H_{44}O_2$
Isobutyl stearate
For textile fiber lubricants, metalworking oils. mp = 20°. *Union Camp.* Name unverified.

32949 Uniflex® TEG-810
capric/caprylate esters with triethylene glycol
Plasticizer for nitrile rubber. *Union Camp.* Name unverified.

32950 Uniflood
Refinery and oilfield chemicals. *Universal-Matthey Products Ltd.* Name unverified.

32951 Uniflot SP 100 Series
Sulfhydril
Copper sulfide collectors for copper/zinc flotation operations. *Hart Chem. Ltd.*

32952 Unifog
Insecticide formulation.

32953 Unifroth G
Glycol blend; frother for flotation operations where selectivity is required. *Hart Chem. Ltd.*

32954 Unifume
137-42-8 6016 205-293-0
$C_2H_4NNaS_2$
methylcarbamodithioic acid sodium salt
metham sodium; methyldithiocarbamic acid sodium salt; sodium methyldithiocarbamate; N-methylaminodithioformic acid sodium salt; sodium N-methylaminodithioformate; sodium metham; metam sodium; sodium metam; SMDC; carbathione; trimaton; VPM. Metam-sodium 40% w/v aqueous solution; for use as a soil sterilant in agriculture and horticulture. soluble in H_2O (722 g/l), alcohols, insoluble in other organic solvents; LD_{50} (rat orl)= 820 mg/kg. *Universal Crop Protection Ltd.*

32955 Unigel
55-63-0 6704 200-240-8
nitroglycerin. A nitroglycerin dynamite; for construction and building industry, explosives, mining, petroleum, and related industries. *Hercules.* Discontinued.

32956 Unihib®
Organic phosphates; for use as corrosion inhibitors in cooling towers and boilers. *Lonza AG.*

32957 Unihib® 106
2809-21-4 3908 220-552-8
$C_2H_8O_7P_2$
1-hydroxyethylidene-1,1-diphosphonic acid
etridonic acid; (1-hydroxyethylidene)diphosphonic acid; ethane-1-hydroxy-1,1-diphosphonic acid; EHDP. Dispersant, scale inhibitor for cooling tower, boiler treatment, oilfield applications. Used medically as a calcium regulator. Soluble in H_2O (69%), insoluble in AcOH; *Lonza Inc.*

32958 Unihib® 305-LC
6419-19-8 229-146-5
Aminotrimethylene phosphonic acid
Dispersant, corrosion and scale inhibitor for cooling tower, boiler treatment, oilfield applications. *Lonza Inc.*

32959 Unihib® 314
2235-43-0 218-791-8
Pentasodium amino tri(methylene phosphonate)
Dispersant, corrosion and scale inhibitor for cooling tower, boiler treatment, oilfield applications. *Lonza Inc.*

32960 Unihib® 905
Diethylene triamine pentamethylene phosphonic acid
Dispersant, corrosion and scale inhibitor for cooling tower, boiler treatment, oilfield applications. *Lonza Inc.*

32961 Unihib® 1324
Diammonium salt of cocamine-di-(methylene phosphonic acid); dispersant, corrosion and scale inhibitor for cooling tower boiler treatment, oilfield applications. *Lonza Inc.*

32962 Unihib® 1704
Bis-hexamethylene triamine phosphonic acid
Dispersant, corrosion and scale inhibitor for cooling tower, boiler treatment, oilfield applications. *Lonza inc.*

32963 Unilab Surgibone
Surgibone. Prosthetic aid. *Unilab Inc.* Name unverified.

32964 Unilin® 425 Alcohol
C_{20}-C_{40} alcohols
Functional polymer for modification of PP, PVC, polyethylene, PS, and highperformance engineering resins; acts as antioxidant, heat stabilizer, uv stabilizer, or viscosity depressant; promotes emollient protective films onto the skin; in cosmetic creams and lotions; used in hot-melt and solvent-based

coatings; textile/leather lubricants and finishes; chemical intermediate; defoamer for pulp/paper processing. *Petrolite*.

32965 Unilink® 4100, 4200
Aromatic diamine
Chain extender for PU elastomers. *UOP*.

32966 Unilink® 450
4,4'-bis(sec-butylamino) diphenyl methane
Used as a chain extender in the production of specialty polyurethane plastics. *UOP Inc*.

32967 Unimate® DCA
dicapryl adipate
Ester used in cosmetics. *Union Camp*. Name unverified.

32968 Unimate® DIPA
6938-94-9 248-299-9
$C_{12}H_{22}O_4$
Diisopropyl adipate
Hexanedioic acid, bis(1-methylethyl) ester; Adipic acid, diisopropyl ester; Ceraphyl 230; Crodamol Da; Beta Dia; Diisopropyl adipate; Iso-adipate 2/043700; Isopropyl adipate; Prodipate; Schercemol Dia; Standamul DIPA; Tegester 504-D; Wickenol 116. Cosmetic ester. *Union Camp*. Name unverified.

32969 Unimate® DIPS
7491-02-3 231-306-4
$C_{16}H_{30}O_4$
Diisopropyl sebacate
Decanedioic acid, bis(1-methylethyl) ester; Sebacic acid, diisopropyl ester; Diisopropyl sebacate; Wickenol 117. Skin emollient, solubilizer, coupler, spreading agent for creams, lotions, bath oils, aerosol toilet preparations. *Union Camp*. Name unverified.

32970 Unimate® EHP
29806-73-3 249-862-1
$C_{24}H_{48}O_2$
Ethylhexyl palmitate
Hexadecanoic acid, 2-ethylhexyl ester; Palmitic acid, 2-ethylhexyl ester; Ceraphyl 368; 2-Ethylhexyl palmitate; Octyl Palmitate; Wickenol 155. Skin emollient for lotions and creams. *Union Camp*. Name unverified.

32971 Unimate® IPM
110-27-0 5234 203-751-4
$C_{17}H_{34}O_2$
isopropyl myristate
tetradecanoic acid 1-methylethyl ester; Estergel. IPM; skin emollient, lubricant, carrier; used in lotions, creams, antiperspirants, bath oils. mp = 3°; bp_{20} = 192.6°; d^{20} = 0.8532; n_D^{25}= 1.432-1.434; insoluble in H_2O, EtOH, soluble in most organic solvents. *Union Camp*. Name unverified.

32972 Unimate® IPP
142-91-6 205-571-1
$C_{19}H_{38}O_2$
isopropyl palmitate
hexadecanoic acid isopropyl ester; palmitic acid isopropyl ester. IPP; skin emollient, lubricant, carrier; used in lotions, creams, antiperspirants, bath oils. mp = 11-13°; d = 0.852; n_D^{20}= 1.010; insoluble in H_2O, EtOH, soluble in most organic solvents. *Union Camp*. Name unverified.

32973 Unimin Series
Blend; selective iron sulfide depressant for use in base metal sulfide separation. *Hart Chem. Ltd*.

32974 Unimite
Ammonia dynamite
Used for wet hole blasting in hard rock for operations such as quarrying and coal stripping. *Hercules*. Discontinued.

32975 Unimoll®
Phthalates plasticizers. *Bayer plc*.

32976 Unimoll® 66 M
84-61-7 201-545-9
$C_{20}H_{26}O_4$
Dicyclohexyl phthalate
Plasticizer for use in formulation of delayed tack heat sealable coatings. mp = 64-66°; insoluble in H_2O, EtOH, soluble in most organic solvents. *Bayer; Miles; Polysar*.

32977 Unimoll® BB
85-68-7 201-622-7
$C_{19}H_{20}O_4$
Benzyl butyl phthalate
Monomeric plasticizer. d = 1.100; n_D^{20} = 1.5400; insoluble in H_2O, EtOH, soluble in most organic solvents. *Bayer AG*.

32978 Unimoll® DB
84-74-2 1622 201-557-4
$C_{16}H_{22}O_4$
dibutyl phthalate
1,2-benzenedicarboxylic acid dibutyl ester; phthalic acid dibutyl ester; DBP. Monomeric plasticizer. Also used as an insect repellent. mp = -35°; bp = 340°; d^{20} = 1.0459; n_D^{20} = 1.4900; insoluble in H_2O (1 part in 2500), very soluble in organic solvents; maximum single does tolerated by rats, >8 g/kg. *Bayer AG*. Discontinued.

32979 Unimoll® DM
131-11-3 3304 205-011-6
$C_{10}H_{10}O_4$
Dimethyl phthalate
1,2-benzenedicarboxylic acid dimethyl ester; phthalic acid dimethyl ester; methyl phthalate; dimethyl 1,2-benzenedicarboxylate acid; DMP; Palatinol M; Fermine; Avolin; Mipax. Fixative for perfumes, plasticizer for cellulose acetate and acetate butyrate. bp = 282°; d = 1.1900; n+ss20D = 1.5150; insoluble in H_2O, soluble in all organic solvents; LD_{50} (rat orl) = 8.2 g/kg. *Bayer AG*. Discontinued.

32980 Union Carbide® A-137
2943-75-1 220-941-2
$C_{14}H_{32}O_3Si$
Octyltriethoxysilane
triethoxy-n-octylsilane. Coupling agent, crosslinking agent providing durability, gloss, hiding power to coatings. bp_2 = 98-99°; d = 0.878. *Union Carbide*.

32981 Union Carbide® A-151
78-08-0 201-081-7
$C_8H_{18}O_3Si$
vinyltriethoxysilane
triethoxyvinylsilane. Coupler promoting bonding between organic and inorganic components of a system; crosslinking agent; provides durability, gloss, hiding power to coatings. bp = 160-161°; d = 0.903; n+ss20D = 1.3960. *Union Carbide*.

32982 Union Carbide® A-162
2031-67-6 217-983-9
$C_7H_{18}O_3Si$
methyltriethoxysilane
triethoxymethylsilane. Coupling agent, crosslinking agent providing durability, gloss, hiding power to coatings. bp = 141-143°; d = 0.897; n+ss20D = 1.3835. *Union Carbide*.

32983 Union Carbide® A-163
1185-55-3 214-685-0
$C_4H_{12}O_3Si$
methyltrimethoxysilane
trimethoxymethylsilane. Coupling agent, crosslinking agent providing durability, gloss, hiding power to coatings. bp = 102-103°; d = 0.955; n_D^{20} = 1.3700. *Union Carbide*.

32984 Union Carbide® A-171
2768-02-7 220-449-8
$C_6H_{12}Si$
Vinyltrimethoxysilane
trimethoxyvinylsilane. Coupling agent, crosslinking agent providing gloss, hiding power to coatings. *Union Carbide*.

32985 Union Carbide® A-172
1067-53-4 213-934-0
$C_{11}H_{24}O_6Si$
Vinyl tris-(2-methoxyethoxy) silane
tris-(2-methoxyethoxy)vinylsilane. Coupling agent, crosslinking agent providing gloss, hiding power to coatings. mp = -30°; bp = 285°; d = 1.0300; n_D^{20}= 1.4284. *Union Carbide*.

32986 Union Carbide® A-174
2530-85-0 219-785-8
$C_{10}H_{20}O_5Si$
γ-Methacryloxypropyltrimethoxysilane
3-(trimethoxsilyl)propyl methacrylate. Coupling agent, crosslinking agent providing gloss, durability, hiding power, adhesion promotion to coatings. bp = 190°; d = 1.0450; n_D^{20}= 1.4313. *Union Carbide*.

32987 Union Carbide® A-186
3388-04-3 222-217-1
$C_{11}H_{22}O_4Si$
β-(3,4-Epoxycyclohexyl) ethyltrimethoxy silane
trimethoxy[2-(7-oxabicyclo[4.1.0]hept-3-yl)ethyl]silane. Coupling agent, crosslinking agent, adhesion promoter for coatings; provides durability. bp = 310°; d = 1.065; n+ss20D = 1.5000. *Union Carbide*.

32988 Union Carbide® A-187
2530-83-8 219-784-2
γ-Glycidoxypropyltrimethoxy silane
Coupling agent, crosslinking agent, adhesion promoter for coatings; provides durability. bp_2 = 120°; d = 1.070; n_D^{20}= 1.4290. *Union Carbide*.

32989 Union Carbide® A-189
4420-74-0 224-588-5

$C_6H_{16}O_3SSi$
γ-Mercaptopropyltrimethoxy silane
trimethoxymercaptopropylsilane. Coupling agent, crosslinking agent adhesion promoter for coatings; provides durability; features active hydrogen reaction, chain transfer, end blocking. bp = 198°; d = 1.039; n_D^{20} = 1.4440. *Union Carbide.*

32990 Union Carbide® A-1100
919-30-2 213-048-4
$C_9H_{23}NO_3Si$
γ-Aminopropyltriethoxy silane
triethoxyaminopropylsilane; 3-(triethoxysilyl)propylamine. Coupling agent, crosslinking agent, adhesion promoter for coatings; provides durability; features active hydrogen reaction, end blocking. bp = 217°; d = 0.942; n_D^{20} = 1.4210. *Union Carbide.*

32991 Union Carbide® A-1106
58160-99-9 261-145-5
$C_3H_{11}NO_3Si$
Silanetriol, (3-aminopropyl)-
σ-Aminopropyl)trihydroxysilane. Aminoalkyl silicone solution; finish for woven fiberglass; coupling agent for glass fiber sizes; filler treatments; additive to water-soluble/dispersion resins include vinyl and acrylic latex, epoxies, and phenolic binder dispersions. for coatings, adhesi *Union Carbide.*

32992 Union Carbide® A-1110
13822-56-5 237-511-5
$C_9H_{23}NO_3Si$
γ-Aminopropyltrimethoxysilane
trimethoxyaminopropylsilane; 3-(triethoxysilyl)propylamine. Coupling agent, crosslinking agent, adhesion promoter for coatings; features active hydrogen reaction. *Union Carbide.*

32993 Union Carbide® A-1120
1760-24-3 217-164-6
$C_8H_{22}N_2O_3Si$
N-β-(Aminoethyl) γ-aminopropyltrimethoxy silane
N-[3-(trimethoxysilyl)propyl]ethylenediamine. Coupling agent, crosslinking agent, adhesion promoter for coatings; features active hydrogen reaction. bp_{15} = 146°; d = 1.019; n_D^{20} = 1.4450. *Union Carbide.*

32994 Union Carbide® A-1160
116912-64-2 245-876-7
γ-Ureidopropyltriethoxysilane
[3-[tri(ethoxy/methoxy)silyl]propyl]urea. Coupling agent, crosslinking agent, adhesion promoter for coatings; features active hydrogen reaction. d = 0.910; n_D^{20} = 1.3900. *Union Carbide.*

32995 Union Carbide® XLP-57D11
Polyester modifier; low profile additive giving improved deep color pigmentation when used in conjunction with orthophthalic-modified rigid polyester resins. *Union Carbide.*

32996 Union Carbide® Y-11343
Low molecular weight silicone/aminofunctional silane coupling agent; crosslinker, adhesion promoter for one- and two-pack RTV silicone sealant systems. *Union Carbide.*

32997 Unipel
Pelleted fertilizer. *Monsanto (Solaris).* Name unverified.

32998 Uniperol® EL
Oxyethylated vegetable oil; surfactant, dispersant, leveling agent, emulsifier for fatty acids, fatty oils, waxes in textile dyeing and printing; stable to hard water. *BASF/Fibers; BASF AG.*

32999 Uniperol® O
Fatty alcohol-polyglycol ether; emulsifier, dyeing auxiliary, leveling and wetting agent, dispersant, detergent used in textile processing. *BASF/Fibers; BASF AG.*

33000 Uniperol® W, W Flakes
Aliphatic ethoxylate
Surfactant, dispersant, leveling agent, protective colloid; for wool and synthetic fibers; sl. wetting action, no detergency; stable to acids, alkalies, water-hardening salts, heavy metal ions, electrolytes. *BASF/Fibers; BASF AG.*

33001 Unipertan P-24, P-242
Hydrolyzed animal collagen, tyrosine, riboflavin; raw material for cosmetics, toiletries, and pharmaceuticals. *Lipo.*

33002 Unipertan P-242
Hydrolyzed animal collagen with tyrosine and adenosine triphosphate. A raw material for toiletries, cosmetics and pharmaceuticals. *Lipo.*

33003 Uniplast
A proprietary trade name for phenol-formaldehyde molding compound. No manufacturer.

33004 Uniplex 80
77-93-0 201-070-7
$C_{12}H_{20}O_7$
Triethyl citrate
ethyl citrate, citric acid triethyl ester. Plasticizer for food packaging bp = 294sg; d+ss20= 1.137; n_D^{20} = 1.4455; soluble in H_2O (7%), more soluble in organic solvents. *Unitex.*

33005 Uniplex 82
77-89-4 201-066-5
$C_{14}H_{22}O_8$
Acetyl triethyl citrate
triethyl O-acetylcitrate. Plasticizer for food packaging bp_{100} = 228-229°; d = 1.136; n_D^{20} = 1.4390. *Unitex.*

33006 Uniplex 84
77-90-7 201-067-0
$C_{22}H_{34}O_8$
Acetyl tributyl citrate
tributyl O-acetylcitrate. Plasticizer for indirect and direct food contact applications; milling lubricant for aluminum foil or sheet steel for use in cans for beverage and food products; in PVC toys, cellulose nitrate films, aerosol hair sprays, and dairy product cartons. d = 1.050; n_D^{20} = 1.4430. *Unitex.*

33007 Uniplex 108
80-39-7 201-275-1
$C_9H_{13}NO_2S$
N-ethyl *p*-toluene sulfonamide
Plasticizer for nylon, shellac, cellulose acetate, protein materials, PVAc adhesives, and nitrocellulose lacquers. mp = 63-65°; bp_{245} = 208°; *Unitex.*

33008 Uniplex 108
1077-56-1 214-073-3
$C_9H_{13}NO_2S$
N-Ethyl *o*-toluene sulfonamide
Plasticizer for nylon, shellac, cellulose acetate, protein materials, PVAc adhesives, and nitrocellulose lacquers. *Unitex.*

33009 Uniplex 110
131-11-3 3304 205-011-6
dimethyl phthalate
Solv. and plasticizer for celiulose acetate butyrate compositions; solv. for organic catalysts; plasticizer and solv. for aerosol hair sprays. *Unitex.*

33010 Uniplex 150
84-74-2 1622 201-557-4
Dibutyl phthalate
Solv./plasticizer in fingernail polish, nail polish remover, hair sprays, organic peroxide catalysts, adhesives, coatings; compatible with cellulosics, methacrylate, PS, PVB, vinyl chloride, urea-formaldehyde, melamine-formaldehye, phenolics. *Unitex.*

33011 Uniplex 155
84-69-5 201-553-2
Diisobutyl phthalate
Plasticizer for thermoplastic and thermoset resins; solv./plasticizer in cellophane, resin coated sand for foundry casting, organic peroxides. *Unitex.*

33012 Uniplex 173
70-55-3 200-741-1
p-toluene sulfonamide
Plasticizer for thermoplastic and thermoset resins; imparts gloss and wetting to melamine, urea and phenolic resins. *Unitex.*

33013 Uniplex 225
35325-02-1 252-512-0
N-(2-hydroxypropyl) benzenesulfonamide
Plasticizer for polyamide, polyurethane, polyacrylic, cellulose ester; excellent antistatic properties and resis. to extraction by water and dry-cleaning solvs.; for paints, lacquers; stabilizer for pigmented unsaturated polyester resins. *Unitex.*

33014 Uniplex 250
84-61-7 201-545-9
$C_{20}H_{26}O_4$
Dicyclohexyl phthalate
dicyclohexyl benzene-1,2-dicarboxylate. Heat activated plasticizer for heat seal applications such as food wrappers/labels, pharmaceutical labels and other applications where delayed heat activated adhesive is required; for printing ink formulations for paper, vinyl, textiles, and other substra mp = 64-65°. *Unitex.*

33015 Uniplex 260
614-33-5 210-379-6
$C_{24}H_{20}O_6$
Glyceryl tribenzoate
Polymer modifier; plasticizer; for heat seal applications, lacquers, films, in PVAc-based adhesives, cellophane coatings, nitrocellulose coatings, nail lacquer formulations, printing inks, polishes; extrusion and injection molding processing aid. mp = 68-72°. *Unitex.*

33016 Uniplex 270
1459-93-4 215-951-9
$C_{10}H_{10}O_4$
Dimethylisophthalate
dimethyl benzene-1,3-dicarboxylate. Modifies clarity and melting pt. of PET resins used in films, blow-molded bottles, etc. mp = 66-67°; bp_{12} = 124° *Unitex.*

33017 Uniplex 310
71486-48-1 275-521-1
Cyclohexyl isooctyl phthalate
Unitex.

33018 Uniplex 552
4196-86-5 224-079-8
$C_{33}H_{28}O_8$
Pentaerythritol tetrabenzoate
Plasticizer for adhesives intended for heat seal applications mp = 102-104°. *Unitex.*

33019 Uniplex 600
1338-51-8 8872 215-667-5
Toluenesulfonamide-formaldehyde resin; modifier and adhesion promoter for synthetic and natural resins used in adhesives and coatings applications; extender in polymide resins. *Unitex.*

33020 Uniplex 680
1338-51-8 8872 215-667-5
o,p-toluenesulfonamideformaldehyde resin in butyl acetate; used in nail polish formulations in conjuction with nitrocellulose to improve durability and gloss; also in PVAc adhesives to impart quick tack and green strength. Used in formulations to bond cellophane. *Unitex.*

33021 Uniplex 809
9004-93-7
PEG di-2-ethylhexoate
Plasticizer with low volatility and excellent heat resistance; for polyester and polyamide engineering plastics; improves mold release. *Unitex.*

33022 Unipor
Pour point depressants. *Universal-Matthey Products Ltd.* Name unverified.

33023 Uniquat
Alkyl imidazoline benzyl quaternary ammonium compounds; germicidal applications in hospitals, institutions, and industrial water treatment. *Lonza AG.*

33024 Unique
A smokeless powder; designed for use in light through heavy shotshell loads; can also be used in handgun loads. *Hercules.* Discontinued.

33025 Uni-Rez® 221
Glyceryl rosinate
Used in clear and pigmented lacquers, sealers, cold-cut additives, aluminum paints, white and light-tint interior paints and enamels. *Union Camp.* Name unverified.

33026 Uni-Rez® 709
Maleic resin; used in aqueous flexographic inks, emulsion floor polishes, varnishes with good color and gloss retention. *Union Camp.* Name unverified.

33027 Uni-Rez® 1039
Limed rosin solution with 5% calcium oxide; additive to low cost paints, varnishes, gloss enamels; as letdown vehicle. *Union Camp.* Name unverified.

33028 Uni-Rez® 1502
Polyamide resin in xylene solution; curing agent; provides good adhesion to metals, concrete, many plastics, good film resiliency, toughness, and chemical resistance; used in epoxy coatings formulated to meet air pollution regulations. *Union Camp.* Name unverified.

33029 Uni-Rez® 1548
Polymide resin; used in flexographic printing inks for plastic substrates. *Union Camp.* Name unverified.

33030 Uni-Rez® 1552
Polymide adhesive resin; designed for bonding porous substrates. *Union Camp.* Name unverified.

33031 Uni-Rez® 2100
Reactive polyamide (condensation product of polymerized fatty acids with polyalkyl polyamines); viscosity resin used with epoxy resins to formulate coatings and adhesives, and impart outstanding adhesion, flexibility, impact resistance, and chemical resistance to coatings; reacts with epoxy resins at room temperature imparting resiliency and toughness to the cured resin; can be used in 1:1 proportions with epoxy resins. *Union Camp.* Name unverified.

33032 Uni-Rez® 2355
Reactive polyamide; epoxy curing agent for adhesives, coatings, and floor toppings. *Union Camp.* Name unverified.

33033 Uni-Rez® 2620
Polyamide adhesive resin; used for hot melt bonding of leather and other porous substrates. *Union Camp.* Name unverified.

33034 Uni-Rez® 2800
Low viscosity amidoamine; used in highly filled systems include adhesives for concrete, grouting compounds, concrete floor toppings, high solids coatings; gives medium cure rates with liquid epoxy resin. *Union Camp.* Name unverified.

33035 Uni-Rez® 7003
Maleic resin; used in nitrocellulose lacquers, printing inks; promotes hardness, high gloss, and adhesion. *Union Camp.* Name unverified.

33036 Uni-Rez® 7705
Aqueous resin solution; used in aqueous gravure inks for use on paper and board substrates. *Union Camp.* Name unverified.

33037 Uni-Rez® 9002
Phenolic resin, rosin-modified; used in sheet-fed lithographic inks, hot melt adhesives, and surface coatings. *Union Camp.* Name unverified.

33038 Uni-Rez® A-800 Light
Maleic glyceryl rosinate; used in lacquers and soft oil varnishes; gives excellent hardness, gloss, and fast dry. *Union Camp.* Name unverified.

33039 Uniscrub®
18472-51-0 2140 242-354-0
chlorhexidine gluconate
4% chlorhexidine gluconate solution; antiseptic cleansing solution. *Seton Healthcare Group plc.*

33040 Unisept® Solution
18472-51-0 2140 242-354-0
0.5% Chlorhexidine gluconate solution; antibacterial agent for general antiseptic purposes. *Seton Healthcare Group plc.*

33041 Uniset® D-124F
SMD adhesive for dielectric applications; for pin transfer or syringe dispense; high strength, IR cure. *Emerson & Cuming Polymer Group.*

33042 Unisiv® 3A
Molecular sieve in carrier oil; water scavenger in polyurethane coatings, adhesives, elastomers, and other moisture-sensitive systems. *UOP.*

33043 Unisize HA-70
Polyvinyl alcohol, tackified; derived from super hydrolyzed grades; tackified grade for specialty paper applications. *Air Prods & Chem/Polymers.*

33044 Unislip 1753
112-84-5 204-009-2
$C_{22}H_{43}NO$
erucamide
erucylamide; *cis*-13-docosenoamide. Slip and antiblock agent; mold release for PP. mp= 79°. *Unichema.*

33045 Unislip 1759
301-02-0 206-103-9
$C_{18}H_{35}NO$
Oleamide
9-Octadecenamide, (Z)-; Adogen 73; Armoslip CP; Octadecene amide; Oleamide; Oleic acid amide; Oleyl amide; Slip-eze. Lubricant, slip and antiblock agent. *Unichema.*

33046 Unisol
Textile auxiliary chemicals. *ICI Chem & Polymers Ltd.* Discontinued.

33047 Unispar 40
Feldspar
Filler, flatting agent with low oil absorption; for traffic paint, interior/exterior architectural coatings, protective, maintenance and marine coatings, mastics and adhesives. *Unimin.*

33048 Unisperse-E
Pigment pastes for emulsion paints. *Ciba plc.*

33049 Unisperse-P
Aqueous pigment dispersions for wallpaper printing mass coloration of paper and paper coating. *Ciba plc.*

33050 Unistab D-33
1,3,5-Trimethyl-2,4,6-tris(3,5-di-t-butyl-4-hydroxybenzyl) benzene
Supplied as a 40% solution in polymeric organosilicon; antioxidant for polyolefin processing. *Union Carbide.* Name unverified.

33051 Uni-Tac® 70
Noncrystallizing rosin; tackifier for SBR, natural rubber, butyl rubber, ethylene-vinyl acetate, and other polymers; for water and solv.-based construction adhesives, pressure-sensitive, sealant, hot melt and rubber compounding. *Union Camp.* Name unverified.

33052 Uni-Tac® R85
Glyceryl rosinate; tackifier for use in EVA, SBS, SIS and other hot melts, pressure-sensitive adhesives, in rubber compounding, sealants, coatings; stabilized to provide good heat and aging stability. *Union Camp.* Name unverified.

33053 Uni-Tac® R99
8050-26-8 232-479-9
Pentaerythrityl rosinate; tackifier for EVA, SBS, SIS, other hot melts, construction adhesives, in rubber compounding, sealants, coatings; stabilized to provide good heat and aging stability. *Union Camp.* Name unverified.

33054 Unite
Chemically modified polyolefins containing maleic anhydride functionality; compatibilizers in polymer blends and alloys (include recycled plastics), as polymeric coupling agent in reinforced and filled polymers (especially PP), as adhesive agent for bonding polyolefins to various substrates. *Aristech.*

33055 United
7440-44-0 1855 231-153-3
Carbon black. *Cabot Carbon Ltd.*

33056 Unitex
9006-04-6 232-689-0
Centrifuged ammoniated NR latex; used for latex foam, dipped goods, surface coatings, adhesives, extruded thread, molded and cast products; produces very light colored latex films. *Guthrie Latex.*

33057 Unithane
Urethane polymers. *Cray Valley Ltd.*

33058 Unithox 450
C_{20}-C_{40} pareth-10; component for water-based PU mold release agents; metalworking additive. *Petrolite.*

33059 Unitreat
Refinery and oilfield chemicals. *Universal Matthey Products Ltd.* Name unverified.

33060 Univadine
Dyeing and printing assistant. *Ciba plc.* Name unverified.

33061 Unival DMDA-6200
Blow molding resin for household and industrial chemicals containers. *Union Carbide.*

33062 Univan
A nickel-vanadium steel.

33063 universal balsam
Consists of camphor (1 part), lead acetate (6 parts), beeswax (16 parts) and rape oil (48 parts).

33064 Univest
A stereospecific, low-molecular weight polybutadiene; a universal binder for almost all types of sand materials; neither cement nor water needed. *Hüls AG.*

33065 Univol U304
A proprietary trade name for a mixture of distilled C_{20}-C_{22} acids. *UOP Inc.* Name unverified.

33066 Univol U312
A proprietary trade name for a mixture of caprylic and capric acids. *UOP Inc.* Name unverified.

33067 Uniwax 1747
3061-75-4 1051 221-304-1
$C_{22}H_{45}NO$
Behenamide
docsoamide; behenic acid amide. Lubricant, mold release agent for EPDM and PVC processing. mp = 111-112°. *Unichema.*

33068 Uniwax 1750
124-26-5 204-693-2
$C_{18}H_{27}NO$
Stearamide
octadecamide. Antiblock additive for polyolefin films; lubricant, release agent for rubber compounding, and PVC processing. mp = 98-102°; bp_{12} = 250°. *Unichema.*

33069 Uniwax 1760
110-30-5 203-755-6
$C_{38}H_{76}N_2O_2$
Ethylene bis-stearamide
N,N'-ethylenebisstearamide. Lubricant, slip and antiblock agent, release agent for polyolefins, chloroprene rubber, and PVC. mp = 144-146°. *Unichema.*

33070 Unlloy Chrome Steels
A proprietary trade name for alloys containing 4-6% chromium, 0.1-0.25% carbon, up to 0.6% manganese and 0.4-0.6% molybdenum and 1.0-1.25% tungsten. No manufacturer.

33071 Unna's stain
A microscopic stain. It contains 0.15 g methyl green, 0.5 g pyronin, 5 ml 95% alcohol, 20 ml glycerin, and the whole made up to 100 ml with 2% carbolic acid solution.

33072 Unocal 76 Res 701
PVDC emulsion polymer; laminating adhesive, excellent adhesion, solvent urethane replacement, flame retardant textile backcoating. *Unocal/Polymers.*

33073 Unocal 76 Res 777
PVDC emulsion polymer; latex vehicle for interior flat and semigloss flame retardant coatings. *Unocal/Polymers.*

33074 Unocal 76 Res 1026
Styrene acrylic emulsion polymer; for elastomeric roof coatings with excellent cold temperature flexibility. *Unocal/Polymers.*

33075 Unocal 76 Res 1300
Styren acrylic emulsion polymer; extremely tough and firm-flexible high performance polymer with outstanding water resistance for textile applications. *Unocal/Polymers.*

33076 Unocal 76 Res 3114
Acrylic emulsion polymer; for textile finishing, nonwoven binder, flocking, and laminating adhesive. *Unocal/Polymers.*

33077 Unocal 76 Res 3512
Acrylic emulsion polymer; self-cross-linking acrylic exhibiting good adhesion to nylon, PP, and polyester films; nonblocking primer for films. *Unocal/Polymers.*

33078 Unocal 76 Res 4305
Carboxylated SBR emulsion polymer; binder for textiles and nonwovens. *Unocal/Polymers.*

33079 Unocal 76 Res 6004
Acrylic emulsion polymer; excellent flexibility and wet adhesion in exterior and interior coatings. *Unocal/Polymers.*

33080 Unocal 76 Res 6206
108-05-4 10130 203-545-4
Vinyl acetate homopolymer emulsion; for adhesives, wood glue. *Unocal/Polymers.*

33081 Unocal 76 Res 6255
Vinyl acrylic emulsion polymer; for compounding high peel, high tack pressure-sensitive adhesives. *Unocal/Polymers.*

33082 Unocal 76 Res 6272
Acrylic emulsion polymer; compounding base for repositionable high peel and tack pressure sensitive adhesives. *Unocal/Polymers.*

33083 Unocal 76 Res 6931
Vinyl acrylic emulsion polymer; soft binder with good tensile properties for textiles and nonwovens. *Unocal/Polymers.*

33084 Unocal 76 Res 7800
Vinyl acrylic emulsion polymer; for papercoating applications. *Unocal/Polymers.*

33085 Unocal 76 Res 9410
Carboxylated SBR emulsion polymer; for compounding low temperature pressure-sensitive tapes and label systems. *Unocal/Polymers.*

33086 Unocal 76 Res S-55
108-05-4 10130 203-545-4
Vinyl acetate homopolymer emulsion; for textile finishing and adhesives. *Unocal/Polymers.*

33087 UOP
Antioxidants. *Universal-Matthey Products Ltd.* Name unverified.

33088 UOP
Catalysts *Universal-Matthey Products Ltd.* Name unverified.

33089 UOP
Antiozonants. *Siba Hegner Ltd.*

33090 UOP 288
N,N'-Bis-(1-methyl/peptyl)-p-phenylene diamine
A proprietary antioxidant. *UOP Inc.* Name unverified.

33091 UP 13600
1317-38-0 2713 215-269-1
Precipitated cupric oxide
Am. Chemet.

33092 Upgrade
Soluble concentrate containing 360 g/l chlormequat and 180 g 2-chloroethylphosphoric acid per liter; plant growth regulator for winter wheat. *Rhône-Poulenc Crop Protection Ltd.*

33093 Upilex
Film. *Imperial Chemical Industries plc.*

33094 Upixon
110-85-0 7617 203-808-3
$C_4H_{10}N_2$
piperazine
hexahydropyrazine; piperazidine; diethylenediamine; Lumbrical; Worm-Away; Wurmirazin; Uvilon. An anthelmintic drug. Also used as a chemical intermediate. mp = 108-110°; bp = 145-146°; soluble in H_2O, organic solvents. *Bayer AG.*

33095 Uplees powder
An explosive containing 62-65% ammonium nitrate, 12 1/2-14 1/2% sodium nitrate, 4-6% trinitro-toluene, 13 1/2% ammonium chloride, and 2-4% starch.

33096 Up-Start
Plant starter. *Monsanto (Solaris).* Name unverified.

33097 UR-20CF/000
Polyurethane, 20% carbon fiber-reinforced. *Compounding technical.* Name unverified.

33098 UR-20GF/000
Polyurethane, 20% glass fiber-reinforced. *Compounding Tech.* Name unverified.

33099 UR-30CF/000 Foamed
Polyurethane foam, 30% carbon fiber-reinforced. *Compounding technical.* Name unverified.

33100 Urac
A proprietary trade name for urea-formaldehyde derivatives. No manufacturer.

33101 uracil
66-22-8 9985 200-621-9
$C_4H_4N_2O_2$
2,4(1H,3H)-pyrimidinedione
2,4-dioxypyrimidine; 2,4-dihydroxypyrimidine; 2,4-dioxopyrimidine; 2,4-pyrimidinediol; 4-hydroxy-2(1H)-pyrimidinone; 2-hydroxy-4(1H)-pyrimidinone; 2-hydroxy-4(3H)-pyrimidinone. Biochemical research. mp = 335°; λ_m = 202.5, 259.5 (ε 9200, 8200, pH 7); soluble in H_2O, insoluble in organic solvents. *Aldrich Ltd; PCR; Penta Mfg.; Schweizerhall.*

33102 Uracryl
A trademark for a range of acrylic synthetic resins in emulsion form. *Unilever.* Name unverified.

33103 Uradex
Active ingredients; diurex plus uragan; selective pre-emergence herbicide mixture. *Agan Chemical Manufacturers Ltd.*

33104 Uradil
A proprietary range of resins dispersible in water. Uradil 580/585 used for air-drying and storing. Uradil 587/588 are acrylic resins cross-linked by water-thinnable amino resins. Uradil 503 and 415 are nonoxidizing oil-free polyesters. No manufacturer.

33105 Uragan
314-40-9 1402 206-245-1
$C_9H_{13}BrN_2O_2$
5-bromo-6-methyl-3-(1-methylpropyl)-2,4(1H,3H)pyrimidinedione
5-bromo-3-*sec*-butyl-6-methyluracil; 5-bromo-6-methyl-3-(1-methylpropyl)uracil; Borea; Bromax; Croptex Onyx; Du Pont Herbicide 976; Hyvar X; Nalkil; Rokar X; Rout; Staa-Free; Urox B. Active ingredient; bromacil; 5-bromo-3-*sec*-butyl-6-methyluracil; versatile herbicide for control of established annual and perennial broadleaf weeds and grasses and brush. mp = 158-169°; d^{25} = 1.55; soluble in H_2O (815 mg/l), less soluble in organic solvents; LD_{50} (rat orl)= 5200 mg/kg. *Agan Chemical Manufacturers Ltd.*

33106 Uralane® 5774-A/B
Two-component urethane adhesive; high strength, self-extinguishing adhesive curing at room or elevated temps. to form tough, impact resistance bonds to thermoplastics or metal substrates; for aircraft industry. *Ciba-Geigy/Furane.*

33107 Uralite
A proprietary trade name for urea-formaldehyde. No manufacturer.

33108 Uralite 3103
Ester PU; non-MDA, room temperature/elevated temperature curing compound. *Hexcel.*

33109 Uralite 3111
Two-component urethane elastomer; room temperature-curing urethane elastomer; low viscosity, long pot life for easy vacuuming, soft, flexible, yet tough; reproduces the finest detail; will flex to release from moderate undercuts without tearing; for makin *Hexcel.*

33110 Uralite 3150, 3152
Two-component urethane elastomer; low moisture sensitivity casting elastomer; high abrasion and impact resistance, fast cure and quick demold, optimum hardness for cold set boxes, good adhesion to many substrates; contains no TDI or MOCA; used in cold set *Hexcel.*

33111 Uralite 3530
PU; adhesive; mix ratio 100:47 *Hexcel.*

33112 Uralloy® Hybrid Polymer LP-2035
PU polymer in styrene monomer; low profile additive designed for use in fiber-reinforced unsaturated polyester composites; controls shrinkage, provides automotive Class A surface appearance, offers easier processing. *Olin.* Discontinued.

33113 uramil
118-78-5 9987 204-277-0
$C_4H_5N_3O_3$
5-amino-2,4,6(1H,3H,5H)-pyrimidinetrione
Amidomalonylurea; dialuramide; 5-amino-2,4,6-pyrimidinetriol; Murexan. mp > 400°; insoluble in cold H_2O, slightly soluble in hot H_2O; insoluble in organic solvents.

33114 Uramon
A proprietary trade name for a fertilizer containing 43% nitrogen in the form of urea or similar compounds. No manufacturer.

33115 uranine
518-47-8 4194 208-253-0
$C_{20}H_{10}N_2O_5$
3',6'-dihydroxyspiro[isobenzofuran-1(3H), 9'[9H]xanthen]-3-one
fluoresceine sodium; 9-(o-carboxyphenyl)-6-hydroxy-3H-xanthen-3-one; 3',6'-dihydroxyfluoran; 3',6'-fluorandiol; 9-(o-carboxyphenyl)-6-hydroxy-3-isoxanthenone; resorcinolphthalein; C.I. Solvent Yellow 94; C.I. 45350:1; D&C Yellow no. 7;. The sodium or potassium salt of fluoresceine, dyes silk and wool yellow. Used as a diagnostic aid in corneal examination. mp = 314-316°; insoluble in H_2O, organic solvents; λ_m = 460, 493.5 nm.

33116 Uranox
A high nickel stainless steel.

33117 urea and thiourea resins
carbamide; carbonyl diamide. Resins obtained by the reaction between urea or thiourea and formaldehyde; used in fertilizers, animal feed, plastics, as a stabilizer in explosives, medicine, adhesives, pharmaceuticals, cosmetics, dentifrices, sulfamic acid production; viscosity modifier.

33118 urea glue
A glue formed from the condensate of urea and formaldehyde; used in conjunction with a hardener.

33119 urea-bromine
$(CH_4N_2O)_4{\cdot}CaBr_2$
A combination of urea and calcium bromide, containing 36% bromine.

33120 urea-formaldehyde resin
Amino resin; thermosetting resin; pigment-grinding medium; aids adhesion and toughness of coatings; wet strength resin in paper treatment; in automotive enamels and primers, metal decorating finishes; modifier for water-sol. polymers. *Akzo; Am. Cyanamid; Bakelite GmbH; Cargill; DSM UK; Georgia-Pacific; Hercules; Sybron.*

33121 ureaphos
A fertilizer containing phosphate of ammonia and urea.

33122 urease
9002-13-5 10009 232-656-0
urea amidohydrolase
Urastat. Enzyme which hydrolyzes urea to ammonium carbonate; in determination of urea in urine, blood, and other body fluid. Solble in H_2O, λ_m= 278.5 nm. *Accurate Chem. & Scientific; Schweizerhall; Worthington Biochemical.*

33123 Urecoll®
Urea-formaldehyde resins; used without hardener to produce adhesives for paper processing; with hardener to produce binders for granular materials, bonding fiber webs. *BASF AG; BASF plc.*

33124 Ureflex 6005
Two-component polyurethane elastomer systems; elastomer formulated for high abrasion resistance, tear strength and resistance to oxygen and common industrial chemicals; used in gaskets, o-rings, potted components, custom mold work. *Flexible Prods.*

33125 Ureflex 6011
Two-component polyurethane elastomer system; re-enterable potting compound for gaskets, o-rings, potted components, custom mold work. *Flexible Prods.*

33126 Ureka
A mixture of diphenylguanidine and mercaptobenzothiazole. used as a rubber vulcanization accelerator.

33127 Ureka B
A blend similar to Ureka with a portion of the diphenylguanidine replaced by Guantal. *General Electric.* Name unverified.

33128 Ureka C
Benzothiazylthiobenzoate
Rubber vulcanization accelerator. *General Electric.* Name unverified.

33129 Ureol® 6414A/5117B
Polyurethane resin/hardener system; for pattern making, core boxes, hammer form tools, explosion molding dies. *Ciba-Geigy Plastics UK.* Name unverified.

33130 Urepan®
PU elastomer; for building, furniture, shoes, automotive and mech. engineering, sports surfacing, sporting goods, textiles, electrical industry, domestic appliances. *Bayer AG.*

33131 Ureresolve
Glycol-based solvent; used hot (175-205°) for removing heavily encrusted urethanes. *Dynaloy.*

33132 Uresamin
Etherified urea/formaldehyde resins; used for stoving finishes and acid curing lacquers. *Resinous Chemicals Ltd.*

33133 Uresolve 411
Flammable solvent for urethane coatings, RTV and silicone oils; for electronic use (conformal coating removal, cable assembly repair, depotting). *Dynaloy.*

33134 Ureth HALL® 2050
Polyester polyol; for urethanes *C.P. Hall.*

33135 urethane
51-79-6 10013 200-123-1
$C_3H_7NO_2$
carbamic acid ethyl ester
ethyl carbamate; ethyl urethane; ethyl urethan. Used as an intermediate for pharmaceuticals and pesticides; in biochemical research and medicine. mp = 48-50°; soluble in H_2O (2 g/ml), less soluble in organic solvents; MLD (mus ip) = 2.1 g/kg.

33136 Urethon
Plastic film for sunblinds. *May & Baker Ltd.* Name unverified.

33137 Uretix
A proprietary urea formaldehyde molding material. *Nisshin Boseki.* Name unverified.

33138 Urexpan
Polyurethane sealant
Self-leveling sealant for caulking dead-level horizontal joints subject to heavy foot and vehicular traffic. *Pecora Corporation.*

33139 uric acid
69-93-2 10014 200-720-7
$C_5H_4N_4O_3$
7,9-dihydro-1H-purine2,6,8(3H)-trione
2,6,8-trihydroxypurine; 2,6,8-trioxypurine; 8-hydroxyxanthine; purine-2,6,8-triol; purine-3,6,8(1H,3H,9H)-trione; lithic acid; uric oxide. Organic synthesis. d = 1.89; poorly soluble in H_2O, organic solvents. *Biosynth AG; Burlington Bio-Medical; ICI Spec.; Spectrum Chem. Mfg.*

33140 uridine
58-96-8 10016 200-407-5
$C_9H_{12}N_2O_6$
1-β-D-ribofuranosyluracil
uracil-1-β-D-ribofuranoside; D-ribosyl uracil. Used in biochemical research. mp = 165°; $[\alpha]_D^{25}$ = 4° (c = 2); λ_m = 261, 205 nm (ε 10100, 9800); soluble in H_2O. *Am. Biorganics; R.W. Greeff; Penta Mfg.*

33141 Urine-Pak
Test combinations for determining uric acid, cholesterol, HDL cholesterol and urinary proteins in clinical chemistry. *Bayer AG.*

33142 Uristix
A proprietary test strip impregnated with a citrate buffer, tetrabromophenol blue, glucose oxidase, peroxidase and *o*-toluidine; used for the detection of protein and glucose in urine. *B. C. Ames.* Name unverified.

33143 Urlite 3109/3741
Ether PU, non-MDA, room temperature/elevated temperature curing compound. *Hexcel.*

33144 Urobilistix
A proprietary test strip impregnated with p-dimethylaminobenzaldehyde in an acid buffer; used to detect urobiliogen in urine. *B. C. Ames.* Name unverified.

33145 urokinase
9039-53-6 10024 232-917-9
urokinase
WIN 22005; Abbokinase; Actosolv; Breokinase; Persolv; Purochin; Ukidan; Uronase; Win-Kinase. A plasminogen activator isolated from human urine.

33146 Uromat PE
A proprietary trade name for an aqueous dispersion based on titanium dioxide used for delustering synthetic fiber fabrics. *Ciba plc.* Name unverified.

33147 Uromiro(n)
440-58-4 5031 207-125-1
Iodamide
Diagnostic aid. *Bracco Industria Chimica SpA.* Name unverified.

33148 Uroplas®
A range of urea molding compounds; used for electrical field, sanitary field, electrotechnical industry and pottery. *AMC SPREA S.p.A.*

33149 Urotuf®
Synthetic resins and resin solutions. *Reichhold.* Discontinued.

33150 Urovist, Urovist Sodium 300
737-31-5 3040 212-004-1
diatrizoate meglumine
Diagnostic aid. *Berlex Laboratories Inc.* Name unverified.

33151 Ursol®
Oxidation bases; for dyeing furs. *BASF AG.*

33152 Urtal
An acrylonitrile-butadiene-styrene resin.

33153 Urtenol
A combination of oils for the textile trades possessing penetration and detergent properties.

33154 Usacert
High strength water soluble powder food colors certified by FDA (USA); used for coloring of foodstuffs and pharmaceuticals. *Morton Int'l. Ltd.* Discontinued.

33155 Usagran
Granular food colors, FDA certified. *Morton Int'l. Ltd.* Discontinued.

33156 Usalake
Water insoluble, aluminum lake food colors; used for coloring of foodstuffs and pharmaceuticals. *Morton Int'l. Ltd.* Discontinued.

33157 USI in Oval
64-17-5 3806 200-578-6
Ethyl alcohol
Quantum Chemical Corp.

33158 USI in Oval
Ethyl ether
Quantum Chemical Corp.

33159 USI in Oval
Polyolefin resins. *Quantum Chemical Corp.*

33160 Usol Copper Green
Wood preservative formulated from copper naphthenate for use in solvent or water reducible bases; used for brushing, dipping, soaking or mopping of liquid to various wood species. *Standard Tar Products Co Inc.*

33161 Usol Organiclear Stain & Sealer
Proprietary wood preservatives to prevent wood rot, decay, mold and termite attack; used for brushing, dipping, soaking or mopping of liquid to various wood species. *Standard Tar Products Co Inc.*

33162 Usol Zinclear
Zinc naphthenate wood preservative formulated in solvent or water reducible bases; used for brushing, dipping, soaking or mopping of liquid to various wood species. *Standard Tar Products Co Inc.*

33163 USP®
Organic peroxide catalyst. *Witco Corporation.*

33164 USP® -90MD
15667-10-4 239-741-1
80% solution of 1,1-di(*t*-amyl peroxy)cyclohexane in odorless mineral spirits; initiator for heat-cured polyester resin cures and acrylates. *Witco/Argus.* Discontinued.

33165 USP® -240
37187-22-7 253-384-9
Ketone peroxide solution from acetylacetone; used for rapidly curing polyester resins with low levels of cobalt accelerators; DOT organic peroxide label is not required. *Witco Corporation.*

33166 USP® -245
13052-09-0 235-935-5
$C_{24}H_{46}O_6$
2,5-dimethyl-2,5-di(2-ethyl hexanoyl peroxy) hexane
Hexaneperoxoic acid, 2-ethyl-, 1,1,4,4-tetramethyl-1,4-butanediyl ester; Peroxyhexanoic acid, 2-ethyl-, 1,1,4,4-tetramethyltetramethylene ester; 2,5-Di(2-ethylhexanoylperoxy)-2,5-dimethylhexane; 2,5-Dimethyl-2,5-bis(2-ethylhexanoylperoxy)hexane; 2,5-Dimethyl-2,5-bis(2-ethyl-1-hexanoylperoxy)hexane; 2,5-Dimethyl-2,5-di(2-ethylhexanoylperoxy)hexane; 2,5-Dimethyl-2,5-hexanediol bis(2-ethylperoxyhexanoate); 2,5-Dimethylhexane 2,5-diper-2-ethylhexanoate; Lupersol 256; USP 245. Catalyst for heated curing of polyester resin systems; features rapid cures and outstanding surface finishes. *Witco Corporation.*

33167 USP® -355M
3851-87-4 223-356-0
$C_{18}H_{34}O_4$
Peroxide, bis(3,5,5-trimethyl-1-oxohexyl)
bis(3,5,5-trimethylhexanoyl) peroxide; Peroxide, bis(3,5,5-trimethylhexanoyl); 3,5,5-Trimethylhexanoyl peroxide. 75% Solution of 3,5,5-trimethyl hexanoyl peroxide in odorless mineral spirits; initiator. *Witco Corporation.*

33168 USP® -400P
3006-86-8 221-111-2
$C_{14}H_{28}O_4$
Peroxide, cyclohexylidenebis[(1,1-dimethylethyl)
Peroxide, cyclohexylidenebis[tert-butyl; 1,1-Bis(tert-butyl peroxy)cyclohexane; Chaloxyd P 1250AL; Cyclohexane, 1,1-bis(tert-butyldioxy)-; 1,1-Di(tert-butylperoxy)cyclohexane; Lupersol 331; Lupersol 331-80B; Trigonox 22B50. 80% solution of 1,1-di (t-butylperoxy) cyclohexane in butyl benzyl phthalate;

initiator useful in heat-cured polyester resin systems where improved flow and pot life are critical. *Witco Corporation.*

33169 USP® -690
t-butyl peroxy-2-ethyl hexanoate [3006-82-4] and 1,1-di (*t*-butyl peroxy) cyclohexane [3006-86-8] in a phthalate plasticizer; single initiator for elevated temperature curing of unsaturated polyester resins and compounds over a broad temperature range. *Witco Corporation.*

33170 USP® -800
75-91-2 1604 200-915-7
$C_4H_{10}O_2$
tert-butyl hydroperoxide
1,1-dimethylethylhydroperoxide. 70% *t*-butyl hydroperoxide with water as major diluent; initiator. mp= 5°; bp_{15} = 37°; d = 0.9400; n_D^{20}= 1.3840; soluble in organic solvents. *Witco Corporation.*

33171 Uspulun
hydroxymercurichlorophenol sulfates
A material containing sodium sulfate, sodium hydroxide, aniline, and mercury-chloro-phenol; used as a fungicide. Seed dressing for control of fungal diseases on cereals, rice, cotton, and vegetables. *Bayer AG.*

33172 Uspulun
hydroxymercurichlorophenol sulfate
Seed dressing for control of fungal diseases on cereals, rice, cotton and vegetables. Used as a fungicide. *Bayer AG.*

33173 Ustinex®
Products with combinations of different herbicidal compounds (aminotriazole, diuron, methabenzthiazuron, phenoxies); herbicides used for control of weeds on paths, open spaces, parks and sports grounds. *Bayer AG.*

33174 U-T-C
Colorant dispersions; for coloring of nonaqueous coating compositions. *Hüls Am.*

33175 Utica Steel
A proprietary trade name for a die steel; it contains 1.4% tungsten, 1.25% carbon, 0.4% chromium and 0.2% vanadium. No manufacturer.

33176 Utocyl
Antibiotic sulfonamide veterinary ethical. *Ciba plc.* Name unverified.

33177 Uval
4065-45-6 9157 223-772-2
$C_{14}H_{12}O_6S$
5-benzoyl-4-hydroxy-2-methoxybenzenesulfonic acid
Sulisobenzone; 3-benzoyl-4-hydroxy-6-methoxybenzenesulfonic acid; 2-benzoyl-5-methoxy-1-phenol-4-sulfonic acid; 2-hydroxy-4-methoxybenzophenone-5-sulfonic acid; benzophenone-4; NSC-60584; Spectra-Sorb UV-284; Sungard; Uvinul MS-40. Ultraviolet screen. mp = 145°; soluble in H_2O (250 mg/ml), less soluble in organic solvents. *Dorsey Pharmaceuticals.* Name unverified.

33178 Uvaseb 770
52829-07-9 258-207-9
$C_{28}H_{52}N_2O_4$
Bis(2,2,6,6-tetramethyl-4-piperidyl) sebacate
Decanedioic acid, bis(2,2,6,6-tetramethyl-4-piperidinyl) ester; Bis(2,2,6,6-tetramethyl-4-piperidinyl)sebacate; Bis(2,2,6,6-tetramethyl-4-piperidyl)sebacate; HALS 1; Tinuvin 770. Light stabilizer for polyolefins, ABS, PUR acrylics, PS and styrene copolymers, thermoplastic elastomers, ethylene-propylene copolymers, polyamides. *Enichem Synthesis SpA.*

33179 Uvasil 125
Polymethyl propyl 3-oxy-[4(2,2,6,6-tetramethyl) piperidinyl] siloxane - SiO_2
Polymethyl propyl 3-oxy-[4(2,2,6,6-tetramethyl) piperidinyl] siloxane absorbed on 30% fumed silica; uv-stabilizer for PP, PE, PS, ABS, polyamides, TPU, etc. *Enichem Synthesis SpA.*

33180 Uvasil 299
102089-33-8
Polymethyl propyl 3-oxy-[4(2,2,6,6-tetramethyl) piperidinyl] siloxane
UV-stabilizer for PP, PE, PS, ABS, polyamides, TPU, etc. *Enichem Synthesis SpA.*

33181 Uvazol 236
3896-11-5 223-445-4
C17H18ClN3O
2(2'-Hydroxy-3'-t-butyl-5'-methylphenyl)-5-chlorobenzotriazole
2-*tert*-butyl-6-(5-chloro-2H-benzotriazol-2-yl)-4-methylphenol; Phenol, 2-(5-chloro-2H-benzotriazol-2-yl)-6-(1,1-dimethylethyl)-4-methyl-; Benazol PBKh; Bumetrizole; 2-*tert*-Butyl-6-(5-chloro-2H-benzotriazol-2-yl)-*p*-cresol; 2-(5-Chloro-2H-benzotriazol-2-yl)-6-(1,1-dimethylethyl)-4-methylphenol; 5-Chloro-2-(3-*tert*-butyl-2-chloro-5-methylphenyl)-2H-benzotriazole; *p*-Cresol,; 2-*tert*-butyl-6-(5-chloro-2H-benzotriazol-2-yl)-; Geigy industrial chemical; 2-(2'-Hydroxy-3'-*tert*-butyl-5'-methylphenyl)-5-chlorobenzotriazole; Tinuvin 326,. Strong uv absorber protecting polyolefins, styrenics, acrylics, PVC, unsaturated polyesters, elastomers, PC, and PU. mp = 144-147°. *Enichem Synthesis SpA.*

33182 Uvazol 237
3864-99-1 223-383-8
$C_{20}H_{24}ClN_3O$
2-(3',5'-Di-*t*-butyl-2'-hydroxyphenyl)-5-chlorobenzotriazole
2,4-di-*tert*-butyl-6-(5-chloro-2H-benzotriazol-2-yl)phenol. Strong uv absorber protecting polyolefins, PVC, unsaturated polyester, acrylics, ABS. mp = 150-153°. *Enichem Synthesis SpA.*

33183 Uvazol 311
3147-75-9 221-573-5
$C_{20}H_{25}N_3O$
2(2'-Hydroxy-5'-t-octylphenyl) benzotriazole
Phenol, 2-(2H-benzotriazol-2-yl)-4-(1,1,3,3-tetramethylbutyl)-; 2-(2H-Benzotriazol-2-yl)-4-(1,1,3,3-tetramethylbutyl) phenol; Cyasorb 5411; 2-(2-Hydroxy-5-*tert*-octylphenyl)benzotriazole; Octrizole. Strong uv absorber protecting PNMA, PC, PS, ABS, unsaturated polyesters. *Enichem Synthesis SpA.*

33184 Uvazol P
2440-22-4 3503 219-470-5
$C_{13}H_{11}N_3O$
2(2'-Hydroxy-5'-methylphenyl) benzotriazole
2-(2H-benzotriazol-2-yl)-4-methylphenol; 2-(2H-benzotriazol-2-yl)-*p*-cresol; 2-(2'-hydroxy-5'-methylphenyl)benzotriazole; Tinuvin® P;. Strong uv absorber protecting styrenics, acrylics, polyesters, PC, PVC, unsaturated elastomers. mp = 131-133°; soluble in organic solvents. *Enichem Synthesis SpA.*

33185 UV-Chek
For uv stabilizers and absorbers for polyolefins and related polymers, in Int'l. Class 1. *Ferro.*

33186 Uvecryl®
UV and electron beam curable materials. *UCB Chemical Sector.*

33187 Uvecryl® P 36
Acrylated derivative of benzophenone; reactive photoinitiator for low odor, uv-cured coatings. *UCB Radcure.* Discontinued.

33188 Uvekol
Ultraviolet curable resins for glass laminating. *UCB Chemical Sector.*

33189 Uvi-Nox 1494
4306-88-1 224-320-7
$C_{23}H_{40}O$
Diisobutyl nonyl phenol.
Phenol, 2,6-bis(1,1-dimethylethyl)-4-nonyl-; Phenol, 2,6-di-tert-butyl-4-nonyl-; 2,6-Di-tert-butyl-4-nonylphenol. *Rhône-Poulenc Surf.*

33190 Uvinul®
UV protecting compounds. *BASF plc.*

33191 Uvinul® 400
131-56-6 1138 205-029-4
$C_{13}H_{10}O_3$
(2,4-dihydroxyphenyl)phenylmethanone
benzoresorcinol; Benzophenone-1; 2,4-dihydroxybenzophenone; 4-benzoylresorcinol; resbenzophenone. Uv absorber; used for polyester, acrylics, PS, in outdoor paints and coatings, varnishes, colored liquid toiletries and cleaning agents, filters for photographic color films and prints, and rubber-based adhesives. mp = 144-145°; insoluble in H_2O, soluble in organic solvents. *BASF.* Name unverified.

33192 Uvinul® 408
1843-05-6 6838 217-421-2
$C_{21}H_{26}O_3$
[2-hydroxy-4-(octyloxy)phenyl]phenylmethanone
octabenzone; 2-Hydroxy-4-n-octoxy-benzophenone; 2-hydroxy-4-(octyloxy) benzophenone; benzophenone-12; Spectra-Sorb UV 531. Uv absorber and stabilizer for polyethylene, PP, plasticized and rigid PVC, and other polymers; offers good compatibility, max. protection, and min. color and as a low order of toxicity. mp = 45-46°. *BASF.* Name unverified.

33193 Uvinul® 490
1341-54-4
Benzophenone-11
UV absorber; used in NC lacquer, fluorescent paint, inks, and for protecting furniture woods, colored liquid toiletries and cleaning agents, isocyanate systems, and butyrate metal lacquers. *BASF.* Name unverified.

33194 Uvinul® D-49
131-54-4 1130 205-027-3
$C_{15}H_{14}O_5$
bis(2-hydroxy-4-methoxyphenyl)methanone
Benzophenone-6; 2,2'-dihydroxy-4,4'-dimethoxybenzophenone. Most economical of the near-uv absorbers; greater heat stability and more sol. (in chlorinated and aromatic solvs.); gives broad protection to plastics, coatings, textiles such as PVC, chlorinated polyesters, epoxies, acrylics, urethanes, cellulosics, and mp = 139-140°; λ_m = 284, 340 nm (log ε 4.12, 4.12). *BASF.* Name unverified.

33195 Uvinul® D-50
131-55-5 205-028-9
$C_{13}H_{10}O_5$
2,2',4,4'-tetrahydroxybenzophenone
Benzophenone-2. Commercial uv absorber with the broadest uv absorp. spectrum; retards fading of pigments and dyestuffs; prolongs the life of polymeric materials; photostabilizes cosmetic formulations; and minimizes discoloration of synthetic rubber of plastic latices. mp = 200-203°. *BASF.* Name unverified.

33196 Uvinul® DS-49
3121-60-6 221-498-8
$C_{15}H_{14}NaO_8S$
Benzophenone-9
Benzenesulfonic acid, 4-hydroxy-5-(2-hydroxy-4-methoxybenzoyl)-2-methoxy-, monosodium salt; Benzophenone-9; General aniline & film; Sodium 2,2'-dihydroxy-4,4'-dimethoxy-5-sulfobenzophenone; Uvinul DS 49. Sulfonated deriv. of Uvinul D-49; uv absorber in cosmetic formulations to prevent fading of colors and viscosity changes caused by uv light; in textiles and water-based paints. *BASF.* Name unverified.

33197 Uvinul® M-40
131-57-7 7088 205-031-5
$C_{14}H_{12}O_3$
(2-hydroxy-4-methoxyphenyl)phenylmethanone
oxybenzone; Benzophenone-3; 2-hydroxy-4-methoxybenzophenone; 4-methoxy-2-hydroxybenzophenone; MOB; Spectra-Sorb UV 9. UV absorber; similar to Uvinul 400 except for higher sol. in aromatic solvs.; good weather resistance in resins and plastics; stabilizes PVC and polyesters against uv-light degradation; used in NC lacquers, varnishes, and oil-based paints. mp = 66°; LD_{50} (rat orl) > 12.8 g/kg. *BASF.* Name unverified.

33198 Uvinul® M-493
1341-54-4
Benzophenone-11
BASF. Name unverified.

33199 Uvinul® MS-40
4065-45-6 9157 223-772-2
Benzophenone-4
Sulfonated, water-sol. deriv. of Uvinul M-40; uv absorber in sunscreen products and in hair sprays and shampoos for dyed and tinted hair; for leather and textile fibers. *BASF.* Name unverified.

33200 Uvinul® N-35
5232-99-5 226-029-0
$C_{18}H_{15}NO_2$
2-propenoic acid, 2-cyano-3,3-diphenyl-, ethyl ester
Etocrylene; acrylic acid, 2-cyano-3,3-diphenyl-, ethyl ester; acrylonitrile, β,β-bis(cyclopropyl)-α-carbethoxy-; acrylonitrile, 3,3-dicyclopropyl-2-(ethoxycarbonyl)-; α-carbethoxy-β,β-biscyclopropyl acrylonitrile; CE 2; α-Cyano-β-phenylcinnamic acid, ethyl ester; Ethyl 2-cyano-3,3-diphenylacrylate; Ethyl α-cyano-β,β-diphenylacrylate; Ethyl 2-cyano-3,3-diphenyl-2-propenoate; Ethyl (diphenylmethylene)cyanoacetate; Etocrilene; Etocrylene; USAF A-15972; UV Absorber-2; Uvinul N-35; Uvinul N 35,. Noncolor contributing uv absorber; does not contain aromatic hydroxyl groups; effective under varying pH conditions; for NC lacquers and PVC; used in alkaline systems such as urea-formaldehyde and epoxyamine formulations, and in cosmetics. *BASF.* Name unverified.

33201 Uvinul® N-539
6197-30-4 228-250-8
$C_{24}H_{27}NO_2$
2-ethylhexyl-2-cyano-3,3-diphenylacrylate
Octocrylene. Uv absorber; in flexible and rigid PVC; used in NC lacquers, varnishes, vinyl flooring, and oil-based paints; in aerosol and oil-based suntan lotions; nonreactive with metallic driers. mp = -10°; $bp_{1.5}$ = 218°; d = 1.051; n+ss20$_D$ = 1.5670. *BASF.* Name unverified.

33202 Uvinul® O-18
118-60-5 204-263-4
$C_{15}H_{22}O_3$
2-ethylhexyl salicylate
octyl salicylate. bp_{21} = 189-190°; d = 1.014; n_D^{20} = 1.5020. *BASF.* Name unverified.

33203 Uvinul® P 25
15716-30-0
PEG-25 PABA
UV absorber for sunscreen products. *BASF.* Name unverified.

33204 Uvinul® T 150
Octyl triazone
UV filter. *BASF.* Name unverified.

33205 Uvistat 12, 24, 247, 2211
A proprietary trade name for a series of additives for protecting plastics against uv light; they have the general formula R_1,C_6H_5-CO LTD-C_6H_4OH-OR; Uvistat 247 is particularly effective for the stabilization of polyolefins. *Octel Chemicals Ltd.* Discontinued.

33206 Uvistat Aftersun
A moisturizer. *Windsor Healthcare Ltd.*

33207 Uvistat Babysun Aftersun
A proprietary preparation containing calamine; a moisturizer. *Windsor Healthcare Ltd.*

33208 Uvistat Sun Cream, Sun Block, Ultrablock, Lotion, Lipscreen, Babysun
Proprietary preparations containing combinations of 2-ethyl hexyl p-methoxycinnamate (uvb sunscreen), 4-5-butyl-1-4·g-methoxydibenzoylmethane (uva sunscreen), 2-hydroxy-4-methoxy-4·g-methylbenzophenone (uva sunscreen); and titanium dioxide (a physical blocker) *Windsor Healthcare Ltd.*

33209 Uvitex
A proprietary trade name for a series of fluorescent brighteners for incorporation in soap-based and synthetic detergents; Uvitex SFC is a stilbenic derivative giving high intensity whites on cellulosic fibers; Uvitex SK is a benzoxazole derivative effect *Ciba plc.* Name unverified.

33210 Uvitex MA
A proprietary trade name for an imidazole derivative fluorescent brightener for acrylic fibers; for application in the dope before spinning. *Ciba plc.* Name unverified.

33211 Uvitex MP
A proprietary trade name for a heterocyclic-stilbene type fluorescent brightener for polyamide fibers. *Ciba plc.* Name unverified.

33212 V.M. and P. Naphtha
ligroin; refined solvent naphtha; Benzolin; Canadol. Varnish-maker's and painter's naphtha, a deodorized petroleum product.

33213 V-1065
RTV silicone; silicone rubber recommended when using polyester, castable PU resins, waxes and low-melt alloys (pewter); excellent for print pads. *Perma-Flex Mold.*

33214 V-19
35634-74-3
2-Phenylazo-4-methoxy-2,4-dimethylvaleronitrile
Polymerization initiator. *Wako Pure Chem. Ind.; Wako Chem. USA.* Discontinued.

33215 V-30
10288-28-5 233-638-5
1-[(1-Cyano-1-methylethyl)azo] formamide
Polymerization initiator. *Wako Pure Chem. Ind.; Wako Chem. USA.* Discontinued.

33216 V-40
2094-98-6 218-254-8
$C_{14}H_{20}N_4$
1,1·g-Azobis(cyclohexane-1-carbonitrile)
Polymerization initiator. mp = 115-118° *Wako Pure Chem. Ind.; Wako Chem. USA.*

33217 V-59
13472-08-7 236-740-8
2,2·g-Azobis(2-methylbutyronitrile)
Polymerization initiator. *Wako Pure Chem. Ind.; Wako Chem. USA.*

33218 V-60
78-67-1 949 201-132-3
$C_8H_{12}N_4$
2,2·g-Azobis(2-methylpropionitrile)
2,2·g-Azobisisobutyronitrile; AIBN; 2,2·g-dicyano-2,2·g-azopropane; Porofor-57. Polymerization initiator. mp = 103-105°; uv max (ethanol) = 345 nm *Wako Pure Chem. Ind.; Wako Chem. USA.*

33219 V-601
2589-57-3 219-976-6
Dimethyl 2,2·g-azobis (2-methylpropionate)
Polymerization initiator. *Wako Pure Chem. Ind.; Wako Chem. USA.*

33220 V-65
4419-11-8 224-583-8
2,2·g-Azobis(2,4-dimethylvaleronitrile)
Polymerization initiator. *Wako Pure Chem. Ind.; Wako Chem. USA.*

33221 V-70
15545-97-8 239-593-8
2,2·g-Azobis(4-methoxy-2,4-dimethylvaleronitrile)
Polymerization initiator. *Wako Pure Chem. Ind.; Wako Chem. USA.*

33222 V-90®
7758-23-8 1740 231-837-1
$CaH_4O_8P_2$
Monocalcium phosphate, anhydrous
calcium dihydrogen phosphate; calcium phosphate, monobasic; acid calcium phosphate; calcium biphosphate; monocalcium orthophospate; monocalcium

phosphate; primary calcium phosphate. Chiefly used in fertilizers. Also leavening agent for baking, cereal. dec at 200°; d_4^{18}= 2.220 *Rhône-Poulenc Food Ingreds.*

33223 V-9415
8007-18-9
NiSbTi
Pigment for thermoplastic and thermoset resins, especially high temp. engineering resins, PVC siding and profile, and industrial finishes *Ferro/Color.*

33224 VA/crotonates copolymer
25609-89-6
Luviset CA 66; Resyn® 28-1310. Vinyl acetate/crotonic acid copolymer; hair fixative for aerosols, hair sprays, setting lotions, conditioners. *BASF; Nat'l. Starch.* Name unverified.

33225 VA-044
27776-21-2 248-655-3
2,2·g-Azobis[2-(2-imidazolin-2-yl)propane]dihydrochloride
Polymerization initiator. *Wako Pure Chem. Ind.; Wako Chem. USA.*

33226 VA-058
102843-39-0
2,2·g-Azobis[2-(3,4,5,6-tetrahydropyrimidin-2-yl) propane] dihydrochloride
Polymerization initiator. *Wako Pure Chem. Ind.; Wako Chem. USA.* Discontinued.

33227 VA-060
118585-13-0
2,2·g-Azobis[2-[1-(2-hydroxyethyl)-2-imidazolin-2-yl] propane] dihydrochloride
Polymerization initiator. *Wako Pure Chem. Ind.; Wako Chem. USA.* Discontinued.

33228 VA-080
104222-32-4
2,2·g-Azobis[2-methyl-N-[1,1-bis(hydroxymethyl)-2-hydroxyethyl]propionamide]
Polymerization initiator. *Wako Pure Chem. Ind.; Wako Chem. USA.* Discontinued.

33229 VA-086
61551-69-7
2,2·g-Azobis[2-methyl-N-(2-hydroxyethyl)-propionamide]
Polymerization initiator. *Wako Pure Chem. Ind.; Wako Chem. USA.* Discontinued.

33230 VA-088
3682-94-8
2,2·g-Azobis(2-methylpropionamide) dihydrate
Polymerization initiator. *Wako Pure Chem. Ind.; Wako Chem. USA.* Discontinued.

33231 VA-545
88684-42-8
2,2·g-Azobis(2-methyl-N-phenylpropionamidine) dihydrochloride
Polymerization initiator. *Wako Pure Chem. Ind.; Wako Chem. USA.* Discontinued.

33232 VA-545
124960-38-9
2,2·g-Azobis[N-(4-chlorophenyl)-2-methylpropionamidine]dihydrochloride
Polymerization initiator. *Wako Pure Chem. Ind.; Wako Chem. USA.* Discontinued.

33233 Vacrel®
Soldermask; for the electrical industry. *DuPont UK.*

33234 Vacsol
Organic solvent wood preservative; for pressure treatment for joinery and carcassing timber used above ground contact. *Hickson & Welch Ltd.*

33235 Vactran
High grade chemicals for vacuum evaporation. *British Drug Houses.* Name unverified.

33236 vacuum silicon iron
Alloys containing about 0.15 or 3.4% silicon, made by melting *in vacuo* ; contain about 0.01% carbon and have remarkable magnetic properties.

33237 Vaderm
66734-13-2 213 266-464-3
$C_{28}H_{37}ClO_7$
Alclometasone dipropionate
Aclosone; Aclovate; Almeta; Delonal; Ecoderm; Legederm; Miloderm; Modrasone; Perderm; Vaderm; Sch 22219; 7-Chloro-17,21-dipropionate-11-hydroxy-16-methylpregna-1,4-diene-3,20-dione. Anti-inflammatory. mp = 212-216°; $[\alpha]_D^{26}$ = +42.6° (c = 0.3 in DMF); uv max (methanol) = 242 nm *Schering Corp.* Discontinued.

33238 Valadol
103-90-2 45 203-157-5
$C_8H_9NO_2$
Acetaminophen
4-acetamidophenol; 4·g-hydroxyacetanilide, paracetamol; p-hydroxyacetanilide; p-acetamidophenol; p-acetaminophenol; p-acetylaminophenol; N-acetyl-p-aminophenol; Abensanil; Acamol; Acetalgin; Alpiny; Amadil; Anaflon; Anhiba; Apamide; APAP; Ben-u-ron; Bickie-mol; Calpol; Captin; Cetadol; Dafalgan; Datril; Dial-a-gesic; Dirox; Disprol; Doliprane; Dolprone; Dymadon; Enelfa; Eneril; Eu-Med; Exdol; Febrilix; Finimal; Gelocatil; Hedex; Homoolan; Korum; Lyteca; Momentum; Naprinol; Nobedon; Ortensan; Pacemo; Paldesic; panadol; Panaleve; Panasorb; Panets; Panex; Panofen; Parelan; Paraspen; Parmol; Pasolind; Pasolind N; Salzone; Tabalgin; Tapar; Temlo; Tempra; Tralgon; Tylenol,. Analgesic; antipyretic. mp = 169-170.5°; d_4^{21}= 1.293; uv max (ethanol) = 250 nm *Bristol-Myers Squibb Co Inc.* Name unverified.

33239 Valanone B
3-β-m-4-Oxo-2,2,7,7-tetramethyltricyclo[6,2,1,0,3,8] undecane
Bush Boake Allen Ltd.

33240 Valclene®
Drycleaning fluid. *DuPont UK.*

33241 valerian
8057-49-6 232-501-7
The dried rhizome of *Valeriana officinalis* containing starch, a resinous substance, and essential oil of valerian.

33242 n-valeric acid
109-52-4 203-677-2
$C_5H_{10}O_2$
Valeric acid, Normal
valerianic acid; n-pentanoic acid; propylacetic acid. Intermediate for flavors and perfumes, ester-type lubricants, plasticizers, pharmaceuticals, and vinyl stabilizers. 10042 d_4^{20}= 0.939; mp = -34.5°; bp = 186-187°; n_D^{20} = 1.4086 *Aldrich; Hoechst Celanse; Union Carbide.*

33243 valerone
108-83-8 203-620-1
$C_9H_{18}+O$
diisobutyl ketone
2,6-dimethyl-4-heptanone; isovalerone. mp = -46°; bp = 169 °; d = 0.809; n_D^{20} = 1.4128.

33244 Valfor® 100
1344-00-9 215-684-8
Sodium silicoaluminate
Ion exchange and selective absorp./adsorp.; detergent builder; anticaking agent. *PQ Corp.*

33245 Valinate®
For the agriculture industry. *DuPont UK.*

33246 valonia
The acorn cups of *Quercus oegilops* contain about 35% of tannin and are used in the leather industry.

33247 Valox®
Polybutylene terephthalate
Used for plastic components for automotive, electrical, electronics, lighting, medical, packaging, audio etc. *GE Plastics Ltd.*

33248 Valox® 210HP, 220HP, 230HP, 260HP, 280HP
26062-94-2
PBT polyester resin; engineering thermoplastics for healthcare products; FDA compliance; provides dimensional stability, high surface gloss, chemical resistant to most oils and greases. *GE Plastics.*

33249 Valox® 325
26062-94-2
PBT polyester resin; general purpose injection molding resin with dimensional stability, high surface gloss, chemical resistant, low water absorp., low coefficient of friction and mold release. *GE Plastics.*

33250 Valox® 508
26062-94-2
PBT polyester alloy, 30% glass-reinforced; offers improved flatness, appearance, good flow. *GE Plastics.*

33251 Valox® 701
26062-94-2
PBT polyester resin, 35% glass/mineral-filled; reinforced resin combining superior elec. properties with improved stiffness and flatness; used in business machine components, keyboards and panels, circuit breakers, industrial motor controls, high voltage TV componentry, card guides, and automotive ignition parts (rotors and adaptor rings). *GE Plastics.*

33252 Valox® 780
26062-94-2
PBT polyester resin, 40% glass/mineral reinforced; superior electric performance, improved flatness, UL94 V-O recognition, reduced shrinkage, good flow. *GE Plastics.*

33253 Valox® 815
26062-94-2
PBT polyester alloy, 15% glass-reinforced; offers improved surface

appearance while maintaining elec. performance, superior heat resistance, and dimensional stability of standard glass-reinforced PBT grades; used in decorator lamp fixtures, business machine bezels, mechanical and electrical copier components where chemical resistance to lubricants and toners is required, ribbon cartridges, lighting fixtures for corrosive environments, and some automotive aftermarket products. *GE Plastics.*

33254 Valox® 9215
25038-59-9 7730
PET polyester resin, 15% glass-reinforced; for applications requiring higher heat. *GE Plastics.*

33255 Valox® 9715M
Polycyclohexylene terephthalate, 15% glass-reinforced, flame-retardant; high heat grade. *GE Plastics.*

33256 Valox® DR48
26062-94-2
PBT polyester resin 15% glass-reinforced, flame-retardant; provides low moisture absorp., dimensional stability in harsh environments, excellent mech. strength and stiffness; good mold-fill characteristics and improved flow for molding thin sections. *GE Plastics.*

33257 Valox® FV608
Thermoplastic polyester foam resin; engineering structural foam resin features high dist. temp., chemical and solvent resistance, good fatigue endurance, high flex. strength and modulus, and good moldability; used for electronic enclosures, processing control instruments, and exterior transportation applications. *GE Plastics.*

33258 Valox® HV7075
PBT/PET/IM blend, 68% mineral-reinforced; injection moldable ceramic/ivory/metal replacement with excellent surface gloss, improved impact. *GE Plastics.*

33259 Valray Alloy 1
A trademark for an alloy of 20% chromium and the balance nickel with controlled manganese, carbon and silicon. *Wiggin Alloys Ltd.* Unverified.

33260 Valtox
Granular systemic carbamate insecticide and nematode containing 5% w/w carbofuran; used to control a wide range of soil and seedling pests including cabbot root fly, cabbage stem weevil, flea beetle, early aphids in brassicas, turnip root fly, fruit fly, millipedes, symphilids, beet leaf miner, springtails, wireworms, free-living nematodes, potato cyst eelworms, carrot fly and carrot willow aphid. *Bayer plc.*

33261 valve bronze
An alloy of from 83-89% copper, 4-5% tin, 3-7% zinc, and 3-6% lead.

33262 Valvoline®
A trade name for lubricating oils; they are American petroleum products. No manufacturer.

33263 Vamac®
Ethylene acrylic elastomer. *DuPont UK.*

33264 Vamin Series
Parenteral solutions; crystalline amino acids for intravenous nutrition. *KabiVitrum AB.* Name unverified.

33265 Vamitox
Herbicide. *May & Baker Ltd.* Name unverified.

33266 Van Bac®
Blend of mutated microorganisms; used for waste treatment in paper industry. *R. T. Vanderbilt Co Inc.* Discontinued.

33267 Van Ermengem's stain
Solution a) contains 1 g of osmic acid, 20 g tannin, 150 ml distilled water, and 8 drops glacial acetic acid. Solution b) is a 0.25-0.5% silver nitrate solution. Solution c) contains 6 g tannin, 1 g gallic acid, 20 g sodium acetate, and 700 ml water; a microscopic stain.

33268 Van Gel®
1327-43-1 215-478-8
Processed magnesium aluminum silicate (smectite); used as thickener, viscosity stabilizer, dispersion adjuster and mineral filler for rubber, paint, household products, ceramics and plastics. *R. T. Vanderbilt Co Inc.* Discontinued.

33269 Van Gel® B
Magnesium aluminum silicate (smectite); thickener and visc. stabilizer for dispersions; for industrial and agricultural uses. *R. T. Vanderbilt Co Inc.* Discontinued.

33270 vanadium
7440-62-2 10053 231-171-1
V
vanadium
Metallic element; target material for x-rays, manufacture of alloy steels. mp = 1917°; $d^{18.7}$ = 6.11 *Atomergic Chemetals; Cerac; Noah Chem.*

33271 vanadium alum
$(NH_4)_2SO_4 \cdot V_2(SO_4)_3 \cdot 24H_2O$
An ammonium-vanadium sulfate.

33272 vanadium brass
An alloy of 70% copper, 29.5% zinc, and 0.5% vanadium.

33273 vanadium bronze
HVO_3
Metavanadic acid
Used as a pigment in the place of gold bronze; also the name for an alloy of 61% copper, 38.5% zinc, and 0.5% vanadium.

33274 vanadium chloride (ous)
10213-09-9 10062 233-517-7
$VOCl_2$
vanadous chloride
vanadyl dichloride; vanadium oxydichloride. Strong reducing agent, purification of hydrogen chloride from arsenic. Used as a mordant in printing fabrics. d = 2.88; disproportionates at 384° *Atomergic Chemetals; Noah Chem.*

33275 vanadium pentoxide
1314-62-1 10056 215-239-8
V_2O_5
vanadium (V) oxide
vanadic acid anhydride. Catalyst for oxidation of sulfur dioxide in sulfuric acid manufacture, ferrovanadium, catalyst for organic reactions, ceramic coloring material, vanadium salts, inhibiting uv transmission in glass, photographic developer, textile dyeing. d = 3.35; mp = 690° *Aldrich; Atomergic Chemetals; Cerac; Kerr-McGee; Shinko Chem.*

33276 vanadium steel
An alloy of steel with vanadium.

33277 vanadium-manganese brass
An alloy containing 58.56% copper, 38.54% zinc, 1.48% aluminum, 1% iron, 0.48% manganese, and 0.03% vanadium.

33278 vanadium-molybdenum steels
Alloys containing from 0.1-1.0% carbon, 0.52-6.0% molybdenum and 0.1-1.0% vanadium.

33279 Vanair
A proprietary preparation of benzoyl peroxide and sulfur used in the treatment of acne. *Carter-Wallace Ltd.* Name unverified.

33280 Vanall (K)
Sorbitan stearate, mono and diglycerides, polysorbate 60 hydrated blend with propylene glycol, lactic acid, sodium proprionate preservatives; emulsifier used in the food industry for cake formulations. *Am. Ingredients/Patco.*

33281 Van-Amid
Epoxy curing agents; used as curing agents for Vanoxy resins. *R. T. Vanderbilt Co Inc.* Discontinued.

33282 Vanate®
A series of ethylene diamine tetraacetic acid compounds; chelating agents. *R. T. Vanderbilt Co Inc.* Discontinued.

33283 Vanax®
Broad range of sulfenamide, dithiocarbamate, thiourea, thiadiazine, isophthalate, guanidine, and aldehyde-amine accelerators; used in natural, synthetic, and latex rubbers as both primary and secondary accelerators. *R. T. Vanderbilt Co Inc.* Discontinued.

33284 Vanax® 552
98-77-1 202-698-4
Piperidinium pentamethylene dithiocarbamate
Accelerator for NR, SR, cements, and latexes; peptizer for sulfur-modified G-type neoprenes. *R. T. Vanderbilt Co Inc.* Discontinued.

33285 Vanax® 808 HP
34562-31-7 252-091-3
Butyraldehyde-aniline condensation product; accelerator for NR, SBR, CR, IIR, and latexes; activator for acidic accelerators; also for reclaims, hard rubber stocks, CR cements containing. litharge. *R. T. Vanderbilt Co Inc.*

33286 Vanax® 833
68411-19-8 270-108-2
Butyraldehyde-monobutylamine condensation product; accelerator for NR, SR, latexes, and reclaim; also in self-curing CR cements. *R. T. Vanderbilt Co Inc.* Discontinued.

33287 Vanax® A
103-34-4 3437 203-103-0
$C_8H_{16}N_2O_2S_2$
4,4·g-Dithiodimorpholine
4,4·g-Dithiobis[morpholine]; morpholine, N,N·g-disulfide; 4,4·g-dimorpholine disulphide. 4,4·g-Dithiodimorpholine in paraffin wax; vulcanizing agent; sulfur donor for NR and synthetic rubbers; functions as primary accelerator for NR, IR, SBR, NBR, IIR elastomers, and as primary and secondary accelerator in EPDM; powder Used as a staining p mp = 124-125° *R. T. Vanderbilt Co Inc.* Discontinued.

33288 Vanax® CPA
71172-17-3 275-226-8
Dimethylammonium hydrogen isophthalate

Accelerator for W and T-type neoprenes; used in press cured, injection molded, and LCM stocks. *R. T. Vanderbilt Co Inc.* Discontinued.

33289 Vanax® DOTG
97-39-2 202-577-6
$C_{15}H_{17}N_3$
N,N·g-Di-ortho-tolylguanidine
1,3-Di-o-tolylguanidine. Accelerator for NR and SR; secondary accelerator. mp = 176-178° *R. T. Vanderbilt Co Inc.* Discontinued.

33290 Vanax® DPG
102-06-7 3383 203-002-1
$C_{13}H_{13}N_3$
N,N·g-diphenyl guanidine
1,3-diphenylguanidine; Melaniline; Vulkazit. Accelerator for NR and SR; secondary accelerator. mp = 150°; dec about 170°; d = 1.13 *R. T. Vanderbilt Co Inc.* Discontinued.

33291 Vanax® MBM
3006-93-7 221-112-8
$C_{14}H_8N_2O_4$
m-Phenylendedimaleimide
N,N·g-(1,3-phenylene)dimaleimide; 1,3-dimaleimidobenzene. Accelerator; coagent in peroxide-cured polymers. Also used for cross-linking proteins. mp = 197-200° *R. T. Vanderbilt Co Inc.* Discontinued.

33292 Vanax® NS
95-31-8 202-409-1
$C_{11}H_{14}N_2S_2$
N-t-Butyl-2-benzothiazole-sulfenamide
N-tert-Butyl-2-benzothiazolesulphenamide. Accelerator; delayed action accelerator for natural and synthetic rubbers; lower scorch and faster cure than Durax; used in tires, mechanical and extruded goods. mp = 105-110° *R. T. Vanderbilt Co Inc.* Discontinued.

33293 Vanax® PML
16971-82-7 241-053-1
Di-ortho-tolylguanidine salt of dicatechol borate; accelerator for NR and SR stocks, CR, neoprenes for wire and cable and mechanical goods; activator and mild antioxidant in NR and SBR. *R. T. Vanderbilt Co Inc.* Discontinued.

33294 Vanbeenol
121-32-4 3904 204-464-7
$C_9H_{10}O_3$
Ethyl vanillin
3-Ethoxy-4-hydroxybenzaldehyde; ethylprotocatechuic aldehyde; bourbonal; ethylprotal; vanilal; Ethavan; Ethovan. Flavoring and perfumery mp = 77-78° *Bush Boake Allen Ltd.*

33295 Vanceril
5534-09-8 1047 226-886-0
beclamethasone dipropionate
An antiallergic, antiasthmatic (inhalant) and anti-inflammatory (topical). *Schering Corp.* Discontinued.

33296 Vanchem®
A series of organic compounds; corrosion inhibitors, chemical intermediates, adhesion promoters, and crosslinking agents for petroleum and other industries. *R. T. Vanderbilt Co Inc.* Discontinued.

33297 Vanchem® HM-4346
Aromatic polyisocyanate in toluene; adhesion promotion or primer or crosslinking agent. *R. T. Vanderbilt Co Inc.*

33298 Vanchem® HM-50
Aromatic polyisocyanate in monochlorobenzene; adhesion promotion or primer or crosslinking agent. *R. T. Vanderbilt Co Inc.*

33299 Vancide®
A series of antimicrobial agents and fungicides; industrial preservatives for latex, paint, paper, cutting fluids, coolants, ceramics, plastics, household products, agriculture etc. *R. T. Vanderbilt Co Inc.* Discontinued.

33300 Vancide® 51
Sodium dimethyldithiocarbamate (27.6%) and sodium 2-mercaptobenzothiazole (2.4%); fungicide for use as a preservative in latex and starch paste; bactericide for sol. cutting fluids and coolants; paper mill slimicide; used in petrol. storage tanks, recirculating cooling towers, paper and paperboard, and cotton fabric *R. T. Vanderbilt Co Inc.* Discontinued.

33301 Vancide® 51Z
Zinc dimethyldithiocarbamate (87%) and zinc 2-mercaptobenzothiazole (7.5%); fungicide for use in neoprene compositions; also for preservation of adhesives, for industrial cooling water slime control, in sanitizing cleansing compounds, for textile mildew and bacterial growth inhibition and as mold inhibitor in caulking compounds *R. T. Vanderbilt Co Inc.*

33302 Vancide® 89
133-06-2 1815 205-087-0
$C_9H_8Cl_3NO_2S$
Captan
3a,4,7,7a-Tetrahydro-2-[(trichloromethyl)thio]-1H-isoindole-1,3(2H)dione; N-(trichloromethylthio)-4-cyclohexene-1,2-dicaboximide; N-trichloromethyl-thio-3a,4,7,7a-tetrahydropthalimide; N-trichloromethyl-mercapto-4-cyclohexene-1,2 dicarboximide; ENT 26538; N-(trichloromethylmercapto)-δ^4-tetrahydorphthalimide; SR-406; Merpan; Orthocide-406. Captan; fungicide for natural and synthetic rubber compounds containing susceptible plasticizers; industrial preservative for vinyl, polyethylene, paint, lacquer, soap, wallpaper flour paste. Bacteriostat in soap. mp = 178°; d = 1.74 *R. T. Vanderbilt Co Inc.* Discontinued.

33303 Vancide® MZ-96
137-30-4 10305 205-288-3
$C_6H_{12}N_2S_4Zn$
Zinc dimethyldithiocarbamate
Bis(dimethylcarbamodithioato-S,S·g)zinc; bis(dimethyldithiocarbamato)zinc; dimethyldithiocarbamic acid zinc salt; zinc bis(dimethylthiocarbamoyl) disulfide; methyl cymate; Methasan; Zimate; Zirberk; Karbam White; Corozate; Fuclasin; Fuklasin; Zeralte; Ziram. Antimicrobial, preservative for starch and synthetic latex adhesives and food packaging adhesives. mp = 250°; d_4^{25}= 1.66 *R. T. Vanderbilt Co Inc.* Discontinued.

33304 Vanclay
5162
$Al_2Si_2O_5(OH)_4$
Kaolin
Bolus alba; China clay; porcelin clay; white bole; argilla. Kaolin; absorbent. *R. T. Vanderbilt Co Inc.* Discontinued.

33305 Vancocin Hydrochloride
1404-93-9 10066
$C_{66}H_{76}Cl_3N_9O_{24}$
Vancomycin hydrochloride
Lyphocin; Vancor. Antibacterial. uv max (H_2O) = 282 nm *Eli Lilly & Co.*

33306 Vancorr®
A series of metal salts of alkylaryl sulfonate in solvent; corrosion inhibitor for paint. *R. T. Vanderbilt Co Inc.* Discontinued.

33307 Vancote®
Antimicrobial agent; industrial preservative for coatings. *R. T. Vanderbilt Co Inc.* Discontinued.

33308 Vancure D.A.A
95-33-0 202-411-2
N-Cyclohexyl-2-benzthiazyl sulfonamide
A proprietary accelerator. *K & K Greeff Chemicals Ltd.* Name unverified.

33309 Vandar®
Elastomer modified polybutylene terephthalate. *Hoechst UK.*

33310 Vandar® 2100
26062-94-2
Thermoplastic alloy (elastomer-modified PBT); outstanding ductility, high impact, excellent paintability, chemical resistant, very low water absorp.; designed to withstand strong abuse in service; used for automotive body components and housings, furnitur *Hoechst Celanese.*

33311 Vandex®
7782-49-2 8572 231-957-4
Se
Selenium
Selenium powder; vulcanizing agent for rubber. *R. T. Vanderbilt Co Inc.*

33312 Vandike 7085
A proprietary trade name for a vinyl acetatebutyl acrylate copolymer emulsion used in emulsion paints. *BOC Group plc.* Name unverified.

33313 Vandike 7086
A proprietary trade name for a vinyl acetatebutyl acrylate copolymer emulsion; used in the manufacture of nondrip emulsion paints. *BOC Group plc.* Name unverified.

33314 Vandike P.360
A proprietary trade name for a vinyl acetatedioctyl maleate copolymer emulsion used in water-based adhesives. *BOC Group plc.* Name unverified.

33315 Vandrox®
Carboxymethyl starch
Viscosity increasing agent and water retention agent for coatings. *R. T. Vanderbilt Co Inc.*

33316 vandura silk
(Gelatin Silk). An artificial silk prepared from gelatin and formaldehyde. It is also the name for a silk made from casein and formaldehyde.

33317 Vanease (K)
Hydrogenated polysorbate 80, glyceryl lactyl palmitate, sodium carboxymethyl cellulose, sodium propionate, glyceryl tristearate, sodium benzoate, acetic acid; hydrated emulsifier used in food industry to increase aeration, smoothness and stability of icings and fillings. *Am. Ingredients/Patco.*

33318 Vanesta®
Polyoxyethylene fatty acid ester

Starch dispersion stabilizer for use in coatings and laminating adhesives. *R. T. Vanderbilt Co Inc.* Discontinued.

33319 Vanex
Blended organic and inorganic acids in combination with surfactants, wetting agents, and inhibitors; cleaner for brick in new construction that may contain vanadium or metals. *Nova Chemical Inc.* Name unverified.

33320 Vanfre®
A series of proprietary formulations; processing aids for elastomers. *R. T. Vanderbilt Co Inc.* Discontinued.

33321 Vanfre® DFL
Processing aid for mold lubrication; corrosion inhibitor. *R. T. Vanderbilt Co Inc.* Discontinued.

33322 Vanfre® IL-1
Sodium alkyl sulfates
Internal lubricant for CR, NBR, CSM, and NR; improves flow characteristics of highly loaded compounds; improves release from mill rolls; provides smoother extrusions; disperses easily. *R. T. Vanderbilt Co Inc.* Discontinued.

33323 Vangard
13684-63-4 7384 237-199-0
$C_{16}H_{16}N_2O_4$
Phenmedipham
3-(Methylphenyl)carbamic acid 3-[(methoxycarbonyl)amino]phenyl ester; m-hydroxycarbanilic acid methyl ester m-methylcarbanilate; methyl 3-(m-tolylcarbamoyloxy)phenylcarbamate; Schering 38584; Betanal; Vanguard. 114 g/liter phenmedipham EC; selective herbicide for annual weeds in cultivated beet species and strawberries. mp = 139-142° *Farmers Crop Chemicals Ltd.* Discontinued.

33324 Vanguard
13684-63-4 7384 237-199-0
Emulsifiable concentrate containing 114 g/l of phenmedipham. Used for weed control in beet crops.

33325 Vanguard
13684-63-4 7384 237-199-0
Emulsifiable concentrate containing 114 g/l of phenmedipham. Used for weed control in beet crops.

33326 vanilla
121-33-5 10069 204-465-2
The cured unripe fruit of *vanilla planifolia.* Contains vanillin. Used as an aromatic and flavoring material.

33327 vanillin
121-33-5 10069 204-465-2
$C_8H_8O_3$
benzaldehyde, 4-hydroxy-3-methoxy
4-hydroxy-3-methoxybenzaldehyde; methylprotcatechuic aldehyde; vanillic aldehyde. Substituted aromatic aldehyde; methyl ether of protocatechuic aldehyde; perfumes, flavorings, pharmaceuticals, lab reagent, source of L-dopa. mp = 80-81°; d = 1.056; bp = 285°; bp_{15} = 170° *Penta Mfg.; Schweizerhall; Trafford Chem. Ltd.*

33328 Vanisperse CB
Fractionated sodium salt of oxylignin; high dispersing power; broad range of applications. *Borregaard Ligno Tech.*

33329 Vanitox
Selective weed killer. *May & Baker Ltd.* Name unverified.

33330 Vankalite
A proprietary trade name for a beryllium-copper alloy used for setting diamonds in drills. No manufacturer.

33331 Vanlube®
A full line of antioxidants, antiwear, extreme pressure additives, metal deactivators, and friction reducers for petroleum lubricants; multifunctional additives for lubricants, functional fluids, and fuels. *R. T. Vanderbilt Co Inc.* Discontinued.

33332 Vanobid
1403-17-4 1789 215-763-7
$C_{47}H_{84}N_2O_{18}$ (Candicidin D)
Candicidin
Levorin; Candeptin; Candimon. Candicidin; antifungal. *Merrell Dow Pharmaceuticals Inc.* Name unverified.

33333 Vanodine
Iodophors. *Evans Vanodine International Ltd.*

33334 Vanox®
A full line of amine and phenol rubber antioxidants; antioxidants for rubber, elastomers and polymers. *R. T. Vanderbilt Co Inc.* Discontinued.

33335 Vanox® 12
68411-46-1 202-965-5
p,p·g-Dioctyldiphenylamine
Antioxidant for elastomers used in adhesives, hot melts. *R. T. Vanderbilt Co Inc.* Discontinued.

33336 Vanox® 1290
35958-30-6 252-816-3
$C_{30}H_{46}O_2$
2,2·g-Ethylidene bis(4,6-di-t-butylphenol)
Antioxidant; oxidative inhibitor for polymers; process stabilizer for polyolefins; stabilizer for PU and PS. mp = 162-164° *R. T. Vanderbilt Co Inc.* Discontinued.

33337 Vanox® 1320
17540-75-9 241-533-0
$C_{18}H_{30}O$
2,6-Di-t-butyl-4-sec-butyl phenol
Antioxidant for PU foam; oxidation inhibitor and scorch preventer for mfg. and storage of bun stock; stabilizer for PU foam. mp = 25=sg; bp_{10} = 141-142°; d = 0.902; $[\alpha]^{25}$ = 0° (c = 1, $CHCl_3$) *R. T. Vanderbilt Co Inc.* Discontinued.

33338 Vanox® 3C
101-72-4 202-969-7
N-Isopropyl-N·g-phenyl-p-phenylenediamine
Antioxidant, antiozonant for dk. colored NR and SR. *R. T. Vanderbilt Co Inc.* Discontinued.

33339 Vanox® 6H
101-87-1 202-984-9
N-Cyclohexyl-N·g-phenyl-p-phenylenediamine
Antioxidant, antiozonant for dk. colored NR and SR. *R. T. Vanderbilt Co Inc.*

33340 Vanox® AM
9003-79-6
Diphenylamine and acetone low-temp. reaction product; antioxidant for NR and SR. *R. T. Vanderbilt Co Inc.* Discontinued.

33341 Vanox® AT
68411-20-1 270-109-8
Butyraldehyde-aniline condensation product; antioxidant for CR, NR, SBR, EPDM stocks, and latexes; activator for thiuram and thiazoles. *R. T. Vanderbilt Co Inc.* Discontinued.

33342 Vanox® GT
27676-62-6 248-597-9
Tris(3,5-di-t-butyl-4-hydroxy benzyl) isocyanurate
Antioxidant for PP, polyethylene, PU, and polymers. *R. T. Vanderbilt Co Inc.*

33343 Vanox® MTI
53988-10-6 258-904-8
2-Mercaptotoluimidazole
Antioxidant for NR, SR; synergist with Agerite antioxidants. *R. T. Vanderbilt Co Inc.* Discontinued.

33344 Vanox® NBC
13927-77-0 237-696-2
Nickel di-n-butyldithiocarbamate
Antioxidant-antiozonant for SBR, NBR, CR, CSM, and ECO. *R. T. Vanderbilt Co Inc.* Discontinued.

33345 Vanox® ODP
101-67-7 202-965-5
Dioctylated diphenylamine
General purpose antioxidant for all elastomers. *R.T.Vanderbilt.*

33346 Vanox® PCX
128-37-0 1583 204-881-4
$C_{15}H_{24}O$
2,6-Di-t-butyl-4-methyl-phenol
Butylated hydroxytoluene; 2,6-bis(1,1-dimethyl-ethyl)-4-methylphenol; 2,6-di-tert-butyl-p-cresol; BHT; Antrancine 8; Tenox BHT; Ionol CP; Sustane; Dalpac; Impruvol; Vianol. Oxidation inhibitor in soaps and cosmetics. mp = 69-70°; bp = 264-265°; d_4^{20}= 1.048 *R. T. Vanderbilt Co Inc.* Discontinued.

33347 Vanox® SKT
34137-09-2 251-844-3
3,5-Di-t-butyl-4-hydroxyhydrocinamic acid triester of 1,3,5-tris (2-hydroxyethyl)-s-triazine-2,4,6-(1H,3H,5H)-trione
Antioxidant; stabilizer for polyolefins; hot melt and food packaging applications. *R. T. Vanderbilt Co Inc.* Discontinued.

33348 Vanox® SWP
85-60-9 201-618-5
4,4·g-Butylidene-bis-(6-t-butyl-m-cresol)
Antioxidant for natural and synthetic latexes; nonstaining. *R. T. Vanderbilt Co Inc.* Discontinued.

33349 Vanox® ZMTI
61617-00-3 262-872-0
Zinc 2-mercapto-toluimidazole
Antioxidant for NR and SR, EPDM, nitrile stock; synergist with Agerite antioxidants. *R. T. Vanderbilt Co Inc.* Discontinued.

33350 Vanoxy
Epoxy resins; uses include protective coatings, laminates, adhesives, castings, tooling, flooring, surfacing, potting and encapsulating. *R. T. Vanderbilt Co Inc.* Discontinued.

33351 Vanplast®
Plasticizers and peptizers for rubber. *R. T. Vanderbilt Co Inc.*

33352 Vanplast® 201
Barium dinonylnaphthalene sulfonate
Corrosion inhibitor used in automotive rubber protective coatings. *R. T. Vanderbilt Co Inc.* Discontinued.

33353 Vanquish
Bactericidal detergent. *ICI Chem & Polymers Ltd.*

33354 Vanseal®
A series of sarcosinate type surfactants; surfactants for cosmetic and toiletry applications. *R. T. Vanderbilt Co Inc.* Discontinued.

33355 Vanseal® 35
61791-59-1 263-193-2
Sodium cocoyl sarcosinate
Biodeg. industrial grade surfactant with outstanding mildness, lather building, and conditioning properties; compatible with cationics; used for soaps, bath gels, shampoos, shaving creams, dentifrices textile and leather processing. *R. T. Vanderbilt Co Inc.* Discontinued.

33356 Vanseal® CS
68411-97-2 270-156-4
Cocoyl sarcosine
Biodeg. surfactant, foaming and wetting agent, detergent, foam booster for soaps, bath gels, shampoos, shaving creams, dentifrices, rug shampoos, oven cleaners, dishwash, textile/leather processing; offers tolerance to hard water, mildness. *R. T. Vanderbilt Co Inc.* Discontinued.

33357 Vanseal® LS
97-78-9 202-608-3
Lauroyl sarcosine
Biodeg. surfactant, foaming and wetting agent, detergent, foam booster for soaps, bath gels, shampoos, shaving creams, dentifrices, rug shampoos, oven cleaners, dishwash, textile/leather processing; offers tolerance to hard water, mildness. *R. T. Vanderbilt Co Inc.* Discontinued.

33358 Vanseal® NACS-30
61791-59-1 263-193-2
Sodium cocoyl sarcosinate
Biodeg. surfactant, foaming and wetting agent, detergent, foam booster for soaps, bath gels, shampoos, shaving creams, dentifrices, rug shampoos, oven cleaners, dishwash, textile/leather processing; offers tolerance to hard water, mildness. *R. T. Vanderbilt Co Inc.* Discontinued.

33359 Vanseal® NALS-30
137-16-6 4379 205-281-5
$C_{15}H_{28}NNaO_3$
Sodium lauroyl sarcosinate
N-Methyl-N-(1-oxododecyl)glycine sodium salt; N-lauroylsarcosine sodium salt; sodium N-lauroyl sarcosinate; Medialan LL-99. Biodegradable surfactant, foaming and wetting agent, detergent, foam booster for soaps, bath gels, shampoos, shaving creams, dentifrices, rug shampoos, oven cleaners, dishwash, textile/leather processing; offers tolerance to hard water, mildness. *R. T. Vanderbilt Co Inc.* Discontinued.

33360 Vanseal® OS
110-25-8 203-749-3
Oleoyl sarcosine
Biodeg. surfactant, foaming and wetting agent, detergent, foam booster for soaps, bath gels, shampoos, shaving creams, dentifrices, rug shampoos, oven cleaners, dishwash, textile/leather processing; offers tolerance to hard water, mildness. *R. T. Vanderbilt Co Inc.* Discontinued.

33361 Vanseb®
Sulfur, salicylic acid; dandruff shampoo. *Allergan Inc.*

33362 Vanseb-T®
Coal tar solution, sulfur, salicylic acid; tar dandruff shampoo. *Allergan Inc.*

33363 Vansil
21738-42-1 7051 244-556-4
$C_{14}H_{21}N_3O_3$
Oxamniquine
1,2,3,4-Tetrahydro-2-[[(1-methyl-ethyl)amino]methyl]-7-nitro-6-quinolinemethanol; 1,2,3,4-tetrahydro-2-[(isopropylamino)methyl]-7-nitro-6-quinolinemethanol; 6-hydroxymethyl-2-isopropylaminomethyl-7-nitro-1,2,3,4-tetrahydroquinoline; UK 4271; Mansil. A proprietary preparation of oxamniquine; an antihelmintic. mp = 147-149° *Pfizer International.*

33364 Vansil®
1344-95-2 1749 215-710-8
$CaSiO_3$;Ca_2SiO_4; Ca_3SiO_5
Calcium silicate (wollastonite)
Functional filler, reinforcing agent, bright color used in adhesives, ceramics, elastomers, insulating materials, cosmetics, plastics, paint, resins, sealants, and wallboards. *R. T. Vanderbilt Co Inc.* Discontinued.

33365 Vansil® W-10
13983-17-0 1749 237-772-5
$CaSiO_3$;Ca_2SiO_4; Ca_3SiO_5
Wollastonite
Extender pigment for solvent-thinned and latex paints. *R. T. Vanderbilt Co Inc.* Discontinued.

33366 Vanstay®
Stabilizers for polyvinyl chloride. *R. T. Vanderbilt Co Inc.*

33367 Vantac
Acrylic emulsion pressure sensitive adhesives. *Rhône-Poulenc UK.*

33368 Vantalc®
14807-96-6 9207 238-877-9
$3MgO{\cdot}4SiO_2{\cdot}H_2O$
Talc
Talcum; French chalk; magnesium silicate, hydrous. Hydrous magnesium silicate (industrial talc); extender and filler used primarily in paints; also used in paper, rubber and plastics. mp = 800° *R. T. Vanderbilt Co Inc.*

33369 Vantard®
A series of retarder compounds; retarder to control scorch in processing of elastomers. *R. T. Vanderbilt Co Inc.* Discontinued.

33370 Vantoc
Industrial disinfectants and bactericides. *ICI Chem & Polymers Ltd.* Name unverified.

33371 Vantoc AL
Aqueous blend of higher alkyl trimethyl ammonium bromide; pale, straw colored liquid; bactericide in the brewing and food processing industries. *ICI Chem & Polymers Ltd.*

33372 Vantoc CL
139-07-1 203-351-5
Lauryl dimethyl benzyl ammonium chloride in aqueous solution; used for general disinfection of plants and equipment in brewing, soft drinks and foodstuffs industries. *ICI Chem & Polymers Ltd.*

33373 Vantocil
Biocides and bactericides. *ICI Chem & Polymers Ltd.*

33374 Vantropol
Detergent/sanitizer. *ICI Chem & Polymers Ltd.*

33375 Vanwax®
Protective wax blends; used for sunlight and ozone protection of rubber. *R. T. Vanderbilt Co Inc.*

33376 Vanzak®
Nonionic surfactants
Used in pulp and paper manufacturing for pitch control. *R. T. Vanderbilt Co Inc.* Discontinued.

33377 Vanzyme®
9000-92-4 640
α-Amylase
Starch converting enzymes used in paper manufacturing. *R. T. Vanderbilt Co Inc.*

33378 Vapona
62-73-7 3129 200-547-7
$C_4H_7Cl_2O_4P$
Dichlorvos
Phosphoric acid2,2-dichloroethenyl dimethyl ester; phosphoric acid 2,2-dichlorovinyl dimethyl ester; O,O-dimethyl O-(2,2,-dichlorovinyl) phosphate; dichlorophos; dichlorovos; DDVP; SD 1750; Astrobot; Atgard; Canogard; Dedevap; Dichlorman; Divipan; Equigard; Equigel; Estrosol; Herkol; Nogos; Nuvan; Task; Verdisol. A proprietary preparation of dichlorvos; an insecticide. d_4^{25}= 1.415; bp_{20} = 140°; $bp_{1.0}$ = 84°; $bp_{0.5}$ = 72°; $bp_{0.01}$ = 30°; n_D^{25} = 1.451. No manufacturer.

33379 Vaporole
A proprietary formulation of amylnitrite; indicated for rapid relief of angina pectoris due to coronary artery disease. *The Wellcome Foundation Ltd.*

33380 Varamide® A-7
93-83-4 202-281-7
Oleamide DEA (2:1); rust inhibitor and base for o/w emulsifiers, detergents, anticorrosive cleaners, and thickener for waterless hand cleaners; degreasers; emulsifier for oils in fiber and yarn lubricants; foam suppressor in solvent and dye carrier emulsi *Sherex/Div. of Witco.*

33381 Varamide® C-212
68140-00-1 268-770-2
Cocamide MEA (1:1); foam booster and stabilizer in aqueous systems; degreaser; hair conditioning agent; visc. modifier; for household and industrial applications. *Sherex/Div. of Witco.*

33382 Varamide® LO-1
Linoleamide DEA (1:1); thickener and foam stabilizer for shampoos, baby bath, and hand soap; conditioning agent. *Sherex/Div. of Witco.*

33383 Varamide® MA-1
Refined cocamide DEA (1:1); foam stabilizer and booster, thickener; basic liq. superamide for shampoos, bubble bath, and dishwasher; low cost equivalent

to lauric superamide; does not require melting; gives higher visc. and foam stability than conventiona *Sherex/Div. of Witco.*

33384 Varamide® ML-1
120-40-1 204-393-1
Lauramide DEA (1:1); thickener and foam stabilizer for shampoo, bubble bath, and hand laundry detergent; gives the highest visc., foam level, and stability of the superamides in series. *Sherex/Div. of Witco.*

33385 Varamide® T-55
Tallow MEA ethoxylate
Detergent base for floor and hard-surface cleaners; tolerant to high builder levels and hard water. *Sherex/Div. of Witco.*

33386 Varcum® 1198
Alkyl phenolic resin; heat-reactive resin; curing agent for butyl rubber; modifier for pressure sensitive adhesives based on SBR and natural rubber. *OxyChem/Durez.*

33387 Variamine
101-64-4 6079 202-962-9
$C_{13}H_{14}N_2O$
N-(p-Methoxyphenyl)-p-phenylenediamine
N-(4-Methoxyphenyl)-1,4-benzenediamine; 4-amino-4·g-methoxydiphenylamine; 4-methoxy-4·g-aminodiphenylamine; Variamine Blue Base. Azoic dyestuffs. mp = 102°; bp_{12} = 238° *Hoechst AG.*

33388 Variclene
Green aqueous gel containing 0.5% w/w brilliant green BP, 0.5% w/w lactic acid BP; an aid in the topical treatment of venous and other types of skin ulcers. *Dermal Laboratories Ltd.*

33389 Vari-Cut
Fiberod product approximately 1/8 in. diameter cut to lengths between 1/4 in. and 4 in.; thermoplastic molding compounds for compression, transfer and injection molding. Application areas include automotive, appliances, equipment, sporting, etc. *Polymer Composites.* Unverified.

33390 Varidase
8784
A proprietary preparation streptokinase and streptodornase; a fibrinolytic drug. *Lederle Laboratories.* Name unverified.

33391 Varifoam® SXC
TEA-lauryl sulfate, cocamidopropyl hydrxysultaine, lauramide DEA, methylparaben; high foaming, cost-effective shampoo conc. *Sherex/Div. of Witco.*

33392 Varine C
61791-38-6 263-170-7
Cocohydroxyethyl imidazoline
Emulsifier, anticorrosive, raw material for surfactant product; shampoo base, penetrating oils, antistats, corrosion inhibitors, paints, printing inks, textiles, adhesives. *Sherex/Div. of Witco.*

33393 Varine O Acetate
Oleic imidazoline acetate
Anticorrosive. *Sherex/Div. of Witco.*

33394 Varine O®
Oleyl hydroxyethyl imidazoline
Emulsifier, anticorrosive for automotive body panels, raw material; shampoo base, penetrating oils, antistats, corrosion inhibitors, paints, printing inks, textiles, adhesives. *Sherex/Div. of Witco.*

33395 Varine T
61791-39-7 263-171-2
Tall oil hydroxyethyl imidazoline
Anticorrosive for automobile industry. *Sherex/Div. of Witco.*

33396 Variotin
19504-77-9 7193 243-116-9
$C_{17}H_{25}NO_3$
Pecilocin
1-(8-Hydroxy-6-methyl-1-oxo-2,4,6-dodecatrienyl)-2-pyrrolidinone; Supral. A proprietary trade name for pecilocin; used in the treatment of fungal skin infections. $[\alpha]_D^{28}$= -5.68° *Leo Laboratories.* Name unverified.

33397 Variquat® 50MC
8001-54-5 1066 231-635-3
$[C_6H_5CH_2N(CH_3)_2R]^+Cl^-$; R = C_8H_{17} to $C_{18}H_{37}$
Benzalkonium chloride
alkylbenzyldimethylammonium chloride; Benirol; BTC; Capitol; Cequartyl; Drapolene; Drapolex; Enuclen; Germinol; Germitol; Osvan; Paralkan; Roccal; Rodalon; Zephiran Chloride; Zephirol. Germicide, algicide, disinfectant, sanitizer, deodorant; used in pesticides and mfg. of sanitizers; food processing, dairy, restaurant, industrial and household products. Alkylammonium chlorides of general formula $[C_6H_5CH_2N(CH_3)_2R]^+Cl^-$; R = C_8H_{17} to $C_{18}H_{37}$ *Sherex/Div. of Witco.*

33398 Variquat® 50ME
Dimethyl alkyl (C12-C16) benzyl ammonium chloride (50%), ethyl alcohol (7.5%) in water; specialty quaternary, germicide used for disinfection and sanitizing for hospitals, beautician instruments, food processing plants. *Sherex/Div. of Witco.*

33399 Variquat® 638
70750-47-9 274-846-6
PEG-2 cocomonium chloride
IPA; detergent booster, antistat, emulsifier for hard surface cleaners, other liq. detergents, textiles; plating bath foam blanket; base for hair conditioners, creme rinses, antistats; coemulsifier. *Sherex/Div. of Witco.*

33400 Variquat® B200
56-93-9 200-300-3
$C_{10}H_{16}ClN$
Benzyltrimethyl ammonium chloride
Dispersant, dye leveler and retarder, emulsifier used in textile industry. mp = 238°; n_D^{20}= 1.4440 *Sherex/Div. of Witco.*

33401 Variquat® K1215
61791-10-4
PEG-15 cocomonium chloride
Emulsifier, antistat for personal care products; textile antistat. *Sherex/Div. of Witco.*

33402 Varisoft® 2 TD
68910-56-5 272-746-7
Ditridecyl dimethyl ammonium chloride
aq ethanol. Softener, conditioner, base for hair conditioners and cream rinses; especially for Afro-Amer. hair. *Sherex/Div. of Witco.*

33403 Varisoft® 5 TD
Ethoxylated di C12-18 ammonium chloride in propylene glycol; hair conditioner, antistat, emulsifier, softener for textured hair; good rinseability, excellent manageability and shine. *Sherex/Div. of Witco.*

33404 Varisoft® 110
Dihydrogenated tallowamidoethyl hydroxyethylmonium methosulfate
IPA/water; nonyellowing fabric softener conc, for home and commercial laundries, textile processing. *Sherex/Div. of Witco.*

33405 Varisoft® 110-PG
Dihydrogenated tallowamidoethyl hydroxyethylmonium methosulfate in aqueous propylene glycol; base for hair conditioners and creme rinses; antistat. *Sherex/Div. of Witco.*

33406 Varisoft® 136-100P
Proprietary; quaternary for fabric softeners. *Sherex/Div. of Witco.*

33407 Varisoft® 250, 300, 355
112-02-7 203-928-6
$C_{19}H_{42}NCl$
Cetrimonium chloride
cethyltrimethylammonium chloride; hexadecyltrimethylammonium chloride. Base for hair conditioners/creme rinses; antistat, surfactant, emulsifier; hair grooming aids; imparts softness and manageability without greasiness. d = 0.968; n_D^{20}= 1.3778 *Sherex/Div. of Witco.*

33408 Varisoft® 300, 355
112-02-7 203-928-6
cetrimonium chloride
Base for hair conditioners and cream rinses. Imparts softness and manageability without greasiness. *Sherex/Div. Of Witco.*

33409 Varisoft® 432-100
1812-53-9 217-325-0
Dicetyl dimonium chloride
Coemulsifier; antistat, conditioner for hair care preps., creme rinses. *Sherex/Div. of Witco.*

33410 Varisoft® 445
Methyl (1) hydrogenated tallow amidoethyl (2) hydrogenated tallow imidazolinium methosulfate; fabric softener conc. for home and commercial laundries, textile processing. *Sherex/Div. of Witco.*

33411 Varisoft® 461
61789-18-2 263-038-9
Cocotrimonium chloride
IPA; base for hair conditioners and cream rinses; used in hot oil treatments. *Sherex/Div. of Witco.*

33412 Varisoft® 462
61789-77-3 263-087-6
Dicocodimonium chloride, aqueous IPA; base for hair conditioners and creme rinses. *Sherex/Div. of Witco.*

33413 Varisoft® 470
Ditallowdimonium chloride, aqueous IPA; base for hair conditioners and cream rinses. *Sherex/Div. of Witco.*

33414 Varisoft® 471
Tallowtrimonium chloride, IPA; base for hair conditioners and cream rinses. *Sherex/Div. of Witco.*

33415 Varisoft® 475
86088-85-9 289-151-3
Quaternium-27

IPA; fabric softener conc. for home and commercial laundries, textile processing; hair conditioner. *Sherex/Div. of Witco.*

33416 Varisoft® 3690
Methyl (1) oleylamidoethyl (2) oleyl imidazolinium methosulfate, IPA; quaternary for laundry detergent-softeners. *Sherex/Div. of Witco.*

33417 Varisoft® BT-85
17301-53-0 241-327-0
Behentrimonium chloride
Antistat, suspending agent for body and hand creams and lotions, hair conditioner. *Sherex/Div. of Witco.*

33418 Varisoft® BTMS
Cetearyl alcohol, behenyltrimonium methosulfate
Self-emulsifying wax for hair and skin formulations. *Sherex/Div. of Witco.*

33419 Varisoft® C SAC
Cetearyl alcohol, stearalkonium chloride, PEG-40 castor oil; conc. for formulating cream rinses and conditioners; based on Varisoft SDAC. *Sherex/Div. of Witco.*

33420 Varisoft® CRC
Cetearyl alcohol, stearalkonium chloride, PEG-40 castor oil; conc. formulating cream rinses and conditioners; based on Varisoft SDAC. *Sherex/Div. of Witco.*

33421 Varisoft® CTB-40
57-09-0 2068 200-311-3
$C_{19}H_{42}BrN$
Cetrimonium bromide
N,N,N-Trimethyl-1-hexadecanaminium bromide; hexadecyltrimethylammonium bromide; cetyltrimethylammonium bromide; Bromat; Cetab; Cetavlon; Cetylamine; C.T.A.B.; Lissolamine V; Micol; Quamonium. Base for hair conditioners and creme rinses; antistat; conditioner for skin creams and lotions. mp = 237° - 243° *Sherex/Div. of Witco.*

33422 Varisoft® DHT
68002-59-5 268-072-8
Quaternium-18
IPA; hair conditioner and antistat in cream rinse concs. *Sherex/Div. of Witco.*

33423 Varisoft® LAC
112-00-5 203-927-0
$C_{15}H_{34}ClN$
n-Dodecyl trimethyl amonium chloride
Lauryltrimonium chloride. Base for hair conditioners and cream rinses. mp = 237° *Sherex/Div. of Witco.*

33424 Varisoft® OIMS
Quaternium-81, aqueous IPA; base for hair conditioners, creme rinses, antistats; coemulsifier; curl retention properties. *Sherex/Div. of Witco.*

33425 Varisoft® PIMS
Quaternium 27 in propylene glycol; base for hair conditioners and creme rinses. *Sherex/Div. of Witco.*

33426 Varisoft® SDAC-W
122-19-0 204-527-9
Stearalkonium chloride
Base for hair conditioners; imparts softness, manageability, antistatic properties. to hair. *Sherex/Div. of Witco.*

33427 Varisoft® ST-50, TSC
112-03-8 203-929-1
Steartrimonium chloride
Steartrimonium chloride aqueous alcohol solution; substantive quaternary imprating softness and manageability to hair. Base for hair conditioners. *Sherex/Div. of Witco.*

33428 Varisoft® TA-100
107-64-2 203-508-2
Distearyldimonium chloride
Coemulsifier, antistat, conditioner for hair and skin care products, pigmented cosmetics. *Sherex/Div. of Witco.*

33429 Varisoft® TC-90
52467-63-7
Tricetylmonium chloride
Base for hair conditioners, creme rinses, 2-in-1 shampoos; antistat. *Sherex/Div. of Witco.*

33430 Varisoft® TIMS
68122-86-1 268-531-2
Quaternium-27
Base for creme rinses and conditioners; antistat, substantivity agent, conditioner. *Sherex/Div. of Witco.*

33431 Varisoft® TSC
112-03-8 203-929-1
steartrimonium chloride
Base for hair conditioners. *Sherex/Div. Of Witco.*

33432 Variton
62-97-5 3361 200-552-4
$C_{21}H_{27}NO_4S$
Diphemanil methylsulfate
4-(Diphenylmethylene)-1,1dimethylpiperidinium methyl sulfate; p-(α-phenylbenzylidene)-1,1-dimethylpiperidinium methylsulfate; N,N-dimethyl-4-piperidylidene-1,1-diphenylmethane methylsufate; Demotil; Diphenatil; Nivelona; Prantal. Anticholinergic. mp = 194-195° *Schering Corp.* Discontinued.

33433 Varitox
650-51-1 9756 211-479-2
$C_2Cl_3NaO_2$
Sodium trichloroacetate
trichloracetic acid, sodium salt; Konesta. mp > 300° *May & Baker Ltd.* Name unverified.

33434 varnish
A mixture of boiled oil with various gum resins and oil of turpentine. Used as protective coatings.

33435 Varnodag
A trademark for a varnish made from phenol-formaldehyde synthetic resin with colloidal graphite. No manufacturer.

33436 varnoline
A petroleum distillate used as a lubricant.

33437 Varonic® 32-E20
9004-98-2
$C_{58}H_{116}O_{21}$
polyoxyethylene(20) oleyl ether
Oleth-20; Brij 99®. Emulsifier, stabilizer. mp = 25-30° *Sherex/Div. of Witco.*

33438 Varonic® 63 E20
68439-49-6 7737
Ceteareth-20
Emulsifier stabilizer, emulsifier, solubilizer, moisturizer, and emollient for creams and lotions; surfactant for stick formulations and hair conditioners. *Sherex/Div. of Witco.*

33439 Varonic® BD
Cetearyl alcohol
ceteareth-20. Self-emulsifying wax, visc. modifier for hair conditioners, creams and lotions; emulsifier, solubilizer. *Sherex/Div. of Witco.*

33440 Varonic® BG
112-92-5 8762 204-017-6
$C_{18}H_{38}O$
Stearyl alcohol
ceteareth-20; 1-octadecanol; n-octadecyl alcohol. Self-emulsifying wax, visc. modifier for hair conditioners, creams and lotions; emulsifier, solubilizer. mp = 56-58°; bp = 340-355°; bp_{15} = 210°; *Sherex/Div. of Witco.*

33441 Varonic® DM55
Methyl capped glycol ether; solvent with low toxicity and excellent grease cutting properties for surface cleaners; textile scouring. *Sherex/Div. of Witco.*

33442 Varonic® K202
61791-14-8
PEG-2 cocamine
Emulsifier, antistat; corrosion inhibitor in metal finishing (e.g., as cutting oil additives); detergent, antifouling, antistalling, and deicing agent in gasoline; also in textile lubricants, oil field emulsification. *Sherex/Div. of Witco.*

33443 Varonic® Li-42
51158-08-8
PEG-20 glyceryl stearate; low-irritation detergent, emulsifier, lubricant, solubilizer for household and industrial applications, personal care products. *Sherex/Div. of Witco.*

33444 Varonic® LI-48
PEG-80 glyceryl tallowate; emulsifier, solubilizer, thickener, dispersant, antiirritant surfactant used in household and industrial applications, personal care products. *Sherex/Div. of Witco.*

33445 Varonic® LI-63
68201-46-7
PEG-30 glyceryl cocoate
Low-irritation detergent, emulsifier, solubilizer for soaps specialized lubricants, personal care products. *Sherex/Div. of Witco.*

33446 Varonic® Q202
PEG-2 oleamine
Anticorrosive emulsifier for metalworking, grinding oils. *Sherex/Div. of Witco.*

33447 Varonic® Q-230
PEG-30 oleamine
Compatiblizer or antiprecipitant in dye bath of acid and cationic dyes; mild dye leveler and/or stripping agent for acid dyes; hydrophilic emulsifier. *Sherex/Div. of Witco.*

33448 Varonic® S202
10213-78-2 233-520-3

PEG-2 stearamine
Polymer additive and emulsifier. *Sherex/Div. of Witco.*

33449 Varonic® T202, T220
61791-44-4 263-177-5
PEG-2 tallowamine, PEG-20 tallowamine
T202 - Lubricant, softener, scouring aid, dye leveler and antistat for textiles; in synthetic latex paints; emulsifier for latex, dyes, and oils; acid cleaners; process modifier in polymer industry; raw material for quaternary and amphoteric surfactants. T220 - Acid dye leveler for nylon; antiprecipitant for mixed dye baths; migrating agent for dispersed dyes; wool lubricant and antistat; antistat for spin finishes. *Sherex/Div. of Witco.*

33450 Varonic® T-220
61791-44-4 263-177-5
PEG020 tallowamine
Acid bath leveler for nylon, antiprecipitant for mixed dye baths, migrating agent for dispersed dyes, wool lubricant and antistat, antistat for spin finishes.

33451 Varonic® U-215
61791-26-2
PEG-15 hydrogenated tallowamine
Nylon leveling agent; antiprecipitant for mixed dye baths; migrating agent for dispersed dyes. *Sherex/Div. of Witco.*

33452 Varox®
Organic peroxides; curing agents and crosslinking agents for elastomers and polymers. *R. T. Vanderbilt Co Inc.* Discontinued.

33453 Varox® 365
Lauramine oxide
Detergent, foam booster/stabilizer for anionic surfactants for shampoo and detergent systems, textiles; hypochlorite-stable. *Sherex/Div. of Witco.*

33454 Varox® 1770
68155-09-9 268-938-5
Cocamidopropylamine oxide
Mild detergent, foam booster for liq. soaps, shampoos, textile scouring. *Sherex/Div. of Witco.*

33455 Varsol®
General purpose hydrocarbon solvents. *Exxon Intn'l.; Exxon UK.* Name unverified.

33456 Varsol® 1
74742-48-9 6120
Mineral spirits
Stoddard solvent, Texsolve S. Aliphatic hydrocarbon solvent $d^{15.6}_{15.6}$ = 0.754-0.820; bp = 149-208° *Exxon.* Name unverified.

33457 Varsol® 18
64741-92-0; 64742-48-9
Aliphatic hydrocarbon solvent *Exxon.* Name unverified.

33458 Varsulf® S-1333
Disodium ricinoleamido MEA-sulfosuccinate aqueous disp.; detergent, refatting agent used in dishwash, liq. soaps, and personal care products; anti-irritant for other surfactants. *Sherex/Div. of Witco.*

33459 Varsulf® SBF-12
Disodium lauryl sulfosuccinate
Low-irritation foaming agent for shampoos, bubble bath, body cleansers; some conditioning and moisturizing effects; detergent for fine fabric wash systems. *Sherex/Div. of Witco.*

33460 Varsulf® SBFA-30
Disodium laureth sulfosuccinate aqueous disp.; detergent, refatting agent used in dishwash, fine fabric wash, liq. soaps, and personal care products. *Sherex/Div. of Witco.*

33461 Varsulf® SBL-203
25882-44-4 247-310-4
Disodium laramido MEA-sulfosuccinate aqueous disp.; detergent, refatting agent used in dishwash, light duty detergents, personal care products, rug and upholstery shampoos. *Sherex/Div. of Witco.*

33462 Vasagel
Thickening agent; used to thicken automatic dishwashing detergents. *Production Chemicals Ltd.*

33463 Vascoloy-Ramet D
A proprietary trade name for a corrosion-resisting alloy of 80% tantalum carbide, 20% tungsten and nickel. No manufacturer.

33464 Vasconite BT
A suspension of magnesium compounds and combustion catalysts; a proprietary anticorrosion agent added to poor-quality boiler fuels. *Gamlen Chemical Industries SA.* Name unverified.

33465 Vascuals
59-02-9 10159 200-412-2
$C_{29}H_{50}O_2$
Vitamin E
3,4-Dihydro-2,5,7,8-tetramethyl-2-(4,8,12-trimethyltridecyl)-2H-1-benzopyran-6-ol; 2,5,7,8-tetramethyl-2-(4',8',12'-trimethyltridecyl)-6-chromanol; α-tocopherol; 5,7,8-trimethyltocol; antisterility vitamin; Eprolin-S; Epsilan; Ephynal; Syntopherol; E-Vimin; Evipherol; Etavit; Phytogermine; Profecundin; Tokopharm; Viprimol; Viteolin; Esorb; Vascuals; Covitol; Evion. Vitamin E supplement. d_4^{25}= 0.9500; mp = 2.5-3.5°; $bp_{0.1}$ = 200-220°; n_D^{25}= 1.5045; uv max = 294 nm; $[\alpha]_D^{20}$= +24.00° *USV Pharmaceutical Corp.* Unverified.

33466 Vasocort
A proprietary preparation of hydrocortisone, hydroxyamphetamine hydrobromide and phenylephrine hydrochloride; used as a nasal spray in cases of allergic rhinitis. *SmithKline Beecham.* Name unverified.

33467 Vasogen
A trademark for an ointment vehicle. An oxygenated petroleum. *Parogen.* No manufacturer.

33468 Vasogen®
1636
Zinc oxide
calamine; dimethicone; Eczederm. Used for treatment of diaper rash. *Pharmax Ltd.*

33469 Vassgro DSM
919-86-8 6129 213-052-6
Emulsifiable concentrate containing 500 g demeton-S-methyl per liter; a systemic organophosphorus insecticide. *L W Vass (Agricultural) Ltd.* Discontinued.

33470 Vassgro Flowable Sulphur
7704-34-9 8948 231-722-6
S
Sulfur
Brimstone, sulphur. Suspension concentrate containing 720 g/l sulfur; a protectant fungicide. d = 2.06; mp = 115.21°; bp = 444.6° *L W Vass (Agricultural) Ltd.*

33471 Vassgro Mini Slug Pellets
108-62-3 5983 202-945-6
Pellets containing 6% w/w metaldehyde; snail and slug bait *L W Vass (Agricultural) Ltd.*

33472 Vassgro Spreader
Nonionic wetting agent containing nonylphenol ethylene oxide condensates; for use with herbicide and insecticide sprays. *L W Vass (Agricultural) Ltd.*

33473 Vaucher's alloy
An alloy of 75% zinc, 18% tin, 4.5% lead, and 2.5% antimony.

33474 Vauquelin's salt
A compound obtained by treating palladium chloride with ammonia, $Pd(NH_3)_2]Cl_2 \cdot PdCl_2$.

33475 Vaycron
Elastomer. *Hydro Polymers Ltd.*

33476 Vazo®
Vinyl polymerization catalyst. *DuPont UK.*

33477 VC20
Iodophor concentrate. *Evans Vanodine International Ltd.*

33478 Vebonol
Anabolic steroid veterinary ethical. *Ciba plc.* Name unverified.

33479 Vec
A proprietary trade name for a vinylidene chloride synthetic resin. No manufacturer.

33480 Vecortenol
Steroid veterinary ethical. *Ciba plc.* Name unverified.

33481 Vecortenol-Vioform
Steroid quinoline topical veterinary ethicals. *Ciba plc.* Name unverified.

33482 Vectal
Selective and total herbicide. *Schering Agrochemicals Ltd.* Discontinued.

33483 Vector®
Styrenic block copolymers (SIS, SBS); thermoplastic elastomers for adhesives, asphalt modification, polymer modification, molding, extrusion, health care products. *Dexco Polymers.*

33484 Vectra® A115
Liq. crystal polymer, low glass reinforced; features high tensile, chemical resistant, virtually zero mold shrinkage, excellent elec. properties, easy processing for molding and extrusion; used in electronics, fiber optics, automotive, aircraft/aerospace, chemical processing, industrial, manufacturing fields, encapsulation of electronic components; A115 is general purpose grade with very easy flow. *Hoechst Celanese/Engineering Plastics.*

33485 Vedacoll
Bitumen mastix. *Vedag GmbH.* Name unverified.

33486 Vedacolor
Lacquer with an aqueous solvent and artificial resin base; for roof surface. *Vedag GmbH.* Name unverified.

33487 Vedafix
Skirting rails. *Vedag GmbH.* Name unverified.

33488 Vedaflex
Modified bitumen membrane; for sealing flat roofs and sloped roofs without additional surface protection. *Vedag GmbH.* Name unverified.
33489 Vedaform
Bitumen shingles in different shapes and colors; for steep roofing. *Vedag GmbH.* Name unverified.
33490 Vedag BM
Emulsion for the improvement of cement mortar. *Vedag GmbH.* Name unverified.
33491 Vedagit
Bitumen plus filler; pebble-bedding compound for cold application. *Vedag GmbH.* Name unverified.
33492 Vedagolan
Bitumen emulsions. *Vedag GmbH.* Name unverified.
33493 Vedagully System
PU-integrated hard foam; for roof inlets. *Vedag GmbH.* Name unverified.
33494 Vedagum
Bituminous crevice filler, approved for underlays; for hot application. *Vedag GmbH.* Name unverified.
33495 Vedalith Facade System
Glass fiber-reinforced cement; bracket-mounted, respiratory, heat insulting system facade, shock-resistant, noncombustible. *Vedag GmbH.* Name unverified.
33496 Vedaphalt
Bitumen emulsions; for road surfaces treatment. *Vedag GmbH.* Name unverified.
33497 Vedaphon
Antidrone materials; for motor cars. *Vedag GmbH.* Name unverified.
33498 Vedapor
Hard foam together with insulation strips; for roll form insulation. *Vedag GmbH.* Name unverified.
33499 Vedapurit
PS or PU hard foam, sandwiched or nonsandwiched; for hard foam insulating boards. *Vedag GmbH.* Name unverified.
33500 Vedasin
Universal sealing compound; for cellars, subgrade purposes. *Vedag GmbH.* Name unverified.
33501 Vedastar
Domelight. *Vedag GmbH.* Name unverified.
33502 Vedatect
Bitumen roof sheets with various inserts. *Vedag GmbH.* Name unverified.
33503 Vedatex
Bitumen solvent glue; for steam barriers and heat insulating material. *Vedag GmbH.* Name unverified.
33504 Vedathene
Self-adhesive sheet; for waterproofing systems. *Vedag GmbH.* Name unverified.
33505 Vedatherm
Shingle thermal insulating board; for insulating sloped roofs. EPS with chipboard (V 100 G). *Vedag GmbH.* Name unverified.
33506 Vedril
Polymethacrylates. *Montedison UK Ltd.* Name unverified.
33507 Veecote®
12269-78-2
Hydrated aluminum silicate mineral (pyrophyllite); extender pigment for paints and coatings. *R. T. Vanderbilt Co Inc.* Discontinued.
33508 Veegum®
12199-37-0
Complex colloidal magnesium aluminum silicate derived from natural smectite clays; thickener, visc. modifier, stabilizer for emulsions, suspensions, sol'ns., liqs., creams, and pastes, cosmetics, toiletries, toothpaste, pharmaceuticals, paints, textile finishes; chemical specialities; industrial applications; suspending agent for powders and pigments; binder for inorganic powders and pigments; disintegrating agent in tablets; pigment dispersant; improves spreadability of lotions. *R. T. Vanderbilt Co Inc.* Discontinued.
33509 Veegum® PRO
Tromethamine magnesium aluminum silicate
Emulsion stabilizer and suspending agent for cosmetics, pharmaceuticals, veterinary products, chemical specialties, household products; superior soap and surfactant compatibility. *R. T. Vanderbilt Co Inc.* Discontinued.
33510 veepa oil
(Veppam oil, neem oil). Margosa oil, obtained from the seeds of *Melia Azadirachta* .
33511 Vegamino 30-SF
Vegetable amino acids; moisture binding agent substantive to skin and hair; used in perms and for chemically treated hair. *Brooks Industries.*
33512 vegetable alkali
1310-58-3 7806 215-181-3
KOH
Potassium hydroxide
Potassium hydrate; caustic potash; potassa. Used to manufacture soap; absorbs CO_2; removes paint and varnish; used in organic syntheses. mp = 360°; mp(anhydrous) = 380°.
33513 vegetable black
1815
Carbon, Amorphous
A very light lamp-black containing 99% carbon.
33514 vegetable butter
(Lactine, vegetaline, cocoaline, laureol, nucoline, albene, palmine, cocose). Names for an edible fat prepared from coconut oil and palm nut oil; used in chocolate manufacture as a substitute for cocoa butter.
33515 vegetable calomel
7521
Podophyllum
May apple; mandrake root; Indian apple. The resin of *Podo-phyllum* . Cathartic. Contains 3-6% resin, 0.2-1% podophyllotoxin, picropodophyllin, quercetin, peltatins.
33516 vegetable casein
1892
Legumin
avenin. Legumin, found in leguminous seeds.
33517 vegetable ethiops
A form of charcoal obtained by the incineration of *Fuci* .
33518 vegetable gelatin
9002-18-0 182 232-658-1
$(C_{12}H_{18}O_9)_x$
Agar
Gelose; Japan agar; Bengal isinglass; Ceylon isinglass; Chinese isinglass; Japanese isinglass; Layor Carang. Agar-agar.
33519 vegetable glue
(Aparatine). A glue obtained by treating starch with alkali; an adhesive.
33520 vegetable ivory
(Corajo). Tagua nut, the fruit of *Phytelephas macrocarpa* of South America.
33521 vegetable jelly
9000-69-5 7194 232-553-0
Pectin
Pectin *(qv)* , found in vegetable juices.
33522 vegetable oil
68956-68-3 273-313-5
(Oils, vegetable); Expressed oil of vegetable origin consisting primarily of triglycerides of fatty acids; paints as drying oils, shortening, salad dressings; rubber softeners; dietary supplements. *Arista Industries; Jojoba Growers & Processors; Karlshamns; Lipo; Mendell; A.E. Staley Manufacture.*
33523 vegetable rouge
36338-96-2 1918 252-981-1
$C_{43}H_{42}O_{22}$
Carthamin
6-β-D-Glucopyranosyl-2-[[3-β-D-glucopyranosyl-2,3,4-trihydroxy-5-[3-(4-hydroxyphenyl)-1-oxo-2-propenyl]-6-oxo-1,4-cyclohexadien-1-yl]methylene]-5,6-dihydroxy-4-[3-(4-hydroxyphenyl)-1-oxo-2-propenyl]-4-cyclohexene-1,3-dione; cathamic acid; safflor carmine; safflor red; C.I. Natural Red 26; C.I. 75140. Carthamin, the coloring matter of *Carthamus tinctorius* mixed with French chalk; used as a cosmetic.
33524 vegetable soda
The general name for the ash of soda plants (land plants).
33525 vegetable tallow
The name applied to vegetable fats similar to tallow, such as Chinese tallow and Malabar tallow.
33526 Vegetable Turf and Ornamental Weeder
1861-32-1 2896 217-464-7
$C_{10}H_6Cl_4O_4$
Dacthal
DCPA; 2,3,5,6-tetrachloro-1,4-benzenedicarboxilic acid dimethyl ester; 2,3,5,6-tetrachloroterephthalic acid dimethyl ester; dimethyl 2,3,5,6-tetrachloroterephthalate; chlorthal-methyl; Rid. A preemergence sprayable herbicide that can be used in vegetable gardenson ornamentals and in turf areas; controls spurge. mp = 155-156° *Lawn & Garden Products Inc.*
33527 vegetable wool
A product obtained from green pine and fir cones by processes of fermentation, washing, and disintegration. It is mixed with cotton for the production of yarns.
33528 vegetaline
The name given to a preparation of lactic acid; used in tanning processes for

the removal of lime. It is obtained from the drainage water of preserve manufacture by evaporation and contains from 8.6-9.6% lactic acid.

33529 Vegetoil
Vegetable oil plus emulsifiers; maximizes performance of pesticides. *Draxel Chemical Company*. Unverified.

33530 Velan
A proprietary product; it is a complex organic compound soluble in water which renders fabric fibers water repellant. No manufacturer.

33531 Velcorin®
Dimethyl dicarbonate
An organic preservative for the cold sterilization of soft drinks based on fruit juices. *Bayer AG*.

33532 Velicren
Acrylic fiber; for sewing threads, carpets, and fur fabrics. *SNIA (UK) Ltd*. Discontinued.

33533 Velon
9900
A proprietary trade name for a vinylidene chloride synthetic resin. No manufacturer.

33534 Veloset
Catalysts for sodium silicate bonded sands. *Foseco (F.S.) Ltd.*

33535 Velpak
Seaweed based crop stimulant. *Bayer AG*.

33536 Velpar;
51235-04-2 4734 257-074-4
$C_{12}H_{20}N_4O_2$
Hexazinone
3-Cyclohexyl-6-(dimethylamino)-1-methyl-1,3,5-triazine-2,4(1H,3H)-dione; DPX 3674. Soluble concentrate of 240 g hexazinone per liter; used for control of weeds in forestry plantations. mp = 97-100.5° *Selectokil Ltd (Velpar); DuPont UK (Velpar®).*

33537 Velpar®
51235-04-2 4734 257-074-4
hexazinone
Soluble concentrate of hexazinone (240 g/l). Used for control of weeds in forestry plantations. *Selectokil Ltd.; DuPont UK.*

33538 Velpeau's caustic powder
A caustic consisting of burnt alum and powdered savin tops.

33539 Velsan
75-39-8 38 200-868-2
C_2H_7NO
Acetaldehyde Ammonia
1-Aminoethanol; α-aminoethyl alcohol; aldehyde ammonia. A proprietary rubber vulcanization accelerator; it is aldehyde ammonia. mp = 97°; bp = 110°; d = 0.852. No manufacturer.

33540 Velsan® D8P-16
Cetyl PPG-2 isodeceth-7 carboxylate
Emollient for skin care, sun care, hair care, pigmented products, bath products. *Sandoz.*

33541 Velsan® D8P-3
Isopropyl PPG-2-isodeceth-7-carboxylate
Noncomedogenic emollient that helps to solubilize cosmetic actives; provides emulsion visc., wetting, and softer feel for skin care, sun care, hair care, bath, and pigmented products. *Sandoz.*

33542 Velsan® P8-16
Cetyl C12-15-pareth-9-carboxylate
Emollient for cosmetics providing emulsion visc., wetting, and softer feel to skin care, sun care, bath, and pigmented products. *Sandoz.*

33543 Velsan® P8-3
Isopropyl C12-15 pareth-9-carboxylate
Emollient for cosmetics providing emulsion visc., wetting, and softer feel to skin care, sun care, bath, and pigmented products. *Sandoz.*

33544 Veltol
118-71-8 5752 204-271-8
$C_6H_6O_3$
Maltol
3-Hydroxy-2-methyl-4H-pyran-4-one; 3-hydroxy-2-methyl-4-pyrone; 3-hydroxy-2-methyl-γ-pyrone; larixinic acid; Palatone; Veltol. Flavoring agent mp = 161-162°; *Pfizer Ltd.* Name unverified.

33545 Veltol Plus
Ethyl maltol
Pfizer Ltd.

33546 Velva Coat
Foundry core or mold coating utilized to improve the surface finish of cast metals. *Ashland Chemical Company.* Name unverified.

33547 Velva Dri
Foundry core and mold coatings implying refractories and chlorinated solvents; foundry core or mold coating utilized to improve the surface finish of cast metals. *Ashland Chemical Company.* Name unverified.

33548 Velvalite
Foundry core and mold coatings implying refractories and alcohol; foundry core or mold coating utilized to improve the surface finish of cast metals. *Ashland Chemical Company.* Name unverified.

33549 Velvaplast
Foundry core and mold coatings implying refractories and water solvent; foundry core or mold coating utilized to improve the surface finish of cast metals. *Ashland Chemical Company.* Name unverified.

33550 velvet black
A variety of gas carbon black.

33551 Velvet Veil 310
12001-26-2, 7631-86-9 8440
Mica
Silica. Provides lubricious feel to skin for pressed and loose powds.; light stable; produces soft focus optical blurring of wrinkles and blemishes. *Presperse.*

33552 Velvetex
1333-86-4 1856 231-153-3
A proprietary carbon black (thermatomic carbon) in a soft form used in rubber mixings. No manufacturer.

33553 Velvetex® AB-45
68424-94-2 270-329-4
Coco-betaine
Surfactant, conditioner, emulsifier, solubilizer used in industrial use, liq. detergents, cleansing emulsions, personal care products; visc. builder, gelling agent; lime soap dispersant; frothing agent. *Henkel/Cospha; Henkel Canada.*

33554 Velvetex® BA-35
Cocamidopropyl betaine
Dispersant, foaming and wetting agent, antistat for household and industrial detergents; lime soap dispersant; biodeg.; stable in strong acid and alkaline solution. *Henkel/Cospha; Henkel Canada.*

33555 Velvetex® CDC
Disodium cocoamphodiacetate
Mild, high foaming surfactant for low-irritation shampoos, conditioners, body cleansers. *Henkel/Cospha; Henkel Canada.*

33556 Velvetol 77-19
Blend; for textile applications. *Rhône-Poulenc Surf.*

33557 Venice soap
An olive oil soap.

33558 Venice turpentine
The oleo-resin of the larch, *Pinus larix* . It contains 20-22% essential oil and 74-80% rosin. A substance is often sold under this name consisting of a mixture of rosin oil, rosin, and turpentine; used in the varnish industry.

33559 Venit
Proteinaceous substances used as food or as a food ingredient. *Courtaulds plc.* Discontinued.

33560 Venite
9004-70-0 8195
Pyroxylin
Cellulose nitrate; nitrocellulose; colloidion cotton; soluble gun cotton; collodion wool; colloxylin; xylodin; celloidin; Parlodion. A proprietary pyroxylin product. Used in lacquer coatings, inks, adhesives, explosives ignites at 160-170°. No manufacturer.

33561 Ventilago Madraspanta
(Ouratpatti, Pitti, Lokandi). An Indian dyestuff obtained from a climbing shrub.

33562 Venzar®
2164-08-1 5459 218-499-0
$C_{13}H_{18}N_2O_2$
Lenacil
3-Cyclohexyl-6,7-dihydro-1H-cyclopentapyrimidine-2,4-(3H,5H)-dione; 3-cyclohexyl-5,6-trimethyleneuracil; 3-cyclohexyl-1,5,6,7-tetrahydo-2H-cyclopentapyrimidine-2,4(3H)-dione; du Pont 634. Wettable powder containing 80% lenacil; used for control of annual dicoyledons and meadow grass in beet, fruit herbaceous perennials. mp = 316-317° *DuPont UK.*

33563 VeoVa
Resin intermediate. *Shell UK.*

33564 VeoVa 10 Monomer
A vinyl monomer containing the tertiary versatic structure can be copolymerized with vinyl acetate and used in all types of emulsion paints. *Shell.* Name unverified.

33565 Verafil
Thermoplastic molding compounds. *Ciba plc.* Name unverified.

33566 Veragel® 200
1415-73-2 314 215-808-0
aloin. Aloe vera gel. *Dr. Madis Labs.*

33567 veratrine
62-59-9; 8051-02-3 2025,8280,9856
$C_{32}H_{49}NO_9$
Cevadine
(Z)-4α,9-epoxycevane-3β,4,12,14,16β,17,20-heptol 3-(2-methyl-2-butenoate;)Cevadilla; caustic barley. A mixture of various alkaloids obtained from the seeds of *Veratrum sabadilla* . It causes sneezing and irritation when inhaled. mp = 145-155°(veratrine); cevadine decomposes at 213-214.5°; $[\alpha]_D^{20}$= +12.8° (c = 3.2 in alc).

33568 veratrole
91-16-7 10089 202-045-3
$C_8H_{10}O_2$
1,3-dimethoxybenzene
pyrocatechol dimethylether; 1,2-dimethoxybenzene; o-dimethoxybenzene. Used in medicine as an antiseptic. d_{25}^{25} = 1.084; mp = 22-23°; bp = 206-207°.

33569 Verdalia A
Tricyclodecenyl methyl ether.
Quest Int'l. UK Ltd.

33570 Verdict
A line of herbicides based primarily on haloxyfop. *Dow UK.* Discontinued.

33571 verdigris
142-71-2 2690 205-553-3
$(CH_3CO_2)_2Cu$
copper(II) acetate
cupric acetate. Basic copper accetate; it is usually a mixture of mono-, di- and tri-acetates of copper. Green verdigris consists chiefly of the basic acetate, $2(C_2H_3O_2)_2Cu_2O$. Blue verdigris consists mainly of the basic acetate, $(C_2H_3O_2)_2Cu_2O$. The various forms of verdigris are used in dyeing and calico printing and for the preparation of oil and water colors.

33572 Verdilyn
1-Ethoxy-1-phenylethoxyethane
Quest Int'l. UK Ltd.

33573 Verdinal
3,5,5-Trimethyl hexanal
Quest Int'l. UK Ltd.

33574 verditer blue
12069-69-1 2697 235-113-6
$CH_2Cu_2O_5$
Copper(II) carbonate, basic
cupric carbonate, basic. (Verditer green, bremen green, mineral blue, bremen blue). An anhydrous basic copper carbonate, produced by the addition of sodium carbonate to a hot solution of copper sulfate or nitrate. Verditer green is an intermediate product; used for paper-staining; copper hydrate and copper carbonate are both sold under these names. mp = 200°.

33575 verditers
Highly basic copper carbonates.

33576 Verdiviton
A proprietary preparation of sodium, calcium, potassium, and manganese glycerophosphates, and vitamin B complex. *Bristol-Myers Squibb Pharmaceuticals Ltd.* Name unverified.

33577 Verdiviton Elixir
Multi vitamin and mineral mix in aqueous solution. *Bristol-Myers Squibb Pharmaceuticals Ltd.* Discontinued.

33578 Verdley
Range of composts and soil conditioners based on peat or bark. *ICI Chem & Polymers Ltd.*

33579 Verdone 2
Contains mecoprop and 2,4-D; selective lawn weedkiller. *ICI Garden Products.*

33580 Verdone CDA
94-75-7;7058-19-0 2802,5666 202-361-1;
$C_8H_6Cl_2O_3$;$C_{10}H_{11}ClO_3$
2,4-D; Mecoprop
2,4-D: (2,4-dichlorophenoxy)acetic acid; Hedonal; Trinoxol; Mecoprop: (±)-2-(4-chloro-2-methylphenoxy)-propanoic acid; (±)-2-[(4-chloro-o-tolyl)oxy]-propionic acid; mechlorprop; MCPP; CMPP; RD 4593; Astix CMPP; IsoCornox; Compitox; Compitox Plus; Proponex-Plus. Emulsifiable concentrate containing 6.7% 2,4-D, and 13.3% mecoprop; used to control weeds in grassland. 2,4-D: mp = 138°; $bp_{0.4}$ = 160° Mecoprop: mp = 95-96°; $[\alpha]_D^{25}$= +19 (alcohol) *ICI Agrochemicals Professional Products.*

33581 Verdoracine
4-Isopropyl-1-methyl-2-propenyl benzene.
Quest Int'l. UK Ltd.

33582 Verdoxan
2,2,5,5-Tetramethyl-4-isopropyl-1,3-dioxane
Fragrance raw material for green or fruty notes. *Henkel/Cospha.*

33583 Verilite
An alloy of 96% aluminum, 2.5% copper, 0.7% nickel, 0.4% silicon, and 0.3% manganese.

33584 Verinor
Precious metal dental casting alloy. *Degussa Ltd.*

33585 Veritas
Lube oil. *Monsanto (Solaris).* Name unverified.

33586 Veriviton Elixir
Multi vitamin and mineral mixture in an aqueous solution. *Bristol-Myers Squibb Pharmaceuticals Ltd.*

33587 verjuice
An old name for the very sour juice of unripe green grapes or crabapples. It contains tartaric, racemic and mailc acids.

33588 Vermadax
Combined ovine anthelmintic and fasciolacide. *RMB Animal Health Ltd.*

33589 vermiculite
1318-00-9 10095 406-060-8
Hydrated magnesium-iron-aluminum silicate
Lightweight concrete aggregate, insulation, sound conditioning, fireproofing, plaster, soil conditioner, fertilizer additive, seed bed, refractory, lubricant, oil-well drilling mud, filler in plastics, rubber and paint; absorbent; packing; carrier; animal feed additive *Filter-Media.*

33590 vermilionettes
Red pigments. Combinations of eosin and white lead or zinc white.

33591 Vermox
31431-39-7 5647 250-635-4
$C_{16}H_{13}N_3O_3$
Mebendazole
(5-benzoyl-1H-benzimidazol-2-yl)carbamic acid methyl ester; 5-benzoyl-2-benzimidazolecarbamic acid methyl ester; methyl 5-benzoyl-2-benzimidazolecarbamate; R 17635; Bantenol; Lomper; Mebenvet; Noverme; Ovitelmin; Pantelmin; Telmin; Vermicidin; Vermirax. A proprietary preparation containing mebendazole; antihelmintic. mp = 288.5° *Janssen Pharmaceutical Ltd.* Name unverified.

33592 Vernafine
Organic pigment pastes; for paints. *Colour-Chem Ltd.* Name unverified.

33593 Vernalin
Speciality dyestuffs for dyeing all types of leathers. *Colour-Chem Ltd.* Name unverified.

33594 Vernamine Binders
Binders based on synthetic resin dispersions based on acrylic and other monomers; for leather. *Colour-Chem Ltd.* Name unverified.

33595 Vernaminol Liquors
Synthetic oil derived from wax and hydrocarbons and their derivatives, free from fatty acids; for leather. *Colour-Chem Ltd.* Name unverified.

33596 Vernasein
Extremely fine dispersions of organic pigments in suitable aqueous medium; for leather. *Colour-Chem Ltd.* Name unverified.

33597 Vernasol
Disperse dyes; for polyester and polyester component in blended fibers and fabrics. *Colour-Chem Ltd.* Name unverified.

33598 Vernatan
Synthetic tanning agents, comprising replacement syntans, resin tanning agents and acrylic syntans; for leather. *Colour-Chem Ltd.* Name unverified.

33599 Vernol Liquors
Fat liquors derived from natural vegetable oils, animal fats, and synthetic esters; for leather. *Colour-Chem Ltd.* Name unverified.

33600 Vernol®
106-25-2 6560 203-378-7
$C_{10}H_{18}O$
Nerol
3,7-dimethyl-2,6-octadien-1-ol; 2,6-dimethyl-2,6-octadien-8-ol. Specially pure nerol (cis-3,7-dimethylocta-2,6-dien-1-ol); sweet, fresh, citrus-rose odor used in fragrances. bp_{745} = 224-225°; bp_{25} = 125°; d^{15} = 0.8813 *Bush Boake Allen Ltd.*

33601 Vernonite
Proprietary trade name for acrylic synthetic resins for denture bases, etc. No manufacturer.

33602 Verofix®
A range of reactive dyestuffs; for the dyeing of wool and polyamide. *Bayer AG.*

33603 Versabacs
A polycellular carpet-backing system. *Dow UK.* Discontinued.

33604 Versaflex® 1
Versatate acrylic terpolymer emulsion; emulsion designed to prevent paint blistering and peeling soon after application and before the paint film is leached of its water extractables; used in latex paints for interior flats, primer sealers, exterior mason *W R Grace/Organics.*

33605 Versal 150
1344-28-1 369 215-691-6
Al_2O_3
Alumina
aluminum oxide. High-purity, high-performance catalytic grades; as catalyst support in washcoat slurries, extrudates, spheres, or tablets; also in abrasives, antislip paper coatings, polishes, ceramic binding agents and raw materials, scavengers. d_4^{20}= 4.0; mp about 2000° *La Roche Chem.*

33606 Versalon® 1140
A trade name for a polyamide resin used as an adhesive between plasticized vinyl resins and metal. No manufacturer.

33607 Versamag® DC
1309-42-8 5706 215-170-3
$Mg(OH)_2$
Magnesium hydroxide
magnesium hydrate. Inorg. filler providing flame retardancy and smoke suppression to elastomers, plastics, and thermosets including EPDM, PP, PE, PVC; used in wire and cable compounds, conduit/tubing, film and sheet. *Morton Int'l./Plastics Additives.*

33608 Versamid®
Polyamide resins, solid thermoplastic and liquid reactable. *Cray Valley Ltd.*

33609 Versapen
3511-16-8 4706 222-512-5
$C_{19}H_{23}N_3O_4S$
Hetacillin
6-(2,2-dimethyl-5-oxo-4phenyl-1-imidazolidinyl)-3,3-dimethyl-7-oxo-4-thia-1azabicyclo[3.2.0]heptane-2-carboxylic acid; 6-(2,2-dimethyl-5-oxo-4-phenyl-1-imidazolidinyl)penicillanic acid; phenazacillin; BRL 804; Penplenum; Versatrex. Antibacterial. mp = 189.2-191.0° (dec); $[\alpha]_D^{25}$ = +366° (pyridine) *Bristol Laboratories.* Name unverified.

33610 Versapen K
5321-32-4 4706 226-182-3
$C_{19}H_{22}KN_3O_4S$
Hetacillin potassium
Uropen; Hetacillin K; Natacillin. Antibacterial. *Bristol Laboratories.* Name unverified.

33611 Versatic
Saturated tertiary monocarboxylic acid. *Shell UK.* Name unverified.

33612 Versatint Fugitive tints
Water soluble, biodegradable colorants; tinting for fiber and yarn identification. *Milliken.*

33613 Versa-TL 3
Dispersant, crystal modifier, antiscalant, sludge preventive in boilers and cooling towers; calcium phosphate inhibitor; iron control aid; on-line cleaner; thermally stable. *Hart Chem. Ltd.*

33614 Versa-TL® 4, TL 7
Sulfonated styrene/maleic anhydride
Alco Chemical Corp.

33615 Versatyl-42
Octyl acrylamide/acrylates copolymer; hairspray polymer for systems containing large proportion of hydrocarbon propellant; for aerosol and pump hairsprays, setting lotions, spritzes. *Nat'l. Starch.*

33616 Versa-TL 125
Polyelectrolyte polymer; dispersant; provides electroconductive properties for electrophotography and electrography; antistat for films and fibers. *Hart Chem. Ltd.*

33617 Versene AG
Chelated micronutrients used by growers to provide their crops with these necessary nutrients; zinc, managanese, iron, copper and magnesium; these elements stimulate growth hormones and contribute to plant-crop health and yields. *Dow Cheml Co Ltd, UK & Ireland.*

33618 Versene CA
62-33-9 3555 200-529-9
$C_{10}H_{12}CaN_2Na_2O_8$
Edetate calcium disodium
[[N,N·g-1,2-ethanediylbis[N-(carboxymethyl)glycinato]](4-)-N,N·g,O,O·g,O^N,$O^{N·}$g]calciate(2-)disodium; [(ethylenedinitrilo)tetraacetato]caciate(2-) disodium; ethylenediaminetetraacetic acid calcium disodium chelate; calcium disodium (ethylenedinitrilo)tetraacetate; calcium disodium ethylenediamine tetraacetate; EDTA calcium; edathamil calcium disodium; calcium disodium edetate; edetic acid calcium disodium salt; sodium calciumedetate; Calcitetracemate Disodium; Calcium Disodium Versenate; Ledclair; Mosatil; Antallin; Sormetal. Chelating agent. *Dow Cheml Co Ltd, UK & Ireland.*

33619 Versenol AG
Chelated micronutrients used by growers to provide their crops with these necessary nutrients - zinc, managanese, iron, copper and magnesium; these elements stimulate growth hormones and contribute to plant-crop health and yields. *Dow Cheml Co Ltd, UK & Ireland.*

33620 Versicane
Lignocaine hydrochloride. *May & Baker Ltd.; Rhône-Poulenc Rorer Ltd.* Name unverified.

33621 Versicon Conductive Polymer
Inherently conductive polymer for electromagnetic shielding, cable shielding, electrostatic control, conductive coatings, resistive heating, sensors, and other commercial applications *AlliedSignal.* Discontinued.

33622 Versiflex
A proprietary trade name for a transparent vinyl chloride acetate. No manufacturer.

33623 Versilan
Strong soil penetrating action, detergency, and emulsification; hard surface cleaners; metal degreasing; laundry powders; raw wool scouring. *Henkel Chemicals Ltd.* Name unverified.

33624 Versilan
Formulated surfactants. *Harcros.*

33625 Versilan MX134
EO/PO copolymer, modified; biodeg. moderate foaming, good wetting surfactant for highly built liqs., acid cleaners, steam cleaning, vehicle cleaning, general purpose floor and wall cleaners, descalers, concrete and aluminum cleaners, derusting agents. *Harcros.*

33626 Versilok® 201
Acrylic adhesive; structural adhesive bonding unprepared metals, plastics, and ceramics; excellent environmental and chemical resistant; semiflexible, medium cure, medium visc. *Lord.*

33627 Versilube® F50
Chlorophenyl methyl siloxane
Lubricating fluid with excellent oxidative and thermal stability; for lubricating sliding, rolling, and antifriction mechanisms in hydraulic systems, vacuum pumps, servo motors, instruments, controls, and bearings. *GE Silicones.*

33628 Versilube® G-321
Dimethyl-diphenyl silicone grease
Provides lubrication at temps. from -73 C to 204 C; for ball bearings used in maintenance-neglected locations. *GE Silicones.*

33629 Versilyt
Formulated surfactants. *Harcros.*

33630 Verstarktes chromammonit
An explosive containing 70 parts ammonium nitrate, 10 parts potassium nitrate, 12.5 parts trinitro-toluene, 7 parts chromium-ammonium alum, and 6 parts petroleum jelly.

33631 Vert d'eau
A jeweler's alloy containing 60% gold and 40% silver.

33632 Vertal 92, 200
14807-96-6 9207 238-877-9
$3MgO·4SiO_2·H_2O$
Talc
Magnesium silicate, hydrous; Talcum; French chalk. Platy talc; reinforcement in black PP compounds; dusting/parting agent for rubbers; filler in autobody compounds, caulks, putties, and sealants. mp = 800° *Luzenac Am.*

33633 Vertal 200
14807-96-6 9207 238-877-9
Talc (hydrous magnesium silicate) for general industrial, high impact applications. *Luzenac Am.*

33634 Vertalec
Verticillium lecanii; a fungal parasite used to control aphids and whitefly. *Koppert (UK) Ltd.*

33635 Vertan
Chelating agents; Vertan 600 controls the hardness of boiler feedwater; other Vertan chelating agents are used for removing water hardness deposits and metal oxide scale from industrial process equipment. *Dow UK.* Discontinued.

33636 Vertelon
4-Methyl-2-(1-phenylethyl)-1-3-dioxolane
Quest Int'l. UK Ltd.

33637 Vertifume
Fumigant containing carbon tetrachloride and carbon disulfide plus a fire inhibitor; used for controlling insects infesting stored grains; may be applied as a liquid to the grain mass or via a closed recirculating system; treated grain must be aired thoro *Dow UK.* Discontinued.

33638 Vertocinth
$C_{12}H_{18}O_2$
1-ethoxy-1-(2-phenylethoxy) ethane
Leafy green, floral odor used in fragrances. *Bush Boake Allen Ltd.*

33639 Verton®
Long fiber-reinforced thermoplastic compounds. *LNP; ICI Chemicals & Polymers Ltd.*

33640 Verton® OF-700-10
9016-75-5
PPS, 50% long glass fiber-reinforced; long fiber composite offering optimum mech. and thermal properties for demanding structural components in automotive, ordnance, chemical processing, and office equip. markets. *LNP.*

33641 Verton® RF-700-10 EM HS
32131-17-2
Nylon 6/6, 50% long glass fiber-reinforced. *LNP.*

33642 Verton® RF-7007 HS
32131-17-2
Nylon 6/6, 35% long glass fiber-reinforced; long fiber composite offering optimum mech. and thermal properties for demanding structural components in automotive, ordnance, chemical processing, and office equip. markets. *LNP.*

33643 Vertrel®
For electrical applications. *DuPont UK.*

33644 Verv®
5793-94-2 11, 1711 227-335-7
$(C_{24}H_{44}O_6)_2 \cdot Ca$
octadecanoic acid, 2-(1-carboxyethoxy)-1-methyl-2-oxoethyl ester, calcium salt
Calcium stearoyl-2-lactylate; Stearic acid, ester with lactic acid bimol. ester calcium salt; Artodan CP 80; Calcium $\alpha,\alpha(\alpha$-(stearoyloxy)propionyloxy) propionate; Calcium stearoxyl-2-lactylate; Calcium stearoyl-2-lactylate; Calcium stearoyl-2-lactyllactate; Calcium stearoyllactylate; Calcium stearyl lactylate; Calcium stearyl-2-lactylate; Emalsy B 1-1000; LAA; Lamegin CSL; Pationic 930; Stearyl-2-lactylic acid calcium salt; Verv; Verv-Ca,. Starch and protein complexing agent, softener for use in yeast-leavened bakery products; conditioning agent in dehydrated potatoes. *Am. Ingredients/Patco.*

33645 Vespel®
A proprietary trade name for polyimide in the form of prefabricated parts. Stable at high temperatures. *DuPont UK.*

33646 Vestagon® B 31
Cyclic amidine
High gloss epoxy and hybrid powder coatings. *Hüls Am.*

33647 Vestamelt
A range of thermoplastic copolyesters; hot-melt adhesives for bonding textiles. *Hüls AG.*

33648 Vestamid®
A large range of polyamides and copolyamides; suitable for injection molding, extrusion, thermoforming, blow molding, rotational molding and fluidized bed coating. *Hüls AG.*

33649 Vestamid® D 14
Polyamide 6/12; molding compound. *Hüls AG.*

33650 Vestamid® E40M-S3
Polyamide 12 elastomer; stabilized against hot air and sunlight; more flexible and more impact resistant than corresponding plasticized grades. *Hüls AG.*

33651 Vestamid® L 1600
Polyamide 12; basic grade homopolymer suitable for injection moldings requiring easy melt flow. *Hüls AG.*

33652 Vestamid® L 1833
Polyamide 12, long glass fiber-filled; molding compound. *Hüls AG.*

33653 Vestamid® X 3500
Polyamide 12; antistatic and elec. conductive molding compound. *Hüls AG.*

33654 Vestamid® X 4178
Polyamide 12; self-extinguishing molding compound. *Hüls AG.*

33655 Vestamin® IPD
Epoxy curative for coatings, adhesives, castings and composites. *Hüls Am.*

33656 Vestamin® TMD
2855-13-2 220-666-8
$C_{12}H_{18}N_2O_2$
Cyclohexanemethanamine, 5-amino-1,3,3-trimethyl-
Trimethylhexamethylenediamine; Cyclohexanemethylamine, 5-amino-1,3,3-trimethyl-; Araldite HY 5083; IPD; Isophorone diamine; 1-Amino-3-(aminomethyl)-3,5,5-trimethylcyclohexane; 1-Amino-3,3,5-trimethyl-5-aminomethylcyclohexane; 1,3,3-Trimethyl-1-aminomethyl-5-aminocyclohexane; 3-Aminomethyl-3,5,5-trimethylcyclohexylamine; 3,3,5-Trimethyl-5-aminomethylcyclohexylamine; 5-Amino-1,3,3-trimethylcyclohexanemethanamine; 5-Amino-1,3,3-trimethylcyclohexanemethylamine. Epoxy curative for flexible coatings, adhesives, castings and composites. *Hüls Am.*

33657 Vestanat® IPDI
4098-71-9 223-861-6
$C_{12}H_{18}N_2O_2$
Cyclohexane, 5-isocyanato-1-(isocyanatomethyl)-1,3,3-trimethyl-
Isophorone diisocyanate; Isocyanic acid , methylene(3,5,5-trimethyl-3,1-cyclohexylene) ester; IPDI; Isophorone diamine diisocyanate; Isophorone diisocyanate; 1-(Isocyanatomethyl)-5-isocyanato-1,3,3-trimethylcyclohexane; 1-Isocyanato-3-(isocyanatomethyl)-3,5,5-trimethylcyclohexane; 1-Isocyanato-3,3,5-trimethyl-5-(isocyanatomethyl)cyclohexane; 1-Isocyanato-5-(isocyanatomethyl)-3,3,5-trimethylcyclohexane; 1,3,3-Trimethyl-1-(isocyanatomethyl)-5-isocyanatocyclohexane; 3-(Isocyanatomethyl)-3,5,5-trimethylcyclohexyl isocyanate; 3,3,5-Trimethyl-5-(isocyanatomethyl) cyclohexyl isocyanate; 5-Isocyanato-1-(isocyanatomethyl)-1,3,3-trimethyl cyclohexane. Raw material for mfg. of light stable polyurethanes. *Hüls Am.*

33658 Vestanat® T 1890 L
Polyisocyanate prepolymer; prepolymer for crosslinking of two-component polyurethane paints. *Hüls Am.*

33659 Vestanat® TMDI
Trimethylhexamethylene diisocyanate
Raw material for manufacture of flexible, light stable polyurethanes. *Hüls Am.*

33660 Vestar®
Chemical for treating wearing apparel, threads, yarns and textile fabrics. *Dow Corning.* Discontinued.

33661 Vestenamer®
Polyoctylene
Suitable for the manufacture of rubber blends for injection moldings, extruded products (profiles, hoses), calendered articles and tires. *Hüls AG.*

33662 Vestenamer® 6213
Polyoctenamer
Nonstaining stabilizer; used as a blend component for other rubbers to improve plasticity, enhance filler incorporation and dispersion, improve flowability in extrusion, injection and compression molding, increase surface smoothness in calendering operati *Hüls Am; Hüls AG.*

33663 Vesticoat® UT 647
Polyester-based isocyanate-terminated prepolymer; prepolymer for flexible moisture curing one component PU coatings. *Hüls Am.*

33664 Vestiform
High molecular polymethacrylate for processing polyvinyl chloride; particularly for improving the deepdraw properties of films and sheets. *Hüls AG.*

33665 Vestinol
A range of alkyl phthalates; primary plasticizers for polyvinyl chloride and paints etc. *Hüls AG.*

33666 Vestinol AH
117-81-7 1291 204-211-0
Dioctyl phthalate
A proprietary trade name for a vinyl plasticizer. *Hüls AG.*

33667 Vestodur
polybutylene terephthalate
A range of polybutylene terephthalate with and without various additives; Vestodur without fillers is suitable for injection molding and extrusion and Vestodur containing fillers is suitable for injection molding. *Hüls AG.*

33668 Vestogrip
Polyisoprene rubber. *Hüls AG.*

33669 Vestolen
High density polyethylene and polypropylene. *Hüls AG.* Discontinued.

33670 Vestolen® A3512
9002-88-4 7728
HDPE; narrow molecular weight distribution polyethylene offering lowest dens. and highest stress cracking resistance in series, and high notched impact strength; used for tubular film, injection molded parts, and hollow moldings with high stresscracking resistance; R3512 film grade is available for high strength, shock resistant film having a low permeability and high transparency and suitable for the production of laminated as well as sterilized film. *Hüls AG.* Discontinued.

33671 Vestolen® AS
A trademark for a flame resistant HDPE building material. *Hüls AG.* Discontinued.

33672 Vestolen® AX4304
9002-88-4 7728
HDPE; narrow molecular weight distribution grade. *Hüls AG.* Discontinued.

33673 Vestolen® EM
A range of elastomer-modified polypropylene; wide range of possible uses in the automobile industry: bumper coverings, bumper comers and instrument panels. *Hüls AG.* Discontinued.

33674 Vestolen® P2000
9003-07-0 7741
Polypropylene homopolymer; used for injection molding of packaging. *Hüls AG.* Discontinued.

33675 Vestolen® P2000 CR
9003-07-0 7741
Polypropylene homopolymer; used for extrusion of fibers. *Hüls AG.* Discontinued.

33676 Vestolen® P6000
9003-07-0 7741
Polypropylene homopolymer; used for injection molding of automotive, elec., other tech. articles, packaging, medical goods. *Hüls AG.* Discontinued.

33677 Vestolen® P6700
9003-07-0 7741
Polypropylene block copolymer; used for injection molding of battery cases. *Hüls AG.* Discontinued.

33678 Vestolen® P7032 G
9003-07-0 7741
Polypropylene, reinforced; used for extrusion of sheets, compression molding of sheets, injection molding of automotive, elec., other tech. goods, and household appliances. *Hüls AG.* Discontinued.

33679 Vestolit®
9002-86-2 7746 206-625-7
$(C_2H_3Cl)_x$
Polyvinyl chloride
Ethene, chloro-, homopolymer; 103EP8; 110A; 195J; 238; 309M; AD 254; Airex; AL 30; AL 31; AL 30 (polymer); Amerace A 30; Armodour; Aron 1100; Aron compound HW; Aron NS 1100; Aron TS 1100; Astralon; Atactic poly(vinyl chloride;)Bakelite; Bakelite OYNV; Bakelite QSAH 7; Bakelite QSAN 7; Bakelite QYAC 10; Bakelite QYJV; Bakelite QYJV 1; Bakelite QYOH-1; Bakelite QYSJ; Bakelite QYSL 7; Bakelite QYTO 7; Bakelite UCA 3310; Benvic; Benvic EB 16; Blacar; Blacar 1716; Bolatron 6200; Boltaron; Bonloid; Breon; Breon 107; Breon 113; Breon 121; Breon 125/10; Breon 130/1; Breon 151; Breon 4001; Breon 4121; Breon 111EP; Breon 112EP,. *Hüls AG.*

33680 Vestolit® B 7021
9002-86-2 7746 206-625-7
polyvinyl chloride
Microsuspension PVC; for paste processing; more suitable for dip coating and casting than pastes based on the Vestolit E series; suitable for mixing with grades of the E series for spreadcoating. *Hüls AG.* Name unverified.

33681 Vestolit® B 7090
9003-22-9
Vinyl chloride/vinyl acetate copolymer; for paste processing of floor coverings, carpet backings, adhesives/laminating coatings, and spray coating. *Hüls AG.* Name unverified.

33682 Vestolit® E 6003, E 6503, E 7003
9002-86-2 7746 206-625-7
polyvinyl chloride
Emulsion PVC; resin emulsion preferable used for processing with plasticizer; E 6003 used for calendering of floor coverings; E 6503 used for calendering of film/sheet and floor coverings, and extrusion of sections and hosepipes; E 7003 used for calendering of floor coverings and extrusion of sections and hosepipes. *Hüls AG.* Name unverified.

33683 Vestolit® HI
206-625-7
polyvinyl chloride
High impact polyvinyl chloride; used for window sections, films, profiles and sheets. *Hüls AG.* Name unverified.

33684 Vestolit® LF HI-SP 5735
9002-86-2 7746 206-625-7
polyvinyl chloride
Rigid PVC; easy flow for injection molding. *Hüls AG.* Name unverified.

33685 Vestolit® M 5867
9002-86-2 7746 206-625-7
polyvinyl chloride
Mass PVC; resin for processing without plasticizer and containing no residues of suspension aids; M 5867 is easy-flowing grade used for the production of injection-molded parts and hollow moldings by extrusion of structural foam furniture; and construction sections and thick sheet. *Hüls AG.* Name unverified.

33686 Vestolit® P 1330 K
9002-86-2 7746 206-625-7
polyvinyl chloride
PVC homopolymer; for paste processing for leathercloth, floor coverings, canvas materails, unsupported layers/films, dipping, casting, spray coating. *Hüls AG.* Name unverified.

33687 Vestolit® S 6058
9002-86-2 7746 206-625-7
polyvinyl chloride
Suspension PVC; resin suspension for processing without plasticizer; used for processing by injection molding (e.g., fittings), for the extrusion of thin-walled pipes and sections (e.g., elec. conduits, drinking straws), and for the production of rigid film and sheet. *Hüls AG.* Name unverified.

33688 Vestopal
A range of unsaturated polyester resins; used for hand lay-up and compression molding, industrial moldings, for polymer-concrete and in fiber spraying processes. *Hüls AG.*

33689 Vestoplast
A range of predominantly amorphous olefin copolymers; used in hot melts, adhesives, anticorrosion strips, putties, sealing compounds, and road marking compounds. *Hüls AG.*

33690 Vestopren
Thermoplastic rubber; a rubber concentrate for improving the impact stength of polyolefins, especially polypropylene. *Hüls AG.*

33691 Vestoran® 1100
Polyphenylene ether; molding compound offering reduced shrinkage and warpage; produces dimensionally stable moldings which resist hydrolysis; used for instrument engineering, electrical and electronic engineering, parts with large surface areas in automotive engineering and office equipment industry. *Hüls AG.* Name unverified.

33692 Vestosint
Nylon 12 fluidized bed coating powders. *Hüls AG.*

33693 Vestowax
A range of Fischer-Tropsch waxes; suitable for use in hot melts, printing inks, lacquers and as aids for processing rubber and plastics. *Hüls AG.*

33694 Vesturit
A range of saturated polyesters for lacquers; used as binder components for stoving finishes containing solvents, for 'medium' and 'high' solids paints, for water soluble stoving finishes and as lacquer resins for highly flexible coatings. *Hüls AG.* Name unverified.

33695 Vestypor
A range of expandable polystyrene; suitable for a wide range of molding applications. *Hüls AG.*

33696 Vestyron
A range of polystyrenes; suitable for a wide range of molding applications. *Hüls AG.*

33697 Vestyron 550
Polystyrene containing low residual monomer; a packaging material. *Hüls AG.* Name unverified.

33698 Vestyron 551
Polystyrene packaging materials with low residual monomer and exceptional stress cracking resistance in contact with oils. *Hüls AG.* Name unverified.

33699 Vestyron X984 and X1260AK
Polystyrenes with high impact resistance. *Hüls AG.* Name unverified.

33700 Vesulong
852-19-7 212-707-3
$C_{16}H_{16}N_4O_2S$
4-amino-N-(3-methyl-1-phenyl-1H-pyrazol-5-yl)benzenesulfonamide
sulfamethylphenazole; sulfapyrazole; 3-methyl-1-phenyl-5-(sulfanilamido)pyrazole; sulfamethylphenazole; sulfapyrzole. Antibacterial agents. Sulfonamide veterinary ethicals. mp = 195°. *Ciba plc.* Name unverified.

33701 Vetanabol
An anabolic steroid used as an ethical veterinary medicine. *Ciba plc.* Name unverified.

33702 Vetol
A proprietary pure vegetable oil palm product. No manufacturer.

33703 VF-077
19706-80-0
2,2·g-azobis[2-(hydroxymethyl) propionitrile]
Polymerization initiator. *Wako Pure Chem. Ind.; Wako Chem. USA.* Discontinued.

33704 Via Rasa
It is the calcium salt of p-toluenechlorosulfonamide; an insecticide.

33705 Vialon Fast Dyes
1:2 metal complex dyes for dyeing and printing polyamide fibers. *BASF; BASF plc.*

33706 Vi-Alpha
68-26-8 10150 200-683-7
$C_{20}H_{30}O$
3,7-dimethyl-9-(2,6,6-trimethyl-1-cyclohexen-1-yl)-2,4,6,8-nonatetraen-1-ol
all-*trans*-retinol; Retinol; anti-infective vitamin; antixerophthalmic vitamin; axerophthol; biosterol; lard-factor; oleovitamin A; vitamin A_1; vitamin A

alcohol; Acon; Afaxin; Agiolan; Alphalin; Anatola; Aoral; Apexol; Apostavit; Atav; Avibon; Avita; Avitol; Axerol; Dohyfral A; Epiteliol; Nio-A-Let; Prepalin; Testavol; Vaflol; Vitpex; Vogan; Vogan-Nu; Vi-Dom-A. Vitamin A; antixerophthalmic. mp = 61-63°; bp = 120-125°; *Lederle Laboratories USA.* Name unverified.

33707 Vibatex
Finishing agent. *Ciba plc.* Name unverified.

33708 Vibrabond
Acrylic polymer cement flooring for heavy duty and chemical resistance. *Prodorite Ltd.*

33709 Vibrac
A nickel chrome steel.

33710 Vibrac steel
A nickel-chromium - molybdenum steel.

33711 Vibrathane®
A range of polyether-based and polyesterbased prepolymers; offer high abrasion resistance, chemical resistance, and electrical properties. *Uniroyal.* Name unverified.

33712 Vibrathane® 5004
Polyester PU; millable urethane with excellent resistant to abrasion, fuel, oil, and ozone, low temp. flexibility, high load bearing, high tensile and tear strength, low compression set; used for o-rings, seals, gaskets, for the automotive, aerospace, air *Uniroyal.*

33713 Vibrathane® 8007
Ester-MDI PU prepolymer; non-MBCA, castable; FDA wet food approval; for food contact articles. *Uniroyal.*

33714 Vibrathane® B602
PU prepolymer (polyether-TDI); castable, high resilience; for mining, slurry parts, linings. *Uniroyal.*

33715 Vibratussan
A proprietary preparation of doxycycline hyclate and codeine. An antibiotic and antitussive. *Pfizer International.* Discontinued.

33716 Vibrin® E-010-01
Bisphenol-A epoxy-based vinyl ester resin in styrene; corrosion-resistant resin for hand lay-up, spray-up, filament winding and pultrusion processes; for chemical tanks, pipe, fume handling equip., equip. for desulfurization processes. *Owens-Corning Fiberglas.*

33717 Vibrin® E-085
68002-44-8
Epoxy novolac-based vinyl ester resin in styrene; corrosion-resistant resin for hand lay up, spray-up, filament winding and pultrusion processes; resistant to chems., oxidation and heat; for chemical tanks, pipe, fume handling equipment and equipment for desulfurization processes. *Owens-Coming Fiberglas.*

33718 Vibrin® E-701
Two-stage isophthalic unsaturated polyester resin; corrosion-resistant resin for tanks, vessels, sm. diameter pipe, grating and air handling equipment. *Owens-Corning Fiberglas.*

33719 Vibrin® E-750
Bisphenol-A fumarate-based unsaturated polyester; corrosion-resistant resin for chemical tanks, small diameter pipe, fume handling equipment. subject to high temps. *Owens-Corning Fiberglas.*

33720 VIC®
Polyester and acrylic urethane; high performance, low temperature cure, corrosion-resistant coatings for three-dimensional objects, plastics, metal, heat-sensitive substrates. *Ashland.* Name unverified.

33721 Vicalloy
A proprietary trade name for a high permeability alloy containing iron with 36-62% cobalt and 6-16% vanadium. No manufacturer.

33722 Viclan
Polyvinylidene chloride copolymer resins and latexes. *ICI Chem & Polymers Ltd.*

33723 Vicmos powder
A smokeless 33-grain powder.

33724 Vicol®
Acrylic polymers designed particularly for application to cotton and spun cellulosic yarns and for sizing continuous filament viscose. *Allied Colloids Ltd.*

33725 Vicryl
Polyglactin
Surgical suture material, absorbable. *Ethicon Inc.* Name unverified.

33726 Victor bronze
An alloy of 58.5% copper, 38.5% zinc, 1.5% aluminum, 1% iron, and 0.03% vanadium.

33727 Victor Cream®
7758-16-9 8713 231-835-0
$H_2Na_2O_7P_2$
disodium hydrogen pyrophosphate
sodium acid pyrophosphate. Leavening agent for baking, cereals. Used in baking powder. mp = 220° (dec); d = 1.86 (hexahydrate). *Rhône-Poulenc Food Ingreds.*

33728 Victor metal
An alloy of 50% copper, 34.3% zinc, 15.4% nickel, 0.28% iron. and 0.11% aluminum; used for sand castings and marine work.

33729 Victor Powder
A trademark for a smokeless powder containing nitroammonium nitrate, wood meal, and potassium chloride. No manufacturer.

33730 Victorium
A proprietary trade name for lignin thermoplastic materials. No manufacturer.

33731 Victory®
63231-60-7 264-038-1
Microcrystalline wax; plastic wax offering high ductility, flexibility at very low temps.; provides protective barrier properties against moisture vapor and gases; used in hot-melt adhesives; hot-melt coatings; in antisunchecking agents in rubber goods, electrical insulating agents, leather treating agents, water repellents for textiles, rustproof coatings, cosmetic ingredients, and as a plasticizer for petroleum waxes used in crayons, dental compounds, chewing gum base and candles. *Petrolite.*

33732 Victrex
High temperature polymers such as polyether-etherketone and polyethersulfone. *ICI Chem & Polymers Ltd.*

33733 Victron
A proprietary trade name for polystyrene and vinylite resins. No manufacturer.

33734 Vicure® 10
22499-12-3 245-039-6
$C_{15}H_{14}O_2$
2-methoxy-2-phenylacetophenone
Isobutyl benzoin ether; benzoin isobutyl ether. Photosensitizer for uv curable systems, e.g., coatings, inks, graphic arts. $bp_{0.5}$ = 133°; d = 0.9850; n+ss20$_D$ = 1.5485 *Akzo.*

33735 Vicure® 55
15206-55-0 239-263-3
$C_9H_8O_3$
methyl phenylglyoxalate
methyl benzoylformate. High purity photoinitiator for uv curable systems, especially acrylate-based formulations. bp = 246-248°; d = 1.1550; n_D^{20} = 1.5268. *Akzo.*

33736 Viczsal
An ammoniacal solution of copper and zinc phenolates. A wood preservative.

33737 Vidal's caustic powder
A caustic consisting of burnt alum and powdered savin tops.

33738 Vi-Daylin
A proprietary preparation containing vitamins A, D, C, thiamine, riboflavine, nicotinamide and pyridoxine. *Abbott Laboratories.* Name unverified.

33739 Videne Disinfectant Solution, Videne Disinfectant Tincture, Videne Powder
25655-41-8 7880
$(C_6H_9NO)_n \cdot xI$
1-ethenyl-2-pyrrolidinone polymers complexed with iodine
poly[1-(2-oxo-1-pyrrolidinyl)ethylene]-iodine; polyvinylpyrrolidone-iodine; polyvidone-iodine; P.V.P.I.; RP-143; Kollidon; Periston; Plasdone; Plasmosan; Protagent; Subtosan; Vinisil. Pre-operative skin antiseptics containing povidone-iodine USP. Soluble in H_2O, EtOH, $CHCl_3$, insoluble in Et_2O. *3M Pharmaceuticals.* Discontinued.

33740 Videne Disinfectant Tincture
25655-41-8 7880
povidone-iodine
A preoperative skin antiseptic containing povidone-iodine USP. *3M Pharmaceuticals.*

33741 Videne Surgical Scrub
25655-41-8 7880
Surgical scrub containing povidone-iodine USP. *3M Pharmaceuticals.* Discontinued.

33742 Videobil
41473-08-9 5077
$C_{15}H_{18}I_3NO_5$
2-[[2-[3-(acetylamino)-2,4,6-triiodophenoxy]ethoxy]methyl]butanoic acid
Iopronic acid; (±)-2-[[2-(3-acetamido)-2,4,6-triiodophenoxy]ethoxy]methyl]butyric acid; B-11420; SQ-21983; Bilimiro; Bilmiron; Oravue; Videobil. Radio-opaque medium used as a diagnostic aid. *Bracco Industria Chimica SpA.* Name unverified.

33743 Videocolagio
Iodoxamic acid
Diagnostic aid. *Bracco Industria Chimica SpA.* Name unverified.

33744 Vi-Dom-A
68-26-8 10150 200-683-7
Vitamin A; antixerophthalmic. *Bayer.* Name unverified.

33745 Vienna cement
A metallic cement made from 86% copper and 14% mercury; an imitation gold.

33746 Vienna paste
A mixture of limeand potash.

33747 Viennese tombac
An alloy of 97% copper and 2.8% zinc.

33748 Vigantol® E Comp
Vitamins A, D_3, and H; used against vitamin deficiency diseases in veterinary medicine. *Bayer AG.*

33749 Vigazoo
58095-31-1
$C_9H_{10}N_2O_2S$
(4,5,6,7-tetrahydro-7-oxobenzo[b]thien-4-yl)urea
Sulbenox; CL-206576. Growth stimulant; used in veterinary medicine. mp = 245-246°; LD_{50} (rat orl) > 5000 mg/kg. *Am. Cyanamid.*

33750 Vigilan
8006-54-0 5371 232-348-6
Lanolin
Adeps lane; Agnolin; Agnolin No. 1; Alapurin; Amber lanolin; Anhydrous lanolin; Anhydrous Lanum; Cosmelan; Lanain; Lanalin; Lanesin; Lanichol; Laniol; Lanolin; Lanolin; Lanolin, anhydrous; Lanolin oil; Lanum; Oesipos; Processed lanolin; Viscolan; Wool fat; Wool grease. Lanolin oil; emulsifier; skin and hair substantive emollient and moisturizer; solubilizer for immiscible fluids; oil-water and/or water-oil emulsions; adds elegance, lubricity, and hydration to makeup and facial preparations; spreads rapidly. *Fanning.*

33751 Vigilan AWS
68458-58-8
PPG-12-PEG-65 lanolin oil; oil-water emulsifier; plasticizer, emollient, solubilizer, and wetting agent in personal care products; disperses to form milky emulsion in water used in bath oils. *Fanning.*

33752 Vigorite
A safety explosive for mines. It consists of 30% nitroglycerin, 49% potassium chlorate 7% potassium nitrate, 9% wood pulp and 5% magnesium carbonate.

33753 Vi-Grow
Chemical products used for physical conditioning of soil. *Coutaulds plc.*

33754 Vikane
2699-79-8 220-281-5
SO_2F_2
sulfuryl fluoride
Fumigant based primarily on sulfuryl fluoride is used specifically for the control of drywood termites and wood boring beetles infesting wood in structures, furniture and lumber; odorless, colorless, noncorrosive, and does not react to produce malodors. mp = -136°; bp = -55°; soluble in H_2O (25% v/v), more soluble in organic solvents. *Dow UK. Discontinued.*

33755 Viking
Tires. *Monsanto (Solaris).* Name unverified.

33756 Vikro
Proprietary alloy containing from 63-65% nickel, 13-23% chromium, 0.5-1% silicon, up to 1% manganese and carbon, and the balance iron. No manufacturer.

33757 Vileda
Household cloths. *Carl Freudenberg.* Name unverified.

33758 Viledon Compact
Nonwoven table cover. *Carl Freudenberg.* Name unverified.

33759 Viledon Filter
Nonwoven air filter. *Carl Freudenberg.* Name unverified.

33760 Vilene
Freudenberg nonwovens. *Carl Freudenberg.* Name unverified.

33761 Vilit
A range of soluble vinyl chloride copolymers; used as binders for paints, heat sealable lacquers for aluminum foils, for coatings on metal, concrete and cardboard. *Hüls AG.*

33762 Vimlite
A proprietary trade name for a cellulose acetate plastic. No manufacturer.

33763 Vinac® 1000 DEV
9003-20-7
Polyvinyl acetate homopolymer emulsion; base for wood adhesives. *Air Prods & Chem/Polymers.*

33764 Vinac® XX-210, XX-220, XX-230, XX-240
9003-20-7
Polyvinyl acetate homopolymer emulsion; base for adhesive compounding for carton sealing, white glues, wood veneering. *Air Prods & Chem/Polymers.*

33765 Vinaccia tartar
Tartar obtained from the manufacture of wines. Principally tartaric acid.

33766 vinaconic acid
598-10-7 209-917-2
C_5H_6O+4
propene-3,3-dicarboxylic acid
vinyl-malonic acid; 1,1-Cyclopropanedicarboxylic acid; Cyclopropane-1,1-dicarboxylic acid; Ethylenemalonic acid; Vinaconic acid. mp = 135-140°;.

33767 Vinacron
Polyvinyl chloride plastisol dispersion; used for protective and decorative coatings, rotomolding, spread coating and dip molding. *Loes Enterprises Inc.*

33768 Vinacryl
Acrylic copolymer emulsions. *Vinamul Ltd.*

33769 Vinacryl 4001/B
A proprietary trade name for a 50% concentrated acrylic copolymer emulsion used as a cement additive. *Vinyl Products.* Name unverified.

33770 Vinacryl 4005
A proprietary acrylic copolymer emulsion soluble in alkali. *Vinyl Products.* Name unverified.

33771 Vinacryl 4152, 4500/X, 4501/X
Proprietary trade names for vinyl acrylic copolymer emulsions used as adhesives. *Vinyl Products.* Name unverified.

33772 Vinacryl 4160
A proprietary trade name for an acrylic copolymer emulsion used as a 46.5% concentrate for a binder in paper board manufacture. *Vinyl Products.* Name unverified.

33773 Vinacryl 4260
A proprietary emulsion of poly-2-ethoxyethyl methacrylate. *Vinyl Products.* Name unverified.

33774 Vinacryl 4290
A proprietary polybutyl methacrylate emulsion. *Vinyl Products.* Name unverified.

33775 Vinacryl 4320
A proprietary self-reactive vinyl acrylic copolymer emulsion used in the finishing of textiles. *Vinyl Products.* Name unverified.

33776 Vinacryl 4322
A proprietary self-reactive vinyl acrylic copolymer emulsion. *Vinyl Products.* Name unverified.

33777 Vinacryl 4450
A proprietary trade name for a vinyl-acrylic copolymer; used in crack filling compounds. *Vinyl Products.* Name unverified.

33778 Vinacryl 4512
A proprietary acrylic polymer emulsion. *Cray Valley Ltd.* Discontinued.

33779 Vinacryl 7170, 7172, and 7175
A proprietary range of styrene acrylic copolymer emulsions. *Vinyl Products.* Name unverified.

33780 Vinacryl R3929, R3940
Proprietary trade names for a 55% concentrated vinyl-acrylic copolymer emulsions used for nonfray carpet backings. *Vinyl Products.* Name unverified.

33781 Vinal
A proprietary trade name for a synthetic vinyl resin. No manufacturer.

33782 Vinalak 5150
A proprietary self-reactive acrylic polymer solution in isopropyl acetate. *Vinyl Products.* Name unverified.

33783 Vinamul
Vinyl acetate homopolymer and copolymer emulsions. *Vinamul Ltd.*

33784 Vinamul 3240
A proprietary vinyl acetate-ethylene copolymer emulsion. *Vinyl Products.* Name unverified.

33785 Vinamul 3250
A proprietary vinyl acetate-ethylene copolymer emulsion containing a nonionic emulsifying system. *Vinyl Products.* Name unverified.

33786 Vinamul 6000
A proprietary vinyl acetate emulsion of the unsaturated acid copolymer type, soluble in alkali. *Vinyl Products.* Name unverified.

33787 Vinamul 6050
A proprietary vinyl acetate-vinyl caprate-unsaturated acid terpolymer emulsion, internally plasticized. *Vinyl Products.* Name unverified.

33788 Vinamul 6208
A proprietary trade name for a 50% concentrated vinyl-acrylic copolymer emulsion used for wallpaper grounding, printing and overcoating. *Vinyl Products.* Name unverified.

33789 Vinamul 6275
A proprietary trade name for a 55% concentrated vinyl-acrylic copolymer emulsion used in the production of emulsion paints. *Vinyl Products.* Name unverified.

33790 Vinamul 6705
A proprietary ethylene grafted vinyl acetate copolymer emulsion. *Vinyl Products.* Name unverified.

33791 Vinamul 6888
A proprietary modified vinyl acetate-acrylate copolymer emulsion. *Vinyl Products.* Name unverified.

33792 Vinamul 6930
A proprietary trade name for 52% concentrated vinyl acetate - Veo Va 911 copolymer emulsion used for emulsion paints. *Vinyl Products.* Name unverified.

33793 Vinamul 7700
A proprietary polystyrene emulsion. *Vinyl Products.* Name unverified.

33794 Vinamul 7715
A proprietary polystyrene emulsion containing 15% dibutyl phthalate in a nonvolatile plasticizer. *Vinyl Products.* Name unverified.

33795 Vinamul 8400
A proprietary trade name for a 50% emulsion of polyvinyl acetate; used for adhesives. *Vinyl Products.* Name unverified.

33796 Vinamul 8430
A proprietary polyvinyl acetate emulsion. *Vinyl Products.* Name unverified.

33797 Vinamul 8460 and 9000
Proprietary polyvinyl acetate emulsions. *Vinyl Products.* Name unverified.

33798 Vinapol
Vinyl acetate homopolymer and copolymer powders. *Vinamul Ltd.*

33799 Vinapol 1000
A proprietary polyvinyl acetate powder dispersible in water. *Vinyl Products.* Name unverified.

33800 Vinapol 1030
A proprietary polyvinyl acetate powder plasticized with 10% dibutyl phthalate and dispersible in water. *Vinyl Products.* Name unverified.

33801 Vinapol 1070
A proprietary finely-divided polyvinyl acetate powder. *Vinyl Products.* Name unverified.

33802 Vinapol 1088
A proprietary acrylic processing and used in the processing of rigid PVC compounds. *Vinyl Products.* Name unverified.

33803 Vinapol R.3800, R.3863
A proprietary trade name for a water dispersible polyvinyl acetate powder. *Vinyl Products.* Name unverified.

33804 Vinapol R.3626
A proprietary trade name for a water-dispersible alkali soluble vinyl acetate powder. *Vinyl Products.* Name unverified.

33805 Vinatex
Polyvinyl chloride plastisols
Hydro Chemicals Ltd.

33806 Vinavil
Vinyl acetate homopolymers and copolymers. *Montedison UK Ltd.* Name unverified.

33807 Vinchel 11, Vinchel 20
Proprietary metal-free liquid organic complexes with a stabilizer system of barium and cadmium soaps; used as chelating agents in PVC compounds. *Vinyl Products.* Name unverified.

33808 Vinchel 20
A proprietary metal-free liquid organic complex with a stabilizer system of barium and cadmium soaps and tribasic barium sulfate . Used as a chelating agent in PVC compounds. *Vinyl Products.* Name unverified.

33809 Vinchel 22
A proprietary zinc-based chelating agent for PVC compounds; a basic lead carbonate stabilizing system is used. *Vinyl Products.* Name unverified.

33810 Vinchel 35
A proprietary zinc-based chelating agent for PVC compounds; a stabilizer system of barium and cadmium soaps. *Vinyl Products.* Name unverified.

33811 vinclozolin
50471-44-8 10122 256-599-6
$C_{12}H_9Cl_2NO_3$
3-(3,5-dichlorophenyl)-5-ethenyl-5-methyl-2,4-oxazolidinedione
Ronilan; vinclozalin; Dichlorophenyl)-5-ethenyl-5-methyl-2,4-oxazolidinedione; Ornalin; Oxazolidinedione, 3-(3,5-dichlorophenyl)-5-ethenyl-5-methyl-; Vorlan. Fungicide. mp = 108°; $bp_{0.05}$ = 131°; soluble in H_2O (1 g/l), more soluble in organic solvents; LD_{50} (rat orl) = 10 g/kg.

33812 Vinco 248
A proprietary stabilizer of the liquid barium and cadmium complex type; used with PVC polymers sensitive to zinc. *NL Victor Wolf Ltd.* Name unverified.

33813 Vinco 249C
A proprietary stabilizer for PVC with a liquid barium, cadmium, and zinc base. *NL Victor Wolf Ltd.* Name unverified.

33814 Vinco 265
A proprietary stabilizer for PVC pastes with a liquid barium, cadmium, and zinc base. *NL Victor Wolf Ltd.* Name unverified.

33815 Vinco 99A
A proprietary stabilizer for PVC and PVA of the liquid barium/cadmium/zinc complex type. *NL Victor Wolf Ltd.* Name unverified.

33816 Vinco 99G
A stabilizer similar to Vinco 99A used with pastegrade resins in rotational molding. *NL Victor Wolf Ltd.* Name unverified.

33817 Vinco A183
A proprietary liquid complex used as a stabliser and initiator in the production of expanded PVC. *NL Victor Wolf Ltd.* Name unverified.

33818 Vinco A33
A proprietary liquid potassium/zinc complex used as a stabilizer and initiator in the production of expanded PVC. *NL Victor Wolf Ltd.* Name unverified.

33819 Vindex
Bromoxynil + clopyralid; herbicide mixture for weed control in cereals. *Quadrangle Agrochemicals.*

33820 Vinex 1003
9002-89-5 7745
Polyvinyl alcohol copolymer resin; thermoplastic resin for extrusion into cast or tubular blown film, injection or blow molding into bottles, melt spinning into fiber for industrial or personal care applications *Air Prods & Chems. Inc.*

33821 Vinex 2019
9002-89-5 7745
Polyvinyl alcohol copolymer resins; thermoplastic for melt blown fiber, injection molding, extrusion, lamination. *Air Prods & Chems. Inc.*

33822 Vinex 2025
9002-89-5 7745
Polyvinyl alcohol copolymer resin; thermoplastic for extrusion into cast or tubular blown film, injection or blow molding into bottles, melt spinning into fiber for industrial or personal care articles. *Air Prods & Chems. Inc.*

33823 Vinex 2034
9002-89-5 7745
Polyvinyl alcohol copolymer resins; thermoplastic for cast film, sheet, blown film, melt spun fiber, blow molding. *Air Prods & Chems. Inc.*

33824 Vinex 2144
9002-89-5 7745
Polvyinyl alcohol copolymer resin; thermoplastic resin for extrusion into cast or tubular blown film, injection or blow molding into bottles, melt spinning into fiber for industrial or personal care applications. *Air Prods & Chems. Inc.*

33825 Vinex 5030
9002-89-5 7745
Polvyinyl alcohol copolymer resin; thermoplastic resin for extrusion into cast or tubular blown film, injection or blow molding into bottles, melt spinning into fiber for industrial or personal care applications. *Air Prods & Chems. Inc.*

33826 Vinidur®
Vinyl chloride/acrylate graft copolymers; produces very high to high impact, weather resistant parts preferably by extrusion, but also by calendering, injection molding, e.g., profiles for outdoor applications (window frames), pipes, panels, films. *BASF AG; BASF aplc.*

33827 Vinisil
9003-39-8 7879 294-352-4
$(C_6H_9NO)_n \cdot xI$
1-ethenyl-2-pyrrolidinone polymers
povidone; poly[1-(2-oxo-1-pyrrolidinyl)ethylene]-iodine; polyvinylpyrrolidone-iodine; polyvidone-iodine; P.V.P.I.; RP-143; Kollidon; Periston; Plasdone; Plasmosan; Protagent; Subtosan; Vinisil. Povidone; used as a dispersing and suspending agent in pharmaceutic preparations. mp = 225° (dec). *Abbott Laboratories.* Name unverified.

33828 Vinnapas®
Homopolymer or copolymer, polyvinyl acetate solutions, dispersions, solid resins; used for lacquers and adhesives, paints. *Wacker-Chemie GmbH.*

33829 Vinnapas® A 50
VA homopolymer disp.; binder for fabrics, glass fibers; textile auxiliary and coating; paper coatings. *Wacker-Chemie.* Name unverified.

33830 Vinnapas® EN 426
VA/ethylene copolymer disp. with reactive groups; binder for fabrics, glass fiber; textile auxiliary and coating. *Wacker-Chemie.* Name unverified.

33831 Vinnapas® EP 177
Vinyl acetate/ethylene copolymer disp.; for textile finishing/softening. *Wacker-Chemie.* Name unverified.

33832 Vinnapas® LL 364
Styrene/acrylic disp.; for textile finishing. *Wacker-Chemie.* Name unverified.

33833 Vinnathen
A proprietary trade name for a vinyl acetate/ethylene copolymer which can be crosslinked with peroxides; can be used for cable jackets. No manufacturer.

33834 Vinnol®
A proprietary grade of polyvinyl chloride. *Wacker-Chemie GmbH; Wacker Chemicals Ltd.*

33835 Vinnol® 50
9003-22-9
Vinyl chloride/vinyl acetate copolymer disp.; binder for fabrics, glass fiber; textile auxiliary and coating; and paper coatings. *Wacker-Chemie.* Name unverified.

33836 Vinnol® LL 352
Vinyl chloride/ethylene copolymer dispersion; for paper, carpet, and textile industries. *Wacker-Chemie.* Name unverified.

33837 Vinofan®
Polymer dispersions based on vinyl compounds; for building adhesives and packaging adhesives; binders for paper and board coatings, for bonding fiber webs, and for textile coating. *BASF AG.*

33838 Vinoflex
PVC; used for extrusion (pipes, profiles and cables), calendering (plasticized and UPVC films), injection molding (fittings), blowmolding (bottles). *BASF plc.*

33839 Vinoflex®
9002-86-2 7746 206-625-7
PVC; after stabilization can be processed into rigid parts for extrusion, calendering, and injection molding (film, pipes, profiles, hollow articles, and panels); with addition of plasticizers produces flexible products such as film, cable insulation, cab *BASF AG.*

33840 Vinoflex® 377
A proprietary PVC-emulsion homopolymer used in the production of PVC film. *BASF plc.* Name unverified.

33841 Vinoflex® 516
A proprietary PVC suspension-type homopolymer. It is an easily-flowing powder used in the extrusion of rigid PVC. *BASF plc.* Name unverified.

33842 Vinoflex® 526
A proprietary PVC homopolymer similar to Vinoflex 516. *BASF plc.* Name unverified.

33843 Vinoflex® 534
A proprietary PVC suspension-type homopolymer used in the extrusion of high-quality cables. *Anic Agricultura Spa.* Unverified.

33844 Vinoflex® 535
A proprietary PVC suspension-type homopolymer in the form of an easy-flowing powder with porous particles; used in the making of soft, calendered products. *BASF plc.* Name unverified.

33845 Vinoflex® 719
A proprietary PVC suspension-type polymer used in the making of rigid and tough, weather-resistant products. *BASF plc.* Name unverified.

33846 Vinsalyn
A range of thermoplastic resins; used as binders and/or binder extenders in industrial adhesives and mastics. *Hercules.* Discontinued.

33847 Vinsol
Dark pine resin. *Hercules Ltd.*

33848 Vinsol
A proprietary trade name for the black residue from the extraction of rosin with solvents; used as an insulating varnish. No manufacturer.

33849 Vinsol Emulsion
An emulsion of aliphatic hydrocarbon insoluble resin; used as a modifier of water-based adhesives and coatings. *Hercules.* Discontinued.

33850 Vinsol Resin
Wood rosin derivative. *Langley Smith & Co Ltd.*

33851 Vinuran®
Polymers for modifying PVC; improves PVC properties including impact strength, resistant to heat deformation and decomposition, processing, and deep-drawing behaviour; used for hot water pipes, impact-resistant profiles, films, hollow articles, panels. *BASF AG; BASF plc.*

33852 Vinychlon®
A series of vinylchloride polymers. *Mitsui Co Ltd.* Name unverified.

33853 vinyl acetate
108-05-4 10130 203-545-4
$C_4H_6O_2$
acetic acid ethenyl ester
acetic acid vinyl ester. Unsaturated ester; in polymerized form for plastic masses, films, and lacquers. Intermediate for industrial and consumer products including PVAc to produce paints, adhesives and coatings; PVAL (adhesives, coatings, packaging films); polyvinyl acetals (insulation, safety glass interlayer); EVA copolymers (films, coatings, adhesives). *Exxon Belgium; Hoechst Celanese; Quantum/USI; Union Carbide.*

33854 Vinyl Acetate
108-05-4 10130 203-545-4
$C_4H_6O_2$
Acetic acid ethenyl ester
Everflex® 81L; acetic acid vinyl ester; Plyamul® 40305-00; Unocal 76 Res 6206; Unocal 76 Res S-55; Vinnapas® A 50; Vinyl Acetate Monomer,. Vinyl acetate copolymer emulsion; paint and coating emulsion for use as binder for clay coating of paper and paperboard; factory finishes, ceiling tile, wall board, and textile treatments. MW=86.09; liquid; polimerizes in light to colorless transparent mass; pb=72.7°; sol.in water; misc with alcohol, ether; LD_{50}(rat, orl)=2.92 g/kg *WR Grace/Organics; Reichhold/Emulsion Polymers; Unocal/Polymers; Wacker-Chemie GmbH; Quantum/USI.* Name unverified.

33855 Vinyl Acetate Monomer
108-05-4 10130 203-545-4
$C_4H_6O_2$
acetic acid vinyl ester
Intermediate for industrial and consumer products including polyvinyl acetate, used in paints, adhesives and coatings, polyvinyl alcohol, used in adhesives, coatings and packaging films, polyvinyl acetals, used in insulation and as a safety glass interlayer and ethylene vinyl acetate copolymers, used in films, coatings and adhesives. bp = 71-73°; d= 0.932; n_D^{20}= 1.3950; LD_{50} (rat orl) = 2.92 g/kg. *Quantum/USI.* Name unverified.

33856 vinyl compounds and polymers
A compound having the vinyl grouping (CH_2=CH); basis for varieties of plastics including vinyl chloride, vinyl acetate, etc.

33857 vinyl ester resin
Thermoset used in chemical processing industry, pulp and paper mills, corrosive-resistant applications. *Ashland; Cook Composites & Polymers; Dajac Labs; Union Carbide.*

33858 vinyl pyridine
1337-81-1
C_7H_9N
Pyridine, ethenyl-
Pyridine, vinyl-; Vinylpyridine. Used for adhesives, tire cord, and industrial goods dips. *Penta Mfg.; Rasching; Reilly Industries.*

33859 Vinyl pyridine
1337-81-1
Good-rite® 2528X10; Vinyl pyridine latex. Used for adhesives in tire cord chafer fabric and industrial rubber goods; produces wickproof chafer fabric in one pass; very stable high solids RFL dips; excellent adhesion retention. *BF Goodrich.* Name unverified.

33860 Vinylite
A proprietary trade name for polyvinyl acetate, polyvinyl chloride-acetate and polyvinyl chloride synthetic resins. No manufacturer.

33861 Vinylite V
A proprietary name for an interpolymer of PVC and PVA. No manufacturer.

33862 Vinylite X
A proprietary trade name for polyvinyl chloride acetate. No manufacturer.

33863 Vinyloid
A proprietary trade name for a polyvinyl acetate resin. No manufacturer.

33864 Vinylseal
A proprietary trade name for vinyl acetate resin adhesives. No manufacturer.

33865 vinyltrichlorosilane
75-94-5 200-917-8
$C_2H_3Cl_3Si$
(trichlorosilyl)ethylene
trichlorovinylsilane. Intermediate for silicones; coupling agent in adhesives and bonds. mp = -95°; bp = 90°; d = 1.2700; n_D^{20}= 1.4362 *Hüls Am.; PCR; Schweizerhall; Union Carbide.*

33866 vinyltriethoxysilane
78-08-0 201-081-7
$C_8H_{18}O_3Si$
(triethoxysilyl)ethylene
triethoxyvinylsilane. Intermediate, especially when acidic by-products are undesirable; filler. bp = 160-161°; d = 0.9030; n_D^{20}= 1.3978. *Hüls Am.; PCR; Union Carbide.*

33867 Vinyon
A proprietary trade name for vinyl resins for textile fibers. No manufacturer.

33868 Vinyon Fiber
A proprietary trade name for a material manufactured from polyvinyl chloride-acetate. No manufacturer.

33869 Vinyzene
58-36-6 7095 200-377-3
$C_{24}H_{16}As_2O_3$
10,10'-oxydiphenoxarsine
10,10'-oxybis-10H-phenoxarsine; bis(phenoxarsin-10-yl)ether; bis(10-phenoxarsyl)oxide; bis(10-phenoxarsinyl)oxide; OBPA Vinyzene. Condensation product of epoxidized soy bean oil and 10, 10-g-oxybisphenoxyarsine; used as a fungicide and bactericide. mp= 184-185°; d = 1.41; insoluble in H_2O, soluble in organic solvents; LD_{50} (rat orl) = 35 mg/kg.

33870 Vinyzene® BP-505 DIDP
5% 10,10·g-Oxybisphenoxarsine in diisodecyl phthalate; antimicrobial, bacteriostat, fungistat for PVC, PU, other plastics, synthetic rubber; recommended for film and sheet, extruded profiles, plastisols, molded goods, organosols, fabric coatings. *Morton Int'l./Plastics Additives.*

33871 Vinyzene® IT-3000 DIDP
2-n-octyl-4-isothiazolin-3-one
2-n-Octyl-4-isothiazolin-3-one in diisodecyl phthalate carrier; fungicide for plastics (PVC, PU); recommended for vinyl film, sheeting, extruded profiles, plastisols, molded goods, organosols, fabric coatings, urethane shoe soles, and foams. *Morton Int'l./Plastics Additives.*

33872 Vinyzene® SB-1 EAA
10,10·g-Oxybisphenoxarsine in ethylene-acrylic acid copolymer carrier; fungicide, bactericide for polyolefins; for interior automotive parts, exterior drainage hose, underground cable jacketing, buoyant floats. *Morton Int'l./Plastics Additives.*

33873 Vioflor
A proprietary preparation of volatile hydrocarbons; used to deodorize turpentine substitutes. No manufacturer.

33874 Vioglaze
UV curing coatings; gloss varnish for roller coating paper and board and roller coating for flooring. *Coates Coatings Ltd.*

33875 violet copper
1317-39-1 2734 215-270-7
Cu_2O
cuprous oxide
Red copper oxide. Reduced copper, cuprous oxide, prepared by reducing cupric oxide by heating with copper. Used as a fungicide and antiseptic.

33876 violet phosphorus
7723-14-0 7503 231-768-7
P
phosphorus
red phosphorus; Hittorf's phosphorus. The coarse-grained red variety is metallic or violet phosphorus. Sublimes 416°; d = 2.34; insoluble in organic solvents.

33877 violet powder
Perfumed starch powder.

33878 violet tungsten
Potassium tritungstate
$K_2W_3O_9 \cdot W_2O_5$. Used as a pigment.

33879 violettol
A mixture of 10% ionone and 90% salicylaldehyde; used to strengthen natural violet perfume.

33880 Violetton
Trade names for the commercial ionones; used as perfumes. No manufacturer.

33881 violuric acid
26351-19-9 201-741-4
$C_4H_3N_3O_4$
2,4,5,6(1H,3H)-pyrimidinetetrone-5-oxime
alloxan-5-oxime;5-isonitrosobarbituric acid; 5-(hydroxyimino)barbituric acid. mp = 240-241° (dec); insoluble in H_2O, soluble in EtOH.

33882 Viomycin
An antibiotic produced by certain strains of *Streptomyces griseus var. purpureus.* No manufacturer.

33883 Vionate Powder
Multivitamin and mineral mix for veterinary use. *Ciba plc.* Name unverified.

33884 Vipla
Emulsion polyvinyl chloride emulsion homopolymers. No manufacturer.

33885 Viplan
Polyvinyl chloride paste polymers. *European Vinyls Corporation Ltd.*

33886 Viplex 680-P
64742-04-7 265-103-7
Heavy paraffinic distillate solvent extract; secondary vinyl plasticizer. *Crowley Chem.*

33887 Viplex 885
68477-29-2 270-719-4
Highly aromatic (99%) petrol. oil; plasticizer/extender for epoxy systems, coatings, potting/encapsulation; promotes low system visc., better wetting of fillers and pigments, higher loading levels; modifier for polyesters. *Crowley Chem.*

33888 Viplex 895-BL
64741-81-7 265-082-4
Polynuclear aromatic hydrocarbon
Secondary vinyl plasticizer. *Crowley Chem.*

33889 Vipophan PVC
Shrink film. *Lonza AG.*

33890 Virco
Textile dyeing auxiliaries. *Virkler Chemical Co.* Name unverified.

33891 Vironex
Folpet 40%, cymoxanil 4%; wettable powder used as protective fungicide for foliage application to ornamental and crop plants. *Industrias Quimicas Del Vailes SA.*

33892 Viscalex®
Acrylic-based thickening agents for use in aqueous adhesives, paints, latex compounds, cosmetics, and ceramics. *Allied Colloids Ltd.*

33893 Viscasil®
9006-65-9 3264
Dimethicone
dimethyl polysiloxane; dimeticon; polydimethylsiloxane;. Defoamer, release agent, emollient in cosmetics, polishes, paint additives, and mechanical devices; lubricant in rubber or plastic-to-metal applications; suggested for auto polish, skin care, hair care, textile softeners, antifoams for petrol. refining, r Insoluble in H_2O, EtOH, soluble in $CHCl_3$, Et_2O. *GE Silicones.*

33894 Visc-Ayd® 812
With Visc-Ayd Activator 814 as bodying agent for solvent-thinned coatings including trade sales paints, house paints, enamels, and flat alkyds. *Elementis Specialties.*

33895 Vischem
A series of diester and polyolester synthetic based greases with various thickeners; high temperature greases for all kinds of bearings, worm gears, slides, etc. *Ultrachem Inc.* Name unverified.

33896 viscoid
A mixture of viscose and clay with powdered horn or zinc oxide.

33897 Viscolas®
Viscoelastic urethane material; formulated to absorb and dissipate shock, provide good resiliency, rapid recovery, good damping; for industrial work gloves, tool grips, shoe inserts, modular orthotics, bike seat cushions, exercise equip. *E-A-R.*

33898 viscolith
The hard mass obtaimed when a viscose solution coagulates.

33899 Viscoloid
A proprietary trade name for pyroxylin plastics. No manufacturer.

33900 Viscoplex
Solutions in mineral oil of long chain fatty alcohol/acrylic/methacrylic acid esters; pour point depressants and viscosity index improvers for mineral oils. *Comelius Chemical Co Ltd.* Name unverified.

33901 viscose
The sodium salt of cellulose xanthate, obtained by the action of carbon disulfide and alkali upon cellulose. In thin sheets; it is used as a substitute for glass and celluloid; as a thickening and dressing substance, as a partial substitute for resin glue.

33902 viscose silk
An artificial silk produced when viscose is forced through narrow orifices into ammonium chloride solution.

33903 Viscosin
A proprietary refined oil tar. No manufacturer.

33904 Visco-Stab
Viscosity stabilizer for cuprous oxide-containing paints based on bioMeT 300 Series antifoulant polymers. *Elf Atochem N. Am.*

33905 Visor
Deodorant for industrial processes and products. *CPL Group Ltd.*

33906 Visqueen
Polyethylene film. *ICI Chem & Polymers Ltd.* Discontinued.

33907 Vista LPA, LPA-140, LPA-210
Mixtures of hydrotreated isoparaffins and naphthenes with very low levels of aromatics; solvent for food applications, pesticides, coatings, household products, water, paper, and mining chemicals; used as textile lubricants, chemical process solvents, freezing point depressants and printing ink solvents. *Vista.*

33908 Vista®
For medical Applications. *DuPont UK.*

33909 Vistaflex
Thermoplastic elastomer. *Exxon Int'l.; Exxon UK.*

33910 VistaFlex®
Thermoplastic elastomer
For parts requiring good surface appearance, replacing thermoplastic olefin blends, styrene-block TPEs, and plasticized PVC. *Advanced Elastomer Systems.*

33911 Vistal
Polyethylene films in widths up to 12 m; films based on polyethylene LD, MD, LLD in thicknesses of 10 to 800 microns; shrink and stretch films for packaging, bundling and palletizing, coextruded films, embossed and repellent films for industrial, horticulture and agriculture applications, peelable

films for easy opening packaging, special films for incorporation into laminates, for balloons used in space research, and for building and road construction. *UCB nv Film Sector.* Discontinued.

33912 Vistalon
Polyolefin; ethylene propylene copolymers. *Exxon Int'l.; Exxon UK.* Name unverified.

33913 Vistalon 719
9010-79-1
Ethylene/propylene copolymer; EPM rubber for modifying polyolefins. *Exxon.* Name unverified.

33914 Vistalon 7000
EPDM terpolymer; fast extruding; used in hose and tubing, elec. insulation, thermoplastic polyolefins, weather-strips, molded goods; sulfur-curable; fast curing. *Exxon.* Name unverified.

33915 Vistalon 2200
EPDM terpolymer; used alone or in blends (especially in butyl rubber) where controlled cure rate and easy processing are needed; imparts excellent ozone and heat resistant; excellent processing during extrusion, molding, or calendering. *Exxon.* Name unverified.

33916 Vistalon 3666
EPDM terpolymer, 75 phr paraffinic oil; accepts high loadings of carbon black and oil to yield low cost, fast extruding compounds with low tension set and good snappiness. *Exxon.* Name unverified.

33917 Vistalon 3708
EPDM terpolymer; high green strength, high extrusion rate, extra high tensile strength; maintains good physical properties when highly extended; used for medium and high hardness extrusion compounds, hose, weather-strip, molded goods; sulfur-curable. *Exxon.* Name unverified.

33918 Vista-Marc
Etafilcon A; contact lens material. *Vistakon Inc.* Name unverified.

33919 Vistanex®
Semi-solid polyisobutylene. *Exxon Int'l.; Exxon UK.*

33920 Vistanex® MML-80, MML-100, MML-120, MML-140
9003-27-4
Polyisobutylene used as polymeric additive for rubbers to impart stability, inertness, ozone, and chemical resistant, electrical inertness; also useful in molded and extruded products and coated materials. *Exxon.* Name unverified.

33921 Vistone
Mild extreme pressure oiliness agent. *Exxon UK.* Name unverified.

33922 Vita Zinc
1314-13-2 10279 215-222-5
Low purity zinc oxide
Animal feed supplement. *Manchem Ltd.* Name unverified.

33923 Vita-Cos
Wheat germ glycerides. *CasChem.*

33924 Vita-E, Vita-E Gelucaps
Proprietary preparations of D-α-tocopherol used in the treatment of vascular disorders. *Bioglan Laboratories.* Name unverified.

33925 Vitafeed 101
Soluble powder containing NK 26:26 and trace elements; fertilizer. *Vitax Ltd.*

33926 Vitafeed 102
Soluble powder containing NK 18:36 and trace elements; fertilizer. *Vitax Ltd.*

33927 Vitafeed 103
Soluble powder containing NK 13:43 and trace elements; fertilizer. *Vitax Ltd.*

33928 Vitafeed 111
Soluble powder containing NPK 19:19:19 plus 0.2% Mg and trace elements; fertilizer. *Vitax Ltd.*

33929 Vitafeed 301
Soluble powder containing NK 36:12 plus 0.15% Mg and trace elements; fertilizer. *Vitax Ltd.*

33930 Vitalum
A trademark for materials of the abrasive class and consisting essentially of alumina. No manufacturer.

33931 Vitamalt
A proprietary preparation containing vitamins A, B, C, and D. No manufacturer.

33932 vitamin A
68-26-8 10150 200-683-7
$C_{20}H_{30}O$;
Retinol; vitamin A alcohol; axerophthol; Vogan. Dietary supplement. *Am. Biorganics; BASF; Hoffmann-La Roche; Penta Mfg.*

33933 Vitamin B Complex CLR
Yeast extract with B vitamins in water-alcohol medium; products for treatment of greasy hair, dandruff and oily skin. *Dr. Kurt Richter; Henkel/Cospha.*

33934 vitamin B$_{12}$
68-19-9 10152 200-680-0
$C_{63}H_{88}CoN_{14}O_{14}P$
cyanocobalamin
cyanocon(III) alamin; α-(5,6-dimethylbenzimidazolyl) cyanocobamide; cobalamin. Dietary supplement; deficiency in man causes pernicious anemia and neural degeneration. *EM industries; Hoffmann-La Roche; Roussel Uclaf; Schweizerhall.*

33935 vitamin B$_2$
83-88-5 8367 201-507-1
$C_{17}H_{20}N_4O_6$
7,8-dimethyl-10-(D-*ribo*-2,3,4,5-tetrahydroxypentyl)isoalloxazine
riboflavin; 7,8-dimethyl-10-ribitylisoalloxazine; lactoflavine; vitamin G; Beflavin; Flavaxin. An essential dietary factor. mp = 278-282° (dec); $[\alpha]_D^{25}$= -117° (H_2O); λ_m = 220-225, 266, 371, 444, 475 nm; slightly soluble in H_2O, organic solvents; LD_{50} (rat orl > 10 g/kg.

33936 vitamin B$_5$ calcium salt
137-08-6 7147 205-278-9
$C_{18}H_{32}CaN_2O_{10}$
(R)-N-(2,4-dihydroxy-3,3-dimethyl-1-oxobutyl)-β-alanine calcium salt
calcium pantothenate; Calpanate; Pantholin. An essential dietary factor. mp = 195-196° (dec); $[\alpha]_D^{20}$= 28.2° (c = 5); soluble in H_2O (0.36 g/ml), poorly soluble in organic solvents.

33937 vitamin B$_6$ hydrochloride
58-56-0 8166 200-386-2
$C_8H_{12}ClNO_3 \cdot HCl$
5-hydroxy-6-methyl-3,4-pyridinedimethanol hydrochloride
pyridoxine hydrochloride; pyridoxol hydrochloride; pyridoxinium chloride; Bonasanit; Hexabione hydrochloride; Hexabetalin; Hexavibex; Pyridipca; Pyridox; Benadon; Bécilan; Hexermin; Campoviton 6; Hexobion. An essential dietary factor. Used in medicine and nutrition. mp = 205-212° (dec); λ_m = 290 nm (ε 8400, 0.1N HCl); soluble in H_2O (0.22 g/ml), less soluble in organic solvents. *Napp Laboratories Inc.*

33938 vitamin C
50-81-7 867 200-066-2
$C_6H_8O_6$
ascorbic acid
L-ascorbic acid; L-xyloascorbic acid; 3-oxo-L-gulofuranolactone; L-3-ketothreohexuronic acid lactone; antiscorbutic vitamin; cevitamic acid; Adenex; Allercorb; Ascorin; Ascorteal; Ascorvit; Cantan; Cantaxin; Catavin C; Cebicure; Cebion; Cecon; Cegiolan; Celaskon; Celin; Cenetone; Cereon; Cergona; Cescorbat; Cetamid; Cetebe; Cetemican; Cevalin; Cevatine; Cevex; Cevimin; Ce-Vi-Sol; Cevitan; Cevitex; Cewin; Ciamin; Cipca; Concemin; C-Vimin; Davitamon C; Duoscorb; Hybrin; Laroscorbine; Lemascorb; Planavit C; Proscorbin; Redoxon; Ribena; Scorbacid; Scorbu-C; Testascorbic; Vicelat; Vitacee; Vitacimin; Vitacin; Vitascorbol; Xitix. Deficiency in man causes scurvy. It probably plays a part in the production of collagen in the tissues; used in nutrition, color fixing, flavoring, dietary supplement. mp= 190-192°; d = 1.65; $[\alpha]_D^{20}$= 48° (c = 1 MeOH); λ_m = 265 nm (pH 7); soluble in H_2O (300 mg/l), insoluble in organic solvents.

33939 vitamin C sodium salt
134-03-2 8723 205-126-1
$C_6H_7NaO_6$
sodium ascorbate
vitamin C sodium; Ascorbin; Sodascorbate; Natrascorb; Cenolate; Ascorbicin; Cebitate. mp = 218° (dec); $[\alpha]_D^{2}$= 104°; very soluble in H_2O (620 g/l).

33940 vitamin D$_2$
50-14-6 10156 200-014-9
$C_{28}H_{44}O$
(3β,5Z,7E,22E)-9,10-secoergosta-5,7,10(19)-,22-tetraen-3-ol
oleovitamin D$_2$; viosterol; Condol; Decaps; Dee-Ron; Deltalin; De-Rat Concentrate; Deratol; Detalup; Diactol; Divit-Urto; Drisdol; D-Tracetten; Ergorone; Ertron; Fortodyl; Hi-Deratol; Infron; Metadee; Mina D$_2$; Mulsiferol; Mykostin; Ostelin; Radiostol; ercalciol; calciferol; ergocalciferol; Radstein; Shock-Ferol Sterogyl; Uvesterol-D; Vio-D. A fat soluble vitamin, deficiency of which causes rickets, in children and osteomalacia in adults; $C_{28}H_{44}O$; used as a dietary supplement and as a rodenticide. mp= 115-118°; $[\alpha]_D^{25}$= 83° (c= 2 Me_2CO); λ_m = 264.5 nm ($E^{1\%}_{1cm}$ 459, hexane); insoluble in H_2O, soluble in organic solvents.

33941 vitamin D$_3$
67-97-0 10157 200-673-2
$C_{27}H_{44}O$
(3β,5Z,7E)-9,10-secocholesta-5,7,10(19)-trien-3-ol
oleovitamin D$_3$; cholecalciferol; CC; Duphafral D$_3$ 1000; Delsterol; Deparal; Ebivit; Micro-Dee; Neo Dohyfral D$_3$; Provitina; Ricketon; Trivitan; D$_3$-Vicotrat; Vi-De-3-Hydrosol; Vigantol®; Vigorsan. Vitamin D$_3$ in oily solution; veterinary preparation to prevent rickets and osteomalacia. mp = 85-87°; $[\alpha]_D^{20}$= 85° (c = 1.6, Me_2CO); λ_m = 264.5 nm ($E^{1\%}_{1cm}$ = 470, EtOH). *Bayer AG.*

33942 vitamin D$_4$
511-28-4 10158 208-127-5

$C_{28}H_{46}O$
93β,5Z,7E)-9,10-secoergosta5,7,10(19)-trien-3-ol
22:23-dihydrovitamin D_2; 22,23-dihydroergocalciferol. A synthetic vitamin derived from 22-dihydroergosterol by irridation with uv light. It has less activity than vitamins D_2 and D_3. mp = 96-98°; $[\alpha]_D^{18}$ = 89° (c = 0.47, Me_2CO); λ_m= 265 nm; insoluble in H_2O, organic solvents.

33943 vitamin E
59-02-9 10159 200-412-2
$C_{29}H_{50}O_2$
[2R-2RName unverified.(4RName unverified.,8RName unverified.)]-3,4-dihydro-2,5,7,8-tetramethyl-2-(4,8,12-trimethyltridecyl)-2H-1-benzopyran-6-ol
(+)-α-tocopherol; 2,5,7,8-tetramethyl-2-(4',8',12'-trimethyltridecyl)-6-chromanol; α-tocopherol; 5,7,8-trimethyltocol; antisterility vitamin. A fat soluble vitamin; used as an antioxidant, in meat curing, nutrient. mp = 3°; bp $_{0.1}$ = 200-220°; d^{25}_4 = 0.950; n_D^{25}= 1.5045; λ_m = 294 nm ($E^{1\%}_{1cm}$ 71); $[\alpha]^{25}_{5461}$ = -3° (C_6H_6); insoluble in H_2O, soluble in organic solvents.

33944 Vitamin F Alcohol-Soluble CLR
Complex of essential free fatty acids, hydro-alcohol solubilized; products for treatment of dry skin and hair. *Dr. Kurt Richter; Henkel/Cospha.*

33945 Vitamin F Ethyl Ester CLR
Complex of essential esterified fatty acids in lipophilic medium; products for treatment of dry skin and hair. *Dr. Kurt Richter; Henkel/Cospha.*

33946 Vitamin F Forte CLR
Complex of essential free fatty acids in lipophilic medium; products for treatment of dry skin and hair. *Dr. Kurt Richter; Henkel/Cospha.*

33947 Vitamin F Glyceryl Ester CLR
Complex of essential esterified fatty acids in lipophilic medium; products for treatment of dry skin and hair. *Dr. Kurt Richter; Henkel/Cospha.*

33948 Vitamin F Water-Soluble CLR
Complex of essential free fatty acids, hydro-alcohol solubilized; products for treatment of dry skin and hair. *Dr. Kurt Richter; Henkel/Cospha.*

33949 vitamin K_1
84-80-0 7536 201-564-2
$C_{31}H_{46}O_2$
[R-[RName unverified.,RName unverified.-(E)]]-2-methyl-3-(3-7,11,15-tetramethyl-2-hexadecenyl)-1,4-naphthalenedione
2-methyl-3-phytyl-1,4-naphthoquinone; phytomenadione; phylloquinone; 3-phytylmenadione; phytomenadione; phytonadione; Aqua-Mephytin; Konakion; Mephyton; Mono-Kay; Veda-K_1; Veta-K_1. A fat soluble vitamin; deficiency gives rise to hemorrhage; $[\alpha]_D^{25}$ = -0.28° (dioxane); n_D^{20} = 1.5263; λ_m = 242, 248, 260, 269, 325 nm ($E^{1\%}_{1cm}$ 396, 419, 383, 387, 68, petroleum ether); insoluble in H_2O, soluble in organc solvents.

33950 vitamin K_2
84-81-1
$C_{41}H_{56}O_2$
(all E)-2-(3,7,11,15,19,23-hexamethyl-2,6,10,14,18,22-tetracosahexaenyl)-3-methyl-1,4-naphthalenedione
menaquinone-6; 2-difarnesyl-3-methyl-1,4-naphthoquinone; farnoquinone; MK 6; vitamin $K_{2(30)}$. A fat soluble vitamin. 2-methyl-3-difarnesyl-1,4-napthaquinone; synthesized in the gut by bacteria. mp = 50°; λ_m = 243, 248, 261, 270, 325-328 nm ($E^{1\%}_{1cm}$ 304, 320, 290, 292, 53, petroleum ether).

33951 vitamin M
59-30-3 4253 200-419-0
$C_{19}H_{19}N_7O_6$
N-[4-[[(2-amino-1,4-dihydro-4-oxo-6-pteridinyl)methyl]amino]benzoyl]-L-glutamic acid
folic acid; folacin; PGA; pteroylglutamic acid; folsäure; Cytofol; Folacin; Foldine; Foliamin; Folicet; Folipac; Folettes; Folsan; Folvite; Incafolic; Millafol. Considered a member of the vitamin B complex; used in medicine, nutrition, and as a food additive. mp = 250° (dec); $[\alpha]_D^{25}$= 23° (c = 0.5 in 0.1N NaOH); λ_m = 256, 283, 368 nm (log ε 4.43, 4.40, 3.96, pH 13); poorly soluble in H_2O, organic solvents.

33952 Vitaminets
Proprietary multivitamin preparation containing the B complex vitamins, vitamins A, C, D and E, biotin, nicotinamide, and various minerals or trace elements (Ca, Mg, Mn and P). *Roche Products Ltd.* Name unverified.

33953 Vitapet®
Fish oils, fish liver oils, vegetable oils; dietary supplement for domestic animals and birds conditioning oil. *Seven Seas Ltd.*

33954 Vitaplant CLR Oil-Soluble N
Calendula extract and lipoid extract of pig skins in oil medium; products for aging, damaged and sun-burned skin. *Dr. Kurt Richter; Henkel/Cospha.*

33955 Vitaplant CLR Water-Soluble
Echinacea extract and aloe juice in water-alcohol medium; products for aging, damaged and sun-burned skin. *Dr. Kurt Richter; Henkel/Cospha.*

33956 Vitapup®
Vitamins and minerals, skimmed milk powder; fortified milk feed puppies. *Seven Seas Ltd.*

33957 Vitavax RS Flowable
Carboxin + γ-HCH + thiram; a fungicide and insecticide dressing for oilseed rape. *Uniroyal Chemical Ltd.*

33958 Vitax Micro Gran, Vitax Turf Tonic
10028-22-5 4079 233-072-9
Ferric sulfate
Used for moss control in turf. *Vitax Ltd.*

33959 Vitax Q4
Powder fertilizer containing NPK 5.3:7.5:10 plus 1.8% Mg and trace elements; all-purpose base and top dressing fertilizer. *Vitax Ltd.*

33960 Vitax Turf Tonic
10028-22-5 4079 233-072-9
$Fe_2O_{12}S_3$
Iron (III) sulfate
ferric persulfate; ferric sesquisulfate; ferric tersulfate; Sulfuric Acid, Iron(3+) Salt (3:2); Iron Tersulfate; Ferric sulfate monohydrate; ferric sulfate. Used for moss control in turf. d^{18} = 3.097; slowly soluble in H_2O, EtOH, poorly soluble in organic solvents *Vitax Ltd.*

33961 Vitazyme® A-Plus
Retinyl palmitate polypeptide
Natural protein complexed vitamins for cosmetics use. *Brooks Industries.*

33962 Viten
Vital wheat gluten; used for flour fortification (bread making). *Roquette (UK) Ltd.*

33963 Vitesse®
For agricultural applications. *DuPont UK.*

33964 Vitexol® K
Emulsified mixture of aliphatic hydroxyl compounds and a neutral phosphoric acid ester; defoamer for dyeing of cotton/polyester knitgoods in jet dyeing machines. *BASF.*

33965 Viton®
Fluoroelastomer. *DuPont UK.*

33966 Viton® A
Fluoroelastomer
General purpose, high temp. and fluid resistant polymer for rubber goods which must withstand extremes of heat and corrosive fluids and chemicals; used in roll covers, sol'n. coatings, gaskets, seals, tubing; curable with Diak curing agents. *DuPont; Du Pont (UK) Ltd.* Name unverified.

33967 Viton® A-35
Fluoroelastomer
Low viscosity analog of Viton A which is easier and safer to process; offers vulcanizate properties similar to those of Viton A except for tensile strength and modulus which are slightly lower; used in high-solids sol'n. coatings and for blending with Viton A or Viton AHV for viscosity control. *DuPont; Du Pont (UK) Ltd.* Name unverified.

33968 Viton® AHV
A copolymer similar to Viton A but possessing greater strength at high temperatures; Mooney viscosity 180 (100 s). *DuPont UK.* Name unverified.

33969 Viton® B
Fluoroelastomer
General purpose, high temperature and fluid resistant elastomer with slightly improved thermal stability and fluid resistance over Viton A; used for o-rings, shaft seals, sol'n. coatings, roll covers, tubing, hose lining; may be cured with Diak curing agents. *DuPont.* Name unverified.

33970 Viton® B-50
Fluoroelastomer
Low visc. analog of Viton B which is easier to process; vulcanizate properties are similar to those of Viton B except for tensile strength and elong. which are slightly lower; used in high solids sol'n. coatings and for blending with Viton B for viscosity control *DuPont.* Name unverified.

33971 Viton® C-10
Fluoroelastomer
Very low visc. form of Viton A, especially useful in sol'n. applications such as sealants, high solids coatings, dipped goods, and adhesives; used in dry polymer applications where its low visc. permits easy mixing and molding of highly loaded, high durometer stocks; vulcanizate properties similar to those of Viton A except for tensile strength and modulus, which are slightly lower. *DuPont.* Name unverified.

33972 Viton® Curative No 30
50% Dihydroxy aromatic compound and 50% fluoroelastomer; used in combination with Viton Curative No. 20 as a curing system for Viton fluoroelastomers. *DuPont.* Name unverified.

33973 Viton® Curative No. 40
33% Benzophenone compound and 67% fluoroelastomer; used in combination with Viton Curative No. 20 as a curing system for Viton fluoroestomers. *DuPont.* Name unverified.

33974 Viton® Curative No.20.
33% Organophosphonium salt and 67% fluoroelastomer; used in combination

with Viton Curative No. 30 or No. 40 as a curing system for Viton fluoroelastomer. *DuPont.* Name unverified.

33975 Viton® E-430

A proprietary fluoroelastomer with good processing and storage properties. *DuPont UK.* Name unverified.

33976 Viton® Free Flow HD

50% Fluoropolymer/50% HDPE blend; processing additive for improved performance in HDPE resins. *DuPont.*

33977 Viton® Free Flow TA

50% Fluoropolymer/50% talc partitioning agent blend; processing additive designed for reduced interactions with talc antiblocks. *DuPont.*

33978 Viton® GFLT

Fluoroelastomer

Premium grade offering hydrocarbon fuel and methanol resistant; excellent resistant to hot water, steam, aqueous acids, and solvs.; recommended for o-rings, lip seals, molded shapes; peroxide-curable. *DuPont.* Name unverified.

33979 Viton® LN

A waxy semi-solid fluoroelastomer of the Viton type used as a plasticizer to improve molding and extrusion characteristics. *DuPont UK.* Name unverified.

33980 Vitradur

Ultra high molecular weight high density polyethylene; used for paper and textile accessories, bunker linings, and skating rinks. *Stanley Smith & Co Plastics Ltd.*

33981 Vitrafix

Coordination compound for glass laminates. *ICI Chem & Polymers Ltd.* Discontinued.

33982 Vitrafos®

10124-56-8 8814 233-343-1

Sodium hexametaphosphate

Food additive for baking, cereals, meat, poultry, dairy/cheese, processed foods. *Rhône-Poulenc Food Ingreds.*

33983 Vitrahose

Polyvinyl chloride

Tubes and hoses for many applications. *Stanley Smith & Co Plastics Ltd.* Discontinued.

33984 Vitralen

Ultra high molecular weight high density polyethylene sheet; used for orthopedic splints. *Stanley Smith & Co Plastics Ltd.*

33985 Vitralene

Polypropylene sheet and block; used for anti-acid fabrications, cutting boards. *Stanley Smith & Co Plastics Ltd.*

33986 Vitralex

Acrylonitrile/butadiene/styrene(ABS), polyvinylidene fluoride (PVDF) Sheet and block; used for anti-acid fabrications. *Stanley Smith & Co Plastics Ltd.*

33987 Vitrapad

Polyvinylchloride or polyolefine based sheet and block; used for cutting boards for all industries. *Stanley Smith & Co Plastics Ltd.*

33988 Vitraplas

Polyolefins

Used for general purpose plastics applications. *Stanley Smith & Co Plastics Ltd.*

33989 Vitrathene

Polyethylene sheet, rod, block and massive castings; used for radiation protection, chemically resistant plant, electronic and radar insulation, orthopedics, textile and paper trade accessories, bunker linings, cutting boards and tabletops. *Stanley Smith & Co Plastics Ltd.*

33990 Vitre-colloid

A proprietary cellulose acetate. No manufacturer.

33991 Vitreo-colloid

A proprietary trade name for cellulose acetate. No manufacturer.

33992 vitreon

A porcelain used in the condensing plant for nitric acid.

33993 vitreosil

60676-86-0 262-373-8

SiO_2

silicon dioxide glass

Silica glass. Fused Silica; used as an ablative material in rocket engines, as fibers in reinforced plastics; special camera lenses.

33994 Vitride®

22722-98-1 245-178-2

$C_6H_{16}AlNaO_4$

dihydrobis(2-methoxyethanolato-O,O')aluminate(1-) sodium

sodium bis(2-methoxyethoxy) aluminum hydride; sodium bis(2-methoxyethoxy)aluminum hydride; RED-AL. Reducing agent. d = 1.0360. *Zeeland.*

33995 vitriol stone

Impure ferric sulfate obtained by the oxidation of pyrites; used in the manufacture of fuming sulfuric acid.

33996 Vitrite

Multivitamin syrup; dietary supplement. *Marfleet Refining Co.* Name unverified.

33997 Vitrite

Polyolefin-based sheet and block; used for nuclear shielding and radiation protection. *Stanley Smith & Co Plastics Ltd.*

33998 Vitromail®

A range of inorganic decoration colors; available in the form of silk screen pastes, thermoplastics, and powder for use in the enameling and glass industries. *Bayer AG.*

33999 Vitrone

Polyvinyl chloride sheet; used for thermoformed packaging, chemically resistant fabrications. *Stanley Smith & Co Plastics Ltd.*

34000 Vivol

Cloth and textile oils. *Crosfield Chemicals Ltd.* Discontinued.

34001 Vizor

2164-08-1 5459 218-499-0

$C_{13}H_{18}N_2O_2$

3-cyclohexyl-6,7-dihydro-1H-cyclopentapyrimidine-2,4(3H,5H)-dione

lenacil; 3-cyclohexyl-1,5,6,7-tetrahydrocyclopentapyrimidine-2,4(3H)-dione; 3-cyclohexyl-5,6-trimethyleneuracil; lenacile; Adol; Du Pont 634; Elbatan; Venzar. Vizor is a herbicide containing lenacil, a slective herbicide absorbed by roots; inhibits photosynthesis. Used for control of annual grasses and broad-leaved weeds in fruit and vegetable crops and annual dicotyledons and meadow grass in beet, fruit herbaceous perennials. mp = 315-317°; d = 1.32; soluble in H_2O (6 mg/l), more soluble in organic solvents; LD_{50} (rat orl) > 11,000 mg/kg. *ICI Chem & Polymers Ltd; Chemolimpex; Fahlberg-List; Du Pont; Farm Protection Ltd.*

34002 Vizor

2164-08-1 5459 218-499-0

lenacil

Suspension concentrate containing 440 g/l of lenacil. Used for control of annual docotyledons and meadow grass in beet, fruits and herbaceous perennials. *ICI Chem & Polymers Ltd.; Farm Protection Ltd.*

34003 Vlemasque

Vliem-Dome. Sulfurated lime; a solution lime, sublimed sulfur and water scabicide. *Dermik Laboratories Inc; Miles Pharmaceuticals.* Name unverified.

34004 Vliseline

Nonwovens for the apparel industry. *Carl Freudenberg.* Name unverified.

34005 Vocol

Zinc-O,O-di-n-butylphosphorodithioate

Vulcanization accelerator. *Monsanto Co; Monsanto plc.*

34006 Vogel's alloy

An alloy containing 8 parts copper, 1 part zinc, 2 parts tin, and 1 part lead; used for polishing steel.

34007 Voidform

Cut polystyrene shapes and molded cylinders; for formation of voids within reinforced concrete structures and for providing profiles with cast concrete. *Vencel Resil Ltd.* Discontinued.

34008 Voidmaster

VR Voidmaster. Molded expanded polystyrene panels; used between prestressed concrete beams to provide floor insulation. *Vencel Resil Ltd.*

34009 Voidox® 100%

A modified fatty acid derivative of a substituted phenol; a food grade antioxidant. *Guardian Chemical.* Name unverified.

34010 Volan®

For the chemical industry. *DuPont UK.*

34011 Volathion®

14816-18-3 238-887-3

$C_{12}H_{15}N_2O_3PS$

4-ethoxy-7-phenyl-3,5-dioxa-6-aza-4-phosphaoct-6-ene-8-nitril 4-sulfide

Phoxim; O,O-diethyl α-cyanobenzylideneamino-oxyphosphonothioate; 2-(diethoxyphosphinothioyloxyimino)-2-phenylacetonitrile; phoxime; Bay 5621; Bay 77488; Baythion; Volaton. Foliage- and soil-applied insecticide for control of lepidopterous larvae, beetles and their larvae and locusts on a wide range of crops including cotton maize and vegetables. mp = 5-6°; d^{20} = 1.176; n_D^{20} = 1.5405; soluble in H_2O (7 mg/l), very soluble in organic solvents; LD_{50} (rat orl) = 1976-2170 mg/kg. *Bayer AG.*

34012 volcanite

A mixture of selenium and sulfur found in the Lipari Islands.

34013 Volck

An oil insecticide. *Monsanto (Solaris).* Name unverified.

34014 Volckmann's solution
A solution containing thymol, alcohol, glycerin, and water.

34015 Volclay HPM-20
Sodium bentonite, technical; suspending agent, viscosifier, gellant, and binder for household products, automotive products, aerosols, paints and enamels. *Am. Colloid.*

34016 Vole
A cotton fabric grade of Tufnol industrial laminates. *Tufnol Ltd.*

34017 Volenite
A rubber substitute made from rosin, oil, and some fibrous material. No manufacturer.

34018 Volkite
A molded rubber product used for electrical insulation. No manufacturer.

34019 Volpo
Polyethoxylated fatty alcohols. *Croda Chem. Ltd; Croda Inc.*

34020 Volpo 3
9004-98-2
$(C_2H_4O)_{20}\cdot C_{18}H_{36}O$
Oleth-3
Poly(oxy-1,2-ethanediyl), α-9-octadecenyl-ω-hydroxy-, (Z)-; Glycols, polyethylene, mono-9-octadecenyl ether, (Z)-; Amerox oe-20; Ameroxol EO 2; Ameroxol OE 10; Ameroxol OE-20; Atlas G-3915; Atlas G-3920; BO 2; BO 7; Brij 92; Brij 93; Brij 96; Brij 97; Brij 98; Brij 99; Brij 92((2)-oleyl); Brij 96((10) oleyl); Brij 98((20) oleyl); Decaethoxy oleyl ether; Dehydol 100; EL-620; EL-719; Emalex 515; Emery 6802; Emulgen 408; Emulgen 420; Emulgen 430; Emulgin 010; Emulgin 05; Emulphor; Emulphor ON-870; Emulphor O; Emulphor ON 870; Emulphor Surfactants; Ethoxol 20; Ethyleneoxide-oleyl alcohol adduct; Ethylene oxide-oleyl alcohol condensate; G 3910; G 3920; Genapol 0 100; Genapol O; Genapol O 020; Genapol O 050; Genapol O 080; Genapol O 120; Genpol O 050; Leonil-O; Lipal 2OA; Lipal 20-oa,. Emollient, lubricant, emulsifier, solubilizer for cosmetic application; emulsifier in astringent creams and lotions; clear gel formation; superfatting in shampoos, foaming bath preparations, and Carbopol gels; used in cold waves, depilatories, and hair straighteners; solubilizer for bromo acids in lipsticks and liquid rouge; spreading agent for bath oils. mp = 25-30° *Croda Inc.*

34021 Volpo 5
9004-98-2
Oleth-5
Emollient, lubricant, emulsifier, solubilizer for cosmetic applications. *Croda Inc.*

34022 Volpo 10
9004-98-2
Oleth-10
Emollient, lubricant, emulsifier, solubilizer for cosmetic applications. *Croda Inc.*

34023 Volpo 20
9004-98-2
Oleth-20
Emollient, lubricant, emulsifier, fragrance solubilizer for cosmetic applications. *Croda Inc.*

34024 Volpo 25 D 10
68131-39-5
PEG-10 C12-15 ether; emulsifier, dispersant, wetting agent, gelling agent, scouring and solubilizing agent for industrial and cosmetic applications. *Croda Chem. Ltd.*

34025 Volpo 25 D 3
68131-39-5
PEG-3 C12-15 ether; emulsifier, dispersant, wetting agent, gelling agent, scouring and solubilizing agent for industrial and cosmetic applications. *Croda Chem. Ltd.*

34026 Volpo CS-3
68439-49-6 7737
Ceteareth-3
Emulsifier, dispersant, wetting agent, and gellant, solubilizer for industrial and cosmetic applications. *Croda Chem. Ltd.*

34027 Volpo L4
9002-92-0 7717
Laureth-4
General purpose emulsifier and dispersant. *Croda Chem. Ltd.*

34028 Volpo N3
9004-98-2
Oleth-3
Distilled; emulsifier, dispersant, wetting agent, gelling agent, scouring and solubilizing agent for industrial and cosmetic applications. *Croda Chem. Ltd.*

34029 Volpo O3
9004-98-2
Oleth-3
Emulsifier, dispersant, wetting agent, gelling agent, scouring and solubilizing agent for industrial and cosmetic applications. *Croda Chem. Ltd.*

34030 Volpo S-2/S-10/S-20
9005-00-9
Steareth-2
Emulsifier for oil-water systems, cosmetics; stable over wide pH range. *Croda Inc.*

34031 Volpo T-3
24938-91-8
Trideceth-3
Emulsifier, dispersant, wetting agent, gelling agent, scouring and solubilizing agent for industrial and cosmetic applications. *Croda Chem. Ltd.*

34032 Voltalef
PCTFE; for molded parts. *Elf Atochem.*

34033 Volucon
Standard volumetric concentrates. *May & Baker Ltd.* Name unverified.

34034 Volucon
Volume concentrates for rapid preparation of laboratory solutions. *Rhône-Poulenc Laboratory Prods. Ltd.*

34035 Volunteered
Mixture of dalapon and di-1-p-menthene; a grass weed herbicide. *Mandops (UK) Ltd.*

34036 Volusol
Standard volumetric solutions. *May & Baker Ltd.* Name unverified.

34037 Von Forster powder
A gelatinized nitrocellulose flake powder, with a little calcium carbonate.

34038 Von Vetter's solution
An aqueous solution of glycerin, sugar, and potassium nitrate; used to preserve anatomical specimens.

34039 Vonges dynamite
An explosive containing 75% nitroglycerin, 20.8% randanite (decomposed felspar), 3.8% quartz, and 0.4% magnesium carbonate.

34040 Voran
A proprietary polyether-based polyurethane elastomer crosslinked with diamine. *Dow UK.* Discontinued.

34041 Voranate 3071
Specialty isocyanate; for PU industry for appliances and specialty foams. *Dow Chemical.*

34042 Voranate T-80, Type I, Type II
584-84-9 9668 209-544-5
$C_9H_6N_2O_2$
toluene-2,4-diisocyanate
2,4-diisocyanato-1-methylbenzene; 2,4-diisocyanatotoluene; 2,4-tolylene diisocyanate; TDI; Nacconate 100. TDI; for PU industry for flexible slabstock foam, molded flexible foam. mp = 19.5-21.5°; bp = 251°; d^{20}_4 = 1.2244; reacts with H_2O, soluble in most organic solvents. *Dow Chemical.*

34043 Voranol
Polyether polyols
Used with isocyanates in the formation of urethanes. *Dow Cheml Co Ltd, UK & Ireland.*

34044 Voranol 202, 225
Polyether polyol
For PU industry for appliances and rigid foams. *Dow Chemical.*

34045 Voranol 234-630
Polyether polyol
Triol functionality, high reactivity for PU industry for industrial/ consumer RIM and structural polymers, dynamic elastomers, adhesives, binders, coatings, and sealants. *Dow Chemical.*

34046 Voranol 3741
Polyether polyol; for flexible slab foam. *Dow Chemical.*

34047 Voranol 4148
Polyether polyol; triol functionality for PU industry for molded flexible foam. *Dow Chemical.*

34048 Vorite 63
Urethane prepolymer (TDI-based). *CasChem.* Name unverified.

34049 Vorite 105, 110, 115, 125
Polymerized castor oil; pigment wetting/dispersing agent; plasticizer for resins, gums, polymers; lubricant, penetrant; coupling solvent; adhesion promoter; for cellulose lacquers, inks, adhesives, industrial lubricants, polishes, caulks, leather dressing The different grades indicate different viscosities and pour points. d = 0.975 *CasChem.*

34050 Vossenblue 705-81
Iron blue pigment; for publication gravure inks. *Degussa.*

34051 Vostec
Boiler water treatment. *Grace Dearbom Ltd.*

34052 VPA No.3
Processing aid providing improved mold release for Viton® fluoroelastomers

cured with bisphenol or peroxide systems; for use for o-rings, seals, molded shapes, diaphragms, and tubing. *DuPont.*

34053 VPC®
Polyester urethane
High performance coatings for high speed continuous sheet stock, vinyl, paper, metal, plastic. *Ashland.* Name unverified.

34054 VPI
Vapor phase corrosion inhibitors. *Contract Chemicals (Knowsley) Ltd.*

34055 VR Coving
Expanded polystyrene; coving for domestic properties. *Vencel Resil Ltd.*

34056 VR-110
39198-34-0
2,2·g-Azobis(2,4,4-trimethylpentane)
Polymerization initiator. *Wako Pure Chem. Ind.; Wako Chem. USA.*

34057 VR-160
927-83-3
$C_8H_{18}N_2$
2,2·g-Azobis(2-methylpropane)
azo-*tert*-butane; 2,2'-azo-2,2'-dimethylpropane; Diazene, bis(1,1-dimethylethyl)-; Azoethane, 1,1,1',1'-tetramethyl-; Azo-tert-butane; 2,2'-Azoisobutane; Bis(tert-butyl)diimine; 1,2-Bis(1,1-dimethylethyl)diazene; di-tert-butyldiazene. Polymerization initiator. bp_{80} = 47-48°. *Wako Pure Chem. Ind.; Wako Chem. USA.* Discontinued.

34058 VR-500
Air dry phenolic coatings; thin film coating for structural steel, construction equipment, sewage plants, marine finishes, heating systems. *Heresite Protective Coatings.*

34059 V-thane
Solid polyurethane elastomers *Hallam Polymer Engineering Ltd.* Name unverified.

34060 Vuelite
A proprietary trade name for a cellulose acetate plastic; Vuelite-reinforced is a transparent cellulose acetate sheet reinforced with wire mesh. No manufacturer.

34061 Vuepak
A proprietary trade name for an acetate wrapping material. No manufacturer.

34062 Vulcabest
Incombustible cladding panels. *Vulcan Plastics Ltd.*

34063 Vulcaboard
Vandal resistant panel. *Vulcan Plastics Ltd.*

34064 Vulcabond®
Bonding agent for PVC. *Akzo Chemie UK Ltd.*

34065 Vulcabond® C 10
Rubber to metal adhesion promoter; specially designed to improve adhesion between rubber and steel cord in radial tires *Akzo.* Name unverified.

34066 Vulcabond® E
Polymeric phenolic compound in 3N aqueous ammonia; textile bonding agent for rubber to polyester *Akzo.* Name unverified.

34067 Vulcabond® N 15
Rubber to metal adhesion promoter. *Akzo.* Name unverified.

34068 Vulcabond® TX
Isocyanate compound in solvent rubber to textile bonding agent. *Akzo.* Name unverified.

34069 Vulcabond® VP
25% Solution of an isocyanurate-type polymer of TDI in dibutyl phthalate; bonding agent added to PVC plastisols to improve adhesion to a reinforcing substrate, e.g., nylon and polyester textiles. *Akzo.* Name unverified.

34070 Vulcabrite
Stainless steel faced cladding panels. *Vulcan Plastics Ltd.*

34071 Vulcaflex
A complex substituted secondary amine of the dimethoxy-diphenylamine type; an antioxidant which offers protection against flex-cracking in rubber materials. *Vulnax International Ltd.* Name unverified.

34072 Vulcaid
1317-36-8 5433 215-267-0
Opb
lead oxide
A proprietary litharge of small particle size. mp = 886°; bp = 1470°; d = 9.5300; LD_{40} (rat ip) = 40 mg Pb/100 g). . No manufacturer.

34073 Vulcaid 27
150-88-9 205-777-1
$C_{10}H_{20}O_2S_4Zn$
Zinc butyl xanthate
Carbonodithioic acid, O-butyl ester, zinc salt; Carbonic acid, dithio-, O-butyl ester, zinc salt; Butylxanthate, zinc salt; Butylxanthogenic acid zinc salt; Carbonic acid, dithio-, o-butyl ester, zinc salt; Nocceler Zbx; Zinc, bis(o-butyl carbonodithioato-S,S')-, (t-4)-; Zinc O-butyldithiocarbonate; Zinc butyl xanthate; Zinc dibutylxanthate. A proprietary trade name for a rubber vulcanization accelerator. No manufacturer.

34074 Vulcaid 28
103-49-1 3058 203-117-7
Dibenzylamine
A proprietary trade name for a rubber vulcanization accelerator. No manufacturer.

34075 Vulcaid 33
A liquid amine condensation product; a proprietary trade name for an antioxidant. No manufacturer.

34076 Vulcaid 44
A proprietary trade name for an antioxidant; a naphthol-amine reaction product. No manufacturer.

34077 Vulcaid 55
An acetaldehyde-aniline condensation product; a proprietary trade name for an antioxidant. No manufacturer.

34078 Vulcaid 111
Butyraldehyde-aniline
A proprietary trade name for a rubber vulcanization accelerator. No manufacturer.

34079 Vulcaid 222
97-74-5 202-605-7
$C_6H_{12}N_2S_3$
Tetramethyl-thiuram monosulfide
bis(dimethylthiocarbamyl) sulfide. A proprietary trade name for a rubber vulcanization accelerator. mp = 106-108°. No manufacturer.

34080 Vulcaid 444B
Heptaldehyde-aniline
A proprietary trade name for a rubber vulcanization accelerator. No manufacturer.

34081 Vulcaid DPG
102-06-7 3383 203-002-1
$C_{13}H_{13}N_3$
1,3-diphenylguanidine
N,N'-diphenylguanidine; *sym*-diphenylguanidine; Melaniline; Vulkazit. Proprietary trade name for a rubber vulcanization accelerator. mp = 148-150°; d = 1.13; poorly soluble in H_2O, soluble in organic solvents; MLD (dog iv) = 25 mg/kg. No manufacturer.

34082 Vulcaid LP
Lead pentamethylene dithiocarbamate
A proprietary rubber vulcanization accelerator. No manufacturer.

34083 Vulcaid P
Piperidine pentamethylene dithiocarbamate
Melts at 172 C and is soluble in water, alcohol, and benzene; a rubber vulcanization accelerator. No manufacturer.

34084 Vulcaid ZP
Zinc pentamethylenedithiocarbamate
A proprietary rubber vulcanization accelerator. No manufacturer.

34085 Vulcalap
GRP shiplap weatherboard. *Vulcan Plastics Ltd.*

34086 Vulcalon
GRP faced cladding panels. *Vulcan Plastics Ltd.*

34087 Vulcalucent
GRP Translucent sheets and domelights. *Vulcan Plastics Ltd.*

34088 Vulcamel
Butyraldehyde-ammonia
A proprietary a rubber vulcanization accelerator. No manufacturer.

34089 Vulcamin
Aluminum faced cladding panels. *Vulcan Plastics Ltd.*

34090 Vulcan
Herbicide containing clopyralid and bromoxynil. *ICI Chem & Polymers Ltd.* Discontinued.

34091 Vulcan Bronze
A proprietary trade name for a bearing bronze containing 1.0% silicon with iron and nickel. No manufacturer.

34092 Vulcan powder
An explosive containing 30% nitroglycerin, 52.5% sodium nitrate, 7% sulfur, and 10.5% charcoal.

34093 Vulcan® XC72R
1333-86-4 1856 231-153-3
carbon
Carbon black; for coloring conductive plastics. *Cabot; Cabot Carbon Ltd.*

34094 Vulcanex
A proprietary rubber vulcanization accelerator; it is a Schiff s base. No manufacturer.

34095 Vulcaniline
138-89-6 6736 205-343-1

$C_8H_{10}N_2O$
p-nitroso-N,N-dimethylaniline
N,N-dimethyl-4-nitrosoaniline; Accelerine. A proprietary trade name for a rubber vulcanization accelerator. mp = 85-87°; insoluble in H_2O, soluble in organic solvents. No manufacturer.

34096 Vulcanine
A patented material made from rubber, asbestos, litharge, lime, sulfur, and zinc oxide. No manufacturer.

34097 Vulcanized fiber
Hard fiber; red fiber; grey fiber; vegetable fiber; whalebone fiber; Egyptian fiber; fiberoid; horn fiber. A material made by treating sheets of paper with zinc chloride solution and subjecting the gelatinized sheets to pressure. The paper is sometimes m.

34098 Vulcanol
A proprietary rubber vulcanization accelerator. No manufacturer.

34099 Vulcanox Crack and Joint Sealant
Polyurethane; poured or extruded into cracks and joints as a moisture-proof sealant; cures to a solid with rubber band elasticity. *Metalcrete Mfg Co.*

34100 Vulcaplas
A proprietary trade name for an organic polysulfide synthetic elastic material. No manufacturer.

34101 Vulcaplast
GRP faced, aluminum framed cladding panels. *Vulcan Plastics Ltd.*

34102 Vulcapont
A proprietary rubber vulcanization accelerator; it consists of equal parts of Thionex and Vulcanol. No manufacturer.

34103 vulcasbeston
A mixture of rubber and asbestos; a heat and electrical insulator.

34104 Vulcase
A proprietary preparation of colloidal sulfur. No manufacturer.

34105 Vulcastab® EFA
Formaldehyde, ammonia, ethyl chloride condensate; secondary gelling agent and stabilizer for latex foams *Akzo.* Name unverified.

34106 Vulcastab® LW
Ethylene oxide and fatty alcohol condensate; stabilizer for latex rubber processing *Akzo.* Name unverified.

34107 Vulcastab® T
Ammonium polymethacrylate solution stabilizer and thickener for latex rubber processing *Akzo.* Name unverified.

34108 Vulcasteel
Steel faced cladding panels. *Vulcan Plastics Ltd.*

34109 Vulcastop
Short stoppers used in water emulsion polymerization processes; for synthetic elastomers production. *Vulnax International Ltd.* Name unverified.

34110 Vulcatuf
Bandit resistant panel. *Vulcan Plastics Ltd.*

34111 Vulcawall
Thermally broken curtain wall cladding system. *Vulcan Plastics Ltd.*

34112 Vulcazol
494-47-3 4832 207-790-8
$C_{15}H_{12}N_2O_3$
1-(2-furanyl)-N,N'-bis(2-furanylmethylene)methanediamine
N,N'-difurfurylidene-2-furanmethanediamine; furfuramine. A vulcanization accelerator. mp= 117°; bp = 250° (dec); λ_m = 259, 215 nm (log ε 4.18, 4.16); insoluble in H_2O, souble in organic solvents.

34113 Vulcoferran
A trademark for linings principally of rubber or ebonite for chemical apparatus. No manufacturer.

34114 Vulcogene
Thiocarbanilide
A proprietary trade name for a rubber vulcanization accelerator. No manufacturer.

34115 Vulcogene ND
102-06-7 3383 203-002-1
Diphenylguanidine
A proprietary trade name for a rubber vulcanization accelerator. No manufacturer.

34116 Vulcoid
A proprietary trade name for a phenolic resin impregnated vulcanized fiber. No manufacturer.

34117 Vulconex
Ethylidineaniline
A proprietary trade name for a rubber vulcanization accelerator. No manufacturer.

34118 Vul-Cup 40KE and R
25155-25-3 246-678-3
$C_{20}H_{34}O_4$
bis(*tert*-butyldioxyisopropyl)benzene
peroxide, [phenylenebis(1-methylethylidene)bis(1,1-dimethylethyl)]; (phenylenediisopropylidene)bis(*tert*-butylperoxide). Used as vulcanizing and polymerizing agents. Insoluble in H_2O, soluble in organic solvents. *Hercules.* Name unverified.

34119 Vul-Cup, Vul-Cup 40KE, Vul-Cup R
25155-25-3 246-678-3
$C_{20}H_{34}O_4$
α,α-bis(t-butylperoxy) diisopropylbenzene
bis(*tert*-butylperoxyisopropyl)benzene. Crosslinking agents for rubber and plastics. mp = 44-48°. *Hercules; Hercules Ltd.* Discontinued.

34120 Vulkacit®
Range of accelerators. *Bayer AG; Bayer plc.*

34121 Vulkacit® 470
A rubber vulcanization accelerator; it is a condensation product of homologous acroleins with aromatic bases. *Bayer AG.* Name unverified.

34122 Vulkacit® 576
A condensation product of homologous acroleins aromatic bases; a rubber vulcanization accelerator most suitable for regenerated rubber. *Bayer AG.* Name unverified.

34123 Vulkacit® 774
The dithiocarbamate of cyclohexyl ethylamine; a proprietary trade name for an ultra-accelerator for rubber. *Bayer AG.* Name unverified.

34124 Vulkacit® 1000
93-69-6 202-268-6
$C_9H_{13}N_5$
o-tolylbiguanide
Imidodicarbonimidic diamide, N-(2-methylphenyl)-; Biguanide, 1-*o*-tolyl-; Aliant; Eponoc B; N-(2-Methylphenyl)imidodicarbonimidic diamide; Nocceler Bg; Sopanox; 1-*o*-Tolylbiguanide; *o*-Tolylbiguanide; *o*-Tolyldiguanide; Vulkacit 1000. A rubber vulcanization accelerator. *Bayer AG.* Name unverified.

34125 Vulkacit® A
Aldehyde-ammonia, a vulcanization accelerator. *Bayer AG.* Name unverified.

34126 Vulkacit® BP
A rubber vulcanization accelerator; it is a paste, and consists of a mixture of bases. *Bayer AG.* Name unverified.

34127 Vulkacit® CRV/LG
3-methylthiazolidinethione-2
Accelerator for polychloroprene elastomers, halobutyl rubbers especially chlorobutyl; used for mechanical goods, cables, hoses, membranes, fabric proofings, vulcanizing solutions; beige-brown pellets. *Bayer Corp; Polysar.*

34128 Vulkacit® CT
A proprietary trade name for a rubber vulcanization accelerator. *Bayer AG.* Name unverified.

34129 Vulkacit® CZ/EGC, DZ/EGC
95-33-0 202-411-2
$C_{13}H_{16}N_2S_2$
Benzothiazyl-2-cyclohexyl sulfenamide
2-Benzothiazolesulfenamide, N-cyclohexyl-; Accelerator CZ; Accicure Hbs; Benzothiazyl-2-cyclohexylsulfenamide; CBS; Conac A; Conac S; Curax; 2-(Cyclohexylaminothio)benzothiazole; N-Cyclohexylbenzothiazole-2-sulfenamide; N-Cyclohexyl-2-benzothiazolesulfenamide; N-Cyclohexyl-2-benzothiazolylsulfonamide; N-Cyclohexyl-2-benzothiazylsulfenamide; Delac S; Durax; Ekagom CBS; Nocceler Cz; Pennac Cbs; Rhodifax 16; Royal CBTS; Sanceler cm-po; Santocure; Santocure Vulcanization Accelerator; Soxinol Cz; Sulfenamide Ts; Sulfenax; Sulfenax CB; Sulfenax CB 30; Sulfenax cb/k; Sulfenax TsB; Thiohexam; Vulcafor CBS; Vulcafor Hbs; Vulkacit C; Vulkacit CZ; Vulkacit cz/c; Vulkacit cz/k; Vulkacit CZ/MG/C; Vulkacite CZ,. Fast accelerators giving delayed onset of cure; used for tires, dynamically stressed technical goods, technical moldings and extrudates. *Bayer Corp; Polysar.*

34130 Vulkacit® D/C
102-06-7 3383 203-002-1
Diphenyl guanidine
Accelerator for NR, IR, BR, SBR, NBR, CR, with very slow onset of cure; used alone for bulky goods, e.g., tires, buffers, roll covers; secondary accelerator with mercaptos for mech. goods, extrudates and moldings, footwear, fabric proofings, cable sheathings and insulation. *Bayer Corp; Polysar.*

34131 Vulkacit® DM/C
120-78-5 3435 204-424-9
$C_{14}H_8N_2S_4$
2,2'-dithiobis[benzothiazole]
MBTS; 2,2'-dibenzothiazyl disulfide; benzothiazyl disulfide; dibenzthiazyl disulfide; mercaptobenzthiazyl ether; Thiofide. Dibenzothiazyl disulfide, mineral oil coated; delayed action, semi-ultra accelerator for NR, IR, BR, SBR, NBR, CR, IIR, and chlorosulfonated polyethylene; used for mech.

goods, tires, conveyor belts, cables, hoses, rubber footwear, expanded rubber goods, proofed fabrics, intricate molded goods. Yellowish coated powder, mp = 180°; d = 1.50; insoluble in H_2O, sparingly soluble in organic solvents; *Bayer Corp; Polysar.*

34132 Vulkacit® FP
Methylene-*p*-toluidine
A proprietary trade name for a rubber vulcanization accelerator. *Bayer AG.* Name unverified.

34133 Vulkacit® M
149-30-4 5916 205-736-8
2-mercaptobenzothiazole
A proprietary rubber vulcanization accelerator. *Bayer AG.* Name unverified.

34134 Vulkacit® M, Vulkacit® Merkapto/C
149-30-4 5916 205-736-8
$C_7H_5NS_2$
2(3H)-benzothiazolethione
2-Mercaptobenzothiazole; 2-benzothiazolethiol; MBT, Captax; Drmacid; Mertax; Thiotax. Mineral oil coated; semi-ultra accelerator for NR, IR, BR, SBR, NBR, IIR, and EPDM; used alone in bulky goods or in combination for molded and extruded goods, hoses, conveyor belts, tires, footwear, cables, expanded rubber goods, proofed fabrics and latex articles. mp = 179-182°; d = 1.420; insoluble in H_2O, soluble in organic solvents. *Bayer Corp; Polysar.*

34135 Vulkacit® Merkapto/MGC
155-04-4 205-840-3
$C_{14}H_8NS_4Zn$
2(3H)-benzothiazolethione zinc salt
Bantex. Zinc salt of 2-mercaptobenzothiazole, mineral oil coated. d_4^{25}= 1.70. *Bayer Corp; Polysar.*

34136 Vulkacit® P
Piperidinepiperidyldithioformate, a vulcanization accelerator. *Bayer AG.* Name unverified.

34137 Vulkacit® P Extra
The zinc salt of ethylphenyldithiocarbaminic acid, a rubber vulcanization accelerator. *Bayer AG.* Name unverified.

34138 Vulkacit® TR
A rubber vulcanization accelerator; it is a mixture of the free bases of polyamines of ethylene. *Bayer AG.* Name unverified.

34139 Vulkadur®
Reinforcing resin. *Bayer AG; Bayer plc.*

34140 Vulkalent®
Retarders. *Bayer AG; Bayer plc.*

34141 Vulkalent® TM
136-85-6 205-265-8
$C_7H_7N_3$
5-methyl-1H-benzo-1,2,3-triazole
5-methyl-1H-benzotriazole. Blend of 4- and 5-methylbenzotriazole; prevulcanization retarder for sulfur-modified CR and halobutyl rubbers; also effective in NBR when Vulkacit DM/sulfur curing systems are used; applications including conveyor belting, hose, and molded goods. mp = 80-82°; bp_{12} = 210-212°; *Bayer Corp; Polysar.*

34142 Vulkanol®
Synthetic plasticizers. *Bayer AG; Bayer plc.*

34143 Vulkanol® 88
Dibutyl methylene bisthioglycolate
Plasticizer to improve low-temp. flexibility and elastic behavior of vulcanizates; used for nitrile and chloroprene rubber. *Bayer; Miles; Polysar.*

34144 Vulkanol® 90
Di-2-ethylhexyl ester thiodiglycolate
Plasticizer for natural and synthetic rubber including NBR, NBR/PVC blends, CR, SBR, BR, chlorosulfonated polyethylene, and chlorinated polyethylene; used for hoses, seals, shock absorbers. roll covers, and conveyor belting. *Bayer Corp; Polysar.*

34145 Vulkanol® OT
Ether thioether
Plasticizer to improve low-temp. flexibility and elastic behavior of vulcanizates; used for nitrile and chloroprene rubber; applications including hoses, seals, mechanical goods. *Bayer; Miles; Polysar.*

34146 Vulkanox®
Antioxidants. *Bayer AG; Bayer plc.*

34147 Vulkanox® 4010 NA
101-72-4 202-969-7
$C_{15}H_{18}N_2$
N-isopropyl-N'-phenyl-*p*-phenylene diamine
1,4-Benzenediamine, N-(1-methylethyl)-N'-phenyl-; p-Phenylenediamine, N-isopropyl-N'-phenyl-; 4010 NA; 4-Anilino-N-isopropylaniline; Antigen 3c; Antigene 3c; Antigene 3C; Antioxidant IP; Antioxidant 40NA; Antioxidant 4010 NA; ASM 4010MA; Cyzone; Cyzone IP; Diafen FP; Diaphen FP; Elastozone 34; N-Fenyl-N'-isopropyl-p-fenylendiamin (czech); Flexzone 3c; Flexzone 3C; Ipognox 44; 4-(Isopropylamino)diphenylamine; p-Isopropylaminodiphenyl amine; N-Isopropyl-N'-fenyl-p-fenylendiamin(Czech); N-Isopropyl-N'-phenyl-p-phenylenediamine; NCI-C56304; Nocrack 810NA; Nonox ZA; Orflex PP; Ozonon 3C; Permanax 115; N-Phenyl-N'-isopropyl-1,4-phenylenediamine; N-Phenyl-N'-isopropyl-p-phenylenediamine; N-(2-Propyl)-N'-phenyl-p-phenylenediamine; Santoflex 36; Santoflex IP; S-IP,. Staining and discoloring antioxidant/antiozonant for protection of rubber from ozone attack, oxidation, heat aging, flexcracking, and rubber poisons; suitable for natural and synthetic rubbers; used for dynamically stressed goods, tech. goods, spring comp *Bayer Corp; Polysar.*

34148 Vulkanox® 4020
N-(1,3-dimethylbutyl)-N-g-phenyl-p-phenylene diamine
Staining stabilizer for cold- and hot-polymerized emulsion and solution SBR and BR; protects vulcanizates against heat, oxidation, and flexcracking. *Bayer Corp; Polysar.*

34149 Vulkanox® BKF
119-47-1 204-327-1
$C_{23}H_{32}O_2$
2,2'-Methylene-bis(4-methyl-6-t-butylphenol)
2,2'methylenebis(6-*tert*-4-methylphenol). Nonstaining antioxidant for natural and synthetic rubbers; used in transparent, bathing, surgical tech., and latex goods, fabric proofings; stabilizer for hot-and cold-polymerized emulsion-SBR, BR, and NBR, and solution-BR, IR, and ABS. mp = 123-127°. *Bayer Corp; Polysar.*

34150 Vulkanox® DDA
Styrenated diphenyl amine; staining and discoloring antioxidant for natural and synthetic rubbers, exp. CR, giving protection against oxidation, heat, flexcracking, and rubber poisons; used for inner tubes, sponge rubber, seals/gaskets, soles, latex goods, heat resistant articles, stabilizer for cold- and hot-polymerized NBR and NBR or SBR latexes. *Bayer Corp; Polysar.*

34151 Vulkanox® HS/LG
2,2,4-trimethyl-1,2-dihydroquinoline polymer
Antioxidant for technical goods and rubber products; used in latex. *Bayer Corp; Polysar.*

34152 Vulkanox® KB
128-37-0 1583 204-881-4
$C_{15}H_{24}O$
2,6-di-*t*-butyl-*p*-cresol
butylated hydroxytoluene; 2,6-bis(1,1-dimethylethyl)-4-methylphenol; BHT; Antrancine 8; Tenox BHT; Ionol CP; Sustane; Dalpac; Improvol; Vianol. Nondiscoloring/nonstaining antioxidant for light colored and transparent natural and synthetic rubber goods; used in fabric proofings, toys, bathing, latex, and dipped goods; stabilizer in emulsion and solution-polymerized elastomers (SBR, NBR, BR, IR); costabilizer for ABS. mp = 70°; bp = 265°; d_4^{20}= 1.048; insoluble in H_2O, soluble in organic solvents; LD_{50} (mus orl)= 1040 mg/kg. *Bayer Corp; Polysar.*

34153 Vulkanox® MB-2/MGC
4- and 5-Methylmercaptobenzimidazole, mineral oil coated; nondiscoloring/ nonstaining antioxidant for natural and synthetic rubbers; synergist; protects against rubber poisons; sensitizing agent for latex. *Bayer Corp; Polysar.*

34154 Vulkanox® OCD/SG
Octylated diphenyl amine; weakly staining and discoloring antioxidant for CR, NR, IR, SBR, BR, NBR, protecting against rubber poisons; used for dynamically stressed goods, tires, air hoses, belts, sponge rubber, footwear, and profiles. *Bayer Corp; Polysar.*

34155 Vulkanox® ZMB2/C5
Zinc salt of 4- and 5-methylmercaptobenzimidazole; nonstaining and nondiscoloring antioxidant for natural and synthetic rubber; synergist; used mainly for heat-resistant thiuram-cured goods, white, colored, and transparent goods and latex foam; sensitizer for latexes. *Bayer Corp; Polysar.*

34156 Vulkaresen
Phenolic resins, resole types. *Hoechst AG.*

34157 Vulkazon®
Antioxidants. *BASF AG.*

34158 Vulkazon® AFS/LG
Bis(1,2,3,6-tetrahydrobenzaldehyde) pentaerythritol acetal
Nonstaining antiozonant for CR, IIR, CIIR, BIIR; used for hoses, extruded profiles, cables, sheeting, proofed goods; beige to gray powder. *Bayer Corp; Polysar.*

34159 Vulklor
118-75-2 2121 204-274-4
$C_6Cl_4O_2$
2,3,5,6-tetrachloro-*p*-benzoquinone
p-chloranil; 2,3,5,6-tetrachloro-2,5-cyclohexadiene-1,4-dione; tetrachloro quinone; Spergon. Combines with R6 to form an effective bonding system for compounds featuring steel cord reinforcement; used to activate GMF; also functions as a vulcanizing agent without sulfur; used in natural, SBR, nitrile,

butyl and chlorobutyl rubbers. mp = 290°; insoluble in H_2O, sparingly soluble in organic solvents; LD_{50} (rat orl) = 4.0 g/kg. *Uniroyal.* Name unverified.

34160 Vulkollan®
High quality polyurethane elastomer. *Bayer AG; Bayer plc.*

34161 Vulkollan® 18W
Water-crosslinked PU elastomer; exhibits superior balance of properties including ultimate tensile strength, tear strength, compression set, and high wear resistant under wet abrasion conditions; used in pump diaphragms, coupling elements, solid tires, rollers, pump housings, seals, wipers, bushes and split sections for anti-friction bearings and flexible couplings. *Bayer; Miles.*

34162 vulpinite
A variety of anhydrite (calcium sulfate) mixed with silica.

34163 Vultac® 2
68555-98-6
Alkyl phenol disulfide
NR, NBR, SBR vulcanizer; plasticizer; NBR, SBR tackifier; accelerates SBR. *Elf Atochem.*

34164 Vultex
A trademark for vulcanized rubber latex preserved with ammonia. No manufacturer.

34165 Vybar® 103
8002-74-2 7155 232-315-6
Paraffin Wax, granular
Synthetic wax; ethylenederived hydrocarbon polymer; lubricant, anticaking agent, modifier; used in paraffin; used in candles to replace stearic acid; opacifies the candle and imparts resistance to thermal shock; pigment disp. to wet inorg. pigments and fi mp= 53°; *Petrolite.*

34166 Vybex 22008 BKFR
26062-94-2
PBT, 10% glass fiber-reinforced; flame retardant thermoplastic. *Ferro/Engineering Thermoplastics.*

34167 Vybex 22028 BKFR
26062-94-2
PBT, 15% glass fiber-reinforced; flame retardant thermoplastic *Ferro/Engineering Thermoplastics.*

34168 Vybex 40001 NA
PET/PC alloy; high-impact grade. *Ferro/Engineering Thermoplastics.*

34169 Vybex 40004 NAFC
PET/PC alloy; food contact grade. *Ferro/Engineering Thermoplastics.*

34170 Vycel
Petroleum-derived surfactant; patented surfactant. *Crowley Chem.*

34171 Vydate®
23135-22-0 245-445-3
$C_7H_{13}N_3O_3S$
2-(dimethylamino)-N-[[(methylamino)carbonyl]oxy]-2-oxoethanimidothioic acid methyl ester
oxamyl; N',N'-dimethyl-N-[(methylcarbamoyl)oxy]-1-thiooxamimidic acid methyl ester; N,N-dimethyl-α-methylcarbamoyloxyimino-α-(methylthio)acetamide; methyl 1-(dimethylcarbamoyl)-N-(methylcarbamoyloxy)thioformimidate; thioxamyl; DPX-1410; Blade; DPX 1410. Granules containing 10% w/w oxamyl; contact and systemic insecticide, acaricide and nematicide. Cholinesterase inhibitor. mp = 108-10°; d = 0.97; soluble in H_2O (280 g/l), more soluble in organic solvents; LD_{50} (rat orl) = 5.4 mg/kg. *DuPont UK.*

34172 Vydax®
Fluorocarbons. *DuPont UK.*

34173 Vydon
A proprietary vinyl resin. No manufacturer.

34174 Vydyne
Nylon and nylon copolymer resins including 66.69 and 66/6 copolymers; designed for molding and extrusion applications. *Monsanto Co.* Name unverified.

34175 Vyflex E
Environmentally friendly outdoor thermoplastic coating power; used for traffic engineering, fencing posts and panels. *Plascoat Systems Ltd.*

34176 Vyflex NT80S
A proprietary PVC powder coating material, containing no toxic metals and supporting no microbiological growth; can be safely brought into contact with drinking water. *Plastic Coatings Ltd.* Name unverified.

34177 Vygen
A trademark for PVC resins. No manufacturer.

34178 Vykamol 83G
Sorbitan ester/polysorbate blend. *Croda Surf. Ltd.*

34179 Vyloc
Oven wall catalysts. *DuPont UK.*

34180 Vylor®
Polymer. *DuPont UK.*

34181 Vylox®
Chemicals used in paper/paperboard products. *Reichhold.*

34182 Vynamon
Pigments for polyvinyl chloride. *ICI Chem & Polymers Ltd.*

34183 Vynathene®
Vinyl acetate copolymers in liquid form; industrial chemicals for use as pour point depressant additives in fuels and lubricants. *Quantum Chemical Corp.*

34184 Vynel N
Polymer emulsion; hand builder; imparts full, resilient hand to cotton and blended fabrics. No manufacturer.

34185 Vyram®
Thermoplastic elastomer; general-performance rubber replacement with temperature and oil resistance equivalent to EPDM, SBR, and natural rubber. *Advanced Elastomer Systems.*

34186 W.A. powder
An American smokeless powder. It is a guncotton-nitroglycerin powder, with barium and potassium nitrates.

34187 W.G.S. Hydrogenated Fish Glyceride 117, 128
Triester of long chain fatty acids and glycerin; used in wax compounds, textile softeners and sizes, yarn lubricants, grease sticks, polishes, crayons, candles, leather stuffing, wire drawing compounds, paper coatings, and plastics. *Werner G. Smith.*

34188 W.G.S. R-60 Z-5
Fish oil derivative containing 40% C_{20} and C_{22} acids. Wetter for hot metal surfaces; high water absorptive capacity; easy emulsification; for marine lubricants, Degras replacement in metalworking oils and rust preventatives; scavenger in chlorinat *Werner G. Smith.* Discontinued.

34189 W.G.S. Synaceti 116 NF/USP
8002-23-1 232-302-5
Cetyl esters; emollient for cosmetics; slip aid for inks; gloss/slip aid for varnish; processing aid and lubricant for plastics; in binder formulations for pencils. *Werner G. Smith.* Discontinued.

34190 W180
Ester/quaternary amine blend; softener, rewetting agent for cotton, nylon, and polyester fibers. *Hart Chem. Ltd.*

34191 W2 Beta
A proprietary name for a special Angelo shellac. *Zinsser NV.* Name unverified.

34192 Wachsemulsion 1864
Carnauba/nonionic wax emulsion; manufacturing of cleaning agents with glossing effect. *Zschimmer & Schwarz.*

34193 Wafex, Wafolin
8061-52-7
Calcium/magnesium lignosulfonate
Unfermented calcium lignosulfonate; industrial binder, dispersant, and emulsifier. Pellet binder in animal feeds; contributes some nutritive value. *Borregaard Ligno Tech.*

34194 Wagner's reagent
A solution of 2 g potassium iodide in 100 ml water.

34195 Wakal® A Range
Alginates
Gelling agents for desserts and filling creams. *Grünau.*

34196 Wakal® J Range
9000-40-2 232-541-5
Locust bean gum; thickener and stabilizer for food industry. *Grünau.*

34197 Wakal® K Range
9000-07-1 1914 232-524-2
Carrageenan
Gelling agents for desserts and filling creams. *Grünau.*

34198 Walkover Moss Killer
10028-22-5 4079 233-072-9
Ferrous sulfate
Used for moss control in turf. *Walkover Sprayers Ltd.*

34199 Wallkyd® 11-024
Flat alkyd. *Reichhold.*

34200 Wallpol® 40,136, 40-165
Vinyl acetate acrylic emulsion copolymers; high performance emulsion for interior semigloss enamels; excellent scrub resist and wet adhesion. *Reichhold/Emulsion Polymers.*

34201 walnut oil
Oils, walnut. Oil derived from the nut meats of walnuts, *Juglans* spp.; emollient for makeup, skin, and hair care products. *Arista Industries; Pacific Anchor; Penta manufacturing.*

34202 Walsrode Powder
A proprietary smokeless powder containing 98.6% nitrocotton and 1.4% volatile matter. No manufacturer.

34203 Wando Steel
A proprietary non-shrinking steel containing 1.05% manganese, 0.5% chromium, 0.5% tungsten, and 0.95% carbon. No manufacturer.

34204 Wanin AM
8061-53-8
Unfermented ammonium lignosulfonate; binder, dispersant, and auxiliary tanning agent. *Borregaard Ligno Tech.*

34205 waras
Wars; warrus. A resinous powder which covers the seed pods of *Flemingia congesta* , of India; used in Arabia as a dye.

34206 Warcodet D
Alkylphenol ethoxylate nonionic surfactant as a colorless viscous liquid; detergent and emulsifier with wetting, penetrating and soil suspending properties used for textiles. *Warwick International Ltd.* Name unverified.

34207 Warcodet K54
Sodium alkylaryl sulfonate
Anionic surfactant in the form of a clear golden brown liquid; detergent with wetting, penetrating and soil suspension properties; used in general purpose detergents and in the textile industry e.g. for scouring. *Warwick Chemical Ltd.* Name unverified.

34208 Warcodet V
Phosphate ester in the form of sodium/potassium salts; anionic surfactant for wetting, scouring and washing off. *Warwick International Ltd.* Name unverified.

34209 Warcodye CLP
Low foam cationic surfactant; leveling wool and polyamide with acid reactive dyes. *Warwick International Ltd.* Name unverified.

34210 Warcodye RWL
Ethoxylated amine based amphoteric; leveler used in reactive dyes on wool. *Warwick International Ltd.* Name unverified.

34211 Warcosoft WSC
Cationic surfactant in liquid form; permanent softener for use with acrylics. *Warwick International Ltd.* Name unverified.

34212 Warcowet O
577-11-7 3460 209-406-4
Sodium dioctyl sulfosuccinate in liquid form; wetting agent with good penetration, stable to boiling, most effective at 30-60 C; used in textiles e.g., in scouring, bleaching, package dyeing, piece dyeing and printing. *Warwick International Ltd.* Name unverified.

34213 Wardol
Nonionic emulsifiers *Courtaulds Chemicals, Leek.*

34214 Wareflex® 650
Dibutoxyethoxyethyl adipate
Plasticizer. *Sartomer.*

34215 Warefog
101-21-3 2240 202-925-7
Chlorpropham as a foggable solution; for controlling sprouting in ware potatoes. *Wheatley Chemical Co Ltd; Dean Agrochemicals Ltd.* Name unverified.

34216 Wargonin Compact
Chemically desugared sodium/calcium lignosulfonate, ·5 1% sugar content; water-reducing, strength-increasing, air-excluding additive for concrete. *Borregaard Ligno Tech.*

34217 Wargotan
8061-52-7
Unfermented calcium lignosulfonate with low calcium and iron content; auxiliary tanning agent and dispersant. *Borregaard Ligno Tech.*

34218 Warmaline
Expanded polystyrene veneer 2mm thick; lining for domestic walls for use beneath wallpapers. *Vencel Resil Ltd.*

34219 Warne's metal
An alloy of 26% nickel, 37% tin, 26% bismuth, and 11% cobalt.

34220 Wasc
Nonionic surfactant consisting of nonylphenol ethoxylate in liquid form; liquid cleaners for hard water. *Berol Kemi (UK) Ltd.* Name unverified.

34221 Waspend
Contains pirimiphos-methyl and synergized pyrethrins; pestkiller for flying or crawling pests. *ICI Garden Products.*

34222 Watchmaker's alloy
An alloy of 59% copper, 40% zinc, and 1.2% lead.

34223 Water Brite
10124-56-8 8814 233-343-1
Sodium hexametaphosphate
Used for water treatment. *Flexibulk Ltd.*

34224 water gas
10176
blue gas. A general name for a mixture of gases obtained by the decomposition of steam by incandescent carbon. It usually contains from 43-44% carbon monoxide, 48-49% of hydrogen, 3-4% of carbon dioxide, and 3-4% nitrogen. Used as a source of hydrogen in the manufa.

34225 Water Lock® A-100
Starch/acrylates/acrylamide copolymer. Superabsorbent polymer; able to absorb or immobilize large quantities of aqueous fluids, such as alkalies, dilute acids, and body fluids. *Grain Processing.*

34226 water mica
Clear transparent Muscovite (potash mica).

34227 Waterez®
Synthetic resins. *Reichhold.*

34228 Waterlity
Water-blown polyurethane film based on methylene diphenylene isocyanate. *Imperial Chemical Industries plc.*

34229 Watsonite
A proprietary material used as a mica substitute. It is made from scrap mica with a binding agent. No manufacturer.

34230 wattle bark
mimosa. A tanning material obtained from species of *Acacia* . Contains tannin in amounts varying from 12-49%.

34231 Watt's and Li's solution
A solution used for the electrodeposition of iron. It contains 150 g ferrous sulfate, 75 g ferrous chloride, 120 g ammonium sulfate, and 1,000 ml water.

34232 Wave
Vinyl acrylic emulsion. *Air Prods & Chems. Inc.*

34233 wax butter
Wax oil. A thick oil obtained by the distillation of beeswax. It consists mainly of cerotene $C_{27}H_{54}$, melissine $C_{30}H_{60}$, and palmitic acid; formerly used medicinally externally and internally.

34234 Wax C
N,N'-Distearyl ethylene diamine
A high melting point amide wax; recommended as an internal lubricant in processing ABC, PVC, and PS. *Hoechst UK.* Name unverified.

34235 Waxemul
Wax emulsion. *Tenneco Malros Ltd.* Name unverified.

34236 Waxenol® 801
65591-14-2 265-839-9
Arachidyl propionate
Binder; emollient, solubilizer, and ingredient in cosmetic and toiletries; lubricant for pressed powders; metal working lubricant and corrosion inhibitor for metal surface. *CasChem.*

34237 Waxenol® 810
3234-85-3 221-787-9
$C_{28}H_{46}O_2$
Myristyl myristate
myristyl tetradecanoate. Emollient with unusual afterfeel; especially for stick preps. *CasChem.*

34238 Waxenol® 815
540-10-3 2073 208-736-6
$C_{32}H_{64}O_2$
Cetyl palmitate
hexadecanoic acid hexadecyl ester; hexadecyl palmitate; palmitic acid hexadecyl ester. Emollient additive; internal lubricant and binder in pressed powders; in metalworking lubricant coatings. mp = 54°; d^{20} = 0.989; n_D^{70} = 1.4398; insoluble in H_2O, soluble in EtOH, Et_2O *CasChem.*

34239 Waxenol® 821
Synthetic beeswax
Lipophilic emulsifier, emollient. *CasChem.*

34240 Waxenol® 822
Arachidyl behenate
Emollient for sticks, creams, lotions; barrier properties for enhanced moisturization; gloss and film-forming properties. *CasChem.*

34241 Waxigel
Modified maize starch (pregelatinized). *Roquette (UK) Ltd.* Discontinued.

34242 Waxilys
Maize starch; used for foods, textiles, board and paper. *Roquette (UK) Ltd.*

34243 Waxit
A wetting agent used by gold-and silversmiths in the jewelry and dentistry industries. *Degussa Ltd.*

34244 Waxolan P-5
Propoxylated lanolin wax; enhances lubricity and rigidity of lipsticks. *Amerchol Corp.* Discontinued.

34245 Waxoline
Oil and wax soluble dyes. *ICI Chem & Polymers Ltd.*

34246 Wayfarer
61791-44-4 263-177-5
Cationic wetting agent containing 80% tallow amine ethoxylate; spreader for phosphonoglycine herbicide sprays. *Service Chemicals Ltd.*

34247 Wayfos A
Phosphate acid
Acid form; antistat, emulsifier, wetting agent, detergent, coupling agent, solubilizer, lubricant, corrosion inhibitor; for alkaline built cleaners, textiles, plastics, metal, emulsion polymerization, and agricultural applications. *Olin.*

34248 Wayfos M-60
Aromatic phosphate ester, free acid; corrosion inhibitor, hydrotrope; drycleaning detergent; emulsifier for pesticides, emulsion polymerization. *Olin.*

34249 Wayhib® S
Triethanolamine phosphate ester, sodium salt; detergent builder, sequestrant, corrosion inhibitor, pipeline scale inhibitor, water circulating systems; for air conditioning and boiler treatment compounds. *Olin.*

34250 Weathershield
Decorative range of exterior paint products. *ICI Chem & Polymers Ltd.*

34251 Webas
Bitument emulsion for road surface treatment. *Vedag GmbH.* Name unverified.

34252 Webert Alloy
A proprietary copper silicon alloy containing small amounts of manganese. No manufacturer.

34253 Wecobee® M
68334-28-1 269-820-6
Hydrogenated vegetable oil; used in personal care products, pharmaceuticals, food industry. *Stepan/PVO; Stepan Europe.*

34254 Wedel's oil
5374
Consists of 1 part bergamot (linalyl acetate), 4 parts camphor, and 32 parts oil of almonds. Used in perfumery.

34255 Wedl's stain
A microscopic stain containing 1 g orseilie, 20 ml absolute alcohol, 5 ml 60% acetic acid, and 40 ml water.

34256 Weed and Brushkiller
Emulsifiable concentrate containing 144 g 2,4-D, 32 g dicamba and 144 g mecoprop per liter; an herbicide to control perennial and woody weeds. *Synchemicals Ltd.*

34257 Weed Hoe
2163-80-6 6020 218-495-9
MSMA; post-emergence herbicide on turf to control established crabgrass, dollis grass, and nutselse. *Lawn & Garden Products Inc.*

34258 Weed Stopper
19044-88-3 7015 242-777-0
Oryzalin
A pre-emergence herbicide for use on ornamentals, trees, roses, flower beds, bulbs, and warm season turf; controls annual grasses any many broadleaf weeds; may be tank-mixed with Roundup. *Lawn & Garden Products Inc.*

34259 Weedex S2
122-34-9 8681 204-535-2
Granules containing 2% w/w simazine; a soil-acting herbicide. *Hortichem Ltd.*

34260 Weedkill
Aminotriazole + 2,4-D + diuron + simazine; used for total weed control in non crop areas. *Dermaglen Ltd.*

34261 Weedmaster
1698-60-8 216-920-2
Suspension concentrate containing 430 g chloridazon per liter; a pyridazinone herbicide for beet crops. *Portman Agrochemicals Ltd.*

34262 Weedol
Weedkiller containing paraquat [1910-42-5] and diquat [85-00-7] for gardeners. *ICI Garden Products.*

34263 Weedone® DPC
2,4-DP and 2,4-D. Post-emergent herbicide for control of annual and perennial broadleaf weeds on golf courses and other ornamental turf area. *Rhône-Poulenc/Ag; W.A. Cleary.*

34264 Wegin
Ketone resins and formaldehyde scavengers. *British Traders & Shippers Ltd.*

34265 Weichharz 398A
A proprietary trade name for a nondrying alkyd made from adipic acid and trimethylene glycol. *Vianova Resins.* Discontinued.

34266 Weichmacher 238S, 333A
Proprietary trade names for esters formed from adipic acid and C_4-C_9 synthetic fatty acids with pentaerythritol. *Vianova Resins.* Discontinued.

34267 Weichmacher 90
A polymer from acetylene, glycerin, and ethylene oxide. No manufacturer.

34268 Weigert's stain
A microscopic stain made by dissolving fuchsine and resorcinol in ferric chloride solution.

34269 weighted silk
Silk impregnated with various inorganic and organic substances in order to increase the weight. Tannin and metallic salts are often used.

34270 Weighter Finish 585
Blend of starch with organic weighting agents and humectants; weighter finisher for cellulosic fibers. *Arol Chem. Prods.*

34271 Weisalloy
A proprietary sheet aluminum alloy. No manufacturer.

34272 Weldox
Adhesives compositions. *Hardman.*

34273 Welgum
Alginates for ceramics, electrodes, and water treatment. *Alginate Industries Ltd.* Name unverified.

34274 Welladyne
Iodophor detergent germicide for cleaning. *Ciba plc.* Name unverified.

34275 Wellbrom
Oil well fluid components. *Ethyl Corp.*

34276 Welltex 300 F
8061-52-7
Fermented calcium lignosulfonate; paper sizing agent. *Borregaard Ligno Tech.*

34277 Welmet
A proprietary chromium nickel-molybdenum steel. No manufacturer.

34278 Welvic
A trademark for a range of plasticized and unplasticized polyvinyl chlorides used in the manufacture of cables, flooring, pipes, etc. *ICI Chem & Polymers Ltd.* Discontinued.

34279 WEP® 662P
Unsaturated polyester resin; water-extendable resin for casting plaques, figurines, other decorative items; designed to be extended with water and cured at room temperature. *Ashland.*

34280 Wepco
Caulking, waterproofing, and cementitious materials. *Weatherguard/Marbleloid Products Inc.* Name unverified.

34281 WesBio
A liquid, biodegradable microbiocide containing sodium salts of dimethyl and ethylene dithiocarbamates; used to prevent fermentation in drilling fluid and corrosion in producing wells. *Westbridge Research Group.* Unverified.

34282 Wescodyne
Iodophor disinfectants. *Ciba plc.* Name unverified.

34283 WesLoTemp
A liquid, sodium polyacrylate polymer-based drilling fluid additive used as a dispersant/deflocculant; used in clay-based fresh water drilling fluid systems subject to low temperatures. *Westbridge Research Group.* Unverified.

34284 Wessalith
1344-00-9 215-684-8
Sodium aluminum silicate
Silicic acid, aluminum sodium salt; Aluminosilicic acid, sodium salt; 23P; Aluminum sodium silicate; Alusil ET; AMSR 3; Decalso; Decalso F; Degussa P820; Sasil; Sodium aluminosilicate; Sodium aluminum silicate; Sodium silicoaluminate; Vulkasil A 1; Zeolex; Zeolex 7; Zeolex 7A; Zeolex 100; Zeolex 23A; Zeolex 23P; Zeolex 25; Zeolex 35; Zeolex 45. Phosphate replacement in laundry detergents. *Degussa AG.* Name unverified.

34285 Wessalon, 50, 50S, S
7631-86-9 8637 231-545-4
silica. Amorphous precipitated silica; for pesticides industry. *Degussa; Degussa AG.*

34286 WesScaleStop
A liquid, acrylic homopolymer scale inhibitor; used in fresh water-based drilling fluid systems. *Westbridge Research Group.* Unverified.

34287 Wessell's silver
A nickel silver alloy of 51-65% copper, 19-32% nickel, 12-17% zinc, and 2% silver.

34288 WesSperse
A dry, blended sodium polyacrylate/chrome free lignosulfonate drilling fluid additive used as a dispersant/deflocculant; used in clay-based drilling fluid systems subject to calcium, magnesium, and chloride contamination and high temperature. *Westbridge Research Group.* Unverified.

34289 West African copaiba
Illurin balsam, an oleo-resin, is known by this name; used as a substitute for balsam of copaiba.

34290 West African gum
A gum arabic resembling Senegal gum, obtained from *Acacia nilotica* .

34291 West System Brand Products
Liquid epoxy resin, hardeners and accessories; used for wood encapsulation, bonding, filling and fairing. *Wessex Resins & Adhesives Ltd.*

34292 WesTemp, WesTemp K+
Liquid, sodium (K+ - potassium) polyacrylate polymer-based drilling fluid additives used as dispersant/deflocculants; used in clay-based fresh water drilling fluid systems subject to high temperatures. *Westbridge Research Group.* Unverified.

34293 Westfalite No. 3
An explosive consisting of 58-61% ammonium nitrate, 13-15% potassium nitrate, 4-6% trinitrotoluene, and 20-22% ammonium chloride.

34294 Westhin, Westhin K+
Liquid, sodium (K+ - potassium) polyacrylate polymer-based drilling fluid additives used as dispersant/deflocculants; used in clay-based fresh water drilling fluid systems. *Westbridge Research Group.* Unverified.

34295 Westo-Flocs
Polymer flocculants; for wastewater. *Garvey Chemical Corp.* Name unverified.

34296 Weston®
Processing aids for thermoplastics. *GE Specialty.* Name unverified.

34297 Weston® 399
26523-78-4 247-759-6
Trisnonylphenyl phosphite (contains 0.75% wt. triisopropanolamine). Stabilizer used in epoxies, hot-melt adhesives, PU, polyester, SBR, PP; in molding and extrusion of PP, HDPE, LDPE, HIPS, PC, ABS, PVC, polyesters, in calendering of ABS, PVC; in film app *GE Specialty.*

34298 Weston® 430
36788-39-3 253-211-7
Tris(dipropyleneglycol) phosphite
Stabilizer for hot-melt adhesives, PU, polyesters; used in molding, extrusion, and film applications in PP, HDPE, LDPE, PVC, and polyesters; also useful for PP fiber applications and calendering of PVC. *GE Specialty.*

34299 Weston® 439
93356-94-6
Poly 4,4' isopropylidenediphenol neodol 25 alcohol phosphite
Chelating agent in conjunction with metallic soaps to enhance the clarity and color of PVC formulations; improves color and light stability and clarity of PVC formulations; FDA approved. *GE Specialty.*

34300 Weston® 474
68610-62-8 271-870-9
tris-Neodol-25 phosphite
Stabilizer for hot-melt adhesives, PU, polyesters; used in molding, extrusion, and film applications in PP, HDPE, LDPE, PVC, and polyesters; also useful for PP fiber applications and calendering of PVC. *GE Specialty.*

34301 Weston® 491
80584-87-8 279-500-8
Diphenyl didecyl (2,2,4-trimethyl-1,3-pentanediol) diphosphite
Stabilizer for PU, hot melt adhesives; used in molding and extrusion of PP, HDPE, LDPE, PC, ABS, PVC; used in film applications for HDPE, LDPE, and PP; used in calendering of PVC, molding of PU, and fiber PP applications. *GE Specialty.*

34302 Weston® 494
68133-13-1 268-665-1
Diisooctyl octylphenyl phosphite
Stabilizer for PU, hot melt adhesives; used in molding and extrusion of PP, HDPE, LDPE, PC, ABS, PVC; used in film applications for HDPE, LDPE, and PP; used in calendering of PVC, molding of PU, and fiber PP applications. *GE Specialty.*

34303 Weston® 600
26544-27-4 247-779-5
Diisodecyl pentaerythritol diphosphite
Stabilizer for hot-melt adhesives, PU, polyesters; used in molding, extrusion, and film applications in PP, HDPE, LDPE, PVC, and polyesters; also useful for PP fiber applications and calendering of PVC. *GE Specialty.*

34304 Weston® 618, Weston® 619
3806-34-6 223-276-6
Distearyl pentaerythritol diphosphite
Color, heat stabilizer; improves the processing and thermal stability of PP; synergistic with light stabilizers such as benzophenones, benzotriazoles, and hindered amines; used in olefin polymers, PS, and rubber-modified PS for food contact and packaging. 619 Contains 1.0% wt. triisopropanolamine; used in olefin polymers and polystyrene. *GE Specialty.*

34305 Weston® DHOP
80584-86-7 279-499-4
Poly (dipropyleneglycol) phenyl phosphite
Stabilizer for PU, hot melt adhesives; used in molding and extrusion of PP, HDPE, LDPE, PC, ABS, PVC; used in film applications for HDPE, LDPE, and PP; used in calendering of PVC, molding of PU, and fiber PP applications. *GE Specialty.*

34306 Weston® DLP
21302-09-0 244-325-8
$C_{24}H_{51}O_3P$
Dilauryl phosphite
Phosphonic acid, didodecyl ester; Didodecyl phosphite; Di-n-Dodecyl phosphite; Dilauryl hydrogen phosphite; Dilauryl phosphite. Stabilizer. *GE Specialty.*

34307 Weston® DOPI
36116-84-4 252-873-4
Diisooctyl phosphite
Stabilizer for hot-melt adhesives, PU, polyesters; used in molding, extrusion, and film applications in PP, HDPE, LDPE, PVC, and polyesters; also useful for PP fiber applications and calendering of PVC. *GE Specialty.*

34308 Weston® DPDP
26544-23-0 247-777-4
Diphenyl isodecyl phosphite
Stabilizer for flexible and rigid PVC; reacts principally by chelation of metallic chlorides during PVC compounding; for use in conjunction with primary heat stabilizers. *GE Specialty.*

34309 Weston® DPP
4712-55-4 225-202-8
$C_{12}H_{11}O_3P$
Diphenyl phosphite
Phosphonic acid, diphenyl ester; Diphenoxyphosphine oxide; Diphenyl hydrogen phosphite; Diphenyl phosphite; Diphenyl phosphonate; Phenyl phosphonate ((PhO)2HPO); Phosphorous acid, diphenyl ester. Stabilizer used in epoxies, hot-melt adhesives, PU, polyester, SBR, PP; in molding and extrusion of PP, HDPE, LDPE, HIPS, PC, ABS, PVC, polyesters, in calendering of ABS, PVC; in film applications of PP, PE, PVC; fiber applications of PP, polyesters. *GE Specialty.*

34310 Weston® DSP
19047-85-9 242-784-9
Distearyl phosphite
Stabilizer for hot-melt adhesives, PU, polyesters; used in molding, extrusion, and film applications in PP, HDPE, LDPE, PVC, and polyesters; also useful for PP fiber applications and calendering of PVC. *GE Specialty.*

34311 Weston® DTDP
36432-46-9 253-034-5
Di-tridecyl phosphite
Stabilizer. *GE Specialty.*

34312 Weston® EGTPP, Weston® TPP
101-02-0 202-908-4
$C_{18}H_{15}O_3P$
Triphenyl phosphite
EGTPP has 0.5% w/w triisopropanolamine; reactive diluent for epoxy applications including adhesives, coatings, laminates, potting and soldering compounds, tooling; viscosity reducer. TPP is a stabilizer used in epoxies, hot-melt adhesives, PU, polyester, SBR, PP; in extrusion of PP, HDPE, LDPE, HIPS, PC, ABS, PVC, polyesters, in calendering of ABS, PVC; in film applications of PP, PE, PVC; fiber applications of PP, polyesters. d = 1.180; n^{20}+siD = 1.5910 *GE Specialty.*

34313 Weston® EHDPP
15647-08-2 239-716-5
Ethylhexyl diphenyl phosphite
Color and processing stabilizer in ABS, PC, PU, coatings, PET fiber; secondary stabilizer improving color and heat stability of PVC. *GE Specialty.*

34314 Weston® ODPP
26401-27-4 247-658-7
Diphenyl isooctyl phosphite
Stabilizer for PU, hot melt adhesives; used in molding and extrusion of PP, HDPE, LDPE, PC, ABS, PVC; used in film applications for HDPE, LDPE, and PP; used in calendering of PVC, molding of PU, and fiber PP applications. *GE Specialty.*

34315 Weston® PDDP
25550-98-5 247-098-3
Phenyl diisodecyl phosphite
Stabilizer for PU, hot melt adhesives; used in molding and extrusion of PP, HDPE, LDPE, PC, ABS, PVC; used in film applications for HDPE, LDPE, and PP; used in calendering of PVC, molding of PU, and fiber PP applications. *GE Specialty.*

34316 Weston® PNPG
3057-08-7 221-291-2
Phenyl neopentylene glycol phosphite
Stabilizer for PU, hot melt adhesives; used in molding and extrusion of PP, HDPE, LDPE, PC, ABS, PVC; used in film applications for HDPE, LDPE, and

PP; used in calendering of PVC, molding of PU, and fiber PP applications. *GE Specialty.*

34317 Weston® PTP
13474-96-9 236-753-9
Heptakis (dipropylene-glycol) triphosphite
Stabilizer for hot-melt adhesives; PU, polyesters; used in molding, extrusion, and film applications in PP, HDPE, LDPE, PVC, and polyesters; also useful for PP fiber applications and calendering of PVC. *GE Specialty.*

34318 Weston® TDP
25448-25-3 246-998-3
Triisodecyl phosphite
Stabilizer for hot-melt adhesives, PU, polyesters; used in molding, extrusion, and film applications in PP, HDPE, LDPE, PVC, and polyesters; also useful for PP fiber applications and calendering of PVC. *GE Specialty.*

34319 Weston® THOP
80584-85-6 279-498-9
Tetraphenyl dipropyleneglycol diphosphite
Stabilizer for PU, hot melt adhesives; used in molding and extrusion of PP, HDPE, LDPE, PC, ABS, PVC; used in film applications for HDPE, LDPE, and PP; used in calendering of PVC, molding of PU, and fiber PP applications. *GE Specialty.*

34320 Weston® TIOP
25103-12-2 246-614-4
$C_{24}H_{51}O_3P$
Triisooctyl phosphite
Stabilizer for hot-melt adhesives, PU, polyesters; used in molding, extrusion, and film applications in PP, HDPE, LDPE, PVC, and polyesters; also useful for PP fiber applications and calendering of PVC. d = 0.891; n^{20}_D= 4.4490 *GE Specialty.*

34321 Weston® TLP
3076-63-9 221-356-5
Trilauryl phosphite
Stabilizer for hot-melt adhesives, PU, polyesters; used in molding, extrusion, and film applications in PP, HDPE, LDPE, PVC, and polyesters; also useful for PP fiber applications and calendering of PVC. *GE Specialty.*

34322 Weston® TLTTP
1656-63-9 216-751-4
Trilauryl trithiophosphite
Stabilizer for hot-melt adhesives, PU, polyesters; used in molding, extrusion, and film applications in PP, HDPE, LDPE, PVC, and polyesters; also useful for PP fiber applications and calendering of PVC. *GE Specialty.*

34323 Weston® TNPP
26523-78-4 247-759-6
Trisnonylphenyl phosphite
Stabilizer used in epoxies, hot-melt adhesives, PU, polyester, SBR, PP; in molding and extrusion of PP, HDPE, LDPE, HIPS, PC, ABS, PVC, polyesters, in calendering of ABS, PVC; in film applications of PP, PE, PVC; fiber applications of PP, polyesters. *GE Specialty.*

34324 Weston® TSP
2082-80-6 218-217-6
Tristearyl phosphite
Stabilizer for hot-melt adhesives, PU, polyesters; used in molding, extrusion, and film applications in PP, HDPE, LDPE, PVC, and polyesters; also useful for PP fiber applications and calendering of PVC. *GE Specialty.*

34325 Westoran®
A trade name for a cleaning agent for cotton; it contains emulsified hydrocarbons, and is also used as an insecticide. No manufacturer.

34326 Westphalite I
A safety explosive for mines, consisting of 95% ammonium nitrate and 5% resin.

34327 Westphalite II
Westphalite, Improved. A safety explosive for mines, containing 92% ammonium nitrate, 3% potassium nitrate, and 5% resin.

34328 Westrol®
A trade name for a cleaning liquid for cotton; it contains oils with a solvent; soaps containing trichlorethylene; a degreasing agent. No manufacturer.

34329 Westropol®
A trade name for a cleaning and degreasing agent. No manufacturer.

34330 Westvaco® Resin 90
8050-09-7
Stabilized tall oil rosin. Emulsifier for synthetic rubber emulsion polymerization; tackifier, stabilizer, plasticizer. *Westvaco.*

34331 WesVis
A liquid, ammonium polyacrylate invert emulsion drilling fluid additive used as a viscosifier, bentonite extender, selective flocculant and hole sweep; used in water-based drilling fluid systems. *Westbridge Research Group.* Unverified.

34332 Wet 6
Dental preparation; wetting agent for waxes. *Bayer AG.* Discontinued.

34333 weta material
A porcelain substitute consisting of fine, uniformly distributed carborundum particles with silicates and metals of the iron series, cobalt and nickel, and sinters after firing at 1400 C.

34334 Wetanol
modified sulfated fatty acid ester. A proprietary trade name for a wetting agent for textiles, etc. No manufacturer.

34335 Wetfix
Range of cationic surfactants comprised of amino groups and selected aliphatic hydrocarbon chains; liquid form, heat stable; promotes and retains adhesion between asphalt and aggregate; incorporated into asphalt which will eventually be used for surface dressing, asphaltic macadams and concretes. *Thomas Swan & Co Ltd.* Discontinued.

34336 Wetstrez®
Synthetic resins. *Reichhold.*

34337 Wetter Nobelit B
A gelatinous permitted explosive (group P1). *Dynamit Nobel Wien GmbH.* Name unverified.

34338 Wetter-dynamite
A safety explosive for mines, consisting of 53% nitroglycerin, 14% kieselguhr, and 33% magnesium sulfate.

34339 Wetter-dynammon
An Austrian explosive containing 94% ammonium nitrate, 2% potassium nitrate, and 4% charcoal.

34340 Wetteren powder
A guncotton powder, containing a little calcium carbonate, gelatinized with amyl acetate.

34341 Wetter-Fulminite
An explosive containing ammonium nitrate.

34342 Wetting Agent FCGB
Sodium laureth phosphate
Wetting agent for metal surfaces; useful in cyanide copper plating baths to produce clean finishes. *Henkel/Functional Prods.*

34343 Wettol® D 1
Sodium salt of phenolsulfonic acid condensation product; emulsifier, wetting agent, and dispersant for formulation of wettable powders for crop protection. *BASF AG.*

34344 Wettol® EM 1
Calcium alkylaryl sulfonate
Emulsifier, wetting and dispersing agent for formulation of pesticides. *BASF AG.*

34345 Wetz
Surfactant with antifoam. *Draxel Chemical Company.* Unverified.

34346 Weyl and Zeitler's solution, Wheeler's solution
Pyrogallol in sodium hydroxide solution; used to absorb oxygen.

34347 Whale
A cotton fabric grade of Tufnol industrial laminates. *Tufnol Ltd.*

34348 Wheat Germ Oil, Wheat Germ Oil CLR
8006-95-9
Oil of wheat germ; *triticum aestivum* germ oil. Oil obtained by expression or extraction of wheat germ; a vitamin E concentrate; emollient for skin and hair care products; gloss agent in makeups; moisturizer. Fatty oil of wheat germ, natural vitamin E carrier; products for general skin protection. *Arista Industries; Croda; Penta manufacturing; Provital; Dr. Kurt Richter; Henkel/Cospha.*

34349 Wheat germ oil;
8006-95-9
EmCon W; Lipovol WGO; Super Refined Wheat Germ Oil. Skin/hair conditioner, occlusive solv.; for hair conditioners, shampoos, lipsticks, cleansers, moisturizing creams and lotions, skin care products *Fanning; Lipo.*

34350 Wheat germamidopropyl betaine
133934-09-5
N-(Carboxymethyl)-N,N-dimethyl-3-[(1-oxowheat germ alkyl)amino]-1-propanaminium hydroxides, inner salt
Afaine+tm WGB; Incronam WG-30; Mackham WGB; Schercotaine WOAB; Wheat germ oil amido betaine. Foam booster/stabilizer. Liquid; pH=6.5 *Croda Inc.; McIntyre; Scher.* Discontinued.

34351 Wheat-Pro EN-20, Wheat-Tein NL
100864-25-1
Hydrolyzed wheat protein; substantive protein, film-former, anti-irritant, protectant, moisturizer for skin and hair care products (shampoos, conditioners, styling products, creams, lotions, hand soaps). cosmetic ingredient for skin and hair care products *Brooks Industries; Maybrook.*

34352 whetstone
Oilstone
Honestone. Hard rocks, usually siliceous in character; used for sharpening

tools. Suitable rocks include hornstone, sandstone, slate, lydian stone, schist, etc.

34353 white acid
hydrofluoric acid - ammonium fluoride. Used for etching glass.

34354 white alloy
An alloy of 10% cast iron, 10% copper, and 80% zinc. This name is also applied to alloys containing 49-53% copper, 23-24% zinc, 22-24% nickel, and 2% iron. They are nickel silvers.

34355 white button alloy
A nickel-silver containing from 49-53% copper, 23-24.5% zinc, 22-24% nickel, and 2-2.5% iron.

34356 white cast iron
A good variety of cast iron. It usually contains 97% iron and 3% carbon, mainly in the uncombined state.

34357 white copper
A nickel silver usually containing 70% copper, 18% zinc, and 12% nickel.

34358 white copperas
$Fe_2(SO_4)_3 \cdot 9H_2O$
cinquinolite
Mineral ferric sulfate, The name white copperas is also used for zinc sulfate.

34359 White Cosmetic
A trade name for basic nitrate or mixture of basic nitrates obtained by adding water to bismuth nitrate. No manufacturer.

34360 white dammar
Manila copal resin, obtained from *Vateria indica* .

34361 White Gold
Water soluble polymer for oilfield use. *BP Chemicals Ltd.* Discontinued.

34362 white gold
Various alloys are known by this term. A jeweler's alloy of gold whitened by means of silver, is called white gold. An alloy of 90% gold, and 10% palladium, and a platinum substitute, containing 59% nickel, and 41% gold are both known by this name.

34363 white gunpowder
A mixture of 2 parts potassium chlorate and 1 part each potassium ferrocyanide and sugar. An ingredient of explosives.

34364 White House Cement Paint
Colored cement; decorative finish for stonework, masonry, etc. *Calder Colours (Ashby) Ltd.*

34365 white insect wax
white lac
Arjun wax of India produced by the insect *Ceroplastes ceriferus* .

34366 white lead
1319-46-6 11, 1016 215-290-6
Ceruse
A basic carbonate of lead, the composition of which is variable; used as a pigment.

34367 white metal
An alloy of 54% copper, 24% nickel, and 22% zinc. It is a nickel silver. It is also the name applied to bearing metals.

34368 white oils
Eggs oils. A liniment usually containing turpentine, acetic acid, and eggs. Sometimes ammonia and camphor are added.

34369 white paste
$Cu_2(CNS)_2$
Copper sulfocyanide.

34370 white precipitate
10124-48-8 5927 233-335-8
ClH_2HgN
ammoniated mercury
Lemery's white precipitate; mercury amide chloride; aminomercuric chloride; mercury ammonium chloride; white mercuric precipitate. Mercury ammonium chloride, used for the preparation of cinnabar and in medicine as a topical anti-infective. d = 5.38.

34371 white pyrites
1317-37-9 4106 215-268-6
FeS
ferrous sulfide
iron sulfuretmagnetkies; pyrrhotine; troillite. A variety of iron pyrites. mp = 1194°; d = 4.84; insoluble in H_2O.

34372 white solder
An alloy of 10% nickel, 45% copper, and 45% zinc. A soldering alloy.

34373 white spirit
A turpentine substitute. It is usually a petroleum product, having flash point and degree of evaporation similar to turpentine.

34374 white tellurium
bunsenine; krennerite. A mineral. It contains from 25-29% gold, 2.7-14.6% silver, and 2.5-19.5% lead, as tellurides.

34375 Whitetex
1332-58-7 5294 296-473-8
Coarse particle size calcined kaolin (aluminum silicate); used as a reinforcer and extender in plastics, PVC, rubber, coatings. *Engelhard.*

34376 Whitworth's steel
Steel which has been subjected to high pressures to eliminate blow-holes.

34377 Wiborg phosphate
A German fertilizer made by heating mineral phosphate with soda. It consists mainly of a tetraphosphate.

34378 Wickenol® 101, 105
110-27-0 5234 203-751-4
$C_{17}H_{34}O_2$
tetradecanoic acid 1-methylethyl ester
Tetradecanoic acid, 1-methylethyl ester; Myristic acid, isopropyl ester; Bisomel; component of Sardo Bath Oil; Crodamol I.P.M.; Deltyl Extra; Emcol-Im; Emerest 2314; Estergel; IPM; Isomyst; Isopropyl Myristate; Isopropyl myristate; Isopropyl tetradecanoate; JA-fA IPM; Kessco IPM; Kessco isopropyl myristate; Kesscomir; Plymoutm IPM; Promyr; Revenge; Sinnoester MIP; Starfol IPM; Stepan D-50; Tegester; Tetradecanoic acid, isopropyl; Tetradecanoic acid, isopropyl ester; 1-Tridecanecarboxylic acid, isopropyl ester; Unimate IPM; Wickenol 101. Emollients, solubilizers, and lubricants for use in cosmetic and toilet preparations. mp = 3°; bp_{20} = 193°; d^{20} = 0.8532; n^{20}_D = 1.433; insoluble in H_2O, soluble in most organic solvents. *CasChem.*

34379 Wickenol® 111
142-91-6 205-571-1
$C_{19}H_{38}O_2$
Isopropyl palmitate
Hexadecanoic acid, 1-methylethyl ester; Palmitic acid, isopropyl ester; Crodamol IPP; Deltyl; Deltyl Prime; Emcol-ip; Emerest 2316; Estol 103; Hexadecanoic acid, isopropyl ester; Isopal; Isopalm; Isopropyl Hexadecanoate; Isopropyl n-hexadecanoate; Isopropyl N-hexadecanoate; Isopropyl palmitate; Ja-fa ipp; Ja-fa ippkessco; Kessco IPP; Kessco isopropyl palmitate; Palmitic acid esters; Plymouth ipp; Propal; Revenge; Sinnoester PIT; Starfol ipp; Stepan D-70; Tegester isopalm; Unimate ipp; Usaf ke-5; Wickenol 111. Emollient, solubilizer, and lubricant for use in cosmetic and toilet preparations. mp = 11-13°; d = 0.852; n^{20}_D = 1.4380. *CasChem.*

34380 Wickenol® 127
112-10-7 203-934-9
$C_{21}H_{42}O_2$
Isopropyl stearate
Octadecanoic acid, 1-methylethyl ester; Stearic acid, isopropyl ester; Isopropyl stearate; 1-Methylethyl octadecanoate; Octadecanoic acid, isopropyl ester; Revenge; Wickenol 127. Emollient, cosolvent and lubricant. *CasChem.*

34381 Wickenol® 131
68171-33-5 269-023-3
$C_{21}H_{42}O_2$
Isopropyl isostearate
Isooctadecanoic acid, 1-methylethyl ester; Isopropyl isostearate; Isostearic acid, isopropyl ester; 2-Propyl isooctadecanoate. Lubricant, emollient, solubilizer. *CasChem.*

34382 Wickenol® 136
Isopropyl esters of myristic, palmitic and stearic acids. Emollient, conditioner, moisturizer, solubilizer, vehicle, and solvent for cosmetic and toilet preparations. *CasChem.*

34383 Wickenol® 139
61789-91-1 5279
Synthetic jojoba oil; Hydrogenated jojoba oil; Jojoba oil; Jojoba wax; Waxes and Waxy substances, jojoba; Waxes and Waxy substances, jojoba oil. Emollient, plasticizer, lubricant, coupler for hair and skin care products; replacement for natural jojoba oil. *CasChem.*

34384 Wickenol® 141
110-36-1 203-759-8
$C_{18}H_{36}O_2$
Butyl myristate
butyl hexadecanoate; tetradecanoic acid n-butyl ester; Tetradecanoic acid, butyl ester; Myristic acid, butyl ester; Bumyr; Butyl myristate; Butyl Tetradecanoate; Butyl N-tetradecanoate; Wickenol 141. Emollient, solubilizer, and lubricant for use in cosmetic and toilet preparations; plasticizer. d = 0.860; n^{20}_D= 1.4405. *CasChem.*

34385 Wickenol® 142
Octyldodecyl myristate
Emollient ingredient, plasticizer for cosmetic and pharmaceutical preparations. *CasChem.*

34386 Wickenol® 143
3687-45-4 222-980-4
$C_{36}H_{68}O_2$
Oleyl oleate

9-Octadecenoic acid (Z)-, 9-octadecenyl ester, (Z)-; Oleic acid, (Z)-9-octadecenyl ester; Cetiol; Oleic acid oleyl ester; Oleyl oleate. Lubricant used in cosmetic and toilet preparations; replacement for sperm oil in addition to functioning as cosmetic additive. *CasChem.*

34387 Wickenol® 144

59231-34-4 261-673-6

$C_{28}H_{54}O_2$

Isodecyl oleate

9-Octadecenoic acid (Z)-, isodecyl ester; Ceraphyl 140A; Ceraphyl 140-A; Isodecyl oleate; 9-Octadecenoic acid, isodecyl ester. Emollient, wetting agent and pigment binder for cosmetics. *CasChem.*

34388 Wickenol® 151

42131-25-9

Isononyl isononanoate

Emollient, moisturizer, pigment wetter; silky emolliency and solvent characteristics; for hair care products. *CasChem.*

34389 Wickenol® 152

41395-89-5

Isodecyl isononanoate

Silky emollience and solvent characteristics for skin and hair care products; pigment wetter. *CasChem.*

34390 Wickenol® 153

59231-37-7 261-675-7

$C_{22}H_{44}O_2$

Isotridecyl isononanoate

Hexanoic acid, 3,5,5-trimethyl-, isotridecyl ester; Isotridecyl isononanoate; Isotridecyl 3,5,5-trimethylhexanoate; Wickenol 153. Silky emollience and solvent characteristics for skin and hair care products; pigment wetter, moisturizer. *CasChem.*

34391 Wickenol® 155

29806-73-3 249-862-1

$C_{24}H_{48}O_2$

Octyl palmitate

Hexadecanoic acid, 2-ethylhexyl ester; Palmitic acid, 2-ethylhexyl ester; Ceraphyl 368; 2-Ethylhexyl palmitate; Octyl Palmitate; Wickenol 155. Emollient, moisturizer, pigment wetter/dispersant; increases water vapor porosity of fatty components used in cosmetic and topical pharmaceutical preparations. *CasChem.*

34392 Wickenol® 156

22047-49-0 244-754-0

$C_{26}H_{52}O_2$

Octyl stearate

Octadecanoic acid, 2-ethylhexyl ester; Stearic acid, 2-ethylhexyl ester; 2-Ethylhexyl octadecanoate; 2-Ethylhexyl stearate; Octyl stearate; Wickenol 156. Emollient, moisturizer, pigment wetter/dispersant; increases water vapor porosity of fatty components used in cosmetic and topical pharmaceutical preparations. *CasChem.*

34393 Wickenol® 158

103-23-1 203-090-1

$C_{22}H_{42}O_4$

di-(2-ethylhexyl) adipate

Hexanedioic acid, bis(2-ethylhexyl) ester; Adipic acid, bis(2-ethylhexyl) ester; Adipic acid, di(2-ethylhexyl) ester; Adipol 2EH; Beha; Bis(2-ethylhexyl) adipate; Bis-(2-ethylhexyl)ester kyseliny adipove (czech;)Bisoflex DOA; Deha; Di(2-ethylhexyl) adipate; Dioctyl adipate; DOA; Effemoll DOA; Effomoll DOA; Ergoplast Addo; Flexol A 26; Flexol plasticizer 10-a; Flexol plasticizer a-26; Hexanedioic acid, dioctyl ester; Kemester 5652; Kodaflex DOA; Mollan S; Monoplex DOA; Monsanto DOA; NCI-C54386; Octyl adipate (VAN); Plastomoll DOA; PX-238; Reomol Doa; Rucoflex Plasticizer Doa; Sicol 250; Staflex Doa; Truflex DOA; Uniflex Doa; Vestinol OA; Wickenol 158; Witamol 320; Witcizer 412,. Emollient, moisturizer, pigment wetter/dispersant, cosolvent; increases water vapor porosity of fatty components used in cosmetic and topical pharmaceutical preparations. bp = 166-168°; d = 0.990; n^{20}_D = 1.4470. *CasChem.*

34394 Wickenol® 159

2915-57-3 220-836-1

$C_{20}H_{38}O_4$

Dioctyl succinate

Butanedioic acid, bis(2-ethylhexyl) ester; Bis(2-ethylhexyl) succinate; Di(2-ethylhexyl) butanedioate; Di(2-ethylhexyl) succinate; Dioctyl succinate; Succinic acid, bis(2-ethylhexyl) ester; Wickenol 159. Emollient, moisturizer, pigment wetter/dispersant; increases water vapor porosity of fatty components used in cosmetic and topical pharmaceutical preparations. *CasChem.*

34395 Wickenol® 160

59587-44-9 261-819-9

$C_{17}H_{34}O_2$

2-Ethylhexyl pelargonate

Nonanoic acid, 2-ethylhexyl ester; Octyl pelargonate. Emollient, moisturizer, pigment wetter/dispersant; increases water vapor porosity of fatty components used in cosmetic and topical pharmaceutical preparations; improves stick formulations. *CasChem.*

34396 Wickenol® 161

Mixture of dioctyl adipate, octyl stearate, octyl palmitate. Emollient, moisturizer, pigment wetter/dispersant; increases water vapor porosity of fatty components used in cosmetic and topical pharmaceutical preparations. *CasChem.*

34397 Wickenol® 163

Mixture of dioctyl adipate, octyl stearate, octyl palmitate. Emollient, solubilizer; fragrance enhancer. *CasChem.*

34398 Wickenol® 171

29710-25-6 249-793-7

$C_{26}H_{52}O_3$

Octyl hydroxystearate

Octadecanoic acid, 12-hydroxy-, 2-ethylhexyl ester; 2-Ethylhexyl 12-hydroxystearate; 2-Ethylhexyl oxystearate; Octyl hydroxystearate; Wickenol 171. Emollient, moisturizer, pigment wetter/dispersant; increases water vapor porosity of fatty components used in cosmetic and topical pharmaceutical preparations; refatting agent, counterirritant, cosolvent, solubilizer. *CasChem.*

34399 Wickenol® 174

Myristyl octanoate

Emollient; provides soft satiny skin afterfeel. *CasChem.*

34400 Wickenol® 506

1323-03-1 215-350-1

$C_{17}H_{34}O_3$

Myristyl lactate

Propanoic acid, 2-hydroxy-, tetradecyl ester; Lactic acid, tetradecyl ester; Ceraphyl 50; Myristyl lactate; Tetradecyl lactate. Emollient with smooth, satiny afterfeel. *CasChem.*

34401 Wickenol® 535

Wheat germ glycerides; hydrophilic/hydrophobic emollient, emulsifier, skin lubricant; anti-irritant. *Caschem.*

34402 Wickenol® 545

Glucose glutamate; substantive humectant, skin conditioner, moisturizer, emulsifier, surfactant, thickener used in personal care products; enhances lather in surfactant systems. *CasChem.*

34403 Wickenol® 550

9050-36-6 232-940-4

Maltodextrin

Maltodextrin; Dextrin, malto; Maltodextrin 24DE; Maltodextrin I; Mar Rex 1918; Snowflake. Absorbent for lipophilic materials for powder bath applications; foodgrade carrier for flavors. *CasChem.*

34404 Wickenol® 707

9035-85-2

$(C_3H_6O)_nC_{16}H_{34}O$

Poly[oxy(methyl-1,2-ethanediyl)], α-hexadecyl-ω-hydroxy-

Glycols, polypropylene, monohexadecyl ether; Polyoxypropylene cetyl ether; Polyoxypropylene (30) cetyl ether; Polyoxypropylene hexadecyl ether; Polypropylene glycol (10) cetyl ether; Polypropylene glycol hexadecyl ether; Polypropylene glycol monocetyl ether; Polypropylene glycol monohexadecyl ether; PPG-10 cetyl ether; Procetyl; Procetyl alcohol 30; Procetyl AWS; Propoxylated hexadecyl alcohol; Wickenol 707. PPG-30 cetyl ether; all-purpose fluid, nongreasy emollient with hydroalcoholic compatibility; provides coupling and emulsion stability; foam modifier; enhances sheen and manageability. *CasChem.*

34405 Wickenol® 727

68439-53-2

PPG-30 lanolin alcohol ether

Nongreasy emollient, solubilizer, bodying agent; also low odor derivative. *Caschem.*

34406 Wiegold alloy

A brass containing aluminum and resembling gold in appearance. It consists of 67.73% copper, 32% zinc, and 0.27% aluminum, but some analysts state that it contains 0.25-0.5% lead; a dental alloy.

34407 wild ginger oil

862

Asarum; Canada snakeroot; Indian ginger. The oil of *Asarum canadense*. Contains methyl eugenol.

34408 Wilhelmit

A German explosive containing potassium chlorate and mineral oil.

34409 Wilkinite

1302-78-9 1082 215-108-5

aluminum silicate hydrate

Jelly rock; Bentonite;. A colloidal clay; used as a substitute for China clay as a paper filter.

34410 Wills Metallic 'O' Rings
Metallic sealing systems; extreme pressure and temperature sealing for fluid and vacuum service. *Fothergill Tygaflor Ltd.* Name unverified.

34411 Wilmil
Alloys similar in composition to silumin. No manufacturer.

34412 Wilmot's aluminum solder
An alloy of 86% tin and 14% bismuth.

34413 Wingstay® 100
Mixed diaryl *p*-phenylenediamine; antioxidant/antiozonant for tires, camelback, mechanical goods; stabilizer for SBR, polybutadiene, and neoprene. *Goodyear; R. T. Vanderbilt Co Inc.*

34414 Wingstay® C
Butylated dimethylbenzyl phenol
Stabilizer for nonstaining SBR and nitrile polymers. *Goodyear; R. T. Vanderbilt Co Inc.*

34415 Wingstay® S
Styrenated phenol
Nonstaining, nondiscoloring antioxidant for foam rubber, white sidewalls, kitchen and drug sundries, garden hose, shoe soles; stabilizer for SBR and BR. *Goodyear; R. T. Vanderbilt Co Inc.*

34416 Wingstay® T
A blend of substituted phenols; a proprietary antioxidant. *Goodyear.* Name unverified.

34417 Wingtack® 10
C_5 hydrocarbon resin. Plasticizer/tackifier for pressure sensitives, hot melts and sealants. *Goodyear.* Name unverified.

34418 Wingtack® Extra
C_5 hydrocarbon resin. Tackifier resin for adhesives; superior resistance to uv light and heat aging. *Goodyear.* Name unverified.

34419 Winnofil
Precipitated calcium carbonate surface treated with calcium stearate; used for giving tires high tensile strength. *ICI Chem & Polymers Ltd.*

34420 Witafrol®
Antifoam agents; used in mining industry, food industry, water engineering, and the paper industry. *Hüls AG.*

34421 Witafrol® 7420
26402-26-6 6335 247-668-1
$C_{11}H_{22}O_4$
octanoic acid monoester with 1,2,3-propanetriol;
monooctanoin; α-monocaprylin; caprylic acid monoglyceride; glyceryl monocaprylate; octanoic acid glycerol ester; DL-glyceryl-1-mono-octanoate; caprylic/capric glycerides. Surfactant for pharmaceutical, cosmetic, nutritional fields; as emulsifier, solubilizer, dispersant, plasticizer, lubricant, consistency regulator, skin/mucous membrane protectant, refatting agent, penetrant, carrier, adsorption promoter, antifoaming agent. mp = 39.5-40.5°; *Hüls Am.*

34422 Witafrol® 7440
Hydrophilic fatty acid ester; foam preventer for food industry. *Hüls AG.*

34423 Witafrol® 7456
Mixture of hydrophilic fatty acid esters. Food antifoam; pronounced/spontaneous retardation effect on foam formation caused by proteins. *Hüls AG.*

34424 Witafrol® 7480 N
Hydrophilic fatty acid ester. Defoamer and foam preventive for sugar, dairy industries, jams, fruit flavors/juices, seasonings, and other food products. *Hüls AG.*

34425 Witafrol® 7490
Hydrophilic fatty acid ester. Biodegradable antifoam agent for sugar industry. *Hüls AG.*

34426 Witafrol® 7497 N
Hydrophilic fatty acid ester. Foam preventer for juice extraction/purification, external water circulation systems of sugar factories; biodegradable. *Hüls AG.*

34427 Witamol
Plasticisers; used in PVC processing, imitation leather, heat resistant cables and sheets, special products with high resistance to cold and weather, application to lacquer. *Hüls AG.*

34428 Witarix®
Hard fats for chocolate. *Hüls AG.*

34429 Witarix® 212
68990-82-9 273-627-2
Hydrogenated palm kernel oil; fats for chocolate and confectionery products with soft consistency; also for personal care products. *Hüls Am.*

34430 Witarix® 250
Cocoglycerides
Fats for chocolate and confectionery products with soft consistency; also for personal care applications. *Hüls Am.*

34431 Witarix® 440
8016-70-4 232-410-2
Hydrogenated soybean oil; fats for chocolate and confectionery products with soft consistency; also for personal care applications. *Hüls Am.*

34432 Witarix® 450
Hydrogenated peanut oil; fats for chocolate and confectionery products with soft consistency; also for personal care applications. *Hüls Am.*

34433 Witbreak DGE-128A
Glycol ester;. Base for crude oil demulsifiers. *Witco Corporation.*

34434 Witbreak DPG-15
POE glycol
Oilfield surfactant, demulsifier. *Witco Corporation.*

34435 Witbreak DRA-21
Oxyalkylated phenolic resin; base for crude oil demulsifier. *Witco Corporation.*

34436 Witbreak DTG-62
Polyoxyalkylene glycol
Oilfield surfactant; demulsifier. *Witco Corporation.*

34437 Witbreak RTC-326
Polymeric amine
Oilfield surfactant, reverse demulsifier. *Witco/Organics.* Discontinued.

34438 Witcamide
Surfactant for cosmetics, toiletries, pharmaceutical, processing, agricultural and other industries. *Baxenden Chemicals Ltd.*

34439 Witcamide® 128T
Cocamide DEA
Detergent, foam booster/stabilizer, viscosity modifier, substantive conditioner. *Witco Corporation.*

34440 Witcamide® 5195
120-40-1 204-393-1
$C_{16}H_{33}NO_3$
Lauramide DEA
Diethanolamide lauric acid; Diethanolamine lauric acid amide; Diethanollauramide; N,N-Diethanollauramide; Diethanol lauric acid amide; N,N-Diethanollauric acid amide; N,N-Diethylollauramide; Emid 6511; Emid 6541; Ethylan mld; Hetamide ml; Lauramide, N,N-bis(hydroxyethyl)-; Lauramide DEA; Lauric acid diethanolamide; Lauric acid, diethanolamine; Lauric acid diethanolamine Con(1/1;) Lauric acid diethanolamine con(1/1); Lauric acid diethanolamine condensate; Lauric diethanolamide; Lauroyl diethanolamide; Lauroyldiethanolamine; Lauryl Diethanolamide; Lauryl diethanolamide (ACN); LDA; LDE; Monamide 150LW; Monamid 150-LW,. viscosity modifier, foam stabilizer, lubricant, conditioner, lubricant, emulsifier, wetting agent, thickener, penetrant, dye dispersant, scouring aid, antistat; cosmetics and toiletries; industrial foamer and stabilizer; metal processing; textile surfacta *Witco Corporation.*

34441 Witcamide® 61
111-05-7 203-828-2
$C_{21}H_{41}NO_2$
9-Octadecenamide, N-(2-hydroxypropyl)-, (Z)-
Oleamide, N-(2-hydroxypropyl)-; N-(2-Hydroxypropyl)oleamide; Oleamide MIPA; Oleic monoisopropanolamide; Oleylmonoisopropanolamide; Steinamid IPE 280. Oleamide monoisopropanolamide and isopropanolamine; hair conditioner, emulsifier, lubricant; cosmetics and toiletries. *Witco Corporation.*

34442 Witcamide® 70
111-57-9 203-883-2
$C_{20}H_{41}NO_2$
Stearamide MEA
Octadecanamide, N-(2-hydroxyethyl)-; Clindrol 200-MS; Comperlan HS; Cycloamide SM; N-(2-Hydroxyethyl)octadecanamide; N-(Hydroxyethyl)stearamide; N-(2-Hydroxyethyl)stearamide; Loramine S 280; Marlamid M 18; Monoethanolamine stearic acid amide; Ninol 1301; Onyx Wax EL; Stearamide MEA; Stearamyl; Stearic acid monoethanolamide; Stearic ethanolamide; Stearic ethylolamide; Stearic monoethanolamide; Stearic monoethanolamine; Stearoylethanolamide; N-Stearoylethanolamine; Stearoyl monoethanolamide; Teric CME7. Opacifier, conditioner, lubricant, thickener, gelling agent, mold release agent, binder; cosmetics and toiletries; base for antiperspirant and makeup sticks. *Witco Corporation.*

34443 Witcamide® Coco Condensate
Cocamide DEA
diethanolamine. Detergent, foam stabilizer, lubricant, conditioner. *Witco/Organics.* Discontinued.

34444 Witcamide® MAS
14351-40-7 238-310-5
$C_{38}H_{75}NO_3$
Stearamide MEA-stearate
Octadecanoic acid, 2-[(1-oxooctadecyl)amino]ethyl ester; Stearamide, N-(2-stearoyloxyethyl)-; Stearamidoethyl stearate; 2-Stearamidoethyl stearate. Opacifier, conditioner, lubricant, gelling agent for personal care, household,

and institutional liquid soaps; used as partial or total replacement for vegetable waxes in polishes; coating agent for paper and textiles; mold release agent for industrial pr *Witco Corporation.*

34445 Witcamine® 209
1-(2-Aminoethyl)-2-n-alkyl-2-imidazoline
Emulsifier, lubricity and wetting agent; penetrant, dye dispersant, scouring aid, antistat, corrosion inhibitor intermediate in petrol. industry, metal processing, textiles. *Witco Corporation.*

34446 Witcamine® 211
Mixed 1-(2-aminoethyl)-2-n-alkyl-2-imidazoline. Intermediate for oil or water soluble salts used as corrosion inhibitors, bactericides, softeners; for petrol. industry. *Witco/Organics.* Discontinued.

34447 Witcamine® 6606
61791-44-4 263-177-5
Amines, bis(2-hydroxyethyl)tallow alkyl; N,N-Bis(2-hydroxyethyl)(tallow alkyl)amine; Ethanol, 2,2'-iminobis-, N-tallow alkyl derivs.; Tallow, amines, N,N-bis(2-hydroxyethyl). PEG-15 tallow amine; antistat, dispersant, oil-water emulsifier, lubricant, substantivity and wetting agent. *Witco/Organics.* Discontinued.

34448 Witcamine® 6622
PEG-30 oleamine
Antistat, dispersant, oil-water emulsifier, lubricant, substantivity and wetting agent. *Witco/Organics.* Discontinued.

34449 Witcamine® AL42-12
Cationic surfactant in the form of tall oil imidazoline; antistatic and corrosion inhibitor for use in carwash and wax formulations. *Witco Chemical Ltd.* Discontinued.

34450 Witcamine® E-607
N-(lauroyl colamino formyl methyl) pyridinium chloride. Cationic surfactant used in deodorants, after shave and hair rinses. *Witco Chemical Ltd.* Discontinued.

34451 Witcamine® E-607S
N-(Stearoyl colamine formyl methyl) pyridinium chloride. Cationic surfactant used in hair conditioners and as a nonirritant emollient. *Witco Chemical Ltd.* Discontinued.

34452 Witcamine® RAD 0500
POE rosin amine
Oilfield surfactant; corrosion inhibitor intermediate, scouring agent, softener, dye assistant, emulsifier for acid-stable emulsions; inhibitor for HCl in acid pickling. *Witco Corporation.*

34453 Witcizer 100
142-77-8 205-559-6
$C_{22}H_{42}O_2$
butyl oleate
9-Octadecenoic acid (Z)-, butyl ester; Oleic acid, butyl ester; Advaplast 42; Butyl oleate; Hallco C-503 Plasticizer; Kessco 554; Kesscoflex BO; 9-Octadecenoic acid, butyl ester (Z)-; Plasthall 503; Uniflex Byo; Wilmar Butyl Oleate; Witcizer 100; Witcizer 101. Proprietary trade name for butyl oleate. No manufacturer.

34454 Witcizer 312
117-81-7 1291 204-211-0
$C_{24}H_{38}O_4$
bis(2-ethylhexyl) phthalate
1,2-Benzenedicarboxylic acid, bis(2-ethylhexyl) ester; Phthalic acid, bis(2-ethylhexyl) ester; Behp; 1,2-Benzenedicarboxylic acid, bis(ethylhexyl) ester; o-Benzenedicarboxylic acid, dioctyl ester; Bis(2-ethylhexyl) 1,2-benzenedicarboxylate; Bis-(2-ethylhexyl)ester kyseliny ftalove (czech); Bis(ethylhexyl) phthalate; Bis(2-ethylhexyl)phthalate; Bis(2-ethylhexyl) phthalate; Bisoflex 81; Bisoflex DOP; Celluflex DOP; Compound 889; DAF 68; DEHP; Di(2-ethylhexyl) orthophthalate; Di(ethylhexyl) phthalate; Di(2-ethylhexyl) phthalate; Di-2-ethylhexylphthalate (osha); Dioctyl o-benzenedicarboxylate; Dioctyl phthalate (VAN); Di-sec-octyl phthalate; Di-sec-octyl phthalate (acgih,osha); DOP; Ergoplast Fdo; Ethylhexyl phthalate; 2-Ethylhexyl phthalate; Eviplast 80; Eviplast 81; Fleximel; Flexol DOP; Flexol plasticizer dop; Good-rite gp 264; Hatcol Dop; Hercoflex 260; Kodaflex DOP; Mollan O; NCI-C52733; Nuoplaz Dop,. Plasticizer. Also used in vacuum pumps. No manufacturer.

34455 Witcizer 313
27554-26-3 248-523-5
$C_{40}H_{72}O_4$
diisooctyl phthalate
1,2-Benzenedicarboxylic acid, diisooctyl ester; Corflex 880; Diisooctyl phthalate; DIOP; Flexol Plasticizer Diop; Hexaplas m/o; Hexaplas M/O; Isooctyl phthalate; Phthalic acid, bis(6-methylheptyl) ester; Phthalic acid, diisooctyl ester. Proprietary trade name for diisooctyl phthalate. No manufacturer.

34456 Witco TX
Modified toluene sulfonic acid in liquid form; anionic surfactant used as a catalyst. *Witco Corporation.*

34457 Witco® 1298 HA
Surfactant blend; surfactant for insecticides. *Witco Corporation.*

34458 Witco® 1298
27176-87-0 248-289-4
$C_{18}H_{30}O_3S$
Benzenesulfonic acid, dodecyl-
Arylsulfonat BASF; Biosoft S-100; Bio-soft S 100; Calsoft LAS 99; Conco AAS-985; Conoco 597; Conoco SA 597; DDBSA; Dobanic Acid 83; Dobanic Acid Jn; n-Dodecylbenzenesulfonate; Dodecylbenzene sulfonic acid; Dodecylbenzenesulfonic acid; n-Dodecylbenzenesulfonic acid; N-Dodecylbenzenesulfonic acid; Dodecylbenzenesulphonic Acid; E 7256; Elfan WA sulfonic acid; Elfan Wa Sulphonic Acid; Intermediate 122; Lakeway SA; Laurylbenzenesulfonate; Laurylbenzenesulfonic acid; Marlon as 3; Na2584 (dot); Nacconol 98SA; Nansa 1042p; Nansa SSA; Pilot ABS-99; Retzloff Intermediate No. 122; Retzulfonic DPB; Retzulfonic SD-12; Richonic Acid B; Sulframin 1298; Sulframin acid 1298; Sulframine acid 1298; Tex-Wet 1197; Tex-Wet 1199; Ultrawet 99LS; Witco 1298 sulfonic acid,. Detergent intermediate, oil-water emulsifier, solubilizer, wetting agent, and detergent for household products, metal cleaning; emulsion polymerization surfactant for latex stabilization and pigment dispersion. *Witco Corporation.*

34459 Witcobond
Solvent based polyurethane adhesives for laminating fabrics and foam; solvent based polyurethanes for laminating films and boards; water-based polyurethane coatings for textiles, leather, glass fiber sizing, paints, lacquers, and others. *Baxenden Chemicals Ltd.*

34460 Witcobond® W-160
PU aqueous dispersion; used as high-performance adhesives or coatings; crosslinkable; solventless. *Witco Corporation.*

34461 Witcobond® W-232, W-234
Aliphatic PU aqueous colloidal dispersion; highly functional, protective top finishes for PVC, other plastics, metal, and wood; used in automotive, paint, textile, and other industrial markets; yields coatings with excellent resistance to abrasion, hydrol *Witco Corporation.*

34462 Witcobond® W-240
Self-crosslinking PU aqueous colloidal dispersion; produces nondiscoloring, high-performance, protective top finishes for metal, rigid plastics, and wood; such coatings exhibit excellent resistance to abrasion, hydrolysis, oxidative discoloration, impact, *Witco Corporation.*

34463 Witcobond® XW
25928-94-3
Epoxy aqueous dispersion; used in waterborne coatings; exhibits good noncryst. properties, freeze-thaw stability, low foaming tendencies, good chemical and water resistance *Witco Corporation.*

34464 Witcodet 100
Formulated product; detergent concentrated base for dishwash, carwash, laundery, all-purpose, and shampoo formulations. *Witco Corporation.*

34465 Witcodet 100 and P280
Alkylaryl sulfonate in liquid form; anionic surfactant used for liquid detergents, e.g., windscreen washer. *Witco Chemical Ltd.* Discontinued.

34466 Witcodet AE
Pearlized blend; detergent concentrated for bubble bath, hand and body soaps, shampoo. *Witco Corporation.*

34467 Witcoflex
Solvent based polyurethane textile coatings. *Baxenden Chemicals Ltd.*

34468 Witcolate 1050, Witcolate SE-5
Sodium C12-15 pareth sulfate
Detergent, detergent base, emulsifier, foamer, and wetting agent for the detergent industry. *Witco Corporation.*

34469 Witcolate 6400
151-21-3 8782 205-788-1
$C_{12}H_{26}NaO_4S$
Sodium lauryl sulfate
Sulfuric acid monododecyl ester sodium salt; AI3-00356; Akyposal SDS; Aquarex Me; Aquarex methyl; AS-C12; Avirol 101; Avirol 118 conc; Berol 452; Carsonol sls; Carsonol sls paste B; Carsonol sls special; Conco sulfate wa; Conco sulfate wa-1200; Conco sulfate wa-1245; Conco sulfate wag; Conco sulfate wan; Conco sulfate WAS; Conco sulfate wn; Cycloryl 21; Cycloryl 31; Cycloryl 580; Cycloryl 585N; DDS; Dehydag Sulfate Gl Emulsion; Dehydag Sulphate Gl Emulsion; Detergent 66; Dodecyl alcohol, hydrogen sulfate, sodium salt; Dodecyl sodium sulfate; Dodecyl Sulfate Sodium; n-Dodecyl sulfate sodium; N-Dodecyl sulfate sodium; Dodecyl sulfate, sodium salt; Dodecylsulfuric acid, sodium salt; Dreft; Dupanol; Dupanol-C; Dupanol WAQ; Duponal; Duponal WAQE; Duponol; Duponol C; Duponol Me; Duponol methyl; Duponol QC; Duponol qx; Duponol WA; Duponol WA dry; Duponol

WAE; Duponol WAG,. Detergent, foaming agent, and wetting agent. *Witco Corporation.*

34470 Witcolate 6465
Sodium 2-ethylhexyl sulfate
Detergent, dispersant, wetting agent. *Witco/Organics.* Discontinued.

34471 Witcolate 7093
Sodium deceth sulfate
Industrial detergent, foamer and wetting agent for pressure-spray applications; electrolyte tolerance. *Witco Corporation.*

34472 Witcolate AE, Witcolate LES-60A
67762-19-0
Ammonium laureth sulfate
(C10-C16)Alcohol ethoxylate, sulfated, ammonium salt; (C10-C16)Alkyl alcohol, ethoxylate, sulfuric acid, ammonium salt; (C10-C16) Alkylethoxylate sulfuric acid, ammonium salt; (C13-C16)Alkyl ethoxylate sulfuric acid, ammonium salt; Poly(oxy-1,2-ethanediyl), .alpha.-sulfo-.omega.-hydroxy-, C10-16-; alkyl ethers, ammonium salts SDA 15-067-01. Detergent, foaming agent, wetting agent. *Witco Corporation.*

34473 Witcolate AE-3
Ammonium C12-15 pareth sulfate
Detergent, wetting agent, emulsifier, foamer for the detergent industry. *Witco Corporation.*

34474 Witcolate D-510
Sodium 2-ethylhexyl sulfate
Detergent, wetting agent and penetrant for industrial use and polymerization reactions; dispersant for bleaching powders; lime soap and grease dispersant. *Witco Corporation.*

34475 Witcolate D51-51
Nonoxynol-4 sulfate
Antistat for synthetic fibers and polymer products; surfactant, wetting agent and dispersant; emulsifier for polymerization of acrylics, vinyl acetate, vinyl acrylics, styrene, SAN, styrene acrylic, and vinyl chloride. *Witco Corporation.*

34476 Witcolate D51-53
Nonoxynol-10 sulfate
Emulsifier for acrylic, vinyl acetate, vinyl acrylic, styrene, SAN, styrene acrylic, and vinyl chloride polymerization. *Witco Corporation.*

34477 Witcolate ES-2, Witcolate LES-60C
9004-82-4
$(C_2H_4O)_nC_{12}H_{26}NaO_4S$
Sodium laureth (1) sulfate
Poly(oxy-1,2-ethanediyl), α-sulfo-ω-(dodecyloxy)-, sodium salt; Glycols, polyethylene, mono(hydrogen sulfate), dodecyl ether, sodium salt; Avirol 100E; Conco Sulfate WE; Cycloryl NA; Dodecyl alcohol, ethoxylated and sulfated, sodium salt; Elfan 242; Elfan NS 242; Elfan NS 243; Empicol ESB 3; Empicol ESB 30; Empimin KSN; Empimin KSN 27; Empimin KSN 60; Empimin KSN 70; Etoxon epa:sodium lauryl ether sulfate; α-Lauryl-ω-hydroxypoly(oxyethylene)sulfate, sodium salt; Maprofix 60S; Maprofix ES; Polyethylene glycol sulfate monododecyl ether, sodium salt; Retzolate 60; Rewopol NL-2; Sipon ES; Sipon LES 25; Sodium dodecylpolyethoxysulfate; Sodium dodecylpoly(oxyethylene) sulfate; Sodium lauryl ether sulfate; Sodium (lauryloxypolyethoxy)ethyl sulfate; Sodium laurylpoly(oxyethylene) sulfate; Sodium lauryl sulfate ethoxylate; Sodium poly(oxyethylene) lauryl ether sulfate; Standapol ES 2; Texapon N40; Texapon N 40; Zetesol LES 2,. Detergent, foaming agent, wetting agent. *Witco Corporation.*

34478 Witcolate NH
2235-54-3 218-793-9
$C_{12}H_{29}NO_4S$
Ammonium lauryl sulfate
Sulfuric acid, monododecyl ester, ammonium salt; Akyposal ALS 33; Ammonium dodecyl sulfate; Ammonium n-dodecyl sulfate; Ammonium N-dodecyl sulfate; Ammonium Lauryl Sulfate; Ammonium lauryl sulfate; Dodecyl ammonium sulfate; Dodecyl sulfate, ammonium salt; Lauryl ammonium sulfate; Lauryl sulfate ammonium salt; Maprofix NH; Montopol LA 20; Neopon Lam; Presulin; Richonol AM; Sinopon; Sipex A; Sipon LA 30; Siprol I22; Siprol L22; Stepantan AM; Sterling AM; Sulfuric acid, lauryl ester, ammonium salt; Texapon A 400; Texapon Special. Detergent, foaming agent, and wetting agent. *Witco Corporation.*

34479 Witcolate TLS-500
139-96-8 205-388-7
$C_{12}H_{26}O_4S.C_6H_{15}NO_3$
TEA-lauryl sulfate
Sulfuric acid, monododecyl ester, compd. with 2,2',2''-nitrilotris[ethanol] (1:1); Akyposal TLS; Cycloryl TAWF; Cycloryl Wat; Dodecyl sulfate, triethanolamine salt; Drene; Elfan 4240 T; Emal T; Emersal 6434; Lauryl sulfate ester, triethanolamine salt; Lauryl sulfate, triethanolamine salt; Lauryl sulfuric acid, triethanolamine salt; Maprofix TLS; Maprofix TLS-500; Maprofix TLS-65; Maprofix TSL-65; Melanol lp20t; Melanol LP 20 T; Propaste T; Rewopol TLS-40; Richonol T; Sipon LT; Sipon LT-6; Sipon LT-40; Standapol T; Standapol TLS 40; Steinapol TLS-40; Stepanol WAT; Sterling WAT; Sulfetal LT; Sulfuric acid, dodecyl ester, triethanolamine salt; TEA-Lauryl sulfate; Texapon t-35; Texapon t-42; Texapon T-35; Texapon T-42; Texapon TH; Triethanolamine dodecyl sulfate; Triethanolamine lauryl sulfate; Triethanolamine monododecyl sulfate; Triethanolamine salt of lauryl sulfate; Tylorol LT-50,. Detergent, wetting agent, foaming agent. *Witco Corporation.*

34480 Witconate 605A
Calcium dodecylbenzene sulfonate
Industrial detergent, dispersant, oil-water and w/o emulsifier, lubricant, wetting agent, and demulsifier; used in agriculture and oilfield applications. *Witco Corporation.*

34481 Witconate 60T
TEA-dodecylbenzenesulfonate
Detergent, wetter, emulsifier, foaming agent; and base for household, industrial, and cosmetic/toiletry specialty compounds. *Witco Corporation.*

34482 Witconate 93S
Amine dodecylbenzene sulfonate
Detergent, detergent base, emulsifier, foamer, wetting agent, dispersant, and solubilizer for the detergent industry; biodeg. *Witco Corporation.*

34483 Witconate AOK
68439-57-6 270-407-8
Sodium C_{14}-C_{16} olefin sulfonate
Detergent, foaming agent, wetting agent. *Witco Corporation.*

34484 Witconate D24-25
Calcium alkylaryl sulfonate
Oil soluble emulsifier. *Witco Chemical Ltd.* Discontinued.

34485 Witconate DS
1322-98-1 215-347-5
$C_{16}H_{26}NaO_3S$
Sodium decylbenzenesulfonate
Benzenesulfonic acid, decyl-, sodium salt; Decylbenzene, sodium sulfonate; Decylbenzenesulfonic acid, sodium salt; Santomerse D; Santomerse-O; Sodium Decylbenzenesulfonamide; Sodium decylbenzenesulfonate; Ultrawet DS. Detergent, foaming agent, and wetting agent. *Witco Corporation.*

34486 Witconate LXH
Branched TEA-dodecylbenzene sulfonate
Detergent, foaming agent, wetting agent, and dispersant. *Witco Corporation.*

34487 Witconate NIS
1562-00-1 216-343-6
$C_2H_6NaO_4S$
Sodium isethionate
Ethanesulfonic acid, 2-hydroxy-, monosodium salt; Ethanesulfonic acid, 2-hydroxy-, sodium salt; 2-Hydroxyethanesulfonic acid sodium salt; Isethionic acid, sodium salt; Sodium 2-hydroxyethanesulfonate; Sodium β-hydroxyethanesulfonate; Sodium 2-hydroxyethyl sulfonate; Sodium isethionate. Detergent, foaming agent, and wetting agent. *Witco Corporation.*

34488 Witconate P-1059
Isopropylamine dodecylbenzene sulfonate
Emulsifier, solubilizer, detergent, and wetting agent for oil-based systems; dispersant in oil and waterbased systems; used in dry-cleaning surfactants; hydrotrope for liquid detergents; oilfield demulsifier. *Witco Corporation.*

34489 Witconate PTSA
Toluene sulfonic acid in crystal form; anionic surfactant with hydrotrope properties used as a catalyst. *Witco.* Discontinued.

34490 Witconate SCS 45%
32073-22-6 250-913-5
$C_9H_{12}NaO_3S$
Sodium cumene sulfonate
Benzene, (1-methylethyl)-, monosulfo deriv., sodium salt; Cumenesulfonic acid, sodium salt; Cyclophil SCS40; Sodium cumenesulfonate; Sodium cumolsulfonate. Hydrotrope, solubilizer, coupler and processing aid in detergent manufacturing and industrial processes; antiblocking and anticaking agent in powder products; formulates shampoos, aerosols, cutting oils, glue; textile finishing. *Witco/Organics.* Discontinued.

34491 Witconate STS
Sodium toluene sulfonate in liquid form; anionic surfactant. *Witco Chemical Ltd.* Discontinued.

34492 Witconate SXS 40%
1300-72-7 215-090-9
$C_8H_{10}NaO_3S$
Sodium xylene sulfonate
Benzenesulfonic acid, dimethyl-, sodium salt; Conco SXS; Cyclophil SXS30; Dimethylbenzenesulfonic acid, sodium salt; Eltesol SX 30; Hydrotrope; Naxonate; Naxonate G; NCI-C55403; Sodium dimethylbenzenesulfonate; Sodium Xylenesulfonate; Sodium xylenesulfonate; Stepanate X; Surco Sxs; Ultrawet 40SX; Xylenesulfonic acid, sodium salt. Hydrotrope, solubilizer, coupler and processing aid in detergent manufacturing and industrial processes; antiblocking and anticaking agent in powder products; formulates

shampoos, aerosols, cutting oils, glue; textile finishing. *Witco/Organics.* Discontinued.

34493 Witconate TX Acid
Modified toluene sulfonic acid; wetting agent, hydrotrope, coupler and solubilizer for liquid detergents; anticaking aid in dry neutralization; catalyst in organic reactions. *Witco/Organics; Witco SA.*

34494 Witconol
Emollient; for bath oils, hair preparations; cosmetic preparations. *Witco Corporation.*

34495 Witconol 14
9007-48-1
Polyglyceryl-4 oleate
1,2,3-Propanetriol, homopolymer, (Z)-9-octadecenoate; Demal 14; Emcol 12-14-18; Oleic acid, ester with 1,2,3-propanetriol homopolymer (1:1); Oleic acid polyglyceride; Plurol Oleique; Polyglycerol monooleate; Polyglycerol oleate; Polyglyceryl oleate. Water-oil and oil-water emulsifier, lubricant, emollient, spreader, sticker, and antifoamer for industrial use, aerosols. *Witco/Organics; Witco SA.*

34496 Witconol 2301
112-62-9 6965 203-992-5
$C_{19}H_{36}O_2$
Methyl oleate
methyl *cis*-9-octadecenoate, oleic acid methyl ester. Defoamer, lubricant, moisture barrier. *Witco Corporation.*

34497 Witconol 2326
123-95-5 1625 204-666-5
$C_{22}H_{44}O_2$
Butyl stearate
octadecanoic acid butyl ester; butyl octadecanoate. Lubricant. mp = 27°; bp = 343°; d_{25}^{25} = 0.855-0.875; slightly soluble in H_2O, soluble in EtOH, Et_2O. *Witco/Organics.* Discontinued.

34498 Witconol 2380
1323-39-3 215-354-3
$C_{21}H_{42}O_3$
Propylene glycol stearate
Octadecanoic acid, monoester with 1,2-propanediol; Stearic acid, monoester with 1,2-propanediol; Atlas G 924; Cerasynt PA; Cerasynt PN; Crill 26; Dragil-P; Emcol ps-50 rhp; Emcol PS-50 RHP; Emerest 2381; Monosteol; Monosteol TG; Noca; Nonex 32; Pegosperse PS; 1,2-Propanediol monostearate; Propylene glycol monostearate; 1,2-Propylene glycol monostearate; Propylene glycol stearate; Propylene glycol stearic acid, ester; Prostearin; Tegin P; USAF KE-13. Oil-water emulsifier, lubricant, and opacifier. *Witco/Organics.* Discontinued.

34499 Witconol 2400, Witconol MST, Witconol RHT
31566-31-1 4498 250-705-4
$C_{21}H_{42}O_4$
Glyceryl stearate
octadecanoic acid monoester with 1,2,3-propanetriol; monostearin. Oil-water emulsifier, lubricant. Emulsifier for cosmetic, pharmaceutical, aerosol formulations; internal lubricant, plasticizer, and emulsifier in industrial applications; flow control agent for polymerization reactions; dispersant. Lubricant, plasticizer, oil-water emulsifier used in industrial applications. *Witco/Organics; Witco SA.*

34500 Witconol 2421
111-03-5 203-827-7
$C_{21}H_{40}O_4$
Glyceryl oleate
9-Octadecenoic acid (Z)-, 2,3-dihydroxypropyl ester; Olein, 1-mono-; Aldo HMO; Aldo MO; Glycerin 1-monooleate; Glycerol 1-monooleate; Glycerol .alpha.-monooleate; Glycerol .alpha.-cis-9-octadecenate; Glyceryl Monooleate; 1-Glyceryl oleate; Monoolein; 1-Monoolein; .alpha.-Monoolein; 1-Monooleoylglycerol; 1-Oleoylglycerol; 1-Oleylglycerol. Defoamer, oil-water emulsifier, lubricant, moisture barrier. *Witco/Organics.* Discontinued.

34501 Witconol 2500
1338-43-8 8872 215-665-4
$C_{24}H_{44}O_6$
Sorbitan oleate
Sorbitan, mono-9-octadecenoate, (Z)-; Sorbitan, monooleate; Anhydrosorbitol monooleate; Arlacel 80; Arlacel A; Armotan MO; Emsorb 2500; Glycomul O; Ionet S-80; Liposorb o-20; Liposorb O; ML 33F; MI 55F; MO 55F; Monodehydrosorbitol monooleate; Montan 80; Nikkol SO-10; Nikkol SO-15; Nikkol SO-30; Nikkol SO 10; Nikkol so-15; Nikkol so-30; Nonion op80r; Nonion OP 80R; O 250; Oleic acid, monosorbitan ester; Radiasurf 7155; Sorbester P 17; Sorbitan, ester, mono-9-octadecenoate, (Z)-; Sorbitan monooleic acid ester; Sorbitan O; Sorbitan oleate; Sorgen 40; Span 80. Water-oil emulsifier, lubricant, coupling agent. *Witco/Organics.* Discontinued.

34502 Witconol 2503
26266-58-0 8872 247-569-3
$C_{60}H_{108}O_8$
Sorbitan trioleate
Sorbitan, tri-9-octadecenoate, (Z,Z,Z)-; Sorbitan, trioleate; Anhydrosorbitol trioleate; Aracel 85; Arlacel 85; Crill 5; Emasol 430; Emsorb 2503; Glycomul TO; Ionet S-85; Ionet S 85; Liposorb To; Nissan nonion OP-85; Nissan nonion op 85; Nissan nonion op 85r; Nonion OP 85R; Op 85r; Protachem Sto; Rheodol sp 030; Sorbitan, esters, tri-9-octadecenoate, (Z,Z,Z)-; Sorbitan, tris(9-octadecenoate), (Z)-; Span 85; TE-33. Water-oil emulsifier, lubricant, coupling agent. *Witco Corporation.*

34503 Witconol 2620
9004-81-3
PEG-4 laurate
Oil-water emulsifier, lubricant. *Witco Corporation.*

34504 Witconol 2622
9005-02-1
PEG-4 dilaurate
Oil-water emulsifier, lubricant. *Witco Corporation.*

34505 Witconol 2640
9004-99-3
PEG-8 stearate
Oil-water emulsifier, lubricant. *Witco/Organics.* Discontinued.

34506 Witconol 2642
9005-08-7
PEG-8 distearate
Oil-water emulsifier, lubricant. *Witco/Organics.* Discontinued.

34507 Witconol 2648
9005-07-6
PEG-8 dioleate
Oil-water emulsifier, lubricant, defoamer. *Witco Corporation.*

34508 Witconol 2720
9005-64-5 8872
Polysorbate 20
Oil-water emulsifier, dispersant, viscosity modifier, coupling agent. *Witco/Organics.* Discontinued.

34509 Witconol 2722
9005-65-6 7742
Polysorbate 80
Oil-water emulsifier, dispersant, viscosity modifier, coupling agent. *Witco/Organics.* Discontinued.

34510 Witconol 5906
61791-12-6
PEG-30 castor oil
Dispersant, oil-water emulsifier, lubricant. *Witco Corporation.*

34511 Witconol 6903
9005-70-3 8872
Polysorbate 85
Oil-water emulsifier, dispersant, lubricant, and coupling agent. *Witco Corporation.*

34512 Witconol APM
63793-60-2
PPG-3 myristyl ether; lubricant, emulsifier, wetting agent, penetrant, dye dispersant, scouring aid, antistat, solvent coupler; metal processing; textile surfactant; emollient oil for cosmetics and toiletries, solubilizer. *Witco/Organics; Witco SA.*

34513 Witconol APS
25231-21-4
PPG-11 stearyl ether; lubricant, emulsifier, wetting agent, penetrant, dye dispersant, scouring aid, antistat, solvent coupler; synthetic oils for personal care products; metal processing; textile surfactant; emollient oil for cosmetics and toiletries, so *Witco Corporation.*

34514 Witconol EGMS
Glycol stearate
Opacifier, conditioner. *Witco Corporation.*

34515 Witconol H31A
9004-96-0
PEG-8 oleate
Lubricant and plasticizer in oils and polymers; oil-water emulsifier for mineral and vegetable oils, and solvents, cosmetic and industrial applications; improves flow and leveling of coatings, increases spreadability of personal care products; defoamer. *Witco/Organics; Witco SA; Witco Israel.*

34516 Witconol H33
9005-07-6
PEG-8 dioleate
Industrial detergent; defoamer, oil-water emulsifier, lubricant. *Witco Corporation.*

34517 Witconol NP-100
9016-45-9 6772
Nonoxynol-10

Pigment dispersant, emulsifier, latex stabilizer, and leveling agent for polymerization reactions. *Witco Corporation.*

34518 Witconol NS-108LQ
Polyalkoxylated alkyphenol
Wetting agent, dispersant for agriculture aqueous suspension concs. *Witco/Organics; Witco SA.*

34519 Witcor 3630
Imidazoline
Oilfield surfactant, corrosion inhibitor. *Witco Corporation.*

34520 Witcor CI-1
Complex surfactant; oilfield surfactant, corrosion inhibitor. *Witco Corporation.*

34521 Witcor PC200
Amine alkylaryl sulfonate
Oilfield surfactant for paraffin inhibition. *Witco/Organics.* Discontinued.

34522 Witepsol®
Neutral hard fats based on mixtures of triglycerides; used for preparation of suppositories. *Hüls AG.*

34523 Witepsol® E75, E76, E85, H5, H12, H15, H19, H32, H35, H37, H39, H42, H175, H185
hydrogenated cocoglycerides
Suppository bases for hydrophilic and lipophilic drugs. *Hüls Am.*

34524 Witflow 60
Alkylaryl polyether alcohol
Pigment wetting agent, spreading agent, flow modifier for paints/coatings. *Witco Corporation.*

34525 Witflow 901
Sodium diester sulfosuccinate
Pigment wetting agent, dispersant, flow modifier for paints/coatings. *Witco Corporation.*

34526 Witflow 916
9004-96-0
PEG-400 oleate
Leveling agent, flow modifier, and defoamer for paints/coatings. *Witco/Organics.* Discontinued.

34527 Witflow 950
Polypropoxy quaternary ammonium chloride
Pigment wetting, grinding and suspension aid for oilbased paints/coatings. *Witco Corporation.*

34528 Witflow 990
9005-64-5 8872
Polysorbate 20
Oil-water emulsifier, antistatic finishing agent for paints/coatings. *Witco/Organics.* Discontinued.

34529 Witflow 991
9005-65-6 7742
Polysorbate 80
Pigment dispersant for paints/coatings. *Witco/Organics.* Discontinued.

34530 Withnell powder
An explosive containing 88-92% ammonium nitrate 4-5% trinitrotoluene, and 4-6% flour.

34531 Witocan®
Fats based on coconut and palm kernel oils; hard fat with neutral flavor; melts very rapidly in the mouth without leaving any unpleasant or greasy taste; prepares products which are easy to cut and bite and remain firm. *Hüls AG.*

34532 Witsol
Aliphatic ink solvents. *Witco Corporation.*

34533 Wittenburg weather dynamite
An explosive, consisting of 25% nitro-glycerin, 34% potassium nitrate, 38.5% rye meal, 1% wood meal, 1% barium nitrate and 0.5% sodium bicarbonate.

34534 Wittol wax
A wax possessing similar properties to beeswax. It is a proprietary material, and is suitable for acid-proof linings. No manufacturer.

34535 woad
A dark, clay-like preparation made from the leaves of the woad plant, *Isatis tinctoria* ; used for the purpose of exciting fermentation in the indigo vat.

34536 Wolf N Lamid IG
A proprietary polyamide resin soluble in alcohol; used as a base for varnishes. *NL Victor Wolf Ltd.* Name unverified.

34537 Wolfaid
Saturated and unsaturated polyesters (liquid and solid); processing aids for PVC flooring, pigment dispersing aid for unsaturated polyester. *NL Victor Wolf Ltd.* Name unverified.

34538 Wolfamid
Nonreactive polyamide resins; for packaging inks, cold seal release lacquers, thermographic systems, thixotropic alkyd resins. *NL Victor Wolf Ltd.* Name unverified.

34539 Wolfert
A rubber substitute consisting of felt which has been impregnated with a vulcanized oil.

34540 Wolfin 18
Thermoplastic, bitumen proof insulating foil for structures; for flat roofs, foundation insulation, protection against leakage oil, protection against seepage and pressure water. *Degussa AG.* Name unverified.

34541 Wolfkur
Reactive polyamide resins, polyamino amides, and amine adducts; epoxy curing agents for coatings and adhesives. *NL Victor Wolf Ltd.* Name unverified.

34542 Wolflex
Saturated polyesters; polymeric plasticizers for PVC plastics. *NL Victor Wolf Ltd.* Name unverified.

34543 Wolfol
Saturated polyesters with hydroxyl groups; polyester polyols for polyurethanes used in elastomers and adhesives. *NL Victor Wolf Ltd.* Name unverified.

34544 Wolframium
An alloy of 98% aluminum, 1.4% antimony, 0.4% copper, 0.1% tin, and 0.04% tungsten.

34545 Wollastocoat®
13983-17-0 1749 237-772-5
H_2O_3CaSi
calcium silicate
Wollastonite; Aedelforsite; Cab-o-lite; Cab-O-Lite; Cab-o-lite 100; Cab-o-lite 130; Cab-o-lite 160; Cab-o-lite F 1; Cab-O-Lite F 1; Cab-O-Lite P 4; Calcium silicate; Casiflux; Casiflux vp 413-004; Dab-o-lite P 4; F 1; FW 325; FW 50; Fw 200 (mineral;)Gillebachite; NCI-C55470; Nyad; Nyad 10; Nyad 325; Nyad G; Nycor; Nycor 200; Nycor 300; Okenite; Rivaite; Schalstein; Tabular Spar; Tremin; Vansil; Vansil W 10; Vansil W 20; Vansil W 30; Vilnite; Wollastokup; Wollastonite; Wollastonite calcium silicates. Family of surface-modified wollastonite products; high aspect ratio grades, milled grades and fine particle size grades; $CaSiO_3$; used in polymer composites, coatings, adhesives, elastomers and friction products. *NYCO® Minerals Inc.*

34546 Wollastokup®
13983-17-0 1749 237-772-5
calcium silicate
Chemically coupled wollastonite, high aspect ratio and fine particle sizes: CaO, SiO_2, Fe_2O_3, Al_2O_3, MnO, MgO, and TiO_2; used for polymer composites, high performance coatings, adhesives, elastomers, and friction products. *Malvern Minerals.*

34547 wollastonite
13983-17-0 1749 237-772-5
Calcium metasilicate
Inert filler, extender pigment for paints. *Am. Colloid; Cornelius Chemical Co Ltd; Nyco Min.; R. T. Vanderbilt Co Inc.*

34548 Wollaston's cement
Consists of 1 part beeswax, 4 parts resin, and 5 parts plaster of Paris; used for fossils.

34549 Wolle
An abbreviation for collodium-wolle nitrocellulose in various viscosities.

34550 wood flour
Wood meal; wood flock. Finely powdered wood, usually white pine; used as a rubber, linoleum, or soap filler; as a filler in dynamite; in fur cleaning; in polishing agents.

34551 wood oil
The final fractions obtained in the distillation of wood spirit, containing high boiling ketones.

34552 wood-apple gum
(Katbel-ki-gond, velampishin, kapithamia piscum). The gum of *Feronia elephantum* .

34553 wood-cloth
Strips of wood treated with sulfurous acid or alkaline bisulfite, making the fiber stronger.

34554 Wood's alloys
Low-melting alloys of bismuth, tin and lead, usually containing cadmium. One alloy contains 50% bismuth, 27% lead, 13% tin and 10% cadmium; another contains 50% bismuth, 25% lead, and 25% tin; a third, contains 50% bismuth, 25% lead, 12.5% tin, and 12.5%.

34555 Woodthane WTP 311-6
Rigid polyurethane systems; for dispensing froth foams through low or high pressure pour machines or hand mix; for furniture, mirror and picture frames, decorative molding, and carvings. *Flexible Prods.*

34556 Woodtones
Woodstain for interior use. *ICI Chem & Polymers Ltd.*

34557 Worm Ender
A natural occurring biological insecticide used for the control of worms and caterpillars on vegetables, ornamentals, fruit trees, shade trees, etc.; may be used up to the day of harvest. *Lawn & Garden Products Inc.*

34558 wormwood
absinthium. The dried leaves and flowering-tops of *Artemisia absinthium* . Contains absinthin. Used as a flavoring agent in alcoholic beverages such as vermouth.

34559 wort
Malt is crushed and heated with water until the starch is converted into sugar by the diastase in the malt. The resulting liquid is known as wort.

34560 Wovco SP
Trademark for a range of polyethylene plastics reinforced with carbon fiber. *Worcester Valve Co, Haywards Heath.* Name unverified.

34561 WR Base
Fatty melamine condensate; used to make waxmelamine textile water repellents. *Clark.*

34562 WRA 1000 Varnish
Liquid polyurethane resin and hardener; used for varnishing directly onto wood or on top of epoxy coatings; contains a uv inhibitor. *Wessex Resins & Adhesives Ltd.* Discontinued.

34563 WRA Epoxy Resin Underwater Series
Liquid and putty epoxies; specifically formulated for underwater use; for general bonding, filling and fairing. *Wessex Resins & Adhesives Ltd.*

34564 WRA System 100 Laminating Composition
Liquid epoxy resin and hardener; laminating resin specifically for use with glass cloth, carbon fiber, aramid and hybrids. *Wessex Resins & Adhesives Ltd.* Discontinued.

34565 WRA System 17
Thixotropic resin and hardener; general purpose adhesive (bonding, laminating), bonding wood, concrete, most metals, stone, china, GRP, unglazed ceramics, rubber. *Wessex Resins & Adhesives Ltd.*

34566 WRA System 80
PVA crosslinking liquid; general purpose wood glue; waterproof; complies with DIN68602 Section B and BS4071. *Wessex Resins & Adhesives Ltd.* Discontinued.

34567 Wresacryl
Hydroxy acrylic resins for use with polyisocyanates; used for automotive paints and industrial finishes. *Resinous Chemicals Ltd.*

34568 Wresilac
Etherified melamine/formaldehyde resins; used for stoving finishes and acid curing lacquers. *Resinous Chemicals Ltd.*

34569 Wresinate
Metal resinates; for marine antifoulings; improves gloss and drying in alkyd finishes. *Resinous Chemicals Ltd.*

34570 Wresinite
Maleinized rosin esters; used for nitrocellulose lacquers, printing inks and to improve gloss in air drying paints. *Resinous Chemicals Ltd.*

34571 Wresinol
Alkyd resins (oil modified polyesters); used for air drying decorative paints, air drying, and stoving industrial finishes. *Resinous Chemicals Ltd.*

34572 Wresinyl
Alkyl phenol/formaldehyde resins, terpene phenolic resins; used for oil varnishes and adhesives. *Resinous Chemicals Ltd.*

34573 Wresipol
Unsaturated polyester resin in styrene; used for glass fiber reinforced laminate, casting and potting. *Resinous Chemicals Ltd.*

34574 Wright's stain
A microscopic stain for white blood corpuscles consisting of 1 g methylene blue eosin mixture in 600 ml methyl alcohol.

34575 Wukonil
Paraffin wax emulsions. *Suddeutsche Emulsions GmbH.*

34576 Wurster's blue
100-22-1 9367 202-831-6
$C_{10}H_{16}N_2$
tetramethyl-*p*-phenylenediamine
N,N,N',N'-tetramethyl-1,4-phenylenediamine; Wurster's reagent;. An oxidation product of tetramethyl-p-phenylenediamine; used as an indicator. mp = 51-52°; bp = 260°; soluble in H_2O, freely soluble in organic solvents.

34577 wurtzillite
Tabbyite; egenite; eonite. An asphaltic mineral, soluble in hot water.

34578 Wytox® 312
26523-78-4 247-759-6
$C_{45}H_{69}O_3P$
Tris (nonylphenyl) phosphite
Phenol, nonyl-, phosphite (3:1); Irgafos TNPP; JP 351; Mark 1178; Mark 829; Nonylphenyl phosphite; P 3; P 3 (antioxidant); Phosphorous acid, tris(nonylphenyl) ester; Tri(mononony lphenyl) phosphite; Trinonylphenol phosphite; Tri(nonylphenyl) phosphite; Tris(nonyl phenyl) phosphite; UVI-NOX 3100. Antioxidant for PE, PP, PS, PVC, ABS, nylon, food packaging; processing, and color stabilizer. A non staining, nondiscoloring, low volatility antioxidant for polyolefins, vinyl chloride polymers, high impact polystyrenes, etc. *Uniroyal; National Polychemicals.*

34579 Wytox® 335
A modified polymeric phosphite stabilizer for emulsion type styrene-butadiene polymers; outstanding suppression of gel build-up; exceptionally resistant to hydrolysis. *National Polychemicals.* Unverified.

34580 Wytox® ADP
Alkylated diphenylamine
An antioxidant for rubber for protection against heat aging and flex cracking. *National Polychemicals. Unverified.*

34581 Wytox® BHT
Alkylated *p*-cresol
A nonstaining antioxidant for plastics. *National Polychemicals.* Unverified.

34582 Wytox® LT
123-28-4 204-614-1
Dilauryl thiodipropionate
An antioxidant suitable for use in plastics of the polyolefine and ABS type in contact with food. *National Polychemicals.* Unverified.

34583 X.L
High boiling tar acids. *Coalite Fuels & Chemicals Ltd.* Name unverified.

34584 X50-S
Si69 and N330 carbon black (1:1); reinforcing filler for rubber industry. *Degussa.*

34585 X-743
Aromatic plasticizer; aromatic, chemically inert, nonsaponifiable plasticizer exhibiting low reactivity; used in adhesives (mastic, pressure sensitive), rubber (cements, mechanical and molded goods, tires), and caulking compounds. *Neville.*

34586 Xanco-Frac®
Heteropolysaccharide product; gum for use as a viscosifier in oil field hydraulic fracturing fluids; suspending agent. *Kelco/Oil Field Prods.*

34587 Xanflood®
11138-66-2 10191 234-394-2
polysaccharide B-1459; Kelrol F; Kelzan; xanthan gum; Corn sugar gum. Industrial-grade xanthan gum; foam stabilizer, flocculant, suspending, gelling agent, rheology modifier, lubricant for industrial applications especially for secondary and tert. oil recovery. High molecular weight hetero polysaccharide gum produced by a pure-culture fermentation of a carbohydrate with xanthomonas campestris; thickening and suspending agent; in drilling fluids, ore flotation, foods, and pharmaceuticals mw > 10^6; *Kelco; Kelco Int'l; Meer.*

34588 Xantalgin®, Xanthano®
Alginate impression materials; dental preparation. *Bayer AG.*

34589 xanthene
92-83-1 202-194-4
$C_{13}H_{10}O$
dibenzopyran, tricyclic
9H-Xanthene; Xanthene; Dibenzo[a,e]pyran; Dibenzopyran, tricyclic; 10H-9-Oxaanthracene; Xanthane. Used in organic synthesis and as a fungicide. mp= 101-102° *Aldrich; Schweizerhall.*

34590 xanthine
69-89-6 10193 200-718-6
$C_5H_4N_4O_2$
3,7-dihydro-1H-purine-2,6-dione
dioxopurine; 2,6-dihydroxypurine. Organic synthesis, medicine. Soluble in H_2O (1 g/14 l); less soluble in EtOH *Am. Biorganics; Dajac Labs; U.S. Biochemical.*

34591 xanthophyll
127-40-2 10197 204-840-0
$C_{40}H_{56}O_2$
β,ε-carotene-3,3'-diol
carotenoid; lutein; vegetable lutein; vegetable luteol; Bo-Xan. The yellow pigment occurring in green vegetation and some animal products. mp = 190°; $[\alpha]^{18}_{Cd}$ = 165° (c = 0.7, C_6H_6); λ_m = 481, 453, 429, 333, 268 nm (ε 142000, 152000, 100000, 15500, 35000); insoluble in H_2O, soluble in organic solvents.

34592 Xanthopicrin
A yellow coloring matter from the bark of *Xanthoxylum caribgum* .

34593 Xantopren®, Xantopren® Function; Xantopren® Plus, Xantygen®
Precision impression materials on an elastomer basis in several different viscosities; used in dentistry. Xantopren® Plus is a silicone based material; Xantygen® is a thermoplastic. *Bayer AG.*

34594 XAS 10961.01
Polybutylene oxide polyol
High hydrophobicity polyol for formultion of improved moisture barrier polyurethane adhesives, sealants and elastomers in construction, marine and electronics markets. *Dow Plastics.*

34595 XEA 9361
Two-component adhesive; room-temperature curing adhesive with high shear and peel strength and flexibility; for general purpose bonding and applications requiring high elong., e.g., sealing and cryogenic applications *Hysol Aerospace Prods.*

34596 Xenacryl®
Acrylic resin; used for surface coatings, adhesives and sealants. *Baxenden Chemicals Ltd.*

34597 Xenalak®
Acrylic polymer in solution; used for surface coatings. *Baxenden Chemicals Ltd.*

34598 Xenith
Blend of polymers, amine, and solvents; zinc electroplating additive. *Taskem Inc.* Name unverified.

34599 xenon
7440-63-3 10206 231-172-7
Xeneisol 133A; Xenomatic. A heavy inert gaseous element present in minute quantities in the atmosphere; luminescent tubes, flash lamps in photography, fluorimetry, lasers, tracer studies, and anesthesia. Xeneisol 133A, Xenon Xe 133-VSS and Xenomatic are proprietary trade names for the 133 isotope of xenon *Air Prods & Chem; Electrochem Ltd; Liquid Air Corp.; Mallinckrodt Inc.*

34600 Xenoy®
Polymer alloys; used for plastic components for automotive, electrical electronics, lighting, medical, packaging, audio etc. *GE Plastics Ltd.*

34601 Xenoy® 2230, 5220, 6120
Thermoplastic alloys, PC-based. Impact modified; alloy offering dimensional stability, mechanical performance. *GE Plastics.*

34602 Xenoy® 5770
PC/polyester alloy, 20% glass/mineral-rein-forced; impact-modified grade with chemical resistance, good surface and low warpage; for lawn and garden applications. *GE Plastics.*

34603 Xenoy® DX2735
Thermoplastic alloy; unreinforced blend offering good chemical and heat resistance, excellent low temperature impact performance. *GE Plastics.*

34604 Xenoy® DX6125
Thermoplastic engineering structural foam resin; used for material handling, industrial, and recreational applications requiring toughness, chemical resistance, heat resistance, and load carrying capabilities. *GE Plastics.*

34605 Xeroder® S100, L67
Silicone-based water repellent agent, especially for chrome leathers and chrome suede; used for leather industry. *Bayer AG; Bayer plc.*

34606 X-Ite
A proprietary alloy containing 37-39% nickel and 17-19% chromium with iron. No manufacturer.

34607 XL-All Insecticide
54-11-5 6611 200-193-3
$C_{10}H_{14}N_2$
3-(1-methyl-2-pyrollidinyl)pyridine
Nicotine; 1-methyl-2-(3-pyridiyl)pyrrolidine; β-pyridyl-α-N-methylpyrrolidine. Component of leaves of *Nictoiana tabacum* and *N. rustica.* An alkaloid insecticide. bp = 243-248°; d = 1.017; n^{20}_D = 1.5265; $[\alpha]^{20}_D$ = -168±5° (c = 5, H_2O); miscible with H_2O, soluble in organic solvents. *Synchemicals Ltd.*

34608 XP Pastes
Plasticizer dispersions of organic pigments. *Reckitts Colours Ltd.*

34609 XP-4
Phosphate animal feed supplement. *FMC.* Discontinued.

34610 XR 7
Semipermanent mold release agent; forms a tough, durable film with excellent adhesion to mold surface; gives easy and multiple releases with most thermosetting rubber and plastic materials. *Releasomers.*

34611 XSA 80
25321-41-9 246-839-8
Xylene sulfonic acid
Catalyst for resins. *Hart Chem. Ltd.*

34612 XT® Polymer 250, XT® Polymer 375
Acrylic multipolymers; polymer sheet for thermoformed packaging applications requiring transparency and toughness at an economical price; suited for the food market where grease-proofness, good gas barrier characteristics, and freedom from odor or taste transfer are mandatory, can be hot-filled, contains no plasticizers to leach out; used in tubs for margarine and dairy products, lids, portion packs, candy trays, meat trays etc.; FDA-approved for food contact use; 250 grade has lower impact strength but is more rigid than 375. *Cyro Industries.*

34613 X-Tan® Special C
Textile auxiliary which suppresses chlorine release in chlorite bleaching; prevents metal corrosion. *Sybron.*

34614 Xtol
61790-12-3 263-107-3
Tall oil fatty acid. *Georgia-Pacific.*

34615 XXX-1
Lubricant for food processing conveyors. *CasChem.*

34616 Xydar® G-540 LCP
Liquid crystal polymer, glass-reinforced; injection molding resin for electrical and electronic applications; high strength and stiffness, high heat deflection temperature, inherent flame resistance, low warpage, excellent weld line strength. *Amoco Chemical Co.*

34617 Xylan 1052
A proprietary lubricant based on PTFE, for use under extreme pressures and in extremes of temperature. *Whitfield Plastics.* Name unverified.

34618 Xylan 330
A proprietary aerosol form of polytetrafluoroethylene used as a mold-releasing agent in plastics processing. *Whitfield Plastics.* Name unverified.

34619 xylazine
7361-61-7 10213 230-902-1
$C_{12}H_{16}N_2S\cdot HCl$
N-(2,6-dimethylphenyl)-5,6-dihydro-4H-1,3-thiazin-2-amine hydrochloride
Narcoxyl; Rompun; Xylapan; Xylasol. Sedative, analgesic, muscle relaxant, used in veterinary medicine. (free base) mp = 140-142°; insoluble in H_2O, soluble in organic solvents; LD_{50} (rat orl) = 130 mg/kg.

34620 xylene
1330-20-7 10214 215-535-7
C_8H_{10}
methyl toluene
dimethylbenzene; xylol. Aromatic compound.; commercial mixture of 3 isomers; *o-, m-, p*-xylene; solvent; raw material for production of benzoic acid, phthalic anhydride, dyes, other organics; aviation gasoline; protective coating; solvent for alkyd resins, lacquers, enamels bp = 137-144°F; d = 0.860; n^{20}_D = 1.4970. Ashland; Crowley; Exxon; Fina; Mallinckrodt; Mitsubishi Petrochem.; Mitsui Petrochem. Ind.; Mobil; Shell; Texaco; Unocal.

34621 Xylene sulfonic acid
25321-41-9 246-839-8
Eltesol® XA; Xylene sulfonic acid , aqueous solution; Manro XSA; Reworyl® X; XSA 80. Catalyst in foundry and chemical industries; hydrotrope for agricultural formulations. *Albright & Wilson UK; Manro Products Ltd; Rewo Chemicals Ltd; Hart Chem. Ltd.*

34622 *m*-xylene
108-38-3 203-576-3
C_8H_{10}
m-xylene
1,3-dimethylbenzene. Solvent and raw mateial for preparation of organic compounds. bp = 138-139°; d = 0.866; n^{20}_D= 1.4970.

34623 *o*-xylene
95-47-6 202-422-2
C_8H_{10}
o-xylene
1,2-dimethylbenzene. Solvent and raw mateial for preparation of organic compounds. mp = -25°; bp= 143-145°; d= 0.877; n^{20}_D= 1.5050;.

34624 *p*-xylene
106-42-3 203-396-5
C_8H_{10}
p-xylene
1,4-dimethylbenzene. Solvent and raw mateial for preparation of organic compounds.

34625 xylenol
1300-71-6 10215 215-089-3
$C_8H_{10}O$
dimethylphenol
hydroxydimethylbenzene; cresylic acid; dimethylhydroxybenzene. Mixture of 6 possible isomers. Disinfectants, solvents, pharmaceuticals, insecticides, fungicides, plasticizers, rubber chemicals, additives to lubricants and gasoline, manufacture of polyphenylene oxide, wetting agents, and dyestuffs. *Coalite Chem. Div.; Crowley; Crowley Tar Prods.*

34626 Xyligen®
Active ingredients for wood preservative formulations; protective agents against wood-destructive fungi; additives to glues to protect wood-based materials. *BASF AG.*

34627 Xylite
A proprietary rubber vulcanization accelerator; It is a tarry diphenylguanidine. No manufacturer.

34628 Xylock 225
A proprietary condensation product of phenols with an aryl alkyl ether; used as a high-performance, heat-stable molding resin. *Albright & Wilson Ltd.* Name unverified.

34629 Xylok
High performance resins. *Albright & Wilson Ltd., Phosphates & Speciality Business.* Name unverified.

34630 xylol
Commercial xylene. It consists of a mixture of about 60% *m*-xylene, 10-25% *o- and p*-xylene, ethylbenzene, and small quantities of trimethyl-benzene, paraffin, and thioxene.

34631 Xylon FR
A proprietary nylon containing a flame-retarding additive. *Dart Industries Inc.* Name unverified.

34632 Xylose
58-86-6 10220 200-400-7
$C_5H_{10}O_5$
D-(+)-xylose
Wood sugar; Xylomed; Xylopfan. Dyeing, tanning, diabetic food, source of ethanol. mp = 144-145°; d^{20}_4 = 1.525; $[\alpha]^{20}_D$ = 92° → 18.6° (c = 10, 16 hours); soluble in H_2O, poorly soluble in organic solvents. *Aldrich; Am. Biorganics; Penta Mfg.*

34633 Xyron®
Modified polyphenylene ether resin. *Ashahi Chem. Industry.*

34634 yaba bark
The bark of *Andira excelsa.*

34635 Yalloy
An aluminum alloy containing 4% copper, 2% nickel, and 1.5% magnesium with small amounts of iron and silicon. It has a specific gravity of 2.8; used for die-cast pistons, etc.

34636 Yaltox®
1563-66-2 1851 216-353-0
A free flowing granule containing 5% w/w carbofuran; soil-applied insecticide for control of soil and seedling pests including cabbage root fly, cabbage stem weevil, flea beetle, cabbage stem flea beetle, early aphids in brassicas, turnip root fly, frit fly, millipedes, symphilids, beet leaf miner, springtails, wireworms, free living nematodes, potato cyst eelworm, carrot fly and carrot willow aphid. *Bayer AG; Bayer plc.*

34637 yama-mai silk
A silk produced by the Japanese oak caterpillar of Japan, China, and India.

34638 Yarmor
Pine oil terpene liquids; antiskinning agents in protective coatings, for pigment grinding aids and as an antifoam agent. *Hercules.*

34639 yeast
68876-77-7
(Barm) Fungus with unicellular growth form; fermentation of sugar, molasses, cereals for alcohol; brewing; baking; food supplement; protein biosynthesis; source of vitamins, enzymes, nucleic acids; biochemical research. *Champlain Industries; Gist-Brocades Food Ingreds.*

34640 yeast extract
8013-01-2 232-387-9
Saccharomycetacae
Extract of yeast. Used in fermentation of sugars, etc., brewing, baking, food supplement. *Atomergic Chemetals; Champlain Industries; Gist-Brocades Food Ingreds.*

34641 Yeast Lactase L-50,000
Yeast lactase; enzyme for hydrolyzing lactose in dairy products (milk, whey, cheese, yogurt), pharmaceuticals (digestive aids). *Solvay Enzymes.*

34642 Yellow 201
Iron oxides, bismuthoxychloride, inorganic colorant. *Presperse.* Discontinued.

34643 yellow bark
The bark of *Cinchona calisaya.*

34644 yellow carmine
Italian pink; yellow lake. Pigments perpared by precipitating the glucoside quarcitrin with alumina.

34645 yellow gold
An alloy of 53% gold, 25% silver, and 22% copper.

34646 yellow liquors
The drainage from alkali waste heaps.

34647 yellow prussiate of potash
14459-95-1 7793 237-323-3
$C_6FeK_4N_6 \cdot 3H_2O$
potassium ferrocyanide trihydrate
Ferro prussiate of potassium; potassium ferrocyanide; potassium hexkis(cyano-C)ferrate(4-) trihydrate; potassium hexacyanoferrate(II;)Anhydrous form, RN 13943-58-3. Used for tempering steel, engraving, and as a lab reagent. d = 1.85.

34648 yellow prussiate of soda
13601-19-9 8760 237-081-9
$C_6FeN_6Na_4 \cdot 10H_2O$
sodium ferrocyanide decahydrate
tetrasodium hexakis(cyano-C)ferrate(4-); sodium hexacyanoferrate(II); sodium prussiate yellow. Used in ore flotation, photography, as anti-caking agent; catalyst in emulsion polymerization. mp = 81.5°, dec 435°; soluble in H_2O (14%), insoluble in organic solvents.

34649 yellow soda ash
A soda ash (sodiumcarbonate) containing traces of ferric oxide.

34650 Yellow sulphur
7704-34-9 9142 231-722-6
Brimstone; sulphur. Sulfur powder; garden fungicide. Used to manufacture sulfuric acid. mp = 114°; bp = 445°; d = 2.0700; *Vitax Ltd.*

34651 yellow wax
A viscous, semi-solid, involatile substance obtained by the distillation of the still residues of petroleum oil. It contains polycyclic aromatic hydrocarbons such as anthracene.

34652 Yellowstone
1302-78-9 1082 215-108-5
$Al_2O_3 \cdot 4SiO_2 \cdot H_2O$
Wyoming sodium bentonite
hydrated aluminum silicate; Wilkinite; montmorillonite. Used for steel production and foundries. *Bromhead & Denison Ltd.*

34653 yenshee
The dregs and carbonized opium which remains after smoking opium. Contains 1-10% morphine.

34654 Yeoman
Lanolin BP. *Croda Chem. Ltd.*

34655 yerba mate
Paraguay tea; Jesuit's tea; St. Batholomew's tea. Leaves of a tree found in Paraguay, containing caffeine and tannin.

34656 Yieldmaster®
Fibers. *DuPont UK.*

34657 ylang-ylang oil
10233
Cananga oil. Essential oil distilled from the flowers of *Conanga odorata* ; used in perfumery. Consists primarily of geraniol and linalool esterified with acetic and benzoic acids. d^{20}_{20} = 0.930-0.950; $[\alpha]_D$= -27 - -50°.

34658 Yodoxin
83-73-8 5063 201-497-9
$C_9H_5I_2NO$
5,7-diiodo-8-quinolinol
Iodoquinol; diiodohydroxyquin; diiodo-oxyquinoline; SS 578; Diodoquin; Di-Quinol; Disoquin; Floraquin; Dyodin; Dinoleine; Searlequin; Diodoxylin; Meobiquin; Rafamebin; Ioquin; Direxiode; Stanquinate; Quniadome; Zoaquin; Enterosep; Embequin. Anti-amebic. mp = 200-215°; insoluble in H_2O, slightly soulbe in organic solvents *Glenwood Inc.*

34659 Yoloy
A proprietary alloy; it is a steel containing 1% copper, 2% nickel, and up to 0.2% carbon. No manufacturer.

34660 Yonckite
A Belgian explosive consisting of ammonium perchlorate, ammonium nitrate, sodium nitrate, and trinitrotoluene or nitronaphthalene.

34661 Yoracryl Dyes
Cationic dyes for acrylic, modified acrylic and other cationic dyeable fibers such as cationic dyeable polyester and nylon. No manufacturer.

34662 York Krystal Kleer Castor Oil
Castor oil. Used in lipsticks as a dye solvent and to impart gloss and emollience. *United Catalysts.*

34663 York USP Castor Oil
Castor oil USP. Used in lipsticks as a dye solvent and to impart gloss and emollience. *United Catalysts.*

34664 Yorkshire grease
Wakefield grease. The recovered fatty acids from wool grease.

34665 Y-Pof
Solvent based graffiti remover. *Kalon Chemicals Ltd.*

34666 YSE-Cure B-001
Amine adduct; epoxy curing agent for adhesives and coatings. *Ajinomoto.*

34667 YSE-Cure B-002
Amine adduct. Used as an epoxy curing agent for coating, flooring and casting. *Ajinomoto.*

34668 YSE-Cure B-002, B-002W, B-003
Amine adduct; epoxy curing agent for coating, flooring, casting; B-002W is an

anti-crystallization version of B-002; B-003 is a clear colorless version of B-002. *Ajinomoto.*

34669 YSE-Cure B-003
Amine adduct. Used as an epoxy curing agent for coating, flooring and casting. Clear, colorless version of B-002. *Ajinomoto.*

34670 YSE-Cure C-002, F-100
Amine adduct; epoxy curing agent for lining and adhesives; acid resist. F-100 serves as raw material for epoxy curing agents and imide compounds. *Ajinomoto.*

34671 YSE-Cure F-100
Amine. Raw material for epoxy curing agents and imide compounds. *Ajinomoto.*

34672 YSE-Cure LX-1N
Modified amine. Epoxy curing agent for lining and potting. Resistantto heat and chemicals. *Ajinomoto.*

34673 YSE-Cure N-001, N-002, LX-1N
Amine adduct; epoxy curing agent elec. potting applications; LX-1N is an epoxy curing agent for adhesives, grout and coating; flexible. *Ajinomoto.*

34674 YSE-Cure PX-3
Modified amine. Inexpensive epoxy curing agent for grout and lining (tar-epoxy). *Ajinomoto.*

34675 YSE-Cure PX-3, QX-2, QX-3, RX-2, RX-3
Epoxy curing agents for lining, adhesives, and putty; fast curing. PX is a modified amine; QX is a thiourea condensate; RX-2, 3 are accelerated modified amines. *Ajinomoto.*

34676 YSE-Cure RX-2
Accelerated modified amine. Fast acting epoxy curing agentfor lining and adhesives. *Ajinomoto.*

34677 YSE-Cure RX-3
Accelerated modified amine. Fast acting epoxy curing agent for lining and flooring. *Ajinomoto.*

34678 YSE-Cure S-002
Amine adduct solution; epoxy curing agent; primer for metal, concrete, mortar, and wood; excellent adhesion. *Ajinomoto.*

34679 Y-Tack
Polyglycol ether and polyacrylic copolymerides in aqueous solution; warp and weft treatment chemicals which provide a combination of film forming yarn cover, intra fiber bonding, lubrication and static control. *Thomas Swan & Co Ltd.* Discontinued.

34680 ytterbium
7440-64-4 10239 231-173-2
Yb
Ytterbium
Metallic element, valencies 2 and 3; used in lasers, dopant for garnets, portable x-ray source, chemical research. d = 6.977; has face-centered or body-centered cubic structure *Aldrich; Atomergic Chemetals; Cerac; Rhône-Poulenc Basic.*

34681 ytterbium oxide
1314-37-0 10239 215-234-0
O_3Yb_2
Ytterbium oxide
ytterbia. Special alloys, dielectric ceramics, carbon rods for industrial lighting, catalyst, and special glasses. soluble in dilute acids *Atomergic Chemetals; Noah Chem.; Rhône-Poulenc Basic.*

34682 Ytterbium Yb-169 DTPA
12111-24-9 7265 235-169-1
Pentetate calcium trisodium Yb 169
Radioactive-tagged chelating agent. *3M.* Discontinued.

34683 yttrium
7440-65-5 10240 231-174-8
Y
Yttrium
Metallic element; nuclear technology, iron and other alloys, deoxidizer for vanadium and other nonferrous metals, microwave ferrites, coating on high-temperature alloys, and special semiconductors. mp = 1509°; bp ·e 3000°; d = 4.472 *Atomergic Chemetals; Cerac; Noah Chem.; Rhône-Poulenc Basic.*

34684 yttrium oxide
1314-36-9 10240 215-233-5
Y_2O_3
yttria
Phosphors for color TV tubes, yttrium-iron garnets for microwave filters, stabilizer for high-temperature service materials. d = 5.03; LD_{50} (rat ip) = 500 mg/kg *Atomergic Chemetals; New Metals & Chems. Ltd; Noah Chem.; Rhône-Poulenc Basic; Shin-Etsu Chem.*

34685 Yukalon
Trademark for a proprietary grade of polyethylene. *Mitsubishi Petrochem.* Name unverified.

34686 Z Span Spansule Capsule
7446-19-7 10293
zinc sulfate monohydrate
Sustained release preparation of zinc sulfate for use where inadequate diet calls for supplementary zinc and where treatment of zinc deficiency is indicated. *SmithKline Beecham.*

34687 Z.P.D
The zinc salt of pentamethylene-dithiocarbamic acid; an ultra-rubber vulcanization accelerator.

34688 zakin rubber
A rubber-like substance prepared from glue or similar material.

34689 Zam Metal
A proprietary alloy of zinc with aluminium and magnesium. No manufacturer.

34690 Zamak Alloys
Proprietary alloys of zinc with aluminium and sometimes small amounts of copper and magnesium; copper-aluminium-zinc alloys, suitable for die castings contain: 3.9-4.3% aluminium, 0.9-2.9% copper, 0.003-0.06% magnesium, and the rest zinc; SG 6.64-6.7. No manufacturer.

34691 Zam-Buk
Antiseptic ointment. *Fisons plc, Pharmaceuticals Div.* Name unverified.

34692 Zanil
2277-92-1 7092 218-904-0
$C_{13}H_6Cl_5NO_3$
2,3,5-trichloro-N-(3,5-dichloro-2-hydroxyphenyl)-6-hydroxybenzamide
3,3',5,5',6-pentachloro-2'-hydroxysalicylanilide; oxyclozanide. A proprietary preparation of oxyclozanide; a veterinary anthelmintic. mp = 209-211°. No manufacturer.

34693 Zap-Oglobin, Zaponin
Reagents which destroy red blood cells to leave white blood cells in suspension for analyzis; used for white blood cell and hemoglobin determination. *Coulter Electronics Ltd.*

34694 Zapon®
Metal complex dyes; for air-drying, acid-hardening, amine-hardening transparent lacquers, baking finishes, polyurethane lacquers, peroxide-hardening polyester lacquers, and wood stains. *BASF AG.*

34695 Zaponin
Reagent which destroys red blood cells leaving white cells in suspension for analysis. Used for white blood cell determination with semi-automatic cell counters. *Coulter Electronics Ltd.* Discontinued.

34696 Zauberin
A material containing Chloramine T; a detergent and bleaching agent. No manufacturer.

34697 ZB2335
12447-61-9
$B_6O_{11}Zn_2 \cdot xH_2O$
zinc borate
boron zinc oxide hydrate. Zinc borate for flame retardancy. *Borax Europe Ltd.* Discontinued.

34698 Zedox
1314-23-4 10313 215-227-2
O_2Zr
Zirconium oxide
zirconia; zirconium dioxide; zirconic anhydride; baddleyite;. mp = 2680°; bp_{760} = 4300°; d = 5.85; insoluble in H_2O, soluble in mineral acids; *Anzon Ltd.* Name unverified.

34699 Zeeospheres® 200
Silica-alumina ceramic
Strong, hard, inert, thick-walled hollow spheres for use as filler for a variety of plastic resins in injection molding, extrusion, SMC, BMC, RTM, compression molding, potting/encapsulating, adhesives, tooling, casting, flooring, grouting, sealants, mastics, coatings, films, and other applications. *Zeelan.*

34700 Zeese
Honey substitute for diabetics containing sorbitol, fructose, water, citric acid, permitted color E150 and flavoring. *LAB Ltd.* Unverified.

34701 Zeiodelite
A mixture obtained by stirring 24 parts powdered glass into 20 parts melted sulfur; used as a cement, and for taking casts.

34702 Zeise's salt
$[Pt(C_2H_4)Cl_3]K$
A salt, formed when potassium chloride is added to a solution of platinous chloride saturated with ethylene.

34703 Zeklan
Chemical products for use in industry. *Courtaulds plc.*

34704 Zelco metal
An alloy of 83% zinc, 15% aluminum, and 2% copper.

34705 Zelcon
Fabric conditioner. *DuPont UK.*

34706 Zelec
Antistatic agent. *DuPont UK.*
34707 Zellwonet
Proprietary viscose packing material. No manufacturer.
34708 Zelulone
A proprietary artificial yarn made from wood pulp. No manufacturer.
34709 Zemdrain®
Fibers. *DuPont UK.*
34710 Zemid® 610
9002-88-4 7728
Toughened polyethylene resin, mineral-reinforced; balanced impact resistance and stiffness even at low temperatures; for industrial and consumer products, e.g., equipment shields, shrouds, lawn mower decks, snow shovels, instrument cases, swimming pool equipment, snowboards. *DuPont.*
34711 Zemid® 650
9002-88-4 7728
Toughened HDPE resin, min.-reinforced; good processability in blow molding, high stiffness and toughness. *DuPont.*
34712 Zenith
Organic brightener system; used for bright nickel electroplating (high leveling, fast brightening). *Engelhard Technologies Ltd.*
34713 Zenker's fluid
A solution containing 2.5 g potassium chromate, 1 g sodium sulfate, 5 g mercuric chloride, 5 ml glacial acetic acid, and 100 ml water.
34714 Zennapron
Mixture of 2,4-D and mecoprop. Used to control weeds in grassland. *BP Oil Ltd.* Discontinued.
34715 Zentralin
611-92-7 210-283-4
$C_{15}H_{16}N_2O$
1,3-dimethyl-1,3-diphenylurea
Zentralit I; Zentralit II. Used for explosives. mp = 120-122°;.
34716 Zeolex
Synthetic sodium aluminum silicate; flow promoter, pigment extender. *Cornelius Chemical Co Ltd.* Name unverified.
34717 zeolite synthetic
68989-22-0 10250
Water softener, detergent builder, cracking catalyst, adsorbents, desiccants, in solar collectors (as heating and cooling agents). *Crosfield Chemie BV; Ethyl; Miles; PQ; Sybron.*
34718 zeolites
1318-02-1 10250 215-283-8
Hydrated aluminum silicates containing alkali or alkaline earth metals occurring naturally and used in ion-exchangers, absorbents, dessicants, and solar collectors.
34719 Zeolum
Crystalline hydrous alumino-silicate of alkali metals or alkaline earth metals; molecular sieve showing strong selective adsorption capacity; for petrochemical industry, drying of gases and liqs., separation and purification processes. *Tosoh.* Name unverified.
34720 Zeonet A
108-80-5 2767 203-618-0
$C_3H_3N0_3$
Isocyanuric acid
1,3,5-Triazine-2,4,6(1H,3H,5H)-trione; s-Triazine-2,4,6(1H,3H,5H)-trione; Cyanuric Acid;
Cyanuric acid; Isocyanurate acid; Isocyanuric acid; Kyselina kyanurova; Pseudocyanuric acid; 1,3,5-Triazine-2,4,6-triol; s-Triazine-2,4,6-triol; S-Triazine-2,4,6-triol; s-Triazinetriol; s-2,4,6-Triazinetriol; S-2,4,6-Triazinetriol; Sym-Triazinetriol; 2,4,6-Triazinetrione; s-Triazinetrione; s-Triazine-2,4,6-trione; S-Triazine-2,4,6(1H,3H,5H)-trione; Tricarbimide; Tricyanic acid; Trihydroxycyanidine; 2,4,6-Trihydroxy-1,3,5-triazine. Curing agent for Nipol AR and HyTemp 4050 series polyacrylate elastomers. *Zeon.*
34721 Zeonet B
1120-02-1 214-294-5
$C_{21}H_{46}BrN$
Trimethyl octadecyl ammonium bromide
1-Octadecanaminium, N,N,N-trimethyl-, bromide; Ammonium, trimethyloctadecyl-, bromide; Morpan O; Octadecyltrimethylammonium bromide; n-Octadecyltrimethylammonium bromide; Softex; Stearyltrimethylammonium bromide; N,N,N-Trimethyl-1-octadecanaminium bromide; Trimethyloctadecylammonium bromide. Accelerator for Nipol AR and HyTemp 4050 series polyacrylate elastomers. *Zeon.*
34722 Zeonet U
102-07-8 1829 203-003-7
$C_{13}H_{12}N_2O$
N,N'-Diphenylurea
diphenylcarbamide; carbanilide; 1,3-diphenylurea; *sym*-diphenylurea; Centralite. Retarder to slow down the cure of Nipol AR and HyTemp 4050 series elastomers. mp = 238°; bp_{760} = 260° (dec); d = 1.239; slightly soluble in H_2O (0.15 g/l), Me_2CO, EtOH, $CHCl_3$; soluble in Et_2O, pyridine *Zeon.*
34723 Zeonex 280
Amorphous polyolefin
Offers low water absorp., transmittance, high heat resist. for optical applications (optical disks, plastic lenses). *Nippon Zeon/Zeonex.*
34724 Zeospan 303, 306
Polyether elastomer
Oil-resistant, high-elasticity rubber with excellent balance of ozone resistance and cold resistance. *Zeon.*
34725 Zeothix
A ground silica; thickening agent. *Cornelius Chemical Co Ltd.* Name unverified.
34726 Zeotokol
A coarse dolerite (an igneous rock composed essentially of labradorite and anorthite, with augite and sometimes olivine) ground up; and used as a fertilizer.
34727 Zepel®
Fabric fluoridizer. *DuPont UK.*
34728 Zephiran Chloride
8001-54-5 1086 231-635-3
Benzalkonium chloride
Benirol; BTC; Capitol; Cequartyl; Drapolene; Drapolex; Enuclen; Germinol; Germitol; Osvan; Paralkan; Roccal; Rodalon; Chloride; Zephirol. Cationic surface agent. Germicide. Pharmaceutic aid. A preparation for the disinfection of hands and instruments. *Sterling Drug Inc; Bayer.* Name unverified.
34729 Zephron®
For the chemical industry. *DuPont UK.*
34730 Zephyr
Inks used in the screen printing process; for bill-boards, signs and displays. *AlliedSignal Inc/Sinclair and Valentine Division.* Name unverified.
34731 Zerex
A proprietary trade name for a polyvinyl alcohol antifreeze compound. No manufacturer.
34732 Zerofil
A proprietary rock wool which has been treated with asphalt. No manufacturer.
34733 Zerogen® 10, Zerogen® 60
1309-42-8 5706 215-170-3
H_2O_2Mg
Magnesium hydroxide
magnesium hydrate, Marinco H. Halogen-free magnesium hydroxide composition; flame retardant/smoke suppressant for thermoplastics and elastomeric formulations. insoluble in H_2O, aqueous suspension has pH 9.5-10.5 *J.M. Huber/Solem.*
34734 Zerol
Refrigeration fluid. *Monsanto (Solaris).* Name unverified.
34735 Zerone
A proprietary trade name for an antifreeze product composed of methanol and polyvinyl alcohol. No manufacturer.
34736 Zeronox
Catalyst for the reduction of nitrogen oxides. *Hüls AG.*
34737 Zerotherm
Self drying coatings for sand molds and cores. *Foseco (F.S.) Ltd.*
34738 Zerox
302-01-2 4809 206-114-9
H_4N_2
Hydrazine
Hydrazine solutions (35% in water); used for water treatment. mp = 1.4°; bp_{760} = 113.5°; d = 1.021; n^{20}_D = 1.4700; LD_{50} (mus orl) = 59 mg/kg *Schering Industrial Products Ltd.* Discontinued.
34739 Zetabon
Plastic coated steel and aluminum used in wire and cable for armor, corrosion and lightning protection. *Dow Cheml Co Ltd, UK & Ireland.*
34740 Zetag®
High molecular weight cationic polyelectrolytes for sewage and industrial effluent treatment supplied in microbead, powder or liquid form; typical applications include primary sedimentation processes, sludge dewatering, sludge thickening, phosphate removal processes. *Allied Colloids Ltd.*
34741 Zetax®
155-04-4 205-840-3
$C_{14}H_{10}N_2S_4Zn$
Zinc 2-mercaptobenzothiazole
2(3H)-Benzothiazolethione, zinc salt; 2-Benzothiazolethiol, zinc salt (2:1);

Bis(2-benzothiazolylthio)zinc; Bis(mercaptobenzothiazolato)zinc; Bis(mercaptobenzthiazolato)zinc; Hermat Zn-MBT; Mercaptobenzothiazole, zinc salt; 2-Mercaptobenzothiazole, zinc salt; OXAF; Pennac ZT; Tisperse MB-58; Usaf gy-7; Vulkacit ZM; Zenite; Zenite Special; Zetax; Zinc benzothiazolethiolate; Zinc 2-benzothiazolethiolate; Zinc mercaptobenzothiazole salt; Zinc salt of 2-mercaptobenzothiazole; ZMBT; ZMBT, Waxed; ZnMB; Zn 2-Mercaptobenzothiazole. Secondary accelerator in latex foam curing systems. *R. T. Vanderbilt Co Inc.*

34742 Zetesap
Base for soap-free soap tablets. *Surfachem Ltd.*

34743 Zetesol 100
MIPA-laureth sulfate-laureth-4-cocamide DEA
Detergent and emulsifier for personal care products; for foam bath preps. with high oil concs. *Zschimmer & Schwarz.*

34744 Zetesol 856 T
MIPA C12-15 pareth sulfate
Detergent for cosmetics, shampoos, bath preparations, and liquid synthetic soap. *Zschimmer & Schwarz.*

34745 Zetesol AP
Ammonium C12-15 pareth sulfate
Propylene glycol; detergent for cosmetics, shampoos, bath preps., liquid hand cleaners, dishwash, household cleaners. *Zschimmer & Schwarz.*

34746 Zetesol NL
Sodium laureth sulfate
Detergent for personal care, household and industrial cleaners. *Zschimmer & Schwarz.* Discontinued.

34747 Zetpol® 1010
Hydrogenated nitrile rubber; elastomer for fuel hose, fuel diaphragm, Freon packing, in-tank insulator applications *Zeon.*

34748 Zetpol® 2000L
Acrylonitrile-ethylene-butadiene terpolymer
Elastomer for o-rings, packings, gaskets, oil seals, oil hose, well head seals, blow out preventor, and water pump seal applications. *Zeon.*

34749 Zetpol® 2020
Hydrogenated nitrile rubber; elastomer for oil seals, synchronous belt, rolls, oil hose, and drilling pipe protector applications. *Zeon.*

34750 Zetpol® ZSC-2295
Nitrile rubber, zinc oxide-methacrylic acid modified; used for very high durometer, high abrasion resist. compounds. *Zeon.*

34751 Zettnow's stain
A microscopic stain made up of solution a (10 g tannic acid and 30 ml of a 5% solution of tartar emetic) and solution b (1g silver sulfate in 250 ml water). Take 50 ml and add ethylamine until precipitate redissolves.

34752 Zeus
An alloy of 20% silver and 80% copper; used for fuse wire.

34753 Zewa powder
Sodium lignin sulfate obtained by evaporation of waste sulfite lyes. It has detergent and water-softeninig properties.

34754 Zewa SL 2
8061-51-6
lignosulfonic acid, sodium salt
Virtually desugared sodium lignosulfonate (·5 1% sugar content); Water-reducing concrete additive; dispersant in pesticides. *Borregaard LignoTech.*

34755 Zewakol
Modified lignosulfonates; extenders for U/F resins in particle board mfg. *Borregaard Ligno Tech.*

34756 Zewalon FN
8061-51-6
Sodium lignosulfonate
Auxiliary tanning agent. *Borregaard Ligno Tech.*

34757 Zewaphosphate
A phosphate fertilizer.

34758 Zidanit
12122-67-7 10300 235-180-1
zineb
A fungicide and insecticide. *Makhteshim Chemical Works Ltd.* Discontinued.

34759 Zienam
Imipenem and cilastatin sodium; broad-spectrum β-lactam antibiotic. *Merck & Co Inc.*

34760 Zilloy
A proprietary zinc alloy containing zinc with 1% copper, 0.01% magnesium, and lead and cadmium in addition; the rolled sheets are suitable for building purposes. No manufacturer.

34761 Zimalium
Zincalium. Alloys containing 74-93.5% aluminum, 2.8-14.8% zinc, and 3.7-11.2% magnesium.

34762 Zimate®
137-30-4 10305 205-288-3
$C_6H_{12}N_2S_4Zn$
bis(dimethylcarbamodithioato-S,S')zinc
bis(dimethyldithiocarbamat)zinc; zinc dimethyldithiocarbamate; dimethyldithiocarbamic acid zinc salt; methyl cymate; Methasan; Zirberk; Karbam White; Crorzate; Fuclasin; Fuklasin; Zerlate; ziram. Zinc diamyl, dibutyl, diethyl and dimethyl dithiocarbamates; rubber accelerators for NR and latex. mp = 250°; d^{25}_4 = 1.66; insoluble in H_2O, soluble in organic solvents; LD_{50} (rat orl) = 1.4 g/kg *R. T. Vanderbilt Co Inc.* Discontinued.

34763 Zimco
121-33-5 10069 204-465-2
$C_8H_8O_3$
4-hydroxy-3-methoxybenzaldehyde
Vanillin; methylprotocatechuic aldehyde; vanillic aldehyde; 3-methoxy-4-hydroxybenzaldehyde. Pharmaceutical aid (flavor). mp = 81-83°; bp = 285°, bp_{15} = 170°; d = 1.056; soluble in H_2O (1%), organic solvents; LD_{50} (rat orl)= 1580 mg/kg *Sterwin Chemicals Inc.* Name unverified.

34764 Zinar®
Rosin-based resinate; tackifier resin for adhesives; *Arizona.* Discontinued.

34765 zinc
7440-66-6 10255 231-175-3
Zn
Zinc
Metallic element; Zn; alloys, galvanizing iron and other metals, and fungicides. mp = 419.5°; bp = 908°; d^{25} = 7.14; *Aldrich; Cerac; Cuproquim; Ferro/Bedford; Pasminco Europe; Zinc Corp. of Am.*

34766 zinc ammonium chloride
14639-98-6 573 238-687-6
$Cl_5H_{12}N_3Zn$
ammonium pentachlorozincate
Welding, soldering flux, dry batteries, and galvanizing. sublimes 340°; d = 1.81; soluble in H_2O *Blythe, William Ltd; Du Pont.*

34767 zinc anhydride
A variety of lithopone, also known as zinc barytes and consisting of a mixture of calcium and barium sulfates and zinc oxide.

34768 zinc bende
(Black Jack; blende; rosinjack; pseudo-galena). A mineral that is zinc sulfide, ZnS; used as a pigment.

34769 zinc borate
1332-07-6 215-566-6
B_2O_3
boric acid
zinc salt; Zinc Borate 2335. Medicine, fireproofing textiles, fungistat, mildew inhibitor. Zinc Borate 2335 (Borax Consolidated Ltd) is a specialty flame retardant additive to plasticized PVC and other polymers to reduce afterglow and smoke. *BA Chem. Ltd; Climax Performance; U.S. Borax & Chem.; Borax Consolidated Ltd.*

34770 zinc bromide
7699-45-8 10258 231-718-4
$ZnBr_2$
zinc bromide anhydrous
Photographic emulsions, manufacture of rayon, radiation viewing shield. mp = 394°; bp = 697°; d = 4.22; soluble in H_2O, organic solvents *Atomergic Chemetals; Cerac; Ethyl; Great Lakes; Hoechst Celanese; Ryvan Ltd.*

34771 zinc carbonate
3486-35-9 10260 222-477-6
$ZnCO_3$
carbonic acid, zinc salt (1:1)
Smithsonite. Inorganic salt; accelerator-activator for transparent natural and synthetic rubber goods, adhesives; as pigment; fireproofing filler for rubber and plastics; topical antiseptics, cosmetics, and lotions. poorly soluble in H_2O *Allchem Industries; Harcros Durham; Nihon Kagaku Sangyo; Spectrum Chem. Mfg.*

34772 zinc chloride
7646-85-7 10261 231-592-0
$ZnCl_2$
Zinc chloride
butter of zinc; zinc trace. Catalyst, dehydrating and condensing agent in organic synthesis, fireproofing, preserving food, electroplating, antiseptic denaturant for alcohol. An astringent and dentin desensitizer. mp =290°; bp = 732°; d^{25} = 2.907; soluble in H_2O (432 g/100 g), less soluble in organic solvents; LD_{50} (rat iv) = 60-90 mg/kg AlliedSignal; Elf Atochem N. Am.; Blythe, William Ltd; EM Industries; Mallinckrodt; Armour Pharmaceutical Co.

34773 zinc chromate
13530-65-9 10262 236-878-9
$H_2O_6CrZn_2$
Zinc chromate(VI) hydroxide

zinc yellow; buttercup yellow; C.I. Pigment Yellow 36; Citron yellow; zinc chrome. Yellow pigments. slightly soluble in H_2O *Colores Hispania SA; Landers-Segal Color.*

34774 zinc fluoride
7783-49-5 10265 232-001-9
F_2Zn
Zinc fluoride
Phosphors, ceramic glazes, wood preservative, electroplating, and organic fluorination. mp = 872°; bp = 1500°; d^{25} = 5.00; anhydrous form is poorly soluble in H_2O *Elf Atochem N. Am; Atomergic Chemetals; Cerac; Hoechst Celanese; Noah Chem.*

34775 zinc formosul
24887-06-7 246-515-6
A basic zinc-formaldehyde-sulfoxylate; used in fat-splitting.

34776 Zinc gluconate
4468-02-4 224-736-9
Gluconal® ZN. Pharmaceutical/food grade mineral source for human and veterinary pharmaceutical preps., dietary supplements, fortified foods and animal feed. *Akzo Chemie; Atomergic Chemetals.*

34777 zinc grey
The name originally used for zinc dust employed for painting on iron. The term is now used for finely ground zinc blende. A mixture of zinc oxide with finely divided charcoal is sold under this name. It is produced in the manufacture of zinc.

34778 zinc hydroxide
20427-58-1 243-814-3
H_2O_2Zn
collozine; colloidal zinc hydroxide. d = 3.050.

34779 Zinc hydroxystannate
12027-96-2
$ZnSn(OH)_6$
Flamtard H. Flame retardant for PVC, polychloroprene, chlorosulfonated polyethylene, other halopolymers. White powder; soluble in strong acids and bases; mw=286.12; sp gr=3.4; decomp pt at 180°; LD_{50}=rat oral>5000 mg/kg; rat dermal>2466 mg/kg *Alcan; Joseph Storey.*

34780 zinc lactate
16039-53-5 10271 240-178-9
$C_6H_{10}O_6Zn$
Zinc lactate
lactic acid zinc salt. soluble in H_2O (2%) *Patco.*

34781 zinc naphthenate
12001-85-3 234-409-2
$Zn(C_6H_5COO)_2$
Drier and wetting agent in paints, varnishes, resins; insecticide, fungicide, mildew preventive; wood preservative; waterproofing textiles; and insulating materials. *King Industries.*

34782 zinc nitrate
10196-18-6 10273 231-943-8
N_2O_6Zn>H_2O
zinc nitrate hexahydrate
Acidic catalyst, latex coagulant, reagent, intermediate, mordant. mp = 36°; d= 2.065; soluble in water (50%), alcohol *Blythe, William Ltd; Mallinckrodt; Nihon Kagaku Sangyo.*

34783 Zinc Omadine
13463-41-7 8178 236-671-3
bis-(1-hydroxy-2(1H)-pyridinethionato-O,S)zinc
zinc pyrithione; zinc pyridine thione; bis(2-pyridylthio)zinc 1,1'-dioxide; Pyrithione zinc; De-Squaman; Vancide ZP; Zincon Dandruff Shampoo; zinc 2-pyridinethiol-1-oxide; Zink Pyrion. An antibacterial; antifungal; and antiseborrheic. In Head and Shoulders. Antimicrobial inhibiting growth of Gram-negative and Gram-positive bacterial, fungi, mold and yeast; used in aqueous metal coolant and cutting fluids, PVC plastics. *E R Squibb & Sons Inc; Olin; Lederle Laboratories USA; Allchem Industries; Ruetgers-Nease.*

34784 Zinc Omadine®, 48% Fine Particle Disp
13463-41-7 8178 236-671-3
zinc pyrithione
Antimicrobial. Inhibits growth of Gram-begative and Gram-positive bacteria, fungi, molds and yeasts. Used in aqueous metal coolants and cutting oils. *Olin.*

34785 Zinc Omadine®, Powd
13463-41-7 8178 236-671-3
zinc pyrithione
Antidandruff agent for shampoos. Inhibits growth of Gram-negative and Gram-positive bacteria and fungi. Used as a cosmetic preservative. *Olin.*

34786 zinc oxide
1314-13-2 10279 215-222-5
ZnO
zinc oxide
Flowers of zinc, Philosopher's wool; Zinox; Zinc powder; C.I. Pigment White 4; C.I. 77947; Chinese white; zinc white; Tertiary zinc oxide; Zinc Oxide No. 185; Zinc Oxide No. 318. UV absorber; accelerator activator; pigment in white paints, cosmetics, driers, dental cements; mold inhibitor in paints; in manufacture of opaque glass, enamels, tires, printing inks, porcelains; reagent in analytical chemistry. Also used as a flame retardant. d = 5.67; n_D = 2.0041; insoluble in H_2O *Am. Chemet; Asarco; Eagle Zinc; General Chem.; Harcros Durham; Mallinckrodt; Zinc Corp. of Am.*

34787 Zinc Oxide No. 185
1314-13-2 10279 215-222-5
Tertiary zinc oxide. An accelerator activator, pigment and reinforcing agent for rubber and applications not requiring a high degree of purity. Also for production of zinc compounds. *Eagle Zinc.*

34788 Zinc Oxide No. 318
1314-13-2 10279 215-222-5
zinc oxide
American process zinc oxide. Lead-free. Slow curing fine particle size pigment and reinforcing agent for the rubber industry. Also used in the manufacture of various zinc compounds.

34789 Zinc Oxide Transparent
1314-13-2 10279 215-222-5
Highly disperse precipitated zinc oxide; vulcanization accelerator activator for transparent rubber goods, in natural and syn. rubbers; acid acceptor in polychloroprene adhesives; light-colored reinforcing filler; used for vulcanizates. *Bayer AG; Miles; Polysar.*

34790 zinc phosphate
7779-90-0 10284 231-944-3
$O_8P_2Zn_3$
Zinc phosphate
(tetrahydrate) hopeite. Used in dental cements, phosphors. Insoluble in H_2O, alcohol Calgon; Colores Hispania SA; Hammond Lead Prods.; Pasminco Europe; G. Whitfield Richards; Witco/Allied-Kelite.

34791 zinc pyrithione
13463-41-7 8178 236-671-3
$C_{10}H_8N_2O_2S_2Zn$
1-hydroxy-2(1H)-pyridinethione zinc salt
zinc pyrithione; zinc pyridinethione; Desquaman; Zink Pyrion. Fungicide and bactericide. Ingredient in Head and Shoulders.

34792 Zinc stannate
12036-37-2
$ZnSnO_3$
Flamtard S. Flame retardant for PVC, polychloroprene, chlorosulfonated polyethylene, polypropylene, other halopolymers. Density=4.25; decomp pt>570° *Alcan; Atomergic Chemetals; Wm. Blythe Ltd; Joseph Storey.*

34793 zinc stearate
557-05-1 10292 209-151-9
$C_{36}H_{70}O_4Zn$
zinc octadecanoate
octadecanoic acid zinc salt, zinc octadecanoate;. Zinc salt; zinc salt of stearic acid; in cosmetics, pharmaceuticals, lacquers, ointments, tablet manufacture; mold release agent for plastic; filler, antifoamer; flatting agent in lacquers; as a drying lubricant and dusting agent for rubber; waterproofing agent for concrete, paper and textiles. mp = 120°; insoluble in H_2O, EtOH, Et_2O, soluble in C_6H_6 *Allchem Industries; Ferro/Grant; Magnesia GmbH; Mallinckrodt; Norac; Stave; Syn. Prods.*

34794 zinc sulfate
7446-20-0 10293
$ZnSO_4 \cdot 7H_2O$
zinc sulfate heptahydrate
white vitriol; white copperas; zinc vitriol. Rayon, manufacture, dietary supplement, animal feeds, mordant, wood preservative, analytical reagent. mp = 40°; d = 1.9570 *Aldrich; Mallainckrodt.*

34795 zinc sulfate, monohydyrate
7446-19-7 10293
$ZnSO_4 \cdot H_2O$
Rayon manufacture, agricultural sprays, chemical intermediate, dyestuffs, electroplating. soluble in H_2O, insoluble in EtOH *Allchem Industries; Blythe, William Ltd; EM Industries.*

34796 zinc sulfide
1314-98-3 10294 215-251-3
SZn
Inorganic salt
zinc blende; wurtzite; sphalerite. Pigment; in white and opaque glass, plastics, dyeing, paints, linoleum, leather, dental rubber; fungicide; anhydrous in x-ray screens, and TV screens. insoluble in H_2O *Cerac; Chemson GmbH; Noah Chem.; Ore & Chem. Corp.; Sachtleben GmbH.*

34797 zinc sulfide grey
Calamine white. A pigment that is a dense zinc oxide used for painting iron, and is artificially made by tinting lithopone with ochers and charcoal.

34798 zinc undecenoate
557-08-4 9983 209-155-0
$C_{22}H_{38}O_4Zn$
zinc 10-undecenoate
zinc 10-hendecenoate; zinc undecylenate. Base for ointments. mp= 115-116°.

34799 zinc vitriol
7733-02-0 10293 231-793-3
O_4SZn
zinc sulfate
Used in rayon manufacture, in animal feeds, as a wood preservative and analytical reagent.

34800 zinc yellow
13530-65-9 10262 236-878-9
CrO_4Zn
zinc chromate
zinc chrome yellow; buttercup yellow; lemon yellow. Used as a pigment in paints, varnishes, oil colors, linoleum, rubber etc.

34801 Zincazol
zinc α-phenylbiguanide
A proprietary rubbber vulcanization accelerator. No manufacturer.

34802 zincocalcite
A calcite (calcium carbonate) containing zinc carbonate.

34803 Zincofol
A fungicide. *Monsanto (Solaris).* Name unverified.

34804 Zincov™
Inhibitor. *Calbiochem. Corp.*

34805 Zincrex
Flux for zinc and alloys. *Foseco (F.S.) Ltd.*

34806 Zinctrace
7646-85-7 10261 231-592-0
zinc chloride
An astringent and dentin desensitizer. *Armour Pharmaceutical Co.* Name unverified.

34807 Zineb
12122-67-7 10300 235-180-1
$(C_4H_6N_2S_4Zn)_x$
[[1,2-ethanediylbis[carbamodithioato]](2-)]zinc
zinc ethylenebis(dithiocarbamate) (polymeric); zinebe, Zidan; Aaphytora; Acuprex; Aspor; Dipher; Dithane Z 78; Diiner; Ditiozin; Enozin; Hexaphane; Kypzin; Lonacol; Parzate; Permilan; Phytox; Polyram-Z; Sepineb; Tiezene; Tritoftorol; Zinosan; Zinugec;. Foliar fungicide with protective action, insecticide. Repellent to birds and rodents. Used for control of fungi in fruits, vines, vegetables and ornamentals. Controls scab in apples and pears. dec 157°; soluble in H_2O 10 mg/l, insoluble in organic solvents; LD_{50} (rat orl) > 5200 mg/kg Agrimont; Diachem; Bayer; Pennwalt Holland; Makhteshim Chemical Works Ltd; Rhône Poulenc, Visplant; Rohm & Haas.

34808 zineb
12122-67-7 10300 235-180-1
zinc dithiocarbamate
Insecticide and fungicide.

34809 Zink Pyrion
13463-41-7 8178 236-671-3
zinc pyrithione
Used as an anti-dandruff agent, preservative, antibacterial and antimicrobial. *Ruetgers-Nease.*

34810 Zinkan
A proprietary combination of aluminium coated with zinc; obtained by rolling at elevated temperatures. No manufacturer.

34811 Zinkgrau
Inexpensive, off-color zinc oxide pigment.

34812 Zinkoxyd Activ®
1314-13-2 10279 215-222-5
ZnO
zinc oxide
Highly disperse precipitated zinc oxide; vulcanization accelerator activator for rubber goods based on natural and synthetic elastomers and latex applications; suitable at high levels for dynamically stressed articles, e.g., buffers and rollers, and in low concentrations in transparent and translucent goods. *Bayer AG; Miles; Polysar.*

34813 Zinnal
A proprietary dual metal consisting of aluminium sheet coated on both sides with tin. No manufacturer.

34814 Zinol
Powder flux for treating dross on hot-dip Galvanizing baths. *Foseco (F.S.) Ltd.* Discontinued.

34815 Zinox
1314-13-2 10279 215-222-5
Ozn
zinc oxide
Precipitated zinc oxide. Used as a paint pigment. *Am. Chemet.*

34816 Zinsser's insulating wax
a) Consists of beeswax; b) consists of shellac, rosin, and oxide of iron.

34817 Zintox
zinc arsenate
A proprietary trade name for an agricultural spray containing basic zinc arsenate. No manufacturer.

34818 Zip Grip
Cyanoacrylate
For bonding closely matted surfaces and maintenance, production, and prototype bonding. *ITW Devcon.*

34819 Ziploc
Brand plastic storage bags. *Dow UK.*

34820 Zippo
A trade name for an aluminium solder for joining aluminum to itself, to copper, zinc, tin, or brass. No manufacturer.

34821 Zircomplex
Ammoniacal zirconium complex; thixotrope in emulsion paints. *Rhône-Poulenc UK.*

34822 zirconium boride
12045-64-6 234-963-5
ZrB_2
zirconium diboride
Refractory for aircraft and rocket applications, thermocouple protection tubes, high-temp. electrical conductor, cutting-tool component, coating tantalum, cathode in high-temp. electrochemical systems; oxidation-resistant composites. *Atomergic Chemicals; Boride Ceramics & Composites Ltd; Cerac; Noah Chem.*

34823 zirconium hydride
7704-99-6 10309 231-727-3
ZrH_2
Vacuum tube getter, powder metallurgy, source of hydrogen, metal-foaming agent, nuclear moderator, reducing agent, and hydrogenation catalyst. *Atomergic Chemetals; Cerac; Degussa; Morton Int'l.*

34824 zirconium oxide
1314-23-4 10313 215-227-2
ZrO_2
zirconia
zirconium dioxide; zirconic anhydride. Unstabilized: prod. of piezoelectric crystals, high-frequency induction coils, ceramic glazes, glasses, heatresistant fibers; hydrous: odor absorbent, and poison ivy treatment. mp = 2700°; bp = 4300°; d = 5.8500; insoluble in H_2O Atomergic Chemetals; Ferro/Transelco; Hüls Am.; Magnesium Elektron Ltd; TAM Ceramics; Zircar Prods.

34825 zirconium silicate
14940-68-2 10314 233-252-7
$ZrSiO_4$
silicic acid, zirconium salt (1:1)
zircon; zirconium orthosilicate. Glaze opacifier; stabilizes color shades; used in white and colored glazes for sanitary ware, wall tile, glazed brick, structural tile, stoneware, dinnerware, special porcelains, refractory compositions, epoxy formulations, and encapsulating resins. mp = 2550° *Elf Atochem N. Am.; Atomergic Chemetals; Du Pont; TAM Ceramics.*

34826 zirconium tetrachloride
10026-11-6 10307 233-058-2
$ZrCl_4$
zirconium (IV) chloride
Source of the pure metal, analytical chemistry, water repellents for textiles, tanning agent, zirconium compounds, special catalysts (Friedel-Crafts, Ziegler). mp = 437°; d = 2.8030; LD_{50} (rat orl) = 1588 mg/kg *Atomergic Chemetals; Noah Chem.*

34827 zirconium tetrafluoride
7783-64-4 10308 232-018-1
ZrF_4
zirconium fluoride
Component of molten salts for nuclear reactors. mp = 640°; bp = 905°; d^{16} = 4.6; LD_{50} (mus orl) = 98 mg/kg *Elf Atochem N. Am.; Atomergic Chemetals; Cerac.*

34828 Zircosil
10101-52-7 10314 233-252-7
O_4SiZr
zirconium orthosilicate
Zirconium silicates. Used in refractories, ceramics, cements and coatings for casting molds. Also in jewelry. mp = 2550° *Anzon Ltd.* Name unverified.

34829 Zircosol P
Recycled paper additive. *Magnesium Elektron Ltd.*
34830 Zircotan
Synthetic tanning agent. *Rohm & Haas.*
34831 Zirex®
Rosin-based resinate; tackifier resin for adhesives; *Arizona.* Discontinued.
34832 Zirgel®
Thixotropic gelling agent. *Magnesium Elektron Ltd.*
34833 Zirkonal
Zirconium aluminum compounds. *Giulini Corp.*
34834 Zirmax®
Zirconium hardener. *Magnesium Elektron Ltd.*
34835 Zirmel®
Zirconium, magnesium, rare-earth and other chemicals, powders and alloys; for use in industry, including metalworking. *Magnesium Elektron Ltd.*
34836 Ziscon
Ziskon. An alloy of 60% aluminum and 40% zinc.
34837 Zisium
An alloy of 60% aluminum and 40% zinc.
34838 Zisnet F-PT
638-16-4 211-322-8
$C_3H_3N_3S_3$
2,4,6-Trimercapto-s-triazine
trithiocyanuric acid; *s*-triazine-2,4,6-trithiol. Curing agent for epichlorohydrin rubber; used in place of ethylene thiourea and red lead; gives improved heat resist., less mold fouling, reduced toxicity; oil treated to reduce dusting. mp > 300° *Zeon.*
34839 Zithate®
1000-90-4 213-680-0
Zinc isopropyl xanthate
Accelerator for vulcanization of rubber. *R. T. Vanderbilt Co Inc.* Discontinued.
34840 Zitrilon® 10%
Zinc chelate with 10% Zn; for foliar application; prevents and cures zinc deficiency. *BASF AG.*
34841 Zitro®
Rosin-based resinate; tackifier resin for adhesives; *Arizona.* Discontinued.
34842 ZMBT
155-04-4 205-840-3
$C_{14}H_8N_2S_4Zn$
2-mercaptobenzothiazole zinc salt
Bantex. Semi-ultra accelerator. Also used as a fungicide. d^{25}_4 = 1.70 *Akrochem.*
34843 Z-M-L
Zinc salt of mercaptobenzothiazole with laurex; a proprietary rubber vulcanizing accelerator. No manufacturer.
34844 Zoamix
148-01-6 3321 205-706-4
$C_8H_7N_3O_5$
2-methyl-3,5-dinitrobenzamide
3,5-dinitro-*o*-toluamide; 3,5-dinitro-2-methylbenzamide; dinitolmide; Zoalene. Coccidiostat; often used on an exchange program with Coyden (clopidol); the main purpose in rotating Zoamix coccidiostat with Coyden coccidiostat is to prevent poultry from developing a resistance to the latter product. mp = 181° *Dow UK.* Discontinued.
34845 Zodiac
Organic brightener system; used for bright nickel electroplating. *Harshaw Chemicals Ltd.*
34846 Zoflora
Concentrated floral disinfectant. *Thornton & Ross Ltd.*
34847 Zohar 60 SD(L), Zohar Automat SD, Zohar KAL SD
Sodium alkylbenzene sulfonate and builders; (60 SD) spray-dried concentrate for detergent powder manufacturing; SD spray-dried built powder for machine washing. *Zohar Detergent Factory.*
34848 Zohar Automat SD
Sodium alkylbenzenesulfonate and builders. Spray-dried built powder for machine washing. *Zohar Detergent Factory.*
34849 Zohar EGMS
Glycol stearate
Coemulsifier for o/w emulsions; opacifier, pearlescent, emollient, superfatting agent for mfg. of shampoos, liquid toilet soaps, bath products. *Zohar Detergent Factory.*
34850 Zohar GLST
31566-31-1 4498 250-705-4
Glyceryl stearate
octadecanoic acid monoester with 1,2,3-propanetriol; monostearin. Emulsifier, thickener, superfatting agent. *Zohar Detergent Factory.*
34851 Zohar KAL SD
Sodium alkylbenzenesulfonate and builders. Spray-dried built powder for machine washing. *Zohar Detergent Factory.*
34852 Zoharconc A.D
Aircraft detergent concentrate. *Zohar Detergent Factory.*
34853 Zoharconc Dead Sea
Foaming bath concentrate (with Dead Sea minerals). *Zohar Detergent Factory.*
34854 Zoharconc DIS
Disinfectant liquid concentrate. *Zohar Detergent Factory.*
34855 Zoharconc FC
Floor cleaner concentrate without wax. *Zohar Detergent Factory.*
34856 Zoharconc FS
Domestic fabric softener concentrate. *Zohar Detergent Factory.*
34857 Zoharconc J-Super
Textile softener concentrate for denim. *Zohar Detergent Factory.*
34858 Zoharconc RA
Rinse aid concentrate. *Zohar Detergent Factory.*
34859 Zoharex A-10
Fatty acid polyglycol ester; detergent for laundry liquids. *Zohar Detergent Factory.*
34860 Zoharfoam
Foaming agent for drilling use. *Zohar Detergent Factory.*
34861 Zoharlab
Linear sodium alkylbenzene sulfonate
Raw material for manufacturing of detergents. *Zohar Detergent Factory.*
34862 Zoharpon DT-80
Alkanolamine lauryl sulfate
Raw material for shampoo and foam baths. *Zohar Detergent Factory.*
34863 Zoharpon ETA 27
9004-82-4
Sodium lauryl ether sulfate (2 EO)
Raw material for personal care products, detergents. *Zohar Detergent Factory.*
34864 Zoharpon LAA
2235-54-3 218-793-9
Ammonium lauryl sulfate
Raw material for clear body and hair shampoos. *Zohar Detergent Factory.*
34865 Zoharpon LAD
DEA lauryl sulfate
Raw material for shampoos, cosmetics, and light duty detergents. *Zohar Detergent Factory.*
34866 Zoharpon LAEA 253
67762-19-0
Ammonium lauryl ether sulfate (3 EO)
Raw material for shampoos, cosmetics, and light duty detergents. *Zohar Detergent Factory.*
34867 Zoharpon LAM
MEA-lauryl sulfate
Raw material for shampoos and foam baths. *Zohar Detergent Factory.*
34868 Zoharpon LAS Special
Sodium lauryl sulfate
Raw material for shampoos. *Zohar Detergent Factory.*
34869 Zoharpon LAT
139-96-8 205-388-7
TEA lauryl sulfate
Raw material for mild shampoos and skin care preps. *Zohar Detergent Factory.*
34870 Zoharpon LMT42
Sodium lauryl methyl taurate
Mild detergent, foamer, dispersant used in soap, syndet toilet bars, shampoos, and bubble baths. *Zohar Detergent Factory.*
34871 Zoharpon MgES
67702-21-4 221-450-6
Magnesium lauryl ether sulfate (2 EO)
Raw material for mild shampoos and cosmetic preps. *Zohar Detergent Factory.*
34872 Zoharpon MgS
Magnesium lauryl sulfate
Raw material for mild shampoos and cosmetic preps. *Zohar Detergent Factory.*
34873 Zoharquat 50
8001-54-5 1086 231-635-3
Benzalkonium chloride
Disinfectant, fungicide, bacteriocide and algicide. *Zohar Detergent Factory.*

34874 Zoharsoft 90
Quaternary imidazoline derivative. Fabric softener base for domestic and commercial laundry use. *Zohar Detergent Factory.*

34875 Zoharsoft DAS
Fatty acid-amine derivative. Softener base for textile industry for all fibers (cotton, rayon, acetates, wool, nylon). *Zohar Detergent Factory.*

34876 Zoharsyl L-30
137-16-6 4379 205-281-5
$C_{15}H_{28}NNaO_3$
N-methyl-N-(1-oxododecyl)glycine sodium salt
Sodium lauroyl sarcosinate; Gardol; Medialan LL-99;. Raw material for manufacturing of hair shampoos, conditioners, toothpastes, carpet and upholstery shampoos; anticorrosive properties. *Zohar Detergent Factory.*

34877 Zohartaine AB
683-10-3 211-669-5
Lauryl betaine. Foam booster, mild ingredient for shampoos and detergents; industrial foamer. *Zohar Detergent Factory.*

34878 Zohartaine TM
Dihydroxyethyltallow glycinate
Thickener and anticorrosive agent for tech. acid formulations, industrial cleaners; component of mild shampoos. *Zohar Detergent Factory.*

34879 Zoharteric D-2
Disodium cocoamphodiacetate
sodium trideceth sulfate. Component of mild, nonirritating conditioning shampoos, bubble baths, baby products. *Zohar Detergent Factory.*

34880 Zoharteric DJ
Disodium lauroamphodiacetate
Component of mild, nonirritating conditioning shampoos, bubble baths; detergent for specialty cleaners. *Zohar Detergent Factory.*

34881 Zoharteric D-SF
Disodiumcocoamphopropionate
Component of extra mild, nonirritating shampoos, bubble baths; detergent for heavy-duty household and industrial cleaners with tolerance for alkalies and electrolytes. *Zohar Detergent Factory.*

34882 Zoharteric LF
13039-35-5 235-907-2
Sodium capryloamphoacetate
Surfactant and wetting agent for low-foam heavy-duty household and industrial cleaners; tolerance for alkalies and electrolytes. *Zohar Detergent Factory.*

34883 Zoharteric LF-SF
68815-55-4 272-383-4
Disodium capryloamphodipropionate
Surfactant and wetting agent for lowfoam heavy-duty household and industrial cleaners; high tolerance for alkalies and electrolytes. *Zohar Detergent Factory.*

34884 Zoharteric M
Sodium cocoamphoacetate
Component for personal care products; detergent for specialty cleaners. *Zohar Detergent Factory.*

34885 Zoharteric M-2
Sodium cocoamphoacetate
sodium trideceth sulfate. Component of mild, nonirritating conditioning shampoos, bubble baths, and baby products. *Zohar Detergent Factory.*

34886 Zoldine® MS-52
137796-06-6
$C_{11}H_{23}NO$
4-Ethyl-2-methyl-2-(3-methylbutyl)-1,3-oxazolidine
Oxazolidine, 4-ethyl-2-methyl-2-(3-methylbutyl)-. Moisture scavenger; urethane crosslinker. *Angus.*

34887 Zoldine® ZT-55
6542-37-6 229-457-6
$C_6H_{11}NO_3$
Oxazolidine
1H,3H,5H-Oxazolo[3,4-c]oxazole-7a(7H)-methanol; component of Nuosept 95; GDUE; 5-Hydroxymethyl-1-aza-3,7-dioxabicyclo[3.3.0]octane; 5-(Hydroxymethyl)-1-aza-3,7-dioxabicyclo(3.3.0)octane ; M 3; 5-Methylol-1-aza-3,7-dioxabicyclo[3.3.0]octane; M 3 (heterocycle); Nuosept 95; 1H,3H,5H-Oxazolo(3,4-C)oxazole-7a(7H)-methanol. Cross-linking agent for resorcinol phenol-formaldehyde or proteinbased resin systems; raw material for synthesis; used in hair care products. *Angus.*

34888 Zolone
2310-17-0 7489 218-996-2
$C_{12}H_{15}ClNO_4PS_2$
S-[(6-chloro-2-oxo-3(2H)-benzoxazoly)methyl] O,O-diethyl phosphorodithioate
S-6-chloro-2,3-dihydro--2-oxobenzoxazol-3-ylmethyl phosphorodithioate; O,O-diethyl phosphorodithioate S-ester with 6-chloro-3-(mercaptomethyl)-2-benzoxazolinone; ENT 27163; 11974 RP; Azofene; Rubitox;. Emulsifiable concentrate containing 350 g/l phosalone; an organophosphorus insecticide and acaricide. Used in control of insects and spider mites on fruit and vegetables. mp = 45-48°; soluble in H_2O (10 mg/l), readily soluble in organic solvents; LD_{50} (rat orl) = 120-175 mg/kg *Hortichem Ltd; Rhône Poulenc Crop Protection Ltd; Voltas.*

34889 Zolyse
9004-07-3 2320 232-671-2
chymotrypsin
Proteolytic enzyme from pancreas. *Alcon Laboratories Inc.* Name unverified.

34890 Zonarez® 7115, 7115 LITE, Alpha 25
Polyterpene resin produced (7115) from dipinene, (Alpha 25) from α-pinene; bright, clear, pale-colored, low molecular weight polymers imparting high levels of tack and adhesion to elastomeric and polymeric materials. Used in manufacture of pressure-sensitive adhesives, rubber cements, hot melt adhesives and coatings, can sealants, caulking, inks, paints, varnishes, chewing gum bases, moisture-resistant soft gelatin capsules of ascorbic acid and its salts. *Arizona.* Discontinued.

34891 Zonarez® Alpha 25
Polyterpene resin based on α-pinene. Provides clear, light-colored thermoplastic polymers imparting tack and strength to adhesives; increases hard resin loading; co-tackifier for the manufacture of pressure-sensitivie adhesives, solvent-based adhesives, emulsion adhesives and hot-melt adhesives/coatings. Especially designed as co-tackifier for adhesive systems based on thermoplastic block copolymers such as Kraton®, plasticizing and softening properties. FDA compliant. *Arizona.* Discontinued.

34892 Zonatac®
Modified polyterpene resins. Tackifying resins for adhesives and sealants. *Arizona.* Discontinued.

34893 Zonatac® 105 LITE, 501 LITE
Modified terpene hydrocarbon resin; light-colored, low odor thermoplastic resins for use in hot-melt adhesives and coatings, hot-melt pressure-sensitive adhesives, and solvent-based thermoplastic elastomer pressure-sensitive adhesive systems; FDA compliance *Arizona.* Discontinued.

34894 Zonatac® 105, 501
Modified terpene hydrocarbon resin; light-colored, low odor thermoplastic resins for use in hotmelt adhesives and coatings, hot-melt pressure-sensitive adhesives, and solvent-based thermoplastic elastomer pressure-sensitive adhesive systems; FDA compliance. *Arizona.* Discontinued.

34895 Zonester® 55
Glycerol ester of Arizona DR-24 disproportionated tall oil rosin; thermoplastic resin ester used as tackifier for difficult-to-tackify elastomers such as SBR as well as natural and syn. latexes; used in pressure-sensitive adhesives, contact cements, latex adhesives where good flexibility at low temperatures is desirable. *Arizona.* Discontinued.

34896 Zonester® 65
8050-26-8 232-479-9
Pentaerythritol ester of disproportionated tall oil rosin; thermoplastic resin ester used in SBR latex, natural rubber, and neoprene adhesives. *Arizona.* Discontinued.

34897 Zonester® 85
Glyceryl ester of tall oil rosin; thermoplastic resin ester used in chewing gums, hot-melt adhesives, hot-melt coatings, mastic adhesives, contact cements, and pressure-sensitive adhesives. *Arizona.* Discontinued.

34898 Zonolite®
A proprietary trademark for a mineral product made by the heat treatment of vermiculite, a mineral similar to a crude mica. Used in the manufacture of building materials and as high temperature insulation. No manufacturer.

34899 Zonyl®
Fluorochemical surfactant. *DuPont UK.*

34900 Zonyl® FSA, FSE, UR
Fluorochemical surfactant; wetting agent, emulsifier, dispersant, corrosion inhibitor, leveling agent for adhesives, agricultural, polishes, polymerization, pigment grinding, cleaners, coatings and paints, fire fighting, *DuPont.*

34901 Zonyl® FSE
Fluorochemical surfactant, leveling agent for emulsions, pigment dispersant, wetting and foaming agent for corrosive media. *DuPont.*

34902 Zonyl® UR
Fluorochemical surfactant. Wetting agent, antifoam, corrosion inhibitor and leveling agent for floor polishes. *DuPont.*

34903 zootic acid
74-90-8 4836 200-821-6
HCN
hydrocyanic acid
hydrogen cyanide; prussic acid; blausäure; formonitrile; zyklon. Used in the manufacture of acrylonitrile, acrylates, dyes, and pesticides. mp = -13.4°; bp = 25.6°; d = 0.941 (air = 1); LC_{50} (rat inh 5 min) = 544 ppm.

34904 Zopaque
13463-67-7 9612 236-675-5
titanium dioxide
A proprietary form of titanium dioxide for rubber mixing. No manufacturer.

34905 Zoramide CM
68140-00-1 268-770-2
Cocamide MEA
Foam booster, thickener, superfatting agent. *Zohar Detergent Factory.*

34906 Zoramox
Coconut amido alkyl amine oxide
Wetting agent, viscosity booster/stabilizer for shampoos, bubble baths; viscosity builder for low pH shampoos, other liquid detergents. *Zohar Detergent Factory.*

34907 Zorbax®
For the medical industry. *DuPont UK.*

34908 Zorite
A proprietary alloy containing 35% nickel, 15% chromium, 1.75% manganese, and 0.5% manganese, and 0.5% carbon with iron. No manufacturer.

34909 Zs-04-xxyyAU
Anisotropically conductive flip chip adhesives; microelectronic adhesives curable at moderate temps. *Zymet.*

34910 Z-Siloxide
A trade name for a silica glass obtained by fusing silica with 0.1-2.0% zirconium dioxide. No manufacturer.

34911 Zumisite
Yeast food. *ABM Chemicals Ltd.* Name unverified.

34912 Zusolat 1004
Fatty alcohol polyglycol ether (4EO)
Washing and cleansing agent. *Zschimmer & Schwarz.* Discontinued.

34913 Zusolat 1005/85
Fatty alcohol polyglycol ether (5 EO)
Dispersant, emulsifier, wetting agent for manufacture. of washing and cleaning agents, dishwash, household/industrial and metal cleaners, textile, leather, and paper auxiliary. *Zschimmer & Schwarz.* Discontinued.

34914 Zusomin C 108
61791-14-8
PEG-8 cocamine
Basic material and component for dyeing and textile auxiliaries; intermediate for quaternization; metal cleaners. *Zschimmer & Schwarz.* Discontinued.

34915 Zusomin O 105
PEG-5 oleamine
Basic material for textile and dyeing auxiliaries, intermediate for quaternization, metal cleaners. *Zschimmer & Schwarz.* Discontinued.

34916 Zusomin S 110
26635-92-7
PEG-10 stearamine
Basic material for dyeing and textile auxiliary; intermediate for quaternization, metal cleaners. Zschimmer & Schwarz. Discontinued .

34917 Zusomin TG 102
61791-44-4 263-177-5
PEG-2 tallowamine
Basic material and component for dyeing and textile auxiliary; intermediate for quaternization and metal cleaners. *Zohar Detergent Factory.*

34918 Zylar® 90-495, Zylar® 93-541
Methyl methacrylate butadiene styrene terpolymer
Features excellent gamma ray recovery, high clarity, abuse/scratch resist.; for medical devices, office accessories, small appliances, and toys. *Novacor.*

34919 Zylar® 91
Acrylic
Clear impact acrylic for extrusion with excellent gamma ray recovery, high clarity, abuse/scratch resistance; for medical devices, thermoformed packaging. *Novacor.*

34920 Zylar® 93-541
Methyl methacrylate-butadiene-styrene terpolymer. A clear impact acrylic characterized by exceptional moldability and abuse resistance. Used for displays, medical devices, office accessories, small appliances and toys. *Novacor.*

34921 Zylar® ST 94-561
Alloy; clear impact resistant alloy with high clarity, strong and toughness and superior molding characteristics; for medical devices, safety devices, and toys. *Novacor.*

34922 zymase
An alcohol-producing enzyme secreted by yeast cells that decomposes grape sugar. No manufacturer.

34923 zymin
Permanent yeast. A product obtained by partially drying ordinary yeast, immersing it in acetone for 15 minutes, which kills yeast, drying on filter paper, and washing with ether. It produces alcohol from grape sugar. No manufacturer.

34924 zymocasein
A phospho-protein obtained from yeast. No manufacturer.

34925 Zymogen
A commercial product consisting of a nitrogenous substance that provides food for yeast in fermentation. No manufacturer.

PART II

THESAURUS

Chemical and Trade Names

Chemical and Trade Names

238, *See* Vestolit®, 33679
249, *See* Maneb, 18526
612, *See* Ethohexadiol, 10968
08-union carbide, *See* Rhodapon® BOS, 26632
103EP8, *See* Vestolit®, 33679
11,561 RP, *See* Carbetamex, 5189
110A, *See* Vestolit®, 33679
11974 RP, *See* phosalone, 23691
1358F, *See* Dapsone, 7751
1540 DS, *See* PEG-150 distearate series, 23037
195J, *See* Vestolit®, 33679
200 + 235 mesh barium sulfate, *See* Nycor®Barytes, 21882
200 + 325 mesh barium sulfate, *See* Nycor® Celestite, 21883
200 DS, *See* PEG-150 distearate series, 23037
23P, *See* Wessalith, 34284
2b, *See* Sumquat® 6210, 30126
2B, *See* stearalkonium chloride, 29592
309M, *See* Vestolit®, 33679
35SL, *See* Nacconol® 40G, 20647
3CF, *See* Genomoll P, 12838
40% Solids methyl methacrylate polymer in MEK, *See* Acryloid® A-101, 425
4010 NA, *See* Vulkanox® 4010 NA, 34147
4H, *See* Trilon® BS, 32327
522 Rubber accelerator, *See* Accelerator 2P,187
550 saccharine, *See* saccharin, 27195
58 12 315, *See* Propoxur, 25198
6000 DS, *See* PEG-150 distearate series, 23037
6000-DS Mapeg®, *See* PEG-150 distearateseries, 23037
61 Tandem 5K, *See* Ethylan® GES6, 11108
611C55, *See* thenium closylate, 31568
8 Tween® 60, 60K, 61, *See* Ethylan® GES6,11108
80WP, *See* cyanazine, 7472
96-H-60, *See* Haloxon, 13567
A, *See* α-alanine, 1333
A 361, *See* atrazine, 3394
A 1114, *See* prometryn, 25140
A 1530, *See* Timonox, 31806, 31807
A 1866, *See* terbutryn, 31230
A 2079, *See* simazine, 28464
A 10846, *See* DMSO, 8762
A 40% solution of acrylamide., *See* **Acrylo-40**,420
A Metal, 1
A-Fax®, 920
A.B.S. 87%, 2
A.T.S, 3
A-0020, 4
A-1, A-1 Thiocarbanilide, 5
A-17, 7; *See* Butane, 4734
A-108, 9
A1530, *See* Octoguard FR-10, 22000
A1582, *See* Octoguard FR-10, 22000
a1588 lp, *See* Octoguard FR-10, 22000
A-2, 6; *See* Akrochem® Thio No. 1, 1181
A2 Monobel, *See* Monobel, 20219
A-31, 8
A-3823A, *See* monensin, 20208
A-4696, *See* Kamoran, 15708
A-5MP, *See* adenosine monophosphate, 680
A-625/641ABS 301K, ABS 500FR-1, 10
A7 Vapam, *See* metam-sodium, 19444
AA, 11
AA Standard, AA USP, 12
AA#2 Lime Additive, 13
AA-9, AA-10, *See* Nacconol® 40G, 20647; sodium dodecylbenzenesulfonate, 28760
AA-10, *See* Nacconol® 40G, 20647; sodium dodecylbenzenesulfonate, 28760
A-acid, 14
AAmlube V, 15
Aaphytora, *See* Zineb, 34807
AAprotect, 16; *See* Accelerator MZ Powder, 209
AAtack, *See* Rezifilm, 26521
Aaterra WP, 17
Aatex, 18
Aatrex, *See* atrazine, 3394
Aatrex 4l, *See* atrazine, 3394
Aatrex 80W, *See* atrazine, 3394
Aatrex Nine-O, *See* atrazine, 3394
Aavolex, *See* AAprotect, 16
Aazira, *See* AAprotect, 16
AB, 19
AB1000-F, 20
AB3500, 3500AR, 21
abaca, 22
Abalon, 23
Abalyn, 24
aba-odo, 25
Abate 1-SG, 2-CG, 4-E, 5CG, 26
Abatia, 27
Abavit B, 28
Abavit S, 29
Abbalide, 30
Abbarome, 31
Abbavert, 32
Abbcite No. 2, 33
Abbocillin-DC, *See* Penicillin V, 23111
Abbokinase, *See* urokinase, 33145
ABC Trieb, 34
Abcure S-40-25, 35
Abelite, 36
Abel's reagent, 37
Abensanil, *See* Valadol, 33238
Abequito, 38
Abeson Nam, *See* Nacconol® 40G, 20647, sodium dodecylbenzenesulfonate, 28760
Abex® 12S, 39
Abex® 23S, 40
Abex® EP-110, 41, *See* Ammonium nonoxynol-4sulfate, 2246
Abex® JKB, 42
Abex® VA 50, 43
Abex®LIV/30, 44
abietic acid, 45
Abietic anhydride, 46
abietinic acid, *See* abietic acid, 45
Abietyl alcohol, dihydro-, *See* Abitol® E, 71
Abil® AV 20-1000, *See* Phenyl trimethicone, 23635
Abil® AV 8853, *See* Phenyl trimethicone, 23635
Abil® AV 8853, 20-1000, 47
Abil® B 8839, 48
Abil® B 8851, 8852, 49
Abil® B 9950, 50
Abil® B 88183, 88184, 51
Abil® EM-90, 97, 52
Abil® K 4, 53
Abil® OS12, OS13, *See* Abil® OS5, 54
Abil® OS5, 54
ABIL® OSW 12, OSW 13, 55
Abil® S201, 56
Abil® S255, 57
Abil® Wax 2434, 58
Abil® Wax 2434, 59
Abil® Wax 2440, 60
Abil® Wax 9800, 61
Abil® Wax 9801, 62
Abil® Wax 9809, 63
Abil® Wax 9814, 64
Abil® WE 09, 65
Abil® WS 08, 66
Abil®10-10000, 67
Abil®-Quat 3270, 3272, 68
Abiol, 69; *See* Imidazolidinyl urea, 14896
Abisol, 70
Abitol® E, 71
Ablefilm® 550, 72
Ablufoam HT, 73
Ablufoam SAE, 74
Abluhide DS, 75
Abluhide F Series, 76
Ablumide CDE, 77
Ablumide CME, 78
Ablumide LDE, 79
Ablumide LME, 80
Ablumide SDE, 81
Ablumide SME, 82
Ablumine 08, 83
Ablumine 08, 10, 12, 1214, 3500, *See* Benzalkonium chloride, 3958
Ablumine 230, 84
Ablumine 280, 85
Ablumine D10, 86
Ablumine DHT75, 87
Ablumine DT, 88
Ablumox C-7, 89
Ablumox CAPO, 90
Ablumox LO, 91; *See* lauramine oxide, 16832
Ablumox T-15, 92
Ablumul AG-306, 93
Ablumul EP, 94
Ablunol 200ML, 95
Ablunol 200MO, 96
Ablunol 200MS, 97
Ablunol CO 10, 98
Ablunol DEGMS, 99
Ablunol EGMS, 100
Ablunol GML, 101
Ablunol GMO, 102
Ablunol GMS, 103
Ablunol LA-3, 104
Ablunol LMO, 105
Ablunol NP4, 106
Ablunol OA-6, 107; *See* oleth-8, 22126
Ablunol S-20, 108
Ablunol S-40, 109
Ablunol S-60, 110
Ablunol S-80, 111
Ablunol S-85, 112
Ablunol SA-7, 113; *See* steareth series, 29596
Ablunol T-20, 114
Abluphat AP Series, 115
Abluphat LP Series, 116
Ablupol AF, 117
Ablusoft A, 118
Ablusoft C-70, 119
Ablusoft ES, 120
Ablusoft PE, 121
Ablusol C-78, *See* dioctyl sodium sulfosuccinate, 8550
Ablusol DA, 122
Ablusol DBC, 123
Ablusol DBD, 124
Ablusol DBM, 125; *See* Ammonium dodecylbenzene sulfonate, 2238
Ablusol DBT, 126
Ablusol LAE, *See* disodium lauryl sulfosuccinate, 8649
Ablusol LDE, 127
Ablusol LMS, 128
Ablusol ML, 129
Ablusol PM, 130
Abluter BE, 131
Abluter BE, CPB, *See* Cocamidopropyl betaine,6551
Abluter DCM-2, *See* disodium cocoamphodiacetate, 8643
Abluter GL Series, 132
Abluton A, 133
Abluton CAT, 134
Abluton CMN, 135
Abluton CTP, 136
Abluton MK9, 137
Abluton SR, 138
Abluton T30, 139
Abluwax EBS, 140
ABM 5C Chlormequat Plus, 141
ABM Chlormequat 40, 72.5, 142
Abocast, 143
Abocrete, 144
Abocure, 145
Abol, 146; *See* pirimicarb, 23879
abomasal enzyme, *See* rennet, 26110
Abopon, 147
Aboseal, 148
Abracol®, 149
Abradux, 150; *See* A-2, 6
Abramant, 151; *See* A-2, 6
Abramax, 152

Chemical and Trade Names

Chemical and Trade Names

Chemical and Trade Names

Adaptinol, 627
Adarola, 628
ADC, 629
Adcora A3, 630
Adcora P6, 631
Adcora SP, 632
Adcora V, 633
Adcortyl with Graneodin Cream, 634
Adcote, 635
Add F, *See* formic acid, 12271
Add It To Oil, 636
Addabond, 637
Addacoat, 638
Addacol, 639
Addaflex, 640
Addaflor, 641
Addagrout, 642
Addalevel, 643
Addamortar, 644
Addapitch, 645
Addaprime, 646
Addaseal, 647
Addasure, 648
Adder, 649
Adder, 650
Add-H, 651
Addipast, 652
Additin 30, 653, Akrochem® Antioxidant PANA,1161, Antioxidant PAN, 2590, Naugard® PANA, 20798
Additive-A, 654, *See* calcium lignosulfonate, 4956, 4957
Additol, 655, 656
Add-M, 657
Adehyde-amine, *See* Accelerator A50, 194
Adeka Catioace DM, PD Series, 658
Adeka CR-5, 659
Adeka Dipropylene Glycol, *See* Dipropylene glycol, 8611
Adeka ED-505, 660
Adeka EP-4100, 661
Adeka Estol, 662
Adeka GH-200, 663
Adeka Glycilol ED-505, *See* Adeka ED-505, 660
Adeka Hypote, 664
Adeka Kiku-Lube, 665
Adeka Lub E-500, 666
Adeka Optomer KR Series, 667
Adeka P-400, 668
Adeka PR-3007, 669
Adeka Sakura-Lube 100, 670
Adeka Sole CO, 671
Adeka Sole YA, 672
Adekacol CS, PS, TS, 673
Adekacol EC Series, 674
Adekamine E Series, 675
Adekanol, 676
Adekatol DES, DS, HAN, LS, TR, SAN, YES, 677
Adekatol LA, LO, NP, OA, PC, SO Series, 678
Adenex, *See* vitamin C, 33938
Adenine, 679
Adeno, *See* 5'-Adenylic Acid, 683
adenosine 5'-(dihydrogen phosphate), *See* 5'-adenylic Acid, 683
adenosine monophosphate, 680, *See* AMP, 2301,My-B-Den, 20547
adenosine phosphate, *See* 5'-adenylic Acid, 683
adenosine-5'-phosphoric acid, *See* adenosine monophosphate, 680
adenosine, 5'-(tetrahydrogen) triphosphate, *See* Adenosine triphosphate, 681
adenosine triphosphate, 681
Adenotriphos, 682
5'-adenylic Acid, 683, *See* adenosine monophosphate, 680, My-B-Den, 20547
t-adenylic acid, *See* adenosine monophosphate, 680
Adeps lane, *See* Vigilan, 33750
Adepsine oil, *See* Kaydol, 15810, paraffin, liquid., 22744
Adequan, 684
Adergon, *See* gentian violet, 12845
Adermin hydrochloride, *See* pyridoxinehydrochloride, 25483
Adermykon, *See* Chlorphenesin, 6161
Adetol, *See* Adenosine triphosphate, 681
ADF-600, 685
ADF-610, 686
Adflex, 687
Adheso, 688
Adimoll® BO, 689
Adimoll® DB, 690
Adimoll® DH, 691
Adimoll® DN, 692
Adimoll® DO, 693
Adimoll® DO, *See* dioctyl adipate, 8548
Adinol, 694, 695
Adinol CT, 696, 697
Adion, *See* permethrin, 23384, 23385, 23386
Adipic acid, 698, *See* Adi-pure®, 706
adipic acid bis(2-ethylhexyl) ester, *See* Kodaflex® DOA, 16272, Plasthall® DOA,24002, Uniflex® DOA, 32945, Wickenol® 158,34393, Unimate® DIPA, 32968
Adipocere, 699
Adipol 2EH, *See* dioctyl adipate, 8548, Plasthall® DOA, 24002, PX-238, 25440, Wickenol® 158, 34393
Adipon, 700
Adiprene®, 701
Adiprene® BL-16, *See* Adiprene®, 701
Adiprene® L-100, 702
Adiprene® L-42, 703
Adiprene® LW-570, 704
Adiprene® M-400, 705
Adi-pure®, 706, *See* Adipic acid, 698
Adirondackite, 707
Adjab fat, *See* Njave butter, 21392
Adjunct B, 708
Adjust 4, 709
ADK CIZER C-8, *See* Jayflex® TOTM, 15559
ADK CIZER E-500, *See* Chlorinated paraffin,6105
ADK CIZER O-130P, 710
ADK CIZER O-180A, 711
ADK STAB 144, 712
ADK STAB 465, 713
ADK STAB 466, 714
ADK STAB 1292, 715
ADK STAB 1413, 716
ADK STAB 1500, 717
ADK STAB 2335, 718
ADK STAB AC-122, 719
ADK STAB AC-133, 720
ADK STAB AC-169, 721
ADK STAB AP-536, 722
ADK STAB BT-11, 723, *See* dibutyltin laurate, 8371
ADK STAB BT-31, 724, *See* Dibutyltin maleate, 8372
ADK STAB BT-53A, *See* Dibutyltin maleate, 8372
ADK STAB BT-83, 725
ADK STAB EC-14, 726
ADK STAB FL-21, 727
ADK STAB GR-16, 728
ADK STAB LA-32, 729
ADK STAB LA-57, 730
ADK STAB LS-2, 731, *See* Dibutyltin maleate, 8372
ADK STAB LS-8, 732, *See* butyl stearate, 4781
ADK STAB NA-11, 733
ADK STAB OF-14, 734
ADK STAB OT-1, 735
ADK STAB OT-9, 736
ADK STAB RUP-9, 737
ADM, 738
ADM-407, 407C, 739
Adma®, 740
ADMA-2, *See* Onamine 12, 22164
Adma® 8, 10, 12, 741
Adma® 12, *See* dimethyl lauramine, 8507
Adma® 14, 742
Adma® 16, 743, *See* N,N-dimethylhexadecylamine, 8524
Adma® 18, 744
Adma® 246-451, 745
Adma® 1214, 746
Adma® WC, 747
Admerol®, 748
Admerol® 75-M-70, 749
Admex®, 750
Admiral® FPS Type 3089 FS, 751
Admiralty brass, 752
Admiralty gun metal, 753
Admiralty nickel, *See* Adnic, 762
Admiralty nickel, *See* Adnic, 762
Admiralty white metal, 754
Admire, 755, *See* Confidor, 6725
Admos Alloys, 756
Admox®, 757
Admox® 14-85, 758
Admox® 18-85, 759
Admox® 1214, 760
Admul, 761
Adnic, 762
Adobacillin, *See* Ampicillin, 2361
Adogen®, *See* tallowamine, 30748
Adogen® 66, 763
Adogen 73, *See* Unislip 1759, 33045
Adogen® 137, 764
Adogen® 170, 765
Adogen® 185, 766
Adogen® 412, 767
Adogen® 417, 768, Arquad® S-50, 3120
Adogen® 442, 769
Adogen® 442-P100, 770
Adogen® 461, 771
Adogen® 462, 772
Adogen® 470, 773
Adogen® 471, 774
Adogen® 477, 775
Adogen® MA-108 SF, 776
Adogen® MA-112 SF, 777
Adogen® S-18 V, 778
Adogen® TA-100, 779,
Adogen® TA-100, *See* Adogen® TA-101, 780,
Adogen® TA-100, TA-101, *See* Distearyldimonium chloride, 8714, Genamin DSAC, 12771
Adogen® TA-101, 781
Adogenen 142, *See* Amine 18-90, 2131, SteamfilmFG, 29589, Armeen® 18,2938, Armid® HTD, 2963
Adol, *See* lenacil, 16984, Vizor, 34001
Adol® 52 NF, 782
Adol® 62 NF, 783
Adol® 64, 784
Adol® 66, 785
Adol® 85, 786
Adox 3125, 787
Adpro AP 2112-GP, 788
Adpro AP 8210-HS, 789, Adpro AP 2112-GP, 788
Ad-Pro-MTS, 790
Adroit, 791
Adronal, 792
Adronal acetate, 793
Adsee® 775, 794
Adsee® 799, 795
Adtac™ LV, 796
Adurol, 797
Aduvex®, 798
Advagum, 799
Advance, 800
Advantage, 801, *See* carbosulfan, 5245
Advantage 52-B, 802
Advantage 70DYX, 803
Advantage 136 Defoamer, 805
Advantage 1007B Defoamer, 804
Advantage CP, 806
Advantage DF 110, 807
Advantage M104 Defoamer, 808
Advantage M1251 Production Aid, 809
Advantage™ 10 Defoamer, 810
Advantage® 101M, 811
Advantage® 124, 812
Advapak® ML-1325, 813

Chemical and Trade Names

Chemical and Trade Names

aldehyde ammonia, *See* Velsan, 33539
aldehyde C_1, *See* formaldehyde, 12258
aldehyde C_{14}, 1536
Alder bark, 1537
Alderton's solution, 1538
Aldesan, *See* glutaraldehyde, 13062, Sonacide,29073
aldifen, *See* Dinitra, 8540
Aldo HMO, *See* Ablunol GMO, 102, Radiasurf®7150, 25801, Witconol 2421, 34500
Aldo MO, *See* Radiasurf® 7150, 25801, Witconol 2421, 34500
Aldo® HMS KFG, 1539
Aldo® MLD, 1540
Aldo® MO, 1541
Aldo® MR, 1542
Aldo® MS FG, 1543
Aldo® PGHMS KFG, 1544
Aldobond, 1545
Aldocit, *See* Aldrin, 1558
Aldocoat, 1546
Aldogen, 1547
Aldol 62, *See* Adol® 62 NF, 783, Stearal, 29591
Aldol-α-naphthylamine resin, *See* AntioxidantRES, 2592
Aldomax GA-100, 1548
Aldones, 1549
D-Aldonolactone, *See* GDL, 12705
D-*threo*-Aldono-1,5-lactone, *See* GDL, 12705
Aldosperse, 1550
Aldosperse® 40/60 FG, 1551
Aldosperse® ML 23, 1552
Aldosperse® MO-50, 1553
Aldosperse® MS-20 FG, 1554
Aldosperse® O-20 KFG, 1555
Aldosperse® TS-40 KFG, 1556
Aldrex, *See* Aldrin, 1558
Aldrey, 1557
Aldrin, 1558
Aldrin, *See* Aldrin Dust, 1559
Aldrin Dust, 1559
aldrine, *See* Aldrin, 1558
Aldrin-R, *See* Aldrin, 1558
Aldrite, *See* Aldrin, 1558
Aldron, *See* Aldrin, 1558
Aldrosol, *See* Aldrin, 1558
Aldrox, *See* aluminum hydroxide, 1905
Aldur, 1560
Al-dur-ba, 1561
Aldydale, 1562
Aldyl®, 1563
Alecra, 1564
Alegro, *See* phenmedipham, 23595, 23596,23597, 23598
Alepol, 1565
Alexandrian laurel oil, *See* laurel nut oil., 16833
Alexis, 1566
Alexis Antibloom, 1567
Alexite, 1568, *See* aluminum oxide, 1919
alfa, 1569
Alfa, Aquilite, *See* Thiovit, 31724
Alfa®, 1570
Alfacron, 1571, *See* azamethiphos, 3517, Iodofenphos, 15213
Alfacron 10WP, 1572, *See* azamethiphos, 3517
Alfadex, 1573
Alfaprostol, 1574
Alfavet, *See* Alfaprostol, 1574
Alfenide, *See* nickel silvers, 21132
Alfenol 8, *See* Ablunol NP4, 106
Alfenol 18, *See* Ablunol NP4, 106
Alfenol 22, *See* Ablunol NP4, 106
Alfenol 28, *See* Ablunol NP4, 106
Alfenol 710, *See* Ablunol NP4, 106
Alferium, 1575
Alferric, 1576, *See* antimony trichloride, 2573
Alficron, *See* azamethiphos, 3517
Alfol, 1577
Alfol 18, *See* Adol® 62 NF, 783, Stearal, 29591
Alforder, 1578
Alfralat, 1579
Alfrax B301, 1580, aluminum oxide, 1919
Alftalat, 1581
Algae Treat, 1582
Algafen, 1583, *See* Anthiphen, 2520
Algalex 104, 1584
Algalith, 1585
algaroba, *See* locust bean gum, 17545
Algarobilla, 1586
Algarobillin, 1587
Algarovilia, 1588
algeldrate, *See* aluminum hydroxide, 1905
Alger metal, 1589
Algerite, *See* Akrochem® Antioxidant PANA,1161, Antioxidant PAN, 2590 Naugard® PANA, 20798
Algier's metal, 1590
Algin, 1591 *See* Kelcosol®, 15832, Kelgin,15841, Kelgin® F, 15842, kelgin® HV, LV, MV,15843, Kelgin® QH, QL, QM, 15844, Kelgin® XL, 15845, Kelgum, 15846, Keltex®, 15878, Keltex® HV, 15879, Keltex® S, 15880, Keltone®, 15883, Keltone® HV, LV, 15884, Mozanon®, 20386
Alginade MR, MRE, 1592
Alginate, *See* Cocoloid®, 6565, Concentrated Dariloid®, 6709, Concentrated Dariloid® XL,6711, Dariloid® Q, 7775, Dariloid® QH, 7776, Wakal® A Range, 34195
alginic acid, 1593, *See* Kelacid®, 15824, Saltialgine H8, 27297, Ammonium alginate,2228,
alginic acid, ester with 1,2-propanediol, *See*propylene glycol alginate, 25216
Alginic acid NF, *See* Satialgine H8, 27481
Alginic acid potassium salt, 1594, *See* Kelmar®, 15856, potassiumalginate,24735, Alginic acid sodium salt, *See* Kelcosol®, 15832, Kelgin, 15841, Kelgin® F, 15842, kelgin® HV,LV, MV, 15843, Kelgin® QH, QL, QM, 15844, Kelgin® XL, 15845, Kelgum, 15846, Keltex®,15878, Keltex® HV, 15879, Keltex® S, 15880, Keltone®, 15883, Keltone® HV, LV, 15884
Alginoid iron, *See* Algiron, 1596
Alginoplast®, 1595
Algiron, 1596
Algisium-C, 1597
Algistat, 1598
Algitox®, 1599
Algodon, 1600
Algodon de Seda, 1601
Algofen, 1602, *See* Anthiphen, 2520
Algoflon®, 1603
algofrene type 2, *See* dichlorodifluoromethane,8388
Algol, 1604
Algol®, 1605
Algon 100, 1606
Algrain, *See* ethyl alcohol, 11064
Algran, *See* Aldrin, 1558
Algulose, 1607
Algylen, *See* trichloroethylene, 32217
Alhydex, *See* glutaraldehyde, 13062, Sonacide,29073
Aliant, *See* Vulkacit® 1000, 34124
alibated iron, 1608
Alibi, 1609
Alicep®, 1610
Ali-Clean, 1611
Aliette, 1612, 1613
Aliette Extra, 1614
Alimemazine, *See* trimeprazine,
Alimet, 1615, *See* methionine hydroxy analog,19486
Alipa®, 1616
Aliphatic alcohol, *See* propylene glycol, 25214
Aliphatic alcohol alkoxylate, *See* Burco LAF-6,4690
Aliphatic amide, *See* oleamide, 22107
Aliphatic amine, *See* Diethanolamine, 8434, diisopropanolamine, 8466,Norcure 131, 21535,
Aliphatic amine tetrol, *See* Amicure® CL-485,2076
aliphatic dialdehyde, *See* Highlink 40®, 80®,14085
Aliphatic diol, *See* ethylene glycol, 11123
Aliphatic ethoxylate, *See* Uniperol® W, WFlakes, 33000
Aliphatic ethoxylates, *See* Intratex® CA-2,15179
Aliphatic hydrocarbon, *See* n-heptane, 13790
Aliphatic hydrocarbon with oxygenerated hydrocarbons, *See* Actrel, 562
aliphatic ketone, *See* Dihydroxyacetone, 8459
Aliphatic phosphate ester, *See* ChemphosTC-444, 6012
Aliphatic polyurethane, *See* Baymod® PU, 3816
Aliphatic resin, stabilized with 0.05%antioxidant, *See* Adtac™ LV, 796
Aliphatic urethane acrylate, *See* CN 953, 6478
Aliphatic urethane acrylate resin, *See* CN 962,6480, CN 966, 6481
Alirox, *See* EPTC, 10729
Aliso, 1617, *See* aluminum isopropoxide, 1909
Alisol®, 1618
Alistell, 1619, 1620
Alizarin, 1621
Alizarin CAF, *See* flavopurpurin, 11904
Alizarin DCA, *See* flavopurpurin, 11904
Alizarin FA, *See* flavopurpurin, 11904
Alizarin GB, *See* flavopurpurin, 11904
Alizarin GD, RF, RT, RX, SC, SSA, SX, SX Extra, WG, *See* anthrapurpurin, 2531
Alizarin GI, *See* flavopurpurin, 11904
Alizarin JCA, *See* flavopurpurin, 11904
Alizarin No. 10CA, *See* flavopurpurin, 11904
Alizarin oil, *See* Turkey red oils., 32607
Alizarin RG, *See* flavopurpurin, 11904
Alizarin SDG, *See* flavopurpurin, 11904
Alizarin VAR, *See* flavopurpurin, 11904
Alizarin VCA, *See* flavopurpurin, 11904
Alizarin X, *See* flavopurpurin, 11904
Alizarine, 1622
Alka, 1623
Alkaflo®, 1624
Alkafoam D, 1625
Alkagel, 1626
Alkagel I.S., *See* aluminum hydroxide, 1905
alkali cellulose, 1627
Alkali Surfactant NM, 1628
Alkalit, 1629
Alkalsite, 1630
Alkamide® 101 CG, 1631
Alkamide® 200 CGN, 1632
Alkamide® 327, 1634, *See* Alkamide® LE, 1646
Alkamide® 2104, 1633
Alkamide® C-212, 1635
Alkamide® C-5, 1636
Alkamide® CDE, 1637
Alkamide® CP-1255, 1638
Alkamide® DC-212/MP, 1639
Alkamide® DIN-295/S, 1640
Alkamide® DO-280, 1641
Alkamide® DS-280/S, 1642
Alkamide® HTDE, 1643
Alkamide® L-203, 1644
Alkamide® L7DE, 1645
Alkamide® LE, 1646
Alkamide® LIPA/C, 1647
Alkamide® OIP, 1648
Alkamide® R-280, 1649
Alkamide® S-280, 1650
Alkamide® SDO, 1651
Alkamide® STED, *See* Advawax® 290, 820
Alkamide® STEDA, 1652
Alkamide® WRS 1-66, 1653
Alkaminox® C-2, 1654
Alkamuls SML, *See* Kemester® S20, 15966, Radiasurf® 7125, 25797, Sorbon S-20, 29117
Alkamuls SMO, *See* Kemester® S80, 15970, Radiasurf® 7155, 25802, Sorbon S-80, 29121
Alkamuls SMS, *See* Radiamuls® Sorb 2145, Kemester® S60, Sorbon S-60, 29119
Alkamuls® 14/R, 1655
Alkamuls® 400-DO, 1656
Alkamuls® 400-DO, 600-DO, *See* PEG-2 dioleate series, 23040
Alkamuls® 400-MO, 1657
Alkamuls® 600-DO, 1658
Alkamuls® A, 1659
Alkamuls® AG-900, 1660

Chemical and Trade Names

Alkylolamides, *See* Ninol®, 21231
Alkylphenol, *See* Antioxydant NV3, 2596
Alkylphenol alkoxylate, *See* Makon® NI10, NI20, NI30, 18429
Alkylphenol ethoxylate, *See* Basopon® LN,3725
Alkylphenol oxyethylate, *See* Emulan® PO,10491
Alkylphenol polyglycol ether, *See* Marlophen® X, 18889
Alkylphenoxy POE acid phosphate, *See* Fosterge LF, 12335
Alkylsulfamido carboxylic acid, *See* BohrmittelHoechst, 4376
Alkynol®, 1735
all-*trans*-retinol, *See* Vi-Alpha, 33706
all-*trans*-squalene, *See* squalene, 29358
Allabond Twenty/Twenty Adhesive, 1736
Allabond Twenty/Twenty NM, 1737
Allactol, 1738
Allantoin, 1739, *See* Fancol TOIN, 11453
Allegheny 33, 44, 55, 66, 1740
Allegheny Metal, 1741
Allenite, *See* pentahydrite,
Allenoy, 1742
Allen's metal, 1743
Allercorb, *See* vitamin C, 33938
Alleron, *See* parathion, 22807
allethrin, 1744
allethrine, *See* allethrin, 1744
allethrins, *See* D-Trans allethrin, 9070
Alleviate, *See* allethrin, 1744, piperonyl butoxide, 23875
Allguard, 1745
Allie, Brushoff, *See* metsulfuron-methyl, 19606
Allied Whiting, *See* CP Filler, 6928
Alligator Pearl oil, *See* Avocado Oil, 3498
Alligator wood, 1746
Allisan, 1747, *See* dicloran, 8395, Fumite Dicloran, 12459
All-O, 1748
Alloid, *See* Kelcosol®, 15832 Kelgin, 15841 Kelgin® F, 15842 kelgin® HV, LV, MV, 15843 Kelgin® QH, QL, QM, 15844 Kelgin® XL,15845 Kelgum, 15846 Keltex®, 15878 Keltex® HV, 15879 Keltex® S, 15880 Keltone®, 15883 Keltone® HV, LV, 15884
allomaleic acid, *See* Fumaric acid, 12453
Allophanic acid, 4,4'-*o*-phenylenebis[3-thio-, dimethyl ester, *See* thiophanate methyl, 31706
Alloprene, 1749
Allose, *See* Kelcosol®, 15832 Kelgin,15841,Kelgin® F, 15842 kelgin® HV, LV, MV, 15843 Kelgin® QH, QL, QM, 15844 Kelgin® XL, 15845 Kelgum, 15846 Keltex®, 15878 Keltex® HV, 15879 Keltex® S, 15880 Keltone®, 15883 Keltone® HV, LV, 15884
Alloxan, 1750
alloxan-5-oxime, *See* violuric acid, 33881
Alloxydim-sodium, *See* Clout, 6446
Alloy 24, *See* Alloy Y, 1779
Alloy 109, 1752
Alloy 122, 1753
Alloy 142, 1754
Alloy 145, 1755
Alloy 195, 1756
Alloy 2129, 1757
Alloy 2L5, 1758
Alloy 2L8, 1759
Alloy 39, 1751
Alloy 3L11, 1760
Alloy AM4-4, 1761
Alloy AM7-4, 1762
Alloy AMF, 1763
Alloy AP33, 1764
Alloy JL, 1765
Alloy L10, 1766
Alloy L11, 1767
Alloy L24, *See* Alloy Y., 1779
Alloy L5, 1768
Alloy L7, 1769
Alloy L8, 1770
Alloy MG7, 1771
Alloy N, 1772
Alloy NCT3, 1773
Alloy of copper, tin, lead and zinc, *See* acidbronze, 329
Alloy RR, 1774
Alloy Steel, 1775
Alloy T, 1776
Alloy W.7.1, 1777
Alloy W.9, 1778
Alloy Y, 1779
Alloys Wm, 1780
Alluman, 1781
Ally, *See* metsulfuron-methyl, 19606
Ally®, 1782
Allyl 2,4,6-Trobromophenyl ether, *See* tribromophenyl allyl ether, 32203
allyl alcohol, *See* propenol, 25181
allyl alcohol dibromide, *See* 2,3-dibromo-1-propanol, 8360, dibromopropanol, 8361
Allyl alcohol ethoxylate, *See* Sipomer® AAE,28530
allyl catechol methylene ether, *See* safrol,27229
Allyl chloride, *See* Barchlor, 3626
Allyl diglycol carbonate, *See* CR-39, 6953
Allyl-1,2-dimethoxybenzene, *See* methyl eugenol, 19534
4-allyl-1,2-dimethoxybenzene, *See* methyl eugenol, 19534
Allyl-2,3-epoxypropyl ether, *See* Ageflex AGE,965, Allyl glycidyl ether, 1783, Sipomer® AGE, 28531
1-allyl-2,3-epoxypropane, *See* Ageflex AGE, 965, Allyl glycidyl ether, 1783, Sipomer® AGE, 28531
allyl glucosinolate, *See* sinigrin, 28500
Allyl glycidyl ether, 1783
allyl glycidyl ether-ethylene glycol prepolymer(18/1), *See* Ageflex AGE, 965, Allyl glycidyl ether, 1783, Sipomer® AGE, 28531
Allyl methacrylate, 1784, *See* Ageflex AMA, 966, Sipomer® AM, 28532
4-allyl-1-methoxybenzene, *See* Estragole, 10897
4-allyl-1,2-methylenedioxybenzene, *See* safrol,27229
Allyl Phthalate, *See* diallyl phthalate, 8283, Nonflammable Decobest DA, 21448, RX®1-501N, 27154
4-allyl veratrole, *See* methyl eugenol, 19534
Allyl veratrole, *See* methyl eugenol, 19534
4-allylanisole, *See* Estragole, 10897
allyldioxybenzene methylene ether, *See* safrol,27229
Allylic alcohol, *See* propenol, 25181
Allylmethacrylate, *See* SR-201, 29360
1-allyloxy-2,3-epoxy-propane, *See* Ageflex AGE,965, Allyl glycidyl ether, 1783, Sipomer® AGE, 28531
***p*-Allylphenol**, *See* chavicol, 5861
Allylphenylcinchoninic ester, *See* atoquinol, 3380
***m*-allylpyrocatechin methylene ether**, *See* safrol, 27229
Allyltrimethoxy silane, *See* CA0567, 4847
Allyltrimethyl silane, *See* CA0570, 4848
Almacarb, *See* Alumina hydrate, 1892, aluminumhydroxide, 1905
Almadina, *See* potato gum, 24789
Almag, 1785
Almagel, *See* Alumina hydrate, 1892
Alman, *See* Kelcosol®, 15832, Kelgin, 15841, Kelgin® F, 15842, kelgin® HV, LV, MV, 15843, Kelgin® QH, QL, QM, 15844, Kelgin® XL, 15845, Kelgum, 15846, Keltex®, 15878, Keltex® HV,15879, Keltex® S, 15880, Keltone®, 15883, Keltone® HV, LV, 15884
Almasilium, 1786
Almederm, *See* hexachlorophene, 13992
Almelec, 1787
Almen's reagent, 1788
Almeta, *See* Vaderm, 33237
Almo Steel, 1789
almond artificial essential oil, *See* Benzaldehyde, 3957, Ethereal Oil of Bitter Almonds,10944
Almondamide DEA (1:1), *See* Incromide Ald,14981
Almondamidopropalkonium chloride, *See* Incroquat AL-85, 15029
Almondamidopropyl betaine, *See* IncronamAL-30, 15017
Almondamidopropyl dimethylamine lactate, *See* Incromate ALL, 14965
Almondamidopropylamine oxide, *See* Incromine Oxide AL, 15001
Almora, 1790
Almora, *See* magnesium gluconate, 18362
Almstab, 1791
Alnovol, 1792, 1793
Alocrom, 1794
Aloe vera gel, *See* Aloe Vera Powd.200XXXExtract-Microfine, 1795
Aloe vera gel-lecithin, *See* Dermasome® V,8114
Aloe Vera Powd. 200XXXExtract-Microfine,1795
aloin, *See* Veragel® 200, 33566
Alomite, 1796
Alon®, 1797
Alox, 1798
Alox® 111, 1799
Alox® 152, 1800
Alox® 318F, 1801
Alox® 350, 1802
Alox® 436A, 1803
Alox® 488, 1804
Alox® 575, 1805
Alox® 606, 606-55, 606-70, 1806
Alox® 904, 1807
Alox® 1680, 1808
Alox® 2000, 1809
Alox® 2028L, 1810
Alox® 2211Y, 1811
Alox® 2301, 1812
Aloxicoll, 1813
Aloxite, 1814, *See* aluminum oxide, 1919
Alpacca, 1815, *See* nickel silvers, 21132
Alpakka, *See* Alpacca., 1815
Alpen, *See* Ampicillin, 2361
Alperox-F, 1816
Alpex, 1817, 1818
Alpfa, 1819
Alpha Chymar, 1820
Alpha Daphnone, 1821
Alphacept, *See* Alfaprostol, 1574
Alphachloralose, 1822
Alphachroic, 1823
Alphacypermethrin, *See* Fastac, 11500
Alphadim® 90AB, 1824
Alphadim® 90LC, 1825
Alphadim® 90NLK, 1826
Alphadim® 90SBK, 1827
Alphalin, *See* Vi-Alpha, 33706
Alphamint, 1828
alpha-monoacetin, *See* monacetin, 20095
Alphanol, 1829, 1830
Alphanol®, 1831
Alpha-Ruvite, 1832
Alphasol IB, *See* Rewopol® SBDB 45, 26371
Alphasol OT, 1833
Alpha-Step® MC-48, 1834
Alpha-Step® ML-40, 1835
Alphatex®, 1836, *See* aluminum silicate, 1924
Alphenate, 1837
Alphenol 8, *See* Ablunol NP4, 106
Alphide, 1838
Alphogen, 1839
Alphol, 1840
Alphoxat O 105, 1841
Alphoxat S 110, 1842
Alphozone, *See* Alphogen, 1839
Alpiny, *See* Valadol, 33238
Alplate, 1843
Alpolit, 1844
Alprokyds, 1845
alquifon, 1846
Alreco, 1847
Alresat, 1848
Alresen, 1849
Alromin Ru 1000, 1850
Alrosperse® 100, 1851
Alscoap AF Series, 1852

Chemical and Trade Names

aluminum tin bronze, 1931
aluminum trifluoride, *See* Alcan AluminumFluoride, 1412
aluminum trihydrate, *See* Polarite® 880E(W),24302, aluminum hydroxide, 1905
aluminum trihydroxide, *See* Alumina hydrate,1892
aluminum trioxide, *See* A-2, 6
aluminum, tris(5-oxo-L-prollinato-), *See*aluminum PCA, 1921
aluminum tristearate, *See* aluminum stearate,1929, Novogel® ST, 21729, Synpro® 404,30593, Synpro® 404, 30593
aluminum zirconium tetrachlorohydrexglycine, *See* Reach® AZP-701, AZP-703, 25920, Rezal® 36GP, 26518
aluminum/zirconium chlorohydrate, *See* DowCorning® AZG-368, 8872
aluminum/zirconium glycine, *See* Dow Corning® AZG-370, 8873
aluminum-boro-tartrate, *See* Boral, 4409
aluminum-formaldehyde-sulfite, *See* Moronal,20342
aluminum-magnesium hydroxy carbonate, *See* L-55R® Acid Neutralizer,16529
aluminum-magnesium-sodium silicate, *See* Perkasil® KS 207, 23316
aluminum-nickel-titanium, 1932
Alumite, 1933, *See* Activated alumina, 495,aluminum oxide, 1919
Alundum, *See* Activated alumina, 495, aluminum oxide, 1919
Alundum®, 1934
Aluni, 1935
Aluphos, *See* aluminum phosphate, 1922
Alusec, 1936
Alusil, 1937, *See* aluminum silicate, 1924
Alusil ET, 1938, *See* aluminum silicate, 1924 Wessalith, 34284
Alu-Tab™, *See* Alumina hydrate, 1892, aluminum hydroxide, 1905
Alutyl, *See* cinchophen, 6307
Alveograf, 1939
Alvex, 1940
Alytol, 1941
Alzen, 1942
AM B 50, *See* Cocamidopropyl betaine, 6551
Amadil, *See* Valadol, 33238
amalgam, 1943
Amangan, *See* Maneb, 18526
Amargosite, 1944
Amarin, 1945
Amasil®, 1946
Amasil® P, 1947
amatin, *See* Julin's chloride, 15632
Amatols, 1948
Amax, *See* Accelerator MF, 208 Amax®, AmaxNo 1, 1950 Delac MOR, 8000 OBTS, 21970 OMTS, 22161 Perkacit® MBS, 23284
Amax XLP, 1949
Amax®, Amax No 1, 1950
Amaze®, 1951
Amazin®, 1952
Ambazyme, 1953
amber, 1954
Amber, *See* Propachlor, 25163
amber acid, *See* succinic acid, 29915
Amber lanolin, *See* Vigilan, 33750
Amberdeen, *See* Bakelite., 3589
Amberglow, 1955
ambergris, 1956
amber-guaiacum resin, 1957
Ambergum® 721, 1958
Amberite, 1959
Amberlac®, 1960
Amberlac® 13-801, 1961
Amberlite®, 1962
Amberlite® IRA-68, 1963
Amberlite® IRP-64, 1964
Amberoid, *See* nickel silvers, 21132
Amberol, 1965
Ambersil, 1966
Ambiflo, 1967
Ambiteric, 1968
Ambiteric D, 1969
Ambitrol, 1970
Amblosin, Amfipen, *See* Ampicillin, 2361
Amborate, 1971
Amborol, 1972
Amboryl Acetate, 1973
Ambra, 1974
Ambrac, *See* nickel silvers, 21132
Ambrac Metal, 1975
Ambraloys, 1976
Ambramicina, *See* tetracycline, 31365
Ambramycin, *See* tetracycline, 31365
Ambrasite, *See* Bakelite, 3589
Ambrene, 1977
ambrite, 1978
ambroid, 1979
Ambrol, 1980
Ambrottolide, *See* musk ambrette, 20532
ambroxan, 1981
Ambush, 1982, *See* Kafil, 15655, permethrin,23384, 23385, 23386
Ambush C, 1983, *See* Topclip Parasol, 32037
Ambushfog, *See* permethrin, 23384, 23385,233866
Amcar CL, 1984
Amcar OCP, 1985
Amchlor, *See* Ammonium chloride, 2235, Katapone VV-328, 15785
Amcide, 1986, *See* Ammonium sulfamate, 2254
Amcron, 1987
Amdye PH-12, 1988
AME-100, *See* Acetamide MEA, 283
AME 4000®, 1989
Amea 100, *See* Polyflex, 24435
Ameen, 1990
Ameenex 70 WS, 1991
Ameenex C-18, 1992
Ameenex Polymer, 1993
AMEO, *See* Aktisil AM, 1201, *See* CA0750,4852, Prosil® 220, 25237
Amephyt, *See* ametryn, 2051
Amerace A 30, *See* Vestolit®, 33679
Amercell Polymer HM-1500, 1994
Amerchol L 101, *See* Solulan, 28998
Amerchol Polysorbate, 1995
Amerchol Polysorbate1, *See* Amerchol® L-500, 2002
Amerchol® 400, 1996, *See* Amerchol® L-500, 2002
Amerchol® BL, 1997
Amerchol® C, 1998, *See* Amerchol® L-500, 2002
Amerchol® CAB, 1999
Amerchol® H-9, 2000
Amerchol® L-10, *See* Amerchol® L-500, 2002
Amerchol® L-101, 2001, 2003
Amerchol® L-500, 2002
Amerchol® L-99, *See* Amerchol® L-101, 2003, Amerchol® L-500, 2002
Amerchol® RC, 2004, *See* Amerchol® L-500, 2002
Amercoat 635, *See* bioMeT TBTF, 4171
Amercor® 8730, 2005
Amerfloc Plus® 5270, 2006
Amerfloc® 2, 2007
Amerfloc® 275, 2008, *See* Amerfloc Plus® 5270,2006
Amergel® 100, 2009
Amergize, 2010
Amergy® 5400, 2011
Ameribond, 2012, *See* calcium lignosulfonate,4956, 4957
Ameribond 2000, 2013
American Cyanamid 12880, *See* Dimethoate, 8496
American mace butter, *See* otoba butter, 22375
American silver, *See* nickel silvers, 21132
American vermilion, *See* chrome red, 6209
Ameripol Synpol 1009, 2014
Ameripol Synpol 1013/8000, 2015
Amerite, 2016
Amerlate® LFA, 2017
Amerlate® P, 2018
Amerlate® WFA, 2019, *See* Amerlate® LFA, 2017
Amerol, 2020
Amerone, 2021
Amerox OE-20, *See* Ablunol OA-6, 107, Volpo 3,34020
Ameroxol EO 2, *See* Ablunol OA-6, 107, Volpo 3,34020
Ameroxol OE 10, *See* Ablunol OA-6, 107 Volpo 3, 34020
Ameroxol OE-20, *See* Ablunol OA-6, 107, Volpo 3, 34020
Ameroxol® OE-2, 2022
Ameroxol® OE-2, OE-5, *See* oleth-8, 22126
Ameroxol® OE-5, 2023
Ameroyal, 2024
Amerplex® 605, 2025
Amerscan MDP Kit, 2026
Amerscent 86, 2027
Amerscreen, 2028
Amersep® MP-3, 2029
Amersil® DMC-287, DMC-357, 2030
Amersil® L-45 Grades, 2031
Amersil® ME-358, 2032
Amersil® simethicone, 2033
Amersil® VS-7207, 2034
Amersite® 2, 2035, 2036
Amersperse 1200, 2037
Amerstat®, 2038
Amerstat® 233, 2039
Amerstat® 250, 2040
Amerstat® 251, 2041
Amerstat® 272, 2042
Amerstat® 282, 2043
Amerstat® 294, 2044
Amerstat® 300, 2045
Amertrol, 2046
Amerzine®, 2047
amesite, 2048
Amethopterin, *See* Methotrexate,
Ametoterina, *See* Aminitrazole, 2164
Ametox, 2049
Ametrex, 2050, *See* ametryn, 2051, Amigan, 2117
Ametryn, 2051, *See* Ametrex, 2050
ametryne, *See* ametryn, 2051
Ametryn-terbutryn, *See* Amigan, 2117
Amfaid, 2052
Amfix, 2053
Amfix FRL, 2054
Amgard TBEP, 2055
Amgard® CPC 452, 2056
Amgard®TOF, *See* trioctyl phosphate, 32349
Amianth, *See* Amianthus, 2057
Amianthus, 2057
Amical® 48, 2058, *See* Amical® Flowable, 2061
Amical® 85, 2059
Amical® 101, 2060
Amical® Flowable, 2061
Amical® SC, 2062
amicarbalide, 2063
Amichrome, 2064
Amicide, *See* Ammonium sulfamate, 2254
Amicon® C-860-4, 2065
Amicon® C-940-4, 2066
Amicon® CT-4042-5, 2067
Amicon® ECT-86, 2068
Amicon® ME-868, 2069
Amicon® SC-220, 2070
Amicon® SC-2634A/B, 2071
Amicon® SC-3613, 2072
Amicon® TG-86, 2073
Amicure® 352, 2074
Amicure® AEP, 2075
Amicure® CL-485, 2076
Amicure® DBU, 2077
Amicure® PACM, 2078
Amicure® SA, 2079
Amicure® TEDA, 2080
Amicure® TMR 30, 2081
Amidan, 2082
amide C_1, *See* formamide, 12261
amide C_1, *See* formamide, 12261
amide C_{18}, *See* stearamide, 29593

amide C_{18}, *See* stearamide, 29593
Amide CMA-2, 2083
Amide RMA-2, 2084
Amide sulfonates, *See* Emkapon 4S, DS, SS,TS, 10249
Amide wax, *See* Acrawax® B, 382
Amides, coco, N-(hydroxyethyl), *See* Ablumide CME, 78
Amides, coco, N-[3-(dimethylamino)propyl],N-oxide, *See* Ablumox CAPO, 90
Amidex 1285, 2085
Amidex AME, 2086, *See* Acetamide MEA, 283
Amidex C, 2087
Amidex CE, 2088
Amidex CIPA, 2089, *See* Cocamide MIPA, 6550
Amidex CME, 2090
Amidex CP, 2091
Amidex KD, 2092
Amidex KME, 2093
Amidex L-9, 2094, *See* Alkamide® LE, 1646
Amidex LD, 2095, *See* Alkamide® LE, 1646
Amidex LIPA, 2096
Amidex LMMEA, 2097
Amidex LN, 2098
Amidex O, 2099
Amidex PK, 2100
Amidex RC, 2101
Amidex S, 2102
Amidex SME, 2103
Amidex TD, 2104
Amidinoquanidine, *See* Biquanide, 4218
Amido betaine C, 2105
Amido betaine C, C-45, *See* Cocamidopropyl betaine, 6551
Amido Betaine-L, *See* lauramidopropyl betaine, 16830
***ortho*-amidobenzoic acid**, *See* anthranilic acid, 2529
Amidocid®, 2106
Amido-G Acid, *See* Salt, Amido-G, β-naphthylamine-6,8-disulfonic acid, 27294
Amidogene, 2107
Amido-G-Salt, *See* Amido-G-Acid., 27294
amidol, 2108
Amidomalonylurea, *See* uramil, 33113
Amido-R-Acid, *See* β-naphthylamine-3,6- disulfonic acid, 27295
amidosulfonic acid, *See* sulfamic acid, 29963
Amidox®, 2109
Amidox® C-2, 2110
Amidox® L-2, 2111
Amidozid, 2112
Amiema MA-OD, 2113
Amiema MA-OL, 2114
Amiesite, 2115
Amietol, 2116
Amigan, 2117
Amigel, 2118
Amigen, 2119
Amihope LL-11, 2120
Amikapron, 2121
Amil peroxyethylhexanoate, *See* Esperox®570, 10854
Amilan, *See* Cabelec® 1015, 4861
amilee, *See* paraffin, liquid., 22744
Amilperoxy pivalate, 2122, *See* Aztec® *t*-Amyl peroxypivalate-75 OM, 3540
Aminacrine hydrochloride, *See* Monacrin, 20097
Amine, 2123
Amine 0, 2124
Amine 10, 2126
Amine 12, 2127
Amine 12-98D, 2128
Amine 14D, 2129
Amine 16D, 2130, *See* Armeen® 16D, 2936
Amine 18-90, 2131
Amine 220®, *See* Marlowet® 5440, 18928
Amine 740, 2145
Amine 760, 2146
Amine 780, 2147
Amine 2HBG, 2132, *See* Amine2 VT, 2163
Amine 2M1214D, 2133
Amine 2M1218D, 2134
Amine 2M12D, 2135, *See* dimethyl lauramine, 8507
Amine 2M14-50D, 2136
Amine 2M14D, 2137
Amine 2M16D, 2138
Amine 2M18D, 2139
Amine 2M810D, 2140
Amine 2MBGD-M, 2141
Amine 2MHBGD, 2142
Amine 2MKKD, 2143
Amine 2MOLD, 2144
Amine 8 D, 2125
Amine Acetate HBG, 2148
Amine Acetate KK, 2149
Amine alkylaryl sulfonate, *See* Eccoterge ASB,9566, Witcor PC200, 34521
Amine B11, 2150
Amine BG, 2151, *See* tallowamine, 30748
Amine borates, *See* Crodinhib, 7178
Amine C_4, 2152
Amine carboxylate, *See* Actracor 1987, 522
Amine CS-1135®, 2153
Amine CS-1246, 2154
Amine D, 2155
Amine dodecylbenzene sulfonate, *See* Hetsulf60T, 13980, Marlon® AMX, 18876, Ninate® 411, 21228, Rhodacal® 2283, 26578, Witconate 93S, 34482
Amine ethoxylate, *See* Sandoxylate® NT-15,27359
Amine ethoxylates, *See* Empilan® AM Series,10386
Amine HBG, 2156
Amine HBGD, 2157
Amine hydrochloride, *See* Catalyst CC, 5434
Amine KK, 2158
Amine M2HBG, 2161
Amine M210D, 2159, *See* Armeen® M2-10D, 2953, didecyl methylamine, 8415
Amine M218, 2160
Amine neutralized sulfonic acid, *See* Marvanol® Penetrant 35, 18984
Amine OL, 2162
Amine oxide, *See* Ninox®, 21240, Tomah AO-728 Special, 31964
Amine oxides, *See* Casamox, 5370, Rewomin,26331, Rewominox, 26332
Amine2 VT, 2163
Amines, bis(2-hydroxyethyl)tallow alkyl, *See* Witcamine® 6606, 34447
Amines, coco alkyldimethyl, N-oxides, *See* Barlox® 12, 3649
Amines, n-cocoalkyltrimethylenedi-, *See* JetAmine DC, 15582
Amines, n-tallow alkyltrimethylenedi-, *See* JetAmine DT, 15590
aminic acid, *See* formic acid, 12271
Aminitrazol, *See* Aminitrazole, 2164
Aminitrazole, 2164
Aminitrozol, *See* Aminitrazole, 2164
Aminitrozole, *See* Aminitrazole, 2164
Aminitrozolum, *See* Aminitrazole, 2164
Amino acid, *See* glycine, 13095
Amino Acid Gelatinization Agent, 2165
1-Amino-3-(aminomethyl)-3,5,5-trimethylcyclohexane, *See* Vestamin® TMD, 33656
2-Amino-5-[(4-amino-3-sulfophenyl)(4-imino-3-sulfo-2,5-cyclohexadien-1-ylidene)methyl]-3-methylbenzenesulfonic acid disodiumsalt, *See* acid fuchsine, 333
***p*-aminobenzenesulfonic acid**, *See* sulfanilic acid, 29966
2-amino-1-butanol, *See* AB, 19
Amino-6-tert-butyl-3-(methylthio)-s-triazin-5(4H)-one, *See* metribuzin, 19595
4-amino-6-tert-butyl-3-(methylthio)-as-triazin-5(4H)-one, *See* metribuzin, 19595, 19596
1-amino-2-carboxybenzene, *See* anthranilic acid, 2529
1-amino-1,2-carboxyethane, *See* Aspartic acid,3234
5-Amino-4-chloro-2-phenylpyridazin-3(2H)-one, *See* chloridazon, 6103
5-Amino-4-chloro-2-phenyl-3(2H)-pyridazinone, *See* chloridazon, 6103
***o*-amino cinnamic acid lactam**, *See* carbostyril,5244
[(4-amino-3,5-dichloro-6-fluoro-2-pyridinyl)oxy]acetic acid, *See* fluroxypyr, 12117
[(4-amino-3,5-dichloro-6-fluoro-2-pyridinyl)oxy] acetic acid 1-methylheptyl ester, *See* fluroxypyr 1-methylheptyl ester, 12118
N-[4-[[(2-amino-1,4-dihydro-4-oxo-6-pteridinyl) methyl]amino]benzoyl]-L-glutamic acid, *See* **vitamin M**, 33951
2-amino-1,7-dihydro-6H-purin-6-one, *See* guanine, 13443
Amino-2,2-dimethylethanol, *See* 2-amino-2-methyl-1 propanol, 2182
Amino-6-(1,1-dimethylethyl)-3-(methylthio)-1,2,4-triazin-5(4H)-one, *See* metribuzin, 19595, 19596
4-amino-6-(1,1-dimethylethyl)-3-(methylthio)-1,2,4-triazin-5(4H)-one, *See* metribuzin, 19595, 19596
4-Amino-N-(4,6-dimethyl-2-pyrimidinyl)benzenesulfonamide, *See* sulfamethazine, 29962
1-amino-2-ethylhexane, *See* Armeen® L8D,2952
2-amino-N-ethyl-N-phenyl-benzamine, *See* Fumite Dicloran, 12459
2-amino-2-ethyl-1,3-propanediol, 2170, *See* AEPD®, 832, AEPD® 85, 833, AMPD, 2304
3-amino-α-ethyl-2,4,6-triiodobenzeneprop-anoic acid, *See* Telepaque, 31057
3-amino-α-ethyl-2,4,6-triiodohydrocinnamicacid, *See* Telepaque, 31057
Amino Gluten MG, 2166
1-amino hexadecane, *See* Armeen® 16D,2936, Crodamine 1.16D, 7101
α-amino-β-hydroxybutyric acid, *See* L-threonine,
4-amino-1-hydroxybenzene, *See* *p*-aminophenol, 2185, Takatol, 30725, Unal,32908
2-amino-1-hydroxybutane, *See* AB, 19
2-amino-2'-hydroxydiethylamine, *See* Aminoethylethanolamine, 2172
2-amino-2-hydroxymethylpropanediol, *See* Tris Amino®, 32405
2-amino-2-(hydroxymethyl)-1,3-propanediol, *See* THAM, 31557, tris (hydroxymethyl)aminomethane, 32403, Tris Amino®, 32405
5-Amino-4-hydroxynaphthalene-1,7-disul-fonic acid, *See* K-acid, 15652
[6R-[6α,7β(R*)]]-7-[[Amino-(4-hydroxyphenyl) acetyl]amino]-3methyl-8-oxo-5-thia-1-azabicyclo-[4.2.0]oct-2-ene-2-carboxylic acid monohydrate, *See* Cefadroxil, 5547
N-(*p*-(((2-amino-4-hydroxy-6-pteridinyl)-methyl)amino)benzoyl)-L-glutamic acid, *See* folic acid, 12196
N-(*p*-amino-4-hydroxy-6-pteridinyl)-N'-ethyl-N'-phenylformamidine, *See* Ethoxy carbonyl phenylethyl phenylformamidine, 11042
α-amino-β-hydroxy-propionic acid, *See* L-serine, 17705, 28035
2-amino-3-hydroxypropionic acid, *See* L-serine, 17705, 28035
L-2-amino-3-hydroxypropionic acid, *See* L-serine, 17705, 28035
2-amino-6-hydroxypurine, *See* guanine, 13443
α-amino-β-imidazolepropionic acid, *See* histidine,
1-α-amino-3-indolepropionic acid, *See* DL-tryptophan,
(+)-2-amino-3-mercaptopropionic acid, *See* L-cysteine, 7602
1-amino-2-methoxybenzene, *See* *o*-anisidine,2473
1-amino-4-methoxybenzene, *See* *p*-anisidine,2474
3-amino-N-(α-methoxycarbonylphenethyl) succinamic acid, *See* Aspartame, 3233
4-amino-4'-methoxydiphenylamine, *See* Variamine, 33387
4-Amino-N-(3-methoxypyrazinyl)benzene-sulfonamide, *See* Kelfizina, 15838
***o*-Amino methyl benzoate**, *See* methyl anthranilate, 19513
2-amino-4-(methylmercapto)butyric acid, *See* DL-methionine, 19487
2-amino-3-methylpentanoic acid, *See* L-isoleucine,
4-amino-N-(3-methyl-1-phenyl-1H-pyrazol-5-yl)benz

Chemical and Trade Names

Chemical and Trade Names

3003, 3478, Carsonol® SES-A, 5336, Empicol® EAA, 10312, Empicol® EAB, 10313, Empicol® EAC, 10314, NonasolN4AS, 21439, Rhodapex® AB-20, 26622, Rhodapex® EAY, 26627, Standapol® EA-1,29472, Steol® CA-460, 29661, Sulfochem EA-1, 29998, Sulfochem EA-2, EA-3, 29999, Sulfochem EA-60, 30000, Texapon EA-1,31457, Witcolate AE, Witcolate LES-60A,34472

Ammonium laureth sulfate (4 EO), *See* Polystep® B-11, 24591

Ammonium laureth-3 sulfate, *See* Nutrapon AL 60, 21814

Ammonium lauroyl sarcosinate, 2241, *See* Hamposyl® AL-30, 13597

Ammonium lauryl ether (1) sulfate, *See* Texapon PNA-127, 31474

Ammonium lauryl ether (3 EO) sulfate, *See*Ungerol AM3-75, 32920

Ammonium lauryl ether sufate, *See*Ammonium laureth sulfate, 2240, Calfoam NEL-60, 4985, Sulfotex OT, 30035, Texapon NA,31468

Ammonium lauryl ether sulfate (3 EO), *See*Zoharpon LAEA 253, 34866

Ammonium lauryl sulfate, 2242, *See* AkyposalALS 33, 1302 Avirol® A, 3477, Carsonol® ALS-R, 5332, DeSonol A, 8186, Empicol® AL30,10306, Marlinat® DFN 30, 18853, Nutrapon PP3563, 21824, Octosol ALS-28, 22024, Perlankrol® DAF25, 23325, Polystep® B-7,24597, Rhodapon® L-22, L-22/C, 26635, Standapol® A, 29467, Štepanol® AM, 29693, Sulfochem ALS, 29996, Texapon ALS, 31453, Ufarol Am 30, 32742, Witcolate NH, 34478, Zoharpon LAA, 34864

Ammonium lauryl sulfosuccinate, *See* Monamate LNT-40, 20107

Ammonium lauryl triethoxy sulfate, *See* Empicol® ALL, 10307

Ammonium laurylbenzenesulfonate, *See* Ablusol DBM, 125

Ammonium lignosulfonate, 2243, *See* LignosolTS, 17240, Tembind A 002, 31077

Ammonium metavanadate, *See* KMAmmonium Metavanadate, 16235

Ammonium molybdate, 2244

ammonium molybdate-potassium ferrocyanide, *See* Huber's reagent, 14455

ammonium muriate, *See* Ammonium chloride,2235

Ammonium myreth sulfate, *See* Standapol®7088, 29465

Ammonium naphthalene sulfonate, *See*Emery® 5366 (Lomar PWA Llq.), 10176

Ammonium naphthalene-formaldehydesulfonate, *See* Dehscofix 929, 7928

Ammonium nitrate, 2245, *See* Ansax, 2488,Old Plantation, 22106

Ammonium nonoxynol-4 sulfate, 2246, *See* Abex® EP-110, 41, Aerosol® NPES 458, 883, Polystep® B-1, 24590, Rhodapex® CO-415,26623, Rhodapex® CO-436, 26625

Ammonium nonoxynol-9 sulfate, *See* Sulfochem436, 29995

Ammonium nonyl phenoxy polyethoxysulfate, *See* Nutrol S-60 5350, 21848

Ammonium nonylphenol ethoxy sulfate, *See* Fenopon EP, 11606

Ammonium octyl/decyl-sulfate, *See* TensatilDA120, 31152

Ammonium oxalate, 2247

Ammonium paramolybate, *See* Ammonium molybdate, 2244

ammonium paratungstate, *See* Ammoniumtungstate, 2258

Ammonium pentaborate, 2248

ammonium pentachlorozincate, *See* zincammonium chloride, 34766

Ammonium pentadecafluorooctanoate, *See* Fluorad® FC-118, 12058

Ammonium perchlorate, *See* KM AmmoniumPerchlorate, 16236

Ammonium perfluoroalkyl sulfonate, *See*Fluorad® FC-93, 12064

ammonium perfluoroctanate, *See* Fluorad® FC-118, 12058

Ammonium perfluorooctanoate, *See* Fluorad® FC-118, 12058

Ammonium Peroxodisulfate, *See* Ammoniumpersulfate, 2249, thioxydant lumire, 31725

Ammonium persulfate, 2249, *See* thioxydantlumire, 31725

Ammonium phosphate dibasic, 2250

Ammonium phosphate monobasic, 2251, *See*Firesaife, 11836

ammonium picrate, *See* Explosive D, 11314

ammonium picrate-potassium nitrate, *See* Brugrepowder, 4623

ammonium picronitrate, *See* Explosive D, 11314

Ammonium polyacrylate, 2252, *See* Alcogum9639, 1468

Ammonium polyacrylate and sodium laureth-6carboxylate, *See* Akypogene KTS, 1259

ammonium polymannuronate, *See* Ammoniumalginate, 2228

ammonium polyphosphate, *See* Phos-chek,23694

Ammonium propionate, *See* Luprosil® NC,17920

ammonium purpurate, *See* murexide, 20506

ammonium reineckate, *See* Reinecke's salt,26033

ammonium reineckate trihydrate, *See* Reinecke'ssalt, 26033

Ammonium rhodanide, *See* Ammoniumthiocyanate, 2256

Ammonium Rhodantate, *See* Ammoniumthiocyanate, 2256

Ammonium Rhodonide, *See* Ammoniumthiocyanate, 2256

ammonium salt, *See* Ammonium alginate, 2228

Ammonium salt of sulfated nonylphenoxy POEethanol, *See* Ammonium nonoxynol-4 sulfate,2246

Ammonium stearate, 2253

Ammonium sulfamate, 2254, *See* Amcide, 1986

Ammonium sulfate, 2255, *See* Nipro (ii), 21304

Ammonium sulfate (2:1), *See* Ammonium sulfate,2255

ammonium sulfobituminate, *See* Perichthol,23270

Ammonium sulfocyanate, *See* Ammoniumthiocyanate, 2256

Ammonium Sulfocyanide, *See* Ammoniumthiocyanate, 2256

ammonium tetrathiocyanodiaminochromate, *See* Reinecke's salt, 26033

Ammonium thiocyanate, 2256

Ammonium thiosulfate, 2257

Ammonium tungstate, 2258

Ammonium tungstate pentahydrate, *See*Ammonium tungstate, 2258

ammonium uriate, *See* Katapone VV-328, 15785

Ammonium Vanadate (V), *See* KM AmmoniumMetavanadate, 16235

ammonium wolframate, *See* Ammonium tungstate, 2258

Ammonium xylene sulfonate, 2259, *See* Eltesol® AX 40, 9865, HartotropeAXS, 13691, Naxonate® 4AX, 20822, Stepanate® AXS, 29672

Ammonium, (3-chloro-2-hydroxy propyl) trimethyl-, chloride, *See* Reagens-CF2, 25928

Ammonium, benzyldimethyloctadecyl-,chloride, *See* Ablumine 280, 85

Ammonium, diallyldimethyl-, chloride,polymers, *See* Agefloc WT-40, 1010

Ammonium, trimethyloctadecyl-, bromide, *See*Zeonet B, 34721

Ammonium/triethanolamine lauryl sulfate, *See*Sulphatol B6, 30064

Ammonium-β-tetralin-sulfonate, *See*Majammonium, 18421

Ammonium-stannic-chloride, *See* pink salt,23863

ammonocarbonous acid, *See* hydrocyanic acid,14584

Ammonyx 4, *See* Sumquat® 6210, 30126,Cycloton® SCS, 7542, octadecylbenzene methanaminium chloride, 21985, stearalkonium chloride, 29592, Ablumine 280, 85

Ammonyx 485, *See* Ablumine 280, 85, Cycloton®SCS, 7542, octadecyl benzenemethanaminium chloride, 21985, stearalkonium chloride, 29592,Sumquat® 6210, 30126,

Ammonyx 490, *See* Ablumine 280, 85, Cycloton®SCS, 7542,octadecylbenzenemethanaminium chloride,21985, stearalkonium chloride, 29592,Sumquat® 6210, 30126

Ammonyx 4002, *See* Ablumine 280, 85,Cycloton® SCS, 7542,octadecylbenzenemethanaminium chloride,21985, stearalkonium chloride, 29592,Sumquat® 6210, 30126

Ammonyx AO, *See* Ablumox LO, 91, Ammonyx® LO, 2264, Empigen®OB, 10375, Rhodamox® LO, 26620

Ammonyx CA Special, *See* Ablumine 280, 85,Cycloton® SCS, 7542, octadecylbenzene methanaminium chloride, 21985, stearalkonium chloride, 29592,Sumquat® 6210, 30126

Ammonyx DME, *See* Sumquat® 6020, 30121

Ammonyx G, *See* Sumquat® 6050, 30124

Ammonyx LO, *See* Ablumox LO, 91, Ammonyx® LO, 2264, Empigen®OB, 10375, Rhodamox® LO, 26620

Ammonyx T, *See* Sumquat® 6050, 30124

Ammonyx® 4, 4B, 485, 4002, 2260

Ammonyx® CETAC, CETAC-30, 2261, *See*Arquad® 16-50, 3108, Cetrimonium chloride,5814

Ammonyx® CO, 2262

Ammonyx® DMCD-40, *See* lauramine oxide,16832

Ammonyx® KP, 2263

Ammonyx® LO, 2264, *See* lauramine oxide,16832

Ammonyx® OAO, 2265

Amo Vitrax®, 2266

Amo® Balanced Salt Solution, 2267

Amoco® 1012, 2268, *See* Adpro AP 2112-GP, 788

Amoco® 1016, 2269, *See* Adpro AP 2112-GP,788

Amoco® 1246, 2270, *See* Adpro AP 2112-GP, 788

Amoco® 4018, 2271, *See* Adpro AP 2112-GP,788

Amoco® 5016, 2272, *See* Adpro AP 2112-GP,788

Amoco® 6114, 2273, *See* Adpro AP 2112-GP,788

Amoco® 6400p, 2274, *See* Adpro AP 2112-GP,788

Amoco® 7234, 2275, *See* Adpro AP 2112-GP,788

Amoco® 7239, 2276, *See* Adpro AP 2112-GP,788

Amoco® 7728, 2277, *See* Adpro AP 2112-GP,788

Amoco® 8244, 2278, *See* Amoco® 8410, 2279

Amoco® 8410, 2279

Amoco® 9119, 2280, *See* Adpro AP 2112-GP,788

Amoco® BR-310, 2281

Amoco® CI-500, 2282

Amoco® H-15, 2283

Amoco® H2R, 2284

Amoco® H3E, 2285

Amoco® L-14, 2286

Amoco® PIA, 2287

Amoco® R1, 2288

Amoco® R5, 2289

Amodel® A-1115HS, A-1145HS, 2290

Amodel® A-1340HS, 2291

Amodel® AF-1115VO, AF-1133VO, AF-1145VO,2292

Amodel® ET-1000, 2293

Amodimethicone, *See* SM 2059, 28626

Amollan® A, 2294, 2295

Amoloid HV, *See* Ammonium alginate, 2228

Amoloid HV, LV, 2296

Amoloid LV, *See* Ammonium alginate, 2228

Amonyl 380 BA, 2297

Amonyl 675 SB, 2298

Amonyl DM, 2299

Amonyl® 380BA, 440 NI, *See* Cocamidopropyl betaine, 6551

Amorphous mineral silicate, *See* Micro PExtender, 19652

Amorphous nylon, *See* Ashlene® 870, 3222

Amorphous polyolefin, *See* Zeonex 280, 34723

AMP, 2300, 2301, *See* adenosinemonophosphate, 680, 5'-adenylic Acid, 683, 2-amino-2-methyl-1 propanol, 2182

AMP isostearoyl hydrolyzed keratin, *See* OleoKeratin

Anchor-bac®, 2427
Anchorite, 2428
Anchorlube G-771, 2429
Anchred, 2430
anchusa acid, *See* Alkanet, 1695
Anchusin, *See* Alkanet, 1695
Ancor® CR-538, CR-539, 2431
Ancor® LB-503, LB-504, 2432
Ancor® OW-1, 2433
Ancor® OW-9, 2434
Ancrack, 2435
andalusite, 2436, *See* Tisyn®, 31857
Andaria®, 2437
Anderol®, 2438
Anderol® Premium Plus, 2439
Andersil, 2440
andersonite, 2441
Andrade indicator, *See* acid fuchsine, 333
Andrez 8000, 2442
Androx 3961, 2443
Andur, 2444
Anedco ADM-407, 2445
Anedco ADM-407C, 2446
Anedco AF-800, 2447
Anedco AF-801, 2448
Anedco AW-395, 2449
Anedco AW-396, 2450
Anedco AW-397, 2451
Anedco DF-6002, 2452
Anedco DF-6031, 2453
Anedco DF-6130, 2454
Anema, *See* Telone, 3 1069
Anesthol, *See* Anestile,
Anesthyl, *See* Anestile,
Anethole, *See* Arizole® Anethole Extra, 2885
aneurine hydrochloride, *See* thiaminehydrochloride, 31665
Angarite, 2455
angel red, *See* Indian Red, 15068
Angelite, *See* aluminum phosphate, 1922
Angio-Conray, 2456
Angiografin, *See* Reno-M, 26113
Angiovist 282, 2457
Anhiba, *See* Valadol, 33238
Anhydrite, 2458, *See* calcium sulfate(anhydrous),4971
anhydrite (natural form), *See* calcium sulfate(anhydrous), 4971
anhydrohexitol diisostearate, *See* Sorbitan diisostearate, 29109
Anhydrol, 2459, *See* ethyl alcohol, 11064
Anhydrol Forte, *See* aluminum chloride,anhydrous, 1902
Anhydrone, 2460, *See* Dehydrite®, 7963
Anhydrosorbitol monolaurate, *See* Ablunol S-20, 108
Anhydrosorbitol monooleate, *See* Ablunol S-80, 111,Witconol 2500, 34501
Anhydrosorbitol monopalmitate, *See* AblunolS-40, 109
Anhydrosorbitol monostearate, *See* Ablunol S-60, 110
Anhydrosorbitol trioleate, *See* Ablunol S-85,112, Witconol 2503, 34502
Anhydrosorbitol tristearate, *See* sorbitantristearate, 29111
Anhydrosorbitols, *See* Anidrisorb, 2464
anhydro-*o*-sulfaminebenzoic acid, *See*saccharin, 27195
Anhydrous Borax, *See* Dehybor, 7942
Anhydrous calcium chloride, *See* Liquical,17459
Anhydrous chromium oxychloride, *See*Etard'sreagent, 10901
anhydrous gypsum, *See* calcium sulfate(anhydrous), 4971
anhydrous hydrofluoric acid, *See* hydrogenFluoride, 14589
Anhydrous lanolin, *See* Anhydrous LanolinP9SRA, 2463, Vigilan, 33750
Anhydrous Lanolin HP-2050, 2461
Anhydrous Lanolin P80, 2462
Anhydrous Lanolin P9SRA, 2463
Anhydrous Lanum, *See* Vigilan, 33750
Anhydrous monocalcium phosphate, *See* Py-Ran, 25468
Anhydrous sodium dimetaborate, *See* Pyrobor, 25494
anhydrous sodium potassium aluminosilicate, *See* Minex 2, 19828
anhydrous sodium sulfate, *See* Crimidesa, 7052
anhydrous stannous chloride, *See* Fascat®2004, 11496
anhydrous sulfate of lime, *See* calcium sulfate(anhydrous), 4971
anhydrous tin tetrachloride, *See* Fascat® 4400, 11497
Anicon kombi, *See* MCPA, 19183
Anicon m, *See* MCPA, 19183
Anidrisorb, 2464
Anilazine, 2465, *See* Dairene®, 7698, Dyrene®,9437
aniline, *See* phenylamine, 23638
aniline black, 2476
Aniline black in paste, *See* aniline black, 2476
aniline oil, *See* phenylamine, 23638
aniline violet, *See* gentian violet, 12845
aniline violet pyoktanine, *See* gentian violet,12845
aniline-m-sulfonic acid, *See* metanilic acid,19445
m-aniline sulfonic acid, *See* metanilic acid,19445
4-Anilino-N-isopropylaniline, *See* Vulkanox®4010 NA, 34147
anilinonaphthalene, *See* Antioxidant PBN, 2591
1-anilinonaphthalene, *See* Akrochem®Antioxidant PANA, 1161, Antioxidant PAN, 2590
2-anilinonaphthalene, *See* Antioxidant PBN,2591, Stabilator A.R, 29400
8-Anilino-1-naphthalenesulfonic acid, *See*phenyl-peri acid, 23647
Anilino-phenyl methacrylamide, *See* PolystayAA-1, 24583
***p*-anilinesulfonic acid**, *See* sulfanilic acid, 29966
anilotic acid, *See* nitrosalicylic acid,
Animal bone charcoal, *See* Octojet 104, 22004
animal charcoal, 2466
animal coniine, *See* Cadaverine, 4883
animal glycerin, 2467
Animal oil, *See* Bone oil, 4402
animal starch, 2468, *See* Liver sugar, 17527
anime resin, 2469
Anionic dyes, *See* Basacid® Dyes, 3678
Anionic surfactant, *See* Manro HCS, 18594
Anionic surfactant as pale cream flakes, *See*Arylan S, 3155
Anionic/nonionic emulsifiers (Agrimul 26-B,Agrimul 70-A, Agrimul-A-300, Agrimul N-300), *See* Agrimul, 1069
***p*-anisaldehyde**, 2470
Aniscol, 2471
anise camphor, *See* Arizole® Anethole Extra,2885
anisic aldehyde, *See* *p*-anisaldehyde,
***m*-anisidine**, 2472, *See* *m*-anisidine, 2472
***o*-anisidine**, 2473
***ortho*-Anisidine**, *See* *o*-anisidine, 2473
***p*-anisidine**, 2474
***para*-Anisidine**, *See* *p*-anisidine, 2474
2-Anisidine, *See* *o*-anisidine, 2473
Anisoline, *See* Rhodamine 3B,
***o*-anisylamine**, *See* *o*-anisidine, 2473
***p*-anisylamine**, *See* *p*-anisidine, 2474
Aniyaline, *See* Anilazine, 2465
Anka steel, 2475
Anlonyx® 12S, 2477
annaline, *See* gypsum, 13518
Annamene, *See* styrol, 29875
annatto, 2478
annotto, *See* annatto, 2478
annulene, *See* benzene, 3961
Annulex BHT, *See* Oxyguard, 22457
anode mud, 2479
Anode slime, *See* anode mud, 2479
Anodyne, *See* Antipyrine,
anodynon, *See* ethyl chloride, 11067
Anofex, *See* DDT, 7820
***p*-anol**, 2480
Anonaid TH, 2481
anone, *See* cyclohexanol, 7521, cyclohexanone,7522
Anotex, 2482
anotto, *See* annatto, 2478
Anox® 20, 2483
Anox® 70, 2484
Anox® IC-14, 2485
Anox® PP 18, 2486
anprolene, *See* ethylene oxide, 11131
anproline, *See* ethylene oxide, 11131
1,8-ANS, *See* phenyl-peri acid, 23647
Ansa, 2487
Ansar, *See* MSMA, 20417, Neoarsycodyl, 20871
Ansar 170, *See* Neoarsycodyl, 20871
Ansar 529, *See* Neoarsycodyl, 20871
Ansar 6.6, *See* Neoarsycodyl, 20871
Ansax, 2488, *See* Ammonium nitrate, 2245
Anscor®, 2489
anserine, 2490
Ansol, 2491
Ansol A, 2492
Ansol B, 2493
Ansol E-121, 2494
Ansol E-181, 2495
Anstex AK-25, 2496
ansul ether 121, *See* monoglyme, 20227
Ant Flip, 2497, *See* Alcan Aluminum SulfateLiquid, 1413
Ant Gun!, 2498
Ant Killer, 2499
Ant Killer Dust, 2500
ant oil, *See* oil of ants,
Antabuse, *See* Disulfiram, 8725
Antadix, *See* Disulfiram, 8725
Antage AW, *See* Polyflex, 24435
Antage DP, *See* Agerite® DPPD, 1016, N,N'-diphenyl-*p*-phenylenediamine, 8603, JZF,15638, Naugard® J, 20796, PermanaxDPPD, 23361
Antak, 2501
Antallin, *See* Versene CA, 33618
Antara®, 2502
Antaron ET-201, 2503
Antarox, *See* Alkasurf® NP-4, 1715, Triton® N-57, 32430
antarox bl-344, *See* Triton® N-57, 32430
Antarox Bl-344, *See* Alkasurf® NP-4, 1715
Antarox CO, *See* Ablunol NP4, 106
Antarox CO 430, *See* Ablunol NP4, 106
Antarox CO 530, *See* Ablunol NP4, 106
Antarox CO 630, *See* Ablunol NP4, 106
Antarox CO 730, *See* Ablunol NP4, 106
Antarox CO 850, *See* Ablunol NP4, 106
Antarox CO 880, *See* Ablunol NP4, 106
Antarox CO 970, *See* Ablunol NP4, 106
Antarox® 17-R-2, 2504, *See* Antarox® E-100,2509
Antarox® 461/P, 2505
Antarox® 497/P, 2506
Antarox® B-10, 2507
Antarox® BL-214, 2508
Antarox® E-100, 2509
Antarox® L-61, 2510
Antarox® L-61, *See* Antarox® E-100, 2509
Antarox® LA-EP 15, 2511
Antarox® PGP 23-7, 2512
Antec Farm Fluid S, 2513
Antec Longlife 250 S, 2514
Antec OO-Cide, 2515
Antec Virkon S, 2516
Antelope, 2517
Antene, *See* AAprotect, 16
Antepan, 2518
Anthion, 2519
Anthipen, *See* Dichlorophen, 8392
Anthiphen, 2520
Anthium Dioxcide, 2521
anthocyanins, 2522
Anthosin®, 2523
Anthoxan, 2524
anthracene, *See* Sterilite Hop Defoliant, 29734
9,10-Anthracenedione, *See* Anthraquinone,2532
9,10-Anthracenedione, 2,6-dihydroxy-, *See*anthraflavic

mercaptobenzimidazole, 19334
AOR/GR, 2638
Aoral, See Vi-Alpha, 33706
Aosoft, 2639
AP 50, See Octoguard FR-10, 22000
Apache, See benodanil, 3942
Apagallin, 2640
Apamide, See Valadol, 33238
APAP, See Valadol, 33238
Apavap, See dichlorvos, 8394
Apec®, 2641
Apec® DP9-9330, 2642
Apec® HT DP9-9350, 2643
Apec® HT KU 1-9350, 2644
Apec® KL 1-9306, 2645
Apec® KU 1-9309, 2646
Apex 400, 2647
Apexior, 2648
Apexol, See Vi-Alpha, 33706
APG® 225 Glycoside, 2649
APG® 300 CS, 2650
APG® 300 Glycoside, 2651
APG® 600 CS, 2652
APG® 600 Glycoside, 2653
Aphamite, See parathion, 22807
Aphit, See nickel silvers, 21132
Aphox, 2654
Aphox, See Abol, 146, pirimicarb, 23879, Rapid,25884
Aphrogene, 2655
Aphthite, 2656
Aphtite, 2657
Apiezon, 2658
Apifac, 2659
Apifil, 2660
apiol, See parsley camphor, 22865
apioline, See parsley camphor, 22865
Apl-Luster, See thiabendazole, 31661
APM, See Aspartame, 3233
Apollo 50C, 2661
Apolloy, 2662
aposafranine, 2663
Aposet 707, 2664
Apostavit, See Vi-Alpha, 33706
Appeel®, 2665
Apperitive Saffron of Iron, 2666
apple acid, 2667, See hydroxysuccinic acid,14671, Malic Acid, 18483
Appretan, 2668
Appretan Ant, 2669
Appretan CPF, 2670
Appretan GM, 2671
Appretan TN, 2672
Appreteen, 2673
APR®, 2674
Apricot kernel oil PEG-6 esters, 2675, See Labrafil M 1944 CS, 16537
Apricotamide DEA, See Schercomid SAP, 27656
Apricotamidopropyl betaine, See SchercotaineAPAB, 27694
Apricotamidopropyl ethyldimonium ethosulfate, See Schercoquat APAS, 27672
Aprocarb, See Baygon®, 3790, propoxur, 25201
Apron, See metalaxyl, 19430
Apron Combi, 2676
Apron T, 2677
APV, See Diethoxol, 8435, ethyl di-lcinol, 11068
aq ethanol, See Varisoft® 2 TD, 33402
aqua fortis, See nitric acid, 21332
Aqua Gro G Granular, 2678
Aqua Magic, 2679
Aqua Mer, 2680
Aqua Paste™, 2681
aqua regia, 2682
Aqua Thix, 2683
Aquabase, 2684, 2685
Aquabloc, 2686
Aquabrome, 2687
Aqua-Chem®, 2688
Aquacide, See Katalon, 15782
Aquacillin, See Penicillin V, 23111
Aquacoat, 2689
Aquadag, 2690, See Acheson's DeflocculatedGraphite, 322
Aquafil I, See Ablunol 200ML, 95
Aquafil II, See Ablunol 200ML, 95
Aquaflex, 2691
Aquaflim, 2692
Aquafloc, 2693
Aquaforte, 2694
Aquagel, 2695
Aqua-Gro L Liquid, 2696
Aqua-Gro S Spreadable, 2697
Aqua-kleen, See Planotox, 23914
Aqualac, 2698
Aqualease 2802, 2699
Aqualease 6102, 2700
Aqualipid 95, 2701, See Alcolec® 439-C, 1482
Aqualite, 2702
Aqualon® Cellulose Gum, 2703, See Aqualon® CMC-T, 2705
Aqualon® CMC-T, 2704, 2705
Aqualon® CMHEC-37L, 2706
Aqualon® EHEC, 2707
Aqualose, 2708
Aqualose L30, 2709
Aqualose LL100, 2710
Aqualose W20, 2711
Aqualox® 225-100, 225A-100, 2712
Aqualox® 232, 2713
Aqualox® 2268, 2714
Aqualube, 2715
Aqua-Mephytin, See vitamin K+71, 33949
AquaMephyton, See Konakion, 16322
Aquamet M, 2716
Aquamollin, See Kalex 220 Crystal, 15668,Kalex Liq. 50%, 15671, tetrasodium EDTA,31391
Aquamollin®, 2717
Aquanol, 2718
Aquapel®, 2719, 2720
Aquapel® 360XC, See Aquapel®, 2720
Aquaperle, 2721
Aquaphil K, 2722
Aquaplex, 2723
Aquapol, 2724
Aquaprint, 2725
Aquaresin, 2726
Aquarex Me, See sodium lauryl sulfate, 28781,Witcolate 6400, 34469
Aquarex methyl, See sodium lauryl sulfate,28781, Witcolate 6400, 34469
Aquarite, 2727
Aquaseal, 2728
Aquaseal Aquaflex Liquid Felt, 2729
Aquaseal Firmafix, 2730
Aquaseal Reflect, 2731
Aquaseal Weatherwise Standard, 2732
Aquasil, 2733
Aquasil®, 2734
Aquasoft®, 2735
Aquasol, 2736
Aquasorb® A250, 2737, See Aqualon® CMC-T,2705
Aquasperse, 2738
Aquastab PA 48, 2739
Aquastab PH 502, See oxalyl bis(benzylidenehydrazide), 22418
Aquastore, 2740
Aquasun, 2741
Aquasuspen, See Penicillin V, 23111
Aquatac® 5527, 2742
Aquatac® 6085, 2743
Aquatac® 8005, 2744
Aquatec, 2745
Aqua-Tein C, 2746
Aquathane Series, 2747
Aquathene AQ 120-000, 2748
Aquathene MP, 2749
Aquathene MV, 2750
Aquatherm, 2751
Aquatreat AR-225-D, 2752
Aquatreat AR-232, 2753
Aquatreat AR-626, 2754
Aquatreat AR-648, 2755
Aquatreat AR-7-H, 2756
Aquatreat AR-900, 2757
Aquatreat DNM-30, 2758
Aquatreat DNM-9, DNM-25, DMN-360, 2759
Aquatreat KM, 2760
Aquatreat SDM, 2761
Aquatrend®, 2762
Aqua-Trete®, 2763
Aquazine, See simazine, 28464
Aquazym, 2764
arabic gum, See gum arabic, 13480
Arabin, See gum arabic, 13480
D-araboascorbic acid, See isoascorbic acid, 15338
araboascorbic acid, monosodium salt, D-, See Eribate, 10753, Neo-Cebitate®, 20878, sodiumerythorbate, 28762
Aracast, 2765
Aracel 40, See Ablunol S-40, 109
Aracel 85, See Ablunol S-85, 112, Witconol 2503,34502
Arachidyl behenate, See Waxenol® 822, 34240
Arachidyl propionate, See Waxenol® 801, 34236
Arachidyl-behenyl 1,3-propylenediamine, See Kemamine® D-190, 15911
arachis oil, See Katchung oil, 15788, peanut oil,22969
Aracnol F, See cyhexatin, 7568
Aracnol K, See Tedion V-18, 30919, tetradifon,31366
Aragonite, See calcium carbonate, 4940
Aragran, See Terbufos, 31226
Arakote® 3000, 2766
Araldite 502, See dibutyl phthalate, 8366
Araldite HT, See Dapsone, 7751
Araldite HY 5083, See Vestamin® TMD, 33656
Araldite RD-1, See Ageflex BGE, 967
Araldite®, 2767
Araldite® 2001, 2768
Araldite® CY 225, 2769
Araldite® ECN 1235, 2770
Araldite® GT 6060, 2771
Araldite® GZ 540 X-90, 2772
Araldite® LT 8052, 2773
Araldite® LY 8047, 2774
Araldite® PT 810, 2775
Araldite® PY 306, 2776
Araldite® XD 4955, 2777
Araldite® XD 897, 2778
Araldite® XU GY 358, 2779
Aranox®, 2780
Arasan, See Rezifilm, 26521, Thiram, 31726,Thiurad, 31730
arasan 70, See Rezifilm, 26521
arasan-m, See Rezifilm, 26521
Arassist APH, 2781
Arassist HKM, 2782
Aratan, See Agallol, 954, Atiran, 3286, Ceresan,5745
Aratex, 2783
Arathane, See Karathane, 15759
Aratronic® 5001, 2784
Aratronic® 5040, 2785
Aravite® 3001, 2786
Arazate®, 2787
Arazete®, See Akrochem® Z.B.E.D, 1183
Arbeflex, 2788
Arbestab, 2789
Arbo, 2790
Arbocaulk, 2791
Arbocel, 2792
Arbocrylic, 2793
Arboflex, 2794
Arbofoam, 2795
Arbogard, 2796
Arbokol, 2797
Arbolite, 2798
Arbomast, 2799
Arborsan®, 2800
Arboseal, 2801

Arbosil, 2802
Arbostrip, 2803
Arbotect, *See* thiabendazole, 31661
Arbuz, *See* papain, 22698
Arbyl, 2804
Arbylen, 2805
ARcare® On/OFF 7810, 2806
Arcel Moldable Polyethylene Copolymers, 2807
Archil, 2808
Arcolloy, 2809
Arcoloy, 2810
Arconate® Propylene Carbonate, 2811
Arcosolv® DPM, 2812
Arcosolv® DPMA, 2813
Arcosolv® PM, 2814
Arcosolv® PMA, 2815
Arcosolv® TPM, 2816
Arcosolve DPM, *See* Icinol DPM, 14815, Poly-Solv® DPM, 24555
Arctite Injection Mortar, 2817
Arctite Quickbinder, 2818
Arctite Slurry 200 B, 2819
Arctite Tanking Mortar 500, 2820
Arcton, 2821
Arcton, *See* chlorotrfluoromethane, 6153
Arcton 6, *See* dichlorodifluoromethane, 8388
Arcton 12, *See* dichlorodifluoromethane, 8388
Arcton 114, *See* Cryofluorane, 7278
Ardeer Powder, 2822
Ardel® D-100, 2823
Ardenite, 2824
Ardent, 2825
Ardmorite, 2826
Ardux, 2827
arecaidine, 2828
arecaine, *See* arecaidine, 2828
arecaline, *See* arecoline, 2829
arecholine, *See* arecoline, 2829
arecoline, 2829
Arecoline-acetarsol, *See* Cestarsol, 5779
Aredion, *See* tetradifon, 31366
Arelon, 2830
Aremco-Bond, 2831
Aremco-Cast, 2832
Aremco-Coat, 2833
Aremsol A, 2834
Aremsol MA, 2835
Arena, *See* Tecgran, 30886, tecnazene, 30898
Arenka, 2836
Arenolite, 2837
Aresenid, 2838, *See* arsenic acid, 3137
Aresin, 2839
Areskap, 2840
Aresket, 2841
Aresklene, 2842
Aretan, 2843, *See* Agallol, 954, Atiran, 3286
Aretan 6, *See* Agallol, 954, Atiran, 3286,Ceresan, 5745
Aretone 270, 2844
Argal, *See* argol,
Argentai, 2845
Argentalium, 2846
Argentan, 2847, *See* nickel silvers, 21132
Argentan solder, *See* nickel silvers, 21132
Argentin, *See* nickel silvers, 21132
Argentoproteinum, *See* protargentum, 25251
argentous fluoride, *See* silver fluoride, 28423,tachiol, 30686
argentous oxide, *See* silver(I) oxide, 28432
Argezin, *See* atrazine, 3394
Argidone®, 2848, *See* arginine PCA, 2849
argilla, *See* kaolin, 15725, Kayphobe-ABO,15816, Vanclay, 33304
arginine PCA, 2849, *See* Argidone®, 2848
Argiroide, See nickel silvers, 21132
Argo Brand Corn Starch, *See* corn starch, 6837
Argobase, 2850
Argobase 125, 2851
Argobase EU, 2852
Argonol, 2853
Argonol 1SO, 2854
Argonol 40, 2855
Argonol 50 Pharmaceutical, 2856
Argonol ACE5, 2857
Argowax, 2858
Argowax Standard, 2859
Argozie, 2860
Argozoil, *See* Argozie, 2860
Arguad® 16-29, 16-50, *See* Cetrimonium chloride, 5814
Argus DLTDP, 2861
Argus DMTDP, 2862
Argus DSTDP, 2863
Argus DSTDP, *See* Distearyl thiodipropionate, 8713
Argus DTDTDP, 2864
Arguzoid, 2865, *See* Argozie, 2860
Arguzoil, *See* Argozie, 2860
Argylene, 2866
Argyroide, *See* nickel silvers, 21132
Argyrolith, 2867, *See* nickel silvers, 21132
Ariabel, 2868
Ariagran, 2869
Arianor, 2870
Ariavit, 2871
Aricel, 2872
Aricyl, 2873
Aridex, 2874
Aridry B, 2875
Aries Antox, *See* Polyflex, 24435
Arigal PMP, 2876
Arigran, 2877
Arikrome S, 2878
Arimid®, 2879
Ariotox, *See* metaldehyde, 19431
Aripol, 2880
Aristar, 2881
Aristoflex, 2882
Aristol A, 2883
Aristonate H, 2884
Arizole® Anethole Extra, 2885
Arizole® Pine Oil, 2886
Arizona 208, 2887
Arizona DR-22, 2888
Arizona DR-24, 2889
Arizona DRS-40, 2890
Arizona DRS-43, 2891
Arizona DRS-50, 2892
Arizona FA-7001, 2893
Arizona shellac, *See* Sonora gum, 29076
Arklone, 2894
Arko metal, 2895
Arkopal, 2896
Arkopal N, 2897, *See* Ablunol NP4, 106
Arkopal N-040, *See* Ablunol NP4, 106
Arkopal N-060, *See* Ablunol NP4, 106
Arkopal N-090, *See* Ablunol NP4, 106, Alkasurf®NP-4, 1715, Triton® N-57, 32430
Arkopal N 100, *See* Ablunol NP4, 106
Arkopal N 110, *See* Ablunol NP4, 106
Arkopal N 150, *See* Ablunol NP4, 106
Arkopal N 300, *See* Ablunol NP4, 106
Arkopon T, 2898
Arkotine, *See* DDT, 7820
Arlacel 20, *See* Ablunol S-20, 108, Kemester®S20, 15966, Radiasurf® 7125, 25797, Sorbon S-20, 29117
Arlacel 40, *See* Ablunol S-40, 109
Arlacel 60, *See* Ablunol S-60, 110, Radiamuls®Sorb 2145, Kemester® S60, 15968, Sorb 2161,Sorb 2166, 25783,Sorbon S-60, 29119
Arlacel 80, *See* Ablunol S-80, 111, Kemester®S80, 15970, Radiasurf® 7155, 25802, SorbonS-80, 29121, Witconol 2500, 34501
Arlacel 85, *See* Ablunol S-85, 112, Witconol 2503,34502
Arlacel A, *See* Witconol 2500, 34501
Arlacel A (VAN), *See* Ablunol S-80, 111
Arlacel® 20, 2899
Arlacel® 40, 2900
Arlacel® 60, 2901
Arlacel® 80, 2902
Arlacel® 83, 2903
Arlacel® 85, 2904
Arlacel® 165, 2905
Arlacel® 186, 2906
Arlacide A, 2907, *See* chlorhexidine acetate,6100
Arlagard, 2908
Arlamol, 2909
Arlamol® E, 2910
Arlamol® ISML, 2911
Arlasolve® 200, 2912
Arlasolve® DMI, 2913
Arlatone® B, 2914
Arlatone® G, 2915
Arlatone® T, 2916
Arlin, 2917
Arlinflex, 2918
Arlon®, 2919
Arloy®, 2920
Armac®, 2921
Armco, 2922
Armco Ingot Iron, 2923
Armeen, 2924
armeen 16d, *See* Armeen® 16, 2935, Armeen®16D, 2936, Crodamine 1.16D, 7101, Amine16D, 2130
Armeen 1180, *See* Amine 18-90, 2131, Armeen®18, 2938, Armid® HTD, 2963, Crodamine1.18D, 7102, Steamfilm FG, 29589
Armeen DM-12D, *See* Onamine 12, 22164
Armeen DM 16D, *See* Adma® 16, 743,Crodamine 3.A16D, 7106, N,N-dimethylhexadecylamine, 8524
Armeen DM 18D, *See* Crodamine 3.A18D, 7107,Dymanthine, 9327
Armeen®, 2925
Armeen® 2-10, 2926
Armeen® 2-18, 2927
Armeen® 2C, 2928
Armeen® 2HT, 2929, *See* Amine2 VT, 2163
Armeen® 2T, 2930
Armeen® 3-12, 2931
Armeen® 3-16, 2932
Armeen® 12, 2933
Armeen® 12D, 2934
Armeen® 16, 2935
Armeen® 16D, 2936
Armeen® 16D, 2937
Armeen® 18, 2938
Armeen® 18D, 2939
Armeen® C, 2940
Armeen® CD, 2941
Armeen® DM12D, 2942, *See* dimethyllauramine, 8507
Armeen® DM16D, 2943
Armeen® DM18D, 2944
Armeen® DMCD, 2945
Armeen® DMHTD, 2946
Armeen® DMOD, 2947, *See* Jet Amine DMOD,15586
Armeen® DMSD, 2948
Armeen® DMTD, 2949, *See* Jet Amine DMTD,15588
Armeen® HT, 2950
Armeen® HTD, 2951
Armeen® L8D, 2952
Armeen® M2-10D, 2953, *See* didecylmethylamine, 8415
Armeen® M2C, 2954
Armeen® M2HT, 2955
Armeen® OL, 2956
Armeen® OLD, 2957
Armeen® T, 2958
Armeen® T, TD, *See* tallowamine, 30748
Armeen® TD, 2959
Armenian bole, *See* Indian Red, 15068
Armenian cement, 2960
Armid® E, 2961
Armid® HT, 2962
Armid® HTD, 2963
Armid® O, 2964
Armide, *See* Camite, 5054
Armillatox®, 2965
Armite, 2966
Armix 146, 2967
Armix 176, 2968
Armodour, *See* Vestolit®, 33679

Armofilm, 2969
Armoflo®, 2970
Armofog, 2971
Armogard, 2972
Armogloss, 2973
Armohib®, 2974
Armohib® 18, 28, 2975
Armol, *See* Metol, 19584
Armor-Kote, 2976
Armor-ply, 2977
Armoslip CP, *See* Unislip 1759, 33045
Armoslip®, 2978
Armostat®, 2979
Armotan ML, *See* Ablunol S-20, 108
Armotan MO, *See* Ablunol S-80, 111, Witconol2500, 34501
Armotan MS, *See* Ablunol S-60, 110
Armotan PMO-20, *See* Alkamuls® T-80, 1693
Armoteric LB, 2980
Armoteric SB, 2981
Armourcote, 2982
Armowax, 2983
Armowax, 2984
Armowax EBS, 2985, *See* Advawax® 290, 820
Armstrong acid, *See* Schaeffer's acid, 27558
Armul 17, 2986
Armul 22, 88, 2987
arnatto, *See* annatto, 2478
Arneel DN, 2988
Arneel HF, 2989
Arneel S, 2990
Arneel TOD, 2991
Arneel® OD, 2992
Arnica Oil CLR, 2993
Arnica Yellow, 2994
Arnite A.K.U, 2995
Arnite G, 2996
Arnold's Base, *See* Michler's Base, 19644
arnotto, *See* annatto, 2478
ARO, 2997
Arobleach HW, 2998
Arobleach MX, 2999
Arobon, *See* carob flour, 5303
Arochem, 3000
Aroclean MC-4, 3001
Aroclear, 3002
Arodet AA-350, 3003
Arodet AN-100, 3004
Arodet AN-160, 3005
Arodet BLN Special, 3006
Arodet BN-100, 3007
Arodet E-15, 3008
Arodet HCS, 3009
Arodet MKD, 3010
Arodet N-100, 3011
Arodet TA-8, 3012
Arofene, 3013
Arofix F-6, 3014
Arofix SRN, 3015
Aroflat®, 3016
Aroflat® 3113-P-30, 3017
Aroflint®, 3018
Aroflint® 202-A6X-60, 3019
Aroflint® 303-X-90, 3020
Arofoam, 3021
Arofoam SNI, 3022
Arofos 200 Conc, 3023
Arofos 326, 3024
Aroful BV-50, 3025
Arogrip, 3026
Arol Biodet, 3027
Arol Defoamer NA2X, 3028
Arol Woolbrite, 3029
Arolev ADL-30, 3030
Arolev CDD, 3031
Arolev MTR-7, 3032
Arolon®, 3033
Arolon® 580-W-42, 3034
Arolterge 100M, 3035
Arolube MIT-1, 3036
Aromabator PC-80, 3037
Aromabator PC-88, 3038
Aromaplas, 3039
Aromasol, 3040
Aromasol 17, 3041
Aromatic alkoxylates, *See* Agrisol PX401, 1071
Aromatic diamine, *See* Unilink® 4100, 4200,32965
Aromatic diisocyanate, *See* Baymidur® K88,3812
Aromatic Oil 745, 3042
Aromatic phosphate ester, *See* Chemphos TC-227, 6008
Aromatic phosphate ester of ethoxylatedphenol, *See* Rhodafac® BP-769, 26594
Aromatic phosphate ester type surfactants, *See* Adekacol CS, PS, TS, 673
Aromatic polyglycol ether, *See* Emulvin® W,10531
Aromatic Solvent 150, 3043
Aromatic sulfonate, *See* Agrilan® WP101, 1062,Sandogen, 27332
Aromatic urethane acrylate resin, *See* CN 970,6482
Aromatic urethane methacrylate, *See* CN 974,6483
Aromax® DMC, *See* Cocamine oxide, 6557
Aromax® DMC-W, *See* Cocamine oxide, 6557
Aromex, 3044
Aromix, 3045, 3046
Aromox DMMC-W, *See* Ablumox LO, 91,Ammonyx® LO, 2264, Empigen® OB, 10375,Rhodamox® LO, 26620
Aromox® C/12-W, 3047
Aromox® DMMC-W, *See* lauramine oxide, 16832
Aron α, 3048
Aron 1100, *See* Vestolit®, 33679
Aron compound HW, *See* Vestolit®, 33679
Aron M 1, *See* methyl ethyl ketoxime, 19532
Aron NS 1100, *See* Vestolit®, 33679
Aron TS 1100, *See* Vestolit®, 33679
Aroplax, *See* Arochem, 3000
Aroplaz®, 3049
Aroplaz® 3667-Z-80, 3050
Aroplaz® 6820-100, 3051
Aropol, 3052
Aropol 2036, 3053
Aropol 7020, 3054
Aropol 7240T-15, 3055
Aropol 7710, 3056
Aropol 8321, 3057
Aropol 8420, 3058
Aropol Phase α, 3059
Aropol Phase II, 3060
Aropol WEP, 3061
Aroquest 100, 3062
Aroquest 120, 3063
Aroquest M Special, 3064
Aroquest MLC, 3065
Aroset®, 3066
Arosoft Base LCS-2, 3067
Arosoft GSE-D, 3068
Arosoft LC-15, 3069
Arosol, *See* Igepal® Cephene Distilled, 14850,Rewopal® MPG 10, 26344
Arosolve 570-HF, 3071
Arosolve 9D5R, 3070
Arosolve MN-LF, 3072
Arosolve RCB, 3073
Arosolve XNF-1, 3074
Arosulf SBO-65, 3075
Arosulf SCO-75%, 3076
Arosurf 1855E40, *See* Ablunol DEGMS, 99
Arosurf® 66-E2, 3077
Arosurf® 66-PE12, 3078
Arosurf® AA 23, 3079
Arosurf® MG-70, 3080
Arosurf® TA-100, 3081, *See* Adogen® TA-101,780, Genamin DSAC, 12771
Arotap®, 3082
Arotech, 3083
Arotex, 3084
Arothix, 3085
Arothix 4000-P-40, 3086
Arotran 50437-8, 3087
Arova 16, 3088
Arowet 70 E, 3089
Arowet ODA, 3090
Arowet SC-75, 3091
Arozyme TD, 3092
ARP®, 3093
Arpak 4322, 3094
Arpal Non Selex, 3095
Arpocox, 3096
arporcarb, *See* Propoxur, 25198, 25199, 25200
Arprinocid, *See* Arpocox, 3096
Arpro 3313, 3097, *See* Adpro AP 2112-GP, 788
Arprocarb, *See* Baygon®, 3790
Arpylen, 3098
Arquad, 3099
Arquad 10, *See* BTC® 1010-80, 4638, Querton210Cl-50, 25651, Radiaquat 6412, 25790
Arquad DM18b-90, *See* Ablumine 280, 85,Cycloton® SCS, 7542,octadecylbenzenemethanaminium chloride,21985, stearalkonium chloride, 29592,Sumquat® 6210, 30126,
Arquad®, 3100
Arquad® 2C-70 Nitrite, 3101
Arquad® 2C-75, 3102, *See* Adogen® 462, 772
Arquad® 2HT-75, 3103
Arquad® 2T-75, 3104
Arquad® 12-37W, 3105, *See* Arquad® 12-37W,3105
Arquad® 12-50, 3106
Arquad® 16-29, 3107, *See* Arquad® 16-50,3108
Arquad® 16-50, 3108
Arquad® 18-50, 3109
Arquad® 210-50, 3110
Arquad® 218-100, 3111, *See* Adogen® TA-101,780
Arquad® 218-75, 3112, *See* Adogen® TA-101,780
Arquad® 218-75, 218-100, *See* GenaminDSAC, 12771
Arquad® 316(W), 3113
Arquad® C-33W, 3114
Arquad® C-50, 3115
Arquad® DM14B-90, *See* JAQ PowderedQuaternary, 15517
Arquad® DMCB-80, 3116
Arquad® DMHTB-75, 3117
Arquad® DMMCB, *See* Benzalkonium chloride,3958
Arquad® HTL8(W) MS-85, 3118
Arquad® M2HTB-80, 3119
Arquad® S-50, 3120
Arquad® T-27W, 3121, tallow trimonium chloride,30747
Arquad® *See* Arquad® T-50, 3122
Arquad® T-50, 3122
Arquard® B-100, 3123
Arrconox S.P, 3124
Arrcorez 16, 3125
Arrcorez 17, 3126
Arresin, 3127, *See* Aresin, 2839, Atlas Linuron,3319
arrhenal, 3128
Arrivo, *See* cypermethrin, 7588, Topclip Parasol,32037
Arroconox AHT, DNL and DNP, 3129
Arrow Tool Steels, 3130
Arsan600, 3131
Arsanote, *See* Neoarsycodyl, 20871
Arsenal, 3132, *See* imazapyr, 14893
Arsenal
Arsenal XL, 3133
Arsenal®, 3134
Arsenal® XL, 3135
arsenic, 3136
arsenic (III) chloride, *See* arsenic trichloride, 3141
arsenic (III) oxide, *See* arsenic trioxide, 3143
arsenic (III) trioxide, *See* arsenic trioxide, 3143
arsenic (V) oxide, *See* arsenic pentoxide, 3140
arsenic (white), *See* arsenic trioxide, 3143
arsenic acid, 3137, *See* Aresenid, 2838, arsenicpentoxide, 3140
arsenic acid anhydride, *See* arsenic pentoxide,3140
arsenic acid, monopotassium salt, *See* potassiumarsenate, 24737
arsenic anhydride, *See* arsenic pentoxide, 3140
arsenic bronze, 3138

arsenic butter, *See* arsenic trichloride, 3141
arsenic chloride, *See* arsenic trichloride, 3141
arsenic disulfide, *See* realgar, 25930, red algar,25950
arsenic hydride, *See* arsine, 3144
arsenic orange, *See* realgar, 25930
arsenic oxide, *See* arsenic pentoxide, 3140
arsenic oxide (3), *See* arsenic trioxide, 3143
arsenic oxide (As_2O_3), *See* arsenic trioxide, 3143
arsenic pentasulfide, 3139
arsenic pentoxide, 3140, *See* arsenic pentoxide,3140
arsenic sesquioxide, *See* arsenic trioxide, 3143
arsenic sesquioxide (As_2O_3), *See* arsenic trioxide,3143
arsenic trichloride, 3141, *See* arsenic trichloride,3141
arsenic trifluoride, 3142
arsenic trihydride, *See* arsine, 3144
arsenic trioxide, 3143, *See* arsenic trioxide, 3143
arsenic trisulfide, *See* orpiment, 22317
arsenicum album, *See* arsenic trioxide, 3143
arsenious acid, *See* arsenic trioxide, 3143
arsenious acid copper(2+) salt(1:1), *See* cupricarsenite, 7376, Scheele's green, 27562
arsenious chloride, *See* arsenic trichloride, 3141
arsenious fluoride, *See* arsenic trifluoride, 3142
arsenious oxide, *See* arsenic trioxide, 3143,poison flour, 24286
arsenite, *See* arsenic trioxide, 3143
arseniuretted hydrogen, *See* arsine, 3144
arsenolite, *See* arsenic trioxide, 3143
arsenous acid, *See* arsenic trioxide, 3143
arsenous acid anhydride, *See* arsenic trioxide,3143
arsenous anhydride, *See* arsenic trioxide, 3143
arsenous chloride, *See* arsenic trichloride, 3141
arsenous hydride, *See* arsine, 3144
arsenous oxide, *See* arsenic trioxide, 3143
arsenous oxide anhydride, *See* arsenic trioxide,3143
arsenous trichloride, *See* arsenic trichloride,3141
arsine, 3144
Arsinette, 3145
arsinyl, *See* arrhenal, 3128
arsodent, *See* arsenic trioxide, 3143
Arsonate, *See* Neoarsycodyl, 20871
Arsonate Liquid, *See* MSMA, 20417
arsonic acid copper(2+) salt (1:1), *See* cupricarsenite, 7376, Scheele's green, 27562
Art Bronze, 3146
Artam, *See* cinchophen, 6307
Artic, 3147, *See* methyl chloride, 19522
Artic Mist, 3148, *See* Altalc 200 USP, 1870
artifical barite, *See* Barium sulfate, 3644
artifical essential oil of almond, *See*Benzaldehyde, 3957
artificial almond oil, *See* Benzaldehyde, 3957,Ethereal Oil of Bitter Almonds, 10944
Artificial Bitter Almond Oil, *See* Benzaldehyde,3957, Ethereal Oil of Bitter Almonds, 10944
artificial chappe, *See* staple fiber, 29518
artificial cinnabar, *See* mercuric sulfide, red andblack, 19350
Artificial essential oil of almond, *See* EtherealOil of Bitter Almonds, 10944
artificial gum, *See* dextrin, 8235
artificial oil of ants, *See* furfural, 12491
artificial rubber, *See* Thinoline, 31675
artificial sweetening substance Gendorf 450, *See* saccharin sodium, 27196
artificial wool, *See* staple fiber, 29518
Artisil®, 3149
Artodan CP 80, *See* Verv®, 33644
Artodan SP 55 Kosher, 3150
Arton F, 3151
Arton G, 3152
ARTZ, *See* hyaluronic acid sodium salt, 14483
Arubren®, 3153
Arvest, *See* Cerone, 5756
Arvetane, 3154
Aryl polyglycol ether, *See* EW-POL 8021,11280, Luprintol® PE, 17915
Aryl polyoxyether, *See* Syn Fac® 222, 30513,Syn Fac® 334, 30514
Arylan PWS, *See* Ablusol DBM, 125
Arylan S, 3155
Arylan SBC, *See* Nacconol® 40G, 20647,sodium dodecylbenzenesulfonate, 28760
Arylan®, 3156
Arylan® CA, 3157
Arylan® PWS, 3158
Arylan® SBC Acid, 3159
Arylan® SC15, 3160
Arylan® SNS, 3161
Arylan® SP, 3162
Arylan® SX, 3163
Arylan® TE/C, 3164
Arylene M40, 3165
Arylene M60, *See* dioctyl sodium sulfosuccinate,8550
Arylethoxy phosphate potassium salt, *See*Phospholan® KPE4, 23718
Arylmate®, 3166
Arylon® LP 401 NC10, 3167
Arylsulfonamidocarboxylic acid, *See* HostacorH Liq. N, 14330
Arylsulfonat BASF, *See* Witco® 1298, 34458
AS, 3168
AS-10GF, 3169, *See* Acrylonitrile-butadiene-styrene, 438, 439
AS-15CF/000, 3170, *See* Acrylonitrile-butadiene-styrene, 438, 439
ASA, 3171
ASA copolymer, *See* Blendex® 975, 4344
Asa Dulcis, *See* benzoin, 3975
Asadene®, 3172
Asaflex®, 3173
Asahi Aji®, 3174
Asaprene®, 3175
asaprol, 3176, *See* Abrastol, 155
asaprol-etrasol, *See* Abrastol, 155, asaprol,3176
Asarum, *See* wild ginger oil, 34407
asbestine pulp, *See* Agalite, 953
asbestos, *See* Amianthus, 2057
Asbesto-Wet, 3177
AS-C12, *See* Witcolate 6400, 34469
ASC-4, *See* Tuasol 100, 32543
Ascabin, *See* Benzyl benzoate, 3987
Ascabiol, *See* Benzyl benzoate, 3987
Ascaridil, *See* Tramisol®, 32134
Asceptichrome, *See* mercurochrome, 19354
Ascinin® P, R, Special, 3178
ascorbic acid, *See* vitamin C, 33938
L-ascorbic acid, *See* vitamin C, 33938
L-Ascorbic acid, monosodium salt, *See*sodium ascorbate, 28734
L-(+)-ascorbic acid sodium salt, *See* sodiumascorbate, 28734
Ascorbicin, *See* sodium ascorbate, 28734,vitamin C sodium salt, 33939
Ascorbin, *See* sodium ascorbate, 28734,vitamin C sodium salt, 33939
Ascorbosilane C, 3179
Ascorbyl methylsilanol pectinate, *See*Ascorbosilane C, 3179
Ascorin, *See* vitamin C, 33938
Ascorteal, *See* vitamin C, 33938
Ascorvit, *See* vitamin C, 33938
Ascot, 3180
Ascurit, *See* prochloraz, 25057
Asepsin, *See* Antisepsin, 2614
Aseptisil, 3181
Aseptoform, *See* methylparaben, 19577
aseptoform butyl, *See* Nipabutyl, 21249
Aseptoform P, *See* Nipasol M, 21282
Aseptoforms, 3182
Ashberry metal, 3183
Ashbury Metal, *See* Ashberry metal, 3183
Ashlade 4% At Gran, 3184, *See* Ashlade Atrazine 50 FL, 3188
Ashlade 4-60 CCC, 700 CCC, 3185
Ashlade 5C, 3186
Ashlade Adjuvant Oil, 3187
Ashlade Atrazine 50 FL, 3188
Ashlade Blight Fungicide, 3189
Ashlade Cosmic FL, 3190
Ashlade CP, 3191
Ashlade D-Moss, 3192
Ashlade Flotin, 3193
Ashlade Flotin, *See* Farmatin, 11480
Ashlade Halt, 3194
Ashlade Linuron, 3195
Ashlade M, 3196
Ashlade Mancarb FL, 3197, *See* Ashlade M, 3196
Ashlade SMC, 3198
Ashlade TCNB, 3199
Ashland Hi-Sol 10, 3200
Ashland Hi-Sol 15, 3201
Ashland Kwik-Dri, 3202
Ashland Lacolene, 3203
Ashlene®, 3204
Ashlene® 61-2M, 3205
Ashlene® 520, 3206
Ashlene® 520-13G, 3207
Ashlene® 520MS, 3208
Ashlene® 525-13G, 3209
Ashlene® 527, 3210
Ashlene® 527LD-13G, 3211
Ashlene® 528BR-WO, 3212
Ashlene® 528L-13G, 3213
Ashlene® 541, 3214
Ashlene® 541S, 3215
Ashlene® 630-33G, 3216
Ashlene® 735, 3217
Ashlene® 830L, 3218, 3219
Ashlene® 840, 3220
Ashlene® 858, 3221
Ashlene® 870, 3222
Ashlene® 980L, 3223
Ashlene® 980LS-40G, 3224
Ashlene® 981S, 3225
Asiaticoside, *See* Madecassol, 18291
Askure, 3226
ASM 4010MA, *See* Vulkanox® 4010 NA, 34147
ASM MB, *See* Anti-Oxidant MB, 2589, 2-mercaptobenzimidazole, 19334
Asmer, 3227
***as*-methylphenylethylene**, *See* α-methylstyrene,19579
Asn, *See* Asparagine, 3232
A-Sol, 3228
Asp, 3229, *See* Aspartic acid, 3234
ASP®, 3230
Aspac®, 3231
L-Asparagic acid, *See* Aspartic acid, 3234
Asparagine, 3232
L-β-asparagine, *See* Asparagine, 3232
asparaginic acid, *See* aspartic acid, 3234
Asparamide, *See* Asparagine, 3232
Aspartame, 3233, *See* Equal, 10731
Aspartic acid, 3234
Aspartic acid β amide, *See* Asparagine, 3232
L-Aspartic acid, *See* Aspartic acid, 3234
L-aspartyl-L-phenylalanine methyl ester, *See*Aspartame, 3233
Aspect® TPPE, 3235
Asphalt, *See* Karetnja, 15765
asphaltenes, 3236
Asphaltum, *See* Karetnja, 15765
Asplit, 3237
Asplosal, *See* Aspirin,
Aspon, 3238
Aspor, *See* Zineb, 34807
Aspro, *See* Aspirin,
Aspulum, 3239
Aspumit AP, 3240
Aspumit SDM, 3241
Assaf, 3242
Assam white, *See* gutta-susu, 13511
Assault, *See* imazapyr, 14893
Asset, 3243
Assugrin, *See* sodium cyclamate, 28754
Assure, *See* quizalofop-ethyl, 25697
Assurgrin Feinsüss, *See* sodium cyclamate,28754

Assurgrin Vollsüss, *See* sodium cyclamate,28754
Astacin® Finish PUD, 3244
A-stage resin, *See* phenolic resin, 23607
Astemisan, *See* Hismanal, 14132
Asterite, 3245
Astick, 3246
Astingol, 3247
Astix, 3248
Astix CMPP, *See* Verdone CDA, 33580
Aston 123, 3249
Aston RC, 3250
Astra®, 3251
Astradur® A and T, 3252
Astraflex®, 3253
Astragal®, 3254
Astralex, 3255
Astralon, *See* Vestolit®, 33679
Astralon®, 3256
Astratone, *See* Emeressence® 1150, 10077,ethylene brassylate, 11121
Astrazon®, 3257
Astringen, *See* Aloxicoll, 1813, aluminumchlorohydrate, 1903, Reach® 101, 201, 501,25919
Astro Floctite, 3258
Astro Mel, 3259
Astrobot, *See* dichlorvos, 8394, Vapona, 33378
Astrol, 3260
Astrolith, 3261
Astrolok, 3262
Astroplax, 3263
Astroturf, 3264
Astrowet 0-70-PG, 3265
Astrowet 0-75, 3266
Astryl, 3267
Astryn® 63A6-2, 3268, *See* Adpro AP 2112-GP,788
Astryn® 63F4-2, 3269
Astryn® 65F4-4, 3270 *See* Adpro AP 2112-GP,788
Astryn® 65F5-4, 3271, *See* Adpro AP 2112-GP,788
Astryn® 734-2, *See* Adpro AP 2112-GP, 788
Astryn® 73F4-2, 3272
Astryn® 73F5-2, 3273, *See* Adpro AP 2112-GP,788
Astryn® 78F4-2, 3274, *See* Adpro AP 2112-GP,788
Astryn® BA16G, 3275, *See* Adpro AP 2112-GP,788
Astryn® SD068-4, 3276, *See* Adpro AP 2112-GP, 788
Asugryn, *See* sodium cyclamate, 28754
Asulam, 3277, *See* Asulox, 3278, 3279
Asulox, 3278, 3279, Asulam, 3277
Asuntol, *See* Perizin®, 23276
Asuntol®, 3280
AT-7, *See* hexachlorophene, 13992
AT-20GF, 3282, *See* Acetal, 280
AT 1806M; AT 4030M, 3281
Atactic poly(vinyl chloride), *See* Vestolit®,33679
Atalco C, *See* Adol® 52 NF, 782, Cachalot® C-50 NF, 4878
Atalco S, *See* Adol® 62 NF, 783, Stearal, 29591
Atar Phenol, 3283
Atasorb, *See* Attapulgus, 3405
Atav, *See* Vi-Alpha, 33706
Atazinax, *See* atrazine, 3394
ATC, *See* Ammonium thiocyanate, 2256
Ateri, *See* nickel silvers, 21132
Aterite, 3284
Atgard, 3285, *See* dichlorvos, 8394, Vapona,33378
Atgard C, *See* dichlorvos, 8394
Atgard V, *See* dichlorvos, 8394
Athado, *See* terbutryn, 31230
Athenon, *See* glycine, 13095
Atiran, 3286, *See* Agallol, 954, Atiran, 3286,Ceresan, 5745
Atlac®, 3287
Atlac® 382-05A, 3288
Atlac® 797CT, 3289
Atlacide, 3290, *See* KM Sodium Chlorate,16244, sodium chlorate, 28747
Atlacide Extra, 3291
Atlacide, AMS, Ammate, *See* Amcide, 1986
Atladox HI, 3292
Atlas 10 Bronze, 3293
Atlas 5C Chlormequat, 3294
Atlas 89, 3295
Atlas 90, 3296
Atlas Adherbe, 3297
Atlas Adherbe®, 3298
Atlas Adjuvant Oil, 3299
Atlas Brown, 3300, 3301
Atlas Chlormequat 46, 700, 3302
Atlas CIPC 40, 3303
Atlas D, 3304
Atlas Defoamer AFC, 3305
Atlas Electrum, 3306
Atlas EM-2, 3307
Atlas EMJ-2, 3308
Atlas G 924, *See* Witconol 2380, 34498
Atlas G-2127, *See* Ablunol 200ML, 95
Atlas G-2129, *See* Ablunol 200ML, 95
Atlas G-2142, *See* Ablunol 200MO, 96
Atlas G-2144, *See* Ablunol 200MO, 96
Atlas G-3915, *See* Ablunol OA-6, 107, Volpo 3,34020
Atlas G-3920, *See* Ablunol OA-6, 107, Volpo 3,34020
Atlas Gold, 3309, 3310
Atlas Indigo, 3311, 3312
Atlas JG#1, 3313
Atlas Leather Odor, 3314
Atlas Libsorb, 3315
Atlas Lignum, 3316
Atlas Lignum Granules, 3317
Atlas Linuron, 3318, 3319
Atlas M 130, 3320
Atlas MCPA, 3321
Atlas Orange, 3322
Atlas Pink C, 3323, 3324
Atlas Protrum® K, 3325
Atlas Red, 3326, 3327
Atlas Sheriff, 3328
Atlas Sheriff®, 3329
Atlas Silver, 3330, 3331
Atlas Solan, 3332
Atlas Steel, 3333
Atlas Steward, 3334
Atlas Steward®, 3335
Atlas Tanked Cod Oil, 3336
Atlas Tecgran, 3337
Atlas Total A, Total S, 3338
Atlas WA-100, *See* dioctyl sodium sulfosuccinate,8550
Atlas®, 3339
Atlasbeam #1, 3340
Atlasol 103, 3341, *See* sodium decyl sulfate, 28755
Atlasol 118-U, 3342
Atlasol 170, 3343
Atlasol 177, 3344
Atlasol 178, 3345
Atlasol BSC, 3346
Atlasol CSN, 3347
Atlasol KAD, 3348
Atlasol KMM, 3349
Atlastan AR, 3350
Atlastan LC, 3351
Atlavar, 3352
Atlazin, 3353
Atlox, 3354
Atlox 1285, *See* Ablunol CO 10, 98
Atlox 1300, *See* Ablunol CO 10, 98
Atlox 4862, *See* Ablusol ML, 129
Atlox 5000, *See* Ablunol DEGMS, 99
Atmer®, 3355
Atmer® 100, 3356
Atmer® 103, 3357
Atmer® 105, 3358
Atmer® 106, 3359
Atmer® 121, 3360
Atmer® 122, 3361
Atmer® 1007, 3362
Atmer® 7001, 3363
Atmer® 8112, 3364
Atmido, 3365
Atmos, 3366
Atmos 150, 3367
Atmos 300, 3368
Atmos 659K, 3369
ATMP, *See* Fostex AMP, 12337
Atmul 124, 3371
Atmul 2622K, 3372
Atmul 80, 3370
Atocin, *See* cinchophen, 6307
Atolex ASL/C, 3373
Atolex ASL/C100, 3374
Atolex DA/25, 3375
Atolex Polythene Emulsions, 3376
Atolex QE, 3377
Atomit, *See* CP Filler, 6928
Atomite, 3378, *See* CP Filler, 6928
Atomite®, 3379
Atonin, *See* aspirin,
Atophan, *See* cinchophen, 6307
atoquinol, 3380
Atoxan, *See* Carbaryl, 5181
ATP, *See* Adenosine triphosphate, 681
ATP Nucleotides, 3381
Atpet, 3382
Atpet 300, 3383
Atpet 400, 3384
Atpet 600, 3385
Atprime®, 3386
Atraflow, 3387, *See* Ashlade Atrazine 50 FL,3188
Atraflow
Atraflow Plus, 3388, 3389
Atraflow Plus, *See* Ashlade Atrazine 50 FL, 3188
Atragan, 3390
Atral, *See* Acaprin®, 181, *See* 1,3-di-6-quinolylurea, 8618
Atramentum Stone, 3391
Atramet Combi, 3392
Atranex, 3393, *See* Ashlade Atrazine 50 FL,3188, atrazine, 3394
Atrasine, *See* atrazine, 3394
Atrataf, *See* atrazine, 3394
Atratol, *See* atrazine, 3394
Atratol A, *See* atrazine, 3394
atrazine, 3394
Atrazine, *3394*, *See* Ashlade 4% At Gran, 3184,Ashlade Atrazine 50 FL, 3188, Atraflow, 3387, Atranex, 3393, Borocil A, 4430, Boroflow A, 4434, Chlorea, 6094, Gesaprim 500FW, 12922, Herbazin Total, 13805, Laddok®, 16593, Mascot Gauntlet, 19002, Primatol AA, 24969
Atrazine + 2,4-D + sodium chlorate, *See*Atlavar, 3352
Atrazine 4l, *See* atrazine, 3394
Atrazine 80W, *See* atrazine, 3394
Atrazine, boromacil and diuron, *See* BorocilExtra, 4431
atrazine-aminotriazole, *See* Atlazin, 3353,Boroflow A/ATA, 4435
atrazine-cyanazine, *See* Holtox, 14280
Atrazines, *See* atrazine, 3394
atrazine-terbuthylazine, *See* Gardoprim A500FW, 12669
Atred, *See* atrazine, 3394
Atrex, *See* atrazine, 3394
Atrinal, 3395
Atriphos, *See* Adenosine triphosphate, 681
Atrixo, 3396
Atroban, *See* permethrin, 23384, 23385, 23386
atropine sulfate, *See* Isopto Atropine, 15414
atroxindol, 3397
Atrust, 3398
Attaclay, 3399, *See* Attapulgus, 3404, 3405
Attacote, 3400, *See* Attapulgus, 3404, 3405
Attaflow, 3401, *See* Attapulgus, 3404, 3405
Attagel, *See* Attapulgus, 3404, 3405
Attane, 3402
Attane 4601, 4802, 3403
Attapulgite, 3404, *See* Attapulgus, 3405 Fuller'searth, 12441 Gelsorb B, 12747
Attapulgite clay, *See* Diluex®, FG, 8482
Attapulgus, 3405, *See* Attapulgite, 3404, 3405
Attasorb, 3406, *See* Attapulgite, 3404, 3405
Attrex, *See* atrazine, 3394

Atul Fast Yellow R, *See* Butter yellow, 4759
Atul orange R, *See* Sudan I, 29938
ATZ, *See* atrazine, 3394
aubepine, *See p*-anisaldehyde, 2470
Audrey, 3407
Auel solder, 3408
Augsburg metal, 3409
Augus Hot Rod, *See* sodium hydroxide, 28772
Aules, *See* Rezifilm, 26521
Auracryl, 3410
Aurantine, 3411
Aurantlol®, 3412
Aurasperse, 3413
auric chloride, *See* gold trichloride, 13175
Aurocyanase, 3414
Auromet55, 3415
Aurum, *See* gold, 13171
Aurum 400, 450, 500, 3416
Aurum JAF 3040, 3417
Aurum JCN 3030, 3419
Aurum JCN 6030, 3418
Aurum JGN 3030, 3420
Aurum JNF 3010, 3421
Aurum JNF 3020, 3422
Aurum JQF 3025, 3423
Aurum JRF 3025, 3424
Aurum JRN 3015, 3425
Aurum paradoxum, *See* Telloy®, 31058, tellurium, 31062
Aurum Series, 3426
austenite, 3427
Australian gold, 3428
Austrapen, *See* Ampicillin, 2361
Austrapol, 3429
Austratex, 3430
Austrian cinnabar, *See* chrome red, 6209
Austrostab, 3431
Austrox, 3432
AuSub, 3433
Autan, *See* diethyl toluamide, 8438
Autan®, 3434
Auto Command®, 3435
Autofroth, 3436
Autogal, 3437
Automate, 3438
automolite, 3439
Autopak, 3440
Autopoon NI, 3441
Autopour, 3442
Autopur WK 4121, 3443
Autovisuel®, 3444
Autoworm, *See* oxfendazole, 22426
Autumn Kite, 3445
Autumn Lawn Food, 3446
Auxiliary PR-10BT, 3447
AV-290, *See* Avoparcin, 3502
Ava-ava, *See* kava-kava resin, 15805
Avabond, 3448
Avadex, 3449
Avadex BE, *See* triallate, 32190
Avadex BW, *See* triallate, 32190
Avadex® BW, 3450
Avadyne AV1200/CA100, 3451
Avalon, 3452
Avamid, 3453
Avamid 150, 3454
Avanel® S-30, 3455
avantin, *See* isopropanol, 15405
Avantine, 3456
Avaunt®, 3457
Avazyme, 3458, *See* α-Chymotrypsins, 6286
Avecolite, 3459
Avenge 2, 3460
avenin, *See* vegetable casein, 33516
Aventox, *See* trietazine, 32238
Aventox SC, 3461
Aveonal, 3462
Aversin, 3463
Avgard, 3464, 3465
Avialite, 3466
Aviamide-6, 3467
Aviashine, 3468
Aviawash, 3469
Avibon, *See* Vi-Alpha, 33706
Avicel, 3470
Avicell-RC, 3471
Avicol, *See* quintozene, 25692
Aviester, 3472
Avilon, 3473
Avional D, 3474
Avirol, 3475, *See* monopol soap, 20252
Avirol ® A, *See* Ammonium lauryl sulfate, 2242
Avirol 100E, *See* Witcolate ES-2, Witcolate LES-60C, 34477
Avirol 101, *See* Witcolate 6400, 34469, Rhodapon® 101-10, 26631, sodium lauryl sulfate, 28781
avirol 118, *See* Rhodapon® 101-10, 26631, sodium lauryl sulfate, 28781
Avirol 118 conc, *See* Witcolate 6400, 34469
Avirol® 125E, 3476
Avirol® A, 3477
Avirol® AE 3003, 3478, *See* Ammonium laurethsulfate, 2240
Avirol® FES 996, 3479, *See* Avirol® SE 3002,3484
Avirol® SA 4106, 3480
Avirol® SA 4108, 3481
Avirol® SA 4110, 3482, *See* sodium decyl sulfate,28755
Avirol® SA 4113, 3483
Avirol® SE 3002, 3484
Avirol® T 40, 3485, *See* Akyposal TLS42, 1322
Avisol, 3486
Avisol® G, 3487
Avistin®, 3488
Avistin® PN, 3489
Avita, *See* Vi-Alpha, 33706
Avitex, 3490
Avitige®, 3491
Avitol, *See* Vi-Alpha, 33706
Avitone®, 3492
Avitone® A, 3493
Avitrol, 3494
Avivage, 3495
Avivan, 3496
Avivan SO 6, *See* Ablunol SA-7, 113
Avloprocil, *See* Penicillin V, 23111
Avlosulfon, *See* Dapsone, 7751
Avobenzone, *See* Parsol® 1789, 22866
Avocado fatty oil, *See* Avocado Oil, 3497
Avocado Oil, 3497, 3498, *See* Avocado Oil CLR,3499
Avocado Oil CLR, 3499
Avocado Oil CLR, *See* Avocado Oil, 3497, 3498
Avoilefin, 3500
Avolan®, 3501
Avolin, *See* dimethyl phthalate, 8508, Kemester®DMP, 15956, Kodaflex® DMP, 16271, Unimoll®DM, 32979
Avoparcin, 3502
Avotan, *See* Avoparcin, 3502
Avron, 3503
AVT-75, 3504, *See* 2-amino-2-methyl-1 propanol,2182
Award, *See* Topas 100, 32026
AX 363, *See* CP Filler, 6928
Axall, 3505
Axarel®, 3506
Axelcon, 3507
Axelglo, 3508
Axerol, *See* Vi-Alpha, 33706
axerophthol, *See* Vi-Alpha, 33706, vitamin A,33932
Axin, *See* Age, 960
Axiquel, 3509
Axite, 3510
Axol® C 62, 3511
Axol® E 61, 3512, *See* Acetylated hydrogenatedlard glyceride, 315
Axol® L 61, L62, 3513
Axuris, *See* gentian violet, 12845
Ay 6108, *See* Ampicillin, 2361
AY-6608, *See* pentagastrin, 23146
Ayrtol, 3514
AZ, 3515
5-aza-10-arsenaanthracene chloride, *See* Adamsite, 624
azabenzene, *See* pyridine, 25481
1-azabicyclo[2.2.2]octan-3-one hydrochloride, *See* 3-quinuclidinone hydrochloride, 25693
azacyclohexane, *See* piperidine, 23874
azacyclotridecane-2-one polyamide, *See* nylon 12, 21912
Azadieno, *See* Taktic, 30729
azadirachtin, 3516
3-azaindole, *See* Benzimidazole, 3966
azamethiphos, 3517, *See* Alfacron 10WP, 1572
3-azapentane-1,5-diamine, *See* D.E.H. 20,7610
azaperone, 3518
Azelaic acid, 3519, *See* anchoic acid, 2422, Emerox® 1110, 10115, Emery's L-110, 10186
4-azido-N-(1-methylethyl)-6-methylthio-1,3,5-triazin-2-amine, *See* Brasoran 50 WP, 4484
azimidobenzene, *See* 1H-benzotriazole, 3982, benzotriazole1H-benzotriazole, 3982, Preventol® Cl 8, 24930
aziminobenzene, *See* benzotriazole1H-benzotriazole, 3982, Preventol® Cl 8, 24930
azindole, *See* Benzimidazole, 3966
Azine, *See* pyridine, 25481
Azinos, *See* Azinphos-ethyl, 3520, Gusathion®A, 13502
Azinotox 500, *See* atrazine, 3394
Azinphos-ethyl, 3520, *See* Cotnion-Ethyl, 6882, Gusathion® A, 13502
Azinphosmethyl, *See* Cotnion-Methyl, 6884
Azinphos-methyl, *See* Gusathion®, 13501
azinphos-methyl, *See* Guthion®, 13505
azinphos-methyl-azinphos-ethyl, *See* Cotnion-Ethyl-Methyl, 6883
azinphos-methyl-demeton-S-methyl sulfone, *See* Gusathion® MS, 13503
Azinugec E, *See* Azinphos-ethyl, 3520
Aziplex, 3521
Aziprotryne, *See* Brasoran 50 WP, 4484
Azo diisobutyronitrile, *See* Poly-zole AZDN,24674
Azo dyestuffs, *See* Cotonerol®, 6886
azo-*tert*-butane, *See* VR-160, 34057
Azoamine Red 2H, *See p*-nitroaniline, 21345
1,1'-azobiscarbamide, *See* Azodicarbonamide,3524, Kempore® 60/14FF, 15987, Santechem21-21, 27382
Azobiscarbonamide, *See* Azodicarbonamide,3524, Kempore®60/14FF, 15987, Santechem 21-21, 27382
Azobiscarboxamide, *See* Azodicarbonamide,3524, Kempore®60/14FF, 15987, Santechem 21-21, 27382
2,2'-Azobis[N-(4-chlorophenyl)-2-methylpropion amidine]dihydrochloride, *See* VA-545, 33232
1,1'-Azobis(cyclohexane-1-carbonitrile), *See* V-40, 33216
2,2'-Azobis(2,4-dimethylvaleronitrile), *See* V-65, 33220
1,1'-azobisformamide, *See* Azodicarbonamide,3524
1,1'-azobisformamide, *See* Santechem 21-21,27382
1,1'-azobisformamide, *See* Kempore® 60/14FF, 15987
2,2'-Azobis[2-[1-(2-hydroxyethyl)-2-imidazolin-2-yl] propane] dihydrochloride, *See* VA-060, 33227
2,2'-azobis[2-(hydroxymethyl) propionitrile], *See* VF-077, 33703
2,2'-Azobis[2-(2-imidazolin-2-yl)propane]dihydrochloride, *See* VA-044, 33225
Azobisisobutyronitrile, *See* Poly-zole AZDN,24674
2,2'-Azobisisobutyronitrile, *See* V-60, 33218
2,2'-azo-bis(isobutyronitrile), *See* Poly-zoleAZDN, 24674
2,2'-Azobis(4-methoxy-2,4-dimethylvaleronitrile), *See* V-70, 33221
2,2'-Azobis[2-methyl-N-[1,1-bis(hydroxymethyl)-2-hydroxyethyl]propionamide], *See* VA-080,33228
2,2'-Azobis(2-methylbutyronitrile), *See* V-59,33217
2,2'-Azobis[2-methyl-N-(2-hydroxyethyl)-propionamide], *See* VA-086, 33229
2,2'-Azobis(2-methyl-N-phenylpropionamidine)dihydrochloride, *See* VA-545, 33231
2,2'-Azobis(2-methylpropane), *See* VR-160,34057

barbasco, *See* rotenone, 27000
Barberite, 3624
Barbouze's Alloy, 3625
Barchlor, 3626
Bardac 22, *See* BTC® 1010-80, 4638, Radiaquat6412, 25790, Querton 210Cl-50, 25651
Bardac 2250/2280, *See* Radiaquat 6410, 6412,25789, Radiaquat 6412, 25790, Rewoquat B10, 26389
Bardew, 3627
Bardyne, 3628
Barex® 210, 3629
Barfoed's Reagent, 3630
Baridol, *See* Barium sulfate, 3644, Radiopaque,25830
Bario metal, 3631
Barisol Super BRM, 3632
Baritop, *See* Radiopaque, 25830
barium, 3633
barium acetate, 3634
barium binoxide, *See* barium peroxide, 3642
barium bromide, 3635
barium carbonate, 3636, *See* Durex white, 9208
barium chloride, 3637
barium chromate, 3638
barium chromate(VI), *See* barium chromate,3638
barium dibromide, *See* barium bromide, 3635
barium dichloride, *See* barium chloride, 3637
barium dinitrate, *See* barium nitrate, 3639
barium dinonylnaphthalene sulfonate, *See*Vanplast® 201, 33352
barium dioxide, *See* barium peroxide, 3642
barium fluosilicate, *See* Flosol, 12026
barium hexafluorosilicate, *See* Flosol, 12026
barium hydroxide, *See* Caustic baryta, 5468
barium mercuric iodide, *See* Rohrbach'ssolution, 26915
barium metatitanate, *See* barium titanate, 3646
barium monoxide, *See* barium oxide, 3641,baryta, 3675
barium nitrate, 3639, 3640
barium nitrite, *See* Stickstoffoxydbaryt, 29762
barium oxide, 3641, *See* baryta, 3675
barium perchlorate, *See* Desicchlora, 8139
barium peroxide, 3642
barium protoxide, *See* barium oxide, 3641
barium protoxide, calcined baryta, *See* baryta,3675
barium stearate, 3643, *See* Synpro® bariumStearate, 30596
barium sulfate, 3644, *See* Albaryt, 1353, BlancFixe Micro, 4313, Citobaryum, 6332,Esophotrast, 10835, Ewo, 11279, fixed white,11858, Huberbrite 1, 14452, Oratrast, 22271,Radiopaque, 25830, Sachtoperse® HU,27202, terra ponderosa, 31325
barium sulfonate, 3645
barium superoxide, *See* barium peroxide, 3642
barium titanate, 3646
barium titanate(IV), *See* barium titanate, 3646
barium titanium oxide, *See* barium titanate,3646
barium/cadmium, *See* Lankromark® LC68,16693, Synpron 1800, 30615
barium/cadmium, *See*
barium/calcium lignosulfonate, *See* Traffaid 30B, 32132
barium/zinc, *See* Interstab® LT-4805R, 15165, Lankromark® DP6404Z, Lankromark® LZ121, 16702, Lankromark® LZ187, 16703,Lankromark® LZ616, 16705, Lankromark®LZ693, 16706
barium-lead, *See* Haro® Mix YK-110, 13648
Barkite B, 3647
Barlene 125, *See* Onamine 12, 22164
Barlene® 12S, *See* dimethyl lauramine, 8507
Barlene® 18S, *See* Dymanthine, 9327
Barlox, 3648
Barlox® 12, 3649, *See* Cocamine oxide, 6557
Barlox® 14, 3650
Barlox® 16S, 3651
Barlox® 18S, 3652
Barlox® C, 3653
barm, *See* yeast, 34639
Barnon, *See* Power Flame, 24824
Barnon, *See* flamprop-isopropyl, 11887, Gunner,13493
barolub fta, *See* stearic acid, 29599
Baros, 3657
Baros camphor, *See* DL-borneol, 4428
Barosperse, *See* barium sulfate, 3644,Radiopaque, 25830
Barotrast, *See* Barium sulfate, 3644
Barquat sb-25, *See* Ablumine 280, 85, Cycloton® SCS, 7542, octadecylbenzene methanaminiumchloride, 21985, stearalkonium chloride, 29592,Sumquat® 6210, 30126
Barquat®, 3665
Barquat® 50-28, 80-28, *See* Benzalkoniumchloride, 3958
Barquat® CME-35, 3666
Barquat® CT-29, 3667, *See* Cetrimonium chloride,5814
Barquat® MX-50, MX-80 BTC® 824, *See* JAQPowdered Quaternary, 15517
Barra Super, *See* Ablusol ML, 129
Barras, 3668
Barrialon® CX, 3669
Barrialon® S, 3670
Barrialon® SF, 3671
Barricade, *See* Topclip Parasol, 32037
Barricade, Cymbush, *See* cypermethrin, 7588
Barrier, 3672
Barrier-Guard®, 3673
Barsilowsky's base, 3674
Barsito, *See* Barium oxide, 3641
barwood, *See* redwoods, 25987
baryta, 3675
baryta yellow, *See* barium chromate, 3638
barytes, natural, *See* barium sulfate, 3644
Barytes, natural, *See* Barium sulfate, 3644
BAS-083, *See* mepiquat chloride, 19327
BAS 089-00E, *See* Ablusol DBD, 124
BAS 351H, *See* Bentazon, 3947
BAS 438, 3676
BAS-517H, *See* Cycloxydim, 7544
BAS 3170F, *See* benodanil, 3942
BAS 3460, *See* Carbendazim, 5187, 5188, Derosal,16327
BAS 3460F, *See* Carbendazim, 5187, 5188
BAS 30000F, *See* nitrothal-isopropyl, 21380
BAS 32500F, *See* thiophanate methyl, 31706
BAS 46402F, 3677
BAS 47900H, *See* metazachlor, 19465, Track,32129
BAS-67054, *See* Derosal, 16327
BAS 67054F, *See* Carbendazim, 5187, 5188
BAS-85559X, *See* mepiquat chloride, 19327
BAS 90520H, *See* Sethoxydim, 28099
Basacid® Dyes, 3678
Basacryl Salt, 3679
Basacryl/Baflxan, 3680
Basacryl® Dyes, 3681
Basacryl® Salt NB-KU, 3682
Basacryl® Salt TX-412, 3683
Basagran 4E, *See* Bentazon, 3947
Basagran®, *See* Bentazon, 3947
Basamid, *See* Dazomet, 7806, N-521® Biocide,20640
Basamid P, *See* Dazomet, 7806, N-521® Biocide,20640
Basamid-fluid, *See* metam-sodium, 19444
Basamid-puder, *See* Dazomet, 7806, N-521®Biocide, 20640
Basammon® Extra 25, 3684
Basantol® Dyes, 3685
Basathrin, *See* cypermethrin, 7588, TopclipParasol, 32037
Basazol®, 3686
Bascal®, 3687
Bascal® S, 3688
Base 10-L, 3689
Base 36, 3690
Base 75, 3691
Base 104, 3692
Base 500-A, 3693
Base 865, 3694
Base 7800, 3695
Base HS, 3696
Base ML, 3697
Base MT, 3698
Base Nacrante 2078, 3699
Base Nacrante 6030 CP, 3700
base oil, *See* blown oils, 4354
Base Wax 36-AG, 3701
Basensol®, 3702
Basex, 3703
BASF Reactive Resist Liquid, 3704
BASF Ursol P base, *See* *p*-aminophenol, 2185
Basfapon, *See* dalapon, 7705
Basfapon B, *See* Dowpon, 8943
Basfapon/N, *See* dalapon, 7705
Basfapon®, 3705
Basfoliar® 34, 3707
Basfoliar® 6-12-6, 3706
Basic Alumina, *See* aluminum oxide, 1919
basic aluminum chloride, *See* aluminumchlorohydrate, 1903, Reach® 101, 201, 501,25919
basic aluminum tannate, *See* tannal, 30802
basic bismuth chloride, *See* bismuthoxychloride, 4241
basic bismuth salicylate, *See* Bismuthsubsalicylate, 4244
basic chloride, *See* powder of Algaroth, 24813
basic copper carbonate ($Cu_2(OH)_2CO_3$), *See*cupric carbonate, 7378
basic copper chloride, *See* Cupravit®, 7370
basic cupric carbonate, *See* cupric carbonate,7378, malachite, 18458
basic cupric chloride, *See* Cupravit®, 7370
Basic Fuchsin, *See* pararosaniline, 22798
basic lead acetate, *See* subacetate of lead, 29902
basic lead carbonate, *See* Ceruse, 5769,Kremser White, 16399
basic lead phthalate, *See* EPlthal 120, 10665
basic magnesium hypochlorite, *See* Mangol,18561
Basic Red 1, *See* Rhodamine G and G Extra,26617
Basic Red 9, monohydrochloride, *See*pararosaniline, 22798
Basic Violet 3, *See* gentian violet, 12845
Basic Violet 10, *See* Rhodamine B, 26616
Basic Violet BN, *See* gentian violet, 12845
Basilen®, 3708
Basilex, 3709
Basimid G, *See* Dazomet, 7806, N-521®Biocide, 20640
Basinetto silk, *See* Galettame silk, 12603
Basinex, *See* dalapon, 7705
Basocoll® CM, 3710
basofor, *See* barium sulfate, 3644
Basoform®, 3711
Basogal® C, 3712
Basojet® PEL 200%, 3713
Basokol® NB-S, 3714
Basol® WS, 3715
Basolan® F, 3716
Basomol®, 3717
Basonat®, 3718
Basonyl® Dyes, 3719
Basopal®, 3720
Basophen® M, 3721
Basophen® RA, 3722
Basophob®, 3723
Basoplast®, 3724
Basopon® LN, 3725
Basopon® TX-110, 3726
Basopor®, 3727
Basoset® 162, 3728
Basosoft®, 3729
Basotect®, 3730
Basotol®, 3731
Basotrope® W, 3732
Basovit® Dyes, 3733
Bastamol®, 3734
Bastanet, 3735
bastose, 3736
Basudin, *See* Diazinon Liquid, 8344
Basudin, Neocidol, *See* Diazinon Liquid, 8344
Basyntyn®, 3737
Batana oil, *See* patava oil, 22883
Batazina, *See* simazine, 28464
Batchite, 3738

Bath metal, 3739
Bath Wax, *See* Sta-Nut EE, 29508
Battal, 3740
Battal FL, *See* Battal, 3740
battery acid, *See* sulfuric acid, 30055
battery copper, 3741
Baudoin's metal, 3742
Baum's Acid, *See* Schaeffer's acid, 27558
Bavistin, *See* Derosal, 16327
Bavistin 3460, *See* Carbendazim, 5187, 5188
BAX-1526, *See* Chymopapain, 6284
Baxan, *See* Cefadroxil, 5547
Baxton®, 3743
Bay 5024, *See* methiocarb, 19484
Bay 5621, *See* Sebacil®, 27796, Volathion®, 34011
Bay 5712a, *See* Euparen® M, 11198
Bay 9002, *See* Rametin, 25856
Bay 9010, *See* propoxur, 25201
Bay 9010 Baygon, *See* Baygon®, 3790
Bay 9026, *See* methiocarb, 19484
Bay 15922, *See* trichlorfon, 32212
Bay 16259, *See* Azinphos-ethyl, 3520
Bay 18436, *See* demeton-S-methyl, 8055
Bay 19149, *See* dichlorvos, 8394
Bay 23129, *See* thiometon, 31700
Bay 25/154, *See* demeton-S-methyl, 8055
Bay 25141, *See* Terracur® P, 31329
Bay 25634, *See* Racumin®, 25713
Bay 29493, *See* Tiguvon®, 31788
Bay 30130, *See* Propanil, 25174
Bay 33172, *See* fuberidazole, 12428
Bay 36205, *See* quinomethionate, 25686
Bay 37344, *See* methiocarb, 19484
Bay 39007, *See* Propoxur, 25198, 25199, 25200,25201
Bay 45432, *See* Omethoate, 22155
Bay 49854, *See* Euparen® M, 11198
Bay 70143, *See* Carbodan, 5197, Rampart, 25860
Bay 71628, *See* methamidophos, 19474,Tamaron®, 30760
Bay 72483, *See* methabenzthiazuron, 19472
Bay 77488, *See* Sebacil®, 27796, Volathion®,34011
Bay 78418, *See* Edifenphos, 9618
Bay 78537, *See* Carbodan, 5197, Rampart, 25860
Bay 94337, *See* metribuzin, 19595, 19596
Bay 105807, *See* isoprocarb, 15404
Bay BUE 1452, *See* Peropal®, 23401
Bay DRW 1139, *See* Metamitron, 19443
Bay ene 11183 B, *See* Racumin®, 25713
Bay h 5757, *See* Rintal®, 26774
Bay HWG 1608, *See* Raxil, 25908
Bay KWG 0599, *See* Bitertanol, 4288
Bay L13/59, *See* trichlorfon, 32212
Bay MEB 6447, *See* Triadimefon, 32184
Bay NTN 19701, *See* pencycuron, 23097, 23098,23099
Bay NTN 8629, *See* Tokuthion®, 31932
Bay SLJ 0312, *See* Cropotex®, 7211
Bay Vh 5757, *See* Rintal®, 26774
Bayblend®, 3744
Bayblend® DP2-1448, 3745
Bayblend® FR 1439, 3746
Bayblend® T 44, 3747
Bayblend® T 45 MN, 3748
Bayblend® T 88-2N, 3749
Baybond®, 3750
Bayboran®, 3751
BAY-BUE 1452, *See* azocyclotin, 3523
Baycast, 3752
Baycid, *See* Tiguvon®, 31788
Baycidal, 3753
Bayclin, 3754
Bayco®, 3755
Baycoll®, 3756
Baycoll® 17, 3757
Baycoll® MD 3040, 3758
Baycor®, *See* Bitertanol, 4288
Baycryl®, 3759
Bayderm® A, 3760
Bayderm® Colours B-TO, 3761
Bayderm® Colours C-TO, 3762
Bayderm® KF, 3763
Bayderm® Lacquers Auxiliaries, 3764
Baydur®, 3765
Baydur® STR, 3766
Bayer 21/199, *See* Perizin®, 23276
Bayer 73, *See* clonitrilide, 6432
Bayer 205, *See* Naganol, 20660
Bayer 2353, *See* niclosamide, 21146
Bayer 5072, 3767
Bayer 6159H, *See* metribuzin, 19595, 19596
Bayer 19149, *See* dichlorvos, 8394
Bayer 25634, *See* Racumin®, 25713
Bayer 25648, *See* clonitrilide, 6432
Bayer 25820, *See* Rametin, 25856
Bayer 37344, *See* methiocarb, 19484
Bayer 39007, *See* Baygon®, 3790
Bayer 47531, *See* Dichlofluanid, 8379
Bayer B 5122, *See* Baygon®, 3790
Bayer Base Plates Glass-Clear, 3768
Bayer CM, 3769
Bayer Perlon®, 3770
Bayer SBR Latex 200 C, 3771
Bayer UV Absorber 325, 340, 3772
BAY-FCR 1272, *See* cyfluthrin, 7563
Bayferon®, 3773
Bayfidan, *See* triadimenol, 32185
Bayfill®, 3774
Bayfit®, 3775
Bayflex®, 3776
Bayfol®, 3777
Bayfol® CR 6-2, 3778
Bayfolan®, 3779
Bayfresh®, 3780
Baygal®, 3781
Baygal® K30, 3782
Baygal® K115, 3783
Baygal® K190, 3784
Baygal® K390, 3785
Baygard®, 3786, 3787,
Baygenal®,3788
Bayglaze, 3789
Baygon, *See* Propoxur, 25198, 25199, 25200,25201
Baygon®, 3790
Baygon® MEB Spray, 3791
Bayguard, 3792
Bayhydrol, 3793
Bayhydrol 140 AQ, 3794
Baykanol® AK, HLX, SL, 3795
Baykanol® Liquor TN, 3796
Baykisol®, 3797
Baylan®, 3798
Baylectrol®, 3799
Bayleton, *See* Triadimefon, 32184
Bayleton® 5, 3800
Bayleton® BM, 3801
Bayleton® CF, 3802
Baylith®, 3803
Baylube®, 3804
Bayluscid, *See* clonitrilide, 6432, Mansonil®,18611
Bayluscide, *See* clonitrilide, 6432
Bayluscide®, 3805
Baymat®, 3806
Baymer®, 3807
Baymetex®, 3808
Baymicron®, 3809
Baymid, 3810
Baymidur®, 3811
Baymidur® K88, 3812
Baymin®, 3813
Baymod®, 3814
Baymod® 50, 90/92, AKU3-2086, *See*Acrylonitrile-butadiene-styrene, 438
Baymod® A, 3815
Baymod® A KU3-2086, *See* Acrylonitrile-butadiene-styrene, 439
Baymod® PU, 3816
Baymoflex A, 3817
Baymoflex A KU3-2069.A, 3818
Baymol® A, D, 3819
Baymosthrin, 3820
Baynat®, 3821
Bayo 9867, *See* ciprofloxacin, 6317
Bayofly®, 3822
Bayothrin, 3823
Bayovac, 3824
Bayplast®, 3825
Baypreg®, 3826
Baypren Latex B, *See* Baypren® 110, 3828
Baypren Latex GK, *See* Baypren® 110, 3828
Baypren Latex L 200A, *See* Baypren® 110,3828
Baypren Latex L345, *See* Baypren® 110, 3828
Baypren Latex MKB, *See* Baypren® 110, 3828
Baypren Latex SK, *See* Baypren® 110, 3828
Baypren Latex T, *See* Baypren® 110, 3828
Baypren®, 3827
Baypren® 110, 3828
Baypren® 110 VSC, 3829
Baypren® 216, 3830
Baypren® 310, 3831
Baypren® AT-H, AT-M, AT-S, 3832
Baypren® EM1, 3833
Baypren® Latex KA 8348, 3834
Baypren® Latex L 200A, 3835
Baypren® M1, 3836
Bayprint®, 3837
Bayrena®, 3838
Bayrusil, *See* quinalphos, 25679
Bayrusil®, 3839
Baysan®, 3840
Bayscript®, 3841
Baysical®, 3842
Baysilex®, 3843
Baysilone®, 3844
Baysin®, 3845
Baysolvex®, 3846
Baysport®, 3847
Baystal®, 3848
Baysynthol®, 3849
Baytan, 3850, *See* Ceresan, 5745, triadimenol,32185
Baytan®, 3851
Baytan® IM, 3852
Baytec®, 3853
Baytex, *See* Tiguvon®, 31788
Baytherm®, 3854
Baythion, *See* Sebacil®, 27796, Volathion®, 34011
Baythroid, *See* cyfluthrin, 7563
Baythroid 2, *See* cyfluthrin, 7563
Baythroid H, *See* cyfluthrin, 7563
Baytigan® AR, 3855
Baytroid®, 3856
BAY-V1 1704, *See* cyfluthrin, 7563
Bayvap®, 3857
Bayvarol, 3858
Bazak, 3859
Ba-Zn, *See* ADK STAB AC-122, 719, ADK STABAC-169, 721, ADK STAB AP-536, 722, ADKSTAB OF-14, 734, ADK STAB RUP-9, 737
Ba-Zn salt, *See* Mark® 4700, 18728
BB Accelerator, 3860
BB10GF/15T, 3861
BBP, *See* Butyl benzyl phthalate, 4764
BBS, 3862
BBTS, 3863
BCDMH, *See* Dantoin® GSD-550, 7733, Halobrom,13560, Quesbrom, 25657
CDMH, *See*
BCF, 3864
BCM, *See* Carbendazim, 5187, 5188 Konker®,16327
BDMA, *See* Dabco® B-16, 7646, Pentamin BDMAetc, 23163
BDO, *See* 1,4-butanediol, 4735
Buffer, 3865
Be Square® 185, 3866, *See* Microcrystalline wax,19666
Bécilan, *See* vitamin B_6 hydrochloride, 33937
Beanfeast, 3867
Bear, 3868
Bearflex® LAO, 3869
Bearium, 3870

Beaudouin's reagent, 3871
Beaver steel, 3872
Becalin, *See* Unasyn® IM/IV, 32913
Bechilite, *See* pricite, 24943
Beckacite® 110, 115, 3873
Beckacite® 425, 3874
Beckacite® 4900, 3875
Beckamine®, 3876
Beckamine® 21-500, 3877
Beckol, 3878
Beckolin®, 3879
Beckolloid, 3880
Beckopol®, 3881
Beckosol, 3882
Beckosol 13-400, 3883
Becksol, 3884
Beckton white, *See* lithopone, 17520
beclamethasone dipropionate, *See* Vanceril,33295
Beclovent Inhaler, *See* octyl methoxycinnamate,22039
Becosal, 3885
Becxopox, 3886
Bedesol, 3887
BEE, *See* Planotox, 23914
beef tallow, *See* tallow, 30739
Beeswax, 3888, *See* Cirami No. 1, 6318
beet molasses, *See* molasses, 20043
beet sugar, *See* sucrose, 29934
Beetafil, 3889
Beetafin, 3890
Beet-Kleen, *See* Chlorpropam, 6162, Propham,25185
Beetle, 3891
Beetle Resin BT 333, 3892
Beetle Resin BT 334, 3893
Beetle Resin W69, 3894
Beetomax, *See* phenmedipham, 23595, 23596,23597, 23598
Beetup, *See* phenmedipham, 23595, 23596,23597, 23598
Beflavin, *See* vitamin B+72, 33935
Begasses oil, *See* pyrene oil, 25477
BEHA, *See* dioctyl adipate, 8548, Plasthall® DOA,24002, PX-238, 25440
Beha, *See* Wickenol® 158, 34393
Behenalkonium chloride, *See* Incroquat B65C,15030, Kemamine® BQ-2802C, 15906
Behenalkonium methosulfate, *See* IncroquatBehenyl TMS, 15035
Behenamide, *See* Kemamide® B, 15894, Uniwax1747, 33067
Behenamide DEA (1:1), 3895 , *See* IncromideBED, 14983
Behenamide MEA (1:1), *See* Incromide BEM,14984
Behenamidopropyl betaine, *See* Behenylbetaine, 3900
Behenamidopropyl dimethylamine, 3896, *See* Chemidex B, 5961, Incromine BB, 14997, Mackine 601, 18222, Schercodine B, 27635
Behenamidopropyl ethyldimonium ethosulfate, *See* Schercoquat BAS, 27673
Behenamidopropyl PG-dimonium chloride, *See* Lexquat® AMG-BEO, 17175
Behenamine oxide, 3897
Beheneth-10, *See* Mulsifan CB, 20452
Beheneth-30, *See* Emalex BHA-30, 9931
Behenic acid, *See* Hystrene® 5522, 14771,Prifrac 2989, 24956
Behenic acid (90%), *See* Hystrene® 9022, 14765
behenic acid amide, *See* Uniwax 1747, 33067
Behenoxy dimethicone, *See* Abil® Wax 2440, 60
Behenoyl PG-trimonium chloride, *See* Akypoquat 131, 1271
Behentrimonium chloride, 3898, *See* GenaminKDM-F, 12772, Varisoft® BT-85, 33417
Behenyl alcohol, 3899, *See* Dehydag Wax 22(Lanette), 7946
Behenyl behenate, *See* Starfol® BB, 29534
Behenyl betaine, 3900, *See* Incronam B-40,15018
Behenyl erucate, *See* Kemester® BE, 15954,Crodamol BE, 7117, Schercemol BE, 27585
Behenyl polyethoxy ethylmethacrylate, *See* CDS-1801, 5526, DV-1801, 9295
BEHP, *See* Bis(2-ethylhexy) Phthalate, 4228,dioctyl phthalate, 8549, Reomol DCP, 26130,Witcizer 312, 34454
Belclene, 3901
Belco, 3902
Beldex, 3903
Belfasin 320 Crushed, 3904
Belgard, 3905
Belgian slag, *See* Thomas slag, 31743
Beligno Seeds, *See* olives of Java, 22141
Belite, 3906
Bell Mine Pulverized Limestone, *See* CP Filler,6928
bell pepper, 3907
Bellasol, 3908
Bellauxine, 3909
Bellclo, 3910
Bellite, 3911
Belloid, 3912
Belmac Straight, 3913
Belmark, *See* fenvalerate, 11620
Belpro, 3914
Belro, 3915
Belsil ADM 6041E, 3916
Belsil DM 0.65, 3917
Belsil DMC 6031, 3918
Belsil PDM 20, 3919
Belsil SDM 6021, 3920
Belsoft, 3921
Belsol, 3922
Beltherm, 3923
Belzak AC, 3924
Belzak BL-50, 3925
Bemal, 3926
Bemberg®, 3927
Bemillese®, 3928
BEN-30, *See* benazolin, 3935
Benadon, *See* pyridoxine hydrochloride, 25483,vitamin B_6 hydrochloride, 33937
Ben-A-Gel®, 3929 , *See* Bentonite, 3954
Benalaxyl-mancozeb, *See* Galben M, 12600
Benalite, 3930
Benaloid, 3931
Benaqua, 3932
Benathix, 3933
Benazalox, 3934, *See* Dow Shield, 8887
Benazol PBKh, *See* Uvazol 236, 33181
benazolin, 3935
benazolin-bromoxynil-ioxynil-mecoprop, *See* Jaguar, 15493
Ben-cornox, *See* benazolin, 3935
Bendalite, 3936
Bendiocarb, 3937, Ficam, 11760, Garvox 3G,12685, Seedox SC, 27840
Bendioxide, *See* Bentazon, 3947
Benecel® Methylcellulose, 3938
Benedict Plate, 3939, *See* nickel silvers, 21132
Benefex, *See* Benfluralin, 3940
Benefin, *See* Benfluralin, 3940
Benefit, *See* benodanil, 3942
Benfluralin, 3940
Benfos, *See* dichlorvos, 8394
Benfuracarb, *See* Oncol 10G, 22173
Bengal catechu, *See* cutch,
Bengal isinglass, *See* agar-agar, 955, vegetable gelatin, 33518
Beni oil, *See* gingelly oil, 12946
Benirol, *See* Variquat® 50MC, 33397, ZephiranChloride, 34728
BeniSeed oil, *See* gingelly oil, 12946, Sesame oil, 28075
Benlate, *See* Benomyl, 3944
Bennatate, 3941
Benne oil, *See* gingelly oil, 12946, Sesame oil,28075
Benodanil, 3942, *See* Calirus, 5009, MascotClearing, 18999
Benol, 3943
Benomyl, 3944
Benopan, *See* benazolin, 3935
Benoxyl, *See* Abcure S-40-25, 35, AcetOxyl 2.5and 5, 305, Benzoyl peroxide, 3983
Bensapol, 3945
Bensecal, *See* benazolin, 3935
Bentalan, *See* betamethasone 21 phosphate, 4060
Bentalol, 3946
Bentazon, 3947, *See* Galaxy®, 12599, Laddok®,16593, Storm®, 29797
Bentazone, *See* Bentazon, 3947
Bentazone + MCPA + MCPB, *See* Acumen, 573
bentazon-isoproturon-dichlorprop, *See* Herbatox®, 13800
bentazon-isoproturon-dichlorprop, *See* Graminon® Plus, 13249
Bentiromide, *See* Chymex, 6283
Bentobrite® 770, 3948
Bentokol, 3949
Bentone, 3950
Bentone Gel, 3951
Bentone SD, 3952
Bentone® EW, MA, *See* Hectorite, 13752
Bentone-34, 3953
Bentonite, 3954, *See* Ben-A-Gel®, 3929,BentoPharm, 3955, Brebent, 4496, Bregel, 4501, Gadorgel, 12563, Korthix, 16362, Korthix H-NF, 16363, Montigel, 20299, Wilkinite, 34409
Bentonite USP/NF, *See* Polargel® HV, 24295
BentoPharm, 3955, *See* Bentonite, 3954
Ben-u-ron, *See* Valadol, 33238
Benvic, 3956, *See* Vestolit®, 33679
Benvic EB 16, *See* Vestolit®, 33679
Benylate, *See* Benzyl benzoate, 3987
Benzac, *See* Benzoyl peroxide, 3983
Benzac B, *See* Benzoyl peroxide, 3983
1,2-Benzacenaphthene, *See* Idryl, 14841
Benzagel, *See* Benzoyl peroxide, 3983
Benzagel 10, *See* Abcure S-40-25, 35, AcetOxyl 2.5and 5, 305, Benzoyl peroxide, 3983
Benzaknen, *See* Abcure S-40-25, 35, AcetOxyl2.5 and 5, 305, Benzoyl peroxide, 3983
benzalacetone, *See* benzylidene acetone, 3992
benzalacetophenone, *See* Chalkone, 5849
benzaldehyde, 3957, *See* Essential Oil of BitterAlmonds, 10864, Ethereal Oil of Bitter Almonds, 10944
benzaldehyde, 4-hydroxy-3-methoxy, *See* vanillin, 33327
Benzalkonium chloride, 3958, *See* Alkaquat® DMB-451-50, DMB-451-80, 1707, *See* Arquard® B-100, 3123, *See* Empigen® BCB50, 10364, *See* Exameen 3500 3714, 11286, *See* FMB 500-15 U.S.P, 12133, *See* Germ-i-Tol, 12898, *See* Jordaquat® 350, 15621, *See* Lutensit® K-OC, 17976, *See* Paramos, 22776, *See* Protectol® KLC 50, 80, 25263, *See* Quatrene CB-50, 25613, *See* Rewoquat B 50, 26390, *See* Variquat® 50MC, 33397, *See* Zephiran Chloride, 34728, *See* Zoharquat 50, 34873 NF, *See* Empigen® BAC50, 10362, dibromide, *See* Callusolve, 5015-dimethicone, *See* Conotrane, 6730
Benzamine Blue, *See* Niagara blue, 21091
Benzamine®, 3959
benzaminoacetic acid, *See* Hippuric acid, 14126
benzamizole, *See* Knot Out, 16257, Ratio, 25897
Benzanil, 3960
Benzar, *See* benazolin, 3935
Benzathonium Chloride, *See* benzethoniumchloride, 3963
Benzelmin, *See* oxfendazole, 22426
Benzenamine, N-(3-phenyl-4,5-bis((trifluoromethyl)imino)-2-thiazolidinylidene)-, *See* Cropotex®, 7211
Benzenamine, N,N'-methanetetraylbis-, *See* Stabilizer 2013-P®, 29406
Benzene, 3961
benzene azimide, *See* benzotriazole1H-benzotriazole, 3982, Preventol® CI 8, 24930
benzene carboxaldehyde, *See* Benzaldehyde,3957, Ethereal Oil of Bitter Almonds, 10944
benzene carboxylic acid, *See* benzoic acid, 3973
benzene chloride, *See* Abluton T30, 139,chlorobenzene, 6119

Chemical and Trade Names

Chemical and Trade Names

benzo-furane resin, *See* paracoumarone resin,22729

7-Benzofuranol, 2,3-dihydro-2,2-dimethyl,methylcarbamate, *See* Carbodan, 5197,Rampart, 25860

Benzofuroline, *See* resmethrin, 26218

benzoglycolic acid, *See* mandelic acid, 18518

Benzoglyoxaline, *See* Benzimidazole, 3966

benzohydrol, *See* Benzhydrol, 3965

Benzoic acid, 3972, 3973, *See* flowers ofBenjamin, 12031

benzoic acid butyl ester, *See* Butyl benzoate,4763, n-Butyl benzoate, 4790

benzoic acid, 2-hydroxy-, *See* salicylic acid,27270

Benzoic acid, 2-hydroxy-, butyl ester, *See*Brunol, 4625

benzoic acid, 4-hydroxy-, phenylmethyl ester, *See* Nipabenzyl, 21248

Benzoic Acid K, 3974

benzoic acid methyl ester, *See* methyl benzoate, 19515

Benzoic acid phenylmethyl ester, *See* Benzylbenzoate, 3987

benzoic acid potassium salt, *See* potassiumbenzoate, 24738

benzoic acid potassium salt trihydrate, *See*potassium benzoate, 24738

benzoic acid sodium salt, *See* sodium benzoate, 28735

benzoic aldehyde, *See* Benzaldehyde, 3957,Ethereal Oil of Bitter Almonds, 10944

benzoic sulfimide, *See* saccharin, 27195

o-benzoic sulfimide, *See* saccharin, 27195

benzoimidazole, *See* Benzimidazole, 3966

Benzoin, 3975

Benzoin ether, *See* Glocure, 12997

benzoin isobutyl ether, *See* Vicure® 10, 33734

benzol, *See* Benzene, 3961

Benzolin, *See* Ligroin, 17244, V.M. and P.Naphtha, 33212

Benzonitrile, 3976

benzoperoxide, *See* benzoyl peroxide, 3983

Benzophenone, *See* ADK STAB 1413, 716

Benzophenone-1, *See* Uvinul® 400, 33191

Benzophenone-2, 3977, *See* Uvinul® D-50,33195

Benzophenone-3, *See* Cyasorb® UV 9, 7495,Escalol® 567, 10789, Neo Heliopan® BB,20867, Ritaphenone 3, 26823, Rhodialux A,26675, Syntase® 62, 30621, Uvinul® M-40,33197

Benzophenone-4, *See* Rhodialux S, 26676,Syntase® 230, 30622, Uval, 33177, Uvinul® MS-40,33199

Benzophenone-6, 3978, *See* Uvinul® D-49, 33194

Benzophenone-6,

Benzophenone 8, *See* Cyasorb® UV 24, 7496,Spectra-Sorb UV 24, 29230

Benzophenone-9, 3979, *See* Uvinul® DS-49,33196

Benzophenone-11, 3980, *See* Uvinul® 490, 33193Uvinul® M-493, 33198

Benzophenone-12, *See* Hostavin® ARO 8, 14418,Lowilite® 22, 17642, Spectra-Sorb UV 531, 29232, Uvinul® 408, 33192

2H-1-Benzopyran-2-one, 4-hydroxy-3-(1,2,3,4-tetrahydro-1-naphthalenyl)-, *See* Racumin®,25713

Benzoresorcin, *See* benzophenone-1,

Benzoresorcinol, 3981, *See* Benzoresorcinol,3981, Uvinul® 400, 33191

benzosulfimide, *See* saccharin, 27195

o-benzosulfimide, *See* saccharin, 27195

benzosulfinide, *See* saccharin, 27195

1,2-benzothiazol-3(2H)-one 1,1-dioxide, *See*saccharin, 27195

Benzothiazole, 2-(4-morpholinyldithio)-, *See*Morfax, 20314

2(3H)-benzothiazolethione, *See* 2-mercaptobenzothiazole, 19334

2-benzothiazolethiol, *See* 2-mercaptobenzothiazole, 19334

2-Benzothiazolesulfenamide, N-cyclohexyl-, *See* Vulkacit® CZ/EGC, DZ/EGC, 34129

4-Benzothiazole-2-sulfenyl morpholine, *See*Santocure MOR/MOR90, 27401

2-Benzothiazolethiol, *See* Accelerator Mercapto,207, mercaptobenzothiazole, 19335, Thiotax,31719, Vulkacit® M, Vulkacit® Merkapto/C,34134

2-benzothiazolethiol, sodium salt, *See* Nacap®,20643

2-Benzothiazolethiol, zinc salt (2:1), *See*Zetax®, 34741

benzothiazolethiol, sodium salt, *See* Nacap®,20643

Benzothiazolethiol, zinc salt, *See* OctocureZMBT-50, 21998, Oxaf, 22408

2(3H)-Benzothiazolethione, *See* AcceleratorMercapto, 207, Thiotax, 31719, Vulkacit® M,Vulkacit® Merkapto/C, 34134

2(3H)-benzothiazolethione sodium salt, *See*Nacap®, 20643, Nuodex 84, 21779

2(3H)-benzothiazolethione zinc salt, *See*Octocure ZMBT-50, 21998, Oxaf, 22408,Vulkacit® Merkapto/MGC, 34135, Zetax®,34741

benzothiazolethione, sodium salt, *See* Nacap®,20643

Benzothiazolethione, zinc salt, *See* OctocureZMBT-50, 21998

(Benzothiazolyl)-N,N'-dimethylurea, *See* methabenz thiazuron, 19472

(benzothiazol-2-yl)-1,3-dimethylurea, *See* methabenz thiazuron, 19472

(Benzothiazolyl)-1,3-dimethylurea, *See* methabenzthiazuron, 19472

2-Benzothiazolyl-N-morpholinosulfide, *See*Amax®, Amax No 1, 1950, Delac MOR, 8000,OBTS, 21970, OMTS, 22161

2-(2-benzothiazolyloxy)-N-methyl-N-phenylacetamide, *See* mefenacet, 19229

Benzothiazyl 1-2-dicyclohexyl sulfenamide, *See* Akrochem® DCBS Granules, 1166

Benzothiazyl disulfide, *See* M B T S, 18100,MBTS, 19172, Vulkacit® DM/C, 34131

Benzothiazyl-2-cyclohexyl sulfenamide, *See*Vulkacit® CZ/EGC, DZ/EGC, 34129

2-Benzothiazyl-N-morpholine disulfide, *See*Accelerator MF, 208

Benzothiazylthiobenzoate, *See* Ureka C, 33128

Benzotriazole, *See* Preventol® CI 8, 24930

1,2,3-Benzotriazole, *See* Preventol® CI 8, 24930

1H-benzotriazole, *See* Preventol® CI 8, 24930

1,2,3-Benzotriazole, *See* benzotriazole1H-benzotriazole, 3982

1H-benzotriazole, 3982, *See* ADK STAB LA-32,729

2-(2H-benzotriazol-2-yl)-4-methylphenol, *See*Topanex 100BT, 32019, Uvazol P, 33184

2-(2H-benzotriazol-2-yl)-*p*-cresol, *See* Topanex100BT, 32019, Uvazol P, 33184

2-(2H-Benzotriazol-2-yl)-4-(1,1,3,3-tetramethylbutyl) phenol, *See* Uvazol 311,33183

4,4'-(3H,2,1-benzoxathiol-3-ylidene)bisphenolS,S-dioxide, *See* phenolsulfonphthalein,23611

4,4'-(3H-2,1-Benzoxathiol-3-ylidene)bis(2-methylphenol) S,S-dioxide, *See* cresol red,6996

Benzoyl alcohol, *See* benzyl alcohol, 3986

Benzoyl cyanide-O-(diethoxyphosphino thioyl)oxime, *See*Sebacil®,27796

Benzoyl hydride, *See* Benzaldehyde, 3957,Ethereal Oil of Bitter Almonds, 10944

Benzoyl methide, *See* acetophenone, 300

benzoyl peroxide, 3983, *See* Abcure S-40-25,35, Aztec® Benzoyl Peroxide-70-77, 3553,Aztec® Benzoyl Peroxide-Dry, 3554, Cadet®BPO-70W, 4891, Cadox® 40E, 4906, Cadox® BPO-W40,4908, Cadox® BS, 4909, Cadox® BTA, 4910, Cadox® BTW-50, 4911, Clear by Design, 6385, Dermoxyl®, 8132, Florox, 12025,Lucidol, 17758, Lucidol RM, 17761, Lucidol-78, 17762, Luperco AFR-250, 17834, Luzidol,18054, Benzoyl peroxide, 3983, Benzoyl peroxide, 3983

4-benzoyl resorcinol, *See* Benzoresorcinol,3981

o-benzoyl sulfimide, *See* saccharin, 27195

Benzoyl superoxide, *See* Abcure S-40-25, 35,AcetOxyl 2.5 and 5, 305, benzoyl peroxide,3983

benzoylaminoacetic acid, *See* Hippuric acid,14126

(±)-4-(benzoylamino)-5-(dipropylamino)-5-oxopentanoic acid, *See* proglumide, 25116

5-benzoyl-2-benzimidazolecarbamic acidmethyl ester, *See* Mebendazole, 19198,Vermox, 33591

(5-Benzoyl-1H-benzimidazol-2-yl)-carbamicacid methyl ester, *See* Mebendazole, 19198,Telmin, 31066, Vermox, 33591

benzoylglycin, *See* Hippuric acid, 14126

N-Benzoyl-glycinamide, *See* Hippuryl Amide,14127

benzoylglycocoll, *See* Hippuric acid, 14126

3-benzoyl-4-hydroxy-6-methoxybenzenesulfonic acid, *See* Uval,33177, Rhodialux S, 26676

5-Benzoyl-4-hydroxy-2-methoxybenzenesulfonic acid, *See* Rhodialux S, 26676, Uval, 33177

2-benzoyl-5-methoxyphenol, *See* NeoHeliopan® BB, 20867, Rhodialux A, 26675,Cyasorb® UV 9, 7495

2-benzoyl-5-methoxy-1-phenol-4-sulfonic acid, *See* Uval, 33177, Rhodialux S, 26676

4-Benzoyl-4-methyl diphenyl sulfide, *See*Speedcure BMDS, 29260

4-benzoyl resorcinol, *See* Uvinul® 400, 33191

1,2-Benzphenanthrene, *See* Chrysene, 6252

Benzthiazyl disulfide, *See* Thiofide, 31685

Benztrimonium chloride, *See* benzyl trimethylammonium chloride, 3989

Benzyl, 3984

benzyl acetate, 3985

benzyl alcohol, 3986, *See* Bentalol, 3946

Benzyl alcohol, 2,4-dichloro- α-(chloromethylene)-, diethyl phosphate, *See* Chlorfenvinphos, 6098

N-Benzylamine, *See* Sumine® 2005, 30110

Benzyl benzoate, 3987, *See* Peruscabin, 23440

benzyl butyl phthalate, *See* butyl benzylphthalate, 4764, Unimoll® BB, 32977

Benzyl carbinol, *See* 2-phenylethyl alcohol,23642

Benzyl cellulose, *See* Benzex, 3964

benzyl chlorocarbonate, *See* benzylchloroformate, 3988

benzyl chloroformate, 3988

benzyl dimethyl ammonium C_{14} benzyl dimethylammonium chloride, *See* JAQ Powdered Quaternary, 15517

benzyl dimethyl dodecyl ammonium chloride, *See* Dehyquart LDB, 7986

Benzyl dimethyl tetradecyl ammoniumchloride, *See* JAQ Powdered Quaternary, 15517

benzyl dimethylamine, *See* Dabco® B-16, 7646,Pentamin BDMA etc, 23163

N-benzyldimethylamine, *See* Pentamin BDMAetc, 23163

benzyl dimethyloctadecylammonium chloride, *See* Cycloton® SCS, 7542

benzyl dimethylstearylammonium chloride, *See*Cycloton® SCS, 7542

Benzyl iodide, *See* Fraissite, 12374

benzyl n-butyl phthalate, *See* Butyl benzylphthalate, 4764

benzyl stearyldimethylammonium chloride, *See*Cycloton® SCS, 7542

Benzyl sulfoxide, *See* Preventol® CI 5, 24928

Benzyl triethyl ammonium chloride, *See*Sumquat® 2355, 30120

Benzyl trimethyl ammonium chloride, 3989, *See* Hipochem Migrator J, 14114, Migrassist® D,19746

Benzyl Tuex®, 3990

benzylamine, 3991

Benzylbenzenecarboxylate, *See* Benzyl benzoate,3987

Benzylcarbonyl chloride, *See* benzylchloroformate, 3988

benzylchlorophenol, *See* Nipacide® BCP, 21252

o-benzyl-*p*-chlorophenol, *See* Nipacide® BCP,21252

Benzyldimethyl ketal, *See* Escacure® KB1, 10785

benzyldimethylamine N-benzyldimethylamine, *See* Dabco® B-16, 7646, Pentamin BDMA etc,23163

benzyldimethylamine N-benzyl-N,N-dimethylamine, *See* Dabco® B-16, 7646,Pentamin BDMA etc, 23163

benzyldimethyloctadecylammonium chloride, *See* Ablumine 280, 85, octadecylbenzene methanaminium chloride, 21985

benzyldimethylstearylammonium chloride, *See*octadecylbenzenemethanaminium chloride,21985

5-Benzylfurfuryl chrysanthemate, *See*resmethrin, 26218

5-benzyl-3-furylmethyl-(+)-*trans*-chrysanthemate, *See*

Reslin, 26216
Benzylhemiformal, *See* Akyposept B, 1323
benzylidene acetone, 3992
Benzylideneacetone, *See* benzylidene acetone,3992
benzylideneacetophenone, *See* Chalkone, 5849
N-Benzyl-N-isopropylpivalamide, *See* tebutam,30879
N-Benzyl-N-isopropyltrimethylacetamide, *See*tebutam, 30879
Benzyl-N-isopropyltrimethylacetamide, *See*Comodor, Comodor 600, 6667
Benzyloctyl adipate, *See* Adimoll® BO, 689
Benzylparaben, *See* Nipabenzyl, 21248
benzylpenicillin potassium, *See* Penicillin Gpotassium, 23109, 23110
Benzylpenicillin procaine, *See* Penicillin V,23111
benzylpenicillinic acid potassium salt, *See*Penicillin G potassium, 23109, 23110
benzylstearyldimethylammonium chloride, *See*octadecylbenzenemethanaminium chloride,21985
Benzyltriethyl ammonium chloride, 3993
Benzyltrimethyl ammonium chloride, *See*Variquat® B200, 33400
Benzytol, *See*chloroxylenol, 6159, Nipacide®MX, 21257
Beosit, *See* thiodan, 31680
Bepanthen, *See* d-Panthenol, 22690
Bercema NMC 50, *See* Carbaryl, 5181
Bercotox, 3994
Berelex, *See* Regulex, 26019
Bergamol, *See* Phanteine, 23569
Bergauf, 3995
Berger Colorizer - Full Gloss/Vinyl Matte/VinylSilk, 3996
Berger Cuprinol Woodpaints and Woodstains,3997
Berger mixture, 3998
Berkatekt, 3999
Berkflam B 10E, *See* Decabromodiphenyl oxide,7837, Octoguard FR-01, 21999, Saytex® 102E,27514
Berkstop, 4000
Berlin Brown, 4001
Berlin red, *See* Indian Red, 15068
Bernel® Ester 168, 4002
Bernel® Ester 2014, 4003
Bernel® Ester CO, 4004, *See* cetyl octanoate,5821
Bernel® Ester DID, 4005
Bernel® Ester DISM, 4006
Bernel® Ester DOM, 4007
Bernel® Ester EHP, 4008
Bernel® Ester NPDC, 4009
Bernel® Ester TOC, 4010
Bernel® OPG, 4011
Bernit, 4012
Berol, 4013
Berol 08, *See* Ablunol SA-7, 113
Berol 09, 4014
Berol 26, 4015
Berol 108, 4016
Berol 191, 4017
Berol 195, 4018
Berol 198, 4019
Berol 199, 4020
Berol 259, 4021
Berol 260, 4022
Berol 267, 4023
Berol 272 and 716, 4024
Berol 278, 281, 282, 291 and 292, 4025, *See*Berol 09, 4014
Berol 302, 4026
Berol 307, 4027, *See* Cocamide DEA, 6547
Berol 381, 4028
Berol 386, 4029
Berol 391, 4030
Berol 397, 4031
Berol 452, *See* Rhodapon® 101-10, 26631,sodium lauryl sulfate, 28781, Witcolate 6400,34469
Berol 455, 4032
Berol 456, 4033
Berol 457, 4034
Berol 474, 4035
Berol 475, 4036
Berol 484, 4037
Berol 490, 4038
Berol 496, 4039
Berol 513 and 525, 4040
Berol 518, 4041
Berol 521, 4042
Berol 563, 4043
Berol 594, 4044
Berol 733, 4045
Berol 784, 4046
Berol 822, 4047
Berol 829, 4048
Berol WASC, 4049
Berpak, 4050
Bersch bearing metal, 4051
Berthier's alloy, 4052
Bertholiet salt, *See* potassium chlorate, 24746
Berylla, 4053
beryllium, 4054
beryllium bronze, 4055
beryllium oxide, *See* Berylla, 4053, Glucina,13035
Besconus, 4056
Besiege®, 4057
Beta, *See* phenmedipham, 23595, 23596,23597, 23598
Beta 2 Vangard, *See* Phenmedipham, 23595,23596, 23597, 23598
Beta Dia, *See* Unimate® DIPA, 32968
Beta Lite® 3503, 4058
Beta Plus, 4059
Betacide P, *See* Nipasol M, 21282
Betacyclodextrin, *See* Kleptose®, 16222
Betaflow, *See* phenmedipham, 23595, 23596,23597, 23598
Betaine, *See* Oxyneurine, 22463
Betaine amphoterics, *See* Lonzaine®, 17590
Betaine hydrazide hydrochloride, *See* Girard'sreagent T, 12950
betaines, coco alkyldimethyl, *See* AccobetaineCL, 222
Betalion, *See* phenmedipham, 23595, 23596,23597, 23598
betamas, *See* betamaze,
betamase, *See* betamaze,
betamaz, *See* betamaze,
betamethasone 21 phosphate, 4060
betamethasone 21 phosphate disodium salt, *See*betamethasone 21 phosphate, 4060
Betamix, *See* phenmedipham, 23595, 23596,23597, 23598
Betanal, *See* phenmedipham, 23595, 23596,23597, 23598
Betanal, *See* Vangard, 33323
Betanal, *See* phenmedipham, 23595, 23596,23597, 23598
Betanal, *See* phenmedipham, 23595, 23596,23597, 23598
betanal E, 4061
Betanal E, *See* phenmedipham, 23595, 23596,23597, 23598
betanal Tandem, 4062
Betanal Tandem, *See* Phenmedipham, 23595,23596, 23597, 23598
betapal Concentrate, 4063
betaprene® 253, 4064
betaprone, 4065
betaseal® 43518, 4066
betaseal® 43520A, 4067
betaseal® 43555, 4068
betaseal® 58702, 4069
betasol Ot-A, 4070
betathane, 4071
betazole hydrochloride, 4072
BET-C-30, *See* Cocamidopropyl betaine, 6551
Betnersol Injectable, *See* betamethasone 21phosphate, 4060
betol, *See* naphthalol, 20712
Betosip, *See* phenmedipham, 23595, 23596,23597, 23598
Betricing, 4073
Betrkake, 4074
Betrox, 4075
Betsolan, *See* betamethasone 21 phosphate,4060
Better, *See* chloridazon, 6103
Better Flowable, 4076
betula oil, *See* Teaberry Oil, 30872
BET-W, *See* Cocamidopropyl betaine, 6551
Beutene, 4077
Bevacid, 4078
Bevaloid, 4079
Bevaloid 35, *See* Ablusol ML, 129
Bevaloid 35 and 36, 4080
Bevaloid 211, 4081
Bevaloid 1299, 4082
Bevaloid 6423, 4083
Bevaloid 6522, 4084
Bevaloid 6703, 4085
Bevaloid 6744, 4086
Bevaloid DA 6805, 4087
Beviros, 4088
Bevitack Resins, 4089
Bewoid, 4090
Bewopac, 4091
Bex, 4092
Bexane, *See* MCPB, 19184
Bexfilm, 4093
Bexloy®, 4094
Bexphane, 4095
Bexton, 4096, *See* Propachlor, 25163, Ramrod,25861
Bexton 4L, *See* Propachlor, 25163, Ramrod,25861
Beycopon, 4097
Beycostat, 4098
Beycostat 148 K, 4099
Beycostat 231, 4100
Beycostat 256A, 4101
Beycostat 273 P, 4102
Beycostat 319 A, 4103
Beycostat 656 A, 4104
Beycostat B 070 A, 4105
Beycostat B 706 A, 4106
Beycostat B 706 E, 4107
Beycostat LP 4 A, 4108
Beycostat NE, 4109
Beycostat QA, 4110
BF 200, *See* CP Filler, 6928
BFP 64K, 4111
BFP 65, 4112
BFP 74, 4113
BFP 75, 4114
BGE, *See* Ageflex BGE, 967, Ageflex TBGE, 998
BH 2,4-D Ester 40, 4115
BH CMPP/2,4-D, 4116
BH Dalapon, 4117, *See* dalapon, 7705
BH Dockmaster, 4118
BH MCPA, *See* MCPA, 19183
BH MCPA 75, 4119
BH Prefix D, 4120
BHA, 4121, *See* Butylated hydroxyanisole, 4787,Nipantiox, 21276, Sustane® 1-F, 30432
Bhimsaim camphor, *See* DL-Borneol, 4428
BHT, 4122, *See* Anti-Oxydant Bayer, 2595,Butylated hydroxytoluene, 4788, Deenax, 7874, Ionol, 15226, Lowinox® BHT, 17664, Nipanox® BHT, 21273, Oxyguard, 22457, Ralox® BHT food grade, 25845, Spectratech® CM 11340, KM 11264, 29240, Vanox® PCX, 33346, Vulkanox® KB, 34152
BHT, Antrancine 8, *See* Tenamine 3, 31092
BHT-lecithin-soybean oil-mineral oil, *See*Banox ES, 3619
Bi Ammonium Phosphate, 4123
Bi(2-ethylhexyl)trimellitate ester, *See* ReomolDCP, 26130
biacetyl, *See* Diacetyl, 8250
Biactol, 4124
Biakmetals, 4125
biantimony trioxide, *See* Octoguard FR-10,22000
Biasbeston, 4126
bibenzal, *See* stilbene, 29764
bibenzene, *See* diphenyl, 8592
bibenzylidene, *See* stilbene, 29764

Bibesol, *See* dichlorvos, 8394
Bicarbonate of soda, *See* sodium bicarbonate,28736
Bicep, *See* metolachlor, 19585
Bicep 6L, *See* metolachlor, 19585
Bichloride of Mercury, *See* mercuric chloride,19347
Bickie-mol, *See* Valadol, 33238
Bicor®, 4127
Bicor® 220 AB, 250 AB, 310 AB, 380 AB, 420HS, 4129
Bicor® 240 B, 306 B, 420 B, 470 B, 4130
Bicor® 318 ASB, 252 ASB, 4131
Bicor® 70 PXS, 4128
Bicyclo[3.1.1]heptane, 6,6-dimethyl-2-methylene-, *See* β-l-pinene, 23858
Bideron, *See* Tokuthion®, 31932
Bidisin, 4132
Bidocef, *See* Cefadroxil, 5547
Biebrich Scarlet Red, *See* Sudan IV, 29941
Bielzite, 4133
Bife, 4134
bifenox, 4135
bifenox-chlorotoluron, *See* Dicurane Duo495FW, 8410
bifenox-isoproturon, *See* Invicta Duo 495FW,15201, 15202
bifenox-isoproturon-mecoprop, *See* Foxstar,12349
biformal, *See* glyoxal 40%, 13146
biformyl, *See* glyoxal 40%, 13146
biformylchlorazin, *See* triforine, 32254
Biguanide, 1-*o*-tolyl-, *See* Vulkacit® 1000,34124
Biju®, 4136
BIK, 4137
Bikini Cream, 4138
Bikorit, 4139
Bilevon, *See* hexachlorophene, 13992
Bilevon® -Solution, 4140
Bilgen bronze, 4141
Bilimiro, *See* Videobil, 33742
Bilimiron, 4142
Bi-Lite®, 4143
Bilitrast, *See* keraphen, 16055
Bilivist, 4144
Bilmiron, *See* Videobil, 33742
Bilopaque, 4145
Bilston, 4146
Bilt-Cote®, 4147
Bilt-Plates®, 4148
Bilt-Rex®, 4149
Bima's redwood, *See* redwoods, 25987
Binab T, 4150
binapacryl, *See* Morocide, 20340
Binarite, *See* marcasite, 18700
Bind, 4151
Bi-nell®, 4152
Binotal, *See* Ampicillin, 2361
Bio Flydown, *See* permethrin, 23384, 23385,23386
Bio Terge®, 4153
Bioacid, 4154
Bio-add, 4155
bioallethrin-permethrin, *See* Insektigun, 15145
Bioban CS 1135, *See* Amine CS-1135®, 2153,Canguard® 327, 5088, Oxaban®-A, 22404
Bioban P 1487, *See* 4,4'-(2-Ethyl-2-nitrotrimethylene)-dimorpholine, 11141,Nitrobutylmorpholine, 21351
Bioban® BNPD-40, 4156, *See* Broponol, 4616
Bioban® CS-1135, 4157
Bioban® GK, 4158
Bioban® N-95, 4159
Bioban® P-1487®, 4160
Biobenzylfuroline, *See* Reslin, 26216
Biobor, 4161
BioCare® Polymer HA-24, 4162
BioCare® SA, 4163
Biocide, 4164, *See* Piror, 23881
Biocides, *See* Protectol®, 25259
Bio-clave, *See* Nipacide® BCP, 21252
Biocolina, *See* choline chloride, 6171
Biocyde, 4165
Biodeg, *See* Arol Biodet, 3027
Biodynes® TRF Ultra-5, 4166
Bio-Feed, 4167
Bioguard, *See* thiabendazole, 31661
Biomart, 4168
Biomate, 4169
bioMeT 14, 4170
Biomet 33, *See* methyl isothiocyanate, 19543
bioMeT TBTF, 4171
bioMeT TBTO, 4172
Biomin® Cinque, 4173
Biomin® Marine, 4174
Bionex, *See* Azinphos-ethyl, 3520, Gusathion® A,13502
Biopal NR-20, *See* Biopal® NR-20, 4175
Biopal® NR-20, 4175
Biopal® NR-20W, 4176
Biopal® VRO-20, 4177
Biophos 35, 4178
Bioplex RNA, 4179
Biopol, 4180
Bio-Pol® OE, 4181
Biopol® TE, 4182
Bio-Pruf®, 4183
Biopure 100, *See* Abiol, 69, Germall® 115,12893
Bioques, 4184
Bioques Q, 4185
Bioques Z, 4186
Bioresmethrin (D-*trans* isomer), *See* resmethrin,26218
Biorion 450 Super, 4187
Biosil, 4188
Biosint Supra, 4189
Biosoft C100, 4190
Biosoft D, 4191
Bio-soft D-35x, *See* Nacconol® 40G, 20647,sodium dodecylbenzenesulfonate, 28760
Bio-soft D-40, *See* sodiumdodecylbenzenesulfonate, 28760
Bio-soft D-60, *See* sodiumdodecylbenzenesulfonate, 28760
Bio-soft D-62, *See* sodiumdodecylbenzenesulfonate, 28760
Biosoft N-300, 4192
Bio-soft S 100, *See* Witco® 1298, 34458
Biosoft S and D-35X, 4193
Biosoft S-100, *See* Witco® 1298, 34458
Biosoft S-100 and JN, 4194
Bio-Soft®, 4195
Bio-Soft® 9283, 4196
Bio-Soft® E-400, 4197
Bio-Soft® EA-4, 4198
Bio-Soft® FF 400, 4199, *See* Deceth-6, 7848
Bio-Soft® LD-95, 4200
Bio-Soft® MT 40, 4201, *See* Foam-Coll 4CT,12151
Bio-Soft® N-300, 4202
Bio-Soft® S-100, 4203
Bio-Soft® TD 400, *See* Trideceth series, 32232
Bio-Soft® TD400, 4204
Biosol, 4205
Biosperse®, 4206
Biosperse® 240, 4207
Biosperse® 250, 4208
Biostat A.1, 4209
biosterol, *See* Vi-Alpha, 33706
Biosulphur Powder, 4210
Bio-Surf I-20, 4211
Bio-Terge® AS-40, 4212
Bio-Terge® AS-40 and AS-90F, 4213
Bio-Terge® PAS-8S, 4214
Bio-Tetra, *See* tetracycline, 31365
Biothane System 228, 4215
biotin, *See* Ritatin, 26834
Biotrol, *See* Bactospeine, 3582
Bioxone, *See* methazole, 19481
Biphenyl, 4216, *See* diphenyl, 8592
1,1'-biphenyl, *See* diphenyl, 8592
biphenyl-2-ol, *See* *o*-phenylphenol, 23648
Biphenylol, *See* *o*-phenylphenol, 23648
(1,1-Biphenyl)-2-ol, *See* *o*-phenylphenol, 23648
1,1'-Biphenyl-2-ol, *See* *o*-phenylphenol, 23648
2-biphenylol, *See* Nipacide® OPP, 21258, *o*-phenylphenol, 23648
[1,1'-Biphenyl]-2-ol, sodium salt, *See* DowicideA, 8932
Biphenylol, sodium salt, *See* Dowicide A, 8932
2-Biphenylol, Sodium Salt, *See* Dowicide A,8932
3,3',4,4'-Biphenyltetramine, *See* DAB, 7639
biphenyl-4-yl-1,2,3,4-tetrahydro-1-naphthyl)-4-hydroxy-1(2H)-benzopyran-2-one, *See*difenacoum, 8449, Neosorexa, 20967
3-(3-(4-Biphenylyl)-1,2,3,4-tetrahydro-1-naphthyl)-4-hydroxycoumarin, *See* Ratak,25896
3-[3-(1,1'-Biphenyl)-4-yl-1,2,3,4-tetrahydro-1-naphthalenyl]-4-hydroxy-2H-1-benzopyran-2-one, *See* Ratak, 25896
3-(3-Biphenyl-4-yl-1,2,3,4-tetrahydro-1-naphthyl)-4-hydroxycoumarin, *See* Ratak,25896
(2-biphenylyloxy)sodium, *See* Dowicide A,8932
β-[1,1'-Biphenyl)-4-yloxy]- α-(1,1-dimethylethyl)-1H-1,2,4-triazol-1-ethanol, *See* Bitertanol, 4288
Bi-play®, 4217
Biquanide, 4218
Bird Manure, *See* guano, 13444
Birgin, *See* Propham, 25185, 25186
Birgin®, 4219
Birlan, *See* Chlorfenvinphos, 6098
Birlane, 4220, 4221, *See* Chlorfenvinphos, 6098
Birlane 10G, *See* Chlorfenvinphos, 6098
Birmabright, 4222
Birmasil alloy, 4223
Birmidium, 4224
Birmingham platina, 4225
Birmite, *See* Burmite, 4709
Birox®, 4226
Birutan, *See* rutin trihydrate, 27150
Bis-2, 4229
Bisabolol, *See* Camilol, 5053
α-bisabolol, *See* Camilol, 5053, Hydagen® B,14525
α-(-)-Bisabolol, *See* Kamillosan, 15707
bis(aceto)dihydroxytrilead, *See* subacetate oflead, 29902
bis(aceto)tetrahydroxytrilead, *See* subacetateof lead, 29902
bis(acetato)tetrahydroxytrilead, *See*subacetateoflead, 29902
bis(acetato-O)tetrahydroxytrilead, *See*subacetate of lead, 29902
3,5-bis(acetylamino)-2,4,6-triiodobenzoic acidsodium salt, *See* diatrizoate sodium, 8337
Bis-acrylamide, *See* methylene bisacrylamide,19567
bis(acyloxyethyl) hydroxyethylmethylammonium methosulfate, *See*Dehyquart AU-36, 7981
bis(allyl ether) of tetrabromobisphenol A, *See*Great Lakes BE-51, 13299
bis(β-amino-β-carboxyethyl) disulfide, *See*cystine, 7603
bis(2-aminoethyl) amine, *See* D.E.H. 20, 7610,diethylenetriamine, 8446
bis(beta-aminoethyl)amine, *See* D.E.H. 20,7610
N,N'-bis(2-aminoethyl)-1,2-ethanediamine, *See* Texacure EA-24, 31439, Texlin® 300, 31507, triethylenetetramine, 32245
Bis(4-aminophenyl)sulfone, *See* Dapsone, 7751 4,4'-diaminodiphenyl sulfone, 8302
bis(2-benzothiazolylthio)zinc, *See* Zetax®, 34741
N,N-Bis(2-(bis-(carboxymethyl)amino)ethyl)-glycine, *See* Aroquest M Special, 3064, Pentaquest OPAC 0201, 23170
bis(2-(bis(carboxymethyl)amino)ethyl)glycinepentasodium salt, *See* Cheelox® 80, 5864,Pentaquest Extra 0685, 23169, Trilon® C Liq, 32328
N-[2-[Bis(carboxymethyl)amino]ethyl]-N-(2-hydroxyethyl)glycine trisodium salt, *See* Kalex OH, 15672, Trilon® D Liq, 32329
bis(2-butoxyethyl) phthalate;, *See* Plasthall® 200, 23979
4,4'-bis(sec-butylamino) diphenyl methane, *See* Unilink® 450, 32966
bis(4-t-butylcyclohexyl) peroxydicarbonate, *See* Perkadox® 16, 23301
Bis(tert-butyl)diimine, *See* VR-160, 34057
bis(*tert*-butyldioxyisopropyl)benzene, *See*Perkadox®

14, 23300, Vul-Cup 40KE and R,34118
1,1-Bis(*tert*-butyl peroxy)cyclohexane, *See*USP® -400P, 33168
α-α-Bis (*t*-butylperoxy) diisopropylbenzene, *See* Luperox 802, 17841, Retilox® F 40 MG,26274
Bisbutyl peroxy diisopropyl benzene, 4230
bis(*tert*-butylperoxyisopropyl)benzene, *See*Vul-Cup, Vul-Cup 40KE, Vul-Cup R, 34119
bis(t-butylperoxyisopropyl) benzene, *See*Polyvel CR-5P, 24658
Bisbutyl peroxy trimethylcyclohexane, *See*Aztec® 1,1-Bis(*t*-Butylperoxy)-3,3,5-TrimethylCyclohexane, 3549
2,6-Bis(*t*-butyl)phenol, *See* 2,6-dibutylphenol,8370
N,N-bis(carboxymethyl)glycine, *See*nitrilotriacetic acid, 21339, TG Buffer, 31547
Bis(carboxymethyl)glycine, trisodium salt, *See*Cheelox® NTA-Na3, 5869
1,2-Bis(Chlorodimethylsilyl)-ethane, *See*CT2015, 7313
7-[[bis(2-chloroethoxy)phosphinyl]oxy]-3-chloro-4-methyl-2H-1-benzopyran-2-one, *See*Haloxon, 13567
2,2-Bis(2-chlorophenyl-4-chlorophenyl)-1,1-dichloroethane, *See* Lysoform, 18095
N,N'-bis (4-chlorophenyl)-3,12-diimino-2,4,11,13-tetraazatetradecane-diimidamidecompound with D-gluconic acid, *See*chlorhexidine digluconate, 6101
1,1-bis(*p*-chlorophenyl)-2,2,2-trichloroethanol, *See* Acarin, 183, Dicofol, 8398, Kelthane, 15881
bis(5-chloro-2-hydroxyphenyl)methane, *See*Nuophene, 21783
Biscumylphenyl trimellitate, *See* Kenplast® ESI,16012
bis(2,3-dibromopropyl ether) oftetrabromo bisphenol A, *See* Great Lakes PE-68, 13309
bis(dibutyldithiocarbamate)nickel complex, *See* Naugard® NBC, 20797, NBC, 20831,Perkacit® NDBC, 23287
bis(2,4-di-t-butylphenyl) pentaerythritoldiphosphite, *See* Ultranox® 626, 32850,32851
Bisdibutylphenyl pentaerythritol diphosphite, *See* Alkanox® P-24, 1705
bis(2,4-di-t-butylphenyl)pentaerythritoldiphosphite, *See* Lowinox® 243, 17656
bis(dibutylthiocarbamoyl) disulfide, *See*Akrochem® TBUT, 1178
bis(diethylamino)xanthylium chloride 3,6-bis(diethylamino)xanthylium chloride, *See*pyronine B, 25522
bis(diethylcarbamodithioato-S,S') Zinc, *See*Ancazate ET, 2415, Octocure ZDE-50, 21996,Perkacit® ZDEC, 23296
bis(diethyldithiocarbamate)cadmium complex, *See* Cadmate®, 4892, ethyl cadmate, 11065
bis(diethyldithiocarbamate)zinc complex, *See*Ancazate ET, 2415, Octocure ZDE-50, 21996,Perkacit® ZDEC, 23296
bis(diethylthiocarbamyl) disulfide, *See*Akrochem® TETD, 1180
bis(dimethylamino)dimethylsilane, *See*CB2100, 5473
4,4'-Bis(dimethylamino)benzhydrol, *See*Michler's hydrol, 19645
4,4'-bis(dimethylamino)benzophenone, *See*ketone base, 16140
***p,p'*-bis(dimethylamino)diphenylmethane,** *See*Michler's Base, 19644
4,4'-Bis(dimethylamino)diphenyl carbinol, *See*Michler's hydrol, 19645
70% bis(dimethylaminoethyl) ether, 30%dipropylene glycol, *See* Dabco® BL-11, 7648
***p,p'*-bis(N,N-dimethylaminophenyl)methane,** *See* Michler's Base, 19644
N-[4-[bis[4-(dimethylamino)phenyl]methylene]-2,5-cyclohexadien-1-ylidene]-N-methylmethanaminium chloride, *See*gentian violet, 12845
bis(*p*-(N,N-dimethylamino)phenyl)methane, *See*Michler's Base, 19644
bis[4-(dimethylamino)phenyl]methanone, *See*ketone base, 16140, Michler's ketone, 19646
bis(dimethylamino)xanthylium chloride 3,6-bis(dimethylamino)xanthylium chloride, *See*pyronine Y, 25523
4,4'-Bis (α,α-dimethyl-benzyl) diphenylamine, *See* Naugard® 445, 20790
bis(dimethylbenzyl) ether, *See* KP 555, 16380
bis(α,α-dimethylbenzyl) peroxide, *See* Aztec®DCP-R, 3556, Dicumylperoxide, 8407, Percumyl D, 23230, Peroximon® DC-40, 23409, Polyvel PCL-20, 24661
bis(dimethylcarbamodithioato-S,S') Copper, *See* Akrochem® Cu.D.D, 1165, copperdimethyldithiocarbamate, 6779, Cumate®,7351
bis(dimethylcarbamodithioato-S,S')zinc, *See*AAprotect, 16, Accelerator MZ Powder, 209,Vancide® MZ-96, 33303, Zimate®, 34762
bis(dimethyldithiocarbamat)zinc, *See*Zimate®, 34762
bis(dimethyldithiocarbamato) nickel complex, *See* methyl niclate®, 19553
bis(dimethyldithiocarbamato)lead, *See*Ledate®, 16949, methyl ledate, 19547
bis(dimethyldithiocarbamato)zinc, *See*Accelerator MZ Powder, 209, Vancide® MZ-96,33303
bis(dimethylthiocarbamoyl) disulfide, *See*Thiurad, 31730
bis(dimethylthiocarbamyl) disulfide, *See*Akrochem® TMTD, 1182, Thiram, 31726
bis(dimethylthiocarbamyl) sulfide, *See*Ancazide IS, 2420, Monex, 20210, TMTM,31906, Unads®, 32907, Vulcaid 222, 34079
2,5-Bis(1,1-dimethylethyl)-1,4-benzenediol, *See* Dibutylhydroquinone, 8368
1,2-Bis(1,1-dimethylethyl)diazene, *See* VR-160, 34057
3,5-bis(1,1-dimethylethyl)-4-hydroxy benzenepropanoic acid octadecylester, *See* Naugard® 76, 20789
2,6-bis (1,1-dimethylethyl)-4-methylphenol, *See* BHT, 4122 Butylated hydroxytoluene, 4788, Ralox® BHTfood grade, 25845, Tenamine 3, 31092,Vanox® PCX, 33346, Vulkanox® KB, 34152
bis(1,1-dimethylethyl)peroxide, *See* Aztec® Di-*t*-Butyl Peroxoide, 3557, 2,6-di-*t*-butylphenol, 4799, di-*t*-butyl peroxide, 8369, Trigonox® B,32295
N,N'-Bis(1,4-dimethylpentyl)-*p*-phenylenediamine, *See* Naugard® I-2, 20792
1,5-bis(2,4-dimethylphenyl)-3-methyl-1,3,5-triazapenta-1,4-diene, *See* Taktic, 30729
bis(p-ethylbenzylidene) sorbitol, *See* NC-4,20832
bis(2-ethylhexanoate)tin, *See* Metacure® T-9,19416
2,6-bis(2-ethylhexyl)hexahydro-7a-methyl-1H-imidazo[1,5-c]imidazole, *See* hexedine, 14027
bis(2-ethylhexyl) 1,2-benzenedicarboxylate, *See* Witcizer 313, 34455
Bis(2-ethylhexyl) adipate, *See* Wickenol® 158,34393
bis(2-ethylhexyl) decanedioate, *See* Plasthall®DOS, 24005, Reomol DOS, 26131
bis(2-Ethylhexyl) Ester Decanedionic Acid, *See* Plasthall® DOS, 24005
bis(2-Ethylhexyl) Ester Sebacic Acid, *See*Plasthall® DOS, 24005, Reomol DOS, 26131
bis(2-ethylhexyl) hexanedioate, *See* dioctyladipate, 8548
bis(2-ethylhexyl) hydrogen phosphite, *See*Syn-O-Ad® P-310, 30560
bis(2-ethylhexyl) peroxydicarbonate, *See*Trigonox® EHP, 32299
bis(2-ethylhexyl) Phthalate, 4228, Bisoflex 81,4249, Bisoflex 82, 4252, Kodaflex® DOP,16273, Witcizer 312, 34454
bis(2-ethylhexyl) sebacate, *See* Plasthall® DOS, 24005, Reomol DOS, 26131, Wickenol® 159, 34394
bis(2-ethylhexyl) sulfosuccinate, sodium salt, *See* Empimin® OP70, 10434, Octowet 40, 22035
bis(2-ethylhexyl) terephthalate, *See* Kodaflex®DOTP, 16274
bis-(2-ethylhexyl)-1,2-benzenedicarboxylate, *See* Reomol DCP, 26130
bis(2-ethylhexyl)adipate, *See* Uniflex® DOA,32945
Bis-(2-ethylhexyl)ester, *See* Wickenol® 158,34393
Bis-(2-ethylhexyl)ester kyseliny ftalove, *See*Witcizer 312, 34454
bis(2-ethylhexyl)ester phthalic acid, *See* Reomol DCP, 26130
bis(2-ethylhexyl)hexanedioate, *See* PX-238,25440
bis(2-ethylhexyl)phthalate, *See* Bis(2-ethylhexy) Phthalate, 4228, Reomol DCP, 26130, Witcizer312, 34454
bis(2-ethylhexyl)sodium sulfosuccinate, *See*Empimin® OP70, 10434, Octowet 40, 22035
bis(2-ethylhexyl)-S-sodium sulfosuccinate, *See*dioctyl sodium sulfosuccinate, 8550
Bis(ethylhexyl) phthalate, *See* Witcizer 312, 34454
bis(D-gluconato) copper, *See* copper gluconate,6781
bis(glycidoxypropyl) tetramethyldisiloxane, *See*CB2405, 5474
Bis-hexamethylene triamine phosphonic acid, *See* Unihib® 1704, 32962
Bis-hydrogenated tallowalkyl-2-hydroxypropylamines, *See* Propomeen 2HT-11, 25193
bis(4-hydroxy-5-*tert*-butyl-2-methylphenyl)sulfide, *See* Rutenol, 27147
1,3-Bis(2-hydroxyethyl)-5,5-dimethyl-2,4-imidazolidinedione, *See* DEDM hydantoin, 7872
bis(2-hydroxyethyl ether) oftetrabromo bisphenol A, *See* Great Lakes BA-50,13297
bis(2-hydroxyethyl), *See* Tomah E-19-2, 31974
bis(hydroxyethyl)amine, *See* Diethanolamine,8434
bis(hydroxyethyl) aminopropyltriethoxy silane, *See* CB2408, 5475
bis(2-hydroxyethyl)ammoniumdodecyl benzenesulfonate, *See* Ablusol DBD,124
N,N-Bis (2-hydroxyethyl) coco amine, *See*Chemstat® 273-C, 6036
bis(2-hydroxyethyl)decanamide, *See* AmidexCP, 2091, Standamid® CD, 29454
Bishydroxyethyl dihydroxypropylstearaminiumchloride, *See* Monaquat TG,20135
N,N-Bis(2-hydroxyethyl)dodecanamide, *See*Ablumide LDE, 79, lauramide DEA, 16828
N,N-Bis (2-hydroxyethyl)-N-(3-dodecyloxy-2-hydroxypropyl) methyl ammoniummethosulfate, *See* Cyastat® 609, 7504
bis(2-hydroxyethyl) isodecyloxypropylamine, *See* Tomah E-14-2, 31969
Bishydroxyethylisodecyloxypropylamineoxide, *See* Tomah AO-14-2, 31963
bis(2-hydroxyethyl)isotridecyloxypropylamine, *See* Tomah E-18-2, 31971
N,N-Bis(2-hydroxyethyl)octadecanamide, *See*Ablumide SDE, 81
N,N-Bis(hydroxyethyl)octadecanamide, *See*Ablumide SDE, 81
bis(2-hydroxyethyl) octadecyloxypropylamine, *See* Tomah E-18-2, 31972
bis(2-hydroxyethyl) octyl methylammonium *p*-toluene sulfonate, *See* Chemstat® 106G/90,6032
N,N-Bis(β-hydroxyethyl)stearamide, *See*Ablumide SDE, 81
N,N-Bis(2-hydroxyethyl)stearamide, *See*Ablumide SDE, 81
N,N-Bis (2-hydroxyethyl) stearyl amine, *See*bishydroxyethyl-N,N-Bis(2-hydroxyethyl) stearylamine, 4231, Chemstat®273-E, 6037
N,N-Bis(2-hydroxyethyl) stearyl amine, 4231
bis(2-hydroxyethyl) sulfide, *See* thiodiglycol,31682
N,N-Bis(2-hydroxyethyl)(tallow alkyl)amine, *See* Witcamine® 6606, 34447
2,2-Bis(4,4'-Hydroxyphenyl)propane, *See*Bisphenol A, 4283
2,2-bis(4-Hydroxyphenyl)propane, *See*Bisphenol A, 4283
Bis[2-hydroxy-5-*t*-octyl-3-(benzotriazol-2-yl)phenyl, *See* Mixxim® BB/100, 19989
bis(2-hydroxy-4-methoxyphenyl) methanone, *See* benzophenone-6, 3978, Uvinul® D-49, 33194
bis(4-hydroxyphenyl) dimethylmethane, *See*Bisphenol A, 4283
bis(4-hydroxyphenyl)propane, *See* Bisphenol A,4283

bis(2-hydroxypropyl)amine, *See* diisopropanolamine, 8466
bis(1-hydroxy-2(1H)-pyridinethionato-O,S)zinc, *See* Zinc Omadine, 34783
bis(2-hydroxy-3,5,6-trichlorophenyl)methane, *See* hexachlorophene, 13992
bis(1,4-isocyanatophenyl)methane, *See* Rubinate® LF-168, 27108
2,4-bis-(isopropylamino)-6-methylthio-1,3,5-triazine, *See* Prometrex, 25139
Bis-isotridecyl hydrogen phosphite, *See* diisotridecyl hydrogen phosphite, 8471
bis(lauroyloxy)di(n-butyl)stannane, *See* Dabco® T-12, 7673
Bismaleimide resin, *See* Fiberite 986, 11731
Bismate®, 4232 , *See* Bismuthdimethyldithiocarbamate, 4238
Bismet, 4233
bis(mercaptobenzothiazolato)zinc, *See* Octocure ZMBT-50, 21998, Zetax®, 34741
bis(mercaptobenzthiazolato)zinc, *See* Zetax®,34741
Bismet, *See* Bismuth dimethyldithiocarbamate,4238
2,2-bis(methacryloxymethyl)butyl methacrylate, *See* Perkalink® 400, 23314
1,2-bis(methacryloyoxy) ethane, *See* ethyleneglycol dimethacrylate, 11125
o-bis(3-methoxycarbonyl-2-thioureidobenzene, *See* thiophanate methyl, 31706
2,2-bis (*p*-methoxyphenol)-1,1,1-trichloroethane, *See* methoxychlor, 19504
bis-(2-Methoxypropyl) ether, *See* Icinol DPM,14815, Poly-Solv® DPM, 24555
bis(1-methylamyl) Sodium Sulfosuccinate,4227
bis(methylamyl) Sodium Sulfosuccinate, *See* Lankropol® KMA, 16713
Bis-(N-methylbenzamide)ethoxymethyl silane, *See* CB2409.5, 5476
bis(methylethyl)-1,1'-biphenyl, *See* NusolvABP-103, 21801
bis(1-methylethyl)-1,1'-Biphenyl, *See* NusolvABP-103, 21801
N,N'-bis(1-methylethyl)-6-methylthio-1,3,5-triazine-2,4-diamine, *See* prometryn, 25141
bis(1-methylethyl) 5-nitro-1,3-benzenedicarboxylate, *See* nitrothal-isopropyl, 21380
N-[2-[[2,6-bis(1-methylethyl)phenyl]amino]2-oxoethyl]-N-(carboxymethyl)glycine, *See* Hepatolite, 13789
N,N'-Bis-(1-methyl/peptyl)-p-phenylene diamine, *See* UOP 288, 33090
Bis(1-methyl-1-phenylethyl) peroxide, *See* Dicumyl peroxide, 8407
N,N'-bis(2-methylphenyl)thiourea, *See* Accelerator A22, 192
N,N'-bis(1-methylpropyl)-1,4-benzenediamine, *See* Naugalube® 403, 20782
bis(1-methyl-1-phenylethyl) peroxide, *See* Aztec® DCP-R, 3556, Percumyl D, 23230,Peroximon® DC-40, 23409, PolyvelPCL-20, 24661
bis(2-methylpropyl) hexanedioate, *See* diisobutyl adipate, 8460
Bismica 46, 4234, *See* Bismuth chlorideoxide,4237
bismite, *See*, , *See* Bismuth trioxide, 4245
Bismogenol, *See* Bismuth subsalicylate, 4244
bismuth, 4235
bismuth (III) nitrate, basic, *See* Bismuthnitrate,4239
Bismuth(III) oxychloride, *See* Bismuthoxychloride, 4241
bismuth boro-phenate, *See* markasol, 18732
bismuth bronze, 4236, *See* nickel silvers, 21132
bismuth carbonate basic, *See* bismuthsubcarbonate, 4242
bismuth chloride oxide, 4237, *See* bismuthoxychloride, 4241
bismuth dimethyldithiocarbamate, 4238, *See* Bismate®, 4232, Bismet, 4233
bismuth hydroxide nitrate oxide, *See* Bismuthsubnitrate, 4243
bismuth nitrate, 4239
bismuth nitrate, basic, *See* bismuthsubnitrate,4243
bismuth oxide, 4240, *See* Bismuth trioxide,4245, flowers of bismuth, 12032
bismuth oxycarbonate, *See* Bismuthsubcarbonate, 4242
bismuth oxychloride, 4241, *See* Biju®,4136,Bismuth chloride oxide, 4237, Mearlite®GBU,19192, Pearl I, II, III,22979,Pearl Supreme UVS,22981,Pearl white, 22982, Pearl-Glo®, 22984
bismuth oxychloride pearls, *See* Perlex, 23332
bismuth oxynitrate, *See* Bismuth subnitrate,4243
bismuth salicylate basic, *See* Bismuthsubsalicylate, 4244
bismuth subcarbonate, 4242
bismuth subchloride, *See* bismuth oxychloride,4241
bismuth subnitrate, 4243, *See* magister ofbismuth, 18320
bismuth subsalicylate, 4244
bismuth trinitrate, *See* bismuth nitrate, 4239
bismuth trioxide, 4245, *See* Bismuth oxide, 4240
bismuth vanadate, *See* pacherite, 22536
bismuth vanadate, *See* pacherite, 22536
bismuth violet, *See* gentian violet, 12845
bismuth yellow, *See* Bismuth oxide, 4240,Bismuth trioxide, 4245
bismuth subchloride, *See* Bismuth chlorideoxide, 4237
bismuth, tris(dimethylcarbamodithioato-S,S')-(OC-6-11)-, *See* Bismate®, 4232,Bismet, 4233, Bismuth dimethyldithio carbamate, 4238
bismuth, tris(dimethylcarbamodithioato-S,S')-(OC-6-11)-, *See*
bismuthoxy nitrate, *See* Bismuth nitrate, 4239
bismuthoxychloride, *See* U Pink 113, UViolet 109, 32696
bismuthyl chloride, *See* Bismuth chlorideoxide, 4237
bismuthyl nitrate, *See* Bismuth nitrate, 4239
bismuthyl nitrate, Bismuth white, *See* Bismuth subnitrate, 4243
1,2-Bis(octadecanamido)ethane, *See* Abluwax EBS, 140
Bisoflex, 4246
Bisoflex 8N, 4254
Bisoflex 799, 4247
Bisoflex 79A, 4248
Bisoflex 81, 4249, *See* Bis(2-ethylhexy)Phthalate, 4228 dioctyl phthalate, 8549 Reomol DCP, 26130 Witcizer 312, 34454
Bisoflex 81, *See* Bis(2-ethylhexy) Phthalate, 4228
Bisoflex 82, 4252, *See* Bis(2-ethylhexy)Phthalate, 4228
Bisoflex 88, 4253
Bisoflex 799, 4247
Bisoflex 810, 4250
Bisoflex 819, 4251
Bisoflex BP9, 4255
Bisoflex DEP, 4256
Bisoflex DMP, 4257
Bisoflex DNA, 4258
Bisoflex DOA, *See* dioctyl adipate, 8548, Plasthall® DOA, 24002, PX-238, 25440, Wickenol® 158, 34393
Bisoflex DOP, 4259, *See* Bis(2-ethylhexy) Phthalate, 4228, dioctyl phthalate, 8549, Reomol DCP, 26130, Witcizer 312, 34454
bisoflex DOS, *See* Plasthall® DOS, 24005, Reomol DOS, 26131
Bisoflex DUP, 4260
Bisoflex L711P, 4261
Bisoflex L79, 4262
Bisoflex L79P, 4263
Bisoflex L9, 4264
Bisoflex L911, 4265
Bisoflex L911P, 4266
Bisoflex ODN, 4267
Bisol, 4268
Bisolene, 4269
Bisolite, 4270
Bisolube, 4271
Bisomel, *See* Wickenol® 101, 105, 34378
Bisomer, 4272
Bisomer 2HEMA, 4273, *See* hydroxyethyl methacrylate, 14662
Bisomer 2HPMA, 4274
Bisomer D10M, 4275
Bisomer DALP, 4276
Bisomer DAM, 4277
Bisomer DBF, 4278
Bisomer DBM, 4279, *See* dibutyl fumarate, 8364
Bisomer DNM, 4280
Bisomer DOM, 4281
Bisoprufe, 4282
Bisoxyl, *See* Bismuth oxychloride, 4241
bis(pentabromophenyl) ether, *See* Octoguard FR-01, 21999, Saytex® 102E, 27514, Thermoguard® 505, 31632
bis(pentamethylene)thiuram tetrasulfide, *See* Sulfads®, 29960
Bisphenol A, 4283, *See* Parabis, 22717
4,4'-bisphenol A, *See* Bisphenol A, 4283
Bisphenol A ethoxylate, *See* Igepal® BPA-6, 14845
Bisphenol-A polyols, *See* Syn Fac®, 30512
bis(phenoxarsin-10-yl)ether, *See* Vinyzene, 33869
bis(10-phenoxarsinyl)oxide, *See* Vinyzene, 33869
bis(10-phenoxarsyl)oxide, *See* Vinyzene, 33869
***p*-bis(phenylamino)benzene,** *See* Agerite® DPPD, 1016
bis(2-pyridylthio)zinc 1,1'-dioxide, *See* Zinc Omadine, 34783
Bispyrithione, *See* Omadine® MDS, 22152
bis(6-quinolyl)urea bis methosulfate, *See* Acaprin®, 181, 1,3-di-6-quinolylurea, 8618
Bis-stearamide, *See* Foamkill® CMP, 12172
Bissy nuts, *See* kola nut, 16303
bis(tetrabromophthalimido)ethane 1,2-bis(tetrabromophthalimido)ethane, *See* Saytex® BT-93®, 27520
Bistetramethyl piperidinyl sebacate, 4284
bis(1,2,3,6-tetrahydrobenzaldehyde) pentaerythritol acetal, *See* Vulkazon® AFS/LG, 34158
Bis (2,2,6,6-tetramethyl-4-piperidinyl) decanedioate, *See* Bistetramethyl piperidinyl sebacate, 4284, Lowilite® 77, 17645
bis(2,2,6,6-tetramethyl-4-piperidinyl) sebacate, *See* Bistetramethyl piperidinyl sebacate, 4284, Lowilite® 77, 17645, Tinuvin® 770, 31836, Uvaseb 770, 33178
bis(2,2,6,6-tetramethyl-4-piperidyl) sebacate, *See* Uvaseb 770, 33178
1,3-bis(*o*-tolyl)-2-thiourea, *See* Accelerator A22, 192
bis(tribromophenoxy) ethane, *See* Great Lakes FF 680, 13305
bis(tributyltin)oxide, *See* bioMeT TBTO, 4172
bis(tri-n-butyl)oxide, *See* Fungitrol Tinox, 12482
bis(tri-n-butyltin) oxide, *See* Keycide® X-10, 16153
bistrichloromethylsulfone, 4285, *See* Amerstat® 294, 2044
bis(trichloromethyl) sulfone, *See* bistrichloromethylsulfone, 4285 Stauffer N-1386®, 29580
bis(3-(triethoxysilyl)propyl) tetrasulfane, *See* Aktisil PF 216, 1204, Si 69, 28190
Bis[3-(triethoxysilyl)propyl]tetrasulfide, *See* Aktisil PF 216, 1204, CB2494, 5478
bis(trimethoxysilylpropyl) ethylene diamine, *See* CB2493, 5477
bis(3,5,5-trimethylhexanoyl) peroxide, *See* Trigonox® 36-C75, 32270, USP® -355M, 33167
Bistrimethylsilyl acetamide, 4286, *See* Bistrimethylsilyl acetamide, 4286, CB2500, 5479, Dynasylan® BSA, 9399
N,O-Bis(trimethylsilyl)acetamide, *See* Bistrimethylsilyl acetamide, 4286, CB2500, 5479, Dynasylan® BSA, 9399
bis(trimethylsilyl) amine, *See* hexamethyldisilazane, 14003
N,O-Bis(trimethylsilyl)acetamide, *See* Bistrimethylsilyl acetamide, 4286
bis(trimethylsilyl)acetamide, *See* Dynasylan® BSA, 9399
N,N-bis(2,4-xylyliminomethyl)methylamine, *See* Taktic, 30729
bis[tris(2-methyl-2-phenylpropyl)tin] oxide, *See*

Torque, 32077
Bi-Tarco, 4287
Bitemol, *See* simazine, 28464
Bitemol S 50, *See* simazine, 28464
Bitertanol, 4288
Bitran, 4289
Bitran H, 4290
Bitrex®, 4291
Bitter almond-oil camphor, *See* Benzoin, 3975
Bitter ash, *See* Hoppit, 14304
Bitter salt, *See* magnesium sulfate, 18368
Bitter wood, *See* Hoppit, 14304
bittern, 4292
Bitulan, *See* Perichthol, 23270
Bitumastic, 4293
Bitumen, 4294, *See* Karetnja, 15765
bituminol, *See* Perichthol, 23270
bitumol, *See* Perichthol, 23270
Bitumuls, 4295
Bitusize, 4296
Bituvar, 4297
Bivelon, *See* hexachlorophene, 13992
Biverm, *See* phenothiazine, 23618
Bivert, 4298
B-K, *See* sodium hypochlorite, 28773
BK Powder, *See* Repak, 26141
BL 3, 4299
BL-60®, 4300
BLA, *See* subacetate of lead, 29902
Blacar, *See* Vestolit®, 33679
Blacar 1716, *See* Vestolit®, 33679
Black 103, 4301
Black and White Bleaching Cream, 4302, *See* Tecquinol® Tech. Grade, 30910
Black catechu, *See* cutch,
black copper oxide, *See* cupric oxide, 7386
Black Diamond, *See* Carbon, 5217
Black Grip, 4303
Black hypo, *See* burnt hypo, 4712
black iron, *See* pyrolignite of iron, 25515
black iron liquor, *See* pyrolignite of iron, 25515
black lead, *See* graphite, 13277
Black Lead Ore, *See* alquifon, 1846
Black Leaf Lawn & Garden Fungicide, *See* Chlorothalonil, 6150, Repulse, 26151
black liquor, *See* pyrolignite of iron, 25515
black manganese oxide, *See* KM Manganese Dioxide, 16238
black mordant, *See* pyrolignite of iron, 25515
Black Out® Black, 4304
black oxide of iron, *See* magnetic iron ore, 18374
Black Pearls® 1100, 4305
black powder, *See* gunpowder, 13495
Black-Out®, 4306
Blackox, 4307
Bladafum®, 4308
Bladan M, *See* Parathion-methyl, 22808
Bladan®, 4309
Blade, 4310, *See* Vydate®, 34171
Bladex, *See* cyanazine, 7472
Bladex-B, *See* Planotox, 23914
Blagden Resins, 4311
Blagdenite, 4312
blanc d'argent, *See* flake white, 11871
blanc de perle, *See* Bismuth oxychloride, 4241, pearl white, 22982
blanc d'Espagne, *See* bismuth oxychloride, 4241
Blanc fixe, *See* Radiopaque, 25830
Blanc Fixe Micro, 4313, *See* Barium sulfate, 3644
Blanc fixe, articifical percipitate, *See* Barium sulfate, 3644
blanc fixe, artificial, precipitated, *See* Barium sulfate, 3644
blanchite, *See* hydrosulfite, 14643
Blancol, 4314, *See* Ablusol ML, 129
Blancol C, *See* Photine, 23756
Blancol Dispersant, *See* Ablusol ML, 129
Blancol N, 4315
Blancorol®, 4316
Blandofen CAZ, 4317
Blandofen CT, 4318, *See* Adogen® TA-101, 780, Genamin DSAC, 12771
Blandofen FA, 4319
Blandol®, 4320
Blanket Adhesive H-98, 4321
blankit, *See* hydrosulfite, 14643
Blankit®, 4322
Blanko-Blech, 4323
Blankophor R, *See* Photine, 23756
Blankophor®, 4324
Blanose, 4325
blasting gelatin, 4326
Blastoff, *See* Edifenphos, 9618
Blatt gold, 4327
Blatt silver, 4328
Blattanex, *See* Baygon®, 3790, Propoxur, 25198, 25199, 25200, 25201
Blattanex®, 4329
Blattanex® 20, 4330
Blattanex® Residual Spray, 4331
Blattosep, *See* Baygon®, 3790
blausäure, *See* zootic acid, 34903
Blaxon LT, 4332
Blazer, *See* acifluorfen, 343
Blazer®, 4333
Blazon, 4334
BLE, 4335
BLE® 25, 4336
bleach, *See* sodium hypochlorite, 28773
bleach liquor, 4337
bleached wood pulp, *See* cellulose, 5626
Bleachit® 1A, 4338
bleiweiss, *See* Kremser White, 16399
Blend of benzalkonium chloride and phenyl phenoxide, *See* Acticide 50X®, 463
Blend of chlorinated and nonchlorinated methyl isothiazolones with nonmetal salts stabilizing system, *See* Acticide LG®, 468
Blend of glycols, fatty acids, and nonionic surfactants in a hydrocarbon base, *See* Actrafoam A, B, C, S, 525
Blend of liquid hydrocarbons, hydrophobic silica, synthetic copolymers, and nonionic emulsifiers, *See* Agitan 281, 1032
Blend of liquid hydrocarbons, polyglycols, and amorphous silica, *See* Agitan P 800, 1035
Blend of modified organo polysiloxanes with nonionic alkoxylated comps, *See* Agitan VP 725, 1036
Blend of vegetable oils, modified solids, nonionic emulsifiers, and silicone defoamer, *See* Agitan 301, 1033
blenda, *See* oil white, 22086
Blenderm, 4339
Blendex® 101, 4340, *See* Acrylonitrile-butadiene-styrene, 439
Blendex® 131, 336, 338, 467, *See* Acrylonitrile-butadiene-styrene, 438
Blendex® 310, 4341, *See* Acrylonitrile-butadiene-styrene, 439
Blendex® 340, 4342
Blendex® 586, 4343
Blendex® 975, 4344
Blendex® HPP 801, 4345
Blendmax 322, 4346
Blendur®, 4347
Blendur® KU 3-4513, 4348
bleu d'azure, *See* smalt, 28638
bleu de saxe, *See* smalt, 28638
Blex, 4349
Blitex, *See* ronnel, 26951
Blitox, *See* copper oxychloride, 6784, Cupravit®, 7370
Blitox 50, *See* Cupravit®, 7370
BLO®, 4350 , *See* Butyrolactone, 4803
Block-Out A-SF, 4351
blood meal, 4352
blood sugar, *See* glucose, 13052
Bloodit, 4353
Blotic, *See* Propetamphos, 25183
Blo-trol, *See* acetyl tributyl citrate, 313
blown oils, 4354
Blown rapeSeed oil, *See* Actralube 7142, 538
BLS™ 1770, *See* Bistetramethyl piperidinyl sebacate, 4284, Lowilite® 77, 17645
BL-S578, *See* Cefadroxil, 5547
Blue asbestos, *See* Crocidolite, 7060
Blue Basic Lead Sulphate, 4355
Blue copperas, *See* cupric sulfate, 7387
Blue Dot, 4356
blue gas, *See* water gas, 34224
blue gold, 4357
Blue Mold, 4358
Blue Powder, 4359
blue silver, *See* Niello silver, 21167
Blue stone, *See* cupric sulfate, 7387
Blue Tetrazolium, *See* tetrazolium chloride, 31395
Blue Tetrazolium chloride, *See* tetrazolium chloride, 31395
Blue vitriol, *See* cupric sulfate, 7387, Kocide® Copper Sulfate Pentahydrate Crystals, 16262
Blueminster Resin Emulsions, 4360
bluestone, *See* Kocide® Copper Sulfate Pentahydrate Crystals, 16262
Bluish eosin, *See* erythrosin, 10778
BMA, *See* Butyl methacrylate, 4774
BMC, *See* Konker®, 16327 Carbendazim, 5187, 5188
BMC 100, 102, 4361
BMC 200, 4362
BMC 800, 4363
BMC 1000, 4364
BMK, *See* Carbendazim, 5187, 5188
BN, *See* Boron nitride, 4443
B-Nine, 4365
B-NINE, *See* daminozide, 7723
B-O, *See* Boric anhydride, 4423
BO 2, *See* Ablunol OA-6, 107, Volpo 3, 34020
BO 7, *See* Ablunol OA-6, 107, Volpo 3, 34020
BOA, *See* Nipantiox, 21276, Sustane® 1-F, 30432
Bobierre metal, 4366
BOC anhydride, *See* Di-*t*-butyl dicarbonate, 8367
Bodenstein, 4367
Bohemian earth, 4368
Bohemian topaz, 4369
Bohnalite B, 4370
Bohnalite J, 4371
Bohnalite S43, 4372
Bohnalite S51, 4373
Bohnalite U, 4374
Bohnalite Y, 4375
Bohrmittel Hoechst, 4376
boiled oil, 4377
Boileezers, *See* aluminum oxide, 1919
boiler plug alloy, 4378
Boiler-Aid, 4379
Boisambrene Forte, 4380
Bolatron 6200, *See* Vestolit®, 33679
Bolda, 4381
Bolda FL, 4382
Boldo, *See* bromadiolone, 4565, Rentokil Deadline, 26116
bole, *See* Indian Red, 15068
boletic acid, *See* Fumaric acid, 12453
Bolfo, *See* Baygon®, 3790
Bolfo®, 4383
Bolstar, 4384
bolster silver, 4385
Boltaron, *See* Vestolit®, 33679
Bolus alba, *See*
Bolus alba, *See* ASP®, 3230, Bilt-Cote®, 4147, Kayphobe-ABO, 15816, kaolin, 1 5725, Peerless®, 23023, SPS, 29349, Vanclay, 33304
Bombardier, 4386, *See* Chlorothalonil, 6150, Repulse, 26151
Bombay catechu, *See* cutch,
Bombay gum, *See* East India Gum, 9460
Bonapicillin, *See* Ampicillin, 2361
Bonasanit, *See* pyridoxine hydrochloride, 25483,

vitamin B_6 hydrochloride, 33937
Bond-A-Tint, 4387
Bonder/Gardobond, 4388
Bonder/Gardoclean, 4389
Bonderite, 4390
Bonderlube/Gardolube, 4391
Bonding Agent 2001, 4392
Bonding Agent M 3, 4393
Bonding Agent P 1, 4394
Bonding Agent R6, 4395
Bonding Agent TN, 4396
Bondogen, 4397
bondou, *See* red water bark, 25973
Bond-Plus, 4398
Bond-Plus HM, 4399
BondTint, 4400
Bondur, 4401
bone ash, *See* calcium phosphate (tribasic), 4967
Bone charcoal, *See* animal charcoal, 2466
Bone oil, 4402
Bone tar, *See* bone oil, 4402
Bonide, *See* Chlorothalonil, 6150, Repulse, 26151
Bonloid, *See* Vestolit®, 33679
Bonner L-894, 4403
Bonoform, *See* Acetosol, 304
Bonomold Op, *See* Nipasol M, 21282
Bonosol, 4404
Bonotex, 4405
Bonsai, *See* paclobutrazol, 22541
Bontex, 4406
Bonzi, 4407, *See* paclobutrazol, 22541
Boost, 4408
Bora-Care, *See* Polybor®, 24369
Boracic acid, *See* Boric Acid, 4422
Boral, 4409
Borascu, 4410
Borate derivs, *See* Crodinhib RT70, RT70S, 7179
Borateem, *See* Optibor Boric Acid, 22230
Borates, *See* Optibor Boric Acid, 22230
Borax, 4411, *See* sodium borate anhydrous, 28739
Borax decahydrate, *See* Borascu, 4410
Borax Glass, 4412
Boraxo, *See* Boric oxide, 4424
Boraxusta, 4413
Borcher's metal, 4414
Bordeaux turpentine, *See* French turpentine, 12398
Bordeaux/cufraneb, *See* Macuprax, 18286
Borden, 4415
Borderland Black, 4416
Bordermaster, *See* MCPA, 19183
Borea, *See* bromacil, 4564, Uragan, 33105
Borester, 4417, *See* Tim-bor industrial, 31798
Borester 7, 4418
Boresters, 4419
Borethyl, 4420
Borfax, 4421
Boria, *See* Boric anhydride, 4423
Boric Acid, 4422, *See* Ant Flip, 2497, Homberg's salt, 14285, sal sedativus, 27254
boric acid, *See* zinc borate, 34769
boric acid (HBO_2), sodium salt, *See* sodium metaborate, 28784
Boric acid, anhydride, 4423, *See* Boric anhydride, 4423
Boric acid, disodium salt, tetrahydrate, *See* Polybor®, 24369
boric acid, zinc salt, *See* zinc borate, 34769
Boric oxide, 4424, 4425, *See* Boric anhydride, 4423
Borite, 4426
Borium, 4427
***endo*-2-bornanol**, *See* DL-Borneol, 4428
Borneo camphor, *See* DL-Borneol, 4428
Borneo rubber, *See* gutta-susu, 13511
Bor-Nitrophoska® 13-13-21+0.1B, 4429
bornyl alcohol, *See* DL-Borneol, 4428
Borocil, *See* Tim-bor Professional, 31799
Borocil A, 4430
Borocil Extra, 4431
Borocil K, 4432
Borofax, *See* Boric Acid, 4422
Boroflow, 4433
Boroflow A, 4434
Boroflow A/ATA, 4435
Boroflow S/ATA, 4436, 4437
Boroflux, 4438
Borogard® ZB, 4439
boroglyceride, 4440
boroglycerine, *See* boroglyceride, 4440
Borolon, 4441
Boron, 4442
boron bromide, *See* Boron tribromide, 4444
boron carbide, *See* Tetrabor®, 31357
boron chloride, *See* Boron trichloride, 4445
boron fluoride, *See* Boron trifluoride, 4446
boron nitride, 4443
Boron oxide, *See* Boric anhydride, 4423
boron oxide (B_2O_3), *See* Boric anhydride, 4423
boron sesquioxide, *See* Boric anhydride, 4423
boron tribromide, 4444, *See* Trona Boron Tribromide, 32452
boron trichloride, 4445, *See* Trona Boron Trichloride, 32453
boron trifluoride, 4446
boron trioxide, *See* Boric anhydride, 4423
boron zinc oxide hydrate, *See* ZB2335, 34697
borosalicylic acid, 4447
Boroxo, *See* Firebrake ZB, 11825
Borrebond, 4448, *See* calcium lignosulfonate, 4956, 4957
Borrechel, 4449
Borresperse CA/CAF, 4450, *See* calcium lignosulfonate, 4956
Borrewell C, 4451
Borrewell FC, 4452
Borrewell FE, 4453
Bortin 45, 4454
bortran, *See* Fumite Dicloran, 12459
Bortrysan, *See* Anilazine, 2465
Borvicote, 4455
Bos MH, 4456
Bosan Supra, *See* DDT, 7820
Boselon, 4457
Bostik, 4458
Bostik 1100FS, 4459
Botran, *See* dicloran, 8395, Fumite Dicloran, 12459
Botran 75W, *See* dicloran, 8395, Fumite Dicloran, 12459
Botrilex, 4460
Botrilex, *See* quintozene, 25692
Botryogen, *See* red vitriol, 25971
Bouchardt's reagent, 4461
Bounty, *See* paclobutrazol, 22540, 225411
bourbonal, *See* Ethavan, 10937, ethyl vanillin, 11087, Rhodiarome, 26678, Vanbeenol, 33294
Bourbonne's metal, 4462
Bourbouze aluminum solder, 4463
Bourbouze solder, 4464
Bovanide, *See* Ranide, 25868
Bovidermol, *See* DDT, 7820
Bovinal-20, 4465
Bovine catalase, *See* Catalase L, 5429
Bovinox, 4466
Bovizole, *See* Tecto, Tecto 60%, 30913, thiabendazole, 31661
Bowhill's stain, 4467
Bow-wire brass, *See* brass, 4485
Bo-Xan, *See* xanthophyll, 34591
Boxite, 4468
Boxolon, 4469
Boygon, *See* Baygon®, 3790
Bozefloc, 4470
Bozetol, 4471
Bozzle, 4472
BP LDPE, 4473
BP Mycocide, 4474
BP polystrene, 4475
BPA polycarbonates, *See* Lexan®, 17102
BPO, *See* Benzoyl peroxide, 3983
BR, *See* polybutadiene, 24372
BR 55N, *See* Decabromodiphenyl oxide, 7837; Octoguard FR-01, 21999; Saytex® 102E, 27514
BR Destral, 4476, *See* Solubor DF, 28991
Brabant PCNB, 4477
Bradophen, 4478
Bradsil, 4479
Bradsyn, 4480
Braemer's reagent, 4481
Brakol, 4482
Branched dodecylbenzene sulfonic acid, *See* Nansa® SBA, 20692
Branched polyether polyol, *See* Baygal® K30, 3782
Branched sodium alkylbenzene sulfonate, *See* Nansa® SB30, 20691
Branched TEA-dodecylbenzene sulfonate, *See* Witconate LXH, 34486
brasilazina oil orange, *See* Sudan I, 29938
Brasivol, 4483, *See* aluminum oxide, 1919
Brasoran 50 WP, 4484
brass, 4485
brass, iron, 4488
brass, iserlohn, 4486
brass, leaded, 4487
Brassica campestris oil, *See* rapeSeed oil, 25874
Brassicol, *See* quintozene, 25692
Bras-sicol, 4489
Brassix, *See* Trifluralin, 32252
Brassix, Digermin, *See* Treflan, 32162
Bravo, *See* Chlorothalonil, 6150, Repulse, 26151
Bravo 500, 4490
Bravocarb, 4491
Bravo-w-75, *See* Repulse, 26151
Braxo, *See* Fertibor, 11706
Brazil wax, *See* carnauba, 5296, Carnauba wax, 5297
Brazil wood, *See* Fustic, 12522, redwoods, 25987
Brazilian Elemi, 4492
brazing solder, 4493
brazing solder, *See* nickel silvers, 21132
Breaker F, 4494
Breaxit, 4495
Brebent, 4496, *See* Bentonite, 3954
Brebond, 4497
Breecht's double salt, 4498
Breedervac 1 Plus, II Plus, III Plus, IV plus, 4499
Breedervac-Reo-Plus, RN Plus, 4500
Bregel, 4501, *See* Bentonite, 3954
Brek, *See* chloridazon, 6103
brellin, *See* Regulex, 26019
Bremen blue, *See* cupric carbonate, 7378
Bremen green, *See* cupric carbonate, 7378
Breokinase, *See* urokinase, 33145
Breon, 4502, *See* Korogel, 16354, Koroplate, 16357, Vestolit®, 33679
Breon 107, *See* Vestolit®, 33679
Breon 111EP, *See* Vestolit®, 33679
Breon 112EP, *See* Vestolit®, 33679
Breon 113, *See* Vestolit®, 33679
Breon 121, *See* Vestolit®, 33679
Breon 125/10, *See* Vestolit®, 33679
Breon 130/1, *See* Vestolit®, 33679
Breon 151, *See* Vestolit®, 33679
Breon 4001, *See* Vestolit®, 33679
Breon 4121, *See* Vestolit®, 33679
Breox, 4503
Breslle wood, *See* redwoods, 25987
Bresin 2, 2E, 4504
Brestan, 4505
Bretol, *See* Sumquat® 6020, 30121
Bretol®, 4506
bretonite, 4507
Brevinyl, *See* dichlorvos, 8394
Brevinyl E50, *See* dichlorvos, 8394
Breviol, 4508
Brifur, *See* Propoxur, 25198, 25199, 25200
Brightray Alloy B, 4509
Brightray Alloy C, 4510
Brightray Alloy F, 4511
Brightray Alloy H, 4512
Brightray Alloy S, 4513
Brij 35, *See* Akyporox RLM 160, 1290

Plus, 4529
Bromoxynil-chlorsulfuron-ioxynil, *See* Glean® TP, 12983
Bromoxynil-ioxynil-isoproturon, *See* Astrol, 3260
bromoxynil-oxynil, *See* Deloxil, 8030
Bromsalans, *See* Tuasol 100, 32543
Brom-tetragnost, *See* Bromeikon, 4568
Bronchodiagnostin, *See* Propyliodone, 25225, 25226
Bronchoselectan, *See* acetrizoate sodium, 307
Bronco, 4591, *See* glyphosate, 13148
bronidiol, *See* Bronopol-Boots, 4596
Bronidox L, 4592
Bronkhodiagnostin, *See* Propyliodone, 25225, 25226
Bron-Newcavac-M, 4593
Bronocot, 4594, *See* Bronopol-Boots, 4596
Bronopol, 4595, *See* Bronopol-Boots, 4596, Myacide® AS Plus, 20544, Myacide® S-1, S-2, 20545
Bronopol-Boots, 4596
Bronosol, *See* Bronopol, 4595, Bronopol-Boots, 4596, Broponol, 4616
Bronotabs, 4597
Bronox, 4598
Bronze, 4599
Bronze A, 4600
bronze acetate, 4601
bronze bearing metals, 4602
bronze powder, *See* copper, 6777
bronze wire, 4603
bronze, Durbar, *See* bronze, 4599
bronze, Eclipse, *See* bronze, 4599
bronze, Olympic, *See* bronze, 4599
bronze, Phono, *See* bronze, 4599
bronze, Vulcan, *See* bronze, 4599
bronzing liquids, 4604
bronzing powder, *See* mosaic gold, 20355
bronzing solder, 4605
bronzite, 4606
brookite, *See* titanium dioxide, 31867
Brookosome® EFA, 4607
Brookosome® EPO, 4608
Brookosome® Fucus, 4609
Brookosome® TRF, 4610
Brookswax D, 4611
Brookswax P, 4612
Brophos 5C10, 4613
Brophos OL-3, 4614, *See* oleth-phosphate Series (2-20), 22127
Brophos OL-3N, 4615
Broponol, 4616
Brosombra, *See* Propyliodone, 25225, 25226
Brown 208, 4617
brown acetate of lime, *See* calcium acetate, 4937
brown acetate of lime, gray acetate of lime, *See* calcium acetate, 4936
Brown barberry gum, *See* Morocco gum, 20339
brown coal, *See* Lignite, 17231
Brown Copp, 4618
Brown lead ore, 4619
Brown manganese ore, *See* manganite, 18555
Brown ore, 4620
Brown oxide of tungsten, 4621
Brown precipitate, 4622
Brozgerite, *See* Clevite,
Brugre powder, 4623
Brunner's salt, 4624
Brunol, 4625
Brunswick black, 4626
Brunswick blue, 4627
Brush Buster, 4628
Brush killer 64, *See* Planotox, 23914
Brush wire, 4629
Brush-B-Gon, 4630
Brushcrete, 4631
Brussels System, 4632
BSA, *See* Bistrimethylsilyl acetamide, 4286, CB2500, 5479, Dynasylan® BSA, 9399
BSWL 202, 4634, *See* lead silicate, 16915
BT, *See* tetrazolium chloride, 31395
BTC, *See* Variquat® 50MC, 33397, Zephiran Chloride, 34728
BTC® 50NF, 65NF, 835, 8358, *See* Benzalkonium chloride, 3958
BTC® 99, 4633
BTC® 818, 4635
BTC® 824, 4636
BTC® 824, 2565, *See* Exameen 824 3724, 11284
BTC® 885 P40, 4637
BTC 1010, *See* Radiaquat 6410, 6412, 25789, Radiaquat 6412, 25790, Rewoquat B 10, 26389
BTC® 1010-80, 4638
BTC® 2565, 4639
BTEAC, *See* Benzyltriethyl ammonium chloride, 3993
BTG alloy, 4640
BTS 27419, *See* Taktic, 30729
BTS 40542, *See* prochloraz, 25057
BTS-7693, *See* benazolin, 3935
Bubber Shet, 4641
Bubble Breaker® 3056A, 4642
Bubble Breaker® 748, 4643
Bubble Breaker® DMD-1, 4644
Bubblefil, 4645
Buca, 4646
Bucarpolate, 4647
Buckland's cement, 4648
Buckroid, 4649
BUCS, *See* 2-butoxyethanol, 4753
Buctril, 4650, *See* Bromoxynil, 4590
Budale, 4651
Budene® 1207, 4652
Budene® 1254, 4653
Budex 5130, *See* Fostex AMP, 12337
Bueno, *See* MSMA, 20417, Neoarsycodyl, 20871
Bueno 6, *See* Neoarsycodyl, 20871
Bufa, 4654
bufexamac, 4655, *See* Parfenac, 22830
Bufexamic acid, *See* bufexamac, 4655
Bufferight, 4656
Bug Check® BF, 4657
Bug Check® SRB, 4658
Bug Gun!, 4659
Bu-Gas, 4660, *See* Butane, 4734
Bug-Geta, 4661
Bulbold, *See* Kemstrene® 96.0%, 15991
Bull metal, 4662
Bulldock, 4663
Bullet Brass, *See* brass, 4485
Bullion®, 4664
Bullseye, 4665, 4666
Bullseye CDA, 4667
Bumal, 4668
Bumetran, *See* Taktic, 30729
Bumetrizole, *See* Uvazol 236, 33181
Bumper, 4669
Bumyr, 4670, *See* Butyl myristate, 4775, Wickenol® 141, 34384
Buna Hüls AP, 4671
Buna Hüls EM, 4672
Buna S, 4673
Buna SL, 4674
Buna SS, 4675
Buna VI, 4676
Buna® CB, 4677
Buna® CB 11, 4678
Buna® CB 22, 4679
bunamidine hydrochloride, 4680, *See* Scolaban, 27745
Bunatex, 4681
Bunge Biscuit Flakes, *See* hydrogenated soybean oil, 14596
bunsenine, *See* white tellurium, 34374
bunsenite, *See* nickel oxide, 21130
Bunt-cure, *See* Julin's chloride, 15632
Bunte's salt, 4682
Bunt-no-more, *See* Julin's chloride, 15632
Bur-A-Loy® 3873, 4683
Bur-A-Loy® 3874, 4684
Bur-A-Loy® 5915, 4685
Bur-A-Loy® 7130, 4686
Burco Anionic APS, 4687
Burco CS-LF, 4688
Burco DFE-45, 4689
Burco LAF-6, 4690
Burco NCS-80, 4691
Burco NPS-225, 4692
Burco TME, 4693
Burcofac 1060, 4694
Burcol BP-181, 4695
Burcop, 4696
Burcosperse AP Liq, 4697
Burcoterge DG-40, 4698
Burcotreat 900-A, 4699
Burcowet TM-LF, 4700
Burez, 4701
Burgess 30-P, 4702
Burgess 2211, 4703
Burgess KE, 4704
Burgess solder, 4705
Burgess-Hambuechen solution, 4706
Burgundy Lake, 4707
Burkeite, 4708
Burma black varnish, *See* thitsi, 31729
Burmite, 4709
Burmol®, 4710
burnt alum, *See* aluminum potassium sulfate, 1923
burnt carmine, 4711
burnt hypo, 4712
burnt iron, 4713
burnt lime, *See* KM Pebble Lime, 16240, lime, 17265
burnt magnesia, 4714
burnt nickel, 4715
burnt pyrites, 4716
burnt topaz, 4717
burnt umber, 4718
Buro-sol Concentrate, 4719, *See* aluminum acetate solution, 1898
Burow's solution, 4720, *See* novacetoform, 21648
Burr Brass, *See* brass, 4485
Burrow's Solution, *See* aluminum acetate solution, 1898
Bursoline, 4721
Burtolin, 4722, *See* maleic hydrazide, 18477
Burtonite V-7-E, *See* Prinza® Range, 25010
Busan 85, *See* Aquatreat KM, 2760
Busan 110, *See* Amerstat® 282, 2043, N-948® Biocide, 20641
Busan 1020, *See* metam-sodium, 19444
Busan® 1506, *See* Bioban® GK, 4158
Bush metal, 4723
Bushwacker, 4724
but-2-yne-1,4-diol, 4725, *See* but-2-yne-1,4-diol, 4725
Butac®, 4726
Butachlor, 4727, *See* Butanex, 4736, Machete, 18124
Butacide, *See* piperonyl butoxide, 23875
Butacite®, 4728
Butaclor, 4729
Butadiene, *See* Acralen® BS, 377
1,3-butadiene, 2-methyl-, homopolymer, *See* Natsyn® 2200, 20763
Butadiene rubber, *See* polybutadiene, 24372
butadiene/styrene copolymer, 4730
Butadiene-acrylonitrile-N-(4-anilinophenyl) methacrylamide terpolymer, *See* Chemigum® HR662, 5974
butadione, *See* Diacetyl, 8250
2,3-butadione, *See* Diacetyl, 8250
Butakon A2554, 4731
Butakon ML 57711, 4732
Butal, *See* Butyraldehyde, 4801
butaldehyde, *See* Butyraldehyde, 4801
butalyde, *See* Butyraldehyde, 4801
butam, *See* tebutam, 30879
butamisole hydrochloride, 4733, *See* Styquin, 29859
n-butanal, *See* Butyraldehyde, 4801
1-Butanamine, *See* Amine C4, 2152
n-butanenitrile, *See* nitrile C_4, 21338
1-butanol, *See* Tebol 88, 99, 30878
n-butanol, *See* Tebol 88, 99, 30878
***t*-butanol-diacetone alcohol, *See* Glassclad® 18, 12970**

1,2-butanolide, *See* Butyrolactone, 4803
1,2,3,4-butane tetracarboxylate, *See* Mixxim® HALS 63, 19992
1,4-butanediamine, *See* putrescine, 25423
1,4-butanolide, *See* Butyrolactone, 4803
2-butanone peroxide, 4738, *See* 2-butanone peroxide, 4738, methyl ethyl ketone peroxide, 19531
2-butanone oxime, *See* methyl ethyl ketoxime, 19532
2-Butanone, 4737, *See* methyl ethyl ketone, 19530
2,3-butanedione, *See* Diacetyl, 8250
2,3-Butanedionedioxime, *See* dimethylglyoxime, 8523
2,3-butanolone, *See* acetyl methyl carbinol, 312
butan-2-one, *See* methyl ethyl ketone, 19530
Butanal, *See* Butyraldehyde, 4801
n-Butane, *See* A-17, 7, Bu-Gas, 4660, Butane, 4734
butane dioxime, *See* dimethylglyoxime, 8523
butane-1,4-diol, *See* 1,4-butanediol, 4735, Dabco® BDO, 7647
butane-2,3-dione, *See* Diacetyl, 8250
butanedioic acid mono(2,2-dimethylhydrazide), *See* daminozide, 7723
butanedioic anhydride, *See* succinic anhydride, 29916
Butanedioic acid, *See* succinic acid, 29915
Butanedioic acid, 2,3-dihydroxy-, [R-(R*,R*)]-, disodium salt, *See* sodium tartrate, 28825
Butanedioic acid, sulfo-, 1,4-dihexyl ester, sodium salt, *See* Octosol HA-80, 22025
Butanedioic acid, bis(2-ethylhexyl) ester, *See* Wickenol® 159, 34394
butanediol, *See* 1,4-butanediol, 4735, Dabco® BDO, 7647
Butanediol diglycidyl ether, *See* Heloxy® 67, 13774
Butanedione, *See* Diacetyl, 8250
Butanenitrile, 4-(trichlorosilyl)-, *See* CC3555, 5495
Butanenitrile, *See* butyronitrile, 4805
***tert*-butanethiol**, *See t*-butyl mercaptan, 15434
Butanex, 4736, *See* Butachlor, 4727
butanimide, *See* succinimide, 29917
butanol-iso, *See* isobutanol, 15344
butanone oxime, *See* methyl ethyl ketoxime, 19532
Butanone, *See* methyl ethyl ketone, 19530
Butanox, 4739, *See* Esperfoam® FR, 10842
Butarez, 4740
Butasan, *See* Butyl Zimate®, 4783, Octocure ZDB-50, 21995, Perkacit® ZDBC, 23295
Butasan Vulcanization Accelerator, 4741
Butazate, 4742, *See* Butyl Zimate®, 4783, Octocure ZDB-50, 21995, Perkacit® ZDBC, 23295
Butazate 50D, 4743
Butazin, *See* Butyl Zimate®, 4783, Octocure ZDB-50, 21995, Perkacit® ZDBC, 23295
1-butene, homopolymer, *See* polybutene, 24373
2-Butenedioic acid (Z)-, dioctyl ester, *See* PX-538, 25444
***cis*-butenedioic acid**, *See* maleic acid, 18475
2-Butenedioic acid (E)-, dibutyl ester, *See* dibutyl fumarate, 8364
(E)-2-Butenedioic acid, *See* Fumaric acid, 12453
(Z)-Butenedioic acid, *See* maleic acid, 18475
2-Butenedioic acid (Z)-, dioctyl ester, *See* Octomer DOM, 22011
2-Butenedioic acid (Z)-, dibutyl ester, *See* Bisomer DBM, 4279, dibutyl maleate, 8365, Octomer DBM, 22008, PX-504, 25443
2-butenedioic acid, dibutyl ester, *See* dibutyl maleate, 8365
***cis*-butenedioic anhydride**, *See* maleic anhydride, 18476
2-butenedioic acid, dibutyl ester, *See* dibutyl maleate, 8365
***trans*-2-butenoic acid**, *See* Crotonic Acid, 7260
2-butenoic acid, *See* crotonic acid, 7260
Buteth-2 carboxylic acid, *See* Akypo LF 5, 1222, Akypo MB 2528S, 1225
Buteth-2 carboxylic acid and capryleth-9 carboxylic acid, *See* Akypo LF 6, 1223
Butex, 4744, 4745, *See* Exxon® Butyl 077, 11355
Butirex, *See* 2,4-DB, 7810
Butisan, 4746
Butisan S, *See* metazachlor, 19465, Track, 32129
Butisan® S, 4747
Butoben, *See* Butylparaben, 4797, Nipabutyl, 21249
butocide, *See* piperonyl butoxide, 23875
Butofan®, 4748
Butofan® D, 4749
Butoflin, *See* Thripstick®, 31766
Butonal®, 4750
Butormone, *See* 2,4-DB, 7810
Butox, 4751, *See* Thripstick®, 31766
2-butoxime, *See* methyl ethyl ketoxime, 19532
Butoxone, 4752, *See* 2,4-DB, 7810
butoxycarbonyl methyl butyl phthalate, *See* Reomol 4PG, 26126
butoxydiethylene glycol, *See* Butyl Carbitol®, 4766
Butoxydiglycol, *See* Ektasolve® DB, 9662, Butyl Carbitol®, 4766
1-butoxy-2,3-epoxypropane, *See* Ageflex BGE, 967
4-Butoxy-4'-(dimethylamino)thiocarbanilide, *See* thiambutosine, 31663
Butoxy(ethoxy)ethyl ester of piperonylic acid, *See* Bucarpolate, 4647
Butoxyethanol, *See* Butyl Cellosolve®, 4767, Ektasolve® EB, 9668
2-butoxyethanol, *4753*, *See* Butyl Cellosolve®, 4767, Ektasolve® EB, 9668
2-n-Butoxyethanol, *See* 2-butoxyethanol, 4753
(Butoxyethoxy)ethanol, *See* Butyl Carbitol®, 4766
Butoxyethanol acetate, *See* diethylene glycol butyl ether acetate, 8443, Ektasolve® EB Acetate, 9669
2-butoxyethanol acetate, 4754, *See* 2-butoxyethanol acetate, 4754
Butoxyethanol ester of 2,4-D, *See* Planotox, 23914
2-butoxy-ethanol phosphate (3:1), *See* Amgard TBEP, 2055, tributoxyethyl phosphate, 32206
2-(2-Butoxyethoxy)ethanol, *See* Butyl Carbitol®, 4766, Butyl Di-icinol, 4769, Ektasolve® DB, 9662
2-(2-n-Butoxyethoxy)ethanol, *See* Butyl Carbitol®, 4766
2-(2-Butoxyethoxy)ethyl acetate, *See* diethylene glycol butyl ether acetate, 8443
Butoxy ethyl acetate, *See* Sensolve BEA, 27953
2-n-Butoxyethyl 4-(dimethylamino) benzoate, *See* Speedcure BEDB, 29259
Butoxyethyl oleate, *See* Plasthall® 325, 23987
Butoxyethyl stearate, 4755
2-Butoxyethyl stearate, *See* Butoxyethyl stearate, 4755
4-butoxy-N-hydroxybenzeneacetamide, *See* bufexamac, 4655
N-(butoxymethyl)-2-chloro-2',6'-diethylacetanilide, *See* Butachlor, 4727, Butanex, 4736
N-(Butoxymethyl)acrylamide, *See* Cylink NBMA, 7573
Butoxymethyl oxirane, *See* Ageflex BGE, 967, Ageflex TBGE, 998
(butoxymethyl)oxirane, *See* Ageflex BGE, 967
(*tert*-butoxymethyl)oxirane, *See* Ageflex TBGE, 998
N-(4-butoxyphenyl)-N'-[4-(dimethylamino) phenyl]urea, *See* thiambutosine, 31663
Butoxyne 497, 4756
Butoxynol-5 carboxylic acid, *See* Akypo TBP 40, 1246
Butoxynol-19 carboxylic acid, *See* Akypo TBP 180, 1245
***para*-Butoxyphenylacetohydroxamic acid**, *See* bufexamac, 4655
Butter color, *See* annatto, 2478
Butter of arsenic, *See* arsenic trichloride, 3141
Butter of paraffin, 4757
Butter of sulfur, 4758
butter of zinc, *See* zinc chloride, 34772
Butter yellow, 4759, *See* Butter yellow, 4759
buttercup yellow, *See* zinc chromate, 34773, zinc yellow, 34800
Button metal, 4760
Button solder, 4761
Butvar, 4762
***tert*-Butyl Acetate**, 4785
butyl acetone, *See* methyl n-amyl ketone, 19552
Butyl acetoxystearate, *See* Paricin® 6, 22840
Butyl acetyl ricinoleate, *See* B.A.R, 3561, Flexricin® P-6, 11960
n-butyl acid phosphate, *See* Syn-O-Ad® P-417, 30568
Butyl acrylate, *See* n-Butyl Acrylate, 4786
n-Butyl Acrylate, 4786
n-butyl alcohol, *See* Tebol 88, 99, 30878
butyl alcohol, *See* Tebol 88, 99, 30878
i-butyl alcohol, *See* isobutanol, 15344
Butyl Aldehyde, *See* Butyraldehyde, 4801
n-butyl aldehyde, *See* Butyraldehyde, 4801
Butyl amine, *See* Amine C4, 2152
n-butylamine, *See* Amine C4, 2152
***sec*-butylamine**, *See* CSC 2-aminobutane, 7308
[1-[(Butylamino)carbonyl]-1H-benzimidazol-2-yl]carbamic acid methyl ester, *See* Benomyl, 3944
2-*t*-butylamino-4-chloro-6-ethylamino-1,3,5-triazine, *See* Terbuthylazine, 31227
***tertiary*-Butylaminoethyl methacrylate**, *See* Ageflex FM-4, 984
N-(*tert*-Butylamino)ethyl methacrylate, *See* Ageflex FM-4, 984
2-*tert*-butylamino-4-ethylamino-6-methylthio-1,3,5-triazine, *See* Terbutrex, 31229
Butyl ammonium caprylate, *See* Rewocoros BAC, 26304
N-butylated-*p*-aminophenol, *See* Tenamine 1, 31090
Butylated dimethylbenzyl phenol, *See* Wingstay® C, 34414
Butylated hydroxyanisole, 4787, *See* Anti-Oxydant Bayer, 2595, BHA, 4121, BHT, 4122, Oxyguard, 22457, Nipanox® BHT, 21273, Nipantiox, 21276 ,Spectratech® CM 11340, KM 11264, 29240, Tenamine 3, 31092, Vanox® PCX, 33346, Vulkanox® KB, 34152
Butylated triphenyl phosphate, *See* Syn-O-Ad® 8485, 30559
N,N-butyl benzene sulfonamide, *See* Plasthall® BSA, 23993
n-Butylbenzene sulfonamide, *See* Butylbenzene sulfonamide, 4789, Dellatol® BBS, 8026, Plastomoll® BMB, 24056
Butylbenzene sulfonamide, 4789
4-n-butylbenzene sulfonamide, *See* Dellatol® BBS, 8026
N-butyl-benzenesulfonamide, *See* Dellatol® BBS, 8026
N-n-butylbenzene sulfonamide, *See* Dellatol® BBS, 8026, Plastomoll® BMB, 24056
4-n-Butylbenzenesulfonamide n-Nonanoic Acid, *See* Plastomoll® BMB, 24056
Butyl benzoate, 4763, *See* n-Butyl benzoate, 4790, Hipochem B-3-M, 14106, Marvanol® Carrier BB, 18980
N-*t*-Butyl-*o*-benzothiazole-2-sulfenamide, *See* Butylbenzothiazole sulfenamide, 4791, Delac NS, 8001
Butylbenzothiazole sulfenamide, 4791
N-*t*-Butyl-2-Benzothiazolesulfenamide, *See* Delac NS, 8001
N-tert-Butyl-2-benzothiazolesulphenamide, *See* Vanax® NS, 33292
N-*t*-Butyl-2-benzothioazole sulfenamide, *See* BBTS, 3863
Butyl 2-benzothiazole sulfenamide, *See* Perkacit® TBBS, 23289
Butyl benzyl phthalate, 4764, *See* Santicizer 160, 27393
n-Butyl Benzyl Phthalate, *See* Butyl benzyl phthalate, 4764
Butyl bisbutyl peroxy valerate, 4765
n-Butyl-4,4-bis (*t*-butylperoxy) valerate, *See* Butyl bisbutyl peroxy valerate, 4765, Lupersol 230, 17855
n-Butyl 4,4-di(*t*-butylperoxy) valerate, *See* Trigonox® 17, 32264
n-butylcarbinol, *See* Amyl alcohol, 2381
Butyl Carbitol, 4766, *See* Butyl Carbitol®, 4766, Ektasolve® DB, 9662
Butyl Carbitol®, 4766
Butyl Carbitol® Acetate, *See* diethylene glycol butyl ether acetate, 8443
n-Butyl Carbitol, *See* Butyl Carbitol®, 4766

Butyl carbitol acetate, *See* diethylene glycol butyl ether acetate, 8443
Butyl Cellosolve, *See* 2-butoxyethanol, 4753, Butyl Icinol, 4771, Dowanol® EB, 8894, Ektasolve® EB, 9668
Butyl Cellosolve Acetate, *See* 2-butoxyethanol acetate, 4754
n-Butyl Cellosolve Acetate, *See* 2-butoxyethanol acetate, 4754
Butyl Cellosolve®, 4767
Butyl Cellosolve™, *See* Butyl Cellosolve®, 4767
Butyl Cellosolve™ Acetate, *See* 2-butoxyethanol acetate, 4754
Butyl Chemosept, *See* Butylparaben, 4797, Nipabutyl, 21249
2-*tert*-butyl-6-(5-chloro-2H-benzotriazol-2-yl)-4-methylphenol, *See* Uvazol 236, 33181
2-*tert*-butyl-6-(5-chloro-2H-benzotriazol-2-yl)-*p*-cresol, *See* Uvazol 236, 33181
α-butyl-α-(4-chlorophenyl)-1H-1,2,4-triazole-1-propanenitrile, *See* myclobutanil, 20549
***p*-Butylcresol**, 4792
di-*tert*-butyl-*p*-cresol;, *See* Oxyguard, 22457
2-*t*-Butyl-*m*-cresol, *See* Lowinox® MBMC, 17668
2-*t*-Butyl-*p*-cresol, *See* *p*-Butylcresol, 4792, Lowinox® MBPC, 17669
butyl-5-chloro-6-methyluracil, *See* terbacil, 31220
3-*tert*-butyl-5-chloro-6-methyluracil, *See* terbacil, 31220
Butyl cumyl peroxide, 4768
***t*-butyl cumyl peroxide**, *See* Trigonox® T, 32308
2-*t*-Butyl cyclohexyl acetate, *See* Ortholate, 22332
Butyl 2-cyanoacrylate, *See* Enbucrilate, 10537
N-butyl-N'-(3,4-dichlorophenyl)-N-methylurea, *See* neburon, 20841
1-Butyl-3-(3,4-dichlorophenyl)-1-methylurea, *See* neburon, 20841
Butyl-3-(3,4-dichlorophenyl)-1-methylurea, *See* neburon, 20841
Butyl-N'-(3,4-dichlorophenyl)-N-methylurea, *See* neburon, 20841
N-Butyl-N'-(3,4-dichlorophenyl)-N-methylurea, *See* Kloben®, 16225
Butyldiglycol acetate, *See* diethylene glycol butyl ether acetate, 8443
6-butyl-1,4-dihydro-4-oxo-7-(phenylmethoxy)-3-quinolinecarboxylic acid methyl ester, *See* methyl benzoquate, 19516
4-tert-Butyl-2,6-dimethyl-3,5-dinitroacetophenone, *See* acetyl-dinitro-butyl-xylene, 317
tert-Butyldimethyldinitroactophenone 4-tert-Butyl-2,6-dimethyl-3,5-dinitroactophenone, *See* ketone musk, 16141
4'-tert-Butyl-2',6'-dimethyl-3',5'-dinitroacetophenone, *See* acetyl-dinitro-butyl-xylene, 317
5-butyl-2-(dimethylamino)-6-methyl-4(1H)pyrimidinone, *See* dimethirimol, 8495
***t*-Butyldimethylchloro silane**, *See* CB2790, 5481
2-butyl-4,6-dimethyldihydropyran, *See* Gyrane, 13520
4-t-Butyl-2,6-dimethyl-3,5-dinitroacetophenone, *See* acetyl-dinitro-butyl-xylene, 317
***t*-Butyldiphenylchlorosilane**, *See* CB2805, 5482
O-butyl diethylene glycol, *See* Butyl Carbitol®, 4766
butyl digol, *See* Butyl Carbitol®, 4766, Ektasolve® DB, 9662
butyl diicinol, *See* Butyl Carbitol®, 4766, Ektasolve® DB, 9662
Butyl Di-icinol, 4769
n-Butyldimethylchlorosilane, *See* CB2785, 5480
5-*t*-butyl-1,3-dinitro-4-methoxy-2-methylbenzene, *See* musk ambrette, 20532
butyl dioxitol, *See* Butyl Carbitol®, 4766
Butyl dithiocarbamate, *See* Ancazate XX, 2418
Butyl Eight®, 4770
1,4-butylene bromide, *See* 1,4-dibromobutane, 8357
butylene glycol, *See* 1,4-butanediol, 4735
butylene glycol, *See* Dabco® BDO, 7647
1,4-Butylene glycol, *See* 1,4-butanediol, 4735, Dabco® BDO, 7647
1,3-Butylene glycol diacrylate, *See* SR-212, 29367
1,3-Butylene glycol dimethacrylate, *See* Ageflex 1,3 BGDMA, 964
butyl-2,3-epoxypropyl ether, *See* Ageflex BGE, 967
n-Butyl 2,3-Epoxypropyl Ether, *See* Ageflex BGE, 967
butylethylacetic acid, *See* 2-ethylhexoic acid, 11137
butylethylene, *See* Neodene® 6, 20886
butyl ether, *See* PPG butyl ether series, 24841
Butyl ethyl acetic acid, *See* 2-ethylhexoic acid, 11137
N-butyl-N-ethyl-2,6-dinitro-4-trifluoromethylaniline, *See* Benfluralin, 3940
Butyl ethylene, *See* 1-hexene, 14029
Butyl glycidyl ether, *See* Ageflex BGE, 967, Epirez 501, 10654, Heloxy® 61, 13770
O,O-*t*-butyl O-(2-ethylhexyl) monoperoxycarbonate, *See* Lupersol TBEC, 17878
***tert*-Butyl Glycidyl Ether**, *See* Ageflex TBGE, 998
Butyl glycol acetate, *See* 2-butoxyethanol acetate, 4754
n-butyl glycol phthalate, *See* Dibutoxyethyl phthalate, 8363
butyl hexadecanoate, *See* Wickenol® 141, 34384
Butyl hydroperoxide, *See* Aztec® *t*-Butyl Hydroperoxide-70, Aq, 3541
***t*-Butyl hydroperoxide**, *See* Aztec® *t*-Butyl Hydroperoxide-70, Aq, 3541, TBHP-70, 30853, Trigonox® A-80, 32291, USP® -800, 33170
***tertiary*-Butyl hydroperoxide**, *See* Aztec® *t*-Butyl Hydroperoxide-70, Aq, 3541
***t*-Butylhydroquinone**, 4793
2-tert-Butylhydroquinone, *See* Sustane® TBHQ, 30437
***tert*-Butylhydroquinone**, *See* Eastman® MTBHQ, 9471
2-*tert*-Butylhydroquinone, *See* Eastman® MTBHQ, 9471
2(3)-tert-butyl-4-hydroxyanisole, *See* Sustane® 1-F, 30432
2-*tert*-butyl-4-hydroxyanisole, *See* Nipantiox, 21276
Butyl Hydroxyanisole, *See* Nipantiox, 21276, Sustane® 1-F, 30432
***tert*-butyl-4-hydroxyanisole**, *See* Sustane® 1-F, 30432
3-*tert*-butyl-4-hydroxyanisole, *See* BHA, 4121, Butylated hydroxyanisole, 4787
***tert*-Butyl *p*-hydroxyanisole**, *See* Sustane® 1-F, 30432
butyl *p*-hydroxybenzoate, *See* Butylparaben, 4797
n-butyl *p*-hydroxybenzoate, *See* Butylparaben, 4797
Butyl *p*-hydroxybenzoate, *See* butylparaben, 4797
2-(3,5 di-tert-butyl-4-hydroxybenzylidene) malononitrile, *See* Malonoben, 18492
5-*tert*-butyl-4-hydroxy-2-methylphenyl sulfide, *See* Ultranox® 236, 32846
butyl-4-hydroxybenzoate, *See* Nipabutyl, 21249
Butyl α-Hydroxypropionate, *See* Butyl lactate, 4773
Butyl 2-hydroxypropanoate, *See* Butyl lactate, 4773
butylhydroxytoluene, *See* Oxyguard, 22457
Butylideneaniline, *See* Grasselerator 508, 13290
4,4' -Butylidenebis(6-t-butyl-m-cresol), *See* Santowhite, 27431, Vanox® SWP, 33348
4,4'-Butylidene-bis(2-*t*-butyl-5-methylphenol, *See* Lowinox® 44B25, 17648
4,4-butylidenebis[2-(1,1-dimethylethyl)-5-methylphenol, *See* Santowhite, 27431
Butyl Icinol, 4771, *See* Butyl Cellosolve®, 4767
Butylite, 4794
Butyl Kamate®, 4772
Butyl lactate, 4773
n-butyl lactate, *See* Butyl lactate, 4773
***t*-butyl mercaptan**, 15434
Butyl methacrylate, 4774
n-Butyl Methacrylate, *See* Butyl methacrylate, 4774
butyl 2-methacrylate, *See* Butyl methacrylate, 4774
butyl methoxydibenzoylmethane, *See* Parsol® 1789, 22866
4-*tert*-butyl-4'-methoxy-dibenzoylmethane, *See* Parsol® 1789, 22866
***tert*-butyl-4-methoxyphenol**, *See* Sustane® 1-F, 30432
***p*-*t*-butyl-α-methyldihydrocinnamic aldehyde**, *See* Lilestralis, 17253
6-*t*-butyl-3-methyl-2,4-dinitroanisole, *See* musk ambrette, 20532
***tert*-butyl methyl ketone**, *See* pinacolone, 23848
2-*tert*-Butyl-4-methylphenol, *See* Cao, 5098
butyl 2-methyl-2-propenoate, *See* Butyl methacrylate, 4774
Butyl-2-Methyl-2-Propenate, *See* Butyl methacrylate, 4774
Butyl-2-propenoate, *See* n-Butyl Acrylate, 4786
Butyl myristate, 4775, *See* Bumyr, 4670, Wickenol® 141, 34384
Butyl Namate®, 4776
Butylnaphthalene sodium sulfonate, *See* Emkal BNS, 10227
Butylnaphthalene sulfonic acid, *See* Emkal BNS Acid, 10228
butyl norate, *See* Dabco® T-12, 7673
butyl octadecanoate, *See* butyl stearate, 4781, Witconol 2326, 34497
n-butyl octadecanoate, *See* butyl stearate, 4781
Butyl 9-octadecenoate, *See* butyl oleate, 4778
2-butyl-1-octanol, *See* 2-butyloctanol, 4795
2-butyloctanol, 4795
Butyl octyl phthalate, 4777, *See* PX-914, 25446
Butylol, 4796
Butyl oleate, 4778, *See* Butyl Oleate C-914, 4779, Emerest® 2328, 10089, Kemester® 4000, 15943, Plasthall® 503, 23989, Priolube 1405, 25014, Radia® 7040, 25717, Uniflex® BYO, 32940, Witcizer 100, 34453
n-butyl oleate, *See* Butyl Oleate C-914, 4779
Butyl Oleate C-914, 4779
Butyl Oleate C-914, *See* Butyl oleate, 4778
Butyl oxitol, *See* 2-butoxyethanol, 4753
Butylparaben, 4797, *See* Lexgard B, 17143, LiquaPar® Oil, 17451
n-butyl-paraben, *See* Nipabutyl, 21249
Butyl Parasept, *See* Butylparaben, 4797, Nipabutyl, 21249
***t*-butyl peracetate**, *See* Aztec® *t*-butyl peracetate-50 OMS, 60 OMS, 75 OMS, 3542, Trigonox® F-C50, 32300
***tertiary*-Butyl peracetate**, *See* Aztec® *t*-butyl peracetate-50 OMS, 60 OMS, 75 OMS, 3542
***tert*-butyl perbenzoate**, *See* Aztec® *t*-Butyl Perbenzoate, 3543, Polyvel CR-5T, 24659, Trigonox® 93, 32276, Trigonox® C, 32297
***t*-butyl perbenzoate**, 4798, *See* Polyvel CR-5T, 24659, Trigonox® C, 32297
***tert*-Butylperoxide**, *See* di-*t*-butyl peroxide, 8369
***tert*-Butylperoxide**, *See* Aztec® Di-*t*-Butyl Peroxoide, 3557
***t*-butyl peroxide bis(1,1-di-methylethyl)peroxide**, *See* di-*t*-butyl peroxide, 8369
***t*-butyl peroxyacetate**, *See* Aztec® *t*-butyl peracetate-50 OMS, 60 OMS, 75 OMS, 3542
***t*-butyl peroxy acetate**, *See* Esperox® 12MD, 10845
Butylperoxy crotonate, *See* Esperox® 13M, 10846
1,1-Di(tert-butylperoxy)cyclohexane, *See* USP® -400P, 33168
α,α-bis(t-butylperoxy) diisopropylbenzene, *See* Vul-Cup, Vul-Cup 40KE, Vul-Cup R, 34119
***t*-Butyl peroxy 2-ethyl hexanoate**, *See* Esperox® 28, 10847
Butylperoxy isobutyrate, *See* Aztec® *t*-Butyl Peroxyisobutyrate-75 OMS, 3546
***t*-Butyl peroxy maleate**, *See* Esperox® 41-25A, 10850
***t*-butyl peroxy 2-methylbenzoate**, *See* Esperox® 497M, 10851
Butylperoxy neodecanoate, *See* Aztec® *t*-Butyl Peroxyneodecanoate-50 OMS, 75 OMS, 3547
1,1-bis(*t*-butylperoxy)-3,3,5-trimethylcyclohexane, *See* Aztec®, 3539
Butyl peroctoate, *See* Aztec® *t*-Butyl Peroctoate, 3544
***t*-Butyl peroctoate**, *See* Trigonox® 21, 32265
***t*-butyl peroxybenzoate**, *See* *t*-butyl perbenzoate, 4798, Esperox® 10, 10844, Trigonox® 93, 32276
***tert*-butyl peroxybenzoate**, *See* Aztec® *t*-Butyl Perbenzoate, 3543
***t*-butyl peroxycrotonate**, *See* Esperox® 13M, 10846
***tert*-butyl peroxy-2-ethylhexanoate**, *See* Aztec® *t*-Butyl Peroctoate, 3544, Aztec® *t*-Butyl Peroctoate-50 OMS,

3545, Trigonox® 21, 32265
t-butyl peroxy-2-ethylhexyl carbonate, *See* Esperox® C-496, 10860
t-butyl peroxyisobutyrate, *See* Trigonox® 41-C75, 32271
t-butyl peroxy-2-methylbenzoate, *See* Trigonox® 97-C75, 32277
t-Butyl peroxyneodecanoate, *See* Aztec® t-Butyl Peroxyneodecanoate-50 OMS, 75 OMS, 3547, Esperox® 33M, 10849, Lupersol 10, 17842, Trigonox® 23, 32267
t-butyl peroxyneoheptanoate, *See* Esperox® 750M, 10857
t-Butyl peroxypivalate, *See* Aztec® t-Butyl Peroxypivalate-75 OMS, 3548
tert-butyl peroxypivalate, *See* Aztec® t-Butyl Peroxypivalate-75 OMS, 3548, Esperox® 31M, 10848, Trigonox® 25-C75, 32268
t-butyl peroxy-3,5,5-trimethyl hexanoate, *See* Trigonox® 42, 32272, Trigonox® 42 PR, 32273, Trigonox® 42S, 32274
Butylphen, *See* 4-t-butyl phenol, 4800, Terbutol, 31228
p-tert-butylphenol, *See* Terbutol, 31228
t-Butylphenol, *See* Lowinox® 070, 17650
p-t-butylphenol, *See* Terbutol, 31228
4-t-butyl phenol, 4800
Butyl phenol formaldehyde, *See* Albertol 142-R, 1369
tert-Butylphenyl diphenyl phosphate, *See* Syn-O-Ad® 8478, 30555
tert-Butylphenyl diphenyl phosphate, *See* Syn-O-Ad® 8485, 30559
p-tert-Butyl phenyl glycidyl ether, *See* Heloxy® 65, 13773
N,N'-di-sec-butyl-p-phenylenediamine, *See* Tenamine 2, 31091
butyl phenylmethyl 1,2-benzenedicarboxylate, *See* Butyl benzyl phthalate, 4764
Butyl phosphate, *See* tributyl phosphate, 32207
n-Butyl phosphate ester, *See* Marlophor® DS-Acid, 18892
Butyl phosphorotrithioate, *See* Def®, 7876
Butyl phthalate, *See* dibutyl phthalate, 8366
n-butyl phthalate, *See* Kodaflex® DBP, 16264
n-Butylphthalate, *See* dibutyl phthalate, 8366
Butyl phthalyl butyl glycolate, *See* Reomol 4PG, 26126
Butyl phthalyl butyl glycollate, *See* Reomol 4PG, 26126
n-Butyl phthalyl-n-butyl glycolate, *See* Morflex 190, 20319
Butyl ricinoleate, *See* Flexricin® P-3, 11958
butyl rubber, 4780, *See* Kalar® 5214, 15662
Butyl salicylate, *See* Brunol, 4625
Butyl sebacate, *See* Plasthall® DBS, 23994, Reomol DBS, 26129
n-butyl sebacate, *See* Plasthall® DBS, 23994, Reomol DBS, 26129
butyl stearate, 4781, *See* ADK STAB LS-8, 732, Emerest® 2325, 10088, Kemester® 5510, 15948, Kessco® BS, 16103, Lexolube® BS-Tech, 17163, Oleo-Coll LP, 22118, Radia® 7051, 25718, Uniflex® BYS-Tech, Unimate® BYS, 32941, Witconol 2326, 34497
n-butyl stearate, *See* Priolube 1451, 25020
Butyl Stearate C-895, *See* butyl stearate, 4781
Butyl Tegosept, *See* Nipabutyl, 21249
Butyl Tetradecanoate, *See* Wickenol® 141, 34384
Butyl N-tetradecanoate, *See* Wickenol® 141, 34384
n-butyl tetradecanoate, *See* Butyl myristate, 4775
S-t-butylthio-methyl-O,O-diethyl phosphorodithioate, *See* Terbufos, 31226
S-tert-butylthiomethyl O,O-diethyl phosphorodithioate, *See* Terbufos, 31226
Butyl thiotin, *See* Lankromark® BT050, 16687
Butyltin carboxylate, *See* Lankromark® BM271, 16686, Okstan M 62, 22098
Butyltin carboxylate, *See*
Butyltin mercaptide, *See* Okstan M 69 S, 22099
Butyl toluene, 4782
p-t-Butyl toluene, *See* Butyl toluene, 4782, Lowinox® PTBT, 17673
p-tert-butyl toluene, *See* Butyl toluene, 4782
butyl 2-[4-(5-trifluoromethyl-2-pyridyloxy)phenoxy]propionate, *See* Fluazifop-butyl, 12038
2-butyl-4,4,6-trimethyl-1,3-dioxane, *See* Herborane, 13807
Butyl Tuads®, *See* Akrochem® TBUT, 1178
Butyl zimate, *See* Octocure ZDB-50, 21995, Perkacit® ZDBC, 23295
Butyl Zimate®, 4783
Butylxanthate, zinc salt, *See* Vulcaid 27, 34073
Butylxanthogenic acid zinc salt, *See* Vulcaid 27, 34073
Butyl-m-xylene, 4784
t-Butyl-m-xylene, *See* Butyl-m-xylene, 4784, Lowinox® TBMX, 17675
5-tert-butyl-m-xylene, *See* Butyl-m-xylene, 4784
1,4-Butynediol, *See* but-2-yne-1,4-diol, 4725
2-butyne-1,4-diol, *See* but-2-yne-1,4-diol, 4725
Butynorate, *See* dibutyltin laurate, 8371
Butyrac, *See* 2,4-DB, 7810
Butyraldehyde, 4801
n-Butyraldehyde, *See* Butyraldehyde, 4801
Butyraldehyde-ammonia, *See* Vulcamel, 34088
Butyraldehyde-aniline, *See* Beutene, 4077, Vulcaid 111, 34078
Butyric acid nitrile, *See* Butyronitrile, 4805
butyric acid sodium salt, *See* sodium butyrate, 28742
Butyric aldehyde, *See* Butyraldehyde, 4801
butyro flavine, *See* Butter yellow, 4759
Butyroin, 4802
butyrolactam, *See* 2-pyrrolidone, 25536
γ-butyrolactam, *See* Soluphor® P, 29023
Butyrolactone, 4803 Agsol Ex BLO, 1090, Agrisynth BLO, 1074, FoamFlush™, 12154
γ-Butyrolactone, *See* Agrisynth BLO, 1074, BLO®, 4350, Butyrolactone, 4803, Dynasolve 699, 9389, GBL, 12704
Butyrone, 4804
Butyronitrile, 4805
n-Butyryl tri-n-hexyl citrate, *See* Citroflex B-6, 6346
N-butyryl-p-aminophenol, *See* Suconox-4®, 29923
Bu-White, 4806
BWF, 4807
BX 310, 4808
BXA, 4809
B-X-A, 4810
BXA Flake, 4811
BXT, 4812
BY-59-18, 4813
Byacin, 4814
Byakisol 30, 4815
Byatran, 4816, 4817
Byco A, C, O, 4818
Byco A,C,O, *See* gelatin, 12722
Bygran, *See* Tecgran, 30886, tecnazene, 30898
Bygran F, 4819, 4820
Bygran S, 4821
BYK® 020, 4822
BYK® 024, 4823
BYK® 045, 4824
BYK® 151, 4825
BYK® 156, 4826
BYK® 307, 4827
BYK® A500, 4828
BYK® -Catalyst 450, 4829
BYK® ES 80, 4830
BYK® P104, 4831
Bykanol® -N, 4832
Byketol® OK, 4833
Bykumen®, 4834
Bynel®, 4835
BZF-60, *See* Benzoyl peroxide, 3983
BZI, *See* Benzimidazole, 3966
BZT, *See* benzethonium chloride, 3963
C, *See* Bachite, 3575, Neospectra, 20969
C 70, *See* Hystrene® 1835, 14755
C 550, *See* Nacconol® 40G, 20647
C 1983, *See* chloroxuron, 6158, Tenoran, 31119
C 2242, *See* Talisman, 30735, Tolurex, 31958
C 3126, *See* Metobromuron, 19583
C 7019, *See* Brasoran 50 WP, 4484
C 8949, *See* Chlorfenvinphos, 6098
C isoparafin series, *See* Exxsol® D-40, D-60, D-80, D-110, D-130, 11359
C. I. 75470, *See* carminic acid, 5294
C. I. Basic Red 9.HCl, *See* pararosaniline, 22798
C. I. Natural Red 4, *See* carminic acid, 5294
C. I. Solvent Yellow 94, *See* Fluorescein, 12080
C.E. powders, 4836
C.I, *See* Kremser White, 16399
C.I. 1037, *See* oxyalizarin, 22438
C.I. 10305, *See* reflorit, 26003
C.I. 11020, *See* Butter yellow, 4759
C.I. 11270, *See* Chrysoidine, 6256
C.I. 12055, *See* Sudan I, 29938
C.I. 12140, *See* Sudan II, 29939
C.I. 19140, *See* tartrazine, 30835
C.I. 23850, *See* Niagara blue, 21091
C.I. 26100, *See* Sudan III, 29940
C.I. 26105, *See* Sudan IV, 29941
C.I. 37035, *See* p-nitroaniline, 21345
C.I. 40600, *See* Photine, 23756
C.I. 40850, *See* Canthaxanthin, 5095
C.I. 42500, *See* pararosaniline, 22798
C.I. 42555, *See* gentian violet, 12845
C.I. 42590, *See* methyl green, 19540
C.I. 42685, *See* acid fuchsine, 333
C.I. 44050, *See* Additin 30, 653
C.I. 44050, *See* Naugard® PANA, 20798
C.I. 44050, *See* Akrochem® Antioxidant PANA, 1161
C.I. 44050, *See* Antioxidant PAN, 2590
C.I. 45010, *See* pyronine B, 25522
C.I. 45170, *See* Rhodamine B, 26616
C.I. 45350, *See* Fluorescein disodium salt, 12081
C.I. 45350:1, *See* Fluorescein, 12080, uranine, 33115
C.I. 45440, *See* rose bengal, 26963
C.I. 58000, *See* Alizarin, 1621
C.I. 58050, *See* quinizarin, 25683
C.I. 58205, *See* oxyalizarin, 22438
C.I. 73000, *See* Indigo, 15074
C.I. 73015, *See* Indigo Carmine, 15075
C.I. 75140, *See* vegetable rouge, 33523
C.I. 75410, *See* oxyalizarin, 22438
C.I. 75500, *See* Juglone, 15631
C.I. 75530, *See* Alkanet, 1695
C.I. 75670, *See* quercetin, 25647
C.I. 76076, *See* Aminogen II, 2177
C.I. 76505, *See* resorcinol, 26230
C.I. 76515, *See* pyrogallol, 25512
C.I. 76515, *See* pyrogallic acid, 25510
C.I. 76550, *See* p-aminophenol, 2185
C.I. 76630, *See* 1,6-naphthalenediol, 20709
C.I. 77007, *See* ultramarine, 32821
C.I. 77103, *See* Barium chromate, 3638
C.I. 77231, *See* Compactrol®, 6670
C.I. 77357, *See* Fischer's salt, 11848
C.I. 77402, *See* cuprous oxide, 7405
C.I. 77410, *See* cupric acetoarsenite, 7374
C.I. 77410, *See* mountain green, 20369
C.I. 77491, *See* Jeweller's rouge, 15611
C.I. 77491, *See* Ferroxide, 11694
C.I. 77578, *See* red lead, 25960
C.I. 77597, *See* Kremser White, 16399
C.I. 77600, *See* Chrome yellow, 6212
C.I. 77800, *See* ruthenium red, 27148
C.I. 77891, *See* Kronos®, 16439
C.I. 77891, *See* Kronos®, 16440
C.I. 77891, *See* Kronos® 1000, 16441
C.I. 77891, *See* Kronos® 2020, 16442
C.I. 77891, *See* Kronos® 2073, 16443
C.I. 77891, *See* Kronos® 3020, 16444
C.I. 77891, *See* titanium dioxide, 31867
C.I. 77947, *See* zinc oxide, 34786
C.I. 77947, *See* Ken-Zinc®, 16050
C.I. 77947, *See* K-Zinc, 16527
C.I. 77947, *See* Activox, 507
C.I. 450095, *See* pyronine Y, 25523

C.I. Acid Blue 74, *See* Indigo Carmine, 15075
C.I. Acid Red 94, *See* rose bengal, 26963
C.I. Acid Violet 19, *See* acid fuchsine, 333
C.I. Acid Yellow 23, *See* tartrazine, 30835
C.I. Acid Yellow 73, *See* Fluorescein disodium salt, 12081
C.I. Azoic Diazo Component 114, *See* Aminogen I, 2176
C.I. Azoic Diazo Component 37, *See* *p*-nitroaniline, 21345
C.I. Basic Red 1, monohydrochloride, *See* Rhodamine G and G Extra, 26617
C.I. Basic Red 9, *See* pararosaniline, 22798
C.I. Basic Red 9 monohydrochloride, *See* pararosaniline, 22798
C.I. Direct Blue 14, *See* Niagara blue, 21091
C.I. Fluorescent Brightener 30, *See* Photine, 23756
C.I. Food Red15, *See* Rhodamine B, 26616
C.I. Food Yellow 4, *See* tartrazine, 30835
C.I. Mordant Red 11, *See* Alizarin, 1621
C.I. Natural Brown 7, *See* Juglone, 15631
C.I. Natural Red 1, *See* quercetin, 25647
C.I. Natural Red 8, *See* oxyalizarin, 22438
C.I. Natural Red 16, *See* oxyalizarin, 22438
C.I. Natural Red 20, *See* Alkanet, 1695
C.I. Natural Red 26, *See* vegetable rouge, 33523
C.I. Natural Yellow 10, *See* quercetin, 25647
C.I. Natural Yellow 10 & 13, *See* quercetin, 25647
C.I. Natural Yellow 11, *See* maclurin, 18235
C.I. Oxidation Base 32, *See* pyrogallol, 25512
C.I. Oxidation Base 32, *See* pyrogallic acid, 25510
C.I. Oxidation Base 6A, *See* *p*-aminophenol, 2185
C.I. Pigment Blue 29, *See* ultramarine, 32821
C.I. Pigment Blue 66, *See* Indigo, 15074
C.I. Pigment Green 21, *See* cupric acetoarsenite, 7374
C.I. Pigment Green 21, *See* mountain green, 20369
C.I. Pigment Red 104, *See* Molybdate Red, 20074
C.I. Pigment Red 105, *See* red lead, 25960
C.I. Pigment Red 83, *See* Alizarin, 1621
C.I. Pigment White 11, *See* Timonox, 31807
C.I. Pigment White 25, *See* Compactrol®, 6670
C.I. Pigment White 4, *See* zinc oxide, 34786
C.I. Pigment White 4, *See* Ken-Zinc®, 16050
C.I. Pigment White 4, *See* K-Zinc, 16527
C.I. Pigment White 5, *See* Lithopone, 17520
C.I. Pigment White 6, *See* Kronos®, 16439
C.I. Pigment White 6, *See* Kronos®, 16440
C.I. Pigment White 6, *See* Kronos® 1000, 16441
C.I. Pigment White 6, *See* Kronos® 2020, 16442
C.I. Pigment White 6, *See* Kronos® 2073, 16443
C.I. Pigment White 6, *See* Kronos® 3020, 16444
C.I. Pigment White 6, *See* titanium dioxide, 31867
C.I. Pigment Yellow 13, *See* Irgalite Yellow BGW, 15277
C.I. Pigment Yellow 31, *See* Barium chromate, 3638
C.I. Pigment Yellow 34, *See* Chrome yellow, 6212
C.I. Pigment Yellow 36, *See* zinc chromate, 34773
C.I. Pigment Yellow 40, *See* Fischer's salt, 11848
C.I. Solvent Orange 7, *See* Sudan II, 29939
C.I. Solvent Red 24, *See* Sudan IV, 29941
C.I. Solvent Yellow 2, *See* Butter yellow, 4759
C.I. Solvent Yellow 14, *See* Sudan I, 29938
C.I. Solvent Yellow 94, *See* uranine, 33115
C.I. Vat Blue 1, *See* Indigo, 15074
C.I.. Developer 17, *See* *p*-nitroaniline, 21345
C.P.D, 4837
C.P.R, 4838
C.T.A.B, *See* Acetoquat CTAB, 303, Cetavlex, Cetavlon, 5787, Cetrimonium bromide, 5813, Sumquat® 6030, 30122, Varisoft® CTB-40, 33421
C_5 alkane, *See* pentanen-pentane, 23164
C_5 hydrocarbon resin, *See* Wingtack® 10, 34417, Wingtack® Extra, 34418
C_6 linear alpha olefin, *See* 1-hexene, 14029
C_6-C_{10} aliphatic alcohol phthalate, *See* Bisoflex L79P, 4263
C_6-C_{10} trimellitates, *See* Bisoflex L911P, 4266
C_6-C_8 alpha olefin, *See* Neodene® 6/8, 20888
C_7-C_8 isoparaffin, *See* Isopar® C, 15388
C_8 alkyl acetate, *See* Exxate® 800, 11348
C_8 fatty alcohol phosphate ester, *See* Beycostat 256A, 4101
C_8, C_{10} alkyl polyglycoside, *See* Glucopon 225, 13049
C_8-C_{10} aliphatic phthalate, *See* Bisoflex 810, 4250
C_8-C_{10} triglycerides, *See* Delios®, 8025
C_8-C_{10} trimellitates, *See* Bisoflex 819, 4251
C_8-C_9 isoparaffin, *See* Isopar® E, 15389
C_9 alkyl acetate, *See* Exxate® 900, 11349
C_9 oxo-alcohol polyglycol ether carboxylic acid, *See* Marlowet® 4539, 18921
C_9-C_{11} alcohols, *See* Neodol® 91, 20917
C_9-C_{11} ethoxylate (2 EO), *See* Teric 9A2, 31254
C_9-C_{11} pareth-3 (2.5 EO), *See* Neodol® 91-2.5, 20918
C_9-C_{11} pareth-6, *See* Rhodasurf® 91-6, 26647
C_9-C_{13} alkyl benzene sulfonate, sodium salt, *See* Teepol CM44, 30929
C_9-C_{13} primary alcohol sulfate, sodium salt, *See* Teepol HB6, 30932
$C_{8\text{-}10}$ alkyl polysaccharide ether, *See* APG® 225 Glycoside, 2649
C_{10} alkyl acetate, *See* Exxate® 1000, 11350
C_{10}-C_{11} isoparaffin, *See* Isopar® G, 15390
C_{10}-C_{12} pareth-3, *See* Surfonic® L12-3, 30367
$C_{10}H_{21}O_4S.Na$, *See* sodium decyl sulfate, 28755
C_{11}-C_{13} isoparaffin, *See* Isopar® L, 15391
C_{11}-oxo alcohol ethoxylate (3 EO), *See* Marlipal® 011/30, 18859
C_{12} alcohol polyglycol ether, *See* Marlowet® BL, 18930
C_{12} linear primary alcohol, *See* lauryl alcohol, 16864
C_{12}-alkylbenzyldimethylammonium chloride, *See* Dehyquart LDB, 7986
C_{12}-C_{13} alcohols, *See* Neodol® 23, 20911
C_{12}-C_{13} alcohols, ethoxylated, *See* Neodol® 23-6.5, 20912
C_{12}-C_{14} alcohol polyglycol ether carboxylic acid, *See* Marlowet® 1072, 18918
C_{12}-C_{14} alkyl dimethylamine oxide, *See* Lilaminox M24, 17251
C_{12}-C_{14} pareth-2, *See* Surfonic® L24-2, 30368
C_{12}-C_{14} pareth-2.9, *See* Genapol® 24-L-3, 12781
C_{12}-C_{14} pareth-3, *See* Genapol® 42-L-3, 12783
C_{12}-C_{15} alcohols, *See* Neodol® 25, 20913
C_{12}-C_{15} pareth-12, *See* Mulsifan RT 203/80, 20457
C_{12}-C_{15} pareth-3, *See* Rhodasurf® LA-3, 26657
C_{12}-C_{15} pareth-7, *See* Rhodasurf® 25-7, 26646
C_{12}-C_{15} pareth-7 carboxylic acid, *See* Sandopan® DTC Linear P Acid, 27340
C_{12}-C_{16} alkyl polyglycoside, *See* Glucopon 600, 13051
C_{12}-C_{16} pareth-1, *See* Genapol® 26-L-1, 12782
C_{12}-C_{18} fatty acid triglyceride, *See* Softenol® 3100, 28847
$C_{12\text{-}14}$ coconut fatty alcohol, *See* Philcohol 1214, 23665
$C_{12\text{-}15}$ alcohols octanoate, *See* Hetester FAO, 13894
$C_{12\text{-}15}$ ether amine, *See* Adogen® 185, 766
$C_{12\text{-}15}$ pareth-3, *See* Bio-Soft® E-400, 4197
C_{13} alcohol ethoxylates, *See* Dehscoxid 730/740 Series, 7941
C_{13} alkyl acetate, *See* Exxate® 1300, 11351
C_{13} fatty alcohol phosphate ester, *See* Beycostat B 706 A, 4106
C_{13} oxo-alcohol polyglycol ether carboxylic acid, *See* Marlowet® 4538, 18920
C_{13}-C_{14} isoparaffin, *See* Isopar® M, 15392, Sepigel 305, 27984
C_{13}-oxo alcohol ethoxylate (2 EO), *See* Marlipal® 013/20, 18860
C_{14} fatty acid triglyceride, *See* Softenol® 3114, 28848
C_{14}-alkylbenzyldimethylammonium chloride, *See* JAQ Powdered Quaternary, 15517
C_{14}-C_{15} pareth-7, *See* Neodol® 45-7, 20916
C_{14}-C_{15} alcohol, *See* Neodol® 45, 20915
C_{14}-C_{16} pareth-7, *See* Surfonic® L46-7, 30369
C_{14} alcohols, *See* Philcohol 1400, 23666
C_{16} alcohols, *See* Philcohol 1600, 23667
$C_{16\text{-}18}$ alcohols, *See* Philcohol 1618, 23668
C_{18} alcohol, *See* Philcohol 1800, 23669
C_{16} linear primary alcohol, *See* cetyl alcohol, 5816
C_{16}-C_{18} alcohol polyglycol ether, *See* Marlowet® PW, 18936
C_{18}-C_{16} alkyl polyglycoside, *See* Glucopon 425, 13050
C_{18}-C_{36} acid, *See* Syncrowax AW1-C, 30530
C_{18}-C_{36} acid glycol ester, *See* Syncrowax ERL-C, 30532
C_{18}-C_{36} acid triglyceride, *See* Syncrowax HGL-C, 30533
C_{20}-C_{40} alcohols, *See* Unilin® 425 Alcohol, 32964
C_{21} dicarboxylic amido alkyl amine, *See* Indulin® MQK, 15092
C-1, 4839
C-1, *See* aluminum oxide, 1919
C-10015, *See* Chlorfenvinphos, 6098
$C_{10\text{-}16}$-Alkylbenzenesulfonic acid, calcium salt, *See* Nansa® 1042, 20681
C-108, *See* Hystrene® 1835, 14755
(C_{10}-C_{16}) Alkylethoxylate sulfuric acid, ammonium salt, *See* Witcolate AE, Witcolate LES-60A, 34472
(C_{10}-C_{16})Alkyl alcohol, ethoxylate, sulfuric acid, ammonium salt, *See* Witcolate AE, Witcolate LES-60A, 34472
C10R,C17,C20, C20R, *See* Jaguar® C-13S, C-14S, 15496
C-110, *See* Hystrene® 1835, 14755
$C_{12\text{-}14}$ alkyl dimethylamine oxide, 4840, *See* C12-14 alkyl dimethylamine oxide, 4840
$C_{12\text{-}14}$ fatty alcohol glycoside, *See* Plantaren 600 CS UP, 23918
$C_{12\text{-}14}$ pareth-3, *See* Laureth-2, 16852
$C_{12\text{-}16}$ alkyl polyglycoside, *See* Plantaren 1200, 23916
$C_{12\text{-}16}$ pareth-1, 4841
C-1297, *See* Philacid 1200, 23660
C_{13} linear primary alcohol, *See* tridecyl alcohol, 32233
(C_{13}-C_{16})Alkyl ethoxylate sulfuric acid, ammonium salt, *See* Witcolate AE, Witcolate LES-60A, 34472
(C_{14}-C_{18}) and (C_{16}-C_{18}) Unsatured trialkylcarboxylic acid, trimethylolpropane ester, *See* Radia® 7370, 25740
(C_{14}-C_{18}) and (C_{16}-C_{18}) Unsatured trialkylcarboxylic acid, trimethylolpropane ester, *See* Radia® 7371, 25741
C_{16}-alkyltrimethylammonium chloride, *See* Rhodaquat® M242C/29, 26644
C_{18} linear alcohol, *See* stearyl alcohol, 29604
C-2, 4842
C2292-59a, *See* Rampart, 25860
C-254, *See* Avoparcin, 3502
C44 dimer acid, *See* Pripol 1004, 25029
C-715u, 4843
C-7819B, *See* narasin, 20741
$C_{8\text{-}10}$ fatty alcohol glycoside, *See* Plantaren CG 60, 23919
$C_{8\text{-}18}$ alkylamido betaine, *See* Amphosol® CB3, 2342
C-84, 4844
$C_{9\text{-}11}$ pareth-4, *See* Berol 260, 4022
C-920u, 4845
C-9491, *See* Iodofenphos, 15213
Ca, *See* calcium, 4935
CA, *See* cellulose acetate, 5627
CA 70203, *See* triforine, 32254
Ca/Zn, *See* Bäropan MC 8046 SP, 3654
CA0397, 4846
CA0567, 4847
CA0570, 4848
CA0699, 4849
CA0700, 4850
CA0700, *See* Aminoethylaminopropyltrimethoxy silane, 2171
CA0742, 4851
CA0750, 4852
CA0880, 4853
CA0880, *See* Aminopropyltrimethoxysilane, 2187
CA0900, 4854
CA-25, 4855
CA-394-60S, 4856
CA750, *See* γ-aminopropyltriethoxysilane, 2186
CAB, *See* cellulose acetate butyrate, 5629
CAB, *See* CAB-171-15S, 4857
CAB, *See* Cocamidopropyl betaine, 6551
CAB-171-15S, 4857, *See* Cellulose acetate butyrate, 5628
CAB-381-0.1, 381-0.5, 381-2, 381-20, 4858

Chemical and Trade Names

calcium dihydrogen phosphate, *See* V-90®, 33222
calcium dioxide, *See* calcium peroxide, 4964, Oxy-Gro, 22456
calcium disodium (ethylenedinitrilo)tetraacetate, *See* Versene CA, 33618
calcium disodium edetate, *See* Versene CA, 33618
calcium disodium ethylenediamine tetraacetate, *See* Versene CA, 33618
calcium disodium versenate, 4946, *See* Versene CA, 33618
calcium dodecylbenzene sulfonate, *See* Ablusol DBC, 123, Agrilan® X98, 1063, Arylan® CA, 3157, Casul® 70 HF, 5421, Kowet 12, 16373, Nansa® EVM50, 20686, Rhodacal® 70/B, 26575, Rhodacal® CA, 70%, 26580, Tensaryl SB Ca, 31148, Witconate 605A, 34480
calcium dodecylbenzene sulfonate/oleyl alcohol ethoxylate, *See* Emulsogen IC, 10523
calcium fluoride, 4947
calcium formate, 4948
calcium glubionate, *See* Neo-Calglucon, 20877
calcium gluconate, *See* Neo-Calglucon, 20877
calcium gluconate anhydrous, *See* Gluconal® CA A, CAM, 13038
calcium gummate, *See* gum arabic, 13480
calcium hydrate, *See* calcium hydroxide, 4952
calcium hydride, 4949, *See* Hydrolete, 14609, hydrolith, 14611
calcium hydrochlorphosphate, 4950
calcium hydrogen orthophosphate, *See* calcium hydrogen phosphate, 4951
calcium hydrogen phosphate, 4951
calcium hydrosulfite, *See* Fredo, 12382
calcium hydroxide, 4952, *See* slaked lime, 28591
calcium hydroxyapatite, *See* Aluminum Grade Bone Ash, BCP 600, 1904, Natural Bone Ash, BCP 400, 20764, Synthetic Bone Ash, 30642
calcium hypochloride, *See* Repak, 26141
calcium hypochlorite, 4953, *See* chloride of lime, 6104, HTH, 14443, Hyporit, 14720, Inductor, 15089, Pittabs, 23888, Pittclor, 23890, Prestochlor, 24919, Pulsar, 25384, Stellos, 29641, Swim clear, 30460
calcium iodate, 4954
calcium lactophosphate, 4955
calcium lignosulfonate, 4956, *See* Borrebond, 4448, Borresperse CA/CAF, 4450, Darvan® No. 404, 7785, Goulac, 13215, Lignorosin, 17233, Lignosite®, 17234, Lignosol B, 17236, Lignosol SF, 17239, Marasperse GFC, 18688, Norlig 11 DA, 21563, Norlig A, 21565, Ultrazine CA, 32884
calcium lignosulfonates, 4957, *See* Glutrin, 13065
calcium metasilicate, *See* Microcal, 19657, Silasorb, 28268, Wwollastonite, 34547
calcium molybdate, 4958
calcium molybdate(VI), *See* calcium molybdate, 4958
calcium monohydrogen phosphate, *See* Calbrite, 4928
calcium monohydrogen phosphate dihydrate, *See* Emcompress®, 10064
calcium montanate, 4959
calcium nitrate, 4960, *See* lime nitrate, 17267 nitrocalcite, 21352 saltpeter rot, 27299
calcium octadecanoate, *See* calcium stearate, 4970
calcium oleate, 4961
calcium orthosilicate, *See* Limeolivine, 17272
calcium oxide, 4962, *See* Caloxol CP2, 5023, KM Pebble Lime, 16240, lime, 17265
calcium oxychloride, *See* calcium hypochlorite, 4953, Repak, 26141
calcium pantothenate, *See* Pantholin, 22693, Ritapan CAP, 26817, vitamin B_5 calcium salt, 33936
calcium pectolith, *See* calcium silicate, 4969
calcium pelargonate, *See* Synpro® calcium Pelargonate, 30598
calcium permanganate, 4963, *See* Acerdol, 275, Monol, 20229
calcium peroxide, 4964, *See* Fertilox, 11707, Oxy-Gro, 22456
calcium petroleum sulfonate, *See* calcium petronate 25H, 25C and 300, 4965
calcium petronate 25H, 25C and 300, 4965
calcium phosphate, *See* Cefkaphos®, 5549
calcium phosphate (monobasic), 4966, *See* H.T, 13524, Ibex, 14805, V-90®, 33222
calcium phosphate, dibasic, *See* Calbrite, 4928
calcium phosphate (tribasic), 4967, *See* Ephos, 10628
calcium phosphate hydroxide, *See* Alveograf, 1939
calcium phosphide, *See* Photophor®, 23763
calcium polysulfide, *See* Orthorix, 22337
calcium propionate, 4968, *See* Luprosil® Salt, 17921
calcium propionate and calcium formate, *See* Amasil® P, 1947
calcium pyroborate, *See* Meyerhofferite, 19617
calcium salt: MHA, *See* methionine hydroxy analog, 19486
calcium silicate, 4969, *See* Cecasol, 5536, Extrusil, 11331, Micro-Cel® T-38, 19663, Nyad®, 21860, Wollastocoat®, 34545, 34546
calcium silicate (wollastonite), *See* Vansil®, 33364
calcium stearate, 4970, *See* Afco-Chem CS, 921, Hallcote® CSD, 13555, Petrac® CP-11, 23463, Synpro® 15F, 30590
calcium stearoxyl-2-lactylate, *See* Verv®, 33644
calcium stearoyl lactylate, *See* Radiamuls® CSL 2980, 25772
calcium stearoyl-2-lactylate, *See* Pationic CSL, 22906, Verv®, 33644
calcium stearoyl-2-lactyllactate, *See* Verv®, 33644
calcium stearoyllactylate, *See* Verv®, 33644
calcium stearoyloxy calcium α,α(α-(stearoyloxy)propionyloxy)propionate, *See* Verv®, 33644
calcium stearyl lactylate, *See* Verv®, 33644
calcium stearyl-2-lactylate, *See* Verv®, 33644
calcium stelate, *See* Crolactil CSL, 7184
calcium sulfate, *See* calcium sulfate (anhydrous), 4971, Clay Breaker, 6376, Drierite, 9027, Gypsum, 13518, pearl-hardening, 22985
calcium sulfate (anhydrous), 4971, *See* Anhydrite, 2458
calcium sulfate (dihydrate), 4972
calcium sulfate dihdrate p.a, *See* calcium sulfate (dihydrate), 4972
calcium sulfate dihydrate NF, *See* calcium sulfate (dihydrate), 4972, Compactrol®, 6670
calcium sulfite, *See* katarsit, 15786
calcium sulfonate, 4973, *See* Lubrizol® 2152, 17744
calcium superoxide, *See* calcium peroxide, 4964
calcium superphosphate, *See* acid calcium phosphate, 330, calcium phosphate (monobasic), 4966
calcium tetraborate, *See* Meyerhofferite, 19617
calcium titanate, 4974
calcium zirconate, 4975
calcium(II) carbonate (1:1), *See* CP Filler, 6928
calcium/magnesium lignosulfonate, *See* Durabond, 9136
calcium/magnesium lignosulfonate, *See* Wafex, Wafolin, 34193
calcium/zinc, *See* Lankromark® LZ495, 16704
calcium-β-naphthol-γ-sulfonate, *See* Abrastol, 155, asaprol, 3176
calcium-zinc, *See* Haro® Mix ZC-028, ZC-029, ZC-030, 13649
Calcogas M, *See* Sudan I, 29938
Calcosan, *See* calcium chloride, 4943, Sure-Step, 30334
Calcozine Magenta N, *See* pararosaniline, 22798
calcozine red 6g, *See* Rhodamine G and G Extra, 26617
calcozine violet 6BN, *See* gentian violet, 12845
calcozine violet C, *See* gentian violet, 12845
Cal-C-Vita, 4976
Caldiox, 4977
Caldo Bordeles Valles, *See* cupric sulfate, basic, 7388
Caldura, 4978
Caledon, 4979
Calendula Oil CLR, 4980
calendulin, *See* Calendula Oil CLR, 4980
Calester, 4981
Calfax 10L-45, 4982
Calfax DB-45, 4983
Calfoam ES-30, 4984
Calfoam NEL-60, 4985, *See* Ammonium laureth sulfate, 2240
Calfoam SLS-30, 4986
Calgene C-100-S, *See* stearyl hydroxyethyl imidazoline, 29606
Calgene DOSS-70, *See* dioctyl sodium sulfosuccinate, 8550
Calglucon, *See* Neo-Calglucon, 20877
Calgon, 4987
Calgon® GW 12x40, *See* carbon, activated, 5225
Calgon® GW 12x40, *See* Carbon, 5213
Calgon® Type BPL 4x10, 6x16, 12x30, *See* Carbon, 5214
Calgon® Type HGR, *See* carbon, activated, 5225
Calgon® Type RB, *See* Carbon, 5215
Calgon® Type SGL, 4988, *See* Carbon, 5216, carbon, activated, 5225
Calgonite, 4989
Cal-Grid, 4990
Calibre 200-4, 4991
Calibre 2060-4, 4997
Calibre 302-E, 4992
Calibre 400-10, 4993
Calibre 510, 550, 4994
Calibre 700-4, 4995
Calibre 1001CD, 4996
Calibre LG2010, 4998
Calibre®, 4999
Calibrite, 5000
Calido, 5001
Calido brass, 5002
Calido-elalco, 5003
Califlux® 90, 5004
Calight RPO, 5005
Caliment, 5006, *See* Calbrite, 4928
Calimulse PRS, 5007
Calipharm, 5008
Calirus, 5009, *See* benodanil, 3942
Calite, 5010
Calixin®, 5011
Calkleen, 5012
Callaway 4000 Series, 5013
Calloseal, Callotek, 5014
Callusolve, 5015
Calmathion, *See* malathion, 18460
Calnox, *See* sodium polyacrylate, 28811
Calofil, 5016
Calofort®, 5017
Calomel, 5018
calomic, 5019
Calopake®, 5020
Caloreen, 5021
Caloride, *See* calcium chloride, 4943
calorite, 5022
Calorose, *See* Invert Sugar, 15200
Caloxol CP2, 5023
Calpanate, *See* vitamin B+75 calcium salt, 33936
Calpol, *See* Valadol, 33238
Calprona K, 5024
Calquat, 5025
Calsan, 5026
Calsil, 5027
Calsoft AOS-40, 5028
Calsoft F-90, 5029, *See* sodium dodecylbenzenesulfonate, 28760
Calsoft L-40, *See* sodium dodecylbenzenesulfonate, 28760
Calsoft L-60, *See* Nacconol® 40G, 20647, sodium dodecylbenzenesulfonate, 28760
Calsoft LAS-99, 5030, *See* Witco® 1298, 34458
Calsoft SXS 96, *See* Eltesol® SX 30, 9873, Naxonate® 4L, 20824, Pilot SXS-40, 23843
Calsoft T-60, 5031
Calsol, *See* Kalex 220 Crystal, 15668, Kalex Liq. 50%, 15671
Calsol 510, 5032
Calsol 804, 5033
Calsol S, *See* tetrasodium EDTA, 31391

Calsolene, 5034
Calstrip, 5035
Calsuds 81, 5036
Calsun Bronze, 5037
Caltaine C35, *See* Cocamidopropyl betaine, 6551
Calthane, 5038
Calthane ND 1100, 5039
Calthane NF 0710, 5040
Calthane NF 1900, 5041
Cal-Tint, 5042
Calurea, 5043
calx, *See* calcium oxide, 4962, KM Pebble Lime, 16240
Camadil, *See* CP Filler, 6928
Cambe wood, *See* redwoods, 25987
Cambendazole, *See* Equiben, 10733, Novazole, 21711
Cambilene, 5044
Cambrelle, 5045
Cambrite, 5046
Camel-CAL®, Camel-CARB®, Camel-FIL, Camel-FINE, 5047, *See* calcium carbonate, 4942
Camelia metal, 5048
Camel-TEX®, CamelWITE®, 5049, *See* calcium carbonate, 4942
Cameo®, 5050
Camflex, 5051
Camie 300, 5052
Camilol, 5053, *See* Hydagen® B, 14525
Camite, 5054
Campanuline, *See* muscarine, 20523
Campbell's CIPC 40%, 5055 , *See* Chlorpropam, 6162
Campbell's DB Straight, 5056
Campbell's Destox, 5057
Campbell's Dioweed 50, 5058
Campbell's DSM, 5059
Campbell's Field Marshal, Grassland Herbicide, 5060
Campbell's Linuron 45%, 5061
Campbell's MC Flowable, 5062, 5063
Campbell's MCPA 25, 50, 5064, 5065
Campbell's Nabam Soil Fungicide, 5066
Campbell's New Camppex, 5067
Campbell's Nico-Soap, 5068
Campbell's Rapier, 5069
Campbell's Redipon, 5070
Campbell's Redipon Extra, 5071
Campbell's Redlegor, 5072
Campbell's Sugar Beet Herbicide, 5073
Campbell's Trifluron, 5074
Campbell's X-Spor, 5075
***endo*-2-camphanol**, *See* DL-Borneol, 4428
camphor, *See* DL-Borneol, 4428
D-Camphor-10-sulfonic acid, *See* d-camphorsulfonic acid, 5076
d-camphorsulfonic acid, 5076
Camphortar, *See* Sanoscent, 27374
camphostyl, *See* d-camphorsulfonic acid, 5076
Camphre de Persil, *See* parsley camphor, 22865
Campoviton 6, *See* pyridoxine hydrochloride, 25483, vitamin B_6 hydrochloride, 33937
camsylate, *See* d-camphorsulfonic acid, 5076
Camtex, 5077
Camwood, 5078
Canacert, 5079
Canada snakeroot, *See* wild ginger oil, 34407
Canadian Certicol, 5080
Canadien 2000, *See* bromadiolone, 4565, Rentokil Deadline, 26116
Canadium, 5081
Canadol, *See* V.M. and P. Naphtha, 33212
Canadol, *See* petroleum ether, 23497, Ligroin, 17244
Canagel 75, 5082
cananga oil, *See* ylang-ylang oil, 34657
Canarin, 5083
candelilla wax, 5084, *See* Cirami No. 1, 6318
Candeptin, *See* Vanobid, 33332
Canderel, *See* Aspartame, 3233
Candex, *See* atrazine, 3394
Candex®, 5085
Candicidin, *See* Vanobid, 33332
Candimon, *See* Vanobid, 33332
Candle pitch, *See* stearin pitch, 29600
candle tar, *See* stearin pitch, 29600
candlenut oil, *See* Kekuna oil, 15823
cane sugar, *See* sucrose, 29934
Canesten®, 5086
Canfelzo, 5087
Canguard® 327, 5088
Canguard® 409, 5089
Canguard® 454, 5090, *See* Bioban® GK, 4158
Canilep, 5091
Canilep D.D, 5092
Canogard, *See* dichlorvos, 8394, Vapona, 33378
Canopar, 5093
Cantamega 1000, 2000, 5094
Cantan, *See* vitamin C, 33938
Cantaxin, *See* vitamin C, 33938
Canthaxanthin, 5095
Cantreece®, 5096
Cantrex, *See* Kantrex, 15718, Klebcil, 16209
Can-Trol, *See* MCPB, 19184
Canzler Wire, 5097
Cao, 5098, *See* *p*-Butylcresol, 4792
CaO, *See* calcium oxide, 4962
Caocobre, *See* cuprous oxide, 7405
caoutchouc, 5099
C-A-P, 5100
CAP, *See* Cellulose acetate propionate, 5630
Cap copper, 5103
C-A-P Enteric Coating Polymer, 5101
CA-P20, 5102
CAP-482-0.5, *See* Cellulose acetate propionate, 5630
CAP-482-20, *See* Cellulose acetate propionate, 5630
Capa, 5104
Caparol, *See* prometryn, 25141
Caparol 4L, *See* prometryn, 25141
Caparol 80W, *See* prometryn, 25141
Capasal, 5105
Capcure Emulsifier 37S, 5106
cape blue, *See* Crocidolite, 7060
Capella®, 5107
Caperase, *See* Catalase L, 5429
Capexco, 5108
Capital 170, 5109
Capitol, *See* Variquat® 50MC, 33397, Zephiran Chloride, 34728
Capmul O, *See* Kemester® S80, 15970, Radiasurf® 7155, 25802, Sorbon S-80, 29121
Capmul POE-O, *See* Alkamuls® T-80, 1693
Capmul® EMG, 5110
Capmul® GDL, 5111, *See* glyceryl dilaurate, 13079
Capmul® GMO, 5112
Capmul® GMS, 5113
Capmul® MCM, 5114
Capmul® O, 5115
Capmul® POE-L, 5116
Capmul® POE-O, 5117
Capmul® POE-S, 5118, *See* Ethylan® GES6, 11108
Capmul® S, 5119
Caposil, 5120
Cappagh brown, 5121
Capramide DEA, *See* Amidex CP, 2091
Capramide DEA 1:1 Capramide DEA, *See* Monamid® 150-CW, 20114
Capran® 77C, 5122
Capran® Unidraw®, 5123
Capran® Emblem, 5124
Capri blue gon, 5125
capric acid, *See* n-Capric acid, 5126, Prifrac 2906, 24951
n-Capric acid, 5126
capric acid chloride, *See* Decanoyl chloride, 7844
capric alcohol, *See* Exxal® 10, 11338
capric diethanolamide, *See* Amidex CP, 2091, Standamid® CD, 29454
capric glycerides, *See* Softigen® 767, 28855
capric/caprylate esters with triethylene glycol, *See* Uniflex® TEG-810, 32949
caproic acid, 5127
n-caproic acid acid, *See* caproic acid, 5127
Caproic aldehyde 1,2-ethanediol cyclic acetal, *See* Abbavert, 32
Caprol® 10G100, 5131, *See* Polygliceryl-10 decaoleate, 24453
Caprol® 10G10S, 5128
Caprol® 10G2O, 5129
Caprol® 10G40, 5130
Caprol® 2G4S, 5132
Caprol® 3GO, 5133
Caprol® 3GS, 5134 *See* Polyglyceryl stearate series, 24457
Caprol® 6G20, 5136
Caprol® 6G2S, 5135
Caprol® ET, 5137
Caprolactam, *See* Nipro (i), 21303
ϵ-Caprolactone, *See* ϵ-caprolactone monomer, 5138
6-caprolactone monomer, *See* ϵ-caprolactone monomer, 5138
ϵ-caprolactone monomer, 5138
Caprolan, *See* Cabelec® 1015, 4861
Caprolin, *See* Carbaryl, 5181
Capron, 5139
Capron 8200, 5140
Capron 8202C, 8203C, 8202CQ, 5141
Capron 8206 S, 5142
Capron 8230G, 5143
Capron 8231G, 8323G, 8233G, 5144
Capron 8253, 8350, 8351, 8352, 5145
Capron 8267 G HS, 5146
Capron 8270 HS, 5147
Capron 8331G, 8332G, 8333G, 8334G, 5148
Capron® 8202, 5149
Capron® 8203C HS, 5150
Capron® 8232G HS FR, 5151
Capron® 8233G HS, 5152
Capron® 8253, 5153
Capron® 8259, 5154
Capron® 8266G HS, 5155
Capron® 8280, 5156
capryl alcohol, 5157, *See* 1-octanol, 21988
Capryl hydroxyethyl imidazoline, *See* Mackazoline CY, 18199, Monazoline CY, 20194
Caprylene, *See* octene-1, 21991
Capryleth-4 carboxylic acid, *See* Akypo MB 1614/1, 1224
Capryleth-6 carboxylic acid, *See* Akypo LF 1, 1217
Capryleth-9 carboxylic acid, *See* Akypo LF 2, 1218
Capryleth-9 carboxylic acid and hexeth-4 carboxylic acid, *See* Akypo LF 4, 1220
caprylic acid, 5158, *See* Emery® 657, 10136, Prifrac 2901, 24950
caprylic acid monoglyceride, *See* Witafrol® 7420, 34421
caprylic acid sodium salt, *See* sodium caprylate, 28743
caprylic acid triglyceride, *See* Captex® 8000, 5170, Emalex O.T.G, 9976, Miglyol® 808, 19736
Caprylic acid-capric acid, 5160, *See* Industrene® 365, 15105
caprylic alcohol, *See* 1-octanol, 21988
Caprylic imidazoline, 5159, *See* Crodazoline Cy, 7160
Caprylic/capric acid (C_{8-10}), *See* Philacid 0810, 23658
caprylic/capric acid triglyceride, 5161, *See* Emalex K.T.G, 9967
Caprylic/capric carboxylic propionate, *See* Monateric 810-A-50, 20142
Caprylic/capric glycerides, *See* Imwitor® 742, 14945 Witafrol® 7420, 34421
Caprylic/capric triglyceride, *See* Captex® 300, 5168, Lexol GT 855, GT 865, 17149, Liponate GC, 17377, Mazol® 1400, 19135, Miglyol® 812, 19737, Myritol 318, 20580, Neobee® M-5, 20875, Tegosoft® CT, 31018
Caprylic/capric/diglyceryl succinate, *See* Miglyol® 829, 19739
Caprylic/capric/linoleic triglyceride, *See* Captex® 810B, 5171, Miglyol® 818, 19738
Caprylic/capric/stearic triglyceride, *See* Emalex SG-37, 9991
Caprylin, *See* Captex® 8000, 5170, Emalex O.T.G,

9976, Miglyol® 808, 19736
Capryloamphopropionate, *See* Miranol® JAS-50, 19904
Caprylyl pyrrolidone, *See* Surfadone LP-100, 30345
Capsicum annuum l, *See* Acetron®, Acetron® GP; Acetron® NS, 308
capsule metal, 5162
Captafol, 5163, *See* Difolatan, 8452, Merpafol, 19368, Sanspor, 27379
Captan, *See* Hormone Rooting Powder, 14310, Merpan, 19369, Vancide® 89, 33302
Captan + fosetyl-aluminum + thiabendazole, *See* Aliette Extra, 1614
Captan Granular, 5164
Captan, Captan-Col, Captan-50, Captan-83P, Captan Granular, 5165
captan-lindane, *See* Gammalex, 12634
Captax, 5166, *See* Accelerator Mercapto, 207, Thiotax, 31719
Captex® 200, 5167
Captex® 300, 5168, *See* caprylic/capric acid triglyceride, 5161
Captex® 800, 5169
Captex® 8000, 5170, *See* tricaprylin, 32211
Captex® 810B, 5171
Captin, *See* Valadol, 33238
Captor, 5172
Caradate 30, *See* Isonaphthol, 15377, MDI, 19188, Rubinate® LF-168, 27108
Caradate, Caradol, 5173
Caramba, 5174
Caramba Felgenglanz, 5175
Caramba Felgenneu, 5175
Caramba Lackkrone, 5176
Caramba Perlglanz, 5176
Carbadox, 5177, *See* Mecadox, 19200
Carbagel, 5178
carbam, *See* metam-sodium, 19444
Carbam, sodium salt, *See* metam-sodium, 19444
carbamaldehyde, *See* formamide, 12261
carbamic acid ethyl ester, *See* urethane, 33135
carbamic acid, (3-chlorophenyl)-, 1-methylethyl ester, *See* Chlorpropam, 6162
carbamic acid, (5-butyl-1H-benzimidazol-2-yl)-, methyl ester, monohydrochloride, *See* Delsene® 50 DF, 8041
carbamic acid, [1,2-phenylenebis(iminocarbono thioyl)] bis-, dimethyl ester, *See* thiophanate methyl, 31706
carbamic acid, 1H-benzimidazol-2-yl, methyl ester, *See* Carbendazim, 5187, 5188
carbamic acid, dimethyldithio-, compound with dimethylamine (1:1), *See* Ultra-DMC, 32791
carbamic acid, dimethyldithio-, dimethylamine salt (1:1), *See* Ultra-DMC, 32791
carbamic acid, dimethyldithio-, potassium salt, hydrate, *See* Aquatreat KM, 2760
carbamic acid, ethylenebis[dithio-, manganese salt, *See* Maneb, 18526
carbamic acid, methyl-, 1-naphthyl ester, *See* Carbaryl, 5181
carbamic acid, methyl, 2,2-dimethyl-2,3-dihydro-7-benzofuranyl ester, *See* Carbodan, 5197, Rampart, 25860
carbamic acid, methyl-, 2,2-dimethyl-2,3-dihydrobenzofuran-7-yl ester, *See* Carbodan, 5197, Rampart, 25860
carbamic acid, methyl, 2,3-dihydro-2,2-dimethyl-7-benzofuranyl ester, *See* Carbodan, 5197, Rampart, 25860
carbamic acid, methyl-, 2,3-dihydro-2,2-dimethyl-7-benzofuranyl ester, *See* Carbodan, 5197, Rampart, 25860
carbamic acid, methyl-, 4-(methylthio)-3,5-xylyl ester, *See* methiocarb, 19484
carbamic acid, methyl-, isopropoxyphenyl carbamic acid, methyl-,o-isopropoxyphenyl ester, *See* Baygon®, 3790
carbamic acid, N-methyl,1-naphthyl ester, *See* Carbaryl, 5181
carbamic acid, N-methyl-1-naphthyl-, *See* Carbaryl, 5181
carbamic acid, phenyl-, ethyl ester, *See* carbanilic ether, 5179
carbamide, *See* urea and thiourea resins, 33117
carbamine, *See* Carbaryl, 5181
4-carbaminophenyl arsonic acid, *See* Leucarsone, 17017
carbamodithioic acid,, *See* metiram, 19582
carbamodithioic acid, dimethyl-, compd. with N-methylmethanamine (1:1), *See* Ultra-DMC, 32791
Carbanilic acid, *m*-chloro- isopropyl ester, *See* Chlorpropam, 6162
Carbanilic acid, ethylcarbamoyl ethyl ester Carbanilic acid, (1-ethylcarbamoyl)ethyl ester, D-(-)-, *See* Carbetamex, 5189
carbanilic ether, 5179
carbanilide, 5180, *See* Zeonet U, 34722
carbanilide, 4-butoxy-4'-(dimethylamino)thio-, *See* thiambutosine, 31663
carbanilide, dimethylthio carbanilide, 2,2'-dimethylthio-, *See* Accelerator A22, 192
Carbaril, *See* Carbaryl, 5181
Carbarilo, *See* Carbaryl, 5181
Carbarilum, *See* Carbaryl, 5181
Carbaryl, 5181, *See* Carylderm, 5364, Microcarb, 19659, Murvin 85, 20522, Tornado, 32071
Carbaryl + pyrethins, *See* Microcarb T, 19660
Carbate Flowable, 5182, 5183
Carbatene, *See* metiram, 19582
Carbathene, 5184
Carbathiin, *See* carboxin, 5255
carbathione, *See* Unifume, 32954
Carbatox, *See* Carbaryl, 5181
Carbatox 75, *See* Carbaryl, 5181
Carbatox-60, *See* Carbaryl, 5181
Carbavur, *See* Carbaryl, 5181
Carbazinc, *See* AAprotect, 16
Carbazole Blue, 5185
Carbazole Violet, 5186
carbazotic acid, *See* picric acid, 23821, reflorit, 26003
Carbendazim, 5187, 5188, *See* Battal, 3740, Derosal WDG, 8133, Hinge, 14095, Konker®, 16327, Maxim, 19082, Stempor DG, 29648
carbendazim + mancozeb, *See* Kombat, 16315
carbendazim + flusilazole, *See* Punch® C, 25397
carbendazim + tecnazene, *See* Hortag Carbotec, 14314
carbendazime, *See* Carbendazim, 5187, 5188
carbendazim-flusilazole, *See* Early Impact, 9453
carbendazim-maneb, *See* Headland Dual, 13729
carbendazim-maneb, *See* Hispor 45WP, 14135
carbendazim-maneb, *See* Delsene® M Flowable, 8042
carbendazim-tecnazene, *See* Hickstor 6 +m2 MBC, 14063
carbendazim-tecnazene, *See* Hortag Tecnacarb Dust, 14315
Carbendazol, *See* Carbendazim, 5187, 5188
carbendazole, *See* Konker®, 16327
Carbendazole, *See* Carbendazim, 5187, 5188
Carbendazym, *See* Carbendazim, 5187, 5188
Carbetamex, 5189
Carbetamid, *See* Carbetamex, 5189
Carbetamide, *See* Carbetamex, 5189
Carbethamide, *See* Carbetamex, 5189
Carbethoxy Malathion, *See* malathion, 18460
α-carbethoxy- β,β-biscyclopropyl acrylonitrile, *See* Uvinul® N-35, 33200
Carbetovur, *See* malathion, 18460
Carbetox, *See* malathion, 18460
Carbide 6-12, *See* ethyl hexanediol, 11073
Carbidopa, *See* Lodosin, 17549
Carbilys, 5190
carbinol, *See* methyl alcohol, 19511
Carbital®, 5191 , *See* calcium carbonate, 4942
Carbital® 35, 5192, *See* calcium carbonate, 4941
Carbital® 50, 5193, *See* calcium carbonate, 4941
carbitol, *See* Diethoxol, 8435, Ektasolve® DE, DE-HG, 9664, Ethoxydiglycol, 11044, ethyl di-lcinol, 11068, Transcutol, 32137
carbitol cellosolve, *See* Diethoxol, 8435, ethyl di-lcinol, 11068
Carbitol®, 5194 , *See* Ethoxydiglycol, 11044
Carbitol® Low Gravity, *See* Ethoxydiglycol, 11044
Carbo Alumina, 5195
Carbobenzoxy Chloride, *See* benzyl chloroformate, 3988
Carbo-corundum, 5196
Carbodan, 5197
Carbodicyclohexylimide, *See* dicyclohexyl carbodiimide, 8412
Carbodiimide, *See* Stabilizer 2013-P®, 29406
Carboform, 5198
Carbofos, *See* malathion, 18460
Carbofrax, 5199
Carbofuran, *See* Carbodan, 5197, Curaterr®, 7414, Nex, 21079, Rampart, 25860, Sipcam UK Carbosip 5G, 28515
Carbofurane, *See* Rampart, 25860
carbo-gel, 5200
Carbogran, 5201
Carbogran E, 5202
Carbogran UF, 5201
Carbokaylene, 5202
carbol xylene, 5203
Carbolan, 5204
Carbolfuchsine, 5205
Carbolic Acid, 5206
Carbolic oils, *See* middle oils, 19722
Carbolon, 5207
Carboloy, 5208
Carbomang, 5209
Carbomant, 5210
Carbomer, 5211
Carbomer 208, *See* Antil® 208, 2559
Carbomethoxyaniline, *See* methyl anthranilate, 19513
***o*-carbomethoxyaniline**, *See* methyl anthranilate, 19513
Carbomix® 1605, 1609, 3651, 5212
Carbon, 5213,, 5215, 5216, 5217
Carbon, *See* Addipast, 652, Calgon® Type SGL, 4988, Carbon, 5216, Carbon, 5217, Carboraffin, 5236, Degusorb, 7916, Derussole, 8135, Hydraffin, 14533, Hydrodarco, 14585, Metasil DA, 19454, Metasil W/2, 19459, Microsperse, 19699, Sorbonorit, 29128, Special Black 100 Powd, 29216, Statex, 29552, Tack, 30689, Vulcan® XC72R, 34093
Carbon 4E, 5218
carbon bisulfide, 5219, *See* carbon disulfide, 5223
Carbon bisulfuret, *See* carbon disulfide, 5223
Carbon black, 5220, *See* Black Pearls® 1100, 4305, Continex® LH-10, 6741, Continex® N351, 6743, Furnex, 12496, Mogul® L, 20036, Octojet 104, 22004, Acticarbone, 461
Carbon Bronze, 5221
Carbon decolorizing, *See* Acticarbone, 461
carbon dioxide, 5222, *See* Cardice, 5259 Dricold, 9026 Drikold, 9032 dry ice, 9057
carbon disulfide, 5223, *See* carbon bisulfide, 5219
Carbon Lampblack, *See* Octojet 104, 22004
Carbon nickel-chromium iron alloy, *See* Adamite, 623
carbon oxychloride, *See* phosgene, 23701
Carbon S, *See* Aquatreat SDM, 2761, methyl namate®, 19551, Octopol SDM-40, 22014, Perkacit® SDMC, 23288, sodium dimethyldithiocarbamate, 28758, Thiostop N, 31717
Carbon soot, *See* Octojet 104, 22004
Carbon sulfide, *See* carbon disulfide, 5223
Carbon tetrachloride, 5224, *See* Katharin, 15791, phenoxin, 23623, seretin, 28026
Carbon trifluoride, *See*
Carbon trifluoride, *See* Fluoroform, 12099, Genetron® HFC23, 12829
carbon, activated, 5225, *See* Acticarbone, 461
Carbon, Amorphous, *See* vegetable black, 33513
Carbonado, *See* Carbon, 5217, carbon, activated, 5225
Carbonato(2-(O:O'))dihydroxydicopper, *See* cupric carbonate, 7378
(carbonato(2-))dihydroxydicopper, *See* cupric

carbonate, 7378
Carbondale silver, 5226, *See* nickel silvers, 21132
Carbonet, 5227
carbonic acid, *See* sodium carbonate, 28744
carbonic acid 1,1-methylethyl2-(1-methylpropyl)-4,6-dinitrophenyl ester, *See* Acrex, 384
carbonic acid 1-methylethyl 2-(1-methylpropyl)-4,6-dinitrophenyl ester, *See* Dinobuton, 8543
carbonic acid calcium salt, *See* Kalvan, 15701, Kotamite®, 16370
carbonic acid diphenyl ester, *See* Phenol Carbonate, 23604
carbonic acid disodium salt, compd. with hydrogen peroxide (H2O2) (2:3), *See* sodium percarbonate, 28807
carbonic acid gas, *See* carbon dioxide, 5222
carbonic acid magnesium salt, *See* magnesium carbonate, 18359
carbonic acid, ammonium copper salt, *See* Croptex Fungex, 7215
carbonic acid, barium salt (1:1), *See* Barium carbonate, 3636
carbonic acid, calcium salt, *See* calcium carbonate, 4940
carbonic acid, dipotassium salt, *See* potassium carbonate, 24744
carbonic acid, disodium salt, *See* sodium carbonate monohydrate, 28745
carbonic acid, dithio-, O-butyl ester, zinc salt, *See* Vulcaid 27, 34073
carbonic acid, lead(2+) salt (1:1), *See* Halcarb 20, 13545
carbonic acid, magnesium salt (1:1), *See* Elastocarb Tech Light, Heavy, 9691
carbonic acid, monoammonium salt), *See* Ammonium bicarbonate, 2231
carbonic acid, nickel salt, *See* nickel carbonate, 21123
carbonic acid, nickel(2+) salt (1:1), *See* nickel carbonate, 21123
carbonic acid, sodium salt (2:3), *See* S-Carb, 27549, sodium sesquicarbonate, 28816
carbonic acid, zinc salt (1:1), *See* Akrochem® 9930 Zinc Oxide Transparent, 1152, zinc carbonate, 34771
carbonic anhydride, *See* carbon dioxide, 5222
carbonic dichloride, *See* phosgene, 23701
Carbonin, 5228
Carbonochloridic acid 1-methylethyl ester, *See* isopropyl chloroformate, 15407
Carbonodithioic acid, O-(1-methylethyl) ester, sodium salt, *See* Aero 343 Xanthate, 842
Carbonodithioic acid, O-butyl ester, zinc salt, *See* Vulcaid 27, 34073
carbonyl chloride, *See* phosgene, 23701
carbonyl diamide, *See* urea and thiourea resins, 33117
carbonyl dichloride, *See* phosgene, 23701
8,8'-[carbonylbis[imino-3,1-phenylenecarbonyl imino(4-methyl-3,1-phenylene)carbonylimino]]bis-1,3,5-naphthalenetrisulfonic acid hexasodium salt, *See* Naganol, 20660
3,3'-(carbonyldiimino)bisbenzenecarboximidamide, *See* amicarbalide, 2063
Carbophenothion, *See* Trithion, 32422
Carbophos, *See* malathion, 18460
Carboplastic, 5229
Carbopol® 613, 614, 5230
Carbopol® 910, 5231
Carbopol® 934, 934P, 940, 5232
Carbopol® 941, 5233
Carbo-Pulbit®, 5234
Carbora, 5235
Carboraffin, 5236
Carborex, 5237
Carborundum, 5238, *See* Meccarb, 19202
Carbosal, 5239
Carbose D, *See* carboxymethylcellulose, 5257, carboxymethylcellulose sodium, 5258, Cekol, 5558
Carboset 514A, 5240
Carboset 531, 5241
Carbosorb, 5242
Carbostat 2203, 5243
carbostyril, 5244
carbosulfan, 5245, *See* Marshal 10G, 18955
Carbotex, 5246
Carbothialdine, *See* Dazomet, 7806, N-521® Biocide, 20640
Carbowax 200, *See* Alkapol PEG 300, 1706
Carbowax 1000 monostearate, *See* Ablunol DEGMS, 99
Carbowax 4000 monostearate, *See* Ablunol DEGMS, 99
Carbowax PEG 400, *See* Alkapol PEG 300, 1706
Carbowax PEG 8000, *See* Alkapol PEG 300, 1706
Carbowax® Compound 20M, 5247
Carbowax® MPEG 350, 5248
Carbowax® PEG 200, 5249
Carbowax® PEG 540 Blend, 5250
Carbowax® PEG 8000, 5251
Carbowax® Sentry, 5252
carbox metal, 5253
Carboxide, 5254
carboxin, 5255
Carboxin sulfone, *See* oxycarboxin, 22441, Ringmaster, 26771
Carboxine, *See* carboxin, 5255
carboxin-imazalil-thiabendazole, *See* Cerevax, Cerevax Extra, 5750
carboxyacetic acid, *See* malonic acid, 18491
Carboxyaniline, *See* anthranilic acid, 2529
***o*-carboxyaniline**, *See* anthranilic acid, 2529
Carboxybenzene, *See* Benzoic acid, 3973
β-Carboxyethyl acrylate, *See* Sipomer® β-CEA, 28527
(2-Carboxyethyl)-N-dodecyl- β-alanine, disodium salt, *See* disodium lauryliminodipropionate, 8650
N-(2-Carboxyethyl)-N-(tallow acyl)- β-alanine, *See* disodium tallowiminodipropionate, 8655
3-carboxy-5-hydroxy-1-*p*-sulfophenyl-4-*p*sulfophenylazopyrazole trisodium salt, *See* tartrazine, 30835
Carboxylated NBR, *See* Chemigum® NX-775, 5978
[3-[[[(3-carboxylato-1-oxopropyl)amino]carbonyl]amino]-2-methoxypropyl]hydroxymercurate(1-) sodium compound with 3,7-dihydro-1,3-dimethyl-1H-purine-2,6-dione, *See* meralluride, 19329
carboxylic acid (C_{10}), *See* n-Capric acid, 5126
Carboxylic acid amine salt, *See* Actracor 129, 856, 519
carboxylic acid C_7, *See* heptanoic acid, 13791
N-(Carboxymethyl)-N,N,-dimethyl-3-[(1-oxooctadecenyl)amino]-1-propanaminium hydroxide, inner salt, *See* Oleamidopropyl betaine, 22108
N-(Carboxymethyl)-N,N-dimethyl-1-docosanaiminum hydroxide, inner salt, *See* Behenyl betaine, 3900
N-(Carboxymethyl)-N,N-dimethyl-3-[(1-oxowheat germ alkyl)amino]-1-propanaminium hydroxides, inner salt, *See* Wheat germamidopropyl betaine, 34350
N-(Carboxymethyl)-N,N-dimethyl-9-octadecen-1-aminium hydroxide, inner salt, *See* oleyl betaine, 22132
carboxymethyl ether cellulose sodium salt, *See* carboxymethylcellulose, 5257, carboxymethylcellulose sodium, 5258, Cekol, 5558
Carboxymethyl hydroxyethyl cellulose, *See* Aqualon® CMHEC-37L, 2706
Carboxymethyl mercaptosuccinic acid, 5256, *See* Evanacid® 3CS, 11251
Carboxymethyl starch, *See* Vandrox®, 33315
Carboxymethyl)imino)bis(ethylenenitrilo))tetraacetic acid, pentasodium salt, *See* Cheelox® 80, 5864, Trilon® C Liq,
carboxymethylcellulose, 5257, *See* Aquasorb® A250, 2737
carboxymethylcellulose sodium, 5258, *See* Aqualon® Cellulose Gum, 2703, Aqualon® CMC-T, 2705, Blanose, 4325
carboxymethylcellulose sodium, technical grades, *See* Aqualon® CMC-T, 2704
N-(Carboxymethyl)N,N-dimethyl-3-[(1-oxododecyl)amino]-1-propanaminium hydroxide, inner salt, *See* lauramidopropyl betaine, 16830
(Carboxymethyl)-N'-(2-hydroxyethyl)-N,N'-ethylenediglycine, trisodium salt, *See* Trilon® D Liq, 32329
N-(carboxymethyl)-N'-(2-hydroxyethyl)-N,N'-ethylenediglycine trisodium salt, *See* Kalex OH, 15672
[[(Carboxymethyl)imino]bis(ethylenenitrilo)]-tetra-acetic acid, *See* Aroquest M Special, 3064, Pentaquest OPAC 0201, 23170
Carboxymethylmethylcellulose, *See* Benecel® Methylcellulose, 3938
(Carboxymethyl)trimethylammonium hydroxide inner salt, *See* Oxyneurine, 22463
9-(2-carboxyphenyl)-3,6-bis(diethylamino)xanthylium chloride, *See* Rhodamine B, 26616
9-(*o*-carboxyphenyl)-6-hydroxy-3-isoxanthenone, *See* uranine, 33115
9-(ocarboxyphenyl)-6-hydroxy-3H-xanthen-3-one, *See* uranine, 33115
Cardice, 5259, *See* carbon dioxide, 5222
Cardinal, 5260
Cardinal Red J, 5261
Cardiografin, *See* Reno-M, 26113
Cardio-Green, 5262
Cardipol® LP, 5263
Cardis® 36, 5264
Cardol, *See* acajou balsam, 180
Cardolite, 5265
Cardolite® NC-507, 5267
Cardolite® NC-511, 5268
Cardolite® NC-1307, 5266
Cardomec, *See* Ivermectin, 15454
Cardotek-30, *See* Ivermectin, 15454
Cardox, 5269
Cardura, 5270
Cardura E, 5271
Cardura E-10 (monoglycidyl ester), *See* glycidyl decanoate, 13092, Glydexx N-10, 13141
Cariflex, 5272
Cariflex Butadiene Rubber (BR), 5273
Cariflex Isoprene Rubbers (IR), 5274
Cariflex S, 5275
Cariflex Styrene-Butadiene Rubbers (SBR), 5276
Cariflex Thermoplastic Rubbers (TR), 5277
carin, *See* hexamine, 14008
Carina, 5278
Carinex, Carinex SB41, SI73, 5279
Carletti's indicator, 5280
Carlisle Wax 280, *See* Abluwax EBS, 140
Carlona 460, 462, 463, 5281
Carlona 55-004, 60-010, 60-060, 60-120, 5282
Carlona LB 157, LF 456, LF 459, 5283
Carlona P PLZ 532, 5284
Carlona P PY 61, 5285
carloon bronze, 5286
carmalum, 5287
Carmargo® White, 5288
Carmethose, *See* carboxymethylcellulose, 5257, carboxymethylcellulose sodium, 5258, Cekol, 5558
Carminaph, *See* Sudan I, 29938
Carmine, 5289
Carmine 40, 5290
Carmine 224, 5291
Carmine alum lake, *See* Carmine, 5289
carmine lake, 5292
carmine red, 5293
carminic acid, 5294
carnallite, 5295
Carnauba, 5296
Carnauba Spray 200, *See* Carnauba wax, 5297
Carnauba wax, 5297
carnidazole, 5298, *See* Spartrix, 29212
Carnot's reagent, 5299
Carnoy's fluid, 5300
Caro bronze, 5301
Caroat, 5302
Carob bean gum, *See* Industrial gum, 15114
carob flour, 5303

carob Seed gum, *See* carob flour, 5303
Carolid® MN-1, 5304
Caromax, 5305
Carophyll, *See* Canthaxanthin, 5095
Carophyll Red, *See* Canthaxanthin, 5095
Caro's acid, 5306
Caro's reagent, 5307
Carosella, *See* calcium molybdate, 4958
Carotaben, *See* carotene, 5308
Carotaben Plus, *See* Canthaxanthin, 5095
carotene, 5308
β-Carotene, *See* carotene, 5308, Lucarotin® 10%, 17753, Solatene, 28900
β,ε-carotene-3,3'-diol, *See* xanthophyll, 34591
β,β-Carotene-4,4'-dione, *See* Canthaxanthin, 5095
carotenoid, *See* xanthophyll, 34591
Caroubier, 5309
Carovax, 5310
Carp, 5311
Carp Brand, 5312
Carpene, *See* Dodine FL, WP, 8785, Radspor FT, 65WP, 25833
Carpenter 22Cr-13Ni-5Mn, 5313
Carpenters Wood Glue, 5314
Carpolin, *See* Carbaryl, 5181
Carrageen, *See* carrageenan, 5315
Carrageenan, 5315, *See* Wakal® K Range, 34197
Carrageenin, *See* carrageenan, 5315
Carrel-Dakin solution, *See* sodium hypochlorite, 28773
Carriant Series, 5316
Carrisorb, 5317, *See* Attapulgite, 3404
Carrot Oil CLR, 5318
Carsamide®, 5319
Carsamide® AMEA, 5320, *See* Acetamide MEA, 283
Carsamide® CA, 5321
Carsamide® CMEA, 5322, *See* Cocamide MEA (1:1), 6549
Carsamide® SAL-7, 5323
Carset, 5324
Carsil, 5325
Carsilon, 5326
Carsofoam® 1618, 5327
Carsofoam® BS-I, 5328
Carsofoam® MSP, 5329
Carsofoam® T-60-L, 5330
Carsoft® T-90, *See* Incrosoft T-90, 15051
Carsonal® DLS, *See* DEA-lauryl sulfate, 7827
Carsonal® SES-A, *See* Ammonium laureth sulfate, 2240
Carsonol sls, *See* Rhodapon® 101-10, 26631, sodium lauryl sulfate, 28781, Witcolate 6400, 34469
Carsonol sls paste B, *See* Rhodapon® 101-10, 26631, sodium lauryl sulfate, 28781, Witcolate 6400, 34469
Carsonol sls special, *See* Rhodapon® 101-10, 26631, sodium lauryl sulfate, 28781, Witcolate 6400, 34469
Carsonol®, 5331
Carsonol® ALS-R, 5332, *See* Ammonium lauryl sulfate, 2242
Carsonol® AOS, 5333
Carsonol® DLS, 5334
Carsonol® MLS, 5335, *See* magnesium lauryl sulfate, 18364
Carsonol® SES-A, 5336
Carsonol® SES-S, 5337
Carsonol® SHS, 5338
Carsonol® SLS Paste B, 5339
Carsonol® TLS, 5340
Carsonon N-9, *See* Alkasurf® NP-4, 1715, Triton® N-57, 32430
Carsonon®, 5341
Carsonon® 144-P, 5342, *See* PPG-3 myristyl ether, 24844
Carsonon® 169-P, 5343, *See* PPG-30 cetyl ether series, 24845
Carsonon® N-4, 5344
Carsoquat SDQ-25, *See* Ablumine 280, 85 Cycloton® SCS, 7542, octadecylbenzenemethanaminium chloride, 21985, stearalkonium chloride, 29592, Sumquat® 6210, 30126
Carsoquat®, 5345
Carsoquat® 816-C, 5346
Carsoquat® 868 E, 5347
Carsoquat® CB, 5348
Carsoquat® CT-429, 5349, *See* Cetrimonium chloride, 5814, cetyl trimethyl ammonium chloride, 5826
Carsoquat® SDQ-25, 5350
Carsosoft®, 5351
Carsosoft® CFI-90, 5352
Carsosoft® S-90, 5353
Carsosoft® T-90, 5354
Carsosulf SXS, *See* Eltesol® SX 30, 9873, Naxonate® 4L, 20824, Pilot SXS-40, 23843
Carspray#2, CW, 5355
Carstab® DLTDP, 5356
Carstab® DSTDP, 5357, *See* Distearyl thiodipropionate, 8713
Cartaretin F-4, 5358
Carterite, 5359
Carthamin, *See* vegetable rouge, 33523
carthamus tinctorius oil, *See* Neobee® 18, 20872, safflower oil, 27222
Cartolac, 5360
Cartose, 5361
carvacrol, 5362, *See* oxycymol, 22445
(+)-carvene, *See* Limonene, 17280
carvol, *See* carvone, 5363
carvone, 5363
Carylderm, 5364, *See* Carbaryl, 5181
caryophyllic acid, *See* eugenol,
Casabet, 5365
Casabet 655, 5365
Casahib, 5366
Casamer, 5367
Casamid, 5368
Casamine, 5369
Casamox, 5370
Casaquat, 5371
Casateric, 5372
Casathane, 5373
casca bark, *See* red water bark, 25973
Cascade, 5374, 5375
Cascamite, 5376
Casco, 5377
Cascogel TM, 5378
Cascomelt, 5379
Cascophen, 5380
Cascophen Resorcinol Resin RS 216/RXS-8, 5381
Casco-resin, Casco Resin TM, 5382
Cascorez TM, 5383, 5384, 5385
casein, 5386, *See* kalzose, 15703
casein glue, 5387
casein magnesia, 5388
casein paints, 5389
casein silk, 5390
casein sodium glycerophosphate, *See* Sanatogen, 27318
casein sodium salt, *See* sodium caseinate, 28746
caseogum, 5391
Cashmilon®, 5392
Casiflux, *See* Wollastocoat®, 34545
Casiflux vp 413-004, *See* Wollastocoat®, 34545
Casilan, 5393
Casoron, 5394
Casoron G, 5395
Casoron G4, 5396
Cassappret®, 5397
Cassastat®, 5398
Cassatan®, 5399
cassava, 5400
cassava meal, 5401
Cassella's acid, 5402
Cassella's Acid F, *See* Delta acid, 8044
Casselmam's green, 5403
cassic acid, *See* rhein, 26534
cassiterite, *See* tin(IV) oxide, 31817
Cassulfon®, 5404
Cassurit®, 5405
Cast acrylic, *See* Polypenco® Cast Acrylic Rod, 24509
cast brass, 5406
cast yellow brass, 5407
Castaldo, 5408
Castaloy, 5409
Castaway Plus, 5410
Castellanos powder, 5411
Castethane, 5412
casting copper, 5413
castiron D2, 5414
Casto-Magic, 5415
Castomer, 5416
castor oil, 5417, *See* AA Standard, AA USP, 12, Cosmetol® X, 6871, DB Oil, 7807
castor oil ethoxylate, *See* Emulan® EL, 10489, Sandoxylate® C-32, 27358
castor oil, ethoxylated, *See* Ablunol CO 10, 98
castor oil, ethylene glycol polymer, *See* Ablunol CO 10, 98
castor oil, hydrogenated, *See* hydrogenated castor oil, 14592
castor oil, polyethoxylated, *See* Ablunol CO 10, 98
castor oil, sodium salt, *See* Soricinol 40, 29143
castor oil, sulfated, sodium salt, *See* oleite, 22114
castor/organoclay complex, *See* Advitrol 8-10, 825
castorwax, *See* hydrogenated castor oil, 14592
Castrix Grains, *See* Crimidine, 7053
Castrol GTX, 5418
Castrol Turbomax, 5419
Castung 103 G-H, 5420
Casul® 70 HF, 5421
Caswell Adhesives, 5422
CAT, *See* simazine, 28464
CAT (herbicide), *See* simazine, 28464
C-A-T Enteric Coating Polymer, 5423
Catabond, 5424
Cata-Chek, 5425
Cataflot, 5426
Catafor, 5427
Cataid, 5428
Catalase, *See* Catalase L, 5429
Catalase L, 5429
Catalazuli, 5430
Catalex, 5431
Catalin CAO-3, *See* Oxyguard, 22457
Catalpo, 5432
Catalyst 9, 5433
Catalyst CC, 5434
Catalyst RD Liq, 5435
Catalyst ZA, 5436
Catalyzed hydrazine, *See* Scav-Ox® II, 27553
Catapol SR, 5437
Catarase, *See* α-Chymotrypsins, 6286
Catarase®, 5438
Catavar, 5439
Catavat Black N-JBB, 5440
Catavin C, *See* vitamin C, 33938
CatCO 600, CatCO 610ST, 5441
catechol, *See* pyrocatechin, 25497, kachin, 15651
Catechu, *See* gum catechu, 13483
Catex, 5442
Cat-floc TL, *See* Agefloc WT-40, 1010
cathamic acid, *See* vegetable rouge, 33523
Catigene 4513, 5443
Catigene BR 80 B, 5444
Catigene DC/100, 5445
Catigene Red-brown, 5446
Catigene SR, 5447
Catigene T80, 5448
Catigene® 50 USP, 5450
Catigene® 818, 5451
Catigene® 1011, 5449
Catigene® B 50, 5452
Catigene® CA 56, 5453
Catigene® CETAC 30, 5454
Catigene® D80, T50, T80, *See* Benzalkonium chloride, 3958
Catigene® DC 100, 5455, *See* Exameen 824 3724, 11284
Catinal CB-50, 5456
Catinal HTB-70, 5457, *See* Cetrimonium bromide, 5813

Chemical and Trade Names

Catinal OB-80E, 5458
Cationaster-C, 5459, *See* 3-Chloro-2-hydroxypropyl trimethylammonium chloride (50% aq solution), 6127
Cationic Collagen Polypeptides, 5460
Cationic dyestuffs, *See* Astrazon®, 3257, Panacryl, 22654
Cationic Guar C-261, *See* Jaguar® C-13S, C-14S, 15496
Cationic polyelectrolyte, *See* Daxad® CP-2, 7804
Cationic Softener X Concentrate, 5461
Catisol AO 100, 5462
Catomer VA, 5463, *See* Everflex® SP-1084, 11266
Catosal®, 5464
Cat's gold, 5465, *See* mosaic gold, 20355
Cat's silver, *See* Cat's gold, 5465
Caucho Blanco, 5466
Caust X, 5467
caustic barley, *See* veratrine, 33567
caustic baryta, 5468
caustic lime, *See* calcium hydroxide, 4952, lime, 17265
caustic potash, *See* potassium hydroxide, 24756, vegetable alkali, 33512
caustic soda, *See* sodium hydroxide, 28772
Causul, 5469
cavi-trol, *See* sodium fluoride, 28764
Caytur 21 & 22, 5470
Caytur 4, 5471
Cazin, 5472
Ca-Zn, *See* ADK STAB GR-16, 728
CB, *See* Cocamidopropyl betaine, 6551
CB 313, *See* Lysoform, 18095
CB/M, *See* Cocamidopropyl betaine, 6551
CB2100, 5473
CB2405, 5474
CB2408, 5475
CB2409.5, 5476
CB2493, 5477
CB2494, 5478
CB2500, 5479
CB2785, 5480
CB2790, 5481
CB2805, 5482
CB-4-34, 5483
CBC, *See* Cocamidopropyl betaine, 6551
CBR Monoteric ADA, *See* Cocamidopropyl betaine, 6551
CBS, *See* Delac S, 8002, Naugard® I-2, 20792, Perkacit® CBS, 23277, Santocure, 27400, Vulkacit® CZ/EGC, DZ/EGC, 34129
CBTS, 5484, *See* Delac S, 8002, Naugard® I-2, 20792, Perkacit® CBS, 23277, Santocure, 27400
CC, *See* vitamin D_3, 33941
CC-103, 5485
CC3005, 5486, *See* 2-chloroethylmethyldichlorosilane, 6122
CC3270, 5487, *See* chloromethyldimethylchlorosilane, 6129
CC3275, 5488, *See* chloromethyldichloromethylsilane, 6128
CC3285, 5489, *See* chloromethyltrimethyl silane, 6130
CC3290, 5490, *See* 3-chloropropylmethyldimethoxysilane, 6137
CC3291, 5491, *See* chloropropyltrichlorosilane, 6139
CC3292, 5492, *See* 3-chloropropyltriethoxysilane, 6140
CC3300, 5493, *See* chloropropyltrimethoxysilane, 6138, 3-chloropropyltrimethoxysilane, 6142
CC3433, 5494, *See* 2-cyanoethyltriethoxysilane, 7478
CC3555, 5495, *See* 3-cyanopropyltrichlorosilane, 7482
CCA Type C, *See* arsenic acid, 3137
CCA Type C Wood Preservative 50-60%, 5496
CCC, *See* CCC 700, 5497
CCC 700, 5497
CCN52, *See* cypermethrin, 7588, Topclip Parasol, 32037
Cd, *See* cadmium, 4893
CD480, 5498
CD492, 5499
CD3770, 5500
CD3780, 5501
CD4153, 5502
CD4153, *See* Di-*t*-butoxydiacetoxysilane, 8362
CD4368, 5503
CD4450, 5504
CD5400, 5505
CD5430, 5506
CD5470, 5507
CD5600, 5508
CD5600, *See* EXP-49, 11305
CD5605, 5509
CD5610, 5510
CD5635, 5511
CD5636, 5512
CD5950, 5513
CD6000, 5514
CD6010, 5515
CD6150, 5516
CD6210, 5517
CD6220, 5518
CDA Dicotox Extra, 5519
CDA Mildothane, 5520
CDA Roval, 5521
CDA Simflow Plus, 5522
CDA Viper, 5523
CDB 90, 5524
CDB Clearon, 5525
CDNA, *See* dicloran, 8395, Fumite Dicloran, 12459
CDS-1801, 5526
CDT, *See* simazine, 28464
CE 2, *See* Uvinul® N-35, 33200
Cérite, 5527
CE6250, 5528
CE6345, 5529, *See* ethyltriacetoxysilane, 11146
CE6350, 5530, *See* ethyltrichlorosilane, 11147
Ceanel, 5531
ceara wax, *See* Carnauba wax, 5297
Cebicure, *See* vitamin C, 33938
Cebion, *See* vitamin C, 33938
Cebion®, 5532
Cebitate, *See* sodium ascorbate, 28734, vitamin C sodium salt, 33939
Cecagel, 5533
Cecaperl, 5534
Cecarbon, 5535
Cecasil, *See* calcium silicate, 4969, Extrusil, 11331
Cecasol, 5536
Ce-Cobalin, 5537
Cecolene 1, 5538
Cecolene 2, 5539
Cecon, *See* vitamin C, 33938
Cedarite, *See* Chemawinite, 5889
Cedepal FA-406, 5540
Cedepal FS-406, 5541
Cedepal TD-403, 5542
Cedepal® Range, 5543
Cedephos® FA600, 5544
Cederan® P 23, 5545
Ceepree® C200, 5546
Ceepryn, *See* Acetoquat CPC, 302, cetyl pyridinium chloride, 5823
Cefa-Drops, *See* Cefadroxil, 5547
Cefadroxil, 5547
Cefadyl, *See* Cephapirin sodium,
Cefamox, *See* Cefadroxil, 5547
Cefkanat®, 5548
Cefkaphos®, 5549
Ceforal, *See* Cefadroxil, 5547
Cefracycline suspension, *See* tetracycline, 31365
Cegemett® Range, 5550
Cegepal® Range, 5551
Cegepaot® Range, 5552
Cegeskin® Range, 5553
Cegesol® Range, 5554
Cegesterin® Range, 5555
Cegiolan, *See* vitamin C, 33938
Ceistran® N66G30-01-4, 5556
Cekas, 5557
Cekherbex, *See* MCPA, 19183
Cekiuron, *See* diuron, 8736
Cekol, 5558
Ceku C.B, *See* Julin's chloride, 15632
Cekufon, *See* trichlorfon, 32212
Cekugib, *See* Regulex, 26019
Cekumethion, *See* Parathion-methyl, 22808
Cekupropanil, *See* Propanil, 25174
Cekusan, *See* dichlorvos, 8394, simazine, 28464
Cekusil Universal C, *See* Atiran, 3286, Agallol, 954
Cekuzina-S, *See* simazine, 28464
Cekuzina-T, *See* atrazine, 3394
CELA 50, *See* triforine, 32254
Cela W-524, *See* triforine, 32254
Celacol, 5559
Celafuse, 5560
Celanese, 5561
Celanese® Nylon 1000-1, 5563
Celanese® Nylon 1003-1, 5564, *See* Celanese® Nylon 1000-1, 5563
Celanese® Nylon 1500-1, 5565, *See* Celanese® Nylon 1000-1, 5563
Celanese® Nylon 6/6, 5562
Celanese® Nylon 7420, 7423, 5566, *See* Celanese® Nylon 1000-1, 5563
Celanese® Nylon N-186, 5567, *See* Celanese® Nylon 1000-1, 5563
Celanex®, 5568
Celanex® 1300A, 1600A, 2000K, 5569
Celanex® 3310, 5300, J600, 5570
Celanex® J600, 5571
Celaskon, *See* vitamin C, 33938
Celastic, 5572
Celasyl, 5573
Celatene Colors, 5574
Celatom, 5575
Celcon® EC90+xg, 5576
Celcon® GB25, 5577
Celcon® LW90, 5578
Celcon® M25, 5579
Celcon® M270, 5580
Celcon® U10, 5581
Celcon® UV25Z, 5582
Celenex® 1300A, 1600A, 2000K, 3310, 5300, J600, *See* Grilpet® XE3060, 13402
Celestial blue, *See* Brunswick blue, 4627
Celestol, 5583
Celestols, 5584
Celestron, 5585
Celevac, *See* Glutolin, 13064
Celin, *See* vitamin C, 33938
Celite, *See* Celite®, 5586, Infusorial earth, 15125
Celite®, 5586
Celite® 110, 5587
Celite® 209, 5588
Celite® 270, 5589
Celite® HSC, 5611
Celite® R-625, 5590
Celite® Snow Floss, 5591
Celkate T-21, 5592
Cellacephate, 5593
Cellamine PAD, 5594
Cellanite, 5595
Cellastine, 5596
Cellasto®, 5597
Cellestren, 5598
Cellidor, 5599
Cellit, 5600
Cellitazol® STN, 5601
Celliton® Dyes, 5602
Cellmore, 5603
Cellobond, 5604
cellodin, 5605
celloidin, *See* Kodaloid, 16283, Venite, 33560
Cellokyd®, Cellokyd®-2708, 5606
Cellolax, *See* carboxymethylcellulose, 5257, carboxymethylcellulose sodium, 5258, Cekol, 5558
Cellolyn, 5607
Cellon, *See* Acetosol, 304
Cellophane, 5608
Cellosize, 5609
Cellosize® HEC QP Grades, 5610

Cellosolve, *See* EE Solvent, 9636, Ethoxyethanol, 11045, ethyl icinol, 11074
Cellosolve acetate, *See* EE Acetate, 9635
Cellosolve solvent, *See* ethyl icinol, 11074
Cellothyl, *See* Glutolin, 13064
Cellthane, 5612
Celluclast, 5613
Cellucon, *See* Glutolin, 13064
Cellucraft, 5614
Celluflex 179, *See* Syn-O-Ad® 8484, 30558, TCP, 30857
Celluflex CEF, *See* Genomoll P, 12838
Celluflex DOP, *See* Witcizer 312, 34454
Celluflex DPB, *See* dibutyl phthalate, 8366
Celluflex M179, 5615
Celluflex TPP, *See* triphenyl phosphate, 32391
Celluflow C-25, 5616
Celluflow TA-25, 5617
Cellufluor, 5618
Cellulase, 5619, 5620, *See* Celluzyme® 2400 T, 5638
Cellulase 4000, 5621
Cellulase AC, 5622
cellulith, 5623
Celluloid, 5624
celluloid-caoutchouc, 5625
Cellulose, 5626, *See* Celluflow C-25, 5616, Fibra-Cel®, 11748, Solka-Floc® BW-40, BW-100, BW-200, BW-2030, UF-900-FCC & NF, 28949
Cellulose 2-hydroxypropyl ether, *See* Klucel®, 16229, Klucel® E, G, H, J, L, M, 16230, Klucel® EF, 16231
cellulose acetate, 5627, *See* CA-394-60S, 4856, Courtaulds, 6914, Kodapak®, 16284, Rotuba H, 27007, Sericose, 28031, Tenite® 105-MS, 31104, Tenite® Cellulosic Acetate, 31108
Cellulose acetate butyrate, 5628, 5629, *See* CAB-171-15S, 4857, CAB-381-0.1, 381-0.5, 381-2, 381-20, 4858, CAB-500-5, 4859, Cellulose acetate butyrate, 5628, Tenite® 264-MH, 31106
cellulose acetate ester, *See* cellulose acetate, 5627
Cellulose acetate phthalate, *See* C-A-P, 5100, C-A-P Enteric Coating Polymer, 5101
Cellulose acetate propionate, 5630, *See* Cellulose acetate propionate, 5630, Tenite® 360-H2, 31107
Cellulose acetate propionate ester, 5631, *See* Cellulose acetate propionate, 5630, Cellulose acetate propionate ester, 5631
Cellulose acetate trimellitate, *See* C-A-T Enteric Coating Polymer, 5423
cellulose acetobutyrate, *See* Cellulose acetate butyrate, 5629
Cellulose butyrate, *See* Tenite® Cellulosic Butyrate, 31109
Cellulose ethyl ether, *See* ethyl cellulose, 11066
cellulose gum (CTFA), *See* carboxymethylcellulose sodium, 5258
cellulose methyl ether, *See* methylcellulose, 19566, Glutolin, 13064
Cellulose NF, *See* Elcema® F150, G250, P100, 9721
Cellulose nitrate, *See* Kodaloid, 16283, Venite, 33560
cellulose pitch, 5632
Cellulose propionate, *See* Tenite® Cellulosic Propionate, 31110
cellulose triacetate, 5633, *See* Celluflow TA-25, 5617
Cellulose turpentine oil, *See* sulfite turpentine oil, 29994
cellulose, acetate butanoate, *See* Cellulose acetate butyrate, 5629
cellulose, acetate propanoate, *See* Cellulose acetate propionate, 5630
Cellulose, 2-hydroxyethyl ether, *See* Hetastarch, 13891
cellulose, nitrate, *See* nitrocellulose, 21353
Cellulosic, *See* cellulose acetate, 5627
Cellulosine, 5634
Cellulysin®, 5635
Cellumeth, *See* Glutolin, 13064
celluphos 4, *See* tributyl phosphate, 32207
Celluvarno, 5636
Celluzyme®, 5637
Celluzyme® 2400 T, 5638, *See* Cellulase, 5620
Celmar®, 5639
Celmer, *See* Ceresan, 5745
Celmontite, 5640
Cel-O-Brandt, *See* carboxymethylcellulose, 5257, carboxymethylcellulose sodium, 5258, Cekol, 5558
Celogen®, 5641
Celogen® AZ 120, 130, 150, 180, 199, 5642, *See* Azodicarbonamide, 3524
Celogen® OT, 5643
Celogen® RA, 5644
Celogen® TSH, 5645
Celogen® XP-100, 5646
Celosen AZ, *See* Azodicarbonamide, 3524
Celphos, *See* Talunex, 30757
Celquat® H-100, SC-240, 5647
Celsit, 5648
Cel-Soft #2, 5649
Celstran® ACG40-01-4, 5650
Celstran® N6G30-01-4, N66G30-01-4, 5651
Celstran® PBTG30-01-4, 5652, *See* Grilpet® XE3060, 13402
Celstran® PCG30-01-4, 5653
Celstran® PETG30-01-4, 5654, *See* Polyethylene terephthalate, 24426
Celstran® PPG30-01-4, 5655
Celstran® PPSG30-01-4, 5656
Celstran® PUG30-01-4, 5657
Celstran® SMAG30-01-4, 5658
Celtex, 5659
Celthion, *See* malathion, 18460
Celtite, 5660
Celulon, 5661
cement copper, 5662
cement mortar, 5663
Cement Prodor, 5664
cement, adamantine, 5665
cement, American, 5666
Cementation copper, *See* cement copper, 5662
cementation steel, 5667
cementite, 5668
cementite, independent, 5669
Cemerin, *See* ametryn, 2051
Cemset, 5670
Cemulsol 1050, *See* Ablunol 200MO, 96
Cemulsol A, *See* Ablunol 200MO, 96
Cemulsol C 105, *See* Ablunol 200MO, 96
Cemulsol D-8, *See* Ablunol 200MO, 96
Ceneg, 5671
Cenegen® 7, 5672
Cenegen® CJB, 5673
Cenekol® 1141, 5674
Cenekol® FT Supra, 5675
Cenetone, *See* vitamin C, 33938
Cenolate, *See* sodium ascorbate, 28734, vitamin C sodium salt, 33939
Centa™, 5676
Centari®, 5677
Centelase Dermatologico, *See* Madecassol, 18291
Centex, 5678
Centimide, *See* Rhodaquat® M242B/99, 26643
Centralite, 5679, *See* Zeonet U, 34722
centrallasite, *See* calcium silicate, 4969
Centrex® 811, 5680
Centrex® 833, 5681
Centrifugal Syrup, 5682
Centrocap® 162SS, 162US, 5683
Centrol® 2FSB, 2FUB, 3FSB, 3FUB, *See* Centrolex® C, 5686, Centrophase® C, 5688, Centrophase® HR, 5689, Centrophase® HR2B, HR2U, 5690, Centromix® CPS, 5687
Centrol® CA, 5684
Centrolene® A, S, 5685
Centrolex® C, 5686
Centromix® CPS, 5687
Centrophase® C, 5688
Centrophase® HR, 5689
Centrophase® HR2B, HR2U, 5690 *See* Centrolex® C, 5686, Centrophase® C, 5688, Centrophase® HR, 5689, Centrophase® HR2B, HR2U, 5690, Centromix® CPS, 5687
Centrophil® K, 5691
century 1210, *See* stearic acid, 29599
century 1220, *See* stearic acid, 29599
century 1230, *See* stearic acid, 29599
century 1240, *See* stearic acid, 29599
Cenwax® G, 5692
Cenwax® ME, 5693
Cepacol, *See* Acetoquat CPC, 302, cetyl pyridinium chloride, 5823
Cephos, *See* Cefadroxil, 5547
Cephreine, 5694
Cephrol, 5695, *See* citronellol, 6348
Cequartyl, *See* Zephiran Chloride, 34728, Variquat® 50MC, 33397
Cera alba, *See* beeswax, 3888
Cerabond 18, 5696
Cerabrit, 5697
Cerachem®, 5698
Cerachrome®, 5699
Ceracolor, 5700
Cerafiber®, 5701
Ceralan®, 5702
Ceralan® Hartolan, *See* Lanolin alcohol, 16741
Ceralumin C, 5703
Ceramabond, 5704
Ceramcel, 5705
Ceramer® 67, 5706
Ceramitalc, 5707
Ceramite, 5708
Ceramol, 5709
Ceramtex, 5710
Ceranine PN Base, 5711
Ceraphyl 28, *See* cetyl lactate, 5820
Ceraphyl 50, *See* Wickenol® 506, 34400
Ceraphyl 140A, *See* Wickenol® 144, 34387
Ceraphyl 140-A, *See* Wickenol® 144, 34387
Ceraphyl 230, *See* Unimate® DIPA, 32968
Ceraphyl 368, *See* Unimate® EHP, 32970, Wickenol® 155, 34391
Ceraphyl 791, *See* isocetyl stearoyl stearate, 15349
Ceraphyl ICA, *See* Exxal® 16, 11341
Ceraphyl® 28, 5712, *See* cetyl lactate, 5820
Ceraphyl® 31, 5713
Ceraphyl® 41, 5714
Ceraphyl® 45, 5715
Ceraphyl® 50, 5716
Ceraphyl® 55, 5717
Ceraphyl® 60, 5718
Ceraphyl® 65, 5719
Ceraphyl® 70, 5720
Ceraphyl® 85, 5721
Ceraphyl® 140, 5722, *See* decyl oleate, 7869
Ceraphyl® 140-A, 5723
Ceraphyl® 230, 5724
Ceraphyl® 368, 5725
Ceraphyl® 375, 5726
Ceraphyl® 424, 5727
Ceraphyl® 791, 5728
Ceraphyl® 847, 5729
Ceraphyl® GA, 5730
Ceraphyl® ICA, 5731
Ceraphyl® IPL, 5732
Cerasine Yellow GG, *See* Butter yellow, 4759
Cerasynt 660, *See* Ablunol DEGMS, 99
Cerasynt PA, *See* Witconol 2380, 34498
Cerasynt PN, *See* Witconol 2380, 34498
Cerasynt® 303, 5733
Cerasynt® 840, 5734
Cerasynt® D, 5735
Cerasynt® M, PA, 5736
Cerasynt® WM, 5737
Cercobin, 5738
Cercobin M, *See* thiophanate methyl, 31706
Cercobin methyl, *See* thiophanate methyl, 31706
Cereclor, 5739, *See* Chlorcosane, 6092
Cereflo, 5740
Cerelose, 5741
Ceremix, 5742

PEG-2 dioleate series, 23040
Chemax PEG 400 DO, 5923
Chemax PEG 600 DO, 5924
Chemax SBO, 5925
Chemax SCO, 5926
Chemax SCO, SCO/75, *See* Eureka 102, 11207
Chemax TO-8, *See* PEG tallate series, 23035
Chemax TO-10, 5927
Chemax TO-10, *See* PEG tallate series, 23035
Chemax TO-16, 5928
Chemax TO-16, *See* PEG tallate series, 23035
Chemax TO-B, 5929
Chembetaine BW, 5930
Chembetaine BW, CB, *See* Coco betaine, 6561
Chembetaine C, CGF, 5931
Chembetaine CAS, 5932, *See* Cocamidopropyl hydroxysultaine, 6555
Chembetaine CB, 5933
Chembetaine L, 5934, *See* lauramidopropyl betaine, 16830
Chembetaine OL, *See* oleyl betaine, 22132
Chembetaine OL-30, 5935, *See* oleyl betaine, 22132
Chembetaine S, 5936
Chembetaine TG, 5937
Chembetaine® C, CGF, CL, *See* Cocamidopropyl betaine, 6551
Chem-Calk® 500, 5938
Chemcaulk, 5939
Chemcoat, 5940
Chemcogen AC, 5941
Chemdur, 5942
Chemeen 18-2, 5943
Chemeen 18-2, *See* PEG stearamine series, 23026
Chemeen C-2, 5944
Chemeen DT-3, 5945
Chemeen HT-5, 5946
Chemeen O-30, O-30/80, 5947
Chemeen S-2, 5948, *See* PEG soyamine series, 23033
Chemeen T-2, 5949
Chemester 300-OC, *See* Ablunol 200MO, 96
Chemet, *See* MPI DMSA Kidney Reagent, 20393
Chemetron Fire Shield, *See* Octoguard FR-10, 22000, Timonox, 31807
Chemfac NC-0910, 5950
Chemfac PA-080, PB-082, 5951
Chemfac PB-184, 5952, *See* oleth-phosphate Series (2-20), 22127
Chemfac PC-006, 5953
Chemfac PD-600, 5954
Chemfac PN-322, PX-322, 5955
Chemfac RD-1200, 5956
Chemfax 5AM-100, 5957
Chem-Fish, *See* rotenone, 27000
Chemglaze®, Chemglaze® Z-004, 5958
Chem-Hoe, *See* Propham, 25186
Chemical 39 Base, 5959
Chemical Base 6532, 5960
Chemical Mace, *See* chloroacetophenone, 6112
chemical red, *See* Indian Red, 15068
Chemidex B, 5961, *See* Behenamidopropyl dimethylamine, 3896
Chemidex C, 5962, *See* Cocamidopropyl dimethylamine, 6552
Chemidex L, 5963, *See* lauramidopropyl dimethylamine, 16831
Chemidex M, 5964
Chemidex O, 5965, *See* oleamidopropyl dimethylamine, 22109
Chemidex P, 5966, *See* Palmitamidopropyl dimethylamine, 22630
Chemidex R, 5967
Chemidex S, 5968
Chemidex SE, 5969
Chemidex SI, 5970
Chemidex SO, 5971
Chemidex T, 5972
Chemidex WC, 5973, *See* Cocamidopropyl dimethylamine, 6552
chemifluor, *See* sodium fluoride, 28764
Chemigum® HR662, 5974
Chemigum® Latex 260, 5975
Chemigum® N318B, N917, 5976
Chemigum® N5, N7, 5977
Chemigum® NX-775, 5978
Chemigum® P7-D, 5979
Chemigum® P83, 5980
Chemigum® TPE 03050, 5981
Chemlease 55, 5982
Chemlease 88, 5983
Chemlease 158R, 5985
Chemlease 906E, 5984
Chemlease SP 40, 5986
Chemline, 5987
Chemlock, 5988
Chemlok® 205, 5989
Chemlok® 220, 5990
Chemlok® 459, 5991
Chemlok® 607, 5992
Chemlube, 5993
chem-mite, *See* rotenone, 27000
Chemocide PK, *See* Nipasol M, 21282
Chemocin, *See* Cupravit®, 7370
Chemonite Part A, *See* arsenic acid, 3137
Chemoside PK, *See* Nipasol M, 21282
Chemoxide CAW, 5994, *See* Cocamidopropylamine oxide, 6556
Chemoxide L, 5995, *See* lauramidopropyl betaine, 16830
Chemoxide LM-30, 5996, *See* lauramine oxide, 16832
Chemoxide O, 5997
Chemoxide O1, 5998
Chemoxide SAO, 5999
Chemoxide ST, 6000
Chemoxide T, 6001
Chemoxide TAO, 6002
Chemoxide WC, 6003
Chemoxide® WC, *See* Cocamine oxide, 6557
Chemphonate AMP, 6004, *See* Aminotrimethylene phosphonic acid, 2191
Chemphonate AMP-S, 6005
Chemphonate HEDP, 6006
Chemphonate NP, 6007
Chemphos TC-227, 6008
Chemphos TC-231S, 6009
Chemphos TC-337, 6010
Chemphos TC-341, 6011
Chemphos TC-444, 6012
Chemphos TDAP, 6013
Chemphos TR-414W, 6014
Chemphos TR-505, 6015
Chemphos TR-505, TR-515, TR-541, *See* oleth-phosphate Series (2-20), 22127
Chemphos TR-505D, 6016
Chemphos TR-510, 6017
Chemphos TR-510S, 6018
Chemphos TR-515, 6019
Chemphos TR-515D, 6020
Chemphos TR-541, 6021
Chemprene 50, 75, 6022
Chemprene R-10, 6023
Chemquat 12-50, 6024
Chemquat 16-50, 6025
Chemquat 16-50, *See* Cetrimonium chloride, 5814
Chemquat C/33W, 6026, *See* Coco trimethyl ammonium chloride, 6562
Chemraz®, 6027
Chem-Rez, 6028
Chemsalan NLS 30, 6029
Chemsalan RLM 28, 6030
Chemset, 6031
Chemstat® 106G/90, 6032
Chemstat® 122, 6033
Chemstat® 172, 6034
Chemstat® 192/NCP, 6035
Chemstat® 273-C, 6036, *See* Cocamide DEA, 6547
Chemstat® 273-E, 6037, *See* bishydroxyethyl-N,N-Bis(2-hydroxyethyl) stearyl amine, 4231, PEG stearamine series, 23026
Chemstat® 9820A, 6038
Chemstat® AF-906, 6039
Chemstat® HTSA#1, 6040
Chemstat® HTSA#3, 6041
Chemstat® P-400, 6042
Chemstat® PS-101, 6043
Chemsulf S2EH-Na, 6044
Chemsulf SBO/65, 6045
Chemsulf SCO/75, 6046
Chemtac 20, 35, 6047
Chemtan A 60, *See* Amine CS-1246, 2154, Oxaban®-E, 22405
Chemtech Cypermethrin, 6048
Chem-Trete®, 6049
Chemzoline 1411, 6050
Chemzoline C-22, 6051
Chemzoline T-11, T-33, 6052
Chemzoline T-44, 6053
Chenzinsky-Plehn's solution, 6054
Ches® 500, 6055
Chesguar C10,, *See* Jaguar® C-13S, C-14S, 15496
Chesguar HP4, *See* hydroxypropyl guar, 14666
Chesguar HP4R, *See* hydroxypropyl guar, 14666
Chesguar HP6, *See* hydroxypropyl guar, 14666
Chia oil, 6056
Chian turpentine, 6057
Chicle, 6058
Chierite, 6059
Chierol, 6060
Chilcote, 6061
Childion, 6062
Chile saltpeter, *See* sodium nitrate, 28800
Chili niter, *See* Chili saltpeter, 6063
Chili saltpeter, 6063
Chilisa FE®, 6064
Chiltern Cropspray 11E, 6065
Chiltern Cyperkill 10, 6066
Chiltern Fazor, 6067
Chiltern IPU, 6068
Chiltern Kocide 101, 6069, *See* cupric hydroxide, 7382
Chiltern Ole, 6070, *See* Chlorothalonil, 6150
Chiltern Pyrazol, 6071
Chim, *See* Terrathion, 31343
Chimassorb 90, *See* Cyasorb® UV 9, 7495, Neo Heliopan® BB, 20867, Rhodialux A, 26675
Chimassorb® 119FL, 6072
Chimin IMB, *See* disodium cocoamphodiacetate, 8643
Chimin LX, *See* lauramidopropyl betaine, 16830
China clay, *See* ASP®, 3230, Bilt-Cote®, 4147 kaolin, 15725, Kayphobe-ABO, 15816, Peerless®, 23023, SPS, 29349, Vanclay, 33304
China silver, *See* Argyrolith, 2867, nickel silvers, 21132
chinaldine, *See* quinaldine, 25678
Chinese bean oil, *See* soya bean oil, 29169
Chinese bronze, 6073
Chinese glue, 6074
Chinese grass, *See* ramie, 25857
Chinese insinglass, *See* Agar (Agar-agar), 955, vegetable gelatin, 33518
Chinese red, *See* chrome red, 6209
Chinese silver, 6075
Chinese tallow, 6076
Chinese vermilion, *See* scarlet vermillon, 27550
Chinese wax, 6077
Chinese white, 6078, *See* Activox, 507, zinc oxide, 34786
Chinese white copper, 6079
Chinese wood oil, 6080, *See* tung oil, 32580
Chinese yellow, *See* ocher, 21976
chinomethionat, *See* quinomethionate, 25686
Chinosol, *See* Quinosol, 25687
Chinufur, *See* Rampart, 25860
Chio turpentine, *See* Chian turpentine, 6057
Chios turpentine, *See* Chian turpentine, 6057
Chip calcium chloride, *See* Sure-Step, 30334
Chipco Thiram 75, *See* Rezifilm, 26521
Chiptox, *See* MCPA, 19183
Chisso-rite, 6081
Chitan, *See* Kytamer® PC, 16522

5-Chloro-3-(1,1-dimethylethyl)-6-methyl-2,4-(1H,3H)-pyrimidinedione, *See* terbacil, 31220
N'-(3-chloro-4-(5-(1,1-dimethylethyl)-2-oxo-1,3,4-oxadiazol-3(2H)-yl)phenyl)-N,N-dimethylurea, *See* Pradone Plus, 24850
4-Chloro-3,5-dimethylphenol, *See* chloroxylenol, 6159, Nipacide® MX, 21257
Chloro-3,5-dimethylphenol, *See* chloroxylenol, 6159
2-chloro-N-(2,6-dimethylphenyl)-N-(1H-pyrazol-1-ylmethyl)acetamide, *See* metazachlor, 19465, Track, 32129
2-chloro-N-(2,6-dimethylphenyl)-N-(tetrahydro-2-oxo-3-furanyl)acetamide, *See* Patafol, 22881
Chlorodimethylsilane, *See* CD5470, 5507
7-Chloro-17,21-dipropionate-11-hydroxy-16-methylpregna-1,4-diene-3,20-dione, *See* Vaderm, 33237
4-Chloro-N-1,3-dithietan-2-ylidene-2-methylbenzeneamine, *See* Abequito, 38
Chlorodyne, *See* Embedyne, 10023
1-chloro-2,3-epoxypropane, *See* Epichlorhydrin, 10632
Chloroethane, *See* ethyl chloride, 11067
2-Chloroethanephosphonic acid, *See* Cerone, 5756
Chloroethanoic acid, *See* monochloroacetic acid, 20221
2-chloro-ethanol Phosphate (3:1), *See* Genomoll P, 12838
2-chloroethanolphosphate, *See* Genomoll P, 12838
chloroethene homopolymer, *See* polyvinyl chloride, 24668
2-chloroethyl phosphonic acid, *See* Cerone, 5756
(Chloroethyl)phosphonic acid, *See* Cerone, 5756
2-chloro-4-ethylamino-6-isopropylamino-1,3,5-triazine, *See* Atranex, 3393
N-(2-chloroethyl)dimethylamine hydrochloride, *See* dimethylaminoethyl chloride hydrochloride, 8516
chloroethylene polymer, *See* Korogel, 16354, Koroplate, 16357, polyvinyl chloride, 24668
2-chloroethylmethyldichlorosilane, 6122, *See* CC3005, 5486
6-chloro-N-ethyl-N'-(1-methylethyl)-1,3,5-triazine-2,4-diamine, *See* atrazine, 3394
2-chloro-N-(2-ethyl-6-methylphenyl)-N-(2-methoxy-1-methylethyl)acetamide, *See* metolachlor, 19585
(2-chloroethyl)trimethylammonium chloride, *See* CCC 700, 5497, Chlormequat chloride, 6110, Cycogan, 7545
chlorofenvinphos, *See* Chlorfenvinphos, 6098
Chlorofin 42, 6123
Chloroflo® 40, 42, 6124
chloroform, 6125
Chloroformic acid benzyl ester, *See* benzyl chloroformate, 3988
chloroformyl chloride, *See* phosgene, 23701
chlorofos, *See* trichlorfon, 32212
Chlorohexidine Diacetate, *See* chlorhexidine acetate, 6100
Chlorohydric acid, *See* hydrochloric acid, 14575
2-Chloro-5-hydroxy-1,3-dimethylbenzene, *See* chloroxylenol, 6159
5-chloro-2-hydroxydiphenylmethane, *See* Nipacide® BCP, 21252
3-chloro-7-hydroxy-4-methyl-2H-1-benzopyran-2-one bis(chloroethyl) phosphate, *See* Haloxon, 13567
3-chloro-7-hydroxy-4-methylcoumarin bis(2-chloroethyl) phosphate, *See* Haloxon, 13567
3-Chloro-2-hydroxypropyl trimethylammonium chloride (50% aq solution), 6127, *See* Cationaster-C, 5459, CHPTA 65%, 6176, Reagens-CF2, 25928, Quab 188, 25562
2-chloro-5-hydroxytoluene, *See* Raschit, 25892
6-chloro-3-hydroxytoluene, *See* Raschit, 25892
2-Chloro-5-hydroxy-*m*-xylene, *See* chloroxylenol, 6159
Chloro-5-hydroxy-*m*-xylene, *See* chloroxylenol, 6159
Chloro-IFK, *See* Chlorpropam, 6162
Chloro-IPC, *See* Chlorpropam, 6162
2-Chloro-N-(1-methylethyl)-N-phenylacetamide, *See* Ramrod, 25861
α-Chloro-N-isopropylacetanilide, *See* Ramrod, 25861
2-Chloro-N-isopropyl-N-phenylacetamide, *See* Ramrod, 25861
2-Chloro-N-isopropylacetanilide, *See* Ramrod, 25861
2-Chloro-N-isopropyl-acetanilide, *See* Prolex, 25128
2-chloro-N-isopropylacetanilideCP 31393, *See* Propachlor, 25163
***o*-(chloromercuri)phenol**, *See* mercufenol chloride, 19345
chloromethane, *See* methyl chloride, 19522
Chloromethyl chloroformate, *See* Palite, 22610
5-Chloro-2-methyl isothiazolones, *See* Kathon® 886, 15793
Chloromethyl menthyl ether, *See* forman, 12263
Chloromethyl methyldichlorosilane, *See* CC3275, 5488
chloromethyl oxirane, *See* Epichlorhydrin, 10632
N-Chloro-4-methylbenzenesulfonamide sodium salt, *See* Chloramine T, 6086, Ketjensept, 16133
O-(3-chloro-4-methyl-2-oxo-2H-1-benzopyran-7-yl) O,O-diethyl phosphorothioate, *See* Perizin®, 23276
N-Chlorobenzenesulfonamide sodium salt, *See* Chloramine B, 6085
chloromethyldichloromethylsilane, 6128
chloromethyldimethylchlorosilane, 6129
Chloromethyldimethylchlorosilane, *See* CC3270, 5487, chloromethyldimethylchlorosilane, 6129
(Chloromethyl)dimethylsilane, *See* chloromethyldimethylchlorosilane, 6129
4-chloro-α-(1-methylethyl)benzeneacetic acid cyano(3-phenoxyphenyl)methyl ester, *See* fenvalerate, 11620
2-Chloro-N-(1-methylethyl)-N-phenylacetamide, *See* Propachlor, 25163, 25164, Ramrod, 25861
5-Chloro-2-methyl-4-isothiazolin-3-one and 2-methyl-4-isothiazolin-3-one, *See* Biosperse® 250, 4208
Chloromethylmethyldichlorosilane, *See* CC3275, 5488, chloromethyldichloromethylsilane, 6128
(Chloromethyl)methyldichlorosilane, *See* CC3275, 5488, chloromethyldichloromethylsilane, 6128
Chloromethyloxirane, *See* Epichlorhydrin, 10632
4-chloro-3-methylphenol, *See* Raschit, 25892
4-(4-Chloro-2-methylphenoxy) butyric acid, *See* MCPB, 19184
(4-Chloro-2-methylphenoxy)acetic acid, *See* MCPA, 19183
4-(4-chloro-2-methylphenoxy)butanoate, *See* MCPB, 19184
(±)-2-(4-chloro-2-methylphenoxy)propanoic acid, *See* Mecoprop, 19204, Verdone CDA, 33580
N-((4-chlorophenyl)methyl)-N-cyclopentyl-N'-phenylurea, *See* pencycuron, 23097, 23098, 23099
N-[[(4-chlorophenyl)amino]carbonyl]-2,6-difluorobenzamide, *See* Diflubenzuron, 8450
3-(3-chloro-4-methylphenyl)-1,1-dimethylurea, *See* Talisman, 30735
chloro-4-(methylphenyl)-2-methylpentamide, *See* Solan, 28887
N-(3-Chloro-4-methylphenyl)-2-methylpentanamide, *See* pentanochlor, 23165
N'-(3-chloro-4-methylphenyl)-N,N-dimethylurea, *See* Talisman, 30735, Tolurex, 31958
3'-Chloro-2-methyl-toluidide, *See* pentanochlor, 23165
chloromethyltrimethyl silane, 6130, *See* CC3285, 5489, chloromethyltrimethyl silane, 6130
3-chloro-4-methylumbelliferone di(2-chloroethyl) phosphate, *See* Haloxon, 13567
Chloroneb 65W Fungicide, 6131
Chloroneb Systemic Flowable Fungicide, 6132
Chloro-N-isopropylacetanilide I, *See* Propachlor, 25163
chloronitrous acid, *See* aqua regia, 2682
S-[(6-chloro-2-oxo-3(2H)-benzoxazolyl)methyl] O,O-diethyl phosphorodithioate, *See* phosalone, 23691, Zolone, 34888
S-[(6-chloro-2-oxooxazolo[4,5-b]pyridin-3(2H)-yl)methyl] O,O-dimethyl phosphorothioate, *See* azamethiphos, 3517
4-chloro-2-oxo-3(2H)-benzothiazoleacetic acid, *See* benazolin, 3935
Chloroperoxyl, *See* Anthium Dioxcide, 2521
Chlorophacinone, 6133, *See* Drat, 8972, Ridene, 26746
chlorophene, *See* Nipacide® BCP, 21252
Chlorophenothane, *See* DDT, 7820
Chlorophenotoxum, *See* DDT, 7820
β-(4-chlorophenoxy)- α-(1,1-dimethylethyl)-1H-1,2,4-Triazole-1-ethanol, *See* triadimenol, 32185
1-(4-chlorophenoxy)-3,3-dimethyl-1-(1H-1,2,4-triazol-1-yl) butan-2-one, *See* Triadimefon, 32184
[1S-[1α(Z),2β(R*), 3α,5α]]-7-[2-[[3-(3-chlorophenoxy)-2-hydroxypropyl]thio]-3,5-dihydroxycyclopentyl]-5-heptenoic acid, *See* Reprodin®, 26149
3-[*p*-(*p*-chlorophenoxy)phenyl]-1,1-dimethylurea, *See* chloroxuron, 6158, Tenoran, 31119
3-[4-(4-chlorophenoxy)phenyl]-1,1-dimethylurea, *See* chloroxuron, 6158, Tenoran, 31119
N-[4-(4-chlorophenoxy)phenyl]-N,N-dimethylurea, *See* chloroxuron, 6158, Tenoran, 31119
3-(4-Chlorophenoxy)-1,2-propanediol, *See* Chlorphenesin, 6161
***p*-chlorophenyl chloride**, *See* *p*-dichlorobenzene, 8386
N,5-Bis(4-chlorophenyl)-3,5-dihydro-3-[(1-methylethyl)imino]-2-phenazinamine, *See* Clofazimine, 6427
N,N-bis(4-chlorophenyl)-3,12-diimino-2,4,11,13-tetraazatetradecanediimidamide diacetate, *See* chlorhexidine acetate, 6100
***p*-chlorophenyl- α-glyceryl ether**, *See* Chlorphenesin, 6161
Chlorophenyl methyl siloxane, *See* SF1250, 28130, Versilube® F50, 33627
(3-chlorophenyl)carbamic acid, 1-methylethyl ester, *See* Chlorpropam, 6162
2-(*o*-Chlorophenyl)-2-(*p*-chlorophenyl)-1,1-dichloroethane, *See* Lysoform, 18095
N-(4-chlorophenyl)-N'-(3,4-dichlorophenyl)urea, *See* TCC, 30856
N,N'-bis (4-chlorophenyl)-3,12-diimino-2,4,11,13-tetraazatetradecane-diimidamide compound with D-gluconic acid, *See* chlorhexidine digluconate, 6101
1-(4-Chlorophenyl)-2-[[3-(10,11-dihydro-5H-dibenz[b,f]azepin-5-yl)propyl]methylamino]ethanone, *See* Lofepramin Hydrochloride, 17552
(RS)-1-(4-chlorophenyl)-4,4-dimethyl-3-(1H-1,2,4-triazol-1-ylmethyl)pentan-3-ol, *See* Raxil, 25908
2-Chloro-1-phenylethanone, *See* chloroacetophenone, 6112
(±)-α-[2-(4-chlorophenyl)ethyl]- α-(1,1-dimethylethyl)-1H-1,2,4-triazole-1-ethanol, *See* Raxil, 25908
α-(2-chlorophenyl)- α-(4-fluorophenyl)-5-pyrimidinemethanol, *See* nuarimol, 21762
N'-(4-chlorophenyl)-N-methoxy-N-methylurea, *See* Aresin, 2839
β-((4-chlorophenyl)methyl)- α-(1,1-dimethylethyl)-1H-1,2,4-triazole-1-ethanol, (R*,R*)-(+-)-, *See* paclobutrazol, 22542
[(chlorophenyl)-methyl]-N-cyclopentyl-N'-phenylurea, *See* pencycuron, 23097, 23098, 23099
(R*,R*)-(±)-β-[(4-chlorophenyl)methyl]- α-(1,1dimethylethyl)-1H-1,2,4-triazole-1-ethanol, *See* paclobutrazol, 22540, 22541
1-[2-[[(4-Chlorophenyl)methyl]-thio]-2-(2,4-dichlorophenyl)ethyl]-1H-imidazole, *See* Sulconazole, 29958
2-Chloro-6-phenylphenol, *See* Dowicide 3, 8928
6-Chloro-2-phenylphenol, *See* Dowicide 3, 8928
Chloro-*o*-phenylphenol, *See* Dowicide 3, 8928
Chloro-6-phenylphenol, *See* Dowicide 3, 8928
2-[(4-chlorophenyl)phenylacetyl]-1H-Indene-1,3(2H)-dione, *See* Chlorophacinone, 6133, Ridene, 26746
2-[(*p*-chlorophenyl)phenylacetyl]-1,3-indandione, *See* Karate, 15758
(Chlorophenyl)phenylacetyl-1,3-indandione, *See* Karate, 15758, Ridene, 26746
2-((p-Chlorophenyl)phenylacetyl)-1,3-indandione, *See* Karate, 15758, Ridene, 26746

Chemical and Trade Names

2-[(4-Chlorophenyl)phenylacetyl]-1H-indene-1,3(2H)-dione, *See* Karate, 15758, Ridene, 26746
***p*-chlorophenyl 2,4,5-trichlorophenyl sulfone**, *See* Tedion V-18, 30919, tetradifon, 31366
4-chlorophenyl-2,4,5-trichlorophenyl sulfone, *See* Tedion V-18, 30919, tetradifon, 31366
chlorophyll, 6134, *See* Amplex, 2362
chloropicrin, 6135
Chloropolyethylene, *See* Solpolac, 28963
Chloropotassuril, *See* Kay Ciel, 15807, K-Contin, 15818, K. Tab, 15642, Kaon-Cl, 15729, K-Lor, 16226, K-Lyte/C1, 16234, KM Potassium Chloride, 16242, Potassium chloride, 24748, potassium chloride, 24749
chloroprene, 6136, *See* Adcora P6, 631, Baypren® 110, 3828
Chloropropene, *See* Barchlor, 3626
3-chloroprene, *See* Barchlor, 3626
α-Chloroprene, *See* chloroprene, 6136
β-Chloroprene, *See* chloroprene, 6136
Chloroprene rubber, *See* Baypren® 110 VSC, 3829, Baypren® 216, 3830, Baypren® 310, 3831, Baypren® AT-H, AT-M, AT-S, 3832, Baypren® EM1, 3833, Baypren® M1, 3836, polychloroprene, 24386, Adcora P6, 631
γ-chloropropylene-oxide, *See* Epichlorhydrin, 10632
1-chloro-2-propene, *See* Barchlor, 3626
3-chloro-1-propene, *See* Barchlor, 3626
3-Chloropropene, *See* Barchlor, 3626
3-chloropropene-1, *See* Barchlor, 3626
Chloropropham, *See* Spud-Nic®, 29350, Chlorpropam, 6162
3-Chloropropyl trimethoxy silane, *See* Dow Corning® Z-6076, 8883
α-chloropropylene, *See* Barchlor, 3626
3-chloro-1-propylene, *See* Barchlor, 3626
3-chloropropylene, *See* Barchlor, 3626
chloropropylene oxide, *See* Epichlorhydrin, 10632
chloropropylene oxide elastomer, *See* Epichlorohydrin elastomer, 10633
3-chloropropylmethyldimethoxysilane, 6137, *See* CC3290, 5490
chloropropyltrichlorosilane, 6139, *See* CC3291, 5491
3-Chloropropyltrichlorosilane, *See* CC3291, 5491
3-chloropropyltriethoxysilane, 6140, 6141, *See* CC3292, 5492, Dynasylan® CPTEO, 9402
(3-Chloropropyl)triethoxysilane, *See* 3-chloropropyltriethoxysilane, 6141
3-(chloropropyl)trimethoxysilane, *See* 3-chloropropyltrimethoxysilane, 6143
3-chloropropyltrimethoxysilane, 6142, 6143, CC3300, 5493, CC3555, 5495, chloropropyltrimethoxysilane, 6138, Dynasylan® CPTMO, 9403
(3-Chloropropyl)trimethoxy-silane, *See* Dynasylan® CPTMO, 9403
2-chloro-N-(pyrazol-1-ylmethyl)acet-2',6'-xylidide, *See* Track, 32129
2-chloropyridine, 6144, *See* 2-chloropyridine, 6144
1-[(6-Chloro-3-pyridinyl)methyl]-4,5-dihydro-N-nitro-1H-imidazol-2-amine, *See* Confidor, 6725
1-(6-Chloro-3-pyridiylmethyl)-N-nitroimidazolidin-2-ylideneamine, *See* Confidor, 6725
Chloropyrifos, *See* Chlorpyrifos, 6163
Chloropyrifos-methyl, *See* Chlorpyrifos-methyl, 6164
Chloropyriphos, *See* Chlorpyrifos, 6163
Chloropyripos, *See* Chlorpyrifos, 6163
1-Chloro-2,5-pyrrolidinedione, *See* N-chlorosuccinimide, 6148
2-[4-[(6-chloro-2-quinoxalinyl)oxy]phenoxy] propanoic acid ethyl ester, *See* quizalofop-ethyl, 25697
4-chlororesorcinol, 6145
Chloros, 6146, *See* sodium hypochlorite, 28773
Chlorosoda, 6147
Chlorosuccinimide, *See* N-chlorosuccinimide, 6148
N-chlorosuccinimide, 6148
Chlorosulfonated paraffins, *See* Cloparten, Cloparten Z, 6434
Chlorotex, 6149
Chlorothal, *See* dimethyl tetrachloroterephthalate, 8512
Chlorothalonil, 6150, *See* Bombardier, 4386, Bravo 500, 4490, Chiltern Ole, 6070, Chlorothalonil, 6150, Contact 75, 6738, Daconil Turf, 7681, Groutcide 75, 13432, Jupital, 15633, Nopcocide® N-40-D, 21487, Power Chlorothalonil 50, 24818, Sipcam UK Rover 500, 28516
chlorothalonil-fenpropimorph, *See* Corbel® CL, Corbel® Star, 6803
chlorothalonil-flutriafol, *See* Impact Excel, 14909
chlorothalonil-metalaxyl, *See* Folio 575FW, 12206
Chlorothanonil, *See* Chlorothalonil, 6150
Chlorothanonil, *See* Repulse, 26151
Chlorothene, 6151, *See* trichloroethane, 32216
Chlorothene NU, *See* Chlorothene, 6151
Chlorothene VG, *See* Chlorothene, 6151
2-chlorothiophene, 6152
***o*-chlorotoluene**, *See* Halso® 99, 13570
(N-chloro-*p*-toluenesulfonamido)sodium, *See* Chloramine T, 6086, Ketjensept, 16133
chlorotoluron, *See* Dicurane 500 FW, 8409, Ludorum, 17778, Talisman, 30735, Toro, 32074
3-(3-chloro-*p*-tolyl)-1,1-dimethylurea, *See* Talisman, 30735, Tolurex, 31958
(Chloro-o-tolyloxy)acetic acid, *See* MCPA, 19183
4-(4-chloro-o-tolyloxy)butyric acid, *See* MCPB, 19184
(±)-2-[(4-chloro-o-tolyl)oxy]-propionic acid, *See* Verdone CDA, 33580
(RS)-2-(4-Chloro-*o*-tolyloxypropionic acid, *See* Mecoprop, 19204
chlorotrfluoromethane, 6153, *See* Genetron® 13, 12819
6-chloro-N,N,N'-triethyl-1,3,5-triazine-2,4-diamine, *See* trietazine, 32238
1-chloro-2,2,2-trifluoroethyl ether, *See* Isoflurane, 15363
2-chloro-N-[[[4-(trifluoromethoxy)phenyl]amino] carbonyl]benzamide, *See* triflumuron, 32248
5-[2-chloro-4-(trifluoromethyl)phenoxy]-2-nitrobenzoic acid, *See* acifluorfen, 343
3-(2-Chloro-3,3,3-trifluoro-1-propenyl)-2,2-dimethylcycloproanecarboxylic acid cyano(3-phenoxyphenyl)methyl ester, *See* Karate, 15757
2-Chloro-N,N,N-trimethylethanaminium chloride, *See* CCC 700, 5497, Chlormequat chloride, 6110
2-chloro-N,N,6-trimethyl-4-pyrimidinamine, *See* Crimidine, 7053
Chlorotrimethylsilane, *See* CT2950, 7325
chlorotrimethylsilicane, *See* CT2950, 7325
chlorous acid sodium salt, *See* sodium chlorite, 28749
β-Chlorovinyldichloroarsine, *See* Lewisite, 17089
Chlorovis 150A, 6154
Chlorowax, 6155
Chlorowax 40, *See* Electrofine® S-70, 9782
Chlorowax 40, 50, 70, LV, 6156
Chlorowax 500c, *See* Electrofine® S-70, 9782
Chlorox, *See* sodium hypochlorite, 28773
chloroxethose, 6157
chloroxuron, 6158, *See* Tenoran, 31119
chloroxyfenidim, *See* chloroxuron, 6158
chloroxylenol, 6159, *See* Ayrtol, 3514, chloroxylenol, 6159
***p*-chloro-*m*-xylenol**, *See* Nipacide® MX, 21257
2-Chloro-*m*-xylenol, *See* chloroxylenol, 6159
4-Chloro-3-xylenol, *See* chloroxylenol, 6159
***para*-chloro-*meta*-xylenol**, *See* chloroxylenol, 6159
Chloro-3,5-xylenol, *See* chloroxylenol, 6159
Chloropyriphos, *See* Chlorpyrifos, 6163
Chlorozone, 6160
chlorphencarb, *See* chloroxuron, 6158
Chlorphenesin, 6161
Chlorpromazine hydrochloride, *See* Largactil, 16786
Chlorpropam, 6162
chlorpropham, *See* Atlas CIPC 40, 3303 Campbell's Chlorpropam, 6162 CIPC 40%, 5055 Mirvale, 19939 MSS CIPC, 20428 Residuren, 26166
Chlorprophame, *See* Chlorpropam, 6162
chlorpropham-fenuron, *See* Croptex Chrome, 7214
Chlorpyrifos, 6163, *See* Crossfire, 7239, Dowco 179, 8902, *See* Dursban®, 9272, Dursban® 14G, 9273, Dursban® 2E, 9274, Spannit, 29198, Talon, 30751
Chlorpyrifos-dimethoate, *See* Atlas Sheriff, 3328, Salut®, 27303
chlorpyrifos-ethyl, *See* Talon, 30751
Chlorpyrifos-methyl, 6164, *See* Chlorpyrifos-methyl, 6164 Cooper Graincote, 6759 Reldan 50, 26037
Chlorpyrofos, *See* Chlorpyrifos, 6163
Chlorpyrophos, *See* Chlorpyrifos, 6163
Chlor-Tabs, 6165
Chlorthal-Dimethyl, *See* dimethyl tetrachloroterephthalate, 8512
chlorthal-dimethyl-propachlor, *See* Decimate, 7851
Chlorthal-methyl, *See* dimethyl tetrachloroterephthalate, 8512, Vegetable Turf and Ornamental Weeder, 33526
Chlortiepin, *See* thiodan, 31680
Chlortoluron, *See* Tolurgan, 31959, Talisman, 30735, Tolurex, 31958
Chlorvinphos, *See* dichlorvos, 8394
chloryl, *See* ethyl chloride, 11067
chloryl anesthetic, *See* ethyl chloride, 11067
Chlorylen, *See* trichloroethylene, 32217
Chlumin, 6166
Cholaxine, *See* A-625/641ABS 301K, ABS 500FR-1, 10
Cholebrine, 6167
cholecalciferol, *See* vitamin D+73, 33941
Cholegnostyl, *See* Bromeikon, 4568
Cholepulvis, *See* keraphen, 16055
Cholest-5-en-3 β-ol, *See* cholesterol, 6168, Kathro, 15795
cholesterin, *See* cholesterol, 6168, Kathro, 15795
cholesterol, 6168, *See* Dastar, 7788, Fancol CH, 11435, Kathro, 15795
Cholestrophane, 6169
Choletec, 6170
Choleth-5, *See* Emalex CS-5, 9937
Choleth-24, *See* Fancol CH-24, 11436, Forlan C-24, 12249, Solulan® C-24, 29005
Choleth-24-ceteth-24, *See* Fancol CH-24, 11436
cholic acid-hexamine, *See* Felamine, 11584
choline chloride, 6171, *See* Chlormequat chloride, 6110
choline dichloride, *See* CCC 700, 5497, Chlormequat chloride, 6110
choline hydrochloride, *See* choline chloride, 6171
cholinechloride, *See* Terpal® CC, M, 31310
Chlografin Meglumine, 6172
Cholumbrin, *See* keraphen, 16055
Chondroitin sulfate, *See* Cromoist CS, 7199
Chondrus, *See* carrageenan, 5315
Chopper, *See* imazapyr, 14893
CHP, 6173
CHP-5, 6174
CHP-158, 6175
CHPTA 65%, 6176 , *See* (3-Chloro-2-hydroxy)*o*-propyltrimethyl ammonium chloride, 6126
Christolit, 6177
Chrogo U42, 6178
Chroma-Cal®, Chroma-Chem, 6179
Chromaflo®, 6180
Chromagan, 6181
Chromagel, 6182
Chromagen, 6183
Chromaguard, 6184
Chromalay, 6185
chromaline, 6186
Chroma-Lite®, 6187
chromaloy, 6188
chromaluminum, 6189
Chroman B, 6190
Chroman Co, 6191
chromar, *See* isopropanol, 15405
Chromargans, 6192
Chromargyre, *See* mercurochrome, 19354
Chromaset DF-100, 6193
Chromasist 1487A, 6194
Chromastral, 6195
Chromate of soda, *See* sodium chromate, 28750
chromate(1-), diaminetetrakis(thiocyanato-n)-, ammonium, (OC-6-11)-, *See* Reinecke's salt, 26033
chromatized gelatin, 6196

Chromatogram, 6197
chromax, 6198
chromax bronze, 6199
Chrombral, 6200
chrome, 6201
chrome alum, *See* chromic potassium sulfate, 6224
chrome amalgam, 6202
chrome black, 6203
chrome bronze, 6204, *See* chromous trioxide, 6247
chrome carmine, chinese scarlet, *See* chrome red, 6209
chrome cement, *See* chromatized gelatin, 6196
chrome cinnabar, *See* chrome red, 6209
chrome complex solution, *See* Arikrome S, 2878
chrome emerald green, *See* chromium green, 6232
Chrome Fast Cyanine B, BN, 6205
chrome garnet, *See* chrome red, 6209
chrome glue, *See* chromatized gelatin, 6196
chrome green, *See* Accrox, 258, chromic oxide, 6223, chromium green, 6232, M100, 18107
Chrome Green 106, 6206, *See* chromic oxide, 6223
chrome iron, 6207
chrome lignosulfonate, *See* Borrewell C, 4451
chrome ocher, *See* M100, 18107
chrome ochre, *See* Accrox, 258
chrome orange, *See* chrome red, 6209
chrome oxide, *See* chromic oxide, 6223, M100, 18107
chrome oxide green, *See* Accrox, 258, chromic oxide, 6223, M100, 18107
chrome prune, 6208
chrome red, 6209
chrome ruby, *See* chrome red, 6209
chrome steels, 6210
chrome steels, high, 6211
chrome yellow, 6212
Chromeduol, 6213
chromels, 6214
chrome-nickel steel, high, 6215
Chrometan, 6216
Chrome-tin Pink, 6217
Chrometrace, 6218
Chromglaserite, 6219
chromia, *See* Accrox, 258, chromic oxide, 6223, chromium oxide (ic), 6236, M100, 18107
chromic acetate, *See* chromium acetate (ic), 6229
chromic acid, 6220, *See* chromous trioxide, 6247
chromic acid, diammonium salt, *See* Ammonium dichromate, 2237
chromic acid ($H_2Cr_2O_7$), diammonium salt, *See* Ammonium dichromate, 2237
chromic acid, disodium salt, *See* sodium chromate, 28750
chromic acid (H_2CrO_4), disodium salt, *See* sodium chromate, 28750
chromic anhydride, *See* chromic acid, 6220, chromous trioxide, 6247
chromic chloride, 6221, Chrometrace, 6218
chromic chloride hexahydrate, *See* Chrometrace, 6218
chromic fluoride, 6222
chromic hydroxide, *See* Guignet's Green, 13456
chromic oxide, 6223, *See* Accrox, 258, chromium oxide (ic), 6236, M100, 18107, red oxide of chromium, 25965, chrome bronze, 6204, M100, 18107
chromic oxide pigment, *See* chromic oxide, 6223, M100, 18107
Chromic Oxide Sesquioxide, *See* chromic oxide, 6223, M100, 18107
Chromic Oxide Sesquioxide, *See*
chromic potassium sulfate, 6224
Chromic potassium sulfate dodecahydrate, *See* chromic potassium sulfate, 6224
chromic sulfate, *See* chromium sulfate, 6237
chromidium, 6225
Chromiform, 6226
Chromitan®, 6227
chromium, 6228
chromium acetate (ic), 6229
chromium bronze, 6230
chromium chloride(III) anhydrous, *See* chromic chloride, 6221
chromium copper, 6231
chromium disodium oxide, *See* sodium chromate, 28750
chromium green, 6232
chromium hydroxide-bismuth oxychloride oxide, *See* H Chrome Green 105, 13521
chromium manganese, 6233
chromium molybdenum, 6234
chromium nickel, 6235
chromium oxide (ic), 6236
chromium sesquichloride, *See* chromic chloride, 6221
chromium sesquioxide, *See* Accrox, 258, chromic oxide, 6223, chromium oxide (ic), 6236
chromium sodium oxide (Na_2CrO_4), *See* sodium chromate, 28750
chromium sulfate, 6237
chromium trichloride, *See* chromic chloride, 6221
chromium trioxide, *See* chromic acid, 6220, chromous trioxide, 6247
chromium (III) chloride, *See* chrometrace, 6218
chromium (III) oxide hydrate, *See* M100, 18107
chromium (III) oxide hydrate, *See* chromic oxide, 6223
chromium (III) oxide, *See* chromic oxide, 6223
chromium (III) oxide, *See* chromic oxide, 6223
chromium (III) oxide, *See* M100, 18107
chromium Acetate, *See* chromium acetate (ic), 6229
chromium chloride (3), *See* chrometrace, 6218
chromium chloride, *See* chrometrace, 6218
chromium fluoride, *See* fluorchrome, 12072
chromium hydroxide, *See* chromic oxide, 6223
chromium hydroxide-bismuth oxychloride, *See* chrome Green 106, 6206
chromium oxide (Cr+72O+73), *See* chromic oxide, 6223, M100, 18107
chromium oxide green pigments, *See* chromic oxide, 6223, M100, 18107
chromium sesquioxide, *See* chromic oxide, 6223, M100, 18107
chromium triacetate, *See* chromium acetate (ic), 6229
chromium trifluoride, *See* chromic fluoride, 6222
chromium trioxide, *See* chrome bronze, 6204, red oxide of chromium, 25965
chromium(III) fluoride, *See* chromic fluoride, 6222
chromium(III) sulfate hydrate, *See* chromium sulfate, 6237
chromium(III) acetate, *See* chromium acetate (ic), 6229
chromium(III) acetate, basic, *See* chromium acetate (ic), 6229
chromium(III) acetate n-hydrate, *See* chromium acetate (ic), 6229
chromium(III) fluoride tetrahydrate, *See* chromic fluoride, 6222
chromium(II) chloride anhydrous, *See* chromous chloride, 6245
chromium(VI) oxide, *See* chromous trioxide, 6247
chromium-molybdenum steel, 6238
chromium-vanadium steel, 6239
chromium-vanadium-molybdenum steels, 6240
chromogen C, *See* chromotrope acid, 6244
chromogen I, *See* chromotrope acid, 6244
chromol, 6241
chromo-nitric acid, *See* Perenyl's fluid, 23241
chromosal®, 6242
chromospun, 6243
chromotrope acid, 6244
chromous chloride, 6245, *See* chromous chloride, 6245
chromous sulfate, 6246
chromous sulfate pentahydrate, *See* chromous sulfate, 6246
chromous trioxide, 6247
Chromovan Steel, 6248
Chromozin, *See* atrazine, 3394
chromule, *See* chlorophyll, 6134
chronin, 6249
chronite, 6250
N-(chrysanthemoxymethyl)-1-cyclohexene-1,2-dicarboximide, *See* Killgerm® Py-Kill W, 16182
chrysazin-3-carboxylic acid, *See* rhein, 26534
chrysazol, 6251
Chrysene, 6252
Chrysocale, 6253
Chrysochalk, 6254
Chrysoform, 6255
Chrysoidine, 6256
chrysoidine orange, *See* Chrysoidine, 6256
chrysoidine Y, *See* Chrysoidine, 6256
Chrysoidine crystal, *See* Chrysoidine, 6256
Chryson, *See* resmethrin, 26218
Chrysorin, 6257
Chrysron, *See* resmethrin, 26218
Chryzoplus, Chryzopon, Chryzosan, Chryzotek, 6258
CHT Activator NB, 6259
CHT Antifoam MI, 6260
CHT Biavin 109, 6261
CHT Carrier GR-A, 6262
CHT Contavan ALR, 6263
CHT Cotoblanc HTD-N, 6264
CHT Defoamer SC, 6265
CHT Egasol SP, 6266
CHT Felosan TAK-NO, 6267
CHT Heptol NWS, 6268
CHT Intensol TH-B, 6269
CHT Lavotan DS, 6270
CHT Lustraffin BA, 6271
CHT Meropan BRE, 6272
CHT Prisulon SNP-113S, 6273
CHT Rapidoprint M-4, 6274
CHT Retinol M, 6275
CHT Rewin MRT, 6276
CHT Sarabid DLO Conc, 6277
CHT Subitol HLF Conc, 6278
CHT Thickener 8300E, 6279
CHT Tubingal 220A, 6280
CHT Tubiprint PERL C, 6281
CHT Viscavin DMS, 6282
Chugaev's reagent, *See* dimethylglyoxime, 8523
Chwastox, *See* MCPA, 19183
Chymar, *See* α-Chymotrypsins, 6286
Chymetin, *See* α-Chymotrypsins, 6286
Chymex, 6283
Chymodiactin, *See* Chymopapain, 6284
Chymolase, *See* α-Chymotrypsins, 6286
Chymopapain, 6284, *See* Discase, 8624
Chymoral, 6285
chymosin, *See* rennet, 26110
Chymotrypsin, *See* Alpha Chymar, 1820, Avazyme, 3458, Enzeon, 10604, Zolyse, 34889
α-Chymotrypsins, 6286
δ-chymotrypsin, *See* Deanase D.C, 7828
CI 58205, *See* oxyalizarin, 22438
CI 77052, *See* Octoguard FR-10, 22000
CI pigment white 11, *See* Octoguard FR-10, 22000
CI-2, 6287
CI77713, *See* magnesium carbonate, 18359
CI7810, 6288, 6289
Cialit, 6290
Ciamin, *See* vitamin C, 33938
Ciba 1906, 6291, *See* thiambutosine, 31663
Ciba 3126, *See* Metobromuron, 19583
Ciba, Cibanone, 6292
Cibacet, Cibacrolan; Cibacron; Cibalan, 6293
Cibacoll™, 6294
Cibamin, 6295
Cibanite, 6296
Cibanold, 6297
Cibaphasol 6042, 6298
Cibatex 248, 6299
Cibatex PA, 6300
Ciclopiroxolamine ethanolamine salt (1:1), *See* Loprox, 17604
Cidex, *See* glutaraldehyde, 13062, Sonacide, 29073, Ucarcide® 225, 32711
Cidorel, *See* nuarimol, 21762
cidrase, 6301
Ciflox, *See* ciprofloxacin, 6317
CI-IPC, *See* Chlorpropam, 6162
CIL, 6302

Cilbond, 6303
Cilcast, 6304
Cilicaine, *See* Penicillin V, 23111
Cimcool wafers, *See* S.S.T® Sump Saver Tablets, 27180, Tris Nitro®, 32408
Cimet, 6305
Cimexan, *See* malathion, 18460
Cimfix 606, Cimpact 699, 710, 6306
Cinartc 200, *See* Dowicil® 75, 8937
cinchophen, 6307
cinchophen allyl ester, *See* atoquinol, 3380
Cinconal, *See* cinchophen, 6307
Cindal, 6308
Cindumix, 6309
cinene, *See* Achilles Dipentene, 323, Dipentene, 8584
cinereine, 6310
cinnamaldehyde, 6311
cinnamein, 6312
cinnamene, *See* styrol, 29875
cinnamenol, *See* styrol, 29875
cinnamic acid, 6313
cinnamic alcohol, *See* cinnamyl alcohol, 6315
cinnamic aldehyde, *See* cinnamaldehyde, 6311
cinnamol, *See* styrol, 29875
cinnamoyl chloride, 6314
cinnamyl alcohol, 6315, *See* Peruvin, 23443, styrone, 29888
cinnamyl cinnamate, *See* styracin, 29860
cinnamyl β-phenyl acrylate, *See* styracin, 29860
cinnamylic acid, *See* cinnamic acid, 6313
(E)-N-cinnamyl-N-methyl-1-naphthalenemethylamine, *See* Naftifine, 20650
Cinquasia, 6316
cinquinolite, *See* white copperas, 34358
CIPA, *See* Ramrod, 25861
CIPC, *See* Chlorpropam, 6162
Cipca, *See* vitamin C, 33938
Ciprobay, *See* ciprofloxacin, 6317
Ciprofloxacin, 6317
Ciproxan, *See* Ciprofloxacin, 6317
Ciproxin, *See* Ciprofloxacin, 6317
Ciram, *See* AAprotect, 16
Cirami No. 1, 6318
Circacid, 6319
Circadet, 6320
Circaline MK 11, 6321
Circosan, 6322
Cire De Lanol CTO, 6323
Cirol, 6324
Cirrasol, 6325
Cirrasol TCS, *See* Ablunol 200ML, 95
Cirtoxin, *See* Clopyralid, 6436
***cis*-9,10-octadecenoamide**, *See* oleamide, 22107, Polydis® TR 121, 24411
***cis*-citral**, *See* Neral, 20996
Cisdene® 1203, 6326
Cislin, *See* Thripstick®, 31766
Cismollan® BH, 6327
Cistobil, *See* Telepaque, 31057
Cithrol, 6328
Cithrol 10MS, *See* Ablunol DEGMS, 99
Cithrol EGMR N/E, *See* glycol ricinoleate, 13104
Cithrol GMS A/S, 6329
Cithrol PGMR N/E, *See* Propylene glycol ricinoleate, 25222
Cithrol PO, *See* Ablunol 200MO, 96
Cithrol PS, *See* Ablunol DEGMS, 99
Citmol 316, 6330
Citmol 320, 6331
Citobaryum, 6332
Citobaryum, *See* Radiopaque, 25830
Citol, *See* *p*-aminophenol, 2185, Unal, 32908, Takatol, 30725
Citotray, 6333
Citowett, 6334
Citox, *See* DDT, 7820
Citraclean, 6335
citral (*cis* and *trans*), 6336
Citral A (*cis*-3,7-dimethyl-2,6-octadienal), *See* Neral, 20996
Citralka, 6337
Citranaxanthine, *See* Lucantin® CX, 17751
Citranova, 6338
Citranox®, 6339
Citrest, 6340
citric acid, 6341, *See* BFP 74, 4113, Citraclean, 6335
citric acid triethyl ester, *See* Hydagen® C.A.T, 14526, Uniplex 80, 33004
citric acid, tributyl ester, acetate, *See* acetyl tributyl citrate, 313
Citrical, *See* calcium carbonate, 4940, Kalvan, 15701, Kotamite®, 16370
Citridic acid, *See* equisetic acid, 10735
Citrobaryum, *See* Barium sulfate, 3644
Citroflex 2, *See* triethyl citrate, 32242
Citroflex A, *See* acetyl tributyl citrate, 313
Citroflex A 4, *See* acetyl tributyl citrate, 313
Citroflex A-2, 6342
Citroflex A-4, 6343, *See* Estaflex, 10874
Citroflex A-6, 6344
Citroflex A-8, 6345
Citroflex B-6, 6346
Citron yellow, *See* zinc chromate, 34773
citronellal, 6347, 6348, *See* Cephrol, 5695
β-citronellol, *See* citronellol, 6348
R-(+)-β-citronellol, *See* Cephrol, 5695
citronellyl acetate, *See* Cephreine, 5694
Citrozone, 6349
Citrus distillates, *See* Stoner A500, 29787
Citrus pectin, *See* pectin, 23017
Citrylated mono-diglycerides, *See* Radiamuls® Citrem 2931, 2932, 25771
Ckuper, *See* copper oxychloride, 6784
CL 5,279, *See* Aminitrazole, 2164
CL 7521, *See* Dodine FL, WP, 8785
CL-81588, *See* Avoparcin, 3502
CL 252,925, *See* imazapyr, 14893
CL-206214, *See* butamisole hydrochloride, 4733
CL-206576, *See* Vigazoo, 33749
Cladex®, 6350
Clairolite L, 6351
Clairsol, 6352
Clampdown, 6353
Clar+lon® A410P, 6366
Claradex CH-540, 6354, *See* Acrylonitrile-butadiene-styrene, 439
Clar-Apel, 6355
Clarase® 5,000, 40,000, 6356
Clarcel, 6357
Clarcel Flo, 6358
Clarex® L, 6359
Clarfina, 6360
Clarifex 800, 6361
Clarifloc, 6362, 6363
Clarifoil, 6364
Clarion®, 6365
Clarit, 6367
Clarital®, 6368
Clarite, 6369
Clark's patent alloy, 6370
Clar-O-Cel, 6371
Clarosan, *See* terbutryn, 31230
Clarosan 1FG, 6372
Clarstabil, 6373
Clarus metal, 6374
Clarvin, 6375
claudelite, *See* arsenic trioxide, 3143
Clay Breaker, 6376
Claymaster, 6377
Claysil, 6378
CLD 2, 6379
Clean Crop MSMA 6 Plus, *See* Neoarsycodyl, 20871
Clean Crop MSMA 6.6, *See* Neoarsycodyl, 20871
Clean Wiz, 6380
Cleanacres CMPP, 6381
Cleanacres PDR 675, 6382
Clean-Up, 6383
Cleapact TI-100, TI-100S, 6384
Clear by Design, 6385
Clear Conc. 7174, 6386
Clear Tint®, 6387
Clearam, 6388
Clearate G, *See* Ablunol DEGMS, 99
Clearbreak TEB, 6389
Clearcol, 6390
Cleargum®, 6391
Clearon, 6392
Clearon®, 6393
Clearpol, 6394
Clearsite, 6395
Clearsol®, 6396
Clear-Stat AS401, 6397
Clearteck, *See* Kaydol, 15810, Klearol, 16208
Cleartuf Series, *See* Polyethylene terephthalate, 24426
Clearway, 6398
Clearys Waterless Hand Cleaner, 6399
Cleaval, 6400
Clebrium alloys, 6401
Cleensheen, *See* malathion, 18460
Cleland regent, *See* 1,4-Dithiothreitol, 8730
Clelands Reagent, 6402, *See* 1,4-Dithiothreitol, 8730
Clenecorn, 6403
Clenesco, 6404
Clerici solution, 6405
Clerit, 6406
Clerite, 6407
Clermait®, 6408
Cleroxide, 6409
1,6-Cleve's acid, *See* Cleve's β acid, 6410
1,7-Cleve's acid, *See* Cleve's ω-Acid or J-acid, 6412, Delta acid, 8044
β acid, 6410
γ-Acid, 6411
γ-acid, *See* Cleve's γ-Acid, 6411
ω-Acid or J-acid, 6412
ω- or δ-acid, 6413
Cliché metal, 6414
Clifton Chlormequat 46, 6415, *See* Chlormequat chloride, 6110
Clifton CMPP Amine 60, 6416
Clifton Glyphosate Additive, 6417
Clifton Wetter, 6418
Climacel, 6419
Climax, 6420
Climax 193, 6421
Clinafarm, 6422, *See* Enilconazole, 10582, imazalil, 14892
Clindrol 101cg, *See* Ablumide LDE, 79
Clindrol 200, *See* Alkamide® 327, 1634
Clindrol 200 L, *See* Ablumide LDE, 79
Clindrol 200CGN, *See* Active #2, 501
Clindrol 200-MS, *See* Ablumide SME, 82, Witcamide® 70, 34442
Clindrol 200-S, *See* Ablumide SDE, 81
Clindrol 202CGN, *See* Active #2, 501
Clindrol 203cg, *See* Ablumide LDE, 79
Clindrol 210cgn, *See* Ablumide LDE, 79
Clindrol 868, *See* Ablumide SDE, 81
Clindrol Superamide 100CG, *See* Active #2, 501
Clindrol Superamide 100L, *See* Ablumide LDE, 79
Clinifeed 400, Favour, ISO, Protein Rich, 6423
Clinistix, 6424
Clinitest, 6425
Clipper, 6426, *See* paclobutrazol, 22540, 22541
Clipper, Parlay, *See* paclobutrazol, 22542
CIK, *See* Potassium chloride, 24748
Clofazimine, 6427
Clofenotane, *See* DDT, 7820
Clofentezine, *See* Apollo 50C, 2661
Clofenvineosum, *See* Chlorfenvinphos, 6098
Clofenvinfos, *See* Chlorfenvinphos, 6098
Clofenvinphos, *See* Chlorfenvinphos, 6098
Cloisonné®, 6428
Clonevac D-78, 6429
Clonevac-30, 30T, 6430
Clonevac-30-Ma5, 6431
clonitrilide, 6432

Clonitrilide, *See* Bayluscide®, 3805
Clonotox, *See* Mecoprop, 19204
Cloparin, Cloparol, 6433
Cloparten, Cloparten Z, 6434
Clophen, 6435
Clopiralid, *See* BH 2,4-D Ester 40, 4115
Cloprop, *See* Bidisin, 4132
Clopyralid, 6436, *See* Agrichem, 1045, Agricorn D, 1051, Campbell's Destox, 5057, Dow Shield, 8887, Format, 12267, Hadranol, 13533, Shield, 28173, silvapron, 28409
Clopyralid + propyzamide, *See* Matrikerb, 19063
Clopyralid-cyanazine, *See* Coupler SC, 6907
Clopyralid-fluroxypyr-ioxynil, *See* Hotspur, 14425
Clopyralid-triclopyr, *See* Grazon 90, 13296
Clorafin, 6437
Cloran, 6438
cloretilo, *See* ethyl chloride, 11067
Clorina, *See* Chloramine T, 6086
clorius, *See* methyl benzoate, 19515
clorofene, *See* Nipacide® BCP, 21252
Clorox, *See* sodium hypochlorite, 28773
Clortex, 6439
ClortoCaffaro, *See* Chlorothalonil, 6150, Repulse, 26151
Clortokem, *See* Talisman, 30735, Tolurex, 31958
Clortol, 6440
Clortosip, *See* Chlorothalonil, 6150, Repulse, 26151
Closantel, 6441, *See* Flukiver, 12051
Clostridiopeptidase A, *See* Collagenase, 6615
Closyl 30 2089, 6442
Closyl LA 3584, 6443, *See* Gardol®, 12668
Cloth oil, 6444
Clotrimazole, *See* Canesten®, 5086, Gyne-Lotrimin, 13513, Lotrimin, 17631, Mono-Baycuten®, 20217
Cloustonite, 6445
Clout, 6446
Clovacorn Extra, 6447
Clovean, 6448
Clovotox, 6449
Clowes' solution, 6450
CLSP 499, 6451
Club, 6452, *See* paclobutrazol, 22540, 225411
Clysar®, 6453
CM8450, 6454
CM8500, 6455
CM8550, 6456
CM8620, 6457
CM8645, 6458
CM8650, 6459
CM8750, 6460
CM8930, 6461
CM8980, 6462
CM9000, 6463
CM9050, 6464
CM9100, 6465
CM9160, 6466
CM9220, 6467
CMA, *See* pentanochlor, 23165
CMC, *See* carboxymethylcellulose, 5257, carboxymethylcellulose sodium, 5258, Cekol, 5558
CMD 834, 6468
CMDMCS, *See* CC3270, 5487
CMMP, *See* pentanochlor, 23165, Solan, 28887
CMP acetate, *See* MCPA, 19183
CMPP, *See* Verdone CDA, 33580
CN, *See* chloroacetophenone, 6112
CN 103, 6469
CN 104 A80, 6470
CN 104 B80, 6471
CN 104 C75, 6472
CN 104 D80, 6473
CN 104 F50, 6474
CN 111, 6475
CN 300, 6477
CN 953, 6478
CN 960, 6479
CN 962, 6480
CN 966, 6481
CN 970, 6482
CN 974, 6483
CN112 C60, 6476
CNA, *See* dicloran, 8395, Fumite Dicloran, 12459
CNC, *See* copper naphthenate, 6783
CNC Antifoam 30-FG, 6484
CNC Antifoam A-107, 6485
CNC Defoamer 12, 34, 407, 6486
CNC Defoamer 69, 97, 6487
CNC Detergent E, 6488
CNC Dispersant WB Series, 6489
CNC Dispersion PE, 6490
CNC Foam Assist AA, AA-100, 6491
CNC Gel Series, 6492
CNC Inhibitor 30, 6493
CNC Leveler JH, 6494
CNC PAL 100, 200, 300, 6495
CNC PAL V-8 Supra, 6496
CNC Product PW, 6497
CNC Product ST, 6498
CNC Soft C-1, 6499
CNC Sol 72-N Series, 6500
CNC Sol BD, CNC Sol XNN #11, 6501
CNC Solv 809, 6502
CNC Wet CP Conc, 6503
CNComerse IMP, 6504
CO, *See* Cocamidopropyl betaine, 6551
CO 1895F, *See* Adol® 62 NF, 783, Stearal, 29591
CO-1214 Natural, *See* Ablumide LDE, 79
CO-1670, *See* Adol® 52 NF, 782, Cachalot® C-50 NF, 4878
CO9745, 6505
CO9750, 6506
CO9800, 6507
CO9810, 6508
CO9816, 6509
CO9817, 6510
CO9819, 6511
CO9830, 6512
CO9835, 6513
COAB Naxaine® C, *See* Cocamidopropyl betaine, 6551
COAB Ralufon® 414, *See* Cocamidopropyl betaine, 6551
Coagulant CHA, 6514
Coagulant WS, 6515
coal oil, *See* Kendex OCTG, 16002
coal tar acids, *See* cresylic acid, 7013
coal tar creosote, *See* Creosote, 6991
coal tar cresols, *See* cresylic acid, 7013
coal tar naphtha, *See* naphtha, 20705
coal tar phenols, *See* cresylic acid, 7013
Coalatex, 6516
Coaldet, 6517
Coalite N.T.P, 6518
Coaltec, 6519
coarse metal, 6520
Coasol, 6521
Coate®, 6522
Coatmaster® K580, 6523
cobalamin, *See* vitamin B_{12}, 33934
cobalt, 6524
cobalt, *See* cobalt, 6524
cobalt(II) bromide;, *See* Mancobride Mancanese, 18515
cobalt(III) oxide, *See* cobaltic oxide, 6532
cobalt 254, 6525
cobalt black, *See* cobaltic oxide, 6532
cobalt boroacylate, *See* Manobond, 18569
cobalt brass, 6526
cobalt bromide, *See* Mancobride Mancanese, 18515
cobalt bronze, 6527
cobalt chloride (ous), 6528
cobalt dibromide, *See* Mancobride Mancanese, 18515
cobalt gluconate, *See* Gluconal® CO, 13040
cobalt hydroxide oxide, *See* cobaltic oxide monohydrate, 6533
cobalt oxide, *See* cobaltic oxide, 6532
cobalt oxide-molybdenum oxide, *See* Comox, 6669
cobalt steel, 6529
cobalt(II) nitrate, *See* cobaltous nitrate, 6539
cobalt(II) oxide, *See* cobaltous oxide, 6540
cobalt(II) sulfate, *See* cobaltous sulfate, 6541
cobalt trifluoride, *See* cobaltic fluoride, 6531, cobaltic trifluoride, 6534
cobalt-chromium-molybdenums steel, 6530
cobaltic fluoride, 6531
cobaltic hydroxide, *See* cobaltic oxide monohydrate, 6533
cobaltic oxide, 6532
cobaltic oxide monohydrate, 6533, *See* cobaltic oxide, 6532
cobaltic trifluoride, 6534
cobaltous acetate tetrahydrate, 6535
cobaltous carbonate, 6536
cobaltous carbonate basic, *See* cobaltous carbonate, 6536
cobaltous chloride, 6537
cobaltous chloride $_{60}$Co, *See* Cobatope-60, 6543
cobaltous chloride anhydrous, *See* cobalt chloride (ous), 6528, cobaltous chloride, 6537
cobaltous hydroxide carbonate, *See* cobaltous carbonate, 6536
cobaltous monoxide, *See* cobaltous oxide, 6540
cobaltous naphthenate, 6538
cobaltous nitrate, 6539
cobaltous oxide, 6540, *See* prepared cobalt oxide, 24897
cobaltous sulfate, 6541
Cobaltron Steel Alloy, 6542
Coban, *See* monensin, 20208
Cobatope-60, 6543
Cobex, 6544
Cobox, *See* copper oxychloride, 6784
Cobox® L, 6545
Cobratec#99, *See* benzotriazole1H-benzotriazole, 3982, Preventol® Cl 8, 24930
Cobratec 99, *See* benzotriazole1H-benzotriazole, 3982, Preventol® Cl 8, 24930
Cobratec TT-100, *See* Preventol® Cl7-100, 24929, tolyltriazole, 31962
Cobrol, 6546
Cocamide DEA, 6547, *See* Accomid C, 230, Active #2, 501, Alkamide® 101 CG, 1631, Alkamide® 200 CGN, 1632, Alkamide® CDE, 1637, Amidex KD, 2092, Aminol KDE, 2179, Comperlan COD, 6674, Comperlan KD, 6675, Crillon CDY, 7048, Emalex N-83, 9972, Ethox COA, 11021, Ethylan® LD, 11114, Hymolon CWC, 14687, Incromide CA, 14985, Marlamid® DF 1218, 18740, Monaterge 85 HF, 20140, Norfox® X, 21557, Oramide DL 200 AF, 22265, Purton CFD, 25419, Stamid HT 3901, 29440, Standapol® AP Blend, 29468, Tohol N-220, 31929, Witcamide® 128T, 34439, Witcamide® Coco Condensate, 34443
Cocamide DEA (1:1), *See* Carsamide® CA, 5321, Empilan® 2502, 10384, Empilan® CDE, 10389, Mackamide C, 18153, Monamid® 7-100, 20111, Schercomid SCE, 27657, Standamid® KD, 29455
Cocamide DEA (2:1), *See* Comperlan PKDA, 6678, Laurel SD-101, 16839, Mazamide® 68, 19106
Cocamide DEA and DEA-dodecylbenzenesulfonate, *See* Active #4, 502
Cocamide DEA superamide, *See* Calamide C, O, 4925
Cocamide/DEA (1:1), *See* Ablumide CDE, 77
cocamide-DEA (1:2), *See* Adeka Sole CO, 671
Cocamide MEA, 6548, *See* Amidex KME, 2093, Cocamide MEA, 6548, Comperlan P 100, 6677, Empilan® CME, 10391, Incromide CM, 14986, Mackamide CMA, 18154, Marlamid® M 1218, 18744, Marlamid® PG 20, 18746, Monamide, 20123, Ninol® CMP, 21235, Ninol® CNR, 21236, Rewomid® C 212, 26320, Zoramide CM, 34905, Standapol® 7088, 29465, Standapol® Pearl Conc. 7130, 29476
Cocamide MEA (1:1), 6549 , *See* Alkamide® C-212, 1635, Carsamide® CMEA, 5322, Cocamide MEA (1:1), 6549, Foamole M, 12178, Mazamide® CFAM, 19109
Cocamide MEA 1:1 Cocamide MEA, *See* Monamid® CMA, 20118
Cocamide-MEA, *See* Adeka Sole YA, 672

Chemical and Trade Names

Cocamide-MEA (1:1), *See* Ablumide CME, 78
Cocamide MIPA, 6550, *See* Empilan® CIS, 10390, Rewomid® IPP 240, 26327
Cocamidopropyl betaine, 6551, *See* Abluter BE, 131, Amonyl 380 BA, 2297, Ampholyt JB 130, 2333, Amphosol® CA, 2341, Amphotensid B4 F, 2347, Aremsol A, 2834, Chembetaine C, CGF, 5931, Dehyton® KE, 7992, Dehyton® PK, 7995, Incronam 30, 15016, Lexaine® C, 17091, Lonzaine® C, 17592, Mackam 35, 18136, Mackam J, 18145, Mafo® C, 18299, Mirataine® BET-C-30, 19923, Monateric ADA, 20152, Norfox® Coco Betaine, 21547, Nutrol Betaine MD 3863, 21846, Nutrol Betaine OL 3798, 21847, Ralufon® 414, 25847, Rewoteric® AM B-13, 26417, Rewoteric® AM B-14LS, 26418, Rewoteric® AM B-15, 26419, Schercotaine CAB, 27695, Standapol® AP Blend, 29468, Tego Betain CK D, 30960, Tego® -Betaine C, 30961, Tego® -Betaine E, 30962, Tego® -Betaine L-7, 30964, Tego® -Betaine S, 30966, Velvetex® BA-35, 33554, Lumorol K 28, 17814
Cocamidopropyl betaine, cocamide DEA and glycol stearate, *See* Base Nacrante 2078, 3699
Cocamidopropyl betaine, glycol distearate, cocamide DEA, cocamide MEA, *See* Tego® -Pearl B-48, 30985
Cocamidopropyl dimethyl amine oxide, *See* Empigen® OS/A, 10378
Cocamidopropyl dimethyl glycine, *See* Cocamidopropyl betaine, 6551
Cocamidopropyl dimethylamine, 6552, *See* Chemidex C, 5962, Chemidex WC, 5973, Incromine CB, 14998, Lexamine C-13, 17097, Mackine 101, 18215, Mazeen® DAPL, 19123, Mazeen® SHCFA, 19127, Miramine® CODI, 19868, Schercodine C, 27636
Cocamidopropyl dimethylamine lactate, 6553
Cocamidopropyl dimethylamine oxide, *See* Patogen AO-30, 22914, Rhodamox® CAPO, 26619, Standamox PCAW, 29461
Cocamidopropyl dimethylamine propionate, 6554, *See* Emcol® 1655, 10046, Incromate CDP, 14967
Cocamidopropyl dimethyl-amine propionate, water, *See* Foamid 117, 12155
Cocamidopropyl ethyl dimonium ethosulfate, *See* Schercoquat CAS, 27674
Cocamidopropyl hydroxysultaine, 6555, *See* Crosultaine C-50, 7241, Mafo® CSB, 18301, Mirataine® BSC, 19926, Mirataine® CBS, CBS Mod, 19927, Rewoteric® AM CAS, 26421, Rewoteric® AM CAS-15, 26422, Schercotaine SCAB, 27700
Cocamidopropyl lauryl ether, *See* Marlamid® KL, 18742, Marlamid® KLP, 18743
Cocamidopropyl PG-dimonium chloride, *See* Lexquat® AMG-WC, 17179
Cocamidopropyl PG-dimonium chloride phosphate, *See* Phospholipid PTC, 23722
Cocamidopropyl)dimethylamine oxide, *See* Ablumox CAPO, 90
Cocamidopropylamine oxide, 6556, *See* Ablumox CAPO, 90, Aminoxid WS 35, 2196, Amyx CDO 3599, 2392, Barlox® C, 3653, Chemoxide CAW, 5994, Cocamidopropylamine oxide, 6556, Mackamine CAO, 18168, Mazox® CAPA, 19149, Monalux CAO, 20103, Ninox® FCA, 21241, Rewominox B 204, 26333, Schercamox C-AA, 27572, Standamox CAW, 29460, Varox® 1770, 33454
Cocamidopropylhydroxysultaine, *See* Amonyl 675 SB, 2298, Chembetaine CAS, 5932, Cocamidopropyl hydroxysultaine, 6555, Lexaine® CSB-50, 17092
Cocamine, *See* Amine KK, 2158, Armeen® C, 2940, Jet Amine PC, 15592
Cocamine (primary amine), *See* Armeen® CD, 2941
Cocamine acetate, *See* Acetamin 24, 284, Amine Acetate KK, 2149
n-Cocamine acetate, *See* Acetamin C, 286
Cocamine oxide, 6557, *See* Barlox® 12, Chemoxide WC, 6003, Genaminox KC, 12779, Karox AO-30, 15774, Mackamine CO, 18169, Naxide 1230, 20817, Schercamox DMC, 27574
Cocamino hydroxy sulfobetaine, *See* Empigen® 5509, 10357
Cocaminopropionic acid, *See* Ampholyte KKDP-60, 2334, Mackam 151C, 18137
Cocamphohydroxypropyl sulfonate, *See* Sandoteric CFL, 27354
Cocbetaine Nutrol betaine MD 3863, *See* Cocamidopropyl betaine, 6551
Coceth-27, 6558, *See* Dehydol LT 3, 7953
Coceth-5, *See* Genapol® C-050, 12787, Genapol® GC-050, 12789
cochineal, *See* carminic acid, 5294
Cochineal extract, *See* Carmine, 5289
cochrome, 6559
Coclor, 6560
Coco alkyl aminopropionic acid, *See* Ampholyte KKE-70, 2335
Coco alkyl dimethyl benzyl quaternary amine, *See* Anedco AW-395, 2449
Coco benzyl imidazolinium chloride, *See* Quatrene C-5-6, 25612
Coco betaine, 6561, *See* Chembetaine BW, 5930, Coco betaine, 6561, Lonzaine® 12C, 17591, Mafo® CB 40, 18300
Coco-1,3-diaminopropane, *See* Duomeen® CD, 9113
N-Coco-1,3-diaminopropane, *See* Cocodiamine, 6564, Duomeen® C, 9112
Coco diethanolamide, *See* Ablumide LDE, 79
Coco dimethylamine oxide, *See* Cocamine oxide, 6557, Empigen® 5083, 10354, Genaminox CS, 12778
Coco fatty acid MEA ethoxylate, *See* Mulsifan RT 72, 20459
Coco fatty acids, *See* Hystrene® 1835, 14755
Coco hydroxyethyl imidazoline, *See* Miramine® C, 19867, Unamine® C, 32911
Coco imidazoline betaine, *See* Empigen® CDR10, 10368
Coco imidazoline dicarboxylate, *See* Amphoterge® NX, 2354
Coco propylenediamine, *See* Cocodiamine, 6564
n-Coco propylene diamine, *See* Diamine KKP, 8297
Coco trimethyl ammonium chloride, 6562, *See* Chemquat C/33W, 6026, Coco trimethyl ammonium chloride, 6562, Jet Quat C-50, 15603
cocoa butter, 6563
Cocoalkonium chloride, *See* Mariazin® KC 21/50, 18715
Cocoalkyl dimethyl benzyl ammonium chloride, *See* Arquad® DMCB-80, 3116
Cocoalkyldimethylbenzyl ammonium chloride, *See* Querton KKBCl-50, 25656
1-Cocoalkylguanidinium chlorides, *See* Querton GCl-50, 25655
Cocoalkyltrimethylenediamine, *See* Jet Amine DC, 15582
Cocoamido-3-propyldimethylamine oxide, *See* Ablumox CAPO, 90
Cocoamidopropyl betaine, *See* Amphoteen BCA-30, 2344
3-Cocoamidopropyl dimethylamine oxide, *See* Ablumox CAPO, 90
Cocoamidopropylhydroxy sultaine, *See* Sandobet SC, 27327
Cocoamidopropylhydroxysultaine, *See* Lonzaine® CS, 17593
Cocoamphoacetate, *See* Empigen® CDR40, 10369
Cocoamphocarboxyglycinate, *See* Dehyton® G, 7990, disodium cocoamphodiacetate, 8643
Cocoamphocarboxypropionate acid, *See* Ampholak YCO-40, 2328
Cocoamphodiacetate, *See* Monateric 805, 20141, Monateric CDL, 20157
Cocoamphodipropionic acid, *See* Miranol® C2M Anhyd. Acid, 19886
Cocoamphoglycinate, *See* Empigen® CDR40, 10369
Cocobetainamido amphopropionate, *See* Rewoteric® QAM 50, 26432
Cocobetaine, *See* Amphoteen BCM-30, 2345, Chembetaine CB, 5933, Incronam CD-30, 15020, Mackam CB-35, 18140, Velvetex® AB-45, 33553
Cocodiamine, 6564
Cocodimonium hydroxyethyl cellulose, *See* Crodacel QM, 7066
Cocodimonium hydroxypropyl hydrolyzed keratin, *See* Kera-Quat WKP, 16056
Coco-EDTA-amide, *See* Rewopol® CHT 12, 26360
Cocoglycerides, *See* Novata 299, A, AB, B, BBC, BC, BCF, BD, C, D, E, 21704, Witarix® 250, 34430
Coco-hydrolyzed soy protein, *See* Oleo-Soy C, 22122
Cocohydroxyethyl imidazoline, *See* Varine C, 33392
Cocoiminodiglycinate, *See* Ampholak XCE, 2320
Cocoiminodipropionate, *See* Ampholak YCE, 2327
Cocoiminodipropionate half sodium salt, *See* Ampholak YCA/P, 2326
Cocoloid®, 6565
Cocomonoethanolamide, *See* Ablumide LME, 80
Coconut acid, *See* Hystrene® 1835, 14755
Coconut Acid, Diethanolamide, *See* Active #2, 501
Coconut alcohol ethoxylate, *See* Genapol® 2299, 12784
Coconut amido alkyl amine oxide, *See* Zoramox, 34906
coconut butter, 6566
Coconut cake, *See* poonac, 24688
Coconut fatty acid amine condensate, *See* CNC Detergent E, 6488
Coconut fatty acid, monoethanolamide, *See* Ablumide CME, 78
coconut milk, *See* coconut butter, 6566
Coconut monoisopropanolamide, *See* Cocamide MIPA, 6550
Coconut oil, *See* coconut butter, 6566, Konut, 16334
Coconut Oil Acids Diethanolamide, *See* Active #2, 501
Coconut oil amide of diethanolamine, *See* Ablumide LDE, 79
Coconut Oil Diethanolamine, *See* Active #2, 501
Coconut oil fatty acid ethanolamide, *See* Ablumide CME, 78
Coconut oil fatty acids, ethanolamine condensate, *See* Ablumide CME, 78
Coconut oil fatty acids, monoethanolamide, *See* Ablumide CME, 78
Coconut oil propane diamine, *See* Radiamine 6560, 25766
Coconut oil, monoethanolamide, *See* Ablumide CME, 78
Coconut oils acids, *See* Hystrene® 1835, 14755
Coconut Oils® 76, 92, 110, 6567, *See* Esi-Terge 40% Coconut Oil Soap, 10825
Coconut poly-diethanolamide, *See* Alkamide® DC-212/MP, 1639
N-Coconut 1,3-propylenediamine, *See* Kemamine® D-650, 15912
Cocopropionate, *See* Alkateric® AP-C, 1720
cocotrimethylammonium chloride, *See* Adogen® 461, 771
Cocotrimonium chloride, *See* Adogen® 461, 771, Arquad® C-33W, 3114, Jet Quat C-50, 15603, Marlazin® KC 30/50, 18753, Varisoft® 461, 33411
Cocotrimonium chloride, IPA, *See* Arquad® C-50, 3115
Cocotripropylene tetraamine, *See* Amine 780, 2147
Cocoyl hydroxyethyl imidazoline, *See* Mackazoline C, 18198, Monazoline C, 20192, Schercozoline C, 27712
N-Cocoyl-N-methyl glycine, *See* Cocoyl sarcosine, 6568
N-Cocoyl Sarcosinate, *See* Crodasinic L, 7143
Cocoyl sarcosine, 6568, *See* Hamposyl® C, 13598, Vanseal® CS, 33356
cod liver, *See* cod liver oil, 6569
cod liver oil, 6569
Codacide Oil, 6570
Codal, *See* metolachlor, 19585
Code 321, 6571, *See* hydrogenated soybean oil, 14596
β-codeine, *See* neopine, 20939
Codite, 6572
Codoil, *See* retinol, 26277
Coeruleoactite, *See* aluminum phosphate, 1922
Cofill, 6573
Co-Gell® A2/B270, 6574
Cohedur®, 6575

Cohedur® RK, 6576
Coherex®, 6577
COHRlastic® 400, 6578
COHRlastic® 1010, 6579
COHRlastic® 1867, 6580
COHRlastic® 3320, 6581
COHRlastic® 8016, 6582
COHRlastic® 9041, 6583
COHRlastic® F12, 6584
COHRlastic® FR17, 6585
COHRlastic® R10450, 6586
COHRlastic® R10490, 6587
COHRlastic® TC100, 6588
cohydrol, 6589
coinage bronze, 6590
Coke, 6591
Cola, *See* kola nut, 16303
Colace, *See* Empimin® OP70, 10434, Octowet 40, 22035
Colacryl, 6592
Colamine, 6593
Colanyl, 6594
Colasta, 6595
Colastex, 6596
Colcar®, 6597
Colcolor, 6598
colcothar, *See* Indian Red, 15068
Cold polymer, *See* Chemigum® NX-775, 5978
cold varnishes, 6599
Coldcide-25 microbiocide, *See* glutaraldehyde, 13062, Sonacide, 29073
Colepax, *See* Telepaque, 31057
Coles solder, 6600
Col-Evac, 6601, *See* sodium bicarbonate, 28736
Colex 900 BP, 6602
Colex 1000 FR, 6603
Colex 2000, 2200, 6604
Colex 4300, 6605
Colex G, *See* calcium lignosulfonate, 4956
Colfite, 6606
colfosceril palmitate, 6607
Colgne yellow, *See* Chrome yellow, 6212
Colidosan, *See* Soricinol 40, 29143
Colla -Gel AC, *See* gelatin, 12722
Collacral®, 6608
Collafix®, 6609
collagen, 6610
Collagen 15K, 6611
Collagen amino acids, *See* Amino-Collagen-25,-40, 2168, Collamino 25, 6616
Collagen amino acids-acetamide MEA-propylene glycol, *See* Aqua-Tein C, 2746
Collagen CLR, 6612
Collagen fiber, *See* collagen, 6610
Collagen glycerides, *See* Colla-Moist CG, 6617
Collagen Hydrolyzate Cosmetic 55, 6613, *See* Extiat®, 11323
Collagen Native Extra 1%, 6614
Collagenase, 6615
Collamino 25, 6616, *See* Extiat®, 11323
Colla-Moist CG, 6617
Colla-Moist WS, 6618
Collasol, 6619
Collatex, 6620, *See* Ammonium alginate, 2228
collaurin, 6621
Collavin, *See* Propham, 25185, 25186
collene, 6622
Collet steel, 6623
Collex G, 6624, *See* calcium lignosulfonates, 4957
Collid 488T, *See* Algin, 1591
Collidine, 6625
γ-Collidine, *See* Collidine, 6625
2,4,6-Collidine, *See* Collidine, 6625
collodion, 6626
collodion cotton, 6627, *See* guncotton, 13492, Kodaloid, 16283, Venite, 33560
collodion wool, *See* Kodaloid, 16283, Venite, 33560
Colloid 106, 6628
Colloid 111, 111D, 6629
Colloid 202, 6630
Colloid 815M, *See* Ablunol 200MO, 96
colloidal arsenic, *See* arsenic, 3136
colloidal clay, *See* Bentonite, 3954
colloidal selenium, *See* selenium, 27861
colloidal silicate, *See* Agrosil® LR, 1082
colloidal zinc hydroxide, *See* zinc hydroxide, 34778
colloidal zinc oxide, *See* Activox B, 508
Colloidox, *See* Cupravit®, 7370
Collokit, 6631
Collone, 6632
Colloresin D, 6633
Colloresin DK, 6634
Collosol Argentum, 6635
colloxylin, *See* Kodaloid, 16283, Venite, 33560
collozine, *See* zinc hydroxide, 34778
Collunosol, *See* 2,4,5-trichlorophenol, 32221
Collys, 6636
Colmonoy, 6637
Colmonoy No.6, 6638
Cologel, *See* Glutolin, 13064
Cologne spirit, *See* ethyl alcohol, 11064
Cologne spirits (alcohol), *See* ethyl alcohol, 11064
Colona Steel, 6639
colophony, *See* rosin, 26976
Color Seal, 6640
Colorado silver, 6641
Colorlok, 6642
Colormatch®, 6643
Color-Max, 6644
Colorol 70, 6645
Colorol Rust Binder, 6646
Colorol Standard, 6647
Colortrend, 6648
Col-o-tex, 6649
Colour-Chem, 6650
Col-o-vin, 6651
Colsol, 6652
Colstar®, 6653
Colturiet, 6654
Colugel, 6655
Colugel, *See* Alumina hydrate, 1892
Columbium, *See* niobium, 21245
Colvinal®, 6656
Colza oil, *See* rapeSeed oil, 25874
Comac Bordeaux Plus, 6657
Comac Macuprax, 6658
Comac Parasol, 6659, *See* cupric hydroxide, 7381, 7382, 7383, Schweitzer's reagent, 27738
Combined Seed Dressing, 6660
combi-schutz, *See* isopropanol, 15405
Combismalt®, 6661
Comboloob 0609, 6662
Combovac-30, 6663
Comet metal, 6664
Commando, 6665, *See* Flamprop-M-isopropyl, 11888
Commercial varieties of edible olive oil, *See* Aix Oil, 1138
Commodore, *See* λ-cyhalothrin, 7567
Common Degras, 6666
common salt, *See* sodium chloride, 28748
Comodor, *See* tebutam, 30879
Comodor, Comodor 600, 6667
Comoflastic®, 6668
Comox, 6669
compact bitumen, *See* Bitumen, 4294
Compactrol®, 6670 , *See* calcium sulfate (dihydrate), 4972
Compak, 6671
Compalox, 6672, *See* aluminum oxide, 1919
Compass, 6673
Compazine, *See* prochlorperazine dimaleate,
Compd 3-120, *See* sulfaquinoxaline, 29968
compd 469, *See* Isoflurane, 15363
Compd. 4072, *See* Chlorfenvinphos, 6098
Compd. 79891, *See* narasin, 20741
Comperlan COD, 6674
Comperlan HS, *See* Ablumide SME, 82, Witcamide® 70, 34442
Comperlan KD, 6675, *See* Active #2, 501
Comperlan LD, 6676, *See* Ablumide LDE, 79
Comperlan LM, *See* Ablumide LME, 80
Comperlan LS, *See* Active #2, 501
Comperlan P 100, 6677, *See* Cocamide MEA, 6548
Comperlan PD, *See* Active #2, 501
Comperlan PKDA, 6678
Compimide, 6679
Compitox, *See* Mecoprop, 19204, Verdone CDA, 33580
Compitox Extra, 6680
Compitox Plus, *See* Verdone CDA, 33580
Complemix, 6681
Complemix® 100, *See* dioctyl sodium sulfosuccinate, 8550
Complex aliphatic hydroxyl compound phosphate ester, *See* Actrafos 110, 110A, 527
Complex amine and thioureas, *See* Crown L-60B Acid Inhibitor, 7271
Complex diester, *See* Actralube Syn-147, 540
Complex phosphate ester of nonionic ethoxylates, *See* CNC Dispersion PE, 6490
Complex phosphate ester, free acid, *See* Agrilan® F546, 1059
Complex polyglycol ester, *See* Inversol 140, 15199
complex with different cupric acetate:cupric hydroxide:water ratios, *See* cupric acetate, basic, 7373
Complexone, *See* Kalex 220 Crystal, 15668, Kalex Liq. 50%, 15671, tetrasodium EDTA, 31391
Complexone(I), *See* TG Buffer, 31547
Complement, 6682
component of Nuosept 95, *See* Zoldine® ZT-55, 34887
component of Sardo Bath Oil, *See* Wickenol® 101, 105, 34378
Composibor, 6683
Compound 42, *See* Killgerm® Sewarin P, 16184, Kypfarin, 16519
Compound 118, *See* Aldrin, 1558
Compound 403/401, 6684
Compound 889, *See* Bis(2-ethylhexy) Phthalate, 4228, dioctyl phthalate, 8549 Reomol DCP, 26130, Witcizer 312, 34454
Compound-4018, *See* Cumate®, 7351
Compound 4049, *See* malathion, 18460
Compound 7744, *See* Carbaryl, 5181
Compound G-11, *See* hexachlorophene, 13992
Compounded polymeric emulsion, *See* Adflex, 687
Compressed polyether acoustical foam, *See* Aerofonic, 856
Compritol 888, 6685
Compron, 6686
Comtek, 6687
Com-Trol, 6688
Conac A, *See* Vulkacit® CZ/EGC, DZ/EGC, 34129
Conac S, *See* Vulkacit® CZ/EGC, DZ/EGC, 34129
Conacure, 6689
Conap® UC-21, 6690
Conap® UC-28 (formerly Conap DPUC-11898), 6691
Conapoxy® FR-1270, 6692
Conapoxy® TE-1257/Conacure® EA-08, 6693
Conastic® AD-20, 6694
Conastic® ST-115, 6695
Conathane® CE-1155, 6696
Conathane® CE-1163, 6697
Conathane® EN-20, 6698
Conathane® EN-2521, 6699
Conathane® EN-4, 6700
Conathane® RN-1501, 6701
Conathane® RN-1558, 6702
Conathane® RN-1570, 6703
Conathane® TU-400, 6704
Conathane® TU-4010, 6705
Conathane® TU-50A, 6706
Conathane® UC-17, 6707
Conathane® UC-34, 6708
Conc. polyphosphate, *See* Drewgard® 120, 8994
Concemin, *See* vitamin C, 33938
Concentrated Dariloid®, 6709
Concentrated Dariloid® KB, 6710

Concentrated Dariloid® XL, 6711
Conco AAS-35, *See* sodium dodecylbenzenesulfonate, 28760
Conco AAS 35H, *See* Nacconol® 40G, 20647, sodium dodecylbenzenesulfonate, 28760
Conco AAS-90, *See* Nacconol® 40G, 20647
Conco AAS-985, *See* Witco® 1298, 34458
Conco Emulsifier K, *See* Active #2, 501
Conco NI, *See* Alkasurf® NP-4, 1715
conco ni, *See* Triton® N-57, 32430
conco ni-90, *See* Triton® N-57, 32430
Conco Ni-90, *See* Alkasurf® NP-4, 1715
Conco sulfate wa, *See* Rhodapon® 101-10, 26631, sodium lauryl sulfate, 28781, Witcolate 6400, 34469
Conco sulfate wa-1200, *See* sodium lauryl sulfate, 28781, Rhodapon® 101-10, 26631, Witcolate 6400, 34469
Conco sulfate wa-1245, *See* Rhodapon® 101-10, 26631, Witcolate 6400, 34469
Conco sulfate wag, *See* Rhodapon® 101-10, 26631, Witcolate 6400, 34469
Conco sulfate wan, *See* Rhodapon® 101-10, 26631, Witcolate 6400, 34469
Conco sulfate WAS, *See* Rhodapon® 101-10, 26631, Witcolate 6400, 34469
Conco sulfate WE, *See* Witcolate ES-2, Witcolate LES-60C, 34477
Conco sulfate wn, *See* Rhodapon® 101-10, 26631, Witcolate 6400, 34469
Conco SXS, *See* Eltesol® SX 30, 9873, Naxonate® 4L, 20824, Pilot SXS-40, 23843, Witconate SXS 40%, 34492
Conco XAL, *See* Ammonyx® LO, 2264, Empigen® OB, 10375, Rhodamox® LO, 26620
Conco XALDDNO, *See* Ablumox LO, 91
Concurat® L, 6712
Condens-Aid, 6713
Condensate PL, *See* Ablumide LDE, 79
Condensed ammonium naphthalene sulfonate, *See* Lomar® PWA, 17568
Condensed potassium naphthalene sulfonate, *See* Lomar® HP, 17566
Condensed sodium naphthalene sulfonate, *See* Lomar® LS, 17567
condenser foil, 6714
Condensol®, 6715
Conditioner Base, 6716
Condol, *See* vitamin D+72, 33940
condor, *See* oil white, 22086
Conductive Nylon 12, 6717
Conductive silicone rubber, *See* COHRlastic® 1867, 6580
conductivity bronze, 6718
Conducto-Lube®, 6719
Conductomer ABS-22, 6720, *See* Acrylonitrile-butadiene-styrene, 439
Conductomer HDC-22HLMI-M, 6721
Conducto-Wrap, 6722
Condux, 6723
Condy's fluid, 6724
Confidor, 6725
Congo Blue, *See* Niagara blue, 21091
Conifer and Shrub Fertiliser, 6726
Conjugated glycopolypeptides, *See* Acticulum, 474
Conlex, 6727
Connettivina, *See* hyaluronic acid sodium salt, 14483
Conn's stain, 6728
Conoco 597, *See* Witco® 1298, 34458
Conoco C-50, *See* Nacconol® 40G, 20647, sodium dodecylbenzenesulfonate, 28760
Conoco C-60, *See* sodium dodecylbenzenesulfonate, 28760
Conoco C-550, *See* sodium dodecylbenzenesulfonate, 28760
Conoco SA 597, *See* Ablusol DBM, 125, Witco® 1298, 34458
Conoco SD 40, *See* sodium dodecylbenzenesulfonate, 28760
Conol 1675, *See* Adol® 62 NF, 783, Stearal, 29591
Conol 30F, *See* Adol® 62 NF, 783, Stearal, 29591
Conol 30F, *See*
Conotrane, 6730, *See* hydrargaphen, 14539
Conrex, 6731
Consal, 6732
Constab, 6733
Constantan, 6734
Constantin, 6735
Constaphyl, *See* Dicloxacillin sodium, 8396
constructal, 6736
Construction 1200®, 6737
Contact 75, 6738
Contain, *See* imazapyr, 14893
Contaverm, *See* phenothiazine, 23618
Contavern, *See* phenothiazine, 23618
Contex, 6739
Continental® Clay, 6740
Continex® LH-10, 6741
Continex® LH-10, N-351, 6742
Continex® LH-10, N-351, *See* carbon black, 5220
Continex® N351, 6743
Contrac, *See* bromadiolone, 4565, Rentokil Deadline, 26116
contracid, 6744
Contractors 1000®, 6745
Contradet, 6746
Contralin, *See* Disulfiram, 8725
Contraqua LF, 6747
Contrastol W, 6748
Contraven, *See* Terbufos, 31226
contraverm, *See* phenothiazine, 23618
Control 1-100, 6749, *See* corn starch, 6837
Controx® KS, 6750
Controx® Range, 6751
Cook's alloys, 6752
Cooksons, 6753, *See* antimony trioxide, 2574
Cool-Amp, 6754
Coolanol, 6755
Cool-Treet, 6756
Coomassie, 6757
co-op hexa, *See* Julin's chloride, 15632
Coop turbex, *See* Jet Amine DC, 15582
Cooper Coopex, 6758, *See* Permethrin, 23385
Cooper Graincote, 6759
Coopercote, 6760
Cooperite, 6761
Cooper's gold, 6762
Cooper's pen metal, 6763
Cooper's speculum metal, 6764
Coopex, *See* permethrin, 23384, 23385, 23386
Copac, 6765
Copac® E, 6766
Copaiba oil, *See* Jesuit's balsam, 15580
Copaifera langsdorffi oil, *See* Jesuit's balsam, 15580
copal oils, 6767
Copaloy, 6768
Copel alloy, 6769
Copene, 6770
Coperflex BR-45, 6771
Coperflex SSBR-B18 4525, 6772
Copernick, 6773
Copharcilin, *See* Ampicillin, 2361
Copisil, 6774
Copo® 1500, Copo® 1712, 6775
Copolyamide, *See* Swiss Polyamid Grilon, 30463
Copolyester, *See* Kodar® A150 Copolyester, 16288
Copolyester 13339, *See* Polyethylene terephthalate, 24426
Copolymer 186, 6776
Copolymeric carboxylate, *See* Sokalan® PM 10, 28879
copper, 6777, *See* casting copper, 5413
copper(I) bromide, *See* cuprous bromide, 7402
copper(I) iodide, *See* cuprous iodide, 7404
copper(I) oxide, *See* Cuprox, 7409
copper(I) thiocyanate, *See* cuprous thiocyanate, 7407
copper(II) acetate, *See* verdigris, 33571
copper(II) arsenate, *See* cupric arsenate, 7375
copper(II) carbonate, basic, *See* verditer blue, 33574
copper(II) chloride dihydrate, *See* cupric chloride dihydrate, 7379
copper(II) chloride oxide hydrate, *See* copper oxychloride, 6784
copper(II) hydroxide, *See* Schweitzer's reagent, 27738
copper(II) hydroxide carbonate, *See* cupric carbonate, 7378
copper(II) naphthenate, *See* copper naphthenate, 6783
copper(II) sulfate pentahydrate, *See* Kocide® copper Sulfate Pentahydrate Crystals, 16262
copper acetate arsenite, *See* cupric acetoarsenite, 7374
copper acetyl tyrosinate methylsilanol, *See* Tyrosilane C, 32688
copper ammonium carbonate, *See* Croptex Fungex, 7215
Copper Antracol®, 6778
copper arsenate, *See* cupric arsenate, 7375
copper bromide, *See* cuprous bromide, 7402
copper bronze, *See* copper, 6777
copper carbonate hydroxide, *See* cupric carbonate, 7378
copper carbonate, basic, *See* Croptex Fungex, 7215
copper, [+lm-[carbonato(2-)-O:O']]dihydroxydi-, *See* cupric carbonate, 7378
copper chloride, *See* cuprous chloride, 7403
copper chloride ($CuCl_2$), mixt. with copper oxide (CuO), hydrate, *See* Cupravit®, 7370
copper chloride oxide, *See* Cupravit®, 7370
copper count N, *See* Croptex Fungex, 7215
copper dihydroxide, *See* cupric hydroxide, 7381, Schweitzer's reagent, 27738
copper dimethyldithiocarbamate
copper dimethyldithiocarbamate, 6779, *See* Akrochem® Cu.D.D, 1165, copper dimethyldithiocarbamate, 6779, Cumate®, 7351, Perkacit® CDMC, 23278
Copper Euparen®, 6780
copper formate, *See* cupric formate, 7380
copper froth, *See* Tyrolite, 32687
copper gluconate, 6781, *See* Gluconal® CU, 13041
copper green, 6782
copper hydrate, *See* cupric hydroxide, 7381, 7382, Schweitzer's reagent, 27738
copper hydroxide, *See* Comac Parasol, 6659, cupric hydroxide, 7381, 7382, Schweitzer's reagent, 27738
copper hydroxide (ic), *See* cupric hydroxide, 7381
copper hydroxide sulfate, *See* cupric sulfate, basic, 7388
copper iodide, *See* cuprous iodide, 7404
copper naphthenate, 6783
copper oxalate, *See* cupric oxalate, 7385
copper oxide, *See* cupric oxide, 7386, Cupridan, 7389
copper oxide copper(I) oxide, *See* cuprous oxide, 7405
copper oxide hydrated, *See* cupric hydroxide, 7381
copper oxychloride, 6784, 6785, copper oxychloride, 6784, Cupravit®, 7370, Cuprokylt, 7395, Cuprosana, 7399, Curenox-50, 7419, FS Dricol 50, 12426, Headland Inorganic Liquid copper, 13731
copper PCA methylsilanol, *See* Silhydrate C, 28312
copper precipitate, *See* cement copper, 5662
copper rust, *See* malachite, 18458
copper silumin, 6786
copper solder, 6787
copper steel, 6788
copper sulfate, *See* cupric sulfate, 7387
copper sulfate monohydrate, *See* Cusamon, 7428
copper sulfate pentahydrate, *See* cupric sulfate, 7387
copper sulfocyanide, *See* white paste, 34369
copper sulphate pentahydrate copper(II) sulphate pentahydrate, *See* Kocide® copper Sulfate Pentahydrate Crystals, 16262
copper uversol, *See* copper naphthenate, 6783
copper, aluminum, tin alloy, *See* Calsun Bronze, 5037
copper, bis(dimethylcarbamodithioato-S,S')-, (SP-4-1)-, *See* Akrochem® Cu.D.D, 1165, Cumate®, 7351, Perkacit® CDMC, 23278
copper, bisdimethylcarbamodithioato copper, bis(dimethylcarbamodithioato-S,S')-, (SP-4-1)-, *See* copper dimethyldithiocarbamate, 6779
copper, bis(D-gluconato), *See* copper gluconate, 6781

copperas, *See* iron sulfate (ous), 15321, Iron vitriol, 15322
copperas, *See* ferrous sulfate, 11690
copper-Count, 6789
copper-Count-N, *See* Monterey Liqui-Cop, 20293
copper-dichlorfluanid, *See* copper Euparen®, 6780
copper-Lonacol, 6790
copper-nap-all, *See* copper naphthenate, 6783
copper-nickel, 6791
copperone, *See* cupferron, 7365
copper-Sandoz, *See* cuprous oxide, 7405
coppertox, 6792
coppertrace, 6793, *See* cupric chloride dihydrate, 7379
copper-Zinc alloy, *See* Abyssinian gold, 167
copper-zineb, *See* Copper-Lonacol, 6790
Coppesan, 6794
copra, *See* coconut butter, 6566
Copramyl, *See* Ablumide LME, 80, Rewomid® L 203, 26328
Coprantex, 6795
Coprantol, *See* copper oxychloride, 6784
Coprol, 6796
Cops 1, 6797
Coptal, 6798
Coptal WA OSN, *See* dioctyl sodium sulfosuccinate, 8550
Coptox, *See* copper oxychloride, 6784
Corafilm, Coravol, 6799
Corail, *See* Raxil, 25908
Co-Ral, *See* Perizin®, 23276
Corasole, 6800
Co-Rax, *See* Killgerm® Sewarin P, 16184, Kypfarin, 16519
Corban, 6801
Corbel®, 6802
Corbel® CL, Corbel® Star, 6803
Corbel® Duo, 6804
Corbrite, 6805
Corcert, 6806
Cordetec 100, 6807
Cordianine, *See* Allantoin, 1739
Cordite, 6808
Cordite MD, *See* Cordite, 6808
Cordocel, *See* Nuophene, 21783
Cordura®, 6809
Corephen® 10, 6810
Coresize 630, 6811
Corexit, 6812
Corfast, 6813
Corfix, 6814
Corflex 880, *See* Witcizer 313, 34455
Corgran, 6815
Coriacide, 6816
Corial Primer, 6817
Corial®, 6818
Corialbinder®, 6819
Corian®, 6820
corichrome, 6821
Corid, *See* Amprolium hydrochloride, 2365
coridine, 6822
Corilene, 6823
Corinal, 6824
Corindite, 6825
corioflavines, 6826
Coripact, 6827
coriphosphines, 6828
Coriumine, 6829
Corlake, 6830
Corlar®, 6831
Cormix, 6832
Cormul, 6833
corn oil, 6834
corn oils, *See* corn oil, 6834
Corn oil PEG-6 esters, 6835, *See* Labrafil M 2125 CS, 16540
corn starch, 6836, 6837, *See* Control 1-100, 6749
corn sugar, *See* glucose, 13052
Corn sugar gum, *See* Xanflood®, 34587
corn syrup, 6838
Cornalith, *See* Gallatite, 12608
Cornish bronze, 6839
Cornox, *See* benazolin, 3935
Cornox-m, *See* MCPA, 19183
Corn-Pro 35, 6840
Cornuite, 6841
Corolox, Corotox, Corowalt, 6842
Corona, 6843
Corona Corozate, *See* AAprotect, 16
Coronet, 6844
Coronium, 6845
Corox, 6846
Corozate, *See* AAprotect, 16, Accelerator MZ Powder, 209, Vancide® MZ-96, 33303
Corozo, 6847
Corrigast, *See* propantheline bromide, 25176
Corro-Guard, 6848
Corrolite®, 6849
Corronel 220, 6850
Corronel Alloy 230, 6851
corronil, 6852
Corrosalloy, 6853
corrosiron, 6854
Corrosist, 6855
corrosive mercury chloride, *See* mercuric chloride, 19347
Corrosive Sublimate, *See* mercuric chloride, 19347
Corry's Moss Remover, 6856
Corry's slug death, *See* metaldehyde, 19431
Corsair, *See* Kafil, 15655, permethrin, 23384, 23385, 23386
Corseal, 6857
Corsodil, *See* chlorhexidine digluconate, 6101
Corsodyl, *See* chlorhexidine digluconate, 6101, Rotersept, 27002
Corten, 6858
Corton A1, 6859
Cortone, 6860
Cortymol® LP, 6861
Corubin, 6862
Corundite, 6863
Corundum, *See* aluminum oxide, 1919, Bikorit, 4139
Corvic, 6864
Cosan, *See* Thiovit, 31724
Cosbiol®, 6865
Cosiderm Collagen Masks, 6866
Coslettized steel, 6867
Cosmedia Guar C-261 N, 6868
Cosmedia Guar® C-14-S, *See* Jaguar® C-13S, C-14S, 15496
Cosmedia Polymer HSP 1180, 6869
Cosmelan, *See* Vigilan, 33750
Cosmetic Gelatin, *See* gelatin, 12722
Cosmetic Lanolin USP, 6870
Cosmetol® X, 6871
Cosmic, 6872
Cosmic® FL, 6873
Cosmica®, 6874
Cosmocil, 6875
Cosmopen, *See* Penicillin G potassium, 23109, 23110
Cosmos Alloy, 6876
Cosmowax, 6877
Cosmowax K, 6878
Cosmowax P, 6879
Cotazym, 6880
Cotazyme, *See* Accelerase, 185
Cothias metal, 6881
Cotnion-Ethyl, 6882, *See* Azinphos-ethyl, 3520, Gusathion® A, 13502
Cotnion-Ethyl-Methyl, 6883, *See* Azinphos-ethyl, 3520
Cotnion-Methyl, 6884
Cotofilm, *See* hexachlorophene, 13992
Cotolan Fast, 6885
Cotonerol®, 6886
Cotopa, 6887
Cottestren, 6888
Cottestren® C Dyes, 6889
Cottoclarin, 6890
cotton black, 6891
cotton blue, 6892
cotton fiber, *See* cellulose, 5626
cotton wax, 6893
Cottonex, 6894
Cotton-Pro, *See* prometryn, 25141
Cotton-Pro®, 6895 , *See* prometryn, 25140
Cotton*Seed* oil, hydrogenated, *See* Sta-Nut EE, 29508
Cotton*Seed* oil, partially hydrogenated, *See* Sta-Nut EE, 29508
Couch and Grass Killer, 6896
Cougar, 6897
Couloscope, 6898
Coulter Clenz, 6899
Coulter Clone, 6900
coumalic acid, 6901
Coumalux®, 6902
Coumaphos, *See* Asuntol®, 3280, Muscatox, 20525, Perizin®, 23276
Coumarin, 4-hydroxy-3-(1,2,3,4-tetrahydro-1-naphthyl)-, *See* Racumin®, 25713
Coumarin, 3-(3-(4-biphenylyl)-1,2,3,4-tetrahydro-1-naphthyl)-4-hydroxy-, *See* Ratak, 25896
coumarone resin, *See* paracoumarone resin, 22729
coumarone-indene resin, 6903
4-Coumaryl alcohol, *See* 4-hydroxycoumarin, 14661
Coumatetralyl, *See* Racumin®, 25713
Coumetralyl, *See* Racumin®, 25713
Coumou oil, *See* patava oil, 22883
Countdown, 6904
Counter, *See* Terbufos, 31226
Country Fresh Disinfectant, 6905
Coupler, 6906
Coupler SC, 6907
Coupsil VP 6109, 6908
Courcel®, 6909, 6910
Courlene, *See* A-C® Polyethylene 6, 6A, 7, 7A, 8, 8A, 9, 9A, 617, 617A, 175, Cabelec® 1017, 4862
Courline, 6911
Cournova, 6912
court plaster, 6913
Courtaulds, 6914
Courtek, 6915
Courthene, 6916
Courtochrome, 6917
Covar, 6918
Coveral, 6919
Coverite, 6920
Covermark, 6921
Covexin, 6922
Covitol, *See* tocopherol, 31918, Vascuals, 33465
Covon, 6923
Cow Gum, 6924
Cowles' aluminum bronze, 6925
coxcomb pyrites, *See* marcasite, 18700
Coxistac, *See* Salinomycin, 27271
Coyden, 6926
Cozirc, 6927
Cozyme, *See* d-Panthenol, 22690
CP 1044 J3, *See* bufexamac, 4655
CP 23426, *See* triallate, 32190
CP 31393, *See* Ramrod, 25861
Cp 31393, *See* Propachlor, 25163
CP Filler, 6928
CP0110, 6929
CP0156, 6930
CP0160, 6931
CP0280, 6932
CP0320, 6933
CP0330, 6934
CP0800, 6935
CP0810, 6936
CP-105, *See* SR-379, 29382
CP-105, *See* glycidyl methacrylate 2,3-epoxypropyl methacrylate, 13094
CP-153-2 (25% in xylene), 6937
CP-343-1(100%), 6938
CP-12009-18, *See* morantel tartrate, 20307
CP-14445-16, *See* Telopar, 31070
CP-50144, *See* Alachlor, 1327

CP-53619, *See* Butachlor, 4727
CPE, 6939
C-petroleum naphtha, 6940
CPH-211-N, 6943, *See* PEG-2 dioleate series, 23040
CPH-250-SE, 6944
CPH-43-N, 6941
CPH-52-SE, 6942
CPPD, *See* Anti-oxidant 4010, 2581
CPS 034, 6945
CPS 076, 6946
CPS 120, 6947
CPS 130, 6948
CPS 140, 6949
CPS 340, 6950
CPS 925, 6951
CPS 9120, 6952
CPTMO, *See* CC3300, 5493, chloropropyltrimethoxy silane, 6138, Dynasylan® CPTMO, 9403
CR-39, 6953
CR-242, *See* proglumide, 25116
CR-1639, *See* Karathane, 15759
CR 3029, *See* Maneb, 18526
Crab grass killer, *See* arsenic acid, 3137
Crackdown, *See* Thripstick®, 31766
Cradocap, *See* Cetrimonium bromide, 5813
Crafol AP-16, 6954
Crafol AP-31, 6958
Crafol AP-53, 6959
Crafol AP-64, 6960
Crafol AP-201, 6955
Crafol AP-260, 6956
Crafol AP-262, 6957
Crag, *See* Dazomet, 7806, N-521® Biocide, 20640
Crag 85W, *See* Dazomet, 7806, N-521® Biocide, 20640
Crag 974, *See* Dazomet, 7806, N-521® Biocide, 20640
Craig gold, 6961, *See* nickel silvers, 21132
Cranco, 6962
Crastine®, 6963 , *See* Grilpet® XE3060, 13402
Crastine® S 600, Crastine® SG 625, 665 FR, 653, XB 3035, 6964
Crastine® SG 625, 6965
Crastine® SG 665 FR, 6966
Crastine® SO 653, 6967
Crastine® XB 3035, 6968
Crastine® XMB 1068, 6969, *See* Polyethylene terephthalate, 24426
Cravenette EFC, 6970
Craymer, 6971
Crayvallac, 6972
Cream E45, 6973, *See* Synwax, 30659
Cream of tartar, *See* potassium bitartrate, 24740
Creamalin, *See* aluminum hydroxide, 1905
Creamtex, 6974
Credo, *See* sodium fluoride, 28764
Crelan®, 6975
Cremalys, 6976
Cremba, 6977
Cremerol HMG, 6978
Cremnitz, *See* flake white, 11871
cremnitz white, *See* flake white, 11871
Cremophor A, *See* Ablunol DEGMS, 99
Cremophor El, *See* Ablunol CO 10, 98
Cremophor® A 11, 6979
Cremophor® A 11, A 25, *See* Ceteareth-11, 5789
Cremophor® A 25, 6980
Cremophor® EL, 6981
Cremophor® NP 10, 6982
Cremophor® NP 14, 6983
Cremophor® RH 40, 6984
Cremophor® RH 60, 6985
Cremophor® S 9, 6986
Cremophor® WO 7, 6987
Cremorin, *See* aluminum hydroxide, 1905
crems white, *See* flake white, 11871
Crenette, 6988
Creolin®, 6989
creosol, 6990
Creosote, 6991
Crepetrol® 190, 6992
Cresamol, *See* Kresamin, 16401
Cresatin, *See* *m*-cresyl acetate, 7011, Kresatin, 16402
Cresatin Metacresylacetate, *See* Kresatin, 16402
Cresatin-Sulzberger, *See* Kresatin, 16402
Cresavon, 6993
Cresol, 6994, *See* oxytoluol, 22472
Cresol (All Isomers), *See* cresylic acid, 7013
Cresol (mixed isomers), *See* cresylic acid, 7013
***m*-cresol**, 6997
***o*-cresol**, 6998
***p*-cresol**, 6999
***m*-Cresol acetate**, *See* Kresatin, 16402
***m*-cresol acetic acid ester**, *See* Kresatin, 16402
cresol carboxylic acids, *See* cresotic acids, 7001
cresol dicyclopentadiene butylated reaction product, *See* Akrochem® Antioxidant 12, 1157
cresol diphenyl phosphate, *See* Disflamoll® DPK, TPK, 8641
***m, p*-Cresol hydrophobe**, *See* Hetoxide MPC, 13926
***m-,p*-cresol mixture**, *See* cresylic acid, 7013
cresol purple, 6995
cresol red, 6996
cresol, o-epoxypropyl ether, *See* DY 023, 9300
Cresolox®, 7000
cresols, *See* cresylic acid, 7013
***m*-cresolsulfonphthalein**, *See* cresol purple, 6995
***o*-Cresolsulfonphthalein**, *See* cresol red, 6996
Cresopur, *See* benazolin, 3935
cresotic acids, 7001
Cressylite, 7002
Crestalan, 7003
Crester KZ, 7004
crestmoreite, *See* calcium silicate, 4969
Crestolan NF, Crestosolve 630, 7005
Crestomer®, 7006
Crestomer® 1066A, 7007
Crestomer® 1080, 7008
Crestopene 5X, 7009
Crestophen®, 7010
***m*-cresyl acetate**, 7011, *See* *m*- cresyl acetate, 7011, Kresatin, 16402
***p*-cresyl acetate**, *See* Narceol, 20742
Cresyl diphenyl phosphate, *See* diphenylcresyl phosphate, 8601, Disflamoll® DPK, TPK, 8641
Cresyl glycidyl ether, *See* DY 023, 9300
***o*-Cresyl glycidyl ether**, *See* DY 023, 9300, Heloxy® 62, 13771
cresyl phenyl phosphate, *See* Disflamoll® DPK, TPK, 8641
***o*-cresyl phosphate**, *See* Plastic X, 24026
Cresylene, 7012
***o*-cresylic acid**, *See* *o*-cresol, 6998
***p*-cresylic acetate**, *See* Narceol, 20742
cresylic acid, 7013, *See* Productol, 25084 xylenol, 34625
***m*-cresylic acid**, *See* *m*-cresol, 6997
***p*-cresylic acid**, *See* *p*-cresol, 6999
Cresylic acids, *See* E.C.A, 9447
cresylol, *See* Cresol, 6994
Cresyl-p-toluenesulfonate, *See* Mittel KP, 19979
***Creta proeparata* B.P.**, *See* prepared chalk, 24896
Creto, 7014
Crex, 7015
Crexathix, 7016
Cri-Line FDA-612, APC-718-75, 7017
Cri-Line FDA-715, 7018
Cri-Line GP-715, 7019
Cri-Line HF-618-65, 7020
Cri-Line IF-612, 7021
Cri-Line LC-508-55, 7022
Cri-Line LC-612-THK, 7023
Cri-Line LC-708, 7024
Cri-Line SP-508, 7025
Cri-Line SP-815, 7026
Criliprint 788FYN, 7027
Crill 1, 7028
Crill 2, 7029, *See* Ablunol S-40, 109
Crill 3, 7030, *See* Ablunol S-60, 110
Crill 4, 7031
Crill 5, *See* Ablunol S-85, 112, Witconol 2503, 34502
Crill 6, 7032, *See* Sorbitan isostearate, 29110
Crill 20,21,22,23, *See* Ablunol DEGMS, 99
Crill 26, *See* Witconol 2380, 34498
Crill 35, 7033
Crill 43, 7034
Crill 45, 7035
Crillet, 7036
Crillet 1, 7037
Crillet 2, 7038
Crillet 3, 7039
Crillet 3, 31, *See* Ethylan® GES6, 11108
Crillet 4, 7040
Crillet 6, 7041
Crillet 11, 7042
Crillet 31, 7043
Crillet 35, 7044
Crillet 35, *See* Polysorbate 65, 24571
Crillet 41, 7045
Crillet 45, 7046
Crill-K-3, *See* Ablunol S-60, 110
Crillon, 7047
Crillon CDY, 7048
Crillon L.D.E, *See* Ablumide LDE, 79
Crillon L.M.E, *See* Ablumide LME, 80
Crillon LDE, 7049
Crillon LME, 7050, *See* Rewomid® L 203, 26328
Crillon ODE, 7051
Crimidesa, 7052
Crimidine, 7053
Crimson Lake, 7054
Crinovyl, *See* Korogel, 16354, Koroplate, 16357
Crisalin, *See* Trifluralin, 32252
Crisamina, *See* atrazine, 3394
Crisapon, *See* dalapon, 7705
Crisatrina, *See* atrazine, 3394
Crisazina, *See* atrazine, 3394
Crisazine, *See* atrazine, 3394
Criscobre, *See* cupric hydroxide, 7381, copper oxychloride, 6784, Schweitzer's reagent, 27738
Criseociclina, *See* tetracycline, 31365
Crisfuran, *See* Rampart, 25860
Cristal, *See* Kemstrene® 96.0%, 15991
Cristalomicina, *See* Kantrex, 15718, Klebcil, 16209
Cristapen, *See* Penicillin G potassium, 23109, 23110
Cristite, 7055
Crisuron, *See* diuron, 8736
Criterion, 7056
Crittox MZ, *See* mancozeb, 18517
Crllitex H-50, 7057
Croak, 7058
Crocase, *See* saffron sugar, 27226
Crocell, 7059
Crocidolite, 7060
crocus, 7061
Crocus Martius, 7062
Croda Bath Oil Disperant, 7063
Croda Fluid, 7064
Crodacel QL, 7065
Crodacel QM, 7066
Crodacel QS, 7067
Crodacid, *See* Philacid 1400, 23662
Crodacol C70, 7068, *See* cetyl alcohol, 5817
Crodacol C95NF, 7069
Crodacol CS50, 7070
Crodacol S, *See* Adol® 62 NF, 783, Stearal, 29591
Crodacol S70, 7071
Crodacol S95NF, 7072
Crodacol-cas, *See* Adol® 52 NF, 782, Cachalot® C-50 NF, 4878
Crodacreme, 7073
Crodafos, 7074
Crodafos 25 D2 Acid, 7075
Crodafos CAP, 7076
Crodafos CDP, 7077, *See* cetyl diethanolamine phosphate, 5818
Crodafos CS2 Acid, 7078
Crodafos N3 Acid, 7081
Crodafos N3 Neutral, 7082

Crodafos N5 Acid, 7083
Crodafos N10 Acid, 7079
Crodafos N10 Acid, N3-Acid, N5 Acid, O2 AcidEmpicol® 0216, *See* oleth-phosphate Series (2-20), 22127
Crodafos N10 Neutral, 7080
Crodafos O2 Acid, 7084
Crodafos O2 TEA, 7085
Crodafos SG, 7086
Crodafos T2 Acid, 7087
Crodafos T2 Acid, *See* Trideceth phosphate series, 32231
Crodakyd, 7088
Crodalan AWS, 7089
Crodalan C24, 7090
Crodalan IPL, 7091
Crodalan LA, 7092
Crodamer, 7093
Crodamet Series, 7094
Crodamide, 7095
Crodamide E, ER, 7096
Crodamide O, OR, 7097
Crodamide S, SR, 7098
Crodamine 1, 7099
Crodamine 1.0, 1.0D, 7100
Crodamine 1.16D, 7101
Crodamine 1.18D, 7102
Crodamine 1.HT, 7103
Crodamine 1.T, 7104, *See* tallowamine, 30748
Crodamine 2.C, 2.S and 2.HT, 7105
Crodamine 3.A16D, 7106
Crodamine 3.A18D, 7107
Crodamine 3.AED, 7108
Crodamine 3.AOD, 7109, *See* Jet Amine DMOD, 15586
Crodamine 3A, 7110
Crodamine 3ABD, 7111
Crodamine 3AED, 7112
Crodamine 3AHRD, 3ARD, 7113
Crodamine 3AOD, 7114
Crodamol, 7115
Crodamol 1PM, 7116
Crodamol BE, 7117
Crodamol BM, *See* Butyl myristate, 4775
Crodamol CAP, 7118, *See* Cetearyl octanoate, 5791
Crodamol CP, 7119, *See* cetyl palmitate, 5822
Crodamol CSP, 7120
Crodamol Da, *See* Unimate® DIPA, 32968
Crodamol GTC/C, 7121
Crodamol I.P.M, *See* Wickenol® 101, 105, 34378
Crodamol IPP, *See* Wickenol® 111, 34379
Crodamol MM, 7122
Crodamol PETS, *See* Pentaerythrityl tetrastearate, 23143
Crodamol PMP, 7123
Crodamol PTC, 7124
Crodamol PTIS, 7125
Crodamol SS, 7126, *See* cetyl esters, 5819
Crodamol TC, *See* triethyl citrate, 32242
Crodamol W, 7127
Crodapearl, 7128
Crodapearl Liq, 7129
Crodapearl NI Liquid, 7130
Crodaplast, 7131
Crodapol, 7132
Crodapur, 7133
Crodarom Avocadin, 7134
Crodarom Calendula O, 7135
Crodarom Carrot O, 7136
Crodarom Chamomile A, 7137
Crodarom Chamomile EO, O, 7138
Crodarom Nut O, 7139
Crodarom St. John's Wort O, 7140
Crodascoop, 7141
Crodasinic, 7142
Crodasinic C, *See* Cocoyl sarcosine, 6568
Crodasinic L, 7143, *See* lauroyl sarcosine, 16861
Crodasinic LS30, 7144
Crodasinic LS30, LS35, *See* Gardol®, 12668
Crodasinic LS35, 7145
Crodasinic LT40, 7146
Crodasinic O, 7148, *See* Oleoyl sarcosine, 22123
Crodasinic OS35, 7147
Crodasone W, 7149
Crodasub, 7150
Crodatem, 7151
Crodateric, 7152
Crodateric C, 7153
Crodateric Cy, 7154
Crodateric O, 0.100, 7155
Crodateric S, 7156
Crodax, 7157
Crodax DP 100, 7158
Crodax DP 50, 7159
Crodazoline Cy, 7160, *See* Caprylic imidazoline, 5159
Crodazoline O, 7161, 7162
Crodazoline S, 7163, *See* stearyl hydroxyethyl imidazoline, 29606
Croderol G7000, 7164
Crodesta, 7165
Crodesta DKS F10, 7166
Crodesta DKS F110, 7167, *See* Grilloten PSE141G, 13378
Crodesta F-10, 7168
Crodesta F-160, 7169, *See* Grilloten PSE141G, 13378
Crodesta SL-40, 7170
Crodet C10, 7171
Crodet L4, 7172
Crodet O 6, *See* Ablunol 200MO, 96
Crodet O4, 7173
Crodet S4, 7174
Crodex A, 7175
Crodex C, 7176
Crodex N, 7177
Crodinhib, 7178
Crodinhib RT70, RT70S, 7179
Crodol, 7180
Croduret 10, 7181
Crodyne BY19, 7182, *See* gelatin, 12722
Crofilcon A, *See* CSI, 7310
Croian, 7183
Crolactil CSL, 7184
Crolactil SISL, 7185
Crolactil SSL, 7186
Crolastin, 7187
Crolastin 10 Powder, 7188
Crolec 4135, 7189
Croloy, 7190
Cromal, 7191
Cromalit, 7192
Cromalit 150, 7193
Cromaloy II, 7194
Cromaloy III, 7195
Cromaloy IV, 7196
Cromeen, 7197
Cromo Steel, 7198
Cromoist CS, 7199
Cromoist HYA, 7200
Cromoist O25, 7201
Cromophtal, 7202
Cromophytal C-20, M-20, 7203
Cromosan, *See* magnesium acetate, 18357
Cromul 0685, 7204
Cronectin, 7205
Cronetal, *See* Disulfiram, 8725
Croneton®, 7206
Cronite, 7207
Cropepsol, *See* Extiat®, 11323
Cropepsol, Cropeptone, 7208
Cropeptide W, 7209
Cropeptone, *See* Extiat®, 11323
Cropol 60, 7210
Cropotex, *See* Cropotex®, 7211
Cropotex®, 7211
Croptex, *See* Propachlor, 25163
Croptex Amber, 7212
Croptex Bronze, 7213, *See* pentanochlor, 23165
Croptex Chrome, 7214, *See* Chlorpropam, 6162
Croptex Fungex, 7215
Croptex Onyx, *See* bromacil, 4564, Uragan, 33105
Croptex Pewter, 7216
Cropure Apricot Kernel, 7217
Cropure Avocado, 7218
Cropure Babassu, 7219
Cropure EPO, 7220
Cropure Orange Roughy, 7221
Cropure Wheat Germ, 7222
Croquat HH, 7223
Croquat L, 7224
Croquat M, 7225
Croquat S, 7226
Croquat WKP, 7227
Crorzate, *See* Zimate®, 34762
Croscarmellose sodium, *See* Ac-Di-Sol, 267, CLD 2, 6379
Croscolor, 7228
Croscour, 7229
Crosdurn, 7230
Crosil, 7231
Crosilk, 7232
Crosilk 10,000, 7233
Crosilk Liq, 7234
Crosilk Powder, 7235, *See* Fibro-Silk Powd, 11754
Crosilkquat, 7236
Croslube, 7237
Crosoft, 7238
Crospovidone, *See* Kollidon® CL, 16313
Crossbow, *See* Timbrel, 31800
Crossfire, 7239 *See* Chlorpyrifos, 6163, resmethrin, 26218
Crosslinkable ethylene-propylene-diene rubbers, *See* AEI Compound 505/401, 830
Crosslinked acrylic emulsion copolymer, *See* Acrysol® ASE-60, 443
Crosterene, 7240
Crosultaine C-50, 7241
Crosultaine E-30, 7242
Crosultaine T-30, 7243
Crotein A, C, O, 7244, *See* Extiat®, 11323
Crotein AD Anhyd, 7245
Crotein ADW, 7246
Crotein ASC, 7247
Crotein ASK, 7248
Crotein CAA, 7249, *See* Extiat®, 11323
Crotein CAA/CF, O, SPA, *See* Extiat®, 11323
Crotein CAA/SF, 7250
Crotein HKP, 7251
Crotein HKP Powd, 7252
Crotein HWE, 7253
Crotein IP, 7254
Crotein K, WKP, 7255
Crotein O, 7256
Crotein Q, 7257
Crotein SPA, 7258
Crothix, 7259
α-Crotonic Acid, *See* Crotonic Acid, 7260
Crotonic Acid, 7260
Crotorite, 7261
Crotothane, *See* Karathane, 15759
Crottle, *See* Cudbear, 7342
Cro-tung, 7262
Crovol A40, 7263
Crovol M40, 7264
Crovol PK40, 7265
Crow, 7266
Crow Chex, *See* cupric oxalate, 7385
Crown Acid Aid X, 7267
Crown Anti-Foam, 7268
Crown Foamer 20, 7269
Crown L-1011 Acid Inhibitor, 7270
Crown L-60B Acid Inhibitor, 7271
Crown solder, 7272
Croysulfone, *See* Dapsone, 7751
Cru Fax P9N, 7273
Crude Arsenic, *See* arsenic trioxide, 3143
Crude dinitrodichlorobenzene, *See* Parazol, 22814
Crude oil, *See* Kendex OCTG, 16002
Crude Sicilian sulfur, *See* Greggio, 13348

Crude sodium sulfate, *See* salt cake, 27281
Cruverlite, 7274
Crylene, 7275
CryOfine® Butyl, 7276, *See* Exxon® Butyl 077, 11355
CryOfine® EPDM, 7277
Cryofluorane, 7278
Cryolite, 7279, *See* Kryalith, 16448
Cryolite
Cryoseal, 7280
Cryptone, 7281
Cryptonol, 7282
Crystal, 7283, 7284
Crystal 1000, 7285
crystal glass, 7286
Crystal Inhibitor #5, 7287
Crystal Inhibitor No. 5 EI-620, *See* Ablunol CO 10, 98
Crystal Polystyrene, 7288, *See* Polystyrene 101, 24602, Polystyrene 220, 24603
Crystal PS, *See* Mobil 1240, 19996, Mobil 2120, 19997
crystal violet, *See* gentian violet, 12845
crystal violet 10B, *See* gentian violet, 12845
crystal violet 5BO, *See* gentian violet, 12845
crystal violet 6B, *See* gentian violet, 12845
crystal violet 6BO, *See* gentian violet, 12845
crystal violet AO, *See* gentian violet, 12845
crystal violet AON, *See* gentian violet, 12845
crystal violet base, *See* gentian violet, 12845
crystal violet BPC, *See* gentian violet, 12845
crystal violet FN, *See* gentian violet, 12845
crystal violet HI2, *See* gentian violet, 12845
crystal violet O, *See* gentian violet, 12845
crystal violet SS, *See* gentian violet, 12845
Crystal® O, Crystal® Crown, 7289
Crystalite, 7290
crystalline, *See* paraffin, 22742
Crystalline aluminosilicate, *See* HSZ-320NAA, 14441
Crystalline aluminum oxide, *See* Dycron, 9309
Crystalline calcium carbonate, *See* Iceland spar, 14813
Crystallized copper sulfate, *See* Sulfato de Cobre Valles, 29974
crystallized verdigris, *See* cupric acetate, 7372
crystallose, *See* saccharin sodium, 27196
Crystalor BC-1, 7291
Crystalor DC-6, 7292
Crystamet 1020, 7293
Crystapen, *See* Penicillin G potassium, 23109, 23110
Crystex®, Crystex® Regular, 7294
Crysthion, *See* Gusathion® A, 13502
Crystic® 199, 7297
Crystic® 2-414PA, 7295
Crystic® 39PA, 7296
Crystic® 471PALV, 7298
Crystic® 581PA, 7299
Crystic® Fireguard, 7300
Crystic® Fireguard 75PA, 7301
Crystic® Impel, 7302
Crystic® Impreg, 7303
Crystic® Prefil F, 7304
Crystic® Pregel 17, 7305
Crysticillin, *See* Penicillin V, 23111
Crystolon®, 7306
Crysylol, *See* cresylic acid, 7013
CS1590, 7307
CSA, *See* d-camphorsulfonic acid, 5076
CSAC, *See* EE Acetate, 9635
CSC 2-aminobutane, 7308
CSE-6000 Series, 7309
CSI, 7310, *See* CT2950, 7325
CT1750, 7311
CT1800, 7312
CT2015, 7313
CT2030, 7314
CT2050, 7315
CT2090, 7316
CT2500, 7317
CT2507, 7318
CT2520, 7319
CT2523, 7320
CT2902, 7321
CT2910, 7322
CT2925, 7323
CT2928, 7324
CT2950, 7325
CT2970, 7326
CT2970, *See* EXP-51, 11306
CT3250, 7327
CT3254, 7328
CT3600, 7329
CT3610, 7330
CT3795, 7331
CT-690, CT-700, 7332
CTA, *See* Sumquat® 6020, 30121
CTAB, *See* Rhodaquat® M242B/99, 26643
CTABr, *See* Rhodaquat® M242B/99, 26643
CTC-3300, 7333, *See* Celanese® Nylon 1000-1, 5563
CTR 6669, *See* Carbendazim, 5187, 5188, Derosal, 16327
CTW, 7334
CTX-308, 7335
CTX-312, 7336
CTXC-020, 7337
Cuba black, 7338
Cuba orange, 7339
Cuba wood, *See* Fustic, 12522
cubanite, 7340
cube root, *See* rotenone, 27000
cubic niter, *See* sodium nitrate, 28800
cubic niter, Nitratine, *See* Chili saltpeter, 6063
cubic saltpeter, soda niter, *See* Chili saltpeter, 6063
Cubor, *See* rotenone, 27000
Cu-Ca-Zn-Sn, *See* Admiralty brass, 752
Cuclat, 7341
Cudbear, 7342, *See* Archil, 2808
Cudrox, *See* cupric hydroxide, 7381, Schweitzer's reagent, 27738
Cufenium, 7343
Cufor, *See* cupric formate, 7380
cufraneb, 7344
cufranebe, *See* cufraneb, 7344
Cuidrox, *See* cupric hydroxide, 7381, Schweitzer's reagent, 27738
Cuidrox, *See*
Cuite, 7345
Cuivre poli, *See* Abyssinian gold, 167
Culminal® Hydroxypropylmethylcellulose, 7346
Culminal® Methylcellulose, 7347
Cultar, 7348, *See* paclobutrazol, 22540, 22541 22542
Cu-lyt, *See* cuprous chloride, 7403
Cumal, 7349, *See* cuminaldehyde, 7354
Cuman, *See* AAprotect, 16
Cuman L, *See* Bisphenol A, 4283
cumar, *See* paracoumarone resin, 22729
cumar gum, *See* paracoumarone resin, 22729
Cumar resin, *See* paracoumarone resin, 22729
Cumar®, Cumar® P-10, R-1, 7350
Cumate, *See* Akrochem® Cu.D.D, 1165, copper dimethyldithiocarbamate, 6779, Cumate®, 7351, Perkacit® CDMC, 23278
Cumate®, 7351 **,** *See* copper dimethyldithiocarbamate, 6779
Cumatetralyl, *See* Racumin®, 25713
cumene, 7352
α-cumene hydroperoxide, *See* Aztec® CHP-80, 3555; CHP-158, 6175
cumene hydroperoxide, *See* Aztec® CHP-80, 3555, CHP-158, 6175, HPC-9, 14438, Trigonox® 239A, 32290, Trigonox® K-95, 32303
cumene peroxide, *See* Aztec® DCP-R, 3556, Dicumyl peroxide, 8407, Percumyl D, 23230, Peroximon® DC-40, 23409, Polyvel PCL-20, 24661
cumene sulfonic acid, 7353 *See* Eltesol® CA 65, 9866, Reworyl® C, 26403
cumenesulfonic acid, sodium salt, *See* Witconate SCS 45%, 34490
cumenyl hydroperoxide, *See* Aztec® CHP-80, 3555, CHP-158, 6175
cumenyl N-methylcarbamate, *See* Etrofolan®, 11156
cuminaldehyde, 7354
cumol, *See* cumene, 7352
α-cumyl hydroperoxide, *See* Aztec® CHP-80, 3555, CHP-158, 6175
cumyl Hydroperoxide, *See* Aztec® CHP-80, 3555, CHP-158, 6175
cumyl peroxide, *See* Aztec® DCP-R, 3556, Dicumyl peroxide, 8407, Percumyl D, 23230, Peroximon® DC-40, 23409, Polyvel PCL-20, 24661
cumyl peroxyneoheptanoate, *See* Esperox® 740M, 10855, Trigonox® 99-B75, 32278
cumyl phenol, 7355
cumylperoxy neodecanoate, *See* Esperox® 939M, 10858
4-cumylphenol, *See* cumyl phenol, 7355
cumyl-phenyl acetate, *See* Kenplast® ES-2, 16010
cumyl-phenyl benzoate, *See* Kenplast® ESB, 16011
cumylphenyl neodecanoate, *See* Kenplast® ESN, 16013
Cunilate 2174-NO, 7356
Cuniloy, 7357
Cuniphen, *See* Nuophene, 21783
Cuniphen 2778-1, 7358
Cunitex, 7359
Cupalit, 7360
Cupar, *See* cupric arsenate, 7375
cuperatin, 7361
Cupertine, 7362
Cupertine Folpet, 7363
Cupertine Super, 7364
cupferron, 7365
Cupolloy, 7366
cupral, *See* Octopol SDE-25, 22013
cuprammonium silk, 7367
Cuprammonium sulfate, *See* Krystallazurin, 16468
cupranium, 7368
Cuprase, 7369
cuprate silk, *See* cuprammonium silk, 7367
Cupravit, *See* copper oxychloride, 6784
Cupravit blue, *See* cupric hydroxide, 7381, Schweitzer's reagent, 27738
Cupravit®, 7370 **,** *See* copper oxychloride, 6784, 6785
Cuprenox, *See* copper oxychloride, 6784
Cupreous pyrites, *See* Sachtocup, 27198
Cuprex, 7371
cupric acetate, 7372, *See* verdigris, 33571
cupric acetate, basic, 7373
cupric acetoarsenite, 7374, *See* mountain green, 20369
cupric ammonium carbonate, *See* Croptex Fungex, 7215
cupric arsenate, 7375
cupric arsenite, 7376, *See* Scheele's green, 27562
cupric bromide, 7377
cupric carbonate, 7378
cupric carbonate basic, *See* mountain blue, 20366, verditer blue, 33574
cupric carbonate, basic, *See*
cupric chloride, *See* Coclor, 6560
cupric chloride dihydrate, 7379, *See* Coppertrace, 6793
upric chloride dihydrate
cupric chromite(III), *See* chrome black, 6203
cupric formate, 7380
cupric gluconate, *See* Gluconal® CU, 13041
cupric hydroxide, 7381, 7382, 7383
cupric hydroxide, *See* Chiltern Kocide 101, 6069, cupric hydroxide, 7383, 7382
cupric nitrate, 7384
cupric nitrate trihydrate, *See* cupric nitrate, 7384
cupric oxalate, 7385
cupric oxide, 7386
cupric subacetate, *See* cupric acetate, basic, 7373
cupric subcarbonate, *See* cupric carbonate, 7378
cupric sulfate, 7387
cupric sulfate, ammoniated, *See* Krystallazurin, 16468
cupric sulfate, basic, 7388
Cupric sulfate, pentahydrate, *See* Kocide® Copper Sulfate Pentahydrate Crystals, 16262
Cupric sulfate/cupric hydroxide, *See* Casselmam's green, 5403

β-cycloamylose, *See* cyclodextrins, 7516
Cyclochem, 7515
Cyclodan, *See* thiodan, 31680
4-cyclodecyl-2,6-dimethylmorpholine acetate, *See* dodemorph-acetate, 8782
cyclodextrins, 7516
4-cyclododecyl-2,6-dimethylmorpholine acetate, *See* dodemorph-acetate, 8782
Cyclo-Flo, 7517
Cyclofor, 7518
cycloglucan, *See* cyclodextrins, 7516
Cyclogol, 7519
2,5-Cyclohexadiene-1,4-dione, bis(O-benzoyloxime), *See* Dibenzo GMF, 8353
cyclohexanamine, *See* cyclohexylamine, 7525
Cyclohexanamine, N-cyclohexyl-, nitrite, *See* Dichan 100, 8377, dicyclohexylamine nitrite, 8414
cyclohexane, 7520
1,4-Cyclohexane dimethanol divinyl ether, *See* Rapi-Cure CHVE, 25879
Cyclohexane dimethanol, monovinyl ether, *See* Rapi-Cure CHMVE, 25878
Cyclohexane, 5-isocyanato-1-(isocyanatomethyl)-1,3,3-trimethyl-, *See* Vestanat® IPDI, 33657
Cyclohexane, 1,1-bis(tert-butyldioxy)-, *See* USP® - 400P, 33168
cyclohexanediamine, *See* 1,2-diaminocyclohexane, 8301
1,2-cyclohexanediamine, *See* Diaminocyclohexane, 8300, 8301
5-amino-1,3,3-trimethyl-, *See* Vestamin® TMD, 33656
cyclohexanesulfamic acid, *See* Cyclamic acid, 7509
cyclohexanesulfamic acid, monosodium salt, *See* sodium cyclamate, 28754
cyclohexanol, 7521, *See* Adronal, 792, Hexalin, 13998, Hydralin, 14536
Cyclohexanol acetate, *See* Adronal acetate, 793
cyclohexanone, 7522, *See* Sextone, 28118
Cyclohexatriene, *See* Benzene, 3961
3-cyclohexy-6-(dimethylamino)-1-methyl-1,3,5-triazine-2,4(1H,3H)-dione, *See* hexazinone, 14020, 14021
Cyclohexyl acetate, *See* H.A. Solvent, 13522
Cyclohexyl acrylate (monomer), *See* SR-220, 29368
Cyclohexyl amine acetate, *See* Coagulant CHA, 6514
cyclohexyl benzothiazole sulfenamide, 7523
N-Cyclohexyl-2-benzothiazole sulfenamide, *See* CBTS, 5484, cyclohexyl benzothiazole sulfenamide, 7523, Delac S, 8002, Durax®, 9181, P Furbac, 12490, erkacit® CBS, 23277, Santocure, 27400
N-Cyclohexyl-2-benzthiazyl sulfonamide, *See* Vancure D.A.A, 33308
cyclohexyl chloride, 7524
Cyclohexyl isooctyl phthalate, *See* Uniplex 310, 33017
Cyclohexyl methacrylate, *See* Ageflex CHMA, 968, SR-208, 29364
Cyclohexyl methacrylate, monomer, *See* Ageflex CHMA, 968
Cyclohexyl vinyl ether, *See* Rapi-Cure CVE, 25880
cyclohexylamine, 7525
2-(Cyclohexylaminothio)benzothiazole, *See* Vulkacit® CZ/EGC, DZ/EGC, 34129
N-Cyclohexyl-2-benzothiazole sulfenamide, *See* Delac S, 8002
N-Cyclohexyl-2-benzothiazyl sulphenamide, *See* Delac S, 8002
N-Cyclohexylbenzothiazyl sulphenamide, *See* Delac S, 8002
N-Cyclohexyl-2-benzothiazolylsulfonamide, *See* Vulkacit® CZ/EGC, DZ/EGC, 34129
N-Cyclohexylbenzothiazole-2-sulfenamide, *See* CBTS, 5484, Perkacit® CBS, 23277, Vulkacit® CZ/EGC, DZ/EGC, 34129
N-Cyclohexylbenzothiazyl sulfenamide, *See* Naugard® I-2, 20792
Cyclohexylbenzothiazyl sulphenamide, *See* Perkacit® CBS, 23277
N-cyclohexylcyclohexanamine nitrite, *See* Dichan 100, 8377, dicyclohexylamine nitrite, 8414
N-cyclohexylcyclohexanamine nitrite, *See* 3-cyclohexyl-6,7-dihydro-1H-cyclopentapyrimidine-2,4(3H,5H)-dione, *See* lenacil, 16984, Venzar®, 33562, Vizor, 34001
3-Cyclohexyl-6-(dimethylamino)-1-methyl-1,3,5-triazine-2,4(1H,3H)-dione, *See* Velpar;, 33536
[1R-[1α(Z),2β(S*),3α,5α]]-7-[2-(5-Cyclohexyl-3-hydroxy-1-pentynyl)-3,5-dihydroxycyclopentyl]-5-heptenoic acid methyl ester, *See* Alfaprostol, 1574
Cyclohexylmethane, *See* methyl cyclohexane, 19525
N-Cyclohexyl-N-phenyl-4-phenylenediamine, *See* Anti-oxidant 4010, 2581, Vanox® 6H, 33339
cyclohexyl-2-pyrrolidone, *See* CHP, 6173
N-Cyclohexyl-2-pyrrolidone, *See* CHP, 6173
N-2-Cyclohexyl-2-pyrrolidone, *See* Agsol Ex 6C, 1088
Cyclohexylsulfamic acid, *See* Cyclamic acid, 7509
cyclohexylsulfamic acid, monosodium salt, *See* sodium cyclamate, 28754
3-cyclohexyl-1,5,6,7-tetrahydo-2H-cyclopenta pyrimidine-2,4(3H)-dione, *See* Venzar®, 33562
3-cyclohexyl-1,5,6,7-tetrahydrocyclopentapyrimidine-2,4(3H)-dione, *See* Vizor, 34001
N-(Cyclohexylthio)phthalimide, *See* Santogard PVI, 27411
3-cyclohexyl-5,6-trimethyleneuracil, *See* Venzar®, 33562, Vizor, 34001
Cyclolac, 7526
Cyclolube® 62, NN-1, 7527
Cyclolube® 85, 7528
cyclomaltoheptaose, *See* cyclodextrins, 7516
Cyclomatic, 7529
Cyclomatic Dur, 7530
Cyclomethicone, *See* Abil® B 8839, 48 ABIL® K 4, 53, Amersil® VS-7207, 2034, Dow Corning® 344 Fluid, 345 Fluid, 8861, Masil® SF-V, 19024, Sentry Cyclomethicone, 27958, SF1173, 28128, SWS-03314, 30473
Cyclomethicone, dimethiconol, dimethicone, *See* ABIL® OSW 12, OSW 13, 55
Cyclomide, 7531
Cyclomide LM, *See* Ablumide LME, 80
Cyclomide SD, *See* Ablumide SDE, 81
Cyclomycin, *See* tetracycline, 31365
Cyclon, *See* hydrocyanic acid, 14584
Cycloneopentyl, cyclo (dimethylaminoethyl) pyrophosphato zirconate, *See* KZ TPPJ, 16526
Cyclonette, 7532
Cyclonox, 7533
1,2,3,4,5-cyclopentanepentol, *See* acorn sugar, 367
cyclopentimine, *See* piperidine, 23874
Cyclophil SCS40, *See* Witconate SCS 45%, 34490
Cyclophil SXS30, *See* Eltesol® SX 30, 9873, Naxonate® 4L, 20824, Pilot SXS-40, 23843, Witconate SXS 40%, 34492
Cyclophos, 7534
Cyclopol, 7535
Cyclopropane-1,1-dicarboxylic acid, *See* vinaconic acid, 33766
Cyclopropanecarboxylic acid, 3-(2,2-dichloro ethenyl)-2,2-dimethyl-, cyano (3-phenoxyphenyl) methyl ester, *See* cypermethrin, 7588
1,1-Cyclopropanedicarboxylic acid, *See* vinaconic acid, 33766
1-Cyclopropyl-6-fluoro-1,4-dihydro-4-oxo-(1-piperazinyl)-3-qunioline carboxylic acid, *See* ciprofloxacin, 6317
Cyclops metal, 7536
Cyclorans, 7537
Cyclorubbers, 7538
Cycloryl, 7539
Cycloryl 21, *See* Rhodapon® 101-10, 26631, Witcolate 6400, 34469
Cycloryl 21 d, *See* sodium lauryl sulfate, 28781
Cycloryl 31, *See* Witcolate 6400, 34469
cycloryl 31, *See* Rhodapon® 101-10, 26631
Cycloryl 580, *See* Rhodapon® 101-10, 26631, Witcolate 6400, 34469
Cycloryl 580 and 585N, 7540
Cycloryl 585N, *See* Rhodapon® 101-10, 26631, Witcolate 6400, 34469
Cycloryl NA, *See* Witcolate ES-2, Witcolate LES-60C, 34477
Cycloryl TAWF, *See* Perlankrol® ATL40, 23324, Rhodapon® LT-6, 26638, Standapol® T, 29481, Witcolate TLS-500, 34479
Cycloryl Wat, *See* Witcolate TLS-500, 34479
Cycloteric, 7541
Cycloton® SCS, 7542
Cyclovertal, 7543
Cycloxydim, 7544, *See* Focus®, 12184, Stratos®, 29806
Cycocel, *See* CCC 700, 5497, *See* Chlormequat chloride, 6110
Cycocel®, *See* Chlormequat chloride, 6110
Cycogan, 7545, *See* CCC 700, 5497, Chlormequat chloride, 6110
Cycolac®, 7546
Cycolac® GPM4700, *See* Acrylonitrile-butadiene-styrene, 439
Cycolac® GPX2800, *See* Acrylonitrile-butadiene-styrene, 439
Cycolac® KCS, *See* Acrylonitrile-butadiene-styrene, 439
Cycolac® KJM, *See* Acrylonitrile-butadiene-styrene, 439
Cycolac® X-11, *See* Acrylonitrile-butadiene-styrene, 439
Cycolac® CKM1, 7547, *See* Acrylonitrile-butadiene-styrene, 439
Cycolac® DH, 7548, *See* Acrylonitrile-butadiene-styrene, 439
Cycolac® GPM4700, 7549
Cycolac® GPX2800, 7550
Cycolac® KCS, 7551
Cycolac® KJM, 7552
Cycolac® X-11, 7553
Cycolin® GCM1900, GCM2900, 7554
Cycoloy® C1110, 7555
Cycoloy® C2800, 7556
Cycoloy® MC8100, 7557
Cycom, 7558
Cycom® MCG Fiber, 7559
Cydril, 7560
Cy-Ex, 7561
Cyfloc 6000, 7562
cyfluthrin, 7563, *See* Baytroid®, 3856 Solfac, 28930
β-cyfluthrin, *See* Bulldock, 4663
Cyfoxylate, *See* cyfluthrin, 7563
Cyglas®, 7564
Cygna, 7565
Cygon, *See* Dimethoate, 8496
Cy-Guard, 7566
λ-cyhalothrin, 7567, *See* Hallmark, 13557, Karate, 15757
cyhexatin, 7568
Cykelin, 7569
Cylence, 7570
Cylink ISOBU-M-AMD, 7571
Cylink M.B.A, 7572
Cylink NBMA, 7573
Cylink NM-AMD, 7574
Cylok® GM, 7575
Cylok® R, 7576
Cymag, 7577, *See* sodium cyanide, 28753
Cymate, *See* AAprotect, 16
Cymbal metal, 7578
Cymbilide, 7579
Cymbush, 7580, *See* Topclip Parasol, 32037
cymenol, *See* oxycymol, 22445
2-*p*-cymenol, *See* carvacrol, 5362
3-*p*-cymenol, *See* thyme camphor, 31775
Cymogran, 7581
cymoxanil, 7582
cymoxanil-mancozeb, *See* Fytospore, 12545
cymoxanil-mancozeb, *See* Curzate® M, 7427
Cymperator, 7583, *See* cypermethrin, 7588, Topclip Parasol, 32037
Cymyl orange, 7584
Cynoff, *See* cypermethrin, 7588, Topclip Parasol, 32037
Cynorex, 7585
Cypan, 7586
cypentil, *See* piperidine, 23874

Chemical and Trade Names

Dalapon, 7705, *See* Couch and Grass Killer, 6896
Dalapon 85, *See* dalapon, 7705
Dalapon sodium salt, *See* Dowpon, 8943
Dalapon-sodium, *See* Basfapon®, 3705
Dal-E-Rad, *See* MSMA, 20417, Neoarsycodyl, 20871
Dalf dust, *See* Baygon®, 3790
Dalfratex, 7706
Dalpac, *See* Anti-Oxydant Bayer, 2595, Butylated hydroxytoluene, 4788, Oxyguard, 22457, Ralox® BHT food grade, 25845, Tenamine 3, 31092, Vanox® PCX, 33346, Vulkanox® KB, 34152
Dalpac, *See*
Dalpad, 7707
Daltocel, 7708
Daltoflex, 7709
Daltofoam, 7710
Daltogard, 7711
Daltolac, 7712
Daltomold, 7713
Daltoped, 7714
Daltorez, 7715
Daltorol, 7716
Dalysep, *See* Kelfizina, 15838
Dama® 1010, 7718, *See* didecyl methylamine, 8415
Dama® 810, 7717
Damar, *See* gum Cowrie, 13484
Damar bronze, *See* bronze, 4599, Damascus bronze, 7720
Damascenized steel, 7719
Damascus bronze, 7720
Damfin, 7721
Damiana, 7722
Daminozide, 7723, *See* Alar, 1334, B-Nine, 4365, Dazide, 7805
Dammar, *See* Dammar Resin, 7724, gum Cowrie, 13484
Dammar Resin, 7724
DAMO, *See* CA0700, 4850, Dynasylan® DAMO, 9404, Prosil® 3128, 25239
Damoil, 7725
Damox, 7726
Danazol, *See* Danol, 7728
Danex, 7727, *See* trichlorfon, 32212
Danicut, *See* Taktic, 30729
Danitol, *See* fenpropathrin, 11615
Danol, 7728
Danol diols, 7729
Dantion DMDMH 55, *See* DMDM hydantoin, 8754, Glydant®, 13140, Mackstat® DM, 18234
Dantochlor, *See* 1,3-Dichloro-5,5-dimethyl hydantoin, 8389
Dantocol® DHE, 7730, *See* DEDM hydantoin, 7872
Dantoest® DHE DL, *See* DEDM hydantoin dilaurate, 7873
Dantogard®, 7731
Dantoguard, *See* DMDM hydantoin, 8754, Glydant®, 13140, Mackstat® DM, 18234, Nipaguard® DMDMH, 21269
Dantoin DMDMH, *See* DMDM hydantoin, 8754, Glydant®, 13140, Mackstat® DM, 18234, Nipaguard® DMDMH, 21269
Dantoin DMDMH 55, *See* Nipaguard® DMDMH, 21269
Dantoin® BCDMH, *See* Bromochloro dimethyl hydantoin, 4574
Dantoin® DCDMH, 7732, *See* 1,3-Dichloro-5,5-dimethyl hydantoin, 8389
Dantoin® GSD-550, 7733, *See* Bromochloro dimethyl hydantoin, 4574
Dantoin® MDMH, 7734
Daotan, 7735
DAP, *See* Ammonium phosphate dibasic, 2250, diallyl phthalate, 8283, Nonflammable Decobest DA, 21448, RX® 1-501N, 27154
Daphnetin, 7736
Dapon 35, 7737, *See* diallyl phthalate, 8283, Nonflammable Decobest DA, 21448, RX® 1-501N, 27154
Dapon M, 7738
Dapon R, *See* diallyl phthalate, 8283, Nonflammable Decobest DA, 21448, RX® 1-501N, 27154
Daponite Sheet, 7739, *See* diallyl phthalate, 8283
Dappol, 7740
Dapro 5005, 7741
Dapro Defoamer NA 1621, 7742
Dapro DF 1181, 7745
Dapro DF 880, 7743
Dapro DF 900, 7744
Dapro S-65, 7746
Dapro U-99, 7747
Dapro W-77, 7748
Dapro-7, 7749
Dapsetyn, 7750
Dapsone, 7751
dapsone, *See* 4,4'-diaminodiphenyl sulfone, 8302
Dapsone, *See* Dapsone, 7751
Dapsonum, *See* Dapsone, 7751
Dapsyvet, 7752
Dapsyvet, *See* Dapsetyn, 7750
Daracide, 7753
Daraclean, 7754
Daradefoam, 7755
Darafloc, 7756
Daramin, *See* sucramine, 29928
Darammon, *See* Katapone VV-328, 15785
Darammon, *See* Ammonium chloride, 2235
Daran® 229, 7757
Daran® 8350, 7758
Darasperse, 7759
Daraspray, 7760
Daratak® 89L, 7761
Daratak® MX, 7762
Daratak® RP2000, 7763
Daratak® RP2000, SP1011, *See* Everflex® SP-1084, 11266
Daratak® SP1011, 7764
Daratak® XB-3631, 7765
Darathane® WB-4000, 7766
D'Arcet's alloy, 7767
dar-chem 14, *See* stearic acid, 29599
Darco, 7768
Darex® 110L, 7769
Darex® 165L, 7770
Darex® 5281L, 7771
Darex® 550L, 7772
Darex® 636L, 7773
Dariloid® 100, 7774
Dariloid® 100, *See* diglyme, 8455
Dariloid® Q, 7775
Dariloid® Q, QH, *See* Algin, 1591
Dariloid® QH, 7776
Darilold®, 7777
Darmex, 7778
Darmex Plus, 7779
Darmycel Agarifume Smoke, 7780
Darmycel Agarifume Smoke, *See* Permethrin, 23385
Dartex®, 7781
Darvan 1, *See* Ablusol ML, 129
Darvan no. 1, *See* Ablusol ML, 129
Darvan®, 7782
Darvan® No. 1, 7783
Darvan® No. 2, 7784
Darvan® No. 404, 7785, *See* calcium lignosulfonate, 4956
Darvan® No.404, *See* calcium lignosulfonates, 4957
Darvan® SMO, 7786
Dasag, *See* Surophosphate, 30421
Dasanit, *See* Terracur® P, 31329
Dasanit®, 7787
Dassitox, *See* Diazinon Liquid, 8344
Dastar, 7788
Datac, 7789
Datagel, 7790
Datamuls, 7791
Datem, 7792
Datril, *See* Valadol, 33238
Daubond DC-9200-A/DC-9200-B, 7793
Daubond DC-9300, 7794
Daudelin solder, 7795
Davainex, *See* dibutyltin laurate, 8371
Davey's gray, 7796
Davis metal, 7797
Davitamon C, *See* vitamin C, 33938
Davpon, *See* dalapon, 7705
Dawe's Nutrigard, *See* Polyflex, 24435
Dawson bronze, 7798
Daxad 11, *See* Ablusol ML, 129
Daxad 15, *See* Ablusol ML, 129
Daxad 18, *See* Ablusol ML, 129
Daxad no. 11, *See* Ablusol ML, 129
Daxad® 11, 7799
Daxad® 19L-40, 7800
Daxad® 23, 7801
Daxad® 30-30, 7802
Daxad® 37LN10-35, 7803
Daxad® CP-2, 7804
Dazide, 7805
Dazide, *See* daminozide, 7723
Dazomet, 7806
Dazomet, *See* Amerstat® 233, 2039
Dazomet, *See* N-521® Biocide, 20640
Dazzel, *See* Diazinon Liquid, 8344
DB, *See* 2,4-DB, 7810
2,4-DB, 7810, *See* Butoxone, 4752, Campbell's DB Straight, 5056, Embutox, 10039
2,4-DB [94-82-6] - MCPA [94-74-6], *See* Campbell's Redlegor, 5072
2,4-DB-MCPA, *See* Farmon 2,4-DB Plus, 11483
DB Acetate, *See* diethylene glycol butyl ether acetate, 8443
DB Oil, 7807
DB-1 Defoamer, 7808
DB-19 Antifoam Compd, 7809
Dba, *See* diethylene glycol butyl ether acetate, 8443
DBA Accelerator, 7811
DBCM, *See* chlorobromoform, 6121
DBDPE, *See* Saytex® 102E, 27514
DBDPO, *See* Octoguard FR-01, 21999, Decabromodiphenyl oxide, 7837
DBE-4, *See* dimethyl succinate, 8509
DBE-IB, *See* diisobutyl adipate, 8460
DBEP, *See* Dibutoxyethyl phthalate, 8363
DBM, *See* dibutyl maleate, 8365, Dibutyltin maleate, 8372
DBNPA, *See* Amerstat® 300, 2045, Biosperse® 240, 4207
DBNPD, *See* Agerite® White, 1024
DBNPG, *See* dibromoneopentyl glycol, 8358
DBP, 7812, *See* 2,4-dibromophenol, 8359, dibutyl phthalate, 8366, Kodaflex® DBP, 16264, Unimoll® DB, 32978
DBPC, 7813, *See* BHT, 4122, Butylated hydroxytoluene, 4788, Oxyguard, 22457, Tenamine 3, 31092
DBTL, *See* Dabco® T-12, 7673
DBU, *See* Diazabicycloundecene, 8341
DC 150, 7814
DC Cristobalite, 7815
DC700, 7816
DCB, *See* *p*-dichlorobenzene, 8386
DCC, *See* dicyclohexyl carbodiimide, 8412
DCCI, *See* dicyclohexyl carbodiimide, 8412
DCH-99, *See* Diaminocyclohexane, 8300
DCHP, *See* dicyclohexyl phthalate, 8413
DCI-3, 7817
DCMO, *See* carboxin, 5255
DCMOO, *See* oxycarboxin, 22441, Ringmaster, 26771
DCMU, *See* diuron, 8736
DCMX, *See* Nipacide® DX, 21255
DCNA, *See* dicloran, 8395, Fumite Dicloran, 12459
DCP, 7818, *See* Dicapryl phthalate, 8375, dimethyl tetrachloroterephthalate, 8512
DCP-0, *See* calcium phosphate, dibasic,
DCPA, *See* Dacthal, 7685, dimethyl tetrachloroterephthalate, 8512, Vegetable Turf and Ornamental Weeder, 33526
DCPA, *See* dimethyl tetrachloroterephthalate, 8512
DCP-O, *See* calcium hydrogen phosphate, 4951
DCR 736, *See* methiocarb, 19484
DCT, *See* simazine, 28464

D-D92, *See* Telone, 31069
D-D95, *See* Telone, 31069
DDBS 100, 7819
DDBSA, *See* dodecylbenzene sulfonic acid, 8779, Witco® 1298, 34458
DDD, *See* Perthane, 23435
2,4'-DDD, *See* Lysoform, 18095
DDDM, *See* Nuophene, 21783
DDH, *See* 1,3-Dichloro-5,5-dimethyl hydantoin, 8389
DDM, *See* n-dodecyl mercaptan,
DDNO, *See* Ammonyx® LO, 2264, Empigen® OB, 10375, Rhodamox® LO, 26620
DDOA, *See* dimethoxane, 8498
DDS, *See* Dapsone, 7751, Witcolate 6400, 34469
4,4'-DDS, *See* Dapsone, 7751, 4,4'-diaminodiphenyl sulfone, 8302
DDT, 7820, *See* De De Tane, 7821
***p,p'*-DDT**, *See* DDT, 7820
DDTC, *See* Octopol SDE-25, 22013
DDVP, *See* Baygon®, 3790, Dedevap®, 7870, dichlorvos, 8394, Vapona, 33378
DE 83R, *See* Decabromodiphenyl oxide, 7837, Octoguard FR-01, 21999, Saytex® 102E, 27514
De De Tane, 7821
De Han salt, 7822
De Rossi's stain, 7823
DEA, *See* diethanolamine, 8434, diethylamine, 8439
DEA dodecylbenzene sulfonate, *See* Ablusol DBD, 124, Marlopon® ADS 50, 18902
DEA lauryl sulfate, *See* Carsonol® DLS, 5334, Zoharpon LAD, 34865
DEA oleate, *See* Mackamide ODM, 18161
DEA oleth-3 phosphate, *See* Brophos OL-3N, 4615
DEA-acrylinoleate and DEA-dodecylbenzene sulfonate, *See* Monaterge 85 HF, 20140
DEA-cetyl phosphate, *See* Amphisol, 2308, Surfagene FGD 600, 30354
De-Acidite, 7824
Deacylated gellan gum, *See* Gellan gum, 12736
Deacylated PS-60, *See* Gellan gum, 12736
dead silver, 7825
Deadline, 7826
DEAE, *See* 2-diethylaminoethanol, 8440
DEA-lauryl sulfate, 7827, *See* Empicol® 0031/T, 10298, Empicol® DA, 10310, Empicol® DLS, 10311, NutraponDE3796, 21831, Standapol® DEA, 29471, Sulfochem DLS, 29997, Texapon DEA, 31456
DEA-lauryl sulfate, DEA-lauramino-propionate, sodium lauraminopropionate and propylene glycol, *See* Miracare® XL, 19860
Deanase D.C, 7828
Deanol, 7829, *See* Alkanolamine, 1701, Dabco® DMEA, 7669, dimethylethanolamine, 8520, Tegoamin® DMEA, 30995, Texacat® DME, 31428
Deanox, 7830
DEA-oleth-10 phosphate, *See* Chemphos TR-505D, 6016, Crodafos N10 Neutral, 7080
DEA-oleth-3 phosphate, *See* Chemphos TR-515D, 6020, Crodafos N3 Neutral, 7082
Dearcide, 7831
DEAS Base, 7832
Deasol, 7833
Debenal®, 7834
Debron 711, 7835
Debroxide, *See* Abcure S-40-25, 35, AcetOxyl 2.5 and 5, 305, Benzoyl peroxide, 3983
Debut®, 7836
Dec, *See* Decalin®, 7840
DECA, *See* Decabromodiphenyl oxide, 7837
decabromo biphenyloxide, *See* Great Lakes DE-83R, 13304
decabromobiphenyl ether, *See* Decabromodiphenyl oxide, 7837, Great Lakes DE-83R, 13304, Octoguard FR-01, 21999, Saytex® 102E, 27514, Thermoguard® 505, 31632
Decabromodiphenyl oxide, 7837, *See* FR-1210, 12365, Saytex® 102E, 27514
decabromodiphenyl oxide-antimony trioxide, *See* Flame Out #44, 11875
Decabromodiphenyloxide, *See* Thermoguard® 505, 31632
Decabromophenyl ether, *See* Decabromodiphenyl oxide, 7837
1,1',2,2',3,3',4,4',5,5'-decachlorobi-2,4-cyclopentadien-1-yl, *See* dienochlor, 8429
Decadex, 7838
Decaethoxy oleyl ether, *See* Ablunol OA-6, 107, Volpo 3, 34020
Decaglycerol decaoleate, *See* Hodag PGO-1010 (formerly Hodag SVO-10107), 14217, Polygliceryl-10 decaoleate, 24453
Decaglyceryl decastearate, *See* Hodag PGS-1010, 14226
Decaglyceryl dioleate, *See* Hodag PGO-102, 14218
Decaglyceryl distearate, *See* Hodag PGS-102, 14227
Decaglyceryl monolaurate, *See* Hodag PGL-101, 14212
Decaglyceryl monooleate, *See* Hodag PGO-101, 14216
Decaglyceryl monostearate, *See* Hodag PGS-101, 14225
Decaglyceryl octastearate, *See* Hodag PGS-108, 14230
Decaglyceryl tetraoleate, *See* Mazol® PGO-104, 19141
Decaglyceryl tetrastearate, *See* Hodag PGS-104, 14229
Decaglyceryl trioleate, *See* Hodag PGO-103, 14219
Decaglyceryl tristearate, *See* Hodag PGS-103, 14228
Decaglycryl octaoleate, *See* Hodag PGO-108, 14221
[1R-(1α,4aβ,4bα,10aα)]-1,2,3,4,4a,4b,5,6,10,10a-decahydro-1,4a-dimethyl-7-(1-methylethyl)-1-phenanthrenecarboxylic acid, *See* abietic acid, 45
Decahydronaphthalene, *See* Decalin®, 7840
Decalex, 7839
Decalin®, 7840
Decalso, *See* Wessalith, 34284
Decalso F, *See* Wessalith, 34284
Decaltal®, 7841
Decamethylcyclopentasiloxane, *See* CD3770, 5500, D3770, 7635
1,1'-decamethylenebis[4-aminoquinaldinium chloride], *See* Dequalinium Chloride, 8077
Decamethyltetrasiloxane, *See* CD3780, 5501, D3780, 7636
decamine, *See* Dequalinium Chloride, 8077
decamphorized oil of turpentine, 7842
Decan-1-ol, *See* Exxal® 10, 11338
1-Decanamine, *See* Amine 10, 2126, Armeen® 2-10, 2926
1-Decanaminium, N-decyl-N,N-dimethyl-, chloride, *See* BTC® 1010-80, 4638
Decanaminium, N-decyl-N,N-dimethyl-, chloride, *See* Radiaquat 6412, 25790
1,10-decanedioic acid, *See* sebacic acid, 27795
decanedioic acid, *See* sebacic acid, 27795
n-decanedioic acid, *See* sebacic acid, 27795
decanedioic acid dibutyl ester, *See* Plasthall® DBS, 23994, Reomol DBS, 26129
Decanedioic acid, bis(1-methylethyl) ester, *See* Unimate® DIPS, 32969
Decanedioic acid, bis(2,2,6,6-tetramethyl-4-piperidinyl) ester, *See* Bistetramethyl piperidinyl sebacate, 4284, Lowilite® 77, 17645, Uvaseb 770, 33178
1,1'-(1,10-Decanediyl)-bis-[4-amino-2-methylquinolinium chloride], *See* Dequalinium Chloride, 8077
n-decanenitrile, *See* Nitrile 10 D, 21335
n-decanoic acid, *See* n-Capric acid, 5126
Decanol, *See* decyl alcohol, 7867, Epal® 10, 10617, Exxal® 10, 11338
1-Decanol, *See* Exxal® 10, 11338
decanonitrile, *See* Nitrile 10 D, 21335
Decanox-F, 7843
Decanoyl chloride, 7844
Decanoyl peroxide, *See* Decanox-F, 7843
Decap, 7845, *See* dimethyl sulfoxide, 8511
Decaps, *See* vitamin D+72, 33940
Decaris, *See* Tramisol®, 32134
Decatylen, *See* Dequalinium Chloride, 8077
Deccox, 7846
Deccozil, *See* imazalil, 14892
Decelox, 7847
1-decene, *See* Neodene® 10, 20891
dec-1-ene, *See* Neodene® 10, 20891
decene-1, *See* Neodene® 10, 20891
n-Decenyl succinic anhydride, *See* Milldride® nDSA, 19787
Deceth-2, *See* Bio-Soft® FF 400, 4199
Deceth-4, *See* Chemal DA-4, Desonic® DA-4, 8182, Genapol® DA-040, 12788, Iconol DA-4, 14819, Marlipal® 1012/4, 18865, Oxetal D 104, 22421, Prox-onic DA-1/04, 25313, Trycol® 5950, 32502
Deceth-4 phosphate, *See* Cedephos® FA600, 5544, Monafax 1214, 20100, Rhodafac® RA-600, 26603
Deceth-6, 7848, *See* Desonic® DA-6, 8183, Iconol DA-6, 14820, Marlipal® KF, 18868
Deceth-9, *See* Iconol DA-9, 14821
dechan, *See* Dichan 100, 8377, dicyclohexylamine nitrite, 8414
Dechlorane A-O, 7849, *See* antimony trioxide, 2574, Octoguard FR-10, 22000, Timonox, 31807
Dechlorane® Plus 25, Plus 35, Plus 515, 7850
Decimate, 7851
Decis, 7852, *See* Thripstick®, 31766
Decisquick, 7853
Deck Seal-PD, 7854
Declar®, 7855
Declid, *See* undecylenic acid, 32916
Deco Board P, Deco Poly, 7856
Decoart, 7857
Decol, 7858
Decola Back sheet, 7859
Decola Excel, New Marine, F, FG, MA, MF, 7860
Decola PFC, 7861
Decolamide, 7862
Deconate, *See* Neoarsycodyl, 20871
Deconyl, 7863
Decopress, 7864
Decoquinate, *See* Deccox, 7846
Decorpa, *See* Prinza® Range, 25010
Decothane® SP, 7865
Decrolin®, 7866
decroline, *See* hydrosulfite, 14643
De-Cut, *See* maleic hydrazide, 18477
decyl alcohol, 7867, *See* Emery® 3323, 10167, Exxal® 10, 11338, Lorol C10, 17610
n-decyl alcohol, *See* decyl alcohol, 7867
n-decyl Alcohol, *See* Exxal® 10, 11338
decyl bromide, 7868
decylbromide, *See* 1-Bromodecane, 4578
n-decyl bromide, *See* 1-Bromodecane, 4578, decyl bromide, 7868
decyl carbinol, *See* Neodol® 1, 20908, Neoflex® 11, 20921
N-decyl-N,N-dimethyldecanaminium chloride, *See* Querton 210Cl-50, 25651
Decyl diphenyl ether disulfonic acid, *See* Poly-Tergent® 3B2 Acid, 24631
N-Decyl diphenyloxide disulfonate, *See* Dowfax 3BO, 8912
Decyl isostearate, *See* Schercemol DEIS, 27591
decyl octyl phthalate, *See* Good-rite® GP-265, 13205
n-decyl n-octyl phthalate, *See* Good-rite® GP-265, 13205
decyl oleate, 7869, *See* Ceraphyl® 140, 5722, Cetiol® V, 5806, Schercemol DO, 27598, Tegosoft® DO, 31019
Decyl polyglucose, *See* APG® 300 CS, 2650, APG® 300 Glycoside, 2651, Plantaren 2000, 23917
Decyl sodium sulfate, *See* Atlasol 103, 3341
n-decyl trimellitate, *See* Plasthall® 8-10 TM-E, 23991
Decyl(sulfophenoxy)benzenesulfonic acid, disodium salt, *See* Calfax 10L-45, 4982, Poly-Tergent® 3B2, 24630
Decylamine, *See* Amine 10, 2126
n-Decylamine, *See* Amine 10, 2126
Decylbenzene, sodium sulfonate, *See* Witconate DS,

34485
Decylbenzenesulfonic acid, sodium salt, *See* Witconate DS, 34485
N-decyldecanamine, *See* Armeen® 2-10, 2926
N-decyl-N,N-dimethyl-1-decanaminium chloride, *See* Radiaquat 6410, 6412, 25789
Decylsulfuric acid sodium salt, *See* Atlasol 103, 3341
DEDC, *See* Octopol SDE-25, 22013
Dedelo, *See* DDT, 7820
Dedevap, *See* dichlorvos, 8394 Vapona, 33378
Dedevap®, 7870 , *See* dichlorvos, 8394
Dedico 5981, 7871
Dedisol, *See* Telone, 31069
DEDK, *See* Octopol SDE-25, 22013
DEDM hydantoin, 7872, *See* Dantocol® DHE, 7730
DEDM hydantoin dilaurate, 7873
DeDTC, *See* Octopol SDE-25, 22013
Ded-weed, *See* MCPA, 19183
deeline, *See* paraffin, liquid, 22744
Deenax, 7874, *See* Anti-Oxydant Bayer, 2595
Deep Feed, 7875
Dee-Ron, *See* vitamin D+72, 33940
DEET, *See* diethyl toluamide, 8438
DEF, *See* Def®, 7876
Def®, 7876
Defend, *See* Dimethoate, 8496
Defirust, 7877
Deflavit® ZA, 7878
Defoamer 1713, 7879
Defoamer A 50, 7880
Defoamer B 90, 7881
Defoamer C5B, 7882
Defoamer DF-160-L, 7883
Defoamer KCE/S, 7884
Defoamer NXZ, 7885
Defoamer S, 7886
Defoamer SF, 7887
Defoamer WB Series, 7888
Defol, 7889
Defol, *See* KM Sodium Chlorate, 16244, sodium chlorate, 28747
Defolia, 7890
Defomax, 7891
Degadur, 7892
Degalan, 7893
Degalex, 7894
Degament, 7895
Degapas, 7896
Degaplast, 7897
Degaroute, 7898
Degaser, 7899
DEGEE, *See* Ethoxydiglycol, 11044
Deglas, 7900
Degopol, 7901
Degopur, 7902
Degras, 7903
Degras Special, 7904
De-Green, *See* Def®, 7876
Degreez, 7905
Degressal® SD 20, 7906
Degressal® SNC, 7907
Degubond, 7908
Degucast, 7909
Degudent, 7910
Deguflex, 7911
Deguform, 7912
Degulor, 7913
Deguphos, 7914
Degupress, 7915
Degusorb, 7916
Degussa methyl isothiocyanate, *See* methyl isothiocyanate, 19543
Degussa P820, *See* Wessalith, 34284
Degutron, 7917
Deguvest, 7918
DEHA, *See* dioctyl adipate, 8548, Plasthall® DOA, 24002, PX-238, 25440, Wickenol® 158, 34393
Dehesive®, 7919
Dehesive® 920, 7920
DEHP, *See* Bis(2-ethylhexy) Phthalate, 4228, dioctyl phthalate, 8549, Reomol DCP, 26130, Witcizer 312, 34454
DEHS, *See* Plasthall® DOS, 24005, Reomol DOS, 26131
Dehscofix 904, 7921
Dehscofix 911, 7922
Dehscofix 912, 7923
Dehscofix 916, 7924
Dehscofix 917, 7925
Dehscofix 918, 7926
Dehscofix 923, 7927
Dehscofix 929, 7928
Dehscofix CO Series, 7929
Dehscotex, 7930
Dehscotex BA Series, 7931
Dehscotex DT 809, 7932
Dehscotex DT Series, 7933
Dehscotex DY Series, 7934
Dehscotex FW Series, 7935
Dehscotex MC Series, 7936
Dehscotex SN Series, 7937
Dehscotex VP-PF, 7938
Dehscotex WA Series, 7939
Dehscoxid 700 Series, 7940
Dehscoxid 730/740 Series, 7941
Dehybor, 7942
Dehydag Sulfate GI Emulsion, *See* Witcolate 6400, 34469
Dehydag Sulphate GI Emulsion, *See* Witcolate 6400, 34469
Dehydag Wax 14, 7943
Dehydag Wax 16, 7944
Dehydag Wax 18, 7945
Dehydag Wax 22 (Lanette), 7946
Dehydag Wax E, 7947
Dehydag Wax N, 7948
Dehydag Wax O, 7949
Dehydag Wax SX, 7950
Dehydag Wax W, 7951
Dehydol 100, *See* Ablunol OA-6, 107, Volpo 3, 34020
Dehydol LS 2, 7952
Dehydol LT 3, 7953, *See* Coceth-27, 6558
Dehydol PCS 6, 7954, *See* Ceteareth-6, 5788
Dehydol PID 6, 7955
Dehydol PLS 1, 7956
Dehydol PTA 7, 7957
Dehydol TA 11, 7958
Dehydran 1019, 7960
Dehydran 1293, 7961
Dehydran 520, 7959
Dehydran P 12, 7962
Dehydrite®, 7963
Dehydroabietylamine, *See* Amine D, 2155
Dehydroacetic acid, 7964
Dehydrophen 65, 7965
Dehydrophen PNP 4, 7966
Dehydrophen POP 4, 7967
Dehydrothio-*p*-toluidinesulfonic acid, *See* D.T.S, 7626
Dehymuis SSO, 7968
Dehymuls, 7969
Dehymuls E, 7970
Dehymuls F, 7971
Dehymuls HRE 7, 7972
Dehymuls K, 7973
Dehymuls SML, 7974
Dehymuls SMO, 7975
Dehymuls SMS, 7976
Dehypon Conc, 7977
Dehypon LS-24, 7978
Dehypon LT 054, 7979
Dehyquart A, 7980, *See* Cetrimonium chloride, 5814
Dehyquart AU-36, 7981
Dehyquart C Crystals, 7982, *See* laurylpyridinium chloride, 16867
Dehyquart CDB, 7983
Dehyquart D, 7984
Dehyquart DAM, 7985, *See* Adogen® TA-101, 780, Distearyldimonium chloride, 8714, Genamin DSAC, 12771
Dehyquart LDB, 7986
Dehyquart LT, 7987
Dehyquart SP, 7988
Dehyquart STC-25, *See* Ablumine 280, 85, Cycloton® SCS, 7542, octadecylbenzenemethanaminium chloride, 21985, stearalkonium chloride, 29592, Sumquat® 6210, 30126
Dehyton AB 30, *See* Accobetaine CL, 222
Dehyton® AB-30, 7989
Dehyton® G, 7990, *See* disodium cocoamphodiacetate, 8643
Dehyton® K, 7991
Dehyton® KE, 7992
Dehyton® PAB-30, 7993
Dehyton® PG, 7994, *See* disodium cocoamphodiacetate, 8643
Dehyton® PK, 7995
Dehyton® PK, 3016 B, K, KE, KE 3016, PK 45, *See* Cocamidopropyl betaine, 6551
Dehyton® PLG, 7996
Dehyton®W, *See* disodium cocoamphodiacetate, 8643
DEK, *See* diethyl ketone, 8436
Dekadin, *See* Dequalinium Chloride, 8077
DeKalin, *See* Decalin®, 7840
Dekamin, *See* Dequalinium Chloride, 8077
Dekelmin, *See* metyridine, 19609
Dekol®, 7997
Dekol® N, 7998
Dekryll, 7999
Dekrysil, *See* 4,6-dinitrocresol, 8542
Delac MOR, 8000
Delac NS, 8001, *See* Butylbenzothiazole sulfenamide, 4791
Delac S, 8002, *See* cyclohexyl benzothiazole sulfenamide, 7523, Vulkacit® CZ/EGC, DZ/EGC, 34129
Delafila, 8003
Delaglas® A, 8004
Delalot's alloy, 8005
Delan, *See* Dithianone, 8729
Delan-Col, 8006, *See* Dithianone, 8729
Delanium, 8007
Delaphos, 8008
Delapon, *See* Basfapon®, 3705
Delaprism®, 8009
Delaville, 8010
Delchowyte, 8011
Deleaf defoliant, *See* Def®, 7876
Delegol®, Delegol-T, 8012
Delevap, *See* dichlorvos, 8394
Delf® Clene, 8013
Delf® Drape, 8014
Delf® Fabric Protector, 8015
Delf® HD Aerosol Adhesive, 8016
Delf® MP Aerosol Adhesive, 8017
Delf® Silicone Aerosol, 8018
Delfloc® 50, 8019
Delft blue, 8020
Delhi rustless iron, 8021
Delial®, 8022
Delicron, 8023
Delight, 8024
Delios®, 8025
Deliquescent potassium carbonate, *See* oil of tartar, 22076
Dellatol® BBS, 8026, *See* Butylbenzene sulfonamide, 4789
Delnet, 8027
Delo, 8028
Delonal, *See* Vaderm, 33237
Deloxan, 8029
Deloxil, 8030
Delpet®, 8031
***m*-Delphene**, *See* diethyl toluamide, 8438
Delrin®, 8032
Delrin® 100, 500, 8033, *See* Acetal, 280
Delrin® 100AF, 500AF, 8034
Delrin® 100ST, 500T, 8035
Delrin® 107, 507, 8036, *See* Acetal, 280

8168
Desmophen®, 8170
Desmorapid®, 8171
Desmotherm®, 8172
Desodora, 8173
Desogrip, 8174
Desomeen, 8175
DeSonate 50-S, 8176
DeSonate 60-S, 8177
DeSonate AOS, 8178
DeSonate SA, 8179
DeSonate SA-H, 8180
Desonic® 20N, 8184
Desonic® 30C, 8181
Desonic® DA-4, 8182, *See* Deceth-6, 7848
Desonic® DA-6, 8183, *See* Deceth-6, 7848
Desonic® S Series, 8185
DeSonol A, 8186, *See* Ammonium lauryl sulfate, 2242
DeSonol AE, 8187, *See* Ammonium laureth sulfate, 2240
DeSonol S, 8188
DeSonol SE, 8189
DeSonol SE-2, 8190
DeSonol T, 8191
Desophos®, 8192
Desoplas, 8193
DeSotan® SMO, 8194
DeSotan® SMO-20, 8195
DeSotan® SMT, 8196
DeSotan® SMT-20, 8197
Desoxon 1, *See* Oxymaster, 22460
Despacilina, *See* Penicillin V, 23111
Desquaman, *See* zinc pyrithione, 34791
De-Squaman, *See* Zinc Omadine, 34783
Dessicant L-10, *See* arsenic acid, 3137
Dessin, *See* Acrex, 384, Dinobuton, 8543
Desson, *See* chloroxylenol, 6159
Destral, 8198, *See* Hi-bor, 14055
Destral BR, *See* Composibor, 6683
Destun, *See* Perfluidone, 23248
Desugared hardwood calcium/magnesium lignosulfonate, *See* Curbeton 0550, 7417
Desulfex, 8199
DET, *See* diethyl toluamide, 8438
DETA, *See* D.E.H. 20, 7610, diethylenetriamine, 8446
***m*-DETA**, *See* diethyl toluamide, 8438
Detac, 8200
Detaclad®, 8201
DeTAINE CAPB-35, CAPB-35HV, *See* Cocamidopropyl betaine, 6551
Detal, *See* 4,6-dinitrocresol, 8542
Detalup, *See* vitamin D+72, 33940
Detamide, *See* diethyl toluamide, 8438
Detarex, 8202
Detarex PY, *See* Cheelox® 80, 5864, Pentaquest Extra 0685, 23169, Trilon® C Liq, 32328
Detarol, 8203
Deterflo A 210, 8204
Detergent 8®, 8205
Detergent 66, *See* Rhodapon® 101-10, 26631 sodium lauryl sulfate, 28781 Witcolate 6400, 34469
Detergent HD-90, *See* Nacconol® 40G, 20647, sodium dodecylbenzenesulfonate, 28760
Detergyl, 8206
Deterlon, *See* Nacconol® 40G, 20647
deterlon ultrawet 99ls, *See* sodium dodecylbenzenesulfonate, 28760
Deterpal, 8207
Deterpal 832, 8208
Deterpal LC, 8209
Dethlac, 8210
Dethmor, 8211
Detia, *See* Talunex, 30757
Detmol MA, *See* malathion, 18460
Detmol U.A, *See* Chlorpyrifos, 6163
Det-O-Jet®, 8212
Detox, *See* DDT, 7820
Detoxan, *See* DDT, 7820
Dettol, 8213, *See* chloroxylenol, 6159, Nipacide® MX, 21257
Deuterated N-Methyl-2-Pyrrolidone, *See* NMP, 21406
Deva, 8214
Devarda's alloy, 8215
Developer P, *See p*-nitroaniline, 21345
Developer, part B, *See* sodium gluconate, 28769
Devermine, *See* Mansonil®, 18611
Devikol, *See* dichlorvos, 8394
Devisulphan, *See* thiodan, 31680
Devithion, *See* Parathion-methyl, 22808
Devol Red GG, *See p*-nitroaniline, 21345
Devoton, *See* methyl acetate, 19506
Devrinol, 8216
Devrinol T, 8217
Deward Steel, 8218
Dewitt Deadline, 8219
Dewrance metal, *See* Durance's metal, 9164
DEWT L, 8220
dexamethasone-neomycin sulfate, *See* Dexamist Ear Spray, 8221
Dexamist Ear Spray, 8221
Dexel, 8222
Dexil, 8223
Dexine 521, 8224
Dexine 656, 8225
Dexine 687, 8226
Dexine 759, 8227
Dexine 779, 8228
Dexlar®, 8229
Dexol, *See* sodium perborate, 28806
Dexol Fungicide Containing Daconil, *See* Repulse, 26151
Dexon, 8230
Dexonite, 8231
Dexoplas, 8232
dexpanthenol, *See* d-Panthenol, 22690
Dextran, 8233, 8234
Dextraven, *See* Dextran, 8233
dextrin, 8235
Dextrin, malto, *See* Wickenol® 550, 34403
dextrine, *See* dextrin, 8235
Dextrin-maltose, *See* glucose syrup, 13054
Dextrinozole, *See* ozole, 22484
Dextroform, 8236
Dextrol OC-20, 8237, 8238
Dextrone X, 8239
dextronic acid, *See* gluconic acid, 13048
Dextropur, *See* glucose, 13052
Dextrose, *See* glucose, 13052
Dextrose monohydrate, *See* Roferose, 26904
Dextrosol, *See* glucose, 13052
Dextrostix®, 8240
Dextrozyme, 8241
Dexuron, 8242
DFF, *See* Diflufenican, 8451
DFFD, *See* Agerite® DPPD, 1016, N,N'-diphenyl-*p*-phenylenediamine, 8603, JZF, 15638, Naugard® J, 20796, Permanax DPPD, 23361
DHA, *See* Dehydroacetic acid, 7964, dihydroxyacetone, 8459
DHAA, *See* Dehydroacetic acid, 7964
Dhak gum, *See* kino, Bengal, 16198
di(sec-butyl) peroxydicarbonate, *See* Lupersol 225, 17854
di(*p*-chlorophenyl)trichloromethylcarbinol, *See* Acarin, 183, Kelthane, 15881
di-(*tert*-butylperoxy)phthalate, *See* Trigonox® 111-B40, 32280
di(α-amino-β-thiol)propionic acid, *See* cystine, 7603
di(2-ethylhexyl) adipate, *See* dioctyl adipate, 8548, Palatinol® DOA, 22590, Uniflex® DOA, 32945, Wickenol® 158, 34393
di(2-ethylhexyl) butanedioate, *See* Wickenol® 159, 34394
di(2-ethylhexyl) peroxydicarbonate, *See* Lupersol 223, 17852
di(2-ethylhexyl) phthalate, *See* dioctyl phthalate, 8549, Kodaflex® DOP, 16273, Palatinol® DOP, 22591, Witcizer 312, 34454
di(2-ethylhexyl) succinate, *See* Wickenol® 159, 34394
di(2-ethylhexyl)sulfosuccinic acid, sodium salt, *See* Empimin® OP70, 10434, Octowet 40, 22035
di(2-hydroxyethyl) amine, *See* diethanolmine, 8434
di(2-Hydroxy-n-propyl) amine, *See* diisopropanolamine, 8466
di-(2-methoxyethyl) phthalate, *See* Kodaflex® DMEP, 16270
di(2-methylbenzoyl) peroxide, *See* Perkadox® 20, 23303
di-(2-octyl)phthalate, *See* Dicapryl phthalate, 8375
di(2-octyldodecyl) N-stearoyl-L-glutamate, *See* Amiter SG-2000, 2201
di-(2-t-butylperoxyisopropyl)benzene, *See* Perkadox® 14, 23300
di-(4-t-butylcyclohexyl) peroxydicarbonate, *See* Perkadox® 16-W40-GB5, 23302
di(dimethylcyclohexyl)oxalate, *See* Barkite B, 3647
di(ethylhexyl) phthalate, *See* Witcizer 312, 34454
di(hardened tallow)dimethylammonium chloride, *See* Adogen® TA-100, 779, Prepagen WK, 24894, Sumquat® 6045, 30123
di(hydrogenated tallowalkyl) (2-hydroxy-2-methylethyl) quaternary ammonium chloride, *See* Propoquad® 2HT/11, 25196
di(n-propyl) peroxydicarbonate, *See* Lupersol 221, 17851
di(phenylmercuric) dodecenyl succinate, *See* Super AD-IT, 30164
di(s-butyl) peroxydicarbonate, *See* Trigonox® SBP, 32307
di-*o*-tolyl-ethylenediamine, *See* Stabilite Alba, 29403
di-*o*-tolylguanidine, *See* Accelerator DT, 201, Akrochem® DOTG, 1167, D.O.T.G, 8813
di-*o*-tolylthiourea, *See* D.O.T.T, 7623
di-*p*-methoxydiphenylamine, *See* Thermoflex, 31630
di-*t*-butoxydiacetoxysilane, 8362, *See* CD4153, 5502, Dynasylan® BDAC, 9398
di-*t*-butyl dicarbonate, 8367
di-*t*-butyl peroxide, 8369, *See* Aztec® Di-*t*-Butyl Peroxide, 3557, Trigonox® B, 32295
di-*t*-butyl peroxyphthalate, *See* Trigonox® 111-B40, 32280
di-*t*-butyl-*para*-cresol, *See* DBPC, 7813
di-*tert*-butyl peroxide, *See* Aztec® Di-*t*-Butyl Peroxoide, 3557
di-*tertiary*-butyl peroxide, *See* Aztec® Di-*t*-Butyl Peroxoide, 3557
di-α-cumyl peroxide, *See* Aztec® DCP-R, 3556, Dicumyl peroxide, 8407, Percumyl D, 23230, Polyvel PCL-20, 24661
di-β-naphthyl-*p*-phenylenediamine, *See* Anti-Oxidant DNP, 2586
di-1,2-propylene glycol, *See* Dipropylene glycol, 8611
di-2-ethylhexyl adipate, *See* Monoplex® DOA, 20247, Morflex 310, 20322, Plastomoll® DOA, 24058
di-2-ethylhexyl azelate, *See* Morflex 410, 20325, Priplast 3018, 25026
di-2-ethylhexyl ester thiodiglycolate, *See* Vulkanol® 90, 34144
di-2-ethylhexyl peroxydicarbonate, *See* Espercarb® 840, 10841
di-2-ethylhexyl phthalate, *See* Plasthall® DOP, 24004
di-2-ethylhexyl sebacate, *See* Monoplex® DOS, 20248, Reomol DOS, 26131
di-2-ethylhexylphthalate, *See* NLA-20, 21396
di-2-ethylhexylphthalate (OSHA), *See* Witcizer 312, 34454
DI-43, 8243
Diabase Developer®, 8244
diabetin, *See* levulose, 17085
Diablack® A, 8245, *See* Derussole, 8135
Diacelliton Dye®, 8246
Diacetazotol, 8247
Diacetin, 8248
Diacetone, *See* Diacetone alcohol, 8249
Diacetone alcohol, 8249
Diacetotoluide, *See* Diacetazotol, 8247
1,2-diacetoxyethane, *See* ethylene glycol diacetate, 11124

Diarsenic oxide, *See* arsenic trioxide, 3143
diarsenic pentasulfide, *See* arsenic pentasulfide, 3139
Diarsenic Pentoxide, *See* arsenic pentoxide, 3140
Diarsenic Trioxide, *See* arsenic trioxide, 3143
diastase, 8331
diastatic enzyme, 8332
Diastix®, 8333
Diater, *See* diuron, 8736
Diaterr-fos, *See* Diazinon Liquid, 8344
Diathol Grounder®, 8334
diatomaceous earth, 8335, 8336, *See* Celite®, 5586, Celite® 209, 5588, Celite® 270, 5589
Diatomaceous silica, Flux calcined, *See* Celite® 110, 5587
diatomite, *See* diatomaceous earth, 8335
Diatrizoate meglumine, *See* Hypaque Meglumine, 14705, Reno-M, 26113, Urovist, Urovist Sodium 300, 33150
diatrizoate methylglucamine, *See* Reno-M, 26113
diatrizoate sodium, 8337, *See* MD 50, 19186
diaveridine, 8338
Diavite, 8339
Diax, 8340
1,4-diazabicyclo[2.2.2]octane, *See* TEDA-L33, 30916
Diazabicycloundecene, 8341
1,8-diazabicyclo[5.4.0]undec-7-ene, *See* Diazabicycloundecene, 8341
1,3-diaza-2,4-cyclopentadiene, *See* Imidazole, 14895
1,3-diazaindene, *See* Benzimidazole, 3966
2,3-diazaindole, *See* benzotriazole1H-benzotriazole, 3982; Preventol® CI 8, 24930
Diazajet, *See* Diazinon Liquid, 8344
Diazamine, 8342
Diazene, bis(1,1-dimethylethyl)-, *See* VR-160, 34057
Diazenedicarboxamide, *See* Azodicarbonamide, 3524, Kempore® 60/14FF, 15987, Santechem 21-21, 27382
Diazenedicarboxylic Acid Diamide, *See* Azodicarbonamide, 3524
Diazide, *See* Diazinon Liquid, 8344
Diazil, *See* sulfamethazine, 29962
Diazine, 8343
1,3-diazine, *See* pyrimidine, 25484
diazinon, *See* Agridin 60, 1053, Antlak, 2629, Diazitol, Diazitol Liquid, 8345, Diazol, 8347, Diziktol, 8745
Diazinon Liquid, 8344
Diazitol, *See* Diazinon Liquid, 8344
Diazitol, Diazitol Liquid, 8345
Diazo Fast Red GG, *See* *p*-nitroaniline, 21345
Diazol, 8346, 8347, *See* Diazinon Liquid, 8344
1,3-diazole, *See* Imidazole, 14895
Diazolidinyl urea, 8348, *See* Germall® II, 12894
Diazolidinyl urea-propylene glycol-methylparaben-propylparaben, *See* Germaben® II, 12891
Diazolidinylurea, *See* Diazolidinyl urea, 8348
Diazone, 8349
Diazopon SS-837, 8350
diazoresorcin, 8351, *See* Resazurin, 26161
diazoresorcinol, *See* Resazurin, 26161
diazoting salts, *See* sodium nitrite, 28801
Diazyme® L-200, 8352
DIBA, *See* diisobutyl adipate, 8460
Dibam, *See* Aquatreat SDM, 2761, methyl namate®, 19551, Octopol SDM-40, 22014, Perkacit® SDMC, 23288, sodium dimethyldithiocarbamate, 28758, Thiostop N, 31717
Dibam A, *See* Aquatreat SDM, 2761, methyl namate®, 19551, Octopol SDM-40, 22014, Perkacit® SDMC, 23288, sodium dimethyldithiocarbamate, 28758, Thiostop N, 31717
Dibasic calcium phosphate dihydrate, *See* Emcompress®, 10064
Dibasic lead carbonate, *See* lead subcarbonate, 16918
Dibasic lead phosphite, *See* Haro® Chem PDF, 13643
Dibehenyldimonium chloride, *See* Kemamine® Q-2802C, 15923
Dibehenyldimonium methosulfate, *See* Incroquat DBM-90, 15038
Dibenzo GMF, 8353
Dibenzo[a,e]pyran, *See* xanthene, 34589
dibenzo-1,4-thiazine, *See* phenothiazine, 23618
Dibenzocycloheptadienone, *See* Dibenzosuberone, 8354
Dibenzo-p-thiazine, *See* phenothiazine, 23618
Dibenzopyran, tricyclic, *See* xanthene, 34589
Dibenzosuberone, 8354
Dibenzothiazine, *See* phenothiazine, 23618
Dibenzothiazole disulfide, *See* Perkacit® MBTS, 23286
2,2'-dibenzothiazyl disulfide, *See* Vulkacit® DM/C, 34131
Dibenzoyl peroxide, *See* AcetOxyl 2.5 and 5, 305, Benzoyl peroxide, 3983
Dibenzoyl resorcinol, *See* Dow DBR, 8884
Dibenzoyl-*p*-quinone-dioxime, *See* Dibenzo GMF, 8353
Dibenzthiazyl disulfide, *See* Ancatax, 2412, Vulkacit® DM/C, 34131
Dibenzyl azelate, *See* Plasthall® DBZZ, 23995
Dibenzyl ether, *See* Erganol, 10751
Dibenzylamine, *See* Accelerator DBA, 200, Vulcaid 28, 34074
Dibenzyldithiocarbamic acid zinc salt, *See* Arazate®, 2787
Dibenzylidene sorbitol, *See* Millithix® 925, 19794
Dibenzylsulfoxide, *See* Preventol® CI 5, 24928
Dibenzyltoluene, *See* Lipinol T, 17333
Dibexin, 8355
DIBK, *See* diisobutyl ketone, 8461
Diboron trioxide, *See* Boric anhydride, 4423
Dibovan, *See* DDT, 7820
Dibrom, 8356
Dibromo-*o*-cresolsulfonphthalein, *See* Bromcresol purple, 4567
Dibromo/tribromo salicylanilide, *See* Temasept I, 31075
dibromo-1-propanol, *See* 2,3-dibromo-1-propanol, 8360, dibromopropanol, 8361
Dibromo-2-carbamoylacetonitrile, *See* Amerstat® 300, 2045, Biosperse® 240, 4207
Dibromo-3-nitrilopropionamide, *See* Amerstat® 300, 2045, Biosperse® 240, 4207
3,5-dibromo-N-(4-bromophenyl)-2-hydroxybenzamide, *See* Temasept IV, 31076, Tuasol 100, 32543
1,4-dibromobutane, 8357
dibromochloromethane, *See* chlorobromoform, 6121
5,5'-dibromo-*o*-cresolsulfonphthalein, *See* Bromcresol purple, 4567
2,2-Dibromo-2-cyanoacetamide, *See* Amerstat® 300, 2045, Biosperse® 240, 4207
1,2-dibromo-4-(1,2-dibromoethyl)cyclohexane, *See* Saytex® BCL-462, 27518
(2',7'-dibromo-3',6'-dihydroxy-3-oxospiro[isobenzofuran-1(3H),9'-[9H]xanthen]-4'-yl)hydroxymercury sodium salt, *See* mercurochrome, 19354
Dibromodiiodohexamethylene-tetramine, *See* Chrysoform, 6255
Dibromoethyldibromocyclohexane, *See* Saytex® BCL-462, 27518
Dibromogallic acid, *See* Gallobromol, 12616
β-dibromohydrin, *See* 2,3-dibromo-1-propanol, 8360, dibromopropanol, 8361
3,5-dibromo-4-hydroxybenzonitrile, *See* Bromoxynil, 4590
dibromohydroxymercurifluorescein disodium salt, *See* mercurochrome, 19354
dibromoneopentyl glycol, 8358, *See* FR-522, 12355
2,2-Dibromo-3-nitrilo propionamide, *See* Amerstat® 300, 2045, Biosperse® 240, 4207
Dibromopentaerythritol, *See* dibromoneopentyl glycol, 8358
2,4-dibromophenol, 8359
2,4-dibromophenol, *See* FR-612, 12356
dibromopropanol, 8361, 2,3-dibromo-1-propanol, 8360, dibromopropanol, 8361
2,3-dibromopropanol, *See* 2,3-dibromo-1-propanol, 8360, dibromopropanol, 8361
Dibromostyrene, *See* Great Lakes DBS, 13301
Dibutoxy ethyl phthalate, *See* Plasthall® 200, 23979
Dibutoxyethoxyethyl adipate, *See* Plasthall® 226, 23986, Wareflex® 650, 34214
Dibutoxyethoxyethyl glutarate, *See* Plasthall® 224, 23985
Dibutoxyethoxyethyl phthalate, *See* Plasthall® 220, 23984
Dibutoxyethyl adipate, *See* Plasthall® 203, 23981
Dibutoxyethyl azelate, *See* Plasthall® 205, 23982
Dibutoxyethyl glutarate, *See* Plasthall® 201, 23980
Dibutoxyethyl phthalate, 8363
Dibutoxyethyl phthalate, *See* Plasthall® 200, 23979
Dibutoxyethyl sebacate, *See* Plasthall® 207, 23983
Dibutyl adipate, *See* Adimoll® DB, 690, Cetiol® B, 5797
Dibutyl ammonium oleate, *See* Activator 1102, 498
2,4-di-*tert*-butyl-6-(5-chloro-2H-benzotriazol-2-yl)phenol, *See* Uvazol 237, 33182
2,6-di-*t*-butyl-*p*-cresol, *See* BHT, 4122, Butylated hydroxytoluene, 4788
2,6-di-*tert*-butyl-*p*-cresol, *See* Ralox® BHT food grade, 25845, Vanox® PCX, 33346
Dibutyl decanedioate, *See* Plasthall® DBS, 23994, Reomol DBS, 26129
dibutyl fumarate, 8364
Di-n-butyl fumarate, *See* dibutyl fumarate, 8364
N,N-dibutyl-4-(hexyloxy)-1-naphthalene carboximidamide hydrochloride, *See* bunamidine hydrochloride, 4680
2,6-di-*tert*-butyl-1-hydroxy-4-methylbenzene, *See* Oxyguard, 22457
Dibutyl ketone, *See* diisobutyl ketone, 8461
dibutyl maleate, 8365, *See* Bisomer DBM, 4279, Octomer DBM, 22008, PX-504, 25443
Di-n-butyl Maleate, *See* dibutyl maleate, 8365
Dibutyl methylene bisthioglycolate, *See* Vulkanol® 88, 34143
6,6'-di-*tert*-butyl-2,2'-methylenedi-*p*-cresol, *See* Cyanox® 2246, 7489
***tert*-Dibutyl peroxide**, *See* di-*t*-butyl peroxide, 8369
Di-*tertiary*-butyl peroxide, *See* di-*t*-butyl peroxide, 8369
Dibutyl peroxy cyclohexane-butylbenzyl phthalate, *See* Aztec® 1,1-Bis(*t*-Butylperoxy)Cyclohexane-80 BBP, 3550
Dibutylphenylene N,N'-Di-s-butyl-*p*-phenylene diamine, *See* Kerobit® BPD, 16083
Dibutyl phosphite, *See* Syn-O-Ad® P-316, 30562
dibutyl phthalate, 8366, *See* Bufa, 4654, DBP, 7812, Kodaflex® DBP, 16264, Kodaflex® DBP, 16264, Morflex 240, 20321, Palatinol® C, 22585, PX-104, 25432, Unimoll® DB, 32978, Uniplex 150, 33010
Di-n-Butyl Phthalate, *See* dibutyl phthalate, 8366, Kodaflex® DBP, 16264
Di-*t*-butyl pyrocarbonate, *See* Di-*t*-butyl dicarbonate, 8367
Di-*tert*-butyl pyrocarbonate, *See* Di-*t*-butyl dicarbonate, 8367
Dibutyl sebacate, *See* Kodaflex® DBS, 16265, Plasthall® DBS, 23994, Reomol DBS, 26129, Uniflex® DBS, 32942
N,N-dibutylthiourea, *See* Pennzone B 0685, 23120
N,N'-di-N-butylthiourea, *See* Pennzone B 0685, 23120
Dibutyl tin diacetate, *See* Metacure® T-1, 19414, Metacure® T-12, 19417
Dibutyl-*o*-Phthalate, *See* dibutyl phthalate, 8366
dibutyl-1,2-benzene dicarboxylate, *See* dibutyl phthalate, 8366
dibutylated hydroxytoluene, *See* Oxyguard, 22457
dibutyl 1,2-benzenedicarboxylate, *See* dibutyl phthalate, 8366
dibutylbis[(1-oxodecyl)oxy]stannane, *See* dibutyltin laurate, 8371
dibutylbis(lauroyloxy) tin, *See* Dabco® T-12, 7673
dibutylbis[(1-oxododecyl)oxy]stannane, *See* Dabco® T-12, 7673
2,6-Di-*t*-butyl-4-*sec*-butyl phenol, *See* Vanox® 1320, 33337
2,6-Di-*tert*-butyl-4-*sec*-butylphenol, *See* Isonox® 132, 15384
Di-*t*-butyl-4-butylphenol, *See* Isonox® 132, 15384

Chemical and Trade Names

dibutylcarbamodithioic acid, sodium salt, *See* Octopol NB-47, 22012
2,6-di-*t*-butyl-*p*-cresol, *See* BHT, 4122, Oxyguard, 22457, Tenamine 3, 31092, Vulkanox® KB, 34152
dibutyldiacetoxystannane, *See* Metacure® T-1, 19414
Di-n-Butyldiacetoxytin, *See* Metacure® T-1, 19414
di-tert-butyldiazene, *See* VR-160, 34057
Di-n-butyldilauryltin, *See* Dabco® T-12, 7673
2,6-Di-*tert*-butyl- α-dimethylamino-*p*-cresol, *See* Ethanox® 703, 10935
2,2-Dibutyl-1,3,2-dioxastannepin-4,7-dione, *See* Dibutyltin maleate, 8372
Dibutylglycol phthalate, *See* Palatinol® K, 22592
Dibutylhydroquinone, 8368
2,5-Di-*t*-butylhydroquinone, *See* Dibutylhydroquinone, 8368, Eastman® DTBHQ, 9466
3,5-di-*tert*-butyl-4-hydroxyhydrocinnamate, *See* Naugard® 76, 20789
3,5-Di-*t*-butyl-4-hydroxybenzoic acid n-hexadecyl ester, *See* Cyasorb® UV 2908, 7501
3,5-Di-t-butyl-4-hydroxyhydrocinamic acid triester of 1,3,5-tris (2-hydroxyethyl)-s-triazine-2,4,6-(1H,3H,5H)-trione, *See* Vanox® SKT, 33347
2,2-Bis[4-(2-(3,5-di-t-butyl-4-hydroxyhydrocinnamoyl oxy)) ethoxyphenyl] propane, *See* Topanol® 205, 32022
2-(3',5'-Di-t-butyl-2'-hydroxyphenyl)-5-chloro benzotriazole, *See* Uvazol 237, 33182
2,6-di-*t*-butyl 4-methylphenol, *See* Deenax, 7874, Vanox® PCX, 33346
2,6-Di-tert-butyl-4-nonylphenol, *See* Uvi-Nox 1494, 33189
Dibutylperoxy dicarbonate, *See* Espercarb® 438M-60, 10840
2,2-di-(*t*-butylperoxy) butane, *See* Trigonox® D-E50, 32298
1,1-Di-t-butylperoxycyclohexane in butyl benzyl phthalate, *See* Trigonox® 22-BB80, 32266
1,1-Di-(*tert*-butylperoxy)-3,3,5-trimethyl cyclohexane, *See* Aztec® 1,1-Bis(*t*-Butylperoxy)-3,3,5-Trimethyl Cyclohexane, 3549, Lupersol 231, 17856, Peroximon® S-164/40P, 23410, Trigonox® 29, 32269
6,6'-di-*tert*-butyl-2,2'-methylenedi-*p*-cresol, *See* Antioxidant 2246, 2580
2,6-di-*tert*-butyl-4-methylphenol, *See* Ralox® BHT food grade, 25845, Tenamine 3, 31092
***tert*-Dibutyl peroxide**, *See* Aztec® Di-*t*-Butyl Peroxoide, 3557
Di-*t*-butyl peroxyazelate, *See* Trigonox® 169-OP50, 32289
1,1-Di-(*tert*-butylperoxy)-3,3,5-trimethyl cyclohexane, *See* Aztec®, 3539
2,6-dibutylphenol, 8370
2,6-di-*t*-butylphenol, 4799, *See* 2,6-dibutylphenol, 8370, Isonox® 103, 15382, Lowinox® 001, 17646
2,6-Di-*tert*-Butylphenol, *See* 2,6-di-*t*-butylphenol, 4799
N,N'-di-*s*-butyl-*p*-phenylenediamine, *See* Naugalube® 403, 20782
N,N'-di-*sec*-butyl-*p*-phenylenediamine, *See* Naugalube® 403, 20782
Dibutylphthalate, *See* NLA-10, 21395
di-*t*-butyl pyrocarbonate, *See* di-t-butyl dicarbonate,
1,3-dibutyl-2-thiourea, *See* Pennzone B 0685, 23120
1,3-dibutylthiourea, *See* Pennzone B 0685, 23120
Di-n-butyltin diacetate, *See* Metacure® T-1, 19414
Dibutyltin didodecanoate, *See* Dabco® T-12, 7673
dibutyltin dilaurate, *See* ADK STAB BT-11, 723, Dabco® T-12, 7673, Synpron 1009, 30611
Di-n-butyltin Dilaurate, *See* Dabco® T-12, 7673
Dibutyltin dillaurate, *See* TEDA-T411, 30917
dibutyltin laurate, 8371
Dibutyltin maleate, 8372, *See* ADK STAB BT-31, 724
Dibutyltin mercaptide, *See* ADK STAB 1292, 715, Lankromark® BT120A, 16688
dicacodyl, *See* Cacodyl, 4882
dicalcium phosphate, *See* Caliment, 5006 Calipharm, 5008,, Lucaphos® 40, 48, 17752
dicalcium phosphate (CTFA), *See* calcium phosphate, dibasic,
Dicalcium phosphate dihydrate, *See* Emcompress®, 10064
dicalcium phosphate, anhydrous, *See* calcium hydrogen phosphate, 4951
Dicalite 14, 14B, and 14W, 8373
Dicamba, 8374, *See* Tracker, 32130
dicamba-MCPA-mecoprop, *See* Dock-Ban, 8774, Docklene, 8775, Headland Relay, 13733, Herrisol, 13883, Hyprone, 14723, Hysward, 14772
dicamba-mecoprop, *See* Di-Farmon, 8448 Endox, 10552,, Farmon Condox, 11486, Hyban, 14487, Hygrass, 14678
dicamba-mecoprop-triclopyr, *See* Fettel, 11710
dicamba-paclobutrazol, *See* Holdfast D, 14274
1,2-Dicaproyl-*sn*-glycero(3) phosphatidylcholine, *See* Phospholipon® CC, 23727
Dicapryl adipate, *See* Uniflex® DCA, Unimate® DCA, 32967 Unimate® DCA, 32943
Dicapryl phthalate, 8375, *See* DCP, 7818, Reomol DCP, 26130, Uniflex® DCP, 32944
Dicapryl/dicaprylyl dimonium chloride, *See* FMB 302-8, 12132
Dicapryl/dicaprylyl dimonium chloride-myristalkonium chloride, *See* FMB 504-5, 12134
Dicarbam, *See* Carbaryl, 5181
Dicarbonic acid, bis(1,1-dimethylethyl) ester, *See* Di-*t*-butyl dicarbonate, 8367
Dicarboxylic acid, *See* succinic acid, 29915
dicarboxylic acid C_3, *See* malonic acid, 18491
dicarboxylic acid C_6, *See* Adipic acid, 698
dicarboxylic acid C_8, *See* suberic acid, 29903
dicarboxymethane, *See* malonic acid, 18491
Dicarburetted hydrogen, *See* olefiant gas,
Dicestal, 8376, *See* Dichlorophen, 8392, Nuophene, 21783
Diceteareth-10 phospate, *See* Marlophor® T10-Acid, 18900
Dicetyl dilinoleate, *See* Liquiwax DC-EFA/SS, 17486
Dicetyl dimonium chloride, *See* Carsoquat® 868 E, 5347, Varisoft® 432-100, 33409
Dicetyldimonium chloride-myristalkonium chloride, *See* FMB 1210-5, 1210-8, 12136
Dichan 100, 8377, *See* dicyclohexylamine nitrite, 8414
Dichevrol, 8378
Dichlobenil, *See* BH Prefix D, 4120, Casoron, 5394, Casoron G, 5395, Casoron G4, 5396, Fydulan, Fydumas, Fydusit, 12532
Dichlofluanid, 8379, *See* Elvaron®, 9893 Euparen®, 11197
Dichlofluanid-methyl, *See* Euparen® M, 11198
Dichlofuanide, 8380
Dichlone, 8381
dichloracetylene, *See* Acetylene dichloride, 320, Dioform, 8553
Dichloramine-M, 8383
Dichloramine T, 8382
Dichloraminet, *See* Dichloramine T, 8382
Dichloran, *See* dicloran, 8395, Fumite Dicloran, 12459
***p*-Dichlorbenzene**, *See* Kaydox, 15811
dichloricide, *See* *p*-dichlorobenzene, 8386
Di-chloricide, *See* Kaydox, 15811
Dichloricide Aerosol, *See* Strobane, 29818
Dichloricide Mothproofer, *See* Strobane, 29818
Dichlorman, *See* dichlorvos, 8394, Vapona, 33378
dichloro-1,3,5-triazinetrione, potassium salt, 8384
dichloro-1,3,5-triazinetrione, sodium salt, 8385
Dichloro-4-methylphenyl-O,O-dimethyl phosphorothioate, *See* Risolex, 26782
Dichloro-4-nitroaniline, *See* Fumite Dicloran, 12459
3,6-dichloro-*o*-anisic acid, *See* Dicamba, 8374
***o*-Dichlorobenzene**, *See* Chloroben, 6117, Dizene, 8744
***p*-dichlorobenzene**, 8386, *See* Paradow, 22738
1,4-dichlorobenzene, *See* *p*-dichlorobenzene, 8386
1,4-Dichlorobenzene paste, *See* Kaydox, 15811
2,4-dichlorobenzenemethanol, *See* 2,4-dichlorobenzyl alcohol, 8387
***p*-dichlorobenzol**, *See* *p*-dichlorobenzene, 8386
2,4-Dichlorobenzoyl peroxide, *See* Cadox® TDP, 4915, Cadox® TS-50S, 4916
2,4-dichlorobenzyl alcohol, 8387, *See* Myacide® SP, 20546
1,1-Dichloro-2-(o-chlorophenyl)-2-(*p*-chlorophenyl)ethane, *See* Lysoform, 18095
Dichlorocide, *See* *p*-dichlorobenzene, 8386
1,1-Dichloro-2,2-bis(2,4'-dichlorophenyl)ethane, *See* Lysoform, 18095
dichlorodifluoromethane, 8388, *See* Genetron® 12, 12818
1,4-dichloro-2,5-dimethoxybenzene, *See* Terraneb SP Turf Fungicide, 31337
1,3-Dichloro-5,5-dimethylhydantoin, 8389, *See* Dantoin® DCDMH, 7732; Hydan, 14528
1,3-Dichloro-5,5-dimethyl-2,4-imidazolidinedione, *See* 1,3-Dichloro-5,5-dimethyl hydantoin, 8389
1,1-dichloro-N-((dimethylamino)sulfonyl)-1-fluoro-N-(4-methylphenyl)methanesulfonamide, *See* Euparen® M, 11198
1,1-Dichloro-N-[(dimethylamino)-sulfonyl]-1-fluoro-N-phenylmethanesulfenamide, *See* Dichlofluanid, 8379
2,4-dichloro-3,5-dimethylphenol, *See* Nipacide® DX, 21255
3,5-Dichloro-N-(1,1-dimethyl-2-propynyl)benzamide, *See* Kerb 50W, 16064, 16065, Kerb Propyzamide 50, 16066, Rapier, 25889
2,4'-Dichlorodiphenyldichloroethane, *See* Lysoform, 18095
***p*-dichlorodiphenylmethane**, *See* Dichloroditane, 8390
dichlorodiphenylsilane, *See* CD5950, 5513
Dichlorodiphenyltrichlorethane, *See* DDT, 7820
Dichloroditane, 8390
1,2-dichloroethane, *See* ethylene dichloride, 11122
1,2-dichloroethylene, *See* Acetylene dichloride, 320, Dioform, 8553
1,2-Dichloroethene (mixed isomers), *See* Dioform, 8553
3-(2,2-Dichloroethenyl)-2,2-dimethylcyclopropanecarboxylic acid (3-phenoxyphenyl)methyl ester, *See* Kafil, 15655, permethrin, 23384, 23385, 23386
β,β-dichloroethyl sulfide, *See* mustard gas, 20538
Dichloroethylene, *See* Dioform, 8553
***sym*-Dichloroethylene**, *See* Dioform, 8553
1,2-Dichloroethylene, *See* Dioform, 8553
1,2-Dichloroethylene (mixture), *See* Dioform, 8553
***cis* & *trans* 1,2-dichloroethylene**, *See* Dioform, 8553
1,2-Dichloroethylene (*cis* & *trans*), *See* Dioform, 8553
Dichloroethylenes, *See* Dioform, 8553
Dichlorofen, *See* Nuophene, 21783
Dichlorofluoroethane, 8391
1,1-Dichloro-1-fluoroethane, *See* Dichlorofluoroethane, 8391
Dichlorofluoromethyl(thio)phthalimide, *See* Fluorfolpet, 12087
N-(Dichlorofluoromethylthio)phthalimide, *See* Fluorfolpet, 12087
2,3-dichloro-N-4-fluorophenylmaleimide, *See* fluoromide, 12108, Sparticide®, 29211
dichloroisocyanuric acid potassium salt, *See* potassium dichloroisocyanurate, 24753
dichloroisocyanuric acid sodium salt, *See* sodium dichloroisocyanurate,
dichloromercury, *See* mercuric chloride, 19347
dichloromethane, *See* Distillex DS3, 8719, methylene chloride, 19568
3,6-dichloro-2-methoxybenzoic acid, *See* Dicamba, 8374
3-[2,4-Dichloro-5-(1-methylethoxy)phenyl]-5-(1,1-dimethylethyl)-1,3,4-oxadiazol-2(3H)-one, *See* oxadiazon, 22406
O-(2,6-dichloro-4-methylphenyl) O,O-dimethyl phosphorothioate, *See* Risolex, 26782
2,3-Dichloro-1,4-naphthalenedione, *See* Dichlone, 8381
Dichloronaphthoquinone, *See* Dichlone, 8381
2,3-dichloro-1,4-naphthoquinone, *See* Dichlone, 8381
2,6-Dichloro-4-nitroaniline, *See* Fumite Dicloran, 12459
Dichloro-4-nitroaniline, *See* dicloran, 8395

2,6-dichloro-4-nitrobenzeneamine, *See* dicloran, 8395
2',5-dichloro-4'-nitrosalicylanilide compound with 2-aminoethanol (1:1), *See* clonitrilide, 6432
Dichlorophen, 8392, *See* Algafen, 1583, Algofen, 1602, Anthiphen, 2520, Dicestal, 8376, Ecco MP® 2004, 9513, Fungo®, 12483, Mascot Moss Killer, 19004, Nuophene, 21783, Panacide, 22652, Super Mosstox, 30184
Dichlorophen-ferrous sulfate, *See* Aither's Lawn Sand Plus, 1137
dichlorophen-mecoprop-dichlorprop-dicamba-benazolin, *See* Green Up Lawn Feedn Weed Plus Moss Killer, 13331
4-(2,4-dichlorophenoxy)butanoic acid, *See* 2,4-DB, 7810
(2,4-dichlorophenoxy)acetic acid, *See* Verdone CDA, 33580
(2,4-dichlorophenoxy)acetic acid 2-butoxyethyl ester, *See* Planotox, 23914
5-(2,4-dichlorophenoxy)-2-nitrobenzoic acid methyl ester, *See* bifenox, 4135
O-2,4-dichlorophenyl O-ethyl S-propyl phosphorodithioate, *See* Tokuthion®, 31932
dichlorophenyl urea, *See* diuron, 8736
(Dichlorophenyl)-5-ethenyl-5-methyl-2,4-oxazolidinedione, *See* vinclozolin, 33811
1-(3',4'-dichlorophenyl)-3-(4'-chlorophenyl)urea, *See* TCC, 30856
3-(3,4-dichlorophenyl)-1,1-dimethylurea, *See* Diurex, 8734, diuron, 8736
N'-(3,4-Dichlorophenyl)-N,N-dimethylurea, *See* Karmex, 15771 Karmex®, 15772
3-(3,5-dichlorophenyl)-5-ethenyl-5-methyl-2,4-oxazolidinedione, *See* vinclozolin, 33811
O-(2,4-dichlorophenyl) O-ethyl S-propyl phosphorodithioate, *See* Tokuthion®, 31932
N'-(3,4-Dichlorophenyl)-N-methoxy-N-methylurea, *See* Atlas Linuron, 3319
3-(3,4-dichlorophenyl)-1-methoxy-1-methylurea, *See* Linurex, 17316
3-(3,4-dichlorophenyl)-1-methyl-1-n-butylurea, *See* Kloben®, 16225
[2S-(2α,5α,6β)]-6-[[[3-(2,6-dichlorophenyl)-5-methyl-4-isoxazolyl]carbonyl]amino]-3,3-dimethyl-7-oxo-4-thia-1-azabicyclo[3.2.0]heptane-2-carboxylic acid, *See* Dicloxacillin sodium, 8396
2-(3,4-dichlorophenyl)-4-methyl-1,2,4-oxadiazolidine-3,5-dione, *See* methazole, 19481
1-[2-(2,4-dichlorophenyl)pentyl]-1H-1,2,4-triazole, *See* Topas 100, 32026
N-(3,4-dichlorophenyl)propanamide, *See* Propanil, 25174
(±)-1-[2-(2,4-dichlorophenyl)-2-(2-propenyloxy)ethyl]-1H-imidazole, *See* imazalil, 14892
1-[2-(2,4-Dichlorophenyl)-2-(2-propenyloxy)ethyl]-1H-imidazole, *See* Enilconazole, 10582
1-[[2-(2,4-dichlorophenyl)-4-propyl-1,3-dioxolan-2-yl]methyl]-1H-1,2,4-triazole, *See* Radar, Radar Propiconazole, 25714
(±)-2-(3,5-dichlorophenyl)-2-(2,2,2-trichloroethyl)oxirane, *See* Tandem, 30796
(RS)-2-(3,5-dichlorophenyl)-2-(2,2,2-trichloroethyl)oxirane, *See* Tandem, 30796
Dichlorophos, *See* dichlorvos, 8394, Vapona, 33378
1,2-dichloropropane, *See* propylene dichloride, 25213
dichloropropane, *See* propylene dichloride, 25213
α,β-dichloropropane, *See* propylene dichloride, 25213
dichloropropanes, *See* propylene dichloride, 25213
2,2-dichloropropanoic acid, *See* dalapon, 7705
2,2-dichloropropanoic acid, sodium salt, *See* Dowpon, 8943
1,3-dichloro-1-propene, *See* Telone, 31069
1,3-dichloropropene, *See* Telone, 31069
2,2-dichloropropionic acid, *See* Couch and Grass Killer, 6896
Dichloropropionic acid, sodium salt, *See* Dowpon, 8943
3,6-Dichloropyridine-2-carboxylic acid, *See* Clopyralid, 6436
dichlorotetraaquochromium chloride dihydrate, *See* Chrometrace, 6218
***sym*-Dichlorotetrafluoroethane**, *See* Cryofluorane, 7278
Dichlorotetrafluoroethane, *See* Cryofluorane, 7278, Genetron® 114, 12822
1,2-Dichlorotetrafluoroethane, *See* Cryofluorane, 7278
1,2-Dichloro-1,1,2,2-tetrafluoroethane, *See* Cryofluorane, 7278
Dichloro-1,1,2,2-tetrafluoroethane, *See* Cryofluorane, 7278
3,5-Dichlorotetrahydro-2,4,6-trioxo-*s*-triazin-1(2H)-yl potassium, *See* ACL 56, 59, 60, 66, 354, dichloro-1,3,5-triazinetrione, potassium salt, 8384
1,3-Dichloro-1,1,3,3-tetraisopropyldisiloxane, *See* CD4368, 5503
1,3-Dichlorotetraisopropyldisiloxane, *See* CD4368, 5503
1,3-Dichlorotetraisopropylsiloxane, *See* CD4368, 5503
1,3-Dichloro-1,3,5-triazine2,4,6-[1H,3H,5H]-trione potassium salt, *See* ACL 56, 59, 60, 66,, 354
1,3-dichloro-s-triazine-2,4,6(1H,3H,5H)-trione, potassium salt, *See* dichloro-1,3,5-triazinetrione, potassium salt, 8384
1,3-dichloro-1,3,5-triazine-2,4,6(1H,3H,5H)-trione sodium salt, *See* dichloro-1,3,5-triazinetrione, sodium salt, 8385
***s*-triazin-2,4,6(1H,3H,5H)trione potassium**, *See* dichloro-1,3,5-triazinetrione, potassium salt, 8384
4,4'-dichloro-α-(trichloromethyl)-benzhydrol, *See* Kelthane, 15881
Dichlorotrifluoroethane, 8393 *See* Genetron® 123, 12823
1,1-Dichloro-2,2,2-trifluoroethane, *See* Dichlorotrifluoroethane, 8393, Genetron® 123, 12823
2,2-dichloro-1,1,1-trifluoroethane, *See* Dichlorotrifluoroethane, 8393, Genetron® 123, 12823
N, N-Dichloro-*p*-toluenesulfonamide, *See* Dichloramine T, 8382
dichlorovos, *See* Vapona, 33378
2,4-Dichloro-*m*-xylenol, *See* Nipacide® DX, 21255
2,4-di-4-chloro-3,5-xylenol, *See* Nipacide® PX, 21260
dichlorprop-MCPA, *See* Farmon 2,4-DP+MCPA, 11484, Hemoxone, 13784
Dichlorprop-P, *See* Duplosan® DP, 9131
Dichlor-Stapenor, *See* Dicloxacillin sodium, 8396
Dichlorvos, 8394, *See* Atgard, 3285, Dedevap®, 7870, Dethlac, 8210, Divipan, 8740, Mafu®, 18305, Nogos, 21420, Vapona, 33378
Dichlosale, *See* Mansonil®, 18611
dichromic acid, ($H_2Cr_2O_7$), disodium salt, dihydrate, *See* sodium dichromate, 28756
Dichromic acid, diammonium salt, *See* Ammonium dichromate, 2237
dichromium trioxide, *See* chromic oxide, 6223, M100, 18107
dickite, *See* Tisyn®, 31857
Dicköl varnish, *See* stand oil, 29450
Dickül varnish, *See* stand oil, 29450
Diclofop-Methyl, *See* Hoegrass, 14271
dicloran, 8395 *See* Allisan, 1747 Fumite Dicloran, 12459
Dicloron, *See* dicloran, 8395, Fumite Dicloran, 12459
Dicloxacillin sodium, 8396, *See* Dycill, 9307
Dicloxin, *See* Dicloxacillin sodium, 8396
Dicocamine, *See* Armeen® 2C, 2928, Radiamine 6260, 25759
Dicoco dimethyl ammonium chloride, *See* Dye Retarder #1, 9310
Dicoco dimonium chloride - isopropanol, *See* Tomah Q-2C, 31989
Dicoco nitrite, *See* Arquad® 2C-70 Nitrite, 3101
Dicocodimethylammonium chloride, *See* Radiaquat 6462, 25793
Dicocodimonium chloride, *See* Accoquat 2C-75, 2C-75H, 242, Adogen® 462, 772, Arquad® 2C-75, 3102
Dicocomethylamine, *See* Armeen® M2C, 2954, Jet Amine M2C, 15591, Radiamine 6360, 25764
Dicofen, 8397
Dicofol, 8398, *See* Fumite Dicofol, 12460, Kelthane, 15881
Dicontal® New, 8399
dicophane, *See* DDT, 7820
dicopper chloride trihydroxide, *See* copper oxychloride, 6784
Dicopur-M, *See* MCPA, 19183
Dicosal, 8400
Dicotox, 8401
Dicotox Extra, 8402
Dicrodamine, 8403
Dicron 45Sc, 8404
Dicrylan, 8405
Dicrylan 270, 8406
Dicumyl peroxide, 8407, *See* Aztec® DCP-R, 3556, Di-Cup, 8408, Esperal® 115, 10836, Luperox 500R, 17840, Percumyl D, 23230, Perkadox® BC, 23307, Peroximon® DC-40, 23409, Polyvel PCL-20, 24661
di-α-cumyl peroxide, *See* Dicumyl peroxide, 8407
Di-Cup, 8408, *See* Aztec® DCP-R, Percumyl D, 23230, Polyvel PCL-20, 24661, Peroximon® DC-40, 23409, Dicumyl peroxide, 8407
Dicuran, *See* Talisman, 30735, Tolurex, 31958
Dicurane, *See* Talisman, 30735, Tolurex, 31958
Dicurane 500 FW, 8409, *See* Ludorum, 17778
Dicurane Duo 495FW, 8410
Dicyandiamide, 8411
2,2'-Dicyano-2,2'-azopropane, *See* Poly-zole AZDN, 24674, V-60, 33218
1,3-dicyano-2,4,5,6-tetrachlorobenzene, *See* Repulse, 26151
Dicyclohexyl adipate, *See* Sipalin AOC, 28512
Dicyclohexyl ammonium nitrite, *See* Dichan 100, 8377, dicyclohexylamine nitrite, 8414
dicyclohexyl benzene-1,2-dicarboxylate, *See* Uniplex 250, 33014
N,N'-Dicyclohexyl-2-benzothiazole sulfenamide, *See* Perkacit® DCBS, 23279
dicyclohexyl carbodiimide, 8412
dicyclohexyl phthalate, 8413, *See* KP 201, 16379, Morflex 150, 20317, Unimoll® 66 M, 32976, Uniplex 250, 33014
Dicyclohexyl sodium sulfosuccinate, *See* Aerosol® A-196-85, 887, Rewopol® SBDC 40, 26372
dicyclohexylamine nitrite, 8414, *See* Dichan 100, 8377, dicyclohexylamine nitrite, 8414
dicyclohexylaminonitrite, *See* Dichan 100, 8377, dicyclohexylamine nitrite, 8414
N,N'-dicyclohexylcarbodiimide, *See* dicyclohexyl carbodiimide, 8412
Dicyclopentenyl acrylate, *See* Sipomer® DCPA, 28533
dicyclopentenyl methacrylate, *See* Sipomer® DCPM, 28534
dicysteine, *See* cystine, 7603
didecyl adipate, *See* Good-rite® GP-236, 13203, Morflex 330, 20324
Didecyl dimethyl ammonium chloride, *See* Catigene® 818, 5451, Catigene® 1011, 5449, BTC® 1010-80, 4638, Radiaquat 6412, 25790
Didecyl dimethyl ammonium methosulfate, *See* Ablumine D10, 86
didecyl dimethylammonium chloride, *See* Querton 210Cl-50, 25651
Didecyl dimonium chloride, *See* BTC® 99, 4633
didecyl methylamine, 8415, *See* Armeen® M2-10D, 2953, Dama® 1010, 7718
Didecyl phthalate, *See* Good-rite® GP-266, 13206
Didecylamine, *See* Armeen® 2-10, 2926
Di-n-decylamine, *See* Armeen® 2-10, 2926
Didecyldimethylammonium chloride, *See* Querton 210Cl-50, 25651, Radiaquat 6410, 6412, 25789, Radiaquat 6412, 25790, Rewoquat B 10, 26389
Didecyldimonium chloride, *See* Arquad® 210-50, 3110, FMB 210-8, 210-15, 12131, Rewoquat B 10, 26389
Didecylphthalate, *See* NLA-40, 21398
8,14-didehydro-4,5 α-epoxy-3-methoxy-17-methylmorphinan-6 α-ol, *See* neopine, 20939
Didi-Col, 8416
Didigam, *See* DDT, 7820
Didigram, 8417

Didimac, 8418, *See* DDT, 7820
1,2-Di-(dimethylamino)-ethane, *See* Propamine D, 25170
Didin Fluid, 8419
Didocyl phosphite, *See* dilauryl phosphite, 8475
didodecyl 3,3'-thiodipropionate, *See* dilauryl thiodipropionate, 8476
Didodecyl phosphite, *See* dilauryl phosphite, 8475, Weston® DLP, 34306
Di-n-dodecyl phosphite, *See* dilauryl phosphite, 8475, Weston® DLP, 34306
Didodecyl 3,3-thiodipropionate, *See* Argus DLTDP, 2861, Carstab® DLTDP, 5356, Cyanox® LTDP, 7491
DIDP, *See* diisodecyl phthalate, 8462
Didpex® A40 and N40, 8420
Didronel, 8421
Didroxan, *See* Nuophene, 21783
Didroxane, *See* Dichlorophen, 8392
die-casting alloys, 8422
Di-el, 8423
Dieline, 8424
Dielmoth, 8425
Dieltamid, *See* diethyl toluamide, 8438
Diene®, 8426
Diene® 35AC, 55AC, 8427
Diene® 70AC, 8428
dienochlor, 8429
Dienol, 8430
Diepoxy, 8431
Dieselect, 8432
Diester quaternary ammonium methyl sulfate, *See* Stepantex® VS 90, 29712
Diesterex N, 8433
diethamine, *See* diethylamine, 8439
N,N-Diethanollauramide, *See* Ablumide LDE, 79
Diethanol lauric acid amide, *See* Ablumide LDE, 79, Witcamide® 5195, 34440
Diethanolamide lauric acid, *See* Ablumide LDE, 79, Witcamide® 5195, 34440
Diethanolamides of the Fatty Acids of Coconut Oil, *See* Active #2, 501
Diethanolamine, 8434, *See* Dabco® DEOA-LF, 7667, Witcamide® Coco Condensate, 34443
Diethanolamine dodecylbenzenesulfonate, *See* Ablusol DBD, 124
diethanolamine lauric acid amide, *See* Ablumide LDE, 79, lauramide DEA, 16828, Witcamide® 5195, 34440
Diethanolamine lauryl sulfate, *See* DEA-lauryl sulfate, 7827
Diethanolamine stearic acid amide, *See* Ablumide SDE, 81
Diethanolamine-dodecylbenzenesulfonic acid adduct, *See* Ablusol DBD, 124
Diethanolamine-dodecylbenzenesulfonic acid salt, *See* Ablusol DBD, 124
Diethanollauramide, *See* Ablumide LDE, 79, Witcamide® 5195, 34440
N,N-Diethanollauramide, *See* Witcamide® 5195, 34440
N,N-diethanollauric acid amide, *See* Alkamide® 327, 1634, Witcamide® 5195, 34440
Diethanolstearamide, *See* Ablumide SDE, 81
Diethchinalphion, *See* Bayrusil®, 3839
1,3-Diethenyl-1,1,3,3-tetramethyl-disiloxane, *See* CD6210, 5517
(SP-5-13)-[7,12-diethenyl-3,8,13,17-tetramethyl-21H,23H-porphine-2,8-dipropanoato(4-)-N_{21},N_{22},N_{23},N_{24}]hydroxyferrate(2-) dihydrogen, *See* Hematin, 13780
Diethoxol, 8435
diethoxy, nitro-phenoxy phosphorothioate, *See* parathion, 22807
1,1-Diethoxyacetal, *See* Acetron®, Acetron® GP; Acetron® NS, 308
diethoxydimethylsilane, *See* CD5600, 5508
1,1-Diethoxyethane, *See* Acetal, 280, Acetron®, Acetron® GP; Acetron® NS, 308
α-[[(diethoxyphosphinothioyl)oxy]imino] benzeneacetonitrile, *See* Sebacil®, 27796
Diethoxyphosphinothioyl)oxy)imino)benzeneacetonitrile, *See* Sebacil®, 27796
2-(diethoxyphosphinothioyloxyimino)-2-phenylacetonitrile, *See* Volathion®, 34011
2-diethoxyphosphinylimino-4-methyl-1,3-dithiolane, *See* mephosfolan, 19326
2-[(Diethoxyphosphinyl)oxy]-1H-benz[de]isoquinoline-1,3(2H)-dione, *See* Rametin, 25856
Diethquinalphion, *See* quinalphos, 25679
diethyl [(dimethoxyphosphinothioyl)thio]butanedioate, *See* malathion, 18460
diethyl 1,2-benzenedicarboxylate, *See* diethyl phthalate, 8437
Diethyl 2-(dimethoxyphosphinothioyl-thio)succinate, *See* Malathion 60, 18461
Diethyl acetal, *See* Acetron®, Acetron® GP; Acetron® NS, 308
diethyl 1,2-benzenedicarboxylate, *See* diethyl phthalate, 8437
O,O-diethyl α-cyanobenzylideneamino-oxyphosphonothioate, *See* Volathion®, 34011
O,O-diethyl O-(alpha-cyano benzylidene amino)phosphorothioate, *See* Sebacil®, 27796
Diethyl Dimpylatum, *See* Diazinon Liquid, 8344
diethyl ester, *See* diethyl phthalate, 8437
diethyl ester, oxalic acid, *See* oxalic ether, 22416
diethyl ether, *See* ethyl ether, 11069, sulfuric ether, 30056
Diethyl hexyl phthalate, *See* Morflex 210, 20320
diethyl ketone, 8436
O,O-diethyl O-4-methylsulfinylphenyl phosphorothioate, *See* Terracur® P, 31329
O,O-diethyl O-[4-(methylsulfinyl)phenyl] phosphorothioate, *See* Terracur® P, 31329
Diethyl N,N-bis(2-hydroxyethyl) aminomethyl phosphonate, *See* Fyrol® 6, 12536
O,O-diethyl O-naphthaloximide phosphate, *See* Rametin, 25856
O,O-Diethyl-O-(4-nitrophenyl) phosphorothioate, *See* ethyl parathion, 11079
O,O-diethyl O-*p*-nitrophenyl phosphorothioate, *See* parathion, 22807
O,O-diethyl S-[(4-oxo-1,2,3-benzotriazin-3(4H)-yl) methyl] phosphorodithioate, *See* Gusathion® A, 13502
O,O-diethyl S-[(4-oxo-1,2,3-benzotriazin-3(4H)-yl)methyl] phosphorodithioate, *See* Azinphos-ethyl, 3520
O,O-diethyl O-2-quinoxalinyl phosphorothioate, *See* quinalphos, 25679
Diethyl O-(2-quinoxalyl) phosphorothioate, *See* quinalphos, 25679
Diethyl O-(quinoxalin-2-yl) thiophosphate, *See* quinalphos, 25679
Diethyl O-2-quinoxalinyl phosphorothioate, *See* quinalphos, 25679
Diethyl O-quinoxalin-2-yl thionophosphate, *See* quinalphos, 25679
Diethyl Oxide, *See* ethyl ether, 11069
diethyl phthalate, 8437, *See* Kodaflex® DEP, 16266, Palatinol® A, 22583
Diethyl phthalimidophosphonothioate, *See* Plondrel, 24160
diethyl sulfide 2,2'-dicarboxylic acid, *See* 3,3'-thiodipropionic acid, 31684
diethyl toluamide, 8438
Diethyl toluene diamine, *See* Ethacure® 100, 10906
diethyl-2-thiourea, *See* Pennzone E 0686, 23121
α-diethylaminoaceto-2,6-xylidide, *See* lidocaine, 17216
diethylamine, 8439
diethylamine N,N-diethylamine, *See* diethylamine, 8439
α-Diethylaminoaceto-2,6-xylidide, *See* lidocaine, 17216
Diethylaminoethanol, *See* Pennad 150, 23113
β-diethylaminoethanol, *See* 2-diethylaminoethanol, 8440
2-diethylaminoethanol, 8440
N,N-diethylaminoethanol, *See* 2-diethylaminoethanol, 8440
N,N-Diethylaminoethyl acrylate, *See* Ageflex FA-2, 976
Diethylaminoethyl acrylate, *See* Ageflex FA-2, 976
2-(Diethylamino)-ethyl acrylate, *See* Ageflex FA-2, 976
Diethylaminoethyl acrylate dimethyl sulfate, *See* Ageflex FA-2Q50DMS, 973
N,N-Diethylaminoethyl acrylate Q-Salt, methosulfate, *See* Ageflex FA-2Q50DMS, 973
2-diethylaminoethyl alcohol, *See* 2-diethylaminoethanol, 8440
Diethylaminoethyl stearate, *See* Cerasynt® 303, 5733
7-Diethylamino-4-methyl-2H-benzopyran-2-one, *See* Coumalux®, 6902
Diethylaminotrimethylsilane, *See* CD4450, 5504
N,N-Diethylaminotrimethyl silane, *See* CD4450, 5504
Di(2-ethylbutyl)phthalate, *See* Jayflex® DHP, 15549
Diethylcarbamazine Citrate, *See* Francide, 12378
Diethylcarbamodithioic acid, sodium salt, *See* Octopol SDE-25, 22013
O,O-diethyl (1,3-dihydro-1,3-dioxo-2H-isoindol-2-yl)phosphonothioate, *See* Plondrel, 24160
diethyldiphenyl dichloroethane, 8441, *See* diethyldiphenyldichloroethane, 8441, Perthane, 23435
Diethyldithiocarbamic acid sodium salt, *See* Octopol SDE-25, 22013
diethyldithiocarbamic acid sodium salt trihydrate, *See* Thiostop E, N, 31716
diethyldithiocarbamic acid tellurium salt, *See* Akrochem® TDEC, 1179, ethyl tellurac®, 11084
diethylene DB, *See* Butyl Carbitol®, 4766
diethylene diamine, *See* piperazine, 23873
1,4-diethylene dioxide, *See* 1,4-dioxane, 8574
Diethylene dioxide, *See* Dioxane, 8574
diethylene ether, *See* 1,4-dioxane, 8574
diethylene glycol, 8442
diethylene glycol butyl ether, *See* Butyl Carbitol®, 4766
diethylene glycol butyl ether acetate, 8443
diethylene glycol cetyl ether, *See* Akyporox RC 200, 1289
diethylene glycol diabietate, *See* Flexalyn, 11921
diethylene glycol dibenzoate, *See* Benzoflex 2-45, 3970
diethylene glycol dimethacrylate, 8444, *See* Ageflex DEGDMA, 962, SR-231, 29369
diethylene glycol dimethyl ether, *See* diglyme, 8455
diethylene glycol dodecyl ether, *See* Laureth-2, 16853
diethylene glycol ethyl ether, *See* Diethoxol, 8435 Ethoxydiglycol, 11044, ethyl di-Icinol, 11068
diethylene glycol laurate, *See* Mapeg® DGLD, 18640, PEG-2 laurate, 23041
Diethylene glycol monobutyl ether, *See* Butyl Carbitol®, 4766
Diethylene glycol monoethyl ether, *See* Carbitol®, 5194, Diethoxol, 8435, Dioxitol, 8576, Ektasolve® DE, DE-HG, 9664, Ethoxydiglycol, 11044, ethyl di-Icinol, 11068
diethylene glycol monolaurate, *See* Glaurin, 12978
diethylene glycol monomethyl ether, *See* Ektasolve® DM, 9666, Poly-Solv® DM, 24554, Transcutol, 32137
diethylene Glycol Mono-n-butyl Ether, *See* Butyl Carbitol®, 4766
diethylene glycol monooleate, *See* PEG-2 oleate, 23042
diethylene glycol monopropyl ether, *See* diethylene glycol propyl ether, 8445, Ektasolve® DP, 9667
diethylene glycol n-butyl ether, *See* Butyl Carbitol®, 4766, Dowanol® DB, 8890
diethylene glycol oleate, *See* Radiasurf® 7400, 25807
diethylene glycol propyl ether, 8445
diethylene glycol stearate, *See* Ablunol DEGMS, 99, PEG-2 stearate, 23043, Radiasurf® 7410, 25811
diethylene glycol, monobutyl ether, *See* Butyl Carbitol®, 4766
diethylene oxide, *See* 1,4-dioxane, 8574, Dioxane, 8574, tetrahydrofuran, 31370
diethylene oximide, *See* morpholine, 20346
diethylene triamine pentaacetic acid adduct, *See* Aroquest M Special, 3064
diethylenetriamine pentamethylene phosphonic acid,

See Unihib® 905, 32960
diethylenediamine, *See* Upixon, 33094
1,4-diethylene dioxide, *See* Dioxane, 8574
diethyleneglycol, *See* Her, 13798
1,4-Diethyleneoxide, *See* Dioxane, 8574
diethylenetriamine, 8446, *See* D.E.H. 20, 7610
Diethylenetriamine adduct, *See* D.E.H. 52, 7612
Diethylenetriamine-N,N,N',N'',N''-pentaacetic acid, *See* Aroquest M Special, 3064, Pentaquest OPAC 0201, 23170
Diethylenetriaminepentaacetic acid, *See* Aroquest M Special, 3064, Pentaquest OPAC 0201, 23170
Diethylenetriaminepentaacetic acid, pentasodium salt, *See* Cheelox® 80, 5864, Trilon® C Liq, 32328
Diethylenetriamine-pentakis(methylene phosphonic acid), *See* Briquest® 543-45AS, 4535
(Diethylenetrinitrilo)pentaacetic acid, pentasodium salt, *See* Cheelox® 80, 5864, Trilon® C Liq, 32328
diethylenimide oxide, *See* morpholine, 20346
diethylethanolamine, *See* 2-diethylaminoethanol, 8440
2,2-Diethyl-ethanolamine, *See* 2-amino-2-methyl-1 propanol, 2182
N,N-diethylethanolamine, *See* 2-diethylaminoethanol, 8440
O,O-diethyl S-[(ethylthio)methyl]phosphorodithioate, *See* Terrathion, 31343
2,5-Di(2-ethylhexanoylperoxy)-2,5-dimethylhexane, *See* USP® -245, 33166
Di-(2-ethylhexyl)adipate, *See* Kodaflex® DOA, 16272, Wickenol® 158, 34393
Di(2-ethylhexyl) orthophthalate, *See* Witcizer 312, 34454
Diethylhexyl peroxy dicarbonate, *See* Espercarb® 840, 10841
Diethylhexyl phthalate, *See* Bis(2-ethylhexy) Phthalate, 4228, Reomol DCP, 26130
di(2-ethylhexyl phthalate, *See* Bis(2-ethylhexy) Phthalate, 4228
Di(ethylhexyl)azelate, *See* Plasthall® DOZ, 24007
di(2-ethylhexyl)orthophthalate, *See* Reomol DCP, 26130
di-(2-ethylhexyl)sebacate, *See* Uniflex® DOS, 32946
N,N-diethyl-2-hydroxyethylamine, *See* 2-diethylaminoethanol, 8440
N,N-diethyl-N-(β-hydroxyethyl)amine, *See* 2-diethylaminoethanol, 8440
N,N-Diethylhydroxylamine, *See* Pennstop® 1866, 23118
diethylin, 8447
N,N-Diethyl-3-methylbenzamide, *See* diethyl toluamide, 8438
O,O-diethyl-O-(6-methyl-2-(1-methylethyl)-4-pyrimidinyl)phosphorothioate, *See* Diazinon Liquid, 8344
o-dimethoxybenzene, *See* veratrole, 33568
Diethylol dimethyl hydantoin, *See* DEDM hydantoin, 7872
Diethylol dimethyl hydantoin dilaurate, *See* DEDM hydantoin dilaurate, 7873
Diethylolamine, *See* Diethanolamine, 8434
N,N-Diethylollauramide, *See* Witcamide® 5195, 34440
N-(2,6-Diethylphenylcarbamoylmethyl)iminodiacetic acid, *See* EHIDA Kit, 9649
O,O-diethyl O-(1-phenyl-1H-1,2,4-triazol-3-yl) phosphorothioate, *See* triazophos, 32196
O,O-Diethyl phosphoroamidothioate, *See* Depat, 8074
O,O-Diethyl phosphorochlorodithioate, *See* EP-2, 10612
O,O-diethyl phosphorodithioate S-ester with 6-chloro-3-(mercaptomethyl)-2-benzoxazolinone, *See* Zolone, 34888
O,O-Diethyl phosphorodithioic acid, *See* EP-1, 10611
Diethylphosphorodithioate, *See* EP-1, 10611
O,O-diethylphosphorodithioate, *See* EP-1, 10611
O-O-diethylphosphorodithioic acid, *See* EP-1, 10611
N,N-diethyl-N-[2-[4-(1,1,3,3-tetramethyl butyl) phenoxy]ethyl]benzenemethanaminium chloride, *See* octaphen, 21989
N,N'-diethylthiocarbamide, *See* Pennzone E 0686, 23121
diethylthiourea, *See* Pennzone E 0686, 23121
1,3-diethyl-2-thiourea, *See* Pennzone E 0686, 23121
1,3-Diethylthiourea, *See* Pennzone E 0686, 23121
N,N-diethyl-*m*-toluamide, *See* diethyl toluamide, 8438
O,O-diethyl-O-3,5,6-trichloro-2-pyridyl phosphorothioate, *See* Talon, 30751
N,N-Diethyltrimethylsilylamine, *See* CD4450, 5504
N,N-Diethyl-1,1,1-trimethylsilylamine, *See* CD4450, 5504
Di-Farmon, 8448
2-difarnesyl-3-methyl-1,4-naphthoquinone, *See* vitamin K+72, 33950
difenacoum, 8449
Difenacoum, *See* Neosorexa, 20967, Ratak, 25896
Difenakum, *See* Ratak, 25896
Difenphos, *See* Abate 1-SG, 2-CG, 4-E, 5CG, 26
Difentan, *See* Nuophene, 21783
Difenzoquat methyl sulfate, *See* Avenge 2, 3460
Diffu-K, *See* K. Tab, 15642, Kaon-Cl, 15729, Kay Ciel, 15807, K-Contin, 15818, K-Lor, 16226, K-Lyte/C1, 16234, KM Potassium Chloride, 16242, potassium chloride, 24749
Diflon S-3, *See* MS-180 Freon® TF Solv, 20415
Diflubenzuron, 8450, *See* Dimilin, 8529
Diflufenican, 8451
diflufenicanil, *See* Diflufenican, 8451
diflufenican-isoproturon, *See* Javelin, 15540
difluorodichloromethane, *See* dichlorodifluoromethane, 8388
difluoromethane-pentafluoroethane, *See* Genetron® Refrigerant 32/125, 12830
difluorophenyltrifluoromethyl N-(2,4-difluorophenyl)-2-[3-(trifluoromethyl)phenoxy]-3-pyridine carboxamide, *See* Diflufenican, 8451
Difluron, *See* Diflubenzuron, 8450
Difolatan, 8452, *See* captafol, 5163
diformyl, 8453
1,3-diformylpropane, *See* glutaraldehyde, 13062, Sonacide, 29073, Ucarcide® 225, 32711
N,N'-difurfurylidene-2-furanmethanediamine, *See* Vulcazol, 34112
digallic acid, *See* tannic acid, 30807
Digermin, *See* Trifluralin, 32252
Diglyceryl diisostearate, *See* Emalex DISG-2, 9943
Diglyceryl ether tetracetate, *See* Glyakol, 13066
Diglyceryl monostearate, *See* Emalex MSG-2, 9970, Polyglyceryl stearate series, 24457
diglycine, *See* N-glycylglycine, 13139
diglycol, See Diethoxol, 8435, diethylene glycol, 8442, ethyl di-Icinol, 11068
diglycol laurate, *See* PEG-2 laurate, 23041
diglycol methyl ether, *See* diglyme, 8455
diglycol monobutyl ether, *See* Butyl Carbitol®, 4766
Diglycol monobutyl ether acetate, *See* diethylene glycol butyl ether acetate, 8443
Diglycol monoethyl ether, *See* Diethoxol, 8435, Ethoxydiglycol, 11044, ethyl di-Icinol, 11068
diglycol oleate, *See* PEG-2 oleate, 23042
diglycol stearate, *See* PEG-2 stearate, 23043
Diglycolamine® Agent (DGA®), 8454
diglyme, 8455
Dihalo, 8456
Dihexyl phthalate, *See* Jayflex® DHP, 15549
Dihexyl sodium sulfosuccinate, *See* Aerosol® MA-80, 893, Bis(1-methylamyl) Sodium Sulfosuccinate, 4227, Monawet MM-80, 20180
dihydro-2(3H)-furanone, *See* butyrolactone, 4803
dihydro-2,5-diketotetrahydrofuran, *See* succinic anhydride, 29916
dihydro-2,5-dioxofuran, *See* maleic anhydride, 18476
dihydro-2,5-furandione, *See* succinic anhydride, 29916
22,23-Dihydroabamectin, *See* Ivermectin, 15454
Dihydroabietyl alcohol, *See* Abitol® E, 71
22,23-dihydroavermectin B_1, *See* Ivermectin, 15454
1,3-dihydro-2H-benzimidazole-2-thione, *See* Anti-Oxidant MB, 2589
dihydrobis(2-methoxyethanolato-O,O')aluminate(1-) sodium, *See* Vitride®, 33994
dihydrocoumaranyl2,2-Dimethyl-7-coumaranyl N-methylcarbamate, *See* Rampart, 25860
10,11-dihydro-5H-dibenzo[a,d] cyclohepten-5-one, *See* Dibenzosuberone, 8354
10,11-Dihydrodibenzo[a,d]cyclohepten-5-one, *See* Dibenzosuberone, 8354
dihydrodicyclopentadienyl acrylate, *See* Sipomer® DCPA, 28533
9,10-dihydro-4,5-dihydroxy-9,10-dioxo-2-anthracenecarboxylic acid, *See* rhein, 26534
2,3-Dihydro-2,2-dimethyl-7-benzofuranyl methylcarbamate, *See* Rampart, 25860
2,3-dihydro-2,2-dimethyl-7-benzofuranyl [(dibutylamino)thio]methyl carbamate, *See* carbosulfan, 5245
5,10 Dihydro-5,10-dioxonaphtho[2-3-b]-1,4-dithiin-2,3-dicarbonitrile, *See* Dithianone, 8729
6,7-Dihydrodipyrido[1,2-a:2',1'-c]pyrazinediium dibromide, *See* Katalon, 15782
22,23-dihydroergocalciferol, *See* vitamin D_4, 33942
dihydrogen dioxide, *See* hydrogen peroxide, 14590
Dihydrogenated tallow amine, *See* Radiamine 6240, 25758
N,N-Di(hydrogenated tallow) amine, *See* Amine 2HBG, 2132
Dihydrogenated tallow benzylmonium chloride, *See* Arquad® M2HTB-80, 3119
Dihydrogenated tallow dimethyl ammonium chloride, *See* Adogen® 442-P100, 770, Querton 442, 25654, Radiaquat 6475, 6480, 25796
Dihydrogenated tallow dimethyl ammonium methosulfate, *See* Ablumine DHT75, 87
Dihydrogenated tallow dimethyl ammonium methyl sulfate, *See* Adogen® 137, 764
Dihydrogenated tallow methylamine, *See* Amine M2HBG, 2161, Armeen® M2HT, 2955, Kemamine® T-9701, 15931, Radiamine 6343, 25762
Dihydrogenated tallow phthalic acid amide, *See* Stepan TAB® -2, 29670
Dihydrogenated tallowamidoethyl hydroxyethylmonium methosulfate, *See* Varisoft® 110, 33404
Dihydrogenated tallowamine, *See* Amine2 VT, 2163
4,5-dihydroimidazole-2(3H)-thione, *See* Perkacit® ETU, 23283
1,6-dihydro-6-iminopurine, *See* Adenine, 679
3,6-dihydro-6-iminopurine, *See* Adenine, 679
dihydroisophorone, 8457
dihydrojasmone, 8458
1,2-dihydro-2-ketobenzisosulfonazole, *See* saccharin, 27195
4,5-dihydro-2-methyl-1H-imidazole, *See* lysidine, 18087
2-[4,5-dihydro-4-methyl-4-(1-methylethyl)-5-oxo-1H-imidazol-2-yl]-3-pyridine carboxylic acid, *See* Arsenal®, 3134, imazapyr, 14893
5,6-dihydro-2-methyl-N-phenyl-1,4-Oxathiin-3-carboxamide 4,4-dioxide, *See* oxycarboxin, 22441, Ringmaster, 26771
Dihydro-2-methyl-1,4-oxathiin-3-carboxanilide 4,4-dioxide, *See* Ringmaster, 26771
Dihydrooxirene, *See* ethylene oxide, 11131
2,3-dihydro-3-oxobenzisosulfonazole, *See* saccharin, 27195
2-(1,3-Dihydro-3-oxo-2H-indol-2-ylidene)-1,2-dihydro-3H-indol-3-one, *See* Indigo, 15074
4,5-dihydro-5-oxo-1-(4-sulfophenyl)-4-[(4-sulfophenyl)azo]-1H-pyrazole-3-carboxylic acid trisodium salt, *See* tartrazine, 30835
2,5-Dihydroperoxy-2,5-dimethylhexane, *See* Luperox 2,5-2,5, 17839
3,7-dihydro-1H-purine-2,6-dione, *See* xanthine, 34590
7,9-dihydro-1H-purine-2,6,8(3H)-trione, *See* uric acid, 33139
1,7-dihydro-6H-purin-6-one, *See* hypoxanthine, 14721
1,2-dihydropyridazine-3,6-dione, *See* maleic hydrazide, 18477
1,2-Dihydro-3,6-pyridazinedione, *See* maleic hydrazide, 18477
[2R-2R*(4R*,8R*)]-3,4-dihydro-2,5,7,8-tetramethyl-2-

(4,8,12-trimethyltridecyl)-2H-1-benzopyran-6-ol, *See* vitamin E, 33943
3,4-Dihydro-2,5,7,8-tetramethyl-2-(4,8,12-trimethyltridecyl)-2H-1-benzopyran-6-ol, *See* tocopherol, 31918
3,4-Dihydro-2,5,7,8-tetramethyl-2-(4,8,12-trimethyltridecyl)-2H-1-benzopyran-6-ol, *See* Vascuals, 33465
1,2-dihydro-2,2,4-trimethylquinoline homopolymer, *See* Agerite® MA, 1019
1,2-dihydro-2,2,4-trimethylquinoline homopolymer, *See* Akrochem® Antioxidant DQ, 1160
Dihydrotrimethylquinoline polymer, *See* Akrochem® Antioxidant DQ, 1160
22:23-dihydrovitamin D_2, *See* vitamin D_4, 33942
Dihydroxy diphenyl sulfonates, *See* Eltesol® 7200 Series, 9862
Dihydroxyacetone, 8459
2',4'-dihydroxyacetophenone, *See* resacetophenone, 26152
2,4-Dihydroxyacetophenone, *See* resacetophenone, 26152
1,8-Dihydroxyanthracene, *See* chrysazol, 6251
1,5-dihydroxy-9,10-anthracenedione, *See* anthrarufin, 2533
1,4-dihydroxy-9,10-anthracenedione, *See* quinizarin, 25683
Dihydroxyanthracoumarin, *See* styrogallol, 29874
1,2-dihydroxyanthraquinone, *See* Alizarin, 1621
1,4-dihydroxyanthraquinone, *See* quinizarin, 25683
1,5-Dihydroxy-anthraquinone, *See* anthrarufin, 2533
1,8-dihydroxyanthraquinone-3-carboxylic acid, *See* rhein, 26534
2,3-Dihydroxyanthraquinone, *See* Histazarin, 14136
2,6-Dihydroxyanthraquinone, *See* anthraflavic acid, 2526
2,7-Dihydroxyanthraquinone, *See* isoanthraflavic acid, 15337
4,5-dihydroxy-2-anthraquinonecarboxylic acid, *See* rhein, 26534
Dihydroxyanthraquinonecarboxylic acid, *See* Munjistin, 20499
1,2-dihydroxybenzene, *See* pyrocatechol, 25498
2,4-dihydroxybenzophenone, *See* benzophenone-1,
3,4-dihydroxybenzoic acid, *See* protocatechuic acid, 25289
***m*-dihydroxybenzene**, *See* resorcin, 26228, resorcinol, 26230
***p*-dihydroxybenzene**, *See* hydroquinone, 14632, quinol, 25685
1,2-dihydroxybenzene, *See* kachin, 15651
1,3-dihydroxybenzene, *See* resorcin, 26228, resorcinol, 26230
3,4-dihydroxybenzoic acid, *See* protocatechuic acid, 25289
2,4-dihydroxybenzophenone, *See* Benzoresorcinol, 3981, Uvinul® 400, 33191
7,8-dihydroxy-2H-1-benzopyran-2-one, *See* Daphnetin, 7736
1,4-Dihydroxybutane, *See* 1,4-butanediol, 4735, Dabco® BDO, 7647
DL-2,3-dihydroxybutanedioic acid, *See* tartaric acid, 30832
1,8-dihydroxy-3-carboxyanthraquinone, *See* rhein, 26534
1,8-Dihydroxycoumarin, *See* Daphnetin, 7736
7,8-Dihydroxycoumarin, *See* Daphnetin, 7736
dihydroxydiethyl ether, *See* diethylene glycol, 8442
2,2'-dihydroxy-4,4'-dimethoxybenzophenone, *See* benzophenone-6, 3978, Uvinul® D-49, 33194
1,3-dihydroxydimethyl ketone, *See* Dihydroxyacetone, 8459
(R)-N-(2,4-dihydroxy-3,3-dimethyl-1-oxobutyl)- β-alanine calcium salt, *See* vitamin B+75 calcium salt, 33936
2,2-(4,4-dihydroxydiphenyl)propane, *See* Bisphenol A, 4283
4,4'-dihydroxdiphenylpropane, *See* Bisphenol A, 4283
4,4'-dihydroxy-2,2-diphenylpropane, *See* Bisphenol A, 4283
4,4'-dihydroxydiphenyl-2,2-propane, *See* Bisphenol A, 4283
Dihydroxydiphenylsilane, *See* CD6150, 5516
Dihydroxyethyl cocamine oxide, *See* Aromox® C/12-W, 3047, Schercamox CMA, 27573
dihydroxyethyl sulfide, *See* thiodiglycol, 31682
Dihydroxyethyl tallow glycinate, *See* Chembetaine TG, 5937, Mackam TM, 18150, Mirataine® TM, 19932, Monateric 1202, 20150, Rewoteric® AM TEG, 26427
Dihydroxyethyl tallowamine oxide, *See* Chemoxide T, 6001
Di-(2-hydroxyethyl)-5,5-dimethyl hydantoin, *See* DEDM hydantoin, 7872
Di-(2-hydroxyethyl)-5,5-dimethyl hydantoin dilaurate, *See* DEDM hydantoin dilaurate, 7873
Dihydroxyethyltallow glycinate, *See* Zohartaine TM, 34878
3',6'-dihydroxyfluoran, *See* uranine, 33115
1,6-dihydroxyhexane, *See* hexamethylene glycol, 14005
2,2'-dihydroxy-3,5,6,3',5',6'-hexachloro diphenylmethane, *See* hexachlorophene, 13992
Dihydroxy-3,3',5,5',6,6'-hexachlorodiphenylmethane, *See* hexachlorophene, 13992
1,6-dihydroxyhexane, *See* Hexamethylene Glycol, 14005
(S)-5,8-dihydroxy-2-(1-hydroxy-4-methyl-3-pentenyl)-1,4-naphthalenedione, *See* Alkanet, 1695
2,4-dihydroxy-N-(3-hydroxypropyl)-3,3-dimethylbutanamide, *See* d-Panthenol, 22690
2,2-dihydroxy-1H-indene-1,3(2H)-dione, *See* ninhydrin, 21230
2,2'-Dihydroxy-4-methoxybenzophenone, *See* Cyasorb® UV 24, 7496
1,3-dihydroxy-5-methylbenzene, *See* orcinol, 22280
1,1-Di(hydroxymethyl)ethylamine, *See* 2-amino-2-methyl-1,3-propanediol, 2183
1,3-dihydroxy-2-methyl-2-propylamine, *See* 2-amino-2-methyl-1,3-propanediol, 2183
2,4-dihydroxy-5-methylpyrimidine, *See* thymine, 31777
1,3-dihydroxynaphthalene, *See* naphthoresorcin, 20726
1,6-dihydroxynaphthalene, *See* 1,6-naphthalenediol, 20709
1,8-Dihydroxynaphthalene-3,6-disulfonic acid, *See* chromotrope acid, 6244
2,3-dihydroxynaphthalene, *See* 2,3-naphthalenediol, 20710
2,7-dihydroxynaphthalene, *See* 2,7-naphthalenediol, 20711
5,8-dihydroxy-1,4-naphthalenedione, *See* naphthazarin, 20713
4,5-Dihydroxynaphthalene-2,7-disulfonic acid, *See* chromotrope acid, 6244
5,8-dihydroxy-1,4 naphthoquinone, *See* naphthazarin, 20713
β-di-*p*-hydroxyphenylpropane, *See* Bisphenol A, 4283
2-(3,4-dihydroxyphenyl)-4H-1-benzopyran-4-one, *See* quercetin, 25647
(2,4-dihydroxyphenyl)phenylmethanone, *See* Uvinul® 400, 33191
2-(3,4-dihydroxyphenyl)-3,5,7-trihydroxy-4H-1-benzopyran-4-one, *See* quercetin, 25647
(3,4-dihydroxyphenyl)(2,4,6-trihydroxyphenyl)-methanone, *See* maclurin, 18235
1,2-dihydroxypropane, *See* propylene glycol, 25214, 25215
1,3-dihydroxy-2-propanone, *See* dihydroxyacetone, 8459
2,3-dihydroxypropanethiol, *See* Thiovanol®, 31723
2,3-Dihydroxypropyl dodecanoate, *See* Aldo® MLD, 1540, glyceryl monolaurate, 13083, Grindtek ML 90, 13410
2,6-dihydroxypurine, *See* xanthine, 34590
2,4-dihydroxypyrimidine, *See* uracil, 33101
3',6'-dihydroxyspiro[isobenzofuran-1(3H), 9'[9H]xanthen]-3-one, *See* Fluorescein, 12080, uranine, 33115
3',6'-Dihydroxyspiro[isobenzofuran-1(3H),9'-[9H]xanthen]-3-one disodium, *See* Fluorescein disodium salt, 12081
dihydroxysuccinic acid, *See* tartaric acid, 30832
Dihydroxysulfonaphthoic acid, *See* nigrotic acid, 21179
dihydroxytoluene, *See* orcinol, 22280
2,5-dihydroxytoluene, *See* toluhydroquinone, 31951
3,5-dihydroxytoluene, *See* orcinol, 22280
2,4-dihydroxy-1,3,5-trinitrobenzene, *See* styphnic acid, 29858
Diiner, *See* Zineb, 34807
Diiodo-*p*-phenolsulfonic acid, *See* Iodozol, 15221
α-diiodohydrin, *See* iohydrin, 15223
1,3-diiodo-2-hydroxypropane, *See* iohydrin, 15223
diiodohydroxyquin, *See* Yodoxin, 34658
1,3-diiodoisopropyl alcohol, *See* iohydrin, 15223
diiodomethane, *See* methylene iodide, 19569
Diiodomethyl *p*-tolyl sulfone, *See* Amical® 48, 2058
1-((diiodomethyl)sulfonyl)-4-methylbenzene, *See* Amical® 48, 2058
3,5-diiodo-4-oxo-1(4H)-pyridineacetic acid propyl ester, *See* Propyliodone, 25226
diiodo-oxyquinoline, *See* Yodoxin, 34658
3,5-Diiodo-4(1H)-pyridinone, *See* Iopydone, 15235
3,5-Diiodo-4-pyridone, *See* Iopydone, 15235
5,7-diiodo-8-quinolinol, *See* Yodoxin, 34658
3,5-diiodosalicylic acid, *See* Diosal, 8572
diiron trioxide, *See* Ferroxide, 11694, Jeweller's rouge, 15611
Diisoarachidyl dilinoleate, *See* Liquiwax DIEFA, 17489
Diisoarachidyl dodecanedioate, *See* Liquiwax DIADD, 17487
diisobutyl adipate, 8460, *See* Plasthall® DIBA, 23996
Diisobutyl azelate, *See* Plasthall® DIBZ, 23997
diisobutyl hexanedioate, *See* diisobutyl adipate, 8460
diisobutyl ketone, 8461, *See* valerone, 33243
Diisobutyl maleate, *See* Octomer DIBM, 22009
Diisobutyl nonyl phenol, *See* Uvi-Nox 1494, 33189
Diisobutyl phthalate, *See* Kodaflex® DIBP, 16267, Palatinol 1C, 22580, Uniplex 155, 33011
Diisobutyl Sodium Sulfosuccinate, *See* Bevaloid 6423, 4083, Monawet MB-45, 20178
Di-isobutylphenol-formaldehyde, *See* Albertol 237-R, 1371
Diisocetyl adipate, *See* Schercemol DICA, 27593
Diisocetyl dodecanedioate, *See* Liquiwax DICDD, 17488
2,4-diisocyanato-1-methylbenzene, *See* Voranate T-80, Type I, Type II, 34042
4,4'-diisocyanatodiphenylmethane, *See* Rubinate® LF-168, 27108
diisocyanatodiphenylmethane, *See* Rubinate® LF-168, 27108
2,4-diisocyanatotoluene, *See* toluene diisocyanate, 31947, Voranate T-80, Type I, Type II, 34042
Diisodecyl adipate, *See* Kodaflex® DIDA, 16268, Monoplex® DDA, 20245, Plasthall® DIDA, 23998
Diisodecyl glutarate, *See* Plasthall® DIDG, 23999
Diisodecyl pentaerythritol diphosphite, *See* Weston® 600, 34303
diisodecyl phthalate, 8462, *See* Bisoflex BP9, 4255, Jayflex® DIDP, 15550, Kodaflex® DIDP, 16269, Palatinol® Z, 22596, PX-120, 25436
Diisodecyl sodium sulfosuccinate, *See* Rewopol® SBDD 65, 26373
Diisodecylphthalate, *See* NLA-30, 21397, Palatinol® DIDP, 22588
Diisoheptyl phthalate, *See* Jayflex® 77, 15543
Diisohexyl sulfosuccinate, *See* Rewopol® SBMB 80, 26378
diisononyl adipate, *See* Adimoll® DN, 692, Jayflex® DINA, 15551, PX-209, 25439
diisononyl phthalate, 8463 *See* Jayflex® DINP, 15552, Jayflex® DIOP, 15553, Palatinol® DN, 22589, Palatinol® N, 22594, PX-109, 25433
Diisonoyl adipate, *See* Monoplex® DIOA, 20246, Plasthall® DIOA, 24000, Plastomoll® NA, 24059
Diisooctyl dodecanedioate, *See* Plasthall® DIODD, 24001
diisooctyl hydrogen phosphite, 8464

Diisooctyl maleate, *See* Octomer DIOM, 22010
Diisooctyl octylphenyl phosphite, *See* Weston® 494, 34302
diisooctyl phosphite, 8465, *See* Doverphos® DIOP, 8846, Weston® DOPI, 34307
Diisooctyl phthalate, *See* Palatinol® D10, 22586, Witcizer 313, 34455
Diisooctyl sulfosuccinate, *See* Triumphnetzer ZSG, 32435
diisopropanolamine, 8466
Diisopropyl adipate, *See* Schercemol DIA, 27592, Unimate® DIPA, 32968
diisopropyl amine N,N-diisopropyl amine, *See* diisopropylamine, 8469
1,3-diisopropyl benzene, 8467, *See* 1,3-diisopropyl benzene, 8467
Diisopropyl dimer dilinoleate, *See* Bernel® Ester DID, 4005, Schercemol DID, 27594
Diisopropyl naphthalene, *See* KMC, 16247
Diisopropyl sebacate, *See* Schercemol DIS, 27595, Unimate® DIPS, 32969
diisopropylamine, 8469
***m*-diisopropylbenzene**, *See* 1,3-diisopropyl benzene, 8467
***p*-diisopropylbenzene**, *See* 1,4-diisopropylbenzene, 8468
1,4-diisopropylbenzene, 8468
Diisopropylbenzene monohydroperoxide, *See* Trigonox® M-50, 32305
diisopropylbenzene peroxide, *See* Aztec® DCP-R, 3556, dicumyl peroxide, 8407, Percumyl D, 23230, Peroximon® DC-40, 23409, Polyvel PCL-20, 24661
Diisopropylbiphenyl, *See* Nusolv ABP-103, 21801
Diisopropylnaphthalenesulfonic acid, sodium salt, *See* Aerosol® OS, 895
diisostearyl dimer dilinoleate, 8470, *See* Dermol DISD, 8123, Schercemol DISD, 27596
Diisostearyl fumarate, *See* Schercemol DISF, 27597
diisotridecyl hydrogen phosphite, 8471
Dikar, 8472
Dikegulac, *See* Cutlass, 7442
Dikegulac sodium, *See* Atrinal, 3395
2,3-diketobutane, *See* Diacetyl, 8250
Diketone Alcohol, *See* Diacetone alcohol, 8249
2,5-diketopyrrolidine, *See* succinimide, 29917
Dikopan, *See* Basfapon®, 3705
Dikotex, *See* MCPA, 19183
Dilasoft, 8473
Dilatin NA Liquid, 8474
dilauroyl peroxide, *See* Alperox-F, 1816, Laurox®, 16859
Dilauryl acetyl dimonium chloride, *See* Schercoquat ALA, 27671
Dilauryl ester, *See* dilauryl thiodipropionate, 8476
Dilauryl hydrogen phosphite, *See* dilauryl phosphite, 8475, Doverphos® 274, 8844, Weston® DLP, 34306
Dilauryl Peroxide, *See* Alperox-F, 1816
Dilauryl phosphite, 8475, *See* Weston® DLP, 34306
dilauryl thiodipropionate, 8476, *See* Aquastab PA 48, 2739, Argus DLTDP, 2861, Carstab® DLTDP, 5356, Cyanox® LTDP, 7491, Evanstab® 12, 11255, Lankromark® DLTDP, 16689, Wytox® LT, 34582
Dilauryl-3-3'-thiodipropionate, *See* Lowinox® DLTDP, 17666
Dilectene, 8477
Dilecto, 8478
Dilexo, 8479
Dilinoleic acid, 8480
Dilinoyl dimethicone copolyol citrate, *See* Pecosil DAS 36, 23004
Dillex, 8481
diloxanide, *See* Entamide, 10593
Diluex®, FG, 8482, *See* Attapulgite, 3404
Dilurgen, *See* meralluride, 19329
Dilver, 8483
Dimacide, 8484
dimagnesium orthophosphate, *See* magnesium phosphate, dibasic,
1,3-dimaleimidobenzene, *See* Vanax® MBM, 33291
Dimanin, 8485
Dimanin A Special, 8486
dimanthine, *See* Armeen® DM18D, 2944
Dimantine, *See* Crodamine 3.A18D, 7107, Dymanthine, 9327, Kemamine® T-9902, 15933
Dimas, *See* daminozide, 7723
Dimatos, *See* infusorial earth, 15125
Dimazon, 8487, *See* Diacetazotol, 8247
Dimdac, 8488
Dimefilcon A, *See* Gelflex, 12734
dimefuron, *See* Pradone Plus, 24850
Dimeglumine gadopentetate, *See* Magnevist, 18376
Dimension, 8489
Dimension D-9300 BK, 8490
Dimension Master®, 8491
Dimension®, 8492
Dimer acid, *See* Empol® 1010, 10445, Empol® 1061, 10447, Pripol 1009, 25030, dilinoleic acid, 8480
Dimer acids, *See* Craymer, 6971
(R*,S*)-2,3-dimercaptobutanedioic acid, *See* MPI DMSA Kidney Reagent, 20393
(R*,R*)-1,4-Dimercapto-2,3-butanediol, *See* 1,4-Dithiothreitol, 8730
Dimercuridiammonium chloride, *See* infusible white precipitate, 15124
Dimethicone, 8493, *See* Abil®10-10000, 67, Belsil DM 0.65, 3917, CPS 034, 6945, CPS 340, 6950, Dow Corning® 200 Fluid, 8859, Foamkill® 810F, 12170, Masil® SF 1,000,000, 19017, Masil® SF 5, 19018, Masil® SF 500, 19020, Masil® SF 500,000, 19021, PS034, 25359, Sentry Dimethicone, 27959, Viscasil®, 33893, AF 10 FG, 907, Vasogen®, 33468
Dimethicone aqueous emulsion, *See* AF 9020, 914
Dimethicone copolyol, 8494, *See* Abil® B 88183, 88184, 51, Abil® B 8851, 8852, 49, Alkasil® NE58-50, 1709, Belsil DMC 6031, 3918, Dow Corning® 190 Surfactant, 8857, SF1188, 28129, Silwet® L-720, 28451, Silwet® L-7500, 28453, Silwet® L-7600, 28454, Silwet® L-7605, 28455, Tego® Foamex 3062, 30977, Tego® Glide 100, 30980, Tego® Glide 410, 30981, Tego® WetKL 245, 30992
Dimethicone copolyol adipate, *See* Fancorsil SLA, 11461
Dimethicone copolyol phosphate, *See* Pecosil PS-100, 23012
Dimethicone propyl PG-betaine, *See* Abil® B 9950, 50
Dimethicone propylethylenediamine behenate, *See* Pecosil OS-100B, 23006
Dimethicone propylethylenediamine dilinoleate, *See* Pecosil OS-100DA, 23007
Dimethicone propylethylenediamine hydroxystearate, *See* Pecosil OS-100HS, 23008
Dimethicone propylethylenediamine laurate, *See* Pecosil OS-100L, 23009
Dimethicone propylethylenediamine myristate, *See* Pecosil OS-100M, 23010
Dimethicone propylethylenediamine undecylenate, *See* Pecosil OS-100U, 23011
Dimethicone trimethylsiloxysilicate, *See* Dow Corning® QF1-3593A, 8877
Dimethiconol, *See* Masil® SFR 70, 19023
Dimethicream, *See* Sentry Dimethicone, 27959
dimethirimol, 8495
Dimethoate, 8496, *See* Perfekthion®, 23244, Rogor E, 26908, Systemic Insecticide, 30664, Turbair, 32595
Dimethoate Bayer, 8497
dimethoate-met, *See* Omethoate, 22155
dimethoxane, 8498, *See* Givgard DXN, 12951
Dimethoxy neotetrazolium, *See* tetrazolium chloride, 31395
2,6-Dimethoxy phenol, *See* dimethoxyphenol, 8500
dimethoxy tetrahydrofuran, 8499
2,5-dimethoxy tetrahydrofuran, *See* dimethoxy Tetrahydrofuran, 8499
Dimethoxy-4-(2-propenyl)benzene, *See* methyl eugenol, 19534
1,4-Dimethoxyacetylene, *See* but-2-yne-1,4-diol, 4725
3,4-Dimethoxyallyl benzene, *See* methyl eugenol, 19534
4,4'-di(methoxy)-azobenzene, *See* Oxymaster, 22460
1,2-dimethoxybenzene, *See* veratrole, 33568
1,3-dimethoxybenzene, *See* veratrole, 33568
3,3'-(3,3'-Dimethoxy-4,4'-biphenylene)bis(2,5-diphenyl-2H-tetrazolium chloride), *See* tetrazolium chloride, 31395
dimethoxydimethylsilane, *See* CD5605, 5509
Dimethoxyethane, *See* monoglyme, 20227
α,β-dimethoxyethane, *See* monoglyme, 20227
1,2-Dimethoxyethane, *See* Ansol E-121, 2494
dimethoxymethane, *See* methylal, 19564
dimethoxyphenol, 8500
2,2-Dimethoxy-2-phenylacetophenone, *See* Photocure 51, 23757
5-[(3,4-dimethoxyphenyl)methyl]-2,4-pyrimidine diamine, *See* diaveridine, 8338
dimethoxypropane, 8501
2,2-Dimethoxypropane, *See* dimethoxypropane, 8501, DMP, 8757
4,7-dimethoxy-5-(2-propenyl)-1,3-benzodioxole, *See* parsley camphor, 22865
Dimethoxytetrahydrofuran, *See* dimethoxy Tetrahydrofuran, 8499, Protectol® DMT, 25260
Dimethyl myristamine, *See* Amine 2M14D, 2137
dimethyl (2,2,2-trichloro-1-hydroxyethyl)phosphonate, *See* trichlorfon, 32212
dimethyl 1,2-benzenedicarboxylate, *See* dimethyl phthalate, 8508, Kemester® DMP, 15956, Kodaflex® DMP, 16271, Unimoll® DM, 32979
Dimethyl 1,3-benzenedicarboxylate, *See* Morflex 1129, 20330
Dimethyl 2,2'-azobis (2-methylpropionate), *See* V-601, 33219
dimethyl adipate, 8502
dimethyl anthranilate, 8503
dimethyl azelate, 8504, *See* Emery® 2914, 10158
dimethyl behenamine, *See* Adogen® MA-112 SF, 777
dimethyl benzene-1,3-dicarboxylate, *See* Uniplex 270, 33016
dimethyl 1,2-benzene-dicarboxylate, *See* dimethyl phthalate, 8508
C_{14} dimethyl benzyl ammonium chloride, *See* JAQ Powdered Quaternary, 15517
Dimethyl benzyl lauryl ammonium chloride, *See* Dehyquart LDB, 7986
Dimethyl benzylamine, *See* Dabco® B-16, 7646, Pentamin BDMA etc, 23163
dimethyl benzyloctadecylammonium chloride, *See* Cycloton® SCS, 7542
dimethyl bis(*p*-hydroxyphenyl)methane, *See* Bisphenol A, 4283
dimethyl butanedioate, *See* dimethyl succinate, 8509
dimethyl butylphenyl N-(1,3-dimethyl butyl)-N'-phenyl-p-phenylene diamine, *See* Akrochem® Antiozonant PD-2, 1164, Vulkanox® 4020, 34148
N,N-Dimethyl caprylamidecapramide, *See* Hallcomid® M-8-10, 13552
dimethyl carbinol, *See* isopropanol, 15405
Dimethyl Carbitol, *See* diglyme, 8455
Dimethyl Cellosolve, *See* monoglyme, 20227
O,O-Dimethyl chlorophosphorothioate, *See* MP-2, 20388
Dimethyl chlorothiophosphate, *See* MP-2, 20388
Dimethyl cocamine, *See* Armeen® DMCD, 2945
N,N-dimethyl cocamine, *See* Amine 2M1218D, 2134
Dimethyl cocoamine, *See* Amine 2MKKD, 2143, Jet Amine DMCD, 15585
Dimethyl diallyl ammonium chloride, *See* Ageflex mDMDAC, 993, Agefloc PC20HV, 1008, Agequat C505, 1013, Agestat 41, 1028
Dimethyl dicarbonate, *See* Velcorin®, 33531
O,O-dimethyl O-(2,2,-dichlorovinyl) phosphate, *See* Vapona, 33378
Dimethyl dicoconut ammonium chloride, *See* Kemamine® Q-6502C, 15924
O,O-Dimethyl-S-(1,2-di(ethoxycarbonyl)ethyl) phosphorodithioate, *See* Fyfanon, 12533
dimethyl diketone, *See* Diacetyl, 8250
dimethyl distearyl ammonium chloride, *See* Adogen®

TA-100, 779, Prepagen WK, 24894, Sumquat® 6045, 30123
Dimethyl dithiocarbamate dimethylammonium salt, *See* Ultra-DMC, 32791
N,N-Dimethyl dodecyl tetradecylamine, *See* Amine 2M1214D, 2133
Dimethyl erucylamine, *See* Crodamine 3.AED, 7108
Dimethyl ester of tetrachloroterephthalic acid, *See* dimethyl tetrachloroterephthalate, 8512
Dimethyl ethanolamine, *See* Alkanolamine, 1701
dimethyl ether, 8505, *See* methyl ether, 19529
Dimethylethylethylmethylthio N-(1,1-Dimethylethyl)-N'-ethyl-6-(methylthio)-1,3,5-triazine-2,4-diamine, *See* terbutryn, 31230
dimethyl ethynyl carbinol, *See* methyl butynol, 19518
dimethyl formamide, 8506
dimethyl glyoxime, *See* dimethylglyoxime, 8523
dimethyl hexanedioate, *See* dimethyl adipate, 8502
3,5-Dimethyl 1-hexyn-3-ol, *See* Surfynol® 61, 30373
5,5-dimethyl hydantoin, *See* DM hydantoin, 8751
Dimethyl hydrogenated tallow amine, *See* Alphadim® 90LC, 1825, Amine 2MHBGD, 2142, Armeen® DMHTD, 2946, Kemamine® T-9742D, 15932
Dimethyl isophthalate, *See* Morflex 1129, 20330
Dimethyl isosorbide, *See* Arlasolve® DMI, 2913
dimethyl lauramine, 8507, *See* Amine 2M12D, 2135, Armeen® DM12D, 2942, N,N-dimethyldodecylamine, 8519, Empigen® AB, 10358
dimethyl lauramine isostearate, *See* Parapel® HC-85, 22785, Parapel® LIS, 22787
Dimethyl lauramine oxide, *See* Karox LO, 15775
Dimethyl lauryl benzyl ammonium chloride, *See* Retarder N, 26256
Dimethyl methanephosphonate, *See* dimethylmethyl phosphonate, 8525, Fyrol® DMMP, 12539
O,O-Dimethyl-S-methoxycarbonylmethyl phosphorodithioate, *See* MPEM, 20392
O,O-dimethyl S-[2-(methylamino)-2-oxoethyl] phosphorothioate, *See* Omethoate, 22155
O,O-Dimethyl-S-(N-methylcarbamoylmethyl) phosphorodithioate, *See* Dimethoate, 8496
Dimethyl methylphosphonate, *See* dimethylmethyl phosphonate, 8525, Fyrol® DMMP, 12539
O,O-dimethyl O-[3-methyl-4-(methylthio)phenyl]phosphorothioate, *See* Tiguvon®, 31788
Dimethyl myristamine, *See* Empigen® AH, 10361
Dimethyl O-(2,6-dichloro-4-methylphenyl) phosphorothioate, *See* Risolex, 26782
O,O-dimethyl O-(4-nitrophenyl) phosphorothioate, *See* Parathion-methyl, 22808
O,O-Dimethyl-O-(4-nitrophenyl) phosphorothioate, *See* methyl parathion, 19556
O,O-dimethyl O-4-nitro-*m*-tolyl phosphorothioate, *See* Novathion, 21708
Dimethyl octadecylamine, *See* Amine 2M18D, 2139
dimethyl octadecylbenzylammonium chloride, *See* Cycloton® SCS, 7542
***trans*-3,7-dimethyl octa-2,6-dien-1-ol**, *See* Meranol, 19331
R-(+)-3,7-Dimethyl-6-octen-1-ol, *See* Cephrol, 5695
Dimethyl oleamine, *See* Jet Amine DMOD, 15586
Dimethyl oleylamine, *See* Crodamine 3.AOD, 7109, Jet Amine DMOD, 15586
4,4-Dimethyloxazolidine, *See* Amine CS-1135®, 2153, Canguard® 327, 5088, Oxaban®-A, 22404
Dimethyl palmitamine, *See* Amine 2M16D, 2138, Armeen® DM16D, 2943
N-(1,4-Dimethylpentyl)-N'-phenyl-p-phenylenediamine, *See* Naugard® I-3, 20793
N-(2,6-dimethylphenyl)-5,6-dihydro-4H-1,3-thiazin-2-amine hydrochloride, *See* xylazine, 34619
N-(2,6-dimethylphenyl)-N-(methoxyacetyl)-DL-alanine methyl ester, *See* metalaxyl, 19430
N-(2,4-dimethylphenyl)-N-[[(2,4-dimethylphenyl) imino]methyl]-N-methylmethaniminamide, *See* Taktic, 30729
N-(2,6-dimethylphenyl)-2-methoxy-N-(2-oxo-3-oxazolidinyl)acetamide, *See* oxadixyl, 22407
O,O-Dimethyl phosphorochloridothioate, *See* MP-2, 20388
O,O-Dimethyl phosphorodithioic acid, *See* MP-1, 20387
Dimethyl phosphochloridothioate, *See* MP-2, 20388
Dimethyl phosphorochlorodithioate, *See* MP-2, 20388
dimethyl phthalate, 8508, *See* Fermine, 11631, Kemester® DMP, 15956, Kodaflex® DMP, 16271, Palatinol® M, 22593, Unimoll® DM, 32979, Uniplex 110, 33009
dimethyl polysiloxane, *See* dimethicone, 8493, Viscasil®, 33893
Dimethyl polysiloxane aqueous emulsion, *See* AF 60, 909
Dimethyl silicone, *See* Sentry Dimethicone, 27959
Dimethyl silicone fluid, *See* Anedco DF-6130, 2454
Dimethyl soya amine, *See* Jet Amine DMSD, 15587
Dimethyl stearamidopropyl [(2-pyrrolidonyl) methyl] ammonium chloride, *See* Surfadone QSP, 30347
Dimethyl stearamine, *See* Armeen® DM18D, 2944, Kemamine® T-9902, 15933
dimethyl succinate, 8509
dimethyl sulfide, 8510, *See* Exact-S®, 11281, methyl sulfide, 19561
dimethyl sulfoxide, 8511, *See* Demavet, 8053, DMSO, 8762
Dimethyl Sulfur Oxide, *See* DMSO, 8762
Dimethyl tallowamine, *See* Jet Amine DMTD, 15588
dimethyl tetrachloroterephthalate, 8512, *See* Dacthal, 7685
dimethyl 2,3,5,6-tetrachloroterephthalate, *See* Vegetable Turf and Ornamental Weeder, 33526
3,5-Dimethyl tetrahydro-2-H,1,3,5-thiadiazone-2-thione, *See* Amerstat® 233, 2039
Dimethyl thiophosphoryl chloride, *See* MP-2, 20388
dimethyl trichlorophenyl thiophosphate, *See* Korlan, 16351
O,O-dimethyl O-(2,4,5-trichlorophenyl) phosphorothioate, *See* ronnel, 26951
Dimethyl (2,4,5-trichlorophenyl) phosphorothionate, *See* ronnel, 26951
Dimethyl O-(2,4,5-trichlorophenyl) thiophosphate, *See* ronnel, 26951
Dimethyl Yellow, *See* Butter yellow, 4759
Dimethyl Yellow Analar, *See* Butter yellow, 4759
Dimethyl Yellow N,N-dimethylaniline, *See* Butter yellow, 4759
Dimethyl(N-octadecylphenylmethyl)ammonium chloride, *See* Ablumine 280, 85
dimethyl(tetradecyl)amine, *See* Adma® 14, 742, N,N-dimethyltetradecylamine, 8527
Dimethyl(trimethylsilyl)amine, *See* CD5400, 5505
Dimethyl-*m*-dioxan-4-ol acetate, *See* dimethoxane, 8498
Dimethyl-*m*-dioxan-4-yl acetate, *See* dimethoxane, 8498
Dimethyl[(1,2-phenylene)bis(iminocarbonothioyl)]bis[carbamate], *See* thiophanate methyl, 31706
Dimethyl-1,3-dioxan-4-ol acetate, *See* dimethoxane, 8498
Dimethyl-1-hexadecanamine, *See* Adma® 16, 743, N,N-dimethylhexadecylamine, 8524
Dimethyl-1-octadecanamine, *See* Crodamine 3.A18D, 7107, Dymanthine, 9327
Dimethyl-2-hydroxyethylamine, *See* 2-amino-2-methyl-1 propanol, 2182
Dimethyl-3-(2-benzothiazolyl)urea, *See* methabenzthiazuron, 19472
Dimethyl-3-(2-benzthiazolyl)-harnstoff, *See* methabenzthiazuron, 19472
Dimethyl-4-chlorophenol, *See* chloroxylenol, 6159
dimethylacetal formaldehyde, *See* methylal, 19564
dimethylacetic acid, *See* isobutyric acid, 15348
Dimethylacetone, *See* diethyl ketone, 8436
Dimethylamine dimethyldithiocarbamate, *See* D.D.D, 7608, Ultra-DMC, 32791
dimethylamine hydrochloride, 8513
Dimethylamine/epichlorohydrin copolymer, *See* Agefloc B-50LV, 1006
4-(Dimethylamino)benzoic Acid, *See* Escalol® 507, 10787
4-(dimethylamino)benzoic acid isoamyl ester, *See* Spectraban, 29225
4-dimethylaminoazobenzene, *See* Butter yellow, 4759
dimethylaminobenzaldehyde, 8515
***p*-dimethylaminobenzaldehyde**, *See* dimethylaminobenzaldehyde, 8515
4-dimethylaminobenzaldehyde, *See* dimethylaminobenzaldehyde, 8515
Dimethylaminobenzenecarbonal, *See* dimethylaminobenzaldehyde, 8515
***p*-Dimethylamino-benzenediazo sodium sulfonate**, *See* Bayer 5072, 3767
4-(dimethylamino)benzoic acid, 8514
2-Dimethylamino-5,6-dimethylpyrimidin-4-yl dimethylcarbamate, *See* Abol, 146
2-(Dimethylamino)-5,6-dimethyl-4-pyrimidinyl dimethylcarbamate, *See* Abol, 146
2-(dimethylamino)-5,6-dimethyl-4-pyrimidinyl dimethylcarbamate, *See* pirimicarb, 23879
N,N-dimethyl-1-aminododecane, *See* Onamine 12, 22164
dimethylaminoethanol, *See* Alkanolamine, 1701, Dabco® DMEA, 7669, deanol, 7829, dimethylethanolamine, 8520
2-dimethylaminoethanol, *See* Alkanolamine, 1701 Dabco® DMEA, 7669, deanol, 7829, dimethylethanolamine, 8520, Texacat® DME, 31428
2-(dimethylamino)ethanol, *See* Tegoamin® DMEA, 30995
2-(2-Dimethylaminoethoxy)ethanol, *See* Texacat® ZR-70, 31437
N-dimethylaminoethanol, *See* Alkanolamine, 1701, Dabco® DMEA, 7669, deanol, 7829
N,N-Dimethylaminoethanol, *See* Alkanolamine, 1701, Dabco® DMEA, 7669, deanol, 7829
Dimethylaminoethyl acrylate, *See* Ageflex FA-1, 975
N,N-Dimethylaminoethyl acrylate, *See* Adame, 622, Ageflex FA-1, 975
Dimethylaminoethyl acrylate, *See* Adame, 622
2-dimethylaminoethyl acrylate, *See* Adame, 622
Dimethylaminoethyl acrylate dimethyl sulfate, *See* Ageflex FA-1Q80DMS, 972
Dimethylaminoethyl acrylate methyl chloride, *See* Ageflex FA-1Q75MC, 971
β-Dimethylaminoethyl alcohol, *See* Alkanolamine, 1701, Dabco® DMEA, 7669, deanol, 7829, Tegoamin® DMEA, 30995, Texacat® DME, 31428
dimethylaminoethyl chloride hydrochloride, 8516
Dimethylaminoethyl methacrylate, *See* Ageflex FM-1, 983 Sipomer® 2M1M, 28528
Dimethylaminoethyl methacrylate methyl chloride, *See* Ageflex FM-1Q80MC, 982
N,N-Dimethylaminoethyl methacrylate, *See* Ageflex FM-1, 983 Sipomer® 2M1M, 28528
dimethylaminoethylacrylate, *See* Ageflex FA-1, 974
2-(dimethylamino)-N-[[(methylamino)carbonyl]oxy]-2-oxoethanimidothioic acid methyl ester, *See* Vydate®, 34171
2-dimethylamino-2-methyl-1-propanol, 8517, *See* DMAMP-80, 8753
[4S-(4α,4aα,5α,5aα,6α,12aα)]-4-(dimethylamino)-1,4,4a,5,5a,6,11,12a-octahydro-3,5,10,12,12a-pentahydroxy-6-methyl-1,11-dioxo-2-naphthacenecarboxamide monohydrate, *See* Doxycycline hydrochloride, 8946
4-(Dimethylamino)-1,4,4a,5,5a,6,11,12a-octahydro-3,6,10,12,12a-pentahydroxy-6-methyl-1,11-dioxo-2-naphthacenecarboxamide, *See* tetracycline, 31365
N-Dimethylamino-N'-phenyl-N'-(fluorodichlormethylthio) sulfamide, *See* Dichlofluanide, 8380
bis[*p*-(dimethylamino)phenyl]methane, *See* Michler's Base, 19644
4-[[4-(dimethylamino)phenyl][4-(dimethylimino)-2,5-cyclohexadien-1-ylidene]methyl]-N-ethyl-N,N-dimethylbenzeneaminium bromide chloride, *See*

methyl green, 19540
Dimethylaminopropyl behenamide, *See* Behenamidopropyl dimethylamine, 3896
N-[3-Dimethylamino)propyl]coco amides, *See* Cocamidopropyl dimethylamine, 6552
3-(N,N-Dimethylamino)propyl cocoamido amine oxide, *See* Ablumox CAPO, 90
N-[3-(Dimethylamino)propyl]cocoamide lactate, *See* Cocamidopropyl dimethylamine lactate, 6553
N-[3-(Triethoxysilyl)-propyl]-4,5-dihydroimidazole, *See* Dynasylan® IMEO, 9411
N-[3-(Dimethylamino)propyl]docosamide, *See* Behenamidopropyl dimethylamine, 3896
N-[3-(Dimethylamino)propyl]dodecanamide, *See* lauramidopropyl dimethylamine, 16831
N-[3-(Dimethylamino)propyl]hexadecanamide, *See* Palmitamidopropyl dimethylamine, 22630
Dimethylaminopropyl lauramide, *See* lauramidopropyl dimethylamine, 16831
N-[3-Dimethylaminopropyl]-9-octadecenamide, *See* oleamidopropyl dimethylamine, 22109
N-[3-(Dimethylamino)propyl]-9-octadecenamide-N-oxide, *See* oleamidopropylamine oxide, 22110
Dimethylaminopropyl oleamide, *See* oleamidopropyl dimethylamine, 22109
Dimethylaminopropyl palmitamide, *See* Palmitamidopropyl dimethylamine, 22630
Dimethylaminopyrazolone, *See* Permidan, 23391
dimethylaminosuccinamic acid, *See* daminozide, 7723
(4-dimethylamino-*o*-tolyl)phosphonous acid sodium salt, *See* Tonophosphan, 32013
Dimethylaminotrimethylsilane, *See* CD5400, 5505
N,N-Dimethylaminotrimethylsilane, *See* CD5400, 5505
Dimethylammonium chloride, *See* dimethylamine hydrochloride, 8513
Dimethylammonium dimethyldithiocarbamate, *See* Ultra-DMC, 32791
Dimethylammonium hydrogen isophthalate, *See* Vanax® CPA, 33288
dimethylarsinic acid sodium salt, *See* sodium cacodylate,
N,N-Dimethyl-1-behenamine-N-oxide, *See* Behenamine oxide, 3897
α-(5,6-dimethylbenzimidazolyl)cyanocobamide, *See* vitamin B+72, 33935
dimethylbenzene, *See* xylene, 34620
1,2-dimethylbenzene, *See* *o*-xylene, 34623
1,3-dimethylbenzene, *See* *m*-xylene, 34622
1,4-dimethylbenzene, *See* *p*-xylene, 34624
dimethylbenzenesulfonic acid, sodium salt, *See* Eltesol® SX 30, 9873, Naxonate® 4L, 20824, Pilot SXS-40, 23843, Witconate SXS 40%, 34492
α-(5,6-dimethylbenzimidazolyl) cyanocobamide, *See* vitamin B_{12}, 33934
2,2-Dimethyl-1,3-benzodioxol-4-ol methylcarbamate, *See* Bendiocarb, 3937
2,5-Dimethyl-*p*-benzoquinone, *See* phlorone, 23676
α,α-Dimethylbenzyl hydroperoxide, *See* CHP-5, 6174
N,N-dimethylbenzylamine, *See* Dabco® B-16, 7646, Pentamin BDMA etc, 23163
α,α-Dimethylbenzylhydroperoxide, *See* Aztec® CHP-80, 3555, CHP-158, 6175
dimethylbenzyloctadecylammonium chloride, *See* Ablumine 280, 85, octadecylbenzenemethanaminium chloride, 21985, stearalkonium chloride, 29592
3,3'-[(3,3'-dimethyl[1,1'-biphenyl]-4,4'-diyl)bis(azo)]bis[5-maino-4-hydroxy-2,7-naphthalenedisulfonic acid] tetrasodium salt, *See* Niagara blue, 21091
1,1'-dimethyl-4,4'-bipyridinium, *See* paraquat, 22796
2,5-Dimethyl-2,5-bis-(benzoylperoxy) hexane, *See* Luperox 118, 17838
2,5-dimethyl-2,5-bis(*t*-butylperoxy)hexane, *See* dimethyldibutyl peroxyhexane, 8518
2,5-Dimethyl-2,5-bis(2-ethylhexanoyl-peroxy) hexane, *See* Lupersol 256, 17858, USP® -245, 33166
2,5-dimethyl-2,5-bis(*t*-butylperoxy)hexane, *See* Aztec® 2,5-Di, 3551, Trigonox® 101, 32279
3,3-Dimethyl-2-butanone, *See* pinacolone, 23848
3,3-dimethylbutan-2-one, *See* pinacolone, 23848
3,3-Dimethylbutyldimethylchlorosilane, *See* CD5430, 5506
2,4-Dimethyl-6-*t*-butylphenol, *See* Lowinox® 624, 17661
N-1,3-Dimethylbutyl-N'-phenyl-p-phenylene diamine, *See* Permanax 6PPD, 23356
N,N-(Dimethyl)-N',N'- diisopropanol-1,3-propanediamine, *See* Texacat® DPA, 31430
2-(dimethylamino)-5,6-dimethyl-4-pyrimidinyl ester, *See* pirimicarb, 23879, Rapid, 25884
N,N-Dimethylbehenylamine, *See* Crodamine 3ABD, 7111
N,N-dimethylbenzenemethanamine, *See* Dabco® B-16, 7646
N,N-Dimethyl $C_{18\text{-}22}$ amines, *See* Crodamine 3AHRD, 3ARD, 7113
dimethyl-carbamodithioic acid, sodium salt, *See* Aquatreat SDM, 2761, Perkacit® SDMC, 23288
dimethyl-carbamodithioic acid, tetra-anhydrosulfide with orthothioselenious acid, *See* methyl selenac, 19559
Dimethylcarbinol, *See* isopropanol, 15405
Dimethylchlorosilane, *See* CD5470, 5507
N,N-Dimethyl-N-(3-cocamidopropyl)amine oxide, *See* Ablumox CAPO, 90
N,N-Dimethyl-N-[3-(coconut oil alkyl)amidopropyl]amine oxide, *See* Ablumox CAPO, 90
2,5-dimethyl-2,5-Cyclohexadiene-1,4-dione, *See* phlorone, 23676
3,6-Dimethyl-3-cyclohexene-1-carbaldehyde, *See* Cyclovertal, 7543
2,3-Dimethyl-3-cyclohexene-1-carboxy-aldehyde, *See* Ligustral, 17247
β-methyladipate, *See* Sipalin MOM, 28514
Dimethylcyclohexyl adipate, *See* Sipalin AOM, 28513
N,N'-Dimethylcyclohexylamine, *See* Toyocat®-DMCH, 32101
N,N dimethyl cyclohexyl ammonium dibutyl dithiocarbamate, *See* Akrochem® Accelerator CZ-1, 1153
2,3'-dimethyl-4'-(diacetylamino)azobenzene, *See* Diacetazotol, 8247
Dimethyldiallylammonium chloride, *See* Dimdac, 8488
dimethyldibutyl peroxyhexane, 8518
2,5-Dimethyl-2,5-di (*t*-butyl peroxy)hexane, *See* Aztec® 2,5-Di, 3551, Esperal® 120, 10837, Lupersol 101, 17846, Polyvel CR-5F, 24657, Trigonox® 101, 32279
2,5-Dimethyl-2 5-di(t-butylperoxy)hexyne-3, *See* Aztec® 2,5-Tri, 3552, Lupersol 130, 17847, Esperal® 230, 10838, Polyvel CR-L10, 24660, Trigonox® 145, 32286
2,5-Dimethyl-2,5-di(*tertiary*-butylperoxy)-hexane, *See* dimethyldibutyl peroxyhexane, 8518
1,1-dimethyl-3-(3,4-dichlorophenyl)urea, *See* Karmex, 15771, Karmex®, 15772
dimethyldidecylammonium chloride, *See* Radiaquat 6410, 6412, 25789
Dimethyldiethoxy silane, *See* CD5600, 5508, EXP-49, 11305
2,5-dimethyl-2,5-di(2-ethyl hexanoyl peroxy) hexane, *See* USP® -245, 33166
2,2-dimethyl-2,3-dihydrobenzoduranyl-7 N-methylcarbamate, *See* Carbodan, 5197, Rampart, 25860
Dimethyldimethoxysilane, *See* CD5605, 5509
5,6-dimethyl-2-dimethylamino-4-dimethylcarbamoyloxypyrimidine, *See* Rapid, 25884
N, N' -Dimethyl-N, N'-di-(1 methylpropyl)-p-phenylenediamine, *See* Eastozone 32, 9480
dimethyldioctadecylammonium chloride, *See* Adogen® TA-100, 779, Prepagen WK, 24894, Sumquat® 6045, 30123
2,6-dimethyl-*m*-dioxan-4-ol acetate, *See* dimethoxane, 8498
2,6-dimethyl-1,3-dioxan-4-ol acetate, *See* dimethoxane, 8498
2,6-dimethyl-*m*-dioxan-4-yl acetate, *See* dimethoxane, 8498
2,6-dimethyl-*m*-dioxan-4-yl ester acetic acid, *See* dimethoxane, 8498
3,3-Dimethyl-1,5-dioxaspiro[5,5] undecane, *See* Thymoxane, 31779
Dimethyl-diphenyl silicone grease, *See* Versilube® G-321, 33628
2,3-Dimethyl-2,3-diphenylbutane, *See* Perkadox® 30, 23305
3,4-Dimethyl-3,4-diphenylhexane, *See* Perkadox® 58, 23306
Dimethyldiphenylurea, *See* Centralite, 5679
1,3-dimethyl-1,3-diphenylurea, *See* Zentralin, 34715
dimethyldisulfide, *See* Sulfa-Hitech® 0382, 29961
2,5-Dimethyl-2,5-di(*tertiary*-butylperoxy)-hexane, *See* Aztec® 2,5-Di, 3551, Polyvel CR-5F, 24657, Trigonox® 101, 32279
2,5-Dimethyl-2,5-di-(*tert*-butylperoxy)hexyne-3, *See* Aztec® 2,5-Tri, 3552, Polyvel CR-L10, 24660
Dimethyldithiocarbamic acid compd. with dimethylamine (1:1), *See* Ultra-DMC, 32791
Dimethyldithiocarbamic acid dimethylamine salt, *See* Ultra-DMC, 32791
dimethyldithiocarbamic acid sodium salt, *See* Aquatreat SDM, 2761, methyl namate®, 19551, Octopol SDM-40, 22014, sodium dimethyldithiocarbamate, 28758
dimethyldithiocarbamic acid zinc salt, *See* Accelerator MZ Powder, 209, Vancide® MZ-96, 33303, Zimate®, 34762
Dimethyldithiocarbamic acid, dimethylammonium salt, *See* Ultra-DMC, 32791
dimethyldithiocarbamic acid, lead salt, *See* Ledate®, 16949, methyl ledate, 19547
dimethyldithiocarbamic acid, sodium salt, *See* sodium dimethyldithiocarbamate, 28758
N,N-dimethyl-1-dodecanamine, *See* Onamine 12, 22164
N,N-Dimethyl-1-docosanamine-N-oxide, *See* Behenamine oxide, 3897, lauramine oxide, 16832
N,N-dimethyldodecylamine, *See* Onamine 12, 22164
Dimethyldodecylamine, *See* Onamine 12, 22164
N,N-dimethyldodecylamine, 8519
N,N-dimethyldodecylamine oxide, *See* Ablumox LO, 91, Ammonyx® LO, 2264, Empigen® OB, 10375, Rhodamox® LO, 26620
N,N-dimethyldodecylamine-N-oxide, *See* Rhodamox® LO, 26620
Dimethyldodecylamine oxide, *See* Ablumox LO, 91, Ammonyx® LO, 2264, Empigen® OB, 10375, Rhodamox® LO, 26620
dimethyldodecylamine-N-oxide, *See* Ablumox LO, 91, Ammonyx® LO, 2264, Empigen® OB, 10375, Rhodamox® LO, 26620
N,N-dimethyldodecylamine-N-oxide, *See* Ammonyx® LO, 2264 *See* Ablumox LO, 91, Ammonyx® LO, 2264, Empigen® OB, 10375, Rhodamox® LO, 26620
dimethylene oxide, *See* ethylene oxide, 11131
N,N-Dimethylerucylamine, *See* Crodamine 3AED, 7112
N,N-Dimethylethanamine, *See* dimethylethylamine, 8521
N,N-Dimethylethanolamine, *See* Texacat® DME, 31428, Toyocat® -DMA, 32100
dimethylethanolamine, 8520, *See* Alkanolamine, 1701, Dabco® DMEA, 7669, deanol, 7829, Tegoamin® DMEA, 30995
N-[(1,1-dimethylethoxy)carbonyl]- β-alanyl-L-tryptophanyl-L-methionyl-L- α-aspartyl--L-phenylalaninamide, *See* pentagastrin, 23146
((1,1-dimethylethoxy)methyl)oxirane, *See* Ageflex TBGE, 998
Dimethylethoxysilane, *See* CD5635, 5511
Dimethylethyl hydroperoxide, *See* Aztec® *t*-Butyl Hydroperoxide-70, Aq, 3541
1,1-dimethylethyl hydroperoxide, *See* Trigonox® A-80, 32291
4-(1,1-Dimethylethyl)phenol, *See* 4-*t*-butyl phenol, 4800

dimethylethylamine, 8521
N,N-dimethylethylamine, 8522, *See* dimethylethylamine, 8521
2-[(1,1-Dimethylethyl)amino]ethyl 2-methyl-2-propenoate, *See* Ageflex FM-4, 984
2-(1,1-Dimethylethyl)-1,4-benzenediol, *See* t-Butylhydroquinone, 4793
1-[4-(1,1-Dimethylethyl)-2,6-dimethyl-3,5-dinitrophenyl]ethanone, *See* acetyl-dinitro-butyl-xylene, 317
Dimethyl-O-(2-ethyl-4-ethoxy-pyrimidinyl-6)-thionophosphate, *See* etrimfos, 11155
2,5-Dimethyl-2,5-bis(2-ethyl-1-hexanoylperoxy)hexane, *See* USP® -245, 33166
2,5-Dimethyl-2,5-di(2-ethylhexanoylperoxy)hexane, *See* USP® -245, 33166
2,5-Dimethyl-2,5-(2-ethylhexanoylperoxy) hexane, *See* Trigonox® 141, 32285
1,1-dimethylethylhydroperoxide, *See* Aztec® *t*-Butyl Hydroperoxide-70, Aq, 3541, TBHP-70, 30853, USP® -800, 33170
1-(1,1-dimethylethyl)-2-methoxy-4-methyl-3,5-dinitrobenzene, *See* musk ambrette, 20532
(1,1-dimethylethyl)-4-methoxyphenol, *See* BHA, 4121, Butylated hydroxyanisole, 4787
2,6-bis(1,1-dimethylethyl)-4-methylphenol, *See* Oxyguard, 22457
1-(1,1-dimethylethyl)-4-methyl-benzene, *See* Butyl toluene, 4782
bis(1,1-dimethylethyl)-4-methylphenol;, *See* Oxyguard, 22457
1,1-dimethylethyl 1-methyl-1-phenylethyl peroxide, *See* Butyl cumyl peroxide, 4768
4-(1,1-dimethylethyl)phenol, *See* Terbutol, 31228
2,6-(1,1-Dimethylethyl)phenol, *See* 2,6-di-*t*-butylphenol, 4799
(1,1-dimethylethyl)phenyl diphenyl ester phosphoric acid, *See* Syn-O-Ad® 8485, 30559, Syn-O-Ad® 8478, 30555
1-[3-[4-(1,1-dimethylethyl)phenyl]-2-methyl propyl] piperidine, *See* fenpropidin, 11616
Dimethylethylpyridine, *See* parvoline, 22874
S-[[(1,1-dimethylethyl)thio]methyl] O,O-diethyl phosphorodithioate, *See* Terbufos, 31226
N,N-Dimethylformamide, *See* dimethyl formamide, 8506, Dynasolve 100, 9386
dimethylformocarbothialdine, *See* Dazomet, 7806, N-521® Biocide, 20640
dimethylglyoxal, *See* Diacetyl, 8250
dimethylglyoxime, 8523
2,5-dimethyl-4-heptanone, *See* diisobutyl ketone, 8461
2,6-dimethyl-4-heptanone, *See* valerone, 33243
N,N-dimethyl-1-Hexadecanamine, *See* Adma® 16, 743, N,N-dimethylhexadecylamine, 8524
N,N-dimethylhexadecylamine, 8524, *See* Adma® 16, 743, Crodamine 3.A16D, 7106, N,N-dimethylhexadecylamine, 8524, Onamine 16, 22166
2,5-Dimethyl-2,5-hexanediol bis(2-ethylperoxyhexanoate), *See* USP® -245, 33166
2,5-Dimethylhexane 2,5-diper-2-ethylhexanoate, *See* USP® -245, 33166
5,5-dimethyl hydantoin, *See* DM hydantoin, 8751
dimethylhydroxybenzene, *See* xylenol, 34625
1,1-Dimethyl-3-hydroxybutyl peroxy-2-ethylhexanoate, *See* Lupersol 665-M50, 17868
N,N-Dimethyl-2-Hydroxyethylamine, *See* Alkanolamine, 1701, Dabco® DMEA, 7669, deanol, 7829, Tegoamin® DMEA, 30995 Texacat® DME, 31428
N,N-dimethyl-N-(2-hydroxyethyl)amine, *See* Alkanolamine, 1701, Dabco® DMEA, 7669, deanol, 7829
5,5-dimethyl-2,4-imidazolidinedione, *See* DM hydantoin, 8751
Dimethylisophthalate, *See* Uniplex 270, 33016
dimethylketol, *See* acetyl methyl carbinol, 312
dimethylketone, *See* acetone, 297
N,N-dimethyllaurylamine, *See* Onamine 12, 22164
N,N-Dimethyl-N-lauryl-N-(3-sulfopropyl)-ammonium betaine, *See* Ralufon® DL, 25848
N,N-Dimethyl-2-mercaptoethylamine hydrochloride, *See* Thiofluor™, 31686
dimethylmethane, *See* propane, 25173
2,4-Dimethyl-6-(1-methyl cyclohexyl) phenol, *See* Permanax WSL, 23368
3,7-Dimethyl-7-methoxyoctan-2-ol, *See* Osyrol, 22373
dimethylmethyl phosphonate, 8525
4,4'-dimethylmethylenediphenol, *See* Bisphenol A, 4283
N,N-dimethyl-α-methylcarbamoyloxyimino-α-(methylthio)acetamide, *See* Vydate®, 34171
N',N'-dimethyl-N-[(methylcarbamoyl)oxy]-1-thiooxamimidic acid methyl ester, *See* Vydate®, 34171
dimethylmethylene-*itp,p'*ro-diphenol, *See* Bisphenol A, 4283
6,6-dimethyl-2-methylenebicyclo [3.1.1] heptane, *See* β-l-pinene, 23858, β-pinene, 23859
2,2-dimethyl-N-(1-methylethyl)-N-(phenylmethyl)propanamide, *See* Comodor, Comodor 600, 6667 tebutam, 30879
(R*,R*)-α,4-dimethyl-α-(4-methyl-3-pentenyl)-3-cyclohexene-1-methanol, *See* Hydagen® B, 14525
(1R trans)-2,2-dimethyl-3-(2-methyl-1-propenyl)cyclopropanecarboxylic acid [5-(phenylmethyl)-3-furanyl] methyl ester, *See* Reslin, 26216
2,2-Dimethyl-3-(2-methyl-1-propenyl) cyclopropanecarboxylic acid (1,3,4,5,6,7-hexahydro-1,3-dioxo-2H-isoindol-2-yl)methyl ester, *See* tetramethrin, 31384
2,2-dimethyl-3-(2-methyl-1-propenyl) cyclopropanecarboxylic acid (3-phenoxyphenyl)methyl ester, *See* phenothrin, 23619
2,2-Dimethyl-3-(2-methyl-1-propenyl) cyclopropanecarboxylic acid (1,3,4,5,6,7-hexahydro-1,3-dioxo-2H-isoindol-2-yl)methyl ester, *See* Killgerm® Py-Kill W, 16182
2,2-dimethyl-3-(2-methylpropenyl) cyclopropanecarboxylic acid ester with N-(hydroxymethyl)-1-cyclohexene-1,2-dicarboximide, *See* Killgerm® Py-Kill W, 16182
Dimethyl-N-(1-methylethyl)-N-(phenylmethyl)propanamide, *See* Comodor, Comodor 600, 6667
1,2-Dimethyl-5-nitro-1H-imidazole, *See* Dimetridazole, 8528
N,N-dimethyl-4-nitrosoaniline, **See** Vulcaniline, 34095
Dimethyl-n-octadecyl-1-octadecanaminium chloride, *See* Adogen® TA-100, 779
Dimethyl-n-octadecylbenzenemethanaminium chloride, *See* Cycloton® SCS, 7542 octadecylbenzenemethanaminium chloride, 21985 stearalkonium chloride, 29592
3,7-dimethyl-2,6-octadienal, *See* citral (cis and trans), 6336
Z-3,7-dimethyl-2,6-octadiene-1-al, *See* Neral, 20996
N,N-dimethyl-1-octadecanamine, *See* Kemamine® T-9902, 15933
N,N-Dimethyl-1-octadecanamine, *See* Dymanthine, 9327
N,N-dimethyloctadecylamine, *See* Crodamine 3.A18D, 7107, Dymanthine, 9327, Kemamine® T-9902, 15933
Dimethyloctadecylamine hydrochloride, *See* Onamine 18, 22167
N,N-Dimethyloctadecylamine hydrochloride , *See* Onamine 18, 22167
N,N-dimethyl-N-octadecylbenzenemethanaminium chloride, *See* Cycloton® SCS, 7542, stearalkonium chloride, 29592, Sumquat® 6210, 30126
Dimethyloctadecylchloro silane, *See* CD5636, 5512
Dimethyl-n-octadecylchlorosilane, *See* CD5636, 5512
dimethyl-n-octadecyl-1-octadecanaminium chloride, *See* Adogen® TA-100, 779, Prepagen WK, 24894, Sumquat® 6045, 30123
2,6-dimethyl-2,6-octadien-8-ol, *See* Vernol®, 33600
2,6-dimethyl-2-octen-8-ol, *See* Cephrol, 5695, citronellol, 6348
S-(-)-3,7-dimethyl-7-octen-1-ol, *See* rhodinol, 26685
2,7-Dimethyloct-5-en-4-one, *See* Isotagetone, 15424
3,6-Dimethyl-4-octyne-3,6-diol, *See* Surfynol® 82, 30374, Surfynol® 82, 82S, 30375
3,7-dimethyl-1,6-octadien-3-ol, *See* Phantol, 23570
3,7-dimethyl-1,6-octadien-3-yl acetate, *See* Phanteine, 23569
3,7-Dimethyl-2,6-octadienal, *See* citral (*cis* and *trans*), 6336, Neral, 20996
3,7-dimethyl-2,6-octadien-1-ol, *See* Vernol®, 33600
(E)-3,7-Dimethyl-2,6-octadien-1-ol, *See* Meranol, 19331
***trans*-3,7-dimethyl-2,6-octadien-1-ol**, *See* Geraniol, 12885
(E)-3,7-Dimethyl-2,6-octadien-1-ol, *See* Geraniol, 12885
(E)-3,7-dimethyl-2,6-octadien-1-yl acetate, *See* Meraneine, 19330
3,7-dimethyl-6-octenal, *See* citronellal, 6347
3,7-dimethyl-6-octen-1-ol, *See* citronellol, 6348
3,7-Dimethyl-6-octen-1-yl acetate, *See* Cephreine, 5694
Dimethylol ethylene urea, *See* Quecodur AE, 25637
Dimethyloldimethyl hydantoin, **See** DMDM hydantoin, 8754, Glydant®, 13140, Mackstat® DM, 18234
N,N-Dimethyl oleamide, *See* Hallcomid® M-18-OL, 13551
Dimethyloleamine, *See* Armeen® DMOD, 2947
N,N-Dimethyloleylamine, *See* Crodamine 3AOD, 7114
Dimethyl-1-oxa-3-aza-cyclopentane, *See* Amine CS-1135®, 2153, Canguard® 327, 5088, Oxaban®-A, 22404
Dimethyloxazolidine, *See* Amine CS-1135®, 2153, Canguard® 327, 5088, Oxaban®-A, 22404
7,7-dimethyl-2-oxobicyclo[2.2.1]heptane-1-methanesulfonic acid, *See* d-camphorsulfonic acid, 5076
[2S-(2α,5α,6β)]-3,3-Dimethyl-7-oxo-6-[(phenylacetyl)amino]-4-thia-1-azabicyclo-[3.2.0]heptane-2-carboxylic acid compound with 2-(diethylamino)ethyl 4-aminobenzoate(1:1) monohydrate, *See* Penicillin V, 23111
[2S-(2α,5α,6β)]-3,3-Dimethyl-7-oxo-6-[(phenylacetyl)amino]-4-thia-1-azabicyclo[3.2.0]heptane-2-carboxylic acid monopotassium salt, *See* Penicillin G potassium, 23109, 23110
6-(2,2-dimethyl-5-oxo-4-phenyl-1-imidazolidinyl)-3,3-dimethyl-7-oxo-4-thia-1azabicyclo[3.2.0]heptane-2-carboxylic acid, *See* Versapen, 33609
6-(2,2-dimethyl-5-oxo-4-phenyl-1-imidazolidinyl)penicillanic acid, *See* Versapen, 33609
N,N-dimethyl-(3-palmitamidopropyl)-glycine betaine, *See* pendecamaine, 23101
Dimethylparabanic acid, *See* Cholestrophane, 6169
dimethylphenol, *See* xylenol, 34625
dimethylphenol phosphate (3:1), *See* Coalite N.T.P, 6518, Syn-O-Ad® 8475M, 30554
Dimethyl(2-phenoxyethyl)-2-thenylammonium *p*-chlorobenzenesulfonate Canopar, *See* thenium closylate, 31568
N,N-Dimethyl-N-(2-phenoxyethyl)-2-thiophen methanaminium salt with 4-chlorobenzenesulfonic acid (1:1), *See* thenium closylate, 31568
1-[(2,4-dimethylphenyl)azo]-2-Naphthalenol, *See* Sudan II, 29939
1-(2,4-dimethylphenylazo)-2-naphthol, *See* Sudan II, 29939
2,4-Dimethyl-6-phenyldihydropyrane, *See* Fenyrane, 11621, Pelargene, 23076
dimethyl-phenylene-green, 8526
2,4-Dimethyl-4-phenyltetrahydrofuran, *See* Rhubafuran, 26718
O,S-dimethyl phosphoramidothioate, *See* methamidophos, 19474, Tamaron®, 30760
dimethylphosphoric acid 3,5,6-trichloro-2-pyridinyl ester, *See* Torelle, 32058
O,O-dimethylphosphorochloridothioate, *See* MP-2,

20388
Dimethylpiperazine, *See* Lupetazin, 17881
N,N'-dimethyipiperazine, *See* Texacat® DMP, 31429
1,4-dimethylpiperazine, *See* Texacat® DMP, 31429
Dimethylpiperazine tartrate, *See* lycetol, 18063
1,1-dimethylpiperidinium chloride, *See* mepiquat chloride, 19327
N,N-dimethyl-4-piperidylidene-1,1-diphenylmethane methylsufate, *See* Variton, 33432
Dimethylpolysiloxane, *See* Sentry Dimethicone, 27959
2,2-dimethyl-1,3-propanediol, *See* neopentyl glycol, 20937, NPG® Glycol, 21751
2,2-dimethylpropane-1,3-diol, *See* neopentyl glycol, 20937, NPG® Glycol, 21751
2,2-dimethylpropaneperoxoic acid, 1,1-dimethylethyl ester, *See* Trigonox® 25-C75, 32268
N,N-dimethyl-N-2-propenyl-, chloride, *See* Agefloc WT-40, 1010
N,N-dimethyl-N-2-propenyl-2-propen-1-aminium chloride, *See* Ageflex mDMDAC, 993
(dimethylpropyl)phenol, *See* Orthophen® 278, 22336
(2,3-Dimethylpropyl) dimethylchlorosilane, *See* CD5610, 5510
(dimethylpropyl)phenol, *See* Pentaphen® 67, 23168
***p*-(α,α-dimethylpropyl)phenol**, *See* Orthophen® 278, 22336, Pentaphen® 67, 23168
***p*-(1,1-dimethylpropyl)phenol**, *See* Orthophen® 278, 22336, Pentaphen® 67, 23168
2,6-dimethylpyridine, *See* 2,6-Lutidine, 17995
N_1-(4,6-dimethyl-2-pyrimidinyl)sulfanilamide, *See* sulfamethazine, 29962
7,8-dimethyl-10-ribitylisoalloxazine, *See* vitamin B_2, 33935
7,8-dimethyl-10-(1'-d-ribityl)isoalloxazine, *See* riboflavin, 26722
7,8-dimethyl-10-(D-*ribo*-2,3,4,5-tetrahydroxypentyl)isoalloxazine, *See* vitamin B+72, 33935
6,8-dimethyl-5,7-dihydroxy-4'-methoxy-flavanone, *See* matteucinol, 19068
Dimethylsilanol hyaluronate, *See* D.S.H.C, 7625
dimethylstearamine, *See* Adogen® MA-108 SF, 776
N,N-Dimethyl-N-stearic acid-amidopropyl-N-(3-sulfopropyl)-ammonium betaine, *See* Ralufon® TA, 25851
N,N-dimethyl stearylamine, *See* Onamine 18, 22167
Dimethylstearylamine, *See* Crodamine 3.A18D, 7107, Dymanthine, 9327
dimethylsulfide-α,α'-dicarboxylic acid, *See* thiodiglycolic acid, 31683
Dimethylsulfoxide, *See* DMSO, 8762
Dimethyltallowamine, *See* Armeen® DMTD, 2949
N,N-Dimethyl tallowamine, *See* Amine 2MBGD-M, 2141, Amine 2MOLD, 2144
N,N-Dimethyl-N-tallow-N-(3-sulfopropyl)ammonium betaine, *See* Ralufon® DT, 25849
N,N-dimethyltetradecylamine, 8527, *See* Amine 2M14-50D, 2136, Onamine 14, 22165
N,N-Dimethyl-N-tetradecylbenzene-methanaminium chloride, *See* JAQ Powdered Quaternary, 15517
N,N-dimethyl-N-[2-[2-[4-(1,1,3,3-tetramethyl butyl)phenoxy]ethoxy]ethyl]benzenemethaminium chloride, *See* benzethonium chloride, 3963
2,7-dimethylthianthrene, *See* Odylen®, 22046
2,2'-dimethylthiocarbanilide, *See* Accelerator A22, 192
dimethyltrichlorophenylthiophosphate, *See* ronnel, 26951
N-[2,4-dimethyl-5-[[(trifluoromethyl) sulfonyl]amino]phenyl]acetamide, *See* mefluidide, 19230
1,1-dimethyl-3-(α,α,α-trifluoro-m-tolyl)urea, *See* Cottonex, 6894
3,7-dimethyl-9-(2,6,6-trimethyl-1-cyclohexen-1-yl)-2,4,6,8-nonatetraen-1-ol, *See* Vi-Alpha, 33706
dimethyltrimethylene glycol, *See* neopentyl glycol, 20937, NPG® Glycol, 21751
N,N-Dimethyltrimethylsilylamine, *See* CD5400, 5505
dimethyrimol, *See* dimethirimol, 8495
dimeticon p, *See* Viscasil®, 33893
Dimetridazole, 8528, *See* Emtryl, 10482
Dimexide, *See* DMSO, 8762
Dimezathine, *See* sulfamethazine, 29962
Dimidim-R, *See* sulfamethazine, 29962
S-Dimidine, *See* sulfamethazine, 29962
Dimilin, 8529, *See* Diflubenzuron, 8450
Dimitone, *See* Dapsone, 7751
Dimodan LS Kosher, 8530
Dimodan O Kosher, 8531
Dimodan PM, 8532
Dimodan PV, PV 300 Kosher, 8533
Dimodan PVP Kosher, 8534
Dimodan S, 8535
4,4'-(2-ethyl-2-nitrotrimethylene), *See* Fuelsaver®, 12432
dimorpholine, *See* Fuelsaver®, 12432
4,4'-dimorpholine disulphide, *See* Vanax® A, 33287
Dimorpholine N,N'-disulfide, *See* Akrochem® Accelerator R, 1154, Naugex SD-1, 20801, Sulfasan, 29970
2,2'-Dimorpholinodiethylether, *See* Texacat® DMDEE, 31427
Dimpylate, *See* Diazinon Liquid, 8344
Dimul DDM K, 8536
Dimundite, *See* Camite, 5054
Dimycin, 8537
1,2-Dimyristoyl-*sn*-glycero(3) phosphatidylcholine, *See* Phospholipon® MC, 23728
Dimyristyl peroxydicarbonate, *See* Perkadox® 26-fl, 23304
Dimyristyl thiodipropionate, *See* Argus DMTDP, 2862, Cyanox® MTDP, 7492, Evanstab® 14, 11257
Dinamene, 8538
Dinaphthalene-methane sulfonate sodium salt, *See* Rhodacal® RM/77-D, 26591
N,N'-Di-β-naphthyl-*p*-phenylenediamine, *See* Agerite® White, 1024
di-n-butyl hydroxytoluene, *See* Oxyguard, 22457
Di-n-butyl phosphite, *See* Syn-O-Ad® P-316, 30562
Di-n-Butyl(maleate)tin, *See* Dibutyltin maleate, 8372
Di-n-butyldithiocarbamic Acid Zinc Salt, *See* Octocure ZDB-50, 21995, Perkacit® ZDBC, 23295
Di-n-butylphthalate, *See* Palatinol® DBP, 22587
Di-n-decylamine, *See* Radiamine 6310, 25761
Dingler's green, 8539
Di-n-hexyl adipate, *See* Adimoll® DH, 691
Di-n-hexyl azelate, *See* Priplast 3013, 25025
Diniternal, *See* Dinitra, 8540
dinitolmide, *See* Zoamix, 34844
Dinitra, 8540
dinitramine, *See* Cobex, 6544
Dinitro-*t*-butyl-*m*-cresol-methyl ether, *See* Ambrene, 1977
Dinitroanthraquinone, *See* Fritzsche's reagent, 12415
Dinitrobenzoic acid, 8541
3,5-dinitrobenzoic acid, *See* Dinitrobenzoic acid, 8541
dinitrobenzoic acid 3,5-dinitrobenzoic acid, 21349
Dinitrocresol, *See* 4,6-dinitrocresol, 8542
4,6-dinitrocresol, 8542, *See* Dekryll, 7999
Dinitro-*o*-cresol, *See* 4,6-dinitrocresol, 8542
3,5-dinitro-N_4,N_4-dipropylsulfanilamide, *See* oryzalin, 22347
2,6-dinitro-N,N-dipropyl-4-(trifluoromethyl) benzeneamine, *See* Treflan, 32162
dinitrogen monoxide, *See* Nitral, 21318
dinitrogen tetroxide, *See* nitrous gas or air, 21381
3,5-Dinitro-2-Hydroxytoluene, *See* 4,6-dinitrocresol, 8542
Dinitrol, *See* 4,6-dinitrocresol, 8542
2,6-dinitro-3-methoxy-4-*tert*-butyltoluene, *See* musk ambrette, 20532
3,5-dinitro-2-methylbenzamide, *See* Zoamix, 34844
α-dinitrophenol, *See* Dinitra, 8540
2,4-Dinitrophenol, *See* Dinitra, 8540
Dinitrophenols, *See* Dinitra, 8540
dinitrosol, *See* 4,6-dinitrocresol, 8542
Dinitro-*tert*-butylxylyl methyl ketone, *See* ketone musk, 16141
3,5-dinitro-o-toluamide, *See* Zoamix, 34844
Dinitrotoluene, *See* D.N.T, 7622
2,4-Dinitrotoluene, *See* D.N.T, 7622
Dinitrotoluol, *See* D.N.T, 7622
2,4-dinitrotoluol, *See* D.N.T, 7622
Di-n-nonyl phthalate, *See* Ceneg, 5671
Dinobuton, 8543, *See* Acrex, 384
Dinocap, *See* Karathane, 15759
Di-n-octyltin dilaurate, *See* ADK STAB OT-1, 735
Di-n-octyltin maleate, *See* ADK STAB OT-9, 736
di-n-octyltin mercaptide, *See* ADK STAB 465, 713
Dinofan, *See* Dinitra, 8540
Dinoleine, *See* Yodoxin, 34658
Dinonyl adipate, *See* Bisoflex DNA, 4258
Dinonyl phenol, 8544
Dinonyl phthalate, *See* Ceneg, 5671
Dinonylphenol ethoxylate, *See* Berol 272 and 716, 4024
Dinonylphthalate, *See* Ceneg, 5671
dinopol 235, *See* Good-rite® GP-265, 13205
Dinoram, 8545
Dinoramox, 8546
Dinoseb, *See* Dynamyte, 9361, Premerge, 24888
DINP, *See* Jayflex® DINP, 15552
Di-n-propyl ketone, *See* Butyrone, 4804
Dioctadecyl disulfide, *See* Hostanox® SE 10, 14376
dioctadecyl ester, *See* Distearyl thiodipropionate, 8713
3,3'-dioctadecyl thiodipropionate, *See* Distearyl thiodipropionate, 8713
Dioctadecyl 3,3'-thiodipropionate, *See* Alkanox® 240-3T, 1704, Evanstab® 18, 11258
Dioctadecylamine, *See* Armeen® 2-18, 2927
Dioctanoyl peroxide, *See* Perkadox® SE-8, 23310
Di-n-octanoyl peroxide, *See* Perkadox® SE-8, 23310
Dioctrahedral smectite, *See* Attapulgite, 3404
Dioctyl, 8547
Dioctyl (2-ethylhexyl) phosphite, *See* Syn-O-Ad® P-310, 30560
dioctyl adipate, 8548, *See* Adimoll® DO, 693, Bisoflex DMP, 4257, Good-rite® GP-223, 13201, Jayflex® DOA, 15554, Kodaflex® DOA, 16272, Plasthall® DOA, 24002, Polycizer DOA, 24391, Uniflex® DOA, 32945, PX-238, 25440, Plasthall® DOA, 24002, Wickenol® 158, 34393
dioctyl azelate, *See* Kodaflex® DOZ, 16275, Plasthall® DOZ, 24007
dioctyl dilinoleate, *See* Kemester® 3681, 15942
dioctyl dimethyl ammonium chloride, *See* Catigene® 818, 5451
dioctyl diphenylamine, *See* Naugalube® 438, 20783
dioctyl dodecanedioate dioate, *See* Plasthall® DODD, 24003
dioctyl maleate, *See* Bernel® Ester DOM, 4007 Ceraphyl® 45, 5715 PX-538, 25444 Octomer DOM, 22011
dioctyl o-benzenedicarboxylate, *See* Witcizer 312, 34454
dioctyl phthalate, 8549, *See* Bis(2-ethylhexy) Phthalate, 4228, Kodaflex® DOP, 16273, Octoil, 22002, Plasticizer 28P, 24030, Polycizer DOP, 24390, PX-138, 25438, Vestinol AH, 33666, Reomol DCP, 26130, Witcizer 312, 34454
di-*sec*-octyl phthalate, *See* dioctyl phthalate, 8549
Dioctyl sebacate, *See* Octoil S, 22003 Plasthall® DOS, 24005, Uniflex® DOS, 32946
dioctyl sodium sulfosuccinate, 8550, *See* Aerosol® OT-75%, 897, Arowet SC-75, 3091, Arylene M40, 3165, Coprol, 6796, Dioctyl, 8547, Discol DFW, 8631, Disponil SUS IC 8, 8709, Empimin® OP70, 10434, Hodag DOSS-70, 14178, Mackanate DOS-40, 18186, Marlinat® DF 8, 18850, Mazawet® DOSS 70, 19118, Monawet MO-65-150, 20183, Monawet MO-70, 20184, Monawet MO-84R2W, 20185, Rewopol® SBDO 75, 26374, Schercopol DOS-70, 27666, Triton® GR-5M, 32428, Octowet 40, 22035
Dioctyl sodium sulfosuccinate, USP/BP, *See* Complemix, 6681
Dioctyl succinate, *See* Wickenol® 159, 34394
Dioctyl sulfosodiumsuccinate, *See* dioctyl sodium sulfosuccinate, 8550

Dioctyl sulfosuccinate, *See* Ablusoft C-70, 119, Amwet DOSS, 2378
Dioctyl terephthalate, *See* Kodaflex® DOTP, 16274
Dioctylated diphenylamine, *See* Agerite® HP-S, 1018, Vanox® ODP, 33345
Dioctylated diphenylamine 65% Dioctylated diphenylamine, 35% diphenyl-*p*-phenylenediamine, *See* Agerite® Hipar®T, 1017
Dioctylcyclohexane, *See* Cetiol® S, 5803
4,4'-dioctyldiphenylamine, *See* Agerite® Stalite, 1022, Naugalube® 438, 20783
Dioctyldodeceth-2 lauroyl glutamate, *See* Amiter LGOD-2, 2199
Dioctyldodecyl lauroyl glutamate, *See* Amiter LGOD, 2198
3,5-diodo-1-propoxycarbonyl-methylpyrid-4-one, *See* Propyliodone, 25225
Diodoquin, *See* Yodoxin, 34658
Diodoxylin, *See* Yodoxin, 34658
Diofan®, 8551
Diofan® D, 8552
Dioform, 8553, *See* Acetylene dichloride, 320
dioform, *See* Dioform, 8553
diokan, *See* Dioxane, 8574
diol 14b, *See* 1,4-butanediol, 4735, Dabco® BDO, 7647
Diolane, *See* hexylene glycol, 14042
Dioleyl hydrogen phosphite, 8554
Dioleyl imidazoline methosulfate, *See* Rewoquat W 3690, W 3690 PG, 26398
Dioleyl tocopheryl methylsilanol, *See* Liposiliol C, 17403
Diolpate®, 8555
Dioltech 311, 8556
Di-On, 8557, *See* diuron, 8736
Dion®, 8558
Dion® Cor-Res, 8559
Dion® FR 6308, 8560
Dion® FR6657, 8561
Dion® VER, 8562
Dion® VER 9100 NP, 8563
Dional 11,113, 8564
Dionil®, 8565
Dionil® OC, 8566
Dionil® OC, *See* PEG-3 oleamide, 23048
Dionil® SD, 8567
Dion-Iso®, 8568
Dionosil, 8569, *See* Propyliodone, 25225, 25226
DIOP, *See* Witcizer 313, 34455
Diorez®, 8570
Diorez® SC, 8571
Diosal, 8572
Diostril, *See* Propyliodone, 25225, 25226
Diothyl, *See* pyrimithate, 25485
Di-o-tolyl thiourea, *See* Accelerator A22, 192
1,3-Di-o-tolylguanidine, *See* Vanax® DOTG, 33289
Diox, *See* Dioxane, 8574
Diox DR 22, 8573
dioxacarb, *See* Famid, 11422
1,4-Dioxacycloheptadecane-5,17-dione, *See* Emeressence® 1150, 10077, ethylene brassylate, 11121
1,4-Dioxacyclohexadecane-5,16-dione, *See* Arova 16, 3088
1,4-Dioxacyclohexane, *See* Dioxane, 8574
3,6-Dioxa-1-decanol, *See* Butyl Carbitol®, 4766
3,6-Dioxadecanol, *See* Butyl Carbitol®, 4766
2,5-Dioxahexane, *See* monoglyme, 20227
Dioxan-4-ol, 2,6-dimethyl-, acetate, *See* dimethoxane, 8498
Dioxane, 8574
1,4-dioxane, *See* Dioxane, 8574
3,6-dioxaoctan-1-ol, *See* Diethoxol, 8435, ethyl di-Icinol, 11068
3,6-dioxa-1-octanol, *See* Diethoxol, 8435, ethyl di-Icinol, 11068
1,5-Dioxaspiro[5.5]undecane 3,3-dicarboxylic acid, bis(2,2,6,6-tetramethyl-4-piperidinyl)ester, *See* Topanex 500H, 32020
3,16-Dioxa-8,9,10,11-tetrathia-4,15-disilaoctacane, *See* Aktisil PF 216, 1204, CB2494, 5478, Si 69, 28190
1,1-Diox-1,2-benzisothiazol-3-one, *See* saccharin, 27195
1,1-dioxide-1,2-benzisothiazol-3(2H)-one, *See* saccharin, 27195
Dioxine, 8575
Dioxitol, 8576, *See* Diethoxol, 8435, ethyl di-Icinol, 11068
Dioxitol-Low Gravity, *See* Ethoxydiglycol, 11044
9,10-dioxoanthracene, *See* Anthraquinone, 2532
2,3-dioxobutane, *See* Diacetyl, 8250
4,4'-Dioxo-β-carotene, *See* Canthaxanthin, 5095
dioxogen, 8577
2,5-dioxo-4-imidazolidinyl urea, *See* Allantoin, 1739
1,3-dioxolan-2-one, *See* Texacar® EC, 31424
Dioxopentane, *See* glutaraldehyde, 13062, Sonacide, 29073
5,5'-[(1,3-dioxo-1,3-propanediyl)bis (methylimino)] bis[N,N'-bis[2,3-dihydroxy-1-(hydroxymethyl) propyl]-2,4,6-triiodo-1,3-benzenedicarboxamide, *See* iotrolan, 15244
dioxopurine, *See* xanthine, 34590
2,4-dioxopyrimidine, *See* uracil, 33101
2,5-dioxotetrahydrofuran, *See* succinic anhydride, 29916
dioxybenzone, *See* Cyasorb® UV 24, 7496, Spectra-Sorb UV 24, 29230
4,4'-Dioxydiphenyl, *See* Anti-Oxidant DOD, 2587
Dioxyethylene ether, *See* Dioxane, 8574
2,4-dioxypyrimidine, *See* uracil, 33101
DIPA, *See* diisopropanolamine, 8466
DIPA, *See* diisopropylamine, 8469
DIPA Commercial Grade, *See* diisopropanolamine, 8466
DIPA Low Freeze Grade 85, *See* diisopropanolamine, 8466
DIPA Low Freeze Grade 90, *See* diisopropanolamine, 8466
DIPA NF, *See* diisopropanolamine, 8466
Dipalitmoyletyl hydroxyethylmonium methosulfate, *See* Stepanquat® 6585, 29700
Dipalmitoyl hydroxy proline, 8578, *See* Dipalmitoyl hydroxy proline, 8578, Lipacide DPHP, 17321
dipalmitoyl phosphatidylcholine, *See* colfosceril palmitate, 6607
1,2-Dipalmitoyl-*sn*-glycero(3) phosphatidylcholine, *See* Phospholipon® PC, 23729
Dipan, *See* diphenyl acetonitrile, 8593, diphenylacetonitrile, 8600
Dipanol, 8579
Dipel®, 8580
Dipentaerythritol, 8581
Dipentaerythrityl hexacaprylate/hexacaprate, 8582, *See* Dipentaerythrityl hexacaprylate/hexacaprate, 8582, Liponate DPC-6, 17375, Lipovol MOS-130, 17430, Lipovol MOS-350, 17431
Dipentamethylene thiuram disulfide, *See* Accelerator 4P, 188
Dipentamethylene thiuram hexasulfide, *See* DPTT, 8952
Dipentamethylenethiuram tetrasulfide, *See* Perkacit® DPTT, 23282
Dipentek, 8583, *See* Dipentaerythritol, 8581
Dipentene, 8584
Dipentene No.122, 8585
2,5-di-*t*-pentylhydroquinone, *See* Santovar, 27428
Di-Petronate Series, 8586
Dipex, 8587
***m*-Diphar**, *See* Maneb, 18526
Diphasol, 8588
Diphemanil methylsulfate, *See* Variton, 33432
Diphen 60-B, 8589
Diphenacoum;, *See* Ratak, 25896
Diphenal, 8590
Diphenamid, *See* Enide, 10581
Diphenasone, *See* Dapsone, 7751
Diphenatil, *See* Variton, 33432
diphenatrile, *See* diphenyl acetonitrile, 8593, diphenylacetonitrile, 8600
Diphenoxyphosphine oxide, *See* Weston® DPP, 34309
Diphenthane 70, *See* Nuophene, 21783
Di-phenthane-70, *See* Dichlorophen, 8392
diphenyl, 8591, 8592 *See* biphenyl, 4216
diphenyl acetonitrile, 8593, *See* diphenylacetonitrile, 8600
diphenylacetonitrile, 8600, *See* diphenyl acetonitrile, 8593
diphenylacetylene, *See* Tolan, 31933
Diphenylamine, *See* OA-505, 21954
Diphenylamine-acetone, *See* Agerite® Superflex®, 1015
N,N-Diphenyl-1,4-benzenediamine, *See* N,N'-diphenyl-*p*-phenylenediamine, 8603
diphenylcarbamide, *See* carbanilide, 5180, Zeonet U, 34722
Diphenylcarbamyl dimethyldithiocarbamate, *See* O.N.V, 21943
Diphenylcarbinol, *See* Benzhydrol, 3965
Diphenyl carbonate, *See* Phenol Carbonate, 23604
diphenyl cresol phosphate, *See* diphenylcresyl phosphate, 8601, Disflamoll® DPK, TPK, 8641
diphenylcresyl phosphate, 8601, *See* Disflamoll® DPK, TPK, 8641
Diphenyldichlorosilane, *See* CD5950, 5513
Diphenyldiethoxy silane, *See* CD6000, 5514
Diphenyldimethoxy silane, *See* CD6010, 5515
N,N'-diphenyl-1,4-diaminobenzene, *See* Agerite® DPPD, 1016
Diphenyl didecyl (2,2,4-trimethyl-1,3-pentanediol) diphosphite, *See* Weston® 491, 34301
Diphenyl ether, *See* diphenyl oxide, 8598, geranium crystals, 12887, phenyl ether, 23633
Diphenyl-2-ethylhexyl phosphate, *See* diphenyl octyl phosphate, 8597
***trans*-1,2-diphenylethylene**, *See* toluylene, 31960
diphenylethyne, *See* Tolan, 31933
N,N'-diphenyl guanidine, *See* Vanax® DPG, 33290
N,N'-Diphenylguanidine, *See* Akrochem® DPG, 1168, Dynamine, 9355, Perkacit® DPG, 23281
Diphenyl guanidine, *See* Akrochem® DPG, 1168, D.P.G, 7624, Dynamine, 9355, melaniline, 19259, Phenaldine, 23585, Vulcogene ND, 34115, Vulkacit® D/C, 34130
***sym*-diphenylguanidine**, *See*
***sym*-diphenylguanidine**, *See* Akrochem® DPG, 1168, Dynamine, 9355, Perkacit® DPG, 23281, Vanax® DPG, 33290, Vulcaid DPG, 34081
1,3-Diphenylguanidine, *See* Akrochem® DPG, 1168, Dynamine, 9355, Perkacit® DPG, 23281, Vanax® DPG, 33290, Vulcaid DPG, 34081
N,N'-diphenylguanidine, *See* Akrochem® DPG, 1168, Dynamine, 9355, Perkacit® DPG, 23281, Vanax® DPG, 33290, Vulcaid DPG, 34081
Diphenyl-guanidine salt of mercaptobenzothiazole, *See* Accelerator W80, 216
Diphenyl guanidine/dibenzyl dithiocarbaminic acid, *See* Accelerator W29, 215
Diphenyl hydrogen phosphite, *See* Weston® DPP, 34309
diphenyl isodecyl phosphite, 8594, *See* Doverphos® 8, 8837, Weston® DPDP, 34308
Diphenyl isodecyl-phosphate, *See* Syn-O-Ad® 8479, 30556
diphenyl isooctyl phosphite, 8595, *See* Doverphos® DPIOP, 8848, Weston® ODPP, 34314
diphenyl isotridecyl phosphite, 8596
4,4'-diphenylmethane diisocyanate, *See* Isonaphthol, 15377, Rubinate® LF-168, 27108
diphenylmethane-4', 4-diisocyanate, *See* MDI, 19188
Diphenyl methanol, *See* Benzhydrol, 3965
Diphenylmethyl Diisocyanate;, *See* Rubinate® LF-168, 27108
4-(Diphenylmethylene)-1,1dimethylpiperidinium methyl sulfate, *See* Variton, 33432
Diphenyl mono *o*-xylenyl phosphate, *See* Dow Plasticizer No. 5, 8885, Dow Plasticizer No. 55, 8886
diphenyl mono(*p-tert*-butylphenyl)phosphate, *See* Syn-O-Ad® 8478, 30555

Chemical and Trade Names

mono(*p-tert*-butylphenyl)phosphate, *See* Syn-O-Ad® 8485, 30559
diphenyl octyl phosphate, 8597, *See* diphenyl octyl phosphate, 8597
2,2-di(4-phenylol)propane, *See* Bisphenol A, 4283
diphenylolpropane, *See* Bisphenol A, 4283
diphenyloxazole, 8602
2,5-diphenyloxazole, *See* diphenyloxazole, 8602
diphenyl oxide, 8598, *See* phenyl ether, 23633
diphenyloxide-4,4'-disulfohydrazide, *See* Celogen® OT, 5643
1,4-diphenyl-3-(phenylamino)-1H-1,2,4-triazolium inner salt, *See* Nitron, 21367
Diphenyl phenylene diamine, *See* N,N'-diphenyl-*p*-phenylenediamine, 8603
Diphenyl-*p*-phenylene diamine, 65:35 ratio, *See* Agerite® HP-S, 1018
N,N'-Diphenyl-*p*-phenylene diamine, 8603, *See* Agerite® DPPD, 1016, N,N'-diphenyl-*p*-phenylenediamine, 8603, JZF, 15638, Naugard® J, 20796, Permanax DPPD, 23361
N,N'-Diphenyl-*p*-phenylenediamine, 8603, *See* Agerite® DPPD, 1016, N,N'-diphenyl-*p*-phenylenediamine, 8603, JZF, 15638, Naugard® J, 20796, Permanax DPPD, 23361
diphenyl phosphite, 8599, *See* Doverphos® 213, 8843, Doverphos® DPP, 8849, Weston® DPP, 34309
Diphenyl phosphonate, *See* Weston® DPP, 34309
Diphenyl PPD, *See* Agerite® DPPD, 1016, N,N'-diphenyl-*p*-phenylenediamine, 8603, JZF, 15638, Naugard® J, 20796, Permanax DPPD, 23361
1,3-Diphenyl propenone, *See* Chalkone, 5849
1,3-Diphenyl-2-propen-1-one, *See* Chalkone, 5849
Diphenylsilanediol, *See* CD6150, 5516
10% Diphenylstibine 2-ethyl hexoate solution in dioctylphthalate, *See* bioMeT 14, 4170
N,N-diphenylsulfourea, *See* A-1, A-1 Thiocarbanilide, 5, Akrochem® Thio No. 1, 1181
N,N'-diphenylthiocaramide, *See* diphenylthiourea, 8604
S-diphenylthiocarbamide, *See* diphenylthiourea, 8604
Diphenylthiocarbazone, *See* dithizone, 8731
diphenylthiourea, 8604, *See* Stabilizer C, 29407
***sym*-diphenylthiourea**, *See* diphenylthiourea, 8604, dithizone, 8731
N,N-diphenylthiourea, *See* A-1, A-1 Thiocarbanilide, 5, diphenylthiourea, 8604
1,2-diphenyl-2-thiourea, *See* Akrochem® Thio No. 1, 1181
1,3-diphenylthiourea, *See* diphenylthiourea, 8604
1,3-diphenyl-2-thiourea, *See* A-1, A-1 Thiocarbanilide, 5, diphenylthiourea, 8604
Diphenylurea, *See* carbanilide, 5180
***sym*-diphenylurea**, *See* carbanilide, 5180, Zeonet U, 34722
N,N'-Diphenylurea, *See* carbanilide, 5180, Zeonet U, 34722
1,3-diphenylurea, *See* carbanilide, 5180, Zeonet U, 34722
diphenyl tolyl ester phosphoric acid, *See* Disflamoll® DPK, TPK, 8641
diphenyl tolyl phosphate, *See* Disflamoll® DPK, TPK, 8641
Dipher, *See* Zineb, 34807
Diphone, 8605, *See* Dapsone, 7751
diphosgene, *See* Perstoff, 23430, Superpalite, 30260
Diphosphoric acid, *See* sodium pyrophosphate, 28814
diphosphoric acid disodium salt, *See* sodium acid pyrophosphate, 28727
diphosphoric acid tetrapotassium salt, *See* Tetrakal, 31375
diphosphorus pentoxide, *See* phosphorus pentoxide, 23746
Di-p-hydroxy-triphenylmethane, *See* Leucobenzaurin, 17022
Diphyl® T, 8606
Dipicrylamine, 8607, *See* Hexil, 14031
Dipirartril-tropico, *See* DMSO, 8762
Diplast® N, *See* Jayflex® DINP, 15552
Diplast® R, *See* diisodecyl phthalate, 8462
Diplast® TM, *See* Jayflex® TOTM, 15559
Diplast® TM8, *See* Jayflex® TOTM, 15559
Dipofene, *See* Diazinon Liquid, 8344
Dipolymer, 8608
dipotassium carboate, *See* potassium carbonate, 24744, 24745
dipotassium disulfite, *See* potassium metabisulfite, 24761
dipotassium glycyrrhizinate, *See* Ritamectant K2, 26815
dipotassium hexafluorotitanate, *See* Aflammit TI, 939
dipotassium hexafluorozirconate, *See* Aflammit ZR, 941
dipotassium hydrogen orthophosphate, *See* potassium phosphate, dibasic, 24773
dipotassium monotitanium hexafluoride, *See* Aflammit TI, 939
dipotassium orthophosphate, *See* potassium phosphate, dibasic, 24773
dipotassium peroxydisulfate, *See* Anthion, 2519
dipotassium phosphate, *See* potassium phosphate, dibasic, 24773
Dipotassium titanium hexafluoride, *See* Aflammit TI, 939
Dipotassium zirconium hexafluoride, *See* Aflammit ZR, 941
Dippel's animal oil, *See* bone oil, 4402
Dippel's oil, *See* bone oil, 4402
Dipping metal, 8609
Diprane®, 8610
dipropyl-2,2'-dihydroxy-amine, *See* diisopropanolamine, 8466
N,N-Di-n-propyl-2,6-dinitro-4-trifluoromethylaniline, *See* Trifluralin, 32252
Dipropyl ketone, *See* Butyrone, 4804
O,O-Di-n-propyl phosphorodithioic acid, *See* NPP-1, 21752
N,N-Dipropyl-4-trifluoromethyl-2,6-dinitroaniline, *See* Trifluralin, 32252
4-(dipropylamino)-3,5-dinitrobenzenesulfonamide, *See* oryzalin, 22347
Dipropylene glycol, 8611, *See* Larostat® 377 DPG, 16796
Dipropylene glycol n-butyl ether, *See* Dowanol® DPnB, 8893
Dipropylene glycol cocamine, *See* Propomeen C/12, 25194
Dipropylene glycol diacrylate, *See* Laromer®, 16790
Dipropylene glycol dibenzoate/diethylene glycol dibenzoate 1/1 blend, *See* Benzoflex 2-45, 3970
Dipropylene glycol methyl ether, *See* Dowanol® PPM, 8899, Icinol DPM, 14815
Dipropylene glycol methyl ether acetate, *See* Arcosolv® DPMA, 2813
Dipropylene glycol monomethyl ether, *See* Icinol DPM, 14815 Poly-Solv® DPM, 24555
Dipropylene glycol salicylate, *See* Dipsal, 8615
Dipropylene glycol tallowamine, *See* Propomeen T/12, 25195
Diprosin A-100, 8612
Diprosin K-80, 8613
Diprosin N-70, 8614
Dipsal, 8615
dipterex, *See* trichlorfon, 32212
Dipterex®, 8616
Dipterex® 80, 8617
Diquanide, *See* Biquanide, 4218
Diquat dibromide, *See* Katalon, 15782, Midstream, 19723, Power Diquat, 24820, Reglone, Reglox, 26017
diquat-paraquat, *See* Dukatalon, 9088, Farmon PDQ, 11490
diquat-paraquat, *See*
Di-Quinol, *See* Yodoxin, 34658
6,6'-diquinolinylurea bis methosulfate, *See* Acaprin®, 181, 1,3-di-6-quinolylurea, 8618
N,N'-di-6-quinolinylurea bis methosulfate, *See* Acaprin®, 181, 1,3-di-6-quinolylurea, 8618
1,3-di-6-quinolylurea, 8618
***sym*-di-(6-quinolyl)urea bis methosulfate,** *See* Acaprin®, 181, 1,3-di-6-quinolylurea, 8618
1,3-Di-6-quinolylurea bismethosulfate, *See* Acaprin®, 181, 1,3-di-6-quinolylurea, 8618
diresorcinolphthalein, *See* fluorescein, 12080, 12081, 12082
Diresul, 8619
Direx, *See* Anilazine, 2465
Direx 4L, *See* diuron, 8736
Direx 80W, *See* diuron, 8736
Direx® 4L, 8620
Direxiode, *See* Yodoxin, 34658
Direz, *See* Anilazine, 2465
Dirigold, *See* oranium bronze, 22269
Dirimal, *See* oryzalin, 22347
Diroval, 8621
Dirox, *See* Valadol, 33238
Dirubin, 8622, *See* aluminum oxide, 1919
Disaccharide, *See* α-D-Lactose, 16589
Disadine, 8623
N,N'-Disalicylidene-1,2-propane diamine, *See* Keromet MD 60, MD 80, 16092, KMD-50, 16248, Metal Deactivator S, 19428
Discase, 8624, *See* Chymopapain, 6284
Discelite, *See* Diatomaceous silica, 8336
Discharge Agent DP, 8625
Discharge Lake R and RR, *See* paranitraniline red, 22780
Disco 727, 8626
Discodye 1148, 8627
Discofix DBA, 8628
Discol 715, 8629
Discol 1457, 8630
Discol DFW, 8631, *See* dioctyl sodium sulfosuccinate, 8550
Discolite, 8632
Discoloc 70-A, 8633
Discolube 473-A, 8634
Discopen 216, 8635
Discosoft 1043-S, 8636
Discoterge 326-D, 8637
Discozone MAC, 8638
Di-sec-octyl phthalate, *See* Witcizer 312, 34454
Di-sec-octyl phthalate (acgih,osha), *See* Witcizer 312, 34454
Disfico, 8639
Disflamol®TOF, *See* trioctyl phosphate, 32349
Disflamoll DPK, *See* Disflamoll® DPK, TPK, 8641
Disflamoll TCA, *See* Genomoll P, 12838
Disflamoll®, 8640
Disflamoll® DPK, *See* diphenylcresyl phosphate, 8601
Disflamoll® DPK, TPK, 8641
Disflamoll® DPO, *See* diphenyl octyl phosphate, 8597
Disflamoll® TOF, *See* trioctyl phosphate, 32349
Disflamoll® TP, 8642
Disilyn, *See* benzethonium chloride, 3963
Disinfection oil, *See* saprol, 27448
Disodium 2,2'-dihydroxy-4,4'-dimethoxy-5,5'-disulfobenzophenone, *See* Benzophenone-9, 3979
Disodium 3,3'-(dodecylimino) dipropionate, *See* disodium lauryliminodipropionate, 8650
Disodium alkyl amidopolyethoxy sulfosuccinate, *See* Aerosol 200, 880
Disodium alkyl sulfosuccinate, *See* Aerosol® 501, 884
disodium aminobenzoyl *sym*-disodium-*m*-amino-benzoyl-*m*-amino-*p*-methyl-benzoyl-1-naphthyl-amino-4,6,8-trisulfonate-urea, *See* Poulenc 309, 24803
Disodium C_{12}-C_{15} pareth sulfosuccinate, *See* Emcol® 4300, 10049
Disodium caproamphodipropionate, *See* Miranol® S2M-SF Conc, 19910
Disodium capryloamphodiacetate, *See* Ampholak XJO, 2322, Amphoterge® J-2, 2349, Mackam 2CY, 18133, Miranol® J2M Conc, 19902, Miranol® JB, 19905, Schercoteric CY-2, 27705
Disodium capryloamphodipropionate, *See* Amphoterge® KJ-2, 2352, Miranol® J2M-SF Conc, 19903, Miranol® JBS, 19906, Monateric 811, 20143, Monateric 1000, 20147, Zoharteric LF-SF, 34883

disodium carbonate, compound with hydrogen peroxide (2:3), *See* sodium percarbonate, 28807
disodium carbonate, hydrogen peroxide (2:3), *See* sodium percarbonate, 28807
Disodium cetearyl sulfosuccinate, *See* Empicol® STT, 10339
Disodium cetyl stearyl sulfosuccinamate, *See* Empimin® MK/B, 10430
disodium chromate, *See* sodium chromate, 28750
Disodium cocamido MEA-sulfosuccinate, *See* Mackanate CM, 18182, Rewopol® SBC 212, 26370, Schercopol CMS-Na, 27665
Disodium cocamido MIPA sulfosuccinate, *See* Mackanate CP, 18183, Monamate C-1142, 20104, Monateric 805, 20141
Disodium cocoamphodiacetate, 8643, *See* Amphoterge® W-2, 2356, Dehyton® PG, 7994, Mackam 2C, 18131, Miranol® 2CIB, 19884, Miranol® FB-NP, 19895, Monateric CDTD, 20158, Monateric CDX-38, 20160, Monateric CSH-32, 20165, Rewoteric® AM 2C NM, 26414, Schercoteric MS-2, 27708, Velvetex® CDC, 33555, Zoharteric D-2, 34879
Disodium cocoamphodiacetate, propylene glycol, *See* Miranol® C2M Conc. NP-PG, 19887
Disodium cocoamphodiacetate, sodium lauryl sulfate and hexylene glycol, *See* Miracare® 2MCA, 19849
Disodium cocoamphodipropionate, *See* Amphoterge® K-2, 2351, Miranol® C2M-SF 70%, 19888, Miranol® FBS, 19896, Monateric CEM-38, 20161, Rewoteric® AM 2C SF, 26415
Disodium cocoamphodipropionate, sodium lauryl sulfate and hexylene glycol, *See* Miracare® 2MCA-SF, 19850
Disodium cocoyl glutamate, *See* Acylglutamate CS-21, 598
Disodium cocoyl sulfosuccinamate, *See* Empimin® MH, 10428
Disodium cocoyl/tallowyl glutamate, *See* Acylglutamate GS-21, 602
Disodium deceth-6 sulfosuccinate, *See* Aerosol® A-102, 885, Mackanate A-102, 18179, Monawet TD-30, 20190
disodium dichromate dihydrate, *See* sodium dichromate, 28756
Disodium difluoride, *See* sodium fluoride, 28764
disodium dihydrogen ethylene diamine tetraacetetate, *See* disodium EDTA, 8644, edetate disodium, 9613
disodium dihydrogen ethylenediaminetetraacetate, *See* sodium versenate, 28831
disodium dihydrogen pyrophosphate, *See* Taterfos®, 30844
disodium dihydrogen pyrophosphate hexahydrate, *See* sodium acid pyrophosphate, 28727
Disodium dimethicone copolyol sulfosuccinate, *See* Mackanate DC-30, 18184
Disodium distyrylbiphenyl disulfonate, *See* tinopal® CBS-X, 31828
disodium disulfite, *See* sodium metabisulfite, 28783
disodium dithionite, *See* sodium hydrosulfite, 28771
disodium edetate, *See* disodium EDTA, 8644, edetate disodium, 9613
disodium edetate USP, *See* Sequestrene® NA2 Edetate USP, 27998
disodium EDTA, 8644, *See* edetate disodium, 9613, Questal DI 0770, 25661, Trilon® BD, 32325
Disodium EDTA dihydrate, *See* Sequestrene® NA2, 27997
disodium ethylenediaminetetraacetate, *See* sodium versenate, 28831
disodium hexafluorosilicate, *See* SSF, 29390
disodium hydrogen phosphate, *See* disodium phosphate, 8652, tasteless salts, 30843
disodium hydrogen phosphate, dihydrate, *See* disodium phosphate, dihydrate, 8653
disodium hydrogen pyrophosphate, *See* Victor Cream®, 33727
disodium IMP, *See* disodium inosinate, 8645
disodium inosinate, 8645
disodium 5'-inosinate, *See* disodium inosinate, 8645
disodium isodecyl sulfosuccinate, *See* Aerosol® A-268, 888
disodium laneth-5 sulfosuccinate, 8646, *See* Incrosul LAFS, 15053, Rewolan® 5, 26314
disodium lauramido MEA-sulfosuccinate, 8647, *See* Geropon® SBL-203, 12912, Incrosul LMS, 15055, Mackanate LM-40, 18189, Marlinat® SRN 30, 18856, Rewopol® SBL 203, 26377
Disodium laureth sulfosuccinate, *See* Geropon® ACR/4, 12903, Geropon® SBFA-30, 12911, Incrosul LTS, 15058, Mackanate EL, 18187, Rewopol® SBFA 30, 26376, Schercopol LPS, 27668, Setacin 103 Spezial, 28081
Disodium laureth-3 sulfosuccinate, *See* Surfagene S 30, 30361
Disodium laurethsulfosuccinate, *See* Lumorol K 28, 17814, Stepan-Mild® SL3, 29682, Texapon SB-3, 31476, Stepan-Mild® LSB, 29681
Disodium lauroamphodiacetate, *See* Amphoterge® L Special, 2353, Empigen® CDL60, 10367, Mackam 2L, 18134, Miranol® BM Conc, 19885, Miranol® H2M Conc, 19898, Monateric 951A, 20145, Rewoteric® AM 2L-40, 26416, Zoharteric DJ, 34880
Disodium lauroamphodiacetate and sodium trideceth sulfate, *See* Miracare® BT, 19853
Disodium lauroamphodipropionate, *See* Miranol® H2M-SF Conc, 19899
Disodium lauryl beta-iminodipropionate, *See* Monateric 1188M, 20149
disodium lauryl ethoxy sulfosuccinate, 8648, *See* Empicol® SDD, 10337
disodium lauryl sulfosuccinate, 8649, *See* Empicol® SLL, 10338, Geropon® LSS, 12908, Incrosul LS, 15056, Mackanate LO, 18190, Monamate LA-100, 20106, Rewopol® SBF 12, 26375, Setacin F Spezial Paste, 28082, Varsulf® SBF-12, 33459
disodium lauryl sulfosuccinate - propylene glycol, *See* Rhodaterge® SSB, 26668
disodium lauryliminodipropionate, 8650
disodium molybdate dihydrate, *See* sodium Molybdate, dihydrate, 28796
Disodium mono-and didodecyl diphenyl oxide disulfonate, *See* Aerosol® DPOS-45, 881
Disodium monofluorophosphate, *See* Albaphos Dental Na 211, 1348
Disodium Monosilicate, *See* sodium metasilicate, 28788
Disodium N-[3-(dodecyloxy)propyl]sulfosuccinamate, *See* Octosol A-1, 22021
Disodium N-lauryl β-iminodipropionate, *See* disodium lauryliminodipropionate, 8650, Deriphat 160, 8090
Disodium N-lauryl iminodipropionate, *See* Miranol® H2C-HA, 19897
Disodium N-octadecyl sulfosuccinamate, *See* Octosol A-18, 22022
Disodium nonoxynol-10 sulfosuccinate, *See* Aerosol® A-103, 886, Mackanate A-103, 18180, Monawet 1240, 20177
Disodium N-tallow aminodipropionate, *See* Mirataine® A2P-TS-30, 19921
Disodium N-tallow- β iminodipropionate, *See* Deriphat 154, 8089, disodium tallowiminodipropionate, 8655
Disodium octaborate, *See* Polybor®, 24369
Disodium octaborate tetrahydrate, *See* Polybor®, 24369, Tim-Bor®, 31797
Disodium octylphenoxy sulfosuccinate, *See* Geropon® S-1585, 12910
Disodium oleamide PEG-2 sulfosuccinate, *See* Monamate OPA-100, 20109, Texapon SH 100, 31479
Disodium oleamido MEA-sulfosuccinate, *See* Incrosul OMS, 15059, Mackanate OM, 18193, Schercopol OMS-Na, 27669
disodium oleamido MIPA sulfosuccinate, 8651, *See* Emcol® 416L, 10045, Mackanate OP, 18194, Sole Terge 8, 28914
Disodium oleamido PEG-2 sulfosuccinate, *See* Anlonyx® 12S, 2477, Mackanate OD-35, 18192, Standapol® SH-100, 29479
Disodium oleamido PEG-2 sulfosuccinate and sodium lauryl sulfate, *See* Standapol® SHC-101, 29480
Disodium oleamido-MIPA sulfosuccinate, *See* Emcol® K8300, 10062
Disodium oleth sulfosuccinate, *See* Mackanate O-3, 18191
Disodium oleth-3 sulfosuccinate, *See* Incrosul OTS, 15060
disodium orthophosphate, *See* disodium phosphate (anhydrous), 8652, tasteless salts, 30843
Disodium PEG-4 cocamido MIPA-sulfosuccinate, *See* Rewopol® SBZ, 26379
disodium peroxodisulfate, *See* sodium persulfate, 28809
Disodium phosphate, *See* Adjunct B, 708, tasteless salts, 30843
disodium phosphate (anhydrous), 8652
disodium phosphate, dihydrate, 8653
disodium pyrophosphate, *See* sodium acid pyrophosphate, 28727
disodium ricinoleamido MEA-sulfosuccinate, 8654, *See* Geropon® SBR-3, 12913, Mackanate RM, 18195, Monamate RMEA-40, 20110, Rewoderm® S 1333, 26311
disodium salt, *See* disodium EDTA, 8644, edetate disodium, 9613, sodium carbonate, 28744
disodium salt pyrosulfurous acid, *See* sodium metabisulfite, 28783
Disodium sitostereth-14 sulfosuccinate, *See* Rewoderm® SPS, 26312
Disodium stearoyl glutamate, *See* Acylglutamate HS-21, 604, 605
disodium stearyl sulfosuccinamate, *See* Aerosol® 18, 878
disodium sulfate, *See* sodium sulfate, 28821
disodium sulfide, *See* sodium sulfide, 28822
disodium sulfite, *See* sodium sulfite, 28823
Disodium tallow betaiminodipropionate, *See* Monateric TDB-35, 20175
Disodium tallow sulfosuccinamate, *See* Rewopol® B 1003, 26358
disodium tallowiminodipropionate, 8655, *See* Mirataine® T2C-30, 19931
disodium tartrate, *See* sodium tartrate, 28825
Disodium tetraborate pentahydrate, *See* Neobor, 20876
Disodium tridecyl sulfosuccinate, *See* Incrosul TS, 15061
Disodium undecylenamido MEA-sulfosuccinate, *See* Mackanate UM, 18196, Rewocid® SBU 185 P, 26297, Surfagene MB 1705, 30360
Disodium wheat germamido PEG-2 sulfosuccinate, *See* Mackanate WGD, 18197
Disodium wheat germamphodiacetate, *See* Mackam 2W, 18135
Disodium-calcium EDTA dihydrate, *See* Sequestrene® NA2Ca, 27999
Disodiumcocoamphopropionate, *See* Zoharteric D-SF, 34881
Disodiumtallamphodipropionate, *See* Miranol® L2M-SF Conc, 19908
Disolite, 8656
Disoquin, *See* Yodoxin, 34658
Dispargen, 8657
Disparit B, 8658
Disperbyk®, 8659
Disperbyk®-181, 8660
Dispercab, 8661, 8662
Dispercap, *See* Cellulose acetate propionate, 5630
Dispercel, 8663
Dispercoll C-74, 8664
Dispercryl, 8665
Disperfin, 8666
Dispergator NF, *See* Ablusol ML, 129
Disperkyd, 8667
Dispermid, 8668
Dispersant 1084, 8669
Dispersant LF-88, 8670

Disperse mb-61, *See* Rutenol, 27147
Disperse Yellow 60, *See* Abluwax EBS, 140
Disperse-Ayd, 8671
Disperse-Ayd 1, 8672
Disperse-Ayd 6, 8673
Disperse-Ayd 15, 8674
Disperse-Ayd W-22, 8675
Disperser NF, *See* Ablusol ML, 129
Dispersing agent NF, *See* Ablusol ML, 129
Dispersite, 8676
Dispersogen A, 8677
Dispersogen SL, 8678
Dispersol, 8679, 8680
Dispersol 103, 105, 8681
Dispersol Aca, *See* Ablusol ML, 129
Dispersol yellow PP, *See* Sudan I, 29938
Disperstat, 8682
Dispersyd, 8683
Dispervyn, 8684
Dispex®, 8685
Dispex® G40 and GA40, 8686
Displasol DP, 8687
Disponil AAP 307, 8688
Disponil AEP 5300, 8689
Disponil AES 13, 8690
Disponil FES 32, 8691
Disponil FES 92E, 8692
Disponil G 200, 8693
Disponil O 5, 8694
Disponil SML 100 F1, 8695
Disponil SML 104 F1, 8696
Disponil SML 120 F1, 8697
Disponil SMO 100 F1, 8698
Disponil SMO 120 F1, 8699
Disponil SMP 100 F1, 8700
Disponil SMP 120 F1, 8701
Disponil SMS 100 F1, 8702
Disponil SMS 120 F1, 8703
Disponil SMS 120 F1, *See* Ethylan® GES6, 11108
Disponil SSO 100 F1, 8704
Disponil STO 100 F1, 8705
Disponil STO 120 F1, 8706
Disponil STS 100 F1, 8707
Disponil STS 120 F1, 8708
Disponil STS 120 F1, *See* Polysorbate 65, 24571
Disponil SUS IC 8, 8709
Disponil® SUS 65, *See* disodium lauryl sulfosuccinate, 8649
Disponyl® SUS 87 Special, *See* dioctyl sodium sulfosuccinate, 8550
Dispray, 8710
Disprol, *See* Valadol, 33238
Disproportionated rosin, *See* Arizona DR-24, 2889
Dissolvine, 8711
Dist. methyl canolate, *See* Emery® 2231, 10150
Distaquaine, *See* Penicillin V, 23111
Distreareth-2 lauroyl glutamate, *See* Amiter LGS-2, 2200
1,2-Distearoyl-*sn*-glycero(3)phosphatidylcholine, *See* Phospholipon® SC, 23730
Distearyl dimethyl ammonium chloride, *See* Dehyquart DAM, 7985, Distearyldimonium chloride, 8714
Distearyl dimonium chloride, *See* Adogen® TA-100, 779, Adogen® TA-101, 780, Arosurf® TA-100, 3081, Arquad® 218-100, 3111
distearyl ester, *See* Distearyl thiodipropionate, 8713
N,N'-Distearyl ethylene diamine, *See* Wax C, 34234
Distearyl hydrogen phosphite, *See* Doverhos® 251, 8827
Distearyl methylamine, *See* Radiamine 6346, 25763
Distearyl pentaerythritol diphosphite, 8712
Distearyl pentaerythritol diphosphite, *See* Weston® 618, Weston® 619, 34304
Distearyl phosphite, *See* Weston® DSP, 34310
Distearyl phthalate, *See* Radia® 7505, 25744, Sicolub® DSP, 28218
Distearyl thiodipropionate, 8713, *See* Alkanox® 240-3T, 1704, Argus DSTDP, 2863, Carstab® DSTDP, 5357, Cyanox® STDP, 7493, Evanstab® 18, 11258, Lankromark® DSTDP, 16691
distearyl-3,3'-thiodipropionate, *See* Lowinox® DSTDP, 17667
distearyldimethylammonium chloride;, *See* Prepagen WK, 24894, Sumquat® 6045, 30123
Distearyldimonium chloride, 8714, *See* Adogen® TA-100, 779, Genamin DSAC, 12771, Genamin KSE, 12773, Varisoft® TA-100, 33428, Prepagen WK, 24894, Sumquat® 6045, 30123
Distearylpentaerythritol diphosphite, *See* Doverphos® S-680, 8850
Distec, 8715
Distillase® L-200, 8716, *See* Diazyme® L-200, 8352
Distillates (petroleum), hydrotreated light naphthenic, *See* Jayflex® 210, 15544
Distilled hydrogenated animal glyceride, *See* Myverol® 18-00, 20619
Distillex DS1, 8717
Distillex DS2, 8718
Distillex DS3, 8719
Distillex DS4, 8720
Distillex DS5, 8721
Distillex DS6, 8722
Distillex DS7, 8723
Distol 8, *See* tetrasodium EDTA, 31391
Distol 8 Irgalon, *See* Kalex 220 Crystal, 15668, *See* Kalex Liq. 50%, 15671
Distoline, 8724
Distyryl-phenyl, *See* Ecco White® FW-5, 9517
Disulfide, bis(dibutylthiocarbamoyl), *See* Akrochem® TBUT, 1178
Disulfiram, 8725, *See* Akrochem® TETD, 1180
disulfoton, *See* Disyston® FE-10, 8726
disulfuric acid, *See* fuming sulfuric acid, 12457
disulfurous acid, dipotassium salt, *See* potassium metabisulfite, 24761
disulfurous acid, disodium salt, *See* sodium metabisulfite, 28783
Disulone, *See* Dapsone, 7751
Disyston® FE-10, 8726
Ditallow amidoammonium methosulfate, *See* Rewoquat W 222 LM, 26397
Ditallow diamido methosulfate, *See* Incrosoft T-90, 15051
Ditallow dimethyl ammonium methosulfate, *See* Ablumine DT, 88
Ditallow dimonium chloride, *See* Arquad® 2T-75, 3104
Ditallow imidazoline methyl sulfate;, *See* Quaternium-27, 25607
Ditallowamine, *See* Armeen® 2T, 2930, Radiamine 6270, 25760
Ditallowdimethylammonium chloride, *See* Radiaquat 6470, 25794
Ditallowdimonium chloride, *See* Adogen® 470, 773
Ditensamine C, O and S, 8727
Diterdodecyl pentasulfide, *See* TPS 32, 37, 32115
Diterdodecyl trisulfide, *See* TPS 20, 32113
Diternonyl trisulfide, *See* TPS 27, 32114
Diterpene, *See* menthol terpine hydrate, 19321
Ditertio nonyl pentasulfide, *See* TPS 327, 32116
Ditetrazolium chloride, *See* tetrazolium chloride, 31395
Dithane, 8728
Dithane 945, *See* mancozeb, 18517
Dithane M 22, *See* Maneb, 18526
Dithane M 22 special, *See* Maneb, 18526
Dithane M-45, *See* Karamate, 15749, mancozeb, 18517, Maneb, 18526
Dithane S-31, *See* Maneb, 18526
Dithane Z 78, *See* Zineb, 34807
2,3-dithiabutane, *See* Sulfa-Hitech® 0382, 29961
Dithianone, 8729, *See* Delan-Col, 8006
2,2'-dithiobis(pyridine) 1,1'-dioxide, *See* Omadine® MDS, 22152
2,2'-dithiobis[benzothiazole], *See* Vulkacit® DM/C, 34131
β,β'-dithiobisalanine, *See* cystine, 7603
3,3'-dithiobis(2-aminopropionic acid), *See* cystine, 7603
4,4'-Dithiobis[morpholine], *See* Akrochem® Accelerator R, 1154, Sulfasan, 29970, Vanax® A, 33287
2,2'-Dithiobis(pyridine-N-oxide), *See* Omadine® MDS, 22152
Dithiocarb, *See* Octopol SDE-25, 22013
Dithiocarbamate, *See* Accelerator R5, 213
dithiocarbamate sodium, *See* Octopol SDE-25, 22013
Dithiocarbonic anhydride, *See* carbon bisulfide, 5219, carbon disulfide, 5223
dithiocyanatomethane, *See* Amerstat® 282, 2043, N-948® Biocide, 20641, Tolcide MBT, 31934
β,β'-dithiodialanine, *See* cystine, 7603
4,4'-Dithiodimorpholine, *See* Akrochem® Accelerator R, 1154, Naugex SD-1, 20801, Sulfasan, 29970, Vanax® A, 33287
2,2-Dithiodipyridine-1,1'-dioxide, *See* Omadine® MDS, 22152
dithiometon, *See* thiometon, 31700
dithionous acid, disodium salt, *See* sodium hydrosulfite, 28771
dithiooxamide, *See* hydrogen rubeanide, 14591
Dithiothreitol, *See* Clelands Reagent, 6402, 1,4-Dithiothreitol, 8730
1,4-Dithiothreitol, 8730
1,4-dithio-L-threitol, *See* 1,4-Dithiothreitol, 8730
dithizone, 8731
Dithymol diiodide, *See* Iosol, 15237
ditiocarb sodium, *See* Octopol SDE-25, 22013
Ditiozin, *See* Zineb, 34807
Ditolyguanidine, 8732
Di-o-tolylguanidine, *See* Ditolyguanidine, 8732
N,N-Di-*o*-tolylguanidine, *See* Perkacit® DOTG, 23280
N,N'-Di-*ortho*-tolylguanidine, *See* Vanax® DOTG, 33289
1,3-Di-*o*-tolylguanidine, *See* Akrochem® DOTG, 1167, Ditolyguanidine, 8732
N,N'-Di-*o*-tolylthiourea, *See* Accelerator A22, 192
Ditranil, *See* dicloran, 8395, Fumite Dicloran, 12459
Ditridecyl adipate, *See* Afilan TDA, 934, Kemester® 5654, 15949, Lexolube® 2X-109, 17158
Ditridecyl dimethyl ammonium chloride, *See* Varisoft® 2 TD, 33402
Di-tridecyl phosphite, *See* Weston® DTDP, 34311
ditridecyl phthalate, *See* Bisoflex L711P, 4261, Jayflex® DTDP, 15555, PX-126, 25437
Ditridecyl sodium sulfosuccinate, *See* Aerosol® TR-70, 899, Monawet MT-70, 20187
Ditridecyl thiodipropionate, *See* Argus DTDTDP, 2864, Cyanox® 711, 7486, Evanstab® 13, 11256
Ditrifon, *See* trichlorfon, 32212
Ditrimethylolpropane tetraacrylate, 8733
Ditrosol, *See* 4,6-dinitrocresol, 8542
ditungsten pentoxide, *See* tungsten blue, 32584
Diundecyl linear phthalate, *See* Jayflex® DUP, 15556
Diundecyl phthalate, *See* Jayflex® DUP, 15556, PX-111, 25435
Diurex, 8734
Diurex, *See* diuron, 8736, Karmex, 15771, Karmex®, 15772
Diurol, 8735
diuron, 8736, *See* Direx® 4L, 8620 Karmex, 15771 Karmex®, 15772
Diuron 4L, *See* diuron, 8736
Diuron 80, *See* diuron, 8736
Diuron Bayer, 8737
diuron-paraquat, *See* Dexuron, 8242
Divergan® F, R, 8738
Diver's liquid, 8739
divinylene sulfide, *See* thiophene, 31707
Divinylenimine, *See* Pyrrole, 25535
Divinyltetramethyldisiloxane, *See* CD6210, 5517
1,3-Divinyltetramethyldisiloxane, *See* CD6210, 5517
Divipan, 8740, *See* dichlorvos, 8394, Vapona, 33378
Divit-Urto, *See* vitamin D_2, 33940
Diwatex 30, 40, 8741
Dixie, *See* Catalpo, 5432
Dixie 5 and Dixie Special 102, 8742
Dixie Clay®, 8743
Dizene, 8744

Chemical and Trade Names

Diziktol, 8745
Dizinon, *See* Diazinon Liquid, 8344
DK-Ester, 8746
DKP, *See* potassium phosphate, dibasic, 24773
DL-3117, *See* iotrolan, 15244
DLG-10, 20, 8747
DLPA 375, 8748
DLS Base, 8749
D-Lube, 8750
DM, *See* Adamsite, 624
DM hydantoin, 8751
DM-2, 8752
DMAB, *See* Butter yellow, 4759
DMAE, *See* Alkanolamine, 1701, dimethylethanolamine, 8520, Dabco® DMEA, 7669, deanol, 7829
DMAEMA, *See* Ageflex FM-1, 983, Sipomer® 2M1M, 28528
DMAMP-80, 8753
DMASA, *See* daminozide, 7723
DMC, *See* dimethylaminoethyl chloride hydrochloride, 8516
DMCS, *See* CD5470, 5507
DMDK, *See* Aquatreat SDM, 2761, sodium dimethyldithiocarbamate, 28758, methyl namate®, 19551, Octopol SDM-40, 22014, Perkacit® SDMC, 23288, Thiostop N, 31717
DMDM hydantoin, 8754, *See* DMDM hydantoin, 8754, Nipaguard® DMDMH, 21269
DMDMH, *See* DMDM hydantoin, 8754, Glydant®, 13140, Mackstat® DM, 18234, Nipaguard® DMDMH, 21269
DMDS, *See* Sulfa-Hitech® 0382, 29961
DMDT, *See* methoxychlor, 19504
DME, *See* monoglyme, 20227
DMF, *See* dimethyl formamide, 8506
DMF (insect repellent), *See* dimethyl phthalate, 8508
DMFA, *See* dimethyl formamide, 8506
DMI-689, 8755
DMMP, *See* dimethylmethyl phosphonate, 8525, Fyrol® DMMP, 12539
DM-Nitrophen™, 8756
DMOC, *See* carboxin, 5255
DMP, 8757, 8758, *See* dimethoxypropane, 8501, dimethyl phthalate, 8508, Kemester® DMP, 15956, Kodaflex® DMP, 16271, Unimoll® DM, 32979
DMP-30, *See* Actiron NX 3, 492
DMPA, *See* Photocure 51, 23757
DMR-503, 8759
DMR-504, 8760
DMS, *See* MPI DMSA Kidney Reagent, 20393
DMS-4-828, 8761
DMS-70, *See* dimethyl sulfoxide, 8511, DMSO, 8762
DMS-90, *See* dimethyl sulfoxide, 8511, DMSO, 8762
DMSA, *See* MPI DMSA Kidney Reagent, 20393
DMSO, 8762, *See* Decap, 7845, dimethyl sulfoxide, 8511
DMSP, *See* Terracur® P, 31329
DMTT, *See* N-521® Biocide, 20640
DMU, *See* diuron, 8736
DNC, *See* 4,6-dinitrocresol, 8542
DNOC, *See* 4,6-dinitrocresol, 8542
4,6-DNOC, *See* 4,6-dinitrocresol, 8542
DNOCP, *See* Karathane, 15759
2,4-DNP, *See* Dinitra, 8540
DNPD, *See* Agerite® White, 1024
DNPDA, *See* Agerite® White, 1024
Dnsbp, *See* DDT, 7820
DNT, *See* D.N.T, 7622
2,4-DNT, *See* D.N.T, 7622
DNTP, *See* parathion, 22807
DO-160, 8763
DOA, *See* dioctyl adipate, 8548, Plasthall® DOA, 24002, PX-238, 25440, Wickenol® 158, 34393
Dobane (Detergent Alkylate), 8764
Dobanic Acid 83, *See* Witco® 1298, 34458
Dobanic Acid Jn, *See* Witco® 1298, 34458
Dobanic Acids JN and 83, 8765
Dobanol, 8766, 8767
Dobanol ethoxylates, 8768
Dobanol ethoxysulfates, 8769
Dobanox, 8770
Dobatex, 8771
Dobbin's reagent, 8772
Dobell solution, 8773
Dobendan, *See* Acetoquat CPC, 302, cetyl pyridinium chloride, 5823
Dock-Ban, 8774
Docklene, 8775
1-Docosanamine, N,N-dimethyl-,N-oxide, *See* Behenamine oxide, 3897
Docosanoic acid, 1,2,3-propanetriyl ester, *See* Syncrowax HR-C, 30534
1-docosanol, *See* Behenyl alcohol, 3899
13-docosenamide, *See* erucamide, 10770
***cis*-13-docosenamide**, *See* erucamide, 10770
(Z)-13-Docosenamide, *See* Armid® E, 2961, erucamide, 10770, Petrac® Eramide®, 23464
***cis*-13-docosenoamide**, *See* Unislip 1753, 33044
(Z)-13-docosenoic acid, *See* erucic acid, 10771
docsoamide, *See* Uniwax 1747, 33067
Doctor metal, 8776
docusate calcium, *See* Surfak, 30362
docusate potassium, *See* Dialose, 8284
docusate sodium, *See* Aerosol OT, 877, Aerosol® GPG, 882, Alcopol O, 1499, Anonaid TH, 2481, Bevaloid 1299, 4082, Cropol 60, 7210, dioctyl sodium sulfosuccinate, 8550, Elfanol® 883, 9828, Empimin® OP70, 10434, Modane Soft, 20015, Octowet 40, 22035
DODA-hydrochloride, *See* Onamine 18, 22167
1,2,3,4,7,8,9,10,13,13,14,14-dodecachloro-1,4,4a,5,6,6a,7,10,10a,11,12,12a-dodecahydro-1,4:7,10-dimethanodibenzo[a,e]cyclooctene, *See* Dechlorane® Plus 25, Plus 35, Plus 515, 7850
dodecahydrophenylamine nitrite, *See* Dichan 100, 8377, dicyclohexylamine nitrite, 8414
dodecahydrosqualene, *See* squalane, 29357
Dodecamethylpentasiloxane, *See* D6219.5, 7637
dodecamolybdophosphoric acid, *See* phosphomolybdic acid, 23731
Dodecanamide, N,N-bis(2-hydroxyethyl)-, *See* Ablumide LDE, 79, 80
Dodecanamide, N-(2-hydroxyethyl)-, *See* Ablumide LME, 79, 80
2-dodecanamidoethanol, *See* Ablumide LME, 80, Rewomid® L 203, 26328
1-Dodecanamine, *See* Armeen® 12, 2933, Amine 12, 2127
i-dodecanamine, N,N-dimethyl, *See* Onamine 12, 22164
1-Dodecanamine, N,N-dimethyl-, N-oxide, *See* Ablumox LO, 91, Rhodamox® LO, 26620
Dodecanamine, acetate, *See* Acetamin 24, 284
1-Dodecanamine, N,N-dimethyl-, N-oxide, *See* Ammonyx® LO, 2264, Empigen® OB, 10375
Dodecanaminium, N,N,N-trimethyl-, chloride, *See* Adogen® 412, 767, Arquad® 12-50, 3106
dodecanenitrile, *See* Nitrile 12, 21336
1-dodecanethiol, *See* n-dodecyl mercaptan,
n-dodecanoate, *See* Philacid 1200, 23660
dodecanoic acid, *See* lauric acid, 16856, Philacid 1200, 23660
n-dodecanoic acid, *See* lauric acid, 16856
dodecanoic acid, 2,3-dihydroxypropyl ester, *See* Ablunol GML, 101, glyceryl monolaurate, 13083
dodecanoic acid 1,2-ethanediyl ester, *See* Emalex EG-di-L, 9949, glycol dilaurate, 13100, Kemester® EGDL, 15958
dodecanoic acid 2-(2-hydroxyethoxy)ethyl ester, *See* PEG-2 laurate, 23041
dodecanoic acid methyl ester, *See* methyl laurate, 19545
dodecanoic acid, 1,2,3-propanetriyl ester, *See* Trilaurin, 32317
dodecanoic acid, monoester with 1,2,3-propanetriol, *See* glyceryl monolaurate, 13083
dodecanol, *See* Emery® 3326, 10169, Epal® 12, 10618
1-Dodecanol, 8777, *See* Emery® 3326, 10169 lauryl alcohol, 16864
Dodecan-1-ol, *See* Emery® 3326, 10169
Dodecanonitrile, *See* Nitrile 12, 21336
dodecanoyl chloride, *See* lauroyl chloride, 16860
dodecanoyl peroxide, *See* Alperox-F, 1816
bis(dodecanoyloxy)di-n-butylstannane, *See* Dabco® T-12, 7673
1-dodecene, *See* Neodene® 12, 20895
dodec-1-ene, *See* Neodene® 12, 20895
dodecene-1, 8778, *See* Neodene® 12, 20895
dodecenyl succinic anhydride, *See* Milldride® DDSA, 19782
n-Dodecenyl succinic anhydride (C12-linear), *See* Milldride® nDDSA, 19786
dodecoic acid, *See* lauric acid, 16856, Philacid 1200, 23660
dodecycldimethylamine oxide, *See* Ablumox LO, 91, Ammonyx® LO, 2264, Empigen® OB, 10375, Rhodamox® LO, 26620
dodecyl acrylate, *See* lauryl acrylate, 16863
Dodecyl alcohol, *See* 1-dodecanol, 8777, Emery® 3326, 10169, Exxal® 12, 11339, lauryl alcohol, 16864
Dodecyl alcohol, ethoxylated and sulfated, sodium salt, *See* Witcolate ES-2, Witcolate LES-60C, 34477
Dodecyl alcohol, hydrogen sulfate, sodium salt, *See* Witcolate 6400, 34469
dodecyl alcohol, monoether with polyethylene glycol, hydrogen sulfate sodium salt, *See* sodium laureth sulfate, 28778
dodecyl alcohol, monoether with polyethylene glycol, hydrogen sulfate sodium salt, *See* Standapol® AP Blend, 29468
Dodecyl ammonium sulfate, *See* Witcolate NH, 34478
Dodecyl benzene sulfonic acid, *See* Manro HA, 18593
Dodecyl dimethyl benzyl ammonium chloride, *See* Dehyquart LDB, 7986
Dodecyl dimethylamine (40%), tetradecyl dimethylamine (50%), hexadecyl dimethylamine (10%), *See* Adma® 246-451, 745
Dodecyl diphenyl ether disulfonic acid, *See* Poly-Tergent® 2A1 Acid, 24628
Dodecyl diphenyloxide disulfonic acid, *See* Dowfax 2A0, 8907
dodecyl ester of sulfuric acid, ammonium salt, *See* Octosol ALS-28, 22024, Rhodapon® L-22, L-22/C, 26635, Standapol® A, 29467
Dodecyl gallate, *See* Progallin LA, 25110
dodecyl methacrylate, *See* Ageflex FM-12, 986
n-Dodecyl methacrylate, *See* Ageflex FM-12, 986
Dodecyl 2-methyl-2-propenoate, *See* Ageflex FM-12, 986
N-dodecyl-2-pyrrolidone, *See* Agsol Ex 12, 1089
Dodecyl sodium sulfate, *See* Witcolate 6400, 34469
Dodecyl Sulfate Sodium, *See* Witcolate 6400, 34469
Dodecyl sulfate, sodium salt, *See* Witcolate 6400, 34469
Dodecyl sulfate, ammonium salt, *See* Witcolate NH, 34478
dodecyl sulfate, magnesium salt, *See* Standapol® MG, 29475
Dodecyl sulfate, triethanolamine salt, *See* Perlankrol® ATL40, 23324, Rhodapon® LT-6, 26638, Standapol® T, 29481, Witcolate TLS-500, 34479
n-Dodecyl sulfate sodium, *See* Witcolate 6400, 34469
dodecyl trichlorosilane, *See* CD6220, 5518
Dodecyl 3,4,5-trihydroxybenzoate, *See* Progallin LA, 25110
n-dodecyl trimethylammonium chloride, *See* Adogen® 412, 767, Arquad® 12-50, 3106, Octosol 562, 22020, Varisoft® LAC, 33423
Dodecyl/tetradecyl ether amine, *See* Jet Amine PE 1214, 15593
dodecylamine, *See* Amine 12, 2127, Armeen® 12, 2933, Armeen® 12D, 2934
n-dodecylamine, *See* Amine 12, 2127, Radiamine 6164, 25754
n-Dodecylamine, *See* Armeen® 12, 2933, Armeen® 12D, 2934
Dodecylbenzene, *See* Naxel DDB 500, 20815
Dodecylamine acetate, *See* Acetamin 24, 284
n-dodecylbenzene, *See* Alkylate 215, 1732

dodecylbenzene sodium sulfonate, *See* sodium dodecylbenzenesulfonate, 28760
dodecylbenzene sulfonate, *See* Tensaryl SB, 31147
dodecylbenzene sulfonate, isopropylamine salt, *See* Rhodacal® IPAM, 26586
n-Dodecylbenzenesulfonate, *See* Witco® 1298, 34458
n-dodecylbenzene sulfonic acid, *See* Ufacid K, 32736
n-Dodecylbenzenesulfonic acid, *See* Witco® 1298, 34458
dodecylbenzene sulfonic acid, 8779, *See* DeSonate SA, 8179, Emka DDBSA, 10210, Lumosäure A, 17815, Manro BA and NA, 18588, Maranil DBS, 18679, Marlon® AS3, 18878, Nansa® SSA, 20695, Pentine Acid 5431, 23182, Reworyl® K, 26404, Rhodacal® LA Acid, 26587, Stepantan® H-100, 29707, Witco® 1298, 34458
dodecylbenzene sulfonic acid (linear), *See* Calsoft LAS-99, 5030, Polystep® A-13, 24586
Dodecylbenzene sulfonic acid, calcium salt, *See* Nansa® 1042, 20681
Dodecylbenzenesulfonic acid, *See* Naxel AAS-98S, 20813, Rhodacal® ABSA, 26579, Witco® 1298, 34458
Dodecylbenzenesulfonic acid ammonium salt, *See* Ablusol DBM, 125
Dodecyl-benzenesulfonic acid calcium salt, *See* Ablusol DBC, 123
Dodecylbenzenesulfonic acid diethanolamine salt, *See* Ablusol DBD, 124
dodecylbenzenesulfonic acid, sodium salt, *See* sodium dodecylbenzenesulfonate, 28760
dodecylbenzenesulfonyl titanate, *See* LICA 09, 17196
Dodecylbenzenesulphonic Acid, *See* Witco® 1298, 34458
Dodecylbenzyl trimethyl ammonium chloride, *See* Country Fresh Disinfectant, 6905
Dodecylbenzyltrimonium chloride, *See* Country Fresh Disinfectant, 6905
n-dodecyldimethylamine, *See* Onamine 12, 22164
dodecyldimethylamine, *See* dimethyl lauramine, 8507, Onamine 12, 22164
n-dodecyldimethylamine oxide, *See* Ammonyx® LO, 2264, Empigen® OB, 10375, Rhodamox® LO, 26620
dodecyldimethylamine oxide, *See* Ablumox LO, 91
dodecyldimethylamine-tetradecyldimethylamine (65:35), *See* Adma® 1214, 746
α-dodecylene, *See* dodecene-1, 8778
dodecylguanidine monoacetate, *See* Dodine FL, WP, 8785, Radspor FT, 65WP, 25833
Dodecylisoquinolinium bromide, *See* Isothan Q-75, 15427
2-[2-(Dodecyloxy)ethoxy]ethanol, *See* Laureth-2, 16853
4-dodecyloxy-2-hydroxybenzophenone, 8780
1-dodecylpyridinium chloride, *See* Dehyquart C Crystals, 7982, laurylpyridinium chloride, 16867
Dodecylsulfuric acid, sodium salt, *See* Witcolate 6400, 34469
Dodecyl-3,3'-thiodipropionate, *See* Cyanox® LTDP, 7491
dodecylthioethanol, 8781
2-(Dodecylthio)ethanol, *See* dodecylthioethanol, 8781, DV-1936, 9296
Dodecyltrichlorosilane, *See* CD6220, 5518
n-Dodecyltrichlorosilane, *See* CD6220, 5518
Dodecyltrimethylammonium chloride, *See* Adogen® 412, 767, Arquad® 12-50, 3106
dodemorph, *See* F-238, 11376
dodemorph-acetate, 8782, *See* Meltatox®, 19313
Dodicor 2565, 8783
Dodigen, 8784
Dodigen 1881, *See* Radiaquat 6410, 6412, 25789 Radiaquat 6412, 25790 Rewoquat B 10, 26389
Dodine, *See* Dodine FL, WP, 8785, Radspor FT, 65WP, 25833
Dodine FL, WP, 8785
Dodoxynol series, 8786
Dodoxynol-4, *See* Rexol 65/4, 26470
Dodoxynol-5, *See* T-Det® DD-5, 30863
Dodoxynol-6, *See* Dodoxynol series, 8786
Dodoxynol-6, *See* Igepal® RC-520, 14874
Dodoxynol-9, *See* Prox-onic DDP-09, 25314
Dodoxynol-10, *See* Igepal® RC-620, 14875
Doff, 8787
Dog Off, 8788
doguadine, *See* Dodine FL, WP, 8785
DOHP, *See* Dioleyl hydrogen phosphite, 8554
Dohyfral A, *See* Vi-Alpha, 33706
Doktacillin, *See* Ampicillin, 2361
Dolan, Dolanit, 8789
Dolasol TF, 8790
Doler Brass, 8791
Dolicur, *See* dimethyl sulfoxide, 8511, *See* DMSO, 8762
Doligur, *See* DMSO, 8762
Doliprane, *See* Valadol, 33238
Dolofil, 8792
dolomite, chalk, *See* Calbux, 4929, calcium carbonate, 4941
Dolomol, 8793
Dolprone, *See* Valadol, 33238
Domba oil, *See* laurel nut oil, 16833
Domeboro, 8794, *See* aluminum acetate solution, 1898, novacetoform, 21648
Domestos, 8795
Dominate, 8796
Domolite, *See* CP Filler, 6928
Domoso, *See* dimethyl sulfoxide, 8511, DMSO, 8762
Donarit1, Donarit2, Donarit3, 8797
Donarite, 8798
Dontalol®, 8799
Donut Pyro®, 8800
Doom, 8801
doom barkteli, *See* red water bark, 25973
door plate brass, *See* brass, 4485
Doowpon, *See* Basfapon®, 3705
DOP, *See* Bis(2-ethylhexy) Phthalate, 4228, dioctyl phthalate, 8549, Reomol DCP, 26130, Witcizer 312, 34454
Dope, 8802
Dorallin A.R, *See* Penicillin V, 23111
Dorcolor®, 8803
Dore silver, 8804
Dorin, 8805
Dorindan, 8806, *See* Dorin, 8805
Dorix®, 8807
Dorlastan®, 8808
Dorlone II, *See* Telone, 31069
Dormakil, 8809
Dormone, 8810
Dorox, 8811
Doruplant, *See* ametryn, 2051
Dorvicide A, *See* Dowicide A, 8932
DOS, *See* Plasthall® DOS, 24005, Reomol DOS, 26131
Dosaflo, 8812
DOTT, *See* Accelerator A22, 192
Double Bond, 8814
Double Green S.F, *See* methyl green, 19540
double nickel salt, 8815
Double Shield, 8816
Double White, 8817
Double/Bubble, 8818
Doublet, 8819
Doublet Twitchell reagent, 8820
Doucil, 8821
Dovenix, *See* nitroxynil, 21384
Doverchlor 10, 8822
Doverguard® 152, *See* Paroil® 152, 22857
Doverguard® 170,700,152,700-S, *See* Chlorinated paraffin, 6105
Doverguard® 700, 8823
Doverguard® 5761, *See* Paroil® 5761, 22858
Doverguard® 8133, 8824
Doverguard® 8207-A, 8825
Doverguard® 8410, 8826
Doverhos® 251, 8827
Doverlub 8136, 8828
Doverlub 8506, 8829
Doverlub 8527, 8830
Doverlub 8531, 8831
Doverlub 8621, 8832
Doverphos®, 8833
Doverphos® 10, 8838, *See* Doverphos® 10, 8838
Doverphos® 10-HR, 8839
Doverphos® 11, 8840
Doverphos® 12, 8841
Doverphos® 213, 8843, *See* diphenyl phosphite, 8599
Doverphos® 253, *See* Dioleyl hydrogen phosphite, 8554
Doverphos® 269, *See* diisotridecyl hydrogen phosphite, 8471
Doverphos® 271L, *See* dilauryl phosphite, 8475
Doverphos® 274, 8844, *See* dilauryl phosphite, 8475
Doverphos® 298, *See* diisooctyl hydrogen phosphite, 8464
Doverphos® 4, 4-HR, *See* Trisnonylphenyl phosphite, 32411
Doverphos® 4-HR, 8845, *See* Doverphos®4, 8834
Doverphos® 6, 8835
Doverphos® 7, 8836
Doverphos® 8, 8837
Doverphos® 8 DPDP, *See* diphenyl isodecyl phosphite, 8594
Doverphos® 53, 8842
Doverphos® 75, *See* diphenyl isotridecyl phosphite, 8596
Doverphos® DIOP, 8846, *See* diisooctyl phosphite, 8465
Doverphos® DPGDP, 8847
Doverphos® DPIOP, 8848, *See* diphenyl isooctyl phosphite, 8595
Doverphos® DPP, 8849, *See* diphenyl phosphite, 8599
Doverphos® S-680, 8850, *See* Distearyl pentaerythritol diphosphite, 8712
Doverphos® S-686, S-687, *See* Distearyl pentaerythritol diphosphite, 8712
Doverphos® TIOP, 8851
Doverphos® TLP, 8852
Doverphos®4, 8834
Dow 276-V2, 8853
Dow Corning® 1-2531 Release Coating, 8854
Dow Corning® 7 Compound, 8855
Dow Corning® 24 Emulsion, 8856
Dow Corning® 29, 54, Q2-5211 Superwetting Agent, Q2-5220 Surfactant, Q4-3667 Fluid, *See* dimethicone copolyol, 8494
Dow Corning® 28, *See* dimethicone copolyol, 8494
Dow Corning® 190 Surfactant, 8857
Dow Corning® 197 Surfactant, 8858
Dow Corning® 200 Fluid, 8859
Dow Corning® 203 Fluid, 8860
Dow Corning® 344 Fluid, 345 Fluid, 8861
Dow Corning® 556 Fluid, 8862
Dow Corning® 929, 8863
Dow Corning® 1500 Compd, 8864
Dow Corning® 3225C Formulation Aid, 8865
Dow Corning® 7224, 8866
Dow Corning® ACH-303, 8867
Dow Corning® ACH7-308, 8868
Dow Corning® Antifoam 1410, 8869
Dow Corning® Antifoam A, 8870
Dow Corning® Antifoam C, 8871
Dow Corning® AZG-368, 8872
Dow Corning® AZG-370, 8873
Dow Corning® FF-400, 8874
Dow Corning® FS-1265 Fluid, 8875
Dow Corning® Q1-6106, 8876
Dow Corning® QF1-3593A, 8877
Dow Corning® Z-6020, 8878, *See* Aminoethylamino propyltrimethoxy silane, 2171
Dow Corning® Z-6030, 8879
Dow Corning® Z-6032, 8880
Dow Corning® Z-6040, 8881
Dow Corning® Z-6075, 8882
Dow Corning® Z-6076, 8883, *See* chloropropyl trimethoxysilane, 6138
Dow DBR, 8884
Dow Plasticizer No. 5, 8885
Dow Plasticizer No. 55, 8886
Dow Shield, 8887

24453
Drewpol® 10-4-O, 9014
Drewpol® 3-1-O, 9015
Drewpone® 60K, 9016, *See* Ethylan® GES6, 11108
Drewpone® 65K, 9017, *See* Polysorbate 65, 24571
Drewpone® 80K, 9018
Drewsorb 60, *See* Ablunol S-60, 110
Drewsorb® 60K, 9019
Drewsperse® 611, 9020
Drewsperse® S-825, 9021
Drewtrol® 6955S, 9022
Drexar, *See* MSMA, 20417
Drexar 530, 9023, *See* Neoarsycodyl, 20871
Drexel diuron 4L, *See* diuron, 8736
Drexel MSMA 6 Plus, *See* Neoarsycodyl, 20871
Drexel MSMA 6.6, *See* Neoarsycodyl, 20871
Dri Film® DF1040, 9024
Dricoid® 200, 9025, *See* diglyme, 8455
Dricold, 9026
dried egg white, *See* Albumen, 1402
drierite, 9027
Driers, 9028
Drift Proof, 9029
Driftol, 9030
Drikalite®, 9031
dri-kil, *See* rotenone, 27000
Drikold, 9032
drill rod brass, *See* brass, 4485
Drimarene®, 9033
Drimax®, 9034
Drimix®, 9035
Drip Syrup, *See* golden syrup, 13184
Drisdol, *See* vitamin D+72, 33940
Dri-Sil, 9036
Drisorb, 9037
Drisoy®, 9038
Dritan, 9039
Driton, 9040
Drittel silver, 9041
Drivanil, 9042
Driverit, 9043
Driverol MPL, 9044
Driverol OMM, 9045
Driveron, 9046
Drivolan, 9047
Drivosol, 9048
Driwal, *See* Ceresit, 5747
Drize, *See* paclobutrazol, 22541
Drmacid, *See* Vulkacit® M, Vulkacit® Merkapto/C, 34134
Drometrizole, *See* Topanex 100BT, 32019
Dromisol, *See* dimethyl sulfoxide, 8511, DMSO, 8762
Droncit®, 9049
Drontal, 9050
Drontal Plus, 9051
drop chalk, *See* calcium carbonate, 4942
Drossa, 9052
Drott, 9053
Droxarol, *See* bufexamac, 4655
Droxaryl, *See* bufexamac, 4655
Droxol 200, 9054
Droxychrome, 9055
Drupina 90, *See* AAprotect, 16
dry and clear, *See* Benzoyl peroxide, 3983
Dry Flo®, 9056 *See* aluminum starch octenyl succinate, 1928
dry ice, 9057
Dry Lightning, 9058
Dry Pexol® 200, 9059
Dry Pexol® 243, 9060
Dry Seed TRIGGRR, 9061
Dry Size XL20C, 9062
Dry-Blend® NCG Fiber, 9063
Drymax®, 9064
Drymet® 59, 9065
Dryobalanops camphor, *See* DL-Borneol, 4428
Dryspersion®, 9066
Drytech, 9067
DS 60, *See* Nacconol® 40G, 20647
DSM, *See* demeton-S-methyl, 8055
DSMA, *See* arrhenal, 3128
D-sorbite, *See* sorbitol, 29113
DSP, *See* disodium phosphate (anhydrous), 8652, tasteless salts, 30843
DSP-2, *See* disodium phosphate, dihydrate, 8653
DSP-O, *See* disodium phosphate (anhydrous), 8652
DSS, *See* Dapsone, 7751, dioctyl sodium sulfosuccinate, 8550
DSTDP, *See* Distearyl thiodipropionate, 8713
D-steel, 9068
DSX 1514, 9069
DTA, *See* Dithianone, 8729
DTBP, *See* Aztec® Di-*t*-Butyl Peroxide, 3557, di-*t*-butyl peroxide, 8369, Trigonox® B, 32295
2DTBP, *See* 2,6-dibutylphenol, 8370
2,6 DTBP, *See* 2,6-dibutylphenol, 8370
(2,6 DTBP), *See* 2,6-di-*t*-butylphenol, 4799
DTC, *See* Octopol SDE-25, 22013
DTDP, *See* Jayflex® DTDP, 15555
DTMC, *See* Acarin, 183, Dicofol, 8398, Kelthane, 15881
DTPA, *See* Aroquest M Special, 3064, Pentaquest OPAC 0201, 23170
D-Tracetten, *See* vitamin D+72, 33940
D-Trans, 9070
L-DTT, *See* 1,4-Dithiothreitol, 8730
DU 112307, *See* Diflubenzuron, 8450
du Pont 634, *See* lenacil, 16984, Venzar®, 33562, Vizor, 34001
Du Pont Adjuvant, 9071
Du Pont Enrich®, 9072
Du Pont Herbicide 976, *See* bromacil, 4564, Uragan, 33105
Du Pont Linuron 50, 4L, 9073
Du Pont Pakwrap®, 9074
Du Pont Permissible No. 1, 9075
Dual, *See* metolachlor, 19585
Dual 8E, *See* metolachlor, 19585
Dual 25G, *See* metolachlor, 19585
Dualite M6001AE, M6017AE, 9076
Duallor, 9077
Duasyn, 9078
Dubbin, 9079
dublofix, *See* ethyl chloride, 11067
Dubox®, 9080
Dubronax, *See* Dapsone, 7751
Duco, 9081
Ducobee-Hy, 9082
Dudley metal, 9083
Duet®, 9084
Duette, 9085
Duffi-K, *See* Potassium chloride, 24748
Dufox, 9086
Duhnul-balasan, *See* Mecca balsam, 19201
DUK-880, 9087
Dukatalon, 9088
Duke's metal, 9089
Dulceta, 9090
Dulcin, 9091
dulcine, *See* Dulcin, 9091
dulcite, *See* galactitol, 12592
dulcitol, *See* galactitol, 12592, Melampyrite, 19258
dulcose, *See* galactitol, 12592
Dulenza, 9092
Dullray, 9093
Dulux, 9094
Dulzor-Etas, *See* sodium cyclamate, 28754
Dumacene C13, NP707, NP7710 and NPX10, 9095
Dumet, 9096
Dumitone, *See* Dapsone, 7751
Dunclad CE, 9097
Dunclad VN, 9098
dunkelgelb, *See* Sudan I, 29938
Dunlop 6593, 9099
Dunlop Grade 6167, 9100
Dunlop PL, 9101
Dunnite, 9102
Dunova®, 9103
Duocid, *See* Unasyn® IM/IV, 32913
Duocrome®, 9104
duodecyclic acid, *See* Philacid 1200, 23660
duodex, *See* Nacap®, 20643, Nuodex 84, 21779
Duofas, *See* Ranide, 25868
Duofol T, 9105
Duo-kill, *See* dichlorvos, 8394
Duolite, 9106
Duolith, *See* Cryptone, 7281
Duomac®, 9107
Duomat, 9108
Duomatic, 9109
Duomeen, 9110
Duomeen®, 9111
Duomeen® C, 9112, *See* Cocodiamine, 6564
Duomeen® CD, 9113, *See* Cocodiamine, 6564
Duomeen® OL, 9114
Duomeen® OTM, 9115
Duomeen® T, 9116, *See* Diamine BG, 8295
Duomeen® TDO, 9117
Duomeen® TTM, 9118
Duoquad®, 9119
Duoquad® O-50, 9120
Duoquad® T-50, 9121
Duoscorb, *See* vitamin C, 33938
Duoteric, 9122
Duothane, 9123
Duovac-C, 9124
Duovac-M, 9125
Duovac-Ma5, 9126
Dupanal, *See* Rhodapon® 101-10, 26631, sodium lauryl sulfate, 28781
Dupanol, *See* Witcolate 6400, 34469
Dupanol WAQ, *See* Witcolate 6400, 34469
Dupanol-C, *See* Witcolate 6400, 34469
Duphacid, *See* Diflubenzuron, 8450
Duphafral D+73 1000, *See* vitamin D+73, 33941
Duphar, *See* tetradifon, 31366
Dupical, 9127
Duplosan New System CMPP, 9128
Duplosan®, 9129
Duplosan® CMPP, 9130
Duplosan® DP, 9131
Duponal WAQE, *See* Rhodapon® 101-10, 26631, sodium lauryl sulfate, 28781, Witcolate 6400, 34469
Duponol, 9132, *See* Rhodapon® 101-10, 26631, sodium lauryl sulfate, 28781
Duponol C, *See* sodium lauryl sulfate, 28781, Witcolate 6400, 34469
Duponol LS, *See* Supralated LS, 30310
Duponol ME, *See* Supralated ME, 30311, Witcolate 6400, 34469
Duponol methyl, *See* Witcolate 6400, 34469
Duponol QC, *See* Witcolate 6400, 34469
Duponol qx, *See* Witcolate 6400, 34469
Duponol WA, *See* Supralated WA, 30312, Witcolate 6400, 34469
Duponol WA dry, *See* Witcolate 6400, 34469
Duponol WAE, *See* Witcolate 6400, 34469
Duponol WAG, *See* Witcolate 6400, 34469
Dupont 1179, *See* methomyl, 19502
DuPont Zonyl FSA Fluorinated Surfactants, *See* isopropanol, 15405
DuPont Zonyl FSJ Fluorinated Surfactants, *See* isopropanol, 15405
DuPont Zonyl FSN Fluorinated Surfactants, *See* isopropanol, 15405
DuPont Zonyl FSO-100 Fluorinated Surfactants, *See* Dioxane, 8574
DuPont Zonyl FSP Fluorinated Surfactants, *See* isopropanol, 15405
Dupranin CR, 9133
Dupranin W, 9134
Duprene, 9135
Durabetason, *See* betamethasone 21 phosphate, 4060
Durabond, 9136
Durabond 650, 655, 9137
Duracef, *See* Cefadroxil, 5547
Duracillin, *See* Penicillin V, 23111
Duracore, 9138
Durad, 9139

Durad 110 hydraulic fluid, *See* Fyrquel® EHC, 12544, triaryl phosphate, 32194
Duradene® 706, 9140
Duradene® 710, *See* Steron 210, 29751
Duradene® 711, 9141
Duradene® 750, 9142
Duradene® 755, 9143
Duradiene®, 9144
Duraflex® 8410, 9145
Durafoam, 9146
Duraform, 9147
Durafur Brown RB, *See* *p*-aminophenol, 2185
Duraguard, 9148
Dural, 9149, *See* aluminum oxide, 1919, Duralumin, 9160
Duralac, 9150
Duralam®, 9151
Duralcon, 9152
Duralit, 9153
Duralium, 9154
Duralkan K Concentrate, 9155
Duralloy, 9156
Duralon, 9157
Duraloy, 9158
Duralum, 9159
Duralumin, 9160
Duramite®, 9161
Durana metal, 9162
Duranalium, 9163
Durance's metal, 9164
Durand's metal, 9165
Durango, *See* Guayale, 13454
Duranic, 9166
Duranit, 9167
Duranite, 9168
Duranox, 9169
Duranthrene® Dyes, 9170
Durapatite, *See* Alveograf, 1939, Periograf, 23274
duraphat, *See* sodium fluoride, 28764
Duraphos™ AP-230, *See* dilauryl phosphite, 8475
Duraphos™ AP-240, *See* Dioleyl hydrogen phosphite, 8554
Duraplex®, 9171
Duraplus® 1, 9172
DuraSoft, 9173
Durasol Acid Blue B, 9174
Durasorb, *See* DMSO, 8762
Durastat® AS-5760, 9175
Durastat® AS-5814-2, 9176
Durastat® AS-5903-3, 9177
Durastic, 9178
Durastrength 200, 9179
Duratears, *See* Synwax, 30659
Duratex, 9180
Duratox, *See* demeton-S-methyl, 8055
Duravos, *See* dichlorvos, 8394
Durax, *See* CBTS, 5484, Delac S, 8002, Durax®, 9181, Perkacit® CBS, 23277, Santocure, 27400, Vulkacit® CZ/EGC, DZ/EGC, 34129
Durax®, 9181, *See* cyclohexyl benzothiazole sulfenamide, 7523
Durax® Rodform, *See* cyclohexyl benzothiazole sulfenamide, 7523
Durazol, 9182
Durbar Bronze, 9183
Durbar Hard Bronze, 9184
Durcoton, 9185
Durecol, 9186
Durehete 900, 9187
Durehete 950, 9188
Durehete 1050, 9189
Durel, 9190
Durelast, 9191
Durelast®, 9192
Dur-Em® 117, 9193
Dur-Em® GMO, 9194
Durene, 9195
Duresco, *See* lithopone, 17520
Durethan®, 9196
Durethan® A 30 S, 9197
Durethan® B 30 S, B 31 SK, 9198
Durethan® B 35 F, B 38 F, B 40 F, 9199
Durethan® BKV, 9200
Durethan® BKV 115, 9201
Durethan® BKV 30 H, 9202
Durethan® BM 30 X, 9203
Durethan® KL1-2402/30, 9204
Durethan® RM KU 2-2501/30, 9205
Durethane, 9206
Durex, 9207
Durex white, 9208, *See* Barium carbonate, 3636
Durez® 115, 9209
Durez® 123, 9210
Durez® 152, 9211
Durez® 18420, 9212
Durez® 24150, 9213
Durez® 25000, 9214
Durez® 32633, 9215
Durfax® 60, 9216
Durfax® 60, *See* Ethylan® GES6, 11108
Durfax® 65, 9217
Durfax® 65, *See* Polysorbate 65, 24571
Durfax® 80, 9218
Durferrit, 9219
Durham's stain, 9220
Duricef, *See* Cefadroxil, 5547
Durichlor, 9221
Duridine, 9222
Durifan AR30, BK 30, 9223
Durimet Alloys, 9224
Durine, 9225
Duriron, 9226
Durisol, 9227
Durkex, *See* hydrogenated soybean oil, 14596
Durkex 100DS SF18-350, *See* AF 10 FG, 907
Durkex 500, 9228
Durkex Durola, 9229
Durko, 9230
Durlac® 100W, 9231
Durlite F, 9232
Durlite Gold MBN II, 9233
Dur-Lo®, 9234
Duro cement, 9235
Durocide, 9236
Durodi Steel, 9237
Durofer, 9238
Duroftal, 9239
Duroglass, 9240
Duroil, 9241
Durol, *See* Durene, 9195
Durola Select, 9242
Durolastik, 9243
Duroloy®, 9244
Durolube, 9245
Duromel, 9246
Duromel B108, 9247, *See* Duratex, 9180
Duronze, 9248
Durophen, 9249
Durophen 127-B, 9250
Durophen 170W, 9251
Durophen 218V, 9252
Durophen 287W, 9253
Durophen 309W, 9254
Durophen 330V, 9255
Duroplaz 610, 810, 911, 9256
Duroprene®, 9257
Duroseal, 9258
Durosehl, 9259
Durosil, 9260
Duroslip, 9261
Durosoft, 9262
Durosol, 9263
Durostabe, 9264
Duroterm, 9265
Durotex, 9266
Durotex 7603, 9267
Durotint, 9268
Durowynd, 9269
Durox, 9270
Duroxyn, 9271
Dursban, *See* Chlorpyrifos, 6163, Talon, 30751
Dursban 4E, *See* Chlorpyrifos, 6163
Dursban F, *See* Chlorpyrifos, 6163
Dursban®, 9272
Dursban® 2E, 9274
Dursban® 14G, 9273
Durtan 60, *See* Ablunol S-60, 110
Durtan® 60, 9275
Dusantox IPPD, *See* N-Isopropyl-N'-phenyl-*p*-phenylene diamine, 15413
Dustallay, 9276
Dustex, 9277
Dutch camphor, 9278
Dutch liquid, *See* ethylene dichloride, 11122
Dutch metal, 9279
Dutch oil, *See* ethylene dichloride, 11122
Dutch pink, 9280
Dutch varnish, 9281
Dutch white, 9282
Dutch Yellow, *See* Persian berry carmine, 23418
Du-Ter, 9283
Du-Ter®, 9284 , *See* Farmatin, 11480
Duthane, 9285
Dutral, 9286
Dutral-Co, 9287
Dutral-Ter, 9288
Dutrex, 9289
Dutrex 20, 25, 9290
Dutrex Process and Extender Oils, 9291
Duxalid, 9292
Duxlte, 9293
Duxol, 9294
DV-1801, 9295
DV-1936, 9296, *See* dodecylthioethanol, 8781
DV-2301, 9297
D-Visor, 9298
DW 3418, *See* cyanazine, 7472
D-Wax, 9299
Dwell, *See* Aaterra WP, 17
DY 023, 9300
Dyafac PEG 6DO, 9301, *See* PEG-2 dioleate series, 23040
Dyamul, 9302
Dyapol, 9303
Dybenal, *See* 2,4-dichlorobenzyl alcohol, 8387
Dybin®, 9304
Dycarb, *See* Bendiocarb, 3937
Dycastal, 9305
Dy-Chek, 9306
Dycill, 9307, *See* Dicloxacillin sodium, 8396
Dycote, 9308
Dycron, 9309, *See* aluminum oxide, 1919
Dye Retarder #1, 9310
Dyeset® 100 Conc, 9311
Dyetone®, 9312
Dyeweld SUPR, 9313
Dyflor 2000, 9314
Dyflor L90, 9315
Dyfonate, 9316
Dykol, *See* DDT, 7820
Dykor 204, 9317
Dylark® 132, 9318
Dylene, 9319
Dylite® D195B, 9320, *See* Dylite® R2595B EPS, 9321
Dylite® R2595B EPS, 9321
Dylon, 9322
Dylonite, 9323
Dylox, *See* trichlorfon, 32212
Dylox®, 9324
Dylux®, 9325
Dymacryl, 9326
Dymadon, *See* Valadol, 33238
Dymanthine, 9327, *See* Adma® 18, 744, Adogen® MA-108 SF, 776, Crodamine 3.A18D, 7107, Kemamine® T-9902, 15933
Dymanthine hydrochloride, *See* Onamine 18, 22167
Dymanthine, N,N-dimethyl octadecanamine, *See*

Onamine 18, 22167
Dymax® Light-Weld® Adhesives, 9328
Dymax® Light-Welder™, 9329
Dymax® Multi-Cure® Adhesives, 9330
Dymel®, 9331
Dymel®, *See* Disflamoll® TP, 8642
dymerex, 9332
Dymetrol®, 9333
Dymos, *See* azamethiphos, 3517
Dymsol® 2031, 9335
Dymsol® 38C, 9334
Dymsol® PA, 9336
Dynacal, 9337
Dyna-Carbyl, *See* Carbaryl, 5181
Dynacast, 9338
Dynacerin®, 9339
Dynacerin® 660, 9340
Dynacet®, 9341
Dynacoll, 9342
Dynaflex, 9343
Dynaflock, 9344
Dynaflush, 9345
Dyna-Form, 9346
Dynaglaze, 9347
Dynagrout, 9348
Dynagunit, 9349
Dynamag, 9350
Dynamar® Brand Specialities, 9351
Dynamar® FC, 9352
Dynamar® PPA-790, 9353
Dynamask, 9354
Dynamine, 9355
dynamite, 9356, *See* Kieselguhr dynamite, 16177
dynamite acid, 9357
dynamite glycerin, 9358
Dynamites, 9359
Dynamullit, 9360
Dynamyte, 9361
Dynapen, *See* Dicloxacillin sodium, 8396
Dynapol® H, 9362
Dynapol® L, 9363 9364
Dynapol® L 205, 9365
Dynapol® LH, 9366, 9367
Dynapol® P, 9368, 9369
Dynapol® S, 9370, 9371
Dynapor, 9372
Dynasan®, 9373
Dynasan® 110, 9374
Dynasan® 112, 9375
Dynasan® 114, 9376
Dynasan® 116, 9377
Dynasan® 118, 9378
Dynasan® P60, 9379
Dynasil P, *See* CT2090, 7316
Dynasil®, 9380
Dynasil® 40, 9381
Dynasil® A, 9382
Dynasil® CA, 9383
Dynasil® CM, 9384
Dynasil® M, 9385
Dynasolve 100, 9386, *See* dimethyl formamide, 8506
Dynasolve 150, 9387
Dynasolve 165, 9388
Dynasolve 699, 9389
Dynasolve M-30, 9390
Dynasperse A, 9391
Dynaspinell, 9392
Dynastite, 9393
Dynasylan 1411, *See* CA0699, 4849
Dynasylan 1505, *See* CA0742, 4851
Dynasylan 1506, *See* CA0742, 4851
Dynasylan 2201, *See* CT2507, 7318
Dynasylan AMEO, *See* Aktisil AM, 1201, CA0750, 4852, Prosil® 220, 25237
Dynasylan AMMO, *See* CA0880, 4853, Dynasylan® AMMO, 9397
Dynasylan BSA, *See* Bistrimethylsilyl acetamide, 4286, CB2500, 5479, Dynasylan® BSA, 9399
Dynasylan DAMO, *See* CA0700, 4850, Dynasylan® DAMO, 9404, Prosil® 3128, 25239
Dynasylan GLYMO, *See* CG6720, 5836, Prosil® 5136, 25240, Aktisil EM, 1202
Dynasylan HMDS, *See* hexamethyldisilazane, 14003, Prosil® HMDS, 25242
Dynasylan IBTMO, *See* CI7810, 6288, Prosil® 178, 25235
Dynasylan MEMO, *See* Dow Corning® Z-6030, 8879, Prosil® 248, 25238
Dynasylan MTES, *See* Dynasylan® MTES, 9413
Dynasylan MTMO, *See* Aktisil MM, 1203, Prosil® 196, 25236
Dynasylan OCTEO, *See* Prosil® 9202, 25241
Dynasylan PTMO, *See* CP0810, 6936
Dynasylan TRIAMO, *See* CT2910, 7322
Dynasylan VTEO, *See* CV-4910, 7451
Dynasylan VTMO, *See* CV-4917, 7452
Dynasylan VTMOEO, *See* Aktisil VM, 1205, CV-5000, 7453
Dynasylan®, 9394
Dynasylan®, 9395
Dynasylan® AMEO, AMEO-P, *See* γ-aminopropyltriethoxysilane, 2186
Dynasylan® AMEO, Dynasylan® AMEO-P, 9396
Dynasylan® AMMO, 9397, *See* Aminopropyltrimethoxysilane, 2187
Dynasylan® BDAC, 9398, *See* Di-*t*-butoxydiacetoxysilane, 8362
Dynasylan® BSA, 9399
Dynasylan® BSM, 9400, 9401
Dynasylan® CPTEO, 9402, 9403 *See* chloropropyltrimethoxysilane, 6138
Dynasylan® DAMO, 9404
Dynasylan® DAMO-P, 9405, *See* Aminoethylaminopropyltrimethoxy silane, 2171
Dynasylan® DAMO-T, 9406, *See* Aminoethylaminopropyltrimethoxy silane, 2171
Dynasylan® ETAC, 9407, *See* ethyltriacetoxysilane, 11146
Dynasylan® GLYMO, 9408
Dynasylan® HMDS, 9409
Dynasylan® IBTMO, 9410
Dynasylan® IMEO, 9411
Dynasylan® MEMO, 9412
Dynasylan® MTES, 9413
Dynasylan® MTMO, 9414
Dynasylan® MTMS, 9415
Dynasylan® OCTEO, 9416
Dynasylan® TCS, 9417
Dynasylan® TRIAMO, 9418
Dynasylan® VTC, 9419
Dynasylan® VTEO, 9420
Dynasylan® VTMO, 9421
Dynasylan® VTMOEO, 9422
Dynat W, 9423
Dynatex GTZ, 9424, *See* Polyisoprene, 24471
Dynatherm, 9425
Dynatred, 9426
Dynatrol, 9427
Dynaweld, 9428
Dynazirkon, 9429
Dyne, 9430
Dynemate 200, 9431
Dynex, *See* diuron, 8736
Dynobel, 9432
Dynomel, 9433
Dyodin, *See* Yodoxin, 34658
Dyphene, 9434
Dyprin, *See* DL-methionine, 19487
Dypur, 9435
Dyquex®, 9436
Dyrene, *See* Anilazine, 2465
Dyrene 50W Triazine, *See* Anilazine, 2465
Dyrene®, 9437 , *See* Anilazine, 2465
dyslytite, *See* schreibersite, 27733
Dysoid, 9438
Dytek® A, 9439
Dytel®, 9440
Dytherm®, 9441
DYTOL F-11, *See* Cachalot® C-50 NF, 4878, Adol® 52 NF, 782
Dytron® XL, 9442
Dyvax®, 9443
Dyzol, *See* Diazinon Liquid, 8344
DZ910, 9444
E 7256, *See* Witco® 1298, 34458
E alloy, 9445
E.B Golden Glitter, Neutral Glitter, 9446
E.C.A, 9447
E.O.D, *See* 2-octyldodecyl erucate, 22043
E2, *See* Ablunol 200MO, 96
E-3810, 9448
E-3824, 9449
E45 Cream, 9450
E-600, *See* paraoxon, 22783
E-605, *See* parathion, 22807
E-607, *See* Lapyrium Chloride, 16782
E-9405, 9451
eaklite, *See* calcium silicate, 4969
Earex®, 9452
Early Impact, 9453
earth archil, 9454
earth wax, *See* ceresin, 5746, ozokerite, 22483
Earthcide, *See* quintozene, 25692
earthnut oil, *See* Katchung oil, 15788
Earwig bait, *See* sodium silicofluoride, 28818, SSF, 29390
Ease Release 200 Series, 9455
Ease Release 2040 Series, 9456
Ease Release 2191, 9457
Easigel, 9458
Easisperse, 9459
East India Gum, 9460
East Indian Balsam of Copaiba, 9461
Eastbond, 9462
Eastman® 910, 9463
Eastman® AQ-38S, 9464
Eastman® C-11 Ketone, 9465
Eastman® DB Acetate, *See* diethylene glycol butyl ether acetate, 8443
Eastman® DE, *See* Ethoxydiglycol, 11044
Eastman® DTBHQ, 9466, *See* Dibutylhydroquinone, 8368
Eastman® EP, *See* ethylene glycol propyl ether, 11130
Eastman® HQMME, 9467, *See* hydroquinone monomethyl ether, 14633
Eastman® Inhibitor DOBP, *See* 4-dodecyloxy-2-hydroxybenzophenone, 8780
Eastman® Inhibitor OABH, *See* oxalyl bis (benzylidenehydrazide), 22418
Eastman® Inhibitor OPS, 9468
Eastman® Inhibitor Poly TDP 2000, 9469
Eastman® Inhibitor RMB, 9470, *See* Resorcinol benzoate, 26231
Eastman® MTBHQ, 9471, *See* *t*-Butylhydroquinone, 4793
Eastman® P4C5B-030, 9472
Eastman® Poly TDP 2000, 9473
Eastman® Yellow, 9474
Eastobrite® OB-1, 9475, *See* Ethenediyl diphenylene bisbenzoxazole, 10940
Eastoflex B1020, 9476
Eastoflex E1003, 9477
Eastoflex P1010, 9478
Eastotac H-100, 9479
Eastozone, *See* Naugard® I-2, 20792
Eastozone 32, 9480
Eastozone 33, *See* Naugard® I-2, 20792
Easy Cleen, 9481
Easy off-D, *See* Def®, 7876
Easy-Flo, 9482
Easypoxy, 9483
Easytec, *See* Tecgran, 30886, tecnazene, 30898
eau de Brouts, 9484
eau de Goudron, 9485
Eau grison, *See* Orthorix, 22337
EB1500-1AR, 9486
EB3000-2, 9487

Ebal, 9488, *See* ethyl-benzaldehyde, 11118
EBDC, manganese salt, *See* Maneb, 18526
Ebecryl®, 9489
Ebecryl® 110, 9490
Ebecryl® 150, 9491
Ebecryl® 600, 9492
Ebecryl® 629, 9493
Ebecryl® 1360, 9494
Ebert and Merz's α-acid, 9495
Ebert and Merz's β-acid, 9496, *See* 2,6-naphthalenedisulfonic acid, 20708
Ebivit, *See* vitamin D_3, 33941
Ebner's fluid, 9497
Ebonestos, 9498
Ebonite, 9499, *See* Dexonite, 8231
Ebonized monel, 9500
Ebontex, 9501
Ebony black, 9502, *See* gas black, 12686
Eborex, 9503
E-BR® 8405, 9504
E-BR® 8471, 9505
Ebrok, 9506
Ebucin, *See* Neo-Calglucon, 20877
EC-25®, -25K, 9507
Eca, 9508
Ecco Defoamer Heavy, 9509
Ecco Defoamer NSD, 9510
Ecco Defoamer S, 9511
Ecco Fast Binder 1500, 9512
Ecco MP® 2004, 9513
Ecco Resin 234, 9514
Ecco Rez 3070, 9515
Ecco Rez M-300-7, 9516
Ecco White® FW-5, 9517
Eccoblanc W-55-Q, 9518
Eccobond®, 9519
Eccobond® 114, 9520
Eccobond® Adhesive Special #2, 9521
Eccobond® Paste 99, 9522
Eccobond® SF40, 9523
Eccobrite RB, 9524, *See* stilbene, 29764
Eccoclean CR-46, 9525
Eccocoat®, 9526
Eccodye Colors, 9527
Eccofix 101-40, 9528
Eccoflo® HiK, 9529
Eccofloat, 9530
Eccofloat EG35, 9531
Eccofloat Encapsulant 1421, 9532
Eccofloat HG452, 9533
Eccofloat PC61, 9534
Eccofloat PP22 and 24, 9535
Eccofloat SP 12, 20, 9536
Eccofloat SS40, 9537
Eccofloat UG 36, 9538
Eccofloat US 35, 9539
Eccofoam® FPH, 9540
Eccofoam® PP, 9541
Eccoful DL Conc, 9542
Eccogel F, 9543
Eccolube L-54, 9544
Eccopel 10, 9545
Eccopuff, 9546
Eccoro®, 9547
Eccoscour CB, 9548
Eccoscour D-7, 9549
Eccoseal®, 9550
Eccoshield®, 9551
Eccosil®, 9552
Eccosil® 1776, 9553
Eccosil® 2CN, 9554
Eccosoft C-2000, 9555
Eccosol 150, 9556
Eccosolv C-14, 9557
Eccosorb® 269E, 9558
Eccosorb® AN, 9559
Eccosorb® Coating 268E, 9560
Eccosorb® MF, 9561
Eccospheres, 9562
Eccostat, 9563
Eccostat C, 9564
Eccoterge 200, 9565
Eccoterge ASB, 9566
Eccoterge MV Conc, 9567
Eccotherm® TC-11, 9568
Eccowax UL-100, 9569
Eccowet® W-50, 9570
Ecdel® 9965, 9571
Ecdel® 9967, 9572
Ecepox® PB1 and PB2, 9573
Ecetate disodium, *See* Trilon® BD, 32325
Echappe silk, 9574
Echicaoutchin, 9575
Echlomezol, *See* Aaterra WP, 17
Echo, 9576, 9577, *See* mefluidide, 19230
echurin, 9578
Eclipse, 9579
Ecobinder®, 9580
Ecobond, 9581
Ecoderm, *See* Vaderm, 33237
Ecolac, 9582
Ecolo, 9583
Ecolotec, 9584
Econocat, 9585
Econogel, 9586
Economy Flor-Dri, 9587
Economy Flor-Dri, *See* Attapulgite, 3404
Econopred, 9588
Ecoro, 9589
Ecosyl, 9590
Ecothene EC 101, 9591
ECP-170, 9592
ecru silk, 9593
Ectiban, *See* permethrin, 23384, 23385, 23386
Ectiban, *See* Kafil, 15655
Ectodex, *See* Taktic, 30729
Ectoral, *See* ronnel, 26951
Ectorl, *See* Korlan, 16351
Eczederm, *See* Vasogen®, 33468
Edaplan LA 400, 9594
Edaplan LA 411, 9595
Edaplan VP LA 420, 9596
Edasil, 9597
edathamil, *See* edetic acid, 9615, Kalex Acids, 15669
edathamil calcium disodium, *See* Versene CA, 33618
edathamil disodium, *See* Trilon® BD, 32325
EDC, *See* ethylene dichloride, 11122
EDDP, *See* Edifenphos, 9618
Edecrin®, 9598
Edelfeka, 9599
Edelresanol, 9600
Edelweiss, *See* oil white, 22086
Edelwit, 9601
Edenol 74, 9602
Edenol 302, 9603
Edenol 888, *See* Plasthall® DOS, 24005, Reomol DOS, 26131
Edenol B35, 9605
Edenol B316, 9604
Edenol D72, 9606
Edenol D82, 9607
Edenol HS 235, 9608
Edenor ITS, 9609
Edenor PTO, 9610, *See* Pentaerythrityl tetraoleate, 23141
Eder's solution, 9611
Edeta®, 9612
edetate calcium disodium, *See* Versene CA, 33618
edetate disodium, 9613, *See* disodium EDTA, 8644, Endrate, 10553, Hamp-Ene® Na2, 13588, Questal DI 0770, 25661, sodium versenate, 28831
edetate sodium, *See* Aroquest 100, 3062, Hamp-Ene® 100, 13586, Kalex 220 Crystal, 15668, Kalex Liq. 50%, 15671, tetrasodium EDTA, 31391
edetate tetrasodium, *See* Hamp-Ene® Na4, 13590
edetate trisodium, 9614, *See* Hamp-Ene® Na3 Liq, 13589
edetic acid, 9615, *See* Hamp-Ene® Acid, 13587, Kalex Acids, 15669, Trilon® BS, 32327
edetic acid calcium disodium salt, *See* Versene CA, 33618
edetic acid tetrasodium salt, *See* Kalex 220 Crystal, 15668, Kalex Liq. 50%, 15671
edetic acid trisodium salt, *See* edetate trisodium, 9614
Edicol, 9616
Edifas, 9617
Edifenphos, 9618, *See* Hinosan®, 14097
Edimet, 9619
Edinol, 9620
Edistir®, 9621
Edistir® FA, 9622
Edistir® N 1280, N 1281, 9623
Edistir® RC, 9624
Edistir® RK, 9625
Edistir® RKV, 9626
Edistir® RV 8, 9627
Edistir® SR 550, SRL 550, 9628
Edistir® UT/1, 9629
Edistir® UT/SF, 9630
Edit®, 9631
Edolan®, 9632
Edrizar, *See* Taktic, 30729
EDTA, *See* Kalex Acids, 15669, Sequestrene® AA, 27996, Trilon® BS, 32327
EDTA (CTFA), *See* edetic acid, 9615
EDTA calcium, *See* Versene CA, 33618
EDTA free acid, *See* Trilon® BS, 32327
EDTA Na4, *See* tetrasodium EDTA, 31391
EDTA tetrasodium, *See* tetrasodium EDTA, 31391, Kalex 220 Crystal, 15668, Kalex Liq. 50%, 15671
EDTA trisodium, *See* edetate trisodium, 9614
EDTA, free base, *See* Trilon® BS, 32327
EDTA, iron (III) derivative, sodium salt, *See* sodium iron edetate, 28776
EDTP, *See* tetra(2-hydroxypropyl) ethylenediamine, 31356
Edunine, 9633
Edward's speculum, 9634
EE, *See* ethoxyethanol, 11045
2EE, *See* ethyl icinol, 11074
EE Acetate, 9635
EE Solvent, 9636
EEA, *See* EE acetate, 9635, Ethoxyethanol acetate, 11046
2EEA, *See* EE Acetate, 9635
Eel antifriction metal, 9637
EF-689, *See* fluroxypyr, 12117
EFED, *See* Doverphos® 10, 8838
Efetaal, 9638
Effemoll DOA, *See* dioctyl adipate, 8548, Plasthall® DOA, 24002, PX-238, 25440, Wickenol® 158, 34393
Effesay, 9639
Effomoll DOA, *See* dioctyl adipate, 8548, Plasthall® DOA, 24002, PX-238, 25440, Wickenol® 158, 34393
Effusan, *See* 4,6-dinitrocresol, 8542
Efica, 9640
Efuzin, *See* Dodine FL, WP, 8785
Efweko, 9641
Egalex, 9642
Egalisal, 9643
Egalon Colours, Auxiliaries, Thinners, 9644
EGDMA, *See* ethylene glycol dimethacrylate, 11125, Perkalink® 401, 23315
EGDME, *See* monoglyme, 20227
EGDS, *See* glycol distearate, 13102
EGEE, *See* ethyl icinol, 11074
egenite, *See* wurtzillite, 34577
egg albumin, *See* Albumen, 1402
Egg oil, 9645
Egg white Solids Type P-20, P-11, P-18G, P-19, P-21, P-25, P-39 P-110, PF-1, *See* Albumen, 1402
Eggs oils, *See* white oils, 34368
Eglantine, 9646
EGO-4, 9647
Egyptian fiber, *See* Vulcanized fiber, 34097
Egyptianized clay, 9648
EH 2-EH, *See* 2-ethylhexanol, 11136

EH diol, *See* ethyl hexanediol, 11073
EH2-EH, *See* 2-ethylhexanol, 11136
EHA, *See* octyl acrylate, 22037
ehag sulfate gl emulsion, *See* sodium lauryl sulfate, 28781
EHD, *See* ethyl hexanediol, 11073
EHDP, *See* Unihib® 106, 32957
EHIDA Kit, 9649
Ehrhard's metal, 9650
Ehrlich-Biondi stain, 9651
Ehrlich's diazo reagent, 9652
Ehrlich's hematoxylin, 9653
Ehrlich's Reagent, *See* dimethylaminobenzaldehyde, 8515
ethylene glycol monobutyl ether, *See* Ektasolve® EB, 9668
EI 47470, *See* mephosfolan, 19326
eicos-1-ene, *See* Neodene® 20, 20900
Eicosyl docosylamine, *See* Amine B11, 2150
Eisler's bronze, 9654
Ekagom CBS, *See* Vulkacit® CZ/EGC, DZ/EGC, 34129
Ekagom GS, *See* Perkacit® MBTS, 23286
Ekagom tb, *See* Rezifilm, 26521
Ekaland CBS, *See* cyclohexyl benzothiazole sulfenamide, 7523
Ekaland DOTG, *See* Ditolyguanidine, 8732
Ekaland DPPD, *See* Agerite® DPPD, 1016, N,N'-diphenyl-*p*-phenylenediamine, 8603, JZF, 15638, Naugard® J, 20796, Permanax DPPD, 23361
Ekaland TETD, *See* Akrochem® TETD, 1180
Ekaline, 9655
Ekaline G 80, *See* Ablunol SA-7, 113
Ekalux, *See* quinalphos, 25679
Ekamet, *See* etrimfos, 11155
Ekamet G, *See* etrimfos, 11155
Ekamet ULV, *See* etrimfos, 11155
Ekanda rubber, 9656
Ekatin, 9657, *See* thiometon, 31700
E-Kote 3042, 9658
Eksmin, *See* permethrin, 23384, 23385, 23386
Eksmin, *See* Kafil, 15655
Ektapro® EEP Solvent, 9659
Ektar® FB PG003, 9660
Ektasolve DB, *See* Butyl Carbitol®, 4766
Ektasolve DE, *See* ethyl di-Icinol, 11068, Diethoxol, 8435
Ektasolve EB, *See* 2-butoxyethanol, 4753
Ektasolve EB Acetate, *See* 2-butoxyethanol acetate, 4754
Ektasolve EB solvent, *See* 2-butoxyethanol, 4753
Ektasolve EE, *See* ethyl icinol, 11074
Ektasolve EE acetate solvent, *See* EE Acetate, 9635
Ektasolve™ EB, *See* Butyl Cellosolve®, 4767
Ektasolve™ EB Acetate, *See* 2-butoxyethanol acetate, 4754
Ektasolve®, 9661
Ektasolve® DB, 9662
Ektasolve® DB Acetate, 9663
Ektasolve® DE, *See* Ethoxydiglycol, 11044
Ektasolve® DE Acetate, 9665
Ektasolve® DE, DE-HG, 9664
Ektasolve® DM, 9666
Ektasolve® DP, 9667, *See* diethylene glycol propyl ether, 8445
Ektasolve® EB, 9668
Ektasolve® EB Acetate, 9669
Ektasolve® EEH, 9670
Ektasolve® EP, 9671, *See* ethylene glycol propyl ether, 11130
Ektasolve® PM Acetate, 9672
Ektogan, 9673
Ektogen, *See* Ektogan, 9673
EL-107, *See* Knot Out, 16257, Ratio, 25897
EL-110, *See* Benfluralin, 3940
EL-119, *See* oryzalin, 22347
EL-228, *See* nuarimol, 21762
EL-620, *See* Ablunol OA-6, 107, Volpo 3, 34020
EL-719, *See* Ablunol OA-6, 107, Volpo 3, 34020
Elacid CLR, 9674
Elacid Richter, 9675
elainic acid, *See* oleic acid, 22112
elaldehyde, *See* paraldehyde, 22759
Elancolan, *See* Treflan, 32162, Trifluralin, 32252
Elaol, 9676
Elaol, *See* dibutyl phthalate, 8366
Elaol 1, 9677
Elaol 2, 9678
Elaol 3, 9679
Elaol 4, 9680
Elaol VI, 9681
Elargol, 9682
Elastalloy® 6713, 9683
Elastan®, 9684
Elas-Tein AS-20, 9685
elastic asbestos, *See* mountain cork, 20368
elastic bitumen, *See* elaterite, 9713
Elasti-glass, 9686
Elastinhydrolysate, Liquid, 9687
Elastite, 9688
Elastoblend 8480, 9689
Elastobond, 9690
Elastocarb Tech Light, Heavy, 9691
Elastocarb Tech Light, Tech Heavy, *See* magnesium carbonate, 18359
Elastocarb UF, 9692
Elastocell®, 9693
Elastocoat®, 9694
Elastoflex®, 9695
Elastofoam® I, 9696
Elastoid 1300, 9697
Elastolac, 9698
Elastolit® D, 9699
Elastolith, 9700
Elastollan® 1154D, 9701
Elastollan® C-59D, 9702
Elastollan® S-60D, 9703
Elastomag® 100, 9704
Elastopal®, 9705
Elastopan®, 9706
Elastopor®, 9707
Elastopreg®, 9708
Elastorid®, 9709
Elastosil®, 9710
Elastosil® LR 3001, 9711
Elastosil® LR 3003/20, 9712
Elastozone 34, *See* Flexzone 3C, 11967, Permanax IPPD, 23364, Vulkanox® 4010 NA, 34147
elaterite, 9713, *See* Aeonite, 831
Elaterite, Artificial, 9714
Elayl, *See* olefiant gas,
Elbanil, *See* Chlorpropam, 6162
Elbasol, 9715
Elbatan, *See* lenacil, 16984, Vizor, 34001
Elbelan, 9716
Elbelene, 9717
Elbenyl, 9718
Elbeplast, 9719
Elbestret, 9720
Elcema® F150, G250, P100, 9721
Elcomet, 9722
Eldopaque, 9723, *See* quinol, 25685, Tecquinol® Tech. Grade, 30910
Eldoquin, *See* quinol, 25685, Tecquinol® Tech. Grade, 30910
eldrin, *See* rutin trihydrate, 27150
Electrafil®, *See* Grilpet® XE3060, 13402
Electrafil® F-1700/CF/10/A, 9724, *See* Polyphenylene oxide, 24518
Electrafil® F-4/CN/40, 9725
Electrafil® G-1/SS/5, 9726
Electrafil® G-1204/SS/3, 9728, *See* Acrylonitrile-butadiene-styrene, 439
Electrafil® G-1704/SS/5, 9729, *See* Polyphenylene oxide, 24518
Electrafil® G-1854/SS/7, 9730
Electrafil® G-50/SS/10, 9727
Electrafil® J-1/30/CF/7/H, 9731
Electrafil® J-2/CF/30, 9732, 9733, *See* nylon 6/10, 21911
Electrafil® J-4/CF/30, 9734
Electrafil® J-7/20/EC, 9735
Electrafil® J-30/CF/20, 9736
Electrafil® J-50/20/CF/10, 9737
Electrafil® J-60/CF/30, 9738
Electrafil® J-80/CF/10/TF/10, 9739
Electrafil® J-80/CF/10/TF/10, *See* Acetal, 280
Electrafil® J-100/CF/30, 9740
Electrafil® J-1100/CF/30, 9741
Electrafil® J-1105/CF/30, 9742
Electrafil® J-1106/CF/30, 9743, *See* Polyetherimide, 24423
Electrafil® J-1200/CF/10, 9744, *See* Acrylonitrile-butadiene-styrene, 439
Electrafil® J-1300/CF/30/TF/15, 9745
Electrafil® J-1400/CF/20, 9746, 9747, *See* Polysulfone, 24619
Electrafil® J-1700/CF/10, 9748, *See* Polyphenylene oxide, 24518
Electrafil® J-1701/CF/10/FR, 9749
Electrafil® J-1800/CF/30, 9750, 9751, *See* Polyethylene terephthalate, 24426
Electrafil® JM-61/CF/10, 9752
Electrafil® M-1526/EC, 9753
Electrafil® PC-50/EC, 9754
Electrafil® PE-90/EC, 9755
Electrafil® PP-60/CC/20/EC, 9756
Electrafil® TR-1900/EC, 9757, *See* Polytetrafluoroethylene, 24642
Electran, 9758
Electrathane, 9759
Electraurol, 9760
Electric bronze, 9761
electric metal, *See* telegraph bronze, 31056
Electricidal, 9762
Electriridol, 9763
Electrisil, 9764
Electrisil 758, 9765
Electrisil 9025, 9766
Electrit, 9767
electro-cf 12, *See* dichlorodifluoromethane, 8388
Electroclear, 9768
Electrocuprol, 9769
Electrodag®, 9770
Electrodag® 112, 9771
Electrodag® 415, 9772
Electrodag® 415C, 9773
Electrodag® 4371, 9774
Electrodag® 438, 9775
Electrodag® 439, 9776
Electrodag® 442, 9777
Electrodag® 550, 9778
Electrodag® 24501, 9779
Electrodyn, 9780
electro-filtros, 9781
Electrofine® S-70, 9782, *See* Chlorinated paraffin, 6105
Electro-fused Cement, *See* fused cement, 12517
electro-granodized iron and steel, 9783
Electrolon, *See* Carbora, 5235
Electrolyilc chlorogen (E.C.), 9784
electrolyte acid, *See* sulfuric acid, 30055
Electromagnesia, *See* Dynatherm, 9425
Electromagnesia (magnesium oxide), *See* Dynamag, 9350
Electromartiol, 9785
Electromate, 9786
Electromercurol, 9787
Electronite, 9788
Electronite No. 2, 9789
Electropalladiol, 9790
electropate, *See* Argyrolith, 2867 nickel silvers, 21132
Electroplatinol, 9791
Electrorhodiol, 9792
Electrorubin, 9793
Electrose, 9794
Electroselenium, 9795
Electrotype metal, 9796
Electro-Wash®, 9797
Electrox, 9798

Electrozone, 9799
Electrum, *See* nickel silvers, 21132
Electrundum, 9800
Elefac I-205, 9801
Elektra, 9802
Elektron®, 9803 , 9804
Elemite, 9805
Elephant bronze, 9806
Elephant-S bronze, 9807
Eleudron-Solution, 9808
Eleuthera bark, *See* sweet bark, 30453
Eleven, *See* hexachlorophene, 13992
Elexar®, 9809
Elexar® 8421, 9810
Elfacos®, 9811
Elfan 242, *See* Witcolate ES-2, Witcolate LES-60C, 34477
Elfan 4240 T, *See* Witcolate TLS-500, 34479
Elfan 4240 T, *See* Standapol® T, 29481
Elfan 4240 T, *See* Rhodapon® LT-6, 26638
Elfan 4240 T, *See* Perlankrol® ATL40, 23324
Elfan NS 242, *See* Witcolate ES-2, Witcolate LES-60C, 34477
Elfan NS 243, *See* Witcolate ES-2, Witcolate LES-60C, 34477
Elfan WA, *See* Nacconol® 40G, 20647
elfan wa sandet 60, *See* sodium dodecylbenzenesulfonate, 28760
Elfan WA Sulfonic acid, *See* Witco® 1298, 34458
Elfan WA Sulphonic Acid, *See* Witco® 1298, 34458
Elfan® 200, 9812
Elfan® 240 and 240S, 9813
Elfan® 240M and 240M/S, 9814
Elfan® 240T and 240T/S, 9815
Elfan® 280, 9816
Elfan® 680, 9817
Elfan® 2240 Mg, *See* magnesium lauryl sulfate, 18364
Elfan® A432, 9818
Elfan® KT550, 9819
Elfan® NS 242, 243S, 252 S, 9820
Elfan® NS 243 S Mg, 9821, *See* magnesium laureth sulfate, 18363
Elfan® NS 682 KS, 9822
Elfan® OS 46, 9823
Elfan® WA Series, 9824
Elfanol® 510, 9825
Elfanol® 616, 9826, *See* disodium lauryl sulfosuccinate, 8649
Elfanol® 850, 9827
Elfanol® 883, 9828
Elfapur® N50, 9829
Elfapur® N70, 9830
Elfapur® N90, N120 and N150, 9831
Elftex® 675, 9832
Elfugin, 9833
Elgetol, *See* 4,6-dinitrocresol, 8542
Elgetol 30, *See* 4,6-dinitrocresol, 8542
Elgetox, *See* 4,6-dinitrocresol, 8542
Elhuyarite, 9834
Elianite I, 9835
Elianite II, 9836
Eliminal, 9837
Elimite®, 9838 , *See* Permethrin, 23385
Elintaal, 9839
Elinvar, 9840
Elite, *See* Raxil, 25908
Elite Fast, 9841
Eljon, 9842
Elkalub, 9843
Elkem Microsilica, 9844
Elkonite, 9845
Ellagitannin, 9846
ellagite, 9847
Elliott's Lawn Sand, 9848, *See* ferric sulfate, 11652
Elliott's Moss Killer, 9849, *See* ferric sulfate, 11652
Elmarid, 9850
Elner's German silver, 9851, *See* nickel silvers, 21132
Elocril, 9852, *See* Iodofenphos, 15213
Elon, *See* Metol, 19584
Elosal, *See* Thiovit, 31724
Elotex, 9853
Eloxal, 9854
ELP-3, 9855
Elromid KD 80, *See* Active #2, 501
Elsner's reagent, 9856
Eltaga®, 9857
Eltesol SX 30, 9873, *See* Naxonate® 4L, 20824, Pilot SXS-40, 23843, Witconate SXS 40%, 34492
Eltesol SX93, 9873, *See* Naxonate® 4L, 20824, Pilot SXS-40, 23843
Eltesol®, 9858
Eltesol® 4009, 4018, 9859
Eltesol® 4402, 4403, FDA 55/8, 9860
Eltesol® 5400 Series, 9861
Eltesol® 7200 Series, 9862
Eltesol® AC60, 9863, *See* Ammonium curmene sulfonate, 2236
Eltesol® ACS 60, 9864
Eltesol® AX 40, 9865, *See* Ammonium xylene sulfonate, 2259
Eltesol® CA 65, 9866, *See* Cumene sulfonic acid, 7353
Eltesol® MGX, 9867
Eltesol® PSA 65, 9868, *See* Phenosulfonic acid, 23617
Eltesol® PT 93, 9869
Eltesol® PX 40, PX 93, 9870
Eltesol® SC 93, 9871, *See* sodium cumene sulfonate, 28752
Eltesol® ST 40, 9872
Eltesol® SX 30, 9873, *See* sodium xylene sulfonate, 28832
Eltesol® TA 65, 9874
Eltesol® TPA, 9875
Eltesol® TSX, 9876
Eltesol® XA, 9877, *See* Xylene sulfonic acid, 34621
Eltex, 9878
Eltex P, 9879
Eludril Mouthwash, 9880
Eludril Spray, 9881
Elugent™, 9882
Elvace®, 9883
Elvace® 1870, 9884
Elvacite®, 9885
Elvaloy®, 9886
Elvaloy® EP-4043, HP441, 9887
Elvamide®, 9888
Elvanol®, 9889
Elvanol® 20-25, 9890
Elvanol® 71-30, 9891
Elvanol® 90-50, 9892
Elvaron®, 9893 , *See* Dichlofluanid, 8379
Elvax®, 9894
Elvax® 40-W, 9895
Elvax® 260, 9896
Elvax® 310, 9897
Elvax® 550, 560, 9898
Elvax® 4260, 9899
Elvax® 4320, 9900
Elvax® D, 9901
Elverite, 9902
Elveron®, Elvon®, 9903
Em-1, 9904
EM-550, 9905
EM-600, 9906, *See* PEG tallate series, 23035
EM-980, 9907
EMA, 9908, *See* ethyl methacrylate, 11077
Ema Resins, 9909
Emac SP 2205, 9910
Emaillit, 9911
emal 10, *See* sodium lauryl sulfate, 28781, Rhodapon® 101-10, 26631
Emal o, *See* Rhodapon® 101-10, 26631, sodium lauryl sulfate, 28781
EMAL T, *See* Rhodapon® LT-6, 26638, Perlankrol® ATL40, 23324, Standapol® T, 29481, Witcolate TLS-500, 34479
Emalex 103, 9912
Emalex 200 di-IS, 9915
Emalex 200 di-L, 9916, *See* PEG-4 dilaurate series, 23050
Emalex 200 di-O, 9917, *See* PEG-2 dioleate series, 23040
Emalex 200 di-S, 9918
Emalex 218, 9919
Emalex 300 di-IS, 9920
Emalex 400 di-IS, 9921
Emalex 400A, 9922
Emalex 508, 9923, *See* oleth-8, 22126
Emalex 515, *See* Ablunol OA-6, 107, Volpo 3, 34020
Emalex 600 di-IS, 9924
Emalex 640, 9925, *See* steareth series, 29596
Emalex 709, 9926
Emalex 805, 9927
Emalex 1605, 9913
Emalex 1805, 9914
Emalex 2505, 9928
Emalex 6300 Di-ST, 9929, *See* PEG-150 distearate series, 23037
Emalex 6300 M-ST, 9930
Emalex BHA-30, 9931
Emalex C-20, 9932
Emalex CC-10, 9933
Emalex CC-16, 9934
Emalex CC-18, 9936
Emalex CC-168, 9935
Emalex CC-168, *See* cetyl octanoate, 5821
Emalex CS-5, 9937
Emalex CWS-3, 9938
Emalex DEG-di-IS, 9939
Emalex DEG-di-L, 9940, *See* PEG-4 dilaurate series, 23050
Emalex DEG-di-O, 9941, *See* PEG-2 dioleate series, 23040
Emalex DEG-m-S, 9942
Emalex DISG-2, 9943
Emalex DISG-3, 9944
Emalex DSG-2, 9945
Emalex EG-2854-IS, 9946
Emalex EG-2854-O, 9947
Emalex EG-2854-S, 9948
Emalex EG-di-L, 9949, *See* glycol dilaurate, 13100
Emalex EG-di-O, 9950
Emalex EG-di-S, 9951
Emalex EGS-A, 9952
Emalex ET-2020, 9953
Emalex ET-8020, 9954
Emalex ET-8040, 9955
Emalex GM-5, 9956
Emalex GMS-55FD, 9957
Emalex GMS-A, 9958
Emalex GMS-ASE, 9959
Emalex GWIS-115, 9960
Emalex GWIS-303, 9961
Emalex GWO-303, 9962
Emalex GWS-204, 9963
Emalex GWS-303, 9964
Emalex HC-5, 9965
Emalex J.J O-V, 9966
Emalex K.T.G, 9967, *See* caprylic/capric acid triglyceride, 5161
Emalex LWIS-2, 9968
Emalex LWS-3, 9969
Emalex MSG-2, 9970, *See* Polyglyceryl stearate series, 24457
Emalex MTS-30E, 9971 *See* Phenyl trimethicone, 23635
Emalex N-83, 9972
Emalex NN-15, 9973
Emalex NN-7, 9974
Emalex NP-2, 9975
Emalex O.T.G, 9976
Emalex O.T.G, *See* tricaprylin, 32211
Emalex OD-5, 9977
Emalex OE-6, 9978
Emalex OP-25, 9979
Emalex PEIS-3, 9980
Emalex PG-di-IS, 9981
Emalex PG-di-L, 9982
Emalex PG-di-O, 9983, *See* Propylene glycol dioleate,

25217
Emalex PG-di-S, 9984
Emalex PGML, 9985
Emalex PGMS, 9986
Emalex PGO, 9987, *See* Propylene glycol oleate, 25221
Emalex RWIS-105, 9988
Emalex RWIS-305, 9989
Emalex RWL-120, 9990
Emalex SG-37, 9991
Emalex SPE-100S, 9992
Emalex SPE-150S, 9993
Emalex SPIS-100, 9994, *See* Sorbitan isostearate, 29110
Emalex SPIS-150, 9995
Emalex SPO-100, 9996
Emalex SPO-150, 9997
Emalex SWS-4, 9998
Emalex TEG-di-IS, 9999
Emalex TEG-di-L, 10000, *See* PEG-4 dilaurate series, 23050
Emalex TPIS-303, 10001
Emalex TPM-303, 10002
Emalex TPS-203, 10003
Emalex TPS-303, 10004
Emaline, 10005
Emalsy B 1-1000, *See* Verv®, 33644
Emanil, *See* Kerecid, 16067
Emanon 1112, *See* Ablunol 200ML, 95
Emanon 3113, *See* Ablunol DEGMS, 99
Emanon 4115, *See* Ablunol 200MO, 96
Emasol 110, *See* Ablunol S-20, 108
Emasol 430, *See* Ablunol S-85, 112, Witconol 2503, 34502
Emaweld®, 10006
Embacel, 10007
Embacide, 10008
Embacoid, 10009
Embadot, 10010
Embaflx, 10011
Embafume, 10012
Embalith, 10013
Embamix, *See* Potassium Iodide, 24759
Embanox, 10014, *See* Butylated hydroxyanisole, 4787, Nipantiox, 21276, Sustane® 1-F, 30432
Embanox BHT, *See* Oxyguard, 22457
Embanox®, 10015
Embaphase, 10016
Embark, 10017, *See* mefluidide, 19230
Embaspeed, 10018
Embatex, 10019
Embathion, 10020
Embatype, 10021
Embazin, 10022
Embedyne, 10023
Embephen, *See* Nuophene, 21783
Embequin, 10024, *See* Yodoxin, 34658
Embesafe, 10025
Embesol, 10026
Emblet® M, 10027
Embond 55, 10028
Embond 66, 10029
Embond 125, 10030
Embond 168, 10031
Embond 169, 10032
Embond 212, 10033
Embond 401, 10034
Embond 560, 10035
Embond Surface Tackifier, 10036
embrithite, 10037
Embrol, 10038
Embutone, *See* 2,4-DB, 7810
Embutox, 10039
Emcast 1510, 1511, 10040
Emcast 1550, 1551, 10041
Emcepan, *See* MCPA, 19183
Emcocel® 90M, 10042
Emcol, 10043
Emcol 12-14-18, *See* Witconol 14, 34495
Emcol d 5-10, *See* Niaproof® Anionic Surfactant 08, 21092, Rhodapon® BOS, 26632
Emcol E-607, *See* Lapyrium Chloride, 16782
Emcol H 31a, *See* Ablunol 200MO, 96
Emcol H 35-A, *See* Ablunol DEGMS, 99
Emcol H-2A, *See* Ablunol 200MO, 96
Emcol PS-50 RHP, *See* Witconol 2380, 34498
Emcol®, 10048
Emcol® 4, 10044
Emcol® 416L, 10045, *See* disodium oleamido MIPA sulfosuccinate, 8651
Emcol® 1655, 10046, *See* Cocamidopropyl dimethylamine propionate, 6554
Emcol® 3780, 10047, *See* stearamidopropyl dimethylamine lactate, 29595
Emcol® 4300, 10049
Emcol® 4350, 10050
Emcol® 4403, *See* disodium lauryl sulfosuccinate, 8649
Emcol® 4500, 10051
Emcol® 4560, 4500, *See* dioctyl sodium sulfosuccinate, 8550
Emcol® 4600, 10052
Emcol® 4776, 10053
Emcol® 4910, 10054
Emcol® 5430, 10055
Emcol® 6613, *See* isostearamidopropyl dimethylamine lactate, 15420
Emcol® 6748, Coco betaine, *See* Cocamidopropyl betaine, 6551
Emcol® CC-42, 10056 *See* PPG-40 diethylmonium chloride, 24846
Emcol® CC-55, 10057
Emcol® DG, NA30, *See* Cocamidopropyl betaine, 6551
Emcol® DMCD-40, *See* lauramine oxide, 16832
Emcol® DOSS, 10058
Emcol® E-607L, 10059, *See* Lapyrium Chloride, 16782
Emcol® E-607S, 10060
Emcol® ISML, 10061
Emcol® K8300, 10062
Emcol® L, *See* lauramine oxide, 16832
Emcol® LO, 10063, *See* lauramine oxide, 16832
Emcol®CMCD, *See* Empigen® CDR40, 10369
Emcol-Im, *See* Wickenol® 101, 105, 34378
Emcol-ip, *See* Wickenol® 111, 34379
Emcompress®, 10064
EmCon E-5, 10065, *See* Egg oil, 9645
EmCon Limnanthes Alba, 10066
EmCon TEA TREE, 10067
EmCon W, 10068, *See* Wheat germ oil;, 34349
Emcor, 10069, *See* Attapulgite, 3404
Emcosoy®, 10070 , *See* Soy Flour, 29168
EMD-34946, *See* Reprodin®, 26149
Emdex®, 10071
Emdite, 10072
Emdithene, 10073
Emerald, 10074
Emerald bronze, 10075
emerald green, *See* chromium green, 6232, cupric acetoarsenite, 7374, mountain green, 20369
Emercide® 1199, 10076
Emeressence 1160, *See* Igepal® Cephene Distilled, 14850, Rewopal® MPG 10, 26344
Emeressence® 1150, 10077, *Sèe* ethylene brassylate, 11121
Emeressence® 1151, 10078
Emeressence® 1160 Rose Ether, 10079 *See* 2-Phenoxyethanol, 23627
Emeressence® 1174 Fir Balsam, 10080
Emerest 2314, *See* Wickenol® 101, 105, 34378
Emerest 2316, *See* Wickenol® 111, 34379
Emerest 2381, *See* Witconol 2380, 34498
Emerest 2640, *See* Ablunol DEGMS, 99
Emerest 2646, *See* Ablunol 200MO, 96
Emerest 2660, *See* Ablunol 200MO, 96
Emerest® 2301, 10081, *See* methyl oleate, 19554
Emerest® 2302, 10082
Emerest® 2308, 10083
Emerest® 2310, 10084
Emerest® 2314, 10085
Emerest® 2316, 10086, *See* isopropyl palmitate, 15410
Emerest® 2324, 10087
Emerest® 2325, 10088, *See* butyl stearate, 4781
Emerest® 2328, 10089, *See* Butyl oleate, 4778
Emerest® 2350, 10090
Emerest® 2355, 10091
Emerest® 2380, 10092
Emerest® 2384, 10093, *See* Propylene glycol isostearate, 25218
Emerest® 2388, 10094
Emerest® 2400, 10095
Emerest® 2410, 10096
Emerest® 2419, 10097, *See* glyceryl dioleate, 13080
Emerest® 2421, 10098
Emerest® 2423, 10099
Emerest® 2452, 10100
Emerest® 2485, 10101
Emerest® 2610, 10102
Emerest® 2617, 10103
Emerest® 2620, 10104
Emerest® 2622, 10105, *See* PEG-4 dilaurate series, 23050
Emerest® 2624, 10106
Emerest® 2625, 10107
Emerest® 2630, 10108
Emerest® 2634, 10109
Emerest® 2636, 10110
Emerest® 2647, 10111
Emerest® 2660, 10112
Emerest® 2704, 10113, *See* PEG-4 dilaurate series, 23050
Emerlube® 5919, 10114
Emerox® 1110, 10115, *See* Azelaic acid, 3519
Emersol 120, *See* stearic acid, 29599
Emersol 132, *See* stearic acid, 29599
Emersol 150, *See* stearic acid, 29599
Emersol 153, *See* stearic acid, 29599
Emersol 6349, *See* stearic acid, 29599
Emersal 6400, *See* Rhodapon® 101-10, 26631, sodium lauryl sulfate, 28781, Witcolate TLS-500, 34479
Emersal 6465, *See* Niaproof® Anionic Surfactant 08, 21092, Rhodapon® BOS, 26632
Emersist 7210, 10116
Emersoft 7700, 10117
Emersol® 110, 10118
Emersol® 143, 10119, *See* Palmitic Acid, 22635
Emersol® 210, 10120
Emersol® 210, 6333 NF, 7021, *See* oleic Acid, 22112
Emersol® 315, 10121
Emersol® 871, 10122, *See* isostearic acid, 15421
Emersol® 6333 NF, 10123
Emersol® 6349, 10124
Emersol® 7021, 10125
Emerstat® 6660, 10126
Emerwax® 1251, 10127
Emerwax® 1253, 10128
Emerwax® 1266, 10129
Emery 655, *See* Philacid 1400, 23662
Emery 5791, *See* 2-mercaptoethanol, 19336
Emery 6705, *See* Igepal® Cephene Distilled, 14850
emery 6705, *See* Rewopal® MPG 10, 26344
Emery 6802, *See* Volpo 3, 34020
Emery 6802, *See* Ablunol OA-6, 107
Emery 15393, *See* Ablunol DEGMS, 99
Emery® 400, 10130
Emery® 515, 10131
Emery® 610, 10132
Emery® 621, 10133
Emery® 621, 622, 626, 627, *See* Hystrene® 1835, 14755
Emery® 650, 10134
Emery® 650, *See* lauric acid, 16856
Emery® 654, 10135
Emery® 654, *See* Myristic acid, 20574
Emery® 657, 10136
Emery® 912, 10137
Emery® 1202, 10138, *See* Pelargonic acid, 23078
Emery® 1650, 10139
Emery® 1730, 10140
Emery® 2203, 10141
Emery® 2204, 10142

Emery® 2209, 10143, *See* methyl caprylate-caprate, 19520
Emery® 2214, 10144, *See* methyl myristate, 19550
Emery® 2216, 10145, *See* methyl palmitate, 19555
Emery® 2218, 10146, *See* methyl stearate, 19560
Emery® 2219, 10147
Emery® 2224, 10148
Emery® 2230, 10149
Emery® 2231, 10150
Emery® 2232, 10151
Emery® 2253, 10152, *See* methyl coconate, 19523
Emery® 2255, 10153
Emery® 2270, 10154
Emery® 2301, 10155
Emery® 2301, 2219, *See* methyl oleate, 19554
Emery® 2895 Foamaster Soap L, 10156, *See* sodium tallowate, 28824
Emery® 2900, 10157
Emery® 2914, 10158, *See* dimethyl azelate, 8504
Emery® 2957, 10159
Emery® 3304, 10160
Emery® 3310, 10161
Emery® 3312, 10162
Emery® 3317, 10163
Emery® 3320, 10164
Emery® 3321, 10165
Emery® 3322, 10166
Emery® 3323, 10167
Emery® 3324, 10168
Emery® 3326, 10169, *See* 1-dodecanol, 8777
Emery® 3332, 10170
Emery® 3334, 10171
Emery® 3336, 10172
Emery® 3343, 10173
Emery® 3357, 10174
Emery® 5353 Lomar PW, 10175
Emery® 5366 (Lomar PWA Liq.), 10176
Emery® 5370 Sellogen W, 10177
Emery® 6220, 10178
Emery® 6221 Monolan 2500, 10179
Emery® 6358, *See* caprylic/capric acid, 5160
Emery® 6686, 10180
Emery® 6701, 10181
Emery® 6717, 10182
Emery® 6744, 10183
Emery® 6744, *See* Cocamidopropyl betaine, 6551
Emery® 6750 Nopcosperse AD-6 Liq, 10184
Emery® 9331, *See* 2,4-dibromophenol, 8359
Emery® 9332, *See* 2,4,6-tribromophenol, 32202
Emery® 9336, *See* dibromoneopentyl glycol, 8358
Emery® HP-2050, 10185
Emery®5412, *See* Empigen® CDR40, 10369
Emery's L-110, 10186
Emery's L-114, 10187
Emgard® 2033, 10188
Emgard® 2063, 10189
EMI-24, 10190
Emid 6511, *See* Witcamide® 5195, 34440
Emid 6541, *See* Witcamide® 5195, 34440
Emid® 6500, 10191
Emid® 6515, 10192
Emid® 6519, 10193
Emid® 6545, 10194
Emid® 6590, 10195
Emisan 6, *See* Ceresan, 5745
EMI-X®, 10196
EMI-X®, *See* Grilpet® XE3060, 13402
EMI-X® DC-1008, 10197
EMI-X® OC-1008, 10198
EMI-X® PC-1008, 10199
EMI-X® PDX-83393, 10200
EMI-X® PDX-A-88128, 10201 *See* Acrylonitrile-butadiene-styrene, 439
EMI-X® PDX-D-87815, 10202
EMI-X® PDX-O-91074, 10203
EMI-X® PDX-P-90305, 10204
EMI-X® PDX-R-89496, 10205
EMI-X® PDX-W-88341, 10206
EMI-X® RC-1008, 10207
EMI-X® WC-1008, 10208
Emka Catalyst P-35, 10209
Emka DDBSA, 10210
Emka Defoam AA, 10211
Emka Defoam BC, NC, 10212
Emka Defoam DP, 10213
Emka Graphite Remover, 10214
Emka Transfer Remover, 10215
Emkabase CA, 10216
Emkabase ODC-2, 10217
Emkabond UR, 10218
Emkacide GS-2, 10219
Emkadixol, 10220
Emkafix RXC, 10221
Emkafol D, 10222
Emkafume FA, 10223
Emkagen 49, 10224
Emkagen 49AM, 10225
Emkagen BT, 10226
Emkal BNS, 10227
Emkal BNS Acid, 10228
Emkal NNS, NNS Acid, 10229
Emkal NOBS, 10230
Emkalane WL, 10231
Emkalar Base C50L, 10232
Emkalite BAC, 10233
Emkalon KLA, 10234
Emkalon TN, 10235
Emkalube F-11, 10236
Emkane Acid, 10237
Emkane HAD, 10238
Emkane HAX, 10239
Emkanet B, 10240
Emkanol NC, NCD 25, 35, 45, 55, 10241
Emkanyl 85, 10242
Emkanyl BRX, 10243
Emkapel DE, 10244
Emkapene AV, AVX, 10245
Emkapene RW, 10246
Emkaperm, 10247
Emkapol 200, *See* Alkapol PEG 300, 1706
Emkapol PO-18, 10248, *See* Potassium oleate, 24764
Emkapon 4S, DS, SS, TS, 10249
Emkapon BC, 10250
Emkapon DAC, DAC-50, 10251
Emkapon Jel 500 Conc, 10252
Emkapon ML, 10253
Emkapruf ABR, FL, 10254
Emkarate, 10255
Emkaron GA-1, 10256
Emkarox, 10257
Emkasan QA-50, 10258
Emkasene 800, 10259
Emkaset, 10260
Emkasize CF, 10261
Emkasol DE, 10262
Emkasorb, 10263
Emkastat MLT, 10264
Emkatan K, 10265
Emkatard, 10266
Emkaterge B, 10267
Emkatex 11, 21, 10268
Emkatex 49-P, 10269
Emkatex AA, 10270
Emkatex DX, DXP, 10271
Emkatex NE, 10272
Emkatint BRN, 10273
Emkatol M, 10274
Emkawate AS, 10275
Emkazyme, 10276
Emmatos, *See* malathion, 18460
Emol Keleet, 10277
Empal, 10278, *See* MCPA, 19183
EM-PB, 10279
Empee® FR 42 LM, 10280
Empee® PE-112, 10281
Empee® PE-113, 10282
Empee® PO Conc. 61, 10283
Empee® PP Conc.33, 10284
Empee® PP-301, 10285
Empee® PP-459, 10286
Empee® PP-560, 10287
Empee® PS-921, 10288
Emperor alloy, 10289
Emperor brass, 10290
Empetal, 10291
Emphos, 10292
Emphos CS-136, 10293
Emphos CS-1361, 10294
Emphos PS-21A, 10295
Emphos TS-230, 10296
Empicol, 10297
Empicol ESB 3, *See* Witcolate ES-2, Witcolate LES-60C, 34477
Empicol ESB 30, *See* Witcolate ES-2, Witcolate LES-60C, 34477
Empicol lpz, *See* Rhodapon® 101-10, 26631, sodium lauryl sulfate, 28781
Empicol Is 30, *See* Rhodapon® 101-10, 26631, sodium lauryl sulfate, 28781
Empicol lx 28, *See* sodium lauryl sulfate, 28781, Rhodapon® 101-10, 26631
Empicol ML30, *See* magnesium lauryl sulfate, 18364
Empicol® 0031/T, 10298, *See* DEA-lauryl sulfate, 7827
Empicol® 0045, 10299
Empicol® 0045V, 10300
Empicol® 0303, 10301
Empicol® 0585/A, 10302
Empicol® 0758, 10303, *See* sodium decyl sulfate, 28755
Empicol® 0775, 10304
Empicol® 9060X, 10305
Empicol® AL30, 10306
Empicol® ALL, 10307
Empicol® BSD 52, 10308
Empicol® CHC 30, 10309
Empicol® DA, 10310, *See* DEA-lauryl sulfate, 7827
Empicol® DLS, 10311
Empicol® EAA, 10312, *See* Ammonium laureth sulfate, 2240
Empicol® EAA, EAB, EAC, *See* Ammonium laureth sulfate, 2240
Empicol® EAB, 10313, *See* Ammonium laureth sulfate, 2240
Empicol® EAC, 10314, *See* Ammonium laureth sulfate, 2240
Empicol® EGB, EGC, 10315, *See* magnesium laureth sulfate, 18363
Empicol® EGB, EGC, EGC70, *See* magnesium laureth sulfate, 18363
Empicol® EL, 10316
Empicol® EMB, 10317
Empicol® ESA, 10318
Empicol® ESB, 10319
Empicol® ESC/AU, 10320
Empicol® ETB, 10321
Empicol® HL25, 10322
Empicol® L Series, 10323
Empicol® LM, 10324
Empicol® LMV/T, 10325
Empicol® LQ33/T, 10326
Empicol® LS30, 10327
Empicol® LX, 10328
Empicol® LXV, 10329
Empicol® LY28/S, 10330
Empicol® LZ, 10331
Empicol® LZG 30, 10332
Empicol® LZP, 10333
Empicol® LZV, 10334
Empicol® MD, 10335
Empicol® ML 26/F, 10336, *See* magnesium lauryl sulfate, 18364
Empicol® SDD, 10337, *See* disodium lauryl ethoxy sulfosuccinate, 8648, disodium lauryl sulfosuccinate, 8649
Empicol® SLL, 10338, *See* disodium lauryl sulfosuccinate, 8649
Empicol® STT, 10339
Empicol® TA40, 10340

Empicol® TAS30, 10341
Empicol® TL40, 10342
Empicol® XC35, 10343
Empicol® XM 17, 10344
Empicol® XPA, 10345
Empicryl®, 10346
Empicryl® 6045, 10347
Empicryl® APD, 10348
Empicryl® DH122, 10349
Empicryl® PPT38, 10350
Empicryl® PT1334, 10351
Empicyl, 10352
Empigen, 10353
Empigen ® BS/AU, BS/F, BS/FA, BS/H, BS/P, *See* Cocamidopropyl betaine, 6551
Empigen® 5083, 10354
Empigen® 5089, 10355
Empigen® 5107, 10356
Empigen® 5509, 10357
Empigen® AB, 10358
Empigen® AF, 10359
Empigen® AG, 10360
Empigen® AH, 10361
Empigen® BAC50, 10362
Empigen® BB, 10363
Empigen® BCB50, 10364, *See* Benzalkonium chloride, 3958
Empigen® BCB50, BAC50/BP, BAC 80, BCB50, *See* Benzalkonium chloride, 3958
Empigen® BS, 10365
Empigen® BS/H, 10366
Empigen® CDL60, 10367
Empigen® CDR10, 10368
Empigen® CDR40, 10369
Empigen® CDR60, *See* Empigen® CDR40, 10369
Empigen® CHB40, 10370
Empigen® CM, 10371
Empigen® FKC75L, 10372
Empigen® FKH75L, 10373
Empigen® FRC75S, 10374
Empigen® FRC90S, *See* Quaternium-27, 25607
Empigen® OB, 10375, *See* lauramine oxide, 16832
Empigen® OB/AU, *See* Cocamine oxide, 6557
Empigen® OC, 10376
Empigen® OH25, 10377
Empigen® OS/A, 10378
Empigen® OY, 10379, *See* PEG-3 lauramine oxide, 23047
Empigen® XDR302, 10380
Empigen®ABE, *See* dimethyl lauramine, 8507
Empilan AP-100, *See* Ablunol 200ML, 95
Empilan BP-100, *See* Ablunol 200MO, 96
Empilan BQ-100, *See* Ablunol 200MO, 96
Empilan CDE, *See* Active #2, 501
Empilan CP-100, *See* Ablunol DEGMS, 99
Empilan CQ-100, *See* Ablunol DEGMS, 99
Empilan®, 10381
Empilan® 0004, 10382
Empilan® 2020, 10383
Empilan® 2502, 10384
empilan® 7132, 10385
Empilan® AM Series, 10386
Empilan® BD, 10387
Empilan® BQ 100, 10388
Empilan® CDE, 10389
Empilan® CIS, 10390, *See* Cocamide MIPA, 6550
Empilan® CME, 10391
Empilan® EGMS, 10392
Empilan® GMS LSE32, 10393
Empilan® GMS NSE32, 10394
Empilan® K Series, 10395
Empilan® KA10/80, 10396
Empilan® KB 2, 10397
Empilan® KB 2, *See* Laureth-2, 16853
Empilan® KCA Series, 10398
Empilan® KCB Series, 10399
Empilan® KCL Series, 10400
Empilan® KCMP 0703/F, 10401
Empilan® KCP Series, 10402
Empilan® KCXSeries, 10403
Empilan® KI Series, 10404
Empilan® KL 6, 10405
Empilan® KM 11, 10406
Empilan® KS Series, 10407
Empilan® LDE, 10408
Empilan® LIS, 10409
Empilan® LME, 10410
Empilan® LP10, 10411
Empilan® MAA, 10412
Empilan® NP9, 10413
Empilan® OPE9.5, 10414
Empilan® P7061, 10415
Empilan® SM Series, 10416
Empimin KSN, *See* Witcolate ES-2, Witcolate LES-60C, 34477
Empimin KSN 27, *See* Witcolate ES-2, Witcolate LES-60C, 34477
Empimin KSN 60, *See* Witcolate ES-2, Witcolate LES-60C, 34477
Empimin KSN 70, *See* Witcolate ES-2, Witcolate LES-60C, 34477
Empimin®, 10417
Empimin® 3060, 10418
Empimin® 3116, 10419
Empimin® BMA, 10420
Empimin® BMB, 10421
Empimin® BMC, 10422
Empimin® KSN27, 10423
Empimin® LAM30/AU, 10424
Empimin® LR28, 10425
Empimin® LSM30, 10426
Empimin® MA, 10427
Empimin® MH, 10428
Empimin® MHH, 10429
Empimin® MK/B, 10430
Empimin® MKK, 10431
Empimin® MSS, 10432
Empimin® MTT, 10433
Empimin® OP70, 10434, *See* dioctyl sodium sulfosuccinate, 8550
Empimin® OT, 10435
Empimin® SDS, 10436, *See* sodium decyl sulfate, 28755
Empimin® SQ25, 10437
Empiphos, 10438
Empiphos DF Series, 10439
Empiquaternary, 10440
Empire, *See* Chlorpyrifos, 6163
Empiwax, 10441
Empiwax SK, 10442
Emplets Potassium Chloride, *See* Potassium chloride, 24748
Emplex, 10443
Empol® 1004, 10444
Empol® 1010, 10445
Empol® 1016, *See* Dilinoleic acid, 8480
Empol® 1020, *See* Dilinoleic acid, 8480
Empol® 1022, 1004, 1026, *See* Dilinoleic acid, 8480
Empol® 1040, 10446
Empol® 1045, *See* Trilinoleic acid, 32320
Empol® 1061, 10447
EMQ, *See* Polyflex, 24435
EMQ,, *See* ethoxyquin, 11049
Emralon, 10448
Emralon 304, 10449
Emralon 8301-01, 10450
Emralon 8301-01, 304, *See* Polytetrafluoroethylene, 24642
EMS 209, 10451
Emsac Concrete Additive, 10452
Emsodur, 10453
Emsodur Micro, 10454
Emsorb 2500, *See* Ablunol S-80, 111, Kemester® S80, 15970, Radiasurf® 7155, 25802, Sorbon S-80, 29121, Witconol 2500, 34501
Emsorb 2503, *See* Ablunol S-85, 112, Witconol 2503, 34502
Emsorb 2505, *See* Ablunol S-60, 110
Emsorb 2510, *See* Ablunol S-40, 109
Emsorb 2515, *See* Ablunol S-20, 108, Kemester® S20, 15966, Radiasurf® 7125, 25797, Sorbon S-20, 29117
Emsorb 6900, *See* Alkamuls® T-80, 1693, Radiasurf® 7157, 25804
Emsorb® 2500, 10455
Emsorb® 2502, 10456
Emsorb® 2503, 10457
Emsorb® 2505, 10458
Emsorb® 2507, 10459
Emsorb® 2510, 10460
Emsorb® 2515, 10461
Emsorb® 2516, 10462, *See* Sorbitan isostearate, 29110
Emsorb® 2518, 10463, *See* Sorbitan diisostearate, 29109
Emsorb® 2720, 10464
Emsorb® 2721, 10465
Emsorb® 2722, 10466
Emsorb® 2726, 10467
Emsorb® 2728, 10468, *See* Ethylan® GES6, 11108
Emsorb® 2728, 6906, 6909, *See* Ethylan® GES6, 11108
Emsorb® 6900, 10469
Emsorb® 6901, 10470
Emsorb® 6903, 10471
Emsorb® 6906, 10472, *See* Ethylan® GES6, 11108
Emsorb® 6908, 10473
Emsorb® 6909, 10474, *See* Ethylan® GES6, 11108
Emsorb® 6913, 10475
Emsorb® 6915, 10476
Emsorb® 6917, 10477
Emthox® 2730, 10478
Emthox® 5882, 10479
Emthox® 5885, 10480
Emthox® 5967, 10481
Emtryl, 10482, *See* Dimetridazole, 8528
EMU® Powd. 120 FD, 10483
Emulamid TO-21, 10484
Emulan, 10485
Emulan®, 10486
Emulan® A, 10487
Emulan® AF, 10488
Emulan® EL, 10489
Emulan® OC, 10490
Emulan® PO, 10491, 10492
Emulcid, 10493
Emuldan HV 40 Kosher, HV 52 Kosher, 10494
Emulgade 1000 NI, 10495
Emulgade C, 10496
Emulgade EO-10, 10497
Emulgade F, 10498
Emulgator E 2149, 10499
Emulgator E 2155, 10500
Emulgator E 2568 SE, *See* Teginacid C, 30949
Emulgator U4, 10501
Emulgeant 710, 10502
Emulgeen P, *See* Potassium ricinoleate, 24778
emulgen, 10503
Emulgen 320P, *See* Ablunol SA-7, 113
Emulgen 408, *See* Ablunol OA-6, 107, Volpo 3, 34020
Emulgen 420, *See* Ablunol OA-6, 107, Volpo 3, 34020
Emulgen 430, *See* Ablunol OA-6, 107, Volpo 3, 34020
Emulgin 010, *See* Ablunol OA-6, 107, Volpo 3, 34020
Emulgin 05, *See* Ablunol OA-6, 107, Volpo 3, 34020
Emulphogene®, *See* Rhodasurf®,
Emulphopal HC, 10504
Emulphor, *See* Ablunol OA-6, 107, Volpo 3, 34020
Emulphor O, *See* Ablunol OA-6, 107, Volpo 3, 34020
Emulphor A, *See* Ablunol 200MO, 96
Emulphor EL 719, *See* Ablunol CO 10, 98
Emulphor EL-620, *See* Ablunol CO 10, 98
Emulphor ON-870, *See* Ablunol OA-6, 107, Volpo 3, 34020
Emulphor Surfactants, *See* Volpo 3, 34020
Emulphor UN-430, *See* Ablunol 200MO, 96
Emulphor vn 430, *See* Ablunol 200MO, 96
Emulphor vt-650, *See* Ablunol DEGMS, 99
Emulpon, 10505
Emulsamin, 10506
Emulsene, 10507
emulsepr (obsolete), *See* Lapyrium Chloride, 16782

Emulsifier 4, 10508
Emulsifier 632/90%, 10509
Emulsifier Component L, *See* Ablunol CO 10, 98
Emulsifier Component M G-1284, *See* Ablunol CO 10, 98
Emulsifier K 30 40%, 10510
Emulsifier L.W, 10511
Emulsifier L-32, *See* Ablunol 200MO, 96
Emulsifier no. 104, *See* Rhodapon® 101-10, 26631, sodium lauryl sulfate, 28781
Emulsifier WHC, 10512
Emulsil, 10513
Emulsilac S, 10514
Emulsin, 10515
Emulsion 212, *See* Fyrol® FR-2, 12540
Emulsion C-340, 10516
Emulsi-Phos, 10517
Emulsogen, 10518
Emulsogen 2144, 10519
Emulsogen CP 136, 10520
Emulsogen EL-050, 10521
Emulsogen HEL-050, 10522
Emulsogen IC, 10523
Emulsogen M, 10524
Emulsynt GDL, 10525, *See* glyceryl dilaurate, 13079
Emultex, 10526
Emultex 307, 328, 10527
Emultex AC431, 10528
Emultex®, 10529
Emulvin®, 10530
Emulvin® W, 10531
Emulzome, 10532
Emunon 3115, *See* Ablunol DEGMS, 99
Emvelop®, 10533 , *See* hydrogenated cotton*Seed* oil, 14593
Emzyaml No. 1, 10534
EN-27766, *See* Rapid, 25884
EN-40, *See* Extiat®, 11323
EN-55, *See* Extiat®, 11323
EN-55X, *See* Extiat®, 11323
enamel white, *See* lithopone, 17520
Enameled iron-granite, *See* Agate ware, 957
Enanth, *See* Nylon 7,
enanthic acid, *See* heptanoic acid, 13791
Enanthic acid triglyceride, *See* Special Oil 107, 29220
enanthic alcohol, *See* 1-heptanol, 13792, heptyl alcohol, 13797
enanthic anhydride, *See* heptanoic acid, 13791
enanthyl alcohol, *See* 1-heptanol, 13792, heptyl acohol, 13797
Enathene® EA 705-009, 10535
Enathene® EA 720-009, 10536
Enbucrilate, 10537
Encapsulated MgO, 10538
Encapsulation, 10539
Encelac®, 10540
Encem Steel, 10541
Enceprint®, 10542
Encore, 10543
Endanil, 10544
Endcor, 10545
Endegal, 10546
Endermol, 10547
Endobil, 10548
Endocel, *See* thiodan, 31680
Endocrocine, 10549
Endomirabil, 10550
Endosol, *See* thiodan, 31680
Endosulfan, *See* thiodan, 31680
endotryptase, 10551
Endox, 10552, *See* Racumin®, 25713
endran, *See* Baygon®, 3790
Endrate, 10553
Endrate Disodium, *See* edetate disodium, 9613, Trilon® BD, 32325
Endrate Tetrasodium, *See* Kalex 220 Crystal, 15668 Kalex Liq. 50%, 15671 tetrasodium EDTA, 31391
Endrocid, *See* Racumin®, 25713
Endrocide, *See* Racumin®, 25713
Endspray, 10554
Endura, 10555
Endura®, 10556
Enduracrete, 10557
Enduraflex, 10558
Enduraflor, 10559
Enduragloss, 10560
Enduraguard EP, 10561
Endurakote, 10562
Enduralay, 10563
Enduralith, 10564
Enduratop, 10565
En-Dur-Lon, 10566
Enduro Alloys, 10567
Endurol, 10568
Ene 11183 B, *See* Racumin®, 25713
Enelchem Products, 10569
Enelfa, *See* Valadol, 33238
Enerade® 3045, 10570
Enerade® 7101, 7102, 10571
Energex, 10572
Energol, *See* *p*-aminophenol, 2185
Eneril, *See* Valadol, 33238
Enervite®, 10573
English bearing metal, 10574
English blue, *See* mountain blue, 20366
English iron oxide red, *See* Ferroxide, 11694, Jeweller's rouge, 15611
English metal, 10575
English powder, *See* powder of Algaroth, 24813
English red, *See* Indian Red, 15068
English white, *See* calcium carbonate, 4940
English white bearing metal, 10576
engobe, 10577
engravers acid, *See* nitric acid, 21332
Engravers acid, *See* nitric acid, 21332
Enhance, 10578, 10579
Enhance®, 10580
Enheptin, *See* 2-amino-5-nitrothiazole, 2184
Enheptin premix, *See* 2-amino-5-nitrothiazole, 2184
Enheptin-A, *See* Aminitrazole, 2164
Enheptin-T, *See* 2-amino-5-nitrothiazole, 2184
Enheptyne, *See* 2-amino-5-nitrothiazole, 2184
enial orange I, *See* Sudan I, 29938
Enial Yellow 2G, *See* Butter yellow, 4759
Enide, 10581
Enilconazole, 10582, *See* Imaverol, 14891, imazalil, 14892
Enisyl, 10583, *See* L-(+)-lysine hydrochloride, 18089
Enkolan, *See* Cabelec® 1015, 4861
Enmag, 10584
Enovit methyl, *See* thiophanate methyl, 31706
Enovit-Supper, *See* thiophanate methyl, 31706
Enozin, *See* Zineb, 34807
ENR 25, 10585, *See* Polyisoprene, 24471
ENSA-6, 10586
EN-SD, *See* Extiat®, 11323
Enseal, *See* K. Tab, 15642, Kaon-Cl, 15729, Kay Ciel, 15807, K-Contin, 15818, K-Lor, 16226, K-Lyte/C1, 16234, KM Potassium Chloride, 16242, Potassium chloride, 24748, 24749
Ensecote S, 10587
Enso DTO 10-30, 10588
Enso Rosin, 10589
Ensol2, 10590
Ensoline, 10591
Ensoline 203 AS, 10592
ENT 987, *See* Thiram, 31726
ENT 987, *See* Thiurad, 31730
ENT 1025, *See* Acelan A, 268
ENT 1716, *See* methoxychlor, 19504
ENT 14250, *See* piperonyl butoxide, 23875
ENT 14875, *See* Maneb, 18526
ENT 15108, *See* parathion, 22807
ENT 15949, *See* Aldrin, 1558
ENT 17292, *See* Parathion-methyl, 22808
ENT 17510, *See* allethrin, 1744
ENT 17957, *See* Perizin®, 23276
ENT 18060, *See* Chlorpropam, 6162
ENT 19763, *See* trichlorfon, 32212
ENT 20218, *See* diethyl toluamide, 8438
ENT 22014, *See* Azinphos-ethyl, 3520
ENT 23648, *See* Acarin, 183, Dicofol, 8398, Kelthane, 15881
ENT 23737, *See* tetradifon, 31366
ENT 23979, *See* thiodan, 31680
ENT 24727, *See* Karathane, 15759
ENT 24945, *See* Terracur® P, 31329
ENT 25540, *See* Tiguvon®, 31788
ENT 25567, *See* Rametin, 25856
ENT 25606, *See* quinomethionate, 25686
ENT 25670, *See* isoprocarb, 15404
ENT 25671, *See* Baygon®, 3790, Propoxur, 25198
ENT 25726, *See* methiocarb, 19484
ENT 25991, *See* mephosfolan, 19326
ENT 26538, *See* Captan, Captan-Col, Captan-50, Captan-83P, Captan Granular, 5165, Vancide® 89, 33302
ENT 27163, *See* phosalone, 23691, Zolone, 34888
ENT 27164, *See* Rampart, 25860
ENT 27311, *See* Chlorpyrifos, 6163, Talon, 30751
ENT 27395, *See* cyhexatin, 7568
ENT 27396, *See* methamidophos, 19474
ENT 27408, *See* Iodofenphos, 15213
ENT 27520, *See* Chlorpyrifos-methyl, 6164
ENT 27738, *See* Torque, 32077
ENT 27766, *See* Abol, 146, pirimicarb, 23879
ENT 27920, *See* Terbufos, 31226
ENT 27967, *See* Taktic, 30729
ENT 28203, *See* Trifluralin, 32252
ENT 29054, *See* Diflubenzuron, 8450
ENT 29106, *See* Abequito, 38
ENT 51142, *See* simazine, 28464
Entamide, 10593
Entarex, 10594
Enterokanacin, *See* Kantrex, 15718, Klebcil, 16209
Enterosep, *See* Yodoxin, 34658
Entex, *See* Tiguvon®, 31788
Entprol, *See* Mazamide® 1214, 19107, tetra(2-hydroxypropyl) ethylenediamine, 31356
Entramin, 10595, *See* 2-amino-5-nitrothiazole, 2184
Entrox, 10596
Enuclen, *See* Variquat® 50MC, 33397, Zephiran Chloride, 34728
Enusin Colours, 10597
Envex® 1001, 10598
Envilon, *See* Korogel, 16354, Koroplate, 16357
Envirez, 10599
Envirocats, 10600
Envirosafe, 10601
Enviroseal, 10602
Enviroset, 10603
Enzactin, *See* Kodaflex® Triacetin, 16281, triacetin, 32182
enzaknen, *See* Benzoyl peroxide, 3983
Enzeon, 10604, *See* α-Chymotrypsins, 6286
Enzypan, 10605
EO/PO alkylphenol block polymer, *See* Antarox® 461/P, 2505, Antarox® 497/P, 2506
EO/PO block copolymer bis-mono-phosphate ester, *See* Hoe S 3618, 14259
EO/PO copolymer complex, *See* Agrilan® AEC178, 1055
eonite, *See* wurtzillite, 34577
EO-PO block copolymer, *See* Adekanol, 676
Eosin, 10606
eosin 3J, *See* Eosin, 10606
eosin 4J extra, *See* Eosin, 10606
Eosin A, *See* Eosin, 10606
eosin A extra, *See* Eosin, 10606
eosin B, *See* Eosin, 10606
eosin Blush, *See* erythrosin, 10778
eosin C, *See* Eosin, 10606
eosin DH, *See* Eosin, 10606
eosin G, *See* Eosin, 10606
eosin G Extra, *See* Eosin, 10606
eosin GGF, *See* Eosin, 10606
eosin J, *See* eosin, 10606

eosin JJS, *See* Eosin, 10606
eosin KS, *See* Eosin, 10606
eosin yellowish, *See* Eosin, 10606
Eosin YS, 10607
Eosolate, 10608
EP, *See* ethylene glycol propyl ether, 11130
EP Lead, 10609
EP Pastes, 10610
EP-1, 10611
EP-2, 10612
EP-452, *See* phenmedipham, 23597
Epal®, 10614
Epal® 6, 10615
Epal® 6, *See* 1-hexanol, 14011
Epal® 8, 10616
Epal® 10, 10617
Epal® 12, 10618
Epal® 14, 10619
Epal® 16NF, 10626
Epal® 18NF, 10627
Epal® 20+, 10613
Epal® 108, 10620
Epal® 610, 10621
Epal® 1012, 10622
Epal® 1214, 10623
Epal® 1412, 10624
Epal® 1618, 10625
Ephos, 10628
Ephynal, 10629, *See* tocopherol, 31918, Vascuals, 33465
Epibond® 1217-A/B, 10630
Epibond® 1544-A/B, 10631
Epichlorhydrin, 10632
dl-α-epichlorohydrin, *See* Epichlorhydrin, 10632
Epichlorohydrin elastomer, 10633
epi-clear, *See* Benzoyl peroxide, 3983
Epiclon, 10634
EPICS, 10635
Epi-Cure 8515, 10636
Epi-Cure® 87, 10637
Epidermin in Oil, 10638
Epidermin Water-Soluble, 10639
Epiflex, 10640
Epiglaubite, 10641
Epigon, *See* permethrin, 23384, 23385, 23386
Epihydrin, 10642
Epikem, 10643
Epikote, 10644
Epikote DX-209-B-80, DX-210-B-80, 10645
Epikote DX-231-B-91, 10646
Epikure 3400, 10647
Epilink, 10648
Epilok, 10649
Epilon, 10650
Epiphassol, 10651
Epiphen, 10652
Epires, 10653
Epirez 501, 10654
Epirez 502, 10655
Epirez 520C, 10656
Epirolin-S, *See* tocopherol, 31918
Episol, 10657, 10658
Epitar, Epitate, 10659
Epiteliol, *See* Vi-Alpha, 33706
Epitone, 10660
Epivax, 10661
Eplink, 10662
EPlstatic® 100, 10663
EPlstatic® 110, 10664
EPlthal 120, 10665, *See* lead phthalate basic, 16914
EPM rubber, 10666
Epocap®, 10667
Epocap® 16129 A/B, 10668
Epocap® 16358 A/B, 10669
Epocast 1610, 10670
Epocrete, 10671
Epocure, 10672
Epodil, 10673
Epodur, 10674
Epok, 10675
Epolast, 10676
Epolene®, 10677
Epolene® C-10, 10678
Epolene® C-13, 10679
Epolene® C-16, 10680
Epolene® E-14, 10681
Epolene® N-15, 10683
Epolene® N-20, N-21, 10684
Epoleon® N-100, 10685
Epoleon® N-7C, 10682
Epolite 1301, 10686
Epolite 1302, 10687
Epolite 2300, 10688
Epolite 2315, 10689
Epolite 3300, 10690
Epolite 5363, 10691
Epomarine, 10692
Epon®, 10693
Epon® 8280, 10694
Epon® Resin DPL-1911, 10695
Eponc®, 10696
Eponoc B, *See* Vulkacit® 1000, 34124
Epophen, 10697
Eporal, *See* Dapsone, 7751
Eposet, 10698, 10699
Eposolve 299-R, 10700
Epotal® 181 D, 10701
Epo-Tek® E-3081, 10702
Epotuf®, 10703
Epotuf® 38-690, 10704
Epotuf® Hardener 37-612, 10705
Epotuf® Hardener 37-621, 10706
Epoweld®, 10707
Epoweld® 19157, 10708
Epoxidized glycol dioleate, *See* Monoplex® S-75, 20251
Epoxidized linseed oil, *See* ADK CIZER O-180A, 711
Epoxidized octyl tallate, *See* Monoplex® S-73, 20250
Epoxidized oleate, *See* Priplast 1431, 25023
Epoxidized soybean oil, *See* Drapex® 6.8, 8969
Epoxidized vegetable oils, *See* Agrilan® FS101, 1060
Epoxidized X-70 and X-75, 10709
Epoxol 7-4, 10710
Epoxol 8-2B, 10711
Epoxol 9-5, 10712
Epoxol G-5, 10714
Epoxol 80, 130, 10713
Epox-S, 10715
Epoxy, *See* Norcast 3258, 21532
Epoxy acrylate, *See* CN 103, 6469
Epoxy acrylate/Di-isodecyl adipate, *See* CN 104 F50, 6474
Epoxy acrylate/GPTA, *See* CN 104 D80, 6473
Epoxy acrylate/HDDA, *See* CN 104 B80, 6471
Epoxy acrylate/Trimethylolpropane triacrylate, *See* CN 104 C75, 6472
Epoxy acrylate/Tripropylene glycol diacrylate, *See* CN 104 A80, 6470
Epoxy Adhesive, 10716
epoxy modifier, *See* Heloxy® 61, 13770
Epoxy mortar, *See* Addamortar, 644
Epoxy novolak acrylate/TMPTA, *See* CN112 C60, 6476
Epoxy Plus, 10717
Epoxy Putty Pack (EP-3/EHP-12), 10718
epoxy resin, 10719
Epoxy resin, *See* Amicon® C-860-4, 2065, Fiberite 7669, 11733, Fiberite 7701, 11734, Stycast® 1090, 29848, Stycast® 1210, 29849, Stycast® 1266, 29850, Stycast® 1467, 29851, Stycast® 2850-FT, 29852
Epoxy top coat, *See* Addaflor, 641
1,2-Epoxy-3-Allyloxypropane, *See* Ageflex AGE, 965, Allyl glycidyl ether, 1783, Sipomer® AGE, 28531
1,2-epoxy-3-butoxypropane, *See* Ageflex BGE, 967
(Z)-4α,9-epoxycevane-3 β,4,12,14,16 β,17,20-heptol 3-(2-methyl-2-butenoate), *See* veratrine, 33567
2-(3,4-Epoxycyclohexyl) ethyltriacetoxysilane, *See* CE6250, 5528
β-(3,4-Epoxycyclohexyl) ethyl trimethoxy silane, *See* CE6250, 5528, Union Carbide® A-186, 32987
epoxyethane, *See* ethylene oxide, 11131
1,2-epoxyethane, *See* ethylene oxide, 11131
1,8-epoxy-*p*-menthane, *See* eucalyptol,
1,2-epoxypropane, *See* propylene oxide, 25223
2,3-epoxypropyl methacrylate, *See* glycidyl methacrylate,
Epoxyprene 50, 10720
epoxypropane, *See* propylene oxide, 25223
2,3-epoxypropane, *See* propylene oxide, 25223
2,3-epoxypropyl butyl ether, *See* Ageflex BGE, 967
epoxypropyl methacrylate 2,3-epoxypropyl methacrylate, *See* SR-379, 29382
2,3-epoxypropyl neodecanoate, *See* glycidyl decanoate, 13092, Glydexx N-10, 13141
(2,3-Epoxypropyl)trimethyl ammonium chloride, *See* Ogtac 85 V, 22052
2,3-epoxypropyltrimethylammonium chloride, *See* Quab 151, 25561
EPR, *See* EPM rubber, 10666
Eprofil, *See* Tecto, Tecto 60%, 30913, thiabendazole, 31661
Eprolin, 10721
Eprolin-S, *See* Vascuals, 33465
Eprylac, 10722
Epsilan, *See* tocopherol, 31918, Vascuals, 33465
Epsilan-M, 10723
epsom salts, *See* magnesium sulfate, 18368
EPsyn® 40-A, 10724
EPsyn® 55, 10725
EPsyn® 5508, 10726
EPsyn® P-557, 10727
EPTAC 1, *See* AAprotect, 16
Eptam, *See* EPTC, 10729
Eptam 6E, 10728
EPTC, 10729
Epurite, 10730
EQ, *See* Polyflex, 24435
Equal, 10731, *See* Aspartame, 3233
equalized guano, 10732
Equiben, 10733
Equigand, *See* dichlorvos, 8394
Equigard, *See* dichlorvos, 8394, Vapona, 33378
Equigel, *See* dichlorvos, 8394
Equigel, *See* Vapona, 33378
Equiguard, *See* dichlorvos, 8394
Equilase, *See* Catalase L, 5429
Equionic, 10734
equisetic acid, 10735
Equity, *See* Chlorpyrifos, 6163
Equivurm Plus, 10736, *See* Mebendazole, 19198
Equi-Vurm Plus, *See* Telmin, 31066
Equizole, *See* Tecto, Tecto 60%, 30913, thiabendazole, 31661
Equron, *See* hyaluronic acid sodium salt, 14483
Eqvalan, 10737, *See* Ivermectin, 15454
Era 147, 10738
Era 164, 10739
Era CR1, 10740
Era CR15 (CB), 10741
Era metal, 10742
Era® Dyes, 10743
Eraclene, 10744
Eradex, *See* Chlorpyrifos, 6163
Eranol, 10745
Erbanil, *See* Propanil, 25174
erbia, *See* erbium oxide, 10747
Erbium, 10746
Erbium oxide, 10747
ercalciol, *See* vitamin D+72, 33940
ercerhinol, 10748
Ercotina, *See* propantheline bromide, 25176
Ercusol®, 10749
Erdmann's reagent, 10750
ergadenylic acid, *See* adenosine monophosphate, 680
Ergamisol, *See* Tramisol®, 32134
Erganol, 10751
ergocalciferol, *See* vitamin D+72, 33940
Ergol, 10752

Ergoplast ADDO, *See* dioctyl adipate, 8548, Plasthall® DOA, 24002, PX-238, 25440, Wickenol® 158, 34393
Ergoplast FDO, *See* Bis(2-ethylhexy) Phthalate, 4228, dioctyl phthalate, 8549, Reomol DCP, 26130, Witcizer 312, 34454
Ergoplast SNO, *See* Plasthall® DOS, 24005, Reomol DOS, 26131
Ergorone, *See* vitamin D+72, 33940
5α-Ergosta-6,8,22E-trien-3 β-ol, *See* Fungisterol, 12479
Eribate, 10753, *See* sodium erythorbate, 28761
Ericon, 10754
erinitrit, *See* sodium nitrite, 28801
Erinofort, 10755
Erinoid, 10756
Erio, 10757
Eriochrome, 10758
Erioclarite, 10759
Erional, 10760, 10761
Eriopon, 10762
Eriosept, *See* Dequalinium Chloride, 8077
Erkantol®, 10763
Erlickl's solution, 10764
Ermite, 10765
Erostabil, *See* Sembonit, 27935
Ertalon® LFX, 10766
Ertalon® PETP, 10767
Ertalyte®, 10768 , *See* Polyethylene terephthalate, 24426
Ertefon, *See* trichlorfon, 32212
Ertilen, 10769
Ertron, *See* vitamin D+72, 33940
Erucalkonium chloride, *See* Kemamine® BQ-2982B, 15907
erucamide, 10770
Erucamide, *See* Armid® E, 2961, Crodamide E, ER, 7096, Kemamide® E, 15895, Polydis® TR 131, 24412, Unislip 1753, 33044
Erucamide TEA (2:1), *See* Laurel SD-520T, 16841
Erucamidopropyl hydroxysultaine, *See* Crosultaine E-30, 7242
erucic acid, 10771, *See* Prifrac 2990, 24957
erucic acid amide, *See* erucamide, 10770
Erucyl arachidate, *See* Kemester® JO, 15964
Erucyl erucamide, *See* Kemamide® E-221, 15897
Erucyl erucate, *See* Kemester® EE, 15957, Schercemol EE, 27599
Erucyl stearamide, *See* Kemamide® S-221, 15902
Erucylamide, *See* Armid® E, 2961, erucamide, 10770, Petrac® Eramide®, 23464, Unislip 1753, 33044
Ervamine, 10772
Ervamix, 10773
Ervol, 10774
erycorbin, *See* isoascorbic acid, 15338
Erythorbic acid, 10775, *See* isoascorbic acid, 15338
Erythorbic Acid Monosodium Salt, *See* Eribate, 10753
Erythrin, 10776
erythrin methyl eosin, *See* Erythrin, 10776
Erythritol, *See* lichen sugar, 17204
D-erythro-hex-2-enonic acid, *See* Erythorbic acid, 10775
D-erythro-hex-2-enonic acid γ-lactone, *See* isoascorbic acid, 15338
D-erythro-hex-2-enonic acid γ-lactone monosodium salt, *See* sodium erythorbate, 28762
D-erythro-3-ketohexonic acid lactone, *See* isoascorbic acid, 15338
Erythromycin, *See* A.T.S, 3
Erythrosiderite, 10777
erythrosin, 10778
Erythrosin A, 10779
Erythrosin B, *See* erythrosin, 10778
erythrosin D, *See* erythrosin, 10778
Erythrosin G, 10780
Erythrosine, *See* Iodesin, 15205, iodoeosin, 15212
ES-7CF/000, 10781
Esaflon, 10782
Esbenite, 10783, 10784
Esberiven, *See* Melilot, 19279
Escacure® KB1, 10785
Escaid, 10786
Escalo 506, *See* Spectraban, 29225
Escalol 507, *See* Padimate O, 22549
Escalol 507, *See* Solarchem® O, 28898
Escalol 557, *See* Neo Heliopan® AV, 20866
Escalol 557, *See* Parsol® MCX, 22867
Escalol 567, *See* Cyasorb® UV 9, 7495
Escalol 567, *See* Neo Heliopan® BB, 20867
Escalol 567, *See* Rhodialux A, 26675
Escalol® 507, 10787
Escalol® 557, 10788
Escalol® 557, *See* octyl methoxycinnamate, 22039
Escalol® 567, 10789
Escalol® 587, 10790
Escane, 10791
Eschel, 10792
Eschka mixture, 10793
Esco Extract, 10794
Escomer, 10795
Escopol® R-020, 10796
Escor, 10797
Escorene, 10798
Escorez, 10799
Escort, *See* metsulfuron-methyl, 19606
Escorto, 10800
Escoweld, 10801
Esdeform, 10802
Esdesol, 10803
Esdogen, 10804
esdragol, *See* Estragole, 10897
esdragon, *See* Estragole, 10897
E-Series® Electronics Cleaner/Degreaser 2000, 10805
E-Series® Freez-It®, 10806
E-Series® Freez-It® Antistat, 10807
E-Series® Ultrajet®, 10808
Eshalit, 10809
Esi-Cryl 1E10N, 10810
Esi-Cryl 20/20, 10811
Esi-Cryl 40, 10816
Esi-Cryl 246, 10812
Esi-Cryl 325N, 10813
Esi-Cryl 1540A, 10814
Esi-Cryl Respond I, 10815
Esi-Det 21M, 10817
Esi-Det CDA, 10818
Esi-Det EP-20, 10819
Esi-Graph 743, 10822
Esi-Graph 745, 10823
Esi-Graph 1045, 10824
Esi-Terge 10, 10820
Esi-Terge 320, 10821
Esi-Terge 40% Coconut Oil Soap, 10825
Esi-Terge B-15, 10826
Esi-Terge HA-20, 10827
Esi-Terge L-75, 10828
Esi-Terge N-100, 10829
Esi-Terge S-10, 10830
Esi-Terge SXS, 10831
Esi-Terge SXS, *See* sodium xylene sulfonate, 28832
Esi-Terge T-60, 10832
Eskacillin, *See* Penicillin G potassium, 23109, 23110
Eskamel, *See* resorcinol, 26230
Eskimo, 10833
eskimon 12, *See* dichlorodifluoromethane, 8388
Esmaillite, 10834
Esophotrast, 10835, *See* Radiopaque, 25830
Esorb, *See* tocopherol, 31918, Vascuals, 33465
Espadol, *See* chloroxylenol, 6159
Esperal, *See* Disulfiram, 8725
Esperal® 115, 10836
Esperal® 120, 10837
Esperal® 230, 10838
Esperase® 16.0L, 10839
Espercarb® 438M-60, 10840
Espercarb® 840, 10841
Esperfoam® FR, 10842
Esperox®, 10843
Esperox® 10, 10844
Esperox® 12MD, 10845, *See* Lupersol 70, 17844
Esperox® 13M, 10846
Esperox® 28, 10847
Esperox® 31M, 10848
Esperox® 33M, 10849
Esperox® 41-25A, 10850
Esperox® 497M, 10851
Esperox® 5100, 10859
Esperox® 545M, 10852
Esperox® 551M, 10853
Esperox® 551M, *See* Amilperoxy pivalate, 2122
Esperox® 570, 10854
Esperox® 740M, 10855
Esperox® 747M, 10856
Esperox® 750M, 10857
Esperox® 939M, 10858
Esperox® C-496, 10860
Esrakon, 10861
Essar (W), 10862
Essence of Bigarade, 10863
Essence of mirbane, *See* nitrobenzene, 21348
Essence of Myrbane, *See* nitrobenzene, 21348
essence of niobe, *See* methyl benzoate, 19515
Essential Oil of Bitter Almonds, 10864
essential salt of urine, *See* microcosmic salt, 19665
Essex powder, 10865
Esshete 1250, 10866
Esshete CML, 10867
Esshete CRM2, 10868
Esshete CRM5, 10869
Esskol®, 10870
Estabex, 10871
Estabex 2307, 2349, 10872
Estabex 2386, 10873
Estaflex, 10874
Estalan JB, *See* Cetearyl octanoate, 5791
Estaloc 61000 Series, 10875
Estalol®, 10876
Estane®, 10877
Estane® 5701 F1, 10878
Estane® 58092, 10879
Estane® 58300, 10880
Estasol, 10881
Ester, *See* Advapak® ML-1325, 813
Ester 14, *See* Ablunol 200ML, 95
Ester 25, *See* paraoxon, 22783
Ester Copal, 10882
Ester ethoxylate, *See* Glytex® 513, 13156
Ester of cetyl alcohol and stearic acid, *See* cetyl stearate, 5825
Estergel, 10883, *See* Radia® 7190, 25729, Unimate® IPM, 32971, Wickenol® 101, 105, 34378
Esterified, *See* Loobwax 0761, 17598
Esterifilcon A, *See* Revlen, 26288
Esterkem, 10884
Estermone, 10885
Esterol, 10886
Esterolane, 10887
Esterox, 10888
Esterpol, 10889
Estol, 10890
Estol 103, *See* Wickenol® 111, 34379
Estol 1407, 10891
Estol 1468, 10892
Estol 1476, 10893
Estol 1481, 10894
Estolan, 10895
Estonate, *See* DDT, 7820
Estoral, 10896
Estragole, 10897
Estron, 10898
Estrosel, *See* dichlorvos, 8394
Estrosol, *See* dichlorvos, 8394, Vapona, 33378
Esurol, *See* methiocarb, 19484
Etabus, *See* Disulfiram, 8725
Etadurin 31, 10899
Etapuron FT, 10900
Etard's reagent, 10901
Etavit, *See* tocopherol, 31918, Vascuals, 33465
Etazin, *See* Ancazate ET, 2415

ETCMTB, *See* Aaterra WP, 17
Eteleen, 10902
Etephon, *See* Florel Fruit Eliminator, 12012
EternaBrite, 10903
Eternite, 10904
Ethacol, 10905
Ethacure® 100, 10906
Ethacure® 300, 10907
Ethafoam, 10908
Etha-Keratin ISO, 10909
ethal
Ethal, 10910, *See* Adol® 52 NF, 782, Cachalot® C-50 NF, 4878, cetyl alcohol, 5815, 5816, 5817, Rita CA, 26788
Ethal 326, 10911
Ethal 368, 10912
Ethal 926, 10913
Ethal 3328, 10914
Ethal BPA-6, 10915
Ethal CSA-25, 10916
Ethal DA-4, 10917, *See* Deceth-6, 7848
Ethal DDP-7, 10918
Ethal DNP-8, 10919
Ethal EH-2, 10920
Ethal LA-4, 10921
Ethal NP-1.5, 10922
Ethal NP-6, 10923
Ethal OA-10, 10924
Ethal OA-10, *See* oleth-8, 22126
Ethal TDA-3, 10926
Ethal TDA-6, 10927
Ethal TDA-6, TDA-3, TDA-18, *See* Trideceth series, 32232
Ethal TDA-18, 10925
Ethana®, 10928
ethanal, *See* Acetaldehyde, 281
Ethanaminium, N,N,N-trimethyl-2-[(1-oxo-2-propenyl)oxy]-, chloride, *See* Ageflex FA-1Q75MC, 971
ethandial, *See* glyoxal 40%, 13146
ethandionic acid, dihydrate, *See* oxalic acid dihydrate, 22414
Ethane, *See* Lysoform, 18095
Ethane, 2-(*o*-chlorophenyl)-2-(*p*-chlorophenyl)-1,1-dichloro-, *See* Lysoform, 18095
Ethanedial, *See* glyoxal 40%, 13146
1,2-Ethanediamine, *See* ethylenediamine, 11133
ethane, 1,2-dichloro, *See* ethylene dichloride, 11122
ethane dioic acid, *See* oxalic acid, 22413
ethanedioic acid, *See* oxalic acid, 22413
Ethanedioic acid copper salt, *See* cupric oxalate, 7385
Ethanedioic acid diammonium salt monohydrate, *See* Ammonium oxalate, 2247
ethane-1,2-dioic acid, *See* oxalic acid, 22413
ethanedioic acid, diethyl ester, *See* oxalic ether, 22416
ethanedioic acid, dihydrate, *See* oxalic acid dihydrate, 22414
1,2-ethanediol, *See* ethylene glycol, 11123, glycol, 13099
1,2-ethanediol, dimethyl ether, *See* monoglyme, 20227
1,2-ethanediol dimethacrylate, *See* ethylene glycol dimethacrylate, 11125
1,2-ethanedione, *See* glyoxal 40%, 13146
Ethane-1,2-dione, *See* glyoxal 40%, 13146
ethanedionic acid, *See* oxalic acid, 22413
ethanedionic acid dihydrate, *See* oxalic acid dihydrate, 22414
ethanedithioamide, *See* hydrogen rubeanide, 14591
(ethanediyl) derivatives, *See* polysorbate 40, 24567
1,1'-(1,2-ethanediyl)bis[benzene], *See* stilbene, 29764, *cis*-stilbene, 29765, Tolan, 31933, toluylene, 31960
N,N'-1,2-ethanediylbis [N-(carboxymethyl glycine], *See* edetic acid, 9615
N,N'-1,2-Ethanediylbis[N-(carboxymethyl)glycine] disodium salt, *See* edetate disodium, 9613
N,N'-1,2-ethanediylbis[N-(carboxymethyl)glycine] disodium salt dihydrate, *See* Trilon® BD, 32325
ethanediylbis(N-(carboxymethyl)glycine) disodium salt, *See* sodium versenate, 28831
N,N'-1,2-ethanediylbis[N-(carboxmethyl)glycine] tetrasodium salt, *See* tetrasodium EDTA, 31391
N,N'-1,2-Ethanediylbis[N-(carboxymethyl)glycine] trisodium salt, *See* edetate trisodium, 9614
[[1,2-Ethanediylbis[carbamodithioato]](2-)manganese mxture with [1,2-ethanediylbis[carbamodithioato]](2-)]zinc, *See* mancozeb, 18517
[[1,2-Ethanediylbis(carbamodithioato)](2-)]manganese mixt. With [[1,2-ethanediylbis (carbamodithioato)](2-)]zinc, *See* Karamate, 15749
[[1,2-ethanediylbis[carbamodithioato]](2-)]zinc, *See* Zineb, 34807
[[N,N'-1,2-ethanediylbis[N-(carboxymethyl)glycinato]](4-)-N,N',O,O',O$_N$,O$_N$] calciate(2-)disodium, *See* Versene CA, 33618
N,N'-1,2-Ethanediylbis-(N-(carboxymethyl)glycine), *See* Kalex Acids, 15669, Trilon® BS, 32327
N,N'-1,2-Ethanediylbis[N-(carboxymethyl)glycine] tetrasodium salt, *See* Kalex 220 Crystal, 15668
[[N,N'-ethanediylbis[N-(carboxymethyl)glycinato]](4-)]-N,N',O,O',O:ssN:ks,O:ssN':ks-ferrate(1-) sodium, *See* sodium iron edetate, 28776
2,2'-(1,2-ethanediyl)bis[4,5,6,7-tetrabromo-1H-isoindole-1,3(2H)-dione, *See* Saytex® BT-93®, 27520
ethane-1-hydroxy-1,1-diphosphonic acid, *See* Unihib® 106, 32957
Ethane nitrile, *See* acetonitrile, 299
ethanenitrile, *See* acetonitrile, 299
ethane pentachloride, *See* pentachlorethane, 23134
ethaneperoxoic acid, *See* Oxymaster, 22460
Ethaneperoxoic acid, 1,1-dimethylethyl ester, *See* Aztec® *t*-butyl peracetate-50 OMS, 60 OMS, 75 OMS, 3542
Ethanesulfonic acid, 2-hydroxy-, monosodium salt, *See* Witconate NIS, 34487
Ethanesulfonic acid, 2-hydroxy-, sodium salt, *See* Witconate NIS, 34487
ethane, 1,1,1-trichloro-, *See* 1,1,1-trichloroethane, 32215, trichloroethane, 32216
Ethanimidic acid, N-(trimethylsilyl)-, trimethylsilyl ester, *See* Bistrimethylsilyl acetamide, 4286, CB2500, 5479, Dynasylan® BSA, 9399
Ethanite, 10929
ethanoic acid, *See* Acetic acid, 292
ethanoic anhydrate, *See* acetic anhydride, 293
ethanoic anhydride, *See* acetic anhydride, 293
ethanol, *See* ethyl alcohol, 11064, Lankropol® KMA, 16713, Lankropol® KO2, 16715, Punctilious® Ethyl Alcohol, 25398
Ethanol, *See*
ethanol 200 proof, *See* ethyl alcohol, 11064
ethanol, 2-butoxyphosphate (3:1), *See* tributoxyethyl phosphate, 32206
Ethanol, 2-(2-ethoxyethoxy), *See* Ethoxydiglycol, 11044
Ethanol, 2,2'-iminobis-, N-tallow alkyl derivs, *See* Witcamine® 6606, 34447
Ethanol, 2,2'-oxybis-, dibenzoate, *See* Benzoflex 2-45, 3970
ethanol, 2,2'-oxybis-, monoethyl ether, *See* ethyl di-lcinol, 11068
Ethanol-2-phenoxy, *See* Igepal® Cephene Distilled, 14850, Rewopal® MPG 10, 26344
Ethanol, 2-propoxy, *See* ethylene glycol propyl ether, 11130
ethanol, silent spirit, *See* spirits of wine, 29304
Ethanol, 2-(2,4,5-trichlorophenoxy)-, hydrogen sulfate, *See* Garlon, 12672
ethanolamine, 10930
Ethanolamine alkylbenzene sulfonate, *See* Textol 80 (L), 31540
Ethanolamine salt of niclosamide, *See* Bayluscide®, 3805
Ethanolamine sesquisulfite, *See* Measac, 19195
Ethanolamine sulfite, *See* Thioset® M, 31714
Ethanolamine-borate ester, *See* Actracor M, 523
1-ethanol-2-thiol, *See* 2-mercaptoethanol, 19336
ethanone, 1-(2,4-dihydroxyphenyl)-, *See* resacetophenone, 26152
Ethanone, 1-[4-(1,1-dimethylethyl)-2,6-dimethyl-3,5-dinitrophenyl, *See* acetyl-dinitro-butyl-xylene, 317
Ethanone, 1-(2-pyridinyl)-, *See* 2-Acetyl pyridine, 321
ethanonitrile, *See* acetonitrile, 299
Ethanox, *See* Ethion, 10955
Ethanox® 323, 10931
Ethanox® 330, 10932
Ethanox® 398, 10933
Ethanox® 702, 10934
Ethanox® 703, 10935
ethanoyl chloride, *See* acetyl chloride, 311
Ethasan, *See* Ancazate ET, 2415
Etha-Soy ISO, 10936
Ethavan, 10937, *See* ethyl vanillin, 11087, Rhodiarome, 26678, Vanbeenol, 33294
Ethazate, 10938, *See* Ancazate ET, 2415, Octocure ZDE-50, 21996
Ethazate 50D, 10939
Ethazol, *See* Aaterra WP, 17
Ethazole, *See* Aaterra WP, 17
Ethazole (fungicide), *See* Aaterra WP, 17
Ethene, *See* olefiant gas,
ethene homopolymer, *See* A-C® Polyethylene 6, 6A, 7, 7A, 8, 8A, 9, 9A, 617, 617A, 175, polyethylene, 24424
Ethene oxide, *See* ethylene oxide, 11131
Ethene, chloro-, homopolymer, *See* Vestolit®, 33679
Ethene, tetrafluoro-, homopolymer, *See* polytetrafluoroethylene, 24640
Ethenediyl diphenylene bisbenzoxazole, 10940
2,2'-(1,2-ethenediyl)bis[5-[[(phenylamino)carbonyl] amino]benzenesulfonic acid] disodium salt, *See* Photine, 23756
2,2'-(1,2-Ethenediyldi-4,1-phenylene)bisbenzoxazole, *See* Eastobrite® OB-1, 9475, Ethenediyl diphenylene bisbenzoxazole, 10940
ethenol, homopolymer, *See* polyvinyl alcohol, 24666
ethenyl acetate, *See* vinyl acetate, 33854
ethenyl acetate, homopolymer, *See* polyvinyl acetate homopolymer, 24665
ethenylbenzene, *See* styrol, 29875
ethenylbenzene, homopolymer, *See* polystyrene, 24601
1-(ethenyloxy)butane, *See* Shostakovsky Balsam, 28187
1-ethenyl-2-pyrrolidinone, homopolymer, *See* PVP, 25429
1-ethenyl-2-pyrrolidinone, polymer with acetic acid ethenyl ester, *See* Agrimer VA 6, 1067
1-ethenyl-2-pyrrolidinone polymer with 1 eicosene, *See* PVP/eicosene copolymer, 25430
1-ethenyl-2-pyrrolidinone polymers, *See* Agrimer 15L, 1065, Tears Plus®, 30876, Vinisil, 33827
1-ethenyl-2-pyrrolidinone polymers complexed with iodine, *See* Videne Disinfectant Solution, Videne Disinfectant Tincture, Videne Powder, 33739
e1-ethenyl-2-pyrrolidinone, homopolymer, *See* PVP, 25429
Ethenyltriacetate-silanetriol, *See* CV-4800, 7449, Dow Corning® Z-6075, 8882
Ethenyltriethyloxy-silane, *See* CV-4910, 7451
Ethephon, *See* Cerone, 5756, Ethrel C, 11051, Ethrel-E, 11052, Ethrel-R, 11053
ether, *See* ethyl ether, 11069
ether alcohol, *See* Butyl Cellosolve®, 4767
ether carboxylate, *See* Hostawet TDC, 14420
ether chloratus, *See* ethyl chloride, 11067
ether hydrochloric, *See* ethyl chloride, 11067
ether muriatic, *See* ethyl chloride, 11067
ether thioether, *See* Vulkanol® OT, 34145
ether, methylated, 10941
Etheramine 13, 10942
Etherdiamine 13, 10943
Ethereal Oil of Bitter Almonds, 10944
Etherin, *See* olefiant gas,
Ethfac 104, 10946
Ethfac 142W, 10947
Ethfac 391, 10948
Ethfac 1018, 10945

Ethfac NP-110, 10949
Ethfac PB-1, 10950
Ethfac PD-6, 10951
Ethfac PP-16, 10952
Ethidium, 10953
ethine, *See* acetylene, 318
ethinyl trichloride, *See* trichloroethylene, 32217
Ethiodan, 10954, *See* Iophendylate, 15234
Ethiofencarb, *See* Arylmate®, 3166, Croneton®, 7206
Ethiol, *See* Ethion, 10955
Ethiolacar, *See* malathion, 18460
Ethion, 10955
ethirimol-flutriafol-thiabendazole, *See* Ferrax, 11647
ethocel, *See* ethyl cellulose, 11066
Ethocel Standard Premium, 10956, *See* ethyl cellulose, 11066
Ethoduomeen, 10957
Ethoduomeen® T/13, 10958
Ethoduomeen® T/20, 10959
Ethoduomeen® T/25, 10960
Ethoduoquad® T/15-50, 10961
Ethofat 0/15, *See* Ablunol 200MO, 96
Ethofat 60/15, *See* Ablunol DEGMS, 99
Ethofat 60/20, *See* Ablunol DEGMS, 99
Ethofat 60/25, *See* Ablunol DEGMS, 99
Ethofat®, 10962
Ethofat® 18/14, 10963
Ethofat® 242/25, 10964
Ethofat® 242/25, *See* PEG tallate series, 23035
Ethofat® O/20, 10965
Ethofoil, 10966
ethofumesate, 10967
ethofumesate-phenmedipham, *See* betanal Tandem, 4062
Ethohexadiol, 10968, *See* ethyl hexanediol, 11073
Ethokem, 10969
Ethokem C/12, 10970
Ethokem C/12, *See* Cocamide DEA, 6547
Ethol, *See* Adol® 52 NF, 782, Cachalot® C-50 NF, 4878, cetyl alcohol, 5815, 5816, 5817, Rita CA, 26788
Ethomeen®, 10971
Ethomeen® 18/12, 10972
Ethomeen® 18/12, *See* PEG stearamine series, 23026
Ethomeen® 18/15, 10973
Ethomeen® 18/20, 10974
Ethomeen® 18/25, 10975
Ethomeen® 18/60, 10976
Ethomeen® C/12, 10977, *See* Cocamide DEA, 6547
Ethomeen® C/15, 10978
Ethomeen® C/20, 10979
Ethomeen® C/25, 10980
Ethomeen® S/12, 10981
Ethomeen® S/12, S/15,S/25, *See* PEG soyamine series, 23033
Ethomeen® S/15, 10982
Ethomeen® S/20, 10983, *See* PEG soyamine series, 23033
Ethomeen® S/25, 10984
Ethomeen® T/12, 10985
Ethomeen® T/15, 10986
Ethomeen® T/25, 10987
Ethomid®, 10988
Ethomid® HT/23, *See* PEG hydrogenated tallow series, 23029
Ethomid® HT/23, HT/60, *See* PEG hydrogenated tallow series, 23029
Ethomid® HT/60, 10989
Ethomid® HT/60, 10990
Ethomid® O/17, 10991
Ethomulsion, 10992
ethopabate, 10993
Ethopropazine hydrochloride, *See* Lysivane, 18091
Ethoquad®, 10994
Ethoquad® 18/12, 10995
Ethoquad® 18/25, 10996
Ethoquad® C/12, 10997
Ethoquad® C/12 Nitrate, 10998, *See* PEG-2 cocomonium nitrate, 23039
Ethoquad® C/25, 10999
Ethoquad® CB/12, 11000
Ethoquad® CB/12, *See* PEG-2 cocobenzonium chloride, 23038
Ethoquad® O/12, 11001
Ethoquad® T/12, 11002, *See* PEG-2 tallowmonium chloride, 23044
Ethoquad® T/13-50, 11003
Ethosperse®, 11004
Ethosperse® CA-2, 11005
Ethosperse® G-26, 11006, *See* Glycereth series, 13069
Ethotal, 11007
Ethovan, *See* ethyl vanillin, 11087, Rhodiarome, 26678, Vanbeenol, 33294
Ethox, *See* ethylene oxide, 11131
Ethox 1122, 11009, *See* PEG-9 pelargonate, 23054
Ethox 1212, 11010
Ethox 1358, 11011
Ethox 1372, 11012
Ethox 2156, 11013
Ethox 2423, 11014
Ethox 25-R-8, 11008
Ethox 2610, 11015
Ethox 2659, 11016
Ethox 2684, 11017
Ethox 3113, 11018
Ethox CAM-2, 11019
Ethox CO-5, 11020
Ethox COA, 11021
Ethox DL-5, 11022, *See* PEG-4 dilaurate series, 23050
Ethox DO-2, *See* PEG-2 dioleate series, 23040
Ethox DO-9, 11023
Ethox DT-15, 11024
Ethox DTO-9A, 11025, *See* PEG ditallate series, 23028
Ethox HCO-16, 11026
Ethox HO-50, 11027, *See* PEG sorbitan hexaoleate series, 23032
Ethox L-61, 11028
Ethox LF-1226, 11029
Ethox MA-8, 11030
Ethox MI-9, 11031, *See* PEG isostearate series, 23030
Ethox ML-5, 11032
Ethox MO-9, 11033
Ethox MS-8, 11034, *See* PEG stearate series, 23034
Ethox OAM-308, 11035
Ethox PPG 1025 DTO, 11036
Ethox SAM-10, 11037
Ethox SO-9, 11038
Ethox TAM-2, 11039
Ethox TO-8, 11040, *See* PEG tallate series, 23035
Ethoxol, 11041
Ethoxol 20, *See* Volpo 3, 34020
ethoxy benzene, *See* phenetole, 23590
Ethoxy carbonyl phenylethyl phenylformamidine, 11042, 11043
ethoxy diglycol, *See* Diethoxol, 8435, ethyl di-Icinol, 11068
2-Ethoxy etheracetate, *See* EE Acetate, 9635
Ethoxy ethyl acetate, *See* Sensolve EEA, 27954
O-(6-ethoxy-2-ethyl-4-pyrimidinyl) O,O-dimethyl phosphorothioate, *See* etrimfos, 11155
Ethoxy propyl acetate, *See* Sensolve EPA, 27955
4-ethoxyaniline, *See* Phenetidine, 23589
4-ethoxybenzenamine, *See* Phenetidine, 23589
N_2-(4-Ethoxy-carbonylphenyl)-N'-methyl-N'-phenylformamidine, *See* Ethoxy carbonyl phenylmethyl phenylformamidine, 11043, Givsorb® UV-1, 12952
Ethoxydiglycol, 11044, *See* ethyl di-Icinol, 11068, Simulsol 5719, 28486
(±)-2-ethoxy-2,3-dihydro-3,3-dimethyl-5-benzofuranylmethanesulfonate, *See* ethofumesate, 10967
6-ethoxy-1,2-dihydro-2,2,4-trimethylquinoline, *See* ethoxyquin, 11049
1-Ethoxy-1-(3,7-dimethyl-1,6-octadienyloxy)-ethane, *See* Elintaal, 9839
ethoxydimethylsilane, *See* CD5635, 5511
2-ethoxyethanol acetate, *See* ethoxyethanol acetate, 11046
Ethoxyethane, *See* ethyl ether, 11069
Ethoxyethanol, 11045, *See* ethyl icinol, 11074
2-ethoxyethanol, *See* 110 EE Solvent, 9636, ethyl icinol, 11074, Oxitol, 22432, 45 Poly-Solv® EE, 24556
2-ethoxyethanol, ester with acetic acid, *See* EE Acetate, 9635
Ethoxyethanol acetate, 11046 *See* EE Acetate, 9635
2-ethoxyethanol acetate, *See* EE Acetate, 9635, Ethoxyethanol acetate, 11046
(Ethoxyethoxy)ethanol, *See* Diethoxol, 8435
2-(β-ethoxyethoxy)ethanol, *See* Diethoxol, 8435
2-(2-Ethoxyethoxy) ethanol, *See* Carbitol®, 5194, Diethoxol, 8435, Ektasolve® DE, DE-HG, 9664, Ethoxydiglycol, 11044, ethyl di-Icinol, 11068, Transcutol, 32137
(Ethoxyethoxy)ethanol, *See* ethyl di-Icinol, 11068
2-(β-ethoxyethoxy)ethanol, *See* ethyl di-Icinol, 11068
2-(2-Ethoxyethoxy)ethanol, *See* Diethoxol, 8435
2-(ethoxyethoxy)ethanol, *See* ethyl di-Icinol, 11068
2-(2-Ethoxyethoxy)-ethyl acrylate, *See* SR-256, 29372
Ethoxyethyl acetate, *See* EE Acetate, 9635
2-ethoxyethyl acetate, *See* EE Acetate, 9635, ethoxyethanol acetate, 11046
β-ethoxyethyl acetate, *See* EE Acetate, 9635
1-Ethoxy-1-hexoxyethane, *See* Lilvert, 17257
3-ethoxy-4-hydroxybenzaldehyde, *See* Ethavan, 10937, ethyl vanillin, 11087, Rhodiarome, 26678, Vanbeenol, 33294
(±)-(EZ)-2-(1-ethoxyiminobutyl)-5-[2-(ethylthio)propyl]-3-hydroxycyclohex-2-enone, *See* Sethoxydim, 28099
2-(1-(ethoxyimino)butyl)-3-hydroxy-5-(tetrahydro-2H-thiopyran-3-yl)-2-cyclohexene-1-one, *See* Cycloxydim, 7544
Ethoxylan® 1685, 11047, *See* PEG lanolin series, 23025
Ethoxylate, *See* Arodet HCS, 3009
Ethoxylate, *See* Capcure Emulsifier 37S, 5106
Ethoxylated (40 mol) lanolin, *See* Lanogel® 31, 16730
Ethoxylated (75 mol) lanolin, *See* Lanogel® 41, 16731
Ethoxylated (85 mol) lanolin, *See* Lanogel® 61, 16732
Ethoxylated 1,2-ethanediol, *See* Alkapol PEG 300, 1706
Ethoxylated alcohol, *See* Amwet DAD, 2377, Arodet BN-100, 3007, Scherpol LSB, 27718
Ethoxylated alkanolamides, *See* Amidox®, 2109
Ethoxylated alkyl guanidine-amine complex, *See* Aerosol® C-61, 890
Ethoxylated amide, *See* Schercoterge 140, 27704
Ethoxylated amine, *See* Syn Fac® TEA-97, 30516
ethoxylated and propoxylated octyl/decyl alcohols, *See* Antarox® BL-214, 2508
Ethoxylated C_{12-18} phosphate (8 EO) sodium salt, *See* Crafol AP-64, 6960
Ethoxylated C_{12}-C_{14} fatty alcohol (1 EO), *See* Dehydol PLS 1, 7956
Ethoxylated C_{16} phosphate ester, *See* Crafol AP-16, 6954
ethoxylated castor oil, *See* Ricinion, 26732
Ethoxylated cocamide, *See* Empilan® LP10, 10411
Ethoxylated cocamine (2-25 EO), *See* Genamin C Grades, 12769
Ethoxylated coco amine, *See* Chemstat® 122, 6033
Ethoxylated decyl alcohol sulfosuccinate monoester, *See* Ablusol DA, 122
Ethoxylated derivative, *See* Marvanol® GAW, 18982
Ethoxylated dodecylmercaptan, *See* Burco TME, 4693
Ethoxylated fatty amine, *See* Hartonyl L531, 13684
Ethoxylated fatty amine oxides, *See* Noxamine, 21747
Ethoxylated lanolin, *See* Lantrol® PLN, 16770
Ethoxylated lauryl alcohol, *See* AE-1, 827
Ethoxylated naphthol sulfonic acid, *See* ENSA-6, 10586
Ethoxylated nonyl phenol, *See* AD-749, 613, Triton® N-57, 32430
Ethoxylated nonylphenol phosphate, *See* Stepfac® 8170, 29713
Ethoxylated nonylphenols, *See* Desonic® 20N, 8184
Ethoxylated oleic acid, *See* Eumulgin PLT 4, 11184
Ethoxylated oleo-cetyl alcohol (30 EO), *See* Mergital

OC 30E, 19361
Ethoxylated oleyl amine, *See* Chemstat® 172, 6034
Ethoxylated oleyl/cetyl alcohol, *See* Eumulgin EP .5.2L, 11174
Ethoxylated phosphate, *See* Merpol® A, 19375
Ethoxylated polystyrylphenol, *See* Soprophor® 37, 29084
Ethoxylated stearyl amine, *See* Chemstat® 192/NCP, 6035
Ethoxylated sulfonate, *See* Burco Anionic APS, 4687
Ethoxylated sulfosuccinate, *See* Schercopon 2WD, 27670
ethoxylated tallow amine, *See* Team, 30875
Ethoxylated trimethylol propane, *See* Sipomer® TMPEO, 28542
Ethoxylated trimethylolpropane triacrylate, *See* Ageflex EOTMPTA, 963
Ethoxylated tristyrylphenol phosphate, *See* Soprophor® 3D33, 29085
Ethoxylated, propoxylated lauryl alcohol, *See* Dehypon LS-24, 7978
Ethoxylated/propoxylated fatty alcohols, *See* Antispumin ZU, 2617
ethoxylated-2,4,7,9-tetramethyl-5-decyne-4,7-diol, *See* Surfynol® 420, 30384, 30385, 30386, 30387
2-ethoxynaphthalene, *See* nerolin, 21007
Ethoxyol® 1707, 11048
4-Ethoxy-7-phenyl-3,5-dioxa-6-aza-4-phosphaoct-6-ene-8-nitrile 4-sulfide, *See* Sebacil®, 27796, Volathion®, 34011
1-ethoxy-1-(2-phenylethoxy) ethane, *See* Vertocinth, 33638
1-Ethoxy-1-phenylethoxyethane, *See* Efetaal, 9638, Fiorivert, 11820, Verdilyn, 33572
4-ethoxyphenylurea, *See* Dulcin, 9091
(4-Ethoxyphenyl)urea, *See* Dulcin, 9091
Ethoxyquin, 11049, *See* Santoquin, 27420
***p*-Ethoxy-quinaldine-*p*-methoxy-quinolineethyl-cyanin-bromide**, *See* pinachrom, 23847
5-Ethoxy-3-trichloromethyl-1,2,4-thiadiazole, *See* Aaterra WP, 17
6-Ethoxy-2,2,4-trimethyl-1,2-dihydroquinoline, *See* Polyflex, 24435, Raluquin®, 25852
ethoxytrimethylsilane, *See* CT2970, 7326
Ethrel, 11050, *See* Cerone, 5756
Ethrel C, 11051
Ethrel-E, 11052
Ethrel-R, 11053
Ethsorbox L-20, 11054
Ethsorbox O-20, 11055
Ethsorbox S-20, 11056, *See* Ethylan® GES6, 11108
Ethsorbox TO-20, 11057
Ethsorbox TS-20, 11058, *See* Polysorbate 65, 24571
Ethulon, 11059
Ethulose, 11060
Ethyl (diphenylmethylene)cyanoacetate, *See* Uvinul® N-35, 33200
ethyl β-naphthyl ether, *See* nerolin, 21007
ethyl 10-(*p*-iodophenyl)undecylate, *See* Pantopaque, 22695
Ethyl 2-cyano-3,3-diphenyl-2-propenoate, *See* Uvinul® N-35, 33200
ethyl 2-methyl-2-propenoate, *See* ethyl methacrylate, 11077
ethyl 2-naphthyl ether, *See* nerolin, 21007
ethyl 3-ethoxypropionate, 11062, *See* Ektapro® EEP Solvent, 9659
ethyl *p*-hydroxybenzoate, 11061
Ethyl 4-iodophenylundec-10-enoate, *See* Ethiodan, 10954
ethyl acetate, 11063
Ethyl acetic ester, *See* ethyl acetate, 11063
ethyl acetone, *See* methyl propyl ketone, 19557
Ethyl acetyl glycolate, *See* EE Acetate, 9635
ethyl alcohol, 11064, *See* ethylol, 11142, grain alcohol, 13244, spirits of wine, 29304, USI in Oval, 33157
ethyl aldehyde, *See* Acetaldehyde, 281
ethyl-*p*-aminobenzoate hydrochloride, *See* benzocaine,
Ethyl-2-*t*-butylcyclohexylcarbonate, *See* Floramat, 12005
7-Ethyl bicyclooxazolidine, *See* Oxaban®-E, 22405
ethyl cadmate, 11065, *See* Cadmate®, 4892, cadmium diethyldithiocarbamate, 4896, ethyl cadmate, 11065
ethyl carbamate, *See* urethane, 33135
ethyl carbanilate, *See* carbanilic ether, 5179, Phenylurethane, 23652
ethyl carbinol, *See* propanol, 25175 propylan-propyl alcohol, 25211
ethyl Carbitol, *See* Diethoxol, 8435, ethyl di-Icinol, 11068
Ethyl Cellosolve, *See* ethyl icinol, 11074
Ethyl Cellosolve Acetate, *See* EE Acetate, 9635
ethyl cellulose, 11066
ethyl cetab, *See* Sumquat® 6020, 30121
ethyl chloride, 11067
ethyl citrate, *See* Hydagen® C.A.T, 14526 triethyl citrate, 32242 Uniplex 80, 33004
Ethyl α-cyano-β,β-diphenylacrylate, *See* Uvinul® N-35, 33200
Ethyl 2-cyano-3,3-diphenylacrylate, *See* Uvinul® N-35, 33200
ethyl cymate, *See* Octocure ZDE-50, 21996
***p,p'*-ethyl DDD**, *See* diethyldiphenyldichloroethane, 8441
ethyl dehydrocyclogeranate, *See* ethyl safranate, 11080
ethyl diethylene glycol, *See* Diethoxol, 8435, ethyl di-Icinol, 11068
ethyl digol, *See* Carbitol®, 5194, Diethoxol, 8435, Ektasolve® DE, DE-HG, 9664, ethyl di-Icinol, 11068
ethyl di-Icinol, 11068, *See* Ethoxydiglycol, 11044
ethyl dimethyl methane, *See* isopentane, 15394
ethyl dimethylamine oxides, *See* Admox®, 757
3-Ethyl-2,4-dioxaspiro (5.5) undec-8-ene, *See* Spiroflor, 29306
S-ethyl dipropylcarbamothioate, *See* EPTC, 10729
ethyl ether, 11069, *See* ethyl ether, 11069, ethyl ether Anhydrous A.C.S, 11070, ethyl ether USP/ACS, 11071, USI in Oval, 33158
ethyl ether Anhydrous A.C.S, 11070
ethyl ether USP/ACS, 11071
ethyl-*o*-(6-(ethylamino)-3-(ethylimino)-2,7-dimethyl-3H-xanthen-9-yl)benzoate monohydrochloride, *See* Rhodamine G and G Extra, 26617
Ethyl Glycol, *See* ethyl icinol, 11074
Ethyl Green, 11072, *See* methyl green, 19540
Ethyl Guthion, *See* Azinphos-ethyl, 3520, Gusathion® A, 13502
2-ethyl hexanal 1,2-ethanediol cyclic acetal, *See* Abbavert, 32
ethyl hexanediol, 11073, *See* Ethohexadiol, 10968, ethyl hexanediol, 11073, Lanoquat® 1756, 16744
Ethyl-1,3-hexanediol, *See* ethyl hexanediol, 11073
2-ethyl-1,3-hexanediol, *See* ethyl hexanediol, 11073
2-ethyl-1,3-hexanediol, *See* Ethohexadiol, 10968
2-Ethyl-1,3-hexanediol, *See* ethyl hexanediol, 11073
2-ethyl hexanediol, *See* ethyl hexanediol, 11073
2-Ethyl-1,2-Hexanediol, *See* ethyl hexanediol, 11073
Ethyl-1,3-Hexane Diol-2, *See* ethyl hexanediol, 11073
2-ethyl-1-hexanol, *See* isooctyl alcohol, 15385
2-Ethyl-1-hexanol, *See* Aerofroth 88, 859, 2-ethylhexanol, 11136, isooctyl alcohol, 15385
2-Ethyl-1-hexanol ester with diphenyl phosphate, *See* diphenyl octyl phosphate, 8597
2-ethyl-1-hexanol hydrogen sulfate sodium salt, *See* Rhodapon® BOS, 26632
2-ethyl hexanol phosphate, *See* Base 104, 3692
2-ethyl-1-hexanol sulfate sodium salt, *See* Rhodapon® BOS, 26632
ethyl Hexyl Acetate, *See* 2-ethylhexyl acetate, 11138
ethyl(2)-Hexyl Acetate, *See* 2-ethylhexyl acetate, 11138
Ethyl-2-hexylacetoacetate, *See* Jessate, 15579
ethyl hexyl acrylate, *See* octyl acrylate, 22037
ethyl hexyl phthalate, *See* Bis(2-ethylhexy) Phthalate, 4228, Reomol DCP, 26130
ethyl hexylene glycol, *See* Ethohexadiol, 10968, ethyl hexanediol, 11073
1-ethyl-2-hydroxybenzene, *See* phlorol, 23675
ethyl-4-hydroxybenzoate, *See* ethyl *p*-hydroxybenzoate, 11061
ethyl hydroxyethylcellulose, *See* Aqualon® EHEC, 2707
ethyl hydroxymethyl oleyl oxazoline, *See* Alkaterge® E, 1717
2-ethyl-2-hydroxymethyl-1,3-propanediol, *See* trimethylolpropane, 32338
2-Ethyl-2-(hydroxymethyl)-1,2,3-propanediol, *See* Ageflex TMPTA, 1001
2-ethyl-2-hydroxymethyl-1,3-propanediol, *See* trimethylolpropane, 32338
ethyl icinol, 11074
ethyl 10-(*p*-iodophenyl)hendecanoate, *See* Iophendylate, 15234
ethyl 4-iodophenylundec-10-enoate, *See* Iophendylate, 15234
ethyl iodophenylundecylate, *See* Myodil, 20569
ethyl 10-(*p*-iodophenyl)undecylate, *See* Iophendylate, 15234
ethyl isothiocyanate, 11075, *See* mustard oil, 20539
ethyl Ketone, *See* diethyl ketone, 8436
ethyl lactate, 11076
ethyl L-lactate, *See* ethyl lactate, 11076
ethyl L-(-)-lactate, *See* ethyl lactate, 11076
ethyl maltol, *See* Veltol Plus, 33545
ethyl methacrylate, 11077
ethyl 2-methacrylate, *See* ethyl methacrylate, 11077
ethyl α-methylacrylate, *See* ethyl methacrylate, 11077
ethyl-α-methyl acrylate, *See* ethyl methacrylate, 11077
N-ethyl-N'-(1-methylethyl)-6-(methylthio)-1,3,5-triazine-2,4-diamine, *See* ametryn, 2051
ethyl methyl ketone, *See* methyl ethyl ketone, 19530
ethyl methyl ketone peroxide, *See* 2-butanone peroxide, 4738, methyl ethyl ketone peroxide, 19531
ethyl methyl ketoxime, 19532
Ethyl-2-methyl pentanoate, *See* Manzanate, 18623
ethyl-2-methyl-2-propenoate, *See* ethyl methacrylate, 11077
N-[3-(1-ethyl-1-methylpropyl)-5-isoxazolyl]-2,6-dimethoxybenzamide, *See* Ratio, 25897
ethyl methyl sulfide, *See* methyl ethyl sulfide, 19533
ethyl 2-methyl-2-propenoate, *See* ethyl methacrylate, 11077
N-ethyl morpholine, 11140
Ethyl morrhuate, *See* Liponate EM, 17376
ethyl mustard oil, *See* ethyl isothiocyanate, 11075
ethyl namate®, 11078
ethyl 1-naphthalene acetate, *See* Tre-Hold, 32165
ethyl 2-(1-naphthyl) acetate, *See* Tre-Hold, 32165
ethyl-1-naphthyl acetate, *See* Tre-Hold, 32165
ethyl nitrate, *See* nitric ether, 21333
ethyl nitrile, *See* acetonitrile, 299
Ethyl-(2-nitrotrimethylene)dimorpholine, *See* 4,4'-(2-Ethyl-2-nitrotrimethylene)-dimorpholine, 11141
ethyl octadecanoate, *See* Radia® 7185, 25727
ethyl oleate, *See* Radia® 7187, 25728
ethyl orthosilicate, *See* ethyl silicate, 11083
ethyl oxalate, *See* oxalic ether, 22416
ethyl oxide, *See* ethyl ether, 11069
ethyl pabate, *See* ethopabate, 10993
Ethyl Parasept, *See* ethyl *p*-hydroxybenzoate, 11061
ethyl parathion, 11079
N-ethyl perfluorooctanesulfonamide, *See* Finitron, 11815
ethyl phenylcarbamate, *See* Phenylurethane, 23652
ethyl N-phenylcarbamate, *See* carbanilic ether, 5179
ethyl N-phenylcarbanilate, *See* Phenylurethane, 23652
ethyl phosphate, *See* triethyl phosphate, 32243
ethyl phthalate, *See* diethyl phthalate, 8437, Kodaflex® DEP, 16266
ethyl polysilicate, *See* Silbond® 40, 28284
Ethyl-3-propyl-1,3-propanediol, *See* ethyl hexanediol, 11073
ethyl protal, *See* Ethavan, 10937, Rhodiarome, 26678
ethyl protocatechualdehyde, *See* Rhodiarome, 26678
ethyl proto-catechualdehyde-3-ethyl ether, *See* Ethavan, 10937
ethyl safranate, 11080

ethyl salicylate, 11081, *See* Sal-ethyl, 27264
ethyl selenac, 11082, *See* Selazate, 27853
ethyl silicate, 11083, *See* Silester, 28303
ethyl silicon trichloride, *See* CE6350, 5530
ethyl stearate, *See* Radia® 7185, 25727
ethyl tellurac®, 11084, *See* Akrochem® TDEC, 1179
Ethyl Tellurac®, *See* Akrochem® TDEC, 1179
Ethyl Tellurac®Rodform, *See* Akrochem® TDEC, 1179
S-[2-(ethylthio)ethyl] O,O-dimethyl phosphorodithioate, *See* demeton-S-methyl, 8055, thiometon, 31700
ethyl thiophamate, *See* thiophanate methyl, 31706
ethyl toluenesulfonamide, *See* ethyltoluene sulfonamide, 11145
N-Ethyl *o*-toluene sulfonamide, *See* Uniplex 108, 33008
N-Ethyl o/p toluenesulfonamide, *See* Rit-Cizer #8, 26839, Sibercizer C6, 28194
ethyl-*p*-toluenesulfonate, 11089, *See* Mittel AEP, 19978
N-ethyl-*p*-toluenesulfonamide, *See* ethyltoluene sulfonamide, 11145, Uniplex 108, 33007
ethyl tuads®, 11085 , *See* Disulfiram, 8725
Ethyl Tuads® Rodform, *See* Akrochem® TETD, 1180
ethyl tuex, 11086
ethyl urethan, *See* urethane, 33135
ethyl urethane, *See* urethane, 33135
ethyl vanillin, 11087, *See* Ethavan, 10937 Rhodiarome, 26678 Vanbeenol, 33294
Ethyl zimate, *See* Ancazate ET, 2415, Octocure ZDE-50, 21996, Perkacit® ZDEC, 23296
ethyl zimate®, 11088
Ethyl Ziram, *See* Octocure ZDE-50, 21996, Perkacit® ZDEC, 23296
D-N-Ethylacetamide carbanilate, *See* Carbetamex, 5189
2-ethylamino, *See* trietazine, 32238
2-Ethylamino-4-isopropylamino-6-methylthio-s-triazine, *See* ametryn, 2051
(E)-3-[[(ethylamino)methoxyphosphinothioyl]oxy]-2-butenoic acid 1-methylethyl ester, *See* Propetamphos, 25183
Ethylan, 11090, *See* diethyldiphenyldichloroethane, 8441
Ethylan A15, *See* Active #2, 501
Ethylan L, *See* Ablunol 200ML, 95
Ethylan L 3, *See* Ablunol 200ML, 95
Ethylan LD, *See* Active #2, 501
Ethylan mld, *See* Witcamide® 5195, 34440
Ethylan® 44, 11092
Ethylan® 172, 11091
Ethylan® A2, 11094
Ethylan® A10, 11093
Ethylan® BD10, 11095
Ethylan® BV, 11096
Ethylan® CD109, 11097
Ethylan® CD123, 11098
Ethylan® CD913, 11099
Ethylan® CDP2, 11100
Ethylan® D252, 11101
Ethylan® FO30, 11102, 11103
Ethylan® GEL2, 11104
Ethylan® GEO8, 11105
Ethylan® GEO81, 11106
Ethylan® GEP4, 11107
Ethylan® GES6, 11108
Ethylan® GL20, 11109
Ethylan® GO80, 11110
Ethylan® GPS85, 11111
Ethylan® GS60, 11112
Ethylan® GT85, 11113
Ethylan® LD, 11114
Ethylan® NP1, 11115
Ethylan® TN-10, 11116
Ethylan® TT-15, 11117
ethyl-benzaldehyde, 11118
4-ethylbenzaldehyde, *See* Ebal, 9488, ethyl-benzaldehyde, 11118
ethylbutylcarbinol, 11119
2-ethylcapronic acid, *See* 2-ethylhexoic acid, 11137
α-ethylcaproic acid, *See* 2-ethylhexoic acid, 11137
1-(Ethylcarbamoyl)ethyl phenylcarbamate, *See* Carbetamex, 5189
D-(-)-1-(Ethylcarbamoyl)ethyl phenylcarbamate, *See* Carbetamex, 5189
Ethylcellulose, *See* Hercules® K, 13860
Ethylcellulose, *See* Hercules® N, 13862
Ethylcellulose, *See* Hercules® T, 13866
ethylcyanine, 11120
Ethylcyclohex-3-ene carboxylate, *See* Ginsene, 12947
ethyldibutoxy phthalate, *See* dibutoxyethyl phthalate, 8363
ethyldiethylperoxide, *See* Aztec® *t*-Butyl Hydroperoxide-70, Aq, 3541
Ethyldiglycol, *See* Transcutol, 32137
ethyldigol, *See* Transcutol, 32137
O-ethyldigol, *See* Diethoxol, 8435
Ethyldihydro-1H,3H,5H-oxazolo(3,4-c)oxazole, *See* Amine CS-1246, 2154, Oxaban®-E, 22405
Ethyldimethylamine, *See* dimethylethylamine, 8521
N-Ethyldimethylamine, *See* dimethylethylamine, 8521 N,N-dimethylethylamine, 8522
ethylene alcohol, *See* ethylene glycol, 11123, glycol, 13099
Ethylene bis[dithiocarbamic acid], manganese salt, *See* Maneb, 18526
N,N'-Ethylene bis 12-hydroxystearamide, *See* Paricin® 285, 22839
N,N'- Ethylene bisstearamide, *See* Glycowax® 765, 13137
Ethylene bis(stearamide), *See* Abluwax EBS, 140, Acrawax® C, 383, Advawax® 290, 820, Alkamide® STEDA, 1652, Armowax EBS, 2985, Glyco, 13137, Nopcowax 22-DS, 21513
ethylene bisstearamide, *See* Nopcowax 22-DS, 21513
ethylene bis-stearamide, *See* Advawax® 290, 820, Alkamide® STEDA, 1652, Radia® 7506, 25745, Uniwax 1760, 33069
Ethylene bis-tetrabromophthalimide, *See* Saytex® BT-93®, 27520
ethylene brassylate, 11121, *See* Emeressence® 1150, 10077
Ethylene carbonate, *See* Texacar® EC, 31424
Ethylene carboxamide, *See* Acrylamide, 410
ethylene chloride, *See* ethylene dichloride, 11122
Ethylene copolymer [9010-77-9] and aluminum stearate [300-92-5], *See* ACuflow AF-1, 571
ethylene diacetate, *See* ethylene glycol diacetate, 11124
Ethylene diamine, *See* ethylenediamine, 11133
Ethylene diamine distearyl amide, *See* Sicolub® EDS, 28220
ethylene diamine tetraacetic acid, *See* edetic acid, 9615
ethylene diamine tetraacetic acid, sodium salt, *See* tetrasodium EDTA, 31391
ethylene dichloride, 11122
Ethylene diglycol monoethyl ether, *See* Diethoxol, 8435, Ethoxydiglycol, 11044, ethyl di-Icinol, 11068
ethylene dilaurate, *See* Emalex EG-di-L, 9949, Kemester® EGDL, 15958
ethylene dimethacrylate, *See* ethylene glycol dimethacrylate, 11125
ethylene dimethyl ether, *See* monoglyme, 20227
Ethylene dioleamide, *See* Kemamide® W-20, 15903
Ethylene distearamide, *See* Kemamide® W-39, 15904
2,2-Ethylene dithiodiethanol, *See* Tegochrome® 22, 30998
Ethylene dodecanedioate, *See* Arova 16, 3088, Emeressence® 1151, 10078
ethylene glycol, 11123, *See* glycol, 13099, Ilexan E, 14883
ethylene glycol bisthioglycolate, *See* glycol dimercaptoacetate, 13101
ethylene glycol butyl ether, *See* Butyl Cellosolve®, 4767
Ethylene glycol butyl ether acetate, *See* 2-butoxyethanol acetate, 4754,
ethylene glycol diacetate, 11124
Ethylene glycol dilaurate, *See* glycol dilaurate, 13100
ethylene glycol dimethacrylate, 11125, *See* Ageflex EGDMA, 969, ethylene glycol dimethacrylate, 11125, Perkalink® 401, 23315, SR-206, 29363
ethylene glycol dimethyl ether, *See* Ansol E-121, 2494, monoglyme, 20227
Ethylene glycol dioleate, *See* Emalex EG-di-O, 9950
Ethylene glycol distearate, *See* Emalex EG-di-S, 9951, glycol distearate, 13102, Radia® 7266, 25735, Rewopal® PG 280, 26348, Rita EDGS, 26789
ethylene glycol distearate VA, 11126
ethylene glycol ethyl ether, *See* Ethoxyethanol, 11045, ethyl icinol, 11074
ethylene glycol ethyl ether acetate, *See* EE Acetate, 9635, Ethoxyethanol acetate, 11046
ethylene glycol ethylene ether, *See* Dioxane, 8574
Ethylene glycol hydroxystearate, *See* Paricin® 15, 22835
Ethylene glycol methacrylate, *See* Bisomer 2HEMA, 4273, Sipomer® HEM-D, 28536
ethylene glycol methyl ether, 11127, *See* methyl cellosolve®, 19521
Ethylene glycol monethyl ether acetate, *See* EE Acetate, 9635
Ethylene glycol mono phenyl ether, *See* Igepal® Cephene Distilled, 14850, Rewopal® MPG 10, 26344
ethylene glycol monobutyl ether, 11128, *See* Butyl Cellosolve®, 4767, Ektasolve® DB, 9662
Ethylene glycol monobutyl ether acetate, *See* 2-butoxyethanol acetate, 4754, EE Acetate, 9635
Ethylene Glycol Mono-n-butyl Ether Acetate, *See* 2-butoxyethanol acetate, 4754
Ethylene glycol monoethyl ether, *See* EE Solvent, 9636, ethyl icinol, 11074, Poly-Solv® EE, 24556
Ethylene glycol monomethyl ether, *See* methyl cellosolve®, 19521, Poly-Solv® EM, 24557
ethylene glycol monomethyl ether acrylate, *See* Ageflex MEA, 994
Ethylene glycol monopalmitate, *See* glycol palmitate, 13103
ethylene glycol monophenyl ether, *See* 2-Phenoxyethanol, 23627
Ethylene glycol monopropyl ether, *See* Ektasolve® EP, 9671, ethylene glycol propyl ether, 11130
Ethylene glycol monoricinoleate, *See* glycol ricinoleate, 13104
Ethylene glycol monostearate, *See* Ablunol EGMS, 100, Alkamuls® EGMS/C, 1664, Alkamuls® SEG, 1684, Emalex EGS-A, 9952
ethylene glycol monostearate VA, 11129
Ethylene glycol n-butyl ether, *See* 2-butoxyethanol, 4753 Dowanol® EB, 8894
Ethylene glycol octyl phenyl ether, *See* Triton® X-15, 32433
Ethylene glycol phenyl ether, *See* Dowanol® EPh, 8895, Igepal® Cephene Distilled, 14850, Rewopal® MPG 10, 26344
ethylene glycol propyl ether, 11130
Ethylene glycol stearate, *See* lauramide EG, 16829, Radiasurf® 7270, 25806
Ethylene glycon dilaurate, *See* Emalex EG-di-L, 9949
Ethylene Methacrylate, *See* ethylene glycol dimethacrylate, 11125 Perkalink® 401, 23315
Ethylene methyl acrylate copolymer, *See* Emac SP 2205, 9910
ethylene oxide, 11131
Ethylene oxide and carbon dioxide, *See* T-gas, 31549
Ethylene oxide-oleyl alcohol condensate, *See* Volpo 3, 34020
ethylene succinic acid, *See* succinic acid, 29915
ethylene tetrachloride, *See* perchloroethylene, 23222
ethylene thiourea, 11132, *See* Perkacit® ETU, 23283
Ethylene undecane dicarboxylate, *See* Emeressence® 1150, 10077, ethylene brassylate, 11121
Ethylene/acrylic acid copolymer, *See* A-C® Copolymer 580, 172
ethylene/acrylic acid copolymer, Mg ionomer, *See* ACIyn®250, 262, 296, 361
Ethylene/acrylic acid/vinyl acetate copolymer, *See* ACter 1450, 1450A, 460
ethylene/chlorotrifluoroethylene copolymer (ECTFE),

See Halar® 300, 13542
Ethylene/diethylene glycol 2-ethythexyl ether, *See* Ektasolve® EEH, 9670
Ethylene/propylene oxide adducts, *See* Miravon B12DF, 19934
Ethylene/VA copolymer, *See* Airflex® 323, 1100
Ethylenebis dibromonorbornane dicarboximide, *See* Saytex® BN-451, 27519
ethylenebis(dithiocarbamate) complexed with 8.15% Zn, 8.05% Mn, 5.5% Cu, 1.0% Fe, *See* cufraneb, 7344
Ethylenebis(dithiocarbamic acid), polymer with ammonia complex of zinc ebdc, *See* metiram, 19582
ethylenebis(iminodiacetic acid) tetrasodium salt, *See* Kalex 220 Crystal, 15668, Kalex Liq. 50%, 15671
N,N'-Ethylenebisoleamide, *See* Kemamide® W-20, 15903
N,N'-Ethylene bis-ricinoleamide, *See* Flexricin® 185, 11956
Ethylenebis[stearamide], *See* Abluwax EBS, 140
N,N'-Ethylene bisstearamide, *See* Acrawax® C, 383, Advawax® 290, 820, Kemamide® W-39, 15904, Uniwax 1760, 33069, Kemwax, 15993
N,N'-Ethylenebis[stearamide], *See* Abluwax EBS, 140
ethylenebis(dithiocarbamic acid) manganese zinc complex, *See* Karamate, 15749
ethylenecarboxylic acid, *See* acrylic acid, 412
ethylenediamine, 11133
ethylenediamine tetraacetic acid, *See* Trilon® BS, 32327
ethylenediamine tetraacetic acid tetrasodium salt, *See* tetrasodium EDTA, 31391
ethylenediaminetetraacetic acid, *See* disodium EDTA, 8644, edetate disodium, 9613, Kalex Acids, 15669
ethylenediaminetetraacetic acid calcium disodium chelate, *See* Versene CA, 33618
Ethylenediamine-N,N,N',N'-tetraacetic acid, *See* Trilon® BS, 32327
ethylenediaminetetraacetic acid tetrasodium salt, *See* Kalex 220 Crystal, 15668, Kalex Liq. 50%, 15671, Trilon® B Powd, 32324
ethylenediaminetetraacetic acid trisodium salt, *See* edetate trisodium, 9614
***cis*-1,2-ethylenedicarboxylic acid**, *See* maleic acid, 18475
trans-1,2-ethylenedicarboxylic acid, *See* Fumaric acid, 12453
(ethylenedinitrilo)tetraacetic acid trisodium salt;, *See* edetate trisodium, 9614
[(ethylenedinitrilo)tetraacetato]caciate(2-) disodium, *See* Versene CA, 33618
ethylenedinitrilotetraacetic acid, *See* Trilon® BS, 32327
(ethylenedintrilo)tetraacetic acid, *See* Kalex Acids, 15669
(Ethylenedinitrilo) tetraacetic acid tetrasodium, *See* Trilon® B Powd, 32324
(ethylenedintrilo)tetraacetic acid tetrasodium salt, *See* Kalex 220 Crystal, 15668, Kalex Liq. 50%, 15671
(ethylenedinitrilo)tetra-2-propanol, *See* tetra(2-hydroxypropyl) ethylenediamine, 31356
1,1'-ethylene-2,2'-dipyridylium dibromide, *See* Katalon, 15782
N,N'-ethylenedi(stearamide), *See* Acrawax® C, 383, Advawax® 290, 820, Alkamide® STEDA, 1652, Armowax EBS, 2985, Glyco, 13137, Nopcowax 22-DS, 21513
Ethylenemalonic acid, *See* vinaconic acid, 33766
Ethyleneoxide-oleyl alcohol adduct, *See* Volpo 3, 34020
ethylene-propylene copolymer, *See* EPM rubber, 10666
1,3-ethylene-2-thiourea, *See* Perkacit® ETU, 23283
N,N'-ethylenethiourea, *See* Perkacit® ETU, 23283
ethylene-vinyl acetate copolymer, *See* Bakelite DQD-3269, 3593, Baymod®, 3814
Ethylene-vinyl chloride copolymer latex, *See* Airflex® 4500, 1102, Airflex® 4514, 1103, Airflex® 4530, 1104, Airflex® 4814, 1105
Ethylene-vinyl chloride emulsion, *See* Airflex® RB-35, 1108, Airflex® RB-40, 1109
N-Ethylethanamine, *See* diethylamine, 8439
Ethylethoxy propionate, *See* ethyl 3-ethoxypropionate, 11062
Ethylflo 162, 11134
Ethylflo 180, 11135
ethylformic acid, *See* propionic acid, 25190
Ethylglycol acetate, *See* EE Acetate, 9635
ethylhexadecyldimethylammonium bromide, *See* Sumquat® 6020, 30121
2-Ethylhexane-1,3-diol, *See* ethyl hexanediol, 11073
2-ethylhexane-diol-1,3, *See* ethyl hexanediol, 11073
2-ethylhexanoic acid, *See* 2-ethylhexoic acid, 11137
Ethylhexanoic acid, *See* 2-ethylhexoic acid, 11137
Ethylhexanoic acid zinc salt, *See* Octoate Z, 21992
2-Ethylhexanoic acid, cetyl/stearyl ester, *See* Cetearyl octanoate, 5791
2-ethylhexanol, 11136, *See* Aerofroth 88, 859
2-Ethylhexanol tails, *See* Aerofroth 99, 860
2-ethylhexoic acid, 11137, *See* 2-ethylhexoic acid, 11137, trioctyl phosphate, 32349
2-ethylhexyl acetate, 11138, *See* Phenol Carbonate, 23604
2-ethylhexyl acrylate, *See* octyl acrylate, 22037
2-Ethylhexyl alcohol, *See* Aerofroth 88, 859, 2-ethylhexanol, 11136
2-Ethylhexyl diphenyl ester phosphoric acid, *See* diphenyl octyl phosphate, 8597
2-Ethylhexyl diphenyl phosphate, *See* diphenyl octyl phosphate, 8597
Ethylhexyl diphenyl phosphite, *See* Weston® EHDPP, 34313
2-Ethylhexyl hexadecanoate, *See* octyl palmitate, 22040
2-Ethylhexyl hydrogenated tallowalkyl methosulfate, *See* Arquad® HTL8(W) MS-85, 3118
2-Ethylhexyl 12-hydroxystearate, *See* Wickenol® 171, 34398
2-ethylhexyl methoxycinnamate, *See* Parsol® MCX, 22867
2-Ethylhexyl nitrate, *See* Exchem GO-1, 11294
2-Ethylhexyl octadecanoate, *See* Wickenol® 156, 34392
Ethylhexyl oleate, *See* Radia® 7331, 25736
2-Ethylhexyl oxystearate, *See* Wickenol® 171, 34398
Ethylhexyl palmitate, *See* Unimate® EHP, 32970
2-Ethylhexyl palmitate, *See* octyl palmitate, 22040, Unimate® EHP, 32970, Wickenol® 155, 34391
2-Ethylhexyl pelargonate, *See* Wickenol® 160, 34395
2-Ethylhexyl phosphate, *See* Hostaphat 2122, 14383, Rhodafac® PEH, 26601
2-Ethylhexyl phosphate ester, *See* Marlophor® IH-Acid, 18895
Ethylhexyl phthalate, *See* Witcizer 312, 34454
2-Ethylhexyl phthalate, *See* Reomol DCP, 26130
2-Ethylhexyl p-methoxycinnamate, *See* Neo Heliopan® AV, 20866
2-Ethylhexyl polyphosphoric ester acid anhydride, *See* Strodex® MO-100, 29819
2-ethylhexyl 2-propenoate, *See* octyl acrylate, 22037
Ethylhexyl salicylate, *See* Dermoblock OS, 8116
2-Ethylhexyl salicylate, *See* Dermoblock OS, 8116, Neo Heliopan® OS, 20870, Uvinul® O-18, 33202
2-ethylhexyl sebacate, *See* Plasthall® DOS, 24005, Reomol DOS, 26131
2-Ethylhexyl stearate, *See* Afilan EHS, 931, Lexolube® T-110, 17164, octyl stearate, 22041, Wickenol® 156, 34392
2-Ethylhexyl sulfosuccinate, *See* Penetron OT-30, 23108
2-Ethylhexyl sulfosuccinate sodium, *See* dioctyl sodium sulfosuccinate, 8550
2-Ethylhexyl Trimellitate, *See* PX-338, 25441
2-Ethylhexyl vinyl ether, *See* Rapi-Cure EHVE, 25882
2-Ethylhexyl phthalate, *See* Witcizer 312, 34454
2-ethylhexyl p-methoxy-cinnamate, *See* octyl methoxycinnamate, 22039
ethylhexylamine, *See* Armeen® L8D, 2952
2-Ethylhexylamine, *See* Armeen® L8D, 2952
Ethylhexylcinnamate, *See* Neo Heliopan® AV, 20866, Parsol® MCX, 22867
2-ethylhexyl-2-cyano-3,3-diphenylacrylate, *See* Neo Heliopan® 303, 20865, Uvinul® N-539, 33201
2-ethylhexyl-4-dimethylaminobenzoate, *See* Padimate O, 22549, Solarchem® O, 28898
2-ethyl-1,3-hexylene glycol, *See* ethyl hexanediol, 11073
2-ethylhexyl hexadecanoate, *See* octyl palmitate, 22040
Bis(2-ethylhexyl)hexanedioate, *See* Plasthall® DOA, 24002
2-Ethylhexyl-4-methoxycinnamate, *See* Neo Heliopan® AV, 20866
2-Ethylhexyloxypropylamine, *See* Tomah PA-12EH, 31980
2-ethylhexyl palmitate, *See* octyl palmitate, 22040
2-Ethylhexylpropenoate, *See* octyl acrylate, 22037
ethylibutoxy phthalate, *See* Dibutoxyethyl phthalate, 8363
Ethylic acid, *See* Acetic acid, 292
2,2'-Ethylidene bis(4,6-di-t-butylphenol), *See* Vanox® 1290, 33336
Ethylidene diethyl ether, *See* Acetron®, Acetron® GP; Acetron® NS, 308
Ethylidene-aniline, *See* Accelerator E-A, 203
Ethylidenebis dibutylphenyl fluorophosphonite, *See* Ethanox® 398, 10933
Ethylidenebisdibutylphenol, *See* Isonox® 129, 15383
2,2'-ethylidenebis (4,5-di-*tert*-butylphenol), *See* Isonox® 129, 15383
2,2'-Ethylidenebis(4,6-di-t-butylphenyl) fluorophosphonite, *See* Ethanox® 398, 10933
Ethylidenelactic acid, *See* lactic acid, 16570
Ethylidine aniline/acetaldehyde (2:1), *See* Accelerator A7, 196
Ethylidine diethyl ether, *See* Acetron®, Acetron® GP; Acetron® NS, 308
Ethylidineaniline, *See* Vulconex, 34117
D-N-Ethyllactamide carbanilate, *See* Carbetamex, 5189
D-N-Ethyllactamide carbanilate (ester), *See* Carbetamex, 5189
Ethylmethyl imidazole, *See* 2-ethyl-4-methyl imidazole, 11139
2-ethyl-4-methyl imidazole, 11139, *See* EMI-24, 10190
4-Ethyl-2-methyl-2-(3-methylbutyl)-1,3-oxazolidine, *See* Zoldine® MS-52, 34886
N-[3-(1-Ethyl-1-methylpropyl)-5-isoxazolyl]-2,6-dimethoxybenzamide, *See* Knot Out, 16257
7-ethyl-2-methyl-4-undecanol hydrogen sulfate sodium salt, *See* Niaproof® Anionic Surfactant 4, 21093
ethylmorpholine, *See* N-ethyl morpholine, 11140
4-ethylmorpholine, *See* N-ethyl morpholine, 11140, Texacat® NEM, 31431
N-Ethylmorpholine, *See* Toyocat® -NEM, 32106, Texacat® NEM, 31431
4,4'-(2-Ethyl-2-nitrotrimethylene), 11141
ethylol, 11142
Ethylol Colored, 11143
ethylparaben, *See* ethyl *p*-hydroxybenzoate, 11061, Nipagin A, 21263
ethyl-*p*-dimethyl aminobenzoate, *See* Speedcure EDB, 29261
2-(N-Ethylperfluorooctanesulfonamido) ethyl acrylate, *See* Fluorad® FX-13, 12066
Ethylphenol, *See* phlorol, 23675
***o*-ethylphenol**, *See* phlorol, 23675
1,1-bis(*p*-ethylphenyl)-2,2-dichloroethane, *See* diethyldiphenyldichloroethane, 8441
***p*-Ethylphenyl-thiocarbimide**, *See* phenethyl mustard oil, 23588
N-(1-Ethylpropyl)-3,4-dimethyl-2,6-dinitrobenzenamine, *See* pendimethalin, 23102
2-ethyl-2-propyl-1,3-propanediol, *See* ethyl hexanediol, 11073
ethylprotal, *See* ethyl vanillin, 11087, Vanbeenol, 33294

ethylprotocatechuic aldehyde, *See* ethyl vanillin, 11087, Vanbeenol, 33294
1-ethyl-2-pyrrolidone, *See* Agsol Ex 2, 1087
N-Ethyl-2-pyrrolidone, *See* Agsol Ex 2, 1087, NEP, 20987
N-Ethyltetrahydroquinoline, *See* Kairoline, 15658
2-[5-ethyltetrahydro-5-[tetrahydro-3-methyl-5-[tetrahydro-6-hydroxy-6-(hydroxymethyl)-3,5-dimethyl-2H-pyran-2-yl]-2-furyl]-2-furyl]-9-hydroxy-β-methoxy-α,+lc,2,8-tetramethyl-1,6-dioxaspiro[4.5]decane-7-butyric acid, *See* monensin, 20208
Ethylthiodiglycol, *See* Glyecin, 13143
ethylthiurad, 11144, *See* Disulfiram, 8725
ethyltoluene sulfonamide, 11145
ethyltriacetoxysilane, 11146, *See* CE6345, 5529, Dynasylan® ETAC, 9407
ethyltrianol, *See* Raxil, 25908
ethyltrichlorosilane, 11147, *See* CE6350, 5530
Ethyltrigol, *See* Trioxitol, 32356
Ethylzimate, *See* Octocure ZDE-50, 21996
ethyne, *See* acetylene, 318
Eticol, *See* paraoxon, 22783
Etidronate disodium, *See* Didronel, 8421
Etidronic Acid, *See* Briquest® ADPA-60A, 4536, Fostex P, 12338
Etifenin, *See* EHIDA Kit, 9649
Etilon, *See* parathion, 22807
Etingal® A, 11148
Etingal® L, 11149
Etingal® S, 11150
Etiol, *See* malathion, 18460
ETMT, *See* Aaterra WP, 17
ETO, *See* ethylene oxide, 11131
Etocas, 11151
Etocrilene, *See* Uvinul® N-35, 33200
Etocrylene, *See* Uvinul® N-35, 33200
EtOH, *See* ethyl alcohol, 11064
Etophen 102, 11152
Etophen 114, 11153
E-Toplex, 11154
Etoxon epa:sodium lauryl ether sulfate, *See* Witcolate ES-2, Witcolate LES-60C, 34477
Etridiazol, *See* Aaterra WP, 17
Etridiazole, *See* Aaterra WP, 17
etridonic acid, *See* Unihib® 106, 32957
etrimfos, 11155, *See* Satisfar, 27488
Etrimphos, *See* etrimfos, 11155
Etrofolan, *See* Etrofolan®, 11156, isoprocarb, 15404
Etrofolan®, 11156 , *See* isopropyl phenylmethyl carbamate, 15411
Etrolan, *See* Etrofolan®, 11156
Etrolene, *See* Korlan, 16351, ronnel, 26951
Etronite, 11157
ETU, 11158, *See* ethylene thiourea, 11132, Perkacit® ETU, 23283
Etyl Tuex, *See* Disulfiram, 8725
Eubeco, 11159
Euchrome, *See* Cappagh brown, 5121
Euclorina, *See* Chloramine T, 6086, Ketjensept, 16133
Eucopine, 11160
Euderm®, 11161
euflavine, *See* Gonacrine, 13192
eugenic acid, *See* eugenol,
eugenol methyl ether, *See* methyl eugenol, 19534
Euka-drya, 11162
Eukanol®, 11163
Eukesol® Binder S, 11164
Eukesolar® Dyes, 11165
Eukinase, 11166
Eulan SP, *See* cyfluthrin, 7563
Eulan® 33, 11167
Eulan® BLS, 11168
Eulan® WA, 11169
Eulysin® WP, 11170
Eu-Med, *See* Valadol, 33238
Eumulgin 05, 11171
Eumulgin B2, 11172
Eumulgin C4, 11173
Eumulgin EP .5.2L, 11174
Eumulgin EP2, 11175
Eumulgin HRE 40, 11176
Eumulgin L, 11177
Eumulgin M8, 11178
Eumulgin M8, O5, PWM2, WM5, *See* oleth-8, 22126
Eumulgin PA 10, 11179
Eumulgin PA 12, 11180
Eumulgin PA 30, 11181
Eumulgin PC 2, 11182
Eumulgin PK 23, 11183
Eumulgin PLT 4, 11184
Eumulgin PPG 40, 11185
Eumulgin PRT 36, 11186
Eumulgin PST 5, 11187
Eumulgin PWM2, 11188
Eumulgin RO 40, 11189
Eumulgin SML 20, 11190
Eumulgin SMO 20, 11191
Eumulgin SMS 20, 11192, *See* Ethylan® GES6, 11108
Eumulgin WM5, 11193
Eunasin, *See* benazolin, 3935
Eunatrol, 11194, *See* sodium oleate, 28802
euonymit, *See* galactitol, 12592
euosmite, 11195
Eupad, 11196
Euparen (e), *See* Dichlofluanid, 8379
Euparen®, 11197 , *See* Dichlofluanid, 8379
Euparen® M, 11198
Euperlan®, 11199
Euperlan® MPK 850, 11200
Euperlan® PK 810, 11202
Euperlan® PK 900, 11203
Euperlan® PK 3000, 11201
euphorbia gum, *See* potato gum, 24789
Eupittonic acid, *See* pittaccal, 23889
Eupolen®, 11204
Euranaat LS3, *See* disodium lauryl sulfosuccinate, 8649
Eurecryl, 11205
Euredur, 11206
Eureka 102, 11207
Eureka 392, 11208, *See* Eureka 392, 11208
Eureka 400-R, 11209
Eureka 800, 11210
Eureka 800-R, 11211
Eureka 1014-M, 11212
Eureka 1067-A, 11213
Eureka Alloy, 11214
Eureka compound, *See* burnt hypo, 4712
Eureka E-2, 11215
Eurelon, 11216
Euremelt, 11217
Eurepox, 11218
Euresyst, 11219
Euretek®, 11220
Euretek® 540, 11221
European Elastin 10, 11222, *See* hydrolyzed elastin, 14612
Europolymer, 11223
Europrene, 11224
Europrene AR, 11225
Europrene CIS, 11226
Europrene Lattice, 11227
Europrene N, 11228
Europrene NEOCIS, 11229
Europrene SOL, 11230
Europrene SOL T, 11231
Euroquat C45, CPB, K, LA, LAC, *See* Cocamidopropyl betaine, 6551
Eurotex, 11232
Eurylon, 11233
Eusapon®, 11234
Eusin®, 11235
Eusol, 11236
Eusolex 232, *See* Neo Heliopan® Hydro, 20868
Eusolex 4360, *See* Cyasorb® UV 9, 7495, Neo Heliopan® BB, 20867, Rhodialux A, 26675
Eusolex 6007, *See* Padimate O, 22549, Solarchem® O, 28898
Eusolvan, 11237 *See* ethyl lactate, 11076
Eutannin, 11238
Eutanol G, 11239, *See* octyl dodecanol, 22038
Eutanol G16, 11240, *See* Exxal® 16, 11341
Euthylen®, 11241
Euvinyl® C, 11242
Euviprint®, 11243
Euxyl, 11244
Euxyl K 400, *See* Igepal® Cephene Distilled, 14850, Rewopal® MPG 10, 26344
EVA, 11245
Eval, 11246
Eval® E105, 11247
Eval® E151, 11248
Eval® G115, 11249
Eval® K102, 11250
Evanacid® 3CS, 11251 *See* Carboxymethyl mercaptosuccinic acid, 5256
Evangard® 18MP, 11252, *See* octadecyl mercaptopropionate, 21983
Evanol, 11253
Evans' cement, 11254
Evansite, *See* aluminum phosphate, 1922
Evanstab® 12, 11255
Evanstab® 13, 11256
Evanstab® 14, 11257
Evanstab® 18, 11258, *See* Distearyl thiodipropionate, 8713
Evasol, *See* Dequalinium Chloride, 8077
Evasperse, 11259
Evatane, 11260
Event, 11261
Everbrite, 11262
Eveready Prestone, 11263
Everflex® 81L, *See* Vinyl Acetate, 33854
Everflex® 515L, 11264
Everflex® E, 11265
Everflex® SP-1084, 11266
Evergreen, 11267
Everitt's salt, 11268
Everlastic, 11269
Everlube, 11270
Everseal, 11271
Eversoft Plastex, 11272
Eversoft Sea Mex, 11273
Eversoft Tees Powder, 11274
Eversun, 11275
Evik, *See* ametryn, 2051
Evik 80W, *See* ametryn, 2051
E-Vimin, *See* tocopherol, 31918, Vascuals, 33465
Evion, *See* tocopherol, 31918, Vascuals, 33465
Evipherol, *See* tocopherol, 31918, Vascuals, 33465
Eviplast 80, *See* Bis(2-ethylhexy) Phthalate, 4228, dioctyl phthalate, 8549, Reomol DCP, 26130, Witcizer 312, 34454
Eviplast 81, *See* Bis(2-ethylhexy) Phthalate, 4228, dioctyl phthalate, 8549, Reomol DCP, 26130, Witcizer 312, 34454
Evola, *See* *p*-dichlorobenzene, 8386
Evolve, 11276
Evo-stik 873 Super, 11277
Ewer-Pick acid, 11278
Ewo, 11279
EW-POL 8021, 11280
Exact-S®, 11281 , *See* methyl sulfide, 19561
Exaltone, 11282
Exam, 11283
Exameen 824 3724, 11284
Exameen 2125 M 3704, 11285
Exameen 3500 3714, 11286
Exameen 3500, 3580, *See* Benzalkonium chloride, 3958
EXC-33, 11287
Excalibur, *See* +ll-cyhalothrin, 7567
Excelite, 11288
Excellerex, 11289
Excello, 11290
Excelo, 11291
Excelon, 11292
Excelsior, 11293

Chemical and Trade Names

Exchem GO-1, 11294
Exdol, *See* Valadol, 33238
Exelderm, *See* Sulconazole, 29958
Exell, 11295
Exem, *See* Exam, 11283
Exetor, *See* Timbrel, 31800
Exgraphite, 11296
Exhoran, *See* Disulfiram, 8725
Exitelite, *See* Octoguard FR-10, 22000, Timonox, 31807, UltraFine® II, 32799
Exkin, 11297
Exl-die Steel, 11298
Exobloc BF-1000, 11299
Exocerol® OM, 11300
Exoderil, *See* Naftifine, 20650
Exodin, *See* Diazinon Liquid, 8344
Exofene, *See* hexachlorophene, 13992
Exolit® IFR-23, 11301
Exolon, 11302
Exoryl, 11303
Exotherm, *See* Chlorothalonil, 6150, Repulse, 26151
Exotherm Termil, *See* Chlorothalonil, 6150, Repulse, 26151
EXP-28, 11304
EXP-49, 11305
EXP-51, 11306
EXP-58, 11307
EXP-3864, *See* quizalofop-ethyl, 25697
Expancel® 091, 11308
Expancel® 551 DE, 11309
Expand, *See* Sethoxydim, 28099
expanded graphite, 11310
Expandex® 175, 11311
Expandex® 5PT, 11312
expanding solder, 11313
Expar, *See* permethrin, 23384, 23385, 23386
Explosive D, 11314
explosive gum, 11315
Explotab®, 11316
Exprol, 11317
exsiccated alum, *See* aluminum potassium sulfate, 1923
exsiccated sodium sulfite, *See* sodium sulfite, 28823
Exsyproteines 2%, 11318
Extend®, 11319
Extendopel, 11320
Extendospheres® Metalite® Zinc, 11321
Extendospheres® SG, 11322
Extiat®, 11323
Extir, 11324
Extol, 11325
Extra Bond, 11326
extra white metal, 11327
Extract of *Rosmarinus officinalis*, *See* Pristene R20, 25039
Extract of vermilion, *See* scarlet vermillon, 27550
Extract of yeast, *See* yeast extract, 34640
Extrakt 52, 11328
extrax fish-tox, *See* rotenone, 27000
Extrazine II, *See* atrazine, 3394
Extrox, 11329
Extrudoil®, 11330
Extrusil, 11331
Extrusion-Plus, 11332
Exxal®, 11333
Exxal® 6, 11334
Exxal® 7, 11335
Exxal® 8, 11336
Exxal® 9, 11337
Exxal® 10, 11338, *See* decyl alcohol, 7867
Exxal® 12, 11339
Exxal® 13, 11340
Exxal® 16, 11341
Exxal® 18, 11342
Exxal® 20, 11343
Exxal® 20, *See* octyl dodecanol, 22038
Exxal® 26, 11344
Exxal® L1315, 11345
Exxate®, 11346
Exxate® 600, 11347
Exxate® 800, 11348
Exxate® 900, 11349
Exxate® 1000, 11350
Exxate® 1300, 11351
Exxelor, 11352
Exxon® Bromo XP-50, 11353
Exxon® Butyl 065, 11354
Exxon® Butyl 077, 11355
Exxon® Butyl 268, 11356
Exx-Print, 11357
Exxsol®, 11358
Exxsol® D-40, D-60, D-80, D-110, D-130, 11359
Exxsol® Heptane, 11360, *See* n-heptane, 13790
Exxsol® Hexane, 11361, *See* n-hexane, 14009
Exxsol® Isopentane, 11362, *See* isopentane, 15394
Exxtraflex, 11363
Exzyme, 11364
Eymid, 11365
Eymyd® Prepreg, 11366
Eymyd® Resin L-20N, 11367
Eypel®, 11368
Eypel® A Acoustic Barrier Sheet, 11369
Eypel® F, 11370
E-Z Mix, 11371
EZ Mold Lubricant, 11372
EZA®, 11373
E-Z-Off D, *See* Def®, 7876
E-Z-Paque, *See* Radiopaque, 25830
F 1-F, *See* Butylated hydroxyanisole, 4787
f1-tabs, *See* sodium fluoride, 28764
F 1, *See* Nyad®, 21860, Wollastocoat®, 34545
F 10, *See* Maneb, 18526
F 12, *See* dichlorodifluoromethane, 8388, Freon®, 12400
F-40 F-400, *See* Algin, 1591
F 90, *See* Nacconol® 40G, 20647
F-100 Dried Gel, *See* Alumina hydrate, 1892
F-238, 11376
F-309, 11377
F-310, 11378
F-319, *See* Tachigaren 70, 30685
F-500, -1000,-3600,etc, *See* Alumina hydrate, 1892
F-500, -3600, etc, 11379
F-600 Proctin BUS Protanal 686, *See* Algin, 1591
F 735, *See* carboxin, 5255
F 1358, *See* Dapsone, 7751
F-1000 Dried Gel, 11380
F-1000®, 11381 , *See* Alumina hydrate, 1892
F-1991, *See* Benomyl, 3944
F-2000, 11382
F-2000 Dried Gel, 11383
F-2000 FR, 11384
F-2001, 11385
F-2100 Dried Gel, 11386
F-2200, 11387
F-2200 Dried Gel, 11388
F-2300, 11389
F-2300H, 11390
F-2400, 11391
F-2400E, 11392
F Acid, *See* Cassella's acid, 5402
F.A.S.T. Lube System, 11374
FA, *See* furfuryl alcohol, 12493, Idryl, 14841
FA-1, 11393
FA-8, 11395
FA-14, 11394
Fabelnyl, 11396
Fabrene®, 11397
Fabrethane, 11398
Fabrex, 11399
Fabrifil, 11400
Fabriglide, 11401
Fabrikoid, 11402
Fabroil, 11403
Fabrolite, 11404
Facet®, 11405
F-acid, *See* Delta acid, 8044
Facteka, 11406
Factice, 11407, *See* Crodasub, 7150
factis, 11408
factitious air, *See* Nitral, 21318
Factoprene NS, Z, 11409
Faexin extract, 11410
Fagacid, 11411
Fahlun diamonds, 11412
Fahralloy, 11413
Fair-2, *See* maleic hydrazide, 18477
Fairey Metal, 11414
Fair-Plus, *See* maleic hydrazide, 18477
Fairprene, 11415
Fairy Ring Destroyer, 11416
Faktex, 11417
Faktogel, 11418
FAL Cypermethrin 10, 11419
Falapen, *See* Penicillin G potassium, 23109
Faligruen, *See* Cupravit®, 7370
Falisan, *See* Agallol, 954, Atiran, 3286, Ceresan, 5745
Falithrom, *See* Phenprocoumon, 23630
falitiram, *See* Rezifilm, 26521
Falkaloid, Falkyd, 11420
Fallowmaster, *See* Dicamba, 8374
FAM 30, 11421
Famid, 11422
Famodan MS Kosher, 11423
Famodan TS Kosher, 11424
Famosan, 11425
Famous, 11426, *See* hydrogenated soybean oil, 14596
Fanal, 11427
fanal pink gfk, *See* Rhodamine G and G Extra, 26617
Fancol 707, *See* PPG-30 cetyl ether series, 24845
Fancol Acel, 11428
Fancol ALA, 11429
Fancol ALA-10, 11430
Fancol C, 11431
Fancol CA, 11432
Fancol CAB, 11433
Fancol CB, 11434
Fancol CH, 11435
Fancol CH-24, 11436
Fancol CO-30, 11437
Fancol DL, 11438
Fancol Gingko Extract, 11439
Fancol HCO-25, 11440
Fancol HL, 11441
Fancol HL-20, 11442
Fancol HL-20, *See* PEG-20 hydrogenated lanolin, 23045
Fancol HON, 11443
Fancol HON, *See* Honey, 14295
Fancol Karite Butter, 11444, *See* Shea Butter, 28141
Fancol Karite Extract, 11445, *See* Shea butter extract, 28143
Fancol LA, 11446
Fancol LA-5, 11447
Fancol LAO, 11448
Fancol Menthol, 11449
Fancol OA-95, 11450
Fancol SA, 11451
Fancol SA-15, 11452
Fancol TOIN, 11453
Fancol WGFA, 11454
Fancor D, 11455
Fancor IPL, 11456
Fancor Lanwax, 11457
Fancor LFA, 11458
Fancorsil A, 11459
Fancorsil P, 11460
Fancorsil SLA, 11461
Fancoscour PO, VC, 11462
Fanfare, 11463
Fanghidi Sclofani, 11464
Fantan, 11465
Fantasit, 11466
Fanwax G, 11467
Fanwax P, 11468
Fanwax P, *See* Ethylan® GES6, 11108
FAR Mark I through X, 11469
Farber, *See* Chlorothalonil, 6150, Repulse, 26151
Far-Go, *See* triallate, 32190
Far-Go/Avadex BW, 11470

Fargro Chlormequat, 11471
farina, 11472
farine, 11473
Faringets, 11474
farinose, 11475
Farlite, 11476
Farmacel, 11478
Farmacel, Farmacel 645, 11477
Farmaneb, 11479
Farmatin, 11480
Farmco Atrazine, *See* atrazine, 3394
Farmitalia 204/122, *See* Kelfizina, 15838
Farmon, 11481
Farmon 2,4-D, 11482
Farmon 2,4-DB Plus, 11483
Farmon 2,4-DP+MCPA, 11484
Farmon Blue, 11485
Farmon Condox, 11486
Farmon MCPA 50, 11487
Farmon MCPB Plus, 11488
Farmon Mini Slug Pellets, 11489, *See* metaldehyde, 19431
Farmon PDQ, 11490
Farmon TCA, 11491
Farnesol, 11492
farnoquinone, *See* vitamin K+72, 33950
Farrant's medium, 11493
Farronic, 11494
Fascat® 2000, 11495
Fascat® 2004, 11496
Fascat® 4400, 11497
Fasco WY-HOE, *See* Chlorpropam, 6162
Fascol, 11498
Fasinex, 11499
Fast Garnet B Base, *See* Aminogen I, 2176
Fast Garnet Base B, *See* Aminogen I, 2176
fast oil orange, *See* Sudan I, 29938
Fast Oil Yellow B, *See* Butter yellow, 4759
Fast Red 2G Base, *See* *p*-nitroaniline, 21345
Fast Red 2G Salt, *See* *p*-nitroaniline, 21345
Fast Red Base 2J, *See* *p*-nitroaniline, 21345
Fast Red Base GG, *See* *p*-nitroaniline, 21345
Fast Red GG Base, *See* *p*-nitroaniline, 21345
Fast Red MP Base, *See* *p*-nitroaniline, 21345
Fast Red P Base, *See* *p*-nitroaniline, 21345
Fast Red P Salt, *See* *p*-nitroaniline, 21345
Fast red salt, *See* Azoene, 3526
Fast Red Salt 2J, *See* *p*-nitroaniline, 21345
Fast Red Salt GG, *See* *p*-nitroaniline, 21345
Fast Yellow, *See* Butter yellow, 4759
Fastac, 11500
Fasteeth, 11501
Fasteeth Extra Hold, 11502
Fastex, 11503
Fastusol®, 11504
Fat Ponceau R, *See* Sudan IV, 29941
Fat Yellow, *See* Butter yellow, 4759
Fat Yellow A, *See* Butter yellow, 4759
Fat Yellow AD OO, *See* Butter yellow, 4759
Fat Yellow ES, *See* Butter yellow, 4759
Fat Yellow R, *See* Butter yellow, 4759
Fat Yellow R (8186), *See* Butter yellow, 4759
Fat, Chocolate, *See* vegetable butter, 33514
Fatal, *See* dimethyl tetrachloroterephthalate, 8512
Fatal Flip, 11505
Fats and Glyceridic oils, vegetable, hydrogenated, *See* BBS, 3862
Fatsco, 11506
Fatty acid alkanolamide, *See* Inhibitor RT 212, 15129
Fatty acid alkanolamide polyglycol ether, *See* Marlowet® OCM, 18935
Fatty acid ethoxylate, *See* Secoster® A, 27816
Fatty acid ethoxylate (8 EO), *See* Industrol® TFA-8, 15117
Fatty acid imidazolines, *See* Hartamine Series, 13664
Fatty acid polydiethanolamide, *See* Pentamid C12, 23161
Fatty acid polyglycol ester, *See* Mulsifan RT 1, 20454
Fatty acid/fatty alcohol-based, *See* Advantage™ 10 Defoamer, 810
Fatty acid-amine derivative, *See* Zoharsoft DAS, 34875
Fatty acids C_{18}, unsaturated, trimers, *See* Trilinoleic acid, 32320
Fatty acids, C_{14-18} and C_{16-18} unsaturated, triesters with trimethylolpropane, *See* Radia® 7370, 25740, Radia® 7371, 25741
Fatty acids, C_{16-18} and C_{18}-unsatyrated, iso-butyl esters, *See* Radia® 7230, 25732
Fatty acids, tallow, *See* tallow fatty acid, 30743
Fatty alcohol, *See* cetyl alcohol, 5816, lauryl alcohol, 16864, myristyl alcohol, 20577
Fatty alcohol benzoate, *See* Pentonate DB, 23184
Fatty alcohol ether sulfate, *See* Akypomine® BC 50, 1264, Akyposal RLM 70, 1320
Fatty alcohol oxyethylate, *See* Emulan® AF, 10488, Emulan® OC, 10490
Fatty alcohol polyglycol ether, *See* Dehypon LT 054, 7979
Fatty alcohol polyglycol ether (4EO), *See* Zusolat 1004, 34912
Fatty alcohol polyglycol ether (5 EO), *See* Zusolat 1005/85, 34913
Fatty alcohol sulfate, *See* Akypomine® BC/S, 1265
Fatty alcohol sulfate, anionic, *See* Adipon, 700
Fatty alcohol/PEG methyl ether, *See* Rewopal® MT 65, 26345
Fatty amide, *See* J-Soft 111E, 15626, Schercotarder, 27703
Fatty amine alkoxylate, *See* empilan® 7132, 10385
Fatty ester, *See* Actralube 310, 536, Amsoft FA, 2369
Fatty methyl taurate, *See* Emkapon Jel 500 Conc, 10252
Faturan, *See* Bakelite, 3589
Faversham powder No. 2, 11507
Favlerite No. 1, 11508
Favlerite No. 2, 11509
Favour, 11510
Fax, 11511
Faxola, 11512
Fazor, *See* maleic hydrazide, 18477
FB 48, 11513
FB/2, *See* Katalon, 15782
FBC CMPP, 11514
FBC Fly Dip, 11515
FBC MCPA, 11516
FBC Pirimicarb 50, 11517
FBC Protectant Fungicide, 11518
FBC Slug Destroyer, 11519
FBC Winter Dip, 11520
FBC-32197, *See* quizalofop-ethyl, 25697
FC 12, *See* dichlorodifluoromethane, 8388
FC 113, 11521
FC 133, *See* MS-180 Freon® TF Solv, 20415
FC-123, *See* Dichlorotrifluoroethane, 8393
FC-123, *See* Genetron® 123, 12823
Fc-125, *See* Genetron® HFC 125, 12828
FC-126 strain, *See* Marexine-CA, 18705
FCR 1272, *See* cyfluthrin, 7563
FD&C Blue No. 2, *See* Indigo Carmine, 15075
FD&C Yellow No. 5, *See* tartrazine, 30835
fda 0101, *See* sodium fluoride, 28764
feathered tin, 11522
Feb Brickclean, 11523
Feb Hybit, 11524
Feb Hyseal Slurry, 11525
Feb Oxide, 11526
Feb Supercrete, 11527
Feb Sylane, 11528
Feb Wintamix, 11529
febantel, *See* Rintal®, 26774
Febclean, 11530
Febclear Super, 11531
Febco, 11532
Febcrete AEA, 11533
Febcure, 11534
Febdura Standard, 11535
Febexpan, 11536
Febface, 11537
Febfast PG, 11538
Febflex One Coat, 11539
Febflor, 11540
Febflow Accelerating, 11541
Febflow Retarding, 11542
Febfoam, 11543
Febglaze, 11544
Febgrout, 11545
Febguard Bonding Agent, 11546
Febkol Elastomer 110 and 122, 11547
Febkol Plastomer 555, 11548
Febmast GP, 11549
Febmix Admix, 11550
Febol Standard, 11551
Febond PVA, 11552
Febond SBR, 11553
Febox, 11554
Febpitch, 11555
Febplast Ready Mixed, 11556
Febplate, 11557
Febproof, 11558
Febrail No. 1, 11559
Febrilix, *See* Valadol, 33238
Febrok, 11560
Febseal, 11561
Febset NF, 11562
Febsilicon, 11563
Febspeed, 11564
Febstik, 11565
Febstrike, 11566
Febtex, 11567
Febtile, 11568
Febtite Liquid, 11569
Febtone, 11570
Febweld, 11571
Fecama, *See* dichlorvos, 8394
Fecap, 11572
Fecraloy, 11573
Feculose, 11574
Fecundal, *See* imazalil, 14892
Fedal-telmin, *See* Mansonil®, 18611
Fedralite, 11575
Feedercalc, 11576, 11577
Feedex, 11578
Feedmate, 11579
Feedol, 11580
Feeno, *See* phenothiazine, 23618
Fehlings solution, 11581
Fehling's solution, neutral, 11582
Fekama, *See* dichlorvos, 8394
Fekta® RT, 11583
Felamine, 11584
Feldspar, *See* Unispar 40, 33047
Feliniffa P, 11585
Feliniffa RC, 11586
Felixite, 11587
Felsinosima, 11588
Felspar, 11589
Felton 3T, 11590
Felvinone, 11591
Felzodox, 11592
FEMA No. 2433, *See* ethylene oxide, 11131
Fenafix, 11593
Fenamine, *See* atrazine, 3394
Fenaminosulf, *See* Bayer 5072, 3767
Fenasal, *See* Mansonil®, 18611
Fenatrol, *See* atrazine, 3394
Fenbendazole, 11594, *See* Panacur, 22655
Fenbutatin oxide, *See* Torque, 32077
Fenbutatin oxyde, *See* Torque, 32077
Fenchlorfos, *See* ronnel, 26951
Fenchlorophos, *See* ronnel, 26951
Fenchlorphos, *See* Korlan, 16351
Fenclofos, *See* ronnel, 26951
Fender, *See* phenmedipham, 23595
Fender, Goliath, *See* phenmedipham, 23597
Fenetrazole, *See* Raxil, 25908
Fenitex, 11595
Fenitrothion, *See* Dicofen, 8397, Fenitex, 11595,

Fenitrothion EC, 11596, Folithion®, 12207
Fenitrothion EC, 11596
Fenmedifam, *See* phenmedipham, 23597
Fennite, 11597
Fennosan B 100, *See* Dazomet, 7806, N-521® Biocide, 20640
Fenocil, 11598
Fenoil®, 11599
Fenolac, 11600
Fenolite, 11601
Fenom, *See* cypermethrin, 7588, Topclip Parasol, 32037
Fenopon AC-78, 11602
Fenopon CD, 11603
Fenopon CN-42, 11604
Fenopon CO, 11605
Fenopon EP, 11606
Fenopon SE, 11607
Fenopon T-33 and T-43, 11608
Fenopon T-51, 11609
Fenopon T-77, 11610
Fenopon TC-42, 11611
Fenopon TK32, 11612, *See* sodium methyl tall oil acid taurate, 28793
Fenopon TN-74, 11613, *See* sodium methyl tall oil acid taurate, 28793
fenoprop, *See* Kuron, 16496
Fenotec, 11614
Fenothrin, *See* phenothrin, 23619
Fenoverm, *See* phenothiazine, 23618
fenoxyl, *See* Dinitra, 8540
Fenoxyl Carbon N, *See* Dinitra, 8540
Fenpropathrin, 11615, *See* Meothrin, 19322
fenpropidin, 11616
Fenpropimorph, *See* Corbel®, 6802
Fenpropimorph-Carbendazim, *See* Corbel® Duo, 6804
fensulfothion, *See* Dasanit®, 7787
fentanyl citrate-fluanisone, *See* Hypnorm, 14717
Fenthion, *See* Lebaycid®, 16932, Tiguvon®, 31788
Fentiazine, *See* phenothiazine, 23618
Fentin acetate-aneb, *See* Hytin, 14787
Fentin acetate-maneb, *See* Brestan, 4505
Fentin hydroxide, *See* Du-Ter, 9283, Farmatin, 11480
Fentin hydroxide, *See*
fentin hydroxide-metoxuron, *See* Endspray, 10554
Fenton's metal, 11617
Fenton's reagent, 11618
Fentro, 11619
fenvalerate, 11620
N-Fenyl-N'-isopropyl-p-fenylendiamin (czech), *See* Vulkanox® 4010 NA, 34147
Fenyrane, 11621
Feospan, *See* ferrous sulfate, 11690
F.E.P, 11375
Ferad, 11622
Feraloy, 11623
Fergapol, 11624
Fergatac, 11625
Fer-in-Sol, *See* ferrous sulfate, 11690
Ferlosa, 11626
fermentation alcohol, *See* ethyl alcohol, 11064
fermentation amyl alcohol, *See* Fusel oil, 12518
fermentation butyl alcohol, *See* isobutanol, 15344
Fermenticide, 11627
Fermenzyme® L-200, 11628
Fermet alloy, 11629
fermide, *See* Rezifilm, 26521
Fermide 850, *See* Rezifilm, 26521
Fermin, 11630
Fermine, 11631, *See* dimethyl phthalate, 8508, Kemester® DMP, 15956, Kodaflex® DMP, 16271, Unimoll® DM, 32979
Fernacol, 11632, *See* Rezifilm, 26521
Fernambuco wood, *See* Redwoods, 25987
Fernasan, 11633, *See* Rezifilm, 26521, Thiurad, 31730
Fernasan a, *See* Rezifilm, 26521
Fernasul, 11634
Fernesta, 11635
Fernex, 11636, 11637
Fernico Alloy, *See* Kovar® Alloy, 16372
Fernide, 11638, *See* Rezifilm, 26521
Fernimine, 11639
Fernol, 11640
Fernos, *See* Abol, 146, pirimicarb, 23879, Rapid, 25884
Fernox, 11641
Fernoxone, 11642
Fero-Gradumet, *See* ferrous sulfate, 11690
Ferophosphorus briquettes, *See* Losilphos, 17624
Ferox-Celotex, 11643
Ferozon, 11644
Ferquatac, 11645
Ferralium® Alloy 255, 11646
Ferrate(1-),, *See* sodium iron edetate, 28776
Ferrate(4-), hexakis(cyano-C)-, tetrasodium, (OC-6-11)-, *See* sodium ferrocyanide, 28763
Ferrax, 11647
Ferric ammonium EDTA, *See* Sequestrene® NH4Fe, 28002, Trilon® B-FA, 32326
ferric chloride, 11648
ferric chloride anhydrous, *See* ferric chloride, 11648
ferric chloride-calcium hypochlorite, *See* ferrochlor, 11664
ferric hydroxide, 11649, *See* ferrugo, 11700
ferric hydroxide oxide, *See* ferric hydroxide, 11649
ferric nitrate, 11650
ferric oxide, 11651 *See* Black 103, 4301, Blackox, 4307, Crocus Martius, 7062, Disperfin, 8666, Ferroxide, 11694, Foundrox, 12344, Hydroferrox®, 14586, Indian Ocher, 15067, Jeweller's rouge, 15611, Kroma Red, 16418, Rainbow Custom Colored Mortars, 25836, trip, 32358, Two Cubed Eight, 32638, sal mineral, 27248
ferric oxide red, *See* ferric oxide, 11651, Ferroxide, 11694, Jeweller's rouge, 15611
ferric oxide-linseed oil, *See* iron putty, 15319
ferric persulfate, *See* Elliott's Lawn Sand, 9848, ferric sulfate, 11652, Vitax Turf Tonic, 33960
ferric sesquioxide, *See* Ferroxide, 11694, Kroma Red, 16418
ferric sesquisulfate, *See* Elliott's Lawn Sand, 9848, ferric sulfate, 11652, Vitax Turf Tonic, 33960
ferric sodium edetate, *See* sodium iron edetate, 28776
ferric sodium EDTA, *See* sodium iron edetate, 28776
Ferric subcarbonate, *See* Apperitive Saffron of Iron, 2666
ferric sulfate, 11652, *See* Elliott's Lawn Sand, 9848, Elliott's Moss Killer, 9849, Greenmaster Autumn, 13342, Greenmaster Mosskiller, 13344, iron sulfate (ic), 15320, Maxicrop Moss Killer & Conditioner, 19078, Vitax Micro Gran, Vitax Turf Tonic, 33958, Vitax Turf Tonic, 33960, Monterey 30% Iron, 20286
ferric sulfate monohydrate, *See* Elliott's Lawn Sand, 9848, ferric sulfate, 11652, Vitax Turf Tonic, 33960
ferric tersulfate, *See* Elliott's Lawn Sand, 9848, ferric sulfate, 11652, Vitax Turf Tonic, 33960
ferric trichloride, *See* Ferric chloride, 11648
ferric trisulfate, *See* ferric sulfate, 11652, iron sulfate (ic), 15320
Ferri-Darotin, 11653
Ferrikalite, 11654
Ferriplex, 11655
Ferriplus, 11656
Ferrisul, 11657
ferrite, 11658
Ferro 66, *See* ferrous chloride, 11689
Ferro prussiate of potassium, *See* yellow prussiate of potash, 34647
ferro-aluminum, 11659, *See* aluminum iron, 1906
ferroaluminum silicate, 11660, *See* Ferrosil 14, 11682
ferro-argentan, 11661
ferro-boron, 11662
ferro-carbon-titanium, 11663
ferrochlor, 11664
Ferrochrome, *See* chrome iron, 6207
Ferrochrome lignosulfonate, *See* Borrewell FC, 4452
ferro-chromium, 11665
ferro-cobalt, 11666
ferrocobaltite, 11667
Ferrocrete, 11668
ferro-cupralium, 11669
Ferro-Cure, 11670
ferrodur, 11671
ferroferrite, *See* Magnetic iron ore, 18374
Ferrofloc, 11672, *See* ferrous chloride, 11689
Ferrofos 509, *See* Fostex AMP, 12337
Ferrogen, 11673
ferro-magnesite, 11674
ferro-manganese, 11675
ferro-molybdenum, 11676
ferron, 11677
ferro-nickel, 11678
ferronite, 11679
Ferrophos Pigment, 11680
ferro-phosphorus, 11681
Ferrosil 14, 11682, *See* ferroaluminum silicate, 11660
ferro-silicon, 11683
ferro-silicon-aluminum, 11684
ferrosoferric oxide, *See* Ferric oxide, 11651
Ferrostrane, *See* sodium iron edetate, 28776
Ferrostrene, *See* sodium iron edetate, 28776
Ferrotone, 11685
Ferrotubes, 11686
ferro-tungsten, 11687
ferro-uranium, 11688
ferrous acetate, *See* pyrolignite of iron, 25515
ferrous ammonium sulfate, *See* Mohr's salt, 20039
ferrous chloride, 11689, *See* Ferrofloc, 11672
ferrous fumarate-folic acid, *See* Fecap, 11572
ferrous gluconate, *See* Gluconal® FE, 13042
ferrous oxide, *See* glassite, 12971, glossite, 13013
ferrous sulfate, 11690, *See* green vitriol, 13337, Green-up Mossfree, 13346, Hart Lawn Sand, 13657, Hart Moss Killer, 13658, Iron vitriol, 15322
ferrous sulfate, *See* iron sulfate (ous), 15321, sal Martis, 27247, salt of steel, 27290, Shoemaker's black, 28184, Walkover Moss Killer, 34198
ferrous sulfate anhydrous, *See* iron sulfate (ous), 15321
ferrous sulfate heptahydrate, *See* ferrous sulfate, 11690, Iron vitriol, 15322, Taylors Lawn Sand, 30849
ferrous sulfide, *See* white pyrites, 34371
ferro-vanadium, 11691
ferro-vanadium (No. 205), 11692
Ferrox, 11693
Ferroxide, 11694
ferroxyl reagent, 11695
Ferrozell, 11696
ferro-zirconium, 11697
Ferrozoid, 11698
ferrozone, 11699
ferrugo, 11700
ferrul, 11701
ferrum, 11702
Ferrux, 11703
Ferry Alloy, 11704
Ferry metal, 11705
Fertibor, 11706
Ferti-lome, *See* Chlorothalonil, 6150, Repulse, 26151
Fertilox, 11707
Fervinal, *See* Sethoxydim, 28099
Fesofor, *See* ferrous sulfate, 11690
Fesotyme, *See* ferrous sulfate, 11690
Festoform, 11708
Fetid quartz, *See* stink quartz, 29769
Fetrilon® Combi, 11709
Fettel, 11710
fettorange 4a, lg, r, *See* Sudan I, 29938
Feuille Morte, 11711
Feximac, *See* bufexamac, 4655
Fezudin, *See* Diazinon Liquid, 8344
FFA-5, FFA-9, 11712
Fiba-Bond Cl, W, 11713
Fiber Pare, 11714
Fiber, Egyptian, *See* Vulcanized fiber, 34097
Fiberfil® J-1/30, 11715
Fiberfil® J-4/35, 11717
Fiberfil® J-7/33, 11720
Fiberfil® J-7/33/IT, 11721
Fiberfil® J-17/30/VO, 11716

Fiberfil® J-60/30/E8, 11718
Fiberfil® J-60/30/FR, 11719
Fiberfil® NY-16/MF/40, 11722
Fiberfil® NY-7, 11723
Fiberfil® NY-7/VO, 11724
Fiberfil® PP-60/TC/40, 11725
Fiberfrax® 6000 RPS, 11726
Fiberglas® 101C, 11727
Fiberil® M-1492, 11728
Fiberite, 11729
Fiberite 944, 11730
Fiberite 986, 11731
Fiberite 6070, 11732
Fiberite 7669, 11733
Fiberite 7701, 11734
Fiberite 9002, 11735
Fiberkal, 11736
Fiberlac, 11737
Fiberloc® 803GR10, 11738
Fiberloid, 11739
Fiberlon, 11740
Fiberod, 11741
Fiberoid, 11742, *See* Vulcanized fiber, 34097
Fiberstos, 11743
Fiberstran® G-1/50, 11744
Fiberstran® G-3/50, 11745
Fiberstran® G-4/45, 11746
Fibestos, 11747
Fibra-Cel®, 11748
Fibral, 11749
Fibralda, 11750
Fibravyl, *See* Korogel, 16354, Koroplate, 16357
fibrino-plastic substance, *See* Globulin,
fibrinoplastin, *See* Globulin,
Fibro, 11751
Fibroc, 11752
Fibron, 11753
Fibro-Silk Powd, 11754
Fibrotan, *See* hydrargaphen, 14539
Fibrotex, 11755
Fibrox, 11756
Fibrox 030 SC, 11757
Fibrox 300, 11758
Fibrredux, 11759
Ficam, 11760
FICAM D;, *See* Bendiocarb, 3937
Ficam Plus;, *See* Bendiocarb, 3937
Ficam ULV;, *See* Bendiocarb, 3937
Ficam W, *See* Bendiocarb, 3937
Ficel®, 11761
Ficel® AC2, 11762 , *See* Azodicarbonamide, 3524
Ficel® AZDN-LF, 11763
Fi-Chlor, 11764
fichtelite, 11765
Fi-Clor, 11766
Ficote, 11767
Fi-Cryl, 11768
Fiddle gum, *See* Gum tragacanth, 13490
Fielder®, 11769
Field's orange vermilion, *See* scarlet vermillon, 27550
Fierroso, *See* Acerado, 274
Fi-Gard, 11770
Filamid, 11771
Filastic, 11772
Filcryl, 11773
file bronze, 11774
Filester, 11775
Filex, 11776, *See* Propamocarb hydrochloride, 25171
Fi-Line, 11777
Filitox, *See* methamidophos, 19474 Tamaron®, 30760
Fillite, 11778, 11779
Fillite Hollow Microspheres, 11780
Fillite Solid Microspheres - PFA, 11781
Fillmaster, 11782
Fillpak, 11783
Film Plus, 11784
Filmerine, *See* sodium nitrite, 28801
Filmex, 11785
Filmite, 11786
Filmon®, 11787
Filofin, 11788
Filon, 11789
Filpro, 11790
Filt-char, 11791
Filter-cel, 11792
Filtracarb, 11793
Filtracite, 11794
Filtram, 11795
Filtrasorb® 100, 200, 11796
Filtrez, 11797
Filtrol, 11798
Filtros, 11799
Filtrosol A, 11800
Final Flip, 11801
Final Touch, 11802
Finaplix, 11803, *See* Trenbolone acetate, 32171
finasol osr (sub 2), *See* sodium lauryl sulfate, 28781
fine black, *See* aniline black, 2476
fine gold, 11804
Fine particle rubber, *See* CryOfine® Butyl, 7276
Fine particle rubber filler, *See* CryOfine® EPDM, 7277
fine silver, 11805
Fine-Clad®, 11806
Finesse®, 11807
Finestol, 11808
Finex-25, 11809
Finex-25-020, 11810
Finimal, *See* Valadol, 33238
finings, 11811
Finish®, 11812
Finistrol ESJ Concentrate, 11813
Finistrol GZ, 11814
Finitron, 11815
Finizym, 11816
Finnish turpentine, *See* turpentine, 32610
Finntitan, 11817
Fintex 572, 11818
Fiolax, 11819
Fiorivert, 11820
fir wool oil, 11821
Fire PRF+72 1000 FM, 11822
Fire Retardant FR-8, 11823
fire-armour, 11824
Firebrake ZB, 11825
Firebrake®, *See* Hydroboracite, 14563
Firebrake® ZB, 11826
Firecheck, 11827
fireclay, 11828
Firecol, 11829
Firecrete, 11830
fire-damp, 11831
FireGuard 910, 11832
FireMaster®, 11833
Firemaster® 642, 11834
Firemaster® HP-36, 11835
Firesaife, 11836
FireShield® H, 11837
Fireshield® H, HPM, HPM-UF, L, *See* antimony trioxide, 2574
FireShield® HPM, 11838
FireShield® L, 11839
Firit, 11840
Firmadent, 11841
Firmagral, 11842
Firnis, 11843
First Choice Electroless Palladium, 11844
Firthite, 11845
Fischer-Langbein solution, 11846
Fischer's reagent, 11847
Fischer's salt, 11848
Fischer's yellow, *See* Fisher's salt, 11848
Fischer-Tropsch wax, *See* Synwax, 30659
fisetin, 11849
fish gelatin, *See* isinglass, 15333
Fish oil derivative containing 40% C+72+70 and C+72+72 acids, *See* W.G.S. R-60 Z-5, 34188
Fisons 18-15, MCPB, 11850
Fisons P.C.P, 11851
Fi-Vi, 11852
Fixanal, 11853
Fixaplus, 11854
Fixapret®, 11855
Fixat, 11856
Fixatek, 11857
fixed white, 11858
Fixegal, 11859
Fixin, 11860
Fixinvar, 11861
Fixodent, 11862
Fixogene, 11863
Fixol, 11864
fixopone, *See* oil white, 22086
Fix-Sol, 11865
FK 140, 11866
FK 300DS, 11867
FK 500LS, 11868
FL7P, 11869
flake lead, *See* white lead, 34366
flake lead ceruse, *See* Kremser White, 16399
flake litharge, 11870
flake white, 11871, *See* pearl white, 22982
Flamarret®, 11872
Flamco, 11873
Flame Guard, 11874, *See* aluminum oxide, 1919
Flame Out #44, 11875
Flame Out CO, 11876
Flamegard® 908, 11877
Flamenco®, 11878
Flamenol, 11879
Flaming, 11880
Flammastik, 11881
Flammentin ASN, 11882
Flammentin PS, 11883
Flammex, 11884
Flammocite, 11885
Flamolin, 11886
flamprop-isopropyl, 11887
flamprop-isopropyl, *See* flamprop-isopropyl, 11887, Gunner, 13493, Power Flame, 24824
flamprop-M isopropyl, *See* Commando, 6665
Flamprop-methyl, *See* Lancer, 16653, Perfluidone, 23248
Flamprop-M-isopropyl, 11888, *See* Gunner, 13493
Flamtard H, 11889, *See* Zinc hydroxystannate, 34779
Flamtard S, 11890, *See* Zinc stannate, 34792
Flandrac, 11891
Flat-Ayd, 11892
Flat-Ayd® Bases, 11893
Flatting Agent OK412, 11894
Flavaxin, 11895, *See* vitamin B_2, 33935
Flavaxin, *See* riboflavin, 26722
flavaxin
Flavazol, 11896
Flaveosine, 11897
flavin, *See* riboflavin, 26722
Flavinduline, 11898
flavine, 11899
Flavocents, 11900
flavoline, 11901
Flav-O-Lok, 11902
flavone, 11903
flavopurpurin, 11904
Flavotint, 11905
Flav-R-Keep FP-51, 11906
Flavurol, *See* mercurochrome, 19354
flax Seed, *See* Linseed
flax wax, 11907
Flax*Seed* oil, *See* Linseed oil, 17309
FLC-2, 11908
Flea Flip, 11909
Flea-B-Gon, 11910
Flectol H, 11911
Flectol H, polymer, *See* Agerite® MA, 1019, Akrochem® Antioxidant DQ, 1160
Flectol ODP, 11912
Flectol Pastilles, 11913
Flectron, *See* cypermethrin, 7588, Topclip Parasol,

32037
Flee, *See* permethrin, 23384, 23385, 23386
Fletcher's alloy, 11914
Fletcher's bearing alloys, 11915
Fleur, 11916
Fleurelle®, 11917
Flex Carbon, 11918
Flex®, 11919
Flexade Regular, 11920
Flexalyn, 11921
Flexamine, 11922
Flexamine g, *See* Agerite® DPPD, 1016, N,N'-diphenyl-*p*-phenylenediamine, 8603, JZF, 15638, Naugard® J, 20796
Flexan® 130, 11923
Flexane®, 11924
Flexane® 80, 11925
Flexane® 94, 11926
Flexbond, 11927
Flexchlor, 11928
Flexcote, 11929
Flexcrete, 11930
Flexcryl, 11931
Flexel® 1010, 11932
Flexible epoxy compound, *See* Addaflex, 640
Flexible Fyrex®, 11933
Flexidor, 11934, *See* Knot Out, 16257, Ratio, 25897
Fleximel, *See* Bis(2-ethylhexy) Phthalate, 4228, dioctyl phthalate, 8549, Reomol DCP, 26130, Witcizer 312, 34454
Flexin, 11935
Flexipol® FP-100(M), 11936
Flexipol® FSF-106, 11937
Flexipol® NDTP-311-1.8, 11938
Flexipol® NP-311-2, 11939
Flexipol® NS-322-2, 11940
Flexizone GH, *See* Anti-oxidant 4010, 2581
Flexobond 329, 11941
Flexobond 423, 11942
Flexocel, 11943
Flexol, 11944
Flexol A 26, *See* dioctyl adipate, 8548, Plasthall® DOA, 24002, PX-238, 25440, Wickenol® 158, 34393
Flexol DOP, *See* Bis(2-ethylhexy) Phthalate, 4228, dioctyl phthalate, 8549, Reomol DCP, 26130, Witcizer 312, 34454
Flexol Plasticizer 10-A, *See* dioctyl adipate, 8548, Plasthall® DOA, 24002, PX-238, 25440, Wickenol® 158, 34393
Flexol Plasticizer 3GH, 11945
Flexol Plasticizer 3GO, 11946
Flexol Plasticizer A-26, *See* dioctyl adipate, 8548, Plasthall® DOA, 24002, PX-238, 25440, Wickenol® 158, 34393
Flexol Plasticizer Diop, *See* Bis(2-ethylhexy) Phthalate, 4228, dioctyl phthalate, 8549, Reomol DCP, 26130, Witcizer 312, 34454, Witcizer 313, 34455
Flexomer DFDA-1137 Natural 7, 11947
Flexone 3c, *See* Flexzone 3C, 11967, Permanax IPPD, 23364
Flexonyl, 11948
Flexoresin, 11949
Flexricin® 9, 11950, *See* Propylene glycol ricinoleate, 25222
Flexricin® 13, 11951
Flexricin® 15, 11952, *See* glycol ricinoleate, 13104
Flexricin® 17, 11953
Flexricin® 100, 11954
Flexricin® 115, 11955
Flexricin® 185, 11956
Flexricin® P-1, 11957
Flexricin® P-3, 11958
Flexricin® P-4, 11959, *See* Methyl acetyl ricinoleate, 19508
Flexricin® P-6, 11960
Flexricin® P-8, 11961, *See* glyceryl triacetyl ricinoleate, 13089
Flex-Shield, 11962
Flexsol 43, 11963
Flexthane 610 EXP, 611 EXP, 620 EXP, 11964
Flextron, 11965
Flexzone, 11966
Flexzone 3C, 11967, *See* N-Isopropyl-N'-phenyl-*p*-phenylene diamine, 15413, Vulkanox® 4010 NA, 34147
Flexzone 6-H, 11968
Flicker Flake™, 11969
Fligene CI, *See* cypermethrin, 7588, Topclip Parasol, 32037
Flint, 11970, 11971
flint alloy, 11972
flint glass, 11973
flint metal, 11974
flintcast, 11975
flinty zinc ore, 11976
Fliselina, 11977
Flit, 11978
Flixapret, 11979
Fl-Mo 5BMP, 11996
Fl-Mo 80/20, 11997
Fl-Mo 1082, 11998
Fl-Mo 1093, 11999
Fl-Mo DEL, DEH, 12000
Fl-Mo Lowfoam, 12001
Fl-Mo Suspend, 12002
Flo Chem Extra, 11980
Flo-Aid, 11981
float tin, 11982
floatstone, 11983
Flocculant T-9, 11984
Flochel, 11985
Flo-Cillin Aqueous, *See* Penicillin V, 23111
Flock-Lok® 850, 11986
Flock-Lok® 851, 11987
Floclean 103, 11988
Flo-Con, 11989
Flocon 100, 11990
Flocsil, 11991
Floex, 11992
Flogel, 11993
Flo-Guard, 12034
Flolys®, 11994
Flomac, 11995
Flor Sherry, 12003
Florafoam, 12004
Floramat, 12005
Florane, 12006
Floranid® N 32, 12007
Floranit, 12008
Floraquin, *See* Yodoxin, 34658
Florasan, *See* imazalil, 14892
Floratex, 12009
Florco®, 12010 , *See* Attapulgite, 3404
Florco® -X, 12011, *See* Attapulgite, 3404
Flordimex, *See* Cerone, 5756
Florel, *See* Cerone, 5756
Florel Fruit Eliminator, 12012
Florence lake, 12013
Florentine brown, 12014
Flores martis, *See* ferric chloride, 11648
Florex®, 12015 , *See* Attapulgite, 3404
Florex® Ag-Dri 6/30, *See* Attapulgite, 3404
Florex® Ag-Dri 6/30, LVM 8/16, RVM 8/16, 12016
Floricin, 12017
Florida phosphates, 12018
Floridine, *See* sodium fluoride, 28764
Florigel® H-Y, 12019, *See* Attapulgite, 3404
Florisil, 12020, *See* Celkate T-21, 5592, magnesium silicate, 18366
Florite, 12021
Florizine, *See* Floricin, 12017
Flor-Kleen, 12022, *See* Attapulgite, 3404
Florocid, *See* sodium fluoride, 28764
Florocyclene, 12023
Florosal, 12024
Florox, 12025
Flosol, 12026
Flotox, 12027
flour of sulfur, 12028
Flovan, 12029
flowers of antimony, 12030, *See* Octoguard FR-10, 22000, Timonox, 31807, UltraFine® II, 32799
flowers of Benjamin, 12031
Flowers of Benzoin, *See* flowers of Benjamin, 12031
flowers of bismuth, 12032
flowers of brimstone, *See* flowers of sulfur,
flowers of camphor, *See* Camphor,
flowers of copper, *See* Copper oxide,
flowers of tin, 12033, *See* tin ash, 31811, tin(IV) oxide, 31817
flowers of zinc, *See* Ken-Zinc®, 16050, K-Zinc, 16527, philosopher's wool, 23670, zinc oxide, 34786
flow-powder, 12035
Flowtron mosquito attractant, *See* Morillol®, 20335
Floxan SC-5211, 12036
flozenges, *See* sodium fluoride, 28764
fluates, 12037
Fluazifop-butyl, 12038, *See* Fusilade, 12519
Flube, *See* Ablusol ML, 129
Flubendazole, 12039
Flubenol, *See* Flubendazole, 12039
Flubenzimine, *See* Cropotex®, 7211
flucythrinate, *See* Tomahawk, 32001
Fludor solder, 12040
Fluf® 10-0-10, 12041
Flufenprop-isopropyl, *See* flamprop-isopropyl, 11887, Gunner, 13493, Power Flame, 24824
Flugne 113, 12042
FLUID 4, *See* MCPA, 19183
Fluid EEZ 1000, 12043
Fluid Flex, 12044
fluid gelatin, *See* oil pulp, 22082
Fluidiram, 12045
Fluilan, 12046
Fluilan AWS, 12047, *See* PPG-12-PEG-65 lanolin oil;, 24843
Fluisil® S55K, 12048
Fluisol, 12049
Fluitex, 12050
Flukanide, *See* Ranide, 25868
Flukiver, 12051, *See* Closantel, 6441
Flumoxal, *See* Flubendazole, 12039
Flumoxane, *See* Flubendazole, 12039
Fluocinonide, *See* Lidex, 17215
Fluolite, 12052
fluometuron, *See* Cottonex, 6894, Meturon® 4L, 19607
fluometuron-MSMA, *See* Croak, 7058
Fluon, *See* polytetrafluoroethylene, 24640, 24642
Fluon®, 12053
Fluon® AD1, AD1L, AD1H, 12054
Fluon® AD1, AD1L, AD1H, CDI, G170, *See* Polytetrafluoroethylene, 24642
Fluon® CDI, 12055
Fluon® G170, 12056
fluor, *See* fluorspar, 12113
Fluorad Surfactants, 12057
Fluorad® FC-24, 12059
Fluorad® FC-26, 12060
Fluorad® FC-93, 12064
Fluorad® FC-95, 12065
Fluorad® FC-118, 12058
Fluorad® FC-722, 12061
Fluorad® FC-724, 12062
Fluorad® FC-740, 12063
Fluorad® FX-8, 12067
Fluorad® FX-13, 12066
fluoral, *See* sodium fluoride, 28764
fluoram, 12068
Fluor-Amps, 12069
Fluoranar, 12070
3',6'-fluorandiol, *See* Fluorescein, 12080, uranine, 33115
Fluoranthrene, *See* Idryl, 14841
Fluoraz, 12071
fluorchrome, 12072
Fluorel® FC-2120, 12073
Fluorel® FC-2144, 12074

Fluorel® FC-2173, 12075
Fluorel® FC-2211, 12076
Fluorel® FT-2481, 12077
Fluorel® FX-9038, 12078
Fluoresbrite, 12079
Fluorescein, 12080
Fluorescein disodium salt, 12081
Fluorescein sodium, *See* Fluor-Amps, 12069, Fluorescein disodium salt, 12081, Fluorescite, 12086, Fluor-I-Strip, 12091
Fluoresceine, 12082
fluoresceine sodium, *See* uranine, 33115
Fluorescent, 12083
Fluorescent Blue, 12084
Fluorescent Red 5B, 12085
Fluorescite, 12086
Fluorescite, *See* Fluorescein disodium salt, 12081
Fluorfolpet, 12087
fluorident, *See* sodium fluoride, 28764
fluorigard, *See* sodium fluoride, 28764
Fluorinated sulfonamide-based powder, *See* Acticide APA, 464
Fluorine, 12088
fluorineed, *See* sodium fluoride, 28764
Fluorinert FC72, *See* Flutec PP1, 12121
Fluorinert Liquids, 12089
Fluorinse, 12090, *See* sodium fluoride, 28764
Fluor-I-Strip, 12091, *See* Fluorescein disodium salt, 12081
fluoritab, *See* sodium fluoride, 28764
fluorite, *See* calcium fluoride, 4947, fluorspar, 12113
Fluorl, 12092
Fluoroaliphatic ethoxylate, *See* Fluowet 40 M, 12114
Fluorobenzene, 12093
Fluorobenzenes, *See* Fluorobenzene, 12093
5-(*p*-fluorobenzoyl)-2-benzimidazolecarbamic acid methyl ester, *See* Flubendazole, 12039
[5-(4-Fluorobenzoyl)-1H-benzimidazol-2-yl]carbamic acid methyl ester, *See* Flubendazole, 12039
Fluorocarbon, *See* Aquaflim, 2692, McLube 1700, 19179
Fluorocarbon 12, *See* dichlorodifluoromethane, 8388
Fluorocarbon 11, *See* trichlorofluoromethane,
Fluorocarbon 113, *See* MS-180 Freon® TF Solv, 20415
Fluorocarbon 114, *See* Cryofluorane, 7278
Fluorocarbon 123, *See* Genetron® 123, 12823
Fluorocarbon 123, *See* Dichlorotrifluoroethane, 8393
fluorocid, *See* sodium fluoride, 28764
Fluorocomp® FC-101, 12094
Fluorocomp® FC-101, FC-144, FC-174, FC-182, *See* Polytetrafluoroethylene, 24642
Fluorocomp® FC-144, 12095
Fluorocomp® FC-174, 12096
Fluorocomp® FC-182, 12097
N-(Fluordichloromethylthio) phthalimid, *See* Fluorfolpet, 12087
Fluoroelastomer, *See* Adcora V, 633, Viton®, 33965, Viton® A, 33966, Viton® A-35, 33967, Viton® B, 33969, Viton® B-50, 33970, Viton® C-10, 33971, Viton® GFLT, 33978
Fluoroether Grease 834, 12098
Fluorofolpet, *See* Fluorfolpet, 12087
Fluoroform, 12099, *See* Genetron® HFC23, 12829
Fluoroglide, 12100
fluoroimide, *See* Sparticide®, 29211
fluor-o-kote, *See* sodium fluoride, 28764
fluorol, 12101, *See* sodium fluoride, 28764
Fluorol® Dyes, 12102
Fluorolene®, 12103
Fluorolube® GR-290, GR-362, GR-470, GR-544, GR-660, 12104
Fluorolux, 12105
Fluoromelt®, 12106 , *See* Polytetrafluoroethylene, 24642
Fluoromelt® FP-F-FMX1, 12107
fluoromide, 12108
Fluorophenol, 12109
4-fluorophenol, *See* Fluorophenol, 12109
***p*-fluorophenol**, *See* Fluorophenol, 12109
1-[(4-fluorophenyl)methyl]-N-[1-[2-(4-methoxyphenyl)ethyl]4-piperidinyl]-1H-benzimidazol-2-amine, *See* Hismanal, 14132
1-[[bis-(4-fluorophenyl)methylsilyl]methyl]-1h-1,2,4-triazole, *See* flusilazole, 12119
1-(4-fluorophenyl)-4-[4-(2-pyridinyl)-1-piperazinyl]-1-butanone, *See* azaperone, 3518
Fluorophlogopite, *See* mica, 19626
fluoroplast 4, *See* MS-122, 20413
Fluoropolymer, *See* Cytop, 7605
fluoros, *See* sodium fluoride, 28764
fluorosilicic acid, *See* hydrofluosilicic acid, 14587, keramyl, 16054
Fluorosilicone, *See* CF1-3510, 5833
Fluorosint® 500, 12110, *See* Polytetrafluoroethylene, 24642
Fluorotex, 12111
fluorotrichloromethane, *See* trichlorofluoromethane,
9-fluoro-11 β,17,21-trihydroxypregan-1,4-diene-3,20-dione 21-acetate, *See* isoflupredone acetate, 15362
Fluorotri-n-butyltin, *See* bioMeT TBTF, 4171
FluorSave, 12112
fluorspar, 12113, *See* calcium fluoride, 4947
fluosilicic acid, *See* montanine, 20271
Fluowet 40 M, 12114
Fluowet OL, 12115
Fluphenazine, 12116
Flupirtine maleate, *See* Katadolon, 15780
Flura, *See* sodium fluoride, 28764
flura-gel, *See* sodium fluoride, 28764
flura-loz, *See* sodium fluoride, 28764
flurcare, *See* sodium fluoride, 28764
fluroxypyr, 12117
fluroxypyr 1-methylheptyl ester, 12118
fluroxypyr-meptyl, *See* fluroxypyr 1-methylheptyl ester, 12118
flursol, *See* sodium fluoride, 28764
flusilazole, 12119, *See* Santion, 27396
Flutec, 12120
Flutec PP1, 12121
FLUTEC PP1, *See* Flutec PP1, 12121
Flutec PP2, 12122
Flutec PP3, 12123
FLUTEC PP3102, *See* Flutec PP3, 12123
Flutec PP9, 12124
FLUTEC PP9, *See* Flutec PP9, 12124
flutriafol, *See* Impact, 14908
Fluvermal, *See* Flubendazole, 12039
Fluvia, *See* Jelutong, 15574
Flux-Off®, 12125
Fluxol, 12126
fluzilazol, *See* flusilazole, 12119
Fly Fighter, *See* dichlorvos, 8394
Fly-Die, *See* dichlorvos, 8394
Flypel, *See* diethyl toluamide, 8438
Flytek, *See* methomyl, 19502
Flytrol, *See* Diazinon Liquid, 8344
FM 1132, 12127
FM 3510, 12128
FM 21288, 12129
F-MA 11®, 12130
FMB 65-15 Quat, 65-28 Quat, *See* Exameen 824 3724, 11284
FMB 65-15, 65-28, 12135
FMB 210-8, 210-15, 12131
FMB 302-8, 12132
FMB 302-8 Quat, *See* FMB 302-8, 12132
FMB 451-8 Quat, *See* JAQ Powdered Quaternary, 15517
FMB 500-15 U.S.P, 12133
FMB 504-5, 12134
FMB 6075-5, 6075-8, 12138
FMB 1210-5, 1210-8, 12136
FMB 3328-5, 3328-8, 12137
FMC 1240, *See* Ethion, 10955
FMC 4512, *See* pentanochlor, 23165
FMC 5462, *See* thiodan, 31680
FMC 5488, *See* tetradifon, 31366
FMC 9260, *See* Killgerm® Py-Kill W, 16182, tetramethrin, 31384
FMC 10242, *See* Rampart, 25860
FMC 30980, *See* cypermethrin, 7588
FMC 35001, *See* carbosulfan, 5245
FMC 33297, *See* Kafil, 15655, permethrin, 23384, 23385, 23386
foam tannin, 12139
FMC 45806, *See* cypermethrin, 7588
Foam Tint, 12140
Foamacure, 12141
Foamaster, 12142
Foamaster 340, 12143
Foamaster 371-S, 12144
Foamaster A, 12145
Foamaster AP, 12146
Foamaster NXZ, 12147
Foamaster Soap L, 12148, *See* sodium tallowate, 28824
Foamaster VC, 12149
Foam-Coll 4C, 12150
Foam-Coll 4CT, 12151
Foamer CD, 12152
Foamex AD-50, AD-100, AD-300, J-275, 12153
FoamFlush™, 12154
Foamid 117, 12155, *See* Cocamidopropyl dimethylamine propionate, 6554
Foamid AME-70, *See* Acetamide MEA, 283
Foamid AME-75, *See* Acetamide MEA, 283
Foamid AME-100, *See* Acetamide MEA, 283
Foamkill® 30 Series, 12156
Foamkill® 30HP, 12157
Foamkill® 400A, 12158
Foamkill® 608, 12159
Foamkill® 614, 12160
Foamkill® 618 Series, 12161
Foamkill® 634C, 12162
Foamkill® 639, 12163
Foamkill® 639JOH, 12164
Foamkill® 639Q, 12165
Foamkill® 649, 12166
Foamkill® 654NS, 12167
Foamkill® 663J, 12168
Foamkill® 80J Series, 12169
Foamkill® 810F, 12170
Foamkill® 836B, 12171
Foamkill® CMP, 12172
Foamkill® EFT, 12173
Foamkill® MS, 12174
Foam-Kon 20, 12175
Foamole A, 12176
Foamole B, 12177
Foamole M, 12178
Foamosul, 12179
Foampol LPS, *See* disodium lauryl sulfosuccinate, 8649
Foamquat IAES, 12180
Foam-Soy C, 12181
Foamtaine CAB, CAB-G, *See* Cocamidopropyl betaine, 6551
Foam-Wheat C, 12182
Fob metal, *See* brass, 4485
Focal, 12183
Focus, *See* Cycloxydim, 7544
Focus®, 12184
Fodel®, 12185
FOE 1976, *See* mefenacet, 19229
Foerdite, 12186
Fogard, *See* atrazine, 3394
foil lead, 12187
Foilcote, 12188
Foilgrip, 12189
folacid, *See* folic acid, 12196, vitamin M, 33951
folbal, *See* folic acid, 12196
Folcord, *See* cypermethrin, 7588, Topclip Parasol, 32037
folcysteine, *See* folic acid, 12196
Foldine, *See* vitamin M, 33951
Folettes, *See* folic acid, 12196, vitamin M, 33951
Folex 6EC, *See* Def®, 7876
Folex-P, 12190
Foliac Super Red, 12191
Folia-Feed, 12192

Foliafume, *See* rotenone, 27000
Foliamin, *See* folic acid, 12196, vitamin M, 33951
Foliar 36 Extra, 12193
Foliar Nitrophoska, 12194
Foliar TRIGGRR, 12195
folic acid, 12196, *See* Folvite, 12211 Mission Prenatal, 19945, vitamin M, 33951
Folicet, *See* vitamin M, 33951
Folicote Transpiration Minimizer, 12197
Folicur, 12198, *See* Raxil, 25908
Folidol, *See* parathion, 22807
Folidol® E605, 12199
Folidol® M, 12200
Folidol-M, *See* Parathion-methyl, 22808
Folimat, *See* Omethoate, 22155
Folimat®, 12201
Folin-Dennis solution, 12202
Folin-McEllroy sugar reagents, 12203
Folin's uranium acetate mixture, 12204
Folin's Uric Acid Reagent, 12205
Folio 575FW, 12206
Folipac, *See* folic acid, 12196, vitamin M, 33951
Folithion®, 12207
Folosan, 12208, *See* quintozene, 25692
Folpan, 12209, *See* Folpet, 12210
Folpel, *See* Folpet, 12210
Folpet, 12210, *See* Folpan, 12209
folpet-cupric sulfate, *See* Cupertine Folpet, 7363
Folpex, *See* Folpet, 12210
folsäure, *See* vitamin M, 33951
Folsan, *See* vitamin M, 33951
folsaure, *See* folic acid, 12196
folsav, *See* folic acid, 12196
foluite, *See* folic acid, 12196
Folvite, 12211, *See* vitamin M, 33951
Fomac, 12212, *See* hexachlorophene, 13992
Fomblin, 12213
Fome-Cor, 12214
Fomescol, 12215
fomitine, 12216
Fomox®, 12217
Fomrez, 12218
Fomrez® 50, 12219
Fomrez® 4393, 12220
Fomrez® A1228, 12221
Fomrez® ED400, 12222
Fomrez® EPD28, 12223
Fomrez® ET190, 12224
Fonderma, *See* sodium pyrithione, 28813
Fonganil, *See* Furalaxyl, 12487
Fongarid, 12225, *See* Furalaxyl, 12487
Fonofos, *See* Dyfonate, 9316
Fonoline® White, Yellow, 12226
Fontaine's powder, 12227
Food Blue 1, *See* Indigo Carmine, 15075
Food Orange 8, *See* Canthaxanthin, 5095
foots, 12228
foots oil, *See* foots, 12228
Foraflon® 1000 HD, 12229
Foral 85, 105 and AX, 12246
Forane, *See* Isoflurane, 15363
Foraperle®, 12230
Forate, *See* Terrathion, 31343
Forbes metal, 12231
Forbest 13, 12233
Forbest 50, 12234
Forbest 62B, 12235
Forbest 410, 12236
Forbest 780, 12237
Forbest 1000B, 12232
Forbest MW 23, 12238, *See* Microcrystalline wax, 19666
Forbest WP, 12239
Force, 12240
Forcite, 12241
Fordath Resins, 12242
Fore, *See* mancozeb, 18517
Forene, *See* Isoflurane, 15363
Forest Bark, 12243
For-Ester, 12244
Forex®, 12245
Foriod, *See* keraphen, 16055
Forlan, 12247
Forlan 200, 12248
Forlan C-24, 12249
Forlan C-24, *See* Fancol CH-24, 11436
Forlan L, 12250
Forlan LM, 12251
Forlanit P, 12252
Forlay, 12253
Forlkyl, 12254
Formac 40, 12255
Formacel®, 12256
Formadermine, *See* Guaiacol methylene ether,
formagen, 12257
Formal, *See* methylal, 19564
formaldehyde, 12258, *See* Durine, 9225, Dyna-Form, 9346, Lysoform, 18095
formaldehyde dimethyl acetal, *See* methylal, 19564
Formaldehyde ethyl cyclododecylacetal, *See* Boisambrene Forte, 4380
formaldehyde sodium sulfoxylate, *See* sodium formaldehydesulfoxylate, 28766
formaldehyde-ammonia, *See* hexamine, 14008
formaldehyde-ammonia 6:4, *See* hexamine, 14008
Formaldehyde-aniline, *See* Accelerator A1010, 191
Formaldehyde-anilines, *See* Accelerator A5-10, 195
formaldehydesulfoxylic acid sodium salt, *See* sodium formaldehydesulfoxylate, 28766
Formaldehyde-sulfurous acid, *See* Sulfiformin, 29989
formalin, 12259, *See* formaldehyde, 12258
Formalite, 12260
formalith, *See* formaldehyde, 12258
formamide, 12261
formamidinesulfinic acid, *See* formammidinesulfinic acid, 12262, thiourea dioxide, 31721
formammidinesulfinic acid, 12262
forman, 12263
Formanek's indicator, 12264
formaniline, 12265
Formapex, 12266
Format, 12267, *See* Dow Shield, 8887
Formax, 12268
Formel NF, 12269
formerly Heaveaplus MG, *See* Megapoly, 19237
Formex, 12270
formic acid, 12271, *See* Amasil®, 1946
Formic acid, sodium salt, *See* Formax, 12268, sodium formate, 28767
formic aldehyde, *See* formaldehyde, 12258
Formica, 12272
Formimidic Acid, *See* formamide, 12261
Formin, *See* Grasselerator 102, 13289, hexamine, 14008
Formit, *See* Bakelite, 3589
Formite, *See* Bakelite, 3589
Formitrol, *See* formaldehyde, 12258
Formkote, 12273
Formodac, 12274
formol, *See* formalin, 12259
Formol 55, 12275
formol-chloral, *See* formalin, 12259
formol-chlorl, *See* formalin, 12259
formolide, 12276
formolites, 12277
formolyptol, *See* formaldehyde, 12258
Formon®, 12278
formonitrile, *See* hydrocyanic acid, 14584, prussic acid, 25352, zootic acid, 34903
Formosa camphor, *See* camphor,
formose, 12279
Formosul, *See* Hydrosulfite NF, 14647
Formoxy methyl isolongifolene, *See* Amborate, 1971
Formrez, 12280
Formrez®, 12281
Formrez® 11, 12282
Formrez® L49-28, 12283
Formrez® T-279, T-280, 12284
Formula 111, 12286
Formula 90, 12285
Formula 300, *See* stearic acid, 29599
Formula 405, 12287
Formula AC, 12288
Formula S, 12289
Formusol®, *See* sodium formaldehydesulfoxylate, 28766
Formusol® SA, 12290
Formvar, 12291
N-Formyldimethylamine, *See* dimethyl formamide, 8506
***p*-formyl-N,N-dimethylaniline**, *See* dimethylamino benzaldehyde, 8515
7-formyl-5-isopropyl-2-methyl bicyclo (2.2.2)-oct-2-ene, *See* Maceal, 18119
***p*-formyltoluene**, *See* PTAL, 25375, *p*-tolyl aldehyde, 31961
Fornax, 12292
fornitrol, 12293
Forociben Premix, 12294
Foron, 12295
Forpen, *See* Penicillin G potassium, 23109
Forstan, *See* quinomethionate, 25686
For-Syn, *See* resmethrin, 26218
Fortafix, 12296
Forte, *See* phenothrin, 23619
Fortex, 12297
Fortex®, 1229 , *See* Microcrystalline wax, 19666 8
Fortiflex®, 12299
Fortiflex® A60-70-99, A60-70-119, 12300
Fortiflex® B45-06R-09, 12301
Fortiflex® G36-24-149, 12302
Fortiflex® G38-70C, 12303
Fortiflex® J36-25-142, 12304
Fortiflex® K36-55-122, 12305
Fortiflex® T50-200, 12306
Fortiflex® XF-855, 12307
Fortiflex® XF-855, 12308
Fortigro, *See* carbadox, 5177
Fortilene®, 12309
Fortilene® 1001, 12310
Fortilene® 1602, 12311
Fortilene® 1802, 12312
Fortilene® 2104, 12313
Fortilene® 3151, 12314
Fortilene® 4104, 4109, 12315
Fortilene® 4209, 12316
Fortilene® 5801, 12317
Fortilene® 9000, 12318
Fortimax, 12319
Fortisan, 12320
Fortodyl, *See* vitamin D_2, 33940
Fortol, *See* cyanazine, 7472
Fortress, 12321
Fortrex, 12322
Fortrol, 12323, *See* cyanazine, 7472
Fortron® 0205B4, 12324
Forturf, *See* Chlorothalonil, 6150, Repulse, 26151
Fosalsil, 12325
Foscast, 12326
Foset, 12327
Fosetyl, *See* Aliette, 1612
Fosetyl-aluminum, *See* Aliette, 1613
Fosfakol, *See* paraoxon, 22783
FOS-FALL "A", *See* Def®, 7876
Fosfalugel, *See* aluminum phosphate, 1922
Fosfamid, *See* Dimethoate, 8496
Fosfamide CPD-170, 12328
Fosfamide N, 12329
Fosferno, 12330
Fosferno, *See* parathion, 22807
Fosfil, 12331
Fosfothion, *See* malathion, 18460
foshagite, *See* calcium silicate, 4969
foshallasite, *See* calcium silicate, 4969
Foshell, 12332
Fosoil, 12333
Fospirate, *See* Chlorpyrifos-methyl, 6164, Torelle, 32058
fossil flour, *See* Celite®, 5586, infusorial earth, 15125
fossil salt, *See* Rock salt, 26887
fossil wax, *See* ozokerite, 22483
Fostap, 12334

Fumite Lindane, 12461
Fumite Permethrin, 12462, *See* Permethrin, 23385
Fumite Pirimiphos Methyl Smoke, 12463, 12464
Fumite Propoxur, 12465
Fumite Ronilan, 12466
Fumite TCNB, 12467
Fumite TCNB Smoke, 12468
Fumite Tecnalin, 12469
Fumyl-O-Gas, 12470
Functional polyorganosiloxane, *See* Coagulant WS, 6515
Fungacetin, *See* Kodaflex® Triacetin, 16281, triacetin, 32182
Fungaflor, 12471, *See* imazalil, 14892
Fungal lactase, *See* Fungal Lactase 100,000, 12472
Fungal Lactase 100,000, 12472
Fungal pectinase, *See* Pearex-L®, 22972, Pectinase AT, 23019
Fungal Protease Conc, 12473
Fungalysin™, 12474
Fungamyl, 12475
Fungazil, *See* imazalil, 14892
Fungex, 12476
Fungicide, *See* Albricide, 1387
Fungicide Fx, *See* Nuophene, 21783
Fungicide GM, *See* Nuophene, 21783
Fungicide M, *See* Nuophene, 21783
Fungicides, *See* Fungitrol, 12481
Fungi-Fluor, 12477
Funginex, 12478, *See* triforine, 32254
Fungisterol, 12479
Fungitex 656, 12480
Fungitox, *See* thiophanate methyl, 31706
Fungitrol, 12481
Fungitrol Tinox, 12482
Fungo, *See* thiophanate methyl, 31706
Fungo 50, *See* thiophanate methyl, 31706
Fungo®, 12483
fungol b, *See* sodium fluoride, 28764
Fungostop, *See* AAprotect, 16
Fungus Fighter, 12484
Furac No. 3, 12485
Furacarb, *See* Rampart, 25860
Furadan, *See* Carbodan, 5197, Rampart, 25860
Furadan 3G, *See* Carbodan, 5197, Rampart, 25860
Furadan 4f, *See* Rampart, 25860
Furadan 75 wp, *See* Rampart, 25860
Furadane, *See* Rampart, 25860
Furakem S, 12486
Furalaxyl, 12487
furalaxyl, *See* Fongarid, 12225, Furalaxyl, 12487
2-furaldehyde, *See* furfural, 12491, QO® Furfural, 25554
Furamazone, *See* Nifuraldezone, 21169
Furan, *See* QO® Furan, 25552
furan-2-carboxylic acid, *See* 2-furoic acid, 12497
Furanace, *See* Nifurpirinol, 21170
2-furancarbinol, *See* furfuryl alcohol, 12493
2-furancarboxaldehyde, *See* furfural, 12491
2-furancarboxylic acid, *See* 2-Furoic acid, 12497
Furanculine, 12488
2,5-Furandione, *See* maleic anhydride, 18476
2,5-Furandione, polymer with methoxyethylene, *See* Agrimer VEMA-H-240, 1068
2-furanmethanol, *See* furfuryl alcohol, 12493
2-(2-furanyl)-1H-benzimidazole, *See* fuberidazole, 12428
1-(2-furanyl)-N,N'-bis(2-furanylmethylene) methanediamine, *See* Vulcazol, 34112
Fura-Tone® NC-1012, 12489
Furazolidone, *See* Furoxone, 12499
Furbac, 12490
furfural, 12491, *See* QO® Furfural, 25554
furfural alcohol, *See* furfuryl alcohol, 12493
furfural resins, 12492
furfuramine, *See* Vulcazol, 34112
furfuryl alcohol, 12493, *See* QO® Furfuryl Alcohol (FA®), 25555
Furloe, *See* Chlorpropam, 6162
Furloe 3 EC, *See* Chlorpropam, 6162
Furlong®, 12494
furnace black, *See* carbon black, 5220
furnace-calamine, 12495
Furnex, 12496
7-furocoumarin, *See* Oxsoralen-Ultra, 22437
Furodan, *See* Rampart, 25860
Furoic acid, *See* 2-Furoic acid, 12497
2-Furoic acid, 12497
α-Furoic acid, *See* 2-Furoic acid, 12497
furoica, *See* 2-Furoic acid, 12497
Furotec, 12498
Furoxone, 12499
furpirinol, *See* Nifurpirinol, 21170
Furpyrinol, *See* Nifurpirinol, 21170
Furro P Base;, *See* *p*-aminophenol, 2185
Fursatil CS12, 12500
Fursatil CS15, 12501
Fursatil CS25, 12502
Fursatil CS30, 12503
Fursatil CS40, 12504
Fursatil CS60, CS65, 12505
Fursatil CS71, 12506
Fursatil CS81, 12507
Furuculin, 12508
furyl alcohol, *See* furfuryl alcohol, 12493
α-furylcarbinol, *See* furfuryl alcohol, 12493
2-furylcarbinol, *See* furfuryl alcohol, 12493
2-Furylmethanol, *See* furfuryl alcohol, 12493
Fusabond® MB-110D, 12509
Fusabond® MC-197D, 12510
Fusabond® MZ-109D, 12511
Fusabond® MZ-203D, 12512
Fusarex, 12513, *See* Tecgran, 30886, tecnazene, 30898
Fusariol, 12514
Fuscochlorin, 12515
Fuscorhodin, 12516
Fused calcium oxide, *See* Dynacal, 9337
fused cement, 12517
Fused magnesium oxide, *See* Magotex, 18406
Fused mullite, *See* Dynamullit, 9360
fusel oil, 12518
fusible salt, *See* microcosmic salt, 19665
fusible salt of urine, *See* microcosmic salt, 19665
Fusilade, 12519, *See* Fluazifop-butyl, 12038
Fusion®, 12520
Fussolon, 12521
Fustic, 12522
fustin, 12523
Futura Flex, 12524
Futura Thane, 12525
Futurit, 12526
FW 18, 12527
FW 50, *See* Nyad®, 21860, Wollastocoat®, 34545
Fw 200 (mineral), *See* Nyad®, 21860, Wollastocoat®, 34545
FW 200 Beads and Powd, 12528
FW 293, *See* Acarin, 183, Dicofol, 8398, Kelthane, 15881
FW 325, *See* Wollastocoat®, 34545
FW-734, *See* Propanil, 25174
FX Pastes, 12529
FX-512, 12530
Fyarestor, 12531
FYCOL 8, *See* Cupravit®, 7370
Fydulan, Fydumas, Fydusit, 12532
Fyfanon, 12533, *See* malathion, 18460
Fyran J2K, 12534
Fyraway, *See* antimony trioxide, 2574
Fyrebloc, *See* antimony trioxide, 2574
Fyrex, *See* Ammonium phosphate dibasic, 2250
Fyrex®, 12535
Fyrol CEF, *See* Genomoll P, 12838
Fyrol CF, *See* Genomoll P, 12838
Fyrol® 6, 12536
Fyrol® 38, 12537
Fyrol® CEF, 12538
Fyrol® DMMP, 12539, *See* dimethylmethyl phosphonate, 8525
Fyrol® FR-2, 12540
Fyrol® PBR, 12541
Fyrol® PCF, 12542
Fyrquel® 150, 12543
Fyrquel® EHC, 12544, *See* triaryl phosphate, 32195
Fytic acid, *See* phytic acid, 23777
Fytolan, *See* copper oxychloride, 6784, Cupravit®, 7370
Fytospore, 12545
Fyzol 11E, 12546
G 4, *See* Nuophene, 21783
G-4, *See* Dichlorophen, 8392
G-4 (Compound G4), *See* Nuophene, 21783
G-11, *See* hexachlorophene, 13992
G-623, 12549
G-635, 12550
G 665, *See* Carbendazim, 5187
G-1300, *See* Ablunol CO 10, 98
G 2129, *See* Ablunol 200ML, 95
G-2162, 12551
G-3300, 12552
G 3710, *See* Ablunol SA-7, 113
G 3720, *See* Ablunol SA-7, 113
G-3720-POE, *See* Ablunol SA-7, 113
G 3910, *See* Volpo 3, 34020
G 3920, *See* Volpo 3, 34020
G-4252, 12553
G-4280, 12554
G 27692, *See* simazine, 28464
G 30027, *See* atrazine, 3394
G-34161, *See* prometryn, 25141
G 34162, *See* ametryn, 2051
G 34360, *See* desmetryn, 8142
G 3694POE, *See* Ablunol SA-7, 113
G-24480, *See* Diazinon Liquid, 8344
G-27901, *See* trietazine, 32238
G-30028, *See* propazine, 25179
G-30320, *See* Clofazimine, 6427
GA3, *See* gibberellic acid, 12934, Regulex, 26019
G Resin, 12547
G salt, *See* glycine, 13095
G Varnish, 12548
G.P.V, *See* Penicillin V, 23111
g1v gard dxn, *See* dimethoxane, 8498
Gaardocyclene, 12555
Gabbett's stain, 12556
Gabbro, 12557
Gabian oil, 12558
Gabraster, 12559
Gabrosa, 12560
G-acid, 12561
Gadalan brands, 12562
Gadorgel, 12563
Gadose, 12564
Gad's cement, 12565
Gaduol, 12566
Gafac®, 12567
Gafamide CDD-518, 12568
Gafanol E 200, *See* Alkapol PEG 300, 1706
gafcol eb, *See* 2-butoxyethanol, 4753
Gafen LB-400, LE-500 and LS-500, 12569
Gafen LE-700, LP-700 and LK-500, 12570
Gafen LM-400, 12571
Gafen LM-600, 12572
Gaffix® VC-713, 12573
Gafgard 233 and 233E, 12574
Gafgard 238, 12575
Gafgard 245, 12576
Gafgard 277, 12577
Gafgard 280, 12578
Gafite, 12579
Gafite LW, 12580
Gaflex, 12581
Gafoam AD, 12582
Gafquat® 734, 12583, *See* Polyquaternium-11, 24530
Gafquat® 755, *See* Polyquaternium-11, 24530
Gafquat® 755N, *See* Polyquaternium-11, 24530
Gafquat® HS-100, 12584
Gafstat AD-510 and AE-610, 12585
Gafstat AS-610 and AS-710, 12586
Gaftuf, 12587
gagat, 12588

Chemical and Trade Names

Gala, 12589
galactan, 12590
Galactasol® Guar Derivs, 12591, *See* hydroxypropyl guar, 14666
galactitol, 12592, *See* Melampyrite, 19258
D-galactitol, *See* sorbitol, 29113
4-O-β-D-Galactopyranosyl-D-glucose, *See* α-D-Lactose, 16589
galactosidase, 12593
β-galactosidase, *See* galactosidase, 12593
4-(β-D-galactosido)-D-glucose, *See* α-D-Lactose, 16589
Galag, 12594
galagum, 12595
Galahad A, 12596
Galalith, 12597
Galam butter, 12598
Galaxy, *See* Bentazon, 3947
Galaxy®, 12599
Galben M, 12600
Galden, 12601
Galenite, 12602
Galesan, *See* Diazinon Liquid, 8344
Galettame silk, 12603
Galicar, 12604
Galipot, 12605
Gallal, 12606
Gallant, 12607
Gallatite, 12608
Gallery, *See* Knot Out, 16257, Ratio, 25897
Gallery One, 12609
Gallic acid, 12610
gallic acid propyl ester, *See* propyl gallate, 25204, Sustane® PG, 30436
Gallicin, 12611
Gallion, 12612
Gallipoli oil, 12613
Gallisin, 12614
Gallium, 12615
Gallium citrate-$_{67}$Ga, *See* Neoscan, 20963
Gallobromol, 12616
Gallochrome, *See* mercurochrome, 19354
gallotannic acid, *See* tannic acid, 30807
gallotannin, *See* tannic acid, 30807
Gallotox, *See* acetoxyphenylmercury, 306, Haloxon, 13567, phenylmercury acetate, 23646
Gallstone, 12617
Galorn, 12618
Galoryl, 12619
galoxone, *See* Haloxon, 13567
Galt glass, 12620
Galvanit, 12621
galvanized iron, 12622
Galvano Lac, 12623
Galvoline, 12624
Galvomag, 12625
Galvorod, 12626
Gamanase, 12627
Gamanil, *See* Lofepramin Hydrochloride, 17552
Gamasol 90, *See* dimethyl sulfoxide, 8511 DMSO, 8762
Gambia, *See* Jelutong, 15574
Gambier, *See* cutch,
Gambine R, *See* Dioxine, 8575
Gambir, *See* gum catechu, 13483
Gambir catechu, *See* gum catechu, 13483
Gamboge, 12628
Gamboge butter, 12629
Gamma acid, 12630
Gamma-BHC Dust, 12631
Gamma-Col, 12632
Gamma-HCH Dust, 12633
Gammalex, 12634
Gammalin, 12635
Gammasan, 12636
Gammatox, 12637
Gammatrol, 12638
Gammexane, 12639
Gamonil®, *See* Lofepramin Hydrochloride, 17552
Gamophene, *See* hexachlorophene, 13992
Gandural, *See* nuarimol, 21762
Ganex® Et-201, 12640
Ganex® P-904, 12641
Ganex® V-216, 12642, *See* PVP/hexadecene copolymer, 25431
Ganex® V-220, 12643, *See* PVP/eicosene copolymer, 25430
Ganex® WP-660, 12644
gangue, 12645
Ganicin, 12646
Ganister, 12647
Ganocide, 12648, *See* drazoxolan, 8974
Gansil, *See* Chloramine T, 6086, Ketjensept, 16133
Gant, 12649
Gantrez®, 12650
Gantrez® AN-119, 12651
Gantrez® AN-8194, 12652
Gantrez® B-773, 12653
Gantrez® ES-225, 12654
Gantrez® M-154, 12655, *See* Polyvinyl methyl ether, 24669
Gantrez® S-95, 12656
Gantrez® SP-215, 12657
Gantrez® V-215, 12658
garantose, *See* saccharin, 27195
Garbacryl, 12659
Garbritol, 12660
Garcinia oil, *See* Kokum Butter, 16302
Garcrete, 12661
Gardamide, 12662
Garden Lime, 12663
Gardenal Sodium, *See* Phenobarbitone sodium, 23600
Gardeniol, 12664
Gardentox, *See* Diazinon Liquid, 8344
Gardinol, 12665, *See* Rhodapon® 101-10, 26631, sodium lauryl sulfate, 28781, Teepol CM44, 30929, Teepol FC5, 30930, Teepol GD53, 30931, Teepol HB6, 30932, Teepol PB, 30933
Gardinox, 12666
Gardiquat 12H, 1450, 1480, SV 480, *See* Benzalkonium chloride, 3958
Gardlite, 12667
Gardol, *See* Crodasinic LS30, 7144, Sarkosyl NL30, 27465, sodium lauroyl sarcosinate, 28779, Zoharsyl L-30, 34876
Gardol®, 12668, *See* Maprosyl® 30, 18671, Medialan LD, 19214, Secosyl, 27823
Gardoprim, *See* Terbuthylazine, 31227
Gardoprim A 500FW, 12669
Garganine, 12670
Garj, 12671
Garlon, 12672, *See* Timbrel, 31800, Triclopyr, 32227
Garlon 4, 12674, *See* Triclopyr, 32227
garnet lac, *See* Lemon Lac, 16983
Garnex, 12675
Garoflam, 12676
Garomix, 12677
Garosorb, 12678
Garospers, 12679
Garox, *See* Benzoyl peroxide, 3983
Garozinc, 12680
Garpak, 12681
Garseal, 12682
Gartop, 12683
Gartube, 12684
Garvox, *See* Bendiocarb, 3937
Garvox 3G, 12685
gas black, 12686
Gas Blue, *See* soda blue, 28712
gas oil, 12687
Gasbinda, 12688
Gasil, 12689
Gasil EBC, EBN, 12690
Gaskoid, 12691
gasoline, solene, *See* petroleum ether, 23497
Gastex, 12692
Gastratox 6G Slug Pellets, 12693, *See* metaldehyde, 19431
Gastro Caloreen, 12694
Gastrodiagnost, *See* pentagastrin, 23146
Gastrografin, 12695
Gat 15, 12696
Gatodan 415, 12697
Gatorooter, *See* metam-sodium, 19444
Gaucho, 12698, *See* Confidor, 6725
Gauduin's fluid, 12699
gauging metal, 12700
Gaultheria oil, 12701
Gaultheric acid, 12702
Gauntlet, *See* nuarimol, 21762
Gauslinite, *See* Burkeite, 4708
Gazelle, 12703
GBL, 12704, *See* Butyrone, 4804
GDL, 12705
GDME, *See* monoglyme, 20227
GDUE, *See* Zoldine® ZT-55, 34887
GE 2557, 12706
Geax, 12707
Geblitol, 12708
Gecet F100, 12709
Gechophen, *See* Chlorphenesin, 6161
Gedanite, 12710
Gedeflex®, 12711
Gedelite®, 12712
Gedge's metal, *See* Aich metal, 1095
gedrite, 12713
Gefir, *See* Nuophene, 21783
Geigy 27,692, *See* simazine, 28464
Geigy 30,027, *See* atrazine, 3394
Geigy industrial chemical, *See* Uvazol 236, 33181
Geko, 12714
Gel Flo, 12715
Gel II, 12716, *See* sodium fluoride, 28764
Gel Power, 12717
gel rubber, 12718
Gelamite D, 12719
Gelaprime F, 12720
Gelatase, 12721
gelatin, 12722, *See* Crodyne BY19, 7182, Pharmagel, 23572
gelatin carbonite, 12723
Gelatin NF, *See* Byco A, C, O, 4818
gelatin, vegetable, 12724
Gelatinastralite, 12725
Gelatine Donarit 1,2,3, 12726
Gelatine Donarit 2 E, 12727
Gelatine Donarit S, 12728
Gelatinous ammon dynamites, *See* Gelatine Donarit 1,2,3, 12726
gelato-glycerin, 12729
Gelcharg, 12730
Gelcotar®, 12731
Gelcotar® Liquid, 12732
Geleol, 12733
Gelflex, 12734
Gelfoam, *See* gelatin, 12722, Pharmagel, 23572
Gelkyd, 12735
Gellan gum, 12736, *See* Gelrite®, 12746, Kelco-Gel® Gellan Gum, 15827
Gelline, 12737
Gelobel, 12738
Gelocatil, *See* Valadol, 33238
Gelose, *See* Agar (Agar-agar), 955, galactan, 12590, vegetable gelatin, 33518
Gelosine, 12739
Gelot 64®, 12740
Geloxite, 12741
Geloy® BG10, 12742
Geloy® GY1020, 12743
Geloy® GY1220, 12744
Geloy® XP1001, 12745
Gelrite®, 12746 , *See* Gellan gum, 12736
Gelsorb B, 12747, *See* Attapulgite, 3404
Gelucire 35/10, 12748
gelucystine (L form), *See* cystine, 7603
Gelumina, *See* aluminum hydroxide, 1905
gelution, *See* sodium fluoride, 28764
Gelva, 12749

Gelvatol, 12750
Gemalgene, *See* trichloroethylene, 32217
Gemex, 12751
Gemfibrozil, *See* Lopid, 17599
Gemglo, 12752
Gemlite, 12753
Gemme, 12754
Gemstone, 12755
Gemstone M.1.2, 12756
Gemtex SC-40, *See* dioctyl sodium sulfosuccinate, 8550
Gemtone®, 12757
Genaden TA-080, *See* PEG tallate series, 23035
Genagen C-100, 12758
Genagen CA-050, 12759
Genagen CA 818, CAB, *See* Cocamidopropyl betaine, 6551
Genagen O-090, 12760
Genagen P-070, 12761, *See* PEG-7 palmitate, 23051
Genagen PL-090, 12762, *See* PEG-9 pelargonate, 23054
Genagen S-080, 12763
Genagen TA-080, 12764
Genal P4300-CM, 12765
Genamid®, 12766
Genamid® 151, 12767
Genamin, 12768
Genamin C Grades, 12769
Genamin CTAC, 12770, *See* Cetrimonium chloride, 5814
Genamin DSAC, 12771, *See* Adogen® TA-101, 780, Genamin DSAC, 12771
Genamin KDM-F, 12772, *See* Behentrimonium chloride, 3898
Genamin KSE, 12773
Genamin KSL, 12774
Genamin T-020, 12775
Genamin TA Grades, 12776 *See* tallowamine, 30748
Genamine C-020, 12777
Genaminox CS, 12778
Genaminox KC, 12779, *See* Cocamine oxide, 6557
Genapol 0 100, *See* Volpo 3, 34020
Genapol O, *See* Volpo 3, 34020
Genapol O 020, *See* Volpo 3, 34020
Genapol O 050, *See* Volpo 3, 34020
Genapol O 080, *See* Volpo 3, 34020
Genapol O 120, *See* Volpo 3, 34020
Genapol S, *See* Ablunol SA-7, 113
Genapol S 020, *See* Ablunol SA-7, 113
Genapol S 150, *See* Ablunol SA-7, 113
Genapol®, 12780
Genapol® 2299, 12784
Genapol® 24-L-3, 12781
Genapol® 24-L-3, 42-L-3, *See* Laureth-2, 16852
Genapol® 26-L-1, 12782, *See* C12-16 pareth-1, 4841
Genapol® 42-L-3, 12783
Genapol® AMS, 12785
Genapol® ARO, 12786
Genapol® C-050, 12787, *See* Coceth-27, 6558
Genapol® DA-040, 12788, *See* Deceth-6, 7848
Genapol® GC-050, 12789, *See* Coceth-27, 6558
Genapol® LRO Liq, Paste, 12790
Genapol® O-020, 12791, *See* oleth-8, 22126
Genapol® PF 10, 12792
Genapol® PGM Conc, 12793
Genapol® PL 120, 12794
Genapol® PMs, 12795
Genapol® S-020, 12796
Genapol® T Grades, 12797
Genapol® TS Powd, 12798, *See* PEG-150 distearate series, 23037
Genapol® TSM, 12799
Genapol® UD-030, 12800
Genapol® V 2908, 12801
Genapol® X-040, 12802, See Trideceth series, 32232
Genapol® ZRO Liq, Paste, 12803
Genasco, 12804
Genatosan Skin Bar, 12805
Gen-che, 12806
Genclor, 12807
Gendriv 162, 12808
Gendriv 492S, 12809
Genelit, 12810
Genep, *See* EPTC, 10729
General aniline & film, *See* Uvinul® DS-49, 33196
Generol® 122, 12811
Generol® 122E5, 12812
Generon, 12813
Genesolv A Solvent, 12814
Genesolv D Solvent, 12815
Genetron, *See* chlorotrfluoromethane, 6153
Genetron 12, *See* dichlorodifluoromethane, 8388
Genetron Dry Refrigerants, 12816
Genetron® 11, 12817
Genetron® 12, 12818
Genetron® 13, 12819
Genetron® 22, 12820
Genetron® 113, 12821
Genetron® 114, 12822, *See* Cryofluorane, 7278
Genetron® 123, 12823, *See* Dichlorotrifluoroethane, 8393
Genetron® 134a, 12824
Genetron® 141b, *See* Dichlorofluoroethane, 8391
Genetron® 500, 12825
Genetron® 502, 12826
Genetron® 503, 12827
Genetron® HFC23, 12829
Genetron® HFC 125, 12828
Genetron® Refrigerant 32/125, 12830
Genetron® Refrigerant 125/143a, 12831
Genie®, 12832
Genisol, 12833
Genitox, *See* DDT, 7820
Genitron, 12834, *See* Azodicarbonamide, 3524
Genitron AC, *See* Azodicarbonamide, 3524
Genitron AC 2, *See* Azodicarbonamide, 3524
Genitron AC 4, *See* Azodicarbonamide, 3524
Genkiene, 12835
Genoa oil, 12836
Genochrome, 12837
Genol, *See* Metol, 19584
Genomoll P, 12838
Genotherm, 12839
Genoxide®, 12840
Genpol O 050, *See* Volpo 3, 34020
Gensil, 12841
Genster, 12842
gentersal, *See* gentian violet, 12845
Genthane SR. (GS 338), 12843
gentheivite, 12844
Gentian Violet, 12845, *See* Pyoctanin, 25455
gentianic acid, *See* gentisin, 12847
Gentiannie, 12846
Gentiaverm, *See* gentian violet, 12845
gentisin, 12847
Gentran, *See* Dextran, 8233
Genu®, 12848
Genuzan, 12849
Geolast®, 12850
Geolite, 12851
Geolith, 12852
Geomet, *See* Terrathion, 31343
Geon, *See* Korogel, 16354, Koroplate, 16357
Geon® 8700A, 12853
Geon® 8720, 12854
Geon® 8812, 8813, 12855
Geon® 8896, 12856
Geon® 83457, 12857
Geon® 83718, 12858
Geon® 86100, 86101, and 86103, 12859
Geon® 87239, 87241, 12860
Geon® 87396, 12861
Geon® 87420, 12862
Geon® HTX 92190, 12863
Geon® HTX-6110, 12864
Geon® W015, 12865
Geonter, *See* terbacil, 31220
Geopan SF365, *See* Ablunol CO 10, 98
Georgia Gulf 3131 Clear 02, 12866
Georgia Gulf 5006, 12867
Georgia Gulf 5006 General, 12868
Georgia Gulf 9105, 12869
Georgia Gulf 9151, 12870
Georgia Gulf 9175J, 12871
Georgia Gulf 9202, 12872
Georgia Gulf CL-7049, 12873
Georgia Gulf EH-71L, 12874
Georgia Gulf EX-240, 12875
Georgia Gulf HH-1900, 12876
Georgia Gulf HM-7054, 12877
Georgia Gulf SP-7107, 12878
Georgia Gulf UV-7160, 12879
Geoseal, 12880
Geosol, *See* ferrous sulfate, 11690
Geostone®, 12881
Geostop, 12882
Gepel, 12883
Geracryl, 12884
geranial, *See* citral (*cis* and *trans*), 6336
Geraniol, 12885
geraniol acetate, 12886, *See* Geranyl acetate, 12889, Meraneine, 19330
geranium crystals, 12887
geranium oil, 12888
Geranyl acetate, 12889, *See* geraniol acetate, 12886
geranyl alcohol, *See* Geraniol, 12885
Gerhardt's caustic, 12890
Germaben® II, 12891
Germalgene, 12892, *See* trichloroethylene, 32217
Germall 115, *See* Abiol, 69, Germall® 115, 12893
Germall® 115, 12893, *See* Imidazolidinyl urea, 14896
Germall® II, 12894, *See* Diazolidinyl urea, 8348
Germa-medica, *See* hexachlorophene, 13992
German silver, *See* nickel silvers, 21132
German silver solder, 12895
German turpentine, *See* turpentine, 32610
German yeast, 12896
germane, *See* monogermane, 20225
Germanin, *See* Naganol, 20660
Germanium, 12897
Germanium tetrahydride, *See* monogermane, 20225
Germinol, *See* Variquat® 50MC, 33397, Zephiran Chloride, 34728
Germitol, *See* Variquat® 50MC, 33397, Zephiran Chloride, 34728
Germ-i-Tol, 12898
Germul, 12899
Germul A 735, 12900
Geronol, 12901
Geronol AG-100/200 Series, 12902
Geronol® ACR/4, *See* disodium lauryl sulfosuccinate, 8649
Geropon® ACR/4, 12903
Geropon® ACR/4, SBFA-30, *See* disodium lauryl sulfosuccinate, 8649
Geropon® AS-200, 12904
Geropon® CYA/60, 12905
Geropon® CYA/DEP, 12906
Geropon® DOS, 12907, *See* dioctyl sodium sulfosuccinate, 8550
Geropon® LSS, 12908
Geropon® MLS/A, 12909, *See* sodium methallyl sulfonate, 28790
Geropon® S-1585, 12910
Geropon® SBFA-30, 12911
Geropon® SBL-203, 12912, *See* disodium lauramido MEA-sulfosuccinate, 8647
Geropon® SBR-3, 12913, *See* disodium ricinoleamido MEA-sulfosuccinate, 8654
Geropon® SS-L7DE, 12914
Geropon® T-22/A, 12915
Geropon® T-33, 12916
Geropon® TC-42, 12917, *See* sodium methyl cocoyl taurate, 28791
Geropon® TK-32, 12918, *See* sodium methyl tall oil acid taurate, 28793, 28794
Geropon® WS-25, WS-25-I, 12919, *See* sodium dinonyl sulfosuccinate, 28759

Geropon® X2152, 12920
Gesafid, *See* DDT, 7820
Gesafloc, *See* trietazine, 32238
Gesagard, 12921, *See* prometryn, 25140, 25141
Gesagard 50, *See* prometryn, 25140
Gesagarde 50 Wp, *See* prometryn, 25140
Gesamil, *See* propazine, 25179
Gesamoos, *See* chloroxuron, 6158, Tenoran, 31119
Gesapax, *See* ametryn, 2051
Gesapon, *See* DDT, 7820
Gesaprim, *See* atrazine, 3394
Gesaprim 50, *See* atrazine, 3394
Gesaprim 500, *See* atrazine, 3394
Gesaprim 500FW, 12922
Gesapun, *See* simazine, 28464
Gesarex, *See* DDT, 7820
Gesarol, *See* DDT, 7820
Gesatop, 12923, *See* simazine, 28464
Gesatop-50, *See* simazine, 28464
Gesektin K, *See* ronnel, 26951
Gesilit, 12924
Gesoprim, *See* atrazine, 3394
Gesteins-tremonit V, 12925
Gesteins-Westfalit B and C, 12926
Gesterol 50, 12927
Gestinal®, 12928
Getah wax, *See* Java wax, 15539
Getren® 4/200, 12929
Getren® FD 575, 12930
Geusapon, *See* DDT, 7820
Gevral, 12931
GFS®, 12932
GH, *See* Nuophene, 21783
GH5, 12933
Gibb-3-ene-1,10-dicarboxylic acid, 2,4a,7-trihydroxy-1-methyl-8-methylene-, 1,4a-lactone, (1α,2β,4aα,4bβ,10β)-, *See* Regulex, 26019
gibberellic acid, 12934
Gibberellic acid, *See* Activol, 506, Regulex, 26019
Gibberellin, *See* Regulex, 26019
Gibberellin 1, *See* Regulex, 26019
gibberellin A_3, *See* gibberellic acid, 12934, Regulex, 26019
gibberellin X, *See* gibberellic acid, 12934, Regulex, 26019
gibbrel, *See* Regulex, 26019
Gibb-tabs, *See* Regulex, 26019
Gibrel, *See* Regulex, 26019
gib-sol, *See* Regulex, 26019
gib-tabs, *See* Regulex, 26019
Giemsa's stain, 12935
Gilalgin, 12936
Gildent, 12937
gilding metal, 12938
gilding solutions, 12939
Gillebachite, *See* Wollastocoat®, 34545
Gilsonite, 12940, 12941
Gilsonite and Design, 12942
Gilstone, 12943
Gilumag, 12944
ginal, 12945
Gingelly oil, 12946, *See* Sesame oil, 28075
ginger grass oil, *See* Rose+a7 oil, 26968
Gingili oil, *See* Gingelly oil, 12946, Sesame oil, 28075
Gingivit, *See* Nuophene, 21783
Ginsene, 12947
Gin-shi-bui-chi, 12948
Gippon, 12949
Girard's reagent T, 12950
Githagin, *See* struthiin, 29840
Givgard DXN, 12951
Givsorb® UV-1, 12952, *See* Ethoxy carbonyl phenylmethyl phenylformamidine, 11043
Givsorb® UV-2, *See* Ethoxy carbonyl phenylethyl phenylformamidine, 11042
gjellebaekite, *See* calcium silicate, 4969
glacial acetic acid, *See* Acetic acid, 292
glacial acrylic acid, *See* acrylic acid, 412
Gladiator, 12953, *See* chloridazon, 6103
Glagerite, 12954
Glance pitch, *See* Manjak, 18566
Glanzan PHN Conc, 12955
Glascol®, 12956
Glascol® HN2, 12957
Glascol® HN4, 12958
Glascol® PA6, 12959
Glascol® PA8, 12960
Glascol® PN 8, 12961
Glasdag, 12962
Glaser's salt, 12963
Glasgro, 12964
Glaskyd®, 12965
Glass gall, *See* sandiver, 27326
Glass Guard, 12966
Glass H, 12967
Glass Liquor, *See* soluble glass, 28980
glass silk, 12968
Glass Sponge, 12969
Glass, Water, *See* soluble glass, 28980
Glassclad® 18, 12970
glassite, 12971
glass-maker's soap, 12972
Glassona, 12973
Glassy sodium, *See* Glass H, 12967, Sodaphos, 28721, sodium hexametaphosphate, 28770
Glassy sodium hexametaphosphate-sodium chloride-sodium erythorbate, *See* Freez-Gard® Formula FP-88E, 12386
glassy sodium metaphosphate, *See* Calgon, 4987, Kalex HMP, 15670, Oilfos, 22089
Glauber's salt, 12974
glauconic acids, 12975
Glaucosil, 12976
Glauramine, 12977
Glaurin, 12978
Glazamine, 12979
Glaze N Seal Waterbase Clear Concrete and Brick Sealer, 12980
Glaze 'N Seal Concrete and Masonry Sealer, 12981
Glazier's salt, 12982
Glean® TP, 12983
Gleem, *See* sodium fluoride, 28764
Glekosa, *See* Chloramine T, 6086
Glendion, 12984
Glessite, 12985
glialka, *See* glyphosate, 13148
Glievor bearing metals, 12986
Glifonox, *See* Rodeo, 26899
glimmer, 12987
glimmer glist, *See* mica, 19626
Glissofluid® A 10, A 13, 12988
Glissopal®, 12989
Glissosafe®, 12990
Glissoviscal® B, 12991
glist, *See* glimmer, 12987, mica, 19626
Glitzi, 12992
Glizarin Binder, 12993
Glissolube®, 12994
GLOB®, 12995
Globe Granite, 12996
globol, *See* *p*-dichlorobenzene, 8386
globularicitrin, *See* rutin trihydrate, 27150
Glocure, 12997
Glofoam, 12998
Glokem, 12999
Glokill, 13000
Glokill 77, 13001
Glokill PQ, 13002
Glomeen, 13003
Glo-Mold®, 13004
Glonoine oil, 13005
Glopol, 13006
Glopol 461, 13007
Gloquat 1032, 13008
Gloquat C, *See* Country Fresh Disinfectant, 6905
Gloquaternary, 13009, 13010, 13011
Gloria®, 13012
glossite, 13013
Glossova, 13014
Glover's wool, *See* Tanner's wool, 30804
Glowtein, 13015
Gloy, 13016
Gluadin® AGP, 13017
glucal, 13018
Glucal, *See* Neo-Calglucon, 20877
Glucam® E-10, 13019, *See* methyl gluceth-10, 19535
Glucam® E-20, 13020, *See* methyl gluceth-20, 19536
Glucam® E-20 Distearate, 13021, *See* methyl gluceth-20 distearate, 19537
Glucam® P-10, 13022
Glucam® P-20, 13023
Glucam® P-20 Distearate, 13024
Glucam® P-20, P-10, *See* PPG methyl glucose ether series, 24842
Glucamate® DOE-120, 13025
Glucamate® SSE-20, 13026, *See* PEG-20 methyl glucose sesquistearate, 23046
Glucanal, 13027
β-Glucanase, *See* Glucanex® L-300, 13029
Glucanex, 13028
Glucanex® L-300, 13029
Glucarolactone, *See* GDL, 12705
glucase, 13030
Glucate® DO, 13031, *See* methyl glucose dioleate, 19538
Glucate® IS, 13032
Glucate® SS, 13033, *See* methyl glucose sesquistearate, 19539
glucid, *See* saccharin, 27195
Glucidex, 13034
Glucina, 13035
glucitol, *See* sorbitol, 29113
D-glucitol, *See* A-625/641ABS 301K, ABS 500FR-1, 10, sorbitol, 29113
D-Glucitol, anhydro-, monooctadecanoate, *See* Ablunol S-60, 110
Gluckauf, 13036
Glucoamylase, *See* Agrilan® AEC266, 1056, Diazyme® L-200, 8352, Distillase® L-200, 8716
Glucobiogen, *See* Neo-Calglucon, 20877
Glucodin, 13037
Glucolin, *See* glucose, 13052
Gluconal® CA A, CAM, 13038
Gluconal® CA M B, 13039
Gluconal® CO, 13040
Gluconal® CU, 13041
Gluconal® FE, 13042
Gluconal® K, 13043, *See* Potassium gluconate, 24755
Gluconal® MG, 13044
Gluconal® MN, 13045
Gluconal® NA, 13046
Gluconal® ZN, 13047, *See* Zinc gluconate, 34776
gluconate, *See* chlorhexidine digluconate, 6101
Gluconic Acid, 13048
D-Gluconic acid, *See* Gluconic Acid, 13048
D-gluconic acid calcium salt, *See* calcium gluconate,
D-Gluconic acid, δ-lactone, *See* GDL, 12705
D-gluconic acid magnesium salt, *See* magnesium gluconate, 18362
D-gluconic acid monosodium salt, *See* sodium gluconate, 28769
D-gluconic acid potassium salt, *See* potassium d-gluconate, 24752
Gluconic acid lactone, *See* GDL, 12705
Gluconic acid potassium salt, *See* Gluconal® K, 13043, Katorin, 15798, Potassium gluconate, 24755
gluconic acid sodium salt, *See* sodium gluconate, 28769
Gluconic lactone, *See* GDL, 12705
Glucono delta lactone, *See* GDL, 12705
D-(+)-Glucono-1,5-lactone, *See* GDL, 12705
Gluconolactone, *See* GDL, 12705
δ-Gluconolactone, *See* GDL, 12705
D-Glucono-δ-lactone, *See* GDL, 12705
Glucono δ-lactone, *See* GDL, 12705
Gluconsan K, *See* Gluconal® K, 13043, Katorin, 15798, potassium d-gluconate, 24752

cocoglycerides, *See* Cutina® CBS, 7436
Glyceryl tri(acetoxystearate), *See* Paricin® 8, 22841
Glyceryl triacetate, *See* Kodaflex® Triacetin, 16281, triacetin, 32182
Glyceryl triacetyl hydroxystearate, *See* Hetester HCA, 13895, Naturechem® GTH, 20771
glyceryl triacetyl ricinoleate, 13089, *See* Baker P-8, 3596, Naturechem® GTR, 20772
Glyceryl tribehenate, *See* Syncrowax HR-C, 30534, Syncrowax HRS-C, 30535
Glyceryl tribenzoate, *See* Uniplex 260, 33015
Glyceryl tricapryl caprylate, *See* Crodamol GTC/C, 7121
Glyceryl tricaprylate-caprate, *See* Radiamuls® MCT 2108, 25776
Glyceryl triheptanoate, *See* Radiamuls® GTH 2375, GTH 2376, 25774
Glyceryl tri-laurate, *See* Trilaurin, 32317
glyceryl trinitrate, *See* nitroglycerin,
Glyceryl trioctanoate, *See* Captex® 8000, 5170, Emalex O.T.G, 9976, Miglyol® 808, 19736, tricaprylin, 32211
Glyceryl trioleate, *See* Hodag GTO, 14185, Kemester® 1000, 15938, Priolube 1435, 25019, Radia® 7363, 25739, triolein, 32351
Glyceryl tristearate, *See* Kemester® 5500, 15947, Kemester® 6000, 15952, Kemester® GMS (Powd.), 15962, Kemester® 5500, 15947, Kemester® 6000, 15952
Glyceryl-lactostearate, *See* Atmul 2622K, 3372
DL-glyceryl-1-mono-octanoate, *See* Witafrol® 7420, 34421
Glycidoxy propyl trimethoxysilane, *See* Aktisil EM, 1202
glycidoxypropyl methyl diethoxysilane, 13090
3-Glycidoxypropyl trimethoxysilane, *See* Prosil® 5136, 25240, Dow Corning® Z-6040, 8881
3-Glycidoxypropylmethyldiethoxy silane, *See* CG6710, 5835, glycidoxypropyl methyl diethoxysilane, 13090, GP-137, 13221
γ-Glycidoxypropyltrimethoxy silane, *See* CG6720, 5836,Prosil® 5136, 25240 Union Carbide® A-187, 32988
σ-Glycidoxypropyltrimethoxysilane, *See* Aktisil EM, 1202, CG6720, 5836, Prosil® 5136, 25240
3-Glycidoxypropyltrimethoxysilane, *See* CG6720, 5836, Dynasylan® GLYMO, 9408
glycidyl acrylate, 13091
glycidyl α-methylacrylate, *See* SR-379, 29382
Glycidyl butyl ether, *See* Ageflex BGE, 967, Ageflex TBGE, 998
Glycidyl butyl ether monomer, *See* Ageflex BGE, 967
glycidyl decanoate, 13092, *See* Glydexx N-10, 13141
glycidyl ether, 13093
Glycidyl methacrylate, *See* SR-379, 29382
glycidyl methacrylate 2,3-epoxypropyl methacrylate, 13094, *See* glycidyl methacrylate 2,3-epoxypropyl methacrylate, 13094
glycidyl α-methylacrylate, *See* glycidyl methacrylate 2,3-epoxypropyl methacrylate, 13094
Glycidyl n-butyl ether, *See* Ageflex BGE, 967
glycidyl neodecanoate, *See* glycidyl decanoate, 13092, Glydexx N-10, 13141
Glycidyl trimethyl ammonium chloride, *See* Ogtac 85 V, 22052
glycin, *See* Glyconyl, 13124
glycine, 13095
Glycine betaine, *See* Oxyneurine, 22463
Glycine, N,N-bis(2-(bis(carboxymethyl)amino)ethyl)-, pentasodium salt, *See* Cheelox® 80, 5864, Trilon® C Liq, 32328
Glycine, N,N-bis(carboxymethyl)-, trisodium salt, *See* Cheelox® NTA-Na3, 5869
glycine, free base, *See* glycine, 13095
Glycine, N-(hydroxymethyl)-, monosodium salt, *See* Suttocide® A, 30440
Glycine, N-(phosphonomethyl)-, compd. with 2-propanamine (1:1), *See* Rodeo, 26899
Glycine, N,N'-1,2-ethanediylbis[N-(carboxymethyl)-, disodium salt, *See* sodium versenate, 28831
Glycine, N,N'-1,2-ethanediylbis[N-(carboxymethyl)-, tetrasodium salt, *See* Trilon® B Powd, 32324
glycine, N-methyl-N-(1-oxododecyl)-, sodium salt, *See* sodium lauroyl sarcosinate, 28779
glycine-iron sulfate (1:1), *See* glycine, 13095
glyco, *See* paraffin, liquid, 22744
glyco metal, 13096
glycobrom, 13097
glycocoll, *See* glycine, 13095
Glycoderm, 13098
Glycogen, *See* animal starch, 2468, Liver sugar, 17527
glycogenic acid, *See* gluconic acid, 13048
glycol, 13099
glycol, *See* ethylene glycol, 11123
glycol butyl ether, *See* 2-butoxyethanol, 4753
glycol diacetate, *See* ethylene glycol diacetate, 11124
glycol dilaurate, 13100, *See* Kemester® EGDL, 15958
glycol dimercaptoacetate, 13101, *See* glycol dimercaptoacetate, 13101
Glycol Dimethacrylate, *See* ethylene glycol dimethacrylate, 11125, Perkalink® 401, 23315
Glycol dimethyl ether, *See* monoglyme, 20227
glycol distearate, 13102, *See* Alkamuls® EGDS, 1663, Emerest® 2355, 10091, ethylene glycol distearate VA, 11126, Genapol® PMs, 12795, Kemester® EGDS, 15959, Kessco® EGDS, 16105, Lexemul® EGDS, 17136, Lipo EGDS, 17338, Mapeg® EGDS, 18641, McAlester EGDS, 19176, Pegosperse® 50 DS, 23065, Rewopal® PG 280, 26348, Secoster® DMS, 27817
Glycol ditallowate, *See* Marlamid® PG 20, 18746
Glycol ester, *See* Atlas EM-2, 3307, Witbreak DGE-128A, 34433
glycol ether DB, *See* Butyl Carbitol®, 4766
glycol ether eb, *See* 2-butoxyethanol, 4753
glycol ether eb acetate, *See* 2-butoxyethanol, 4753
glycol ether EE, *See* ethyl icinol, 11074
glycol ether EE acetate, *See* EE Acetate, 9635
Glycol Ether PM, *See* Arcosolv® PM, 2814, Dowanol® PM, 8896, Icinol PM, 14816, propylene glycol monomethyl ether, 25220, Solvenon® PM, Solvent PM, 29050
Glycol ethyl ether, *See* ethyl icinol, 11074
glycol ethylene ether, *See* Dioxane, 8574
Glycol hydroxystearate, *See* Naturechem® EGHS, 20769
Glycol Methacrylate, *See* Bisomer 2HEMA, 4273, Sipomer® HEM-D, 28536
glycol MIPA stearate, *See* Crodapearl Liq, 7129
Glycol monoethyl ether, *See* ethyl icinol, 11074
glycol monomethyl ether acrylate, *See* Ageflex MEA, 994
Glycol monopalmitate, *See* glycol palmitate, 13103
glycol monophenyl ether, *See* Igepal® Cephene Distilled, 14850, Rewopal® MPG 10, 26344
Glycol monoricinoleate, *See* glycol ricinoleate, 13104
glycol palmitate, 13103, *See* glycol palmitate, 13103, Lanol P, 16737
glycol ricinoleate, 13104, *See* Flexricin® 15, 11952
Glycol stearate, *See* Ablunol EGMS, 100, Emerest® 2350, 10090, Empilan® EGMS, 10392, ethylene glycol monostearate VA, 11129, Hodag EGMS, 14179, Kemester® EGMS, 15960, Kessco® EGMS, 16106, Lexemul® EGMS, 17137, Lipo EGMS, 17339, Mackester EGMS, 18211, Mackester IP, 18213, Mapeg® EGMS, 18642, Monthyle, 20298, Pegosperse® 50 MS, 23066, Rita EGMS, GMS, 26790, Schercemol EGMS, 27600, Secoster® EMS, 27819, Witconol EGMS, 34514, Zohar EGMS, 34849, Nutrapon TK 3603, 21826, Standapol® 7092, 29466
Glycol stearate - PEG-2 stearate - trilaneth-4 phosphate, *See* Sedefos 75®, 27833
Glycol, monobutyl ether, *See* Butyl Carbitol®, 4766
Glycol, monoethyl ether acetate, *See* EE Acetate, 9635
Glycol/butylene glycol montanate, *See* Hoechst Wax E, 14263
glycolic acid, 13105
glycolic acid phenyl ether, *See* phenoxyacetic acid, 23626
glycoline, *See* paraffin, liquid, 22744
Glycols, polyethylene, mono(hydrogen sulfate), dodecyl ether, sodium salt, *See* Witcolate ES-2, Witcolate LES-60C, 34477
Glycols, polyethylene, mono(nonylphenyl) ether, *See* Ablunol NP4, 106, Triton® N-57, 32430
Glycols, polyethylene, mono-9-octadecenyl ether, (Z)-, *See* Ablunol OA-6, 107, Volpo 3, 34020
Glycols, polyethylene, monolaurate, *See* Ablunol 200ML, 95
Glycols, polyethylene, monooctadecyl ether, *See* Ablunol SA-7, 113
Glycols, polyethylene, monooleate, *See* Ablunol 200MO, 96
Glycols, polyethylene, monostearate, *See* Ablunol DEGMS, 99
Glycols, polypropylene, monohexadecyl ether, *See* Wickenol® 707, 34404
Glycolube® 100, 13106
Glycolube® 140, 13107
Glycolube® P, 13108
Glycolube® VL, 13109, *See* Paraffin, 22743
Glycomul L, *See* Ablunol S-20, 108 Kemester® S20, 15966 Radiasurf® 7125, 25797 Sorbon S-20, 29117
Glycomul LC, *See* Ablunol S-20, 108
Glycomul O, *See* Ablunol S-80, 111, Kemester® S80, 15970, Radiasurf® 7155, 25802, Sorbon S-80, 29121, Witconol 2500, 34501
Glycomul P, *See* Ablunol S-40, 109
Glycomul S, *See* Ablunol S-60, 110, Kemester® S60, 15968, Radiamuls® Sorb 2145, Sorb 2161, Sorb 2166, 25783, Sorbon S-60, 29119
Glycomul TO, *See* Ablunol S-85, 112, Witconol 2503, 34502
Glycomul®, 13110
Glycomul® L, 13111
Glycomul® O, 13112
Glycomul® P, 13113
Glycomul® S FG, 13114
Glycomul® SOC, 13115
Glycomul® TO, 13116
Glycomul® TS KFG, 13117
glycon dp, *See* stearic acid, 29599
glycon s-70, *See* stearic acid, 29599
glycon s-80, *See* stearic acid, 29599
glycon s-90, *See* stearic acid, 29599
glycon tp, *See* stearic acid, 29599
Glycon® G 100, G 300, 13118
Glycon® P-45, 13119, *See* Palmitic Acid, 22635
Glycon® S-65, 13120, *See* hydrogenated tallow acid, 14598
Glycon® S-90, 13121
Glycon® TP, 13122
glyconic acid, *See* gluconic acid, 13048
Glyconol®, 13123
Glyconyl, 13124
glycophenol, *See* saccharin, 27195
glycosal, 13125
Glycosin, *See* saccharin, 27195
Glycosperse 0-20, *See* Alkamuls® T-80, 1693
Glycosperse 0-20 Veg, *See* Alkamuls® T-80, 1693
Glycosperse 0-20X, *See* Alkamuls® T-80, 1693
Glycosperse®, 13126
Glycosperse® HTO-40, 13127
Glycosperse® L-10, 13128
Glycosperse® L-20, 13129
Glycosperse® O-20 FG, O-20 KFG, 13130
Glycosperse® O-5, 13131
Glycosperse® P-20, 13132
Glycosperse® S-20 FG, S-20 KFG, 13133, *See* Ethylan® GES6, 11108
Glycosperse® TS-20 FG, *See* Polysorbate 65, 24571
Glycosperse® TS-20 FG, TS-20 KFG, 13134
Glycosterine, 13135
glycosthene, *See* glycine, 13095
glycothymoline, 13136
Glycowax® 765, 13137, *See* Advawax® 290, 820
Glycozone, 13138, *See* hydrozone, 14672

glycyl alcohol, *See* glycerin, 13071
N-glycylglycine, 13139
Glydant, *See* DMDM hydantoin, 8754, Glydant®, 13140, Mackstat® DM, 18234, Nipaguard® DMDMH, 21269
Glydant Plus, *See* DMDM hydantoin, 8754, Nipaguard® DMDMH, 21269
Glydant®, 13140 , *See* DMDM hydantoin, 8754
Glydexx N-10, 13141, *See* glycidyl decanoate, 13092
Glydus, 13142
Glyecin, 13143
Glyezin®, 13144
Glykocellon, *See* Cekol, 5558
Glykocellon, *See* carboxymethylcellulose, 5257, carboxymethylcellulose sodium, 5258
Glyme, *See* monoglyme, 20227
Glymin®, 13145
Glymo, *See* Aktisil EM, 1202, CG6720, 5836, Prosil® 5136, 25240
glymol, *See* Kaydol, 15810, paraffin, liquid, 22744
Glyoxal, *See* diformyl, 8453, glyoxal 40%, 13146, Protectol® GL 40, 25262
glyoxal 40%, 13146
glyoxalin, *See* imidazole, 14895
glyoxylaldehyde, *See* glyoxal 40%, 13146
glyoxyldiureide, *See* Allantoin, 1739
Glyphogan, 13147
glyphosate, 13148, *See* Glyphogan, 13147, Roundup, 27011, Spasor, 29213, Sting, 29768
Glyphosate + simazine, *See* Mascot Ultrasonic, 19008
Glyphosate Amine, *See* Rodeo, 26899
Glyphosate isopropylamine salt, *See* Rodeo, 26899
Glyprosol 20, 13149
Glyptal 2557,2559, 13150
glyptal resins, 13151
Glyrol, 13152
Glysantin®, 13153
Glysobuzole, *See* Stabinol, 29413
Glytex, 13154
Glytex® 203, 13155
Glytex® 513, 13156
Glytex® 663, 13157
Glytex® 1085, 13158
Glythermin®, 13159
GMA, *See* Bisomer 2HEMA, 4273, Sipomer® HEM-D, 28536
GMB Prime F-25, *See* Algin, 1591
GMD, 13160
Gmelin's blue, *See* Turnbull's blue, 32609
GMF, 13161
GMS Base, 13162
GMS/SE Base, 13163
Goa butter, *See* Kokum butter, 16302
Gobapur Acide Pur, 13164
Gofrativ, 13165
Gofravik, 13166
Go-Go-San, *See* pendimethalin, 23102
Gohi Iron, 13167
Gohsefimer, 13168
Gohsenol, 13169
Gohseran, 13170
Gold, 13171
gold bronze, *See* copper, 6777, saffron bronze, 27224
gold button brass, *See* brass, 4485
gold chloride, *See* gold trichloride,
Gold Coinage, *See* standard gold, 29485
Gold Guard, 13172
gold size, 13173
Gold solders, 13174
gold trichloride, 13175, *See* gold chloride,
gold-bloom, *See* Calendula Oil CLR, 4980
Gold-copper, *See* Chrysochalk, 6254
golden acorn, 13176
golden antimony sulfide, 13177, *See* antimony pentasulfide, 2570
Golden Bear 102, 13178
Golden Bear 2013-10, 13179
Golden Bear 4013-10, 13180
Golden Dawn, 13181
Golden Dew, *See* Thiovit, 31724
Golden Fleece, 13182
Golden Hermetite, 13183
golden Sulfide of Antimony, *See* golden antimony sulfide, 13177
Golden Sulfuret of Antimony, *See* golden antimony sulfide, 13177
golden syrup, 13184
Golden Wax, 13185
Golden-Pea-Pro EN-15, 13186
goldenseal, 13187
Goliath, 13188, *See* phenmedipham, 23595, 23596, 23597, 23598
Golpanol®, 13189
Golpanol® MBS, 13190
Goltix®, 13191 , *See* Metamitron, 19443
gommelin, *See* dextrin, 8235
gommeline, *See* British gum,
Gonacrine, 13192
Gondang wax, *See* Java wax, 15539
gong metal, 13193
Goniosol, 13194
Gooch and Eddy reagent, 13195
Good Gulf, 13196
GoodLife, 13197
Good-rite GP 264, *See* Bis(2-ethylhexy) Phthalate, 4228, dioctyl phthalate, 8549, Reomol DCP, 26130, Witcizer 312, 34454
Good-rite nix, *See* Aero 343 Xanthate, 842
Good-rite Polyacrylates, 13198
Good-rite® 2528X10, 13199, *See* Vinyl pyridine, 33859
Good-rite® 3150, 13200
Good-rite® GP-223, 13201, *See* dioctyl adipate, 8548
Good-rite® GP-235, 13202
Good-rite® GP-236, 13203
Good-rite® GP-261, 13204
Good-rite® GP-265, 13205, *See* octyldecyl phthalate, 22042
Good-rite® GP-266, 13206
Good-rite® K-702, 13207
Good-rite® K-705BD, 13208
Good-rite® K-752, 13209
Goodyear LPR-6632, 13210
gooroo nuts, *See* kola nut, 16303
Gopmann solder, 13211
Gordon superflex D, 13212
Gossamer®, 13213
gossypose, *See* raffinose, 25835
Gougeon Laminating Epoxy, 13214
Goulac, 13215, *See* calcium lignosulfonate, 4956, 4957
Goulard powder, 13216
Gountdown, *See* Metamitron, 19443
Government Rubber Styrene, *See* KR04 K-Resin Polymer, 16382, 16383, Kraton®, 16386, Kraton® D 1101, 16387, Kraton® D 1116, 16389, Kraton® D 2103, 16391, K-Resin Polymer KR01, 16404, K-Resin Polymer KR03, 16405, K-Resin Polymer KR04, 16406, K-Resin Polymer KR05, 16407, K-Resin Polymer KR10, 16408, Krylene® 606, 16449, Krylene® 608, 16450
GP-4, 13217
GP-4-E, 13218
GP-66 Miracle Cleaner, 13219
GP-71-SS, 13220
GP-137, 13221, *See* glycidoxypropyl methyl diethoxysilane, 13090
GP-180, 13223
GP-187, 13224
GP-209, 13225
GP-210, 13226
GP-217, 13227
GP-227, 13228
GP-262, 13229
GP-310-I, 13230
GP-7000, 13231
GPC-5544, *See* tebutam, 30879
G-P-D, 13232
G-Picoline, *See* γ-picoline, 23817
GP-II, 13233
GR Acid, 13234
Graessorb, 13235
Grafene, 13236
Grafil, 13237
Grafita, 13238
Grafitix (Anti-graffiti), 13239
Grafitix (Baent), 13240
Grafitix (Ravalement), 13241
Grafix, 13242
grahamite, 13243
Grahams salt, *See* Calgon, 4987, Kalex HMP, 15670, sodium hexametaphosphate, 28770
Graham's Salt, Sodium Polymetaphosphate, *See* Glass H, 12967
grain alcohol, 13244, *See* ethyl alcohol, 11064
,grain oil, *See* fusel oil, 12518
,Grain Store Smoke, 13245
grains D'Ambrette, 13246
grains of paradise, 13247
Gramazine, 13248
Gramevin, *See* Basfapon®, 3705, dalapon, 7705
Gramevin, *See*
Graminon® Plus, 13249
Gramisan, *See* Agallol, 954, Atiran, 3286, Ceresan, 5745
Gramixel, 13250
grammite, *See* calcium silicate, 4969
Gramonol, 13251
Gramonol 5, 13252
Gramonol Five, 13253
Gramoxone, 13254, 13255
Gramoxone X, 13256
Grampenil, *See* Ampicillin, 2361
Gramp's solder, 13257
Gram's iodine stain, 13258
Gram's stain, 13259
Gramuron, 13260
Graney bronze, 13261
Granodine, 13262
granodized steel, 13263
granol, 13264
Granolube, 13265
Granosan®, 13266
Granox NM, *See* Julin's chloride, 15632
Granstar, *See* metsulfuron-methyl, 19606
Granstock, 13267
Granubor, 13268
Granuform, 13269
Granugen, *See* Synwax, 30659
Granular Weedkiller, 13270
Granulestin, *See* Kelecin®, 15835
Granulite® BF 6/16, 13271
Granulite® TR-10, 13272
Granulite® WTP-10, 13273
granulose, 13274
Granurex, *See* neburon, 20841
Granutox, *See* Terrathion, 31343
Grape Seed oil, 13275
grape sugar, *See* glucose, 13052
Graphalloy, Silver, 13276
graphite, 13277, *See* Aquadag, 2690 Lonza KS, 17589 Rollit, 26921
Graphite fiber, *See* Mangnamite, 18560
graphite metal, 13278
graphitic carbon, 13279
graphitic temper carbon, 13280
graphitites, 13281
Graphitol, 13282
Graphol, *See* Metol, 19584
Graphsize, 13283
Graphtol®, 13284
Grappier cements, 13285
Grasal Brilliant Yellow, *See* Butter yellow, 4759
grasal orange, *See* Sudan I, 29938
grasan orange r, *See* Sudan I, 29938
Grasidim, *See* Sethoxydim, 28099
Graslam, 13286
Grass Greenizit, 13287
Grasselerator 101, 13288
Grasselerator 102, 13289
Grasselerator 508, 13290

thiocyanate, 13442
guanidine hydrothiocyanate, *See* guanidine thiocyanate, 13442
Guanidine isothiocyanate, *See* Guanidine thiocyanate, 13442
Guanidine mononitrate, *See* Guanidine nitrate, 13441
Guanidine nitrate, 13441
Guanidine thiocyanate, 13442
guanidinium chloride, *See* guanidine hydrochloride,
Guanidinium Isothiocyanate, *See* Guanidine thiocyanate, 13442
guanidinium nitrate, *See* Guanidine nitrate, 13441
guanidinium rhodanide, *See* guanidine thiocyanate, 13442
guanidinium thiocyanate, *See* Guanidine thiocyanate, 13442
guanine, 13443
guaniol, *See* Geraniol, 12885
guano, 13444
Guanylurea Sulfate, *See* Grossmann reagent, 13426
N-Guanylurea sulfate, *See* Grossmann reagent, 13426
guar flour, *See* guar gum, 13445, Prinza® Range, 25010
guar gum, 13445, *See* hydroxypropyl guar, 14666
Guar gum (*cyamopsis tetragonolobus*), *See* Prinza® Range, 25010
Guar gum, 2-hydroxy-3-(trimethylammonio) propyl ether, chloride, *See* Jaguar® C-13S, C-14S, 15496
Guar hydroxypropyl trimethyl ammonium chloride, *See* Jaguar® C-13S, C-14S, 15496
Guar hydroxypropyl trimonium chloride, *See* Cosmedia Guar C-261 N, 6868, Hi-Care® 1000, 14060, Jaguar® C-13S, C-14S, 15496, N-Hance® 3000, 21083
Guara, 13446
Guaran, *See* guar gum, 13445, Prinza® Range, 25010
Guaranine, *See* caffeine,
Guardar, 13447
Guardian, 13448
Guardsep, 13449
Guardsman, 13450, 13451
Guarem, *See* Prinza® Range, 25010
Guartec CAP, 13452, 13453
Guayale, 13454
Guay-azulene sodium sulfonate, *See* Rita AZ, 26787
Guazatine, *See* Rappor, 25890
Guazatine combined with imazalil, *See* Rappor Plus, 25891
Guesapon, *See* DDT, 7820
Guesarol, *See* DDT, 7820
Guettier metal, 13455
Guicitrina, *See* Ampicillin, 2361
Guignet's Green, 13456, *See* chromium green, 6232
Guillaume Alloy, 13457
Guillaume Metal, 13458
Guinea Grains, *See* Grains of Paradise, 13247
Gulf Lite, 13459
Gulf Lubcote, 13460
Gulf No-Rust, 13461
Gulfad-C, 13462
Gulfco, 13463
Gulfcrest, 13464
Gulfcrown, 13465
Gulfcut, 13466
Gulfgem, 13467
Gulfknit, 13468
Gulfleet, 13469
Gulflex, 13470
Gulflube, 13471
Gulfpride, 13472
Gulfspin, 13473
Gulftene, 13474
Gulftex, 13475
Gulftow, 13476
Gulftronic, 13477
Gulfwax, 13478
gulitol, *See* sorbitol, 29113
L-gulitol, *See* A-625/641ABS 301K, ABS 500FR-1, 10, sorbitol, 29113
Gum acacia, *See* gum arabic, 13480
gum Amritsar, 13479
Gum anime, *See* anime resin, 2469
gum arabic, 13480
gum benguela, 13481
gum benjamin, 13482
gum benzoin, *See* gum benjamin, 13482
gum camphor, *See* camphor,
gum catechu, 13483
gum Cowrie, 13484
gum cyamopsis, *See* guar gum, 13445, Prinza® Range, 25010
Gum D, 13485
Gum Damar, *See* Dammar Resin, 7724, gum Cowrie, 13484
gum dragon, *See* gum tragacanth, 13490
Gum E, 13486
gum guar, *See* Prinza® Range, 25010
gum Juniper, *See* gum sandarac,
gum karaya, *See* karaya gum, 15760
gum kino, *See* kino, 16196
gum lac, *See* lac, 16547
gum Lini, 13487
Gum MB, 13488
gum rosin, *See* Abietic anhydride, 46, rosin, 26976
Gum rosin acid, *See* Snowtack 342A, 28678
Gum senegal, *See* gum arabic, 13480
gum thus, 13489, *See* frankincense, 12376
gum tragacanth, 13490, *See* Dragon gum, 8956
gum, fiddle, *See* Gum tragacanth, 13490
gumbo, *See* bentonite, 3954
Gummeline, *See* dextrin, 8235
Gummi gutta, *See* Gamboge, 12628
Gun metal, 13491
guncotton, 13492 *See* nitrocellulose, 21353
Gunner, 13493, *See* Flamprop-M-isopropyl, 11888
Gunning's reagent, 13494
gunpowder, 13495
Guntapite, 13496
Gurdynamite, *See* dynamite, 9356, Kieselguhr dynamite, 16177
Gurjun balsam or oil, 13497
Gurley's metal, 13498
Gurney's bronze, 13499
Gurr, 13500
guru nuts, *See* kola nut, 16303
Gusathion K forte, *See* Azinphos-ethyl, 3520
Gusathion®, 13501
Gusathion® A, 13502, *See* Azinphos-ethyl, 3520
Gusathion® MS, 13503
Gusto, 13504, *See* phenmedipham, 23596, 23597, 23598, 23599
Gutex R 1513, *See* Azinphos-ethyl, 3520
Guthion®, 13505
Guthrie's eutectic alloy, 13506
gutta-gerip, *See* gutta-susu, 13511
gutta-percha, 13507
gutta-shea, 13508
Gutta-Siak, 13509
gutta-Singarip, *See* gutta-susu, 13511
Gutta-sundik, 13510
gutta-susu, 13511
guvacine, 13512
Gyneclorina, *See* Chloramine T, 6086
Gyne-Lotrimin, 13513
gynocardic acid, 13514
Gynochrome, *See* mercurochrome, 19354
Gynofon, *See* Aminitrazole, 2164
Gynol, 13515
gypsite, 13516
Gypsona, 13517
Gypsum, 13518
Gypsum-F, 13519
Gyrane, 13520
gyrolite, *See* calcium silicate, 4969
Gyron, *See* DDT, 7820
H, *See* hypoxanthine, 14721
H 119, *See* chloridazon, 6103
H 1803, *See* simazine, 28464
H Chrome Green 105, 13521
H.A. Solvent, 13522
H.E. cellulose, *See* Hetastarch, 13891
H.E.S. 6-H.E.S, *See* Hetastarch, 13891
H.M.T, *See* hexamine, 14008
H.M.T.D, 13523
H.T, 13524
H.T.S, 13525
H_2O-$_3H$, *See* Tritiotope, 32423
H_2old EP-1, 13526
H_2O_2, *See* hydrogen peroxide, 14590
H-30, 13527
H7250, 13528, *See* Hexamethyldisiloxane, 14004
H7301, 13529
H7310, 13530, *See* Hexamethyldisiloxane, 14004
HA 819, 13531
Hachi-sugar, *See* sodium cyclamate, 28754
H-acid, 13532
Hadranol, 13533
Haemodyn, *See* Tears Plus®, 30876
Haemofort, *See* ferrous sulfate, 11690
Hagafilm, Hagatreat, 13534
Hager's reagent, 13535
Hagevap, 13536
Hahnmann's mercury, 13537
haiari, *See* rotenone, 27000
Haine's solution, 13538
Hair Complex 20/70n, 13539
Hair Complex Aquosum, 13540
Hair salt, *See* magnesium sulfate, 18368
Halamid, *See* Chloramine T, 6086, Ketjensept, 16133
Halane, *See* 1,3-Dichloro-5,5-dimethyl hydantoin, 8389
Halar E-CTFE Film, 13541
Halar® 300, 13542
Halbase 10, 13543
Halberland metal, 13544
Halcarb 20, 13545
Halenol, *See* Nuophene, 21783
Halethazole, *See* Episol, 10657
half-stuff, 13546
Haliborange, 13547
halite, *See* rock salt, 26887
Halizan, *See* metaldehyde, 19431
Hallco C-503 Plasticizer, *See* Witcizer 100, 34453
Hallco CPH 43, *See* Ablunol 200ML, 95
Hallco® C-491, 13548
Hallco® C-918, 13549
Hallcomid®, 13550
Hallcomid® M-18-OL, 13551
Hallcomid® M-8-10, 13552
Hallcote®, 13553
Hallcote® 573, 13554
Hallcote® CSD, 13555
Hallcote® ZS 5050, 13556
Hallmark, 13557, *See* λ-cyhalothrin, 7567
Hal-Lub-D, 13558
Hal-Lub-N, 13559
Halobrom, 13560, *See* Bromochloro dimethyl hydantoin, 4574
Halocarbon 12, *See* dichlorodifluoromethane, 8388
Halocarbon 13, *See* chlorotrfluoromethane, 6153
Halocarbon 13, *See* Genetron® 13, 12819
Halocarbon 23, *See* Fluoroform, 12099, Genetron® HFC23, 12829
Halocarbon 113, *See* MS-180 Freon® TF Solv, 20415
Halocarbon 114, *See* Cryofluorane, 7278
Haloflex, 13561
Halofuginone hydrobromide, 13562, *See* Stenorol, 29650
Halogen compound with metal oxide and binder, *See* Aflamman CN, 937
Halon, 13563, *See* dichlorodifluoromethane, 8388
Halon 122, *See* dichlorodifluoromethane, 8388
Halon 242, *See* Cryofluorane, 7278
Halon 10001, *See* methyl iodide, 19542
Halowax, *See* Nibren wax, 21097
Halowax 1014, 13564 *See* Halowax 1014, 13564
Halowax 4000 B-2, 13565
Haloxil, 13566
Haloxon, 13567

haloxon-oxyclozanide, *See* Haloxil, 13566
haloxyfop, *See* Gallant, 12607
Halphen reagent, 13568
Halphos, 13569
HALS 1, *See* Uvaseb 770, 33178
Halso® 99, 13570
Halstab 30, 13571
Halstab P-1, 13572
Halt, *See* cypermethrin, 7588
Haltex 300, 13573
Halthal, 13574
HaltImperator, *See* Topclip Parasol, 32037
Haltox, *See* methyl bromide, 19517
Halumin, 13575
Halycitrol, 13576
Hamamell tannin, 13577
hambergite, 13578
Hamburg blue, *See* mountain blue, 20366
Hamburg white, 13579
Hamilton metal, 13580
hammer slag, 13581
Hammonia metal, 13582
Hamonite, 13583
Hampamide B, 13584
Hampden Steel, 13585
Hampene, *See* Trilon® BS, 32327
Hamp-Ene® 100, 13586
Hamp-Ene® Acid, 13587
Hamp-Ene® Na2, 13588
Hamp-Ene® Na3 Liq, 13589
Hamp-Ene® Na3 Liq, *See* edetate trisodium, 9614
Hamp-Ene® Na4, 13590
HAMP-EX 80, *See* Cheelox® 80, 5864, Pentaquest Extra 0685, 23169, Trilon® C Liq, 32328
Hamp-Ex® 80, 13591, *See* Pentasodium pentetate, 23172
Hamp-Ex® Acid, 13592
Hampfoam 35, 13593
Hamp-Ol® 120, 13594
Hamp-Ol® Acid, 13595
Hamp-Ol® Crystals, 13596
Hamposyl L 30, *See* sodium lauroyl sarcosinate, 28779
Hamposyl® AL-30, 13597, *See* Ammonium lauroyl sarcosinate, 2241
Hamposyl® C, 13598
Hamposyl® C, CZ, *See* Cocoyl sarcosine, 6568
Hamposyl® C-30, 13599
Hamposyl® L, 13600, *See* lauroyl sarcosine, 16861
Hamposyl® L-30, 13601
Hamposyl® M, 13602
Hamposyl® M-30, 13603
Hamposyl® O, 13604, *See* Oleoyl sarcosine, 22123
Hamposyl® S, 13605
Hamposyl® TL-40, 13606
Hamposyl® TOC-30, 13607
Hampshire® DEG, 13608, *See* sodium dihydroxyethyl glycinate, 28757
Hampshire® EDG, 13609
Hampshire® NTA 150, 13610
Hampshire® NTA Acid, 13611
Hampshire® NTA Na3, 13612
HAN® 857, 13613
Handi-Wrap, 13614
Handyfoam, 13615
Hansa, 13616
Hansa oil, 13617
Hanus' Iodine bromide solution, 13618
Harco Foamstopper®, 13619
hard aluminum, 13620
Hard Cure, 13621
Hard fiber, *See* Vulcanized fiber, 34097
hard jatoba, 13622
hard lead, 13623
hard metal, 13624
hard paraffin, *See* paraffin, 22742, 22743
hard platinum, 13625
hard solder, 13626
hard zinc, 13627
Hardcote, 13628
hardened rosins, 13629
Hardened rubber, *See* Ebonite, 9499
hardenite, 13630
hard-finish plaster, 13631
hard-head, 13632
Hardite, 13633
Hardite X, 13634
Hardset, 13635
Hargus Steel, 13636
Harlington bronze, *See* Harrington Bronze, 13653
Harle's solution, 13637
Harlington bronze, *See* Harrington bronze, 13653
Harmomang A and B, 13638
Harmonia bronze, 13639
Harmony, 13640, *See* thifensulfuron-methyl, 31673
Harmony Extra, *See* thifensulfuron-methyl, 31673
Harmony®, 13641
Haro® Chem P28G, 13642
Haro® Chem PDF, 13643
Haro® Chem PTS-E, 13644
Haro® Mix CE-701, 13645
Haro® Mix CE-701, Mix CK-711, Mix MH-204, *See* lead, 16900
Haro® Mix CK-711, 13646
Haro® Mix MH-204, 13647
Haro® Mix YK-110, 13648
Haro® Mix ZC-028, ZC-029, ZC-030, 13649
Haro® Mix ZT-514, 13650
Haro® Wax L333, 13651
Harrier, 13652
Harrington bronze, 13653
Harringtonite, 13654
Harris' hematoxyiin stain, 13655
Harrison's indicator, 13656
Hart Lawn Sand, 13657, *See* ferric sulfate, 11652
Hart Moss Killer, 13658, *See* ferric sulfate, 11652
Hartamide 9137, 13659
Hartamide AD, 13660
Hartamide LDA, 13661
Hartamide LMEA, 13662
Hartamide OD, 13663
Hartamine Series, 13664
Hartasist 16, 13665
Hartasist 46, 13666
Hartasperse DI-4900 Series, 13667
Hartbreak Series, 13668
Hartenol LAS-30, 13669
Hartenol LES 60, 13670
hartin, 13671
hartite, 13672
Hartmetall, *See* Camite, 5054
Hartofix 2X, 13673
Hartofol 40, 13674
Hartofol 60T, 13675
Hartolan, 13676
Hartolite, 13677, *See* Lanolin alcohol, 16741
Hartolon 5683, 13678
Hartolon NA, 13679
Hartomer 4900, 13680
Hartomer GP 2164, 13681
Hartomer LD 31, 13682
Hartomul PE-30, 13683
Hartonyl L531, 13684
Hartopol 25R2, 13685
Hartopol L42, 13686
Hartopol LF-1, 13687
Hartopol P65, 13688
Hartosoft 171, 13689
Hartosoft S5793, 13690
hartosol, *See* isopropanol, 15405
Hartotrope AXS, 13691, *See* Ammonium xylene sulfonate, 2259
Hartotrope KTS 44, 13692
Hartotrope STS-40, Powd, 13693
Hartotrope SXS 40, Powd, 13694, *See* sodium xylene sulfonate, 28832
Hartowet 5917, 13695
Hartshorn oil, *See* bone oil, 4402
Hartshorn spirit, *See* hartshorn and volatile alkali,
Harvesan, 13696
Harvestra®, 13697
Harvite, 13698
Hascrome, 13699
Hastelloy®, 13700
hatchette brown, *See* Florentine brown, 12014
hatchettine, *See* bitumen, 4294
Hatcol DOP, *See* Bis(2-ethylhexy) Phthalate, 4228, dioctyl phthalate, 8549, Reomol DCP, 26130, Witcizer 312, 34454
Havero-extra, *See* DDT, 7820
Havidote, *See* Kalex Acids, 15669
Hayem's solution, 13701
Haylite No. 1, 13702
Haylite No. 3, 13703
Haynes alloy No. 25, 13704
Haynes metals, 13705
Haynes® 242, 13706
Haysite, 13707
Haysite 14100, 13708
Haysite 24500, 13709
Haysite 42000, 13710
Haysite EHC-P, 13711
Haystellite, *See* Camite, 5054
HB-40, 13712
HBD, *See* tributyltin oxide, 32209
HBR, 13713
HC 200 Concentrate, 13714
HC-913, 13715
HCB, *See* Julin's chloride, 15632
HCFC-123, *See* Dichlorotrifluoroethane, 8393, Genetron® 123, 12823
HCFC-141b, *See* Dichlorofluoroethane, 8391
γ-HCH, *See* Gammasan, 12636, Murfume Grain Store Smoke, 20512, New Kotol, 21067, Unicrop Leatherjacket Pellets, 32931
HCN, *See* hydrocyanic acid, 14584
HCP, *See* hexachlorophene, 13992
HD, *See* mustard gas, 20538
HDDA, *See* Ageflex HDDA, 990
HD-Echelon 45/50, 13717
HD-Echelon 90/95, 13718
HD-Echelon 110/130, 13716
HD-Eutanol, 13719
HDODA, *See* Ageflex HDDA, 990
HDPE, *See* polyethylene, high-density, 24428
HDPE 04352N, 13720
HDPE 25053-P, 13721
HDPE 32060C, 13722
HDPE 35053, 13723
HDPE IP-10, 13724
HDPE resin, *See* Marlex® HXM 50100, 18816
Hégor, 13725
Head and Shoulders, 13726
Headland Charge, 13727
Headland Dephend, 13728, *See* Phenmedipham, 23596
Headland Dual, 13729
Headland Guard, 13730
Headland Inorganic Liquid Copper, 13731, *See* copper oxychloride, 6785
Headland Intake, 13732
Headland Relay, 13733
Headland Spirit, 13734
Headland Swift, 13735
Healon, *See* hyaluronic acid sodium salt, 14483
Healonid, *See* hyaluronic acid sodium salt, 14483
Heartgard 30, *See* Ivermectin, 15454
Heavithane, 13736
heavy carburetted hydrogen, *See* olefiant gas,
Heavy naphtha, *See* HAN® 857, 13613
Hebron Pabracr, 13737
Heckel's solution, 13738
Hecla 35, 13739
Hecla 115, 13740
Hecla 135, 138, 13741
Hecla 138H, 13742
Hecla 155, 13743
Hecla 174, 13744
Hecla 180, 13745

Hecla 306, 13746
Hecla 307, 13747
Hecla powder, 13748
Hectabrite® AW, 13749, *See* Hectorite, 13752
Hectalite® 200, 13750, *See* Hectorite, 13752
hecto violet R, *See* gentian violet, 12845
hectograph violet SR, *See* gentian violet, 12845
Hector bases, 13751
Hectorite, 13752
hectorite laponite, 13753
Hedapur M 52, *See* MCPA, 19183
Hedarex M, *See* MCPA, 19183
hedeoma oil, *See* pennyroyal oil, 23119
hedeoma pulegioides, *See* pennyroyal oil, 23119
Hedex, *See* Valadol, 33238
hedgehog crystals, 13754
Hedonal, *See* Verdone CDA, 33580, MCPA, 19183
Hedonal M, *See* MCPA, 19183
Hedonal®, 13755
Hefilcon A, *See* SoftMate DW, 28862
Hefilcon B, *See* Naturvue, 20780, Toric Contact Lens, 32059
Hegolit 3, 13756
Heidenhain's chrome hematoxyiin, 13757
Heiloy, 13758
Helarion, 13759, *See* metaldehyde, 19431
Helenite, *See* Elaterite, 9713
Heliane, 13760
Helicon, *See* aspirin,
Helio®, 13761
Heliochrysin, 13762
Heliofil, 13763
Heliogen®, 13764
Heliolac, 13765
Heliophan, 13766, *See* Homosalate, 14292, Kemester® HMS, 15963
Heliquat BAC 50, *See* Benzalkonium chloride, 3958
Helizarin®, 13767
helmetina, *See* phenothiazine, 23618
Helmetine, *See* phenothiazine, 23618
Helmiantin, *See* Mansonil®, 18611
helmirone, *See* Haloxon, 13567
Helothion, 13768
Heloxy® 7, 13769
Heloxy® 61, 13770
Heloxy® 62, 13771
Heloxy® 64, 13772
Heloxy® 65, 13773
Heloxy® 67, 13774
HEM, *See* N-ethyl morpholine, 11140
Hemachates, 13775
Hema-Combistix, 13776
Hemalum, 13777
Hemastix, 13778
Hematest, 13779
Hematin, 13780
Hematine paste and powder, *See* logwood, 17557
heme-a, *See* Bisomer 2HEMA, 4273, Sipomer® HEM-D, 28536
Hemicellulase Conc, 13781
Hemicelluloses, 13782
Hemodex, *See* Dextran, 8233
Hemostatic Serum, *See* Hemoplastin,
Hemoterge, 13783
Hemoxone, 13784
Hempel's solution, 13785
1-hendecanol, *See* Neodol® 1, 20908 Neoflex® 11, 20921
10-hendecenoic acid, *See* undecylenic acid, 32916
henna, *See* Lawsone, 16891
Hentax Type P-1800, *See* Albumen, 1402
Hentex Type P-2100;, *See* Albumen, 1402
Hentriacontane, *See* candelilla wax, 5084
HEP, 13786
Hepacholine, *See* choline chloride, 6171
hepar sulfur, 13787
Heparin sodium, *See* Heparin, Sodium Salt, 13788
Heparin, Sodium Salt, 13788
hepatic acid, *See* sulfuretted hydrogen, 30053
Hepatolite, 13789
Heprinar, *See* Heparin, Sodium Salt, 13788
Hepsal, *See* Heparin, Sodium Salt, 13788
1-heptadecanecarboxylic acid, *See* stearic acid, 29599
heptadecanoic acid, *See* isostearic acid, 15421
heptadecanoic acid, 16-methyl-, *See* isostearic acid, 15421
2-Heptadecyl-4,5-dihydro-1H-imidazole, *See* stearyl Hydroxyethyl imidazoline, 29606
2-Heptadecyl-2-imidazoline-1-ethanol, *See* stearyl Hydroxyethyl imidazoline, 29606
2-(8-heptadecenyl)-4,5-dihydro-1H-imidazole-1-ethanol, *See* Marlowet® 5440, 18928, Sovatex IM17H, 29160
Heptadecyl hydroxyethylimidazoline, *See* Atlasol KAD, 3348
Heptadecyl sodium sulfate, *See* Rexowet 77, 26504
heptahydrate; bitter salts, *See* magnesium sulfate, 18368
Heptakis (dipropylene-glycol) triphosphite, *See* Weston® PTP, 34317
Heptaldehyde-aniline, *See* heptene, 13794, Vulcaid 444B, 34080
2,2,4,4,6,8,8-Heptamethylnonane, *See* Permethyl 101A, 23387
Heptan-4-one, *See* Butyrone, 4804
heptanal, *See* Heptoic aldehyde,
n-heptane, 13790, *See* Exxsol® Heptane, 11360
1,7-Heptanedicarboxylic acid, *See* Azelaic acid, 3519
heptanoic acid, 13791
1-heptanol, 13792, *See* heptyl alcohol, 13797
3-heptanol, *See* ethylbutylcarbinol, 11119
2-heptanone, *See* methyl n-amyl ketone, 19552
heptan-2-one, *See* methyl n-amyl ketone, 19552
4-Heptanone, *See* Butyrone, 4804
Hepteen Base, 13793
heptene, 13794
heptenophos, 13795, *See* heptenophos, 13795, Hostaquick, 14407
heptoic acid, *See* heptanoic acid, 13791
n-heptoic acid, *See* heptanoic acid, 13791
Heptokill, 13796
heptyl alcohol, 13797, *See* 1-heptanol, 13792
n-Heptyl alcohol, *See* 1-heptanol, 13792
***sec*-Heptyl alcohol**, *See* ethylbutylcarbinol, 11119
n-heptyl *p*-hydroxybenzoate, *See* Nipaheptyl, 21272
2-Heptyl tetrahydrofuran, *See* Florane, 12006
heptylalcohol, *See* 1-heptanol, 13792
Heptylated diphenylamine, *See* Permanax HD, 23362, Permanax HD (SE), 23363
n-heptylic acid, *See* heptanoic acid, 13791
Her, 13798
Herald, *See* fenpropathrin, 11615
Herapath's salt, 13799
Herbadox, *See* pendimethalin, 23102
Herb-All, *See* Neoarsycodyl, 20871
Herbalt, *See* neburon, 20841
Herbatox®, 13800
Herbax, *See* Propanil, 25174
Herbazin, *See* simazine, 28464
Herbazin 50, 13801, *See* simazine, 28464
Herbazin Plus, 13802
Herbazin Plus SC, 13803
Herbazin Special, 13804
Herbazin Total, 13805
Herbicide, *See* Paraquat + Plus, 22797
Herbicide 732, *See* terbacil, 31220
Herbicide M, *See* MCPA, 19183
Herbodox, *See* pendimethalin, 23102
Herbohn bronze, 13806
Herborane, 13807
Herbrak®, 13808
Herbyl acetate, *See* Jasmacyclene, 15529
Hercat 627, 13809
Herclor, 13810
Herco®, 13811
Herco® Pine Oil, *See* Herco®, 13811
Hercobind DS, 13812
Hercobond® 339, 13813
Hercoflat®, 13814
Hercoflat® Texturing Pigments and Flatting Agent, 13815
Hercoflav, 13816
Hercoflex 260, *See* Bis(2-ethylhexy) Phthalate, 4228, dioctyl phthalate, 8549, Reomol DCP, 26130, Witcizer 312, 34454
Hercoflex® 600, 13817
Hercoflex® 707, 13818
Hercoflex® 707A, 13819
Hercoflex® 900, 13820
Hercoflex® Plasticizer, 13821
Hercofloc, 13822
Hercofroth, 13823
Hercol 2, 13824
Hercol 2X, 13825
Hercolube®, 13826
Hercolube® Synthetic Ester, 13827
Hercolyn®, 13828
Hercolyn® D, 13829
Hercomix, 13830
Hercon® 2, 13831
Hercon® 2X, 13832
Hercon® 32, 13833
Hercon® 40, 48, 13834
Herco-Prills, 13835
Hercoprime, 13836
Hercopruf, 13837
Hercose AP, 13838
Hercose C, 13839
Hercosett®, 13840
Hercosett® 125, 13841
Hercosol, 13842
Hercosol TP-S, 13843
Hercosplit WR, 13844
Hercotac®, 13845
Hercotac® AD, *See* Hercotac®, 13845
Hercotac® LA, 13846
Hercotuf, 13847
Hercules metal, 13848
Hercules P 6 PE 200, *See* pentaerythritol, 23138
Hercules® 4 Defoamer, 13849
Hercules® 247, 13851
Hercules® 356 Defoamer, 13852
Hercules® 752 Size, 13853
Hercules® 1098, 13854
Hercules® 2051GS Defoamer, 13855
Hercules® 37M6-8, 13850
Hercules® 37M6-8, *See* formaldehyde, 12258
Hercules® AR 150, 13856, 13857
Hercules® AS, 13858
Hercules® Improved Tech. PE, 13859
Hercules® K, 13860
Hercules® K, N, T, *See* ethyl cellulose, 11066
Hercules® Mono-PE, 13861, *See* Pentaerythritol, 23139
Hercules® N, 13862
Hercules® RES A-2338, 13863
Hercules® RS, 13864
Hercules® SS, 13865
Hercules® T, 13866
Hercules® Tech Di-PE, *See* Dipentaerythritol, 8581
Hercules® Type AS4, 13867
Hercules® X Dry Size, 13868
Herculine FR, 13869
Herculite, 13870
Herculon, 13871
Herculoy, 13872
Hercures®, 13873
Heresite, 13874
Heritage, *See* Treflan, 32162, Trifluralin, 32252
Herkal, *See* dichlorvos, 8394
Herkol, *See* dichlorvos, 8394, Vapona, 33378
Herkules, 13875
hermal, *See* Rezifilm, 26521
Hermann's fluid, 13876
hermat tmt, *See* Rezifilm, 26521
Hermat ZDK, *See* Octocure ZDE-50, 21996
Hermat Zdm, *See* Bisphenol A, 4283
Hermat ZN-MBT, *See* Octocure ZMBT-50, 21998, Oxaf,

22408, Perkacit® ZMBT, 23298, Zetax®, 34741
Hermes, 13877
hermesetas, *See* saccharin, 27195
Hermite fluid, 13878
Herolith, 13879
Heron, 13880
Hero-Prills, *See* Ammonium nitrate, 2245
Herox®, 13881
Herpes-Gel, *See* Kerecid, 16067
Herpid, *See* dimethyl sulfoxide, 8511
Herplex, *See* Kerecid, 16067
Herrifex DS, 13882
Herrisol, 13883
Herschel's crystals, 13884
Heryl, *See* Rezifilm, 26521
HES, *See* Hetastarch, 13891
Hespan, *See* Hetastarch, 13891
Hespander, *See* Hetastarch, 13891
Hest CSO, *See* Cetearyl octanoate, 5791
Hest MS, 13885
Hestar, *See* Hetastarch, 13891
Hestat, *See* Hetastarch, 13891
Hesthsulphid, 13886
Hestsol, *See* Hetastarch, 13891
Hetacillin, *See* Versapen, 33609
Hetacillin K, *See* Versapen K, 33610
Hetacillin potassium, *See* H-K Mastitis, 14150, Versapen K, 33610
Hetamide MA;, *See* Acetamide MEA, 283
Hetamide ml, *See* Witcamide® 5195, 34440
Hetamine 5L-25, 13887 *See* stearamidopropyl dimethylamine lactate, 29595
Hetan SL, 13888
Hetan SO, 13889
Hetan SS, 13890
Hetastarch, 13891
heteroauxin, *See* Rhizopon A, AA, 26573
Hetester 412, 13892
Hetester 3236S, 13893
Hetester FAO, 13894
Hetester HCA, 13895
Hetester HCP, 13896
Hetester HSS, 13897, *See* isocetyl stearoyl stearate, 15349
Hetester ISS, 13898
Hetester MS, 13899
Hetester PCA, 13900
Hetester PCP, 13901
Hetester PHA, 13902
Hetester PMA, 13903
Hetester SSS, 13904
Hetester TICC, 13905
Hetlan AC, 13906
Hetoxamate 200 DL, 13907, *See* PEG-4 dilaurate series, 23050
Hetoxamate 400 DS, 13908, *See* PEG-150 distearate series, 23037
Hetoxamate FA-5, 13909, *See* PEG tallate series, 23035
Hetoxamate LA-5, 13910
Hetoxamate MO-2, 13911
Hetoxamate SA-5, 13912
Hetoxamide C-4, 13913
Hetoxamine 0-2, 13914
Hetoxamine C-2, 13915
Hetoxamine S-2, 13916, *See* PEG soyamine series, 23033
Hetoxamine ST-5, 13917
Hetoxamine T-2, 13918
Hetoxide BN-13, 13919
Hetoxide BY-1.8, 13920
Hetoxide C-2, 13921
Hetoxide C-200-50%, 13922
Hetoxide DNP-4, 13923, *See* Ethal DNP-8, 10919
Hetoxide G-7, 13924, *See* Glycereth series, 13069
Hetoxide HC-16, 13925
Hetoxide MPC, 13926
Hetoxide NP-4, 13927
Hetoxide P-3, 13928
Hetoxol C-24, 13931
Hetoxol CA-2, 13932
Hetoxol CAWS, 13933
Hetoxol CD-4, 13934
Hetoxol CS, 13935
Hetoxol CS-4, 13936
Hetoxol D, 13937
Hetoxol G, 13938
Hetoxol IS-2, 13939
Hetoxol J, 13940
Hetoxol L-4, 13941
Hetoxol LS-9, 13942
Hetoxol M-3, 13943
Hetoxol MP-3, 13944, *See* PPG-3 myristyl ether, 24844
Hetoxol OA-3 Special, 13945, *See* oleth-8, 22126
Hetoxol OL-2, 13946, *See* oleth-2, 22125
Hetoxol P, 13947
Hetoxol PLA, 13948
Hetoxol SP-15, 13949
Hetoxol STA-2, 13950, *See* steareth series, 29596
Hetoxol TD-3, 13951, *See* Trideceth series, 32232
Hetoxol TDEP-15, 13952
Hetphos OA-3, 13953, *See* oleth-phosphate Series (2-20), 22127
Hetphos SG, 13954
Hetquat S-20, 13955
Hetron®, 13956
Hetoxol 15 CSA, 13929
Hetron® 92, 13957
Hetron® 99P, 13958
Hetron® 197-3, 13959
Hetron® 197AT, 13960
Hetron® 692, 13961
Hetron® 700 DMA, 13962
Hetron® 800, 13963
Hetoxol 916P, 13930
Hetron® 922, 13964
Hetron® FR 991, 13965
Hetsorb L-4, 13968
Hetsorb L-10, 13966
Hetsorb L-20, 13967
Hetsorb L-80-72%, 13969
Hetsorb O-5, 13971
Hetsorb O-20, 13970
Hetsorb P-20, 13972
Hetsorb S-4, 13974
Hetsorb S-20, 13973
Hetsorb S-20, S-4, *See* Ethylan® GES6, 11108
Hetsorb TO-20, 13975
Hetsorb TS-20, 13976, *See* Polysorbate 65, 24571
Hetsulf 40, 40X, 13977
Hetsulf 50A, 13978, *See* Ammonium dodecylbenzene sulfonate, 2238
Hetsulf 60S, 13979
Hetsulf 60T, 13980
Hetsulf Acid, 13981
Hetsulf IPA, 13982
Heucophos™ZPO, *See* Delaphos, 8008
Heucophos™ZPZ, *See*,
Heusler alloy, 13983
Heveacrumb, 13984
Heveasyn Nitrile Latex, *See* Acrylonitrile-butadiene rubber, 437
Heveasyn Polychloroprene latex, *See* Baypren® 110, 3828
Heveatex, 13985
Hevikote, 13986
Hevyteck, *See* Kaydol, 15810, Klearol, 16208
hex, *See* hexamine, 14008
hexa c.b, *See* Julin's chloride, 15632
hexa-2,4-dienoic acid, *See* sorbic acid, 29105
Hexaammonium molybdate, *See* Ammonium molybdate, 2244
Hexabalm, *See* hexachlorophene, 13992
Hexabetalin, 13987, *See* pyridoxine hydrochloride, 25483, vitamin B_6 hydrochloride, 33937
Hexabione hydrochloride, *See* pyridoxine hydrochloride, 25483 vitamin B_6 hydrochloride, 33937
Hexabromocyclododecane, 13988, *See* FR-1206, 12363, Great Lakes CD-75P, 13300, Saytex® HBCD-LM, 27522
1,2,5,6,9,10-Hexabromocyclododecane, *See* Hexabromocyclododecane, 13988, Saytex® HBCD-LM, 27522
,Hexabutyl distannoxane, *See* Keycide® X-10, 16153, tributyltin oxide, 32209
Hexacal, 13989
Hexacarb, 13990
Hexacert, 13991
hexachlornaphthalene, *See* Halowax 1014, 13564
Hexachlorobenzene, *See* Julin's chloride, 15632
Hexachlorodimethyl sulfone, *See* Amerstat® 294, 2044, bistrichloromethylsulfone, 4285, Stauffer N-1386®, 29580
Hexachlorodivinyl ether, *See* chloroxethose, 6157
1,2,3,4,10,10-hexachloro-1 α,4α,4aβ,5α,8α,8aβ-hexahydro1,4:5,8-dimethanonaphthalene, *See* Aldrin, 1558
6,7,8,9,10,10-hexachloro-1,5,5a,6,9,9a-hexahydro-6,9-methano-2,4,3-benzodioxathiepin 3-oxide, *See* thiodan, 31680
hexachlorophene, 13992, *See* Fascol, 11498
Hexacide, 13993
Hexacol, 13994
hexadecanamide, *See* oleyl palmitamide, 22133
1-hexadecanamine, *See* Amine 16D, 2130, Armeen® 16, 2935, Armeen® 16D, 2936, Crodamine 1.16D, 7101
1-Hexadecanamine, N,N-dimethyl-, *See* Crodamine 3.A16D, 7106
Hexadecanaminium, N,N,N-trimethyl-, bromide, *See* Rhodaquat® M242B/99, 26643
Hexadecanaminium, N,N,N-trimethyl-, chloride, *See* Rhodaquat® M242C/29, 26644
1-Hexadecanaminium, N,N,N-trimethyl-, chloride, *See* Rhodaquat® M242C/29, 26644
hexadecanoic acid, *See* palmitic acid, 22631, 22632, 22633, 22634, 22635, Wickenol® 111, 34379
hexadecanoic acid hexadecyl ester, *See* cetyl palmitate, 5822, Kemester® CP, 15955, Kessco® 653, 16100, Radia® 7500, 25742, Waxenol® 815, 34238
Hexadecanoic acid isopropyl ester, *See* Kessco® IPP, 16116, Radia® 7200, 25730, Unimate® IPP, 32972
hexadecanoic acid methyl ester, *See* methyl palmitate, 19555
Hexadecanoic acid, 2-ethylhexyl ester, *See* Unimate® EHP, 32970, Wickenol® 155, 34391
Hexadecanoic acid, isopropyl ester, *See* Wickenol® 111, 34379
Hexadecanol, *See* Exxal® 16, 11341, Adol® 52 NF, 782, Cachalot® C-50 NF, 4878, cetyl alcohol, 5817, Rita CA, 26788, Adol® 52 NF, 782, Cachalot® C-50 NF, 4878, cetyl alcohol, 5815
1-hexadecanol, *See* cetyl alcohol, 5817, 5816
Hexadecan-1-ol, *See* Adol® 52 NF, 782, Cachalot® C-50 NF, 4878
Hexadecanol NF, *See* Epal® 16NF, 10626
1-Hexadecanol, acetate, *See* Acelan A, 268
1-hexadecanol lactate, *See* cetyl lactate, 5820
1-hexadecene, *See* Neodene® 16, 20898
hexadec-1-ene, *See* Neodene® 16, 20898
hexadecene-1, *See* Neodene® 16, 20898
Hexadecenyl succinic anhydride, *See* Milldride® HDSA, 19783
Hexadecyl acetate, *See* Acelan A, 268
Hexadecyl alcohol, *See* Adol® 52 NF, 782, Cachalot® C-50 NF, 4878
n-hexadecyl alcohol, *See* cetyl alcohol, 5815
n-Hexadecyl ethanoate, *See* Acelan A, 268
Hexadecyl 12-hydroxy-9-octadecenoate, *See* cetyl ricinoleate, 5824
Hexadecyl methacrylate, *See* Ageflex FM-68, 988
hexadecyl palmitate, *See* cetyl palmitate, 5822, Kemester® CP, 15955, Kessco® 653, 16100, Radia® 7500, 25742, Rewowax CG, 26434, Waxenol® 815, 34238
n-Hexadecyl pyridinium bromide monohydrate, *See* Acetoquat CPB, 301
Hexadecyl trimethyl ammonium bromide, *See* Catinal

sodium erythorbate, 28762
Hexermin, *See* pyridoxine hydrochloride, 25483, vitamin B$_6$ hydrochloride, 33937
Hexeth-4 carboxylic acid, *See* Akypo LF 3, 1219
Hexetidine, 14030, *See* Sterisil, 29739
Hexide, *See* hexachlorophene, 13992
Hexil, 14031, *See* Dipicrylamine, 8607
hexite, *See* Dipicrylamine, 8607, Hexil, 14031
Hexnitrol, 14032
Hexo, 14033
Hexobion, *See* pyridoxine hydrochloride, 25483, vitamin B$_6$ hydrochloride, 33937
Hexoll, 14034
Hexomax, 14035
hexonic acid, 2-ethyl-, *See* 2-ethylhexoic acid, 11137
hexophene, *See* hexachlorophene, 13992
Hexoran, 14036
Hexoran A15, 14037
Hexosan, *See* hexachlorophene, 13992
Hexsotate, 14038
hexyl, *See* Dipicrylamine, 8607, Hexil, 14031
n-Hexyl acrylate, *See* Ageflex FA-6, 977, Ageflex n-HA, 996
n-hexyl acrylate, *See*
hexyl alcohol, 14039, *See* Emery® 3321, 10165, 1-hexanol, 14011
***sec*-Hexyl Alcohol**, *See* methyl amyl alcohol, 19512
n-Hexyl alcohol, *See* 1-hexanol, 14011
Hexyl alcohol, *See* Exxal® 6, 11334
2-n-Hexyl cyclopentanone, *See* Jasmatone, 15531
Hexyl Jasmat®, 14040
Hexyl laurate, *See* Cetiol® A, 5796
n-Hexyl methacrylate, *See* SR-211, 29366
Hexyl phosphate, *See* Hostaphat HI, 14385
4-Hexyl-1,3-benzenediol compd with 9-acridinamine(1:1), *See* Acrisorcin, 392
hexylene, *See* 1-hexene, 14029, Neodene® 6, 20886
hexylene glycol, 14041, 14042, *See* Isol, 15366
Hexyloxypropylamine, *See* Tomah PA-10, 31979
Hexyltrichlorosilane, *See* CH7332, 5842
n-Hexyltrichlorosilane, *See* CH7332, 5842
3-hexyne, 2,5-dimethyl-2, 5-di(*t*-butylperoxy)-, *See* Aztec® 2,5-Tri, 3552, Polyvel CR-L10, 24660
Heyn's reagent, 14043
HF 120M, *See* Algin, 1591
HFC, 14044
HFC 60, *See* Algin, 1591
HHDN, *See* Aldrin, 1558
Hi Temp EC-1000, 14045
Hi Temp EC-4000, 14046
Hi Temp EC-5000, 14047
Hibbo, 14048
Hibicare, *See* Rotersept, 27002
Hibicare, *See* chlorhexidine digluconate, 6101
Hibiclens, 14049, *See* chlorhexidine digluconate, 6101 Rotersept, 27002
Hibidil, 14050, *See* Rotersept, 27002, chlorhexidine digluconate, 6101
Hibiscrub, 14051, *See* chlorhexidine digluconate, 6101, Rotersept, 27002
Hibisol, 14052, *See* Rotersept, 27002, chlorhexidine digluconate, 6101
Hibispray, 14053
Hibistat, *See* Rotersept, 27002, *See* chlorhexidine digluconate, 6101
Hibitane, 14054, *See* chlorhexidine digluconate, 6101, Rotersept, 27002
Hibitane chlorhexidine gluconate, *See* chlorhexidine digluconate, 6101, Rotersept, 27002
Hi-bor, 14055
Hibosol, 14056
Hibudine, 14057
Hi-Build, 14058
Hi-Carbolon®, 14059
Hi-Care® 1000, 14060
Hi-Cat, 14061
Hickstor, 14062, *See* Tecgran, 30886, tecnazene, 30898
Hickstor 6 + MBC, 14063
Hicond-2000, 14064
Hicore 90, 14065
HiD 9301, 14066
HiD 9602, 14067
HiD 9632, 14068
HiD 9650, 14069
hidaco crystal violet, *See* gentian violet, 12845
hidaco oil orange, *See* Sudan I, 29938
Hi-Deratol, *See* vitamin D+72, 33940
Hidosin, 14070
Hiduminium, 14071
Hi-Ex Foam, 14072
HiFax AB 6023, 14073
HiFax CA 45A, 14074
HiFax CB 17AC, 14075
HiFax ETA 3011, 14076
HiFax ETA 3095, 14077
HiFax ETA 5012, 14078
HiFax RTA 3263E, 14079
Hi-Fibre, 14080
Higalmetox, *See* methoxychlor, 19504
Higaluron, *See* Talisman, 30735, Tolurex, 31958
HiGel, 14081
High acyl gellan gum, *See* Gellan gum, 12736, Kelco-Gel® Gellan Gum, 15827
High density polyethylene, *See* Compound 403/401, 6684
High solids fluidized polymer suspension, *See* Admiral® FPS Type 3089 FS, 751
High Temperature Deodorant #4896, OS, 14082
high tensile brass, 14083
High Trees Mixture B, 14084
Highlink 40®, 80®, 14085
Highly dispersed pyrogenic silica, *See* Aerosil®, 872
high-strength hydrogen peroxide, *See* hydrogen peroxide, 14590
Hightensite, 14086
Highuron, *See* Talisman, 30735, Tolurex, 31958
HiGlass BJ44A, 14087
HiGlass PF062-2, 14088
HiGlass PF072-1, 14089
HiGlass SB 224-2, 14090
Hi-Gloss 1, 14091
Higosan, *See* Agallol, 954, Atiran, 3286, Ceresan, 5745
Hi-heet, 14092
Hiirogane, 14093
Hilda, *See* thiodan, 31680
Hildit, *See* DDT, 7820
hillebrandite, *See* calcium silicate, 4969
Hills-McCanna Alloy No.45, 14094
Hinge, 14095
Hinochloa, *See* mefenacet, 19229
Hinochloa®, 14096
Hinosan, *See* Edifenphos, 9618
Hinosan®, 14097
Hi-oleic safflower oil, *See* Neobee® 18, 20872, safflower oil, 27222
Hiotrol, 14098
Hioxy, *See* hydrogen peroxide, 14590
Hipec®, 14099
Hipercilina, *See* Penicillin G potassium, 23109
Hipernick®, 14100
Hipersil®, 14101
Hipersolv, 14102
Hi-Pflex® 100, 14103
Hi-pHase® 35, 14104
Hi-pHorm 67, 14105
Hipochem B-3-M, 14106, *See* n-Butyl benzoate, 4790
Hipochem Carrier 761, 14107
Hipochem Compatibilizer WMC, 14108
Hipochem EK-18, 14109
Hipochem GM, 14110
Hipochem Jet Dye T, 14111
Hipochem Jet Scour, 14112
Hipochem M-51, 14113
Hipochem Migrator J, 14114
Hipochem MS-BW, 14115
Hipochem MTD, 14116
Hipochem PDO, 14117
Hipochem SRC, 14118
Hipofix 491, 14119
Hi-Point® 90, 14120, *See* Esperfoam® FR, 10842
Hipolon New, 14121
Hiposcour® 3-80, 14122
Hiposcour® BFS, 14123
Hipowet IBS, 14124
Hippodin, *See* Hippuran-^{131}I, 14125, Nephroflow, 20989
Hippuran I-131, *See* Hippuran-^{131}I, 14125
Hippuran-$_{131}$I, 14125
Hippuric acid, 14126
***p*-hippuric acid sodium salt**, *See* sodium *para*-aminohippurate, 28804
Hippuryl Amide, 14127
Hipputope, 14128
Hipputope, *See* Hippuran-^{131}I, 14125
HiPure Liq. Gelatin, Cosmetic Grade, *See* gelatin, 12722
Hirathiol, 14129, *See* Perichthol, 23270
Hi-Selon, 14130
Hi-Sil, 14131
Hismanal, 14132
Hi-Sol®, 14133
Hi-Sol® 10,15,70, 14134
Hispor 45WP, 14135
Histalog, *See* betazole hydrochloride, 4072
Histamen, *See* Hismanal, 14132
Histaminos, *See* Hismanal, 14132
Histazarin, 14136
Histazol, *See* Hismanal, 14132
L-histidine, *See* histidine,
Histimin, *See* betazole hydrochloride, 4072
histoacryl, *See* Enbucrilate, 10537
Histo-Acryl, 14137
Hitac 300, 14138
HiTEC®, 14139
HiTEC® 300, 14140
HiTEC® 800, 14141
HiTEC® 2900, 14142
HiTEC® 4000, 14143
HiTEC® 4400, 14144
HiTEC® 4700, 14145
Hitenso, 14146
HiTint, 14147
Hitox®, 14148
Hittorf's phosphorus, *See* violet phosphorus, 33876
Hi-Yield Dessicant H-10, *See* arsenic acid, 3137
Hi-Zex, 14149
H-K Mastitis, 14150
HM-0230, 14151
HM-0652, 14152
HM-0814, 14153
HM-6300, 14154
HMB, *See* Cyasorb® UV 9, 7495, Neo Heliopan® BB, 20867, Rhodialux A, 26675
HMDS, *See* hexamethyldisilazane, 14003, Prosil® HMDS, 25242
HMDSO, *See* CH7310, 5841, Hexamethyldisiloxane, 14004
HMN, *See* Permethyl 101A, 23387
HMT, *See* Grasselerator 102, 13289
HMTA, *See* Grasselerator 102, 13289, hexamethylene tetramine, 14006
Hobane, 14155, 14156
HOBt, *See* 1-hydroxybenzotriazole, 14659
Hodag 20-L, 14157
Hodag 22-L, 14158, *See* PEG-4 dilaurate series, 23050
Hodag 40-O, 14159
Hodag 40-R, 14160, *See* PEG ricinoleate series, 23031
Hodag 42-0, 14161, *See* PEG-2 dioleate series, 23040
Hodag 150-S, 14162
Hodag 602-S, 14163, *See* PEG-150 distearate series, 23037
Hodag Antifoam CO-350, 14164
Hodag Antifoam F-1, 14165
Hodag Antifoam FD-82, 14166
Hodag C-100-L, 14167, *See* lauryl hydroxyethyl imidazoline, 16865
Hodag C-100-O, 14168
Hodag C-100-S, 14169, *See* stearyl hydroxyethyl

imidazoline, 29606
Hodag C-100-T, 14170
Hodag CC-22, 14171
Hodag CSA-80, 14173
Hodag CSA-101, 14172
Hodag DGL, 14174
Hodag DGO, 14175
Hodag DGS, 14176
Hodag DGS-C, 14177
Hodag DOSS-70, 14178
Hodag EGMS, 14179
Hodag GML, 14180
Hodag GMO, 14181
Hodag GMP, 14182
Hodag GMR, 14183
Hodag GMS, 14184
Hodag GTO, 14185
Hodag Nonionic 1017-R, 14186
Hodag Nonionic 1035-L, 14187
Hodag Nonionic 1044-L, 14188
Hodag Nonionic 1064-L, 14189
Hodag Nonionic 1088-F, 14190
Hodag Nonionic 2017-R, 14191
Hodag Nonionic E-5, 14192
Hodag Nonionic GR-8, 14193
Hodag Nonionic GRH-25, 14194
Hodag Nonionic ID-5, 14195
Hodag Nonionic L-4, 14196
Hodag Nonionic S-2, 14197, *See* steareth series, 29596
Hodag Nonionic TD-15, 14198 , *See* Trideceth series, 32232
Hodag PB-285, 14199, *See* PPG butyl ether series, 24841
Hodag PE-005, 14200
Hodag PE-005-K, 14201
Hodag PE-1803, *See* oleth-phosphate Series (2-20), 22127
Hodag PEG 200, 14202
Hodag PEG 300, 14203
Hodag PEG 400, 14204
Hodag PEG 540, 14205
Hodag PEG 600, 14206
Hodag PEG 1000, 14207
Hodag PEG 1450, 14208
Hodag PEG 3350, 14209
Hodag PEG 8000, 14210
Hodag PGL, 14211
Hodag PGL-101, 14212
Hodag PGML, 14213
Hodag PGMP, 14214
Hodag PGMS, 14215
Hodag PGO-61, 14222
Hodag PGO-62, 14223
Hodag PGO-101, 14216
Hodag PGO-102, 14218
Hodag PGO-103, 14219
Hodag PGO-104 (formerly Hodag SVO-1047), 14220
Hodag PGO-108, 14221
Hodag PGO-1010 (formerly Hodag SVO-10107), 14217 , *See* Polygliceryl-10 decaoleate, 24453
Hodag PGS, 14224
Hodag PGS-61, 14231
Hodag PGS-62, 14232
Hodag PGS-101, 14225
Hodag PGS-101, PGS-61, *See* Polyglyceryl stearate series, 24457
Hodag PGS-1010, 14226
Hodag PGS-102, 14227
Hodag PGS-103, 14228
Hodag PGS-104, 14229
Hodag PGS-108, 14230
Hodag PGSH, 14233
Hodag PGSH-61, 14234
Hodag PGSH-62, 14235
Hodag Polyglycol 5035, 14236
Hodag PSML-20, 14237
Hodag PSMO-20, 14238
Hodag PSMP-20, 14239
Hodag PSMS-20, 14240
Hodag PSMS-20, SVS-18, *See* Ethylan® GES6, 11108
Hodag PSTS-20, 14241, *See* Polysorbate 65, 24571
Hodag SML, 14242
Hodag SMO, 14243
Hodag SMP, 14244
Hodag SMS, 14245
Hodag SMS, *See* Ablunol S-60, 110
Hodag SSO, 14246
Hodag STO, 14247
Hodag STS, 14248
Hodag SVO-1047, 14249
Hodag SVO-629, 14250
Hodag SVO-9, 14251
Hodag SVS-18, 14252
Hoe 2671, *See* thiodan, 31680
HOE 2873, *See* Afugan, 952
HOE 2960, *See* triazophos, 32196
Hoe 02982, *See* heptenophos, 13795
Hoe 17411, *See* Carbendazim, 5187
HOE-17411, *See* Derosal, 16327
Hoe S 1816, 14253
Hoe S 1984 (TP 2279), 14254
Hoe S 2650, 14255
Hoe S 2713, 14256
Hoe S 2721, 14257
Hoe S 2749, 14258
Hoe S 3618, 14259
Hoe S 3680, 14260
HOE-881v, *See* fenbendazole, 11594
Hoechst New Blue, 14261
Hoechst Wax C, 14262
Hoechst Wax E, 14263
Hoechst Wax OP, 14264
Hoechst Wax PE 190, 14265
Hoechst Wax PE 520, 14266
Hoechst Wax PED 191, 14267
Hoechst Wax PED 521, 14268
Hoechst Wax PP 230, 14269
Hoechst Wax S, 14270
Hoechst Wax S, *See* Montan acid wax, 20262
Hoechst-095K, *See* acesulfame potassium, 277
Hoegrass, 14271
Hoenle's cement, 14272
Holcote, 14273
Holdfast, *See* paclobutrazol, 22540, 22541
Holdfast D, 14274
Holfos bronze, 14275
Holite, 14276
holligold, *See* Calendula Oil CLR, 4980
Hollofil®, 14277
holmia, *See* holmium oxide, 14279
holmium, 14278
holmium oxide, 14279
Holtox, 14280
Homac, 14281
Homagenets Aorl, 14282
Homberg's metal, 14283
Homberg's phosphorus, 14284
Homberg's salt, 14285
Hombifine®, 14286
Hombisorp®, 14287
Hombitan®, 14288
Hombitec®, 14289
Homidium chloride, *See* Novidium, 21717
Homo Size 7A, 14290
homoanisic acid, *See* *p*-methoxyphenylacetic acid, 19505
Homocatechol methyl ester, *See* creosol, 6990
Homocatechol monomethyl ether, *See* creosol, 6990
Homokol, 14291
homomenthyl salicylate, *See* homosalate, 14292, Kemester® HMS, 15963
Homoolan, *See* Valadol, 33238
homophan, *See* paratophan,
Homosalate, 14292, *See* Filtrosol A, 11800, Heliophan, 13766, Homosalate, 14292, Kemester® HMS, 15963
Hondostab, 14293
Hondurite, 14294
Honestone, *See* whetstone, 34352
Honey, 14295
Honey Wax, 14296
Honeycat, 14297
hoof oil, *See* neatsfoot oil, 20837
Hopcalite I, 14298
Hopkin's lactic acid reagent, 14299
Hopkin's-Cole reagent, 14300
Hopkin's-Cole Tyrosine C reagent, 14301
Hopp II, 14302
Hopper salt, 14303
Hoppit, 14304
Horco X, 14305
Hordaflex, Hordalub, 14306
Hordamer, 14307
Horizon, *See* Raxil, 25908
Horizon/Horizont, 14308
Horizon®, 14309
Hormodin, *See* 4-indol-3-ylbutyric acid, 15086
hormodin IBA, *See* 4-indol-3-ylbutyric acid, 15086
Hormone Rooting Powder, 14310
Hormotuho, *See* MCPA, 19183
horn fiber, *See* Vulcanized fiber, 34097
Horn O'Plenty, 14311
Horna, 14312
Hortag Aquasulf, 14313
Hortag Carbotec, 14314
Hortag Tecnacarb Dust, 14315
Hortag Tecnazene Plus, 14316
Hortag Thiram, 14317
Hortichem Spraying Oil, 14318
Hortus, 14319
Hospex, *See* glutaraldehyde, 13062
Hospex, *See* Sonacide, 29073
Hostacerin CG, 14320
Hostacerin DGO, 14321
Hostacerin DGS, 14322
Hostacerin O-20, 14323, *See* oleth-8, 22126
Hostacerin T-3, 14324
Hostacerin WO, 14325
Hostacor, 14326
Hostacor 2098, 14327
Hostacor 2125, 14328
Hostacor BBM, 14329
Hostacor H Liq. N, 14330
Hostacor TP 2445, 14331
Hostadrill Brands, 14332
Hostadur, 14333
Hostaflam, 14334
Hostaflex, 14335
Hostaflon®, 14336
Hostaflon® C2, 14337
Hostaflon® ET, 14338
Hostaflon® TF 1101, 14339
Hostaflon® TF 1101, 1620, 2071, 5032, 5537, *See* Polytetrafluoroethylene, 24642
Hostaflon® TF 1620, 14340
Hostaflon® TF 2071, 14341
Hostaflon® TF 5032, 14342
Hostaflon® TF 5537, 14343
Hostaflot L Grades, 14344
Hostaform, 14345
Hostaform C 2521, 14346
Hostaform C 9021 ELS, 14347
Hostaform C 9021 K, 14348
Hostaform S 27076, 14349
Hostalen®, 14350
Hostalen® EP 4450, 14351
Hostalen® G, 14352
Hostalen® GB 6950, 14353
Hostalen® GM 5010 T2, 14354
Hostalen® GM 7745 HP, 14355
Hostalen® GUR 5121, 14356
Hostalen® OO, 14357
Hostalen® PP 927, 14358
Hostalit, 14359
Hostalit Z, 14360
Hostalub®, 14361
Hostalub® CAF 484 SB, 14362
Hostalub® FA 1, 14363

Hostalub® H 4, 14364
Hostalub® VP Ca W 2, 14365
Hostalub® VP Ca W 2, *See* calcium montanate, 4959
Hostalub® WE4, 14366
Hostalub® WE4, *See* glyceryl montanate, 13086
Hostalub® XL 165, 14367
Hostalux KCB, 14368
Hostamer Brands, 14369
Hostamid, 14370
Hostamont, 14371
Hostanox®, 14372
Hostanox® 03, 14373
Hostanox® OSP 1, 14374
Hostanox® PAR 24, 14375
Hostanox® SE 10, 14376
Hostapal 2345, 14377
Hostapal N-040, 14378
Hostaperm, 14379
Hostaphan, 14380
Hostaphane, 14381
Hostaphat, 14382
Hostaphat 2122, 14383
Hostaphat AR K, 14384
Hostaphat HI, 14385
Hostaphat KL 340N, 14386
Hostaphat KO 300, 14387
Hostaphat KO 380, 14388
Hostaphat KW 340 N, 14389
Hostapon CAS, 14390
Hostapon CT Paste, 14391
Hostapon IDC, 14392
Hostapon KA Powd, 14393
Hostapon KTW, *See* sodium lauroyl taurate, 28780
Hostapon KTW New, 14394, *See* sodium lauroyl taurate, 28780
Hostapon SO, 14395
Hostapon STT Paste, 14396, *See* sodium methyl stearoyl taurate, 28792
Hostapon T Powd, 14397
Hostapon TF, 14398
Hostapor, 14399
Hostapren, 14400
Hostaprime® HC 5, 14401
Hostaprint, 14402
Hostapur DOS Hi conc, 14403
Hostapur DTC, 14404
Hostapur OS, 14405
Hostapur SAS, 14406
Hostaquick, 14407
Hostaquick, *See* heptenophos, 13795
Hostarex Grades, 14408
Hostastat®, 14409
Hostastat® FA 14, 14410
Hostastat® System E1956, 14411
Hostatec, 14412
Hostathion, 14413, *See* triazophos, 32196
Hostatint, 14414
Hostatron®, 14415
Hostatron® System P1941, 14416
Hostavin®, 14417
Hostavin® ARO 8, 14418
Hostavin® N 30, 14419
Hostawet TDC, 14420
Hostcyclin, *See* tetracycline, 31365
Hostiren, 14421
Hot Melt Wetnes Indicator®, 14422
Hotbac®, 14423
Hotspur, 14424
Hotspur, 14425
houillite, 14426
House Plant Leaf Shine, 14427
House Plant Liquid Feed, 14428
House Plant Pest Killer, 14429
Houseplant Long Lasting Feed, 14430
Howard's silver, 14431
Howflex, 14432
Howstik, 14433
Howtex, 14434
Howtol, 14435
Hoyle's metals, 14436
Hoyt Metal, 14437
HPC-9, 14438, *See* Esperfoam® FR, 10842
HRS-16 Pentac, *See* dienochlor, 8429
HS, *See* mustard gas, 20538
HS 85S, *See* Nacconol® 40G, 20647
HSB 1900, 14439
H-scale, 14440
HSZ-320NAA, 14441
HT Non-Ionic Wetter, 14442
HTAB, *See* Rhodaquat® M242B/99, 26643
HTH, 14443
HTH, *See* Repak, 26141
HT-Proteolytic Conc, 14444
HTSA #1, 14445
HTSA #3, 14446
HTSA #3, *See* stearyl erucamide, 29605
Hünefeld solution, 14447
Hubel's reagent, 14448
Huber 40C, 14449
Huber 65A, 14450
Huber 95, 14451
Huberbrite 1, 14452
Hubercarb® Q 6-20, 14453
Hubercarb® W 2, 14454
Huber's reagent, 14455
Hudroson, 14456
Hugel A, 14457
Hugel AH, 14458
Hugel B, 14459
Hugel BC 10, 14460
Hugel CH14, 14461
Hulot's solder, 14462
Humber, 14463
Humectant SD-35, 14464
Humectant SD-35, *See* Panthenylethyl ether, 22691
Humectol®, 14465
Humifen, 14466
Humifen BA-77, 14467
Humifen BX-78, 14468
Humifen nbl 85, *See* Ablusol ML, 129
humko industrene r, *See* stearic acid, 29599
Hungazin, *See* atrazine, 3394
Huntsman 201, 14469
Huntsman 240, 14470
Huntsman 312, 14471
Huntsman 331, 14472
Huntsman 351, 14473
Huntsman 474, 14474
Huntsman 765, 14475
Huppert's reagent, 14476
Huron, 14477
Hurr nut, 14478
Husept extra, *See* chloroxylenol, 6159
Husman metal, 14479
HVA-2, 14480
HVP, 14481, *See* hydrolyzed vegetable protein, 14613
HWG 1608, *See* Raxil, 25908
Hyacid, *See* hyaluronic acid sodium salt, 14483
Hyadur, *See* dimethyl sulfoxide, 8511, DMSO, 8762
Hyalase, 14482
Hyalgan, *See* hyaluronic acid sodium salt, 14483
Hyalovet, *See* hyaluronic acid sodium salt, 14483
hyaluronic acid, *See* Cromoist HYA, 7200
Hyaluronic Acid Sodium, *See* Amo Vitrax®, 2266
hyaluronic acid sodium salt, 14483, *See* ActiMoist, 486
Hyamine, *See* benzethonium chloride, 3963
Hyamine 1622, *See* benzethonium chloride, 3963
Hyamine® 10X, 14484
Hyamine® 1622 50%, 14485
Hyasorb, *See* Penicillin G potassium, 23109
Hybaite, 14486
Hyban, 14487
Hyb-lum, 14488
Hybon® 2011, 14489
Hybri-Chem 100, 14490
Hybrin, *See* vitamin C, 33938
Hy-Brite, 14491
Hybrite®, 14492
Hycal, 14493
Hycar 1552, *See* Acrylonitrile-butadiene rubber, 437
Hycar 1561, *See* Acrylonitrile-butadiene rubber, 437
Hycar 1571, *See* Acrylonitrile-butadiene rubber, 437
Hycar 1572, *See* Acrylonitrile-butadiene rubber, 437
Hycar 1572x64, *See* Acrylonitrile-butadiene rubber, 437
Hycar® 1203X17, 14494
Hycar® 1204X5, 14495
Hycar® 1204X9, 14496
Hycar® 1205X3, 14497
Hycar® 1273, 14498
Hycar® 1300X8, 14499
Hycar® 1402 H82, 14500
Hycar® 1402 H83, 14501
Hycar® 1403 H84, 14502
Hycar® 1552, 14503
Hycar® 1577, 14504
Hycar® 2100, 14505
Hycar® 2550H33, 14506
Hycar® 2550H5, 14507
Hycar® 2550H55, 14508
Hycar® 2570H28 and 2570H29, 14509
Hycar® 2570X5, 14510
Hycar® 26345, 14511
Hycar® 2671H49, 14512
Hycar® 4021, 14513
Hycar® 4032, 14514
Hycar® 4043, 14515
Hycar® 4201, 14516
Hycar® ATBN, CTB, CTBN, VTBN, 14517
Hycar® Reactive Liquid Polymer, 14518
Hycathane, 14519
Hycel M Series, 14520
Hy-Chlor, *See* Repak, 26141
Hychlorite, *See* sodium hypochlorite, 28773
Hycol, 14521
Hycolln, 14522
Hycon, 14523
Hycote, 14524
Hydagen DEO, *See* Oxyguard, 22457
Hydagen® B, 14525
Hydagen® C.A.T, 14526
Hydagen® C.A.T, *See* triethyl citrate, 32242
Hydagen® DEO, 14527
Hydan, 14528
Hydan®, 14529
Hydantoin ester, *See* Glytex® 663, 13157
Hydantoin, 3-bromo-1-chloro-5,5-dimethyl-, *See* Bromochloro dimethyl hydantoin, 4574, Dantoin® GSD-550, 7733, Halobrom, 13560, Quesbrom, 25657
Hydecat, 14530
Hydex® 100 Gran.206, 14531
Hydon, 14532
Hydracillin, *See* Penicillin V, 23111
Hydraffin, 14533
Hydra-guard, 14534
Hydral® 710, 14535
Hydralin, 14536
Hydramyl, *See* Pentane, 23164
Hydrangea Colourant, 14537
hydrangin, *See* umbelliferone, 32899
Hydraphen, *See* hydrargaphen, 14539
Hydraphthal, 14538
hydrargaphen, 14539
hydrargyrum, *See* mercury, 19355
Hydrasal®, 14540
Hydrated alumina, *See* Hydral® 710, 14535, Onyx Classica, 22185
Hydrated aluminum oxide, *See* H-30, 13527
Hydrated aluminum silicate, *See* Dixie Clay®, 8743, Langford Clay, 16674, Pyrax® A, 25474, Yellowstone, 34652
Hydrated cellulose, *See* alkali cellulose, 1627
hydrated copper oxide, *See* cupric hydroxide, 7383
hydrated cupric oxide, *See* cupric hydroxide, 7381, 7382
hydrated ferric oxide, *See* ferric hydroxide, 11649
Hydrated lime, *See* calcium hydroxide, 4952, Edelwit, 9601

hydrated magnesium silicate, *See* talc, 30731
Hydrated magnesium-iron-aluminum silicate, *See* vermiculite, 33589
hydrated silica, *See* Akrochem® Rubbersil RS-150/RS-200, 1177, Sipernat® 22, 28519, Sipernat® 50, 28522
Hydrated silicate, *See* Metasil MQC, 19455
hydraulan®, 14541
hydraulic bronze, 14542
hydraulic cements or mortars, 14543
hydraulic limes, 14544
Hydraulic mortar, *See* hydraulic cements, 14543
hydrazine, 14545, *See* Amerzine®, 2047, Ultra Pure, 32780, Zerox, 34738
hydrazine anhydrous, *See* hydrazine, 14545, Ultra Pure, 32780
hydrazine base, *See* hydrazine, 14545
Hydrazine hydrobromide, *See* Hyflux M, 14676
hydrazine monohydrobromide, *See* Hyflux M, 14676
Hydrazine sulfate, 14546
hydrazine yellow, *See* tartrazine, 30835
2-hydrazino-N,N,N-trimethyl-2-oxoethanaminium chloride, *See* Girard's reagent T, 12950
Hydrenol D, 14547
Hydresol®, 14548
Hydrex®, 14549
Hydrholac, 14550
Hydrin® C, 14551
Hydrin® C-CG, 14552
Hydrin® T, 14553
β-hydrindone, *See* 2-indanone, 15063
Hydrine, 14554
Hydriodic acid, 14555
Hydrisan, 14556
Hydro, 14557
Hydro AWC, *See* sodium formaldehydesulfoxylate, 28766
Hydro Paste®, 14558
Hydroabietyl alcohol, *See* Abitol® E, 71
Hydroace Series, 14559
Hydroba, 14560
Hydroblok, 14561
Hydrobol, 14562
Hydroboracite, 14563
Hydrobromic acid, 14564
hydrobuna, 14565
Hydrocarbon, *See* Actracor 800, 521, paraffin, 22742
hydrocarbon black, *See* gas black, 12686
hydrocarbon cement, 14566
Hydrocarbon epoxy, *See* Tactix 556, 30697
hydrocarbon gas black, *See* gas black, 12686
hydrocarbon oil, *See* paraffin, liquid, 22744
Hydrocarbon oil-based defoamer, *See* Advantage 136 Defoamer, 805
Hydrocarbon propellant, *See* propane, 25173
hydrocarbon-aldehyde resins, 14567
Hydrocell YP-30, 14568
Hydrocerol Compound, 14569
Hydrocerol® BIH, 14570
Hydrocerol® CF 70, 14571
Hydrocerol® LC, 14572
Hydrocerol® TAF 50, 14573
Hydro-Chem, 14574
hydrochloric acid, 14575, 14576, *See* marine acid, 18718, soldering acid, 28909
Hydrochloric acid dimethylamine, *See* dimethylamine hydrochloride, 8513
hydrochloric ether, *See* ethyl chloride, 11067
Hydrochloride, *See* hydrochloric acid, 14575
Hydrochlorofluorocarbon 22, *See* Genetron® 22, 12820
hydrocinnamic acid, 14577
Hydrocol®, 14578 , 14579
Hydrocoll AG-SD, PGA, PGB, *See* gelatin, 12722
Hydrocoll AL-50,, *See* Extiat®, 11323
Hydrocoll AL-50, AL-55, EN-40, EN-55, EN-55-X, EN-SD, EN-SD-1M, EN-SD-10M, 14580
Hydrocoll G-40, 14581**,** *See* gelatin, 12722
Hydrocoll G-55, 14582
Hydrocoll™ G-40, G-55, *See* gelatin, 12722
Hydrocolloids, *See* Frigesa® D, F, IC, 12410
hydrocortisone-neomycin, *See* Corton A1, 6859
Hydro-Cure, 14583
hydrocyanic acid, 14584, *See* prussic acid, 25352, zootic acid, 34903
hydrocyanic acid sodium salt, *See* sodium cyanide, 28753
hydrocyanite, *See* cupric sulfate, 7387
Hydrodarco, 14585
Hydroferrox®, 14586
hydrofluoric acid - ammonium fluoride, *See* white acid, 34353
Hydrofluoric acid gas, *See* hydrogen Fluoride, 14589
hydrofluosilicic acid, 14587, *See* keramyl, 16054, sand acid, 27321
hydrofol 1895, *See* stearic acid, 29599
hydrofol acid 150, *See* stearic acid, 29599
hydrofol acid 1255, *See* Philacid 1200, 23660
hydrofol acid 1295, *See* Philacid 1200, 23660
hydrofol acid 1495, *See* Philacid 1400, 23662
Hydrofol Acid 1655, 14588, *See* stearic acid, 29599
hydrofol acid 1855, *See* stearic acid, 29599
hydrogen arsenide, *See* arsine, 3144
hydrogen bromide in acetic acid, *See* hydrobromic acid, 14564
hydrogen carboxylic acid, *See* formic acid, 12271
hydrogen chloride, *See* hydrochloric acid, 14575
Hydrogen chloride (acid), *See* hydrochloric acid, 14575
hydrogen cyanide, *See* zootic acid, 34903
hydrogen dioxide, *See* hydrogen peroxide, 14590
hydrogen fluoride, 14589
hydrogen hexafluorosilicate, *See* hydrofluosilicic acid, 14587, keramyl, 16054
Hydrogen monosulfide, *See* sulfuretted hydrogen, 30053
Hydrogen nitrate, *See* nitric acid, 21332
hydrogen peroxide, 14590, *See* dioxogen, 8577, Genoxide®, 12840, hydrozone, 14672, Oxzone, 22478, ozogen, 22481, Perhydrol, 23268, Peroxal, 23403, Peroxol, 23411, Peroxyl, 23413, Proxy, 25348, Truzone, 32496
hydrogen peroxide sodium carbonate adduct, *See* sodium percarbonate, 28807
hydrogen peroxide-urea, *See* Hyperol, 14712
Hydrogen platinochloride, *See* Lummer's solution, 17809
hydrogen rubeanide, 14591
hydrogen sulfate, *See* sulfuric acid, 30055
hydrogen sulfite sodium, *See* sodium acid sulfite, 28728
Hydrogen telluride, *See* telluretted hydrogen, 31060
Hydrogenated bisphenol A, *See* Millad® HBPA, 19776
hydrogenated castor oil, 14592
hydrogenated cocoglycerides, *See* Softisan® 100, 28857, Witepsol® E75, E76, E85, H5, H12, H15, H19, H32, H35, H37, H39, H42, H175, H185, 34523
Hydrogenated coconut acid, *See* Industrene® 223, 15103
hydrogenated cotton*Seed* oil, 14593, *See* Emvelop®, 10533, Sta-Nut EE, 29508
Hydrogenated dimer acid, *See* Pripol 1025, 25031
Hydrogenated ditallow amine, *See* Kemamine® S-970, 15927
Hydrogenated ditallowamine, *See* Armeen® 2HT, 2929
Hydrogenated jojoba oil, *See* Wickenol® 139, 34383
Hydrogenated lanolin, *See* Fancol HL, 11441, Satulan, 27492
Hydrogenated lecithin, *See* PhosPho H-150, 23710
Hydrogenated palm/palm kernel oil PEG-6 complex, *See* hydrogenated palm/palm kernel oil PEG-6 esters, 14594, Labrafil M 2130 BS, 16541
hydrogenated palm/palm kernel oil PEG-6 esters, 14594 *See* Labrafil M 2130 BS, 16541
hydrogenated polyisobutene, 14595, *See* hydrogenated polyisobutene, 14595, Luvitol HP, 18037, Panalane® L-14E, 22657
Hydrogenated rotenone, *See* rotenone, 27000
Hydrogenated soya lecithin, *See* PhosPho H-00, 23709
hydrogenated soybean oil, 14596, *See* Akorex, 1149
Hydrogenated starch hydrolysate, *See* Lipo Polyol NC, 17344
hydrogenated stripped coconut acid, 14597
hydrogenated tallow acid, 14598, *See* Prifac 9428, 24949
Hydrogenated tallow amine, *See* Amine HBG, 2156, Crodamine 1.HT, 7103, Jet Amine PHT, 15594
(Hydrogenated tallow) amine (primary amine), *See* Armeen® HT, 2950
Hydrogenated tallow amine (tech.), *See* Kemamine® P-970, 15920
hydrogenated tallow amine acetate, *See* Acetamin HT, 287, Radiamac 6149, 25748
Hydrogenated tallow dimethylbenzyl ammonium chloride, *See* Querton 441-BC, 25653
Hydrogenated tallow fatty acid, *See* Glycon® S-65, 13120, hydrogenated tallow acid, 14598
hydrogenated tallow glyceride citrate, 14599, *See* hydrogenated tallow glyceride citrate, 14599, Lamegin® ZE 30, 60, 16620
hydrogenated tallow glyceride lactate, 14600, *See* Axol® L 61, L62, 3513, hydrogenated tallow glyceride lactate, 14600, Lamegin® GLP 20, 16619
Hydrogenated tallow glycerides, *See* Neustrene® 059, 21022
Hydrogenated tallow propanediamine, *See* Radiamine 6540, 25765
N-hydrogenated tallow 1,3-propylenediamine, *See* Kemamine® D-970, 15913
Hydrogenated tallowalkonium chloride, *See* Arquad® DMHTB-75, 3117
Hydrogenated tallowamide, *See* Armid® HT, 2962
Hydrogenated tallowamine, *See* Armeen® HTD, 2951
Hydrogenated tallowamine acetate, *See* Amine Acetate HBG, 2148
Hydrogenated triglycerides, *See* Sett®, 28107
hydrogenated vegetable oil, 14601, *See* Aratex, 2783
hydrogenite, 14602
hydro-giene, *See* cuprous iodide, 7404
hydroiodic acid, *See* hydriodic acid, 14555
Hydrokeratin AL-30, 14603
Hydrokote® 95, 14604
Hydrol 100, 14605
Hydrolact, 14606
Hydrolactin 2500, 14607
Hydrolactol 70, 14608
Hydrolete, 14609, *See* calcium hydride, 4949
Hydrolin, 14610, *See* sodium hydrosulfite, 28771
hydrolith, 14611
Hydrolose,, *See* Glutolin, 13064
Hydrolyzed collagen, *See* Collagen Hydrolyzate Cosmetic 55, 6613, Cropepsol, Cropeptone, 7208, Extiat®, 11323, Lexein® X-250HP, 17131, Polypeptide 10, 24516, Polypeptide 37, 24517
hydrolyzed elastin, 14612 *See* Actigen E, 479, European Elastin 10, 11222, Ritalastin EL-10, 26813, Solu-Lastin 30, 29011
Hydrolyzed fibronectin, *See* Cronectin, 7205
Hydrolyzed keratin, *See* Hydrokeratin AL-30, 14603, Kera-Tein 1000, 16059
Hydrolyzed mucopolysaccharides, *See* Actiglow C, 481
Hydrolyzed polyacrylonitrile, *See* Cypan, 7586
Hydrolyzed reticulin, *See* Reticusol, 26272
Hydrolyzed soya lecithin, *See* Lecithin Water Dispersible CLR, 16941
hydrolyzed vegetable protein, 14613
hydrolyzed wheat protein, 14614
Hydro-Marc, 14615
Hydromethylabietate, *See* Superbrillantoline®, 30222
Hydromilk EN-20, 14616
Hydromol Cream, 14617
Hydromol Emollient, 14618
Hydron Blue, 14619
Hydron®, 14620
Hydronal, 14621
Hydronalium, 14622
hydronaphthol, 14623
hydrone, 14624

Hydronyx, 14625
γ-hydrooxybutyric acid lactone, *See* Butyrolactone, 4803
Hydropalat 535, 14626
Hydropalat® A, 14627
Hydropalat® B, 14628
Hydroperoxide, *See* hydrogen peroxide, 14590
Hydroperoxide,1,1-dimethylethyl, *See* Aztec® *t*-Butyl Hydroperoxide-70, Aq, 3541
2-hydroperoxy-2-methylpropane, *See* Aztec® *t*-Butyl Hydroperoxide-70, Aq, 3541
Hydrophilic fatty acid ester, *See* Witafrol® 7480 N, 34424, Witafrol® 7490, 34425, Witafrol® 7497 N, 34426
Hydrophilic polysaccharide, *See* Algin, 1591
Hydrophilol ISO, 14629, *See* Propylene glycol isostearate, 25218
Hydrophobized silicone, *See* Atlas Defoamer AFC, 3305
Hydrophobol, 14630
Hydropur®, 14631
hydroquinol, *See* hydroquinone, 14632, quinol, 25685, Tecquinol® Tech. Grade, 30910
hydroquinone, 14632, *See* Black and White Bleaching Cream, 4302, Eldopaque, 9723, quinol, 25685
***m*-hydroquinone**, *See* resorcinol, 26230
hydroquinone monomethyl ether, 14633, *See* Eastman® HQMME, 9467
Hydroquinone,1,4-benzenediol, *See* Tecquinol® Tech. Grade, 30910
Hydro-resin A, 14634
Hydros, 14635
Hydros® 1, 14636
Hydros® F, 14637
Hydrosil®, 14638
hydrosilicofluoric acid, *See* keramyl, 16054
hydrosol, 14639, 14640
Hydrosol®, 14641
Hydrosoy 2000, 14642, *See* hydrolyzed vegetable protein, 14613
hydrosulfite, 14643, *See* sodium hydrosulfite, 28771
Hydrosulfite A, 14644
Hydrosulfite A.W, *See* Discolite, 8632
Hydrosulfite AWC, 14645
Hydrosulfite BASF, 14646
Hydrosulfite NF, 14647
Hydrosulfite NF Conc, *See* Hyraldite C EXT, 14730
hydrosulfuric acid, *See* sulfuretted hydrogen, 30053
Hydrosulphit®, 14648
Hydrotek, 14649
hydrotreated light naphthenic distillate, *See* Jayflex® 210, 15544
Hydrotriticum 2000, 14650, *See* hydrolyzed wheat protein, 14614
Hydrotriticum Powd, 14651
Hydrotriticum QL, 14652
Hydrotriticum QM, 14653
Hydrotriticum QS, 14654
Hydrotriticum WAA, 14655
hydrotrope, *See* Eltesol® SX 30, 9873, Naxonate® 4L, 20824, Pilot SXS-40, 23843, Witconate SXS 40%, 34492
Hydrotropes, *See* Stepanate®, 29671
Hydrous magnesium calcium silicate, *See* I T Talc, 14800
Hydrous magnesium silicate, *See* ABT-2500®, 166, talc, 30731
Hydrous silica, *See* Britesorb, 4540
Hydrous sodium polysilicate, *See* Britesil, 4539
hydrous wool fat, *See* lanolin, 16740
2-Hydroxy-4-acryloyloxyethoxy benzophenone, *See* Cyasorb® UV 2098, 7499
2-hydroxy benzoic acid, butyl ester, *See* Brunol, 4625
4-hydroxy-2H-1-benzopyran-2-one, *See* 4-hydroxycoumarin, 14661
7-hydroxy-2H-1-benzopyran-2-one, *See* umbelliferone, 32899
1-Hydroxy-1H-benzotriazole, *See* 1-hydroxybenzotriazole, 14659
3-hydroxy-2-butanone, *See* acetyl methyl carbinol, 312
2-(2'-Hydroxy-3'-*tert*-butyl-5'-methylphenyl)-5-chloro benzotriazole, *See* hydroxybutylmethyl phenylchlorobenzotriazole, 14660, Lowitite® 26, 17677, Uvazol 236, 33181
2-Hydroxy-3-chloropropyltrimethylammonium chloride, *See* Reagens-CF2, 25928
1-(4'-hydroxy-3'-coumarinyl)-1-phenyl-3-butanone, *See* Killgerm® Sewarin P, 16184
1-(4'-hydroxy-3'-coumarinyl)-1-phenyl-3-butanone, *See* Kypfarin, 16519
hydroxy-p-cymene, *See* oxycymol, 22445
2-hydroxy-*p*-cymene, *See* carvacrol, 5362
3-hydroxy-*p*-cymene, *See* thyme camphor, 31775
4-hydroxy-3,5-diiodo-benzenesulfonic acid, *See* Iodozol, 15221
2-hydroxy-3,5-diiodobenzoic acid, *See* Diosal, 8572
Hydroxy-dimercuro ammonium hydroxide, *See* Millon's mase, 19795
2-Hydroxy-1,3-dimethoxybenzene, *See* dimethoxyphenol, 8500
1-hydroxy-3,6-dioxaoctane, *See* Diethoxol, 8435, ethyl di-Icinol, 11068
2-Hydroxy-1,2-diphenylethanone, *See* Benzoin, 3975
2-hydroxy-1-ethanethiol, *See* 2-mercaptoethanol, 19336
hydroxy ether, *See* ethyl icinol, 11074
2-Hydroxy-N-(2-hydroxyethyl)propanamide, *See* Lactamide MEA, 16568
(S)-N,N'-Bis[2-hydroxy-1-(hydroxymethyl)ethyl]-5-[(2-hydroxy-1-oxopropyl)amino]-2,4,6-triiodo-1,3-benzenedicarboxamide, *See* Iopamidol, 15233
α-hydroxy-α,α-bis(4-hydroxy-*m*-tolyl)-o-toluenesulfonic acid γ-sultone, *See* cresol red, 6996
4-hydroxy-3-iodo-5-nitrobenzonitrile, *See* nitroxynil, 21384
8-hydroxy-7-iodoquinoline-5-carboxylic acid, *See* Loretine, 17606
4-Hydroxy-2-keto-4-methylpentane, *See* Diacetone alcohol, 8249
2-hydroxy-4-methoxybenzophenone, *See* Cyasorb® UV 9, 7495, Lankromark® LE296, 16698, Neo Heliopan® BB, 20867, Uvinul® M-40, 33197
2-Hydroxy-4-methoxybenzophenone-5-sulfonic acid, *See* Rhodialux S, 26676 Uval, 33177
4-hydroxy-3-methoxybenzaldehyde, *See* vanillin, 33327, Zimco, 34763
6-hydroxy-7-methoxy-5-benzofuranacrylic acid δ-lactone, *See* Oxsoralen-Ultra, 22437
4-hydroxy-3-methoxy-1-methylbenzene, *See* creosol, 6990
2-hydroxy-4-methoxyphenyl, *See* benzophenone-3,
(2-hydroxy-4-methoxyphenyl)(2-hydroxyphenyl)methanone, *See* Spectra-Sorb UV 24, 29230
(2-hydroxy-4-methoxyphenyl)phenylmethanone, *See* Cyasorb® UV 9, 7495, Rhodialux A, 26675, Uvinul® M-40, 33197
1-hydroxy-4-(2-methyl-2-butyl)benzene, *See* Orthophen® 278, 22336, Pentaphen® 67, 23168
Hydroxy methyl isolongifolene, *See* Amborol, 1972
3-hydroxy-5-methylisoxazole, *See* Tachigaren 70, 30685
1-(8-Hydroxy-6-methyl-1-oxo-2,4,6-dodecatrienyl)-2-pyrrolidinone, *See* Variotin, 33396
4-hydroxy-4-methyl-2-pentanone, *See* diacetone alcohol, 8249
2-Hydroxy-2-methyl-4-pentanone, *See* Diacetone alcohol, 8249
4-hydroxy-4-methylpentan-2-one, *See* Diacetone alcohol, 8249
4-Hydroxy-4-methyl-2-pentanone, *See* Diacetone alcohol, 8249
Hydroxy-4-methyl-2-pentanone, *See* Diacetone alcohol, 8249
(2'-Hydroxy-5'-methylphenyl) benzotriazole, *See* Lowilite® 55, 17643, Topanex 100BT, 32019, Uvazol P, 33184
2-(2'-hydroxy-5'-methylphenyl)benzotriazole, *See* Uvazol P, 33184
3-Hydroxy-2-methyl-4H-pyran-4-one, *See* Veltol, 33544
5-hydroxy-6-methyl-3,4-pyridinedimethanol, *See* pyridoxine hydrochloride, 25483
5-hydroxy-6-methyl-3,4-pyridinedimethanol hydrochloride, *See* vitamin B+76 hydrochloride, 33937
3-hydroxy-2-methyl- γ-pyrone, *See* Veltol, 33544
3-hydroxy-2-methyl-4-pyrone, *See* Veltol, 33544
2-hydroxy-4-(methylthio)-butanoic acid, *See* methionine hydroxy analog, 19486
hydroxy-9-octadecenoic acid, monopotassium salt, *See* Solricin® 135, 28964
2-Hydroxy-1,4-naphthalenedione, *See* Lawsone, 16891
5-Hydroxy-1,4-naphthalenedione, *See* Juglone, 15631
8-Hydroxy-1,4-naphthalenedione, *See* Juglone, 15631
3-hydroxy-2,7-naphthalenedisulfonic acid, *See* R-Acid, 25710
4-hydroxy-1-naphthalenesulfonic acid, *See* Nevile and Winther's acid, 21046
7-Hydroxy-2-naphthalenesulfonic acid, *See* Cassella's acid, 5402
6-hydroxy-1-naphthol, *See* 1,6-naphthalenediol, 20709
2-hydroxy-1,4-naphthoquinone, *See* Lawsone, 16891
5-Hydroxynaphthoquinone, *See* Juglone, 15631
5-Hydroxy-*p*-naphthoquinone, *See* Juglone, 15631
5-Hydroxy-1,4-naphthoquinone, *See* Juglone, 15631
8-hydroxy-1,4-naphthoquinone, *See* Juglone, 15631
N-Hydroxy-N-nitrosobenzenamine ammonium salt, *See* cupferron, 7365
2-Hydroxy-1,2,3-nonadecanetricarboxylic acid, *See* Agaricic, 956
12-hydroxy-(*cis*)-9-octadecenoic acid, *See* P® -10 Acid, 22499, ricinoleic acid, 26734
12-Hydroxy-9-octadecenoic acid ester, *See* cetyl ricinoleate, 5824
12-hydroxy-9-octadecenoic acid methyl ester, *See* methyl ricinoleate, 19558
[R-(Z)]-12-hydroxy-9-octadecenoic acid sodium salt, *See* Soricinol 40, 29143
12-hydroxy-9-octadecenoic acid, monoester with 1,2,3-propanetriol, *See* glyceryl monoricinoleate, 13085
12-Hydroxy-9-octadecenoic acid, monoester with 1,2-propanediol, *See* Propylene glycol ricinoleate, 25222
5-Hydroxy-4-octanone, *See* Butyroin, 4802
2-Hydroxy-4-n-octoxybenzophenone, *See* Cyasorb® UV 531, 7497
2-Hydroxy-4-n-octoxy-benzophenone, *See* Uvinul® 408, 33192
2-hydroxy-4-(octyloxy)benzophenone, *See* Cyasorb® UV 531, 7497, Uvinul® 408, 33192
2-Hydroxy-4-octyloxybenzophenone, *See* Lankromark® LE285, 16697
[2-hydroxy-4-(octyloxy)phenyl]phenylmethanone, *See* Spectra-Sorb UV 531, 29232, Uvinul® 408, 33192
2-(2-Hydroxy-5-*t*-octylphenyl)-benzotriazole, *See* Cyasorb® UV 5411, 7503, Uvazol 311, 33183
γ-hydroxy-β-oxobutane, *See* acetyl methyl carbinol, 312
4-Hydroxy-3-(3-oxo-1-phenylbutyl)-2H-1-benzopyran-2-one, *See* Killgerm® Sewarin P, 16184, Kypfarin, 16519
7-hydroxy-3H-phenoxazin-3-one 10-oxide, *See* diazoresorcin, 8351, Resazurin, 26161
1-Hydroxy-2-phenoxyethane, *See* Igepal® Cephene Distilled, 14850, 2-Phenoxyethanol, 23627, Rewopal® MPG 10, 26344
hydroxy-2-phenyl acetophenone, *See* Benzoin, 3975
α-hydroxy-α-phenylacetophenone, *See* benzoin, 3975
Hydroxy-2-phenylbenzene, *See* o-phenylphenol, 23648
4-Hydroxy-3-(1-phenylpropyl)-2H-1-benzopyran-2-one, *See* Phenprocoumon, 23630
2-hydroxy-1,2,3-propanetricarboxylic acid, *See* citric acid, 6341
2-hydroxy-1,2,3-propanetricarboxylic acid silver salt,

See silver citrate, 28422
2-Hydroxy-1,2,3-propanetricarboxylic acid, calcium salt (2:3), *See* calcium citrate, 4944
2-Hydroxy-1,2,3-propanetricarboxylic acid, monoester with 1,2,3-propanetriol monooctadecanoate, *See* glyceryl stearate citrate, 13088
2-Hydroxy-1,2,3-propanetricarboxylic acid, triethyl ester, *See* triethyl citrate, 32242
4-hydroxy-1-propenylbenzene, *See* *p*-anol, 2480
6-Hydroxy-2H-pyridazin-3-one, *See* maleic hydrazide, 18477
6-hydroxy-3(2H)-pyridazinone, *See* maleic hydrazide, 18477
1-hydroxy-2-pyridine, 14669
2-hydroxypyridine, *See* 1-hydroxy-2-pyridine, 14669
N-Hydroxy-2-pyridinethione, sodium salt, *See* Sodium Omadine® 40% Aq. Sol'n, 28803, sodium pyrithione, 28813
1-hydroxy-2(1H)-pyridinethione zinc salt, *See* zinc pyrithione, 34791
2-hydroxy-4(1H)-pyrimidinone, *See* uracil, 33101
2-hydroxy-4(3H)-pyrimidinone, *See* uracil, 33101
4-hydroxy-2(1H)-pyrimidinone, *See* uracil, 33101
Hydroxy quinoline, *See* 8-hydroxyquinoline, 14670
12-hydroxy stearic acid, *See* Ceroxin GL, 5758
4-Hydroxy-3-(1,2,3,4-tetrahydro-1-naftyl)-cumarine, *See* Racumin®, 25713
4-Hydroxy-3-(1,2,3,4-tetrahydro-1-naphthalenyl)-2H-1-benzopyran-2-one, *See* Racumin®, 25713
4-Hydroxy-3-(1,2,3,4-tetrahydro-1-naphthyl)coumarin, *See* Racumin®, 25713
2-Hydroxy-N,N,N-trimethylethanaminium chloride, *See* choline chloride, 6171
(R)-4-hydroxy-N,N,N-trimethyl-10-oxo-7-[(1-oxahexadecyl)oxy]-3,5,9-trioxa-4-phosphapentacosan-1-aminium inner salt 4-oxide, *See* colfosceril palmitate, 6607
2-hydroxy-1,3,5-trinitrobenzene, *See* reflorit, 26003
4-hydroxy-2,3,5-trinitrobenzene, *See* trinitrocresol, 32347
1-hydroxy-2-[(6-O- β-D-xylopyranosyl- β-D-glucopyranosyl)oxy]-9,10-anthracenedione, *See* rubianic acid, 27103
hydroxyacetic acid, *See* glycolic acid, 13105
4'-hydroxyacetanilide, paracetamol, *See* Valadol, 33238 **Hydroxdiphenyl**, *See* *o*-phenylphenol, 23648
***o*-hydroxyacetophenone**, 14657
***p*-hydroxyacetophenone**, 14658
2'-hydroxyacetophenone, *See* *o*-hydroxyacetophenone, 14657
2-hydroxyacetophenone, *See* *o*-hydroxyacetophenone,
2-Hydroxyacetophenone, *See* *o*-hydroxyacetophenone, 14657
4'-hydroxyacetophenone, *See* *p*-hydroxyacetophenone, 14658
4-hydroxyacetophenone, *See* p-hydroxyacetophenone,
4-hydroxyacetophenone, *See* *p*-hydroxyacetophenone, 14658
Hydroxal, 14656
3-hydroxyalanine, *See* L-serine, 17705, 28035
Hydroxyaluminum distearate, *See* Synpro® 303, 30592
hydroxyaminoiminodiphenyl-sulfide, *See* Thionoline, 31705
4-Hydroxyaniline;, *See* *p*-aminophenol, 2185, Takatol, 30725
***p*-hydroxyaniline**, *See* *p*-aminophenol, 2185, Unal, 32908, Takatol, 30725
hydroxyanisole, *See* Eastman® HQMME, 9467
***o*-hydroxyanisole**, *See* guaiacol,
***p*-Hydroxyanisole**, *See* Eastman® HQMME, 9467, hydroquinone monomethyl ether, 14633
2-hydroxybenzaldehyde, *See* salicylaldehyde, 27269
***o*-hydroxybenzaldehyde**, *See* salicylaldehyde, 27269
4-hydroxybenzaldehyde, *See* p-hydroxybenzaldehyde,
***p*-hydroxybenzene**, *See* Tecquinol® Tech. Grade, 30910
***m*-hydroxybenzenesulfonic acid**, *See* Eltesol® PSA 65, 9868
3-hydroxybenzisothiazole-S,S-dioxide, *See* saccharin, 27195
(2-Hydroxybenzoato-O+b1)-oxobismuth, *See* Bismuth subsalicylate, 4244
2-hydroxybenzoic acid, *See* salicylic acid, 27270
***o*-hydroxybenzoic acid**, *See* salicylic acid, 27270
***p*-hydroxybenzoic acid sodium salt**, *See* Nipacombin SK, 21261
4-hydroxybenzoic acid butyl ester, *See* Butylparaben, 4797, Nipabutyl, 21249
2-hydroxybenzoic acid ethyl ester, *See* ethyl salicylate, 11081
4-hydroxybenzoic acid ethyl ester, *See* ethyl *p*-hydroxybenzoate, 11061
***para*-hydroxybenzoic acid ethyl ester**, *See* ethyl *p*-hydroxybenzoate, 11061
2-hydroxybenzoic acid methyl ester, *See* Teaberry Oil, 30872
2-hydroxybenzoic acid 2-naphthalenyl ester, *See* naphthalol, 20712
2-hydroxybenzoic acid phenyl ester, *See* phenyl salicylate, 23634
4-hydroxybenzoic acid propyl ester, *See* propylparaben, 25231
2-Hydroxybenzoic acid 3,3,5-trimethylcyclohexyl ester, *See* Homosalate, 14292, Kemester® HMS, 15963
hydroxybenzopyridine, *See* 8-hydroxyquinoline, 14670
1-hydroxybenzotriazole, 14659
N-Hydroxybenzotriazole, *See* 1-hydroxybenzotriazole, 14659
α-hydroxybenzyl phenyl ketone, *See* Benzoin, 3975
N,N'-(2-Hydroxybenzylidene)-1,2-propandiamine, *See* Metal Deactivator S, 19428
Hydroxybiphenyl, *See* *o*-phenylphenol, 23648
2-hydroxybiphenyl, *See* Nipacide® OPP, 21258, *o*-phenylphenol, 23648
Hydroxybutanedioic acid, *See* Malic Acid, 18483
4-hydroxybutanoic acid lactone, *See* Butyrolactone, 4803
hydroxybutylmethylphenylchlorobenzotriazole, 14660
4'-hydroxybutyranilide, *See* Suconox-4®, 29923
3-hydroxybutyric acid lactone, *See* Butyrolactone, 4803
4-hydroxybutyric acid, γ-lactone, *See* butyrolactone, 4803
Hydroxybutyl vinyl ether, *See* Rapi-Cure HBVE, 25883
Hydroxybutyl vinyl ether carbonate, *See* Rapi-Cure BHC, 25877
1-Hydroxy-2-butylamine, *See* AB, 19
***endo*-2-hydroxycamphane**, *See* DL-Borneol, 4428
Hydroxocobalamin, *See* Ducobee-Hy, 9082
7-hydroxycoumarin, *See* umbelliferone, 32899
4-hydroxycoumarin, 14661
hydroxydimethylbenzene, *See* xylenol, 34625
2-Hydroxydiphenyl, *See* *o*-phenylphenol, 23648
2-hydroxydiphenyl sodium, *See* Dowicide A, 8932
***o*-Hydroxydiphenyl**, *See* *o*-phenylphenol, 23648
Hydroxydiphenyl, sodium salt, *See* Dowicide A, 8932
Hydroxydodecane, *See* Emery® 3326, 10169
1-Hydroxyethane 1-carboxylic acid, *See* lactic acid, 16570
2-Hydroxyethanesulfonic acid sodium salt, *See* Witconate NIS, 34487
Hydroxyethanoic acid, *See* glycolic acid, 13105
N-(2-Hydroxyethyl)acetamide, *See* Amidex AME, 2086, Mackamide AME-75, AME-100, 18152
1-Hydroxyethyl-2-alkyl-imidazoline, *See* Rewopon® IM OA, 26387
Hydroxyethylbenzene, *See* 2-phenylethyl alcohol, 23642
1-[[(2-hydroxyethyl)carbamoyl]methyl]pyridinium chloride laurate (ester), *See* Lapyrium Chloride, 16782
Hydroxyethylcellulose, *See* Cellosize, 5609, Cellosize® HEC QP Grades, 5610, Hetastarch, 13891, Natrosol® 250, 20758, Natrosol® Hydroxyethylcellulose, 20760, Tylose® H Series, 32660
Hydroxyethyl cetyldimonium phosphate, *See* Luviquat® Mono CP, 18025
Hydroxyethyl coco imidazoline, *See* Chemzoline C-22, 6051
β-hydroxyethyldimethylamine, *See* Alkanolamine, 1701, Dabco® DMEA, 7669, deanol, 7829
1-Hydroxyethyl-1-diphosphonic acid, *See* Chemphonate HEDP, 6006
N-(2-hydroxyethyl)dodecaneamide, *See* Rewomid® L 203, 26328
N-(2-hydroxyethyl)ethylene dinitrilo triacetic acid, *See* Emkasene 800, 10259
N-(2-Hydroxyethyl)ethylenediamine, *See* Aminoethylethanolamine, 2172
N-(2-Hydroxyethyl)ethylenediaminetriacetic acid trisodium salt hydrate, *See* Kalex OH, 15672
N-(2-Hydroxyethyl)ethylenediaminetriacetic acid, trisodium salt, *See* Trilon® D Liq, 32329
(Hydroxyethyl)ethylenediaminetriacetic acid, trisodium salt, *See* Trilon® D Liq, 32329
N-hydroxyethylethylenediaminetriacetic acid trisodium salt, *See* Kalex OH, 15672
2-Hydroxyethyl hexadecanoate, *See* glycol palmitate, 13103
2-Hydroxyethyl 12-hydroxy-9-octadecenoate, *See* glycol ricinoleate, 13104
N (2-Hydroxyethyl) 12-hydroxystearamide, *See* Paricin® 220, 22838
(1-Hydroxyethylidene)biphosphonic acid, *See* Briquest® ADPA-60A, 4536
(1-hydroxyethylidene)diphosphonic acid, *See* Unihib® 106, 32957
1-hydroxyethylidene-1,1-diphosphonic acid, *See* Unihib® 106, 32957
N-(2-hydroxyethyl)lauramide, *See* Rewomid® L 203, 26328
1-Hydroxyethyl-2-lauric imidazoline, *See* Hodag C-100-L, 14167, lauryl hydroxyethyl imidazoline, 16865
2-hydroxyethyl mercaptan, *See* 2-mercaptoethanol, 19336
hydroxyethyl methacrylate, 14662
2-hydroxyethyl methacrylate, *See* Bisomer 2HEMA, 4273, hydroxyethyl methacrylate, 14662, Igepal® Cephene Distilled, 14850, Sipomer® HEM-D, 28536
β-hydroxyethyl methacrylate, *See* Bisomer 2HEMA, 4273, Sipomer® HEM-D, 28536
N-(2-Hydroxyethyl)octadecanamide, *See* Ablumide SME, 82
1-Hydroxyethyl-2-oleic imidazoline, *See* Hodag C-100-O, 14168
2-hydroxyethyl phenyl ether, *See* Rewopal® MPG 10, 26344
β-Hydroxyethyl phenyl ether, *See* Igepal® Cephene Distilled, 14850, Rewopal® MPG 10, 26344
β-hydroxyethylphenyl ether, *See* 2-Phenoxyethanol, 23627
10-[3'-[4-(β-hydroxyethyl)-1-piperazinyl]propyl]-3-trifluoromethylphenothiazine, *See* Fluphenazine, 12116
N-(2-Hydroxyethyl)-2 pyrrolidone, *See* HEP, 13786
N(β-Hydroxyethyl) ricinoleamide, *See* Flexricin® 115, 11955
hydroxyethyl starch, *See* Hetastarch, 13891
N-(2-Hydroxyethyl)stearamide, *See* Ablumide SME, 82, Witcamide® 70, 34442
N-(Hydroxyethyl)stearamide, *See* Witcamide® 70, 34442
Hydroxyethyl stearamide-MIPA, *See* Crodapearl NI Liquid, 7130
1-Hydroxyethyl-2-stearic imidazoline, *See* Hodag C-100-S, 14169, stearyl Hydroxyethyl imidazoline, 29606
1-Hydroxyethyl-2-tall oil imidazoline, *See* Hodag C-100-T, 14170
Hydroxyethyl-*s*-triazine, *See* Bioban® GK, 4158
2-(hydroxyethyl)trimethylammonium chloride, *See* choline chloride, 6171
2β-Hydroxygibberellin 1, *See* Regulex, 26019
1-hydroxyheptane, *See* 1-heptanol, 13792 heptyl

alcohol, 13797
1-hydroxyhexane, *See* 1-hexanol, 14011
6-Hydroxyhexan-6-olide, *See* +le-caprolactone monomer, 5138
5-(hydroxyimino)barbituric acid, *See* violuric acid, 33881
Hydroxyl terminated polybutadiene, *See* Nisso PB, 21312
Hydroxylamine, *See* oxyammonia, 22439
Hydroxylan, *See* hydroxylated lanolin, 14663
hydroxylated lanolin, 14663, *See* OHlan®, 22055, Ritahydrox, 26807
Hydroxylated lecithin, *See* Alcolec® Z-3, 1488, Centrolene® A, S, 5685, PhosPho 642, 23706
Hydroxylated soy lecithin, *See* Lipotin H, 17424
Hydroxylizaric Acid, *See* oxyalizarin, 22438
Hydroxyl-type acrylic polymer, *See* Acryloid® WR-97, 435
Hydroxyl-type acrylic polymer,, *See* Acryloid® AT-51, 429
Hydroxyl-type polymer, *See* Acryloid® AT-63, 430
hydroxymercurichlorophenol sulfate, *See* Uspulun, 33171, 33172
Hydroxymethanesulfinic acid sodium salt, *See* sodium formaldehydesulfoxylate, 28766
hydroxymethanesulfonic acid monosodium salt, *See* Rongalit® C, 26945
5-Hydroxymethoxymethyl-1-aza-3,7-dioxabicylco [3.3.0]octane (24.5%), 5-hydroxymethyl-1-aza-3,7-dioxabicylco[3.3.0]octane (17.7%) and 5-hydroxypoly[methyleneoxyl]methyl-1-aza-3,7-dioxabicyclo[3.3.0] octane (7.8%), *See* Bioban® N-95, 4159
(Hydroxymethyl)-2-nitro-1,3-propanediol, *See* Tris Nitro®, 32408
Hydroxymethyl-2-propylamine, *See* 2-amino-2-methyl-1 propanol, 2182
N-(Hydroxymethyl)acrylamide, *See* NM-AMD, 21404
(Hydroxymethyl)-2-amino-1,3-propanediol, *See* Tris Amino®, 32405
[2-(Hydroxymethyl) amino] ethanol, *See* Troysan® 174, 32477
2-[(Hydroxymethyl) amino]-2-methylpropanol, *See* Troysan® 192, 32478
5-(Hydroxymethyl)-1-aza-3,7-dioxabicyclo(3.3.0)octane, *See* Zoldine® ZT-55, 34887
5-Hydroxymethyl-1-aza-3,7-dioxabicyclo[3.3.0]octane, *See* Zoldine® ZT-55, 34887
(hydroxymethyl)benzene, *See* benzyl alcohol, 3986
1-(hydroxymethyl)-5,5-dimethylhydantoin, *See* Dantoin® MDMH, 7734
1,3-bis (hydroxymethyl)-5,5-dimethyl-2,4-imidazolidine dione, *See* DMDM hydantoin, 8754, Glydant®, 13140, Mackstat® DM, 18234, Nipaguard® DMDMH, 21269
2-hydroxymethylfuran, *See* furfuryl alcohol, 12493
3-hydroxymethyl-n-heptan-4-ol, *See* ethyl hexanediol, 11073
6-hydroxymethyl-2-isopropylaminomethyl-7-nitro-1,2,3,4-tetrahydroquinoline, *See* Vansil, 33363
Tris(hydroxymethyl)methylamine, *See* Tris Amino®, 32405
Hydroxymethyl-n-heptan-4-ol, *See* ethyl hexanediol, 11073
2-hydroxymethyl-2-nitropropanediol, *See* Tris Nitro®, 32408 S.S.T® Sump Saver Tablets, 27180
2-Hydroxymethyl-2-nitro-1,3-propanediol, *See* Tris Nitro®, 32408
(Hydroxymethyl)-2-nitropropane-1,3-diol, *See* Tris Nitro®, 32408
(Hydroxymethyl)-2-nitropropanediol, *See* Tris Nitro®, 32408
1-hydroxymethylpropane, *See* isobutanol, 15344
2,2-bis(hydroxymethyl)-1,3-propanediol, *See* pentaerythritol, 23138
N-(Hydroxymethyl)-2-propenamide, *See* NM-AMD, 21404
1-(Hydroxymethyl)propylamine, *See* AB, 19
hydroxymimetite, 14664
N-hydroxynaphthalimide diethyl phosphate, *See* Rametin, 25856
4-Hydroxynitrobenzene, *See* *p*-nitrophenol, 21368
1-hydroxyoctadecane, *See* Adol® 62 NF, 783, Stearal, 29591
d-12-hydroxyoleic acid, *See* ricinoleic acid, 26734
Hydroxyphenazone, *See* Resorufin, 26232
3-hydroxyphenol, *See* resorcin, 26228, resorcinol, 26230
***p*-hydroxyphenol**, *See* quinol, 25685
***o*-hydroxyphenyl arsenate**, *See* pyrocatechol arsenic acid, 25499
β-p-hydroxyphenylalanine, *See* tyrosine,
1-(2-Hydroxyphenyl)ethanone, *See* o-hydroxyacetophenone, 14657
1-(4-hydroxyphenyl)ethanone, *See* p-hydroxyacetophenone, 14658
***p*-hydroxyphenylglycine**, *See* Glyconyl, 13124
N-(4-Hydroxyphenyl)glycine, *See* Glyconyl, 13124
***o*-hydroxyphenylmercuric chloride**, *See* mercufenol chloride, 19345
γ-(*p*-hydroxyphenyl)- α-propylene, *See* chavicol, 5861
hydroxyprolisilane C, 14665
2-hydroxypropane, *See* isopropanol, 15405
(R)-2-hydroxypropanoic acid, *See* D-lactic acid, 16572
(S)-2-hydroxypropanoic acid, *See* L-lactic acid, 16571
2-hydroxypropanoic acid, *See* lactic acid, 16570, DL-lactic acid, 16573
2-hydroxypropanoic acid monosodium salt, *See* sodium lactate, 28777
2-hydroxypropanoic acid, tetradecyl ester, *See* myristyl lactate, 20578
α-Hydroxypropanoic acid, *See* lactic acid, 16570
2-hydroxypropanoic acid, *See* lactic acid, 16570
3-Hydroxypropene, *See* propenol, 25181
2-hydroxypropionic acid, *See* lactic acid, 16570
3-hydroxypropionic acid β-lactone, *See* betaprone, 4065
3-hydroxypropionic acid lactone, *See* betaprone, 4065
2-Hydroxypropionic acid hexadecyl ester, *See* cetyl lactate, 5820
2-hydroxypropionic acid, *See* lactic acid, 16570
hydroxypropyl alginate, *See* propylene glycol alginate, 25216
N-(2-hydroxypropyl) benzenesulfonamide, *See* Uniplex 225, 33013
Hydroxypropyl bisstearyldimonium chloride, *See* M-Quat® Dimer 18, 20405
Hydroxypropyl ether guar gum, *See* hydroxypropyl guar, 14666
hydroxypropyl guar, 14666, *See* Galactasol® Guar Derivs, 12591 Jaguar® HP 60, 15501 Jaguar® HP 8, 15502, Jaguar® HP-11, 15503
Hydroxypropyl guar gum, *See* hydroxypropyl guar, 14666, Jaguar® HP 60, 15501
Hydroxypropyl guar hydroxypropyl trimonium chloride, *See* Jaguar® C-162, 15498, 15499
Hydroxypropyl hydroxyethylcellulose, *See* Natrovis® Water-Soluble Polymer, 20762
hydroxypropyl methacrylate, 14667
2-Hydroxypropyl methacrylate, *See* Bisomer 2HPMA, 4274
Hydroxypropyl methyl cellulose
hydroxypropyl methylcellulose, 14668, *See* Goniosol, 13194, Lacril®, 16565, Methocel® 40-202, 19489, Methocel® E3 Premium, 19493, Methocel® E4M, 19494, Methocel® E5, 19495, Methocel® F4M, 19496,Methocel® K100MP, 19497, Methocel® K35, 19498,
2-hydroxypropylamine, *See* isopropanolamine, 15406
Hydroxypropylcellulose, *See* Klucel®, 16229, Klucel® E, G, H, J, L, M, 16230, Klucel® EF, 16231
N-(2-Hydroxypropyl)oleamide, *See* Witcamide® 61, 34441
6-hydroxypurine, *See* hypoxanthine, 14721
2-Hydroxyquinoline, *See* carbostyril, 5244
8-hydroxyquinoline, 14670
8-Hydroxyquinoline sulfate, *See* Cryptonol, 7282
Hydroxystearamidopropyl trimonium chloride, *See* Surfactol® Q2, 30342
Hydroxystearamidopropyl trimonium methosulfate, propylene glycol, *See* Surfactol® Q3, 30343
hydroxysuccinic acid, 14671 *See* Malic Acid, 18483
N-hydroxysuccinic acid, *See* apple acid, 2667, Malic Acid, 18483
Hydroxytoluene, *See* benzyl alcohol, 3986, cresylic acid, 7013
4-hydroxytoluene, *See* *p*-cresol, 6999
o-hydroxytoluene, *See* *o*-cresol, 6998
α-Hydroxytoluene, *See* benzyl alcohol, 3986
β-hydroxytricarballylic acid, *See* citric acid, 6341
2-hydroxytriethylamine, *See* 2-diethylaminoethanol, 8440
β-hydroxytriethylamine, *See* 2-diethylaminoethanol, 8440
8-hydroxyxanthine, *See* uric acid, 33139
hydrozone, 14672
Hyflo NS, S, 14673, *See* Polyisoprene, 24471
Hyflo Super-Cel, 14674
Hyflux, 14675
Hyflux M, 14676
Hy-glo Steel, 14677
Hygrass, 14678
hygrol, 14679
Hygroplex HHG, 14680
Hyjet, 14681
hylastic, 14682
Hylemox, *See* Ethion, 10955
Hylene M50, *See* Isonaphthol, 15377, MDI, 19188, Rubinate® LF-168, 27108
Hylene®, 14683
Hylenta, *See* Penicillin G potassium, 23109
Hylite Color-Max, 14684
Hymec, 14685
Hymexazol, *See* Tachigaren 70, 30685
Hymod, 14686
Hymolon CWC, 14687
Hymono, 14688
Hymush, 14689
Hyonate, *See* hyaluronic acid sodium salt, 14483
Hyonic, 14690
Hyonic GL 400, 14691
Hyonic NP-40, 14692
Hyonic OP-7, 14694
Hyonic OP-55, 14693
Hyonic PE-100, 14695
Hyosan, *See* Nuophene, 21783
Hypacel, 14696
Hypalon®, 14697
Hypalon® CP 826, 14698
Hypan® QT 100, 14699
Hypan® SA100H, 14700
Hypan® SR150H, 14701
Hypan® SR150H, SA100H, *See* Acrylic acid acrylonitrogens copolimer, 413
Hypan® SS201, 14702, *See* Ammonium acrylates acrylonitrogens copolymer, 2227
Hypan® SS500V, 14703
Hypan® SS500W, 14704
Hypaque, *See* diatrizoate sodium, 8337
Hypaque Cysto, *See* Reno-M, 26113
Hypaque Meglumine, 14705, *See* Reno-M, 26113
Hypaque Sodium, *See* diatrizoate sodium, 8337
Hypax, 14706
Hyper+Ion 1050, 14707
Hyper+Ion 2050A, 14708
Hyperdrol, *See* Aloxicoll, 1813, aluminum chlorohydrate, 1903, Reach® 101, 201, 501, 25919
Hyperiast®, 14709
Hyperit, *See* hyperol, 14712
Hyperiz, *See* Aztec® CHP-80, 3555, CHP-158, 6175
Hyperkil Bait, 14710
Hypermer, 14711
Hypernick, *See* Hipernick®, 14100
Hyperol, 14712
Hypersal, 14713
Hypersal®, 14714

Hypersol, 14715
hy-phi 1199, *See* stearic acid, 29599
hy-phi 1205, *See* stearic acid, 29599
hy-phi 1303, *See* stearic acid, 29599
hy-phi 1401, *See* stearic acid, 29599
Hy-Phos, *See* Calgon, 4987, Kalex HMP, 15670, sodium hexametaphosphate, 28770
Hypnodil, 14716
Hypnone, *See* acetophenone, 300
Hypnorm, 14717
HYPO, *See* sodium thiosulfate, 28826
Hypochlorous acid, 14718, *See* HyPure A, 14724
Hypochlorous Acid, Calcium Salt, *See* Repak, 26141
hypochlorous acid-calcium biborate-calcium chloride, *See* Eusol, 11236
Hypoiodous acid, 2,2'-dimethyl-5,5'-bis(1-methylethyl)[1,1'-biphenyl]-4,4'-diyl ester, *See* Iosol, 15237
Hypol® FHP 2000, 14719
hyponitrous acid anhydride, *See* Nitral, 21318
Hyporice, *See* sodium thiosulfate, 28826
Hyporit, 14720
hypoxanthine, 14721, *See* sarcine, 27460
Hypromellose, 14722
Hyprone, 14723
HyPure A, 14724, *See* Hypochlorous acid, 14718
HyPure C, 14725, *See* Chloric acid, 6102
HyPure K, 14726, *See* Potassium hypochlorite, 24757
HyPure L, 14727, *See* lithium hypochlorite, 17509
HyPure N, 14728
Hyquat 70, 75, 14729
hyraldite, *See* hydrosulfite, 14643
Hyraldite C Ext, 14730
Hysa, 14731
Hysede, 14732
Hysol® 1C Epoxi-Patch Kit, 14733
Hysol® 6C Epoxi-Patch Kit, 14736
Hysol® 342, 14735
Hysol® 1942, 14738
Hysol® 2000, 14734
Hysol® 7804, 14737
Hysol® EA9460, 14739
Hysol® EE0067/HD3561 and Hysol EE0067/HD3615, 14740
Hysol® EO1016, 14741
Hysol® ES4228, 14742
Hysol® MBI-02, 14743
Hysol® MG1 Series, 14744
Hysol® OSO100, 14745
Hysol® PC18, 14746
Hysol® Polyshot 1X, 14747
Hysol® XC7-W529, 14748
Hysorb, 14749
Hyspray, 14750
Hystar®, 14751
Hystar® 3375, 14752
Hystar® TPF, 14753
Hystor, *See* tecnazene, 30898
Hystor 10, 14754
Hystore, *See* Tecgran, 30886
Hystrene ® 9014, *See* Myristic acid, 20574
Hystrene 80, *See* stearic acid, 29599
Hystrene 4516, *See* stearic acid, 29599
Hystrene 5016, *See* stearic acid, 29599
Hystrene 7018, *See* stearic acid, 29599
Hystrene 9014, *See* Philacid 1400, 23662
Hystrene 9512, *See* Philacid 1200, 23660
Hystrene 9718, *See* stearic acid, 29599
Hystrene s 97, *See* stearic acid, 29599
Hystrene t 70, *See* stearic acid, 29599
Hystrene® 1835, 14755
Hystrene® 3022, 14756
Hystrene® 3675, 14757
Hystrene® 4516, 14758
Hystrene® 5012, 14759, *See* hydrogenated stripped coconut acid, 14597
Hystrene® 5016, 14760
Hystrene® 5460, 14761, *See* Trilinoleic acid, 32320
Hystrene® 5522, 14771
Hystrene® 7018 FG, 14762
Hystrene® 8016, 14763, *See* Palmitic Acid, 22635
Hystrene® 9014, 14764
Hystrene® 9022, 14765
Hystrene® 9512, 14766, *See* lauric acid, 16856
Hystrene® 9514, 14767, *See* Myristic acid, 20574
Hystrene® 9718 NF FG, 14768
HyStretch V-29, 14769
HyStretch V-43, 14770
Hysward, 14772
Hytak, 14773
Hytane, 14774
Hytec, 14775, *See* Tecgran, 30886, tecnazene, 30898
Hytec Super, 14776
Hytemco, 14777
HyTemp 4051, 14778
HyTemp 4052, 14779
HyTemp NPC-50, 14780
Hyten M Steel, 14781
Hyten®, 14782
Hy-ten-sl, 14783
Hytex, 14784
Hythane, 14785
Hytherm, 14786
Hytin, 14787
Hy-TL, 14788
Hytox, 14789, *See* Etrofolan®, 11156, isoprocarb, 15404
Hytrast, *See* lopydone, 15235
Hytrel®, 14790
Hytrol, 14791, 14792
Hytrol O, *See* cyclohexanone, 7522
Hyvar X, 14793, *See* bromacil, 4564, Uragan, 33105
Hyvar® X, 14794
Hy-Vic, 14795
Hy-Vin, 14796
Hyvis, 14797
Hy-Yield H-10, *See* arsenic acid, 3137
Hyzod® AC-1000, 14798
Hyzon, *See* chloridazon, 6103
I T Talc, 14800
I.P.S, 14799
IAA, *See* Rhizopon A, AA, 26573
Iachiol, 14801
IA-IA alloy, 14802
Ial, *See* hyaluronic acid sodium salt, 14483
IBA, *See* 4-indol-3-ylbutyric acid, 15086, isobutanol, 15344
Ibbal, 14803
Ibdu®, 14804
Ibex, 14805
Ibiosuc, *See* sodium cyclamate, 28754
Icdal, 14806
ice colors, 14807
Ice Melt, 14808
Ice No. 2, 14810
Ice No. 12K, 14809, *See* Polysorbate 65, 24571
ice spar, *See* cryolite, 7279, Kryalith, 16448
Iceberg®, 14811
Icecap® K, 14812
Iceland spar, 14813
Iceline, *See* Preservaline, 24902
Ichden, *See* Perichthol, 23270
Ichtammon, *See* Perichthol, 23270
Ichthadone, *See* Perichthol, 23270
Ichthalum, *See* Perichthol, 23270
ichthammol, *See* Perichthol, 23270
ichthammonium, *See* Perichthol, 23270
Ichthium, *See* Perichthol, 23270
Ichthopur, *See* Perichthol, 23270
Ichthosan, *See* Perichthol, 23270
Ichthosauran, *See* Perichthol, 23270
ichthosulfol, *See* Perichthol, 23270
Ichthymall, *See* Perichthol, 23270
Ichthynat, *See* Perichthol, 23270
Ichthyocolla, *See* isinglass, 15333
Ichthyol, *See* Perichthol, 23270
Ichthyopon, *See* Perichthol, 23270
Ichthysalle, *See* Perichthol, 23270
ICI, *See* Dapsone, 7751
ICI-29661, *See* pyrimithate, 25485
ICI-50123, *See* pentagastrin, 23146
ICI-55052, *See* methyl benzoquate, 19516
Icinol, 14814
Icinol DPM, 14815
Icinol PM, 14816
ICI-PP-333, *See* paclobutrazol, 22542
Icomeen® T-2, 14817
Icomeen® T-15, 14818
Icon, *See* +II-cyhalothrin, 7567
Iconol DA-4, 14819, *See* Deceth-6, 7848
Iconol DA-6, 14820, *See* Deceth-6, 7848
Iconol DA-9, 14821, *See* Deceth-6, 7848
Iconol NP-30, 14823
Iconol NP-40, 14824
Iconol NP-50, 14825
Iconol NP-70, 14826
Iconol NP-100, 14822
Iconol OP-10, 14827
Iconol OP-30, 14828
Iconol OP-40, 14829
Iconol TDA-10, 14834
Iconol TDA-3, 14830
Iconol TDA-3, TDA-6, TDA-8, TDA-9, TDA-10;, *See* Trideceth series, 32232
Iconol TDA-6, 14831
Iconol TDA-8, 14832
Iconol TDA-9, 14833
Iconol WA-1, 14835
Iconol WA-4, 14836
Iconsim®, 14837
Iconyl, *See* glycine, 13095
1-icosene, *See* Neodene® 20, 20900
icos-1-ene, *See* Neodene® 20, 20900
ICR, 14838
Ideal alloy, 14839
Idexur, *See* Kerecid, 16067
Idilon, *See* Carbora, 5235
Iditol, 14840
Idoxene, *See* Kerecid, 16067
Idoxuridine, *See* Kerecid, 16067
IDPP, *See* Syn-O-Ad® 8479, 30556
Idryl, 14841
IDU, *See* Kerecid, 16067
Idulea, *See* Kerecid, 16067
IDUR, *See* Kerecid, 16067
Iduridin, *See* Kerecid, 16067
IE-40-A, 14842
IFC, *See* Propham, 25185, 25186
IFK, *See* Propham, 25185, 25186
Igasurine, *See* brucine,
Igelit PCU, 14843
Igepal CO, *See* Alkasurf® NP-4, 1715, Triton® N-57, 32430
Igepal CO-630, *See* Alkasurf® NP-4, 1715, Triton® N-57, 32430
Igepal® 131, 14844
Igepal® BPA-6, 14845
Igepal® CA-210, 14846
Igepal® CA-620 and CA-630, 14847
Igepal® CA-720, 14848
Igepal® CA-897, 14849
Igepal® Cephene Distilled, 14850
Igepal® CO-210, 14851
Igepal® CO-430, 14852
Igepal® CO-520, 14853
Igepal® CO-530, 14854
Igepal® CO-610, 14855
Igepal® CO-630, 14856
Igepal® CO-660, 14857
Igepal® CO-710, 14858
Igepal® CO-720, 14859
Igepal® CO-730, 14860
Igepal® CO-850, 14861
Igepal® CO-880, 14862
Igepal® CO-887, 14863
Igepal® CO-890, 14864
Igepal® CO-970, 14865
Igepal® CO-997, 14866

Igepal® CTA-639W, 14867
Igepal® DM-430, 14868
Igepal® DM-430, DM-530, DM-710, DM-730, DM-970, *See* Ethal DNP-8, 10919
Igepal® DM-530, 14869
Igepal® DM-710, 14870
Igepal® DM-730, 14871
Igepal® DM-970, 14872
Igepal® OD-410, 14873
Igepal® RC-520, 14874, *See* Dodoxynol series, 8786
Igepal® RC-620, 14875
Igepon®, 14876
Igetaleim MA, 14877
Igetaleim UA, 14878
Igewsky's reagent, 14879
Iglodine, 14880
Ignicide, 14881
Igran, *See* terbutryn, 31230
Igran 50, *See* terbutryn, 31230
Igran 500, *See* terbutryn, 31230
Iguafen, 14882
Ikathene, *See* A-C® Polyethylene 6, 6A, 7, 7A, 8, 8A, 9, 9A, 617, 617A, 175
Ikurin, *See* Ammonium sulfamate, 2254
Ilcocillin P, *See* Penicillin V, 23111
Iletin, *See* insulin,
Ilexan E, 14883, *See* ethylene glycol, 11123
Ilexan HT, 14884
Ilexan P, 14885
Ilexan S, 14886
ilixathin, *See* rutin trihydrate, 27150
Illium, 14887
Ilmenite, 14888
Ilopan, *See* d-Panthenol, 22690
ilothion, *See* iohydrin, 15223
Ilozyme, 14889, *See* Accelerase, 185
Imacol, 14890
Imaverol, 14891
Imaverol, *See* Enilconazole, 10582
imazalil, 14892, *See* Clinafarm, 6422, Enilconazole, 10582, Fungaflor, 12471, Magnate, 18349
imazamethabenz-methyl, *See* Dagger, 7693
imazapyr, 14893, *See* Arsenal, 3132, Arsenal®, 3134
Imber, *See* Thiovit, 31724
Imicure EMI-24, 24S, 14894
Imidacloprid, *See* Admire, 755, Confidor, 6725, Gaucho, 12698
Imidazole, 14895
1H-Imidazole, 1-(trimethylsilyl)-, *See* CT3600, 7329
Imidazolidinedione, 3-bromo-1-chloro-5,5-dimethyl-, *See* Bromochloro dimethyl hydantoin, 4574 Dantoin® GSD-550, 7733 Halobrom, 13560 Quesbrom, 25657
Imidazolidinedione, 1-(hydroxymethyl)-5,5-dimethyl-, *See* Dantoin® MDMH, 7734
Imidazolidinethione, *See* Perkacit® ETU, 23283
2-Imidazolidinethione, *See* Akroform® ETU-22 PM, 1191, ethylene thiourea, 11132, ETU, 11158, Perkacit® ETU, 23283
Imidazolidinyl urea, 14896, *See* Abiol, 69, Germall® 115, 12893
Imidazoline, *See* Sylfam 2082, 30480, Witcor 3630, 34519
Imidazoline 18, 14897
Imidazoline SOH, *See* stearyl hydroxyethyl imidazoline, 29606
Imidazoline-2-thiol, *See* ethylene thiourea, 11132, Perkacit® ETU, 23283
Imidazolines/amines, *See* Rewopon®, 26386
Imidazolinium methosulfate, *See* Marlosoft® IQ 75, 18906
Imidiazolidinyl urea NF, *See* Abiol, 69
Imidodicarbonimidic diamide, *See* Biquanide, 4218
Imidodicarbonimidic diamide, N-(2-methylphenyl)-, *See* Vulkacit® 1000, 34124
Imidole, *See* Pyrrole, 25535
2-Imidozolidimethione, *See* Perkacit® ETU, 23283
Imidrol, 14898
Imidurea, *See* Abiol, 69, Germall® 115, 12893
Imidurea NF, *See* Abiol, 69, Germall® 115, 12893
iminazole, *See* Imidazole, 14895
2,2'-iminobisethanol, *See* Diethanolamine, 8434
2,2'-iminobisethylamine, *See* D.E.H. 20, 7610
1,1'-Iminobis-2-propanol, *See* 8466, propanolamine,
2,2'-iminodiethanol, *See* Diethanolamine, 8434
2,2'-Iminodiethylamine, *See* diethylenetriamine, 8446
1,1'-iminodipropan-2-ol, *See* diisopropanolamine, 8466
1,1'-imino-2-propanol, *See* diisopropanolamine, 8466
Iminopropionate, *See* Amphoteric 400, 2358
Imiodid, 14899
Imlar®, 14900
Immadium, 14901
Immedial®, 14902
Immergan® A, 14903
Immetal, 14904
Immuno-bed, 14905
Imogen, 14906
Imogen sulfite, 14907
IMP sodium, *See* disodium inosinate, 8645
Impact, 14908, *See* Fluphenazine, 12116
Impact Excel, 14909, *See* Fluphenazine, 12116
Impact®, 14910
Impad, 14911
Imperator, *See* cypermethrin, 7588
Imperator, *See* permethrin, 23384, 23385, 23386
imperatorin, 14912, *See* peucedanin, 23541
imperial green, *See* cupric acetoarsenite, 7374, mountain green, 20369
Imperial metal, 14913
Impervite, *See* Ceresit, 5747
Impet® 330, 14914, *See* Polyethylene terephthalate, 24426
Implenal®, 14915
Imposil, 14916
Impra®, 14917
Impra® Concentrates, 14918
Impra-blolan, 14919
Impra-color, 14920
Impra-elan, 14921
Imprafix®, 14922
Impralan®, 14923
Impraleum, 14924
Impralit, 14925
Impranil®, 14926
Impregnant, 14927
Impression, 14928
Impressional, 14929
Imprez, 14930
Impriment Black 7101-A, 14931
Imprimus®, 14932
Improved Kelmar®, 14933 , *See* Alginic acid potassium salt, 1594
Impruvol, *See* Anti-Oxydant Bayer, 2595, Butylated hydroxytoluene, 4788, Oxyguard, 22457, Ralox® BHT food grade, 25845, Tenamine 3, 31092, Vanox® PCX, 33346, Vulkanox® KB, 34152
Impsonite, 14934
Imron®, 14935
Imsol, 14937
imsol a, *See* isopropanol, 15405
Imsil® A-10, 14936
Imuthiol, *See* Octopol SDE-25, 22013
Imwitor®, 14938
Imwitor® 191, 14939
Imwitor® 308, 14940, *See* glyceryl caprylate, 13077
Imwitor® 312, 14941
Imwitor® 369, 14942, *See* Monoglyceride citric ester, 20226
Imwitor® 369, 370, *See* Monoglyceride citric ester, 20226
Imwitor® 370, 14943, *See* glyceryl stearate citrate, 13088
Imwitor® 742, 14945
Imwitor® 780 K, 14946, *See* isostearic acid, 15421
Imwitor® 914, 14947
Imwitor® 928, 14948, *See* glyceryl cocoate, 13078
Imwitor® 960, 14949
Imwitor® 988, *See* glyceryl caprylate, 13077
Inakor, *See* atrazine, 3394
Inalium, 14950
incafolic, *See* folic acid, 12196, vitamin M, 33951
incidol, *See* Benzoyl peroxide, 3983
Inco Chrome Nickel, 14951
Incoblend, 14952
Incoloy Alloy 800, 14953
Incoloy Alloy 825, 14954
Incoloy Alloy 901, 14955
Incoloy Alloy DS, 14956
Incomparable, 14957
Inconel, 14958
Inconel Alloy 600, 14959
Inconel Alloy 700, 14960
Inconel Alloy 718, 14961
Inconel Alloy X-750, 14962
incorporation factor, IFP, *See* Kemstrene® 96.0%, 15991
Incrocas 30, 14963
Incrodet TD7-C, 14964
Incromate ALL, 14965
Incromate BAL, 14966
Incromate CDL, *See* Cocamidopropyl dimethylamine lactate
Incromate CDP, 14967, *See* Cocamidopropyl dimethylamine propionate, 6554
Incromate IDL, 14968, *See* isostearamidopropyl dimethylamine lactate, 15420
Incromate ISML, 14969
Incromate Mink L, 14970
Incromate ODL, 14971
Incromate OLL, 14972
Incromate SDL, 14973, *See* stearamidopropyl dimethylamine lactate, 29595
Incromate SEL, 14974
Incromate WGL, 14975
Incromectant AMEA-100, 14976
Incromectant AMEA-100, AMEA-70, *See* Acetamide MEA, 283
Incromectant AQ, 14977
Incromectant LAMEA, 14978
Incromectant LMEA, *See* Lactamide MEA, 16568
Incromectant LMEA-70, 14979
Incromectant LQ, 14980
Incromide ALd, 14981
Incromide BAD, 14982
Incromide BED, 14983, *See* Behenamide DEA (1:1), 3895
Incromide BEM, 14984
Incromide CA, 14985
Incromide CM, 14986
Incromide L-90, 14987
Incromide LA, 14988
Incromide LCL, 14989
Incromide LLT, 14990
Incromide LM-70, 14991
Incromide Mink D, 14992
Incromide OD, 14993
Incromide OLD, 14994
Incromide SED, 14995
Incromide WGD, 14996
Incromine BB, 14997, *See* Behenamidopropyl dimethylamine, 3896
Incromine CB, 14998, *See* Cocamidopropyl dimethylamine, 6552
Incromine IB, 14999
Incromine OPB, 15000, *See* oleamidopropyl dimethylamine, 22109
Incromine OPM, OPB, *See* oleamidopropyl dimethylamine, 22109
Incromine Oxide AL, 15001
Incromine Oxide B, *See* Behenamine oxide, 3897
Incromine Oxide B-30P, *See* Behenamine oxide, 3897
Incromine Oxide B50, *See* Behenamine oxide, 3897
Incromine Oxide BA, 15002
Incromine Oxide C, 15003
Incromine Oxide I, 15004
Incromine Oxide ISMO, 15005
Incromine Oxide L, *See* lauramine oxide, 16832
Incromine Oxide L-40, 15006

Incromine Oxide M, 15007
Incromine Oxide Mink, 15008
Incromine Oxide O, *See* oleamidopropylamine oxide, 22110
Incromine Oxide OD-50, 15009
Incromine Oxide OL, 15010
Incromine Oxide S, 15011
Incromine Oxide SE, 15012
Incromine Oxide WG, 15013
Incromine PB, *See* Palmitamidopropyl dimethylamine, 22630
Incromine SB, 15014
Incronam 1-30, 15015
Incronam 30, 15016, *See* Cocamidopropyl betaine, 6551
Incronam AL-30, 15017
Incronam B-40, 15018, *See* Behenyl betaine, 3900
Incronam BA-30, 15019
Incronam CD-30, 15020
Incronam Mink 30, 15021
Incronam OD-50, 15022, *See* oleyl betaine, 22132
Incronam OL-30, 15023, *See* Oleamidopropyl betaine, 22108
Incronam SE-30, 15024
Incronam WG-30, *See* Wheat germamidopropyl betaine, 34350
Incropol CS-20, 15025
Incropol L-23, 15026
Incroquat 100, 15027
Incroquat 248, 15028
Incroquat AL-85, 15029
Incroquat B65C, 15030
Incroquat BA-85, 15031
Incroquat Behenyl BDQ/P, 15032
Incroquat Behenyl TMC, 15033
Incroquat Behenyl TMC/P, 15034
Incroquat Behenyl TMS, 15035
Incroquat CR Conc, 15036
Incroquat CTC-30, 15037, *See* Cetrimonium chloride, 5814
Incroquat DBM-90, 15038
Incroquat I-85, 15039
Incroquat Mink-85, 15040
Incroquat O-50, 15041
Incroquat OL-85, 15042
Incroquat S-75CG, 15043
Incroquat SDQ-25, 15044
Incroquat SE-85, 15045
Incroquat WG-85, 15046
Incrosoft 100, 15047
Incrosoft 248, 15048
Incrosoft CF1-75, 15049
Incrosoft S-75, 15050, *See* Quaternium-27, 25607
Incrosoft S-75
Incrosoft S-90, *See* Quaternium-27, 25607
Incrosoft S-90M, *See* Quaternium-27, 25607
Incrosoft T-75, *See* Incrosoft T-90, 15051
Incrosoft T-90, 15051
Incrosoft T-90HV, *See* Incrosoft T-90, 15051
Incrosperse, 15052
Incrosul LAFS, 15053, *See* disodium laneth-5 sulfosuccinate, 8646
Incrosul LMA, 15054
Incrosul LMS, 15055, *See* disodium lauramido MEA-sulfosuccinate, 8647
Incrosul LS, 15056
Incrosul LSA, 15057
Incrosul LTS, 15058
Incrosul OMS, 15059
Incrosul OTS, 15060
Incrosul TS, 15061
Indalca, 15062
Indandione, 2-((p-chlorophenyl)phenylacetyl)-, *See* Ridene, 26746
indane, *See* Fumite Lindane, 12461
indanone, *See* 2-indanone, 15063
2-indanone, 15063
Indanthren®, 15064
1,2,3-indantrione monohydrate, *See* ninhydrin, 21230
Indapamide, *See* Lozol, 17688
Indazin, 15065
Indene-1,3(2H)-dione, 2-((4-chlorophenyl) phenyl acetyl)-, *See* Ridene, 26746
India gum, 15066
India tragacanth, *See* karaya gum, 15760
Indian apple, *See* vegetable calomel, 33515
Indian frankincense, *See* olibanum, 22134
Indian ginger, *See* wild ginger oil, 34407
Indian gum, *See* gum arabic, 13480
Indian gutta-percha, *See* pala gum, 22553
Indian lake, *See* carmine lake, 5292
Indian Ocher, 15067
Indian Red, 15068
Indian saffron, *See* turmeric, 32608
Indian tragacanth, *See* Normacol, 21566
Indian Yellow, 15069
Indianol, *See* *p*-aminophenol, 2185
indican, 15070, *See* plant indican, 23915
indicator, universal, 15071
indicolite, 15072
Indigal, 15073
Indigo, 15074
Indigo blue, *See* Indigo, 15074
Indigo Carmine, 15075
Indigo red, *See* indirubin, 15078
Indigotin, *See* Indigo, 15074
Indigotin disulfonate sodium, *See* Indigo Carmine, 15075
Indigotine, *See* Indigo, 15074, Indigo Carmine, 15075
Indilitans, 15076
Indio, 15077
indirubin, 15078
Indisin®, 15079
indium, 15080
indium oxide, 15081
Indocarbon, 15082
Indofast®, 15083
Indoil CPD 142 and CPD 143, 15084
Indol-3-yl acetic acid, *See* Rhizopon A, AA, 26573
indol-3-yl potassium sulfate, *See* indican, 15070
4-(3-indole)butyric acid, *See* 4-indol-3-ylbutyric acid, 15086
α-aminopropionic acid, *See* DL-tryptophan,
Indoleacetate, *See* Rhizopon A, AA, 26573
3-Indoleacetic Acid, *See* Rhizopon A, AA, 26573
β-Indoleacetic Acid, *See* Rhizopon A, AA, 26573
1H-Indole-3-acetic acid, *See* Rhizopon A, AA, 26573
1H-indole-3-butanoic acid, *See* 4-indol-3-ylbutyric acid, 15086
indolebutyric acid, *See* 4-indol-3-ylbutyric acid, 15086
3-indolebutyric acid, *See* 4-indol-3-ylbutyric acid, 15086
1H-Indole-2,3-dione, *See* Isatin, 15331
Indole-2,3-dione, *See* Isatin, 15331
indoline, 15085
2,3-Indolinedione, *See* Isatin, 15331
Indolylacetic acid, *See* Rhizopon A, AA, 26573
(Indol-3-yl)acetic acid, *See* Rhizopon A, AA, 26573
Indolyl-3-acetic acid, *See* Rhizopon A, AA, 26573
4-indol-3-ylbutyric acid, 15086, *See* Chryzoplus, Chryzopon, Chryzosan, Chryzotek, 6258, Seradix, 28005
1H-indol-3-yl-β-D-glucopyranoside, *See* plant indican, 23915
Indonex VG, 15087
Indopol, 15088
Induchlor, *See* calcium hypochlorite, 4953
Induclor, 15089
Indulin® 201, 15090
Indulin® AQS, 15091
Indulin® MQK, 15092
Indulin® SA-L, 15093
Indulin® W-1, 15094
Indulin® XD-70, 15095
Induline, 15096
Induline yellow, *See* Flavinduline, 11898
Indur, 15097
Indurite, 15098
Indusoil, 15099
industrene 5016, *See* stearic acid, 29599
industrene 8718, *See* stearic acid, 29599
industrene 9018, *See* stearic acid, 29599
industrene r, *See* stearic acid, 29599
Industrene® 104, 15100, *See* oleic acid, 22112
Industrene® 120, *See* linolenic acid, 17306
Industrene® 126, 15101
Industrene® 143, 15102, *See* tallow fatty acid, 30743
Industrene® 223, 15103, *See* hydrogenated stripped coconut acid, 14597
Industrene® 224, 15104
Industrene® 325, 328, *See* Hystrene® 1835, 14755
Industrene® 365, 15105, *See* caprylic/capric acid, 5160
Industrene® 4516, 15106, *See* Palmitic Acid, 22635
Industrene® 4518, 15107
Industrene® 5016, 15108
Industrene® 7018 FG, 15109
Industrene® D, 15110, *See* Dilinoleic acid, 8480
Industrene® R, 15111
Industrial Dyne, 15112
Industrial Dynemate, 15113
Industrial gum, 15114
Industrial talc, *See* I T Talc, 14800
Industrial, cosmetic, or platy talc, *See* talc, 30731
Industrol CO-36, *See* Ablunol CO 10, 98
Industrol® DW-5, 15115
Industrol® N3, 15116
Industrol® TFA-8, 15117
Industrol® TO-16, 15118, *See* PEG tallate series, 23035
Inertex, 15119
Inevitan, *See* paclobutrazol, 22540, 22541
Infacare, 15120
Infasoft, 15121
Infavina, 15122
InFilm, 15123
Infiltrina, *See* DMSO, 8762
Infron, *See* vitamin D+72, 33940
infusible white precipitate, 15124
Infusorial earth, 15125, *See* diatomaceous earth, 8335
Ingalite, *See* Gallatite, 12608
Ingotol, 15126
Inhibine, *See* hydrogen peroxide, 14590
Inhibited 1,2-propylene glycol, *See* Ilexan P, 14885
Inhibitor, 15127
Inhibitor 60S, 15128
Inhibitor RT 212, 15129
Inipol, 15130
Initial®, 15131
Initiator BK, 15132
Injacom, 15133
Inklurit®, 15134
Inkovar 335, 15135
Inkovar 617, 15136
Inkrustin®, 15137
INM-6316, *See* thifensulfuron-methyl, 31673
Innovex, 15138
Inochrome, 15139
Inoculin, 15140
Inoderme, 15141
Inopak, 15142
Inorganic acid, *See* potassium carbonate, 24744, sodium carbonate, 28744, sulfuric acid, 30055, zinc sulfide, 34796
Inosine 5'-disodium phosphate, *See* disodium inosinate, 8645
Inositol, *See* meat-sugar, 19196, muscle sugar, 20526, phaseomannite, 23579
Inositol Hexaphosphate, *See* phytic acid, 23777
Inositolhexaphosphoric acid, *See* phytic acid, 23777
Inositol-hexaphosphoric acid, *See* phytic acid, 23777
inositol hexaphosphoric acid hexasodium salt, *See* Rencal, 26105
levo-inositol monomethyl ether, *See* quebrachite, 25635
Inotab, 15143
INPC, *See* Propham, 25185, 25186
Insect repellant, *See* ethyl hexanediol, 11073
Insect Spray for House Plants, 15144
insect wax, *See* Chinese wax, 6077
Insecticide 1179, *See* methomyl, 19502

Insecticide 3960-X14, *See* Strobane, 29818
Insecticide No. 4049, *See* malathion, 18460
Insectigas D, *See* dichlorvos, 8394
Insectophene, *See* thiodan, 31680
6-12 insect repellent, *See* ethyl hexanediol, 11073
Insektigun, 15145
Insol-U & RM, 15146
insoluble saccharin, *See* saccharin, 27195
Insoluble sulfur, *See* Manox, 18577
Instant Gasket, 15147
Instant Ocean, 15148
Instrument bronze, 15149
Insubeta, *See* Invert Sugar, 15200
Insulgard, 15150
Insullac, 15151
Insulmag®, 15152
Insural, 15153
Insurok, 15154
Integuard, 15155
Intercept, 15156
Intercide, 15157
Interferon, 15158
Intergravin-orales, *See* calcium chloride, 4943
Intermediate 122, *See* Witco® 1298, 34458
internally compensated tartaric acid, *See* mesotartaric acid, 19400
internally compensated tartaric acid, *See meso*-tartaric acid, 30831
釋Internol, *See* paraffin, liquid, 22744
Interol, *See* paraffin, liquid, 22744
Interox H48, 15159
Intersept®, 15160
Interstab®, 15161
Interstab® BZ-4828A, 15162
Interstab® BZ-4836, 15163
Interstab® FR930, 15164
Interstab® LT-4805R, 15165
Interwood®, 15166
Intestibar, *See* Radiopaque, 25830
Intexan SB-85, *See* Ablumine 280, 85, Cycloton® SCS, 7542, octadecylbenzenemethanaminium chloride, 21985, stearalkonium chloride, 29592, Sumquat® 6210, 30126
Intexsan SB-85, *See* Ablumine 280, 85, Cycloton® SCS, 7542, octadecylbenzenemethanaminium chloride, 21985, stearalkonium chloride, 29592, Sumquat® 6210, 30126
Intimate Contact, 15167
Intob, 15168
Intracarrier® ATM, 15169
Intradex, *See* Dextran, 8233
Intrafomil® AK, 15170
Intral®, 15171
Intralan® Salt HA, 15172
Intralan® Salt N, 15173
Intrapan, *See* d-Panthenol, 22690
Intraphasol COP, 15174
Intraquest® TA Solution, 15175
Intrasoft® OCN, 15176
Intrassist® LA-LF, 15177
Intratex® A, 15178
Intratex® CA-2, 15179
Intratex® DD, 15180
Intratex® N, 15181
Intravon® JU, 15182
Intravon® SOL-N, 15183
Intrex, 15184
Intrex DW81, 15185
Intrex HA70, 15186
Invaderm, 15187
Invaderm C9B, 15188
Invadine, 15189
Invalon, 15190
Invar, 15191
invariant, 15192
Invaro Steel, 15193
Invasol, 15194
Invephos 20, 15195
Invephos 21C, 15196
Inveres EVH, 15197
Inveres K-82, 15198
Inversol 140, 15199
Invert Sugar, 15200
Invertase, *See* sucrase, 29929
invertin, *See* sucrase, 29929
Invesol, *See* Invert Sugar, 15200
Invicta Duo 495FW, 15201, 15202
Invisi-Gard, *See* Baygon®, 3790, Propoxur, 25198, 25199, 25200
INY-6202, *See* quizalofop-ethyl, 25697
Iodal, 15203
Iodamide, *See* Uromiro(n), 33147
Iodamide meglumine, *See* Renovue-65, Renovue-DIP, 26115
Iodamide, N-Methyl-D-glucamine salt, *See* Jodomiron, 15615
Iodanisol, 15204
Iodantifebrin, *See* Iodoacetanilide,
Iodeikon, *See* keraphen, 16055
iodeosin B, *See* erythrosin, 10778
iodeosin G, *See* Erythrosin G, 10780
Iodesin, 15205
Iodestone, *See* magnetic iron ore, 18374
Iodex, 15206
Iodic acid, sodium salt, *See* sodium iodate, 28774
Iodicyl, *See* Diiodosalicylic acid methyl ester,
Iodin, 15207
Iodin green (Griesbach), *See* methyl green, 19540
Iodine, 15208
iodine-potassium Iodide solution, *See* Lugol's solution, 17788
Iodipamide meglumine, *See* Cholografin Meglumine, 6172
iodized salt, 15209
Iodoacetone, *See* bretonite, 4507
***o*-Iodoanisole**, *See* Iodanisol, 15204
iodobenzene, 15210
N-(2-iodobenzoyl)glycine-$_{131}$I monosodium salt, *See* Hippuran-:ss131:ksI, 14125, Hipputope, 14128
Iodobio 45, 15211
5-Iodo-2'-deoxyuridine, *See* Kerecid, 16067
iodoeosin, 15212
Iodofenfos, *See* Iodofenphos, 15213
Iodofenphos, 15213, *See* Elocril, 9852
iodoform, 15214
Iodognost, *See* keraphen, 16055
***o*-Iodohippurate sodium-$_{131}$I**, *See* Hippuran-:ss131:ksI, 14125, Nephroflow, 20989
Iodohydroxyquinolinesulfonic acid, *See* Loretine, 17606
iodol, 15215
iodomethane, *See* methyl iodide, 19542
iodomethanesulfonic acid sodium salt, *See* Abrodil, 158, Methiodal sodium, 19485
Iodomethylbenzene, *See* Fraissite, 12374
4-Iodo-ι-methylbenzenedecanoic acid ethyl ester, *See* Iophendylate, 15234
(±)-4-(iodo-$_{123}$I)-α-methyl-N-(1-methylethyl) benzene ethanamine hydrochloride, *See* Iofetamine ^{123}I hydrochloride, 17553
Iodopaque, *See* acetrizoate sodium, 307
Iodophene, *See* Kerasol, 16058
Iodophene sodium, *See* keraphen, 16055
2-Iodo-N-phenylbenzamide, *See* benodanil, 3942
Iodophthalein, *See* Kerasol, 16058
Iodophthalein sodium, *See* Apagallin, 2640, keraphen, 16055
3-Iodo-2-propynyl butyl carbamate, *See* Troysan® Polyphase® EC17, 32479
iodoquinol, *See* Embequin, 10024, Yodoxin, 34658
Iodorayorl, *See* keraphen, 16055
Iodosorb, 15216
Iodothymol, *See* Iosol, 15237
Iodotope, *See* Iodotope I-125, 15217, Iodotope I-131, 15218, Iodotope Therapeutic, 15219, Radiocaps-131, 25827
Iodotope I-125, 15217
Iodotope I-131, 15218
Iodotope Therapeutic, 15219
Iodotrimethylsilane, *See* CT3610, 7330
Iodoval, 15220
Iodoxamic acid, *See* Endobil, 10548, Endomirabil, 10550, Videocolagio, 33743
***p*-Iodoxyanisol**, *See* isoform oxiosol, 15365
Iodozol, 15221
Iodron, 15222
Iodtetragnost, *See* keraphen, 16055
iofendylate, *See* Pantopaque, 22695
Iofetamine $_{123}$I hydrochloride, 17553, *See* Perfusamine, 23251
Iohexol, *See* Omnipaque, 22159
iohydrin, 15223
iomapidol, *See* Iopamidol, 15233
Iomesan, *See* Mansonil®, 18611
Iomezan, *See* Mansonil®, 18611
Ionac ECP-88, 15224
Ionet ML-400, *See* Ablunol 200ML, 95
Ionet MS-1000, *See* Ablunol DEGMS, 99
Ionet S 60, *See* Ablunol S-60, 110
Ionet S 85, *See* Ablunol S-85, 112 Witconol 2503, 34502
Ionet S-20, *See* Ablunol S-20, 108
Ionet S-60, *See* Ablunol S-60, 110
Ionet S-80, *See* Ablunol S-80, 111 Witconol 2500, 34501
Ionet S-85, *See* Ablunol S-85, 112 Witconol 2503, 34502
Ionex, 15225
Ionol, 15226
Ionol CP, 15227, *See* Anti-Oxydant Bayer, 2595, Butylated hydroxytoluene, 4788, Oxyguard, 22457, Ralox® BHT food grade, 25845, Tenamine 3, 31092, Vanox® PCX, 33346, Vulkanox® KB, 34152
Ionol CPA-Feed, 15228
Ionol K65, 15229
Ionomer resins, 15230
Ionox, 15231
Ionpure, 15232
Iopamidol, 15233
Iopamiro, *See* Iopamidol, 15233
Iopamiron, *See* Iopamidol, 15233
Iopanoic acid, *See* Telepaque, 31057
Iophendylate, 15234, *See* Ethiodan, 10954, Pantopaque, 22695
Iopronic acid, *See* Bilimiron, 4142, Oravue, 22272, Videobil, 33742
iopropane, *See* iohydrin, 15223
Iopydone, 15235
Ioquin, *See* Yodoxin, 34658
Iosan, 15236
Iosol, 15237
Iotect, 15238
Iotek 7010, 15239
Iotek 8000, 15240
Iothalamate acid sodium salt, *See* Angio-Conray, 2456
Iothion, *See* iohydrin, 15223
Iotox, 15241
Iotril, 15242
Iotrilex, 15243
Iotrol, *See* iotrolan, 15244
iotrolan, 15244
Ioxynil, *See* Actrilawn 10, 566, Totril, 32081
ioxynil octanoate, *See* Iotril, 15242, Iotrilex, 15243
ioxynil-mecoprop, *See* Iotox, 15241
IPA, *See* isopropanol, 15405
IPC, *See* Chlorpropam, 6162, Propham, 25186
IPC-IPPC, *See* Propham, 25185
IPD, *See* Vestamin® TMD, 33656
IPDI, *See* Vestanat® IPDI, 33657
Ipecac, 15245
ipecacuanha, *See* Ipecac, 15245
Ipersan, *See* Treflan, 32162, Trifluralin, 32252
Iphaneine, 15246
IPM, *See* isopropyl myristate, 15409, Wickenol® 101, 105, 34378
IPMC, *See* Baygon®, 3790
Ipodate calcium, *See* Oragrafin Calcium, 22260
Ipodate sodium, *See* Bilivist, 4144, Oragrafin Sodium, 22261
Ipognox 44, *See* Vulkanox® 4010 NA, 34147

ipomic acid, 15247
ipon lsb, *See* sodium lauryl sulfate, 28781
Iporka, 15248
IPP, *See* isopropyl palmitate, 15410
IPPC, *See* Propham, 25186
IPPD, *See* Flexzone 3C, 11967, N-Isopropyl-N'-phenyl-*p*-phenylene diamine, 15413, Permanax IPPD, 23364
Iprit, *See* mustard gas, 20538
Iprodione, *See* CDA Roval, 5521, Roval Dust, 27021, Roval Flo, 27022, Roval Green, 27023, Roval WP, 27024
Ipsilene, 15249
Ipso, 15250
IPX, *See* Nipacide® BIT, 21253
IR, *See* polyisoprene, 24470, 24471
Irabond, 15251
iradicav, *See* sodium fluoride, 28764
Iragcet, 15252
Iranolin, *See* Firnagral, 11842
Irasolve, 15253
Irathane, 15254
Ircogel, 15255
Ircogel® 900, 15256, *See* calcium sulfonate, 4973
Ircogel® 905, 15257
Ircosperse 2170, 15258
Irenat, *See* KM Sodium Perchlorate, 16245
iretol, 15259
Irgaclarol, 15260
Irgacure, 15261
Irgaderm, 15262
Irgaferm BC Champagne, 15263
Irgafin, 15264
Irgafiner, 15265
Irgafos, 15266
Irgafos TNPP, *See* Wytox® 312, 34578
Irgalan, 15267
Irgalevone, 15268
Irgalite, 15269
Irgalite Blue GST, 15270
Irgalite C-20, 15271
Irgalite Dispersed, 15272
Irgalite M-20, 15273
Irgalite MPS, 15274
Irgalite PDS, 15275
Irgalite PR, 15276
Irgalite Yellow BGW, 15277
Irgalite Yellow F4G, 15278
Irgalon, 15279, *See* tetrasodium EDTA, 31391
Irganol, 15280
Irganox, 15281
Irganox 1076, *See* Naugard® 76, 20789
Irgapadol, 15282
Irgaphor, 15283
Irgaplastol M-20, 15284
Irgapyrol, 15285
Irgarol, 15286
Irgasan, 15287
Irgasol®, 15288
Irgasperse, 15289
Irgastab®, 15290
Irgatan, 15291
Irgatron, 15292
Irgawax, 15293
Irgazin, 15294
Irgazin C-20, M-20, 15295
Irgoferm CM Montrachet, 15296
iridin, 15297
Iridite®, 15298
iridium steel, 15299
Iridosmine, 15300
Irigenin, 15301
Irilac®, 15302
Iris blue, *See* Fluorescent Blue, 12084
Irish moss extract, *See* carrageenan, 5315
Irish pearl moss, 15303
Irish Peat Wax, 15304
irisin, *See* iridin, 15297
Irisol®, 15305
Irium, *See* sodium lauryl sulfate, 28781, Texapon K-12, K-1296, L-100, 31461, Rhodapon® 101-10, 26631
Irlux®, 15306
iron, 15307
Iron 30% Iron, *See* Monterey 30% Iron, 20286
Iron sulfate, *See* Monterey 30% Iron, 20286
Iron A, 15308
Iron B, 15309
iron black, 15310
Iron chloride, *See* ferrous chloride, 11689
iron chlorides, *See* ferric chloride, 11648
Iron D, 15311
Iron D2, 15312
iron dichloride, *See* ferrous chloride, 11689
iron flint, 15313
iron froth, 15314
Iron G, 15315
Iron L, 15316
Iron lignosulfonate, *See* Borrewell FE, 4453
iron liquor, *See* pyrolignite of iron, 25515
Iron Man, 15317
iron minium, *See* Indian Red, 15068
Iron nitrate, *See* ferric nitrate, 11650
Iron ore A, hematite type, 15318
Iron (III) oxide, *See* ferric oxide, 11651, Ferroxide, 11694, Jeweller's rouge, 15611
iron oxide, *See* ferric oxide, 11651, Ferroxide, 11694, Jeweller's rouge, 15611
iron oxide, *See* ferric oxide, 11651
Iron oxide (Fe_2O_3), *See* Ferroxide, 11694, Jeweller's rouge, 15611
Iron oxide fume, *See* Ferroxide, 11694, Jeweller's rouge, 15611
Iron oxide red, *See* Ferroxide, 11694, Jeweller's rouge, 15611
Iron oxides, bismuthoxychloride, *See* Red 139, 25948
iron persulfate, *See* iron sulfate (ic), 15320
Iron Protochloride, *See* ferrous chloride, 11689
iron putty, 15319
Iron pyrolignite, *See* pyrolignite of iron, 25515
iron red, *See* Indian Red, 15068
iron saffron, *See* Indian Red, 15068
iron sesquichloride, *See* ferric chloride, 11648
iron sulfuretmagnetkies, *See* white pyrites, 34371
Iron tersulfate, *See* Elliott's Lawn Sand, 9848, ferric sulfate, 11652, iron sulfate (ic), 15320, Vitax Turf Tonic, 33960
Iron vitriol, 15322, *See* ferrous sulfate, 11690, iron sulfate (ous), 15321
Iron(II) chloride, *See* ferrous chloride, 11689
iron(III) chloride, *See* ferric chloride, 11648
iron(III) nitrate, *See* ferric nitrate, 11650
Iron(III) oxide, *See* Kroma Red, 16418, OSO® 440, 22363, OSO® 1905, 22364, sal mineral, 27248
Iron(III) oxide dihydrate, *See* Ferroxide, 11694, Jeweller's rouge, 15611
iron sulfate (ous), 15321
iron sulfate (ic), 15320
Iron (III) sulfate, *See* Elliott's Lawn Sand, 9848, ferric sulfate, 11652, Vitax Turf Tonic, 33960
iron(II) sulfate heptahydrate, *See* sal Martis, 27247
Iron, Armco, *See* Armco Ingot Iron, 2923
Ironac, 15323
iron-andradite, 15324
Ironate, *See* ferrous sulfate, 11690
iron-leucite, 15325
iron-ore cement, 15326
iron-orthoclase, 15327
Irox, 15328
Isanol®, 15329
Isarit, 15330
Isatin, 15331
Isceon, 15332
Isdopher, *See* Bardyne, 3628
Isethionic acid, sodium salt, *See* Witconate NIS, 34487
isinglass, 15333
iskia-c, *See* sodium ascorbate, 28734
Iso Isostearyle WL 3196, 15334
Iso Isotearyle WL 3196, *See* isostearyl isostearate, 15422
iso soap, 15335
ISO.PPC, *See* Propham, 25185
Iso-adipate 2/043700, *See* Unimate® DIPA, 32968
isoamyl acetate, 15336, *See* Jargonelle pear essence, 15521, pear oil, 22971
Isoamyl acid phosphate, *See* Syn-O-Ad® P-415, 30567
Isoamyl ethanoate, *See* isoamyl acetate, 15336, pear oil, 22971
isoamyl salicylate, *See* sanfoin, 27364
β-isoamylene, *See* pental, 23147
isoanethole, *See* Estragole, 10897
isoanthraflavic acid, 15337
Isoarachidyloxypropylamine, *See* Tomah PA-24, 31987
Isoascorbate, *See* Eribate, 10753, Neo-Cebitate®, 20878, sodium erythorbate, 28762
isoascorbic acid, 15338, *See* erythorbic acid, 10775
isoascorbic acid, sodium salt, *See* Eribate, 10753, Neo-Cebitate®, 20878, sodium erythorbate, 28762
isoascorbic acid- γ-lactone, *See* Erythorbic acid, 10775
isobelite, *See* polyhalite, 24463
1,3-isobenzofurandione, *See* phthalic anhydride, 23770, Retarder AK, 26253
1,3-isobenzofurandione, *See*
Isobond, 15339
Isobornyl acrylate, *See* Ageflex IBOA, 991, Sipomer® IBOA, 28537
Isobornyl methacrylate, *See* Sipomer® IBOMA, 28538, Ageflex IBOMA, 992
Isobrite®, 15340
Isobu-M-AMD, 15341
isobutad, 15342
isobutane, 15343, *See* A-31, 8, isobutane, 15343
isobutanol, 15344
Isobutanol-2-amine, *See* 2-amino-2-methyl-1 propanol, 2182
Isobutanolamine, *See* 2-amino-2-methyl-1 propanol, 2182
N-(Isobutoxymethyl)acrylamide, *See* Cylink ISOBU-M-AMD, 7571, Isobu-M-AMD, 15341
isobutyl acetate, 15345
Isobutyl adipate, *See* diisobutyl adipate, 8460
isobutyl alcohol, *See* isobutanol, 15344
Isobutyl benzoin ether, *See* Vicure® 10, 33734
isobutyl formate, 15346
isobutyl o-hydroxybenzoate, *See* orchindone, 22279
Isobutyl ketone, *See* diisobutyl ketone, 8461
isobutyl maleate, *See* Polycare® 509, 24377
Isobutyl methacrylate polymer in VM&P naphtha, *See* Acryloid® B-67, 432
Isobutyl Niclate®, 15347
Isobutyl oleate, *See* Priolube 1414, 25017, Radia® 7230, 25732
Isobutyl phenylacetate, *See* Iphaneine, 15246
Isobutyl salicylate, *See* orchindone, 22279
Isobutyl stearate, *See* Emerest® 2324, 10087, Estol 1476, 10893, Kemester® 5415, 15946, Kessco® IBS, 16113, Radia® 7241, 25734, Uniflex® IBYS, 32948
Isobutyl α-toluate, *See* Iphaneine, 15246
Isobutylated lanolin oil, *See* Argonol 40, 2855
***p*-Isobutylbenzaldehyde**, *See* Ibbal, 14803
Isobutylene isoprene copolymer, *See* Exxon® Butyl 077, 11355
Isobutylene-isoprene elastomer, *See* Exxon® Butyl 077, 11355
2,2'-Isobutylidene-bis(4,6-dimethylphenol), *See* Lowinox® 22IB46, 17651
Isobutylmethylcarbinol, *See* methyl amyl alcohol, 19512
isobutylparaben, *See* LiquaPar® Oil, 17451
N-(5-Isobutyl-1,3,4-thiadiazol-2-yl)-*p*-methoxy benzenesulfonamide, *See* Stabinol, 29413
Isobutyltrimethoxysilane, *See* CI7810, 6288 Dynasylan® IBTMO, 9410 Prosil® 178, 25235
isobutyric acid, 15348
Isobuzol, *See* Stabinol, 29413
Isocarb, *See* Baygon®, 3790
Isoceteth-20, *See* Arlasolve® 200, 2912
Isocetyl alcohol, *See* Eutanol G16, 11240, Exxal® 16,

11341, Michel XO-150-16, 19641
Isocetyl octanoate, *See* Bernel® Ester 168, 4002
Isocetyl salicylate, *See* Dermol ICSA, 8126
Isocetyl stearate, *See* Afilan ICS, 932, Kemester® 5822, 15951, Kessco® ICS, 16114, Schercemol ICS, 27603
isocetyl stearoyl stearate, 15349, *See* Ceraphyl® 791, 5728, Hetester HSS, 13897
Isoclad, 15350
Isocon, 15351
Iso-Cornox, 15352, 15353, *See* Mecoprop, 19204 Verdone CDA, 33580
Isocothane, *See* Karathane, 15759
Isocracking, 15354
Isocreme, 15355
Isoctyl acrylate, *See* Ageflex FA-8, 978, SR-440, 29383
Isocure, 15356
1-Isocyanato-3-(isocyanatomethyl)-3,5,5-trimethylcyclohexane , *See* Vestanat® IPDI, 33657
5-Isocyanato-1-(isocyanatomethyl)-1,3,3-trimethylcyclohexane, *See* Vestanat® IPDI, 33657
3-(Isocyanatomethyl)-3,5,5-trimethylcyclohexyl isocyanate , *See* Vestanat® IPDI, 33657
1-(Isocyanatomethyl)-5-isocyanato-1,3,3-trimethylcyclohexane, *See* Vestanat® IPDI, 33657
γ-Isocyanatopropyl triethoxy Silane, *See* CI7840, 6289
Isocyanatopropyltriethoxy silane, *See* CI7840, 6289
3-Isocyanatopropyltriethoxysilane, *See* CI7840, 6289
1-Isocyanato-3,3,5-trimethyl-5-(isocyanato methyl)cyclohexane , *See* Vestanat® IPDI, 33657
Isocyanic acid, methylene(3,5,5-trimethyl-3,1-cyclohexylene) ester, *See* Vestanat® IPDI, 33657
γ-isocyanopropyltriethoxysilane, *See* CI7840, 6289
Isocyanurate acid, *See* Zeonet A, 34720
isocyanuric acid, 15357, *See* Zeonet A, 34720
isocyanuric acid triallyl ester, *See* triallyl isocyanurate, 32192
isocyanuric acid, dichloropotassium salt, *See* dichloro-1,3,5-triazinetrione, potassium salt, 8384
isocyanuric chloride, *See* Queschlor, 25658, trichloroisocyanuric acid, 32218
Isodamp® C-1002, 15358
Isodeceth-3, *See* Remcopal 273, 26070, Remcopal LO 2B, 26092
Isodeceth-4, *See* Oxetal ID 104, 22422, Rhodasurf® DA-4, 26653, Rhodasurf® DA-530, 26654
Isodeceth-5, *See* Hodag Nonionic ID-5, 14195
Isodeceth-6, *See* Rhodasurf® 860/P, 26648
Isodeceth-9, *See* Ethylan® CD109, 11097
Isodecyl 2-methylpropenoate, *See* isodecyl methacrylate, 15359
isodecyl 2-propenoate, *See* Sipomer® IDA, 28539
Isodecyl acrylate, *See* Ageflex FA-10, 979, Sipomer® IDA, 28539
Isodecyl diphenyl phosphate, *See* Syn-O-Ad® 8479, 30556
Isodecyl isononanoate, *See* Wickenol® 152, 34389
isodecyl methacrylate, 15359, *See* Ageflex FM-10, 985, isodecyl methacrylate, 15359
Isodecyl 2-methyl-2-propenoate, *See* Ageflex FM-10, 985, isodecyl methacrylate, 15359
Isodecyl 2-methylpropenoate, *See* Ageflex FM-10, 985
Isodecyl neopentanoate, *See* Dermol 105, 8118
Isodecyl octanoate, *See* Dermol 108, 8119
Isodecyl oleate, *See* Ceraphyl® 140-A, 5723, Mackester IDO, 18212, Schercemol IDO, 27604, Wickenol® 144, 34387
Isodecyl oxypropyl amine acetate, *See* Tomah PA-14 Acetate, 31983
Isodecyl stearate, *See* Lexolube® B-108, 17161
Isodecyldiphenyl Phosphate, *See* Syn-O-Ad® 8479, 30556
N-Isodecyloxypropyl-1,3-diaminopropane, *See* Tomah DA-14, 31966
N-isododecyloxypropyl-1,3-diaminopropane, *See* Tomah DA-16, 31967
Isodecyloxypropyl dihydroxyethyl methyl ammonium chloride, *See* Tomah Q-14-2, 31990
Isodecyloxypropylamine, *See* Tomah PA-14, 31982
Isoderm®, 15360
isodium hexafluorosilicate, *See* sodium silicofluoride, 28818
Isododecyl alcohol, *See* 2-butyloctanol, 4795, Michel XO-150-12, 19640
Isododecyloxypropylamine, *See* Tomah PA-16, 31984
isodulcit, *See* rhamnose, 26529
Isodurindine, 15361
Isoeicosanol, *See* Disponil G 200, 8693
isoestragole, *See* Arizole® Anethole Extra, 2885
Isofenphos, *See* Amaze®, 1951, Amidocid®, 2106, Oftanol®, 22049, Pryfon, 25355
isoflupredone acetate, 15362, *See* Predef, 24870
Isoflurane, 15363
Isofoam, 15364
isoform oxiosol, 15365
Isoheptyl alcohol, *See* Exxal® 7, 11335
isohexadecanol, *See* Michel XO-150-16, 19641
Isohexadecyl alcohol, *See* Michel XO-150-16, 19641
Isohexyl neopentanoate, *See* Schercemol 65, 27579
isohol, *See* isopropanol, 15405
Isoindole-1,3(2H)-dione, 2-((dichlorofluoromethyl) thio)-, *See* Fluorfolpet, 12087
Iso-Iodeikon, *See* TIP, 31847
Isol, 15366
Isol R, 15367
Isolan® GI 34, 15368
Isolan® GO 33, 15369
Isolan®, Isolan® K, 15370
Isolantite, 15371
Isolaureth-3, *See* Tergitol® TMN-3, 31248
Isolene® 40, 75, 400, 15372
Isolierstahl, *See* Bakelite, 3589
Isolit, 15373
Isoloss® LS, 15374
α-isomethylionone, *See* Alpha Daphnone, 1821
Isomol, 15375
Isomyst, *See* Wickenol® 101, 105, 34378
Isona, *See* Eribate, 10753, Neo-Cebitate®, 20878, sodium erythorbate, 28762
Isonal, 15376
Isonaphthol, 15377
Isonate 125M, *See* Isonaphthol, 15377, MDI, 19188, Rubinate® LF-168, 27108
Isonate 125MF, *See* Isonaphthol, 15377, MDI, 19188, Rubinate® LF-168, 27108
Isonate® 125M, 15378
Isonate® 2125M, 15379
isonicotinic acid, 15380
5-isonitrosobarbituric acid, *See* violuric acid, 33881
Isonol, 15381
Isononanol phosphate ester, *See* Marlowet® 5311, 18925
Isononyl alcohol, *See* Exxal® 9, 11337, Neoflex® 9, 20922
Isononyl isononanoate, *See* Wickenol® 151, 34388
Isononyl stearate, *See* Radia® 7510, 25746
isononyloxypropylamine, *See* Tomah PA-13i, 31981
Isonox® 103, 15382, *See* 2,6-dibutylphenol, 8370
Isonox® 129, 15383
Isonox® 132, 15384
isooctadecanoic acid, *See* isostearic acid, 15421
Isooctadecanoic acid, 1-methylethyl ester, *See* Wickenol® 131, 34381
isooctahexacontane, *See* Permethyl 104A, 23389
isooctanol, *See* isooctyl alcohol, 15385
Isooctyl acrylate, *See* Ageflex FA-8, 978
Isooctyl acrylate (monomer), *See* SR-440, 29383
isooctyl alcohol, 15385, *See* Exxal® 8, 11336
isooctyl mercaptoacetate, *See* isooctyl thioglycolate, 15386
Isooctyl O-acetylcitrate, *See* Citroflex A-8, 6345
Isooctyl phthalate, *See* Witcizer 313, 34455
Isooctyl stearate, *See* Priolube 1458, 25021, Radia® 7131, 25724
Isooctyl tallate, *See* Plasthall® 100, 23978
isooctyl thioglycolate, 15386
isooctylmercaptoacetate, *See* isooctyl thioglyconate, 15386
Isopal, *See* Wickenol® 111, 34379
Isopalm, *See* Wickenol® 111, 34379
Isopar®, 15387
Isopar® C, 15388
Isopar® E, 15389
Isopar® G, 15390
Isopar® L, 15391
Isopar® M, 15392, *See* Exxsol® D-40, D-60, D-80, D-110, D-130, 11359
Isopaste, 15393
isopentane, 15394, *See* Exxsol® Isopentane, 11362
8-Isopentenyloxypsorlene, *See* imperatorin, 14912
Isopentyl Acetate, *See* isoamyl acetate, 15336
isopentyl alcohol, acetate, *See* isoamyl acetate, 15336, pear oil, 22971
isopentyl ester acetic acid, *See* isoamyl acetate, 15336, pear oil, 22971
isophorone, 15395
Isophorone diamine, *See* Vestamin® TMD, 33656
Isophorone diamine diisocyanate, *See* Vestanat® IPDI, 33657
Isophorone diisocyanate, *See* Vestanat® IPDI, 33657
isophthalic acid, *See* Amoco® PIA, 2287
isophthalic acid, di-(2-propenyl) ester, *See* Dapon M, 7738
Isoplac, 15396
Iso-Planotox, 15397
Isoplast, 15398
Isoplast 101, 15399
Isoplast 101LGF40 Nat, 101LGF60 Blk, 15400
Isoplast 302, 15401
Isopoxy, 15402
IsoPPC, *See* Propham, 25186
Isoppc, *See* Propham, 25186
Isoprene rubber, *See* polyisoprene, 24470, 24471
Isoprep®, 15403
isoprocarb, 15404, *See* Etrofolan®, 11156, Mipcin®, 19846
isoprocarbe, *See* isoprocarb, 15404
isopropanol, 15405
isopropanolamine, 15406
(R)-4-isopropenyl-1-methyl-1-cyclohexene, *See* Limonene, 17280
isopropenylbenzene, *See* α-methylstyrene, 19579
isopropenylbenzene, *See* α-methylstyrene, 19579
o-isopropoxyphenyl methyl carbamate, *See* Propoxur, 25198, 25199, 25200
2-isopropoxyphenyl methyl carbamate, *See* Propoxur, 25198, 25199, 25200
1-Isopropoxy-2-propanol, *See* Solvenon® IPP, 29047
2-isopropoxy-1-propanol, *See* Solvenon® IPP, 29047
Isopropyl 0 PPD, *See* Flexzone 3C, 11967
Isopropyl 2,4-dinitro-6-*itsec*ro-butylphenyl carbonate, *See* Dinobuton, 8543
Isopropyl 4-aminobenzenesulfonyl di(dodecylbenzenesulfonyl) titanate (monoalkoxy), *See* Ken-React® 26S (KR 26S), 16018
Isopropyl adipate, *See* Unimate® DIPA, 32968
isopropyl alcohol, *See* Alcowipe®, 1531, Avantine, 3456, I.P.S, 14799, isopropanol, 15405
Isopropyl alcohol 70%, *See* Sterets Pre-Injection Swabs®, 29728
(-)-2,3-o-Isopropyl alcohol, *See* isopropanol, 15405
isopropyl benzene peroxide, *See* Aztec® DCP-R, 3556, Dicumyl peroxide, 8407, Percumyl D, 23230, Peroximon® DC-40, 23409, Polyvel PCL-20, 24661
Isopropyl N-benzoyl-N-(3-chloro-4-fluorophenyl)alanine, *See* flamprop-isopropyl, 11887, Gunner, 13493, Power Flame, 24824
Isopropyl C_{12-15} pareth-9-carboxylate, *See* Velsan® P8-3, 33543
Isopropyl carbanilate, *See* Propham, 25186, 25185
Isopropyl carbanilate, *See* Propham
isopropyl carbinol, *See* isobutanol, 15344
isopropyl carbitol, *See* isobutanol, 15344
isopropyl-2-chloroacetanilide, *See* Propachlor, 25163
N-Isopropyl-2-chloroacetanilide, *See* Ramrod, 25861
N-Isopropyl-α-chloroacetanilide, *See* Ramrod, 25861

isopropyl chlorocarbanilate, *See* Chlorpropam, 6162
isopropyl chloroformate, 15407
isopropyl-o-cresol, *See* oxycymol, 22445
6-isopropyl-*m*-cresol, *See* thymol, 31778
Isopropyl dimethacryl isostearoyl titanate (monoalkoxy), *See* Ken-React® 7 (KR 7), 16015
Isopropyl esters of myristic, palmitic and stearic acids, *See* Wickenol® 136, 34382
Isopropyl Hexadecanoate, *See* Wickenol® 111, 34379
isopropyl n-hexadecanoate, *See* isopropyl palmitate, 15410, Wickenol® 111, 34379
Isopropyl isostearate, *See* Emerest® 2310, 10084 Schercemol 318, 27582 Wickenol® 131, 34381
Isopropyl lanolate, *See* Amerlate® P, 2018, Crodalan IPL, 7091, Fancor IPL, 11456, Lanesta S, 16658, Ritasol, 26832
Isopropyl linoleate, *See* Ceraphyl® IPL, 5732
isopropyl mercaptan, 15408
2-isopropyl-5-methylphenol, *See* thymol, 31778
4-Isopropyl-1-methyl-2-propenyl benzene, *See* Verdoracine, 33581
isopropyl myristate, 15409, *See* Crodamol 1PM, 7116, Emerest® 2314, 10085, Estergel, 10883, Kessco® IPM, 16115, Lexol 3975, 17147, Lexol IPM-NF, 17150, Liponate IPM, 17378, Promyr, 25152, Radia® 7190, 25729, Tegosoft® M, 31024, Tegosoft® Liquid M, 31023, Unimate® IPM, 32971, Wickenol® 101, 105, 34378, Wickenol® 101, 105, 34378
Isopropyl n-hexadecanoate, *See* isopropyl palmitate, 15410
Isopropyl oleate, *See* Radia® 7231, 25733
isopropyl palmitate, 15410, *See* Emerest® 2316, 10086, Kessco® IPP, 16116, Lexol 3975, 17147, Liponate IPP, 17379, Propal, 25168, Radia® 7200, 25730, Tegosoft® P, 31026, Unimate® IPP, 32972, Wickenol® 111, 34379
N-isopropyl-N'-phenyl-*p*-phenylene diamine, *See* Flexzone 3C, 11967, Permanax IPPD, 23364, Vanox® 3C, 33338, Vulkanox® 4010 NA, 34147
Isopropyl phenyl urethane, *See* Propham, 25186
Isopropyl phenylcarbamate, *See* Propham, 25185
isopropyl phenylmethyl carbamate, 15411
Isopropyl phosphate ester, *See* Marlophor® CS-Acid, 18891
Isopropyl PPG-2-isodeceth-7-carboxylate, *See* Velsan® D8P-3, 33541
Isopropyl stearate, *See* Lexol 3975, 17147, Tegosoft® S, 31027, Wickenol® 127, 34380
Isopropyl tetradecanoate, *See* Wickenol® 101, 105, 34378
Isopropyl thioxanthone, *See* Speedcure ITX, 29263
Isopropyl tri(dioctylphosphato) titanate (monoalkoxy), *See* Ken-React® 12 (KR 12), 16017
Isopropyl tri(dioctylpyrophosphato) titanate (monoalkoxy), *See* Ken-React® 38S (KR 38S), 16020
Isopropyl tri(N ethylaminoethylamino) titanate (monoalkoxy), *See* Ken-React® 44 (KR 44), 16023
Isopropyl tridodecylbenzenesulfonyl titante (monoalkoxy), *See* Ken-React® 9S (KR 9S), 16016
isopropyl-*m*-cresol, *See* thyme camphor, 31775
isopropyl-*o*-cresol, *See* carvacrol, 5362
4-Isopropyl-5,5-dimethyl-1,3-dioxane, *See* Anthoxan, 2524
N-Isopropyl-N'-fenyl-p-fenylendiamin, *See* Vulkanox® 4010 NA, 34147
2-Isopropyl-phenyl-N-methylcarbamate, *See* Etrofolan®, 11156, isopropyl phenylmethyl carbamate, 15411
N-Isopropyl-N'-phenyl-p-phenylenediamine, 15413, *See* Vulkanox® 4010 NA, 34147
N-isopropyl-N'-phenyl-4-phenylenediamine, *See* Flexzone 3C, 11967
Isopropylamine alkylbenzene sulfonate, *See* Ufasan IPA, 32748
4-Isopropylamine diphenylamine, *See* Nonox ZA, 21456
Isopropylamine dodecylbenzene sulfonate, *See* Arylan® PWS, 3158, Calimulse PRS, 5007, Kowet 3300, 16374, Nansa® YS94, 20698, Naxel AAS-Special 3, 20814, Pentine 1185 5432, 23181, Rhodacal® 330, 26577, Rhodacal® IPAM, 26586, Witconate P-1059, 34488
Isopropylamine dodecylbenzene sulfonate (branched), *See* Polystep® A-11, 24585
Isopropylamine glyphosate, *See* Rodeo, 26899
Isopropylamino)-4-(methylamino)-6-(methylthio)-s-triazine, *See* desmetryn, 8142
4-(Isopropylamino)diphenylamine, *See* Vulkanox® 4010 NA, 34147
***p*-Isopropylaminodiphenylamine**, *See* N-Isopropyl-N'-phenyl-*p*-phenylene diamine, 15413, Vulkanox® 4010 NA, 34147
Isopropylan® 33, 15412
***p*-Isopropylbenzaldehyde**, *See* Cumal, 7349, cuminaldehyde, 7354
isopropylbenzene, *See* cumene, 7352
isopropylbenzene hydroperoxide, *See* Aztec® CHP-80, 3555, CHP-158, 6175
Isopropylbiphenyl, *See* Nusolv ABP-62, 21802
2-Isopropylbiphenyl, *See* Nusolv ABP-62, 21802
Isopropylcarbinol, *See* isobutanol, 15344
3-isopropylcumene, *See* 1,3-diisopropyl benzene, 8467
3-and 4-isopropylcumene, *See* 1,3-, 1,4-diisopropyl benzene, 8467
4-isopropylcumene, *See* 1,4-diisopropylbenzene, 8468
Isopropylformic acid, *See* isobutyric acid, 15348
***p,p'*-isopropylidenebisphenol**, *See* Bisphenol A, 4283
4,4'-Isopropylidenebis (2,6-dibromophenol), *See* tetrabromobisphenol-A, 31359
(Isopropylidenedioxy)phenyl N-methylcarbamate;, *See* Bendiocarb, 3937
4,4'-isopropylidenediphenol (CTFA), *See* bisphenol A, 4283
***p,p'*-Isopropylidenediphenol**, *See* Bisphenol A, 4283
1-isopropylidene-4-methyl-2-cyclohexanone, *See* pulegone, 25379
Isopropylparaben, *See* LiquaPar® Oil, 17451
Isopropylphenyl diphenyl phosphate, *See* Syn-O-Ad® 8480, 30557
2-Isopropylphenyl methylcarbamate, *See* Mipcin®, 19846
Isopropylthioxanthone, *See* Speedcure ITX, 29262
Isopropylxanthic acid, sodium salt, *See* Aero 343 Xanthate, 842
Isoproturon, *See* Arelon, 2830, Hytane, 14774, Protugan, 25300
isoproturon-isoxaben, *See* Fanfare, 11463, Ipso, 15250
isoproturon-pendimethalin, *See* Encore, 10543
Isopto Atropine, 15414
Isopto Cetamide, 15415
Isopurpurin,, *See* anthrapurpurin, 2531
Isoquassin, *See* picrasmin, 23819
Isoquinolinium, 2-dodecyl-, bromide, *See* Isothan Q-75, 15427
Isorate, 15416
Isoset®, 15417
Isoset® WD3-A322 Emulsion Resin, 15418
Isoset® WD3-CM402 Emulsion Resin, 15419
Isosorbide laurate, *See* Arlamol® ISML, 2911
Isostearamidopropyl alkonium chloride, *See* Schercoquat IB, 27680
Isostearamidopropyl betaine, *See* Incronam 1-30, 15015, Lexaine® IS, 17093, Mackam ISA, 18144, Schercotaine IAB, 27697
Isostearamidopropyl dimethylamine, *See* Chemidex SI, 5970, Incromine IB, 14999, Mackine 401, 18219, Schercodine I, 27637
Isostearamidopropyl dimethylamine glycolate, *See* Katemul IG-70, 15789
isostearamidopropyl dimethylamine lactate, 15420, *See* Incromate IDL, 14968
Isostearamidopropyl epoxypropyl dimonium chloride, *See* Schercoquat IEP, 27681
Isostearamidopropyl ethyldimonium ethosulfate, *See* M-Quat® 522, 20401, Schercoquat IAS, 27679
Isostearamidopropyl laurylacetodimonium chloride, *See* Schercoquat IALA, 27678
Isostearamidopropyl morpholine, *See* Mackine 421, 18220
Isostearamidopropyl morpholine lactate, *See* Emcol® ISML, 10061, Incromate ISML, 14969
Isostearamidopropyl morpholine oxide, *See* Incromine Oxide ISMO, 15005, Mackamine ISMO, 18171
Isostearamidopropyl PG-dimonium chloride, *See* Lexquat® AMG-IS, 17176
Isostearamidopropylamine oxide, *See* Incromine Oxide I, 15004, Mackamine IAO, 18170
Isostearaminopropalkonium chloride, *See* Incroquat I-85, 15039
Isosteareth-2, *See* Arosurf® 66-E2, 3077, Hetoxol IS-2, 13939
Isosteareth-5, *See* Emalex 1805, 9914
Isosteareth-6 carboxylic acid, *See* Sandopan® TA-10, 27348
isostearic acid, 15421, *See* Emersol® 871, 10122, Prisorine 3508, 25036
Isostearic acid and isostearoyl hydrolyzed collagen, *See* Proto-Lan IP, 25295
Isostearic acid, isopropyl ester, *See* Wickenol® 131, 34381
Isostearic acid, sorbitan oleate and cocoyl hydrolyzed keratin, *See* Proto-Lan KT, 25296
Isostearoyl PG-trimonium chloride and behenoyl PG-trimonium chloride, *See* Akypoquat 129, 1270
isostearyl alcohol, *See* Adol® 66, 785, Diadol 18G, 8264, Michel XO-150-1620, 19642
Isostearyl benzyl imidonium chloride, *See* Schercoquat IIB, 27682
Isostearyl diglyceryl succinate, *See* Imwitor® 780 K, 14946
Isostearyl dimethylamidopropyl ethonium ethosulfate, *See* Foamquat IAES, 12180
Isostearyl erucate, *See* Schercemol ISE, 27605
Isostearyl ethylimidonium ethosulfate, *See* Monaquat ISIES, 20134, Schercoquat IIS, 27683
Isostearyl hydroxyethyl imidazoline, *See* Monazoline IS, 20195, Schercozoline I, 27713
isostearyl isostearate, 15422, *See* Iso Isostearyle WL 3196, 15334, Schercemol 1818, 27584, Starfol® IS, 29536
Isostearyl lactate, *See* Patlac® IL, 22909
Isostearyl neopentanoate, *See* Ceraphyl® 375, 5726, Dermol 185, 8120, Schercemol 185, 27581
Isostearyl stearoyl stearate, *See* Hetester ISS, 13898
Isostearyl-erucyl erucate, *See* Scheroba Oil, 27717
Isostearylpropyl dimethylamine, *See* Mazeen® DAPI, 19122
isotachiol, 15423
Isotagetone, 15424
Isotagetone 50, 15425
Isotearamide DEA, *See* Mackamide ISA, 18155
Isoterge, 15426
Isothan Q-75, 15427, *See* Isothan Q-75, 15427
Isothiazolone, 2-octyl-, *See* Skane® M-8, 28563
isothiocyanatoethane, *See* ethyl isothiocyanate, 11075
Isothiocyanatomethane, *See* methyl isothiocyanate, 19543
isothymol, *See* carvacrol, 5362, oxycymol, 22445
Isoton, 15428
isotonic salt solution, *See* physiological salt solution, 23776
Isotox, 15429
Isotridecyl alcohol polyglycol ether carboxylate, *See* Hostapur DTC, 14404
Isotridecyl isononanoate, *See* Wickenol® 153, 34390
Isotridecyl stearate, *See* Sicolub® TDS, 28222
Isotridecyl 3,5,5-trimethylhexanoate, *See* Wickenol® 153, 34390
3-(Isotridecyloxy) 1-propaneamine, *See* Etheramine 13, 10942
Isotridecyloxypropyl dihydroxyethyl methyl ammonium chloride, *See* Tomah Q-17-2, 31991
Isotridecyloxypropylamine, *See* Tomah PA-17, 31985
N-Isotridecyloxypropyl-1,3-diaminopropane, *See* Tomah DA-17, 31968
Isotron 2, *See* dichlorodifluoromethane, 8388

Isoundeceth-3, *See* Prox-onic UA-03, 25347
isovalerone, *See* diisobutyl ketone, 8461, valerone, 33243
Isovie, *See* Iopamidol, 15233
Isovist, *See* iotrolan, 15244
isovitamin C, *See* Erythorbic acid, 10775, isoascorbic acid, 15338
Isovue, *See* Iopamidol, 15233
Isovyl, *See* Korogel, 16354, Koroplate, 16357
Isoxaben, *See* Flexidor, 11934, Knot Out, 16257
isoxaben-methabenzthiazuron, *See* Glytex, 13154
Issolin, 15430
Issolith, *See* Bakelite, 3589
Istambul, *See* Taktic, 30729
Isteropac E.R, *See* Renovue-65, Renovue-DIP, 26115
Italcor, 15431, *See* aluminum oxide, 1919
Italian pink, *See* yellow carmine, 34644
ithiobis(pyridine N-oxide), *See* Omadine® MDS, 22152
ITP®, 15432
itrol, 15433, *See* silver citrate, 28422
IUDR, *See* Kerecid, 16067
Iupital® F10, 15435
Ivaleur, 15436
Ivarbase 98, 15437
Ivarbase 101, 15438
Ivarbase 3210, 15439
Ivarbase 3230, 15440
Ivarbase 3240, 15441
Ivarbase 3250, 15442
Ivarlan 3100, 15443
Ivarlan 3310, 15444, *See* Lanolin alcohol, 16741
Ivarlan 3400, 15445
Ivarlan 3406, 15446
Ivarlan 3420, 15447
Ivarlan 3450, 15448, *See* PEG-20 hydrogenated lanolin, 23045
Ivarlan AWS, 15449, *See* PPG-12-PEG-65 lanolin oil;, 24843
Ivarlan C-24, 15450
Ivarlan HL, 15451, *See* Fancol HL, 11441
Ivarlan Light, 15452
Ivarlan OH, 15453, *See* hydroxylated lanolin, 14663
Ivermectin, 15454, *See* Eqvalan, 10737
Ivomec, *See* Ivermectin, 15454
Ivoran, *See* DDT, 7820
ivoride, 15455
Ivorin-Profalon, 15456
Ivosit, 15457
Iwox, 15458
Ixan, 15459
Ixan®, 15460
Ixef® 1022, 15461
Ixodex, *See* DDT, 7820
Ixol, 15462
Ixolite, 15463
Ixper, 15464
Izal, 15465
J-acid, 15482
J Slip NS-77, 15466
J Soft C 4, *See* Ablumine 280, 85, Cycloton® SCS, 7542, octadecylbenzenemethanaminium chloride, 21985, Sumquat® 6210, 30126, stearalkonium chloride, 29592
J Wet 19A, 15467
J-13, 15468
Jabclad, 15469
Jabdec, 15470
Jabdie, 15471
Jabilite Thermodek, 15473
Jablina Insulating Panels, 15472
Jablite, 15474
Jablite Cavity, 15475
Jablite Flooring, 15476
Jablite Insulation Board, 15477
Jablite Thermacel, 15478
Jablite Thermoclik, 15479
Jablite WallLok, 15480
Jabroc, *See* Permali, 23349
Jacana metal, 15481
Jacksonite, *See* prehnite, 24883
Jacoby metal, 15483
Jacquemart's reagent, 15484
Jacutin®, 15485
JA-fA IPM, *See* Wickenol® 101, 105, 34378
Ja-fa ipp, *See* Wickenol® 111, 34379
Ja-fa ippkessco, *See* Wickenol® 111, 34379
Jaffamine®, 15486
Jagalux, 15487
Jagalyd, 15488
Jagapol, 15489
JagDril CC, 15490
Jagotex, 15491
Jagotex Esi-Cryl, 15492
Jaguar, 15493
Jaguar gum A-20-D, *See* Prinza® Range, 25010
Jaguar no.124, *See* Prinza® Range, 25010
Jaguar plus, *See* Prinza® Range, 25010
Jaguar® 413, 15494
Jaguar® C, 15495
Jaguar® C-13S, C-14S, 15496, 15497
Jaguar® C-162, 15498, 15499
Jaguar® Guar Gum, 15500
Jaguar® HP 8, 15502
Jaguar® HP-11, 15503
Jaguar® HP8, HP60, HP120, HP-11, HP-200, *See* hydroxypropyl guar, 14666
Jaguar® HP 60, 15501
jalcase, 15504
Jamaica wood, *See* logwood, 17557
janthone, 15505
Janus, 15506
Jaon, 15507
Jaon, *See* Potassium gluconate, 24755
Japan Agar, *See* agar-agar, 955, vegetable gelatin, 33518
Japan camphor, 15508
Japan drier, *See* terebine, 31234
Japan earth, *See* cutch,
Japan isinglass, *See* agar-agar, 955
Japan sago, 15509
Japan tallow, 15510
Japan varnishes, 15511
Japan wax, *See* Japan tallow, 15510
Japanese acid clay, 15512
Japanese bell metal, 15513
Japanese brass, *See* Sin-Chu, 28495
Japanese bronze, 15514
Japanese gelatin, *See* agar-agar, 955
Japanese silver, 15515
Japanese wax, *See* Chinese wax, 6077
Japidermic, 15516
JAQ Powdered Quat, *See* Exameen 824 3724, 11284
JAQ Powdered Quaternary, 15517
jara jara, 15518
Jarcal, 15519
Jarfix 391, 15520
Jargonelle pear essence, 15521
Jarofast, 15522
Jarosol, 15523
Jarozyme 491, 15524
Jarytherm, 15525
Jascitile, 15526
Jasilyn, 15527
Jasmacyclat, 15528
Jasmacyclene, 15529
Jasmal, 15530
Jasmatone, 15531
Jasmin acetate, *See* Jasmopyrane, 15533
Jasmolide, 15532
Jasmopyrane, 15533
Jasmorange®, 15534
Jatex, 15535
Jatob duro, 15536
Jatob lagrima, 15537
Jatob resin, 15538
Java olives, *See* olives of Java, 22141
Java wax, 15539
Javelin, 15540
Javelle water, *See* sodium hypochlorite, 28773
Javex, *See* sodium hypochlorite, 28773
Jaydalene, 15541
Jayflex®, 15542
Jayflex® 77, 15543
Jayflex® 210, 15544
Jayflex® 215, 15545
Jayflex® 911, 15546
Jayflex® 3209, 15547
Jayflex® 7911, 15548
Jayflex® DHP, 15549
Jayflex® DIDP, 15550, *See* diisodecyl phthalate, 8462
Jayflex® DINA, 15551
Jayflex® DINP, 15552
Jayflex® DIOP, 15553
Jayflex® DOA, 15554, *See* dioctyl adipate, 8548
Jayflex® DTDP, 15555
Jayflex® DUP, 15556
Jayflex® L11P, *See* Jayflex® DUP, 15556
Jayflex® L9P, 15557
Jayflex® TINTM, 15558
Jayflex® TOTM, 15559
Jayflex® UDP, 15560, *See* undecyl dodecyl phthalate, 32915
Jaysol, *See* ethyl alcohol, 11064
Jaysol s, *See* ethyl alcohol, 11064
Jazz®, 15561
JB-4, 15562
J-Black-20, 15563
Jectoflo, 15564
Jectomag, 15565
Jectothane, 15566
Jeffamine® BuD-2000, 15567
Jeffamine® D-2000, 15568
Jeffamine® DU-700, 15569
Jeffersol DB, *See* Butyl Carbitol®, 4766
Jeffersol EB, *See* 2-butoxyethanol, 4753
Jeffersol EE, *See* ethyl icinol, 11074
Jeffox, 15570
jelly rock, *See* Wilkinite, 34409
Jel-O-Mer®, 15571
Jel-O-Mer® 46-902, 15572
Jelonet, 15573
Jelutong, 15574
Jenner's stain, 15575
Jeppel's oil, *See* bone oil, 4402
Jer Dri WRN, 15576
Jerotex P, 15577
Jersey lily white, 15578
Jersey lily white, *See* Lithopone, 17520
Jessate, 15579
Jesuit's balsam, 15580
Jesuit's tea, *See* yerba mate, 34655
Jet, 15581
Jet Amine DC, 15582
Jet Amine DE 810, 15583
Jet Amine DE-13, 15584
Jet Amine DMCD, 15585
Jet Amine DMOD, 15586
Jet Amine DMSD, 15587
Jet Amine DMTD, 15588
Jet Amine DO, 15589
Jet Amine DT, 15590, *See* Cocodiamine, 6564
Jet Amine M2C, 15591
Jet Amine PC, 15592
Jet Amine PE 1214, 15593
Jet Amine PHT, 15594
Jet Amine PO, 15595
Jet Amine PS, 15596
Jet Amine PT, 15597, *See* tallowamine, 30748
Jet Amine TET, 15598
Jet Amine TP, 15599
Jet Amine TRT, 15600
jet black, *See* gas black, 12686
Jet Jel®, 15601
Jet Quat 2C-75, 15602
Jet Quat C-50, 15603
Jet Quat DT-50, 15604
Jet Quat S-50, 15605

Jet Quat T-50, 15606
Jet Quat T-50, T-27W, *See* tallow trimonium chloride, 30747
Jetfill 700C, 15607
Jet-Flex® 101, 15608
Jet-Lube J-75®, 15609
Jeunite, 15610
Jeweller's rouge, 15611, *See* Ferroxide, 11694, Kroma Red, 16418
Jeyes disinfectant, 15612
Jicwood, *See* Permali, 23349
JL 43155AS, 15613
JL 43176, 15614
Jodairol, *See* Hippuran-^{131}I, 14125, Nephroflow, 20989
jodfenphos, *See* Iodofenphos, 15213
Jodid, *See* Potassium Iodide, 24758, 24759
Jodomiron, 15615, *See* Renovue-65, Renovue-DIP, 26115
Jogral, 15616
Johannisbrotmehl, *See* carob flour, 5303
Jojoba oil, 15617, *See* Emalex J.J O-V, 9966, Wickenol® 139, 34383
Jojoba wax, *See* Wickenol® 139, 34383
Jojobeads, 15618
Jonylon, 15619
Jopamiro, *See* Iopamidol, 15233
Jordapon® CI-50 Disp, 15620
Jordaquat® 350, 15621
JP 351, *See* Wytox® 312, 34578
JR Surfacer, 15622
JR-228, JR-228-1, 15623
J-Red-10, 15624
J-Red-12FS, 15625
J-Soft 111E, 15626
JSR-10, 15627
JSR-12, 15628
JSR-21, 15629
Jubilee®, 15630
Judean pitch, *See* Karetnja, 15765
Juglone, 15631
Julian's carbon chloride, *See* Julin's chloride, 15632
Julin's chloride, 15632
Jupital, 15633, *See* Chlorothalonil, 6150, Repulse, 26151
Jureong, *See* permethrin, 23384, 23385, 23386
Justice®, 15634
justite, 15635
jutahy, *See* jutahycica, 15636
jutahycica, 15636
jutahycica resins, 15637
Juvelith, *See* Bakelite, 3589
JZF, 15638, *See* Agerite® DPPD, 1016 N,N'-diphenyl-*p*-phenylenediamine, 8603 Naugard® J, 20796 Permanax DPPD, 23361
K, *See* lysine, 18088
K 129, 15639, *See* Montmorillonite, 20300
K 129-H, 15640
K de Krizia, 15641
K III, *See* 4,6-dinitrocresol, 8542
K IV, *See* 4,6-dinitrocresol, 8542
K. Tab, 15642
K.A. alloy, 15643
K.L.X, 15644
K.S. magnet steel, 15645
K.S. powder, 15646
K/Zn, *See* Bärostab® KK 47 S, 3659
$K_2W_3O_9.W_2O_5$, *See* violet tungsten, 33878
K-10 Bentonite clay, *See* Fulmont Activated Bleaching Earths, 12448
K-11941, *See* Alfaprostol, 1574
K154, 15647
K285, 15648
kabaite, 15649
Kabikinase, 15650
Kabipenin, *See* Penicillin V, 23111
kachin, 15651
K-acid, 15652
kadaya, *See* karaya gum, 15760, Normacol, 21566
kadaya gum, *See* karaya gum, 15760
Kadel® E-1000, 15653
kaempferol, 15654
Kafil, 15655, *See* cypermethrin, 7588, permethrin, 23384, 23386, 23385
Kafil Super, *See* Topclip Parasol, 32037
Kafylox, *See* K-Lyte, 16233
Kagolin 5.8FG, 15656
Kagoo oil, *See* korung oil, 16364
kainite, 15657
Kairoline, 15658
Kaiserling solution, 15659
Kaladex, 15660
Kalammon, 15661
Kalar® 5214, 15662
Kalar® 5214, 5263, *See* Exxon® Butyl 077, 11355
Kalar® 5263, 15663
Kalbord, 15664
Kalcohl 80, *See* Adol® 62 NF, 783, Stearal, 29591
Kalcrete, 15665
Kalene® 800, 15666, *See* Exxon® Butyl 077, 11355
Kaleoilris, 15667
Kaleorid, *See* Kay Ciel, 15807, K-Contin, 15818, K. Tab, 15642, Kaon-Cl, 15729, K-Lor, 16226, K-Lyte/C1, 16234, KM Potassium Chloride, 16242, potassium chloride, 24748, 24749
Kaleorod, *See* Potassium chloride, 24748, 24749
Kalex, *See* tetrasodium EDTA, 31391
Kalex 220 Crystal, 15668
Kalex Acids, 15669
Kalex HMP, 15670
Kalex Liq. 50%, 15671
Kalex OH, 15672
Kalex Penta, 15673, *See* Pentasodium pentetate, 23172
Kalex® 13361, 15674
Kalex® 15036, 15675
Kalex® 20171, 15676
Kaliammon saltpeter, 15677
Kalidone®, 15678
Kalif, 15679
Kalimozan, *See* Gluconal® K, 13043
Kalimozan, *See* Katorin, 15798
Kalimozan, *See* potassium d-gluconate, 24752, Potassium gluconate, 24755
Kalipol, 15680
Kalipol 18, 15681
Kalitabs, *See See* Kay Ciel, 15807, K-Contin, 15818, K. Tab, 15642, Kaon-Cl, 15729, K-Lor, 16226, K-Lyte/C1, 16234, KM Potassium Chloride, 16242, potassium chloride, 24748, 24749
Kalium, *See* potassium, 24732
Kalium-Duriles, *See See* Kay Ciel, 15807, K-Contin, 15818, K. Tab, 15642, Kaon-Cl, 15729, K-Lor, 16226, K-Lyte/C1, 16234, KM Potassium Chloride, 16242, potassium chloride, 24748, 24749
kalkeisenollvin, 15682
Kalkor, 15683
Kalle's acid, 15684
Kallodent®, 15685
Kallodoc, 15686
Kallte, 15687
Kalluzoto, 15688
Kalmex, 15689
Kalmin, 15690
Kalminex, 15691
Kaloempang beans, *See* olives of Java, 22141
Kalorex, 15692
Kalpack, 15693
Kalpad, 15694
Kalpur, 15695
Kalpur TE, *See* Bioban® GK, 4158, Nipacide® BK, 21254
Kalrez®, 15696
Kalseal, 15697
Kalsert, 15698
Kaltas, 15699
Kaltop, 15700
Kalvan, 15701
Kalzana, 15702
kalzose, 15703
kam 1000, *See* stearic acid, 29599
kam 2000, *See* stearic acid, 29599
kam 3000, *See* stearic acid, 29599
Kamala, *See* Kamela, 15706
Kamax, 15704
Kamax T-150, 15705
Kambara earth, *See* Japanese acid clay, 15512
Kambe wood, *See* redwoods, 25987
kameela, *See* Kamela, 15706
Kamela, 15706
Kamila, *See* Kamela, 15706
Kamillosan, 15707
Kamoran, 15708
Kampstoff Lost, *See* mustard gas, 20538
Kamycin, *See* Kantrex, 15718, Klebcil, 16209
Kamynex, *See* Kantrex, 15718, Klebcil, 16209
Kanabristol, *See* Kantrex, 15718, Klebcil, 16209
Kanacedin, *See* Kantrex, 15718, Klebcil, 16209
Kanamycin A sulfate, *See* Kantrex, 15718, Klebcil, 16209
Kanamytrex, *See See* Kantrex, 15718, Klebcil, 16209
Kanaqua, *See See* Kantrex, 15718, Klebcil, 16209
Kanasig, *See See* Kantrex, 15718, Klebcil, 16209
Kanatrol, *See See* Kantrex, 15718, Klebcil, 16209
kandiset, *See* saccharin, 27195
Kane-Ace, 15709
Kane's salt, 15710
Kanescin, *See* Kantrex, 15718, Klebcil, 16209
Kanfotrex, 15711
Kanga butter, *See* lamy butter, 16643
Kanicin, *See* Kantrex, 15718, Klebcil, 16209
Kanigen, 15712
Kanja butter, *See* lamy butter, 16643
Kankerex, 15713
Kannasyn, 15714 *See* Kantrex, 15718, Klebcil, 16209
Kano, *See* Kantrex, 15718, Klebcil, 16209
Kanten, 15715
Kanthal Alloy, 15716
Kantmelt, 15717
Kantrex, 15718, *See* Klebcil, 16209
Kantrexil, 15719, *See* Kantrex, 15718, Klebcil, 16209
Kantrim, 15720
Kantrox, *See* Kantrex, 15718, Klebcil, 16209
Kantstik Q Powd, 15721
Kaokote, 15722
Kaokote F, 15723
Kaola, 15724
kaolin, 15725, *See* Anhydrol, 2459 ASP®, 3230, Bilt-Cote®, 4147, Bilt-Plates®, 4148, Buca, 4646, Continental® Clay, 6740, kaolin, 15725, Kayphobe-ABO, 15816, Peerless®, 23023, Peerless® No. 1, 23024, SPS, 29349, Vanclay, 33304
kaolin clay, *See* Dixie Clay®, 8743, Langford Clay, 16674
kaolinase, 15726
Kaolinite, *See* ASP®, 3230 Bilt-Cote®, 4147 Peerless®, 23023 SPS, 29349 Tisyn®, 31857
Kaomax, 15727
Kaomel, 15728
Kaon, *See* Gluconal® K, 13043, Katorin, 15798, potassium d-gluconate, 24752
Kaon P, *See* Potassium gluconate, 24755
Kaon-Cl, 15729, *See* Kay Ciel, 15807, K-Contin, 15818, K. Tab, 15642, Kaon-Cl, 15729, K-Lor, 16226, K-Lyte/C1, 16234, KM Potassium Chloride, 16242, Potassium chloride, 24748, 24749
Kaopectate, *See* ASP®, 3230, Bilt-Cote®, 4147, Peerless®, 23023, SPS, 29349
Kaopolite® 1152, 15730
Kaopolite® AB, 15731
Kaopolite® SF, 15732
Kaoprem-E, 15733
Kaorich Beads, 15734
Kaorich Gold, 15735
Kaowool®, 15736
kapak, 15737
Kapazang oil, 15738
Kapex, 15739
kapithamia piscum, *See* wood-apple gum, 34552

Kapitol, 15740
kapok, 15741
Kappadione, *See* menadiol sodium diphosphate, 19319
Kapron, *See* Cabelec® 1015, 4861
Kapsol, 15742
Kapsovit, 15743
Kapton®, 15744
Kapur Kachri, *See* sanna,
Kara Lube Al-Conc, 15745
Karafac 78, 15746
Karakane, 15747
Karalube DKL, 15748
Karamate, 15749, *See* mancozeb, 18517
Karamide 121, 15750
Karamide CO9A, 15751
Karamide ST-DEA, 15752
Karaphos HSPE, 15753
Karaphos SWPE, 15754
Karasoft YB-11, 15755
Karasurf AS-26, 15756
Karate, 15757, 15758, *See* λ-cyhalothrin, 7567
Karathane, 15759
karaya gum, 15760, *See* Normacol, 21566
Karaya Paste, 15761
Karbam White, *See* AAprotect, 16, Accelerator MZ Powder, 209, Vancide® MZ-96, 33303, Zimate®, 34762
Karbation, *See* metam-sodium, 19444
Karbofos, *See* malathion, 18460
Karbofuranu, *See* Rampart, 25860
Karbolite, 15762
Karboresin, 15763
Karbos, 15764
Karetnja, 15765
karidium, *See* sodium fluoride, 28764
karigel, *See* sodium fluoride, 28764
Karion,, *See* A-625/641ABS 301K, ABS 500FR-1, 10
kari-rinse, *See* sodium fluoride, 28764
Karite gum, *See* gutta-shea, 13508
Karlex 12006 BKFR, 12018 BKFR, 15766
Karlex 40002 NA, 15767
Karlex 40003 NA, 15768
Karma metal, 15769
Karmarsch metal, 15770
Karmex, 15771, *See* diuron, 8736
Karmex 80W, *See* diuron, 8736
Karmex DL, *See* diuron, 8736
Karmex®, 15772
Karolith, 15773
Karox AO-30, 15774
Karox LO, 15775
karstenute, *See* calcium sulfate (anhydrous), 4971
Kartacid C 60, *See* Hystrene® 1835, 14755
Karvol, 15776
Kasal®, 15777
Kasil, 15778
Kaskay, *See* Kay Ciel, 15807, K-Contin, 15818, K. Tab, 15642, Kaon-Cl, 15729, K-Lor, 16226, K-Lyte/C1, 16234, KM Potassium Chloride, 16242, Potassium chloride, 24748, 24749
Kastone®, 15779
Katadolon, 15780
Katalabu gum, 15781
Katalon, 15782
Katamine AB, *See* Ablumine 280, 85, Cycloton® SCS, 7542, octadecylbenzenemethanaminium chloride, 21985, stearalkonium chloride, 29592, Sumquat® 6210, 30126
Katanol, 15783
Katapol®, 15784
Katapone VV-328, 15785
katarsit, 15786
Katavel oil, 15787
Katbél - ki - gond, *See* wood-apple gum, 34552
Katchung oil, 15788, *See* peanut oil, 22969
katechu, *See* cutch,
Katemul IG-70, 15789
Katemul IGU-70, 15790
Katharin, 15791
Kathol, *See* p-aminophenol, 2185
Kathon, *See* Skane® M-8, 28563
Kathon 893, *See* Skane® M-8, 28563
kathon lp preservative, *See* Skane® M-8, 28563
kathon sp 70, *See* Skane® M-8, 28563
Kathon®, 15792
Kathon® 886, 15793
Kathon® 893, 15794
Kathro, 15795
Katigen®, 15796
katilo, *See* karaya gum, 15760, Normacol, 21566
Katioran® AF, 15797
Katorin, 15798, *See* Gluconal® K, 13043, Katorin, 15798, potassium d-gluconate, 24752, Potassium gluconate, 24755
Katzenstein bearing metal, *See* bearing metals,
Kau drega, *See* Talotalo gum, 30752
Kauk Catalyst, 15799
Kauramin®, 15800
Kauranat®, 15801
Kauresin®, 15802
Kaurit®, 15803
Kauritil, *See* copper oxychloride, 6784, Cupravit®, 7370
Kauropal®, 15804
kautschin, *See* Achilles Dipentene, 323
Kava, *See* kava-kava resin, 15805
kava-kava resin, 15805
kawa, *See* kava-kava resin, 15805
Kawasaki Hakkinko, 15806
Kay Ciel, 15807, *See* Potassium chloride, 24748
Kayamer, 15808
Kayarad, 15809
Kayazinon, *See* Diazinon Liquid, 8344
Kayazol, *See* Diazinon Liquid, 8344
Kayback, *See* Kay Ciel, 15807, K-Contin, 15818, K. Tab, 15642, Kaon-Cl, 15729, K-Lor, 16226, K-Lyte/C1, 16234, KM Potassium Chloride, 16242, Potassium chloride, 24748, 24749
Kay-Cee-L, *See* Kay Ciel, 15807, K-Contin, 15818, K. Tab, 15642, Kaon-Cl, 15729, K-Lor, 16226, K-Lyte/C1, 16234, KM Potassium Chloride, 16242, Potassium chloride, 24748, 24749
Kaydol, 15810, *See* Klearol, 16208
Kaydox, 15811
Kayexalate, 15812
Kaylene, 15813
Kaylene-ol, 15814
Kaynitro, 15815
Kayphobe-ABO, 15816
KB, *See* Nacconol® 40G, 20647
kb (surfactant), *See* sodium dodecylbenzenesulfonate, 28760
K-Bond, 15817
KC 119, *See* Algin, 1591
KCl+m.MgCl+72+m.6H+72O, *See* carnallite, 5295
K-Contin, 15818
K-Contin, *See* Kay Ciel, 15807, K-Contin, 15818, K. Tab, 15642, Kaon-Cl, 15729, K-Lor, 16226, K-Lyte/C1, 16234, KM Potassium Chloride, 16242, Potassium chloride, 24748, 24749
K-Cop, 15819
KDA, *See* potassium arsenate, 24737
Keene's alloy, *See* nickel silvers, 21132
Keene's Cement, 15820
Keffekilite, 15821
Keffekill, *See* Sepiolite,
Kefroxil, *See* Cefadroxil, 5547
Keical-Ace, 15822, *See* Extrusil, 11331
Kekuna oil, 15823
Kelacid®, 15824
Kelburon, 15825
Kelco® HV, LV, *See* Algin, 1591
Kelco-Crete™, 15826
Kelcogel, *See* Gellan gum, 12736
Kelco-Gel® Gellan Gum, 15827
Kelcoloid® D, 15828
Kelcoloid® DH, DSF, 15829
Kelcoloid® HV, LV, 15830
Kelcoloid® O, S, 15831
Kelcosol®, 15832 , *See* Algin, 1591
Keldax®, 15833
Keleastoi, 15834
Kelecin, *See* Kelecin®, 15835
Kelecin®, 15835
kelene, *See* ethyl chloride, 11067
Kel-F 81 Plastic, 15837
Kel-F 3700 Elastomer, 15836
Kelfizina, 15838
Kelfizine W, *See* Kelfizina, 15838
Kelflo®, 15839
Kelfo®, 15840
Kelgin, 15841, *See* Kelcosol®, 15832, Kelgin, 15841, Kelgin® F, 15842, kelgin® HV, LV, MV, 15843, Kelgin® QH, QL, QM, 15844, Kelgin® XL, 15845, Kelgum, 15846, Keltex®, 15878, Keltex® HV, 15879, Keltex® S, 15880, Keltone®, 15883, Keltone® HV, LV, 15884
Kelgin® F, 15842
Kelgin® F, HV, LV, MV, QL, XL, *See* Algin, 1591
kelgin® HV, LV, MV, 15843
Kelgin® QH, QL, QM, 15844
Kelgin® XL, 15845
Kelgum, 15846
Kelgum®, 15847
Kelig, 15848
Kelig 32, 15849
Kelig FS, 15850
Kelisema Collagen-IMZ Complex, *See* Empigen® CDR40, 10369
Kellin®, 15851
Kellite, 15852
Kel-Lite™ CM, 15853
Kellox®, 15854
Kelly's paint, 15855
Kelmar®, 15856
Kelmer®, 15857
kelp, 15858, *See* Hismanal, 14132
kelp salt, 15859
Kelpak, 15860
kelpchar, 15861
Kelpol®, 15862
kelpol® 835-M-50, 15863
Kelpoxy®, 15864
Kelpoxy® G202-100, 15865
Kelprox, 15866
Kelrinal, 15867
Kelrol F, *See* Kelgum®, 15847, Xanflood®, 34587
Kelset®, 15868, *See* Algin, 1591
Kelsol®, 15869
Kelsol® 3922-G-80, 15870
Kelstar, 15871
Keltan, 15872
Keltan 312, 15873
Keltan 512, 15874
Keltan 4703, 15875
Keltan 4802, 15876
Keltan TP, 15877
Keltex®, 15878
Keltex® HV, 15879
Keltex® S, 15880
Kelthane, 15881, *See* Acarin, 183, Dicofol, 8398
Kelthix, 15882
Keltone®, 15883
Keltone® HV, LV, 15884, *See* Algin, 1591
Keltose®, 15885
Keltose®, *See* Ammonium alginate, 2228
Keltrol F, *See* Kelzan®, 15890, Kelzan® AR, 15891, Kelzan® D, M, XC Polymer, 15892, Kelzan® S, 15893,, Kelflo®, 15839, Kel-Lite™ CM, 15853, Keltrol®, 15886, Rhodigel® EZ, 26683, Rhodopol® 23, XGD, 26696
Keltrol®, 15886 , 15887
Keltrol® 1001-M-60, 15888
Keltrol® F, 15889
Kelvis®, *See* Algin, 1591
Kelzan, *See* Kelgum®, 15847, Kelflo®, 15839, Kel-Lite™ CM, 15853, Keltrol®, 15886, Keltrol® F, 15889, Rhodigel® EZ, 26683, Rhodopol® XGD, 26697, *See*

Kessco IPM, *See* Wickenol® 101, 105, 34378
Kessco IPP, *See* Wickenol® 111, 34379
Kessco isopropyl myristate, *See* Wickenol® 101, 105, 34378
Kessco isopropyl palmitate, *See* Wickenol® 111, 34379
Kessco PTS, *See* Pentaerythrityl tetrastearate, 23143
Kessco X-211, *See* Ablunol DEGMS, 99
Kessco®, 16099
Kessco® 653, 16100
Kessco® 874, 16101
Kessco® 887, 16102
Kessco® BS, 16103, *See* butyl stearate, 4781
Kessco® EGAS, 16104
Kessco® EGDS, 16105
Kessco® EGMS, 16106
Kessco® GDL, 16107, *See* glyceryl dilaurate, 13079
Kessco® GDS 386F, 16108
Kessco® GMC-8, 16109
Kessco® GML, 16110
Kessco® GMO, 16111
Kessco® GMS, 16112
Kessco® IBS, 16113
Kessco® ICS, 16114
Kessco® IPM, 16115
Kessco® IPP, 16116, *See* isopropyl palmitate, 15410
Kessco® Octyl Isononanoate, 16117
Kessco® Octyl Palmitate, 16118
Kessco® PEG 200 DL, 16119
Kessco® PEG 200 DO, 16120, *See* PEG-2 dioleate series, 23040
Kessco® PEG 200 DS, 16121, *See* PEG-150 distearate series, 23037
Kessco® PEG 200 ML, 16122
Kessco® PEG 200 MO, 16123
Kessco® PEG 200 MS, 16124
Kessco® PEG-200DL, *See* PEG-4 dilaurate series, 23050
Kessco® PGML, 16125
Kessco® PGMS, 16126
Kesscoflex BO, *See* Witcizer 100, 34453
Kesscomir, *See* Wickenol® 101, 105, 34378
Kester, 16127
Kester Wax® K 48, 16128
Kestrel, *See* permethrin, 23384, 23385, 23386
Ketjenblack® EC-310 NW, 16129
Ketjencat, 16130
Ketjenflex, 16131
Ketjenlube® 115, 16132
Ketjensept, 16133
Ketjensil® SM 405, 16134
keto resins, 16135
Ketochromin, *See* Dihydroxyacetone, 8459
Keto-Diastix, 16136
α-ketoglutaric acid, 16137, *See* α-ketoglutaric acid, 16137
β-ketoglutaric acid, 16138, *See* ADA, 615
2-ketoglutaric acid, *See* α-ketoglutaric acid, 16137
3-Ketoglutaric acid, *See* ADA, 615, β-ketoglutaric acid, 16138
ketohexamethylene, *See* cyclohexanone, 7522
ketolin-h, *See* Nipacide® BCP, 21252
Ketomax, 16139
ketone base, 16140
Ketone C-7, *See* methyl n-amyl ketone, 19552
ketone moschus, *See* acetyl-dinitro-butyl-xylene, 317, ketone musk, 16141
ketone musk, 16141, *See* acetyl-dinitro-butyl-xylene, 317
Ketone peroxide, *See* Cadox® F-85, 4912
Ketonone, 16142
Ketonone B, 16143
Ketonone E, 16144
Ketonone M, 16145
Ketonone M.O, 16146
α-ketopropionic acid, *See* pyruvic acid, 25538
6-ketosabinane, *See* tanacetone, 30775
Ketostix, 16147
L-3-ketothreohexuronic acid lactone, *See* vitamin C, 33938
γ-ketovaleric acid, *See* levulinic acid, 17084
Ketovite, 16148
Ketrax, 16149
Keviar® 29, 49, 16150
Kevlar®, 16151
key alloy, 16152
Keycide® X-10, 16153
Keydime, 16154
Keykote®, 16155
Keystone, 16156
Keytrol, 16157
KF Polymer® C-1000, 16158
KF Polymer® T-850, 16160
KF Polymer® T-1300, 16159
KF Polymer® U-1000, 16161
KF Polymer® W-1000, 16162
Khaki Yellow C, 16163
Khakl, 16164
khari salt, 16165
KHE 0145, *See* Etrofolan®, 11156
K-IAO, *See* Gluconal® K, 13043, Katorin, 15798, potassium d-gluconate, 24752, Potassium gluconate, 24755
Kidnamin, 16166
kidney cotton, 16167
kiel compound, 16168
kien oll, 16169
Kienmeyer's amalgam, 16170
Kieralon®, 16171
Kieralon® C, 16172
Kieralon® ED, 16173
Kieralon® NB-ED, 16174
Kieralon® TX-199, 16175
Kieralon® TX-410 Conc, 16176
kieselguhr, *See* Celite®, 5586, diatomaceous earth, 8335, infusorial earth, 15125
Kieselguhr dynamite, 16177, *See* dynamite, 9356
Kil, 16178
Kilfoam, 16179
Kilianite, 16180
killed spirits, 16181
Killgerm® Py-Kill W, 16182
Killgerm® Ratak Cut Wheat Rat Bait, 16183
Killgerm® Sewarin P, 16184
Killgerm® Sol Odamask H, 16185
Killgerm® ULV 400, 16186
Killgerm® ULV 500, 16187
Killmaster, *See* Chlorpyrifos, 6163
Kilmet, 16188
Kilnet, 16189
Kilprop, *See* Mecoprop, 19204
Kilsem, *See* MCPA, 19183
Kilumal, *See* fenpropathrin, 11615
Kimitsu Algin I-1, I-2, I-3, *See* Algin, 1591
Kinel 5502, 16190
Kinel 5514, 16191
Kinel 5517, 16192
Kinetic No 12, *See* Freon®, 12400
Kinetite, 16193
king's blue, *See* smalt, 28638
King's yellow, *See* Chrome yellow, 6212
Kingston bronze, 16194
Kinite, 16195
kino, 16196
kino gum, *See* kino, 16198
kino, Australian, 16197
kino, Bengal, 16198
kino, Botany Bay, *See* kino, Australian, 16197
Kino, Madras, *See* kino, Bengal, 16198
Kino-yellow, *See* maclurin, 18235
Kipsin, *See* methomyl, 19502
Kiresuto P, *See* Cheelox® 80, 5864, Pentaquest Extra 0685, 23169, Trilon® C Liq, 32328
Kirnol® Range, 16199
kish, 16200
Kisvax, *See* carboxin, 5255
Kite, 16201
Kiton, 16202
Kittool fiber, 16203
Kival, 16204
Kiwi Lustr 277, *See* dicloran, 8395, Fumite Dicloran, 12459
KL-990, *See* bioMeT TBTF, 4171
Klea, 16205
Klearfac® 870, 16206
Klearfac® AA270, 16207
Klearol, 16208
Klebcil, 16209, *See* Kantrex, 15718
Kleen-Dent, 16210
Kleenite, 16211
Kleenmold, 16212
Kleenup, 16213
Kleerox® HCS, 16214
Klee's salt, 16215
Klegecell R30, 16216
Kleinenberg's fat mixture, 16217
Kleinenberg's fixative, 16218
Klein's reagent, 16219
Klenal, 16220
Klenenberg's stain, 16221
Kleptose®, 16222
Klerat, 16223
Klere-Seal®, 16224
Kloben, *See* Kloben®, 16225, neburon, 20841
Kloben®, 16225
K-Lor, 16226
Klor-Con, *See* Kay Ciel, 15807, K-Contin, 15818, K. Tab, 15642, Kaon-Cl, 15729, K-Lor, 16226, K-Lyte/C1, 16234, KM Potassium Chloride, 16242, potassium chloride, 24748, 24749
Kloro 6001, 16227, *See* Chlorinated paraffin, 6105
Kluberlubrication, 16228
Klucel®, 16229
Klucel® E, G, H, J, L, M, 16230
Klucel® EF, 16231
Klucine, 16232
K-Lyte, 16233
K-Lyte/C1, 16234, *See* Potassium chloride, 24748
KM Ammonium Metavanadate, 16235
KM Ammonium Perchlorate, 16236
KM Fly Ash, 16237
KM Manganese Dioxide, 16238
KM Muriate of Potash, 16239
KM Pebble Lime, 16240
KM Phosphate Rock, 16241
KM Potassium Chloride, 16242
KM Potassium Perchlorate, 16243
KM Sodium Chlorate, 16244
KM Sodium Perchlorate, 16245
KM Vanadium Pentoxide, 16246
KMC, 16247
KMD-50, 16248
K-Monel, 16249
Knapp's solution, 16250
Knauerit 2, 16251
Knauerit S, 16252
Knave, 16253
Kneiss alloy, 16254
Knights patent zinc white, *See* Lithopone, 17520
Knit-Soft 30 NCPM, 16255
Knittex, 16256
K-Norm, *See* Kay Ciel, 15807, K-Contin, 15818, K. Tab, 15642, Kaon-Cl, 15729, K-Lor, 16226, K-Lyte/C1, 16234, KM Potassium Chloride, 16242, potassium chloride, 24748, 24749
Knot Out, 16257
Knox Out, *See* Diazinon Liquid, 8344
Knox Out 2FM, *See* Diazinon Liquid, 8344
Koban, *See* Aaterra WP, 17
Kobu, *See* quintozene, 25692
Kobutol, *See* quintozene, 25692
kochenite, 16258
kochite, *See* Tisyn®, 31857
Kochlin's Bearing Bronze, 16259
Koch's acid, 16260
Kocide, *See* cupric hydroxide, 7381, Schweitzer's reagent, 27738

Kocide 101, *See* cupric hydroxide, 7381, Schweitzer's reagent, 27738
Kocide® 20/20, 16261
Kocide® Copper Sulfate Pentahydrate Crystals, 16262
Kodabond® Copolyester 5116, 16263
Kodaflex DOA, *See* dioctyl adipate, 8548, Plasthall® DOA, 24002, PX-238, 25440, Wickenol® 158, 34393
Kodaflex DOP, *See* Bis(2-ethylhexy) Phthalate, 4228, dioctyl phthalate, 8549, Witcizer 312, 34454
kodaflex DP, *See* Reomol DCP, 26130
Kodaflex TOTM, *See* PX-338, 25441
Kodaflex® DBP, 16264
Kodaflex® DBS, 16265
Kodaflex® DEP, 16266, *See* diethyl phthalate, 8437
Kodaflex® DIBP, 16267
Kodaflex® DIDA, 16268
Kodaflex® DIDP, 16269
Kodaflex® DMEP, 16270
Kodaflex® DMP, 16271
Kodaflex® DOA, 16272, *See* dioctyl adipate, 8548
Kodaflex® DOP, 16273
Kodaflex® DOTP, 16274
Kodaflex® DOZ, 16275
Kodaflex® HS-3, 16276
Kodaflex® HS-4, 16277
Kodaflex® OIDP, 16278
Kodaflex® TEG-EH, 16279
Kodaflex® TOTM, 16280, *See* Jayflex® TOTM, 15559
Kodaflex® Triacetin, 16281
Kodaflex® TXIB, 16282
Kodaloid, 16283
Kodapak®, 16284
Kodapak® 5214A, 16285, *See* Polyethylene terephthalate, 24426
Kodapak® PET, *See* Polyethylene terephthalate, 24426
Kodapak® PET Copolyester 13339, 16286
Kodar®, 16287
Kodar® A150 Copolyester, 16288
Kodar® PETG Copolyester 6763, 16289, *See* Polyethylene terephthalate, 24426
Kodar® Thermx Copolyester 6761, 16290
Kodel, 16291
Kodelon, *See* *p*-aminophenol, 2185
Kodofil, Kodolite, Kodosoff, 16292
Koenig solder, 16293
Koerzit, 16294
Koerzit, I, II, III, 16295
Kogasin III, 16296
Kohacool L-400, 16297
Koka Seki, 16298
Koken, 16299
koko, 16300
kokowal, 16301
kokum butter, 16302
Kola, *See* kola nut, 16303
kola nut, 16303
kola *Seeds*, *See* kola nut, 16303
Kolax, 16304
Kolene, 16305
kol-kol gum, 16306
Kollercast, 16307
Kollercure®, 16308
Kollerdur® L 90, 16309
Kollerdur® MO118, 16310
Kollerdure® M0122, 16311
Kollermox®, 16312
Kollidon, *See* Tears Plus®, 30876, Videne Disinfectant Solution, Videne Disinfectant Tincture, Videne Powder, 33739, Vinisil, 33827
Kollidon® CL, 16313
kolm, 16314
Kolodust, *See* Thiovit, 31724
Kolofog, *See* Thiovit, 31724
Kolospray, *See* Thiovit, 31724
Kombat, 16315
Kombé arrow poison, 16316
Komeen®, 16317
Kommoid, 16318
Kompak, 16319
Komplexon, *See* Kalex 220 Crystal, 15668, Kalex Liq. 50%, 15671, tetrasodium EDTA, 31391
Kompolite, 16320
kon oil, 16321
Konakion, 16322, *See* vitamin K_1, 33949
Konator, 16323
Kondang wax, *See* Java wax, 15539
Konel, 16324
Konesta, *See* Varitox, 33433
Konforme® AR, 16325
konilite, 16326
Konker®, 16327
konnan bark, 16328
Konservan SN, 16329
Konstrastin, 16330
Konstructal, 16331
Kontakt, 16332
Kontrastin, 16333
Konut, 16334, *See* Esi-Terge 40% Coconut Oil Soap, 10825
konzentrole, 16335
Koolkat, 16336
Kopan, *See* Bakelite, 3589
Kopol®, 16337
Koppert Moss Killer, 16338
Koppeschaar solution, 16339
Kopr-Kote®, 16340
Kopsol, *See* DDT, 7820
Kop-Thion, *See* malathion, 18460
Korad A, 16341
Koraid PSM, 16342
Korantin, 16343
Koraton, 16344
Koreforte®, 16345
Koreon, 16346
Koresin®, 16347
Korestab®, 16348
Koretack®, 16349
Korever®, 16350
Korium, *See* Nuophene, 21783
Korlan, 16351, *See* ronnel, 26951
Korlite, 16352
Koro, 16353
Korogel, 16354
Koron, 16355
Koronit, 16356
Koroplate, 16357
Koroseal, 16358
Korpad, 16359
Korspray, 16360
Kortaid, 16361
Korthix, 16362
Korthix H-NF, 16363
Kortofin, *See* Aldrin, 1558
Korum, *See* Valadol, 33238
korung oil, 16364
Kosmos Black, 3XB, BB, and F4, 16365
Kosmos®, 16366
Kosmos® 10, 16367
Kosmos® 21, 16368
Kostil, 16369
Kotamite®, 16370
Kotebond, 16371
K-Otek, *See* Thripstick®, 31766
K-Othrin, *See* Thripstick®, 31766
Kovar® Alloy, 16372
Kowet 12, 16373
Kowet 3300, 16374
KP, *See* Algin, 1591
KP-2, 16375
KP-23, 16376
KP 90, 16377
KP-140, *See* Amgard TBEP, 2055, tributoxyethyl phosphate, 32206
KP-140®, 16378
KP 201, 16379, *See* dicyclohexyl phthalate, 8413
KP 555, 16380
KPM, *See* Algin, 1591
K-Pool, 16381
KR, *See* antimony trioxide, 2574
KR01 K-Resin Polymer, 16382
KR04 K-Resin Polymer, 16383
kraft paper, 16384
Kraft's metal, 16385
Kraton®, 16386
Kraton® D 1101, 16387
Kraton® D 1107, 16388
Kraton® D 1116, 16389
Kraton® D 1320X, 16390
Kraton® D 2103, 16391
Kraton® FG 1901X, 16392
Kraton® G 1650, 16393
Kraton® G 1701X, 16394
Kraton® G 1726X, 16395
Kraton® G 2701, 16396
Kraton® G 7430, 16397
Kraton® RP 6404, 16398
Kraut's reagent, *See* Dragendorf's reagent, 8954
Krecalvin, *See* dichlorvos, 8394
kregasan, *See* Rezifilm, 26521
Kremnitz, *See* flake white, 11871
Kremol, *See* Kaydol, 15810, Klearol, 16208
Krems white, *See* flake white, 11871
Kremser White, 16399
Krenite, 16400
Krennerite, *See* white tellurium, 34374
Kresamin, 16401
Kresatin, 16402
Kreside, 16403
K-Resin Polymer KR01, 16404
K-Resin Polymer KR03, 16405
K-Resin Polymer KR04, 16406
K-Resin Polymer KR05, 16407
K-Resin Polymer KR10, 16408
Kresival, 16409
Kresolin, *See* Kresopolin, 16410
Kresopolin, 16410
Krezone, *See* MCPA, 19183
Kricinol 35, 16411, *See* Potassium ricinoleate, 24778
Kriegr-o-dip, 16412
Kristalex, 16413
Kristel Gold II, 16414
Krist-o-kleer, 16415
Kristol, 16416
Krokoloy, 16417
Kroma Red, 16418
Kromaplast, 16419
Kromax, 16420
Kromore, 16421
Kromosperse, 16422
Kronagold, 16423
Kronaplate, 16424
Krona-Syn, 16425
Kronitex 50 triaryl phosphate, *See* Fyrquel® EHC, 12544, triaryl phosphate, 32194
Kronitex CDP, *See* Disflamoll® DPK, TPK, 8641
Kronitex TCP, *See* Syn-O-Ad® 8484, 30558, *See* TCP, 30857
Kronitex®, 16426
Kronitex® 25, 16427
Kronitex® 25, 50, 100, 200, 1840, *See* triaryl phosphate, 32195
Kronitex® 50, 16428
Kronitex® 100, 16429
Kronitex® 200, 16430
Kronitex® 1840, 16431
Kronitex® 3600, 16432
Kronitex® PB-460, 16433
Kronitex® TBP, 16434
Kronitex® TCP, 16435
Kronitex® TOF, 16436, *See* trioctyl phosphate, 32349
Kronitex® TPP, 16437
Kronitex® TXP, 16438
Kronos®, 16439 , 16440
Kronos® 1000, 16441
Kronos® 2020, 16442
Kronos® 2073, 16443

Kronos® 3020, 16444
Krovar, 16445, *See* diuron, 8736
Krovar®, 16446
Kruppin, 16447
Kryalith, 16448
Krynac® 19.65, 16451
Krynac® 20H35, 16452
Krynac® 27.50, 16453
Krynac® 34.140, 16454
Krynac® 34.35, 34.50, 16455
Krynac® 34.60 SP, 16456
Krynac® 34.80, 16457
Krylene® 606, 16449
Krylene® 608, 16450
Krynac® 823X2, 16458
Krynac® 833, 16459
Krynac® 843, 16460
Krynac® 850, 16461
Krynac® 881 and 882, 16462
Krynac® 882X1, 16463
Krynac® NV 850, 16464
Krynac® PXL 34.17, 16465
Krynac® X 1.46, 16466
Kryolith, *See* cryolite, 7279
Krystalex, 16467
Krystallazurin, 16468
Krystallos, 16469
Krystaltite Film, 16470
Krytox®, 16471
Krytox® GPL 206, 16472
KS-052P, 16473
K-Slag, 16474
KT-012P, 16475
K-Tab, *See* Kay Ciel, 15807, K-Contin, 15818, K. Tab, 15642, Kaon-Cl, 15729, K-Lor, 16226, K-Lyte/C1, 16234, KM Potassium Chloride, 16242, Potassium chloride, 24748, 24749
K-Tea, 16476
KTPP, *See* potassium tripolyphosphate, 24787
Küttner silk, 16477
KUE 13032c, *See* Dichlofluanid, 8379
Kuhne phosphor bronze, 16478
Kukident, 16479
Kukkersite, 16480
kullo, *See* karaya gum, 15760, Normacol, 21566
kullo, *See*
Kumulan®, 16481
Kumulus, *See* Thiovit, 31724
Kumulus® DF, FL, 16482
Kumulus® S, 16483
Kunstharz HW, 16484
Kunststein, 16485
kuoxam, 16486
Kupferdermasan, 16487
Kupricol, *See* Cupravit®, 7370
Kuprikol, *See* Cupravit®, 7370
Kuracap, 16488
Kurade, 16489
Kurchi, 16490
Kuro Bishi®, 16491
Kurofan, 16492
Kurofan D, 16493
Kurom 1, 16494
Kuromoji oil, 16495
Kuron, 16496
Kurrodur, 16497
Kurrol salts, 16498
Kusum oil, *See* kon oil, 16321
kutch, *See* cutch,
kuteera, *See* karaya gum, 15760, Normacol, 21566
Ku-zyme HP, *See* Accelerase, 185
K-Van, 16499
Kwarc, *See* Diflufenican, 8451
K-White, 16500
Kwik Dri, 16501
Kwikfill, 16502
Kwlk-Green, 16503
Kyanol, *See* phenylamine, 23638
Kylar, *See* daminozide, 7723
Kymene®, 16504
Kymene® 109, 16505
Kymene® 435, 16506
Kymene® 557H, 16507
Kymene® 557LX, 16508
Kymo-trypure, *See* α-Chymotrypsins, 6286
Kynar® 301 F, 16509
Kynar® 460, 16510
Kynar® 700 Series, 16511
Kynar® 7200, 7201, 16512
Kynar® Flex® 2800, 2801, 16513
Kynar® Flex® 2850, 16514
Kynar® Flex® 2900, 16515
Kynite, 16516
Kynol, 16517
Kyolox BAT, 16518
Kypfarin, 16519
Kypman, *See* Maneb, 18526
Kypman 80, *See* Maneb, 18526
Kypzin, *See* Zineb, 34807
Kyrock, 16520
Kyselina kyanurova, *See* Zeonet A, 34720
kyseliny adipove (Czech), *See* Wickenol® 158, 34393
Kysite, 16521
Kytamer® PC, 16522
KZ 55, 16523
KZ OPPR, 16524
KZ TPP, 16525
KZ TPPJ, 16526
K-Zinc, 16527
L, *See* Leucine, 17020
L-55R® Acid Neutralizer, 16529
L-66, 16530
L-36352, *See* Treflan, 32162, Trifluralin, 32252
L.A.S, 16528, *See* Labrasol, 16543, PEG-8 caprylic/capric glyceride, 23053
LAA, *See* Verv®, 33644
lab, *See* rennet, 26110
Labilite, *See* Maneb, 18526
Labilite M, *See* thiophanate methyl, 31706
Labitan, 16531
Laboprin, 16532
Labosept, 16533, *See* Dequalinium Chloride, 8077
Labrafac Hydro WL 1219, 16534
Labrafac Lipophile WL 1349, 16535, *See* caprylic/capric acid triglyceride, 5161
Labrafil ISO, 16536
Labrafil M 1944 CS, 16537, *See* Apricot kernel oil PEG-6 esters, 2675
Labrafil M 1969 CS, 16538
Labrafil M 1980 CS, 16539, *See* olive oil PEG-6 esters, 22139
Labrafil M 2125 CS, 16540, *See* Corn oil PEG-6 esters, 6835
Labrafil M 2130 BS, 16541, *See* hydrogenated palm/palm kernel oil PEG-6 esters, 14594
Labrafil M 2735 CS, 16542, *See* Triolein PEG-6 esters, 32352
Labrasol, 16543, *See* PEG-8 caprylic/capric glyceride, 23053
LABS 100/H.V, 16544, *See* tridecylbenzene sulfonic acid, 32236
LABS-100, 16545
Labstix®, 16546
labstone, *See* Kemite, 15977
lac, 16547
Lac lake, *See* carmine lake, 5292
lac, Japanese, 16548
Lacanite, 16549
lacca, *See* lac, 16547
Laccain, 16550
lac-dye, 16551
Lackmoid, 16552
lacmus, 16553
Lacolene, 16554
lacolin, *See* sodium lactate, 28777
Lacorene, 16555
Lacqran, 16556
Lacqrene 550, 16557
Lacqrene 635, 811, 835 and 836, 16558
Lacqrene 740, 16559
Lacqrene E, 16560
Lacqsan 125 and 125L, 16561
Lacqsan E, 16562
Lacqtene 1070 MN20, 1200 MN26, 16563
lacquer, 16564
lacquer orange vg, *See* Sudan I, 29938
Lacril®, 16565 , *See* hydroxypropyl methylcellulose, 14668
Lacri-Lube® NP, S.O.P®, 16566
Lacrinite, 16567
Lacrisert, *See* Klucel®, 16229, Klucel® E, G, H, J, L, M, 16230, Klucel® EF, 16231
β-lactamase, *See* Penicillinase, 23112
β-Lactamase substrate, *See* Padac®, 22546
Lactamide MEA, 16568 *See* Incromectant LMEA-70, 14979, Mackamide LME, 18157, Parapel® LAM-100, 22786, Schercomid LME, 27652
Lactamide, N-ethyl-, carbanilate (ester), D-, *See* Carbetamex, 5189
Lactamidopropyl trimonium chloride, *See* Incromectant LQ, 14980
lactase, *See* galactosidase, 12593, β-galactosidase, 12593
lacteol, 16569
lactic acid, 16570, *See* Patlac® LA, 22910
D(-)-lactic acid, *See* D-lactic acid, 16572
d-lactic acid, *See* L-lactic acid, 16571
DL-lactic acid, 16573
l-lactic acid, *See* D-lactic acid, 16572
L-lactic acid, 16571
Lactic Acid Butyl Ester, *See* Butyl lactate, 4773
lactic acid cetyl ester, *See* cetyl lactate, 5820
lactic acid ethyl ester, *See* ethyl lactate, 11076
lactic acid hexadecyl ester, *See* cetyl lactate, 5820
Lactic acid monoethanolamide, *See* Lactamide MEA, 16568
lactic acid, monosodium salt, *See* sodium lactate, 28777
Lactic acid, tetradecyl ester, *See* Wickenol® 506, 34400
lactic acid zinc salt, *See* zinc lactate, 34780
Lacticol 336, 16574
lactigen, *See* lacteol, 16569
Lactil®, 16575
lactilloids, 16576, *See* lacteol, 16569
Lactimon®, 16577
Lactin, 16578
lactine, *See* Ablunol DEGMS, 99 vegetable butter, 33514
Lactinium, 16579
Lactite, *See* Gallatite, 12608
Lactitis, 16580
Lactobacilline, *See* lacteol, 16569
Lactodan B 30, 16581
lactoflavin, *See* vitamin B_2, 33935
lactoflavine, *See* vitamin B_2, 33935
lactoflavine, Zinvit-G, *See* riboflavin, 26722
lactoform, *See* Gallatite, 12608
Lactoid, 16582
Lactol, 16583
Lactolin, 16584
Lactolith, 16585
Lactoloid, 16586
Lactonaphthol, *See* Lactol, 16583
lactone, *See* lacteol, 16569
Lactoprene, 16587
lactorite, *See* Gallatite, 12608
Lactosan, 16588
Lactose, *See* Lactin, 16578, milk sugar, 19768
α-D-Lactose, 16589
Lactose molasses, 16590
lactovagan, *See* DL-lactic acid, 16573
Lactoyl methylsilanol elastinate, *See* Lasilium C, 16805
Lactozym, 16591
lactylated mono-diglycerides, *See* Radiamuls® Lactem 2950, 25775
Ladalrod, 16592

Laddock, *See* atrazine, 3394
Laddok®, 16593
Ladelloy, 16594
laevo-glucose, *See* levulose, 17085
Laevuflex, 16595
Lafil WL 3254, 16596, *See* Polyglyceryl isostearostearate, 24456
Laguncurin, *See* maclurin, 18235
Lakeland AMA LF, 16597
Lakeland C, 16598
Lakeland CAB, 16599
Lakeland CTA/N, 16600
Lakeland N, 16601
Lakeland PA, PAE, PPE, 16602
Lakewax, 16603
Lakeway SA, *See* Witco® 1298, 34458
Lalicopharsol, 16604
Lalitecsol, 16605
Lamacit CA, *See* Ablunol DEGMS, 99
Lamalgin, 16606
Lamar L-300, *See* Tuasol 100, 32543
Lambrex, 16607
Lambrit (Anfo-explosives), 16608
Lamecreme LPM, 16609
Lamecreme SA 7, 16610
Lamefin, 16611
Lamefix, 16612
Lamefix 680, 16613
Lameform TGI, 16614
Lamefrost® Range, 16615
Lamegin CSL, *See* Verv®, 33644
Lamegin® DW 8000 HW, 16616
Lamegin® DW Range, 16617
Lamegin® EE, 16618, *See* Acetylated hydrogenated tallow glyceride, 316
Lamegin® GLP 10, 20, *See* hydrogenated tallow glyceride lactate, 14600
Lamegin® GLP 20, 16619
Lamegin® ZE 30, 60, 16620, *See* hydrogenated tallow glyceride citrate, 14599
Lamegum, 16621
Lamemul® Range, 16622
Lamephan, 16623
Lamepon, 16624
Lamepon PA-TR, 16625, *See* TEA-abietoyl hydrolyzed collagen, 30871
Lamepon S, 16626, *See* Foam-Coll 4C, 12150
Lamepon ST40, 16627, *See* Foam-Coll 4CT, 12151
Lamepon UD, 16628, *See* Potassium undecylenoyl hydrolyzed collagen, 24788
Lameprint, 16629
Lamequat L, 16630
Lamequick® Range, 16631
Lamesoft LMG, 16632
Lamesorb® Range, 16633
Lamex 173/FR, 16634
Lamex 185, 16635
Lamex 186, 16636
Lamicoid, 16637
Lamictal Tablets, 16638
Laminac EPX-176, 16639
laminated talc, *See* mica, 19626
Lamitex, 16640
Lamol, 16641
Lampblack, *See* Octojet 104, 22004, sugar charcoal, 29953
Lamprene, *See* Clofazimine, 6427
Lampronol, 16642
Lamy butter, 16643
Lanacron, 16644
Lanain, *See* Cosmetic Lanolin USP, 6870, lanolin, 16740, Vigilan, 33750
Lanaire, 16645
Lanalin, *See* Cosmetic Lanolin USP, 6870, lanolin, 16740, Vigilan, 33750
Lanamine®, 16646 , *See* Mixed isopropanolamines myristate, 19984
Lanaperl, 16647
Lanapex, 16648
Lan-Aqua-Sol 100, 16649, *See* PEG-75 lanolin, 23052
Lanasol, 16650
Lanbritol, 16651
Lancare, 16652
Lancer, 16653, *See* Perfluidone, 23248
Lancosol, 16654
Landemul, 16655
Landmaster, *See* glyphosate, 13148
Landromil, 16656
Lanesin, *See* Cosmetic Lanolin USP, 6870, lanolin, 16740, Vigilan, 33750
Lanesta G, 16657, *See* glyceryl lanolate, 13081
Lanesta S, 16658
Lanestren, 16659
Laneth-5, *See* Polychol 5, 24387, Ritawax 5, 26836, Solulan® 5, 29000
Laneth-20, *See* Aqualose W20, 2711
Laneto 27, 16660
Laneto AWS, *See* PEG -12 PEG-50 lanolin, 23027, PPG-12-PEG-65 lanolin oil;, 24843
Lanette 14, 16661, 16662
Lanette 16, 16663
Lanette 18 DEO, 16664
Lanette E, 16665
Lanette N, 16666
Lanette O, 16667
Lanette SX, 16668
Lanette Wax, 16669
Lanette Wax Ester, 16670
Lanette Wax SX, 16671
lanette wax-s, *See* Rhodapon® 101-10, 26631, sodium lauryl sulfate, 28781
Lanexol AWS, 16672, *See* PPG-12-PEG-65 lanolin oil, 24843
Lanfrax® 1776, 16673
Langford Clay, 16674
langite, *See* cupric sulfate, basic, 7388, Cusatrib, 7429
Lanichol, *See* Cosmetic Lanolin USP, 6870, lanolin, 16740, Vigilan, 33750
Laniol, *See* Cosmetic Lanolin USP, 6870, lanolin, 16740, Vigilan, 33750
Lanital, 16675
Lankrocell® D15L, 16676
Lankroflex®, 16677
Lankroflex® ED3, 16678
Lankroflex® ED6, 16679
Lankroflex® GE, 16680
Lankrol, 16681
Lankrolan, 16682
Lankroline, 16683
Lankrolyte, 16684
Lankromark®, 16685
Lankromark® BM271, 16686
Lankromark® BT050, 16687
Lankromark® BT120A, 16688
Lankromark® DLTDP, 16689
Lankromark® DP6404Z, 16690
Lankromark® DSTDP, 16691, *See* Distearyl thiodipropionate, 8713
Lankromark® LC475, 16692
Lankromark® LC68, 16693
Lankromark® LC90, 16694
Lankromark® LE65, 16700
Lankromark® LE109, 16695, *See* Trisnonylphenyl phosphite, 32411
Lankromark® LE230, 16696
Lankromark® LE285, 16697
Lankromark® LE296, 16698
Lankromark® LE296, 16699
Lankromark® LZ121, 16702
Lankromark® LZ187, 16703
Lankromark® LZ495, 16704
Lankromark® LZ616, 16705
Lankromark® LZ693, 16706
Lankromark® LZ1034, 16701
Lankromark® OT450, 16707
Lankromul, 16708
Lankroplast®, 16709
Lankroplast® L542, 16710
Lankroplast® V2012, 16711
Lankropol®, 16712
Lankropol® KMA, 16713, *See* Bis(1-methylamyl) Sodium Sulfosuccinate, 4227
Lankropol® KO Special, 16714
Lankropol® KO2, 16715
Lankropol® KPH70, 16716
Lankropol® KSB 22, 16717
Lankropol® KSG 72, 16718
Lankropol® ODS, 16719
Lankrosol, 16720
Lankrosol HS101, 16721
Lankrosol SXS, 16722
Lankrosperse, 16723
Lankrostat® 16, 16724
Lankrostat® LME, 16725
Lankrothane, 16726
Lannabait, *See* methomyl, 19502
Lannate, *See* methomyl, 19502
Lannate LB, *See* methomyl, 19502
Lannate(R), *See* methomyl, 19502
Lannate®, 16727
Lanocerin®, 16728
Lanogel® 21, 16729
Lanogel® 31, 16730
Lanogel® 41, 16731
Lanogel® 61, 16732
Lanogene®, 16733
Lanoiac, 16734
Lanoid, *See* Mulsoid,
Lanol 14 M, 16735
Lanol 1688, 16736
Lanol 1688, *See* Cetearyl octanoate, 5791
Lanol P, 16737, *See* glycol palmitate, 13103
Lanol S, *See* Adol® 62 NF, 783, Stearal, 29591
Lanolex L-40, 16738
Lanolic Acid, 16739
lanolin, 16740, *See* Amerchol® 400, 1996, Amerchol® C, 1998, Anhydrous Lanolin HP-2050, 2461, Anhydrous Lanolin P80, 2462, Cosmetic Lanolin USP, 6870, Forlan, 12247, Ultra Anhydrous Lanolin HP-2060, 32779, Vigilan, 33750
Lanolin acid, *See* Amerlate® LFA, 2017, Amerlate® WFA, 2019, Ritalafa®, 26808, Skliro Distilled, 28572
Lanolin alcohol, 16741, *See* Amerchol® 400, 1996, Amerchol® L-500, 2002
lanolin alcohol, *See* Amerchol® L-101, 2001
lanolin alcohol, *See* Amerchol® L-101, 2003
lanolin alcohol, *See* Amerchol® C, 1998
Lanolin alcohol BP, *See* Argowax Standard, 2859
Lanolin alcohols, *See* Ceralan®, 5702, Fancol LA, 11446
lanolin alcohol, *See* Forlan, 12247, Hartolan, 13676, Ivarlan 3310, 15444, Oleo-Coll LP, 22118, Solulan, 28998, Ritawax, 26835
lanolin alcohols, *See* Solulan, 28998
Lanolin BP/EP, *See* Corona, 6843
Lanolin fatty acids, *See* Fancor LFA, 11458
Lanolin oil, *See* Lantrol® 1673, 16767, Vigilan, 33750
Lanolin wax, *See* Lanocerin®, 16728
Lanolin, anhydrous, *See* Vigilan, 33750
Lanoline, 16742
Lanoplast, 16743
Lanoplast, CA, *See* cellulose acetate, 5627
Lanoquat® 1756, 16744
Lanosol, 16745
Lanosterol, 16746
Lanotein AWS 30, 16747
Lanox, *See* methomyl, 19502
LANOX 90, *See* methomyl, 19502
Lanox 216, *See* methomyl, 19502
Lanoxicaps, 16748
Lanoxin, 16749
Lanoxine-PG, 16750
Lanpharsol, 16751
Lanpol, 16752
Lanpolamide 5, 16753
Lanstar, 16754
Lanstar NP2 and NP4, 16756

Lanstar NP40, NP50 and NP100, 16757
Lanstar NP100/50, 16755
Lanstar PCH, PC2 and PCO, 16758
Lanstar PS, 16759
Lanstar PSW, 16760
Lantecsol, 16761
Lanthana, *See* lanthanum oxide, 16766
Lanthanol LAL, 16762
lanthanum, 16763
lanthanum chloride, 16764
lanthanum chloride anhydrous, *See* lanthanum chloride, 16764
lanthanum nitrate, 16765
lanthanum nitrate hexahydrate, *See* lanthanum nitrate, 16765
lanthanum oxide, 16766
lanthanum sesquioxide, *See* lanthanum oxide, 16766
lanthanum trioxide, *See* lanthanum oxide, 16766
Lantrol® 1673, 16767
Lantrol® AWS 1692, 16768, *See* PPG-12-PEG-65 lanolin oil, 24843
Lantrol® HP-2073, 16769
Lantrol® PLN, 16770, *See* PEG-75 lanolin, 23052
Lanum, *See* Cosmetic Lanolin USP, 6870, Vigilan, 33750
Lanvis, 16771
Lanxide, 16772
Lapis Smiridis, 16773
Lapis-lazuli blue, *See* ultramarine, 32821
Lapix, 16774
Lapofloc, 16775
Laponite®, 16776
Laponite® B, 16777
Laponite® D, 16778
Laponite® S, 16779
Laponite® XLG, 16780
Lapotan, 16781
Laptran, *See* Plondrel, 24160
Lapyrium Chloride, 16782, *See* Emcol® E-607L, 10059
Laquanol, 16783
Laractone®, 16784
Laraflex®, 16785
Laraflex®, *See* Laraflex®, 16785
Lard glyceride, *See* BFP 74, 4113, Dimodan S, 8535
Lard glycerides, *See*
Lard glycerides-citric acid, *See* BFP 75, 4114
Lard glycerides-TBHQ-citric acid, *See* BFP 65, 4112
Lard methyl ester, *See* methyl ester L, 19527
lard-factor, *See* Vi-Alpha, 33706
Larex, *See* Salmocid, 27277
Largactil, 16786
Largon, *See* Diflubenzuron, 8450
Laricic acid, *See* Agaricic, 956
Laridal, *See* Hismanal, 14132
larixinic acid, *See* Veltol, 33544
larnite, *See* calcium silicate, 4969
Larocin, 16787
Larodur®, 16788
Laroflex®, 16789
Laromer®, 16790
Laromer® POEA, 16791
Laromer® TPGDA, 16792
Laromid®, 16793
Laropal®, 16794
Laroscorbine, *See* vitamin C, 33938
Larostat® 88, 16798
Larostat® 143, 16795
Larostat® 377 DPG, 16796
Larostat® 451, 16797
Larostat® JMR, 16799
Larotid, 16800
Laroxyl, 16801
Larvacide, 16802
Larvacide 100, *See* chloropicrin, 6135
Larvex, 16803
Laser, 1680, *See* Cycloxydim, 7544 4
Lasilium C, 16805
Lasilso, 16806
Lasix, 16807
Lasso, 16808, *See* Alachlor, 1327
Lastex, 16809
Lastil, 16810
Lastilac, 16811
Latekoll®, 16812
Latene, 16813
Latensol AP8, 16814
Latex, *See* Sentry Dimethicone, 27959
Latex Foam, 16815
Lathanol® LAL, 16816
Latibon®, 16817 , *See* calcium formate, 4948
Latkem, 16818
Latol 4, 16819
Latol 1550, 16822
Latol MOD, 16823
laudran, *See* Dabco® T-12, 7673
Laughing gas, *See* Nitral, 21318
Laur 101A, *See* Laur 101B, 16820
Laur 101B, 16820
Laur 676U, 16821
Laur Q-1331, 16824
Laur Red #10 Silicone Pigment, 16825
Laural, 16826
Lauralkonium chloride, *See* Catinal CB-50, 5456, Dehyquart LDB, 7986
lauramide 11, 16827
lauramide DEA, 16828, *See* Amidex L-9, 2094, Amidex LD, 2095, Crillon LDE, 7049, Emalex NN-7, 9974, Incromide L-90, 14987, Incromide LLT, 14990, Incromide LM-70, 14991, Mackamide L10, 18156, Monamid® 150-LMWC, 20115, Ninol® 30-LL, 21234, Witcamide® 5195, 34440
lauramide DEA, *See* Alkamide® 327, 1634, Witcamide® 5195, 34440
Lauramide DEA (1:1), *See* Carsamide® SAL-7, 5323, Emid® 6519, 10193, Empilan® LDE, 10408, Schercomid SL-Extra, 27659, Standamid® KDS, 29456
Lauramide DEA (2:1), *See* Mazamide® 1214, 19107
1:1 Lauramide DEA, *See* Alkamide® LE, 1646
2:1 Lauramide DEA, *See* Mazamide® L-298, 19110
Lauramide DEA sulfosuccinate monoester, *See* Ablusol LDE, 127
Lauramide DEA superamide, *See* Comperlan LD, 6676
lauramide EG, 16829
Lauramide MEA, *See* Ablumide LME, 80, Amidex LMMEA, 2097, Crillon LME, 7050, Empilan® LME, 10410, Incromide LCL, 14989, Mackamide LMM, 18158, Ninol® LMP, 21237, Rewomid® L 203, 26328
Lauramide MEA, *See*
1:1 Lauramide MEA, *See* Monamid® LMA, 20120
Lauramide MEA sulfosuccinate monoester, *See* Ablusol LMS, 128
Lauramide MIPA, *See* Alkamide® LIPA/C, 1647, Amidex LIPA, 2096, Empilan® LIS, 10409, Rewomid® IPL 203, 26326
1:1 Lauramide MIPA, *See* Monamid® LIPA, 20119
Lauramide, N,N-bis(hydroxyethyl)-, *See* Witcamide® 5195, 34440
Lauramide-DEA (1:1), *See* Ablumide LDE, 79
Lauramide-MEA (1:1), *See* Ablumide LME, 80
lauramidopropyl betaine, 16830, *See* Chembetaine L, 5934, Chemoxide L, 5995, lauramidopropyl betaine, 16830, Lexaine® LM, 17094, Lexate BPQ, 17121, Mackam LMB, 18146, Mafo® LMAB, 18302, Mirataine® BB, 19922, Monateric LMAB, 20172, Tego® -Betaine L-90, 30965
lauramidopropyl dimethylamine, 16831, *See* Chemidex L, 5963, lauramidopropyl dimethylamine, 16831, Lexamine L-13, 17098, Mackine 801, 18224, Schercodine L, 27638
Lauramidopropyl PEG-dimonium chloride phosphate, *See* Phospholipid PTD, 23723
Lauramidopropyl PG-dimonium chloride, *See* Lexquat® AMG-M, 17177
(3-Lauramidopropyl) trimethylammonium methyl sulfate, *See* Cyastat® LS, 7505
Lauramidopropylamine oxide, *See* Mackamine LAO, 18172
Lauramine, *See* Amine 10, 2126, Amine 12-98D, 2128, Armeen® 12, 2933, Amine 12, 2127
Lauramine (primary amine), *See* Armeen® 12D, 2934
lauramine oxide, 16832, *See* Ablumox LO, 91, Ammonyx® LO, 2264, Amyx LO 3594, 2394, Chemoxide LM-30, 5996, Emcol® LO, 10063, Empigen® OB, 10375, Incromine Oxide L-40, 15006, Mackamine LO, 18173, Mazox® LDA, 19152, Ninox® L, 21242, Rewominox L 408, 26334, Rewominoxid L 408, 26337, Schercamox DML, 27575, Varox® 365, 33453, Ammonyx® LO, 2264, Empigen® OB, 10375, Rhodamox® LO, 26620, Ablumox LO, 91
Lauraminopropionic acid, *See* Mackam 151L, 18138, Mirataine® HC-Acid, 19929
Laurdimonium hydroxyethyl cellulose, *See* Crodacel QL, 7065
Laurdimonium hydroxypropyl hydrolyzed wheat protein, *See* Hydrotriticum QL, 14652
Laurel Camphor, *See* Japan camphor, 15508
laurel oil, 16833
Laurel PEG 400 DT, 16834, *See* PEG ditallate series, 23028
Laurel PEG 400 MO, 16835
Laurel PEG 400 MT, 16836, *See* PEG tallate series, 23035
Laurel R-50, 16837, *See* Eureka 102, 11207
Laurel SBT, 16838, *See* sulfated butyl tallate, 29972
Laurel SD-101, 16839
Laurel SD-400, 16840
Laurel SD-520T, 16841
Laurel SMR, 16842
Laurel SRO, 16843
Laurel wax, 16844
Laurelphos 39, 16845
Laurelphos 400, 16846, *See* oleth-phosphate Series (2-20), 22127
Laurelphos RH-44, 16847
Laurelterge 837, 1390, 16848
Laureltex 308, 308 LF, 16849
Laurent's acid, 16850
Laurent's aluminum solder, 16851
Laurent's naphthaldinic acid, *See* Laurent's acid, 16850
Laureol, *See* vegetable butter, 33514
Laureth-2, 16852, 16853, Akyporox RLM 22, 1291, Empilan® KB 2, 10397, Marlipal® 24/20, 18862, Oxetal VD 20, 22425, Prox-onic LA-1/02, 25321
Laureth-2 phosphate, *See* Surfagene FDD 402, 30353
Laureth-3, *See* Ablunol LA-3, 104, Ethal 326, 10911, Remcopal 121, 26061, Rewopal® LA 3, 26343, Standapol® CS Paste, 29470
Laureth-3 carboxylic acid, *See* Rewopol® CL 30, 26361
Laureth-3 phosphate, *See* Marlophor® MO 3-Acid, 18898, Rewophat EAK 8190, 26352
Laureth-4, *See* Akyporox RLM 40, 1292, Brij® 30, 4515, Brij® 30SP, 4516, Chemal LA-4, 5882, Ethal LA-4, 10921, Hetoxol L-4, 13941, Hodag Nonionic L-4, 14196, Lipocol L-4, 17354, Macol® LA-4, 18267, Marlipal® 124, 18864, Mulsifan CPA, 20453, Remcopal 4, 26052, Rhodasurf® L-4, 26656, Simulsol P4, 28489, Trycol® 5882, 32499, Volpo L4, 34027
Laureth-4 carboxylic acid, *See* Akypo RLM25, 1237
Laureth-4 phosphate, *See* Chemphos TR-510, 6017, Rhodafac® RD-510, 26605
Laureth-5, *See* Mulsifan RT 23, 20458
Laureth-5 carboxylic acid, *See* Akypo 1690 S, 1214, Akypo RLM 38, 1236, Akypo RLMQ 38, 1238, Marlinat® CM 40, 18849, Sandopan® LA-8, 27345
Laureth-5 carboxylic acid and sodium octyl sulfate, *See* Akypo TFC-S, 1247
Laureth-6, *See* Dehydol PID 6, 7955
Laureth-7, *See* Marlipa® MG, 18857, Sepigel 305, 27984
Laureth-7 phosphate, *See* Akypomine® MW 05, 1266
Laureth-8, *See* Akyporox RLM 80V, 1293
Laureth-9, *See* Emalex 709, 9926, Ethal 926, 10913, Remcopal 258, 26069, Remcopal L9, 26089, Remcopal LP, 26093, Standapol® Pearl Conc. 7130, 29476
Laureth-9, steareth-9, *See* Hetoxol LS-9, 13942

Laureth-10.5, *See* Remcopal L12, 26090
Laureth-11, *See* Remcopal 21411, 26063
Laureth-11 carboxylic acid, *See* Akypo RLM 100, 1232
Laureth-12, *See* Remcopal 21912 AL, 26064
Laureth-14 carboxylic acid, *See* Akypo RLM 130, 1234
Laureth-16, *See* Rhodasurf® B-1, 26650
Laureth-17 carboxylic acid, *See* Akypo RLM 160, 1235
Laureth-20, *See* Remcopal 20, 26056, Sellig LA 1150, 27882
Laureth-23, *See* Incropol L-23, 15026, *See* Ritox 35, 26851
Laureth-2I, *See* Dehydol LS 2, 7952
Laurex®, 16854
Laurex® CS, 16855
Lauric, *See* Ablunol 200ML, 95
lauric acid, 16856, *See* Emery® 650, 10134, Philacid 1200, 23660, Prifrac 2920, 24952
Lauric acid (95%), *See* Hystrene® 9512, 14766
Lauric acid diethanolamide, *See* Alkamide® 327, 1634, Witcamide® 5195, 34440
Lauric acid diethanolamine Con(1/1), *See* Witcamide® 5195, 34440
Lauric acid diethanolamine condensate, *See* Witcamide® 5195, 34440
lauric acid ethanolamide, *See* Ablumide LME, 80, Rewomid® L 203, 26328
Lauric acid, diethanolamine, *See* Witcamide® 5195, 34440
lauric acid, dibutylstannylene derivative, *See* Dabco® T-12, 7673
lauric acid, dibutyltin derivative, *See* Dabco® T-12, 7673
Lauric acid, ethylene oxide adduct, *See* Ablunol 200ML, 95
Lauric acid 1-monoglyceride, *See* Ablunol GML, 101
Lauric acid α-monoglyceride, *See* Ablunol GML, 101
Lauric acid, sorbitan ester, *See* Ablunol S-20, 108
lauric diethanolamide, *See* lauramide DEA, 16828, Witcamide® 5195, 34440
lauric ethylolamide, *See* Rewomid® L 203, 26328
Lauric myristic dimethylethyl ammonium ethosulfate, *See* Larostat® 377 DPG, 16796
Lauric/myristic acid (C_{12-14}), *See* Philacid 1214, 23661
Lauridit, 16857
Lauridit KDG, *See* Active #2, 501
Lauridit LM, *See* Ablumide LME, 80
Laurin, 1-mono-2,3-dihydroxypropyl dodecanoate, *See* Ablunol GML, 101
Laurodin, 16858
laurostearic acid, *See* Philacid 1200, 23660
Laurox®, 16859
N-lauryl acrylate (monomer), *See* SR-335, 29378
N-Lauroyl-p-aminophenol, *See* Suconox-12®, 29925
lauroyl chloride, 16860
N-(lauroyl colamino formyl methyl) pyridinium chloride, *See* Witcamine® E-607, 34450
Lauroyl diethanolamide, *See* Witcamide® 5195, 34440
lauroyl hydrolyzed collagen, *See* Lipo-Peptide AME 30, 17399
Lauroyl lysine, *See* Amihope LL-11, 2120
Lauroyl peroxide, *See* Laurydol, 16862
Lauroyl PG-trimonium chloride and hexylene glycol, *See* Akypoquat 132, 1272
Lauroyl Sarcosinate, *See* Crodasinic L, 7143
lauroyl sarcosine, 16861, *See* Crodasinic L, 7143, Hamposyl® L, 13600, Vanseal® LS, 33357
N-lauroylsarcosine sodium salt, *See* sodium lauroyl sarcosinate, 28779
n-lauroylsarcosine, sodium salt, *See* sodium lauroyl sarcosinate, 28779
N-(lauroylcolaminoformylmethyl)pyridinium chloride, *See* Lapyrium Chloride, 16782
Lauroyldiethanolamine, *See* Witcamide® 5195, 34440
lauroylethanolamide, *See* Rewomid® L 203, 26328
n-lauroylsarcosine, *See* Crodasinic L, 7143
N-lauroylsarcosine sodium salt, *See* Vanseal® NALS-30, 33359
lauroylsarcosine, sodium salt, *See* sodium lauroyl sarcosinate, 28779
Laurtrimonium chloride, *See* Arquad® 12-50, 3106, Chemquat 12-50, 6024, Dehyquart LT, 7987, Empigen® 5089, 10355, Adogen® 412, 767, Arquad® 12-50, 3106
Laur-trimonium chloride, *See* Arquad® 12-37W, 3105
Laurydol, 16862
Laurydone®, *See* lauryl PCA, 16866
lauryl acrylate, 16863, *See* Ageflex FA-12, 980
lauryl alcohol, 16864, *See* Emery® 3332, 10170, Lipocol L, 17353, Lorol C12, 17611, Philcohol 1200, 23664, 1-dodecanol, 8777, Emery® 3326, 10169
Lauryl alcohol(40-48% C_8, 51-59% C_{10}), *See* Lorol C_8-C_{10} Special, 17617
Lauryl alcohol (65-69% C_{12}, 24-28% C_{14}, 4-8% C_{16}), *See* Lorol C_{12}-C_{14}, 17612
Lauryl alcohol (70-75% C_{12}, 24-30% C_{14}), *See* Lorol Special, 17621
Lauryl alcohol ethoxylate, *See* Empilan® 2020, 10383
lauryl amine, *See* Amine 12, 2127, Armeen® 12, 2933, Armeen® 12D, 2934
Lauryl ammonium sulfate, *See* Witcolate NH, 34478
Lauryl betaine, *See* Amphoteen 24, 2343, Dehyton® PAB-30, 7993, Rewoteric® AM DM-35L, 26423, Zohartaine AB, 34877
Lauryl Diethanolamide, *See* Witcamide® 5195, 34440
Lauryl dimethyl benzyl ammonium chloride, *See* Dehyquart LDB, 7986
Lauryl dimethylamine, *See* dimethyl lauramine, 8507
Lauryl dimethylamine oxide, *See* lauramine oxide, 16832
Lauryl gallate, *See* Progallin LA, 25110
Lauryl glucoside, *See* APG® 600 Glycoside, 2653
lauryl hydroxyethyl imidazoline, 16865, *See* Mackazoline L, 18200, Schercozoline L, 27714
Lauryl hydroxysultaine, *See* Rewoteric® AM HC, 26425
Lauryl isoquinolinium bromide, *See* Isothan Q-75, 15427
Lauryl lactate, *See* Ceraphyl® 31, 5713
Lauryl methacrylate, *See* Ageflex FM-12, 986, Ageflex FM-25, 987
n-Lauryl methacrylate, *See* Ageflex FM-12, 986
Lauryl methyl gluceth-10 hydroxypropyldimonium chloride, *See* Glucquat® 125, 13058
Lauryl monoethanolamide, *See* Rewomid® L 203, 26328
lauryl PCA, 16866 , *See* lauryl PCA, 16866
Lauryl phosphate, *See* Crafol AP-201, 6955
Lauryl polyglucose, *See* APG® 600 CS, 2652
Lauryl pyridinium bisulfate, *See* Dehyquart D, 7984
Lauryl pyridinium chloride, *See* Dehyquart C Crystals, 7982
n-Lauryl pyridinium chloride, *See* Dehyquart C Crystals, 7982
Lauryl pyrrolidone, *See* Agsol Ex 12, 1089, Surfadone LP-300, 30346
Lauryl pyrrolidonecarboxylate, *See* lauryl PCA, 16866
Lauryl sulfate ammonium salt, *See* Witcolate NH, 34478
Lauryl sulfate ester, triethanolamine salt, *See* Rhodapon® LT-6, 26638, Standapol® T, 29481, Witcolate TLS-500, 34479
Lauryl sulfate, triethanolamine salt, *See* Witcolate TLS-500, 34479
Lauryl sulfuric acid, triethanolamine salt, *See* Witcolate TLS-500, 34479
Lauryl trimethyl ammonium chloride, *See* Adogen® 412, 767, Arquad® 12-50, 3106, Octosol 562, 22020
Lauryl/stearyl thiodipropionate, *See* Mark® 5095, 18730
laurylamidoethanol, *See* Ablumide LME, 80, Rewomid® L 203, 26328
Laurylamine acetate, *See* Acetamin 24, 284
Laurylbenzenesulfonate, *See* Witco® 1298, 34458
Laurylbenzenesulfonic Acid, *See* dodecylbenzene sulfonic acid, 8779, Pentine Acid 5431, 23182, Witco® 1298, 34458
lauryldimethylamine, *See* Onamine 12, 22164
n-lauryldimethylamine, *See* Onamine 12, 22164
lauryldimethylamine oxide, *See* Ablumox LO, 91, Ammonyx® LO, 2264, Empigen® OB, 10375, Rhodamox® LO, 26620
α-Lauryl-ω-hydroxypoly(oxyethylene)sulfate, sodium salt, *See* Witcolate ES-2, Witcolate LES-60C, 34477
n-lauryl mercaptan, *See* n-dodecyl mercaptan,
Lauryloleylmethylamine soy amino acids, *See* Soy-Amino Quat L/O, 29171
laurylpyridinium chloride, 16867
β-laurylthiopropionate, *See* Seenox 412S, 27843
Lauryltrimonium chloride, *See* Varisoft® LAC, 33423
Lausofan, 16868
Lautal, 16869
Lautarite, *See* calcium iodate, 4954
Lauth's violet, 16870
Lavalloy, 16871
Lavarock, 16872
Lavasul, 16873
Lavender drops, 16874
Laventin® CW, 16875
Lavite, 16876
Lavrex, 16877
Lawinit 100, 16878
Lawn Food, 16879
Lawn Plus, 16880
Lawn Spot Weeder, 16881
Lawn Weed Gun1, 16882
Lawn Weedkiller, 16883
Lawn Weeds Killer, 16884
Lawnsman, 16885
Lawnsman Mosskiller, 16886
Lawnsman Spring Feed, 16887
Lawnsman Weed and Feed, 16888
Lawnsman Winterizer, 16889
Lawnturf®, 16890
Lawsone, 16891
Laxans, *See* Purgen,
Laxatin, *See* Purgen,
Laxatol, *See* Purgen,
Laxatoline, *See* Purgen,
Laxen, *See* Purgen,
Laxiconfect, *See* Purgen,
Laxin, *See* Purgen,
Laxoin, *See* Purgen,
Laxophen, *See* Purgen,
Layor Carang, *See* Agar (Agar-agar), 955, vegetable gelatin, 33518
Laysa, 16892
Laysa Plan, 16893
Laytex, 16894
LCA-1, 16895
LCA-20, 16896
LCA-127, 16897
LCP-20CF/000, 16898
LDA, *See* Ablumide LDE, 79, Alkamide® 327, 1634, Witcamide® 5195, 34440
LDE, *See* Ablumide LDE, 79, Alkamide® 327, 1634, Witcamide® 5195, 34440
LDPE, *See* polyethylene, low-density, 24430
LE 79-519, *See* permethrin, 23384, 23385, 23386
LE 79600, *See* cypermethrin, 7588, Topclip Parasol, 32037
Le Sage cement, 16899
lea-cov, *See* sodium fluoride, 28764
lead, 16900, 16901
lead, *See* Haro® Mix CE-701, 13645, Haro® Mix CK-711, 13646, Haro® Mix MH-204, 13647
lead acetate, 16902 *See* Goulard powder, 13216, Ledac, 16948
lead antimonate(V), *See* Naples yellow, 20733
lead ashes, 16903
lead bronze, 16904
lead carbonate, *See* Halcarb 20, 13545, Ledca, 16950
lead carbonate hydroxide, *See* Ceruse, 5769
lead carbonate, basic Lead(II) carbonate, basic, *See* Ceruse, 5769
lead chamber crystals, *See* nitrosylsulfuric acid, 21379
lead chloride, 16905, *See* Leclo, 16944
lead chromate(VI), *See* Chrome yellow, 6212

lead chromate(VI) oxide, *See* chrome red, 6209
lead chromate + bleaching powder, *See* Acagine, 179
lead diacetate, 16906
lead dimethyldithiocarbamate, *See* Ledate®, 16949, methyl ledate, 19547
lead dioxide, *See* Leadoxe, 16925, lead peroxide, 16913, Lepro, 17005, LP-100, 17691
lead distearate, *See* lead stearate, 16917
lead flake, *See* white lead, 34366
lead fluoborate, 16907
lead formate, 16908, *See* lead formate, 16908, Ledfo, 16959
lead glass crystal, *See* flint glass, 11973
lead metasilicate, *See* lead silicate, 16915
lead molybdate, 16909
lead monosilicate, *See* BSWL 202, 4634
lead monoxide, 16910, *See* lead ocher, 16912
lead nitrate, 16911, *See* Ledni, 16961
lead ocher, 16912
lead orthoplumbate, *See* red lead, 25960
lead oxide, *See* Vulcaid, 34072
lead oxide red, *See* red lead, 25960
lead oxide yellow, *See* lead monoxide, 16910
lead oxide, black, *See* lead oxide,
lead pentamethylene dithiocarbamate, *See* Vulcaid LP, 34082
lead peroxide, 16913
lead phosphite, *See* Halphos, 13569
Lead phosphite dibasic, *See* Halphos, 13569
lead phthalate basic, 16914
lead protoxide, *See* lead monoxide, 16910
lead silicate, 16915, *See* BSWL 202, 4634
lead silicate, potassium silicate, *See* crystal glass, 7286
lead solder, 16916
lead stearate, 16917
lead stearate, *See* Hal-Lub-D, 13558, Hal-Lub-N, 13559, Haro® Chem P28G, 13642, P-51, 22518
lead subacetate, *See* subacetate of lead, 29902
lead subcarbonate, 16918, *See* Ceruse, 5769, Kremser White, 16399
lead suboxide, *See* lead oxide,
lead tetroxide, *See* minium tego, 19832, red lead, 25960
lead thiosulfate, *See* Lethi, 17015
lead tungate, 16919
lead vinegar, 16920
lead water, 16921
lead whites, *See* flake white, 11871
lead(II) acetate trihydrate, *See* lead diacetate, 16906
lead(II) carbonate, *See* Halcarb 20, 13545
lead(II) n-octadecanoate, *See* lead stearate, 16917
lead(II) stearate, *See* lead stearate, 16917
lead(IV) oxide, *See* lead peroxide, 16913
lead, bis(dimethylcarbamodithioato-S,S')-, (T-4)-, *See* Ledate®, 16949, methyl ledate, 19547
Lead, bis[carbonato(2-)]dihydroxytri-, *See* Ceruse, 5769
leaded bronze, 16922
leaded gun metal, 16923
leaded zinc oxides, 16924
Leadoxe, 16925
Leaf green, *See* chlorophyll, 6134, M100, 18107
Leak Detector, 16926
Lean coal, 16927
Leantin, 16928
Leatherlubric, 16929
Lebanon No. 34, 16930
Lebanon No. 48, 16931
Lebaycid, *See* Tiguvon®, 31788
Lebaycid®, 16932
Lebbin Salt, 16933
Lebon 200, *See* Cocamidopropyl betaine, 6551
Lecin, 16934
Lecipon, 16935
Lecitase, 16936
Lecithan, 16937
Lecithcerebrin, 16938
lecithin, 16939, *See* Actiflo® 68, 70, 476, Alcolec® 439-C, 1481, Alcolec® 439-C, 1482, Centrocap® 162SS, 162US, 5683, Centrol® CA, 5684, Centrophase® C, 5688, Centrophase® HR, 5689, Centrophase® HR2B, HR2U, 5690, Centrophil® K, 5691, Dermasome® MT, 8107, Dermasome® TRF, 8112, Kelecin®, 15835, Lecithin L-Range, 16940, Oleo-Coll LP, 22118, PhosPho E-100, 23707, PhosPho F-97, 23708, PhosPho LCN-TS, 23711
Lecithin albuminate, *See* Litalbin, 17497
lecithin-allantoin, *See* Dermasome® A, 8104
lecithin and sodium hyaluronate, *See* Dermasome® H, 8106
Lecithin, butyl stearate, coco-hydrolyzed animal protein, oleoyl sarcosine, sesame oil and lanolin alcohol, *See* Proto-Lan 8, 25294
Lecithin L-Range, 16940
Lecithin Water Dispersible CLR, 16941
Lecithin-Polysorbate 80, *See* Centromix® CPS, 5687
lecithin-soluble collagen, *See* Dermasome® SC, 8110
lecithin-superoxide dismulase, *See* Dermasome® SOD, 8111
lecithin-tocopheryl acetate, *See* Dermasome® E, 8105
lecithmedullan, 16942
lecithol, 16943, *See* Kelecin®, 15835
Leclo, 16944, *See* lead chloride, 16905
Lectricon, 16945
Lectro Cast, 16946
Lecutyl, 16947
Ledac, 16948
Ledate, *See* Ledate®, 16949, methyl ledate, 19547
Ledate®, 16949
Ledca, 16950
Ledclair, *See* Versene CA, 33618
Leddel alloy, 16951
Ledeburite, 16952
Ledebur's metal, 16953
Ledercillin, *See* Penicillin V, 23111
Ledercort, 16954
Lederfen, 16955
Ledermix, 16956
Lederplex, 16957
Lederspan, 16958
Ledfo, 16959, *See* lead formate, 16908
Ledmin LPC, 16960, *See* laurylpyridinium chloride, 16867
Ledni, 16961, *See* lead nitrate, 16911
Ledon 12, *See* dichlorodifluoromethane, 8388
Ledrite Brass, 16962
Leecure B Series, 16963
Leefex, 16964
Leegen®, 16965
Lees, 16966
Leffmann and Beam's glycerol reagent, 16967
Legederm, *See* Vaderm, 33237
Legion, 16968
Legumex, *See* 2,4-DB, 7810
Legumex Extra, 16969
Legumin, *See* vegetable casein, 33516
Legupren®, 16970
Legurame, *See* Carbetamex, 5189
Leguval®, 16971
Leinsaat oils, 16972
Leipzig yellow, *See* Chrome yellow, 6212
Leitch's blue, *See* Cyanine, 7475
Lekutherm®, 16973
Lemarome n, *See* Neral, 20996
Lemarquand's alloy, 16974
Lemascorb, *See* vitamin C, 33938
Lembergite, 16975
Lemberg's solution, 16976
Lemco, 16977
Lemery's white precipitate, *See* white precipitate, 34370
Lemnian earth, 16978
lemoflur, *See* sodium fluoride, 28764
Lem-O-Fos®, 16979
Lemol, 16980
Lemolac, 16981
lemon chrome, *See* barium chromate, 3638
Lemon Delph, 16982
Lemon Lac, 16983
lemon yellow, *See* Barium chromate, 3638, zinc yellow, 34800
lemonene, *See* Biphenyl, 4216
Lemonol, *See* Geraniol, 12885, Meranol, 19331
lenacil, 16984, *See* Venzar®, 33562, Vizor, 34001, 34002
lenacile, *See* lenacil, 16984
lenacil-phenmedipham, *See* DUK-880, 9087
Leneta, 16985
Lenetol, 16986
Lenium, 16987
Lenka, 16988
Lensine, 16989
Lentagran, 16990, *See* Pyridate, 25480
Lentana, 16991
Lenticillin, *See* Penicillin V, 23111
Lentizol, 16992
Lento-Kalium, *See* potassium chloride
Lento-Kalium, *See* Kay Ciel, 15807, K-Contin, 15818, K. Tab, 15642, Kaon-Cl, 15729, K-Lor, 16226, K-Lyte/C1, 16234, KM Potassium Chloride, 16242, Potassium chloride, 24748, 24749
Lentopen, 16993
Lentrek, *See* Chlorpyrifos, 6163
Lenzing P84, 16994
Lenzit, *See* gypsum, 13518
Lenzol, 16995
Leo 640, *See* Lofepramin Hydrochloride, 17552
Leo K, 16996, *See* Potassium chloride, 24748
Leomin AN, 16997
Leomin FANF, 16998
Leona®, 16999
Leonil, *See* Nekal, 20855
Leonil DB Powd, 17000
Leonil EBL, 17001
Leonil OS, 17002
Leonil-O, *See* Volpo 3, 34020
Leophen® BN, 17003
Lepandin, 17004
lepargylic acid, *See* anchoic acid, 2422, Azelaic acid, 3519
Lepidinquinoline ethylcyanine bromide, *See* ethylcyanine, 11120
Lepro, 17005
Lepton®, 17006
Lerbek, 17007
Lerifond, 17008
Leriphen, 17009
Lerite, 17010
Leromoll, 17011
Lesan, 17012
Les-cav, *See* sodium fluoride, 28764
Lethalbine, 17013
Lethane, 17014
Lethelmin, *See* phenothiazine, 23618
Lethi, 17015
Leu, *See* Leucine, 17020
Leucaniline, 17016
Leucarsone, 17017
Leucaurin, 17018
Leuchtol, 17019
Leucine, 17020, *See* l-leucine, 17021
l-leucine, 17021
L-Leucine, *See* Leucine, 17020
1-leucine, *See* Leucine, 17020
Leuco-4, *See* Adenine, 679
leucoargilla, *See* Lithomarge, 17518
Leucobenzaurin, 17022
Leucogen, 17023
Leucol, 17024
Leucoline, *See* Quinoline,
Leucomycin®, 17025
Leuconine, 17026
Leucophor KNR, 17027
Leucophor R, *See* Photine, 23756
Leucopure EGM Powd, 17028
Leucoturic acid, *See* oxalantin, 22411
Leukanol NF, *See* Ablusol ML, 129

leukarion, *See* oil white, 22086
Leukeran Tablets, 17029
Leukochthol, *See* Perichthol, 23270
Leukomycin®, 17030
Leukon, 17031
leukonin, *See* Leuconine, 17026
Leukonöl LBA-2, 17032
Leukotan® 974, 17033
Leukotrop® W, 17034
Leuna Gas, 17035
Leuna M, *See* MCPA, 19183
Leuna saltpeter, 17036
Leunaphos, 17037
Leuprolide acetate, *See* Lupron, 17918
Leutalux, 17038
Levacast®, 17039
Levacell®, 17040
Levacide, *See* Tramisol®, 32134
Levaderm®, 17041
Levadin, *See* Tramisol®, 32134
Levafil, 17042
Levafix®, 17043
Levaflex®, 17044
Levaform®, 17045
Levagard®, 17046
Levagel®, 17047
Levair, 17048
Levair®, 17049
Levalaine, 17050
Levalan N, 17051
Levalin®, 17052
Levamisole, *See* Ketrax, 16149
Levamisole hydrochloride, *See* Tramisol®, 32134
Levanox®, 17053
Levanyl®, 17054
Levapon®, 17055
Levapren®, 17056
Levapren® 400, 17057
Levapren® K, 17058
Levasil®, 17059
Levasint®, 17060
Levasol®, 17061
Levasole, *See* Tramisol®, 32134
Levcarb, 17062
Levegal®, 17063
Levelan P14B, 17064
Levelan P208, 17065
Levelan P307, 17066
Levelan P357, 17067
Levelan® P208, 17068
Levenol PW, *See* Ablunol SA-7, 113
Levepox®, 17069
Levesol, 17070
Levn-Lite, 17071
Levochrom®, 17072
Levodip®, 17073
Levofin®, 17074
Levogel®, 17075
Levogen®, 17076
Levogen® LF, 17077
levomenol, *See* Kamillosan, 15707
Levopress, 17078
Levorin, *See* Vanobid, 33332
Levotan®, 17079
Levotherm®, 17080
Levothroid, 17081
Levothyroxine sodium, *See* Levothroid, 17081
Levovermax, *See* Ketrax, 16149
Levoxin®, 17082
levulic acid, *See* levulinic acid, 17084
Levuline, 17083
levulinic acid, 17084
levulose, 17085, *See* fructose, 12420, levulose, 17085
Lewasorb®, 17086
Lewatit®, 17087
Lewis metal, 17088
Lewisite, 17089
Lewisol 28, 17090
Lexaine® C, 17091
Lexaine® CSB-50, 17092
Lexaine® IS, 17093
Lexaine® LM, 17094, *See* lauramidopropyl betaine, 16830
Lexaine® O, 17095, *See* Oleamidopropyl betaine, 22108
Lexamine 22, 17096, *See* stearamidoethyl diethylamine, 29594
Lexamine C-13, 17097, *See* Cocamidopropyl dimethylamine, 6552
Lexamine L-13, 17098, *See* lauramidopropyl dimethylamine, 16831
Lexamine O-13, 17099, *See* oleamidopropyl dimethylamine, 22109
Lexamine S-13, 17100
Lexamine S-13 Lactate, 17101, *See* stearamidopropyl dimethylamine lactate, 29595
Lexan®, 17102
Lexan® 121, 17103
Lexan® 141L, 17104
Lexan® 144LR, 17105
Lexan® 151, 153, 154, 17106
Lexan® 241, 17107
Lexan® 500, 503, 17108
Lexan® 920, 17109
Lexan® 3412, 17110
Lexan® 8040, 17111
Lexan® FL400, 17112
Lexan® GR1310, 17113
Lexan® HF1110, HF1130, HF1140, 17114
Lexan® HP1, HPS1, 17115
Lexan® OQ1020, 17116
Lexan® PK2040, 17117
Lexan® PPC4501, 17118
Lexan® SP1010, 17119
Lexan® WR1210, WR1240, 17120
Lexate BPQ, 17121
Lexate CRC, 17122
Lexate PX, 17123
Lexate TA, 17124
Lexate TL, 17125
Lexe®, 17126
Lexein® A200, 17127, *See* Myristoyl hydrolyzed collagen, 20576
Lexein® A520, 17128, *See* TEA-abietoyl hydrolyzed collagen, 30871
Lexein® CP-125, 17129
Lexein® S620S/Superpro 5A, 17130
Lexein® X-250HP, 17131, *See* Extiat®, 11323
Lexell, 17132
Lexemul® 503, 17133
Lexemul® 561, 17134
Lexemul® CS-20, 17135
Lexemul® EGDS, 17136
Lexemul® EGMS, 17137
Lexemul® GDL, 17138, *See* glyceryl dilaurate, 13079
Lexemul® P, 17139
Lexemul® PEG-200 DL, 17140, *See* PEG-4 dilaurate series, 23050
Lexemul® PEG-400ML, 17141
Lexemul® T, 17142
Lexgard B, 17143, *See* Butylparaben, 4797
Lexgard bronopol, *See* Bronopol-Boots, 4596
Lexgard M, 17144
Lexgard P, 17145
Lexite Granular Carpet, 17146
Lexol 3975, 17147
Lexol EHP, 17148
Lexol GT 855, GT 865, 17149, *See* caprylic/capric acid triglyceride, 5161
Lexol IPM-NF, 17150
Lexol IPP, 17151, *See* isopropyl palmitate, 15410
Lexol PG 800, 17152
Lexol PG 855, 17153
Lexol PG 900, 17154
Lexol SS, 17155
Lexolube® 2J-237, 17156
Lexolube® 2T-237, 17157
Lexolube® 2X-109, 17158
Lexolube® 3G-310, 17159
Lexolube® 4N-415, 17160
Lexolube® B-108, 17161
Lexolube® B-109, 17162
Lexolube® BS-Tech, 17163, *See* butyl stearate, 4781
Lexolube® T-110, 17164, *See* octyl stearate, 22041
Lexone, *See* metribuzin, 19595, 19596
Lexone 4L, *See* metribuzin, 19595, 19596
Lexone 75DF, *See* metribuzin, 19595, 19596
Lexone DF, *See* metribuzin, 19595, 19596
Lexone®, 17165
Lexorez 1100-25, 17166
Lexorez 1101-50A, 17167
Lexorez 1130-30, 17168
Lexorez 1131-190, 17169
Lexorez 1400-35, 17170
Lexorez 1721-65P, 17171
Lexorez 1821-50, 17172
Lexorez 5901-55, 17173
Lexquat® 2240, 17174
Lexquat® AMG-BEO, 17175
Lexquat® AMG-IS, 17176
Lexquat® AMG-M, 17177
Lexquat® AMG-O, 17178
Lexquat® AMG-WC, 17179
Lexquat® CH, 17180
Ley Cornox, 17181
LF 5/60LF 20, *See* Algin, 1591
LF 20/40, *See* Algin, 1591
LF 60 Sobalg FD 100 Range, *See* Algin, 1591
LG Wax, 17182
LHS 40% Coconut Oil Soap, 17183
Liancare, 17184
Libavius' fuming spirit, 17185
Libfer®, 17186
Libfer® SP, 17187
Libollite, 17188
Libral Range, 17189
Librebor®, 17190
Librel®, 17191
Libreleaf®, 17192
Libsorb®, 17193
Libspray®, 17194
LICA 01, 17195
LICA 09, 17196
LICA 12, 17197
LICA 38, 17198
LICA 44, 17199
LICA 97, 17200
LICA 99, 17201
Lice Rid, *See* malathion, 18460
Licella yarn, 17202
lichen starch, 17203
lichen sugar, 17204
Lichenin, *See* lichen starch, 17203
Lichner's blue, 17205
Lichtenbergs's metal, 17206
Licomer, 17207
Liconite, 17208
Licowet, 17209
Licryfilcon A, *See* Licryl-55, 17210, Licryl-70, 17211
Licryl-55, 17210
Licryl-70, 17211
Licuado Instante, 17212
Lida-Mantle, 17213
Lidamidine hydrochloride, *See* Lidarral, 17214
Lidarral, 17214
Lidex, 17215
lidocaine, 17216, *See* Lida-Mantle, 17213
Lidofilcon B, *See* Sauflon PW, 27496
Lieben solution, 17217
Lieberkuhn's jelly, 17218
Liebmann and Studer's acid, 17219
Life, 17220
Ligantraal, 17221
Ligdynite, 17222
Light carburetted hydrogen, *See* marsh gas, 18954
light ligrin, *See* petroleum ether, 23497
Light spar, *See* gypsum, 13518
Light Water Foam, 17223

Chemical and Trade Names

Lipocol O-2, 17356, *See* oleth-8, 22126
Lipocol S, 17357
Lipocol S-2, 17358, *See* steareth series, 29596
Lipocol S-20, *See* Ablunol SA-7, 113
Lipocol SC-4, 17359
Lipocol TD-3, 17360 , *See* Trideceth series, 32232
Lipocutin®, 17361
Lipodan CDS Kosher, 17362
Lipodan CRE Kosher, 17363
Lipodan SET Kosher, 17364
Lipoderm®, 17365
Lipo-Hepin, *See* Heparin, Sodium Salt, 13788
Lipo-Hepinette, *See* Heparin, Sodium Salt, 13788
Lipolan, 17366, *See* Fancol HL, 11441
Lipolan 31, 17367, *See* PEG-20 hydrogenated lanolin, 23045
Lipolan R, 17368
Lipolase, 17369
Lipo-Lutin, 17370
Lipomulse 165, 17371
Liponate 2-DH, 17372
Liponate CL, 17373
Liponate CRM, 17374, *See* cetyl ricinoleate, 5824
Liponate DPC-6, 17375, *See* Dipentaerythrityl hexacaprylate/hexacaprate, 8582
Liponate EM, 17376
Liponate GC, 17377
Liponate IPM, 17378
Liponate IPP, 17379, *See* isopropyl palmitate, 15410
Liponate MM, 17380
Liponate NPGC-2, 17381
Liponate PB-4, 17382
Liponate PC, 17383
Liponate PE-810, 17384
Liponate PO-4, 17385, *See* Pentaerythrityl tetraoleate, 23141
Liponate PS-4, 17386, *See* Pentaerythrityl tetrastearate, 23143
Liponate SPS, 17387, *See* cetyl esters, 5819
Liponate SS, 17388
Liponate TDS, 17389
Liponate TDTM, 17390, *See* tridecyl trimellitate, 32235
Liponic 70-NC, 17391
Liponic EG-1, 17392, *See* Glycereth series, 13069
Liponic SO-20, 17393
Lipopeg 2-DL, 17394, *See* PEG-4 dilaurate series, 23050
Lipopeg 4-DO, 17395, *See* PEG-2 dioleate series, 23040
Lipopeg 4-DS, 17396, *See* PEG-150 distearate series, 23037
Lipopeg 4-L, 17397, *See* Ablunol 200ML, 95
Lipopeg 6000-DS, 17398
Lipo-Peptide AME 30, 17399
Lipoproteol LCO, 17400
Lipoproteol LK, 17401
Lipoquat R, 17402, *See* Ricinoleamidopropyl ethyldimonium ethosulfate, 26733
Liposiliol C, 17403
Liposorb 0-20, *See* Alkamuls® T-80, 1693
Liposorb L, 17404
Liposorb L-10, 17405, *See* PEG-10 sorbitan laurate, 23036
Liposorb L-20, 17406
Liposorb O, 17407, *See* Ablunol S-80, 111, Witconol 2500, 34501
Liposorb O-20, 17408, *See* Ablunol S-80, 111, Radiasurf® 7157, 25804, Witconol 2500, 34501
Liposorb O-5, 17409
Liposorb P, 17410, *See* Ablunol S-40, 109
Liposorb P-20, 17411
Liposorb S, 17412, *See* Ablunol S-60, 110
Liposorb S-20, 17413, *See* Ablunol S-60, 110
Liposorb S-4, 17414
Liposorb S-4, S-20, *See* Ethylan® GES6, 11108
Liposorb SQO, 17415
Liposorb TO, 17416, *See* Ablunol S-85, 112, Witconol 2503, 34502
Liposorb TO-20, 17417
Liposorb TS, 17418
Liposorb TS-20, 17419, *See* Polysorbate 65, 24571
Liposorb™, 17420
Lipostat Tablets, 17421
Lipotin 100, 100J, SB, 17422
Lipotin A, 17423
Lipotin H, 17424
Lipotril, *See* choline chloride, 6171
Lipovol A, 17425, *See* Avocado Oil, 3498, 3497
Lipovol ALM, 17426, *See* Sweet Almond oil, 30452
Lipovol G, 17427, *See* Grape Seed oil, 13275
Lipovol HS, 17428, *See* hydrogenated soybean oil, 14596
Lipovol J, 17429
Lipovol MOS-70, 17432
Lipovol MOS-130, 17430
Lipovol MOS-350, 17431
Lipovol P, 17433
Lipovol PAL, 17434, *See* Palm oil, 22626
Lipovol SAF, 17435
lipovol saf(lipo), *See* Neobee® 18, 20872, safflower oil, 27222
Lipovol SES, 17436, *See* Sesame oil, 28075
Lipovol SO, 17437
Lipovol SOY, 17438, *See* Soybean oil, 29173
Lipovol SUN, 17439
Lipovol WGO, 17440, *See* Wheat germ oil;, 34349
Lipowax D, 17441
Lipowax G, 17442
Lipowax NI, 17443
Lipowax P, 17444
Lipowax PR, 17445
Lipowax P-SPEC, 17446
Lipowitz's alloy, 17447
Lipoxol® 12000, 17448
Lipozyme, 17449
Liquamar, 17450, *See* Phenprocoumon, 23630
LiquaPar® Oil, 17451
Liquapen, 17452
Liquemin, 17453, *See* Heparin, Sodium Salt, 13788
liqueur de Ferraile, *See* pyrolignite of iron, 25515
Liqueur de van Swieten, 17454
Liquibor, 17455
Liquibor 169, 17456
Liquibor 524, 17457
Liquibrom, 17458
Liquical, 17459
Liqui-Cee, 17460
Liquid 99, 17461
Liquid Absorption Base Type T, 17462
Liquid Bases, 17463
Liquid bleach, *See* sodium hypochlorite, 28773
Liquid bronzes, 17464
Liquid calcium chloride, *See* Liquidow, 17478
Liquid Code XLR, 17465, *See* Potassium pyroantimonate, Acid, 24776
Liquid Copper Fungicide, 17466
Liquid Crystal CN/9, 17467
Liquid drier, 17468
Liquid drier, Japan drier, *See* terebine, 31234
Liquid Feed for Hanging Baskets, 17469
Liquid gold, 17470
Liquid Growmore, 17471
Liquid Latex, 17472, *See* Everflex® SP-1084, 11266
Liquid Lightning, 17473
liquid paraffin, *See* Kaydol, 15810, Klearol, 16208, Mineral oil,
liquid petrolatum, *See* Kaydol, 15810
liquid pitch oil, *See* Oil of tar, 22075
Liquid polyisobutylene, *See* Parapol, 22795
Liquid Q4 Borders and Beds Fertiliser, 17474
Liquid resins, 17475
liquid silver, *See* Mercury, 19355
Liquid storax, *See* storax calamita, 29795
Liquid Tomato Feed, 17476
Liquidambar, 17477
Liquidow, 17478
Liquifilm Forte®, 17479
Liquifilm Tears®, 17480
Liquigel®, 17481 , *See* Alumina hydrate, 1892
Liquigel-AM, 17482
Liquimeth, 17483
Liqui-Nox®, 17484
Liquinure, 17485
Liquiphene, *See* acetoxyphenylmercury, 306, phenylmercury acetate, 23646
Liquiwax DC-EFA/SS, 17486
Liquiwax DIADD, 17487
Liquiwax DICDD, 17488
Liquiwax DIEFA, 17489
Liquor, Glass, *See* soluble glass, 28980
liquorice, 17490
Lisat, 17491
Liskonum Tablets, 17492
Lissamine, 17493
Lissanol, 17494
Lissapol, 17495
Lissatan Ac, *See* Ablusol ML, 129
Lissolamine V, *See* Acetoquat CTAB, 303, Cetavlex, Cetavlon, 5787, Cetrimonium bromide, 5813, Sumquat® 6030, 30122, Varisoft® CTB-40, 33421
Listab, 17496
Litalbin, 17497
Litarol, *See* Bromoxynil, 4590
Litefax, 17498
Litex, 17499
Lithane, 17500
Litharge, 17501, *See* lead monoxide, 16910
Litharge 33, 17502
litharge, leaded, *See* lead oxide,
Lithargrite, 17503
Lithene®, 17504
Lithex, 17505
Lithic acid, 17506, *See* uric acid, 33139
Lithio-mercuric-iodide, *See* mercuricide, 19351
Lithiopiperazine, 17507
lithium, 17508
lithium carbonate, *See* Liskonum Tablets, 17492
lithium hypochlorite, 17509, *See* HyPure L, 14727
lithium iodide, *See* Tenephrol, 31103
lithium lauryl sulfate, *See* Empicol® HL25, 10322
lithium molybdate, 17510
lithium nitrate, 17511
lithium octadecanoate, *See* lithium stearate, 17512
lithium salt, *See* lithium stearate, 17512
lithium stearate, 17512, *See* Afco-Chem LIS, 922
lithium-β-hydroxynaphthalene- α-monosulfonate, *See* Naplithin, 20734
Litho-carbon, 17513
lithoclastite, 17514
Lithocolla, *See* Lithomarge, 17518
Lithoform, 17515
lithographer's varnish, *See* stand oil, 29450
lithographer's varnish, *See* stand oil, 29450
lithographic stone, *See* Calbux, 4929, calcium carbonate, 4941, CP Filler, 6928, limestone, 17273
Lithol, 17516, *See* Perichthol, 23270
Lithol® Pigments, 17517
Lithomarge, 17518
Litho-oil, 17519
Lithopone, 17520
Lithostar, 17521
Lithostat, 17522
Litmus, 17523
litnum bronze, 17524
Littorl, 17525
liver *lactobacillus casei* factor, *See* folic acid, 12196
liver of antimony, 17526
liver of sulfur, *See* sulfurated potash, 30052
liver ore, *See* cinnabar,
liver starch, *See* Liver sugar, 17527
liver sugar, 17527
Liveroid, 17528
Lizetan Spray, 17529
LL, *See* chromotrope acid, 6244
LL-AV290, *See* Avoparcin, 3502
LLDPE, *See* polyethylene, linear low density, 24429
Llinol, *See* Linseed oil, 17309
limonene, Inactive, *See* dipentene, 8584

LSD, 17704
L-serine, 17705
Luaktin®, 17706
Lubafax, 17707
Lubasin®, 17708
Lube-Booster® 1320 II, 17709
Lube-Lok®, 17710
Lubestat, 17711
Lubestine, 17712
Lubit® 64, 17713
Lubix, 17714
Lubolid, 17715
Lubrajel® CG, DV, MS, TW, 17716
Lubran 145, 17717
Lubran AD, 17718
Lubrhophos® HR-719, 17719
Lubrhophos® LB-400, 17720
Lubrhophos® LE-500, 17721
Lubrhophos® LE-700, 17722
Lubrhophos® LF-200, 17723
Lubrhophos® LM-400, 17724
Lubrhophos® LP-700, 17725
Lubrhophos® LS-500, 17726, *See* Trideceth phosphate series, 32231
Lubri-Bond®, 17727
Lubricant EHS, 17728
Lubricated PPS, *See* RTP 1378, 27072
Lubricin 25, 17729
Lubricin N-1, 17730
Lubrico, 17731
Lubricomp®, 17732
Lubricomp® 189, 17733
Lubricomp® DFL-4036, 17734
Lubricomp® DL-4030, 17735
Lubrigel, 17736
Lubrigel, *See* Montmorillonite, 20300
Lubril CAT-XIVC, 17737
Lubril PF-570, 17738
Lubril QC, 17739
Lubrimet® P 600, P 900, 17740
Lubrisol, 17741
Lubritab®, 17742
Lubrite B33, 17743
Lubrizol® 2152, 17744
Lubrizol® 2152, 2064, *See* calcium sulfonate, 4973
Lubrizol® 2153, 17745
Lubrol, 17746
Lubrol 90, 17747
Lubrol EA, *See* Abluwax EBS, 140
Lubrol N13, 17748
Lubrol N5, 17749
Lucalen®, 17750
Lucantin® CX, 17751
Lucaphos® 40, 48, 17752
Lucarotin® 10%, 17753
Lucca oil, 17754
Lucel ADA, *See* Azodicarbonamide, 3524
Lucerno, 17755
Luchem AS-946, 17756
Lucidene, 17757
Lucidol, 17758, *See* Abcure S-40-25, 35, AcetOxyl 2.5 and 5, 305, Benzoyl peroxide, 3983
Lucidol 75FP, 17759
Lucidol GS, 17760
Lucidol RM, 17761
Lucidol-78, 17762
Lucilite, 17763
Lucinite, *See* aluminum phosphate, 1922
Lucipal, 17764
Lucirin® BDK, 17765
Lucitone, *See* Vernonite, 33601
Lucite®, 17766
Lucobit®, 17767
Lucofen S A, 17768
Luconyl®, 17769
Lucovyl, 17770
Lucryl®, 17771 , *See* Polymethylmethacrylate, 24492
Ludigol F, 17772, *See* sodium *m*-nitrobenzene sulfonate, 28795
Ludigol®, 17773
Ludiomil, 17774
Ludlum alloy, 17775
Ludlum No. 602 Steel, 17776
Ludopal®, 17777
Ludorum, 17778, *See* Talisman, 30735, Tolurex, 31958
Ludox®, 17779
Lufibrol® E, 17780
Lufibrol® FW, 17781
Lufibrol® NB-7, 17782
Lufilen®, 17783
Luflbrol®, 17784
Luftseide, 17785
Lugalvan, 17786
Luganil® Dyes, 17787
Lugol's solution, 17788
Lugo's powder, 17789
Luhydran®, 17790
Lukens Bone Wax, 17791
Lulea tar, 17792
Lumarith® EC, 17793
Lumarith® ER, 17794
Lumattin®, 17795
Lumbang oil, 17796
Lumbrical, *See* Upixon, 33094
Lumen Alloy 11-C, 17797
Lumen bronze, 17798
Lumicon 68 Silver Amalgam/Powder, 17799
Lumicon Non Gamma 2, 17800
Lumifor® Gluma, 17801
Lumilux, 17802
Luminal Sodium, 17803
Luminous®, 17804
Lumisol RV, *See* Photine, 23756
Lumite, 17805
Lumiten® E, 17806
Lumiten® I, N, 17807
Lumitol®, 17808
Lummer's solution, 17809
Lumnite cement, 17810
Lumo Stabil S 80, 17811
Lumo WW 75, 17812
Lumorol 4153, 17813
Lumorol K 28, 17814
Lumosäure A, 17815
lump-Lac, *See* lac, 16547
lunac s 20, *See* stearic acid, 29599
lunar caustic, 17816
Lunosol, 17817
Luo-calcite, 17818
Luo-chalybite, 17819
Luo-diallogite, 17820
Luo-magnesite, 17821
Lupeose, 17822
luperco, *See* Benzoyl peroxide, 3983
Luperco 101-P20, 17823, *See* dimethyldibutyl peroxyhexane, 8518
Luperco 230-XL, 17824, *See* Lupersol 230, 17855
Luperco 230-XL, *See* Butyl bisbutyl peroxy valerate, 4765
Luperco 231-XL, 17825
Luperco 233-XL, 17826
Luperco 331-XL, 17827
Luperco 500-40KE, 17828
Luperco 801-XL, 17829, *See* Butyl cumyl peroxide, 4768
Luperco 802-40KE, 17830
Luperco A, 17831
Luperco AC, 17832
Luperco AFR, 17833
Luperco AFR-250, 17834
Luperco AST, 17835
Luperfoam 40, 17836
Luperox, 17837
Luperox 2,5-2,5, 17839
Luperox 118, 17838
Luperox 500R, 17840
Luperox 802, 17841, *See* Bisbutyl peroxy diisopropyl benzene, 4230
luperox fl, *See* Benzoyl peroxide, 3983
Lupersol 10, 17842
Lupersol 11, 17843, *See* Esperox® 31M, 10848
Lupersol 70, 17844
Lupersol 80, 17845
Lupersol 101, 17846
Lupersol 130, 17847
Lupersol 188-M75, 17848
Lupersol 188-M75, 288-M75, *See* Esperox® 939M, 10858
Lupersol 219-M60, 17849
Lupersol 220-D50, 17850
Lupersol 221, 17851
Lupersol 223, 17852, *See* Espercarb® 840, 10841
Lupersol 224, 17853
Lupersol 225, 17854, *See* Espercarb® 438M-60, 10840
Lupersol 230, 17855, *See* Butyl bisbutyl peroxy valerate, 4765
Lupersol 231, 17856
Lupersol 233-M75, 17857
Lupersol 256, 17858, *See* USP® -245, 33166
Lupersol 288-M75, 17859
Lupersol 331, *See* USP® -400P, 33168
Lupersol 331-80B, 17860, *See* USP® -400P, 33168
Lupersol 531-80B, 17861
Lupersol 533-M75, 17862
Lupersol 546-M75, 17863, *See* Esperox® 545M, 10852
Lupersol 553-M75, 17864
Lupersol 554-M50, 554-M75, 17865, *See* Amilperoxy pivalate, 2122
Lupersol 555-M60, 17866
Lupersol 575, 17867, *See* Esperox® 570, 10854
Lupersol 665-M50, 17868
Lupersol 688-T50, 17869
Lupersol DDM, 17870
Lupersol DDM-9, 17871
Lupersol DDM-9, DSW, *See* Esperfoam® FR, 10842
Lupersol DEL, 17872
Lupersol DSW, 17873
Lupersol KDB, 17874
Lupersol P-31, P-33, 17875
Lupersol PDO, 17876
Lupersol TAEC, 17877
Lupersol TBEC, 17878
Lupersol TBIC-M75, 17879
Lupersol® 227, 17880
Lupetazin, 17881
Luphen® D, 17882
Lupinit, 17883
Lupolen, *See* A-C® Polyethylene 6, 6A, 7, 7A, 8, 8A, 9, 9A, 617, 617A, 175 Cabelec® 1017, 4862
Lupolen 804H and 1814H, 17902
Lupolen 1800 H/M/S, 17884
Lupolen 1810E, 17885
Lupolen 1810H, 17886
Lupolen 1812D and 1812EH, 17887
Lupolen 1814E, 17888
Lupolen 1852E/H, 17889
Lupolen 2040EX and 2410DX, 17890
Lupolen 2410S, 17891
Lupolen 2424H and 2425K, 17892
Lupolen 2430H, 17893
Lupolen 2452 E, 17894
Lupolen 3010 S, 17895
Lupolen 3020 D, 17896
Lupolen 3020 KX and 3025 KX, 17897
Lupolen 4261 AX, 17898
Lupolen 5011 K, 17899
Lupolen 5052 C, 17900
Lupolen 6011 K, 17901
Lupolen V-2524EX and V-3510K, 17903
Lupolen®, 17904
Lupolex, 17905
Lupranat®, 17906
Lupranol®, 17907
Lupraphen, 17908
Luprenal®, 17909
Luprimol®, 17910
Luprintan®, 17911
Luprintan® ATP, 17912

Luprintan® DCA, 17913
Luprintol®, 17914
Luprintol® PE, 17915
Luprofil®, 17916
Lupromag®, 17917
Lupron, 17918
Luprosil®, 17919
Luprosil® NC, 17920
Luprosil® Salt, 17921
Luprosil® Sodium Salt, 17922
Lurafix® Dyes, 17923
Luramid®, 17924
Luran®, 17925
Luran® 358 N, 17926
Luran® 378 P G7, 17927
Luran® 757R, 17928
Luran® 776S, 17929
Luran® KR 2517, 17930
Luran® KR 2556, 17931
Luran® S, 17932
Lurantin®, 17933
Luranyl®, 17934
Lurapret®, 17935
Lurazol® Dyes, 17936
Luredox® BP, PO, 17937
Luredur®, 17938
Luresin®, 17939
Lurgi metal, 17940
luride, *See* sodium fluoride, 28764
luride lozi-tabs, *See* sodium fluoride, 28764
luride-sf, *See* sodium fluoride, 28764
Luron® Binder, 17941
Lurotex®, 17942
Lurotex® A-25, 17943
Luscin, 17944
Lusol, 17945
Lusol®, 17946
Lusolvan® FBH, 17947
Lustilac, 17948
Lustra-Cellulose, 17949
Lustral, 17950
Lustralite®, 17951
Lustralite® 44-444, 17952
Lustran, 17953
Lustran ABS, 17954
Lustran SAN, 17955
Lustran Ultra ABS, 17956
Lustranyl, 17957
Lustra-Pearl®, 17958
Lustrasol®, 17959
Lustre, 17960
Lustrex, 17961
Lustron, 17962
Lustrose, 17963
Lusynton® A, 17964
Lutan®, 17965
Lutate, 17966
Lutavit, 17967
Lutecin, *See* nickel silvers, 21132
lutein, *See* xanthophyll, 34591
Lutensit®, 17968
Lutensit® A-BO, 17969
Lutensit® A-ES, 17970
Lutensit® A-LBA, 17971
Lutensit® AN 10, 17972
Lutensit® A-PS, 17973
Lutensit® AS 2230, 2270, 17974
Lutensit® K-LC, 17975
Lutensit® K-OC, 17976
Lutensol®, 17977
Lutensol® A 7, 17978
Lutensol® AO 10, 17979
Lutensol® AO 3, 17980
Lutensol® AP 10, 17981
Lutensol® AP 20, 17982
Lutensol® AP 6, 17983
Lutensol® AT 11, 17984
Lutensol® ED 140, 17985
Lutensol® FA 12, 17986
Lutensol® FSA 10, 17987
Lutensol® FSA 10, *See* PEG-3 oleamide, 23048
Lutensol® LF 220, 221, 223, 224, 17988
Lutensol® ON 30, 17989
Lutensol® TO 3, 17990
luteol, 17991
Lutetia, 17992
Lutexal® TX-401, 17993
Lutexan, 17994
Lutidine, *See* 2,6-Lutidine, 17995
2,6-Lutidine, 17995
α,α'-lutidin, *See* 2,6-Lutidine, 17995
Lutofan®, 17996
Lutonal®, 17997
Lutonal® A, 17998
Lutonal® D, 17999
lutosol, *See* isopropanol, 15405
Lutrabond, 18000
Lutradur, 18001
Lutranal® LC, 18002
Lutrizol®, 18003
Lutrol®, 18004
Lutrol® E 300, 18005
Lutrol® E 400, 18006
Lutrol® E 1500, 18007
Lutrol® E 4000, 18008
Lutrol® E 6000, 18009
Lutrol® OP-2000, 18010
Lutrol® W-3520, 18011
Lutron, 18012, 18013
Lutropur, 18014
Luvican M170, 18015
Luviflex® VBM 35, 18016, *See* Acrylates/PVP copolymer, 411
Luviform® ES 22, 18017
Luviform® ES 42, 18018
Luviform® FA 119, 18019
Luviquat®, 18020
Luviquat® FC 370, 18021, *See* Polyquaternium-16, 24531
Luviquat® FC 550, 18022, *See* Polyquaternium-16, 24531
Luviquat® FC 905, 18023, *See* Polyquaternium-16, 24531
Luviquat® HM 552, 18024, *See* Polyquaternium-16, 24531
Luviquat® Mono CP, 18025
Luviset, 18026
Luviset CA 66, 18027, *See* VA/crotonates copolymer, 33224
Luviset CAP, 18028
Luviset CAP X, 18029
Luviskol®, 18030
Luviskol® K12, K17, K30, K60, 18031
Luviskol® K80, K90, 18032
Luviskol® VA 28 E, 18033
Luviskol® VAP 343 E, 18034
Luvisoft, 18035
Luvitol EHO, 18036
Luvitol HP, 18037
Luvitol HP, *See* hydrogenated polyisobutene, 14595
Luvitol® EHO, *See* Cetearyl octanoate, 5791
Luwax A, 18038
Luwax AF 30, 18039
Luwax AH 3, 18040
Luwax EAS 1, 18041
Luwax EVA 1, 18042
Luwax OA 2, 18043
Luwipal®, 18044 , *See* Melamine formaldehyde resin, 19256
Luxalloy, 18045
Luxate, 18046
Luxene, 18047
Luxene 44, 18048
Luxer®, 18049
Luxol, 18050
Luxon, *See* Haloxon, 13567
Luxor, 18051
Luxor® 1639, *See* disodium inosinate, 8645
Luzenac B170, 18052
Luzerne, 18053
Luzidol, 18054
LV, *See* Algin, 1591
LVC, *See* Algin, 1591
LX, 18055
LX-685® 125, 18056
LX-782®, 18057
lyargol, 18058
Lycadex®, 18059 , *See* Maltodextrin, 18500
Lycal®, 18060
Lycanol, 18061
Lycasin, 18062
Lycedan, *See* adenosine monophosphate, 680, 5'-adenylic Acid, 683
lycetol, 18063
lycine, 18064
Lycopon, 18065
Lycra®, 18066
Lycresse®, 18067
Lyddite, 18068
lyddite, *See* reflorit, 26003
Lydian stone, 18069
Lye, *See* potassium hydroxide, 24756, sodium hydroxide, 28772
Lye glycerin, 18070
lye, caustic, *See* sodium hydroxide, 28772
Lynex®, 18071
Lynite, 18072
Lynx, 18073, *See* Raxil, 25908
Lyocol, 18074
Lyofix, 18075
Lyofix 363, 18076
Lyofix F, 18077
Lyogen, 18078
Lyonore, 18079
Lyons sugar, 18080, *See* sucramine, 29928
Lyphocin, *See* Vancocin Hydrochloride, 33305
Lypsyl, 18081
Lyracamine, 18082
Lyric®, Lyril®, 18083
Lys, *See* lysine, 18088
Lysargine, 18084
Lysase, 18085
Lyse S, S III, S III Diff, 18086
Lysergic acid diethylamide, *See* LSD, 17704
lysidine, 18087
lysine, 18088
Lysine hydrochloride, *See* Enisyl, 10583, L-(+)-lysine hydrochloride, 18089
dl-lysine monohydrochloride, 18090
L-(+)-lysine hydrochloride, 18089
L-Lysine monohydrochloride, *See* L-(+)-lysine hydrochloride, 18089
L-lysine, *See* lysine, 18088
Lysitol, *See* Lysol, 18096
Lysivane, 18091
Lysmeral®, 18092
Lysochlor, 18093
Lysodren, 18094
Lysoform, 18095
Lysol, 18096
Lyteca, *See* Valadol, 33238
Lythol Oil, 18097
Lytor & RM, 18098
Lytron, 18099
Lytron 810, *See* maleic anhydride, 18476
Lytron 820, *See* maleic anhydride, 18476
M 3, *See* Zoldine® ZT-55, 34887
M 3 (heterocycle), *See* Zoldine® ZT-55, 34887
M33, MN3, 18108
M50, 18109
M66, 18110
M 73, *See* clonitrilide, 6432
M-81, *See* thiometon, 31700
M100, 18107
M 176, *See* DMSO, 8762
M 3196, *See* Chlorpyrifos-methyl, 6164
M9030, 18111

Chemical and Trade Names

M B T S, 18100
M Velpar, *See* diuron, 8736
M Violet 112, 18101
M&B 2878, *See* 2,4-DB, 7810
M&B 10064, *See* Bromoxynil, 4590
M&B 38544, *See* Diflufenican, 8451
M.B, 18102
M.B.A, 18103
M.F.C, 18104
M.N.T, 18105
M.O.D, 18106
MA, *See* maleic anhydride, 18476
αsol MA, *See* Bis(1-methylamyl) Sodium Sulfosuccinate, 4227
MA 20, 18112
Maali resin, 18113
Macadamite, 18114
Macaja butter, 18115, *See* mocaya oil, 20008
Macaloid, 18116
Macassar nutmeg butter, 18117
Macassar oil, *See* kon oil, 16321
mace butter, 18118
Mace Butter, American, *See* otoba butter, 22375
Maceal, 18119
Macerase®, 18120
MacFarland's alloy, 18121
Macgill metal, 18122
Machacon juice, 18123
Machete, 18124, *See* Butachlor, 4727
machine bronze, 18125
Mach's metal, 18126
Macht's metal, 18127
M-acid, 18128
Mackalene™ 116, *See* Cocamidopropyl dimethylamine lactate, 6553
Mackalene™ 416, *See* isostearamidopropyl dimethylamine lactate, 15420
Mackam 1C, 18129
Mackam 1L, 18130
Mackam 2C, 18131
Mackam 2CT, 18132
Mackam 2CY, 18133
Mackam 2L, 18134
Mackam 2W, 18135
Mackam 35, 18136
Mackam 151C, 18137
Mackam 151L, 18138
Mackam 160C, 18139
Mackam CB-35, 18140
Mackam CET, 18141
Mackam CSF, 18142
Mackam HV, 18143
Mackam ISA, 18144
Mackam J, 18145
Mackam LMB, 18146
Mackam MLT, 18147
Mackam OB-30, 18148
Mackam RA, 18149
Mackam TM, 18150
Mackam WGB, 18151
Mackam™, *See* Empigen® CDR40, 10369
Mackam™ 2C, *See* disodium cocoamphodiacetate, 8643
Mackam™ 35, 35HP, J, L, *See* Cocamidopropyl betaine, 6551
Mackam™ HV, *See* Oleamidopropyl betaine, 22108
Mackam™ LMB-LS, *See* lauramidopropyl betaine, 16830
Mackam™ OB-30, *See* oleyl betaine, 22132
Mackamide™ AME-75, *See* Acetamide MEA, 283
Mackamide AME-75, AME-100, 18152
Mackamide C, 18153
Mackamide CMA, 18154
Mackamide ISA, 18155
Mackamide L10, 18156
Mackamide LME, 18157
Mackamide LMM, 18158
Mackamide LOL, 18159
Mackamide MO, 18160
Mackamide ODM, 18161
Mackamide OP, 18162
Mackamide PK, 18163
Mackamide PKM, 18164
Mackamide R, 18165
Mackamide S, 18166
Mackamide SMA, 18167
Mackamide™ LME, *See* Lactamide MEA, 16568
Mackamine CAO, 18168
Mackamine CO, 18169
Mackamine IAO, 18170
Mackamine ISMO, 18171
Mackamine LAO, 18172
Mackamine LO, 18173
Mackamine O2, 18174
Mackamine OAO, 18175
Mackamine SAO, 18176
Mackamine SO, 18177
Mackamine WGO, 18178
Mackamine™ CO, *See* Cocamine oxide, 6557
Mackamine™ LO, *See* lauramine oxide, 16832
Mackamine™ OAO, *See* oleamidopropylamine oxide, 22110
Mackanate A-102, 18179
Mackanate A-103, 18180
Mackanate AY-65TD, 18181
Mackanate CM, 18182
Mackanate CP, 18183
Mackanate DC-30, 18184
Mackanate DC-30A, 18185
Mackanate DOS-40, 18186
Mackanate EL, 18187
Mackanate LA, 18188
Mackanate LM-40, 18189
Mackanate LM-40, *See* disodium lauramido MEA-sulfosuccinate, 8647
Mackanate LO, 18190
Mackanate O-3, 18191
Mackanate OD-35, 18192
Mackanate OM, 18193
Mackanate OP, 18194
Mackanate OP, *See* disodium oleamido MIPA sulfosuccinate, 8651
Mackanate RM, 18195
Mackanate UM, 18196
Mackanate WGD, 18197
Mackanate™ SL3, *See* disodium lauryl sulfosuccinate, 8649
Mackanate™ DOS-75, *See* dioctyl sodium sulfosuccinate, 8550
Mackazoline C, 18198
Mackazoline CY, 18199
Mackazoline L, 18200, *See* lauryl hydroxyethyl imidazoline, 16865
Mackazoline O, 18201
Mackechnie, 18202
Mackechnie's bronze, 18203
Mackenzie's amalgam, 18204
Mackenzie's metal, 18205
Mackernium 006, 18206
Mackernium 007, 18207
Mackernium KP, 18208
Mackernium NLE, 18209
Mackernium SDC-25, 18210
Mackester EGMS, 18211
Mackester IDO, 18212
Mackester IP, 18213
Mackester TD-88, 18214
Mackgard DM, *See* DMDM hydantoin, 8754, Glydant®, 13140, Mackstat® DM, 18234, Nipaguard® DMDMH, 21269
Mackham WGB, *See* Wheat germamidopropyl betaine, 34350
Mackine 101, 18215, *See* Cocamidopropyl dimethylamine, 6552
Mackine 201, 18216
Mackine 301, 18217
Mackine 321, 18218
Mackine 401, 18219
Mackine 421, 18220
Mackine 501, 18221
Mackine 601, 18222
Mackine 701, 18223
Mackine 801, 18224, *See* lauramidopropyl dimethylamine, 16831
Mackine 901, 18225
Mackine™ 501, *See* oleamidopropyl dimethylamine, 22109
Mackine® 601, *See* Behenamidopropyl dimethylamine, 3896
Mackpro KLP, 18226
Mackpro MLP, 18227
Mackpro NLP, NLP-Special, 18228
Mackpro NLW, 18229
Mackpro NSP, 18230
Mackpro SLP, 18231
Mackpro WWP, 18232
Mack's cement, 18233
Mackstat® DM, 18234
Mackstat® DM, *See* DMDM hydantoin, 8754
maclurin, 18235
Maclurin, *See* maclurin, 18235
Macol® 1, 18236
Macol® 2, 18237
Macol® 4, 18238
Macol® 8, 18239
Macol® 15, 18240
Macol® 16, 18241
Macol® 18, 18242
Macol® 19, 18243
Macol® 23, 18244
Macol® 27, 18245
Macol® 33, 18246
Macol® 34, 18247
Macol® 35, 18248
Macol® 40, 18249
Macol® 42, 18250
Macol® 44, 18251
Macol® 46, 18252
Macol® 72, 18253
Macol® 77, 18254
Macol® 85, 18255
Macol® 101, 18256
Macol® 108, 18257
Macol® 123, 18258
Macol® 125, 18259
Macol® 300, 18260
Macol® 660, 18261
Macol® CA-2, 18262
Macol® CPS, 18263
Macol® CSA-2, 18264
Macol® DNP-5, 18265, *See* Ethal DNP-8, 10919
Macol® E-200, 18266
Macol® LA-4, 18267
Macol® LF-110, 18268
Macol® NP-4, 18269
Macol® OA-2, 18270, *See* oleth-8, 22126
Macol® OP-3, 18271
Macol® P-500, 18272
Macol® SA-2, 18273, *See* steareth series, 29596
Macol® TD-3, 18274, *See* Trideceth series, 32232
Macol® WSL-2000, 18275
Macor, 18276
Macquer's salt, 18277
Macrogol, *See* Alkapol PEG 300, 1706
Macrogol laurate 600, *See* Ablunol 200ML, 95
macrogol nonylphenyl ether, *See* Alkasurf® NP-4, 1715, Triton® N-57, 32430
macrogol stearate, 18278
macrogols, 18279
Macrolex®, 18280
Macrolide antibiotic, *See* Leucomycin®, 17025, Trubin, 32485
Macromite, 18281
macrose, *See* dextran, 8233, 8234
Macrosorb, 18282
Macrospherical® 95, 18283
Macrynal, 18284
Maculanin, 18285

Macuprax, 18286
Madanite, 18287
Madar fiber, 18288
madder, 18289
Maddrell salts, 18290
Madecassol, 18291
madhurin, *See* saccharin sodium, 27196
madol oil, 18292
Madquat Q-6, 18293
Madurit®, 18294
Mafe®, *See* aluminum oxide, 1919
Mafloc® 700, 18295
Mafloc® 718, 18296
Mafloc® 764, 18297
Mafo® 13, 18298
Mafo® C, 18299
Mafo® CB 40, 18300
Mafo® CSB, 18301
Mafo® LMAB, 18302
Mafo® OB, 18303
Maftec®, 18304
Mafu, *See* dichlorvos, 8394
Mafu Strip, *See* dichlorvos, 8394
Mafu®, 18305
Mafura fat, 18306
Mag-40, 18307
Magadi soda, 18308
Magala® 0.5E, 18309
Magcal, *See* Ken-Mag®, 16007
MagChem® 10B, 18310
MagChem® 20, 18311
MagChem® 1060, 18312
Magcoke, 18313
Magecol, 18314
Magenta, *See* pararosaniline, 22798
Magicote Masonry Paint, 18315
Magicote Non Drip and Liquid Gloss, 18316
Magicote Solid Emulsion, 18317
Magicote Vinyl Matt, 18318
Magicote Vinyl Silk, 18319
magister of bismuth, 18320
magister of sulfur, 18321
magistery of lead, *See* white lead, 34366
magistral, 18322
Magistry of bismuth, *See* Bismuth subnitrate, 4243
Maglite, *See* Ken-Mag®, 16007
Maglite Y, 18323
Maglite® D, 18324
Magmet, 18325
Magna A, 18326
Magna Flow, 18327
Magna Tac, 18328
Magnabrite® F, 18329
Magnacell, 18330
Magnacide, 18331
Magnaclean, 18332
Magnaclear, 18333
Magnacryl, 18334
Magnaflo, 18335
Magnafloc®, 18336
Magnafloc® LT, 18337
Magnakyd, 18338
magnalium, 18339
Magnamite, 18340
Magnaphoscal®, 18341
Magnaplas PA213-BF83, 18342
Magnaset, 18343
Magnasoft, 18344
Magnasol, 18345
Magnasorb, 18346
Magnasperse, 18347
Magnaspheres®, 18348
Magnate, 18349, *See* imazalil, 14892
Magnatreat, 18350
magnesia, 18351, *See* magnesium oxide, Ken-Mag®, 16007
magnesia alba, 18352
magnesia alumina spinel, *See* M66, 18110
magnesia bleaching liquid, 18353
magnesia magma, *See* magnesium hydroxide,
magnesia usta, *See* Ken-Mag®, 16007
magnesia usta, *See*
magnesia white, 18354
magnesia-citrate mixture, 18355
magnesite, *See* Elastocarb Tech Light, Heavy, 9691
magnesium, 18356
magnesium acetate, 18357
magnesium acrylate copolymer, *See* AClyn®246A, 360
magnesium aluminate, *See* Dynaspinell, 9392
magnesium aluminum carbonate, *See* Alcamizer 1, 1409, Alcamizer 2, 1410
magnesium aluminum silicate, *See* Attaclay, 3399
magnesium aluminum silicate NF, *See* Magnabrite® F, 18329
magnesium base alloys, 18358
magnesium borate, *See* Antifungin, 2554
magnesium carbonate, 18359, *See* Elastocarb Tech Light, Heavy, 9691, Elastocarb UF, 9692, Magocarb-33, 18404
magnesium Carbonate Basic, *See* Elastocarb Tech Light, Heavy, 9691
magnesium carbonate hydroxide, *See* Elastocarb Tech Light, Heavy, 9691, magnesia alba, 18352
magnesium(II) carbonate(1:1), *See* magnesium carbonate, 18359
magnesium chloride, 18360
magnesium chloride anhydrous, *See* magnesium chloride, 18360
magnesium dioxide, *See* Novozone®, 21745
magnesium fatty alcohol sulfate, *See* Stepanol Mg, 29686
magnesium fluoride, 18361, *See* Magtran, 18412
magnesium gluconate, 18362, *See* Almora, 1790 Gluconal® MG, 13044
magnesium D-gluconate, *See* magnesium gluconate, 18362
magnesium hydrate, *See* magnesium hydroxide, Versamag® DC, 33607
magnesium hydrate, Marinco H, *See* Zerogen® 10, Zerogen® 60, 34733
magnesium hydrogen phosphate, *See* magnesium phosphate, dibasic,
magnesium hydroxide, *See* FR-20, 12352, Gilumag, 12944, Magnifin® H7A, 18377, Magnifin® H10, 18378, Magnifin® H10C, 18379, Magoh-S, 18405, MGH-93, 19622, Mylanta, 20564, Versamag® DC, 33607, Zerogen® 10, Zerogen® 60, 34733
magnesium laureth sulfate, 18363, *See* Empicol® EGB, EGC, 10315, Texapon MG, 31465
magnesium laureth-11 carboxylate, *See* Akypo® Soft 100 MgV, 1250
magnesium lauryl ether sulfate, *See* magnesium laureth sulfate, 18363
magnesium lauryl ether sulfate (2 EO), *See* Zoharpon MgES, 34871
magnesium lauryl sulfate, 18364, *See* Akyposal MGLS, 1309, Carsonol® MLS, 5335, Empicol® ML 26/F, 10336, Rhodapon® LM, 26637, Standapol® MG, 29475, Stepanol® MG, 29696, Sulfetal MG 30, 29986, Sulfochem MG, 30008, Zoharpon MgS, 34872, Standapol® MG, 29475, Lumorol K 28, 17814
magnesium monododecyl sulfate, *See* magnesium lauryl sulfate, 18364
magnesium nitrate, 18365, *See* Magnisal, 18380
magnesium(II) nitrate, *See* magnesium nitrate, 18365
magnesium nitrate hexahydrate, *See* Magnisal, 18380
magnesium(II) nitrate hexahydrate, *See* Magnisal, 18380
magnesium octadecanoate, *See* magnesium stearate, 18367
magnesium oxide, *See* Burnt magnesia, 4714, Corox, 6846, Elastomag® 100, 9704, Encapsulated MgO, 10538, Flamarret®, 11872, Insulmag®, 15152, Ken-Mag®, 16007, MagChem® 10B, 18310, MagChem® 20, 18311, MagChem® 1060, 18312, Maglite® D, 18324, magnesia, 18351, Magox® 98HR, 18407, Magox® Super Premium, 18408, Scorchguard O, 27754, Scorchguard-bound, 27755, Anscor®, 2489
magnesium oxide sticks, *See* Magrods, 18410
magnesium oxide-zinc oxide, *See* Garomix, 12677
magnesium oxychloride, *See* magnesia bleaching liquid, 18353, Magnocid, 18385
magnesium perchlorate, *See* Anhydrone, 2460
magnesium perchlorate trihydrate, *See* Dehydrite®, 7963
magnesium peroxide, *See* Novozone®, 21745
magnesium phosphate, secondary, *See* magnesium phosphate, dibasic,
magnesium propionate, *See* Lupromag®, 17917
magnesium ricinoleate, *See* Maricol, 18716
magnesium silicate, 18366, *See* Celkate T-21, 5592, Stellar 500, 29638
magnesium silicate, hydrated, *See* Silverline 665, 28439
magnesium silicate, hydrous, *See* Vantalc®, 33368, Vertal 92, 200, 33632
magnesium stearate, 18367
magnesium stearate, *See* Afco-Chem MGS, 923, stearopodis, 29602, Synpro® 90, 30591
magnesium stearate NF, *See* Petrac® MG-20 NF, 23466
magnesium sulfate, 18368, *See* magnesium sulfate, 18368, sal amarum, 27243, sel d'Angleterre, 27850
magnesium sulfonate, 18369
magnesium xylene sulfonate, *See* Eltesol® MGX, 9867
magnesium-Monel, 18370
magnesium-perhydrol, *See* Biogen,
Magnesol, 18371
magnet steel, 18372
Magnetic 6, *See* Thiovit, 31724
Magnetic Black, 18373
magnetic iron ore, 18374
magnetic oxide of iron, *See* Iron ore A, hematite type, 15318
magnetic pyrites, 18375
magnetic sulfide of iron, *See* Iron ore A, hematite type, 15318, magnetic pyrites, 18375
magnetite, *See* Iron ore A, hematite type, 15318, magnetic iron ore, 18374
Magnevist, 18376
Magnifin® H10, 18378
Magnifin® H10C, 18379
Magnifin® H7A, 18377
Magnisal, 18380
Magno, 18381
Magnobond 3, 18382
Magnobond 6504, 18383
Magno-Ceram, 18384
Magnocid, 18385
magnodat, *See* Biogen,
Magnogene, *See* magnesium chloride, 18360
magnolia metal, 18386
Magnolium, 18387
Magnoloop I, 18388
Magnox, 18389
Magnum, 18390, 18391
Magnum 240, 18392, *See* Acrylonitrile-butadiene-styrene, 439
Magnum 275, 18393, *See* Acrylonitrile-butadiene-styrene, 439
Magnum 445 HQ, 18394, *See* Acrylonitrile-butadiene-styrene, 439
Magnum 788HP, 18395, *See* Acrylonitrile-butadiene-styrene, 439
Magnum 2610, 18396, *See* Acrylonitrile-butadiene-styrene, 439
Magnum 3661, 18397, *See* Acrylonitrile-butadiene-styrene, 439
Magnum 4420, 18398, *See* Acrylonitrile-butadiene-styrene, 439
Magnum 9450P, 18399, *See* Acrylonitrile-butadiene-styrene, 439
Magnum FG960, 18400, *See* Acrylonitrile-butadiene-styrene, 439
Magnum® F, 18401
Magnuminium, 18402
Magnum-White, 18403

Magocarb-33, 18404, *See* magnesium carbonate, 18359
Magoh-S, 18405
Magotex, 18406
Magox® 98HR, 18407
Magox® Super Premium, 18408
Magrex, 18409
Magrods, 18410
Magspa, 18411
Magtran, 18412
mahogany acid, 18413
mahogany brown, 18414
Maillechort, 18415
Maincote, 18416
Mainstay, *See* Atlas Tanked Cod Oil, 3336
Mainstay®, 18417
Maitac, *See* Taktic, 30729
Maize gluten amino acids-sodium chloride, *See* Amino Gluten MG, 2166
maize oil, *See* corn oil, 6834
Maizena, 18418
Maizina, *See* atrazine, 3394
Maizolith, 18419
Majamin, 18420
Majammonium, 18421
majolica, 18422
majunga noir, 18423
MAK, *See* methyl n-amyl ketone, 19552
Makalot, 18424
Maki, *See* bromadiolone, 4565, Rentokil Deadline, 26116
Maklurin, *See* maclurin, 18235
Makon, *See* Alkasurf® NP-4, 1715, Triton® N-57, 32430
Makon NP6, NP10 and 4, 8, 12, 14 and 30, 18425
Makon OP6 and OP9, 18426
Makon® 4, 18427
Makon® 8240, 18428
Makon® NI10, NI20, NI30, 18429
Makon® OP-6, 18430
Makroblend®, 18431
Makroblend® DP4-1368, 18432
Makroblend® UT 400, 18433
Makrofol®, 18434
Makrofol® BL 2-2, 18435
Makrofol® DE 1-1, 18436
Makrofol® FR 6-2, 18437
Makrolon®, 18438
Makrolon® 1006 Tint, 18439
Makrolon® 1143, 18440
Makrolon® 2400, 18441
Makrolon® 2600, 2800, 18442
Makrolon® 3100, 3200, 18443
Makrolon® 3208, 18444
Makrolon® 6355, 6455, 18445
Makrolon® 8325, 18446
Makrolon® AL-2647 1068 Tint, 18447
Makrolon® FCR-2405, FCR-2407, FCR-2458, 18448
Makrolon® GV, 18449
Makrolon® HMS-3118, 18450
Makrolon® LQ-2847, LQ-3147, LQ-3187, 18451
Makrolon® LTG-3123, 18452
Makrolon® Rx-2530, 18453
Makrolon® SF-600, 18454
Maktion, 18455
malabar tallow, 18456, *See* piney tallow, 23861
malacca primers, 18457
malachite, 18458, *See* copper green, 6782, cupric carbonate, 7378
Malacide, *See* malathion, 18460
Malagran, *See* malathion, 18460
Malamar, *See* malathion, 18460
Malapaho, 18459
Malaspray, *See* malathion, 18460
Malathion, 18460, *See* Fyfanon, 12533
Malathion 60, 18461
Malathion Dust, 18462
Malathion Liquid, 18463
Malay camphor, *See* DL-Borneol, 4428
Malazide, 18464, *See* maleic hydrazide, 18477
Maldene, 18465
Maldene 285, 18466
Maldene 286, 18467
Maldene 288, 18468
Maldene 289, 18469
Maldene 292, 18470
Maldene 293, 18471
Maldene 300, 18472
Maldene 631, 18473
Maldison, *See* malathion, 18460
male fern, 18474
maleic acid, 18475
maleic acid anhydride, *See* maleic anhydride, 18476
maleic acid hydrazide, *See* maleic hydrazide, 18477
Maleic acid/olefin copolymer, sodium salt, *See* Acusol® 460ND, 586
maleic anhydride, 18476
maleic hydrazide, 18477, *See* Bos MH, 4456, Burtolin, 4722, Chiltern Fazor, 6067, Malazide, 18464, MSS MH18, 20433, Regulox K, 26020, Royal Slo-Gro, 27034
maleinic acid, *See* maleic acid, 18475
Malenite, 18478
malethamer, 18479
Malezafin 55 Plus, 18480
Malezafin 57 LV, 18481
Malezafin LV-4, 18482
Malic Acid, 18483, *See* apple acid, 2667 hydroxysuccinic acid, 14671
Malipuran, *See* bufexamac, 4655
Malix, *See* thiodan, 31680
malladrite, 18484
malleable iron, 18485
malleable nickel, 18486
Mallebrein, 18487
Mallebrin, *See* aluminum chlorate, 1901
Mallet alloy, 18488
Mallet bark, 18489
Malloydium, 18490
malonic acid, 18491
malonic methyl ester nitrile, *See* methyl cyanoacetate, 19524
Malonoben, 18492
Maloran, 18493
Malotte's alloy, 18494
MALP, *See* aluminum phosphate, 1922
Malros, 18495
malt sugar, *See* maltose, 18501
maltase, *See* glucase, 13030
maltha, 18496
malthactite, 18497
Malthenes, *See* petrolenes, 23496
malthite, 18498
Maltisorb, 18499
maltobiose, *See* maltose, 18501
Maltodextrin, 18500, *See* Lycadex®, 18059, Maltodextrin, 18500, Maltrin® M040, 18502, Microduct®, 19670, Mor-Rex® I-920, 20348, Wickenol® 550, 34403
Maltodextrin 24DE, *See* Wickenol® 550, 34403
Maltodextrin I, *See* Wickenol® 550, 34403
Maltol, *See* Veltol, 33544
maltonic acid, *See* Gluconic Acid, 13048
Maltos, *See* maltose, 18501
maltose, 18501
Maltrin® M040, 18502
Maltrin® M040, M510, *See* Maltodextrin, 18500
Maltrin® M200, 18503
Maltrin® M510, 18504
Maltrin® QD M600, 18505
maltyl, 18506
maltzyme, 18507
maluminum, 18508
Malzid, *See* maleic hydrazide, 18477
Mammacillin, *See* Penicillin V, 23111
man oil, *See* bone oil, 4402
Manal, 18509
Manalox, 18510
Manalox AG, 18511
Manalox AS, 18512
Manam, *See* Maneb, 18526
Mancarb Plus, 18513
Manchem, 18514
Mancobride Mancanese, 18515
Mancona bark, *See* red water bark, 25973
Mancopper, 18516
mancozeb, 18517, *See* Dithane, 8728, Karamate, 15749
mancozebe, *See* mancozeb, 18517
mancozeb-metalaxyl, *See* Fubol, 12429
Mancozin, *See* mancozeb, 18517
mandelic acid, 18518
Mandelin's reagent, 18519
Mandops Barleyquat B, 18520
Mandops Bettaquat B, 18521
Mandops Halloween, Hele Stone, 18522
Mandops Narsty, 18523
Mandops Podquaternary, 18524
Mandops Spring Poquaternary, 18525
mandrake root, *See* vegetable calomel, 33515
Maneb, 18526, *See* Campbell's X-Spor, 5075, Headland Spirit, 13734
Maneb 80, *See* Maneb, 18526
Maneba, *See* Maneb, 18526
maneb-cupric sulfate, *See* Cupertine, 7362
manebe, *See* Maneb, 18526
Maneb-R, *See* Maneb, 18526
Maneb-tridemorph-carbendazim, *See* Cosmic® FL, 6873
Manex, 18527, *See* Maneb, 18526
Manfloc, 18528
Mangabeira rubber, 18529
Mangal, 18530
mangaloy, 18531
manganaluminum bronze, 18532
Manganar, 18533
manganated linseed oil, 18534
manganese, 18535, *See* Dienol, 8430, Tripart® Liquid manganese, 32377, Tronamang Electrolytic manganese Metal, 32463
manganese acetate, 18536, *See* Manal, 18509
manganese arsenate, *See* manganar, 18533
manganese binoxide, *See* KM manganese Dioxide, 16238, manganese dioxide, 18541
manganese bister, *See* Mangatrace, 18559
manganese black, *See* manganese dioxide, 18541
manganese borate, *See* siccative, 28200
manganese boron, 18537
manganese brass, 18538
manganese bronze, 18539
manganese(II) carbonate, *See* manganese white, 18551
manganese chloride tetrahydrate, *See* Mangatrace, 18559
manganese cupro nickel, 18540
manganese dichloride, *See* Mangatrace, 18559
manganese dioxide, 18541, *See* glass-maker's soap, 12972, KM manganese Dioxide, 16238, Mangoxe, 18563
manganese ethylenbis(dithiocarbamate) (polymeric) complex with zinc salt, *See* Karamate, 15749
manganese German silver, 18542
manganese gluconate, 18543, *See* Gluconal® MN, 13045
manganese nickel, 18544
manganese nickel brass, 18545
manganese nickel silver, 18546
manganese ore A, 18547
manganese oxide, *See* manganese dioxide, 18541
manganese(IV) oxide, *See* KM manganese Dioxide, 16238
manganese peroxide, *See* manganese dioxide, 18541
manganese peroxide, *See* KM manganese Dioxide, 16238
manganese(II) phosphate, *See* manganese Violet, 18550
manganese sesquioxide, *See* manganite, 18555
manganese silver, *See* manganese German Silver, 18542
manganese steel, 18548
manganese sulfate, 18549, *See* Mansu, 18612
manganese(II) sulfate, *See* manganese sulfate, 18549

manganese superoxide, *See* KM manganese Dioxide, 16238
manganese tool steel containing some tungsten, *See* Air-hardening steel, 1113
manganese velvet brown, *See* umber, 32900
manganese violet, 18550
manganese violet and bismuthoxychloride, *See* M Violet 112, 18101
manganese white, 18551
manganese, [[1,2-ethanediylbis[carbamodithioato]](2-)]-, *See* Maneb, 18526
manganese, ethylenebisdithiocarbamato manganese, [ethylenebis[dithiocarbamato]]-, *See* Maneb, 18526
manganese-aluminum brass, 18552
manganese-zinc ethylenebis(dithiocarbamate), *See* mancozeb, 18517
manganic, 18553
Manganin, 18554
manganite, 18555
Mangan-Neusilber, 18556
Mangano Steel, 18557
mangano-titanium, 18558
manganous chloride, *See* Mangatrace, 18559
manganous sulfate, *See* manganese sulfate, 18549
Mangatrace, 18559
Mangnamite, 18560
Mangol, 18561
Mangonic, 18562
Mangoxe, 18563
Manguard®, 18564
Manhardt's Aluminum Bronze, 18565
Manjak, 18566
Mannheim gold, 18567
Mannitol, *See* hexanhexol, 14010, mushroom sugar, 20529
Mannolit, *See* Chloramine T, 6086
L-mannomethylose, *See* rhamnose, 26529
Mannose, *See* seminose, 27939
Manoblend, 18568
Manobond, 18569
Manocat, 18570
Manofast, 18571
Manofil, 18572
Manomet, 18573
Manosec, 18574
Manosil, 18575
Manosperse, 18576
Manox, 18577, 18578, *See* Maneb, 18526
Manoxol, 18579
Manoxol MA, 18580
Manoxol OT, OT/P and OT/B, 18581
Manoxol OT60, 18582
Manoxolot, 18583
Manplex, 18584
Manqueta, 18585
Manquta, *See* Manqueta, 18585
Manro, 18586
Manro ALS, 18587
Manro BA and NA, 18588
Manro BES, 18589
Manro D Paste, 18590
Manro DL28, 18591
Manro DS 35, 18592
Manro HA, 18593
Manro HCS, 18594
Manro KXS, 18595
Manro MA 35, 18596
Manro ML 33, 18597
Manro NEC, 18598
Manro NP, 18599
Manro PTSA, 18600
Manro SBS, 18601
Manro SDBS, 18602
Manro SIOS, 18603
Manro SLS28, 18604
Manro SLS45, 18605
Manro STS, 18606
Manro SXS, 18607, *See* sodium xylene sulfonate, 28832
Manro TDBS, 18608
Manro TL40, 18609
Manro XSA, 18610, *See* Xylene sulfonic acid, 34621
Manroteric CAB, *See* Cocamidopropyl betaine, 6551
Manroteric CDX38, CLV, CSH-32, *See* disodium cocoamphodiacetate, 8643
Mansil, *See* Vansil, 33363
Mansonil, *See* Mansonil®, 18611
Mansonil®, 18611
Mansu, 18612
Mantin, 18613
Mantrilon®, 18614
Mantrilon® FL, 18615
Manucol DH, DM, DMF, LB, *See* Algin, 1591
Manucol DM, 18616
Manucol Ester E/RK, 18617
Manucol Ester EX/LL, 18618
Manucol, Manucol DH, 18619
Manucreme, 18620
Manugel DJX, *See* Algin, 1591
Manugel DMB GHB, *See* Algin, 1591
Manugel PTJ, 18621
Manutex, 18622
Manzanate, 18623
Manzate, *See* Karamate, 15749
Manzate 200, *See* mancozeb, 18517
Manzate®, 18624
Manzate® 200 DF, 18625
manzeb, *See* Karamate, 15749, mancozeb, 18517
Manzi, *See* Maneb, 18526
Manzin, *See* Karamate, 15749, mancozeb, 18517
Man-Zox, *See* Maneb, 18526
Map$_4$, 18626
Mapeg® 200 DL, 18627
Mapeg® 200 DO, 18628, *See* PEG-2 dioleate series, 23040
Mapeg® 200 DOT, 18629, *See* PEG ditallate series, 23028
Mapeg® 200 DS, 18630
Mapeg® 200 ML, 18631
Mapeg® 200 MO, 18632
Mapeg® 200 MOT, 18633, *See* PEG tallate series, 23035
Mapeg® 200 MS, 18634
Mapeg® 200DL, *See* PEG-4 dilaurate series, 23050
Mapeg® 1500 MS, 18635
Mapeg® 1540 DS, 18636
Mapeg® 6000 DS, 18637
Mapeg® CO-16H, 18638
Mapeg® CO-5, 18639
Mapeg® DGLD, 18640
Mapeg® EGDS, 18641
Mapeg® EGMS, 18642
Mapeg® S-40, 18643
Mapeg® TAO-15, 18644, *See* PEG tallate series, 23035
Maphos® 17, 18645
Maphos® 33, 18646
Maphos® 60A, 18647
Maphos® 66H, 18648
Maphos® 78, 18649
Maphos® 8135, 18650
Maphos® FDEO, 18651
Maphos® JA 60, 18652
Maphos® L 13, 18653
Mapico, 18654
maple sugar sand, 18655
Mapo®, 18656
Maposol, *See* metam-sodium, 19444
Maprenal, 18657
Maprenal®, 18658
Maprofix 60S, *See* Witcolate ES-2, Witcolate LES-60C, 34477
Maprofix 60S and 60N, 18660
maprofix 563, *See* Rhodapon® 101-10, 26631, sodium lauryl sulfate, 28781
Maprofix 563 and LK.USP, 18659
Maprofix ES, *See* Witcolate ES-2, Witcolate LES-60C, 34477
Maprofix ES-2, 18661
Maprofix ESY, 18662
Maprofix MG, 18663
Maprofix NH, *See* Witcolate NH, 34478
Maprofix NH and NHL, 18664
Maprofix NH, NHL, *See* Ammonium lauryl sulfate, 2242
Maprofix TAS, 18665
Maprofix TLS, 18666, *See* Perlankrol® ATL40, 23324, Rhodapon® LT-6, 26638, Standapol® T, 29481, Witcolate TLS-500, 34479
Maprofix TLS-500, *See* Witcolate TLS-500, 34479
Maprofix TLS-65, *See* Witcolate TLS-500, 34479
Maprofix WA, WAC and WAQ, 18667
Maprofix WAC-LA and LCP, 18668
Mapromin, 18669
Mapron®, 18670
Maprosyl 30, *See* sodium lauroyl sarcosinate, 28779
Maprosyl® 30, 18671
maprotiline hydrochloride, *See* Ludiomil, 17774
Maquat LC-12S, MC-1412, MC-1416, MC-6025, *See* Benzalkonium chloride, 3958
Mar Rex 1918, *See* Wickenol® 550, 34403
Marabond 21, 18672
Marabout silk, 18673
Maracarb, 18674
Maracarb N-1, 18675
Maracell XE, 18676
Maracon, 18677
Maramul SS, 18678
Maranil, *See* Nacconol® 40G, 20647, sodium dodecylbenzenesulfonate, 28760
Maranil DBS, 18679
Maranil Powd. A, 18680
maranta, 18681
Maranyl®, 18682
Maranyl® A125, 18683
Maranyl® A175S, 18684
Maranyl® A360, 18685
Maranyl® TA505HS, 18686
Marasperse 52 CP, 18687
Marasperse GFC, 18688, *See* calcium lignosulfonates, 4957
Marasperse N-22, 18689
Marbalettes®, 18690
marble, *See* Calbux, 4929, calcium carbonate, 4941, CP Filler, 6928
Marble (OSHA), *See* CP Filler, 6928
Marbledust, 18691
Marbleloid, 18692
Marblemite, 18693
Marblette, 18694
Marbo, 18695
Marbolith, *See* Bakelite, 3589
Marbon B, 18696
Marbon Latex, 18697
Marbon Resins, 18698
Marbon White, 18699, *See* Lithopone, 17520
Marc brandy oil, *See* fusel oil, 12518
marcasite, 18700
marcasol, *See* markasol, 18732
marchies, *See* margines, 18710
Marchon® DC 1102, 18701
Marcoumar, *See* Phenprocoumon, 23630
marcs, 18702
Mareepa, 18703
Maretin, 18704, *See* Rametin, 25856
Marexine-CA, 18705
Marezzo marble, 18706
Marfanil-Prontalbin®, 18707
Margalite, 18708
Margarite, *See* mica, 19626
margarodite, 18709
margines, 18710
margol, 18711
margosa bark, 18712
Margosan-O, 18713
Margraff alloy, 18714
Mariazin® KC 21/50, 18715
Maricol, 18716
Marignac's salt, 18717
Marigold, *See* Calendula Oil CLR, 4980

Marinco C, *See* magnesia alba, 18352
marine acid, 18718, *See* hydrochloric acid, 14575
marine fiber, 18719
marine oil, 18720
marine plasma extract, 18721
marine salt, 18722
marine soap, 18723
Marisilan, *See* Ampicillin, 2361
Mark 80, 18724
Mark 829, *See* Wytox® 312, 34578
Mark 1178, *See* Wytox® 312, 34578
Mark® 1330, 18725
Mark® 1414, 18726
Mark® 2140, 18727
Mark® 4700, 18728
Mark® 5060, *See* Distearyl pentaerythritol diphosphite, 8712
Mark® 5089, 18729
Mark® 5095, 18730
Mark-A-Leak AW, 18731
markasol, 18732
Marksman, 18733
Markstat® AL-12, 18734
Markus alloy, *See* nickel silvers, 21132
Markwet NR-25, 18735
Markwet WL-12, 18736
Marlamid D 1218, *See* Active #2, 501
Marlamid M 18, *See* Ablumide SME, 82, Witcamide® 70, 34442
Marlamid®, 18737
Marlamid® A 18, 18738
Marlamid® D 1885, 18739
Marlamid® DF 1218, 18740
Marlamid® DF 1818, 18741
Marlamid® KL, 18742
Marlamid® KLP, 18743
Marlamid® M 1218, 18744
Marlamid® M 1618, 18745
Marlamid® PG 20, 18746
Marlate, *See* methoxychlor, 19504
Marlate 2-MR Emulsifiable Insecticide, 18747
Marlate 50 WP, 18748
Marlate 300 Flowable, 18749
Marlate 400 Flowable Concentrate, 18750
Marlate Methoxychlor Insecticide, 18751
Marlazin®, 18752
Marlazin® KC 30/50, 18753
Marlazin® L 10, 18754
Marlazin® OL 2, 18755
Marlazin® S 10, 18756
Marlazin® T 10, 18757
Marlex 1708, 18758
Marlex® BMN 55500, 18759
Marlex® BMN TR-880, 18760
Marlex® CL-50, 18763
Marlex® CL-100, 18761
Marlex® CL-200, 18762
Marlex® EHM 6003, 18764
Marlex® EHM 6006, 18765
Marlex® EHM 6007, 18766
Marlex® EMN TR-885, 18767
Marlex® ER9-0002, 18768
Marlex® ER9-0020, 18769
Marlex® HGH-050, 18770
Marlex® HGL-050-01, 18771
Marlex® HGL-050-01 (Antistatic), 18772
Marlex® HGL-120-01 (Antistatic), 18773
Marlex® HGL-200 (Antistatic), 18774
Marlex® HGL-350 (Antistatic), 18775
Marlex® HGN-020-01, 18776
Marlex® HGN-120-01 (Nucleated), 18777
Marlex® HGN-200 (Nucleated), 18778
Marlex® HGN-200A, 18779
Marlex® HGN-350 (Nucleated), 18780
Marlex® HGX-010, 18781
Marlex® HGX-030, 18782
Marlex® HGX-040, 18783
Marlex® HGX-330 (Controlled Rheology), 18784
Marlex® HGZ-050-02, 18785
Marlex® HGZ-120-02, 18786
Marlex® HGZ-120-04, 18787
Marlex® HGZ-200, 18788
Marlex® HGZ-350, 18789
Marlex® HHM 4903, 18790
Marlex® HHM 5202, 18791
Marlex® HHM TR-130, 18792
Marlex® HHM TR-140, 18793
Marlex® HHM TR-144, 18794
Marlex® HHM TR-210, 18795
Marlex® HHM TR-226, 18796
Marlex® HHM TR-230 Black, 18797
Marlex® HHM TR-232 Black, 18798
Marlex® HHM TR-250 Black, 18799
Marlex® HHM TR-400, 18800
Marlex® HHM TR-418 (Black, Orange), 18801
Marlex® HHM-4515, 18802
Marlex® HLM-020, 18803
Marlex® HLN-120-01, 18804
Marlex® HLN-200 (Antistatic, Nucleated), 18805
Marlex® HLN-350 (Antistatic, Nucleated), 18806
Marlex® HMN 4550, 18808
Marlex® HMN 5060, 18809
Marlex® HMN 54140, 18812
Marlex® HMN 5580, 18810
Marlex® HMN 6060, 18811
Marlex® HMN TR-942, 18813
Marlex® HMN-938, 18807
Marlex® HMX-020-01 (Lubricant), 18814
Marlex® HNS-080, 18815
Marlex® HXM 50100, 18816
Marlex® RGX-020, 18817
Marlex® RGX-020 (Antistat), 18818
Marlex® RMN-020C, 18820
Marlex® RMX-020, 18819
Marlex® TR.610, 18821
Marlex® TR.885, 18822
Marley Bitumen Paint Primer, 18823
Marley Carpet Cleaner, 18824
Marley Cement Accelerator, 18825
Marley Cement Colorant, 18826
Marley Cement Dustproofer, 18827
Marley Cement Plasticiser, 18828
Marley Cement Waterproofer, 18829, *See* Potassium oleate, 24764
Marley Cork Tile Adhesive, 18830
Marley Exterior Water Repellent, 18831
Marley Floor Cleaner, 18832
Marley Floor Gloss, 18833
Marley Floor Primer, 18834
Marley Homelay Adhesive, 18835
Marley Interior Waterproofer, 18836
Marley Mastic, 18837
Marley Patchit®, 18838
Marley Roofbond®, 18839
Marley Roofseal®, 18840
Marley Stick and Lift, 18841
Marley Superwax, 18842
Marley Universal Flooring Adhesive, 18843
Marleybond PVA, 18844
Marlican®, 18845
Marlie's alloy, 18846
Marlinat®, 18847
Marlinat® 242/28, 18848
Marlinat® CM 40, 18849
Marlinat® DF 8, 18850
Marlinat® DFK 30, 18851
Marlinat® DFL 40, 18852
Marlinat® DFN 30, 18853, *See* Ammonium lauryl sulfate, 2242
Marlinat® KT 50, 18854
Marlinat® SL 3/40, 18855
Marlinat® SRN 30, 18856
Marlipa® MG, 18857
Marlipal 1850, *See* Ablunol SA-7, 113
Marlipal®, 18858
Marlipal® 011/30, 18859
Marlipal® 013/20, 18860
Marlipal® 1/12, 18861
Marlipal® 1012/4, 18865, *See* Deceth-6, 7848
Marlipal® 124, 18864
Marlipal® 1618/6, 18866
Marlipal® 1850/5, 18867
Marlipal® 1850/5, *See* oleth-8, 22126
Marlipal® 24/20, 18862, *See* Laureth-2, 16852
Marlipal® 24/939, 18863
Marlipal® KF, 18868
Marlipal® SU, 18869
Marloid® CAS, 18870
Marlon 375a, *See* Nacconol® 40G, 20647
marlon 375a X 2073, *See* sodium dodecylbenzenesulfonate, 28760
Marlon A, *See* Nacconol® 40G, 20647, sodium dodecylbenzenesulfonate, 28760
Marlon A 350, *See* Nacconol® 40G, 20647, sodium dodecylbenzenesulfonate, 28760
Marlon A 375, *See* Nacconol® 40G, 20647, sodium dodecylbenzenesulfonate, 28760
Marlon as 3, *See* Witco® 1298, 34458
Marlon®, 18871
Marlon® A350, 18872
Marlon® A360, A365, A375, 18873
Marlon® A365, 18874
Marlon® A390, A396, ARL, 18875
Marlon® AMX, 18876
Marlon® ARL, 18877
Marlon® AS3, 18878
Marlon® PF 40, 18879
Marlon® PS 30, 18880
Marlophen®, 18881
Marlophen® 81N, 18882
Marlophen® 85, 18883
Marlophen® 810, 18884
Marlophen® 810N, 18885
Marlophen® 830N, 18886
Marlophen® DNP 16, 18887, *See* Ethal DNP-8, 10919
Marlophen® P 1, 18888
Marlophen® X, 18889
Marlophor, 18890
Marlophor® CS-Acid, 18891
Marlophor® DS-Acid, 18892
Marlophor® FC, 18893
Marlophor® HS-Acid, 18894
Marlophor® IH-Acid, 18895
Marlophor® LN-Acid, 18896
Marlophor® MD, 18897
Marlophor® MO 3-Acid, 18898
Marlophor® ND-Acid, 18899
Marlophor® T10-Acid, 18900
Marlopon, 18901
Marlopon® ADS 50, 18902
Marlopon® AMS 60, 18903
Marlopon® AT, 18904
Marlosoft® A 18 M, 18905
Marlosoft® IQ 75, 18906
Marlosoft® IQ 90, 18907
Marlosol®, 18908
Marlosol® 183, 18909
Marlosol® BS, 18910, *See* PEG-150 distearate series, 23037
Marlosol® FS, 18911, *See* PEG-2 dioleate series, 23040
Marlosol® OL2, 18912
Marlosol® R70, 18913
Marlosol® RF3, 18914
Marlosol® TF3, 18915, *See* PEG tallate series, 23035
Marlotherm, 18916
Marlowet®, 18917
Marlowet® 1072, 18918
Marlowet® 4536, 18919
Marlowet® 4538, 18920
Marlowet® 4539, 18921
Marlowet® 4702, 18922
Marlowet® 4800, 18923
Marlowet® 4900, 18924
Marlowet® 5311, 18925
Marlowet® 5324, 18926
Marlowet® 5400, 18927
Marlowet® 5440, 18928

Marlowet® 5459, 18929
Marlowet® BL, 18930
Marlowet® FOX, 18931
Marlowet® ISM, 18932
Marlowet® LVS, 18933
Marlowet® NF, 18934
Marlowet® OCM, 18935
Marlowet® PW, 18936
Marlowet® R 11, 18937
Marlowet® R 40, 18938
Marlowet® WOE, 18939
Marlowet® WOE, *See* oleth-8, 22126
Marlox®, 18940
Marlox® 3000, 18941
Marlox® FK 14, 18942
Marlox® FK 64, 18943
Marlox® L 6, 18944
Marlox® MO 124, 18945
Marlox® MS 48, 18946
marls, 18947
Marmelosin, *See* imperatorin, 14912
Marme's reagent, 18948
Marmite, 18949
Marmo Bardiglio de Bergamo, 18950
Maroxol-50, *See* Dinitra, 8540
Marphos, 18951
Marquat Pigments, 18952
Marseilles soap, 18953
marsh gas, 18954, *See* methane, 19475
Marshal, *See* carbosulfan, 5245
Marshal 10G, 18955
Marshal/suSCon, 18956
Marshal/suXon, 18957
Marshall 10G, *See* carbosulfan, 5245
marshite, *See* cuprous iodide, 7404
Marsipol, 18958
martensite, 18959
Martifin, 18960, *See* Alumina hydrate, 1892
Martin steel, 18961
Martinal, 18962, *See* Alumina hydrate, 1892
Martinal® OL-111 LE, 18963, *See* Alumina hydrate, 1892
Martinal® ON-4608, 18964, *See* Alumina hydrate, 1892
Martinal® OS, 18965, *See* Alumina hydrate, 1892
martinite, 18966
Martino's alloys, 18967
Martin's cement, 18968
Martipol, 18969, *See* aluminum oxide, 1919
Martisorb, 18970
Martos-10, *See* maltose, 18501
Martoxin, 18971
Marttisorb, *See* aluminum oxide, 1919
Marvaloy 750, 18972
Marvanbrite CF, 18973
Marvanfix® ATA, 18974
Marvanfix® C, 18975
Marvanlube® 92, 18976
Marvanlube® BHC, 18977
Marvanol® Aftertreat 2AF, 18978
Marvanol® BAN, 18979
Marvanol® Carrier BB, 18980, *See* Butyl benzoate, 4763, n-Butyl benzoate, 4790
Marvanol® Defoamer AM-2, 18981
Marvanol® GAW, 18982
Marvanol® LSL, 18983
Marvanol® Penetrant 35, 18984
Marvanol® Pretreat GD-P, 18985
Marvanol® RD2-1852, 18986
Marvanol® REAC A-213, 18987
Marvanol® SBO (60%), 18988
Marvanol® SCO (50%), 18989 , *See* Eureka 102, 11207
Marvanol® Scour 2 Base, 18990
Marvanol® SPO (60%), 18991
Marvanquest 1022, 18992
Marvanscour® KW, 18993
Marvanscour® LF, 18994
Marvansoft 1771, 18995
Marvantex RBDS, 18996
Marvex, *See* dichlorvos, 8394
Marvinol, *See* Korogel, 16354, Koroplate, 16357
Marvylan, 18997
Marweld M-17, 18998
Mary-bud, *See* Calendula Oil CLR, 4980
Mascot, *See* triflumuron, 32248
Mascot Clearing, 18999
Mascot Cloverkiller, 19000
Mascot Contact Turf Fungicide, 19001
Mascot Gauntlet, 19002
Mascot Highway, 19003
Mascot Moss Killer, 19004
Mascot Selective, 19005
Mascot Super Selective, 19006
Mascot Systemic Turf Fungicide, 19007
Mascot Ultrasonic, 19008
Masil® 1066C, 19013
Masil® 173, 19009
Masil® 264, 19010
Masil® 280, 19011
Masil® 756, 19012
Masil® EM 100, 19014
Masil® EM 100,000, 19016
Masil® EM 350, 19015
Masil® SF 5, 19018
Masil® SF 201, 19019
Masil® SF 500, 19020
Masil® SF 500,000, 19021
Masil® SF 1,000,000, 19017
Masil® SF-MH, 19022
Masil® SFR 70, 19023
Masil® SF-V, 19024
Maslip® 500, 19025
Masocare, 19026
Masodine, 19027
Masol, 19028
Masonry Stain and Seal, 19029
Masoten®, 19030
Masquol P 320, *See* Fostex AMP, 12337
Massa Estarinum, 19031
Massa Estarinum® CM, 19032
Massaranduba, 19033
massecuite, 19034
Massicot, *See* lead monoxide, 16910, litharge, 17501
Master Bond AC82, 19035
Master Bond EP11HT, 19036
Master Bond EP21HT, 19037
Master Bond EP30HT, 19038
Master Bond EP34CA, 19039
Master Bond EP75, 19040
Master Bond Supreme 11HT, 19041
Master Bond Supreme 3HT, 19042
Master Sil 701, 19043
Masterblok, 19044
Masterbond, 19045
Mastercarb, 19046
Mastercolor, 19047
Masterflam, 19048
Masterwood, 19049
mastic, 19050
Masticillin® C, M, 19051
Masuron, 19052
Matador, *See* λ-cyhalothrin, 7567
Matalex, 19053
matali, 19054
Match, 19055, *See* cyanazine, 7472
Mateflex, 19056
Mater-Bi®, 19057
Matexil, 19058
Mathesius Metal, 19059
matico-camphor, 19060
Matikus, 19061
MATO, *See* Mansonil®, 18611
Matrigon, *See* Clopyralid, 6436
Matrikerb, 19062, 19063
Matrimid® 5292 System, 19064
Matrix alloy, 19065
Matrix®, 19066
Matsuka alcohol, *See* Morillol®, 20335
Matt salt, 19067
Matte, *See* coarse metal, 6520
matteucinol, 19068
Mattheylec, 19069
Mattina®, 19070
maturex, 19071
maucherite, 19072
mawele, 19073
Maxahibit TT-50, 19074
Maxepa®, 19075
Maxhete, 19076
maxi braun®, 19077
Maxicrop Moss Killer & Conditioner, 19078, *See* ferric sulfate, 11652
Maxigard, 19079
Maxilon, 19080
Maxilvry Steel, 19081
Maxim, 19082
MaxiMate, 19083
Maximite, 19084, *See* Cordite, 6808
Maxim-Nordenfelt powder, *See* MN powder, 19994
maxium metal, 19085
Maxon, 19086
May apple, *See* vegetable calomel, 33515
Mayari iron, 19087
Mayari steel, 19088
Mayclene, 19089
Mayco Base 1351, 19090
Maydol, *See* corn oil, 6834
Mayer's albumen, 19091
Mayer's solution, 19092
Mayoquest 1320, *See* Fostex AMP, 12337
Mayphos 45, 19093
Maypon, 19094
Maypon 4C, 19095, *See* Foam-Coll 4C, 12150
Maypon 4CT, 19096, *See* Foam-Coll 4CT, 12151
Maypon UD, 19097, *See* Potassium undecylenoyl hydrolyzed collagen, 24788
maytee, 19098
May-Tein C, 19099, *See* Foam-Coll 4C, 12150
May-Tein CT, 19100, *See* Foam-Coll 4CT, 12151
May-Tein KK, 19101
May-Tein KT, 19102
May-Tein R, 19103
Mazak, 19104
Mazam, 19105
Mazamide® 68, 19106
Mazamide® 1214, 19107
Mazamide® C-2, 19108
Mazamide® CFAM, 19109
Mazamide® L-298, 19110
Mazamide® L-5, 19111
Mazamide® LLD, 19112
Mazamide® O 20, 19113
Mazamide® SMEA, 19114
Mazamide® SS 20, 19115
Mazawax® 163R, 19116
Mazawet® 36, 19117
Mazawet® DOSS 70, 19118
Mazclean EP, 19119
Mazeen® 173, 19120
Mazeen® C-2, 19121
Mazeen® DAPI, 19122
Mazeen® DAPL, 19123, *See* Cocamidopropyl dimethylamine, 6552
Mazeen® S-2, 19126, *See* PEG soyamine series, 23033
Mazeen® S-2, S-15, *See* PEG soyamine series, 23033
Mazeen® S-13, 19124
Mazeen® S-15, 19125, *See* PEG soyamine series, 23033
Mazeen® SHCFA, 19127, *See* Cocamidopropyl dimethylamine, 6552
Mazeen® T-2, 19128
Mazide, *See* maleic hydrazide, 18477
Mazide 25, 19129
Mazide Selective, 19130
Mazin®, 19131
Mazol® 80 MG K, 19132
Mazol® 165C, 19133
Mazol® 300 K, 19134
Mazol® 812, *See* caprylic/capric acid triglyceride, 5161

Mazol® 1400, 19135
Mazol® GMO, 19136
Mazol® GMR, 19137
Mazol® GMS, 19138
Mazol® PETO, 19139, *See* Pentaerythrityl tetraoleate, 23141
Mazol® PGMS, 19140
Mazol® PGO-104, 19141
Mazol® PGO-31 K, 19142
Mazola, *See* corn oil, 6834
Mazon® 18A, 19143
Mazon® 41, 19144
Mazon® 60T, 19145
Mazon® 85, 19146
Mazon® 1045A, 19147
Mazon® RI 4A, 19148
Mazox® CAPA, 19149
Mazox® CDA, 19150
Mazox® KCAO, 19151
Mazox® LDA, 19152, *See* lauramine oxide, 16832
Mazox® MDA, 19153
Mazox® ODA, 19154
Mazox® SDA, 19155
Maztreat® 246, 19156
Maztreat® BOM, 19157
Maztreat® CA Powd, 19158
Mazu® 10 P Mod 11, 19159
Mazu® 43 C, 19160
Mazu® 142, 19161
Mazu® 252, 19162
Mazu® 5118, 19163
Mazu® DF 243, 19167
Mazu® DF 100S, 19164
Mazu® DF 200SP, 19165
Mazu® DF 205SX, 19166
MB 450, 19168
MBA, *See* M.B.A, 18103, methylene bisacrylamide, 19567
MBC, *See* Carbendazim, 5187, 5188, Konker®, 16327
MBI, *See* Isonaphthol, 15377, MDI, 19188, Rubinate® LF-168, 27108
MBR-8251, *See* Perfluidone, 23248
MBR12325, *See* mefluidide, 19230
MBS, 19169
MBT, 19170, *See* Accelerator Mercapto, 207, 2-mercaptobenzothiazole, 19334, Thiotax, 31719
MBT, Captax, *See* Vulkacit® M, Vulkacit® Merkapto/C, 34134
MBT®, 19171
MBTS, 19172, *See* Perkacit® MBTS, 23286, Vulkacit® DM/C, 34131
MBTS rubber accelerator, *See* Perkacit® MBTS, 23286
MC 2508, 19173
MC®, 19174
MC-4379, *See* bifenox, 4135
MC580, 19175
MCA, *See* acetic acid, 292, monochloroacetic acid, 20221
McAlester EGDS, 19176
MCB, *See* Abluton T30, 139
McGill Metal, 19177
M-Cillin, *See* Penicillin G potassium, 23109, 23110
M-Clean D, 19178
McLube 1700, 19179
McLube 1777, 19180, *See* Polytetrafluoroethylene, 24642
McLube 1829, 19181
McNamee Clay®, 19182
MCP, *See* calcium phosphate (monobasic), 4966, MCPA, 19183
MCP ester, *See* MCPA, 19183
MCPA, 19183, *See* Agrichem MCPA-25, 50, 1048, Agricorn 500, 1050, Agritox 50, 1077, Agroxone 50, 1086, Albar-M, 1351, Atlas MCPA, 3321, BH MCPA 75, 4119, Empal, 10278, Farmon MCPA 50, 11487, FBC MCPA, 11516, Phenoxylene Plus, 23629, Star MCPA, 29525
2,4-MCPA, *See* MCPA, 19183
MCPA Ester, *See* MCPA, 19183
MCPA-MCPB, *See* Farmon MCPB Plus, 11488
MCPA-mecoprop, *See* Greenmaster Extra, 13343
MCPB, 19184, 19185 *See* Belmac Straight, 3913, Fisons 18-15, MCPB, 11850
2,4-MCPB, *See* MCPB, 19184
MCP-butyric, *See* MCPB, 19184
MCPP, *See* Astix, 3248, Verdone CDA, 33580
MCT, *See* Captex® 8000, 5170, Emalex O.T.G, 9976, Miglyol® 808, 19736
MD 50, 19186
MD 60, 19187
MDBA, *See* Dicamba, 8374
M-Det, *See* diethyl toluamide, 8438
MDI, 19188, *See* Isonate® 2125M, 15379
M-Diphar, *See* Maneb, 18526
MDM hydantoin, *See* Dantoin® MDMH, 7734
Me f248, *See* Rampart, 25860
MEA, *See* ethanolamine, 10930
MEA laureth (6) phosphate, *See* Crafol AP-262, 6957
MEA laureth sulfate, *See* Aremsol MA, 2835, Empicol® EMB, 10317
MEA lauryl sulfate, sodium PEG-6 cocamide carboxylate and disodium laureth sulfosuccinate, *See* Akypogene HM 8, 1257
MEA-laureth sulfate, *See* Akyposal MLES 35, 1310
MEA-lauryl sulfate, *See* Akyposal MLS 30, 1311, Empicol® EL, 10316, Empicol® LQ33/T, 10326, Rewopol® MLS 30, 26365, Sulfochem MLS, 30009, Zoharpon LAM, 34867
MEA-PPG-6-laureth-7 carboxylate, nonoxynol-2, *See* Akypogene SO, 1260
Mearl Film, 19189
Mearlin, 19190
Mearlin®, 19191
Mearlite® GBU, 19192
Mearlmaid® AA, 19193
Mearlmica® SVA, 19194
Measac, 19195
meat-sugar, 19196
MEB, *See* Maneb, 18526
Mebatreat, 19197, *See* Mebendazole, 19198
Mebatryne, *See* ametryn, 2051
Mebazine, *See* atrazine, 3394
Mebendazole, 19198, *See* Equivurm Plus, 10736, Vermox, 33591
Mebenvet, 19199, *See* Mebendazole, 19198, Telmin, 31066, Vermox, 33591
Mebrofenin, *See* Choletec, 6170
Mecadox, 19200, *See* carbadox, 5177
Mecarzole, *See* Carbendazim, 5187, 5188
Mecca balsam, 19201
Meccarb, 19202
mechlorprop, *See* Verdone CDA, 33580
Meco, 19203
Mecomec, *See* Mecoprop, 19204
meconium, *See* opium,
Mecopex, *See* Mecoprop, 19204
Mecoprop, 19204, *See* Chafer CMPP Super, 5845, Cleanacres CMPP, 6381, Clenecorn, 6403, Clifton CMPP Amine 60, 6416, Clovotox, 6449, Compitox Extra, 6680, FBC CMPP, 11514, Headland Charge, 13727, Herrifex DS, 13882, Hymec, 14685, Iso-Cornox, 15353, Verdone CDA, 33580
mecoprop-3,6-dichloropicolinic acid-ioxynil, *See* Harrier, 13652
mecoprop-P, *See* Duplosan New System CMPP, 9128, Duplosan® CMPP, 9130
Mecoprop-P, *See*
Mecpa, 19205
Mectizan, *See* Ivermectin, 15454
Mecufix, 19206
Meculon, 19207
Med Gel, 19208
medal bronze, 19209
medal metal, 19210
Medang Losoh oil, 19211
Medialan KA, 19212
Medialan KF, 19213
Medialan LD, 19214
Medialan LL-99, *See* sodium lauroyl sarcosinate, 28779, Vanseal® NALS-30, 33359, Zoharsyl L-30, 34876
Mediben, *See* Dicamba, 8374
Mediker, 19215
Medilave, *See* Acetoquat CPC, 302, cetyl pyridinium chloride, 5823
Medipham, *See* phenmedipham, 23595, 23596, 23597, 23598
Meditar, 19216
Medium 7, 19217
Medium 10, 19218
Medium VS, 19219
Medley®, 19220
Medo, 19221
medol, 19222
Medolit, 19223
Medrin, *See* thiometon, 31700
Medronic acid :ss99m:ksTc complex, *See* Amerscan MDP Kit, 2026
Medusa, 19224
Meehanite, 19225
Meena Harma, 19226
meetco, *See* methyl ethyl ketone, 19530
MEF® -LD, 19227
Mefarol®, 19228
mefenacet, 19229, *See* Hinochloa®, 14096
Mefenal, *See* sulfamethazine, 29962
mefluidide, 19230, *See* Echo, 9577, Embark, 10017
Mefranal, 19231
Mefrasol, 19232
Megacillin Tablets, *See* Penicillin G potassium, 23109
Meganite, 19233
Meganox Plus, 19234
Megapen, *See* Penicillin V, 23111
Megaperm 4510, 19235
Megaperm 6510, 19236
Megapoly, 19237
Megapren C 150, 19238
Megapren Si 10, 20, 30, and 60, 19239
Megapren U225, 19240
Megasil, 19241
Megilp, 19242
Meglum, *See* Tramisol®, 32134
Meglumine, 19243
meglumine amidotrizoate, *See* Reno-M, 26113
Meglumine Diatrizoate, *See* Angiovist 282, 2457, MD 60, 19187
Megomit, 19244
Megum, 19245
Mehltaumittel, *See* dodemorph-acetate, 8782
MEHQ, *See* hydroquinone monomethyl ether, 14633
MEK, *See* 2-butanone, 4737, Meketone, 19247, methyl ethyl ketone, 19530
MEK peroxide, *See* 2-butanone peroxide, 4738, Cadox® HBO-50, 4913, Cadox® L-30, 4914, Esperfoam® FR, 10842, methyl ethyl ketone peroxide, 19531, Quickset® Extra, 25672,
Mekad, 19246
Meketone, 19247
Mekon® White, 19248, *See* Microcrystalline wax, 19666
Mekor®, 19249
Mekor® 70, 19250
MEK-oxime, *See* methyl ethyl ketoxime, 19532
MEKP, *See* 2-butanone peroxide, 4738
Mekure T1, T2, 19251
MEL 80-P, 19252, *See* Melamine formaldehyde resin, 19256
Melacos, 19253
meladinin, *See* Oxsoralen-Ultra, 22437
Melafix DM, 19254
Melaleuca alternifolia oil, *See* EmCon TEA TREE, 10067
Melalith, 19255
Melamine formaldehyde resin, 19256
melamine resin, *See* melamine/formaldehyde resin, 19256
Melamit 200, 19257
melampyrin, *See* galactitol, 12592
Melampyrite, 19258, *See* galactitol, 12592

melampyrum, *See* galactitol, 12592
melaniline, 19259, *See* Akrochem® DPG, 1168, Dynamine, 9355, Perkacit® DPG, 23281, Vanax® DPG, 33290, Vulcaid DPG, 34081
melanoid, 19260
melanol cl, *See* Rhodapon® 101-10, 26631, sodium lauryl sulfate, 28781
melanol cl 30, *See* sodium lauryl sulfate, 28781
Melanol LP 20 T, *See* Perlankrol® ATL40, 23324, Rhodapon® LT-6, 26638, Standapol® T, 29481, Witcolate TLS-500, 34479
Melatix, 19261
Melax, 19262
Melclif®, 19263
melco, 19264
Melcril 4079, 19265
Melcril 4083, 19266
Melcril 4085, 19267
Melcril 4087, 19268
Melcril 5919, 19269
Meld, 19270
Meldin® 3000A, 19271
Meldin® 3000D, 19272
Meldin® 3000F, 19273
Meldin® 3000G, 19274
Meldin® 3000H, 19275
meletin, *See* quercetin, 25647
Melhi N, NS and NLM, 19276
Melibiase, 19277
Meligrin, 19278
Melilot, 19279
Melimax, 19280
melin, *See* rutin trihydrate, 27150
Melinar, 19281
Melinex®, 19282
Melinex® 393, 19283
Melinex® 505, 19284
Melinex® 994, 19285
Melinite, 19286, *See* Lyddite, 18068, reflorit, 26003
melinose, 19287
Meliodent®, 19288
melioform, 19289
Melioran, 19290
Melioran F6, 19291
Meliose, 19292
Melit, 19293
Melite®, 19294
melitose, *See* raffinose, 25835
Mellavax, 19295
Mellite, 19296
mellitriose, *See* raffinose, 25835
Mellol, 19297, *See* 2-phenylethyl alcohol, 23642
Melmac, 19298
Melmag®, 19299
Melmex, 19300
Melnox®, 19301
Melocol, 19302
Melolam, 19303
Melolanel, 19304
Melopas, 19305
Meloprufe®, 19306
Melox®, 19307
Meloxide®, 19308
Meloxine, *See* Oxsoralen-Ultra, 22437
Melprex, *See* Dodine FL, WP, 8785, Radspor FT, 65WP, 25833
Melpure®, 19309
Melrasal®, 19310
Melsprea®, 19311
Meltan®, 19312
Meltatox, *See* dodemorph-acetate, 8782
Meltatox®, 19313
Melted lead oxide granules, *See* Austrox, 3432
Meltron®, 19314
Melurac, 19315
Melweld®, 19316
Melwhite®, 19317
MEMA, *See* Panogen M, 22682
MEMC, *See* Agallol, 954, Atiran, 3286, Ceresan, 5745
Memilene, *See* methomyl, 19502
Memilite, *See* Randanite, 25864
MEMO, *See* Dow Corning® Z-6030, 8879, Prosil® 248, 25238
Memosil, 19318
menadiol sodium diphosphate, 19319, *See* Synkavit, 30548, 30549
menaquinone-6, *See* vitamin K_2, 33950
MENCS, *See* methyl isothiocyanate, 19543
menispermin, 19320
***p*-mentha-1,8-diene**, *See* limonene, 17280
***p*-mentha-6,8-dien-2-one**, *See* carvone, 5363
3-*p*-menthanol, *See* menthol terpine hydrate, 19321
$\Delta_{4,8}$-*p*-menthen-3-one, *See* pulegone, 25379
menthol, *See* Fancol Menthol, 11449
l-menthol, *See* menthol terpine hydrate, 19321
menthol terpine hydrate, 19321
menthyl anthranilate, *See* Neo Heliopan® MA, 20869
menthyl borate, *See* Estoral, 10896
Meobiquin, *See* Yodoxin, 34658
Meothrin, 19322, *See* fenpropathrin, 11615
Mephanac, *See* MCPA, 19183
Mephaneine, 19323
Mephetol, 19324
Mephetol Extra, 19325
mephosfolan, 19326, *See* Cytro-Lane, 7606
Mephyton, *See* Konakion, 16322, *See* vitamin K_1, 33949
Mephyton
mepiquat chloride, 19327, *See* Pix® ULV, 23898
Mepro, *See* Mecoprop, 19204
Mepron, 19328
mepyrapone, *See* metyrapone, 19608
mequinol, *See* Eastman® HQMME, 9467
mequinol (INN), *See* Eastman® HQMME, 9467
Mera industries 2MOM3B, *See* sodium hypochlorite, 28773
meralluride, 19329
Meraneine, 19330
Meranol, 19331
Merantine, 19332
Merbron R, 19333
2-mercaptoacetic acid, *See* thioglycolic acid, 31683
mercaptan C_{12}, *See* n-dodecyl mercaptan,
mercaptan C_3, *See* propyl mercaptan,
mercaptoacetic acid, *See* thioglycolic acid, 31688
2-mercaptoacetic acid, *See* thioglycolic acid, 31688
2-mercaptobenzimidazole, 19334, *See* Anti-Oxidant MB, 2589 2-mercaptobenzimidazole, 19334
mercaptobenzothiazole, 1933, *See* Accelerator Mercapto, 207 5
2-Mercaptobenzothiazole, *See* Captax, 5166, MBT, 19170, MBT®, 19171, mercaptobenzothiazole, 19335, Perkacit® MBT, 23285, Rokon, 26918, Rotax®, 26999, Thiotax, 31719, Vulkacit® M, 34133, Vulkacit® M, Vulkacit® Merkapto/C, 34134
Mercaptobenzothiazole, ammonium salt, *See* Accelerator Z88, 217
Mercaptobenzothiazole, zinc salt, *See* Octocure ZMBT-50, 21998, Oxaf, 22408, Zetax®, 34741
2-mercapto-benzothiazole, sodium, *See* Nacap®, 20643
2-mercaptobenzothiazole zinc salt, *See* Zetax®, 34741, ZMBT, 34842
mercaptobenzthiazyl ether, *See* Vulkacit® DM/C, 34131
mercaptodiacetic acid, *See* thiodiglycolic acid, 31683
mercaptodimethur, *See* methiocarb, 19484
mercaptoethanol, *See* 2-mercaptoethanol, 19336
2-mercaptoethanol, 19336, *See* Sipomer® 2ME, 28529
β-mercaptoethanol, *See* 2-mercaptoethanol, 19336
Mercaptoimidazoline, *See* Perkacit® ETU, 23283
2-mercaptoimidazoline, *See* ethylene thiourea, 11132, Perkacit® ETU, 23283
2-Mercapto-4(5)-methyl benzimidazole, Zinc salt, *See* Akrochem® Antioxidant 58, 1159
mercaptophos, *See* Tiguvon®, 31788
1-Mercaptopropane, *See* n-propyl mercaptan, 25228
1-mercaptopropane, *See* n-propyl mercaptan, 25227, 25229
3-mercaptopropane-1,2-diol, *See* Thiovanol®, 31723
γ-mercaptopropyl trimethoxy silane, *See* Aktisil MM, 1203, Prosil® 196, 25236, Union Carbide® A-189, 32989
γ-Methacryloxypropyl trimethoxy silane, *See* Prosil® 248, 25238
3-Mercaptopropyltrimethoxysilane, *See* CM8500, 6455
γ-mercaptopropyltrimethoxy silane, *See* Prosil® 196, 25236
3-Mercaptopropylmethyldimethoxy silane, *See* CM8450, 6454
3-mercaptopropyltrimethoxysilane, *See* Dynasylan® MTMO, 9414
(3-mercaptopropyl)trimethoxysilane, *See* Prosil® 196, 25236
mercaptopyridine-N-oxide sodium salt, *See* Sodium Omadine® 40% Aq. Sol'n, 28803, sodium pyrithione, 28813
2-Mercaptopyridine-N-oxide, sodium salt monohydrate, *See* Sodium Omadine® 40% Aq. Sol'n, 28803
2-Mercaptopyridine-N-oxide, sodium salt, *See* sodium pyrithione, 28813
mercaptosuccinate, *See* malathion, 18460
Mercaptothion, *See* malathion, 18460
2-Mercaptotoluimidazole, *See* Vanox® MTI, 33343
Mercardan, *See* meralluride, 19329
Mercasin, *See* prometryn, 25140
Mercate, *See* isoascorbic acid, 15338
Mercate 20, *See* Eribate, 10753, Neo-Cebitate®, 20878, sodium erythorbate, 28762
Mercazin, *See* prometryn, 25140
Merce Assist ADB, 19337
Mercerisin OR, 19338
mercerized cotton, 19339
Mercers, 19340
Mercer's liquor, 19341
Merchlorate, *See* Agallol, 954, Atiran, 3286, Ceresan, 5745
Merclor D, 19342
Mercol 25, *See* sodium dodecylbenzenesulfonate, 28760
Mercol 30, *See* Nacconol® 40G, 20647, sodium dodecylbenzenesulfonate, 28760
mercolloid, 19343
Mercoloy, 19344
Mercozen, *See* Perkacit® ETU, 23283
mercufenol chloride, 19345
Mercuhydrin, *See* meralluride, 19329
Mercuram, *See* Rezifilm, 26521
Mercuranine, *See* mercurochrome, 19354
Mercuretin, *See* meralluride, 19329
mercuric acetate, 19346
mercuric bichloride, *See* mercuric chloride, 19347
mercuric chloride, 19347
mercuric chloride-ammonium oxalate, *See* Eder's solution, 9611
mercuric diacetate, *See* mercuric acetate, 19346
mercuric oxide, *See* Kankerex, 15713, mercuric oxide, red and yellow, 19348
mercuric oxide, red, *See* red oxide of mercury, 25966
mercuric oxide, red and yellow, 19348
mercuric oxycyanide, *See* Howard's silver, 14431
mercuric potassium iodide, *See* Mayer's solution, 19092
mercuric subsulfate, *See* turpeth mineral, 32612
mercuric sulfate, 19349
mercuric sulfide, *See* mercuric sulfide, red and black, 19350
mercuric sulfide, red and black, 19350
mercuricide, 19351
mercuriocoleols, 19352
mercuriol, 19353
mercurochrome, 19354
Mercurochrome 220 soluble, *See* mercurochrome, 19354
Mercurocol, *See* mercurochrome, 19354
Mercurome, *See* mercurochrome, 19354
Mercurophage, *See* mercurochrome, 19354
mercurous chloride, *See* Calomel, 5018

Metaxon, *See* MCPA, 19183
metazachlor, 19465, *See* Butisan, 4746, Butisan® S, 4747, Pree®, 24873, Track, 32129
metazachlore, *See* metazachlor, 19465
Metazene, *See* Ageflex FM-12, 986
Metazene®, 19466
Metco 450, 19467
metelilachlor, *See* metolachlor, 19585
Meteor, 19468
Meteor Plus, 19469
meteorite, 19470
metepa, *See* Mapo®, 18656
Metglas®, 19471
methabenzthiazuron, 1947, *See* Tribunil®, 32204 2
Methacetone, *See* diethyl ketone, 8436
Methacide, *See* methyl benzene, 19514, toluene, 31946
Methacrifos, *See* Damfin, 7721
Methacrol, 19473
methacrylate, *See* methyl acrylate, 19509, MVP, 20542
methacrylate copolymer,, *See* Acryloid® 702, 424
methacrylate/butadiene styrene, *See* Paraloid® BTA-702, 22762
methacrylatoethyl trimethylammonium chloride, *See* Madquat Q-6, 18293
β-methacrylic acid, *See* Crotonic Acid, 7260
methacrylic acid 2,3-epoxypropyl ester, *See* glycidyl methacrylate 2,3-epoxypropyl methacrylate, 13094
methacrylic Acid ethyl ester, *See* ethyl methacrylate, 11077
methacrylic Acid methyl ester, *See* methyl methacrylate, 19548
methacrylic acid tetrahydrofurfuryl ester, *See* Ageflex THFMA, 999, tetrahydrofurfuryl methacrylate, 31372
methacryloxyethyltrimethyl ammonium chloride, *See* Ageflex FM-1Q80MC, 982
2,2-bis(methacryloxymethyl)butyl methacrylate, *See* Ageflex TM 402, 403, 404, 410, 421, 423, 451, 461, 462, 1000
γ-methacryloxypropyl trimethoxysilane, *See* Dow Corning® Z-6030, 8879, Prosil® 248, 25238, Union Carbide® A-174, 32986
3-Methacryloxypropyl trimethoxysilane, *See* CM8550, 6456, Dow Corning® Z-6030, 8879, Dynasylan® MEMO, 9412, Prosil® 248, 25238
methacryloxypropyltrimethoxy silane, *See* Dow Corning® Z-6030, 8879, Prosil® 248, 25238
2-(Methacryloyloxy)ethyltrimethylammonium methyl sulfate, *See* Ageflex FM-1Q80DMS, 981
metham, *See* metam-sodium, 19444
metham sodium, *See* Unifume, 32954
methamidophos, 19474, *See* Nitofol®, 21314, Prodex, 25069, Tamaron®, 30760
methanal, *See* formaldehyde, 12258
methanamide, *See* formamide, 12261
methanamine, N-methyl-, hydrochloride, *See* dimethylamine hydrochloride, 8513
methanaminium, 1-carboxy-N,N,N-trimethyl-, inner salt, *See* Oxyneurine, 22463
methane, 19475, *See* fire-damp, 11831, marsh gas, 18954, Mixxim® BB/100, 19989
Methane Base Michler's Hydride, *See* Michler's Base, 19644
methane trichloromethane, bis(2,3,5-trichloro-6-hydroxyphenyl), *See* hexachlorophene, 13992
methane, oxybis-, *See* dimethyl ether, 8505
Methanearsonic acid sodium salt, *See* Neoarsycodyl, 20871
Methanebis[N,N'-(5-ureido-2,4-diketotetrahydro imidazole)-N,N-dimethylol], *See* Abiol, 69
methanecarbonitrile, *See* acetonitrile, 299
methanecarboxylic acid, *See* Acetic acid, 292
methanedicarbonic acid, *See* malonic acid, 18491
methanedicarboxylic acid, *See* malonic acid, 18491
N,N'-Methanetetraylbiscyclohexanamine, *See* dicyclohexyl carbodiimide, 8412
methanoic acid, *See* formic acid, 12271
methanol, *See* methyl alcohol, 19511, pyro alcohol, 25488
N-methanolacrylamide, *See* NM-AMD, 21404
Methanone, bis(2,4-dihydroxyphenyl)-, *See* Benzophenone-2, 3977
Methanone, [2-hydroxy-4-(octyloxy)phenyl]phenyl-, *See* Cyasorb® UV 531, 7497
Methaplex, 19476
methapoxide, *See* Mapo®, 18656
Methar 30, 19477
Methasan, 19478, *See* AAprotect, 16, Accelerator MZ Powder, 209, Vancide® MZ-96, 33303, Zimate®, 34762
Methasol, 19479
Methavin, *See* methomyl, 19502
Methazate, 19480, *See* AAprotect, 16
methazole, 19481, *See* Probe, 25048
methbipyranone,, *See* metyrapone, 19608
methenamine, *See* Grasselerator 102, 13289, hexamethylamine, 14000, hexamethylene tetramine, 14006
N,N'-methenyl-o-phenylenediamine, *See* Benzimidazole, 3966
Methibenzuron, *See* methabenzthiazuron, 19472
Methic, 19482
Methicone, *See* Masil® SF-MH, 19022
methicone, *See* Finex-25-020, 11810
methidathion, 19483, *See* Supracide, 30299
Methilonin, *See* DL-methionine, 19487
methiocarb, 19484, *See* Borderland Black, 4416, Club, 6452, Mesurol®, 19401
Methiodal sodium, 19485, *See* Abrodil, 158, Skiodan Sodium, 28570
DL-methionine, 19487
methionine hydroxy analog, 19486, *See* MHA, 19624
Methocel, 19488
Methocel MC, *See* Glutolin, 13064
Methocel® 40-202, 19489
Methocel® A15-LV, 19490
Methocel® A4C, A4M, 19491
Methocel® A4MP, 19492
Methocel® E3 Premium, 19493
Methocel® E4M, 19494
Methocel® E5, 19495
Methocel® F4M, 19496
Methocel® K100MP, 19497
Methocel® K35, 19498
Meth-O-Gas, 19499
Methokill, 19500
Methoklone, 19501
Methomex, *See* methomyl, 19502
methomyl, 19502
Methomyl 5G, *See* methomyl, 19502
methopyrapone, *See* metyrapone, 19608
Methoxa-Dome, *See* Oxsoralen-Ultra, 22437
methoxalen, *See* Oxsoralen-Ultra, 22437
Methoxone, 19503
1-methoxy-2-acetoxypropane, *See* Dowanol® PMA, 8897
2-Methoxy-1-aminobenzene, *See* *o*-anisidine, 2473
3-Methoxy-1-aminobenzene, *See* *m*-anisidine, 2472
4-Methoxy-1-aminobenzene, *See* *p*-anisidine, 2474
4-methoxy-4'-aminodiphenylamine, *See* Variamine, 33387
2-methoxy-*p*-cresol, *See* creosol, 6990
methoxy DDT, *See* methoxychlor, 19504
methoxy ether of propylene glycol, *See* Arcosolv® PM, 2814, Dowanol® PM, 8896, Icinol PM, 14816, Solvenon® PM, Solvent PM, 29050
methoxy ethyl phthalate, *See* Reomol P, 26132
2-methoxy ethyl phthalate, *See* Reomol P, 26132
9-methoxy-7H-furo(3,2-g)benzopyran-7-one, *See* Oxsoralen-Ultra, 22437
8-methoxy-2',3',6,7-furocoumarin, *See* Oxsoralen-Ultra, 22437
8-Methoxy-4',5':6,7-furocoumarin, *See* Oxsoralen-Ultra, 22437
3-methoxy-4-hydroxybenzaldehyde, *See* Zimco, 34763
4-Methoxy-2-hydroxybenzophenone, *See* Cyasorb® UV 9, 7495, Rhodialux A, 26675, Uvinul® M-40, 33197
3-methoxy-4-hydroxytoluene, *See* creosol, 6990
(2-Methoxy-methylethoxy) propanol, *See* Icinol DPM, 14815
(Methoxymethylethoxy)propanol, *See* Icinol DPM, 14815, Poly-Solv® DPM, 24555
2-Methoxy-1-methylethyl acetate, *See* Arcosolv® PMA, 2815 Dowanol® PMA, 8897
2-Methoxy-4-methylphenol, *See* creosol, 6990
***p*-methoxy-β-methylstyrene**, *See* Arizole® Anethole Extra, 2885
2-[[[[(4-methoxy-6-methyl-1,3,5-triazin-2-yl)amino] carbonyl]amino]sulfonyl] benzoic acid methylester, *See* metsulfuron-methyl, 19606
3-[[[[(4-methoxy-6-methyl-1,3,5-triazin-2-yl)amino] carbonyl]amino]sulfonyl]-2-thiophenecarboxylic acid methyl ester, *See* thifensulfuron-methyl, 31673
2-methoxy-2-phenylacetophenone, *See* Vicure® 10, 33734
(±)-1-methoxy-2-propanol, *See* Arcosolv® PM, 2814, Dowanol® PM, 8896, Icinol PM, 14816, Solvenon® PM, Solvent PM, 29050
1-Methoxy 2-propanol, *See* Arcosolv® PM, 2814, Dowanol® PM, 8896, Icinol PM, 14816, Solvenon® PM, Solvent PM, 29050
1-Methoxy-2-propanol Acetate, *See* Arcosolv® PMA, 2815, Dowanol® PMA, 8897
1-methoxy-4-(1-propenyl)benzene, *See* Arizole® Anethole Extra, 2885
1-methoxy-4-propenylbenzene, *See* Arizole® Anethole Extra, 2885
1-methoxy-4-(2-propenyl)benzene, *See* Estragole, 10897
Methoxy-4-propenylbenzene, *See* Arizole® Anethole Extra, 2885
Methoxy propyl acetate, *See* Sensolve MPA, 27956
1-Methoxy-2-propyl Acetate, *See* Arcosolv® PMA, 2815
4'-methoxy-3,3',5-stilbenetriol-3-glucoside, *See* rhapontin, 26530
3-methoxy-2-sulfanilamidopyrazine, *See* Kelfizina, 15838
***p*-methoxy-α-toluic acid**, *See* *p*-methoxyphenylacetic acid, 19505
[[2-[(methoxyacetyl)amino]-4-(phenylthio)phenyl] carbonimidoyl]biscarbamic acid dimethyl ester, *See* Rintal®, 26774
***p*-methoxyallylbenzene**, *See* Estragole, 10897
***o*-methoxyaniline**, *See* *o*-anisidine, 2473
***p*-methoxyaniline**, *See* *p*-anisidine, 2474
3-methoxyaniline, *See* *m*-anisidine, 2472
4-methoxybenzaldehyde, *See* *p*-anisaldehyde, 2470
2-methoxybenzenamine, *See* *o*-anisidine, 2473
4-methoxybenzenamine, *See* *p*-anisidine, 2474
2-(methoxycarbonylamino)benzimidazole) methyl 2-benzimidazolecarbamate, *See* Konker®, 16327
3-[(methoxycarbonyl)amino]phenyl (3-methylphenyl)carbamate, *See* phenmedipham, 23595, 23598
(Methoxycarbonyl)aniline, *See* methyl anthranilate, 19513
2-(methoxycarbonyl)aniline, *See* methyl anthranilate, 19513
Methoxycarbonylethylene, *See* methyl acrylate, 19509
methoxychlor, 19504, *See* Marlate 2-MR Emulsifiable Insecticide, 18747
Methoxydiglycol, *See* Dowanol® DM, 8891
Methoxydiuron, *See* Atlas Linuron, 3319
2-Methoxyethanol, *See* methyl cellosolve®, 19521, methyl icinol, 19546
2-methoxyethanol, acrylate, *See* Ageflex MEA, 994
2-(2-Methoxyethoxy) ethanol, *See* methyl di-Icinol, 19526
2-(methoxyethoxy)ethanol, *See* Ektasolve® DM, 9666
methoxyethyl acrylate, *See* Ageflex MEA, 994
methoxyethyl mercury acetate, *See* Panogen M, 22682
2-Methoxyethylmercury acetate, *See* Panogen M, 22682
methoxyethyl mercury chloride, *See* Agallol, 954, Atiran, 3286
2-Methoxyethyl ether, *See* diglyme, 8455
(β-Methoxyethyl)mercuric chloride, *See* Agallol, 954

Atiran, 3286 Ceresan, 5745
2-methoxyethylmercuric Chloride, *See* Agallol, 954, Atiran, 3286
β-Methoxyethylmercury chloride, *See* Agallol, 954, Atiran, 3286, Ceresan, 5745
2-methoxyethylmercury Chloride, *See* Agallol, 954, Atiran, 3286
2-(β-methoxyethyl)pyridine, *See* metyridine, 19609
8-Methoxy(furano-3',2':6-7-coumarin, *See* Oxsoralen-Ultra, 22437
8-Methoxyfuranocoumarin, *See* Oxsoralen-Ultra, 22437
2-Methoxynaphthalene, *See* jara jara, 15518
***o*-methoxyphenol**, *See* guaiacol,
***p*-methoxyphenol**, *See* Eastman® HQMME, 9467
4-Methoxyphenol, *See* Eastman® HQMME, 9467, hydroquinone monomethyl ether, 14633
***p*-methoxyphenylacetic acid**, 19505
4-methoxyphenylacetic acid, *See* *p*-methoxyphenylacetic acid, 19505
***p*-methoxyphenylamine**, *See* *p*-anisidine, 2474
***o*-methoxyphenylamine**, *See* *o*-anisidine, 2473
N-(4-Methoxyphenyl)-1,4-benzenediamine, *See* Variamine, 33387
N-[2-[[(4-methoxyphenyl)methyl]-2-pyrimidinylamino]ethyl]-N,N-dimethyl-1-hexadecanaminium bromide, *See* Thonzide, 31746
N-(p-Methoxyphenyl)-p-phenylenediamine, *See* Variamine, 33387
1-(*p*-methoxyphenyl)propene, *See* Arizole® Anethole Extra, 2885
3-(4-Methoxyphenyl)-2-propenoic acid 2-ethylhexyl ester, *See* octyl methoxycinnamate, 22039
Methoxypropanol, *See* Arcosolv® PM, 2814, Dowanol® PM, 8896
1-Methoxypropan-2-ol, *See* Arcosolv® PM, 2814, Dowanol® PM, 8896, propylene glycol monomethyl ether, 25220, Solvenon® PM, Solvent PM, 29050
Methoxypropanol, α isomer, *See* Arcosolv® PM, 2814, Dowanol® PM, 8896, Icinol PM, 14816, Solvenon® PM, Solvent PM, 29050
2-(1-Methoxy)propyl acetate, *See* Arcosolv® PMA, 2815 Dowanol® PMA, 8897
8-Methoxypsoralen, *See* Oxsoralen-Ultra, 22437
N1-(3-methoxy-2-pyrazinyl)sulfanilamide, *See* Kelfizina, 15838
Methozin, *See* Antipyrine,
methyl abietate, *See* Abalyn, 24
methyl acetate, 19506
methyl acetic ester, *See* methyl acetate, 19506
methyl acetone, 19507, *See* methyl ethyl ketone, 19530
methyl 12-acetoxy-9-octadecenoate, *See* methyl acetyl ricinoleate, 19508
methyl 12-acetoxyoleate, *See* methyl acetyl ricinoleate, 19508
methyl acetyl ricinoleate, 19508, *See* Flexricin® P-4, 11959, Naturechem® MAR, 20773
methyl acrylate, 19509
methyl acrylate polymer, 19510
methyl adipate, *See* dimethyl adipate, 8502
methyl alcohol, 19511, *See* spirit of wood, 29300
methyl aldehyde, *See* formaldehyde, 12258
methyl alpha-methylacrylate, *See* methyl methacrylate, 19548
methyl aminoacetic acid, *See* methyl-glycocoll, 19570
methyl *o*-aminobenzoate, *See* methyl anthranilate, 19513
methyl-*p*-amino-*m*-cresol sulfate, *See* Metol, 19584
methyl-*o*-aminophenol and hydroquinone (2:1), *See* Ortol, 22342
methyl amyl alcohol, 19512
methyl amyl ketone, *See* methyl n-amyl ketone, 19552
methyl n-amyl ketone, 19552
methyl anthranilate, 19513
methyl aphoxide, *See* Mapo®, 18656
methyl aracidate, *See* Kemester® 2050, 15941
methyl arecaidin, *See* arecoline, 2829
methyl behenate, *See* Kemester® 9022, 15953
methyl benzene, 19514
methyl benzenecarboxylate, *See* methyl benzoate, 19515
5-methyl-1,3-benzenediol, *See* orcinol, 22280
methyl benzimidazol-2-ylcarbamate, *See* Carbendazim, 5187, 5188
methyl 1H-benzimidazol-2-ylcarbamate, *See* Carbendazim, 5187, 5188
methyl benzoate, 19515
methyl benzoquate, 19516
5-methyl-1,2,3-benzotriazole, *See* Preventol® CI7-100, 24929, tolyltriazole, 31962
methyl-1H-benzotriazole, *See* Preventol® CI7-100, 24929, tolyltriazole, 31962
5-methyl-1H-benzotriazole, *See* Retrocure® G, 26279, Vulkalent® TM, 34141
5-methyl-1H-benzo-1,2,3-triazole, *See* Retrocure® G, 26279, Vulkalent® TM, 34141
methyl 5-benzoyl-2-benzimidazolecarbamate, *See* Vermox, 33591
methyl benzoylformate, *See* Vicure® 55, 33735
methyl bromide, 19517, *See* Embafume, 10012, Meth-O-Gas, 19499
methyl bromide-amyl acetate, *See* Fumyl-O-Gas, 12470
2-methyl-1,3-butadiene, homopolymer, *See* polyisoprene, 24471
2-methyl-1,3-butadiene, *See* polyisoprene, 24470
2-methyl-1-butanol, *See* Amyl alcohol, 2381
3-methyl-1-butanol acetate, *See* pear oil, 22971
3-methyl-1-butanol acetate, *See* isoamyl acetate, 15336
2-methyl-2-butene, *See* pental, 23147
3-methyl-2-butene-1-ol, *See* Prenol, 24892
3-methyl-1-butyl Acetate, *See* isoamyl acetate, 15336, pear oil, 22971
2-methyl butylacrylate, *See* Butyl methacrylate, 4774
3-methyl butyl ester acetic acid, *See* isoamyl acetate, 15336, pear oil, 22971
β-methyl butyl acetate, *See* isoamyl acetate, 15336, pear oil, 22971
***p*-methyl-*tert*-butylbenzene**, *See* Butyl toluene, 4782
1-methyl-4-*tert*-butylbenzene, *See* Butyl toluene, 4782
1-methyl-4-*t*-butyl-3-methoxy-2,6-dinitrobenzene, *See* musk ambrette, 20532
4-methyl-2,6-di-t-butyl-phenol, *See* Oxyguard, 22457
2-methyl-3-(4-t-butylphenyl) propanol, *See* Lysmeral®, 18092
methyl *tert*-butyl ether, *See* Driveron, 9046
methyl-*t*-butyl-ketone, *See* pinacolone, 23848
methyl butynol, 19518
2-methyl-3-butyn-2-ol, *See* methyl butynol, 19518
methyl canolate, *See* Emery® 2232, 10151
methyl caprate, 19519
methyl caprinate, *See* methyl caprate, 19519
methyl caprylate-caprate, 19520, *See* Emery® 2209, 10143
methyl carbamic acid 2,3-dihydro-2,2-dimethyl-7-benzofuranyl ester, *See* Rampart, 25860
methyl carbamic acid, 4-(methylthio)-3,5-xylyl ester, *See* methiocarb, 19484
methyl carbitol, *See* Ektasolve® DM, 9666, Poly-Solv® DM, 24554
methyl cellosolve, *See* ethylene glycol methyl ether, 11127, Poly-Solv® EM, 24557
methyl cellosolve®, 19521
methyl cellosolve acrylate, *See* Ageflex MEA, 994
methyl cellulose, *See* Glutolin, 13064
methyl chavicol, *See* Estragole, 10897
methyl chemosept, *See* methylparaben, 19577
methyl chloride, 19522, *See* Artic, 3147
N-methyl-N(4-chlorobenzoylmethyl)-3-(10,11-dihydro-5H-dibenzo[*b,f*]azepin-5-yl)propylamine hydrochloride, *See* Lofepramin Hydrochloride, 17552
methyl Chloroform, *See* Chlorothene, 6151
3-methyl-4-chlorophenol, *See* Raschit, 25892
methyl chlorophenoxy acetic acid, *See* MCPA, 19183
3-methyl-4-[(2-chlorophenyl)hydrazone]-4,5-isoxazoledione, *See* drazoxolan, 8974
methyl chlorpyrifos, *See* Chlorpyrifos-methyl, 6164
methyl cocoate, *See* Radia® 7117, 25722
methyl coconate, 19523, *See* Emery® 2253, 10152
N-methyl-N-(1-coconut alkyl) glycine, *See* Cocoyl sarcosine, 6568
methyl cyanide, *See* acetonitrile, 299
methyl cyanoacetate, 19524
methyl cyclohexane, 19525
methyl cymate, *See* AAprotect, 16, Accelerator MZ Powder, 209, Vancide® MZ-96, 33303, Zimate®, 34762
methyl decanoate, *See* methyl caprate, 19519
methyl decyl-1-amino decane, *See* didecyl methylamine, 8415
methyl demeton, *See* demeton-S-methyl, 8055
2-methyl-1,5-diaminopentane, *See* Dytek® A, 9439
methyl dicocamine, *See* Jet Amine M2C, 15591
methyl dicoconut tert, amine, *See* Kemamine® T-6501, 15929
methyl digol, *See* Ektasolve® DM, 9666
methyl dihydroxyethyl isoarachidaloxypropyl ammonium chloride - isopropanol, *See* Tomah Q-24-2, 31993
methyl di-icinol, 19526
methyl 1-(dimethylcarbamoyl)-N-(methylcarbamoyloxy)thioformimidate, *See* Vydate®, 34171
methyl-N-(2,4-dimethyl cyclohexene-3-ylidenemethyl)-anthranilate, *See* Ligantraal, 17221
1-methyl-4-(1,5-dimethyl-1-hydroxyhex-4(5)-enyl)cyclohexen-1, *See* Kamillosan, 15707
2-methyl-3,5-dinitrobenzamide, *See* Zoamix, 34844
1-methyl-2,4-ddinitrobenzene, *See* D.N.T, 7622
4-methyl-1,3-dioxolan-2-one, *See* Arconate® Propylene Carbonate, 2811, Solvenon® PC, 29049, Texacar® PC, 31425
5-methyl-2,4-dioxypyrimidine, *See* thymine, 31777
N-methyl-N',N'-diphenylurea, *See* Acardite 2, 182
2-methyl-1,3-di-3-pyridiyl-1-propanone, *See* metyrapone, 19608
methyl disulfide, *See* carbon disulfide, 5223
(4-methyl-1,3-dithiolan-2-ylidene)phosphoramidic acid diethyl ester, *See* mephosfolan, 19326
6-methyl-1,3-dithiolo[4,5-b]quinoxalin-2-one, *See* quinomethionate, 25686
methyl docosanoate, *See* Kemester® 9022, 15953, methyl laurate, 19545
3-methyl dodecanonitrile, *See* Frescile, 12401
methyl dodecylate, *See* methyl laurate, 19545
methyl dursban, *See* Chlorpyrifos-methyl, 6164
methyl eicosenate, *See* Kemester® 2050, 15941
methyl eosin, *See* Erythrin, 10776
methyl ester acrylic acid, *See* methyl acrylate, 19509
methyl ester L, 19527
methyl ester S, 19528
methyl ethanoate, *See* methyl acetate, 19506
methyl ether, 19529, *See* dimethyl ether, 8505, Dymel®, 9331
methyl ethyl ketone, 19530, *See* 2-butanone, 4737, Meketone, 19247, methyl acetone, 19507
methyl ethyl ketone hydroperoxide, *See* 2-butanone peroxide, 4738
methyl ethyl ketone oxime, *See* methyl ethyl ketoxime, 19532
methyl ethyl ketone peroxide, 19531, *See* Butanox, 4739, Cadox® HBO-50, 4913, Cadox® L-30, 4914, Esperfoam® FR, 10842
methyl ethyl ketoxime, 19532
methyl ethyl sulfide, 19533
methyl ethylene oxide, *See* propylene oxide, 25223
methyl eugenol, 19534
***o*-methyl eugenol ether**, *See* methyl eugenol, 19534
methyl eugenyl ether, *See* methyl eugenol, 19534
methyl formal, *See* methylal, 19564
methyl gallate, *See* Gallicin, 12611
methyl gluceth-10, 19535, *See* Glucam® E-10, 13019
methyl gluceth-20, 19536, *See* Glucam® E-20, 13020
methyl gluceth-20 distearate, 19537, *See* Glucam® E-20 Distearate, 13021
methyl glucose dioleate, 19538, *See* Glucate® DO, 13031
methyl glucose sesquiisostearate, *See* Glucate® IS,

13032
methyl glucose sesquistearate, 19539, *See* Glucate® SS, 13033
methyl glycocoll, *See* sarcosine, 27463
methyl glycol, *See* propylene glycol, 25214, 25215
methyl glyoxaidine, *See* lysidine, 18087
2-methyl glyoxaline, *See* 2-methyl imidazole, 19572
methyl green, 19540, *See* Ethyl Green, 11072
methyl hexadecanoate, *See* Emery® 2216, 10145, methyl palmitate, 19555
methyl hexahydrophthalic anhydride, *See* Milldride® MHHPA, 19785
methyl hydride, *See* methane, 19475
methyl-1-hydrogenated tallowamidoethyl-2-hydrogenated tallow imidazolinium-methyl sulfate, *See* Accosoft 808HT, 248
methyl 2-hydroxybenzoate, *See* methyl salicylate,
methyl *p*-hydroxybenzoate, *See* methylparaben, 19577, Nigagin, 21171
methyl hydroxycellulose, 19541, *See* Tylose® P, P-x, PS-x, P-z Series, 32662
methyl hydroxyethylcellulose, *See* Tylose® MHB, 32661
methyl 12-hydroxy-9-octadecenoate, *See* methyl ricinoleate, 19558
4-methyl-4-Hydroxy-2-pentanone, *See* Diacetone alcohol, 8249
methyl hydroxystearate, *See* Cenwax® ME, 5693, Naturechem® MHS, 20774, Paricin® 1, 22833
methyl bis (hydrogenated tallow amido ethyl)2-hydroxyethyl ammonium chloride, *See* Incroquat 100, 15027, Incrosoft 100, 15047
methyl bis (hydrogenated tallowamidoethyl) 2-hydroxyethyl ammonium methyl sulfate, *See* Accosoft 440-75, 244
methyl 4-hydroxybenzoate, *See* methylparaben, 19577
methyl icinol, 19546 , *See* methyl cellosolve®, 19521
2-methyl imidazole, 19572
methyl iodide, 19542
methyl iodine, *See* methyl iodide, 19542
methyl isobutyl ketone peroxide, *See* Trigonox® HM, 32302
16-methyl-Isooctadecanoic acid, *See* isostearic acid, 15421
1-methyl-4-isopropenyl- Δ_{6}-cyclohexen-2-one, *See* carvone, 5363
1-methyl-4-isopropylidene-3-cyclohexanone, *See* pulegone, 25379
1-methyl-7-isopropylphenanthrene, *See* retene, 26271
2-methyl-5-isopropyl-phenol, *See* carvacrol, 5362
N-methyl-2-isopropylphenylcarbam, *See* Etrofolan®, 11156
5-methyl-3(2H)-isoxazolone, *See* Tachigaren 70, 30685
methyl isothiocyanate, 19543
methyl lardate, 19544, *See* Base ML, 3697
methyl laurate, 19545, *See* Emery® 2270, 10154, Stepan C-40, 29666
methyl ledate, 19547, *See* Ledate®, 16949
methyl methacrylate, 19548
methyl methacrylate butadiene styrene terpolymer, *See* Zylar® 90-495, Zylar® 93-541, 34918
methyl methacrylate copolymer, *See* Acryloid® B-44, 431, Polymethylmethacrylate, 24492
methyl methacrylate copolymer, xylene, toluene (70/30), *See* Acryloid® B-99, 433
methyl methacrylate polymer, 19549, *See* Acryloid® A-30, 428
methyl methacrylate polymer in toluene/butanol (90/10), *See* Acryloid® A-21, 426
methyl methacrylate polymer in toluene/MEK/butanol (50/40/10), *See* Acryloid® A-21LV, 427
methyl methacrylate resin, *See* methyl methacrylate polymer, 19549
methyl methylacrylate, *See* methyl methacrylate, 19548
N-methyl methyl anthranilate, *See* dimethyl anthranilate, 8503
methyl O-methyl amino benzoate, *See* dimethyl anthranilate, 8503
methyl N-methyl anthranilate, *See* dimethyl anthranilate, 8503
N-methyl methyl anthranilate, *See* dimethyl anthranilate, 8503
S-methyl-N[(methylcarbamoyl) oxy]thioacetimidate, *See* methomyl, 19502
methyl-1 methyl-3-cyclohexene carboxylate, *See* Tachryrate, 30687
6-methyl-2-(4-methyl-3-cyclohexen-1-yl)-5-hepten-2-ol, *See* Kamillosan, 15707
1-methyl-4-(1-methylethenyl)cyclohexene, *See* Achilles Dipentene, 323
2-methyl-5-(1-methylethenyl)-2-cyclohexene-1-one, *See* carvone, 5363
N-methyl-N'-(1-methylethyl)-6-(methylthio)-, *See* desmetryn, 8142
N-methyl-N'-(1-methylethyl)-6-(methylthio)-1,3,5-triazine-2,4-diamine, *See* desmetryn, 8142
methyl-5-(1-methylethyl)phenol, *See* oxycymol, 22445
N-methyl-N-(1-methylethyl)-N-[2-[(9H-xanthen-9-ylcarbonyl)oxy]ethyl]-2-propanaminium bromide, *See* propantheline bromide, 25176
5-methyl-2-(1-methylethyl) cyclohexanol, *See* menthol terpine hydrate, 19321
5-methyl-2-(1-methylethyl) phenol, *See* thyme camphor, 31775, thymol, 31778
4-methyl-1-(1-methylethyl)bicyclo[3.1.0]hexan-3-one, *See* tanacetone, 30775
5α-methyl-2β-(1α-methylethyl)cyclohexanol, *See* menthol terpine hydrate, 19321
1-methyl-7-(1-methylethyl)phenanthrene, *See* retene, 26271
2-methyl-5-(1-methylethyl)phenol, *See* carvacrol, 5362, oxycymol, 22445
methyl-2-methyl pentylidene anthranilate, *See* Mevantraal, 19610
2-methyl-3-(4-methylphenyl) propanal, *See* Jasmorange®, 15534
1-[[2-methyl-4-[(2-methylphenyl)azo]phenyl]azo]-2-naphthalenol, *See* Sudan IV, 29941
methyl-2-methyl-2-Propenoate, *See* methyl methacrylate, 19548
methyl-2-methylpropenoate, *See* methyl methacrylate, 19548
methyl morpholine, *See* *p*-methyl morpholine, 19573
4-methyl morpholine, *See* *p*-methyl morpholine, 19573, Texacat® NMM, 31432
***p*-methyl morpholine**, 19573
N-methyl morpholine, *See* *p*-methyl morpholine, 19573
methyl bis (modified tallowamidoethyl) 2-hydroxyethyl ammonium methyl sulfate, *See* Accosoft 540 HC, 550-90 HHV, 245
methyl mustard oil, *See* methyl isothiocyanate, 19543
methyl myristate, 19550, *See* Emery® 2214, 10144
methyl N-(2,6-dimethylphenyl)-N-(2-furanylcarbonyl)-DL-alanine, *See* Furalaxyl, 12487
methyl namate, *See* Aquatreat SDM, 2761, Octopol SDM-40, 22014, Perkacit® SDMC, 23288, sodium dimethyldithiocarbamate, 28758, Thiostop N, 31717
methyl namate®, 19551
methyl naphthalene, *See* Abluton CMN, 135
2-methyl-1,4-naphthalenediol diphosphoric acid ester tetrasodium salt, *See* menadiol sodium diphosphate, 19319
N-methyl-N-(1-naphthylmethyl)-3-phenylpropen-1-amine, *See* Naftifine, 20650
methyl-N'-methyl-N'-(2-benzothiazolyl)urea, *See* methabenzthiazuron, 19472
methyl niclate®, 19553
[2-(2-methyl-5-nitro-1H-imidazol-1-yl]carbamothioic acid methyl ester, *See* carnidazole, 5298
methyl n-octadecanoate, *See* Kemester® 4516, 15944
methyl n-octadecyl terephthalamate, *See* Pentamid KH, 23162
methyl n-propyl ketone, *See* methyl propyl ketone, 19557
methyl octadecanoate, *See* methyl stearate, 19560
methyl octadecanoate, *See* Emery® 2218, 10146
methyl n-octadecanoate, *See* Emery® 2218, 10146
methyl *cis*-9-octadecenoate, *See* Emery® 2301, 10155, Kemester® 104, 15935, Witconol 2301, 34496
methyl (Z)-9-octadecenoate, *See* Emery® 2218, 10146
methyl 9-octadecenoate, *See* Emery® 2301, 10155
methyl bis(oleylamidoethyl) 2-hydroxyethyl ammonium methyl sulfate, *See* Accosoft 750, 247
methyl oleate, 19554, *See* Emerest® 2301, 10081, Emery® 2219, 10147, Emery® 2301, 10155, Kemester® 104, 15935, Priolube 1400, 25013, Radia® 7060, 25719, Witconol 2301, 34496
methyl oleate/stearate, *See* Stepan C-68, 29668
4-methyl-1-oxa-4-azacyclohexane, *See* *p*-methyl morpholine, 19573
6-methyl-1,2,3-oxathiazin-4(3H)-one 2,2-dioxide, *See* acesulfame potassium, 277, Sunett, 30140
S(-)-methyloxirane, *See* propylene oxide, 25223
methyl-N-(1-oxododecyl)glycine, sodium salt, *See* sodium lauroyl sarcosinate, 28779
N-methyl-N-(1-oxododecyl)glycine sodium salt, *See* Vanseal® NALS-30, 33359
(Z)-2-[methyl(1-oxo-9-octadecenyl)amino]ethane sulfonic acid, sodium salt, *See* Adinol OT, 697, Arkopon T, 2898
2-methyl-4-oxo-3-(2-propenyl)-2-cyclopenten-1-yl-2,2-dimethyl-3-(2-methyl-1-propenyl) cyclopropane carboxylate, *See* allethrin, 1744
N-methyl-N-(1-oxo-9-octadecenyl)glycine, *See* Oleoyl sarcosine, 22123
N-methyl-N-(1-oxododecyl) glycine, *See* Crodasinic L, 7143, lauroyl sarcosine, 16861
N-methyl-N-(1-oxododecyl) glycine, ammonium salt, *See* Ammonium lauroyl sarcosinate, 2241
N-methyl-N-(1-oxododecyl)glycine, sodium salt, *See* sodium lauroyl sarcosinate, 28779, Zoharsyl L-30, 34876
(methyl-2)oxybispropanol;, *See* Dipropylene glycol, 8611
methyl palm kernelate, *See* Emery® 2255, 10153
methyl palmitate, 19555, *See* Emery® 2216, 10145, Radia® 7120, 25723
methyl palmitate-oleate, *See* Stepan C-65, 29667
methyl paraben, *See* Nigagin, 21171
methyl parasept, *See* methylparaben, 19577
methyl parathion, 19556, *See* Parathion-methyl, 22808
methyl PCT, *See* MP-2, 20388
2-methyl-2,4-pentanediol, *See* hexylene glycol, 14042
methyl-2-pentanol, *See* methyl amyl alcohol, 19512
2-methyl-4-pentanol, *See* methyl amyl alcohol, 19512
4-methyl-2-pentanol, *See* methyl amyl alcohol, 19512
2-methyl-3-pentanol-4-one, *See* Diacetone alcohol, 8249
4-(4-methyl-3-penten-1-yl)-3-cyclohexen-1-carboxyaldehyde, *See* Empetal, 10291
5-methyl-2-pentylphenol, *See* Amyl-*m*-cresol, 2383
methyl pentyl ketone, *See* methyl n-amyl ketone, 19552
methyl phenyl diphenyl phosphate, *See* Disflamoll® DPK, TPK, 8641
(E)-N-methyl-N-(3-phenyl-2-propenyl)-1-naphtalenemethanamine, *See* Naftifine, 20650
1-methyl-1-phenylethylhydroperoxide, *See* Aztec® CHP-80, 3555, CHP-158, 6175
4-methyl-2-(1-phenylethyl)-1-3-dioxolane, *See* Vertelon, 33636
1-methyl-1-phenyl ethylene, *See* α-methylstyrene, 19579
methyl phenyl ketone, *See* acetophenone, 300
N-methyl-N-phenyl-2-methyl butyramide, *See* Gardamide, 12662
2-methyl-4-phenylsulfonyltrifluoromethane sulfonanilide, *See* Perfluidone, 23248
3-methyl-1-phenyl-5-(sulfanilamido)pyrazole, *See* Vesulong, 33700
3-methyl-5-phenyl-1-pentanal, *See* Mefranal, 19231, 19232
methyl phenyl polysilocane, *See* phenyl dimethicone, 23632
methyl phenyl polysiloxane, *See* phenyl dimethicone, 23632
methyl phenylacetate, *See* Mephaneine, 19323
***p*-methylphenyl acetate**, *See* Narceol, 20742

methyl phenylglyoxalate, See Vicure® 55, 33735
methyl phosphonic acid dimethyl ester, See Fyrol® DMMP, 12539
methyl phthalate, See Kemester® DMP, 15956, Kodaflex® DMP, 16271, Unimoll® DM, 32979
2-methyl-3-phytyl-1,4-napthoquinone, See Konakion, 16322, vitamin K+71, 33949
2-methyl-1-propanol, See isobutanol, 15344
2-methyl-1,3-propanediol, See MP Diol Glycol, 20391
2-methyl-2-propanethiol, See t-butyl mercaptan, 15434
2-methyl-1-propene, homopolymer, See polyisobutene, 24469
methyl propenoate, See methyl acrylate, 19509
2-methyl-2-Propenoic Acid Butyl Ester, See Butyl methacrylate, 4774
2-methyl-2-propenoic acid 2-(dimethylamino)ethyl ester, See Sipomer® 2M1M, 28528
2-methyl-2-propenoic acid 2-ethyl-2-[[(2-methyl-1-oxo-2-propenyl)oxy]methyl]-1,3-propanediyl ester, See Perkalink® 400, 23314
2-methyl-2-Propenoic Acid Ethyl Ester, See ethyl methacrylate, 11077
2-methyl-2-propenoic acid 3a,4,7,7a-tetrahydro-4,7-methano-1H-indenyl ester, See Sipomer® DCPM, 28534
2-methyl-2-propenoic acid, cyclohexyl ester, See Ageflex CHMA, 968
methyl-2-propenoic acid, dodecyl ester, See Ageflex FM-12, 986
2-methyl-2-propenoic acid, octadecyl ester, See Ageflex FM-68, 988, octadecyl methacrylate, 21984
2-methyl-2-propenoic acid, 2-hydroxyethyl ester, See Bisomer 2HEMA, 4273
2-methyl-2-propenyl 2-propenoate, See Sipomer® AM, 28532
methyl-6-propoxy-2-benzothiazolyl carbamate, See Tiox, 31845
2-methyl-1-propyl formate, See isobutyl formate, 15346
methyl propyl ketone, 19557
2-methyl propyl phenyl acetate, See Iphaneine, 15246
methyl-2-pyridyl ketone, See 2-Acetyl pyridine, 321
methyl-2-pyrone, See dehydroacetic acid, 7964
1-methyl-2-pyrrolidone, See NMP, 21406
2-methyl-2-pyrrolidone, See, See NMP, 21406
N-methyl-2-pyrrolidone, 19578, See FoamFlush™, 12154, Micropure® Ultra, 19694, NMP, 21406
N-methyl-2-pyrrolidone and dipropylene glycol methyl ether, See Printsolve™ Ink Remover, 25008
N-methyl-2-pyrrolidone, butyrolactone, deceth-6, fragrance, See Ship Shape® Resin Cleaner, 28180
1-methylpyrrolidinone, See NMP, 21406
N-methyl-2-pyrrolidinone, See Agsol Ex1, 1091, NMP, 21406
1-methyl-5-Pyrrolidinone, See NMP, 21406
methyl resinate, See Abalyn, 24
methyl ricinoleate, 19558, See Flexricin® P-1, 11957
methyl rosinate, See Abalyn, 24
methyl salicylate, See Gaultheria oil, 12701, Gaultheric acid, 12702, Teaberry Oil, 30872
methyl selenac, 19559
methyl soyate, See Emery® 2224, 10148, Kemester® 226, 15937
methyl stearate, 19560 Emery® 2218, 10146, Kemester® 4516, 15944, Radia® 7110, 25721
methyl styryl ketone, See benzylidene acetone, 3992
methyl succinate, See dimethyl succinate, 8509
methyl sulfide, 19561
methyl sulfide, See dimethyl sulfide, 8510
methyl sulfinylmethane, See dimethyl sulfoxide, 8511
methyl sulfoxide, See dimethyl sulfoxide, 8511, DMSO, 8762
methyl sunflowerate, See Emery® 2230, 10149
methyl tallowate, See Base MT, 3698 Emery® 2203, 10141 Kemester® 143, 15936
methyl bis (tallowamidoethyl) 2-hydroxyethyl ammonium methyl sulfate, See Accosoft 550L-90, 620-90, 246
methyl-1-tallow amido ethyl-2-tallow imidazolinium methyl sulfate, See Quaternium-27, 25607
methyl tetradecanoate, See Emery® 2214, 10144, methyl myristate, 19550
methyl theobromine, See caffeine,
methyl thioether, See dimethyl sulfide, 8510
methyl thiophanate, See thiophanate methyl, 31706
methyl α-toluate, See Mephaneine, 19323
methyl toluene, See xylene, 34620
methyl 3-(m-tolylcarbamoyloxy)phenylcarbamate, See Vangard, 33323
methyl topsin, See thiophanate methyl, 31706
methyl triethoxysilane, See Dynasylan® MTES, 9413
N-methyl-N-trimethylsilyltrifluoroacetamide, See CM9160, 6466
methyl tuads , 19562®, See Akrochem® TMTD, 1182
methyl tuads® Rodoform, See Akrochem® TMTD, 1182
methyl violet 5BNO, See gentian violet, 12845
methyl violet 5BO, See gentian violet, 12845
methyl violet 10B, See gentian violet, 12845
methyl violet 10BD, See gentian violet, 12845
methyl violet 10BK, See gentian violet, 12845
methyl violet 10BN, See gentian violet, 12845
methyl violet 10BNS, See gentian violet, 12845
methyl violet 10BO, See gentian violet, 12845
methyl viologen(2+), See paraquat, 22796
N-methyl-N'-2,4-xylyl-N-(N,2,4-xylylformimidoyl) formamidine, See Taktic, 30729
N-methylbis-(2,4-xylyliminomethyl)-amine, See Amitraz, 2202, Taktic, 30729
methyl yellow, See Butter yellow, 4759
methyl zimate, See AAprotect, 16
methyl zimate®, 19563
methyl zineb, See Bisphenol A, 4283
methyl ziram, See AAprotect, 16
methylacetaldehyde, See propanal, 25172
methylacetic acid, See propionic acid, 25190
p-methylacetophenone, See Melilot, 19279
methylacetopyronone, See Dehydroacetic acid, 7964
β-methylacrylic acid, See Crotonic Acid, 7260
2-methylacrylic acid methyl ester, See methyl methacrylate, 19548
methylal, 19564
methylaminoacetic acid, See sarcosine, 27463
N-methylaminoacetic acid, See sarcosine, 27463
N-methylaminodithioformic acid sodium salt, See Unifume, 32954
methylaminoethanoic acid, See sarcosine, 27463
(methylamino)-4-(isopropylamino)-6-(methythio)-s-triazine, See desmetryn, 8142
2-methylamino methyl benzoate, See dimethyl anthranilate, 8503
methylamino-4-methylthio-6-isopropylamino-1,3,5-triazine, See desmetryn, 8142
N-methylaminopropyltrimethoxysilane, See CM8620, 6457
methylamyl alcohol, See methyl amyl alcohol, 19512
2-methylbenzoic acid, See o-toluic acid, 31954
4-methylbenzenesulfonic acid, See toluene sulfonic acid, 31950
4-methylbenzoic acid, See p-toluic acid, 31955
4-methylbenzoyl chloride, See p-toluoyl chloride, 31957
4-methylbenzaldehyde, See PTAL, 25375, p-tolyl aldehyde, 31961
methylbenzene, See toluene, 31946
methyl-benzene, 19514
4-methylbenzenesulfonic acid, See toluene sulfonic acid, 31950
4-methylbenzenesulfonic acid ethyl ester, See ethyl-p-toluenesulfonate, 11089
methylbenzenesulfonic acid, sodium salt, See sodium p-toluenesulfonate, 28827
methylbenzethonium chloride, See Diaparene, 8326, Hyamine® 10X, 14484
2-methylbenzoic acid, See o-toluic acid, 31954
3-methylbenzoic acid, See m-toluic acid, 31953
4-methylbenzoic acid, See p-toluic acid, 31955
methylbenzol, See methyl benzene, 19514
2-methylbenzophenone, 19565
5-methylbenzotriazole, See Preventol® CI7-100, 24929, tolyltriazole, 31962
4-methylbenzoyl chloride, See p-toluoyl chloride, 31957
methyl 5-benzoyl-2-benzimidazole-carbamate, See Mebendazole, 19198
methyl-bladan, See Parathion-methyl, 22808
2-methylbutane, See isopentane, 15394
9-[(3-methylbut-2-enyl)oxy]-7H-furo[3,2-g][1]benzopyran-7-one, See imperatorin, 14912
3-methylbutyl acetate, See isoamyl acetate, 15336, pear oil, 22971
methylcarbamic acid 2,3-(isopropylidenedioxy)phenyl ester, See Bendiocarb, 3937
methylcarbamic acid, o-cumenyl ester, See Etrofolan®, 11156
methylcarbamodithioic acid sodium salt, See Unifume, 32954
methylcarbinol, See ethyl alcohol, 11064
methylcellulose, 19566, See Culminal® methylcellulose, 7347, Methocel, 19488, Methocel® A15-LV, 19490, Methocel® A4C, A4M, 19491, Methocel® A4MP, 19492
methylchloroform, See trichloroethane, 32216
methylcyclohexyldichlorosilane, See CM8645, 6458
methylcyclohexyldimethoxysilane, See CM8650, 6459
methylcyclooctylcarbonate, See Jasmacyclat, 15528
3-methylcyclopentadecanone, See muscone, 20527
methyldichlorosilane, See CM8750, 6460
methyldidecylamine, See Amine M210D, 2159
N-methyldidecylamine, See didecyl methylamine, 8415
methyldioctadecylamine, See Amine M218, 2160
methyldiphenylmethyl-dichloramine, See Dichloramine-M, 8383
methyldithiocarbamic acid sodium salt, See Unifume, 32954
(methyldithio)methane, See Sulfa-Hitech® 0382, 29961
N',N'-methylene bis acrylamide, See methylene bisacrylamide, 19567
methylene base, See Michler's Base, 19644
methylene bichloride, See methylene chloride, 19568
methylene bis(thiocyanate), See Amerstat® 282, 2043, N-948® Biocide, 20641
methylene bisacrylamide, 19567
methylene bisphenyl isocyanate, See MDI, 19188, Rubinate® LF-168, 27108
methylene bisphosphonic acid :ss99m:ksTc complex, See Amerscan MDP Kit, 2026
methylene chloride, 19568, See Aerothene, 900, Driverit, 9043, Nevolin®, 21053, salesthin, 27263
methylene dichloride, See methylene chloride, 19568
methylene dimethyl ether, See methylal, 19564
4,4'-methylene bis(N,N-dimethyl)benzenamine, See Michler's Base, 19644
4,4'-methylene bis(N,N'-dimethylaniline), See Michler's Base, 19644
4,4'-methylene-bis-(2,6-di-t-butylphenol), See Ralox® 02, 25841
methylene diphenyl diisocyanate, See Isonaphthol, 15377
4,4'-methylene-bis-(phenylcarbanilate), See Bonding Agent P 1, 4394
methylene di-p-phenylene isocyanate, See MDI, 19188
methylenebis(p-phenylene isocyanate), See Rubinate® LF-168, 27108
methylene iodide, 19569 See Mi-Gee Brand, 19732
methylene oxide, See formaldehyde, 12258
methylene-p-toluidine, See Vulkacit® FP, 34132
methylenebisacrylamide, See methylene bisacrylamide, 19567
N,N'-methylenebisacrylamide, See Cylink M.B.A, 7572, M.B.A, 18103
4,4'-methylenebis[benzeneamine], See Tonox® 22, Tonox® R, 32014
methylenebisbutyl methylphenol, See Anti-oxidant 2246, 2580
4,4-methylenebis(2,6-di-t-butylphenol), See Lowinox® 002, 17647
2,2'-methylenebis(6-tert-butyl)-p-cresol, See Anti-oxidant 2246, 2580
2,2'-methylenebis(6-tert-butyl)-para-cresol, See

MTES, 9413, Union Carbide® A-162, 32982
methyltrimethoxysilane, *See* CM9100, 6465, Dynasylan® MTMS, 9415, Union Carbide® A-163, 32983
methyltris(methylethylketoxime)silane, *See* CM9220, 6467
5-methyluracil, *See* thymine, 31777
methylvinyldichlorosilane, *See* CV-4772, 7448
methyridine, *See* metyridine, 19609
Metillure, 19580
Metiloil®, 19581
Metion, *See* DL-methionine, 19487
metiram, 19582
metiram-cymoxanil, *See* Avisol® G, 3487
Metmercapturan, *See* methiocarb, 19484
Metmercapturon, *See* methiocarb, 19484
Metobromuron, 19583, *See* Patoran, 22916, Patoran® FL, 22917
Metodik, *See* Hismanal, 14132
Metol, 19584
metolachlor, 19585
Metolaclor, *See* metolachlor, 19585
Metolat FC 355, 19586
Metolat FC 515, 19587
Metolat LA 524, 19588
Metolat TH 75, 19589
metolhydroquinone, 19590
metol-quinone, *See* metolhydroquinone, 19590
Metopirone, 19591, *See* metyrapone, 19608
Metoprolol tartrate, *See* Lopressor, 17602
metopyrone, *See* metyrapone, 19608
metoxuron, *See* Dosaflo, 8812
metoxuron-simazine, *See* Hermes, 13877
Metprep, 19592
Metral®, 19593
metramine, *See* hexamine, 14008
Metrax, 19594
metribuzin, 19595, 19596, *See* Lexone®, 17165, Sencor®, 27947
metribuzin + chlorimuron, *See* metribuzin, 19595
metriphonate, *See* trichlorfon, 32212
Metrizamide, *See* Amipaque, 2197
Metro Tiles, 19597
Metronidazole, *See* Satric, 27491
Metro-nite, 19598
Metropad, 19599
Metroprione, *See* metyrapone, 19608
Metrosol AZ, 19600
Metrotect, 19601
Metrotex, 19602
Metrotex Colors, 19603
Metso, 19604, 19605
Metso Beads, Drymet, *See* sodium metasilicate, 28788
metsulfuron-methyl, 19606, *See* Ally®, 1782
metsulfuron-methyl-thifensulfuron-methyl, *See* Harmony®, 13641
Meturon® 4L, 19607
metyrapone, 19608
metyridine, 19609
Mevantraal, 19610
Mewlon, 19611
Mexapol, 19612
Mexene, *See* AAprotect, 16
Mexican onyx, 19613
Mexican turpentine, *See* turpentine, 32610
Mexico seeds, 19614
mexide, *See* rotenone, 27000
Mexitil, 19615
Mexphalte, 19616
Meyerhofferite, 19617
Meyer's solution, 19618
Meyprofix® 509 (redesignated Polycare® 509), 19619
Mezene, *See* AAprotect, 16
Mezopur, *See* methazole, 19481
MF-344, *See* Aaterra WP, 17
MFM-416, *See* Ageflex EGDMA, 969
MFM-418, *See* diethylene glycol dimethacrylate, 8444
MG2/MG4, 19620
M-Gard, 19621
MGH-93, 19622
Mgoa rubber, 19623
MH, *See* maleic hydrazide, 18477
MH-30, *See* maleic hydrazide, 18477
MHA, 19624
Mhoromer, *See* Bisomer 2HEMA, 4273, Sipomer® HEM-D, 28536
m-methylcarbanilate, *See* Vangard, 33323
Mianin, *See* Chloramine T, 6086
Mianine, *See* Ketjensept, 16133, Chloramine T, 6086
miazine, 19625, *See* Imidazole, 14895, pyrimidine, 25484
miazole, *See*
MIBC, *See* methyl amyl alcohol, 19512
MIBK peroxide, *See* Trigonox® HM, 32302
mica, 19626, *See* Velvet Veil 310, 33551
mica Cambric, *See* micanite cloth, 19636
Mica platelets coated with titanium dioxide and/or iron oxide, *See* Afflair® Lustre Pigments, 930
mica silk, 19627
Mica, Basic bismuth chloride, *See* Bismuth chloride oxide, 4237
Mica-bismuth oxychloride, *See* Bismica 46, 4234
Micabond, 19628
Micacoat®, 19629
Micafil B, 19630
Micafil G, 19631
Micafil S, 19632
Micafolium, 19633
Mica-Kote, 19634
micanite, 19635
micanite cloth, 19636
Micarta, 19637
micarta folium, 19638
Michel XO-150-12, 19640
Michel XO-150-16, 19641, *See* Exxal® 16, 11341
Michel XO-150-1620, 19642
Michel XO-150-20, 19643, *See* octyl dodecanol, 22038
Michel XO-24, 19639
Michler's Base, 19644
Michler's hydride, *See* Michler's Base, 19644
Michler's hydrol, 19645
Michler's ketone, 19646, *See* ketone base, 16140
Michler's Methane, *See* Michler's Base, 19644
Micofume, *See* Dazomet, 7806, N-521® Biocide, 20640
Mi-Col, 19647
Micol, *See* Cetavlex, Cetavlon, 5787, Cetrimonium bromide, 5813, Sumquat® 6030, 30122, Varisoft® CTB-40, 33421
Micol Quamonium, *See* Acetoquat CTAB, 303
Micracet, 19648
Micral® 855, 19650
Micral® 932, 19651
Micral® 1000, 19649
Micro DDT 75, *See* DDT, 7820
Micro K, *See* K. Tab, 15642, Kaon-Cl, 15729, Kay Ciel, 15807, K-Contin, 15818, K-Lor, 16226, K-Lyte/C1, 16234, P KM Potassium Chloride, 16242, otassium chloride, 24748, 24749
Micro P Extender, 19652
micro-asbestos, 19653
Microbar, *See* Radiopaque, 25830
Microbator PC-78, 19654
Microbicide M-8, *See* Skane® M-8, 28563
Microbiotone, 19655
Microbloc®, 19656
Microcal, 19657, *See* Extrusil, 11331
Microcal ET, 19658, *See* Extrusil, 11331
Microcarb, 19659, *See* Carbaryl, 5181
Microcarb T, 19660
Microcatalase®, 19661
Micro-Cel® A, 19662
Micro-Cel® A, T-38, *See* Extrusil, 11331
Micro-Cel® T-38, 19663
micro-chek 11, *See* Skane® M-8, 28563
micro-chek 11d, *See* Skane® M-8, 28563
micro-chek skane, *See* Skane® M-8, 28563
Micro-Chek® 11, 19664
microcosmic salt, 19665
Microcrystalline cellulose, *See* Avicel, 3470
Microcrystalline paraffin waxes and hydrocarbon waxes, *See* Mekon® White, 19248, Paracol® 404C, 22724
Microcrystalline wax, 19666
Microcult® GG, 19667
Micro-Dee, *See* vitamin D+73, 33941
Microdol (Extra), 19668
Micro-Dry®, 19669
Microduct®, 19670
Microduct®, *See* Maltodextrin, 18500
Micro-fine®, 19671
MicroForm B, 19672
MicroForm BCS, 19673
Microgen Plus, 19674
Microhoba, 19675
Micro-K, 19676
Microlan, 19677
Microlith-A, 19678
Microlube A, 19679, *See* Paraffin, 22743
micromeritol, 19680
Micromet, 19681
Micromid 1022, 19682
Micromite, 19683, *See* Diflubenzuron, 8450
Micromulse WIO, 19684
Micronal®, 19685
Micronized polyethylene, *See* ACumist B-6, B-12, B-18, C-5, C-12, C-18, 581
Micropaque, 19686, *See* Radiopaque, 25830
MicroPflex 1200, 19687
Micropil, 1968, 19689 8
Micropoly 520, 19690
Micropore, 19691
microporite, 19692
Micropur, 19693
Micropure® Ultra, 19694
Microsan, 19695
Microseal, 19696
Microsil, 19697
Microsized, 19698
Microsperse, 19699
Micro-Step® H-301, 19700
Microstix®, 19701
Micro-supplex®, 19702
Microtex GTZ, 19703
Microthene®, 19704
Microthene® FA 150-00, 19705
Microthene® MA 530-060, 19706
Microthene® MN 701-00, 19707
Microthene® MP 625U, 19708
Microthene® MU 760-00, 19709
Microtomic 280, *See* Abluwax EBS, 140
Microtrast, *See* Radiopaque, 25830
Micro-triever®, 19710
Microtuff 1000, 19711
Micro-White® 07 Slurry, 19712
Micro-White® 10 Codex, 19713
Micro-White® 15, 19714
Micro-White® 25, 19715
Micro-White® 40, 19716
Micro-White® 100, 19717
Microx, 19718
Micryston, 19719
Midas Gold®, 19720
Midas®, 19721
middle oils, 19722
Midstream, 19723
Midvale Alloys, 19724
Midvaloy H.R, 19725
Miedzian, *See* Cupravit®, 7370
Miedzian 5, *See* Cupravit®, 7370
Miedzian 50, *See* Cupravit®, 7370
Mifaslug, 19726, *See* metaldehyde, 19431
Mifatox, 19727, *See* demeton-S-methyl, 8055
Migafar, 19728
Migafar AL, 19729
Migassist® NYL, 19730
Migatex, 19731
Mi-Gee Brand, 19732

Migen, 19733
Mighty Soft, 19734
Miglyol®, 19735
Miglyol® 808, 19736
Miglyol® 808, *See* tricaprylin, 32211
Miglyol® 812, 19737
Miglyol® 812, *See* caprylic/capric acid triglyceride, 5161
Miglyol® 818, 19738
Miglyol® 829, 19739
Miglyol® 840, 19740
Miglyol® Gel, 19741
Mignonette Green, *See* May Green,
mignonette-geranium oil, 19742
migra iron, 19743
Migraine Dolviran®, 19744
Migranil 858, 19745
Migrassist® D, 19746
Mikacion Dye®, 19747
Mikawhite®, 19748
Mikolite, 19749
Mikrobin, 19750
Mikron, 19751
Mila, 19752
milam, *See* Bisphenol A, 4283
Milanol, 19753
Milbam, *See* AAprotect, 16
Milban, *See* dodemorph-acetate, 8782
Milcap, 19754
Mil-Col, 19755, *See* drazoxolan, 8974
Milcurb, 19756, *See* dimethirimol, 8495
mild alkali, 19757
mild lime, 19758
Mildew remover X-14, *See* Repak, 26141
Mildex, *See* Karathane, 15759
Mildothane, 19759, *See* thiophanate methyl, 31706
Mildothane Turf Liquid, 19760
Mil-du-rid, *See* Dowicide A, 8932
Mildvac Ma5 (Mild Mass. Type), 19761
Mildvac-C, 19762
Mildvac-M, 19763
milfuram, *See* Patafol, 22881
Milgard, 19764
Milgo, 19765
Milid, *See* proglumide, 25116
Milide, *See* proglumide, 25116
Miliden X-2, *See* Nipacide® BK, 21254
Miliden X-2, *See* Bioban® GK, 4158
milk acid, *See* lactic acid, 16570
Milk amino acids, *See* Milkamino 20, 19770
milk glass, 19766
milk of lime, 19767
milk protein, *See* casein, 5386
milk sugar, 19768, *See* α-D-Lactose, 16589
milk sugar rennet, *See* Pegnin, 23056
milk tree wax, 19769
Milkamino 20, 19770
milkstone, 19771
Mill Creek, 19772
Millad®, 19773
Millad® 3905, 19774
Millad® 5L71-10, 19775
Millad® HBPA, 19776
millafol, *See* folic acid, 12196, vitamin M, 33951
Millaloy, 19777
Millamine® 5260, 19778
Millathane® 66, 19779
Millathane® 88, 19780
Milldride®, 19781
Milldride® DDSA, 19782
Milldride® HDSA, 19783
Milldride® HHPA, 19784
Milldride® MHHPA, 19785
Milldride® nDDSA, 19786
Milldride® nDSA, 19787
Milldride® ODSA, 19788
Milldride® OSA, 19789
Milldride® TDSA, 19790
Miller's fumigrain, *See* Acrylonitrile, 436
Millidet, 19791
Millifoam, 19792
milling silver, 19793
Millithix® 925, 19794
Millon's mase, 19795
Millon's reagent, 19796
Millophyline, 19797
Mills Plastic, 19798
Miloderm, *See* Vaderm, 33237
Milogard, *See* propazine, 25179
Milowite, 19799
Milstem, 19800
Milton, 19801, *See* sodium hypochlorite, 28773
Miltopan, 19802
Milvan Steel, 19803
Milwaloy, 19804
Mimea, *See* momea, 20093
Mimico, *See* Carbora, 5235
mimosa, *See* wattle bark, 34230
Mina D+72, *See* vitamin D+72, 33940
Minadex, 19805
Minamino, 19806
minargent, 19807
minargentatum, 19808
Mindel® A-670, 19809
Mindel® B-310, 19810, *See* Polysulfone, 24619
Mindel® B-322, 19811, *See* Polysulfone, 24619
Mindel® M-800, 19812, *See* Polysulfone, 24619
Mindel® S-1000, 19813, *See* Polysulfone, 24619
Minder, 19814, 19815
Mindust Series, 19816
mineral acid, 19817
mineral black, 19818
mineral blue, *See* mountain blue, 20366
mineral brown, *See* Cappagh brown, 5121, umber, 32900
mineral butters, 19819
mineral caoutchouc, *See* elaterite, 9713
mineral carbon, 19820, *See* graphite, 13277
Mineral Cotton, *See* slag wool, 28589
mineral fat, *See* petroleum jelly,
mineral flour, 19821
mineral glycerin, *See* paraffin, liquid, 22744
mineral green, *See* cupric acetoarsenite, 7374, mountain green, 20369, Scheele's green, 27562
mineral grey, 19822, *See* slate grey, 28594
Mineral Gum, *See* soluble glass, 28980
Mineral iron sulfide (pyrite), *See* Intral®, 15171
Mineral Jelly, *See* petroleum jelly, petrolatum, 23495
Mineral Jelly No. 5, 19823
mineral khaki, 19824
mineral lake, 19825
Mineral oil, *See* Amerchol® L-101, 2001, 2003, Amerchol® L-500, 2002, Kaydol, 15810, Kendex OCTG, 16002, Klearol, 16208
Mineral oil, coco-hydrolyzed collagen, cetyl alcohol, myristyl myristate, ceteth-16 and hydrogenated lanolin, *See* proto-Lan 4R, 25293
mineral orange, *See* red lead, 25960
Mineral phosphates, *See* Florida phosphates, 12018
mineral pitch, *See* bitumen, 4294 Karetnja, 15765
Mineral pulp, *See* Agalite, 953
mineral red, *See* red lead, 25960
mineral rubber, 19826
Mineral spirits, *See* Varsol® 1, 33456
Mineral superphosphate, *See* superphosphate, 30263
mineral syrup, *See* paraffin, liquid, 22744
mineral tallow, *See* bitumen, 4294
mineral umber, *See* umber, 32900
mineral violet, *See* Manganese Violet, 18550
mineral wax, *See* ceresin, 5746, ozokerite, 22483
mineral white, *See* gypsum, 13518
mineral wool, *See* slag wool, 28589
mineral yeast, 19827
Minex 2, 19828
Minflo, 19829
Mini Slugit Pellets, 19830
Minihep, *See* Heparin, Sodium Salt, 13788
Minite, 19831
minium, *See* red lead, 25960
minium tego, 19832
mink oil, 19833
Mink oil, *See* Emulan, 10485
Minkamide DEA, *See* Incromide Mink D, 14992
Minkamidopropalkonium chloride, *See* Incroquat Mink-85, 15040
Minkamidopropyl betaine, *See* Incronam Mink 30, 15021
Minkamidopropyl dimethylamine, *See* Foamole B, 12177
Minkamidopropyl dimethylamine lactate, *See* Incromate Mink L, 14970
Minkamidopropylamine oxide, *See* Incromine Oxide Mink, 15008
Minlon®, 19834
Minofor, 19835
Minol, *See* paraffin, liquid, 22744
Minolite Antigrisouteuse, 19836
Minoxidil, *See* Loniten, 17584
mint camphor, *See* menthol,
Mintacol, *See* paraoxon, 22783
Mintesol, *See* thiabendazole, 31661
Mintezol, *See* Tecto, Tecto 60%, 30913, thiabendazole, 31661
Mintite, 19837
Minugel, 19838, *See* Attapulgite, 3404
Min-U-Gel® 100, 19839
Min-U-Gel® 100, -200,-AR, *See* Attapulgite, 3404
Min-U-Gel® 200, 19840
Min-U-Gel® AR, 19841
Min-U-Gel® CW, 19842
Min-U-Gel® LF, 19843
Min-U-Sil, 19844
Minyak Kerung, 19845
Minzolum, *See* Tecto, Tecto 60%, 30913, thiabendazole, 31661
Miotisal A, *See* paraoxon, 22783
MIPA, *See* isopropanolamine, 15406
MIPA Aliphatic amine, *See* isopropanolamine, 15406
MIPA C_{12-15} pareth sulfate, *See* Zetesol 856 T, 34744
MIPA-dodecylbenzenesulfonate, *See* Hetsulf IPA, 13982
MIPA-laureth sulfate and cocamide DEA, *See* Texapon IES, 31460
MIPA-laureth sulfate-laureth-4-cocamide DEA, *See* Zetesol 100, 34743
MIPA-lauryl sulfate, *See* Sulfetal CJOT 38, 29983
Mipax, *See* dimethyl phthalate, 8508, Kemester® DMP, 15956, Kodaflex® DMP, 16271, Unimoll® DM, 32979
Mipax, *See* dimethyl phthalate, 8508
MIPC, *See* Etrofolan®, 11156, isoprocarb, 15404
Mipcin, *See* Etrofolan®, 11156, isoprocarb, 15404
Mipcin®, 19846
Mipolam, 19847
Mira metal, 19848
Mirabilite, *See* Glauber's salt, 12974
Miracare® 2MCA, 19849
Miracare® 2MCA-SF, 19850
Miracare® ANL, 19851
Miracare® BC-10, 19852
Miracare® BT, 19853
Miracare® CT 100, 19854
Miracare® M1, 19855
Miracare® MHT, 19856
Miracare® MS-1, 19857
Miracare® NWC, 19858
Miracare® SCS, 19859
Miracare® XL, 19860
Miracle Man, 19861
Miraculoy, 19862
Miraflon®, 19863
Mirage, 19864
Mira-Gel® 463, *See* corn starch, 6837
Miralite, 19865
Miramant, 19866
Miramine® C, 19867
Miramine® CODI, 19868, *See* Cocamidopropyl dimethylamine, 6552
Miramine® GS, 19869

Chemical and Trade Names

Miramine® HPS-B, 19870
Miramine® O, 19871
Miramine® SODI, 19872
Miramine® TO, 19873
Miranate® B, 19874
Miranate® LEC, 19875
Miranol, 19876
Miranol C2M, 19877
Miranol C2M-SF, 19878
Miranol CM, 19879
Miranol DM, 19880
Miranol JEM, 19881
Miranol L2M-SF, 19882
Miranol SM, 19883
Miranol® 2CIB, 19884, *See* disodium cocoamphodiacetate, 8643
Miranol® BM Conc, 19885
Miranol® C2M Anhyd. Acid, 19886
Miranol® C2M Conc. NP, *See* disodium cocoamphodiacetate, 8643
Miranol® C2M Conc. NP-PG, 19887
Miranol® C2M-SF 70%, 19888
Miranol® CM Conc. NP, 19889
Miranol® CM-SF Conc, 19890
Miranol® CS Conc, 19891
Miranol® DM Conc. 45%, 19892
Miranol® Ester PO-LM4, 19893
Miranol® FA-NP, 19894
Miranol® FB-NP, 19895
Miranol® FBS, 19896
Miranol® H2C-HA, 19897
Miranol® H2M Conc, 19898
Miranol® H2M-SF Conc, 19899
Miranol® HM Conc, 19900
Miranol® HM-SF Conc, 19901
Miranol® J2M Conc, 19902
Miranol® J2M-SF Conc, 19903
Miranol® JAS-50, 19904
Miranol® JB, 19905
Miranol® JBS, 19906
Miranol® JS Conc, 19907
Miranol® L2M-SF Conc, 19908
Miranol® OS-D, 19909
Miranol® S2M-SF Conc, 19910
Miranol® SM Conc, 19911
Miranol® TBS, 19912
Miranol®CM Conc. NP, *See* Empigen® CDR40, 10369
Mirapol® 1941, 19915
Mirapol® 550, 19914
Mirapol® 9, 95, 175, 19913
Mirapol® A-15, 19916
Mirapol® AD-1, 19917
Mirapol® AZ-1, 19918
Mira-Set® B, *See* corn starch, 6837
Mirasheen® 202, 19919
Mirasol, 19920
Mirataine CDMB, *See* Accobetaine CL, 222
Mirataine® A2P-TS-30, 19921
Mirataine® BB, 19922
Mirataine® BET-C-30, 19923
Mirataine® BET-CS, *See* Mirataine® CBS, 19927
Mirataine® BET-O-30, 19924, *See* Oleamidopropyl betaine, 22108
Mirataine® BET-P-30, 19925
Mirataine® BSC, 19926
Mirataine® CBS, CBS Mod, 19927
Mirataine® H2C-HA, 19928
Mirataine® HC-Acid, 19929
Mirataine® JC-HA, 19930
Mirataine® T2C-30, 19931, *See* disodium tallowiminodipropionate, 8655
Mirataine® T2C-35%;, *See* disodium tallowiminodipropionate, 8655
Mirataine® TM, 19932
Miratiane BD-J, *See* Cocamidopropyl betaine, 6551
Miravon, 19933
Miravon B12DF, 19934
mirbane oil, 19935, *See* nitrobenzene, 21348
Mirion, 19936
Miristalkonium chloride, *See* JAQ Powdered Quaternary, 15517
Mirlon, *See* Cabelec® 1015, 4861
Mirrolac, 19937
mirror bronze, 19938
Mirvale, 19939
Mischzinn, 19940
miscible carbon disulfide, 19941
Misco, 19942
Miscrome, 19943
Missile, 19944
Mission Prenatal, 19945
mistletoe rubber, 19946
Mist-O-Matic, 19947
Mist-o-matic Ferrax, 19948
Mistral, 19949
Mistral CT, 19950
Mistron CB, 19951
Mistron Vapor-RE, 19952
Mistron ZSC, 19953
Mitaban, 19954
Mitac, *See* Amitraz, 2202, Taktic, 30729
Mitac 20, 19955
Mitacid, *See* cyhexatin, 7568
Mitas®, 19956
Mitchalloy A, 19957
Mitec® GP105A, 19958
Mitifon, *See* Tedion V-18, 30919, tetradifon, 31366
Mitigal, *See* Odylen®, 22046
Mitigan, 19959, *See* Acarin, 183, Dicofol, 8398, Kelthane, 15881
mitigated caustic, 19960
Mitin, 19961
Mitine, 19962
Mition, *See* tetradifon, 31366
Mitis green, *See* cupric acetoarsenite, 7374, mountain green, 20369
Mitotan, *See* Lysoform, 18095
Mitotane, *See* Lysodren, 18094, Lysoform, 18095
Mitrelle, 19963
Mitschlich's ammoniacal salt, 19964
Mitsubishi 4300J, 19965
Mitsubishi BT002, 19966
Mitsubishi ET008, 19967
Mitsubishi F101A, 19968
Mitsubishi JS050, 19969
Mitsubishi Kasei GF-PET 6010G15, 19970, *See* Polyethylene terephthalate, 24426
Mitsubishi Kasei PBT 5008, 19971
Mitsubishi Kasei PBT 5008, PBT 5010F1, *See* Grilpet® XE3060, 13402
Mitsubishi Kasei PBT 5010F1, 19972
Mitsubishi Kasei PPS 704G40, 19973
Mitsubishi L300, 19974
Mitsubishi UF421, 19975
Mitsubishi Yuka-ECX, 19976
Mitsubishi Yuka-SPX, 19977
Mittel AEP, 19978
Mittel KP, 19979
Mittel L, 19980
Mittler's green, *See* chromium green, 6232
Mix, 19981
Mixad, 19982
Mixed 1-(2-aminoethyl)-2-n-alkyl-2-imidazoline, *See* Witcamine® 211, 34446
mixed acid, *See* nitrating acid, 21329
Mixed chlorides of magnesium, potassium and sodium, *See* Abraum salts, 156
Mixed cresols, *See* cresylic acid, 7013
Mixed diaryl *p*-phenylene diamine, *See* Akrochem® Antiozonant MPD-100, 1163
mixed ether, 19983
mixed glycerides of oleic (83%), palmitic (9%), linoleic (4%), stearic (2%) and arachidic (1%) acids, *See* olive oil, 22138
Mixed isopropanolamines myristate, 19984
mixed metal, 19985
mixed vitriol, 19986
Mix-Kit, 19987
Mixobar, *See* Radiopaque, 25830
Mixol, 19988
Mixture of Aerosil and alumina in 5:1 ratio, *See* Aerosil COK 84, 870
Mixture of Aerosil with 15% starch, *See* Aerosil Composition, 871
Mixture of ammonium laureth sulfate and cocamide MEA, *See* Stepanol® AEM, 29692
Mixture of ammonium lauryl sulfate, ammonium laureth sulfate, cocamidopropyl betaine and cocamide DEA, *See* Stepanol® AEG, 29691
Mixture of Cetyl alcohol, Glyceryl stearate, dicetyl dimonium chloride, cetrimonium chloride, *See* Carsoquat® CB, 5348
Mixture of DEA lauryl sulfate, DEA lauraminopropionate and sodium lauraminopropionate, *See* Stepanol® LX, 29694
Mixture of dibenzoyl peroxide and calcium sulfate, *See* Cadox® BTA, 4910
Mixture of dioctyl adipate, octyl stearate, octyl palmitate, *See* Wickenol® 161, 34396, Wickenol® 163, 34397
Mixture of fatty acids, *See* Akrochem® Proaid FLOW, 1175
Mixture of formaldehyde and lindane, *See* Microgen Plus, 19674
Mixture of glycol distearate, sodium laureth sulfate and propylene glycol, *See* Standapol® Pearl Conc. 7130, 29476
Mixture of glycol stearate, lauramine oxide, propylene glycol and ceteareth-20, *See* Standapol® CAT, 29469
Mixture of 1-hexene, 1-heptene and 1-octene, *See* Neodene® 6/8, 20888
Mixture of higher alcohols, *See* Aerofroth 76, 858
Mixture of hydrophilic fatty acid esters, *See* Witafrol® 7456, 34423
Mixture of magnesium lauryl sulfate and disodium laureth sulfosuccinate, *See* Texapon MG 3, 31466
Mixture of three isomeric mononitrotoluenes, *See* M.N.T, 18105
Mixture of phenoxyethanol, methylparaben, ethylparaben, propylparaben and butylparaben, *See* Sepicide HB, 27983
Mixture of powdered PTFE and titanium dioxide, *See* PTFE-20, 25376
Mixture of sodium lauryl sulfate, sodium cetyl sulfate and laureth-3, *See* Standapol® CS Paste, 29470
Mixture of sodium lauryl sulfate, sodium laureth sulfate, lauramide MIPA and cocamide MEA, *See* Standapol® S, 29477
Mixture of surfactants, *See* Aerodri 200, 850
Mixxim® BB/100, 19989
Mixxim® BB/50, 19990
Mixxim® HALS 57, 19991
Mixxim® HALS 63, 19992
Mixxim® HALS 68, 19993
MJF-11567, *See* Cefadroxil, 5547
MK 6, *See* vitamin K+72, 33950
MK-23, *See* fluoromide, 12108
MK 360, *See* thiabendazole, 31661
MK-360, *See* thiabendazole, 31661
MK-360, *See* Tecto, Tecto 60%, 30913
MK-933, *See* Ivermectin, 15454
MK-990, *See* Ranide, 25868
MKP, *See* potassium phosphate, 24773
ML 33F, *See* Ablunol S-80, 111, Witconol 2500, 34501
MI 55F, *See* Ablunol S-80, 111, Witconol 2500, 34501
MLT, *See* malathion, 18460
MMA, *See* methyl methacrylate, 19548
MME, *See* methyl methacrylate, 19548
MN powder, 19994
Mo 1202T, *See* molybdenum trioxide, 20081
MO 55F, *See* Ablunol S-80, 111, Witconol 2500, 34501
moac, 19995
MOB, *See* Cyasorb® UV 9, 7495, Neo Heliopan® BB, 20867, Rhodialux A, 26675, Uvinul® M-40, 33197
Mobil 1240, 19996
Mobil 2120, 19997

Mobil 5350, 19998
Mobil 5600, 19999
Mobil 8020, 20000
Mobil MX 4354, 20001
Mobilrap, 20002
Mobilrapper, 20003
Mobilsol®, 20004
Moca®, 20005
Mocap 10G, 20006
Mocasco Iron, 20007
mocaya oil, 20008
mocha-stone, 20009
mock gold, *See* mosaic gold, 20355
mock lead, 20010
mock silver, 20011
mock vermilion, 20012
MOD, *See* Cyasorb® UV 9, 7495, Neo Heliopan® BB, 20867, Rhodialux A, 26675
Mod Acid, 20013
Modaflow, 20014
Modane Soft, 20015
Modar, 20016
Modar 814, 20017
Modar 826HT, 20018
Modar 865, 20019
Modarez APVC 8, 20020
Modarez MFP Powd, 20021
Moddite, 20022
Moderator, 20023
Modic, 20024
Modicol L, 20025
Modicol S, 20026
Modified 2:1 tall oil fatty acid alkanolamide, *See* Actramide 5264, 545
Modified alcohol ethoxylate, *See* Fl-Mo 80/20, 11997
Modified alkanolamide, *See* Monalube 780, 20102
Modified amine, *See* Michel XO-24, 19639
Modified Butacite, 20027
Modified Dakin's solution, *See* sodium hypochlorite, 28773
Modified dioctylsulfosuccinate, *See* Aerodri 100 104, 849
Modified ethoxylate, *See* Sole-Mulse B, 28924
Modified lecithin, *See* Crolec 4135, 7189
Modified melamine formaldehyde resin, *See* Aerotru 23, 901
Modified quaternary ammonium compound, *See* Acryloft, Acryloft Conc, 421
modified soda, 20028
modified sulfated fatty acid ester, *See* Wetanol, 34334
Modified triaryl phosphate, *See* Santicizer 143, 27390
Modified triglyceride, *See* Actralube 1200, 537
Modified Vinylite X, 20029
Modinal T, 20030
Modown, *See* bifenox, 4135
Modown 4 Flowable, *See* bifenox, 4135
Modrasone, *See* Vaderm, 33237
Modulan®, 20031 , *See* Fancol Acel, 11428
Modulex, 20032
Moellon R, 20033
Mofenar, *See* bufexamac, 4655
Mofix, 20034
Mo-Flo, 20035
Mogador gum, *See* rocco gum, Morocco gum, 20339
Mogul® L, 20036
mohair, 20037
Mohawk Steel, 20038
Mohr's salt, 20039
Molaschar, 20040
molascuit, 20041
molasocarb, 20042
Molass, *See* paclobutrazol, 22540, 22541
molasses, 20043
molasses alcohol, *See* ethyl alcohol, 11064
molassine meal, 20044
Molco, 20045
Mold Wiz Ext, 20046
Mold Wiz Int, 20047
Moldabaste® Moldabaster S, 20048
Moldag 200, 20049
Moldano®, 20050
Moldaroc®, 20051
Moldasil, 20052
Moldastone, 20053
Moldasynt, 20054
Moldcote, 20055
Moldensite, 20056
Moldesite®, 20057
Moldex, 20058
MoldPro 613, 20059
MoldPro 759, 20060
MoldPro 830, 20061
moler, 20062
Molera, 20063
Mol-Iron, *See* ferrous sulfate, 11690
Mollan O, *See* Bis(2-ethylhexy) Phthalate, 4228, dioctyl phthalate, 8549, Reomol DCP, 26130, Witcizer 312, 34454
Mollan S, *See* dioctyl adipate, 8548, Plasthall® DOA, 24002, PX-238, 25440, Wickenol® 158, 34393
Mollescal® C Conc, 20064
Mollifex, 20065
Mollin, 20066
Mollit, 20067
Mollit B, 20068
mollphorus, 20069
Mollutox, *See* clonitrilide, 6432
Molochite®, 20070
Moloie, 20071
Molsidolat, 20072
Moltopren®, 20073
Molurame, *See* AAprotect, 16
Molybdate Red, 20074
molybdena, *See* molybdenum trioxide, 20081
molybdenite, *See* molybdenum disulfide, 20077
molybdenum, 20075
molybdenum anhydride, *See* molybdenum trioxide, 20081
molybdenum dioxide, 20076
molybdenum disulfide, 20077, *See* Moldag 200, 20049
molybdenum nickel, 20078
molybdenum orange,, *See* Molybdate Red, 20074
molybdenum(IV) oxide, *See* molybdenum dioxide, 20076
molybdenum(IV) sulfide, *See* molybdenum disulfide, 20077
molybdenum(VI) oxide, *See* molybdenum trioxide, 20081
molybdenum oxide, *See* molybdenum trioxide, 20081
Molybdenum Permalloy, 20079
molybdenum peroxide, *See* molybdenum trioxide, 20081
molybdenum steel, 20080
molybdenum sulfide, *See*
molybdenum sulfide, *See* molybdenum disulfide, 20077, SLA 1261, 28577
molybdenum trioxide, 20081
molybdic acid anhydride, *See* molybdenum trioxide, 20081
molybdic acid calcium salt, *See* calcium molybdate, 4958
molybdic acid hydride, *See* molybdenum trioxide, 20081
molybdic acid, disodium salt, dihydrate, *See* sodium Molybdate, dihydrate, 28796
molybdic anhydride, *See* molybdenum trioxide, 20081
molybdic oxide, *See* molybdenum trioxide, 20081
molybdic sulfide, *See* molybdenum disulfide, 20077
molybdic trioxide, *See* molybdenum trioxide, 20081
molybdophosphoric acid, *See* phosphomolybdic acid, 23731
molybdosodalite, 20082
Molydag, 20083
Molydag 204, 20084
Molydag 206, 20085
Molydag 208, 20086
Molydag 210, 20087
Molydag 211, 20088
Molydag 214, 20089
Molyhibit 100, *See* sodium Molybdate, dihydrate, 28796
Molykote®, 20090
Molyte, 20091
Molyvan, 20092
momea, 20093
Momentum, *See* Valadol, 33238
MON 0573, *See* glyphosate, 13148
MON 2139, *See* glyphosate, 13148
MON 6000, *See* glyphosate, 13148
Mona NF-10, 20094
monacetin, 20095
Monachit, 20096
Monacrin, 20097
monacrin hydrochloride, *See* Monacrin, 20097
Monafax, 20098
Monafax 785, 20099
Monafax 1214, 20100
Monalube 29-78, 20101
Monalube 780, 20102
Monalux CAO, 20103
Monamate C-1142, 20104
Monamate CPA, 20105
Monamate LA-100, 20106
Monamate LNT-40, 20107
Monamate OPA, 20108
Monamate OPA-100, 20109
Monamate RMEA-40, 20110
Monamid 150D, *See* Active #2, 501
Monamid 150DB, *See* Active #2, 501
Monamid 150-LW, *See* Witcamide® 5195, 34440
Monamid® 7-100, 20111
Monamid® 15-70W, 20112
Monamid® 150-ADY, 20113
Monamid® 150-CW, 20114
Monamid® 150-LMWC, 20115
Monamid® 150-MW, 20116
Monamid® 718, 20117
Monamid® CMA, 20118
Monamid® LIPA, 20119
Monamid® LMA, 20120
Monamid® R31-42, 20121
Monamid® S, 20122
Monamide, 20123
Monamide 150LW, *See* Witcamide® 5195, 34440
Monamine, 20124
Monamine 779, 20125
Monamine AA-100, 20126
Monamine ACO-100, 20127
Monamine ADY-100, 20128
Monamine LM-100, 20129
Monamine T-100, 20130
Monamulse 653-C, 20131
Monamulse Cl, 20132
Monaprin, 20133
Monaquat ISIES, 20134
Monaquat TG, 20135
Monarch® 1100, 20136
Monasirup, *See* Arizole® Anethole Extra, 2885
Monastat 1195, 20137
Monastral, 20138
Monate merge 823, *See* Neoarsycodyl, 20871
Monaterge, 20139
Monaterge 85 HF, 20140
Monateric 805, 20141
Monateric 810-A-50, 20142
Monateric 811, 20143
Monateric 811, 20144
Monateric 951A, 20145
Monateric 985A, 20146
Monateric 1000, 20147
Monateric 1000, 20148
Monateric 1188M, 20149, *See* disodium lauryliminodipropionate, 8650
Monateric 1202, 20150
Monateric 1203, 20151
Monateric ADA, 20152
Monateric CA-35, 20153
Monateric CA-35%, 20154

Monateric CAB, 20155
Monateric CAM-40, 20156
Monateric CDL, 20157
Monateric CDTD, 20158
Monateric CDX38, 20159
Monateric CDX-38, 20160
Monateric CEM-38, 20161
Monateric CEM-38%, 20162
Monateric CM-36S, 20163, *See* Empigen® CDR40, 10369
Monateric CSH 32, 20164
Monateric CSH-32, 20165
Monateric CyNa-50, 20166
Monateric CyNa-50%, 20167
Monateric ISA-35, 20168
Monateric ISA-35%, 20169
Monateric L30, 20170
Monateric LF, 20171
Monateric LMAB, 20172, *See* lauramidopropyl betaine, 16830
Monateric LMM-30, 20173
Monateric TA-35, 20174
Monateric TDB-35, 20175, *See* disodium tallowiminodipropionate, 8655
Monatrope 1250, 20176
Monawet 1240, 20177
Monawet MB-45, 20178, 20179
Monawet MM-80, 20180, 20181
Monawet MO, 20182
Monawet MO-65-150, 20183
Monawet MO-70, 20184
Monawet MO-84R2W, 20185
Monawet MT Series, 20186
Monawet MT-70, 20187
Monawet SNO-35, 20188, 20189
Monawet TD-30, 20190, 20191
Monazol, *See* glycine, 13095
Monazoline C, 20192
Monazoline C, CY, O and T, 20193
Monazoline CY, 20194
Monazoline IS, 20195
Monazoline O, 20196
Monazoline S, *See* stearyl hydroxyethyl imidazoline, 29606
Monazoline T, 20197
Monceren, *See* pencycuron, 23097, 23098 23099
Monceren®, 20198
Moncler Derma, 20199
Mond 70 alloy, 20200
Mond gas, 20201
Mondur, 20202
Monece®, 20203
Monel Alloy 400, 20204
Monel Alloy 414, 20205
Monel Alloy K-500, 20206
Monel metal, 20207
monensic acid, *See* monensin, 20208
monensin, 20208, *See* Romensin, 26930
monetite, 20209, *See* Calbrite, 4928
Monex, 20210
Monitan, 20211, *See* Radiasurf® 7157, 25804
Monite, 20212
Monitor, 20213, *See* methamidophos, 19474, Tamaron®, 30760
Monnex, 20214
mono, *See* MSG, 20416
Mono Ammonium Phosphate (Agricultural Grade), 20215
Mono Methyl Ether Hydroquinone, *See* Eastman® HQMME, 9467
Mono Thiurad, 20216
mono(2-ethylhexyl)sulfate sodium salt, *See* Niaproof® Anionic Surfactant 08, 21092, Rhodapon® BOS, 26632
mono-*p*-phenetol-carbamide, *See* Dulcin, 9091
mono-*tert*-butylhydroquinone, *See* *t*-butylhydroquinone, 4793, Eastman® MTBHQ, 9471, Sustane® TBHQ, 30437
monoacetin, *See* monacetin, 20095
1-monoacetin, *See* monacetin, 20095
monoaluminum phosphate, *See* aluminum phosphate, 1922
monoaluminum stearate, *See* Synpro® 404, 30593
monoammonium phosphate, *See* Ammonium phosphate monobasic, 2251, Mono Ammonium Phosphate (Agricultural Grade), 20215
monoammonium phosphate-diammonium phosphate, *See* Flexible Fyrex®, 11933
monoammonium sulfamate, *See* Ammonium sulfamate, 2254
monobasic calcium phosphate, *See* acid calcium phosphate, 330
monobasic lead acetate, *See* subacetate of lead, 29902
Mono-Baycuten®, 20217
Monobed, 20218
Monobel, 20219
Monobutylamine, *See* Amine C4, 2152
monocalcium orthophospate, *See* acid calcium phosphate, 330, calcium phosphate (monobasic), 4966, V-90®, 33222
Monocalcium phosphate, anhydrous, *See* V-90®, 33222
Monocalcium phosphate, monohydrate, *See* Calcium phosphate, monobasic, 4966, Regent® 12XX, 26014
α-monocaprylin, *See* Witafrol® 7420, 34421
Monocast® MC 901, 20220
Monochlorbenzene, *See* Abluton T30, 139
Monochloro and monobromo hydroquinones, *See* Adurol, 797
monochloroacetic acid, 20221
monochlorobenzene, *See* Abluton T30, 139, chlorobenzene, 6119
monochloroethane, *See* ethyl chloride, 11067
Monochloroethanoic acid, *See* monochloroacetic acid, 20221
monochloromethane, *See* methyl chloride, 19522
Monochlorotoluene, *See* Halso® 99, 13570
monochlorotrimethylsilicon, *See* CT2950, 7325
Mono-Coat®, 20222
Monococo amidoamine quaternary, *See* Tomah Q-511, 31995
monocresyl diphenylphosphate, *See* Disflamoll® DPK, TPK, 8641
Monocron, 20223
Monoctylamine, *See* Amine 8 D, 2125
Monocyclic C_{21} dicarboxylic acid, *See* Latol 1550, 16822
Monodehydrosorbitol monooleate, *See* Ablunol S-80, 111, Witconol 2500, 34501
monododecyl sodium sulfate, *See* sodium lauryl sulfate, 28781
monoethanolamine, *See* ethanolamine, 10930
Monoethanolamine borate, *See* MB 450, 19168
Monoethanolamine lactic acid amide, *See* Lactamide MEA, 16568
Monoethanolamine lauryl sulfate, *See* Sulphatol 33 MO, 30063
Monoethanolamine stearic acid amide, *See* Ablumide SME, 82, Witcamide® 70, 34442
monoethanolethylenediamine, *See* Aminoethylethanolamine, 2172
monoethyl ether of diethylene glycol, *See* Diethoxol, 8435, ethyl di-lcinol, 11068
Monoethyl maleate, *See* Sipomer® MEM, 28540
monoethylene glycol dimethyl ether, *See* monoglyme, 20227
mono(2-ethylhexyl)phosphate, *See* Rhodafac® PEH, 26601
Monofilament, *See* Plataril, 24087
monofluorobenzene, *See* Fluorobenzene, 12093
monoformin, 20224
monogen y 100, *See* Rhodapon® 101-10, 26631, sodium lauryl sulfate, 28781
monogermane, 20225
Monogliceride/emulsifier blend, *See* Advitagel, 822
Monoglyceride citric acid ester, *See* Acidan, 337
Monoglyceride citric ester, 20226, *See* Imwitor® 369, 14942
monoglyme, 20227
monohydratekieserite, *See* magnesium sulfate, 18368
(monohydrate) kieserite, *See* magnesium sulfate, 18368
Monoiodoisovalerylurea, *See* Iodoval, 15220
Monoiodomethane, *See* methyl iodide, 19542
monoisopropanolamine, *See* isopropanolamine, 15406, Ninol® M10, 21238
Mono-Kay, 20228, *See* Konakion, 16322, vitamin K_1, 33949
Monol, 20229
Monolan®, 20230
Monolan® 8000 E/80, 20231
Monolan® P222, 20232
Monolan® PT, 20233
Monolastex Smooth, 20234
1-Monolaurin, *See* Ablunol GML, 101
monolauryl dimethylamine, *See* Onamine 12, 22164
Mono-Line, 20235
monolinuron, *See* Aresin, 2839, Arresin, 3127
Monolite®, 20236
Monolithium salt of N-distilled cocoyl-L-glutamic acid, *See* Acylglutamate DL-12, 600
Mono-Lube®, 20237
Monomax AH90 B, 20238
Monomer, *See* lauryl acrylate, 16863
monomethyl benzene, *See* methyl benzene, 19514
monomethylolacrylamide, *See* NM-AMD, 21404
monomethylol-5,5-dimethylhydantoin, *See* Dantoin® MDMH, 7734
Monomuls 90O18, *See* Radiasurf® 7150, 25801
Monomuls® 90-L12, 20239
Monomuls® 90-O18, 20240
Monomuls® Range, 20241
mono-n-butylamine, *See* Amine C4, 2152
mononitrophenol, *See* *p*-nitrophenol, 21368
mononitrotoluenes, *See* M.N.T, 18105
monooctanoin, *See* Witafrol® 7420, 34421
monoolein, *See* glyceryl oleate, 13084, Radiasurf® 7150, 25801, Witconol 2421, 34500
α-monoolein, *See* Ablunol GMO, 102, Radiasurf® 7150, 25801, Witconol 2421, 34500
1-monoolein, *See* Radiasurf® 7150, 25801
1-Monoolein, *See* Witconol 2421, 34500
1-Monooleoylglycerol, *See* Ablunol GMO, 102, Radiasurf® 7150, 25801, Witconol 2421, 34500
Mono-or di-alkyl tetraamonium salts, *See* Adekamine E Series, 675
Monoparin, *See* Heparin, Sodium Salt, 13788
Monopen, *See* Penicillin G potassium, 23109
monopentaerythritol, *See* pentaerythritol, 23139
Monoplas 279, 20242
Monoplex DOA, *See* dioctyl adipate, 8548, Plasthall® DOA, 24002, PX-238, 25440, Wickenol® 158, 34393
Monoplex DOS, *See* Plasthall® DOS, 24005, Reomol DOS, 26131
Monoplex®, 20243
Monoplex® 5, 20244
Monoplex® DDA, 20245
Monoplex® DIOA, 20246
Monoplex® DOA, 20247, *See* dioctyl adipate, 8548
Monoplex® DOS, 20248
Monoplex® NODA, 20249
Monoplex® S-73, 20250
Monoplex® S-75, 20251
monopol oil, *See* Turkey red oils, 32607
monopol soap, 20252
monopotassium orthophosphate, *See* potassium phosphate, monobasic, 24772, potassium phosphate, 24773
Monopoxy, 20253
monopropylene glycol, *See* propylene glycol, 25215
Monopropylene glycol monomethyl ether, *See* Poly-Solv® MPM, 24558
monorhein, *See* rhein, 26534
monoricinolein, *See* glyceryl ricinoleate, 13085, glyceryl monoricinoleate, 13085
Monorotox, *See* Aresin, 2839
Monoset, 20254

monosilane, *See* Silicane, 28318
Monosiliol C, 20255
monosodium ascorbate, *See* sodium ascorbate, 28734
monosodium dihydrogen phosphate, *See* sodium phosphate, 28810
monosodium dihydrogen phosphate dihydrate, *See* sodium phosphate, 28810
Monosodium erythorbate, *See* Eribate, 10753, sodium erythorbate, 28762
monosodium glutamate, *See* MSG, 20416
monosodium L-glutamate monohydrate, *See* MSG, 20416
monosodium metaborate, *See* sodium metaborate, 28784
monosodium methanearsonate, *See* MSMA, 20417
Monosodium methylarsonate, *See* Drexar 530, 9023, MSMA, 20417
monosodium sulfite, *See* sodium acid sulfite, 28728
Monosorb, 20256
Monosoya amidoamine quaternary, *See* Tomah Q-311, 31994
monostearin, *See* glyceryl stearate, 13087, GMS Base, 13162, Radiamuls® MG 2141, MG 2142, MG 2600, MG 2900, 25777, Witconol 2400, Witconol MST, Witconol RHT, 34499, Zohar GLST, 34850
Monostearyl citrate, *See* Morflex MSC, 20331
Monosteol, *See* Witconol 2380, 34498
Monosteol TG, *See* Witconol 2380, 34498
Monosulfiram, 20257
Monosulph, 20258
monothioethylene glycol, *See* 2-mercaptoethanol, 19336
α-monothioglycerol, *See* Thiovanol®, 31723
Monox, 20259
monoxone, *See* sodium monochloracetate, 28797
Monsanto DOA, *See* Wickenol® 158, 34393
Monsanto Salt, 20260
Montaclere®, *See* Alkylated phenol, 1734
Montago, 20261
Montalere® SPH, *See* Alkylated phenol, 1734
Montan 80, *See* Ablunol S-80, 111, Witconol 2500, 34501
Montan acid wax, 20262, *See* Hoechst Wax S, 14270
montan pitch, 20263
Montana gold, 20264
Montana wax, *See* Irish Peat Wax, 15304
Montane 20, 20265
Montane 40, 20266, *See* Ablunol S-40, 109
Montane 60, 20267, *See* Ablunol S-60, 110
Montane 65, 20268
Montane 80, 20269
Montane 481, 20270
montanin wax, *See* Irish Peat Wax, 15304
montanine, 20271
Montanox 20 DF, 20272
Montanox 40 DF, 20273
Montanox 60 DF, 20274, *See* Ethylan® GES6, 11108
Montanox 61, 20275
Montanox 65, 20276, *See* Polysorbate 65, 24571
Montanox 70, 20277
Montanox 80 DF, 20278
Montanox 81, 20279
Montanox 85, 20280
Montax, 20281
Monteban, 20282, *See* narasin, 20741
Monteine LCK-32, 20283, *See* Foam-Coll 4C, 12150
Monteine LCQ, 20284
Monteine LCT, 20285, *See* Foam-Coll 4CT, 12151
Monterey 30% Iron, 20286
Monterey Bayleton, 20287
Monterey Bloom Popper, 20288
Monterey Foliar Nutrient 11-4-6, 20289
Monterey Herbicide Helper, 20290
Monterey Insulate, 20291
Monterey Iron Chelate 10%, 20292
Monterey Liqui-Cop, 20293
Monterey Perc-O-Late Plus, 20294
Monterey Signal, 20295
Monterey Stimulator 12, 20296
Monthier's blue, 20297
Monthyle, 20298
Montigel, 20299
Montmorillonite, 20300, *See* Fulcat Catalysts, 12436, Fulmont Activated Bleaching Earths, 12448, Yellowstone, 34652, Witcolate NH, 34478
Montopol LA 20, *See*
montopol la paste, *See* Rhodapon® 101-10, 26631, sodium lauryl sulfate, 28781
Montothene G50, 20301
Montreal potash, 20302
Moore Floc, 20303
Moorland, 20304
Moorman's medicated RID-EZY, *See* ronnel, 26951
Mopari, *See* dichlorvos, 8394
Moplefan, 20305
Moplen, 20306
mopper blue, *See* mountain blue, 20366
MOR, *See* Morfax, 20314
morantel tartrate, 20307, *See* Paratect Bolus, 22804
Moranyl, *See* Naganol, 20660
Morat white, 20308
Mordant rouge, *See* red liquor, 25961
Moreau marble, 20309
Morell's solution, 20310
Morestan, 20311, *See* quinomethionate, 25686
Morestan®, 20312
Morfast, 20313
Morfax, 20314
Morflex 100, 20315
Morflex 125, 20316
Morflex 125, 175, *See* octyldecyl phthalate, 22042
Morflex 150, 20317, *See* dicyclohexyl phthalate, 8413
Morflex 175, 20318
Morflex 190, 20319
Morflex 210, 20320
Morflex 240, 20321
Morflex 310, 20322
Morflex 325, 20323
Morflex 330, 20324
Morflex 410, 20325
Morflex 510, 20326
Morflex 525, 20327
Morflex 530, 20328
Morflex 560, 20329
Morflex 1129, 20330
Morflex MSC, 20331
Morflex P50, 20332
Morhal resin, 20333
Morhulin, 20334
Morillol, *See* Morillol®, 20335
Morillol®, 20335
Morintannic acid, *See* maclurin, 18235
Morkit, *See* Anthraquinone, 2532
Morkit®, 20336
Morland's salt, 20337
Moroccan olive oil, 20338
Morocco gum, 20339
Morocide, 20340
Moroline, 20341
Moronal, 20342
Morpan, 20343
Morpan BC, 20344
Morpan NBB, 20345
Morpan O, *See* Zeonet B, 34721
morpholine, 20346
Morpholine, 4-(2-benzothiazolylthio)-, *See*
Morpholine, 4-(2-benzothiazolylthio)-, *See* Accelerator MF, 208, Amax®, Amax No 1, 1950, Delac MOR, 8000, OBTS, 21970, OMTS, 22161
Morpholine, 4,4'-(2-ethyl-2-nitro-1,3-propanediyl)bis-, *See* 4,4'-(2-Ethyl-2-nitrotrimethylene)-dimorpholine, 11141
Morpholine, N,N'-disulfide, *See* Akrochem® Accelerator R, 1154, Naugex SD-1, 20801, Sulfasan, 29970, Vanax® A, 33287
Morpholine, N-cyclododecyl-2,6-dimethyl-, acetate, *See* dodemorph-acetate, 8782
2-(morpholinothio)benzothiazole, *See*
2-(morpholinothio)benzothiazole, *See* Accelerator MF, 208, Amax®, Amax No 1, 1950, Delac MOR, 8000, OMTS, 22161, OBTS, 21970, OMTS, 22161, Perkacit® MBS, 23284
4-Morpholinyl-2-benzothiazole disulfide, *See* Morfax, 20314
4-Morpholinyl-2-benzothiazyl disulfide, *See* Morfax, 20314
2-(4-Morpholinylmercapto)benzothiazol, *See* Morfax, 20314
Morpholinylmercaptobenzothiazole, *See* Morfax, 20314
Morplas, 20347
Mor-Rex® I-920, 20348, *See* Maltodextrin, 18500
morrhua oil, *See* cod liver oil, 6569
Morsep, 20349
Morstrip, 20350
Mortegg Emulsion, 20351
Morto, 20352
Morton's fluid, 20353
Morwet EFW, 20354
mosaic gold, 20355, *See* stannic sulfide, 29496
mosaic silver, 20356
Mosatil, *See* Versene CA, 33618
Moss Gunl, 20357
Mosskil, 20358
Mos-Tox, 20359
Mota, 20360
mother of pearl sulfur, 20361
Motilyn, *See* d-Panthenol, 22690
motiorange r, *See* Sudan I, 29938
Motung Steel, 20362
Moudan white, *See* Morat white, 20308
mou-iéou, 20363
Mould Release Agent N 32, 20364
Mouldrite, 20365
mountain blue, 20366
mountain butter, 20367
mountain cork, 20368
mountain flax, *See* amianthus, 2057
mountain flour, *See* infusorial earth, 15125
mountain green, 20369, *See* malachite, 18458
mountain leather, 20370
mountain milk, 20371
mountain paper, *See* mountain leather, 20370
mountain tallow, *See* bitumen, 4294
mountain wood, 20372
Mountford's paint, 20373
Mourey's aluminum solder, 20374
Mouse Killer, 20375
Mouser, 20376
Moussett's alloy, 20377
moutain paper, *See* mountain leather, 20370
Movin B, 20378
Movyl, *See* Korogel, 16354, Koroplate, 16357
Mowchem, 20379, *See* mefluidide, 19230
Mowilith, 20380
Mowilith®, 20381
Mowiol, 20382
Mowiol®, 20383
Mowital, 20384
Mow-It-Less, 20385
Mozanon®, 20386
8-MP, *See* Oxsoralen-Ultra, 22437
MP Diol Glycol, 20391
MP-1, 20387
MP-2, 20388
MP-10CF-4CC/15T, 20389
MP-10-CF-4CC/15T, *See* Polyphenylene oxide, 24518
M-P-A, 20390
MPEM, 20392
MPI DMSA Kidney Reagent, 20393
MPI Indium DTPa ^{111}In, 20394
MPI Indium Oxine ^{111}In, 20395
MPI Krypton ^{81m}Kr Gas Generator, 20396
MPK, *See* methyl propyl ketone, 19557
MPMD, *See* Dytek® A, 9439
MPP, *See* Tiguvon®, 31788
MPS 500, 20397

M-Quat® 32, 20398
M-Quat® 40, 20399
M-Quat® 257, 20400
M-Quat® 522, 20401, *See* Foamquat IAES, 12180
M-Quat® 1033, 20402
M-Quat® 2475, 20403
M-Quat® B-25, 20404
M-Quat® Dimer 18, 20405
M-Quat® JN, 20406, *See* Ricinoleamidopropyl ethyldimonium ethosulfate, 26733
M-Quat® JO-50, 20407
MRV 1000, 20410
MR-1, MR-1A, MR-17A, MR-17B, 20408
MS-4, 20411
MS-26, 20412
MS-33, *See* Ablunol S-60, 110
MS 33F, *See* Ablunol S-60, 110
MR-502K, MR-502P, MR-502Y, 20409
MS-122, 20413
MS-170 1,1,1 Trichloroethane Solvent, 20414
MS-180 Freon® TF Solv, 20415
MSG, 20416
MSMA, 20417
MSMA, *See* Neoarsycodyl, 20871
M-Soft-1, 20418
M-Soft-10, 20419
MSP, 20420, *See* sodium phosphate, 28810
MSS 2,4-D Amine, 20421
MSS 2,4-D Ester, 20422
MSS 2,4-DB + MCPA, 20423
MSS 2,4-DP, 20424
MSS 2,4-DP + MCPA, 20425
MSS Aminotriazole, 20426
MSS Chlormequat 40, 46, 60, 70, 20427
MSS CIPC, 20428
MSS CMPP, 20429
MSS CMPP/DP, 20430
MSS IPC 50, 20431
MSS MCPA 50, 20432
MSS MH18, 20433
MSS Mircam, 20434
MSS Mircam Plus, 20435
MSS Simazine/Aminotriazole 43FL, 20436
MSS Sugar Beet Herbicide, 20437
MTC, *See* methyl isothiocyanate, 19543
MTES, *See* Dynasylan® MTES, 9413
mucara, *See* karaya gum, 15760, Normacol, 21566
Muccocota gum, *See* ocota cocota gum, 21979
Mucicarmin, 20438
Mucogel®, 20439 , *See* Alumina hydrate, 1892
mudar gum, 20440
Mudge's speculum metal, 20441
muga silk, 20442
Mulch-Magic, 20443
muldan, 20444
mule gum, 20445
Mulgofen, 20446
Muller's fluid, 20447
Mullex, 20448
Mullfrax 301, 20449
mullicite, 20450
mullite, 20451, *See* Tisyn®, 31857
Mulsifan CB, 20452
Mulsifan CPA, 20453
Mulsifan RT 1, 20454
Mulsifan RT 23, 20458
Mulsifan RT 72, 20459
Mulsifan RT 141, 20455
Mulsifan RT 146, 20456
Mulsifan RT 203/80, 20457
Mulsiferol, *See* vitamin D_2, 33940
Multaglut, 20460
Multamat, *See* Bendiocarb, 3937
Multex, 20461
Multi Base, 20462
Multibase ABS 3075, 20463, *See* Acrylonitrile-butadiene-styrene, 439
Multibase ABS 3525 CL, 20464
Multibase ABS 3959, 20465
multibrol, 20466
Multicel, 20467
Multicet, 20468
Multicoild, 20469
Multicrack, 20470
Multicrom, 20471
Multicuer, 20472
Multiflow, 20473
Multigreen® II, 20474
MultiGuard, 20475
Multilind, 20476
Multilind Ointment, 20477
Multiluz, 20478
Multimet, 20479, *See* Bendiocarb, 3937
Multionic, 20480
Multisil, 20481
MultiSperse, 20482
Multisperse CP, 20483
Multispray, 20484
Multistix®, 20485
Multiter, 20486
MultiTherm IG-2 Heat Transfer Fuild, 20487
MultiTherm PG-1Heat Transfer Fluid, 20488
Multivac 4, 20489
Multivite, 20490
Multi-W FL, 20491
Multiwax® 180-M, 20492
Multiwax® 180-M, HS, *See* Microcrystalline wax, 19666
Multiwax® HS, 20493
Multronol, 20494
Mulukilivary, 20495
Mumetal, 20496
Municol, 20497
munjeet, 20498
Munjistin, 20499
Muntz metal, 20500
Murac, 20501
Murald, 20502
Murcurite, 20503
Murdiel, 20504
Murex, 20505
Murexan, *See* uramil, 33113
murexide, 20506
Murfite, 20507
Murfixtan, 20508
Murfly, 20509
Murfotox, 20510
Murfume, 20511
Murfume Grain Store Smoke, 20512
muriacite, *See* calcium sulfate (anhydrous), 4971
muriate of potash, 20513
muriate of soda, 20514
Muriatic acid, *See* hydrochloric acid, 14575, 14576
Muriatic Ether, *See* ethyl chloride, 11067
Muritan, 20515
Murlin Premium Ladle Wash, 20516
Murman's alloy, 20517
Murnil®, 20518
Murox, *See* nuarimol, 21762
Murphex, 20519
Murphicol, 20520
Murphos, 20521
Murvin 85, 20522, *See* Carbaryl, 5181
muscarine, 20523, 20524
Muscatox, 20525
muscle adenylic acid, *See* adenosine monophosphate, 680
muscle sugar, 20526
muscone, 20527, *See* Exaltone, 11282, musk, 20531
Muscovite mica, *See* mica, 19626
Mushet steel, 20528
Mushroom alcohol, *See* Morillol®, 20335
mushroom sugar, 20529
Musiv gold, 20530
musk, 20531
Musk amberette, *See* musk ambrette, 20532
musk ambrette, 20532
Musk B, *See* musk baur, 20533
musk baur, 20533
musk C, *See* ketone musk, 16141
Musk R-1, 20534
Musk T, *See* Emeressence® 1150, 10077, ethylene brassylate, 11121
Musketeer, 20535
muskone, *See* muscone, 20527
Musol 20, 20536
mussanin, 20537
mustard gas, 20538
mustard oil, 20539
Mustone, 20540
Mutan, *See* Timbrel, 31800
Muthmann's liquid, 20541
muthol, *See* paraffin, liquid, 22744
Mutoxin, *See* DDT, 7820
mutton tallow, *See* tallow, 30739
MV 119A, *See* Dithianone, 8729
MV NF/FCC W-300FG, *See* Algin, 1591
MVP, 20542
MXM-7500, 20543
Myacide® AS Plus, 20544
Myacide® S-1, S-2, 20545
Myacide® SP, 20546
My-B-Den, 20547, *See* 5'-adenylic Acid, 683
Mybond®, 20548
Mycil, *See* Chlorphenesin, 6161
Mycivin, *See* lincomycin hydrochloride hemihydrate, 17285
myclobutanil, 20549, *See* Systhane, 30665
Mycoban, *See* calcium propionate, 4968
Mycocide, 20550
Mycose, 20551
Mycotal, 20552
Mycovac-L, 20553
Mycozol, *See* thiabendazole, 31661
Mycronil, *See* Bisphenol A, 4283
Mydochrome, 20554
Mydoneg, 20555
Mydoprint, 20556
myelin, 20557
Myer's naphthol green, 20558
Myflam®, 20559
Mykon, 20560
Mykon 817, 20561
Mykonaid, 20562
Mykostin, *See* vitamin D+72, 33940
Mykroy/Mycalex, 20563
Mylanta, 20564
Mylar, 20565, *See* A-C® 6, 168
Mylar®, 20566
Mylipen, *See* Penicillin V, 23111
Mylocon, 20567
Mylofanol, *See* cinchophen, 6307
Mylol, 20568
Mylone, *See* Dazomet, 7806, N-521® Biocide, 20640
Mylone 85, *See* Dazomet, 7806, N-521® Biocide, 20640
Myodil, 20569
myo-Inositol, *See* inositol,
myo-inositol hexakis(dihydrogen phosphate) sodium salt, *See* Rencal, 26105
myo-Inositol hexakisphosphate, *See* phytic acid, 23777
Myo-inositol hexaphosphate, *See* phytic acid, 23777
myo-Inositol,, *See* phytic acid, 23777
Myoston, *See* adenosine monophosphate, 680, 5'-adenylic Acid, 683
Mypolex®, 20570
myrabola oil, 20571
Myras, 20572
Myreth-3, *See* Hetoxol M-3, 13943
Myreth-3 laurate, *See* Schercemol MEL-3, 27606
Myreth-3 myristate, *See* Cetiol® 1414E, 5795, Lanol 14 M, 16735, Schercemol MEM-3, 27607
Myreth-3 palmitate, *See* Schercemol MEP-3, 27608
Myreth-5 carboxylic acid, *See* Akypostat MA 35, 1324
Myrickite, 20573
Myristalkonium chloride, *See* BTC® 824, 4636, BTC® 2565, 4639, Exameen 824 3724, 11284, FMB 65-15, 65-28, 12135, JAQ Powdered Quaternary, 15517

Myristalkonium chloride quaternium-14, *See* Exameen 2125 M 3704, 11285, FMB 3328-5, 3328-8, 12137
Myristamide DEA, *See* Emalex NN-15, 9973
Myristamide DEA 1:1 Myristamide DEA, *See* Monamid® 150-MW, 20116
Myristamidopropyl betaine, *See* Schercotaine MAB, 27698
Myristamidopropyl dimethylamine, *See* Chemidex M, 5964, Schercodine M, 27639
Myristamine, *See* Amine 14D, 2129
Myristamine oxide, *See* Admox® 14-85, 758, Barlox® 14, 3650, Incromine Oxide M, 15007, Mazox® MDA, 19153, Ninox® M, 21243, Schercamox DMM, 27576
Myristic acid, 20574, *See* Emery® 654, 10135, Philacid 1400, 23662, Prifrac 2940, 24953
Myristic acid (90%), *See* Hystrene® 9014, 14764
Myristic acid (95%), *See* Hystrene® 9514, 14767
myristic acid isopropyl ester, *See* isopropyl myristate, 15409
Myristic acid, butyl ester, *See* Wickenol® 141, 34384
Myristic acid, isopropyl ester, *See* Wickenol® 101, 105, 34378
myristica, 20575
Myristoyl hydrolyzed collagen, 20576, *See* Pro-Tein SM-20, 25271
Myristoyl sarcosine, *See* Hamposyl® M, 13602
myristryltrimethylammonium bromide, *See* myrtrimonium bromide, 20589
myristyl alcohol, 20577, *See* Cachalot® M-43 NF, 4879, cetyl alcohol, 5815, Dehydag Wax 14, 7943, Emery® 3334, 10171, Lanette 14, 16662, Lorol C14, 17613
n-Myristyl dimethyl amine oxide, *See* Empigen® OH25, 10377
Myristyl dimethyl benzyl ammonium chloride, *See* Catigene DC/100, 5445, JAQ Powdered Quaternary, 15517
myristyl lactate, 20578, *See* Ceraphyl® 50, 5716, Wickenol® 506, 34400
Myristyl myristate, *See* Alkamuls® MM/M, 1670, Ceraphyl® 424, 5727, Cetiol® MM, 5801, Crodamol MM, 7122, Kemester® MM, 15965, Schercemol MM, 27609, Waxenol® 810, 34237
Myristyl neopentanoate, *See* Schercemol 145, 27580
Myristyl octanoate, *See* Wickenol® 174, 34399
Myristyl propionate, *See* Lonzest® 143-S, 17595, Schercemol MP, 27610
Myristyl stearate, *See* Hest MS, 13885, Kemester® 1418, 15939, Schercemol MS, 27611
myristyl tetradecanoate, *See* Waxenol® 810, 34237
Myristyleicosyl stearate, *See* Hetester 3236S, 13893
myristyltrimethylammonium bromide, *See* Sumquat® 6110, 30125
Myritol, 20579
Myritol 318, 20580, *See* caprylic/capric acid triglyceride, 5161
Myrj 45, *See* Ablunol SA-7, 113
Myrj 52, *See* Ablunol SA-7, 113
Myrj®, 20581
Myrj® 45, 20582
Myrj® 52, 20583
Myrj® 53, 20584
Myrj® 59, 20585
myrmekite, 20586
myrobalans, 20587
myrrh, 20588
myrticolorin, *See* rutin trihydrate, 27150
myrtrimonium bromide, 20589, *See* Mytab®, 20595, Querton 14Br-40, 25649, Sumquat® 6110, 30125, Rhodaquat® M214B/99, 26642
myrystyltrimethylammonium bromide, *See* Rhodaquat® M214B/99, 26642
Mystery gold, 20590
Mystic metal, 20591
mystin, 20592
Mystolene, 20593
Mystox, 20594
Mytab®, 20595
Mytex, 20596
Mytolac, 20597
Myvacet®, 20598
Myvacet® 5-07K, 20599
Myvacet® 7-00, *See* Acetylated hydrogenated lard glyceride, 315
Myvacet® 7-07K, 20600
Myvacet® 9-08K, 20601
Myvacet® 9-40, 20602
Myvacet® 9-45K, 20603
Myvaplex®, 20604
Myvaplex® 600K, 20605
Myvatem® 30, 20606
Myvatex®, 20607
Myvatex® 3-50K, 20608
Myvatex® 8-06K, 20609
Myvatex® 40-06S K, 20610
Myvatex® 90-10K, 20611
Myvatex® 600PK, 20612
Myvatex® Do Control K, 20613
Myvatex® Mighty Soft®, 20614
Myvatex® Monoset® K, 20615
Myvatex® Super DO, 20616
Myvatex® Texture Lite® K, 20617
Myverol®, 20618
Myverol® 18-00, 20619
Myverol® 18-04K, 20620
Myverol® 18-06K, 20621
Myverol® 18-07K, 20622
Myverol® 18-30, 20623
Myverol® 18-35, 20624
Myverol® 18-40, 20625
Myverol® 18-50K, 20626
Myverol® 18-85K, 20627
Myverol® 18-92, 18-92K, 20628
Myverol® 18-98, 20629
Myverol® 18-99K, 20630
Myverol® P-06K, 20631
Myverol® SMG VK, 20632
MZ 2MZ, *See* 2-methyl imidazole, 19572
MZ2MZ, *See* 2-methyl imidazole, 19572
N, 20633, *See* Asparagine, 3232
N 521, *See* Dazomet, 7806,, *See* N-521® Biocide, 20640
N.C.T, 20634
N.E. powder, 20635
N.P.L. alloy, 20636
N.S. fluid, 20637
N.S.B, 20638
N33, 20639
N-521® Biocide, 20640
N-869, *See* metam-sodium, 19444
N-948® Biocide, 20641
Na Frinse, *See* sodium fluoride, 28764
Na Ta, 20642
NA-22, *See* ethylene thiourea, 11132, Perkacit® ETU, 23283
NA-22-D, *See* Perkacit® ETU, 23283
Na2584 (dot), *See* Witco® 1298, 34458
Na2S, *See* sodium sulfide, 28822
NA-8318, *See* Knot Out, 16257, Ratio, 25897
NAA, *See* Rhizopon B, 26574, Tipoff, 31848
naa 173, *See* stearic acid, 29599
Nabac, *See* hexachlorophene, 13992
nabam, *See* Campbell's Nabam Soil Fungicide, 5066
Nabu, *See* Sethoxydim, 28099
Nacap, *See* Nacap®, 20643, Nuodex 84, 21779
Nacap®, 20643
Naccanol SW, *See* Nacconol® 40G, 20647
Nacconal, 20644
Nacconate 100, *See* toluene diisocyanate, 31947, Voranate T-80, Type I, Type II, 34042
Nacconol, *See* Nacconol® 40G, 20647
Nacconol 35SL, 20645, *See* sodium dodecylbenzenesulfonate, 28760
Nacconol 40F, *See* Nacconol® 40G, 20647
Nacconol 90F and 40F, 20646
Nacconol 98SA, *See* Witco® 1298, 34458
Nacconol® 40G, 20647
Na-Cemmix, *See* Ablusol ML, 129
Nacreous sulfur, *See* mother of pearl sulfur, 20361
Nadavin®, 20648
Nadone, *See* cyclohexanone, 7522
NaDOSS, *See* dioctyl sodium sulfosuccinate, 8550
Naetex-LAM, *See* Lactamide MEA, 16568
NaF, *See* sodium fluoride, 28764
nafeen, *See* sodium fluoride, 28764
Nafion®, 20649
nafpak, *See* sodium fluoride, 28764
Naftalofos, *See* Maretin, 18704, Rametin, 25856
Naftam 2, *See* Antioxidant PBN, 2591, Stabilator A.R, 29400
Naftifine, 20650
Naftifine hydrochloride, *See* Naftifine, 20650
naftifungin, *See* Naftifine, 20650
Naftin, *See* Naftifine, 20650
Naftocit, 20651
Naftocit® Thiuram 16, *See* Akrochem® TMTD, 1182
Naftocit® ZBEC, *See* Akrochem® Z.B.E.D, 1183
Naftogran®, 20652
Naftolen, 20653
Naftolen R 100, 510, 530, 550, 570, X413, X414 X10, 134, 20654
Naftomix, 20655
Naftonox, 20656
Naftopast, 20657
Naftopast® Thiuram 16-P, *See* Akrochem® TMTD, 1182
Naftovin, 20658
Naftozin, 20659
Naganol, 20660
Nageli's solution, 20661
Naglusol, 20662
nahcolite, 20663
Nako Brown R;, *See* *p*-aminophenol, 2185
Nalan, 20664
Nalcite, 20665
Nalco D-1994, *See* Amerstat® 282, 2043, N-948® Biocide, 20641
Nalco L-357, *See* Ablunol CO 10, 98
Nalco® 131, 20666
Nalco® 2300, 20667
Nalco® 2340, 20668
Nalco® 8669, 20669
Nalcon 243, *See* N-521® Biocide, 20640, Dazomet, 7806
Naled, *See* Bromex, 4569, Dibrom, 8356
Nalfleet, 20670
Nalfloc, 20671
Nalidone®, 20672
Nalkil, *See* bromacil, 4564, Uragan, 33105
Nalzin, 20673
Namekil, *See* metaldehyde, 19431
Nametal, 20674
nancic acid, *See* lactic acid, 16570
Nandel®, 20675
Nangawhite, 20676
Nankor, 20677, *See* Korlan, 16351, ronnel, 26951
Nansa, 20678
Nansa 1042p, *See* Witco® 1298, 34458
Nansa HF 80, *See* Nacconol® 40G, 20647, sodium dodecylbenzenesulfonate, 28760
Nansa HS 80, *See* Nacconol® 40G, 20647, sodium dodecylbenzenesulfonate, 28760
Nansa HS 85S, *See* Nacconol® 40G, 20647, sodium dodecylbenzenesulfonate, 28760
NANSA HS 85S C 550 KB HS 85S, *See* sodium dodecylbenzenesulfonate, 28760
Nansa SL, *See* Nacconol® 40G, 20647, sodium dodecylbenzenesulfonate, 28760
Nansa SS, *See* Nacconol® 40G, 20647, sodium dodecylbenzenesulfonate, 28760
Nansa SSA, *See* Witco® 1298, 34458
Nansa UC, 20679
Nansa UCA/S and UCP/S, 20680
Nansa® 1042, 20681
Nansa® 1106/P, 20682
Nansa® 1169/P, 20683
Nansa® AS 40, 20684, *See* Ammonium dodecylbenzene sulfonate, 2238
Nansa® BMC, 20685
Nansa® EVM50, 20686

Nansa® HS80S, 20687
Nansa® LES42, 20688
Nansa® LSS38/A, 20689
Nansa® MA30, 20690
Nansa® SB30, 20691
Nansa® SBA, 20692
Nansa® SL 30, 20693
Nansa® SS 30, 20694
Nansa® SSA, 20695
Nansa® TDB, 20696, *See* LABS 100/H.V, 16544, tridecylbenzene sulfonic acid, 32236
Nansa® TS 50, 20697
Nansa® YS94, 20698
Nansen, 20699
napalite, 20701
napalm, 20700
Napelec, 20702
Napgel, 20703
naphalane, 20704
Naphta, *See* Ligroin, 17244
naphtalin, *See* naphthalene, 20706
naphtha, 20705, *See* Hi-Sol® 10,15,70, 14134, Kensol 10, 16036, Kensol 13, 16037, Kensol 80, 16044, Texsolve V, 31532
Naphthalamine, *See* α-naphthylamine, 20729
Naphthalane, *See* Decalin®, 7840
1-naphthalenamine, *See* Aminogen I, 2176, α-naphthylamine, 20729
naphthalene, 20706
Naphthalene dialdehyde, *See* dicyclohexylamine nitrite, 8414
β-naphthalene sulfonic acid, 20707, *See* Dehscofix 918, 7926
naphthalene, 1,6-dihydroxy-, *See* 1,6-naphthalenediol, 20709
naphthalene-1,3-diol, *See* naphthoresorcin, 20726
naphthalene-1-acetic acid, *See* Rhizopon B, 26574
Naphthalene-2-sulfonic acid, *See* β-naphthalene sulfonic acid, 20707
Naphthalene-2-sulfonic acid hydrate, *See* β-naphthalene sulfonic acid, 20707
1-naphthaleneacetic acid, *See* Phyomone, 23775, Rhizopon B, 26574, Tipoff, 31848
1,2-(1,8-naphthalene)benzene, *See* Idryl, 14841
2,5-naphthalenediol, *See* 1,6-naphthalenediol, 20709
1,6-naphthalenediol, 20709
2,3-naphthalenediol, 20710
2,7-naphthalenediol, 20711
1,6-Naphthalene-disulfonic acid, *See* Ewer-Pick acid, 11278
2,6-naphthalenedisulfonic acid, 20708, *See* Ebert and Merz's β-acid, 9496
2,7-Naphthalenedisulfonic acid, *See* Ebert and Merz's α-acid, 9495
1,2-(1,8-naphthalenediyl)benzene, *See* Idryl, 14841
2-Naphthalenesulfonic acid monohydrate, *See* β-naphthalene sulfonic acid, 20707
1-naphthalene-6-sulfonic acid, *See* Cleve's β acid, 6410
4-amino 2-Naphthalenesulfonic acid, 4-amino-, *See* Cleve's +Ig-Acid, 6411
Naphthalenesulfonic acid, polymer with formaldehyde, sodium salt, *See* Ablusol ML, 129 Rhodacal® N, 26589
Naphthalenesulfonic acid-formaldehyde condensate sodium salt, *See* Ablusol ML, 129
1-Naphthalenol, methylcarbamate, *See* Carbaryl, 5181
α-naphthalenyl methylcarbamate, *See* Carbaryl, 5181
naphthalidam, *See* Aminogen I, 2176, α-naphthylamine, 20729
naphthalidine, *See* Aminogen I, 2176, α-naphthylamine, 20729
naphthalin, *See* naphthalene, 20706
naphthalol, 20712
naphthamine, *See* hexamine, 14008
Naphthane, *See* Decalin®, 7840
naphthazarin, 20713
naphthene, *See* naphthalene, 20706
naphthenic acid, 20714
Naphthenic acid, *See* Agenap HMW-H, 1011
Naphthenic acids, copper salts, *See* copper naphthenate, 6783
Naphthenic distillate, *See* Golden Bear 102, 13178
Naphthenic hydrocarbon, *See* MultiTherm PG-1Heat Transfer Fluid, 20488
Naphthenic rubber process oil, *See* Akrochem® Plasticizer LN, 1171
naphthine, *See* Bitumen, 4294
naphthionic acid, 20715, *See* Piria's acid, 23878
naphthite, 20716
Naphthochrome, 20717
Naphthocyanine, 20718
naphthocyanole, 20719
Naphtho(2,3-b)-*p*-dithiin-2,3-dicarbonitrile, 5,10-dihydro-5,10-dioxo-, *See* Dithianone, 8729
naphthoformol, 20720
Naphthol, 20721
α-Naphthol + 8-9 EO, *See* Igepal® 131, 14844
β-naphthol, *See* hydronaphthol, 14623
2-naphthol, *See* Antioxygene BN, 2600
Naphthol Aristol, 20722
α-Naphthol salicylate, *See* Alphol, 1840
α-Naphtholdisulfonic acid, *See* GR Acid, 13234
1-Naphthol-4,8-disulfonic acid, *See* Schollkopf's acids, 27726
2-Naphthol-3,6-disulfonic acid, *See* R-Acid, 25710
Naphtholite, 20723
naphtholith, 20724
β-naphthol-methyl ether, *See* nerolin, 21007
β-Naphthol- δ-monosulfonic F676, *See* Cassella's acid, 5402
β-naphtholsulfonic acid F, *See* Cassella's acid, 5402
α-naphthol-4-sulfonic acid, *See* Nevile and Winther's acid, 21046
1-Naphthol-7-sulfonic acid (1:7), *See* Liebmann and Studer's acid, 17219
2-naphthol-1-sulfonic acid calcium salt, *See* Abrastol, 155, asaprol, 3176
2-naphthol-6-sulfonic acid, *See* Schaeffer's acid, 27558
2-naphthol-7-sulfonic acid, *See* Cassella's acid, 5402
Naphthopone E, 20725
naphthoresorcin, 20726
naphthoresorcinol, *See* naphthoresorcin, 20726
Naphthoride, 20727
naphthosalol, *See* naphthyl salicylate, naphthalol, 20712
naphthosultone, 20728
β-naphthyl butyl ether, *See* Fragarol, 12372
β-naphthyl ethyl ether, *See* nerolin, 21007
β-Naphthyl methyl ether, *See* jara jara, 15518
1-Naphthyl methylcarbamate, *See* Carbaryl, 5181
α-naphthyl N-methylcarbamate, *See* Carbaryl, 5181
1-Naphthyl phenyl amine, *See* Akrochem® Antioxidant PANA, 1161, Antioxidant PAN, 2590
2-naphthyl salicylate, *See* naphthalol, 20712
naphthylacetic acid, *See* Rhizopon B, 26574, Tipoff, 31848
1-Naphthylacetic acid, *See* Rhizopon B, 26574, Tipoff, 31848
α-naphthylacetic acid, *See* Rhizopon B, 26574
1-naphthylamine, *See* Aminogen I, 2176
1-naphthylamine-8-sulfonic acid, *See* Scholikopf's acids, 27726
α-naphthylamine, 20729, *See* Aminogen I, 2176
α-naphthylamine 7-sulfonic acid, *See* Cleve's +Iw-Acid or J-acid, 6412
Naphthylamine Blue, *See* Niagara blue, 21091
1-Naphthylamine-2,7-disulfonic acid, *See* Kalle's acid, 15684
1-Naphthylamine-3,6-disulfonic acid, *See* Freund's acid, 12404
α-naphthylamine-4,6-disulfonic acid, *See* Dahl's acid II, 7694
α-naphthylamine-4,7-disulfonic acid, *See* Dahl's acid III, 7695
2-naphthylamine-3,6-disulfonic acid, *See* Salt, Amido-R, 27295
2-naphthylamine-6,8-disulfonic acid, *See* Salt, Amido-G, 27294
1-Naphthylamine, 3-sulfo-, *See* Cleve's +Ig-Acid, 6411
1-Naphthylamine-3-sulfonic acid, *See* Cleve's +Ig-Acid, 6411
α-naphthylamine 3-sulfonic acid, *See* Cleve's +Ig-Acid, 6411
α-naphthylamine-4-sulfonic acid, *See* Piria's acid, 23878
α-naphthylamine-6-sulfonic acid, *See* Cleve's β acid, 6410
β-Naphthylamine-7-sulfonic acid, *See* Delta acid, 8044
1-Naphthylamine 4-sulfonic acid, *See* naphthionic acid, 20715
1-Naphthyl-amine-5-sulfonic acid, *See* Laurent's acid, 16850
1-Naphthylamine-8-sulfonic acid, *See* Peri acid, 23269
2-naphthylamine-1-sulfonic acid, *See* Tobias acid, 31914
2-Naphthylamine-6-sulfonic acid, *See* Broenner's acid, 4550
1-Naphthylamine-3,6,8-trisulfonic acid, *See* Koch's acid, 16260
N-(1-Naphthyl)aniline, *See* Akrochem® Antioxidant PANA, 1161, Antioxidant PAN, 2590, Antioxidant PBN, 2591, Naugard® PANA, 20798
1-naphthyl N-methylcarbamate, *See* Sevin, 28115
(2-naphthyloxy) acetic acid, *See* betapal Concentrate, 4063
β-naphthylphenylamine, *See* Antioxidant PBN, 2591
2-naphthylphenylamine, *See* Antioxidant PBN, 2591
***sym*-di-β-Naphthyl-*p*-phenylenediamine**, *See* Agerite® White, 1024
N-(2-naphthyl)-N-phenylamine, *See* Antioxidant PBN, 2591
N-β-naphthyl-N-phenylamine, *See* Antioxidant PBN, 2591
Naphtoelan Red GG Base, *See* *p*-nitroaniline, 21345
Naphtopon E, 20730
Naphtopone® E, 20731
Naphuride, *See* Naganol, 20660
Napisan, 20732
Naples red, *See* murexide, 20506
Naples yellow, 20733
Naplithin, 20734
Napliwi, 20735
NAPP, *See* M.B.A, 18103, *See* methylene bisacrylamide, 19567
Naprinol, See Valadol, 33238
napropamide, *See* Devrinol, 8216
napropamide-trifluralin, *See* Devrinol T, 8217
Napryl, 20736
Napsoft FL, 20737
Naptel, 20738
Naptol, 20739
Napvis, 20740
narasin, 20741
Narceol, 20742
narcotile, *See* ethyl chloride, 11067
Narcoxyl, *See* xylazine, 34619
Nargentol, 20743
Nari oil, *See* Njave butter, 21392
Narki, 20744
Narlex EP-2, 20745
Narlex LD 42, 20746
NAS® 10, 20747
NAS® 30, 20748
NAS® 50, 20749
Nasemo, *See* Mansonil®, 18611
NaTa, *See* trichloroacetic acid, 32213
Natac®, 20750
Natacillin, *See* Versapen K, 33610
Natalon, *See* Carbora, 5235
Natene, 20751
Nathin, 20752
Native gellam gum, *See* Gellan gum, 12736, Kelco-Gel® Gellan Gum, 15827
native paraffin, *See* ozokerite, 22483
Natoil AVO, *See* Avocado Oil, 3497
Natopherol, 20753
Natrascorb, *See* vitamin C sodium salt, 33939, sodium

Chemical and Trade Names

ascorbate, 28734
natreen, *See* saccharin, 27195
natrena®, 20754
natri-c, *See* sodium ascorbate, 28734
natriphene, *See* Dowicide A, 8932
Natritope Chloride, 20755
natrium, 20756, *See* sodium, 28725
Natrolith, 20757
Natrosol® 250, 20758
Natrosol® FPS, 20759
Natrosol® Hydroxyethylcellulose, 20760
Natrosol® Plus CS, 20761
Natrovis® Water-Soluble Polymer, 20762
Natrundum, *See* Carbora, 5235
Natsyn® 2200, 20763
natural asphalt, *See* bitumen, 4294
Natural Bone Ash, BCP 400, 20764
Natural calcium carbonate, *See* CP Filler, 6928
Natural Extract AP, 20765
natural molybdite, *See* molybdenum trioxide, 20081
natural rubber, 20766, *See* Polyisoprene, 24471
Naturchem® CR, 20767, *See* cetyl ricinoleate, 5824
Naturechem® CAR, 20768
Naturechem® CR
Naturechem® EGHS, 20769
Naturechem® GMHS, 20770
Naturechem® GTH, 20771
Naturechem® GTR, 20772, *See* glyceryl triacetyl ricinoleate, 13089
Naturechem® MAR, 20773
Naturechem® MHS, 20774
Naturechem® OHS, 20775
Naturechem® PGHS, 20776
Naturechem® PGR, 20777, *See* Propylene glycol ricinoleate, 25222
Naturechem® THS-200, 20778
Nature's Own Spray Helper, 20779
Naturvue, 20780
Naubuc, 20781
Naugalube® 403, 20782
Naugalube® 438, 20783
Naugalube® 470, 20784
Naugalube® PDA, 20785
Nauganlite, 20786
Nauganlite Powder, 20787
Naugard J, *See* Agerite® DPPD, 1016, N,N'-diphenyl-*p*-phenylenediamine, 8603, JZF, 15638, Naugard® J, 20796, Permanax DPPD, 23361
Naugard® 10, 20788
Naugard® 76, 20789
Naugard® 445, 20790
Naugard® 524, 20791
Naugard® I-2, 20792
Naugard® I-3, 20793
Naugard® I-4, 20794
Naugard® I-6, 20795
Naugard® J, 20796
Naugard® J, *See* N,N'-diphenyl-*p*-phenylenediamine, 8603
Naugard® NBC, 20797
Naugard® PANA, 20798
Naugard® XL-1, 20799
Naugawhite, 20800
Naugex MBT, *See* Perkacit® MBTS, 23286
Naugex SD-1, 20801
Nauli gum, 20802 , *See* Arizole® Anethole Extra, 2885
Navac, 20803
naval bronze, 20804
navy green paint, 20805
Naxchem CD-6M, 20806
Naxchem Detergent CNB, 20807
Naxchem Dispersant K, 20808
Naxchem Emulsifier 700, 20809
Naxchem N-Foam 802, 20810
Naxel AAS-40S, 20811
Naxel AAS-60S, 20812
Naxel AAS-98S, 20813
Naxel AAS-Special 3, 20814
Naxel DDB 500, 20815
Naxell, 20816
Naxide 1230, 20817
Naxolate WA-97, 20818
Naxonac, 20819
Naxonac 510, 20820
Naxonate® 4ST, 20821
Naxonate, *See* Eltesol® SX 30, 9873, Naxonate® 4L, 20824, Pilot SXS-40, 23843, Witconate SXS 40%, 34492
Naxonate G, *See* Eltesol® SX 30, 9873, Naxonate® 4L, 20824, Pilot SXS-40, 23843, Witconate SXS 40%, 34492
Naxonate® 4AX, 20822, *See* Ammonium xylene sulfonate, 2259
Naxonate® 4KT, 20823
Naxonate® 4L, 20824, *See* sodium xylene sulfonate, 28832
Naxonate® 5KT, 20825
Naxonate® 45SC, 20826, *See* sodium cumene sulfonate, 28752
Naxonate® SC, 20827, *See* sodium cumene sulfonate, 28752
Naxonate® SX, 20828, *See* sodium xylene sulfonate, 28832
Naxonic NI-40, 20829
Naxonol CO, 20830
nayper b and bo, *See* Benzoyl peroxide, 3983
Na-Zn, *See* ADK STAB FL-21, 727
NBC, 20831
NBR, cold polymer, *See* Chemigum® HR662, 5974, Chemigum® N318B, N917, 5976
NBR, hot polymer, *See* Chemigum® N5, N7, 5977
NBT, *See* Nitro BT, 21342
N-C$_{16-22}$ alkyl propylene diamine, *See* Diamine B11, 8294
NC-4, 20832
NC-302, *See* quizalofop-ethyl, 25697
NC-1667, *See* trietazine, 32238
NC-6897, *See* Bendiocarb, 3937
NC-8438, *See* ethofumesate, 10967
NCI C00443, *See* Trifluralin, 32252
NCI-C52733, *See* Witcizer 312, 34454
NCI-C53894, *See* pentachlorethane, 23134
NCI-C54386, *See* Wickenol® 158, 34393
NCI-C54886, *See* Abluton T30, 139
NCI-C55152, *See* Timonox, 31807
NCI-C55254, *See* chlorobromoform, 6121
NCI-C55403, *See* Witconate SXS 40%, 34492
NCI-C55470, *See* Wollastocoat®, 34545
NCI-C56304, *See* Vulkanox® 4010 NA, 34147
NCI-C96683, *See* quizalofop-ethyl, 25697
NCS, *See* N-chlorosuccinimide, 6148
NDA, *See* Dichan 100, 8377, dicyclohexylamine nitrite, 8414
Ndilo oil, *See* Laurel nut oil, 16833
NE, 20833
Neacid, 20834
Nealpon, *See* pantopon,
Neantina, 20835
Neantine, *See* Kodaflex® DEP, 16266
neatsfoot oil, 20836, 20837, *See* animal glycerin, 2467
Neazina, *See* sulfamethazine, 29962
Nebony® 100, L-55, 20838
Nebulin, 20839, *See* Tecgran, 30886, tecnazene, 30898
Neburex, 20840
Neburex, *See* neburon, 20841
neburon, 20841, *See* Kloben®, 16225, Noruben, 21591
Necatorina, *See* carbon tetrachloride, 5224, Katharin, 15791
Neccanol SW, *See* Nacconol® 40G, 20647, sodium dodecylbenzenesulfonate, 28760
Necco Fire Retardant 2750, 2578, 2762, *See* Ammonium sulfamate, 2254
Necol, 20842
Nectandra bark, 20843
Nedi, 20844
needle bronze, 20845
needle tin ore, 20846
Neem bark, 20847
Neem oil, *See* veepa oil, 33510
Nefomolit, 20848
Nefrafos, *See* dichlorvos, 8394
Nefusan, *See* Dazomet, 7806, *See* N-521® Biocide, 20640
Neganol®, 20849
Negasunt®, 20850
Negex, 20851
Neguvon®, 20852
n-eicosanol, *See* Behenyl alcohol, 3899
Neillite, 20853
Neisser's stain, 20854
Nekal, 20855
Nekal A, *See* Aerosol® OS, 895, Rhodacal® IN, 26585
Nekal® BA-77, *See* Rhodacal® BA-77,
Nekal® BX, 20856
Nekal® BX-78, *See* Rhodacal® BX-78,
Nekal® SBS, 20857
Nekal® WS-25 and WS-25-1, *See* Geropon® WS-25, WS-25-1, 12919
Nekal® WT-27, *See* Geropon® WT-27,
Nekanil® 907, 20858
Nelco, 20859
Nelio Resin, 20860
Nelpon, *See* Tandem, 30796
Nema, 20861
Nemacin, 20862
Nemacur®, 20863
Nemafax, 20864
Nemapan, *See* Tecto, Tecto 60%, 30913, thiabendazole, 31661
Nemasol, *See* metam-sodium, 19444
Nematolyt, *See* papain, 22698
Nemazene, *See* phenothiazine, 23618
nemazine, *See* phenothiazine, 23618
Nemicide, *See* Tramisol®, 32134
Nemispor, *See* Karamate, 15749, mancozeb, 18517
Neo Dohyfral D+73, *See* vitamin D+73, 33941
Neo Heliopan AV, *See* Neo Heliopan® AV, 20866, Parsol® MCX, 22867
Neo Heliopan, Type AV, *See* octyl methoxycinnamate, 22039
Neo Heliopan® 303, 20865
Neo Heliopan® AV, 20866
Neo Heliopan® BB, 20867
Neo Heliopan® Hydro, 20868
Neo Heliopan® MA, 20869
Neo Heliopan® OS, 20870
Neoalkoxy, *See* LICA 01, 1719, LICA 09, 17196, 5 LICA 12, 17197, LICA 38, 17198, LICA 44, 17199, LICA 97, 17200
Neoalkoxy tris (dioctyl) phosphato zirconate, IPA, *See* NZ 12, 21935
Neoalkoxy tris (dioctyl) pyrophosphato zirconate, IPA, *See* NZ 38, 21937
Neoalkoxy tris (dodecyl) benzene sulfonyl zirconate, IPA, *See* NZ 09, 21934
Neoalkoxy tris (ethylene diamino) ethyl zirconate, *See* NZ 44, 21939
Neoarsycodyl, 20871
Neobar, *See* Barium sulfate, 3644, Radiopaque, 25830
Neobee® 18, 20872
Neobee® 20, 20873
Neobee® 62, 20874
Neobee® M-5, 20875, *See* caprylic/capric acid triglyceride, 5161
Neobor, 20876
Neo-Calglucon, 20877
neo-cebitate, *See* Eribate, 10753, Neo-Cebitate®, 20878, sodium erythorbate, 28762
Neo-Cebitate®, 20878, *See* sodium erythorbate, 28761
Neocid, 20879, *See* DDT, 7820
Neocidol, *See* Diazinon Liquid, 8344
Neocidol Veterinary Powder, 20880
Neocosal, 20881
Neocrest, 20882
NeoCryl, 20883, 20884
Neo-Cytamen, 20885
neodecaneperoxoic acid, 1,1-dimethylpropyl ester,

See Trigonox® 123-C75, 32282
Neodecanoic acid oxiranylmethyl ester, *See* glycidyl decanoate, 13092, Glydexx N-10, 13141
Neodene® 6, 20886
Neodene® 6/12, 20887
Neodene® 6/12, *See* dodecene-1, 8778
Neodene® 6/8, 20888
Neodene® 6/8/10, 20889
Neodene® 8, 20890
Neodene® 10, 20891
Neodene® 10/12/1314, 20892
Neodene® 10-11/12-13, 20893
Neodene® 10-11/12-13/14, 20894
Neodene® 12, 20895
Neodene® 12, *See* dodecene-1, 8778
Neodene® 12/1314, 20896
Neodene® 14, 20897
Neodene® 16, 20898
Neodene® 18, 20899
Neodene® 20, 20900
Neodene® 810, 20901
Neodene® 1012, 20902, *See* dodecene-1, 8778
Neodene® 1014, 20903
Neodene® 1112, 20904
Neodene® 1420, 20905
Neodene® 1624, 20906
Neodene® 2024, 20907
Neodol phosphite, *See* Weston® 474, 34300
***tris*-Neodol-25 phosphite**, *See* Weston® 474, 34300
Neodol® 1, 20908
Neodol® 1-3, 20909
Neodol® 5, 20910
Neodol® 23, 20911
Neodol® 23-6.5, 20912
Neodol® 25, 20913
Neodol® 25-3, 20914
Neodol® 45, 20915
Neodol® 45-7, 20916
Neodol® 91, 20917
Neodol® 91-2.5, 20918
Neo-Duroterm® 3, 5 and 7, 20919
Neo-Duroterm® L, 20920
neo-fat 12, *See* Philacid 1200, 23660
neo-fat 12-43, *See* Philacid 1200, 23660
neo-fat 14, *See* Philacid 1400, 23662
neo-fat 18, *See* stearic acid, 29599
neo-fat 18-53, *See* stearic acid, 29599
neo-fat 18-54, *See* stearic acid, 29599
neo-fat 18-55, *See* stearic acid, 29599
neo-fat 18-59, *See* stearic acid, 29599
neo-fat 18-61, *See* stearic acid, 29599
neo-fat 18-s, *See* stearic acid, 29599
Neoflex® 9, 20922
Neoflex® 11, 20921
Neogen, 20923
Neolan, 20924
Neoleukorit, 20925
Neoloid, *See* AA Standard, AA USP, 12, Cosmetol® X, 6871
Neolyn, 20926
Neolysol, 20927
Neomagnol, *See* Chloramine B, 6085
Neo-Metantyl, *See* propantheline bromide, 25176
Neomodelon 100, 20928
Neomycin hexadecanoate, *See* neomycin palmitate, 20929
Neomycin palmitate, 20929
Neonalium, 20930
Neonite, 20931
Neonite® EG60/6mm, EG61/12mm, 20932
Neopac, 20933
Neopelex 05, *See* Nacconol® 40G, 20647
Neopelline, 20934
Neopen SS, 20935
Neopen® Dyes, 20936
Neopentyl (diallyl) oxy, tri(9,10 epoxy stearoyl) zirconate, *See* NZ 39, 21938
Neopentyl (diallyl) oxy, tri(dodecyl)benzene-sulfonyl zirconate, IPA, *See* NZ 90, 21941
Neopentyl (diallyl) oxy, trimercapto-phenyl zirconate, *See* NZ 89, 21940
Neopentyl (diallyl)oxy, *See* LICA 99, 17201
Neopentyl diallyloxy, trimethacryl zirconate, IPA, *See* NZ 33, 21936
Neopentyl dicaprate, *See* Bernel® Ester NPDC, 4009
neopentyl glycol, 20937, *See* NPG® Glycol, 21751
Neopentyl glycol diacrylate (monomer), *See* SR-247, 29371
Neopentyl glycol dicaprate, *See* Schercemol NGDC, 27612
Neopentyl glycol dicaprylate/dicaprate, *See* Liponate NPGC-2, 17381, Lipovol MOS-350, 17431, Lipovol MOS-70, 17432
Neopentyl glycol dilaurate, *See* Schercemol NGDL, 27613
Neopentyl glycol dioctanoate, *See* Schercemol NGDO, 27614
Neophax FA, 20938
neopine, 20939
Neoplen, 20940
Neopolen® E, 20941
Neopolen® P, 20942
Neopon, 20943
Neopon 33, 20944
Neopon LAM, 20945, *See* Ammonium lauryl sulfate, 2242, Witcolate NH, 34478
Neopon LOA/F, 20946
Neopon LOS, LOS/F and LOS/NF, 20947
Neopon LOT/F, 20948
Neopon LS, 20949
Neopon LT, 20950
Neopralac, 20951
Neoprene, *See* Adcora P6, 631, chloroprene, 6136, Baypren® 110, 382, polychloroprene, 24386 8
Neoprene AH, *See* Baypren® 110, 3828
Neoprene latex, *See* NPR 3911, 21753, NPR 5587, 21754
Neoprene Latex 115, 20952, *See* Baypren® 110, 3828
Neoprene Latex 400, 20953
Neoprene Latex 622, *See* Baypren® 110, 3828
Neoprene Latex 654, *See* Baypren® 110, 3828
Neoprene Latex 671A, *See* Baypren® 110, 3828
Neoprene Latex 750, 20954
Neoprene Latex 842-A, *See* Baypren® 110, 3828
Neoprene NPG 6856, 20955
Neoproc, *See* Penicillin V, 23111
Neo-protosli, 20956
Neopybuthrin, 20957
Neo-Pynamin, *See* Killgerm® Py-Kill W, 16182, tetramethrin, 31384
Neorad, 20958
Neoram, *See* copper oxychloride, 6784
Neoresit, 20959
NeoRez, 20960, 20961
Neorode, 20962
neosaccharin, *See* saccharin, 27195
neosaccharin saccharol, *See* saccharin, 27195
Neoscan, 20963
Neoscoa 203C, 20964
Neosept V, *See* hexachlorophene, 13992
Neosobenil, *See* Nipacide® BCP, 21252
Neosolve® AD-1, 20965
Neosone D, *See* Antioxidant PBN, 2591, Stabilator A.R, 29400
Neosorb, 20966
Neosorex;, *See* Ratak, 25896
Neosorexa, 20967
Neosorexa PP580, *See* difenacoum, 8449, Neosorexa, 20967, Ratak, 25896
Neosorexa;, *See* Ratak, 25896
Neosote, 20968
Neospectra, 20969
Neospinol 264, 20970
Neostar, 20971
Neosulfine, *See* Odylen®, 22046
Neosyl®, 20972
Neosyn, 20973
Neotac, 20974
Neotex, 20975
Neothane, 20976
Neothene, 20977
Neotopsin, *See* thiophanate methyl, 31706
NeoVac, 20978
Neovadine, 20979
Neo-Voronit®, 20980
Neozapon® Dyes, 20981
Neozon D, *See* Antioxidant PBN, 2591, Stabilator A.R, 29400
Neozone, *See* Antioxidant PBN, 2591, Stabilator A.R, 29400
Neozone A, 20982, *See* Additin 30, 653, Akrochem® Antioxidant PANA, 1161, Antioxidant PAN, 2590, Naugard® PANA, 20798
Neozone B, 20983
Neozone C, 20984
Neozone D, 20985
Neozone E, 20986
NEP, 20987, *See* Agsol Ex 2, 1087
,NEPD, 20988
Nephroflow, 20989
,Nephrotest, *See* sodium *para*-aminohippurate, 28804
Nepol® PP40, 20990
NEPS, 20991
Nepton EXT, 20992
Neptun® Bases, 20993
Nequinate, *See* methyl benzoquate, 19516
neradol, 20994, 20995
Neral, 20996, *See* citral, 6336
Neramine, 20997
Nercol, 20998
Nercolan, 20999
Nercosol, 21000
Nerfinol, 21001
Nergandin, 21002
Neri silk, *See* Galettame silk, 12603
Nericur, *See* Abcure S-40-25, 35, AcetOxyl 2.5 and 5, 305, Benzoyl peroxide, 3983
Nericur Gel 5, 21003
Nerkol, *See* dichlorvos, 8394
Nerloate, 21004
Nerol, *See* Vernol®, 33600
neroli oil, 21005
Neroli oil, artificial, *See* methyl anthranilate, 19513
Nerolidol, 21006
nerolin, 21007
nerolin old, *See* jara jara, 15518, nerolin, 21007
Nerolin bromelia, *See* nerolin, 21007
Nervan, 21008
Nervan CP, 21009
Nervanaid, 21010
Nervanaid B, *See* Kalex 220 Crystal, 15668, Kalex Liq. 50%, 15671, tetrasodium EDTA, 31391
Nervanzse, 21011
nerve oil, *See* neatsfoot oil, 20837
nervin, 21012
Nesfield's triple tablets, 21013
Neste Polyethylene, 21014
Netazol, *See* MCPA, 19183
netilmicin sulfate, 21015
Nettolin, 21016
Neudorfite, 21017
Neuphor®, 21018
Neuphor® 100, *See* Neuphor®, 21018
Neuphor® 635, 21019
Neuro-Phosphates, 21020
Neustrene® 045, 21021
Neustrene® 059, 21022
Neustrene® 064, 21023
Neutral Acriflavine, *See* Gonacrine, 13192
neutral alum, 21024
Neutral ammonium flouride, *See* Ammonium Fluoride, 2239
Neutral Degras, 21025
neutral lead acetate, normal lead acetate, *See* lead diacetate, 16906
Neutral oils, 21026
Neutral oxalate of potash, *See* potassium oxalate,

24766
neutral phosphate, 21027
neutral red, 21028
neutral sodium chromate, *See* sodium chromate, 28750
neutral tartar, *See* potassium tartrate,
Neutralaleisen, 21029
Neutralite, 21030
neutralized verdigris, *See* cupric acetate, 7372
Neutramag®, 21031
Neutrapen, *See* Penicillinase, 23112
Neutrase, 21032
neutrazyme, *See* Rhodapon® 101-10, 26631, sodium lauryl sulfate, 28781
Neutrichrome, 21033
Neutrigan®, 21034
neutroflavine, *See* Gonacrine, 13192
Neutrogene, 21035
Neutrol® TE, 21036
Neutrolactis, 21037
neutronyx 600, *See* Alkasurf® NP-4, 1715, Triton® N-57, 32430
neutronyx 600's, *See* Alkasurf® NP-4, 1715, Triton® N-57, 32430
Neutronyx®, 21038
Neutronyx® 656, 21039
Nevada silver, *See* nickel silvers, 21132
Nevastain, 21040
Nevastain R.A, 21041
Nevastain® 21, *See* Alkylated phenol, 1734
Nevastain® A, 21042
Nevchem® 70, 21043
Nevex® 100, 21044
Nevidene, 21045
Neviken, *See* Orthorix, 22337
Nevile and Winther's acid, 21046
Nevillac® 10+sg XL, 21047
Neville, 21048
Nevillite, 21049
Nevin, 21050, *See* Lithopone, 17520
Nevindene, 21051
Nevinol, 21052
Nevolin®, 21053
Nevoxy® EPX-L, 21054
Nevpene® 9500, 21055
nevraltein, 21056
nevrosthe+a7nine, 21057
Nevroz® 1420, 21058
Nevtac® 80, 21059
Nevugon, *See* trichlorfon, 32212
Nevyanskite, *See* iridosmine, 15300
New 5C Cycocel, 21060, 21061
new blue, *See* ultramarine, 32821
New Brick, 21062
New Brick (Heavy Duty), 21063
new cacodyl, *See* arrhenal, 3128
New Formulation SBK Brushwood Killer, 21064
New Hickstor 6, 21065
New Hystor, 21066
New Kotol, 21067
New Legumex, 21068
New Murbetex, 21069
New Verdone, 21070
New Zealand dammar, 21071
Newagit, 21072
Newaloy, 21073
Newcavac, -T, 21074
Newcol 60, *See* Ablunol S-60, 110
Newcol 150, *See* Ablunol 200ML, 95
Newdamp Balancing Fluids, 21075
Newloy, 21076
newtonite, *See* Tisyn®, 31857
Newton's alloy, 21077
New-wrap, 21078
Nex, 21079
NF, *See* Ablusol ML, 129
Nf (dispersant), *See* Ablusol ML, 129
NF 44, *See* thiophanate methyl, 31706
Nf-a, *See* Ablusol ML, 129
NFB, 21080
NFT Fertilizer, 21081
Ngai camphor, 21082
N-Hance® 3000, 21083, *See* Jaguar® C-13S, C-14S, 15496
N'hangellite, 21084
NI-20GF, 21085, *See* nylon 6/10, 21911
NIA 5273, *See* piperonyl butoxide, 23875
NIA 5488, *See* tetradifon, 31366
NIA 10242, *See* Carbodan, 5197, Rampart, 25860
Nia proof 08, *See* Niaproof® Anionic Surfactant 08, 21092
nia proof 08, *See* Rhodapon® BOS, 26632
NIA-17370, *See* resmethrin, 26218
NIA-18739, *See* Reslin, 26216
NIA-33297, *See* permethrin, 23384, 23385, 23386
NIA-33297, *See* Kafil, 15655
Niac, 21086
Niacet, 21087
Niacet Calcium Acetate Tech, 21088
Niacet Sodium Acetate Anhyd. Tech, 21089
niacin, 21090
Niagara 10242, *See* Rampart, 25860
Niagara blue, 21091
Niagaral 242, *See* Carbodan, 5197
Niaproof® Anionic Surfactant 08, 21092
Niaproof® Anionic Surfactant 4, 21093
Niax, 21094
Niax 3CF, *See* Genomoll P, 12838
Niax Flame Retardant 3CF, *See* Genomoll P, 12838
Nibiol, 21095
Nibren, 21096
Nibren wax, 21097
Nibrite, 21098
Nicar, 21099
Nicaragua wood, *See* redwoods, 25987
Nicat, 21100
Nicel, *See* Glutolin, 13064
Nicfo, 21101
Ni-chillite, 21102
Nichroloy, 21103
Nichrome®, 21104
Nichrosi, 21105
nickel, 21106
nickel, *See* malleable nickel, 18486
nickel(II) carbonate, *See* nickel carbonate, 21123
nickel(II)carbonate basic, *See* nickel carbonate, 21123
nickel(II) chloride, *See* nickel chloride, 21125
nickel(II) nitrate, *See* nickel nitrate, 21128
nickel(II) oxide, *See* nickel oxide, 21130
nickel(II) sulfamate dihydrate, *See* nickel sulfamate, 21134
nickel(II) sulfate, *See* nickel sulfate, 21135
nickel 200, 21107
nickel 201, 21108
nickel 204, 21109
nickel 205, 21110
nickel 211, 21111
nickel 212, 21112
nickel 213, 21113
nickel 222, 21114
nickel 223, 21115
nickel 229, 21116
nickel 270, 21117
nickel acetate, 21118
nickel acetate tetrahydrate, *See* nickel acetate, 21118
nickel aluminum bronze, 21119
nickel ammonium sulfate, *See* double nickel salt, 8815
nickel Babbitt, 21120
nickel brass, 21121, *See* nickel silvers, 21132
nickel bronze, 21122, *See* nickel silvers, 21132
nickel carbonate, 21123
nickel carbonate, basic, 21124
nickel chloride, 21125
nickel dialkyldithiocarbamate, *See* Niclate®, 21145
nickel dibutyl dithiocarbamate, *See* Naugard® NBC, 20797, NBC, 20831, Perkacit® NDBC, 23287
nickel diisobutyldithiocarbamate, *See* Isobutyl Niclate®, 15347
nickel dimethyldithiocarbamate, *See* methyl niclate®, 19553
nickel di-n-butyldithiocarbamate, *See* Vanox® NBC, 33344
nickel formate, *See* Nicfo, 21101
nickel glance, 21126
nickel iron, *See* ferro-nickel, 11678
nickel manganese bronze, 21127
nickel monosulfate, *See* nickel sulfate, 21135
nickel monoxide, *See* nickel oxide, 21130, Nico, 21147
nickel nitrate, 21128
nickel oreide, 21129
nickel oxide, 21130
nickel protoxide, *See* nickel oxide, 21130
nickel silver solder, 21131
nickel silvers, 21132
nickel steel, 21133
nickel sulfamate, 21134, *See* Nimate, 21201
nickel sulfate, 21135
nickel zirconium, 21136
nickel(2+) salt, sulfuric acid, *See* nickel sulfate, 21135
nickel, bis(dibutylcarbamodithioato-S,S')-, (SP-4-1)-, *See* Naugard® NBC, 20797, NBC, 20831, Perkacit® NDBC, 23287
nickeladium, 21137
nickel-chromium-copper-Austenitic Iron L, *See* Iron L, 15316
nickelene, 21138
nickelin, 21139, *See* nickel silvers, 21132
nickelin
nickel-linnaeite, *See* polydymite, 24414
nickel-manganese-copper, 21140
nickel-molybdenum steels, 21141
nickeloid, 21142
nickelous chloride, *See* nickel chloride, 21125
nickelous chloride hexahydrate, *See* nickel chloride, 21125
nickelous nitrate, *See* nickel nitrate, 21128
nickelous oxide, *See* nickel oxide, 21130
nickelous sulfate, *See* nickel sulfate, 21135
nickeloy, 21143
Niclad, 21144
Niclate®, 21145
Niclocide, *See* niclosamide, 21146
niclosamide, 21146, *See* clonitrilide, 6432, Mansonil®, 18611
niclosamide ethanolamine, *See* clonitrilide, 6432
Nico, 21147
Nico-400, 21148
Nicobid, 21149
Nicocap, 21150
Nicolar, 21151
Nicolle's Carbol-thionin blue, 21152
Nicolmelt®, 21153
Nicon, 21154
Nicor, 21155
Nicoschwab, *See* Uba, 32698
Nicotine, *See* Campbell's Nico-Soap, 5068, XL-All Insecticide, 34607
Nicotine 40% Shreds, 21156
nicotinic acid, *See* niacin, 21090
nicotinyl alcohol, *See* nicotinyl alcohol,
nicouline, *See* rotenone, 27000
Nicral alloys, 21157
nicro-copper, 21158
Nicrolan, 21159
Nicroman, 21160
Nicron 325, 21161
Nicron 665, 21162
Nicron JS 422, 21163
Nicrosil, 21164
Nicrosilal, 21165
Nicu steel, 21166
Nidrin, *See* Alumina hydrate, 1892
Niello silver, 21167
Niflex, *See* Polyflex, 24435
Niflex D, *See* Polyflex, 24435
Niflor, 21168
Nifuraldezone, 21169
Nifurpirinol, 21170

Nigagin, 21171
Night of Olay Nightcare, 21172
nigraniline, 21173
nigre, 21174
nigrol, 21175
Nigrosin, 21176
Nigrosin Bases, 21177
Nigroth Metal, 21178
nigrotic acid, 21179
Ni-hard, 21180
Nikal®, 21181
Nikkol Avacado Oil, *See* Avocado Oil, 3497
Nikkol BM, *See* Butyl myristate, 4775
Nikkol Decaglyn 10-O, *See* Polygliceryl-10 decaoleate, 24453
Nikkol Grapeseed Oil, *See* Grape seed oil, 13275
Nikkol Sarcosinate LH, *See* lauroyl sarcosine, 16861
Nikkol Sarcosinate OH, *See* Oleoyl sarcosine, 22123
Nikkol SL 10, *See* Ablunol S-20, 108
nikkol sls, *See* Rhodapon® 101-10, 26631, sodium lauryl sulfate, 28781
Nikkol SMT, *See* sodium methyl stearoyl taurate, 28792
Nikkol SO 10, *See* Ablunol S-80, 111, Witconol 2500, 34501
Nikkol SO-15, *See* Ablunol S-80, 111, Witconol 2500, 34501
Nikkol SO-30, *See* Ablunol S-80, 111, Witconol 2500, 34501
Nikkol sp10, *See* Ablunol S-40, 109
Nikkol SS 30, *See* Ablunol S-60, 110
Nikkol Sweet Almond Oil, *See* Sweet Almond oil, 30452
Niklad, 21182
Nikrome, 21183
Nikro-trimmer Steel, 21184
Nilex, 21185
Nilfom 2X, 21186
Nilfom DF-155, 21187
Nilo Alloy 36, 21188
Nilo Alloy 42, 21189
Nilo Alloy 48, 21190
Nilo Alloy 51, 21191
Nilo Alloy 475, 21192
Nilo Alloy K, 21193
Nilo Alloy K45, 21194
Nilo Alloy P50, 21195
Nilomag Alloy 48, 21196
Nilomag Alloy 51, 21197
Nilomag Alloy 471, 21198
Nilomag Alloy 475, 21199
Nilomag Alloy K, 21200
Nilox PBNA, *See* Antioxidant PBN, 2591, Stabilator A.R, 29400
Nilverm, *See* Tramisol®, 32134
Nimate, 21201
nimbecetin, *See* kaempferol, 15654
nimeton, *See* thiometon, 31700
Nimidane, *See* Abequito, 38
Nimocast Alloy 242, 21202
Nimocast Alloy 713, 21203
Nimocast Alloy 771, 21204
Nimocast Alloy PE10, 21205
Nimocast Alloy PK24, 21206
Nimol, 21207
Nimonic Alloy 75, 21208
Nimonic Alloy 80A, 21209
Nimonic Alloy 90, 21210
Nimonic Alloy 93, 21211
Nimonic Alloy 105, 21212
Nimonic Alloy 115, 21213
Nimonic Alloy 118, 21214
Nimonic Alloy PE 11, 21215
Nimonic Alloy PE 13, 21216
Nimonic Alloy PK 31, 21217
Nimonic Alloy PK 33, 21218
Nimox, 21219
Nimrod, 21220
Nimrod T, 21221
Ninate 401, 21222
Ninate 411, 21223
Ninate 415, 21224
Ninate®, 21225
Ninate® 401, 21226
Ninate® 401-A, 21227
Ninate® 411, 21228
Ninate® DS 70, 21229
NINE, *See* daminozide, 7723
ninhydrin, 21230
ninitrogen oxide, *See* Nitral, 21318
Ninol 1281, *See* Active #2, 501
Ninol 1301, *See* Ablumide SME, 82, Witcamide® 70, 34442
Ninol 2012E, *See* Active #2, 501
Ninol AA62, *See* Ablumide LDE, 79, Alkamide® 327, 1634
ninol AA62 extra, *See* Philacid 1200, 23660
Ninol P 621, *See* Active #2, 501
Ninol®, 21231
Ninol® 11-CM, 21232
Ninol® 30-LL, 21234
Ninol® 201, 21233
Ninol® CMP, 21235
Ninol® CNR, 21236
Ninol® LMP, 21237
Ninol® M10, 21238
Ninox, 21239
Ninox CZ-1, *See* Ablunol CO 10, 98
Ninox®, 21240
Ninox® FCA, 21241
Ninox® L, 21242
Ninox® M, 21243
Ninox® SO, 21244
Nio-A-Let, *See* Vi-Alpha, 33706
Niobe oil, *See* methyl benzoate, 19515
niobium, 21245
niobium(V) oxide, *See* niobium oxide, 21246
niobium oxide, 21246
niobium pentoxide, *See* niobium oxide, 21246
Niomil, *See* Bendiocarb, 3937
Niopam, *See* Iopamidol, 15233
Nip, *See* Korogel, 16354, Koroplate, 16357
Nipa 49, *See* Sustane® PG, 30436
Nipa salt, 21247
Nipabenzyl, 21248
Nipabutyl, 21249, *See* Nipabutyl, 21249
Nipabutyl Potassium, 21250
Nipabutyl Sodium, 21251
Nipacide PX, *See* chloroxylenol, 6159
Nipacide®, *See* Benzoic acid, 3972
Nipacide® BCP, 21252
Nipacide® BIT, 21253
Nipacide® BK, 21254, *See* Bioban® GK, 4158
Nipacide® DP, *See* dichlorophen, 8392
Nipacide® DX, 21255
Nipacide® F, 21256
Nipacide® MX, 21257
Nipacide® OPP, 21258
Nipacide® PTAP, 21259
Nipacide® PX, 21260
Nipacombin SK, 21261
Nipafax, 21262
Nipagallin P, *See* Sustane® PG, 30436
Nipagin, *See* methylparaben, 19577
Nipagin A, 21263, *See* ethyl *p*-hydroxybenzoate, 11061
Nipagin A
Nipagin A Ethyl, *See* ethyl *p*-hydroxybenzoate, 11061
Nipagin A Potassium, 21264
Nipagin A Sodium, 21265
Nipagin M, 21266, *See* methylparaben, 19577
Nipagin M Potassium, 21267
Nipagin M Sodium, 21268
Nipagin P, *See* Nipasol M, 21282
Nipaguard DMDMH, *See* DMDM hydantoin, 8754, Glydant®, 13140, Mackstat® DM, 18234, Nipaguard® DMDMH, 21269
Nipaguard® DMDMH, 21269
Nipaguard® DME, *See* Nipaguard® TCC, 21270
Nipaguard® MPS, 21271
Nipaguard® TCC, 21270
Nipaheptyl, 21272
Nipanox® BHT, 21273 , *See* Anti-Oxydant Bayer, 2595
Nipanox® S-1, 21274
Nipanox® Special, 21275
Nipantiox, 21276, *See* Butylated hydroxyanisole, 4787
Nipantiox 1-F, 21277 , Nipantiox, 21276, *See* Sustane® 1-F, 30432
NiPar 640, 21278
NiPar S-10, 21279
NiPar S-20, 21280
Nipasept, 21281
Nipasol, *See* Nipasol M, 21282
Nipasol M, 21282, *See* Nipasol M, 21282
Nipasol M Potassium, 21283
Nipasol M Sodium, 21284
Nipasol P, *See* Nipasol M, 21282
Nipastat, 21285
Nipazol, *See* Nipasol M, 21282
Niphen, *See p*-nitrophenol, 21368
Nipol DN-601, *See* Acrylonitrile-butadiene rubber, 437
Nipol® 1000X132, 21286
Nipol® 1001 CG, 21287
Nipol® 1022X59, 21288
Nipol® 1072, 21289
Nipol® 1203F60, 21290
Nipol® 1204X22, 21291
Nipol® 1411, 21292
Nipol® 2782, 21293
Nipol® AR-31, 21294
Nipol® AR-42, 21295
Nipol® AR-74, 21296
Nipol® DP5123P, 21297
Nippon Ant and Crawling Insect Killer, 21298, *See* Permethrin, 23385
Nippon Ant Killer Liquid, 21299
Nippon Ant Killer Powder, 21300, *See* Permethrin, 23385
Nippon Fly Killer Spray, 21301
Nippon Ready For Use Ant and Crawling Insect Killer, 21302, *See* Permethrin, 23385
Nipro (i), 21303
Nipro (ii), 21304
Nipsan, *See* Diazinon Liquid, 8344
Niptan, *See* EPTC, 10729
Niran, *See* parathion, 22807
Ni-resist, 21305
Nirex, 21306
Nirez® 7002, 21307
Nirez® 9007, 21308
Nirez® 9011, 21309
Nirostaguss, 21310
Nispan Alloy C-902, 21311
Nissan Anon BF, *See* Accobetaine CL, 222
Nissan Cation S2-100, *See* Ablumine 280, 85, Cycloton® SCS, 7542, octadecylbenzenemethanaminium chloride, 21985, stearalkonium chloride, 29592, Sumquat® 6210, 30126
Nissan Nonion L 4, *See* Ablunol 200ML, 95
Nissan nonion op 85, *See* Ablunol S-85, 112, Witconol 2503, 34502
Nissan nonion op 85r, *See* Ablunol S-85, 112, Witconol 2503, 34502
Nissan nonion OP-85, *See* Ablunol S-85, 112, Witconol 2503, 34502
Nissan Nonion PP 40, *See* Ablunol S-40, 109
Nissan nonion PP 40r, *See* Ablunol S-40, 109
Nissan Nonion SP-60, *See* Ablunol S-60, 110
Nissan Tert. Amine BB, *See* dimethyl lauramine, 8507
Nisso PB, 21312
Nitacid, *See* Ramrod, 25861
Nital, *See* nitric acid, 21332
niter, *See* potassium nitrate, 24762, saltpeter, saltpeter flour, 27298
niter cake, 21313
niter spirit, *See* spirit of niter,
Niticid, *See* Propachlor, 25163, 25164
Niticid, *See* Ramrod, 25861
Nitofol®, 21314

Nitolac, 21315
Nitoman, 21316
Nitracc, 21317
Nitrador, *See* 4,6-dinitrocresol, 8542
Nitral, 21318
nitralin, 21319
Nitraline, 21320
Nitralloy, 21321
Nitram, 21322, *See* Ammonium nitrate, 2245
Nitramin IDO, *See* 2-amino-5-nitrothiazole, 2184
Nitramine, *See* Tetryl, 31423
Nitrammite, 21323
Nitrammomkalk, 21324
Nitran, *See* Trifluralin, 32252
***p*-nitraniline**, *See p*-nitroaniline, 21345
Nitraniline N, 21325
Nitrapo, 21326
Nitraprill, 21327
nitrate of soda, *See* sodium nitrate, 28800
nitrated oils, 21328
nitrating acid, 21329
nitrazol, *See* paranitraniline red, 22780
Nitrazol CF Extra, *See p*-nitroaniline, 21345
nitre, *See* potassium nitrate, 24762
Nitrene 11230, 21330
Nitrene C Extra, 21331
nitric acid, 21332, *See* dynamite acid, 9357 fuming nitric acid, 12456
nitric acid red fuming, *See* nitric acid, 21332
nitric Acid, barium Salt, *See* Barium nitrate, 3639
nitric acid, magnesium salt, *See* Magnisal, 18380, magnesium nitrate, 18365
nitric acid, sodium salt, *See* sodium nitrate, 28800
nitric ether, 21333
nitrided steel, 21334
Nitrile 10 D, 21335
Nitrile 12, 21336
Nitrile BG, 21337
nitrile C_4, 21338, *See* Butyronitrile, 4805
Nitrile latex, *See* Chemigum® Latex 260, 5975, Darex® 110L, 7769, Hycar® 1552, 14503
Nitrile rubber,, *See* Acrylonitrile-butadiene rubber, 437
Nitrile-based thermoplastic elastomer, *See* Chemigum® TPE 03050, 5981
nitrile-butadiene rubber, *See* Chemigum® P7-D, 5979, Chemigum® P83, 5980
Nitriloacetate trisodium salt, *See* Cheelox® NTA-Na3, 5869
nitriloacetic acid, *See* TG Buffer, 31547
1,1',1-nitrilopropan-2-ol, *See* triisopropanolamine, 32313
Nitrilotriacetamide, *See* Hampamide B, 13584
nitrilotriacetic acid, 21339, *See* Hampshire® NTA Acid, 13611
Nitrilotriacetic Acid Trisodium Salt, *See* Cheelox® NTA-Na3, 5869
Nitrilotrimethanephosphonic acid, *See* Fostex AMP, 12337
Nitrilotris(methylene)triphosphonic acid, *See* Fostex AMP, 12337
Nitrilotris(methylphosphonic acid), *See* Fostex AMP, 12337
2,2',2-nitrilotris(ethanol), *See* triethanolamine, 32240
2,2',2-nitrilotris(ethanol) hydrochloride, *See* triethanolamine hydrochloride, 32241
1,1',1-nitrilotris-2-propanol, *See* triisopropanolamine, 32313
nitrimidazine, *See* nimorazole,
nitrite rubber, 21340
nitro acid sulfite, *See* nitrosylsulfuric acid, 21379
nitro base, 21341
Nitro BT, 21342, *See* Nitro BT, 21342
Nitro BT monohydrate, *See* Nitro BT, 21342
Nitro Fast, 21343
nitro kleenup, *See* Dinitra, 8540
Nitro-26, 21344
5-nitro-2-aminothiazole, *See* 2-amino-5-nitrothiazole, 2184
***p*-nitroaniline**, 21345
4-nitroaniline, *See p*-nitroaniline, 21345
2-nitrobenzaldehyde, *See o*-nitrobenzaldehyde, 21346
***o*-nitrobenzaldehyde**, 21346
4-nitrobenzaldehyde, *See p*-nitrobenzaldehyde, 21347
***p*-nitrobenzaldehyde**, 21347
5-nitrobenzene-1,3-dicarboxylic acid, *See* 5-nitroisophthalic acid, 21361
nitrobenzene, 21348, *See* mirbane oil, 19935
4-nitrobenzoic acid, *See p*-nitrobenzoic acid, 21350
***p*-nitrobenzoic acid**, 21350
nitrobenzol, *See* nitrobenzene, 21348
nitrobromoform, *See* Bromopicrin, 4585
4-(2-Nitrobutyl) morpholine and 4,4-(2-ethyl-2-nitrotrimethylene) dimorpholine, *See* Bioban® P-1487®, 4160
4-(2-Nitrobutyl)morpholine, *See* Fuelsaver®, 12432, Nitrobutylmorpholine, 21351
Nitrobutylmorpholine, 21351
nitrocalcite, 21352
nitrocarbol, *See* nitromethane, 21366
nitrocellulose, 21353, *See* Hercules® AS, 13858, Hercules® RS, 13864, Hercules® SS, 13865, RS Kodaloid, 16283, Nitrocellulose, 27051, SS Nitrocellulose, 29388, Venite, 33560
nitrocellulose varnishes, 21354
Nitro-chalk, 21355
nitrochloroform, *See* chloropicrin, 6135
nitrocotton, *See* guncotton, 13492, nitrocellulose, 21353
nitro-dextrin, 21356
nitrodracrylic acid, *See p*-nitrobenzoic acid, 21350
***p*-nitrodracylic acid**, *See p*-nitrobenzoic acid, 21350
Nitroetan, *See* NE, 20833, nitroethane, 21357
nitroethane, 21357, *See* NE, 20833
2-Nitro-2-ethyl-1,3-propanediol, *See* NEPD, 20988
Nitroferrite, 21358
Nitrofor, *See* Trifluralin, 32252
Nitroform, 21359, *See* tetranitromethane, 31386
Nitrofuel®, 21360
5-nitro-2-furaldehyde semioxamazone, *See* Nifuraldezone, 21169
6-[2-(5-Nitro-2-furanyl)ethenyl]-2-pyridinemethanol, *See* Nifurpirinol, 21170
2-[3-(5-nitro-2-furanyl)-1-[2-(5-nitro-2-furanyl)ethenyl]-2-propenylidene]hydrazinecarboximidamide, *See* nitrovin, 21382
6-[2-(5-nitro-2-furyl)vinyl]-2-pyridine-methanol, *See* Nifurpirinol, 21170
Nitrogen derivs, *See* Kemamine® AS-650, 15905
Nitrogen dioxide, *See* nitrous gas or air, 21381
nitrogen monoxide, *See* nitrous oxide,
nitrogen oxide, *See* Nitral, 21318
nitroglycerin, *See* explosive gum, 11315, Glonoine oil, 13005
Nitroglycerin dynamite, *See* Hercol 2, 13824, Hercol 2X, 13825, Hercon® 2, 13831, Hercon® 2X, 13832
nitrohydrochloric acid, *See* aqua regia, 2682
nitro-2-(hydroxymethyl)-1,3-propanediol, *See* Tris Nitro®, 32408
2-nitro-2-(hydroxymethyl)-1,3-propanediol, *See* S.S.T® Sump Saver Tablets, 27180, Tris Nitro®, 32408
Nitroisobutylglycerol, *See* S.S.T® Sump Saver Tablets, 27180, Tris Nitro®, 32408
4-nitroisophthalic acid, 21361
5-nitroisophthalic acid, 21361
Nitrolac, 21362
nitrolignin, 21363
nitrolim, 21364
Nitrolite, 21365
nitromagnesite, *See* magnesium nitrate, 18365
nitromethane, 21366
Nitromethane, *See* NM, 21402
2-Nitro-2-methyl-1-propanol, *See* NMP, NMP Conc, 21405
Nitromin IDOo, *See* 2-amino-5-nitrothiazole, 2184
nitromuriatic acid, *See* aqua regia, 2682
Nitron, 21367
α-nitronaphthalene-7-sulfonic acid, *See* Cleve's +lw-or +ld-acid, 6413
1-nitropan, *See* 1-nitropropane, 21373
Nitroparaffin, *See* nitroethane, 21357, nitromethane, 21366
4-Nitrophenol, *See p*-nitrophenol, 21368
***p*-nitrophenol**, 21368, *See* P.N.P, 22493
3-(2-nitrophenyl)-2-propynoic acid sodium salt, *See* nitropropiol, 21375
Nitrophos® 20-20-0, 21369
Nitrophoska® 10-15-20, 21370
nitrophosphate, 21371
Nitropore, *See* Azodicarbonamide, 3524
Nitropore® ATA, 21372
1-nitropropane, 21373 *See* NiPar S-10, 21279
2-nitropropane, 21374, *See* NiPar S-20, 21280
n-nitropropane, *See* 1-nitropropane, 21373
***sec*-nitropropane**, *See* 2-nitropropane, 21374
nitropropiol, 21375
Nitrose, *See* nitrosylsulfuric acid, 21379
Nitrosin Saltpetre, 21376
nitrosodihydroxynaphthalene, *See* Dioxine, 8575
***p*-nitrosodimethylaniline**, *See* nitro base, 21341
***p*-nitroso-N,N-dimethylaniline**, *See* nitro base, 21341
***p*-nitrosodimethylaniline/ β-naphthol (1:1)**, *See* Accelerene V 1, 220
***p*-nitroso-N,N-dimethylaniline**, *See* Vulcaniline, 34095
Nitrosophenylhydroxylamine ammonium salt, *See* cupferron, 7365
nitrososulfuric acid, *See* nitrosylsulfuric acid, 21379
nitro-starch, 21377
nitrosulfate, *See* ferric sulfate, 11652
nitrosulfonic acid, *See* nitrosylsulfuric acid, 21379
nitrosyl silver, 21378
nitrosyl sulfate, *See* nitrosylsulfuric acid, 21379
nitrosylsulfuric acid, 21379
nitrothale-isopropyl, *See* nitrothal-isopropyl, 21380
nitrothal-isopropyl, 21380
Nitrothal-isopropyl, sulfur, *See* Kumulan®, 16481
Nitrothal-isopropyl-sulfur mixture, *See* Pallitop® S, 22623
5-Nitro-2-thiazol-amine, *See* 2-amino-5-nitrothiazole, 2184
5-nitro-2-thiazolamine, *See* 2-amino-5-nitrothiazole, 2184
N-(5-Nitro-2-thiazolyl)acetamide, *See* Aminitrazole, 2164
5-nitro-2-thiazolylamine, *See* 2-amino-5-nitrothiazole, 2184
Nitro-tolueneazonitrosalicylic acid, *See* Persian yellow, 23419
nitrotribromomethane, *See* Bromopicrin, 4585
Nitrotris(hydroxymethyl)methane, *See* S.S.T® Sump Saver Tablets, 27180, Tris Nitro®, 32408
nitrous acid barium salt monohydrate, *See* Barium nitrite, 3640
nitrous acid, sodium salt, *See* sodium nitrite, 28801
nitrous gas or air, 21381
nitrous oxide, *See* Nitral, 21318
nitrous sulfuric acid, *See* nitrosylsulfuric acid, 21379
nitrovin, 21382
Nitrox, *See* Parathion-methyl, 22808
Nitroxan, 21383
nitroxanthic acid, *See* picric acid, 23821, reflorit, 26003
nitroxylsulfuric acid, *See* nitrosylsulfuric acid, 21379
nitroxynil, 21384, *See* Trodax, 32439
Nitryl Hydroxide, *See* nitric acid, 21332
Nitto Nitoflon, 21385
Nitto SPV, 21386
nivan, *See* oil white, 22086
Nivar, *See* invar, 15191
Nivelona, *See* Variton, 33432
Nivitin, *See* A-625/641ABS 301K, ABS 500FR-1, 10
Nix, *See* Kafil, 15655
Nix Creme Rinse, 21387, *See* Permethrin, 23385
Nix Dermal Cream, 21388, *See* Permethrin, 23385
Nix Stix L-515, 21389
Nix Stix X-9021, 21390
Nixenoid, *See* Viscoloid, 33899
Nixon C/A, *See* cellulose acetate, 5627
Nixon C/N, *See* cellulose nitrate,

NP 30, 26094
Nonoxynol-30, *See* Akyporox NP 300V, 1280, Chemax NP-30, 5915, Iconol NP-30, 14823, Igepal® CO-880, 14862, Marlophen® 830N, 18886
Nonoxynol-34, *See* Polystep® F-95B, 24600
Nonoxynol-40, *See* Chemax NP-40, 5916, Iconol NP-40, 14824, Igepal® CO-890, 14864, Sellig N 1050, 27884, T-Det® N-40, 30867
Nonoxynol-50, *See* Iconol NP-50, 14825, Igepal® CO-970, 14865, Remcopal 31250, 26081
Nonoxynol-52, *See* Sellig N 50 100, 27893
Nonoxynol-70, *See* Iconol NP-70, 14826
non-pareil metal, 21457
Nonsepara®, 21458
Nonsilicone antifoam, *See* AF GN-11-P, 915, AF HL-36, 916, AF HL-52, 917
nonsoul ok 1, *See* Octosol 449, 22018
Nontoxol, 21459
Nonxynol-4, *See* Chemax NP-4, 5910
Nonxynol-6, *See* Chemax NP-6, 5911
Nonyl acarbinol, *See* Exxal® 10, 11338
nonyl cyanide, *See* Nitrile 10 D, 21335
Nonyl nonoxinol series, *See* Ethal DNP-8, 10919
Nonyl nonoxynol-4, *See* Hetoxide DNP-4, 13923
Nonyl nonoxynol-5, *See* Macol® DNP-5, 18265
Nonyl nonoxynol-7, *See* Igepal® DM-430, 14868
Nonyl nonoxynol-7 phosphate, *See* Rhodafac® RM-410, 26607
Nonyl nonoxynol-8, *See* Prox-onic DNP-08, 25315, Trycol® 6985, 32506
Nonyl nonoxynol-8, -18, *See* Chemax DNP-8, DNP-18, 5894
Nonyl nonoxynol-9, *See* Igepal® DM-530, 14869
Nonyl nonoxynol-10, *See* Sellig DN 10 100, 27879
Nonyl nonoxynol-10 phosphate, *See* Chemphos TC-341, 6011
Nonyl nonoxynol-15, *See* Igepal® DM-710, 14870
Nonyl nonoxynol-16, *See* Marlophen® DNP 16, 18887
Nonyl nonoxynol-22, *See* Sellig DN 22 100, 27880
Nonyl nonoxynol-24, *See* Igepal® DM-730, 14871
Nonyl nonoxynol-150, *See* Igepal® DM-970, 14872
nonyl phenol, 21460
Nonyl phenol ethoxylate, *See* Berol 09, 4014
Nonyl phenol polyglycol ether sulfate, *See* Rewopol® NOOSE 5, 26369
Nonyl phenyl glycidyl ether, *See* Heloxy® 64, 13772
Nonylated diphenylamine, *See* OA-502, 21953
Nonylic acid, *See* Emery® 1202, 10138, Pelargonic acid, 23078
Nonylnaphthalene sodium sulfonate, *See* Emkal NNS, NNS Acid, 10229
Nonylol, *See* Nonanol, 21436
Nonylphenol disulfide, *See* Ethanox® 323, 10931
Nonylphenol ethoxylate, *See* Anedco AW-397, 2451, Berol 259, 4021, Berol 267, 4023, Berol 278, 281, 282, 291 and 292, 4025, Berol WASC, 4049, Indulin® XD-70, 15095, Nonal 206, 21435
Nonylphenol PEG ester, *See* Arodet N-100, 3011
Nonylphenol polyglycol ether, *See* Marlowet® ISM, 18932
Nonylphenol polyglycol ether carboxylate, *See* Hostapal 2345, 14377
Nonylphenol polyglycol ether carboxylic acid, *See* Marlowet® 4536, 18919
Nonylphenol polyglycol ether phosphate, *See* Rewophat NP 90, 26353
Nonylphenolpolyethylene glycol-iodine complex, *See* Biopal® NR-20, 4175
Nonylphenoxy polyethoxy alcohol, *See* Prechem NPX, 24864
Nonylphenoxyacetic acid (4-Nonylphenoxy)-acetic acid, *See* Akypo NP 70, 1226
Nonylphenoxyl polyethoxy ethanol, *See* Atlas EMJ-2, 3308
Nonylphenoxypoly(ethyleneoxy)ethanol, *See* Triton® N-57, 32430
Nonylphenyl phosphite, *See* Wytox® 312, 34578
Nonylphenyl Polyethyleneglycol Ether, Nonionic, *See* Alkasurf® NP-4, 1715
α-(4-nonylphenyl)- ω-hydroxytetra(oxy-1,2-ethanediyl), *See* T-Det® N-4, 30866
Nopalcol, 21461
Nopalcol 1-L, 21463, *See* Ablunol 200ML, 95
Nopalcol 1-TW, 21464
Nopalcol 2-DL, 21465, *See* PEG-4 dilaurate series, 23050
Nopalcol 4-C, 21467
Nopalcol 4-DTW, 21468
Nopalcol 4-L, 21469
Nopalcol 4-0, 21466
Nopalcol 4-S, 21470
Nopalcol 6-DO, 21471, *See* PEG-2 dioleate series, 23040
Nopalcol 6-L, *See* Ablunol 200ML, 95
Nopalcol 6-R, 21472, *See* PEG ricinoleate series, 23031
Nopalcol 10-COH, 21462
Nopalcol 10-L, *See* Ablunol 200ML, 95
Nopalcol 30-TWH, 21473
Nopalcol 200, 21474
Nopalcol 400, 21475
Nopalcol 600, 21476
Nopco polyethoxylated castor oil, *See* Ablunol CO 10, 98
Nopco Worsted Oil 12, 21477
Nopco® 1179, 21478
Nopco® 2031, 21479
Nopco® 2272-R, 21480
Nopco® Colorsperse 188-A, 21481
Nopco® Foamaster, 21482
Nopco® NXZ, 21483
Nopco® PD#1-D, 21484
Nopcocastor, 21485, *See* Eureka 102, 11207
Nopcochex RA, 21486
Nopcocide® N-40-D, 21487
Nopcocide® N-96, 21488
Nopcocide® N-96-S, 21489
Nopcofloc, 21490
Nopcogen, 21491
Nopcogen 14-S, 21492
Nopcogen 16-L, 21493
Nopcogen 22-0, 21494
Nopcolan SHR3, 21495
Nopcolene, 21496
Nopcolube, 21497
Nopcone, 21498
Nopcosant, 21499
Nopcosant L, 21500
Nopcosize, 21501
Nopcosperse 2B-B, 21502
Nopcostat 237, 21503
Nopcosulf CA-60, -70, 21504
Nopcosulf CA-60, -70, CA, *See* Eureka 102, 11207
Nopcosulf TA-30, 21505
Nopcosulph, 21506
Nopcosurf CA, 21507
Nopcotan, 21508
Nopcote, 21509
Nopcote C-104, 21510
Nopcotex, 21511
Nopcowax, 21512
Nopcowax 22-DS, 21513, *See* Abluwax EBS, 140, Advawax® 290, 820
No-pest, *See* dichlorvos, 8394
No-pest Strip, *See* dichlorvos, 8394
nopinene, *See* β-l-pinene, 23858, β-pinene, 23859
Nora, 21514
18-norabietane, *See* fichtelite, 11765
Noraflor, 21515
Noralastic, 21516
Noralen, 21517
Noram, 21518
Noram DMSD, *See* Jet Amine DMTD, 15588
Noramac, 21519
Norament, 21520
Noramid, 21521
Noramium, 21522
Noramium DA-50, *See* Benzalkonium chloride, 3958
Noramox, 21523
Noraplan, 21524
Noratex, 21525
Norbide, 21526
Norbo, 21527
Norcast 142 Systems, 21528
Norcast 154FR, 21529
Norcast 1460-1, 21530
Norcast 3220G-1, 21531
Norcast 3258, 21532
Norcast 3705, 21533
Norcast 4914-1, 21534
Norcure 131, 21535
Norcure 3298, 21536
Nordel®, 21537
Nordel® 2744, 21538
Nordhausen acid, *See* fuming sulfuric acid, 12457
Nordot 101F, 21539
Nordox, 21540
Noreplast, 21541
Norepol, 21542
Noreseal, 21543
Norfemac, *See* bufexamac, 4655
Norfox® 1101, 21544
Norfox® Agent 2A-2S, 21545
Norfox® ALPHA XL, 21546
Norfox® Coco Betaine, 21547
Norfox® DC, 21548, *See* Cocamide DEA, 6547
Norfox® F-221, 21549
Norfox® IM-38, 21550
Norfox® SLES-60, 21551
Norfox® SLS, 21552
Norfox® Sorbo T-60, 21553, *See* Ethylan® GES6, 11108
Norfox® T-60, 21554
Norfox® TLS, 21555
Norfox® Unimulse OW, 21556
Norfox® X, 21557
Norfroth, 21558
Norgine, 21559, *See* alginic acid, 1593, Kelacid®, 15824
Norit A, *See* carbon, activated, 5225
Norit SA II, *See* carbon, activated, 5225
Norit, C, PK, R, RO, 21560
Norithene, 21561
Norlig, 21562
Norlig 11 DA, 21563, *See* calcium lignosulfonate, 4956
Norlig 11 DA, A, *See* calcium lignosulfonates, 4957
Norlig 415, 21564
Norlig A, 21565
Normacol, 21566
Normal lead acetate, *See* salt of Saturn, 27287
normal powder, 21567
Normal salt solution, *See* physiological salt solution, 23776
Nor-Mer 020, 21568
normersan, *See* Rezifilm, 26521
Normet, 21569
No-Roma, 21570
norox bzp-250, *See* Benzoyl peroxide, 3983
norox bzp-c-35, *See* Benzoyl peroxide, 3983
Norox® MCP, 21571
Norpar®, 21572
Norpar® 12, 21573
norphytane, *See* Pristane, 25037
Norplex laminates, 21574
Norprop, 21575
norralamine, *See* Amine C4, 2152
Norsil 1000, 21576
Norsil RTV 811, 21577
Norsil SG 131, 21578
Norsil SG 169, 21579
Norsolene®, 21580
Norsomix®, 21581
Norsophen®, 21582
Norsorex, 21583
Nortech, 21584
Northovan, 21585
Norton, *See* ethofumesate, 10967
Nortran, *See* ethofumesate, 10967
Nortranese, *See* ethofumesate, 10967

Nortron, 21586, *See* ethofumesate, 10967
Nortron Leyclene, 21587
Nortuff RA 1700-MO, 21588
Nortuff RA 7020-KO, 21589
Nortuff RC 1700-MO, 21590
Noruben, 21591
Norunil, 21592
Norust, 21593
Norval, 21594
Norvas, *See* Norvasc,
Norvinyl, 21595
Norvinyl DX 550, 21596
Norwegian saltpeter, *See* Air Saltpeter, 1097
Noryl®, 21597
Noryl® 731, 21598
Noryl® 1402B, 21599
Noryl® BN25, 21600, *See* Polyphenylene oxide, 24518
Noryl® EM5101, 21601, *See* Polyphenylene oxide, 24518
Noryl® EN185, 21602
Noryl® FN150, 21603
Noryl® FN215X, 21604
Noryl® GTX810, 21605, *See* Polyphenylene oxide, 24518
Noryl® HM3020, 21606, *See* Polyphenylene oxide, 24518
Noryl® HS1000X, 21607, *See* Polyphenylene oxide, 24518
Noryl® N190X, 21608, *See* Polyphenylene oxide, 24518
Noryl® PC180X, 21609, *See* Polyphenylene oxide, 24518
Noryl® PN235, 21610, *See* Polyphenylene oxide, 24518
Noryl® PX0722, 21611, *See* Polyphenylene oxide, 24518
Noryl® SE100, 21612, *See* Polyphenylene oxide, 24518
Noryl® SE1GFN2, 21613
Noryl®EN185, *See* Polyphenylene oxide, 24518
Noryl®FN150, *See* Polyphenylene oxide, 24518
Noryl®FN215X, *See* Polyphenylene oxide, 24518
Noryl®SE1GFN2, *See* Polyphenylene oxide, 24518
Nosifeed 40, 21614
Nosiheptide, 21615, 21616
Nosil 711, 21617
Nosophen, *See* Kerasol, 16058
Nosophene sodium, *See* keraphen, 16055
No-Swab, 21618
Notak, 21619
Notaral, *See* Penicillin G potassium, 23109
Nottingham white, *See* flake white, 11871
Noumgou oil, *See* Njave butter, 21392
Nouraid, 21620
Nourycryl, 21621
Nourydrier, 21622
Nourymix, 21623
Nouryset®, 21624
Nouryset® 156, 21625
Nouryset® 200 HV 250, 21626
Nova, *See* myclobutanil, 20549
Nova 2001, 21627
Nova Furnipol, 21628
Nova Highgloss, 21629
Nova Lime-Lite, 21630
Nova Long-Life, 21631
Nova One, 21632
Nova One Plus, 21633
Nova PC-1000BK, 21634
Nova Pine Fluid, 21635
Nova Starbight, 21636
Nova Stripper Super, 21637
Nova Supercote, 21638
Nova Superphaite, 21639
Nova Supratreet, 21640
Nova Tri-Power, 21641
Novablend® 501, 21642
Novablend® 5555, 21643
Novabloo, 21644
Novabold, 21645
Novacarb, 21646
Novacare, 21647
novacetoform, 21648
Novacleer, 21649
Novacorn, 21650
Novacote, 21651
Novacrete, 21652
Novacross, 21653
Novacryl, 21654
novaculite, 21655
Novacut, 21656
Novadelox, 21657, *See* Benzoyl peroxide, 3983
Novadine, 21658
Novafil, 21659
NovaFlo, 21660
Novafrost, 21661
Novagem, 21662
Novagrip, 21663
Novalak resins, 21664
Novalar, 21665
Novalast 5000, 9000, 21666
Novalene, 21667
Novalift, 21668
Novalite, 21669
Novaloy 6521, 21670
Novaloy® 9000, 21671
Novamid® 1010C, 21672
Novamid® 1020VA2, 21673
Novamid® 2020A, 2420A, 21674
Novamyl, 21675
Novantisol, *See* Neo Heliopan® Hydro, 20868
Novanyl, 21676
Novapel, 21677
Novaphalte, 21678
Novapint, 21679
Novaplaste, 21680
Novapol, 21681
Novapol® GF-0218-F, 21682
Novapol® GI-2024-A, 21683
Novapol® HB-L455-A, 21684
Novapol® LC-0517-A, 21685
Novapol® LE-0220-A, 21686
Novapol® LF-0223-B, 21687
Novapol® LF-Y819-D, 21688
Novapol® PF-0118-B, 21689
Novapol® PR-0636-UG, 21690
Novaquik, 21691
Novaract, 21692
Novarex® 7022A, 21693
Novarex® 7025G10, 21694
Novarex® 7025NB, 21695
Novaruca, *See* glutaraldehyde, 13062
Novasan, 21696
Novasheen, 21697
Novashield, 21698
NovaSize, Dark Fortified, 21699
Novasolve, 21700
NovaSperse, 21701
Novastet, 21702
Nova-T, 21703
Novata 299, A, AB, B, BBC, BC, BCF, BD, C, D, E, 21704
Novate SM-40, *See* sodium dimethyldithiocarbamate, 28758
Novatec, 21705
Novatec 240H, 21706
Novatec-AP, 21707
Novathion, 21708
Novatreet, 21709
Novaways, 21710
Novazole, 21711
Noveloid, 21712
Novemol, 21713
Noverme, *See* Mebendazole, 19198, Telmin, 31066, Vermox, 33591
Novester, 21714
Novex, 21715, 21716
Novidium, 21717
novismuth, *See* Bismuth subnitrate, 4243
Novitane, 21718
Novite, 21719
Novocoll® NC222 US, 21720
Novodur®, 21721
Novodur® L3FR, 21722, *See* Acrylonitrile-butadiene-styrene, 439
Novodur® P2H-AT, 21723, *See* Acrylonitrile-butadiene-styrene, 439
Novodur® P2HE, 21724
Novodur® P2HGV, 21725
Novodur® P2T, P2T-AT, 21726
Novodur® PMTM, 21727
Novodur® PMTM RTP 601, *See* Acrylonitrile-butadiene-styrene, 439
Novofix, 21728
Novogel® ST, 21729
Novol, 21730
novolac resin, 21731, *See* Aerolite, 862, phenolic resin, 23607
Novolen®, 21732
Novoline, Novolith, Novomatic, 21733
Novon® 2020-6001, 21734
Novon® 3001, 21735
Novon® M0121, 21736
Novon® M0289, 21737
Novonacco, *See* Aerosol® OS, 895, Rhodacal® IN, 26585
Novonasco, 21738
Novo-Nastizol A, *See* Hismanal, 14132
Novoperm, 21739
Novophone, *See* Dapsone, 7751
Novoplas, 21740
Novoprotin, 21741
Novor, 21742
Novor 950, 21743
Novotak, 21744
Novotox, *See* dichlorvos, 8394
Novozone®, 21745
Novozym, 21746
Noxaben, *See* Dicloxacillin sodium, 8396
Noxal, *See* Disulfiram, 8725
Noxamine, 21747
Noxamine CA 30, *See* Cocamine oxide, 6557
Noxamium, 21748
Noxfire, *See* rotenone, 27000
Noxfish, *See* rotenone, 27000
Nozolex, 21749
NP-10, 21750
NP-55, *See* Sethoxydim, 28099
NPG® Glycol, 21751
NPP-1, 21752
NPR 3911, 21753
NPR 5587, 21754
NRDC 149, *See* cypermethrin, 7588, Topclip Parasol, 32037
NRDC 161, *See* Thripstick®, 31766
NRDC-107, *See* Reslin, 26216
NRDC-143, *See* Kafil, 15655 permethrin, 23384, 23385, 23386
NS®, 21755
NSAE, *See* Aerosol® OS, 895, Rhodacal® IN, 26585
NSAE Powder, 21756
NSC-1771, *See* Thiram, 31726, Thiurad, 31730
NSC-7571, *See* Monacrin, 20097
NSC-71047, *See* 4,5,8-trimethylpsoralen, 32340
NSC-20264, *See* adenosine monophosphate, 680
NSC-21626, *See* betaprone, 4065
NSC-38721, *See* Lysoform, 18095
NSC-60584, *See* Uval, 33177
N-Serve, 21757
NT-15GF/000, 21758
NTA, 21759, *See* nitrilotriacetic acid, 21339, TG Buffer, 31547
NTA, trisodium salt, *See* Cheelox® NTA-Na3, 5869
NTBBTS, *See* BBTS, 3863, Delac NS, 8001, Perkacit® TBBS, 23289
N-t-Butyl-2-benzothiazole-sulfenamide, *See* Vanax® NS, 33292
NTF, *See* Fostex AMP, 12337
ntimonial saffron, *See* golden antimony sulfide, 13177
NTM, *See* dimethyl phthalate, 8508

NTMP, *See* Fostex AMP, 12337
NTN 33 893, *See* Confidor, 6725
NTN 801, *See* mefenacet, 19229
NTPA, *See* Fostex AMP, 12337
Nuact, 21760
Nuade, 21761
nuarimol, 21762
Nuba, 21763
Nu-bait II, *See* methomyl, 19502
Nubex, 21764
Nubrite, 21765
Nubun, 21766
Nucidol, *See* Diazinon Liquid, 8344
nucin, *See* Juglone, 15631
Nucleant, 21767
Nucol, 21768
Nucoline, *See* vegetable butter, 33514
Nucrel®, 21769
Nudrin, *See* methomyl, 19502
nufluor, *See* sodium fluoride, 28764
Nufol, 21770
Nuglas, 21771
Nujol, *See* Klearol, 16208
Nujol, *See* K. Tab, 15642, Kaon-Cl, 15729, Kay Ciel, 15807, Kaydol, 15810, K-Contin, 15818, K-Lor, 16226, K-Lyte/C1, 16234, KM Potassium Chloride, 16242, Potassium chloride, 24748, 24749
Nullapon, *See* Kalex 220 Crystal, 15668, Kalex Liq. 50%, 15671, tetrasodium EDTA, 31391
Nuloid, *See* Bakelite, 3589
Nulomoline, 21772, *See* Invert Sugar, 15200
NuoCide®, 21773
Nuocure, 21774
Nuocure 28, 21775
Nuodex, 21776, *See* Nacap®, 20643, Nuodex 84, 21779
Nuodex 84, 21779
Nuodex 87, 21780
Nuodex 100, 21777
Nuodex 321 Extra, 21778
Nuodex NA, 21781
Nuolate, 21782
Nuophene, 21783
Nuoplaz, 21784
Nuoplaz DOP, *See* Bis(2-ethylhexy) Phthalate, 4228, dioctyl phthalate, 8549, Reomol DCP, 26130, Witcizer 312, 34454
Nuoplaz® 6959, *See* Jayflex® TOTM, 15559
Nuoplaz® DIDP, *See* diisodecyl phthalate, 8462
Nuoplaz® DTDP, *See* Jayflex® DTDP, 15555
Nuoplaz® TOTM, *See* Jayflex® TOTM, 15559
Nuosept, 21785
Nuosept 95, *See* Zoldine® ZT-55, 34887
Nuosperse, 21786
Nuostabe, 21787
Nuostabe 1317, 21788
Nuostabe 1374, 21789
Nuostabe 1605, 21790
Nuosyn, 21791
Nupel®, 21792
Nupyrin, *See* aspirin,
Nurac, 21793
Nuram, 21794
Nurapon AL1, AL 60, *See* Ammonium laureth sulfate, 2240
Nurelle, *See* cypermethrin, 7588, Topclip Parasol, 32037
Nuremberg gold, 21795
Nuremberg violet, *See* Manganese Violet, 18550
Nurolon, 21796
Nusat, 21797
Nu-Seals Aspirin, 21798
Nu-Seals Sodium Salicylate, 21799
Nu-Set, 21800
Nusolv ABP-62, 21802
Nusolv ABP-103, 21801
NuStar, *See* flusilazole, 12119
Nustar®, 21803
Nusyn, *See* rotenone, 27000
Nusyn-Noxfish, *See* rotenone, 27000
nut oil, 21804, *See* Pondicherry oil, 24683
Nutimalt® Range, 21805
nutmeg butter, 21806
Nutralys, 21807
Nutramigen, 21808
Nutramin, 21809
Nutramon, 21810
Nutranel, 21811
Nutraphos, 21812
Nutrapon AL 1, 21813
Nutrapon AL 60, 21814
Nutrapon B 1365, 21815
Nutrapon BM 3960, 21816
Nutrapon DE 3796, *See* DEA-lauryl sulfate, 7827
Nutrapon DL 3891, 21817
Nutrapon ES-60 3568, 21818
Nutrapon ESY 2299, 21819
Nutrapon FA-50 0066, 21820
Nutrapon HA 3841, 21821, *See* Ammonium lauryl sulfate, 2242
Nutrapon KF 3846, 21822
Nutrapon KPC 0156, 21823
Nutrapon PP 3563, 21824, *See* Ammonium lauryl sulfate, 2242
Nutrapon RS 1147, 21825
Nutrapon TK 3603, 21826
Nutrapon TLS-500, 21827
Nutrapon TW 3987, 21828
Nutrapon W 1367, 21829
Nutrapon WAQE 2364, 21830
NutraponDE3796, 21831
NutraSweet, 21832, *See* Aspartame, 3233, Equal, 10731
Nutrifos, 21833
Nutrilan® FPK, H, M, 21834
Nutrilan® FPK, H, M, I-50, *See* Extiat®, 11323
Nutrilan® I-50, 21835
Nutrilan® Keratin W, 21836
Nutrilan® L, 21837
Nutrilife® Range, 21838
Nutrisoft® 55, 100, 21839
Nutrol 100, 21840
Nutrol 600, 21841
Nutrol 611, 21842
Nutrol 622, 21843
Nutrol 640, 21844
Nutrol 656, 21845
Nutrol Betaine MD 3863, 21846
Nutrol Betaine OL 3798, 21847
Nutrol S-60 5350, 21848
Nutrol SXS 5418, 21849, *See* sodium xylene sulfonate, 28832
Nuva, *See* dichlorvos, 8394
Nuvan, 21850, *See* dichlorvos, 8394, Vapona, 33378
Nuvan 7, *See* dichlorvos, 8394
Nuvan 100ec, *See* dichlorvos, 8394
Nuvan Fly Spray, 21851
Nuvan Top Aerosol, 21852
Nuvanol, 21853
Nuvanol N, *See* Iodofenphos, 15213
Nuvapen, *See* Ampicillin, 2361
Nuvis, 21854
Nuxtra, 21855
n-valeric acid, 33242
NY-10GF, 21856
NY-30CF, 21857
Nya G, *See* Nyad®, 21860
Nyacol a 1530, *See* Octoguard FR-10, 22000
Nyacol®, 21858
Nyacol® A-1530, 21859
Nyad, *See* Wollastocoat®, 34545
Nyad 10, *See* Nyad®, 21860, Wollastocoat®, 34545
Nyad 325, *See* Nyad®, 21860, Wollastocoat®, 34545
Nyad G, *See* Wollastocoat®, 34545
Nyad®, 21860
Nyad® Wollastonite, 21861
Nyala, 21862
Nybex 12034 BKFR, 21863
Nybex 12056 BKFR, 21864
Nybex 13001 BKC, 21865
Nybex 15011 NA, 21866
Nybex 17000 NAX, 21867
Nybex 22008 BKUT, 21868
Nybex 42002 BKHS, 21869
Nybex 52000 NA, 21870
Nycoa® 438, 21871
Nycoa® 446, 21872
Nycoa® 500, 21873
Nycoa® 528, 21874
Nycoa® 567, 21875
Nycoa® 714, 21876
Nycoa® 870, 21877
Nycoa® 1417, 21878
Nycoa® 4015, 21879
Nycoa® 5015, 21880
Nycoat®, 21881
Nycor, *See* Wollastocoat®, 34545
Nycor 200, *See* Nyad®, 21860, Wollastocoat®, 34545
Nycor 300, *See* Nyad®, 21860, Wollastocoat®, 34545
Nycor® Barytes, 21882
Nycor® Celestite, 21883
Nycor® R, 21884
Nydur, 21885
Nye Tact 520, 21886
Nyebar, 21887
Nyflake®, 21888
Nyglas®, 21889
Nykon, 21890
Ny-Kon® I, 21891
Ny-Kon® P, 21892
Ny-Kon® Q, 21893, *See* nylon 6/10, 21911
Ny-Kon® R, 21894
Ny-Kon® V, 21895
Nylander's reagent, 21896
Nylatron®, 21897
Nylatron® 1018 HS, 21898
Nylatron® 1024 HS, 21899
Nylatron® GS-63, 21900
Nylatron® NSB-90, 21901
Nylatrrn® GS, 21902
Nylmerate, *See* acetoxyphenylmercury, 306, phenylmercury acetate, 23646
Nylocrom, 21903
Nylofixan, 21904
Nyloflex, 21905
Nylok® 170, 21906
Nylomine, 21907
nylon, 21908, *See* Cadco® Nylon, 4887
Nylon 6, 21909, *See* Ashlene® 858, 3221, Ceistran® N66G30-01-4, 5556, Durethan® B 30 S, B 31 SK, 9198, Durethan® B 35 F, B 38 F, B 40 F, 9199, Novamid® 1020VA2, 21673
Nylon 6 and 66, *See* Akulon, 1206
Nylon 6 copolymer, *See* Capron® 8253, 5153, Capron® 8259, 5154
Nylon 6 homopolymer, *See* Capron® 8203C HS, 5150, Capron® 8280, 5156
Nylon 6/6, 21910, *See* Durethan® A 30 S, 9197, PA-111, 22523, PA-111CF30, 22524
Nylon 6/10, 21911
Nylon 12, 21912
Nylon 12 (polyamide 12), *See* Grilamid®, 13360
Nylon N-012, 21913
Nylon Resist NCO, 21914
Nyloprint, 21915
Nylosan®, 21916
Nyloset Finish, 21917
Nylosolv, 21918
Nylox, 21919
Nyogel, 21920
NyoGel® 744, 21921
NyoSil, 21922
Nypel, 21923
Nypene, 21924
Nyppon Fly Killer Spray, *See* Permethrin, 23385
Nyrim, 21925
Nyspheres®, 21926
NYsyn® 30-5, 21927
NYsyn® 33-5HM, 21928
NYsyn® 305V, 21929

NYsynblak® 9010, 21930
NYsynblak® DN 120, 21931
Nytal® 100, 21932
NZ 01, 21933
NZ 09, 21934
NZ 12, 21935
NZ 33, 21936
NZ 38, 21937
NZ 39, 21938
NZ 44, 21939
NZ 89, 21940
NZ 90, 21941
NZ 97, 21942
O 250, *See* Ablunol S-80, 111, Witconol 2500, 34501
O.N.V, 21943
O.O.D, 21944
O_2Zr, *See* SM 945, 28621, SM 975, 28622, SM 987, 28623, SM 992, 28624, SM 994, 28625
O9810, 21945
O9816, 21946
OA 40-30, 21947
OA-100A, 21948
OA-154, 21949
OA-252, 21950
OA-270, 21951
OA-300, 21952
OA-502, 21953
OA-505, 21954
OA-700, 21955
OA-770, 21956
OA-951, 21957
Oak Draw 720, 728, 830A, 21958
Oak Kool 625, 632A, 648, 21959
oak moss resin, 21960
Oak Oils, 21961
oak red, 21962
Oak Syncrolube, 21963
Oakite Defoamant, 21964
Oakite Ladd, 21965
Obanol 516, 21966
Obazoline 662Y, 21967
OBDPO, *See* Octabromodiphenyl oxide, 21981
Obermayer's reagent, 21968
obesitol, *See* Neobee® 18, 20872, safflower oil, 27222
Obilique, *See* piperonyl butoxide, 23875
OBPA Vinyzene, *See* Vinyzene, 33869
obsidene, 21969
Obsidian, *See* pumice stone, 25394
OBTS, 21970
Obturin, 21971
Occidine, 21972
Occlusin, 21973
Occultest, 21974
Ocenol, 21975
ocher, 21976
ochermatite, 21977
ochran, 21978
ocota cocota gum, 21979
oct-1-ene, *See* octene-1, 21991
OCTA, *See* Octabromodiphenyl oxide, 21981
octabenzone, *See* benzophenone-12, Cyasorb® UV 531, 7497, Hostavin® ARO 8, 14418, Spectra-Sorb UV 531, 29232, Uvinul® 408, 33192
OctaBoost® 620, 21980
octabromodiphenyl ether 2,2',3,3',4,4',5,5'-octabromodiphenyl ether, *See* Saytex® 111, 27515
octabromodiphenyl oxide, 21981, *See* FR-1208, 12364, Great Lakes DE-79, 13303, Saytex® 111, 27515
octabromodiphenyloxide, *See* Great Lakes DE-79, 13303
octabromom-1,1'-oxybisbenzene, *See* Comperlan HS, 27515
9,12-octadecadienoic acid, *See* linoleic acid, 17305
9,12-octadecadienoic acid, dimer, *See* dilinoleic acid, 8480
octadecanamide, *See* Kemamide® S, 15900, stearamide, 29593, Uniwax 1750, 33068
octadecanamide, N-(2-hydroxyethyl)-, *See* Ablumide SME, 82, Witcamide® 70, 34442
octadecanamide, N,N'-1,2-ethanediylbis-, *See* Acrawax® C, 383, Abluwax EBS, 140, Advawax® 290, 820, Armowax EBS, 2985, Glyco, 13137, Nopcowax 22-DS, 21513
octadecanamide, N,N'-ethylenebis-, *See* Abluwax EBS, 140
octadecanamide, N,N-bis(2-hydroxyethyl)-, *See* Ablumide SDE, 81
1-octadecanamine, *See* Armeen® 18, 2938, Armid® HTD, 2963, Amine 18-90, 2131, Crodamine 1.18D, 7102, Steamfilm FG, 29589
1-octadecanamine, acetate, *See* Acetamin 86, 285, *See* Acetamin T, 288
1-octadecanamine, N,N-dimethyl-, *See* Crodamine 3.A18D, 7107, Dymanthine, 9327
1-octadecanaminium, N,N,N-trimethyl-, bromide, *See* Zeonet B, 34721
octadecanoic acid, *See* lithium stearate, 17512
n-octadecanoic acid, *See* stearic acid, 29599
(Z)-9-octadecanoic acid, *See* oleic Acid, 22112
octadecanoic acid aluminum salt, *See* aluminum stearate, 1929, Novogel® ST, 21729, Synpro® 404, 30593
octadecanoic acid, ammonium salt, *See* Ammonium stearate, 2253
octadecanoic acid, barium salt, *See* Barium stearate, 3643, Synpro® Barium Stearate, 30596
octadecanoic acid butyl ester, *See* ADK STAB LS-8, 732, butyl stearate, 4781, Kemester® 5510, 15948, Kessco® BS, 16103, Radia® 7051, 25718, Uniflex® BYS-Tech, Unimate® BYS, 32941, Witconol 2326, 34497
octadecanoic acid, calcium salt, *See* calcium stearate, 4970, Nopcote C-104, 21510
octadecanoic acid, 2-(1-carboxyethoxy)-1-methyl-2-oxoethyl ester, calcium salt, *See* Verv®, 33644
octadecanoic acid, 12-hydroxy-, 2-ethylhexyl ester, *See* Wickenol® 171, 34398
octadecanoic acid, ester with 1,2,3-propanetriol, *See* GMS Base, 13162
octadecanoic acid, ethanediyl ester octadecanoic acid, 1,2-ethanediyl ester, *See* glycol distearate, 13102
octadecanoic acid, 2-ethylhexyl ester, *See* Wickenol® 156, 34392
octadecanoic acid, isopropyl ester, *See* Wickenol® 127, 34380
octadecanoic acid, lithium salt, *See* lithium stearate, 17512
octadecanoic acid, magnesium salt, *See* Afco-Chem MGS, 923, magnesium stearate, 18367
n-octadecanoic acid methyl ester, *See* Emery® 2218, 10146, methyl stearate, 19560
octadecanoic acid, 1-methylethyl ester, *See* Wickenol® 127, 34380
octadecanoic acid monester with propane-1,2,3-triol, *See* Ablunol GMS, 103, Capmul® GMS, 5113 CPH-250-SE, 6944, glyceryl stearate, 13087, Witconol 2400, Witconol MST, Witconol RHT, 34499, Zohar GLST, 34850
octadecanoic acid, monoester with 1,2-propanediol, *See* Witconol 2380, 34498
octadecanoic acid, 2-[(1-oxooctadecyl)amino]ethyl ester, *See* Witcamide® MAS, 34444
octadecanoic acid 1,2,3-propanetriyl ester, *See* Kemester® 5500, 15947, Kemester® 6000, 15952, Kemester® GMS (Powd.), 15962, Neobee® 62, 20874
octadecanoic acid, potassium salt, *See* potassium stearate, 24784
(Z)-9-octadecanoic acid, potassium salt, *See* Octosol 449, 22018
octadecanoic acid, sodium salt, *See* sodium stearate, 28820
octadecanoic acid, trimer 9,12-octadecanoic acid, trimer, *See* Trilinoleic acid, 32320
octadecanoic acid, zinc salt, *See* zinc stearate, 34793
octadecanoic acid zinc salt, zinc octadecanoate, *See* zinc stearate, 34793
1-octadecanol, *See* Adol® 62 NF, 783, Stearal, 29591, stearyl alcohol, 29604, Varonic® BG, 33440
n-octadecanol, *See* Adol® 62 NF, 783, Stearal, 29591, stearyl alcohol, 29604
octadecan-1-ol, *See* Adol® 62 NF, 783, Stearal, 29591
octadec-9-en-1-ol, *See* Adol® 85, 786
octadecanol NF, *See* Epal® 18NF, 10627
1-octadecanol, monoether with polyethylene glycol, *See* Ablunol SA-7, 113
2,2-Bis(octadecanoyloxymethyl)-1,3-propanediyl dioctadecanoate, *See* pentaerythrityl tetrastearate, 23142
octadecatrienoic acid (Z,Z,Z)-9,12,15-octadecatrienoic acid, *See* linolenic acid, 17306
(Z)-9-octadecenamide, *See* oleamide, 22107
9-octadecenamide, (Z)-, *See* oleamide, 22107, Polydis® TR 121, 24411, Unislip 1759, 33045
9-octadecenamide, N-[3-(dimethylamino)propyl]-,N-oxide, *See* oleamidopropylamine oxide, 22110
9-octadecenamide, N,N-bis(2-hydroxyethyl)-, (Z)-, *See* Active #18, 503, Alkamide® DO-280, 1641, Alkamide® WRS 1-66, 1653, Rewomid® DO 280, 26322
9-octadecenamide, N-(2-hydroxypropyl)-, (Z)-, *See* Witcamide® 61, 34441
(Z)-9-octadecen-1-amine, *See* Crodamine 1.0, 1.0D, 7100, Jet Amine PO, 15595
octadec-1-ene, *See* Neodene® 18, 20899
octadecene-1, *See* Neodene® 18, 20899
1-octadecene, *See* Neodene® 18, 20899
octadecene amide, *See* Unislip 1759, 33045
octadecene nitrile, *See* Arneel® OD, 2992
(octadecenoato-O)phenylmercury, *See* phenylmercuric oleate, 23645
9-octadecenoic acid, *See* oleic acid, 22112
9-octadecenoic acid, 12-(acetyloxy)-,1,2,3-propanetriol ester, *See* glyceryl triacetyl ricinoleate, 13089
9-octadecenoic acid, 12-(acetyloxy)-, butyl ester, [R-(Z)]-, *See* B.A.R, 3561
9-octadecenoic acid, aluminum salt, oleic acid aluminum salt, *See* Olminat, 22143
9-octadecenoic acid (Z)-, butyl ester, *See* Butyl Oleate C-914, 4779, Plasthall® 503, 23989, Witcizer 100, 34453
(Z)-9-octadecenoic acid butyl ester, *See* Butyl Oleate C-914, 4779, Plasthall® 503, 23989
9-octadecenoic acid calcium salt, *See* calcium oleate, 4961
9-octadecenoic acid (Z), decyl ester, *See* decyl oleate, 7869
9-octadecenoic acid (Z)-, 2,3-dihydroxypropyl ester, *See* Ablunol GMO, 102, Witconol 2421, 34500
9-octadecenoic acid ester with 2,2-bis(hydroxymethyl)-1,3-propanediol, *See* Radiasurf® 7156, 25803
9-octadecenoic acid (Z)-, ester with 1,2,3-propanetriol, *See* glyceryl monooleate, 13084
Z-9-octadecenoic acid ethyl ester, *See* Radia® 7187, 25728
9-octadecenoic acid, 12-hydroxy-, [R-(Z)]-, *See* P®-10 Acid, 22499, ricinoleic acid, 26734
9-octadecenoic acid, isodecyl ester, *See* Wickenol® 144, 34387
9-octadecenoic acid (Z)-, isodecyl ester, *See* Wickenol® 144, 34387
(Z)-9-octadecenoic acid methyl ester, *See* Emery® 2301, 10155, Radia® 7060, 25719
9-octadecenoic acid (Z)-, methyl ester, *See* Emery® 2301, 10155
9-octadecenoic acid, monoester with 1,2,3-propanetriol, *See* glyceryl monooleate, 13084
9-octadecenoic acid (Z)-, 9-octadecenyl ester, (Z)-, *See* Wickenol® 143, 34386
octadecenoic acid (Z)-, potassium salt, *See* Octosol 449, 22018
9-octadecenoic acid 1,2,3-propanetriyl ester, *See* Kemester® 1000, 15938, Radia® 7363, 25739, triolein, 32351

9-octadecenoic acid, sodium salt, *See* sodium oleate, 28802
9-octadecenoic acid (Z)-, sodium salt, *See* sodium oleate, 28802
octadecenoic acid sodium salt, *See* sodium oleate, 28802
(Z)-9-octadecen-1-ol, *See* Adol® 85, 786, oleyl alcohol, 22131
9-octadecen-1-ol, *See* Adol® 85, 786, oleyl alcohol, 22131
N-9-octadecenyl hexadecanamide, 21982, *See* oleyl palmitamide, 22133
octadecenyl succinic anhydride, *See* Milldride® ODSA, 19788
octadecenylamine, *See* Crodamine 1.0, 1.0D, 7100, Jet Amine PO, 15595
octadecyl 3-(3,5-di-*tert*-butyl-4-hydroxyphenyl)propionate, *See* Naugard® 76, 20789, Uvinul® O-18, 33202
octadecyl 3,5-di-t-butyl-4-hydroxyhydrocinnamate, *See* Ultranox® 276, 32849
N-octadecyl-13-docosenamide, 21986
octadecyl 3-mercaptopropionate, *See* Evangard® 18MP, 11252, octadecyl mercaptopropionate, 21983
octadecyl alcohol, *See* Adol® 62 NF, 783, Stearal, 29591
n-octadecyl alcohol, *See* Varonic® BG, 33440
octadecyl alcohol, ethoxylated, *See* Ablunol SA-7, 113
octadecyl dibutyl hydroxy hydrocinnamate, *See* Anox® PP 18, 2486
octadecyl-3-(3,5-di-*t*-butyl-4-hydroxyphenyl) propionate, *See* Lowinox® PO35, 17671, Ralox® 530, 25843, Ultranox® 276, 32849
octadecyl-3,5-di-*tert*-butyl-4-hydroxyhydrocinnamate, *See* Naugard® 76, 20789
octadecyl diethanol methyl ammonium chloride, *See* M-Quat® 32, 20398
octadecyl dihydroxyethyl methyl ammonium chloride - isopropanol, *See* Tomah Q-18-2, 31992
octadecyl dimethylamine, *See* Adma® 18, 744
octadecyl ether diamine, *See* Jet Amine DE 810, 15583
octadecyl mercaptopropionate, 21983
octadecyl methacrylate, 21984, *See* Ageflex FM-68, 988, Ageflex FM-1620, 989, octadecyl methacrylate, 21984, stearyl methacrylate, 29607
octadecyl trimethyl ammonium chloride, *See* Octosol 474, 22019
octadecylamine, *See* Amine 18-90, 2131, Armeen® 18, 2938, Armid® HTD, 2963, Steamfilm FG, 29589
n-octadecylamine, *See* Amine 18-90, 2131, Armeen® 18, 2938, Armid® HTD, 2963, Crodamine 1.18D, 7102, Steamfilm FG, 29589
1-octadecylamine, *See* Amine 18-90, 2131, Armeen® 18, 2938, Armid® HTD, 2963, Crodamine 1.18D, 7102, Kemamine® P-990, P-990D, 15922, Steamfilm FG, 29589
octadecylamine, N,N-dimethyl-, hydrochloride, *See* Onamine 18, 22167
octadecylamineadogenen 142, *See* Crodamine 1.18D, 7102
octadecylbenzenemethanaminium chloride, 21985
octadecylbenzydimethylammonium N-octadecyl-N-benzyl-N,N-dimethylammonium chloride, *See* stearalkonium chloride, 29592
N-octadecyl-N-benzyl-N,N-dimethylammonium chloride, *See* Cycloton® SCS, 7542
octadecyldimethyl [3-(trimethoxysilyl) propyl] ammonium chloride, *See* CO9745, 6505
octadecyldimethylbenzylammonium chloride, *See* Cycloton® SCS, 7542, stearalkonium chloride, 29592
octadecyldimethylchlorosilane, *See* CD5636, 5512
n-octadecylic acid, *See* stearic acid, 29599
octadecyloxypoly(ethyleneoxy)ethanol, *See* Ablunol SA-7, 113
octadecyltrichlorosilane, *See* CO9750, 6506
octadecyltrimethylammonium chloride, *See* Arquad® 18-50, 3109, Zeonet B, 34721
n-octadecyltrimethylammonium bromide, *See* Zeonet B, 34721
Octaguard FR-01, *See* Decabromodiphenyl oxide, 7837
1R-1α,2,3,4,4aβ,9,10,10aα-octahydro-1,4a-dimethyl-7-(1-methylethyl)-1-phenanthrenemethanamine, *See* Amine D, 2155
2,3,4,6,7,8,9,10-Octahydropyrimido[1,2-a]azepine, *See* Diazabicycloundecene, 8341
Octalene, *See* Aldrin, 1558
octamethylcyclotetrasilazane, *See* CO9800, 6507
octamethylcyclotetrasiloxane, *See* CO9810, 6508
octamethylcyclotetrasiloxane, *See* O9810, 21945
octamethyldiphosphoramide, *See* Schradan, 27731
octamethyltetrasiloxane, *See* O9810, 21945
octamethyltrisiloxane, *See* CO9816, 6509
octamethyltrisiloxane, *See* O9816, 21946
octamine® Flake, Powd, 21987
1-octanamine, *See* Amine 8 D, 2125
1-octanecarboxylic acid, *See* Emery® 1202, 10138
1,8-octanedicarboxylic acid, *See* sebacic acid, 27795
octane-1,8-dioic acid, *See* suberic acid, 29903
octanedioic acid, *See* suberic acid, 29903
n-octanoic acid, *See* caprylic acid, 5158
octanoic acid glycerol ester, *See* Witafrol® 7420, 34421
octanoic acid monoester with 1,2,3-propanetriol;, *See* Witafrol® 7420, 34421
octanoic acid, 1,2,3-propanetriyl ester, *See* Captex® 8000, 5170, Emalex O.T.G, 9976, Miglyol® 808, 19736
n-octanoic acid, *See* caprylic acid, 5158
octanol, *See* Epal® 8, 10616
1-octanol, 21988, *See* capryl alcohol, 5157
5-octanol-4-one, *See* Butyroin, 4802
octaphen, 21989
octaphenylcyclotetrasiloxane, *See* CO9817, 6510
octave, 21990, *See* prochloraz, 25057
1-octene, *See* Neodene® 8, 20890
octene-1, 21991
1-octene-3-ol, *See* Morillol®, 20335
3-octenol, *See* Morillol®, 20335
octen-3-ol, *See* Morillol®, 20335
octenyl succinic anhydride, *See* Milldride® OSA, 19789
octeth-3 carboxylic acid, *See* Defoamer B 90, 7881
octhilinone, *See* Kathon®, 15792, Micro-Chek® 11, 19664, Skane® M-8, 28563
octoate Z, 21992
Octocrilene, *See* Neo Heliopan® 303, 20865
Octocrylene, *See* Neo Heliopan® 303, 20865, Uvinul® N-539, 33201
Octocure 456, 21993
Octocure 553, 21994
Octocure ZBZ-50, *See* Akrochem® Z.B.E.D, 1183
Octocure ZDB-50, 21995, 21996
Octocure ZDM-50, 21997
Octocure ZMBT-50, 21998
Octoguard FR-01, 21999
Octoguard FR-10, 22000, *See* antimony trioxide, 2574
Octoguard FR-15, 22001
octoic acid, *See* caprylic acid, 5158
Octoil, 22002, *See* Bis(2-ethylhexy) Phthalate, 4228, dioctyl phthalate, 8549, Kodaflex® DOP, 16273, Reomol DCP, 26130
Octoil S, 22003, *See* Plasthall® DOS, 24005, Reomol DOS, 26131
Octojet 104, 22004
Octolite 544, 22005
Octolite 561, 22006
Octolite AO-28, 22007
Octomer DBM, 22008
Octomer DIBM, 22009
Octomer DIOM, 22010
Octomer DOM, 22011
Octopol NB-47, 22012
Octopol SDE-25, 22013
Octopol SDM-40, 22014
Octoran, 22015
Octorez, 22016
Octosol, 22017
Octosol 449, 22018, *See* Potassium oleate, 24764
Octosol 474, 22019
Octosol 562, 22020
Octosol A-1, 22021
Octosol A-18, 22022
Octosol A-18-A, 22023
Octosol ALS-28, 22024, *See* Ammonium lauryl sulfate, 2242
Octosol HA-80, 22025
Octosol IB-45, 22026
Octosol SLS, 22027
Octosol TH-40, 22028
Octosperse TS-10, 22029
Octotint 103, 22030
Octotint 138, 22031
Octovit Tablets, 22032
Octowax 321, 22033, *See* Paraffin, 22743
Octowax 518, 22034
Octowet 40, 22035
Octowet 70A, 22036
Octoxinol, *See* Triton® X-15, 32433
Octoxynol-1, *See* Rexol 45/1, 26469, Triton® X-15, 32433
Octoxynol-1.5, *See* Igepal® CA-210, 14846
Octoxynol-3, *See* Chemax OP-3, 5917, Macol® OP-3, 18271
Octoxynol-4, *See* Dehydrophen POP 4, 7967, Sellig O 4 100, 27900, T-Det® O-4, 30868
Octoxynol-5, *See* Marlophen® 85, 18883, Sellig O 5 100, 27901, Teric X5, 31289
Octoxynol-5.5, *See* Remcopal 306, 26071
Octoxynol-6, *See* Makon® OP-6, 18430, Sellig O 6 100, 27902
Octoxynol-7, *See* Chemax OP-7, 5918, Hyonic OP-7, 14694
Octoxynol-8, *See* Sellig O 8 100, 27903
Octoxynol-9, *See* Nutrol 100, 21840, Prox-onic OP-09, 25332, Remcopal O9, 26095, Sellig O 9 100, 27904
Octoxynol-9 carboxylic acid, *See* Akypo OP 190, 1229, Akypo OP 80, 1230, Akyposal OP 80, 1317
Octoxynol-10, *See* Akyporox OP 100, 1284, Iconol OP-10, 14827, Marlophen® 810, 18884
Octoxynol-11, *See* Remcopal O11, 26096, Sellig O 11 100, 27897, Solubilisant Gamma 2420, 28975
Octoxynol-12, *See* Akyporox OP 115 SPC, 1285, Remcopal O12, 26097, Sellig O 12 100, 27898
Octoxynol-20, *See* Akyporox OP 200, 1286, Sellig O 20 100, 27899
Octoxynol-20 carboxylic acid, *See* Akypo OP 190, 1229
Octoxynol-25, *See* Akyporox OP 250V, 1287, Emalex OP-25, 9979
Octoxynol-30, *See* Chemax OP-30/70, 5919, Iconol OP-30, 14828
Octoxynol-33, sodium laurethyl sulfate, *See* Abex® VA 50, 43
Octoxynol-40, *See* Akyporox OP 400V, 1288, Chemax OP-40, 5920, Chemax OP-40/70, 5921, Iconol OP-40, 14829, Igepal® CA-897, 14849, T-Det® O-407, 30869, Trycol® 6956, 32505
Octrizole, *See* Cyasorb® UV 5411, 7503, Spectra-Sorb UV 5411, 29233, Uvazol 311, 33183
n-Octyl, *See* Plasthall® 8-10 TM-E, 23991
octyl (2-ethylhexyl) acid phosphate, *See* Syn-O-Ad® P-412, 30566
octyl acetate, *See* 2-ethylhexyl acetate, 11138
octyl acrylate, 22037
octyl adipate, *See* Plasthall® DOA, 24002, PX-238, 25440, Wickenol® 158, 34393
octyl alcohol, *See* capryl alcohol, 5157, 2-ethylhexanol, 11136, Emery® 3322, 10166, Emery® 3324, 10168, Lorol C8, 17616
n-octyl alcohol, *See* 1-octanol, 21988
octyl amine, *See* Amine 8 D, 2125
n-octyl n-decyl adipate, *See* Morflex 325, 20323
octyl decyl dimethyl ammonium chloride, *See* Catigene® 818, 5451
octyl decyl phthalate, *See* Good-rite® GP-265, 13205
n-octyl n-decyl phthalate, *See* Good-rite® GP-265, 13205, Morflex 125, 20316

octyl dimethyl PABA, *See* Escalol® 507, 10787, Padimate O, 22549, Solarchem® O, 28898
octyl dimethylamine (7%), decyl dimethylamine (6%), dodecyl dimethylamine (53%), tetradecyl dimethylamine (19%), hexadecyl dimethylamine (9%), octadecyl dimethylamine (6%), *See* Adma® WC, 747
octyl dimethylaminobenzoate, *See* Padimate O, 22549, Solarchem® O, 28898
octyl dimethyl-PABA, *See* Padimate O, 22549, Solarchem® O, 28898
octyl diphenyl phosphate, *See* diphenyl octyl phosphate, 8597
octyl dipropionate, *See* Ampholak YJH-40, 2329
octyl dodecanol, 22038
octyl epoxy tallate, *See* Drapex® 4.4, 8968
octyl hydroxystearate, *See* Naturechem® OHS, 20775, Schercemol OHS, 27615, Wickenol® 171, 34398
octyl isononanoate, *See* Dermol 89, 8117, Kessco® octyl Isononanoate, 16117
N-octyl isothiazolones, *See* Kathon® 893, 15794
octyl methoxycinnamate, 22039, *See* Escalol® 557, 10788, Neo Heliopan® AV, 20866, octyl methoxycinnamate, 22039, Parsol® MCX, 22867
octyl octanoate, *See* Tegosoft® EE, 31020
octyl oxystearate, *See* Dermol OO, 8129
octyl palmitate, 22040, *See* Bernel® Ester EHP, 4008, Kessco® octyl Palmitate, 16118, Lexol EHP, 17148, Schercemol OP, 27617, Tegosoft® OP, 31025, Unimate® EHP, 32970, Wickenol® 155, 34391
octyl pelargonate, *See* Bernel® OPG, 4011, Schercemol OPG, 27618, Wickenol® 160, 34395
octyl phenol ethoxylate, *See* Empilan® OPE9.5, 10414, Triton® X-15, 32433
octyl phosphate, *See* trioctyl phosphate, 32349
n-octyl phosphate ester, *See* Marlophor® HS-Acid, 18894
octyl phthalate, *See* Bis(2-ethylhexy) Phthalate, 4228, Reomol DCP, 26130
octyl salicylate, *See* Dermoblock OS, 8116, Escalol® 587, 10790, Uvinul® O-18, 33202
octyl sebacate, *See* Plasthall® DOS, 24005, Reomol DOS, 26131
octyl sodium sulfate, *See* Rhodapon® OLS, 26639
octyl stearate, 22041, *See* Cetiol® 868, 5794, Wickenol® 156, 34392
octyl tallate, *See* Plasthall® R-9, 24014, Uniflex® 192, Uniflex® EHT, 32938, Uniflex® EHT, 32947
octyl thiotin, *See* Lankromark® OT450, 16707
octyl triazone, *See* Uvinul® T 150, 33204
n-octyl, n-decyl adipate, *See* Plasthall® NODA, 24010
n-octyl, n-decyl adipate mixture, *See* Monoplex® NODA, 20249
octyl/decyloxypropylamine, *See* Tomah PA-1214, 31988
3-octylamine, *See* Armeen® L8D, 2952
n-octylamine, *See* Amine 8 D, 2125
octylated diphenylamine, *See* Akrochem® Antioxidant S, 1162, Anton N, 2631, Antox® N, 2632, Flectol ODP, 11912, Lowinox® ODA, 17670, Octamine® Flake, Powd, 21987, Permanax OD, 23365
octylated diphenylamines, *See* Agerite® Stalite, 1022, Agerite® Stalite® S, 1023
octyldecanol, *See* Exxal® 18, 11342
octyldecyl adipate, *See* Good-rite® GP-235, 13202
octyldecyl phthalate, 22042, *See* Good-rite® GP-265, 13205
octyldimethylamine [7378-99-6], decyldimethylamine [1120-24-7] and dodecyl dimethylamine [112-18-5], *See* Adma® 8, 10, 12, 741
octyldimethylchlorosilane, *See* CO9819, 6511
octyldodecanol, 22038, *See* Amerchol® L-500, 2002, Eutanol G, 11239, Exxal® 20, 11343, Michel XO-150-20, 19643
2-octyl-1-dodecanol, *See* Eutanol G, 11239, Michel XO-150-20, 19643
2-octyldodecan-1-ol, *See* Eutanol G, 11239, Michel XO-150-20, 19643
octyldodeceth-5, *See* Emalex OD-5, 9977
octyldodecoxy dimethicone, *See* Pecosil SG-20, 23014
octyldodecyl dimethicone copolyol adipate, *See* Pecosil GSA 36, 23005
2-octyldodecyl erucate, 22043
octyldodecyl myristate, *See* Bernel® Ester 2014, 4003, M.O.D, 18106, Wickenol® 142, 34385
octyldodecyl neopentanoate, *See* Elefac I-205, 9801
octyldodecyl N-myristoryl-N-methyl alanate, *See* Amiema MA-OD, 2113
2-octyldodecyl oleate, *See* O.O.D, 21944
2-octyldodecyl ricinoleate, *See* R.O.D, 25704
octyldodecyl stearate, *See* Starfol® OS, 29538
octyldodecyl stearoyl stearate, *See* Ceraphyl® 847, 5729
octyldodecyl/dimethicone copolyol citrate, *See* Pecosil CAS-36, 23003
octylene, *See* octene-1, 21991
octylene glycol, *See* Ethohexadiol, 10968, ethyl hexanediol, 11073
2-n-octyl-4-isothiazolin-3-one, *See* Micro-Chek® 11, 19664, Skane® M-8, 28563, Vinyzene® IT-3000 DIDP, 33871
2-octyl-4-isothiazolin-3-one, *See* Kathon®, 15792, Kathon® 893, 15794, Skane® M-8, 28563
2-octyl-3(2H)-isothiazolone, *See* Kathon®, 15792, Kathon® 893, 15794, Skane® M-8, 28563
4-octyl-N-(4-octylphenyl)benzenamine, *See* Agerite® Stalite, 1022, Naugalube® 438, 20783
octylphenol ethoxylate, *See* Merpol® 100, 19374
octylphenol ethoxylates, *See* Desonic® S Series, 8185
octylphenoxypolyethoxyethanol, *See* Triton® X-15, 32433
octyltin mercaptide, *See* ADK STAB 466, 714
octyltrichlorosilane, *See* CO9830, 6512, CO9835, 6513, Dynasylan® OCTEO, 9416, Prosil® 9202, 25241, Union Carbide® A-137, 32980
n-octyltriethoxysilane, *See* Dynasylan® OCTEO, 9416, Prosil® 9202, 25241
ocuba wax, 22044
ODIX, *See* glyoxal 40%, 13146
odoripon al 95, *See* Rhodapon® 101-10, 26631, sodium lauryl sulfate, 28781
Odoron, 22045
Odylen®, 22046
oenanthic acid, *See* heptanoic acid, 13791
oenanthylic acid, *See* heptanoic acid, 13791
oesipos, *See* Cosmetic Lanolin USP, 6870, Vigilan, 33750
O-Ethyl O-8-quinolyl phenylphosphonothioate, *See* Quintiofos, 25691
O-ethyl S,S-diphenyl phosphorodithioate, *See* Edifenphos, 9618
O-ethyldigol, *See* ethyl di-Icinol, 11068
OFA, 22047
Off, *See* diethyl toluamide, 8438
OFHC Copper, 22048
Oftanol®, 22049
Oftentral, 22050
OGA, 22051
Ogtac 85 V, 22052
Ogwin, 22053
O-hi-o, 22054
Ohlan OH, *See* hydroxylated lanolin, 14663
OHlan®, 22055
ohm oil, 22056
Ohmal, 22057
ohmlac kapak, 22058
Ohmoid, 22059
Oil asphalt, 22060
oil babuluam, *See* neatsfoot oil, 20836
oil babulum, *See* neatsfoot oil, 20837
Oil blue, 22061
Oil Die, 22062
Oil Gard, 22063
oil green, *See* Accrox, 258, M100, 18107
Oil of Ajowan, *See* Oil of Ptychotis, 22071
oil of aniseed, *See* Arizole® Anethole Extra, 2885
oil of bitter almond, *See* Benzaldehyde, 3957, Ethereal Oil of Bitter Almonds, 10944
Oil of duty, *See* Oil of Rhodium, 22072
oil of geranium, *See* Rose+a7 oil, 26968
oil of Hartshorn, *See* Bone oil, 4402
Oil of Jojoba, *See* jojoba oil, 15617
Oil of liquid pitch, *See* oil of tar, 22075
Oil of mink, 22064, *See* mink oil, 19833
oil of mirbane, *See* mirbane oil, 19935, nitrobenzene, 21348
oil of myrbane, *See* nitrobenzene, 21348
Oil of niobe, *See* methyl benzoate, 19515
Oil of Palma Christi, 22065, *See* castor oil, 5417, Cosmetol® X, 6871
oil of palmarosa, *See* Rose+a7 oil, 26968
oil of pelargonium, *See* Rose+a7 oil, 26968
Oil of Pennyroyal, 22066
Oil of Peter, 22067
Oil of Petitgrain, 22068
Oil of Petre, *See* Oil of Peter, 22067
Oil of Poley, *See* Oil of Pennyroyal, 22066
Oil of Pompillon, 22069
Oil of Portugal, 22070
Oil of Ptychotis, 22071
oil of rapeseed, *See* rapeseed oil, 25874
Oil of Rhodium, 22072
oil of rose-geranium, *See* Rosé oil, 26968
Oil of Spike, 22073
Oil of sweet almond, *See* Sweet Almond oil, 30452
oil of sweet birch, 22074
oil of tar, 22075
oil of tartar, 22076
oil of tea, 22077
oil of turpentine, 22078
oil of verbena, 22079
oil of wax, *See* wax butter, 34233
oil of wheat, 22080
Oil of wheat germ, *See* Wheat Germ Oil, 34348, Wheat Germ Oil CLR, 34348
Oil Orangesolvent yellow 14, *See* Sudan I, 29938
Oil Paalsgaard, 22081
oil pulp, 22082
Oil Red, *See* Sudan IV, 29941
Oil Red BB, *See* Sudan IV, 29941
Oil Scarlet, *See* Sudan III, 29940
oil shale, 22083
oil skin, 22084
oil varnishes, 22085
oil vulcanized, *See* Thinoline, 31675
oil wax, *See* wax butter, 34233
oil white, 22086
Oil yellow, *See* Butter yellow, 4759
Oil Yellow 2G, *See* Butter yellow, 4759
Oil Yellow 20, *See* Butter yellow, 4759
Oil Yellow 2625, *See* Butter yellow, 4759
Oil Yellow 7463, *See* Butter yellow, 4759
Oil Yellow BB, *See* Butter yellow, 4759
Oil Yellow D, *See* Butter yellow, 4759
Oil Yellow DN, *See* Butter yellow, 4759
Oil Yellow FF, *See* Butter yellow, 4759
Oil Yellow FN, *See* Butter yellow, 4759
Oil Yellow G, *See* Butter yellow, 4759
Oil Yellow G-2, *See* Butter yellow, 4759
Oil Yellow GG, *See* Butter yellow, 4759
Oil Yellow GR, *See* Butter yellow, 4759
Oil Yellow II, *See* Butter yellow, 4759
Oil Yellow N, *See* Butter yellow, 4759
Oil Yellow Pel, *See* Butter yellow, 4759
oil, Bari, *See* aix oil, 1138
oil, Riviera, *See* aix oil, 1138
oil, var, *See* aix oil, 1138
Oildag, 22087
oiled silk, 22088
Oilfos, 22089
Oils, *See* cod liver oil, 6569, olive oil, 22138
Oils, avocado, *See* Avocado Oil, 3498
oils, cod liver, *See* cod liver oil, 6569
oils, corn, *See* corn oil, 6834
Oils, grape seed, *See* Grape seed oil, 13275
oils, jojoba, *See* jojoba oil, 15617
oils, linseed, *See* Linol, 17304, linseed oil, 17309

oils, olive, *See* olive oil, 22138
oils, palm, *See* palm oil, 22625, 22626
oils, palm kernel, *See* palm kernel oil, 22624
Oils, pennyroyal, *See* pennyroyal oil, 23119
oils, pine, *See* pine oil, 23856
oils, safflower, *See* safflower oil, 27222
oils, vegetable, *See* vegetable oil, 33522
oils, walnut, *See* walnut oil, 34201
Oilsol, 22090
oil-soluble resins, 22091
Oilstone, *See* stink-stone, 29770, whetstone, 34352
Oil-Treet, 22092
Ointment Base No. 3, 4, 6, 22093
oiticia, 22094
Okenite, *See* calcium silicate, 4969, Wollastocoat®, 34545
Okerin, 22095
Oko, *See* dichlorvos, 8394
Oko®, 22096
Okol, 22097
Okstan M 62, 22098
Okstan M 69 S, 22099
Okstan X3, 22100
Okstan XO, 22101
OL 3798 Proteric CAB, *See* Cocamidopropyl betaine, 6551
Olapon ND-9, SW, 22102
olate, *See* sodium oleate, 28802
Olay Beauty Bar, 22103
Olay Beauty Cleanser, 22104
OLB-50, *See* oleyl betaine, 22132
olcotrop leather, 22105
old fustic, *See* Fustic, 12522
Old Plantation, 22106, *See* Ammonium nitrate, 2245
Ole, *See* Chlorothalonil, 6150, Repulse, 26151
olea europaea oil, *See* olive oil, 22138
Oleal Yellow 2G, *See* Butter yellow, 4759
Olealkonium chloride, *See* Ammonyx® KP, 2263, Incroquat O-50, 15041, Mackernium KP, 18208, M-Quat® JO-50, 20407
oleamide, 22107
Oleamide, *See* Armid® O, 2964, Crodamide O, OR, 7097, Petrac® Slip-Eze, 23468, Polydis® TR 121, 24411, Unislip 1759, 33045
Oleamide DEA, *See* Active #18, 503, Alkamide® DO-280, 1641, Amidex O, 2099, Crillon ODE, 7051, Mackamide ODM, 18161, Marlamid® D 1885, 18739, Norfox® F-221, 21549
Oleamide DEA (1:1), *See* Emid® 6545, 10194, Incromide OD, 14993, Mackamide MO, 18160, Rewomid® DO 280 SE, 26323, Schercomid SO-A, 27661
Oleamide DEA (2:1), *See* Laurel SD-400, 16840
Oleamide DEA 2:1 Oleamide DEA, *See* Mazamide® O 20, 19113
Oleamide MEA, *See* Schercomid OME, 27654
Oleamide MIPA, *See* Alkamide® OIP, 1648, Mackamide OP, 18162, Rewomid® IPE 280, 26325, Schercomid OMI, 27655, Witcamide® 61, 34441
Oleamide, N-(2-hydroxypropyl)-, *See* Witcamide® 61, 34441
Oleamidopropyl betaine, 22108, *See* Lexaine® O, 17095, Mackam HV, 18143, Mirataine® BET-O-30, 19924
Oleamidopropyl dihydroxypropyl dimonium chloride, *See* Lexate BPQ, 17121
Oleamidopropyl dimethyl glycine, *See* Oleamidopropyl betaine, 22108
oleamidopropyl dimethylamine, 22109, *See* Chemidex O, 5965, Incromine OPB, 15000, Lexamine O-13, 17099, Mackine 501, 18221, Schercodine O, 27640
Oleamidopropyl dimethylamine hydrolyzed collagen, *See* Lexein® CP-125, 17129
Oleamidopropyl dimethylamine lactate, *See* Incromate ODL, 14971
Oleamidopropyl dimethylamine oxide, *See* oleamidopropylamine oxide, 22110
Oleamidopropyl PG-dimonium chloride, *See* Lexquat® AMG-O, 17178
oleamidopropylamine oxide, 22110
Oleamidopropylamine oxide, *See* Mackamine OAO, 18175
Oleamine, *See* Amine OL, 2162, Armeen® OL, 2956, Armeen® OLD, 2957, Jet Amine PO, 15595
Oleamine hydroxypropyl bistrimonium chloride-polyacrylamide, *See* Akypomine® P 191, 1267
Oleamine oxide, *See* Ammonyx® OAO, 2265, Chemoxide O, 5997, Mackamine O2, 18174, Mazox® ODA, 19154, Standamox 01, 29459
oleandocyn, 22111
α olefin, *See* Dialen 6, 8281
α olefin sulfonate, *See* Bio Terge®, 4153
oleic Acid, 22112, *See* Distoline, 8724, Emersol® 210, 10120, Emersol® 7021, 10125, Industrene® 104, 15100, Pamolyn® 125, 22648, Priolene 6900, 25012, potassium oleate, 24763
oleic acid (70%), 15% of linolic acid (15%), stearic acid (15%), *See* red oil, 25964
Oleic acid amide, *See* Unislip 1759, 33045
Oleic acid amide ethoxylate, *See* Lutensol® FSA 10, 17987, PEG-3 oleamide, 23048
oleic acid calcium salt, *See* calcium oleate, 4961
Oleic acid dibutylamide, *See* Rewocoros RA 280, 26305
oleic acid diethanolamide, *See* Active #18, 503, Alkamide® DO-280, 1641, Alkamide® WRS 1-66, 1653, Rewomid® DO 280, 26322
oleic acid methyl ester, *See* Emery® 2301, 10155, Kemester® 104, 15935, Radia® 7060, 25719, Wickenol® 143, 34386, Witconol 2301, 34496
Oleic acid oxyethylate, *See* Emulan® A, 10487
Oleic acid polyglyceride, *See* Witconol 14, 34495
oleic acid potassium salt, *See* Octosol 449, 22018, Potassium oleate, 24764
Oleic acid, (Z)-9-octadecenyl ester, *See* Wickenol® 143, 34386
Oleic acid, butyl ester, *See* Witcizer 100, 34453
Oleic acid, ester with 1,2,3-propanetriol homopolymer (1:1), *See* Witconol 14, 34495
Oleic acid, monosorbitan ester, *See* Ablunol S-80, 111, Witconol 2500, 34501
oleic acid, potassium salt, *See* potassium oleate, 24764
oleic acid, sodium salt, *See* sodium oleate, 28802
oleic acid-linoleic acid, *See* Industrene® 224, 15104
oleic diethanol amide, *See* Active #18, 503, Alkamide® DO-280, 1641, Alkamide® WRS 1-66, 1653
oleic diethanolamide, *See* Schercomid ODA, 27653
Oleic imidazoline, *See* Crodazoline O, 7162
Oleic imidazoline acetate, *See* Varine O Acetate, 33393
Oleic monoisopropanolamide, *See* Witcamide® 61, 34441
olein, *See* Kemester® 1000, 15938, Radia® 7363, 25739, triolein, 32351
olein of saponification, 22113
Olein, 1-mono-, *See* Ablunol GMO, 102, Witconol 2421, 34500
oleite, 22114
oleo, 22115
Oleo Keratin ISO, 22116
Oleoamphocarboxyglycinate, *See* Ampholak XO7, 2323, Ampholak XOO-30P, 2324
oleobismuth, 22117
Oleo-Coll LP, 22118
Oleocuprit, *See* Cupravit®, 7370
oleogen, *See* Parogen, 22854
Oleogesaprim, *See* atrazine, 3394
Oleogesaprim 200, *See* atrazine, 3394
oleoguaiacol, 22119
oleoresins, 22120
Oleosol, 22121
Oleo-Soy C, 22122
Oleotripropylenetetraamine, *See* Amine 740, 2145, Jet Amine TET, 15598
oleovitamin A, *See* Vi-Alpha, 33706
oleovitamin D_2, *See* vitamin D_2, 33940
oleovitamin D_3, *See* vitamin D_3, 33941
Oleoyl PG-trimonium chloride-stearoyl PG-trimonium chloride-behenoyl PG-trimonium chloride-palmitoyl PG-trimonium chloride and trideceth-2, *See* Akypoquat 40, 1269
Oleoyl sarcosine, 22123, *See* Hamposyl® O, 13604, Oleo-Coll LP, 22118, Vanseal® OS, 33360
1-Oleoylglycerol, *See* Ablunol GMO, 102, Radiasurf® 7150, 25801, Witconol 2421, 34500
Olepal ISO, 22124
Oleth-2, 22125, *See* Eumulgin PWM2, Genapol® O-020, 12791, Hetoxol OL-2, 13946, Lipocol O-2, 17356, Macol® OA-2, 18270, Ritoleth 2, 26848
Oleth-2 phosphate, *See* Crodafos O2 Acid, 7084
Oleth-3, *See* Hetoxol OA-3 Special, 13945, Volpo 3, 34020, Volpo N3, 34028, Volpo O3, 34029
Oleth-3 carboxylic acid, *See* Akypo RO 20, 1239
Oleth-3 phosphate, *See* Brophos OL-3, 4614, Chemphos TR-515, 6019, Crodafos N3 Acid, 7081, Hetphos OA-3, 13953
Oleth-4, *See* Chemal OA-4, 5885, Prox-onic OA-1/04, 25329
Oleth-4 phosphate, *See* Chemfac PB-184, 5952, Chemphos TR-541, 6021, Rhodafac® RB-400, 26604
Oleth-5, *See* Chemal OA-5, 5886, Eumulgin 05, 11171, Eumulgin WM5, 11193, Marlipal® 1850/5, 18867, Marlowet® WOE, 18939, Solulan® 5, 29000, Volpo 5, 34021
Oleth-5 phosphate, *See* Crodafos N5 Acid, 7083
Oleth-6, *See* Ablunol OA-6, 107
Oleth-6 carboxylic acid, *See* Akypo RO 50, 1240
Oleth-7, *See* Akyporox RTO 70, 1295
oleth-8, 22126, *See* Emalex 508, 9923
Oleth-9, *See* Akyporox RO 90, 1294
Oleth-10, *See* Ethal OA-10, 10924, Volpo 10, 34022
Oleth-10 carboxylic acid, *See* Akypo RO 90, 1241
Oleth-10 phosphate, *See* Chemphos TR-505, 6015, Crodafos N10 Acid, 7079
Oleth-15 OS 20A, *See* Ablunol SA-7, 113
Oleth-20, *See* Chemal OA-20/70CWS, 5884, Rhodasurf® ON-870, 26658, Simulsol 98, 28481, Trycol® 5971, 32503, Volpo 20, 34023, Varonic® 32-E20, 33437
oleth-phosphate Series (2-20), 22127
oleum, *See* fuming sulfuric acid, 12457
oleum spirits, 22128
oleum white, 22129, *See* Lithopone, 17520
oleum;, *See* fuming sulfuric acid, 12457
Olex, 22130
Oley hydroxyethyl imidazoline, *See* Mackazoline O, 18201
oleyl alcohol, 22131, *See* Cachalot® O-15, 4880, Dermaffine®, 8093, Fancol OA-95, 11450, HD-Eutanol, 13719, Lipocol O, 17355, Novol, 21730, Adol® 85, 786
oleyl amide, *See* oleamide, 22107, Unislip 1759, 33045
Oleyl amine ethoxylate, *See* Lutensol® FA 12, 17986
oleyl betaine, 22132
Oleyl betaine, *See* Chembetaine OL-30, 5935, Incronam OD-50, 15022, Mackam OB-30, 18148, Mafo® OB, 18303
N-Oleyl-1,3-diaminopropane, *See* Duomeen® OL, 9114
Oleyl dimethyl glycine, *See* oleyl betaine, 22132
Oleyl dimethylamine, *See* Jet Amine DMOD, 15586
Oleyl dimethylamine oxide, *See* Chemoxide O1, 5998, Incromine Oxide OD-50, 15009
Oleyl erucate, *See* Cetiol® J600, 5799, Dynacerin®, 9339, Dynacerin® 660, 9340
Oleyl ether phosphate, *See* oleth-phosphate Series (2-20), 22127
Oleyl hydroxyethyl imidazoline, *See* Miramine® O, 19871, Monazoline O, 20196, Schercozoline O, 27715, Unamine® O, 32912, Varine O®, 33394
N-Oleyl imidazolinium hydrochloride, *See* Norfox® IM-38, 21550
Oleyl methylaminoethanoic acid, *See* Oleoyl sarcosine, 22123
Oleyl N-myristoyl-N-methyl alanate, *See* Amiema MA-OL, 2114
Oleyl oleate, *See* Cetiol®, 5793, Schercemol OLO, 27616, Starfol® OO, 29537, Wickenol® 143, 34386, Wickenol® 143, 34386

Oracylic acid, *See* Benzoic acid, 3973
Oragrafin Calcium, 22260
Oragrafin Sodium, 22261
Orahesive, 22262
Orahexal, *See* chlorhexidine digluconate, 6101, Rotersept, 27002
Oraldene, 22263
oralith, 22264
Oramide DL 200 AF, 22265
Oramix L, *See* lauroyl sarcosine, 16861
Oramix O, *See* Oleoyl sarcosine, 22123
Orange, *See* Propachlor, 25163
orange 3RA soluble in grease, *See* Sudan I, 29938
orange a l'huile, *See* Sudan I, 29938
Orange III, 22266
orange lac, *See* lemon lac, 16983
orange lead, 22267
Orange mineral, *See* orange lead, 22267
orange r fat soluble, *See* Sudan I, 29938
orange red, *See* orange lead, 22267
orange resenole no. 3, *See* Sudan I, 29938
orange soluble a l'huile, *See* Sudan I, 29938
orange tungsten, 22268
orange vermilion, *See* scarlet vermillon, 27550
Orangelac, *See* Lemon Lac, 16983
Oranit, *See* Nekal, 20855
oranium bronze, 22269
Orasol, 22270
Oratrast, 22271, *See* Radiopaque, 25830
Oravue, 22272, *See* Videobil, 33742
Orbinamon, 22273
orbit, *See* A-625/641ABS 301K, ABS 500FR-1, 10, Radar, Radar Propiconazole, 25714
Orbitol, 22274
Orca, 22275
Orchard Brand Ziram, *See* AAprotect, 16
Orchard Herbide, 22276
orchellin, *See* Archil, 2808
Orchidee, 22277, 22278, *See* sanfoin, 27364
orchil, *See* Archil, 2808
orchindone, 22279
orcinol, 22280
ordeal bark, 22281, *See* red water bark, 25973
ordeal bean, 22282
Ordnance 204, 500, 22283
ordonezite, 22284
Ordoval, 22285
Ordoval G.2G, *See* neradol, 20995
ore-furnace slag, 22286
Oregon balsam, 22287
Oreide, 22288
Orelite, *See* Metalite, 19433
orellin, 22289
Oresmasin, 22290
Orevac® PP-C, 22294
Orevac® 18211, 22292
Orevac® 18302, 22293
Orevac® 9309, 22291
Orflex PP, *See* Vulkanox® 4010 NA, 34147
Orgamide R, 22295
Organelle, 22296
Organic and inorganic pigments, *See* Akrochem® Powder Colors, 1172
Organic calcium compound, *See* Ircogel® 900, 15256
Organic phosphate ester, *See* Actrafos 152A, 528, Actrafos TDA, 533
Organic phosphorus/nitrogen compound, *See* Aflammit P, 938
Organo molybdenum additive, *See* Adeka Sakura-Lube 100, 670
organol orange, *See* Sudan I, 29938
Organol Yellow ADM, *See* Butter yellow, 4759
Organometallic, *See* Amergy® 5400, 2011
Organopol®, 22297
Organosilane Si203, Si208, 22298
Organo-silicone, *See* Foamkill® 639JOH, 12164, Foamkill® 639Q, 12165, Foamkill® 663J, 12168
Organosiloxane, *See* silicone, 28335
Organotin compound, *See* Advastab® LS-203, 814, Advastab® TM-790 Series, 817, Advastab® WS-499, 818, Certincoat, 5765
Orglas, 22299
Orgozon CC 1118, 22300
Orgozon Conc. 0680, 22301
Orient Oil, *See* Butter yellow, 4759
orient oil orange ps, *See* Sudan I, 29938
oriental blue, *See* ultramarine, 32821
Oriental powder, 22302
Oriental sweet gum, *See* storax calamita, 29795
Orimon, *See* phenothiazine, 23618
Oriodide, *See* lodotope I-125, 15217, lodotope I-131, 15218, lodotope Therapeutic, 15219, Radiocaps-131, 25827
Oriodide-131, 22303
Orisan, 22304
Orkan, 22305
Orlean, *See* annatto, 2478
Orlon®, 22306
Ormolu, 22307
Ornalin, *See* vinclozolin, 33811
Ornalith, *See* Bakelite, 3589
Ornamental Weeder, 22308
Ornithite, 22309
Orobronze, *See* Canthaxanthin, 5095
Oroglas DR, 22310
Oromid®, 22311
Oronal LCG, 22312
Oronite, 22313
Oropon, 22314
Orotan, 22315
Orotric, *See* succinimide, 29917
Orovite, 22316
orphenol, *See* Dowicide A, 8932
orpiment, 22317
orris camphor, 22318
orris root, 22319
Orrs white, *See* Lithopone, 17520
Orr's white, *See* lithopone, 17520
orseille, *See* Archil, 2808
orsin, *See* Aminogen II, 2177
Ortegol®, 22320
Ortegol® 204, 22321
Ortensan, *See* Valadol, 33238
Orthene, 22322
Orthenex, 22323
Ortho, 22324
Ortho Brush Killer A, *See* α-methylstyrene, 19579
Ortho Danitol, *See* fenpropathrin, 11615
Ortho grass killer, *See* Propham, 25186
Ortho Metaldehyde 4% Bait, *See* metaldehyde, 19431
Ortho Multi-Purpose Fungicide Daconil 2787, *See* Chlorothalonil, 6150, Repulse, 26151
Ortho phosphate defoliant, *See* Def®, 7876
Ortho Rose Pride Funginex Rose and Shrub Disease Control, *See* triforine, 32254
Ortho St. Augustine Weed and Feed, *See* atrazine, 3394
Orthoarsenic Acid, *See* arsenic acid, 3137
Orthoboric acid, *See* Boric Acid, 4422
Orthochlorotoluene biphenyl, *See* Tanalon® EFA, 30780
Orthochrom, 22325
Orthochrome T, 22326
ortho-chrysotile, 22327
Orthocide, 22328
Orthocide-406, *See* Captan, Captan-Col, Captan-50, Captan-83P, Captan Granular, 5165, Vancide® 89, 33302
Orthoclase, *See* Felspar, 11589
Ortho-Clear, 22329
Ortho-Gro, 22330
orthohydroxydipbenyl, *See* o-phenylphenol, 23648
Ortho-Klor, 22331
Ortholate, 22332
Ortholeum, 22333
Ortholite, 22334
Orthomatic, 22335
Orthoperiodic Acid, *See* periodic acid, 23273
Orthophen® 278, 22336
orthophosphoric acid, *See* phosphoric acid, 23738, 23739
orthophosphorous acid, *See* phosphorus acid, 23741
Orthorix, 22337
Orthosan MB, *See* Cycloton® SCS, 7542, octadecylbenzenemethanaminium chloride, 21985, stearalkonium chloride, 29592, Sumquat® 6210, 30126
Orthosil, 22338, *See* sodium metasilicate, 28788
Orthotrol, 22339
orthotungstic acid, *See* tungstic acid, 32592
Orthoxenol, *See* o-phenylphenol, 23648
Orth's stain, 22340
Ortizon, *See* Hyperol, 14712
Ortol, 22341, 22342
Ortolan, 22343
Ortolan® Dyes, 22344
Ortosol, 22345
Orvus WA, 22346
orvus wa paste, *See* Rhodapon® 101-10, 26631
orvus wa paste, *See* sodium lauryl sulfate, 28781
oryzalin, 22347, *See* Weed Stopper, 34258
Oryze, *See* paclobutrazol, 22540, 22541
Orzan, 22348
Orzol, 22349
os sepiae, 22350
OS-2, 22351
Osadan, *See* Torque, 32077
Osbon AC, *See* Oxymaster, 22460
Oscodal, 22352
Oscrete, 22353
Osimol, 22354
osmic acid, 22355, *See* osmium tetroxide, 22357
osmic acid
Osmiridium, *See* Iridosmine, 15300
osmium, 22356
osmium tetroxide, 22357, *See* osmic acid, 22355
Osmoglyn, 22358
osmo-kaolin, 22359
osmondite, 22360
osmo-sil, 22361
Osnol, 22362
OSO® 440, 22363
OSO® 1905, 22364
osram, 22365
ossalin, *See* sodium fluoride, 28764
ossein, 22366, *See* collagen, 6610
ossein
Ossin, *See* sodium fluoride, 28764
Ossivite, 22367
Ostamer, 22368
Ostan, 22369
Ostelin, 22370, *See* vitamin D_2, 33940
osteofluor, *See* sodium fluoride, 28764
Ostrilan, 22371
Osvan, *See* Variquat® 50MC, 33397, Zephiran Chloride, 34728
Oswego, 22372
osyritrin, *See* rutin trihydrate, 27150
Osyrol, 22373
otasoral, *See* Potassium gluconate, 24755
ote seeds, 22374
otoba butter, 22375
otoba fat, *See* otoba butter, 22375
otoba wax, *See* otoba butter, 22375
Otokalixin, *See* Kantrex, 15718, Klebcil, 16209
Otoryl, 22376
Ottacide, *See* Nipacide® DX, 21255
Ottasept, 22377, *See* chloroxylenol, 6159
Oulu 102, 22378
Oulu 331, 22379
Oulumer 70, 22380
Oulupale XB 100, 22381
Oulures, 22382
Oulutac 105, 22393
Oulutac 20 D, 22383
Oulutac 20 EP, 22384
Oulutac 30, 22385

Oulutac 30 D, 22386
Oulutac 80, 22387
Oulutac 80 D, 22388
Oulutac 80 D/HS, 22389
Oulutac 90, 22390
Oulutac 90 D, 22391
Oulutac 90 D, 22392
ounce metal, 22394
Ouralpatti, *See* Ventilago Madraspanta, 33561
Oust, 22395
Outflank, *See* permethrin, 23384, 233865, 23386
Outflank, *See* permethrin
ouvarovite, 22396
Ovac, 22397
Ovacryl, 22398
Ovasyn, *See* Amitraz, 2202
Ovasyn, *See* Taktic, 30729
Overnite, 22399
Ovicide, 22400
Ovigest, 22401
Ovitelmin, 22402, *See* Mebendazole, 19198, Telmin, 31066, Vermox, 33591
Ovoflavin, *See* riboflavin, 26722
ovolecithin, *See* lecithin, 16939
Ovucire WL 2944, 22403
10H-9-Oxaanthracene, *See* xanthene, 34589
1-Oxa-4-azacyclohexane, *See* morpholine, 20346
Oxaban®-A, 22404
Oxaban®-E, 22405
oxacyclopropane, *See* ethylene oxide, 11131
Oxadiazon, 22406, *See* Ronstar 2G, 26954
oxadixyl, 22407
Oxaf, 22408
OXAF, *See* Octocure ZMBT-50, 21998, Oxaf, 22408, Perkacit® ZMBT, 23298, Zetax®, 34741
Oxaf 50D, 22409
3-oxa-1-heptanol, *See* 2-butoxyethanol, 4753
11-Oxahexadecanolide, *See* Musk R-1, 20534
3-Oxa-1-hexanol, *See* Ektasolve® EP, 9671
OXAL, *See* glyoxal 40%, 13146
Oxalaldehyde, *See* glyoxal 40%, 13146
oxalan, 22410
oxalantin, 22411
oxalate blasting powder, 22412
oxalic acid, 22413
oxalic acid and oxalate salts, *See* Coalatex, 6516
oxalic acid dihydrate, 22414
oxalic acid potassium salt, *See* potassium oxalate, 24766
oxalic acid tin (II) salt, 22415, *See* tin(II) oxalate, 31816
oxalic acid, dipotassium sal, *See* potassium oxalate, 24766
Oxalic acid, dipotassium salt monohydrate, *See* potassium oxalate, 24765
oxalic ether, 22416
oxalumina, 22417
Oxaluramide, *See* oxalan, 22410
oxalyl bis (benzylidenehydrazide), 22418
2,2'-oxamido bis-[ethyl 3-(3,5-di-tert-butyl-4-hydroxyphenyl) propionate, *See* Naugard® XL-1, 20799
Oxamin LO, *See* lauramine oxide, 16832
Oxamniquine, *See* Vansil, 33363
oxamyl, *See* Blade, 4310, Vydate®, 34171
Oxane E.O, *See* ethylene oxide, 11131
Oxanol, *See* Ablunol SA-7, 113
Oxantel embonate, *See* Telopar, 31070
Oxantel pamoate, *See* Telopar, 31070
1-oxa-2-oxocycloheptane, *See* +le-caprolactone monomer, 5138
3-Oxapentane-1,5-diyl dimethacrylate, *See* Ageflex DEGDMA, 962, diethylene glycol dimethacrylate, 8444
1,4-oxathiin, 2,3-dihydro-5-carboxanilido-6-methyl-, *See* carboxin, 5255
Oxathiin-3-carboxamide, 5,6-dihydro-2-methyl-N-phenyl-, 4,4-dioxide, *See* Ringmaster, 26771
Oxathiin-3-carboxanilide, 5,6-dihydro-2-methyl-, 4,4-dioxide, *See* Ringmaster, 26771
Oxathiincarboxamide 1,4-Oxathiin-3-carboxamide, 5,6-dihydro-2-methyl-N-phenyl-, *See* carboxin, 5255
Oxathiincarboxanilide 1,4-Oxathiin-3-carboxanilide, 5,6-dihydro-2-methyl, *See* carboxin, 5255
Oxatin, *See* carboxin, 5255
oxazolidine, 22419, *See* Alkaterge® C, 1716 Zoldine® ZT-55, 34887
oxazolidin-2-one, *See* oxazolidine, 22419
2-oxazolidine, *See* oxazolidine, 22419
Oxazolidine E, *See* Amine CS-1246, 2154, Oxaban®-E, 22405
oxazolidine, 4-ethyl-2-methyl-2-(3-methylbutyl)-, *See* Zoldine® MS-52, 34886
Oxazolidinedione, 3-(3,5-dichlorophenyl)-5-ethenyl-5-methyl-, *See* vinclozolin, 33811
2-oxazolidinone, *See* oxazolidine, 22419
Oxazolo(3,4-c)oxazole, 7 α-ethyldihydro-, *See* Oxaban®-E, 22405
Oxazolo(3,4-c)oxazole, 7a-ethyldihydro-, *See* Amine CS-1246, 2154
1H,3H,5H-Oxazolo(3,4-C)oxazole-7a(7H)-methanol, *See* Zoldine® ZT-55, 34887
oxenol, *See* *o*-phenylphenol, 23648
2-oxepanone, *See* ϵ-caprolactone monomer, 5138
Oxetal 500/85, 22420
Oxetal D 104, 22421, *See* Deceth-6, 7848
Oxetal ID 104, 22422
Oxetal O 108, 22423
Oxetal TG 111, 22424
Oxetal VD 20, 22425
oxetan-2-one, *See* betaprone, 4065
Oxetanone, *See* betaprone, 4065
oxfendazole, 22426, *See* Synanthic, 30520
Oxi-Chek, 22427
Oxicob, *See* Cupravit®, 7370
oxidate le, *See* methyl benzoate, 19515
oxidation black, *See* aniline black, 2476
oxide yellow, *See* ocher, 21976
Oxidised low molecular weight polyethylene, *See* A-C® Polyethylene 629A, 177
Oxidized HDPE homopolymer, *See* A-C® 316, A-C®629, 169
Oxidized hydrocarbon waxes, *See* Rocsol, 26896
oxidized linseed oil, *See* solidified linseed oil, 28940
oxidized oils, *See* blown oils, 4354
Oxidized polyethylene, *See* A-C® Polyethylene 316, 316A, 325, 330, 392, 395, 629, 680, 176, A-C® Polyethylene 6702, 178, Hoechst Wax PED 191, 14267, Loobwax 0761, 17598
Oxidized pyrogallic acid, *See* pyraloxin, 25463
Oxidized/micronized polyethylene, *See* ACumist A-12, A-18, 580
oxidoethane, *See* ethylene oxide, 11131
α,β-oxidoethane, *See* ethylene oxide, 11131
8α-12-oxido-13,14,15,16-tetra-norlabdane, *See* ambroxan, 1981
Oxilube, 22428
Oximony, 22429
oxine, *See* 8-hydroxyquinoline, 14670
oxirane, *See* ethylene oxide, 11131
Oxi-tan, 22430
Oxitex, 22431
Oxitol, 22432, *See* EE Solvent, 9636, ethyl icinol, 11074
Oxitropium bromide, *See* Oxivent, 22433
Oxivent, 22433
Oxivor, *See* Cupravit®, 7370
5-Oxo compound with L-arginine, *See* arginine PCA, 2849
Oxobutane, *See* methyl ethyl ketone, 19530
2-oxobutane, *See* 2-butanone, 4737, methyl ethyl ketone, 19530
2-oxo-1,5-dipentanedioic acid, *See* α-ketoglutaric acid, 16137
Oxogen, *See* Oxzone, 22478
2-Oxoglutaric acid, *See* α-ketoglutaric acid, 16137
3-Oxoglutaric acid, *See* ADA, 615, β-ketoglutaric acid, 16138
3-oxo-L-gulofuranolactone, *See* vitamin C, 33938
3-oxo-L-gulofuranolactone sodium, *See* sodium ascorbate, 28734
Oxo-hexyl acetate, *See* Exxate® 600, 11347
Oxolin, 22434, *See* Oxone, 22435
5-oxol-Proline, *See* Ajidew A-100, 1142
oxomethane, *See* formaldehyde, 12258
Oxone, 22435
Oxonite, 22436
1-[[2-Oxo-2-[(1-oxododecyl)oxy]ethyl]amino]ethyl] pyridinium chloride, *See* Lapyrium Chloride, 16782
2-oxopentanedioic acid, *See* α-ketoglutaric acid, 16137
3-oxopentanedioic acid, *See* ADA, 615, β-ketoglutaric acid, 16138
4-oxopentanoic acid, *See* levulinic acid, 17084
5-Oxoproline, *See* Ajidew A-100, 1142
2-oxopropionic acid, *See* pyruvic acid, 25538
6-oxopurine, *See* hypoxanthine, 14721
2-Oxo-1,2H-pyran-5-carboxylic acid, *See* coumalic acid, 6901
5-Oxopyrrolidine-2-carboxylic acid, *See* Ajidew A-100, 1142
3-β-m-4-Oxo-2,2,7,7-tetramethyltricyclo[6,2,1,0,3,8] undecane, *See* Valanone B, 33239
Oxsoralen, *See* Oxsoralen-Ultra, 22437
Oxsoralen-Ultra, 22437
Oxtoxynol-12.5, *See* Igepal® CA-720, 14848
oxy wash, *See* Benzoyl peroxide, 3983
Oxy-5, *See* Abcure S-40-25, 35, AcetOxyl 2.5 and 5, 305, Benzoyl peroxide, 3983
Oxy-5, Oxy 10, *See* Benzoyl peroxide, 3983
Oxy-10, *See* Benzoyl peroxide, 3983
oxyalizarin, 22438
Oxyalkylated polyol, *See* Enerade® 3045, 10570
oxyammonia, 22439
oxyanthracene, 22440
Oxybenzone, *See* benzophenone-3, Cyasorb® UV 9, 7495, Neo Heliopan® BB, 20867, Rhodialux A, 26675, Spectra-Sorb UV9, 29229, Uvinul® M-40, 33197
oxybenzopyridine, *See* 8-hydroxyquinoline, 14670
1,1'-oxybisbenzene, *See* diphenyl oxide, 8598, phenyl ether, 23633
***p,p*-Oxybisbenzenesulfonyl hydrazide**, *See* Celogen® OT, 5643
1,1'-Oxybisethane, *See* ethyl ether, 11069
(2,2'-[oxybis(2,1-ethanediyloxy)bisethanol), *See* PEG-4, 23049
2,2'-Oxybisethanol, *See* diethylene glycol, 8442
2,2'-oxybisethanol dimethyl ether, *See* diglyme, 8455
oxybismethane, *See* dimethyl ether, 8505, methyl ether, 19529
1,1'-oxybis(2-methoxy)ethane, *See* diglyme, 8455
1,1'-Oxybis[2-methoxyethane], *See* diglyme, 8455
2,2-(oxybis(methylene)bis(2-(hydroxymethyl))-1,3-propanediol, *See* Dipentaerythritol, 8581
1,1'-Oxybis (2,3,4,5,6-pentabromobenzene), *See* Decabromodiphenyl oxide, 7837, Octoguard FR-01, 21999, Saytex® 102E, 27514
oxybisphenoxarsine 10,10'-oxybis-10H-phenoxarsine, *See* Vinyzene, 33869
1,1'-Oxybis-2-propanol, *See* Dipropylene glycol, 8611
Oxybispropanol, Methyl Ether, *See* Icinol DPM, 14815, Poly-Solv® DPM, 24555
oxycarboxin, 22441, *See* Plantvax, 23921, Plantvax 20, 23922, Plantvax 75, 23923
oxychinolin, *See* 8-hydroxyquinoline, 14670
oxychloride, 22442
oxychloride of tin, 22443
Oxyclozanide, 22444
Oxycur, *See* Cupravit®, 7370
oxycymol, 22445
oxydapatit, *See* calcium phosphate (tribasic), 4967
oxydase, 22446
Oxydasine, 22447
oxydiazol, *See* methazole, 19481
Oxydiazon, *See* oxadiazon, 22406
2,2'-oxydiethanol, *See* diethylene glycol, 8442
2,2'-Oxydiethylene dibenzoate, *See* Benzoflex 2-45, 3970
N-oxydiethylene-benzothiazole sulfenamide, *See* Amax®, Amax No 1, 1950, Accelerator MF, 208 Delac MOR, 8000, OBTS, 21970, OMTS, 22161, Perkacit®

MBS, 23284, Vinyzene, 33869
10,10'-oxydiphenoxarsine, *See*
oxydislin, 22448
Oxydol, *See* hydrogen peroxide, 14590
Oxydpech, 22449
Oxydurit®, 22450
Oxyethylated fatty amine, *See* Amollan® A, 2294
Oxyethylated phenol acrylate monomer, *See* Ebecryl® 110, 9490
oxyfume, *See* ethylene oxide, 11131
oxyfume 12, *See* ethylene oxide, 11131
oxygen, 22451
oxygen cubes, 22452
oxygen powder, 22453
oxygen, solid, *See* oxygen cubes, 22452
oxygenated oil, 22454
oxygenated paraffin, *See* Parogen, 22854
oxygenite, 22455
Oxy-Gro, 22456
Oxyguard, 22457
Oxykisvax, *See* oxycarboxin, 22441
Oxy-L, *See* Abcure S-40-25, 35, AcetOxyl 2.5 and 5, 305
Oxyliquit, 22458
oxylite, *See* Benzoyl peroxide, 3983
oxylith, 22459
Oxymaster, 22460
oxymel, 22461
oxymethylene, *See* formaldehyde, 12258
oxymethyleneurea, *See* Salmocid, 27277
oxymuth saca, 22462
Oxyneurine, 22463
Oxynone, 22464
Oxyper, 22465
Oxyphenbutazone, *See* Tandearil, 30795
Oxyphenine, 22466
Oxypon 288, 22467
Oxypon 306, 22468
Oxypon 328, 22469
Oxypon 365, 22470
Oxypon 2145, 22471
oxypropylated cellulose, *See* Klucel®, 16229, Klucel® E, G, H, J, L, M, 16230, Klucel® EF, 16231
oxypsoralen, *See* Oxsoralen-Ultra, 22437
oxyquinoline, *See* 8-hydroxyquinoline, 14670
Oxytetracycline, *See* Biostat A.1, 4209
oxythioquinox, *See* quinomethionate, 25686
oxytol acetate, *See* EE Acetate, 9635
oxytoluol, 22472
Oxytracyl, 22473
oxytri, 22474
Oxytril, 22475
Oxytril CM, 22476
Oxytril P, 22477
Oxzone, 22478
Ozasol, 22479
Ozatec, 22480
ozocerite, *See* ozokerite, 22483
ozogen, 22481
Ozokerine, 22482
ozokerite, 22483
ozole, 22484
Ozonon 3C, *See* Vulkanox® 4010 NA, 34147
P, *See* proline, 25129
P 3, *See* Wytox® 312, 34578
P 3 (antioxidant), *See* Wytox® 312, 34578
P 50, *See* Ampicillin, 2361
P 210-D, 22485
P 506, 22486
P and G Amide 72, *See* Active #2, 501
p and g emulsifier 104, *See* Rhodapon® 101-10, 26631, sodium lauryl sulfate, 28781
***p*-(α-phenylbenzylidene)-1,1-dimethylpiperidinium methylsulfate**, *See* Variton, 33432
***p,p'*-Dioctyldiphenylamine**, *See* Vanox® 12, 33335
P.A.C, 22487
P.B.N, 22488
P.D.A.B, *See* Butter yellow, 4759
β-p.e.a, *See* 2-phenylethyl alcohol, 23642
P.H.D, 22489
P.I.B, 22490
P.M.F, *See* hydrargaphen, 14539
P.M.G.Metal, 22491
P.M.T. Alloy, 22492
P.N.P, 22493
P.O.T.G, 22494
P.P.D, 22495
P.P.S, 22496
P.S.E. No. 15 powder, 22497
P.T.F.E, *See* polytetrafluoroethylene, 24642
P.U.R.E.-CMC, 22498
P.V.P, *See* Agrimer 15L, 1065, Tears Plus®, 30876
P.V.P.I, *See* Vinisil, 33827, Videne Disinfectant Solution, Videne Disinfectant Tincture, Videne Powder, 33739
P3, 22505
P3-Almeco, 22506
P3-Armourbond, 22507
P3-Carclin, 22508
P3-Croni, 22509
P3-Ferroclene, 22510
P3-Ferromede, 22511
P3-Maxan, 22512
P3-Rodine, 22513
P3-Stripalene, 22514
P3-Suncorrite, 22515
P® -10 Acid, 22499, *See* ricinoleic acid, 26734
P® -10 Acid
P-11, 22501
P13 N, 22502
P-51, 22517, 22518
P-56, 22519
P-60-10 and P-60-20 Cold Molding Compounds, 22520
P-80F, 22521
P-85, 22522
P-289, 22504
P-400 Series, 22516
P0820, 22500
P-1011, *See* Dicloxacillin sodium, 8396
P-2003-K and P-2020-T, 22503
P-7138, *See* Nifurpirinol, 21170
***p,p'*-Isopropylidenediphenol**, *See* Bisphenol A, 4283
PA-57, 22531, *See* Polyisoprene, 24471
PA-80, 22532, *See* Polyisoprene, 24471
PA-111, 22523
PA-111CF30, 22524
PA-111G13, 22525
PA-121, 22526
PA-211, 22527
PA-211G13, 22528
PA-211N40, 22529
PA-221, 22530
PAA, *See* Oxymaster, 22460
Paalsgaard oil, *See* schou oil, 27730
Pabagel, 22533
Pabanol, 22534
PAC, *See* chloridazon, 6103
Pace 6L, *See* metolachlor, 19585
Pacemo, *See* Valadol, 33238
Pacer, 22535
***p*-acetamidophenol**, *See* Valadol, 33238
***p*-acetylaminophenol**, *See* Valadol, 33238
pacherite, 22536
Pacific Sea Kelp Glycolic Extract B-1063, 22537
Packfong, 22538
Packman, 22539
paclobutrazol, 22540, 22541, 22542, *See* Bonzi, 4407, Cultar, 7348, Parlay, 22846
paco, 22543
Pacos, *See* paco, 22543
Pacvac, 22544
Pacwet, 22545
Padac®, 22546
Paddox, 22547, 22548
Paddox
Padimate A, *See* Spectraban, 29225
Padimate 0, 22549, *See* Escalol® 507, 10787 Solarchem® O, 28898
Padophene, *See* phenothiazine, 23618
Pafra, 22550
PAG DSTDP, *See* Distearyl thiodipropionate, 8713
PAG DXTDP, *See* Distearyl thiodipropionate, 8713
Pageant, *See* Chlorpyrifos, 6163
Pagid, 22551
paint white, *See* Bismuth subnitrate, 4243
painter's naphtha, 22552
Paka oil, *See* kon oil, 16321
pala gum, 22553
Palacel, *See* Nuophene, 21783
palaite, 22554
Palamid®, 22555
Palamoll®, 22556
Palamoll® 632, 22557
Palamoll® 644 and 646, 22558
Palamoll® 645 and 647, 22559
Palamoll® 855, 22560
Palanil®, 22561
Palanil® Carrier, 22562
Palanil® Dyes, 22563
Palanil® P Dyes, 22564
Palanil® T Dyes, 22565
Palanthrene®, 22566
Palanthrene® T, 22567
palao amarillo, 22568
Palapreg®, 22569
Palatal®, 22570
Palatal® KR 1397, 22571
Palatal® P5, 22572
Palatal® P8, P50T and P52TL, 22573
Palatal® S333, 22574
Palatase, 22575
Palatex® NB-2, 22576
Palatin® Fast Dyes, 22577
palatinit, 22578
Palatinol, 22579
Palatinol 1C, 22580
Palatinol A, *See* Kodaflex® DEP, 16266
Palatinol AH, *See* Bis(2-ethylhexy) Phthalate, 4228, dioctyl phthalate, 8549, Reomol DCP, 26130
Palatinol BB, *See* Butyl benzyl phthalate, 4764
Palatinol C, *See* dibutyl phthalate, 8366
Palatinol M, *See* dimethyl phthalate, 8508, Kemester® DMP, 15956, Kodaflex® DMP, 16271, Unimoll® DM, 32979
Palatinol® 11, 22581
Palatinol® 711, 22582
Palatinol® A, 22583
Palatinol® AH, 22584
Palatinol® C, 22585
Palatinol® D10, 22586
Palatinol® DBP, 22587
Palatinol® DIDP, 22588, *See* diisodecyl phthalate, 8462
Palatinol® DN, 22589
Palatinol® DOA, 22590, *See* dioctyl adipate, 8548
Palatinol® DOP, 22591
Palatinol® K, 22592
Palatinol® M, 22593
Palatinol® N, 22594, *See* Jayflex® DINP, 15552
Palatinol® TOTM, 22595, *See* Jayflex® TOTM, 15559
Palatinol® Z, 22596
Palatone, 22597
Palatone, *See* Veltol, 33544
Palau, 22598
Paldesic, *See* Valadol, 33238
Pale 4, 22599
pale acid, 22600
pale catechu, *See* cutch, gum catechu, 13483
pale oils, 22601
Palegal® A, 22602
Palegal® NB-SF, 22603
Palenine, 22604
Palette 70, 22605
Paliocrom, 22606
Paliogen®, 22607
Paliotan®, 22608
Paliotol®, 22609
Palite, 22610
paliuroside, *See* rutin trihydrate, 27150
palladium, 22611

palladium asbestos, 22612
palladium black, 22613
palladium chloride (ous), *See* palladium(II) chloride, 22617
palladium chloride anhydrous, *See* palladium(II) chloride, 22617
palladium dichloride, *See* palladium(II) chloride, 22617
palladium gold, 22614
palladium monoxide, *See* palladium oxide, 22615
palladium nitrate (ous), *See* palladium(II) nitrate, 22618
palladium oxide, 22615
palladium red, 22616
palladium(II) chloride, 22617
palladium(II) nitrate, 22618
palladous chloride, *See* palladium(II) chloride, 22617
palladous nitrate, *See* palladium(II) nitrate, 22618
pallethrine, *See* allethrin, 1744
Pallgrip, 22619
Palliag, 22620
Pallinal®, 22621
Pallitop®, 22622
Pallitop® S, 22623
palm acidulated soapstock, *See* palm oil, 22625
palm butter, *See* palm oil, 22625, 22626
palm grease, *See* palm oil, 22625, 22626
palm kernel oil, 22624
Palm kernelamide DEA, *See* Accomid PK, 231, Amidex PK, 2100, Mackamide PK, 18163
Palm kernelamide MEA, *See* Mackamide PKM, 18164
palm oil, 22625, 22626
Palm oil
palm pitch, 22627
palm wax, 22628
palmerite, 22629
palmine, *See* vegetable butter, 33514
palmitamidopropyl betaine, *See* Schercotaine PAB, 27699
palmitamidopropyl dimethylamine, 22630, *See* Chemidex P, 5966
Palmitamine, *See* Amine 16D, 2130, Armeen® 16, 2935, Armeen® 16D, 2936, Armeen® 16D, 2937
palmitamine oxide, *See* Ammonyx® CO, 2262, Amyx CO 3764, 2393, Mazox® CDA, 19150
palmitic acid, 22631, 22632, 22633, 22634, 22635, Emersol® 143, 10119, Glycon® P-45, 13119, Hystrene® 8016, 14763, Industrene® 4516, 15106, Prifrac 2960, 24954
Palmitic acid esters, *See* Wickenol® 111, 34379
palmitic acid hexadecyl ester, *See* cetyl palmitate, 5822, Kemester® CP, 15955, Kessco® 653, 16100, Radia® 7500, 25742, Waxenol® 815, 34238
palmitic acid isopropyl ester, *See* Radia® 7200, 25730, Kessco® IPP, 16116, Unimate® IPP, 32972, Wickenol® 111, 34379
Palmitic acid, 2-ethylhexyl ester, *See* Unimate® EHP, 32970, Wickenol® 155, 34391
Palmitic/stearic acid mono/diglycerides, *See* Tegotens 4100, 31035
palmitin, 22636
Palmityl acetate, *See* Acelan A, 268
palmityl alcohol, *See* Adol® 52 NF, 782, Cachalot® C-50 NF, 4878, cetyl alcohol, 5815, 5816, 5817, Rita CA, 26788
Palmityl dimethylamine, *See* Crodamine 3.A16D, 7106
Palmityl trimethyl ammonium chloride, *See* Radiaquat 6444, 25792
palmitylamine, *See* Amine 16D, 2130, Armeen® 16, 2935, Armeen® 16D, 2936, Crodamine 1.16D, 7101
Palomar®, 22637
Palorium, 22638
Palormone, 22639
Palusol, 22640
Palygorskite, *See* Attapulgite, 3404
Pamak, 22641
Pamolyn, 22642, 22643
Pamolyn 100 FGK, 22644
Pamolyn 100, 100 FG, *See* Pamolyn 100 FGK, 22644
Pamolyn 100, 100 FG, 100, FGK, 125, *See* oleic acid, 22112
Pamolyn 125, 22645
Pamolyn 200, 240, *See* Pamolyn 125, 22645
Pamolyn 327B, 22646
Pamolyn 380, 22647
Pamolyn® 125, 22648, *See* oleic acid, 22112
pan scale, 22649
Panabath, 22650
Panablock, 22651
Panacete 800, *See* Captex® 8000, 5170, Emalex O.T.G, 9976, Miglyol® 808, 19736
Panacide, 22652, *See* Nuophene, 21783
Panaclean, 22653
Panacryl, 22654
Panacur, 22655, *See* fenbendazole, 11594
panadol, *See* Valadol, 33238
Panagran, 22656
Panalane® L-14E, 22657
Panaleve, *See* Valadol, 33238
Panama bark, 22658
Panama crimson, 22659
Panaplate, *See* dichlorvos, 8394
Panasand, 22660
Panasol, 22661
Panasorb, *See* Valadol, 33238
Panaspray, 22662
Panastat, 22663
Panatest, 22664
Panazon, *See* nitrovin, 21382
Panazyme, 22665
pancil, *See* Skane® M-8, 28563
Pancil T, 22666, *See* Skane® M-8, 28563
Pancoxin, 22667, 22668, 22669, *See* Amprolium hydrochloride, 2365, ethopabate, 10993
pancreas diastase, 22670
Pancrease, 22671 , *See* Accelerase, 185
Pancreatic lipase, *See* steapsin, 29590
Pancreatic Lipase 250, 22672, *See* Lipase, 17329
pancreatokinase, 22673
Pancrelipase, *See* Accelerase, 185, Cotazym, 6880, Ilozyme, 14889, Pancrease, 22671
Pancreol, 22674
Pancrex V, 22675
Pancrolin, 22676
Pandex, 22677
Panelyte, 22678
Panets, *See* Valadol, 33238
Panex, 22679, *See* Valadol, 33238
Panheprin, *See* Heparin, Sodium Salt, 13788
Panilax®, 22680
Panmycin, *See* tetracycline, 31365
Panodan 235, FDP-K, SD, SD-K, 22681
Panofen, *See* Valadol, 33238
Panogen, *See* Panogen M, 22682
Panogen M, 22682
Pan-O-Lite, 22683
panoram 75, *See* Rezifilm, 26521
PanOxyl, *See* Abcure S-40-25, 35, AcetOxyl 2.5 and 5, 305, Benzoyl peroxide, 3983
Pansorbin®, 22684
Pantal, 22685
Pantalast® 1120, 22686
Pantarol, 22687
Pantelmin, *See* Mebendazole, 19198, Telmin, 31066, Vermox, 33591
Pantene, 22688
Pantene Grooming Lotion, 22689
Pantenyl, *See* d-Panthenol, 22690
Pantheline, *See* propantheline bromide, 25176
d-Panthenol, 22690, *See* Ritapan D, DL, 26818
DL-Panthenol, *See* Fancol DL, 11438
panthenol-lecithin, *See* Dermasome® P, 8108
Panthenyl ethyl ether, 22691, *See* Humectant SD-35, 14464, Panthenylethyl ether, 22691
Panthenyl triacetate, *See* Ritapan TA, 26820
Panther, 22692
Panthoderm, *See* d-Panthenol, 22690
Pantholin, 22693
Pantholin, *See* vitamin B_5 calcium salt, 33936
Pantolit, 22694
Pantopaque, 22695
Pantosept, 22696
pantothenol, *See* d-Panthenol, 22690
pantothenyl alcohol, 22697, *See* d-Panthenol, 22690
pantothenylol, *See* d-Panthenol, 22690, pantothenyl alcohol, 22697
N-pantoyl-3-propanolamine, *See* d-Panthenol, 22690, pantothenyl alcohol, 22697
PAP, *See* *p*-aminophenol, 2185
papain, 22698
Papain Conc, 22699
papayotin, *See* papain, 22698
Paperad, 22700
paper-clay, *See* bentonite, 3954
paperhanger's alum, 22701
Paperine, 22702
Paper-Pac®, 22703
paper-spar, 22704
Papi 27, 22705
Papi 4901, 22706
Papite, 22707
papoid, *See* papayotin, papain, 22698
paposite, 22708
Pappenheim's stain, 22709
paprika, 22710
Papua nutmeg butter, *See* Macassar nutmeg butter, 18117
Papyrus, 22711
Par, 22712
Par Clay®, 22713
PARA, *See* *p*-dichlorobenzene, 8386
Para butter, *See* para palm oil, 22714
para palm oil, 22714
para red, *See* paranitraniline red, 22780
para toner, 22715
paraacetaldehyde, *See* paraldehyde, 22759
Parabar, 22716
Paraben, *See* Nipasol M, 21282
Parabens, *See* Nipasol M, 21282, Preservals®, 24903
Parabis, 22717, *See* Dichlorophen, 8392, Nuophene, 21783
Parable, 22718
Parabolix® 100, 22719
Paracetaldehyde, *See* paraldehyde, 22759
Parachlorocidum, *See* DDT, 7820
parachlorometacresol, *See* Raschit, 25892
Parachoc, *See* Synwax, 30659
Paracide, *See* *p*-dichlorobenzene, 8386, Kaydox, 15811
Paracol, 22720, 22721
Paracol® 403A6, 22723
Paracol® 404C, 22724
Paracol® 404C, *See* Microcrystalline wax, 19666
Paracol® 800N, 22725
Paracol® 800N, 810N, 1886, M161, *See* Paraffin, 22743
Paracol® 810N, 22726
Paracol® 1886, 22722
Paracol® M161, 22727
paracon, 22728
paracoumarone resin, 22729
Paracril® 1880, 1880LM, 22730
Paracril® 2813, 22731
Paracril® BJLT M-30, 22732
Paracril® OZO, 22733
Paracure, 22734
Paradene®, 22735
Paradene® No.2, 22736
Paradept, *See* Nipasol M Sodium, 21284
paraderil, *See* rotenone, 27000
Paraderm, *See* bufexamac, 4655
paradichlorobenzene, *See* Kaydox, 15811
Paradol, 22737, *See* Neradol, 20995
Paradow, 22738, *See* *p*-dichlorobenzene, 8386
Paradura, 22739
Paradyne, 22740
Paraffagar, 22741
paraffin, 22742, 22743, *See* Glycolube® VL, 13109
paraffin oil, *See* Kaydol, 15810, Klearol, 16208, mineral oil,
paraffin wax, *See* Fortex®, 12298, paraffin, 22742,

22743, Synwax, 30659
Paraffin wax (petroleum), *See* Synwax, 30659
Paraffin wax fume, *See* Synwax, 30659
Paraffin wax, chlorinated, *See* Electrofine® S-70, 9782
Paraffin Wax, granular, *See* Vybar® 103, 34165
Paraffin waxes, *See* Synwax, 30659
Paraffin waxes and Hydrocarbon waxes chlorinated, *See* Chlorinated paraffin, 6105, Electrofine® S-70, 9782
paraffin, liquid, 22744
Paraffinic hydrocarbon, *See* MultiTherm IG-2 Heat Transfer Fuild, 20487
Paraffinum Molle, *See* petroleum jelly, , *See* petrolatum, 23495
Parafil, 22745
Parafilm, 22746
Parafix, 22747
Paraflow, 22748
Para-Flux® 4156, 22749
Paraform, 22750, *See* paraformaldehyde, 22751
paraformaldehyde, 22751, *See* Granuform, 13269
Parafuchsine, *See* pararosaniline, 22798
Paraglas, 22752
paraglobin, *See* globulin,
paraglobulin, *See* globulin,
Paragon II, 22753
Paragon Steel, 22754
Paragon-15, 22755
Paragrid, 22756
Paraguay tea, *See* yerba mate, 34655
paragum, *See* jutahycica resins, 15637
Paragutta, 22757
PARAL, *See* paraldehyde, 22759
Paralac, 22758
paralactic acid, *See* L-lactic acid, 16571
paraldehyde, 22759
Paralegin, *See* Hismanal, 14132
Paralene, 22760
Paralink, 22761
Paralkan, *See* Zephiran Chloride, 34728
Paralkan, *See* Variquat® 50MC, 33397
Paraloid® BTA-702, 22762
Paraloid® EXL 2607, 22763
Paraloid® EXL-3330, 22764
Paraloid® HT-510, 22765
Paraloid® K-120N, 22766
Paraloid® K-120N, *See* Polymethylmethacrylate, 24492
Paraloid® KM-318F, 22767
Paraloop, 22768
Paralux, 22769
Paramagenta, *See* pararosaniline, 22798
Paramel, 22770, 22771
(+)-4(8)-*para*-menthen-3-one, *See* pulegone, 25379
Paramet Ester Gum, 22772
Paramid, 22773
paramidophenol, *See* *p*-aminophenol, 2185
Paramin DF, 22774
Paramins, 22775
Paramos, 22776
Paramoth, *See* *p*-dichlorobenzene, 8386, *See* Kaydox, 15811
Paramount, 22777
Paramount B, 22778
Paramul® SAS, 22779
Paranaphthalene, *See* Sterilite Hop Defoliant, 29734
paranitraniline red, 22780, *See* para toner, 22715
paranitrosodimethylanilide, *See* nitro base, 21341
Paranol, 22781, *See* *p*-aminophenol, 2185, Takatol, 30725, Unal, 32908
Paranox, 22782
paranuggets, *See* *p*-dichlorobenzene, 8386
paraoxon, 22783
Parapak, 22784
Parapel® HC-85, 22785
Parapel® LAM-100, 22786, *See* Lactamide MEA, 16568
Parapel® LIS, 22787
Paraphos, *See* parathion, 22807
Paraplast 8100, 22788
Paraplex®, 22789
Paraplex® G.62, 22790
Paraplex® G-25 100%, 22791
Paraplex® G-50, 22792
Paraplex® G-60, 22793
Parapoid, 22794
Parapol, 22795
paraquat, 22796, *See* Dextrone X, 8239, Gramoxone, 13254, 13255, Gramoxone X, 13256
paraquat [1910-42-5]-monolinuron [330-55-2], *See* Gramonol Five, 13253
Paraquat + Plus, 22797
paraquat-diuron, *See* Gramixel, 13250, Gramuron, 13260
paraquat-monolinuron, *See* Gramonol, 13251, Gramonol 5, 13252
pararosaniline, 22798
Pararosaniline Chloride, *See* pararosaniline, 22798
Pararosaniline Hydrochloride, *See* pararosaniline, 22798
Parasept, *See* ethyl *p*-hydroxybenzoate, 11061
Paraset, 22799
Parasiticine, 22800
Parasol 17, 22801
Paraspen, *See* Valadol, 33238
Paratac, 22802
Parataf, *See* Parathion-methyl, 22808
paratartaric acid, 22803, *See* tartaric acid, 30832
Paratect, *See* morantel tartrate, 20307
Paratect Bolus, 22804
Paratemp, 22805, *See* Extrusil, 11331
Paratherm NF, 22806
parathion, 22807, *See* Bladan®, 4309, ethyl parathion, 11079, Folidol® E605, 12199, Fosferno, 12330, Murphos, 20521
Parathion-methyl, 22808, *See* Folidol® M, 12200, Metacide, 19409
Paratie, 22809
Paratol, 22810
Paratone, 22811
Paratox, *See* Parathion-methyl, 22808
Paratulle®, 22812 , *See* Paraffin, 22743
Paraweb, 22813
parawollastonite, *See* calcium silicate, 4969
Para-zene, *See* *p*-dichlorobenzene, 8386, Kaydox, 15811
Parazol, 22814
Parco, 22815
Parco® 58-C-55, 22816
Parcolene, 22817
Parcryl® 250, 22818
Parcryl® 311, 22819
Pardner, *See* Bromoxynil, 4590
pareira, 22820
Parel 58, 22821
Parel®, 22822
Parelan, *See* Valadol, 33238
Parenamine, 22823, *See* Extiat®, 11323
Parenol, 22824
Parenol liquid, 22825
Parentrovite, 22826
Parenzyme, 22827, *See* trypsin, 32530
Parenzymol, *See* trypsin, 32530
Parez, 22828
Parez 631NC, 22829
Parfenac, 22830, *See* bufexamac, 4655
Parfenal, *See* bufexamac, 4655
Parian cement, 22831
parianite, 22832
Paricin® 1, 22833
Paricin® 6, 22840
Paricin® 8, 22841
Paricin® 9, 22842
Paricin® 13, 22834
Paricin® 15, 22835
Paricin® 18, 22836
Paricin® 210, 22837
Paricin® 220, 22838
Paricin® 285, 22839
Parilene, 22843
Paris green, *See* cupric acetoarsenite, methyl green, mountain green, 20369
Paris Lake, *See* Florence lake, 12013
Paris salts, 22844
Paris White, *See* calcium carbonate, 4940
Paris yellow, *See* Chrome yellow, 6212
Paris yellow, lead antimonate, *See* Naples yellow, 20733
Parkerised Steel, 22845
Parker's cement, *See* Roman cement, 26927
Parlay, 22846, *See* paclobutrazol, 22540, 22541
Parlodion, 22847, *See* Kodaloid, 16283, Venite, 33560
Parlon, 22848
Parlon P, 22849
Parmentine, 22850
Parmetol, 22851
Parmol, *See* Valadol, 33238
Parmr, 22852
Paroa-caxy Oil, 22853
Parodyne, *See* antipyrine,
Parogen, 22854
Paroil®, 22855
Paroil® 10, 145, 1061, 22856
Paroil® 152, 22857
Paroil® 5761, 22858
Parol 70, 22859
Par-o-lac, 22860
Paroleine, *See* Kaydol, 15810 paraffin, liquid, 22744
Paroline, *See* paraffin, liquid, 22744
Parolite, 22861
Parraynite, 22862
parrot coal, 22863
parrot green, *See* cupric acetoarsenite, 7374, mountain green, 20369
Parr's alloys, 22864
Parrycop, *See* Cupravit®, 7370
parsley apiole, *See* parsley camphor, 22865
parsley camphor, 22865
Parsol A, *See* Parsol® 1789, 22866
Parsol MCX, *See* Neo Heliopan® AV, 20866, octyl methoxycinnamate, 22039, Parsol® MCX, 22867
Parsol MOX, *See* Neo Heliopan® AV, 20866, octyl methoxycinnamate, 22039, Parsol® MCX, 22867
Parsol® 1789, 22866
Parsol® MCX, 22867
Parsolin, 22868
Parson's Alloy, 22869
Partagon, 22870
Partial glycerides, *See* Tinamul®, 31819
Partially hydrogenated cottonseed oil, *See* Sta-Nut EE, 29508
Partially hydrogenated terphenyl, *See* HB-40, 13712
Partially hydrogenated vegetable oil, *See* BBS, 3862
Partially/fully hydrolyzed polyvinyl acetate, *See* Mowiol®, 20383
Partinium, 22871
Partox, *See* Ridene, 26746
Partron M, *See* Parathion-methyl, 22808
PartsPrep™ Degreaser, 22872
Parvol, 22873
parvoline, 22874
Parylene N, 22875
Parzate, *See* Zineb, 34807
Paseptol, *See* Nipasol M, 21282
Pasilex, 22876
Pasolind, *See* Valadol, 33238
Pasolind N, *See* Valadol, 33238
Passini's solution, 22877
Passow's slag cement, 22878
pastaccio, 22879
Paste, *See* strass, 29802
Pasturol Plus, 22880
Patafol, 22881
Patafol Plus, 22882
patava oil, 22883
patchouli, 22884
Patco® 3, 22885
Patcote® 305, 22886
Patcote® 306, 22887
Patcote® 309, 22888

Patcote® 315, 22889
Patcote® 337, 22890
Patcote® 500, 22891
Patcote® 512, 22892
Patcote® 525, 22893
Patcote® 555K, 22894
Patcote® 803, 22895
Patcote® 811, 22896
patent bark, *See* quercetin, 25647
patent black, 22897
Patent Fustin, *See* maclurin, 18235
patent red, *See* cinnabar,
patent zinc white, 22898
patents, 22899
Pat-Fix RBD, 22900
Path Gun!, 22901
Pathclear, 22902
Pathocil, *See* Dicloxacillin sodium, 8396
patina, 22903
Pationic 122A, 22904
Pationic 138, 22905
Pationic 930, *See* Verv®, 33644
Pationic CSL, 22906
Pationic ISL, 22907
Pationic SSL, 22908
Patlac® IL, 22909
Patlac® LA, 22910
Pat-Lube Series, 22911
Patogen 311, 22912
Patogen 353, 22913
Patogen AO-30, 22914
Patogen P-10 Acid, 22915
Patoran, 22916
Patoran, *See* Metobromuron, 19583
Patoran® FL, 22917
Pat-Quest CS, 22918
Patrol, 22919, *See* fenpropidin, 11616
Patrole, *See* methamidophos, 19474
Patrole, *See* Tamaron®, 30760
Pat-Soft 1442, 22920
pattern metal, 22921
Pattinson's white lead, 22922
Pattonex, 22923, *See* Metobromuron, 19583
Pattrex, 22924
Pattrit, 22925
Pat-Wet LF-55, 22926
Pat-Wet SP, 22927
pavlin, 22928
Pavy's solution, 22929
Paxbestos, 22930
Paxilon, *See* methazole, 19481
Paxolin, 22931
Payne's grey, 22932
Payne's solution, 22933
Pay-off, *See* pendimethalin, 23102
Payze, *See* cyanazine, 7472
Payzone, 22934, *See* nitrovin, 21382
Pazo, 22935
Pb, *See* lead, 16901
PB, *See* piperonyl butoxide, 23875
PBB-PA, *See* Poly (pentabromobenzyl) acrylate, 24342
PBDPO, *See* Pentabromodiphenyl oxide, 23130
PBI Slug Pellets, 22936
PBI Spreader, 22937
PBN, *See* Antioxidant PBN, 2591, Boron nitride, 4443, Stabilator A.R, 29400
PBNA, *See* Antioxidant PBN, 2591, Stabilator A.R, 29400
PB-Nox, *See* rotenone, 27000
PBT-1100, 22938
PBT-1100, 1100G15, 1300, *See* Grilpet® XE3060, 13402
PBT-1100G15, 22939
PBT-1300, 22940
PBT-1700, 22941
PC, *See* Arconate® Propylene Carbonate, 2811, Solvenon® PC, 29049
PC resin, *See* polycarbonate resin, 24375
PC-000/5T, 22942
PC-20CF, 22949
PC-20GF/15T, 22950
PC-1100, 22943
PC-1100G10, 22944
PC-1100H30, 22945
PC-1244, 22946
PC-1344, 22947
PC-1700G10FR, 22948
PCA, *See* Ajidew A-100, 1142, chloridazon, 6103
PCA 301, 22951
PCA 4-1, 22952
PCHO, *See* paraldehyde, 22759
PCI, 22953
PCL Liq.1002/066240, *See* Cetearyl octanoate, 5791
PCMX, 22954, *See* chloroxylenol, 6159
PCNB, *See* quintozene, 25692
PCP, *See* pentachlorophenol, 23135, 23136
PD 185, *See* stearic acid, 29599
PDB, *See* *p*-dichlorobenzene, 8386, Kaydox, 15811
PDCB, *See* *p*-dichlorobenzene, 8386
PDD 6040-I, *See* Diflubenzuron, 8450
PDQ, *See* MCPB, 19184
PDS, 22955
PDX-82427, 22956
PDX-84347, *See* Polyetherimide, 24423
PDX-84367, 22957
PDX-84368, 22958
PDX-84369, 22959, *See* Grilpet® XE3060, 13402
PE, *See* pentaerythritol, 23139
PE 100 228 FH-VP, 22960
PE 1017, 22961
PE 4517, 22962
PE 5222, 22963
PE 5554-H, 22964
PE 5861, 22965
Pénicline, *See* Ampicillin, 2361
Pétrole Hahn, 22966
Pea Pro-Tein BK, 22967
peach black, 22968
peach wood, *See* redwoods, 25987
peanut oil, 22969, *See* Katchung oil, 15788
Peanut oil PEG-6 esters, *See* Labrafil M 1969 CS, 16538
peanut ore, 22970
pear oil, 22971, *See* Amyl acetate, 2380, isoamyl acetate, 15336
Pearex-L®, 22972
Pearistick 46-10/06, 22973
Pearistick 65-05, 22974
Pearl, 22975
pearl alum, 22976
pearl ash, 22977, *See* potassium carbonate, 24743
Pearl Dust®, 22978
Pearl I, II, III, 22979
pearl powder, *See* bismuth oxychloride, 4241
pearl spar, 22980
pearl stearic, *See* stearic acid, 29599
Pearl Supreme UVS, 22981
Pearl white, 22982, *See* Bismuth chloride oxide, 4237, Bismuth oxychloride, 4241
Pearlex GC 0311, 22983
Pearl-Glo®, 22984
pearl-hardening, 22985
pearlite, 22986
Pearlstick, 22987
Pearlstick 45-05/40, 22988
Pearsall, 22989
Pearsol, 22990
Pearson's cerate, 22991
Pearson's solution, 22992
Peat, 22993
peat coal, 22994
Peaweed, 22995
Peb1, *See* DDT, 7820
Pebax®, 22996
Pebax® 2533 SA 00, 2533 SD 00, 2533 SN 00, 22997
Pebax® 4011 MA 00, 22998
Pebax® 4033 SN 70, 5533 SN 70, 22999
Pebax® 6333 SA 00, 6333 SD 00, 6333 SN 00, 23000
pecan oil, 23001
Peceol Isostearique, 23002
Pecilocin, *See* Variotin, 33396
Pecosil CAS-36, 23003
Pecosil DAS 36, 23004
Pecosil GSA 36, 23005
Pecosil OS-100B, 23006
Pecosil OS-100DA, 23007
Pecosil OS-100HS, 23008
Pecosil OS-100L, 23009
Pecosil OS-100M, 23010
Pecosil OS-100U, 23011
Pecosil PS-100, 23012
Pecosil PS-100K, 23013
Pecosil SG-20, 23014
Pecosil SSP, 23015
pectase, 23016
pectin, 23017
Pectin, *See* vegetable jelly, 33521
pectinase, 23018, *See* Clarex® L, 6359
Pectinase AT, 23019
Pectinex, 23020
Pecutrin®, 23021
Pedameth, *See* DL-methionine, 19487
pediaflor, *See* sodium fluoride, 28764
pedident, *See* sodium fluoride, 28764
Peerless alloy, 23022
Peerless®, 23023
Peerless® No. 1, 23024
PEG, *See* Alkapol PEG 300, 1706, Pluracol® E8000, 24184
PEG lanolin series, 23025
PEG stearamine series, 23026
PEG (1-4) lauryl ether sulfate sodium salt, *See* sodium laureth sulfate, 28778
PEG (3) lauryl dimethyl amine oxide, *See* PEG-3 lauramine oxide, 23047
PEG -12 PEG-50 lanolin, 23027
PEG 100 monolaurate, *See* PEG-2 laurate, 23041
PEG 100 monostearate, *See* PEG-2 stearate, 23043
PEG 200, *See* Alkapol PEG 300, 1706, Nopalcol 200, 21474, Teric PEG 200, 31272, PEG-4, 23049
PEG 200 diacrylate, *See* Ageflex T4EGDA, 970, SR-259, 29373
PEG 200 dilaurate, *See* Ethox DL-5, 11022, Nonex DL-2, 21445
PEG 200 laurate, *See* Ablunol 200ML, 95
PEG 200 stearate, *See* Ablunol 200MS, 97
PEG 300, *See* Teric PEG 300, 31273
PEG 300 abietate, *See* Secoster® MA 300, 27820
PEG 300 stearate, *See* Nonex S3E, 21447
PEG 400, *See* Nopalcol 400, 21475, Teric PEG 400, 31274
PEG 400 caprylate/caprate glycerides, *See* Labrasol, 16543, PEG-8 caprylic/capric glyceride, 23053
PEG 400 dioleate, *See* Nonex DO-4, 21446
PEG 400 ester, *See* Eccoterge 200, 9565
PEG 400 isostearate, *See* Trydet 2644, 32508
PEG 400 monostearate, *See* Alkamuls® S-65-8, 1680
PEG 400 oleate, *See* Nonex 04E, 21443, Secoster® MO 400, 27821
PEG 400 oleate, *See*
PEG 500 cocoate, *See* Nonex C5E, 21444
PEG 600, *See* Emery® 6686, 10180, Nopalcol 600, 21476, Teric PEG 600, 31275
PEG 600 dioleate, *See* Secoster® DO 600, 27818
PEG 600 monostearate, *See* Alkamuls® S-65-8, 1680
PEG 600 monotallate, *See* EM-600, 9906, PEG tallate series, 23035
PEG 600 tallate, *See* Actrol 6M25P, 569
PEG 800, *See* Teric PEG 800, 31276
PEG 1000, *See* Alkapol PEG 300, 1706, Teric PEG 1000, 31277
PEG 1000 oleate, *See* Ethylan® A10, 11093, Ethylan® FO30, 11102
PEG 1500, *See* Teric PEG 1500, 31278
PEG 3350, *See* Teric PEG 3350, 31279
PEG 4000, *See* Teric PEG 4000, 31280
PEG 6000, *See* Teric PEG 6000, 31281
PEG 8000, *See* Teric PEG 8000, 31282
PEG 12000, *See* Teric PEG 12000, 31283

PEG-25 hydrogenated castor oil, *See* Fancol HCO-25, 11440
PEG-25 oleate, *See* Sellig AO 25 100, 27877
PEG-25 PABA, *See* Uvinul® P 25, 33203
PEG-25 propylene glycol stearate, *See* G-2162, 12551
PEG-26 jojoba acid, *See* Oxypon 328, 22469
PEG-26 jojoba alcohol, *See* Oxypon 328, 22469
PEG-27 lanolin, *See* Laneto 27, 16660, Lanogel® 21, 16729
PEG-30 castor oil, *See* Alkamuls® EL-620, 1665, Fancol CO-30, 11437, Sellig R 3395 SP, 27907, Witconol 5906, 34510
PEG-30 glyceryl cocoate, *See* Varonic® LI-63, 33445
PEG-30 glyceryl isostearate, *See* Tagat® I, 30706
PEG-30 glyceryl laurate, *See* Tagat® L, 30707
PEG-30 glyceryl oleate, *See* Tagat® O, 30709
PEG-30 glyceryl stearate, *See* Tagat® S, 30714
PEG-30 lanolin, *See* Aqualose L30, 2709
PEG-30 oleamine, *See* Chemeen O-30, O-30/80, 5947, Eumulgin PA 30, 11181, Rhodameen® OA-860, 26611, Trymeen® 6620, 32524, Trymeen® OAM 30/60, 32526, Varonic® Q-230, 33447, Witcamine® 6622, 34448
PEG-30 oleyl amine, *See* Ethox OAM-308, 11035
PEG-30 stearate, *See* Mergital ST 30/E, 19362, Sellig S 30 100, 27911
PEG-30 tetramethyl decynediol, *See* Surfynol® 420, 30384, Surfynol® 440, 30385, Surfynol® 465, 30386, Surfynol® 485, 30387
PEG-32, *See* Hodag PEG 540, 14205 Hodag PEG 1450, 14208
PEG-32 castor oil, *See* Berol 195, 4018, Berol 199, 4020, Sellig R 3395-C435, 27908
PEG-32 distearate, *See* Mapeg® 1540 DS, 18636
PEG-33 castor oil, *See* Sellig R 3395, 27906
PEG-35 castor oil, *See* Cremophor® EL, 6981
PEG-40, *See* Pluracol® E2000, 24176
PEG-40 castor oil, *See* Berol 108, 4016, Sellig R 4095, 27909
PEG-40 glyceryl cocate, *See* Oronal LCG, 22312
PEG-40 hydrogenated castor oil, *See* Akyporox CO 400, 1273, Amyx ST 3837, 2396, Cremophor® RH 40, 6984
PEG-40 sorbitan diisostearate, *See* Emsorb® 2726, 10467
PEG-40 sorbitan hexatallate, *See* Glycosperse® HTO-40, 13127
PEG-40 sorbitan oleate, *See* Emalex ET-8040, 9955
PEG-40 sorbitan peroleate, *See* Arlatone® T, 2916
PEG-40 stearate, *See* Mapeg® S-40, 18643, Myrj® 52, 20583, Pegosperse® 1750 MS, 23062, Ritox 52, 26852
PEG-40 stearate, sorbitan stearate, and silica, *See* AF 72, 911
PEG-44 castor oil, *See* Sellig R 4495, 27910
PEG-50 hydrogenated tallow amide, *See* Ethomid® HT/60, 10989, 10990, Schercomid HT-60, 27651, Sipenol IT-50-46, 28517
PEG-50 lanolin, *See* Lanolex L-40, 16738
PEG-50 sorbitol hexaoleate, *See* Ethox HO-50, 11027, PEG sorbitan hexaoleate series, 23032
PEG-50 stearamine, *See* Ethomeen® 18/60, 10976
PEG-50 stearate, *See* Myrj® 53, 20584
PEG-50 stearyl amine, *See* Trymeen® 6617, 32523, Trymeen® SAM-50, 32527
PEG-55 propylene glycol oleate, *See* Antil® 141 Liquid, 2557, 2558
PEG-60 hydrogenated castor oil, *See* Cremophor® RH 60, 6985
PEG-60 lanolin, *See* Ivarlan 3406, 15446, Solan 50, 28888
PEG-75, *See* Hodag PEG 3350, 14209, Pluracol® E4000, 24180, Rhodasurf® PEG 3350, 26659
PEG-75 lanolin, 23052, *See* Ethoxylan® 1685, 11047, Ivarlan 3400, 15445, Lan-Aqua-Sol 100, 16649, PEG-75 lanolin, 23052, Solan, 28886, Solangel 401, 28891, Solulan® 75, 29002
PEG-75 lanolin, propylene glycol, ceteth-16, hydrolyzed animal protein and lanolin oil, *See* Proto-Lan 20, 25291
PEG-80 glyceryl tallowate, *See* Rewoderm® LI 48, 26310
PEG-80 sorbitan laurate, *See* Emsorb® 2721, 10465, G-4280, 12554, Hetsorb L-80-72%, 13969, T-Maz® 28, 31894
PEG-80 sorbitan laurate, cocamidopropyl betaine, sodium trideceth sulfate, sodium lauroamphoacetate, PEG-150 distearate and sodium laureth-13 carboxylate, *See* Miracare® BC-10, 19852
PEG-80 sorbitan laurate, sodium trideceth sulfate, PEG-150 distearate, disodium lauraminopropionate, cocamidopropyl hydroxysultaine and sodium laureth-13 carboxylate, *See* Miracare® MS-1, 19857
PEG-80 sorbitan laurate-sodium trideceth sulfate-PEG-150 distearate-disodium lauroampho diacetate-cocamidopropylhydroxysultaine-sodium laureth-13 carboxylate, *See* Carsofoam® BS-I, 5328
PEG-80 sorbitan palmitate, *See* G-4252, 12553
PEG-100 stearate, *See* Arlacel® 165, 2905, Myrj® 59, 20585, Simulsol 165, 28484
PEG-120 methyl glucoside dioleate, *See* Glucamate® DOE-120, 13025
PEG-150, *See* Carbowax® PEG 8000, 5251, Pluracol® E6000, 24182
PEG-150 dinonyl phenyl ether, *See* T-Det® D-150, 30862
PEG-150 distearate, *See* Emalex 6300 Di-ST, 9929, Hodag 602-S, 14163, Lipopeg 6000-DS, 17398, Mapeg® 6000 DS, 18637, PEG-150 distearate series, 23037, Rewopal® PEG 6000 DS, 26347
PEG-150 distearate series, 23037
PEG-150 oleate, *See* Emerest® 2617, 10103
PEG-150 stearate, *See* Emalex 6300 M-ST, 9930, Lipowax PR, 17445, Macol® CPS, 18263, Ritachol® 1000, 26803, Ritapeg 150 DS, 26821
PEG-160 castor oil, *See* Berol 198, 4019
PEG-200 castor oil, *See* Berol 191, 4017
PEG-200 hydrogenated castor oil, *See* Chemax HCO-200/50, 5907
PEG-200 oleate, *See* Ablunol 200MO, 96
PEG-200 trihydroxystearin, *See* Naturechem® THS-200, 20778
PEG-350, *See* Carbowax® Compound 20M, 5247
PEG-400 oleate, *See* Witflow 916, 34526
PEG-660 hydroxystearate, *See* Solutol® HS 15, 29034
PEG-660 tallate, *See* Mapeg® TAO-15, 18644
pegmatite, 23055
pegnin, 23056
Pegol® L-10, 23057
Pegosperse PS, *See* Witconol 2380, 34498
Pegosperse®, 23058
Pegosperse® 50 DS, 23065
Pegosperse® 50 MS, 23066
Pegosperse® 100 L, 23059
Pegosperse® 100 O, 23060
Pegosperse® 100 S, 23061
Pegosperse® 200 DL, 23063, *See* PEG-4 dilaurate series, 23050
Pegosperse® 400 DOT, 23064, *See* PEG ditallate series, 23028
Pegosperse® 1750 MS, 23062
Pegosperse® PMS CG, 23067
Pegoterate, *See* Aviester, 3472
PEGs, *See* Sizing Wax PA, PT, SM, 28560
pegu, *See* cutch,
pegu catechu, *See* cutch,
PEI, *See* PDX-84367, 22957
Peka Glas, 23068
Pekafill, 23069
Pekafix, 23070
Pekalux, 23071
Pekatop, 23072
Pekatray®, 23073
Pelagol Grey P Base, *See* *p*-aminophenol, 2185
Pelagol P Base;, *See* *p*-aminophenol, 2185
Pelamag, 23074
Pelamagsalt, 23075
Pelargene, 23076
pelargidenolon 1497, *See* kaempferol, 15654
Pelargone, 23077
Pelargonic acid, 23078, *See* Emery® 1202, 10138, Emery's L-114, 10187
N-Pelargonoyl-*p*-aminophenol, *See* Suconox-9®, 29924
Pelaspan, 23079
Pelaspan 333FR, 23080
Pelaspan GP, 23081
Pelaspan Mold-a-Pac, 23082
Pelaspan PAC, 23083
Pelemol CR, *See* cetyl ricinoleate, 5824
Peligal, *See* Odylen®, 22046
pelionite, 23084
Pellethane®, 23085
Pellethane® 2102-55D, 23086
Pellethane® 2354-45DGA, 23087
Pellethane® 2363-55D, 23088
Pellidol, *See* Diacetazotol, 8247
Pelonit D, 23089
Pelopon A, *See* Nacconol® 40G, 20647
Pels, 23090
Pelt 14, *See* thiophanate methyl, 31706
Peltex, 23091
Penacolite® B-18-S, 23092
Penacolite® R-2170, 23093
Penaquacaine G, *See* Penicillin V, 23111
Penaryl A, 23094
Penaryl B, 23095
Pen-Bristol, *See* Ampicillin, 2361
Penbritin, *See* Ampicillin, 2361
Penbrock, *See* Ampicillin, 2361
Pencal, *See* calcium arsenate, 4938
Penchlor, 23096
penchlorol, *See* pentachlorophenol, 23135, 23136
penconazole, *See* Topas 100, 32026
pencycuron, 23097, 23098, 23099, *See* Trotis, 32473
pendare, 23100
pendecamaine, 23101
pendimethalin, 23102
Peneteck, 23103
Penetek, *See* Pentaerythritol, 23139
Penetral NA 20, 23104
Penetrol, 23105, 23106
Penetrol 2-EHS, 23107
Penetron OT-30, 23108
Pen-Fifty, *See* Penicillin V, 23111
Penicillamine, *See* Trolovol®, 32449
Penicillin G potassium, 23109, 23110, *See* Liquapen, 17452
Penicillin G procaine, *See* Lentopen, 16993
Penicillin V, 23111
Penicillinase, 23112
Pennac Cbs, *See* Vulkacit® CZ/EGC, DZ/EGC, 34129
pennac cra, *See* Perkacit® ETU, 23283
Pennac ZT, *See* Octocure ZMBT-50, 21998, Oxaf, 22408, Perkacit® ZMBT, 23298, Zetax®, 34741
Pennad 150, 23113
Pennant, *See* metolachlor, 19585
Penncap-M, *See* Parathion-methyl, 22808
Pennchem® Mortar, 23114
Penncozeb, 23115, *See* Karamate, 15749, mancozeb, 18517
Pennfloat® 3-2277, 23116
Pennington's Pride Multi-Purpose Fungicide, *See* Chlorothalonil, 6150, Repulse, 26151
Pennodorant® 1013, 23117
Pennstop® 1866, 23118
pennwhite, *See* sodium fluoride, 28764
pennyroyal oil, 23119
Pennzone B 0685, 23120
Pennzone E, *See* Pennzone E 0686, 23121
Pennzone E 0686, 23121
Penotrane, *See* cadmium diethyldithiocarbamate, 4896, hydrargaphen, 14539
Penoxalin, *See* pendimethalin, 23102
Penplenum, *See* Versapen, 33609

Chemical and Trade Names

Pentine 1185 5432, 23181
Pentine Acid 5431, 23182
Pentol, 23183, *See* Radiasurf® 7156, 25803
1-Pentol, *See* Amyl alcohol, 2381
Pentol (emulsifier), *See* Radiasurf® 7156, 25803
Pentonate DB, 23184
Pentonite, 23185
Pentonium 50, 80, 23186
Pentopan, 23187
Pentosalen, *See* imperatorin, 14912
Pentoxyl M, 23188
Pentrex, 23189, *See* Ampicillin, 2361
Pentrexyl, *See* Ampicillin, 2361
Pentrone, 23190
Pentrone ON, 23191, *See* Niaproof® Anionic Surfactant 08, 21092, Rhodapon® BOS, 26632
Pentrone S, 23192
Pentrosan, 23193
pentyl, 23194
pentyl acetate, *See* Amyl acetate, 2380
n-pentyl acetate, *See* Amyl acetate, 2380
pentyl alcohol, *See* Amyl alcohol, 2381
n-pentyl alcohol, *See* Amyl alcohol, 2381
pentyl vintyl carbinol, *See* Morillol®, 20335
pentylcarbinol, *See* 1-hexanol, 14011, hexyl alcohol, 14039
2-Pentyl-3-methyl-2-cyclopenten-1-one, *See* dihydrojasmone, 8458
Pentylol, 23195
Pentyltetrahydropyranyl acetate, *See* Jasmopyrane, 15533
Pentyltrichlorosilane, *See* CA0900, 4854
Penzold's reagent, 23196
Pep, 23197
Pep Set, 23198
Pepper Dust, 23199
peppermint camphor, *See* menthol terpine hydrate, 19321
Pepsamar, *See* aluminum hydroxide, 1905
pepsin, 23200
Pepsinum, *See* pepsin, 23200
Peptavlon, 23201, *See* pentagastrin, 23146
Peptein® 2000®, 23202 , *See* Extiat®, 11323
Peptein® VgW, 23203
Peptizer 566, 23204
Peptizer 7010, 23205
Peptoil, 23206
Peptorub, 23207
Peptrex, 23208
peracetic acid, *See* Oxymaster, 22460, Proxitane, 25308
Peradinol, 23209
Peralfan T Concentrate, 23210
Perapret®, 23211
Perapret® PE-2, 23212
perborax, 23213
perborin, 23214
perborol, *See* perborax, 23213
Perbunan N Latex, *See* Acrylonitrile-butadiene rubber, 437
Perbunan®, 23215
Perbunan® N, 23216
Perbunan® N 1807 NS, 23217
Perbunan® N/VC70, 23218
Perbutyl H, *See* Aztec® *t*-Butyl Hydroperoxide-70, Aq, 3541
Percarbamid, 23219
Perchloracap, *See* KM Potassium Perchlorate, 16243
Perchlorate of barium, *See* Desicchlora, 8139
perchlorethylene, 23220
perchloric acid, 23221, *See* Fraude's reagent, 12381
Perchloric acid, barium salt, *See* Desicchlora, 8139
Perchloric acid, potassium salt, *See* potassium perchlorate, 24767
perchloride of mercury, *See* mercuric chloride, 19347
perchlorobenzene, *See* Julin's chloride, 15632
perchloroethylene, 23222, *See* Perclene, 23225, Perclene TG, 23226, Perklone, 23321
Perchloromethane, *See* carbon tetrachloride, 5224, Katharin, 15791
Perchloron, 23223, *See* Repak, 26141
Percist, 23224
Perclene, 23225
Perclene TG, 23226
Percol®, 23227
Percolaye, 23228
Percresan, 23229
Percumyl D, 23230
Percumyl H, *See* Aztec® CHP-80, 3555, CHP-158, 6175
Perderm, *See* Vaderm, 33237
Perduren, 23231
Perdynamine, 23232
Perecot, 23233
Peregal®, 23234
Peregal® O, 23235
Peregal® OK, 23236
Peregal® ST, 23237
Pereman, 23238
Perenex, *See* cuprous oxide, 7405
Perenol El, 23239
Perenox, 23240
Perenyl's fluid, 23241
Perfecta, 23242
Perfection®, 23243
Perfekthion, *See* Dimethoate, 8496
Perfekthion®, 23244
Perfilcon, *See* Permalens, 23348
Perfix, 23245
Perflex, 23246, 23247
Perfluidone, 23248
Perfluoro alkyl phosphinate/phosphonate, *See* Defoamer SF, 7887
perfluoro-(methylcyclohexane), *See* Flutec PP2, 12122
perfluoro-1,3-dimethylcyclohexane, *See* Flutec PP3, 12123
perfluoro-1-methyldecalin, *See* Flutec PP9, 12124
perfluoroacetic acid, *See* trifluoroacetic acid, 32249
Perfluoro-compound FC-72, *See* Flutec PP1, 12121
perfluoroethene, *See* MS-122, 20413
perfluoroethylene, *See* MS-122, 20413
perfluoroheptanecarboxylic acid, *See* Fluora, 12060
perfluoro-methylcyclohexane, *See* Flutec PP2, 12122
perfluoro-n-hexane, *See* Flutec PP1, 12121
n-perfluorohexane, *See* Flutec PP1, 12121
perfluorooctane sulfonyl fluoride, *See* Fluorad® FX-8, 12067
perfluoro-1-octanesulfonyl fluoride, *See* Fluorad® FX-8, 12067
perfluorooctanoic acid, *See* Fluorad® FC-26, 12060
perfluorooctanoic acid, ammonium salt, *See* Fluorad® FC-118, 12058
Perfluoropolyether fluids, *See* Aflunox, 942
PerforMax® 403, 23249
perfumery oil, 23250
Perfusamine, 23251, *See* Iofetamine ^{123}I hydrochloride, 17553
Perfusamine
Pergacid, 23252
Pergamin, 23253
Pergamyn, *See* Pergamin, 23253
pergantene, *See* sodium fluoride, 28764
Pergantine, 23254
Pergaprint, 23255
Pergascript, 23256
Pergasol, 23257
pergenol, 23258
Perglanz-Konzen-Trat B48 and B30, 23259
Perglazmittel GM 4006, 23260
Perglow, 23261
perglycerol, 23262
Pergopak, 23263, 23264
Pergut®, 23265
Perhydrate, 23266
Perhydrit, 23267
Perhydrol, 23268, *See* hydrogen peroxide, 14590
perhydrol of magnesia, *See* Biogen,
perhydronaphthalene, *See* Decalin®, 7840
perhydrosqualene, *See* Cosbiol®, 6865, squalane, 29357
Peri acid, 23269, *See* Schollkopf's acids, 27726
Perichthol, 23270
Peridex, *See* chlorhexidine digluconate, 6101, Rotersept, 27002
Perigen, *See* permethrin, 23384, 23385, 23386
Perikol, 23271
perilla oil, 23272
periodic acid, 23273
periodic acid dihydrate, *See* periodic acid, 23273
periodic acid potassium salt, *See* potassium periodate, 24768
Periograf, 23274
Periston, *See* Tears Plus®, 30876, Vinisil, 33827, Videne Disinfectant Solution, Videne Disinfectant Tincture, Videne Powder, 33739
Periygel®, 23275
Perizin®, 23276
Perkacit® CBS, 23277, *See* cyclohexyl benzothiazole sulfenamide, 7523
Perkacit® CDMC, 23278
Perkacit® DCBS, 23279
Perkacit® DOTG, 23280, *See* Ditolyguanidine, 8732
Perkacit® DPG, 23281
Perkacit® DPTT, 23282
Perkacit® ETU, 23283
Perkacit® MBS, 23284
Perkacit® MBT, 23285
Perkacit® MBTS, 23286
Perkacit® NDBC, 23287
Perkacit® SDMC, 23288
Perkacit® TBBS, 23289, *See* BBTS, 3863, Butylbenzothiazole sulfenamide, 4791
Perkacit® TDEC, 23290, *See* Akrochem® TDEC, 1179
Perkacit® TETD, 23291, *See* Akrochem® TETD, 1180
Perkacit® TMTD, 23292, *See* Akrochem® TMTD, 1182
Perkacit® TMTM, 23293
Perkacit® ZBEC, 23294
Perkacit® ZDBC, 23295
Perkacit® ZDEC, 23296
Perkacit® ZDMC, 23297
Perkacit® ZMBT, 23298
Perkadox®, 23299
Perkadox® 14, 23300, *See* Bisbutyl peroxy diisopropyl benzene, 4230, Luperox 802, 17841
Perkadox® 16, 23301
Perkadox® 16-W40-GB5, 23302
Perkadox® 20, 23303
Perkadox® 26-fl, 23304
Perkadox® 30, 23305
Perkadox® 58, 23306
Perkadox® BC, 23307
Perkadox® GS, 23308
Perkadox® RM, 23309
Perkadox® SE-8, 23310
Perkaglycerol, 23311
Perkait® TETD, *See* Disulfiram, 8725
Perkalink® 300, 23312
Perkalink® 301, 23313
Perkalink® 400, 23314
Perkalink® 401, 23315, *See* ethylene glycol dimethacrylate, 11125
Perkasil® KS 207, 23316
Perkasil® KS 300, 23317
Perkasil® KS 404, 23318
Perkasil® VP 406, 23319
Perkin's Base, 23320
Perklone, 23321
perlandrol I, *See* Rhodapon® 101-10, 26631, sodium lauryl sulfate, 28781
Perlankrol®, 23322
Perlankrol® ADP3, 23323
Perlankrol® ATL40, 23324
Perlankrol® DAF25, 23325, *See* Ammonium lauryl sulfate, 2242
Perlankrol® DSA, 23326
Perlankrol® ESD, 23327
Perlankrol® ESK32, 23328
Perlankrol® PA Conc, 23329
perlate salt, *See* salt perlate, 27293

Perlatum 400, 410, 410CG, 420, 510, 23330
Perlatum 415, 415 CG, 425, 23331
Perlex, 23332
Perlextra, 23333
Perlit®, 23334
perlite, 23335
Perlon, 23336, *See* Cabelec® 1015, 4861, nylon 6, 21909
Perma Kleer 140, *See* Cheelox® 80, 5864, Pentaquest Extra 0685, 23169, Trilon® C Liq, 32328
Perma Shield, 23337
Perma Sta, 23338
Permabond, 23339
Permador, 23340
Perma-Flex Blak-Stretchy®, 23341
Perma-Flex Blak-Tufy®, 23342
Perma-Flex Blu-Sil, 23343
Perma-Flex Green-Sil, 23344
Permafuse, 23345
permalba, 23346
Perma-Leaf, 23347
Permalens, 23348
Permali, 23349
Permalloy®, 23350
Permalon, 23351
Permalose, 23352
Permalux, 23353
Permalyn® 7085, 23354
Perma-Mold®, 23355
Permanax 6PPD, 23356
Permanax 18, *See* Agerite® DPPD, 1016, N,N'-diphenyl-*p*-phenylenediamine, 8603, JZF, 15638, Naugard® J, 20796, Permanax DPPD, 23361
Permanax 45, *See* Agerite® MA, 1019, Akrochem® Antioxidant DQ, 1160
Permanax 103, *See* Polyflex, 24435
Permanax 115, *See* Vulkanox® 4010 NA, 34147
Permanax BL, BLN, 23357
Permanax BLW, 23358
Permanax CNS, 23359
Permanax CR, 23360
Permanax DPPD, 23361, *See* Agerite® DPPD, 1016, N,N'-diphenyl-*p*-phenylenediamine, 8603, JZF, 15638, Naugard® J, 20796, Permanax DPPD, 23361
Permanax HD, 23362
Permanax HD (SE), 23363
Permanax IPPD, 23364
Permanax OD, 23365
Permanax OZNS, 23366
Permanax TQ, 23367, *See* Agerite® MA, 1019, Akrochem® Antioxidant DQ, 1160
Permanax WSL, 23368
Permanax WSL Pdr, 23369
Permanax WSO, 23370
Permanax WSP, 23371
Permanax WSP (PQ), 23372
Permanax™ IPPD, *See* N-Isopropyl-N'-phenyl-*p*-phenylene diamine, 15413
Permanent, 23373
permanent blue, *See* ultramarine, 32821
Permanent Encapsulant 185N, 23374
permanent green, *See* chromium green, 6232
permanent violet, *See* Manganese Violet, 18550
Permanent yeast, *See* zymin, 34923
permanent yellow, *See* Barium chromate, 3638
Permanex IPPD, *See* Flexzone 3C, 11967, Permanax IPPD, 23364
permanganc acid potassium salt, *See* potassium permanganate, 24770
Permanite, 23375
Permanone
Permaplex, 23376
Permasect, 23377
Permasect, *See* permethrin, 23384
Permasep®, 23378
Perma-Slik, 23379
Permatag, 23380
Permathin, 23381
Permatol A, 23382
Permax, 23383
permethrin, 23384, 23385, 23386, *See* Cooper Coopex, 6758, Darmycel Agarifume Smoke, 7780, Elimite®, 9838, Fumite Permethrin, 12462, Kafil, 15655, Permasect, 23377
Permethyl 99A, 23390
Permethyl 101A, 23387
Permethyl 102A, 23388
Permethyl 104A, 23389
Permetrina, *See* permethrin, 23384, 23385, 23386
Permidan, 23391
Permilan, *See* Zineb, 34807
Perminal, 23392
Perminvars, 23393
Permit, *See* permethrin, 23384, 23385, 23386
Permobel, 23394
Permonite, 23395
Permutite, 23396
Permyl B-100, 23397
Pernambuco Wood, *See* redwoods, 25987
Pernax, 23398
Peroidin, *See* KM Potassium Perchlorate, 16243
Perone, 23399
Peronoid, 23400
Peropal, *See* azocyclotin, 3523
Peropal®, 23401
perosmic acid anhydride, *See* osmic acid, 22355, osmium tetroxide, 22357
perosmic oxide, *See* osmic acid, 22355, osmium tetroxide, 22357
Perovskite, *See* calcium titanate, 4974
Perox, 23402
Peroxal, 23403
Peroxan, *See* hydrogen peroxide, 14590
peroxidase, 23404
peroxide, *See* hydrogen peroxide, 14590
Peroxide RH-2, 23405
Peroxide, bis(3,5,5-trimethyl-1-oxohexyl), *See* USP® -355M, 33167
Peroxide, bis(3,5,5-trimethylhexanoyl), *See* USP® -355M, 33167
Peroxide, bis(1-oxododecyl), *See* Alperox-F, 1816
Peroxide, bis(1-oxooctyl), *See* Perkadox® SE-8, 23310
Peroxide, cyclohexylidenebis[(1,1-dimethylethyl), *See* USP® -400P, 33168
Peroxide, cyclohexylidenebis[tert-butyl], *See* USP® - 400P, 33168
peroxide, [phenylenebis(1-methylethylidene)bis(1,1-dimethylethyl)], *See* Retilox® F 40 MG, 26274, Vul-Cup 40KE and R, 34118
Peroxide, (3,3,5-trimethylcyclohexylidene)bis(1,1-dimethylethyl), *See* Aztec® 1,1-Bis(*t*-Butylperoxy)-3,3,5-Trimethyl Cyclohexane, 3549, Peroximon® S-164/40P, 23410, Trigonox® 29, 32269
Peroxidol®, 23406
Peroximon, 23407
Peroximon® DC-40, 23409
Peroximon® DC 40 MG, 23408
Peroximon® S-164/40P, 23410
Peroxol, 23411
peroxyacetic acid, *See* Oxymaster, 22460
peroxydecanoic acid, *See* Decanox-F, 7843
Peroxydex, *See* Abcure S-40-25, 35, AcetOxyl 2.5 and 5, 305, Benzoyl peroxide, 3983
Peroxydicarbonate, *See* Trigonox® ADC, 32293
Peroxydicarbonic acid, ditetradecyl ester, *See* Perkadox® 26-fl, 23304
peroxydisulfuric acid, diammonium salt, *See* Ammonium persulfate, 2249
peroxydisulfuric acid dipotassium salt, *See* Anthion, 2519, potassium persulfate, 24771
peroxydisulfuric acid, disodium salt, *See* sodium persulfate, 28809
peroxydol, 23412
Peroxyhexanoic acid, 2-ethyl-, 1,1,4,4-tetramethyltetramethylene ester, *See* USP® -245, 33166
Peroxyl, 23413
Perpentol, 23414
Perrindo®, 23415
Persadox, *See* Abcure S-40-25, 35, AcetOxyl 2.5 and 5, 305, Benzoyl peroxide, 3983
Persa-gel, *See* Abcure S-40-25, 35, AcetOxyl 2.5 and 5, 305, Benzoyl peroxide, 3983
Persellig T, 23416
Persian balsam, 23417
Persian berry carmine, 23418
Persian gum, *See* India gum, 15066
Persian red, *See* chrome red, 6209, Indian Red, 15068
Persian yellow, 23419
persia-perazol, *See* *p*-dichlorobenzene, 8386
Persiderm®, 23420
Persiderm® Black, 23421
Persio, *See* Archil, 2808
persionin, 23422
Persistol®, 23423
Persistol® E, 23424
persodine, 23425
Persoftal®, 23426
Persoftal® PE Special, 23427
Persolv, *See* urokinase, 33145
Persoz's reagent, 23428
Perspex, 23429
Perstoff, 23430
Perstorp Phenolic Moulding Compound, 23431
Perstorp Urea Moulding Compound, 23432
Persulfocyanogen yellow, *See* Canarin, 5083
Persulon, 23433
Persyst, 23434
Persyst, *See* demeton-S-methyl, 8055
Perthane, 23435, *See* diethyldiphenyldichloroethane, 8441
Perthrine, *See* permethrin, 23384, 23385, 23386
Pertinit, 23436
Pertite, 23437, *See* Lyddite, 18068, Reflorit, 26003
Pertscan-99m, 23438
Peru saltpeter, *See* Chili saltpeter, 6063
Peru silver, *See* Chinese silver, 6075
Peruol®, 23439
Peruscabin, 23440
Peruvian balsam, 23441
Peruvin, 23443
Peruvlan bark, 23442
Pervon, 23444
Pescola oil, 23445
Pest-B-Gon, 23446
Pestex, 23447
Pestilizer®, 23448
Pestilizer® B Series, 23449
Pestox III, *See* Schradan, 27731
PET, *See* A-C® 6, 168
PETA, *See* SR-444, 29384
Petalin, 23450
Petameth, 23451
Petcat R-9, 23452, *See* antimony trioxide, 2574
Peter-Kal, *See* K. Tab, 15642, Kaon-Cl, 15729, Kay Ciel, 15807, K-Contin, 15818, K-Lor, 16226, K-Lyte/C1, 16234, KM Potassium Chloride, 16242, Potassium chloride, 24748, 24749
Petiole, 23453
Petlon®, 23454
PETP, *See* pentaerythritol, 23138
Petra, 23455
Petra® 130, 23456, *See* Polyethylene terephthalate, 24426
Petra® 130FR, 23457, *See* Polyethylene terephthalate, 24426
Petra® 230, 23458, *See* Polyethylene terephthalate, 24426
Petra® 242, 23459, *See* Polyethylene terephthalate, 24426
Petrac® 165, 23460
Petrac® 215, 23461
Petrac® 270, 23462
Petrac® CP-11, 23463
Petrac® Eramide®, 23464
Petrac® GMS, 23465
Petrac® MG-20 NF, 23466

Petrac® PHTA, 23467, *See* hydrogenated tallow acid, 14598
Petrac® Slip-Eze, 23468
Petrac® Vyn-Eze®, 23469
Petrac® ZN-41, 23470
Petralol, *See* paraffin, liquid, 22744
Petralon, 23471
Petramin, 23472
Petrarch® A0700, A0701, *See* Aminoethylamino propyl trimethoxy silane, 2171
Petrarch® C3300, *See* chloropropyltrimethoxysilane, 6138
Petrasul, 23473
Petre, 23474
Petrex, 23475
Petrex 7-75T, 23476
Petrinex, 23477
Petro, *See* paraffin, liquid, 22744
Petro 11, 23478
Petro 22, 23479
Petro AG Special, 23480
Petro BAF, 23481
Petro P, 23482
Petro S, 23483
Petro ULF, 23484
Petro WP, 23485
Petroacid, 23486
petrobenzol, 23487
Petroclastite, 23488
Petrofibe 201, 210, 215, 235, 23489
Petrofracteur, 23490
Petrogils, 23491
petrohol, *See* isopropanol, 15405
Petroklastite, *See* Petroclastite, 23488
petrol, 23492
petrol orange y, *See* Sudan I, 29938
Petrol Yellow Wt Resinol, *See* Butter yellow, 4759
Petrolagar, 23493
Petrolane, 23494
petrolatum, 23495, *See* Amerchol® 400, 1996, Amerchol® C, 1998, Forlan, 12247, Mineral Jelly No. 5, 19823, Petrofibe 201, 210, 215, 235, 23489
petrolatum amber, *See* petrolatum, 23495
Petrolatum USP, *See* Protopet, 25297
petrolatum white, *See* petrolatum, 23495
Petrolatum, lanolin, sodium PCA, and polysorbate 85, *See* Ritaderm®, 26806
Petrolatum-lanolin alcohol, *See* Forlan 200, 12248
Petrolax, *See* paraffin, liquid, 22744
petrolenes, 23496
Petroleum, *See* Kendex OCTG, 16002
Petroleum benzin, *See* Kensol 80, 16044, Texsolve V, 31532
petroleum benzine, *See* C-petroleum naphtha, 6940
Petroleum distillates, *See* Jayflex® 215, 15545
petroleum ether, 23497
Petroleum hydrocarbons, *See* petrolatum, 23495
Petroleum hydrocarbon-extract, *See* Kendex 0866, 16000
petroleum jelly, *See* petrolatum, 23495
petroleum naphtha, 23498, *See* Kensol 10, 16036, Kensol 13, 16037, naphtha, 20705
petroleum pitch, 23499
petroleum spirit, 23500
petroleum sulfonate, sodium salt, *See* Actrabase PS-470, 517
Petroleum sulfonates, *See* Petromor, 23508, Petrostep, 23515
petroleum wax, 23501, *See* paraffin, 22742, 22743
petroleum wax, crystalline, *See* paraffin, 22743
Petroleum, Liquid, *See* Klearol, 16208
Petrolia, *See* paraffin, liquid, 22744
Petrolig, 23502
petroline, 23503
Petrolit, 23504
Petrolite® C-400, 23505
Petrolite® C-7500, 23506
Petroll, *See* Aerosol® OS, 895, Rhodacal® IN, 26585
petrolum benzin, *See* Kensol 10, 16036, Kensol 13, 16037
petrolum naphtha, *See* Kensol 80, 16044
Petromix 9, 23507
Petromor, 23508
Petronate® L, HL, K, CR and S, 23509
Petronate® RP, 23510
Petronauba® C, 23511
Petrone A4 and A6C, 23512
Petronol, *See* paraffin, liquid, 22744
Petropul, 23513
Petro-Rez 801, 23514
Petrosio, *See* paraffin, liquid, 22744
Petrostep, 23515
Petrostep A-70, 23516
Petrosul® H-50, 23517
Petrosul® M-50, M-60, M-70, 23518
Petrosulpho, *See* Perichthol, 23270
Petrothene®, 23519
Petrothene® GA 501, 23520
Petrothene® GA 564, 23521
Petrothene® GA 808-090, 23522
Petrothene® HD 5903B, 23523
Petrothene® LB 5003-00, 23524
Petrothene® LB 6001-00, 23525
Petrothene® LF 6030-00, 23526
Petrothene® LP 5102-00, 23527
Petrothene® LS 3150-00, 23528
Petrothene® LT 5704-00, 23529
Petrothene® NA 155-000, 23530
Petrothene® NA 204-000, 23531
Petrothene® NA 341-000, 23532
Petrothene® PA 436, 23533
Petrothene® PP 1510-HC, 23534
Petrothene® PP 2004-MR, 23535
Petrothene® PP 7300-KF, 23536
Petrothene® PP 8000-GK, 23537
Petrothene® PP 8770-HU, 23538
Petrothene® XL, 23539
Petrowet® R, 23540
peucedanin, 23541
Pevafix, 23542
Pevalon, 23543
Pevikon, 23544
pewter, 23545
Pexalyn, 23546
Pexate, 23547
Pexite, 23548
Pexol® 245, 23549
Pexol® 50, Dark Fluid, 23550
Peyton powder, 23551
PF-10GF/15T, 23552, *See* Polysulfone, 24619
PF-20GF, 23553, *See* Polysulfone, 24619
PF-38, *See* Fyrol® FR-2, 12540
Pfeilringspalter, 23554
Pferrico, 23555
Pferrisperse, 23556
Pferritan, 23557
Pferrocal, 23558
Pferromet, 23559
Pferrox, 23560
Pfico$_2$-Hop, 23561
Pfico$_2$-Isohop, 23562
Pfico$_2$-Redihop, 23563
Pfiklor, 23564
PfiKlor, *See* K. Tab, 15642, Kaon-Cl, 15729, Kay Ciel, 15807, K-Contin, 15818, K-Lor, 16226, K-Lyte/C1, 16234, KM Potassium Chloride, 16242, Potassium chloride, 24748, 24749
Pfinodal, 23565
Pfizer citroflex A-4, *See* acetyl tributyl citrate, 313
P-Flakes, 23566
p-fuchsin, *See* pararosaniline, 22798
PG, *See* Sustane® PG, 30436
PG 12, *See* propylene glycol, 25215
PGA, *See* folic acid, 12196
PGA p, *See* vitamin M, 33951
PGME, *See* Arcosolv® PM, 2814, Dowanol® PM, 8896, Icinol PM, 14816, propylene glycol monomethyl ether, 25220, Solvenon® PM, Solvent PM, 29050
PGMEA, *See* Arcosolv® PMA, 2815, Dowanol® PMA, 8897
PH-60-40, *See* Diflubenzuron, 8450
Phaltan, 23567, *See* Folpet, 12210
Phalton, *See* Folpet, 12210
Phamosan, 23568
Phanteine, 23569
Phantol, 23570
Phantom, 23571, *See* Abol, 146
Pharmagel, 23572
Pharmasorb, 23573
Pharmaton, 23574
Pharmatone, 23575
Pharmgel, *See* gelatin, 12722
Pharmolin, 23576
Phase Alpha®, 23577
Phase II®, 23578
phaseomannite, 23579
Phathalogen, 23580
PHC S, *See* Baygon®, 3790
Phemeride, *See* benzethonium chloride, 3963
Phemerol, *See* benzethonium chloride, 3963
Phemerol Chloride, 23581, *See* benzethonium chloride, 3963
Phemersol Chloride, *See* benzethonium chloride, 3963
Phemithyn, *See* benzethonium chloride, 3963
Phemox, 23582
Phenac, 23583
Phenachlor, *See* 2,4,6-trichlorophenol, 32222
phenacyl chloride, *See* chloroacetophenone, 6112
phenador-x, *See* Biphenyl, 4216
phenald resins, 23584
Phenaldine, 23585
Phenalgin, *See* Ammonal, 2212
phenamiphos, *See* Nemacur®, 20863
1-Phenanthrenemethanol, dodecahydro-1,4a-dimethyl-7-(1-methylethyl)-, [1R-(1α,4aβ,4bα,10aα)]-, *See* Abitol® E, 71
1-Phenanthrenemethanol, dodecahydro-1,4a-dimethyl-7-(1-methylethyl)-, *See* Abitol® E, 71
Phenarsazine Chloride, *See* Adamsite, 624
Phenasal, *See* Mansonil®, 18611
Phenasal ethanolamine salt, *See* clonitrilide, 6432
phenazacillin, *See* Versapen, 33609
phenazone, *See* antipyrine,
phenedin, *See* phenacetin,
phenegic, *See* phenothiazine, 23618
Phenegol, 23586
Phenester, 23587
phenethyl mustard oil, 23588
phenethylene, *See* styrol, 29875
2-Phenethyltrichlorosilane, *See* CP0110, 6929
Phenetidine, 23589
***p*-phenetidine**, *See* Phenetidine, 23589
***p*-phenetocarbamide**, *See* Dulcin, 9091
phenetole, 23590
***p*-phenetylurea**, *See* Dulcin, 9091
Phenex, 23591
phenic acid, 23592
phenic alcohol, *See* phenic acid, 23592
Phenistix, 23593
Phenitol, 23594
phenmedipham, 23595, 23596, 23597, 23598, *See* betanal E, 4061, Goliath, 13188, Gusto, 13504, Headland Dephend, 13728, Protrum K, 25299, Vangard, 33323
Phenmerzyl Nitrate, 23599
Phenobarbital, *See* Talpheno, 30755
Phenobarbitone sodium, 23600, *See* Luminal Sodium, 17803
phenoctide, *See* octaphen, 21989
Phenodip, 23601
Phenodur, 23602, 23603
Pheno-epoxy novolac, *See* Quatrex 2410, 25619
Phenoform, *See* Bakelite, 3589
Phenol, *See* Carbolic Acid, 5206, phenic acid, 23592, pyrocatechol, 25498, pyrogallol, 25513
Phenol Carbonate, 23604
Phenol ethoxylate (1 EO), *See* Marlophen® P 1, 18888

Phenol formaldehyde, *See* Corephen® 10, 6810
Phenol formaldehyde resins, *See* Aerolite, 862
Phenol polyglycol ether phosphate ester, *See* Marlowet® 5324, 18926
phenol red, 23605
phenol red, P.S.P., *See* phenolsulfonphthalein, 23611
Phenol sulfonate, *See* Stainaway L2B, 29427
Phenol sulfonic acid, *See* Eltesol® PSA 65, 9868, Phenosulfonic acid, 23617, PSA, 25370
phenol trinitrate, *See* reflorit, 26003
Phenol, 2,4,5-trichloro-, O-ester with O,O-dimethyl phosphorothioate, *See* ronnel, 26951
phenol, 2-methyl-5-(1-methylethyl)-, *See* oxycymol, 22445
Phenol, 4,4'-(1-methylethylidene)bis-, *See* Bisphenol A, 4283
Phenol, 4,4'-(3H-2,1-benzoxathiol-3-ylidene)bis[3-methyl-, S,S-dioxide, *See* cresol purple, 6995
Phenol, 4,4'-thiobis 2-(1,1-dimethylethyl) phosphite, *See* Hostanox® OSP 1, 14374
Phenol, 2-(2H-benzotriazol-2-yl)-4-(1,1,3,3-tetramethylbutyl)-, *See* Uvazol 311, 33183
Phenol, 2,4-bis(1,1-dimethylethyl)-, phosphite (3:1), *See* Hostanox® PAR 24, 14375, Naugard® 524, 20791
Phenol, 2,6-bis(1,1-dimethylethyl)-4-nonyl-, *See* Uvi-Nox 1494, 33189
Phenol, 2-(5-chloro-2H-benzotriazol-2-yl)-6-(1,1-dimethylethyl)-4-methyl-, *See* Uvazol 236, 33181
Phenol, 2,6-di-tert-butyl-4-nonyl-, *See* Uvi-Nox 1494, 33189
Phenol, 2-(1,1-dimethylethyl)-4-methyl-, *See* Cao, 5098
Phenol, isopropylated, phosphate (3:1), *See* Fyrquel® EHC, 12544, triaryl phosphate, 32194
Phenol, 2-(1-methylethoxy)-,methylcarbamate, *See* Baygon®, 3790
Phenol, nonyl-, phosphite (3:1), *See* Doverphos®4, 8834, Wytox® 312, 34578
Phenol, 2,4,5-trichloro-, sodium salt, *See* Dowicide B, 8933
Phenol/formaldehyde resins, *See* Aerophen, 867, 866
Phenolene, Phenolene Supra, 23606
Phenol-formaldehyde, *See* phenolic resin, 23607
phenol-formaldehyde (novolak) resin, *See* novolac resin, 21731
phenolic resin, 23607, *See* Fiberite 6070, 11732
Phenolic/bisphenolic, *See* Naugawhite, 20800
Phenoline, 23608
Phenolite, 23609, 23610
phenolsulfonphthalein, 23611, *See* phenol red, 23605
phenoltetraiodothalein sodium, *See* TIP, 31847
Phenonip, 23612
Phenopreg, 23613
phenopyridine, *See* 8-hydroxyquinoline, 14670
Phenoquin, *See* cinchophen, 6307
Phenoro, 23614
phenosalyl, 23615
phenosan, *See* phenothiazine, 23618
Phenosept, 23616
Phenosulfonic acid, 23617
phenothiazine, 23618, *See* PTZ® Phenothiazine Purified, 25378
10H-Phenothiazine, *See* phenothiazine, 23618
phenothrin, 23619
Phenovarm, *See* phenothiazine, 23618
phenoverm, *See* phenothiazine, 23618
phenovis, *See* phenothiazine, 23618
Phenoweld, 23620
Phenox, 23621
Phenoxethol, *See*
Phenoxethol, *See* Igepal® Cephene Distilled, 14850, 2-Phenoxyethanol, 23627, Rewopal® MPG 10, 26344
Phenoxetol, 23622, *See* Igepal® Cephene Distilled, 14850, 2-Phenoxyethanol, 23627, Rewopal® MPG 10, 26344
phenoxin, 23623
phenoxur, *See* phenothiazine, 23618
phenoxy resin, 23624
phenoxyacetic acid, 23626
phenoxybenzene, *See* diphenyl oxide, 8598, phenyl ether, 23633
phenoxybenzyl chrysanthemate, *See* phenothrin, 23619
phenoxybenzyl (1R)-*cis/trans* chrysanthemate, *See* phenothrin, 23619
phenoxybenzyl-(±)-*cis,trans*-chrysanthemate, *See* phenothrin, 23619
***m*-phenoxybenzyl(±)-*cis,trans*-3-(2,2-dichlorovinyl)-2,2-dimethylcyclopropanecarboxylate:,** *See* Kafil, 15655
phenoxybenzyl-D-Z/E-chrysanthemate, *See* phenothrin, 23619
3-phenoxybutyric acid, *See* MCPB, 19185
Phenoxydiglycol, *See* Igepal® OD-410, 14873
phenoxyethanoic acid, *See* phenoxyacetic acid, 23626
Phenoxyethanol, *See* Emeressence® 1160, Rose Ether, 10079, Igepal® Cephene Distilled, 14850, Phenoxetol, 23622, 2-Phenoxyethanol, 23627, Prox-onic PH-01, 25334
2-Phenoxyethanol, 23627, *See* Igepal® Cephene Distilled, 14850, Rewopal® MPG 10, 26344
Phenoxyethanol and methyl, ethyl, propyl, butyl esters of *p*-hydroxybenzoic acid, *See* Phenonip, 23612
β-phenoxyethyl alcohol, *See* Igepal® Cephene Distilled, 14850
Phenoxyethyl acrylate, *See* Ageflex PEA, 997, Laromer® POEA, 16791
2-Phenoxyethyl acrylate, *See* Melcril 4087, 19268, SR-339, 29379
Phenoxyethyl alcohol, *See* Igepal® Cephene Distilled, 14850 Rewopal® MPG 10, 26344
β-phenoxyethyl alcohol, *See* Rewopal® MPG 10, 26344
Phenoxyethyl-acrylate, *See* Melcril 4087, 19268, SR-339, 29379
Phenoxyl ethanol, *See* Rewopal® MPG 10, 26344
Phenoxyl ethanol, *See* Igepal® Cephene Distilled, 14850
Phenoxylene 50, 23628
Phenoxylene Plus, 23629, *See* MCPA, 19183
3-(phenoxyphenyl)methyl (±)-*cis,trans*-3-(2,2-dichloro ethenyl)-2,2-dimethylcyclopropanecarboxylate, *See* Kafil, 15655
1-Phenoxy-2-propanol, 23625
1-phenoxypropan-2-ol, *See* 1-Phenoxy-2-propanol, 23625
Phenoxythrin, *See* phenothrin, 23619
phenoxytol, *See* Igepal® Cephene Distilled, 14850, Rewopal® MPG 10, 26344
Phenprocoumon, 23630, *See* Liquamar, 17450
phentetiothalien sodium, *See* TIP, 31847
phenthiazine, *See* phenothiazine, 23618
Phenyform, 23631
phenyl β-naphthylamine, *See* P.B.N, 22488
phenyl acrolein, *See* cinnamaldehyde, 6311
DL-phenylalanine, *See* DLPA 375, 8748
2-phenyl alkylbenzene sulfonic acid, *See* Ufacid KA, 32737
phenyl allyl cinnamate, *See* styracin, 29860
3-phenyl allyl cinnamate, *See* styracin, 29860
phenyl benzene, *See* biphenyl, 4216
phenyl carboxylic acid, *See* Benzoic acid, 3973
Phenyl cellosolve, *See* Igepal® Cephene Distilled, 14850, 2-Phenoxyethanol, 23627, Rewopal® MPG 10, 26344
phenyl chloride, *See* Abluton T30, 139, chlorobenzene, 6119
phenyl cyanide, *See* Benzonitrile, 3976
phenyl diisodecyl phosphite, *See* Doverphos® 7, 8836, Weston® PDDP, 34315
phenyl dimethicone, 23632, *See* Belsil PDM 20, 3919
phenyl ether, 23633, *See* diphenyl oxide, 8598
phenyl ethyl alcohol, *See* 2-phenylethyl alcohol, 23642
β-phenyl ethyl alcohol, *See* 2-phenylethyl alcohol, 23642
phenyl ethyl ether, *See* phenetole, 23590
phenyl ethyl isopropyl ether, *See* Petiole, 23453
N-phenyl ethylcarbamate, *See* carbanilic ether, 5179
phenyl fluoride, *See* fluorobenzene, 12093
phenyl glycol, *See* styrolyl alcohol, 29878
phenyl glycol ether, *See* LICA 97, 17200
2-phenyl imidazole, 23643
phenyl iodide, *See* iodobenzene, 15210
Phenyl isopropyl carbamate, *See* Propham, 25186
Phenyl mercaptan, *See* thiophenol, 31708
phenyl mercuric Fixtan, *See* hydrargaphen, 14539
Phenyl mercury acetate powder, *See* Acticide PMA 100, 470
phenyl methane, *See* methyl benzene, 19514, toluene, 31946
phenyl methyl acetate, *See* Plastolin I, 24051
phenyl methyl ketone, *See* acetophenone, 300
Phenyl neopentylene glycol phosphite, *See* Weston® PNPG, 34316
phenyl perchloryl, *See* Julin's chloride, 15632
2-phenyl phenol, *See* *o*-phenylphenol, 23648
***o*-phenyl phenol**, *See* Dowicide 1, 8926, Nipacide® OPP, 21258, Preventol® O Extra, 24931
phenyl phosphate, *See* triphenyl phosphate, 32391
phenyl phosphonate $((PhO)_2HPO)$, *See* Weston® DPP, 34309
phenyl salicylate, 23634
phenyl styryl ketone, *See* Chalkone, 5849
phenyl tolyl ketone, *See* 2-methylbenzophenone, 19565
phenyl 2-tolyl ketone, *See* 2-methylbenzophenone, 19565
phenyl trimethicone, 23635, *See* Abil® AV 8853, 20-1000, 47, Emalex MTS-30E, 9971
phenyl urethan, *See* Phenylurethane, 23652
phenyl-α-naphthylamine, *See* Akrochem® Antioxidant PANA, 1161, Antioxidant PAN, 2590, Neozone A, 20982, Neozone D, 20985
phenyl-β-naphthylamine, *See* Antioxidant PBN, 2591, Antioxygene MC, 2603
phenyl-γ acid, 23636
phenyl-1-naphthylamine, *See* Akrochem® Antioxidant PANA, 1161
Phenyl-3,5-dimethyl-4-(methylthio)-, methylcarbamate, *See* methiocarb, 19484
phenylacetic acid, 23637, *See* α-toluic acid, 31952
3-α-phenyl-β-acetylethyl-4-hydroxycoumarin, *See* Killgerm® Sewarin P, 16184, Kypfarin, 16519
β-phenylacrylic acid, *See* cinnamic acid, 6313
***trans*-3-phenylacryloyl propenoyl chloride,** *See* cinnamoyl chloride, 6314
phenylacrylyl chloride, *See* cinnamoyl chloride, 6314
phenylallyl alcohol, *See* cinnamyl alcohol, 6315
γ-phenylallyl alcohol, *See* cinnamyl alcohol, 6315
phenylamine, 23638
2-phenylaminonaphthalene, *See* Antioxidant PBN, 2591
8-(phenylamino)-1-naphthalenesulfonic acid, *See* phenyl-peri acid, 23647
2-Phenylamino 8-naphthol-6-sulfonic acid, *See* phenyl-γ acid, 23636
4-(Phenylazo)-1,3-benzenediamine hydrochloride, *See* Chrysoidine, 6256
2-Phenylazo-4-methoxy-2,4-dimethylvaleronitrile, *See* V-19, 33214
1-phenylazo-2-naphthol, *See* Sudan I, 29938
1-[4-(phenylazo)phenylazo]-2-naphthol, *See* Sudan III, 29940
4-phenylazo-*m*-phenylenediamine hydrochloride, *See* Chrysoidine, 6256
phenylbenzene, *See* diphenyl, 8592
α-phenylbenzeneacetonitrile, *See* diphenyl acetonitrile, 8593, diphenylacetonitrile, 8600
α-Phenylbenzenemethanol, *See* Benzhydrol, 3965
phenylbenzimidazol-5-sulfonic acid, *See* Neo Heliopan® Hydro, 20868
2-phenyl-1H-benzimidazole-5-sulfonic acid, *See* Neo Heliopan® Hydro, 20868
2-Phenylbenzimidazole-5-sulfonic acid, *See* Neo Heliopan® Hydro, 20868
2-phenylbenzopyran-4-one, *See* flavone, 11903

Chemical and Trade Names

Phosal, 23687, 23688
Phosal 25 SB, 23689
Phosal 53 MCT, 23690
phosalone, 23691
Phosbrite®, 23692
Phosbrite® 172, 23693
Phos-chek, 23694
Phosclene, 23695
Phosclere, 23696
Phos-copper, 23697
Phosfetal 201 K, 23698
Phosfetal 205, 23699
Phosfetal 603, 23700
Phosflex 112, *See* Disflamoll® DPK, TPK, 8641
Phosflex 179-C, *See* Plastic X, 24026
Phosflex® 362, *See* diphenyl octyl phosphate, 8597
Phosflex® CDP, *See* diphenylcresyl phosphate, 8601
phos-flur point two, *See* sodium fluoride, 28764
phosgene, 23701
phosphacol, *See* paraoxon, 22783
Phosphalijel, *See* aluminum phosphate, 1922
Phosphalugel, *See* aluminum phosphate, 1922
Phosphalutab, *See* aluminum phosphate, 1922
phospham, 23702
Phosphamide, *See* Dimethoate, 8496
phosphammite, 23703
Phosphanol series, 23704
Phosphate acid, *See* Wayfos A, 34247
Phosphate ester, *See* Actrafos 104, 109, 526, Actrafos SA-216, 529, Actrafos SN-315, 530, Actrafos SP-407, 531, Arodet TA-8, 3012, Beycostat 231, 4100, Beycostat B 070 A, 4105, Beycostat B 706 E, 4107, Hartomer GP 2164, 13681, Phosphanol series, 23704, Phospholan® PTP7, 23720, Tanapon NF-200, 30784
Phosphate ester sodium salt, *See* Abluphat LP Series, 116
Phosphate ester, sodium salt, free acid, *See* Abluphat AP Series, 115
Phosphate ester, potassium salt, *See* Acrilev AM, AM-Special, 388
Phosphate esters, *See* Pestilizer®, 23448, Rewophat, 26351
Phosphate of soda, *See* disodium phosphate (anhydrous), 8652, tasteless salts, 30843
Phosphated alcohol, *See* Strodex® MOK-70, 29820
Phosphated aliphatic ethoxylate acid anhydride, *See* Strodex® SE-100, 29823
Phosphated nonylphenoxy polyethoxy ethanol, *See* Esi-Terge 320, 10821
Phosphatidylcholine, *See* Phosal 25 SB, 23689, Phosal 53 MCT, 23690, Phospholipon® 90/90G, 23726, Kelecin®, 15835
Phosphatidyl-N-trimethylethanolamine, *See* Phosal 25 SB, 23689, Phosal 53 MCT, 23690
Phosphazote, 23705
Phosphite chelator, *See* Synpron 241, 30610
PhosPho 642, 23706
PhosPho E-100, 23707
PhosPho F-97, 23708
PhosPho H-00, 23709
PhosPho H-150, 23710
PhosPho LCN-TS, 23711
PhosPho PL-50, 23712
PhosPho S-85, 23713
PhosPho T-20, 23714
phospho-12-molybdic acid, *See* phosphomolybdic acid, 23731
Phospho-gélose, 23715
Phosphola® ALF5, 23716
Phospholan®, 23717
Phospholan® KPE4, 23718
Phospholan® PHB 14, 23719
Phospholan® PTP7, 23720
phospholeum, *See* polyphosphoric acid, 24520
Phospholipid EFA, 23721
Phospholipid PTC, 23722
Phospholipid PTD, 23723
Phospholipid PTS, 23724
Phospholipid SV, 23725
Phospholipids, *See* PhosPho PL-50, 23712, PhosPho S-85, 23713, PhosPho T-20, 23714
Phospholipon® 90/90G, 23726
Phospholipon® CC, 23727
Phospholipon® MC, 23728
Phospholipon® PC, 23729
Phospholipon® SC, 23730
Phospholutein, *See* lecithin, 16939
N-phosphomethylglycine, *See* glyphosate, 13148
phosphomolybdic acid, 23731
Phosphomort, 23732
phosphonic acid, *See* phosphorus acid, 23741
phosphonic acid, dibutyl ester, *See* Syn-O-Ad® P-316, 30562
phosphonic acid, didodecyl ester, *See* Weston® DLP, 34306
phosphonic acid, diphenyl ester, *See* diphenyl phosphite, 8599, Doverphos® 213, 8843, Doverphos® DPP, 8849, Weston® DPP, 34309
phosphonodithioimidocarbonic acid cyclic propylene P,P-diethyl ester, *See* mephosfolan, 19326
(phosphonomethyl)glycine, isopropylamine salt, *See* Rodeo, 26899
N-(phosphonomethyl)glycine, *See* glyphosate, 13148
phosphonomethyliminoacetic acid, *See* glyphosate, 13148
Phosphonorm, *See* Aloxicoll, 1813, aluminum chlorohydrate, 1903, Reach® 101, 201, 501, 25919
phosphor bronze, 23733
phosphor copper, 23734
phosphor resin, 23735
phosphor steel, 23736
phosphor tin, 23737
phosphoric acid, 23738, 23739, Marphos, 18951, Solklean™ 101, 28950
phosphoric acid (1,1-dimethylethyl)phenyl diphenyl ester, *See* Syn-O-Ad® 8478, 30555
phosphoric acid (1-methylethyl)phenyl diphenyl ester, *See* Syn-O-Ad® 8480, 30557
phosphoric acid 2-chloro-1-(2,4-dichlorophenyl)ethenyl diethyl ester, *See* Chlorfenvinphos, 6098
phosphoric acid bis(2-chloroethyl) 3-chloro-4-methyl-2-oxo-2H-1-benzopyran-7-yl ester, *See* Haloxon, 13567
phosphoric acid 2,2-dichloroethenyl dimethyl ester, *See* Vapona, 33378
phosphoric acid 2,2-dichlorovinyl dimethyl ester, *See* Vapona, 33378
phosphoric acid diethyl 4-nitrophenyl ester, *See* paraoxon, 22783
phosphoric acid (1,1-dimethylethyl)phenyl diphenyl ester, *See* Syn-O-Ad® 8485, 30559
phosphoric acid ester, *See* Etingal® A, 11148, Etingal® S, 11150
phosphoric acid isodecyl diphenyl ester, *See* Syn-O-Ad® 8479, 30556
phosphoric acid methylphenyl diphenyl ester, *See* diphenylcresyl phosphate, 8601, Disflamoll® DpK, TpK, 8641
phosphoric acid tributyl ester, *See* Kronitex® TBp, 16434, tributyl phosphate, 32207
phosphoric acid tri-n-butyl ester, *See* Kronitex® TBp, 16434, tributyl phosphate, 32207
o-phosphoric acid triethyl ester, *See* triethyl phosphate, 32243
phosphoric acid triphenyl ester, *See* Kronitex® Tpp, 16437, triphenyl phosphate, 32391
phosphoric acid tris (methylphenyl) ester, *See* Syn-O-Ad® 8484, 30558, TCp, 30857
phosphoric acid tris(2-methylphenyl) ester, *See* plastic X, 24026
phosphoric acid, aluminum sodium salt, *See* pan-O-Lite, 22683, Stabil-9, 29399
phosphoric acid, calcium salt (2:1), *See* calcium phosphate (monobasic), 4966
phosphoric acid, diammonium salt, *See* Ammonium phosphate dibasic, 2250
phosphoric acid, tri-o-cresyl ether, *See* plastic X, 24026
o-phosphoric acid, triethyl ester, *See* triethyl phosphate, 32243
phosphoric acid, tris(2-chloroethyl)ester, *See* Genomoll p, 12838
phosphoric acid, trisodium salt, *See* trisodium phosphate, anhydrous, 32412
phosphoric acid, trisodium, 12-hydrate, *See* trisodium phosphate, dodecahydrate, 32413
phosphoric anhydride, *See* phosphorus pentoxide, 23745, 23746
phosphoric chloride, *See* phosphorus pentachloride, 23744
phosphoric D, *See* Iron D, 15311
phosphoric ether, 23740, *See* sulfuric ether, 30056
phosphoric oxide, *See* phosphorus pentoxide, 23745, 23746
phosphoric perchloride, *See* phosphorus pentachloride, 23744
phosphorochloridothioic acid O,O-dimethyl ester, *See* Mp-2, 20388
phosphorodithioic acid, O,O'-diethyl ester, *See* Ep-1, 10611
phosphorodithioic acid, S-[(5-methoxy-2-oxo-1,3,4-thiadiazol-3(2H)-yl)methyl O,O-dimethyl ester, *See* methidathion, 19483
phosphorofluoridic acid, disodium salt, *See* Albaphos Dental Na 211, 1348
phosphorothioic acid O-(2,5-dichloro-4-iodophenyl) O,O-dimethyl ester, *See* Iodofenphos, 15213
phosphorothioic acid O,O-diethyl O-(3,5,6-trichloro-2-pyridinyl) ester, *See* Talon, 30751
phosphorothioic acid, O,O-diethyl O-(3,5,6-trichloro-2-pyridyl) ester, *See* Chlorpyrifos, 6163
phosphorothioic acid O-[2-(dimethylamino)-6-methyl--4-pyrimidinyl] O,O-diethyl ester, *See* pyrimithate, 25485
phosphorothioic acid, O,O-dimethyl O-(3,5,6-trichloro-2-pyridinyl) ester, *See* Chlorpyrifos-methyl, 6164
phosphorothioic acid, O,O'-(thiodi-*p*-phenylene) O,O,O',O'-tetramethyl ester, *See* Abate 1-SG, 2-CG, 4-E, 5CG, 26
phosphorotrithioic acid S,S,S-tributyl ester, *See* Def®, 7876
phosphorous acid, diphenyl ester, *See* Weston® Dpp, 34309
phosphorous acid, tributyl ester, *See* Syn-O-Ad® p-312, 30561
phosphorous acid triethyl ester, *See* phosphorus ether, 23742, triethyl phosphite, 32244
phosphorous acid tri-isooctyl ester, *See* triisooctyl phosphite, 32312
phosphorous acid, triphenyl ester, *See* Doverphos® 10, 8838, Syn-O-Ad® p-399, 30564, triphenyl phosphite, 32393
phosphorous acid, tris(2-ethylhexyl) ester, *See* Syn-O-Ad® p-374, 30563
phosphorous acid, tris(nonylphenyl) ester, *See* Wytox® 312, 34578
phosphorous chloride, *See* phosphorus trichloride, 23747
phosphorthioic acid O,O-dimethyl O-(2,4,5-trichlorophenyl)ester, *See* Korlan, 16351
phosphorus, *See* violet phosphorus, 33876
phosphorus acid, 23741
phosphorus chloride, *See* phosphorus oxychloride, 23743, phosphorus trichloride, 23747
phosphorus(III) chloride, *See* phosphorus trichloride, 23747
phosphorus derivative, *See* tributyl phosphite, 32208
phosphorus ether, 23742
phosphorus imidonitride, *See* phospham, 23702
phosphorus oxychloride, 23743
phosphorus pentachloride, 23744
phosphorus pentoxide, 23745, 23746
phosphorus perchloride, *See* phosphorus pentachloride, 23744
phosphorus salt, *See* microcosmic salt, 19665

Chemical and Trade Names

phosphorus trichloride, 23747
phosphorus triphenyl, *See* triphenyl phosphine, 32392
phosphorus(V) oxide, *See* phosphorus pentoxide, 23746
phosphoryl chloride, *See* phosphorus oxychloride, 23743
phosphoteric® T-C6, 23748
phosphotex, *See* sodium pyrophosphate, 28814
phosphothion, *See* malathion, 18460
phosphotungstic acid, 23749, 23750, 23751
phosphotungstic acid hydrate, *See* phosphotungstic acid, 23750, 23751
phosphowolframic acid, *See* phosphotungstic acid, 23750, 23751
Phosteem, 23752
Phostin, 23753
Phostoxin, 23754, *See* Talunex, 30757
Phosvit, *See* dichlorvos, 8394
Photal, 23755
Photine, 23756
Photine R, *See* Photine, 23756
Photobiline, *See* keraphen, 16055
Photocure 51, 23757
Photoglaze, 23758
Photol, *See* Metol, 19584
Photomer™4013, *See* Ageflex T4EGDA, 970
Photomer®, 23759
Photomer® 3005, 23760
Photomer® 5029, 23761
Photomer® 6360, 23762
Photophor, *See* Photophor®, 23763
Photophor®, 23763
Photo-Rex, *See* Metol, 19584
Photosensitive coupler, *See* Coupler, 6906
photoxylin, *See* cellodin, 5605
Photozinc, 23764
Phoxim, *See* Sebacil®, 27796, Volathion®, 34011
phoxime, *See* Volathion®, 34011
phoxim-methyl, 23765
PhPh, *See* biphenyl, 4216
Phrilon, *See* Cabelec® 1015, 4861
phtahlic acid, butyl ester, ester with butyl glycolate, *See* Reomol 4PG, 26126
Phtalofix® FN, 23766
Phtalogen®, 23767
Phtalogen® K, N1, 23768
Phtalotrop® B, 23769
phthalic acid, *See* diethyl phthalate, 8437
phthalic acid dibutyl ester, *See* dibutyl phthalate, 8366, Kodaflex® DBP, 16264, Unimoll® DB, 32978
phthalic acid di-n-butyl ester, *See* Kodaflex® DBP, 16264
phthalic acid diethyl ester, *See* Kodaflex® DEP, 16266
phthalic acid bis(2-ethylhexyl)ester, *See* Kodaflex® DOP, 16273
phthalic acid diisobutyl ester, *See* Kodaflex® DIBP, 16267
phthalic acid bis(2-methoxyethyl) ester, *See* Reomol P, 26132
phthalic acid dimethyl ester, *See* dimethyl phthalate, 8508, Kemester® DMP, 15956, Kodaflex® DMP, 16271, Unimoll® DM, 32979
phthalic acid dinonyl ester, *See* Ceneg, 5671
phthalic acid dioctyl ester, *See* Bis(2-ethylhexy) Phthalate, 4228, Reomol DCP, 26130
phthalic acid methyl ester, *See* dimethyl phthalate, 8508
Phthalic acid, bis(6-methylheptyl) ester, *See* Witcizer 313, 34455
phthalic acid, diethyl ester, *See* diethyl phthalate, 8437
phthalic acid, bis(2-ethylhexyl) ester, *See* Witcizer 312, 34454
Phthalic acid, diisooctyl ester, *See* Witcizer 313, 34455
phthalic anhydride, 23770, *See* Retarder AK, 26253
Phthalofyne, *See* Whipcide,
Phthalol, *See* diethyl phthalate, 8437
Phthalopal®, 23771
phthalthrin, *See* Killgerm® Py-Kill W, 16182, tetramethrin, 31384
phyban, *See* Neoarsycodyl, 20871
Phyban H.C, *See* Neoarsycodyl, 20871
Phycon 15, 18, 25, 23772
Phycon LPH, 23773
***p*-hydroxyacetanilide**, *See* Valadol, 33238
***p*-hydroxybenzoic acid ester**, *See* Solbrol®, 28902
***p*-hydroxy phenylmorpholine**, *See* Solux, 29037
Phygon, *See* Dichlone, 8381
Phygon Paste, *See* Dichlone, 8381
Phygon XL, *See* Dichlone, 8381
Phylatol, 23774
Phylletten, *See* Dequalinium Chloride, 8077
phylloquinone, *See* Mono-Kay, 20228, vitamin K_1, 33949
phylloquinone, *See*
Phyomone, 23775, *See* Rhizopon B, 26574, Tipoff, 31848
physic nut oil, *See* purging nut oil, 25408
physiological salt solution, 23776
Physostigma, *See* ordeal bean, 22282
Phytat D.B, *See* Rencal, 26105
Phytate sodium, *See* Rencal, 26105
phytic acid, 23777
phytinic acid, *See* phytic acid, 23777
Phyt'iod, 23778
Phytoforol, 23779
Phytogermine, *See* tocopherol, 31918, Vascuals, 33465
phytol, 23780
phytomelin, *See* rutin trihydrate, 27150
Phytomenadione, *See* Konakion, 16322, vitamin K_1, 33949
phytonadione, *See* Konakion, 16322, Mono-Kay, 20228,
Phytonomic oil, *See* Damoil, 7725
Phytox, *See* Zineb, 34807
Phytox MZ, *See* mancozeb, 18517
3-phytylmenadione, *See* Konakion, 16322, vitamin K_1, 33949
PI-20GF/000, 23781, *See* Polyetherimide, 24423
PIB, *See* polybutene, 24373
Pibiter, 23782
Picaltal®, 23783
Picamar, 23784
Picco® 5000, 6000, 23785
Piccodiene®, 23786
Piccolastic®, 23787
Piccolastic® D125, D150, 23788
Piccolyte C, 23789
Piccolyte®, 23790
Piccolyte® A, C, 23791
Piccolyte® HM110, 23792
Piccolyte® S, 23793
Piccomer® XX, 23794
Picconol®, 23795
Picconol® A200, 23796
Picconol® A300, 23797
Picconol® A400, 23798
Picconol® A500 A600, 23799
Piccopale®, 23800
Piccopyn, 23801
Piccotac®, 23802
Piccotex®, 23803
Piccotoner, 23804
Piccoumaron, 23805
Piccoumarone Resins, 23806
Piccovar® AB, 23807
Piccovar® AP, 23808
Piccovar® L, 23809
Picfume, *See* chloropicrin, 6135
pichurim camphor, 23810
picked turkey gum, 23811
Picket, 23812, *See* permethrin, 23384, 23385, 23386
pickle alum, 23813
pickle green, 23814
PICl, *See* phosphorus trichloride, 23747
piclorame, *See* Tordon, 32056
2-picoline, *See* α-picoline, 23815
3-picoline, *See* β-picoline, 23816
4-picoline, *See* γ-picoline, 23817
α-picoline, 23815
β-picoline, 23816
γ-picoline, 23817
σ-Picoline, *See* γ-picoline, 23817
3-picolinic acid, *See* nicotinic acid,
4-picolinic acid, *See* isonicotinic acid, 15380
γ-picolinic acid, *See* isonicotinic acid, 15380
Pi-cone, 23818
picrasmin, 23819
picrate powder, 23820
picric acid, 23821, *See* Hager's reagent, 13535, Lyddite, 18068
picric acid-nitro-flavin, *See* echurin, 9578
picric acid-trinitrocresol, *See* Cressylite, 7002
picric powder, 23822
picro-aniline blue, 23823
picrocarmine, 23824
picrocrocin, 23825
picronitric acid, *See* picric acid, 23821, reflorit, 26003
picronigrosine, 23826
picrontric acid, *See* picric acid, 23821
picro-sulfuric acid, 23827
picrylmethylnitramine, *See* Tetryl, 31423
picrylnitromethylamine, *See* Tetryl, 31423
Pictet crystals, 23828
Pictet's fluid, 23829
Pictet's liquid, 23830
Pictol, *See* Metol, 19584
pictolin, 23831
Pidolic acid, *See* Ajidew A-100, 1142
Pidolidone®, 23832
Pielanase, 23833
Pierrot metal, 23834
Pif-Paf, 23835
pig iron, 23836
pig lead, 23837
Pigment red 101, *See* Ferroxide, 11694, Jeweller's rouge, 15611
Pigment White 1, *See* Kremser White, 16399
Pigmentar, 23838
pilasonite, 23839
Piliogrip Adhesive System for Styructural Bonding, 23840
Piliophen, *See* keraphen, 16055
Pillargon, *See* Propoxur, 25198, 25199, 25200
Pillarich, *See* Repulse, 26151
Pillarich Repulse Taloberg Tuffcide, *See* Chlorothalonil, 6150
Pillaron, *See* methamidophos, 19474, Tamaron®, 30760
Pilot, 23841, 23842, *See* quizalofop-ethyl, 25697
Pilot ABS-99, *See* Witco® 1298, 34458
Pilot HD-90, *See* sodium dodecylbenzenesulfonate, 28760
pilot SP-60, *See* Nacconol® 40G, 20647
Pilot SXS-40, 23843, *See* sodium xylene sulfonate, 28832
Pimel®, 23844
pimelic ketone, *See* cyclohexanone, 7522
pimelite, 23845
pimple metal, 23846
pinachrom, 23847
Pinacolin, *See* pinacolone, 23848
pinacoline, *See* pinacolone, 23848
pinacolone, 23848
pinacyanol, 23849
pinaflavol, 23850
pinakol, 23851
pinakon, *See* hexylene glycol, 14041, 14042
Pinaverdol, 23852
pinchbeck, 23853
Pincoffin, 23854
pine gum, 23855
pine oil, 23856, *See* Arizole® Pine Oil, 2886
pine resin, *See* pine gum, 23855
Pine tar, *See* Stockholm, 29775
α-pinene, 23857
β-pinene, 23859
β-l-pinene, 23858
2(10)-Pinene, *See* β-l-pinene, 23858
2-pinene, *See* α-pinene, 23857

plastic metal, 24022
Plastic Plant Product, *See* plastic wood, 24025
Plastic Steel, 24023
plastic sulfur, 24024
plastic wood, 24025
Plastic X, 24026
Plasticalk, 24027
Plasticede, 24028
Plasticizer 9, 24031
Plasticizer 13, 24029
Plasticizer 28P, 24030
Plasticizer CEL, 24032
Plasticizer E, 24033
Plasticizer REO, 24034
Plasticote, 24035
Plastifix, 24036
Plastigel®, 24037
Plastigen® G, 24038
Plastikon, 24039
Plastilit® 3060, 24040
Plastisorb, 24041
Plastisperse®, 24042
Plastite, 24043
Plastitoy, 24044
Plastitube, 24045
Plastodent, 24046
Plastogen, 24047
Plastokyd, 24048
Plastokyd SC, 24049
Plastol®, 24050
Plastolin I, 24051
Plastomag® 170, 24052
Plastomenite, 24053
Plastomoll DOA, *See* dioctyl adipate, 8548, Plasthall® DOA, 24002, PX-238, 25440, Wickenol® 158, 34393
Plastomoll®, 24054
Plastomoll® 34, 24055
Plastomoll® BMB, 24056, *See* Butylbenzene sulfonamide, 4789
Plastomoll® DMA, 24057
Plastomoll® DOA, 24058, *See* dioctyl adipate, 8548
Plastomoll® NA, 24059
Plastomoll® TAH, 24060
Plastomoll® WH, 24061
Plastone, 24062
Plastone A, 24063
Plastone B, 24064
Plastopal®, 24065
Plastopal® 11, 24066
Plastoprene, 24067
plastoresin orange f4a, *See* Sudan I, 29938
Plastorit, 24068
Plastose, 24069
Plastosperse, 24070
Plastosperse 40, 24071
Plastpak, 24072
Plastplate, 24073
Plastrotyl, 24074
Plastules with Liver, 24075
Plastyrol, 24076
Plastyrol E6X, 24077
Plasvita, 24078
Plasvita® TSM, 24079
Plaswite® LL 7014, 24080
Plaswite® LL 7105, 24081
Plaswite® masterbatches, 24082
Plaswite® PS 7174, 24083
Platalargan, 24084
Platamid, 24085
Platamid® Series, 24086
Plataril, 24087
plate pewter, 24088
plate powder, 24089
plate sulfate, 24090
Plath-Lyse, *See* Dichlorophen, 8392, Nuophene, 21783
Platilon, *See* A-C® Polyethylene 6, 6A, 7, 7A, 8, 8A, 9, 9A, 617, 617A, 175, Cabelec® 1017, 4862
platina, 24091
platine, 24092
Platine-autitre, 24093
Platinite, 24094
platinized asbestos, 24095
Platino, 24096
platino-aceto-osmic acid, *See* Hermann's fluid, 13876
platinoid, 24097
Platinol AH, *See* Bis(2-ethylhexy) Phthalate, 4228, dioctyl phthalate, 8549, Reomol DCP, 26130
Platinol DOP, *See* Bis(2-ethylhexy) Phthalate, 4228, dioctyl phthalate, 8549, Reomol DCP, 26130
Platinor, 24098
Platinum, 24099
platinum black, 24100
platinum bronze, 24101
platinum gold, 24102
platinum grey, *See* zinc grey, 34777
platinum iridium, 24103
platinum silver, 24104
platinum solder, 24105
platinum substitute, 24106
platinum tetrachloride, *See* salt of Norton, 27286
platinum yellow, 24107
platinum, soft, *See* soft platinum, 28840
platnam, 24108
platnik, 24109
Platol II, 24110
Platone, 24111
Platy talc, *See* Silverline 200, 28438, Silverline 665, 28439
Plavolex, *See* Dextran, 8233
Pleocide, *See* Aminitrazole, 2164
Plessite, 24112
Plessy's Green, 24113
Plexar, 24114
Plexar®, 24115
Plexar® PX 108, 24116
Plexene D, *See* Cheelox® 80, 5864, Pentaquest Extra 0685, 23169, Trilon® C Liq, 32328
Plexiglas® DR, 24117
Plexiglas® HFI-10, 24118
Plexiglas® V045, 24119
Plexiglas® VS, 24120
Plexiglo, 24121
Plexigum, 24122
Pleximon, 24123
Plexisol, 24124
Plexol, 24125
Plexol 201, *See* Plasthall® DOS, 24005, Reomol DOS, 26131
Plexophor, 24126
Plextol, 24127
Plialite, 24128
Plictran, 24129, 24130, *See* cyhexatin, 7568
Plimmer and Paine's stain, 24131
plinthite, 24132
Pliobond Adhesives, 24133
Pliobond®, 24134
Plio-Caulk, 24135
Pliocord® LVP-4668, 24136
Plioflex, 24137
Plioflex® 1006, 24138
Plioflex® 1028, 24139
Plioflex® 1905, 24140
Plioform, 24141
Pliogrip®, 24142
Pliolite, 24143
Pliolite® 7103, 7104, 24144
Pliolite® AC-80, 24145
Pliolite® LPF-2108, 24146
Pliolite® S-6B, S-6F, 24147
Pliolite® VT, 24148
Pliolite® VTAC, 24149
Plio-Nail, 24150
Plio-Seam, 24151
Plio-Tac, 24152
Plio-Tac 38, 24153
Pliovic® DR-450, DR-453, DR-454, DR-600, DR-602, DR-652, 24154
Pliovic® M-50, M-70, M-70SC, M-90, 24155
Pliovic® WO-1, WO-2, WO-3, WO-S, 24156
Plioway® EC1, 24157
Plitex, 24158
Plombit, 24159
Plondrel, 24160
Plondrel, *See* Plondrel, 24160
plumbago, *See* graphite, 13277
plumbago grease, 24161
plumber's solder, 24162
plumber's white alloy, 24163
plumbocalcite, *See* tarnowitzite, 30826
plumbous oxide, *See* lead monoxide, 16910
plumboxan, 24164
Plumbral, 24165
Plumbrex, 24166
Plumbrit, 24167
plumose mica, 24168
Pluracol® 220, 24169
Pluracol® 355, 24170
Pluracol® 450, 24171
Pluracol® 581, 24172
Pluracol® E200, 24175
Pluracol® E300, 24177
Pluracol® E400, 24178
Pluracol® E400 NF, 24179
Pluracol® E1000, 24173
Pluracol® E1500, 24174
Pluracol® E2000, 24176
Pluracol® E4000, 24180
Pluracol® E4500, 24181
Pluracol® E6000, 24182
Pluracol® E600NF, 24183
Pluracol® E8000, 24184
Pluracol® V-10, 24185
Pluracol® W170, 24186
Pluracol® W660, 24190
Pluracol® W2000, 24187
Pluracol® W3520N, 24188
Pluracol® W5100N, 24189
Plurafac®, 24191
Plurafac® A-38, 24192
Plurafac® A-39, 24193
Plurafac® B-25-5, 24194
Plurafac® D-25, 24195
Plurafac® LF 120, 24196
Plurafac® LF 220, 24197
Plurafac® RA-20, 24198
Pluraflo® E4A E5G, N5G, 24199
Plurasafe, 24200
Plurexid, *See* chlorhexidine digluconate, 6101, Rotersept, 27002
Pluriol E 200, *See* Alkapol PEG 300, 1706
Pluriol®, 24201
Pluriol® E 200, 24202
Pluriol® E 4000, 24203
Pluriol® P 2000, 24204
Pluriol® P 600, 24205
Plurivite, 24206
Plurol Isostearique, 24207
Plurol Oleique, *See* Witconol 14, 34495
Plurol Oleique WL 1173, 24208
Plurol Stearique WL 1009, 24209
Pluronic® 10R5, 24210
Pluronic® 10R8, 24211
Pluronic® 17R1, 24212
Pluronic® 17R2, 24213
Pluronic® 17R4, 24214
Pluronic® 17R8, 24215
Pluronic® 25R1, 24216
Pluronic® 25R2, 24217
Pluronic® 25R4, 24218
Pluronic® 25R5, 24219
Pluronic® 25R8, 24220
Pluronic® 31R1, 24221
Pluronic® 31R2, 24222
Pluronic® 31R4, 24223
Pluronic® F38, 24226
Pluronic® F68, 24227
Pluronic® F68LF, 24228

Pluronic® F77, 24229
Pluronic® F87, 24230
Pluronic® F88, 24231
Pluronic® F98, 24232
Pluronic® F108, 24224
Pluronic® F127, 24225
Pluronic® L31, 24236
Pluronic® L35, 24237
Pluronic® L42, 24238
Pluronic® L43, 24239
Pluronic® L44, 24240
Pluronic® L61, 24241
Pluronic® L62, 24242
Pluronic® L62D, 24243
Pluronic® L62LF, 24244
Pluronic® L63, 24245
Pluronic® L64, 24246
Pluronic® L72, 24247
Pluronic® L81, 24248
Pluronic® L92, 24249
Pluronic® L101, 24233
Pluronic® L121, 24234
Pluronic® L122, 24235
Pluronic® P65, 24254
Pluronic® P75, 24255
Pluronic® P84, 24256
Pluronic® P85, 24257
Pluronic® P94, 24258
Pluronic® P103, 24250
Pluronic® P104, 24251
Pluronic® P105, 24252
Pluronic® P123, 24253
Pluronic® PE 3100, 24259
Plus, 24260
Plusbrite, 24261
Plus-Gas C, 24262
Plus-Pac®, 24263
Pluviusin, 24264
Plyamul®, 24265
Plyamul® 40305-00, 24266, *See* Vinyl Acetate, 33854
Plydax, *See* Terbufos, 31226
Plymouth IPP, *See* Wickenol® 111, 34379
Plymoutm IPM, *See* Wickenol® 101, 105, 34378
Plymul 98-759, 24267
Plyophen, 24268
Plyothene, 24269
Ply-Pro 25, 24270
Plyron, 24271
Plysolene, 24272
Plytron, 24273
PMA, *See* phenylmercury acetate, 23646, phosphomolybdic acid, 23731
PMA 18, 60, 24274
PMAC, *See* acetoxyphenylmercury, 306, phenylmercuric acetate, 23644, 23645, 23646
PMAS, 24275, *See* acetoxyphenylmercury, 306, phenylmercury acetate, 23646
PMDI, *See* Isonaphthol, 15377, MDI, 19188, Rubinate® LF-168, 27108
PMP, *See* Accelerator 2P, 187
PNA, *See* *p*-nitroaniline, 21345
Pneulec Core Gum, 24276
pneumatogen, 24277
Pneumax DM, *See* Perkacit® MBTS, 23286
p-Nitro-benzeneazo beta-naphthol, *See* paranitraniline red, 22780
p-nitrosodimethylaniline, *See* nitro base, 21341
PNP, *See* *p*-nitrophenol, 21368
Poast, *See* Sethoxydim, 28099
Poast®, 24278
Pocan®, 24279 , *See* Grilpet® XE3060, 13402
Pocan® B 1300, 24280
Pocan® B 1305, 24281
Pocan® B 1505, 24282
Pocan® KU1-7033, 24283
Pocan® S 1506, 24284
Podocarpa-8,11,13-trien-15-amine, 13-isopropyl-, *See* Amine D, 2155
Podophyllum, *See* vegetable calomel, 33515
POE, *See* PEG,
POE (1.5) nonyl phenol, *See* Triton® N-57, 32430
POE (3) lauryl dimethyl amine oxide, *See* PEG-3 lauramine oxide, 23047
POE (4) nonylphenol, *See* Triton® N-57, 32430
POE (4) sorbitan monolaurate, *See* polysorbate 21, 24566
POE (4) sorbitan monostearate, *See* polysorbate 61, 24569
POE (4) stearic acid (monoester), *See* Alkamuls® S-65-8, 1680
POE (4) tridecyl alcohol, *See* Trycol® 5874, 32498
POE (5) nonylphenol, *See* Triton® N-57, 32430
POE (5) sorbitan monooleate, *See* polysorbate 81, 24573
POE (6) isolauryl mercaptan, *See* Prox-onic TM-06, 25346
POE (6) nonylphenol, *See* Triton® N-57, 32430
POE (6) tridecyl alcohol, *See* Trycol® 5874, 32498
POE (8) nonylphenol, *See* Triton® N-57, 32430
POE (8) stearic acid (monoester), *See* Alkamuls® S-65-8, 1680
POE (8) tridecyl alcohol, *See* Trycol® 5874, 32498
POE (10) cetyl alcohol, *See* Akyporox RC 200, 1289
POE (10) nonylphenol, *See* Triton® N-57, 32430
POE (10) octylphenol, *See* Triton® X-15, 32433
POE (14) nonylphenol, *See* Triton® N-57, 32430
POE (20) castor oil (ether, ester), *See* Ricinion, 26732
POE (20) cetyl alcohol, *See* Akyporox RC 200, 1289
POE (20) nonylphenol, *See* Triton® N-57, 32430
POE (20) sorbitan monolaurate, *See* polysorbate 20, 24565
POE (20) sorbitan monooleate, *See* polysorbate 80, 24572, Radiasurf® 7157, 25804
POE (20) sorbitan monopalmitate, *See* polysorbate 40, 24567
POE (20) sorbitan monostearate, *See* Alkamuls® T-60, 1692, polysorbate 60, 24568
POE (20) sorbitan trioleate, *See* Alkamuls® T-85, 1694, polysorbate 85, 24574
POE (20) sorbitan tristearate, *See* polysorbate 65, 24570
POE (30) nonylphenol, *See* Triton® N-57, 32430
POE (40) stearic acid (monester), *See* Alkamuls® S-65-8, 1680
POE (50) stearic acid (monoester), *See* Alkamuls® S-65-8, 1680
POE alkyl ether, *See* Neoscoa 203C, 20964
POE alkyl ether phosphate, *See* Chemfac PD-600, 5954
POE alkyl phenol phosphate, *See* Chemfac NC-0910, 5950
POE ethers and special resins, *See* Adsee® 775, 794
POE fatty acid, *See* Emkatex 11, 21, 10268
POE glycol, *See* Witbreak DPG-15, 34434
POE monostearate, *See* Alkamuls® S-65-8, 1680
POE nonyl phenyl ether, *See* Nonipol 20, 21452
POE octylphenol, *See* Triton® X-15, 32433
POE rosin amine, *See* Witcamine® RAD 0500, 34452
POE sorbitan laurate, *See* Ablunol T-20, 114
POE sorbitan oleate, *See* Sorbon T-80, 29125
POE sorbitan palmitate, *See* Sorbon T-40, 29123
POE sorbitan stearate, *See* Sorbon T-60, 29124
POE sorbitol oleate, *See* Sorbon TR 814, 29126, Sorbon TR 843, 29127
POE stearoylether, *See* Lamecreme SA 7, 16610
POE(5.5) Cetyl Ether, *See* Akyporox RC 200, 1289
POE(7) Cetyl Ether, *See* Akyporox RC 200, 1289
POE(10) Cetyl Ether, *See* Akyporox RC 200, 1289
POE(15) Cetyl Ether, *See* Akyporox RC 200, 1289
POE(20) Cetyl Ether, *See* Akyporox RC 200, 1289
POE(23) Cetyl Ether, *See* Akyporox RC 200, 1289
POE(25) Cetyl Ether, *See* Akyporox RC 200, 1289
POE(30) Cetyl Ether, *See* Akyporox RC 200, 1289
POE(40) Cetyl Ether, *See* Akyporox RC 200, 1289
Poilite, 24285
poison flour, 24286
Poivrette, 24287
Pokalon, 24288
Polacaritox, *See* tetradifon, 31366
Polacrilin, *See* Amberlite® IRP-64, 1964
Polacure® 740M, 24289
Polamine® 250, 24290
Polamine® 650, 1000, 2000, 24291
Polaqua, 24292
Polar, 24293
Polar Dynobel, 24294
Polar Monobel No. 2, *See* Polar Dynobel, 24294
Polar Rex, *See* Polar Dynobel, 24294
Polar Saxonite, *See* Polar Dynobel, 24294
Polar Stomonal, *See* Polar Dynobel, 24294
Polar Super Clifite, *See* Polar Dynobel, 24294
Polar Thames Powder, *See* Polar Dynobel, 24294
Polar Viking, *See* Polar Dynobel, 24294
Polargel® HV, 24295
Polargel® NF, 24296
Polarin® Range, 24297
Polaris, 24298
Polarite®, 24299
Polarite® 420E(W), 24300
Polarite® 420G(W), 24301
Polarite® 880E(W), 24302
Polarwhite, 24303
Polathane, 24304
Polathane STE-73D, STE-83A, STE-90A, STE-95A, 24305
Polathane STS-55, 24306
Polathane XPE-10, XPE-20, XPE-30, 24307
Polawax®, 24308
Polawax® A31, 24309
Polectron 430, 24310
Poley oil, *See* oil of pennyroyal, 22066
Policapram, *See* Aviamide-6, 3467, nylon 6, 21909
Policydal, *See* Kelfizina, 15838
Polidene 528F, 24311
Polidene®, 24312
Polidene® 33-001, 24313
Polidene® 33-004, 24314
Polidene® 33-055, 24315
Polidocanol, *See* Akyporox RLM 160, 1290
Polifil® C-10, 24316
Polifil® CAS-40, 24317
Polifil® GFPP-10, 24318
Polifil® GFPPCC-10, 24319
Polifil® M-20, 24320
Polifil® RMC-10, 24321
Polifil® RMT-10, 24322
Polifil® T-10, 24323
Poligen, 24324
Poligen MMV, 24325
Poligen PE, 24326
Poligen WE 1, 24327
Polimex TR, 24328
Polisax, 24329
Polish turpentine, *See* turpentine, 32610
polishing oil, 24330
Polisin, *See* prometryn, 25140
Politarp, 24331
Politec, 24332
Politef, *See* polytetrafluoroethylene, 24640
Politint 1, 2, 24333
Politol®, 24334
Polival, *See* Tecto, Tecto 60%, 30913, thiabendazole, 31661
Pollack's cement, 24335
Pollopas, 24336
Polnac, 24337
Polnoks R, *See* Agerite® MA, 1019, Akrochem® Antioxidant DQ, 1160
Polomyx, 24338
Poloxalene, *See* Antarox® 17-R-2, 2504, Antarox® E-100, 2509, Industrol® N3, 15116, Therabloat, 31570
Poloxalkol, 24339
Poloxamer 101, *See* Macol® 46, 18252, Pluronic® L31, 24236
Poloxamer 105, *See* Hodag Nonionic 1035-L, 14187, Macol® 35, 18248, Pluronic® L35, 24237
Poloxamer 108, *See* Pluronic® F38, 24226, Pluronic®

F68LF, 24228, Pluronic® L62D, 24243, Pluronic® L62LF, 24244
Poloxamer 122, *See* Macol® 42, 18250, Pluronic® L42, 24238
Poloxamer 123, *See* Pluronic® L43, 24239
Poloxamer 124, *See* Hodag Nonionic 1044-L, 14188, Macol® 44, 18251, Pluronic® L44, 24240
Poloxamer 181, *See* Antarox® L-61, 2510, Macol® 1, 18236, Pluronic® L61, 24241
Poloxamer 182, *See* Macol® 2, 18237, Pluronic® L62, 24242
Poloxamer 183, *See* Pluronic® L63, 24245
Poloxamer 184, *See* Hodag Nonionic 1064-L, 14189, Macol® 4, 18238, Pluronic® L64, 24246
Poloxamer 185, *See* Pluronic® P65, 24254
Poloxamer 188, *See* Macol® 8, 18239, Pluronic® F68, 24227
Poloxamer 212, *See* Macol® 72, 18253, Pluronic® L72, 24247
Poloxamer 215, *See* Pluronic® P75, 24255
Poloxamer 217, *See* Macol® 77, 18254, Pluronic® F77, 24229
Poloxamer 231, *See* Pluronic® L81, 24248
Poloxamer 234, *See* Pluronic® P84, 24256
Poloxamer 235, *See* Macol® 85, 18255, Pluronic® P85, 24257
Poloxamer 237, *See* Antarox® PGP 23-7, 2512, Pluronic® F87, 24230
Poloxamer 238, *See* Hodag Nonionic 1088-F, 14190, Pluronic® F88, 24231
Poloxamer 282, *See* Pluronic® L92, 24249
Poloxamer 284, *See* Pluronic® P94, 24258
Poloxamer 288, *See* Pluronic® F98, 24232
Poloxamer 331, *See* Macol® 101, 18256, Pluronic® L101, 24233
Poloxamer 333, *See* Pluronic® P103, 24250
Poloxamer 334, *See* Pluronic® P104, 24251
Poloxamer 335, *See* Pluronic® P105, 24252
Poloxamer 338, *See* Macol® 108, 18257, Pluronic® F108, 24224
Poloxamer 401, *See* Antarox® E-100, 2509, Pluronic® L121, 24234
Poloxamer 402, *See* Pluronic® L122, 24235
Poloxamer 403, *See* Macol® 23, 18244, Pluronic® P123, 24253
Poloxamer 407, *See* Macol® 27, 18245, Pluronic® F127, 24225
Poloxamine 304, *See* Tetronic® 304, 31399
Poloxamine 504, *See* Tetronic® 504, 31400
Poloxamine 701, *See* Tetronic® 701, 31402
Poloxamine 702, *See* Tetronic® 702, 31403
Poloxamine 704, *See* Tetronic® 704, 31404
Poloxamine 707, *See* Tetronic® 707, 31405
Poloxamine 901, *See* Tetronic® 901, 31406
Poloxamine 904, *See* Tetronic® 904, 31407
Poloxamine 908, *See* Tetronic® 908, 31408
Poloxamine 1101, *See* Tetronic® 1101, 31409
Poloxamine 1102, *See* Tetronic® 1102, 31410
Poloxamine 1104, *See* Tetronic® 1104, 31411
Poloxamine 1107, *See* Tetronic® 1107, 31412
Poloxamine 1301, *See* Tetronic® 1301, 31413
Poloxamine 1302, *See* Tetronic® 1302, 31414
Poloxamine 1304, *See* Tetronic® 1304, 31415
Poloxamine 1307, *See* Tetronic® 1307, 31416
Poloxamine 1501, *See* Tetronic® 1501, 31417
Poloxyl Lanolin, 24340
Poly (2-hydroxypropyl-N,N-dimethyl ammonium chloride), *See* Agefloc A-50, 1005
Poly (dipropyleneglycol) phenyl phosphite, *See* Weston® DHOP, 34305
poly (ethylene glycol adipate), 24341
poly (oxy-1,4-butanediyl)- α-hydro-+lw-hydroxy), *See* polytetramethylene ether glycol, 24643
Poly (pentabromobenzyl) acrylate, 24342
Poly 4,4' isopropylidenediphenol neodol 25 alcohol phosphite, *See* Weston® 439, 34299
Poly bd® R-45HT, 24343
Poly Brand Dessicant, *See* arsenic acid, 3137
Poly C4M, 24344
Poly Check, 24345
Poly Ethylene Oxide, *See* Alkapol PEG 300, 1706
poly solv, *See* diglyme, 8455
Poly(α-methylstyrene-styrene acrylonitrile), *See* Blendex® 586, 4343
Poly(acrylic acid), *See* Acumer 1000, 575
Poly(acrylic acid), ammonium salt, *See* Ammonium polyacrylate, 2252
Poly(acrylonitrile-co-butadiene), *See* Krynac® 19.65, 16451, Krynac® 20H35, 16452, Krynac® 34.140, 16454, Krynac® 34.35, 34.50, 16455, Krynac® 34.80, 16457, Krynac® 823X2, 16458, Krynac® 843, 16460, Krynac® PXL 34.17, 16465
Poly(acrylonitrile-co-butadiene-co-styrene), *See* Claradex CH-540, 6354
Poly(caprolactam), *See* Capran®, Capran® 77C, Capran® Emblem, Capran®Unidraw®, 5124, Celstran® N6G30-01-4, N66G30-01-4, 5651, nylon 6, 21909
Poly(dibromostyrene), *See* Great Lakes PDBS-10, 13307, PDBS-80, 13308
Poly(dimethyl diallyl ammonium chloride), *See* Agefloc WT-20, 1009
Poly(dipropylene glycol)phenyl phosphite, *See* Doverphos® 12, 8841
Poly(ethylene glycol-400) distearate, *See* Kessco® PEG 200 DS, 16121, Kodapak® 5214A, 16285, Kodapak® PET Copolyester 13339, 16286, Kodar® PETG Copolyester 6763, 16289
poly(hexamethyleneadipamide), *See* nylon 6/6, 21910
poly(hexamethyleneadipamide) Polyamide, *See* nylon 6/6, 21910
poly(iminocabonylpentamethylene), *See* Cabelec® 1015, 4861, nylon 6, 21909
poly(iminocarbonylpentamethylene) Polyamide, *See* nylon 6, 21909
Poly(isobutylene-co-isoprene), *See* Kalar® 5214, 15662, Kalar® 5263, 15663, Kalene® 800, 15666
poly(isoprene), *cis*, *See* Natsyn® 2200, 20763
poly(laurolactam), *See* nylon 12, 21912
Poly(methylene)wax, *See* Synwax, 30659
poly(N-acetyl-1,4- β-D-glucopyranosamine), *See* chitin, 6082
poly(oxy-1,2-ethanediyl),, *See* Standapol® AP Blend, 29468
Poly(oxy-1,2-ethanediyl), α-sulfo-ω-hydroxy-, C_{10-16}-alkyl ethers, ammonium salts SDA 15-067-01, *See* Witcolate AE, Witcolate LES-60A, 34472
Poly(oxy-1,2-ethanediyl), α-(1-oxo-9-octadecenyl)- ω-hydroxy-, (Z)-, *See* Ablunol 200MO, 96
Poly(oxy-1,2-ethanediyl), α-(1-oxododecyl)- ω-hydroxy-, *See* Ablunol 200ML, 95, Ablunol DEGMS, 99
Poly(oxy-1,2-ethanediyl), α-(nonylphenyl)- ω-hydroxy-, *See* Ablunol NP4, 106
Poly(oxy-1,2-ethanediyl), α-(nonylphenyl)- ω-hydroxy-, branched, *See* Berol 09, 4014
Poly(oxy-1,2-ethanediyl), α,α'-[1,4-dimethyl-1,4-bis(2-methylpropyl)-2-butyne-1,4-diyl]bis[ω-hydroxy-, *See* Surfynol® 420, 30384, Surfynol® 440, 30385, Surfynol® 465, 30386, Surfynol® 485, 30387
Poly(oxy-1,2-ethanediyl), α-9-octadecenyl- ω-hydroxy-, (Z)-, *See* Volpo 3, 34020
Poly(oxy-1,2-ethanediyl), α-octadecyl- ω-hydroxy-, *See* Ablunol SA-7, 113
poly(oxy-1,2-ethanediyl), α-sulfo-ω-(dodecyloxy)-, sodium salt, *See* sodium laureth sulfate, 28778, Witcolate ES-2, Witcolate LES-60C, 34477
Poly(oxy-1,2-ethanediyl), α-9-octadecenyl- ω-hydroxy-, (Z)-, *See* Ablunol OA-6, 107
Poly(oxycarbonylmethylene), *See* polyglycolic acid, 24459
poly(oxyethylene) lauryl ether sulfate sodium salt, *See* sodium laureth sulfate, 28778, Standapol® AP Blend, 29468
Poly(oxyethylene)-*p*-tert-octylphenyl ether, *See* Triton® X-15, 32433
Poly(pentabromobenzyl) acrylate, *See* FR-1025, 12360
poly(podium acrylate), *See* sodium polyacrylate, 28811
Poly(styrene-co-butadiene), *See* KR01 K-Resin Polymer, 16382, KR04 K-Resin Polymer, 16383, Kraton®, 16386, Kraton® D 1101, 16387, Kraton® D 1116, 16389, Kraton® D 2103, 16391, K-Resin Polymer KR01, 16404, K-Resin Polymer KR03, 16405, K-Resin Polymer KR04, 16406, K-Resin Polymer KR05, 16407, K-Resin Polymer KR10, 16408, Krylene® 606, 16449, Krylene® 608, 16450
Poly(vinyl isobutyl ether), hexane, *See* Gantrez® B-773, 12653
Poly(vinylidene fluoride), *See* KF Polymer® C-1000, 16158, KF Polymer® T-1300, 16159, KF Polymer® T-850, 16160, KF Polymer® U-1000, 16161, KF Polymer® W-1000, 16162, Kynar® 301 F, 16509, Kynar® 460, 16510, Kynar® 700 Series, 16511, Kynar® Flex® 2800, 2801, 16513, Kynar® Flex® 2900, 16515
Poly(vinylidene fluoride-co-hexafluoropropylene), *See* Kynar® Flex® 2850, 16514
Poly(vinylpyrrolidone)Plasmosan, *See* Povidone, 24805
Poly/Bed 812, 24346
Poly/Sep 47, 24347
poly[1-(2-oxo-1-pyrrolidinyl)ethylene], *See* Agrimer 15L, 1065, Tears Plus®, 30876
poly[1-(2-oxo-1-pyrrolidinyl)ethylene]-iodine, *See* Videne Disinfectant Solution, Videne Disinfectant Tincture, Videne Powder, 33739, Vinisil, 33827
poly[imino (1,6-dioxo-1,6-hexanediyl) imino-1,6-hexanediyl, *See* nylon 6/6, 21910
poly[imino(1-oxo-1,6-hexanediyl)], *See* Cabelec® 1015, 4861, nylon 6, 21909
poly[methylenedi(hydroxymethyl)urea], *See* Salmocid, 27277
Poly[oxy(methyl-1,2-ethanediyl)], α-hexadecyl- ω-hydroxy-, *See* Wickenol® 707, 34404
Poly-α-methylstyene, *See* Resin 18, 26174
Poly-α-olefin, *See* Ethylflo 162, 11134, Ethylflo 180, 11135
Polyacetal, *See* Kemlex 10007 NAL1, 15980
polyacrylamide, 24348, *See* Diaclear® MK-166, 8253, Sepigel 305, 27984
Polyacrylamidomethylpropane sulfonic acid, *See* Cosmedia Polymer HSP 1180, 6869
polyacrylate, 24349, *See* Hartomer LD 31, 13682, Size CB, 28559
Polyacrylic, *See* Sokalan® PA 13 PN, 28877
Polyacrylic acid, *See* Acrysol® ASE-75, 444, Acusol® 445, 583, Alcogum L-15, L-26, L-28, L-31, L-35, L-36, 1472, Burcotreat 900-A, 4699, Carbopol® 613, 614, 5230, Sokalan® PA 110 S, 28876
polyacrylic acid, potassium salt, *See* potassium polyacrylate, 24775
polyacrylic acid, sodium salt, *See* sodium polyacrylate, 28811
Polyacrylonitrile-co-butadiene, *See* Cycolac® KJM, 7552
Polyadipate, *See* Ultramoll® M, 32843
Polyaldo® 2010 KFG, 24350
Polyaldo® 2P10 KFG, 24351
Polyaldo® 2S6 KFG, 24352
Polyaldo® DGDO KFG, 24353, *See* Polygliceryl-10 decaoleate, 24453
Polyaldo® HGDS KFG, 24354
Polyaldo® TGMS KFG, 24355, *See* Polyglyceryl stearate series, 24457
Polyalk, 24356
Polyalkoxylated aliphatic ether, *See* Macol® LF-110, 18268
Polyalkoxylated alkyphenol, *See* Witconol NS-108LQ, 34518
Polyalkylene glycol ester, *See* Pronal 502, 502A, 25154
Polyalkylene glycols, *See* Icinol, 14814, Poly-G Fluids, 24442
Polyalkylene oxide-modified polymethylsiloxane, *See* Silwet® L-77, 28450
polyamide, 24357
Polyamide (PA), *See* Celanese® Nylon 6/6, 5562

Polyamide 6, *See* Polyloy® 6, 24477, Swiss Polyamid Grilon, 30464
Polyamide 6/6, *See* Polyloy® A, 24478
Polyamide resin dispersion, *See* Micromid 1022, 19682
Polyamine, *See* Acumer C-3, 578
Polyanthrene KS Liq. New, 24358
Polyaromatic ethoxylate phosphate ester, *See* Agrilan® F513, 1058
Polyarylamide, *See* Ixef® 1022, 15461
Polyarylate, *See* Durel, 9190
Polyarylate resin, *See* Ardel® D-100, 2823
Polyaryletherketone, *See* Ultrapek® KR 4176, 32854
Polybar, *See* Radiopaque, 25830
Polybead, 24359
Polybilt, 24360
Polyblack, 24361
Polyblends, 24362
Polybond®, 24363
Polybond® 1000, 24364
Polybond® 1009, 24365
Polybond® 1011, 24366
Polybond® 2005, 24367
Polybond® 2021, 24368
Polybor®, 24369
Polybrominated salicylanilide, *See* Tuasol 100, 32543
Polybut, 24370
1,2-polybutadiene, 24371
polybutadiene, 24372 Budene® 1207, 4652, Budene® 1254, 4653, Diene® 70AC, 8428
***cis*-polybutadiene**, *See* polybutadiene, 24372
Polybutadiene diacrylate, *See* CN 300, 6477
Polybutadiene-acrylonitrile, *See* Ultrmoll NR, 32887
Polybutene, 24373, *See* Amoco® L-14, 2286
Polybutenes, *See* Hyvis, 14797, Indopol, 15088, Napvis, 20740
Polybutylene, *See* Duraflex® 8410, 9145, Polybutene, 24373
Polybutylene oxide polyol, *See* XAS 10961.01, 34594
Polybutylene terephthalate, *See* Grilpet® XE3060, 13402, Valox®, 33247, Vestodur, 33667
Polycaprolactam, *See* Cabelec® 1015, 4861
Polycarbafil®, 24374
Polycarbodiimide, *See* Stabaxol, 29395
polycarbonate resin, 24375
Polycarboxylate AMC 60, 24376
Polycare® 509, 24377
Polycat®, 24378
Polycat® 5, 24380
Polycat® 12, 24379
Polycat® 58, 24381
Polycat® 77, 24382
Polycat® 77, 24383
Polycat® 91, 24384
Polycell, *See* carboxymethylcellulose, 5257, 5258
Polycell, *See* Cekol, 5558
Poly-Chek, 24385
polychloroprene, 24386, *See* Adcora P6, 631, Baypren® 110, 3828, Baypren® Latex KA 8348, 3834, Baypren® Latex L 200A, 3835, Plastifix, 24036
Polychloroprene latex, *See* Daubond DC-9300, 7794, Neoprene Latex 115, 20952
Polychlorotrifluoroethylene, *See* Fluorolube® GR-290, GR-362, GR-470, GR-544, GR-660, 12104
Polychol 5, 24387
polychrome blue of unna, 24388
Polycidine, *See* Dequalinium Chloride, 8077
Polycillin, *See* Ampicillin, 2361
Polycin 12, 24389
Polycizer 162, *See* Polycizer DOP, 24390
Polycizer 332, *See* Polycizer DOA, 24391, dioctyl adipate, 8548
polycizer 532, *See* Good-rite® GP-265, 13205
polycizer 562, *See* Good-rite® GP-265, 13205
Polycizer 962-BPA, *See* PX-126, 25437
Polycizer DBP, *See* dibutyl phthalate, 8366
Polycizer DOA, 24391
Polycizer DOP, 24390
Polyclar® 10, 24392
Polyclear, 24393
Polyclear 32-F, 24394
Polycomp® 139, 24395
Polycomp® 185, 24396
Polycon, 24397
Polycon II, 24398
Polycon S-60 K, 24399
Polycon S-80 K, 24400
Polycon T-60 K, 24401, *See* Ethylan® GES6, 11108
Polycon T-80 K, 24402
Polycote Pedigree, 24403
Polycote Prime, 24404
Polycoupler IMP RFB X-353, 24405
polycroit, 24406
Polycrol, 24407
Polycron, 24408
Polycryl, 24409
polycrylic acid, sodium salt, *See* sodium polyacrylate, 28811
Polycure, 24410
Polycycline, *See* tetracycline, 31365
Polydecene, *See* Synton PAO-100, 30650
Polydiallyldimethylammonium chloride, *See* Agefloc WT-40, 1010
Poly-dibromophenylene oxide, *See* Great Lakes PO-64P, 13314
Polydiethoxysiloxane, *See* CPS 9120, 6952, Dynasil® 40, 9381
poly(1,2-dihydro-2,2,4-trimethylquinoline), *See* Akrochem® Antioxidant DQ, 1160
Polydimethyl diallyl ammonium chloride, *See* Agefloc WT-40, 1010, Praestol® 186K, 24852
Polydimethyldiphenyl siloxane, *See* SF1154, 28127
Polydimethylsiloxane, *See* AF 10 FG, 907, SF96®, 28124
polydimethylsiloxane rubber, *See* B-182, 3569, silicone elastomer, 28337
Polydiol 200, *See* Alkapol PEG 300, 1706
Polydioxanone, *See* PDS, 22955
Polydis® TR 121, 24411
Polydis® TR 131, 24412
Polydur, 24413
polydymite, 24414
Polydyol 30-G, 24415
Polyeite, 24416
Poly-Em, 24417
Poly-epsilon-caprolactam, *See* nylon 6, 21909
Polyester 1606, 24418
Polyester acrylate, *See* Photomer® 5029, 23761
Polyester carbonate, *See* Apec® KL 1-9306, 2645, Apec® KU 1-9309, 2646
Polyester N-95, 24419
Polyester plasticizer, *See* Drapex® P-1, 8971
Polyester polyols, *See* Estolan, 10895, Lupraphen, 17908
Polyester resin, *See* Fiberite 9002, 11735
Polyester urethane, *See* VPC®, 34053
Polyestren®, 24420
Poly-Eth, 24421
Poly-Eth Hi-D, 24422
Polyether elastomer, *See* Zeospan 303, 306, 34724
Polyether glycol, *See* polyethylene glycol, 24425, polytetramethylene ether glycol, 24643
Polyether polyol, *See* Blendur® KU 3-4513, 4348, Multronol, 20494, Pluracol® 220, 24169, Voranol 202, 225, 34044, Voranol 234-630, 34045
Polyether polyol (diol), *See* Poly-G® 20-28, 24445, Poly-G® 20-56, 24446
Polyether polyol (triol), *See* Poly-G® 30-56, 24447
Polyether polyol (VHP diol), *See* Poly-G® 55-28, 24448
Polyether polyol, water-soluble, *See* Adeka PR-3007, 669
Polyether polyols, *See* Lupranol®, 17907, Voranol, 34043
Polyether-etherketone, *See* Arlon®, 2919
Polyetherimide, 24423, *See* Electrafil® J-1106/CF/30, 9743, PI-20GF/000, 23781
Polyether-polyester polyol, *See* Baygal® K115, 3783
Polyethersulfone, *See* Ultrason® E 1010, 32863, Ultrason® E 2010, 32865, Ultrason® E 3010, 32866, Ultrason® E 6010, 32867
Polyethoxy alkyl phenol, *See* Hyonic GL 400, 14691
Polyethoxy ether, *See* Emulgeant 710, 10502
Polyethoxylate, *See* Emkalon TN, 10235
Polyethoxylated lanolin, *See* Solan E, 28889
Polyethoxylated sorbitan esters, *See* Crillet, 7036
polyethylene, 24424, *See* Arcel Moldable Polyethylene Copolymers, 2807, Astroturf, 3264, Cabelec® 1017, 4862, Ecothene EC 101, 9591, Forbest 410, 12236, Fortiflex®, 12299, Fortiflex® A60-70-99, A60-70-119, 12300, Fortiflex® B45-06R-09, 12301, Fortiflex® G36-24-149, 12302, Fortiflex® G38-70C, 12303, Fortiflex® J36-25-142, 12304, Fortiflex® K36-55-122, 12305, Fortiflex® T50-200, 12306, Fortiflex® XF-855, 12307, Fortiflex® XF-855, 12308, Fusabond® MB-110D, 12509, Fusabond® MZ-109D, 12511, Fusabond® MZ-203D, 12512, HDPE 04352N, 13720, HDPE 25053-P, 13721, HDPE 32060C, 13722, HDPE 35053, 13723, HDPE IP-10, 13724, HiD 9301, 14066, HiD 9602, 14067, HiD 9632, 14068, HiD 9650, 14069, Hi-Zex, 14149, Hostalen® G, 14352, Hostalen® GB 6950, 14353, Hostalen® GM 5010 T2, 14354, Hostalen® GM 7745 HP, 14355, Innovex, 15138, Micropoly 520, 19690, Microthene® MA 530-060, 19706, Microthene® MP 625U, 19708, Microthene® MU 760-00, 19709, Nortuff RA 7020-KO, 21589, Polysoft CA, 24551, Tenite® 154DF, 31105
Polyethylene emulsion nonionic, *See* Adalin, 617
polyethylene glycol, 24425, *See* Akrochem® PEG 3350, 1169, Atpet 300, 3383, Atpet 400, 3384, Atpet 600, 3385, Carbowax® Sentry, 5252, Chemstat® P-400, 6042, Ilexan HT, 14884
Polyethylene glycol 1000 diacrylate, *See* Ageflex T4EGDA, 970
Polyethylene glycol 1000 monocetyl ether, *See* Cetomacrogol 1000, 5808
Polyethylene glycol 400, *See* Alkapol PEG 300, 1706
polyethylene glycol 450 nonyl phenyl ether, *See* Triton® N-57, 32430
Polyethylene Glycol 8000, *See* Alkapol PEG 300, 1706
Polyethylene glycol ethers of Cetyl and Stearyl alcohols, *See* Carsofoam® 1618, 5327
Polyethylene glycol mono[4-(1,1,3,3-tetramethylbutyl)phenyl] ether, *See* Macol® OP-3, 18271, Triton® X-15, 32433
Polyethylene glycol monooctadecyl ether, *See* Ablunol SA-7, 113
Polyethylene glycol monostearate, *See* Alkamuls® S-65-8, 1680
Polyethylene glycol sulfate monododecyl ether, sodium salt, *See* Witcolate ES-2, Witcolate LES-60C, 34477
polyethylene glycol, mono(hydrogen sulfate), dodecyl ether, sodium salt, *See* sodium laureth sulfate, 28778, Standapol® AP Blend, 29468
Polyethylene Gylcol, *See* Alkapol PEG 300, 1706
Polyethylene homopolymer, *See* A-C® 6, 168, Polywax® 500, 24671
Polyethylene terephthalate, 24426, *See* Hostaphan, 14380, Hostaphane, 14381, Terylene, 31348,
polyethylene wax, 24427, *See* A-C® Polyethylene 6, 6A, 7, 7A, 8, 8A, 9, 9A, 617, 617A, 175, Epolene® C-10, 10678, Hoechst Wax PE 190, 14265, Hoechst Wax PE 520, 14266, Hoechst Wax PED 521, 14268, Loobwax 0651, 17597
polyethylene, high-density, 24428
polyethylene, linear low density, 24429
polyethylene, low-density, 24430
Polyethyleneglycol 1500, *See* Alkapol PEG 300, 1706
Polyethyleneglycol 4000, *See* Alkapol PEG 300, 1706
Polyethyleneglycol 300 monodecyl ether, *See* Marlipal® 1012/4, 18865
Polyethylenepolyamine, *See* Texlin® 500, 31509
Polyfeed, 24431
Polyfil® WC, 24432
Polyfilm, 24433
Polyfine MF15C, 24434
Polyflex, 24435
Polyflex®, 24436

Chemical and Trade Names

Polyflo, 24437
Polyflo, *See* bioMeT TBTF, 4171
Polyflon, 24438
Polyfon, 24439
Polyfon® F, 24440
Polyfusor Solutions, 24441
Poly-G Fluids, 24442
Poly-G Polyols, 24443
Poly-G® 200, 24444
Poly-G® 20-28, 24445
Poly-G® 20-56, 24446
Poly-G® 30-56, 24447
Poly-G® 55-28, 24448
polygalin, *See* struthiin, 29840
polygallic acid, *See* struthiin, 29840
Polygard, 24449
polygeline, 24450
Polygeline™, 24451
polyglactin, 24452, *See* Vicryl, 33725
Polygliceryl-10 decaoleate, 24453
Polygloss, 24454
Polyglucadyne, 24455
Polyglucin, *See* Dextran, 8233
Polyglycerol esters, *See* Crester KZ, 7004
Polyglycerol monooleate, *See* Witconol 14, 34495
Polyglycerol oleate, *See* Witconol 14, 34495
Polyglycerol polyricinoleate, *See* Radiamuls® Poly 2253, 25782
Polyglyceryl isostearostearate, 24456
Polyglyceryl monostearate, *See* Radiamuls® Poly 2248, 25781
Polyglyceryl oleate, *See* Witconol 14, 34495
Polyglyceryl stearate series, 24457
Polyglyceryl-2 distearate, *See* Emalex DSG-2, 9945
Polyglyceryl-2 isostearate, *See* Apifac, 2659
Polyglyceryl-2 sesquiisostearate, *See* Hoe S 2721, 14257
Polyglyceryl-2 sesquioleate, *See* Hostacerin DGO, 14321
Polyglyceryl-2 tetrastearate, *See* Caprol® 2G4S, 5132
Polyglyceryl-2-PEG-4 stearate, *See* Hostacerin DGS, 14322
Polyglyceryl-3 diisostearate, *See* Emalex DISG-3, 9944, Emerest® 2452, 10100, Lameform TGI, 16614
Polyglyceryl-3 monooleate, *See* Drewpol® 3-1-O, 9015
Polyglyceryl-3 oleate, *See* Caprol® 3GO, 5133, Grindtek PGE 25, 13413, Isolan® GO 33, 15369, Santone® 3-1-SH, 27415
Polyglyceryl-3 stearate, *See* Caprol® 3GS, 5134, Polyaldo® TGMS KFG, 24355, Santone® 3-1-S, 27414
Polyglyceryl-4 isostearate, *See* Isolan® GI 34, 15368
Polyglyceryl-4 isostearate, cetyl dimethicone copolyol, hexyl laurate, *See* Abil® WE 09, 65
Polyglyceryl-4 oleate, *See* Witconol 14, 34495
Polyglyceryl-6 dioleate, *See* Caprol® 6G20, 5136, Plurol Oleique WL 1173, 24208
Polyglyceryl-6 distearate, *See* Caprol® 6G2S, 5135, Plurol Stearique WL 1009, 24209, Polyaldo® 2S6 KFG, 24352, Polyaldo® HGDS KFG, 24354
Polyglyceryl-6 isostearate, *See* Plurol Isostearique, 24207
Polyglyceryl-10 decaoleate, *See* Caprol® 10G100, 5131, Polyaldo® DGDO KFG, 24353
Polyglyceryl-10 decastearate, *See* Caprol® 10G10S, 5128
Polyglyceryl-10 dioleate, *See* Caprol® 10G2O, 5129, Polyaldo® 2010 KFG, 24350
Polyglyceryl-10 dipalmitate, *See* Polyaldo® 2P10 KFG, 24351
Polyglyceryl-10 tetraoleate, *See* Caprol® 10G40, 5130, Drewpol® 10-4-O, 9014
polyglycol, *See* polyethylene glycol, 24425
Polyglycol 1000, *See* Alkapol PEG 300, 1706
Polyglycol B-11-50, 24458
polyglycolic acid, 24459, *See* Dexon, 8230
Polyglycols, *See* Pluriol®, 24201
Polyglyconate, *See* Maxon, 19086
Polygon, 24460
Polygrade, 24461
Polyguide®, 24462
polyhalite, 24463
Polyhall® 21J, 24464
Polyhipe, 24465
Polyhydric alcohol, *See* glycerin, 13071, Maltisorb, 18499
polyhydrite, 24466
Polyimide, *See* Thermid® EL-5010, EL-5512, 31589
Polyimide resin, *See* Fiberite 944, 11730
polyimide, thermoplastic, 24467
polyimide, thermoset, 24468
polyisobutene, 24469, *See* Permethyl 104A, 23389
Polyisobutylene, *See* Oppanol® B, 22222, polyisobutene, 24469, Tree Bug-Lok Adhesive, 32159
Polyisobutylenes, *See* Glissoviscal® B, 12991
Polyisocyanates, *See* Reatane, 25934
polyisoprene, 24470, 24471, *See* ENR 25, 10585
***trans*-polyisoprene**, *See* Natsyn® 2200, 20763
***cis*-1,4-polyisoprene rubber**, *See* polyisoprene, 24470
Polyketone, *See* Kadel® E-1000, 15653
Polylite, 24472, 24473
Polylite®, 24474
Polylite® 32-162, 24475
polylithionite, 24476
Polyloy® 6, 24477
Polyloy® A, 24478
Polylube 1105, 24479
Polylube ASTL, 24480
Polylube DDL, 24481
Polylube Wax, 24482
polymannuronic acid, *See* Kelacid®, 15824, alginic acid, 1593
Polymate, 24483
Polymekon, 24484
Polymekon®, 24485
Polymekon®, *See* Microcrystalline wax, 19666
Polymel #7, 24486
Polymene AZ, 24487
Polymer, *See* Appeel®, 2665
Polymer C, 24488
Polymer of 4-(2-acryloyloxyethoxy)-2-hydroxy benzophenone, *See* Cyasorb® UV 2126, 7500
Polymeric, *See* Ricobond 1031, 1731, 1756, 26735
Polymeric 2,2,4-trimethyl-1,2-dihydroquinoline, *See* Lowinox® ACP, 17662
Polymeric amine, *See* Clearbreak TEB, 6389, Witbreak RTC-326, 34437
Polymeric carboxylic acid, sodium salt, *See* Agesperse 71, 1025, Agesperse 80, 1026
Polymeric MDI, *See* Papi 27, 22705, Papi 4901, 22706
Polymeric plasticizers, *See* Admex®, 750
Polymeric Sealant Gun, 24489
Polymerized β-hydroxy-trimethylene sulfide, *See* oxytri, 22474
Polymerized 1,2-dihydro-2,2,4-trimethylquinoline, *See* Agerite® Resin D®, 1020, Flectol H, 11911, Flectol Pastilles, 11913
Polymerized 2,2,4-trimethyl-1,2-dihydroquinoline, *See* Permanax TQ, 23367, Struktol® T.M.Q, 29838
polymerized oils, *See* blown oils, 4354
Polymet®, 24490
Polymethacrylamidopropyl trimonium chloride, *See* Lexquat® 2240, 17174
Polymethyl methacrylate, *See* Acry-Ace, 402, Edimet, 9619, Polymethylmethacrylate, 24492
Polymethyl propyl 3-oxy-[4(2,2,6,6-tetramethyl) piperidinyl] siloxane, *See* Uvasil 299, 33180
Polymethyl propyl 3-oxy-[4(2,2,6,6-tetramethyl) piperidinyl] siloxane - SiO_2, *See* Uvasil 125, 33179
Polymethyl-3,3,3-trifluoropropyl siloxane, *See* PS181, 25366
Polymethylene polyaniline, *See* Curithane 103, 7423
Polymethylene polyphenyl isocyanate, *See* Desmodur® VKS-2, VKS-4, VKS-18, 8160
Polymethylhydrosiloxane, *See* CPS 120, 6947, Masil® SF-MH, 19022
polymethylmethacrylate, 24491, 24492, *See* Lucryl®, 17771, Polymethylmethacrylate, 24492
Polymethyloctadecylsiloxane, *See* CPS 130, 6948
Polymethyloctylsiloxane, *See* CPS 140, 6949
Polymica 200, 325, 400, 24493
polymignite, 24494
Polymin®, 24495
Polymine D, *See* benzethonium chloride, 3963
Polymines, *See* Laromid®, 16793
Polymist® F-5, 24496, *See* Polytetrafluoroethylene, 24642
Polymoist Mask, 24497
Polymon, 24498
Polymone, 24499
Polymul, 24500
polynoxylin, *See* Salmocid, 27277
polynoxyline, *See* Salmocid, 27277
Polynuclear aromatic hydrocarbon, *See* Viplex 895-BL, 33888
Polyoctadecylmethylsiloxane, *See* PS130, 25364
Polyoctenamer, *See* Vestenamer® 6213, 33662
Polyoctylene, *See* Vestenamer®, 33661
Polyoctylmethylsiloxane, *See* PS140, 25365
Polyoil Hüls 110, 24501
Polyol ester, *See* Glycolube® 100, 13106
Polyolefin, *See* Lockite, 17538, Microthene® FA 150-00, 19705
Polyolefins, *See* Polybond®, 24363, Vitraplas, 33988
Polyox, 24502
Polyox WSR-301, *See* Alkapol PEG 300, 1706
Polyox® WSR 3333, 24503
Polyox® WSR N-10, 24504
Polyoxalkylene glycols, *See* Hartopol LF-1, 13687
Polyoxy alkylated polyalkylene glycols, *See* Supronic B10, B25, B50, B75 and B100, 30325
Polyoxyalkylene alcohol phosphate, *See* Polylube ASTL, 24480
Polyoxyalkylene fatty amine, *See* Ethox 1372, 11012
Polyoxyalkylene glycol, *See* Hartopol 25R2, 13685, Hartopol P65, 13688, Witbreak DTG-62, 34436
Polyoxyalkylene glycol ether, *See* Ethox LF-1226, 11029
Polyoxyalkylene glycol polyol, *See* Pluracol® V-10, 24185
Polyoxyethylated fatty alcohol (nonionic), *See* Peregal® O, 23235
Polyoxyethylated glyceryl isostearate, *See* Tegotens I, 31036
Polyoxyethylated octyl phenol, *See* Triton® X-15, 32433
Polyoxyethylene (1.5) Nonyl Phenol, *See* Alkasurf® NP-4, 1715
Polyoxyethylene (4) Nonylphenol, *See* Alkasurf® NP-4, 1715
polyoxyethylene (4) tridecyl alcohol, *See* Trycol® 5874, 32498
polyoxyethylene (6) tridecyl alcohol, *See* Trycol® 5874, 32498
polyoxyethylene (8) stearic acid (monoester), *See* Alkamuls® S-65-8, 1680
polyoxyethylene (8) tridecyl alcohol, *See* Trycol® 5874, 32498
Polyoxyethylene (9) Nonylphenyl Ether; nonyl phenol ethoxylate, *See* Triton® N-57, 32430
polyoxyethylene (10) cetyl alcohol, *See* Akyporox RC 200, 1289
polyoxyethylene (10) octylphenol, *See* Triton® X-15, 32433
polyoxyethylene (20) castor oil (ether, ester), *See* Ricinion, 26732
polyoxyethylene (20) cetyl alcohol, *See* Akyporox RC 200, 1289
polyoxyethylene (20) sorbitan monooleate, *See* Radiasurf® 7157, 25804
Polyoxyethylene (20) sorbitan monostearate, *See* Alkamuls® T-60, 1692
polyoxyethylene (20) sorbitan trioleate, *See* Alkamuls® T-85, 1694
polyoxyethylene (40) stearic acid (monester), *See* Alkamuls® S-65-8, 1680
polyoxyethylene (50) stearic acid (monoester), *See*

Alkamuls® S-65-8, 1680
Polyoxyethylene 1000, *See* Alkapol PEG 300, 1706
Polyoxyethylene alkylesters, *See* Myrj®, 20581
Polyoxyethylene castor oil, *See* Ablunol CO 10, 98
Polyoxyethylene ether, *See* Ethosperse®, 11004
Polyoxyethylene fatty acid ester, *See* Vanesta®, 33318
Polyoxyethylene glycerol fatty acid esters, *See* Tagat®, 30704
Polyoxyethylene Isocetyl Ether, *See* Emalex 1605, 9913
Polyoxyethylene nonyl phenol, *See* HA 819, 13531
Polyoxyethylene octyl phenyl ether, *See* Triton® X-15, 32433
Polyoxyethylene Sorbitan Monooleate, *See* Alkamuls® T-80, 1693
Polyoxyethylene Sorbitan Monostearate, *See* Alkamuls® T-60, 1692
Polyoxyethylene(20) cetyl ether, *See* Akyporox RC 200, 1289
Polyoxyethylene(20) oleyl ether, *See* Ameroxol® OE-2, 2022, Varonic® 32-E20, 33437
Polyoxyethylene(20) stearyl ether, *See* Macol® SA-2, 18273
Polyoxyethylene(23) lauryl ether, *See* Akyporox RLM 160, 1290
Polyoxyl(10)oleyl ether, *See* Ameroxol® OE-2, 2022
polyoxymethylene, *See* paraformaldehyde, 22751
Polyoxymethylene/polyacetal granules, *See* Ultraform®, 32804
polyoxymethyleneurea, *See* Salmocid, 27277
Polyoxypropylene (30) cetyl ether, *See* Wickenol® 707, 34404
Polyoxypropylene cetyl ether, *See* Wickenol® 707, 34404
Polyoxypropylene glycol, *See* Adeka P-400, 668
Polyoxypropylene hexadecyl ether, *See* Wickenol® 707, 34404
Poly-Pale®, 24505 , 24506
Poly-Pale® Ester 10, *See* Poly-Pale®, 24506
Polypeg-E, 24507
Polypel, 24508
Polypenco® Cast Acrylic Rod, 24509
Polypenco® Nylon 101, 24510
Polypenco® PEEK, 24511
Polypenco® Polycarbonate, 24512
Polypenco® Polysulfone, 24513, *See* Polysulfone, 24619
Polypenco® Q200.5, 24514
Polypentek, 24515
Polypeptide 10, 24516
Polypeptide 10, 37, *See* Extiat®, 11323
Polypeptide 37, 24517
Polyphenyl ether, *See* Santovac, 27427
Polyphenylene ether, *See* PX-10GF/000, 25434
Polyphenylene oxide, 24518, *See* Electrafil® G-1704/SS/5, 9729, PPO, 24847
Polyphenylene sulfide, *See* Fortron® 0205B4, 12324, Techtron™ PPS, 30896
polyphenylmethyl siloxane, *See* phenyl dimethicone, 23632
Polyphlogin, *See* cinchophen, 6307
Polyphos, 24519
Polyphosphate, *See* CHT Heptol NWS, 6268, Quadrafos, 25573
polyphosphoric acid, 24520
Polyphosphorylated surfactant, *See* Arofos 326, 3024
Polyphthalamide, *See* Amodel® A-1115HS, A-1145HS, 2290, Amodel® A-1340HS, 2291, Amodel® AF-1115VO, AF-1133VO, AF-1145VO, 2292, Amodel® ET-1000, 2293
Polyphthalate, *See* Ultramoll® PP, 32844, Ultramoll® TGN, 32845, Ultrmoll PP, 32888, Ultrmoll TGN, 32890
Polyplasdone®, 24521
Polyplasdone® XL, 24522
Polyplate 90, 24523
Polypor, 24524
Polyprene, 24525
Poly-Pro, 24526
polypropene, *See* polypropylene, 24527
Polypropene 25, *See* Avoilefin, 3500
Polypropoxy quaternary ammonium acetate, *See* Emcol® CC-55, 10057
Polypropoxy quaternary ammonium chloride, *See* Witflow 950, 34527
Polypropoxylate, *See* Degressal® SD 20, 7906
polypropylene, 24527, *See* Adpro AP 8210-HS, 789, Amoco® 1012, 2268, Amoco® 1016, 2269, Amoco® 1246, 2270, Amoco® 4018, 2271, Amoco® 5016, 2272, Amoco® 6114, 2273, Amoco® 6400p, 2274, Amoco® 7234, 2275, Amoco® 7239, 2276, Amoco® 7728, 2277, Amoco® 9119, 2280, Cabelec® 3464, 3004, 4867, Fortilene®, 12309, Fortilene® 1001, 12310, Fortilene® 1602, 12311, Fortilene® 1802, 12312, Fortilene® 2104, 12313, Fortilene® 3151, 12314, Fortilene® 4104, 4109, 12315, Fortilene® 4209, 12316, Fortilene® 5801, 12317, Fortilene® 9000, 12318, Hercoflat® Texturing Pigments and Flatting Agent, 13815, Hercotuf, 13847, HiFax AB 6023, 14073, HiFax CA 45A, 14074, HiFax CB 17AC, 14075, HiFax ETA 3011, 14076, HiFax ETA 3095, 14077, HiFax ETA 5012, 14078, HiFax RTA 3263E, 14079, HiGlass BJ44A, 14087, HiGlass PF062-2, 14088, HiGlass PF072-1, 14089, HiGlass SB 224-2, 14090, Hitac 300, 14138, Hostalen® OO, 14357, Hostalen® PP 927, 14358, Moplen, 20306, Stamylan P, 29444, Adpro AP 2112-GP, 788
polypropylene glycol, 24528
Polypropylene glycol, *See* Aerofroth 65, 857
Polypropylene glycol (10) cetyl ether, *See* Wickenol® 707, 34404
Polypropylene glycol alkylphenol ether, *See* Plastilit® 3060, 24040
Polypropylene glycol hexadecyl ether, *See* Wickenol® 707, 34404
polypropylene glycol methyl ether, *See* Arcosolv® PM, 2814, Dowanol® PM, 8896, Icinol PM, 14816, Solvenon® PM, Solvent PM, 29050
Polypropylene glycol monocetyl ether, *See* Wickenol® 707, 34404
Polypropylene glycol monohexadecyl ether, *See* Wickenol® 707, 34404
Polypropylene homopolymer, *See* Adpro AP 2112-GP, 788, Eastman® P4C5B-030, 9472
Polypropylene pulps, *See* Pulpex E and P, 25382
Polypropylene resin, *See* Mitsubishi 4300J, 19965
Polypropylene wax, *See* Hoechst Wax PP 230, 14269
Polyquart H, 24529
Polyquaternium-2, *See* Mirapol® A-15, 19916
Polyquaternium-4, *See* Celquat® H-100, SC-240, 5647
Polyquaternium-6, *See* Agequat 400, 1012, Mackernium 006, 18206, M-Quat® 40, 20399
Polyquaternium-7, *See* Mackernium 007, 18207, Mirapol® 550, 19914
Polyquaternium-10, *See* Ucare® Polymer JR-30M, JR-125, JR-400, LR-30M, LR-400, SR-10, 32712
Polyquaternium-11, 24530, *See* Gafquat® 734, 12583
Polyquaternium-16, 24531, *See* Luviquat® FC 905, 18023, Luviquat® HM 552, 18024, Polyquaternium-16, 24531
Polyquaternium-17, *See* Mirapol® AD-1, 19917
Polyquaternium-18, *See* Mirapol® AZ-1, 19918
Polyquaternium-24, *See* Quatrisoft Polymer LM-200, 25627
Polyquaternium-24-hyaluronic acid, *See* BioCare® Polymer HA-24, 4162
Polyquaternium-27, *See* Mirapol® 9, 95, 175, 19913
Polyquaternium-28, *See* Gafquat® HS-100, 12584
Polyquaternium-29, *See* Lexquat® CH, 17180
Polyquest, 24532
Polyquest 80, 24533, *See* Pentasodium pentetate, 23172
Polyrad, 24534
Polyram, 24535, *See* metiram, 19582
Polyram M, *See* Maneb, 18526
Polyram® -Combi, DF, 24536
Polyram-combi, *See* metiram, 19582
Polyram-Ultra, *See* Rezifilm, 26521
Polyram-Z, *See* Zineb, 34807
Polyran, *See* metiram, 19582
polysaccharide B-1459, *See* Kelflo®, 15839, Kelgum®, 15847, Kel-Lite™ CM, 15853, Keltrol®, 15886, Keltrol® F, 15889, Kelzan®, 15890, Kelzan® AR, 15891, Kelzan® D, M, XC Polymer, 15892, Kelzan® S, 15893, Rhodigel® EZ, 26683, Rhodopol® 23, XGD, 26696, Rhodopol® XGD, 26697, Xanflood®, 3458 7
Polysaccharide S-60, *See* Gellan gum, 12736, Kelco-Gel® Gellan Gum, 15827
Polysalt, 24537
Polysar Bromobutyl 2030, 24538
Polysar Butyl 100, 24539, *See* Exxon® Butyl 077, 11355
Polysar Chlorobutyl 1240, 24540
Polysar EPDM 227, 24541
Polysar EPDM 6463, 24542
Polysar EPDM XG 006, 24543
Polysar EPM XF 004, 24544
Polysar XL 30102, 24545, *See* Exxon® Butyl 077, 11355
Polyseal, 24546
Polyset 100, 24547
Polysil, 24548
Polysiloxane, *See* Antifoam VOL, 2553
Polysiloxane aqueous emulsion, *See* Agitan E 255, 1034
Polysiloxane polyalkyl copolymer, *See* Abil® Wax 2434, 59
Polysiloxane polyether copolymer, *See* Abil® OS5, 54
Polysiloxanes/polymer mixture, *See* BYK® 024, 4823
Polysoft 35, 24549
Polysoft B, 24550
Polysoft CA, 24551
Poly-solv DB, *See* Butyl Carbitol®, 4766
Poly-solv DE, *See* ethyl di-Icinol, 11068
Poly-solv DE ethanol, 2,2'-oxybis-, monoethyl ether, *See* Diethoxol, 8435
poly-solv EB, *See* 2-butoxyethanol, 4753
Poly-solv EE, *See* ethyl icinol, 11074
Poly-solv EE acetate, *See* EE Acetate, 9635
Poly-Solv®, 24552
Poly-Solv® DE, *See* Ethoxydiglycol, 11044
Poly-Solv® DE (High Gravity), 24553
Poly-Solv® DM, 24554
Poly-Solv® DPM, 24555
Poly-Solv® EE, 24556
Poly-Solv® EM, 24557, *See* methyl cellosolve®, 19521
Poly-Solv® MPM, 24558
Poly-Solv® TM, 24559
Poly-Solv® TPM, 24560
Polysolvan E, 24561
Polysolvan O, 24562, 24563
Polysolvan SHS, 24564
Poly-solve MPM, *See* Arcosolv® PM, 2814, Dowanol® PM, 8896, Icinol PM, 14816, propylene glycol monomethyl ether, 25220, Solvenon® PM, Solvent PM, 29050
Polysorbate 20, 24565, Alkamuls® PSMO-20, 1672, Alkamuls® T-20, 1691, Capmul® POE-L, 5116, Crillet 1, 7037, Disponil SML 120 F1, 8697, Emalex ET-2020, 9953, Emsorb® 2720, 10464, Ethylan® GEL2, 11104, Eumulgin SML 20, 11190, Glycosperse® L-20, 13129, Hetsorb L-20, 13967, Hodag PSML-20, 14237, Liposorb L-20, 17406, Montanox 20 DF, 20272, Mulsifan RT 141, 20455, Prox-onic SML-020, 25338, Radiasurf® 7137, 25799, Ritabate 20, 26794, Witconol 2720, 34508, Witflow 990, 34528, Solubilisant Gamma 2420, 28975
Polysorbate 20 NF, *See* Tween® 20, 32625
Polysorbate 21, 24566, *See* Disponil SML 104 F1, 8696, Hetsorb L-4, 13968, Tween® 21, 32626
Polysorbate 40, 24567, *See* Disponil SMP 120 F1, 8701, Ethylan® GEP4, 11107, Glycosperse® P-20, 13132, Hetsorb P-20, 13972, Liposorb P-20, 17411, Montanox 40 DF, 20273, Prox-onic SMP-020, 25341, T-Maz® 40, 31895
Polysorbate 40 NF, *See* Tween® 40, 32627
Polysorbate 60, 24568, *See* Alkamuls® PSMS-20, 1674, Capmul® POE-S, 5118, Disponil SMS 120 F1, 8703, Durfax® 60, 9216, Emsorb® 2728, 10468, Eumulgin SMS 20, 11192, Glycosperse® S-20 FG, S-20 KFG,

Polyvinylpyrrolidone/hexadecene copolymer, *See* PVP/hexadecene copolymer, 25431
Polyvinylpyrrolidone-iodine, *See* Videne Disinfectant Solution, Videne Disinfectant Tincture, Videne Powder, 33739, Vinisil, 33827
Polyviol, 24670
Polywax® 500, 24671
Polywet® ND-2, 24672
Polywet® Z1766, 24673
Polyxyethylated lanolin fatty acids, *See* Lanpol, 16752
Polyxyethylene alkyl ethers, *See* Brij®, 4514
Poly-zole AZDN, 24674
Polyzote, 24675
pomace, 24676
Pomarsal, *See* Rezifilm, 26521
Pomarsol, *See* Rezifilm, 26521, Thiram, 31726, Thiurad, 31730
Pomarsol forte, *See* Rezifilm, 26521
Pomarsol Z Forte, *See* AAprotect, 16
Pomarsol®, 24677
Pomarsolz, *See* AAprotect, 16
pomasol royal, *See* Rezifilm, 26521
Pombe, 24678
Pommetrol M, 24679
Pomoloy, 24680
Pompeian red, *See* Indian Red, 15068
Ponceau fast L salt, *See* Azoene, 3526
Pondermite, 24681
Ponder's stain, 24682
Pondicherry oil, 24683
Pondmaster, *See* glyphosate, 13148
Ponecil, *See* Ampicillin, 2361
ponite, 24684
ponolith, 24685
Ponoxylan, *See* Salmocid, 27277
Pontallor, 24686
Pontamine White BR, *See* Photine, 23756
Pontianac, *See* Jelutong, 15574
Ponticin, *See* rhaponticum, rhapontin, 26530
Pool-Chem, 24687
poonac, 24688
poonahlite, 24689
POP (2) bisphenol, *See* Prox-onic BP-02 P, 25310
POP (15) stearyl alcohol, *See* Prox-onic SA1-015/P, 25337
POP (20) methyl glucoside, *See* Prox-onic MG-020 p, 25323
Pope's solution, 24690
poppy capsules or heads, 24691
populnetin, *See* kaempferol, 15654
porcelain, 24692
porcelain clay, 24693, *See* ASP®, 3230, Bilt-Cote®, 4147, Kayphobe-ABO, 15816, Peerless®, 23023, SPS, 29349
porcelain earth, *See* China clay,
porcelain white, *See* lithopone, 17520
porcelanite, 24694
Porcelave, 24695
porcelin clay, *See* kaolin, 15725, Vanclay, 33304
Porcine Lard oil, *See* Actran Extra, 546
Porocel, 24696
Porofocon A, *See* RX-56, 27157
Porofocon B, *See* GP-II, 13233
Porofor, *See* Azodicarbonamide, 3524
Porofor 57, *See* Ficel® AZDN-LF, 11763
Porofor ADC/R, *See* Azodicarbonamide, 3524
Porofor CHKHZ 21, *See* Azodicarbonamide, 3524
Porofor CHKHZ 21r, *See* Azodicarbonamide, 3524
Porofor®, 24697
Porofor® ADC/E, 24698, *See* Azodicarbonamide, 3524
Porofor-57, *See* Poly-zole AZDN, 24674, V-60, 33218
Porofore 505, *See* Azodicarbonamide, 3524
Poron® 4701-01, 24699
Poron® S2000-80-24031, 24700
Porosil-Clarcel, 24701
porous alum, 24702
porpezite, 24703
porphyry, 24704
porporino, 24705
Portagen, 24706
Portland arrowroot, 24707
Portland cement, 24708
Portland stone, 24709, *See* Calbux, 4929, calcium carbonate, 4941, CP Filler, 6928
Portman 5C Chormequat, 24710
Portman Chlormequat 400, 460, 600, 700, 24711
Portman Isotop, 24712
Portman propachlor 50FL, 24713
Portman Supaquat, 24714
Portsmouth Accelerator No. 3, 24715
Portugallo oil, 24716
Portuguese turpentine, *See* turpentine, 32610
Pos O Print, 24717
Posistac, 24718
Poskydal, 24719
Posse, *See* carbosulfan, 5245
Post-4, 24720
Post-Kite, 24721
pot metal, 24722
potarite, 24723
potash, *See* potassium carbonate, 24744, potassium hydroxide, 24756
potash alum, 24724, *See* Roman alum, 26925
potash bordeaux mixture, 24725
potash chlorate, *See* potassium chlorate, 24746
potash felspar, *See* felspar, 11589
potash glass, 24726
potash salts, *See* Abraum salts, 156
potash water-glass, 24727
pot-ashes, 24728
potash-lead glass, 24729
Potasoral, *See* potassium d-gluconate, 24752
Potasorl, *See* Gluconal® K, 13043, Katorin, 15798
potassa, *See* potassium hydroxide, 24756, vegetable alkali, 33512
potassalumite, 24730
potassic superphosphate, 24731
potassiocuprous cyanide, *See* cuprous potassium cyanide, 7406
potassium, 24732
potassium acetate, 24733, 24734, *See* sal diureticum, 27246
potassium acid arsenate, *See* potassium arsenate, 24737
potassium acid carbonate, *See* K-Lyte, 16233
potassium acid fluoride, *See* Fremy's salt, 12394
potassium acid sulfate, *See* tartarline, 30833
potassium acid tartrate, *See* potassium bitartrate, 24740
potassium alginate, 24735, *See* Alginic acid potassium salt, 1594, Improved Kelmar®, 14933, Sobalg FD 200 Series, 28688
potassium alkyl phosphate ester, *See* Berol 521, 4042
potassium aluminum sulfate, *See* alum, 1883
potassium amalgam, 24736
potassium amyl xanthate, *See* Aero 350 Xanthate, 843
potassium amylate, *See* Maculanin, 18285
potassium antimonate(V), *See* potassium pyroantimonate, Acid, 24776
potassium antimonyl tartrate, *See* antimony potassium tartrate, 2571
potassium arsenate, 24737, *See* Macquer's salt, 18277
potassium benzoate, 24738
potassium benzylpenicillinate, *See* Penicillin G potassium, 23109, 23110
potassium bicarbonate, *See* K-Lyte, 16233
potassium bichromate, 24739
potassium bifluoride, *See* Fremy's salt, 12394
potassium bismuth iodide, *See* Thresh's reagent, 31765
potassium bisulfate, *See* tartarline, 30833
potassium bitartrate, 24740
potassium bromate, 24741, *See* Bromox, 4589
potassium bromide, 24742
potassium butylparaben, *See* Nipabutyl potassium, 21250
potassium cadmium iodide, *See* Marme's reagent, 18948
potassium carbonate, 24743, 24744, 24745, *See* pearl ash, 22977
potassium cetyl phosphate, *See* Amphisol K, 2309
potassium chlorate, 24746
potassium chloride, 24747, 24748, 24749, *See* K. Tab, 15642, Kaon-Cl, 15729, Kay Ciel, 15807, K-Contin, 15818, kelp salt, 15859, K-Lor, 16226, K-Lyte/C1, 16234, KM potassium Chloride, 16242, Micro-K, 19676, muriate of potash, 20513, Pfiklor, 23564, potassium chloride, 24749, Slow-K, 28609
potassium chromic sulfate dodecahydrate, *See* chromic potassium sulfate, 6224
potassium chromium alum dodecahydrate, *See* chromic potassium sulfate, 6224
potassium citrate, 24750
potassium cobaltic nitrite, *See* Fischer's salt, 11848
potassium cocoate, *See* Norfox® 1101, 21544
potassium coco-hydrolyzed collagen, *See* Foam-Coll 4C, 12150
potassium copper cyanide, *See* cuprous potassium cyanide, 7406
potassium cyanide, 24751
potassium cyanocuprate, *See* cuprous potassium cyanide, 7406
potassium dibutyldithiocarbamate, *See* Butyl Kamate®, 4772
potassium dichloro-*s*-triazine-2,4,6-trione, *See* dichloro-1,3,5-triazinetrione, potassium salt, 8384
potassium dichlorocyanurate, *See* dichloro-1,3,5-triazinetrione, potassium salt, 8384
potassium dichloroisocyanurate, 24753, *See* dichloro-1,3,5-triazinetrione, potassium salt, 8384
potassium dichloroisocyanurate dihydrate, *See* ACL 56, 59, 60, 66,, 354
potassium dichloro-s-triazinetrione, *See* triclosene potassium, potassium dichloroisocyanurate, 24753
potassium dichromate, 24754
potassium dichromate(VI), *See* potassium bichromate, 24739, potassium dichromate, 24754
potassium dicyanocuprate(I), *See* cuprous potassium cyanide, 7406
potassium dihydrogen arsenate, *See* potassium arsenate, 24737
potassium dihydrogen orthophosphate, *See* potassium phosphate, monobasic, 24772, potassium phosphate, 24773
potassium dihydroxyethyl cocamine oxide phosphate, *See* Mazox® KCAO, 19151
potassium dimagnesium sulfate, *See* Breecht's double salt, 4498
potassium dimethicone copolyol phosphate, *See* Pecosil PS-100K, 23013
potassium dimethyldithiocarbamate, *See* Aquatreat KM, 2760
potassium disulfato chromate(III) dodecahydrate, *See* chromic potassium sulfate, 6224
potassium dodecylbenzene sulfonate (linear), *See* Polystep® A-15-30K, 24587
potassium ethyl xanthate, *See* Aero 303 Xanthate, 840
potassium ethylparaben, *See* Nipagin A potassium, 21264
potassium Felspar, *See* Felspar, 11589
potassium ferrocyanide, *See* yellow prussiate of potash, 34647
potassium ferrocyanide trihydrate, *See* yellow prussiate of potash, 34647
potassium ferrous ferro-cyanide, *See* Everitt's salt, 11268
potassium fluorozirconate, *See* Aflammit ZR, 941
potassium gluconate, 24755, *See* Gluconal® K, 13043, Jaon, 15507, Katorin, 15798
potassium d-gluconate, 24752
potassium glycol borate, *See* Liquibor 169, 17456
potassium hexacyanoferrate(II), *See* yellow prussiate of potash, 34647
potassium hexadienoate potassium 2,4-hexadienoate, *See* potassium sorbate, 24780
potassium hexafluorotitanate, *See* Aflammit TI, 939
potassium hexafluorozirconate, *See* Aflammit ZR, 941

potassium 2,4-hexadienoate, *See* potassium sorbate, 24780
potassium (E,E)-hexa-2,4-dienoate, *See* potassium sorbate, 24781
potassium hexkis(cyano-C)ferrate(4-) trihydrate, *See* yellow prussiate of potash, 34647
potassium hydrate, *See* potassium hydroxide, 24756, vegetable alkali, 33512
potassium hydrogen carbonate, *See* K-Lyte, 16233
potassium hydrogen difluoride, *See* Fremy's salt, 12394
potassium hydrogen fluoride, *See* Fremy's salt, 12394
potassium hydrogen oxalate, *See* salt of Sorrel, 27289
potassium hydrogen sulfate, *See* tartarline, 30833
potassium hydroquinone monosulfate, *See* Stabilizer, 29404
potassium hydroxide, 24756, *See* vegetable alkali, 33512
potassium hydroxide-sodium silicate-sodium hypochlorite, *See* Det-O-Jet®, 8212
potassium 4-hydroxyphenyl sulfate, *See* Stabilizer, 29404
potassium hypochlorite, 24757, *See* HyPure K, 14726
potassium indoxyl sulfate, *See* indican, 15070
potassium iodide, 24758, 24759
potassium iodobismuthate, *See* Dragendorf's reagent, 8954
potassium lauryl phosphate, *See* Crafol AP-31, 6958
potassium lauryl sulfate, *See* Sulfochem K, 30007
potassium lignosulfonate, 24760
potassium metabisulfite, 24761
potassium metaperiodate, *See* potassium periodate, 24768
potassium metavanadate, *See* K-Van, 16499
potassium methylparaben, *See* Nipagin M potassium, 21267
potassium muriate, *See* potassium chloride, 24749
potassium myronate, *See* sinigrin, 28500
potassium nitrate, 24762, *See* Petre, 23474, prismatic niter, 25035, sal niter, 27250, sal Prunella, 27251, saltpeter, saltpeter flour, 27298
potassium 9-octadecenoate, *See* potassium oleate, 24764, 24763
potassium octoate, *See* Metacure® T-45, 19418
potassium oleate, 24763, 24764, *See* Emkapol PO-18, 10248, Octosol 449, 22018
potassium orthophenyl phenate solution, *See* Acticide 50, 462
potassium oxalate, 24765, 24766
potassium oxalate monohydrate, *See* potassium oxalate, 24765 24766
potassium PCA, *See* Kalidone®, 15678
potassium penicillin G, *See* Penicillin G potassium, 23109, 23110
potassium percarbonate, *See* Antihypo, 2556
potassium perchlorate, 24767, *See* KM potassium Perchlorate, 16243
potassium perchlorate, nitroglycerol, ammonium oxalate, wood meal, and small quantities of collodion cotton, and nitrotoluenes, *See* Ajax powder, 1139
potassium perfluoroalkyl sulfonate, *See* Fluorad® FC-95, 12065
potassium periodate, 24768
potassium Periodate Meta, *See* potassium periodate, 24768
potassium permanganate, 24769, 24770, *See* sin red, 28490
potassium Peroxodisulfate, *See* Anthion, 2519
potassium peroxydisulfate, *See* Anthion, 2519, potassium persulfate, 24771
potassium persulfate, 24771, *See* Anthion, 2519
potassium phosphate ester, *See* Hodag PE-005-K, 14201
potassium phosphate, dibasic, 24773
potassium phosphate, monobasic, 24772, *See* Beycostat 148 K, 4099 Beycostat 273 P, 4102
potassium phosphate, tribasic, 24774
potassium polyacrylate, 24775
potassium polymannuronate, *See* potassium alginate, 24735
potassium propylparaben, *See* Nipasol M potassium, 21283
potassium pyroantimonate, Acid, 24776
potassium pyrophosphate, 24777, *See* Tetrakal, 31375
potassium pyrophosphate, normal, *See* potassium pyrophosphate, 24777
potassium pyrosulfite, *See* potassium metabisulfite, 24761
potassium ricinoleate, 24778, *See* Kricinol 35, 16411 Solricin® 135, 28964
potassium ricinoleate
potassium ricinoleate, *See*
potassium rosinate, *See* Arizona DRS-50, 2892
potassium salt, *See* potassium oleate, 24763
potassium salt of oleic fatty acid, *See* Octosol 449, 22018
potassium salt of tryfac 5552, *See* Tryfac® 5553, 32512
potassium silicate, 24779, *See* Double White, 8817, Kasil, 15778, Potkem, 24795, soluble potash glass, 28983
potassium silicates, *See* Pyramid, 25465
potassium sodium chromate, *See* Chromglaserite, 6219
potassium sodium tartrate tetrahydrate, *See* potassium sodium tartrate,
potassium sorbate, 24780, 24781, *See* Sorbistat K, 29108
potassium stannate, 24782
potassium stannate(IV), *See* potassium stannate, 24782
potassium stearate, 24783, 24784
potassium sulfate, 24785, *See* Glazier's salt, 12982, salt of Lemery, 27285, Trona Sulphate of potash, 32460
potassium sulfate-potassium sulfite, *See* Glaser's salt, 12963
potassium sulfide, *See* hepar sulfur, 13787
potassium sulfite, *See* Stahl's sulfur salt, 29425
potassium tartrate, *See* soluble tartar, 28987
potassium tetroborate, 24786
potassium toluene sulfonate, *See* Eltesol® PT 93, 9869, Hartotrope KTS 44, 13692, Naxonate® 4KT, 20823, Naxonate® 5KT, 20825
potassium trichloroisocyanurate, *See* ACL 56, 59, 60, 66,, 354
potassium triphosphate, *See* potassium tripolyphosphate, 24787
potassium tripolyphosphate, 24787
potassium tritungstate, *See* violet tungsten, 33878
potassium troclosene, *See* ACL 56, 59, 60, 66, 354, dichloro-1,3,5-triazinetrione, potassium salt, 8384
potassium undecylenoyl hydrolyzed collagen, 24788
potassium xylene sulfonate, *See* Eltesol® PX 40, PX 93, 9870, Manro KXS, 18595, Reworyl® KXS, 26405
potassium xylene sulfonate, potassium tallate and potassium cocoate, *See* Akypogene ZA 97 SP, 1263
potassium zirconium hexafluoride, *See* Aflammit ZR, 941
potassium-ammonium-antimonyl-tartrate, *See* Antiluetin, 2560
potassium-magnesium sulfate, *See* schoenite, 27725
potassium-stanno-sulfate, *See* Marignac's salt, 18717
potassuril, *See* potassium d-gluconate, 24752, potassium gluconate, 24755, Gluconal® K, 13043, Katorin, 15798
potato alcohol, *See* ethyl alcohol, 11064
potato gum, 24789
potato oil, *See* fusel oil, 12518
potato oil or spirit, 24790
potato rubber, *See* potato gum, 24789
Potato-Pro EN-15, 24791
Potazote, 24792
potcrate, *See* potassium chlorate, 24746
potentiated acid glutaraldehyde, *See* glutaraldehyde, 13062, Sonacide, 29073
Potentite, *See* Tonite, 32012
Potenzol V, 24793
Potin, 24794
Potinjaune, *See* Potin, 24794
Potkem, 24795
Potosi silver, *See* nickel silvers, 21132
potstone, 24796
Potter's clay, *See* pipeclay, 23872
Potter's Ore, *See* alquifon, 1846
Potting Base, 24797
Pouckpong gum, 24798
Poudre B, 24799
Poudre EF, 24800
Poudre J, 24801
Poudre Pyroxule+a7e, 24802
Poulenc 309, 24803
Pounce, *See* Kafil, 15655, permethrin, 23384, 23385, 23386
Poutet's reagent, 24804
Povidine, *See* Luviskol® K12, K17, K30, K60, 18031, Luviskol® K80, K90, 18032
Povidone, 24805, *See* Agrimer AL-22, 1066, H_2Old EP-1, 13526, Plasdone®, 23935, PVP, 25429, Tears Plus®, 30876, Vinisil, 33827
povidone-iodine, 24806, *See* Disadine, 8623, Videne Disinfectant Tincture, 33740
Powaspray Glymark, 24807
Powax, 24808
Powder 19/04/15H Black 904, 24809
Powder 22/04/00A 400, 24811
Powder 26/04/00, 24812
Powder 215 Natural, 24810
powder of Algaroth, 24813
powder of Algarotti, *See* powder of Algaroth, 24813
Powdered Aloe Vera (1:200) Food Grade, 24814
powdered hydrocyanic acid, 24815
powellite, 24816
Power 64, 640, 700, 24817
Power Chlorothalonil 50, 24818
Power Demo, 24819
Power Diquat, 24820
Power Drive, 24821
Power DSM, 24822
Power Ethephon 48, 24823
Power Flame, 24824, *See* Flamprop-M-isopropyl, 11888
Power Flamprop, 24825, *See* Flamprop-M-isopropyl, 11888
Power Gard, 24826
Power Gro-Stop, 24827
Power MCPA, 24828
Power Non-ionic Wetter, 24829
Power Phosphine Pellets, 24830
Power Platoon, 24831
Power Propiconazole, 24832
Power Spray Save, 24833
Power Swing, 24834
Power Task, 24835
Powers Terebine, 24836
Powerspire, 24837
Powmet, 24838
Pozzolith 400n, *See* Ablusol ML, 129
PP, *See* polypropylene, 24527
PP, *See* polypropylene, 24527
PP 062, *See* Abol, 146
PP 321, *See* λ-cyhalothrin, 7567
PP 333, *See* paclobutrazol, 22540, 22541
PP 383, *See* cypermethrin, 7588, Topclip Parasol, 32037
PP 580, *See* Ratak, 25896
PP Captan 83, 24839
PP-009, *See* Fluazifop-butyl, 12038
PP-062, *See* Rapid, 25884, *See* pirimicarb, 23879
PP-10GF/000, 24840
PP-333, *See* paclobutrazol, 22542
PP-557, *See* Kafil, 15655, permethrin, 23384, 23385, 23386
PP-581, *See* Talon, 30750
PP-781, *See* drazoxolan, 8974
PPG, *See* Pluracol® W3520N, 24188
PPG-2 bisphenol A, *See* Ethox 3113, 11018
PPG-2 methyl ether, *See* Arcosolv® DPM, 2812, Icinol DPM, 14815, Poly-Solv® DPM, 24555

Prifac 9428, *See* hydrogenated tallow acid, 14598
Prifrac 2901, 24950
Prifrac 2906, 24951
Prifrac 2920, 24952
Prifrac 2920, *See* lauric acid, 16856
Prifrac 2940, 24953
Prifrac 2960, 24954
Prifrac 2980, 24955
Prifrac 2989, 24956
Prifrac 2990, 24957
Prifrac 5901, *See* Hystrene® 1835, 14755
Primacor, 24958
Primacor 1320, 24959
Primacor 2912, 24960
Primacor 4990 Dispersion, 24961
Primafloc, 24962
Primagram, *See* metolachlor, 19585
Primal, 24963
Primallor, 24964
Primapel, 24965
primary alcohol ethoxy sulfate, sodium salt, *See* Teepol PB, 30933
primary amyl alcohol, *See* Amyl alcohol, 2381
primary calcium phosphate, *See* acid calcium phosphate, 330, calcium phosphate (monobasic), 4966, V-90®, 33222
primary-n-amyl alcohol, *See* Amyl alcohol, 2381
Primasol® AMK, 24966
Primasol® FP, 24967
Primasol® NB-NF, 24968
Primatol, *See* atrazine, 3394
Primatol A, *See* atrazine, 3394
Primatol AA, 24969
Primatol AD 85WP, 24970
Primatol AP, 24971
Primatol M, *See* Terbuthylazine, 31227
Primatol Q, *See* prometryn, 25140, 25141
Primatol S, *See* simazine, 28464
Primatol SE 500FW, 24972
Primax, 24973, 24974
Primax UH-1060, 24975
Primax UH-1080, 24976
Primax UH-1250, 24977
Primaze, *See* atrazine, 3394
Primazin Fixing Agent RP, 24978
Primazin®, 24979
Prime Flavours, 24980
Primecoat®, 24981
Primene, 24982
Primextra, *See* metolachlor, 19585
Primicid, 24983
Primor, 24984
Primorol 1511, 24985
Primotec, 24986
primrose smokeless, 24987
primrose soluble, *See* erythrosin, 10778
primrose soluble in alcohol, *See* Erythrin, 10776
primross soluble, *See* erythrosin, 10778
primuline base, 24988
Primus, 24989
Primus®, 24990
Prince Rupert's Metal, *See* Prince's metal, 24992
Princep, *See* simazine, 28464
Prince's blue, 24991
Prince's metal, 24992
Prince's metallic, *See* Prince's mineral, 24993
Prince's mineral, 24993
Prinsyl, 24994
Printel, 24995
Printer's acetate, 24996
printer's iron liquor, 24997
Printex, 24998
Printex 25 Beads and Powd, 24999
Printex P, 25000
Printex U Beads and Powd, 25001
Printing Black for Wool, 25002
printing inks, 25003
Printlok 1046, 25004
Printogen, 25005
Printol®, 25006
Printop, *See* simazine, 28464
Printosol, 25007
Printsolve™ Ink Remover, 25008
Printwash, 25009
Prinza® Range, 25010
Prioderm, 25011
Priolene 6900, 25012, *See* oleic acid, 22112
Priolube 1400, 25013, *See* methyl oleate, 19554
Priolube 1405, 25014, *See* Butyl oleate, 4778
Priolube 1407, 25015
Priolube 1409, 25016, *See* glyceryl dioleate, 13080
Priolube 1414, 25017
Priolube 1429, 25018, *See* Propylene glycol dioleate, 25217
Priolube 1435, 25019
Priolube 1451, 25020, *See* butyl stearate, 4781
Priolube 1458, 25021
Priormatt, 25022
Priplast 1431, 25023
Priplast 1562, 25024
Priplast 3013, 25025
Priplast 3018, 25026
Priplast 3157, 25027
Priplast 3191, 25028
Pripol 1004, 25029
Pripol 1009, 25030
Pripol 1017, 1022, 1025, *See* Dilinoleic acid, 8480
Pripol 1025, 25031
Pripol 1040, 25032
Prism, 25033, 25034
prismatic niter, 25035
Prisorine 3508, 25036, *See* isostearic acid, 15421
Pristacin, *See* Acetoquat CPC, 302, cetyl pyridinium chloride, 5823
Pristane, 25037
Pristene 180, 25038
Pristene R20, 25039
Pristerene 4904, 25040
Pristine, 25041
Pro, *See* proline, 25129
Pro Seal, 25042
Pro Weld, 25043
Proaid 9802, 25044
Proaid 9814, 25045
Proaid 9904, 25046
Proban, 25047
Pro-Banthine, *See* propantheline bromide, 25176
Probe, 25048, *See* methazole, 19481
Probe 75 WP, *See* methazole, 19481
proberite, 25049
Probilin, *See* phenolphthalein,
Probimer, 25050
Probimide, 25051
procaine benzylpenicillinate, *See* Penicillin V, 23111
procaine hydrochloride, *See* benzocaine,
Procaine penicillin G, *See* Penicillin V, 23111
Procal, 25052
Procanodia, *See* Penicillin V, 23111
Pro-Care Multi-Purpose Fungicide, *See* Chlorothalonil, 6150, Repulse, 26151
Processed lanolin, *See* Vigilan, 33750
Procetyl, *See* Wickenol® 707, 34404
Procetyl 10, 25053, *See* PPG-30 cetyl ether series, 24845
Procetyl alcohol 30, *See* Wickenol® 707, 34404
Procetyl AWS, 25054, 25055, *See* Wickenol® 707, 34404
Prochinor, 25056
Prochloraz, 25057, *See* Mirage, 19864, Prelude, 24885, Sportak Delta, 29321
prochlorite, 25058
Procilene, 25059
Procinyl, 25060
Procion, 25061
Procol CS-6, *See* Ceteareth-6, 5788
Procol®, 25062
Procom, 25063
proconazole, *See* Radar, Radar Propiconazole, 25714
Procond-101, 25064
Procor 75 AB-X, 25065
procythol, 25066
Prodag, 25067
Prodaram, *See* AAprotect, 16
Prodew 100, 25068
Prodex, 25069
Prodipate, *See* Unimate® DIPA, 32968
Prodoraqua, 25070
Prodorbond, 25071
Prodorcrete GT, 25072
Prodorfilm, 25073
Prodorflor, 25074
Prodorglas, 25075
Prodorglaze, 25076
Prodorguard, 25077
Prodorite, 25078
Prodorlac, 25079
Prodorshield, 25080
Prodox, 25081
Prodox®120, *See* Alkylated phenol, 1734
Product 308, *See* Adol® 52 NF, 782, Cachalot® C-50 NF, 4878
Product AAS 90, 25082
Product MB320, 25083
product no. 75, *See* sodium lauryl sulfate, 28781
product no. 161, *See* sodium lauryl sulfate, 28781
Productol, 25084
Produkt 2058, 25085
Produkt GM 4210, 25086
Produkt GS 5001, 25087
Pro-Etch, 25088
Profalon, 25089
Profam, *See* Propham, 25186
Profax, 25090
Pro-fax® 65F4-4, 25092
Pro-fax® 65F5-4, 25093
Pro-fax® 6323, 25091
Pro-fax® 7523, 25094
Pro-fax® 8523, 25095
Pro-fax® HB-301, 25096
Pro-fax® PC-072PM, 25097
Pro-fax® PD-064, 25098
Pro-fax® PF-101, 25099
Pro-fax® SA-747M, 25100
Pro-fax® SB-242, 25101
Pro-fax® SD-062, 25102
Pro-fax® SE-191, 25103
Pro-fax® SV-256M, 25104
Pro-fax® Z-39S, 25105
Profecundin, *See* tocopherol, 31918, Vascuals, 33465
Proferdex, 25106
Proflan, *See* Treflan, 32162
Progacyl® ADG, 25107
Progacyl® CP-7, 25108
Progacyl® CP-82, 25109
Progallin LA, 25110
Progallin P, 25111, *See* Sustane® PG, 30436
Proganol, *See* protargentum, 25251
Pro-Gas (Gas Disclaimed), 25112
Progasol® COG, 25113
Progene, 25114
Progesterone, *See* Gesterol 50, 12927, Lipo-Lutin, 17370, luteol, 17991
Pro-Gibb, *See* Regulex, 26019
Pro-Gibb Plus, *See* Regulex, 26019
Progilite, 25115
proglumide, 25116
Prograss, *See* ethofumesate, 10967
Progressite, 25117
proidonite, 25118
Proil, 25119
proiodin, 25120
Project® 70 Stainless Type 316, 25121
Prokarbol, *See* 4,6-dinitrocresol, 8542
Prokayvit Oral, 25122
Prolan, *See* Treflan, 32162, Trifluralin, 32252
Prolan®, 25123
Prolaurin, 25124

Chemical and Trade Names

2-propenoic acid, 2-methoxyethyl ester, *See* Ageflex MEA, 994
2-Propenoic acid, 2-methyl-, cyclohexyl ester, *See* Ageflex CHMA, 968
2-Propenoic acid, 2-methyl-, 2-(dimethylamino)ethyl ester, *See* Ageflex FM-1, 983
Propenoic acid, 2-methyl-, dodecyl ester, *See* Ageflex FM-12, 986
2-propenoic acid methyl ester, *See* methyl acrylate, 19509
2-propenoic acid methyl ester, homopolymer, *See* polymethylmethacrylate, 24491
2-Propenoic acid, 2-methyl-, 1,2-ethanediyl ester, *See* ethylene glycol dimethacrylate, 11125
Propenoic acid, (1-methyl-1,2-ethanediyl)bis(oxy(methyl-2,1-ethanediyl)) ester, *See* Ageflex TPGDA, 1003
2-Propenoic acid, 2-methyl-, 2-propenyl ester, *See* Ageflex AMA, 966
2-Propenoic acid, 2-methyl-, 2-propenyl ester, *See* Allyl methacrylate, 1784
2-Propenoic acid octyl ester, *See* octyl acrylate, 22037
2-Propenoic acid oxybis(2,1-ethanediyloxy-2,1-ethanediyl) ester, *See* Ageflex T4EGDA, 970
2-Propenoic acid, 2-phenoxyethyl ester, *See* Ageflex PEA, 997, SR-339, 29379
2-Propenoic acid, 2-phenoxyethyl ester, *See* Melcril 4087, 19268
propenoic acid, sodium carbonate polymer, *See* sodium polyacrylate, 28811
2-Propenoic acid, 2-methyl-, (tetrahydro-2-furanyl)methyl ester, *See* Ageflex THFMA, 999
2-Propenoic acid 3-(trimethoxysilyl)propyl ester, *See* CA0397, 4846
propenol, 25181
1-Propenol-3-ol, *See* propenol, 25181
2-propen-1-ol, *See* propenol, 25181
2-propenol, *See* propenol, 25181
Propenol-3, *See* propenol, 25181
propenonitrile, *See* Acrylonitrile, 436
2-propenyl acrylic acid, *See* sorbic acid, 29105
Propenyl alcohol, *See* propenol, 25181
2-Propenyl Alcohol, *See* propenol, 25181
2-propenyl chloride, *See* Barchlor, 3626
4-(1-propenyl)phenol, *See* *p*-anol, 2480
4-(2-Propenyl)phenol, *See* chavicol, 5861
Propenylanisole, *See* Arizole® Anethole Extra, 2885
***p*-Propenylanisole**, *See* Arizole® Anethole Extra, 2885, Estragole, 10897
***p*-1-propenylanisole,** *See* Arizole® Anethole Extra, 2885
5-(2-propenyl)-1,3-benzodioxole, *See* safrol, 27229
[(2-propenyloxy)methyl]oxirane, *See* Ageflex AGE, 965, Allyl glycidyl ether, 1783, Sipomer® AGE, 28531
***p*-propenylphenol**, *See* *p*-anol, 2480
***p*-propenylphenyl methyl ether,** *See* Arizole® Anethole Extra, 2885
Propetal 241, 25182
Propetamphos, 25183
propezite, 25184
Propham, 25185, 25186, *See* MSS IPC 50, 20431
Propiconazole, *See* Bumper, 4669, Power Propiconazole, 24832, Powerspire, 24837, Radar, Radar Propiconazole, 25714
Propinate, *See* Basfapon®, 3705
Propineb, *See* Antracol®, 2634, Fruvit®, 12423
Propiodone, *See* Propyliodone, 25225, 25226
Propiofan®, 25187
Propiofan® D, 25188
Propiolactone, *See* betaprone, 4065
β-Propiolactone, *See* betaprone, 4065
propiolic acid, 25189
propionaldehyde, *See* propanal, 25172
Propionaldehyde dimethyl acetal, *See* DMP, 8757
Propionamide, N-benzyl-2,2-dimethyl-N-isopropyl-, *See* Comodor, Comodor 600, 6667
Propione, *See* diethyl ketone, 8436
propionic acid, 25190
Propionic acid, *See* Luprosil®, 17919
propionic acid, calcium salt, *See* calcium propionate, 4968
Propionic aldehyde, *See* propanal, 25172
Propionitrile, 2-[[4-chloro-6-(ethylamino)-1,3,5-triazin-2-yl]amino]-2-methyl, *See* cyanazine, 7472
Propionolactone, *See* betaprone, 4065
β-propionolactone, *See* betaprone, 4065
Proplatinum, 25191
Propocon, 25192
Propogon, *See* Propoxur, 25198, 25199, 25200
propol, *See* isopropanol, 15405
Propomeen 2HT-11, 25193
Propomeen C/12, 25194
Propomeen T/12, 25195
Proponex-Plus, *See* Verdone CDA, 33580
Proponic acid, 3,3'-thiodipionate, *See* dilauryl thiodipropionate, 8476
Propoquad® 2HT/11, 25196
Propoquad® T/12, 25197
Propotox, *See* Baygon®, 3790
Propoxur, 25198, 25199, 25200, 25201, *See* Baygon®, 3790, Blattanex®, 4329, Blattanex® 20, 4330, Fumite Propoxur, 12465, Suncide®, 30138, Unden®, Undene®, 32917
Propoxure, *See* Baygon®, 3790
(6-propoxy-2-benzothiazolyl)carbamic acid methyl ester, *See* Tiox, 31845
propoxyethanol, *See* Ektasolve® EP, 9671
2-Propoxyethanol, *See* Ektasolve® EP, 9671, ethylene glycol propyl ether, 11130
2-(2-Propoxyethoxy) ethanol, *See* diethylene glycol propyl ether, 8445
Propoxylated amine, *See* Hexcelcure 160, 14025
Propoxylated hexadecyl alcohol, *See* Wickenol® 707, 34404
Propoxylated trimethylolpropane triacrylate, *See* CD492, 5499
Propoxyol® 1695, 25202
Proprionaldehyde, *See* propanal, 25172
n-propyl 3,4,5-trihydroxybenzoate, *See* Sustane® PG, 30436
N-Propyl 3,5-di-iodo-4-pyridone-N-acetate, *See* Propyliodone, 25225
n-propyl alcohol, *See* propanol, 25175
n-propyl gallate, *See* Sustane® PG, 30436
n-propyl mercaptan, 25227, 25228, 25229
n-Propyllutidine, *See* coridine, 6822
N-propyl-N-[2-(2,4,6-trichlorophenoxy)ethyl]-1H-imidazole-1-carboxamide, *See* Octave, 21990, prochloraz, 25057
n-propyltrichlorosilane, 25232, *See* CP0800, 6935
n-Propyltrimethoxysilane, *See* CP0810, 6936
Propyl Cellosolve, *See* ethylene glycol propyl ether, 11130
propyl *p*-hydroxybenzoate, *See* propylparaben, 25231
propyl acetate, 25203
1-Propyl Acetate, *See* propyl acetate, 25203
propyl alcohol, *See* propanol, 25175
propyl aldehyde, *See* propanal, 25172
Propyl Aseptoform, *See* Nipasol M, 21282
Propyl Butex, *See* Nipasol M, 21282
propyl carbinol, *See* butyl alcohol, Tebol 88, 99, 30878
Propyl cellosolve, *See* Ektasolve® EP, 9671
Propyl Cellosolve®, *See* ethylene glycol propyl ether, 11130
Propyl Chemosept, *See* Nipasol M, 21282
Propyl Chemsept, *See* Nipasol M, 21282
propyl cyanide, *See* butyronitrile, 4805
propyl gallate, 25204, *See* Progallin P, 25111, Sustane® PG, 30436
propyl hydride, *See* propane, 25173
propyl p-hydroxybenzoate, *See* propylparaben, 25231
2-Propyl isooctadecanoate, *See* Wickenol® 131, 34381
propyl methyl ketone, *See* methyl propyl ketone, 19557
propyl oleate, 25205, *See* Emerest® 2302, 10082
propyl parahydroxybenzoate, *See* propylparaben, 25231
Propyl Parasept, *See* Nipasol M, 21282
***sec*-Propyl**, *See* isopropanol, 15405
n-propyl 3,4,5-trihydroxybenzoate, *See* propyl gallate, 25204
Propyl Zithate®, 25206
propylacetate, *See* propyl acetate, 25203
propylacetic acid, *See* n-valeric acid, 33242
Propylan, 25207
Propylan A350, 25208
Propylan G600, 25209
Propylan RF55, 25210
propylan-propyl alcohol, 25211
Propylated triphenyl phosphate, *See* Syn-O-Ad® 8480, 30557
propylene, 25212
1-Propylene, *See* propylene, 25212
Propylene carbonate, *See* Arconate® Propylene Carbonate, 2811, Solvenon® PC, 29049, Texacar® PC, 31425
1,2-Propylene Carbonate, *See* Arconate® Propylene Carbonate, 2811, Solvenon® PC, 29049
propylene dichloride, 25213
α,β-propylene dichloride, *See* propylene dichloride, 25213
propylene epoxide, *See* propylene oxide, 25223
propylene glycol, 25214, 25215, Dowfrost, 8922, Prolugen, 25135, Surfactol® Q1, 30341, Surfactol® Q2, 30342
α-propyleneglycol, *See* propylene glycol, 25214, 25215
propylene glycol 1-methyl ether, *See* Arcosolv® PM, 2814, Dowanol® PM, 8896, Icinol PM, 14816, Solvenon® PM, Solvent PM, 29050
propylene glycol alginate, 25216, *See* Concentrated Dariloid® KB, 6710, Kelcoloid® D, 15828, Kelcoloid® DH, DSF, 15829, Kelcoloid® HV, LV, 15830, Kelcoloid® O, S, 15831, Manucol Ester E/RK, 18617
Propylene glycol capreth-4, *See* Marlox® FK 14, 18942
Propylene glycol ceteth-3 acetate, *See* Hetester PCA, 13900
Propylene glycol ceteth-3 propionate, *See* Hetester PCP, 13901
Propylene glycol dicaprylate/dicaprate, *See* Captex® 200, 5167, Hodag CC-22, 14171, Lexol PG 855, 17153, Miglyol® 840, 19740, Neobee® 20, 20873, Edenol 302, 9603
Propylene glycol diisostearate, *See* Emalex PG-di-IS, 9981
Propylene glycol dilaurate, *See* Emalex PG-di-L, 9982
Propylene glycol dioctanoate, *See* Captex® 800, 5169, Lexol PG 800, 17152
Propylene glycol dioleate, 25217, *See* Emalex PG-di-O, 9983, Priolube 1429, 25018, Radia® 7204, 25731
Propylene glycol dipelargonate, *See* DPPG, 8950, Emerest® 2388, 10094, Lexol PG 900, 17154, Schercemol PGDP, 27619
Propylene glycol distearate, *See* Emalex PG-di-S, 9984
Propylene glycol hydroxystearate, *See* Naturechem® PGHS, 20776, Paricin® 9, 22842
Propylene glycol isoceteth-3 acetate, *See* Hetester PHA, 13902
Propylene glycol isostearate, 25218, *See* Emerest® 2384, 10093
Propylene glycol laurate, *See* Hodag PGML, 14213, Kessco® PGML, 16125, Schercemol PGML, 27620
Propylene glycol methyl ether, *See* Arcosolv® PM, 2814, Dowanol® PM, 8896, Icinol PM, 14816, Solvenon® PM, Solvent PM, 29050
Propylene glycol methyl ether acetate, *See* Arcosolv® PMA, 2815, Dowanol® PMA, 8897, Ektasolve® PM Acetate, 9672
Propylene glycol methyl ether acetate, *See* Dowanol® PMA, 8897
propylene glycol monolaurate, 25219, *See* Emalex PGML, 9985
Propylene glycol monomethacrylate, *See* Bisomer 2HPMA, 4274, hydroxypropyl methacrylate, 14667
propylene glycol monomethyl ether, 25220, *See* Dowanol® PM, 8896, Icinol PM, 14816
α-propylene glycol monomethyl ether, *See* Arcosolv® PM, 2814, Dowanol® PM, 8896, Icinol PM, 14816, Solvenon® PM, Solvent PM, 29050
Propylene Glycol Monomethyl Ether Acetate, *See*

Arcosolv® PMA, 2815
Propylene Glycol Monomethyl Ether Acetate-Methoxy-2-propyl Acetate, *See* Dowanol® PMA, 8897
Propylene glycol monooleate, *See* Emalex PGO, 9987, Propylene glycol oleate, 25221
Propylene glycol monostearate, *See* Aldo® PGHMS KFG, 1544, Emalex PGMS, 9986, Witconol 2380, 34498
1,2-Propylene glycol monostearate, *See* Tegomuls® P 411, 31011, Witconol 2380, 34498
Propylene glycol myristyl ether acetate, *See* Hetester PMA, 13903
Propylene glycol n-butyl ether, *See* Dowanol® PnB, 8898
Propylene glycol oleate, 25221, *See* Radiamuls® PG 2206, 25780
Propylene glycol palmitate, *See* Hodag PGMP, 14214
Propylene glycol ricinoleate, 25222, *See* Flexricin® 9, 11950, Naturechem® PGR, 20777
Propylene glycol stearate, *See* Aldo® PGHMS KFG, 1544, CPH-52-SE, 6942, Emerest® 2380, 10092, Grindtek PGMS 90, 13414, Hodag PGMS, 14215, Hodag PGS, 14224, Kessco® PGMS, 16126, Lipo PGMS, 17343, Mazol® PGMS, 19140, Pegosperse® PMS CG, 23067, Promodan SP, 25144, Schercemol PGMS, 27621, Witconol 2380, 34498, Witconol 2380, 34498
Propylene Glycol Stearate SE, *See* Aldo® PGHMS KFG, 1544, Lexemul® P, 17139
Propylene glycol stearic acid, ester, *See* Witconol 2380, 34498
Propylene glycol, PPG-12-PEG-65 lanolin oil and hydrolyzed animal protein, *See* Proto-Lan 30, 25292
1,2-propylene glycol, *See* propylene glycol, 25215
propylene glycol-behenalkonium chloride, *See* Incroquat Behenyl BDQ/P, 15032, Incroquat Behenyl TMC/P, 15034
propylene oxide, 25223, *See* Epihydrin, 10642
1,2-propylene oxide, *See* propylene oxide, 25223
Propylene oxide - allyl glycidyl ether copolymer, *See* Parel®, 22822
propylene polymer, *See* polypropylene, 24527
propylenephenoxythol, *See* 1-Phenoxy-2-propanol, 23625
Propylex, 25224
propylic aldehyde, *See* propanal, 25172
Propyliodone, 25225, 25226, *See* Dionosil, 8569
Propylol, 25230
propylparaben, 25231, *See* Nipasol M, 21282
N-propylparaben, *See* Nipasol M, 21282
Propylparaben USP, *See* Lexgard P, 17145
N-(2-Propyl)-N'-phenyl-*p*-phenylenediamine, *See* Vulkanox® 4010 NA, 34147
6-Propylpiperonyl Butyl Diethylene Glycol Ether, *See* piperonyl butoxide, 23875
Propylpyrogallol dimethyl ether, *See* Picamar, 23784
n-Propyltrimethoxysilane, *See* CP0810, 6936
Propyltris(trimethylsiloxy)silane, *See* P0820, 22500
2-propynoic acid, *See* propiolic acid, 25189
Propyon, *See* propoxur, 25201
propytal, *See* proponal,
Propyzamide, *See* Campbell's Rapier, 5069, Kerb 50W, 16064, 16065, Kerb Propyzamide 50, 16066, Rapier, 25889
proralone-mop, *See* Oxsoralen-Ultra, 22437
Prosan, 25233
Proscorbin, *See* vitamin C, 33938
Prosil 178, *See* CI7810, 6288, Dynasylan® IBTMO, 9410
Prosil®, 25234
Prosil® 178, 25235
Prosil® 196, 25236
Prosil® 220, 25237, *See* γ-aminopropyltriethoxysilane, 2186
Prosil® 248, 25238
Prosil® 3128, 25239, *See* Aminoethylaminopropyl trimethoxy silane, 2171
Prosil® 5136, 25240
Prosil® 9202, 25241
Prosil® HMDS, 25242
Prosobee, 25243
Prosol 525, 25245
Prosol 4692, 25244
Prosolvin, *See* Reprodin®, 26149
prosopite, 25246
Prosparol, 25247
Prospect, 25248
Prospect®, 25249
Prostar, *See* Propanil, 25174
Prostearin, *See* Witconol 2380, 34498
Prostearyl 15, 25250
Protaben P, *See* Nipasol M, 21282
Protachem 630, *See* Alkasurf® NP-4, 1715, Triton® N-57, 32430
Protachem CER, *See* cetyl ricinoleate, 5824
Protachem Smp, *See* Ablunol S-40, 109
Protachem Sto, *See* Ablunol S-85, 112, Witconol 2503, 34502
Protagent, *See* Agrimer 15L, 1065, Tears Plus®, 30876, Videne Disinfectant Solution, Videne Disinfectant Tincture, Videne Powder, 33739, Vinisil, 33827
Protagon, *See* lecithin, 16939
Protan, *See* Albutannin, 1406
Protanal, *See* Kelcosol®, 15832, Kelgin, 15841, Kelgin® F, 15842, kelgin® HV, LV, MV, 15843, Kelgin® QH, QL, QM, 15844, Kelgin® XL, 15845, Kelgum, 15846, Keltex®, 15878, Keltex® HV, 15879, Keltex® S, 15880, Keltone®, 15883, Keltone® HV, LV, 15884
protargentum, 25251
Protargin, *See* protargentum, 25251
Protargolgranulat, 25252
Protars, 25253
Protasan, 25254
Protasorb O-20, *See* Alkamuls® T-80, 1693
Protavic, 25255
protease, 25256, *See* Fungal Protease Conc, 12473
Protectoid, 25257
Protectol, 25258
Protectol®, 25259
Protectol® DMT, 25260
Protectol® GDA, GT 50, 25261
Protectol® GL 40, 25262
Protectol® KLC 50, 80, 25263
Protectol® TOE, 25264
Protectol®DMT, *See* dimethoxy Tetrahydrofuran, 8499
Protectyl, 25265
Protegin®, 25266
Protegin® W, WX, 25267
Protein concentrates, *See* Cegepaot® Range, 5552
Pro-Tein ES-20, 25268
Protein Grade®, 25269
Protein hydrolysate, *See* Amigen, 2119, Aminosol, 2189, Parenamine, 22823
Pro-Tein SA-20, 25270
Pro-Tein SM-20, 25271, *See* Myristoyl hydrolyzed collagen, 20576
proteinase, 25272, *See* Alcalase® 2,0 T, 1407, Esperase® 16.0L, 10839
protelds, 25273
Proteodermin, 25274
Proteol, 25275
Proteolite, *See* Gallatite, 12608
Proteosilane C, 25276
Prote-pon P 2 EHA-02-Z, 25277
Proteric CDX-38, *See* disodium cocoamphodiacetate, 8643
Proteric CM-36S, *See* Empigen® CDR40, 10369
Prote-sorb SML, 25278
Prote-sorb SMO, 25279
Prote-sorb SMP, 25280
Prote-sorb SMS, 25281
Prote-sorb STO, 25282
Prote-sorb STS, 25283
Protex, 25284
Pro-Tex, 25285
Protexulate, 25286
protheite, 25287
Prothera™, 25288
Prothiofos, *See* Tokuthion®, 31932
Prothiophos, *See* Tokuthion®, 31932
protocatechuic acid, 25289
protogest, 25290, *See* Extiat®, 11323
Proto-Lan 4R, 25293
Proto-Lan 8, 25294
Proto-Lan 20, 25291
Proto-Lan 30, 25292
Proto-Lan IP, 25295 , *See* isostearic acid, 15421
Proto-Lan KT, 25296
Protopet, 25297
(+)-protoquercitol, *See* acorn sugar, 367
Protosol, *See* Dihydroxyacetone, 8459
Protovit, 25298
Protrum K, 25299, *See* phenmedipham, 23595, 23596, 23597, 23598
Protrum K SN 38584, *See* phenmedipham, 23595, 23596, 23597, 23598
Protugan, 25300
Provatene, *See* carotene, 5308
Proventin, 25301
Proventin 7, 25302, 25303
Provil®, 25304
Provisc, *See* hyaluronic acid sodium salt, 14483
proviscol wax, *See* stearic acid, 29599
provitamin A, *See* carotene, 5308
Provitina, *See* vitamin D+73, 33941
Provol, 25305
Provol 50, 25306
Prowl, *See* pendimethalin, 23102
Prowl 3.3E, *See* pendimethalin, 23102
Prowl 4E, *See* pendimethalin, 23102
Proxan, *See* Nipacide® BIT, 21253
Proxan sodium, *See* Aero 343 Xanthate, 842
Proxel, 25307, *See* Nipacide® BIT, 21253
Proxel XL, *See* Nipacide® BIT, 21253
Proxil, *See* Nipacide® BIT, 21253
Proxitane, 25308
Proxol, *See* trichlorfon, 32212
Prox-onic 2EHA-1/02, 25309
Prox-onic BP-02 P, 25310
Prox-onic CC-05, 25311
Prox-onic CSA-1/04, 25312
Prox-onic CSA-1/06, *See* Ceteareth-6, 5788
Prox-onic DA-1/04, 25313, *See* Deceth-6, 7848
Prox-onic DDP-09, 25314, *See* Dodoxynol series, 8786
Prox-onic DNP-08, 25315, *See* Ethal DNP-8, 10919
Prox-onic DT-03, 25316
Prox-onic EP 1090-1, 25317
Prox-onic HR-05, 25318
Prox-onic HRH-05, 25319
Prox-onic L 081-05, 25320
Prox-onic LA-1/02, 25321
Prox-onic MC-02, 25322
Prox-onic MG-020 p, 25323
Prox-onic MHT-015, 25324
Prox-onic MO-02, 25325
Prox-onic MS-05, 25326
Prox-onic MT-02, 25327
Prox-onic NP-04, 25328
Prox-onic OA-1/04, 25329
Prox-onic OA-1/04, *See* oleth-8, 22126
Prox-onic OCA-1/06, 25330
Prox-onic OL-1/05, 25331
Prox-onic OP-09, 25332
Prox-onic PEG-2000, 25333
Prox-onic PH-01, 25334
Prox-onic PPG-900, 25335
Prox-onic SA-1/02, 25336
Prox-onic SA1-015/P, 25337
Prox-onic SML-020, 25338
Prox-onic SMO-05, 25340
Prox-onic SMO-020, 25339
Prox-onic SMP-020, 25341
Prox-onic SMS-020, 25342, *See* Ethylan® GES6, 11108
Prox-onic ST-05, 25343
Prox-onic TA-1/08, 25344, *See* PEG tallate series, 23035

Prox-onic TD-1/03, 25345, *See* Trideceth series, 32232
Prox-onic TM-06, 25346
Prox-onic UA-03, 25347
Prox-onix SA-1/02, *See* steareth series, 29596
Proxy, 25348
Proxyl, 25349
Prozinex, 25350, *See* propazine, 25179
Prozone, 25351
prussic acid, 25352, *See* hydrocyanic acid, 14584, zootic acid, 34903
Pruteen, 25353
Pruv, 25354
Pryfon, 25355
Prystal, 25356
Prystaline, 25357
PS, *See* polystyrene, 24601
PS021, 25358
PS034, 25359
PS071, 25360
PS072, 25361
PS073, 25362
PS-10GF/000, 25363
PS130, 25364
PS140, 25365
PS181, 25366
PS187, 25367
PS3/PS4/PS5, 25368
PS-30GM/000, 25369, *See* Grilpet® XE3060, 13402
PS-60, *See* Gellan gum, 12736, Kelco-Gel® Gellan Gum, 15827
PSA, 25370
PSDF04, 25371
PSDF05, 25372
pseudo-alums, 25373
Pseudocollagen, 25374
Pseudocyanuric acid, *See* Zeonet A, 34720
pseudo-galena, *See* zinc blende, 34767
Pseudopinene, *See* β-l-pinene, 23858
pseudo-wollastonite, *See* calcium silicate, 4969
psoralen-mop, *See* Oxsoralen-Ultra, 22437
Psoralon, *See* Allantoin, 1739
PTAL, 25375
PTAP, *See* Orthophen® 278, 22336, Pentaphen® 67, 23168
pteglu, *See* folic acid, 12196
Pteroylglutamic Acid, *See* folic acid, 12196
L-pteroylglutamic acid, *See* folic acid, 12196
pteroyl-L-glutamic acid, *See* folic acid, 12196
pteroyl-L-monoglutamic acid, *See* folic acid, 12196
pteroylmonoglutamic acid, *See* folic acid, 12196
PTFE, *See* polytetrafluoroethylene, 24640, 24642
PTFE, TFE, *See* polytetrafluoroethylene, 24642
PTFE-19, *See* Polytetrafluoroethylene, 24642
PTFE-20, 25376
PTFET, *See* polytetrafluoroethylene, 24641
PTMEG, *See* polytetramethylene ether glycol, 24643
PTMG, *See* polytetramethylene ether glycol, 24643
***p*-toluquinaldine *p*-toluquinoline-ethylcyanine bromide**, *See* Orthochrome T, 22326
***p*-tolylaminoditolyl-*p*-toluquinone diimine**, *See* Perkin's Base, 23320
PTSA, *See* *p*-toluenesulfonamide, 31949, toluene sulfonic acid, 31950
PTSA 70, 25377
PTZ® Phenothiazine Purified, 25378
PU elastomer, *See* IE-40-A, 14842
Pularin, *See* Heparin, Sodium Salt, 13788
Pulegium oil, *See* pennyroyal oil, 23119
pulegone, 25379
(R)-(+)-Pulegone, *See* pulegone, 25379
D-Pulegone, *See* pulegone, 25379
Pulluzyme, 25380
Pulmolite, 25381
Pulpex E and P, 25382
Pulpzyme, 25383
Pulsar, 25384, 25385
Pulse, 25386
Pulse 600, 25389
Pulse 1310, 25387
Pulse 1735, 25388
Pulse®, 25390
Pultac, 25391
Pulvatex, 25392
Pulvex, *See* Kafil, 15655
Pulvis Conservans, *See* Nipasol M, 21282
Pumice Plus, 25393
pumice stone, 25394
Pumiline, 25395
Pump Repair Putty, 25396
Punch, *See* flusilazole, 12119
Punch® C, 25397
Punctilious® Ethyl Alcohol, 25398
punicin, 25399
Purac, 25400
Purac®, *See* lactic acid, 16570
Puragel, *See* gelatin, 12722, Pharmagel, 23572
Puralin, *See* Rezifilm, 26521
Puralyn, *See* Thiram, 31726
Puraspec, 25401
Puratronic, 25402
Purbeck stone, 25403
Purdox, 25404, *See* aluminum oxide, 1919
Purdurum, *See* Camite, 5054
Pure cellulose, *See* Metasil SA, SB, 19458
Pure Food Powd. Starch 105-A, 131-C, 142-A, *See* corn starch, 6837
Pure Food Powder, *See* corn starch, 6837
Purecal, *See* calcium carbonate, 4940
Pure-Dent® B700, 25405
Pure-Dent® B700, B810, B812, B815, B816, B880, *See* corn starch, 6837
Purez, 25406
Purgatol, 25407
purging nut oil, 25408
Purifloc, 25409
1H-purin-6-amine, *See* Adenine, 679
purine-6-ol, *See* hypoxanthine, 14721
purine-2,6,8-triol, *See* uric acid, 33139
purine-3,6,8(1H,3H,9H)-trione, *See* uric acid, 33139
Purisol, 25410
Purity® 21, 25411, *See* corn starch, 6837
Pur-Oba®, 25412
Purochem, 25413
Purochin, *See* urokinase, 33145
Purocyclina, *See* tetracycline, 31365
Puromix, 25414
Puron, 25415
Purozone, 25416
Purple Copp, 25417
Purple cuprous oxide, *See* Purple Copp, 25417
Purplecopp 97N Premium, 25418
Purpurin, *See* oxyalizarin, 22438
Purpurine, *See* oxyalizarin, 22438
purpuroxanthic acid, *See* Munjistin, 20499
Purree, *See* Indian Yellow, 15069
Purton CFD, 25419, *See* Active #2, 501
Purton SFD, 25420
Purus, *See* Chloramine T, 6086
Purzaust® Catalysts, 25421
Pusher, 25422
putrescine, 25423
putty powder, 25424
puzzuolana, 25425
PV, PV Fast, 25426
PVA, *See* polyvinyl alcohol, 24666
PVAc, *See* polyvinyl acetate homopolymer, 24665
Pvacote, 25427
PVAL, *See* polyvinyl alcohol, 24666
PVB, *See* polyvinyl butyral, 24667
PVC, *See* polyvinyl chloride, 24668, *See* Korogel, 16354, Koroplate, 16357
PVC Deodorant #5417, OS, 25428
PVDF, *See* Dyflor 2000, 9314
PVF2, *See* polyvinyl fluoride,
PVM/MA copolymer, *See* Agrimer VEMA-H-240, 1068
PVOH, *See* polyvinyl alcohol, 24666
PVP, 25429, *See* Tears Plus®, 30876
PVP/eicosene copolymer, 25430, *See* Ganex® V-220, 12643
PVP/ethyl methacrylate/methacrylic acid terpolymer, *See* Stepanhold® Extra, 29679
PVP/hexadecene copolymer, 25431, *See* Ganex® V-216, 12642
PVP/VA copolymer, *See* Agrimer VA 6, 1067
PX 104, *See* dibutyl phthalate, 8366
PX-104, 25432
PX-109, 25433
PX-10GF/000, 25434
PX-111, 25435, *See* Jayflex® DUP, 15556
PX-120, 25436, *See* diisodecyl phthalate, 8462
PX-126, 25437, *See* Jayflex® DTDP, 15555
PX-138, 25438
PX-139, *See* Jayflex® DINP, 15552
PX-209, 25439
PX-238, 25440, *See* dioctyl adipate, 8548, Plasthall® DOA, 24002, PX-238, 25440, Wickenol® 158, 34393
PX-338, 25441, *See* Jayflex® TOTM, 15559
PX-339, 25442, *See* Jayflex® TINTM, 15558
PX-438, *See* Plasthall® DOS, 24005, Reomol DOS, 26131
PX-504, 25443
PX-538, 25444
PX-800, 25445
PX-914, 25446
PX-917, *See* Syn-O-Ad® 8484, 30558, TCP, 30857
Py, *See* pyrimidine, 25484
PY Garden Insect Killer, 25447
PY Garden Insecticide, 25448
PY Powder Garden & Household Insect Killer, 25449
Pybuthrin, 25450, *See* piperonyl butoxide, 23875
Pydrin, *See* fenvalerate, 11620
Pyelokon-R, *See* acetrizoate sodium, 307
Py-Kill, *See* tetramethrin, 31384
Pylen, *See* A-C® Polyethylene 6, 6A, 7, 7A, 8, 8A, 9, 9A, 617, 617A, 175, Cabelec® 1017, 4862
Pylkrome, 25451
Pylumin, 25452
Pynamin, *See* allethrin, 1744
Pynol, 25453
Pynosect, *See* Kafil, 15655, permethrin, 23384 23385 23386, resmethrin, 26218
Pynosect 30, 25454
Pyoctanin, 25455
Pyoktanin, *See* gentian violet, 12845, Pyoctanin, 25455
Pyr, *See* pyridine, 25481
Pyr, *See* pyrimidine, 25484
Pyracur® FL, 25456
Pyradex® T, 25457
Pyradiolin, 25458
Pyradur®, 25459
Pyra-Fog 100, 25460
Pyralin, 25461
Pyralin®, 25462
pyraloxin, 25463
Pyralux®, 25464
Pyramid, 25465
Pyramin, *See* chloridazon, 6103
Pyramin®, 25466
Pyramol, 25467
Py-Ran, 25468
2H-pyran-2,4(3H)-dione, 3-acetyl-6-methyl-, *See* Dehydroacetic acid, 7964
Pyranet, 25469
pyrantimonite, *See* red antimony, 25951
Pyrasteel, 25470
Pyratex, 25471
Pyraton, 25472
Pyrax ABB, *See* Aciculite, 326
Pyrax talcs A and B, 25473
Pyrax® A, 25474
Pyrazine, *See* Antipyrine,
pyrazine hexahydride, *See* piperazine, 23873
Pyrazol, *See* chloridazon, 6103
1H-pyrazole-3-ethanamine dihydrochloride, *See* betazole hydrochloride, 4072
Pyrazoline, *See* Antipyrine,
pyrazon, *See* chloridazon, 6103

Pyrazophos, *See* Afugan, 952, Missile, 19944
Pyre-ML®, 25475
Pyrene, 25476
pyrene oil, 25477
Pyrenone 606, *See* piperonyl butoxide, 23875
Pyresin, *See* allethrin, 1744
Pyretherm, *See* resmethrin, 26218
Pyrethin, *See* Killgerm® ULV 400, 16186, Pyra-Fog 100, 25460
Pyrethins, *See* HC 200 Concentrate, 13714
Pyrethrin, *See* Alfadex, 1573, Dairy Fly Spray, 7701
pyrethrum powder, 25478
Pyrexcel, *See* allethrin, 1744
Pyrgos, *See* Chloramine T, 6086
Pyricit, 25479
Pyridate, 25480
Pyridin, *See* fenvalerate, 11620
pyridine, 25481
pyridine-3-carboxylic acid, *See* nicotinic acid, niacin, 21090
pyridine-4-carboxylic acid, *See* isonicotinic acid, 15380
2,6-pyridinediamine, *See* 2,6-diaminopyridine, 8305, diaminopyridine, 8306
Pyridine, ethenyl-, *See* vinyl pyridine, 33858
2-pyridinol, *See* 1-hydroxy-2-pyridine, 14669
2-pyridinethiol-1-oxide, sodium salt, *See* Sodium Omadine® 40% Aq. Sol'n, 28803, sodium pyrithione, 28813
Pyridine, vinyl-, *See* vinyl pyridine, 33858
Pyridipca, *See* pyridoxine hydrochloride, 25483, vitamin B_6 hydrochloride, 33937
Pyrido rubber, 25482
2-pyridone, *See* 1-hydroxy-2-pyridine, 14669
Pyridox, *See* pyridoxine hydrochloride, 25483, vitamin B_6 hydrochloride, 33937
pyridoxine hydrochloride, 25483, *See* Gravidox, 13294, Hexa-Betalin, 13987, Hexavibex, 14019, vitamin B_6 hydrochloride, 33937
pyridoxinium chloride, *See* vitamin B_6 hydrochloride, 33937
pyridoxol hydrochloride, *See* vitamin B_6 hydrochloride, 33937
β-pyridyl-α-N-methylpyrrolidine, *See* XL-All Insecticide, 34607
Pyrimicarbe, *See* Abol, 146
pyrimidine, 25484, *See* miazine, 19625
2,6-pyridinediamine, *See* 2,6-diaminopyridine, 8305
2,4-pyrimidinediol, *See* uracil, 33101
2,4(1H,3H)-pyrimidinedione, *See* uracil, 33101
2,4,5,6(1H,3H)-pyrimidinetetrone, *See* Alloxan, 1750
2,4,5,6(1H,3H)-pyrimidinetetrone-5-oxime, *See* violuric acid, 33881
pyrimitate, *See* pyrimithate, 25485
pyrimithate, 25485
Pyrinex, 25486, *See* Chlorpyrifos, 6163, Talon, 30751
Pyrisept, *See* Acetoquat CPC, 302, cetyl pyridinium chloride, 5823
pyrites, cockscomb, *See* marcasite, 18700
pyrites, coxcomb, *See* marcasite, 18700
pyrites, radiated, *See* marcasite, 18700
pyrites, white iron, *See* marcasite, 18700
Pyrithione zinc, *See* Head and Shoulders, 13726, Zinc Omadine, 34783, zinc pyrithione, 34791
Pyro, 25487, *See* pyrogallic acid, 25510, pyrogallol, 25512
pyro alcohol, 25488
pyro cotton, 25489
Pyroban G, 25490
pyro-bitumen, 25491
Pyroblak, 25492
Pyrobond, 25493
Pyrobor, 25494
Pyrobrite, 25495
Pyrocast, 25496
pyrocatechin, 25497, *See* , *See* kachin, 15651, pyrocatechol, 25498
pyrocatechol, 25498
Pyrocatechol, *See* kachin, 15651, pyrocatechin, 25497
pyrocatechol arsenic acid, 25499
pyrocatechol dimethylether, *See* veratrole, 33568
pyrocatechol methyl ether, *See* guaiacol,
pyrocatechol monoethyl ether, *See* Ethacol, 10905
Pyro-Chek® 68PB, 25500
Pyro-Chek® LM, 25501
Pyrochlor, 25502
Pyrocide, 25503, *See* allethrin, 1744
Pyroclean, 25504
Pyroclense, 25505
pyrocollodion, 25506
Pyrodialite, 25507
Pyroforane, 25508
pyrofulmin, 25509
pyrogallic acid, 25510, *See* pyrogallol, 25511, 25512, 25513, 25514
pyrogallol, 25511, 25512, 25513, 25514, *See* Piral, 23877, pyrogallic acid, 25510
pyrogallol 1,3-dimethyl ether, *See* dimethoxyphenol, 8500
Pyrogallol dimethyl ether, *See* dimethoxyphenol, 8500
pyroglutamic acid, *See* Ajidew A-100, 1142
L-Pyroglutamic acid, *See* Ajidew A-100, 1142, Pidolidone®, 23832
pyrolignite of iron, 25515
pyrolignite of lime, 25516
Pyrolith, 25517
pyrolytic boron nitride, *See* Boron nitride, 4443
Pyromet® Alloy 625, 25518
pyromic, 25519
pyromorphic phosphorus, 25520
pyromucic acid, *See* 2-Furoic acid, 12497
pyronalorange, *See* Sudan I, 29938
Pyronate, 25521
α-Pyrone-3-carboxylic acid, *See* coumalic acid, 6901
2-Pyrone-5-carboxylic acid, *See* coumalic acid, 6901
Pyronin stain, *See* Pappenheim's stain, 22709
pyronine B, 25522
pyronine G, *See* pyronine Y, 25523
pyronine Y, 25523
Pyronite, *See* Tetryl, 31423
Pyronium, 25524
Pyrophan, 25525
Pyrophosphate, *See* sodium pyrophosphate, 28814
Pyrophosphoric acid, tetrasodium salt, *See* sodium pyrophosphate, 28814
Pyrophyllite, *See* Pyrax® A, 25474, Tisyn®, 31857
Pyroplasmin, *See* Acaprin®, 181, 1,3-di-6-quinolylurea, 8618
pyroracemic acid, *See* pyruvic acid, 25538
pyroretin, 25526
Pyros, 25527
Pyroset, 25528
pyrosin B, *See* erythrosin, 10778
pyrosin G, *See* Erythrosin G, 10780
pyrosine B, *See* erythrosin, 10778
pyrostibnite, *See* red antimony, 25951
pyrosulfuric acid, *See* fuming sulfuric acid, 12457
pyrosulfurous acid, disodium salt, *See* sodium metabisulfite, 28783
Pyroter CPI-40, 25529
Pyroter GPI-25, 25530
Pyrovatex, 25531
Pyrox®, 25532
Pyroxylin, *See* collodion cotton, 6627, guncotton, 13492, Kodaloid, 16283, nitrocellulose, 21353, nitro-starch, 21377, Venite, 33560
Pyrozone, *See* hydrozone, 14672
pyrrhol, 25533
Pyrrhotine, *See* magnetic pyrites, 18375, white pyrites, 34371
pyrrodiazole, 25534
Pyrrole, 25535, *See* pyrrhol, 25533
1H-Pyrrole, *See* pyrrhol, 25533, Pyrrole, 25535
m-pyrrole, *See* NMP, 21406
2-pyrrolidine carboxylic acid, *See* proline, 25129
2,5-pyrrolidinedione, *See* succinimide, 29917
2,5-Pyrrolidinedione, 1-chloro-, *See* N-chlorosuccinimide, 6148
Pyrrolidinedione, 1-chloro-, *See* N-chlorosuccinimide, 6148
2-pyrrolidinone, *See* Soluphor® P, 29023
2-Pyrrolidinone, 1-cyclohexyl-, *See* CHP, 6173
α-pyrrolidone, *See* Soluphor® P, 29023
Pyrrolidone carboxylic acid, aluminum salt, *See* aluminum PCA, 1921
Pyrrolidone carboxylic acid, lauryl ester, *See* lauryl PCA, 16866
pyrrolidone-2, *See* pyrrolidone2-pyrrolidone, 25536, Soluphor® P, 29023
2-pyrrolidone, 25536
2-pyrrolidine carboxylic acid, *See* proline,
5-Pyrrolidone-2-carboxylic acid, *See* Ajidew A-100, 1142
L-2-Pyrrolidone-5-carboxylic acid, *See* Ajidew A-100, 1142
L-5-Pyrrolidone-2-carboxylic acid, *See* Ajidew A-100, 1142
pyrroline, *See* pyrrhol, 25533, pyrrole, 25535
pyrro[*b*]monazole, *See* Imidazole, 14895
pyruvic acid, 25537, 25538
Python, 25539
pyxol, 25540
Q uaternium 12, *See* Querton 210Cl-50, 25651
Q-137, *See* diethyldiphenyldichloroethane, 8441
Q-1300, 25541
Q-1301, 25542
QA-555, 25543
Qamlin, *See* permethrin, 23384, 23385, 23386
Qaulineg, 25544
Qazi-ketcham, *See* ethylene oxide, 11131
Qazul, 25545
Q-Broxin, 25546
QC-8800, 25547
Q-Cast, 25548
Q-Cel® 300, 25549
Q-Crete, 25550
Q-Gum, 25551
QI Damp, *See* Ampicillin, 2361
QO® Furan, 25552
QO® Furcarb®, 25553
QO® Furfural, 25554
QO® Furfuryl Alcohol (FA®), 25555
QO® Polymeg® 650, 25556
QO® Quacorr® Resin/Catalyst Systems, 25557
QO® Tetrahydrofuran (THF), 25558
QO® Tetrahydrofurfuryl Alcohol (THFA®), 25559
QR 819, *See* Ablusol ML, 129
Q-Therm, 25560
Quab 151, 25561
Quab 188, 25562
Quabond® 210, 25563, *See* Everflex® SP-1084, 11266
Quabond® 230, 25564
Quad DSM, 25565
Quad MCPA 50%, 25566
Quad Mini Slug Pellets, 25567
Quad Store, 25568
Quadban, 25569
Quadefome® MAB, 25570
Quad-Fast, 25571
Quad-Keep, 25572
Quadrafos, 25573, 25574
Quadrangle Chlormequat 700, 25575
Quadrangle Cropspray 11E, 25576
Quadrangle Cyper, 25577
Quadrangle Super-Tin 4L, 25578, *See* Farmatin, 11480
Quadrilan® AT, 25579
Quadrilan® BC, 25580
Quadrilan® MY 211, 25581
Quadrol, *See* Mazamide® 1214, 19107, tetra(2-hydroxypropyl) ethylenediamine, 31356
Quadrol®, 25582
Qualamox, 25583
Qualidot, 25584
Qualifix, 25585
Qualitol, 25586
Qualloflex®, Quallofil®, Quallofirm®, 25587
Quamectant AM-50, 25588
Quamilin, 25589, *See* Permethrin, 23385

Quamonium, *See* Cetrimonium bromide, 5813, Cetavlex, Cetavlon, 5787, Sumquat® 6030, 30122, Varisoft® CTB-40, 33421
Quantacure®, 25590
Quantum, 25591, 25592, 25593
Quantum (LOGO), 25594
Quantum®, 25595
Quanyl-quanidine, *See* Biquanide, 4218
Quartex CTAC, *See* Cetrimonium chloride, 5814
quartz, 25596, *See* silica, 28313
quartz glass, 25597
quartzilite, 25598
quarzal, 25599
Quasar, 25600
Quasilan, 25601
Quassia, *See* Dog Off, 8788
Quat Keratin WKP, 25602
Quat-Coll CDMA 40, 25603
Quat-Coll IP10-30, 25604
Quat-Coll QS, 25605
Quatemium-18 - isopropyl alcohol, *See* M-Quat® 257, 20400
Quatermium-22, *See* Ceraphyl® 60, 5718
Quaterium 23, *See* Polyquaternium-11, 24530
Quatermium-26, *See* Ceraphyl® 65, 5719
Quatermium-33, *See* Lanoquat® 1756, 16744
Quatermium-70, *See* Ceraphyl® 70, 5720
Quatermium-83, *See* Rewoquat W 75 H, 26395
Quaternary ammonium chloride, *See* Acid Foamer, 332, Katapone VV-328, 15785
Quaternary ammonium compound, *See* Agrilan® TKA103, 1061
Quaternary ammonium compounds, (carboxymethyl)coco alkyldimethyl, hydroxides, inner salts, *See* Accobetaine CL, 222
Quaternary arylammonium chloride, *See* Dodicor 2565, 8783
Quaternary imidazoline derivative, *See* Zoharsoft 90, 34874
quaternary steels, 25606
Quaternary sulfate, *See* Alacsan T, 1328
Quaternium 12, *See* BTC® 1010-80, 4638, Radiaquat 6412, 25790
Quaternium-18, *See* Adogen® 442, 769, Arquad® 2HT-75, 3103, Kemamine® Q-9702C, 15925, Varisoft® DHT, 33422
Quaternium-24, *See* BTC® 818, 4635
Quaternium-27, 25607, *See* Carsosoft® S-90, 5353, Incroquat S-75CG, 15043, Incrosoft S-75, 15050, Varisoft® 475, 33415, Varisoft® TIMS, 33430
Quaternium-52, *See* Dehyquart SP, 7988
Quaternium-53, *See* Carsosoft® T-90, 5354
Quaternium-72, *See* Incroquat 248, 15028, Incrosoft 248, 15048
Quaternium-80, *See* Abil®-Quat 3270, 3272, 68
Quaternium-82, *See* Amonyl DM, 2299
Quaternium-84, *See* Mackernium NLE, 18209
Quaternized imidazoline, *See* Quatrex 152, 25615, Quatrex 182, 25617
Quaternol 1, *See* Cycloton® SCS, 7542, octadecylbenzenemethanaminium chloride, 21985, stearalkonium chloride, 29592, Sumquat® 6210, 30126
Quat-Pro E, 25608
Quat-Pro S, S-30, 25609
Quatrachlor, *See* benzethonium chloride, 3963
Quatramine, 25610
Quatrene 7670, 25611
Quatrene C-5-6, 25612
Quatrene CB-50, 25613
Quatrex, 25614
Quatrex 152, 25615
Quatrex 162, 25616
Quatrex 182, 25617
Quatrex 1010, 25618
Quatrex 2410, 25619
Quatrex 5010, 25620
Quatrex 6410, 25621
Quatrex CRC, 25622
Quatrex CT-100, 25623
Quatrex CTAC, 25624
Quatrex S, 25625
Quatrex STC-25, 25626
Quatrisoft Polymer LM-200, 25627
Quat-Silk QTM-10, 25628
Quat-Soy CDMA-25, 25629
Quat-Soy LDMA-30, 25630
Quat-Veg Q-30, 25631
Quat-Wheat CDMA-30, 25632
Quat-Wheat QTM-20, 25633
Quat-Wheat SDMA-30, 25634
quebrachite, 25635
quebrachitol, *See* quebrachite, 25635
quebracho, 25636
Quecodur AE, 25637
Quecodur B Granular, 25638
Quecodur CW Conc, 25639
Quecophob HPA, 25640
Queen's metal, 25641
Queen's yellow, *See* turpeth mineral, 32612
Queensland arrowroot, *See* tous-les-mois starch, 32088
Queletox, *See* Tiguvon®, 31788
Quelicin, 25642
Quell Oil, 25643
Quellada, 25644
Quenty®, 25645
Quenty® forty, 25646
quercetin, 25647, *See* Ritacetin, 26799
quercetin 3-rutinoside, *See* rutin trihydrate, 27150
quercetol, *See* quercetin, 25647
Quercitol, *See* acorn sugar, 367
quercitron, 25648
quertine, *See* quercetin, 25647
Querton 14Br-40, 25649
Querton 16Cl-29, 25650, *See* Cetrimonium chloride, 5814
Querton 210Cl-50, 25651
Querton 210CL, *See* Radiaquat 6410, 6412, 25789, Radiaquat 6412, 25790, Rewoquat B 10, 26389
Querton 280, 25652
Querton 441-BC, 25653
Querton 442, 25654
Querton GCl-50, 25655
Querton KKBCl-50, 25656
Quesbrom, 25657, *See* Bromochloro dimethyl hydantoin, 4574
Queschlor, 25658
Quesfloc, 25659
Quesfloc F11283-1, 25660
Questal DI 0770, 25661
Questal Extra Powd. Conc. 0780, 25662
Questal FEC 0800, 25663
Questal Special 0860, 25664
Questex, *See* Kalex 220 Crystal, 15668, Kalex Liq. 50%, 15671, tetrasodium EDTA, 31391
Questric Acid 5286, 25665
Quevenne's iron, 25666
Quiacryl, 25667
Quiana®, 25668
Quick, *See* Chlorophacinone, 6133, Karate, 15758, Ridene, 26746, Rodeo, 26899
Quick Cure®, 25669
quickening liquid, 25670
quicklime, *See* calcium oxide, 4962, KM Pebble Lime, 16240, lime, 17265
Quick-pach, 25671
Quickset® Extra, 25672, *See* Esperfoam® FR, 10842
quicksilver, *See* mercury, 19355
quicksilver vermilion, *See* mercuric sulfide, red and black, 19350
Quickstir, 25673
Quidur, 25674
Quikote, 25675
Quikset, 25676
Quilan, *See* Benfluralin, 3940
Quilastic, 25677
Quimar, *See* α-Chymotrypsins, 6286
Quimoral, *See* α-Chymotrypsins, 6286
Quimotrase, *See* α-Chymotrypsins, 6286
quinaldine, 25678
quinalphos, 25679
Quinclorac, *See* Facet®, 11405
Quindex, 25680
Quindo®, 25681
quindoxin, 25682
Quinine iodo-sulfate, *See* Herapath's salt, 13799
quinizarin, 25683
Quinn's Rubber, 25684
quinofen, *See* cinchophen, 6307
quinofop-ethyl, *See* quizalofop-ethyl, 25697
quinol, 25685
quinol, *See* Tecquinol® Tech. Grade, 30910
Quinol Ed, *See* Polyflex, 24435
2-quinoline, *See* carbostyril, 5244
quinoline blue, *See* Cyanine, 7475
2-quinolinol, *See* carbostyril, 5244
8-quinolinol, *See* 8-hydroxyquinoline, 14670
8-quinolinol sulfate monohydrate, *See* Quinosol, 25687
2(1H)-quinolinone, *See* carbostyril, 5244
quinolor compound, *See* Benzoyl peroxide, 3983
quinomethionate, 25686
quinone dioxime, *See* Accelerator G.M.F, 206
***p*-quinonedioxime**, *See* GMF, 13161
quinophan, *See* cinchophen, 6307
quinosol, 25687
quinovasugar, 25688
quinovitol, *See* Quinovasugar, 25688
quinoxaline-1,4-dioxide copper(II) salt, *See* quindoxin, 25682
quinoxalines, *See* quinomethionate, 25686
(2-quinoxalinylmethylene)hydrazinecarboxylic acid methyl ester N,N'-dioxide, *See* carbadox, 5177
Quinta-Pro Conc, 25689
Quintar, *See* Dichlone, 8381
Quintesse, 25690
Quintiofos, 25691
Quintozene, 25692, *See* Botrilex, 4460, Brabant PCNB, 4477, Bras-sicol, 4489, Folosan, 12208
3-quinuclidinone hydrochloride, 25693
quisqueite, 25694
Quiver, 25695
Quixalud, 25696
quizalofop-ethyl, 25697, *See* Pilot, 23841
Quniadome, *See* Yodoxin, 34658
quolac ex-ub, *See* Rhodapon® 101-10, 26631, sodium lauryl sulfate, 28781
quorn, 25698
Quso® G27, G29, G35, G38, WR55, WR55-FG, WR83, 25699
Quso® WR55-FG, 25700
Q-Vibe, 25701
R 17635, *See* Vermox, 33591
R 23979, *See* imazalil, 14892
R acid, *See* R-Acid, 25710
R Type Solvent®, 25702
R.A.E. 57 alloy, 25703
R.O.D, 25704
R-2 Crystals, 25706
R-12, *See* dichlorodifluoromethane, 8388
R-13, *See* chlorotrfluoromethane, 6153, Genetron® 13, 12819
R-23, *See* Fluoroform, 12099, Genetron® HFC23, 12829
R-40, *See* methyl chloride, 19522
R-114, *See* Cryofluorane, 7278
R-125, *See* Genetron® HFC 125, 12828
R-502, MR-502K, MR-502Y, MR-502P, 25707
R-1007, 25705
R-1608, *See* EPTC, 10729
R-1929, *See* azaperone, 3518
R-7315, *See* Hypnodil, 14716
R-12564, *See* Tramisol®, 32134
R-17635, *See* Mebendazole, 19198, Telmin, 31066
R-17889, *See* Flubendazole, 12039
R-23979, *See* Enilconazole, 10582
R-25831, *See* carnidazole, 5298

R-31520, *See* Closantel, 6441
R-43512, *See* Hismanal, 14132
Rabalon, 25708
Racemethionine, *See* Petameth, 23451
racemic acid, *See* tartaric acid, 30832
racemic menthol, *See* menthol terpine hydrate, 19321
racemic tartaric acid, *See* paratartaric acid, 22803, tartaric acid, 30832
Rachromate-51, 25709
R-Acid, 25710
Rackarock, 25711
Rackarock Special, 25712
Racumin®, 25713
Radapon, *See* Basfapon®, 3705, Dowpon, 8943
Radar, Radar Propiconazole, 25714
Radarsan, *See* Rawstol, 25907
Radazine, *See* atrazine, 3394
raddle, *See* Indian Red, 15068
Radel® A-100, 25715
Radeverm, *See* Mansonil®, 18611
Radex, 25716
Radia® 7040, 25717
Radia® 7051, 25718
Radia® 7060, 25719
Radia® 7108, 25720
Radia® 7110, 25721
Radia® 7117, 25722, *See* methyl coconate, 19523
Radia® 7120, 25723, *See* methyl palmitate, 19555
Radia® 7131, 25724
Radia® 7171, 25725
Radia® 7176, *See* Pentaerythrityl tetrastearate, 23143
Radia® 7176, Radiasurf® 7175, 25726
Radia® 7185, 25727
Radia® 7187, 25728
Radia® 7190, 25729
Radia® 7200, 25730, *See* isopropyl palmitate, 15410
Radia® 7204, 25731, *See* Propylene glycol dioleate, 25217
Radia® 7230, 25732
Radia® 7231, 25733
Radia® 7241, 25734
Radia® 7266, 25735
Radia® 7331, 25736
Radia® 7345, 25737
Radia® 7355, 25738
Radia® 7363, 25739
Radia® 7370, 25740
Radia® 7371, 25741
Radia® 7500, 25742
Radia® 7501, 25743
Radia® 7505, 25744
Radia® 7506, 25745
Radia® 7510, 25746
Radia® 7514, 25747
Radiacid® 631, *See* Hystrene® 1835, 14755
Radiamac 6149, 25748
Radiamac 6169, 25749
Radiamine 6140, 25750
Radiamine 6141, 25751
Radiamine 6160, 25752
Radiamine 6161, 25753
Radiamine 6164, 25754
Radiamine 6170, 25755
Radiamine 6170, 6171, *See* tallowamine, 30748
Radiamine 6171, 25756, *See* tallowamine, 30748
Radiamine 6172, 25757
Radiamine 6240, 25758
Radiamine 6260, 25759
Radiamine 6270, 25760
Radiamine 6310, 25761, *See* didecyl methylamine, 8415
Radiamine 6343, 25762
Radiamine 6346, 25763
Radiamine 6360, 25764
Radiamine 6540, 25765
Radiamine 6560, 25766, *See* Cocodiamine, 6564
Radiamine 6570, 25767
Radiamine 6572, 25768
Radiamox® 6804, *See* lauramine oxide, 16832
Radiamuls® 2602, 25769
Radiamuls® Acetem 2021, 2134, 25770
Radiamuls® Citrem 2931, 2932, 25771
Radiamuls® CSL 2980, 25772
Radiamuls® Datem 2001, 2008, 25773
Radiamuls® GTH 2375, GTH 2376, 25774
Radiamuls® Lactem 2950, 25775
Radiamuls® MCT 2108, 25776
Radiamuls® MG 2141, MG 2142, MG 2600, MG 2900, 25777
Radiamuls® MG 2152, 25778
Radiamuls® PG 2201, 25779
Radiamuls® PG 2206, 25780, *See* Propylene glycol oleate, 25221
Radiamuls® Poly 2248, 25781
Radiamuls® Poly 2253, 25782
Radiamuls® Sorb 2145, Sorb 2161, Sorb 2166, 25783
Radiamuls® Sorb 2147, 25784, *See* Ethylan® GES6, 11108
Radiamuls® Sorb 2157, 25785
Radiamuls® Sorb 2344, Sorb 2345, 25786
Radiamuls® Sorb 2345, 25787
Radiamuls® SSL 2990, 25788
Radiaquat 6410, 6412, 25789
Radiaquat 6412, 25790
Radiaquat 6442, 25791
Radiaquat 6444, 25792, *See* Cetrimonium chloride, 5814
Radiaquat 6462, 25793
Radiaquat 6470, 25794
Radiaquat 6471, 25795
Radiaquat 6475, 6480, 25796
Radiasurf 7125, *See* Ablunol S-20, 108
Radiasurf 7155, *See* Ablunol S-80, 111, Witconol 2500, 34501
Radiasurf® 7125, 25797
Radiasurf® 7135, 25798
Radiasurf® 7137, 25799
Radiasurf® 7145, *See* Radiamuls® Sorb 2145, Sorb 2161, Sorb 2166, 25783
Radiasurf® 7147, 25800, *See* Ethylan® GES6, 11108
Radiasurf® 7150, 25801
Radiasurf® 7155, 25802
Radiasurf® 7156, 25803
Radiasurf® 7157, 25804
Radiasurf® 7175, 25805
Radiasurf® 7270, 25806
Radiasurf® 7400, 25807
Radiasurf® 7402, 25808
Radiasurf® 7403, 25809
Radiasurf® 7404, 25810
Radiasurf® 7410, 25811
Radiasurf® 7414, 25812
Radiasurf® 7417, 25813
Radiasurf® 7423, 25814
Radiasurf® 7443, 25815
Radiasurf® 7444, 25816
Radiasurf® 7453, 25817
Radiasurf® 7454, 25818
Radiasurf® 7473, 25819
Radiasurf® 7600, 25820
Radiasurf® 7900, 25821
radiated pyrites, *See* marcasite, 18700
radicle vinegar, *See* acetic acid, glacial, 292
Radiflam A AE, 25822
Radilon A CP300, 25823
Radilon A, A 32E, 25824
Radilon S BHS200/201, 25825
Radilon S, S 35FL/FLC, 25826
Radiocaps-125, *See* Iodotope I-125, 15217
Radiocaps-131, 25827, *See* Iodotope I-131, 15218, Iodotope Therapeutic, 15219, Radiocaps-131, 25827
Radiographol, *See* Abrodil, 158, Methiodal sodium, 19485
Radio-malt, 25828
Radiometal, 25829
Radiopaque, 25830
Radiose, 25831
Radiostol, *See* vitamin D_2, 33940
Radiostoleum, *See* vitamin A, vitamin D_2, 33932
Radiotetrane, *See* keraphen, 16055
Radizine, *See* atrazine, 3394
Radlasurf® 7140, *See* Radiamuls® MG 2141, MG 2142, MG 2600, MG 2900, 25777
Radmolite, 25832
Radocon, *See* simazine, 28464
Radokor, *See* simazine, 28464
Radspor, *See* Dodine FL, WP, 8785
Radspor FT, 65WP, 25833
Radstein, *See* vitamin D+72, 33940
Radumine, *See* oxalic acid, 22413
Rafamebin, *See* Yodoxin, 34658
raffinate, 25834
raffinose, 25835
rafluor, *See* sodium fluoride, 28764
Rafoxanide, *See* Ranide, 25868
Ragadan, *See* heptenophos, 13795, Hostaquick, 14407
Raid Flying Insect Killer, *See* resmethrin, 26218
Rainbow Custom Colored Mortars, 25836
Rainbow Ware, 25837
RAK®, 25838
Rakel's alloy, 25839
Rakusol®, 25840
Rally, *See* myclobutanil, 20549
Ralothrin, *See* cypermethrin, 7588, Topclip Parasol, 32037
Ralox® 02, 25841
Ralox® 46, 25842
Ralox® 530, 25843
Ralox® 630, 25844
Ralox® BHT food grade, 25845, *See* Anti-Oxydant Bayer, 2595
Ralox® TMQ-R, 25846
Ralufon® 414, 25847
Ralufon® DL, 25848
Ralufon® DT, 25849
Ralufon® N, 25850
Ralufon® TA, 25851
Raluquin®, 25852
Ramasit® KGT, 25853
Ramenti ferri, 25854
Ramet, 25855
Rametin, 25856
ramie, 25857
Ramix, 25858
Ramos, 25859
Rampart, 25860, *See* Terrathion, 31343
Ramrod, 25861, *See* Propachlor, 25163
Ramrod 20G, *See* Propachlor, 25163
Ramrod 65, *See* Ramrod, 25861
Ramrod Flowable, *See* Propachlor, 25163, 25164
Ramrod-atrazine, *See* Propachlor, 25163, 25164
Ramtap, 25862
Rancho, *See* mefenacet, 19229
Rancho®, 25863
Randanite, 25864
Raneoff® S, 25865
Ranestol, 25866
Raney nickel, 25867
Ranger, *See* glyphosate, 13148
Ranide, 25868
Ranotex, 25869
Ransome's stone, 25870
Raolein 131, 25871
RAP, 25872
Rapadex, 25873
rapeseed oil, 25874
Rapeseedamidopropyl epoxypropyl dimonium chloride, *See* Schercoquat ROEP, 27685
Rapeseedamidopropyl ethyldimonium ethosulfate, *See* Schercoquat ROAS, 27684
Rapiblend, 25875
rapic acid, 25876
Rapi-Cure BHC, 25877
Rapi-Cure CHMVE, 25878
Rapi-Cure CHVE, 25879
Rapi-Cure CVE, 25880
Rapi-Cure DVE-3, 25881
Rapi-Cure EHVE, 25882
Rapi-Cure HBVE, 25883

Rapid, 25884, *See* Abol, 146, pirimicarb, 23879
Rapid Purge 2, 25885
Ra-Pid-Gro, 25886
Rapidogen, 25887
Rapidosept®, 25888
Rapier, 25889
Rappor, 25890
Rappor Plus, 25891, *See* imazalil, 14892
Rasayansulfan, *See* thiodan, 31680
Raschit, 25892
Rasorite, *See* sodium metaborate, 28785
Rassamix CDA, 25893
Rassapron, 25894
Rastop;, *See* Ratak, 25896
Rat Flip, 25895
Ratak, 25896, *See* difenacoum, 8449, Neosorexa, 20967
Ratak+, *See* Talon, 30750
Ratimus, *See* bromadiolone, 4565, Rentokil Deadline, 26116
Ratio, 25897
RATO, *See* Captex® 8000, 5170, Emalex O.T.G, 9976, Miglyol® 808, 19736
Ratox, 25898
Rattler 4AS, *See* glyphosate, 13148
Rauxite, 25899
Rauxone, 25900
Rauzene, 25901
Rauzene Ester, 25902
Raven, 25903
Raviac, *See* Chlorophacinone, 6133, Karate, 15758, Ridene, 26746
Ravinil, 25904
Ravolen, 25905
Ravolen 11(T), 25906
Raw palmira root flour, *See* talipot, 30734
raw turkey umber, *See* umber, 32900
raw umber, *See* umber, 32900
Rawstol, 25907
Raxil, 25908
Raybar, 25909
Rayo, 25910
Rayon, 25911
Rayox, 25912
RC 7, 25913
RC Plasticizer DOP, *See* Bis(2-ethylhexy) Phthalate, 4228, dioctyl phthalate, 8549, Reomol DCP, 26130
RC-620, *See* Dodoxynol series, 8786
Rchonate SXS, *See* Naxonate® 4L, 20824
RCR Grey Squirrel Killer Concentrate, 25914
RCRA Waste Number U006, *See* acetyl chloride, 311
RD 4593, *See* Mecoprop, 19204, Verdone CDA, 33580
RD10, 25915
RD-6584, *See* dicloran, 8395, Fumite Dicloran, 12459
RDPE, 25916
Réamur's alloy, 25917
Réboulet's solution, 25918
Reach® 101, 201, 501, 25919
Reach® AZP-701, AZP-703, 25920
Reacrone, 25921
Reactal, 25922
Reactint®, 25923
Reactobond, 25924
Reacton, 25925
Reafor, 25926
Reafree, 25927
Reagens-CF2, 25928
Reakt, 25929
realgar, 25930
realgar, arsenic disulfide, *See* sandaracha, 27325
Realox®, 25931
Reamul, 25932
Rearguard, 25933
Reatane, 25934
Reater, 25935
Reatint, 25936
reaumerite, 25937
Reax® 45A, 25938
Reax® 80C, 25939
Rebelate, *See* Dimethoate, 8496
Reclaim, *See* Clopyralid, 6436
Recoil, 25940, 25941
reconox, *See* phenothiazine, 23618
Recop, *See* copper oxychloride, 6784
Recoura's sulfate, 25942
recovered grease, 25943
Recresal, 25944
Recrete NRC, 25945
Rectified Spirit S. V. R, 25946
Recupex, 25947
Red 139, 25948
Red 2g Base, *See* *p*-nitroaniline, 21345
red acid, 25949
red algar, 25950
red antimony, 25951
red argol, *See* argol,
red arsenic, *See* realgar, 25930
red arsenic glass, *See* realgar, 25930
red bole, *See* Indian Red, 15068
red brass, 25952
red chalk, *See* Indian Red, 15068
red charcoal, 25953
red chromate of potash, *See* potassium dichromate, 24754
red cobalt, 25954
Red Copp, *See* cuprous oxide, 7405
red copper, *See* violet copper, 33875
Red Copper Oxide, *See* cuprous oxide, 7405, violet copper, 33875
Red Dot, 25955
red drops, 25956
Red dye woods, *See* redwoods, 25987
red earth, *See* Indian Red, 15068
red fiber, *See* Vulcanized fiber, 34097
red fuming nitric acid, *See* nitric acid, 21332
red gold, 25957
Red Hermetite, 25958
Red Hot Pellets, 25959
Red iron oxide, *See* Ferroxide, 11694, Jeweller's rouge, 15611
red iron trioxide, *See* ferric oxide, 11651
Red lavender, *See* red drops, 25956
red lavender spirit, *See* spirit of red lavender, 29297
red lead, 25960
red lead oxide, *See* red lead, 25960
Red lead oxide, minium, Paris red, Satum red, *See* red lead, 25960
red liquor, 25961
red metal, 25962
red mordant, *See* red liquor, 25961
red nickel ore, 25963
red ocher, *See* Indian Red, 15068
red oil, 25964, *See* oleic acid, 22112
red orpiment, *See* realgar, 25930
red oxide of chromium, 25965
red oxide of lead, *See* red lead, 25960
red oxide of mercury, 25966
red phosphorus, *See* violet phosphorus, 33876
red potassium chromate, *See* potassium dichromate, 24754
red precipitate, *See* mercuric oxide, red and yellow, 19348, red oxide of mercury, 25966
red prussiate of potash, *See* potassium ferricyanide,
red rudd, *See* Indian Red, 15068
red salts, 25967
red saunderswood, *See* redwoods, 25987
red soda, 25968
Red Star Powder, 25969
red storax, 25970
red vermilion, *See* mercuric sulfide, red and black, 19350
red vitriol, 25971
red wash, 25972
red water bark, 25973
RED-AL, *See* Vitride®, 33994
Redalloy, 25974
Redcopp 97N Premium, *See* cuprous oxide, 7405
Redd Citrus Specialties, 25975
reddingite, 25976
reddle, *See* Indian Red, 15068
Redeem, *See* Timbrel, 31800
Rediclear, 25977
Redicote, 25978
Redisol, *See* vitamin B+71+72, 33934
Redmanol, *See* Bakelite, 3589
redo, *See* hydrosulfite, 14643
Redoxon, *See* vitamin C, 33938
Redray, 25979
Redreid Starch A, B, *See* corn starch, 6837
Reduce® -150, 25980
reduced Michler's ketone, *See* Michler's Base, 19644
reduced turpentine, 25981
reducin, 25982
Reductone, 25983, *See* sodium hydrosulfite, 28771
Redurit, 25984
Redux® 501, 25985
Reduxol Z, 25986
red-water tree bark, *See* red water bark, 25973
redwoods, 25987
Reed C-ABS-17415, 25988
Reed C-NY-261, 25989
Reed C-NY-4892, 25990
Reed C-PBT-1338, 25991
ReedLite C-NY, 25992
ReedLite CPC, 25993
Rees' thionin stain, 25994
Reese's alloy, 25995
Reevon, *See* A-C® Polyethylene 6, 6A, 7, 7A, 8, 8A, 9, 9A, 617, 617A, 175, Cabelec® 1017, 4862
Refagan® N, *See* Refagan,
refikite, 25996
Refine®, 25997
refined silver, 25998
refined solvent naphta, *See* Ligroin, 17244, V.M. and P. Naphtha, 33212
Refined soybean oil, *See* soya bean oil, 29169
Refined undeodorized soybean oil, *See* soya bean oil, 29169
Refinex, 25999
Refkon, 26000
Reflectafoam, 26001
Reflite, 26002
reflorit, 26003
reform phosphate, 26004
Refrax, 26005
Refrigerant 12, *See* dichlorodifluoromethane, 8388
Refrigerant 23, *See* Fluoroform, 12099
Refrigerant 113, *See* MS-180 Freon® TF Solv, 20415
Refrigerant 114, *See* Cryofluorane, 7278
Refrigerant R12, *See* dichlorodifluoromethane, 8388
Refrigerant R40, *See* methyl chloride, 19522
Refrigerant R114, *See* Cryofluorane, 7278
Refuse Trol, 26006
Regal Crown, 26007
Regal® 400R, 26008
Regalite, 26009
Regalox, 26010
RegalStar, 26011
regenerated turpentine, 26012
Regenex, 26013
Regent® 12XX, 26014
regianin, *See* Juglone, 15631
Reginal, 26015
Reginol 2701, 26016
Reglone, 26017, *See* Katalon, 15782
Reglox, *See* Reglone, 26017
Regnis, 26018
Regonal, *See* Prinza® Range, 25010
Regular mineral spirits, *See* Kensol 30, 16038
Regulex, 26019
Regulox, *See* maleic hydrazide, 18477
Regulox K, 26020
Regulus, 26021
Regulus metal, 26022
regulus of antimony, 26023
regulus of Venus, 26024
Regutol, *See* docusate sodium,
Rehydragel® Compressed Gel, 26025, *See* Alumina

hydrate, 1892
Rehydrol®, 26026
Reich's bronze, 26027
Reicolit, 26028
Reillex 202, 26029
Reillex 402 and 425, 26030
Reilline 2200 and 240, 26031
Reilline 4200 and 450, 26032
Reinecke's salt, 26033
Reiset's first base, 26034
Reiset's first chloride, 26035
Reith alloy, 26036
Rekawan, *See* potassium chloride
Rekawan, *See*
Rekawan, *See*
Rekawan, *See* K. Tab, 15642, Kaon-Cl, 15729, Kay Ciel, 15807, K-Contin, 15818, K-Lor, 16226, K-Lyte/C1, 16234, KM Potassium Chloride, 16242, Potassium chloride, 24748, 24749
Reldan, *See* Chlorpyrifos-methyl, 6164
Reldan 50, 26037
Reldan F, *See* Chlorpyrifos-methyl, 6164
Release Agent NL-1, 26038
Release Agent NL-2, 26039
Release Agent NL-10, 26040
Releasil, 26041
Releez, 26042
Relief®, 26043
Relimate®, Relipress®, 26044
Reloder 7, 26045
Relugan GT, 26046
Relugan®, 26047
Rely, *See* Timbrel, 31800
Remafin, 26048
Remazol, 26049
Remcoil, 26050
Remcopal, 26051
Remcopal 4, 26052
Remcopal 6, 26053
Remcopal 10, 26054
Remcopal 18, 26055
Remcopal 20, 26056
Remcopal 25, 26057
Remcopal 29, 26058
Remcopal 40, 26059
Remcopal 40 S3, 26060
Remcopal 121, 26061
Remcopal 207, 26062
Remcopal 220, 26065
Remcopal 229, 26066
Remcopal 234, 26067
Remcopal 238, 26068
Remcopal 258, 26069
Remcopal 273, 26070
Remcopal 306, 26071
Remcopal 334, 26072
Remcopal 349, 26073
Remcopal 666, 26074
Remcopal 3112, 26075
Remcopal 3712, 26076
Remcopal 3820, 26077
Remcopal 4000, 26078
Remcopal 4018, 26079
Remcopal 6110, 26080
Remcopal 21411, 26063
Remcopal 21912 AL, 26064
Remcopal 31250, 26081
Remcopal 33820, 26082
Remcopal D, 26083
Remcopal HC 7, 26084
Remcopal HC 20, 26085
Remcopal HC 33, 26086
Remcopal HC 40, 26087
Remcopal HC 60, 26088
Remcopal L9, 26089
Remcopal L12, 26090
Remcopal L30, 26091
Remcopal LO 2B, 26092
Remcopal LP, 26093
Remcopal NP 30, 26094
Remcopal O9, 26095
Remcopal O11, 26096
Remcopal O12, 26097
Remcopal PONF, 26098
Remedy, *See* Timbrel, 31800
Remelt, *See* ronnel, 26951
Remex, 26099
Remol TRF, *See o*-phenylphenol, 23648
Remsynol, 26100
Remtal, 26101
Renacit® 4, 26102
Renacit® 7, 26103, *See* Akrochem® Peptizer PTP, 1170
Renal AC, *See p*-aminophenol, 2185
Renaleptine, *See* renaglandin,
Renault alloy, 26104
Rencal, 26105
Rendells, 26106
Rendrock, 26107
Renektan, 26108
Renex, 26109
Renex 600's, *See* Alkasurf® NP-4, 1715, Triton® N-57, 32430
Reniten, *See* Renitec,
rennase, *See* rennet, 26110
rennet, 26110
Rennilase, 26111
rennin, *See* rennet, 26110
Renografin, *See* Reno-M, 26113
Renol, 26112
Reno-M, 26113
Renova, 26114
Renovue-65, Renovue-DIP, 26115
Renselin, *See* undecylenic acid, 32916
Rentokil Deadline, 26116
Renyx, 26117
Reochlor (LF and 54), 26118
Reoflam, 26119
Reofos, 26120
Reogen, 26121
Reolube, 26122
Reolube DOS, *See* Plasthall® DOS, 24005, Reomol DOS, 26131
Reolube FAD, 26123
Reomet®, 26124
Reomol, 26125
Reomol 4PG, 26126
Reomol BCF, 26127
Reomol D 79P, *See* Bis(2-ethylhexy) Phthalate, 4228, dioctyl phthalate, 8549, Reomol DCP, 26130
Reomol D79S, 26128
Reomol DBS, 26129
Reomol DCP, 26130
Reomol DOA, *See* dioctyl adipate, 8548, Plasthall® DOA, 24002, PX-238, 25440, Wickenol® 158, 34393
Reomol DOP, *See* Bis(2-ethylhexy) Phthalate, 4228, dioctyl phthalate, 8549, Reomol DCP, 26130
Reomol DOS, 26131
Reomol P, 26132
Reomol PBPS, 26133
Reomol TC9, 26134
Reoplast, 26135
Reoplex, 26136
Reoplex 200, 220, 300, 26137
Reoplex 901, 26138
Reoplex 902, 26139
Reostene, 26140
Repak, 26141
Repeftal, *See* dimethyl phthalate, 8508
Repel, *See* diethyl toluamide, 8438
Repelit, 26142
Repellent 6-12, *See* ethyl hexanediol, 11073
Repello DC, 26143
Repel-O-Tex® QCJ, 26144
Repidose, *See* oxfendazole, 22426
Replay RP 2177, 26145
Replay RP 2236, 26146
Replens, *See* Synwax, 30659
Replicast CS, 26147
Replicast FM, 26148
Repone K, *See* K. Tab, 15642, Kaon-Cl, 15729, Kay Ciel, 15807, K-Contin, 15818, K-Lor, 16226, K-Lyte/C1, 16234, KM Potassium Chloride, 16242, Potassium chloride, 24748, 24749
Reprodin®, 26149
Reproxal, 26150
Repulse, 26151, *See* Chlorothalonil, 6150
resacetophenone, 26152
Resad, 26153
Resamine, 26154
Resan, *See* Bakelite, 3589
Resarit, 26155
Resarix SF, 26156
Resart, 26157
Resartglas GS, 26158
Resartherm, 26159
Resart-PMMA XT, 26160
resazoin, *See* diazoresorcin, 8351, Resazurin, 26161
Resazurin, 26161, *See* diazoresorcin, 8351
resbenzophenone, *See* Uvinul® 400, 33191
Resbuthrin, *See* Reslin, 26216
Rescon, 26162
Rescue, 26163
rescue squad, *See* sodium fluoride, 28764
Resibon, 26164
Resicart, 26165
Residuren, 26166
Residuren Extra, 26167
Resigum, 26168
Resilia, 26169
Resilita, 26170
Resilla, 26171
Resilon, 26172
Resimene, 26173
resin, *See* colophony,
Resin 18, 26174
Resin 164, 26175
Resin 731D, 26176
Resin 885, 3072, 26177
Resin benjamin, *See* gum benjamin, 13482
Resin benzoin, *See* gum benjamin, 13482
resin blende, 26178
resin Damar, *See* Dammar Resin, 7724, gum Cowrie, 13484
resin essence, *See* rosin spirit, 26979
Resin Ether L, 26179
Resin EX, 26180
Resin H, 26181
Resin kino, *See* kino, 16196
resin lutea, 26182
Resin M.S.2, 26183
Resin NC-11, 26184
resin oil, 26185
Resin Release N, 26186
resin spirit, *See* rosin spirit, 26979
Resin WP, 26187
resina animé, *See* jutahycica resins, 15637
Resinall 153, 26188
Resinase, 26189
Resinette, 26190
Resinite, *See* Bakelite, 3589
Resinoid 1324, 26191
Resinoid 2002-4, 26192
Resinoids, *See* Balsamarome®, 3610
resinol, 26193
resinol orange r, *See* Sudan I, 29938
resinous silica, 26194
Resinox, 26195
resins, acrolein, 26196
Resipol, 26197
Resipol DL, 26198
Resipol ML, 26199
Resiren®, 26200
Resisan, *See* dicloran, 8395, Fumite Dicloran, 12459
Resisco, 26201
Resissan, *See* dicloran, 8395, Fumite Dicloran, 12459
Resista, 26202
Resista steel, 26203

Resistac, 26204
Resistal, 26205
resistance bronze, 26206
Resistherm®, 26207
Resistin, 26208
Resistoflex, 26209
Resistolac, 26210
Resistomycin, *See* Kantrex, 15718, Klebcil, 16209
Resistone, 26211
Resistone QD, Resitone QD, 26212
Resistox, 26213
Resithren, 26214
Resitone QD, 26215
Reslin, 26216
Reslin S, 26217
Reslosol, *See* Anusol,
resmethrin, 26218
(+)-*trans*-resmethrin, *See* Reslin, 26216
Resocoton, 26219
Resoform orange g, *See* Sudan I, 29938
Resoform Yellow GGA, *See* Butter yellow, 4759
Resogen® 35 Conc, 26220
Resoglaz, 26221
Resolamin, 26222
resole, *See* phenolic resin, 23607
Resolin®, 26223
Resolin® P, 26224
Resoltex, 26225
resolvable tartaric acid, *See* tartaric acid, 30832
Resonium A, *See* Kayexalate, 15812
Resopol, 26226
Resorband, 26227
resorcin, 26228, *See* resorcinol, 26230
Resorcin blue, *See* Fluorescent Blue, 12084
resorcinal, 26229
resorcinol, 26230, *See* Cohedur® RK, 6576, resorcin, 26228
Resorcinol benzoate, 26231
Resorcinol formaldehyde resins, *See* Aerodux, 852
Resorcinol monobenzoate, *See* Eastman® Inhibitor RMB, 9470, Resorcinol benzoate, 26231
resorcinol phthalein sodium, *See* Fluorescein disodium salt, 12081
Resorcinol/formaldehyde resins, *See* Aerodux, 851
resorcinolphthalein, *See* fluorescein, 12080, Fluoresceine, 12082, uranine, 33115
Resorcylindene, *See* RRV, 27050
Resorufin, 26232
Resovin, 26233
Resovist, *See* Magnevist, 18376
Resovyl, 26234
Responsar, 26235
Respumit®, 26236
Restoration Cleaner, 26237
Restoration Cleaner (Heavy Duty), 26238
Restoration Cleaner (Super Heavy Duty), 26239
Restoration Rinse, 26240
Restor-E (Restoration Chemical Products), 26241
Restore-X Exterior Paint Remover, 26242
Restore-X Weathered Wood Renewer, 26243
Resydrol, 26244
Resyn® 28-1310, 26245, *See* VA/crotonates copolymer, 33224
Resyn® 28-2913, 26246
Retain PE-1001, 26247
Retain PE-5009, 26248
Retain PS-4000, 26249
Retain RP-120, 26250
Retaminol®, 26251
Retard, 26252
Retarded BA, *See* Benzoic acid, 3973
Retarder AK, 26253
Retarder BA, BAX, 26254
Retarder ESEN, 26255
Retarder N, 26256
Retarder OC, 26257
Retarder PX, 26258
Retarder SAFE, 26259
Retarder SAX, 26260
Retarder V-48, 26261
Retardex, *See* Benzoic acid, 3973
Retardine, 26262
Retardit A, 26263
Retardol, 26264
Retargal, 26265
Retariox, 26266
Reten®, 26267
Reten® 157, 26268
Reten® 763, 26269
Reten® 1232, 26270
retene, 26271
Reticusol, 26272
Retilox, 26273
Retilox® F 40 MG, 26274, *See* Bisbutyl peroxy diisopropyl benzene, 4230, Luperox 802, 17841
retin asphalt, *See* retinite, 26276
retinaphtha, *See* methyl benzene, 19514
Retingan® R6, R7, R48, 26275
retinite, 26276
retinol, 26277, *See* Vi-Alpha, 33706, vitamin A, 33932
Retinyl palmitate polypeptide, *See* Vitazyme® A-Plus, 33961
Retnolite, 26278
Retolen, *See* Hismanal, 14132
Retractyl, *See* Korogel, 16354, Koroplate, 16357
Retrocure® G, 26279
Retz alloy, 26280
Retzanol CO-40, *See* Ablunol CO 10, 98
Retzloff Intermediate No. 122, *See* Witco® 1298, 34458
Retzloff Intermediate No. 123, *See* Ablunol CO 10, 98
Retzolate 60, *See* Witcolate ES-2, Witcolate LES-60C, 34477
Retzulfonic DPB, *See* Witco® 1298, 34458
Retzulfonic SD-12, *See* Witco® 1298, 34458
reuniol, *See* roseol, 26973
reussinite, 26281
Re-Vac II, 26282
Revacryl®, 26283
Revalon, 26284
Revatol, 26285
Revatol S, 26286
Revenge, *See* dalapon, 7705, Wickenol® 101, 105, 34378, Wickenol® 111, 34379, Wickenol® 127, 34380
Revertex®, 26287
Revlen, 26288
Revoke, *See* Rodeo, 26899
Revolex, 26289
Revolite, 26290
Revona, 26291
Revuitex®, 26292
Rewagit, 26293, *See* aluminum oxide, 1919
Rewo-amid, 26294
Rewocid®, 26295
Rewocid® DU 185, 26296
Rewocid® SBU 185 P, 26297
Rewocid® U 185, 26298
Rewocid® UTM 185, 26299
Rewocor, 26300
Rewocoros, 26301
Rewocoros B 2045, 26302
Rewocoros B 3032, 26303
Rewocoros BAC, 26304
Rewocoros RA 280, 26305
Rewocoros RAB 90, 26306
Rewocoros TPAC 100, 26307
Rewocors B 3010, 26308
Rewoderm® ES 90, 26309
Rewoderm® LI 48, 26310
Rewoderm® S 1333, 26311
Rewoderm® SPS, 26312
Rewolan®, 26313
Rewolan® 5, 26314, *See* disodium laneth-5 sulfosuccinate, 8646
Rewolan® AWS, 26315
Rewolan® LP, 26316
Rewolub KSM 80, 26317
Rewolub TMP 275, 26318
Rewomid DI 203/S, *See* Alkamide® 327, 1634
Rewomid DL 203/S, *See* Ablumide LDE, 79
Rewomid Dlms, *See* Alkamide® 327, 1634
Rewomid DLMS, *See* Ablumide LDE, 79
Rewomid L 203, *See* Ablumide LME, 80
Rewomid®, 26319
Rewomid® C 212, 26320
Rewomid® DL 203 S, 26321
Rewomid® DO 280, 26322
Rewomid® DO 280 SE, 26323
Rewomid® F, 26324
Rewomid® IPE 280, 26325
Rewomid® IPL 203, 26326
Rewomid® IPP 240, 26327, *See* Cocamide MIPA, 6550
Rewomid® L 203, 26328
Rewomid® R 280, 26329
Rewomid® S 280, 26330
Rewomin, 26331
Rewominox, 26332
Rewominox B 204, 26333
Rewominox L 408, 26334, *See* lauramine oxide, 16832
Rewominox S 300, 26335
Rewominoxid, 26336
Rewominoxid L 408, 26337
Rewominoxid S 300, 26338
Rewomul MG SE, 26339
Rewopal® BN 13, 26340
Rewopal® C 6, 26341
Rewopal® HV 4, 26342
Rewopal® LA 3, 26343
Rewopal® MPG 10, 26344
Rewopal® MT 65, 26345
Rewopal® O 8, 26346
Rewopal® PEG 6000 DS, 26347, *See* PEG-150 distearate series, 23037
Rewopal® PG 280, 26348
Rewopal® RO 40, 26349
Rewopal® TA 11, 26350
Rewophat, 26351
Rewophat EAK 8190, 26352
Rewophat NP 90, 26353
Rewophos EAK 8190, 26354
Rewophos TD40, 26355
Rewophos TD70 and OP80, 26356
Rewopol, 26357
Rewopol NL-2, *See* Witcolate ES-2, Witcolate LES-60C, 34477
Rewopol NLS 30, *See* Rhodapon® 101-10, 26631, sodium lauryl sulfate, 28781
Rewopol TLS-40, *See* Perlankrol® ATL40, 23324, Rhodapon® LT-6, 26638, Standapol® T, 29481, Witcolate TLS-500, 34479
Rewopol® B 1003, 26358
Rewopol® B 2003, 26359
Rewopol® CHT 12, 26360
Rewopol® CL 30, 26361
Rewopol® CLN 100, 26362
Rewopol® CT 65, 26363
Rewopol® DLS, 26364
Rewopol® MLS 30, 26365
Rewopol® NEHS 40, 26366
Rewopol® NL 2-28, 26367
Rewopol® NLS 15 L, 26368
Rewopol® NOOSE 5, 26369
Rewopol® SBC 212, 26370
Rewopol® SBDB 45, 26371
Rewopol® SBDC 40, 26372
Rewopol® SBDD 65, 26373
Rewopol® SBDO 75, 26374
Rewopol® SBF 12, 26375
Rewopol® SBFA 30, 26376, *See* disodium lauryl sulfosuccinate, 8649
Rewopol® SBL 203, 26377, *See* disodium lauramido MEA-sulfosuccinate, 8647
Rewopol® SBMB 80, 26378
Rewopol® SBZ, 26379
Rewopol® SK 275, 26380
Rewopol® SLS, 26381
Rewopol® SMS, 26382
Rewopol® TLS 40, 26383

Ricobond 1031, 1731, 1756, 26735
Ricolite, *See* Bakelite, 3589
Ricon 100, 26736
Ricon 130MA8, 26737
Ricon 159, 26738
Ricon P30/Dispersion, 26739
Ricoroof, 26740
Ricoseal, 26741
Ricotti silk, *See* Galettame silk, 12603
Ricotuff, Ricotuff L.V, 26742
Ric-Syn Wax, 26743
Rid, *See* dimethyl tetrachloroterephthalate, 8512, Vegetable Turf and Ornamental Weeder, 33526
Ridacto®, 26744
Rid-A-Roach, 26745
Ridect PourOn, *See* Kafil, 15655
Ridene, 26746
Ridoline, 26747
Ridomil, *See* metalaxyl, 19430
Ridomil MBC 60WP, 26748
Ridomil Plus 50WP, 26749
Ridosol, 26750
Rifleite, 26751
Rigidex, 26752
Rigidex 3, 26753
Rigidex 9, 26754
Rigidex X4RR, 26755
Rigidite®, 26756
Rigidoll, 26757
Rigillène, 26758
Rigipore, 26759
Rigo's Best Lawn & Garden Fungicide, *See* Chlorothalonil, 6150, Repulse, 26151
Rikemal S 250, *See* Ablunol S-60, 110
rilan wax, 26760
Rilata, 26761
Rilsan®, 26762
Rilsan® BESHVO, BESVO, 26763
Rimflex A/A, 26764
RIMline® 8711B/8700A, 26765
RIMline® GMR-5000, 26766
RIMline® GMR-8711, 26767
Rimplast®, 26768
Rimso-50, *See* dimethyl sulfoxide, 8511, DMSO, 8762
Rimthane, 26769
Ringer solution, 26770
Ringmaster, 26771, *See* oxycarboxin, 22441
Rinoxin, 26772
Rinsan, 26773
Rintal®, 26774
Rintal® Plus, 26775
Rio Resin, 26776
Riogen, *See* phenylmercury acetate, 23646
Riozeb, *See* mancozeb, 18517
Ripcord, *See* cypermethrin, 7588, Topclip Parasol, 32037
Ripercol, 26777, *See* Tramisol®, 32134
Ripping ammonal, 26778
Rippite, 26779
Riselect, *See* Propanil, 25174
Rismavac-CR6, 26780
Riso, 26781
Risolex, 26782
Rissicol, 26783
Ristin, 26784
Riston®, 26785
Risunal, 26786
Rita AZ, 26787
Rita CA, 26788
Rita EDGS, 26789
Rita EGMS, GMS, 26790
Rita HA C-1, 26791
Rita KA, 26792
Rita SA, 26793
Ritabate 20, 26794
Ritabate 60, 26795, *See* Ethylan® GES6, 11108
Ritabate 80, 26796
Ritacet-20, 26797
Ritaceti, 26798, *See* cetyl esters, 5819
Ritacetin, 26799
Ritacetyl®, 26800 , *See* Fancol Acel, 11428
Ritachlor 50%, 26801
Ritachol®, 26802
Ritachol® 1000, 26803
Ritachol® SS, 26804
Ritacollagen BA-1, 26805
Ritaderm®, 26806
Ritahydrox, 26807, *See* hydroxylated lanolin, 14663
Ritalafa®, 26808
Ritalan®, 26809
Ritalan® AWS, 26810, *See* PPG-12-PEG-65 lanolin oil, 24843
Ritalan® C, 26811
Ritalanine, 26812
Ritalastin EL-10, 26813
Ritaloe 1X, 26814
Ritamectant K2, 26815
Ritamectant PCA, 26816
Ritapan CAP, 26817
Ritapan D, DL, 26818
Ritapan NAP, 26819
Ritapan TA, 26820
Ritapeg 150 DS, 26821
Ritapeg 400 DS, 26822, *See* PEG-150 distearate series, 23037
Ritaphenone 3, 26823
Ritaplast, 26824
Ritaplast R, 26825
Ritaplast TN, 26826
Ritapro 100, 26827
Ritapro 165, 26828
Ritaquat Q, 26829
Ritasilk, 26830
Ritasilk Powder, 26831, *See* Fibro-Silk Powd, 11754
Ritasol, 26832
Ritasynt IP, 26833
Ritataine, *See* Cocamidopropyl betaine, 6551
Ritatin, 26834
Ritawax, 26835, *See* Lanolin alcohol, 16741
Ritawax 5, 26836
Ritawax AEO, 26837
Ritawax ALA, 26838
Rit-Cizer #8, 26839
Rit-Cizer #9, 26840
Riteflex®, 26841
Riteflex® 347ZS, 26842
Riteflex® 540, 26843
Riteflex® BP 8929, 26844
Riteflex® BP 9057, 26845
Rit-Ester B-100, 26846
Ritha, 26847
Rititaine B, *See* Cocamidopropyl betaine, 6551
Ritoleth 2, 26848, *See* oleth-8, 22126
Rit-O-Lite MHP-S, 26849
Rit-O-Lite MS-80, 26850
Ritosept, *See* hexachlorophene, 13992
Ritox 35, 26851
Ritox 52, 26852
Ritox 721, 26853, *See* steareth series, 29596
Rivaite, *See* Wollastocoat®, 34545
Rival, 26854
Rivalit P, 26855
riversideite, *See* calcium silicate, 4969
rivet metal, 26856
Riviera Oil, *See* aix oil, 1138
RIX 90149, 26857
Rizolex, 26858, *See* Risolex, 26782
RJ-100, 26859
R-MA 11®, *See* Alumina hydrate, 1892
RMD, 26860
RMR, 26861
Ro-1-7977, *See* MPI DMSA Kidney Reagent, 20393
Ro-12-3049, *See* fenpropidin, 11616
Ro 14-9480/000, *See* tebutam, 30879
Ro-22-9000, *See* Alfaprostol, 1574
Roach salt, *See* sodium fluoride, 28764
Roach Stoppers, 26862
Roachban, 26863
Roanoid, 26864
roaster slag, 26865
Ro-A-Vit, 26866
Robac, 26867
Robac 22, *See* ethylene thiourea, 11132, Perkacit® ETU, 23283
Robac T.B.Z, 26868
Robacure, 26869
Robane®, 26870
Robengatope I-131, 26871
Robert's reagent, 26872
Robertson alloy, 26873
robigenin, *See* kaempferol, 15654
Ro-Bile, 26874
Robond, 26875
Robuoy, *See* Pristane, 25037
Roburite, 26876
Roburite I, 26877
Roburite III, 26878
Rocagel, 26879
Roccal, 26880, *See* Variquat® 50MC, 33397, Zephiran Chloride, 34728
Roccal MC-14, *See* JAQ Powdered Quaternary, 15517
Rocel, 26881
Rochdale salt, *See* Rochelle salt, 26882
Rochelle salt, 26882, *See* Seignette salt, 27847
rock asphalt, 26883
rock cork, 26884, *See* rock wool, 26888
rock crystal, 26885
rock dammar, 26886
rock oil, *See* Kendex OCTG, 16002
rock salt, 26887, *See* sodium chloride, 28748
rock tallow, *See* bitumen, 4294
rock wool, 26888
Rocket® Ultra, 26889
Rocksil, 26890
Rocktex, 26891, *See* rock wool, 26888
Roclys®, 26892
Rocol P.R, 26893
Rocol R.S.7, 26894
rocou, *See* annatto, 2478
Rocryl, 26895
Rocsol, 26896
rod wax, 26897
Rodalon, *See* Variquat® 50MC, 33397, Zephiran Chloride, 34728
Rodea, 26898
Rodentin, *See* Racumin®, 25713
Rodeo, 26899
Rodex, *See* Killgerm® Sewarin P, 16184, Kypfarin, 16519
Rodiatox, *See* parathion, 22807
Rodinal, 26900, *See* *p*-aminophenol, 2185, Takatol, 30725, Unal, 32908
Rodo, 26901
Rody, *See* fenpropathrin, 11615
Roebaryt, 26902
Roesch's aluminum solder, 26903
Roferose, 26904
Roga, 26905
Rogé Cavaills, 26906
Roghan, 26907
Rogor, *See* Dimethoate, 8496
Rogor E, 26908
Rogue, *See* Propanil, 25174
Rohafloc, 26909
Rohagit® SM V, 26910
Rohamere® 8662, 26911
Rohatol® BV 382, 26912
Rohatol® D 362, 26913
Rohn alloys, 26914
Rohrbach's solution, 26915
Roica®, 26916
Rokar X, *See* Uragan, 33105
Rokar Xrout, *See* bromacil, 4564
RokLok® B-3, 26917
RO-KO, *See* rotenone, 27000
Rokon, 26918
Rolafix, 26919

Rolamid CD, *See* Ablumide LDE, 79, Alkamide® 327, 1634
Rolamid CM, *See* Ablumide LME, 80
roll sulfur, 26920
Rollit, 26921
Rollofix X100, 26922
Rol-man Steel, 26923
Rolox, 26924
Roman alum, 26925
Roman bronze, 26926
Roman cement, 26927
Roman vitriol, *See* Kocide® Copper Sulfate Pentahydrate Crystals, 16262
Romane, 26928
Romanite, *See* Roumanite, 27010
Romanium, 26929
Romensin, 26930
Romite, 26931
Rompel's alloy, 26932
romperit G, 26933
Rompun, *See* xylazine, 34619
Rompun®, 26934
Ronabond®, 26935
Ronafix, 26936
Ronaset®, 26937
Rondis, 26938
Rondo, *See* Rodeo, 26899
Ronfalin, 26939
Ronfaloy E, 26940
Ronfaloy V, 26941
Ronfusil Steel, 26942
Rongal®, 26943
Rongalit®, 26944
Rongalit® C, 26945
Rongalite, *See* hydrosulfite, 14643
Rongalite C, 26946, *See* sodium formaldehydesulfoxylate, 28766
Rongalite Conc, *See* Hydrosulfite NF, 14647
Ronia metal, 26947
Ronicol, 26948
Ronilan, *See* vinclozolin, 33811
Ronilan® DF, FL, 26949
Ronilon, 26950
Ronnel, 26951, *See* Korlan, 16351
ronone, *See* rotenone, 27000
Ronoxan, 26952
Ronseal, 26953
Ronstar 2G, 26954, *See* Parogen, 22854
Ronstar 50W, *See* Parogen, 22854
Ronstar TX, 26955
Roofcover LM, 26956
Roo-Pru, *See* metam-sodium, 19444
Root Guard, 26957
Rooting Powder, 26958
Root-Out, 26959
Ropaque, 26960
Ro-Pel, 26961
Rosampline, *See* Ampicillin, 2361
rose B, 26962, *See* erythrosin, 10778
Rose Bengal, 26963
Rose Bengal Extra, *See* Robengatope I-131, 26871
Rose Bengal Sodium ^{125}I, *See* Robengatope I-131, 26871
Rose Bengale B, *See* rose bengal, 26963
Rose Ester, 26964
Rose Ether, *See* Igepal® Cephene Distilled, 14850, Rewopal® MPG 10, 26344
Rose Food, 26965
Rose Plus, 26966
Rose Quartz, 26967
Rosé oil, 26968
Roseclear, 26969
rosein, 26970
Roseline, 26971
Roselle fiber, 26972
roseol, 26973
Rose's metal, 26974
Rosette copper, 26975
Roshé oil, *See* Rosé oil, 26968
rosin, 26976
Rosin amine D, *See* Amine D, 2155
rosin grease, 26977
rosin gum, *See* metribuzin, 19595, rosin, 26976
Rosin modified phenolic resins, *See* Albertol, 1368
rosin oil, 26978, *See* retinol, 26277
rosin spirit, 26979
rosin tin, 26980
Rosinal, 26981
Rosin-glycerol varnish and lacquer resins, *See* Aero, 838
rosinjack, *See* zinc blende, 34767
rosinol, *See* retinol, 26277
Rosintene, 26982
Rosite® 4030FS, 26983
Rosite® ESD Cond. C, 26984
Roskens, 26985
Roskydal®, 26986
Roslte® 3250A, 26987
Roslte® 4010ES, 26988
rosolene, 26989
Ross alloy, 26990
Ross Japan Wax Substitute 525, 26991
Ross Spermaceti Wax Substitute 573, 26992, *See* cetyl esters, 5819
Ross Wax #100, 26993
Ross Wax #145, 165, 26994
Ross's white, *See* Lithopone, 17520
Rota, 26995
Rotacide, *See* rotenone, 27000
Rotal, 26996
Rotalin, 26997, 26998
Rotate, *See* Bendiocarb, 3937
Rotax®, 26999
rotefive, *See* rotenone, 27000
rotefour, *See* rotenone, 27000
rotenone, 27000, *See* Derris Dust, 8134, FS Derris, 12425
(-)-rotenone, *See* rotenone, 27000
Rotenone, hydrogenated, *See* rotenone, 27000
Roter, 27001
Rotersept, 27002
Rotersept, *See* chlorhexidine digluconate, 6101, Rotersept, 27002
rotessenol, *See* rotenone, 27000
rotocide, *See* rotenone, 27000
Rotoval, 27003
Rotoxit, 27004
Rotra bark, 27005
rotten stone, 27006, *See* Tripoli, 32399
Rotuba H, 27007
Rouen white, 27008
rouge, 27009, *See* Ferroxide, 11694, Indian Red, 15068, Jeweller's rouge, 15611
Roumanite, 27010
Roundup, 27011, *See* Rodeo, 26899
Roundup 2.5, *See* glyphosate, 13148
Rousselot Gelatine, 27012, *See* gelatin, 12722
Roussel's solution, 27013
Roussin's black salt, 27014
Roussin's red salt, 27015
Roussin's salts, 27016
Rout, *See* Uragan, 33105
Roux's stain, 27017
Rovace 571, 27018
Rovace 2113, 27019
Rovace 9100, 27020
Rovace® 571, 2113, *See* Everflex® SP-1084, 11266
Roval Dust, 27021
Roval Flo, 27022
Roval Green, 27023
Roval WP, 27024
Rovan, *See* ronnel, 26951
Rovel, 27025
Rovigon, 27026
Rovimix, 27027
Rovisol, 27028
Rovral, 27029
Rovral WP, 27030
Roxanthin Red 10, *See* Canthaxanthin, 5095
Roxion, *See* Dimethoate, 8496
Roxite, 27031
Roxon, 27032
Roxotit, 27033
royal blue, *See* smalt, 28638
Royal CBTS, *See* Vulkacit® CZ/EGC, DZ/EGC, 34129
Royal MBTS, *See* Perkacit® MBTS, 23286
Royal Slo-Gro, 27034
Royalac 133, 27035
Royalac 136, 27036
Royalac 140, 27037
Royalcast® 3105, 27038
Royalene 301-T, 27039
Royalene 306, 27040
Royalene 521, 27041
Royalene 622, 27042
Royaltherm® 1411, 27043
Roydalox, 27044
Roydazide, 27045
Rozol, *See* Chlorophacinone, 6133, Drat, 8972, Karate, 15758, Ridene, 26746
Roztoczol, *See* tetradifon, 31366
Roztoczol Extra, *See* tetradifon, 31366
Roztozol, *See* tetradifon, 31366
RP-50, *See* Benzalkonium chloride, 3958
RP-143, *See* Tears Plus®, 30876, Videne Disinfectant Solution, Videne Disinfectant Tincture, Videne Powder, 33739, Vinisil, 33827
RP-8595, *See* Dimetridazole, 8528
RP 11561, *See* Carbetamex, 5189
RP 11974 RP, *See* Zolone, 34888
RP-17623, *See* Parogen, 22854
RPDE, 27046
RPM, 27047
RP-Thion, *See* Ethion, 10955
RR 5, 27048
RR 53 alloy, 27049
RRV, 27050
RS Nitrocellulose, 27051
RS-8858, *See* oxfendazole, 22426
R-salt, 27052
RT/Duroid® M, 27053, *See* Polytetrafluoroethylene, 24642
RTF 762, 27054
RTP 100 GB 10, 27055
RTP 200FR, 27056
RTP 201A, 27057
RTP 201B, 27058, *See* nylon 6/10, 21911
RTP 201C, 27059
RTP 201D, 27060
RTP 301, 27061
RTP 401, 27062
RTP 501, 27063
RTP 601, 27064
RTP 701, 27065
RTP 801, 27066
RTP 901, 27067, *See* Polysulfone, 24619
RTP 1001, 27068, *See* Grilpet® XE3060, 13402
RTP 1105FR, 27069, *See* Polyethylene terephthalate, 24426
RTP 1201-80D, 27070
RTP 1301, 27071
RTP 1378, 27072
RTP 1401, 27073
RTP 1501, 27074
RTP 2301A, 27075
RTP 2381A, 27076
RTP 3403-3, 27077
RTP 3405-3 TFE 15, 27078
RTP 4001, 27079
RTP 4081, 27080
RTP ESD-300 EM FR, 27081
RTV 11, 27082
RTV 31, 27083
RTV 133, 27084
RTV 511, 27085
RTV 615, 27086
RTV 6156, 27087

RTV silicone, *See* RTV 11, 27082, RTV 31, 27083
RTY-319, *See* Tachigaren 70, 30685
RU 22974, *See* Thripstick®, 31766
RU-19110, *See* halofuginone hydrobromide, 13562
Rubalt, 27088
rubber, 27089, *See* Natsyn® 2200, 20763
Rubber (all-*cis*), *See* Natsyn® 2200, 20763
rubber cements, 27090
rubber formolite, 27091
rubberite, 27092
Rubberlene, 27093
Rubbermakers Sulfur, 27094
rubber-sulfur, 27095
Rubbing Alcohol, *See* isopropanol, 15405
Rubbone, 27096
rubeanic acid, *See* hydrogen rubeanide, 14591
Rubel metal, 27097
rubellan, 27098
rubeosine, 27099
ruberite, 27100
Rub-erok, 27101
Rub-er-red, 27102
ruberythric acid, *See* rubianic acid, 27103
ruberythrinic acid, *See* rubianic acid, 27103
rubian, *See* rubianic acid, 27103
rubianic acid, 27103
rubidium carbonate, 27104
rubidium carbonate, *See* rubidium carbonate, 27104
rubidium chloride, 27105
rubidium iodide, 27106
rubidium sulfate, 27107
Rubidor, *See* azamethiphos, 3517
Rubinate® LF-168, 27108
Rubini's essence, 27109
rubio ore, 27110
Rubitox, *See* phosalone, 23691, Zolone, 34888
Rubmag, 27111
Rubout, 27112
Rubox, 27113
Rubramin PC, 27114
Rubratope-57, 27115
rubrax, 27116
rubrescin, 27117
rubrica, 27118
Rub-tex, 27119
Ruby, *See* niclosamide, 21146
ruby arsenic, *See* realgar, 25930
ruby ore, 27120
ruby powder, 27121
Ruby sulfur, *See* realgar, 25930
ruby tin, 27122
Rucoflex Plasticizer DOA, *See* dioctyl adipate, 8548, Plasthall® DOA, 24002, PX-238, 25440, Wickenol® 158, 34393
Rucoflex® F-2014, 27123
Rucoflex® S-101 Series, 27124
Rucote 102, 27125
Rucothane 2010L, 27126
Rudol, 27127
rufiopin, 27128
Ruge's solution, 27129
Rukseam Tech DDT, *See* DDT, 7820
Rulan®, 27130
Runa, 27131
Runaway, 27132
Ruolz alloys, 27133
Ruselite, 27134
rusma, 27135
Ruspini's solution, 27136
Russian steam glue, *See* steam glue, 29586
Russian tallow, 27137
Russian tula, *See* Niello silver, 21167
Russian turpentine, 27138
Russian white lead, 27139
Russol, *See* paraffin, liquid, 22744
Rustban, 27140
Rustlan Oil, 27141
Rustless steel, *See* stainless steel, 29433
rustlessiIron, 27142
Rust-Tap, 27143
Rutaform, 27144
Rutamod, 27145
Rutasolv DI, 27146
Rutenol, 27147
Rutgers 612, *See* Ethohexadiol, 10968
Rutgers 6-12, *See* ethyl hexanediol, 11073
ruthenium red, 27148
Ruthmol, 27149
rutile, *See* titanium dioxide, 31867
rutin trihydrate, 27150
Rutiox, 27151
rutoside, *See* rutin trihydrate, 27150
Ruvea®, 27152
α-Ruvite, *See* aluminum oxide, 1919
RVPaba Lipstick, 27153
RX® 1-501N, 27154
RX® 3-1-530, 27156
RX® 1906, 27155
RX-56, 27157
RXXL, 27158
Ryax C, 27159
Ryax F, O, 27160
Ryflex, 27161
Rylex, 27162
Rynite®, 27163
Ryoto Ester KA, 27164
Ryoto Ester SP, 27165
Ryoto Sugar Ester LWA-1570, 27166, *See* Grilloten LSE87, 13376
Ryoto Sugar Ester OWA-1570, 27167
Ryoto Sugar Ester P-1570, P-1670, 27168
Ryoto Sugar Ester S-170, 27170
Ryoto Sugar Ester S-570, S-770, 27171
Ryoto Sugar Ester S-1170, 27169, *See* Grilloten PSE141G, 13378
Rytherm, 27172
Ryton, *See* Debron 711, 7835
Ryton® A-200, 27173
Ryton® R-4, 27174
Ryton® V-1, 27175
Ryvin, 27176
Ryzelan, *See* oryzalin, 22347
S, *See* L-serine, 17705, 28035
S 767, *See* Terracur® P, 31329
S 1752, *See* Tiguvon®, 31788
S Monel, 27177
S.21, 27178
S.N.P, *See* parathion, 22807
S.O.S.®, 27179
S.Q, *See* sulfaquinoxaline, 29968
S.S.T® Sump Saver Tablets, 27180
S/B latex, *See* Bayer SBR Latex 200 C, 3771
S-60 RVM, 27183
S-94, *See* Fluphenazine, 12116
S95, *See* dalapon, 7705
S160 Beads and Powder, 27181
S-201, 27182, *See* aluminum oxide, 1919
S-940, *See* Rametin, 25856
S-2539, *See* phenothrin, 23619
S-3151, *See* Kafil, 15655 permethrin, 23384, 23385, 23386
S-3206, *See* fenpropathrin, 11615
S-3349, *See* Risolex, 26782
S-5602, *See* fenvalerate, 11620
S-6876, *See* Omethoate, 22155
S945, 27184
S975, 27185
S987, 27186
S992, 27187
S994, 27188
SA Lawn Ornamental & Vegetable Fungicide, *See* Chlorothalonil, 6150
SAA, *See* succinic anhydride, 29916
Sabalith®, 27189
Sabeco Metal, 27190
Saber, *See* λ-cyhalothrin, 7567
sabion, *See* limo, 17279
Sabre, 27191, *See* Bromoxynil, 4590
Sabre 1628, 27192
Sabulite, 27193
Sabutol, 27194
sacarina, *See* saccharin, 27195
saccharase, *See* sucrase, 29929
saccharimide, *See* saccharin, 27195
saccharin, 27195, *See* Sweeta, 30454
saccharin acid, *See* saccharin, 27195
saccharin ammonium, *See* sucramine, 29928
saccharin insoluble, *See* saccharin, 27195
saccharin sodium, 27196, *See* Sucaryl, 29912
saccharin soluble, *See* saccharin sodium, 27196
saccharinol, *See* saccharin, 27195
saccharinose, *See* saccharin, 27195
Saccharomycetacae, *See* yeast extract, 34640
saccharose, *See* sucrose, 29934
saccharose distearate, *See* Sucro Ester 7, 29931
saccharose monostearate, *See* Grilloten PSE141G, 13378
saccharosonic acid, *See* isoascorbic acid, 15338
saccharum lactis, *See* α-D-Lactose, 16589
Sachsse's solution, 27197
Sachtocup, 27198
Sachtoklar®, 27199
Sachtolen®, 27200
Sachtolith®, 27201
Sachtoperse® HU, 27202
Sachtopyr®, 27203
Sachtosil®, 27204
SACI, 27205
SACI® 200HM, 445a, 450W, 4192, 4194, 4215, 4253, *See* calcium sulfonate, 4973
Sacon®, 27206
sacred bark, 27207
SADH, *See* daminozide, 7723
Sadofos, *See* malathion, 18460
Saduren®, 27208
Safari®, 27209
Safebond 3, 27210
Safe-Break, 27211
Safe-FR, 27212
Safe-Gard, *See* fenbendazole, 11594
Safepak, 27213
safety dynamite, 27214
safety nitro-powder, 27215
safety oil, *See* C-petroleum naphtha, 6940
Safety-Cool, 27216
Safety-Lube®, 27217
Safex, 27218
Safezone® Cleaning Solvent & Flux Remover, 27219
Saffil, 27220
Saffil®, 27221
Saffil® Selexsorb® COS T-1061 T-64 Versal 150, *See* aluminum oxide, 1919
safflor, *See* Neobee® 18, 20872, safflower oil, 27222
safflor carmine, *See* vegetable rouge, 33523
safflor red, *See* vegetable rouge, 33523
safflower oil, 27222, *See* Neobee® 18, 20872
safflower oils, *See* safflower oil, 27222
Safflowеramidopropyl ethyldimonium ethosulfate, *See* Schercoquat FOAS, 27677
saffron, 27223
saffron bronze, 27224
saffron oil, 27225
saffron sugar, 27226
Saflex, 27227
Safoam, 27228
safrol, 27229
Safrotin, *See* Propetamphos, 25183
Saf-T-Side, 27230
Saf-T-Sol, 27231
Sag, 27232
Sagimid, *See* Mansonil®, 18611
Sahli's reagent, 27233
Sahli's stain, 27234
Saisan, 27235, *See* drazoxolan, 8974
Sajji, 27236
sakaloid, 27237
Sakarat, 27238

Sakarat Special, 27239
sakoa oil, 27240
Sakresote, 27241
sal absinthii, 27242, *See* salt of wormwood, 27292
sal amarum, 27243
sal ammoniac, *See* ammonium chloride, 2235, Katapone VV-328, 15785
sal Anglicum, *See* magnesium sulfate, 18368, sal amarum, 27243
sal Catharticum, *See* magnesium sulfate, 18368, sal amarum, 27243
sal chalybis, *See* ferrous sulfate, 11690, iron sulfate (ous), 15321
sal commune, 27244
sal Culinaris, 27245
sal digestnum Sylvii, *See* potassium chloride, 24749
sal diureticum, 27246
sal enixum, *See* tartarline, 30833
sal ethyl, *See* ethyl salicylate, 11081
sal Martis, 27247
sal mineral, 27248
sal Mirabil, 27249
sal niter, 27250
sal Prunella, 27251
sal Rupellensis, 27252
sal Saturni, 27253, *See* lead diacetate, 16906
sal sedativus, 27254, *See* Boric Acid, 4422
sal Seidlitense, *See* magnesium sulfate, 18368, sal amarum, 27243
sal soda, 27255
sal succini, 27256
sal tartar, 27257, *See* sodium tartrate, 28825
sal volatile, 27258
Salacetin, *See* aspirin,
Salachlor, *See* Formax, 12268, sodium formate, 28767
Salaigugl, *See* olibanum, 22134
sal-Alembroth, *See* salt of Alembroth, 27282
Salammonite, *See* Ammonium chloride, 2235
Salantin, *See* aspirin,
Salargyl, 27259
salarmoniac, *See* ammonium chloride, 2232
Salaspin, *See* aspirin,
Salcare®, 27260
salenixon, 27261
saleratus, 27262
salesthin, 27263
Sal-ethyl, 27264
Saletin, *See* aspirin,
Salfax 77, 27265
Salfuride, 27266
Salge metal, 27267
Salhar gum, 27268
Salicyanilide, 3,4,5-tribromo-, *See* Tuasol 100, 32543
salicylal, *See* salicylaldehyde, 27269
salicylaldehyde, 27269
salicylic acid, 27270
salicylic acid 3,3,5-trimethylcyclohexyl ester, *See* Homosalate, 14292, Kemester® HMS, 15963
salicylic acid basic bismuth salt, *See* Bismuth subsalicylate, 4244
salicylic acid ethyl ester, *See* ethyl salicylate, 11081
salicylic aldehyde, *See* salicylaldehyde, 27269
salicylic ether, *See* ethyl salicylate, 11081
Salicyl-salicylate, *See* Salysal, 27306
Salinaphthol, *See* naphthalol, 20712
Salinomycin, 27271, *See* Posistac, 24718
Saliretin resins, *See* Saliretins, 27272
Saliretins, 27272
Saliter, 27273
Salkowski's solution, 27274
Sallit's speculum metal, 27275
Sally Nixon, 27276
Salmiac, *See* Ammonium chloride, 2235, Katapone VV-328, 15785
salmiak, *See* ammonium chloride, 2232
Salmocid, 27277
Salodine, 27278
salol, *See* phenyl salicylate, 23634
salpetersäure, *See* nitric acid, 21332
Salpix, *See* acetrizoate sodium, 307
Salsorb®, 27279
salt, 27280, *See* sodium chloride, 28748
salt cake, 27281
salt of Alembroth, 27282
salt of amber, 27283
salt of chlorhexidine and gluconic acid, *See* chlorhexidine digluconate, 6101
salt of Hartshorn, 27284
salt of Lemery, 27285
salt of Norton, 27286
salt of Saturn, 27287, *See* lead acetate, 16902, lead diacetate, 16906
salt of soda, 27288
salt of Sorrel, 27289
salt of steel, 27290
salt of tartar, *See* pearl ash, 22977 potassium carbonate, 24743, 24744, 24745
salt of tin, 27291
salt of wisdom, *See* salt of Alembroth, 27282
salt of wormwood, 27292
salt perlate, 27293
Salt, Amido-G, 27294
Salt, Amido-R, 27295
Saltex, 27296
Saltialgine H8, 27297
saltpeter, *See* potassium nitrate, 24762
saltpeter rot, 27299
saltpeter superphosphate, 27300
saltpeter, saltpeter flour, 27298
salts of England, *See* magnesium sulfate, 18368
salts of Lemon, *See* salt of Sorrel, 27289
Salts of Sonel, *See* salt of Sorrel, 27289
salufer, 27301
salunol, 27302
Salut®, 27303
Salute, *See* metribuzin, 19595
Saluthion®, *See* Salut®, 27303
Salvarom, 27304
Salvex, 27305
Salvo, *See* Benzoic acid, 3973
Salysal, 27306
Salzburg vitriol, *See* Kocide® Copper Sulfate Pentahydrate Crystals, 16262, mixed vitriol, 19986
Salzone, *See* Valadol, 33238
samarium, 27307
samarium oxide, 27308
Samaron, 27309
Sambarin, 27310
Samite, 27311
samli, 27312
Samson Steel, 27313
Samsonite, 27314
SAN, *See* styrene-acrylonitrile copolymer, 29864
SAN 371F, *See* oxadixyl, 22407
SAN copolymer, *See* styrene-acrylonitrile copolymer, 29864
san nai, *See* sanna,
SAN-322I, *See* Propetamphos, 25183
Sanachlor, 27315
Sanaklenz, 27316
Sanatank, 27317
Sanatogen, 27318
Sanceler cm-po, *See* Vulkacit® CZ/EGC, DZ/EGC, 34129
Sanceler D, *See* Akrochem® DPG, 1168, Dynamine, 9355, Perkacit® DPG, 23281
Sanceller 22, *See* ethylene thiourea, 11132, Perkacit® ETU, 23283
Sanclomycine, *See* tetracycline, 31365
Sancos, *See* pholcodine, menthol, glycerol,
Sanction, *See* flusilazole, 12119
Sanction®, 27319
Sancure® 776, 27320
sand acid, 27321
Sandacid, 27322
sandalwood, 27323
sandarac resin, 27324
sandaracha, 27325
sandelwood, *See* redwoods, 25987
sanderswood, *See* redwoods, 25987
Sandet 60, *See* Nacconol® 40G, 20647
sandiver, 27326
sandix, *See* orange lead, 22267
Sandobet SC, 27327
sandoce, 27328
Sandocryl®, 27329
Sandofan, *See* oxadixyl, 22407
Sandofix, 27330
Sandofluor, 27331
Sandogen, 27332
Sando-K, 27333
Sandol, 27334
Sandolan®, 27335
Sandolube, 27336
Sandopac, 27337
Sandopan®, 27338
Sandopan® DTC Linear P, 27339
Sandopan® DTC Linear P Acid, 27340
Sandopan® DTC-100, 27341
Sandopan® DTC-Acid, 27342
Sandopan® JA-36, 27343
Sandopan® KST, 27344
Sandopan® LA-8, 27345
Sandopan® LS-24, 27346
Sandopan® MA-18, 27347
Sandopan® TA-10, 27348
Sandopan® TS-10, 27349
Sandopur, 27350
Sandorin, 27351
Sandosperse, 27352
Sandostab P-EPQ, 27353
Sandoteric CFL, 27354
Sandoteric TFL Conc, 27355
Sandotex, 27356
Sandoxylate® AC-46, AD-4, 27357
Sandoxylate® C-32, 27358
Sandoxylate® NT-15, 27359
Sandoxylate® SX-408, 27360
Sandozin, 27361
sandscale, 27362
sandstone, 27363
Sanecta, *See* Aspartame, 3233
sanfoin, 27364, *See* Orchidee, 22278
Sangajol, 27365
Saniblanket, Sanifoam, 27366
Saniclor, *See* quintozene, 25692
SaniFoam, *See* Saniblanket, Sanifoam, 27366
Sani-Soil-Set, 27367
Sanitant, 27368
Sanitary 1700, 27369
sanitas, 27370
Sanocide, *See* Julin's chloride, 15632
Sanoform, 27371
Sanogran®, 27372
Sanoleum, 27373
Sanoscent, 27374
Sanoxit, *See* Abcure S-40-25, 35, AcetOxyl 2.5 and 5, 305, Benzoyl peroxide, 3983
Sansalid, 27375
sanse, 27376
sanse oil, *See* sanse, 27376
sansel orange g, *See* Sudan I, 29938
Sansilic 11, 27377
Sansorbin®, 27378
Sanspor, 27379, *See* captafol, 5163
Santac 52, 27380
santalwood, 27381
Santechem 21-21, 27382
Santechem Grey F.R. P.E. Conc, 27383
Santel, 27384
Santiciser, 27385
Santiciser SC, 27386
Santicizer 10, 27387
Santicizer 97, 27388
Santicizer 140, *See* Disflamoll® DPK, TPK, 8641
Santicizer 141, 27389, *See* diphenyl octyl phosphate, 8597

Santicizer 143, 27390
Santicizer 148, 27391
Santicizer 154, 27392
Santicizer 160, 27393, *See* Butyl benzyl phthalate, 4764
Santicizer 711, 27394, *See* Jayflex® DUP, 15556
Santicizer DUP, 27395
Santion, 27396
Santobane, *See* DDT, 7820
Santobrite, 27397, *See* sodium pentachlorphenate, 28805
Santocel, 27398
Santochlor, 27399, *See* *p*-dichlorobenzene, 8386
Santocure, 27400, *See* Delac S, 8002, Naugard® I-2, 20792, Perkacit® CBS, 23277, Vulkacit® CZ/EGC, DZ/EGC, 34129
Santocure MOR/MOR90, 27401
Santocure NS, 27402, *See* Delac NS, 8001, Perkacit® TBBS, 23289
Santocure Vulcanization Accelerator, *See* Vulkacit® CZ/EGC, DZ/EGC, 34129
Santocure®, *See* cyclohexyl benzothiazole sulfenamide, 7523
Santocure® NS, *See* Butylbenzothiazole sulfenamide, 4791
Santocure® TBSI, *See* Butylbenzothiazole sulfenamide, 4791
Santocure®NS, *See* BBTS, 3863
Santoflex, 27403, *See* ethoxyquin, 11049, Polyflex, 24435
Santoflex 1P, 27405
Santoflex 13, 77, 27404
Santoflex 36, *See* Flexzone 3C, 11967, Permanax IPPD, 23364, Vulkanox® 4010 NA, 34147
Santoflex 77, *See* Naugard® I-2, 20792
Santoflex A, 27406, *See* Polyflex, 24435
Santoflex AW, 27407, *See* Polyflex, 24435
Santoflex B, 27408
Santoflex BX, 27409
Santoflex DD. DPA, 27410
Santoflex IP, *See* Flexzone 3C, 11967, Permanax IPPD, 23364
Santoflex IP S-IP, *See* Vulkanox® 4010 NA, 34147
Santoflex® IP, *See* N-Isopropyl-N'-phenyl-*p*-phenylene diamine, 15413
Santogard PVI, 27411
Santolite, 27412
Santomerse, 27413, *See* Nacconol® 40G, 20647
Santomerse 3, *See* sodium dodecylbenzenesulfonate, 28760
Santomerse D, *See* Witconate DS, 34485
Santomerse ME, *See* Nacconol® 40G, 20647, sodium dodecylbenzenesulfonate, 28760
Santomerse no. 85, *See* Nacconol® 40G, 20647
Santomerse-O, *See* Witconate DS, 34485
Santone® 3-1-S, 27414
Santone® 3-1-SH, 27415
Santone® 10-10-O, *See* Polygliceryl-10 decaoleate, 24453
Santone® SDD, *See* Polyglyceryl stearate series, 24457
Santonox, 27416, *See* Rutenol, 27147
Santonox BM, *See* Rutenol, 27147
Santonox R, *See* Rutenol, 27147
Santophen, *See* Nipacide® BCP, 21252
Santophen 1, *See* Nipacide® BCP, 21252
Santophen 20, 27417, *See* pentachlorophenol, 23135, 23136
Santoprene® 181-55, 181-64, 281-55, 281-64, 27418
Santoprene® 281-87, 283-40, 27419
Santoquin, 27420, *See* ethoxyquin, 11049, Polyflex, 24435
Santo-Res, 27421
Santoresin, 27422
santorin earth, 27423
Santosite, 27424
Santosol™ DMS, *See* dimethyl succinate, 8509
Santotan KR, 27425
Santotrac, 27426
Santovac, 27427
Santovar, 27428
Santovar A, *See* Santovar, 27428
Santowax, 27429
Santoweb, 27430
Santowhite, 27431, *See* Rutenol, 27147
sanyan, 27432
Sanylene, 27433
Sapamine, 27434
Sapamine, 27435
sapan wood, *See* redwoods, 25987
Sapecron, 27436, *See* Chlorfenvinphos, 6098
Saphire, 27437
sapin, 27438
sapoform, 27439
Sapogenat, 27440
Sapogenat T, 27441
sapogenin, 27442
sapogenin glycosides, *See* saponins, 27444
saponine, 27443
saponins, 27444
Saponite, *See* Steatite,
Saporin®, 27445
Sapp #4, 27446
sapphire, *See* aluminum oxide, 1919
Saprol, 27447, 27448, *See* triforine, 32254
Sar, *See* sarcosine, 27463
Sar Gel®, 27449
Saran, 27450
Saran 313, 27451
Saran 510, 27452
Saran F-239, F-278, F-310, 27453
Saran Wrap, 27454
Saran® Filament, 27455
Saranex, 27456
Sarapron, 27457
Saratoga Steel, 27458
Sarbox® SB 400, 27459
sarcine, 27460, *See* hypoxanthine, 14721
sarco, 27461
sarcocoll, 27462
sarcolactic acid, *See* L-lactic acid, 16571
sarcosine, 27463, *See* methyl-glycocoll, 19570
Sarcosyl® L, *See* lauroyl sarcosine, 16861
Saret® 500, 515, 27464
sarkin, *See* hypoxanthine, 14721
sarkine, *See* sarcine, 27460
Sarkosyl NL, *See* sodium lauroyl sarcosinate, 28779
Sarkosyl NL30, 27465
Sarkosyl O, 27466
Sarlink 1000, 27467
Sarlink 2000, 27468
Sarlink 3000, 27469
Sarolex, *See* Diazinon Liquid, 8344
Saroul, 27470
Sarpol, 27471
Sartomer, 27472
SAS, *See* sodium alum, 28730
Sascol, 27473
Sasetone, 27474
Sasfroth, 27475
Sasil, *See* Wessalith, 34284
Sasolwaks, 27476, *See* Paraffin, 22743
Sasolwaks M3, 27477, *See* Paraffin, 22743
Sassy bark, *See* ordeal bark, 22281, red water bark, 25973
Satco Metal, 27478
Satecid, *See* Propachlor, 25163, 25164, Ramrod, 25861
Satessa, 27479
Satexlan 20, 27480, *See* PEG-20 hydrogenated lanolin, 23045
Satialgine H8, 27481
Satin gloss black, *See* gas black, 12686
satin rouge, 27482
satin spar, *See* Gypsum, 13518
satin white, 27483
Satina 44, 27484
Satina®, 27485
satin-gloss black, *See* gas black, 12686
satinite, *See* gypsum, 13518
Satintone, 27486
Satinwood, 27487
Satisfar, 27488, *See* etrimfos, 11155
sativic acid, 27489
Satrapol, 27490
Satric, 27491
Satulan, 27492, *See* Fancol HL, 11441
Saturn Glace, 27493
Saturn red, *See* orange lead, 22267, red lead, 25960
Saturseal, 27494
sauconite, 27495
Saucy bark, *See* sassy bark, red water bark, 25973
Sauflon PW, 27496
Saurol, *See* Perichthol, 23270
Savall, 27497, 27498, *See* quinalphos, 25679
Savan, 27499
Savinase, 27500
Savinyl®, 27501
Savloclens, 27502
Savlodil, *See* chlorhexidine,
Savlon, 27503
Savlon Babycare, 27504
savol, 27505
Savona, 27506
savonette oil, 27507
Savoselling REAC 4, 27508
Saxifragin, 27509
Saxin, 27510, *See* saccharin sodium, 27196
Saxol, *See* Kaydol, 15810, paraffin, liquid, 22744
Saxon, 27511
Saxon bark, *See* red water bark, 25973
saxon blue, *See* smalt, 28638
Saxonite, 27512
Saxony blue, *See* smalt, 28638
Saytex 102, *See* Decabromodiphenyl oxide, 7837, Octoguard FR-01, 21999, Saytex® 102E, 27514
saytex 102E, *See* Decabromodiphenyl oxide, 7837, Octoguard FR-01, 21999, Saytex® 102E, 27514
Saytex®, 27513
Saytex® 102E, 27514
Saytex® 111, 27515
Saytex® 115, *See* Pentabromodiphenyl oxide, 23130
Saytex® 120, 27516
Saytex® 8010, 27517
Saytex® 60006L, *See* Hexabromocyclododecane, 13988
Saytex® BCL-462, 27518
Saytex® BCT-610, *See* Hexabromocyclododecane, 13988
Saytex® BN-451, 27519
Saytex® BT-93®, 27520
Saytex® FR-1138, 27521
Saytex® HBCD-LM, 27522, *See* Hexabromocyclododecane, 13988
Saytex® RB-100, 27523
Saytex® RB-49, 27524, *See* tetrabromophthalic anhydride, 31361
Saytex® VBR, 27525
Saytex®111, *See* Octabromodiphenyl oxide, 21981
Sazio, *See* alginic acid, 1593, Kelacid®, 15824
SB 70/52P6, 27526
SB-136, 27527
SB-336, 27528
SB-632, 27529
SBP, 27530
SBP-1390, *See* Reslin, 26216
SBP-1513, *See* Kafil, 15655, permethrin, 23384, 23385, 23386
SBR latex, *See* Goodyear LPR-6632, 13210
SBR Rubber, *See* K-Resin Polymer KR04, 16406, K-Resin Polymer KR05, 16407, KR01 K-Resin Polymer, 16382, KR04 K-Resin Polymer, 16383, Kraton®, 16386, Kraton® D 1101, 16387, Kraton® D 1116, 16389, Kraton® D 2103, 16391, K-Resin Polymer KR01, 16404, K-Resin Polymer KR03, 16405, K-Resin Polymer KR10, 16408, Krylene® 606, 16449, Krylene® 608, 16450
SB-VAC, 27531
SB-VAC Plus Marexine-CA, 27532
SC-10, 27533
SC-17, 27534

SC-53, 27535
SC-18862, *See* Aspartame, 3233
Scabene Lotion, 27536
Scadoplast RA3L, RA350, 27537
Scadoplast RS 20, RS 150, 27538
Scagliola, 27539
S-CAL, 27540
scale, 27541
Scale Cleen, 27542
Scalol, 27543
scandia, 27544
scandium, 27545
scandium oxide, 27546, *See* scandia, 27544
scandium(III) oxide, *See* scandium oxide, 27546
Scarab, 27547
Scarat, 27548
S-Carb, 27549
Scarlet B, *See* Sudan I, 29938
Scarlet Red, *See* Sudan IV, 29941
Scarlet Red Scharlach, *See* Sudan IV, 29941
Scarlet vermillon, 27550
Scav-Ex® 235, 27551
Scav-Ox® 35%, 27552
Scav-Ox® II, 27553
Scent Off, 27554
Scent Sticks, 27555
ScentCap, 27556
Scentinel, 27557
Sch-7056, *See* Acrisorcin, 392
Sch-21480, *See* Tiox, 31845
Sch 22219, *See* Vaderm, 33237
Schaeffer's acid, 27558
Schaeffer's salt, 27559
schallerite, 27560
Schalstein, *See* Wollastocoat®, 34545
Schardinger dextrin, *See* cyclodextrins, 7516
Scharlach b, *See* Sudan I, 29938
Scheele's acid, 27561
Scheele's green, 27562, *See* cupric arsenite, 7376
Scheele's mineral, *See* Scheele's green, 27562
scheeletine, 27563
scheelinite, 27564
Scheelinite, lead tungstate, *See* scheeletine, 27563
scheerite, 27565
Scheiber oil, 27566
Scheibler's reagent, 27567
Scheiderite, 27568
Schellan solution, 27569
Schensand, 27570
Schenvar, 27571
Schercamox C-AA, 27572
Schercamox CMA, 27573
Schercamox DMC, 27574, *See* Cocamine oxide, 6557
Schercamox DML, 27575, *See* lauramine oxide, 16832
Schercamox DMM, 27576
Schercamox DMS, 27577
Schercassist AC, 27578
Schercemol 65, 27579
Schercemol 145, 27580
Schercemol 185, 27581
Schercemol 318, 27582
Schercemol 1688, 27583, *See* Cetearyl octanoate, 5791, cetyl octanoate, 5821
Schercemol 1818, 27584, *See* isostearyl isostearate, 15422
Schercemol BE, 27585
Schercemol CM, 27586
Schercemol CO, 27587, *See* cetyl octanoate, 5821
Schercemol CP, 27588
Schercemol CS, 27589
Schercemol DEGMS, 27590
Schercemol DEIS, 27591
Schercemol DIA, 27592, *See* Unimate® DIPA, 32968
Schercemol DICA, 27593
Schercemol DID, 27594
Schercemol DIS, 27595
Schercemol DISD, 27596
Schercemol DISF, 27597
Schercemol DO, 27598
Schercemol EE, 27599
Schercemol EGMS, 27600
Schercemol GMIS, 27601
Schercemol GMS, 27602
Schercemol ICS, 27603
Schercemol IDO, 27604
Schercemol ISE, 27605
Schercemol MEL-3, 27606
Schercemol MEM-3, 27607, *See* Lanol 14 M, 16735
Schercemol MEP-3, 27608
Schercemol MM, 27609
Schercemol MP, 27610
Schercemol MS, 27611
Schercemol NGDC, 27612
Schercemol NGDL, 27613
Schercemol NGDO, 27614
Schercemol OHS, 27615
Schercemol OLO, 27616
Schercemol OP, 27617
Schercemol OPG, 27618
Schercemol PGDP, 27619
Schercemol PGML, 27620
Schercemol PGMS, 27621
Schercemol SE, 27622
Schercemol TISC, 27623
Schercemol TIST, 27624
Schercemol TT, 27625
Scherco Finish AL, 27626
Scherco Softener #1, 27627
Scherco Softener #2, 27628
Schercoat OE-44, 27629
Schercoat OE-44K, 27630
Schercoat P-110, 27631
Schercoat PC-550, 27632
Schercoat S-220, 27633
Schercoat S-330, 27634
Schercodine B, 27635, *See* Behenamidopropyl dimethylamine, 3896
Schercodine C, 27636, *See* Cocamidopropyl dimethylamine, 6552
Schercodine I, 27637
Schercodine L, 27638, *See* lauramidopropyl dimethylamine, 16831
Schercodine M, 27639
Schercodine O, 27640, *See* oleamidopropyl dimethylamine, 22109
Schercodine P, *See* Palmitamidopropyl dimethylamine, 22630
Schercodine S, 27641
Schercodine T, 27642
Schercolene SB, 27643
Schercolube 707, 27644
Schercomid 304, 27645
Schercomid 1214, 27646
Schercomid AC-S, 27647
Schercomid AME, 27648, *See* Acetamide MEA, 283
Schercomid CDA, 27649, *See* Active #2, 501
Schercomid EAC, 27650
Schercomid HT-60, 27651
Schercomid LME, 27652, *See* Lactamide MEA, 16568
Schercomid ODA, 27653
Schercomid OME, 27654
Schercomid OMI, 27655
Schercomid SAP, 27656
Schercomid SCE, 27657
Schercomid SLE, 27658
Schercomid SL-Extra, 27659
Schercomid SLS, 27660
Schercomid SO-A, 27661
Schercomid SO-T, 27662
Schercomid ST, *See* Ablumide SDE, 81
Schercomid SWG, 27663
Schercomid TO-2, 27664
Schercopol CMS-Na, 27665
Schercopol DOS-70, 27666
Schercopol DS-120, 27667
Schercopol LPS, 27668, *See* disodium lauryl sulfosuccinate, 8649
Schercopol OMS-Na, 27669
Schercopon 2WD, 27670
Schercoquat ALA, 27671
Schercoquat APAS, 27672
Schercoquat BAS, 27673
Schercoquat CAS, 27674
Schercoquat COAS, 27675
Schercoquat DAS, 27676
Schercoquat FOAS, 27677
Schercoquat IALA, 27678
Schercoquat IAS, 27679, *See* Foamquat IAES, 12180
Schercoquat IB, 27680
Schercoquat IEP, 27681
Schercoquat IIB, 27682
Schercoquat IIS, 27683
Schercoquat ROAS, 27684
Schercoquat ROEP, 27685
Schercoquat SAS, 27686
Schercoquat SOAB, 27687
Schercoquat SOAS, 27688
Schercoquat WOAS, 27689
Schercosol DS, 27690
Schercosol NL, 27691
Schercosol P, 27692
Schercosol T, 27693
Schercotaine APAB, 27694
Schercotaine CAB, 27695, *See* Cocamidopropyl betaine, 6551
Schercotaine CAB, CAB-G
Schercotaine CAB-A, 27696
Schercotaine IAB, 27697
Schercotaine MAB, 27698
Schercotaine PAB, 27699
Schercotaine SCAB, 27700
Schercotaine UAB, 27701
Schercotaine WOAB, 27702, *See* Wheat germamidopropyl betaine, 34350
Schercotarder, 27703
Schercoterge 140, 27704
Schercoteric CY-2, 27705
Schercoteric I-AA, 27706
Schercoteric MS, 27707, *See* Empigen® CDR40, 10369
Schercoteric MS-2, 27708, *See* disodium cocoamphodiacetate, 8643
Schercoteric MS-EP, 27709
Schercoteric O-AA, 27710
Schercowet DOS-70, 27711
Schercozoline C, 27712
Schercozoline I, 27713
Schercozoline L, 27714, *See* lauryl hydroxyethyl imidazoline, 16865
Schercozoline O, 27715
Schercozoline S, 27716, *See* stearyl hydroxyethyl imidazoline, 29606
Schering 4072, *See* phenmedipham, 23597
Schering 38584, *See* phenmedipham, 23597, Vangard, 33323
Scheroba Oil, 27717
Scherpol LSB, 27718
Schersoftoil P, 27719
Schiff's reagents, 27720
Schimose, *See* picric acid, 23821
schlempe, 27721
Schlichte®, 27722
Schlippe's salt, 27723
Schneiderite, 27724
schoenite, 27725
Schollkopf's acids, 27726
Schonberg's alloy, 27727
schonite, *See* schoenite, 27725
schorl, 27728
schorl rock, 27729
Schou oil, 27730
Schradan, 27731, *See* Sytam, 30667
schraufite, 27732
schreibersite, 27733
Schultze's reagents, 27734
Schultze's stain, 27735
schungite, 27736
Schutzenberger's salt, 27737

Schweinfurt green, *See* cupric acetoarsenite, 7374, mountain green, 20369
Schweitzer's reagent, 27738
schwelkohle, 27739
Scian turpentine, *See* Chian turpentine, 6057
Scilla, *See* squill, 29359
Scillin, *See* Scillipicrin,
Scillotoxin, *See* Scillipicrin,
Scintillase, 27740
Scintran, 27741
Sclair®, 27742
sclerolac, 27743
Sclerosol, *See* dimethyl sulfoxide, 8511
Sclomo, 27744
Scolaban, 27745, *See* bunamidine hydrochloride, 4680
S-collidine, *See* Collidine, 6625
Scopacron, 27746
Scopacron 50, 75 and 80, 27747
Scopacryl, 27748
Scopasol 550, 27749
Scopol 58M, 58SP, 27750
Scopol 85X, 27751
Scopolux 221SP, 27752
Scorbacid, *See* vitamin C, 33938
Scorbu-C, *See* vitamin C, 33938
Scorchex, 27753
Scorchguard O, 27754
Scorchguard-bound, 27755
Scotch cement, 27756
Scotch foundry pig, 27757
scotch gin, *See* spirit of sweet niter,
scotch soda, 27758
Scotch topaz, 27759
Scotchkote® 213, 214, 27760
Scotchlite, 27761
Scotcil, *See* Penicillin G potassium, 23109
Scotphos, 27762
Scotts Bonus Type S, *See* atrazine, 3394
Scotts OH I, *See* Parogen, 22854
Scour 1161, 27763
Scour KSV, KSV Special, 27764
Scourge, *See* piperonyl butoxide, 23875, resmethrin, 26218
scouring slag, 27765
Scram, 27766
scrap rubber, 27767
Scratch-Guard, 27768
Screen, 27769
Screen Star Photo Emulsion, 27770
Screte, 27771
screw bronze, 27772
Scripset, 27773
Scripset 520, 27774
Scripset 720, 27775
Scuranate, 27776
Scurane V, 27777
Scutl, *See* phenylmercury acetate, 23646
Scutl Riogen, *See* acetoxyphenylmercury, 306
Scuttle, 27778
Scythe, 27779
SD 1750, *See* dichlorvos, 8394, Vapona, 33378
SD 11831, *See* nitralin, 21319
SD 14114, *See* Torque, 32077
SD 15418, *See* cyanazine, 7472
sd alcohol 23-hydrogen, *See* ethyl alcohol, 11064
SD-1, -2, 27780
SD-376, 27781
SD-7859, *See* Chlorfenvinphos, 6098
SD-43775, *See* fenvalerate, 11620
SDBS, *See* Nacconol® 40G, 20647
SDDC, *See* Octopol SDE-25, 22013, sodium dimethyldithiocarbamate, 28758
SDMDTC, *See* Aquatreat SDM, 2761, methyl namate®, 19551, Octopol SDM-40, 22014, Perkacit® SDMC, 23288, sodium dimethyldithiocarbamate, 28758, Thiostop N, 31717
SDS, *See* sodium lauryl sulfate, 28781, Texapon K-12, K-1296, L-100, 31461
SE Wax, 27782
SE-458, 27783
Sea Hawk Biotin, *See* bioMeT TBTF, 4171
sea onion, *See* squill, 29359
SeaBuffer, 27784
SeaCure, 27785
Sea-Gard® Formula FP-91, 27786
SeaGarden, 27787
Seair, 27788
Seal and Heal, 27789
Sealac, 27790
sealite, 27791
Sealum, 27792
Searlequin, *See* Yodoxin, 34658
SeaTest, 27793
sea-water bronze, 27794
sebacic acid, 27795, *See* ipomic acid, 15247
sebacic acid bis(2-ethylhexyl) ester, *See* Uniflex® DOS, 32946
Sebacic acid, diisopropyl ester, *See* Unimate® DIPS, 32969
Sebacil, *See* Sebacil®, 27796
Sebacil®, 27796
Sebase, 27797
Sebastine, 27798
sebkanite, 27799
Sebond, 27800
Sebrite, 27801
***sec*-nitropropane**, *See* 2-nitropropane, 21374
9,10-secocholesta-5,7,10(19)-trien-3-ol, *See* vitamin D_3, 33941
9,10-secoergosta-5,7,10(19)-,22-tetraen-3-ol, *See* vitamin D_2, 33940
9,10-secoergosta-5,7,10(19)-trien-3-ol, *See* vitamin D_4, 33942
Secol, 27802
Secolan S-1, BA-1, BA-1G, 27803
Secolat, 27804
Secomine TA 02, 27805
Secomix® E40, 27806
Secondary acetylenic alcohol, *See* Ancor® OW-1, 2433
secondary ammonium phosphate, *See* Ammonium phosphate dibasic, 2250
secondary calcium phosphate, *See* Calbrite, 4928
secondary sodium phosphate, *See* disodium phosphate (anhydrous), 8652, tasteless salts, 30843
Secosol AL 959, 27807
Secosol AL/MG 50, 27808
Secosol ALL/40, 27809
Secosol DOS/70, 27810
Secosol EA/40, 27811
Secosol® AL 959, 27812
Secosol® ALL40, 27813
Secosol® DOS 70, 27814
Secosov, 27815
Secoster® A, 27816
Secoster® DMS, 27817
Secoster® DO 600, 27818, *See* PEG-2 dioleate series, 23040
Secoster® EMS, 27819
Secoster® MA 300, 27820
Secoster® MO 400, 27821
Secoster® SDG, 27822
Secosyl, 27823
Secosyl, *See* sodium lauroyl sarcosinate, 28779
Secretan, 27824
Secretol, 27825
Securamid®, 27826
Secure, 27827
Securite, 27828
Securitol, 27829
Security Fungi-Gard, *See* Repulse, 26151
Security Fungi-Gard Bravo-W-75, *See* Chlorothalonil, 6150
Security Lime Sulphur, *See* Orthorix, 22337
Securon 540, 27830
Sedanox, 27831
Sedaplant Richter, 27832
Sedatin, *See* antipyrine,
Sedefos 75®, 27833
Sedex, 27834
Sedifloc Flocculant Aids, 27835
Sedipol®, 27836
Sedipur®, 27837
Sedoneural, *See* sodium bromide, 28741
Sedox, *See* Bendiocarb, 3937
Sedral, *See* Cefadroxil, 5547
Sedresan, *See* Agallol, 954, Atiran, 3286, Ceresan, 5745
Seed Base, 27838
seed-lac, 27839
Seedox, *See* Bendiocarb, 3937
Seedox SC, 27840
Seedoxin, *See* Bendiocarb, 3937
Seedrin, *See* Aldrin, 1558
Seedtect, 27841
Seekay Pitch®, 27842
Seekay wax, *See* Nibren wax, 21097
Seenox 412S, 27843
SEF, 27844
Segetan, 27845
Segosin, *See* Abrodil, 158, Methiodal sodium, 19485
Seifert solder, 27846
Seignette salt, 27847
Seignette's salt, *See* Rochelle salt, 26882
Sekawrap, 27848
Sekicel, 27849
sel d'Angleterre, 27850
Seladon green, *See* Bohemian earth, 4368
Selar®, 27851
Selastin EL-10, EL-30, SE EM 95, 27852
Selazate, 27853
Selbana 2001, 27854
Selbax, 27855
Select-A-Sorb, 27856
Selectin, *See* prometryn, 25140, 25141
Selectin 50, *See* prometryn, 25140, 25141
Selective herbicide, *See* Mayclene, 19089
Selectrol, 27857
Select-Trol, 27858
Selek, 27859
Selektin, *See* prometryn, 25140
Selektonon M, *See* MCPA, 19183
Selenac®, 27860
selenious acid, disodium salt, pentahydrate, *See* sodium selenite, 28815
selenious anhydride, *See* selenium dioxide, 27862
selenite, *See* Gypsum, 13518
selenium, 27861, *See* Vandex®, 33311
selenium dialkyldithiocarbamate, *See* Selenac®, 27860
selenium diethyl dithiocarbamate, *See* ethyl selenac, 11082, Selazate, 27853
selenium dimethyldithiocarbamate, *See* methyl selenac, 19559
selenium dioxide, 27862
selenium sulfide, 27863
selenium trisulfide, 27864
selenium, tetrakis(dimethyldithiocarbamate), *See* methyl selenac, 19559
selenous acid anhydride, *See* selenium dioxide, 27862
Seleron, 27865
Selexsorb® COS, 27866, *See* aluminum oxide, 1919
self-hardening steel, *See* Mushet steel, 20528
Selin® O, 27867
Selinon, *See* 4,6-dinitrocresol, 8542
Seliwanoff's reagent, 27868
Seljut, 27869
Sella Fast, *See* Sellaflor, 27871
Sellacron, *See* Sellaflor, 27871
Sellacron, Sella Acid, Sella Fast, Sellaflor, 27870
Sellaflor, 27871
Sellaite, *See* magnesium fluoride, 18361
Sellasol, 27872
Sellifix Helios, 27873
Sellig, 27874
Sellig Antimousse S, 27875
Sellig AO 15 100, 27876
Sellig AO 25 100, 27877

Sellig AO 6100, 27878
Sellig DN 10 100, 27879, *See* Ethal DNP-8, 10919
Sellig DN 22 100, 27880, *See* Ethal DNP-8, 10919
Sellig HR 18 100, 27881
Sellig LA 1150, 27882
Sellig N 4 100, 27891
Sellig N 5 100, 27892
Sellig N 6 100, 27894
Sellig N 8 100, 27895
Sellig N 9 100, 27896
Sellig N 10 100, 27883
Sellig N 11 100, 27885
Sellig N 12 100, 27886
Sellig N 15 100, 27887
Sellig N 20 80, 27889
Sellig N 30 70, 27890
Sellig N 50 100, 27893
Sellig N 1050, 27884
Sellig N 1780, 27888
Sellig O 4 100, 27900
Sellig O 5 100, 27901
Sellig O 6 100, 27902
Sellig O 8 100, 27903
Sellig O 9 100, 27904
Sellig O 11 100, 27897
Sellig O 12 100, 27898
Sellig O 20 100, 27899
Sellig R 20 100, 27905
Sellig R 3395, 27906
Sellig R 3395 SP, 27907
Sellig R 3395-C435, 27908
Sellig R 4095, 27909
Sellig R 4495, 27910
Sellig S 30 100, 27911
Sellig SP 8 100, 27917
Sellig SP 16 100, 27912
Sellig SP 20 100, 27913
Sellig SP 25 50, 27914
Sellig SP 30 100, 27915
Sellig SP 3020, 27916
Sellig Stearo 6, 27918
Sellig SU 4 100, 27922
Sellig SU 18 100, 27919
Sellig SU 25 100, 27920
Sellig SU 30 100, 27921
Sellig SU 50 100, 27923
Sellig T 3 100, 27926, *See* PEG tallate series, 23035
Sellig T 14 100, 27924, *See* PEG tallate series, 23035
Sellig T 1790, 27925, *See* PEG tallate series, 23035
Selligon SP, 27927
Selligor 860 SP, 27928
Sellogen DFL, 27929
Sellogen HR-90, 27930
Sellogen W, *See* Aerosol® OS, 895
Selora, 27931, *See* Potassium chloride, 24748
Seloxone, 27932
Selsun, 27933
seltzers, 27934
Sembonit, 27935
Semeron, 27936, *See* desmetryn, 8142
Semfreeze, 27937
semicoke, 27938
seminose, 27939
Semirit, 27940
semi-steel, 27941
Sempatap, 27942
Sempollan, 27943
Semtol, 27944
senarmontite, *See* Octoguard FR-10, 22000, UltraFine® II, 32799
Senate, 27945, 27946
Sencor, *See* metribuzin, 19595, 19596
Sencor 4L, *See* metribuzin, 19595, 19596
Sencor 75DF, *See* metribuzin, 19595, 19596
Sencor DF, *See* metribuzin, 19595, 19596
Sencor or metribuzin, *See* metribuzin, 19595, 19596
Sencor®, 27947
Sencoral®, *See* metribuzin, 19596
Sencorex, *See* metribuzin, 19595, 19596
Sencorex® WG, 27948
SencorlPreview, *See* metribuzin, 19595, 19596
Sendran, *See* Propoxur, 25198, 25199, 25200
Sendust, 27949
seneca oil, *See* Kendex OCTG, 16002
Senegal gum, 27950, *See* gum arabic, 13480
senegin, *See* struthiin, 29840
Seneprolin, *See* paraffin, liquid, 22744
Senesco, *See* zoloft,
Senfgas, *See* mustard gas, 20538
Sengite, 27951
Sennaar gum, *See* Suakin gum, 29901
Sensitizer, 27952
Sensolve BEA, 27953
Sensolve EEA, 27954
Sensolve EPA, 27955
Sensolve MPA, 27956
Sentinel, 27957, *See* λ-cyhalothrin, 7567, Propachlor, 25163
Sentiphene, *See* Nipacide® BCP, 21252
Sentry Cyclomethicone, 27958
Sentry Dimethicone, 27959
Seoolgen W, *See* Rhodacal® IN, 26585
SEP 55, 27960
Sep 6, 27961
Sepabase, 27962
Sepabase A Grades, 27963
Sepabeads® FP Series, 27964
Sepacid® CE 5209, 27965
Sepacid® CE 5265, 27966
Sepaclear®, 27967
Sepacorr®, 27968
Sepacorr® HT, 27969
Sepaflood®, 27970
Sepaflux®, 27971
Sepakoll®, 27972
Sepapar® P, 27973
Separan, 27974
Separit, 27975
Separol, 27976, 27977
Separol AF 27, 27978
Sepascale®, 27979
Sepasolv® MPE, 27980
Sepawet®, 27981
sepia, 27982
Sepicide HB, 27983
Sepigel 305, 27984
Sepigel A, 27985
Sepimate, *See* Ammonium sulfamate, 2254
Sepineb, *See* Zineb, 34807
Sepisol, *See* Telone, 31069
Sepizin L, *See* Azinphos-ethyl, 3520
Seponver, *See* Closantel, 6441
Sepramar, 27986
Seprate K, 27987
Sept 115, *See* Abiol, 69, Germall® 115, 12893
Septal, 27988
Septalan, *See* Allantoin, 1739
Septiderm-hydrochloride, *See* chloroxylenol, 6159
Septiphene, *See* Nipacide® BCP, 21252
Septisol, *See* hexachlorophene, 13992
Septofen, *See* hexachlorophene, 13992
Septotan, *See* hydrargaphen, 14539
Seqlene® 270, 27989
Sequalog, 27990
Sequenase, 27991
Sequest-All, 27992
Sequestrene, *See* Kalex 220 Crystal, 15668, Kalex Liq. 50%, 15671
Sequestrene NA 2, *See* edetate disodium, 9613, Trilon® BD, 32325
Sequestrene®, 27993
Sequestrene® 220, 27995
Sequestrene® 30A, 27994
Sequestrene® AA, 27996
Sequestrene® NA2, 27997
Sequestrene® NA2 Edetate USP, 27998
Sequestrene® NA2Ca, 27999
Sequestrene® NA3, 28000, *See* edetate trisodium, 9614
Sequestrene® NAFe 13% Fe, 28001
Sequestrene® NH4Fe, 28002
Sequestrene® Tetraammonium, 28003
Sequestrine, *See* tetrasodium EDTA, 31391
Sequion 20H45, *See* Fostex AMP, 12337
Sequion OA, *See* Fostex AMP, 12337
Ser, *See* L-serine, 17705, 28035
Seracelle, 28004
Seradix, 28005, *See* 4-indol-3-ylbutyric acid, 15086
Sera-Pak®, 28006
Seraphos, *See* Propetamphos, 25183
Serdet DCK, 28007
Serdet DFK, 28008
Serdet DFK 30, *See* sodium decyl sulfate, 28755
Serdet DFL, DFM, and DFN, 28009
Serdet DM and DMK, 28010
Serdet DML, 28011
Serdet DNK, 28012
Serdet DPK, 28013
Serdet DSK, 28014
Serdolamide, 28015
Serdox, 28016
Serdox NNP4, 28017
Serdox NNP5 and NNP6, 28018
Serdox NNP7, NNP8.5 and NNP9, 28019
Serdox NNP10, NNP12, 28020
Serdox NNP15, NNP20, NNP25, 28021
Serdox NNP30, NNP30/70, 28022
Serdox NNPQ 7/11, 28023
Serdox NOP 30/70, 28024
Serdox NOP9, 28025
seretin, 28026
Serfene, 28027
sericine, 28028
Sericite PHN, 28029
Sericite SL-012, 28030
Sericose, 28031
Serilan, 28032
Serilene Dyes, 28033
Serilube Series, 28034
(S)-(-)-serine, *See* L-serine, 17705, 28035
L-(-)-serine, *See* L-serine, 17705, 28035
L-serine, 28035, *See* L-serine, 17705, 28035
Serinyl, 28036
Seriplast Dyes, 28037
Seripol Series, 28038
Seriprint Dyes, 28039
Serisol Dyes, 28040
Seritox, 28041, 28042
Serizyme, 28043
Sermag®, 28044
Sermix, 28045
Sermul, 28046
Sermul EA54, EA151 and EA146, 28052
Sermul EA 88, 28047
Sermul EA129, 28048, *See* Ammonium lauryl sulfate, 2242
Sermul EA150, 28049
Sermul EA176, 28050
Sermul EA188, EA136 and EA205, 28051
SeroClear®, 28053
Serotulle®, 28054
Serramix CDA, 28055
Serseal, 28056
Serum albumin, *See* Bovinal-20, 4465
serum-casein, *See* globulin,
Serumpro EN-10, 28057
Servamine KAC 422, 28058
Servamine KEP 4527, 28059
Servamine KET 350, 28060
Servamine KET 4542, 28061
Servamine KOO 330, 28062
Servamine KOO 330B, 28063
Servamine KOO 360, 28064
Servil® Range, 28065, 28066
Servo Ampholyt (B) JA110, 28067
Servo Ampholyt (B) JA140, 28068
Servo Ampholyt (B) JB130, 28069
Servo Brilliant Oil B AZ 75, 28070

Servoxyl VLA 2170, 28071
Servoxyl VLB 1123, 28072
Servoxyl VLE 1159, 28073
Servoxyl VP, 28074
Sesagard, *See* prometryn, 25140
Sesame oil, 28075
Sesamide DEA (1:1), *See* Incromide SED, 14995
Sesamidopropalkonium chloride, *See* Incroquat SE-85, 15045
Sesamidopropyl betaine, *See* Incronam SE-30, 15024
Sesamidopropyl dimethylamine lactate, *See* Incromate SEL, 14974
Sesamidopropylamine oxide, *See* Incromine Oxide SE, 15012
Seseal, 28076
Seseal 8, 28077
Sesolvan® L, 28078
Sestrip, 28079
Setac, 28080
Setacin 103 Spezial, 28081, *See* disodium lauryl sulfosuccinate, 8649
Setacin F Spezial Paste, 28082
Setacure, 28083
Setafix, 28084
Setair, 28085
Setal, 28086
Setalana, 28087
Setalin, 28088
Setalux, 28089
Setamine US, 28090
Setamol®, 28091
Setarol, 28092
Setatack A, 28093
Setatack AF, 28094
Setatack LP, 28095
Setatack P, 28096
Setatack T, 28097
Sethotope, 28098
Sethoxydim, 28099, *See* Checkmate, 5863, Poast®, 24278
sethoxydime, *See* Sethoxydim, 28099
Setic, 28100
Setilon, 28101
Setilose, 28102
Setilthe, 28103
Setirene, 28104
Setit®, 28105
Setreat, 28106
Sett®, 28107
Setter 33, 28108
Sevacarb, 28109
Sevamine KOV 4342B, 28110
Sevefilm 20, 28111
Sevelyte K, 28112
Seven Seas®, 28113
Sevestat ML 300, 28114
Sevin, 28115
Sevin® Brand SL, 28116
Sevinon, *See* undecylenic acid, 32916
Sewarin, 28117
Sewer gas, *See* sulfuretted hydrogen, 30053
Sextone, 28118
Seymourite, 28119
SF18, 28120
SF-20CF, 28121
SF69, 28122
SF81, 28123
SF99, 28125
SF1023, 28126
SF1154, 28127
SF1173, 28128
SF1188, 28129
SF1250, 28130
SF96®, 28124
S-Flakes, 28131, *See* hydrogenated soybean oil, 14596
SH-437, *See* iotrolan, 15244
Shadeacrete, 28132
Shadocol, *See* keraphen, 16055
Shadow, 28133
Shaku-do, 28134
shale, 28135
shale oil, 28136
Shamrox Vacate, *See* MCPA, 19183
Sharstop 204, *See*
Sharstop 204, *See* Aquatreat SDM, 2761, methyl namate®, 19551, Octopol SDM-40, 22014, Perkacit® SDMC, 23288, sodium dimethyldithiocarbamate, 28758, Thiostop N, 31717
Shatah, 28137
Shawinigan Black, 28138, *See* acetylene black, 319
Shawinigan's Black, *See* acetylene black, 319
Shawplas, 28139
Shcherbokov's solder, 28140
Shea Butter, 28141, 28142, *See* Cirami No. 1, 6318, Fancol Karite Butter, 11444
Shea butter extract, 28143
Sheathing bronze, *See* sea-water bronze, 27794
Shebu WS, 28144
Shebu, Refined, 28145, *See* Shea Butter, 28141
Sheduri, *See* sanna,
sheet-lac, *See* lac, 16547
Shell 5A18Z, 28146
Shell 5A95, 28147
Shell 5C64, 28148
Shell 6A01K, 28149
Shell 6C20S, 28150
Shell 7C55H, 28151
Shell D 50, 28152
Shell DS 6C46L, Shell DS 7C04N, 28153
Shell DS 7C04N, 28154
Shell Elexar, 28155
Shell JF 6100, 28156
shell limestone, 28157
Shell PDC 1120, 28158
Shell WRS 6-198, 28159
Shell WRS 6-205, 28160
shellac, *See* lac, 16547
shellackose, 28161
Shellflex Process and Extender Oils, 28162
Shellite, 28163
Shellsol D40, D60, D70, 28164
Shellsol E, A, AB, R, 28165
shellsol T, 28166
Shellswim 11T, 28167
Shellswim 5X, 28168
Shellvis 50 (SAP 150), 28169
Shelspra, *See* Mexphalte, 19616
Sherbelizer®, 28170
Sheriff, *See* carbosulfan, 5245
Sherpa, 28171, *See* cypermethrin, 7588, Topclip Parasol, 32037
Sherwood oil, *See* petroleum ether,
Shibu-ichi, 28172
Shield, 28173, *See* Androx 3961, 2443
Shield Stinger, *See* Clopyralid, 6436
shikimole, 28174
shikon, 28175
Shilajatu, 28176
Shimose, *See* Lyddite, 18068, reflorit, 26003
Shimosite, 28177
shinnamu, 28178
Shinnippon Fast Red GG Base, *See* *p*-nitroaniline, 21345
shio liao, 28179
Ship Shape® Resin Cleaner, 28180
Shipley's solutions, 28181
Shiro Bishi®, 28182
SHL Lawn Sand Plus, 28183
SHL-451A, *See* Magnevist, 18376
SHMP, *See* sodium hexametaphosphate, 28770
Shock-Ferol Sterogyl, *See* vitamin D+72, 33940
Shoemaker's black, 28184
Shoemaker's paste, 28185
Shokusen SE, 28186
shorle, *See* schorl, 27728
short oil varnishes, *See* long oil varnishes, 17580
Shortstop, *See* terbutryn, 31230
Shortstop E, *See* terbutryn, 31230
Shostakovsky Balsam, 28187
shot lead, *See* shot metal, 28188
Shot lead, bullet metal, *See* shot metal, 28188
shot metal, 28188
Shotgun, *See* Acetic acid, 292
Showchlon, *See* sodium hypochlorite, 28773
Showdown, *See* triallate, 32190
Shur-Coal® FCA, 28189
Si 264, 28191
Si 69, 28190
sialonite, 28192
Siapton, 28193
Sibercizer C6, 28194
Siberez, 28195
Sibley alloy, 28196
Sibor, 28197
Sibutol, *See* Bitertanol, 4288
Sibutol®, 28198
Sical, 28199
sicalite, *See* Gallatite, 12608
siccative, 28200
Siccatol, 28201
Siccolam, 28202
Sickle, 28203
SIC-L(TM), *See* CT1800, 7312
Siclor, 28204
Sico®, 28205
Sicocab®, 28206
Sicodop®, 28207
Sicoflex 80, 85, 28208
Sicoflex 85, 28209
Sicoflex 90, 28210
Sicoflex 93, 28211
Sicoflex 95, 28212
Sicoflex 99, 28213
Sicoflex MBS, 28214
Sicoflush® A, 28215
Sicol 150, *See* Bis(2-ethylhexy) Phthalate, 4228, dioctyl phthalate, 8549, Reomol DCP, 26130
Sicol 160, *See* Butyl benzyl phthalate, 4764
Sicol 250, *See* dioctyl adipate, 8548, Plasthall® DOA, 24002, PX-238, 25440, Wickenol® 158, 34393
Sicolen®, 28216
Sicolub®, 28217
Sicolub® DSP, 28218
Sicolub® E, 28219
Sicolub® EDS, 28220
Sicolub® OA2, OA4, 28221
Sicolub® TDS, 28222
Sicomet®, 28223
Sicomin®, 28224
Sicomix®, 28225
Sicopal®, 28226
Sicopharm®, 28227
Sicoplast®, 28228
Sicopos®, 28229
Sicopur®, 28230
Sicopurol®, 28231
Sicorin®, 28232
Sicostab®, 28233
Sicostyren®, 28234
Sicotan®, 28235
Sicotherm®, 28236
Sicotrans®, 28237
Sicoversal®, 28238
Sicovinyl®, 28239
Sicovit®, 28240
Sicromo Steel, 28241
Sidanyl, 28242
Sident 15, 28243
Sident 22LS, 22S, 28244
Sideraphthite, 28245
sidero cement, 28246
Sidot's blende, 28247
sidum radioiodide, *See* Radiocaps-131, 25827
Siemensite, 28248
sienna, 28249
Sierra Leone butter, *See* Lamy butter, 16643
Sifbronze, 28250

Silk Pro-Tein, 28362
silk rubber, *See* sericine, 28028
silk size, *See* sericine, 28028
silk wadding, 28363
Silk, Anaphe, 28364
Silkin, 28365
Silkiol, 28366
Silksoft® Supreme, 28367
Sillikolloid P 87, 28368
sillimanite, *See* Tisyn®, 31857
Sillimanith, *See* Pressolith, 24915
Sillitin N 82, 28369
Sillitin-Aktisil, 28370
Sillman bronze, 28371
Silm, 28372
Silman Steel, 28373
Silmar® 901R, 28374
Silmar® S249, 28375
Silmar® S585, 28376
Silmar® S957, 28377
Silmar® S958, 28378
Silmod 20A, 28379
Silmod SAT-30, 28380
Sil-o-cel, 28381
Siloid®, 28382
Silopren®, 28383
Silopren® HV, 28384
silosuper pink b, *See* Rhodamine G and G Extra, 26617
Silotras orange tr, *See* Sudan I, 29938
Silotras Yellow T2G, *See* Butter yellow, 4759
Sil-O-Wet™, 28385
Siloxanes and Silicones, di-Me, (octadecyloxy)-terminated, *See* Abil® Wax 2434, 58
siloxicon, 28386
siloxide, 28387
Silpruf® 2000, 28388
Silquat Q-100, 28389
Silres® KX, 28390
Silres® MP 42 E, 28391
Silres® MSE 100, 28392
Silres® REN 50, 28393
Silrifos, *See* Chlorpyrifos, 6163
Silso®, 28394
Silsoft, 28395, 28396
Silteg, 28397
Siltek, 28398
Siltek® L Polymer, 28399
Siltem® STM1300, 28400
Siltex, 28401
Siltouch Cotton Plus, 28402
Silumin, 28403
Siluminite, 28404
Silumin-Y, 28405
silundum, 28406
Silva, 28407
Silvacide, *See* Ammonium sulfamate, 2254
Silvacur, 28408
silvapron, 28409
Silvatol, 28410
Silvaz, 28411
Silvel, 28412
silver, 28413, *See* chloridazon, 6103
silver acetate, 28414
silver acetylguaiacoltrisulfonate, *See* Eosolate, 10608
silver alum, 28415
silver amalgam, 28416
silver bell metal, 28417
Silver Bond 30, 28418
silver bronze, 28419
silver carbonate, 28420
silver chloride, 28421
silver citrate, 28422, *See* itrol, 15433
silver fluoride, 28423, *See* lachiol, 14801, tachiol, 30686
silver foil, 28424
silver grain, 28425
silver grey, 28426, *See* slate grey, 28594
silver hyponitrite, *See* nitrosyl silver, 21378
silver ink, 28427
silver iodide, 28428
silver leaf, 28429
silver metal, 28430, *See* aluminum silver, 1927
silver monofluoride, *See* silver fluoride, 28423, tachiol, 30686
silver nitrate, 28431, *See* lunar caustic, 17816
silver nucleate, *See* protargentum, 25251
silver nucleinate, *See* protargentum, 25251
silver protein, *See* protargentum, 25251
silver proteinate, *See* lyargol, 18058, protargentum, 25251
silver saltpeter, *See* silver nitrate, 28431
silver silicofluoride, *See* isotachiol, 15423
silver solder, 28433
silver sulfate, 28434
silver sulfate normal, *See* silver sulfate, 28434
silver sulfide, 28435
silver white, *See* Flake White, 11871
silver(I) fluoride, *See* tachiol, 30686
silver(I) oxide, 28432
Silver, China, *See* Aygyrolith,
silver, colloidal, *See* silver, 28413
silverine, 28436
silvering solutions, 28437
silverite, *See* nickel silvers, 21132
Silverline 200, 28438
Silverline 665, 28439
Silveroid, 28440, *See* nickel silvers, 21132
silver-salt, 28441
Silverstone Supra®, 28442
Silverstone®, 28443
silvestrite, 28444
Silvet, 28445
Silvex, *See* Kuron, 16496, Silvet, 28445
silvisar 550, *See* Neoarsycodyl, 20871
Silvital, 28446
Silvoline, 28447
Silwax® S, 28448
Silwet®, 28449
Silwet® L-720, 28451
Silwet® L-7200, 28452
Silwet® L-7200, L-7210, L-7230, L-7622, *See* dimethicone copolyol, 8494
Silwet® L-7500, 28453
Silwet® L-7600, 28454
Silwet® L-7605, 28455
Silwet® L-7607, 28456
Silwet® L-7622, 28457
Silwet® L-77, 28450
Silylene, oxo-, *See* Monox, 20259
silylium, trimethyl-, chloride, *See* CT2950, 7325
silzin bronze, 28458
Simadex, 28459
Simanex, 28460
Simapron, 28461
Simask®, 28462
Simax, 28463
simazine, 28464, *See* Boroflow, 4433, Gesatop, 12923, Herbazin 50, 13801, Simadex, 28459, Sinazine, 28493
simazine-aminotriazole, *See* Herbazin Plus, 13802
Simazol, 28465
Simchin WS, 28466
Simchin, Natural, 28467
Simethicone, *See* Amersil® simethicone, 2033, Dow Corning® Antifoam A, 8870, Dow Corning® Antifoam C, 8871, Hodag Antifoam F-1, 14165, Mazu® DF 200SP, 19165, Sentry Dimethicone, 27959, Tego® Foamex N, 30979
Simetite, 28468
Simfix, 28469
Simflex, 28470
Simflow, 28471
Simflow Plus, 28472
Similor, 28473
Simmering, 28474
Simoniz, 28475
Simplex, 28476
Simplex Steel, 28477
Simply White, 28478
Simrax, 28479
Simrit, 28480
Simulsol 58, 28483
Simulsol 98, 28481, *See* oleth-8, 22126
Simulsol 165, 28484
Simulsol 989, 28482
Simulsol 1292, 28485
Simulsol 5719, 28486
Simulsol CS, 28487
Simulsol M 45, 28488
Simulsol P4, 28489
sin red, 28490
Sinapoline, 28491
Sinatron, 28492
Sinazine, 28493
Sinbar, *See* terbacil, 31220
Sinbar 80W, *See* terbacil, 31220
Sinbar®, 28494
Sin-Chu, 28495
Sindanyo, 28496
Sinflouran, *See* Treflan, 32162, Trifluralin, 32252
single muriate of tin, 28497
single nickel salt, *See* nickel sulfate, 21135
Single Purpose, 28498, *See* phenylmercury acetate, 23646
Singlex CIP, 28499
sinigrin, 28500
sinigroside, *See* sinigrin, 28500
Sinnoester MIP, *See* Wickenol® 101, 105, 34378
Sinnoester PIT, *See* Wickenol® 111, 34379
sinnopon ls 95, *See* Rhodapon® 101-10, 26631
sinnopon ls 100, *See* sodium lauryl sulfate, 28781
Sinnozon, *See* Nacconol® 40G, 20647, sodium dodecylbenzenesulfonate, 28760
Sinodor, 28501
Sinopon, *See* Witcolate NH, 34478
Sinotex CDB, *See* Benzalkonium chloride, 3958
Sinox, *See* 4,6-dinitrocresol, 8542
sintapon l, *See* Rhodapon® 101-10, 26631, sodium lauryl sulfate, 28781
Sinter-corundum, 28502
Sinterit, 28503
Sinterloy, 28504
Sintrex®, 28505
Sinvabond, 28506
Sinvaset, 28507
Siogel, 28508
Sionit, *See* A-625/641ABS 301K, ABS 500FR-1, 10
Sionon, *See* A-625/641ABS 301K, ABS 500FR-1, 10
sionon®, 28509
Siopel, 28510
Sioplas, 28511
Sipalin AOC, 28512
Sipalin AOM, 28513
Sipalin MOM, 28514
Sipaxo, *See* pendimethalin, 23102
Sipcam UK Carbosip 5G, 28515
Sipcam UK Rover 500, 28516
Sipenol IT-50-46, 28517
Siperin, *See* cypermethrin, 7588, Topclip Parasol, 32037
Sipernat®, 28518
Sipernat® 22, 28519
Sipernat® 22S, 28520
Sipernat® 44, 28521
Sipernat® 50, 28522
Sipernat® 283LS, 28523
Sipernat® D17, 28524
Sipex 30, 28525
Sipex A, *See* Witcolate NH, 34478
Sipex BOS, *See* Niaproof® Anionic Surfactant 08, 21092, Rhodapon® BOS, 26632
Sipex DS, 28526
Sipex OP, *See* Rhodapon® 101-10, 26631, sodium lauryl sulfate, 28781
Sipex SB, *See* sodium lauryl sulfate, 28781
Sipilite, *See* Bakelite, 3589
Sipol S, *See* Adol® 62 NF, 783, Stearal, 29591
Sipomer™ AM, *See* Ageflex AMA, 966, Allyl methacrylate, 1784

Sipomer™ TMPTMA, *See* Ageflex TMPTMA, 1002
Sipomer® β-CEA, 28527
Sipomer® 2M1M, 28528
Sipomer® 2ME, 28529, *See* 2-mercaptoethanol, 19336
Sipomer® AAE, 28530
Sipomer® AGE, 28531, *See* Allyl glycidyl ether, 1783
Sipomer® AM, 28532
Sipomer® DCPA, 28533
Sipomer® DCPM, 28534
Sipomer® HEM-5, 28535
Sipomer® HEM-D, 28536, *See* hydroxyethyl methacrylate, 14662
Sipomer® IBOA, 28537
Sipomer® IBOMA, 28538
Sipomer® IDA, 28539
Sipomer® MEM, 28540
Sipomer® TATM, 28541
Sipomer® TMPEO, 28542
Sipomer® TMPTA, 28543
Sipon ES, *See* Witcolate ES-2, Witcolate LES-60C, 34477
Sipon LA 30, *See* Witcolate NH, 34478
Sipon LES 25, *See* Witcolate ES-2, Witcolate LES-60C, 34477
Sipon LS, *See* Rhodapon® 101-10, 26631, sodium lauryl sulfate, 28781
Sipon LS 100 S, *See* sodium lauryl sulfate, 28781
Sipon IT, *See* Perlankrol® ATL40, 23324, Rhodapon® LT-6, 26638, Standapol® T, 29481
Sipon LT, *See* Witcolate TLS-500, 34479
Sipon LT-6, *See* Witcolate TLS-500, 34479
Sipon LT-40, *See* Witcolate TLS-500, 34479
Sipon®, 28544
Siponate DS 10, *See* Nacconol® 40G, 20647, sodium dodecylbenzenesulfonate, 28760
Siponol S, *See* Adol® 62 NF, 783, Stearal, 29591
Siponol SC, *See* Adol® 62 NF, 783, Stearal, 29591
Siprol L22, *See* Witcolate NH, 34478
SIR 8514, *See* triflumuron, 32248
Sirdate®, 28545
Siriene, 28546
Sirius® Sirius Supra, Sirius Supra LL, 28547
sirlene, *See* propylene glycol, 25215
Sirtan, 28548
Sisellig, 28549
Sistan, *See* metam-sodium, 19444
Sistan®, 28550
Sitilan, 28551
Sitol, 28552
Sitren, 28553
Sit-ruti, *See* sanna,
Sivacur, *See* Raxil, 25908
Siverslice, 28554
Sivex, 28555
Sivex F, 28556
Size, 28557, 28558
Size CB, 28559
Sizing Wax PA, PT, SM, 28560
SKA, 28561
Skane, 28562
Skane HQ, *See* Skane® M-8, 28563
Skane M-8, *See* Skane® M-8, 28563
Skane® M-8, 28563
Skaterpax, 28564
Skatole carboxylic acid, *See* Rhizopon A, AA, 26573
SKB, 28565
Skelleftea, 28566
Skellite®, 28567
Skellysolve, 28568
skimmetin, *See* umbelliferone, 32899
Skin Wiz, 28569
SKINO #2, *See* methyl ethyl ketoxime, 19532
Skinoren, *See* Azelaic acid, 3519
Skiodan, *See* Abrodil, 158, Methiodal sodium, 19485
Skiodan Sodium, 28570
Skleron, 28571
Skliro Distilled, 28572
Skybond, 28573
Skydrol, 28574
Skyllex, 28575
SLA 1208, 1261, 28576
SLA 1261, 28577
SLA 1262, 1275, 28578
SLA 1275, 28579
SLA 1286, 28580
SLA 1611, 1612, 28581, *See* Polytetrafluoroethylene, 24642
SLA 1612, 28582
SLA 2208, 28583
SLA 2239, 28584
Slab Dip AC699, 28585
slack wax, 28586
Slag A, 28587
slag sand, 28588
slag wool, 28589
slagbestos, 28590
slaked lime, 28591, *See* calcium hydroxide, 4952
slate black, 28592
slate filler, *See* slate fust, 28593
slate fust, 28593
slate grey, 28594
slate lime, 28595
Slax, 28596
Slaymor, 28597
SLCC-D, 28598
S-lec, 28599
Slick, 28600
Slick Slide®, 28601
slicker solder, *See* plumber's solder, 24162
Slimicide 508, *See* Amerstat® 300, 2045, Biosperse® 240, 4207
Slimicide DE-488, *See* Aztec® *t*-Butyl Hydroperoxide-70, Aq, 3541
Slimicide MC, *See* Amerstat® 282, 2043, N-948® Biocide, 20641
Slip-Ayd, 28602
Slip-Ayd® Surface Conditioners, 28603
Slip-eze, *See* Unislip 1759, 33045
Slipicone, 28604
Slix, 28605
SLJ 0312, *See* Cropotex®, 7211
Slo-Gro, *See* maleic hydrazide, 18477
slop wax, 28606
S-lost, *See* mustard gas, 20538
Slow-Fe, 28607
Slow-FE Folic, 28608
Slow-K, 28609, *See* K. Tab, 15642, Kaon-Cl, 15729, Kay Ciel, 15807, K-Contin, 15818, K-Lor, 16226, K-Lyte/C1, 16234, KM Potassium Chloride, 16242, Potassium chloride, 24748, 24749
Slow-Sodium, 28610
SLPE, 28611
SLS sodium monolauryl sulfate, *See* Rhodapon® 101-10, 26631
sludge acid, 28612
Slue, 28613
Slug, 28614
Slug Destroyer, 28615
Slug Pellets, 28616
Slug Snail Killer, 28617
Slug-Geta, 28618
Slugit Liquid, 28619
Slugoids, 28620
Slug-Tox, *See* metaldehyde, 19431
SM 2059, 28626
SM 2061, 28627
SM 2133,2135, 28628
SM 2140, 28629
S-M 5731 Process-Type Silicone Sealant, 28630
SM 945, 28621
SM 975, 28622
SM 987, 28623
SM 992, 28624
SM 994, 28625
SMA, 28631, *See* sodium monochloracetate, 28797
SMA 3840, 28633
SMA 5500, 28634
SMA 17352 A, 28632
SMA® 1000, 28635
SMA® 1440, 28636
SMA® 2625, 28637
smalt, 28638
Smaragdine, 28639
Smarect, *See* paclobutrazol, 22540, 22541
S-Maz® 20, 28640
S-Maz® 40, 28641
S-Maz® 60K, 28642
S-Maz® 65K, 28643
S-Maz® 80, 28644
S-Maz® 83R, 28645
S-Maz® 85, 28646
S-Maz® 90, 28647
S-Maz® 95, 28648
SMCA, *See* sodium monochloracetate, 28797
SMDC, *See* metam-sodium, 19444, Unifume, 32954
S-Mez, *See* sulfamethazine, 29962
SMFP, *See* sodium monofluorophosphate, 28798
smithite, 28649
Smithol 22LD, 28650
Smithol PEG Adipate, 28651
Smithsonite, *See* zinc carbonate, 34771
Smitter-Lenian, 28652
Smoke, 28653
smoke black, 28654
Smoke Brown G, *See* oxyalizarin, 22438
smokeless diamond powder, 28655
smoking salts, 28656
smoky quartz, 28657
Smoothex, 28658
Smooth-On, 28659
Smut-Go, *See* Julin's chloride, 15632
SN-30GF/000 FR, 28660
SN 4075, *See* phenmedipham, 23597
SN 5870, *See* Acaprin®, 181
SN 5870, *See* 1,3-di-6-quinolylurea, 8618
SN-38584, *See* phenmedipham, 23597
SN 81742, *See* Sethoxydim, 28099
Snac-Kote, 28661
Snapper CDA, 28662
Sniafil, 28663
Sniafoam, 28664
Sniamid®, 28665
Sniamid® ADS 40 I, 28666
Sniamid® ASN 27T, 28667
Sniamid® SSD 300 EP 021, 28668
Sniamid® SSD AF, 28669
Sniasan®, 28670
Sniatal®, 28671
Sniater®, 28672
Snieciotox, *See* Julin's chloride, 15632
Snip, *See* azamethiphos, 3517
Snomelt, 28673
snow white, *See* zinc white, oil white, 22086
Snow White 200 Mica, 28674
Snowflake, *See* Wickenol® 550, 34403
Snowflake P. E, 28675
Snowflake White, 28676
Snowtack, 28677
Snowtack 342A, 28678
Snowtack SE 325 A, 28679
S-nyl, 28680
So/San 30M, 28681
Soa, 28682
Soap, *See* Actralube Syn-153, 541
soap bark, 28683
soap clay, *See* bentonite, 3954
Soap sulfonate, *See* Actrabase 31-A, 514
Soapearl®, 28684
Soapstone, *See* ABT-2500®, 166, mica, 19626, Olympic, 22146, talc, 30731
Soarblen, 28685
Soarnol, 28686
Sobag FD 300 Series, *See* Ammonium alginate, 2228
Sobalg FD 100 Series, 28687
Sobalg FD 200 Series, 28688
Sobalg FD 300 Series, 28689
Sobalg FD 460, 28690

Sobee, 28691
Sobral 12-101, 28693
Sobral 72-625D, 28694
Sobral 1321, 28695
Sobral 2911, 28696
Sobral 9257, 28697
Sobral AD-002, 28698
Sobral AN-001, 28699
Sobral EE-632, 28700
Sobral L90-20A, 28701
Sobral®, 28692
Sobrom, 28702
Socal, 28703
Socci 3500, 3500-WP, 28704
Sochamine A 271, 28705
Sochamine A 7525, 28706
Sochamine A 7527, 28707, *See* disodium cocoamphodiacetate, 8643
Sochamine A 8955, 28708
Sochamine OX 30Empigen® 5083, *See* Cocamine oxide, 6557
soda, 28709
soda, *See* soda ash, 28710
soda alum, *See* porous alum, 24702, sodium alum, 28730
soda ash, 28710, *See* sodium carbonate, 28744, sodium carbonate monohydrate, 28745
Soda Ash Blocks, 28711
soda blue, 28712
soda Bordeaux mixture, 28713
soda glass, *See* soda-lime glass, 28719
soda glass, soluble, *See* soluble glass, 28980
soda lye, 28714, *See* sodium hydroxide, 28772
soda niter, *See* sodium nitrate, 28800
soda pulp, 28715
soda salt peter, *See* Chili saltpeter, 6063
soda tar, 28716
soda ultramarine, *See* ultramarine, 32821
soda water glass, 28717
soda, calcined, *See* soda ash, 28710
Sodagrain, 28718
soda-lime glass, 28719
sodalumite, *See* Kasal®, 15777
soda-olein, 28720
Sodaphos, 28721
sodascorbate, *See* sodium ascorbate, 28734, vitamin C sodium salt, 33939
Sodasorb, 28722
Sodastraw, 28723
Sodatol, 28724
Sodiformasal, *See* Formasal,
sodium aluminum phosphate, *See* Levair®, 17049
sodium, 28725, *See* Lipoproteol LK, 17401, natrium, 20756
sodium acetate, 28726, *See* Niacet sodium Acetate Anhyd. Tech, 21089
sodium acetate anhydrous, *See* sodium acetate, 28726
sodium acid carbonate, *See* Baking soda, 3600, sodium bicarbonate, 28736
sodium acid methanearsonate, *See* Neoarsycodyl, 20871
sodium acid pyrophosphate, 28727, *See* B.P. Pyro®, 3564, Donut Pyro®, 8800, Perfection®, 23243, Sapp #4, 27446, Taterfos®, 30844, Victor Cream®, 33727
sodium acid sulfite, 28728, *See* Abisol, 70, sodium bisulfite, 28738
sodium acrylic acid/maleic acid copolymer, *See* Acusol® 479N, 587, Acusol® 479ND, 588
Sodium Aerofloat Promoter, 28729
sodium alginate, *See* Algin, 1591, Antimigrant C-45, 2564, Kelcosol®, 15832, Kelgin, 15841, Kelgin® F, 15842, kelgin® HV, LV, MV, 15843, Kelgin® QH, QL, QM, 15844, Kelgin® XL, 15845, Kelgum, 15846, Keltex®, 15878, Keltex® HV, 15879, Keltex® S, 15880, Keltone®, 15883, Keltone® HV, LV, 15884, Manucol DM, 18616, Manucol, Manucol DH, 18619, Manutex, 18622, Mozanon®, 20386, Municol, 20497, Sobalg FD 100 Series, 28687
sodium alginate HV NF/FCC, *See* Algin, 1591
sodium alkane sulfonate, *See* Mersolat® H, 19387
sodium alkanoate, *See* Monatrope 1250, 20176
sodium alkyl diaryl sulfonate, *See* Alkanol® ND, 1698
sodium alkyl diphenyl ether disulfonate, *See* Poly-Tergent® 4C3, 24632
sodium alkyl diphenyl oxide disulfonate, *See* Calfax DB-45, 4983
sodium alkyl ether sulfate, *See* Avirol® 125E, 3476, Empimin® LSM30, 10426, Stepanflote® 85L, 29676
sodium alkyl naphthalene sulfonate, *See* Sellogen DFL, 27929, Sellogen HR-90, 27930, Emery® 5370 Sellogen W, 10177
sodium alkyl sulfate, *See* Melioran, 19290
sodium alkyl sulfates, *See* Vanfre® IL-1, 33322
sodium alkyl sulfonate, *See* Alkanol® 189-S, 1697, Avitone® A, 3493, Petrowet® R, 23540
sodium alkyl triethoxysulfate, *See* Empimin® SQ25, 10437
sodium alkylallyl sulfosuccinate, *See* Trem-LF-40, 32167
sodium alkylaryl polyether sulfonate, *See* Fenopon SE, 11607
sodium alkylaryl sulfonate, *See* Eccoscour CB, 9548, Hexoran A15, 14037, Rexopene, 26483, Udet 950, 32732, Warcodet K54, 34207
sodium alkylbenzene sulfonate, *See* Alkanol® WXN, 1699, Nacconal, 20644
sodium alkylnaphthalene sulfonate, *See* Alkanol® XC, 1700, NSAE Powder, 21756
sodium 1-allyloxy-2-hydroxy-propane sulfonate, *See* Cops 1, 6797
sodium alpha olefin (C_{14}, C_{16}) sulfonate, *See* Polystep® A-18, 24588
sodium alpha olefin sulfonate, *See* Stepantan® AS-12, 29704
sodium alum, 28730
sodium aluminate, 28731, *See* Amerfloc® 2, 2007, Dynaflock, 9344, Dynagrout, 9348, Manfloc, 18528
sodium aluminosilicate, *See* Wessalith, 34284
sodium aluminum chlorhydroxy lactate, 28732, *See* Chloracel®, 6083
sodium aluminum dioxide, *See* sodium aluminate, 28731
sodium aluminum fluoride, *See* cryolite, 7279, Kryalith, 16448
sodium aluminum phosphate, 28733, *See* BL-60®, 4300, Levn-Lite, 17071, Pan-O-Lite, 22683, Stabil-9, 29399
sodium aluminum phosphate acidic, *See* sodium aluminum phosphate, 28733
sodium aluminum silicate, *See* Wessalith, 34284
sodium aluminum sulfate, *See* porous alum, 24702
sodium amidotrizoate, *See* diatrizoate sodium, 8337
sodium amino tris(methylene phosphonate), *See* Chemphonate AMP-S, 6005
sodium *para*-aminohippurate, 28804
sodium 4-aminohippurate hydrate, *See* sodium *para*-aminohippurate, 28804
sodium-ammonium-hydrogen Phosphate, *See* microcosmic salt, 19665
sodium anthraquinone monosulfonate, *See* silver-salt, 28441
sodium D-araboascorbate, *See* sodium erythorbate, 28762
sodium arsenate, *See* Fatsco, 11506
sodium arsenite, *See* Harle's solution, 13637
sodium ascorbate, 28734, *See* Cevalin, 5827 Liqui-Cee, 17460 vitamin C sodium salt, 33939
sodium L-ascorbate, *See* sodium ascorbate, 28734
sodium L-(+)-ascorbate, *See* sodium ascorbate, 28734
sodium benzene-sulfochloroamide, *See* Chloramine B, 6085
sodium benzensulfochloramine, *See* Chloramine B, 6085
sodium benzoate, 28735
sodium benzosulfimide, *See* saccharin sodium, 27196
sodium *o*-benzosulfimide, *See* saccharin sodium, 27196
sodium benzothiazolethiolate, *See* Nacap®, 20643
sodium benzothiazole-2-thiolate, *See* Nacap®, 20643
sodium, (2-benzothiazolylthio)-, *See* Nacap®, 20643
sodium benzothiazol-2-yl sulfide, *See* Nacap®, 20643
sodium biborate, *See* sodium borate anhydrous, 28739
sodium bicarbonate, 28736, *See* Baking soda, 3600, Bufferight, 4656, Col-Evac, 6601
sodium bichromate, *See* sodium dichromate, 28756
sodium biphosphate, *See* sodium phosphate, 28810
sodium bisulfide, 28737
sodium bisulfite, 28738, *See* Abisol, 70, Amersite® 2, 2035, 2036, sodium acid sulfite, 28728
sodium borate, *See* antipyoninum, 2612, Dehybor, 7942, FB 48, 11513, Solubor, 28990
sodium borate ($NaBO_2$), *See* sodium metaborate, 28784
sodium borate anhydrous, 28739
sodium borofluoride, *See* sodium fluoroborate, 28765
sodium borohydride, 28740
sodium bromate, *See* Dyetone®, 9312
sodium bromide, 28741
sodium butanoate, *See* sodium butyrate, 28742
sodium butoxyethoxy acetate, *See* Miranate® B, 19874
sodium butyl oleate, *See* Tetranol, 31387
sodium butylparaben, *See* Nipabutyl sodium, 21251
sodium *sec*-butyl xanthate, *See* Aero 301 Xanthate, 839
sodium butyrate, 28742
sodium n-butyrate, *See* sodium butyrate, 28742
sodium C_{13}-C_{17} alkane sulfonate, *See* Marlon® PF 40, 18879, Marlon® PS 30, 18880
sodium C_{12-15} alkoxypropyl iminodipropionate, *See* Amphoteric N, 2360
sodium C_{12}-C_{15} alkyl ether sulfate (3 EO), *See* T-Det® 25-3S, 30860
sodium C_{10}-C_{18} alkyl sulfonate, *See* Chemstat® PS-101, 6043
sodium C_{8-16} isoalkyl succinyl lactoglobulin sulfonate, *See* Bio-Pol® OE, 4181
sodium C_{12-14} olefin sulfonate, *See* Marlinat® SRN 30, 18856
sodium C_{14}-C_{16} olefin sulfonate, *See* Bio-Terge® AS-40, 4212, Carsonol® AOS, 5333, Nansa® LSS38/A, 20689, Rhodacal® 301-10, 26576, Witconate AOK, 34483
sodium C_{14-16} alpha olefin sulfonate, *See* Norfox® ALPHA XL, 21546
sodium C_{14}-C_{16} olefin sulfonate, glycol distearate, cocamidopropyl betaine, sorbitan laurate, *See* Tego® -Pearl S-33, 30986
sodium C_{14-16} olefin sulfonate, sodium laureth sulfate and lauramide DEA, *See* Miracare® ANL, 19851
sodium C_{12}-C_{15} pareth-6 carboxylate, *See* Sandopan® DTC Linear P, 27339
sodium C_{12-13} pareth sulfate, *See* Akyposal 23 ST 70, 1297
sodium C_{12-15} pareth sulfate, *See* Witcolate 1050, Witcolate SE-5, 34468
sodium C_{12-13} pareth sulfate, sodium PEG-6 cocamide carboxylate, disodium laureth sulfosuccinate and trideceth-2 carboxamide MEA, *See* Akypogene HM 12, 1256
sodium C_{12-15} pareth-3 sulfonate, *See* Avanel® S-30, 3455
sodium calciumedetate, *See* Versene CA, 33618
sodium caproamphoacetate, *See* Miranol® SM Conc, 19911
sodium capryl lactylate, *See* Pationic 122A, 22904
sodium caprylate, 28743
sodium capryleth-9 carboxylate and sodium hexeth-4 carboxylate, *See* Akypo LF 4N, 1221
sodium capryloamphoacetate, *See* Rewoteric® AM V, 26428, Zoharteric LF, 34882
sodium capryloamphohydroxypropylsulfonate, *See* Miranol® JS Conc, 19907
sodium capryloamphopropionate, *See* Monateric CyNa-50, 20166
sodium carbonate, 28744, *See* mild alkali, 19757, sal soda, 27255, salt of soda, 27288, Trona Soda Ash, 32459, Tronalight Light Soda Ash, 32462

sodium carbonate anhydrous, *See* soda ash, 28710
sodium carbonate hydrated, *See* Consal, 6732
sodium carbonate monohydrate, 28745, *See* sodium carbonate monohydrate, 28745
sodium carbonate peroxyhydrate, *See* Oxyper, 22465
sodium carboxymethyl cellulose, *See* Cekol, 5558, carboxymethylcellulose sodium, 5258
sodium caseinate, 28746
sodium cellulose glycolate, *See* carboxymethylcellulose sodium, 5258, Cekol, 5558
sodium cetearyl sulfate, *See* Dehydag Wax E, 7947, Lanette E, 16665
sodium ceteth-13 carboxylate, *See* Sandopan® KST, 27344
sodium cetyl oleyl sulfate, *See* Empicol® CHC 30, 10309
sodium cetyl/stearyl sulfate, *See* Rhodapon® EC111, 26634
sodium chlorate, 28747, *See* Arpal Non Selex, 3095, Atlacide, 3290, Centex, 5678, Defol, 7889, Granular Weedkiller, 13270, KM sodium Chlorate, 16244
sodium chloride, 28748, *See* Alberger® Natural Flake, 1361, Betrox, 4075, Hopper salt, 14303, marine salt, 18722, muriate of soda, 20514, rock salt, 26887, sal commune, 27244, sal Culinaris, 27245, Superior® Granulated, 30244, Tru-Flake® Salt, 32490
sodium chloride ^{22}Na, *See* Natritope Chloride, 20755
sodium chloride solution, *See* Saltex, 27296
sodium chloride-calcium chloride-silver chloride-potassium bitartrate, *See* Cool-Amp, 6754
sodium chlorite, 28749, *See* Adox 3125, 787, C-2, 4842, Chloritane, 6109
sodium chloroacetate, *See* sodium monochloracetate, 28797
sodium-*p*-chlorobenzoate, *See* Mikrobin, 19750
sodium-chloro-*o*-phenylphenate, *See* Dowicide C, 8934
sodium chromate, 28750
sodium chromate(VI), *See* sodium chromate, 28750
sodium chromate(VI) ^{51}Cr, *See* Rachromate-51, 25709
sodium citrate dihydrate, *See* sodium citrate,
sodium citrotartrate, 28751
sodium CMC, *See* carboxymethylcellulose sodium, 5258, Tylose® C, C-p, CB, CB-p, 32659
sodium coceth sulfate, *See* Oronal LCG, 22312
sodium coceth-2 sulfate and triethylene glycol distearate, *See* Base Nacrante 6030 CP, 3700
sodium cocoamphoacetate, *See* Ampholak XCO-30, 2321, Amphoterge® W, 2355, Empigen® CDR40, 10369, Mackam 1C, 18129, Miranol® CM Conc. NP, 19889, Miranol® FA-NP, 19894, Monateric CM-36S, 20163, Schercoteric MS, 27707, Zoharteric M, 34884, Zoharteric M-2, 34885
sodium cocoamphohydroxypropylsulfonate, *See* Miranol® CS Conc, 19891, Schercoteric MS-EP, 27709
sodium cocoamphopropionate, *See* Amphoterge® K, 2350, Mackam CSF, 18142, Miranol® CM-SF Conc, 19890, Monateric CA-35, 20153, Monateric CAM-40, 20156
sodium N-coconut acid N-methyl taurate, *See* Fenopon TC-42, 11611
sodium coconut fatty alcohol sulfate (C_{12}-C_{18}), *See* Elfan® 280, 9816
sodium cocosulfate, *See* Marlinat® KT 50, 18854
sodium cocoyl glutamate, *See* Acylglutamate CS-11, 597
sodium cocoyl isethionate, *See* Hostapon KA Powd, 14393, Jordapon® CI-50 Disp, 15620
sodium cocoyl isethionate-coconut acid-stearic acid, *See* Geropon® AS-200, 12904
sodium N-Cocoyl-N-Methyl Taurate, *See* Adinol CT, 696
sodium cocoyl sarcosinate, *See* Hampfoam 35, 13593, Hamposyl® C-30, 13599, Medialan KA, 19212, Nutrapon RS 1147, 21825, Vanseal® 35, 33355, Vanseal® NACS-30, 33358
sodium-N-cocoyl sarcosinate, *See* Closyl 30 2089, 6442
sodium cumene sulfonate, 28752, *See* Eltesol® SC 93, 9871, Naxonate® 45SC, 20826, Naxonate® SC, 20827, Reworyl® NCS, 26406, Stepanate® SCS, 29673, Witconate SCS 45%, 34490
sodium cumenesulfonate, *See* Witconate SCS 45%, 34490
sodium cumolsulfonate, *See* Witconate SCS 45%, 34490
sodium cyanide, 28753, *See* Cymag, 7577
sodium cyclamate, 28754
sodium cyclohexanesulfamate, *See* sodium cyclamate, 28754
sodium cyclohexyl amidosulfate, *See* sodium cyclamate, 28754
sodium N-cyclohexyl-N-palmitoyl taurate, *See* Fenopon CN-42, 11604
sodium cyclohexyl sulfamate, *See* sodium cyclamate, 28754
sodium cyclohexylsulfamidate, *See* sodium cyclamate, 28754
sodium dalapon, *See* Basfapon®, 3705
sodium D-araboascorbate, *See* Eribate, 10753, Neo-Cebitate®, 20878
sodium deceth sulfate, *See* Cedepal FS-406, 5541, Witcolate 7093, 34471
sodium deceth-2 carboxylate and sodium capryleth-2 carboxylate, *See* Akypo OCD 10 NV, 1228
sodium decylbenzenesulfonate, *See* Witconate DS, 34485
sodium decyl diphenyl ether disulfonate, *See* Poly-Tergent® 3B2, 24630
sodium n-decyl diphenyloxide disulfonate, *See* Dowfax 3B2, 8911
sodium decyl sulfate, 28755, *See* Atlasol 103, 3341, Empicol® 0758, 10303, Empimin® SDS, 10436, Polystep® B-25, 24593
sodium n-decyl sulfate, *See* Atlasol 103, 3341, Avirol® SA 4110, 3482, Sulfotex 110, 30030
sodium dedt, *See* Octopol SDE-25, 22013
sodium di(2-ethylhexyl) sulfosuccinate, *See* docusate sodium, dioctyl sodium sulfosuccinate, 8550
sodium di-*s*-butyl dithiophosphate, *See* Aerofloat 238 Promoter, 855
sodium diaminonaphtholsulfonate, *See* Imogen, 14906
sodium dibutyl naphthalene sulfonate, *See* Supragil® NK, 30308
sodium dibutyl sulfosuccinate, *See* Aerosol® IB-45, 892, Rewopol® SBDB 45, 26371
sodium dibutyldithiocarbamate, *See* Octopol NB-47, 22012
sodium dicarboxyethylcoco phosphoethyl imidazoline, *See* Phosphoteric® T-C6, 23748
sodium dichloro isocyanurate, *See* dichloro-1,3,5-triazinetrione, sodium salt, 8385
sodium dichlorocyanurate, *See* dichloro-1,3,5-triazinetrione, sodium salt, 8385
sodium dichloroisocyanurate dihydrate, *See* CDB Clearon, 5525
sodium 2,2-dichloropropionate, *See* BH Dalapon, 4117
sodium α,α-dichloropropionate, *See* Basfapon®, 3705
sodium dichloro-*s*-triazine-2,4,6-trione, *See* dichloro-1,3,5-triazinetrione, sodium salt, 8385
1-sodium-3,5-dichloro-*s*-triazine-2,4,6-trione, *See* dichloro-1,3,5-triazinetrione, sodium salt, 8385
sodium dichromate, 28756
sodium dichromate dihydrate, *See* sodium dichromate, 28756
sodium dichromate dihydrate ($Na_2Cr_2O_7.2H_2O$), *See* sodium dichromate, 28756
sodium dicloxacillin monohydrate, *See* Dicloxacillin sodium, 8396
sodium dicyclohexyl sulfosuccinate, *See* Octosol TH-40, 22028
sodium diester sulfosuccinate, *See* Witflow 901, 34525
sodium diethyl and sodium di-*sec.* butyl dithiophosphate mixture, *See* Aerofloat 208 Promoter, 853
sodium diethyl dithiophosphate, *See* sodium Aerofloat Promoter, 28729
sodium diethyldithiocarbamate, *See* ethyl namate®, 11078, Octopol SDE-25, 22013
sodium N,N-diethyldithiocarbamate, *See* Octopol SDE-25, 22013
sodium N,N-diethyldithiocarbamate trihydrate, *See* Octopol SDE-25, 22013
sodium N,N-dimethyl dithiocarbamate, *See* methyl namate®, 19551, Octopol SDM-40, 22014, Perkacit® SDMC, 23288, sodium dimethyldithiocarbamate, 28758
sodium dihexyl sulfosuccinate, *See* Bis(1-methylamyl) sodium Sulfosuccinate, 4227, Empimin® MA, 10427, Lankropol® KMA, 16713, Octosol HA-80, 22025
sodium 2,2'-dihydroxy-4,4'-dimethoxy-5-sulfobenzophenone, *See* Uvinul® DS-49, 33196
sodium dihydroxyethyl glycinate, 28757
sodium dihydroxyethylglycinate, *See* Hampshire® DEG, 13608, sodium dihydroxyethyl glycinate, 28757
sodium diisoamyldithiophosphate, *See* Aero 3501 Promoter, 845
sodium diisobutyl dithiophosphate, *See* Aero 3477 Promoter, 844
sodium diisobutyl naphthalene sulfonate, *See* Leonil DB Powd, 17000
sodium diisobutyl sulfosuccinate, *See* Aerosol® IB-45, 891, Octosol IB-45, 22026
sodium diisobutyldithiophosphinate, *See* Aerophine 3418A, 868
sodium diisooctyl sulfosuccinate, *See* Geropon® CYA/DEP, 12906, Lankropol® KO Special, 16714, Manoxol OT60, 18582
sodium diisopropyl dithiophosphate, *See* Aerofloat 211 Promoter, 854
sodium diisopropyl naphthalene sulfonate, *See* Dehscofix 916, 7924, Supragil® WP, 30309
sodium diisopropylnaphthalene sulfonate, *See* Aerosol® OS, 895, Rhodacal® IN, 26585
sodium dimethyl arsenate, *See* sodium cacodylate,
sodium dimethyl dithiocarbamate, *See* Thiostop N, 31717
sodium dimethyl naphthalene-formaldehyde sulfonate, *See* Dehscofix 923, 7927
sodium dimethylbenzenesulfonate, *See* Eltesol® SX 30, 9873, Naxonate® 4L, 20824, Pilot SXS-40, 23843, Witconate SXS 40%, 34492
sodium dimethylcarbamodithioate, *See* Aquatreat SDM, 2761, methyl namate®, 19551, Octopol SDM-40, 22014, sodium dimethyldithiocarbamate, 28758
sodium dimethyldithiocarbamate, 28758, *See* Aquatreat SDM, 2761, methyl namate®, 19551, Octopol SDM-40, 22014, Perkacit® SDMC, 23288
N,N-dimethyldithiocarbamate, *See* Aquatreat SDM, 2761, sodium dimethyldithiocarbamate, 28758
sodium dimethyldithiocarbamates, *See* Amersep® MP-3, 2029
sodium di-n-butyl naphthalene sulfonate, *See* Dehscofix 917, 7925
sodium dinonyl sulfosuccinate, 28759, *See* Geropon® WS-25, WS-25-I, 12919
sodium dioctyl sulfosuccinate, *See* Astrowet 0-70-PG, 3265, Chemax DOSS/70, 5895, Drewfax® S-700, 8991, Drewfex® 0007, 8992, Emcol® 4500, 10051, Emcol® DOSS, 10058, Empimin® OP70, 10434, Empimin® OT, 10435, Geropon® CYA/60, 12905, Geropon® DOS, 12907, Hipochem EK-18, 14109, Lankropol® KO2, 16715, Leonil OS, 17002, Manoxol OT, OT/P and OT/B, 18581, Ninate® DS 70, 21229, Pentasol, 23174, Rexowet ASG-81, 26505, Schercowet DOS-70, 27711, Secosol® DOS 70, 27814, Thorowet G-40 3230, 31755, docusate sodium, , dioctyl sodium sulfosuccinate, 8550, Octowet 40, 22035
sodium dioctyl sulfosuccinate, propylene glycol/water, *See* Aerosol® OT-70 PG, OT-S, 896
sodium dioxide, *See* Oxone, 22435
sodium diphosphate, *See* sodium pyrophosphate, 28814

sodium diphosphate ($Na_4P_2O_7$), *See* sodium pyrophosphate, 28814
sodium disulfite, *See* sodium metabisulfite, 28783
sodium dithionite, *See* Blankit®, 4322, Luredox® BP, PO, 17937, hydrosulfite, 14643, Hydrosulphit®, 14648, Hydros® F, 14637, Reductone, 25983
sodium dithionite hydrate, *See* sodium hydrosulfite, 28771
sodium-N-dodecanoyl-N-methylglycinate, *See* sodium lauroyl sarcosinate, 28779
sodium dodecyl benzene sulfonate, *See* sodium dodecylbenzenesulfonate, 28760
sodium dodecyl diphenyl ether disulfonate, *See* Poly-Tergent® 2A1-L, 24629
sodium dodecyl diphenyloxide disulfonate, *See* Dowfax 2A1, 2EP, 8909, Rhodacal® DSB, 26584
sodium dodecyl sulfate, *See* sodium lauryl sulfate, 28781, Texapon K-12, K-1296, L-100, 31461
sodium dodecylated oxydibenzene disulfonate, *See* Dowfax 2A1, 8908
sodium dodecylbenzene sulfonate, 28760, *See* Akyposal NAF, 1313, Arylan® SC15, 3160, Calsoft F-90, 5029, DeSonate 50-S, 8176, DeSonate 60-S, 8177, Hartofol 40, 13674, Hetsulf 40, 40X, 13977, Hetsulf 60S, 13979, Hetsulf Acid, 13981, Hoe S 2713, 14256, Lumo Stabil S 80, 17811, Lumo WW 75, 17812, Maranil Powd. A, 18680, Marlon® A365, 18874, Nacconol® 40G, 20647, Nansa® HS80S, 20687, Nansa® SL 30, 20693, Nansa® SS 30, 20694, Naxel AAS-40S, 20811, Reworyl® NKS 100, 26407, Rhodacal® DDB-40, 26582, Stepantan® DS-40, 29705, Sul-fon-ate AA-10, 30020, Sulfotex LAS-90, 30031, Tensaryl DF90, 31142, Tensaryl DX54Sp. and DX62, 31143, Tensaryl L48, 31145, Tensopol 30E, LDS, 31188, Tensaryl SB85P, 31149
sodium dodecylbenzene sulfonate (linear), *See* Polystep® A-7, 24589
sodium dodecylbenzenesulfonate, 28760
sodium dodecylpoly(oxyethylene) sulfate, *See* Witcolate ES-2, Witcolate LES-60C, 34477
sodium dodecylpolyethoxysulfate, *See* Witcolate ES-2, Witcolate LES-60C, 34477
sodium dodecylbenzene sulfonate, *See* Nansa® 1106/P, 20682
sodium edetate, *See* Kalex 220 Crystal, 15668, Kalex Liq. 50%, 15671, Trilon® B Powd, 32324
sodium eicosyloxypropyliminodipropionate, *See* amphoteric 300, 2357
sodium erythorbate, 28761, 28762, *See* Eribate, 10753, Neo-Cebitate®, 20878
sodium etasulfate, *See* Niaproof® Anionic Surfactant 08, 21092, Rhodapon® BOS, 26632
sodium ethasulfate, *See* Niaproof® Anionic Surfactant 08, 21092, Rhodapon® BOS, 26632
sodium ethylhexyl sulfate, *See* Empicol® 0585/A, 10302
sodium 2-ethylhexyl sulfate, *See* Avirol® SA 4106, 3480, Carsonol® SHS, 5338, Chemsulf S2EH-Na, 6044, Niaproof® Anionic Surfactant 08, 21092, Rhodapon® BOS, 26632, Tensatil DEH120, 31154, Witcolate 6465, 34470, Witcolate D-510, 34474
sodium 1,4-bis(2-ethylexyl) sulfosuccinate, *See* dioctyl sodium sulfosuccinate, 8550
sodium ethylparaben, *See* Nipagin A sodium, 21265
sodium-ethyl thiosulfate, *See* Bunte's salt, 4682
sodium fatty alcohol ether sulfate, *See* Ungerol, 32919
sodium feredetate, *See* sodium iron edetate, 28776
sodium ferrocyanide, 28763
sodium ferrocyanide decahydrate, *See* sodium ferrocyanide, 28763, yellow prussiate of soda, 34648
sodium fluoride, 28764, *See* Fluorinse, 12090, Fluorl, 12092, fluorol, 12101, Gel II, 12716
sodium fluoride cyclic dimer, *See* sodium fluoride, 28764
sodium fluoroborate, 28765, *See* Pyricit, 25479
sodium fluorophosphate, *See* Albaphos Dental Na 211, 1348
sodium fluorosilicate, *See* sodium silicofluoride, 28818
sodium fluosilicate, *See* malladrite, 18484, sodium silicofluoride, 28818, salufer, 27301, SSF, 29390
sodium formaldehyde bisulfite, *See* Rongalit® C, 26945
sodium formaldehyde sulferylate, *See* Foamosul, 12179
sodium formaldehyde sulfoxylate, 28766, *See* Hyraldite C Ext, 14730, sodium formaldehydesulfoxylate, 28766
sodium formate, 28767, *See* Formax, 12268
sodium formate, hydrate, *See* Formax, 12268, sodium formate, 28767
sodium formate, hydrated, *See* Formax, 12268, sodium formate, 28767
sodium glucoheptonate, 28768
α-sodium glucoheptonate dihydrate, *See* Belzak AC, 3924
β sodium glucoheptonate, *See* Belzak BL-50, 3925
sodium gluconate, 28769, *See* Asahi Aji®, 3174, Gluconal® NA, 13046
sodium glutamate, *See* MSG, 20416
sodium glycereth-1 polyphosphate, *See* Surfagene FPG 50, 30358
sodium *p*-glycollylarsanilate, *See* Astroplax, 3263, Astryl, 3267
sodium hexacyanoferrate, *See* sodium ferrocyanide, 28763
sodium hexacyanoferrate(II), *See* yellow prussiate of soda, 34648
sodium n-hexadecyl diphenyloxide disulfonate, *See* Dowfax 8390, 8914, Dowfax XDS 8390.00, 8920
sodium hexafluoroaluminate, *See* Synkrolith, 30550
sodium hexafluorosilicate, *See* salufer, 27301, sodium silicofluoride, 28818, SSF, 29390
sodium hexametaphosphate, 28770, *See* Glass H, 12967, Hexaphos, 14013, Kalex HMP, 15670, Metagon, 19425, Polyphos, 24519, Sodaphos, 28721, Vitrafos®, 33982, Water Brite, 34223, Glass H, 12967, Calgon, 4987
sodium hexametaphosphates, *See* Limex G, 17276
sodium hexeth-4 carboxylate and trideceth-2, *See* Akypo TPR, 1248
sodium hexyl diphenyloxide disulfonate, *See* Dowfax XDS 8292.00, 8919
sodium hyaluronate, *See* hyaluronic acid, hyaluronic acid sodium salt, 14483, Pronova, 25159, Rita HA C-1, 26791
sodium hyaluronate solution, *See* ActiMoist, 486
sodium hydrate, *See* sodium hydroxide, 28772
sodium hydrofluoride, *See* sodium fluoride, 28764
sodium hydrogen carbonate, *See* Baking soda, 3600, saleratus, 27262, sodium bicarbonate, 28736
sodium hydrogen sulfide, *See* sodium bisulfide, 28737
sodium hydrogen sulfite, *See* Abisol, 70, Leucogen, 17023, sodium acid sulfite, 28728, sodium bisulfite, 28738
sodium hydrogenated tallow dimethyl glycinate, *See* Monateric 1203, 20151
sodium hydrogenated tallow glutamate, *See* Acylglutamate HS-11, 603
sodium hydrogenated tallow glutamate, sodium cocoyl glutamate, *See* Acylglutamate GS-11, 601
sodium hydrosulfide, *See* sodium bisulfide, 28737
sodium hydrosulfite, 28771, *See* Geblitol, 12708, Hybaite, 14486, Hybrite®, 14492, Hydrolin, 14610, Hydronyx, 14625, Hydros, 14635, Hydros® F, 14637, hydrosulfite, 14643, Hydrosulfite AWC, 14645, Hydrosulphit®, 14648, Reductone, 25983, Schutzenberger's salt, 27737
sodium hydrosulfite-sodium pyrophosphate, *See* Hydros® 1, 14636
sodium hydroxide, 28772, *See* Pels, 23090
sodium 2-hydroxyethanesulfonate, *See* Witconate NIS, 34487
sodium 2-hydroxyethyl sulfonate, *See* Witconate NIS, 34487
sodium β-hydroxyethanesulfonate, *See* Witconate NIS, 34487
sodium hydroxymethane sulfonate, *See* Rongalit® C, 26945, sodium formaldehydesulfoxylate, 28766
sodium hydroxymethyl glycinate, *See* Suttocide® A, 30440
sodium hydroxymethylamino acetate, *See* Suttocide® A, 30440
sodium 12-hydroxy-(*cis*)-9-octadecenoate, *See* Soricinol 40, 29143
sodium hypochlorite, 28773, *See* Adeka Hypote, 664, Chloros, 6146, Chlorosoda, 6147, Domestos, 8795, HyPure N, 14728
sodium hypochlorite pentahydrate, *See* sodium hypochlorite, 28773
sodium hypochlorite-sodium perborate, *See* Dakin's solution, 7704
sodium hypophosphite monohydrate, *See* Sofibex, 28835
sodium hyposulfite, *See* sodium hydrosulfite, 28771, sodium thiosulfate, 28826
sodium inosinate, *See* disodium inosinate, 8645
sodium 5-inosinate, *See* disodium inosinate, 8645
sodium iodate, 28774
sodium iodide, 28775
sodium iodide ^{125}I, *See* Iodotope I-125, 15217
sodium iodide ^{131}I, *See* Iodotope I-131, 15218, Iodotope Therapeutic, 15219, Oriodide-131, 22303, Radiocaps-131, 25827
sodium iodine, *See* sodium iodide, 28775
sodium iodomethanesulfonate, *See* Abrodil, 158, Methiodal sodium, 19485
sodium iron edetate, 28776
sodium isethionate, *See* Witconate NIS, 34487
sodium isoascorbate, *See* sodium erythorbate, 28762
sodium D-isoascorbate, *See* Eribate, 10753, sodium erythorbate, 28762
sodium isobutyl xanthate, *See* Aero 317 Xanthate, 841
sodium isodecyl sulfate, *See* Rhodapon® CAV, 26633
sodium isooctyl sulfate, *See* Sulfetal 4105, 29981, Sulfetal FA 40, 29984
sodium O-isopropyl dithiocarbonate, *See* Aero 343 Xanthate, 842
sodium isopropyl naphthalene sulfonate, *See* Rhodacal® IN, 26585
sodium isopropylxanthate, *See* Aero 343 Xanthate, 842
sodium isopropylxanthogenate, *See* Aero 343 Xanthate, 842
sodium isosteareth-6 carboxylate, *See* Sandopan® TS-10, 27349
sodium isostearoamphopropionate, *See* Monateric ISA-35, 20168, Schercoteric I-AA, 27706
sodium isostearoyl lactylate, *See* Crolactil SISL, 7185
sodium isostearoyl-2-lactylate, *See* Pationic ISL, 22907
sodium lactate, 28777
sodium-dl-lactate, *See* sodium lactate, 28777
sodium lauramido DEA sulfosuccinate, *See* Geropon® SS-L7DE, 12914
sodium (lauryloxypolyethoxy)ethyl sulfate, *See* Witcolate ES-2, Witcolate LES-60C, 34477
sodium lauramphoacetate, *See* Mackam 1L, 18130
sodium laureth carboxylate series, *See* Akypo RLM 100 NV, 1233
sodium laureth 11-carboxylate, *See* Akypo® Soft 100 NV, 1251
sodium laureth (1) sulfate, *See* Witcolate ES-2, Witcolate LES-60C, 34477
sodium laureth (2) sulfate, *See* DeSonol SE-2, 8190, Marlinat® 242/28, 18848, Texapon N 25, Texapon NSE, 31467
sodium laureth (3) sulfate, *See* DeSonol SE, 8189, Nutrapon ES-60 3568, 21818, Perlankrol® ADP3, 23323, Rhodapex® ES, 26628
sodium laureth phosphate, *See* Forlanit P, 12252
sodium laureth phosphate, *See* Wetting Agent FCGB, 34342
sodium laureth (10) phosphate, *See* Crafol AP-260, 6956
sodium laureth sulfate, 28778, *See* Akyposal 9278 R, 1301, Akyposal DS 28, 1305, Akyposal DS 56, 1306, Akyposal EO 20 MW, 1307, Akyposal MS SPC, 1312, Akyposal RLM 56 S, 1319, Avirol® FES 996, 3479,

Avirol® SE 3002, 3484, Calfoam ES-30, 4984, Carsonol® SES-S, 5337, Chemsalan RLM 28, 6030, Crodapearl Liq, 7129, Disponil FES 32, 8691, Empicol® ESA, 10318, Empicol® ESB, 10319, Empicol® ESC/AU, 10320, Genapol® ARO, 12786, Genapol® LRO Liq, Paste, 12790, Genapol® ZRO Liq, Paste, 12803, Hartenol LES 60, 13670, Nonasol N4SS, 21440, Rewopol® NL 2-28, 26367, Standapol® 7092, 29466, Standapol® AP Blend, 29468, Standapol® ES-1, 29473, Steol® 4N, 29660, Steol® CS-130, 29663, Steol® OS 28, 29664, Texapon ES-1, 31458, Texapon NSE, 31469, Zetesol NL, 34746, Euperlan® PK 900, 11203, Marlamid® KLP, 18743, Marlon® PF 40, 18879, Monateric CDL, 20157, Texapon QLV, 31475
sodium laureth sulfate (2 EO), *See* Ungerol N2-28, 32921
sodium laureth sulfate (3 EO), *See* Empimin® KSN27, 10423
sodium laureth sulfate (4 EO), *See* Polystep® B-12, 24592
sodium laureth sulfate and disodium laureth sulfosuccinate, *See* Texapon SBN, 31477
sodium laureth sulfate and magnesium laureth-16 sulfate, *See* Akyposal HF 28, 1308
sodium laureth sulfate, cocamide DEA, TEA-lauryl sulfate, *See* Miracare® NWC, 19858
sodium laureth sulfate, MEA laureth-6 carboxylate, cocamide DEA and sodium dodecylbenzene sulfonate, *See* Akypogene WSW-W, 1262
sodium laureth sulfate, mixed with magnesium laureth sulfate, sodium laureth-8 sulfate, sodium oleth sulfate and magnesium oleth sulfate, *See* Texapon ASV, 31454
sodium laureth sulfate, PEG-8, cocamide MEA, glycol disterate, and glycerin, *See* Texapon SG, 31478
sodium laureth sulfate-Cocamide DEA-glycol distearate, *See* Akyposal 2010 S, 1299
sodium laureth sulfate-disodium laureth sulfosuccinate-laureth-6 carboxylic acid-cocamidopropyl betaine-ammonium chloride, *See* Dermalcare® 1673, 8096
sodium laureth sulfosuccinate, *See* Secosol® ALL40, 27813
sodium laureth-1 sulfate, *See* Nutrapon ESY 2299, 21819, Sulfochem ES-1, 30002
sodium laureth-11 carboxylate, *See* Rewopol® CLN 100, 26362
sodium laureth-11 carboxylate, iodine, *See* Akypogene Jod F, 1258
sodium laureth-12 sulfate, *See* Disponil FES 92E, 8692
sodium laureth-13 carboxylate, *See* Miranate® LEC, 19875, Sandopan® LS-24, 27346
sodium laureth-2 sulfate, *See* Sulfochem ES-2, 30003, Sulfochem ES-70, 30006, Texapon PLT-227, 31473
sodium laureth-3 sulfate, *See* Nutrapon BM 3960, 21816, Nutrapon KF 3846, 21822, Nutrapon KPC 0156, 21823, Sulfochem ES-3, 30004, Sulfochem ES-60, 30005
sodium laureth-4 phosphate, *See* Chemphos TR-510S, 6018, Rhodafac® MC-470, 26597
sodium laureth-6 carboxylate, *See* Akypo NTS, 1227, Akypo® Soft 45 NV, 1253
sodium laurethyl sulfate, *See* Abex® 23S, 40
sodium lauriminodipropionate, *See* Mackam 160C, 18139, Mirataine® H2C-HA, 19928
sodium lauroamphoacetate, *See* Ampholyt JA 140, 2332, Dehyton® PLG, 7996, Miranol® HM Conc, 19900, Monateric 985A, 20146, Monateric L30, 20170, Monateric LMM-30, 20173
sodium lauroamphoacetate and sodium trideceth sulfate, *See* Miracare® MHT, 19856
sodium lauroamphopropionate, *See* Miranol® HM-SF Conc, 19901
sodium lauroyl lactylate, *See* Pationic 138, 22905
sodium lauroyl sarcosinate, 28779, *See* Closyl LA 3584, 6443, Gardol®, 12668, Hamposyl® L-30, 13601, Medialan LD, 19214, Sarkosyl NL30, 27465, Vanseal® NALS-30, 33359, Zoharsyl L-30, 34876
sodium lauroyl taurate, 28780, *See* Hostapon KTW New, 14394
sodium lauroylsarcosinate, *See* sodium lauroyl sarcosinate, 28779
sodium N-lauroyl sarcosinate, *See* Crodasinic LS30, 7144, Maprosyl® 30, 18671, Secosyl, 27823, sodium lauroyl sarcosinate, 28779, Vanseal® NALS-30, 33359
sodium lauryl benzene sulfonate, *See* sodium dodecylbenzenesulfonate, 28760
sodium lauryl ether (2) sulfate, *See* Texapon NSO, 31471
sodium lauryl ether sulfate, *See* Sulfotex LMS-E, 30033, sodium laureth sulfate, 28778, Witcolate ES-2, Witcolate LES-60C, 34477
sodium lauryl ether sulfate (2 EO), *See* Zoharpon ETA 27, 34863
sodium lauryl ether sulfate (n = 1-4), *See* sodium laureth sulfate, 28778
sodium-N-lauryl β-iminodipropionate, *See* Deriphat 160C, 8091
sodium lauryl methyl taurate, *See* Zoharpon LMT42, 34870
sodium lauryl sulfate, 28781, *See* Akyposal NLS, 1314, Alscoap LN-40, LN-90, 1853, Calfoam SLS-30, 4986, Carsonol® SLS Paste B, 5339, Chemsalan NLS 30, 6029, DeSonol S, 8188, Dreft, 8976, Drene, 8977, Elfan® 200, 9812, Empicol® 0045, 10299, Empicol® 0045V, 10300, Empicol® LM, 10324, Empicol® LMV/T, 10325, Empicol® LS30, 10327, Empicol® LX, 10328, Empicol® LXV, 10329, Empicol® LY28/S, 10330, Empicol® LZ, 10331, Empicol® LZG 30, 10332, Empicol® LZP, 10333, Empicol® LZV, 10334, Empimin® LR28, 10425, Hartenol LAS-30, 13669, Marlinat® DFK 30, 18851, Norfox® SLS, 21552, Nutrapon DL 3891, 21817, Nutrapon RS 1147, 21825, Nutrapon TK 3603, 21826, Nutrapon W 1367, 21829, Nutrapon WAQE 2364, 21830, Octosol SLS, 22027, Orvus WA, 22346, Perlankrol® DSA, 23326, Polystep® B-3, 24596, Rewopol® NLS 15 L, 26368, Rhodapon® 101-10, 26631, Rhodapon® LCP, 26636, Rhodaterge® SSB, 26668, Standapol® WAQ-LC, 29482, Stepanol® ME Dry, 29695, Stepanol® WA Extra, 29698, Sulfetal C 38, 29982, Sulfochem SAC, 30010, Sulfochem SLC, 30011, Sulfochem SLN, 30012, Sulfochem SLP-95, 30013, Sulfochem SLS, 30014, Sulfochem SLX, 30015, Sulfopon 101, 101 Special, 30023, Sulfopon 101/POL, 30024, Sulfopon P-40, 30027, Sulfotex LCX, 30032, Sulphatol 33, 30062, Supralated ME, 30311, Texapon K-12, K-1296, L-100, 31461, Texapon K-1296, 31462, Texapon VHC Needles, ZHC Needles, 31481, Texapon ZHC Needles, 31482, Texapon ZHC Needles, 31483, Ufarol Na-30, 32743, Witcolate 6400, 34469, Zoharpon LAS Special, 34868, Nutrapon TW 3987, 21828, Rewopol® SLS, 26381
sodium lauryl sulfate (C_{12}-C_{14}), *See* Texapon LS Highly Conc, 31464
sodium lauryl sulfate BP, *See* Empicol® 0303, 10301
sodium lauryl sulfate C_{12}-C_{18}, *See* Texapon OT Highly Conc. Needles, 31472
sodium lauryl sulfate ethoxylate, *See* Witcolate ES-2, Witcolate LES-60C, 34477
sodium lauryl sulfate mixed with sodium laureth sulfate; lauramide MIPA; cocamide MEA; glycol stearate, *See* Texapon EVR, 31459
sodium lauryl sulfate mixed with sodium myristyl sulfate, sodium cetyl sulfate, sodium stearyl sulfate and laureth-10, *See* Texapon CS Paste, 31455
sodium lauryl sulfate USP, BP, *See* Naxolate WA-97, 20818
sodium lauryl sulfate, stearamide MEA, glycol stearate and cocamide MEA, *See* Miracare® M1, 19855
sodium lauryl sulfate-lauryl alcohol, *See* Supralated WA, 30312
sodium lauryl sulfoacetate, *See* Lathanol® LAL, 16816, Stepan-Mild® LSB, 29681
sodium lauryl sulfosuccinate, *See* Secosol® AL 959, 27812
sodium lauryl/propoxy sulfosuccinate, *See* Emcol® 4910, 10054
sodium lauryl/tallow sulfate, *See* Empicol® 0775, 10304
sodium lauryliminodipropionate, *See* Rewoteric® AMLP, 26431
sodium laurylpoly(oxyethylene) sulfate, *See* Witcolate ES-2, Witcolate LES-60C, 34477
sodium lignate, *See* Indulin® SA-L, 15093, Politol®, 24334
sodium lignosulfonate, *See* Darvan® No. 2, 7784, Dynasperse A, 9391, Dyquex®, 9436, Lignosol DXD, 17237, Maracarb N-1, 18675, Marasperse N-22, 18689, Polyfon® F, 24440, Temsperse S 001, 31086, Zewalon FN, 34756
sodium linear decyl diphenyl oxide disulfonate, *See* Calfax 10L-45, 4982
sodium magnesium fluorosilicate, *See* Laponite® B, 16777, Laponite® S, 16779
sodium/magnesium laureth sulfate, *See* Empicol® BSD 52, 10308
sodium magnesium lithium fluoro silicate, *See* hectorite laponite, 13753
sodium magnesium silicate, 28782, *See* Laponite® D, 16778, Laponite® XLG, 16780
sodium 2-mercaptobenzothiolate, *See* Nacap®, 20643
sodium 2-mercaptobenzothiazolate, *See* Nacap®, 20643
sodium 2-mercaptobenzothiazole, *See* Nacap®, 20643
sodium metabisulfite, 28783, *See* Hydros® F, 14637, metabisulfite, 19404
sodium metaborate, 28784, 28785
sodium metaborate ($NaBO_2$), *See* sodium metaborate, 28784
sodium metam, *See* Unifume, 32954
sodium metaperiodate, 28786
sodium metaphosphate, 28787, *See* Sodaphos, 28721
sodium metaphosphate, *See* sodium metaphosphate, 28787
sodium metaphosphate
sodium metaphosphate ($Na_6(PO_3)_6$), *See* sodium hexametaphosphate, 28770
sodium metasilicate, 28788, *See* Lasilso, 16806
sodium metasilicate pentahydrate, 28789
sodium metasilicate, anhydrous, *See* sodium metasilicate, 28788
sodium metasilicate, pentahydrate, *See* Crystamet 1020, 7293
sodium metasilicate-sodium carbonate-POE ester of mixed fatty and resin acids, *See* Alcojet®, 1479
sodium metavanadate, *See* Northovan, 21585
sodium methallyl sulfonate, 28790, *See* Geropon® MLS/A, 12909
sodium metham, *See* Unifume, 32954
sodium methanalsulfoxylate, *See* sodium formaldehydesulfoxylate, 28766
sodium methanearsonate, *See* Neoarsycodyl, 20871
sodium bis(2-methoxyethoxy) aluminum hydride, *See* Vitride®, 33994
sodium N-methylaminodithioformate, *See* Unifume, 32954
sodium N-methyl-N-cocoyl taurate, *See* Geropon® TC-42, 12917, sodium methyl cocoyl taurate, 28791
sodium methyl cocoyl taurate, 28791, *See* Adinol CT, 696, Hostapon CT Paste, 14391, Somepon T25, 29071
sodium methyl naphthalene sulfonate condensate, *See* Supragil® MNS/90, 30307
sodium N-methyl N-palmitoyl taurate, *See* Fenopon TN-74, 11613
sodium N-methyl N-tall oil acid taurate, *See* Fenopon TK32, 11612
sodium methyl oleoyl taurate, *See* Adinol OT, 697, Arkopon T, 2898, Fenopon T-33 and T-43, 11608, Geropon® T-22/A, 12915, Hostapon SO, 14395, Hostapon T Powd, 14397
sodium N-methyl N-oleoyl taurate, *See* Adinol OT, 697,

Geropon® T-33, 12916
sodium N-methyl N-oleyl taurate, *See* Fenopon T-51, 11609, Fenopon T-77, 11610
sodium N-methyl-N-oleoyltaurate, *See* Adinol OT, 697, Arkopon T, 2898
sodium N-methyl-N-stearoyl taurate, *See* sodium methyl stearoyl taurate, 28792
sodium N-methyl-N-tallowyl taurate, *See* Geropon® TK-32, 12918, sodium methyl tall oil acid taurate, 28794
sodium 2-methylprop-2-ene-1-sulfonate, *See* Geropon® MLS/A, 12909
sodium methyl stearoyl taurate, 28792, *See* Hostapon STT Paste, 14396
sodium methyl tall oil acid taurate, 28793, 28794
sodium methylarsinate, *See* arrhenal, 3128
sodium methyldithiocarbamate, *See* Unifume, 32954
sodium methyldithiocarbamodithioate, *See* metam-sodium, 19444
sodium 2,2'-methylene bis-(4,6-di-*t*-butylphenyl) phosphate, *See* ADK STAB NA-11, 733
sodium methylparaben, *See* Nipagin M sodium, 21268
sodium molybdate dihydrate, *See* sodium molybdate, dihydrate, 28796
sodium molybdate (VI), *See* sodium molybdate, dihydrate, 28796
sodium molybdate (VI) dihydrate, *See* sodium Molybdate, dihydrate, 28796
sodium monochloracetate, 28797
sodium monofluoride, *See* sodium fluoride, 28764
sodium monofluorophosphate, 28798
sodium monoiodide, *See* sodium iodide, 28775
sodium monosulfide, *See* sodium sulfide, 28822
sodium montmorillonite, *See* Bentonite, 3954
sodium morrhuate, 28799
sodium muriate, *See* sodium chloride, 28748
sodium myristoyl glutamate, *See* Acylglutamate MS-11, 607
sodium myristoyl sarcosinate, *See* Hamposyl® M-30, 13603
sodium α-naphthalenesulfonate, *See* Aerosol® NS, 894
sodium naphthalene formaldehyde sulfonate, *See* Darvan® No. 1, 7783, Dispersogen A, 8677
sodium naphthalene sulfonate, *See* Aerosol® NS, 894, Chromasist 1487A, 6194, Lomar® D, 17565, Tamol® L Conc, 30767
sodium 1-Naphthalenesulfonate, *See* Aerosol® NS, 894
sodium naphthalene sulfonate formaldehyde condensate, *See* Ablusol ML, 129
sodium naphthaleneformaldehyde sulfonate, *See* Dehscofix 912, 7923
sodium-22 neoprene accelerator, *See* Perkacit® ETU, 23283
sodium nitrate, 28800, *See* Saliter, 27273
sodium (I) nitrate, *See* sodium nitrate, 28800
sodium nitrite, 28801
sodium *m*-nitrobenzene sulfonate, 28795, *See* Ludigol F, 17772
sodium-o-nitro-phenyl-propiolate, *See* nitropropiol, 21375
sodium nonoxynol-4 sulfate, *See* Polystep® B-27, 24594, Rhodapex® CO-433, 26624, Steol® COS 433, 29662
sodium nonoxynol-6 phosphate, *See* Rhodafac® LO-529, 26596, Surfagene FAD 106, 30351
sodium nonoxynol-6 sulfate, *See* Akyposal NPS 60, 1316
sodium nonoxynol-9 phosphate, *See* Chemphos TC-231S, 6009, Emphos CS-1361, 10294
sodium nonoxynol-10 sulfate, *See* Akyposal NPS 100, 1315
sodium nonylbenzene sulfonate, *See* Emkal NOBS, 10230
sodium octadecanoate, *See* sodium stearate, 28820
sodium 9-octadecenoate, *See* sodium oleate, 28802
sodium 1-octane sulfonate, *See* Bio-Terge® PAS-8S, 4214
sodium n-octanoate, *See* sodium caprylate, 28743
sodium octoxynol-3 sulfate, *See* Polystep® C-OP3S, 24598
sodium octoxynol-6 sulfate, *See* Akyposal BD, 1304
sodium octoxynol-9 sulfate, *See* Akyposal OPS 85, 1318
sodium octylphenol ethoxy sulfate, *See* Rhodapex® F-85/SD, 26630
sodium octyl sulfate, *See* Polystep® B-29, 24595, Rewopol® NEHS 40, 26366, Rhodapon® OLS, 26639, Standapol® LF, 29474, Stepantex® B-29, 29709, Sulfotex OA, 30034
sodium octylsulfate, *See* Tensatil D100, 31151
sodium n-octyl sulfate, *See* Avirol® SA 4108, 3481
sodium oleoamphohydroxypropylsulfonate, *See* Miranol® OS-D, 19909, Sandoteric TFL Conc, 27355
sodium oleoamphopropionate, *See* Schercoteric O-AA, 27710
sodium oleate, 28802, *See* Eunatrol, 11194
sodium α olefin sulfonate, *See* Calsoft AOS-40, 5028, DeSonate AOS, 8178
sodium α-olefine sulfonate (C_{14}/C_{16}), *See* Elfan® OS 46, 9823
sodium oleth-7 phosphate, *See* Rhodafac® GB-520, 26595
sodium oleyl sulfate, *See* Rhodapon® OS, 26640, Sulfopon O 680, 30026
sodium oleyl-cetyl alcohol sulfate, *See* Elfan® 680, 9817
sodium oleoyl glutamate - sodium cocoyl glutamate, *See* Acylglutamate AS-12, 596
sodium N-oleoyl sarcosinate, *See* Crodasinic OS35, 7147
sodium omadine, *See* sodium Omadine® 40% Aq. Sol'n, 28803, sodium pyrithione, 28813
Sodium Omadine® 40% Aq. Sol'n, 28803
sodium orthophosphate, *See* trisodium phosphate, anhydrous, 32412
sodium orthosilicate, *See* Acsil, 454, N, 20633, sodium silicate, 28817
sodium-2-[(1-oxododecyl) amino] ethanesulfonate, *See* sodium lauroyl taurate, 28780
sodium oxychloride, *See* sodium hypochlorite, 28773
sodium N-palmityl N-cyclohexyl taurine, *See* Hostapon IDC, 14392
sodium pantothenate, *See* Ritapan NAP, 26819
sodium PCA, *See* Ajidew N-50, 1143, Ritamectant PCA, 26816
sodium PEG-3 lauramide carboxylate, *See* Akypo AD 100 SPC, 1215
sodium PEG-6 cocamide carboxylate, *See* Akypo® Soft KA 250 BV, 1254
sodium PEG-6 cocamide carboxylate, glycol distearate, *See* Akyposal 2010 SD, 1300
sodium pentaborate, *See* Borax Glass, 4412
sodium pentachlorophenate, 28805
sodium pentachlorophenoxide, *See* sodium pentachlorphenate, 28805
sodium perborate, 28806, *See* perborax, 23213, perborin, 23214, peroxydol, 23412
sodium perborate anhydrous, *See* sodium perborate, 28806
sodium percarbonate, 28807
sodium perchlorate, *See* KM sodium Perchlorate, 16245
sodium periodate, *See* sodium metaperiodate, 28786
sodium *m*-periodate, *See* sodium metaperiodate, 28786
sodium permutite, 28808
sodium peroxide, *See* Oxone, 22435, oxygen powder, 22453
sodium peroxydisulfate, *See* sodium persulfate, 28809
sodium persulfate, 28809
sodium pertechnitate $_{99m}$Tc, *See* Pertscan-99m, 23438
sodium petrol. sulfonate, *See* Petrosul® H-50, 23517
sodium petroleum sulfonate, *See* Aristonate H, 2884, Petrosul® M-50, M-60, M-70, 23518
sodium *o*-phenylphenolate, *See* Dowicide A, 8932
sodium *o*-phenylphenoxide, *See* Dowicide A, 8932
sodium *ortho*-phenylphenate, *See* Dowicide A, 8932
sodium phenyl sulfonate, *See* Supragil® GN, 30306
sodium phosphate, 28810, *See* disodium phosphate, dihydrate, 8653, salt perlate, 27293, Sodaphos, 28721, Sorensen's salt, 29136, trisodium phosphate, anhydrous, 32412, trisodium phosphate, dodecahydrate, 32413
sodium phosphate ($Na_6(PO_3)_6$), *See* sodium hexametaphosphate, 28770
sodium phosphate ($Na_4P_2O_7$), *See* sodium pyrophosphate, 28814
sodium phosphate 12-Water, *See* trisodium phosphate, dodecahydrate, 32413
sodium phosphate glass, *See* sodium hexametaphosphate, 28770
sodium phosphate tribasic dodecahydrate, *See* trisodium phosphate, dodecahydrate, 32413
sodium phosphate, dibasic, *See* disodium phosphate (anhydrous), 8652, tasteless salts, 30843
sodium phosphate, dibasic dihydrate, *See* disodium phosphate, dihydrate, 8653
sodium phosphate, monobasic, *See* Aspon, 3238, Recresal, 25944, sodium phosphate, 28810
sodium phosphate, tribasic, *See* trisodium phosphate, anhydrous, 32412
sodium phosphate, tribasic dodecahydrate, *See* trisodium phosphate, dodecahydrate, 32413
sodium phosphates, *See* Caust X, 5467
sodium phosphotungstate, *See* Scheibler's reagent, 27567
sodium phytate, *See* Rencal, 26105
sodium POE(3) tridecyl ether acetate, *See* Akypo ITD 30 N, 1216
sodium POE(6) tridecyl ether acetate, *See* Akypo ITD 30 N, 1216
sodium poly PG-propyl dimethicone thiosulfate, *See* Abil® S201, 56
sodium poly(oxyethylene) lauryl ether sulfate, *See* Witcolate ES-2, Witcolate LES-60C, 34477
sodium polyacrylate, 28811, *See* Acrysol® GS, 447, Acrysol® HV-1, 448, Acusol® 410N, 582, Acusol® 445N, 584, Acusol® 445ND, 585, Acusol® 860N, 594, Alkasperse® A-20, 1712, Drewsperse® 611, 9020, Good-rite® K-705BD, 13208
sodium polycarboxylate, *See* Bevaloid 211, 4081
sodium polymannuronate, *See* Algin, 1591, Kelcosol®, 15832, Kelgin, 15841, Kelgin® F, 15842, kelgin® HV, LV, MV, 15843, Kelgin® QH, QL, QM, 15844, Kelgin® XL, 15845, Kelgum, 15846, Keltex®, 15878, Keltex® HV, 15879, Keltex® S, 15880, Keltone®, 15883, Keltone® HV, LV, 15884
sodium polymetaphosphate, *See* Calgon, 4987, Kalex HMP, 15670, Polyphos, 24519, sodium hexametaphosphate, 28770
sodium polymethacrylate, *See* Alkasperse® M-10, 1713, Tamol® 850, 30766
sodium polynaphthalene sulfonate, *See* Rhodacal® Liquid, Rhodacal® N, 26588
sodium polystyrene sulfonate, *See* Flexan® 130, 11923, Kayexalate, 15812
sodium-potassium tartrate, *See* sal Rupellensis, 27252
sodium propionate, 28812, *See* Luprosil® sodium Salt, 17922, Spac, 29182
sodium propyl paraben, *See* Nipasol M sodium, 21284
sodium prussiate yellow, *See* yellow prussiate of soda, 34648
sodium pyrithione, 28813, *See* sodium Omadine® 40% Aq. Sol'n, 28803, sodium pyrithione, 28813
sodium pyroborate, *See* sodium borate anhydrous, 28739
sodium pyrophosphate, 28814
sodium pyrophosphate (4:1), *See* sodium pyrophosphate, 28814
sodium pyrophosphate ($Na_4P_2O_7$), *See* sodium pyrophosphate, 28814
sodium pyrophosphate, normal, *See* sodium pyrophosphate, 28814
sodium pyrosulfite, *See* sodium metabisulfite, 28783
sodium ricinolate, *See* Soricinol 40, 29143
sodium ricinoleate, *See* Soricinol 40, 29143

sodium rosinate, *See* Arizona DRS-43, 2891
sodium saccharide, *See* saccharin sodium, 27196
sodium saccharin, *See* saccharin sodium, 27196
sodium saccharinate, *See* saccharin sodium, 27196
sodium salt of N,N-diethyldithiocarbamic acid, *See* Octopol SDE-25, 22013
sodium salt of turkey red oil, *See* oleite, 22114
(sodium salt) scifluorfen, *See* acifluorfen, 343
sodium selenite, 28815
sodium selenite pentahydrate, *See* sodium selenite, 28815
sodium sesquicarbonate, 28816
sodium sesquisilicate, *See* Acsil, 454
sodium silicate, 28817
sodium silicate, *See* Acsil, 454, Gasbinda, 12688
sodium silicate glass, *See* Acsil, 454, N, 20633, sodium silicate, 28817
sodium silicate solution, *See* isinglass, 15333
sodium silicates, *See* Crystal, 7284
sodium silicoaluminate, *See* Valfor® 100, 33244, Wessalith, 34284
sodium silicofluoride, 28818, *See* salufer, 27301, SSF, 29390
sodium silico-fluoride, *See* SSF, 29390
sodium stannate, 28819
sodium stannate(IV), *See* preparing salt, 24898, sodium stannate, 28819
sodium starch glycolate, *See* Explotab®, 11316
sodium stearate, 28820
sodium steareth-4 phosphate, *See* Surfagene FHD 704 NV, 30356
sodium stearoamphoacetate, *See* Miranol® DM Conc. 45%, 19892
sodium stearoyl lactylate, *See* Artodan SP 55 Kosher, 3150, Emplex, 10443, Emulsilac S, 10514, Grindtek FAL 1, 13409, Radiamuls® SSL 2990, 25788
sodium stearoyl 2-lactylate, *See* Pationic SSL, 22908
sodium stearoyl lactylates, *See* Crolactil SSL, 7186
sodium stearoyl methyl taurate, *See* sodium methyl stearoyl taurate, 28792
sodium N-stearoyl-N-methyl taurate, *See* sodium methyl stearoyl taurate, 28792
sodium stearyl fumarate, *See* Pruv, 25354
sodium subsulfite, *See* sodium thiosulfate, 28826
sodium sucaryl, *See* sodium cyclamate, 28754
sodium sulfantimonate, *See* Schlippe's salt, 27723
sodium sulfate, 28821, *See* sal Mirabil, 27249
sodium sulfate, anhydrous, *See* sodium sulfate, 28821
sodium sulfate decahydrate, *See* Glauber's salt, 12974
sodium sulfhydrate, *See* sodium bisulfide, 28737
sodium sulfide, 28822, *See* sodium bisulfide, 28737
sodium sulfide (Na_2S), *See* sodium sulfide, 28822
sodium sulfite, 28823, *See* sodium acid sulfite, 28728
sodium sulfite (Na_2SO_3), *See* sodium bisulfite, 28738
sodium sulfite-benzoic acid, *See* Heckel's solution, 13738
sodium α-sulfomethyl cocoate, *See* Alpha-Step® MC-48, 1834
sodium sulfonate-soap derivative, *See* Base 75, 3691
sodium sulforicinoleate, *See* oleite, 22114, Türkischrotöl 100%, 32541
sodium-sulfo-ricinoleate, *See* oleite, 22114
sodium sulfoxylate, *See* Reductone, 25983, sodium hydrosulfite, 28771
sodium sulfuret, *See* sodium sulfide, 28822
sodium superoxide, *See* Oxone, 22435
sodium tallamphodipropionate, *See* Miranol® TBS, 19912
sodium tallamphopropionate, *See* Monateric TA-35, 20174
sodium tallow sulfate, *See* Empicol® TAS30, 10341, Marlinat® KT 50, 18854
sodium tallowate, 28824, *See* Emery® 2895 Foamaster Soap L, 10156, Foamaster Soap L, 12148
sodium tartrate, 28825
sodium tetraborate, *See* sodium borate anhydrous, 28739
sodium tetraborate decahydrate, *See* Borax, 4411
sodium tetrabromophenolphthalein, *See* Bromeikon, 4568
sodium tetrabromophthalate, *See* Great Lakes FR-756, 13306
sodium tetrachlorophenate, *See* Dowicide F, 8935
sodium tetradecyl sulfate, *See* Niaproof® Anionic Surfactant 4, 21093
sodium tetrafluoroborate, *See* sodium fluoroborate, 28765
sodium tetrahydridoborate, *See* sodium borohydride, 28740
sodium tetrahydroborate, *See* sodium borohydride, 28740
sodium-β-tetralin sulfonate, *See* Majamin, 18420
sodium tetrapropylene benzene sulfonate, *See* Tensaryl S30P and S70P, 31146
sodium thioantimonate(V), *See* Schlippe's salt, 27723
sodium thiosulfate, 28826, *See* Ametox, 2049, green mordant, 13321, Lycopon, 18065
sodium thiosulfate pentahydrate, *See* sodium thiosulfate, 28826
sodium tin oxide, *See* sodium stannate, 28819
sodium *p*-toluenesulfochloramide, *See* Chloramine T, 6086
sodium *p*-toluenesulfonate, 28827, *See* Eltesol® ST 40, 9872
sodium toluene sulfonate, *See* Eltesol® ST 40, 9872, Hartotrope STS-40, Powd, 13693, Manro STS, 18606, Naxonat® 4ST, 20821, Reworyl® NTS, 26408
sodium *p*-toluenesulfonchloramide, *See* Ketjensept, 16133
sodium tolyltriazole, *See* Maxahibit TT-50, 19074
sodium trichloroacetate, *See* Varitox, 33433
sodium trichlorophenate, *See* Dowicide B, 8933
sodium 2,4,5-trichlorophenate, *See* Dowicide B, 8933
sodium trideceth sulfate, *See* Akyposal BA 28, 1303, Cedepal TD-403, 5542, Monateric 985A, 20146, Monateric CDTD, 20158, Rhodapex® 674/C, 26621, Zoharteric D-2, 34879, Zoharteric M-2, 34885
sodium trideceth-3 carboxylate, *See* Akypo ITD 30 N, 1216
sodium trideceth-6 carboxylate, *See* Akypo ITD 30 N, 1216
sodium trideceth-7 carboxylate, *See* Sandopan® DTC-100, 27341
sodium trideceth-3 carboxylic acid, *See* Akypo ITD 30 N, 1216
sodium tridecyl sulfate, *See* Avirol® SA 4113, 3483, Rhodapon® TDS, 26641
sodium trideth sulfate, *See* Rhodapex® EST-30, 26629
sodium triphosphate, *See* sodium tripolyphosphate, 28828
sodium triphosphate, tripoly, *See* sodium tripolyphosphate, 28828
sodium tripolyphosphate, 28828, *See* Curafos® STP, 7411, Flav-R-Keep FP-51, 11906, Hysorb, 14749, Lem-O-Fos®, 16979, Polygon, 24460, Rhodia-Phos, 26677, Sea-Gard® Formula FP-91, 27786
sodium tungstate, 28829
sodium tungstate dihydrate, *See* sodium tungstate, 28829
sodium tungstate(VI) dihydrate, *See* sodium tungstate, 28829
sodium tungsten oxide, *See* sodium tungstate, 28829
sodium tungstophosphate, *See* Scheibler's reagent, 27567
sodium undecylenate, 28830
sodium vanadate, *See* Northovan, 21585
sodium vanadate (meta), *See* Northovan, 21585
sodium versenate, 28831, *See* edetate disodium, 9613, Trilon® BD, 32325
sodium vinyl sulfonate, *See* Hartomer 4900, 13680
sodium wolframate, *See* sodium tungstate, 28829
sodium xylene sulfonate, 28832, *See* Eltesol® SX 30, 9873, Esi-Terge SXS, 10831, Hartotrope SXS 40, Powd, 13694, Manro SXS, 18607, Naxonate® 4L, 20824, Nutrol SXS 5418, 21849, Pilot SXS-40, 23843, Reworyl® NXS 40, 26409, Stepanate® SXS, 29674, Sulfotex SXS-40, 30036, Witconate SXS 40%, 34492
sodium *m*-xylene sulfonate, *See* Eltesol® SX 30, 9873, Naxonate® 4L, 20824, Pilot SXS-40, 23843, Witconate SXS 40%, 34492
sodusec, 28833
Sofac, *See* cyfluthrin, 7563
Sofanate, 28834
Sofibex, 28835
Soflens, 28836
So-flo, *See* sodium fluoride, 28764
Sofnol Soda-lime G, 28837
Sofnolite, *See* Sofnol Soda-lime G, 28837
Sofnon 105G, 28838
soft amber, *See* Gedanite, 12710
soft copal, 28839
soft platinum, 28840
Soft Resin P 65, 28841
Soft Touch 1052, 28842
Soft-Clad®, 28843
Softcon, 28844
Softenol®, 28845 , 28846
Softenol® 3100, 28847
Softenol® 3114, 28848
Softenol® 3408, 28849
Softenol® 3829, 28850
Softenol® 3991, 28851
Softex, 28852, 28853, *See* Zeonet B, 34721
Softigen® 701, 28854
Softigen® 767, 28855
Softisan®, 28856
Softisan® 100, 28857
Softisan® 378, 28858
Softisan® 601, 28859
Softisan® 649, 28860
Softisan® Gel, 28861
SoftMate DW, 28862
Softrite, 28863
Softyne H, 28864
Sohnhofen stone, *See* CP Filler, 6928
Soil Fungicide 1823, *See* Terraneb SP Turf Fungicide, 31337
Soil Pests Killer, 28865
Soil TRIGGRR, 28866
soilime, 28867
S-oils, 28868
Soiltex®, 28869
soja bean oil, *See* soya bean oil, 29169
Sokalan® CP 2, 28870
Sokalan® CP 5, 28871
Sokalan® CP 7, 28872
Sokalan® HP 22, 28873
Sokalan® HP 50, 28874
Sokalan® HP 53, 28875
Sokalan® PA 13 PN, 28877
Sokalan® PA 15, 28878
Sokalan® PA 110 S, 28876
Sokalan® PM 10, 28879
Sokoff, 28880
Sokolan, 28881
sol rubber, 28882
Solactol, 28883
Soladox, 28884
Soladox 112, 28885
Solamin, *See* benzethonium chloride, 3963
Solan, 28886, 28887, *See* Croptex Bronze, 7213, pentanochlor, 23165, Solan, 28887
Solan 50, 28888
Solan E, 28889
Solane, 28890
Solangel 401, 28891
Solanine, *See* benzethonium chloride, 3963
Solanthrene, 28892
Solar, 28893, 28894
Solar 90, *See* Nacconol® 40G, 20647, sodium dodecylbenzenesulfonate, 28760
Solar NP, *See* Alkasurf® NP-4, 1715, Triton® N-57, 32430
solar oil, 28895
solar salt, 28896
Solar Steel, 28897
solar winter ban, *See* propylene glycol, 25215

Solarchem® O, 28898, *See* 4-(dimethylamino)benzoic acid, 8514
Solasan 500, *See* metam-sodium, 19444
Solaskil, *See* Tramisol®, 32134
Solasol, 28899
Solatene, 28900, *See* carotene, 5308
Solatol, 28901
Solbrol A, *See* ethyl *p*-hydroxybenzoate, 11061
Solbrol B, *See* Nipabutyl, 21249
Solbrol P, *See* Nipasol M, 21282
Solbrol®, 28902
Solcod, 28903
Solcornol, 28904
Soldaflux, 28905
Soldamoll, 28906
solder, 28907
Solderel®, 28908
soldering acid, 28909
soldering salt, 28910
soldering solution, 28911
Soldis, 28912
soldo, 28913
Sole Tege TS-25, *See* Niaproof® Anionic Surfactant 08, 21092, Rhodapon® BOS, 26632
Sole Terge 8, 28914, *See* disodium oleamido MIPA sulfosuccinate, 8651
Soledon, 28915
Solef®, 28916
Solef® 1008, 28917
Solef® 1010, 28918
Solef® 5008, 28919
Solef® 6010, 28920
Solef® 8808, 28921
Solef® 11008/0003, 28922
Solef® 11010, 28923
Sole-Mulse B, 28924
Solene, *See* petroleum ether,
Solenhofen stone, 28925
Solenite, 28926
Solester, 28927
Solfa, 28928, 28929, *See* Thiovit, 31724
Solfac, 28930, *See* cyfluthrin, 7563
Solflex® 1216, 28931
Solfo Black 2B Supra, *See* Dinitra, 8540
Solfo Black B, *See* Dinitra, 8540
Solfo Black BB, *See* Dinitra, 8540
Solfo Black G, *See* Dinitra, 8540
Solfo Black SB, *See* Dinitra, 8540
Solgen, 28932
Solgen 50, *See* Ablunol S-60, 110
Solgen 90, *See* Ablunol S-20, 108
Solicum, 28933
solid alcohol, 28934
Solid organic peroxides, *See* Perkadox® RM, 23309
Solid photopolymer, *See* AFP®, 945
Solid storax, *See* red storax, 25970
Solidarol, 28935
Solidegal®, 28936
Solidermin®, 28937
Solidex, 28938
solidified alcohol, 28939
solidified linseed oil, 28940
Solidite, 28941
Solidogen LT-13, 28942
Solidogen®, 28943
Solidokoil®, 28944
Soligen, 28945
Solimide® Foam, 28946
Solintor, 28947
Solinure, 28948
Solka-Floc® BW-40, BW-100, BW-200, BW-2030, UF-900-FCC & NF, 28949
Solklean™ 101, 28950
Solkote™ Hi/Sorb™ -II, 28951
Sollacaro's aluminum solder, 28952
Solmed 100, 28953
Solnhofen stone, *See* Calbux, 4929, calcium carbonate, 4941
Solo, 28954
Solochrome, 28955
Solok®, 28956
Solon Conc, 28957
Solon Fe Special, 28958
Solophenyl, 28959
Solosil, 28960
Solox, 28961
Solozone, 28962, *See* Oxone, 22435
Solpolac, 28963
Solricin® 135, 28964, *See* Potassium ricinoleate, 24778
Solricin® 235, 28965
Solricin® 285, 28966
Solricin® 435, 28967
Solsaf-T-Solv™ 403, 28968
solsol needles, *See* Rhodapon® 101-10, 26631
Solsolv™ 301, 28969
Solsperse, 28970
Solstar, 28971
Soltair, 28972
Soltrol, 28973
Solu Kera-Tein M, 28974
Solubacter, *See* TCC, 30856
Solubilisant Gamma 2420, 28975
Solubilisant Gamma 2428, 28976
Solu-Biloptin, 28977
soluble algin, 28978
soluble animal collagen, *See* collagen, 6610
soluble castor oils, *See* blown oils, 4354
Soluble collagen, *See* Actigen C, 478, Collagen Native Extra 1%, 6614
soluble cream of tartar, 28979
soluble fluorescein, *See* Fluorescein disodium salt, 12081
Soluble glass, 28980, *See* Acsil, 454, N, 20633, sodium silicate, 28817
soluble gluside, *See* saccharin sodium, 27196
soluble gun cotton, *See* Kodaloid, 16283, Venite, 33560
soluble indigo blue, *See* Indigo Carmine, 15075
Soluble iodophthalein, *See* keraphen, 16055
soluble oil, *See* Turkey red oils, 32607
soluble phenyle, 28981
soluble pitch, *See* pix solubilis, 23897
Soluble Plant Feed, 28982
soluble potash glass, 28983, *See* Kasil, 15778, potassium silicate, 24779
soluble potash water glass, *See* Kasil, 15778, potassium silicate, 24779
soluble primrose, *See* erythrosin, 10778
Soluble Rose Feed, 28984
soluble saccharin, *See* saccharin sodium, 27196
soluble salumin, 28985
soluble soda glass, *See* soluble glass, 28980
soluble starch, 28986
soluble tartar, 28987
Soluble Tomato Feed, 28988
Solublon, 28989
Solubor, 28990
Solubor DF, 28991
Solu-Coll, 28992
Solu-Coll P, 28993
Solucryl, 28994
Soluene 100 and 350, 28995
Solufeed, 28996
Soluglaucit, *See* paraoxon, 22783
Soluhoba, 28997
Solulan, 28998
Solulan 5, 28999
Solulan® 5, 29000
Solulan® 16, 29001
Solulan® 75, 29002
Solulan® 97, 29003
Solulan® 98, 29004
Solulan® C-24, 29005
Solulan® PB-2, 29007
Solulan® PB-5, 29009
Solulan® PB-10, 29006
Solulan® PB-20, 29008
Solu-Lastin 10, 29010
Solu-Lastin 30, 29011
Solulys®, 29012
Solu-Mar EN-30, 29013
Solu-Mar Native, 29014
Solumin, 29015
Solumin F, 29016
Solumin PFN, 29017
Solumin PV27, 29018
Solumin T45S, 29019
Solumin V27SD, 29020
so-luminum, 29021
Solumix, 29022
Soluphor® P, 29023
Solu-Silk 25, 29024
Solu-Silk Protein, 29025
Solu-Soy EN-25, 29026
Sol-U-Tein 6861, 29027
Sol-U-Tein EA, 29028
Sol-U-Tein EA Type PF-1;, *See* Albumen, 1402
Sol-U-Tein FS-1000, 29029
Sol-U-Tein PS-1000, 29030
Sol-U-Tein VG, 29031
Solutene, 29032
Solution forms containing 1,10-phenanthroline, *See* Activ-8®, Activ-8 in Hexylene Glycol, 494
Solutol, 29033
Solutol® HS 15, 29034
Solutrast, *See* Iopamidol, 15233
Solu-Veg EN-35, 29035, *See* hydrolyzed vegetable protein, 14613
Soluvit Richter, 29036
Solux, 29037
Solva, 29038
Solvanom, *See* dimethyl phthalate, 8508
Solvaperm, 29039
Solvarone, *See* dimethyl phthalate, 8508
Solvatone, 29040
Solvay® Soda, 29041
Solvene, 29042
Solvenol 1, *See* Solvenol 2, 226, 29043
Solvenol 2, 226, 29043
Solvenon® BB, 29044
Solvenon® DIP, 29045
Solvenon® DPM, 29046
Solvenon® IPP, 29047
Solvenon® I, 29048
Solvenon® PC, 29049
Solvenon® PM, Solvent PM, 29050
Solvenon® PP, 29051
Solvent 78, 29052
Solvent 111, *See* Chlorothene, 6151
Solvent 401, 29053
Solvent APV Spec, *See* Ethoxydiglycol, 11044
Solvent GC, 29054
Solvent naphta, *See* Ligroin, 17244
solvent naphtha, 29055
Solvent Orange 7, *See* Sudan II, 29939
Solvent Red 23, *See* Sudan III, 29940
Solvent Red 24, *See* Sudan IV, 29941
Solvent Scour 25/27, 29056
Solveol, 29057
Solvesso, 29058
Solvetek, 29059
Solvethane, 29060
Solvic, 29061
Solvifog (N.R.I.), 29062
Solvigran, 29063
Solvoclarin, 29064
Solvol, 29065
Solvolsol, *See* Diethoxol, 8435, ethyl di-Icinol, 11068
Solvtext, 29066
Somacount, 29067
Somafix, 29068
Somali gum, 29069
Somalia orange i, *See* Sudan I, 29938
Somaton, 29070
Somepon T25, 29071
Sometam, *See* metam-sodium, 19444
Somipront, *See* dimethyl sulfoxide, 8511, DMSO, 8762
Somon, 29072

Sonac, *See* phosphoric acid, 23738, 23739
Sonacide, 29073, *See* glutaraldehyde, 13062
Sonic, *See* glyphosate, 13148
Sonnenschein's reagent, 29074
Sonojell, 29075
Sonora gum, 29076
Sonostat 1111, 29077
Sonostat NTL, 29078
Sontara®, 29079
Sontex, 29080
Sontique®, 29081
Sopanox, 29082, *See* Vulkacit® 1000, 34124
sophoretin, *See* quercetin, 25647
sophorin, *See* rutin trihydrate, 27150
sophorine, *See* ulexine, 32761
SOPP, *See* Dowicide A, 8932
Soprodac, 29083
Soprophor® 37, 29084
Soprophor® 3D33, 29085
Soprophor® BSU, 29086
Sorane, 29087
Sorban, 29088
Sorbax HO-40, 29089
Sorbax HO-50, 29090
Sorbax PML-20, 29091
Sorbax PMO-20, 29092
Sorbax PMO-5, 29093
Sorbax PMP-20, 29094
Sorbax PMS-20, 29095, *See* Ethylan® GES6, 11108
Sorbax PTO-20, 29096
Sorbax PTS-20, 29097, *See* Polysorbate 65, 24571
Sorbax SML, 29098
Sorbax SMO, 29099
Sorbax SMP, 29100
Sorbax SMS, 29101
Sorbax STO, 29102
Sorbax STS, 29103
Sorbelite C, 29104
Sorbester P 17, *See* Witconol 2500, 34501
Sorbeth-20, *See* Liponic SO-20, 17393
sorbic acid, 29105, *See* Sorbistat, 29107
***trans,trans*-sorbic acid**, *See* sorbic acid, 29105
α-*trans*-σ-*trans*-sorbic acid, *See* sorbic acid, 29105
sorbic acid, potassium salt, *See* potassium sorbate, 24780, 24781
Sorbicolan, *See* A-625/641ABS 301K, ABS 500FR-1, 10
Sorbilande, *See* A-625/641ABS 301K, ABS 500FR-1, 10
sorbimacrogol laurate 300, *See* polysorbate 20, 24565
sorbimacrogol oleate 300, *See* Alkamuls® T-80, 1693, polysorbate 80, 24572
sorbimacrogol palmitate 300, *See* polysorbate 40, 24567
sorbimacrogol stearate 300, *See* polysorbate 60, 24568
sorbimacrogol trioleate 300, *See* polysorbate 85, 24574
sorbimacrogol tristearate 300, *See* polysorbate 65, 24571, 24570
sorbin, *See* sorbose, 29129
sorbinose, *See* sorbose, 29129
Sorbismal, 29106
Sorbistat, 29107, *See* sorbic acid, 29105
Sorbistat K, 29108
D-sorbit, *See* sorbitol, 29113
Sorbitan diisostearate, 29109, *See* Emsorb® 2518, 10463
sorbitan distearate, *See* Sorbon S-66, 29120
Sorbitan ester, *See* Atmer® 103, 3357
Sorbitan isostearate, 29110, *See* Crill 6, 7032, Emalex SPIS-100, 9994, Emsorb® 2516, 10462
Sorbitan laurate, *See* Ablunol S-20, 108, Alkamuls® SML, 1685, Arlacel® 20, 2899, Atmer® 100, 3356, Crill 1, 7028, Dehymuls SML, 7974, Disponil SML 100 F1, 8695, Emsorb® 2515, 10461, Ethylan® GL20, 11109, Glycomul® L, 13111, Hetan SL, 13888, Kemester® S20, 15966, Liposorb L, 17404, Montane 20, 20265, Prote-sorb SML, 25278, Radiasurf® 7125, 25797, S-Maz® 20, 28640, Sorbax SML, 29098, Sorbon S-20, 29117
Sorbitan laurate NF, *See* Span® 20, 29187
Sorbitan ML, *See* Ablunol S-20, 108
sorbitan monododecanoate, *See* S-Maz® 20, 28640
Sorbitan monolaurate, *See* Alkamuls® SML, 1685, Kemester® S20, 15966, Radiasurf® 7125, 25797, S-Maz® 20, 28640, Sorbon S-20, 29117
Sorbitan monooleate, *See* Alkamuls® SMO, 1686, DeSotan® SMO, 8194, Kemester® S80, 15970, Sorbon S-80, 29121
Sorbitan monooleic acid ester, *See* Witconol 2500, 34501
Sorbitan monopalmitate, *See* Kemester® S40, 15967, Sorbon S-40, 29118
sorbitan monostearate, *See* Alkamuls® S-60, 1677, Kemester® S60, 15968, Radiamuls® Sorb 2145, Sorb 2161, Sorb 2166, 25783, Sorbon S-60, 29119
Sorbitan monotallate, *See* DeSotan® SMT, 8196
Sorbitan O, *See* Witconol 2500, 34501
Sorbitan oleate, *See* Ablunol S-80, 111, Alkamuls® S-80, 1681, Alkamuls® SMO, 1686, Arlacel® 80, 2902
sorbitan oleate, *See* Atmer® 105, 3358, Capmul® O, 5115, Crill 4, 7031, Dehymuls SMO, 7975, Disponil SMO 100 F1, 8698, Emalex SPO-100, 9996, Ethylan® GO80, 11110, Glycomul® O, 13112, Hetan SO, 13889, Kemester® S80, 15970, Liposorb O, 17407, Montane 80, 20269, Montane 481, 20270, Polycon S-80 K, 24400, Prote-sorb SMO, 25279, Radiasurf® 7155, 25802, S-Maz® 80, 28644, Sorbax SMO, 29099, Sorbon S-80, 29121, Witconol 2500, 34501, Alkamuls® SMO, 1686
Sorbitan oleate NF, *See* Span® 80, 29191
Sorbitan palmitate, *See* Ablunol S-40, Arlacel® 40, 2900, Crill 2, 7029, Disponil SMP 100 F1, 8700, Emsorb® 2510, 10460, Kemester® S40, 15967, Liposorb P, 17410, Montane 40, 20266, Prote-sorb SMP, 25280, Radiasurf® 7135, 25798, S-Maz® 40, 28641, Sorbax SMP, 29100, Sorbon S-40, 29118
Sorbitan palmitate NF, *See* Span® 40, 29188
Sorbitan sesquiisostearate, *See* Emalex SPIS-150, 9995
Sorbitan sesquioleate, *See* Arlacel® 83, 2903, Crill 43, 7034, Dehymuis SSO, 7968, Disponil SSO 100 F1, 8704, Emalex SPO-150, 9997, Emsorb® 2502, 10456, Glycomul® SOC, 13115, Liposorb SQO, 17415, S-Maz® 83R, 28645
Sorbitan sesquistearate, *See* Emalex SPE-150S, 9993
Sorbitan stearate, *See* Ablunol S-60, 110, Alkamuls® S-60, 1677, Alkamuls® SMS, 1687, Arlacel® 60, 2901, Capmul® S, 5119, Crill 3, 7030, Dehymuls SMS, 7976, Emalex SPE-100S, 9992, Emsorb® 2505, 10458, Ethylan® GS60, 11112, Famodan MS Kosher, 11423, Grindtek SMS, 13415, Hetan SS, 13890, Hetsorb S-4, 13974, Kemester® S60, 15968, Liposorb S, 17412, Montane 60, 20267, Polycon S-60 K, 24399, Prote-sorb SMS, 25281, Radiamuls® Sorb 2145, Sorb 2161, Sorb 2166, 25783, S-Maz® 60K, 28642, Sorbax SMS, 29101, Sorbon S-60, 29119,
Sorbitan stearate NF, *See* Span® 60, 60K, 29189
Sorbitan tallate, *See* S-Maz® 90, 28647
Sorbitan trioctadecanoate, *See* sorbitan tristearate, 29111
Sorbitan trioleate, *See* Ablunol S-85, 112, Alkamuls® S-85, 1682, Alkamuls® STO, 1689, Arlacel® 85, 2904, Atmer® 106, 3359, Crill 45, 7035, Disponil STO 100 F1, 8705, Emsorb® 2503, 10457, Ethylan® GT85, 11113, Kemester® S85, 15971, Liposorb TO, 17416, Prote-sorb STO, 25282, Radia® 7355, 25738, S-Maz® 85, 28646, Sorbax STO, 29102, Span® 85, 29192, Witconol 2503, 34502
Sorbitan tristearate, 29111, *See* Alkamuls® S-65, 1678, Alkamuls® STS, 1690, Crill 35, 7033, Emsorb® 2507, 10459, Grindtek STS, 13416, Kemester® S65, 15969, Liposorb TS, 17418, Montane 65, 20268, Prote-sorb STS, 25283, Radia® 7345, 25737, Radiamuls® Sorb 2344, Sorb 2345, 25786, *See* Radiamuls® Sorb 2345, 25787, S-Maz® 65K, 28643, Sorbax STS, 29103, Span® 65, 29190
Sorbitan tritallate, *See* S-Maz® 95, 28648
Sorbitan, ester, mono-9-octadecenoate, (Z)-, *See* Witconol 2500, 34501
Sorbitan, esters, tri-9-octadecenoate, (Z,Z,Z)-, *See* Witconol 2503, 34502
Sorbitan, mono-9-octadecenoate, *See* Ablunol S-80, 111, Alkamuls® SMO, 1686, Witconol 2500, 34501
Sorbitan, monododecanoate, *See* Ablunol S-20, 108, Alkamuls® SML, 1685, Sorbon S-20, 29117
Sorbitan, monohexadecanoate, *See* Ablunol S-40, 109, Sorbon S-40, 29118
Sorbitan, monolaurate, *See* Ablunol S-20, 108
Sorbitan, monooctadecanoate, *See* Ablunol S-60, 110, Alkamuls® S-60, 1677
sorbitan, monohexadecanoate, poly(oxy-1,2-ethanediyl) derivatives, *See* polysorbate 40, 24567
Sorbitan, monooctadecanoate, poly(oxy-1,2-ethanediyl) derivs, *See* Alkamuls® T-60, 1692
Sorbitan, monooleate, *See* Ablunol S-80, 111, Radiasurf® 7155, 25802, Witconol 2500, 34501
Sorbitan, monopalmitate, *See* Ablunol S-40, 109
Sorbitan, monostearate, *See* Ablunol S-60, 110
Sorbitan, tri-9-octadecenoate, (Z,Z,Z)-, *See* Ablunol S-85, 112, Alkamuls® S-85, 1682, Witconol 2503, 34502
Sorbitan, tri-9-octadecenoate, poly(oxy-1,2-ethanediyl) derivs, (Z,Z,Z)-, *See* Alkamuls® T-85, 1694
Sorbitan, trioleate, *See* Ablunol S-85, 112, Witconol 2503, 34502
Sorbitan, tris(9-octadecenoate), (Z)-, *See* Witconol 2503, 34502
sorbite, 29112, *See* sorbitol, 29113
D-sorbite, *See* sorbitol, 29113
sorbitol, 29113, *See* A-625/641ABS 301K, ABS 500FR-1, 10, Hydex® 100 Gran.206, 14531, Liponic 70-NC, 17391, Sorban, 29088, Sorbo®, 29115
Sorbitol (EGIC), 29114
D-sorbitol, *See* A-625/641ABS 301K, ABS 500FR-1, 10, sorbitol, 29113
Sorbo, *See* A-625/641ABS 301K, ABS 500FR-1, 10
Sorbo®, 29115
Sorbol, *See* A-625/641ABS 301K, ABS 500FR-1, 10
Sorbolene®, 29116
Sorbon S-20, 29117
Sorbon S-40, 29118
Sorbon S-60, 29119
Sorbon S-66, 29120
Sorbon S-80, 29121
Sorbon T-20, 29122
Sorbon T-40, 29123
Sorbon T-60, 29124
Sorbon T-80, 29125
Sorbon TR 814, 29126
Sorbon TR 843, 29127
Sorbonorit, 29128
sorbose, 29129
L-(-)-sorbose, *See* sorbose, 29129
Sorbosil, 29130
Sorbostryl, *See* A-625/641ABS 301K, ABS 500FR-1, 10
Sorbothane®, 29131
Sorbsil, 29132
Soreflon, 29133
Sorel Cement, 29134
Sorel's gutta-percha substitutes, 29135
Sorensen's salt, 29136
Sorensen's sodium phosphate, *See* disodium phosphate, dihydrate, 8653
Sorenson's phosphate, *See* disodium phosphate, dihydrate, 8653
Sorethytan (20) Mono-oleate, *See* Alkamuls® T-80, 1693
Sorex Golden Fly Bait, 29137
Sorex Super Fly Spray, 29138
Sorex Wasp Nest Destroyer, 29139
Sorexa CD, 29140
Sorexa Plus, 29141
Sorgan, 29142
Sorgen 40, *See* Witconol 2500, 34501
Sorgen 90, *See* Ablunol S-20, 108
Soricin, *See* Soricinol 40, 29143
Soricinol 40, 29143
Sorlate, 29144, *See* Alkamuls® T-80, 1693, Radiasurf® 7157, 25804

Sormetal, *See* Versene CA, 33618
Soromin®, 29145
Soromine AT, 29146
Sorot, *See* Dequalinium Chloride, 8077
Sorpol 320, 29147
Sorpur®, 29148
Sorrel's alloy, 29149
Soruken 90, *See* Ablunol S-20, 108
Sorvall®, 29150
Sotradecol, *See* Niaproof® Anionic Surfactant 4, 21093
Soubieran's ammonical salt, 29151
Soucol, 29152
Soudan coffee, *See* kola nut, 16303
Soudan i, *See* Sudan I, 29938
Souesite, 29153
Souframine, *See* phenothiazine, 23618
soufre, *See* Thiovit, 31724
Soulan's cement, 29154
Sour gas, *See* sulfuretted hydrogen, 30053
South American spp. of *copaifera l.* oil, *See* Jesuit's balsam, 15580
Southalite, 29155
Sovatex C1, 29156
Sovatex EP 5288, 29157
Sovatex IM12H, 29158
Sovatex IM12N, 29159
Sovatex IM17H, 29160
Sovatex IM17N, 29161
Sovatex MP/1, 29162
Sovatex WA, 29163
Sovermol POL 1008, 29164
Sovermol POL 1012, 29165
Sovprene, 29166
Soxhlet's solution, 29167
Soxinal PZ, *See* AAprotect, 16
Soxinol BZ, *See* Butyl Zimate®, 4783, Octocure ZDB-50, 21995, Perkacit® ZDBC, 23295
Soxinol CZ, *See* Vulkacit® CZ/EGC, DZ/EGC, 34129
Soxinol D, *See* Akrochem® DPG, 1168, Dynamine, 9355, Perkacit® DPG, 23281
Soxinol EZ, *See* Ancazate ET, 2415
Soxinol PZ, *See* AAprotect, 16
Soy Flour, 29168
soy oil, *See* soya bean oil, 29169
Soy polysaccharides, *See* Emcosoy®, 10070
Soy polysaccharides, *See* Soy Flour, 29168
Soya acid, *See* Industrene® 126, 15101
soya bean oil, 29169
Soya ethyidimonium ethosulfate, *See* M-Quat® 1033, 20402
Soya methyl ester, *See* methyl ester S, 19528
Soya oil, *See* soya bean oil, 29169
Soya sterol, *See* Generol® 122, 12811
Soya trimethyl ammonium chloride, *See* Adogen® 417, 768, Jet Quat S-50, 15605
Soya trimethyl ammonium chloride - isopropanol, *See* Tomah Q-S, 31999
Soyaalkyl dimethylamine, *See* Armeen® DMSD, 2948
Soyafluff® 200 W, 29170, *See* Soy Flour, 29168
Soyamide DEA, *See* Alkamide® SDO, 1651, Amidex S, 2102, Marlamid® DF 1818, 18741, Stamid LS 5487, 29441
Soyamide DEA (1:1), *See* Mackamide S, 18166, Schercomid SLS, 27660
Soyamidopropalkonium chloride, *See* Quatrex S, 25625
Soyamidopropyl benzyldimonium chloride, *See* Schercoquat SOAB, 27687
Soyamidopropyl betaine, *See* Chembetaine S, 5936
Soyamidopropyl dimethylamine, *See* Chemidex SO, 5971, Mackine 901, 18225
Soyamidopropyl ethyldimonium ethosulfate, *See* Schercoquat SOAS, 27688
Soyamine, *See* Jet Amine PS, 15596
Soy-Amino Quat L/O, 29171
Soyaminopropylamine, *See* Kemamine® D-999, 15916
Soyarich® 115 W, 29172, *See* Soy Flour, 29168
Soybean acidulated soapstock, *See* soya bean oil, 29169
Soybean deodorizer distillate, *See* soya bean oil, 29169
Soybean oil, 29173, *See* soya bean oil, 29169
Soybean oil bleaching, *See* soya bean oil, 29169
Soybean oil deodorization, *See* soya bean oil, 29169
Soybean oil epoxide, *See* ADK CIZER O-130P, 710
Soybean oil fatty acids, glycerol triester, *See* soya bean oil, 29169
Soybean oil hydrogenated, *See* hydrogenated soybean oil, 14596
Soybean oil, bleached, *See* soya bean oil, 29169
Soybean oil, degummed, *See* soya bean oil, 29169
Soybean oil, deodorized, *See* soya bean oil, 29169
Soybean vegetable oil, winter fraction, *See* soya bean oil, 29169
Soy-che, 29174
Soy-Quat C, 29175
Soy-Tein NL, 29176
Soytrimonium chloride, *See* Arquad® S-50, 3120, Jet Quat S-50, 15605
sozoiodolic acid, *See* Iodozol, 15221
SP-33, 29177
SP-731, 29178
SP-1103, *See* Killgerm® Py-Kill W, 16182, tetramethrin, 31384
SP-2205, 29179
SP-2207, 29180
SP-6700, 29181
Spac, 29182
SpaceRite S-11, 29183
Spalerite, 29184
Spallshield®, 29185
Span 20, *See* Ablunol S-20, 108, Alkamuls® SML, 1685, Kemester® S20, 15966, Radiasurf® 7125, 25797, Sorbon S-20, 29117
Span 40, *See* Kemester® S40, 15967, Sorbon S-40, 29118
Span 60, *See* Kemester® S60, 15968, Radiamuls® Sorb 2145, Sorb 2161, Sorb 2166, 25783, Sorbon S-60, 29119
Span 80, *See* Alkamuls® SMO, 1686, Kemester® S80, 15970, Radiasurf® 7155, 25802, Sorbon S-80, 29121, Witconol 2500, 34501
Span 85, *See* Alkamuls® S-85, 1682, Kemester® S85, 15971, Witconol 2503, 34502
Span®, 29186
Span® 20, 29187
Span® 40, 29188
Span® 60, 60K, 29189
Span® 65, 29190
Span® 80, 29191
Span® 85, 29192
Spandofoam, 29193
Spandra Transparent Dressing, 29194
Spangite, 29195
Spanish oxide, 29196, *See* Indian Red, 15068
Spanish saffron, *See* crocus, 7061
Spanish soap, 29197
Spanish turpentine, *See* turpentine, 32610
Spanish white, *See* Bismuth subnitrate, 4243
Span-K, *See* Kaon-Cl, 15729, Kay Ciel, 15807, K-Contin, 15818, K. Tab, 15642, K-Lor, 16226, K-Lyte/C1, 16234, KM Potassium Chloride, 16242, potassium chloride, 24748, 24749
Spannit, 29198, *See* Chlorpyrifos, 6163
Spanscour EFS, 29199
Spanscour GR, 29200
Spanscour N20, 29201
Sparkaloy, 29202
Spark-L® HPG, 29203
Sparkle Silver®, 29204
Sparkle Silver® Premier (SSP), 29205
Sparkle Silvet, 29206
Sparkle Silvex, 29207
Sparkolac, 29208
Sparmite, 29209
Spartakon, *See* Tramisol®, 32134
Spartase, 29210
Sparticide, *See* fluoromide, 12108
Sparticide®, 29211
Spartrix, 29212, *See* carnidazole, 5298
Spasor, 29213, *See* glyphosate, 13148
spathic zinc ore, *See* calamine,
Spauldite, 29214
SPB-1382, *See* resmethrin, 26218
spear pyrites, *See* marcasite, 18700
Specflex, 29215
Special Black 100 Powd, 29216
Special Black 4, 4A Beads and Powd, 29217
Special Extender, 29218
Special Fat 42/44, 29219
Special Oil 107, 29220
Special Oil 619, 29221
Special phosphite, *See* ADK STAB 1500, 717
Specpure, 29222
Spectamine, *See* lofetamine :ss123;ksl hydrochloride, 17553
Spectra, 29223
Spectra®, 29224
Spectraban, 29225
Spectracide, *See* Diazinon Liquid, 8344
Spectracote, 29226
Spectraguard, 29227
Spectra-Pearl®, 29228
spectrar, *See* isopropanol, 15405
Spectra-Sorb UV 9, 29229, *See* Cyasorb® UV 9, 7495, Neo Heliopan® BB, 20867, Rhodialux A, 26675, Uvinul® M-40, 33197
Spectra-Sorb UV 24, 29230
Spectra-Sorb UV 284, 29231, *See* Rhodialux S, 26676, Uval, 33177
Spectra-Sorb UV 531, 29232, *See* Uvinul® 408, 33192
Spectra-Sorb UV 5411, 29233
Spectratech® CM 10540, 29234
Spectratech® CM 10608, 10634, 10778, 10779, 11013, 11056, 11513, 29235
Spectratech® CM 10777, 11045, 11638, 77242, 29236
Spectratech® CM 11014, 11126, 11172, 11174, 11194, 29237
Spectratech® CM 11053, 11489, 11591, 29238
Spectratech® CM 11246, 29239
Spectratech® CM 11340, KM 11264, 29240, *See* Anti-Oxydant Bayer, 2595
Spectratech® CM 11357, 11367, 29241
Spectratech® CM 11698, 29242
Spectratech® FM 1035H, 29243
Spectratech® FM 1150H, 29244
Spectratech® FM 1776H, 29245
Spectrathene, 29246
Spectraveil, 29247
Spectrim, 29248
Spectrim 5, 29249
Spectrim MM 310, 29250
Spectrim Polyurea HF85, 29251
Spectroflux, 29252
Spectromel, 29253
Spectron, 29254
Spectrosol, 29255
Spectro-Sorb UV 24, *See* Cyasorb® UV 24, 7496
Specular hematite, *See* specularite, 29256
specularite, 29256
speculum metal, 29257
Speed X Accelerator, 29258
Speedcure BEDB, 29259
Speedcure BMDS, 29260
Speedcure EDB, 29261
Speedcure ITX, 29262, 29263
Speedway, 29264
SpeeDye, 29265
Speetan SB60, 29266
spelter, 29267
Spenbond®, 29268
Spence metal, 29269
Spenkel®, 29270
Spenkel® F18-M-60, 29271
Spenlite®, 29272
Spenlite® M22-X-40, 29273
Spensol®, 29274

Steapyrium chloride, *See* Emcol® E-607S, 10060
Stearal, 29591
Stearalkonium bentonite, *See* Tixogel VZ, 31885
stearalkonium chloride, 29592, *See* Ammonyx® 4, 4B, 485, 4002, 2260, Amyx A-25-S 0040, 2391, Amyx ST 3837, 2396, Carsoquat® SDQ-25, 5350, *See* Cycloton® SCS, 7542, Emcol® 4, 10044, Incroquat SDQ-25, 15044, Mackernium SDC-25, 18210, M-Quat® B-25, 20404, octadecylbenzene methanaminium chloride, 21985, Quatrex STC-25, 25626, Rhodaquat® M270C/18, 26645, *See* stearalkonium chloride, 29592, Stedbac®, 29609, Sumquat® 6210, 30126, Varisoft® SDAC-W, 33426
stearamide, 29593, *See* Crodamide S, SR, 7098, Kemamide® S, 15900, Petrac® Vyn-Eze®, 23469, Uniwax 1750, 33068
Stearamide DEA, *See* Ablumide SDE, 81, Alkamide® HTDE, 1643, Amldex SE, 2204, Monamid® 718, 20117
Stearamide MEA, *See* Ablumide SME, 82, Amidex SME, 2103, Mackamide SMA, 18167, Rewomid® S 280, 26330, Witcamide® 70, 34442
Stearamide MEA (1:1), *See* Ablumide SME, 82, Mazamide® SMEA, 19114
Stearamide MEA 1:1 Stearamide MEA, *See* Monamid® S, 20122
Stearamide MEA-stearate, *See* Cerasynt® D, 5735, Witcamide® MAS, 34444
Stearamide, N-(2-stearoyloxyethyl)-, *See* Witcamide® MAS, 34444
Stearamide-DEA (1:1), *See* Ablumide SDE, 81
stearamidoethyl diethylamine, 29594, *See* Chemical Base 6532, 5960, Chemidex SE, 5969, Lexamine 22, 17096
Stearamidoethyl ethanolamine, *See* Avistin® PN, 3489, Chemical 39 Base, 5959, Marlamid® A 18, 18738
Stearamidoethyl stearate, *See* Witcamide® MAS, 34444
2-Stearamidoethyl stearate, *See* Witcamide® MAS, 34444
Stearamidopropyl dimethyl-2-hydroxyethyl ammonium dihydrogen phosphate, *See* Cyastat® SP, 7507
Stearamidopropyl dimethyl-2-hydroxyethyl ammonium nitrate, *See* Cyastat® SN, 7506
stearamidopropyl dimethylamine, *See* Adogen® S-18 V, 778, Chemidex S, 5968, Incromine SB, 15014, Lexamine S-13, 17100, Lexate CRC, 17122, Lipamine SPA, 17328, Mackine 301, 18217, Mazeen® S-13, 19124, Miramine® SODI, 19872, Schercodine S, 27641, TEGO Amid S18, 30957
stearamidopropyl dimethylamine lactate, 29595, *See* Emcol® 3780, 10047, Hetamine 5L-25, 13887, Incromate SDL, 14973, Lexamine S-13 Lactate, 17101
Stearamidopropyl ethyl dimonium ethosulfate, *See* Schercoquat SAS, 27686
Stearamidopropyl morpholine, *See* Mackine 321, 18218
Stearamidopropyl PG-dimonium chloride phosphate, *See* Phospholipid PTS, 23724
Stearamidopropyl pyrrolidonylmethyl dimonium chloride, *See* Surfadone WSP, 30348
Stearamidopropylamine oxide, *See* Chemoxide SAO, 5999, Mackamine SAO, 18176
Stearamidopropylamine oxide, *See*
Stearamine, *See* Armeen® 18, 2938, Armeen® 18D, 2939
Stearamine acetate, *See* Acetamin 86, 285
Stearamine oxide, *See* Admox® 18-85, 759, Amyx SO 3734, 2395, Chemoxide ST, 6000, Incromine Oxide S, 15011, Mackamine SO, 18177, Mazox® SDA, 19155, Ninox® SO, 21244, Rewominox S 300, 26335, Rewominoxid S 300, 26338, Schercamox DMS, 27577
Stearamyl, *See* Ablumide SME, 82, Witcamide® 70, 34442
Steardimonium hydroxyethyl cellulose, *See* Crodacel QS, 7067
steareth series, 29596
Steareth-2, *See* Brij® 72, 4518, Brij® 700 S, 4520, Brij® 721, 4521, Hodag Nonionic S-2, 14197, Lipocol S-2, 17358, Macol® SA-2, 18273, Prox-onic SA-1/02, 25336, Rhodasurf® S-2, 26661, Volpo S-2/S-10/S-20, 34030
Steareth-5, *See* Solulan® 5, 29000
Steareth-7, *See* Ablunol SA-7, 113, Emulgator E 2149, 10499
Steareth-7 carboxylic acid, *See* Akypo RS 60, 1243
Steareth-11 carboxylic acid, *See* Akypo RS 100, 1242
Steareth-20, *See* Lipowax PR, 17445, Macol® CPS, 18263, Ritachol® 1000, 26803, Trycol® 5888, 32500
Steareth-21, *See* Ritox 721, 26853
steareth-25, *See* Tego® Care 150, Care 300, 30967
Steareth-40, *See* Emalex 640, 9925, steareth series, 29596
Steareth-200, *See* Rhodasurf® TB-970, 26663
Stearex, 29597
steargillite, 29598
stearic acid, 29599, *See* Emersol® 110, 10118, Emersol® 6349, 10124, Emery® 400, 10130, Glycon® S-90, 13121, Glycon® TP, 13122, Hydrofol Acid 1655, 14588, Hystrene® 4516, 14758, Hystrene® 7018 FG, 14762, Industrene® 4518, 15107, Industrene® 5016, 15108, Industrene® 7018 FG, 15109, Industrene® R, 15111, Petrac® 270, 23462, Prifrac 2980, 24955, Pristerene 4904, 25040
Stearic acid aluminum salt, *See* aluminum stearate, 1929, Synpro® 404, 30593
stearic acid amide, *See* stearamide, 29593
stearic acid, calcium salt, *See* calcium stearate, 4970
Stearic acid diamide, *See* Trymeen® 6657, 32525
Stearic acid diethanolamide, *See* Ablumide SDE, 81
stearic acid methyl ester, *See* Kemester® 4516, 15944
Stearic acid monoethanolamide, *See* Witcamide® 70, 34442
Stearic acid NF (92%), *See* Hystrene® 9718 NF FG, 14768
stearic acid potassium salt, *See* potassium stearate, 24783
stearic acid sodium salt, *See* sodium stearate, 28820
Stearic acid, ester with lactic acid bimol. ester calcium salt, *See* Verv®, 33644
stearic acid, ester with lactate of lactic acid, calcium salt, *See* Crolactil CSL, 7184
Stearic acid, 2-ethylhexyl ester, *See* Wickenol® 156, 34392
Stearic acid, isopropyl ester, *See* Wickenol® 127, 34380
stearic acid, lead salt, *See* lead stearate, 16917
Stearic acid, monoester with 1,2-propanediol, *See* Witconol 2380, 34498
stearic acid, potassium salt, *See* potassium stearate, 24784
stearic acid, sodium salt, *See* sodium stearate, 28820
Stearic diethanolamide, *See* Ablumide SDE, 81
Stearic ethanolamide, *See* Ablumide SME, 82, Witcamide® 70, 34442
Stearic ethylolamide, *See* Ablumide SME, 82, Witcamide® 70, 34442
Stearic imidazoline, *See* Crodazoline O, 7161, Crodazoline S, 7163
Stearic monoethanolamide, *See* Witcamide® 70, 34442
stearin, *See* Kemester® 5500, 15947, Kemester® 6000, 15952, Kemester® GMS (Powd.), 15962
stearin pitch, 29600
Stearite, 29601, *See* Olympic, 22146
stearix orange, *See* Sudan I, 29938
Stearol, *See* Stearal, 29591
Stearol S, *See* Adol® 62 NF, 783
stearone, *See* Amerchol® 400, 1996
stearophanic acid, *See* stearic acid, 29599
stearopodis, 29602
stearosan, 29603
Stearoxydimethicone, *See* Abil® Wax 2434, 58, Belsil SDM 6021, 3920
N-(Stearoyl colamine formyl methyl) pyridinium chloride, *See* Witcamine® E-607S, 34451
Stearoyl Diethanolamide, *See* Ablumide SDE, 81
Stearoyl lactylates, *See* Softex, 28853
Stearoyl monoethanolamide, *See* Ablumide SME, 82, Witcamide® 70, 34442
Stearoyl sarcosine, *See* Hamposyl® S, 13605
stearoylamide, *See* stearamide, 29593
N-Stearoyl-*p*-amino phenol, *See* Suconox-18®, 29926
N-Stearyl 12-hydroxystearamide, *See* Paricin® 210, 22837
N-Stearoylethanolamine, *See* Ablumide SME, 82
Stearoylethanolamide, *See* Ablumide SME, 82, Witcamide® 70, 34442
N-Stearoylethanolamine, *See* Witcamide® 70, 34442
N-Stearoyl diethanolamine, *See* Ablumide SDE, 81
Steartrimonium chloride, *See* Arquad® 18-50, 3109, Tomah Q-ST-50, 32000, Varisoft® ST-50, TSC, 33427, Varisoft® TSC, 33431
Stearyl 12-hydroxystearate, *See* Paricin® 18, 22836
stearyl alcohol, 29604, *See* Crodacol S70, 7071, Dehydag Wax 18, 7945, Emery® 3343, 10173, Fancol SA, 11451, Lanette 18 DEO, 16664, Lipocol S, 17357, Lorol C18, 17615, Rita SA, 26793, Stearal, 29591, Varonic® BG, 33440, Emulgator E 2149, 10499, Adol® 62 NF, 783, Stearal, 29591, Tego® Care 150, Care 300, 30967
Stearyl alcohol and ceteareth-20, *See* Macol® 125, 18259
Stearyl alcohol and cetrimonium bromide, *See* Miracare® CT 100, 19854
Stearyl alcohol NF, *See* Crodacol S95NF, 7072
Stearyl alcohol polyglycol ether, *See* Genapol® S-020, 12796
Stearyl alcohol USP, *See* Cachalot® S-56, 4881
Stearyl alcohol, cetrimonium chloride, *See* Quatrex CT-100, 25623
Stearyl alcohol-ceteareth-20, *See* Cosmowax K, 6878, Fanwax G, 11467
stearyl citrate, *See* Morflex MSC, 20331
Stearyl dimethicone, *See* Abil® Wax 9800, 61
Stearyl dimethyl amine, *See* Dymanthine, 9327
Stearyl dimethyl amine oxide, *See* Barlox® 18S, 3652
Stearyl dimethyl benzyl ammonium chloride, *See* Ablumine 280, 85, Catinal OB-80E, 5458
Stearyl dimethylamine, *See* Crodamine 3.A18D, 7107
stearyl dimethylbenzylammonium chloride, *See* octadecylbenzenemethanaminium chloride, 21985
stearyl erucamide, 29605, *See* Chemstat® HTSA#3, 6041, HTSA#3, 14446, Kemamide® E-180, 15896, N-octadecyl-13-docosenamide, 21986
Stearyl erucate, *See* Schercemol SE, 27622
Stearyl glucoside, *See* Tego® Care 450, 30968
Stearyl heptanoate, *See* Crodamol W, 7127, Tegosoft® SH, 31028
stearyl Hydroxyethyl imidazoline, 29606, *See* Miramine® GS, 19869
Stearyl imidazoline, *See* Schercozoline S, 27716, stearyl Hydroxyethyl imidazoline, 29606
Stearyl lactyl-2-lactylates, *See* Lisat, 17491
stearyl methacrylate, 29607, *See* Ageflex FM-68, 988, Ageflex FM-1620, 989, octadecyl methacrylate, 21984
Stearyl methicone, *See* Abil® Wax 9809, 63
stearyl stearamide, *See* Kemamide® S-180, 15901
Stearyl stearate, *See* Alkamuls® SS, 1688, Emalex CC-18, 9936, Hetester 412, 13892, Lexol SS, 17155, Liponate SS, 17388, Radia® 7501, 25743, Ritachol® SS, 26804
Stearyl stearoyl stearate, *See* Hetester SSS, 13904
Stearyl Trimethyl Ammonium Chloride, *See* Arquad® 18-50, 3109
stearyl/lauryl thiodipropionate, 29608
Stearyl-2-lactylic acid calcium salt, *See* Verv®, 33644
Stearylamine, *See* Armid® HTD, 2963, Crodamine 1.18D, 7102, Amine 18-90, 2131, Steamfilm FG, 29589, Kemamine® P-990, P-990D, 15922, Armeen® 18, 2938, Armid® HTD, 2963
Stearylamphopolycarboxyglycinate, *See* Ampholak 7TX/C, 2314
Stearyldimethylammonium chloride, *See* Onamine 18, 22167
stearyldimethylbenzylammonium chloride, *See*

Cycloton® SCS, 7542, stearalkonium chloride, 29592
Stearyldimethylethyl ammonium ethosulfate, *See* Larostat® 451, 16797
Stearyltrimethylammonium bromide, *See* Zeonet B, 34721
Steatite, *See* ABT-2500®, 166, talc, 30731
Stebac, *See* Cycloton® SCS, 7542, octadecylbenzenemethanaminium chloride, 21985, stearalkonium chloride, 29592, Sumquat® 6210, 30126
Steclin, *See* tetracycline, 31365
Stedbac, *See* Cycloton® SCS, 7542, octadecylbenzenemethanaminium chloride, 21985, stearalkonium chloride, 29592, Sumquat® 6210, 30126
Stedbac®, 29609
steel, 29610
Steel 01, 29611
Steel A2, 29612
Steel B4, 29613
Steel bronze, *See* Uchatius bronze, 32719
Steel C, 29614
Steel E, 29615
Steel F, 29616
Steel Guard, 29617
Steel H, 29618
Steel I, 29619
Steel M, 29620
Steel N, 29621
Steel N1, 29622
Steel O, 29623
steel ore, 29624
Steel P, 29625
Steel R, 29626
Steel S1, 29627
Steel T, 29628
Steel U, 29629
Steel V, 29630
Steel V2A, 29631
Steel W, 29632
Steel W2, 29633
steelite, 29634
Steinamid DC 2129, *See* Active #2, 501
Steinamid DC 2129e, *See* Active #2, 501
Steinamid DI 203 S, *See* Alkamide® 327, 1634
Steinamid DL 203 S, *See* Ablumide LDE, 79
Steinamid IPE 280, *See* Witcamide® 61, 34441
Steinamid L 203, *See* Ablumide LME, 80
steinapol NLS 90, *See* Rhodapon® 101-10, 26631
Steinapol TLS-40, *See* Witcolate TLS-500, 34479
Steinaryl NKS 50, *See* Nacconol® 40G, 20647, sodium dodecylbenzenesulfonate, 28760
Steinazid SBU 185, 29635
Steinbühl yellow, *See* Barium chromate, 3638
Steladone, *See* Chlorfenvinphos, 6098
Stelex, 29636
Stellak, 29637
Stellar 500, 29638
Stellite®, 29639
Stellited metal, 29640
Stellos, 29641
Stellox 380EC, 29642
Stelogen, 29643
Stelopack, 29644
Stelorit, 29645
Stelotol, 29646
Stelpur, 29647
Stempor DG, 29648
Stenol, *See* Adol® 62 NF, 783, Stearal, 29591
Stenol 1618, 29649
Stenorol, 29650, *See* halofuginone hydrobromide, 13562
Stentor Steel, 29651
Steol 3OS, 29652
Steol 4N, 29653
Steol 7T, 29654
Steol CA-460 and KA-460, 29655
Steol CS-460 and KS-460, 29656
Steol CS-760 and 7N, 29657
Steol FA, 29658
Steol®, 29659
Steol® 4N, 29660
Steol® CA-460, 29661, *See* Ammonium laureth sulfate, 2240
Steol® COS 433, 29662
Steol® CS-130, 29663
Steol® OS 28, 29664
Stepan, 29665
Stepan C-40, 29666
Stepan C-65, 29667
Stepan C-68, 29668
Stepan D-50, *See* Wickenol® 101, 105, 34378
Stepan D-70, *See* Wickenol® 111, 34379
Stepan DS 60, *See* Nacconol® 40G, 20647, sodium dodecylbenzenesulfonate, 28760
Stepan Pearl Series, 29669
Stepan TAB® -2, 29670
Stepanate X, *See* Eltesol® SX 30, 9873, Naxonate® 4L, 20824, Pilot SXS-40, 23843, Witconate SXS 40%, 34492
Stepanate®, 29671
Stepanate® AXS, 29672, *See* Ammonium xylene sulfonate, 2259
Stepanate® SCS, 29673, *See* sodium cumene sulfonate, 28752
Stepanate® SXS, 29674, *See* sodium xylene sulfonate, 28832
Stepanflo, 29675
Stepanflote® 85L, 29676
Stepanform® 1440, 29677
Stepanform® 1750, 29678
Stepanhold® Extra, 29679
Stepanhold® R-1, 29680
Stepan-Mild® LSB, 29681
Stepan-Mild® SL3, 29682
STEPAN-MILD® SL3, *See* disodium lauryl sulfosuccinate, 8649
Stepanol AM, 29683
Stepanol DEA, 29684
Stepanol ME, 29685, *See* Rhodapon® 101-10, 26631
Stepanol MG, 29686
Stepanol WA, WAC, WAQ, 29687
Stepanol WA-100, 29688
Stepanol WAT, 29689, *See* Witcolate TLS-500, 34479
Stepanol®, 29690
Stepanol® AEG, 29691
Stepanol® AEM, 29692
Stepanol® AM, 29693, *See* Ammonium lauryl sulfate, 2242
Stepanol® LX, 29694
Stepanol® ME Dry, 29695
Stepanol® MG, 29696, *See* magnesium lauryl sulfate, 18364
Stepanol® SPT, 29697
Stepanol® WA Extra, 29698
Stepanon CG, 29699
Stepanquat® 6585, 29700
Stepantan A, 29701
Stepantan AM, *See* Witcolate NH, 34478
Stepantan NP 80, 29702
Stepantan®, 29703
Stepantan® AS-12, 29704
Stepantan® DS-40, 29705
Stepantan® DT-60, 29706
Stepantan® H-100, 29707
Stepantex Q90B, 29708
Stepantex® B-29, 29709
Stepantex® CO-30, 29710
Stepantex® TD14, 29711
Stepantex® VS 90, 29712
Stepfac® 8170, 29713
Stepfac® 8171, 29714
Stepfac® PN 10, 29715
Step-Flow 21, 29716
Steposol®, 29717
Steposol® CA-207, 29719
Steposol® CA-60H, 29718
Stepsperse® DF-100, 29720
Stepwet® DF-60, 29721
Steraffine, *See* Stearal, 29591
Steral, *See* hexachlorophene, 13992
Steraskin, *See* hexachlorophene, 13992
Sterbon, *See* Carbora, 5235
Stercofuge, *See* Alginic acid potassium salt, 1594, Kelmar®, 15856, potassium alginate, 24735
Sterculia gum, *See* karaya gum, 15760
Sterculia Kernals, *See* olives of Java, 22141
Sterculia kernels, *See* olives of Java, 22141
Stereon® 840A, 29722
Stereon® 881, 29723
Stereon® 900, 29724
stereotype plate, 29725
steresol, 29726
Sterethox, 29727
Sterets Pre-Injection Swabs®, 29728
Steribath, 29729
Steridex, 29730
Steriflux, 29731
SteriLine 200, 29732
SteriLine 665, 29733
Sterilite Hop Defoliant, 29734
Sterilite®, 29735
Sterillium, 29736
Sterisafe, 29737
Steriseal #40, *See* Aquatreat SDM, 2761, methyl namate®, 19551, Octopol SDM-40, 22014, Perkacit® SDMC, 23288, sodium dimethyldithiocarbamate, 28758, Thiostop N, 31717
Sterisheen, 29738
Sterisil, 29739
Sterisol, *See* hexedine, 14027
sterisol hand disinfectant, *See* isopropanol, 15405
Steritile® Plus, 29740
Steriwipe®, 29741
Sterline, 29742
Sterling, 29743
Sterling AM, *See* Witcolate NH, 34478
sterling gold, *See* standard gold, 29485
sterling silver, *See* standard silver, 29486
sterling solder, 29744
sterling wa paste, *See* Rhodapon® 101-10, 26631
Sterling wat, *See* Perlankrol® ATL40, 23324, Rhodapon® LT-6, 26638, Standapol® T, 29481, Witcolate TLS-500, 34479
Sterlite, 29745
Sterlith, 29746
Sternite, 29747, 29748
Stero WW, 29749
Sterocoll, 29750
Steron 210, 29751
Sterotabs, 29752
Sterotex, *See* hydrogenated soybean oil, 14596
Sterotex®, 29753
Sterotex® HM, 29754
Sterox, 29755, *See* Alkasurf® NP-4, 1715, Triton® N-57, 32430
Sterox DF, DJ, 29756
Steroxin-Hydrocortisone, 29757
Steroxol, 29758
Sterpon, 29759
Sterretite, *See* aluminum phosphate, 1922
Sterro metal, *See* Aich metal, 1095
Ster-Zac, 29760
STFF, *See* sodium tripolyphosphate, 28828
Stibic anhydride, *See* Timonox, 31806
stibium, 29761, *See* antimony, 2567
Stick-lac, *See* lac, 16547
Stickstoffoxydbaryt, 29762
Stik-It, 29763
stilbene, 29764, *See* Eccobrite RB, 9524
***cis*-stilbene**, 29765
***trans*-stilbene**, *See* toluylene, 31960
Stillingia oil, 29766
Stimufol, 29767
Sting, 29768, *See* glyphosate, 13148
Stink Damp, *See* sulfuretted hydrogen, 30053
stink quartz, 29769
stink-stone, 29770

Stipolac, *See* keraphen, 16055
Stir & Sperse®, *See* corn starch, 6837
Stirene, 29771
Stiresol®, 29772
Stirling's gentian violet, 29773
Stirrup-A/WF, *See* Farnesol, 11492
Stirrup-CRW, *See* Farnesol, 11492
Stirrup-H, *See* Farnesol, 11492
Stirrup-HB, *See* Farnesol, 11492
Stirrup-TPW, *See* Farnesol, 11492
Stockade, *See* cypermethrin, 7588, Topclip Parasol, 32037, permethrin, 23384, 23385, 23386
Stockalite, 29774
Stockholm, 29775
Stockholm pitch, 29776
Stoco, 29777
Stoddard Solvent, 29778, *See* Kensol 30, 16038
Stoddard solvent, Texsolve S, *See* Varsol® 1, 33456
Stoffertite, 29779
stoic metal, 29780
Stoke's reagent, 29781
Stoller Flowable Sulphur, 29782
Stomahesive, 29783
Stomoxin, *See* permethrin, 23384, 23385, 23386
Stomp, 29784, *See* pendimethalin, 23102
Stomp H, 29785
stone green, *See* Bohemian earth, 4368
Stone grey, *See* slate grey, 28594
stone wax, 29786
stone, green, *See* Bohemian earth, 4368
Stoner A500, 29787
Stoner E800, 29788
Stoner K206, 29789
stone's bronze, 29790
Stonite, 29791
Stoodite, 29792
Stopetyl, *See* Disulfiram, 8725
Stopmold B, *See* Dowicide A, 8932
Stop-scald, *See* Polyflex, 24435
Stora, 29793
Storaid Dust, Storite SS, 29794
Storalon, *See* Carbora, 5235
storax calamita, 29795
Storite, 29796
Storm, *See* Bentazon, 3947
Storm®, 29797
Stortex, 29798
Stowite, 29799
Stoxil, *See* Kerecid, 16067
Stipend, *See* Chlorpyrifos, 6163
STPP, *See* sodium tripolyphosphate, 28828
Straight chain dodecylbenzene sulfonic acid, *See* Arylan® SBC Acid, 3159
Straight-chain dodecylbenzene, *See* Marlican®, 18845
Strandex, 29800
Strandol, 29801
strass, 29802
Strassburg turpentine, 29803
Strasser solder, 29804
Strata-Fire, 29805
Stratos, *See* Cycloxydim, 7544
Stratos®, 29806
Stratton, 29807
Stratyl, 29808
Strawlink, 29809
Strazine, *See* atrazine, 3394
Strel, *See* Propanil, 25174
Strelax, 29810
Strenes Metal, 29811
Strepsils, 29812
Streptase, *See* Kabikinase, 15650
Streptococcal fibrinolysin, *See* Kabikinase, 15650
Streptokinase, *See* Kabikinase, 15650
streptomycin, dihydrostreptomycin, *See* Dimycin, 8537
Stresnil, 29813, *See* azaperone, 3518
strewing smalt, 29814
Striadine, *See* Adenosine triphosphate, 681
Strim, 29815
Stripcote, 29816
stripping salt, *See* Abraum salts, 156
Strite, 29817
Strobane, 29818
Strodex® MO-100, 29819
Strodex® MOK-70, 29820
Strodex® P-100, 29821
Strodex® PK-90, 29822
Strodex® SE-100, 29823
Strodex® Super V-8, 29824
Stronscan-85, 29825
strontia, 29826
strontium, 29827
strontium carbonate, 29828
strontium chloride, 29829
strontium chloride hexahydrate, *See* Stronscan-85, 29825, strontium chloride, 29829
strontium chloride-$_{85}$Sr, *See* Stronscan-85, 29825
strontium monoxide, *See* strontia, 29826
strontium nitrate, 29830
Strontium nitrate $_{85}$Sr, *See* Strotope, 29832
Strontium oxide, *See* strontia, 29826
strontium titanate, 29831
strontium titanium oxide, *See* strontium titanate, 29831
Strotope, 29832
Struktol® 40 MS Flakes, 29833
Struktol® Activator 73, 29834
Struktol® HP 55, 29835
Struktol® PE H-100, 29836
Struktol® SU 109, 29837
Struktol® T.M.Q, 29838
Struktol® WB 300, 29839
struthiin, 29840
Stryden Forte Rapid, 29841
Stryene-butadiene rubber, *See* A-0020, 4
STS, *See* sorbitan tristearate, 29111
ST-Size, 29842
stucco, 29843
studafluor, *See* sodium fluoride, 28764
Stuk, 29844
stupp, 29845
Sturcal®, 29846
Sturcarb, 29847
Stycast® 1090, 29848
Stycast® 1210, 29849
Stycast® 1266, 29850
Stycast® 1467, 29851
Stycast® 2850-FT, 29852
Stycond-109, 29853
Stygene Series, 29854
Stylac® ABS, 29855, *See* Acrylonitrile-butadiene-styrene, 439
Stylac® AS, 29856
Stylex 72001 NA, *See* Polyphenylene oxide, 24518
Styphen I, 29857
styphnic acid, 29858
Styquin, 29859, *See* butamisole hydrochloride, 4733
styracin, 29860
Styrafil®, 29861
Styraloy 22, 22A, 29862
Styramic H.T. and M.T, 29863
Styrenated diphenylamine, *See* Lowinox® SDA, 17674
Styrenated phenol, *See* Agerite® Spar, 1021, Wingstay® S, 34415
styrene, *See* Acralen® BS, 377, styrol, 29875
styrene glycol, *See* styrolyl alcohol, 29878
styrene latex, *See* Acralen® ATR, 376
styrene monomer, *See* styrol, 29875
styrene polymer, *See* polystyrene, 24601
Styrene-acrylic, *See* Ucar® Vehicle 451, 32710
styrene-acrylonitrile copolymer, 29864
styrene-butadiene rubber, *See* KR01 K-Resin Polymer, 16382, KR04 K-Resin Polymer, 16383, Kraton®, 16386, Kraton® D 1101, 16387, Kraton® D 1116, 16389, Kraton® D 2103, 16391, K-Resin Polymer KR01, 16404, K-Resin Polymer KR03, 16405, K-Resin Polymer KR04, 16406, K-Resin Polymer KR05, 16407, K-Resin Polymer KR10, 16408, Krylene® 606, 16449, Krylene® 608, 16450
Styresol 13-031, 29865
Styrid, 29866
Styrocell, 29867
Styrochrom®, 29868
Styrodur®, 29869
Styrofan®, 29870
Styrofill®, 29871
Styroflex, 29872
Styrofoam Brand Insulation, 29873
styrogallol, 29874
styrol, 29875
styrolene, *See* styrol, 29875
Styrolit®, 29876
Styrolux®, 29877
styrolyl alcohol, 29878
Styromol, 29879
Styron, 29880, *See* styrol, 29875
Styron 421, 29881
Styron 478, 29882
Styron 479, 29883
Styron 697, 29884
Styron 6075, 29885
Styron 6087 SF, 29886
Styronal®, 29887
styrone, 29888, *See* cinnamyl alcohol, 6315
Styroplus®, 29889
Styropol, *See* styrol, 29875
Styropor, 29890, *See* styrol, 29875
Styropor® F, 29891
Styropor® FH, 29892
Styropor® P, 29893
Styrothane 5329/5330, 29894
styryl alcohol, *See* styrolyl alcohol, 29878
Styrylcarbinol, *See* cinnamyl alcohol, 6315
3-(N-Styrylmethyl-2-aminoethylamino) propyltrimethoxy silane hydrochloride, *See* CS1590, 7307
Styvex 22000 NA, 29895
Styvex 32000 BK, 29896
Styvex 40007 BKL2, 29897
Styvex 40007 BKL2, *See* Acrylonitrile-butadiene-styrene, 439
Styvex 42023 NAFR, 29898
Styvex 42023 NAFR, *See* Acrylonitrile-butadiene-styrene, 439
Styvex 72001 NA, 29899
Styxol, *See* Uba, 32698
Su 1906, *See* thiambutosine, 31663
Südflock, 29900
SU-4885, *See* metyrapone, 19608
Suakin gum, 29901
subacetate of lead, 29902
Subdue, *See* metalaxyl, 19430
suberic acid, 29903
Subitol, *See* Perichthol, 23270
Sublaprint, 29904
sublimate, *See* mercuric chloride, 19347
sublimed blue lead, 29905, 29906
sublimed white lead, 29907
sublimoform, 29908
subox, 29909
substitute of tartar, *See* superargol, 30217
Substituted alkylamine derived lanolin acids, *See* Cromeen, 7197
Substituted benzophenone, *See* Lankromark® LE230, 16696
Subtosan, *See* Tears Plus®, 30876, Videne Disinfectant Solution, Videne Disinfectant Tincture, Videne Powder, 33739, Vinisil, 33827
Suburban Propane (and Design), 29910
Sub-Vitralen, 29911
Sucaryl, 29912, 29913
succaril, *See* saccharin sodium, 27196
Succimer, *See* MPI DMSA Kidney Reagent, 20393
Succinchlorimide, *See* N-chlorosuccinimide, 6148
Succinellite, 29914
succinic acid, 29915
Succinic acid, *See* acid of amber, 334, sal succini, 27256, salt of amber, 27283, Succinellite, 29914

succinic acid anhydride, *See* succinic anhydride, 29916
Succinic acid, bis(2-ethylhexyl) ester, *See* Wickenol® 159, 34394
succinic anhydride, 29916
succinimide, 29917, *See* Lubrizol® 2153, 17745
succinite, 29918
succinol, 29919
Succinoxate, *See* Alphogen, 1839
Succinyl peroxide, *See* Alphogen, 1839, succinic anhydride, 29916
Succinylcholine chloride, *See* Quelicin, 25642, Sucostrin, 29927
succinyloxide, *See* succinic anhydride, 29916
Suchar, 29920
Sucker Plucker Concentrate, 29921
Sucker Stuff, 29922
Sucline, *See* sucramine, 29928
sucol b, *See* 1,4-butanediol, 4735, Dabco® BDO, 7647
Suconox-4®, 29923
Suconox-9®, 29924
Suconox-12®, 29925
Suconox-18®, 29926
Sucostrin, 29927
Sucra, *See* saccharin sodium, 27196
sucramine, 29928
sucrase, 29929
sucrate of hydrocarbonate of lime, 29930, *See* sucro-carbonate of lime, 29933
sucre edulcor, *See* saccharin, 27195
Sucrene, *See* dulcine, Dulcin, 9091
sucrette, *See* saccharin, 27195
Sucro Ester 7, 29931, *See* Grilloten PSE141G, 13378
Sucro Ester 15, 29932, *See* Grilloten PSE141G, 13378
sucro-carbonate of lime, 29933
sucrol, *See* dulcine, Dulcin, 9091
sucro-levulose, *See* levulose, 17085
Sucrosa, *See* sodium cyclamate, 28754
sucrose, 29934
Sucrose cocoate, *See* Crodesta SL-40, 7170, Grilloten LSE87K, 13377
Sucrose distearate, *See* Crodesta F-10, 7168, Ryoto Sugar Ester S-570, S-770, 27171
Sucrose di, tristearate, *See* Ryoto Sugar Ester S-170, 27170
Sucrose laurate, *See* Grilloten LSE87, 13376, Ryoto Sugar Ester LWA-1570, 27166
Sucrose mono/di/tri palmitic/stearic acid, *See* Crodesta DKS F110, 7167
Sucrose monolaurate, *See* Ryoto Sugar Ester LWA-1570, 27166
Sucrose monostearate, *See* Crodesta F-160, 7169
sucrose octa-acetate, 29935
D-(+)-sucrose octa-acetate, *See* sucrose octa-acetate, 29935
Sucrose oleate, *See* Ryoto Sugar Ester OWA-1570, 27167
Sucrose palmitate, *See* Ryoto Sugar Ester P-1570, P-1670, 27168
Sucrose ricinoleate, *See* Grilloten ZT40, ZT80, PSE 141G, LSE 87, LSE 87K, 13379
Sucrose stearate, *See* Grilloten PSE141G, 13378, Ryoto Sugar Ester S-1170, 27169
Sucrun 7, *See* sodium cyclamate, 28754
suction gum, 29936
Suction powder, *See* suction gum, 29936
Sudafed Nasal Spray, 29937
Sudan I, 29938
Sudan II, 29939
Sudan III, 29940
Sudan IV, 29941
Sudan j, *See* Sudan I, 29938
Sudan orange r, *See* Sudan I, 29938
Sudan Red, *See* Sudan III, 29940
Sudan Red III, *See* Sudan III, 29940
Sudan®, 29942
Sudermo, *See* Odylen®, 22046
Sudol®, 29943
Sudranol, 29944
Suessette, *See* sodium cyclamate, 28754
Suestamin, *See* sodium cyclamate, 28754
Sufatone SCS/B, 29945
Sufatone SCS/CL, 29946
Sufatone SMC/L, 29947
Sufatone SMC/W, 29948
Sufenax CB, *See* CBTS, 5484, Delac S, 8002, Durax®, 9181, Perkacit® CBS, 23277, Santocure, 27400
Suffa, 29949
Suffa, *See* Thiovit, 31724
Suffix BW, *See* flamprop-isopropyl, 11887, Gunner, 13493, Power Flame, 24824
SufuSorb 8, 29950
sugamo, 29951
sugar, *See* sucrose, 29934
sugar cane wax, 29952
sugar charcoal, 29953
sugar house black, 29954
sugar of lead, 29955, *See* lead diacetate, 16906, lead acetate, 16902
Sugar of Saturn, *See* salt of Saturn, 27287
Sugarin, *See* sodium cyclamate, 28754
Sugaron, *See* sodium cyclamate, 28754
Sugartab®, 29956
Suhler white copper, 29957
Suicalm, *See* azaperone, 3518
Suiminth, *See* morantel tartrate, 20307
Sulbenox, *See* Vigazoo, 33749
Sulconazole, 29958
sulfabenzpyrazine, *See* sulfaquinoxaline, 29968
sulfacetamide sodium, *See* Isopto Cetamide, 15415
Sulfacide, 29959
sulfadimerazine, *See* sulfamethazine, 29962
sulfadimethylpyrimidine, *See* sulfamethazine, 29962
sulfadimidine, *See* sulfamethazine, 29962
Sulfadine, *See* sulfamethazine, 29962
Sulfads®, 29960
sulfa-hitech, *See* Sulfa-Hitech® 0382, 29961
Sulfa-Hitech® 0382, 29961
Sulfalene, *See* Kelfizina, 15838
sulfamethazine, 29962
sulfamethopyrazine, *See* Kelfizina, 15838
sulfamethoxypyrazine, *See* Kelfizina, 15838
sulfamethylphenazole, *See* Vesulong, 33700
sulfamexathine, *See* sulfamethazine, 29962
sulfamic acid, 29963
Sulfamic acid, ammonium salt, *See* Ammonium sulfamate, 2254
Sulfamic Acid, Monoammonium Salt, *See* Ammonium sulfamate, 2254
Sulfamic acid, nickel(2+) salt (2:1), *See* nickel sulfamate, 21134, Nimate, 21201
Sulfamidic acid, *See* sulfamic acid, 29963
sulfamidine, *See* sulfamethazine, 29962
Sulfamin, 29964
sulfammonium, 29965
2-sulfanilamido-3-methoxypyrazine, *See* Kelfizina, 15838
sulfanilic acid, 29966
***m*-sulfanilic acid**, *See* metanilic acid, 19445
Sulfanol, 29967
Sulfanona-mae, *See* Dapsone, 7751
sulfapol, *See* sodium dodecylbenzenesulfonate, 28760, Nacconol® 40G, 20647
sulfapyrazinemethoxyine, *See* Kelfizina, 15838
sulfapyrazole, *See* Vesulong, 33700
sulfaquinoxaline, 29968
sulfaquinoxaline sodium, *See* Embazin, 10022
Sulfarine, 29969
Sulfasan, 29970
Sulfasan R, *See* Akrochem® Accelerator R, 1154, Naugex SD-1, 20801, Sulfasan, 29970
Sulfate alkyl phenol ethoxylate, *See* Eccoscour D-7, 9549
sulfate pulp, 29971
Sulfate Resin, *See* Liquid resins, 17475
Sulfate, Ammonium, *See* Ammonium sulfate, 2255
Sulfated blend of oils, sodium neutralized, *See* Actrasol KAP, 552
Sulfated butyl oleate, *See* Chemax SBO, 5925, Chemsulf SBO/65, 6045
sulfated butyl tallate, 29972, *See* Laurel SBT, 16838, sulfated butyl tallate, 29972
Sulfated castor oil, *See* Eureka 102, 11207
Sulfated caster oil, sodium neutralized, *See* Actrasol 167A, 547, Actrasol C-50, C-75, C-85, 549
Sulfated ester, *See* Duofol T, 9105
Sulfated glyceryl trioleate, sodium neutralized, *See* Actrasol EO, 551
Sulfated methyl ester of soya fatty acid, sodium neutralized, *See* Actrasol MY-75, 553
sulfated oils, *See* Turkey red oils, 32607
Sulfated oleic acid ammonium neutralized, *See* Actrasol SR75, 559
Sulfated oleic acid potassium neutralized, *See* Actrasol SRK 75, 560
Sulfated rapeseed oil, *See* Actrasol 6092, 548, Laurel SRO, 16843
Sulfated ricinoleic acid, potassium neutralized, *See* Actrasol PSR, 555
Sulfated soyabean oil, sodium neutralized, *See* Actrasol OY-75, 554
Sulfated tall oil, *See* Actrasol SS, 561
Sulfated tall oil, potassium neutralized, *See* Actrasol SP 175K, 558
Sulfatine, 29973
Sulfato de Cobre Valles, 29974
Sulfatol CL, 29975
Sulfatol E3, 29976
Sulfatol LS3, 29977
Sulfatol LX/B, 29978
Sulfatol PD/B, 29979
Sulfatol TL/B, 29980
Sulfated butyl oleate, sodium neutralized, *See* Actrasol SBO, 556
Sulfenamide Ts, *See* Vulkacit® CZ/EGC, DZ/EGC, 34129
Sulfenax, *See* Vulkacit® CZ/EGC, DZ/EGC, 34129
Sulfenax CB, *See* Vulkacit® CZ/EGC, DZ/EGC, 34129
Sulfenax CB 30, *See* Vulkacit® CZ/EGC, DZ/EGC, 34129
Sulfenax cb/k, *See* Vulkacit® CZ/EGC, DZ/EGC, 34129
Sulfenax TsB, *See* Vulkacit® CZ/EGC, DZ/EGC, 34129
Sulferrous, *See* ferrous sulfate, 11690
Sulfetal 4105, 29981
Sulfetal C 38, 29982
Sulfetal CJOT 38, 29983
Sulfetal FA 40, 29984
Sulfetal KT 400, 29985
sulfetal l 95, *See* Rhodapon® 101-10, 26631
Sulfetal LT, *See* Witcolate TLS-500, 34479
Sulfetal MG 30, 29986, *See* magnesium lauryl sulfate, 18364
Sulfetal TC 50, 29987
Sulfex, *See* Thiovit, 31724
Sulfhydril, *See* Uniflot SP 100 Series, 32951
sulfide dyestuffs, 29988
sulfide white, *See* lithopone, 17520
Sulfiformin, 29989
Sulfil®, 29990
Sulfinylbis (methane), *See* dimethyl sulfoxide, 8511
Sulfinylbismethane, *See* dimethyl sulfoxide, 8511, DMSO, 8762
Sulfiolinic acid, *See* Loretine, 17606
Sulfirol 8, *See* Niaproof® Anionic Surfactant 08, 21092, Rhodapon® BOS, 26632
Sulfisoxazole acetyl, *See* Lipo Gantrisin, 17340
sulfite carbon, 29991
sulfite pulp, 29992
sulfite turpentine, 29993
sulfite turpentine oil, 29994
Sulflox, *See* Thiovit, 31724
***o*-sulfobenzimide**, *See* saccharin, 27195
***o*-sulfobenzoic acid imide**, *See* saccharin, 27195
sulfobenzoic imide, sodium salt, *See* saccharin sodium, 27196
2-sulfobenzoicimide, *See* saccharin, 27195
Sulfobutanedioic acid 1,4-bis(1-methylpentyl) ester sodium salt, *See* Bis(1-methylamyl) Sodium

Sulfosuccinate, 4227, Lankropol® KMA, 16713
Sulfo-butanedioic acid 1,4-bis(2-ethylhexyl)ester sodium salt, *See* Empimin® OP70, 10434, Octowet 40, 22035
Sulfo-butanedioic acid 1,4-bis(2-methylpropyl)ester sodium salt, *See* Aerosol® IB-45, 892, Rewopol® SBDB 45, 26371
Sulfobutanedioic acid 4-[2-[2-[2-(dodecyloxy)ethoxy]ethoxy]ethyl]ester, disodium salt, *See* disodium lauryl sulfosuccinate, 8649
Sulfobutanedioic acid, 4-isodecyl ester, sodium salt, *See* disodium laneth-5 sulfosuccinate, 8646
sulfocarbanilide, *See* diphenylthiourea, 8604
Sulfocarbolic acid, *See* Eltesol® PSA 65, 9868
sulfocarbonic anhydride, *See* carbon disulfide, 5223
Sulfochem 436, 29995, *See* Ammonium nonoxynol-4 sulfate, 2246
Sulfochem ALS, 29996, *See* Ammonium lauryl sulfate, 2242
Sulfochem DLS, 29997, *See* DEA-lauryl sulfate, 7827
Sulfochem EA-1, 29998, *See* Ammonium laureth sulfate, 2240
Sulfochem EA-1, EA-2, EA-3, EA-60, EA-70, *See* Ammonium laureth sulfate, 2240
Sulfochem EA-2, EA-3, 29999
Sulfochem EA-60, 30000
Sulfochem EA-70, 30001
Sulfochem ES-1, 30002
Sulfochem ES-2, 30003
Sulfochem ES-3, 30004
Sulfochem ES-60, 30005
Sulfochem ES-70, 30006
Sulfochem K, 30007
Sulfochem MG, 30008, *See* magnesium lauryl sulfate, 18364
Sulfochem MLS, 30009
Sulfochem SAC, 30010
Sulfochem SLC, 30011
Sulfochem SLN, 30012
Sulfochem SLP-95, 30013
Sulfochem SLS, 30014
Sulfochem SLX, 30015
Sulfochem TLS, 30016
Sulfochem® MgLES, *See* magnesium laureth sulfate, 18363
α-sulfo-ω-(dodecyloxy)-, sodium salt, *See* Standapol® AP Blend, 29468
sulfoform, 30017
sulfogenol, 30018, *See* Perichthol, 23270
Sulfokyl DAS, 30019
α-sulfomethyl laurate, *See* Calester, 4981
Sulfona, *See* Dapsone, 7751
Sulfona-Mae, *See* Dapsone, 7751
Sulfonamide, *See* Bayrena®, 3838, Eleudron-Solution, 9808
Sul-fon-ate AA-10, 30020
Sul-fon-ate OA-5R, 30021
Sulfonate soap, *See* Base 7800, 3695
Sulfonated aliphatic hydrocarbon, *See* Statexan® K1, 29554
sulfonated castor oil, *See* Standapol® SCO, 29478
Sulfonated castor oils, monopol oil, soluble castor oil, sulfated oil, red oil, oleine, *See* Turkey red oils, 32607
Sulfonated dioctyl succinate, *See* Rexowet 500, 26503
Sulfonated ester, *See* Arowet 70 E, 3089
sulfonated oils, *See* Turkey red oils, 32607
Sulfonated oleic acid, *See* Sul-fon-ate OA-5R, 30021
Sulfonated soyabean oil, sodium neutralized, *See* Actrasol CS-75, 550
Sulfonated styrene/maleic anhydride, *See* Versa-TL® 4, TL 7, 33614
Sulfonated tall oil fatty acid, sodium neutralized wet process phosphoric acid defoamer, *See* Actrasol SP, 557
Sulfonates, *See* Emkanol NC, NCD 25, 35, 45, 55, 10241
o-sulfonbenzoic acid imide sodium salt, *See* saccharin sodium, 27196
Sulfone, *See* tetradifon, 31366
Sulfone UCB, *See* Dapsone, 7751
Sulfone, 2,4,4',5-tetrachlorodiphenyl, *See* tetradifon, 31366
sulfonic acid, 30022
Sulfon-mere, *See* Dapsone, 7751
Sulfonphthal, *See* phenolsulfonphthalein, 23611
Sulfonyl hydrazide, *See* Celogen® XP-100, 5646
sulfonylbis(trichloromethane), *See* Amerstat® 294, 2044, bistrichloromethylsulfone, 4285, Stauffer N-1386®, 29580
4,4'-Sulfonylbisbenzeneamine, *See* Dapsone, 7751
Sulfonyldianiline, *See* Dapsone, 7751
3,3'-sulfonyldianiline, *See* 3,3'-diaminodiphenylsulfone, 8303
4,4'-sulfonyldianiline, *See* Dapsone, 7751, 4,4' -diaminodiphenyl sulfone, 8302
Sulfopon 101, 101 Special, 30023
Sulfopon 101/POL, 30024
Sulfopon LS, 30025
Sulfopon O 680, 30026
Sulfopon P-40, 30027
sulfopon wa 1, *See* Rhodapon® 101-10, 26631
Sulfosept oil, 30028
Sulfosoft, 30029
Sulfospor, *See* Thiovit, 31724
Sulfosuccinate, *See* Basophen® RA, 3722
sulfosuccinic acid 1,4-bis(2-ethylhexyl) ester, *See* Aerosol® OT-MSO, 898
Sulfotep, *See* Bladafum®, 4308
Sulfotex 110, 30030, *See* sodium decyl sulfate, 28755
Sulfotex LAS-90, 30031
Sulfotex LCX, 30032
Sulfotex LMS-E, 30033
Sulfotex OA, 30034
Sulfotex OT, 30035, *See* Ammonium laureth sulfate, 2240
Sulfotex SXS-40, 30036, *See* sodium xylene sulfonate, 28832
Sulfotex T-65, 30037
sulfotex wa, *See* Rhodapon® 101-10, 26631
sulfourea, 30038, *See* thiourea, 31720
sulfoxylate, *See* sodium hydrosulfite, 28771
Sulframin, 30039
Sulframin 14-16 AOS, 30040
Sulframin 33, 30041
Sulframin 85, *See* Nacconol® 40G, 20647, sodium dodecylbenzenesulfonate, 28760
Sulframin 1238, *See* Nacconol® 40G, 20647, sodium dodecylbenzenesulfonate, 28760
Sulframin 1238 slurry, *See* Nacconol® 40G, 20647, sodium dodecylbenzenesulfonate, 28760
Sulframin 1240, *See* Nacconol® 40G, 20647
Sulframin 1250, 30042
Sulframin 1250 slurry, *See* Nacconol® 40G, 20647, sodium dodecylbenzenesulfonate, 28760
Sulframin 1298, *See* Witco® 1298, 34458
Sulframin acid 1298, *See* Witco® 1298, 34458
Sulframine acid 1298, *See* Witco® 1298, 34458
sulftech, *See* sodium sulfite, 28823
Sulfur, 30043, *See* Biosulphur Powder, 4210, Flotox, 12027, flour of sulfur, 12028, Gofrativ, 13165, Gofravik, 13166, Green Sulphur, 13324, Hortag Aquasulf, 14313, Kumulus® DF, FL, 16482, Kumulus® S, 16483, roll sulfur, 26920, Rubbermakers Sulfur, 27094, Solfa, 28928, Solfa, 28929, Stoller Flowable Sulphur, 29782, Struktol® SU 109, 29837, Suffa, 29949, sulfur, 30043, Sulfur-F, 30054, Thiovit, 31724, Vassgro Flowable Sulphur, 33470
sulfur dioxide, 30044
sulfur gold, 30045
sulfur hexafluoride, 30046, *See* Esaflon, 10782
sulfur hydride, *See* sulfuretted hydrogen, 30053
sulfur hydroxide, *See* sulfuretted hydrogen, 30053
sulfur hypochlorite, 30047
Sulfur mustard, *See* mustard gas, 20538
sulfur olive oils, 30048
sulfur rich amino acid concentrate, *See* Aminodermin CLR, 2169
sulfur soap, 30049
sulfur waste, 30050
sulfur, insoluble, *See* Crystex®, Crystex® Regular, 7294
sulfur, sulphur, *See* roll sulfur, 26920
sulfurated antimony, 30051
sulfurated potash, 30052
sulfuretted hydrogen, 30053
Sulfur-F, 30054
sulfuric acid, 30055, *See* spirit of alum, 29295, spirit of vitriol, 29299
Sulfuric acid diammonium, *See* Ammonium sulfate, 2255
sulfuric acid, aluminum ammonium salt (2:1:1) dodecahydrate, *See* Ammonium alum, 2229
Sulfuric Acid, Aluminum Salt (3:2), *See* aluminum sulfate, 1930
Sulfuric Acid, Ammonium Iron (2+) Salt (2:2:1), *See* Mohr's salt, 20039
Sulfuric acid, ammonium iron(2+) salt, *See* Mohr's salt, 20039
Sulfuric acid, disodium salt, *See* sodium sulfate, 28821
Sulfuric acid, dodecyl ester, triethanolamine salt, *See* Witcolate TLS-500, 34479
Sulfuric Acid, Iron(3+) Salt (3:2), *See* ferric sulfate, 11652
Sulfuric Acid, Iron(3+) Salt (3:2), *See* Vitax Turf Tonic, 33960
Sulfuric Acid, Iron(3+) Salt (3:2), *See* Elliott's Lawn Sand, 9848
Sulfuric acid, lauryl ester, ammonium salt, *See* Witcolate NH, 34478
sulfuric acid, magnesium salt (1:1), *See* magnesium sulfate, 18368
Sulfuric Acid, Mercury Salt, *See* mercuric sulfate, 19349
Sulfuric acid, mercury(2+)salt(1:1), *See* mercuric sulfate, 19349
Sulfuric acid, mono(2-ethylhexyl) ester, sodium salt, *See* Rhodapon® BOS, 26632
sulfuric acid monodecyl ester sodium salt, *See* Texapon K-12, K-1296, L-100, 31461
Sulfuric acid, monododecyl ester, ammonium salt, *See* Ammonium lauryl sulfate, 2242, Standapol® A, 29467, Witcolate NH, 34478
Sulfuric acid, monododecyl ester, compd. with 2,2',2''-nitrilotris(ethanol) (1:1), *See* Rhodapon® LT-6, 26638, Standapol® T, 29481, TEA-lauryl sulfate, 30873, Witcolate TLS-500, 34479
sulfuric acid monododecyl ester magnesium salt, *See* Standapol® MG, 29475
sulfuric acid monododecyl ester, sodium salt, *See* sodium lauryl sulfate, 28781, Witcolate 6400, 34469
Sulfuric acid, sodium salt, *See* sodium sulfate, 28821
sulfuric ether, 30056, *See* ethyl ether, 11069
Sulfuril 50, *See* Nacconol® 40G, 20647
sulfuril 50 F 90, *See* sodium dodecylbenzenesulfonate, 28760
sulfurite, 30057
Sulfurized fatty acid, *See* OA-252, 21950
Sulfurized methyl ester, *See* OA-270, 21951
Sulfurized oleic acid, *See* Base 36, 3690
sulfurous acid, disodium salt, *See* sodium sulfite, 28823
sulfurous acid, monoammonium salt, *See* Ammonium bisulfite, 2233
sulfurous acid, monosodium salt, *See* sodium acid sulfite, 28728, sodium bisulfite, 28738
sulfurous acid, sodium salt (1:2), *See* sodium bisulfite, 28738, sodium sulfite, 28823
sulfurous anhydride, *See* sulfur dioxide, 30044
sulfurous oxide, *See* sulfur dioxide, 30044
sulfuryl fluoride, *See* Vikane, 33754
Sulisobenzone, *See* benzophenone-4, Rhodialux S, 26676, Spectra-Sorb UV 284, 29231, Sungard, 30143, Uval, 33177
Sulmet, 30058, *See* sulfamethazine, 29962
Sul-Perm® 10, 30059

Sul-Perm® 110, 30060
Sul-Perm® C, 30061
Sulphadione, *See* Dapsone, 7751
Sulphatol 33, 30062
Sulphatol 33 MO, 30063
Sulphatol B6, 30064
Sulphix, *See* sulfamethazine, 29962
Sulphol, 30065
Sulphol High Fast, 30066
Sulphonic Acid LS, 30067
Sulphonol, 30068
Sulphophone, 30069
Sulphosol, 30070
Sulphotox, *See* Thiovit, 31724
Sulphramin B and TPB, 30071
sulphur, *See* sulfur, 30043, Kumulus® DF, FL, 16482, Kumulus® S, 16483, Yellow sulphur, 34650
Sulprofos, *See* Bolstar, 4384, Helothion, 13768
Sulqui, *See* Mansonil®, 18611
Sulquin, *See* sulfaquinoxaline, 29968
Sulsol, 30072
Sululan, *See* Lanolin alcohol, 16741
Sulveol DC, 30073
Sumac, *See* sumach, 30075
sumacel, 30074
sumach, 30075
sumach wax, *See* Japan tallow, 15510
sumalban, 30076
sumaphos, 30077
Sumatra camphor, *See* DL-Borneol, 4428
Sumatra wax, *See* Java wax, 15539
Sumet Processed lead, 30078
Sumibond PA, 30079
Sumicidin, 30080, *See* fenvalerate, 11620
Sumicool, 30081
Sumicure S, *See* Dapsone, 7751
Sumiflex, 30082
Sumikon, 30083
Sumikon AM, 30084
Sumikon EM, EME, 30085
Sumikon IM, 30086
Sumikon PM, 30087
Sumikon TM, 30088
Sumikon VM, 30089
Sumilac PC, 30090
Sumilite CEL, 30091
Sumilite EI, 30092
Sumilite EL, 30093
Sumilite ELC, 30094
Sumilite FS, 30095
Sumilite IL, 30096
Sumilite ILC, 30097
Sumilite ILI, 30098
Sumilite NS, 30099
Sumilite PL, 30100
Sumilite PLC, 30101
Sumilite Resin PR, 30102, 30103, 30104
Sumilite STS, 30105
Sumilite TFC, 30106
Sumilite TFP, 30107
Sumilite VSL, 30108
Sumilite VSS, 30109
sumilizer wx, *See* Rutenol, 27147
sumilizer wx-r, *See* Rutenol, 27147
Sumine® 2005, 30110
Sumine® 2015, 30111
Suminet, 30112
Sumipipe, 30113
Sumitac EA, 30114
Sumitac GA, 30115
Sumitac VA, 30116
Sumithrin, 30117, *See* phenothrin, 23619
Summetrin, *See* papain, 22698
Summit, 30118, 30119, *See* triadimenol, 32185
Summit 1906, *See* thiambutosine, 31663
Sumquat® 2355, 30120
Sumquat® 6020, 30121
Sumquat® 6030, 30122
Sumquat® 6045, 30123, *See* Adogen® TA-101, 780
Sumquat® 6045
Sumquat® 6050, 30124
Sumquat® 6110, 30125
Sumquat® 6210, 30126
sun bronze, 30127
Sun gold, *See* Heliochrysin, 13762
Sun Wrap®, 30128
Sunaptic Acids, 30129
Sunaptol, 30130
Sunaptol NP55, 30131
Sunaptol NP65, NP70, 30132
Sunaptol NP80, NP95, 30133
Sunaptol NP100, 30134
Sunaptol NP140, 30135
Sunaptol NP350, 30136
Sunarome O, *See* Neo Heliopan® OS, 20870, Dermoblock OS, 8116
Sunarome WMO, *See* Dermoblock OS, 8116, Neo Heliopan® OS, 20870
Sunbrite, 30137
Sun-Bugger #4, *See* resmethrin, 26218
Suncide, *See* Baygon®, 3790, Propoxur, 25198, 25199, 25200, 25201
Suncide®, 30138
Sundora, 30139
Sundown, *See* Padimate O, 22549
Sunett, 30140
Sunette, *See* acesulfame potassium, 277
sunflower seed oil, 30141
Sunfort®, 30142
Sungard, 30143, *See* Rhodialux S, 26676, Uval, 33177
Sunimac ECR, 30144
Sunimac GCR, 30145
Sunnis, 30146
Sunnol, 30147
Sunolith, 30148, *See* ponolith, 24685
Sunoxol, *See* Quinosol, 25687
Sunproof, 30149
Sunproofing Wax 1343, 30150
Sunscreen UV 15, *See* Cyasorb® UV 9, 7495, Neo Heliopan® BB, 20867, Rhodialux A, 26675
SunShade 18-89, 30151
Sunshine, 30152
Suntec® -HD, 30153
Suntec® -LD, 30154
suntei tallow, 30155
Sunveil®, 30156
Sunvex®, 30157
Supaclean, 30158
Suparamin 30, 120, 30159
Suparen, 30160
Suparex, 30161
Supec® G401, 30162
Super A, 30163
Super AD-IT, 30164
Super Alkyd® 574-75TK, 30165
Super Amide L-9A, *See* Ablumide LDE, 79, Alkamide® 327, 1634
Super Amide L-9C, *See* Ablumide LDE, 79, Alkamide® 327, 1634
Super Arsonate, *See* MSMA, 20417
Super Beckacite® 2000, 30166
Super Beckamine®, 30167
Super Beckosol®, 30168
Super Bowl, 30169
super bronze, 30170
Super Cat, 30171
super cement, 30172
super cobalt, *See* cobalt, 6524
Super Corona, 30173
Super D, 30174, *See* Atlas Tanked Cod Oil, 3336
Super Die, 30175
Super Glue, 30176
Super Green, 30177
Super Hartolan, 30178, *See* Lanolin alcohol, 16741
Super Lacolene, 30179
Super Lubestine, 30180
Super Lubracon, 30181
Super Moss Killer & Lawn Fungicide, 30182
Super Mosstox, 30183, 30184, *See* Nuophene, 21783
Super Nevtac® 99, 30185
Super Nickel, 30186
Super Refined Almond NF, 30187
Super Refined Avocado Oil, *See* Avocado Oil, 3497
Super Refined Coconut Oil, 30188, *See* Esi-Terge 40% Coconut Oil Soap, 10825
Super Refined Crossential EPO, 30189
Super Refined Grapeseed Oil, 30190
Super Refined Menhaden, 30191
Super Refined Mink Oil, 30192
Super Refined Olive, 30193
Super Refined Peanut, 30194
Super Refined Safflower USP, 30195
Super Refined Sesame, 30196
Super Refined Sesame Oil, *See* Sesame oil, 28075
Super Refined Shark, 30197
Super Refined Soybean, 30198
Super Refined Soybean Oil, *See* Soybean oil, 29173
Super Refined Sunflower Oil, 30199
Super Refined Wheat Germ Oil, *See* Wheat germ oil;, 34349
Super Refined™ Avocado oil, *See* Avocado Oil, 3498
Super Refined™ Grapeseed oil, *See* Grape seed oil, 13275
Super Six, *See* Thiovit, 31724
Super Solan, 30200
Super Solvitax®, 30201
Super Sta-Tac® 80, 30202
Super Sta-Tac® 100, 30203
Super Sterol Ester, 30204
Super Sulfur No. 1, 30205
Super Sulfur No. 2, 30206
Super Thane® 975-70, 30207
Super Tin® 4L, 30208
Super Tin® 4L, *See* Farmatin, 11480
Super Verdone, 30209
Super Vilex, 30210
Super Weedex, 30211
Super Wet, 30212
Super White Fonoline®, 30213
Super-A, 30214
Superadoplast, 30215
Superam, 30216
superargol, 30217
Super-ascoloy, 30218
Superb, *See* hydrogenated soybean oil, 14596
Superba, 30219
superbasique metal, 30220
Superbond®, 30221
Superbrillantoline®, 30222
Supercadoplast, 30223
Super-caid, *See* bromadiolone, 4565, Rentokil Deadline, 26116
Super-Cel, *See* Celite®, 5586
Super-Ceram, 30224
Superchlon, 30225
Superclear 80-N, 30226
Super-cliffite, 30227
Supercoat®, 30228
Supercol U, *See* Prinza® Range, 25010
Supercol® Guar Gum, 30229
Supercore® S13F, 30230
Supercore® S13F, *See* corn starch, 6837
Supercut, 30231
super-dent, *See* sodium fluoride, 28764
Super-De-Sprout, *See* maleic hydrazide, 18477
Super-excellite, 30232
Superfast Power Pack, 30233
Superfiltchar, 30234
Superfine Lanolin USP, 30235
Superfloc, 30236
Superfloc C507, 30237
Superfloc C521, 30238
Superfloc C567, C573, C577, C581, 30239
Superforcite, 30240
Supergreen, 30241
Superinone, 30242
Superior alloy, 30243

Suva®, 30441
suXon, 30442
Swale, 30443
Swale powder, 30444
Swalite, 30445
Swan, 30446
Swanol AM-3130N, *See* Cocamidopropyl betaine, 6551
Swardsman, 30447
swartziol, *See* kaempferol, 15654
swascol 3l, *See* Rhodapon® 101-10, 26631
Swedelec, 30448
Swedex AR58P-15AC, 30449
Swedex SSL-5AC, 30450
Swedish factory tar, 30451
Swedish green, *See* Scheele's green, 27562
Swedish liquid resin, *See* talloel, 30738
Swedish turpentine, *See* turpentine, 32610
Sweep, *See* Chlorothalonil, 6150, Repulse, 26151
Sweet Almond oil, 30452
Sweet Almond Oil BP 73, *See* Sweet Almond oil, 30452
sweet bark, 30453
sweet birch oil, *See* methyl salicylate, Teaberry Oil, 30872
sweet clover, *See* Melilot, 19279
Sweet wood bark, *See* sweet bark, 30453
Sweeta, 30454, *See* saccharin sodium, 27196
Sweetex, 30455
Sweetex Plus, 30456
Sweetrex®, 30457
sweet-water, 30458
Sweetzyme, 30459
Swim clear, 30460
Swipe 560 EC, 30461
Swirl, 30462
Swiss Polyamid Grilon, 30463, 30464
Swiss Polyester Grilene, 30465
SWP, *See* Santowhite, 27431
SWS-101, 30466
SWS-290, 30467
SWS-725, 30468
SWS-03314, 30473
SWS-06545u, 30474
SWS-7532u, 30469
SWS-7655u, 30470
SWS-7675u, 30471
SWS-7865u, 30472
Sybol, 30475
Sybron, *See* sodium iron edetate, 28776
sycorin, *See* saccharin, 27195
sycose, *See* saccharin, 27195
Sydex, 30476
Syford, 30477
sykose, *See* saccharin, 27195, saccharin sodium, 27196
Syl, 30478
Sylade, 30479
Sylfam 2082, 30480
Sylfan 20, 30481
Sylfat® D-1, 30482
Sylfat® DX, MM, 30483
Sylfat® RD-1, 30484
Sylgard® 170, 30485
Sylgard® 182, 184, 30486
Syl-off®, 30487
Syloid 72, 30488
Sylphane S, 30489
Sylphrap, 30490
Syltherm®, 30491
Syltherm® 444, 30492
Syltherm® XLT, 30493
Sylvacote® K, 30494
Sylvadym® M-35, 30495
Sylvamid®, 30496
Sylvan, 30497
Sylvania Cellophane, 30498
Sylvaros® 20, 30499
Sylvaros® 315, 30500
Sylvaros® R, 30501
Sylvatac® AC, 30502
sylvic acid, 30503, *See* abietic acid, 45
Sylvid®, 30504
Symalit GM 20 PP, 30505
Symax, 30506, 30507
Symclosene, *See* ACL 85, 90 Plus, 355, Fi-Clor, 11766, Queschlor, 25658, trichloroisocyanuric acid, 32218
symdichloroethane, *See* ethylene dichloride, 11122
***sym*-Diphenylthiourea**, *See* Akrochem® Thio No. 1, 1181
Symel, 30508, 30509
Sympathy®, 30510
Symphony®, 30511
Syn Fac®, 30512
Syn Fac® 222, 30513
Syn Fac® 334, 30514
Syn Fac® 8009, 8017, 30515
Syn Fac® TEA-97, 30516
Syn Lube, 30517
Synaceti 116 NF/USP, *See* cetyl esters, 5819
Synacid, *See* hyaluronic acid sodium salt, 14483
Synacril, 30518
Synacto, 30519
Synanthic, 30520, *See* oxfendazole, 22426
Synaqua, 30521
Synasol, 30522, *See* ethyl alcohol, 11064
syncal, *See* saccharin, 27195
Syncelose, *See* Glutolin, 13064
Syn-Chek 1203, 30523
Synchemicals Dalapon, 30524
Synchemicals Total Weed Killer, 30525
Synclyst, 30526
Syncol®, 30527
Syncrolube, 30528
Syncrowax, 30529
Syncrowax AW1-C, 30530
Syncrowax BB4, 30531
Syncrowax ERL-C, 30532
Syncrowax HGL-C, 30533
Syncrowax HR-C, 30534
Syncrowax HRS-C, 30535
Syndane, 30536
Syndite, 30537
Syndraw, 30538
Synektan, 30539
Synergistic blend of aliphatic nitrogen and heterocyclic sulfur containing compounds, *See* Acticide AZ®, 465
Synergistic mixture of FDA approved chlorinated phenols, *See* Acticide CPC, 466
Synergol, 30540
Syn-Fab DC-1, 30541
Synfluid, 30542
Syngran, 30543
Synkad® 100, 30544
Synkad® 200, 30545
Synkad® 303, 30546
Synkad® 6000, 30547
Synkavit, 30548, *See* menadiol sodium diphosphate, 19319
Synkayvit, *See* menadiol sodium diphosphate, 19319
Synkayvite, 30549
Synkrolith, 30550
Synmold, 30551, 30552
Syn-O-Ad® 8412, 30553
Syn-O-Ad® 8475M, 30554
Syn-O-Ad® 8478, 30555
Syn-O-Ad® 8479, 30556
Syn-O-Ad® 8480, 30557
Syn-O-Ad® 8484, 30558
Syn-O-Ad® 8485, 30559
Syn-O-Ad® P-310, 30560
Syn-O-Ad® P-312, 30561
Syn-O-Ad® P-316, 30562
Syn-O-Ad® P-374, 30563
Syn-O-Ad® P-399, 30564
Syn-O-Ad® P-408, 30565
Syn-O-Ad® P-412, 30566
Syn-O-Ad® P-415, 30567
Syn-O-Ad® P-417, 30568
Synocryl, 30569
Synocure, 30570
Synogist, 30571
Synolac, 30572
Synolide, 30573
Synotex 800, 30574
Synotol L-60, *See* Ablumide LDE, 79, Alkamide® 327, 1634
Synouryn, 30575
Synova, 30576
Synox, 30577
Synpenin, *See* Ampicillin, 2361
Synperonic, 30578
Synperonic 3S27 and 3S60S, 30579
Synperonic 3S60A, 30580
Synperonic N, NX, NXP and NDB, 30581
Synperonic NP10 and NP12, 30582
Synperonic NP13 and NP15, 30583
Synperonic NP20 and NP30, 30584
Synperonic NP4, NP5 and NP6, 30585
Synperonic NP8 and NP9, 30586
Synperonic OP, 30587
Synpren, *See* rotenone, 27000
Synpren-Fish, *See* piperonyl butoxide, 23875
Synpro®, 30588
Synpro® 8, 30589
Synpro® 15F, 30590
Synpro® 90, 30591
Synpro® 303, 30592
Synpro® 404, 30593
Synpro® 505 USP, 30594
Synpro® Aluminum Octoate, 30595
Synpro® Barium Stearate, 30596
Synpro® Cadmium Stearate, 30597
Synpro® Calcium Pelargonate, 30598
Synpro® Stannous Stearate, 30599
Synprol, 30600
Synprol Sulphate, 30601
Synprolam, 30602
Synprolam 35, 30603
Synprolam 35 BQC, 30604
Synprolam 35 DM, 30605
Synprolam 35 DMA, 30606
Synprolam 35 N3, 30607
Synprolam 35A, 30608
Synpron, 30609
Synpron 241, 30610
Synpron 1009, 30611
Synpron 1027, 30612
Synpron 1032, 1033, 30613
Synpron 1538, 30614
Synpron 1800, 30615
Synpro-Ware, 30616
Synresin RD 461, 30617
Synsoft, 30618
Synsolve, 30619
Synstryp, 30620
syntapon, *See* Rhodapon® 101-10, 26631
Syntarpen, *See* Dicloxacillin sodium, 8396
Syntase 62, *See* Cyasorb® UV 9, 7495, Neo Heliopan® BB, 20867, Rhodialux A, 26675
Syntase® 62, 30621
Syntase® 230, 30622
Syntergent 55-A, 30623
Syntes 12a, *See* Kalex 220 Crystal, 15668, Kalex Liq. 50%, 15671
Syntex, 30624
Syntexan, *See* dimethyl sulfoxide, 8511, DMSO, 8762
Synteze, 30625
Synthacalk, 30626
Synthacryl, 30627
Synthamel, 30628
Synthamica, 30629
Synthane, 30630
Synthappret®, 30631
Synthaprufe, 30632
Syntharesin®, 30633
Synthasil, 30634
Synthawax, 30635
synthecite, 30636

Synthemul®, 30637
Synthemul® 40-422, 30638
Synthemul® 40-425, 30639
Synthemul® 40850-00, 30640
synthe-plastic, 30641
Synthetic alcohol ethoxylates, *See* Dehscoxid 700 Series, 7940
Synthetic alcohol polyalkylene oxide deriv, *See* Teric BL8, 31264
Synthetic barytes, *See* Barium sulfate, 3644
Synthetic beeswax, *See* Waxenol® 821, 34239
Synthetic Bone Ash, 30642
synthetic cryolite, *See* Synkrolith, 30550
Synthetic ester, *See* Actralube 100, 535
Synthetic hydrocarbons, *See* Santotrac, 27426
Synthetic jojoba oil, *See* Wickenol® 139, 34383
Synthetic lubricant, *See* Krytox® GPL 206, 16472
synthetic oil of bitter almond, *See* benzaldehyde, 3957
synthetic Peru balsam, *See* Perugen,
Synthetic polymer, *See* Akypopress DB, 1268
Synthetic rubber, *See* Acrylonitrile-butadiene rubber, 437
Synthetic Rutile, 30643
Synthetic wax, *See* Ross Wax #100, 26993
synthin, 30644
Synthite, 30645
synthocarbone, 30646
synthol, 30647
Syntholvar, 30648
Synthrapol KB, *See* Deceth-6, 7848
Synthrin, *See* resmethrin, 26218
Syntol K77 and N77, 30649
Synton PAO-100, 30650
Syntopherol, *See* tocopherol, 31918, Vascuals, 33465
Syntopon 8, 30651
Syntopon A, B, C and D, 30652
Syntopon F, G and N, 30653
Syntox, *See* resmethrin, 26218
Syntox Total Weed Killer, 30654
Syntroil, 30655
Syntron, 30656
Syntron C, *See* Cheelox® 80, 5864, Pentaquest Extra 0685, 23169, Trilon® C Liq, 32328
Synvaren, 30657
Synvarol, 30658
Synwax, 30659, *See* Paraffin, 22743
S-yperite, *See* mustard gas, 20538
Syrian asphalt, 30660
Syringol, *See* dimethoxyphenol, 8500
Sys Tec® 1998, 30661
Systamex, 30662, *See* oxfendazole, 22426
Systemic Insecticide, 30663, 30664
Systhane, 30665, *See* myclobutanil, 20549
Systol M, 30666
Sytam, 30667, *See* Schradan, 27731
Sytasol, *See* Acrex, 384, Dinobuton, 8543
Sytobex, 30668
Syton, 30669
Sytron, 30670
Szklarniak, *See* dichlorvos, 8394
2,4,6 T, *See* 2,4,6-trichlorophenol, 32222
T hydro, *See* Aztec® *t*-Butyl Hydroperoxide-70, Aq, 3541
T metal, 30671
T.A.M, 30672
T.I.P.P.S, *See* keraphen, 16055
T-64, 30673
T-64, *See* aluminum oxide, 1919
T-95, *See* Trideceth series, 32232
T-1061, 30674, *See* aluminum oxide, 1919
T1750, 30675
T1920, 30676
T4250, 30677
Tabalgin, *See* Valadol, 33238
Tabbyite, *See* wurtzillite, 34577
Tabilin, *See* Penicillin G potassium, 23109
table spate, *See* calcium silicate, 4969
table sugar, *See* sucrose, 29934
TABS, 30678
Tabular alumina, *See* Activated alumina, 495
Tabular Spar, *See* Wollastocoat®, 34545
Tacamahac oil, *See* laurel nut oil, 16833
Tacamahac resin, 30679
TACC 104, 30680
TACC 524, 30681
TACC 700-82, 30682
TACC AR-1001, 30683
TACC CR-3200, 30684
Tachigaren 70, 30685
tachiol, 30686
Tachryrate, 30687
tachyiite, 30688
tachyol, *See* tachiol, 30686
Tack, 30689
Tackidex, 30690
tackol, 30691
Tacolyn, 30692
Tactel, 30693
Tactesse, 30694
Tactix, 30695
Tactix 123, 30696
Tactix 556, 30697
Tactix 742, 30698
Tactix H31, 30699
Taeniatol, See Nuophene, 21783
Tafasan, *See* Agallol, 954, Atiran, 3286, Ceresan, 5745
Tafasan 6W, *See* Agallol, 954, Atiran, 3286, Ceresan, 5745
taffy, 30700
Tafigel PUR 40, 30701
Taflite 900, 30702
Tag, 30703
Tag Fungicide, *See* acetoxyphenylmercury, 306, phenylmercury acetate, 23646
Tag HL-331, *See* acetoxyphenylmercury, 306, phenylmercury acetate, 23646
Tagat®, 30704
Tagat® 12, 30705
Tagat® I, 30706
Tagat® L, 30707
Tagat® L2, 30708
Tagat® O, 30709
Tagat® O2, 30710
Tagat® R40, 30711
Tagat® R60, 30712
Tagat® R63, 30713
Tagat® S, 30714
Tagat® S2, 30715
Tagat® TO, 30716
Tagit®, 30717
ta-hong, 30718
TAHP-80, 30719
TAIC, *See* triallyl isocyanurate, 32192
Taifun, 30720
tailor's chalk, 30721
Tak, 30722
Taka-Sweet®, 30723
Taka-Therm® L-340, 30724
Takatol, 30725
takineocol, *See* isopropanol, 15405
takizolit, 30726
takizolite, *See* Tisyn®, 31857
ta-kong, 30727
Taktene 220, 30728
Taktic, 30729, *See* Amitraz, 2202
Talatrol, *See* THAM, 31557
Talbor's powder, 30730, *See* cinchona bark,
talc, 30731, *See* ABT-2500®, 166, Altalc 200 USP, 1870, Artic Mist, 3148, Ceramitalc, 5707, Olympic, 22146, Super Lubestine, 30180, Vantalc®, 33368, Vertal 92, 200, 33632
Talc MS, 30732
talca gum, *See* Suakin gum, 29901
Talcord, *See* permethrin, 23384, 23385, 23386
Talcum, *See* talc, 30731, Silverline 665, 28439, Vantalc®, 33368, Vertal 92, 200, 33632
Talent, 30733
talide, *See* tungsten carbide,
talipot, 30734
Talisman, 30735, *See* Tolurex, 31958
talite, 30736
talka gum, *See* Suakin gum, 29901
tall oil, 30737
2:1 Tall oil fatty acid alkanolamide, *See* Actramide 202, 543
Tall oil fatty acids, *See* Oulu 102, 22378
Tall oil hydroxyethyl imidazoline, *See* Miramine® TO, 19873, Monazoline T, 2019, Varine T, 33395 7
Tall oil rosin, *See* Acintol® R Type SFS, 350
Tallamide DEA (1:1), *See* Schercomid SO-T, 27662
Tallamidopropyl dimethylamine, *See* Schercodine T, 27642
Talleol, *See* Liquid resins, 17475
talloel, 30738
tallow, 30739
Tallow acid, *See* Industrene® 143, 15102, Prifac 7920, 24946
Tallow alkyl dimethylamine, *See* Jet Amine DMTD, 15588
N-Tallowalkyl dipropylene triamine, *See* Triameen T, 32193
Tallow alkyl imidazoline methosulfate, *See* Empigen® FRC75S, 10374
N-Tallowalkyl-1,3-propanediamine, *See* Jet Amine DT, 15590
n-Tallow amine acetate, *See* Acetamin T, 288
N-Tallow-1,3-propanediamine dioleate, *See* Duomeen® TDO, 9117
N-Tallow-1,3-propylenediamine, *See* Jet Amine DT, 15590
n-Tallow-propylene diamine, *See* Diamine BG, 8295
tallow amine, 30740, *See* Amine BG, 2151, Crodamine 1.T, 7104, Mulgofen, 20446, Radiamine 6170, 25755, tallowamine, 30748
Tallow amine ethoxylate, *See* Secomine TA 02, 27805
tallow benzyldimethyl ammonium chloride, *See* Cycloton® SCS, 7542, octadecylbenzene methanaminium chloride, 21985, stearalkonium chloride, 29592
tallow clays, 30741, *See* zinc silicate,
tallow diamine, 30742, *See* Jet Amine DT, 15590, tallow diamine, 30742
Tallow diamine diammonium dichloride, *See* Adogen® 477, 775
Tallow dimethylamine, *See* Jet Amine DMTD, 15588
Tallow dipropylene triamine, *See* Lilamin LSP 33, 17250
tallow fatty acid, 30743
Tallow fatty acid amine, *See* Genamin TA Grades, 12776, tallowamine, 30748
Tallow imidazolinium methosulfate, *See* Quaternium-27, 25607
Tallow MEA ethoxylate, *See* Varamide® T-55, 33385
Tallow nitrile, *See* Nitrile BG, 21337
Tallow pentamine, *See* Jet Amine TP, 15599
Tallow propanediamine, *See* Radiamine 6570, 25767
tallow seed oil, 30744
tallow succinamate, *See* Adogen® 170, 765
tallow tetramine, 30745
tallow triamine, 30746, *See* Jet Amine TRT, 15600
Tallow trimethyl ammonium chloride, *See* Jet Quat T-50, 15606, tallow trimonium chloride, 30747
Tallow trimethyl ammonium methosulfate, *See* Empigen® CM, 10371
tallow trimonium chloride, 30747, *See* Arquad® T-27W, 3121, Arquad® T-50, 3122, Jet Quat T-50, 15606
Tallow trimonium chloride, IPA, *See* Adogen® 471, 774
Tallow tripropylene tetramine, *See* Amine 760, 2146
Tallow, amines, N,N-bis(2-hydroxyethyl), *See* Witcamine® 6606, 34447
Tallow-1,3-diaminopropane, *See* Duomeen® T, 9116
Tallowalkonium chloride, *See* Kemamine® BQ-9742C, 15910
Tallowalkylmethyl-bis(2-hydroxy-2-methylethyl) quaternary ammonium methylsulfates, *See* Propoquad® T/12, 25197

N-(Tallowalkyl)trimethylenediamine, *See* Jet Amine DT, 15590
Tallowamide DEA, *See* Amidex TD, 2104
Tallowamide MEA, *See* Marlamid® M 1618, 18745
N -Tallowamido-polyamino-polygincate, *See* Ampholak XTP, 2325
Tallowamidopropyl dimethylamine, *See* Chemidex T, 5972
Tallowamidopropyl hydroxysultaine, *See* Crosultaine T-30, 7243
Tallowamidopropylamine oxide, *See* Chemoxide TAO, 6002
tallowamine, 30748, *See* Jet Amine PT, 15597
Tallowamine (primary amine), *See* Armeen® T, 2958
Tallowamine, distilled, *See* Armeen® TD, 2959
Tallowaminopropylamine, *See* Kemamine® D-974, 15914
Tallowamphopolycarboxyglycinate, *See* Ampholak 7TX, 2313, Ampholak 7TX-SD 55, 2315, Ampholak 7TX-T, 2316
Tallowamphopolycarboxypropionic acid, *See* Ampholak 7TY, 2317
Talloweth-7, *See* Dehydol PTA 7, 7957
Talloweth-7 carboxylic acid, *See* Akypo RT 60, 1244
Talloweth-11, *See* Dehydol TA 11, 7958, Oxetal TG 111, 22424, Rewopal® TA 11, 26350
Tallowtrimethylammonium chloride, *See* Radiaquat 6471, 25795
Tallowtrimonium chloride, *See* Kemamine® Q-9743C, 15926
Tally® 100 Plus, 30749
Talmi gold, *See* Abyssinian gold, 167
Taloberg, *See* Chlorothalonil, 6150, Repulse, 26151
Talon, 30750, 30751, *See* Chlorpyrifos, 6163
Talotalo gum, 30752
ta-lou, 30753
Talpex, 30754
Talpheno, 30755, *See* phenobarbital,
Talstar, 30756
Talunex, 30757
talwaan, 30758
Tam, *See* methamidophos, 19474, Tamaron®, 30760
Tamanox, *See* methamidophos, 19474, Tamaron®, 30760
tamarac, 30759
Tamaron, *See* methamidophos, 19474
Tamaron®, 30760
tambookie grass, 30761
Tamclad 7200, 30762
Tame, *See* fenpropathrin, 11615
Tamguard 840, 840H and 840S, 30763
Tamogam, *See* bromadiolone, 4565, Rentokil Deadline, 26116
Tamol®, 30764
Tamol® 731-25%, 30765
Tamol® 850, 30766
Tamol® L Conc, 30767
Tamolan®, 30768
tampicin, 30769
Tamraghol, *See* Cupravit®, 7370
Tamsil 8, 30770
Tamsil Gold Bond, 30771
Tamtam, 30772
Tanabond STA, 30773
Tanabron W, 30774
tanacetone, 30775
Tanafresh HFO, 30776
Tanal, 30777
Tanalev® 221, 30778
Tanalith, 30779
Tanalon® EFA, 30780
Tanalube® RF, 30781
Tanapal® LD-3, LD-3T, 30782
Tanapel® 54, 30783
Tanapon NF-200, 30784
Tanapure® AC, 30785
Tanasoft® PNL, 30786
Tanassist® JCR, 30787
Tanastat® PH, 30788
Tanaterge® SCP, 30789
Tanatex® Nostick, 30790
tanatol, *See* uba, 32698
Tanavol® URC, 30791
Tanawet® AR, 30792
Tanawet® RCN, 30793
Tanbase®, 30794
Tandearil, 30795
Tandem, 30796
Tandem 5K, 8, 30798
Tandem 11H K, 30797
Tanderil, *See* Tandearil, 30795
tanekaha, 30799
Tangantangan oil, *See* AA Standard, AA USP, 12, castor oil, 5417, Cosmetol® X, 6871
Tanigan®, 30800
tanked oil, 30801
tannal, 30802
tannaline films, 30803
Tannalum, *See* tannal, 30802
tanner's wool, 30804
Tannesco, 30805
Tannex® MGP, 30806
tannic acid, 30807
Tannic acid concentrate, *See* Actan SP, 457
tannin, *See* tannic acid, 30807
Tanolin, 30808
Tanret's reagent, 30809
Tansel, 30810
Tansul, 30811
Tansul-7, 30812
tantalic acid anhydride, *See* tantalum oxide, 30815
tantalum, 30813
tantalum carbide, 30814
tantalum oxide, 30815
tantalum pentoxide, *See* tantalum oxide, 30815
tantcopper, 30816
tantiron, 30817
Tap Aid, 30818
tap cinder, 30819
Tapar, *See* Valadol, 33238
Taquence, 30820
tar camphor, *See* naphthalene, 20706
tar oil, *See* oil of tar, 22075
tara, 30821
tarapon k 12, *See* Rhodapon® 101-10, 26631
Tardex, 30822
Tardex 100, *See* Decabromodiphenyl oxide, 7837, Octoguard FR-01, 21999, Saytex® 102E, 27514
Tardomyocel®, Tardomyocel L, 30823
Tarene, *See* Treflan, 32162, Trifluralin, 32252
Targa, *See* quizalofop-ethyl, 25697
Target MSMA, *See* Neoarsycodyl, 20871
tari, *See* white tan,
Tarimyl, *See* Dapsone, 7751
Tarmac, 30824
Tarmex, 30825
tarnovicite, *See* tarnowitzite, 30826
tarnowitzite, 30826
tarola, 30827
Taroma, 30828
Tarot®, 30829
tarragon, *See* Estragole, 10897
Tarslag, 30830
tartar, *See* argol,
tartar cake, *See* superargol, 30217
tartar emetic, *See* antimony potassium tartrate, 2571
tartar emetic substitute, *See* tartar emetic powder,
tartar substitute, *See* superargol, 30217
tartaric acid, 30832
***meso*-tartaric acid**, 30831
dl-tartaric acid anhydrous, *See* tartaric acid, 30832
tartaric acid disodium salt, *See* sodium tartrate, 28825
tartarline, 30833
tartars, 30834
tartrate lime, *See* limo, 17279
tartrated antimony, *See* antimony potassium tartrate, 2571
tartrated soda, *See* Rochelle salt, 26882
tartrazine, 30835
Tarvia, 30836
tarwar, 30837
tasar silk, *See* tussar silk, 32617
Tasian®, 30838
Task, *See* dichlorvos, 8394, Vapona, 33378
Task Tabs, *See* dichlorvos, 8394
Taski TR101, 30839
Taslan®, 30840
Tasnon®, 30841
Tasprin, 30842
tasteless salts, 30843
Taterfos®, 30844
Tattoo, *See* Bendiocarb, 3937
Tatuzinho, *See* Aldrin, 1558
taurine, 30845
Taylor, 30846
Taylor Oil, 30847
Taylor solder, 30848
Taylorite, *See* bentonite, 3954
Taylors Lawn Sand, 30849, *See* ferric sulfate, 11652
Tayssato, *See* Ceresan, 5745
Tazoline, 30850
2,3,6-TBA, *See* Benzoyl peroxide, 3983
TBAB, 30851, *See* tetrabutyl ammonium bromide, 31362
TBBS, *See* Butylbenzothiazole sulfenamide, 4791
TBDPE, *See* tetrabromodipentaerythritol, 31360
TBE, *See* Muthmann's liquid, 20541
TBEP, 30852, *See* Amgard TBEP, 2055, tributoxyethyl phosphate, 32206
T-BGE, *See* Ageflex TBGE, 998
TBHP, *See* Aztec® *t*-Butyl Hydroperoxide-70, Aq, 3541, TBHP-70, 30853
TBHP-70, 30853
TBHQ, *See* *t*-Butylhydroquinone, 4793, Eastman® MTBHQ, 9471, Sustane® TBHQ, 30437
TBNPA, *See* tribromoneopentyl alcohol, 32201
TBP, *See* Aztec® Di-*t*-Butyl Peroxoide, 3557, di-*t*-butyl peroxide, 8369, 2,4,6-tribromophenol, 32202, tributyl phosphate, 32207
TBP-AE, *See* tribromophenyl allyl ether, 32203
TBS, *See* Temasept IV, 31076, Tuasol 100, 32543
TBT, *See* Butyl toluene, 4782
TBTO, *See* tributyltin oxide, 32209
Tbz, *See* thiabendazole, 31661
TBZ, *See* Tecto, Tecto 60%, 30913
2,4,5-TC, *See* Kuron, 16496
Tc 99m Lungaggregate, 30854
TCA, *See* trichloroacetic acid, 32213
T-Carb, 30855
TCBA, *See* Benzoyl peroxide, 3983
TCC, 30856
1,1,1-TCE, *See* Chlorothene, 6151
TCEP, *See* Genomoll P, 12838
TCIN, *See* Repulse, 26151, Chlorothalonil, 6150
TCLP extraction fluid 2, *See* Acetic acid, 292
TCP, 30857, *See* calcium phosphate (tribasic), 4967, Plastic X, 24026, Syn-O-Ad® 8484, 30558, 2,4,6-trichlorophenol, 32222, tricresyl phosphate, 32230
***m*-TCPN,** *See* Chlorothalonil, 6150, Repulse, 26151
TCPP, *See* Fyrol® FR-2, 12540
TCTH, *See* cyhexatin, 7568
TDA-1, 30858
TDA-6 Teric 13A5, *See* Trideceth series, 32232
TDCPP, *See* tris (2,3-dibromopropyl) phosphate, 32402
T-Det® 25-3A, 30859
T-Det® 25-3S, 30860
T-Det® C-40, 30861
T-Det® D-150, 30862, *See* Ethal DNP-8, 10919
T-Det® DD-5, 30863
T-Det® DD-5, *See* Dodoxynol series, 8786
T-Det® EPO-61, 30864
T-Det® N-1007, 30865
T-Det® N-4, 30866
T-Det® N-40, 30867
T-Det® O-4, 30868
T-Det® O-407, 30869
T-Det® RQ1, 30870
T-DET-N, *See* Triton® N-57, 32430

Chemical and Trade Names

T-det-n, *See* Alkasurf® NP-4, 1715
TDI, *See* toluene diisocyanate, 31947, Voranate T-80, Type I, Type II, 34042
TDI-ester polyurethane prepolymer, *See* Airthane® PST-80A; PST-90A, 1119
TDI-PTMEG polyurethane prepolymer, *See* Airthane® PET-60D; PET-70D; PET-75D; PET-80A; PET-90A; PET-93A; PET-95A, 1116
TDQP, *See* Agerite® MA, 1019, Akrochem® Antioxidant DQ, 1160
TE-33, *See* Witconol 2503, 34502
TEA, *See* triethanolamine, 32240
TEA alkylbenzene sulfonate, *See* Nansa® TS 50, 20697
TEA cocoate, *See* Akypogene FP 35 T, 1255
TEA coco-hydrolyzed collagen, *See* Foam-Coll 4CT, 12151
TEA dodecylbenzene sulfonate, *See* Ablusol DBT, 126, Carsofoam® T-60-L, 5330, Esi-Terge T-60, 10832, Hartofol 60T, 13675, Hoe S 2749, 14258, Mazon® 60T, 19145, Naxel AAS-60S, 20812, Norfox® T-60, 21554, Reworyl® TKS 90/L, 26411, Rhodacal® DOV, 26583, Rhodacal® T, 26592, Stepantan® DT-60, 29706
TEA hydro-iodide, *See* Iodobio 45, 15211
TEA lauroyl glutamate, *See* Acylglutamate LT-12, 606
TEA lauroyl sarcosinate, *See* Crodasinic LT40, 7146, Hamposyl® TL-40, 13606
TEA lauryl sulfate, *See* Avirol® T 40, 3485, DeSonol T, 8191, Marlinat® DFL 40, 18852, Neopon LT, 20950, Norfox® TLS, 21555, Perlankrol® ATL40, 23324, Stepanol® SPT, 29697, Texapon T 42, 31480, Ufarol TA-40, 32744, Zoharpon LAT, 34869
TEA phosphate ester, *See* Surfagene FPT, 30359
Tea tree oil, *See* EmCon TEA TREE, 10067
TEA-abietoyl hydrolyzed collagen, 30871
TEA-acrylates/acrylonitrogens copolymer, *See* Hypan® SS500W, 14704
Teaberry Oil, 30872
TEAC, *See* Sumquat® 2355, 30120
TEA-coco-hydrolyzed collagen, *See* Lexate BPQ, 17121
TEA-cocoyl-glutamate, *See* Acylglutamate CT-12, 599
TEA-dodecylbenzene sulfonate, *See* Bio-Soft® N-300, 4202, Calsoft T-60, 5031, Marlopon® AT, 18904, Rhodacal® DDB 60T, 26581, Sulfotex T-65, 30037, Witconate 60T, 34481
teal oil, *See* gingelly oil, 12946, Sesame oil, 28075
TEA-laureth sulfate, *See* Empicol® ETB, 10321
TEA-lauroyl hydrolyzed keratin amino acids, *See* Lipoproteol LK, 17401
TEA-lauryl sulfate, 30873, *See* Akyposal TLS42, 1322, Carsonol® TLS, 5340, Empicol® TA40, 10340, Empicol® TL40, 10342, Nutrapon TLS-500, 21827, Nutrapon TW 3987, 21828, Rewopol® TLS 40, 26383, Rhodapon® LT-6, 26638, Sulfetal KT 400, 29985, Sulfochem TLS, 30016, Witcolate TLS-500, 34479, Standapol® T, 29481, Witcolate TLS-500, 34479
TEA-lauryl sulfate-cocamide DEA-cocamidopropyl betaine, *See* Carsofoam® MSP, 5329
tea-lead, 30874
Tealox®, *See* aluminum oxide, 1919
Team, 30875
TEA-oleth-2 phosphate, *See* Crodafos O2 TEA, 7085
TEA-PEG-3 cocamide sulfate, *See* Genapol® AMS, 12785
Tears Plus®, 30876
Teatcote Plus, 30877
Tebithiuron, *See* Bushwacker, 4724
Tebol 88, 99, 30878
Tebuconazole, *See* Folicur, 12198, Horizon/Horizont, 14308, Silvacur, 28408, Raxil, 25908
tebutam, 30879, *See* Comodor, Comodor 600, 6667
tebutame, *See* tebutam, 30879
Tebuzate, *See* thiabendazole, 31661
Tec, 30880
Tecagg, 30881
Tecali onyx, *See* onyx of Tecali, 22187
Tecane, 30882
Teccel, 30883
Tec-Char, 30884
Tecfil, 30885
Tecgran, 30886
Techmate, 30887
Techne Coll, 30888
TechneScan MAA, 30889
TechneScan PYP, 30890
TechneScan SSC, 30891
Technetium $_{99m}$Tc salt of Disofenin, *See* Hepatolite, 13789
Technetium-$_{99m}$Tc sulfur colloid, *See* Tesuloid, 31353
Technical grade oxides and zeolites, *See* Baylith®, 3803
Technyl, 30892
Techroline, 30893
Techron, 30894
Techster, 30895
Techtron™ PPS, 30896
Teclam®, 30897
tecnazene, 30898, *See* Ashlade TCNB, 3199, Bygran S, 4821, Fumite TCNB, 12467, Fumite TCNB Smoke, 12468, Fusarex, 12513, Hickstor, 14062, Hystor 10, 14754, Hytec, 14775, Tripart® Arena 6, 32360
tecnazene-thiabendazole, *See* Hortag Tecnazene Plus, 14316, Hytec Super, 14776
Tecnocin, 30899
Tecnoflon®, 30900
Tecnoflon® FOR-45C2/R, 30901
Tecnoflon® FOR-65BI/R, 30902
Tecnoflon® FOR-LHF, 30903
Tecnoflon® NH, NM, NMB, NML, NMLB, 30904
Tecnoflon® P-1, P-2, P-2HV, P-40, 30905
Tecnoflon® TN-LATEX, 30906
Tecnoprene, 30907
Tecoflex Polyurethane, 30908
Tecpril, 30909
Tecquinol, *See* quinol, 25685
Tecquinol® Tech. Grade, 30910
Tecsol, *See* ethyl alcohol, 11064
Tecsol®, 30911 , *See* ethyl alcohol, 11064
Tectilon, 30912
Tecto, *See* thiabendazole, 31661
Tecto 10P, *See* thiabendazole, 31661
Tecto 40F, *See* thiabendazole, 31661
Tecto 60, *See* thiabendazole, 31661
Tecto RPH, *See* thiabendazole, 31661
Tecto, Tecto 60%, 30913
Tectrode, 30914
TEDA-D007, 30915
TEDA-L33, 30916
TEDA-T411, 30917
Tedimon, 30918
Tedion, *See* tetradifon, 31366
Tedion V-18, 30919, *See* tetradifon, 31366
Tedlar®, 30920
Tedur®, 30921
Tedur® KU1-9510-1, 30922
Tedur® KU1-9511, 30923
Tedur® KU1-9530, 30924
Tedur® KU1-9552, 30925
Tedur® KU1-9561, 30926
Teebrix, 30927
TeEDC, *See* Akrochem® TDEC, 1179
Teefroth, 30928
teel oil, *See* gingelly oil, 12946, Sesame oil, 28075
Teepol CM44, 30929
Teepol FC5, 30930
Teepol GD53, 30931
Teepol HB6, 30932
Teepol PB, 30933
teerlack, 30934
Tefaire®, 30935
Teflon, *See* polytetrafluoroethylene, 24640, 24642
Teflon®, 30936
Teflon® F.E.P, 30937
Tefose® 63, 30938
Tefose® 1500, 30939
Tefose® 2000, 30940
Tefose® 2561, 30941
Tefzel® 200, 30942
Tegamine® 18, *See* TEGO Amid S18, 30957
Tegamine® Oxide WS-35, *See* Aminoxid WS 35, 2196
Tegester, *See* Wickenol® 101, 105, 34378
Tegester 504-D, *See* Unimate® DIPA, 32968
Tegester isopalm, *See* Wickenol® 111, 34379
Tegiloxan®, 30943
Tegin P, *See* Witconol 2380, 34498
Tegin®, 30944
Tegin® 4011, 30945
Tegin® C-62 SE, *See* hydrogenated tallow glyceride citrate, 14599
Tegin® C-63 SE, *See* hydrogenated tallow glyceride citrate, 14599
Tegin® E-61, 61 NSE, *See* Acetylated hydrogenated lard glyceride, 315
Tegin® L61, L 62, *See* hydrogenated tallow glyceride lactate, 14600
Tegin® O, 30946
Tegin® V, 30947
Teginacid, 30948
Teginacid C, 30949
Teglac, 30950
TegMeR® 703, 30951
TegMeR® 803, 30952
TegMeR® 804 Special, 30953
TegMeR® 903, 30954
TEGO Amid S18, 30957
Tego Betain CK D, 30960
Tego Pearl N100, Tego Pearl N 300, *See* Tego® -Pearl S-33, 30986
Tego® Airex 900, 30955
Tego® Airex 960, 30956
Tego® -Antiflamm® N, 30958
Tego® -Antifoam, 30959
Tego® -Betaine C, 30961
Tego® betaine C, E, F, F50, L-7, L-7F, L-5351, S, ZF, *See* Cocamidopropyl betaine, 6551
Tego® -Betaine E, 30962
Tego® -Betaine HS, 30963
Tego® -Betaine L-5351, *See* Tego® -Betaine E, 30962
Tego® -Betaine L-7, 30964
Tego® -Betaine L-90, 30965
Tego® -Betaine S, 30966
Tego® Care 150, Care 300, 30967
Tego® Care 450, 30968
Tego® Dispers 610, 30969
Tego® Dispers 630, 30970
Tego® Effect L 104, 30971
Tego® Emulsion 3454, 30972
Tego® Emulsion ASL, 30973
Tego® Emulsion PK, 30974
Tego® Flow 425, 30975
Tego® Foamex 3062, 30977
Tego® Foamex 800, 30976
Tego® Foamex KS 10, 30978
Tego® Foamex N, 30979
Tego® Glide 100, 30980
Tego® Glide 410, 30981
Tego® Hammer 300000, 30982
Tego® Heat Conductive Paste Z, 30983
Tego® IMR 918, 30984
Tego® -Pearl B-48, 30985
Tego® -Pearl S-33, 30986
Tego® Phobe 1030, 30987
Tego® Phobe L 1004, 30988
Tego® Release Agent M 379, 30989
Tego® Silicone Acrylate RC, 30990
Tego® Silicone Paste A, 30991
Tego® WetKL 245, 30992
Tego®-Betaine L-10S, *See* lauramidopropyl betaine, 16830
Tegoamin®, 30993
Tegoamin® 33, 30994
Tegoamin® DMEA, 30995
Tegoamin® PMD, 30996
Tegoamin® SMP, 30997
Tego-betain BL 158, *See* Accobetaine CL, 222

Tegochrome® 22, 30998
Tegocoll®, 30999
Tegocolor®, 31000
Tegodont®, 31001
Tegodor, 31002
Tegoglätte, 31003
Tegoglas, 31004
Tegold, 31005
Tegomag, 31006
Tegoman, 31007
Tegomuls®, 31008
Tegomuls® 19, 31009
Tegomuls® B, 31010
Tegomuls® P 411, 31011
Tegopren®, 31012
Tegosept B, *See* Butylparaben, 4797, Nipabutyl, 21249
Tegosept butyl, *See* Nipabutyl, 21249
Tegosept M, *See* methylparaben, 19577
Tegosept P, *See* Nipasol M, 21282
Tegosil®, 31013
Tegosipon®, 31014
Tegosivin® HL 250, 31015
Tegosoft® 622LD, *See* caprylic/capric acid triglyceride, 5161
Tegosoft® CI, 31016
Tegosoft® CO, 31017, *See* cetyl octanoate, 5821
Tegosoft® CT, 31018
Tegosoft® DO, 31019
Tegosoft® EE, 31020
Tegosoft® GC, 31021
Tegosoft® Liquid, 31022, *See* Cetearyl octanoate, 5791
Tegosoft® Liquid M, 31023
Tegosoft® M, 31024
Tegosoft® OP, 31025
Tegosoft® P, 31026, *See* isopropyl palmitate, 15410
Tegosoft® S, 31027
Tegosoft® SH, 31028
Tegostab®, 31029
Tegostab® B 1048, 31030
Tegostab® B 2219, 31031
Tegostab® B 4900, 31032
Tegostab® B 8406, 31033
Tegostab® BF 2270, 31034
Tegotain D, *See* Tego Betain CK D, 30960
Tegotens 4100, 31035
Tegotens I, 31036
Tegotrenn® LH 157 A, 525, 31037
Tegovakon, 31038
Tegul, 31039
Tegula, 31040
Teka Oil, 31041
Tekblend, 31042
Tekemail, 31043
Tekfoam, 31044
Teknol, 31045
Tekpak-Tekbent, 31046
Tekpak-Tekcem, 31047
Tekstim 8504, 31048
Tekstim 8741, 31049
Tekwaisa, *See* Parathion-methyl, 22808
Telconax, 31050
Telconite, 31051
Telconstan, 31052
Telcothene®, 31053
Telcovin®, 31054
Telebar, *See* Radiopaque, 25830
Teleblock, 31055
telegraph bronze, 31056
telegraph metal, *See* telegraph bronze, 31056
Telepaque, 31057
Teletrast, *See* Telepaque, 31057
Telloy®, 31058
tellurac, *See* Akrochem® TDEC, 1179, ethyl tellurac®, 11084, Perkacit® TDEC, 23290
Tellurac®, 31059
telluretted hydrogen, 31060
Tellurit, 31061
tellurium, 31062, *See* Telloy®, 31058
Tellurium(IV) diethyldithiocarbamate, *See* Akrochem® TDEC, 1179, ethyl tellurac®, 11084, Perkacit® TDEC, 23290, Tellurac®, 31059
tellurium dioxide, *See* tellurium oxide, 31064
tellurium lead, 31063
tellurium oxide, 31064
Tellurium Tubes, 31065
Tellurium, tetrakis(diethylcarbamodithioato-S,S')-, (DD-8-1,1,1",1",1',1',1"',1"')-, *See* ethyl tellurac®, 11084
tellurium, tetrakis(diethyldithiocarbamate)-, *See* Akrochem® TDEC, 1179, ethyl tellurac®, 11084
tellurous acid anhydride, *See* tellurium oxide, 31064
Telmin, 31066 Mebendazole, 19198, Vermox, 33591
Telogen, 31067
Telon®, 31068
Telone, 31069
Telone 2000, *See* Telone, 31069
Telone II, *See* Telone, 31069
Telopar, 31070
Telsit, 31071
Teluran, 31072
Telvar®, 31073
Temadex, 31074
Temasept, *See* Tuasol 100, 32543
Temasept I, 31075
Temasept IV, 31076, *See* Tuasol 100, 32543
Tembind A 002, 31077, *See* Ammonium lignosulfonate, 2243, Lignosol TS, 17240
Temephos, *See* Abate 1-SG, 2-CG, 4-E, 5CG, 26
Temlo, *See* Valadol, 33238
Temlock, 31078
Tempaloy, 31079
temper, 31080
Tempered lead, *See* Noheet metal, 21421
Temperite Alloys, 31081
Tempo, 31082, 31083, 31084
Tempo 2, *See* cyfluthrin, 7563
Tempra, *See* Valadol, 33238
Tempro, 31085
Temsperse S 001, 31086
Tem-Tuf, 31087
Tenac, *See* dichlorvos, 8394
Tenac®, 31088
Tenacite, *See* Bakelite, 3589
Tenacite, *See* Bakelite, 3589
Tenacity, 31089
Tenamene, *See* Naugard® I-2, 20792
Tenamene 2, *See* Naugalube® 403, 20782
Tenamine 1, 31090
Tenamine 2, 31091
Tenamine 3, 31092
Tenasco, 31093
Tenase® 1200, L-340, L-1200, 31094
Tenatine, 31095
Tenax metal, 31096
Tenax Wax, 31097
Tenaxatex VA 632, 31098
Tenaxatex VA 956, 31099
Tenazit, 31100
Tendex, *See* Propoxur, 25198, 25199, 25200, Baygon®, 3790
Tendrelle, 31101
Tendril®, 31102
Tenephrol, 31103
Teniathane, *See* Dichlorophen, 8392, Nuophene, 21783
Teniatol, *See* Dichlorophen, 8392, Nuophene, 21783
Teniotol, *See* Nuophene, 21783
Tenite® PET 9902, 31111, *See* Polyethylene terephthalate, 24426
Tenite® 105-MS, 31104
Tenite® 154DF, 31105
Tenite® 264-MH, 31106
Tenite® 360-H2, 31107, *See* Cellulose acetate propionate, 5630
Tenite® Cellulosic Acetate, 31108
Tenite® Cellulosic Butyrate, 31109
Tenite® Cellulosic Propionate, 31110, *See* Cellulose acetate propionate, 5630
Tenite® Polyethylene, 31112
Tennafast, 31113
Tennafast, Tennalaks, 31114
Tennal, 31115
Tenncol, 31116
Tennessee phosphates, 31117
Tennplas, *See* Benzoic acid, 3973
Tenoban, 31118
Tenoran, 31119, *See* chloroxuron, 6158
Tenox, *See* Nipantiox, 21276
Tenox BHA, *See* Butylated hydroxyanisole, 4787, Sustane® 1-F, 30432
Tenox BHT, *See* Anti-Oxydant Bayer, 2595, Butylated hydroxytoluene, 4788, Oxyguard, 22457, Ralox® BHT food grade, 25845, Tenamine 3, 31092, Vanox® PCX, 33346, Vulkanox® KB, 34152
Tenox PG, *See* Sustane® PG, 30436
Tenox® 2, 31120
Tenox® TBHQ, *See* *t*-Butylhydroquinone, 4793
Tenrez, 31121
Tensabit, 31122
Tensactol, 31123
Tensadal, 31124
Tensagex, 31125
Tensagex BV, 31126
Tensagex DMY, 31127
Tensagex DP24, 31128
Tensagex EOC, 31129
Tensagex SPDL, 31130
Tensami 1/05, 31131
Tensami 3/06, 31132
Tensami 4/07, 31133
Tensami 8/09, 31134
Tensami 10/06, 31135
Tensamina, 31136
Tensamine C, O, S, and SH, 31137
Tensaminox, 31138
Tensarane SBTE, 31139
Tensaryl, 31140
Tensaryl 40CC, 50B, 80B, and 82F, 31141
Tensaryl DF90, 31142
Tensaryl DX54Sp. and DX62, 31143
Tensaryl KD, 31144
Tensaryl L48, 31145
Tensaryl S30P and S70P, 31146
Tensaryl SB, 31147
Tensaryl SB Ca, 31148
Tensaryl SB85P, 31149
Tensaryl SBD, 31150
Tensatil D100, 31151
Tensatil DA120, 31152
Tensatil DB120, 31153
Tensatil DEH120, 31154
Tensiamix, 31155
Tensianol, 31156
Tensibet 50, 31157
Tensibet 55, 31158
Tensidef, 31159
Tensidye, 31160
Tensilac 39, 31161
Tensilac 40, 31162
Tensilac 41, 31163
Tensilite, 31164
Tensimul, 31165
Tensiofix, 31166
Tensioquat C50, 31167
Tensioquat C75, 31168
Tensiorex, 31169
Tensiostat, 31170
Tensipar, 31171
Tensitex, 31172
Tensloy, 31173
Tensocide, 31174
Tensol, 31175, 31176, 31177
Tensoleate, 31178
Tensoline, 31179
Tensomel, 31180
Tensomin, 31181
Tensopac, 31182
Tensopane D, 31183

Chemical and Trade Names

Tensophene 2D30, 31184
Tensophene D12, D15, D18, 31185
Tensophene H10, I10, DT, D36, D42EC, D45, D60, D90, 31186
Tensopol 12A, 12P, 31187
Tensopol 30E, LDS, 31188
Tensopol A, 7, USP, 31189
Tensopol ACL, PCL, 31190
Tensopol AG, MG, 31191
Tensopol DX85, FL, 31192
Tensopol LT, 31193
Tensopol N, 31194, *See* Ammonium lauryl sulfate, 2242
Tensopol SPK, 31195
Tensopol VAL, 31196
Tensoprene, 31197
Tensostat, 31198
Tensovax, 31199
Tensovyl, 31200
Tensuccin D8, 31201
Tensuccin H724, H925, 31202
Tensuccin HS40, 31203
Tensuccin ML, MO, MS, 31204
Tensyl 30, 31205
Tensynvac, 31206
TEOS, *See* Silester, 28303
TEP, *See* triethyl phosphate, 32243
TEPA, *See* tetraethylenepentamine, 31367
Tephal, 31207
Tepperite, 31208
teraffine, *See* Adol® 62 NF, 783
Teralan, 31209
Teranol, *See* tetradifon, 31366
Teraprint, 31210
Terasil, 31211
Terate 101, 131, 31212
Terate 202, 203, 204, 31213
Terate 203, 31214, 31215
Terate 204, 31216, 31217
Terathane, 31218
Terathane® 650, 31219
terbacil, 31220
Terbalin, 31221
terbia, *See* terbium oxide, 31223
terbium, 31222
terbium oxide, 31223
Terblend®, 31224
Terblend® S, 31225
terbuconazole, *See* Raxil, 25908
Terbufos, 31226
Terbuthylazine, 31227, *See* Tyllanex, 32658
Terbutol, 31228, *See* 4-*t*-butyl phenol, 4800
terbutrazole, *See* Raxil, 25908
Terbutrex, 31229, *See* terbutryn, 31230
terbutryn, 31230
terbutryne, *See* Terbutrex, 31229, terbutryn, 31230
Terbytex, 31231
Tercoton, 31232
terebene, 31233
terebine, 31234
Terechloroterephthalic acid dimethyl ester, *See* dimethyl tetrachloroterephthalate, 8512
Teremec, *See* Terraneb SP Turf Fungicide, 31337
Terephane, 31235, 31236
Terephthalic acid, 2,3,5,6-tetrachloro-, dimethyl ester, *See* dimethyl tetrachloroterephthalate, 8512
Terethane®, 31237
Tereton, *See* methyl acetate, 19506
Terfenol, 31238
Terg-A-Zyme®, 31239
Tergemist, *See* Niaproof® Anionic Surfactant 08, 21092, Rhodapon® BOS, 26632
Tergenol 1122, 31240
tergimist, *See* Rhodapon® BOS, 26632
Tergitex KW, 31241
Tergitol 08, *See* Niaproof® Anionic Surfactant 08, 21092, Rhodapon® BOS, 26632
Tergitol 4, *See* Niaproof® Anionic Surfactant 4, 21093
Tergitol anionic 08, *See* Niaproof® Anionic Surfactant 08, 21092, Rhodapon® BOS, 26632
Tergitol NP, *See* Alkasurf® NP-4, 1715, Triton® N-57, 32430
Tergitol NP-14, *See* Alkasurf® NP-4, 1715, Triton® N-57, 32430
Tergitol NP-27, *See* Alkasurf® NP-4, 1715, Triton® N-57, 32430
Tergitol NP-33, *See* Alkasurf® NP-4, 1715
Tergitol NP-35, *See* Alkasurf® NP-4, 1715, Triton® N-57, 32430
Tergitol NP-40, *See* Alkasurf® NP-4, 1715, Triton® N-57, 32430
Tergitol NPX, *See* Alkasurf® NP-4, 1715, Triton® N-57, 32430
Tergitol TP-9, *See* Alkasurf® NP-4, 1715, Triton® N-57, 32430
Tergitol TP-9 (non-ionic), *See* Alkasurf® NP-4, 1715, Triton® N-57, 32430
Tergitol®, 31242
Tergitol® 15-S-3, 31243
Tergitol® 24-L-45, 31244, *See* Laureth-2, 16852
Tergitol® 26-L-3, 31245
Tergitol® D-683, 31246
Tergitol® Min-Foam 1X, 31247
Tergitol® TMN-3, 31248
Tergitol® XD, 31249
Tergitol® XH, 31250
Tergolix, 31251
Tergraf, 31252
Tergum, 31253
Teric 12A2, 31255
Teric 12M2, 31256
Teric 13A5, 31257
Teric 15A11, 31258
Teric 16A16, 31259
Teric 16M2, 31260, *See* PEG soyamine series, 23033
Teric 17A2, 31261
Teric 17M2, 31262
Teric 18M5, 31263
Teric 9A2, 31254
Teric BL8, 31264
Teric CME7, *See* Ablumide SME, 82, Witcamide® 70, 34442
Teric DD5, 31265, *See* Dodoxynol series, 8786
Teric G12A4, 31266
Teric G9A5, 31267
Teric LA4, 31268
Teric N2, 31269
Teric OF4, 31270
Teric PE61, 31271
Teric PEG 200, 31272
Teric PEG 300, 31273
Teric PEG 400, 31274
Teric PEG 600, 31275
Teric PEG 800, 31276
Teric PEG 1000, 31277
Teric PEG 1500, 31278
Teric PEG 3350, 31279
Teric PEG 4000, 31280
Teric PEG 6000, 31281
Teric PEG 8000, 31282
Teric PEG 12000, 31283
Teric PPG 400, 31284
Teric PPG 1000, 31285
Teric PPG 1650, 31286
Teric PPG 2250, 31287
Teric PPG 4000, 31288
Teric X5, 31289
Terinda, 31290
Terlac, 31291
Terlan, 31292
Terluran, 31293
Terluran 846 L, 31294
Terluran 8760 Galvano, 31296
Terluran 886, 31295
Terluran®, 31297, *See* Acrylonitrile-butadiene-styrene, 439
Terlux®, 31298 , *See* Acrylonitrile-butadiene-styrene, 439
Termamyl®, 31299
Termamyl® 120L, 31300
Termex, 31301, 31302
termiertie, *See* Tisyn®, 31857
Termil, *See* Chlorothalonil, 6150, Repulse, 26151
Terminate, 31303
Term-X, 31304
ternary steels, 31305
Terne metal, 31306
Terne plate, 31307
Terohane, 31308
teroylglutamic acid, *See* vitamin M, 33951
Terpal®, 31309
Terpal® CC, M, 31310
Terpanol, 31311
Terpenato, 31312
Terpene hydrocarbon, *See* α-pinene, 23857, β-pinene, 23859
Terpene phenolic, *See* Super Beckacite® 2000, 30166
terpene polychlorinates, *See* Strobane, 29818
terpene resin, 31313
terpestrol, 31314
Terpex D, K-3, S, 31315
Terphane®, 31316
Terpigol, 31317
terpine, 31318
Terpineol, *See* Lily of valley, artificial, 17258
Terpinoxo, 31319
Terposol No. 3, 31320
Terposol No. 8, 31321
Terpurile, 31322
terra alba, *See* gypsum, 13518
terra catechu, *See* cutch,
terra fullonica, 31323, *See* Fuller's earth, 12441
terra Japonica, *See* cutch, gum catechu, 13483
terra Merita, *See* turmeric, 32608
Terra Nova, 31324
terra ponderosa, 31325
terra verte, 31326
Terraclor, *See* quintozene, 25692
Terracote, 31327
terra-cotta, 31328
Terracur® P, 31329
Terradust, 31330
Terrafen®, 31331
Terragloss, 31332
Terraklene, 31333
Terralacke, 31334
Terram, 31335
Terramix CDA, 31336
Terraneb SP Turf Fungicide, 31337
Terranox, 31338
Terrapaint, 31339
Terrapowder, 31340
terrar, 31341
Terra-Systam, 31342
Terrathion, 31343
Terravest 801, 31344
Terrazan, *See* quintozene, 25692
terre verte, *See* Bohemian earth, 4368
Terr-O-Gas, 31345
Terr-O-Gas 100, *See* methyl bromide, 19517
terroline, *See* paraffin, liquid, 22744
Terry's stain, 31346
Tersan, *See* Rezifilm, 26521, Thiram, 31726, Thiurad, 31730
Tersan 75, *See* Rezifilm, 26521
Tersan SP, *See* Terraneb SP Turf Fungicide, 31337
Tersaseptic, *See* hexachlorophene, 13992
Terset, 31347
tertiary acetylenic alcohol, *See* methyl butynol, 19518
tertiary amine, *See* Texacat® DD, 31426
tertiary amine salts, *See* Amicure® SA, 2079
tertiary calcium phosphate, *See* calcium phosphate (tribasic), 4967
tertiary zinc oxide, *See* zinc oxide, 34786
Tertral P Base, *See* *p*-aminophenol, 2185
Tertrogras orange sv, *See* Sudan I, 29938
tertrosulfur black pb, *See* Dinitra, 8540
tertrosulfur pbr, *See* Dinitra, 8540

Terylene, 31348, *See* A-C® 6, 168
Tesal, 31349
Tescol®, 31350
testalin, 31351
Testascorbic, *See* vitamin C, 33938
Testavol, *See* Vi-Alpha, 33706
testifas oil, 31352
Testo, *See* thiabendazole, 31661
Tesuloid, 31353
TETA, *See* triethylenetetramine, 32245
Tetanol, 31354, *See* calcium hevulinate,
TETD, *See* Akrochem® TETD, 1180
tetiothalein sodium, *See* keraphen, 16055
tetjamer, 31355
Tetra, *See* tetraphosphate, 31390
Tetra (2,2 diallyloxymethyl) butyl, di(ditridecyl) phosphito zirconate, *See* KZ 55, 16523
Tetra (2-hydroxypropyl) ethylenediamine, *See* Quadrol®, 25582
tetra olive n2g, *See* Sterilite Hop Defoliant, 29734
tetra(2-hydroxypropyl) ethylenediamine, 31356
Tetraamminecopper sulfate, *See* Krystallazurin, 16468
Tetraammonium EDTA, *See* Sequestrene® Tetraammonium, 28003
Tetraamonium salt type polymer, *See* Adeka Catioace DM, PD Series, 658
Tetraazatetradecanediimidamide, N,N''-bis(4-chlorophenyl)-3,12-diimino-, diacetate, *See* chlorhexidine acetate, 6100
Tetraazatetradecanediimidamide, N,N''-bis(4-chlorophenyl)-3,12-diimino, di-D-gluconate, *See* Rotersept, 27002
Tetra-Base, *See* Michler's Base, 19644
Tetra-base-paper, *See* tetra-paper, 31389
N,N,N',N'-Tetrabenzylthiuram disulfide, *See* Benzyl Tuex®, 3990
Tetrabon, *See* tetracycline, 31365
Tetrabor®, 31357
Tetrabrom, *See* Bromeikon, 4568
tetrabromo phthalatediol, 31358
tetrabromo-*m*-cresolsulfonphthalein, *See* Bromocresol green, 4575
tetrabromoacetylene, *See* Muthmann's liquid, 20541
tetrabromobisphenol A, 31359, *See* FR-1524, 12367, Great Lakes BA-59P, 13298, Saytex® RB-100, 27523
tetrabromobisphenol-A allyl ether, *See* FR-2124, 12368
tetrabromodipentaerythritol, 31360, *See* FR-1034, 12361
tetrabromoethane, *See* Muthmann's liquid, 20541
***s*-Tetrabromoethane**, *See* Muthmann's liquid, 20541
***sym*-Tetrabromoethane**, *See* Muthmann's liquid, 20541
1,1,2,2-tetrabromoethane, *See* Muthmann's liquid, 20541
1,1,2,2-tetrabromoethylene, *See* Muthmann's liquid, 20541
4,5,6,7-tetrabromo-1,3-isobenzofurandione, *See* Great Lakes PHT4, 13312, Saytex® RB-49, 27524
2,2',6,6'-tetrabromo-4,4'-isopropylidene phenol, *See* Saytex® RB-100, 27523
Tetrabromophthalatediol, *See* Great Lakes PHT4-Diol, 13313, tetrabromo phthalatediol, 31358
tetrabromophthalic anhydride, 31361, *See* Great Lakes PHT4, 13312
3,4,5,6-tetrabromophthalic anhydride, *See* Great Lakes PHT4, 13312, Saytex® RB-49, 27524
tetrabutoxypropyl methicone, *See* Masil® 756, 19012
tetrabutoxysilane, *See* CT1750, 7311
tetrabutyl ammonium bromide, 31362, *See* TBAB, 30851
tetrabutyl orthosilicate, *See* CT1750, 7311, T1750, 30675
tetrabutyl silicate, *See* CT1750, 7311
tetrabutyl thiuram disulfide, *See* Akrochem® TBUT, 1178
tetrabutylthioperoxydicarbonic diamide, *See* tetrabutylthiuram disulfide, 31363
tetrabutylthiuram disulfide, 31363
tetracaprate, *See* Liponate PE-810, 17384
Tetracarnit, 31364
tetracemate disodium, *See* edetate disodium, 9613, Trilon® BD, 32325
tetracemate tetrasodium, *See* Kalex 220 Crystal, 15668, Kalex Liq. 50%, 15671
tetracemin, *See* Kalex 220 Crystal, 15668, Kalex Liq. 50%, 15671
tetrachloro-1,3-dicyanobenzene, *See* Repulse, 26151
2,4,5,6-Tetrachloro-1,3-benzenedicarbonitrile, *See* Repulse, 26151
2,3,5,6-tetrachloro-1,4-benzenedicarboxylic acid dimethyl ester, *See* Vegetable Turf and Ornamental Weeder, 33526
2,3,5,6-tetrachloro-1,4-benzenedicarboxylate, *See* dimethyl tetrachloroterephthalate, 8512
2,3,5,6-Tetrachloro-1,4-benzenedicarboxylic acid dimethyl ester, *See* dimethyl tetrachloroterephthalate, 8512
2,3,5,6-tetrachloro-*p*-benzoquinone, *See* Vulklor, 34159
2,3,5,6-tetrachloro-2,5-cyclohexadiene-1,4-dione, *See* chloranil, 6087, Vulklor, 34159
2,4,5,6-tetrachloro-1,3-dicyanobenzene, *See* Repulse, 26151
4,5,6,7-tetrachloro-3',6'-dihydroxy-2',4',5',7'-tetraiodospiro[isobenzofuran-1(3H),9'-[9H]xanthen]-3-one-disodium salt, *See* Robengatope I-131, 26871, rose bengal, 26963
2,4,4',5-tetrachlorodiphenyl sulfone, *See* Tedion V-18, 30919
Tetrachloroethane, *See* Acetosol, 304
***sym*-tetrachloroethane**, *See* Acetosol, 304
1,1,2,2-tetrachloroethane, *See* Acetosol, 304
tetrachloroethene, *See* perchloroethylene, 23222
tetrachloroethylene, *See* Distillex DS4, 8720, Dowper, 8942, Nema, 20861, perchlorethylene, 23220, 23222
2,4,5,6-tetrachloroisophthalonitrile, *See* Repulse, 26151
4,5,6,7-tetrachloro-1,3-isobenzofurandione, *See* Tetrathal, 31393
Tetrachloroisophthalonitrile, *See* Nopcocide® N-96, 21488, NuoCide®, 21773, Repulse, 26151, Siclor, 28204
2,4,5,6-Tetrachloroisophthalonitrile, *See* Nopcocide® N-96-S, 21489
tetrachloromethane, *See* carbon tetrachloride, 5224, Katharin, 15791
1,2,4,5-tetrachloro-3-nitrobenzene, *See* Tecgran, 30886, tecnazene, 30898
2,3,5,6-tetrachloronitrobenzene, *See* Tripart® Arena 6, 32360
tetrachlorophenol, *See* Dowicide 6, 8930
2,3,4,5-tetrachlorophenol, *See* Dowicide 6, 8930
Tetrachlorophthalic anhydride, *See* Tetrathal, 31393
***m*-tetrachlorophthalodinitrile,** *See* Repulse, 26151
***m*-tetrachlorophthalonitrile,** *See* Repulse, 26151
tetrachloroquinone, *See* Vulklor, 34159
tetrachlorosilane, *See* CT1800, 7312 silicon tetrachloride, 28333
tetrachlorosilicon, *See* CT1800, 7312
2,3,5,6-tetrachloroterephthalic acid dimethyl ester, *See* Vegetable Turf and Ornamental Weeder, 33526
4,5,6,7-Tetrachloro-2',4',5',7'-tetraiodofluorescein potassium derivative potassium salt, *See* rose bengal, 26963
Tetrachlorquinone, *See* chloranil, 6087
tetracycline, 31365
Tetracyn, *See* tetracycline, 31365
tetradec-1-ene, *See* Neodene® 14, 20897
tetradecabromodiphenoxy benzene, *See* Saytex® 120, 27516
tetradecafluorohexane, *See* Flutec PP1, 12121
[1S-(1α,4aα,4bβ,7β,8aα,10aβ)-tetradecahydro-1,4a-dimethyl-7-(1-methylethyl)phenanthrene, *See* fichtelite, 11765
1-tetradecanamine, *See* Amine 14D, 2129
tetradecanaminium, N,N,N-trimethyl-, bromide, *See* Rhodaquat® M214B/99, 26642, Sumquat® 6110, 30125
1-tetradecanol, *See* myristyl alcohol, 20577
tetradecanoic acid, *See* myristic acid, 20574, Philacid 1400, 23662
n-tetradecanoic acid, *See* Philacid 1400, 23662
tetradecanoic acid n-butyl ester, *See* Wickenol® 141, 34384
tetradecanoic acid 1-methylethyl ester, *See* isopropyl myristate, 15409, Kessco® IPM, 16115, Radia® 7190, 25729, Unimate® IPM, 32971, Wickenol® 101, 105, 34378
tetradecanoic acid, methyl ester, *See* methyl myristate, 19550
tetradecanol, *See* Epal® 14, 10619
1-tetradecanol, *See* myristyl alcohol, 20577
1-tetradecene, *See* Neodene® 14, 20897
tetradecene-1, *See* Neodene® 14, 20897
tetradecenyl succinic anhydride, *See* Milldride® TDSA, 19790
Tetradecin, *See* tetracycline, 31365
n-tetradecoic acid, *See* Philacid 1400, 23662
tetradecyl alcohol, *See* myristyl alcohol, 20577
tetradecyl dimethyl benzyl ammonium chloride, *See* JAQ Powdered Quaternary, 15517
tetradecyl dimethylamine oxide, *See* Lilaminox M4, 17252
tetradecyl lactate, *See* Wickenol® 506, 34400
tetradecylamine, *See* Amine 14D, 2129
1-tetradecylamine, *See* Amine 14D, 2129
n-tetradecylamine, *See* Amine 14D, 2129
tetradecyl 2-hydroxypropanoate, *See* myristyl lactate, 20578
tetradecyltrimethylammonium bromide, *See* myrtrimonium bromide, 20589, Rhodaquat® M214B/99, 26642, Sumquat® 6110, 30125
tetradifon, 31366
tetradine, *See* lycetol, 18063
Tetradine, *See* Disulfiram, 8725
Tetradiphon, *See* tetradifon, 31366
tetradonium bromide, *See* myrtrimonium bromide, 20589, Sumquat® 6110, 30125, Rhodaquat® M214B/99, 26642
tetraethoxysilane, *See* Dynasil® A, 9382, Silester, 28303
O,O,O',O'-Tetraethyl-S,S'-methylene di(phosphoro dithioate), *See* Ethion, 10955
tetraethyl orthosilicate, *See* ethyl silicate, 11083, Silester, 28303
tetraethylene glycol diacrylate, *See* Ageflex T4EGDA, 970
tetraethylene glycol dicocoate, *See* Lexolube® 2J-237, 17156
tetraethylene glycol dimethacrylate, *See* SR-209, 29365
tetraethylene glycol dimethyl ether, *See* Ansol E-181, 2495
tetraethylenepentamine, 31367, *See* Texlin® 400, 31508
tetraethylrhodamine hydrochloride, *See* Rhodamine B, 26616
Tetraethylthioperoxydicarbonic diamide, *See* Disulfiram, 8725
Tetraethylthiuram disulfide, *See* Ancazide ET, 2419, Disulfiram, 8725, ethyl tuads®, 11085, ethyl tuex, 11086, ethylthiurad, 11144, Perkacit® TETD, 23291
Tetra-ethylthiuram monosulfide, *See* Monosulfiram, 20257
Tetraetil, *See* Disulfiram, 8725
Tetrafidon, *See* tetradifon, 31366
Tetrafilcon A, *See* Aosoft, 2639, Aquaflex, 2691, Permathin, 23381, Superthin, 30287
Tetraflon, 31368
tetrafluorethene, *See* MS-122, 20413
tetrafluorethylene telomer, *See* MS-122, 20413
tetrafluoroethane, *See* Genetron® 134a, 12824
tetrafluoroethene homopolymer, *See* Polytetrafluoroethylene, 24641, 24642
tetrafluoroethylene polymer, *See* polytetrafluoroethylene, 24641, 24642
tetrafluoroethylene resin, *See* polytetrafluoroethylene,

24640
tetrafluorosilane, *See* silicon tetrafluoride, 28334
tetraform, 31369
tetraglyme, *See* Ansol E-181, 2495
(tetrahydrate) hopeite, *See* zinc phosphate, 34790
tetrahydro-*p*-dioxin, *See* Dioxane, 8574
tetrahydro-1,4-dioxin, *See* Dioxane, 8574
tetrahydro-1,4-oxazine, *See* morpholine, 20346
tetrahydro-2,5-dimethoxyfuran, *See* Protectol® DMT, 25260
tetrahydro-2,5-dioxofuran, *See* succinic anhydride, 29916
tetrahydro-2-furancarbinol, *See* tetrahydrofurfuryl alcohol, 31371
tetrahydro-2-furanmethanol, *See* tetrahydrofurfuryl alcohol, 31371
tetrahydro-2-furylmethanol, *See* tetrahydrofurfuryl alcohol, 31371
tetrahydro-2H-1,4-oxazine, *See* morpholine, 20346
tetrahydro-3,5-dimethyl-2H-1,3,5-thiadiazine-2-thione, *See* Dazomet, 7806, N-521® Biocide, 20640
(S)-2,3,5,6-tetrahydro-6-phenylimidazo[2,1-b]thiazole, *See* Ketrax, 16149
(2R,2α,6aα,12aα)-1,2,12,12a-tetrahydro-8,9-dimethoxy-2-(1-methylethenyl)[1]benzopyrano[3,4-b]furo[2,3-H][1]benzopyran-6(6aH)-one, *See* rotenone, 27000
tetrahydrofuran, 31370, *See* Dynasolve 150, 9387, QO® Tetrahydrofuran (THF), 25558
tetrahydrofurfuryl acrylate, *See* SR-285, 29375
tetrahydrofurfuryl alcohol, 31371, *See* QO® Tetrahydrofurfuryl Alcohol (THFA®), 25559 THFA, 31660
tetrahydrofurfuryl methacrylate, 31372, *See* Ageflex THFMA, 999 SR-203, 29361
tetrahydrofurfuryl oleate, *See* Kemester® THFO, 15972
tetrahydrofuryl carbinol, *See* tetrahydrofurfuryl alcohol, 31371
[2S-(2α,4β,5α)]-tetrahydro-4-hydroxy-N,N,N,5-tetramethyl-2-furanmethanaminium, *See* muscarine, 20524
1,2,3,4-Tetrahydro-2-[[(1-methyl-ethyl)amino]methyl]-7-nitro-6-quinolinemethanol, *See* Vansil, 33363
(E)-1,4,5,6-tetrahydro-1-methyl-2-[2-(3-methyl-2-thienyl)ethenyl]pyrimidine tartrate, *See* morantel tartrate, 20307
1,2,5,6-tetrahydro-1-methyl-3-pyridinecarboxylic acid, *See* arecaidine, 2828
1,2,5,6-tetrahydro-1-methyl-3-pyridinecarboxylic acid methyl ester, *See* arecoline, 2829, Cestarsol, 5779
1,2,5,6-tetrahydropyridine-3-carboxylic acid, *See* guvacine, 13512
(E)-3-[2-(1,4,5,6-tetrahydro-1-methyl-2-pyrimidinyl)ethenyl]phenol pamoate, *See* Telopar, 31070
tetrahydronaphthalene, 31373
1,2,3,4-tetrahydronaphthalene, *See* tetrahydronaphthalene, 31373
1,2,3,4-tetrahydro-2-naphthalenol, *See* Tetralol, 31381
***ac*-tetrahydro-β-naphthol**, *See* Tetralol, 31381
tetrahydro-naphthol acetate, *See* Solvol, 29065
3-(1,2,3,4-Tetrahydro-1-naphthyl)-4-hydroxycoumarin, *See* Racumin®, 25713
1,2,5,6-tetrahydronicotinic acid, *See* guvacine, 13512
(4,5,6,7-tetrahydro-7-oxobenzo[b]thien-4-yl)urea, *See* Vigazoo, 33749
(S)-2,3,5,6-tetrahydro-6-phenylimidazo[2,1-b]thiazole hydrochloride, *See* Tramisol®, 32134
N-(3,4,5,6-tetrahydrophthalimide)methyl-*cis,trans*-chrysanthemate, *See* Killgerm® Py-Kill W, 16182
1,2,3,4-tetrahydro-2-[(isopropylamino)methyl]-7-nitro-6-quinolinemethanol, *See* Vansil, 33363
3a,4,7,7a-tetrahydro-2-[(1,1,2,2-tetrachloroethyl)thio]-1H-isoindole-1,3(2H)-dione, *See* captafol, 5163
tetrahydrothiophene, 31374, *See* Pennodorant® 1013, 23117
3a,4,7,7a-Tetrahydro-2-[(trichloromethyl)thio]-1H-isoindole-1,3(2H)dione, *See* Vancide® 89, 33302
Tetrahydroxyanthraquinone, *See* rufiopin, 27128
2,2',4,4'-tetrahydroxy benzophenone, *See* benzophenone-2, 3977, Uvinul® D-50, 33195
2,2,4,4-Tetrahydroxybenzophenol, *See* Benzophenone-2, 3977
3,4',5,7-tetrahydroxyflavone, *See* kaempferol, 15654
tetrahydroxymethylanthraquinone, *See* fisetin, 11849
2,2,6,6,-tetra(hydroxymethyl)-4-oxaheptane-1,7-diol, *See* Dipentaerythritol, 8581
tetrahydroxypropyl ethylenediamine, *See* Mazeen® 173, 19120, Neutrol® TE, 21036
Tetraiode, *See* keraphen, 16055
tetraiodophenolphthalein, *See* Apagallin, 2640, Kerasol, 16058
tetraiodophenolphthalein sodium, *See* keraphen, 16055
4,5,6,7-tetraiodophenolphthalein sodium, *See* TIP, 31847
tetraiodophthalein, *See* Kerasol, 16058
tetraiodophthalein sodium, *See* keraphen, 16055
tetraiodopyrrole, *See* iodol, 15215
1,1,3,3-tetraisopropyl-1,3-dichlorodisiloxane, *See* CD4368, 5503
tetrakal, 31375
tetrakis (2,2,6,6-tetramethyl-4-piperidyl)-1,2,3,4-butane tetracarboxylate, *See* Mixxim® HALS 57, 19991
tetrakis (2,4-di-t-butylphenyl) 4,4'-biphenylylene diphosphonite, *See* Alkanox® 24-44, 1702
tetrakis (2,4-di-tertbutylphenyl) 4,4-biphenylenediphosphonite, *See* Sandostab P-EPQ, 27353
tetrakis (2-ethoxyethoxy)silane, *See* Dynasil® CA, 9383
tetrakis (2-ethylbutoxy) silane, *See* T1920, 30676
tetrakis (2-methoxyethoxy) silane, *See* Dynasil® CM, 9384
tetrakis [methylene (3,5-di-*t*-butyl-4-hydroxyhydrocinnamate)] methane, *See* Ralox® 630, 25844
tetrakis(diethylcarbamodithioato-S,S')tellurium, *See* ethyl tellurac®, 11084
tetrakis(hydroxymethyl)methane, *See* pentaerythritol, 23138
tetrakis[methylene(3,5-di-*tert*-butyl-4-hydroxyhydrocinnamate)methane], *See* Naugard® 10, 20788
N,N',N'',N'''-tetrakis(4,6-bis(butyl-(N-methyl-2,2,6,6-tetramethylpiperidin-4-yl)-amino)triazin-2-yl)-4,7-diazadecane-1,10-diamine, *See* Chimassorb® 119FL, 6072
tetrakis(diethylcarbamodithioato-S,S')tellurium, *See* Akrochem® TDEC, 1179
N,N,N',N'-tetrakis(2-hydroxypropyl)ethylenediamine, *See* tetra(2-hydroxypropyl) ethylenediamine, 31356
Tetralex Plus, 31376
Tetralide®, 31377
tetralin, *See* tetrahydronaphthalene, 31373
Tetralin Extra, 31378
tetraline, 31379, *See* tetrachloroethane,
(α-tetralinyl)-4-hydroxycoumarin, *See* Racumin®, 25713
tetralitbenzol, 31380
Tetralite, *See* Tetryl, 31423
Tetralol, 31381
***ac*-β-tetralol**, *See* Tetralol, 31381
Tetralon®, 31382
3-(α-Tetral)-4-oxycoumarin, *See* Racumin®, 25713
3-(α-Tetralyl)-4-hydroxycoumarin, *See* Racumin®, 25713
3-(D-Tetralyl)-4-hydroxycoumarin, *See* Racumin®, 25713
tetramer, *See* metaldehyde, 19431
Tetramet-125, 31383
tetramethoxysilane, *See* Dynasil® M, 9385
tetramethrin, 31384, *See* Killgerm® Py-Kill W, 16182
tetramethyl decynediol, *See* Surfynol® 104, 30376
tetramethyl silane, *See* CT2050, 7315
tetramethyl silicane, *See* CT2050, 7315
tetramethylthiuram disulfide, *See* Rezifilm, 26521
tetramethyl thiuram disulfide (2/3) + 2-mercaptobenzthiazole (1/3), *See* Accelerator 108, 190
tetramethyl thiuram monosulfide, *See* Ancazide IS, 2420, TMTM, 31906
tetramethyl-*p*-phenylenediamine, *See* Wurster's blue, 34576
tetramethyl-1,3,5,7-tetroxocane, *See* metaldehyde, 19431
tetramethylammonium hydroxide, 31385
tetramethylammonium oxalate, *See* Albiogen, 1376
tetramethylaniline, *See* Isodurindine, 15361
1,2,4,5-tetramethylbenzene, *See* Durene, 9195
(1,1,4,4-tetramethyl-1,4-butanediyl)bis[(1,1-dimethylethyl) peroxide, *See* Aztec® 2,5-Di, 3551, dimethyldibutyl peroxyhexane, 8518, Polyvel CR-5F, 24657, Trigonox® 101, 32279
***p*-(1,1,3,3-tetramethylbutyl)phenol, polymer with ethylene oxide and formaldehyde**, *See* Superinone, 30242
α-[4-(1,1,3,3-tetramethylbutyl)phenyl]-ω-hydroxypoly(oxy-1,2-ethanediyl), *See* Triton® X-15, 32433
2,4,7,9-tetramethyl-5-decyn-4,7-diol, *See* Surfynol® 104E, 30379
***p,p'*-tetramethyldiamindiphenylmethane**, *See* Michler's Base, 19644
tetramethyldiaminobenzophenone, *See* ketone base, 16140 Michler's ketone, 19646
4,4'-Tetramethyldiaminodiphenylmethane, *See* Michler's Base, 19644
N,N,N',N'-tetramethyl-1,2-diaminoethane, *See* Propamine D, 25170
N,N,N',N'-tetramethyl-p,p'-diaminodiphenylmethane, *See* Michler's Base, 19644
tetramethyldiarsine, *See* Cacodyl, 4882
1,1,4,4-tetramethyldichlorodisilethylene, *See* CT2015, 7313
tetramethyldisiloxane, *See* CT2030, 7314
1,1,3,3-tetramethyl-1,3-divinyldisiloxane, *See* CD6210, 5517
1,4-tetramethylene, *See* 1,4-butanediol, 4735, Dabco® BDO, 7647
tetramethylene 1,4-diol, *See* 1,4-butanediol, 4735, Dabco® BDO, 7647
tetramethylene dibromide, *See* 1,4-dibromobutane, 8357
tetramethylene glycol, *See* 1,4-butanediol, 4735, Dabco® BDO, 7647
1,4-tetramethylene glycol, *See* 1,4-butanediol, 4735, Dabco® BDO, 7647
tetramethylene oxide, *See* tetrahydrofuran, 31370
tetramethylene sulfide, *See* tetrahydrothiophene, 31374
tetramethylene sulphide, *See* tetrahydrothiophene, 31374
tetramethylenediamine, *See* putrescine, 25423
N,N,N',N'-tetramethylenediamine, *See* Toyocat® -TE, 32108
N,N,N',N'-tetramethyl-1,2-ethanediamine, *See* Propamine D, 25170
N,N,N',N'-tetramethylethenediamine, *See* Propamine D, 25170
tetramethylethylenediamine, *See* Propamine D, 25170
3,7,11,15-tetramethyl-2-hexadecen-1-ol, *See* phytol, 23780
N,N,N',N'-tetramethylhexanediamine, *See* Toyocat® -MR, 32105
2,2,5,5-Tetramethyl-4-isopropyl-1,3-dioxane, *See* Verdoxan, 33582
tetramethylolmethane, *See* pentaerythritol, 23139
2,6,10,14-tetramethylpentadecane, *See* Pristane, 25037
tetramethylphenylamine, *See* Duridine, 9222
N,N,N',N'-tetramethyl-1,4-phenylenediamine, *See* Wurster's blue, 34576
2,2,6,6-Tetramethyl-4-piperidyl/ β,β,β',β'-tetramethyl-3,9-(2,4,8,10-tetraoxaspiro(5,5) undecane) diethyl -1,2,3,4-butane tetracarboxylate, *See* Mixxim® HALS 68, 19993

2,4,6,8-tetramethyl-1,3,5,7-tetraoxacyclooctane, *See* metaldehyde, 19431
R-2,C-4,C-6,C-8-tetramethyl-1,3,5,7-tetroxocane
2,4,6,8-tetramethyl-1,3,5,7-tetroxocane, *See* metaldehyde, 19431
tetramethylthioperoxydicarbonic diamide, *See* Rezifilm, 26521, Thiram, 31726, Thiurad, 31730
tetramethylthiuram disulfide, *See* Akrochem® TMTD, 1182, methyl tuads®, 19562, Perkacit® TMTD, 23292, Rezifilm, 26521, Thiram, 31726, Thiurad, 31730, Tuex, 32553
tetramethylthiuram monosulfide, *See* Ancazide IS, 2420, Monex, 20210, Mono Thiurad, 20216, Perkacit® TMTM, 23293, Thionex, 31703, Unads®, 32907, Vulcaid 222, 34079
2,5,7,8-tetramethyl-2-(4',8',12'-trimethyltridecyl)-6-chromanol, *See* Vasculals, 33465, vitamin E, 33943
tetran, *See* cyhexatin, 7568 polytetrafluoroethylene, 24640, 24641, 24642
tetra-n-butoxysilane, *See* CT1750, 7311, T1750, 30675
tetra-N-butylammonium bromide, *See* tetrabutyl ammonium bromide, 31362
2,3,4,6-tetranitro-aniline, *See* Tetranyl, 31388
tetranitromethane, 31386, *See* Nitroform, 21359
Tetranol, 31387, *See* Tedion V-18, 30919
tetra-n-propoxysilane, *See* CT2090, 7316
Tetranyl, 31388
2,5,7,10-tetraoxa-6-silaundecane, 6-ethenyl-6-(2-methoxyethoxy)-, *See* Aktisil VM, 1205, CV-5000, 7453, 31389
tetraphenyl dipropylene glycol diphosphite, *See* Doverphos® 11, 8840, Doverphos® DPGDP, 8847, Weston® THOP, 34319
1,1,5,5-tetraphenyl-1,3,3,5-tetramethyltrisiloxane, *See* PSDF04, 25371
tetraphosphate, 31390
tetraphosphoric acid, *See* polyphosphoric acid, 24520
Tetrapom, *See* Rezifilm, 26521
tetrapotassium pyrophosphate, *See* potassium pyrophosphate, 24777
tetrapropoxysilane, *See* CT2090, 7316
tetrapropyl orthosilicate, *See* CT2090, 7316
tetrapropyl silicate, *See* CT2090, 7316
tetrapropylene, *See* dodecene-1, 8778
tetrapropylene benzene sulfonate, *See* Tensaryl KD, 31144
tetrasipton, *See* Rezifilm, 26521
tetrasodium (ethylenedinitrilo)tetraacetate, *See* Trilon® B Powd, 32324
tetrasodium dicarboxyethyl stearyl sulfosuccinamate, *See* Aerosol® 22, 879, Monawet SNO-35, 20188, Rewopol® B 2003, 26359
tetrasodium diphosphate, *See* sodium pyrophosphate, 28814
tetrasodium edetate, *See* tetrasodium EDTA, 31391, Kalex 220 Crystal, 15668, Kalex Liq. 50%, 15671
tetrasodium EDTA, 31391, *See* Cheelox® 100, 5865, Intraquest® TA Solution, 15175, Questal Extra Powd. Conc. 0780, 25662, Questal Special 0860, 25664, Sequestrene® 30A, 27994, Solon Conc, 28957, Thorquest 39, 31756, Trilon® B Powd, 32324
tetrasodium EDTA dihydrate, *See* Sequestrene® 220, 27995
tetrasodium ethylenbis(iminodiacetate), *See* Kalex Liq. 50%, 15671, Kalex 220 Crystal, 15668
tetrasodium ethylene diaminetetraacetate, *See* tetrasodium EDTA, 31391
tetrasodium ethylenediaminetetraacetate, *See* Kalex 220 Crystal, 15668, Kalex Liq. 50%, 15671
tetrasodium hexacyanoferrate, *See* sodium ferrocyanide, 28763
tetrasodium hexakis(cyano-C)ferrate(4-), *See* yellow prussiate of soda, 34648
tetrasodium pyrophosphate, *See* Laponite® S, 16779, sodium pyrophosphate, 28814, TSPP, 32540
tetrasodium salt, *See* sodium pyrophosphate, 28814
Tetraterge D-101, NFF, 31392
Tetrathal, 31393
1,1'-(tetrathiodicarbonothioyl)-bis-Piperidine, *See* Perkacit® DPTT, 23282, Sulfads®, 29960
Tetravos, *See* dichlorvos, 8394
Tetrawet DWN, 31394
tetrazolium blue, *See* tetrazolium chloride, 31395
tetrazolium chloride, 31395
tetrazolium salt, *See* tetrazolium chloride, 31395
2H-tetrazolium, 3,3'-(3,3'-dimethoxy[1,1'-biphenyl]-4,4'-diyl)bis[2-(4-nitrophenyl)-5-phenyl-, dichloride, *See* Nitro BT, 21342
Tetrine, *See* Kalex 220 Crystal, 15668, Kalex Liq. 50%, 15671, tetrasodium EDTA, 31391
Tetron, 31396, *See* sodium pyrophosphate, 28814
Tetrone, 31397
Tetronic® 50R1, 31401
Tetronic® 150R1, 31398
Tetronic® 304, 31399
Tetronic® 504, 31400
Tetronic® 701, 31402
Tetronic® 702, 31403
Tetronic® 704, 31404
Tetronic® 707, 31405
Tetronic® 901, 31406
Tetronic® 904, 31407
Tetronic® 908, 31408
Tetronic® 1101, 31409
Tetronic® 1102, 31410
Tetronic® 1104, 31411
Tetronic® 1107, 31412
Tetronic® 1301, 31413
Tetronic® 1302, 31414
Tetronic® 1304, 31415
Tetronic® 1307, 31416
Tetronic® 1501, 31417
Tetronic® 1502, 31418
Tetronic® 1504, 31419
Tetronic® 1508, 31420
Tetrosan 3,4 D, 31421
Tetroxone, 31422
Tetryl, 31423
tetryl formate, *See* isobutyl formate, 15346
teturamin, *See* Disulfiram, 8725
Tevilon, *See* Korogel, 16354, Koroplate, 16357
Texacar® EC, 31424
Texacar® PC, 31425
Texacat® DD, 31426
Texacat® DMDEE, 31427
Texacat® DME, 31428
Texacat® DMP, 31429
Texacat® DPA, 31430
Texacat® NEM, 31431
Texacat® NMM, 31432
Texacat® TD-33, 31433
Texacat® ZF-10, 31434
Texacat® ZF-22, 31435
Texacat® ZR-50, 31436
Texacat® ZR-70, 31437
Texaco BQ, 31438
Texacure EA-20, *See* diethylenetriamine, 8446
Texacure EA-24, 31439
Texadril 2010, 31440
Texalon 1000A, 31442
Texalon 1200A BK-11, 31443
Texalon 1200A HR-2 BK-16, 31444
Texalon 1308 A, 31445
Texalon 1600A Nat, 31446, *See* nylon 6/10, 21911
Texalon 600A NU, 31441
Texalon GF 600A (6-33), 31447
Texalon GF 1200A (13-40), 31448
Texalys, 31449
Texamine 84(L), 31450
Texanol® Ester-Alcohol, 31451
Texaphor 277, 31452
Texapon A 400, *See* Witcolate NH, 34478
Texapon ALS, 31453, *See* Ammonium lauryl sulfate, 2242
Texapon ASV, 31454
Texapon CS Paste, 31455
Texapon DEA, 31456
texapon DL, *See* Rhodapon® 101-10, 26631
Texapon EA-1, 31457
Texapon EA-1, NA, *See* Ammonium laureth sulfate, 2240
Texapon ES-1, 31458
Texapon EVR, 31459
Texapon IES, 31460
Texapon K-12, K-1296, L-100, 31461
Texapon K-1296, 31462
Texapon L20C, 31463
Texapon LS Highly Conc, 31464
Texapon MG, 31465
Texapon MG 3, 31466
Texapon N 25, Texapon NSE, 31467
Texapon N 40, *See* Witcolate ES-2, Witcolate LES-60C, 34477
Texapon NA, 31468
Texapon NSE, 31469
Texapon NSF, 31470
Texapon NSO, 31471
Texapon OT Highly Conc. Needles, 31472
Texapon PLT-227, 31473
Texapon PNA-127, 31474
Texapon QLV, 31475
Texapon SB-3, 31476
Texapon SBN, 31477
Texapon SG, 31478
Texapon SH 100, 31479
Texapon Special, *See* Witcolate NH, 34478
Texapon T 42, 31480
Texapon T-35, *See* Perlankrol® ATL40, 23324, Rhodapon® LT-6, 26638, Standapol® T, 29481, Witcolate TLS-500, 34479
Texapon T-42, *See* Witcolate TLS-500, 34479
Texapon TH, *See* Witcolate TLS-500, 34479
Texapon VHC Needles, ZHC Needles, 31481
Texapon ZHC Needles, 31482, 31483
Texapon® 1030, *See* sodium decyl sulfate, 28755
Texapon® DEA, *See* DEA-lauryl sulfate, 7827
Texapon® MGLS, *See* magnesium lauryl sulfate, 18364
Texapon® NDS, *See* sodium decyl sulfate, 28755
Texapon® SB-3, *See* disodium lauryl sulfosuccinate, 8649
Texapon®SB-3KC, *See* disodium lauryl sulfosuccinate, 8649
Texapret®, 31484
Texgas, 31485, 31486
Texi TD, 31487
Texicote®, 31488
Texicote® 03-001, 31489
Texicote® 03-019, 31490
Texicote® 1000, 1050, 31491
Texicryl®, 31492
Texicryl® 13-002, 31493
Texicryl® 13-011, 31494
Texicryl® 13-030, 31495
Texicryl® Additive 87-1280, 31496
Texicryl® Ecobinder, 31497
Texicryl® Hyperbinder, 31498
Texigel®, 31499
Texileather, 31500
Texin 3203, 3215, 4203, 4206, 4210, 4215, 31501
Texin 480-A, 31502
Texin 985-A, 31503
Texin 5286, 31504
Texipol®, 31505
Texipol® 63-002, 31506
Texlin® 300, 31507
Texlin® 400, 31508
Texlin® 500, 31509
Texoderm, 31510
Texofor, 31511
Texofor A and B, 31512
Texofor C, 31513
Texofor D, 31514
Texofor E and ED, 31515
Texofor FN, FP and FX, 31516
Texofor G, 31517
Texofor J4, 31518
Texofor M, 31519

Thiersch's antiseptic solution, 31671
thiet-sie, 31672
Thifensulfuron Me, *See* thifensulfuron-methyl, 31673
thifensulfuron-methyl, 31673
Thifor, *See* thiodan, 31680
Thilaven, *See* Perichthol, 23270
thillate, *See* Rezifilm, 26521
Thimer, *See* Rezifilm, 26521
Thimet, *See* Terrathion, 31343
Thimul, *See* thiodan, 31680
Thinners, 31674
Thinoline, 31675
Thinsec, 31676, *See* Carbaryl, 5181
Thinsol, 31677
thioacetamide, 31678
thioalkofen bm4, *See* Rutenol, 27147
thioalkofen bmch, *See* Rutenol, 27147
thioalkofen mbch, *See* Rutenol, 27147
thioalkophene bm-4, *See* Rutenol, 27147
2,2'-thiobis[acetic acid], *See* thiodiglycolic acid, 31683
4,4'-thiobis(6-*t*-butyl-*m*-cresol), *See* Santonox, 27416
4,4'-thiobis (2-*t*-butyl-5-methylphenol), *See* Lowinox® 44S36, 17649, Ultranox® 236, 32846
4,4'-thiobis(2-*tert*-butyl-5-methylphenol), *See* Rutenol, 27147
4,4-thiobis-(6-*t*-butyl-3-methyl-phenol), *See* Rutenol, 27147
1,1'-thiobis[2-chloroethane], *See* mustard gas, 20538
thiobismethane, *See* dimethyl sulfide, 8510, methyl sulfide, 19561
2,2'-thiobis (4-*tert*-octylphenolato)-n-butylamine nickel II, *See* Cyasorb® UV 1084, 7498
3,3'-thiobispropanoic acid, *See* Distearyl thiodipropionate, 8713
3,3'-thiobis[propanoic acid], *See* 3,3'-thiodipropionic acid, 31684
thiobutan-2-one, O-(methylcarbamoyl)oxime, *See* methomyl, 19502
thiocamf, 31679
thiocarbamide, *See* thiourea, 31720
thiocarbamyl sulfenamide, *See* Cure-Rite® 18, 7420
thiocarbanilide, *See* Activit, 505, Akrochem® Thio No. 1, 1181, Anchoracel, 2426, diphenylthiourea, 8604, Vulcogene, 34114
thiocarbonyl chloride, *See* thiophosgene, 31711
3-thiocyanatopropyltriethoxy silane, *See* Si 264, 28191
thiocyanic acid, ammonium salt, *See* Ammonium thiocyanate, 2256
thiocyanic acid, compd. with guanidine (1:1), *See* Guanidine thiocyanate, 13442
thiocyanic acid, methylene ester, *See* Amerstat® 282, 2043, N-948® Biocide, 20641
thiodan, 31680
thiodet, 31681
2,2'-thiodiacetic acid, *See* thiodiglycolic acid, 31683
thiodicarbonic diamide, tetramethyl-, *See* Ancazide IS, 2420, Monex, 20210
2,2'-thiodiethanol, *See* thiodiglycol, 31682
thiodiethylene bisdibutyl hydroxy hydrocinnamate, *See* Anox® 70, 2484
thiodiethylene glycol, *See* thiodiglycol, 31682
thiodiglycol, 31682
thiodiglycolic acid, 31683
thiodihydracrylic acid, *See* 3,3'-thiodipropionic acid, 31684
thiodiphenylamine, *See* phenothiazine, 23618
O,O'-(thiodi-4,1-phenylene)phosphorothioic acid O,O,O',O'-tetramethyl ester, *See* Abate 1-SG, 2-CG, 4-E, 5CG, 26
thiodipropionic acid, *See* dilauryl thiodipropionate, 8476
β,β-thiodipropionic acid, *See* 3,3'-thiodipropionic acid, 31684
3,3'-thiodipropionic acid, 31684
thiodipropionic acid dilauryl ester, *See* dilauryl thiodipropionate, 8476
thiodipropionic acid, distearyl ester, *See* distearyl thiodipropionate, 8713
2-thioethanol, *See* 2-mercaptoethanol, 19336
thioethylene glycol, *See* 2-mercaptoethanol, 19336
Thiofide, 31685, *See* Perkacit® MBTS, 23286, Vulkacit® DM/C, 34131
Thiofluor™, 31686
thiofuran, *See* thiofurfuran, 31687, thiophene, 31707
thiofurfuran, 31687, *See* thiophene, 31707
1-thio-β-D-glucopyranose 1-[N-(sulfoxy)-3-butenimidate] monopotassium salt, *See* sinigrin, 28500
thioglycerin, *See* Thiovanol®, 31723
1-thioglycerol, *See* Thiovanol®, 31723
thioglycol, *See* 2-mercaptoethanol, 19336
thioglycolic acid, 31688, *See* thioglycolic acid, 31688, Thiovanic® Acid, 31722
Thiohexam, *See* Vulkacit® CZ/EGC, DZ/EGC, 34129
Thioknock, *See* Rezifilm, 26521
Thiokol® 2135, 31689
Thiokol® 2153, 2157, 31690
Thiokol® FEC-2232, 31691
Thiokol® FES-2258, 31692
Thiokol® LP, 31693
Thiokol® MC-2027, 31694
Thiokol® R-2100, 31695
Thiokol® RLP-2078, 31696
thiolane, *See* tetrahydrothiophene, 31374
2-thiol-dihydroglyoxaline, *See* Perkacit® ETU, 23283
thiole, *See* thiophene, 31707
Thiolim, 31697
Thiolin, *See* Perichthol, 23270
Thiolite, 31698
Thiolux, *See* Thiovit, 31724
Thiolyte®, 31699
thiometon, 31700, *See* Ekatin, 9657
thiomonoglycol, *See* 2-mercaptoethanol, 19336
Thion, *See* Thiovit, 31724
Thionalide, 31701
Thionex, 31702, 31703, *See* thiodan, 31680
thionine, *See* Lauth's violet, 16870
Thionol®, 31704
Thionoline, 31705
thioperoxydicarbonic diamide, tetrabutyl-, *See* tetrabutylthiuram disulfide, 31363
thiophanate methyl, 31706, *See* CDA Mildothane, 5520, Cercobin, 5738, Nemafax, 20864, Sys Tec® 1998, 30661
Thiophanate-methyl dimethyl, *See* thiophanate methyl, 31706
thiophane, *See* tetrahydrothiophene, 31374
thiophene, 31707, *See* Hopkin's lactic acid reagent, 14299, thiofurfuran, 31687
thiophenol, 31708
Thiophor Bronze 5G, 31709
Thiophor Indigo, 31710
Thiophos, *See* parathion, 22807
thiophosgene, 31711
Thioprene® -48, 31712
thiopropane, *See* dimethyl sulfide, 8510
2-thiopropane, *See* dimethyl sulfide, 8510
Thiosan, *See* Rezifilm, 26521, Thiram, 31726
thiosept oil, 31713
Thioset® M, 31714
Thiosolucin-dihydrostreptomycin preparation, *See* Agavin, 959
Thiostab, 31715, *See* sodium thiosulfate, 28826
Thiostop E, N, 31716
Thiostop N, 31717, *See* Aquatreat SDM, 2761, methyl namate®, 19551, Octopol SDM-40, 22014, Perkacit® SDMC, 23288, sodium dimethyldithiocarbamate, 28758
thio-sul, *See* Ammonium thiosulfate, 2257
thiosulfan, *See* thiodan, 31680
thiosulfuric acid ($H_2S_2O_3$), diammonium salt, *See* Ammonium thiosulfate, 2257
thiosulfuric acid ($H_2S_2O_3$), disodium salt, *See* sodium thiosulfate, 28826
Thiotan, 31718
Thiotax, 31719, *See* Accelerator Mercapto, 207, Vulkacit® M, Vulkacit® Merkapto/C, 34134
thiotetrole, *See* thiophene, 31707
Thiotex, *See* Rezifilm, 26521
thiotox, *See* Rezifilm, 26521
thiourea, 31720, *See* sulfourea, 30038
thiourea dioxide, 31721, *See* formammidinesulfinic acid, 12262, Manofast, 18571
Thiourea, N-(4-butoxyphenyl)-N'-[4-(dimethylamino)phenyl]-, *See* thiambutosine, 31663
Thiourea, N,N'-dibutyl-, *See* Pennzone B 0685, 23120
thiourea, N,N-diethyl-, *See* Pennzone E 0686, 23121
thiourea, N,N'-bis(2-methylphenyl)-, *See* Accelerator A22, 192
Thiovanic® Acid, 31722, *See* thioglycolic acid, 31688
Thiovanol®, 31723
Thiovit, 31724
thioxamyl, *See* Vydate®, 34171
thioxydant lumire, 31725
Thiozin, *See* Perichthol, 23270
Thiprazole, *See* thiabendazole, 31661
Thiram, 31726, *See* Agrichem Flowable Thiram, 1047, Ancazide ME, 2421, Fernide, 11638, FS Thiram 15% Dust, 12427, Hortag Thiram, 14317, Pomarsol®, 24677
Thiram 75, *See* Rezifilm, 26521
Thiram b, *See* Rezifilm, 26521
Thiramad, *See* Rezifilm, 26521
Thirasan, *See* Rezifilm, 26521
This, *See* Thiovit, 31724
Thisol, 31727
Thissirol, 31728
thitsi, 31729
thiulix, *See* Rezifilm, 26521
Thiurad, 31730, *See* Rezifilm, 26521, Thiram, 31726
Thiuram, *See* Rezifilm, 26521
Thiuram disulfide, *See* Akrochem® TMTD, 1182
Thiuramin, *See* Rezifilm, 26521
Thiuramyl, *See* Rezifilm, 26521, Thiram, 31726
Thixatrol, 31731
Thixokon, *See* acetrizoate sodium, 307
Thixolan, 31732
Thixomen, 31733
Thixon® 300, 31734
Thixon® 511-T, 31735
Thixon® 753, 31736
Thixon® 957, 31737
Thixon® 2000, 31738
Thixon® OSN-2, 31739
Thixon® P-15, 31740
Thixotropes, *See* Crayvallac, 6972
Thixotropic alkyds, *See* Gelkyd, 12735
Thixseal, 31741
THME, *See* pentaerythritol, 23138
Thomas meal, 31742
Thomas phosphate, *See* Thomas slag, 31743
Thomas slag, 31743
Thomasite, 31744
Thompsons, 31745
Thonzide, 31746
Thonzonium bromide, *See* Thonzide, 31746
Thoran, 31747, 31748
thoria, *See* thorium dioxide, 31749
thorium anhydride, *See* thorium dioxide, 31749
thorium dioxide, 31749
thorium oxide, *See* thorium dioxide, 31749
Thornel® Carbon Fiber T600/50C 12K, 31750
Thoroclear, 31751
thoron, 31752
Thorosheen®, 31753
Thorotrast, 31754
Thorowet G-40 3230, 31755
Thorowet ML-3 0532, *See* disodium lauryl sulfosuccinate, 8649
Thorquest 39, 31756
Thor-stabilizator BF, 31757
Thorstat ASA, 31758
Thoulet's solution, 31759
Thovaline, 31760
Thowless solder, 31761
Three Elephant Boric Acid, 31762
Three Elephant Pyrobor Dehydrated Borax, 31763

Three Elephant V-Bor Refined Pentahydrate Borax, 31764
Three-part pack epoxy-based grouting material, *See* Addagrout, 642
Thresh's reagent, 31765
Thripstick®, 31766
thrombase, 31767
Thrombo-Hepin, *See* Heparin, Sodium Salt, 13788
Thrombophob, *See* Heparin, Sodium Salt, 13788
thsing-hoa-liao, 31768
3-thujanone, *See* tanacetone, 30775
thujone, *See* tanacetone, 30775
thulia, *See* thulium oxide, 31769
thulium oxide, 31769
thulium(III) oxide, *See* thulium oxide, 31769
Thurcide, *See* Bactospeine, 3582
Thuricide HP, 31770
Thurmalox, 31771
Thurston's alloy, 31772
Thwaites' solution, 31773
T-Hydro, 31774
Thylate, *See* Rezifilm, 26521, Thiram, 31726, Thiurad, 31730
Thylose, *See* carboxymethylcellulose sodium, 5258, Cekol, 5558
thyme camphor, 31775
thymene, 31776
thymic acid, *See* thyme camphor, 31775
thymine, 31777
thymol, 31778, *See* thyme camphor, 31775
thymol iodide, *See* losol, 15237
Thymoxane, 31779
Thynon, *See* Dithianone, 8729
Thyodene, 31780
thyol, 31781
Thyroblock, *See* Potassium Iodide, 24758, 24759
thyroid-stimulating hormone, *See* Thytropar, 31782
Thyrojod, *See* potassium iodide, 24758, 24759
thyrotropic hormone, *See* Thytropar, 31782
Thyroxine 125I, *See* Riamat, 26720 Tetramet-125, 31383
Thytropar, 31782
Ti, *See* titanium, 31865
Ti tree oil, *See* EmCon TEA TREE, 10067
Ti(catalyst), *See* Metacure® T-1, 19414
Tiabenda, *See* thiabendazole, 31661
Tiazon, *See* Dazomet, 7806, N-521® Biocide, 20640
Tibond, 31783
Ticevite, 31784
Ticlatone, *See* Landromil, 16656
Tico, 31785
Tidolith, *See* Cryptone, 7281
Tiers argent, 31786
Tiezene, *See* Zineb, 34807
Tiform, 31787
Tiguvon®, 31788
Til, 31789
til oil, *See* gingelly oil, 12946
til oil, beni oil, *See* Gingelly oil, 12946, Sesame oil, 28075
Tilcarex, *See* quintozene, 25692
Tilcom, 31790
tile ore, 31791, *See* cuprous oxide, 7405
Tilite, 31792
Tillantina, 31793
Tilt, *See* Radar, Radar Propiconazole, 25714
Tilt Turbo, 31794
Timail®, 31795
Timang Steel, 31796
Tim-Bor, *See* Polybor®, 24369
Tim-Bor industrial, 31798
Tim-Bor Professional, 31799
Tim-Bor®, 31797
Timborised, *See* Ammonium biborate, 2230
Timbrel, 31800
Timbrelle, 31801
Time Bomb, 31802
Timelit, *See* Lofepramin Hydrochloride, 17552
Timica, 31803
Timica®, 31804
Timolate, 31805
timolol maleate, *See* Timolate, 31805
Timonox, 31806, 31807, *See* antimony trioxide, 2574, Octoguard FR-10, 22000
Timonox Blue Star, 31808, *See* antimony trioxide, 2574
tin, 31809, *See* Haro® Mix ZT-514, 13650
tin amalgam, 31810
tin ash, 31811
Tin brilliants, *See* Fahlun diamonds, 11412
tin bronze, 31812, *See* mosaic gold, 20355, stannic sulfide, 29496
tin chloride, *See* stannic chloride, 29494
tin (II) chloride anhydrous, *See* stannous chloride, 29502
tin (IV) chloride, *See* stannic chloride, 29494
tin (IV) chloride anhydrous, *See* stannic chloride, 29494
tin(IV) chromate(VI), *See* stannic chromate, 29495
tin crystals, *See* stannous chloride, 29502 tin salts, 31814
tin,dibutyl-,diacetate, *See* Metacure® T-1, 19414
tin dibutyl dilaurate, *See* Dabco® T-12, 7673
tin dichloride, *See* stannous chloride, 29502
tin dichloride, *See* stannous chloride, 29502, tin ash, 31811
tin disulfide, *See* stannic sulfide, 29496
tin(IV) hydroxide, *See* tin white, 31815
tin ore, 31813
tin oxalate, *See* oxalic acid tin (II) salt, 22415
tin(II) oxalate, 31816, *See* oxalic acid tin (II) salt, 22415
tin(IV) oxide, 31817, *See* Superlite, 30252
tin perchloride, *See* stannic chloride, 29494
tin peroxide, *See* tin ash, 31811
tin protochloride, *See* stannous chloride, 29502
tin salt, *See* stannous chloride, 29502
tin salts, 31814
tin(IV) sulfide, *See* stannic sulfide, 29496
tin tetrachloride, *See* stannic chloride, 29494
tin white, 31815
Tinaderm, 31818
Tinamul®, 31819
tincal, 31820
ting-yu, *See* tse-leou, 32534
Tinkal, *See* tincal, 31820
Tinman, 31821
Tinoclarite, 31822
Tinofil, 31823
Tinopal®, 31824
Tinopal® 5BM-GX, AMS-GX, 31825
Tinopal® AMS-GX, 31826, 31827
tinopal® CBS-X, 31828
Tinorex, 31829
Tinosol, 31830
Tinostat, *See* Dabco® T-12, 7673, dibutyltin laurate, 8371
Tinovetin, 31831
Tintacrete®, 31832
Tint-Ayd, 31833
Tintophen X, *See* Photine, 23756
Tinuvin 326, *See* Uvazol 236, 33181
Tinuvin 770, *See* Uvaseb 770, 33178
Tinuvin P, *See* Topanex 100BT, 32019
Tinuvin®, 31834
Tinuvin® 622LD, 31835
Tinuvin® 770, 31836, *See* Bistetramethyl piperidinyl sebacate, 4284, Lowilite® 77, 17645
Tinuvin® P, 31837, *See* Uvazol P, 33184
Tioga Adhesion Promoter 30-6-600, 31838
Tiolene, *See* Thiovit, 31724
Tiona®, 31839
Tiona® HSS, 31840
Tiona® RCL-4, 31841
Tiona® RCL-535, 31842
Tiona® RCS-P, 31843
Tioveil, 31844
Tiox, 31845
Tioxidazole, *See* Tiox, 31845
Tioxide, 31846
TIP, 31847
TIPA, *See* triisopropanolamine, 32313
TIPA lauryl sulfate, *See* Akyposal TIPA 45, 1321, Rewopol® TLS 90 L, 26384
TIPA-laureth sulfate, *See* Akyposal 100 DAL, 1298
Tipoff, 31848, *See* Tre-Hold, 32165
TIPSCI, *See* CD4368, 5503
Tipula, *See* Aldrin, 1558
Ti-pure®, 31849
Ti-Pure® R-103, 31850
Tirade, *See* fenvalerate, 11620
Tirampa, *See* Rezifilm, 26521
Tirucalli gum, 31851
Tisco Steel, 31852
Tisept® Solution, 31853
Tisperse MB-2X, *See* Agerite® White, 1024
Tisperse MB-58, *See* Octocure ZMBT-50, 21998, Oxaf, 22408, Perkacit® ZMBT, 23298, Zetax®, 34741
Ti-Sphere AB-15155A, 31854
Tissalys, 31855
Tissier's metal, 31856
Tisyn®, 31857
Titan, 31858, 31859
Titan cements, 31860
Titan Design, 31861
titanate(2-), hexafluoro-, dipotassium, (OC-6-11)-, *See* Aflammit TI, 939
titania, *See* titanium dioxide, 31867
titanic acid anhydride, *See* titanium dioxide, 31867
titanic anhydride, *See* titanium dioxide, 31867
titanicilron, *See* Ilmenite, 14888
titaniferous iron, *See* ilmenite, 14888
titaniferous iron ore, *See* Ilmenite, 14888
Titanital, 31862
Titanite, 31863
Titanite No. 1, 31864
titanium, 31865
titanium alloy, 31866
titanium dioxide, 31867, *See* Covermark, 6921, Diox DR 22, 8573, Finntitan, 11817, Hitox®, 14148, Hombitan®, 14288, Kronos®, 16439, Kronos®, 16440, Tiona® RCL-535, 31842, Tioxide, 31846, Ti-Pure® R-103, 31850, titanium dioxide, 31867, Titanium Dioxide P25, 31869, Tronox Titanium Dioxide Pigments, Chloride Process, 32465, Zopaque, 34904
94% titanium dioxide, *See* Synthetic Rutile, 30643
titanium dioxide [13463-67-7] and bismuthoxychloride [7787-59-9], *See* Titanium Dioxide 110, 31868
titanium dioxide [13463-67-7]/silica [7631-86-9], *See* Ti-Sphere AB-15155A, 31854
titanium dioxide 110, 31868
titanium dioxide P25, 31869
titanium dioxide/mica, *See* Mearlin®, 19191
titanium dioxide; anatase; titanox, *See* Kronos® 1000, 16441
titanium dioxide; rutile; titanox 2020, *See* Kronos® 2020, 16442
titanium dioxide; rutile; titanox 2073, *See* Kronos® 2073, 16443
titanium dioxide; titanox 3020, *See* Kronos® 3020, 16444
titanium lactates, *See* corichrome, 6821
titanium oxide, *See* titanium dioxide, 31867
titanium potassium fluoride, *See* Aflammit TI, 939
Titanium Putty, 31870
Titanium(IV)Potassium Fluoride, *See* Aflammit TI, 939
titanoferrite, 31871
Titanox, 31872
Titanox Design, 31873
Titanox RA-39, 31874
Titanweiss C, Extra T, Standard T, Standard A, 31875
Titite, 31876
Title®, 31877
Ti-tone, 31878
Titriplex III, *See* edetate disodium, 9613, Trilon® BD, 32325
Ti-trol, *See* EmCon TEA TREE, 10067
Titus®, 31879
Tiuramil, *See* Thiram, 31726

Tiuramyl, See Rezifilm, 26521
Tixit, See Propham, 25186
Tixogel, 31880
Tixogel LAN, 31881
Tixogel OMS, 31882
Tixogel VP, 31883
Tixogel VSP, 31884
Tixogel VZ, 31885
Tixoton, 31886
Tizit, 31887
TKP, See potassium phosphate, tribasic, 24774
TKPP, See potassium pyrophosphate, 24777
TL 314, See Acrylonitrile, 436
TL 898, See mercuric chloride, 19347
TL 1163, See CT2950, 7325
TL 4190, 31888
TLA-227, 31890
TLA-256, 31891, See calcium sulfonate, 4973, Lubrizol® 2152, 17744
TLA-111B, 31889
T-lim, 31892
TM-95, See Rezifilm, 26521
T-Maz 80, See Radiasurf® 7157, 25804
T-Maz® 20, 31893
T-Maz® 28, 31894
T-Maz® 40, 31895
T-Maz® 60, 31896, See Ethylan® GES6, 11108
T-Maz® 61, 31897
T-Maz® 65, 31898, See Polysorbate 65, 24571
T-Maz® 80, 31899
T-Maz® 81, 31900
T-Maz® 85, 31901
T-Maz® 90, 31902
T-Maz® 95, 31903
TMBAC, See benzyl trimethyl ammonium chloride, 3989
TMEDA, See Propamine D, 25170
TMP, 31904
TMPD® Glycol, 31905
TMPTA, See Ageflex TMPTA, 1001
TMS, See CT2050, 7315
TMSDEA, See CD4450, 5504
TMTD, See Akrochem® TMTD, 1182, Rezifilm, 26521, Thiram, 31726, Thiurad, 31730
TMTD Sadoplon, See Rezifilm, 26521
TMTDS, See Rezifilm, 26521
TMTM, 31906, See Ancazide IS, 2420, Monex, 20210
T-Mulz® 66H, 31907
T-Mulz® 596, 31908
T-Mulz® 1158, 31909
T-Mulz® AO2, 31910
T-Mulz® Mal 5, 31911
Tnegal, 31912
TNPP, See Trisnonylphenyl phosphite, 32411
TNT, See Trotyl, 32474
TNX, 31913, See tetranitroxylene,
tobermorite, See calcium silicate, 4969
Tobias acid, 31914
Tobin bronze, 31915
Tochlorine, 31916, See chloramine-T, 6086, Ketjensept, 16133
Tocopherex, 31917
tocopherol, 31918, See vitamin E, 33943
α-tocopherol, See Vascuals, 33465, vitamin E, 33943
tocopherol oil CLR, 31919
(+)-α-tocopherol, See vitamin E, 33943
TOCP, See Plastic X, 24026
TOF, See trioctyl phosphate, 32349
TOF™, See trioctyl phosphate, 32349
Toffix, 31920
Toffix®, 31921
TOFK, See Plastic X, 24026
Togocoll, 31922
Togoplast, 31923
Togotec, 31924
Togotherm, 31925
Toho Me-PEG Series, 31926
Toho PEG Series, 31927
Toho Salt A-5, 31928
Tohol N-220, 31929
toile micanite, See micanite cloth, 19636
Toisin's solution, 31930
Tokiocillin, See Ampicillin, 2361
Tokopharm, See tocopherol, 31918, Vascuals, 33465
Toku Bishi®, 31931
Tokuthion®, 31932
Tol, See methyl benzene, 19514
Tolamine, See Chloramine T, 6086, Ketjensept, 16133
Tolan, 31933
Tolazamide, See Tolinase, 31936
Tolcide MBT, 31934
Tolclofos-methyl, See Basilex, 3709 Rizolex, 26858
toldimfos sodium, See Tonophosphan, 32013
Tolgard, 31935
Tolinase, 31936, See tolanase
Tolite, See Trotyl, 32474
Tolkan, 3193, 31938 7
Tolkan 500, 31939
Tollen's reagent, 31940
Tolochrome, 31941
Tolonate, 31942
Toloy 45, 31943
Tolplaz, 31944
tolu balsam, 31945
p-tolualdehyde, See PTAL, 25375, p-tolyl aldehyde, 31961
toluene, 31946, See Dracyl, 8953, methyl benzene, 19514
toluene (77.5%), 2-butanone (10.9%) and carbon black (2.2%), See Black Out® Black, 4304
o-toluene-azo-β-naphthol, See Sudan IV, 29941
toluene diisocyanate, 31947, See Scuranate, 27776
toluene 2,4-diisocyanate, See toluene diisocyanate, 31947, Voranate T-80, Type I, Type II, 34042
toluene diisocyanate and polymethyl diisocyanate, See Mondur, 20202
3,5-toluenediol, See orcinol, 22280
toluene hexahydride, See methyl cyclohexane, 19525
o-toluene sulfonamide, 31948, See o-toluenesulfonamide, 31948
p-toluene sulfonamide, 31949, See Uniplex 173, 33012
o/p toluene sulfonamide, See Rit-Cizer #9, 26840
toluene sulfonic acid, 31950, See Manro PTSA, 18600
p-toluene sulfonic acid, See PTSA 70, 25377, Reworyl® T, 26410, toluene sulfonic acid, 31950
p-toluene sulfonic acid monohydrate BP, See Eltesol® TSX, 9876
p-toluene sulfonylhydrazide, See Celogen® TSH, 5645
toluene, 4-(diiodomethylsulfonyl)-, See Amical® 48, 2058
toluene-4-sulfonamide, See p-toluenesulfonamide, 31949
p-toluenesulfonyl semicarbazide, See Celogen® RA, 5644
p-(p-toluenesulfonyl amido) diphenylamine, See Aranox®, 2780
p-toluenesulphonamide, See p-toluenesulfonamide, 31949
α-toluenol, See benzyl alcohol, 3986
toluhydroquinone, 31951
toluic a, See diethyl toluamide, 8438
m-toluic acid, 31953
o-toluic acid, 31954
p-toluic acid, 31955
α-toluic acid, 31952, See phenylacetic acid, 23637
m-toluic acid diethylamide, See diethyl toluamide, 8438
p-toluidine, α,α,α-trifluoro-2,6-dinitro-N,N-dipropyl-, See Trifluralin, 32252
toluol, 31956, See methyl benzene, 19514, toluene, 31946
p-toluoyl chloride, 31957
Tolurane, See Talisman, 30735, Tolurex, 31958
Tolurex, 31958, See Talisman, 30735
Tolurgan, 31959, See Tolurex, 31958
tolu-sol, See methyl benzene, 19514
tolutriazole, See Preventol® CI7-100, 24929, tolyltriazole, 31962
toluylene, 31960
toluylene red, See neutral red, 21028
o-toluylic acid, See o-toluic acid, 31954
m-toluylic acid, See m-toluic acid, 31953
p-toluylic acid, See p-toluic acid, 31955
m-tolyl acetate, See m-cresyl acetate, 7011, Kresatin, 16402
p-tolyl aldehyde, 31961
tolyl diiodomethyl sulfone, See Amical® 48, 2058
tolyl diphenyl phosphate, See diphenylcresyl phosphate, 8601, Disflamoll® DPK, TPK, 8641
o-tolyl epoxypropyl ether, See DY 023, 9300
o-tolyl ester phosphoric acid, See Plastic X, 24026
o-tolyl glycidyl ether, See DY 023, 9300
o-tolyl phosphate, See Plastic X, 24026
tris(o-tolyl) phosphate, See Plastic X, 24026
tolyl triazole, See Preventol® CI7-100, 24929
p-tolylacetate, See Narceol, 20742
4"-(o-tolylazo)-o-diacetotoluidide, See Diacetazotol, 8247
o-tolylbiguanide, See Vulkacit® 1000, 34124
1-o-tolylbiguanide, See Vulkacit® 1000, 34124
o-tolylbiguanide, See Vulkacit® 1000, 34124
2,4-tolylene diisocyanate, See toluene diisocyanate, 31947, Voranate T-80, Type I, Type II, 34042
Tolylfluanid, See Euparen® M, 11198
tolyltriazole, 31962, See Preventol® CI7-100, 24929
Tomado, See Carbaryl, 5181
Tomah AO-14-2, 31963
Tomah AO-728 Special, 31964
Tomah BExM-1, 31965
Tomah DA-14, 31966
Tomah DA-16, 31967
Tomah DA-17, 31968
Tomah E-14-2, 31969
Tomah E-14-5, 31970
Tomah E-18-2, 31971, 31972
Tomah E-18-5, 31973
Tomah E-19-2, 31974
Tomah E-24-2, 31975
Tomah E-DT-3, 31976
Tomah E-S-2, 31977, See PEG soyamine series, 23033
Tomah E-T-2, 31978
Tomah PA-10, 31979
Tomah PA-1214, 31988
Tomah PA-12EH, 31980
Tomah PA-13i, 31981
Tomah PA-14, 31982
Tomah PA-14 Acetate, 31983
Tomah PA-16, 31984
Tomah PA-17, 31985
Tomah PA-19, 31986
Tomah PA-24, 31987
Tomah Q-14-2, 31990
Tomah Q-17-2, 31991
Tomah Q-18-2, 31992
Tomah Q-24-2, 31993
Tomah Q-2C, 31989
Tomah Q-311, 31994
Tomah Q-511, 31995
Tomah Q-C-15, 31996, See Ethoquad® C/25, 10999
Tomah Q-D-T, 31997
Tomah Q-DT-HG, 31998
Tomah Q-S, 31999
Tomah Q-ST-50, 32000
Tomahawk, 32001
Tomaset, 32002
Tomato Setting Spray, 32003
Tombac, 32004
Tombasil, 32005
Tombel, 32006
Tomophan, 32007
Tomorite, 32008
ton, See Tisyn®, 31857
Toncan, 32009
Toncas metal, 32010
Tone, 32011
Tonite, 32012
Tonofosfan, See Tonophosphan, 32013
Tonophosphan, 32013
Tonox® 22, Tonox® R, 32014

tonquinol, *See* musk baur, 20533
Tonsil, 32015
tonsil, *See* Frankonite, 12377
tonsil L80, *See* Bentonite, 3954
Tonsillosan, *See* DL-lactic acid, 16573
Tony Red, *See* Sudan III, 29940
Tool Life, 32016
Toolife, 32017
Top 7 Mosaic and Pebble, 32018
Top Form Wormer, *See* Tecto, Tecto 60%, 30913
Topane, *See* Dowicide A, 8932
Topanex 500H, 32020
Topanex 100BT, 32019
Topanol, *See* Oxyguard, 22457
Topanol OC and 0, *See* Oxyguard, 22457
Topanol®, 32021
Topanol® 205, 32022
Topanol® CA, 32023
Topanol® LVT 600, 32024
Topanol® M, 32025
Topas 100, 32026
Topas C 50WP, 32027
Topaz, 32028, *See* Topas 100, 32026
Topaze, *See* Topas 100, 32026
Topbraun®, 32029
Topclip Dridress, 32030
Topclip Fly and Scab Dip, 32031
Topclip Foot Rot Aerosol, 32032
Topclip Formalin, 32033
Topclip Gold Shield, 32034
Topclip Marker Aerosols, 32035
Topclip Marker Fluid, 32036
Topclip Parasol, 32037
Topclip Scab Dip, 32038
Topclip Sheep Dip, 32039
Topclip Vaccines, 32040
Topclip Wormer, 32041
Top-Cop, 32042
Topex, 32043, *See* Benzoyl peroxide, 3983
Topexane, 32044
Topfix, 32045
Tophet, 32046
Tophet A, 32047
Tophet C, 32048
Tophol, *See* cinchophen, 6307
Tophosan, *See* cinchophen, 6307
Topitox, *See* Ridene, 26746
Topline T, *See* Taktic, 30729
Toppel, 32049, 32050, *See* cypermethrin, 7588, Topclip Parasol, 32037
Topshot, 32051
Topsin WP methyl, *See* thiophanate methyl, 31706
Topsol, 32052
Topsym, *See* DMSO, 8762
Topup, 32053
Topusyn, *See* desmetryn, 8142
Torapron, 32054
torbanite, 32055
Torch, *See* Bromoxynil, 4590
Tordon, 32056
Tordon 22K, 32057
Torelle, 32058
Toric Contact Lens, 32059
Torlon® 4203L, 32060
Torlon® 4275, 32061
Torlon® 4301, 32062
Torlon® 4347, 32063
Torlon® 5030, 32064
Torlon® 7130, 32065
Torlon® 7330, 32066
tormentil, 32067
Tormol, 32068
Tornac, 32069
Tornac B 3850, 32070
Tornade, *See* permethrin, 23384, 23385, 23386
Tornado, 32071, *See* Carbaryl, 5181
Tornesit, 32072, 32073
Toro, 32074, *See* Ludorum, 17778, Talisman, 30735, Tolurex, 31958
toron, 32075
Torpedo, *See* permethrin, 23384, 23385, 23386
Torqseal, 32076
Torque, 32077
Torrax, 32078
Torsite, *See* *o*-phenylphenol, 23648
tosylic acid, *See* toluene sulfonic acid, 31950
Totablan, 32079
Totacillin, *See* Ampicillin, 2361
Totacol, 32080
Totalciclina, *See* Ampicillin, 2361
Totalene, *See* trichlorfon, 32212
Totalon, *See* Ketrax, 16149
Totapen, *See* Ampicillin, 2361
TOTM, *See* Jayflex® TOTM, 15559
TOTP, *See* Plastic X, 24026
Totril, 32081
Totroxocane, 2,4,6,8-tetramethyl-, *See* metaldehyde, 19431
Toucas metal, 32082
Touchpong gum, *See* Pouckpong gum, 24798
touchstone, *See* Lydian stone, 18069
tough copper, 32083
Tough Gel, 32084
toughened caustic, 32085
Tournant Oil, 32086
Tournay's metal, 32087
tournesol, *See* litmus, 17523
tous-les-mois starch, 32088
Toval®, 32089
Tower Brick, 32090
Tower Treat, 32091
toxilic acid, *See* maleic acid, 18475
toxilic anhydride, *See* maleic anhydride, 18476
Toximul®, 32092
Toximul® 600, 32099
Toximul® 8240, 32093
Toximul® 8320, 32094
Toximul® D, 32095
Toximul® H-HF, 32096
Toximul® SEE-340, 32097
Toximul® TA-2, 32098
Toyo oil orange, *See* Sudan I, 29938
Toyocat® -DMA, 32100
Toyocat® -DMCH, 32101
Toyocat® -DT, 32102
Toyocat® -ET, 32103
Toyocat® -HPW, 32104
Toyocat® -MR, 32105
Toyocat® -NEM, 32106
Toyocat® -NP, 32107
Toyocat® -TE, 32108
Toyocerin®, 32109
TP-35 Solvent, 32110
TPG, 32111, *See* triphenylguanidine,
TPL, 32112
TPN, *See* Chlorothalonil, 6150, Repulse, 26151
TPP, *See* triphenyl phosphate, 32391, triphenyl phosphite, 32393
TPS 20, 32113
TPS 27, 32114
TPS 32, 37, 32115
TPS 327, 32116
TPX-80CNI, 32117
Tra-Bond 2151, 32118
Tra-Bond F113, 32119
Trabuk, 32120
Tra-Cast 3103, 32121
Tracey B, 32122
Tracey C, 32123
Tracey M Plus, 32124
Tracey MG, 32125
Tracey SSS, 32126
Trachine, 32127
trachyte, 32128
Track, 32129, *See* metazachlor, 19465
Tracker, 32130, *See* Dicamba, 8374
Tra-Duct 2902, 32131
Traffaid 30 B, 32132
Tragasol, 32133, *See* Industrial gum, 15114
Tralgon, *See* Valadol, 33238
Tramat, *See* ethofumesate, 10967
Trametan, *See* Rezifilm, 26521
Tramisol®, 32134
tranexamic acid, *See* Amikapron, 2121
Trans Gard, 32135
Trans-aid, *See* Ammonium thiocyanate, 2256
Transclene, 32136
Transcutol, 32137, *See* Diethoxol, 8435, Ethoxydiglycol, 11044, ethyl di-lcinol, 11068
Transjojoba, 32138
Translink®, 32139
Translink® 37, 32140
Translink® 555, 32141
Transol, 32142
Transoxide, 32143
Transpafill, 32144
Transpalene®, 32145
transpar, 32146
Transpex, 32147
Trans-Vert, *See* Neoarsycodyl, 20871
Trapex, *See* methyl isothiocyanate, 19543
Trapoc resin, *See* Jatob lagrima, 15537
Traseolide, 32148
Tra-Shield 2867, 32149
trass, 32150
Trastan LS, 32151, *See* calcium lignosulfonate, 4956, 4957
Trasulphane, *See* Perichthol, 23270
Traton®, 32152
Traubensäure, *See* tartaric acid, 30832
Travamin, *See* Extiat®, 11323
Travase, 32153
Travert, *See* Invert Sugar, 15200
travertine, 32154
Traytuf Ultra-Clear, 32155
Traytuf Ultra-Clear Ultralen® SP 3700 S, *See* Polyethylene terephthalate, 24426
trazophos, *See* MSS CMPP, 20429
Trédémine, *See* niclosamide, 21146
Tread-Brite, 32156
Treadfast®, 32157
treble superphosphate, 32158
Tredemine, *See* Mansonil®, 18611
Tree Bug-Lok Adhesive, 32159
tree copal, 32160
tree gum, 32161
tree wax, *See* Chinese wax, 6077
Trefanocide, *See* Trifluralin, 32252
Treficon, *See* Trifluralin, 32252
Treflam, *See* Trifluralin, 32252
Treflan, 32162, *See* Trifluralin, 32252
Treflanocide, *See* Trifluralin, 32252
Trefsin®, 32163
Trefsin® 3201-50, 3201-60, 32164
Trehalose, *See* Mycose, 20551
α,α'-Trehalose, *See* Mycose, 20551
Tre-Hold, 32165, *See* Rhizopon B, 26574, Tipoff, 31848
Trelit, 32166
Tremin, *See* Wollastocoat®, 34545
Trem-LF-40, 32167
Tremvac, 32168
Tremvac-FP, 32169
Trenamine D 200, *See* Octosol 449, 22018
Trenamine D 201, *See* Octosol 449, 22018
Trenbest 500, 32170
Trenbolone acetate, 32171, *See* Finaplix, 11803
Trenchmaster, 32172
Trench's flameless explosive, 32173
Trend®, 32174
Trenyline, 32175
trepenol wa tvm 474, *See* Rhodapon® 101-10, 26631
Trepolate F 40, *See* Nacconol® 40G, 20647, sodium dodecylbenzenesulfonate, 28760
Trepolate F 95, *See* Nacconol® 40G, 20647, sodium dodecylbenzenesulfonate, 28760
Trethylene, *See* trichloroethylene, 32217
Tretobond®, 32176

Tret-o-Lite, 32177
Trevin, *See* thiophanate methyl, 31706
Trevira, 32178
TRH-Roche, 32179
Tri, 32180
tri (*m*-amino) phenyl titanate, *See* LICA 97, 17200
tri-(2-butoxyethyl) phosphate, *See* Amgard TBEP, 2055
tri(β-chloroethyl)phosphate, *See* Fyrol® CEF, 12538
tri(β-chloroisopropyl)phosphate, *See* Fyrol® PCF, 12542
tri(β,β'-dichloroisopropyl)phosphate, *See* Fyrol® 38, 12537, Fyrol® FR-2, 12540
tri(dimethylphenyl)phosphate, *See* Coalite N.T.P, 6518, Syn-O-Ad® 8475M, 30554
tri (dioctylphosphato) titanate, *See* LICA 12, 17197
tri (dioctylpyrophosphato) titanate, *See* LICA 38, 17198
tri (N ethylaminoethylamino) titanate, *See* LICA 44, 17199
tri-2-ethyl hexyl trimellitate, *See* Morflex 510, 20326
tri (2-ethylhexyl) trimellitate, *See* Jayflex® TOTM, 15559, Palatinol® TOTM, 22595
tri(2-ethylhexyl)trimellitate ester, *See* PX-338, 25441
tri (hydroxymethyl) aminomethane, *See* Tris Amino®, 32405
tri(nonylphenyl) phosphite, *See* Wytox® 312, 34578
tri(n-octyl n-decyl) trimellitate, *See* Morflex 525, 20327
Tri-4, *See* Treflan, 32162, Trifluralin, 32252
Triabon® 16-8-12-4, 32181
Tri-Abrodil, *See* acetrizoate sodium, 307
triacetin, 32182, *See* Kodaflex® Triacetin, 16281
triacetoxy(vinyl)silane, *See* CV-4800, 7449, Dow Corning® Z-6075, 8882
triacetyl cellulose, *See* cellulose triacetate, 5633
triacetyl glycerin, *See* triacetin, 32182
triacetyl glycerine, *See* Kodaflex® Triacetin, 16281
triacetyl glycerol, *See* triacetin, 32182
triacylglycerol lipase, *See* lipase, 17329
Tri-Ad, 32183
Triadenyl, *See* Adenosine triphosphate, 681
Triadimefon, 32184, *See* Bayleton® 5, 3800, Monterey Bayleton, 20287
triadimenol, 3218, *See* Summit, 30118 5
triadimenol-fuberidazole, *See* Baytan®, 3851
triadimenol-tridemorph, *See* Dorin, 8805, Dorindan, 8806
Triadine® 3, 32186, *See* Bioban® GK, 4158
Triadine® 10, 32187
Triafol, 32188
Triagran®, 32189
triallate, 32190, *See* Avadex® BW, 3450, Far-Go/Avadex BW, 11470
triallyl cyanurate, 32191, *See* Perkalink® 300, 23312
triallyl isocyanurate, 32192, *See* Perkalink® 301, 23313
triallyl trimellitate, *See* Sipomer® TATM, 28541
1,3,5-triallylisocyanurate, *See* triallyl isocyanurate, 32192
1,3,5-triallylisocyanuric acid, *See* triallyl isocyanurate, 32192
2,4,6-Triallyloxy-1,3,5-triazine, *See* Perkalink® 300, 23312, triallyl cyanurate, 32191
triallyl-*s*-triazine-2,4,6(1H,3H,5H)-trione, *See* triallyl isocyanurate, 32192
1,3,5-triallyl-s-triazine-2,4,6(1H,3H,5H)-trione, *See* triallyl isocyanurate, 32192
Triameen T, 32193
triamino-diphenyl-tolyl methane, *See* Leucaniline, 17016
triaminoresorcinol, *See* reducin, 25982
Triamino-triphenyl-carbinol hydrochloride, *See* pararosaniline, 22798
triaryl phosphate, 32194, 32195, *See* Fyrquel® EHC, 12544, Kronitex® 25, 16427, Kronitex® 50, 16428, Kronitex® 100, 16429, Kronitex® 200, 16430, Kronitex® 1840, 16431, Santicizer 154, 27392
Triaryl phosphate ester, *See* Fyrquel® 150, 12543
Triasym, *See* Anilazine, 2465
Triasyn, *See* Anilazine, 2465
Triatix, *See* Amitraz, 2202
Triatox, *See* Amitraz, 2202, Taktic, 30729
Triatrix, *See* Amitraz, 2202
Triaza-1-azoniatricyclo[3.3.1.1(3,7)]decane, 1-(3-chloro-2-propenyl)-, chloride, *See* Dowicil® 75, 8937
Triaza-1-azoniaadamantane, 1-(3-chloroallyl)-, chloride, *See* Dowicil® 75, 8937
3,5,7-Triaza-1-azoniatricyclo[3.3.1.13,7]decane, 1-(3-chloro-2-propenyl)-, chloride, *See* Dowicil® 75, 8937
1,2,3-triaza-1H-indene, *See* benzotriazole1H-benzotriazole, 3982, Preventol® CI 8, 24930
1,2,3-triazaindene, *See* benzotriazole1H-benzotriazole, 3982, Preventol® CI 8, 24930
Triazid, *See* Amitraz, 2202
Triazine A 1294, *See* atrazine, 3394
***s*-Triazine, 2-(*tert*-butylamino)-4-(ethylamino)-6-(methylthio)-**, *See* terbutryn, 31230
***s*-triazine, 2-chloro-4,6-bis(ethylamino)-**, *See* simazine, 28464
***s*-triazine, 2-chloro-4-(ethylamino)-6-(1-cyano-1-methyl)(ethylamino)-**, *See* cyanazine, 7472
1,3,5-Triazine-2,4-diamine, 6-chloro-N,N'-diethyl, *See* simazine, 28464
1,3,5-Triazine-2,4-diamine, N(1,1-dimethylethyl)-N'-ethyl-6-(methylthio)-, *See* terbutryn, 31230
1,3,5-Triazine-2,4-diamine, N,N'-bis(1-methylethyl)-6-(methylthio)-, *See* prometryn, 25140
***s*-triazine, 2,4-bis(ethylamino)-6-chloro-**, *See* simazine, 28464
Triazine, 4-(isopropylamino)-4-(methylamino)-6-(methylthio)-, *See* desmetryn, 8142
***s*-triazine, 2,4-bis(isopropylamino)-6-methylmercapto-**, *See* prometryn, 25140
***s*-triazine, 2,4-bis(isopropylamino)-6-methylthio-**, *See* prometryn, 25140
1,3,5-Triazine, 2,4,6-tris(2-propenyloxy)-, *See* Perkalink® 300, 23312, triallyl cyanurate, 32191
1,3,5-Triazine-2,4,6-triamine, *See* Chimassorb® 119FL, 6072
***s*-Triazine-1,3,5(2H,4H,6H)-triethanol,** *See* Bioban® GK, 4158
1,3,5-Triazine-1,3,5(2H,4H,6H)-triethanol, *See* Bioban® GK, 4158, Nipacide® BK, 21254
s-Triazinetriol, *See* Zeonet A, 34720
***sym*-Triazinetriol**, *See* Zeonet A, 34720
***s*-triazine-2,4,6-triol**, *See* isocyanuric acid, 15357, Zeonet A, 34720
1,3,5-Triazine-2,4,6-triol, *See* Zeonet A, 34720
***s*-2,4,6-Triazinetriol**, *See* Zeonet A, 34720
2,4,6-Triazinetrione, *See* Zeonet A, 34720
***s*-Triazinetrione**, *See* Zeonet A, 34720
***s*-Triazine-2,4,6-trione**, *See* Zeonet A, 34720
***s*-Triazine-2,4,6(1H,3H,5H)-trione**, *See* Zeonet A, 34720
1,3,5-Triazine-2,4,6(1H,3H,5H)-trione, *See* cyanuric acid, 7494, Zeonet A, 34720
Triazin-5(4H)-one, 4-amino-6-tert-butyl-3-(methylthio)-, *See* metribuzin, 19595
triazine, 2,4,6(1H,3H,5H)-trione, 1,3-dichloro-, sodium salt, *See* dichloro-1,3,5-triazinetrione, sodium salt, 8385
triazine 2,4,6(1H,3H,5H)trione, 1,3-dichloro, potassium salt, *See* dichloro-1,3,5-triazinetrione, potassium salt, 8384
Triazine-2,4,6(1H,3H,5H)-trione, 1,3,5-trichloro-, *See* Queschlor, 25658
***s*-triazine-2,4,6-trithiol**, *See* Zisnet F-PT, 34838
1,1',1-(1,3,5-Triazine-2,4,6-triyltris ((cyclohexylimino)-2,1-ethanediyl)tris(3,3,5,5-tetramethylpiperazinone), *See* Good-rite® 3150, 13200
1,2,4-1H-Triazole, *See* pyrrodiazole, 25534
1H-1,2,4-triazole, *See* pyrrodiazole, 25534
1,2,4-Triazol-3-ylamine, *See* Azolan, 3530
triazophos, 32196, *See* Hostathion, 14413
Triazotion, *See* Azinphos-ethyl, 3520
Tribac, *See* Benzoyl peroxide, 3983
Tribase, 32197
Tribasic copper sulfate, *See* Cusatrib, 7429
Tribasic lead sulfate, *See* Haro® Chem PTS-E, 13644
Tribonol, 32198
Tri-Borne, 32199
Tribovax, 32200
tribromo-*t*-butyl alcohol, *See* acetone bromoform, 298
tribromo phenyl allyl ether, *See* Great Lakes PHE-65, 13311
tribromoborane, *See* Boron tribromide, 4444
tribromomethane, *See* Bromoform, 4579
1,1,1-tribromo-2-methyl-2-propanol, *See* acetone bromoform, 298
tribromoneopentyl alcohol, 32201, *See* FR-513, 12354, FR-1360, 12366
tribromonitromethane, *See* Bromopicrin, 4585
tribromophenol, *See* Bromol, 4582, 2,4,6-tribromophenol, 32202
2,4,6-tribromophenol, 32202, *See* Bromol, 4582, FR-613, 12357, Great Lakes PH-73, 13310
tribromophenyl allyl ether, 32203, *See* FR-913, 12359, Great Lakes PHE-65, 13311
tribromophenyl maleimide, *See* Actimer FR-1033, 485
tribromosalicylanilide, *See* Temasept IV, 31076
3,4,5-Tribromosalicylanilide, *See* Tuasol 100, 32543
3,4',5-tribromosalicylanilide, *See* Temasept IV, 31076
Tribromostyrene, *See* Actimer FR-803, 483
tribromsalan, *See* Temasept IV, 31076, Tuasol 100, 32543
Tribufos, *See* Def®, 7876
Tribunil, *See* methabenzthiazuron, 19472
Tribunil®, 32204
Tribuphos, *See* Def®, 7876
Tribute, 32205
tributoxyethyl phosphate, 32206, *See* Amgard TBEP, 2055, KP-140®, 16378, TBEP, 30852, tributoxyethyl phosphate, 32206
tributyl 2-(acetyloxy)-1,2,3-propanetricarboxylate, *See* acetyl tributyl citrate, 313
tributyl acetylcitrate, *See* acetyl tributyl citrate, 313, Citroflex A-4, 6343
tributyl cellosolve phosphate, *See* tributoxyethyl phosphate, 32206
tributyl citrate acetate, *See* acetyl tributyl citrate, 313
tributyl O-acetylcitrate, *See* acetyl tributyl citrate, 313, Uniplex 84, 33006
tributyl phosphate, 32207, *See* Antifoam T, 2551, Kronitex® TBP, 16434, Phos-Ad 100, 23686
tributyl phosphite, 32208, *See* Syn-O-Ad® P-312, 30561
S,S,S-tributyl phosphorotrithioate, *See* Def®, 7876
S,S,S-tributyl phosphorotrithioate, *See* Def®, 7876
S,S,S-Tributyltrithiophosphate, *See* Def®, 7876
tributylfluorostannane, *See* bioMeT TBTF, 4171
tributylphosphorotrithioate, *See* Def®, 7876
tributyltin fluoride, *See* bioMeT TBTF, 4171
tributyltin oxide, 32209, *See* Keycide® X-10, 16153
tricalcium arsenate, *See* calcium arsenate, 4938
tricalcium citrate, *See* calcium citrate, 4944
tricalcium orthophosphate, *See* calcium phospate, tribasic, 4967
tricalcium phosphate, *See* calcium phosphate, tribasic, 4967, Synthetic Bone Ash, 30642
Tricap, 32210
tricaprin, *See* Dynasan® 110, 9374
tricaprylic glyceride, *See* Captex® 8000, 5170, Emalex O.T.G, 9976, Miglyol® 808, 19736
tricaprylin, 32211, *See* Captex® 8000, 5170, Emalex O.T.G, 9976, Miglyol® 808, 19736
tricapryloylglycerol, *See* Captex® 8000, 5170, Emalex O.T.G, 9976, Miglyol® 808, 19736
tricaprylyl glycerin, *See* Captex® 8000, 5170, Emalex O.T.G, 9976, Miglyol® 808, 19736
Tricarbamix Z, *See* AAprotect, 16
tricarbimide, *See* cyanuric acid, 7494, isocyanuric acid, 15357, Zeonet A, 34720
1,2,4-tricarboxybenzene, *See* trimellitic acid, 32335
triceteareth-4 phosphate, *See* Hostaphat KW 340 N, 14389
tricetylmonium chloride, *See* Varisoft® TC-90, 33429
trichloracetic acid, sodium salt, *See* Varitox, 33433

Chemical and Trade Names

Trichlorad, *See* Aminitrazole, 2164
Trichloral, *See* Aminitrazole, 2164
Trichloren, *See* trichloroethylene, 32217
1,1,1-trichlorethane, *See* Cut Aid, 7432, Chlorothene, 6151, Distillex DS1, 8717, Ethana®, 10928, MS-170 1,1,1 Trichloroethane Solvent, 20414, trichloroethane, 32216
trichlorethylene, *See* Disparit B, 8658, Germalgene, 12892
Trichlorfon, 32212, *See* Dipterex®, 8616, Dipterex® 80, 8617, Dylox®, 9324, Masoten®, 19030, Tugon, Tugon OKO, Tugon sp 80, 32570
Trichlormetaphos, *See* ronnel, 26951
Trichlormethylfos, *See* Chlorpyrifos-methyl, 6164
trichloro(3-chloropropyl)silane, *See* CC3291, 5491
trichloro-3-cyanopropylsilane, *See* 3-cyanopropyltrichlorosilane, 7482
trichloroacetic acid, 32213, *See* Farmon TCA, 11491, Na Ta, 20642
trichloroacetyl chloride, *See* Superpalite, 30260
trichloroallyl diisopropylthiocarbamate, *See* Far-Go/Avadex BW, 11470
trichloroarsine, *See* arsenic trichloride, 3141
trichlorobenzene, *See* Abluton CTP, 136
1,2,4-trichlorobenzene, 32214, *See* Hipochem GM, 14110
***unsym*-trichlorobenzene**, *See* 1,2,4-trichlorobenzene, 32214
trichlorobenzene/biphenyl mixture, *See* Tanavol® URC, 30791
trichlorobenzene-butyl benzoate, *See* Arosolve 9D5R, 3070
trichlorobenzoic acid, *See* Benzoyl peroxide, 3983
trichloroborane, *See* Boron trichloride, 4445
3,4,4'-Trichlorocarbanilide, *See* TCC, 30856
1,2,4-trichloro-5-[(4-chlorophenyl)sulfonyl]benzene, *See* Tedion V-18, 30919
trichlorocyanuric acid, *See* trichloroisocyanuric acid, 32218
2,3,5-trichloro-N-(3,5-dichloro-2-hydroxyphenyl)-6-hydroxybenzamide, *See* Oxyclozanide, 22444, Zanil, 34692
trichloro(dodecyl)silane, *See* CD6220, 5518
1,1,1-trichloroethane, *See* trichloroethane, 32216
trichloroethane, 32216, *See* Dabco® CS90, 7653, Solvethane, 29060
α-trichloroethane, *See* Chlorothene, 6151
1,1,1-trichloroethane, 32215
1,1,1-trichloroethane-surfactants-butane-propane, *See* Delf® Fabric Protector, 8015
trichloroethene, *See* trichloroethylene, 32217
trichloroethylene, 32217, *See* Altene DG, 1873
1,1,1-trichloroethylene, *See* Distillex DS2, 8718
1,1'-(2,2,2,-trichloroethylidene)bis[4-chlorobenzene], *See* DDT, 7820
1,1'-(2,2,2-trichloroethylidene)bis[4-methoxy benzene], *See* methoxychlor, 19504
trichloroethylsilane, *See* CE6350, 5530, ethyltrichlorosilane, 11147
trichlorofluoromethane, *See* Distillex DS6, 8722, Genesolv A Solvent, 12814, Genetron® 11, 12817
trichloro-2-hydroxybenzene, *See* 2,4,6-trichlorophenol, 32222
trichloroiminocyanuric acid, *See* ACL 85, 90 Plus, 355, trichloroisocyanuric acid, 32218
1,3,5-trichloroisocyanuric acid, *See* Queschlor, 25658
trichloroisocyanuric acid, 32218, *See* ACL 85, 90 Plus, 355, CDB 90, 5524
N,N',N-trichloroisocyanuric acid, *See* trichloroisocyanuric acid, 32218
N,N',N" -trichloroisocyanuric acid, *See* trichloroisocyanuric acid, 32218
Trichlorometafos, *See* ronnel, 26951
trichloromethane, *See* chloroform, 6125
trichloromethyl chloroformate, *See* Superpalite, 30260
trichloromethyl phenyl carbinyl acetate, *See* Rose Ester, 26964
trichloromethyl(thio)phthalimide, *See* Folpet, 12210
1,1,1-trichloro-2-methyl-2-propanol, *See* chlorobutanol,
α-trichloromethylbenzyl acetate, *See* Rose Ester, 26964
trichloromethyl-chloroformate, *See* Perstoff, 23430
N-(trichloromethylmercapto)- δ_4-tetrahydro phthalimide, *See* Vancide® 89, 33302
N-(trichloromethylmercapto)- Δ_4-tetrahydro phthalimide, *See* Captan, Captan-Col, Captan-50, Captan-**N-trichloromethyl-mercapto-4-cyclohexene-1,2 dicarboximide,** *See* Vancide® 89, 33302
N-(trichloromethylmercapto)phthalimide, *See* Folpet, 12210
2-[(Trichloromethyl)thio]-1H-isoindole-1,3(2H)-dione, *See* Folpet, 12210
N-(trichloromethylthio)-4-cyclohexene-1,2-dicarboximide, *See* Captan, Captan-Col, Captan-50, Captan-83P, Captan Granular, 5165, Vancide® 89, 33302
N-(trichloromethylthio)phthalimide, *See* Folpet, 12210 83P, Captan Granular, 5165
N-trichloromethylthio-3a,4,7,7a-tetrahydrophthalimide, *See* Captan, Captan-Col, Captan-50, Captan-83P, Captan Granular, 5165, Vancide® 89, 33302
N-trichloromethylthiotetrahydrophthalimide, 32219, *See* Captan, Captan-Col, Captan-50, Captan-83P, Captan Granular, 5165
N-Trimethoxysilylpropyl-N,N,N-trimethyl ammonium chloride, *See* CT2925, 7323
trichloromonosilane, *See* trichlorosilane, 32223
1,2,8-trichloronaphthalene, *See* Nibren wax, 21097
trichloronitromethane, *See* chloropicrin, 6135
trichloropentylsilane, *See* CA0900, 4854
trichlorophene, *See* hexachlorophene, 13992
2,3,6-trichlorophenol, 32220, *See* 2,3,6-trichlorophenol, 32220
2,4,5-trichlorophenol, 32221, *See* Dowicide 2, 8927
2,4,6-trichlorophenol, 32222
trichlorophenol, sodium salt, *See* Dowicide B, 8933
(trichlorophenoxy)ethyl hydrogen sulfate, *See* Garlon, 12672
2-(2,4,5-Trichlorophenoxy)propionic acid, *See* Kuron, 16496
trichlorophenylsilane, *See* phenyltrichlorosilane, 23650
S-(2,3,3-trichloro-2-propenyl) bis(1-methylethyl)carbamothioate, *See* triallate, 32190
trichloropropylsilane, *See* CP0800, 6935, n-propyltrichlorosilane, 25232
[(3,5,6-trichloro-2-pyridinyl)oxy]acetic acid, *See* Timbrel, 31800
3,5,6-trichloro-2-pyridyloxyacetic acid, *See* Timbrel, 31800
trichlorosilane, 32223, *See* Dynasylan® TCS, 9417
(trichlorosilyl)ethylene, *See* vinyltrichlorosilane, 33865
4-(trichlorosilyl)butyronitrile, *See* 3-cyanopropyltrichlorosilane, 7482
trichloro-*s*-triazinetrione, *See* Queschlor, 25658, trichloroisocyanuric acid, 32218
trichloro-*t*-butyl alcohol, *See* chlorobutanol,
1,3,5-trichloro-s-triazine-2,4,6-trione, *See* trichloroisocyanuric acid, 32218
1,3,5-trichloro-1,3,5-triazine-2,4,6(1H,3H,5H)-trione, *See* Queschlor, 25658
1,3,5-Trichloro-S-triazine-2,4,6-trione, *See* Queschlor, 25658, trichloroisocyanuric acid, 32218
1-Trichloro-s-triazinetrione, *See* Queschlor, 25658
1,3,5-trichloro-1,3,5-triazine-2,4,6(1H,3H,5H)-trione, *See* ACL 85, 90 Plus, 355, Queschlor, 25658
1,3,5-Trichloro-1-triazine-2,4,6(1H,3H,5H)-trione, *See* Queschlor, 25658
1,2,2-Trichlorotrifluoroethane, *See* MS-180 Freon® TF Solv, 20415
trichlorotrifluoroethane, 32224, *See* Arklone, 2894, Distillex DS5, 8721, Genesolv D Solvent, 12815, Genetron® 113, 12821, MS-180 Freon® TF Solv, 20415
Trichloro 1,2,2-trifluoroethane, *See* MS-180 Freon® TF Solv, 20415
1,1,2-trichloro-1,2,2-trifluoroethane, *See* trichlorotrifluoroethane, 32224
trichlorovinylsilane, *See* CV-4900, 7450, vinyltrichlorosilane, 33865
trichlorphon, *See* Danex, 7727, trichlorfon, 32212
Trichlorpyrphos, *See* Chlorpyrifos, 6163
Trichocid, *See* Aminitrazole, 2164
Trichoderma viride, *See* Binab T, 4150
Trichoman, *See* Aminitrazole, 2164
Trichorad, *See* Aminitrazole, 2164
Trichoral, *See* Aminitrazole, 2164
Trichrome, 32225
Triclene, 32226
Tri-Clene, *See* trichloroethylene, 32217
Triclofenol piperazine, *See* Ranestol, 25866
Triclopyr, 32227, *See* Garlon, 12672
Triclosene potassium, *See* ACL 56, 59, 60, 66, 354, dichloro-1,3,5-triazinetrione, potassium salt, 8384
Tricoid, 32228
Tricontanyl PVP, *See* Ganex® WP-660, 12644
Tricotyl trimellitate, *See* Plasthall® TOTM, 24016
tricresol, 32229, *See* Cresol, 6994, cresylic acid, 7013
tricresyl phosphate, 32230, *See* Kronitex® TCP, 16435, Plastic X, 24026, Syn-O-Ad® 8484, 30558, TCP, 30857
Tricresylmercuroacetate, *See* Cerere, 5743
tricresylphosphate, *See* Plastic X, 24026
tricyanic acid, *See* cyanuric acid, 7494, Zeonet A, 34720
tricyanide, *See* cyanuric acid, 7494, isocyanuric acid, 15357
tricyclodecenyl acetate, *See* Jasmacyclene, 15529
tricyclodecenyl dimethyl propanoate, *See* Pivaloxycyclene, 23894
tricyclodecenyl isobutyrate, *See* Gaardocyclene, 12555
tricyclodecenyl methyl ether, *See* Verdalia A, 33569
tricyclodecenyl propionate, *See* Florocyclene, 12023
4-tricyclo(5.2.1.0. 2,6)-decylidene-8)-butanal, *See* Dupical, 9127
tricyclohexylhydroxystannane, *See* cyhexatin, 7568
tricyclohexylstannoltricyclohexyltin hydroxide, *See* cyhexatin, 7568
1-(tricyclohexylstannyl)-1H-1,2,4-triazole, *See* azocyclotin, 3523, Peropal®, 23401
tricyclotin, *See* azocyclotin, 3523
Tridal, *See* nuarimol, 21762
Tridecan-1-ol, *See* tridecyl alcohol, 32233
1-tridecanecarboxylic acid, *See* Philacid 1400, 23662
1-Tridecanecarboxylic acid, isopropyl ester, *See* Wickenol® 101, 105, 34378
tridecanol, *See* tridecyl alcohol, 32233
1-tridecanol, phthalate, *See* PX-126, 25437
Trideceth phosphate series, 32231
Trideceth series, 32232
Trideceth-2 carboxamide MEA, *See* Aminol A-15, 2178
Trideceth-2 phosphate, *See* Crodafos T2 Acid, 7087
Trideceth-2, sodium dodecylbenzene sulfonate, *See* Akypogene VSM-N, 1261
Trideceth-3, *See* Bio-Soft® TD400, 4204, Chemal TDA-3, 5887, Hetoxol TD-3, 13951, Iconol TDA-3, 14830, Lipocol TD-3, 17360, Macol® TD-3, 18274, Prox-onic TD-1/03, 25345, Rhodasurf® BC-420, 26651, Volpo T-3, 34031
Trideceth-3 phosphate, *See* Surfagene FMD-12-03 KAM, 30357
Trideceth-4, *See* Genapol® X-040, 12802
Trideceth-5, *See* Teric 13A5, 31257
Trideceth-6, *See* Iconol TDA-6, 14831, Rhodasurf® TDA-6, 26664, Trycol® 5940, 32501
Trideceth-6 phosphate, *See* Rhodafac® RS-610, 26608
Trideceth-7 carboxylic acid, *See* Incrodet TD7-C, 14964, Rewopol® CT 65, 26363, Sandopan® DTC-Acid, 27342
Trideceth-8, *See* Iconol TDA-8, 14832
Trideceth-9, *See* Iconol TDA-9, 14833
Trideceth-10, *See* Iconol TDA-10, 14834, Rhodasurf® T-95, 26662
Trideceth-14, *See* Trycol® 5874, 32498
Trideceth-15, *See* Hodag Nonionic TD-15, 14198
Trideceth-19 carboxylic acid, *See* Sandopan® JA-36,

Trixene, 32436
Trixidin, 32437
Trixsalen, See 4,5,8-trimethylpsoralen, 32340
Trixylenyl phosphate, See Kronitex® TXP, 16438, Syn-O-Ad® 8475M, 30554
Trizma, See THAM, 31557. Tris Amino®, 32405
trizma base, See Tris Amino®, 32405
troclosene potassium, See Clearon, 6392
troclosene sodium, See dichloro-1,3,5-triazinetrione, sodium salt, 8385
Trocor, 32438
Trodax, 32439
Trogamid T, 32440
Trogamid T G35, 32441
troillite, See white pyrites, 34371
Trojan, See chloridazon, 6103
Trojan SC, 32442
trolamine, See triethanolamine, 32240
Trolene, 32443, See Korlan, 16351, ronnel, 26951
Trolit F, 32444
Trolit S and Special, 32445
Trolit W, 32446
Trolite, 32447
Trolon, 32448
Trolovol®, 32449
Troluoil, 32450
Tromasin, See papain, 22698
Trombavar, See Niaproof® Anionic Surfactant 4, 21093
Trombovar, See Niaproof® Anionic Surfactant 4, 21093
trometamol, See THAM, 31557, Tris Amino®, 32405
Tromethamine, See THAM, 31557, Tris Amino®, 32405
tromethamine (CTFA), See THAM, 31557, tris (hydroxymethyl) aminomethane, 32403
Tromethamine acrylates/acrylonitrogens copolymer, See Hypan® SS500V, 14703
Tromethamine magnesium aluminum silicate, See Veegum® PRO, 33509
tromethane, See THAM, 31557, Tris Amino®, 32405
trona, See S-Carb, 27549
Trona Anhydrous Sodium Sulphate, 32451
Trona Boron Tribromide, 32452
Trona Boron Trichloride, 32453
Trona Elemental Boron, 32454
Trona Muriate of Potash, 32455
Trona Potassium Chloride, 32456, See Potassium chloride, 24748
Trona Potassium Sulphate, 32457
Trona Salt Cake, 32458
Trona Soda Ash, 32459
Trona Sulphate of Potash, 32460
Tronacarb Sodium Bicarbonate, 32461
Tronalight Light Soda Ash, 32462
Tronamang Electrolytic Manganese Metal, 32463
Tronamang-75, 85 Manganese Aluminum Briquettes, 32464
Tronox Titanium Dioxide Pigments, Chloride Process, 32465
Trooper, See Dicamba, 8374
troostite, 32466
Trophysan, 32467
Tropotox, 32468, See MCPB, 19184
Tropotox Plus, 32469
Trosiplast, 32470
Trosiplast M, 32471
Trosiplast S, 32472
Trotis, 32473, See pencycuron, 23097, 23098, 23099
trotter oil, See Neatsfoot oil, 20837
Trotyl, 32474
Troykyd anti-skin b, See methyl ethyl ketoxime, 19532
Troykyd® D44, 32475
Troykyd® Perma-Dry, 32476
Troysan, See copper naphthenate, 6783
Troysan 142, See Dazomet, 7806, N-521® Biocide, 20640
Troysan® 174, 32477
Troysan® 192, 32478
Troysan® Polyphase® EC17, 32479
Troysol AFL, 32480
Troysol LAC, 32481
Troysol S366, 32482
Troythix A, 32483
TRPGDA, See Ageflex TPGDA, 1003
TRS Rubber, 32484
Trubin, 32485
Trucal, 32486
Trucarb, 32487
Truchem Quintex, 32488
Tru-Color, 32489
Tru-Flake® Salt, 32490
Truflex DOA, See dioctyl adipate, 8548, Plasthall® DOA, 24002, PX-238, 25440, Wickenol® 158, 34393
Truflex DOP, See Bis(2-ethylhexy) Phthalate, 4228, dioctyl phthalate, 8549, Reomol DCP, 26130
Truflex DTDP, See PX-126, 25437
Trufree, 32491
Trugreen 0-0-2, 32492
Trulime, 32493
Trump, 32494
Trustan®, 32495
Truzone, 32496
Trycite, 32497
Trycol® 5874, 32498, See Trideceth series, 32232
Trycol® 5874, 5940, See Trideceth series, 32232
Trycol® 5882, 32499
Trycol® 5888, 32500, See steareth series, 29596
Trycol® 5940, 32501
Trycol® 5950, 32502, See Deceth-6, 7848
Trycol® 5971, 32503, See oleth-8, 22126
Trycol® 6940, 32504
Trycol® 6956, 32505
Trycol® 6985, 32506, See Ethal DNP-8, 10919
Trydet 2610, 32507
Trydet 2644, 32508
Trydet 2676, 32509
Trydet 2682, 32510, See PEG tallate series, 23035
Trydil, See Ammonium pentaborate, 2248
Tryfac® 5552, 32511
Tryfac® 5553, 32512
Tryfac® 5556, 32513
Tryfac® 5573, 32514
Trylon® 6735, 32515
Trylox® 5900, 32516
Trylox® 5921, 32517
Trylox® 6746, 32518
Trylox® 6753, 32519
Trylube 7602, 32520
Trymeen® 6601, 32521
Trymeen® 6606, 32522
Trymeen® 6617, 32523
Trymeen® 6620, 32524
Trymeen® 6657, 32525
Trymeen® OAM 30/60, 32526
Trymeen® SAM-50, 32527
Trymeen® TAM-15, 32528
Trymer, 32529
Trypaflavine, See Gonacrine, 13192
Trypsin, 32530
Tryptar, See trypsin, 32530
Trypure, See trypsin, 32530
TSAR, See Propetamphos, 25183
TSE Mold Release®, 32531
TSE-2000, 32532
tse-hong, 32533
tse-leou, 32534
TSH, See Thytropar, 31782
tsiklomitsin, See tetracycline, 31365
T-siloxide, 32535
TSIM, See CT3600, 7329
Tsimat, See AAprotect, 16
tsing-lieu, 32536
T-Size, 32537
TSP dodecahydrate, See trisodium phosphate, dodecahydrate, 32413
TSP-12, 32538, See trisodium phosphate, dodecahydrate, 32413
TSP-O, 32539, See trisodium phosphate, anhydrous, 32412
TSPP, 32540
TSPP, See sodium pyrophosphate, 28814
TsS 21, See Ablunol SA-7, 113
t-stuff, See hydrogen peroxide, 14590
TTC, See tetrazolium chloride, 31395
TTD, See Akrochem® TETD, 1180, Disulfiram, 8725
TTEGDA, See Ageflex T4EGDA, 970
TTH, See Thytropar, 31782
Türkischrotöl 100%, 32541
Tuads, See Rezifilm, 26521, Thiram, 31726, Thiurad, 31730
Tuads®, 32542
Tuasal, See Tuasol 100, 32543
Tuasol 100, 32543, See Temasept IV, 31076
Tubania, 32544
Tubatoxin, See rotenone, 27000
Tubazole, 32545, 32546
Tubazole M, 32547
Tubercuprose, See cupric formate, 7380
Tubergran, 32548, See quintozene, 25692
Tubodust, See Tecgran, 30886, tecnazene, 30898
Tubodust, Tubostore, 32549, See Tecgran, 30886, tecnazene, 30898
Tubotin, 32550
Tubotox, 32551
tubotoxin, See rotenone, 27000
Tuc-tur metal, 32552
Tudy, See Taktic, 30729
Tuex, 32553
TUEX, See Rezifilm, 26521
Tuf Stuf, 32554
Tufcote, 32555
Tufcote E-50SM, 32556
Tufdene®, 32557
Tuf-Draw, 32558
Tuff Stuff, 32559
Tuffcide, See Chlorothalonil, 6150, Repulse, 26151
Tufflake™, 32560
Tufftride, 32561
Tuflin HS-7066 NT7, 32562
Tuf-Lube, 32563
Tufnol, 32564
Tufprene®, 32565
Tufseal, 32566
Tufset, 32567
Tuftane, 32568
Tuftec®, 32569
Tugen, See Propoxur, 25198, 25199, 25200
Tugon, See trichlorfon, 32212
Tugon, Tugon OKO, Tugon sp 80, 32570
Tula metal, 32571
Tulisan, See Rezifilm, 26521, Thiram, 31726
Tullanox HM-250, 32572
Tumbleblite, 32573
Tumblebug, 32574
Tumblemoss, 32575
Tumbleslug, 32576
tumbleweed, See glyphosate, 13148
Tumbleweed Gel, 32577
Tumenol, See Perichthol, 23270
Tumescal OPE, See o-phenylphenol, 23648
Tumil-K, See Gluconal® K, 13043
Tumil-K, See Katorin, 15798, potassium d-gluconate, 24752, Potassium gluconate, 24755
Tuncast, 32578
Tundak, 32579
tung oil, 32580, See Chinese wood oil, 6080
tunga resin, 32581
Tungophen® B, 32582
Tungstate, decaammonium, See Ammonium tungstate, 2258
tungstate, disodium, (T-4)-, See sodium tungstate, 28829
tungsten, 32583
tungsten (VI) fluoride, See tungsten hexafluoride, 32587
tungsten blue, 32584
tungsten brass, 32585
tungsten bronze, 32586
tungsten carbide, See Camite, 5054
tungsten dioxide, See Brown oxide of tungsten, 4621

tungsten hexafluoride, 32587
tungsten steel, 32588
tungsten steel, high, 32589
tungsten steel, low, 32590
tungsten trioxide, 32591
tungsten-sodium bronze, *See* saffron bronze, 27224
tungstic acid, 32592
tungstic acid anhydride, *See* tungsten trioxide, 32591
tungstic oxide, *See* tungsten trioxide, 32591
tungstic(VI) acid, *See* tungstic acid, 32592
tungstophosphoric acid, *See* phosphotungstic acid, 23749, 23750, 23751
Tunic, *See* methazole, 19481
Tunsten, *See* tungsten, 32583
turacine, 32593
turbadium bronze, 32594
Turbair, 32595, *See* Permethrin, 23385
Turbair Grain Store Insecticide, 32596
Turbair Permethrin, 32597
Turbair Roval, 32598
Turbex, 32599, *See* Jet Amine DC, 15582
turbiston bronze, 32600
Turbith mineral, *See* turpeth mineral, 32612
Turbo, *See* metolachlor, 19585
Turboclean, 32601
Turbo-Grass, 32602
Turbonit, 32603
Turcam, *See* Bendiocarb, 3937
Turfcide, *See* quintozene, 25692
Turfclear, 32604
Turflon, *See* Timbrel, 31800
Turgex, *See* hexachlorophene, 13992
turgolds, 32605
Turgum® S, 32606
Turkey Red Oil, *See* Standapol® SCO, 29478
Turkey Red Oil, sodium salt, *See* oleite, 22114
Turkey Red Oils, 32607
Turkey red oxide, *See* Indian Red, 15068
turkey umber, *See* umber, 32900
Turkish geranium oil, *See* Rose+a7 oil, 26968
turmeric, 32608
Turnbull's blue, 32609
turpenteen, *See* turpentyne, 32611
turpentine, 32610
turpentyne, 32611
turpeth mineral, 32612
Turpex, 32613
Turpinal MD2, *See* Fostex AMP, 12337
turps, *See* oil of turpentine, 22078
Turrisin, 32614
Tursione, 32615
Tusadin, 32616
tussar silk, 32617
tutane, *See* Amine C4, 2152
Tutania, 32618
Tutenag, 32619
Tutenague, *See* Tutenag, 32619
Tutenay, *See* Tutenag, 32619
Tutia, *See* tutty powder, 32623
Tutol No. 2, 32620
Tutol®, 32621
Tutor, 32622
tutty powder, 32623
Tween 60, *See* Alkamuls® T-60, 1692
Tween 80, *See* Alkamuls® T-80, 1693, Radiasurf® 7157, 25804
Tween(R) 60, *See* Alkamuls® T-60, 1692
Tween(R) 80, *See* Alkamuls® T-80, 1693
Tween®, 32624
Tween® 20, 32625
Tween® 21, 32626
Tween® 40, 32627
Tween® 60, 60K, 32628
Tween® 61, 32629
Tween® 65, 65K, 32630
Tween® 65, 65K, *See* Polysorbate 65, 24571
Tween® 80, 80K, 32631
Tween® 81, 32632
Tween® 85, 32633
Twent®, 32634
Twinspan, 32635
Twin-Tak, 32636
Twitchells reagent, 32637
Two Cubed Eight, 32638
two step, *See* phenolic resin, 23607
two-stage resin, *See* phenolic resin, 23607
Twosward, 32639
Tybrite, 32640
Tycel®, 32641
Tycel® 7000 Series, 32642
Tychem®, 32643
Tyclarosol, *See* tetrasodium EDTA, 31391, Kalex 220 Crystal, 15668, Kalex Liq. 50%, 15671
Tygacell, 32644
Tygadure, 32645
Tygafion, 32646
Tygaflor, 32647
Tygafluor, 32648
Tygalam, 32649
Tygan, 32650
Tygatape, 32651
Tygavac, 32652
Tyglas, 32653
Tygon F, 32654
Tylac®, 32655
Tylac® 68151-00, 32656
Tylac® 68202-00, 32657
Tylan, *See* tylosin, 32663
Tylenol, *See* Valadol, 33238
Tyllanex, 32658, *See* Terbuthylazine, 31227
Tylon, *See* tylosin, 32663
Tylorol LT-50, *See* Perlankrol® ATL40, 23324, Rhodapon® LT-6, 26638, Standapol® T, 29481, Witcolate TLS-500, 34479
Tylose, *See* Glutolin, 13064
Tylose MGA, *See* carboxymethylcellulose, 5257, carboxymethylcellulose sodium, 5258, Cekol, 5558
Tylose® C, C-p, CB, CB-p, 32659
Tylose® H Series, 32660
Tylose® MHB, 32661
Tylose® P, P-x, PS-x, P-z Series, 32662
tylosin, 32663
Tyloxapol, *See* Superinone, 30242
Tymelyt, *See* Lofepramin Hydrochloride, 17552
Tynex®, 32664
Tyox®, 32665
Typar®, 32666
Type 798 Roving, 32667
type metal, 32668
typewriter metal, 32669
Typly, 32670
Ty-Ply BN, 32671
Typophor®, 32672
Typro®, 32673
Tyranton, *See* Diacetone alcohol, 8249
Tyrenka, 32674
Tyril 125, 32675
Tyril 880, 32676
Tyril 880B, 32677
Tyril 1011, 32678
Tyrin, 32679
Tyrin CM 0136, 32680
Tyrin CM 566, 32682
Tyrin CM 0836, 32681
Tyrin CM 3615, 32683
Tyrite® 7412, 32684
Tyrite® 7660, 32685
Tyrol-2, 32B, 6, CEP, 32686
Tyrolean earth, *See* Bohemian earth, 4368
Tyrolite, 32687
Tyropanoate sodium, *See* Bilopaque, 4145
Tyrosilane C, 32688
tysonite, 32689
Tytanpol, 32690
Tyvek Practik®, Protech®, 32691
Tyvek®, 32692
Tyzor®, 32693
U 46 KV Fluid, *See* Mecoprop, 19204
U 46®, 32694
U Blue 104, 32695
U Pink 113, U Violet 109, 32696
U35, 32697
U-2069, *See* dicloran, 8395, Fumite Dicloran, 12459
U-4224, *See* dimethyl formamide, 8506
U-6013, *See* isoflupredone acetate, 15362
U-6233, *See* benzotriazole1H-benzotriazole, 3982, Preventol® Cl 8, 24930
U-6233, *See*
Uba, 32698
UBOB®, 32699
UBS, 32700
UC-5500 Series, 32701
UCAR amyl phenol 4t, *See* Pentaphen® 67, 23168
UCAR Solvent IM, *See* Dowanol® PM, 8896
UCAR solvent LM, *See* Arcosolv® PM, 2814, Icinol PM, 14816, propylene glycol monomethyl ether, 25220, Solvenon® PM, Solvent PM, 29050
Ucar® Acrylic 503, 32702
Ucar® Latex 100, 32703
Ucar® Latex 130, 32704, *See* Everflex® SP-1084, 11266
Ucar® Latex 148, 32705
Ucar® Latex 351, 32706
Ucar® Latex 405, 32707
Ucar® Phenoxy Resin PKHM-301, 32708
Ucar® Vehicle 435, 32709
Ucar® Vehicle 451, 32710
Ucarcide, *See* glutaraldehyde, 13062, Sonacide, 29073
Ucarcide® 225, 32711
Ucare® Polymer JR-30M, JR-125, JR-400, LR-30M, LR-400, SR-10, 32712
Ucarsil® FR-1A, FR-1B, 32713
UCC 974, *See* Dazomet, 7806, N-521® Biocide, 20640
Ucecoat, 32714
Ucecryl, 32715
Ucefix, 32716
Uceflex, 32717
Ucelone, 32718
Uchatius bronze, 32719
Ucicline, Ucipol, 32720
Ucinite, 32721
Ucipol, 32722
Ucon®, 32723
Ucon® 50-HB-400, 32724
Ucon® Fluid AP, 32725, *See* PPG butyl ether series, 24841
Ucon® Hydrolube HP-5046, 32726
Ucrete, 32727
Ucuhuba fat, 32728
Udel, 32729
Udel® GF-110, 32730, *See* Polysulfone, 24619
Udel® P-1700, 32731, *See* Polysulfone, 24619
Udet 950, 32732
Udikral, 32733
Udilo oil, *See* laurel nut oil, 16833
Udolac, *See* Dapsone, 7751
UDVF, *See* dichlorvos, 8394
UF 3, *See* Cyasorb® UV 9, 7495, Neo Heliopan® BB, 20867, Rhodialux A, 26675
Ufablend DC, 32734
Ufacem, 32735
Ufacid K, 32736
Ufacid KA, 32737
Ufanon K-80, 32738
Ufapol, 32739
Ufapore GP, 32740
Ufarol, 32741
Ufarol Am 30, 32742, *See* Ammonium lauryl sulfate, 2242
Ufarol Na-30, 32743
Ufarol TA-40, 32744
Ufaryl, 32745
Ufaryl DB80, 32746
Ufasan 35, 32747
Ufasan IPA, 32748
Ufasan TEA, 32749
Ufasoft 75, 32750
Uffelmann's reagent, 32751

Ufomite® 27-802, *See* Melamine formaldehyde resin, 19256
Uformite®, 32752
Uformite® 21-806, 32753
Uformite® 27-802, 32754
Ufoxane 2, 32755
Ugikral RA, RB and SN, 32756
uhligite, 32757
UHMW polyethylene, *See* Hostalen® EP 4450, 14351, Hostalen® GUR 5121, 14356
UK 4271, *See* Vansil, 33363
Ukidan, *See* urokinase, 33145
Ulcatite, 32758
Ulco, 32759
Ulcocid, *See* aluminum phosphate, 1922
Ulcogant®, *See* sucralfate,
Ulcony, 32760
ulexine, 32761
Ulmal, 32762
ulmite, 32763
Ulon, 32764
Ultara®, 32765
Ultem®, 32766
Ultem® 1000, 32767, *See* Polyetherimide, 24423
Ultem® 2100, *See* Polyetherimide, 24423
Ultem® 2100, 2112, 32768
Ultem® 2212, 32769, *See* Polyetherimide, 24423
Ultem® 4000, 32770, *See* Polyetherimide, 24423
Ultem® 6000, 32771, *See* Polyetherimide, 24423
Ultem® CRS5001, *See* Polyetherimide, 24423
Ultem® FXU100, *See* Polyetherimide, 24423
Ultem® CRS5001, 32772
Ultem® CRS5011, 32773
Ultem® FXU100, 32774
Ultem® LTX100A, 32775
Ultimem, 32776
Ulto Accelerator, 32777
Ultra 206, 32778
Ultra Anhydrous Lanolin HP-2060, 32779
Ultra Pure, 32780
ultra sulfate sl-1 WAQE, *See* Rhodapon® 101-10, 26631
Ultra Touch, 32781
Ultra Zinc DMC, 32782
Ultra®, 32783
Ultrabase, 32784
Ultrabion, *See* Ampicillin, 2361
Ultrablend®, 32785
Ultrablend® S, 32786
Ultracast PE-35, PE-60, 32787
Ultracef, *See* Cefadroxil, 5547
Ultracene, 32788
Ultrachem Assembly Fluid 1, 32789
Ultracid, *See* methidathion, 19483
Ultracut, 32790
Ultra-DMC, 32791
Ultradur®, 32792
Ultradur® B 2550, 32793
Ultradur® B 2550, B 4300 G10, B 4500, *See* Grilpet® XE3060, 13402
Ultradur® B 4300 G10, 32794
Ultradur® B 4500, 32795
Ultradur® KR 4001, 32796
Ultrafast® 830 Liq, 32797
ultraferran, 32798
UltraFine® II, 32799, *See* antimony trioxide, 2574
Ultra-Flat, 32800
Ultraflex®, 32801, *See* Microcrystalline wax, 19666
Ultraflo, 32802
Ultrafloc, *See* carrageenan, 5315
Ultraform, 32803
Ultraform®, 32804
Ultraform® H 2320, 32805
Ultraform® N 2200 C4X, 32806
Ultraform® N 2200 G5, 32807
Ultraform® N 2211 PVX, 32808
Ultraform® N 2320, 32809
Ultraform® N 2320 BK 11005 MO, 32810
Ultraform® W 2320, 32811
Ultraglaze® 4400, 32812
Ultrahold 8, 32813
Ultrahold, Ultrahold 8, 32814
Ultralen® SP 3700 S, SP 3705, 32815
Ultralen® SP 3705, 32816, *See* Polyethylene terephthalate, 24426
ultra-light alloys, 32817
Ultralin, 32818
Ultralog, 32819
Ultralumin, 32820
ultramarine, 32821
Ultramarine ash, 32822
Ultramarine blue, *See* U Blue 104, 32695
ultramarine green, *See* Accrox, 258, chromic oxide, 6223, M100, 18107
Ultramarine pink, *See* U Pink 113, U Violet 109, 32696
Ultramarine soda, *See* ultramarine, 32821
ultramarine yellow, *See* barium chromate, 3638
ultramarine, soda, *See* ultramarine, 32821
Ultra-Mg, *See* magnesium gluconate, 18362
Ultramid®, 32823
Ultramid® A3, 32824
Ultramid® A3G5, 32825
Ultramid® A3K, 32826
Ultramid® A3X1G10, 32827
Ultramid® A4, 32828
Ultramid® B3, 32829
Ultramid® B35G3, 32830
Ultramid® B3EG10, B3WG10, 32831
Ultramid® B3G10, 32832
Ultramid® B3K, 32833
Ultramid® B5, 32834
Ultramid® BS 3300, 32835
Ultramid® BS 400 S, 32836
Ultramid® BS 416, 32837
Ultramid® C35, 32838
Ultramid® KR 4205, 32839
Ultramid® S3, 32840, *See* nylon 6/10, 21911
Ultramite, 32841
Ultramoll® I, II, III, 32842
Ultramoll® M, 32843
Ultramoll® PP, 32844
Ultramoll® TGN, 32845
Ultranox® 236, 32846
Ultranox® 246, 32847
Ultranox® 257, 32848
Ultranox® 276, 32849
Ultranox® 626, 32850
Ultranox® 627A, 32851
Ultranyl®, 32852
Ultrapas, 32853
Ultrapek® KR 4176, 32854
Ultrapek® KR 4177 G4, 32855
Ultra-Pflex®, 32856
Ultraphan, 32857
Ultraphor® CW, 32858
Ultraphor® SFG liquid, SFR Liq, 32859
Ultrapole H, *See* Ablumide LME, 80
Ultraprene®, 32860
Ultrasil® VN3SP, 32861
Ultrasol, 32862
Ultrason® E 1010, 32863
Ultrason® E 1010 G2, 32864
Ultrason® E 2010, 32865
Ultrason® E 3010, 32866
Ultrason® E 6010, 32867
Ultrason® KR 4101, 32868
Ultrason® S 1010, 32869, *See* Polysulfone, 24619
Ultrastyr, 32870
Ultratard, *See* Ultralente Iletin,
Ultratechnekow, *See* Pertscan-99m, 23438
Ultratex, 32871
Ultrathane, 32872
Ultrathene®, 32873
Ultrathene® UE 630-000, 32874
Ultrathene® UE 632-000, 32875
Ultrathene® UE 659-04, 32876
Ultravist, 32877
Ultravon, 32878
Ultravon An, 32879
Ultrawet, 32880
Ultrawet 1t, *See* Nacconol® 40G, 20647, sodium dodecylbenzenesulfonate, 28760
Ultrawet 40SX, *See* Eltesol® SX 30, 9873, Naxonate® 4L, 20824, Pilot SXS-40, 23843, Witconate SXS 40%, 34492
ultrawet 60k, *See* sodium dodecylbenzenesulfonate, 28760
Ultrawet 99lS, *See* Nacconol® 40G, 20647
Ultrawet 99LS, *See* Witco® 1298, 34458
Ultrawet DS, 32881, *See* Witconate DS, 34485
ultrawet k, *See* Nacconol® 40G, 20647, sodium dodecylbenzenesulfonate, 28760
Ultrawet K and AOK, 32882
Ultrawet KX, *See* sodium dodecylbenzenesulfonate, 28760
Ultrawet SK, *See* sodium dodecylbenzenesulfonate, 28760
Ultrawet XK, *See* Nacconol® 40G, 20647
Ultrax®, 32883
Ultrazine CA, 32884, *See* calcium lignosulfonate, 4956, 4957
Ultrazine NA, 32885
Ultrazym, 32886
Ultrmoll NR, 32887
Ultrmoll PP, 32888
Ultrmoll PU, 32889
Ultrmoll TGN, 32890
Ultroil, 32891
Ultrol®, 32892
Ultryl 6010, 32893
Ultryl 6500, 32894
Ultryl 6800, 32895
Ultryl 7100, 32896
Ultryl 7150, 32897
Ulvio Cocoa, 32898
umbelliferone, 32899
umber, 32900
Umber brown, mineral brown, velvet brown, chestnut brown, manganese velvet brown, burnt umber, *See* umber, 32900
umbrathor, *See* thorium dioxide, 31749
umbrenal, 32901
Umbrite A, 32902
Umbrite B, 32903
umburana seed, 32904
umea tar, 32905
UN 1134, *See* Abluton T30, 139
UN 1332, *See* metaldehyde, 19431
UN 1669, *See* pentachlorethane, 23134
UN 1717 (DOT), *See* acetyl chloride, 311
UN-28 and UN-32, 32906
Unacid, *See* Unasyn® IM/IV, 32913
UNADS, *See* Ancazide IS, 2420, Monex, 20210
Unads®, 32907
Unal, 32908, *See* *p*-aminophenol, 2185, Takatol, 30725
Unamide J-56, *See* Ablumide LDE, 79, Alkamide® 327, 1634
Unamide S, *See* Ablumide SDE, 81
Unamide® C-5, 32909
Unamide® C-72-3, 32910
Unamine® C, 32911
Unamine® O, 32912
Unamine® S, *See* stearyl hydroxyethyl imidazoline, 29606
Unasyn® IM/IV, 32913
Unasyna, *See* Unasyn® IM/IV, 32913
Unasyne, *See* Unasyn® IM/IV, 32913
Unburn, 32914
Undecalactone, *See* aldehyde C14, 1536
undecan-1-ol, *See* Neodol® 1, 20908, Neoflex® 11, 20921
undecane-1-carboxylic acid, *See* Philacid 1200, 23660
1-undecanecarboxylic acid, *See* Philacid 1200, 23660
1-undecanol, *See* Neodol® 1, 20908, Neoflex® 11, 20921
undecenoic acid, *See* undecylenic acid, 32916
9-undecenoic acid, *See* undecylenic acid, 32916

10-undecenoic acid, *See* undecylenic acid, 32916
10-undecenoic acid, sodium salt, *See* sodium undecylenate, 28830
Undeceth-3, *See* Neodol® 1-3, 20909
undecyl alcohol, *See* Neodol® 1, 20908
Undecyl carbinol, *See* Emery® 3326, 10169
Undecyl cyanide, *See* Nitrile 12, 21336
n-undecyl cyanide, *See* Nitrile 12, 21336
undecyl dodecyl phthalate, 32915, *See* Jayflex® DTDP, 15555
Undecyl pentadecanol, *See* Exxal® 26, 11344
Undecyl phthalate, *See* Palatinol® 11, 22581
Undecylenamidopropyl betaine, *See* Schercotaine UAB, 27701
Undecylenamidopropyltrimonium methosulfate, *See* Rewocid® UTM 185, 26299
undecylenic acid, 32916
Undecylpentadecanol, *See* Exxal® 26, 11344
Undeen, *See* Propoxur, 25198, 25199, 25200, 25201
Unden, *See* Baygon®, 3790, Propoxur, 25198, 25199, 25200, 25201
Unden®, Undene®, 32917
Undene, *See* Propoxur, 25198, 25199, 25200, 25201
Unger A-50, 32918
Ungerol, 32919
Ungerol AM3-75, 32920, *See* Ammonium lauryl sulfate, 2240, 2242
Ungerol N2-28, 32921
Unguentum, 32922
Uni G, 32923
Unibaryt, *See* Barium sulfate, 3644, Radiopaque, 25830
Unibetaine 2C, *See* disodium cocoamphodiacetate, 8643
Unibetaine GC-88, *See* Empigen® CDR40, 10369
Unibind Series, 32924
Uni-Cal 66, 32925
Unicast, 32926
Unicell, 32927
Unichem CALCHLOR, *See* Jarcal, 15519
Unicide U-13, *See* Abiol, 69, Germall® 115, 12893
Unicoat, 32928
Unicor, 32929
Unicrop 6% Mini Slug Pellets, 32930
Unicrop Leatherjacket Pellets, 32931
Unicrop Thianosan, *See* Thiram, 31726
Unicrop Zineb, 32932
Unicrylic, 32933
Unidem, 32934
Unidene, 32935
Unidri A-74, 32936
Unidri M-75, 32937
Unidron, *See* diuron, 8736
Uniflex Byo, *See* Witcizer 100, 34453
Uniflex DOA, *See* dioctyl adipate, 8548, Plasthall® DOA, 24002, PX-238, 25440, Wickenol® 158, 34393
Uniflex TEC, *See* triethyl citrate, 32242
Uniflex® 192, Uniflex® EHT, 32938
Uniflex® 300, 32939
Uniflex® BYO, 32940, *See* Butyl oleate, 4778
Uniflex® BYS-Tech, *See* butyl stearate, 4781
Uniflex® BYS-Tech, Unimate® BYS, 32941
Uniflex® DBS, 32942
Uniflex® DCA, Unimate® DCA, 32943
Uniflex® DCP, 32944, *See* Dicapryl phthalate, 8375
Uniflex® DOA, 32945, *See* dioctyl adipate, 8548
Uniflex® DOS, 32946
Uniflex® EHT, 32947
Uniflex® IBYS, 32948
Uniflex® TEG-810, 32949
Uniflex® TOTM, *See* Jayflex® TOTM, 15559
Uniflood, 32950
Uniflot SP 100 Series, 32951
Uniflow, *See* Thiovit, 31724
Unifoam, *See* Azodicarbonamide, 3524
Unifoam AZ, *See* Azodicarbonamide, 3524
Unifog, 32952
Uniform AZ, *See* Azodicarbonamide, 3524
Unifos, *See* dichlorvos, 8394
Unifos 50 EC, *See* dichlorvos, 8394
Unifroth G, 32953
Unifume, 32954
Unigel, 32955
Unihep, *See* Heparin, Sodium Salt, 13788
Unihib®, 32956
Unihib® 106, 32957
Unihib® 1324, 32961
Unihib® 305-LC, 32958, *See* Aminotrimethylene phosphonic acid, 2191
Unihib® 314, 32959
Unihib® 905, 32960
Unihib® 1704, 32962
Unilab Surgibone, 32963
Unilin® 425 Alcohol, 32964
Unilink® 450, 32966
Unilink® 4100, 4200, 32965
Unimate IPM, *See* Wickenol® 101, 105, 34378
Unimate IPP, *See* Wickenol® 111, 34379
Unimate® BYS, *See* butyl stearate, 4781, Uniflex® BYS-Tech, Unimate® BYS, 32941
Unimate® BYS, *See*
Unimate® DCA, 32967
Unimate® DIPA, 32968
Unimate® DIPS, 32969
Unimate® EHP, 32970
Unimate® IPM, 32971
Unimate® IPP, 32972, *See* isopropyl palmitate, 15410
Unimer U-15, *See* PVP/eicosene copolymer, 25430
Unimer U-151, *See* PVP/hexadecene copolymer, 25431
Unimin Series, 32973
Unimite, 32974
Unimoll BB, *See* Butyl benzyl phthalate, 4764
Unimoll DM, *See* dimethyl phthalate, 8508
Unimoll®, 32975
Unimoll® 66M, 32976, *See* dicyclohexyl phthalate, 8413
Unimoll® BB, 32977, *See* Butyl benzyl phthalate, 4764
Unimoll® DB, 32978
Unimoll® DM, 32979
Unimox LO, *See* lauramine oxide, 16832
Unimycin, *See* oxytetracycline hydrochloride,
Union Carbide 08-Union Carbide, *See* Niaproof® Anionic Surfactant 08, 21092
Union Carbide Flexol 380, *See* Bis(2-ethylhexy) Phthalate, 4228, dioctyl phthalate, 8549, Reomol DCP, 26130
Union Carbide® A-137, 32980
Union Carbide® A-151, 32981
Union Carbide® A-162, 32982
Union Carbide® A-163, 32983
Union Carbide® A-171, 32984
Union Carbide® A-172, 32985
Union Carbide® A-174, 32986
Union Carbide® A-186, 32987
Union Carbide® A-187, 32988
Union Carbide® A-189, 32989
Union Carbide® A-1100, 32990, *See* γ-aminopropyltriethoxysilane, 2186
Union Carbide® A-1106, 32991
Union Carbide® A-1110, 32992, *See* Aminopropyltrimethoxysilane, 2187
Union Carbide® A-1120, 32993, *See* Aminoethylaminopropyltrimethoxy silane, 2171
Union Carbide® A-1160, 32994
Union Carbide® XLP-57D11, 32995
Union Carbide® Y-11343, 32996
Unipel, 32997
Uniperol® EL, 32998
Uniperol® O, 32999
Uniperol® W, W Flakes, 33000
Unipertan P-24, P-242, 33001
Unipertan P-242, 33002
Unipex 108, *See* ethyltoluene sulfonamide, 11145
Uniplast, 33003
Uniplex 80, 33004
Uniplex 82, 33005
Uniplex 84, 33006, *See* Estaflex, 10874
Uniplex 108, 33007, 33008
Uniplex 110, 33009
Uniplex 150, 33010
Uniplex 155, 33011
Uniplex 171, *See* o-toluenesulfonamide, 31948
Uniplex 173, 33012
Uniplex 214, *See* Butylbenzene sulfonamide, 4789
Uniplex 225, 33013
Uniplex 250, 33014, *See* dicyclohexyl phthalate, 8413
Uniplex 260, 33015
Uniplex 270, 33016
Uniplex 310, 33017
Uniplex 552, 33018
Uniplex 600, 33019
Uniplex 680, 33020
Uniplex 809, 33021
Unipol CIM-40, *See* Empigen® CDR40, 10369
Unipol DEA, *See* DEA-lauryl sulfate, 7827
Unipol MGLS, *See* magnesium lauryl sulfate, 18364
Unipon, *See* dalapon, 7705
Unipor, 33022
Uniquart COSM GUAR, *See* Jaguar® C-13S, C-14S, 15496
Uniquat, 33023
Uniquat® QAC-50, QAC-80, *See* Benzalkonium chloride, 3958
Unique, 33024
Uni-Rez® 221, 33025
Uni-Rez® 709, 33026
Uni-Rez® 1039, 33027
Uni-Rez® 1502, 33028
Uni-Rez® 1548, 33029
Uni-Rez® 1552, 33030
Uni-Rez® 2100, 33031
Uni-Rez® 2355, 33032
Uni-Rez® 2620, 33033
Uni-Rez® 2800, 33034
Uni-Rez® 7003, 33035
Uni-Rez® 7705, 33036
Uni-Rez® 9002, 33037
Uni-Rez® A-800 Light, 33038
Uniscrub®, 33039
Unisept, *See* chlorhexidine digluconate, 6101, Rotersept, 27002
Unisept® Solution, 33040
Uniset® D-124F, 33041
Unisiv® 3A, 33042
Unisize HA-70, 33043
Unislip 1753, 33044
Unislip 1759, 33045
Unisol, 33046
Unispar 40, 33047
Unisperse-E, 33048
Unisperse-P, 33049
Unistab D-33, 33050
Uni-Tac® 70, 33051
Uni-Tac® R85, 33052
Uni-Tac® R99, 33053
Unitane, *See* Kronos®, 16439 , 16440, Kronos® 1000, 16441, Kronos® 2020, 16442, Kronos® 2073, 16443, Kronos® 3020, 16444, titanium dioxide, 31867
Unite, 33054
United, 33055
Unitex, 33056, *See* Polyisoprene, 24471
Unithane, 33057
Unithox 450, 33058
Unitolate PGO-1010, *See* Polygliceryl-10 decaoleate, 24453
Unitreat, 33059
Univadine, 33060
Unival DMDA-6200, 33061
Univan, 33062
universal balsam, 33063
Univest, 33064
Univol U 3165, *See* Philacid 1400, 23662
Univol U304, 33065
Univol U312, 33066
Univol U320, *See* Myristic acid, 20574
Univol U332, *See* palmitic acid, 22631
Uniwax 1747, 33067
Uniwax 1750, 33068
Uniwax 1760, 33069, *See* Advawax® 290, 820
Unizeen OA, *See* oleamidopropyl dimethylamine, 22109

Unizole, *See* Dimetridazole, 8528
Unlloy Chrome Steels, 33070
Unna's stain, 33071
Unocal 76 Res 701, 33072
Unocal 76 Res 777, 33073
Unocal 76 Res 1026, 33074
Unocal 76 Res 1300, 33075
Unocal 76 Res 3114, 33076
Unocal 76 Res 3512, 33077
Unocal 76 Res 4305, 33078
Unocal 76 Res 6004, 33079
Unocal 76 Res 6206, 33080, *See* Vinyl Acetate, 33854
Unocal 76 Res 6255, 33081
Unocal 76 Res 6272, 33082
Unocal 76 Res 6931, 33083
Unocal 76 Res 7800, 33084
Unocal 76 Res 9410, 33085
Unocal 76 Res S-55, 33086, *See* Vinyl Acetate, 33854
unresolvable tartaric acid, *See* mesotartaric acid, 19400, *meso*-tartaric acid, 30831
UOP, 33087, 33088, 33089
UOP 288, 33090
UP 13600, 33091
Upamide ACMEA, *See* Acetamide MEA, 283
Upamide LACAMEA, *See* Lactamide MEA, 16568
Upgrade, 33092
Upilex, 33093
Upixon, 33094
Uplees powder, 33095
Up-Start, 33096
UR-20CF/000, 33097, 33098
UR-30CF/000 Foamed, 33099
Urac, 33100
Uracid, *See* aluminum hydroxide, 1905
uracil, 33101
uracil-1-β-D-ribofuranoside, *See* uridine, 33140
Uracryl, 33102
Uradex, 33103
Uradil, 33104
Uragan, 33105, *See* bromacil, 4564
Uralane® 5774-A/B, 33106
Uralite, 33107
Uralite 3103, 33108
Uralite 3111, 33109
Uralite 3150, 3152, 33110
Uralite 3530, 33111
Uralloy® Hybrid Polymer LP-2035, 33112
uramil, 33113
uramine t 80, *See* NM-AMD, 21404
Uramon, 33114
uranin(e), *See* Fluorescein disodium salt, 12081
uranine, 33115
uranine yellow, *See* Fluorescein disodium salt, 12081
Uranox, 33116
Urao, *See* S-Carb, 27549
Urastat, *See* urease, 33122
Urea, *See* Bubber Shet, 4641
urea amidohydrolase, *See* urease, 33122
urea and thiourea resins, 33117
urea glue, 33118
Urea, 1-(2-benzothiazolyl)-1,3-dimethyl-, *See* methabenzthiazuron, 19472
Urea, (aminoiminomethyl)-, sulfate (2:1), *See* Grossmann reagent, 13426
Urea, N,N''-methylenebis[N'-[1-(hydroxymethyl)-2,5-dioxo-4-imidazolidinyl]-, *See* Abiol, 69
Urea, N-(1,3-bis(hydroxymethyl)-2,5-dioxo-4-imidazolidinyl)-N,N'-bis(hydroxymethyl)-, *See* Diazolidinyl urea, 8348
urea/calcium nitrate, *See* Calurea, 5043
urea/formaldehyde resins, 33120, *See* Aerolite, 861
urea-bromine, 33119
urea-lecithin, *See* Dermasome® U, 8113
ureaphos, 33121
urease, 33122
Urecoll®, 33123
Uredofos, *See* Sansalid, 27375
Ureflex 6005, 33124
Ureflex 6011, 33125
5-ureidohydrantoin, *See* Allantoin, 1739
Ureidopropyltriethoxysilane, *See* CT2507, 7318
γ-Ureidopropyltriethoxysilane, *See* Union Carbide® A-1160, 32994
Ureit, *See* urea-formaldehyde resin, 33120
Ureka, 33126
Ureka B, 33127
Ureka C, 33128
Ureol® 6414A/5117B, 33129
Urepan®, 33130
Ureresolve, 33131
Uresamin, 33132
Uresolve 411, 33133
Ureth HALL® 2050, 33134
urethane, 33135
Urethane acrylate resin, *See* CN 960, 6479
Urethane polyol, *See* Polycin 12, 24389
Urethane rubber, *See* Adiprene® L-100, 702, Adiprene® L-42, 703, Adiprene® LW-570, 704
Urethane rubber, non-MBCA, *See* Adiprene® M-400, 705
Urethon, 33136
Uretix, 33137
Urexpan, 33138
uric acid, 33139, *See* Lithic acid, 17506
uric oxide, *See* uric acid, 33139
uridine, 33140
Urimeth, *See* DL-methionine, 19487
urinary indican, *See* indican, 15070
Urine-Pak, 33141
urisol, *See* hexamine, 14008
Uristix, 33142
Uritone, *See* Grasselerator 102, 13289, hexamine, 14008
Urlite 3109/3741, 33143
Urobilistix, 33144
Urocit-K, *See* potassium citrate, 24750
uroformine, *See* hexamine, 14008
urografic acid methylglucamine salt, *See* Reno-M, 26113
urokinase, 33145
Urokon Sodium, *See* acetrizoate sodium, 307
Uromat PE, 33146
urometin, *See* hexamine, 14008
Uromiro, *See* Renovue-65, Renovue-DIP, 26115
Uromiro(n), 33147
Uronase, *See* urokinase, 33145
Uropen, *See* Versapen K, 33610
Uroplas®, 33148
Uropon, *See* dalapon, 7705
Urotropin, *See* Grasselerator 102, 13289
urotropine, *See* hexamethylene tetramine, 14006, hexamine, 14008
Urotuf®, 33149
Urovist, *See* Reno-M, 26113
Urovist Sodium 300, 33150, *See* diatrizoate meglumine,
Urovist, Urovist Sodium 300, 33150
Urox B, *See* bromacil, 4564, Uragan, 33105
Urox D, *See* Aminogen II, 2177, diuron, 8736, Karmex, 15771, Karmex®, 15772
Ursol P, *See* *p*-aminophenol, 2185, Takatol, 30725, Unal, 32908
Ursol P Base, *See* *p*-aminophenol, 2185
Ursol®, 33151
Urtal, 33152
Urtenol, 33153
Urupan, *See* d-Panthenol, 22690
Usacert, 33154
USAF A-15972, *See* Uvinul® N-35, 33200
USAF EK-1651, *See* Accelerator A22, 192
USAF GY-7, *See* Zetax®, 34741
USAF KE-5, *See* Wickenol® 111, 34379
USAF KE-13, *See* Witconol 2380, 34498
Usagran, 33155
Usalake, 33156
USI in Oval, 33157, 33158, 33159
Usol Copper Green, 33160
Usol Organiclear Stain & Sealer, 33161
Usol Zinclear, 33162
usoline, *See* paraffin, liquid, 22744
USP 245, *See* USP® -245, 33166
USP®, 33163
USP® -90MD, 33164
USP® -240, 33165
USP® -245, 33166
USP® -355M, 33167
USP® -400P, 33168
USP® -690, 33169
USP® -800, 33170
Uspulun, 33171, 33172
USR 604, *See* Dichlone, 8381
Ustaad, *See* cypermethrin, 7588, Topclip Parasol, 32037
Ustinex®, 33173
U-T-C, 33174
Utica Steel, 33175
Utocyl, 33176
UV Absorber-2, *See* Uvinul® N-35, 33200
Uvadex, *See* Oxsoralen-Ultra, 22437
Uval, 33177, *See* Rhodialux S, 26676
Uvaseb 770, 33178, *See* Bistetramethyl piperidinyl sebacate, 4284, Lowilite® 77, 17645
Uvasil 125, 33179
Uvasil 299, 33180
Uvazol 236, 33181
Uvazol 237, 33182
Uvazol 311, 33183
Uvazol P, 33184
UV-Chek, 33185
UV-Chek® AM-320, *See* 4-dodecyloxy-2-hydroxybenzophenone, 8780
Uvecryl®, 33186
Uvecryl® P 36, 33187
Uvekol, 33188
Uvesterol-D, *See* vitamin D+72, 33940
uvic acid, *See* tartaric acid, 30832
Uvilon, *See* Upixon, 33094
UVI-NOX 1494, 33189
UVI-NOX 3100, *See*
Uvinul 9, *See* Cyasorb® UV 9, 7495, Neo Heliopan® BB, 20867, Rhodialux A, 26675
Uvinul D49, *See* Benzophenone-6, 3978
Uvinul DS 49, *See* Uvinul® DS-49, 33196
Uvinul M-40, *See* Cyasorb® UV 9, 7495, Neo Heliopan® BB, 20867, Rhodialux A, 26675
Uvinul MS-40, *See* Rhodialux S, 26676, Uval, 33177
Uvinul N 35, *See* Uvinul® N-35, 33200
Uvinul®, 33190
Uvinul® 400, 33191
Uvinul® 408, 33192
Uvinul® 490, 33193
Uvinul® D-49, 33194
Uvinul® D-50, 33195
Uvinul® DS-49, 33196
Uvinul® M-40, 33197
Uvinul® M-493, 33198
Uvinul® MS-40, 33199
Uvinul® N-35, 33200
Uvinul® N-539, 33201
Uvinul® O-18, 33202
Uvinul® P 25, 33203
Uvinul® T 150, 33204
Uvistat 1121, *See* Rhodialux S, 26676
Uvistat 12, 24, 247, 2211, 33205
Uvistat 24, *See* Cyasorb® UV 9, 7495, Neo Heliopan® BB, 20867, Rhodialux A, 26675
Uvistat Aftersun, 33206
Uvistat Babysun Aftersun, 33207
Uvistat Sun Cream, Sun Block, Ultrablock, Lotion, Lipscreen, Babysun, 33208
Uvitex, 33209
Uvitex MA, 33210
Uvitex MP, 33211
Uvon, *See* prometryn, 25140, 25141
V.M. and P. Naphtha, 33212
V.M. & P, *See* Ligroin, 17244
V-1065, 33213
V-18, *See* Tedion V-18, 30919, tetradifon, 31366
V-19, 33214

Varamide® LO-1, 33382
Varamide® MA-1, 33383
Varamide® ML-1, 33384
Varamide® T-55, 33385
Varcum® 1198, 33386
varech, *See* vraic,
Variamine, 33387
Variamine Blue Base, *See* Variamine, 33387
Variclene, 33388
Vari-Cut, 33389
Varidase, 33390
Varifoam® SXC, 33391
Varine C, 33392
Varine O Acetate, 33393
Varine O®, 33394
Varine T, 33395
Varion Cdg, *See* Accobetaine CL, 222
Variotin, 33396
Variquat® 50MC, 33397
Variquat® 50MC, 60LC, 80MC, 80ME, *See* Benzalkonium chloride, 3958
Variquat® 50ME, 33398
Variquat® 638, 33399
Variquat® B200, 33400
Variquat® K 1215, 33401, *See* Ethoquad® C/25, 10999
Varisoft SDC, *See* Cycloton® SCS, 7542, octadecylbenzenemethanaminium chloride, 21985, stearalkonium chloride, 29592, Sumquat® 6210, 30126
Varisoft® 2 TD, 33402
Varisoft® 5 TD, 33403
Varisoft® 110, 33404
Varisoft® 110-PG, 33405
Varisoft® 136-100P, 33406
Varisoft® 250, 300, 355, 33407, *See* Cetrimonium chloride, 5814
Varisoft® 300, 355, 33408
Varisoft® 432-100, 33409
Varisoft® 445, 33410
Varisoft® 461, 33411
Varisoft® 462, 33412
Varisoft® 470, 33413
Varisoft® 471, 33414
Varisoft® 475, 33415
Varisoft® 3690, 33416
Varisoft® BT-85, 33417, *See* Behentrimonium chloride, 3898
Varisoft® BTMS, 33418
Varisoft® C SAC, 33419
Varisoft® CRC, 33420
Varisoft® CTB-40, 33421
Varisoft® DHT, 33422
Varisoft® LAC, 33423
Varisoft® OIMS, 33424
Varisoft® PIMS, 33425
Varisoft® SDAC-W, 33426
Varisoft® ST-50, TSC, 33427
Varisoft® TA-100, 33428, *See* Adogen® TA-101, 780
Varisoft® TC-90, 33429
Varisoft® TIMS, 33430, *See* Quaternium-27, 25607
Varisoft® TSC, 33431
Variton, 33432
Varitox, 33433
varnish, 33434
Varnodag, 33435
varnoline, 33436
Varonic® 32-E20, 33437, *See* oleth-8, 22126
Varonic® 63 E20, 33438
Varonic® BD, 33439
Varonic® BG, 33440
Varonic® DM55, 33441
Varonic® K202, 33442
Varonic® Li-42, 33443
Varonic® LI-48, 33444
Varonic® LI-63, 33445
Varonic® Q202, 33446
Varonic® Q-230, 33447
Varonic® S202, 33448
Varonic® T202, T220, 33449
Varonic® T-220, 33450
Varonic® U-215, 33451
Varox®, 33452
Varox® 270, *See* lauramine oxide, 16832
Varox® 365, 33453, *See* lauramine oxide, 16832
Varox® 375, *See* lauramine oxide, 16832
Varox® 1770, 33454
Varsol 1, *See* Kensol 30, 16038
Varsol®, 33455
Varsol® 1, 33456
Varsol® 18, 33457
Varsulf® S-1333, 33458
Varsulf® SBF-12, 33459
Varsulf® SBFA-30, 33460, *See* disodium lauryl sulfosuccinate, 8649
Varsulf® SBL-203, 33461, *See* disodium lauramido MEA-sulfosuccinate, 8647
Varsulf®
Vasagel, 33462
Vascoloy-Ramet D, 33463
Vasconite BT, 33464
Vasculs, 33465, *See* tocopherol, 31918
Vasheegyite, *See* aluminum phosphate, 1922
Vasocort, 33466
Vasogen, 33467, *See* Parogen, 22854
Vasogen®, 33468
vasoliment, *See* Parogen, 22854
vasoxyl, *See* methoxamine hydrochloride,
Vassgro DSM, 33469
Vassgro Flowable Sulphur, 33470
Vassgro Mini Slug Pellets, 33471
Vassgro Spreader, 33472
Vatrolite, *See* sodium hydrosulfite, 28771
Vatsol OS, *See* Aerosol® OS, 895, Rhodacal® IN, 26585
Vaucher's alloy, 33473
Vauquelin's salt, 33474
Vaycron, 33475
Vazo®, 33476
VC20, 33477
VCN, *See* Acrylonitrile, 436
VCS-438, *See* methazole, 19481
Vebonol, 33478
Vec, 33479
Vecortenol, 33480
Vecortenol-Vioform, 33481
Vectal, 33482, *See* atrazine, 3394
Vectal SC, *See* atrazine, 3394
Vector®, 33483
Vectra® A115, 33484
Vectral SC, *See* atrazine, 3394
Vectrin, *See* resmethrin, 26218
Vedacoll, 33485
Vedacolor, 33486
Vedafix, 33487
Vedaflex, 33488
Vedaform, 33489
Vedag BM, 33490
Vedagit, 33491
Vedagolan, 33492
Vedagully System, 33493
Vedagum, 33494
Veda-K_1, *See* Konakion, 16322, vitamin K_1, 33949
Vedalith Facade System, 33495
Vedaphalt, 33496
Vedaphon, 33497
Vedapor, 33498
Vedapurit, 33499
Vedasin, 33500
Vedastar, 33501
Vedatect, 33502
Vedatex, 33503
Vedathene, 33504
Vedatherm, 33505
Vedril, 33506
Veecote®, 33507
Veegum®, 33508
Veegum® PRO, 33509
veepa oil, 33510
Veetids, *See* V-Cil-K,
Vegamino 30-SF, 33511
vegetable alkali, 33512
vegetable black, 33513
vegetable butter, 33514
vegetable calomel, 33515
vegetable casein, 33516
vegetable ethiops, 33517
vegetable fiber, *See* Vulcanized fiber, 34097
vegetable gelatin, 33518
vegetable glue, 33519
vegetable gum, *See* British gum, dextrin, 8235
vegetable ivory, 33520
vegetable jelly, 33521
vegetable lutein, *See* xanthophyll, 34591
vegetable luteol, *See* xanthophyll, 34591
vegetable oil, 33522
vegetable oil, hydrogenated, *See* hydrogenated vegetable oil, 14601
vegetable pepsin, *See* papain, 22698
vegetable rouge, 33523
vegetable soda, 33524
vegetable spermacetl, *See* Chinese wax, 6077
vegetable tallow, *See* Chinese tallow, 6076
Vegetable Turf and Ornamental Weeder, 33526
vegetable wax, *See* Japan tallow, 15510
vegetable wool, 33527
vegetaline, 33528
Vegetoil, 33529
Vegfru Foratox, *See* Terrathion, 31343
Vegfru-Fosmite, *See* Ethion, 10955
velampishin, *See* wood-apple gum, 34552
Velan, 33530
Velardon, *See* papain, 22698
Velcorin®, 33531
Velicren, 33532
Velmonit, *See* ciprofloxacin, 6317
Velon, 33533
Veloset, 33534
Velpak, 33535, *See* hexazinone, 14020, 14021
Velpar, 33536
Velpar®, 33537
Velpeau's caustic powder, 33538
Velsan, 33539
Velsan® D8P-3, 33541
Velsan® D8P-16, 33540
Velsan® P8-3, 33543
Velsan® P8-16, 33542
Velsicol 58-CS-11, *See* Dicamba, 8374
Veltol, 33544
Veltol Plus, 33545
Velva Coat, 33546
Velva Dri, 33547
Velva Wash, *See* Velvaplast, 33549
Velvalite, 33548
Velvaplast, 33549
velvet black, 33550
velvet brown, *See* umber, 32900
Velvet Veil 310, 33551
Velvetex, 33552
Velvetex AB 45, *See* Accobetaine CL, 222
Velvetex® AB-45, 33553
Velvetex® BA-35, 33554, *See* Cocamidopropyl betaine, 6551
Velvetex® CDC, 33555, *See* disodium cocoamphodiacetate, 8643
Velvetex® GC-88, *See* Empigen® CDR40, 10369
Velvetex® OLB-30, *See* oleyl betaine, 22132
Velvetex® OLB-50, *See* oleyl betaine, 22132
Velvetol 77-19, 33556
Venceweed, *See* 2,4-DB, 7810
Vendex, *See* Torque, 32077
Venetian bole, *See* Indian Red, 15068
Venetian red, *See* Indian Red, 15068
Venice soap, 33557
Venice turpentine, 33558
Venit, 33559
Venite, 33560
Ventilago Madraspanta, 33561
Ventox, *See* Acrylonitrile, 436

Chemical and Trade Names

Venturol, *See* Dodine FL, WP, 8785
Venzar, *See* lenacil, 16984, Vizor, 34001
Venzar®, 33562
Venzoate, *See* Benzyl benzoate, 3987
VeoVa, 33563
VeoVa 10 Monomer, 33564
veppam oil, *See* veepa oil, 33510
Veracillin, *See* Dicloxacillin sodium, 8396
Verafil, 33565
Veragel® 200, 33566
Verantin, *See* oxyalizarin, 22438
veratrine, 33567
veratrole, 33568
veratrole methyl ether, *See* methyl eugenol, 19534
Verdalia A, 33569
Verdican, *See* dichlorvos, 8394
Verdict, 33570
verdigris, 33571, *See* cupric acetate, basic, 7373
verdigris green, *See* verdigris, 33571
Verdilyn, 33572
Verdinal, 33573
Verdipor, *See* dichlorvos, 8394
Verdisol, *See* dichlorvos, 8394, Vapona, 33378
verditer blue, 33574
verditers, 33575
Verdiviton, 33576
Verdiviton Elixir, 33577
Verdley, 33578
Verdone 2, 33579
Verdone CDA, 33580
Verdoracine, 33581
Verdox, *See* Ortholate, 22332
Verdoxan, 33582
Verdyl acetate, *See* Jasmacyclene, 15529
Veridian, *See* chromium green, 6232
Verilite, 33583
Verinor, 33584
Veritas, 33585
Veriviton Elixir, 33586
verjuice, 33587
Vermadax, 33588
Vermicidin, *See* Mebendazole, 19198, Telmin, 31066, Vermox, 33591
vermiculite, 33589
vermilion, *See* cinnabar, mercuric sulfide, red and black, 19350
vermilionettes, 33590
Vermirax, *See* Telmin, 31066, Vermox, 33591
Vermithana, *See* Nuophene, 21783
Vermitid, *See* Mansonil®, 18611
Vermitin, *See* phenothiazine, 23618
Vermizym, *See* papain, 22698
Vermox, 33591, *See* Telmin, 31066 Mebendazole, 19198
Vernafine, 33592
Vernalin, 33593
Vernamine Binders, 33594
Vernaminol Liquors, 33595
Vernasein, 33596
Vernasol, 33597
Vernatan, 33598
Vernol Liquors, 33599
Vernol®, 33600
Vernonite, 33601
Verofix®, 33602
Verol, *See* Metol, 19584
Veronese earth, *See* Bohemian earth, 4368, green earth, 13318
Versabacs, 33603
Versaflex® 1, 33604
Versal 150, 33605, *See* aluminum oxide, 1919
Versalon® 1140, 33606
Versamag® DC, 33607
Versamid®, 33608
Versapen, 33609
Versapen K, 33610
Versar, *See* MSMA, 20417
Versatic, 33611
Versatint Fugitive tints, 33612
Versa-TL 3, 33613
Versa-TL 125, 33616
Versa-TL® 4, TL 7, 33614
Versatrane, *See* hydrargaphen, 14539
Versatrex, *See* Versapen, 33609
Versatyl-42, 33615
Versene, *See* Kalex 220 Crystal, 15668, Kalex Liq. 50%, 15671, tetrasodium EDTA, 31391, Trilon® BS, 32327
versene acid, *See* edetic acid, 9615, Kalex Acids, 15669
Versene AG, 33617
Versene CA, 33618
Versene disodium salt, *See* edetate disodium, 9613, sodium versenate, 28831, Trilon® BD, 32325
Versene-9, *See* edetate trisodium, 9614
Versenex 80, *See* Cheelox® 80, 5864, Pentaquest Extra 0685, 23169, Trilon® C Liq, 32328
Versen-Ol, *See* Trilon® D Liq, 32329
Versenol 120, *See* Trilon® D Liq, 32329
Versenol AG, 33619
Versen-Ol®, *See* Kalex OH, 15672
Versicane, 33620
Versicon Conductive Polymer, 33621
Versiflex, 33622
Versilan, 33623, 33624
Versilan MX134, 33625
Versilok® 201, 33626
Versilube® F50, 33627
Versilube® G-321, 33628
Versilyt, 33629
Verstarktes chromammonit, 33630
Vert d'eau, 33631
Vertac, *See* Nipantiox, 21276, Sustane® 1-F, 30432
Vertac MSMA 400, *See* Neoarsycodyl, 20871
Vertac MSMA 600, *See* Neoarsycodyl, 20871
Vertac MSMA 660, *See* Neoarsycodyl, 20871
Vertal 92, 200, 33632
Vertal 200, 33633
Vertalec, 33634
Vertan, 33635
Vertelon, 33636
Vertifume, 33637
Vertigon Spansule Capsule, *See* vertigon,
Vertocinth, 33638
Vertolan, *See* sulfamethazine, 29962
Verton®, 33639
Verton® OF-700-10, 33640
Verton® RF-700-10 EM HS, 33641
Verton® RF-7007 HS, 33642
Vertrel®, 33643
Verucasep, *See* glutaraldehyde, 13062, Ucarcide® 225, 32711
Veruca-sep, *See* Sonacide, 29073
Verv, *See* Verv®, 33644
Verv®, 33644
Verv-Ca, *See* Crolactil CSL, 7184, Verv®, 33644
vesaloin, *See* hexamine, 14008
vesalvine, *See* hexamine, 14008
Vesamin, *See* acetrizoate sodium, 307
Vespel®, 33645
Vesta phosphate, *See* Rhenania phosphate, 26537
Vestagon® B 31, 33646
Vestalin, *See* Ferrozoid, 11698
Vestamelt, 33647
Vestamid®, 33648
Vestamid® D 14, 33649
Vestamid® E40M-S3, 33650
Vestamid® L 1600, 33651
Vestamid® L 1833, 33652
Vestamid® X 3500, 33653
Vestamid® X 4178, 33654
Vestamin® IPD, 33655
Vestamin® TMD, 33656
Vestanat® IPDI, 33657
Vestanat® T 1890 L, 33658
Vestanat® TMDI, 33659
Vestar®, 33660
Vestenamer®, 33661
Vestenamer® 6213, 33662
Vesticoat® UT 647, 33663
Vestiform, 33664
Vestinol, 33665
Vestinol AH, 33666, *See* Bis(2-ethylhexy) Phthalate, 4228, dioctyl phthalate, 8549, Reomol DCP, 26130
Vestinol OA, *See* dioctyl adipate, 8548, Plasthall® DOA, 24002, PX-238, 25440, Wickenol® 158, 34393
Vestodur, 33667
Vestogrip, 33668
Vestolen, 33669
Vestolen® A3512, 33670
Vestolen® AS, 33671
Vestolen® AX4304, 33672
Vestolen® EM, 33673
Vestolen® P2000, 33674
Vestolen® P2000 CR, 33675
Vestolen® P6000, 33676
Vestolen® P6700, 33677
Vestolen® P7032 G, 33678
Vestolit®, 33679
Vestolit® B 7021, 33680
Vestolit® B 7090, 33681
Vestolit® E 6003, E 6503, E 7003, 33682
Vestolit® HI, 33683
Vestolit® LF HI-SP 5735, 33684
Vestolit® M 5867, 33685
Vestolit® P 1330 K, 33686
Vestolit® S 6058, 33687
Vestopal, 33688
Vestoplast, 33689
Vestopren, 33690
Vestoran® 1100, 33691
Vestosint, 33692
Vestowax, 33693
Vesturit, 33694
Vestypor, 33695
Vestyron, 33696
Vestyron 550, 33697
Vestyron 551, 33698
Vestyron X984 and X1260AK, 33699
Vesulong, 33700
Veta-K_1, *See* Konakion, 16322, vitamin K_1, 33949
Vetanabol, 33701
Vetkelfizina, *See* Kelfizina, 15838
Vetol, 33702
Vetrmirax, *See* Mebendazole, 19198
Vex, *See* piperonyl butoxide, 23875
VF-077, 33703
Via Rasa, 33704
Vialon Fast Dyes, 33705
Vi-Alpha, 33706
Vianin, *See* gentian violet, 12845
Vianol, *See* Anti-Oxydant Bayer, 2595, Butylated hydroxytoluene, 4788, Oxyguard, 22457, Ralox® BHT food grade, 25845, Tenamine 3, 31092, Vanox® PCX, 33346, Vulkanox® KB, 34152
Vibalt, *See* cyanocobalamin,
Vibatex, 33707
Vibazine, *See* buclizine hydrochloride,
Vibrabond, 33708
Vibrac, 33709
Vibrac steel, 33710
Vibrathane®, 33711
Vibrathane® 5004, 33712
Vibrathane® 8007, 33713
Vibrathane® B602, 33714
Vibratussan, 33715
Vibrin® E-010-01, 33716
Vibrin® E-085, 33717
Vibrin® E-701, 33718
Vibrin® E-750, 33719
VIC®, 33720
Vi-Cad, *See* cadmium chloride, 4895
Vicalloy, 33721
Viccilin, *See* Ampicillin, 2361
Vicelat, *See* vitamin C, 33938
Vichy salt, *See* sodium bicarbonate, 28736
Viclan, 33722
Vicmos powder, 33723
Vicol®, 33724
Vicron, *See* calcite,

Vicryl, 33725
Victor bronze, 33726
Victor Cream®, 33727
Victor metal, 33728
Victor Powder, 33729
Victor TSPP, *See* sodium pyrophosphate, 28814
Victoria red, *See* chrome red, 6209
Victoria Silver, *See* nickel silvers, 21132
Victorium, 33730
Victory®, 33731 , *See* Microcrystalline wax, 19666
Victrex, 33732
Victron, 33733
Vicure® 10, 33734
Vicure® 55, 33735
Viczsal, 33736
Vidal's caustic powder, 33737
Vi-Daylin, 33738
Vi-De-3-Hydrosol, *See* vitamin D+73, 33941
Videne Disinfectant Solution, Videne Disinfectant Tincture, Videne Powder, 33739
Videne Disinfectant Tincture, 33740
Videne Surgical Scrub, 33741
Videobil, 33742
Videocolagio, 33743
Videophel, *See* keraphen, 16055
Vi-Dom-A, 33744, *See* Vi-Alpha, 33706
Vidox, *See* zinc oxide, 34786
Vieille powder, *See* Poudre B, 24799
Vienna caustic, Vienna paste, *See* potassium hydroxide, calcium oxide, 24756
Vienna cement, 33745
Vienna green, *See* cupric acetoarsenite, 7374
Vienna Lake, *See* Florence lake, 12013
Vienna paste, 33746
Vienna red, *See* chrome red, 6209
Viennese tombac, 33747
Vifilcon, *See* Softcon, 28844
Vigantol®, *See* vitamin D+73, 33941
Vigantol® E Comp, 33748
Vigazoo, 33749
Vigilan, 33750
Vigilan AWS, 33751, *See* PPG-12-PEG-65 lanolin oil;, 24843
Vigorite, 33752, *See* Bakelite, 3589
Vigorsan, *See* vitamin D_3, 33941
Vi-Grow, 33753
Vikane, 33754
Viking, 33755
Vikro, 33756
Vileda, 33757
Viledon Compact, 33758
Viledon Filter, 33759
Vilene, 33760
Vilit, 33761
Villiaumite, *See* sodium fluoride, 28764
Vilnite, *See* Wollastocoat®, 34545
Viluite, *See* Grossular,
Vimlite, 33762
Vin Clad Super Vinge, *See* bioMeT TBTF, 4171
Vinac® 1000 DEV, 33763
Vinac® 1000 DEV, XX-210, XX-220, XX-230, XX-240;, *See* Everflex® SP-1084, 11266
Vinac® XX-210, XX-220, XX-230, XX-240, 33764
Vinaccia tartar, 33765
vinaconic acid, 33766
Vinacron, 33767
Vinacryl, 33768
Vinacryl 4001/B, 33769
Vinacryl 4005, 33770
Vinacryl 4152, 4500/X, 4501/X, 33771
Vinacryl 4160, 33772
Vinacryl 4260, 33773
Vinacryl 4290, 33774
Vinacryl 4320, 33775
Vinacryl 4322, 33776
Vinacryl 4450, 33777
Vinacryl 4512, 33778
Vinacryl 7170, 7172, and 7175, 33779
Vinacryl R3929, R3940, 33780
Vinal, 33781
Vinalak 5150, 33782
Vinamul, 33783
Vinamul 3240, 33784
Vinamul 3250, 33785
Vinamul 6000, 33786
Vinamul 6050, 33787
Vinamul 6208, 33788
Vinamul 6275, 33789
Vinamul 6705, 33790
Vinamul 6888, 33791
Vinamul 6930, 33792
Vinamul 7700, 33793
Vinamul 7715, 33794
Vinamul 8400, 33795
Vinamul 8430, 33796
Vinamul 8460 and 9000, 33797
Vinapol, 33798
Vinapol 1000, 33799
Vinapol 1030, 33800
Vinapol 1070, 33801
Vinapol 1088, 33802
Vinapol R.3626, 33804
Vinapol R.3800, R.3863, 33803
Vinatex, 33805
Vinavil, 33806
Vincaron, *See* Vincapront,
Vinchel 11, Vinchel 20, 33807
Vinchel 20, 33808
Vinchel 22, 33809
Vinchel 35, 33810
vinclozalin, *See* vinclozolin, 33811
vinclozolin, 33811, *See* Fumite Ronilan, 12466, Konker®, 16327, Mascot Contact Turf Fungicide, 19001, Power Drive, 24821, Ronilan® DF, FL, 26949, Ronilon, 26950
Vinco 99A, 33815
Vinco 99G, 33816
Vinco 248, 33812
Vinco 249C, 33813
Vinco 265, 33814
Vinco A33, 33818
Vinco A183, 33817
Vindex, 33819
vinegar, *See* Acetic acid, 292
Vinegar acid, *See* Acetic acid, 292
vinegar naphtha, *See* ethyl acetate, 11063
vinegar salts, *See* calcium acetate, 4937
Vinex 1003, 33820
Vinex 2019, 33821
Vinex 2025, 33822
Vinex 2034, 33823
Vinex 2144, 33824
Vinex 5030, 33825
Vinicizer 80, *See* Bis(2-ethylhexy) Phthalate, 4228, dioctyl phthalate, 8549, Reomol DCP, 26130
Vinidur®, 33826
Vinisil, 33827, *See* Agrimer 15L, 1065, Tears Plus®, 30876, Videne Disinfectant Solution, Videne Disinfectant Tincture, Videne Powder, 33739
Vinnapas®, 33828
Vinnapas® A 50, 33829, *See* Vinyl Acetate, 33854
Vinnapas® EN 426, 33830
Vinnapas® EP 177, 33831
Vinnapas® LL 364, 33832
Vinnathen, 33833
Vinnol®, 33834
Vinnol® 50, 33835
Vinnol® LL 352, 33836
Vinofan®, 33837
Vinoflex, 33838
Vinoflex®, 33839
Vinoflex® 377, 33840
Vinoflex® 516, 33841
Vinoflex® 526, 33842
Vinoflex® 534, 33843
Vinoflex® 535, 33844
Vinoflex® 719, 33845
Vinsalyn, 33846
Vinsol, 33847, 33848
Vinsol Emulsion, 33849
Vinsol Resin, 33850
Vinstop, *See* Aquatreat SDM, 2761, Octopol SDM-40, 22014, Perkacit® SDMC, 23288, sodium dimethyldithiocarbamate, 28758, Thiostop N, 31717
Vinuran®, 33851
Vinychlon®, 33852
vinyl acetal polymers, butyrals, *See* polyvinyl butyral, 24667
vinyl acetate, 33853, 33854, *See* Polycare® 509, 24377
Vinyl Acetate Monomer, 33855, *See* Vinyl Acetate, 33854
Vinyl acetate/butyl maleate/isobornyl acrylate copolymer, ethanol SDA-40B, *See* Advantage CP, 806
Vinyl acetate/crotonic acid copolymer, *See* Resyn® 28-1310, 26245
Vinyl acetate/crotonic acid/vinyl neodecanoate copolymer, *See* Resyn® 28-2913, 26246
Vinyl acetate-ethylene emulsion, *See* Airflex® RB-8, 1110
Vinyl acetate-ethylene latex, *See* Airflex® CA-50, 1106, Airflex® TL-30, 1111
Vinyl acetate-ethylene latex emulsion, *See* Airflex® RB-11, 1107
Vinyl acetate-ethylene-vinyl chloride terpolymer latex, *See* Airflex® 456, 1101
vinyl acetyl polymers, butyrates, *See* polyvinyl butyral, 24667
Vinyl acrylic, *See* Printlok 1046, 25004
vinyl amide, *See* Acrylamide, 410
vinyl benzene, *See* styrol, 29875
N-[2-(Vinylbenzylamino)-ethyl]-3-aminopropyl trimethoxysilane, *See* Dow Corning® Z-6032, 8880
vinyl bromide, *See* Saytex® VBR, 27525
vinyl butyl ether, *See* Shostakovsky Balsam, 28187
vinyl carbinol, *See* propenol, 25181
Vinyl chloride disp, *See* Lutofan®, 17996
vinyl compounds and polymers, 33856
vinyl cyanide, *See* Acrylonitrile, 436
vinyl ester resin, 33857
Vinyl hexanol, *See* Morillol®, 20335
Vinyl pentyl carbinol, *See* Morillol®, 20335
vinyl pyridine, 3385, 33859 8
Vinyl pyridine latex, *See* Good-rite® 2528X10, 13199, Vinyl pyridine, 33859
Vinyl pyrrolidone/vinyl acetate copolymer, *See* Agrimer VA 6, 1067
Vinyl triethoxysilane, *See* Dynasylan® VTEO, 9420
Vinyl tris (methoxyethoxy) silane, *See* Dynasylan® VTMOEO, 9422
Vinyl tris (methylethylketoxime) silane, *See* CV-5050, 7454
Vinyl tris-(2-methoxyethoxy) silane, *See* Union Carbide® A-172, 32985
Vinyl tris(trimethylsiloxy) silane, *See* CV-5100, 7455
Vinyl/resin lacquer, *See* Plasticote, 24035
vinylbenzol, *See* styrol, 29875
Vinyldimethylchlorosilane, *See* CV-4720, 7447
vinylformic acid, *See* acrylic acid, 412
Vinylite, 33860
Vinylite V, 33861
Vinylite X, 33862
vinyl-malonic acid, *See* vinaconic acid, 33766
Vinylmethyldichlorosilane, *See* CV-4772, 7448
vinyl-n-butyl ether, *See* Shostakovsky Balsam, 28187
Vinylofos, *See* dichlorvos, 8394
Vinyloid, 33863
Vinylophos, *See* dichlorvos, 8394
Vinylpyridine, *See* vinyl pyridine, 33858
1-vinyl-2-pyrrolidinone polymers, *See* Agrimer 15L, 1065, Tears Plus®, 30876
Vinylpyrrolidone/dimethylaminoethyl methacrylate copolymer/diethyl sulfate reaction product, *See* Polyquaternium-11, 24530
Vinylseal, 33864
Vinyltriacetoxy silane, *See* CV-4800, 7449 Dow Corning® Z-6075, 8882

Volkite, 34018
Volomite, *See* tungsten carbide,
Volphor, *See* Terrathion, 31343
Volpo, 34019
Volpo 3, 34020, *See* oleth-8, 22126
Volpo 5, 34021
Volpo 5, 10, 20, N3, O3, *See* oleth-8, 22126
Volpo 10, 34022
Volpo 20, 34023
Volpo 25 D 3, 34025
Volpo 25 D 10, 34024
Volpo CS-3, 34026
Volpo L4, 34027
Volpo N3, 34028
Volpo O3, 34029
Volpo S-2, *See* steareth series, 29596
Volpo S-2/S-10/S-20, 34030
Volpo T-3, 34031, *See* Trideceth series, 32232
Voltalef, 34032
Voltarol, *See* Voltaren,
Voltarol Retard, *See* Voltaren,
Voltoids®, *See* ammonium chloride, 2232
Volucon, 34033, 34034
Volunteered, 34035
Volusol, 34036
vomiting salt, *See* zinc sulfate, 34794
Von Forster powder, 34037
Von Vetter's solution, 34038
Vondalhyd, *See* maleic hydrazide, 18477
Vondozeb, *See* Karamate, 15749
Vondozeb Plus, *See* mancozeb, 18517
Vonduron, *See* diuron, 8736
Vonges dynamite, 34039
Voran, 34040
Voranate 3071, 34041
Voranate T-80, Type I, Type II, 34042
Voranol, 34043
Voranol 202, 225, 34044
Voranol 234-630, 34045
Voranol 3741, 34046
Voranol 4148, 34047
Vorite 63, 34048
Vorite 105, 110, 115, 125, 34049
Vorlan, *See* vinclozolin, 33811
Voronit, *See* fuberidazole, 12428
Vossenblue 705-81, 34050
Vostec, 34051
VPA No.3, 34052
VPC®, 34053
VPI, 34054
VPM, *See* metam-sodium, 19444, Unifume, 32954
VR Claymaster, *See* Claymaster, 6377
VR Coving, 34055
VR Fillmaster, *See* Fillmaster, 11782
VR Voidmaster, *See* Voidmaster, 34008
VR-110, 34056
VR-160, 34057
VR-500, 34058
vraic, *See* kelp, 15858
VTC, *See* CV-4900, 7450
VTEO, *See* CV-4910, 7451
V-thane, 34059
VTMO, *See* CV-4917, 7452
Vuelite, 34060
Vuepak, 34061
Vulcabest, 34062
Vulcaboard, 34063
Vulcabond®, 34064
Vulcabond® C 10, 34065
Vulcabond® E, 34066
Vulcabond® N 15, 34067
Vulcabond® TX, 34068
Vulcabond® VP, 34069
Vulcabrite, 34070
Vulcacid D, *See* Akrochem® DPG, 1168, Dynamine, 9355, Perkacit® DPG, 23281
Vulcacure, *See* AAprotect, 16, Octocure ZDE-50, 21996
Vulcacure ZE, *See* Octocure ZDE-50, 21996
Vulcacure ZM, *See* AAprotect, 16
Vulcaflex, 34071
Vulcafor CBS, *See* Vulkacit® CZ/EGC, DZ/EGC, 34129
Vulcafor HBS, *See* Vulkacit® CZ/EGC, DZ/EGC, 34129
Vulcafor MBTS, *See* Perkacit® MBTS, 23286
Vulcafor TMTD, *See* Rezifilm, 26521
Vulcaid, 34072
Vulcaid 27, 34073
Vulcaid 28, 34074
Vulcaid 33, 34075
Vulcaid 44, 34076
Vulcaid 55, 34077
Vulcaid 111, 34078
Vulcaid 222, 34079
Vulcaid 444B, 34080
Vulcaid DPG, 34081
Vulcaid LP, 34082
Vulcaid P, 34083
Vulcaid ZP, 34084
Vulcalap, 34085
Vulcalon, 34086
Vulcalucent, 34087
Vulcamel, 34088
Vulcamin, 34089
Vulcan, 34090
Vulcan Bronze, 34091
Vulcan powder, 34092
Vulcan® XC72R, 34093
Vulcanex, 34094
Vulcaniline, 34095
Vulcanine, 34096
Vulcanite, *See* Ebonite, 9499
Vulcanized fiber, 34097
Vulcanized oils, *See* Thinoline, 31675
vulcanized paper, *See* vulcanized fiber, 34097
Vulcanol, 34098
Vulcanox Crack and Joint Sealant, 34099
Vulcaplas, 34100
Vulcaplast, 34101
Vulcapont, 34102
vulcasbeston, 34103
Vulcase, 34104
Vulcastab® EFA, 34105
Vulcastab® LW, 34106
Vulcastab® T, 34107
Vulcasteel, 34108
Vulcastop, 34109
Vulcatuf, 34110
Vulcawall, 34111
Vulcazol, 34112
Vulcoferran, 34113
Vulcogene, 34114
Vulcogene ND, 34115
Vulcoid, 34116
Vulconex, 34117
Vul-Cup, *See* Bisbutyl peroxy diisopropyl benzene, 4230, Luperox 802, 17841
Vul-Cup 40KE and R, 34118, *See* Bisbutyl peroxy diisopropyl benzene, 4230, Luperox 802, 17841
Vul-Cup, Vul-Cup 40KE, Vul-Cup R, 34119
vulkacit, *See* Rezifilm, 26521
Vulkacit 1000, *See* Vulkacit® 1000, 34124
Vulkacit C, *See* Vulkacit® CZ/EGC, DZ/EGC, 34129
Vulkacit CZ, *See* Vulkacit® CZ/EGC, DZ/EGC, 34129
Vulkacit CZ/C, *See* Vulkacit® CZ/EGC, DZ/EGC, 34129
Vulkacit CZ/EG, *See* cyclohexyl benzothiazole sulfenamide, 7523
Vulkacit CZ/EGC, *See* cyclohexyl benzothiazole sulfenamide, 7523
Vulkacit CZ/K, *See* Vulkacit® CZ/EGC, DZ/EGC, 34129
Vulkacit CZ/MG/C, *See* Vulkacit® CZ/EGC, DZ/EGC, 34129
Vulkacit D/C, *See* Akrochem® DPG, 1168, Dynamine, 9355, Perkacit® DPG, 23281
Vulkacit DM, *See* Perkacit® MBTS, 23286
Vulkacit DM/C, *See* Perkacit® MBTS, 23286
Vulkacit DZ/EGC, *See* cyclohexyl benzothiazole sulfenamide, 7523
Vulkacit LDA, *See* Octocure ZDE-50, 21996
Vulkacit MTIC, *See* Rezifilm, 26521
Vulkacit NPV/C, *See* ethylene thiourea, 11132, Perkacit® ETU, 23283
Vulkacit NPV/C2, *See* Perkacit® ETU, 23283
Vulkacit NZ, *See* BBTS, 3863, Delac NS, 8001, Perkacit® TBBS, 23289
Vulkacit NZ/EG, *See* BBTS, 3863, Butylbenzothiazole sulfenamide, 4791
Vulkacit ZBEC, *See* Akrochem® Z.B.E.D, 1183
Vulkacit ZDK, *See* Octocure ZDE-50, 21996
Vulkacit ZM, *See* Octocure ZMBT-50, 21998, Oxaf, 22408, Perkacit® ZMBT, 23298, Zetax®, 34741
Vulkacit®, 34120
Vulkacit® 470, 34121
Vulkacit® 576, 34122
Vulkacit® 774, 34123
Vulkacit® 1000, 34124
Vulkacit® A, 34125
Vulkacit® BP, 34126
Vulkacit® CA, *See* thiocarbanilide,
Vulkacit® CRV/LG, 34127
Vulkacit® CT, 34128
Vulkacit® CZ/EGC, DZ/EGC, 34129
Vulkacit® D/C, 34130
Vulkacit® DM/C, 34131
Vulkacit® DZ/EGC, *See* Vulkacit® CZ/EGC, DZ/EGC, 34129
Vulkacit® FP, 34132
Vulkacit® H, *See* hexamine, 14008
Vulkacit® M, 34133
Vulkacit® M, Vulkacit® Merkapto/C, 34134
Vulkacit® Merkapto/MGC, 34135
Vulkacit® P, 34136
Vulkacit® P Extra, 34137
Vulkacit® Thiuram, *See* thiuram,
Vulkacit® TR, 34138
Vulkacite CZ, *See* Vulkacit® CZ/EGC, DZ/EGC, 34129
Vulkacite L, *See* AAprotect, 16
Vulkadur®, 34139
Vulkalent®, 34140
Vulkalent® TM, 34141
Vulkanol®, 34142
Vulkanol® 88, 34143
Vulkanol® 90, 34144
Vulkanol® OT, 34145
Vulkanox 4010 NA, *See* Flexzone 3C, 11967, Permanax IPPD, 23364
Vulkanox 4030, *See* Naugard® I-2, 20792
Vulkanox PBN, *See* Antioxidant PBN, 2591, Stabilator A.R, 29400
Vulkanox®, 34146
Vulkanox® 410NA, *See* N-Isopropyl-N'-phenyl-*p*-phenylene diamine, 15413
Vulkanox® 4010 NA, 34147
Vulkanox® 4020, 34148
Vulkanox® BKF, 34149
Vulkanox® DDA, 34150
Vulkanox® HS/LG, 34151
Vulkanox® KB, 34152, *See* Anti-Oxydant Bayer, 2595
Vulkanox® MB-2/MGC, 34153
Vulkanox® OCD/SG, 34154
Vulkanox® SP, *See* Alkylated phenol, 1734
Vulkanox® ZMB2/C5, 34155
Vulkaresen, 34156
Vulkasil A 1, *See* Wessalith, 34284
Vulkasil®, *See* silica, 28313
Vulkazit, *See* Akrochem® DPG, 1168, Dynamine, 9355, Perkacit® DPG, 23281, Vanax® DPG, 33290, Vulcaid DPG, 34081
Vulkazon®, 34157
Vulkazon® AFS/LG, 34158
Vulklor, 34159
Vulkollan®, 34160
Vulkollan® 18W, 34161
Vulnopol NM, *See* Aquatreat SDM, 2761, methyl namate®, 19551, Octopol SDM-40, 22014, Perkacit® SDMC, 23288, sodium dimethyldithiocarbamate, 28758, Thiostop N, 31717
vulpinite, 34162
Vultac® 2, 34163

Vultex, 34164
Vulvan, *See* Virormone,
Vybar® 103, 34165, *See* Paraffin, 22743
Vybex 22008 BKFR, 34166
Vybex 22008 BKFR, 22028 BKFR, *See* Grilpet® XE3060, 13402
Vybex 22028 BKFR, 34167
Vybex 40001 NA, 34168
Vybex 40004 NAFC, 34169
Vycel, 34170
Vydate®, 34171
Vydax®, 34172
Vydon, 34173
Vydyne, 34174
Vyflex E, 34175
Vyflex NT80S, 34176
Vygen, 34177
Vykamol 83G, 34178
Vyloc, 34179
Vylor®, 34180
Vylox®, 34181
Vynamon, 34182
Vynathene®, 34183
Vynel N, 34184
Vyram®, 34185
W VII/117, *See* fuberidazole, 12428
W.A. powder, 34186
W.G.S. Hydrogenated Fish Glyceride 117, 128, 34187
W.G.S. R-60 Z-5, 34188
W.G.S. Synaceti 116 NF/USP, 34189
W180, 34190
W2 Beta, 34191
W-5219, *See* proglumide, 25116
Wachsemulsion 1864, 34192
Wafex, *See* calcium lignosulfonate, 4956, 4957
Wafex, Wafolin, 34193
Wagner's reagent, 34194
Wakal® A Range, 34195
Wakal® J Range, 34196, *See* Locust bean gum, 17545
Wakal® K Range, 34197
Wakefield grease, *See* Yorkshire grease, 34664
Walchowite, *See* retinite, 26276
Walker's earth, *See* Fuller's earth, 12441
Walkover Moss Killer, 34198, *See* ferric sulfate, 11652
Wall saltpeter, *See* nitrocalcite, 21352
Wallkyd® 11-024, 34199
Wallpol® 40,136, 40-165, 34200
walnut oil, 34201
Walsrode Powder, 34202
Wando Steel, 34203
Wanin AM, 34204, *See* Ammonium lignosulfonate, 2243, Lignosol TS, 17240
waras, 34205
Warcodet D, 34206
Warcodet K54, 34207
Warcodet V, 34208
Warcodye CLP, 34209
Warcodye RWL, 34210
Warcosoft WSC, 34211
Warcowet O, 34212
Wardol, 34213
Warecure C, *See* Perkacit® ETU, 23283
Wareflex® 650, 34214
Warefog, 34215
WARF compound 42, *See* Killgerm® Sewarin P, 16184, Kypfarin, 16519
Warfarin, *See* Dethmor, 8211, Killgerm® Sewarin P, 16184, Kypfarin, 16519, Sakarat, 27238, Sewarin, 28117
Wargonin Compact, 34216
Wargotan, 34217, *See* calcium lignosulfonate, 4956, 4957
Warmaline, 34218
Warne's metal, 34219
warrus, *See* waras, 34205
Wars, *See* waras, 34205
Waruzol, *See* Hismanal, 14132
Wasc, 34220
Wasp Destroyer, *See* Carbaryl, 5181
Waspend, 34221
Waspex, *See* lodofenphos, 15213
Watchmaker's alloy, 34222
Water Brite, 34223
water gas, 34224
water glass, *See* Acsil, 454, N, 20633, sodium silicate, 28817, soluble glass, 28980
Water Lock® A-100, 34225
water mica, 34226
water soluble eosin, *See* Eosin, 10606
Water treatment polymers, *See* Acumer, 574
Water-based all-purpose defoamer, *See* Advantage 1007B Defoamer, 804
Waterez®, 34227
Waterglass, *See* sodium metasilicate, 28788, soluble glass, 28980
Waterlity, 34228
Watsonite, 34229
wattle bark, 34230
Watt's and Li's solution, 34231
Wave, 34232
Wavellite, *See* aluminum phosphate, 1922
wax butter, 34233
Wax C, 34234
Wax extract, *See* Synwax, 30659
Wax oil, *See* wax butter, 34233
Waxed, *See* Zetax®, 34741
Waxemul, 34235
Waxenol® 801, 34236
Waxenol® 810, 34237
Waxenol® 815, 34238
Waxenol® 821, 34239
Waxenol® 822, 34240
Waxes and Waxy substances, jojoba oil, *See* Wickenol® 139, 34383
Waxigel, 34241
Waxilys, 34242
Waxit, 34243
Waxolan P-5, 34244
Waxoline, 34245
Way Up, *See* pendimethalin, 23102
Wayfarer, 34246
Wayfos A, 34247
Wayfos M-60, 34248
Wayhib® S, 34249
Waynecomycin, *See* lincomycin hydrochloride hemihydrate, 17285
WBA 8107, *See* difenacoum, 8449, Neosorexa, 20967, Ratak, 25896
WBA-8119, *See* Talon, 30750
Weathershield, 34250
Webas, 34251
Weber's acid, *See* nitrosylsulfuric acid, 21379
Webert Alloy, 34252
Wecobee® M, 34253
wecoline 1295, *See* Philacid 1200, 23660
Wedel's oil, 34254
Wedl's stain, 34255
weed 108, *See* Neoarsycodyl, 20871
Weed and Brushkiller, 34256
weed drench, *See* propenol, 25181
Weed Hoe, 34257
Weed Stopper, 34258
Weedar, *See* MCPA, 19183
Weedar MCPA, *See* MCPA, 19183
Weed-E-Rad, *See* MSMA, 20417
Weedex, *See* atrazine, 3394
Weedex A, *See* atrazine, 3394
Weedex S2, 34259
Weed-Hoe, *See* Neoarsycodyl, 20871, MSMA, 20417
Weedkill, 34260
Weedmaster, 34261
Weedol, 34262
Weedone, *See* MCPA, 19183
Weedone LV 4, *See* Planotox, 23914
Weedone MCPA Ester, *See* MCPA, 19183
Weedone® DPC, 34263
Weed-rhap, *See* MCPA, 19183
Weed-s-rad, *See* Neoarsycodyl, 20871
Weed-Stopper, *See* oryzalin, 22347
weeviltox, *See* carbon disulfide, 5223
Wegin, 34264
Weichharz 398A, 34265
Weichmacher 238S, 333A, 34266
Weichmacher 90, 34267
Weigert's stain, 34268
weighted silk, 34269
Weighter Finish 585, 34270
dl-Weinsäure, *See* tartaric acid, 30832
Weisalloy, 34271
Weiss-Kupfer, *See* nickel silvers, 21132
weisspiessglanz, *See* Octoguard FR-10, 22000, UltraFine® II, 32799
Weldox, 34272
Welgum, 34273
Welladyne, 34274
Wellbrom, 34275
Welltex 300 F, 34276, *See* calcium lignosulfonate, 4956, 4957
Welmet, 34277
Welvic, 34278, *See* Korogel, 16354, Koroplate, 16357
WEP® 662P, 34279
Wepco, 34280
WesBio, 34281
Wescodyne, 34282
WesLoTemp, 34283
Wespuril, *See* Dichlorophen, 8392, Nuophene, 21783
Wessalith, 34284
Wessalon, 50, 50S, S, 34285
WesScaleStop, 34286
Wessell's silver, 34287
WesSperse, 34288
West African copaiba, 34289
West African gum, 34290, *See* Senegal gum, 27950
West Indian amine resin, *See* Tacamahac resin, 30679
West System Brand Products, 34291
WesTemp, WesTemp K+, 34292
Westfalite No. 3, 34293
Westhin, Westhin K+, 34294
Westo-Flocs, 34295
Weston®, 34296
Weston® 399, 34297 , *See* Trisnonylphenyl phosphite, 32411
Weston® 430, 34298
Weston® 439, 34299
Weston® 474, 34300
Weston® 491, 34301
Weston® 494, 34302
Weston® 600, 34303
Weston® 618, *See* Distearyl pentaerythritol diphosphite, 8712
Weston® 618, Weston® 619, 34304, *See* Distearyl pentaerythritol diphosphite, 8712
Weston® 618F, *See* Distearyl pentaerythritol diphosphite, 8712
Weston® 619, *See* Distearyl pentaerythritol diphosphite, 8712
Weston® DHOP, 34305
Weston® DLP, 34306, *See* dilauryl phosphite, 8475
Weston® DOPI, 34307
Weston® DPDP, 34308, *See* diphenyl isodecyl phosphite, 8594
Weston® DPP, 34309, *See* diphenyl phosphite, 8599
Weston® DSP, 34310
Weston® DTDP, 34311
Weston® EGTPP, Weston® TPP, 34312
Weston® EHDPP, 34313
Weston® ODPP, 34314, *See* diphenyl isooctyl phosphite, 8595
Weston® PDDP, 34315
Weston® PNPG, 34316
Weston® PTP, 34317
Weston® TDP, 34318
Weston® THOP, 34319
Weston® TIOP, 34320
Weston® TLP, 34321
Weston® TLTTP, 34322
Weston® TNPP, 34323

X-Ite, 34606
Xitix, *See* vitamin C, 33938
XL-50, *See* phenothiazine, 23618
XL-All Insecticide, 34607
***o*-Xonal**, *See o*-phenylphenol, 23648
xonaltite, *See* calcium silicate, 4969
XP Pastes, 34608
XP-4, 34609
XR 7, 34610
XSA 80, 34611, *See* Xylene sulfonic acid, 34621
XT® Polymer 250, XT® Polymer 375, 34612
X-Tan® Special C, 34613
Xtol, 34614
XXX-1, 34615
Xydar® G-540 LCP, 34616
xylamide, *See* proglumide, 25116
Xylan 330, 34618
Xylan 1052, 34617
xylanase, *See* Pulpzyme, 25383
Xylapan, *See* xylazine, 34619
Xylasol, *See* xylazine, 34619
xylazine, 34619, *See* Rompun®, 26934
xylene, 34620
***m*-xylene**, 34622
***o*-xylene**, 34623
***p*-xylene**, 34624
Xylene sulfonic acid, 34621, *See* Manro XSA, 18610, Reworyl® X, 26412, XSA 80, 34611
Xylene sulfonic acid, aq.solution, *See* Xylene sulfonic acid, 34621
xylenesulfonic acid, sodium salt, *See* Eltesol® SX 30, 9873, Naxonate® 4L, 20824, Pilot SXS-40, 23843, Witconate SXS 40%, 34492
xylenol, 34625
Xylenol, 4-chloro-, *See* chloroxylenol, 6159
Xyligen®, 34626
Xylite, 34627
L-xyloascorbic acid, *See* vitamin C, 33938
Xylock 225, 34628
L-xylo-hexulose, *See* sorbose, 29129
xyloidin, *See* Kodaloid, 16283, nitro-starch, 21377, Venite, 33560
Xylok, 34629
xylol, 34630, *See* xylene, 34620
Xylomed, *See* Xylose, 34632
Xylomucine, *See* carboxymethylcellulose sodium, 5258, Cekol, 5558
Xylo-Mucine, *See* carboxymethylcellulose, 5257
Xylon FR, 34631
Xylopfan, *See* Xylose, 34632
***p*-xyloquinone**, *See* phlorone, 23676
Xylose, 34632
D-(+)-xylose, *See* Xylose, 34632
Xyron®, 34633
yaba bark, 34634
Yalloy, 34635
Yaltox, *See* Carbodan, 5197, Rampart, 25860
Yaltox®, 34636
yama-mai silk, 34637
yara-yara, *See* jara jara, 15518, Nerolin, 21007
Yarmor, 34638, *See* pine oil, 23856
yeast, 34639
yeast extract, 34640
Yeast Lactase L-50,000, 34641
Yellow 201, 34642
yellow bark, 34643
yellow carmine, 34644
yellow catechu, *See* cutch,
yellow cross gas, *See* mustard gas, 20538
Yellow Cuprocide, *See* cuprous oxide, 7405
yellow earth, *See* ocher, 21976
Yellow GG, *See* Butter yellow, 4759
yellow gold, 34645
Yellow GR, *See* Butter yellow, 4759
yellow lake, *See* yellow carmine, 34644
yellow liquors, 34646
yellow melilot, *See* Melilot, 19279
Yellow ocher, *See* ocher, 21976
yellow precipitate, *See* mercuric oxide, red and yellow, 19348, red oxide of mercury, 25966
yellow prussiate of potash, 34647
yellow prussiate of soda, 34648, *See* sodium ferrocyanide, 28763
yellow soda ash, 34649
Yellow sulfide of arsenic, *See* orpiment, 22317
Yellow sulphur, 34650
yellow sweet clover, *See* Melilot, 19279
yellow wax, 34651
yellow wood, *See* Fustic, 12522
Yellowstone, 34652
yenshee, 34653
Yeoman, 34654
yerba mate, 34655
Yieldmaster®, 34656
Ylanat ortho, *See* Ortholate, 22332
ylang-ylang oil, 34657
yntes 12a, *See* tetrasodium EDTA, 31391
Yodoxin, 34658
yolk powder, *See* lecithin, 16939
Yoloy, 34659
Yomesan, *See* clonitrilide, 6432, niclosamide, 21146
Yonckite, 34660
Yoracryl Dyes, 34661
York Krystal Kleer Castor Oil, 34662
York USP Castor Oil, 34663
Yorkshire grease, 34664
Yperite, *See* mustard gas, 20538
Y-Pof, 34665
YSE-Cure B-001, 34666
YSE-Cure B-002, 34667
YSE-Cure B-002, B-002W, B-003, 34668
YSE-Cure B-003, 34669
YSE-Cure C-002, F-100, 34670
YSE-Cure F-100, 34671
YSE-Cure LX-1N, 34672
YSE-Cure N-001, N-002, LX-1N, 34673
YSE-Cure PX-3, 34674
YSE-Cure PX-3, QX-2, QX-3, RX-2, RX-3, 34675
YSE-Cure RX-2, 34676
YSE-Cure RX-3, 34677
YSE-Cure S-002, 34678
Y-Tack, 34679
ytterbia, *See* ytterbium oxide, 34681
ytterbium, 34680
ytterbium oxide, 34681
Ytterbium Yb-169 DTPA, 34682
yttria, *See* yttrium oxide, 34684
yttrium, 34683
yttrium oxide, 34684
Yukalon, 34685
Yunihomu AZ;, *See* Azodicarbonamide, 3524
Z 11, *See* Aero 343 Xanthate, 842
Z 75, *See* AAprotect, 16
Z Span Spansule Capsule, 34686
Z.P.D, 34687
zaccatila, *See* cochineal,
zaffer, *See* smalt, 28638
zaffre, *See* smalt, 28638
zaharina, *See* saccharin, 27195
zakin rubber, 34688
Zam Metal, 34689
Zamak Alloys, 34690
Zam-Buk, 34691
Zanil, 34692, *See* Oxyclozanide, 22444
zanthotoxin, *See* Oxsoralen-Ultra, 22437
Zap-Oglobin, Zaponin, 34693
Zapon®, 34694
Zaponin, 34695
zapoto gum, *See* chicle, 6058
Zarlate, *See* Bisphenol A, 4283
Zauberin, 34696
ZB2335, 34697
ZBC, *See* Butyl Zimate®, 4783, Octocure ZDB-50, 21995, Perkacit® ZDBC, 23295
ZBeDC, *See* Akrochem® Z.B.E.D, 1183
ZC, *See* AAprotect, 16
Z-C Spray, *See* AAprotect, 16
ZDA, *See* Ageflex ZDA, 1004, Akrochem® ZDA Powd, 1184
ZDC, *See* Ancazate ET, 2415
ZDEC, *See* Ancazate ET, 2415
Zea mays oil, *See* corn oil, 6834
Zeaphos, *See* atrazine, 3394
Zeapos, *See* atrazine, 3394
Zeazin, *See* atrazine, 3394
Zedox, 34698
Zeeospheres® 200, 34699
Zeese, 34700
Zeidan R50, *See* DDT, 7820
Zeidane, *See* DDT, 7820
Zeiodelite, 34701
Zeise's salt, 34702
Zeklan, 34703
Zelan, *See* MCPA, 19183
Zelco metal, 34704
Zelcon, 34705
Zelec, 34706
Zellwonet, 34707
Zeltivar, *See* trichlorfon, 32212
Zeltoxone, *See* Treflan, 32162, Trifluralin, 32252
Zelulone, 34708
Zemdrain®, 34709
Zemid® 610, 34710
Zemid® 650, 34711
Zendium, *See* sodium fluoride, 28764
Zenite, *See* Octocure ZMBT-50, 21998, Oxaf, 22408, Perkacit® ZMBT, 23298, Zetax®, 34741
Zenite Special, *See* Zetax®, 34741
Zenith, 34712
Zenker's fluid, 34713
Zennapron, 34714
Zentralin, 34715
Zentralit I, *See* diethyldiphenylurea, Zentralin, 34715
Zentralit II, *See* dimethyldiphenylurea, Zentralin, 34715
Zeolex, 34716
Zeolex, *See* Wessalith, 34284
Zeolex 7, *See* Wessalith, 34284
Zeolex 7A, *See* Wessalith, 34284
Zeolex 23A, *See* Wessalith, 34284
Zeolex 23P, *See* Wessalith, 34284
Zeolex 25, *See* Wessalith, 34284
Zeolex 35, *See* Wessalith, 34284
Zeolex 45, *See* Wessalith, 34284
Zeolex 100, *See* Wessalith, 34284
zeolite synthetic, 34717
zeolites, 34718
Zeolum, 34719
Zeonet A, 34720
Zeonet B, 34721
Zeonet U, 34722
Zeonex 280, 34723
Zeospan 303, 306, 34724
Zeothix, 34725
Zeotokol, 34726
Zepel®, 34727
Zepharovicht, *See* aluminum phosphate, 1922
Zephiramine, *See* JAQ Powdered Quaternary, 15517
Zephiran Chloride, 34728, *See* Variquat® 50MC, 33397
Zephirol, *See* Variquat® 50MC, 33397, Zephiran Chloride, 34728
Zephirol®, *See* benzalkonium chloride, 3958
Zephron®, 34729
Zephyr, 34730
Zeralte, *See* Vancide® MZ-96, 33303
Zerdane, *See* DDT, 7820
Zerex, 34731
Zerlate, *See* AAprotect, 16, Accelerator MZ Powder, 209, Zimate®, 34762
Zerofil, 34732
Zerogen® 10, Zerogen® 60, 34733
Zerol, 34734
Zerone, 34735
Zeronox, 34736
Zerotherm, 34737
Zerox, 34738
Zertell, *See* Chlorpyrifos-methyl, 6164
Zetabon, 34739

Chemical and Trade Names

PART III

INDEXES

1. CAS Registry Numbers

50-00-0: Durine, 9225; Dyna-Form, 9346; formaldehyde, 12258; Hercules® 37M6-8, 13850;Lysoform, 18095;
50-06-6: Talpheno, 30755
50-14-6: vitamin D+72, 33940
50-21-5: lactic acid, 16570; Patlac® LA, 22910
50-24-8: prednisolone, 24872
50-29-3: DDT, 7820; De De Tane, 7821; Neocid, 20879
50-34-0: propantheline bromide, 25176
50-36-2: Coke, 6591
50-37-3: LSD, 17704
50-65-7: Mansonil®, 18611; niclosamide, 21146
50-70-4: A-625/641ABS 301K, ABS 500FR-1, 10; Hydex® 100 Gran.206, 14531; Liponic 70-NC, 17391; Sorban, 29088; Sorbelite C, 29104; sorbitol, 29113; Sorbitol (EGIC), 29114; Sorbo®, 29115
50-71-5: Alloxan, 1750
50-78-2: Nu-Seals Aspirin, 21798; Tasprin, 30842
50-81-7: vitamin C, 33938
50-99-7: Candex®, 5085; Cerelose, 5741; Emdex®, 10071; Flolys®, 11994; glucose, 13052; Roferose, 26904
51-03-6: piperonyl butoxide, 23875
51-17-2: Benzimidazole, 3966
51-28-5: Dinitra, 8540
51-48-9: Riamat, 26720
51-79-6: urethane, 33135
52-01-7: Laractone®, 16784
52-21-1: Econopred, 9588
52-51-7: Bioban® BNPD-40, 4156; Bronopol, 4595; Bronopol-Boots, 4596; Broponol, 4616; Canguard® 409, 5089; Myacide® AS Plus, 20544; Myacide® S-1, S-2, 20545
52-67-5: Trolovol®, 32449
52-68-6: Danex, 7727; Dipterex®, 8616; Dipterex® 80, 8617; Dylox®, 9324; Masoten®, 19030; Neguvon®, 20852; trichlorfon, 32212; Tugon, Tugon OKO, Tugon sp 80, 32570
52-89-1: cysteine hydrochloride, 7601
52-90-4: L-cysteine, 7602
53-19-0: Lysodren, 18094
54-11-5: Campbell's Nico-Soap, 5068; XL-All Insecticide, 34607
54-21-7: Nu-Seals Sodium Salicylate, 21799
54-31-9: Lasix, 16807
54-36-4: Metopirone, 19591; metyrapone, 19608
54-42-2: Kerecid, 16067
54-64-8: Merthiolate, 19391
55-03-8: Levothroid, 17081
55-22-1: isonicotinic acid, 15380
55-38-9: Lebaycid®, 16932; Tiguvon®, 31788
55-48-1: Isopto Atropine, 15414
55-55-0: Metol, 19584
55-56-1: Hibidil, 14050; Hibiscrub, 14051; Hibisol, 14052; Hibispray, 14053; Savloclens, 27502; Savlon Babycare, 27504; Superspray, 30282
55-63-0: Cascade, 5375; explosive gum, 11315; Glonoine oil, 13005; Hercol 2, 13824; Hercol 2X, 13825; Hercon® 2, 13831; Hercon® 2X, 13832; Hercosplit WR, 13844; Unigel, 32955
55-68-5: Phenmerzyl Nitrate, 23599
56-03-1: Biquanide, 4218
56-23-5: carbon tetrachloride, 5224; Katharin, 15791; phenoxin, 23623; seretin, 28026; tetraform, 31369; Thawpit, 31564
56-35-9: bioMeT TBTO, 4172; Keycide® X-10, 16153; tributyltin oxide, 32209
56-37-1: Benzyltriethyl ammonium chloride, 3993; Sumquat® 2355, **30120**
56-38-2: Bladan®, 4309; ethyl parathion, 11079; Folidol® E605, 12199; Fosferno, 12330; Murphos, 20521; parathion, 22807
56-40-6: glycine, 13095
56-41-7: α-alanine, 1333; Ritalanine, 26812
56-45-1: L-serine, 17705; 28035
56-65-5: Adenosine triphosphate, 681
56-72-4: Asuntol®, 3280; Muscatox, 20525; Perizin®, 23276
56-75-7: Leukomycin®, 17030
56-81-5: Croderol G7000, 7164; dynamite glycerin, 9358; Emery® 912, 10137; glycerin, 13071; Glycon® G 100, G 300, 13118; Glyrol, 13152; Kemstrene® 96.0%, 15991; Lye glycerin, 18070; Osmoglyn, 22358; Pricerine 9071, 24942; Shur-Coal® FCA, 28189; Star, 29521; Superol, 30258
56-87-1: lysine, 18088
56-89-3: cystine, 7603
56-93-9: Benzyl trimethyl ammonium chloride, 3989; Hipochem Migrator J, 14114; Variquat® B200, 33400
56-95-1: Arlacide A, 2907; chlorhexidine acetate, 6100
57-09-0: Acetoquat CTAB, 303; Bromat®, 4566; Catinal HTB-70, 5457; Cetavlex, Cetavlon, 5787; Cetrimide BP, 5812; Cetrimonium bromide, 5813; Rhodaquat® M242B/99, 26643; Sumquat® 6030, 30122; Varisoft® CTB-40, 33421
57-10-3: Emersol® 143, 10119; Glycon® P-45, 13119; Hystrene® 8016, 14763; Industrene® 4516, 15106; palmitic acid, 22631, 22632, 22633, 22634, 22635; Prifrac 2960, 24954
57-11-4: Emersol® 110, 10118; Emersol® 6349, 10124; Emery® 400, 10130; Glycon® S-90, 13121; Glycon® TP, 13122; Hydrofol Acid 1655, 14588; Hystrene® 4516, 14758; Hystrene® 5016, 14760; Hystrene® 7018 FG, 14762; Hystrene® 9718 NF FG, 14768; Industrene® 4518, 15107; Industrene® 5016, 15108; Industrene® 7018 FG, 15109; Industrene® R, 15111; Petrac® 270, 23462; Prifrac 2980, 24955; Pristerene 4904, 25040; Stearex, 29597; stearic acid, 29599; Stearite, 29601
57-13-6: Bubber Shet, 4641; Superprill, 30265
57-30-7: Luminal Sodium, 17803; Phenobarbitone sodium, 23600
57-39-6: Mapo®, 18656
57-48-7: fructose, 12420; levulose, 17085
57-50-1: sucrose, 29934; Sugartab®, 29956
57-55-6: Chilisa FE®, 6064; Dowfrost, 8922; Ilexan P, 14885; Prolugen, 25135; propylene glycol, 25214, 25215
57-57-8: betaprone, 4065
57-68-1: sulfamethazine, 29962
57-74-9: Ortho-Klor, 22331; Syndane, 30536
57-83-0: Gesterol 50, 12927; Lipo-Lutin, 17370; luteol, 17991
57-88-5: cholesterol, 6168; Dastar, 7788; Fancol CH, 11435; Kathro, 15795; Liquid Crystal CN/9, 17467
58-36-6: Vinyzene, 33869
58-46-8: Nitoman, 21316
58-55-9: liquorice, 17490
58-56-0: Gravidox, 13294; Hexa-Betalin, 13987; Hexavibex, 14019; pyridoxine hydrochloride, 25483; vitamin B_6 hydrochloride, 33937
58-85-5: Ritatin, 26834
58-86-6: Xylose, 34632
58-89-9: Atlas Steward, 3334; Fumite Lindane, 12461; Gamma-BHC Dust, 12631; Gamma-Col, 12632; Gamma-HCH Dust, 12633; Gammalin, 12635; Gammasan, 12636; Gammexane, 12639; Jacutin®, 15485; Lorexane, 17607; Murfume Grain Store Smoke, 20512; New Kotol, 21067; Scabene Lotion, 27536
58-96-8: uridine, 33140
59-02-9: Ephynal, 10629; Eprolin, 10721; Epsilan-M, 10723; E-Toplex, 11154; Natopherol, 20753; Phytoforol, 23779; Tocopherex, 31917; tocopherol, 31918; Vascuals, 33465; vitamin E, 33943
59-06-3: ethopabate, 10993; Pancoxin, 22669
59-30-3: folic acid, 12196; Folvite, 12211; Mission Prenatal, 19945; vitamin M, 33951
59-31-4: carbostyril, 5244
59-40-5: Pancoxin, 22667; sulfaquinoxaline, 29968
59-43-8: thiamine, 31664
59-50-7: Lysochlor, 18093; Raschit, 25892
59-51-8: DL-methionine, 19487; Petameth, 23451
59-66-5: Diamox, 8316
59-67-6: Niac, 21086; niacin, 21090; Nico-400, 21148; Nicobid, 21149; Nicocap, 21150; Nicolar, 21151
60-00-4: edetic acid, 9615; Hamp-Ene® Acid, 13587; Kalex Acids, 15669; Questric Acid 5286, 25665; Sequestrene® AA, 27996; Trilon® BS, 32327
60-11-7: Butter yellow, 4759
60-12-8: 2-phenylethyl alcohol., 23642; Mellol, 19297
60-24-2: 2-mercaptoethanol, 19336; Sipomer® 2ME, 28529
60-29-7: ethyl ether, 11069; ethyl ether Anhydrous A.C.S., 11070; ethyl ether USP/ACS, 11071
60-33-3: Emersol® 315, 10121; linoleic acid, 17305; Pamolyn, 22643; Pamolyn 125, 22645; Pamolyn 380, 22647
60-51-5: Dimethoate, 8496; Dimethoate Bayer, 8497; Perfekthion®, 23244; Systemic Insecticide, 30664; Turbair, 32595
60-54-8: tetracycline, 31365
60-57-1: Murdiel, 20504

61-19-8: 5'-adenylic Acid, 683; adenosine monophosphate, 680; AMP, 2301; My-B-Den, 20547
61-82-5: Aminotriazole Bayer, 2190; Azolan, 3530; Boroflow S/ATA, 4437; MSS Aminotriazole, 20426
61-90-5: Leucine, 17020; l-leucine, 17021
62-23-7: *p*-nitrobenzoic acid, 21350
62-33-9: calcium disodium versenate, 4946; Versene CA, 33618
62-38-4: acetoxyphenylmercury, 306 Acticide PMA 100, 470; Agrosan GN, 1081; Ceresol, 5748; Merpectogel, 19370; Phenitol, 23594; phenylmercuric acetate, 23644, 23645, 23646; PMA 18, 60, 24274; Single Purpose, 28498
62-53-3: phenylamine, 23638
62-54-4: Calac, 4924; calcium acetate, 4936, 4937; grey acetate, 13351; Niacet Calcium Acetate Tech, 21088; pyrolignite of lime, 25516
62-55-5: thioacetamide, 31678
62-56-6: sulfourea, 30038; thiourea, 31720
62-73-7: Atgard, 3285; Dedevap®, 7870; Dethlac, 8210; dichlorvos, 8394; Divipan, 8740; Mafu®, 18305; Nogos, 21420; Nuvan, 21850; Nuvan Top Aerosol, 21852; Nuvanol, 21853; Vapona, 33378
62-97-5: Variton, 33432
63-25-2: Carbaryl, 5181; Carylderm, 5364; Microcarb, 19659; Murvin 85, 20522; Sevin, 28115; Sevin® Brand SL, 28116; Thinsec, 31676; Tornado, 32071
63-42-3: Lactin, 16578; milk sugar, 19768
63-68-3: Mepron, 19328
63-75-2: arecoline, 2829
63-89-8: colfosceril palmitate, 6607
64-02-8: Aroquest 100, 3062; Cheelox® 100, 5865; Hamp-Ene® 100, 13586; Hamp-Ene® Na4, 13590; Intraquest® TA Solution, 15175; Kalex 220 Crystal, 15668; Kalex Liq. 50%, 15671; Questal Extra Powd. Conc. 0780, 25662; Questal Special 0860, 25664; Sequestrene® 220, 27995; Sequestrene® 30A, 27994; Solon Conc, 28957; tetrasodium EDTA, 31391; Thorquest 39, 31756; Trilon® B Powd, 32324
64-17-5: ethyl alcohol, 11064; ethylol, 11142; grain alcohol, 13244; Punctilious® Ethyl Alcohol, 25398; Pyro, 25487; spirits of wine, 29304; Synasol, 30522; USI in Oval, 33157
64-18-6: Amasil®, 1946; formic acid, 12271
64-19-7: Acetic acid, 292
65-71-4: thymine, 31777
65-85-0: Benzoic acid, 3972, 3973; flowers of Benjamin, 12031; Retarder BA, BAX, 26254
66-22-8: uracil, 33101
66-71-7: Activ-8®, Activ-8 in Hexylene Glycol, 494
67-03-8: thiamine hydrochloride, 31665
67-43-6: Aroquest M Special, 3064; Chel DTPA, 5873; Hamp-Ex® Acid, 13592; Pentaquest OPAC 0201, 23170
67-45-8: Furoxone, 12499
67-48-1: choline chloride, 6171
67-56-1: methyl alcohol, 19511; pyro alcohol, 25488; spirit of wood, 29300
67-63-0: Alcowipe®, 1531; Avantine, 3456; I.P.S., 14799; isopropanol, 15405; Sterets Pre-Injection Swabs®, 29728
67-64-1: acetone, 297; Sasetone, 27474
67-66-3: chloroform, 6125
67-68-5: Decap, 7845; Demavet, 8053; dimethyl sulfoxide, 8511; DMSO, 8762
67-97-0: vitamin D_3, 33941
68-11-1: thioglycolic acid, 31688; Thiovanic® Acid, 31722
68-12-2: dimethyl formamide, 8506; Dynasolve 100, 9386
68-19-9: Fermin, 11630; Rubramin PC, 27114; Sytobex, 30668; vitamin B_{12}, 33934
68-26-8: Anatola, 2403; A-Sol, 3228; Homagenets Aorl, 14282; retinol, 26277; Ro-A-Vit, 26866; Super A, 30163; Vi-Alpha, 33706; Vi-Dom-A, 33744; vitamin A, 33932
68-94-0: hypoxanthine, 14721; sarcine, 27460
69-09-0: Largactil, 16786
69-53-4: Acillin, 346; Ampicillin, 2361; Unasyn® IM/IV, 32913
69-65-8: hexanhexol, 14010; mushroom sugar, 20529
69-72-7: Retarder SAX, 26260; salicylic acid, 27270
69-79-4: maltose, 18501
69-89-6: xanthine, 34590;
69-93-2: Lithic acid, 17506; uric acid, 33139
70-30-4: Fascol, 11498; hexachlorophene, 13992; Ster-Zac, 29760
70-47-3: Asparagine, 3232
70-53-1: dl-lysine monohydrochloride, 18090
70-55-3: *p*-toluenesulfonamide, 31949; Uniplex 173, 33012
71-23-8: Pro-Gas (Gas Disclaimed), 25112; propanol, 25175; propylan-propyl alcohol, 25211
71-27-2: Quelicin, 25642; Sucostrin, 29927
71-36-3: Tebol 88, 99, 30878
71-41-0: Amyl alcohol, 2381
71-43-2: Benzene, 3961
71-55-6: Chlorothene, 6151; Cut Aid, 7432; Dabco® CS90, 7653; Delf® Fabric Protector, 8015; Distillex DS1, 8717; Ethana®, 10928; MS-170 1,1,1 Trichloroethane Solvent, 20414; Solvethane, 29060; trichloroethane, 32216; Tri-Ethane, 32239
72-17-3: perglycerol, 23262; sodium lactate, 28777
72-43-5: Marlate 2-MR Emulsifiable Insecticide, 18747; Marlate 300 Flowable, 18749; Marlate 400 Flowable Concentrate, 18750; Marlate 50 WP, 18748; Marlate Methoxychlor Insecticide, 18751; methoxychlor, 19504
72-48-0: Alizarin, 1621; Pincoffin, 23854
72-56-0: diethyldiphenyldichloroethane, 8441; Perthane, 23435
72-57-1: Niagara blue, 21091
73-24-5: Adenine, 679
73-40-5: guanine, 13443
74-31-7: Agerite® DPPD, 1016; JZF, 15638; N,N'-diphenyl-*p*-phenylenediamine, 8603; Naugard® J, 20796; Permanax DPPD, 23361
74-82-8: fire-damp, 11831; marsh gas, 18954; methane, 19475
74-83-9: Embafume, 10012;Meth-O-Gas, 19499; methyl bromide, 19517
74-86-2: acetylene, 318
74-87-3: Artic, 3147; methyl chloride, 19522
74-88-4: methyl iodide, 19542
74-90-8: hydrocyanic acid, 14584; prussic acid, 25352; zootic acid, 34903
74-98-6: A-108, 9; propane, 25173
75-00-3: ethyl chloride, 11067
75-05-8: acetonitrile, 299
75-07-0: Acetaldehyde, 281
75-09-2: Aerothene, 900; Distillex DS3, 8719; Driverit, 9043; M-Clean D, 19178; Methoklone, 19501; methylene chloride, 19568; Nevolin®, 21053; salesthin, 27263
75-11-6: methylene iodide, 19569; Mi-Gee Brand, 19732
75-12-7: formamide, 12261
75-15-0: carbon bisulfide, 5219; carbon disulfide, 5223
75-18-3: dimethyl sulfide, 8510; Exact-S®, 11281; methyl sulfide, 19561
75-20-7: calcium carbide, 4939
75-21-8: ethylene oxide, 11131
75-25-2: Bromoform, 4579
75-28-5: A-31, 8; isobutane, 15343
75-33-2: isopropyl mercaptan, 15408
75-36-5: acetyl chloride, 311
75-39-8: Aldamine, 1534; Velsan, 33539
75-44-5: phosgene, 23701
75-46-7: Fluoroform, 12099; Genetron® HFC23, 12829
75-47-8: iodoform, 15214
75-52-5: Nitrofuel®, 21360; nitromethane, 21366; NM, 21402; NM-55®, 21403
75-54-7: CM8750, 6460
75-56-9: Epihydrin, 10642; propylene oxide, 25223
75-59-2: tetramethylammonium hydroxide, 31385
75-66-1: *t*-butyl mercaptan, 15434
75-69-4: Distillex DS6, 8722; Genesolv A Solvent, 12814; Genetron® 11, 12817
75-71-8: dichlorodifluoromethane, 8388; Genetron® 12, 12818; Sterethox, 29727
75-72-9: chlorotrfluoromethane, 6153; Genetron® 13, 12819
75-76-3: CT2050, 7315
75-77-4: CT2950, 7325
75-79-6: CM9000, 6463
75-89-8: 2,2,2-trifluoroethanol, 32250
75-91-2: Aztec® *t*-Butyl Hydroperoxide-70, Aq, 3541; TBHP-70, 30853; T-Hydro, 31774; Trigonox® A-80, 32291; USP® -800, 33170
75-94-5: CV-4900, 7450; Dynasylan® VTC, 9419; vinyltrichlorosilane, 33865
75-97-8: pinacolone, 23848
75-99-0: BH Dalapon, 4117; Couch and Grass Killer, 6896; dalapon, 7705; Synchemicals Dalapon, 30524
76-01-7: pentachlorethane, 23134; pentaline,

23154
76-02-8: Superpalite, 30260
76-03-9: Farmon TCA, 11491; Na Ta, 20642; trichloroacetic acid, 32213;
76-05-1: trifluoroacetic acid, 32249
76-06-2: chloropicrin, 6135; Larvacide, 16802
76-08-4: acetone bromoform, 298
76-13-1: Arklone, 2894; Distillex DS5, 8721; Genesolv D Solvent, 12815; Genetron® 113, 12821; MS-180 Freon® TF Solv, 20415; trichlorotrifluoroethane, 32224
76-14-2: Cryofluorane, 7278; Genetron® 114, 12822
76-22-2: Japan camphor, 15508
76-39-1: NMP, NMP Conc, 21405
76-60-8: Bromocresol green, 4575
76-87-9: Ashlade Flotin, 3193; Du-Ter, 9283; Du-Ter®, 9284; Farmatin, 11480; Quadrangle Super-Tin 4L, 25578; Super Tin® 4L, 30208
77-06-5: Activol, 506; gibberellic acid, 12934; Regulex, 26019
77-58-7: ADK STAB BT-11, 723; Dabco® T-12, 7673; dibutyltin laurate, 8371; Metacure® T-12, 19417; Synpron 1009, 30611; TEDA-T411, 30917
77-71-4: DM hydantoin, 8751
77-86-1: THAM, 31557; tris(hydroxymethyl) aminomethane, 32403; Tris Amino®, 32405; Tris Amino® Molecular Biology Grade, 32406; Tris Amino® Ultra Pure Standard, 32407
77-89-4: Citroflex A-2, 6342; Uniplex 82, 33005
77-90-7: acetyl tributyl citrate, 313; Citroflex A-4, 6343; Estaflex, 10874; Uniplex 84, 33006
77-92-9: Citraclean, 6335; citric acid, 6341;
77-93-0: Hydagen® C.A.T, 14526; triethyl citrate, 32242; Uniplex 80, 33004
77-99-6: trimethylolpropane, 32338
78-04-6: Dibutyltin maleate, 8372
78-05-7: octaphen, 21989
78-08-0: CV-4910, 7451; Dynasylan® VTEO, 9420; Union Carbide® A-151, 32981; vinyltriethoxysilane, 33866
78-10-4: Dynasil® A, 9382; ethyl silicate, 11083; Silbond® Condensed, 28285; Silester, 28303
78-11-5: Penthrit, 23180
78-21-7: Barquat® CME-35, 3666
78-30-8: Plastic X, 24026; tricresyl phosphate, 32230
78-40-0: phosphoric ether, 23740; triethyl phosphate, 32243
78-42-2: trioctyl phosphate, 32349
78-48-8: Def®, 7876
78-51-3: Amgard TBEP, 2055; KP-140®, 16378; TBEP, 30852; tributoxyethyl phosphate, 32206
78-59-1: isophorone, 15395
78-62-6: CD5600, 5508; EXP-49, 11305
78-63-7: Aztec® 2,5-Di, 3551; dimethyldibutyl peroxyhexane, 8518; Esperal® 120, 10837; Luperco 101-P20, 17823; Lupersol 101, 17846; Polyvel CR-5F, 24657; Trigonox® 101, 32279
78-66-0: Surfynol® 82, 30374; Surfynol® 82, 82S, 30375
78-67-1: Ficel® AZDN-LF, 11763; Poly-zole AZDN, 24674; V-60, 33218
78-70-6: Phantol, 23570
78-78-4: Exxsol® Isopentane, 11362; isopentane, 15394
78-83-1: isobutanol, 15344
78-87-5: propylene dichloride, 25213
78-91-1: isopropanolamine, 15406
78-93-3: 2-butanone, 4737; Meketone, 19247; methyl acetone, 19507; methyl ethyl ketone, 19530
79-01-6: Altene DG, 1873; Disparit B, 8658; Distillex DS2, 8718; Germalgene, 12892; trichloroethylene, 32217; Triclene, 32226; trieline, 32237; Triklone, 32315
79-06-1: Acrylamide, 410
79-09-4: Luprosil®, 17919; propionic acid, 25190
79-10-7: acrylic acid, 412
79-11-8: monochloroacetic acid, 20221
79-14-1: glycolic acid, 13105
79-20-9: methyl acetate, 19506
79-21-0: Oxymaster, 22460; Proxitane, 25308
79-24-3: NE, 20833; nitroethane, 21357
79-27-6: Muthmann's liquid, 20541
79-31-2: isobutyric acid, 15348
79-33-4: L-lactic acid, 16571
79-34-5: Acetosol, 304
79-40-3: hydrogen rubeanide, 14591
79-46-9: 2-nitropropane, 21374; NiPar S-20, 21280
79-57-2: Biostat A.1, 4209
79-74-3: Lowinox® AH25, 17663; Santovar, 27428
79-94-7: FR-1524, 12367; Great Lakes BA-59P, 13298; Saytex® RB-100, 27523; tetrabromobisphenol-A, 31359
80-05-7: Bisphenol A, 4283; Millad® HBPA, 19776; Parabis, 22717
80-08-0: 4,4'-diaminodiphenyl sulfone, 8302; Dapsone, 7751
80-10-4: CD5950, 5513
80-15-9: Aztec® CHP-80, 3555; CHP-158, 6175; CHP-5, 6174; HPC-9, 14438; Trigonox® 239A, 32290; Trigonox® K-95, 32303
80-39-7: ethyltoluene sulfonamide, 11145; Uniplex 108, 33007
80-40-0: ethyl-*p*-toluenesulfonate, 11089; Mittel AEP, 19978
80-43-3: Aztec® DCP-R, 3556; Dicumyl peroxide, 8407; Di-Cup, 8408; Esperal® 115, 10836; Luperco 500-40KE, 17828; Luperox 500R, 17840; Percumyl D, 23230; Perkadox® BC, 23307; Peroximon® DC 40 MG, 23408; Peroximon® DC-40, 23409; Polyvel PCL-20, 24661
80-46-6: Nipacide® PTAP, 21259; Orthophen® 278, 22336; Pentaphen® 67, 23168
80-51-3: Celogen® OT, 5643
80-56-8: pineneα-pinene, 23857
80-62-6: methyl methacrylate, 19548
81-07-2: saccharin, 27195; Sweeta, 30454
81-13-0: Fancol DL, 11438; panthend-Panthenol, 22690; pantothenyl alcohol, 22697; Ritapan D, DL, 26818
81-14-1: acetyl-dinitro-butyl-xylene, 317; ketone musk, 16141
81-16-3: Tobias acid, 31914
81-54-9: oxyalizarin, 22438
81-64-1: quinizarin, 25683
81-81-2: Dethmor, 8211; Killgerm® Sewarin P, 16184; Kypfarin, 16519; Ratox, 25898; RCR Grey Squirrel Killer Concentrate, 25914; Sakarat, 27238; Sewarin, 28117; Sorexa Plus, 29141
81-88-9: Rhodamine B, 26616
82-68-8: Botrilex, 4460; Brabant PCNB, 4477; Bras-sicol, 4489; Folosan, 12208; quintozene, 25692; Trim, 32330; Tubergran, 32548
82-71-3: styphnic acid, 29858
82-75-7: Peri acid, 23269; Schollkopf's acids, 27726
82-76-8: phenyl-peri acid, 23647
83-63-6: Diacetazotol, 8247; Dimazon, 8487
83-66-9: musk ambrette, 20532
83-72-7: Lawsone, 16891
83-73-8: Embequin, 10024; Yodoxin, 34658
83-79-4: Derris Dust, 8134; FS Derris, 12425; pyrethrum powder, 25478; rotenone, 27000
83-86-3: phytic acid, 23777
83-88-5: Flavaxin, 11895; riboflavin, 26722; vitamin B_2, 33935
84-60-6: anthraflavic acid, 2526
84-61-7: dicyclohexyl phthalate, 8413; KP 201, 16379; Morflex 150, 20317; Unimoll® 66 M, 32976; Uniplex 250, 33014
84-65-1: Anthraquinone, 2532; Morkit®, 20336
84-66-2: diethyl phthalate, 8437; Kodaflex® DEP, 16266; Palatinol® A, 22583
84-69-5: Kodaflex® DIBP, 16267; Palatinol 1C, 22580; Uniplex 155, 33011
84-74-2: Bufa, 4654; DBP, 7812; dibutyl phthalate, 8366; Kodaflex® DBP, 16264; Morflex 240, 20321; NLA-10, 21395; Palatinol® C, 22585; Palatinol® DBP, 22587; PX-104, 25432; Unimoll® DB, 32978; Uniplex 150, 33010
84-76-4: Ceneg, 5671
84-77-5: NLA-40, 21398
84-78-6: Butyl octyl phthalate, 4777; PX-914, 25446
84-80-0: Konakion, 16322; Mono-Kay, 20228; vitamin K_1, 33949;
84-81-1: vitamin K_2, 33950
84-86-6: naphthionic acid, 20715; Piria's acid, 23878
84-87-7: Nevile and Winther's acid, 21046
85-00-7: Katalon, 15782; Midstream, 19723; Power Diquat, 24820; Reglone, Reglox, 26017
85-44-9: phthalic anhydride, 23770; Retarder AK, 26253; Retarder ESEN, 26255; Retarder PX, 26258
85-60-9: Santowhite, 27431; Vanox® SWP, 33348
85-68-7: Butyl benzyl phthalate, 4764; Santicizer 160, 27393; Unimoll® BB, 32977
85-70-1: Morflex 190, 20319; Reomol 4PG, 26126
85-74-5: Dahl's acid II, 7694

85-75-6: Dahl's acid III, 7695
85-83-6: Sudan IV, 29941
85-86-9: Sudan III, 29940
85-91-6: dimethyl anthranilate, 8503
85-97-2: Dowicide 3, 8928
86-29-3: diphenyl acetonitrile, 8593; diphenylacetonitrile, 8600
86-50-0: Cotnion-Methyl, 6884; Gusathion®, 13501; Guthion®, 13505
86-65-7: Salt, Amido-G, 27294
86-87-3: Phyomone, 23775; Planofix, 23913; Rhizopon B, 26574; Rooting Powder, 26958; Tipoff, 31848
87-08-1: Penicillin V, 23111
87-10-5: Temasept IV, 31076; Tuasol 100, 32543
87-19-4: orchindone, 22279
87-20-7: Orchidee, 22278; sanfoin, 27364
87-51-4: Rhizopon A, AA, 26573
87-58-1: iodol, 15215
87-66-1: Piral, 23877; pyrogallic acid, 25510; pyrogallol, 25511, 25512, 25513, 25514
87-79-6: sorbose, 29129
87-83-2: FR-705, 12358; pentabromotoluene, 23131; Pentabromotoluene, 23132
87-86-5: Dowicide 7, 8931; pentachlorophenol, 23135; pentachlorophenol, 23136; Santophen 20, 27417
87-89-8: meat-sugar, 19196; muscle sugar, 20526; phaseomannite, 23579
87-90-1: ACL 85, 90 Plus, 355; CDB 90, 5524; Fi-Clor, 11766; Queschlor, 25658; trichloroisocyanuric acid, 32218
88-04-0: Ayrtol, 3514; chloroxylenol, 6159; Nipacide® MX, 21257
88-05-1: Mesidine, 19398
88-06-2: 2,4,6-trichlorophenol, 32222
88-14-2: 2-Furoic acid, 12497
88-19-7: *o*-toluenesulfonamide, 31948
88-24-4: Anti-oxidant 425, 2576; Cyanox® 425, 7485
88-41-5: Ortholate, 22332
88-58-4: Dibutylhydroquinone, 8368; Eastman® DTBHQ, 9466
88-85-7: Dynamyte, 9361; Premerge, 24888; Tubotox, 32551
88-89-1: Hager's reagent, 13535; Lyddite, 18068; Pertite, 23437; picric acid, 23821; reflorit, 26003
89-04-3: Jayflex® TOTM, 15559
89-65-6: Erythorbic acid, 10775; isoascorbic acid, 15338
89-78-1: Fancol Menthol, 11449; menthol terpine hydrate, 19321
89-82-7: pulegone, 25379
89-83-8: thyme camphor, 31775; thymol, 31778
89-84-9: resacetophenone, 26152
90-00-6: phlorol, 23675
90-02-8: salicylaldehyde, 27269
90-03-9: mercufenol chloride, 19345
90-04-0: *o*-anisidine, 2473
90-12-0: α-methylnaphthalene, 19574
90-17-5: Rose Ester, 26964
90-30-2: Additin 30, 653; Akrochem® Antioxidant PANA, 1161; Antioxidant PAN, 2590; Naugard® PANA, 20798; Neozone A, 20982; Neozone D, 20985
90-43-7: *o*-phenylphenol, 23648; Dowicide 1, 8926; Nipacide® OPP, 21258; Preventol® O Extra, 24931
90-64-2: mandelic acid, 18518
90-72-2: Actiron NX 3, 492
90-80-2: GDL, 12705
90-94-8: ketone base, 16140; Michler's ketone, 19646
91-01-0: Benzhydrol, 3965
91-10-1: dimethoxyphenol, 8500
91-16-7: veratrole, 33568
91-17-8: Decalin®, 7840
91-20-3: naphthalene, 20706
91-22-5: Leucol, 17024
91-53-2: ethoxyquin, 11049; Polyflex, 24435; Santoquin, 27420
91-56-5: Isatin, 15331
91-57-6: β-methylnaphthalene, 19575
91-63-4: quinaldine, 25678
91-95-2: DAB, 7639
92-28-4: Salt, Amido-R, 27295
92-32-0: pyronine Y, 25523
92-40-0: Cassella's acid, 5402
92-41-1: Ebert and Merz's α-acid, 9495
92-44-4: 2,3-naphthalenediol, 20710
92-52-4: Biphenyl, 4216; diphenyl, 8592
92-71-7: diphenyloxazole, 8602
92-83-1: xanthene, 34589
92-84-2: phenothiazine, 23618; PTZ® Phenothiazine Purified, 25378
93-01-6: Schaeffer's acid, 27558
93-04-9: jara jara, 15518; nerolin, 21007
93-15-2: methyl eugenol, 19534
93-23-2: Isothan Q-75, 15427
93-46-9: Agerite® White, 1024
93-51-6: creosol, 6990
93-56-1: styrolyl alcohol, 29878
93-58-3: methyl benzoate, 19515
93-65-2: Astix, 3248; Duplosan New System CMPP, 9128; Duplosan® CMPP, 9130
93-69-6: Vulkacit® 1000, 34124
93-72-1: Kuron, 16496
93-76-5: Trioxone, 32357
93-82-3: Ablumide SDE, 81; Alkamide® DS-280/S, 1642; Alkamide® HTDE, 1643; Amidex SE, 2204; Karamide ST-DEA, 15752; Monamid® 718, 20117; Nopcogen 14-S, 21492
93-83-4: Active #18, 503; Alkamide® DO-280, 1641; Alkamide® WRS 1-66, 1653; Amidex O, 2099; Crillon ODE, 7051; Emid® 6545, 10194; Hartamide 9137, 13659; Incromide OD, 14993; Laurel SD-400, 16840; Mackamide MO, 18160; Marlamid® D 1885, 18739; Mazamide® O 20, 19113; Ninol® 201, 21233; Norfox® F-221, 21549; Rewomid® DO 280, 26322; Rewomid® DO 280 SE, 26323; Schercomid ODA, 27653; Schercomid SO-A, 27661; Varamide® A-7, 33380
94-13-3: Lexgard P, 17145; Nipasol M, 21282; propylparaben, 25231
94-16-6: sodium *para*-aminohippurate, 28804
94-17-7: Cadox® TDP, 4915; Cadox® TS-50S, 4916
94-18-8: Nipabenzyl, 21248
94-26-8: Butylparaben, 4797; Lexgard B, 17143; Nipabutyl, 21249
94-28-0: Kodaflex® TEG-EH, 16279
94-36-0: Abcure S-40-25, 35; AcetOxyl 2.5 and 5, 305; Aztec® Benzoyl Peroxide-70-77, 3553; Aztec® Benzoyl Peroxide-Dry, 3554; Benzoyl peroxide, 3983; Cadet® BPO-70W, 4891; Cadox® 40E, 4906; Cadox® BPO-W40, 4908; Cadox® BS, 4909; Cadox® BTA, 4910; Cadox® BTW-50, 4911; Clear by Design, 6385; Dermoxyl®, 8132; Florox, 12025; Lucidol, 17758; Lucidol 75FP, 17759; Lucidol GS, 17760; Lucidol RM, 17761; Lucidol-78, 17762; Lucilite, 17763; Lucipal, 17764; Luperco A, 17831; Luperco AC, 17832; Luperco AFR, 17833; Luperco AFR-250, 17834; Luperco AST, 17835; Luzidol, 18054; Nericur Gel 5, 21003; Novadeiox, 21657; Periygel®, 23275; Topex, 32043
94-41-7: Chalkone, 5849
94-59-7: safrol, 27229; shikimole, 28174
94-74-6: Agrichem MCPA-25, 50, 1048; Agricorn 500, 1050; Agritox 50, 1077; Agroxone 50, 1086; Albar-M, 1351; Atlas MCPA, 3321; BH MCPA 75, 4119; Campbell's MCPA 25, 50, 5064; Campbell's MCPA 25, 50, 5065; Chafer MCPA 675, 5846; Empal, 10278; Farmon MCPA 50, 11487; FBC MCPA, 11516; MCPA, 19183; Mecpa, 19205; MSS MCPA 50, 20432; Phenoxylene 50, 23628; Phenoxylene Plus, 23629; Power MCPA, 24828; Quad MCPA 50%, 25566; Star MCPA, 29525
94-75-7: Albar-40, 1349; Albar-Super, 1352; Atlas D, 3304; Campbell's Dioweed 50, 5058; Dicotox Extra, 8402; Dormone, 8810; Farmon 2,4-D, 11482; Fernimine, 11639; Fernoxone, 11642; For-Ester, 12244; MSS 2,4-D Amine, 20421; MSS 2,4-D Ester, 20422; Shell D 50, 28152; Syford, 30477
94-81-5: Belmac Straight, 3913; Fisons 18-15, MCPB, 11850; MCPB, 19184, 19185; Tropotox, 32468
94-82-6: 2,4-DB, 7810; Butoxone, 4752; Campbell's DB Straight, 5056; Embutox, 10039
94-91-7: Keromet MD 60, MD 80, 16092; Metal Deactivator S, 19428
94-96-2: Ethohexadiol, 10968; ethyl hexanediol, 11073
95-05-6: Monosulfiram, 20257
95-14-7: ADK STAB LA-32, 729; benzotriazole1H-benzotriazole, 3982; Preventol® CI 8, 24930
95-19-2: Atlasol KAD, 3348; Crodazoline O, 7161; Crodazoline S, 7163; Hodag C-100-S, 14169; Miramine® GS, 19869; Schercozoline S, 27716; stearyl Hydroxyethyl imidazoline, 29606
95-31-8: BBTS, 3863; Delac NS, 8001; Perkacit® TBBS, 23289; Santocure NS, 27402; Vanax® NS, 33292
95-32-9: Morfax, 20314
95-33-0: CBTS, 5484; cyclohexyl

benzothiazole sulfenamide, 7523; Delac S, 8002; Durax®, 9181; Furbac, 12490; Perkacit® CBS, 23277; Santocure, 27400; Vancure D.A.A, 33308; Vulkacit® CZ/EGC, DZ/EGC, 34129
95-38-5: Crodazoline O, 7162; Marlowet® 5440, 18928; Schercozoline O, 27715; Sovatex IM17H, 29160
95-45-4: dimethylglyoxime, 8523
95-47-6: *o*-xylene, 34623
95-48-7: *o*-cresol, 6998
95-50-1: Chloroben, 6117; Dizene, 8744
95-51-2: *o*-chloroaniline, 6114
95-54-5: *o*-phenylenediamine, 23640; Octolite 544, 22005
95-71-6: toluhydroquinone, 31951
95-88-5: 4-chlororesorcinol, 6145
95-92-1: oxalic ether, 22416
95-93-2: Durene, 9195
95-95-4: 2,4,5-trichlorophenol, 32221; Dowicide 2, 8927
96-05-9: Ageflex AMA, 966; Allyl methacrylate, 1784; Sipomer® AM, 28532; SR-201, 29360
96-13-9: 2,3-dibromo-1-propanol, 8360; dibromopropanol, 8361
96-20-8: AB, 19
96-22-0: diethyl ketone, 8436
96-26-4: Dihydroxyacetone, 8459
96-27-5: Thiovanol®, 31723
96-29-7: methyl ethyl ketoxime, 19532
96-33-3: methyl acrylate, 19509
96-43-5: chlorothiophen2-chlorothiophene, 6152
96-45-7: Akroform® ETU-22 PM, 1191; ethylene thiourea, 11132; ETU, 11158; Perkacit® ETU, 23283
96-48-0: Agrisynth BLO, 1074; Agsol Ex BLO, 1090; BLO®, 4350; Butyrolactone, 4803; Dynasolve 699, 9389; GBL, 12704
96-49-1: Texacar® EC, 31424
96-69-5: Lowinox® 44S36, 17649; Rutenol, 27147; Ultranox® 236, 32846
96-83-3: Telepaque, 31057
97-23-4: Algafen, 1583; Algofen, 1602; Anthiphen, 2520; Dicestal, 8376; Dichlorophen, 8392; Ecco MP® 2004, 9513; Fungo®, 12483; Mascot Moss Killer, 19004; Nuophene, 21783; Panacide, 22652; Preventol®, 24926; Super Moss Killer & Lawn Fungicide, 30182; Super Mosstox, 30183, 30184
97-39-2: Accelerator DT, 201; Akrochem® DOTG, 1167; D.O.T.G, 8813; Ditolyguanidine, 8732; Perkacit® DOTG, 23280; Vanax® DOTG, 33289
97-59-6: Allantoin, 1739; Fancol TOIN, 11453
97-63-2: ethyl methacrylate, 11077
97-64-3: ethyl lactate, 11076; Solactol, 28883
97-74-5: Ancazide IS, 2420; Monex, 20210; Mono Thiurad, 20216; Perkacit® TMTM, 23293; Thionex, 31703; TMTM, 31906; Unads®, 32907; Vulcaid 222, 34079
97-77-8: Akrochem® TETD, 1180; Ancazide ET, 2419; Disulfiram, 8725; ethyl tuads®, 11085; ethyl tuex, 11086; ethylthiurad, 11144; Perkacit® TETD, 23291
97-78-9: Crodasinic L, 7143; Hamposyl® L, 13600; lauroyl sarcosine, 16861; Vanseal® LS, 33357
97-88-1: Butyl methacrylate, 4774
97-90-5: Ageflex EGDMA, 969; ethylene glycol dimethacrylate, 11125; Perkalink® 401, 23315
97-94-9: Borethyl, 4420
97-99-4: QO® Tetrahydrofurfuryl Alcohol (THFA®), 25559; tetrahydrofurfuryl alcohol, 31371; THFA, 31660
98-00-0: furfuryl alcohol, 12493; QO® Furfuryl Alcohol (FA®), 25555
98-01-1: furfural, 12491; QO® Furfural, 25554
98-10-2: Benzenesulfonamide, 3962
98-13-5: C P 0 2 8 0, 6 9 3 2; phenyltrichlorosilane, 23650
98-19-1: Butyl-*m*-xylene, 4784; Lowinox® TBMX, 17675;
98-51-1: Butyl toluene, 4782; Lowinox® PTBT, 17673
98-54-4: 4-*t*-butyl phenol, 4800; Lowinox® 070, 17650; Terbutol, 31228
98-55-5: Lily of valley, artificial, 17258
98-77-1: Accelerator 2P, 187; Akrochem® P.P.D, 1173; Vanax® 552, 33284
98-79-3: Ajidew A-100, 1142; Pidolidone®, 23832
98-82-8: cumene, 7352
98-83-9: α-methylstyrene, 19579
98-86-2: acetophenone, 300
98-95-3: mirbane oil, 19935; nitrobenzene, 21348
99-04-7: *m*-toluic acid, 31953
99-05-8: *m*-Aminobenzoic Acid, 2167
99-20-7: Mycose, 20551
99-24-1: Gallicin, 12611
99-30-9: Allisan, 1747; dicloran, 8395; Fumite Dicloran, 12459
99-34-3: 3,5-dinitrobenzoic acid, 21349; Dinitrobenzoic acid, 8541
99-49-0: carvone, 5363
99-50-3: protocatechuic acid, 25289
99-62-7: 1,3-diisopropyl benzene, 8467
99-68-3: Carboxymethyl mercaptosuccinic acid, 5256; Evanacid® 3CS, 11251
99-76-3: Lexgard M, 17144; methylparaben, 19577; Nigagin, 21171; Nipagin M, 21266
99-79-6: Ethiodan, 10954; Iophendylate, 15234; Myodil, 20569; Pantopaque, 22695
99-93-4: *p*-hydroxyacetophenone, 14658
99-94-5: *p*-toluic acid, 31955
100-01-6: *p*-nitroaniline, 21345
100-02-7: *p*-nitrophenol, 21368; P.N.P, 22493
100-10-7: dimethylaminobenzaldehyde, 8515
100-18-5: 1,4-diisopropylbenzene, 8468
100-22-1: Wurster's blue, 34576
100-37-8: 2-diethylaminoethanol, 8440; Pennad 150, 23113
100-42-5: styrol, 29875
100-46-9: benzylamine, 3991; Sumine® 2005, 30110
100-47-0: Benzonitrile, 3976
100-51-6: Bentalol, 3946; benzyl alcohol, 3986
100-52-7: Benzaldehyde, 3957; Ethereal Oil of Bitter Almonds, 10944
100-74-3: N-ethyl morpholine, 11140; Texacat® NEM, 31431; Toyocat® -NEM, 32106
100-88-9: Cyclamic acid, 7509; hexamic Acid, 14007
100-97-0: Grasselerator 102, 13289; hexamethylamine, 14000; hexamethylene tetramine, 14006; hexamine, 14008
101-02-0: Doverphos® 10, 8838; Doverphos® 10-HR, 8839; Lankromark® LE65, 16700; Syn-O-Ad® P-399, 30564; triphenyl phosphite, 32393; Weston® EGTPP, Weston® TPP, 34312
101-05-3: Anilazine, 2465; Dairene®, 7698; Dyrene®, 9437
101-10-0: Bidisin, 4132
101-20-2: Nipaguard® TCC, 21270; TCC, 30856
101-21-3: Atlas CIPC 40, 3303; Campbell's CIPC 40%, 5055; Chlorpropam, 6162; Mirvale, 19939; MSS CIPC, 20428; Residuren, 26166; Spud-Nic®, 29350; Warefog, 34215
101-34-8: Baker P-8, 3596; Flexricin® P-8, 11961; glyceryl triacetyl ricinoleate, 13089; Naturechem® GTR, 20772
101-37-1: Perkalink® 300, 23312; Perkalink® 301, 23313; triallyl cyanurate, 32191
101-41-7: Mephaneine, 19323
101-43-9: Ageflex CHMA, 968
101-54-2: UBOB®, 32699
101-61-1: Michler's Base, 19644
101-64-4: Variamine, 33387
101-67-7: Naugalube® 438, 20783; Vanox® ODP, 33345
101-68-8: Desmodur® VKS-2, VKS-4, VKS-18, 8160; Isonaphthol, 15377; Isonate® 125M, 15378; Isonate® 2125M, 15379; MDI, 19188; Rubinate® LF-168, 27108
101-72-4: Flexzone 3C, 11967; N-Isopropyl-N'-phenyl-*p*-phenylene diamine, 15413; Permanax IPPD, 23364; Vanox® 3C, 33338; Vulkanox® 4010 NA, 34147
101-77-9: Tonox® 22, Tonox® R, 32014
101-84-8: diphenyl oxide, 8598; geranium crystals, 12887; phenyl ether, 23633
101-87-1: Vanox® 6H, 33339
101-91-7: Suconox-4®, 29923
101-96-2: Kerobit® BPD, 16083; Naugalube® 403, 20782; Tenamine 2, 31091; Topanol® M, 32025;
101-99-5: carbanilic ether, 5179; Phenylurethane, 23652
102-06-7: Akrochem® DPG, 1168; D.P.G, 7624; Dynamine, 9355; melaniline, 19259; Perkacit® DPG, 23281; Phenaldine, 23585; Vanax® DPG, 33290; Vulcaid DPG, 34081; Vulcogene ND, 34115; Vulkacit® D/C, 34130
102-07-8: carbanilide, 5180; Zeonet U, 34722
102-08-9: A-1, A-1 Thiocarbanilide, 5; Activit, 505; Akrochem® Thio No. 1, 1181; Anchoracel, 2426;

diphenylthiourea, 8604; dithizone, 8731; Stabilizer C, 29407
102-09-0: Phenol Carbonate, 23604
102-13-6: Iphaneine, 15246
102-60-3: Mazeen® 173, 19120; Neutrol® TE, 21036; Quadrol®, 25582; tetra(2-hydroxypropyl) ethylenediamine, 31356
102-62-5: Hallco® C-491, 13548
102-71-6: triethanolamine, 32240
102-76-1: Kodaflex® Triacetin, 16281; triacetin, 32182
102-77-2: Accelerator MF, 208; Amax®, Amax No 1, 1950; Delac MOR, 8000; OBTS, 21970; OMTS, 22161; Perkacit® MBS, 23284
102-85-2: Syn-O-Ad® P-312, 30561; tributyl phosphite, 32208
102-87-4: Armeen® 3-12, 2931
102-92-1: cinnamoyl chloride, 6314
103-09-3: 2-ethylhexyl acetate, 11138
103-11-7: octyl acrylate, 22037
103-23-1: Adimoll® DO, 693; dioctyl adipate, 8548; Good-rite® GP-223, 13201; Jayflex® DOA, 15554; Kodaflex® DOA, 16272; Monoplex® DOA, 20247; Palatinol® DOA, 22590; Plasthall® DOA, 24002; Plastomoll® DOA, 24058; Polycizer DOA, 24391; PX-238, 25440; Uniflex® DOA, 32945; Wickenol® 158, 34393
103-24-2: Kodaflex® DOZ, 16275; Plasthall® DOZ, 24007; Priplast 3018, 25026
103-30-0: toluylene, 31960
103-34-4: Akrochem® Accelerator R, 1154; Naugex SD-1, 20801; Sulfasan, 29970; Vanax® A, 33287
103-41-3: cinnamein, 6312
103-44-6: Rapi-Cure EHVE, 25882
103-49-1: Accelerator DBA, 200; Vulcaid 28, 34074
103-73-1: phenetole, 23590
103-82-2: α-toluic acid, 31952; phenylacetic acid, 23637
103-83-3: Dabco® B-16, 7646; Pentamin BDMA etc, 23163
103-88-8: Antisepsin, 2614
103-90-2: Valadol, 33238
104-01-8: *p*-methoxyphenylacetic acid, 19505
104-15-4: Eltesol® TSX, 9876; Manro PTSA, 18600; Mod Acid, 20013; PTSA 70, 25377; toluene sulfonic acid, 31950
104-29-0: Chlorphenesin, 6161
104-46-1: Arizole® Anethole Extra., 2885
104-54-1: cinnamyl alcohol, 6315; Peruvin, 23443; styrone, 29888
104-55-2: cinnamaldehyde, 6311
104-60-9: phenylmercuric oleate, 23645
104-74-5: Dehyquart C Crystals, 7982; laurylpyridinium chloride, 16867; Ledmin LPC, 16960
104-75-6: Armeen® L8D, 2952
104-76-7: 2-ethylhexanol, 11136; Aerofroth 88, 859
104-87-0: *p*-tolyl aldehyde, 31961; PTAL, 25375
104-94-9: *p*-anisidine, 2474
105-34-0: methyl cyanoacetate, 19524
105-55-5: Pennzone E 0686, 23121
105-57-7: Acetal, 280; Acetron®, Acetron® GP; Acetron® NS, 308; AT-20GF, 3282; Cadco® Acetal, 4884; Delrin® 100, 500, 8033; Delrin® 100AF, 500AF, 8034; Delrin® 100ST, 500T, 8035; Delrin® 107, 507, 8036; Delrin® 150A, 550SA, 8037; Delrin® 570, 8038; Delrin® 900, 8039; Delrin® AF Blend, 8040; Electrafil® J-80/CF/10/TF/10, 9739; Thermocomp® KB-1008, 31616
105-60-2: Nipro (i), 21303
105-62-4: Emalex PG-di-O, 9983
105-74-8: Alperox-F, 1816; Laurox®, 16859; Laurydol, 16862
105-75-9: Bisomer DBF, 4278; dibutyl fumarate, 8364
105-76-0: Bisomer DBM, 4279; dibutyl maleate, 8365; Octomer DBM, 22008; PX-504, 25443
105-87-3: Meraneine, 19330
105-95-3: Emeressence® 1150, 10077; ethylene brassylate, 11121
105-97-5: Uniflex® DCA, Unimate® DCA, 32943
105-99-7: Adimoll® DB, 690; Cetiol® B, 5797
106-11-6: Alkamuls® SDG, 1683 PEG-2 stearate, 23043;
106-12-7: PEG-2 oleate, 23042
106-14-9: Ceroxin GL, 5758
106-16-1: Alkamide® R-280, 1649; Rewomid® R 280, 26329
106-17-2: Flexricin® 15, 11952; Glycol ricinoleate, 13104
106-22-9: Cephrol, 5695; citronellol, 6348
106-23-0: citronellal, 6347
106-24-1: Geraniol, 12885; Meranol, 19331
106-25-2: Vernol®, 33600
106-42-3: *p*-xylene, 34624
106-44-5: *p*-cresol, 6999
106-46-7: *p*-dichlorobenzene, 8386; Kaydox, 15811; Paradow, 22738; Santochlor, 27399
106-47-8: *p*-chloroaniline, 6115, 6116
106-50-3: *p*-phenylenediamine, 23641; Aminogen II, 2177
106-58-1: Texacat® DMP, 31429
106-65-0: dimethyl succinate, 8509
106-89-8: Epichlorhydrin, 10632
106-90-1: glycidyl acrylate, 13091
106-91-2: glycidyl methacrylate 2,3-epoxypropyl methacrylate, 13094; SR-379, 29382
106-92-3: Ageflex AGE, 965; Allyl glycidyl ether, 1783; Sipomer® AGE, 28531
106-97-8: A-17, 7; Bu-Gas, 4660; Butane, 4734
107-03-9: n-propyl mercaptan, 25227, 25228, 25229
107-05-1: Barchlor, 3626
107-06-2: ethylene dichloride, 11122
107-13-1: Acrylonitrile, 436
107-15-3: ethylenediamine, 11133
107-18-6: propenol, 25181
107-21-1: ethylene glycol, 11123; glycol, 13099; Ilexan E, 14883
107-22-2: diformyl, 8453; glyoxal 40%, 13146; Protectol® GL 40, 25262
107-35-7: taurine, 30845
107-41-5: hexylene glycol, 14041, 14042; Isol, 15366
107-43-7: Oxyneurine, 22463
107-46-0: CH7310, 5841; H7310, 13530; Hexamethyldisiloxane, 14004
107-51-7: CO9816, 6509; O9816, 21946
107-54-0: Surfynol® 61, 30373
107-64-2: Adogen® TA-100, 779; Adogen® TA-101, 780; Adogen® TA-101, 781; Arosurf® TA-100, 3081; Arquad® 218-100, 3111; Arquad® 218-75, 3112; Blandofen CT, 4318; Dehyquart DAM, 7985; Distearyldimonium chloride, 8714; Genamin DSAC, 12771; Prepagen WK, 24894; Sumquat® 6045, 30123; Varisoft® TA-100, 33428
107-71-1: Aztec® *t*-butyl peracetate-50 OMS, 60 OMS, 75 OMS, 3542; Esperox® 12MD, 10845; Lupersol 70, 17844; Trigonox® F-C50, 32300
107-72-2: CA0900, 4854
107-87-9: methyl propyl ketone, 19557
107-93-7: Crotonic Acid, 7260
107-97-1: methyl-glycocoll, 19570; sarcosine, 27463
107-98-2: Arcosolv® PM, 2814; Dowanol® PM, 8896; Icinol PM, 14816; Poly-Solv® MPM, 24558; propylene glycol monomethyl ether, 25220; Solvenon® PM, Solvent PM, 29050
108-01-0: Alkanolamine, 1701; Dabco® DMEA, 7669; deanol, 7829; dimethylethanolamine, 8520; Tegoamin® DMEA, 30995; Texacat® DME, 31428; Toyocat® -DMA, 32100
108-03-2: 1-nitropropane, 21373; NiPar S-10, 21279
108-05-4: Plyamul® 40305-00, 24266; Unocal 76 Res 6206, 33080; Unocal 76 Res S-55, 33086; vinyl acetate, 33853 33854; Vinyl Acetate Monomer, 33855
108-11-2: methyl amyl alcohol, 19512
108-18-9: diisopropylamine, 8469
108-23-6: isopropyl chloroformate, 15407
108-24-7: acetic anhydride, 293
108-30-5: succinic anhydride, 29916
108-31-6: maleic anhydride, 18476
108-32-7: Arconate® Propylene Carbonate, 2811; Solvenon® PC, 29049; Texacar® PC, 31425
108-38-3: *m*xylene, 34622
108-39-4: *m*-cresol, 6997
108-42-9: *m*-chloroaniline, 6113
108-45-2: *m*-phenylenediamine, 23639
108-46-3: Cohedur® RK, 6576; resorcin, 26228; resorcinol, 26230
108-48-5: 2,6-Lutidine, 17995
108-62-3: Farmon Mini Slug Pellets, 11489; Gastratox 6G Slug Pellets, 12693; Meta, 19403; metaldehyde, 19431; Mifaslug, 19726; PBI Slug Pellets, 22936; Unicrop 6% Mini Slug Pellets, 32930; Vassgro Mini Slug Pellets, 33471
108-65-6: Arcosolv® PMA, 2815; Dowanol® PMA, 8897; Ektasolve® PM Acetate, 9672
108-73-6: phloroglucinol, 23674
108-75-8: Collidine, 6625
108-80-5: cyanuric acid, 7494; isocyanuric acid, 15357; Zeonet A, 34720

108-83-8: diisobutyl ketone, 8461; valerone, 33243
108-87-2: methyl cyclohexane, 19525
108-88-3: methyl benzene, 19514; toluene, 31946
108-89-4: picoline γ-picoline, 23817
108-90-7: Abluton T30, 139; chlorobenzene, 6119
108-91-8: cyclohexylamine, 7525
108-93-0: Adronal, 792; cyclohexanol, 7521; Hexalin, 13998; Hydralin, 14536
108-94-1: cyclohexanone, 7522; Sextone, 28118
108-95-2: Carbolic Acid, 5206; phenic acid, 23592
108-98-5: thiophenol, 31708
108-99-6: picolineβ-picoline, 23816
109-02-4: *p*-methyl morpholine, 19573; Texacat® NMM, 31432
109-06-8: picolineα-picoline, 23815
109-09-1: 2-chloropyridine, 6144
109-13-7: Aztec® *t*-Butyl Peroxy isobutyrate-75 OMS, 3546; Lupersol 80, 17845; Trigonox® 41-C75, 32271
109-28-4: Chemidex O, 5965; Incromine OPB, 15000; Lexamine O-13, 17099; Mackine 501, 18221; oleamidopropyl dimethylamine, 22109; Schercodine O, 27640
109-31-9: Priplast 3013, 25025
109-38-6: Butoxyethyl stearate, 4755; KP-23, 16376
109-43-3: Kodaflex® DBS, 16265; Plasthall® DBS, 23994; Reomol DBS, 26129; Uniflex® DBS, 32942
109-46-6: Pennzone B 0685, 23120
109-52-4: valericn-valeric acid, 33242
109-60-4: propyl acetate, 25203
109-66-0: pentanen-pentane, 23164
109-73-9: Amine C4, 2152
109-74-0: Butyronitrile, 4805; nitrile C_4, 21338
109-86-4: ethylene glycol methyl ether, 11127; methyl cellosolve®, 19521; methyl icinol, 19546; Poly-Solv® EM, 24557
109-87-5: methylal, 19564
109-89-7: diethylamine, 8439
109-97-7: pyrrhol, 25533; Pyrrole, 25535
109-99-9: Dynasolve 150, 9387; QO® Tetrahydrofuran (THF), 25558; tetrahydrofuran, 31370
110-00-9: QO® Furan, 25552; Pennodorant® 1013, 23117; tetrahydrothiophene, 31374
110-02-1: Hopkin's lactic acid reagent, 14299; thiofurfuran, 31687; thiophene, 31707
110-05-4: Aztec® Di-*t*-Butyl Peroxide, 3557; di-*t*-butyl peroxide, 8369; Trigonox® B, 32295
110-15-6: acid of amber, 334; sal succini, 27256; salt of amber, 27283; Succinellite, 29914; succinic acid, 29915
110-16-7: maleic acid, 18475
110-17-8: Fumaric acid, 12453
110-18-9: Propamine D, 25170
110-19-0: isobutyl acetate, 15345
110-25-8: Crodasinic O, 7148; Hamposyl® O, 13604; Oleoyl sarcosine, 22123; Vanseal® OS, 33360
110-26-9: M.B.A, 18103; methylene bisacrylamide, 19567
110-27-0: Crodamol 1PM, 7116; Emerest® 2314, 10085; Estergel, 10883; isopropyl myristate, 15409; Kessco® IPM, 16115; Lexol IPM-NF, 17150; Liponate IPM, 17378; Promyr, 25152; Radia® 7190, 25729; Tegosoft® M, 31024; Unimate® IPM, 32971; Wickenol® 101, 105, 34378
110-30-5: Abluwax EBS, 140; Acrawax® C, 383; Advawax® 290, 820; Alkamide® STEDA, 1652; Armowax EBS, 2985; Glycowax® 765, 13137; Kemamide® W-39, 15904; Kemwax, 15993; Nopcowax 22-DS, 21513; Uniwax 1760, 33069
110-31-6: Advawax® 240, 819; Kemamide® W-20, 15903
110-33-8: Adimoll® DH, 691
110-36-1: Bumyr, 4670; Butyl myristate, 4775; Wickenol® 141, 34384
110-42-9: methyl caprate, 19519
110-43-0: methyl n-amyl ketone, 19552
110-44-1: sorbic acid, 29105; Sorbistat, 29107
110-52-1: 1,4-dibromobutane, 8357
110-54-3: Exxsol® Hexane, 11361; n-hexane, 14009
110-60-1: putrescine, 25423
110-63-4: 1,4-butanediol, 4735; Dabco® BDO, 7647
110-65-6: but-2-yne-1,4-diol, 4725
110-71-4: Ansol E-121, 2494; monoglyme, 20227
110-80-5: EE Solvent, 9636; Ethoxyethanol, 11045; ethyl icinol, 11074; Poly-Solv® EE, 24556
110-82-7: cyclohexane, 7520
110-85-0: Antepan, 2518; piperazine, 23873; Tasnon®, 30841; Upixon, 33094
110-86-1: pyridine, 25481
110-89-4: piperidine, 23874
110-91-8: morpholine, 20346
110-97-4: diisopropanolamine, 8466
110-98-5: Dipropylene glycol, 8611
111-01-3: Cosbiol®, 6865; Dermane, 8103; Robane®, 26870; Romane, 26928; squalane, 29357
111-02-4: squalene, 29358; Supraene, 30301; Supraene®, 30302
111-03-5: Ablunol GMO, 102; Aldo® MO, 1541; Capmul® GMO, 5112; Kemester® 2000, 15940; Kessco® GMO, 16111; Mazol® 300 K, 19134; Mazol® GMO, 19136; Monomuls® 90-O18, 20240; Radiasurf® 7150, 25801; Witconol 2421, 34500
111-05-7: Alkamide® OIP, 1648; Mackamide OP, 18162; Rewomid® IPE 280, 26325; Schercomid OMI, 27655; Witcamide® 61, 34441
111-14-8: heptanoic acid, 13791
111-15-9: EE Acetate, 9635; Ethoxyethanol acetate, 11046; Sensolve EEA, 27954
111-17-1: 3,3'-thiodipropionic acid, 31684
111-20-6: ipomic acid, 15247; sebacic acid, 27795
111-27-3: 1-hexanol, 14011; Epal® 6, 10615; Exxal® 6, 11334; hexyl alcohol, 14039
111-30-8: glutaraldehyde, 13062; Protectol® GDA, GT 50, 25261; Relugan GT, 26046; Sonacide, 29073; Ucarcide® 225, 32711
111-34-2: Shostakovsky Balsam, 28187
111-40-0: D.E.H. 20, 7610; D.E.H. 52, 7612; diethylenetriamine, 8446
111-41-1: Aminoethylethanolamine, 2172
111-42-2: Dabco® DEOA-LF, 7667; Diethanolamine, 8434
111-46-6: diethylene glycol, 8442; Her, 13798
111-48-8: thiodiglycol, 31682
111-55-7: ethylene glycol diacetate, 11124
111-57-9: Ablumide SME, 82; Alkamide® S-280, 1650; Amidex SME, 2103; Mackamide SMA, 18167; Mazamide® SMEA, 19114; Monamid® S, 20122; Rewomid® S 280, 26330; Witcamide® 70, 34442
111-58-0: Schercomid OME, 27654
111-60-4: Ablunol EGMS, 100; Alkamuls® EGMS/C, 1664; Alkamuls® SEG, 1684; Emalex EGS-A, 9952; Monthyle, 20298; Schercemol EGMS, 27600
111-61-5: Radia® 7185, 25727
111-66-0: Neodene® 8, 20890; octene-1, 21991
111-70-6: 1-heptanol, 13792; heptyl alcohol, 13797
111-76-2: 2-butoxyethanol, 4753; Butyl Cellosolve®, 4767; Butyl Icinol, 4771; Dowanol® EB, 8894; Ektasolve® EB, 9668
111-77-3: Dowanol® DM, 8891; Ektasolve® DM, 9666; methyl di-Icinol, 19526; Poly-Solv® DM, 24554
111-82-0: methyl laurate, 19545
111-86-4: Amine 8 D, 2125
111-87-5: 1-octanol, 21988; capryl alcohol, 5157; Emery® 3322, 10166; Emery® 3324, 10168; Epal® 8, 10616; Lorol C8, 17616
111-90-0: Carbitol®, 5194; Diethoxol, 8435; Dioxitol, 8576; Ektasolve® DE, DE-HG, 9664; Ethoxydiglycol, 11044; ethyl di-Icinol, 11068; Transcutol, 32137
111-96-6: diglyme, 8455
112-00-5: Adogen® 412, 767; Arquad® 12-37W, 3105; Arquad® 12-50, 3106; Chemquat 12-50, 6024; Dehyquart LT, 7987; Empigen® 5089, 10355; Octosol 562, 22020; Varisoft® LAC, 33423
112-02-7: Ammonyx® CETAC, CETAC-30, 2261; Arquad® 16-29, 3107; Arquad® 16-50, 3108; Barquat® CT-29, 3667; Carsoquat® CT-429, 5349; Cetrimonium chloride, 5814; cetyl trimethyl ammonium chloride, 5826; Chemquat 16-50, 6025; Dehyquart A, 7980; Genamin CTAC, 12770; Incroquat CTC-30, 15037; Quatrex CTAC, 25624; Querton 16Cl-29, 25650; Radiaquat 6444, 25792; Rhodaquat® M242C/29, 26644;

Varisoft® 250, 300, 355, 33407; Varisoft® 300, 355, 33408
112-03-8: Arquad® 18-50, 3109; Octosol 474, 22019; Tomah Q-ST-50, 32000; Varisoft® ST-50, TSC, 33427; Varisoft® TSC, 33431
112-04-9: CO9750, 6506
112-05-0: Emery® 1202, 10138; Emery's L-114, 10187; Pelargonic acid, 23078
112-07-2: 2-butoxyethanol acetate, 4754; Ektasolve® EB Acetate, 9669
112-10-7: Tegosoft® S, 31027; Wickenol® 127, 34380
112-13-0: Decanoyl chloride, 7844
112-16-3: lauroyl chloride, 16860
112-18-5: Armeen® DM12D, 2942; dimethyl lauramine, 8507; N,N-dimethyldodecylamine, 8519; Onamine 12, 22164
112-24-3: Texacure EA-24, 31439; Texlin® 300, 31507; triethylenetetramine, 32245
112-29-8: 1-Bromodecane, 4578; decyl bromide, 7868
112-30-1: Epal® 10, 10617; Exxal® 10, 11338;Lorol C10, 17610;
112-34-5: Butyl Carbitol®, 4766; Butyl Di-icinol, 4769; Dowanol® DB, 8890; Ektasolve® DB, 9662
112-35-6: Poly-Solv® TM, 24559
112-38-9: undecylenic acid, 32916
112-39-0: Emery® 2216, 10145; methyl palmitate, 19555; Radia® 7120, 25723
112-41-4: dodecene-1, 8778; Neodene® 12, 20895
112-42-5: Neodol® 1, 20908; Neoflex® 11, 20921
112-53-8: 1-dodecanol, 8777; Emery® 3326, 10169; Emery® 3332, 10170; Emery® 3357, 10174; Epal® 12, 10618; Exxal® 12, 11339; lauryl alcohol, 16864; Lipocol L, 17353; Lorol C12, 17611; Lorol C12-C14, 17612; Lorol C8-C10 Special, 17617; Lorol Special, 17621; Philcohol 1200, 23664
112-57-2: tetraethylenepentamine, 31367; Texlin® 400, 31508
112-60-7: PEG-4, 23049
112-61-8: Emery® 2218, 10146; Kemester® 4516, 15944; methyl stearate, 19560
112-62-9: Emerest® 2301, 10081; Emery® 2219, 10147; Emery® 2301, 10155; Kemester® 104, 15935; methyl oleate, 19554; Priolube 1400, 25013; Witconol 2301, 34496
112-69-6: Adma® 16, 743; Armeen® DM16D, 2943; Crodamine 3.A16D, 7106; N,N-dimethylhexadecylamine, 8524
112-70-9: tridecyl alcohol, 32233
112-72-1: Cachalot® M-43 NF, 4879; Dehydag Wax 14, 7943; Emery® 3334, 10171; Epal® 14, 10619; Lanette 14, 16661, 16662; Lorol C14, 17613; myristyl alcohol, 20577; Philcohol 1400, 23666
112-75-4: Adma® 14, 742; Empigen® AH, 10361; N,N-dimethyltetradecyl amine, 8527; Onamine 14, 22165
112-80-1: Distoline, 8724; Emersol® 210, 10120; Emersol® 6333 NF, 10123; Emersol® 7021, 10125; Industrene® 104, 15100; oleic Acid, 22112; Pamolyn 100 FGK, 22644; Pamolyn® 125, 22648; Priolene 6900, 25012; red oil, 25964
112-84-5: Armid® E, 2961; Crodamide E, ER, 7096; erucamide, 10770; Kemamide® E, 15895; Petrac® Eramide®, 23464; Polydis® TR 131, 24412; Unislip 1753, 33044
112-85-6: Hystrene® 9022, 14765; Hystrene® 5522, 14771; Prifrac 2989, 24956
112-86-7: erucic acid, 10771; Prifrac 2990, 24957
112-88-9: Neodene® 18, 20899
112-90-3: Amine OL, 2162; Armeen® OL, 2956; Armeen® OLD, 2957; Crodamine 1.0, 1.0D, 7100; Jet Amine PO, 15595; Kemamine® P-989D, 15921; Radiamine 6172, 25757
112-92-5: Adol® 62 NF, 783; Cachalot® S-56, 4881; Crodacol S70, 7071; Crodacol S95NF, 7072; Dehydag Wax 18, 7945; Emery® 3343, 10173; Epal® 18NF, 10627; Fancol SA, 11451; Lanette 18 DEO, 16664; Lipocol S, 17357; Lorol C18, 17615; Loxiol VPG 1354, 17684; Philcohol 1800, 23669; Rita SA, 26793; Stearal, 29591; stearyl alcohol, 29604; Varonic® BG, 33440
112-99-2: Armeen® 2-18, 2927
113-98-4: Liquapen, 17452; Penicillin G potassium, 23109; Penicillin G potassium, 23110
114-26-1: Blattanex®, 4329; Blattanex® 20, 4330; Fumite Propoxur, 12465; Propoxur, 25198, 25199, 25200, 25201; Suncide®, 30138; Unden®, Undene®, 32917
114-91-0: metyridine, 19609; Promintic, 25143
115-07-1: propylene, 25212
115-10-6: dimethyl ether, 8505; Dymel®, 9331; methyl ether, 19529
115-19-5: methyl butynol, 19518
115-21-9: CE6350, 5530; ethyltrichlorosilane, 11147
115-29-7: thiodan, 31680; Thionex, 31702
115-32-2: Acarin, 183; Dicofol, 8398; Fumite Dicofol, 12460; Kelthane, 15881; Mitigan, 19959
115-40-2: Bromcresol purple, 4567
115-44-6: Lotusate, 17635
115-69-5: 2-amino-2-methyl-1,3-propanediol, 2183; AMPD, 2304
115-70-8: 2-Amino-2-ethyl-1,3-propanediol, 2170; AEPD®, 832; AEPD® 85, 833
115-77-5: Hercules® Improved Tech. PE, 13859; Hercules® Mono-PE, 13861; pentaerythritol, 23138, 23139; Pentek, 23176
115-83-3: Liponate PS-4, 17386; pentaerythrityl tetrastearate, 23142; Radia® 7176, Radiasurf® 7175, 25726
115-86-6: Altal, 1869; Disflamoll® TP, 8642; Kronitex® TPP, 16437; triphenyl phosphate, 32391
115-90-2: Dasanit®, 7787; Terracur® P, 31329;
115-95-7: Phanteine, 23569
115-96-8: Genomoll P, 12838
116-14-3: MS-122, 20413
116-25-6: Dantoin® MDMH, 7734
116-29-0: Tedion V-18, 30919; tetradifon, 31366
117-08-8: Tetrathal, 31393
117-12-4: anthrarufin, 2533
117-18-0: Ashlade TCNB, 3199; Bygran S, 4821; Fumite TCNB, 12467; Fumite TCNB Smoke, 12468; Fusarex, 12513; Hickstor, 14062; Hystor 10, 14754; Hytec, 14775; Nebulin, 20839; New Hickstor 6, 21065; New Hystor, 21066; Quad Store, 25568; Quad-Keep, 25572; Tecgran, 30886; tecnazene, 30898; Tripart® Arena 6, 32360; Tripart® Arena Granules, 32362; Tubodust, Tubostore, 32549
117-39-5: quercetin, 25647; Ritacetin, 26799
117-80-6: Dichlone, 8381
117-81-7: Bis(2-ethylhexy) Phthalate, 4228; Bisoflex 81, 4249; Bisoflex 82, 4252; dioctyl phthalate, 8549; Kodaflex® DOP, 16273; Morflex 310, 20322; Morflex 410, 20325; NLA-20, 21396; Octoil, 22002; Palatinol® DOP, 22591; Plasthall® DOP, 24004; Plasticizer 28P, 24030; Polycizer DOP, 24390; PX-138, 25438; Reomol DCP, 26130; Vestinol AH, 33666; Witcizer 312, 34454
117-82-8: Kodaflex® DMEP, 16270; Reomol P, 26132
117-83-9: Dibutoxyethyl phthalate, 8363; Plasthall® 200, 23979
118-52-5: 1,3-Dichloro-5,5-dimethyl hydantoin, 8389; Dantoin® DCDMH, 7732; Hydan, 14528;
118-55-8: phenyl salicylate, 23634
118-56-9: Filtrosol A, 11800; Heliophan, 13766; Homosalate, 14292; Kemester® HMS, 15963
118-60-5: Dermoblock OS, 8116; Escalol® 587, 10790; Neo Heliopan® OS, 20870; Uvinul® O-18, 33202
118-61-6: ethyl salicylate, 11081; Sal-ethyl, 27264
118-71-8: Veltol, 33544
118-74-1: Julin's chloride, 15632
118-75-2: chloranil, 6087; Vulklor, 34159
118-78-5: uramil, 33113
118-79-6: 2,4,6-tribromophenol, 32202; Bromol, 4582; FR-613, 12357; Great Lakes PH-73, 13310;
118-90-1: *o*-toluic acid, 31954
118-92-3: AA, 11; anthranilic acid, 2529
118-93-4: *o*-hydroxyacetophenone, 14657
118-96-7: Trotyl, 32474
119-06-2: Jayflex® DTDP, 15555; PX-126, 25437
119-07-3: Good-rite® GP-265, 13205; Morflex 125, 20316; Morflex 175, 20318; octyldecyl phthalate, 22042

119-28-8: Cleve's ω-Acid or J-acid, 6412; Delta acid, 8044
119-36-8: Gaultheria oil, 12701; Gaultheric acid, 12702; Teaberry Oil, 30872
119-47-1: Anti-oxidant 2246, 2580; Cyanox® 2246, 7489; Lowinox® 22M46, 17652, 17653; Ralox® 46, 25842; Ultranox® 246, 32847; Vulkanox® BKF, 34149
119-53-9: Benzoin, 3975; Persian balsam, 23417
119-58-4: Michler's hydrol, 19645
119-61-9: ADK STAB 1413, 716
119-64-2: tetrahydronaphthalene, 31373
119-79-9: Cleve's β acid, 6410
120-12-7: Sterilite Hop Defoliant, 29734
120-18-3: β-naphthalene sulfonic acid, 20707; Dehscofix 918, 7926
120-23-0: betapal Concentrate, 4063
120-32-1: Nipacide® BCP, 21252
120-36-5: Campbell's Redipon, 5070; MSS 2,4-DP, 20424
120-40-1: Ablumide LDE, 79; Alkamide® 327, 1634; Alkamide® LE, 1646; Amidex L-9, 2094; Amidex LD, 2095; Carsamide® SAL-7, 5323; Comperlan LD, 6676; Crillon LDE, 7049; Emalex NN-7, 9974; Emid® 6519, 10193; Empilan® LDE, 10408; Hartamide LDA, 13661; Incromide L-90, 14987; Incromide LLT, 14990; Incromide LM-70, 14991; lauramide DEA, 16828; Mackamide L10, 18156; Mazamide® 1214, 19107; Mazamide® L-298, 19110; Monamid® 150-LMWC, 20115; Ninol® 30-LL, 21234; Schercomid SL-Extra, 27659; Standamid® KDS, 29456; Varamide® ML-1, 33384; Witcamide® 5195, 34440
120-47-8: ethyl *p*-hydroxybenzoate, 11061; Nipagin A, 21263
120-51-4: Benzyl benzoate, 3987; Peruscabin, 23440
120-52-5: Dibenzo GMF, 8353
120-54-7: DPTT, 8952; Perkacit® DPTT, 23282; Sulfads®, 29960
120-55-8: Benzoflex 2-45, 3970
120-78-5: Altax®, 1872; M B T S, 18100; MBTS, 19172; Perkacit® MBTS, 23286; Vulkacit® DM/C, 34131
120-80-9: kachin, 15651; pyrocatechin, 25497; pyrocatechol, 25498
120-82-1: 1,2,4-trichlorobenzene, 32214; Hipochem GM, 14110
121-14-2: D.N.T, 7622
121-25-5: Amprol, 2364; Amprolium hydrochloride, 2365; Pancoxin, 22668
121-32-4: Ethavan, 10937; ethyl vanillin, 11087; Rhodiarome, 26678; Vanbeenol, 33294
121-33-5: vanilla, 33326, 33327; Zimco, 34763
121-47-1: metanilic acid, 19445
121-54-0: benzethonium chloride, 3963; Hyamine® 1622 50%, 14485; Phemerol Chloride, 23581
121-57-3: sulfanilic acid, 29966
121-66-4: 2-amino-5-nitrothiazole, 2184; Entramin, 10595
121-75-5: Fyfanon, 12533; malathion, 18460, 18461; Prioderm, 25011
121-79-9: Progallin P, 25111; propyl gallate, 25204; Sustane® PG, 30436
122-00-9: Melilot, 19279
122-03-2: Cumal, 7349; cuminaldehyde, 7354
122-14-5: Dicofen, 8397; Fenitex, 11595; Fenitrothion EC, 11596; Folithion®, 12207; Micromite, 19683; Novathion, 21708
122-18-9: Sumquat® 6050, 30124
122-19-0: Ablumine 280, 85; Ammonyx® 4, 4B, 485, 4002, 2260; Amyx A-25-S 0040, 2391; Carsoquat® SDQ-25, 5350; Catinal OB-80E, 5458; Cycloton® SCS, 7542; Emcol® 4, 10044; Hetquat S-20, 13955; Incroquat SDQ-25, 15044; Mackernium SDC-25, 18210; Miracare® SCS, 19859; M-Quat® B-25, 20404; octadecylbenzene methanaminium chloride, 21985; Quatrex STC-25, 25626; Rhodaquat® M270C/18, 26645; stearalkonium chloride, 29592; Stedbac®, 29609; Sumquat® 6210, 30126; Varisoft® SDAC-W, 33426
122-20-3: triisopropanolamine, 32313
122-32-7: Emerest® 2423, 10099; Kemester® 1000, 15938; Priolube 1435, 25019; Radia® 7363, 25739; triolein, 32351
122-34-9: Boroflow, 4433; Gesatop, 12923; Herbazin 50, 13801; Simadex, 28459; Simanex, 28460; Simapron, 28461; simazine, 28464; Simflow, 28471; Sinazine, 28493; Syngran, 30543; Weedex S2, 34259
122-39-4: OA-505, 21954
122-42-9: MSS IPC 50, 20431; Propham, 25185, 25186
122-46-3: *m*- cresyl acetate, 7011; Kresatin, 16402
122-52-1: phosphorus ether, 23742; triethyl phosphite, 32244
122-57-6: benzylidene acetone, 3992
122-59-8: phenoxyacetic acid, 23626
122-62-3: Monoplex® DOS, 20248; Octoil S, 22003; Plasthall® DOS, 24005; Reomol DOS, 26131; Uniflex® DOS, 32946
122-69-0: styracin, 29860
122-87-2: Glyconyl, 13124
122-99-6: Emeressence® 1160 Rose Ether, 10079; Igepal® Cephene Distilled, 14850; Igepal® OD-410, 14873; Phenoxetol, 23622; 2-Phenoxyethanol, 23627; Prox-onic PH-01, 25334; Rewopal® MPG 10, 26344
123-03-5: Acetoquat CPC, 302; cetyl pyridinium chloride, 5823
123-11-5: *p*-anisaldehyde, 2470
123-19-3: Butyrone, 4804
123-28-4: Argus DLTDP, 2861; Carstab® DLTDP, 5356; Cyanox® LTDP, 7491; dilauryl thiodipropionate, 8476; Evanstab® 12, 11255; Lankromark® DLTDP, 16689; Lowinox® DLTDP, 17666; Wytox® LT, 34582
123-30-8: *p*-aminophenol, 2185; Rodinal, 26900; Takatol, 30725; Unal, 32908
123-31-9: Black and White Bleaching Cream, 4302; Eldopaque, 9723; hydroquinone, 14632; quinol, 25685; Tecquinol® Tech. Grade, 30910
123-33-1: Bos MH, 4456; Burtolin, 4722; Chiltern Fazor, 6067; Malazide, 18464; maleic hydrazide, 18477; Mazide 25, 19129; MSS MH18, 20433; Regulox K, 26020; Royal Slo-Gro, 27034; Sucker Stuff, 29922
123-38-6: propanal, 25172
123-42-2: Diacetone alcohol, 8249; Pyraton, 25472
123-46-6: Girard's reagent T, 12950
123-54-6: acetylacetone, 314
123-56-8: Lubrizol® 2153, 17745; succinimide, 29917
123-63-7: paraldehyde, 22759
123-72-8: Butyraldehyde, 4801
123-76-2: levulinic acid, 17084
123-77-3: Azodicarbonamide, 3524; Celogen® AZ 120, 130, 150, 180, 199, 5642; Ficel® AC2, 11762; Genitron, 12834; Kempore® 60/14FF, 15987; Porofor® ADC/E, 24698; Santechem 21-21, 27382
123-81-9: glycol dimercaptoacetate, 13101
123-88-6: Agallol, 954; Atiran, 3286; Baytan®, 3851;Ceresan, 5745;
123-91-1: Dioxane, 8574
123-92-2: isoamyl acetate, 15336; Jargonelle pear essence, 15521;pear oil, 22971;
123-93-3: thiodiglycolic acid, 31683
123-94-4: glyceryl stearate, 13087
123-95-5: ADK STAB LS-8, 732; butyl stearate, 4781; Emerest® 2325, 10088; Kemester® 5510, 15948; Kessco® BS, 16103; Lexolube® BS-Tech, 17163; Oleo-Coll LP, 22118; Priolube 1451, 25020; Radia® 7051, 25718; Uniflex® BYS-Tech, Unimate® BYS, 32941; Witconol 2326, 34497
123-99-9: Azelaic acid, 3519; Emerox® 1110, 10115; Emery's L-110, 10186
124-03-8: Bretol®, 4506; Sumquat® 6020, 30121
124-04-9: Adipic acid, 698; Adi-pure®, 706
124-07-2: caprylic acid, 5158; Emery® 657, 10136; Prifrac 2901, 24950
124-10-7: Emery® 2214, 10144; methyl myristate, 19550
124-17-4: diethylene glycol butyl ether acetate, 8443
124-22-1: Amine 12, 2127; Armeen® 12, 2933; Armeen® 12D, 2934; Radiamine 6164, 25754
124-26-5: Crodamide S, SR, 7098; Kemamide® S, 15900; Petrac® Vyn-Eze®, 23469; stearamide, 29593; Uniwax 1750, 33068
124-28-7: Adma® 18, 744; Adogen® MA-108 SF, 776; Amine 2M18D, 2139; Armeen® DM18D, 2944;

Crodamine 3.A18D, 7107; Dymanthine, 9327; Kemamine® T-9902, 15933; Onamine 18, 22167
124-30-1: Amine 18-90, 2131; Armeen® 18, 2938; Armeen® 18D, 2939; Armid® HTD, 2963; Crodamine 1.18D, 7102; Kemamine® P-990, P-990D, 15922; Steamfilm FG, 29589
124-38-9: carbon dioxide, 5222; Cardice, 5259; Dricold, 9026; Drikold, 9032; dry ice, 9057
124-48-1: chlorobromoform, 6121
124-68-5: 2-amino-2-methyl-1 propanol, 2182; AMP, 2300; AMP-95, 2302; AVT-75, 3504
124-70-9: CV-4772, 7448
124-94-7: Ledercort, 16954
126-06-7: Bromochloro dimethyl hydantoin, 4574; Dantoin® GSD-550, 7733; Halobrom, 13560; Quesbrom, 25657
126-11-4: S.S.T® Sump Saver Tablets, 27180; Tris Nitro®, 32408
126-14-7: Soa, 28682; sucrose octa-acetate, 29935
126-30-7: neopentyl glycol, 20937; NPG® Glycol, 21751
126-31-8: Abrodil, 158; Methiodal sodium, 19485; Skiodan Sodium, 28570
126-45-4: itrol, 15433; silver citrate, 28422
126-58-9: Dipentaerythritol, 8581
126-72-7: tris (2,3-dibromopropyl) phosphate, 32402
126-73-8: Antifoam T, 2551; Kronitex® TBP, 16434; Phos-Ad 100, 23686; Syn-O-Ad® 8412, 30553; tributyl phosphate, 32207
126-80-7: CB2405, 5474
126-86-3: Surfynol® 104, 30376; Surfynol® 104A, 30377; Surfynol® 104BC, 30378; Surfynol® 104E, 30379; Surfynol® 104H, 30380; Surfynol® 104PA, 30381; Surfynol® 104PG, 30382; Surfynol® 104S, 30383; Surfynol® DF-110D, DF-110L, 30392; Surfynol® PC, 30403; Surfynol® PG-50, 30404; Surfynol® TG, 30407; Surfynol® TG-E, 30408
126-92-1: Empicol® 0585/A, 10302; Niaproof® Anionic Surfactant 08, 21092; Rhodapon® BOS, 26632; Sulfetal 4105, 29981
126-99-8: Baypren® 110, 3828; Baypren® 110 VSC, 3829; Baypren® 216, 3830; Baypren® 310, 3831; Baypren® AT-H, AT-M, AT-S, 3832; Baypren® EM1, 3833; Baypren® Latex KA 8348, 3834; Baypren® Latex L 200A, 3835; Baypren® M1, 3836; chloroprene, 6136; Daubond DC-9300, 7794; Neoprene Latex 115, 20952; Neoprene NPG 6856, 20955; polychloroprene, 24386
127-08-2: potassium acetate, 24733, 24734; sal diureticum, 27246
127-09-3: Niacet Sodium Acetate Anhyd. Tech, 21089; sodium acetate, 28726
127-17-3: pyruvic acid, 25537, 25538; Distillex DS4, 8720;
127-18-4: Dowper, 8942; Nema, 20861; perchlorethylene, 23220, 23222; Perclene, 23225; Perclene TG, 23226; Perklone, 23321
127-20-8: Basfapon®, 3705; Dowpon, 8943
127-25-3: Abalyn, 24
127-39-9: Aerosol® IB-45, 891, 892; Bevaloid 6423, 4083; Geropon® CYA/DEP, 12906; Monawet MB-45, 20178, 20179; Octosol IB-45, 22026; Rewopol® SBDB 45, 26371
127-40-2: xanthophyll, 34591
127-52-6: Chloramine B, 6085
127-65-1: Antibacterin, 2537; Chloramine T, 6086; Chlorozone, 6160; Ketjensept, 16133; Tochlorine, 31916
127-68-4: Ludigol F, 17772; sodium m-nitrobenzene sulfonate, 28795
127-91-3: β-l-pinene, 23858, 23859
128-03-0: Aquatreat KM, 2760
128-04-1: Aquatreat SDM, 2761; methyl namate®, 19551; Octopol SDM-40, 22014; Perkacit® SDMC, 23288; sodium dimethyldithiocarbamate, 28758; Thiostop N, 31717
128-09-6: N-chlorosuccinimide, 6148
128-37-0: Anti-Oxydant Bayer, 2595; BHT, 4122; Butylated hydroxytoluene, 4788; DBPC, 7813; Deenax, 7874; Ionol, 15226; Lowinox® BHT, 17664; Nipanox® BHT, 21273; Oxyguard, 22457; Ralox® BHT food grade, 25845; Spectratech® CM 11340, KM 11264, 29240; Tenamine 3, 31092; Vanox® PCX, 33346; Vulkanox® KB, 34152
128-39-2: 2,6-di-*t*-butylphenol, 4799; 2,6-dibutylphenol, 8370; Isonox® 103, 15382; Lowinox® 001, 17646
128-44-9: saccharin sodium, 27196
128-49-4: Surfak, 30362
129-16-8: mercurochrome, 19354
129-20-4: Tandearil, 30795
129-46-4: Naganol, 20660
129-63-5: acetrizoate sodium, 307
130-14-3: Aerosol® NS, 894
130-17-6: D.T.S., 7626
131-11-3: dimethyl phthalate, 8508; Fermine, 11631; Kemester® DMP, 15956; Kodaflex® DMP, 16271; Palatinol® M, 22593; Unimoll® DM, 32979; Uniplex 110, 33009
131-13-5: menadiol sodium diphosphate, 19319; Synkavit, 30548; Synkayvite, 30549
131-15-7: DCP, 7818; Dicapryl phthalate, 8375
131-17-9: Dapon 35, 7737; Daponite Sheet, 7739; diallyl phthalate, 8283; Nonflammable Decobest DA, 21448; RX® 1-501N, 27154; RX® 3-1-530, 27156
131-49-7: Angiovist 282, 2457; Hypaque Meglumine, 14705; MD 60, 19187; Reno-M, 26113
131-53-3: Cyasorb® UV 24, 7496; Spectra-Sorb UV 24, 29230
131-54-4: Benzophenone-6, 3978; Uvinul® D-49, 33194
131-55-5: Benzophenone-2, 3977; Uvinul® D-50, 33195
131-56-6: Benzoresorcinol, 3981; Uvinul® 400, 33191
131-57-7: Cyasorb® UV 9, 7495; Escalol® 567, 10789; Lankromark® LE296, 16698, 16699; Neo Heliopan® BB, 20867; Rhodialux A, 26675; Ritaphenone 3, 26823; Spectra-Sorb UV9, 29229; Syntase® 62, 30621; Uvinul® M-40, 33197
131-58-8: 2-methylbenzophenone, 19565
131-73-7: Dipicrylamine, 8607; Hexil, 14031;
131-74-8: Explosive D, 11314
132-27-4: Dowicide A, 8932
132-60-5: cinchophen, 6307
132-86-5: naphthoresorcin, 20726
133-06-2: Captan Granular, 5164; Captan, Captan-Col, Captan-50, Captan-83P, Captan Granular, 5165; Hormone Rooting Powder, 14310; Merpan, 19369; Orthocide, 22328; PP Captan 83, 24839; Vancide® 89, 33302
133-07-3: Folpan, 12209; Folpet, 12210; Phaltan, 23567
133-17-5: Hippuran-:ss131:ksl, 14125; Hipputope, 14128; Nephroflow, 20989
133-32-4: 4-indol-3-ylbutyric acid, 15086; Chryzoplus, Chryzopon, Chryzosan, Chryzotek, 6258; Seradix, 28005
133-37-9: paratartaric acid, 22803; sal tartar, 27257; tartaric acid, 30832
133-49-3: Akrochem® Peptizer PTP, 1170; Renacit® 7, 26103
133-53-9: Nipacide® DX, 21255; Nipacide® PX, 21260
133-90-4: Naptol, 20739
133-91-5: Diosal, 8572
134-03-2: Cevalin, 5827; Liqui-Cee, 17460; sodium ascorbate, 28734; vitamin C sodium salt, 33939
134-09-8: Neo Heliopan® MA, 20869
134-20-3: methyl anthranilate, 19513
134-31-6: Cryptonol, 7282; Quinosol, 25687
134-32-7: α-naphthylamine, 20729; Aminogen I, 2176
134-50-9: Monacrin, 20097
134-54-3: Cleve's γ-Acid, 6411
134-62-3: diethyl toluamide, 8438
135-19-3: Antioxygene BN, 2600; hydronaphthol, 14623
135-20-6: cupferron, 7365
135-58-0: Odylen®, 22046
135-88-6: Antioxidant PBN, 2591; Antioxygene MC, 2603; P.B.N, 22488; Stabilator A.R, 29400
136-23-2: Accelerator BZ Powder, 198; Butasan Vulcanization Accelerator, 4741; Butazate, 4742; Butazate 50D, 4743; Butyl Zimate®, 4783; Octocure ZDB-50, 21995; Perkacit® ZDBC, 23295
136-26-5: Amidex CP, 2091; Monamid® 150-CW, 20114; Standamid® CD, 29454
136-30-1: Butyl Namate®, 4776; Octopol NB-47, 22012
136-32-3: Dowicide B, 8933;

136-36-7: Eastman® Inhibitor RMB, 9470; Resorcinol benzoate, 26231
136-53-8: Octoate Z, 21992
136-60-7: Butyl benzoate, 4763; Hipochem B-3-M, 14106; Marvanol® Carrier BB, 18980; n-Butyl benzoate, 4790
136-63-0: Bucarpolate, 4647
136-85-6: Retrocure® G, 26279; Vulkalent® TM, 34141
136-99-2: Hodag C-100-L, 14167; lauryl hydroxyethyl imidazoline, 16865; Mackazoline L, 18200; Schercozoline L, 27714
137-08-6: Pantholin, 22693; Ritapan CAP, 26817; vitamin B_5 calcium salt, 33936
137-16-6: Closyl LA 3584, 6443; Crodasinic LS30, 7144; Crodasinic LS35, 7145; Gardol®, 12668; Hamposyl® L-30, 13601; Maprosyl® 30, 18671; Medialan LD, 19214; Sarkosyl NL30, 27465; Secosyl, 27823; sodium lauroyl sarcosinate, 28779; Tensyl 30, 31205; Vanseal® NALS-30, 33359; Zoharsyl L-30, 34876
137-18-8: phlorone, 23676
137-20-2: Adinol OT, 697; Arkopon T, 2898; Fenopon T-33 and T-43, 11608; Fenopon T-51, 11609; Fenopon T-77, 11610; Geropon® T-22/A, 12915; Geropon® T-33, 12916; Hostapon SO, 14395; Hostapon T Powd, 14397
137-26-8: Agrichem Flowable Thiram, 1047; Akrochem® TMTD, 1182; Ancazide ME, 2421; Fernide, 11638; FS Thiram 15% Dust, 12427; Hortag Thiram, 14317; methyl tuads®, 19562; Perkacit® TMTD, 23292; Pomarsol®, 24677; Rezifilm, 26521; Spotret 75 WDG, 29329; Thiram, 31726; Thiurad, 31730; Tuads®, 32542; Tuex, 32553
137-29-1: Akrochem® Cu.D.D, 1165; copper dimethyldithiocarbamate, 6779; Cumate®, 7351; Perkacit® CDMC, 23278
137-30-4: AAprotect, 16; Accelerator MZ Powder, 209; Ancazate ME, 2416; Methasan, 19478; Methazate, 19480; methyl zimate®, 19563; Octocure ZDM-50, 21997; Perkacit® ZDMC, 23297; Ultra Zinc DMC, 32782; Vancide® MZ-96, 33303; Zimate®, 34762
137-40-6: Luprosil® Sodium Salt, 17922; sodium propionate, 28812; Spac, 29182
137-42-8: metam-sodium, 19444; Sistan®, 28550; Unifume, 32954
137-58-6: Lida-Mantle, 17213; lidocaine, 17216
137-97-3: Accelerator A22, 192
138-22-7: Butyl lactate, 4773
138-32-9: Cetats®, 5786
138-55-6: picrocrocin, 23825
138-86-3: Achilles Dipentene, 323; Dipentene, 8584
138-89-6: nitro base, 21341; Vulcaniline, 34095
138-92-1: betazole hydrochloride, 4072
139-05-9: sodium cyclamate, 28754
139-07-1: Catinal CB-50, 5456; Dehyquart LDB, 7986; Retarder N, 26256; Vantoc CL, 33372
139-08-2: BTC® 824, 4636; BTC® 2565, 4639; Catigene DC/100, 5445; Catigene® DC 100, 5455; Exameen 824 3724, 11284; FMB 65-15, 65-28 , 12135; JAQ Powdered Quaternary, 15517
139-13-9: Hampshire® NTA Acid, 13611; nitrilotriacetic acid, 21339; TG Buffer, 31547
139-33-3: disodium EDTA, 8644; edetate disodium, 9613; Endrate, 10553; Hamp-Ene® Na2, 13588; Questal DI 0770, 25661; Sequestrene® NA2, 27997; Sequestrene® NA2 Edetate USP, 27998; sodium versenate, 28831; Trilon® BD, 32325
139-40-2: propazine, 25179; Prozinex, 25350
139-41-3: Hampshire® DEG, 13608; sodium dihydroxyethyl glycinate, 28757
139-44-6: Rheocin, 26557
139-88-8: Niaproof® Anionic Surfactant 4, 21093
139-89-9: Cheelox® 120, 5866; Cheelox® HE-24, 5868; Chel DM-41, 5872; Emkasene 800, 10259; Hamp-Ol® 120, 13594; Hamp-Ol® Crystals, 13596; Kalex OH, 15672; Questal FEC 0800, 25663; Trilon® D Liq, 32329
139-96-8: Akyposal TLS42, 1322; Avirol® T 40, 3485; Carsonol® TLS, 5340; DeSonol T, 8191; Elfan® 240T and 240T/S, 9815; Empicol® TL40, 10342; Lorol TA and TAR, 17622; Maprofix TLS, 18666; Marlinat® DFL 40, 18852; Neopon LT, 20950; Norfox® TLS, 21555; Nutrapon TLS-500, 21827; Perlankrol® ATL40, 23324; Rewopol® TLS 40, 26383; Rhodapon® LT-6, 26638; Standapol® T, 29481; Stepanol® SPT, 29697; Sulfetal KT 400, 29985; Sulfochem TLS, 30016; TEA-lauryl sulfate, 30873; Tensopol LT, 31193; Texapon T 42, 31480; Ufarol TA-40, 32744; Witcolate TLS-500, 34479; Zoharpon LAT, 34869
140-01-2: Cheelox® 80, 5864; Chel DTPA-41, 5874; Hamp-Ex® 80, 13591; Kalex Penta, 15673; Pentaquest Extra 0685, 23169; Pentaquest OPNA 0256, 23171; Pentasodium pentetate, 23172; Polyquest 80, 24533; Trilon® C Liq, 32328
140-03-4: Flexricin® P-4, 11959; Methyl acetyl ricinoleate, 19508; Naturechem® MAR, 20773
140-04-5: B.A.R, 3561
140-11-4: benzyl acetate, 3985; Plastolin I, 24051
140-31-8: Aminoethylpiperazine, 2173
140-39-6: Narceol, 20742
140-40-9: Aminitrazole, 2164
140-56-7: Bayer 5072, 3767
140-67-0: Estragole, 10897
140-72-7: Acetoquat CPB, 301
140-93-2: Aero 343 Xanthate, 842
141-04-8: diisobutyl adipate, 8460; Plasthall® DIBA, 23996
141-08-2: Aldo® MR, 1542; Flexricin® 13, 11951; glyceryl monoricinoleate, 13085; Hodag GMR, 14183; Mazol® GMR, 19137; Softigen® 701, 28854
141-20-8: Glaurin, 12978; PEG-2 laurate, 23041
141-21-9: Avistin® PN, 3489; Chemical 39 Base, 5959; Marlamid® A 18, 18738
141-22-0: Flexricin® 100, 11954; P®-10 Acid, 22499; ricinoleic acid, 26734
141-23-1: Cenwax® ME, 5693; Naturechem® MHS, 20774; Paricin® 1, 22833
141-24-2: Flexricin® P-1, 11957; methyl ricinoleate, 19558
141-32-2: n-Butyl Acrylate, 4786
141-43-5: Colamine, 6593; ethanolamine, 10930
141-53-7: Formax, 12268; sodium formate, 28767
141-57-1: CP0800, 6935; n-propyltrichlorosilane, 25232
141-62-8: CD3780, 5501; D3780, 7636
141-78-6: ethyl acetate, 11063
141-82-2: malonic acid, 18491
141-86-6: 2,6-diaminopyridine, 8305; diaminopyridine, 8306
141-94-6: Hexetidine, 14030; Oraldene, 22263; Sterisil, 29739
142-08-5: 1-hydroxy-2-pyridine, 14669
142-17-6: calcium oleate, 4961
142-18-7: Ablunol GML, 101; Aldo® MLD, 1540; glyceryl monolaurate, 13083; Grindtek ML 90, 13410; Hodag GML, 14180; Imwitor® 312, 14941; Kessco® GML, 16110; Monomuls® 90-L12, 20239
142-26-7: Acetamide MEA, 283; Amidex AME, 2086; Carsamide® AMEA, 5320; Incromectant AMEA-100, 14976; Lipamide MEAA, 17326; Mackamide AME-75, AME-100, 18152; Schercomid AME, 27648
142-31-4: Rhodapon® OLS, 26639
142-47-2: MSG, 20416
142-48-3: Hamposyl® S, 13605
142-54-1: Alkamide® LIPA/C, 1647; Amidex LIPA, 2096; Empilan® LIS, 10409; Monamid® LIPA, 20119; Rewomid® IPL 203, 26326
142-55-2: Emalex PGML, 9985
142-59-6: Campbell's Nabam Soil Fungicide, 5066
142-62-1: caproic acid, 5127
142-71-2: cupric acetate, 7372; verdigris, 33571
142-72-3: magnesium acetate, 18357
142-77-8: Butyl oleate, 4778; Butyl Oleate C-914, 4779; Emerest® 2328, 10089; Kemester® 4000, 15943; Plasthall® 503, 23989; Priolube 1405, 25014; Uniflex® BYO, 32940; Witcizer 100, 34453
142-78-9: Ablumide LME, 80; Alkamide® L-203, 1644; Amidex LMMEA,

2097; Crillon LME, 7050; Empilan® LME, 10410; Hartamide LMEA, 13662; Incromide LCL, 14989; Mackamide LMM, 18158; Monamid® LMA, 20120; Ninol® LMP, 21237; Rewomid® L 203, 26328
142-82-5: Exxsol® Heptane, 11360; n-heptane, 13790
142-87-0: Atlasol 103, 3341; Empicol® 0758, 10303
142-90-5: Ageflex FM-12, 986
142-91-6: Emerest® 2316, 10086; isopropyl palmitate, 15410;Kessco® IPP, 16116; Lexol IPP, 17151; Liponate IPP, 17379; Propal, 25168; Radia® 7200, 25730; Tegosoft® P, 31026; Unimate® IPP, 32972; Wickenol® 111, 34379
143-07-7: Emery® 650, 10134; Hystrene® 9512, 14766; lauric acid, 16856; Philacid 1200, 23660; Prifrac 2920, 24952
143-18-0: Emkapol PO-18, 10248; Marley Cement Waterproofer, 18829; Octosol 449, 22018; potassium oleate, 24763, 24764
143-19-1: Eunatrol, 11194; sodium oleate, 28802
143-24-8: Ansol E-181, 2495
143-27-1: Amine 16D, 2130; Armeen® 16, 2935; Armeen® 16D, 2936, 2937; Crodamine 1.16D, 7101; Kemamine® P-880D, 15919
143-28-2: Adol® 85, 786; Cachalot® O-15, 4880; Dermaffine®, 8093; Emery® 3312, 10162; Emery® 3317, 10163; Fancol OA-95, 11450; HD-Echelon 90/95, 13718; HD-Eutanol, 13719; Lipocol O, 17355; Novol, 21730; oleyl alcohol, 22131
143-33-9: Cymag, 7577; sodium cyanide, 28753
143-74-8: phenol red, 23605; phenolsulfonphthalein, 23611
144-19-4: TMPD® Glycol, 31905
144-21-8: arrhenal, 3128; Arsinette, 3145; Methar 30, 19477
144-34-3: methyl selenac, 19559
144-55-8: Baking soda, 3600; Bufferight, 4656; Col-Evac, 6601; saleratus, 27262; sodium bicarbonate, 28736; Tronacarb Sodium Bicarbonate, 32461
144-62-7: oxalic acid, 22413
144-87-6: Astryl, 3267
147-73-9: *meso*-tartaric acid, 30831; mesotartaric acid, 19400
147-85-3: proline, 25129
148-01-6: Zoamix, 34844
148-18-5: ethyl namate®, 11078; Octopol SDE-25, 22013
148-24-3: 8-hydroxyquinoline, 14670
148-25-4: chromotrope acid, 6244
148-75-4: R-Acid, 25710
148-79-8: Hymush, 14689; Nemacin, 20862; Storite, 29796; Tecto, Tecto 60%, 30913; thiabendazole, 31661; Thibenzole, 31668; Tubazole, 32545
149-30-4: Accelerator Mercapto, 207; Captax, 5166; MBT, 19170; MBT®, 19171; mercaptobenzothiazole, 19335; Perkacit® MBT, 23285; Rokon, 26918; Rotax®, 26999; Thiotax, 31719; Vulkacit® M, 34133; Vulkacit® M, Vulkacit® Merkapto/C, 34134
149-32-6: lichen sugar, 17204
149-39-3: Hostapon STT Paste, 14396; sodium methyl stearoyl taurate, 28792
149-44-0: Hyraldite C Ext, 14730; sodium formaldehydesulfoxylate, 28766
149-57-5: 2-ethylhexoic acid, 11137
149-74-6: CM8930, 6461
149-91-7: Gallic acid, 12610
150-13-0: Pabagel, 22533; Pabanol, 22534; RVPaba Lipstick, 27153
150-30-1: DLPA 375, 8748
150-38-9: edetate trisodium, 9614; Hamp-Ene® Na3 Liq, 13589; Sequestrene® NA3, 28000
150-76-5: Eastman® HQMME, 9467; hydroquinone monomethyl ether, 14633
150-84-5: Cephreine, 5694
150-86-7: phytol, 23780
150-88-9: Vulcaid 27, 34073
151-06-4: Lucofen S A, 17768
151-21-3: Akyposal NLS, 1314; Alscoap LN-40, LN-90, 1853; Calfoam SLS-30, 4986; Carsonol® SLS Paste B, 5339; Chemsalan NLS 30, 6029; Dreft, 8976; Drene, 8977; Empicol® 0303, 10301; Empicol® LS30, 10327; Empicol® LXV, 10329; Empicol® LY28/S, 10330; Empicol® LZG 30, 10332; Empicol® LZP, 10333; Empimin® LR28, 10425; Hartenol LAS-30, 13669; Marlinat® DFK 30, 18851; Naxolate WA-97, 20818; Norfox® SLS, 21552; Nutrapon DL 3891, 21817; Nutrapon W 1367, 21829; Nutrapon WAQE 2364, 21830; Octosol SLS, 22027; Orvus WA, 22346; Perlankrol® DSA, 23326; Polystep® B-3, 24596; Rewopol® NLS 15 L, 26368; Rhodapon® 101-10, 26631; Rhodapon® LCP, 26636; sodium lauryl sulfate, 28781; Standapol® WAQ-LC, 29482; Stepanol® ME Dry, 29695; Stepanol® WA Extra, 29698; Sulfetal C 38, 29982; Sulfochem SLN, 30012; Sulfochem SLP-95, 30013; Sulfochem SLS, 30014; Sulfochem SLX, 30015; Sulfopon 101, 101 Special, 30023; Sulfopon 101/POL, 30024; Sulfopon P-40, 30027; Sulfotex LCX, 30032; Supralated ME, 30311; Texapon K-12, K-1296, L-100, 31461; Texapon K-1296, 31462; Texapon LS Highly Conc, 31464; Texapon OT Highly Conc. Needles, 31472; Texapon VHC Needles, ZHC Needles, 31481; Texapon ZHC Needles, 31482; Texapon ZHC Needles, 31483; Ufarol Na-30, 32743; Witcolate 6400, 34469
151-38-2: Panogen M, 22682
151-50-8: potassium cyanide, 24751
151-73-5: betamethasone 21 phosphate, 4060
152-16-9: Schradan, 27731; Sytam, 30667
152-47-6: Kelfizina, 15838
152-84-1: madder, 18289; rubianic acid, 27103
153-18-4: rutin trihydrate, 27150
154-42-7: Lanvis, 16771
155-04-4: Octocure ZMBT-50, 21998; Oxaf, 22408; Perkacit® ZMBT, 23298; Vulkacit® Merkapto/MGC, 34135; Zetax®, 34741; ZMBT, 34842
155-58-8: rhapontin, 26530
156-43-4: Phenetidine, 23589
156-54-7: sodium butyrate, 28742
156-62-7: lime nitrogen, 17268
206-44-0: Idryl, 14841
218-01-9: Chrysene, 6252
280-57-9: Dabco® 33-LV, 7640; Dabco® Crystalline, 7652; TEDA-L33, 30916; Tegoamin® 33, 30994; Texacat® TD-33, 31433
288-32-4: Imidazole, 14895
288-88-0: pyrrodiazole, 25534
289-95-2: miazine, 19625; pyrimidine, 25484
298-00-0: Folidol® M, 12200; Metacide, 19409; methyl parathion, 19556; Parathion-methyl, 22808
298-02-2: Terrathion, 31343
298-04-4: Disyston® FE-10, 8726; Solvigran, 29063
298-06-6: EP-1, 10611
298-14-6: K-Lyte, 16233
298-81-7: Oxsoralen-Ultra, 22437
298-83-9: Nitro BT, 21342
299-27-4: Gluconal® K, 13043; Jaon, 15507; Katorin, 15798; potassium d-gluconate, 24752; Potassium gluconate, 24755
299-28-5: Gluconal® CA A, CAM, 13038; Neo-Calglucon, 20877
299-29-6: Gluconal® FE, 13042
299-84-3: Korlan, 16351; Nankor, 20677; ronnel, 26951; Trolene, 32443
300-54-9: muscarine, 20524
300-76-5: Bromex, 4569; Dibrom, 8356
300-92-5: Synpro® 303, 30592
301-02-0: Armid® O, 2964; Crodamide O, OR, 7097; Kemamide® O, 15898; oleamide, 22107; Petrac® Slip-Eze, 23468; Polydis® TR 121, 24411; Unislip 1759, 33045
301-04-2: Goulard powder, 13216; lead acetate, 16902; Ledac, 16948; sal Saturni, 27253; salt of Saturn, 27287; sugar of lead, 29955
301-10-0: Metacure® T-9, 19416
301-12-2: Metasystox R, 19461; Metasystox® R, 19463
301-13-3: Syn-O-Ad® P-374, 30563
302-01-2: Amerzine®, 2047; hydrazine, 14545; Scav-Ox® 35%, 27552; Scav-Ox® II, 27553; Ultra Pure, 32780; Zerox, 34738
304-55-2: MPI DMSA Kidney Reagent, 20393
304-59-6: Rochelle salt, 26882; sal Rupellensis, 27252
305-03-3: Leukeran Tablets, 17029
306-83-2: Dichlorotrifluoroethane, 8393;

Genetron® 123, 12823
306-92-3: Flutec PP9, 12124
307-35-7: Fluorad® FX-8, 12067
309-00-2: Aldrin, 1558; Aldrin Dust, 1559; Murald, 20502
311-45-5: paraoxon, 22783
314-40-9: bromacil, 4564; Hyvar X, 14793; Hyvar® X, 14794; Uragan, 33105
321-55-1: Haloxon, 13567; Loxon, 17687
328-50-7: α-ketoglutaric acid, 16137
330-54-1: Direx® 4L, 8620; Diurex, 8734; diuron, 8736; Diuron Bayer, 8737; Karmex, 15771; Karmex®, 15772
330-55-2: Afalon, 918; Ashlade Linuron, 3195; Atlas Linuron, 3318, 3319; Campbell's Linuron 45%, 5061; Linex® 4L, 17294; Linurex, 17316; Linuron, 17317; Linuron 15, 17318; Linuron 450 FL, 17319; Norunil, 21592; Rotalin, 26997; Rotalin, 26998; Du Pont Linuron 50, 4L, 9073
333-41-5: Agridin 60, 1053; Antlak, 2629; Diazinon Liquid, 8344; Diazitol, Diazitol Liquid, 8345; Diazol, 8347; Diziktol, 8745; Neocidol Veterinary Powder, 20880; Root Guard, 26957
334-48-5: n-Capric acid, 5126; Prifrac 2906, 24951
335-27-3: Flutec PP3, 12123
335-67-1: Fluorad® FC-26, 12060;
338-98-7: isoflupredone acetate, 15362; Predef, 24870
339-44-6: Lycanol, 18061
353-59-3: BCF, 3864
354-33-6: Genetron® HFC 125, 12828
355-02-2: Flutec PP2, 12122
355-42-0: Flutec PP1, 12121
356-12-7: Lidex, 17215
371-41-5: Fluorophenol, 12109
373-02-4: nickel acetate, 21118
379-79-3: Lingraine, 17296; Migraine Dolviran®, 19744
386-17-4: Kerasol, 16058
409-21-2: Carbogran, Carbogran E, Carbogran UF, 5201; Carbomant, 5210; Carborex, 5237; Carsilon, 5326; Exolon, 11302; Lonsicar, 17585; Meccarb, 19202; Resilon, 26172; silicon carbide, 28329; silundum, 28406; Simax, 28463
431-03-8: Diacetyl, 8250
435-97-2: Liquamar, 17450; Phenprocoumon, 23630
437-50-3: gentisin, 12847
440-58-4: Jodomiron, 15615; Uromiro (n), 33147
443-48-1: Satric, 27491
461-58-5: Dicyandiamide, 8411
462-06-6: Fluorobenzene, 12093
462-94-2: Cadaverine, 4883
463-40-1: linolenic acid, 17306
463-71-8: thiophosgene, 31711
464-10-8: Bromopicrin, 4585
467-14-1: neopine, 20939
470-90-6: Birlane, 4220, 4221; Chlorfenvinphos, 6098; Sapecron, 27436;Sedanox, 27831;
471-25-0: propiolic acid, 25189
471-34-1: Aeromatt, 863; Albacar, 1342; Amical® 101, 2060; Amical® 85, 2059; Amical® SC, 2062; Atomite, 3378; Calofil, 5016; Calofort®, 5017; Calopake®, 5020; Camel-CAL®, Camel-CARB®, Camel-FIL, Camel-FINE, 5047; Camel-TEX®, CamelWITE®, 5049; Carbital®, 5191; Carbital® 35, 5192; Carbital® 50, 5193; CC-103, 5485; CP Filler, 6928; Drikalite®, 9031; Duramite®, 9161; Hallcote® 573, 13554; Hi-Pflex® 100, 14103; Hubercarb® Q 6-20, 14453; Hubercarb® W 2, 14454; Iceland spar, 14813; Kalvan, 15701; Kotamite®, 16370; Macromite, 18281; Marbledust, 18691; Marblemite, 18693; Micro-White® 07 Slurry, 19712; Micro-White® 10 Codex, 19713; Micro-White® 100, 19717; Micro-White® 15, 19714; Micro-White® 25, 19715; Micro-White® 40, 19716; mild lime, 19758; No. 1 White, 21410; No. 3 White, 21411; Opacicoat, 22195; Opacimite, 22196; sandscale, 27362; SC-53, 27535; Slab Dip AC699, 28585; Snowflake P. E., 28675; Snowflake White, 28676; Socal, 28703; statuary marble, 29575; Sturcal®, 29846; Super-Pflex® 100, 30261; T-Carb, 30855; Trucal, 32486; Trucarb, 32487; Ultramite, 32841; Ultra-Pflex®, 32856
471-35-2: Cacodyl, 4882
473-34-7: Dichloramine T, 8382
475-38-7: naphthazarin, 20713
478-43-3: rhein, 26534
479-45-8: Tetryl, 31423
481-39-0: Juglone, 15631
482-44-0: imperatorin, 14912; peucedanin, 23541
482-89-3: Indigo, 15074
483-65-8: retene, 26271
485-31-4: Morocide, 20340
485-35-8: ulexine, 32761
485-47-2: ninhydrin, 21230
486-35-1: Daphnetin, 7736
486-54-4: Kalle's acid, 15684
487-60-5: plant indican, 23915
487-94-5: indican, 15070
488-73-3: acorn sugar, 367
494-47-3: Vulcazol, 34112
495-69-2: Hippuric acid, 14126
496-77-5: Butyroin, 4802
497-19-8: Consal, 6732; mild alkali, 19757; Oxyper, 22465; sal soda, 27255; salt of soda, 27288; scotch soda, 27758; soda ash, 28710; sodium carbonate, 28744; Solvay® Soda, 29041; Trona Soda Ash, 32459; Tronalight Light Soda Ash, 32462
497-25-6: oxazolidine, 22419
498-96-4: guvacine, 13512
499-04-7: arecaidine, 2828
499-12-7: equisetic acid, 10735
499-75-2: carvacrol, 5362; oxycymol, 22445
500-05-0: coumalic acid, 6901
500-89-0: Ciba 1906, 6291; thiambutosine, 31663
501-24-6: Cardolite® NC-507, 5267
501-52-0: hydrocinnamic acid, 14577
501-53-1: benzyl chloroformate, 3988
501-65-5: Tolan, 31933
501-92-8: chavicol, 5861
502-44-3: ε-caprolactone monomer, 5138
503-38-8: Perstoff, 23430
504-15-4: orcinol, 22280
504-24-5: Avitrol, 3494
505-48-6: suberic acid, 29903
505-60-2: mustard gas, 20538
506-59-2: dimethylamine hydrochloride, 8513
506-68-3: cyanogen bromide, 7479
506-87-6: baker's salt, 3598; sal volatile, 27258; salt of Hartshorn, 27284
506-93-4: Guanidine nitrate, 13441
507-70-0: DL-Borneol, 4428
511-28-4: vitamin D+74, 33942
512-69-6: raffinose, 25835
513-35-9: pental, 23147
513-77-9: Barium carbonate, 3636; Durex white, 9208
513-79-1: cobaltous carbonate, 6536
513-86-0: acetyl methyl carbinol, 312
514-10-3: abietic acid, 45; sylvic acid, 30503
514-78-3: Canthaxanthin, 5095
515-69-5: Camilol, 5053; Hydagen® B, 14525
516-18-7: Abrastol, 155; asaprol, 3176
517-25-9: Nitroform, 21359; tetranitromethane, 31386
517-88-4: Alkanet, 1695
518-47-8: Fluor-Amps, 12069; Fluorescein disodium salt, 12081; Fluorescite, 12086; Fluor-I-Strip, 12091; uranine, 33115
519-34-6: maclurin, 18235
519-73-3: tritane, 32420
520-18-3: kaempferol, 15654
520-45-6: Dehydroacetic acid, 7964
522-51-0: Dequalinium Chloride, 8077; Labosept, 16533
523-80-8: parsley camphor, 22865
524-30-1: pavlin, 22928
524-34-5: atoquinol, 3380
525-37-1: Ewer-Pick acid, 11278
525-82-6: flavone, 11903
526-95-4: Gluconic Acid, 13048
527-07-1: Asahi Aji®, 3174; Gluconal® NA, 13046; Naglusol, 20662; sodium gluconate, 28769
527-09-3: copper gluconate, 6781; Gluconal® CU, 13041
528-44-9: trimellitic acid, 32335
529-28-2: Iodanisol, 15204
529-86-2: oxyanthracene, 22440
530-91-6: Tetralol, 31381
532-05-8: 1,3-di-6-quinolylurea, 8618; Acaprin®, 181
532-27-4: chloroacetophenone, 6112
532-32-1: sodium benzoate, 28735
532-82-1: Chrysoidine, 6256
533-74-4: Amerstat® 233, 2039; Dazomet, 7806; N-521® Biocide, 20640
533-96-0: S-Carb, 27549; sodium sesquicarbonate, 28816
534-08-7: iohydrin, 15223
534-16-7: silver carbonate, 28420
534-17-8: cesium carbonate, 5772
534-22-5: Sylvan, 30497
534-26-9: lysidine, 18087
534-52-1: 4,6-dinitrocresol, 8542; Dekryll, 7999

535-89-7: Crimidine, 7053
536-90-3: *m*-anisidine, 2472
537-65-5: Oxynone, 22464
538-23-8: Captex® 8000, 5170; Miglyol® 808, 19736; Radiamuls® MCT 2108, 25776; tricaprylin, 32211
538-24-9: Dynasan® 112, 9375; Trilaurin, 32317
538-75-0: dicyclohexyl carbodiimide, 8412
539-12-8: *p*-anol, 2480
540-10-3: cetyl palmitate, 5822; Crodamol CP, 7119; Cutina® CP, 7437; Emalex CC-16, 9934; Kemester® CP, 15955; Kessco® 653, 16100; Precifac ATO, 24865; Radia® 7500, 25742; Rewowax CG, 26434; Schercemol CP, 27588; Starfol® CP, 29535; Waxenol® 815, 34238
540-59-0: acetylene dichloride, 93; Dioform, 8553;
540-88-5: *tert*-Butyl Acetate, 4785
541-02-6: CD3770, 5500; D3770, 7635
541-05-9: CH7260, 5839
541-91-3: Exaltone, 11282; muscone, 20527; musk, 20531
542-05-2: β-ketoglutaric acid, 16138; ADA, 615
542-18-7: cyclohexyl chloride, 7524
542-55-2: isobutyl formate, 15346
542-75-6: Telone, 31069
542-85-8: ethyl isothiocyanate, 11075; mustard oil, 20539
543-80-6: Barium acetate, 3634
544-17-2: calcium formate, 4948; Latibon®, 16817
544-19-4: cupric formate, 7380
544-63-8: Emery® 654, 10135; Hystrene® 9014, 14764; Hystrene® 9514, 14767; Myristic acid, 20574; Philacid 1400, 23662; Prifrac 2940, 24953
546-56-5: CO9817, 6510
546-80-5: tanacetone, 30775
546-88-3: Lithostat, 17522
546-93-0: Elastocarb Tech Light, Heavy, 9691; Elastocarb UF, 9692; Magocarb-33, 18404
547-91-1: Loretine, 17606
548-62-9: gentian violet, 12845; Pyoctanin, 25455
548-76-5: Irigenin, 15301
549-18-8: Laroxyl, 16801; Lentizol, 16992
550-82-3: diazoresorcin, 8351; Resazurin, 26161
550-97-0: Alphol, 1840
551-92-8: Dimetridazole, 8528; Emtryl, 10482
552-22-7: Iosol, 15237
552-89-6: *o*-nitrobenzaldehyde, 21346
552-94-3: Salysal, 27306
553-08-2: Thonzide, 31746
554-13-2: Liskonum Tablets, 17492; Lithane, 17500
554-71-2: Iodozol, 15221
554-95-0: trimesic acid, 32337
555-16-8: *p*-nitrobenzaldehyde, 21347
555-31-7: Aliso, 1617; Aluminum isopropoxide, 1909
555-37-3: Kloben®, 16225; Neburex, 20840; neburon, 20841; Noruben, 21591
555-43-1: Dynasan® 118, 9378; Kemester® 5500, 15947; Kemester® 6000, 15952; Kemester® GMS (Powd.), 15962; Neobee® 62, 20874
555-44-2: Dynasan® 116, 9377
555-45-3: Dynasan® 114, 9376
556-50-3: N-glycylglycine, 13139
556-61-6: methyl isothiocyanate, 19543
556-67-2: CO9810, 6508; O9810, 21945
556-82-1: Prenol, 24892
557-04-0: Afco-Chem MGS, 923; magnesium stearate, 18367; Petrac® MG-20 NF, 23466; Synpro® 90, 30591
557-05-1: Afco-Chem ZNS, 924; Antidust 2, 2541; DLG-10, 20, 8747; Hallcote® ZS 5050, 13556; Petrac® ZN-41, 23470; Synpro® 8, 30589; zinc stearate, 34793
557-08-4: zinc undecenoate, 34798
563-12-2: Ethion, 10955
563-63-3: silver acetate, 28414
569-61-9: pararosaniline, 22798
569-65-3: Longifene, 17581
575-44-0: 1,6-naphthalenediol, 20709
575-75-7: Tonophosphan, 32013
577-11-7: Aerosol OT, 877; Aerosol® GPG, 882; Aerosol® OT-70 PG, OT-S, 896; Aerosol® OT-75%, 897; Aerosol® OT-MSO, 898; Agrilan® AEC266, 1056; Alcopol O, 1499; Anonaid TH, 2481; Arowet SC-75, 3091; Arylene M40, 3165; Astrowet 0-70-PG, 3265; Astrowet 0-75, 3266; Bevaloid 1299, 4082; Chemax DOSS/70, 5895; Complemix, 6681; Coprol, 6796; Cropol 60, 7210; Dioctyl, 8547; Discol DFW, 8631; Disponil SUS IC 8, 8709; Drewfax® S-700, 8991; Drewfex® 0007, 8992; Elfanol® 883, 9828; Emcol® 4500, 10051; Emcol® DOSS, 10058; Empimin® OP70, 10434; Empimin® OT, 10435; Geropon® CYA/60, 12905; Geropon® DOS, 12907; Hipochem EK-18, 14109; Hodag DOSS-70, 14178; Lankropol® KO2, 16715; Lankropol® KPH70, 16716; Leonil OS, 17002; Mackanate DOS-40, 18186; Manoxol OT, OT/P and OT/B, 18581; Manoxol OT60, 18582; Manoxolot, 18583; Marlinat® DF 8, 18850; Mazawet® DOSS 70, 19118; Modane Soft, 20015; Monawet MO, 20182; Monawet MO-65-150, 20183; Monawet MO-70, 20184; Monawet MO-84R2W, 20185; Ninate® DS 70, 21229; Norval, 21594; Octowet 40, 22035; Pentasol, 23174; Rewopol® SBDO 75, 26374; Rexowet ASG-81, 26505; Schercopol DOS-70, 27666; Schercowet DOS-70, 27711; Secosol DOS/70, 27810; Secosol® DOS 70, 27814; Servoxyl VLA 2170, 28071; Tensuccin D8, 31201; Thorowet G-40 3230, 31755; Triton® GR-5M, 32428; Triumphnetzer ZSG, 32435; Warcowet O, 34212
578-94-9: Adamsite, 624
579-38-4: Entamide, 10593
581-64-6: Lauth's violet, 16870
581-75-9: 2,6-naphthalenedisulfonic acid, 20708; Ebert and Merz's β-acid, 9496
582-17-2: 2,7-naphthalenediol, 20711
582-25-2: potassium benzoate, 24738
583-39-1: 2-mercaptobenzimidazole, 19334; Anti-Oxidant MB, 2589
583-91-5: methionine hydroxy analog, 19486; MHA, 19624
584-08-7: Montreal potash, 20302; pearl ash, 22977; Pearl Dust®, 22978; potassium carbonate, 24743, 24744, 24745; sal absinthii, 27242; salt of wormwood, 27292
584-09-8: rubidium carbonate, 27104
584-79-2: allethrin, 1744
584-84-9: Scuranate, 27776; toluene diisocyanate, 31947; Voranate T-80, Type I, Type II, 34042
584-85-0: anserine, 2490
587-61-1: Dionosil, 8569; Propyliodone, 25225, 25226
588-59-0: Eccobrite RB, 9524; stilbene, 29764
589-82-2: ethylbutylcarbinol, 11119
589-97-9: Antihypo, 2556
590-00-1: Sorbistat K, 29108
591-01-5: Grossmann reagent, 13426
591-50-4: iodobenzene, 15210
592-01-8: calcium cyanide, 4945; Cyanolime, 7480; powdered hydrocyanic acid, 24815
592-41-6: 1-hexene, 14029; Neodene® 6, 20886
593-29-3: potassium stearate, 24783, 24784
593-60-2: Saytex® VBR, 27525
593-84-0: Guanidine thiocyanate, 13442
597-09-1: NEPD, 20988
598-10-7: vinaconic acid, 33766
598-56-1: dimethylethylamine, 8521; N,N-dimethylethylamine, 8522
598-62-9: magnesium carbonate, 18359; manganese white, 18551
598-63-0: lead subcarbonate, 16918; Ledca, 16950
598-64-1: Ultra-DMC, 32791
598-82-3: DL-lactic acid, 16573
599-61-1: 3,3'-diaminodiphenylsulfone, 8303
599-64-4: cumyl phenol, 7355
603-35-0: triphenyl phosphine, 32392
608-66-2: galactitol, 12592; Melampyrite, 19258
611-92-7: Zentralin, 34715
613-78-5: naphthalol, 20712
614-33-5: Mollit B, 20068; Plastic A, 24019; Uniplex 260, 33015
614-45-9: *t*-butyl perbenzoate, 4798; Aztec® *t*-Butyl Perbenzoate, 3543; Esperox® 10, 10844; Polyvel CR-5T, 24659; Trigonox® 93, 32276; Trigonox® C, 32297
615-13-4: 2-indanone, 15063
615-58-7: 2,4-dibromophenol, 8359; FR-612, 12356
616-45-5: pyrrolidone2-pyrrolidone, 25536; Soluphor® P, 29023
617-48-1: apple acid, 2667
617-86-7: CT2523, 7320
618-88-2: 5-nitroisophthalic acid, 21361
619-84-1: 4-(dimethylamino)benzoic acid,

8514
620-05-3: Fraissite, 12374
621-08-9: Preventol® CI 5, 24928
621-71-6: Dynasan® 110, 9374
621-82-9: cinnamic acid, 6313
622-16-2: Stabilizer 2013-P®, 29406
622-45-7: H.A. Solvent, 13522
624-04-4: Emalex EG-di-L, 9949; glycol dilaurate, 13100; Kemester® EGDL, 15958
624-89-5: methyl ethyl sulfide, 19533
624-92-0: Sulfa-Hitech® 0382, 29961
625-58-1: nitric ether, 21333
627-83-8: Alkamuls® EGDS, 1663; Emalex EG-di-S, 9951; Emerest® 2355, 10091; ethylene glycol distearate VA, 11126; Genapol® PMs, 12795; glycol distearate, 13102; Kemester® EGDS, 15959; Kessco® EGDS, 16105; Lexemul® EGDS, 17136; Lipo EGDS, 17338; Mapeg® EGDS, 18641; McAlester EGDS, 19176; Pegosperse® 50 DS, 23065; Rewopal® PG 280, 26348; Rita EDGS, 26789; Secoster® DMS, 27817
627-93-0: dimethyl adipate, 8502
628-63-7: Amyl acetate, 2380
628-86-4: Howard's silver, 14431
629-11-8: Hexamethylene Glycol, 14005
629-70-9: Acelan A, 268
629-73-2: Neodene® 16, 20898
629-76-5: Neodol® 5, 20910
630-56-8: Lutate, 17966
631-61-8: Ammonium acetate, 2226
632-69-9: rose bengal, 26963
632-79-1: Great Lakes PHT4, 13312; Saytex® RB-49, 27524; tetrabromophthalic anhydride, 31361
635-78-9: Resorufin, 26232
637-12-7: Aluminum stearate, 1929; Duroseal, 9258; HiGel, 14081; LoGel, 17556; Med Gel, 19208; Metasap® 537, 19449; Novogel® ST, 21729; Synpro® 404, 30593
637-39-8: triethanolamine hydrochloride, 32241
638-16-4: Zisnet F-PT, 34838
638-38-0: Manal, 18509; manganese acetate, 18536
640-15-3: Ekatin, 9657; thiometon, 31700
645-49-8: *cis*-stilbene, 29765
646-13-9: Emerest® 2324, 10087; Estol 1476, 10893; Kemester® 5415, 15946; Kessco® IBS, 16113; Uniflex® IBYS, 32948
650-51-1: Varitox, 33433
657-27-2: Enisyl, 10583; L-(+)-lysine hydrochloride, 18089
657-84-1: Eltesol® ST 40, 9872; Naxonat® 4ST, 20821; Reworyl® T, 26410; sodium *p*-toluenesulfonate, 28827
661-19-8: Behenyl alcohol, 3899; Dehydag Wax 22 (Lanette), 7946; Emery® 3304, 10160; Loxiol VPG 1451, 17685;
661-61-6: Akyposal TIPA 45, 1321; Rewopol® TLS 90 L, 26384
666-99-9: Agaricic, 956
667-83-4: Humectant SD-35, 14464; Panthenylethyl ether, 22691
670-96-2: 2-phenyl imidazole, 23643
681-84-5: Dynasil® M, 9385
682-01-9: CT2090, 7316
683-10-3: Dehyton® PAB-30, 7993; Zohartaine AB, 34877
686-31-7: Esperox® 570, 10854; Lupersol 575, 17867; Trigonox® 121, 32281;
688-37-9: Olminat, 22143
693-33-4: Alkateric® PB, 1721; Mackam CET, 18141; Mirataine® BET-P-30, 19925
693-36-7: Alkanox® 240-3T, 1704; Argus DSTDP, 2863; Carstab® DSTDP, 5357; Cyanox® STDP, 7493; Distearyl thiodipropionate, 8713; Evanstab® 18, 11258; Lankromark® DSTDP, 16691; Lowinox® DSTDP, 17667
693-98-1: 2-methyl imidazole, 19572
694-83-7: 1,2-diaminocyclohexane, 8301; Diaminocyclohexane, 8300
696-59-3: dimethoxy Tetrahydrofuran, 8499; Protectol® DMT, 25260
709-98-8: Propanil, 25174; Sorpur®, 29148; Surcopur, 30331
719-96-0: Fluorfolpet, 12087
731-27-1: Euparen® M, 11198
737-31-5: diatrizoate sodium, 8337; MD 50, 19186; Urovist, Urovist Sodium 300, 33150
756-79-6: dimethylmethyl phosphonate, 8525; Fyrol® DMMP, 12539
757-84-1: Manro STS, 18606
759-94-4: Eptam 6E, 10728; EPTC, 10729
762-16-3: Perkadox® SE-8, 23310
762-72-1: CA0570, 4848
763-69-9: Ektapro® EEP Solvent, 9659; ethyl 3-ethoxypropionate, 11062
768-33-2: CP0160, 6931
770-35-4: phenoxy1-Phenoxy-2-propanol, 23625
780-69-8: CP0320, 6933
786-19-6: Trithion, 32422
793-24-8: Akrochem® Antiozonant PD-2, 1164
804-36-4: nitrovin, 21382
811-54-1: lead formate, 16908; Ledfo, 16959
813-94-5: calcium citrate, 4944
814-91-5: cupric oxalate, 7385
814-94-8: tin(II) oxalate, 31816
816-94-4: Phospholipon® SC, 23730
822-16-2: sodium stearate, 28820
828-00-2: dimethoxane, 8498; Givgard DXN, 12951
834-12-8: Ametrex, 2050; ametryn, 2051
842-07-9: Sudan I, 29938
849-99-0: Sipalin AOC, 28512
852-19-7: Vesulong, 33700
860-07-1: Bromeikon, 4568
860-22-0: Indigo Carmine, 15075
866-84-2: potassium citrate, 24750
868-14-4: potassium bitartrate, 24740; soluble cream of tartar, 28979
868-18-8: sodium tartrate, 28825
868-77-9: Bisomer 2HEMA, 4273; hydroxyethyl methacrylate, 14662; Sipomer® HEM-D, 28536
870-72-4: Rongalit® C, 26945
871-37-4: Chembetaine OL-30, 5935; Incronam OD-50, 15022; Mackam OB-30, 18148; Mafo® OB, 18303; oleyl betaine, 22132
872-05-9: Neodene® 10, 20891
872-50-4: Agsol Ex1, 1091; Micropure® Ultra, 19694; N-methyl-2-pyrrolidone, 19578; NMP, 21406; PartsPrep+sn Degreaser, 22872
873-94-9: dihydroisophorone, 8457
874-60-2: *p*-toluoyl chloride, 31957
877-66-7: Celogen® TSH, 5645
886-50-0: Clarosan 1FG, 6372; Prebane 500, 24861; Terbutrex, 31229; terbutryn, 31230
900-77-6: Cestarsol, 5779
919-30-2: Aktisil AM, 1201; CA0750, 4852; Dynasylan® AMEO, Dynasylan® AMEO-P, 9396; γ-aminopropyltriethoxysilane, 2186; Prosil® 220, 25237; Union Carbide® A-1100, 32990
919-86-8: Azotoz 580, 3538; demeton-S-methyl, 8055; Demetox, 8056; Metasystox 55, 19460; Metasystox® I, 19462; Mifatox, 19727; Power DSM, 24822; Quad DSM, 25565; Vassgro DSM, 33469
921-53-9: soluble tartar, 28987
922-80-5: Aerosol® AY-65, 889
924-42-5: NM-AMD, 21404
927-07-1: Aztec® *t*-Butyl Peroxypivalate-75 OMS, 3548; Esperox® 31M, 10848; Lupersol 11, 17843; Trigonox® 25-C75, 32268
927-83-3: VR-160, 34057
928-24-5: Emalex EG-di-O, 9950
928-65-4: CH7332, 5842
929-06-6: Diglycolamine® Agent (DGA®), 8454
929-77-1: Kemester® 9022, 15953
931-36-2: 2-ethyl-4-methyl imidazole, 11139; EMI-24, 10190
933-75-5: 2,3,6-trichlorophenol, 32220
940-41-0: CP0110, 6929
944-22-9: Dyfonate, 9316
947-42-2: CD6150, 5516
950-10-7: Cytro-Lane, 7606; mephosfolan, 19326
950-37-8: methidathion, 19483; Supracide, 30299; Suprathion, 30318
957-51-7: Enide, 10581
973-21-7: Acrex, 384; Dinobuton, 8543
989-38-8: Rhodamine G and G Extra, 26617
994-30-9: CT2520, 7319
995-33-5: Butyl bisbutyl peroxy valerate, 4765; Luperco 230-XL, 17824; Lupersol 230, 17855
996-50-9: CD4450, 5504
998-30-1: CT2500, 7317
999-21-3: diallyl maleate, 8282
999-81-5: ABM Chlormequat 40, 72.5, 142; Ashlade 4-60 CCC, 700 CCC, 3185; Atlas Chlormequat 46, 700, 3302; CCC 700, 5497; Chlormequat chloride, 6110; Cleanacres PDR 675, 6382; Clifton Chlormequat 46, 6415; Cycogan, 7545; Fargro Chlormequat, 11471; Farmacel, 11478; Farmacel, Farmacel 645, 11477; Headland Swift, 13735; Hyquat 70, 75, 14729; Mandops Barleyquat B, 18520;

Mandops Bettaquat B, 18521; Mandops Spring Poquaternary, 18525; MSS Chlormequat 40, 46, 60, 70, 20427; Portman Chlormequat 400, 460, 600, 700, 24711; Power 64, 640, 700, 24817; Quadrangle Chlormequat 700, 25575; Standup, 29487; Star Chlormequat, 29523; Terpal® CC, M, 31310; Titan, 31859; Tripart® Brevis, 32366; Tripart® Chlormequat 460, 32367

999-97-3: Dynasylan® HMDS, 9409; H7301, 13529; hexamethyldisilazane, 14003;Prosil® HMDS, 25242

1000-50-6: CB2785, 5480

1000-90-4: Propyl Zithate®, 25206; Zithate®, 34839

1002-89-7: Ammonium stearate, 2253

1009-93-4: CH7250, 5838; H7250, 13528; Hexamethylcyclotrisilazane, 14001

1014-69-3: desmetryn, 8142; Semeron, 27936

1025-15-6: triallyl isocyanurate, 32192

1055-55-6: bunamidine hydrochloride, 4680; Scolaban, 27745

1066-30-4: chromium acetate (ic), 6229

1066-33-7: ABC Trieb, 34; Ammonium bicarbonate, 2231

1066-35-9: CD5470, 5507

1067-25-0: CP0810, 6936

1067-33-0: Metacure® T-1, 19414

1067-53-4: Aktisil VM, 1205; CV-5000, 7453; Dynasylan® VTMOEO, 9422; Union Carbide® A-172, 32985

1068-27-5: Aztec® 2,5-Tri, 3552; Esperal® 230, 10838; Lupersol 130, 17847; Polyvel CR-L10, 24660; Trigonox® 145, 32286

1071-27-8: 3-cyanopropyltrichlorosilane, 7482; CC3555, 5495

1071-83-6: Glyphogan, 13147; glyphosate, 13148; Roundup, 27011; Spasor, 29213; Sting, 29768

1072-35-1: Hal-Lub-N, 13559; Haro® Chem P28G, 13642; lead stearate, 16917; P-289, 22504; P-51, 22518

1076-38-6: 4-hydroxycoumarin, 14661

1077-56-1: Uniplex 108, 33008

1085-12-7: Nipaheptyl, 21272

1085-98-9: Dichlofluanid, 8379; Elvaron®, 9893; Euparen®, 11197

1087-21-4: Dapon M, 7738

1094-08-2: Lysivane, 18091

1111-67-7: cuprous thiocyanate, 7407

1112-39-6: CD5605, 5509

1113-02-6: Folimat®, 12201; Omethoate, 22155

1119-97-7: myrtrimonium bromide, 20589; Mytab®, 20595; Querton 14Br-40, 25649; Rhodaquat® M214B/99, 26642; Sumquat® 6110, 30125

1120-02-1: Zeonet B, 34721

1120-28-1: Kemester® 2050, 15941

1120-36-1: Neodene® 14, 20897

1120-49-6: Armeen® 2-10, 2926; Radiamine 6310, 25761

1122-62-9: 2-Acetyl pyridine, 321

1128-08-1: dihydrojasmone, 8458

1143-38-0: anthranol, 2530

1151-11-7: Oragrafin Calcium, 22260; Solu-Biloptin, 28977

1156-19-0: Tolinase, 31936

1163-19-5: Decabromodiphenyl oxide, 7837; FR-1210, 12365; Great Lakes DE-83R, 13304; Octoguard FR-01, 21999; Saytex® 102E, 27514; Thermoguard® 505, 31632

1166-52-5: Progallin LA, 25110

1185-53-1: tris (hydroxymethyl) aminomethane hydrochloride, 32404

1185-55-3: CM9100, 6465; Dynasylan® MTMS, 9415; Union Carbide® A-163, 32983

1187-56-0: Sethotope, 28098

1190-63-2: cetyl stearate, 5825; Schercemol CS, 27589;

1193-65-3: 3-quinuclidinone hydrochloride, 25693

1194-65-6: BH Prefix D, 4120; Casoron, 5394; Casoron G, 5395; Casoron G4, 5396; Fydulan, Fydumas, Fydusit, 12532; Prefix D, 24876

1197-18-8: Amikapron, 2121

1210-35-1: Dibenzosuberone, 8354

1221-56-3: Bilivist, 4144; Oragrafin Sodium, 22261

1225-20-3: Angio-Conray, 2456

1229-29-4: Curatin, 7415

1241-94-7: diphenyl octyl phosphate, 8597

1260-17-9: carminic acid, 5294

1300-71-6: xylenol, 34625

1300-72-7: Eltesol® SX 30, 9873; Esi-Terge SXS, 10831; Hartotrope SXS 40, Powd, 13694; Manro SXS, 18607; Naxonate® 4L, 20824; Naxonate® SX, 20828; Nutrol SXS 5418, 21849; Pilot SXS-40, 23843; Reworyl® NXS 40, 26409; sodium xylene sulfonate, 28832; Stepanate® SXS, 29674; Sulfotex SXS-40, 30036; Witconate SXS 40%, 34492

1302-42-7: Amerfloc® 2, 2007; Dynaflock, 9344; Dynagrout, 9348; Manfloc, 18528; sodium aluminate, 28731

1302-78-9: Ben-A-Gel®, 3929; Bentonite, 3954; BentoPharm, 3955; Brebent, 4496; Bregel, 4501; Fulbent, 12434; Fulbond, 12435; Gadorgel, 12563; K 129-H, 15640; Korthix, 16362; Korthix H-NF, 16363; MicroForm B, 19672; MicroForm BCS, 19673; Montigel, 20299; Polargel® HV, 24295; Polargel® NF, 24296; Wilkinite, 34409; Yellowstone, 34652

1303-28-2: arsenic pentoxide, 3140

1303-32-8: realgar, 25930; red algar, 25950; sandaracha, 27325

1303-33-9: orpiment, 22317

1303-34-0: arsenic pentasulfide, 3139

1303-86-2: Boric anhydride, 4423

1303-96-4: Borax, 4411

1304-28-5: Barium oxide, 3641; baryta, 3675

1304-29-6: Barium peroxide, 3642

1304-56-9: Berylla, 4053; Glucina, 13035

1304-76-3: Bismuth oxide, 4240; Bismuth trioxide, 4245; flowers of bismuth, 12032

1304-85-4: Bismuth subnitrate, 4243; flake white, 11871; magister of bismuth, 18320

1305-62-0: calcium hydroxide, 4952; Edelwit, 9601; lime water, 17270; Red Hot Pellets, 25959; slaked lime, 28591; Trulime, 32493

1305-78-8: calcium oxide, 4962; Caloxol CP2, 5023; Dynacal, 9337; KM Pebble Lime, 16240; lime, 17265

1305-79-9: calcium peroxide, 4964; Fertilox, 11707; Oxy-Gro, 22456

1305-99-3: Photophor®, 23763

1306-06-5: Alveograf, 1939; Periograf, 23274

1306-19-0: cadmium oxide, 4901

1307-52-4: ruthenium red, 27148

1307-96-6: cobaltous oxide, 6540; prepared cobalt oxide, 24897

1308-04-9: cobaltic oxide, 6532

1308-14-1: Guignet's Green, 13456

1308-38-9: Accrox, 258; chromic oxide, 6223; chromium oxide (ic), 6236; M100, 18107

1309-37-1: Black 103, 4301; Blackox, 4307; Crocus Martius, 7062; Deanox, 7830; Disperfin, 8666; ferric oxide, 11651; Ferroxide, 11694; Foundrox, 12344; Hydroferrox®, 14586; Indian Ocher, 15067; Jeweller's rouge, 15611; Kroma Red, 16418; Magnox, 18389; Mapico, 18654; No Vein Compound, 21409; ocher, 21976;Octotint 103, 22030; OSO® 1905, 22364; OSO® 440, 22363; Pferrico, 23555; Pferrisperse, 23556; Pferrox, 23560; Rainbow Custom Colored Mortars, 25836; Red 139, 25948; rouge, 27009; sal mineral, 27248; sienna, 28249; specularite, 29256; Transoxide, 32143; trip, 32358; Two Cubed Eight, 32638

1309-42-8: FR-20, 12352; Gilumag, 12944; Magnaspheres®, 18348; Magnifin® H10, 18378; Magnifin® H10C, 18379; Magnifin® H7A, 18377; Magoh-S, 18405; MGH-93, 19622; Mylanta, 20564; Neutramag®, 21031; Versamag® DC, 33607; Zerogen® 10, Zerogen® 60, 34733

1309-48-4: Burnt magnesia, 4714; Corox, 6846; Dynamag, 9350; Dynatherm, 9425; Elastomag® 100, 9704; Encapsulated MgO, 10538; Flamarret®, 11872; Insulmag®, 15152; Ken-Mag®, 16007; MagChem® 1060, 18312; MagChem® 10B, 18310; MagChem® 20, 18311; Maglite Y, 18323; Maglite® D, 18324; magnesia, 18351; Magotex, 18406; Magox® 98HR, 18407; Magox® Super Premium, 18408; Magrods, 18410; Plastomag® 170, 24052; Rhenomag, 26551; Scorchex, 27753; Scorchguard O, 27754; Scorchguard-bound, 27755; Sermag®, 28044; Tanbase®, 30794

1309-60-0: lead peroxide, 16913; Lepro, 17005 LP-100, 17691;

1309-64-4: antimony trioxide, 2574; Crystic® Prefil F, 7304; Dechlorane A-O,

7849; FireShield® H, 11837; FireShield® HPM, 11838; FireShield® L, 11839; flowers of antimony, 12030; Octoguard FR-10, 22000; Petcat R-9, 23452; Thermoguard® L, 31636; UltraFine® II, 32799
1310-58-3: potassium hydroxide, 24756; vegetable alkali, 33512
1310-73-2: Pels, 23090; soda lye, 28714; sodium hydroxide, 28772
1312-03-4: turpeth mineral, 32612
1312-43-2: indium oxide, 15081
1312-73-8: hepar sulfur, 13787
1312-76-1: Double White, 8817; Kasil, 15778; potassium silicate, 24779; Potkem, 24795; soluble potash glass, 28983
1312-81-8: lanthanum oxide, 16766
1313-13-9: glass-maker's soap, 12972; KM Manganese Dioxide, 16238; manganese dioxide, 18541; Mangoxe, 18563
1313-27-5: molybdenum trioxide, 20081
1313-60-6: Oxone, 22435; oxygen powder, 22453; Solozone, 28962
1313-82-2: Hesthsulphid, 13886; sodium sulfide, 28822
1313-96-8: niobium oxide, 21246
1313-99-1: nickel oxide, 21130; Nico, 21147
1314-08-5: palladium oxide, 22615
1314-11-0: strontia, 29826;
1314-13-2: Activox, 507; Activox B, 508; Canfelzo, 5087; Decelox, 7847; Denzox, 8068; Electrox, 9798; Entrox, 10596; Extrox, 11329; Felzodox, 11592; Finex-25, 11809; Finex-25-020, 11810; Fotofax, 12339; Garozinc, 12680; Ken-Zinc®, 16050; K-Zinc, 16527; Microx, 19718; Octocure 553, 21994; philosopher's wool, 23670; Photozinc, 23764; Rubox, 27113; Vita Zinc, 33922; zinc oxide, 34786; Zinc Oxide No. 185, 34787; Zinc Oxide No. 318, 34788; Zinc Oxide Transparent, 34789; Zinkoxyd Activ®, 34812; Zinox, 34815
1314-20-1: thorium dioxide, 31749
1314-23-4: Dynazirkon, 9429; Kontrastin, 16333; S945, S975, S987, S992, S994, 27184; SM 945, 28621; SM 975, 28622; SM 987, 28623; SM 992, 28624; SM 994, 28625; Zedox, 34698; zirconium oxide, 34824
1314-35-8: tungsten trioxide, 32591
1314-36-9: yttrium oxide, 34684
1314-37-0: ytterbium oxide, 34681
1314-41-6: minium tego, 19832; red lead, 25960
1314-56-3: phosphorus pentoxide, 23745; phosphorus pentoxide, 23746;Superphosphate, 30262;
1314-60-9: Nyacol® A-1530, 21859; Thermoguard® FR, 31635; Timonox, 31806; TPL, 32112
1314-61-0: tantalum oxide, 30815
1314-62-1: KM Vanadium Pentoxide, 16246; vanadium pentoxide, 33275
1314-98-3: Sachtolith®, 27201; Spalerite, 29184;zinc sulfide, 34796;
1315-01-1: stannic sulfide, 29496
1315-04-4: antimony pentasulfide, 2570; golden antimony sulfide, 13177; sulfur gold, 30045
1317-33-5: Moldag 200, 20049; molybdenum disulfide, 20077; Molydag, 20083; Molydag 204, 20084; Molydag 206, 20085; Molydag 208, 20086; Molydag 210, 20087; Molydag 211, 20088; Molydag 214, 20089; Molykote®, 20090; SLA 1208, 1261, 28576; SLA 1261, 28577; SLA 1286, 28580; SLA 2208, 28583
1317-34-6: manganite, 18555
1317-36-8: lead monoxide, 16910; Litharge, 17501; Vulcaid, 34072
1317-37-9: white pyrites, 34371
1317-38-0: cupric oxide, 7386; UP 13600, 33091
1317-39-1: Brown Copp., 4618; Cupridan, 7389; cuprous oxide, 7405; Cuprox, 7409; Lolotint 97, 17562; Nordox, 21540; Perenox, 23240; Purple Copp, 25417; Purplecopp 97N Premium, 25418; ruby ore, 27120; violet copper, 33875
1317-65-3: Atomite®, 3379; Calbux, 4929; calcium carbonate, 4940, 4941, 4942; Garden Lime, 12663; limestone, 17273; Portland stone, 24709; Supercoat®, 30228; Supermite®, 30256
1317-95-9: Tripoli, 32399;
1317-97-1: ultramarine, 32821; umbelliferone, 32899
1318-00-9: Mikolite, 19749; vermiculite, 33589
1318-02-1: zeolites, 34718
1318-72-5: kainite, 15657
1318-93-0: Fulcat Catalysts, 12436; Fulmont Activated Bleaching Earths, 12448; K 129, 15639; Lubrigel, 17736; Montmorillonite, 20300
1319-41-1: Sepigel A, 27985
1319-46-6: Ceruse, 5769; Halcarb 20, 13545; Kremser White, 16399; white lead, 34366
1319-77-3: cresylic acid, 7013; tricresol, 32229
1321-12-6: M.N.T, 18105
1321-65-9: Nibren wax, 21097;
1321-94-4: Arosolve MN-LF, 3072
1322-31-4: Amianthus, 2057
1322-93-6: Aerosol® OS, 895; Dehscofix 916, 7924; Rhodacal® IN, 26585; Supragil® WP, 30309
1322-98-1: Witconate DS, 34485
1323-03-1: Ceraphyl® 50, 5716; myristyl lactate, 20578; Wickenol® 506, 34400
1323-38-2: castor oil, 5417; Surfactol® 13, 30339
1323-39-3: Aldo® PGHMS KFG, 1544; Cerasynt® M, PA, 5736; CPH-52-SE, 6942; Emalex PGMS, 9986; Emerest® 2380, 10092; Grindtek PGMS 90, 13414; Hodag PGMS, 14215; Hodag PGS, 14224; Kessco® PGML, 16125; Kessco® PGMS, 16126; Lipo PGMS, 17343; Mazol® PGMS, 19140; Pegosperse® PMS CG, 23067; Promodan SP, 25144; Radiamuls® PG 2201, 25779; Schercemol PGMS, 27621; Witconol 2380, 34498
1323-42-8: Naturechem® GMHS, 20770; Paricin® 13, 22834
1323-65-5: Dinonyl phenol, 8544
1323-83-7: Kessco® GDS 386F, 16108
1327-33-9: Cooksons, 6753; Thermoguard® UF, 31637; Timonox, 31807; Timonox Blue Star, 31808
1327-41-9: Aloxicoll, 1813; Aluminum chlorohydrate, 1903; Dow Corning® ACH-303, 8867; Locron, 17543; Micro-Dry®, 19669; Reach® 101, 201, 501, 25919; Ritachlor 50%, 26801
1327-43-1: Van Gel®, 33268
1327-53-3: arsenic trioxide, 3143; poison flour, 24286
1330-20-7: xylene, 34620
1330-43-4: antipyoninum, 2612; Dehybor, 7942; FB 48, 11513; sodium borate anhydrous, 28739; Solubor, 28990
1330-61-6: Ageflex FA-10, 979; Sipomer® IDA, 28539
1330-76-3: Bisomer D10M, 4275;Octomer DIOM, 22010;
1330-78-5: Syn-O-Ad® 8484, 30558; TCP, 30857
1330-80-9: Emalex PGO, 9987; Prolein, 25126; Propylene glycol oleate, 25221; Radiamuls® PG 2206, 25780
1330-85-4: Country Fresh Disinfectant, 6905
1330-86-5: Monoplex® DIOA, 20246; Plasthall® DIOA, 24000
1331-61-9: Ablusol DBM, 125; Ammonium dodecylbenzene sulfonate, 2238; Hetsulf 50A, 13978; Nansa® AS 40, 20684
1332-07-6: zinc borate, 34769
1332-14-5: cupric sulfate, basic, 7388; Cusatrib, 7429
1332-40-7: copper oxychloride, 6784 6785; Cupravit®, 7370; Cuprokylt, 7395; Cuprosana, 7399; Curenox-50, 7419; FS Dricol 50, 12426; Headland Inorganic Liquid Copper, 13731
1332-58-7: Anhydrol, 2459; ASP®, 3230; Bilt-Cote®, 4147; Bilt-Plates®, 4148; Buca, 4646; Catalpo, 5432; Continental® Clay, 6740; Dixie Clay®, 8743; Fiberfrax® 6000 RPS, 11726; Fiberkal, 11736; Huber 40C, 14449; Huber 65A, 14450; Huber 95, 14451; kaolin, 15725; Kaowool®, 15736; Kayphobe-ABO, 15816; Langford Clay, 16674; McNamee Clay®, 19182; Par Clay®, 22713; Peerless®, 23023; Peerless® No. 1, 23024; Pharmolin, 23576; Polyplate 90, 24523; Satintone, 27486; SP-33, 29177; SPS, 29349; takizolit, 30726; Translink®, 32139; Whitetex, 34375
1333-39-7: Eltesol® PSA 65, 9868; Phenosulfonic acid, 23617
1333-82-0: chrome bronze, 6204; chromous trioxide, 6247; red oxide of

chromium, 25965

1333-86-4: acetylene black, 319; Addipast, 652; Black Pearls® 1100, 4305; carbon black, 5220; Continex® LH-10, 6741; Continex® LH-10, N-351, 6742; Continex® N351, 6743; Derussole, 8135; Diablack® A, 8245; Efweko, 9641; Elftex® 675, 9832; Flex Carbon, 11918; Furnex, 12496; FW 18, 12527; FW 200 Beads and Powd, 12528; Ketjenblack® EC-310 NW, 16129; Modulex, 20032; Mogul® L, 20036; Monarch® 1100, 20136; Printex, 24998; Printex 25 Beads and Powd, 24999; Printex P, 25000; Printex U Beads and Powd, 25001; S160 Beads and Powder, 27181; Velvetex, 33552; Vulcan® XC72R, 34093

1335-26-8: Novozone®, 21745

1335-32-6: subacetate of lead, 29902

1335-87-1: Halowax 1014, 13564

1337-33-3: Morflex MSC, 20331

1337-76-4: Attaclay, 3399; Attacote, 3400; Attaflow, 3401; Attapulgite, 3404; Attapulgus, 3405; Attasorb, 3406; Carrisorb, 5317; Diluex® , FG, 8482; Economy Flor-Dri, 9587; Emcor, 10069; Florco®, 12010; Florco®-X, 12011; Florex®, 12015; Florex® Ag-Dri 6/30, LVM 8/16, RVM 8/16, 12016; Florigel® H-Y, 12019; Flor-Kleen, 12022; Gelsorb B, 12747; Minugel, 19838; Min-U-Gel® 100, 19839; Min-U-Gel® 200, 19840; Min-U-Gel® AR, 19841; Min-U-Gel® CW, 19842; Min-U-Gel® LF, 19843; Pharmasorb, 23573; QA-555, 25543; Refinex, 25999; S-60 RVM, 27183

1337-81-1: Good-rite® 2528X10, 13199; vinyl pyridine, 33858, 33859

1338-02-9: copper naphthenate, 6783

1338-23-4: 2-butanone peroxide, 4738; Butanox, 4739; Cadox® HBO-50, 4913; Cadox® L-30, 4914; Hi-Point® 90, 14120; Lupersol DDM-9, 17871; Lupersol DSW, 17873; methyl ethyl ketone peroxide, 19531; Quickset® Extra, 25672; Sprayset® MEKP, 29338

1338-24-5: Agenap HMW-H, 1011; naphthenic acid, 20714

1338-39-2: Ablunol S-20, 108; Alkamuls® S-20, 1676; Alkamuls® SML, 1685; Arlacel® 20, 2899; Atmer® 100, 3356; Crill 1, 7028; Dehymuls SML, 7974; Disponil SML 100 F1, 8695; Emsorb® 2515, 10461; Ethylan® GL20, 11109; Glycomul® L, 13111; Hetan SL, 13888; Hodag SML, 14242; Kemester® S20, 15966; Liposorb L, 17404; Montane 20, 20265; Prote-sorb SML, 25278; S-Maz® 20, 28640; Sorbax SML, 29098; Sorbon S-20, 29117; Span® 20, 29187

1338-41-6: Ablunol S-60, 110; Alkamuls® S-60, 1677; Alkamuls® SMS, 1687; Arlacel® 60, 2901; Capmul® S, 5119; Crill 3, 7030; Dehymuls SMS, 7976; Disponil SMS 100 F1, 8702; Drewsorb® 60K, 9019; Durtan® 60, 9275; Emalex SPE-100S, 9992; Emsorb® 2505, 10458; Ethylan® GS60, 11112; Famodan MS Kosher, 11423; Glycomul® S FG, 13114; Grindtek SMS, 13415; Hetan SS, 13890; Hodag SMS, 14245; Kemester® S60, 15968; Liposorb S, 17412; Montane 60, 20267; Polycon S-60 K, 24399; Prote-sorb SMS, 25281; Radiamuls® Sorb 2145, Sorb 2161, Sorb 2166, 25783; S-Maz® 60K, 28642; Sorbax SMS, 29101; Sorbon S-60, 29119; Span® 60, 60K, 29189

1338-43-8: Ablunol S-80, 111; Alkamuls® S-80, 1681; Alkamuls® SMO, 1686; Arlacel® 80, 2902; Atmer® 105, 3358; Capmul® O, 5115; Crill 4, 7031; Dehymuls SMO, 7975; DeSotan® SMO, 8194; Disponil SMO 100 F1, 8698; Emalex SPO-100, 9996; Emsorb® 2500, 10455; Ethylan® GO80, 11110; Glycomul® O, 13112; Hetan SO, 13889; Hodag SMO, 14243; Kemester® S80, 15970; Liposorb O, 17407; Montane 481, 20270; Montane 80, 20269; Polycon S-80 K, 24400; Prote-sorb SMO, 25279; Radiasurf® 7155, 25802; S-Maz® 80, 28644; Sorbax SMO, 29099; Sorbon S-80, 29121; Span® 80, 29191; Witconol 2500, 34501; Uniplex 600, 33019; Uniplex 680, 33020

1341-49-7: Ammonium bifluoride, 2232; fluoram, 12068; Matt salt, 19067

1341-54-4: benzophenone-11, 3980; Uvinul® 490, 33193; Uvinul® M-493, 33198

1343-88-0: Celkate T-21, 5592; magnesium silicate, 18366

1343-98-2: Akrochem® Rubbersil RS-150/RS-200, 1177; Flexin, 11935

1344-00-9: Ketjensil® SM 405, 16134; Valfor® 100, 33244; Wessalith, 34284

1344-09-8: Acsil, 454; Crystal, 7284; Gasbinda, 12688; Marley Cement Dustproofer, 18827; N., 20633; sodium silicate, 28817

1344-28-1: A-2, 6; Abradux, 150;Abramant, 151; Abramax, 152; Abrarex, 153; Abrasit, 154; Activated alumina, 495; Adamant, 621; Alcan AA-100, 1411; Alcan C-70, C-71, C-72, C-73, C-75, 1414; Alcan FRF 5, 10, 20, 30, 40, 60, 80, 85, 1415; Alcan FRF LV2, LV4, LV5, LV6, LV7, LV8, LV9, 1416; Alcan H-10, 1429; Alexite, 1568; Alfrax B301, 1580; Aloxite, 1814; Alpha-Ruvite, 1832; Alufrit, 1880; alumina, 1891; Aluminum oxide, 1919; Aluminum oxide C, 1920; Alumite, 1933; Alundum®, 1934; Bikorit, 4139; Borite, 4426; Borolon, 4441; Boxite, 4468; Brasivol, 4483; C-1, 4839; Carbo Alumina, 5195; Carbo-corundum, 5196; Compalox, 6672; Corolox, Corotox, Corowalt, 6842; Corubin, 6862; Corundite, 6863; D-201, 7632; Dirubin, 8622; Dural, 9149; Dycron, 9309; Flame guard, 11874; H-30, 13527; Haltex 300, 13573; Italcor, 15431; Maftec®, 18304; Martipol, 18969; Martisorb, 18970; Martoxin, 18971; Purdox, 25404; Realox®, 25931; Rewagit, 26293; Rex, 26435; S-201, 27182; Saffil®, 27221; SB-136, 27527; SB-336, 27528; SB-632, 27529; Selexsorb® COS, 27866; T-1061, 30674; T-64, 30673; Versal 150, 33605

1344-40-7: lead phthalate basic, 16914

1344-48-5: mercuric sulfide, red and black, 19350

1344-81-6: Orthorix, 22337

1344-95-2: calcium silicate, 4969; Vansil®, 33364

1345-05-7: Lithopone, 17520

1345-25-1: glassite, 12971; glossite, 13013

1345-44-6: antimony trisulfide, 2575

1369-66-3: dioctyl sodium sulfosuccinate, 8550

1390-65-4: Carmine, 5289; Carmine 40, 5290

1393-92-6: Litmus, 17523

1398-61-4: chitin, 6082; Kytamer® PC, 16522

1400-62-0: Archil, 2808

1401-55-4: quebracho, 25636; tannic acid, 30807

1401-69-0: tylosin, 32663

1403-17-4: Vanobid, 33332

1404-93-9: Vancocin Hydrochloride, 33305

1406-65-1: Amplex, 2362; chlorophyll, 6134

1415-73-2: Veragel® 200, 33566

1420-04-8: Bayluscide®, 3805; clonitrilide, 6432

1446-61-3: Amine D, 2155

1450-14-2: CH 7280, 5840; Hexamethyldisilane, 14002

1459-93-4: Morflex 1129, 20330; Uniplex 270, 33016

1462-54-0: Mirataine® HC-Acid, 19929

1462-55-1: dodecylthioethanol, 8781; DV-1936, 9296

1491-41-4: Maretin, 18704; Rametin, 25856

1493-13-6: Fluorad® FC-24, 12059; trifluoromethanesulfonic acid, 32251

1528-49-0: Morflex 560, 20329

1533-45-5: Eastobrite® OB-1, 9475; Ethenediyl diphenylene bisbenzoxazole, 10940

1538-09-6: Diapen, 8327

1558-33-4: CC 3275, 5488; chloromethyldichloromethylsilane, 6128

1561-92-8: Geropon® MLS/A, 12909; sodium methallyl sulfonate, 28790

1562-00-1: Witconate NIS, 34487

1563-66-2: Carbodan, 5197; Curaterr®, 7414; Nex, 21079; Rampart, 25860; Sipcam UK Carbosip 5G, 28515; Tripart® Nex, 32382; Yaltox®, 34636

1582-09-8: Agriphlan 24, 1070; Ornamental Weeder, 22308; Treflan, 32162;

Trifluralin, 32252; Trigard, 32256; Trilin® 10G, 32319; Trimaran, 32333; Tripart® Trifluralin 48 EC, 32387; Tristar, 32415
1592-23-0: Afco-Chem CS, 921; calcium stearate, 4970; Hallcote® CSD, 13555; Nopcote C-104, 21510; Petrac® CP-11, 23463; Synpro® 15F, 30590
1596-84-5: Alar, 1334; B-Nine, 4365; daminozide, 7723; Dazide, 7805
1600-27-7: mercuric acetate, 19346
1633-05-2: strontium carbonate, 29828
1634-02-2: Akrochem® TBUT, 1178; tetrabutylthiuram disulfide, 31363
1634-04-4: Driveron, 9046
1642-54-2: Franocide, 12378
1643-19-2: TBAB, 30851; tetrabutyl ammonium bromide, 31362
1643-20-5: Ablumox LO, 91; Ammonyx® LO, 2264; Empigen® OB, 10375; lauramine oxide, 16832
1643-20-5: Rhodamox® LO, 26620; Schercamox DML, 27575
1649-18-9: azaperone, 3518; Stresnil, 29813
1656-63-9: Weston® TLTTP, 34322
1669-59-3: Isobu-M-AMD, 15341
1680-21-3: SR-272, 29374
1689-83-4: Actrilawn 10, 566; Totril, 32081
1689-84-5: Bromoxynil, 4590; Buctril, 4650
1689-89-0: nitroxynil, 21384;Trodax, 32439;
1698-60-8: Alicep®, 1610; Atlas Silver, 3330, 3331; Better Flowable, 4076; Chiltern Pyrazol, 6071; chloridazon, 6103;Gladiator, 12953; Paramin DF, 22774; Pyramin®, 25466; Starter Flowable, 29545; Tripart® Gladiator, 32372; Trojan SC, 32442; Weedmaster, 34261
1702-17-6: Agrichem, 1045; Agricorn D, 1051;BH 2,4-D Ester 40, 4115; Campbell's Destox, 5057; CDA Dicotox Extra, 5519; Clopyralid, 6436; Dow Shield, 8887; Format, 12267; Hadranol, 13533; Lontrel, 17586; Shield, 28173; silvapron, 28409
1704-62-7: Texacat® ZR-70, 31437
1715-40-8: Alugan, 1881; Bromocyclene, 4576; Bromodan, 4577
1717-00-6: Dichlorofluoroethane, 8391
1719-57-9: CC3270, 5487; chloromethyldimethylchlorosilane, 6129
1719-58-0: CV-4720, 7447
1732-10-1: dimethyl azelate, 8504; Emery® 2914, 10158
1733-12-6: cresol red, 6996
1746-81-2: Aresin, 2839; Arresin, 3127
1758-73-2: formammidinesulfinic acid, 12262; Manofast, 18571; thiourea dioxide, 31721
1760-24-3: Aminoethylaminopropyltrimethoxy silane, 2171; CA0700, 4850; Dow Corning® Z-6020, 8878; Dynasylan® DAMO, 9404; Dynasylan® DAMO-P, 9405; Dynasylan® DAMO-T, 9406; Prosil® 3128, 25239; Union Carbide® A-1120, 32993
1762-95-4: Ammonium thiocyanate, 2256
1777-82-8: 2,4-dichlorobenzyl alcohol, 8387; Myacide® SP, 20546
1783-96-6: Aspartic acid, 3234
1801-72-5: Sinapoline, 28491
1809-19-4: Syn-O-Ad® P-316, 30562
1812-53-9: Varisoft® 432-100, 33409
1825-62-3: CT2970, 7326; EXP-51, 11306
1843-03-4: Lowinox® CA 22, 17665; Topanol® CA, 32023
1843-05-6: Cyasorb® UV 531, 7497; Hostavin® ARO 8, 14418; Lankromark® LE285, 16697; Lowilite® 22, 17642; Spectra-Sorb UV 531, 29232; Uvinul® 408, 33192
1847-55-8: Rhodapon® OS, 26640; Sulfopon O 680, 30026; Supralated LS, 30310
1854-23-5: 4,4'-(2-Ethyl-2-nitrotrimethylene)-dimorpholine, 11141
1861-32-1: Dacthal, 7685; dimethyl tetrachloroterephthalate, 8512; Vegetable Turf and Ornamental Weeder, 33526
1861-40-1: Benfluralin, 3940
1871-22-3: tetrazolium chloride, 31395
1889-67-4: Perkadox® 30, 23305;
1897-45-6: Bombardier, 4386; Bravo 500, 4490; Chiltern Ole, 6070; Chlorothalonil, 6150; Contact 75, 6738; Daconil Turf, 7681; Groutcide 75, 13432; Jupital, 15633; Nopcocide® N-40-D, 21487; Nopcocide® N-96, 21488; Nopcocide® N-96-S, 21489; NuoCide®, 21773; Power Chlorothalonil 50, 24818; Repulse, 26151; Siclor, 28204; Sipcam UK Rover 500, 28516; Tripart® Faber, 32371; Tripart® Ultrafaber, 32388
1910-42-5: Gramoxone, 13254, 13255; Gramoxone X, 13256; Scythe, 27779; Speedway, 29264
1912-24-9: Ashlade 4% At Gran, 3184; Ashlade Atrazine 50 FL, 3188; Atraflow, 3387; Atranex, 3393; atrazine, 3394; Borocil A, 4430; Boroflow A, 4434; Boroflow A/ATA, 4435; Chlorea, 6094; Gesaprim 500FW, 12922; Herbazin Total, 13805; Mascot Gauntlet, 19002; Primatol AA, 24969
1912-26-1: trietazine, 32238
1918-00-9: Dicamba, 8374; Tracker, 32130
1918-02-1: Tordon, 32056; Tordon 22K, 32057
1918-16-7: Albras Propachlor, 1384; Albrass, 1385; Atlas Orange, 3322; Bexton, 4096; Croptex Amber, 7212; Portman propachlor 50FL, 24713; Prolex, 25128; Propachlor, 25163, 25164; Ramrod, 25861; Tripart® Granular, 32373; Tripart® Sentinel, 32385
1921-70-6: Pristane, 25037
1929-73-3: Planotox, 23914
1931-62-0: Esperox® 41-25A, 10850
1934-21-0: tartrazine, 30835
1948-33-0: *t*-Butylhydroquinone, 4793; Eastman® MTBHQ, 9471; Sustane® TBHQ, 30437
1975-78-6: Nitrile 10 D, 21335
1982-47-4: chloroxuron, 6158; Tenoran, 31119
1983-10-4: bioMeT TBTF, 4171
1984-06-1: sodium caprylate, 28743
2002-29-1: Locorten, 17541
2016-42-4: Amine 14D, 2129
2016-56-0: Acetamin 24, 284
2016-57-1: Amine 10, 2126; Amine 12-98D, 2128
2030-63-9: Clofazimine, 6427
2031-67-6: CM9050, 6464; Dynasylan® MTES, 9413; Union Carbide® A-162, 32982
2032-65-7: Borderland Black, 4416; Club, 6452; Draza®, 8973; Mesurol®, 19401; methiocarb, 19484
2044-56-6: Empicol® HL25, 10322
2051-90-3: Dichloroditane, 8390
2052-14-4: Brunol, 4625
2082-79-3: Anox® PP 18, 2486; Lowinox® PO35, 17671; Naugard® 76, 20789; Ralox® 530, 25843; Ultranox® 276, 32849
2082-80-6: Weston® TSP, 34324
2083-91-2: CD5400, 5505
2094-98-6: V-40, 33216
2116-84-9: Abil® AV 8853, 20-1000, 47; Emalex MTS-30E, 9971; phenyl dimethicone, 23632; Phenyl trimethicone, 23635
2122-70-5: Tre-Hold, 32165
2150-48-3: pyronine B, 25522
2155-71-7: Trigonox® 111-B40, 32280
2156-97-0: Ageflex FA-12, 980; lauryl acrylate, 16863;SR-335, 29378;
2163-42-0: MP Diol Glycol, 20391
2163-80-6: Drexar 530, 9023; MSMA, 20417; Neoarsycodyl, 20871; Weed Hoe, 34257
2164-08-1: lenacil, 16984; Venzar®, 33562; Vizor, 34001, 34002
2164-17-2: Cottonex, 6894; Meturon® 4L, 19607
2167-23-9: Trigonox® D-E50, 32298
2190-04-7: Acetamin 86, 285; Acetamin T, 288
2210-79-9: DY 023, 9300; Heloxy® 62, 13771
2217-44-9: Apagallin, 264 keraphen, 16055; 0
2218-94-2: Nitron, 21367
2221-95-6: fichtelite, 11765
2223-93-0: Synpro® Cadmium Stearate, 30597
2224-33-1: CV-5050, 7454
2224-44-4: Nitrobutylmorpholine, 21351
2227-17-0: dienochlor, 8429; Pentac Aquaflow, 23133
2235-43-0: Unihib® 314, 32959
2235-54-3: Akyposal ALS 33, 1302; Ammonium lauryl sulfate, 2242; Avirol® A, 3477; Carsonol® ALS-R, 5332; DeSonol A, 8186; Lorol NH, 17620; Maprofix NH and NHL, 18664; Marlinat® DFN 30, 18853; Neopon LAM, 20945; Nutrapon HA 3841, 21821; Nutrapon PP 3563, 21824; Octosol ALS-28, 22024; Perlankrol® DAF25, 23325; Polystep® B-7, 24597; Rhodapon® L-22, L-22/C, 26635; Sermul EA129, 28048; Standapol® A, 29467; Stepanol® AM, 29693; Sulfochem ALS, 29996; Tensopol N, 31194; Texapon ALS, 31453; Ufarol Am 30, 32742; Witcolate NH,

34478; Zoharpon LAA, 34864
2244-21-5: ACL 56, 59, 60, 66, 354; CDB Clearon, 5525;Clearon, 6392; dichloro-1,3,5-triazinetrione, potassium salt, 8384; potassium dichloroisocyanurate, 24753
2277-92-1: Oxyclozanide, 22444; Zanil, 34692
2303-01-7: cresol purple, 6995
2303-16-4: Avadex, 3449
2303-17-5: Avadex® BW, 3450; Far-Go/Avadex BW, 11470; triallate, 32190
2307-68-8: Croptex Bronze, 7213; pentanochlor, 23165; Solan, 28887
2310-17-0: phosalone, 23691; Zolone, 34888
2315-02-8: Sudafed Nasal Spray, 29937
2321-07-5: Fluorescein, 12080; Fluoresceine, 12082
2344-80-1: CC3285, 5489; chloromethyltrimethyl silane, 6130
2358-84-1: Ageflex DEGDMA, 962; diethylene glycol dimethacrylate, 8444
2372-21-6: Trigonox® BPIC, 32296
2398-96-1: Tinaderm, 31818
2409-55-4: *p*-Butylcresol, 4792; Cao, 5098; Lowinox® MBPC, 17669
2422-91-5: Desmodur® RE, 8157
2423-66-7: quindoxin, 25682
2425-06-1: captafol, 5163; Merpafol, 19368; Sanspor, 27379
2425-79-8: Heloxy® 67, 13774
2426-08-6: Ageflex BGE, 967; Epirez 501, 10654; Heloxy® 61, 13770
2426-54-2: Ageflex FA-2, 976
2437-25-4: Nitrile 12, 21336
2438-72-4: bufexamac, 4655; Parfenac, 22830
2439-01-2: Morestan, 20311; Morestan®, 20312; quinomethionate, 25686
2439-10-3: Dodine FL, WP, 8785; Radspor FT, 65WP, 25833
2439-35-2: Adame, 622; Ageflex FA-1, 974; Ageflex FA-1, 975
2440-22-4: Lowilite® 55, 17643; Topanex 100BT, 32019; Uvazol P, 33184
2455-24-5: Ageflex THFMA, 999; tetrahydrofurfuryl methacrylate, 31372
2489-77-2: Thiate E, 31666
2492-26-4: Nacap®, 20643; Nuodex 84, 21779;
2495-25-2: Ageflex FM-25, 987;
2495-35-4: Melcril 4085, 19267
2499-95-8: Ageflex FA-6, 977; Ageflex n-HA, 996
2500-83-6: Jasmacyclene, 15529
2524-03-0: MP-2, 20388
2530-83-8: Aktisil EM, 1202; CG6720, 5836; Dow Corning® Z-6040, 8881; Dynasylan® GLYMO, 9408; Prosil® 5136, 25240; Union Carbide® A-187, 32988
2530-85-0: CM8550, 6456; Dow Corning® Z-6030, 8879; Dynasylan® MEMO, 9412; Prosil® 248, 25238; Union Carbide® A-174, 32986
2530-87-2: 3-chloropropyltrimethoxysilane, 6142, 6143; CC3300, 5493; chloropropyltrimethoxysilane, 6138; Dow Corning® Z-6076, 8883; Dynasylan® CPTMO, 9403
2550-06-3: CC3291, 5491; chloropropyltrichlorosilane, 6139
2551-62-4: Esaflon, 10782; sulfur hexafluoride, 30046
2551-83-9: CA0567, 4847
2553-19-7: CD6000, 5514
2571-88-2: Admox® 18-85, 759; Amyx SO 3734, 2395; Barlox® 18S, 3652; Chemoxide ST, 6000; Incromine Oxide S, 15011; Mackamine SO, 18177; Mazox® SDA, 19155; Ninox® SO, 21244; Rewominox S 300, 26335; Rewominoxid S 300, 26338; Schercamox DMS, 27577
2589-57-3: V-601, 33219
2592-95-2: 1-hydroxybenzotriazole, 14659
2593-15-9: Aaterra WP, 17
2599-01-1: Schercemol CM, 27586
2606-93-1: Photine, 23756
2627-95-4: CD6210, 5517
2631-40-5: Etrofolan®, 11156; isoprocarb, 15404; isopropyl phenylmethyl carbamate, 15411; Mipcin®, 19846
2634-33-5: Nipacide® BIT, 21253
2642-71-9: Azinphos-ethyl, 3520; Cotnion-Ethyl, 6882; Gusathion® A, 13502
2644-64-6: Phospholipon® PC, 23729
2673-22-5: Aerosol® TR-70, 899; Monawet MT Series, 20186
2675-77-6: Chloroneb 65W Fungicide, 6131; Chloroneb Systemic Flowable Fungicide, 6132; Terraneb SP Turf Fungicide, 31337
2687-91-4: Agsol Ex 2, 1087;NEP, 20987;
2687-96-9: Agsol Ex 12, 1089
2694-54-4: Sipomer® TATM, 28541
2699-79-8: Vikane, 33754
2724-58-5: Emersol® 871, 10122; isostearic acid, 15421; Prisorine 3508, 25036; Proto-Lan IP, 25295
2754-27-0: CT3254, 7328
2768-02-7: CV-4917, 7452; Dynasylan® VTMO, 9421; Union Carbide® A-171, 32984
2778-96-3: Alkamuls® SS, 1688; Emalex CC-18, 9936; Hetester 412, 13892; Lexol SS, 17155; Liponate SS, 17388; Radia® 7501, 25743; Ritachol® SS, 26804
2807-30-9: Ektasolve® EP, 9671; ethylene glycol propyl ether, 11130
2809-21-4: Briquest® ADPA-60A, 4536; Fostex P, 12338; Unihib® 106, 32957
2855-13-2: Vestamin® TMD, 33656
2857-97-8: CT2928, 7324
2867-47-2: Ageflex FM-1, 983; Sipomer® 2M1M, 28528
2893-78-9: dichloro-1,3,5-triazinetrione, sodium salt, 8385
2897-60-1: CG6710, 5835; glycidoxypropyl methyl diethoxysilane, 13090; GP-137, 13221
2915-53-9: Octomer DOM, 22011; PX-538, 25444
2915-57-3: Wickenol® 159, 34394
2921-88-2: Chlorpyrifos, 6163; Crossfire, 7239; Dowco 179, 8902; Dursban®, 9272; Dursban® 14G, 9273; Dursban® 2E, 9274; Lorsban, 17623; Pyrinex, 25486; Spannit, 29198; suSCon® Blue, 30425; suSCon® Green, 30426; Talon, 30751
2943-75-1: CO9835, 6513; Dynasylan® OCTEO, 9416; Prosil® 9202, 25241; Union Carbide® A-137, 32980
2996-92-1: CP0330, 6934; phenyltrimethoxysilane, 23651
3006-15-3: Aerosol® MA-80, 893; Empimin® MA, 10427; Octosol HA-80, 22025
3006-82-4: Aztec® *t*-Butyl Peroctoate, 3544; Aztec® *t*-Butyl Peroctoate-50 OMS, 3545; Esperox® 28, 10847; Trigonox® 21, 32265
3006-86-8: Aztec® 1,1-Bis(*t*-Butylperoxy)Cyclohexane -80 BBP, 3550; Luperco 331-XL, 17827; Lupersol 331-80B, 17860; Trigonox® 22-BB80, 32266; USP® -400P, 33168
3006-93-7: HVA-2, 14480; Vanax® MBM, 33291
3026-63-9: Avirol® SA 4113, 3483; Rhodapon® TDS, 26641
3033-77-0: Ogtac 85 V, 22052
3034-79-5: Perkadox® 20, 23303
3051-09-0: murexide, 20506
3052-70-8: Lupersol 553-M75, 17864
3055-93-4: Akyporox RLM 22, 1291
3055-95-6: Mulsifan RT 23, 20458; Oxetal VD 20, 22425
3057-08-7: Weston® PNPG, 34316
3060-89-7: Metobromuron, 19583; Patoran, 22916; Patoran® FL, 22917; Pattonex, 22923
3061-75-4: Kemamide® B, 15894; Uniwax 1747, 33067
3064-70-8: Amerstat® 294, 2044; bistrichloromethylsulfone, 4285; Stauffer N-1386®, 29580
3068-76-6: CP0156, 6930
3069-25-8: CM8620, 6457
3069-29-2: CA0699, 4849
3076-63-9: Doverphos® 53, 8842; Doverphos® TLP, 8852; Weston® TLP, 34321
3088-31-1: Akyposal RLM 56 S, 1319
3094-87-9: pyrolignite of iron, 25515
3097-08-3: Akyposal MGLS, 1309; magnesium lauryl sulfate, 18364; Rhodapon® LM, 26637; Standapol® MG, 29475
3101-60-8: Heloxy® 65, 13773
3115-49-9: Akypo NP 70, 1226
3118-97-6: Sudan II, 29939
3121-60-6: Benzophenone-9, 3979; Uvinul® DS-49, 33196
3121-61-7: Ageflex MEA, 994
3129-91-7: Dichan 100, 8377; dicyclohexylamine nitrite, 8414
3144-16-9: d-camphorsulfonic acid, 5076
3147-75-9: Cyasorb® UV 5411, 7503; Spectra-Sorb UV 5411, 29233; Uvazol 311, 33183
3179-56-4: Trigonox® ACS-M28, 32292
3179-76-8: CA0742, 4851
3179-80-4: Chemidex L, 5963; lauramidopropyl dimethylamine, 16831; Lexamine L-13, 17098; Mackine 801, 18224; Schercodine L, 27638
3179-81-5: Cerasynt® 303, 5733
3194-55-6: Saytex® HBCD-LM, 27522
3234-85-3: Alkamuls® MM/M, 1670; Ceraphyl® 424, 5727; Cetiol® MM, 5801;

Crodamol MM, 7122; Kemester® MM, 15965; Liponate MM, 17380; Schercemol MM, 27609; Waxenol® 810, 34237
3244-88-0: acid fuchsine, 333
3251-23-8: cupric nitrate, 7384
3270-71-1: Nifuraldezone, 21169
3277-26-7: CT2030, 7314
3278-89-5: FR-913, 12359; Great Lakes PHE-65, 13311; tribromophenyl allyl ether, 32203
3290-92-4: Ageflex TM 402, 403, 404, 410, 421, 423, 451, 461, 462, 1000; Ageflex TMPTMA, 1002; Perkalink® 400, 23314; SR-350, 29380
3296-90-0: dibromoneopentyl glycol, 8358; FR-522, 12355
3319-31-1: Kodaflex® TOTM, 16280; Palatinol® TOTM, 22595; Plasthall® TOTM, 24016; PX-338, 25441
3322-93-8: Saytex® BCL-462, 27518
3327-22-8: (3-Chloro-2-hydroxy)-n-propyltrimethyl ammonium chloride, 6126; 3-Chloro-2-hydroxypropyl trimethylammonium chloride (50% aq solution), 6127; Cationmaster-C, 5459; CHPTA 65%, 6176; Quab 188, 25562; Reagens-CF2, 25928
3332-27-2: Admox® 14-85, 758; Barlox® 14, 3650; Empigen® OH25, 10377; Incromine Oxide M, 15007; Lilaminox M4, 17252; Mazox® MDA, 19153; Ninox® M, 21243; Schercamox DMM, 27576
3333-67-3: nickel carbonate, 21123
3337-71-1: Asulam, 3277; Asulox, 3278, 3279
3347-22-6: Delan-Col, 8006; Dithianone, 8729
3349-06-2: Nicfo, 21101
3383-96-8: Abate 1-SG, 2-CG, 4-E, 5CG, 26
3388-04-3: CE6250, 5528; Union Carbide® A-186, 32987
3391-86-4: Morillol®, 20335
3398-33-2: sodium undecylenate, 28830
3425-61-4: TAHP-80, 30719; Trigonox® TAHP-W85, 32309
3436-44-0: Phospholipon® CC, 23727
3452-07-1: Neodene® 20, 20900
3452-97-9: Nonanol, 21436
3457-61-2: Butyl cumyl peroxide, 4768; Luperco 801-XL, 178 Trigonox® T, 32308; 29
3458-28-4: seminose, 27939
3459-96-9: amicarbalide, 2063
3483-12-3: 1,4-Dithiothreitol, 8730; Clelands Reagent, 6402
3486-35-9: Akrochem® 9930 Zinc Oxide Transparent, 1152; zinc carbonate, 34771
3511-16-8: Versapen, 33609
3521-84-4: Cholografin Meglumine, 6172
3524-68-3: pentaerythrityl triacrylate, 23144; SR-444, 29384
3567-08-6: Stabinol, 29413
3615-41-6: rhamnose, 26529
3622-84-2: Butylbenzene sulfonamide, 4789; Dellatol® BBS, 8026; Plasthall® BSA, 23993; Plastomoll® BMB, 24056
3632-91-5: Almora, 1790; magnesium gluconate, 18362
3648-20-2: Jayflex® DUP, 15556; PX-111, 25435
3655-00-3: Deriphat 160, 8090; disodium lauryliminodipropionate, 8650; Miranol® H2C-HA, 19897; Monateric 1188M, 20149
3658-48-8: Syn-O-Ad® P-310, 30560
3682-94-8: VA-088, 33230
3687-45-4: Cetiol®, 5793; Schercemol OLO, 27616; Starfol® OO, 29537; Wickenol® 143, 34386
3687-46-5: Ceraphyl® 140, 5722; Cetiol® V, 5806; decyl oleate, 7869; Schercemol DO, 27598; Tegosoft® DO, 31019
3689-24-5: Bladafum®, 4308
3691-35-8: Chlorophacinone, 6133; Drat, 8972; Karate, 15758;Ridene, 26746; Sakarat Special, 27239; Skaterpax, 28564
3696-28-4: Omadine® MDS, 22152
3710-84-7: Pennstop® 1866, 23118
3734-33-6: Bitrex®, 4291
3736-26-3: Trigonox® M-50, 32305
3768-58-9: CB2100, 5473
3775-90-4: Ageflex FM-4, 984
3806-34-6: Distearyl pentaerythritol diphosphite, 8712; Doverphos® S-680, 8850; Weston® 618, Weston® 619, 34304
3811-04-9: potassium chlorate, 24746
3811-73-2: Sodium Omadine® 40% Aq. Sol'n, 28803; sodium pyrithione, 28813
3813-05-6: benazolin, 3935
3825-26-1: Fluorad® FC-118, 12058
3851-87-4: Trigonox® 36-C75, 32270; USP® -355M, 33167
3864-99-1: Uvazol 237, 33182
3878-19-1: fuberidazole, 12428
3896-11-5: hydroxybutylmethylphenylchlorobenzotriazole, 14660; Lowitite® 26, 17677; Uvazol 236, 33181
3902-71-4: 4,5,8-trimethylpsoralen, 32340
3913-02-8: Michel XO-150-12, 19640; 2-butyloctanol, 4795
3926-62-3: sodium monochloracetate, 28797; Somon, 29072
3944-37-4: Solvenon® IPP, 29047
3952-98-5: sinigrin, 28500
3990-03-2: Sipomer® MEM, 28540
4008-48-4: Nibiol, 21095
4065-45-6: Rhodialux S, 26676; Spectra-Sorb UV 284, 29231; Sungard, 30143; Syntase® 230, 30622; Uval, 33177; Uvinul® MS-40, 33199
4070-80-8: Pruv, 25354
4075-81-4: Amasil® P, 1947; calcium propionate, 4968; Luprosil® Salt, 17921
4080-31-3: Dowicil® 75, 8937
4088-22-6: Amine M218, 2160; Radiamine 6346, 25763
4098-71-9: Vestanat® IPDI, 33657
4130-08-9: CV-4800, 7449; Dow Corning® Z-6075, 8882
4171-13-5: Axiquel, 3509
4196-86-5: Uniplex 552, 33018
4219-49-2: glycol palmitate, 13103; Lanol P, 16737
4253-34-3: CM8980, 6462
4292-10-8: lauramidopropyl betaine, 16830; Lexaine® LM, 17094
4304-40-9: Ancaris, 2411; Canopar, 5093; thenium closylate, 31568
4306-88-1: Uvi-Nox 1494, 33189
4369-14-6: CA0397, 4846
4390-04-9: Permethyl 101A, 23387
4419-11-8: V-65, 33220
4420-74-0: Aktisil MM, 1203; CM8500, 6455; Dynasylan® MTMO, 9414; Prosil® 196, 25236; Union Carbide® A-189, 32989
4468-02-4: Gluconal® ZN, 13047; Zinc gluconate, 34776
4484-72-4: CD6220, 5518
4485-12-5: Afco-Chem LIS, 922; lithium stearate, 17512
4511-39-1: Esperox® 5100, 10859; Trigonox® 127, 32284
4584-46-7: dimethylaminoethyl chloride hydrochloride, 8516
4602-84-0: Farnesol, 11492
4658-28-0: Brasoran 50 WP, 4484
4685-14-7: Dextrone X, 8239; paraquat, 22796;Paraquat + Plus, 22797;
4691-65-0: disodium inosinate, 8645
4706-78-9: Sulfochem K, 30007; Tensopol SPK, 31195
4712-55-4: diphenyl phosphite, 8599; Doverphos® 213, 8843; Doverphos® DPP, 8849; Weston® DPP, 34309
4719-04-4: Nipacide® BK, 21254
4722-98-9: Akyposal MLS 30, 1311; Empicol® LQ33/T, 10326
4726-14-1: nitralin, 21319; Planavin, 23909
4744-10-9: DMP, 8757
4748-78-1: Ebal, 9488; ethyl-benzaldehyde, 11118
4766-57-8: CT1750, 7311; T1750, 30675
4810-50-8: Kalidone®, 15678
4862-18-4: Hampamide B, 13584
5026-62-0: Nipagin M Sodium, 21268
5039-78-1: Ageflex FM-1Q80MC, 982; Madquat Q-6, 18293
5064-31-3: Cheelox® NTA-Na3, 5869; Hampshire® NTA 150, 13610; Hampshire® NTA Na3, 13612; NTA, 21759; Trilon® A, A-92, 32323
5089-70-3: 3-chloropropyltriethoxysilane, 6140, 6141; CC3292, 5492; Dynasylan® CPTEO, 9402
5131-24-8: Plondrel, 24160
5221-49-8: pyrimithate, 25485
5221-53-4: dimethirimol, 8495; Milcurb, 19756
5232-99-5: Uvinul® N-35, 33200
5234-68-4: carboxin, 5255
5259-88-1: oxycarboxin, 22441; Plantvax, 23921; Plantvax 20, 23922; Plantvax 75, 23923; Ringmaster, 26771
5274-68-0: Akyporox RLM 40, 1292; Hetoxol L-4, 13941; Mulsifan CPA, 20453
5283-66-9: CO9830, 6512
5306-85-4: Arlasolve® DMI, 2913
5321-32-4: H-K Mastitis, 14150; Versapen K, 33610
5323-95-5: Solricin® 435, 28967; Soricinol 40, 29143

5329-14-6: sulfamic acid, 29963
5333-42-6: Eutanol G, 11239; Exxal® 20, 11343; Michel XO-150-20, 19643; octyl dodecanol, 22038
5355-16-8: diaveridine, 8338
5356-84-3: CV-5100, 7455
5370-01-4: Mexitil, 19615
5392-40-5: citral (*cis* and *trans*), 6336; Neral, 20996
5397-31-9: Tomah PA-12EH, 31980
5422-34-4: Lactamide MEA, 16568; Mackamide LME, 18157; Parapel® LAM-100, 22786; Schercomid LME, 27652
5434-57-1: Schercemol 65, 27579
5456-28-0: ethyl selenac, 11082; Selazate, 27853
5466-77-3: Escalol® 557, 10788; Neo Heliopan® AV, 20866; octyl methoxycinnamate, 22039; Parsol® MCX, 22867
5534-09-8: Vanceril, 33295
5534-95-2: pentagastrin, 23146; Peptavlon, 23201
5578-42-7: CM8645, 6458
5579-93-1: Iopydone, 15235
5598-13-0: Chlorpyrifos-methyl, 6164; Cooper Graincote, 6759; Reldan 50, 26037;
5598-52-7: Torelle, 32058
5611-51-8: Lederspan, 16958
5707-69-7: drazoxolan, 8974; Ganocide, 12648; Mil-Col, 19755; Saisan, 27235
5714-82-9: Ranestol, 25866
5743-34-0: Gluconal® CA M B, 13039
5793-94-2: Crolactil CSL, 7184; Pationic CSL, 22906; Radiamuls® CSL 2980, 25772; Verv®, 33644
5809-08-5: Trigonox® TMPH, 32310
5836-29-3: Racumin®, 25713
5888-33-5: Ageflex IBOA, 991; Sipomer® IBOA, 28537
5892-10-4: Bismuth subcarbonate, 4242
5902-51-2: Sinbar®, 28494; terbacil, 31220
5915-41-3: Terbuthylazine, 31227; Tyllanex, 32658
5968-11-6: sodium carbonate monohydrate, 28745
5980-31-4: hexedine, 14027
5989-27-5: Limonene, 17280
5989-81-1: α-D-Lactose, 16589
6001-97-4: Bis(1-methylamyl) Sodium Sulfosuccinate, 4227; Lankropol® KMA, 16713
6009-70-7: Ammonium oxalate, 2247
6080-56-4: lead diacetate, 16906
6130-64-9: Lentopen, 16993
6144-28-1: Dilinoleic acid, 8480
6147-53-1: cobaltous acetate tetrahydrate, 6535
6153-56-6: oxalic acid dihydrate, 22414
6164-87-0: Ronicol, 26948
6179-44-8: Schercotaine IAB, 27697
6182-11-2: Emalex PG-di-S, 9984
6197-30-4: Neo Heliopan® 303, 20865; Uvinul® N-539, 33201
6209-17-2: Isopto Cetamide, 15415
6221-95-0: Lonzest® 143-S, 17595; Schercemol MP, 27610
6272-74-8: Emcol® E-607L, 10059; Lapyrium Chloride, 16782
6281-42-1: DV-2301, 9297
6283-92-7: Ceraphyl® 31, 5713
6284-40-8: Meglumine, 19243
6317-18-6: Amerstat® 282, 2043; N-948® Biocide, 20641; Tolcide MBT, 31934
6381-77-7: Eribate, 10753; Neo-Cebitate®, 20878; sodium erythorbate, 28761, 28762
6419-19-8: Aminotrimethylene phosphonic acid, 2191; Chemphonate AMP, 6004; Fostex AMP, 12337; Unihib® 305-LC, 32958
6422-86-2: Kodaflex® DOTP, 16274
6440-58-0: DMDM hydantoin, 8754; Glydant®, 13140; Mackstat® DM, 18234; Nipaguard® DMDMH, 21269
6484-52-2: Ammonium nitrate, 2245; Ansax, 2488; Herco-Prills, 13835; Nitram, 21322; Nitrammite, 21323; Old Plantation, 22106
6487-48-5: potassium oxalate, 24765, 24766
6542-37-6: Zoldine® ZT-55, 34887
6620-60-6: proglumide, 25116
6674-22-2: Diazabicycloundecene, 8341
6683-19-8: Anox® 20, 2483; Lowinox® PP35, 17672; Naugard® 10, 20788; Ralox® 630, 25844
6731-36-8: Aztec®, 3539; Aztec® 1,1-Bis(*t*-Butylperoxy)-3,3,5-Trimethyl Cyclohexane, 3549; Luperco 231-XL, 17825; Lupersol 231, 17856; Peroximon® S-164/40P, 23410; Trigonox® 29, 32269
6804-07-5: carbadox, 5177; Mecadox, 19200
6812-78-8: rhodinol, 26685
6834-92-0: Crystamet 1020, 7293; Drymet® 59, 9065; Lasilso, 16806; Metso, 19604; sodium metasilicate, 28788
6837-24-7: CHP, 6173
6843-66-9: CD6010, 5515
6846-50-0: Kodaflex® TXIB, 16282
6865-35-6: Barium stearate, 3643; Synpro® Barium Stearate, 30596
6868-66-2: Miranol® BM Conc, 19885
6881-94-3: diethylene glycol propyl ether, 8445; Ektasolve® DP, 9667
6891-44-7: Ageflex FM-1Q80DMS, 981
6915-15-7: hydroxysuccinic acid, 14671; Malic Acid, 18483
6923-22-4: Monocron, 20223
6938-94-9: Ceraphyl® 230, 5724; Schercemol DIA, 27592; Unimate® DIPA, 32968
6988-21-2: Famid, 11422
7005-47-2: 2-dimethylamino-2-methyl-1-propanol, 8517; DMAMP-80, 8753
7060-74-4: oleandocyn, 22111
7085-19-0: Chafer CMPP Super, 5845; Cleanacres CMPP, 6381; Clenecorn, 6403; Clifton CMPP Amine 60, 6416; Clovotox, 6449; Compitox Extra, 6680; FBC CMPP, 11514; Headland Charge, 13727; Hymec, 14685; Iso-Cornox, 15353; Mascot Cloverkiller, 19000; Mecoprop, 19204
7123-62-8: Diamine OL, 8298
7128-91-8: Ammonyx® CO, 2262; Amyx CO 3764, 2393; Barlox® 16S, 3651; Arquad® 210-50, 3110; BTC® 1010-80, 4638; BTC® 99, 4633; Catigene® 1011, 5449; FMB 210-8, 210-15, 12131; Querton 210Cl-50, 25651; Radiaquat 6410, 6412, 25789; Radiaquat 6412, 25790; Rewoquat B 10, 26389
7173-62-8: Duomeen® OL, 9114; Jet Amine DO, 15589; Kemamine® D-989, 15915
7179-49-9: Lincocin, 17284; lincomycin hydrochloride hemihydrate, 17285; Linocin, 17302
7205-52-9: Rencal, 26105
7212-44-4: Nerolidol, 21006
7235-40-7: carotene, 5308; Lucarotin® 10%, 17753; Solatene, 28900
7246-21-1: Bilopaque, 4145
7281-04-1: Catigene BR 80 B, 5444
7287-19-6: Cotton-Pro®, 6895; Gesagard, 12921; Prometrex, 25139; prometryn, 25140, 25141
7320-34-5: potassium pyrophosphate, 24777; Tetrakal, 31375
7360-38-5: Emalex O.T.G, 9976
7361-61-7: xylazine, 34619
7396-58-9: Amine M210D, 2159; Armeen® M2-10D, 2953; Dama® 1010, 7718; didecyl methylamine, 8415
7398-69-8: Ageflex mDMDAC, 993; Ageflex NB-50, 995; Agefloc PC20HV, 1008; Agequat C505, 1013; Agestat 41, 1028
7414-83-7: Didronel, 8421
7429-90-5: Aluminum, 1897
7439-89-6: iron, 15307; malleable iron, 18485
7439-91-0: lanthanum, 16763
7439-92-1: Haro® Mix CE-701, 13645; Haro® Mix CK-711, 13646; Haro® Mix MH-204, 13647; lead, 16900, 16901
7439-93-2: lithium, 17508
7439-95-4: magnesium, 18356
7439-96-5: Dienol, 8430; manganese, 18535; Tripart® Liquid Manganese, 32377; Tronamang Electrolytic Manganese Metal, 32463
7439-97-6: mercury, 19355
7439-98-7: molybdenum, 20075
7440-02-0: malleable nickel, 18486; nickel, 21106; Nickel 200, 21107; Nickel 201, 21108; Nickel 204, 21109; Nickel 205, 21110; Nickel 211, 21111; Nickel 212, 21112; Nickel 213, 21113; Nickel 222, 21114; Nickel 223, 21115; Nickel 229, 21116; Nickel 270, 21117; Raney nickel, 25867
7440-03-1: niobium, 21245
7440-04-2: osmium, 22356
7440-05-3: palladium, 22611; palladium black, 22613
7440-06-4: Platinum, 24099; platinum black, 24100
7440-09-7: potassium, 24732
7440-16-6: rhodium, 26686
7440-19-9: samarium, 27307
7440-20-2: scandium, 27545
7440-21-3: silicon, 28326
7440-22-4: Collosol Argentum, 6635; silver, 28413

7440-23-5: Metallic Sodium, 19436; Nametal, 20674; natrium, 20756; sodium, 28725
7440-24-6: strontium, 29827
7440-25-7: tantalum, 30813
7440-27-9: terbium, 31222
7440-28-0: thallium, 31552
7440-31-5: Haro® Mix ZT-514, 13650; tin, 31809
7440-32-6: titanium, 31865
7440-33-7: tungsten, 32583
7440-36-0: antimony, 2567; grey antimony, 13352; stibium, 29761; Thermoguard® CPA, 31634
7440-38-2: arsenic, 3136
7440-39-3: Barium, 3633
7440-41-7: Beryllium, 4054
7440-42-8: Boron, 4442; Trona Elemental Boron, 32454
7440-43-9: cadmium, 4893
7440-44-0: Bachite, 3575; Bastanet, 3735; Batchite, 3738; Calgon® Type SGL, 4988; Carbon, 5213; Carbon, 5214, 5215, 5216, 5217; carbon, activated, 5225; Carboraffin, 5236; Cecarbon, 5535; Darco, 7768; Degusorb, 7916; Hydraffin, 14533; Hydrodarco, 14585; Metasil DA, 19454; Metasil W/2, 19459; Microsperse, 19699; Neospectra, 20969; Neotex, 20975; Norit, C, PK, R, RO, 21560; Norithene, 21561; Novacarb, 21646; Octojet 104, 22004; Purac, 25400; Raven, 25903; Regal® 400R, 26008; Sevacarb, 28109; Sorbonorit, 29128; Special Black 100 Powd, 29216; Special Black 4, 4A Beads and Powd, 29217; Statex, 29552; Sterling, 29743; sulfite carbon, 29991; Superba, 30219; Superjet, 30246; Tack, 30689; Thermax® Floform N-990, 31581; Thermax® Stainless, 31582; Thermax® Stainless Floform N-907, 31583; United, 33055
7440-45-1: cerium, 5754
7440-46-2: cesium, 5770
7440-47-3: chromium, 6228
7440-48-4: cobalt, 6524
7440-50-8: casting copper, 5413; copper, 6777
7440-52-0: Erbium, 10746
7440-55-3: Gallium, 12615
7440-56-4: Germanium, 12897
7440-57-5: Gold, 13171
7440-60-0: holmium, 14278
7440-62-2: vanadium, 33270
7440-63-3: xenon, 34599
7440-64-4: ytterbium, 34680
7440-65-5: yttrium, 34683
7440-66-6: Delaville, 8010; zinc, 34765
7440-69-9: Bismuth, 4235
7440-70-2: calcium, 4935
7440-74-6: indium, 15080
7446-07-3: tellurium oxide, 31064
7446-08-4: selenium dioxide, 27862
7446-09-5: sulfur dioxide, 30044
7446-14-2: Haro® Chem PTS-E, 13644
7446-19-7: Z Span Spansule Capsule, 34686; zinc sulfate, monohydrate, 34795
7446-20-0: Op-Thal-Zin, 22229; zinc sulfate, 34794
7446-70-0: Aluminum chloride, anhydrous, 1902; Praestol® K2001, 24854
7447-39-4: Coclor, 6560; Coppertrace, 6793; cupric chloride dihydrate, 7379
7447-40-7: K. Tab, 15642; Kaon-Cl, 15729; Kay Ciel, 15807; K-Contin, 15818; kelp salt, 15859; K-Lor, 16226; K-Lyte/C1, 16234; KM Potassium Chloride, 16242; Leo K, 16996; Micro-K, 19676; muriate of potash, 20513; Pfiklor, 23564; potassium chloride, 24747, 24748, 24749; Sando-K, 27333; Selora, 27931; Slow-K, 28609; Trona Muriate of Potash, 32455; Trona Potassium Chloride, 32456
7487-88-9: magnesium sulfate, 18368; sal amarum, 27243; sel d'Angleterre, 27850
7487-94-7: mercuric chloride, 19347
7488-54-2: rubidium sulfate, 27107
7491-02-3: Schercemol DIS, 27595; Unimate® DIPS, 32969
7491-09-0: Dialose, 8284
7491-14-7: Dipsal, 8615
7492-30-0: Kricinol 35, 16411; Potassium ricinoleate, 24778; Solricin® 135, 28964
7527-91-5: Acrisorcin, 392
7534-94-3: Ageflex IBOMA, 992; Sipomer® IBOMA, 28538
7545-23-5: Emalex NN-15, 9973; Monamid® 150-MW, 20116
7553-56-2: Iodine, 15208
7558-79-4: Adjunct B, 708; disodium phosphate (anhydrous), 8652; salt perlate, 27293; tasteless salts, 30843
7558-80-7: Aspon, 3238; Monosorb, 20256; MSP, 20420; Oilfos, 22089; Recresal, 25944; sodium phosphate, 28810
7585-39-9: cyclodextrins, 7516
7601-54-9: Avgard, 3464; trisodium phosphate, anhydrous, 32412; TSP-O, 32539
7601-89-0: KM Sodium Perchlorate, 16245
7601-90-3: Fraude's reagent, 12381; perchloric acid, 23221
7617-78-9: Tomah PA-14, 31982
7631-46-1: Herapath's salt, 13799
7631-86-9: Aerosil® 130, 150, 873; Aerosil® 200, 874; Aerosil® 875; Cab-O-Sil® L-90, 4868; Cab-O-Sil® TS-530, 4869; Celatom, 5575; Celite®, 5586; Celite® 110, 5587; Celite® 209, 5588; Celite® 270, 5589; Celite® R-625, 5590; Celite® Snow Floss, 5591; Celite® HSC, 5611; Clarcel, 6357; diatomaceous earth, 8335; flint, 11971; Flocsil, 11991; Flo-Guard, 12034; Glaucosil, 12976; Hi-Sil, 14131; Hyflo Super-Cel, 14674; Imsil® A-10, 14936; J Slip NS-77, 15466; Krystallos, 16469; Micropil, 19689; Milowite, 19799; Neosyl®, 20972; rock crystal, 26885; Santocel, 27398; Silantox, 28266; silica, 28313; Silica FK 320 DS, 28315; Silver Bond 30, 28418; Siogel, 28508; Sipernat® 22S, 28520; Sipernat® 283LS, 28523; Sipernat® 50, 28522; Sipernat® D17, 28524; Tamsil 8, 30770; Tamsil Gold Bond, 30771; Wessalon, 50, 50S, S, 34285
7631-90-5: Abisol, 70; Amersite® 2, 2035; Amersite® 2, 2036; Leucogen, 17023; sodium acid sulfite, 28728; sodium bisulfite, 28738
7631-97-2: sodium monofluorophosphate, 28798
7631-99-4: Chili saltpeter, 6063; Saliter, 27273; sodium nitrate, 28800
7632-00-0: sodium nitrite, 28801
7632-04-4: perborax, 23213; perborin, 23214; peroxydol, 23412; sodium perborate, 28806;
7637-07-2: Boron trifluoride, 4446; Leecure B Series, 16963
7646-78-8: Fascat® 4400, 11497; Libavius' fuming spirit, 17185; stannic chloride, 29494
7646-79-9: cobalt chloride (ous), 6528; cobaltous chloride, 6537; Cobatope-60, 6543
7646-85-7: zinc chloride, 34772; Zinctrace, 34806
7646-93-7: tartarline, 30833
7647-01-0: hydrochloric acid, 14575 14576; marine acid, 18718; soldering acid, 28909; spirit of salt, 29298
7647-10-1: palladium(II) chloride, 22617
7647-14-5: Alberger® Natural Flake, 1361; Betrox, 4075; Hopper salt, 14303; marine salt, 18722; muriate of soda, 20514;Natritope Chloride, 20755; rock salt, 26887; sal commune, 27244; sal Culinaris, 27245; Saltex, 27296; Slow-Sodium, 28610; sodium chloride, 28748; Superior® Granulated, 30244; Tru-Flake® Salt, 32490
7647-15-6: sodium bromide, 28741
7647-17-8: cesium chloride, 5773
7647-18-9: antimony pentachloride, 2568
7651-02-7: Adogen® S-18 V, 778; Chemidex S, 5968; Incromine SB, 15014; Lexamine S-13, 17100; Lipamine SPA, 17328; Mackine 301, 18217; Mazeen® S-13, 19124; Miramine® SODI, 19872; Schercodine S, 27641; TEGO Amid S18, 30957
7664-38-2: Marphos, 18951; NFB, 21080; phosphoric acid, 23738, 23739; Solklean+sn 101, 28950
7664-39-3: hydrogen Fluoride, 14589
7664-41-7: ammonia, 2219; spirit of Hartshorn, 29296
7664-93-9: spirit of alum, 29295; spirit of vitriol, 29299; sulfuric acid, 30055
7665-72-7: Ageflex TBGE, 998
7681-11-0: potassium iodide, 24758, 24759
7681-49-4: Fluorinse, 12090; Fluorl, 12092; fluorol, 12101; Gel II, 12716; sodium fluoride, 28764
7681-52-9: Adeka Hypote, 664; Chlorasol®, 6089; Chloros, 6146; Chlorosoda, 6147; Dakin's solution, 7704; Domestos, 8795; HyPure N, 14728; Merclor D, 19342; oxychloride,

22442; salunol, 27302; sodium hypochlorite, 28773
7681-55-2: sodium iodate, 28774
7681-57-4: metabisulfite, 19404; sodium metabisulfite, 28783
7681-65-4: cuprous iodide, 7404
7681-82-5: sodium iodide, 28775
7696-12-0: Killgerm® Py-Kill W, 16182; tetramethrin, 31384
7697-37-2: dynamite acid, 9357; fuming nitric acid, 12456; Nitraline, 21320; nitric acid, 21332
7699-45-8: zinc bromide, 34770
7702-01-4: Schercoteric CY-2, 27705
7704-34-9: Biosulphur Powder, 4210; Crystex®, Crystex® Regular, 7294; Flotox, 12027; flour of sulfur, 12028; Gofrativ, 13165; Gofravik, 13166; Green Sulphur, 13324; Greggio, 13348; Hortag Aquasulf, 14313; Kumulus® DF, FL, 16482; Kumulus® S, 16483; Manox, 18577; mother of pearl sulfur, 20361; Octocure 456, 21993; roll sulfur, 26920; Rubbermakers Sulfur, 27094; Solfa, 28928, 28929; Stoller Flowable Sulphur, 29782; Struktol® SU 109, 29837; Suffa, 29949; sulfur, 30043; Sulfur-F, 30054; Thiovit, 31724; Tripart® Imber, 32374; Vassgro Flowable Sulphur, 33470; Yellow sulphur, 34650
7704-99-6: zirconium hydride, 34823
7705-08-0: ferric chloride, 11648
7718-54-9: nickel chloride, 21125
7719-12-2: phosphorus trichloride, 23747
7720-78-7: green vitriol, 13337; Green-up Mossfree, 13346; iron sulfate (ous), 15321
7722-64-7: potassium permanganate, 24769, 24770; sin red, 28490
7722-76-1: Ammonium phosphate monobasic, 2251; Firesaife, 11836; Mono Ammonium Phosphate (Agricultural Grade), 20215; Phosal, 23687
7722-84-1: dioxogen, 8577; Genoxide®, 12840; hydrogen peroxide, 14590; hydrozone, 14672; Oxzone, 22478; Percarbamid, 23219; Perhydrol, 23268; Perone, 23399; Peroxal, 23403; Peroxol, 23411; Peroxyl, 23413; Proxy, 25348; Truzone, 32496
7722-86-3: Caro's acid, 5306
7722-88-5: Nutrifos, 21833; sodium pyrophosphate, 28814; Tetron, 31396; TSPP, 32540
7723-14-0: violet phosphorus, 33876
7726-95-6: Bromine, 4572
7727-21-1: Anthion, 2519; potassium persulfate, 24771
7727-43-7: Albaryt, 1353; Barium sulfate, 3644; Blanc Fixe Micro, 4313; Citobaryum, 6332; Esophotrast, 10835; Ewo, 11279; fixed white, 11858; Huberbrite 1, 14452; Micropaque, 19686; Oratrast, 22271; Radiopaque, 25830; Raybar, 25909; Roebaryt, 26902; Sachtoperse® HU, 27202; Sacon®, 27206; Sparrnite, 29209; terra ponderosa, 31325
7727-54-0: Ammonium persulfate, 2249; thioxydant lumire, 31725
7727-73-3: Glauber's salt, 12974
7732-18-5: Tritiotope, 32423
7733-02-0: zinc vitriol, 34799
7738-94-5: chromic acid, 6220
7747-35-5: Amine CS-1246, 2154; Oxaban®-E, 22405
7757-79-1: Petre, 23474; potassium nitrate, 24762; prismatic niter, 25035; sal niter, 27250; sal Prunella, 27251; saltpeter, saltpeter flour, 27298
7757-82-6: Crimidesa, 7052; sal Mirabil, 27249; salt cake, 27281; sodium sulfate, 28821; Trona Anhydrous Sodium Sulphate, 32451; Trona Salt Cake, 32458;
7757-83-7: Santosite, 27424; sodium sulfite, 28823
7757-93-9: Calbrite, 4928; calcium hydrogen phosphate, 4951; Caliment, 5006; Calipharm, 5008; Dentplus Special, 8067; Lucaphos® 40, 48, 17752; monetite, 20209
7758-01-2: Bromox, 4589; potassium bromate, 24741
7758-02-3: potassium bromide, 24742
7758-11-4: potassium phosphate, dibasic, 24773
7758-16-9: Antelope, 2517; B.P. Pyro®, 3564; Donut Pyro®, 8800; Perfection®, 23243; Sapp #4, 27446; sodium acid pyrophosphate, 28727; Taterfos®, 30844; Victor Cream®, 33727;
7758-19-2: Adox 3125, 787; C-2, 4842; Chloritane, 6109; sodium chlorite, 28749
7758-23-8: acid calcium phosphate, 330; calcium phosphate (monobasic), 4966; Cefkaphos®, 5549; H.T, 13524; Ibex, 14805; Py-Ran, 25468; Regent® 12XX, 26014; V-90®, 33222
7758-29-4: Curafos® STP, 7411; Flav-R-Keep FP-51, 11906; Hysorb, 14749; Lem-O-Fos®, 16979; Polygon, 24460; Rhodia-Phos, 26677; Sea-Gard® Formula FP-91, 27786
7758-87-4: calcium phosphate (tribasic), 4967; Ephos, 10628; precipitated phosphate, 24866
7758-89-6: Cuproid, 7394; cuprous chloride, 7403
7758-94-3: Ferrofloc, 11672; ferrous chloride, 11689
7758-95-4: Leclo, 16944; lead chloride, 16905
7758-97-6: Chrome yellow, 6212; mock vermilion, 20012
7758-98-7: cupric sulfate, 7387; Mackechnie, 18202; Sulfato de Cobre Valles, 29974
7758-99-8: Kocide® Copper Sulfate Pentahydrate Crystals, 16262
7761-88-8: lunar caustic, 17816; silver nitrate, 28431
7772-98-7: Ametox, 2049; green mordant, 13321; Lycopon, 18065; sodium thiosulfate, 28826
7772-99-8: Fascat® 2004, 11496; single muriate of tin, 28497; stannous chloride, 29502; tin salts, 31814
7773-01-5: Mangatrace, 18559
7773-06-0: Amcide, 1986; Ammonium sulfamate, 2254; Root-Out, 26959
7775-09-9: Arpal Non Selex, 3095; Atlacide, 3290; Centex, 5678; Defol, 7889; Granular Weedkiller, 13270; KM Sodium Chlorate, 16244;sodium chlorate, 28747; **7775-11-3:** sodium chromate, 28750
7775-14-6: Blankit®, 4322; Geblitol, 12708; Hybaite, 14486; Hybrite®, 14492; Hydrolin, 14610; Hydronyx, 14625; Hydros, 14635; Hydros® F, 14637; hydrosulfite, 14643; Hydrosulfite AWC, 14645; Hydrosulfite BASF, 14646; Hydrosulphit®, 14648; Luredox® BP, PO, 17937; Reductone, 25983; sodium hydrosulfite, 28771
7775-19-1: sodium metaborate, 28784, 28785
7775-27-1: sodium persulfate, 28809
7775-41-9: lachiol, 14801; silver fluoride, 28423; tachiol, 30686
7778-18-9: Anhydrite, 2458; calcium sulfate (anhydrous), 4971; Clay Breaker, 6376; drierite, 9027; pearl-hardening, 22985
7778-39-4: Aresenid, 2838; arsenic acid, 3137
7778-41-8: cupric arsenate, 7375
7778-43-0: Fatsco, 11506
7778-44-1: Calcars, 4930; calcium arsenate, 4938
7778-50-9: potassium bichromate, 24739; potassium dichromate, 24754
7778-53-2: potassium phosphate, tribasic, 24774
7778-54-3: calcium hypochlorite, 4953; chloride of lime, 6104; HTH, 14443; Hyporit, 14720; Inductor, 15089; Perchloron, 23223; Pittabs, 23888; Pittclor, 23890; Prestochlor, 24919; Pulsar, 25384; Repak, 26141; Stellos, 29641; Swim clear, 30460
7778-66-7: HyPure K, 14726; Potassium hypochlorite, 24757
7778-74-7: KM Potassium Perchlorate, 16243; potassium perchlorate, 24767
7778-77-0: Beycostat 148 K, 4099; Beycostat 273 P, 4102; potassium phosphate, monobasic, 24772
7778-80-5: glazier's salt, 12982; potassium sulfate, 24785; Trona Potassium Sulphate, 32457; Trona Sulphate of Potash, 32460
7779-90-0: Delaphos, 8008; Diroval, 8621; zinc phosphate, 34790
7782-41-4: Fluorine, 12088
7782-42-5: Aquadag, 2690; Dag 137, 7686; Dag 154, 7687; Dag 197, 7689; graphite, 13277; Lonza KS, 17589; mineral carbon, 19820; Prodag, 25067; Rollit, 26921; SLA 1262, 1275, 28578; SLA 1275, 28579; SLA 2239, 28584
7782-44-7: oxygen, 22451
7782-49-2: selenium, 27861; Vandex®, 33311
7782-50-5: chlorine, 6107
7782-63-0: ferrous sulfate, 11690; Iron vitriol, 15322; sal Martis, 27247; salt of

steel, 27290; Shoemaker's black, 28184; Taylors Lawn Sand, 30849
7782-65-2: monogermane, 20225
7782-76-5: Manganese Violet, 18550; palaite, 22554
7782-78-7: nitrosylsulfuric acid, 21379
7783-03-1: tungstic acid, 32592
7783-06-4: sulfuretted hydrogen, 30053
7783-09-7: telluretted hydrogen, 31060
7783-18-8: Ammonium thiosulfate, 2257
7783-20-2: Ammonium sulfate, 2255; Nipro (ii), 21304
7783-28-0: Ammonium phosphate dibasic, 2250
7783-35-9: mercuric sulfate, 19349
7783-40-6: magnesium fluoride, 18361; Magtran, 18412
7783-49-5: zinc fluoride, 34774
7783-61-1: silicon tetrafluoride, 28334
7783-64-4: zirconium tetrafluoride, 34827
7783-70-2: antimony pentafluoride, 2569
7783-82-6: tungsten hexafluoride, 32587
7783-90-6: silver chloride, 28421
7783-96-2: silver iodide, 28428
7784-18-1: Alcan Aluminum Fluoride, 1412
7784-25-0: Ammonium alum, 2229; Curb, 7416; Guardsman, 13450; Mandops Narsty, 18523
7784-27-2: Aluminum nitrate, 1917
7784-30-7: Aluminum phosphate, 1922
7784-34-1: arsenic trichloride, 3141
7784-35-2: arsenic trifluoride, 3142
7784-41-0: Macquer's salt, 18277; potassium arsenate, 24737
7784-42-1: arsine, 3144
7784-46-5: Harle's solution, 13637
7785-87-7: manganese sulfate, 18549; Mansu, 18612
7785-88-8: Pan-O-Lite, 22683; sodium aluminum phosphate, 28733; Stabil-9, 29399
7786-30-3: magnesium chloride, 18360
7786-81-4: nickel sulfate, 21135
7787-55-9: U Pink 113, U Violet 109, 32696
7787-59-9: Biju®, 4136; Bismuth oxychloride, 4241; Mearlite® GBU, 19192; Pearl I, II, III, 22979; Pearl Supreme UVS, 22981; pearl white, 22982; Pearl-Glo®, 22984; Perlex, 23332
7787-69-1: cesium bromide, 5771
7787-70-4: cuprous bromide, 7402
7787-85-1: 2-chloroethylmethyldichlorosilane, 6122; CC3005, 5486
7788-97-8: chromic fluoride, 6222; fluorchrome, 12072
7789-04-0: Plessy's Green, 24113
7789-09-5: Ammonium dichromate, 2237
7789-12-0: sodium dichromate, 28756
7789-17-5: cesium iodide, 5776
7789-18-6: cesium nitrate, 5777
7789-29-9: Fremy's salt, 12394
7789-38-0: Dyetone®, 9312
7789-43-7: Mancobride Mancanese, 18515
7789-45-9: cupric bromide, 7377
7789-75-5: calcium fluoride, 4947
7789-77-7: Emcompress®, 10064
7789-78-8: calcium hydride, 4949; Hydrolete, 14609; hydrolith, 14611
7789-80-2: calcium iodate, 4954
7789-82-4: calcium molybdate, 4958
7790-21-8: potassium periodate, 24768
7790-26-3: Iodotope I-125, 15217; Iodotope I-131, 15218; Iodotope Therapeutic, 15219; Oriodide-131, 22303; Radiocaps-131, 25827
7790-28-5: sodium metaperiodate, 28786
7790-29-6: rubidium iodide, 27106
7790-69-4: lithium nitrate, 17511
7790-79-6: cadmium fluoride, 4897
7790-80-9: cadmium iodide, 4899
7790-92-3: Hypochlorous acid, 14718; HyPure A, 14724
7790-93-4: Chloric acid, 6102; HyPure C, 14725
7790-98-9: KM Ammonium Perchlorate, 16236
7791-11-9: rubidium chloride, 27105
7791-12-0: thallium chloride, 31554
7803-49-8: oxyammonia, 22439
7803-55-6: KM Ammonium Metavanadate, 16235
7803-62-5: silane, 28265; Silicane, 28318
8000-46-2: geranium oil, 12888
8001-17-0: Egg oil, 9645; EmCon E-5, 10065
8001-20-5: tung oil, 32580
8001-21-6: Lipovol SUN, 17439; sunflower seed oil, 30141; Super Refined Sunflower Oil, 30199
8001-22-7: Lipovol SOY, 17438; soya bean oil, 29169; Soybean oil, 29173; Super Refined Soybean, 30198
8001-23-8: Lipovol SAF, 17435; Neobee® 18, 20872; safflower oil, 27222; saffron oil, 27225; Super Refined Safflower USP, 30195
8001-25-0: Lucca oil, 17754; olive oil, 22138; Super Refined Olive , 30193
8001-26-1: Linol, 17304; Linseed oil, 17309
8001-30-7: corn oil, 6834
8001-31-8: coconut butter, 6566; Coconut Oils® 76, 92, 110, 6567; Esi-Terge 40% Coconut Oil Soap, 10825; Konut, 16334; Super Refined Coconut Oil, 30188
8001-54-5: Morpan BC, 20344; Quadrilan® BC, 25580; Rewoquat B 50, 26390; Roccal, 26880; Variquat® 50MC, 33397; Zephiran Chloride, 34728; Zoharquat 50, 34873
8001-58-9: Creosote, 6991
8001-61-4: Jesuit's balsam, 15580
8001-69-2: Atlas Tanked Cod Oil, 3336; cod liver oil, 6569; Mainstay®, 18417; Super D, 30174
8001-75-0: ceresin, 5746
8001-78-3: hydrogenated castor oil, 14592
8001-79-4: AA Standard, AA USP, 12; Cosmetol® X, 6871; Crystal® O, Crystal® Crown, 7289
8001-85-2: Bone oil, 4402
8002-03-7: Katchung oil, 15788; peanut oil, 22969; Super Refined Peanut , 30194
8002-09-3: pine oil, 23856
8002-13-9: Actralube 7142, 538; rapeseed oil, 25874
8002-16-2: rosin oil, 26978
8002-23-1: cetyl esters, 5819; Crodamol SS, 7126; Dermalcare® SPS, 8101; Liponate SPS, 17387; Ritaceti, 26798; Ross Spermaceti Wax Substitute 573, 26992; spermaceti, 29276; Starfol® Wax CG, 29539; W.G.S. Synaceti 116 NF/USP, 34189
8002-26-4: tall oil, 30737
8002-31-1: cocoa butter, 6563; Fancol CB, 11434
8002-33-3: Actrasol C-50, C-75, C-85, 549; Chemax SCO, 5926; Chemsulf SCO/75, 6046; Eureka 102, 11207; Laurel R-50, 16837; Marvanol® SCO (50%), 18989; Nopcocastor, 21485; Nopcosulf CA-60, -70, 21504; Nopcosurf CA, 21507; Standapol® SCO, 29478
8002-43-5: Actiflo® 68, 70, 476; Alcolec® 439-C, 1481; Alcolec® 439-C, 1482; Alcolec® 440-WD, 1483; Alcolec® 495, 1484; Alcolec® BS, 1485; Alcolec® F-100, 1486; Alcolec® S, 1487; Aqualipid 95, 2701; Blendmax 322, 4346; Centrocap® 162SS, 162US, 5683; Centrol® CA, 5684; Centrolex® C, 5686; Centromix® CPS, 5687; Centrophase® C, 5688; Centrophase® HR, 5689; Centrophase® HR2B, HR2U, 5690; Centrophil® K, 5691; Crolec 4135, 7189; Dermasome® MT, 8107; Dermasome® TRF, 8112; Kelecin®, 15835; lecithin, 16939; Lecithin L-Range, 16940; Lecithin Water Dispersible CLR, 16941; Lipotin 100, 100J, SB, 17422; Phosal, 23688; PhosPho E-100, 23707; PhosPho F-97, 23708; PhosPho LCN-TS, 23711; PhosPho S-85, 23713; PhosPho T-20, 23714
8002-48-0: Nutimalt® Range, 21805
8002-50-4: Super Refined Menhaden , 30191
8002-53-7: Sicolub® E, 28219
8002-64-0: neatsfoot oil, 20836, 20837; Paracol+so M161, 22727; paraffin, 22742, 22743; Paratulle®, 22812; petroleum wax, 23501; Ross Wax #100, 26993; Ross Wax #145, 165, 26994; Sasolwaks, 27476; Sasolwaks M3, 27477; Synwax, 30659; Vybar® 103, 34165
8002-75-3: Lipovol PAL, 17434; palm oil, 22625, 22626
8003-20-6: kerosene, 16096
8006-13-1: Aluminum Acetate Solution, 1898; Buro-sol Concentrate, 4719; Domeboro, 8794; novacetoform, 21648; Printer's acetate, 24996
8006-44-8: candelilla wax, 5084
8006-54-0: Anhydrous Lanolin HP-2050, 2461; Anhydrous Lanolin P80, 2462; Anhydrous Lanolin P9SRA, 2463; Corona, 6843; Coronet, 6844; Cosmetic Lanolin USP, 6870; Fluilan, 12046; Lantrol® HP-2073, 16769; Ultra Anhydrous Lanolin HP-2060, 32779; Vigilan, 33750
8006-64-2: turpentine, 32610
8006-95-9: Cropure Wheat Germ , 7222; EmCon W, 10068; Lipovol WGO, 17440; Wheat Germ Oil, Wheat Germ Oil CLR, 34348; Wheat germ oil;, 34349

8007-12-3: nutmeg butter, 21806
8007-18-9: V-9415, 33223
8007-24-7: Cardolite® NC-511, 5268
8007-43-0: Arlacel® 83, 2903;Crill 43, 7034; Dehymuis SSO, 7968; Disponil SSO 100 F1, 8704; Emalex SPO-150, 9997; Emsorb® 2502, 10456; Glycomul® SOC, 13115; Hodag SSO, 14246; Liposorb SQO, 17415; S-Maz® 83R, 28645
8007-44-1: Oil of Pennyroyal, 22066; pennyroyal oil, 23119
8007-56-5: aqua regia, 2682
8007-69-0: Lipovol ALM, 17426; Sweet Almond oil, 30452
8008-20-6: Pearl, 22975
8008-74-0: Gingelly oil, 12946;Lipovol SES, 17436; Sesame oil, 28075; Super Refined Sesame, 30196
8012-95-1: Balneol, 3609; Emulzome, 10532; Kaydol, 15810
8013-01-2: yeast extract, 34640
8013-05-6: Solricin® 235, 28965
8013-07-8: ADK CIZER O-130P, 710; Drapex® 6.8, 8969; Edenol D82, 9607; Epoxol 7-4, 10710; Hoe S 3680, 14260; Lankroflex® GE, 16680; Paraplex® G.62, 22790; Paraplex® G-60, 22793; Plasthall® ESO, 24008; PX-800, 25445
8013-17-0: Invert Sugar, 15200
8014-95-7: fuming sulfuric acid, 12457
8015-86-9: Carnauba, 5296; Carnauba wax, 5297; Emulsion C-340, 10516
8016-28-2: Actran Extra, 546
8016-35-1: oiticia, 22094
8016-70-4: Akorex, 1149; Code 321, 6571; Diamond D 31, 8310; Famous, 11426; hydrogenated soybean oil, 14596; Lipovol HS, 17428; S-Flakes, 28131; Witarix® 440, 34431
8017-16-1: polyphosphoric acid, 24520
8018-01-7: Dithane, 8728; Karamate, 15749; mancozeb, 18517; Manplex, 18584; Manzate® 200 DF, 18625; Penncozeb, 23115
8020-84-6: Degras Special, 7904
8021-55-4: ozokerite, 22483
8022-00-2: Campbell's DSM, 5059; D S M, 7607; Persyst, 23434; Star DSM, 29524; Tripart® Systemic Insecticide, 32386
8023-79-8: palm kernel oil, 22624
8024-22-4: Grape seed oil, 13275; Lipovol G, 17427; Super Refined Grapeseed Oil, 30190
8024-32-6: Avocado Oil, 3497, 3498; Avocado Oil CLR, 3499; Cropure Avocado, 7218; Lipovol A, 17425
8024-37-1: turmeric, 32608
8027-33-6: Argobase 125, 2851; Argowax Standard, 2859; Ceralan®, 5702; Fancol LA, 11446; Hartolan, 13676; Hartolite, 13677; Ivarlan 3310, 15444; Lanolin alcohol, 16741; Ritawax, 26835; Solulan, 28998; Super Hartolan, 30178
8028-66-8: Fancol HON, 11443; Honey, 14295
8028-98-6: Acetulan, 309
8029-43-4: corn syrup, 6838
8029-44-5: Myverol® 18-85K, 20627
8029-68-3: Perichthol, 23270
8029-76-3: Alcolec® Z-3, 1488; Centrolene® A, S, 5685; Lipotin H, 17424; PhosPho 642, 23706
8029-91-2: Acetylated hydrogenated lard glyceride, 315; Axol® E 61, 3512
8029-92-3: Myvacet® 9-40, 20602
8030-30-6: Aromatic Solvent 150, 3043; Hi-Sol® 10,15,70, 14134; naphtha, 20705; petroleum naphtha, 23498; Texsolve V, 31532
8030-55-5: East Indian Balsam of Copaiba, 9461
8030-78-2: Arquad® T-27W, 3121; Arquad® T-50, 3122; Jet Quat T-50, 15606; tallow trimonium chloride, 30747
8031-09-2: sodium morrhuate, 28799
8031-44-5: Fancol HL, 11441; Ivarlan HL, 15451; Lipolan, 17366; Satulan, 27492; Super-Sat, 30270
8032-30-2: kerogen, 16086
8032-32-4: Ligroin, 17244; petroleum spirit, 23500
8038-93-5: sodium aluminum chlorhydroxy lactate, 28732
8039-09-6: Lan-Aqua-Sol 100, 16649; Lantrol® PLN, 16770; PEG-75 lanolin, 23052
8042-47-5: Drakeol 10, 8962; Drakeol 19, 8963; Drakeol 5, 8960; Drakeol 7, 8961; Paratherm NF, 22806; Parol 70, 22859; Peneteck, 23103
8048-52-0: Gonacrine, 13192
8049-47-6: Pancrease, 22671; Pancrex V, 22675
8050-09-7: rosin, 26976; Westvaco® Resin 90, 34330
8050-26-8: Oulutac 90 D, 22391; Uni-Tac® R99, 33053; Zonester® 65, 34896
8050-81-5: Amersil® simethicone, 2033; Dow Corning® Antifoam A, 8870; Dow Corning® Antifoam C, 8871; Hodag Antifoam F-1, 14165; Mazu® DF 200SP, 19165; Tego® Foamex N, 30979
8051-30-7: Empilan® 2502, 10384
8052-42-4: Bitumen, 4294; petroleum pitch, 23499
8052-48-0: Emery® 2895 Foamaster Soap L, 10156; Foamaster Soap L, 12148; sodium tallowate, 28824
8057-49-6: valerian, 33241
8061-51-6: Darvan® No. 2, 7784; Dynasperse A, 9391;Kelig, 15848; Lignosol DXD, 17237; Maracarb N-1, 18675; Maracell XE, 18676; Marasperse 52 CP, 18687; Marasperse N-22, 18689; Polyfon, 24439; Polyfon® F, 24440; Reax® 45A, 25938; Reax® 80C, 25939; Temsperse S 001, 31086; Ufoxane 2, 32755; Ultrazine NA, 32885; Zewa SL 2, 34754; Zewalon FN, 34756
8061-52-7: Additive-A, 654; Ameribond, 2012; Borrebond, 4448; Borresperse CA/CAF, 4450; calcium lignosulfonate, 4956; calcium lignosulfonates, 4957; Collex G, 6624; Darvan® No. 404, 7785; Glutrin, 13065; Goulac, 13215; Lignorit, 17232; Lignosite®, 17234; Lignosol B, 17236; Lignosol SF, 17239; Marasperse GFC, 18688; Norlig 11 DA, 21563; Norlig A, 21565; Trastan LS, 32151; Ultrazine CA, 32884; Wafex, Wafolin, 34193; Wargotan, 34217; Welltex 300 F, 34276
8061-53-8: Ammonium lignosulfonate, 2243; Lignosol TS, 17240; Tembind A 002, 31077; Wanin AM, 34204
8067-32-1: Precirol ATO, 24867; Precirol WL 2155, 24868
8067-69-4: Quixalud, 25696
8069-64-5: meralluride, 19329
9000-01-5: gum arabic, 13480; picked turkey gum, 23811
9000-05-9: gum benjamin, 13482
9000-07-1: carrageenan, 5315; Clarifloc, 6363; Stamere®, 29437; Wakal® K Range, 34197
9000-16-2: Dammar Resin, 7724
9000-30-0: guar gum, 13445; Jaguar® C, 15495; Jaguar® Guar Gum, 15500; Prinza® Range, 25010; Progacyl® CP-7, 25108; Supercol® Guar Gum, 30229
9000-36-6: karaya gum, 15760
9000-40-2: Locust bean gum, 17545; Wakal® J Range, 34196
9000-65-1: Dragon gum, 8956; gum tragacanth, 13490
9000-69-5: pectin, 23017; vegetable jelly, 33521
9000-70-8: Byco A, C, O, 4818; Crodyne BY19, 7182; gelatin, 12722; Hydrocoll G-40, 14581; Pharmagel, 23572; Pronel Capsules, 25158; Rousselot Gelatine, 27012; SPG Gelatine, 29281
9000-71-9: kalzose, 15703
9000-92-4: Arozyme TD, 3092; Clarase® 5,000, 40,000, 6356; diastase, 8331; Taka-Therm® L-340, 30724; Tenase® 1200, L-340, L-1200, 31094; Termamyl®, 31299; Termamyl® 120L, 31300
9001-05-2: Catalase L, 5429; Microcatalase®, 19661
9001-09-6: Chymopapain, 6284; Discase, 8624
9001-12-1: Collagenase, 6615
9001-19-8: Nervanzse, 21011
9001-57-4: sucrase, 29929
9001-62-1: Lipase, 17329; Pancreatic Lipase 250, 22672
9001-73-4: papain, 22698; Scintillase, 27740
9001-74-5: Penicillinase, 23112
9001-75-6: pepsin, 23200
9001-98-3: rennet, 26110
9002-01-1: Kabikinase, 15650
9002-07-7: Parenzyme, 22827; trypsin, 32530
9002-13-5: urease, 33122
9002-18-0: Agar (Agar-agar), 955; vegetable gelatin, 33518
9002-60-2: Cortymol® LP, 6861
9002-62-0: Tylose® H Series, 32660
9002-71-5: Thytropar, 31782
9002-83-9: Fluorolube® GR-290, GR-362, GR-470, GR-544, GR-660, 12104
9002-84-0: Electrafil® TR-1900/EC, 9757; Emralon 304, 10449; Emralon

8301-01, 10450; Fluon® AD1, AD1L, AD1H, 12054; Fluon® CDI, 12055; Fluon® G170, 12056; Fluorocomp® FC-101, 12094; Fluorocomp® FC-144, 12095; Fluorocomp® FC-174, 12096; Fluorocomp® FC-182, 12097; Fluorolene®, 12103; Fluoromelt®, 12106; Fluorosint® 500, 12110; Hostaflon® TF 1101, 14339; Hostaflon® TF 1620, 14340; Hostaflon® TF 2071, 14341; Hostaflon® TF 5032, 14342; Hostaflon® TF 5537, 14343; McLube 1777, 19180; Polymist® F-5, 24496; polytetrafluoroethylene, 24640, 24641. 24642; RT/Duroid® M, 27053; SLA 1611, 1612, 28581; SLA 1612, 28582

9002-86-2: Advex 91025, 821; Forex®, 12245; Geon® 83457, 12857; Geon® 83718, 12858; Geon® 86100, 86101, 86103, 12859; Geon® 8700A, 12853; Geon® 8720, 12854; Geon® 87239, 87241, 12860; Geon® 87396, 12861; Geon® 87420, 12862; Geon® 8812, 8813, 12855; Geon® 8896, 12856; Geon® HTX 92190, 12863; Geon® W015, 12865; Georgia Gulf 3131 Clear 02, 12866; Georgia Gulf 5006, 12867; Georgia Gulf 5006 General, 12868; Georgia Gulf 9105, 12869; Georgia Gulf 9151, 12870; Georgia Gulf 9175J, 12871; Georgia Gulf 9202, 12872; Georgia Gulf CL-7049, 12873; Georgia Gulf EH-71L, 12874; Georgia Gulf EX-240, 12875; Georgia Gulf HH-1900, 12876; Georgia Gulf HM-7054, 12877; Georgia Gulf SP-7107, 12878; Georgia Gulf UV-7160, 12879; Klegecell R30, 16216; Korogel, 16354; Koroplate, 16357; Lutofan®, 17996; Norvinyl DX 550, 21596; Novablend® 501, 21642; Novablend® 5555, 21643; polyvinyl chloride, 24668; Saran, 27450; Saran 313, 27451; Saran 510, 27452; Saran F-239, F-278, F-310, 27453; Saran Wrap, 27454; Saran® Filament, 27455; Ultraprene®, 32860; Vestolit®, 33679; Vestolit® B 7021, 33680; Vestolit® E 6003, E 6503, E 7003, 33682; Vestolit® LF HI-SP 5735, 33684; Vestolit® M 5867, 33685; Vestolit® P 1330 K, 33686; Vestolit® S 6058, 33687; Vinoflex®, 33839

9002-88-4: A-C® 6, 168; A-C® Polyethylene 6, 6A, 7, 7A, 8, 8A, 9, 9A, 617, 617A, 175; ACtone® P, 513; ACumist A-12, A-18, 580; ACumist B-6, B-12, B-18, C-5, C-12, C-18, 581; Akrowax PE, 1199; Alathon® H5234, 1336; Alathon® L5440, 1337; Alathon® M5560, 1338; Alathon® M6062, 1339; Amsoft MDH-20, 2370; Arcel Moldable Polyethylene Copolymers, 2807; Arpak 4322, 3094; Attane 4601, 4802, 3403; Cabelec® 1017, 4862; Cabelec® 1017, 4863; Cabelec® 3172, 4865; Cabot® PE 9007, 4873; Cadco® UHMW, 4889; Compound 403/401, 6684; Conductomer HDC-22HLMI-M, 6721; Dowlex 2032 2035, 2042, 2500, 3010, 8939; Ecothene EC 101, 9591; Electrafil® PE-90/EC, 9755; Empee® FR 42 LM, 10280; Empee® PE-112, 10281; Empee® PE-113, 10282; Epolene® C-10, 10678; Epolene® C-13, 10679; Epolene® N-20, N-21, 10684; Esi-Cryl 1540A, 10814; Esi-Cryl 1E10N, 10810; Esi-Cryl 325N, 10813; Forbest 410, 12236; Fortiflex®, 12299; Fortiflex® A60-70-99, A60-70-119, 12300; Fortiflex® B45-06R-09, 12301; Fortiflex® G36-24-149, 12302; Fortiflex® G38-70C, 12303; Fortiflex® J36-25-142, 12304; Fortiflex® K36-55-122, 12305; Fortiflex® T50-200, 12306; Fortiflex® XF-855, 12307; Fortiflex® XF-855, 12308; Fusabond® MB-110D, 12509; Hartolon 5683, 13678; HDPE 04352N, 13720; HDPE 25053-P, 13721; HDPE 32060C, 13722; HDPE 35053, 13723; HDPE IP-10, 13724; HiD 9301, 14066; HiD 9602, 14067; HiD 9632, 14068; HiD 9650, 14069; Hi-Zex, 14149; Hoechst Wax PE 190, 14265; Hoechst Wax PE 520, 14266; Hoechst Wax PED 191, 14267; Hoechst Wax PED 521, 14268; Hostalen® EP 4450, 14351; Hostalen® G, 14352; Hostalen® GB 6950, 14353; Hostalen® GM 5010 T2, 14354; Hostalen® GM 7745 HP, 14355; Hostalen® GUR 5121, 14356; Innovex, 15138; Loobwax 0651, 17597; Lupolen®, 17904; Luwax A, 18038; Luwax AF 30, 18039; Luwax AH 3, 18040; Luwax EAS 1, 18041; Luwax EVA 1, 18042; Marlex® CL-200, 18762; Marlex® HHM 4903, 18790; Marlex® HHM TR-140, 18793; Marlex® HHM TR-400, 18800; Marlex® HMN 4550, 18808; Marlex® HXM 50100, 18816; Micropoly 520, 19690; Microthene® MA 530-060, 19706; Microthene® MN 701-00, 19707; Microthene® MP 625U, 19708; Microthene® MU 760-00, 19709; Mitsubishi BT002, 19966; Mitsubishi ET008, 19967; Mitsubishi F101A, 19968; Mitsubishi JS050, 19969; Mitsubishi L300, 19974; Mitsubishi UF421, 19975; Neopolen® E, 20941; Nortuff RA 7020-KO, 21589; Novapol® GF-0218-F, 21682; Novapol® GI-2024-A, 21683; Novapol® HB-L455-A, 21684; Novapol® LC-0517-A, 21685; Novapol® LE-0220-A, 21686; Novapol® LF-0223-B, 21687; Novapol® LF-Y819-D, 21688; Novapol® PF-0118-B, 21689; Novapol® PR-0636-UG, 21690; Octowax 518, 22034; Orevac® 18302, 22293; PE 1017, 22961; PE 4517, 22962; PE 5554-H, 22964; PE 5861, 22965; Petrothene® GA 501, 23520; Petrothene® GA 564, 23521; Petrothene® GA 808-090, 23522; Petrothene® HD 5903B, 23523; Petrothene® LB 5003-00, 23524; Petrothene® LB 6001-00, 23525; Petrothene® LF 6030-00, 23526; Petrothene® LP 5102-00, 23527; Petrothene® LS 3150-00, 23528; Petrothene® LT 5704-00, 23529; Petrothene® NA 155-000, 23530; Petrothene® NA 204-000, 23531; Petrothene® NA 341-000, 23532; Petrothene® PA 436, 23533; Poligen PE, 24326; Polybond® 1009, 24365; polyethylene, 24424; Polysoft B, 24550; Polywax® 500, 24671; Primax UH-1060, 24975; Primax UH-1080, 24976; Primax UH-1250, 24977; Retain PE-1001, 26247; Retain PE-5009, 26248; Retain Rp-120, 26250; Rexene® PE 6010, 26443; Rexene® PE6076, 26444; Ribiene, 26721; Rigidex, 26752; Rigidex 3, 26753; Rigidex 9, 26754; Rigidex X4RR, 26755; Rigillene, 26758; RTP 701, 27065; Spectratech® FM 1150H, 29244; Stat-Kon® FE, 29563; Struktol® PE H-100, 29836; Suntec® -HD, 30153; Suntec® -LD, 30154; Tenite® 154DF, 31105; Tenite® Polyethylene, 31112; Tuflin HS-7066 NT7, 32562; Vestolen® A3512, 33670; Vestolen® AX4304, 33672; Zemid® 610, 34710; Zemid® 650, 34711

9002-89-5: Airvol® 103, 107, 1121; Airvol® 125, 1122; Airvol® 165, 1123; Airvol® 321, 325, 350, 1127; Airvol® MH-82, MM-14, MM-51, MM-81, 1134; Airvol® SH-72, SM-73, 1135; Elvanol® 20-25, 9890; Elvanol® 71-30, 9891; Elvanol® 90-50, 9892; polyvinyl alcohol, 24666; Release Agent NL-2, 26039; Vinex 1003, 33820; Vinex 2019, 33821; Vinex 2025, 33822; Vinex 2034, 33823; Vinex 2144, 33824; Vinex 5030, 33825

9002-91-9: Helarion, 13759; Mini Slugit Pellets, 19830; Mota, 20360; Quad Mini Slug Pellets, 25567; Slug Destroyer, 28615; Slug Pellets, 28616; Slugit Liquid, 28619; Slugoids, 28620; Tripart® Mini Slug Pellets, 32381

9002-92-0: Akyporox RLM 160, 1290; Marlipa® MG, 18857; Marlipal® 124, 18864;

Marlowet® BL, 18930; Volpo L4, 34027
9002-92-4: Vanzyme®, 33377
9002-93-1: Akyporox OP 100, 1284; Akyporox OP 115 SPC, 1285; Akyporox OP 200, 1286; Akyporox OP 250V, 1287; Akyporox OP 400V, 1288; Chemax OP-3, 5917; Chemax OP-30/70, 5919; Chemax OP-40, 5920; Chemax OP-40/70, 5921; Chemax OP-7, 5918; Dehydrophen POP 4, 7967; Hyonic OP-7, 14694; Iconol OP-10, 14827; Iconol OP-30, 14828; Iconol OP-40, 14829; Igepal® CA-210, 14846; Igepal® CA-897, 14849; Macol® OP-3, 18271; Makon® OP-6, 18430; Marlophen® 810, 18884; Marlophen® 85, 18883; Nutrol 100, 21840; Prox-onic OP-09, 25332; Remcopal 306, 26071; Remcopal O11, 26096; Remcopal O12, 26097; Remcopal O9, 26095; Rexol 45/1, 26469; Sellig O 11 100, 27897; Sellig O 12 100, 27898; Sellig O 20 100, 27899; Sellig O 4 100, 27900; Sellig O 5 100, 27901; Sellig O 6 100, 27902; Sellig O 8 100, 27903; Sellig O 9 100, 27904; T-Det® O-4, 30868; T-Det® O-407, 30869; Teric X5, 31289; Triton® X-15, 32433; Trycol® 6956, 32505
9002-97-5: Akypopress DB, 1268
9003-01-4: Acrysol® ASE-75, 444; Acumer 1000, 575; Acusol® 445, 583; Alcogum L-15, L-26, L-28, L-31, L-35, L-36, 1472;Burcotreat 900-A, 4699; Carbopol® 613, 614, 5230; Good-rite® K-702, 13207; Good-rite® K-752, 13209; Sokalan® PA 110 S, 28876
9003-03-6: Acrysol® G-110, 446; Alcogum 9639, 1468; Ammonium polyacrylate, 2252; BYK® 156, 4826; Size CB, 28559
9003-04-7: Acrysol® GS, 447; Acrysol® HV-1, 448; Acusol® 410N, 582; Acusol® 445N, 584; Acusol® 445ND, 585; Acusol® 480N, 589; Acusol® 860N, 594; Alcosperse 107, 124, 149, 157, 1516; Alkasperse® A-20, 1712; Burcosperse AP Liq, 4697; Colloid 202, 6630; Daxad® 37LN10-35, 7803; Drewsperse® 611, 9020; sodium polyacrylate, 28811; Sokalan® CP 7, 28872; Sokalan® PA 15, 28878
9003-05-8: polyacrylamide, 24348; Polyhall® 21J, 24464
9003-07-0: Acctuf 3045, 259 Adpro AP 2112-GP, 788 Adpro AP 8210-HS, 789 Amoco® 1012, 2268 Amoco® 1016, 2269 Amoco® 1246, 2270 Amoco® 4018, 2271 Amoco® 5016, 2272 Amoco® 6114, 2273 Amoco® 6400p, 2274 Amoco® 7234, 2275 Amoco® 7239, 2276 Amoco® 7728, 2277 Amoco® 9119, 2280 Arpro 3313, 3097 Astryn® 63A6-2, 3268 Astryn® 63F4-2, 3269 Astryn® 65F4-4, 3270 Astryn® 65F5-4, 3271 Astryn® 73F4-2, 3272 Astryn® 73F5-2, 3273 Astryn® 78F4-2, 3274 Astryn® BA16G, 3275 Astryn® SD068-4, 3276 Bicor® 240 B, 306 B, 420 B, 470 B, 4130 Cabelec® 3004, 4864 Cabelec® 3464, 4866 Cabelec® 3464, 3004, 4867 Celstran® PPG30-01-4, 5655 Eastman® P4C5B-030, 9472 Electrafil® J-60/CF/30, 9738 Electrafil® JM-61/CF/10, 9752 Electrafil® PP-60/CC/20/EC, 9756 Empee® PP Conc.33, 10284 Empee® PP-301, 10285 Empee® PP-459, 10286 Empee® PP-560, 10287 Epolene® N-15, 10683 Fiberfil® J-60/30/E8, 11718 Fiberfil® J-60/30/FR, 11719 Fiberfil® PP-60/TC/40, 11725 Fiberil® M-1492, 11728 Fortilene®, 12309 Fortilene® 1001, 12310 Fortilene® 1602, 12311 Fortilene® 1802, 12312 Fortilene® 2104, 12313 Fortilene® 3151, 12314 Fortilene® 4104, 4109, 12315 Fortilene® 4209, 12316 Fortilene® 5801, 12317 Fortilene® 9000, 12318 Fusabond® MZ-109D, 12511 Fusabond® MZ-203D, 12512 Hercoflat® Texturing Pigments and Flatting Agent, 13815 Hercotuf, 13847 HiFax AB 6023, 14073 HiFax CA 45A, 14074 HiFax CB 17AC, 14075 HiFax ETA 3011, 14076 HiFax ETA 3095, 14077 HiFax ETA 5012, 14078 HiFax RTA 3263E, 14079 HiGlass BJ44A, 14087 HiGlass PF062-2, 14088 HiGlass PF072-1, 14089 HiGlass SB 224-2, 14090 Hitac 300, 14138 Hoechst Wax PP 230, 14269 Hostalen® OO, 14357 Hostalen® PP 927, 14358 Marlex® HGX-030, 18782 Marlex® HGZ-120-02, 18786 Marlex® HNS-080, 18815 Marlex® RMX-020, 18819 Metallyte 70-2, 70-4, 70-U, 80-2, 80-4, 80-U, 140-2, 140-4, 140-U, 19439 Mitsubishi 4300J, 19965 Neopolen® P, 20942 Nepol® PP40, 20990 Nortuff RA 1700-MO, 21588 Nortuff RC 1700-MO, 21590 OPPalyte® 233 TW, 278 TW, 350 TW, 22219 Orevac® PP-C, 22294 Petrothene® PP 1510-HC, 23534 Petrothene® PP 2004-MR, 23535 Petrothene® PP 7300-KF, 23536 Petrothene® PP 8000-GK, 23537 Petrothene® PP 8770-HU, 23538 Plastech+sn EP 8126, 23971 Plastech+sn PP 3344, 23972 Polifil® C-10, 24316 Polifil® CAS-40, 24317 Polifil® GFPP-10, 24318 Polifil® GFPPCC-10, 24319 Polifil® M-20, 24320 Polifil® RMC-10, 24321 Polifil® RMT-10, 24322 Polifil® T-10, 24323 polypropylene, 24527 PP-10GF/000, 24840 PPX-30GF/000 HC, 24849 Procond-101, 25064 Pro-fax® 6323, 25091 Pro-fax® 65F4-4, 25092 Pro-fax® 65F5-4, 25093 Pro-fax® 7523, 25094 Pro-fax® 8523, 25095 Pro-fax® HB-301, 25096 Pro-fax® PC-072PM, 25097 Pro-fax® PD-064, 25098 Pro-fax® PF-101, 25099 Pro-fax® SA-747M, 25100 Pro-fax® SB-242, 25101 Pro-fax® SD-062, 25102 Pro-fax® SE-191, 25103 Pro-fax® SV-256M, 25104 Pro-fax® Z-39S, 25105 Rexene® 11S12, 26438 Rexene® 13S10A, 26439 Rexene® 14S4A, 26440 Rexene® 18C3A, 26441 Rexene® PP 12R10A, 26445 Rexene® PP 23S2, 26446 Rexene® PP 9234, 26447 Rexene® PP41E2, 26448 RTP 100 GB 10, 27055 SD-376, 27781 Shell 5A18Z, 28146 Shell 5A95, 28147 Shell 5C64, 28148 Shell 6A01K, 28149 Shell 6C20S, 28150 Shell 7C55H, 28151 Shell DS 6C46L, Shell DS 7C04N, 28153 Shell DS 7C04N, 28154 Shell JF 6100, 28156 Shell PDC 1120, 28158 Shell WRS 6-198, 28159 Shell WRS 6-205, 28160 Stat-Kon® M-1 HI, 29564 Symalit GM 20 PP, 30505 Thermocomp® MF-1002, 31617 Transpalene®, 32145 Vestolen® P2000, 33674 Vestolen® P2000 CR, 33675 Vestolen® P6000, 33676 Vestolen® P6700, 33677 Vestolen® P7032 G, 33678
9003-08-1: Basocoll® CM, 3710; Ecco Rez M-300-7, 9516; Luwipal®, 18044; MEL 80-P, 19252; Melamine formaldehyde resin, 19256; Uformite® 27-802, 32754
9003-09-2: Gantrez® M-154, 12655; Polyvinyl methyl ether, 24669
9003-11-6: Antarox® 17-R-2, 2504; Antarox® E-100, 2509; Antarox® L-61, 2510; Antarox® PGP 23-7, 2512; Hodag Nonionic 1017-R, 14186; Hodag Nonionic 1035-L, 14187; Hodag Nonionic 1044-L, 14188; Hodag Nonionic 1064-L, 14189; Hodag Nonionic 1088-F, 14190; Hodag Nonionic 2017-R, 14191; Industrol® N3, 15116; Macol® 1, 18236; Macol® 101, 18256; Macol® 108, 18257; Macol® 15, 18240; Macol® 16, 18241; Macol® 18, 18242; Macol® 19, 18243; Macol® 2, 18237; Macol® 23, 18244; Macol® 27, 18245; Macol® 33, 18246; Macol® 34, 18247; Macol® 35, 18248; Macol® 4, 18238; Macol® 40, 18249; Macol® 42, 18250; Macol® 44, 18251; Macol® 46, 18252; Macol® 72, 18253; Macol® 77, 18254;

Macol® 8, 18239; Macol® 85, 18255; Monolan® 8000 E/80, 20231; Pluronic® 10R5, 24210; Pluronic® 10R8, 24211; Pluronic® 17R1, 24212; Pluronic® 17R2, 24213; Pluronic® 17R4, 24214; Pluronic® 17R8, 24215; Pluronic® 25R1, 24216; Pluronic® 25R2, 24217; Pluronic® 25R4, 24218; Pluronic® 25R5, 24219; Pluronic® 25R8, 24220; Pluronic® 31R1, 24221; Pluronic® 31R2, 24222; Pluronic® 31R4, 24223; Pluronic® F108, 24224; Pluronic® F127, 24225; Pluronic® F38, 24226; Pluronic® F68, 24227; Pluronic® F68LF, 24228; Pluronic® F77, 24229; Pluronic® F87, 24230; Pluronic® F88, 24231; Pluronic® F98, 24232; Pluronic® L101, 24233; Pluronic® L121, 24234; Pluronic® L122, 24235; Pluronic® L31, 24236; Pluronic® L35, 24237; Pluronic® L42, 24238; Pluronic® L43, 24239; Pluronic® L44, 24240; Pluronic® L61, 24241; Pluronic® L62, 24242; Pluronic® L62D, 24243; Pluronic® L62LF, 24244; Pluronic® L63, 24245; Pluronic® L64, 24246; Pluronic® L72, 24247; Pluronic® L81, 24248; Pluronic® L92, 24249; Pluronic® P103, 24250; Pluronic® P104, 24251; Pluronic® P105, 24252; Pluronic® P123, 24253; Pluronic® P65, 24254; Pluronic® P75, 24255; Pluronic® P84, 24256; Pluronic® P85, 24257; Pluronic® P94, 24258

9003-13-8: Hodag PB-285, 14199; PPG butyl ether series, 24841; Ucon® Fluid AP, 32725

9003-17-2: Budene® 1207, 4652; Budene® 1254, 4653; Cisdene® 1203, 6326; Diene® 35AC, 55AC, 8427; Diene® 70AC, 8428; E-BR® 8405, 9504; E-BR® 8471, 9505; Poly bd® R-45HT, 24343; polybutadiene, 24372; Polyoil Hu+a5ls 110, 24501; Ricon 159, 26738; Taktene 220, 30728

9003-18-3: Krynac® 19.65, 16451; Krynac® 20H35, 16452; Krynac® 34.140, 16454; Krynac® 34.35, 34.50., 16455; Krynac® 34.80, 16457; Krynac® 823X2, 16458; Krynac® 843, 16460; Krynac® PXL 34.17, 16465; Nipol® 1000X132, 21286; Nipol® 1001 CG, 21287; Nipol® 1022X59, 21288; Nipol® 1072, 21289; Nipol® 1411, 21292; Tylac® 68151-00, 32656

9003-20-7: Catomer VA, 5463; Daratak® RP2000, 7763; Daratak® SP1011, 7764; Everflex® SP-1084, 11266; Liquid Latex, 17472; polyvinyl acetate homopolymer, 24665; Quabond® 210, 25563; Rovace 2113, 27019; Rovace 571, 27018; Ucar® Latex 130, 32704; Vinac® 1000 DEV, 33763; Vinac® XX-210, XX-220, XX-230, XX-240, 33764

9003-22-9: Vestolit® B 7090, 33681; Vinnol® 50, 33835

9003-24-1: Darvan® No. 1, 7783

9003-27-4: Vistanex® MML-80, MML-100, MML-120, MML-140, 33920

9003-28-5: Actipol® E6, 489; Amoco® BR-310, 2281; Amoco® CI-500, 2282; Amoco® H-15, 2283; Amoco® L-14, 2286; Duraflex® 8410, 9145; Polybutene, 24373; Santac 52, 27380

9003-29-6: Permethyl 104A, 23389

9003-31-0: Natsyn® 2200, 20763; natural rubber, 20766; polyisoprene, 24470

9003-39-8: Agrimer 15L, 1065; Agrimer AL-22, 1066; H+72old EP-1, 13526; Luviskol® K12, K17, K30, K60, 18031; Luviskol® K80, K90, 18032; Peregal® ST, 23237; Plasdone®, 23935; Plasdone® C, 23936; Plasdone® K, 23937; Plasmosan, 23964; Povidone, 24805; PVP, 25429; Sokalan® HP 50, 28874; Sokalan® HP 53, 28875; Tears Plus®, 30876; Vinisil, 33827

9003-53-6: Amoco® H2R, 2284; Amoco® H3E, 2285; Amoco® R1, 2288; Amoco® R5, 2289; Dylene, 9319; Dylite® D195B, 9320; Dylite® R2595B EPS, 9321; Edistir® FA, 9622; Edistir® N 1280, N 1281, 9623; Edistir® RC, 9624; Edistir® RK, 9625; Edistir® RKV, 9626; Edistir® RV 8, 9627; Edistir® SR 550, SRL 550, 9628; Edistir® UT/1, 9629; Edistir® UT/SF, 9630; Electrafil® J-30/CF/20, 9736; Empee® PS-921, 10288; Huntsman 201, 14469; Huntsman 240, 14470; Huntsman 312, 14471; Huntsman 331, 14472; Huntsman 351, 14473; Huntsman 474, 14474; Huntsman 765, 14475; Mobil 1240, 19996; Mobil 2120, 19997; Mobil 5350, 19998; Mobil 5600, 19999; Mobil 8020, 20000; Mobil MX 4354, 20001; Polypenco® Q200.5, 24514; polystyrene, 24601; Polystyrene 101, 24602; Polystyrene 220, 24603; Polystyrene 410, 24604; Polystyrene P 2122, 24605; Replay RP 2177, 26145; Replay RP 2236, 26146; Retain PS-4000, 26249; Rigipore, 26759; RTP 401, 27062; Styrodur®, 29869; Styrofoam Brand Insulation, 29873; Styron 421, 29881; Styron 478, 29882; Styron 479, 29883; Styron 6075, 29885; Styron 6087 SF, 29886; Styron 697, 29884; Styropor® F, 29891; Styropor® FH, 29892; Styropor® P, 29893; Styvex 22000 NA, 29895; Thermocomp® CF-1004, 31609; Thermocomp® NF-1004, 31618

9003-54-7: Darex® 165L, 7770; Luran® 358 N, 17926; Luran® 378 P G7, 17927; Luran® KR 2556, 17931; RTP 501, 27063; SN-30GF/000 FR, 28660; Stylac® AS, 29856; styrene-acrylonitrile copolymer, 29864; Styvex 32000 BK, 29896; Thermocomp® BF-1006, 31607; Thermocomp® BF-1006FR, 31608; Tyril 1011, 32678; Tyril 125, 32675; Tyril 880, 32676; Tyril 880B, 32677

9003-55-8: Bayer SBR Latex 200 C, 3771; Baygon®, 3790; KR01 K-Resin Polymer, 16382; KR04 K-Resin Polymer, 16383; Kraton®, 16386; Kraton® D 1101, 16387; Kraton® D 1116, 16389; Kraton® D 2103, 16391; K-Resin Polymer KR01, 16404; K-Resin Polymer KR03, 16405; K-Resin Polymer KR04, 16406; K-Resin Polymer KR05, 16407; K-Resin Polymer KR10, 16408; Krylene® 606, 16449; Krylene® 608, 16450; Pliolite® S-6B, S-6F, 24147; Styrolux®, 29877; Styroplus®, 29889; Tylac® 68202-00, 32657

9003-56-9: ABS 124ESG, 236F, 236MA, 160; ABS 301K, ABS 500FR-1, 161; Acrylonitrile-butadiene-styrene, 4 3 8 ; Acrylonitrile-butadiene-styrene, 439; AS-10GF, 3169; AS-15CF/000, 3170; Blendex® 101, 4340; Blendex® 310, 4341; Claradex CH-540, 6354; Conductomer ABS-22, 6720; Cycolac® CKM1, 7547; Cycolac® DH, 7548; Cycolac® GPM4700, 7549; Cycolac® GPX2800, 7550; Cycolac® KCS, 7551; Cycolac® KJM, 7552; Cycolac® X-11, 7553; Electrafil® G-1204/SS/3, 9728; Electrafil® J-1200/CF/10, 9744; EMI-X® PDX-A-88128, 10201; Magnum 240, 18392; Magnum 2610, 18396; Magnum 275, 18393; Magnum 3661, 18397; Magnum 4420, 18398; Magnum 445 HQ, 18394; Magnum 788HP, 18395; Magnum 9450P, 18399; Magnum FG960, 18400; Multibase ABS 3075, 20463; Multibase ABS 3525 CL, 20464; Multibase ABS 3959, 20465; Novodur® L3FR, 21722; Novodur® P2H-AT, 21723; Novodur® P2HE, 21724; Novodur® P2HGV, 21725; Novodur® P2T, P2T-AT, 21726; Novodur® PMTM, 21727; RTP 601, 27064; Stat-Kon® AC-1003, 29559; Stat-Kon® AS, 29560; Stylac® ABS, 29855; Styvex 40007 BKL2, 29897; Styvex 42023 NAFR, 29898; Terluran®, 31297; Terlux®, 31298; Thermocomp® AF-1004, 31606

9003-59-2: Kayexalate, 15812

9003-79-6: Aminox® Flake, Powd, 2193; BXA Flake, 4811; Vanox® AM, 33340

9003-99-0: peroxidase, 23404

9004-07-3: α-Chymotrypsins, 6286; Alpha Chymar, 1820; Zolyse, 34889

9004-32-4: Aqualon® Cellulose Gum, 2703; Aqualon® CMC-T, 2704; Aqualon® CMC-T, 2705; Aquasorb® A250, 2737; Blanose, 4325; carboxymethylcellulose, 5257; carboxymethylcellulose sodium, 5258; Cekol, 5558; Tylose® C, C-p, CB, CB-p, 32659
9004-34-6: Celluflow C-25, 5616; cellulose, 5626; CF 31,000 C Coarse, 5830; CF 42,500T Medium, 5831; CF 70,000WDK, Ex. Superfine, 5832; Elcema® F150, G250, P100, 9721; Emcocel® 90M, 10042; Fibra-Cel®, 11748; Solka-Floc® BW-40, BW-100, BW-200, BW-2030, UF-900-FCC & NF, 28949
9004-35-7: CA-394-60S, 4856; Celluflow TA-25, 5617; cellulose acetate, 5627; Courtaulds, 6914; Kodapak®, 16284; Rotuba H, 27007; Tenite® 105-MS, 31104; Tenite® Cellulosic Acetate, 31108
9004-36-8: CAB-171-15S, 4857; CAB-381-0.1, 381-0.5, 381-2, 381-20, 4858; CAB-500-5, 4859; Cabulite, 4877; Cellulose acetate butyrate, 5628; Cellulose acetate butyrate, 5629; Dispercab, 8661; Tenite® 264-MH, 31106
9004-39-1: Cellulose acetate propionate, 5630; Cellulose acetate propionate ester, 5631; Dispercap, 8662; Tenite® 360-H2, 31107; Tenite® Cellulosic Propionate, 31110
9004-53-9: dextrin, 8235
9004-54-0: Dextran, 8233
9004-57-3: Ethocel Standard Premium, 10956; ethyl cellulose, 11066; Hercules® K, 13860; Hercules® N, 13862; Hercules® T, 13866
9004-58-4: Aqualon® EHEC, 2707
9004-61-9: Cromoist HYA, 7200
9004-62-0: Cellosize, 5609; Cellosize® HEC QP Grades, 5610; Hetastarch, 13891; Natrosol® 250, 20758; Natrosol® Hydroxyethylcellulose, 20760
9004-64-2: Klucel®, 16229; Klucel® E, G, H, J, L, M, 1623 Klucel® EF, 16231; 0
9004-65-3: C u l m i n a l ® Hydroxypropylmethylcellulose, 7346; Goniosol, 13194; hydroxypropyl methylcellulose, 14668; Lacril®, 16565; Methocel® 40-202, 19489; Methocel® E3 Premium, 19493; Methocel® E4M, 19494; Methocel® E5, 19495; Methocel® F4M, 19496; Methocel® K100MP, 19497; Methocel® K35, 19498
9004-67-5: Culminal® Methylcellulose, 7347; Glutolin, 13064; Methocel, 19488; Methocel® A15-LV, 19490; Methocel® A4C, A4M, 19491; Methocel® A4MP, 19492; methylcellulose, 19566
9004-70-0: AS, 3168; cellodin, 5605; collodion cotton, 6627; Dispercel, 8663; Hercules® AS, 13858; Hercules® RS, 13864; Hercules® SS, 13865; Kodaloid, 16283; nitrocellulose, 21353; Pyralin, 25461; RS Nitrocellulose, 27051; SS Nitrocellulose, 29388; Venite, 33560
9004-73-3: Masil® SF-MH, 19022
9004-74-4: Carbowax® MPEG 350, 5248
9004-81-3: Ablunol 200ML, 95; Alkamuls® L-9, 1669; Chemax E-200 ML, 5896; Chemax E-400 ML, 5899; Chemax E-600 ML, 5902; CPH-43-N, 6941; Crodet L4, 7172; Emerest® 2620, 10104; Emerest® 2630, 10108; Ethox ML-5, 11032; Hetoxamate LA-5, 13910; Hodag 20-L, 14157; Hodag DGL, 14174; Kessco® PEG 200 ML, 16122; Lexemul® PEG-400ML, 17141; Lipo DGLS, 17337; Lipopeg 4-L, 17397; Mapeg® 200 ML, 18631; Mapeg® DGLD, 18640; Nopalcol 1-L, 21463; Nopalcol 4-L, 21469; Pegosperse® 100 L, 23059; Witconol 2620, 34503
9004-82-4: Abex® 23S, 40; Akyposal 9278 R, 1301; Akyposal EO 20 MW, 1307; Akyposal MS SPC, 1312; Akyposal RLM 70, 1320; Avirol® FES 996, 3479; Avirol® SE 3002, 3484; Calfoam ES-30, 4984; Carsonol® SES-S, 5337; Chemsalan RLM 28, 6030; Disponil FES 32, 8691; Empicol® ESB, 10319; Empimin® KSN27, 10423; Genapol® ARO, 12786; Genapol® LRO Liq., Paste, 12790; Genapol® ZRO Liq., Paste, 12803; Hartenol LES 60, 13670; Marlinat® 242/28, 18848; Nonasol N4SS, 21440; Norfox® SLES-60, 21551; Nutrapon ES-60 3568, 21818; Nutrapon KPC 0156, 21823; Polystep® B-12, 24592; Rewopol® NL 2-28, 26367; Rhodapex® ES, 26628; sodium laureth sulfate, 28778; Standapol® AP Blend, 29468; Steol® 4N, 29660; Steol® OS 28, 29664; Sulfochem ES-2, 30003; Sulfochem ES-70, 30006; Sulfotex LMS-E, 30033; Texapon N 25, Texapon NSE, 31467; Texapon NSE, 31469; Texapon NSO, 31471; Texapon PLT-227, 31473; Ungerol N2-28, 32921; Witcolate ES-2, Witcolate LES-60C, 34477; Zoharpon ETA 27, 34863
9004-83-5: Alcodet® 218, 1459
9004-84-9: soluble starch, 28986
9004-92-2: Hetoxol OL-2, 13946; oleth-2, 22125
9004-93-7: Uniplex 809, 33021
9004-94-3: Acconon 1300, 235
9004-94-8: Genagen P-070, 12761; PEG-7 palmitate, 23051
9004-95-9: Akyporox RC 200, 1289; Brij® 52, 4517; Cetomacrogol 1000, 5808; Disponil O 5, 8694; Emalex 103, 9912; Emalex 1605, 9913; Ethosperse® CA-2, 11005; Hetoxol CA-2, 13932; Lipocol C-2, 17352; Macol® CA-2, 18262; Rhodasurf® C-2, 26652; Simulsol 58, 28483
9004-96-0: Ablunol 200MO, 96; Acconon 400-MO, 234; Alkamuls® 400-MO, 1657; Alkamuls® A, 1659; Alphoxat O 105, 1841; Chemax E-1000 MO, 5905; Chemax E-200 MO, 5897; Chemax E-400 MO, 5900; Chemax E-600 MO, 5903; Crodet O4, 7173; Emalex 218, 9919; Emalex OE-6, 9978; Emerest® 2617, 10103; Emerest® 2624, 10106; Emerest® 2660, 10112; Empilan® BQ 100, 10388; Emulan® A, 10487; Ethofat® O/20, 10965; Ethox MO-9, 11033; Ethylan® A10, 11093; Ethylan® A2, 11094; Ethylan® FO30, 11102; Eumulgin PLT 4, 11184; Genagen O-090, 12760; Hetoxamate MO-2, 13911; Hodag 40-O, 14159; Hodag DGO, 14175; Kessco® PEG 200 MO, 16123; Laurel PEG 400 MO, 16835; Mapeg® 200 MO, 18632; Marlosol® OL2, 18912; Mulsifan RT 1, 20454; Nonex 04E, 21443; Nopalcol 4-0, 21466; Pegosperse® 100 O, 23060; Prox-onic OL-1/05, 25331; Remcopal 207, 26062; Remcopal 6, 26053; Secoster® MO 400, 27821; Sellig AO 15 100, 27876; Sellig AO 25 100, 27877; Sellig AO 6100, 27878; Teric OF4, 31270; Trydet 2676, 32509; Witconol H31A, 34515; Witflow 916, 34526
9004-97-1: Hodag 40-R, 14160; Nopalcol 6-R, 21472; PEG ricinoleate series, 23031
9004-98-2: Ablunol OA-6, 107; Akyporox RO 90, 1294; Akyporox RTO 70, 1295; Ameroxol® OE-2, 2022; Ameroxol® OE-5, 2023; Brij® 93, 4519; Chemal OA-20/70CWS, 5884; Chemal OA-4, 5885; Chemal OA-5, 5886; Emalex 508, 9923; Ethal OA-10, 10924; Eumulgin 05, 11171; Eumulgin M8, 11178; Eumulgin PWM2, 11188; Eumulgin WM5, 11193; Genapol® O-020, 12791; Hetoxol OA-3 Special, 13945; Hostacerin O-20, 14323; Lipocol O-2, 17356; Macol® OA-2, 18270; Marlipal® 1850/5, 18867; Marlowet® WOE, 18939; oleth-8, 22126; Prox-onic OA-1/04, 25329; Rhodasurf® ON-870, 26658; Ritoleth 2, 26848; Simulsol 98, 28481; Trycol® 5971, 32503; Varonic® 32-E20, 33437; Volpo 10, 34022; Volpo 20, 34023; Volpo 3, 34020; Volpo 5, 34021; Volpo N3, 34028; Volpo O3, 34029
9004-99-3: Ablunol 200MS, 97; Ablunol DEGMS, 99; Acconon 200-MS, 400-MS, 233; Alkamuls® S-65-40, 1679; Alkamuls® S-65-8, 1680; Alphoxat S 110, 1842; Cerasynt® 840, 5734; Chemax E-1000 MS, 5906; Chemax E-200 MS, 5898; Chemax E-400 MS, 5901; Chemax

E-600 MS, 5904; Cremophor® S 9, 6986; Crodet S4, 7174; Emalex 400A, 9922; Emalex 6300 M-ST, 9930; Emalex 805, 9927; Emalex DEG-m-S, 9942; Emerest® 2610, 10102; Emerest® 2636, 10110; Ethofat® 18/14, 10963; Ethox MS-8, 11034; Eumulgin PST 5, 11187; Genagen S-080, 12763; Hetoxamate SA-5, 13912; Hodag 150-S, 14162; Hodag DGS, 14176; Hydrine, 14554; Kemester® 5221SE, 15945; Kessco® PEG 200 MS, 16124; Mapeg® 1500 MS, 18635; Mapeg® 200 MS, 18634; Mapeg® S-40, 18643; Marlosol® 183, 18909; Mergital ST 30/E, 19362; Myrj® 45, 20582; Myrj® 52, 20583; Myrj® 53, 20584; Myrj® 59, 20585; Nonex S3E, 21447; Nopalcol 4-S, 21470; PEG stearate series, 23034; Pegosperse® 100 S, 23061; Pegosperse® 1750 MS, 23062; Pegosperse® 50 MS, 23066; Polystate C, 24582; Prox-onic ST-05, 25343; Ritapeg 150 DS, 26821; Ritox 52, 26852; Schercemol DEGMS, 27590; Sellig S 30 100, 27911; Sellig Stearo 6, 27918; Simulsol M 45, 28488; Superpolystate, 30264; Tefose® 1500, 30939; Trydet 2610, 32507; Witconol 2640, 34505

9005-00-9: Ablunol SA-7, 113; Brij® 700 S, 4520; Brij® 72, 4518; Brij® 721, 4521; Emalex 640, 9925; Hetoxol STA-2, 13950; Hodag Nonionic S-2, 14197; Lipocol S-2, 17358; Macol® SA-2, 18273; Prox-onic SA-1/02, 25336; Rhodasurf® S-2, 26661; Rhodasurf® TB-970, 26663; Ritox 721, 26853; steareth series, 29596; Trycol® 5888, 32500; Volpo S-2/S-10/S-20, 34030

9005-02-1: Acconon 200-DL, 232; Emalex 200 di-L, 9916; Emalex DEG-di-L, 9940; Emerest® 2622, 10105; Emerest® 2704, 10113; Ethox DL-5, 11022; Hetoxamate 200 DL, 13907; Hodag 22-L, 14158; Kessco® PEG 200 DL, 16119; Lexemul® PEG-200 DL, 17140; Lipopeg 2-DL, 17394; Mapeg® 200 DL, 18627; Nonex DL-2, 21445; Nopalcol 2-DL, 21465; PEG-4 dilaurate series, 23050; Pegosperse® 200 DL, 23063; Witconol 2622, 34504

9005-07-6: Alkamuls® 400-DO, 1656; Alkamuls® 600-DO, 1658; Chemax PEG 200 DO, 5922; Chemax PEG 400 DO, 5923; Chemax PEG 600 DO, 5924; CPH-211-N, 6943; Dyafac PEG 6DO, 9301; Emalex 200 di-O, 9917; Ethox DO-9, 11023; Hodag 42-0, 14161; Kessco® PEG 200 DO, 16120; Lipopeg 4-DO, 17395; Mapeg® 200 DO, 18628; Marlosol® FS, 18911; Nonex DO-4, 21446; Nopalcol 6-DO, 21471; PEG-2 dioleate series, 23040; Secoster® DO 600, 27818; Witconol 2648, 34507; Witconol H33, 34516

9005-08-7: Emalex 200 di-S, 9918; Emalex 6300 Di-ST, 9929; Genapol® TS Powd, 12798; Hetoxamate 400 DS, 13908; Hodag 602-S, 14163; Kessco® PEG 200 DS, 16121; Lipopeg 4-DS, 17396; Lipopeg 6000-DS, 17398; Mapeg® 1540 DS, 18636; Mapeg® 200 DS, 18630; Mapeg® 6000 DS, 18637; Marlosol® BS, 18910; PEG-150 distearate series, 23037; Rewopal® PEG 6000 DS, 26347; Ritapeg 400 DS, 26822; Witconol 2642, 34506

9005-25-8: Ceraphyl® 847, 5729; Control 1-100, 6749; corn starch, 6836,6837; D12F, 7629; Pure-Dent® B700, 25405; Purity® 21, 25411; Supercore® S13F, 30230

9005-32-7: alginic acid, 1593; Kelacid®, 15824; Saltialgine H8, 27297; Satialgine H8, 27481

9005-34-9: Ammonium alginate, 2228; Amoloid HV, LV, 2296; Collatex, 6620; Sobalg FD 300 Series, 28689; Superloid®, 30254

9005-35-0: Sobalg FD 460, 28690

9005-36-1: Alginic acid potassium salt, 1594; Improved Kelmar®, 14933; Kelmar®, 15856; potassium alginate, 24735; Sobalg FD 200 Series, 28688

9005-37-2: Concentrated Dariloid® KB, 6710; Kelcoloid® D, 15828; Kelcoloid® DH, DSF, 15829; Kelcoloid® HV, LV, 15830; Kelcoloid® O, S, 15831; Manucol Ester E/RK, 18617; Manucol Ester EX/LL, 18618; propylene glycol alginate, 25216; Sherbelizer®, 28170

9005-38-3: Algin, 1591; Dariloid® Q, 7775; Dariloid® QH, 7776; Darilold®, 7777; Kelcosol®, 15832; Kelgin, 15841; Kelgin® F, 15842; kelgin® HV, LV, MV, 15843; Kelgin® QH, QL, QM, 15844; Kelgin® XL, 15845; Kelgum, 15846; Keltex®, 15878; Keltex® HV, 15879; Keltex® S, 15880; Keltone®, 15883; Keltone® HV, LV, 15884; Mozanon®, 20386; Municol, 20497

9005-46-3: casein, 5386; sodium caseinate, 28746

9005-49-6: Heparin, Sodium Salt, 13788; Liquemin, 17453

9005-64-5: Accosperse 20, 249; Alkamuls® PSML-20, 1671; Alkamuls® T-20, 1691; Capmul® POE-L, 5116; Crillet 1, 7037; Crillet 11, 7042; Disponil SML 104 F1, 8696; Disponil SML 120 F1, 8697; Emalex ET-2020, 9953; Emsorb® 2720, 10464; Emsorb® 2721, 10465; Emsorb® 6915, 10476; Ethsorbox L-20, 11054; Ethylan® GEL2, 11104; Eumulgin SML 20, 11190; G-4280, 12554; Glycosperse® L-10, 13128; Glycosperse® L-20, 13129; Hetsorb L-10, 13966; Hetsorb L-20, 13967; Hetsorb L-4, 13968; Hetsorb L-80-72%, 13969; Hodag PSML-20, 14237; Liposorb L-10, 17405; Liposorb L-20, 17406; Montanox 20 DF, 20272; Mulsifan RT 141, 20455; PEG-10 sorbitan laurate, 23036; polysorbate 20, 24565; polysorbate 21, 24566; Prox-onic SML-020, 25338; Radiasurf® 7137, 25799; Sorbax PML-20, 29091; Sorbon T-20, 29122; T-Maz® 20, 31893; T-Maz® 28, 31894; Tween® 20, 32625; Tween® 21, 32626; Witconol 2720, 34508; Witflow 990, 34528

9005-65-6: Accosperse 80, 251; Alkamuls® PSMO-20, 1672; Alkamuls® PSMO-5, 1673; Alkamuls® T-80, 1693; Capmul® POE-O, 5117; Crillet 4, 7040; Crillet 41, 7045; DeSotan® SMO-20, 8195; Disponil SMO 120 F1, 8699; Drewpone® 80K, 9018; Durfax® 80, 9218; Emalex ET-8020, 9954; Emsorb® 2722, 10466; Emsorb® 6900, 10469; Emsorb® 6901, 10470; Ethsorbox O-20, 11055; Ethylan® GEO8, 11105; Ethylan® GEO81, 11106; Eumulgin SMO 20, 11191; Glycosperse® O-20 FG, O-20 KFG, 13130; Glycosperse® O-5, 13131; Hetsorb O-20, 13970; Hetsorb O-5, 13971; Hodag PSMO-20, 14238; Hodag SVO-9, 14251; Liposorb O-20, 17408; Liposorb O-5, 17409; Montanox 80 DF, 20278; Montanox 81, 20279; Mulsifan RT 146, 20456; Polycon T-80 K, 24402; polysorbate 80, 24572; polysorbate 81, 24573; Prox-onic SMO-020, 25339; Prox-onic SMO-05, 25340; Radiamuls® Sorb 2157, 25785; Radiasurf® 7157, 25804; Ritabate 80, 26796; Sorbax PMO-20, 29092; Sorbax PMO-5, 29093; Sorlate, 29144; T-Maz® 80, 31899; T-Maz® 81, 31900; Tween® 80, 80K, 32631; Tween® 81, 32632; Witconol 2722, 34509; Witflow 991, 34529

9005-66-7: Crillet 2, 7038; Disponil SMP 120 F1, 8701; Ethylan® GEP4, 11107; G-4252, 12553; Glycosperse® P-20, 13132; Hetsorb P-20, 13972; Hodag PSMP-20, 14239; Liposorb P-20, 17411; Montanox 40 DF, 20273; polysorbate 40, 24567; Prox-onic SMP-020, 25341; Sorbax PMP-20, 29094; T-Maz® 40, 31895; Tween® 40, 32627

9005-67-8: Accosperse 60, 250; Alkamuls® PSMS-20, 1674; Alkamuls® T-60, 1692; Capmul® POE-S, 5118; Crillet 3, 7039; Crillet 31, 7043; Disponil SMS 120 F1, 8703;

Drewpone® 60K, 9016; Durfax® 60, 9216; Emsorb® 2728, 10468; Emsorb® 6906, 10472; Emsorb® 6909, 10474; Ethsorbox S-20, 11056; Ethylan® GES6, 11108; Eumulgin SMS 20, 11192; Glycosperse® S-20 FG, S-20 KFG, 13133; Hetsorb S-20, 13973; Hetsorb S-4, 13974; Hodag PSMS-20, 14240; Hodag SVS-18, 14252; Liposorb S-20, 17413; Liposorb S-4, 17414; Montanox 60 DF, 20274; Montanox 61, 20275; Norfox® Sorbo T-60, 21553; Polycon T-60 K, 24401; polysorbate 60, 24568; polysorbate 61, 24569; Prox-onic SMS-020, 25342; Radiamuls® Sorb 2147, 25784; Radiasurf® 7147, 25800; Ritabate 60, 26795; Sorbax PMS-20, 29095; Tandem 5K, 8, 30798; T-Maz® 60, 31896; T-Maz® 61, 31897; Tween® 60, 60K, 32628; Tween® 61, 32629

9005-70-3: Alkamuls® PSTO-20, 1675; Alkamuls® T-85, 1694; Crillet 45, 7046; Disponil STO 120 F1, 8706; Emsorb® 6903, 10471; Emsorb® 6913, 10475; Ethsorbox TO-20, 11057; Ethylan® GPS85, 11111; Hetsorb TO-20, 13975; Liposorb TO-20, 17417; Montanox 85, 20280; polysorbate 85, 24574; Sorbax PTO-20, 29096; T-Maz® 85, 31901; Tween® 85, 32633; Witconol 6903, 34511

9005-71-4: Crillet 35, 7044; Disponil STS 120 F1, 8708; Drewpone® 65K, 9017; Durfax® 65, 9217; Ethsorbox TS-20, 11058; Glycosperse® TS-20 FG, TS-20 KFG, 13134; Hetsorb TS-20, 13976; Hodag PSTS-20, 14241; Ice No. 12K, 14809; Liposorb TS-20, 17419; Montanox 65, 20276; polysorbate 65, 24570; Polysorbate 65, 24571; Sorbax PTS-20, 29097; T-Maz® 65, 31898; Tween® 65, 65K, 32630

9005-79-2: Liver sugar, 17527

9006-03-5: chlorinated rubber, 6106

9006-04-6: DPNR, 8949; Dynatex GTZ, 9424; ENR 25, 10585; Hyflo NS, S, 14673; PA-57, 22531; PA-80, 22532; Polyisoprene, 24471; rubber, 27089; Unitex, 33056

9006-42-2: metiram, 19582; Polyram® -Combi, DF, 24536

9006-50-2: Albumen, 1402

9006-59-1: Sol-U-Tein EA, 29028

9006-65-9: AF 10 FG, 907; CPS 034, 6945; dimethicone, 8493; PS034, 25359; Viscasil®, 33893

9007-28-7: Cromoist CS, 7199

9007-34-5: Actigen C, 478; Cationic Collagen Polypeptides, 5460; Clearcol, 6390; collagen, 6610; Collagen CLR, 6612; Collagen Native Extra 1%, 6614; Collasol, 6619; Desamidocollagen, 8136; Polygeline+sn, 24451; Polymoist Mask, 24497; Ritacollagen BA-1, 26805; Secolan S-1, BA-1, BA-1G, 27803; Solu-Coll, 28992

9007-48-1: Caprol® 3GO, 5133; Drewpol® 3-1-O, 9015; Grindtek PGE 25, 13413; Hodag PGO-101, 14216; Hodag PGO-61, 14222; Isolan® GO 33, 15369; Mazol® PGO-31 K, 19142; Santone® 3-1-SH, 27415; Witconol 14, 34495

9008-04-7: Good-rite® K-705BD, 13208

9008-66-6: Electrafil® J-2/CF/30, 9732; NI-20GF, 21085; Ny-Kon® Q, 21893; nylon 6/10, 21911; RTP 201B, 27058; Texalon 1600A Nat, 31446; Thermocomp® QF-1006FR, 31621; Ultramid® S3, 32840

9008-99-8: Pea Pro-Tein BK, 22967

9009-32-9: Emalex MSG-2, 9970

9009-54-5: Aspect® TPPE, 3235; Autofroth, 3436; Autopak, 3440; Autopour, 3442; Desmopan® 150, 8165; Desmopan® 385, 8166; Desmopan® 585, 8167; Desmopan® 786, 8168; Desmopan® KA 8333, 8169; Electrafil® J-100/CF/30, 9740; Estane® 5701 F1, 10878; Estane® 58092, 10879; Estane® 58300, 10880; Isoplast 101, 15399; Isoplast 101LGF40 Nat, 101LGF60 Blk, 15400; Isoplast 302, 15401; Pearistick 46-10/06, 22973; polyurethane foam, 24655; Resin 164, 26175; RTP 2301A, 27075; RTP 2381A, 27076; Texin 480-A, 31502; Texin 5286, 31504; Texin 985-A, 31503; Thermocomp® TF-1004, 31624

9009-86-3: ricin, 26731

9009-99-8: Crosilk Powder, 7235; Fibro-Silk Powd, 11754; Ritasilk Powd, 26831

9010-77-9: A-C® Copolymer 5180, 174; A-C® Copolymer 540, 540A, 580, 5120, 5180, 171; A-C® Copolymer 580, 172; A-C® Copolymer 580, 5120, 5180, 173; AClyn®250, 262, 296, 361; ACuflow AF-1, 571; Primacor 4990 Dispersion, 24961

9010-79-1: Amoco® 8244, 2278; Amoco® 8410, 2279; EPM rubber, 10666; Siltek® L Polymer, 28399; Vistalon 719, 33913

9010-85-9: Butex, 4744; CryOfine® Butyl, 7276; Exxon® Butyl 077, 11355; Kalar® 5214, 15662; Kalar® 5263, 15663; Kalene® 800, 15666; Polysar Butyl 100, 24539; Polysar XL 30102, 24545

9010-98-4: Sovprene, 29166

9011-05-6: Salmocid, 27277

9011-13-6: Celstran® SMAG30-01-4, 5658; Dylark® 132, 9318; Scripset 520, 27774; Scripset 720, 27775; SMA® 1000, 28635; SMA® 1440, 28636; SMA® 2625, 28637; Stapron S SG340, 29519; Stapron S SM300, 29520

9011-14-7: Acryloid® A-30, 428; Acryloid® B-44, 431; Lucryl®, 17771; methyl methacrylate polymer, 19549; Paraloid® K-120N, 22766; polymethylmethacrylate, 24491; Polymethylmethacrylate, 24492

9011-16-9: Agrimer VEMA-H-240, 1068; Gantrez® S-95, 12656; Luviform® FA 119, 18019

9011-17-0: Kynar® Flex® 2850, 16514

9012-09-3: cellulose triacetate, 5633; Triafol, 32188

9012-54-8: Cellulase, 5619; Cellulase, 5620; Cellulase 4000, 5621; Cellulase AC, 5622; Celluzyme® 2400 T, 5638

9014-01-1: AFP 2000, 946; protease, 25256

9014-85-1: Surfynol® 420, 30384;Surfynol® 440, 30385; Surfynol® 465, 30386;Surfynol® 485, 30387;

9014-90-8: Akyposal NPS 100, 1315; Akyposal NPS 60, 1316; Polystep® B-27, 24594; Serdet DNK, 28012; Steol® COS 433, 29662

9014-92-0: Dodoxynol series, 8786; Igepal® RC-520, 14874; Igepal® RC-620, 14875; Prox-onic DDP-09, 25314; Rexol 65/4, 26470; T-Det® DD-5, 30863; Teric DD5, 31265

9014-93-1: Chemax DNP-8, DNP-18, 5894; Ethal DNP-8, 10919; Hetoxide DNP-4, 13923; Igepal® DM-430, 14868; Igepal® DM-530, 14869; Igepal® DM-710, 14870; Igepal® DM-730, 14871; Igepal® DM-970, 14872; Macol® DNP-5, 18265; Marlophen® DNP 16, 18887; Prox-onic DNP-08, 25315; Sellig DN 10 100, 27879; Sellig DN 22 100, 27880; T-Det® D-150, 30862; Trycol® 6985, 32506

9015-51-4: protargentum, 25251

9015-54-7: Collagen Hydrolyzate Cosmetic 55, 6613; Collamino 25, 6616; Cropepsol, Cropeptone, 7208; Crotein A, C, O, 7244; Crotein CAA, 7249; Crotein CAA/SF, 7250; Crotein O, 7256; Crotein SPA, 7258; Extiat®, 11323; Hydrocoll AL-50, AL-55, EN-40, EN-55, EN-55-X, EN-SD, EN-SD-1M, EN-SD-10M, 14580; Lexein® X-250HP, 17131; Nutrilan® FPK, H, M, 21834; Nutrilan® I-50, 21835; Nutrilan® L, 21837; Parenamine, 22823; Peptein® 2000®, 23202; Polypeptide 10, 24516; Polypeptide 37, 24517; protogest, 25290

9016-00-6: Masil® SF 1,000,000, 19017; Masil® SF 5, 19018; Masil® SF 500, 19020; Masil® SF 500,000, 19021; Sentry Dimethicone, 27959

9016-45-9: Ablunol NP4, 106; Akyporox NP 105, 1274; Akyporox NP 1200V, 1275; Akyporox NP 150, 1277; Akyporox NP 200, 1278; Akyporox NP 30, 1279; Akyporox NP 300V, 1280; Akyporox NP 40, 1281; Akyporox NP 90, 1282; Akyporox NP 95, 1283; Alkamuls® AG-900, 1660; Alkasurf® NP-4, 1715; Carsonon® N-4, 5344; Chemax NP-1.5, 5909; Chemax NP-10,

5913; Chemax NP-15, 5914; Chemax NP-30, 5915; Chemax NP-4, 5910; Chemax NP-40, 5916; Chemax NP-6, 5911; Chemax NP-9, 5912; Cremophor® NP 10, 6982; Cremophor® NP 14, 6983; Emalex NP-2, 9975; Empilan® NP9, 10413; Emulgator U4, 10501; Ethal NP-1.5, 10922; Ethal NP-6, 10923; Ethylan® 44, 11092; Ethylan® BV, 11096; Ethylan® NP1, 11115; Etophen 102, 11152; Etophen 114, 11153; Gynol, 13515; Hetoxide NP-4, 13927; Hodag Nonionic E-5, 14192; Hostapal N-040, 14378; Hyonic NP-40, 14692; Hyonic PE-100, 14695; Iconol NP-100, 14822; Iconol NP-30, 14823; Iconol NP-40, 14824; Iconol NP-50, 14825; Iconol NP-70, 14826; Igepal® CO-210, 14851; Igepal® CO-430, 14852; Igepal® CO-520, 14853; Igepal® CO-530, 14854; Igepal® CO-610, 14855; Igepal® CO-630, 14856; Igepal® CO-660, 14857; Igepal® CO-710, 14858; Igepal® CO-720, 14859; Igepal® CO-730, 14860; Igepal® CO-850, 14861; Igepal® CO-880, 14862; Igepal® CO-887, 14863; Igepal® CO-890, 14864; Igepal® CO-970, 14865; Igepal® CO-997, 14866; Indulin® XD-70, 15095; Intercept, 15156; Lutensol® AP 10, 17981; Lutensol® AP 20, 17982; Macol® NP-4, 18269; Makon® 4, 18427; Marlophen® 810N, 18885; Marlophen® 830N, 18886; Naxonic NI-40, 20829; Neutronyx® 656, 21039; Nonal 206, 21435; Nonipol 20, 21452; Nutrol 600, 21841; Nutrol 611, 21842; Nutrol 622, 21843; Nutrol 640, 21844; Nutrol 656, 21845; Polystep® F-1, 24599; Polystep® F-95B, 24600; Poly-Tergent® B-150, 24633; Prox-onic NP-04, 25328; Remcopal 29, 26058; Remcopal 3112, 26075; Remcopal 31250, 26081; Remcopal 334, 26072; Remcopal 33820, 26082; Remcopal 349, 26073; Remcopal 3712, 26076; Remcopal 3820, 26077; Remcopal 6110, 26080; Remcopal 666, 26074; Remcopal L30, 26091; Remcopal NP 30, 26094; Remcopal PONF, 26098; Rewopal® HV 4, 26342; Rexol 25/10, 26466; Rexol 25/4, 26467; Sellig N 10 100, 27883; Sellig N 1050, 27884; Sellig N 11 100, 27885; Sellig N 12 100, 27886; Sellig N 15 100, 27887; Sellig N 1780, 27888; Sellig N 20 80, 27889; Sellig N 30 70, 27890; Sellig N 4 100, 27891; Sellig N 5 100, 27892; Sellig N 50 100, 27893; Sellig N 6 100, 27894; Sellig N 8 100, 27895; Sellig N 9 100, 27896; Surfonic® HDL, 30365; Surfonic® N-10, 30371; T-Det® N-1007, 30865; T-Det® N-4, 30866; T-Det® N-40, 30867; Teric N2, 31269; Triton® N-57, 32430; Trycol® 6940, 32504; Witconol NP-100, 34517

9016-75-5: Celstran® PPSG30-01-4, 5656; Debron 711, 7835; Electrafil® J-1300/CF/30/TF/15, 9745; EMI-X® OC-1008, 10198; EMI-X® PDX-O-91074, 10203; Fortron® 0205B4, 12324; Lubricomp® 189, 17733; Mitsubishi Kasei PPS 704G40, 19973; P.P.S, 22496; Plaslube® J-1300/30/TF/15, 23947; Plaslube® J-1300/CF/20/MS/10/TF/15, 23948; Polycomp® 139, 24395; Polycomp® 185, 24396; RTP 1301, 27071; RTP 1378, 27072; Ryton® A-200, 27173; Ryton® R-4, 27174; Ryton® V-1, 27175; SF-20CF, 28121; Stat-Kon® OC-1006, 29565; Supec® G401, 30162; Techtron+sn PPS, 30896; Tedur® KU1-9510-1, 30922; Tedur® KU1-9511, 30923; Tedur® KU1-9530, 30924; Tedur® KU1-9552, 30925; Tedur® KU1-9561, 30926; Thermocomp® OC-1006, 31619; Verton® OF-700-10, 33640

9031-11-2: galactosidase, 12593

9032-08-0: Agidex, 1030; Aldomax GA-100, 1548; Diazyme® L-200, 8352; Distillase® L-200, 8716; Fermenzyme® L-200, 11628

9032-42-2: Tylose® MHB, 32661

9032-75-1: Clarex® L, 6359; Pearex-L®, 22972; pectinase, 23018; Pectinase AT, 23019; Spark-L® HPG, 29203

9035-85-2: Carsonon® 169-P, 5343; PPG-30 cetyl ether series, 24845; Procetyl 10, 25053; Wickenol® 707, 34404

9038-95-3: Pluracol® W3520N, 24188; Tergitol® XD, 31249; Tergitol® XH, 31250

9039-53-6: urokinase, 33145

9040-38-4: Aerosol® A-103, 886

9046-01-9: Crodafos T2 Acid, 7087; Lubrhophos® LS-500, 17726; Rhodafac® RS-610, 26608; Trideceth phosphate series, 32231

9048-46-8: Bovinal-20, 4465

9050-36-6: Lycadex®, 18059; Maltodextrin, 18500; Maltrin® M040, 18502; Maltrin® M510, 18504; Microduct®, 19670; Mor-Rex® I-920, 20348; Wickenol® 550, 34403

9051-57-4: Abex® EP-110, 41; Aerosol® NPES 458, 883; Ammonium nonoxynol-4 sulfate, 2246; Polystep® B-1, 24590; Rhodapex® CO-415, 26623; Rhodapex® CO-436, 26625; Sulfochem 436, 29995

9056-38-6: nitro-starch, 21377

9063-38-1: Explotab®, 11316

9064-14-6: Marlox® L 6, 18944

9067-32-7: ActiMoist, 486; Amo Vitrax®, 2266; Dermasome® H, 8106; hyaluronic acid sodium salt, 14483; Pronova, 25159; Rita HA C-1, 26791

9076-43-1: Emcol® CC-42, 10056; PPG-40 diethylmonium chloride, 24846

9084-06-4: Ablusol ML, 129; Chromasist 1487A, 6194; Daxad® 11, 7799; Dehscofix 912, 7923; Dispersogen A, 8677; Emery® 5370 Sellogen W, 10177; Lomar® D, 17565; Lomar® LS, 17567; Nopcosant, 21499; Rhodacal® Liquid, Rhodacal® N, 26588; Rhodacal® N, 26589; Sellogen DFL, 27929; Sellogen HR-90, 27930; Tamol® L Conc, 30767

9087-53-0: Procetyl AWS, 25054

9087-61-0: Aluminum starch octenyl succinate, 1928; Dry Flo®, 9056

10004-44-1: Tachigaren 70, 30685

10022-31-8: Barium nitrate, 3639

10024-97-2: Nitral, 21318

10025-69-1: salt of tin, 27291

10025-73-7: Chrometrace, 6218; chromic chloride, 6221

10025-78-2: Dynasylan® TCS, 9417; trichlorosilane, 32223

10025-87-3: phosphorus oxychloride, 23743

10025-91-9: antimony trichloride, 2573

10026-04-7: CT1800, 7312; silicon tetrachloride, 28333

10026-06-9: opacite, 22197

10026-11-6: zirconium tetrachloride, 34826

10026-13-8: phosphorus pentachloride, 23744

10026-18-3: cobaltic fluoride, 6531; cobaltic trifluoride, 6534

10028-22-5: Elliott's Lawn Sand, 9848; Elliott's Moss Killer, 9849; ferric sulfate, 11652; Greenmaster Mosskiller, 13344; Greenmaster Autumn, 13342; Hart Lawn Sand, 13657; Hart Moss Killer, 13658; iron sulfate (ic), 15320; Maxicrop Moss Killer & Conditioner, 19078; Vitax Micro Gran, Vitax Turf Tonic, 33958; Vitax Turf Tonic, 33960; Walkover Moss Killer, 34198;

10028-24-7: disodium phosphate, dihydrate, 8653

10034-81-8: Anhydrone, 2460; Dehydrite®, 7963

10034-85-2: Hydriodic acid, 14555

10034-92-2: thoron, 31752

10034-93-2: Hydrazine sulfate, 14546

10035-10-6: Hydrobromic acid, 14564

10039-53-9: Rachromate-51, 25709

10039-56-2: Sofibex, 28835

10042-76-9: strontium nitrate, 29830; Strotope, 29832

10043-01-3: Alcan Aluminum Sulfate Liquid, 1413; Alferrlc, 1576; Aluminoferric, 1894; Aluminum sulfate, 1930; Clar+Ion® A410P, 6366; Lapotan, 16781; mountain butter, 20367; paperhanger's alum, 22701; pearl alum, 22976; pickle alum, 23813

10043-11-5: Boron nitride, 4443

10043-35-3: Ant Flip, 2497; Boric Acid, 4422; Homberg's salt, 14285; sal sedativus, 27254

10043-52-4: Cal Plus, 4923; calcium chloride,

4943; Dowflake, 8921; Homberg's phosphorus, 14284; Huppert's reagent, 14476; Ice Melt, 14808; Jarcal, 15519; Liquical, 17459; Liquidow, 17478; Marley Cement Accelerator, 18825; Snomelt, 28673; Sure-Step, 30334
10043-67-1: alum, 1883; aluminum potassium sulfate, 1923; potash alum, 24724
10045-89-3: Mohr's salt, 20039
10048-99-4: Rohrbach's solution, 26915
10049-04-4: Anthium Dioxcide, 2521; chlorine dioxide, 6108
10049-05-5: chromous chloride, 6245
10049-07-7: rhodium chloride, 26687
10081-67-1: Naugard® 445, 20790
10090-54-7: Liquid Code XLR, 17465; Potassium pyroantimonate, Acid, 24776
10094-45-8: Chemstat® HTSA#3, 6041; HTSA #3, 14446; Kemamide® E-180, 15896; Kemamide® E-221, 15897; N-octadecyl-13-docosenamide, 21986; stearyl erucamide, 29605
10097-28-6: Monox, 20259
10099-58-8: lanthanum chloride, 16764
10099-59-9: lanthanum nitrate, 16765
10099-74-8: lead nitrate, 16911; Ledni, 16961
10099-76-0: BSWL 202, 4634; lead silicate, 16915
10101-39-0: Cecasol, 5536; Extrusil, 11331; Keical-Ace, 15822; Microcal, 19657; Microcal ET, 19658; Micro-Cel® A, 19662; Micro-Cel® T-38, 19663; Paratemp, 22805; Silasorb, 28268; Silene, 28300
10101-41-4: calcium sulfate (dihydrate), 4972; Compactrol®, 6670
10101-52-7: Zircosil, 34828
10101-89-0: trisodium phosphate, dodecahydrate, 32413; TSP-12, 32538
10102-05-3: palladium(II) nitrate, 22618
10102-40-6: sodium Molybdate, dihydrate, 28796
10102-44-0: nitrous gas or air, 21381
10102-71-3: Kasal®, 15777; porous alum, 24702; sodium alum, 28730
10108-22-2: propylene glycol monolaurate, 25219
10108-64-2: Caddy, 4890; cadmium chloride, 4895
10112-91-1: Calomel, 5018
10117-38-1: Stahl's sulfur salt, 29425
10118-76-0: Acerdol, 275; calcium permanganate, 4963; Monol, 20229
10124-37-5: calcium nitrate, 4960; lime nitrate, 17267;nitrocalcite, 21352;
10124-43-3: cobaltous sulfate, 6541
10124-48-8: white precipitate, 34370
10124-56-8: Kalex HMP, 15670; Limex G, 17276; Sodaphos, 28721; sodium hexametaphosphate, 28770; Vitrafos®, 33982; Water Brite, 34223
10141-00-1: chromic potassium sulfate, 6224
10141-05-6: cobaltous nitrate, 6539
10161-34-9: Finaplix, 11803; Trenbolone acetate, 32171
10163-15-2: Albaphos Dental Na 211, 1348
10190-55-3: lead molybdate, 16909
10192-30-0: Ammonium bisulfite, 2233
10192-93-5: Perkadox® 58, 23306
10196-18-6: zinc nitrate, 34782
10213-09-9: vanadium chloride (ous), 33274
10213-78-2: bishydroxyethyl-N,N-Bis(2-hydroxyethyl) stearyl amine, 4231; Chemeen 18-2, 5943; Chemstat® 273-E, 6037; Ethomeen® 18/12, 10972; PEG stearamine series, 23026; Varonic® S202, 33448
10213-79-3: sodium metasilicate pentahydrate, 28789
10222-01-2: Amerstat® 300, 2045; Biosperse® 240, 4207
10257-55-3: katarsit, 15786
10265-92-6: methamidophos, 19474; Monitor, 20213; Nitofol®, 21314; Prodex, 25069; Tamaron®, 30760
10287-53-3: Speedcure EDB, 29261
10288-28-5: V-30, 33215
10290-12-7: cupric arsenite, 7376; pickle green, 23814; Scheele's green, 27562
10294-26-5: silver sulfate, 28434
10294-33-4: Boron tribromide, 4444; Trona Boron Tribromide, 32452
10294-34-5: Boron trichloride, 4445; Trona Boron Trichloride, 32453
10294-40-3: Barium chromate, 3638
10294-42-5: cerium sulfate, 5755
10294-54-9: cesium sulfate, 5778
10326-41-7: D-lactic acid, 16572
10347-81-6: Ludiomil, 17774
10361-03-2: sodium metaphosphate, 28787
10361-37-2: Barium chloride, 3637
10361-46-3: Bismuth nitrate, 4239
10377-51-2: Tenephrol, 31103
10377-60-3: magnesium nitrate, 18365; Magnisal, 18380
10401-55-5: cetyl ricinoleate, 5824; Liponate CRM, 17374; Naturchem® CR, 20767
10416-59-8: Bistrimethylsilyl acetamide, 4286; CB2500, 5479; Dynasylan® BSA, 9399
10421-48-4: ferric nitrate, 11650
10450-60-9: periodic acid, 23273
10453-86-8: resmethrin, 26218
10476-85-4: Stronscan-85, 29825; strontium chloride, 29829
10525-14-1: Lanamine®, 16646; Mixed isopropanolamines myristate, 19984
10525-29-0: Schercemol NGDL, 27613
10552-74-6: nitrothal-isopropyl, 21380
10553-31-8: Barium bromide, 3635
10591-85-2: Benzyl Tuex®, 3990
10595-72-9: Argus DTDTDP, 2864; Cyanox® 711, 7486; Evanstab® 13, 11256
10605-21-7: Battal, 3740; Carbate Flowable, 5182; Carbate Flowable, 5183; Carbendazim, 5187; Carbendazim, 5188; Hinge, 14095; Konker®, 16327; Mascot Systemic Turf Fungicide, 19007; Maxim, 19082; Stempor DG, 29648; Tripart® Defensor FL, 32370; Turfclear, 32604
11006-75-0: saponins, 27444
11094-60-3: Caprol® 10G100, 5131; Hodag PGO-1010 (formerly Hodag SVO-10107), 14217; Polyaldo® DGDO KFG, 24353; Polygliceryl-10 decaoleate, 24453
11096-18-7: cufraneb, 7344
11096-42-7: Biopal® NR-20, 4175
11097-59-9: Alcamizer 1, 1409
11099-07-3: Ablunol GMS, 103; Aldo® HMS KFG, 1539; Aldo® MS FG, 1543; GMS Base, 13162
11104-88-4: phosphomolybdic acid, 23731
11111-34-5: Tetronic® 1101, 31409; Tetronic® 1102, 31410; Tetronic® 1104, 31411; Tetronic® 1107, 31412; Tetronic® 1301, 31413; Tetronic® 1302, 31414; Tetronic® 1304, 31415; Tetronic® 1307, 31416; Tetronic® 1501, 31417; Tetronic® 1502, 31418; Tetronic® 1504, 31419; Tetronic® 1508, 31420; Tetronic® 304, 31399; Tetronic® 504, 31400; Tetronic® 701, 31402; Tetronic® 702, 31403; Tetronic® 704, 31404; Tetronic® 707, 31405; Tetronic® 901, 31406; Tetronic® 904, 31407; Tetronic® 908, 31408
11113-70-5: M50, 18109
11120-25-5: Ammonium tungstate, 2258
11125-95-4: CCA Type C Wood Preservative 50-60%, 5496
11138-66-2: Dariloid® 100, 7774; Dricoid® 200, 9025; Kelflo®, 15839; Kelgum®, 15847; Kel-Lite+sn CM, 15853; Keltrol®, 15886; Keltrol® F, 15889; Kelzan®, 15890; Kelzan® AR, 15891; Kelzan® D, M, XC Polymer, 15892; Kelzan® S, 15893; Rhodicare XC, 26682; Rhodigel® EZ, 26683; Rhodigel®; Rhodigel® EZ, 26684; Rhodopol® 23, XGD, 26696; Rhodopol® XGD, 26697; Xanflood®, 34587
11140-78-6: Rewoteric® AM DM-35L, 26423
11141-17-6: azadirachtin, 3516; Margosan-O, 18713
11174-62-0: Crotein Q, 7257; Quat-Coll QS, 25605; Quat-Pro S, S-30, 25609
12001-26-2: Alsibronz, 1855; Kemira Phlogopite Mica, 15976; Mearlmica® SVA, 19194; mica, 19626; Micacoat®, 19629; moac, 19995; Nyflake®, 21888; Polymica 200, 325, 400, 24493; Sericite PHN, 28029; Snow White 200 Mica, 28674
12001-85-3: zinc naphthenate, 34781
12002-03-8: cupric acetoarsenite, 7374; mountain green, 20369
12004-14-7: satin white, 27483
12007-56-6: Meyerhofferite, 19617
12007-92-0: Borax Glass, 4412
12016-80-7: cobaltic oxide monohydrate, 6533
12018-10-9: chrome black, 6203
12027-67-7: Ammonium molybdate, 2244
12027-96-2: Flamtard H, 11889; Zinc hydroxystannate, 34779
12033-89-5: Roydazide, 27045
12036-37-2: Flamtard S, 11890; Zinc stannate, 34792

24787
13927-77-0: Naugard® NBC, 20797;NBC, 20831; Perkacit® NDBC, 23287; Vanox® NBC, 33344;
13952-84-6: CSC 2-aminobutane, 7308
13983-17-0: Nyad® , 21860; Nyad® Wollastonite, 21861; Nycor® R, 21884; Vansil® W-10, 33365; Wollastocoat®, 34545; Wollastokup®, 34546; wollastonite, 34547
13997-19-8: methyl benzoquate, 19516; Statyl, 29578
14087-96-6: Cimfix 606, Cimpact 699, 710, 6306
14156-10-6: Decanox-F, 7843
14234-82-3: Octomer DIBM, 22009
14235-86-0: hydrargaphen, 14539
14239-68-0: Cadmate®, 4892; cadmium diethyldithiocarbamate, 4896; ethyl cadmate, 11065
14283-05-7: Krystallazurin, 16468
14324-55-1: Accelerator EZ Powder, 204; Ethazate, 10938; Ancazate ET, 2415; ethyl zimate®, 11088; Octocure ZDE-50, 21996; Perkacit® ZDEC, 23296
14350-96-0: Dehyton® PLG, 7996
14350-97-1: Rewoteric® AM 2L-40, 26416
14351-40-7: Cerasynt® D, 5735; Witcamide® MAS, 34444
14351-50-9: Ammonyx® OAO, 2265; Chemoxide O, 5997; Incromine Oxide OD-50, 15009; Mackamine O2, 18174; Mazox® ODA, 19154; Standamox 01, 29459
14450-05-6: Afilan PP, 933; Emerest® 2485, 10101
14459-95-1: yellow prussiate of potash, 34647
14481-60-8: Aerosol® 18, 878; Empimin® MKK, 10431; Octosol A-18, 22022
14639-98-6: zinc ammonium chloride, 34766
14643-87-9: Ageflex ZDA, 1004; Akrochem® ZDA Powd, 1184
14726-36-4: Akrochem® Z.B.E.D, 1183; Arazate®, 2787; Perkacit® ZBEC, 23294
14727-68-5: Jet Amine DMOD, 15586
14769-73-4: Ketrax, 16149
14807-96-6: ABT-2500®, 166; Altalc 200 USP, 1870; Artic Mist, 3148; Ceramitalc, 5707; I T Talc, 14800; J-13, 15468; Jetfill 700C, 15607; Lubestine, 17712; Luzenac B170, 18052; Microbloc®, 19656; MicroPflex 1200, 19687; Microtuff 1000, 19711; Mistron CB, 19951; Mistron Vapor-RE, 19952; Mistron ZSC, 19953; Nicron 325, 21161; Nicron 665, 21162; Nicron JS 422, 21163; Nytal® 100, 21932; Olympic, 22146; PolyTalc 445, 24624; Select-A-Sorb, 27856; Silverline 200, 28438; Silverline 665, 28439; Special Extender, 29218; Stellar 500, 29638; SteriLine 200, 29732; SteriLine 665, 29733; Super Lubestine, 30180; Supra EF, 30295; Suprafino A, 30303; talc, 30731; Talc MS, 30732; Vantalc®, 33368; Vertal 200, 33633; Vertal 92, 200, 33632
14808-60-7: quartz, 25596; rose quartz, 26967
14816-18-3: Sebacil®, 27796; Volathion®, 34011
14855-76-6: methyl green, 19540
14857-34-2: CD5635, 5511
14882-18-9: Bismuth subsalicylate, 4244
14940-68-2: zirconium silicate, 34825
14960-06-6: Mirataine® H2C-HA, 19928
15086-94-9: Eosin, 10606
15096-52-3: cryolite, 7279; Kryalith, 16448
15206-55-0: Vicure® 55, 33735
15244-38-9: chromium sulfate, 6237; Santotan KR, 27425
15299-99-7: Devrinol, 8216; Devrinol T, 8217
15305-07-4: Q-1301, 25542
15310-01-7: benodanil, 3942; Calirus, 5009; Mascot Clearing, 18999
15317-78-9: Isobutyl Niclate®, 15347
15477-33-5: Aluminum chlorate, 1901; Mallebrein, 18487
15489-90-4: Hematin, 13780
15520-10-2: Dytek® A, 9439
15520-11-3: Perkadox® 16, 23301; Perkadox® 16-W40-GB5, 23302
15521-65-0: methyl niclate®, 19553
15535-29-2: Thioset® M, 31714
15535-69-0: ADK STAB BT-31, 724; ADK STAB LS-2, 731
15545-48-9: Dicurane 500 FW, 8409; Ludorum, 17778; Toro, 32074; Tripart® Ludorum 700, 32378
15545-58-9: Talisman, 30735; Tolurex, 31958; Tolurgan, 31959
15545-97-8: V-70, 33221
15578-26-4: TechneScan PYP, 30890
15599-36-7: Episol, 10657
15625-89-5: Ageflex TMPTA, 1001;Sipomer® TMPTA, 28543; SR-351, 29381
15630-89-4: sodium percarbonate, 28807
15647-08-2: Weston® EHDPP, 34313
15667-10-4: USP® -90MD, 33164
15708-41-5: sodium iron edetate, 28776
15716-30-0: Uvinul® P 25, 33203
15829-53-5: Hahnmann's mercury, 13537
15845-66-6: Aliette, 1612
15972-60-8: Alachlor, 1327; Alagan, 1329; Alanex, 1331; Alazine, 1340
16029-98-4: CT3610, 7330
16034-77-8: Cholebrine, 6167
16039-53-5: zinc lactate, 34780
16066-38-9: Lupersol 221, 17851
16111-62-9: Espercarb® 840, 10841; Lupersol 223, 17852; Trigonox® EHP, 32299
16118-49-3: Carbetamex, 5189
16230-35-6: CB2409.5, 5476
16260-09-6: Chemstat® HTSA#1, 6040;HTSA #1, 14445; Kemamide® P-181, 15899; oleyl palmitamide, 22133
16409-44-2: geraniol acetate, 12886; Geranyl acetate, 12889
16423-68-0: erythrosin, 10778; Iodesin, 15205; iodoeosin, 15212
16545-54-3: Argus DMTDP, 2862; Cyanox® MTDP, 7492; Evanstab® 14, 11257
16580-06-6: Trigonox® 169-OP50, 32289
16595-80-5: Tramisol®, 32134
16672-87-0: Cerone, 5756; Ethrel, 11050; Ethrel C, 11051; Ethrel-E, 11052; Ethrel-R, 11053; Power Ethephon 48, 24823
16693-53-1: Crodasinic LT40, 7146
16721-80-5: sodium bisulfide, 28737
16731-55-8: potassium metabisulfite, 24761
16752-77-5: Lannate®, 16727; methomyl, 19502; Sorex Golden Fly Bait, 29137
16789-79-5: Radiamine 6240, 25758
16830-15-2: Madecassol, 18291
16841-14-8: Kemamine® BQ-2802C, 15906; Chemical Base 6532, 5960; Lexamine 22, 17096; stearamidoethyl diethylamine, 29594
16893-85-9: malladrite, 18484; salufer, 27301; sodium silicofluoride, 28818; SSF, 29390; Uba, 32698
16919-27-0: Aflammit TI, 939
16923-95-8: Aflammit ZR, 941
16940-66-2: sodium borohydride, 28740
16961-83-1: sand acid, 27321
16961-83-4: hydrofluosilicic acid, 14587; keramyl, 16054; montanine, 20271
16971-82-7: Vanax® PML, 33293
17090-79-8: monensin, 20208; Romensin, 26930
17109-49-8: Edifenphos, 9618; Hinosan®, 14097
17125-80-3: Flosol, 12026
17194-00-2: Caustic baryta, 5468
17230-88-5: Danol, 7728
17301-53-0: Behentrimonium chloride, 3898; Genamin KDM-F, 12772; Varisoft® BT-85, 33417
17342-21-1: Dehyquart D, 7984
17540-75-9: Isonox® 132, 15384; Vanox® 1320, 33337
17661-50-6: Hest MS, 13885; Hetester MS, 13899; Kemester® 1418, 15939; Schercemol MS, 27611
17671-27-1: Starfol® BB, 29534
17673-56-2: Cetiol® J600, 5799; Dynacerin®, 9339; Dynacerin® 660, 9340
17689-77-9: CE6345, 5529; Dynasylan® ETAC, 9407; ethyltriacetoxysilane, 11146
17796-82-6: Santogard PVI, 27411
17804-35-2: Benomyl, 3944
17831-71-9: Ageflex T4EGDA, 970
17865-32-6: CM8650, 6459
17932-62-6: 2-cyanoethyltriethoxysilane, 7478; CC3433, 5494
18039-42-4: 5-phenyltetrazole, 23649
18156-74-6: CT3600, 7329
18162-48-6: CB2790, 5481
18162-84-0: CO9819, 6511
18171-19-2: 3-chloropropylmethyldimethoxysilane, 6137; CC3290, 5490
18181-70-9: Elocril, 9852; Iodofenphos, 15213
18194-24-6: Phospholipon® MC, 23728
18265-54-8: TIP, 31847
18282-10-5: flowers of tin, 12033; Superlite, 30252; tin ash, 31811; tin(IV) oxide, 31817
18312-32-8: Crodamol BE, 7117; Kemester® BE, 15954; Schercemol BE, 27585
18395-30-7: CI7810, 6288; Dynasylan®

IBTMO, 9410; Prosil® 178, 25235
18407-94-8: Dynasil® CA, 9383
18448-65-2: Ethoquad® O/12, 11001
18454-12-1: chrome red, 6209
18467-77-1: Cutlass, 7442
18472-51-0: Hibiclens, 14049; Hibitane, 14054; Phisomed, 23671; Rotersept, 27002; Uniscrub®, 33039; Unisept® Solution, 33040
18507-89-6: Deccox, 7846
18641-57-1: Syncrowax HR-C, 30534
18643-08-8: CD5636, 5512
18656-21-8: Renovue-65, Renovue-DIP, 26115
18691-97-9: methabenzthiazuron, 19472; Tribunil®, 32204
18769-78-3: M9030, 18111
18868-43-4: molybdenum dioxide, 20076
18871-14-2: Jasmopyrane, 15533
18917-91-4: Fixin, 11860
19010-66-3: Ledate®, 16949; methyl ledate, 19547; Super Sulfur No. 2, 30206
19044-88-3: oryzalin, 22347; Weed Stopper, 34258
19047-85-9: Weston® DSP, 34310
19321-40-5: Edenor PTO, 9610; Liponate PO-4, 17385; Mazol® PETO, 19139; Pentaerythrityl tetraoleate, 23141
19504-77-9: Variotin, 33396
19666-30-9: oxadiazon, 22406; Ronstar 2G, 26954
19706-80-0: VF-077, 33703
19910-65-7: Espercarb® 438M-60, 10840; Lupersol 225, 17854; Trigonox® ADC, 32293; Trigonox® SBP, 32307
19937-59-8: Dosaflo, 8812
20018-09-1: Amical® 48, 2058
20324-33-8: Arcosolv® TPM, 2816
20344-49-4: ferric hydroxide, 11649; ferrugo, 11700
20354-26-1: methazole, 19481; Probe, 25048
20427-58-1: zinc hydroxide, 34778
20427-59-2: Chiltern Kocide 101, 6069; Comac Parasol, 6659; cupric hydroxide, 7381, 7382, 7383; Schweitzer's reagent, 27738
20566-35-2: Great Lakes PHT4-Diol, 13313; tetrabromo phthalatediol, 31358
20624-25-3: Thiostop E, N, 31716
20667-12-3: silver(I) oxide , 28432
20816-12-0: osmic acid, 22355; osmium tetroxide, 22357
20830-75-5: Lanoxicaps, 16748; Lanoxin, 16749; Lanoxine-PG, 16750
20859-73-8: Phostoxin, 23754; Power Phosphine Pellets, 24830; Talunex, 30757
20941-65-5: Akrochem® TDEC, 1179; ethyl tellurac®, 11084; Perkacit® TDEC, 23290
21041-95-2: cadmium hydroxide, 4898
21087-64-9: Lexone®, 17165; metribuzin, 19595; metribuzin, 19596; Sencor®, 27947; Sencorex® WG, 27948
21142-28-9: Sulfetal CJOT 38, 29983
21245-01-2: Spectraban, 29225
21245-02-3: Escalol® 507, 10787; Padimate O, 22549; Solarchem® O, 28898;
21260-46-8: Bismate®, 4232; Bismet, 4233; Bismuth dimethyldithiocarbamate, 4238
21293-20-9: picrasmin, 23819
21302-09-0: dilauryl phosphite, 8475; Weston® DLP, 34306
21351-79-1: cesium hydroxide, 5775
21548-73-2: silver sulfide, 28435
21645-51-2: Alugel, 1882; Aluminum hydroxide, 1905; Colugel, 6655; Dialume, 8287; F-1000 Dried Gel, 11380; F-1000®, 11381; F-2000, 11382; F-2000 Dried Gel, 11383; F-2100 Dried Gel, 11386; F-2200 Dried Gel, 11388; F-500, -3600, etc, 11379; Hydral® 710, 14535; Hydroxal, 14656; Martifin, 18960; Martinal, 18962; Martinal® OL-111 LE, 18963; Martinal® ON-4608, 18964; Martinal® OS, 18965; Onyx Classica , 22185; Polarite® 880E(W), 24302; Rehydragel® Compressed Gel, 26025; SpaceRite S-11, 29183
21652-27-7: Miramine® O, 19871
21725-46-2: cyanazine, 7472; Fortrol, 12323; Match, 19055
21738-42-1: Vansil, 33363
21810-39-9: Ageflex FA-2Q50DMS, 973
21908-53-2: Kankerex, 15713; mercuric oxide, red and yellow, 19348; red oxide of mercury, 25966;
22023-23-0: Jet Amine DE-13, 15584
22042-96-2: Briquest® 543-45AS, 4535
22047-49-0: Afilan EHS, 931; Cetiol® 868, 5794; Lexolube® T-110, 17164; octyl stearate, 22041; Wickenol® 156, 34392
22128-62-7: Palite, 22610
22204-53-1: Laraflex®, 16785
22224-92-6: Nemacur®, 20863
22313-62-8: Esperox® 497M, 10851; Trigonox® 97-C75, 32277
22499-12-3: Vicure® 10, 33734
22662-39-1: Ranide, 25868
22722-98-1: Vitride®, 33994
22766-82-1: Starfol® OS, 29538
22781-23-3: Bendiocarb, 3937; Ficam, 11760; Garvox 3G, 12685; Seedox SC, 27840
22788-19-8: Emalex PG-di-L, 9982
22801-45-2: O.O.D, 21944
22839-47-0: Aspartame, 3233; Equal, 10731; NutraSweet, 21832
22882-95-7: Ceraphyl® IPL, 5732
22984-54-9: CM9220, 6467
23047-25-8: Lofepramin Hydrochloride, 17552
23089-26-1: Kamillosan, 15707
23103-98-2: Abol, 146; Aphox, 2654; FBC Pirimicarb 50, 11517; Phantom, 23571; pirimicarb, 23879; Pirimor, 23880; Power Demo, 24819; Rapid, 25884
23135-22-0: Blade, 4310; Vydate®, 34171
23184-66-9: Butachlor, 4727; Butanex, 4736; Machete, 18124
23288-60-0: Pertscan-99m, 23438
23386-52-9: Aerosol® A-196-85, 887; Octosol TH-40, 22028; Rewopol® SBDC 40, 26372
23474-91-1: Esperox® 13M, 10846
23505-41-1: Fernex, 11636, 11637; Primicid, 24983; Primotec, 24986
23560-59-0: heptenophos, 13795; Hostaquick, 14407
23564-05-8: CDA Mildothane, 5520; Cercobin, 5738; Mildothane, 19759; Mildothane Turf Liquid, 19760; Sys Tec® 1998, 30661; thiophanate methyl, 31706
23564-06-9: Nemafax, 20864
23593-75-1: Canesten®, 5086; Gyne-Lotrimin, 13513; Lotrimin, 17631; Mono-Baycuten®, 20217
23779-32-0: CT2507, 7318
23947-60-6: Milgo, 19765; Milstem, 19800
23950-58-5: Campbell's Rapier, 5069; Kerb 50W, 16064; Kerb 50W, 16065; Kerb Propyzamide 50, 16066; Rapier, 25889
24017-47-8: Hostathion, 14413; Methoxone, 19503; MSS CMPP, 20429; triazophos, 32196
24304-00-5: Aluminum nitride, 1918
24307-26-4: mepiquat chloride, 19327; Pix® ULV, 23898
24390-14-5: Doxycycline hydrochloride, 8946
24424-99-5: Di-*t*-butyl dicarbonate, 8367
24589-78-4: CM9160, 6466
24602-86-6: Calixin®, 5011
24634-61-5: potassium sorbate, 24780, 24781
24650-42-8: Photocure 51, 23757
24800-44-0: tripropylene glycol, 32400
24801-88-5: CI7840, 6289
24817-92-3: Citroflex A-6, 6344
24887-06-7: Decrolin®, 7866; Parolite, 22861; Reduxol Z, 25986; zinc formosul, 34775
24937-78-8: Airflex® 323, 1100; AT 1806M; AT 4030M, 3281; Fusabond® MC-197D, 12510; Hysol® 1942, 14738; Hysol® 342, 14735
24937-79-9: Dykor 204, 9317; Foraflon® 1000 HD, 12229; KF Polymer® C-1000, 16158; KF Polymer® T-1300, 16159; KF Polymer® T-850, 16160; KF Polymer® U-1000, 16161; KF Polymer® W-1000, 16162; Kynar® 301 F, 16509; Kynar® 460, 16510; Kynar® 700 Series, 16511; Kynar® Flex® 2800, 2801, 16513; Kynar® Flex® 2900, 16515; Solef® 1008, 28917; Solef® 1010, 28918; Solef® 11008/0003, 28922; Solef® 11010, 28923; Solef® 5008, 28919; Solef® 6010, 28920; Solef® 8808, 28921
24938-91-8: Bio-Soft® TD400, 4204; Chemal TDA-3, 5887; Ethal TDA-18, 10925; Ethal TDA-3, 10926; Ethal TDA-6, 10927; Genapol® X-040, 12802; Hetoxol TD-3, 13951; Hodag Nonionic TD-15, 14198; Iconol TDA-10, 14834; Iconol TDA-3, 14830; Iconol TDA-6, 14831; Iconol TDA-8, 14832; Iconol TDA-9, 14833; Lipocol TD-3, 17360; Macol® TD-3, 18274; Merpol® SH, 19378; Prox-onic TD-1/03, 25345; Rhodasurf® BC-420, 26651;

Rhodasurf® T-95, 26662; Rhodasurf® TDA-6, 26664; Teric 13A5, 31257; Trideceth series, 32232; Trycol® 5874, 32498; Trycol® 5940, 32501; Volpo T-3, 34031

24969-11-7: Penacolite® R-2170, 23093

25013-16-5: BHA, 4121; Butylated hydroxyanisole, 4787; Nipantiox, 21276; Nipantiox 1-F, 21277; Sustane® 1-F, 30432; Sustane® BHA, 30435

25034-71-3: Trilene® 65, 32318

25035-04-5: Rilsan® BESHVO, BESVO, 26763; RTP 201C, 27059; Thermocomp® HF-1006, 31613

25038-54-4: Ashlene® 630-33G, 3216; Ashlene® 735, 3217; Ashlene® 830L, 3218; Ashlene® 830L, 3219; Ashlene® 840, 3220; Ashlene® 858, 3221; Cabelec® 1015, 4861; Cadco® Cast Nylon, 4886; Capran® 77C, 5122; Capran® Unidraw®, 5123; Capran®, Capran® 77C, Capran® Emblem, Capran®Unidraw®, 5124; Capron® 8202, 5149; Capron® 8203C HS, 5150; Capron® 8232G HS FR, 5151; Capron® 8233G HS, 5152; Capron® 8253, 5153; Capron® 8259, 5154; Capron® 8266G HS, 5155; Capron® 8280, 5156; Celstran® N6G30-01-4, N66G30-01-4, 5651; CTX-312, 7336; Durethan® B 30 S, B 31 SK, 9198; Durethan® B 35 F, B 38 F, B 40 F, 9199; Durethan® BKV 115, 9201; Durethan® BKV 30 H, 9202; Durethan® BM 30 X, 9203; Durethan® KL1-2402/30, 9204; Durethan® RM KU 2-2501/30, 9205; Electrafil® J-3/CF/30, 9733; EMI-X® PC-1008, 10199; EMI-X® PDX-P-90305, 10204; Fiberfil® J-7/33, 11720; Fiberfil® J-7/33/IT, 11721; Fiberfil® NY-7, 11723; Fiberfil® NY-7/VO, 11724; Fiberstran® G-3/50, 11745; Grilon®, 13380; Grilon® A23GM, 13381; Grilon® BT40Z, 13383; Grilon® PV-15H, 13389; Grilon® PVN-15H, 13390; Grilon® R47HW, 13391; MC®, 19174; Novamid® 1010C, 21672; Novamid® 1020VA2, 21673; NY-10GF, 21856; NY-30CF, 21857; Nybex 12034 BKFR, 21863; Nybex 12056 BKFR, 21864; Nybex 13001 BKC, 21865; Nybex 15011 NA, 21866; Nybex 17000 NAX, 21867; Nybex 42002 BKHS, 21869; Nycoa® 1417, 21878; Nycoa® 4015, 21879; Nycoa® 438, 21871; Nycoa® 446, 21872; Nycoa® 567, 21875; Nycoa® 714, 21876; Nycoa® 870, 21877; Ny-Kon® P, 21892; Nylatron® GS-63, 21900; nylon 6, 21909; PA-211, 22527; PA-211G13, 22528; PA-211N40, 22529; PA-221, 22530; Plaslube® G-3/40/MS/5, 23942; Plaslube® J-3/30/MS/5, 23949; RTP 201A, 27057; Sniamid® ADS 40 I, 28666; Sniamid® ASN 27T, 28667; Stat-Kon® P, 29566; Texalon 1000A, 31442; Texalon 600A NU, 31441; Texalon GF 600A (6-33), 31447; Thermocomp® PC-1006, 31620; Ultramid® B3, 32829; Ultramid® B35G3, 32830; Ultramid® B3EG10, B3WG10, 32831; Ultramid® B3G10, 32832; Ultramid® B3K, 32833; Ultramid® B5, 32834; Ultramid® BS 3300, 32835; Ultramid® BS 400 S, 32836; Ultramid® BS 416, 32837

25038-59-9: Celstran® PETG30-01-4, 5654; Crastine® XMB 1068, 6969; Electrafil® J-1800/CF/30, 9750; Ertalyte®, 10768; Grilpet® EV-30, 13401; Impet® 330, 14914; Kodapak® 5214A, 16285; Kodapak® PET Copolyester 13339, 16286; Kodar® PETG Copolyester 6763, 16289; Mitsubishi Kasei GF-PET 6010G15, 19970; Petra® 130, 23456; Petra® 130FR, 23457; Petra® 230, 23458; Petra® 242, 23459; Polyethylene terephthalate, 24426; RTP 1105FR, 27069; Tenite® PET 9902, 31111; Traytuf Ultra-Clear, 32155; Ultralen® SP 3700 S, SP 3705, 32815; Ultralen® SP 3705, 32816; Valox® 9215, 33254

25038-74-8: nylon 12, 21912

25054-76-6: Lexaine® O, 17095; Mackam HV, 18143; Mirataine® BET-O-30, 19924; Oleamidopropyl betaine, 22108

25057-89-0: Bentazon, 3947

25066-20-0: Chemoxide SAO, 5999; Mackamine SAO, 18176

25068-26-2: Crystalor DC-6, 7292

25086-62-8: Alkasperse® M-10, 1713

25086-89-9: Agrimer VA 6, 1067; Luviskol® VA 28 E, 18033

25103-09-7: isooctyl thioglycolate, 15386

25103-12-2: triisooctyl phosphite, 32312; Weston® TIOP, 34320

25134-01-4: Electrafil® F-1700/CF/10/A, 9724; Electrafil® G-1704/SS/5, 9729; Electrafil® J-1700/CF/10, 9748; MP-10CF-4CC/15T, 20389; Noryl® 731, 21598; Noryl® BN25, 21600; Noryl® EM5101, 21601; Noryl® EN185, 21602; Noryl® FN150, 21603; Noryl® FN215X, 21604; Noryl® GTX810, 21605; Noryl® HM3020, 21606; Noryl® HS1000X, 21607; Noryl® N190X, 21608; Noryl® PC180X, 21609; Noryl® PN235, 21610; Noryl® PX0722, 21611; Noryl® SE100, 21612; Noryl® SE1GFN2, 21613; Polyphenylene oxide, 24518; Stat-Kon® ZC-1003, 29572; Styvex 72001 NA, 29899; Thermocomp® ZF-1004, 31629

25135-51-3: Electrafil® J-1500/CF/20, 9747; Mindel® B-310, 19810; Mindel® B-322, 19811; Mindel® M-800, 19812; Mindel® S-1000, 19813; PF-10GF/15T, 23552; PF-20GF, 23553; Polypenco® Polysulfone, 24513; polysulfone, 24618, 24619; RTP 901, 27067; Thermocomp® GF-1004, 31612; Udel® GF-110, 32730; Udel® P-1700, 32731; Ultrason® S 1010, 32869

25154-52-3: nonyl phenol, 21460

25155-18-4: Diaparene, 8326; Hyamine® 10X, 14484

25155-23-1: Coalite N.T.P, 6518; Kronitex® TXP, 16438; Syn-O-Ad® 8475M, 30554

25155-25-3: Bisbutyl peroxy diisopropyl benzene, 4230; Luperox 802, 17841; Perkadox® 14, 23300; Retilox® F 40 MG, 26274; Vul-Cup 40KE and R, 34118; Vul-Cup, Vul-Cup 40KE, Vul-Cup R, 34119

25155-30-0: DeSonate 50-S, 8176; DeSonate 60-S, 8177; Nacconol® 40G, 20647; Nansa® HS80S, 20687; Rhodacal® DDB-40, 26582; sodium dodecylbenzenesulfonate, 28760; Sul-fon-ate AA-10, 30020

25159-40-4: Mackamine OAO, 18175; oleamidopropylamine oxide, 22110

25167-32-2: Aerosol® DPOS-45, 881

25167-83-3: Dowicide 6, 8930

25168-05-2: Halso® 99, 13570

25168-73-4: Crodesta DKS F110, 7167; Crodesta F-160, 7169; Grilloten PSE141G, 13378; Ryoto Sugar Ester S-1170, 27169; Sucro Ester 15, 29932; Sucro Ester 7, 29931

25213-24-5: Airvol® 203, 205, 1124; Airvol® 205S, 523S, 540S, 1125; Airvol® 205S, 523S, 540S, 1126; Airvol® 425, WS 42, 1128; Airvol® 425, WS 42, 1129; Airvol® 523, 540, 1130; Airvol® 523, 540, 1131; Airvol® 705, 723, 740, 1132; Airvol® 705, 723, 740, 1133

25231-21-4: Acconon E, 238; Arlamol® E, 2910; Fancol SA-15, 11452; Hetoxol SP-15, 13949; Prostearyl 15, 25250; Prox-onic SA1-015/P, 25337; Witconol APS, 34513

25265-77-4: 2,2,4-trimethyl-1,3-pentanediol monoisobutyrate, 32339; Texanol® Ester-Alcohol, 31451

25301-02-4: Superinone, 30242

25311-71-1: Amaze®, 1951; Amidocid®, 2106; Oftanol®, 22049; Pryfon, 25355

25321-41-9: Eltesol® XA, 9877; Manro XSA, 18610; Reworyl® X, 26412; XSA 80, 34611; Xylene sulfonic acid, 34621

25322-68-3: Alkapol PEG 300, 1706; Atpet 300, 3383; Atpet 400, 3384; Atpet 600, 3385; Carbowax® Compound 20M, 5247;

Carbowax® PEG 200, 5249; Carbowax® PEG 540 Blend, 5250; Carbowax® PEG 8000, 5251; Carbowax® Sentry, 5252; Chemstat® P-400, 6042; Droxol 200, 9054; Emery® 6686, 10180; Hodag PEG 1000, 14207; Hodag PEG 1450, 14208; Hodag PEG 200, 14202; Hodag PEG 300, 14203; Hodag PEG 3350, 14209; Hodag PEG 400, 14204; Hodag PEG 540, 14205; Hodag PEG 600, 14206; Hodag PEG 8000, 14210; Ilexan HT, 14884; Lipoxol® 12000, 17448; Lutrol® E 1500, 18007; Lutrol® E 300, 18005; Lutrol® E 400, 18006; Lutrol® E 4000, 18008; Lutrol® E 6000, 18009; Macol® E-200, 18266; Nopalcol 200, 21474; Nopalcol 400, 21475; Nopalcol 600, 21476; Pluracol® E1000, 24173; Pluracol® E1500, 24174; Pluracol® E200, 24175; Pluracol® E2000, 24176; Pluracol® E300, 24177; Pluracol® E400, 24178; Pluracol® E400 NF, 24179; Pluracol® E4000, 24180; Pluracol® E6000, 24182; Pluracol® E600NF, 24183; Pluracol® E8000, 24184; polyethylene glycol, 24425; Poly-G® 200, 24444; Polyox® WSR 3333, 24503; Polyox® WSR N-10, 24504; Polytrix, 24649; Prox-onic PEG-2000, 25333; Rhodasurf® E 400, 26655; Rhodasurf® PEG 3350, 26659; Rhodasurf® PEG 400, 26660; Teric PEG 1000, 31277; Teric PEG 12000, 31283; Teric PEG 200, 31272; Teric PEG 300, 31273; Teric PEG 400, 31274; Teric PEG 4000, 31280; Teric PEG 600, 31275; Teric PEG 6000, 31281; Teric PEG 800, 31276; Teric PEG 8000, 31282

25339-09-7: Afilan ICS, 932; Kemester® 5822, 15951; Kessco® ICS, 16114; Schercemol ICS, 27603

25339-99-5: Grilloten LSE87, 13376; Ryoto Sugar Ester LWA-1570, 27166

25383-99-7: Artodan SP 55 Kosher, 3150; Crolactil SSL, 7186; Emplex, 10443; Emulsilac S, 10514; Grindtek FAL 1, 13409; Pationic SSL, 22908; Radiamuls® SSL 2990, 25788; Swedex SSL-5AC, 30450

25389-94-0: Kantrex, 15718; Kantrim, 15720; Klebcil, 16209

25395-31-7: Diacetin, 8248

25417-20-3: Dehscofix 917, 7925; Supragil® NK, 30308

25446-78-0: Rhodapex® 674/C, 26621; Rhodapex® EST-30, 26629

25448-25-3: Doverphos® 6, 8835; Weston® TDP, 34318

25496-01-9: LABS 100/H.V., 16544; Nansa® TDB, 20696; tridecylbenzene sulfonic acid, 32236

25496-72-4: Tegin® O, 30946

25550-98-5: Doverphos® 7, 8836; Weston® PDDP, 34315

25606-41-1: Filex, 11776; Propamocarb hydrochloride, 25171

25608-12-2: potassium polyacrylate, 24775

25609-89-6: Luviset CA 66, 18027; Resyn® 28-1310, 26245; VA/crotonates copolymer, 33224

25637-84-7: Emerest® 2419, 10097; glyceryl dioleate, 13080; Priolube 1409, 25016

25637-99-4: FR-1206, 12363; Great Lakes CD-75P, 13300; Hexabromocyclododecane, 13988

25640-78-2: Nusolv ABP-62, 21802

25655-41-8: Disadine, 8623; povidone-iodine, 24806; Videne Disinfectant Solution, Videne Disinfectant Tincture, Videne Powder, 33739; Videne Disinfectant Tincture, 33740; Videne Surgical Scrub, 33741

25704-18-1: Afcolene, 926

25736-86-1: Sipomer® HEM-5, 28535

25812-30-0: Lopid, 17599

25882-44-4: disodium lauramido MEA-sulfosuccinate, 8647; Geropon® SBL-203, 12912; Incrosul LMS, 15055; Mackanate LM-40, 18189; Rewopol® SBL 203, 26377; Varsulf® SBL-203, 33461

25928-94-3: Allabond Twenty/Twenty Adhesive, 1736; Allabond Twenty/Twenty NM, 1737; Amicon® C-860-4, 2065; Amicon® CT-4042-5, 2067; Amicon® ECT-86, 2068; Amicon® ME-868, 2069; Amicon® TG-86, 2073; Araldite® 2001, 2768; Araldite® CY 225, 2769; Araldite® ECN 1235, 2770; Araldite® GT 6060, 2771; Araldite® GZ 540 X-90, 2772; Araldite® PT 810, 2775; Araldite® PY 306, 2776; Araldite® XD 4955, 2777; Araldite® XD 897, 2778; Araldite® XU GY 358, 2779; Aratronic® 5001, 2784; Aratronic® 5040, 2785; Aroflint® 303-X-90, 3020; Basoset® 162, 3728; C-84, 4844; CI-2, 6287; Conapoxy® FR-1270, 6692; Conapoxy® TE-1257/Conacure® EA-08, 6693; D.E.R. 317, 7616; D.E.R. 362, 7617; D.E.R. 383, 7618; D.E.R. 642U, 7620; D.E.R. 732, 7621; E-3810, 9448; E-3824, 9449; E-9405, 9451; Emcast 1510, 1511, 10040; Emcast 1550, 1551, 10041; Epocap® 16129 A/B, 10668; Epocap® 16358 A/B, 10669; Epolite 1301, 10686; Epolite 1302, 10687; Epolite 2315, 10689; Epolite 3300, 10690; Epolite 5363, 10691; Epon® Resin DPL-1911, 10695; Epoweld® 19157, 10708; epoxy resin, 10719; FA-1, 11393; FA-14, 11394; FA-8, 11395; FFA-5, FFA-9, 11712; Fiberite 7669, 11733; Fiberite 7701, 11734; Hysol® 1C Epoxi-Patch Kit, 14733; Hysol® 6C Epoxi-Patch Kit, 14736; Hysol® EO1016, 14741; Hysol® MG1 Series, 14744; LCA-1, 16895; LCA-127, 16897; LCA-20, 16896; Master Bond EP11HT, 19036; Master Bond EP21HT, 19037; Master Bond EP30HT, 19038; Master Bond EP34CA, 19039; Master Bond EP75, 19040; Master Bond Supreme 11HT, 19041; Master Bond Supreme 3HT, 19042; Neonite® EG60/6mm, EG61/12mm, 20932; Norcast 142 Systems, 21528; Norcast 1460-1, 21530; Norcast 154FR, 21529; Norcast 3220G-1, 21531; Norcast 3258, 21532; Norcast 3705, 21533; Norcast 4914-1, 21534; Nor-Mer 020, 21568; P-11, 22501; P-51, 22517; P-56, 22519; P-80F, 22521; P-85, 22522; Quatrex 1010, 25618; Redux® 501, 25985; RX® 1906, 27155; Scotchkote® 213, 214, 27760; Sobral EE-632, 28700; Stycast® 1090, 29848; Stycast® 1210, 29849; Stycast® 1266, 29850; Stycast® 1467, 29851; Stycast® 2850-FT, 29852; Supertherm 2003, 30286; Tactix 123, 30696; Tactix 556, 30697; Tactix 742, 30698; Thermoset 100, 31647; Thermoset 300/No. 65 Hardener, 31648; Thermoset 310, 31649; Thermoset DC-232, 31650; Thiokol® FES-2258, 31692; Tra-Bond 2151, 32118; Tra-Bond F113, 32119; Tra-Cast 3103, 32121; Tra-Duct 2902, 32131; Tra-Shield 2867, 32149; Witcobond® XW, 34463

25954-13-6: Krenite, 16400

26002-80-2: phenothrin, 23619; Sumithrin, 30117

26027-37-2: Dionil® OC, 8566; Ethomid® O/17, 10991

26062-79-3: Agefloc WT-20, 1009; Agefloc WT-40, 1010; Agequat 400, 1012; Mackernium 006, 18206; M-Quat® 40, 20399

26062-94-2: Celanex® 1300A, 1600A, 2000K, 5569; Celanex® 3310, 5300, J600, 5570; Celanex® J600, 5571; Celstran® PBTG30-01-4, 5652; Crastine® S 600, Crastine® SG 625, 665 FR, 653, XB 3035, 6964; Crastine® SG 625, 6965; Crastine® SG 665 FR, 6966; Crastine® SO 653, 6967; Crastine® XB 3035, 6968; Electrafil® G-1854/SS/7, 9730; Electrafil® J-1850/CF/30, 9751; EMI-X® PDX-W-88341, 10206; Grilpet® XE3060, 13402; Mitsubishi Kasei PBT 5008, 19971; Mitsubishi Kasei PBT 5010F1, 19972; PBT-1100,

22938; PBT-1100G15, 22939; PBT-1300, 22940; PBT-1700, 22941; PDX-84369, 22959; Pocan® B 1300, 24280; Pocan® B 1305, 24281; Pocan® B 1505, 24282; Pocan® KU1-7033, 24283; Pocan® S 1506, 24284; PS-30GM/000, 25369; RTP 1001, 27068; Stat-Kon® W, 29571; Thermocomp® WC-1006, 31626; Ultradur® B 2550, 32793; Ultradur® B 4300 G10, 32794; Ultradur® B 4500, 32795; Ultradur® KR 4001, 32796; Valox® 210HP, 220HP, 230HP, 260HP, 280HP, 33248; Valox® 325, 33249; Valox® 508, 33250; Valox® 701, 33251; Valox® 780, 33252; Valox® 815, 33253; Valox® DR48, 33256; Vandar® 2100, 33310; Vybex 22008 BKFR, 34166; Vybex 22028 BKFR, 34167

26097-80-3: Equiben, 10733; Novazole, 21711

26112-07-2: Nipagin M Potassium, 21267

26155-31-7: morantel tartrate, 20307; Paratect Bolus, 22804

26183-52-8: Bio-Soft® FF 400, 4199; Chemal DA-4, 5881; Deceth-6, 7848; Desonic® DA-4, 8182; Desonic® DA-6, 8183; Ethal DA-4, 10917; Genapol® DA-040, 12788; Iconol DA-4, 14819; Iconol DA-6, 14820; Iconol DA-9, 14821; Marlipal® 1012/4, 18865; Oxetal D 104, 22421; Prox-onic DA-1/04, 25313; Rhodasurf® DA-4, 26653; Trycol® 5950, 32502

26225-79-6: ethofumesate, 10967; Nortron, 21586;

26256-79-1: Deriphat 160C, 8091

26264-05-1: Arylan® PWS, 3158; Kowet 3300, 16374; Rhodacal® 330, 26577; Rhodacal® IPAM, 26586

26264-06-2: Ablusol DBC, 123; Kowet 12, 16373

26264-58-4: Dehscofix 923, 7927; Rhodacal® RM/210, 26590; Supragil® MNS/90, 30307

26266-57-9: Ablunol S-40, 109; Arlacel® 40, 2900; Crill 2, 7029; Disponil SMP 100 F1, 8700; Emsorb® 2510, 10460; Glycomul® P, 13113; Hodag SMP, 14244; Kemester® S40, 15967; Liposorb P, 17410; Montane 40, 20266; Prote-sorb SMP, 25280; Radiasurf® 7135, 25798; S-Maz® 40, 28641; Sorbax SMP, 29100; Sorbon S-40, 29118; Span® 40, 29188

26266-58-0: Ablunol S-85, 112; Alkamuls® S-85, 1682; Alkamuls® STO, 1689; Arlacel® 85, 2904; Atmer® 106, 3359; Crill 45, 7035; Disponil STO 100 F1, 8705; Emsorb® 2503, 10457; Ethylan® GT85, 11113; Glycomul® TO, 13116; Hodag STO, 14247; Kemester® S85, 15971; Liposorb TO, 17416; Prote-sorb STO, 25282; Radia® 7355, 25738; S-Maz® 85, 28646; Sorbax STO, 29102; Span® 85, 29192; Witconol 2503, 34502

26266-77-3: Abitol® E, 71

26351-19-9: violuric acid, 33881

26401-27-4: diphenyl isooctyl phosphite, 8595; Doverphos® DPIOP, 8848; Weston® ODPP, 34314

26402-26-6: glyceryl caprylate, 13077; Imwitor® 308, 14940; Imwitor® 742, 14945; Witafrol® 7420, 34421

26402-31-3: Flexricin® 9, 11950; Naturechem® PGR, 20777; Propylene glycol ricinoleate, 25222

26426-80-2: Tamol® 731-25%, 30765

26444-49-5: diphenylcresyl phosphate, 8601; Disflamoll® DPK, TPK, 8641

26446-35-5: Hallco® C-918, 13549; monacetin, 20095

26446-38-8: Ryoto Sugar Ester P-1570, P-1670, 27168

26447-10-9: Ammonium xylene sulfonate, 2259; Eltesol® AX 40, 9865; Hartotrope AXS, 13691; Naxonate® 4AX, 20822; Stepanate® AXS, 29672

26483-35-2: Behenamine oxide, 3897

26523-78-4: Doverphos® 4-HR, 8845; Doverphos® 4, 8834; Lankromark® LE109, 16695; Lowinox® TNPP, 17676; Trisnonylphenyl phosphite, 32411; Weston® 399, 34297; Weston® TNPP, 34323; Wytox® 312, 34578

26530-20-1: Kathon®, 15792; Kathon® 893, 15794; Micro-Chek® 11, 19664; Pancil T, 22666; Skane® M-8, 28563

26544-23-0: diphenyl isodecyl phosphite, 8594; Doverphos® 8, 8837; Weston® DPDP, 34308

26544-27-4: Weston® 600, 34303

26545-53-9: Ablusol DBD, 124; Marlopon® ADS 50, 18902

26589-26-4: Acrylates/PVP copolymer, 411; Luviflex® VBM 35, 18016; Stepanhold® Extra, 29679; Stepanhold® R-1, 29680

26590-05-6: Mackernium 007, 18207; Mirapol® 550, 19914

26597-36-4: Kemamine® Q-2802C, 15923

26635-75-6: Amidox® L-2, 2111

26635-92-7: Ethomeen® 18/15, 10973; Ethomeen® 18/20, 10974; Ethomeen® 18/25, 10975; Ethomeen® 18/60, 10976; Ethox SAM-10, 11037; Hetoxamine ST-5, 13917; Marlazin® S 10, 18756; Prox-onic MS-05, 25326; Teric 18M5, 31263; Trymeen® 6617, 32523; Trymeen® SAM-50, 32527; Zusomin S 110, 34916

26635-93-2: Marlazin® OL 2, 18755

26635-93-8: Marlowet® 5400, 18927

26644-46-2: Fairy Ring Destroyer, 11416; Funginex, 12478; Saprol, 27447; triforine, 32254

26658-19-5: Alkamuls® S-65, 1678; Alkamuls® STS, 1690; Crill 35, 7033; Disponil STS 100 F1, 8707; Emsorb® 2507, 10459; Famodan TS Kosher, 11424; Glycomul® TS KFG, 13117; Grindtek STS, 13416; Hodag STS, 14248; Kemester® S65, 15969; Liposorb TS, 17418; Montane 65, 20268; Prote-sorb STS, 25283; Radia® 7345, 25737; Radiamuls® Sorb 2344, Sorb 2345, 25786; Radiamuls® Sorb 2345, 25787; S-Maz® 65K, 28643; Sorbax STS, 29103; sorbitan tristearate, 29111; Span® 65, 29190

26675-46-7: Isoflurane, 15363

26680-54-6: Milldride® OSA, 19789

26741-53-7: Alkanox® P-24, 1705; Blendex® 340, 4342; Ultranox® 626, 32850; Ultranox® 627A, 32851

26748-38-9: Esperox® 750M, 10857

26748-41-4: Aztec® *t*-Butyl Peroxyneodecanoate-50 OMS, 75 OMS, 3547; Esperox® 33M, 10849; Lupersol 10, 17842; Trigonox® 23, 32267

26748-47-0: Esperox® 939M, 10858; Lupersol 188-M75, 17848; Lupersol 288-M75, 17859; Trigonox® 99-B75, 32278

26761-45-5: glycidyl decanoate, 13092; Glydexx N-10, 13141

26780-96-1: Agerite® MA, 1019; Agerite® Resin D®, 1020; Akrochem® Antioxidant DQ, 1160; Flectol H, 11911; Flectol Pastilles, 11913

26787-78-0: Larocin, 16787; Larotid, 16800; Qualamox, 25583

26807-65-8: Lozol, 17688

26850-24-8: Dantocol® DHE, 7730; DEDM hydantoin, 7872

26921-17-5: Timolate, 31805

26952-21-6: Exxal® 8, 11336; isooctyl alcohol, 15385

26970-82-1: sodium selenite, 28815

27176-87-0: dodecylbenzene sulfonic acid, 8779; Pentine Acid 5431, 23182; Rhodacal® ABSA, 26579; Witco® 1298, 34458

27177-77-1: Polystep® A-15-30K, 24587

27178-16-1: Kodaflex® DIDA, 16268; Monoplex® DDA, 20245; Plasthall® DIDA, 23998

27194-74-7: Schercemol PGML, 27620

27195-16-0: Crodesta DKS F10, 7166; Crodesta F-10, 7168; Ryoto Sugar Ester S-570, S-770, 27171

27233-00-7: Hetester HCA, 13895; Naturechem® GTH, 20771

27247-96-7: Exchem GO-1, 11294

27306-78-1: Silwet® L-77, 28450

27306-79-2: Hetoxol M-3, 13943

27306-90-7: Akypo RLM 160, 1235

27321-96-6: Forlan C-24, 12249

27503-81-7: Neo Heliopan® Hydro, 20868

27554-26-3: Morflex 100, 20315; Palatinol® D10, 22586; Witcizer 313, 34455

27607-77-8: CT3795, 7331

27638-00-2: Capmul® GDL, 5111; Emulsynt GDL, 10525; glyceryl dilaurate, 13079; Kemester® GDL, 15961; Kessco® GDL, 16107; Lexemul®

GDL, 17138
27640-89-7: Kemester® EE, 15957; Schercemol EE, 27599
27668-52-6: CO9745, 6505
27676-62-6: Anox® IC-14, 2485; Vanox® GT, 33342
27776-21-2: VA-044, 33225
27790-37-0: Marignac's salt, 18717
27813-02-1: Bisomer 2HPMA, 4274; hydroxypropyl methacrylate, 14667
27841-06-1: Schercemol NGDC, 27612
27883-12-1: Alkamide® CP-1255, 1638
27986-36-3: Akyporox NP 15, 1276
28061-69-0: Armeen® DMOD, 2947; Crodamine 3.AOD, 7109
28108-99-8: Syn-O-Ad® 8480, 30557
28211-18-9: Ganex® V-220, 12643; PVP/eicosene copolymer, 25430
28212-44-4: Sandopan® MA-18, 27347
28434-01-7: Reslin, 26216
28510-23-8: Schercemol NGDO, 27614
28519-02-0: Calfax DB-45, 4983; Chemcogen AC, 5941; Rhodacal® DSB, 26584
28631-35-8: Duplosan® DP, 9131
28631-63-2: Cumene sulfonic acid, 7353; Eltesol® CA 65, 9866; Reworyl® C, 26403
28724-32-5: Ethoquad® 18/25, 10996
28772-56-7: bromadiolone, 4565; Rentokil Deadline, 26116; Slaymor, 28597
28860-95-9: Lodosin, 17549
28874-51-3: Ajidew N-50, 1143; Nalidone®, 20672; Ritamectant PCA, 26816
28905-71-7: trinitrocresol, 32347
28961-43-5: Ageflex EOTMPTA, 963
29051-57-8: Pationic 122A, 22904
29091-05-2: Cobex, 6544
29232-93-7: Actellic, 458; Actellifog, 459; Blex, 4349; Fumite Pirimiphos Methyl Smoke, 12463; Fumite Pirimiphos-Methyl Smoke, 12464; Sybol, 30475
29240-17-3: Amilperoxy pivalate, 2122; Aztec® *t*-Amyl peroxypivalate-75 OM, 3540; Esperox® 551M, 10853; Lupersol 554-M50, 554-M75, 17865; Trigonox® 125-C75, 32283
29297-22-0: Luviquat® FC 370, 18021; Luviquat® FC 550, 18022; Luviquat® FC 905, 18023; Luviquat® HM 552, 18024; Polyquaternium-16, 24531
29317-52-0: Tomah PA-13i, 31981
29381-93-9: Marlopon® AT, 18904
29385-43-1: Preventol® CI7-100, 24929; tolyltriazole, 31962
29406-96-0: polybutad1,2-polybutadiene, 24371
29590-42-9: Ageflex FA-8, 978; SR-440, 29383
29598-76-3: Seenox 412S, 27843
29710-25-6: Dermol OO, 8129; Schercemol OHS, 27615; Wickenol® 171, 34398
29761-21-5: Syn-O-Ad® 8479, 30556
29806-73-3: Bernel® Ester EHP, 4008; Ceraphyl® 368, 5725; Kessco® Octyl Palmitate, 16118; Lexol EHP, 17148; octyl palmitate, 22040; Schercemol OP, 27617; Tegosoft® OP, 31025; Unimate® EHP, 32970; Wickenol® 155, 34391
29860-47-7: Sipomer® TMPEO, 28542
29911-28-2: Dowanol® DPnB, 8893
29964-84-9: Ageflex FM-10, 985; isodecyl methacrylate, 15359
29973-13-5: Arylmate®, 3166; Croneton®, 7206
30007-47-7: Bronidox L, 4592
30286-75-0: Oxivent, 22433
30342-62-2: Crodateric S, 7156
30364-51-3: Hamposyl® M-30, 13603
30416-77-4: Perlankrol® PA Conc, 23329
30525-89-4: Granuform, 13269; paraformaldehyde, 22751
30526-22-8: Hartotrope KTS 44, 13692; Naxonate® 4KT, 20823
30560-19-1: Orthene, 22322
30657-38-6: lauryl PCA, 16866
31001-77-1: CM8450, 6454; Colortrend, 6648
31112-62-6: Amipaque, 2197
31127-82-9: Endobil, 10548
31218-83-4: Propetamphos, 25183
31394-71-5: Hodag CSA-80, 14173; Lutrol® OP-2000, 18010
31430-15-6: Flubendazole, 12039
31431-39-7: Equivurm Plus, 10736; Mebatreat, 19197; Mebendazole, 19198; Mebenvet, 19199; Ovitelmin, 22402; Telmin, 31066; Vermox, 33591
31556-45-3: Emerest® 2308, 10083; Kemester® 5721, 15950; Lexolube® B-109, 17162; Liponate TDS, 17389
31566-31-1: Alkamuls® GMS/C, 1667; Alkamuls® GMS/C, 1668; Capmul® GMS, 5113; Cerasynt® WM, 5737; CPH-250-SE, 6944; Cutina® GMS, Cutina® KD16, Cutina® MD-A, 7440; Geleol, 12733; Imwitor® 191, 14939; Radiamuls® MG 2141, MG 2142, MG 2600, MG 2900, 25777; Tegin® 4011, 30945; Tego® Care 150, Care 300, 30967; Witconol 2400, Witconol MST, Witconol RHT, 34499; Zohar GLST, 34850
31570-04-4: Alkanox® 240, 1703; Hostanox® PAR 24, 14375; Lowinox® 242, 17655; Naugard® 524, 20791
31621-91-7: Ethox 1122, 11009; Genagen PL-090, 12762; PEG-9 pelargonate, 23054
31692-79-2: Masil® SFR 70, 19023
31694-55-0: Acconon ETG, 239; Ethosperse® G-26, 11006; Glycereth series, 13069; Hetoxide G-7, 13924; Liponic EG-1, 17392
31717-87-0: dodemorph-acetate, 8782; F-238, 11376; Meltatox®, 19313
31778-15-1: Evangard® 18MP, 11252; octadecyl mercaptopropionate, 21983
31799-71-0: Lutensol® FSA 10, 17987; PEG-3 oleamide, 23048
32057-14-0: Peceol Isostearique, 23002
32073-22-6: Eltesol® SC 93, 9871; Naxonate® 45SC, 20826; Naxonate® SC, 20827; Reworyl® NCS, 26406; sodium cumene sulfonate, 28752; Stepanate® SCS, 29673; Witconate SCS 45%, 34490
32131-17-2: Ashlene® 520, 3206; Ashlene® 520-13G, 3207; Ashlene® 520MS, 3208; Ashlene® 525-13G, 3209; Ashlene® 527, 3210; Ashlene® 527LD-13G, 3211; Ashlene® 528BR-WO, 3212; Ashlene® 528L-13G, 3213; Ashlene® 541, 3214; Ashlene® 541S, 3215; Ashlene® 61-2M, 3205; Ceistran® N66G30-01-4, 5556; Celanese® Nylon 1000-1, 5563; Celanese® Nylon 1003-1, 5564; Celanese® Nylon 1500-1, 5565; Celanese® Nylon 7420, 7423, 5566; Celanese® Nylon N-186, 5567; CTC-3300, 7333; Durethan® A 30 S, 9197; Electrafil® J-1/30/CF/7/H, 9731; EMI-X® PDX-R-89496, 10205; EMI-X® RC-1008, 10207; Fiberfil® J-1/30, 11715; Fiberfil® J-17/30/VO, 11716; Fiberfil® NY-16/MF/40, 11722; Fiberstran® G-1/50, 11744; Grilon® T300GM, 13393; Grilon® TV-15H, 13394; Leona®, 16999; Maranyl® A125, 18683; Maranyl® A175S, 18684; Maranyl® A360, 18685; Maranyl® TA505HS, 18686; NN-10GF, 21407; NN-20CF, 21408; NT-15GF/000, 21758; Nybex 22008 BKUT, 21868; Nybex 52000 NA, 21870; Nycoa® 500, 21873; Nycoa® 5015, 21880; Nycoa® 528, 21874; Ny-Kon® R, 21894; Nylatron® 1018 HS, 21898; Nylatron® 1024 HS, 21899; Nylatron® NSB-90, 21901; nylon 6/6, 21910; Nylon N-012, 21913; PA-111, 22523; PA-111CF30, 22524; PA-111G13, 22525; PA-121, 22526; Plaslube® G-1/30/SI/2, 23941; Plaslube® J-1/30/MS/5, 23944; Plaslube® J-1/33/TF/13/SI/2, 23945; Plaslube® J-1/CF/15/TF/20, 23946; Plaslube® J-77/30/TF/15, 23956; Plaslube® NY-1/MS/5/TF/30, 23959; Plaslube® NY-1/SI/5, 23960; Plaslube® NY-1/TF/10, 23961; Polypenco® Nylon 101, 24510; RTP 200FR, 27056; Sniamid® SSD 300 EP 021, 28668; Sniamid® SSD AF, 28669; Stat-Kon® PDX-84440, 29567; Stat-Kon® R, 29568; Stat-Kon® RC-1002, 29569; Stat-Kon® RF-15, 29570; Texalon 1200A BK-11, 31443; Texalon 1200A HR-2 BK-16, 31444; Texalon 1308 A, 31445; Texalon GF 1200A (13-40), 31448; Thermocomp® RC-1002, 31622; Ultramid® A3, 32824; Ultramid®

A3G5, 32825; Ultramid® A3K, 32826; Ultramid® A3X1G10, 32827; Ultramid® A4, 32828; Verton® RF-700-10 EM HS, 33641; Verton® RF-7007 HS, 33642
32360-05-7: Ageflex FM-1620, 989; Ageflex FM-68, 988; octadecyl methacrylate, 21984; stearyl methacrylate, 29607
32426-11-2: BTC® 818, 4635
32440-50-9: Ganex® V-216, 12642; PVP/hexadecene copolymer, 25431
32456-28-6: Crodateric O, 0.100, 7155
32534-81-9: FR-1205, 12362; Fyrol® PBR, 12541; Great Lakes DE-71, 13302; pentabromodiphenyl oxide, 23129, 23130
32536-52-0: FR-1208, 12364; Great Lakes DE-79, 13303; Octabromodiphenyl oxide, 21981; Saytex® 111, 27515
32588-76-4: Saytex® BT-93®, 27520
32612-48-9: Texapon PNA-127, 31474
32647-67-9: Millithix® 925, 19794
32954-43-1: Schercotaine PAB, 27699
33089-61-1: Amitraz, 2202; Mitaban, 19954; Mitac 20, 19955; Taktic, 30729
33113-08-5: Croptex Fungex, 7215
33665-90-6: Sunett, 30140
33703-08-1: Adimoll® DN, 692; Jayflex® DINA, 15551; Plastomoll® NA, 24059; PX-209, 25439
33907-46-9: Naturechem® EGHS, 20769; Paricin® 15, 22835
33907-47-0: Naturechem® PGHS, 20776; Paricin® 9, 22842
33939-64-9: Akypo RLM 100 NV, 1233; Akypo® Soft 160 NV, 1252; Rewopol® CLN 100, 26362; Sandopan® LS-24, 27346
33939-65-0: Sandopan® KST, 27344
34014-18-1: Bushwacker, 4724
34123-59-6: Arelon, 2830; Chiltern IPU, 6068; Hytane, 14774; Portman Isotop, 24712; Power Swing, 24834; Protugan, 25300; Sabre, 27191; Tolkan, 31937; Tolkan, 31938; Tolkan 500, 31939
34137-09-2: Vanox® SKT, 33347
34205-21-5: Pradone Plus, 24850
34316-64-8: Cetiol® A, 5796
34360-00-4: Tomah E-14-2, 31969
34424-98-1: Caprol® 10G40, 5130; Drewpol® 10-4-O, 9014; Hodag PGO-104 (formerly Hodag SVO-1047), 14220; Hodag SVO-1047, 14249; Mazol® PGO-104, 19141
34443-12-4: Esperox® C-496, 10860
34562-31-7: Vanax® 808 HP, 33285
34590-94-8: Arcosolv® DPM, 2812; Icinol DPM, 14815; Poly-Solv® DPM, 24555
34643-46-4: Tokuthion®, 31932
34745-96-5: Texacat® DD, 31426
34938-91-8: Dehydol PID 6, 7955
34962-91-9: Dermol 108, 8119
35141-30-1: CT2910, 7322; Dynasylan® TRIAMO, 9418
35256-85-0: Comodor, Comodor 600, 6667; tebutam, 30879
35274-05-6: Ceraphyl® 28, 5712; cetyl lactate, 5820; Liponate CL, 17373
35285-68-8: Nipagin A Sodium, 21265
35285-69-9: Nipasol M Sodium, 21284
35325-02-1: Uniplex 225, 33013
35367-38-5: Diflubenzuron, 8450; Dimilin, 8529
35400-43-2: Bolstar, 4384; Helothion, 13768
35512-33-9: Lentagran, 16990; Pyridate, 25480
35545-57-4: Hetoxide BN-13, 13919
35554-44-0: Enilconazole, 10582; Fungaflor, 12471; Imaverol, 14891; imazalil, 14892; Magnate, 18349
35575-96-3: Alfacron, 1571; Alfacron 10WP, 1572; azamethiphos, 3517
35634-74-3: V-19, 33214
35913-09-8: chlorobenzaldehyde, 6118
35958-30-6: Isonox® 129, 15383; Vanox® 1290, 33336
36116-84-4: diisooctyl phosphite, 8465; Doverphos® DIOP, 8846; Weston® DOPI, 34307
36311-34-9: Ceraphyl® ICA, 5731; Eutanol G16, 11240; Michel XO-150-16, 19641
36330-85-5: Lederfen, 16955
36338-96-2: vegetable rouge, 33523
36409-57-1: Empicol® SLL, 10338; Geropon® LSS, 12908
36432-46-9: Weston® DTDP, 34311
36445-71-3: Calfax 10L-45, 4982; Dowfax 3B2, 8911; Poly-Tergent® 3B2, 24630
36457-19-9: Nipagin A Potassium, 21264
36457-20-2: Nipabutyl Sodium, 21251
36483-57-5: FR-513, 12354; tribromoneopentyl alcohol, 32201
36521-89-8: Sorbon S-66, 29120
36631-30-8: Morflex 530, 20328
36653-82-4: Adol® 52 NF, 782; Cachalot® C-50 NF, 4878; Cetaffine®, 5781; Cetal, 5783; cetyl alcohol, 5815; cetyl alcohol, 5816; cetyl alcohol, 5817; Crodacol C70, 7068; Crodacol C95NF, 7069; Dehydag Wax 16, 7944; Emery® 3336, 10172; Epal® 16NF, 10626; ethal, 10910; Exxal® 16, 11341; Fancol CA, 11432; Lanette 16, 16663; Lipocol C, 17351; Lorol C16, 17614; Loxlol VPG 1743, 17686; Philcohol 1600, 23667; Rita CA, 26788
36734-19-7: CDA Roval, 5521; Roval Dust, 27021; Roval Flo, 27022; Roval Green, 27023; Roval WP, 27024; Turbair Roval, 32598
36788-39-3: Weston® 430, 34298
37067-27-9: Stabilizer, 29404
37106-97-1: Chymex, 6283
37139-99-4: Ammonyx® KP, 2263; Incroquat O-50, 15041; Mackernium KP, 18208; M-Quat® JO-50, 20407
37187-22-7: USP® -240, 33165
37199-81-8: Colloid 111, 111D, 6629
37205-87-1: Marlowet® ISM, 18932
37220-82-9: Emerest® 2421, 10098; glyceryl monooleate, 13084
37251-69-7: Tergitol® D-683, 31246
37294-49-8: Aerosol® A-268, 888
37309-58-3: Synton PAO-100, 30650
37318-14-2: Radiasurf® 7423, 25814
37332-99-3: Avoparcin, 3502
37340-60-6: Rhodafac® LO-529, 26596
37349-34-1: Caprol® 3GS, 5134; Hodag PGS-101, 14225; Hodag PGS-61, 14231; Polyaldo® TGMS KFG, 24355; Polyglyceryl stearate series, 24457; Santone® 3-1-S, 27414
37350-58-6: Lopresoretic, 17601
37354-45-5: disodium lauryl ethoxy sulfosuccinate, 8648; Empicol® SDD, 10337
37475-88-0: ACS 60, 453; Ammonium curmene sulfonate, 2236; Eltesol® AC60, 9863; Reworyl® ACS, 26402
37478-68-5: Caprylic imidazoline, 5159; Crodazoline Cy, 7160; Mackazoline CY, 18199; Monazoline CY, 20194
37893-02-0: Cropotex®, 7211
37924-13-3: Lancer, 16653; Perfluidone, 23248
38260-54-7: etrimfos, 11155; Satisfar, 27488
38304-91-5: Loniten, 17584
38411-30-3: Marlon® A365, 18874
38455-77-5: stannic chromate, 29495
38517-23-6: Acylglutamate HS-11, 603
38566-94-8: Nipabutyl Potassium, 21250
38613-77-3: Alkanox® 24-44, 1702; Sandostab P-EPQ, 27353
38641-94-0: Rodeo, 26899
38720-61-5: Akypostat MA 35, 1324
38916-42-6: Aerosol® 22, 879
39148-24-8: Aliette, 1613
39198-34-0: VR-110, 34056
39236-46-9: Abiol, 69; Germall® 115, 12893; Imidazolidinyl urea, 14896
39255-32-8: Manzanate, 18623
39300-45-3: Karathane, 15759
39354-45-5: Aerosol® A-102, 885; Schercopol LPS, 27668; Setacin 103 Spezial, 28081; Surfagene S 30, 30361
39354-47-5: Emcol® 4300, 10049
39407-03-9: Marlophor® HS-Acid, 18894
39409-82-0: magnesia alba, 18352
39421-75-5: Galactasol® Guar Derivs, 12591; hydroxypropyl guar, 14666; Jaguar® 413, 15494; Jaguar® HP 60, 15501; Jaguar® HP 8, 15502; Jaguar® HP-11, 15503
39464-64-7: Chemphos TC-341, 6011; Lubrhophos® LM-400, 17724; Rhodafac® RM-410, 26607
39464-66-9: Akypomine® MW 05, 1266; Chemphos TR-510, 6017; Marlophor® MO 3-Acid, 18898; Rewophat EAK 8190, 26352; Rhodafac® RD-510, 26605;
39464-67-0: Lubrhophos® LF-200, 17723
39464-69-2: Brophos OL-3, 4614; Chemfac PB-184, 5952; Chemphos TR-505, 6015; Chemphos TR-515, 6019; Chemphos TR-541, 6021; Crodafos N10 Acid, 7079; Crodafos N3 Acid, 7081; Crodafos N5 Acid, 7083; Crodafos O2 Acid, 7084; Hetphos

OA-3, 13953; Laurelphos 400, 16846; Lubrhophos® LB-400, 17720; oleth-phosphate Series (2-20), 22127; Rhodafac® RB-400, 26604
39464-70-5: Chemfac PC-006, 5953; Lubrhophos® LP-700, 17725; Marlowet® 5324, 18926; Rhodafac® BP-769, 26594
39515-41-8: fenpropathrin, 11615; Meothrin, 19322
39529-26-5: Caprol® 10G10S, 5128; Hodag PGS-1010, 14226
39669-97-1: Chemidex P, 5966; Palmitamidopropyl dimethylamine, 22630
40372-72-3: Aktisil PF 216, 1204; CB2494, 5478; Si 69, 28190
40487-42-1: pendimethalin, 23102; Stomp, 29784; Stomp H, 29785
40716-42-5: Amidex RC, 2101; Mackamide R, 18165; Rodea, 26898
40754-60-7: disodium ricinoleamido MEA-sulfosuccinate, 8654; Geropon® SBR-3, 12913
41083-11-8: azocyclotin, 3523; Clermait®, 6408; Peropal®, 23401
41183-64-6: Neoscan, 20963
41205-21-4: fluoromide, 12108; Sparticide®, 29211
41340-25-4: Lodine, 17546
41394-05-2: Countdown, 6904; Goltix®, 13191; Metamitron, 19443
41395-83-9: DPPG, 8950; Emerest® 2388, 10094; Lexol PG 900, 17154;Schercemol PGDP, 27619;
41395-89-5: Wickenol® 152, 34389
41473-08-9: Bilimiron, 4142; Oravue, 22272; Videobil, 33742
41483-43-6: Nimrod, 21220
41484-35-9: Anox® 70, 2484
41621-49-2: Loprox, 17604
41637-38-1: CD480, 5498
41669-30-1: Iso Isostearyle WL 3196, 15334; isostearyl isostearate, 15422; Schercemol 1818, 27584; Starfol® IS, 29536
41672-81-5: Dipalmitoyl hydroxy proline, 8578; Lipacide DPHP, 17321
41890-92-0: Osyrol, 22373
42116-76-7: carnidazole, 5298; Spartrix, 29212
42131-25-9: Wickenol® 151, 34388
42131-28-2: Patlac® IL, 22909
42576-02-3: bifenox, 4135
42612-52-2: Chemphos TR-510S, 6018; Rhodafac® MC-470, 26597
42808-36-6: Laurel SBT, 16838; sulfated butyl tallate, 29972
43121-43-3: Bayleton® 5, 3800; Monterey Bayleton, 20287; Triadimefon, 32184
43154-85-4: disodium oleamido MIPA sulfosuccinate, 8651; Emcol® 416L, 10045; Emcol® K8300, 10062; Mackanate OP, 18194; Sole Terge 8, 28914
43210-67-9: fenbendazole, 11594; Panacur, 22655
43222-48-6: Avenge 2, 3460
44992-01-0: Ageflex FA-1Q75MC, 971
45267-19-4: Chemidex M, 5964; Schercodine M, 27639
48145-04-6: Ageflex PEA, 997; Melcril 4087, 19268; SR-339, 29379; Laromer® POEA, 16791
50291-21-9: Robengatope I-131, 26871
50327-22-5: Stanyl® TE200F6, 29511; Stanyl® TE300, 29512; Stanyl® TE350, 29513; Stanyl® TQ200F6, 29514; Stanyl® TW300, 29515
50435-25-1: Abequito, 38
50471-44-8: Fumite Ronilan, 12466; Mascot Contact Turf Fungicide, 19001; Power Drive, 24821; Ronilan® DF, FL, 26949; Ronilon, 26950; vinclozolin, 33811
50594-66-6: acifluorfen, 343
50643-20-4: Brophos 5C10, 4613; Crodafos SG, 7086; Hetphos SG, 13954
50813-16-6: Glass H, 12967; Hexaphos, 14013; Metagon, 19425; Polyphos, 24519
50975-76-3: CT2902, 7321
51158-08-8: Aldosperse® MS-20 FG, 1554; Capmul® EMG, 5110; Cutina® E24, 7438; Emalex GM-5, 9956; Mazol® 80 MG K, 19132; Tagat® S, 30714; Tagat® S2, 30715; Varonic® Li-42, 33443
51178-59-7: Sipomer® DCPM, 28534
51192-09-7: Tagat® O, 30709; Tagat® O2, 30710
51200-87-4: Amine CS-1135®, 2153; Canguard® 327, 5088; Oxaban®-A, 22404
51218-45-2: metolachlor, 19585
51235-04-2: hexazinone, 14020, 14021; Velpar;, 33536; Velpar®, 33537
51248-32-9: Tagat® L, 30707; Tagat® L2, 30708
51258-15-2: Adeka GH-200, 663; ADK STAB 465, 713
51312-42-6: Scheibler's reagent, 27567
51338-27-3: Hoegrass, 14271
51630-58-1: fenvalerate, 11620; Sumicidin, 30080
51811-79-1: Dextrol OC-20, 8237; Lubrhophos® LE-500, 17721; Lubrhophos® LE-700, 17722; Tryfac® 5556, 32513
51812-80-7: Ceraphyl® 60, 5718
52019-36-0: Chemfac PD-600, 5954
52229-50-2: Gantrez® AN-119, 12651
52292-17-8: Arosurf® 66-E2, 3077; Emalex 1805, 9914; Hetoxol IS-2, 13939
52315-07-8: Cymbush, 7580; Cymperator, 7583; Cyperkill, 7587; cypermethrin, 7588; Cypersect, 7589; Cypertox, 7590
52315-75-0: Amihope LL-11, 2120
52467-63-7: Varisoft® TC-90, 33429
52503-64-7: cupric acetate, basic, 7373
52504-24-2: Softigen® 767, 28855
52508-35-7: Atrinal, 3395
52556-42-0: Cops 1, 6797
52558-73-3: Hamposyl® M, 13602
52581-71-2: Provol 50, 25306
52623-95-7: Laurelphos RH-44, 16847
52645-53-1: Ambush, 1982; Cooper Coopex, 6758; Darmycel Agarifume Smoke, 7780; Elimite®, 9838; Fumite Permethrin, 12462; Kafil, 15655; Nippon Ant Killer Powder, 21300; Nippon Ready For Use Ant and Crawling Insect Killer, 21302; Nix Creme Rinse, 21387; Nix Dermal Cream, 21388; Permasect, 23377; permethrin, 23384; Permethrin, 23385; permethrin, 23386; Picket, 23812; Quamilin, 25589; Turbair Permethrin, 32597
52668-97-0: Marlowet® 4702, 18922
52756-22-6: flamprop-isopropyl, 11887; Flamprop-M-isopropyl, 11888; Gunner, 13493; Power Flame, 24824; Power Flamprop, 24825
52794-79-3: Mackamide ISA, 18155
52820-00-5: Decis, 7852
52829-07-9: Lowilite® 77, 17645; Tinuvin® 770, 31836; Uvaseb 770, 33178;
52907-07-0: Saytex® BN-451, 27519
52918-63-5: Thripstick®, 31766
53003-10-4: Posistac, 24718; Salinomycin, 27271
53043-14-4: Amyl-*m*-cresol, 2383
53092-90-3: Azocoll®, 3522
53220-22-7: Perkadox® 26-fl, 23304
53260-54-1: Fungisterol, 12479
53320-86-8: Laponite® D, 16778; Laponite® XLG, 16780; sodium magnesium silicate, 28782
53370-45-9: Estoral, 10896
53449-39-8: Chlorez® 700-DF, Chlorez® 760, 6096; Doverguard® 700, 8823
53608-75-6: Accelerase, 185
53610-02-9: Akypo® Soft 45 NV, 1253
53633-54-8: Gafquat® 734, 12583; Polyquaternium-11, 24530
53716-50-0: oxfendazole, 22426; Synanthic, 30520; Systamex, 30662
53780-34-0: Echo, 9577; Embark, 10017; mefluidide, 19230; Mowchem, 20379
53879-54-2: CD492, 5499
53894-23-8: Jayflex® TINTM, 15558; Plasthall® TIOTM, 24015; PX-339, 25442
53988-10-6: Vanox® MTI, 33343
54045-08-8: Marlipal® SU, 18869
54392-26-6: Emalex SPIS-100, 9994
54400-62-3: butamisole hydrochloride, 4733; Styquin, 29859
54546-26-8: Herborane, 13807
54667-43-5: Polamine® 650, 1000, 2000, 24291
54982-83-1: Arova 16, 3088
55134-13-9: Monteban, 20282; narasin, 20741
55179-31-2: Bitertanol, 4288
55219-65-3: Spinnaker, 29292; Summit, 30118; triadimenol, 32185
55268-74-1: Droncit®, 9049
55285-14-8: carbosulfan, 5245; Marshal 10G, 18955; Marshal/suSCon, 18956
55298-68-5: neomycin palmitate, 20929
55335-06-3: Timbrel, 31800
55353-21-4: Resyn® 28-2913, 26246
55589-62-3: acesulfame potassium, 277
55635-13-7: Clout, 6446
55779-18-5: Arpocox, 3096
55799-16-1: Nalzin, 20673
55819-53-9: Emcol® 3780, 10047; Hetamine

5L-25, 13887; Incromate SDL, 14973; Lexamine S-13 Lactate, 17101; stearamidopropyl dimethylamine lactate, 29595
55852-13-6: Mackine 321, 18218
55852-15-8: Incromate IDL, 14968; isostearamidopropyl dimethylamine lactate, 15420
56002-14-3: Emalex PEIS-3, 9980;Emerest® 2625, 10107;Ethox MI-9, 11031;Olepal ISO, 22124;PEG isostearate series, 23030;Trydet 2644, 32508
56073-07-5: difenacoum, 8449; Killgerm® Ratak Cut Wheat Rat Bait, 16183; Neosorexa, 20967; Ratak, 25896
56073-10-0: Matikus, 19061; Mouser, 20376; Talon, 30750
56235-92-8: Ceraphyl® 45, 5715
56265-06-6: Argidone®, 2848; arginine PCA, 2849
56388-43-3: Anlonyx® 12S, 2477; Mackanate OD-35, 18192; Monamate OPA-100, 20109; Standapol® SH-100, 29479; Texapon SH 100, 31479
56392-17-7: Lopressor, 17602; Lopressor SR, 17603
56519-71-2: Captex® 800, 5169; Lexol PG 800, 17152
56803-37-3: Syn-O-Ad® 8478, 30555; Syn-O-Ad® 8485, 30559
56863-02-6: Schercomid SLE, 27658
56995-20-1: Katadolon, 15780
57018-04-9: Basilex, 3709; Risolex, 26782
57171-56-9: Ethox HO-50, 11027; PEG sorbitan hexaoleate series, 23032; Trylox® 6746, 32518
57569-76-3: Dermol GL-7A, 8125
57635-48-0: Akypo RO 50, 1240; Akypo RO 90, 1241
57646-30-7: Fongarid, 12225; Furalaxyl, 12487
57808-65-8: Closantel, 6441; Flukiver, 12051
57834-33-0: Ethoxy carbonyl phenylmethyl phenylformamidine, 11043; Givsorb® UV-1, 12952
57837-19-1: metalaxyl, 19430
57966-95-7: cymoxanil, 7582
58068-97-6: Dynasylan® IMEO, 9411
58069-11-7: Dehyquart SP, 7988
58095-31-1: Vigazoo, 33749
58138-08-2: Tandem, 30796
58160-99-9: Union Carbide® A-1106, 32991
58229-88-2: Okstan XO, 22101
58306-30-2: Rintal®, 26774
58353-68-7: Octosol A-1, 22021
58479-61-1: CB2805, 5482
58767-50-3: Chemstat® 106G/90, 6032
58810-48-3: Patafol, 22881
58855-63-3: Brophos OL-3N, 4615; Chemphos TR-505D, 6016; Chemphos TR-515D, 6020; Crodafos N10 Neutral, 7080; Crodafos N3 Neutral, 7082
58958-60-4: Ceraphyl® 375, 5726; Dermol 185, 8120; Schercemol 185, 27581
58965-66-5: Saytex® 120, 27516
59070-56-3: Aldosperse® ML 23, 1552
59130-69-7: Bernel® Ester CO, 4004; Cetearyl octanoate, 5791; cetyl octanoate, 5821; Emalex CC-168, 9935; Schercemol 1688, 27583; Schercemol CO, 27587; Tegosoft® CO, 31017; Tegosoft® Liquid, 31022
59186-41-3: Dehydag Wax E, 7947; Lanette E, 16665; Rhodapon® EC111, 26634
59231-34-4: Ceraphyl® 140-A, 5723; Mackester IDO, 18212; Schercemol IDO, 27604; Wickenol® 144, 34387
59231-37-7: Wickenol® 153, 34390
59272-84-3: Schercotaine MAB, 27698
59355-61-2: Empigen® OY, 10379; PEG-3 lauramine oxide, 23047
59447-55-1: Actimer FR-1025M, 484
59587-44-9: Bernel® OPG, 4011; Schercemol OPG, 27618; Wickenol® 160, 34395
59686-68-9: Cetiol® 1414E, 5795; Lanol 14 M, 16735; Schercemol MEM-3, 27607
59789-51-4: Actimer FR-1033, 485
59792-81-3: Aludone®, 1878; Aluminum PCA, 1921
60166-93-0: Iopamidol, 15233
60207-90-1: Bumper, 4669; Power Propiconazole, 24832; Powerspire, 24837; Radar, Radar Propiconazole, 25714
60209-82-7: Dermol 105, 8118
60270-33-9: Behenamidopropyl dimethylamine, 3896; Chemidex B, 5961; Incromine BB, 14997; Mackine 601, 18222; Schercodine B, 27635
60676-86-0: quartz glass, 25597; vitreosil, 33993
60828-78-6: Tergitol® TMN-3, 31248
61105-31-5: Crocidolite, 7060
61128-46-9: Electrafil® J-1106/CF/30, 9743; PDX-84367, 22957; PI-20GF/000, 23781; Polyetherimide, 24423; Thermocomp® EC-1006, 31611; Ultem®, 32766; Ultem® 1000, 32767; Ultem® 2100, 2112, 32768; Ultem® 2212, 32769; Ultem® 4000, 32770; Ultem® 6000, 32771; Ultem® CRS5001, 32772; Ultem® CRS5011, 32773; Ultem® FXU100, 32774
61318-91-0: Sulconazole, 29958
61368-34-1: Actimer FR-803, 483
61551-69-7: VA-086, 33229
61570-90-9: Tiox, 31845
61617-00-3: Vanox® ZMTI, 33349
61632-57-3: Magala® 0.5E, 18309
61682-73-3: Liponate PB-4, 17382
61693-08-1: hydrogenated polyisobutene, 14595; Luvitol HP, 18037; Panalane® L-14E, 22657
61699-38-5: Jasmacyclat, 15528
61778-68-9: Actrasol 6092, 548
61788-40-7: Hypan® SA100H, 14700
61788-44-1: Akrochem® Antioxidant 16, 1158; Alkylated phenol, 1734
61788-45-2: Amine HBG, 2156; Amine HBGD, 2157; Armeen® HT, 2950; Armeen® HTD, 2951; Crodamine 1.HT, 7103; Jet Amine PHT, 15594; Kemamine® P-970, 15920; Radiamine 6140, 25750; Radiamine 6141, 25751
61788-46-3: Amine KK, 2158; Armeen® C, 2940; Armeen® CD, 2941; Jet Amine PC, 15592; Kemamine® P-650D, 15918; Radiamine 6160, 25752; Radiamine 6161, 25753
61788-48-5: Acelan L, 269; Acetadeps, Acelan A, 279; Acylan, 595;Fancol Acel, 11428; Modulan®, 20031; Ritacetyl®, 26800
61788-49-6: Hetlan AC, 13906
61788-59-8: Emery® 2253, 10152; methyl coconate, 19523; Radia® 7117, 25722
61788-62-3: Armeen® M2C, 2954; Jet Amine M2C, 15591;Kemamine® T-6501, 15929; Radiamine 6360, 25764
61788-63-4: Amine M2HBG, 2161; Armeen® M2HT, 2955; Kemamine® T-9701, 15931; Radiamine 6343, 25762
61788-85-0: Akyporox CO 400, 1273; Alkamuls® COH-5, 1662; Arlatone® G, 2915; Chemax HCO-200/50, 5907; Chemax HCO-5, 5908; Cremophor® RH 40, 6984; Cremophor® RH 60, 6985; Cremophor® WO 7, 6987; Croduret 10, 7181; Dehymuls HRE 7, 7972; Emalex HC-5, 9965; Emulsogen HEL-050, 10522; Ethox HCO-16, 11026; Eumulgin HRE 40, 11176; Fancol HCO-25, 11440; Hetoxide HC-16, 13925; Hodag Nonionic GRH-25, 14194; Mapeg® CO-16H, 18638; Nopalcol 10-COH, 21462; Prox-onic HRH-05, 25319; Remcopal HC 20, 26085; Remcopal HC 33, 26086; Remcopal HC 40, 26087; Remcopal HC 60, 26088; Remcopal HC 7, 26084; Simulsol 1292, 28485; Simulsol 989, 28482; Tagat® R40, 30711; Tagat® R60, 30712; Tagat® R63, 30713; Trylox® 5921, 32517
61788-90-7: Barlox® 12, 3649; Chemoxide WC, 6003; Cocamine oxide, 6557; Empigen® 5083, 10354; Genaminox CS, 12778; Genaminox KC, 12779; Karox AO-30, 15774; Mackamine CO, 18169; Naxide 1230, 20817; Schercamox DMC, 27574
61788-91-8: Armeen® DMSD, 2948; Jet Amine DMSD, 15587
61788-93-0: Amine 2M1218D, 2134; Amine 2MKKD, 2143; Armeen® DMCD, 2945; Jet Amine DMCD, 15585; Kemamine® T-6502D, 15930
61788-95-2: Amine 2MHBGD, 2142; Armeen® DMHTD, 2946
61789-05-7: glyceryl cocoate, 13078; Imwitor® 928, 14948
61789-08-0: Myverol® 18-50K, 20626
61789-10-4: Alphadim® 90LC, 1825; Myverol® 18-40, 20625; Tegomuls® 19,

31009
61789-13-7: Myverol® 18-30, 20623
61789-18-2: Adogen® 461, 771; Arquad® C-33W, 3114; Arquad® C-50, 3115; Chemquat C/33W, 6026; Coco trimethyl ammonium chloride, 6562; Jet Quat C-50, 15603; Marlazin® KC 30/50, 18753; Varisoft® 461, 33411
61789-30-8: Norfox® 1101, 21544
61789-40-0: Amphotensid B4 F, 2347;Cocamidopropyl betaine, 6551;Dehyton® PK, 7995;Incronam 30, 15016;Rewoteric® AM B-13, 26417;Schercotaine CAB, 27695;Tego® -Betaine C, 30961;Tego® -Betaine E, 30962;Tego®-Betaine L-7, 30964
61789-68-2: Ethoquad® CB/12, 11000; PEG-2 cocobenzonium chloride, 23038
61789-71-7: Alkaquat® DMB-451-50, DMB-451-80, 1707; Arquad® DMCB-80, 3116; Mariazin® KC 21/50, 18715; Querton GCI-50, 25655; Querton KKBCI-50, 25656
61789-72-8: Arquad® DMHTB-75, 3117; Kemamine® BQ-9702C, 15909; Querton 441-BC, 25653
61789-73-9: Arquad® M2HTB-80, 3119; Kemamine® BQ-9701C, 15908
61789-75-1: Kemamine® BQ-9742C, 15910
61789-76-2: Armeen® 2C, 2928; Radiamine 6260, 25759
61789-77-3: Accoquat 2C-75, 2C-75H, 242; Adogen® 462, 772; Arquad® 2C-75, 3102;Dye Retarder #1, 9310;Jet Quat 2C-75, 15602;Kemamine® Q-6502C, 15924;M-Quat® 2475, 20403;Radiaquat 6462, 25793;Tomah Q-2C, 31989;Varisoft® 462, 33412
61789-79-5: Amine 2HBG, 2132; Amine2 VT, 2163; Armeen® 2HT, 2929; Kemamine® S-970, 15927
61789-80-8: Adogen® 442, 769; Adogen® 442-P100, 770; Arquad® 2HT-75, 3103; Kemamine® Q-9702C, 15925; M-Quat® 257, 20400; Querton 442, 25654; Radiaquat 6442, 25791; Radiaquat 6475, 6480, 25796
61789-86-4: calcium sulfonate, 4973; Lubrizol® 2152, 17744; TLA-256, 31891
61789-91-1: Emalex J.J O-V, 9966; jojoba oil, 15617; Lipovol J, 17429; Simchin, Natural, 28467; Wickenol® 139, 34383
61789-97-7: tallow, 30739
61790-12-3: Acintol® 736, 2122, D25LR, D30E, DFA, EPG, FA-1, FA-2, R Type3A, LO-3A, SB, SM4, 348; Acofor, 364; Oulu 102, 22378; Pamolyn, 22642; Xtol, 34614
61790-18-9: Jet Amine PS, 15596
61790-31-6: Armid® HT, 2962
61790-33-8: Adogen® 170, 765; Amine BG, 2151; Armeen® T, 2958; Armeen® TD, 2959; Crodamine 1.T, 7104; Genamin TA Grades, 12776; Jet Amine PT, 15597; Radiamine 6170, 25755; Radiamine 6171, 25756; tallow amine, 30740; tallowamine, 30748
61790-35-0: Actrasol SS, 561; Eureka 392, 11208
61790-37-2: Industrene® 143, 15102; tallow fatty acid, 30743
61790-38-3: Glycon® S-65, 13120; hydrogenated tallow acid, 14598; Petrac® PHTA, 23467; Prifac 9428, 24949
61790-41-8: Adogen® 417, 768; Arquad® S-50, 3120; Jet Quat S-50, 15605; Tomah Q-S, 31999
61790-50-9: Arizona DRS-40, 2890; Arizona DRS-50, 2892; Diprosin K-80, 8613
61790-51-0: Arizona DRS-43, 2891; Diprosin N-70, 8614
61790-53-2: Infusorial earth, 15125
61790-57-6: Acetamin C, 286; Amine Acetate KK, 2149; Radiamac 6169, 25749
61790-59-8: Acetamin HT, 287; Amine Acetate HBG, 2148; Radiamac 6149, 25748
61790-63-4: Marlamid® DF 1218, 18740
61790-64-5: Akypogene FP 35 T, 1255
61790-81-6: Aqualose L30, 2709; Ethoxylan® 1685, 11047; Ivarlan 3400, 15445; Ivarlan 3406, 15446; Laneto 27, 16660; Lanogel® 21, 16729; Lanolex L-40, 16738; PEG lanolin series, 23025; Solan, 28886; Solan 50, 28888; Solangel 401, 28891; Solulan® 75, 29002; Super Solan, 30200
61790-85-0: Ethoduomeen® T/13, 10958; Ethoduomeen® T/20, 10959; Ethoduomeen® T/25, 10960
61791-00-2: Aconol X6, 365; Actrol 6M25P, 569; Chemax TO-10, 5927; Chemax TO-16, 5928; Chemax TO-B, 5929; EM-600, 9906; Ethofat® 242/25, 10964; Ethox TO-8, 11040; Genagen TA-080, 12764; Hetoxamate FA-5, 13909; Industrol® TO-16, 15118; Laurel PEG 400 MT, 16836; Mapeg® 200 MOT, 18633; Mapeg® TAO-15, 18644; Marlosol® TF3, 18915; PEG tallate series, 23035; Prox-onic TA-1/08, 25344; Sellig T 14 100, 27924; Sellig T 1790, 27925; Sellig T 3 100, 27926; Trydet 2682, 32510
61791-01-3: Ethox DTO-9A, 11025; Laurel PEG 400 DT, 16834; Mapeg® 200 DOT, 18629; PEG ditallate series, 23028; Pegosperse® 400 DOT, 23064
61791-08-0: Alkamide® C-5, 1636; Amide CMA-2, 2083; Amidox® C-2, 2110; Empilan® LP10, 10411; Empilan® MAA, 10412; Eumulgin C4, 11173; Eumulgin PC 2, 11182; Genagen CA-050, 12759; Hetoxamide C-4, 13913; Rewopal® C 6, 26341; Unamide® C-5, 32909
61791-10-4: Ethoquad® C/25, 10999; Tomah Q-C-15, 31996; Variquat® K1215, 33401
61791-12-6: Ablunol CO 10, 98; Acconon CA-5, CA-9, CA-15, 236; Akyporox RZO 30, 1296; Alkamuls® 14/R, 1655; Alkamuls® B, 1661; Alkamuls® EL-620, 1665; Berol 108, 4016; Berol 191, 4017; Berol 195, 4018; Berol 198, 4019; Berol 199, 4020; Berol 829, 4048; Chemax CO-16, 5891; Chemax CO-5, 5892; Cremophor® EL, 6981; Emalex C-20, 9932; Emulsogen EL-050, 10521; Ethox CO-5, 11020; Eumulgin PRT 36, 11186; Eumulgin RO 40, 11189; Fancol CO-30, 11437; Hetoxide C-2, 13921; Hetoxide C-200-50%, 13922; Hodag Nonionic GR-8, 14193; Incrocas 30, 14963; Makon® 8240, 18428; Mapeg® CO-5, 18639; Marlosol® R70, 18913; Marlowet® R 11, 18937; Marlowet® R 40, 18938; Polymene AZ, 24487; Prox-onic HR-05, 25318; Remcopal 40, 26059; Remcopal 40 S3, 26060; Remcopal 4000, 26078; Remcopal 4018, 26079; Ricinion, 26732; Sellig HR 18 100, 27881; Sellig R 20 100, 27905; Sellig R 3395, 27906; Sellig R 3395 SP, 27907; Sellig R 3395-C435, 27908; Sellig R 4095, 27909; Sellig R 4495, 27910; Stepantex® CO-30, 29710; Surfactol® 318, 30340; T-Det® C-40, 30861; Toximul® 8240, 32093; Trylox® 5900, 32516; Witconol 5906, 34510
61791-13-7: Coceth-27, 6558; Dehydol LT 3, 7953; Genapol® C-050, 12787; Genapol® GC-050, 12789
61791-14-8: Ablumox C-7, 89; Accomeen C2, C5, C10, C15, 224; Alkaminox® C-2, 1654; Berol 397, 4031; Chemeen C-2, 5944; Ethomeen® C/12, 10977; Ethomeen® C/15, 10978; Ethomeen® C/20, 10979; Ethomeen® C/25, 10980; Ethox CAM-2, 11019; Ethylan® TN-10, 11116; Eumulgin PA 12, 11180; Hetoxamine C-2, 13915; Mazeen® C-2, 19121; Prox-onic MC-02, 25322; Teric 12M2, 31256; Trymeen® 6601, 32521; Varonic® K202, 33442; Zusomin C 108, 34914
61791-20-6: Aqualose W20, 2711; Fancol LA-5, 11447; Polychol 5, 24387; Ritawax 5, 26836
61791-24-0: Accomeen S2, S10, S15, 225; Chemeen S-2, 5948; Ethomeen® S/12, 10981; Ethomeen® S/15, 10982; Ethomeen® S/20, 10983; Ethomeen® S/25, 10984; Hetoxamine S-2, 13916; Mazeen® S-15, 19125; Mazeen®

S-2, 19126; PEG soyamine series, 23033; Teric 16M2, 31260; Tomah E-S-2, 31977
61791-25-1: Mirataine® TM, 19932; Rewoteric® AM TEG, 26427
61791-26-2: Berol 381, 4028; Berol 386, 4029; Berol 391, 4030; Berol 457, 4034; Chemeen HT-5, 5946; Ethomeen® T/15, 10986; Ethomeen® T/25, 10987; Prox-onic MHT-015, 25324; Rhodameen® PN-430, 26613; Rhodameen® T-5, 26614; Rhodameen® VP-532/SPB, 26615; Sipenol IT-50-46, 28517; Varonic® U-215, 33451
61791-28-4: Dehydol PTA 7, 7957; Dehydol TA 11, 7958;Oxetal TG 111, 22424; Rewopal® TA 11, 26350
61791-29-5: Crodet C10, 7171; Eumulgin PK 23, 11183; Genagen C-100, 12758; Nonex C5E, 21444; Nopalcol 4-C, 21467; Prox-onic CC-05, 25311
61791-31-9: Chemstat® 273-C, 6036; Cocamide DEA, 6547; Ethokem C/12, 10970; Norfox® DC, 21548; Purton CFD, 25419
61791-32-0: Rewoteric® AM 2C NM, 26414
61791-38-6: Chemzoline C-22, 6051; Mackazoline C, 18198; Monazoline C, 20192; Schercozoline C, 27712; Sovatex EP 5288, 29157; Sovatex IM12H, 29158; Unamine® C, 32911; Varine C, 33392
61791-39-7: Chemzoline T-44, 6053; Hodag C-100-T, 14170; Miramine® HPS-B, 19870; Miramine® TO, 19873; Monazoline T, 20197; Textamine T-1, 31534; Varine T, 33395
61791-41-1: Fenopon TK32, 11612; Geropon® TK-32, 12918; sodium methyl tall oil acid taurate, 28793, 28794
61791-42-2: Geropon® TC-42, 12917; sodium methyl cocoyl taurate, 28791
61791-44-4: Ablumox T-15, 92; Accomeen T2, T5, T15, 226; Agrisorb, 1072; Berol 455, 4032; Berol 456, 4033; Chemeen T-2, 5949; Clifton Glyphosate Additive, 6417; Ethomeen® T/12, 10985; Ethox TAM-2, 11039; Ethylan® TT-15, 11117; Eumulgin PA 10, 11179; Exell, 11295; Frigate, 12408; Genamin T-020, 12775; Hetoxamine T-2, 13918; Hyspray, 14750; Icomeen® T-15, 14818; Icomeen® T-2, 14817; Jogral, 15616; Lo-Dose, 17548; Marlazin® T 10, 18757; Mazeen® T-2, 19128; Power Spray Save, 24833; Prox-onic MT-02, 25327; Team, 30875; Teric 17M2, 31262; Tomah E-T-2, 31978; Topup, 32053; Toximul® TA-2, 32098; Trymeen® 6606, 32522; Trymeen® TAM-15, 32528; Varonic® T202, T220, 33449; Varonic® T-220, 33450; Wayfarer, 34246; Witcamine® 6606, 34447; Zusomin TG 102, 34917
61791-46-6: Chemoxide T, 6001
61791-47-7: Aromox® C/12-W, 3047; Schercamox CMA, 27573
61791-53-5: Duomeen® TDO, 9117
61791-55-7: Diamine BG, 8295; Duomeen® T, 9116; Jet Amine DT, 15590; tallow diamine, 30742
61791-56-8: Deriphat 154, 8089; disodium tallowiminodipropionate, 8655; Mirataine® T2C-30, 19931; Monateric TDB-35, 20175
61791-57-9: Jet Amine TRT, 15600; tallow triamine, 30746; Triameen T, 32193
61791-59-1: Closyl 30 2089, 6442; Hampfoam 35, 13593; Hamposyl® C-30, 13599; Medialan KA, 19212; Vanseal® 35, 33355; Vanseal® NACS-30, 33358
61791-63-7: Cocodiamine, 6564; Diamine KKP, 8297; Duomeen® C, 9112; Duomeen® CD, 9113; Jet Amine DC, 15582; Radiamine 6560, 25766
61792-31-2: Mackamine LAO, 18172
61827-42-7: Rhodasurf® DA-530, 26654
61840-27-5: Cartaretin F-4, 5358
61849-72-7: Glucam® P-10, 13022;Glucam® P-20, 13023; PPG methyl glucose ether series, 24842; Prox-onic MG-020 p, 25323;
61901-02-8: Miranol® HM-SF Conc, 19901
62180-77-2: Esi-Cryl Respond I, 10815
62449-33-6: Norfox® IM-38, 21550
62476-59-9: Blazer®, 4333
62610-77-9: Damfin, 7721
63123-11-5: Eastman® Inhibitor Poly TDP 2000, 9469; Eastman® Poly TDP 2000, 9473
63148-55-0: Alkasil® NE58-50, 1709
63148-56-1: PS181, 25366
63148-57-2: CPS 120, 6947
63148-62-9: Crystal 1000, 7285
63217-13-0: Geropon® WS-25, WS-25-I, 12919; sodium dinonyl sulfosuccinate, 28759
63231-60-7: Be Square® 185, 3866; Forbest MW 23, 12238; Fortex®, 12298; Mekon® White, 19248; Microcrystalline wax, 19666; Multiwax® 180-M, 20492; Multiwax® HS, 20493; Paracol® 404C, 22724; Polymekon®, 24485; Starwax® 100, 29546; Ultraflex®, 32801; Victory®, 33731
63245-28-3: EHIDA Kit, 9649
63284-71-9: nuarimol, 21762; Triminol, 32344
63393-82-8: Neodol® 25, 20913
63393-93-1: Amerlate® P, 2018; Fancor IPL, 11456; Lanesta S, 16658; Ritasol, 26832
63394-02-5: B-182, 3569; C-715u, 4843; C-920u, 4845; COHRlastic® 1867, 6580; COHRlastic® 3320, 6581; COHRlastic® 400, 6578; COHRlastic® 9041, 6583; COHRlastic® R10450, 6586; COHRlastic® TC100, 6588; D1-SEA 210 Silicone, 7631; Elastosil® LR 3001, 9711; Elastosil® LR 3003/20, 9712; Laur 101B, 16820; Laur 676U, 16821; Laur Q-1331, 16824; LIM6045, 17259; Norsil RTV 811, 21577; Pensil® 100, 23127; Rhodorsil® RS 44, RS 48, 26703; RTF 762, 27054; RTV 511, 27085; RTV 615, 27086; Silastic® 21145, 28269; Silastic® GP-30, 28270; Silastic® GP-950+xg, 28271; Silastic® HE-26, 28272; Silastic® LCS-740, 28273; Silastic® LT-40, 28274; Silastic® NPC-40, 28275; Silastic® SPG-30, 28276; Silastic® TR-55, 28277; Silastic® WC-50, 28278; Silastic+sp HS-30, 28279; silicone elastomer, 28337; SWS-06545u, 30474; SWS-7532u, 30469; SWS-7655u, 30470; SWS-7675u, 30471; SWS-7865u, 30472
63428-83-1: nylon, 21908
63449-39-8: Chlorcosane, 6092; Chlorinated paraffin, 6105; Electrofine® S-70, 9782
63451-23-0: Crodateric Cy, 7154
63782-90-1: Commando, 6665
63793-60-2: Carsonon® 144-P, 5342; Hetoxol MP-3, 13944; PPG-3 myristyl ether, 24844; Promyristyl PM-3, 25153; Witconol APM, 34512
64051-29-2: Dioleyl hydrogen phosphite, 8554
64365-11-3: Acticarbone, 461
64628-44-0: Alsystin, 1868; Starycide, 29547; triflumuron, 32248
64665-57-2: Maxahibit TT-50, 19074
64741-81-7: Viplex 895-BL, 33888
64742-04-7: Viplex 680-P, 33886
64742-06-9: HAN® 857, 13613
64742-14-9: Jayflex® 215, 15545; Penreco 2251 Oil, 23123
64742-47-8: Exxsol® D-40, D-60, D-80, D-110, D-130, 11359; Isopar® M, 15392
64742-48-9: Isopar® C, 15388; Isopar® E, 15389; Isopar® G, 15390; Isopar® L, 15391
64742-53-6: Jayflex® 210, 15544
64743-02-8: Neodene® 1420, 20905; Neodene® 1624, 20906; Neodene® 2024, 20907
64771-72-8: Norpar® 12, 21573
64924-67-0: halofuginone hydrobromide, 13562; Stenorol, 29650
65009-35-0: Lidarral, 17214
65381-09-1: caprylic/capric acid triglyceride, 5161; Captex® 300, 5168; Emalex K.T.G, 9967; Labrafac Lipophile WL 1349, 16535; Lexol GT 855, GT 865, 17149; Liponate GC, 17377; Mazol® 1400, 19135; Miglyol® 812, 19737; Myritol 318, 20580; Neobee® M-5, 20875; Tegosoft® CT, 31018
65447-77-0: Tinuvin® 622LD, 31835
65473-14-5: Naftifine, 20650
65497-29-2: Cosmedia Guar C-261 N, 6868; Hi-Care® 1000, 14060; Jaguar® C-13S, C-14S, 15496; Jaguar®

C-13S, C-14S, 15497; N-Hance® 3000, 21083
65591-14-2: Waxenol® 801, 34236
65717-97-7: Hepatolite, 13789
65816-20-8: Ethoxy carbonyl phenylethyl phenylformamidine, 11042
65876-95-1: Penacolite® B-18-S, 23092
65996-61-4: CF 1500, 5829
65997-15-1: Portland cement, 24708
66009-41-4: Crodamol W, 7127; Tegosoft® SH, 31028
66063-05-6: Monceren®, 20198; pencycuron, 23097, 23098. 23099; Trotis, 32473
66085-00-5: Imwitor® 780 K, 14946; Schercemol GMIS, 27601
66108-95-0: Omnipaque, 22159
66197-78-2: Crafol AP-53, 6959
66246-88-6: Topas 100, 32026
66441-23-4: Cheetah R, 5870
66455-14-9: Neodol® 23-6.5, 20912
66455-15-0: Marlipal® KF, 18868
66455-17-2: Neodol® 91, 20917
66455-29-6: Amphoteen 24, 2343; Empigen® BB, 10363
66592-87-8: Cefadroxil, 5547
66734-13-2: Vaderm, 33237
66794-58-9: Crillet 6, 7041; Montanox 70, 20277
66841-24-5: Ambush C, 1983; Ashlade Halt, 3194; Chemtech Cypermethrin, 6048; Chiltern Cyperkill 10, 6066; FAL Cypermethrin 10, 11419; Quadrangle Cyper, 25577; Topclip Parasol, 32037; Toppel, 32049; Toppel, 32050
66988-04-3: Crolactil SISL, 7185; Pationic ISL, 22907
67129-08-2: Butisan, 4746; Butisan® S, 4747; metazachlor, 19465; Pree®, 24873; Track, 32129
67306-00-7: fenpropidin, 11616; Patrol, 22919
67306-03-0: Corbel®, 6802; Mistral, 19949; Power Task, 24835
67375-30-8: Fastac, 11500
67567-23-1: Lupersol 533-M75, 17862
67633-57-2: Monaquat ISIES, 20134; Schercoquat IIS, 27683
67633-59-4: Incroquat I-85, 15039
67633-63-0: Foamquat IAES, 12180; M-Quat® 522, 20401; Schercoquat IAS, 27679
67700-98-5: Amine 2M12D, 2135; Empigen® AB, 10358;
67701-00-2: Armeen® 3-16, 2932
67701-05-7: Hystrene® 1835, 14755
67701-26-2: Softenol® 3100, 28847
67701-27-3: Neustrene® 059, 21022
67701-32-0: Alphadim® 90NLK, 1826
67701-33-1: Alphadim® 90AB, 1824
67702-21-4: Empicol® EGB, EGC, 10315; magnesium laureth sulfate, 18363; Texapon MG, 31465; Zoharpon MgES, 34871
67747-09-5: Mirage, 19864; Octave, 21990; Prelude, 24885; prochloraz, 25057; Sporak, 29316; Sporgon, 29317; Sportak, 29319; Sportak Delta, 29321
67762-19-0: Ammonium laureth sulfate, 2240; Avirol® AE 3003, 3478; Calfoam NEL-60, 4985; Carsonol® SES-A, 5336; DeSonol AE, 8187; Empicol® EAA, 10312; Empicol® EAB, 10313; Empicol® EAC, 10314; Nonasol N4AS, 21439; Nutrapon AL 1, 21813; Nutrapon AL 60, 21814; Polystep® B-11, 24591; Rhodapex® AB-20, 26622; Rhodapex® EA, 26626; Rhodapex® EAY, 26627; Standapol® EA-1, 29472; Steol® CA-460, 29661; Sulfochem EA-1, 29998; Sulfochem EA-2, EA-3, 29999; Sulfochem EA-60, 30000; Sulfochem EA-70, 30001; Sulfotex OT, 30035; Texapon EA-1, 31457; Texapon NA, 31468; Ungerol AM3-75, 32920; Witcolate AE, Witcolate LES-60A, 34472; Zoharpon LAEA 253, 34866
67762-36-1: caprylic/capric acid, 5160; Industrene® 365, 15105; Philacid 0810, 23658
67762-38-3: Radia® 7060, 25719
67762-39-4: Emery® 2209, 10143; methyl caprylate-caprate, 19520
67762-41-8: Exxal® L1315, 11345
67762-96-3: Tego® Foamex 3062, 30977
67774-74-7: Marlican®, 18845
67784-77-4: Ethoquad® T/12, 11002; PEG-2 tallowmonium chloride, 23044
67784-87-6: Myverol® 18-04K, 20620
67784-90-1: Miramine® C, 19867
67799-04-6: Chemidex SI, 5970; Mackine 401, 18219; Mazeen® DAPI, 19122; Schercodine I, 27637
67846-16-6: Schercoquat SAS, 27686
67892-37-9: Schercoteric O-AA, 27710
67923-14-2: CPS 340, 6950
67938-21-0: Emalex DISG-2, 9943
67990-17-4: Miranate® B, 19874
67998-94-1: Sul-fon-ate OA-5R, 30021
68002-44-8: Vibrin® E-085, 33717
68002-59-5: Varisoft® DHT, 33422
68002-61-9: Radiaquat 6471, 25795
68002-71-1: Neustrene® 064, 21023
68002-72-2: Neustrene® 045, 21021
68002-79-9: Radia® 7370, 25740; Radia® 7371, 25741
68002-97-1: Empilan® KB 2, 10397; Laureth-2, 16853; Rhodasurf® L-4, 26656
68003-46-3: Ammonium lauroyl sarcosinate, 2241; Hamposyl® AL-30, 13597
68037-49-0: Chemstat® PS-101, 6043
68037-87-6: CPS 925, 6951
68037-92-3: Amine B11, 2150
68037-93-4: Amine 2M16D, 2138
68037-97-8: Radiamine 6572, 25768
68039-13-4: Lexquat® 2240, 17174
68071-35-2: Crodafos 25 D2 Acid, 7075
68081-96-9: Empicol® AL30, 10306
68081-97-0: Empicol® ML 26/F, 10336
68083-14-7: Silikophen® P 40/W, 28356
68085-85-8: Karate, 15757
68092-28-4: Emulamid TO-21, 10484
68122-86-1: Varisoft® TIMS, 33430
68128-59-6: Octosol A-18-A, 22023
68130-24-5: Dipentaerythrityl hexacaprylate/hexacaprate, 8582; Liponate DPC-6, 17375
68130-43-8: Akypomine® BC 50, 1264
68130-47-2: Rhodafac® RA-600, 26603
68131-37-3: Maltrin® M200, 18503; Maltrin® QD M600, 18505
68131-39-5: Bio-Soft® E-400, 4197; Mulsifan RT 203/80, 20457; Neodol® 25-3, 20914; Rhodasurf® 25-7, 26646; Rhodasurf® LA-3, 26657; Teric 12A2, 31255; Teric G12A4, 31266; Teric LA4, 31268; Volpo 25 D 10, 34024; Volpo 25 D 3, 34025
68131-40-8: Tergitol® 15-S-3, 31243
68133-13-1: Weston® 494, 34302
68139-30-0: Amonyl 675 SB, 2298; Chembetaine CAS, 5932; Cocamidopropyl hydroxysultaine, 6555; Crosultaine C-50, 7241; Lexaine® CSB-50, 17092; Lonzaine® CS, 17593; Mafo® CSB, 18301; Rewoteric® AM CAS, 26421; Rewoteric® AM CAS-15, 26422; Sandobet SC, 27327; Schercotaine SCAB, 27700
68140-00-1: Ablumide CME, 78; Adeka Sole YA, 672; Alkamide® C-212, 1635; Amidex CME, 2090; Amidex KME, 2093; Carsamide® CMEA, 5322; Cocamide MEA, 6548; Cocamide MEA (1:1), 6549; Comperlan P 100, 6677; Emid® 6500, 10191; Empilan® CME, 10391; Foamole M, 12178; Incromide CM, 14986; Mackamide CMA, 18154; Marlamid® M 1218, 18744; Mazamide® CFAM, 19109; Monamid® CMA, 20118; Monamide, 20123; Ninol® CMP, 21235; Ninol® CNR, 21236; Rewomid® C 212, 26320; Varamide® C-212, 33381; Zoramide CM, 34905
68140-01-2: Chemidex C, 5962; Chemidex WC, 5973; Cocamidopropyl dimethylamine, 6552; Incromine CB, 14998; Lexamine C-13, 17097; Mackine 101, 18215; Mazeen® DAPL, 19123; Mazeen® SHCFA, 19127; Miramine® CODI, 19868; Schercodine C, 27636
68140-08-9: Amidex TD, 2104
68140-98-7: Alkaterge® E, 1717
68153-28-6: Promax® HV, 25136; Promosoy® 20/60, 25147
68153-32-2: Radiaquat 6470, 25794
68153-63-9: Marlamid® M 1618, 18745
68153-64-0: Nopalcol 1-TW, 21464
68154-36-9: Radiasurf® 7125, 25797
68154-97-2: Marlox® FK 64, 18943; Propetal 241, 25182
68155-09-9: Ablumox CAPO, 90; Aminoxid WS 35, 2196; Amyx CDO 3599, 2392; Barlox® C, 3653; Chemoxide CAW, 5994; Cocamidopropyl amine oxide, 6556; Empigen® OS/A, 10378; Incromine Oxide C, 15003; Mackamine CAO, 18168; Mazox® CAPA, 19149; Monalux

CAO, 20103; Ninox® FCA, 21241; Patogen AO-30, 22914; Rewominox B 204, 26333; Rhodamox® CAPO, 26619; Schercamox C-AA, 27572; Standamox CAW, 29460; Standamox PCAW, 29461; Varox® 1770, 33454

68155-12-2: Accomid PK, 231

68155-20-4: Monamine T-100, 20130; Schercomid SO-T, 27662; Schercomid TO-2, 27664

68155-24-8: Ethomid® HT/60, 10989; Ethomid® HT/60, 10990; PEG hydrogenated tallow series, 23029

68171-33-5: Schercemol 318, 27582; Wickenol® 131, 34381;

68171-38-0: Emerest® 2384, 10093; Hydrophilol ISO, 14629; Propylene glycol isostearate, 25218

68173-73-7: Texlin® 500, 31509

68184-04-3: Akyposal MLES 35, 1310; Empicol® EL, 10316

68186-34-5: Rhodafac® GB-520, 26595

68187-29-1: Acylglutamate CT-12, 599

68187-32-6: Acylglutamate CS-11, 597

68187-76-8: oleite, 22114; Türkischrotöl 100%, 32541

68188-30-7: Chemidex SO, 5971; Mackine 901, 18225

68201-46-7: Cetiol® HE, 5798; Rewoderm® ES 90, 26309; Tegosoft® GC, 31021; Varonic® LI-63, 33445

68201-49-0: Albalan, 1344; Fancor Lanwax, 11457; Lanfrax® 1776, 16673; Lanocerin®, 16728

68238-35-7: Crotein HKP, 7251; Keramino 25, 16053; Kera-Tein AA, 16060

68238-87-9: Emsorb® 2518, 10463; Sorbitan diisostearate, 29109

68239-42-9: Glucam® E-10, 13019; methyl gluceth-10, 19535; methyl gluceth-20, 19536

68239-43-0: Glucam® E-20, 13020

68299-16-1: Esperox® 545M, 10852; Lupersol 546-M75, 17863; Trigonox® 123-C75, 32282

68299-17-2: Rhodapon® CAV, 26633

68308-22-5: calcium montanate, 4959; Hostalub® VP Ca W 2, 14365

68308-34-9: shale oil, 28136

68308-67-8: Larostat® 88, 16798; M-Quat® 1033, 20402

68309-95-5: Bacote®, 3578

68310-73-6: Duoquad® O-50, 9120

68333-79-9: Phos-chek, 23694

68334-00-9: C-Flakes, 5834; Duratex, 9180; Duromel, 9246; Duromel B108, 9247; Emvelop®, 10533; hydrogenated cottonseed oil, 14593; Lubritab®, 17742; Sta-Nut EE, 29508

68334-21-4: Empigen® CDR40, 10369; Schercoteric MS, 27707

68334-28-1: Aratex, 2783; BBS, 3862; Cirol, 6324; Creamtex, 6974; Durkex 500, 9228; Durko, 9230; Durlite F, 9232; hydrogenated vegetable oil, 14601; Hydrokote® 95, 14604; K.L.X., 15644; Kaokote F, 15723; Kaola, 15724; Kaomel, 15728; Kaoprem-E, 15733; Kaorich Beads, 15734; Kaorich Gold, 15735; Lipo SS, 17345; Lipodan SET Kosher, 17364; Magna A, 18326; Optima 23B, 22238; Sterotex®, 29753; Sterotex® HM, 29754; Wecobee® M, 34253

68359-37-5: Bulldock, 4663; cyfluthrin, 7563; Baytroid®, 3856; Solfac, 28930

68389-70-8: Glucamate® SSE-20, 13026; PEG-20 methyl glucose sesquistearate, 23046

68391-01-5: Arquard® B-100, 3123

68391-03-7: Querton 280, 25652

68391-07-1: Kemamine® T-9742D, 15932

68411-00-7: Neodene® 1112, 20904

68411-19-8: Vanax® 833, 33286

68411-20-1: Vanox® AT, 33341

68411-46-1: Agerite® Stalite® S, 1023; Vanox® 12, 33335

68411-97-2: Cocoyl sarcosine, 6568; Hamposyl® C, 13598; Vanseal® CS, 33356

68412-53-3: Rhodafac® PE-510, 26600; Rhodafac® RE-610, 26606

68412-54-4: Berol 09, 4014; Berol 278, 281, 282, 291 and 292, 4025; Berol WASC, 4049

68424-43-1: Amerlate® LFA, 2017; Amerlate® WFA, 2019; Fancor LFA, 11458; Ritalafa®, 26808; Skliro Distilled, 28572

68424-59-9: Fancol Karite Extract, 11445; Shea butter extract, 28143

68424-60-2: Cetiol® SB45, 5804; Fancol Karite Butter, 11444; Lipex 102, 17331; Shea Butter, 28141; Shebu, Refined, 28145

68424-66-8: hydroxylated lanolin, 14663; Ivarlan OH, 15453; OHlan®, 22055; Ritahydrox, 26807

68424-94-2: Accobetaine CL, 222; Amphoteen BCM-30, 2345; Chembetaine BW, 5930; Chembetaine CB, 5933; Coco betaine, 6561; Incronam CD-30, 15020; Lonzaine® 12C, 17591; Mackam CB-35, 18140; Mafo® CB 40, 18300; Velvetex® AB-45, 33553

68424-95-3: FMB 302-8, 12132

68425-37-6: Philcohol 1214, 23665

68425-42-3: Cocamidopropyl dimethylamine lactate, 6553

68425-43-4: Cocamidopropyl dimethylamine propionate, 6554; Emcol® 1655, 10046; Foamid 117, 12155; Incromate CDP, 14967

68425-47-8: Alkamide® DIN-295/S, 1640; Alkamide® SDO, 1651; Amidex S, 2102; Mackamide S, 18166; Marlamid® DF 1818, 18741; Purton SFD, 25420; Schercomid SLS, 27660; Stamid LS 5487, 29441

68425-50-3: Chemidex T, 5972

68439-39-4: Marlophor® DS-Acid, 18892

68439-45-2: Empilan® KA10/80, 10396

68439-46-3: Berol 260, 4022; Neodol® 91-2.5, 20918; Oxetal 500/85, 22420; Rhodasurf® 91-6, 26647; Teric 9A2, 31254; Teric G9A5, 31267

68439-49-6: Acconon W230, 241; Ceteareth-11, 5789; Ceteareth-20 BP, 5790; Ceteareth-6, 5788; Cetomacrogol 1000 BP, 5809; Cremophor® A 11, 6979; Cremophor® A 25, 6980; Dehydol PCS 6, 7954; Empilan® KM 11, 10406; Emthox® 5885, 10480; Eumulgin B2, 11172; Hetoxol 15 CSA, 13929; Hetoxol CS-4, 13936; Hostacerin T-3, 14324; Incropol CS-20, 15025; Lipocol SC-4, 17359; Macol® CSA-2, 18264; Marlipal® 1618/6, 18866; Marlowet® 4800, 18923; Marlowet® FOX, 18931; Marlowet® PW, 18936; Plurafac® A-38, 24192; Plurafac® A-39, 24193; Prox-onic CSA-1/04, 25312; Remcopal 229, 26066; Remcopal 238, 26068; Rhodasurf® A-1P, 26649; Ritacet-20, 26797; Sellig SU 18 100, 27919; Sellig SU 25 100, 27920; Sellig SU 30 100, 27921; Sellig SU 4 100, 27922; Sellig SU 50 100, 27923; Simulsol CS, 28487; Teginacid C, 30949; Teric 16A16, 31259; Varonic® 63 E20, 33438; Volpo CS-3, 34026

68439-50-9: Genapol® 24-L-3, 12781; Genapol® 42-L-3, 12783; Marlipal® 24/20, 18862; Surfonic® L24-2, 30368; Tergitol® 24-L-45, 31244

68439-51-0: Marlox® MO 124, 18945

68439-53-2: Hetoxol PLA, 13948; Solulan® PB-2, 29007; Wickenol® 727, 34405

68439-57-6: Bio-Terge® AS-40, 4212; Carsonol® AOS, 5333; Nansa® LSS38/A, 20689; Norfox® ALPHA XL, 21546; Polystep® A-18, 24588; Rhodacal® 301-10, 26576; Witconate AOK, 34483

68439-70-3: Amine 2M14D, 2137

68439-73-6: Kemamine® D-974, 15914; Radiamine 6570, 25767

68440-05-1: Cocamide MIPA, 6550; Empilan® CIS, 10390

68440-66-4: Silwet® L-7500, 28453

68440-90-4: CPS 140, 6949; PS140, 25365

68458-58-8: Fluilan AWS, 12047; Ivarlan AWS, 15449; Lantrol® AWS 1692, 16768; PPG-12-PEG-65 lanolin oil;, 24843; Ritalan® AWS, 26810; Vigilan AWS, 33751

68458-88-8: Ivarlan 3420, 15447; Lanexol AWS, 16672; PEG -12 PEG-50 lanolin, 23027

68476-03-9: Hoechst Wax S, 14270; Montan acid wax, 20262

68476-38-0: glyceryl montanate, 13086; Hostalub® WE4, 14366

68476-78-8: molasses, 20043

68477-29-2: Viplex 885, 33887

68479-64-1: Schercopol OMS-Na, 27669

68484-43-1: Fancol WGFA, 11454

68511-41-1: Jet Amine PE 1214, 15593

68513-95-1: Emcosoy®, 10070; Soy Flour, 29168; Soyafluff® 200 W, 29170; Soyarich® 115 W, 29172
68515-47-9: Jayflex® UDP, 15560; undecyl dodecyl phthalate, 32915
68515-48-0: diisononyl phthalate, 8463; Jayflex® DINP, 15552; Jayflex® DIOP, 15553; Palatinol® DN, 22589; Palatinol® N, 22594; PX-109, 25433
68515-49-1: Bisoflex BP9, 4255; diisodecyl phthalate, 8462; Jayflex® DIDP, 15550; Kodaflex® DIDP, 16269; NLA-30, 21397; Palatinol® DIDP, 22588; Palatinol® Z, 22596; PX-120, 25436
68515-50-4: Jayflex® DHP, 15549
68515-65-1: Mackanate CP, 18183; Monamate C-1142, 20104
68515-73-1: APG® 225 Glycoside, 2649; Burco NPS-225, 4692; Glucopon 225, 13049; Glucopon 425, 13050
68516-06-3: Propomeen C/12, 25194
68526-84-1: Exxal® 9, 11337
68526-85-2: decyl alcohol, 7867
68526-86-3: Exxal® 13, 11340
68527-05-9: Neoflex® 9, 20922
68527-24-2: Escopol® R-020, 10796
68551-12-2: C12-16 pareth-1, 4841; Genapol® 26-L-1, 12782
68551-13-3: Empilan® KCMP 0703/F, 10401
68551-14-4: Tergitol® Min-Foam 1X, 31247
68554-09-6: Propoquad® 2HT/11, 25196
68554-53-0: Abil® Wax 2434, 58; Belsil SDM 6021, 3920
68554-65-4: Silwet® L-720, 28451
68555-36-2: Mirapol® A-15, 19916
68555-98-6: Vultac® 2, 34163
68584-22-5: Nansa® 1042, 20681; Nansa® SSA, 20695
68584-24-7: Nansa® YS94, 20698
68584-25-8: Nansa® TS 50, 20697
68585-05-7: Stasoft J, 29549
68585-34-2: Empicol® ESA, 10318; Empicol® ESC/AU, 10320; Empimin® SQ25, 10437
68585-44-4: DEA-lauryl sulfate, 7827; Empicol® 0031/T, 10298; Empicol® DA, 10310
68585-47-4: Empicol® LX, 10328
68586-07-2: MB 450, 19168
68603-25-8: Antarox® BL-214, 2508
68603-42-9: Active #2, 501; Adeka Sole CO, 671; Alkamide® 101 CG, 1631; Alkamide® CDE, 1637; Aminol KDE, 2179; Schercomid CDA, 27649; Schercomid SCE, 27657
68603-64-5: Diamine HBG, 8296; Kemamine® D-970, 15913; Radiamine 6540, 25765
68604-44-4: Radia® 7171, 25725
68604-71-7: Miranol® C2M-SF 70%, 19888; Miranol® FBS, 19896
68604-73-9: Miranol® CS Conc, 19891; Sandoteric CFL, 27354; Schercoteric MS-EP, 27709
68607-29-4: Duoquad® T-50, 9121; Jet Quat DT-50, 15604
68607-75-0: Abil® Wax 9809, 63; CPS 130, 6948; PS130, 25364
68608-61-7: Miranol® SM Conc, 19911
68608-63-9: Miranol® DM Conc. 45%, 19892
68608-64-0: Ampholak XJO, 2322; Miranol® J2M Conc, 19902; Miranol® JB, 19905
68608-65-1: Ampholak XCO-30, 2321; Miranol® CM Conc. NP, 19889; Miranol® FA-NP, 19894
68608-66-2: Empigen® CDL60, 10367; Miranol® H2M Conc, 19898; Miranol® HM Conc, 19900
68608-88-8: Nansa® SBA, 20692
68610-26-4: Tomah PA-19, 31986
68610-38-8: Miranol® OS-D, 19909; Sandoteric TFL Conc, 27355
68610-39-9: Miranol® JS Conc, 19907
68610-43-5: Miranol® H2M-SF Conc, 19899
68610-51-5: Akrochem® Antioxidant 12, 1157; Ultranox® 257, 32848
68610-62-8: Weston® 474, 34300
68630-96-6: Monateric ISA-35, 20168; Schercoteric I-AA, 27706
68647-16-5: Smithol PEG Adipate, 28651
68647-53-0: Dehyton® G, 7990; disodium cocoamphodiacetate, 8643; Miranol® 2CIB, 19884
68647-73-4: EmCon TEA TREE, 10067
68647-77-8: Chemoxide TAO, 6002
68648-27-1: Fancol HL-20, 11442; Ivarlan 3450, 15448; Lipolan 31, 17367; PEG-20 hydrogenated lanolin, 23045; Satexlan 20, 27480; Super-Sat AWS-4, 30271
68648-66-8: Ceraphyl® GA, 5730
68648-87-3: Alkylate 215, 1732
68649-29-6: Rhodafac® PL-620, 26602
68650-39-5: Dehyton® PG, 7994; Miranol® C2M Conc. NP-PG, 19887; Miranol® FB-NP, 19895; Schercoteric MS-2, 27708
68650-79-3: Schercodine T, 27642
68715-87-7: Duomeen® OTM, 9115
68783-22-2: Rhodameen® HT-50, 26610; Schercomid HT-60, 27651
68783-24-4: Armeen® 2T, 2930; Radiamine 6270, 25760
68783-25-5: Duomeen® TTM, 9118
68783-63-1: Acconon TGH, 240
68783-78-8: Arquad® 2T-75, 3104
68784-08-7: Schercopol CMS-Na, 27665
68797-65-9: Schercoquat BAS, 27673
68813-55-8: Telopar, 31070
68814-69-7: Amine 2MBGD-M, 2141; Amine 2MOLD, 2144; Armeen® DMTD, 2949; Jet Amine DMTD, 15588
68815-23-6: Neutral Degras, 21025
68815-45-2: Miranol® S2M-SF Conc, 19910
68815-55-4: Amphoterge® KJ-2, 2352; Miranol® J2M-SF Conc, 19903; Miranol® JBS, 19906; Monateric 1000, 20147; Monateric 811, 20143; Zoharteric LF-SF, 34883
68815-56-5: Geropon® SBFA-30, 12911
68844-77-9: Hismanal, 14132
68876-77-7: yeast, 34639
68877-55-4: Miranol® JAS-50, 19904
68890-92-6: disodium laneth-5 sulfosuccinate, 8646; Incrosul LAFS, 15053; Rewolan® 5, 26314
68891-17-8: Akypo ITD 30 N, 1216; Sandopan® DTC-100, 27341
68891-21-4: Berol 272 and 716, 4024
68891-38-3: Berol 474, 4035
68891-39-4: Rhodapex® CO-433, 26624
68901-05-3: Ageflex TPGDA, 1003; Laromer® TPGDA, 16792
68908-44-1: Empicol® TA40, 10340
68910-56-5: Varisoft® 2 TD, 33402
68911-61-5: Epal+20+m2, 10613
68911-79-5: Amine 740, 2145; Jet Amine TET, 15598; tallow tetramine, 30745
68915-25-3: Croquat HH, 7223; Croquat WKP, 7227; Kera-Quat WKP, 16056; Quat Keratin WKP, 25602
68918-77-4: Lamepon PA-TR, 16625; Lexein® A520, 17128; TEA-abietoyl hydrolyzed collagen, 30871
68919-40-4: Ampholak YCO-40, 2328; Miranol® C2M Anhyd. Acid, 19886
68919-41-5: Amphoterge® K, 2350; Mackam CSF, 18142; Miranol® CM-SF Conc, 19890; Monateric CA-35, 20153 Monateric CAM-40, 20156;
68920-65-0: Foam-Coll 4C, 12150; Lamepon S, 16626; Maypon 4C, 19095; May-Tein C, 19099; Monteine LCK-32, 20283
68920-66-1: Eumulgin EP .5.2L, 11174
68921-45-9: Agerite® Stalite, 1022
68921-83-5: Ceraphyl® 70, 5720
68936-95-8: Glucate® SS, 13033; methyl glucose sesquistearate, 19539
68937-41-7: Fyrquel® EHC, 12544; Kronitex® 100, 16429; Kronitex® 1840, 16431; Kronitex® 200, 16430; Kronitex® 25, 16427; Kronitex® 50, 16428; triaryl phosphate, 32194; triaryl phosphate, 32195
68937-54-2: Tego® Flow 425, 30975; Tego® Foamex 800, 30976
68937-55-3: Abil® B 88183, 88184, 51; Abil® B 8851, 8852, 49; Silwet® L-7200, 28452
68937-90-6: Hystrene® 5460, 14761; Trilinoleic acid, 32320
68938-15-8: hydrogenated stripped coconut acid, 14597; Hystrene® 5012, 14759; Industrene® 223, 15103
68938-54-5: Silwet® L-7600, 28454; Silwet® L-7605, 28455; Silwet® L-7622, 28457
68951-72-4: Propomeen T/12, 25195
68951-89-3: Crotein ASC, 7247; Pro-Tein ES-20, 25268
68951-92-8: Lamepon UD, 16628; Maypon UD, 19097; Potassium undecylenoyl hydrolyzed collagen, 24788
68952-15-8: Pro-Tein SA-20, 25270
68952-16-9: Bio-Soft® MT 40, 4201; Foam-Coll 4CT, 12151; Lamepon ST40, 16627; Maypon 4CT, 19096; May-Tein CT, 19100; Monteine LCT, 20285
68952-35-2: Kronitex® TCP, 16435
68952-98-7: Akwilox 133, 1213
68953-11-7: Foamole B, 12177
68953-58-2: Tixogel VP, 31883
68953-64-0: Ceraphyl® 65, 5719
68953-96-8: Nansa® EVM50, 20686
68954-89-2: Akypo RCS 60, 1231; Akypo RLM 130, 1234; Akypo RLM25, 1237;

Akypo RLMQ 38, 1238; Akypo RS 100, 1242; Akypo RS 60, 1243; Akypo RT 60, 1244
68955-19-1: Empicol® LM, 10324; Empicol® LMV/T, 10325; Empicol® LZ, 10331; Empicol® LZV, 10334
68955-20-4: Empicol® TAS30, 10341
68955-45-3: Radia® 7506, 25745
68956-08-1: Kessco® 887, 16102
68956-68-3: vegetable oil, 33522
68957-18-6: Akyposal DS 28, 1305; Akyposal DS 56, 1306
68958-58-7: Emalex GWIS-115, 9960
68958-64-5: Tagat® TO, 30716
68966-38-1: Monazoline IS, 20195; Schercozoline I, 27713
68987-89-3: Akypo NTS, 1227; Akypo® Soft 100 NV, 1251
68989-00-4: Benzalkonium chloride, 3958; Empigen® BCB50, 10364
68989-03-7: Rewoquat CPEM, 26391
68989-22-0: zeolite synthetic, 34717
68990-05-6: Axol® C 62, 3511
68990-06-7: hydrogenated tallow glyceride lactate, 14600; Lamegin® GLP 20, 16619
68990-58-9: Acetylated hydrogenated tallow glyceride, 316; Lamegin® EE, 16618
68990-59-0: Acidan N 12, 338; hydrogenated tallow glyceride citrate, 14599; Lamegin® ZE 30, 60, 16620
68990-63-6: Super Refined Shark, 30197
68990-82-9: Paramount B, 22778; Witarix® 212, 34429
68991-88-8: Miranol® L2M-SF Conc, 19908; Miranol® TBS, 19912
69009-90-1: Nusolv ABP-103, 21801
69011-06-9: Halthal, 13574
69011-84-3: Akyposal BD, 1304
69013-18-9: Surfonic® LF-17, 30370
69227-20-9: Mulsifan CB, 20452
69227-21-0: Marlox® MS 48, 18946
69304-37-6: CD4368, 5503
69331-39-1: Amphisol, 2308; Surfagene FGD 600, 30354
69377-81-7: fluroxypyr, 12117
69430-24-6: Abil® B 8839, 48; ABIL® K 4, 53; ABIL® OSW 12, OSW 13, 55; Amersil® VS-7207, 2034; Dow Corning® 344 Fluid, 345 Fluid, 8861; Fancorsil A, 11459; Masil® SF-V, 19024; Sentry Cyclomethicone, 27958; SF1173, 28128; SWS-03314, 30473
69430-36-0: Crotein ASK, 7248; Crotein K, WKP, 7255; Hydrokeratin AL-30, 14603; Kera-Tein 1000, 16059; Nutrilan® Keratin W, 21836
69468-44-6: Tagat® 12, 30705; Tagat® I, 30706
69537-28-8: Akypoquat 131, 1271
69633-04-1: Garlon, 12672; Garlon 2, 12673; Garlon 4, 12674; Triclopyr, 32227
69806-50-4: Fluazifop-butyl, 12038; Fusilade, 12519; Gallant, 12607
70024-77-0: Ampholak XOO-30P, 2324
70084-87-6: hydrolyzed wheat protein, 14614; Hydrotriticum 2000, 14650
70084-94-5: Soy-Tein NL, 29176
70124-77-5: Tomahawk, 32001
70161-44-3: Suttocide® A, 30440
70191-76-3: Poly-Tergent® 4C3, 24632
70225-05-7: Liponate TDTM, 17390; tridecyl trimellitate, 32235
70288-86-7: Eqvalan, 10737; Ivermectin, 15454
70321-78-7: Lubrhophos® HR-719, 17719
70331-94-1: Naugard® XL-1, 20799
70356-09-1: Parsol® 1789, 22866
70496-39-8: Behenamide DEA (1:1), 3895; Incromide BED, 14983
70609-66-4: Hostapon KTW New, 14394; sodium lauroyl taurate, 28780
70632-06-3: Miranate® LEC, 19875; Sandopan® DTC Linear P, 27339
70693-04-8: Michel XO-150-1620, 19642
70693-05-9: Exxal® 26, 11344
70693-32-2: Liponate NPGC-2, 17381
70729-87-2: Parapel® LIS, 22787
70750-47-9: Ethoquad® C/12, 10997; Variquat® 638, 33399
70851-07-9: Amphoteen BCA-30, 2344; Mirataine® BET-C-30, 19923
70851-08-0: Mirataine® BSC, 19926; Mirataine® CBS, CBS Mod, 19927
70914-02-2: Oxypon 288, 22467; Oxypon 365, 22470
70914-20-4: Exxal® 7, 11335
71010-52-1: Gellan gum, 12736; Gelrite®, 12746; Kelco-Gel® Gellan Gum, 15827
71060-61-2: Propomeen 2HT-11, 25193
71060-72-5: Arquad® 316(W), 3113
71172-17-3: Vanax® CPA, 33288
71289-18-1: Mazox® CDA, 19150
71329-50-5: Jaguar® C-162, 15498; Jaguar® C-162, 15499
71486-48-1: Uniplex 310, 33017
71487-00-8: Ethoquad® C/12 Nitrate, 10998; PEG-2 cocomonium nitrate, 23039
71487-01-9: Arquad® 2C-70 Nitrite, 3101
71566-49-9: Dermol 89, 8117; Kessco® Octyl Isononanoate, 16117
71786-47-5: magnesium sulfonate, 18369
71864-46-5: CD5610, 5510
71888-89-6: Jayflex® 77, 15543
71902-01-7: Crill 6, 7032; Emsorb® 2516, 10462; Sorbitan isostearate, 29110
72160-13-5: Akypo OP 190, 1229; Akypo OP 80, 1230
72162-46-0: Tomah DA-14, 31966
72319-06-3: Lexein® A200, 17127; Myristoyl hydrolyzed collagen, 20576; Pro-Tein SM-20, 25271
72347-89-8: Caprol® 2G4S, 5132
73049-73-7: European Elastin 10, 11222; hydrolyzed elastin, 14612
73138-79-1: Standapol® 1610, 29464
73250-68-7: Hinochloa®, 14096; mefenacet, 19229
73334-07-3: Ultravist, 32877
73523-00-9: Reprodin®, 26149
73790-28-0: Clinafarm, 6422
74051-80-2: Checkmate, 5863; Poast®, 24278; Sethoxydim, 28099
74115-24-5: Apollo 50C, 2661
74176-31-1: Alfaprostol, 1574
74223-64-6: Ally®, 1782; metsulfuron-methyl, 19606
74381-53-6: Lupron, 17918
74623-31-7: Pluracol® W170, 24186; Pluracol® W2000, 24187; Pluracol® W5100N, 24189; Pluracol® W660, 24190
74742-48-9: Varsol® 1, 33456
75422-21-0: Empimin® LAM30/AU, 10424
75782-86-4: Neodol® 23, 20911
75782-87-5: Neodol® 45, 20915
76483-21-1: Marlophor® CS-Acid, 18891
76578-14-8: Pilot, 23841; quizalofop-ethyl, 25697
76674-21-0: Fluphenazine, 12116; Impact, 14908
76738-62-0: Bonzi, 4407; Cultar, 7348; paclobutrazol, 22540, 22541, 22542; Parlay, 22846
77732-09-3: oxadixyl, 22407
78266-06-5: Choletec, 6170
78330-12-8: Aristonate H, 2884
78350-78-4: Trigonox® ADC-NS60, 32294
78491-02-8: Diazolidinyl urea, 8348; Germall® II, 12894
79277-27-3: thifensulfuron-methyl, 31673
79533-80-5: corichrome, 6821
79770-24-4: iotrolan, 15244
79770-97-1: Propoquad® T/12, 25197
80584-85-6: Doverphos® 11, 8840; Doverphos® DPGDP, 8847; Weston® THOP, 34319
80584-86-7: Doverphos® 12, 8841; Weston® DHOP, 34305
80584-87-8: Weston® 491, 34301
81131-70-6: Lipostat Tablets, 17421
81334-34-1: Arsenal, 3132; Arsenal®, 3134; imazapyr, 14893
81405-85-8: Dagger, 7693
81406-37-3: fluroxypyr 1-methylheptyl ester, 12118; Starane, 29528
82204-94-2: Amiter LGOD, 2198
82469-79-2: Citroflex B-6, 6346
82558-50-7: Flexidor, 11934; Knot Out, 16257; Ratio, 25897; Tripart® Ratio, 32383
82560-54-1: Oncol, 22172; Oncol 10G, 22173
82657-04-3: Talstar, 30756
83164-33-4: Diflufenican, 8451
83601-81-4: Ashlade M, 3196; Ashlade Mancarb FL, 3197; Delsene® 50 DF, 8041; Derosal WDG, 8133
83682-78-4: Phospholipid PTD, 23723
83933-91-3: methyl glucose dioleate, 19538
84082-44-0: Behenyl betaine, 3900; Incronam B-40, 15018
84087-01-4: Facet®, 11405
84501-49-5: Empimin® SDS, 10436; sodium decyl sulfate, 28755
84507-84-1: Lamictal Tablets, 16638
84539-90-2: Radia® 7514, 25747
84604-14-8: Pristene R20, 25039
84605-09-4: Schercemol ISE, 27605
84605-13-0: Schercemol MEL-3, 27606
84605-14-1: Schercemol MEP-3, 27608
84605-15-2: Schercoquat IEP, 27681
84812-94-2: Ampholyte KKDP-60, 2334; Ampholyte KKE-70, 2335; Mackam 151C, 18137
84930-16-5: Nipasol M Potassium, 21283
84988-74-9: Radia® 7040, 25717
84988-79-4: Priolube 1414, 25017; Radia® 7230, 25732

CAS Registry Numbers

85005-47-6: Argonol 40, 2855
85005-55-6: Marlophen® 81N, 18882
85009-19-9: Santion, 27396
85049-34-9: Priolube 1429, 25018; Propylene glycol dioleate, 25217; Radia® 7204, 25731
85049-36-1: Radia® 7187, 25728
85049-37-2: Radia® 7331, 25736
85068-76-4: Iofetamine :ss123;ksl hydrochloride, 17553; Perfusamine, 23251
85116-87-6: Radia® 7231, 25733
85116-93-4: Radiasurf® 7175, 25805
85116-97-8: Radiasurf® 7410, 25811
85209-91-2: ADK STAB NA-11, 733
85251-77-0: Radiasurf® 7600, 25820
85408-49-7: C12-14 alkyl dimethylamine oxide, 4840; Lilaminox M24, 17251
85409-09-2: Radia® 7108, 25720
85509-19-9: flusilazole, 12119
85536-07-8: L.A.S, 16528; Labrasol, 16543
85536-08-9: Corn oil PEG-6 esters, 6835; Labrafil M 2125 CS, 16540
85536-14-7: Marlon® AS3, 18878
85536-23-8: Aminol N, 2180; Aminol TEC N, 2181
85586-21-6: Radia® 7110, 25721
85632-63-9: Lilamin LSP 33, 17250
85721-33-1: ciprofloxacin, 6317
85736-49-8: Radiasurf® 7402, 25808; Radiasurf® 7403, 25809; Radiasurf® 7404, 25810; Radiasurf® 7443, 25815; Radiasurf® 7444, 25816
85865-69-6: Radia® 7241, 25734
86050-77-3: Magnevist, 18376
86088-85-9: Incrosoft S-75, 15050; Quaternium-27, 25607; Varisoft® 475, 33415
86089-12-5: Mackam RA, 18149; Rewoteric® AM R40, 26426
86438-78-0: Mirataine® BB, 19922
87616-36-2: Ceraphyl® 85, 5721
88103-59-7: 2-octyldodecyl erucate, 22043
88497-58-9: Sandopan® DTC Linear P Acid, 27340
88671-89-0: myclobutanil, 20549; Systhane, 30665
88684-42-8: VA-545, 33231
88917-22-0: Arcosolv® DPMA, 2813
89339-41-3: Ultracast PE-35, PE-60, 32787
90193-76-3: Radia® 7505, 25744
90268-48-7: Rewopol® B 1003, 26358
90283-04-8: Schercoquat ALA, 27671
90388-14-0: cetyl diethanolamine phosphate, 5818; Crodafos CDP, 7077
90453-59-1: Akypo® Muls 400, 1249
90529-57-0: Schercoquat SOAS, 27688
90624-75-2: Mirapol® AD-1, 19917
90624-76-3: Mirapol® AZ-1, 19918
90640-45-2: Diamine B11, 8294
90730-68-0: Kemamine® BQ-2982B, 15907
91031-31-1: Radia® 7266, 25735
91031-48-0: Priolube 1458, 25021; Radia® 7131, 25724
91031-57-1: Radia® 7510, 25746
91052-47-0: Radiasurf® 7900, 25821
91080-18-1: Exsyproteines 2%, 11318
91465-08-6: λ-cyhalothrin, 7567; Hallmark, 13557
91723-32-9: Dynacet®, 9341
91723-55-6: Rewoquat W 75 H, 26395
91744-38-6: glyceryl stearate citrate, 13088; Imwitor® 369, 14942; Imwitor® 370, 14943; Monoglyceride citric ester, 20226
91824-88-3: Isolan® GI 34, 15368
91995-05-0: Ampholak YCA/P, 2326
93356-94-6: Weston® 439, 34299
93455-78-8: Radiasurf® 7400, 25807
93572-63-5: Rewoquat DQ 35, 26393
93685-79-1: Permethyl 102A, 23388
93820-52-1: Rewoteric® AMKSF 40, 26430
94109-05-4: Incromide BEM, 14984
94441-92-6: Ampholak YJH-40, 2329
94552-41-7: Schercoquat ROAS, 27684
95706-86-8: Alkaterge® -T-IV, 1718
95823-35-1: Mark® 2140, 18727
96328-09-1: Mark® 5089, 18729
96690-41-4: Crosilk 10,000, 7233; Ritasilk, 26830; Silk Pro-Tein, 28362; Solu-Silk Protein, 29025
96726-23-9: Miranol® Ester PO-LM4, 19893
97281-23-7: Radiasurf® 7270, 25806; Radiasurf® 7414, 25812; Radiasurf® 7417, 25813; Radiasurf® 7453, 25817; Radiasurf® 7454, 25818; Radiasurf® 7473, 25819
97338-28-8: Ceraphyl® 791, 5728; Hetester HSS, 13897; isocetyl stearoyl stearate, 15349
97404-50-7: glyceryl lanolate, 13081; Lanesta G, 16657
97488-62-5: Ampholak 7TY, 2317
97488-91-0: Apricot kernel oil PEG-6 esters, 2675; Labrafil M 1944 CS, 16537
97593-29-8: Dimodan PVP Kosher, 8534; Dimodan S, 8535
97593-31-2: Acidan, 337
97659-50-2: Ampholak YCE, 2327
97659-51-3: Ampholak XCE, 2320
97659-53-5: Ampholak 7TX, 2313; Ampholak 7TX/C, 2314; Ampholak 7TX-SD 55, 2315; Ampholak 7TX-T, 2316; Ampholak XO7, 2323
97692-58-5: T.A.M, 30672
97808-04-3: Amine 780, 2147
97999-44-5: Marlowet® 5311, 18925
98073-10-0: Glucam® E-20 Distearate, 13021; methyl gluceth-20 distearate, 19537
99330-44-6: Akypo® Soft 100 MgV, 1250
100085-10-7: Actigen E, 479
100085-64-1: Rewoteric® QAM 50, 26432
100864-25-1: Wheat-Pro EN-20, Wheat-Tein NL, 34351
101205-02-1: CO9800, 6507; Cycloxydim, 7544; Focus®, 12184; Stratos®, 29806
102089-33-8: Uvasil 299, 33180
102523-96-6: Abil® B 9950, 50
102843-39-0: VA-058, 33226
103213-19-0: Schercemol DISD, 27596
103213-20-3: Bernel® Ester DID, 4005; Schercemol DID, 27594
103213-22-5: Schercemol TIST, 27624
103597-45-1: Mixxim® BB/100, 19989
103819-46-1: Labrafil M 1980 CS, 16539; olive oil PEG-6 esters, 22139
104222-32-4: VA-080, 33228
104909-82-2: Akypo TBP 180, 1245
105391-15-9: Akypo LF 3, 1219; Akypo LF 5, 1222; Akypo MB 2528S, 1225
105827-78-9: Admire, 755; Confidor, 6725; Gaucho, 12698
106392-12-5: Dowfax 30C05, 30C10, 50C15, 8910
106436-39-9: Ceraphyl® 55, 5717
106990-43-6: Chimassorb® 119FL, 6072
107397-59-1: Tetronic® 150R1, 31398
107534-96-3: Folicur, 12198; Horizon/Horizont, 14308; Raxil, 25908; Silvacur, 28408
107600-33-9: Akypo LF 2, 1218
107600-36-2: Akyposal 100 DAL, 1298
107628-03-5: Akypo® Soft KA 250 BV, 1254
107987-23-5: Quatrisoft Polymer LM-200, 25627
108419-32-5: Exxate® 800, 11348
108419-33-6: Exxate® 900, 11349
108419-34-7: Exxate® 1000, 11350
108419-35-8: Exxate® 1300, 11351
108818-88-8: Rhodafac® BG-510, 26593
109075-72-1: Dehypon LT 054, 7979
109464-53-1: Sokalan® HP 22, 28873
109678-33-3: FR-1034, 12361; tetrabromodipentaerythritol, 31360
110152-58-4: Indulin® W-1, 15094
110615-47-9: APG® 600 Glycoside, 2653; Glucopon 600, 13051
111019-03-5: Crodafos CAP, 7076
111174-64-2: Quat-Pro E, 25608
111905-55-6: Schercoquat DAS, 27676
112324-11-5: Schercoquat ROEP, 27685
112324-16-0: Lipoquat R, 17402; M-Quat® JN, 20406; Ricinoleamidopropyl ethyldimonium ethosulfate, 26733; Surfactol® Q4, 30344
113431-53-1: Schercemol DISF, 27597
113431-54-2: Schercemol TISC, 27623
113492-03-8: Schercoquat CAS, 27674
113492-04-9: Schercoquat FOAS, 27677
113976-90-2: APG® 300 Glycoside, 2651
115047-92-2: CDS-1801, 5526; DV-1801, 9295
115340-78-8: Schercoquat APAS, 27672
115340-80-2: Schercoquat WOAS, 27689
116912-64-2: Union Carbide® A-1160, 32994
117272-76-1: Silwet® L-7607, 28456
118337-09-0: Ethanox® 398, 10933
118585-13-0: VA-060, 33227
118777-77-8: Katemul IG-70, 15789
123754-28-9: Ammonium acrylates acrylonitrogens copolymer, 2227; Hypan® SS201, 14702
124046-39-5: Schercomid SWG, 27663
124960-38-9: VA-545, 33232
125740-36-5: Tomah Q-14-2, 31990
125804-12-8: Dermol G-76, 8124
125804-13-9: Dermol L45, 8127
127358-81-0: Dermol DISD, 8123; diisostearyl dimer dilinoleate, 8470
128952-18-1: D.S.H.C, 7625
128973-71-7: Algisium-C, 1597
128973-73-9: Theophyllisilane C, 31569
129541-36-2: Katemul IGU-70, 15790
130097-36-8: Esperox® 740M, 10855
130124-24-2: Carsosoft® T-90, 5354; Incrosoft T-90, 15051
130986-04-8: Liposiliol C, 17403
131044-77-4: Tyrosilane C, 32688

131044-78-5: Silhydrate C, 28312
131081-39-5: Lasilium C, 16805
133101-79-8: Proteosilane C, 25276
133798-12-6: Schercotaine UAB, 27701
133934-08-4: Schercotaine APAB, 27694
133934-09-5: Mackam WGB, 18151; Schercotaine WOAB, 27702; Wheat germamidopropyl betaine, 34350
134112-42-8: Schercoquat IALA, 27678
136097-97-7: Kester Wax® K 48, 16128
136505-00-5: Hypan® SR150H, 14701
137796-06-6: Zoldine® MS-52, 34886
138208-68-1: Dermol ICSA, 8126
138314-11-1: Dermolan GLH, 8130
144610-95-5: Schercemol 145, 27580
151066-66-5: Acusol® 820, 591; Acusol® 830, 592
185123-36-8: Schercomid SAP, 27656
270028-82-6: Empicol® ETB, 10321
594477-57-3: FR-1025, 12360; Poly (pentabromobenzyl) acrylate, 24342
617788-68-9: Laurel SRO, 16843
689990-06-7: Axol® L 61, L62, 3513
977039-11-4: Quat-Soy CDMA-25, 25629; Soy-Quat C, 29175
977053-96-5: Rewomul MG SE, 26339
977059-33-8: Hydrosoy 2000, 14642
977067-77-8: Empicol® EMB, 10317
977077-71-6: Crosilk Liq, 7234

2. EINECS Numbers

200-001-8: Durine, 9225;Dyna-Form, 9346; formaldehyde, 12258; Hercules® 37M6-8, 13850; Lysoform, 18095; paraformaldehyde, 22751;
200-007-0: Talpheno, 30755;
200-014-9: vitamin D_2, 33940;
200-018-0: lactic acid, 16570; Patlac® LA, 22910;
200-021-7: prednisolone, 24872;
200-024-3: DDT, 7820; De De Tane, 7821; Neocid, 20879;
200-030-6: propantheline bromide, 25176;
200-032-7: Coke, 6591;
200-033-2: LSD, 17704;
200-056-8: Mansonil®, 18611; niclosamide, 21146;
200-061-5: A-625/641ABS 301K, ABS 500FR-1, 10; Hydex® 100 Gran.206, 14531; Liponic 70-NC, 17391; Sorban, 29088; Sorbelite C, 29104; sorbitol, 29113; Sorbitol (EGIC), 29114; Sorbo®, 29115;
200-062-0: Alloxan, 1750;
200-064-1: Nu-Seals Aspirin, 21798; Tasprin, 30842;
200-066-2: vitamin C, 33938;
200-075-1: Candex®, 5085; Cerelose, 5741; Emdex®, 10071; Flolys®, 11994; glucose, 13052; Roferose, 26904;
200-076-7: piperonyl butoxide, 23875;
200-081-4: Benzimidazole, 3966;
200-087-7: Dinitra, 8540;
200-101-1: Riamat, 26720;
200-123-1: urethane, 33135;
200-133-6: Laractone®, 16784;
200-134-1: Econopred, 9588;
200-143-0: Bioban® BNPD-40, 4156; Bronopol, 4595; Bronopol-Boots, 4596; Broponol, 4616; Canguard® 409, 5089; Myacide® AS Plus, 20544; Myacide® S-1, S-2, 20545;
200-148-8: Trolovol®, 32449;
200-149-3: Danex, 7727; Dipterex®, 8616; Dipterex® 80, 8617; Dylox®, 9324; Masoten®, 19030; Neguvon®, 20852; trichlorfon, 32212; Tugon, Tugon OKO, Tugon sp 80, 32570;
200-157-7: cysteine hydrochloride, 7601;
200-158-2: L-cysteine, 7602;
200-166-6: Lysodren, 18094;
200-193-3: Campbell's Nico-Soap, 5068; XL-All Insecticide, 34607;
200-198-0: Nu-Seals Sodium Salicylate, 21799;
200-203-6: Lasix, 16807;
200-206-2: Metopirone, 19591; metyrapone, 19608;
200-207-8: Kerecid, 16067;
200-210-4: Merthiolate, 19391;
200-221-4: Levothroid, 17081;
200-228-2: isonicotinic acid, 15380;
200-231-9: Lebaycid®, 16932; Tiguvon®, 31788;
200-235-0: Isopto Atropine, 15414;
200-237-1: Metol, 19584;
200-238-7: Hibidil, 14050; Hibiscrub, 14051; Hibisol, 14052; Hibispray, 14053; Savloclens, 27502; Savlon Babycare, 27504; Superspray, 30282;
200-240-8: Cascade, 5375; explosive gum, 11315; Glonoine oil, 13005; Hercol 2, 13824; Hercol 2X, 13825; Hercon® 2, 13831; Hercon® 2X, 13832; Hercosplit WR, 13844; Unigel, 32955;
200-242-9: Phenmerzyl Nitrate, 23599;
200-251-8: Biquanide, 4218;
200-262-8: carbon tetrachloride, 5224; Katharin, 15791; phenoxin, 23623; seretin, 28026; tetraform, 31369; Thawpit, 31564;
200-268-0: bioMeT TBTO, 4172; Keycide® X-10, 16153; tributyltin oxide, 32209;
200-270-1: Benzyltriethyl ammonium chloride, 3993; Sumquat® 2355, 30120;
200-271-7: Bladan®, 4309; ethyl parathion, 11079; Folidol® E605, 12199; Fosferno, 12330; Murphos, 20521; parathion, 22807;
200-272-2: glycine, 13095;
200-273-8: α-alanine, 1333; Ritalanine, 26812;
200-274-3: L-serine, 17705, 28035;
200-283-2: Adenosine triphosphate, 681;
200-285-3: Asuntol®, 3280; Muscatox, 20525; Perizin®, 23276;
200-287-4: Leukomycin®, 17030;
200-289-5: Croderol G7000, 7164; dynamite glycerin, 9358; Emery® 912, 10137; glycerin, 13071; Glycon® G 100, G 300, 13118; Glyrol, 13152; Kemstrene® 96.0%, 15991; Lye glycerin, 18070; Osmoglyn, 22358; Pricerine 9071, 24942; Shur-Coal® FCA, 28189; Star, 29521; Superol, 30258;
200-294-2: lysine, 18088;
200-296-3: cystine, 7603;
200-300-3: Benzyl trimethyl ammonium chloride, 3989; Hipochem Migrator J, 14114; Variquat® B200, 33400;
200-302-4: Arlacide A, 2907; chlorhexidine acetate, 6100;
200-311-3: Acetoquat CTAB, 303; Bromat®, 4566; Catinal HTB-70, 5457; Cetavlex, Cetavlon, 5787; Cetrimide BP, 5812; Cetrimonium bromide, 5813; Rhodaquat® M242B/99, 26643; Sumquat® 6030, 30122; Varisoft® CTB-40, 33421;
200-312-9: Emersol® 143, 10119; Glycon® P-45, 13119; Hystrene® 8016, 14763; Industrene® 4516, 15106; palmitic acid, 22631; palmitic acid, 22632; palmitic acid, 22633; palmitic acid, 22634; Palmitic Acid, 22635; Prifrac 2960, 24954;
200-313-4: Emersol® 110, 10118; Emersol® 6349, 10124; Emery® 400, 10130; Glycon® S-90, 13121; Glycon® TP, 13122; Hydrofol Acid 1655, 14588; Hystrene® 4516, 14758; Hystrene® 5016, 14760; Hystrene® 7018 FG, 14762; Hystrene® 9718 NF FG, 14768; Industrene® 4518, 15107; Industrene® 5016, 15108; Industrene® 7018 FG, 15109; Industrene® R, 15111; Petrac® 270, 23462; Prifrac 2980, 24955; Pristerene 4904, 25040; Stearex, 29597; stearic acid, 29599; Stearite, 29601;
200-315-5: Bubber Shet, 4641; Superprill, 30265;
200-322-3: Luminal Sodium, 17803; Phenobarbitone sodium, 23600;
200-326-5: Mapo®, 18656;
200-333-3: fructose, 12420; levulose, 17085;
200-334-9: sucrose, 29934; Sugartab®, 29956;
200-338-0: Chilisa FE®, 6064; Dowfrost, 8922; Ilexan P, 14885; Prolugen, 25135; propylene glycol, 25214; propylene glycol, 25215;
200-340-1: betaprone, 4065;
200-346-4: sulfamethazine, 29962;
200-349-0: Ortho-Klor, 22331; Syndane, 30536;
200-350-6: Gesterol 50, 12927; Lipo-Lutin, 17370; luteol, 17991;
200-353-2: cholesterol, 6168; Dastar, 7788; Fancol CH, 11435; Kathro, 15795; Liquid Crystal CN/9, 17467;
200-377-3: Vinyzene, 33869;
200-383-6: Nitoman, 21316;
200-385-7: liquorice, 17490;
200-386-2: Gravidox, 13294; Hexa-Betalin, 13987; Hexavibex, 14019; pyridoxine hydrochloride, 25483; vitamin B_6 hydrochloride, 33937;
200-399-3: Ritatin, 26834;
200-400-7: Xylose, 34632;
200-401-2: Atlas Steward, 3334; Fumite Lindane, 12461; Gamma-BHC Dust, 12631; Gamma-Col, 12632; Gamma-HCH Dust, 12633; Gammalin, 12635; Gammasan, 12636; Gammexane, 12639; Jacutin®, 15485; Lorexane, 17607; Murfume Grain Store Smoke, 20512; New Kotol, 21067; Scabene Lotion, 27536;
200-407-5: uridine, 33140;
200-412-2: Ephynal, 10629; Eprolin, 10721; Epsilan-M, 10723; E-Toplex, 11154; Natopherol, 20753; Phytoforol, 23779; Tocopherex, 31917; tocopherol, 31918; Vascuals, 33465; vitamin E, 33943;
200-414-3: ethopabate, 10993; Pancoxin, 22669;
200-419-0: folic acid, 12196; Folvite, 12211; Mission Prenatal, 19945; vitamin M, 33951;
200-420-6: carbostyril, 5244;
200-423-2: Pancoxin, 22667; sulfaquinoxaline, 29968;
200-425-3: thiamine, 31664;
200-431-6: Lysochlor, 18093; Raschit, 25892;
200-432-1: DL-methionine, 19487; Petameth, 23451;
200-440-5: Diamox, 8316;
200-441-0: Niac, 21086; niacin, 21090; Nico-400, 21148; Nicobid, 21149; Nicocap, 21150; Nicolar, 21151;
200-449-4: edetic acid, 9615; Hamp-Ene® Acid, 13587; Kalex Acids, 15669; Questric Acid 5286, 25665; Sequestrene® AA, 27996; Trilon® BS, 32327;
200-455-7: Butter yellow, 4759;
200-456-2: 2-phenylethyl alcohol., 23642; Mellol, 19297;
200-464-6: 2-mercaptoethanol, 19336; Sipomer® 2ME, 28529;
200-467-2: ethyl ether, 11069; ethyl ether Anhydrous A.C.S., 11070; ethyl ether USP/ACS, 11071;
200-470-9: Emersol® 315, 10121; linoleic acid, 17305; Pamolyn, 22643; Pamolyn 125, 22645; Pamolyn 380, 22647;
200-480-3: Dimethoate, 8496; Dimethoate Bayer, 8497; Perfekthion®, 23244; Systemic Insecticide, 30664; Turbair, 32595;
200-481-9: tetracycline, 31365;
200-484-5: Murdiel, 20504;
200-500-0: 5'-adenylic Acid, 683; adenosine monophosphate, 680; AMP, 2301;

My-B-Den, 20547;
200-521-5: Aminotriazole Bayer, 2190; Azolan, 3530; Boroflow S/ATA, 4437; MSS Aminotriazole, 20426;
200-522-0: Leucine, 17020; l-leucine, 17021;
200-526-2: *p*-nitrobenzoic acid, 21350;
200-529-9: calcium disodium versenate, 4946; Versene CA, 33618;
200-532-5: acetoxyphenylmercury, 306; Acticide PMA 100, 470; Agrosan GN, 1081; Ceresol, 5748; Merpectogel, 19370; Phenitol, 23594; phenylmercuric acetate, 23644; phenylmercury acetate, 23646; PMA 18, 60, 24274; Single Purpose, 28498;
200-539-3: phenylamine, 23638;
200-540-9: Calac, 4924; calcium acetate, 4936, 4937; grey acetate, 13351; Niacet Calcium Acetate Tech, 21088; pyrolignite of lime, 25516;
200-541-4: thioacetamide, 31678;
200-543-5: sulfourea, 30038; thiourea, 31720;
200-547-7: Atgard, 3285; Dedevap®, 7870; Dethlac, 8210; dichlorvos, 8394; Divipan, 8740; Mafu®, 18305; Nogos, 21420; Nuvan, 21850; Nuvan Top Aerosol, 21852; Nuvanol, 21853; Vapona, 33378;
200-552-4: Variton, 33432;
200-555-0: Carbaryl, 5181; Carylderm, 5364; Microcarb, 19659; Murvin 85, 20522; Sevin, 28115; Sevin® Brand SL, 28116; Thinsec, 31676; Tornado, 32071;
200-559-2: Lactin, 16578; α-D-Lactose, 16589; milk sugar, 19768;
200-562-9: Mepron, 19328;
200-565-5: arecoline, 2829;
200-567-6: colfosceril palmitate, 6607;
200-573-9: Aroquest 100, 3062; Cheelox® 100, 5865; Hamp-Ene® 100, 13586; Hamp-Ene® Na4, 13590; Intraquest® TA Solution, 15175; Kalex 220 Crystal, 15668; Kalex Liq. 50%, 15671; Questal Extra Powd. Conc. 0780, 25662; Questal Special 0860, 25664; Sequestrene® 220, 27995; Sequestrene® 30A, 27994; Solon Conc, 28957; tetrasodium EDTA, 31391; Thorquest 39, 31756; Trilon® B Powd, 32324;
200-578-6: ethyl alcohol, 11064; ethylol, 11142; grain alcohol, 13244; Punctilious® Ethyl Alcohol, 25398; Pyro, 25487; spirits of wine, 29304; Synasol, 30522; USI in Oval, 33157;
200-579-1: Amasil®, 1946; formic acid, 12271;
200-580-7: Acetic acid, 292;
200-616-1: thymine, 31777;
200-618-2: Benzoic acid, 3972, 3973; flowers of Benjamin, 12031; Retarder BA, BAX, 26254;
200-621-9: uracil, 33101;
200-629-2: Activ-8®, Activ-8 in Hexylene Glycol, 494;
200-641-8: thiamine hydrochloride, 31665;
200-652-8: Aroquest M Special, 3064; Chel DTPA, 5873; Hamp-Ex® Acid, 13592; Pentaquest OPAC 0201, 23170;
200-653-3: Furoxone, 12499;
200-655-4: choline chloride, 6171;
200-659-6: methyl alcohol, 19511; pyro alcohol, 25488; spirit of wood, 29300;
200-661-7: Alcowipe®, 1531; Avantine, 3456; I.P.S., 14799; isopropanol, 15405; Sterets Pre-Injection Swabs®, 29728;
200-662-2: acetone, 297; Sasetone, 27474;
200-663-8: chloroform, 6125;
200-664-3: Decap, 7845; Demavet, 8053; dimethyl sulfoxide, 8511; DMSO, 8762;
200-673-2: vitamin D+73, 33941;
200-677-4: thioglycolic acid, 31688; Thiovanic® Acid, 31722;
200-679-5: dimethyl formamide, 8506; Dynasolve 100, 9386;
200-680-0: Fermin, 11630; Rubramin PC, 27114; Sytobex, 30668; vitamin B_{12}, 33934;
200-683-7: Anatola, 2403; A-Sol, 3228; Homagenets Aorl, 14282; retinol, 26277; Ro-A-Vit, 26866; Super A, 30163; Vi-Alpha, 33706; Vi-Dom-A, 33744; vitamin A, 33932;
200-697-3: hypoxanthine, 14721; sarcine, 27460;
200-701-3: Largactil, 16786;
200-709-7: Acillin, 346; Ampicillin, 2361; Unasyn® IM/IV, 32913;
200-711-8: hexanhexol, 14010; mushroom sugar, 20529;
200-712-3: Retarder SAX, 26260; salicylic acid, 27270;
200-716-5: maltose, 18501;
200-718-6: xanthine, 34590;
200-720-7: Lithic acid, 17506; uric acid, 33139;
200-733-8: Fascol, 11498; hexachlorophene, 13992; Ster-Zac, 29760;
200-739-0: dl-lysine monohydrochloride, 18090;
200-741-1: *p*-toluenesulfonamide, 31949; Uniplex 173, 33012;
200-746-9: Pro-Gas (Gas Disclaimed), 25112; propanol, 25175; propylan-propyl alcohol, 25211;
200-747-4: Quelicin, 25642; Sucostrin, 29927;
200-751-6: Tebol 88, 99, 30878;
200-752-1: Amyl alcohol, 2381;
200-753-7: Benzene, 3961;
200-756-3: Chlorothene, 6151; Cut Aid, 7432; Dabco® CS90, 7653; Delf® Fabric Protector, 8015; Distillex DS1, 8717; Ethana®, 10928; MS-170 1,1,1 Trichloroethane Solvent, 20414; Solvethane, 29060; trichloroethane, 32216; Tri-Ethane, 32239;
200-772-0: perglycerol, 23262; sodium lactate, 28777;
200-779-9: Marlate 2-MR Emulsifiable Insecticide, 18747; Marlate 300 Flowable, 18749; Marlate 400 Flowable Concentrate, 18750; Marlate 50 WP, 18748; Marlate Methoxychlor Insecticide, 18751; methoxychlor, 19504;
200-782-5: Alizarin, 1621; Pincoffin, 23854;
200-785-1: diethyldiphenyldichloroethane, 8441; Perthane, 23435;
200-786-7: Niagara blue, 21091;
200-796-1: Adenine, 679;
200-799-8: guanine, 13443;
200-806-4: Agerite® DPPD, 1016; JZF, 15638; N,N'-diphenyl-*p*-phenylenediamine, 8603; Naugard® J, 20796; Permanax DPPD, 23361;
200-812-7: fire-damp, 11831; marsh gas, 18954; methane, 19475;
200-813-2: Embafume, 10012; Meth-O-Gas, 19499; methyl bromide, 19517;
200-816-9: acetylene, 318;
200-817-4: Artic, 3147; methyl chloride, 19522;
200-819-5: methyl iodide, 19542;
200-821-6: hydrocyanic acid, 14584; prussic acid, 25352; zootic acid, 34903;
200-827-9: A-108, 9; propane, 25173;
200-830-5: ethyl chloride, 11067;
200-835-2: acetonitrile, 299;
200-836-8: Acetaldehyde, 281;
200-838-9: Aerothene, 900; Distillex DS3, 8719; Driverit, 9043; M-Clean D, 19178; Methoklone, 19501; methylene chloride, 19568; Nevolin®, 21053; salesthin, 27263;
200-841-5: methylene iodide, 19569; Mi-Gee Brand, 19732;
200-842-0: formamide, 12261;
200-843-6: carbon bisulfide, 5219; carbon disulfide, 5223;
200-846-2: dimethyl sulfide, 8510; Exact-S®, 11281; methyl sulfide, 19561;
200-848-3: calcium carbide, 4939;
200-849-9: ethylene oxide, 11131; Ilexan HT, 14884;
200-854-6: Bromoform, 4579;
200-857-2: A-31, 8; isobutane, 15343;
200-861-4: isopropyl mercaptan, 15408;
200-865-6: acetyl chloride, 311;
200-868-2: Aldamine, 1534; Velsan, 33539;
200-870-3: phosgene, 23701;
200-872-4: Fluoroform, 12099; Genetron® HFC23, 12829;
200-874-5: iodoform, 15214;
200-876-6: Nitrofuel®, 21360; nitromethane, 21366; NM, 21402; NM-55®, 21403;
200-877-1: CM8750, 6460;
200-879-2: Epihydrin, 10642; propylene oxide, 25223;
200-882-9: tetramethylammonium hydroxide, 31385;
200-890-2: *t*-butyl mercaptan, 15434;
200-892-3: Distillex DS6, 8722; Genesolv A Solvent, 12814; Genetron® 11, 12817;
200-893-9: dichlorodifluoromethane, 8388; Genetron® 12, 12818; Sterethox, 29727;
200-894-4: chlorotrfluoromethane, 6153; Genetron® 13, 12819;
200-899-1: CT2050, 7315;
200-900-5: CT2950, 7325;
200-902-6: CM9000, 6463;
200-913-6: 2,2,2-trifluoroethanol, 32250;
200-915-7: Aztec® *t*-Butyl Hydroperoxide-70, Aq, 3541; TBHP-70, 30853; T-Hydro, 31774; Trigonox® A-80, 32291; USP® -800, 33170;
200-917-8: CV-4900, 7450; Dynasylan® VTC, 9419; vinyltrichlorosilane, 33865;
200-920-4: pinacolone, 23848;
200-923-0: BH Dalapon, 4117; Couch and Grass Killer, 6896; dalapon, 7705; Synchemicals Dalapon, 30524;
200-925-1: pentachlorethane, 23134; pentaline, 23154;
200-926-7: Superpalite, 30260;
200-927-2: Farmon TCA, 11491; Na Ta, 20642; trichloroacetic acid, 32213;
200-929-3: trifluoroacetic acid, 32249;
200-930-9: chloropicrin, 6135; Larvacide, 16802;

200-931-4: acetone bromoform, 298;
200-936-1: Arklone, 2894; Distillex DS5, 8721; Genesolv D Solvent, 12815; Genetron® 113, 12821; MS-180 Freon® TF Solv, 20415; trichlorotrifluoroethane, 32224;
200-937-7: Cryofluorane, 7278; Genetron® 114, 12822;
200-945-0: Japan camphor, 15508;
200-972-8: Bromocresol green, 4575;
200-990-6: Ashlade Flotin, 3193; Du-Ter, 9283; Du-Ter®, 9284; Farmatin, 11480; Quadrangle Super-Tin 4L, 25578; Super Tin® 4L, 30208;
201-001-0: Activol, 506; gibberellic acid, 12934; Regulex, 26019;
201-039-8: ADK STAB BT-11, 723; Dabco® T-12, 7673; dibutyltin laurate, 8371; Metacure® T-12, 19417; Synpron 1009, 30611; TEDA-T411, 30917;
201-051-3: DM hydantoin, 8751;
201-056-0: dimethoxypropane, 8501;
201-064-4: THAM, 31557; tris (hydroxymethyl) aminomethane, 32403; Tris Amino®, 32405; Tris Amino® Molecular Biology Grade, 32406; Tris Amino® Ultra Pure Standard, 32407;
201-066-5: Citroflex A-2, 6342; Uniplex 82, 33005;
201-067-0: acetyl tributyl citrate, 313; Citroflex A-4, 6343; Estaflex, 10874; Uniplex 84, 33006;
201-069-1: Citraclean, 6335; citric acid, 6341;
201-070-7: Hydagen® C.A.T, 14526; triethyl citrate, 32242; Uniplex 80, 33004;
201-074-9: trimethylolpropane, 32338;
201-077-5: Dibutyltin maleate, 8372;
201-078-0: octaphen, 21989;
201-081-7: CV-4910, 7451; Dynasylan® VTEO, 9420; Union Carbide® A-151, 32981; vinyltriethoxysilane, 33866;
201-083-8: Dynasil® A, 9382; ethyl silicate, 11083; Silbond® Condensed, 28285; Silester, 28303;
201-084-3: Penthrit, 23180;
201-094-8: Barquat® CME-35, 3666;
201-103-5: Plastic X, 24026;
201-114-5: phosphoric ether, 23740; triethyl phosphate, 32243;
201-116-6: trioctyl phosphate, 32349;
201-120-8: Def®, 7876;
201-122-9: Amgard TBEP, 2055; KP-140®, 16378; TBEP, 30852; tributoxyethyl phosphate, 32206;
201-126-0: isophorone, 15395;
201-127-6: CD5600, 5508; EXP-49, 11305;
201-128-1: Aztec® 2,5-Di, 3551; dimethyldibutyl peroxyhexane, 8518; Esperal® 120, 10837; Luperco 101-P20, 17823; Lupersol 101, 17846; Polyvel CR-5F, 24657; Trigonox® 101, 32279;
201-131-8: Surfynol® 82, 30374; Surfynol® 82, 82S, 30375;
201-132-3: Ficel® AZDN-LF, 11763; Poly-zole AZDN, 24674; V-60, 33218;
201-134-4: Phantol, 23570;
201-142-8: Exxsol® Isopentane, 11362; isopentane, 15394;
201-148-0: isobutanol, 15344;
201-152-2: propylene dichloride, 25213;
201-156-4: isopropanolamine, 15406;
201-159-0: 2-butanone, 4737; Meketone, 19247; methyl acetone, 19507; methyl ethyl ketone, 19530;
201-167-4: Altene DG, 1873; Disparit B, 8658; Distillex DS2, 8718; Germalgene, 12892; trichloroethylene, 32217; Triclene, 32226; trieline, 32237; Triklone, 32315;
201-173-7: Acrylamide, 410;
201-176-3: Luprosil®, 17919; propionic acid, 25190;
201-177-9: acrylic acid, 412;
201-178-4: monochloroacetic acid, 20221;
201-180-5: glycolic acid, 13105;
201-185-2: methyl acetate, 19506;
201-186-8: Oxymaster, 22460; Proxitane, 25308;
201-188-9: NE, 20833; nitroethane, 21357;
201-191-5: Muthmann's liquid, 20541;
201-195-7: isobutyric acid, 15348;
201-196-2: L-lactic acid, 16571;
201-203-9: hydrogen rubeanide, 14591;
201-209-1: 2-nitropropane, 21374; NiPar S-20, 21280;
201-212-8: Biostat A.1, 4209;
201-222-2: Lowinox® AH25, 17663; Santovar, 27428;
201-236-9: FR-1524, 12367; Great Lakes BA-59P, 13298; Saytex® RB-100, 27523; tetrabromobisphenol-A, 31359;
201-245-8: Bisphenol A, 4283; Millad® HBPA, 19776; Parabis, 22717;
201-248-4: 4,4'-diaminodiphenyl sulfone, 8302; Dapsone, 7751;
201-251-0: CD5950, 5513;
201-254-7: Aztec® CHP-80, 3555; CHP-158, 6175; CHP-5, 6174; HPC-9, 14438; Trigonox® 239A, 32290; Trigonox® K-95, 32303;
201-275-1: ethyltoluene sulfonamide, 11145; Uniplex 108, 33007;
201-276-7: ethyl-*p*-toluenesulfonate, 11089; Mittel AEP, 19978;
201-279-3: Aztec® DCP-R, 3556; Dicumyl peroxide, 8407; Di-Cup, 8408; Esperal® 115, 10836; Luperco 500-40KE, 17828; Luperox 500R, 17840; Percumyl D, 23230; Perkadox® BC, 23307; Peroximon® DC 40 MG, 23408; Peroximon® DC-40, 23409; Polyvel PCL-20, 24661;
201-280-9: Nipacide® PTAP, 21259; Orthophen® 278, 22336; Pentaphen® 67, 23168;
201-286-1: Celogen® OT, 5643;
201-297-1: methyl methacrylate, 19548;
201-321-0: saccharin, 27195; Sweeta, 30454;
201-327-3: Fancol DL, 11438; d-Panthenol, 22690; pantothenyl alcohol, 22697; Ritapan D, DL, 26818;
201-328-9: acetyl-dinitro-butyl-xylene, 317; ketone musk, 16141;
201-331-5: Tobias acid, 31914;
201-359-8: oxyalizarin, 22438;
201-368-7: quinizarin, 25683;
201-377-6: Dethmor, 8211; Killgerm® Sewarin P, 16184; Kypfarin, 16519; Ratox, 25898; RCR Grey Squirrel Killer Concentrate, 25914; Sakarat, 27238; Sewarin, 28117; Sorexa Plus, 29141;
201-383-9: Rhodamine B, 26616;
201-435-0: Botrilex, 4460; Brabant PCNB, 4477; Bras-sicol, 4489; Folosan, 12208; quintozene, 25692; Trim, 32330; Tubergran, 32548;
201-436-6: styphnic acid, 29858;
201-437-1: Peri acid, 23269; Schollkopf's acids, 27726;
201-438-7: phenyl-peri acid, 23647;
201-490-0: Diacetazotol, 8247; Dimazon, 8487;
201-493-7: musk ambrette, 20532;
201-497-9: Embequin, 10024; Yodoxin, 34658;
201-501-9: Derris Dust, 8134; FS Derris, 12425; pyrethrum powder, 25478; rotenone, 27000;
201-506-6: phytic acid, 23777;
201-507-1: Flavaxin, 11895; riboflavin, 26722; vitamin B_2, 33935;
201-544-3: anthraflavic acid, 2526;
201-545-9: dicyclohexyl phthalate, 8413; KP 201, 16379; Morflex 150, 20317; Unimoll® 66 M, 32976; Uniplex 250, 33014;
201-549-0: Anthraquinone, 2532; Morkit®, 20336;
201-550-6: diethyl phthalate, 8437; Kodaflex® DEP, 16266; Palatinol® A, 22583;
201-553-2: Kodaflex® DIBP, 16267; Palatinol 1C, 22580; Uniplex 155, 33011;
201-557-4: Bufa, 4654; DBP, 7812; dibutyl phthalate, 8366; Kodaflex® DBP, 16264; Morflex 240, 20321; NLA-10, 21395; Palatinol® C, 22585; Palatinol® DBP, 22587; PX-104, 25432; Unimoll® DB, 32978; Uniplex 150, 33010;
201-559-5: Jayflex® DHP, 15549;
201-560-0: Ceneg, 5671;
201-561-6: NLA-40, 21398;
201-562-1: Butyl octyl phthalate, 4777; PX-914, 25446;
201-564-2: Konakion, 16322; Mono-Kay, 20228; vitamin K_1, 33949;
201-567-9: naphthionic acid, 20715; Piria's acid, 23878;
201-568-4: Nevile and Winther's acid, 21046;
201-579-4: Katalon, 15782; Midstream, 19723; Power Diquat, 24820; Reglone, Reglox, 26017;
201-607-5: phthalic anhydride, 23770; Retarder AK, 26253; Retarder ESEN, 26255; Retarder PX, 26258;
201-618-5: Santowhite, 27431; Vanox® SWP, 33348;
201-622-7: Butyl benzyl phthalate, 4764; Santicizer 160, 27393; Unimoll® BB, 32977;
201-624-8: Morflex 190, 20319; Reomol 4PG, 26126;
201-629-5: Dahl's acid II, 7694;
201-630-0: Dahl's acid III, 7695;
201-635-8: Sudan IV, 29941;
201-638-4: Sudan III, 29940;
201-644-7: Dowicide 3, 8928;
201-662-5: diphenyl acetonitrile, 8593; diphenylacetonitrile, 8600;
201-676-1: Cotnion-Methyl, 6884; Gusathion®, 13501; Guthion®, 13505;
201-689-2: Salt, Amido-G, 27294;
201-705-8: Phyomone, 23775; Planofix, 23913; Rhizopon B, 26574; Rooting Powder, 26958; Tipoff, 31848;
201-722-0: Penicillin V, 23111;
201-723-6: Temasept IV, 31076; Tuasol 100, 32543;
201-729-9: orchindone, 22279;
201-730-4: Orchidee, 22278; sanfoin, 27364;
201-741-4: violuric acid, 33881;

201-748-2: Rhizopon A, AA, 26573;
201-754-5: iodol, 15215;
201-762-9: Piral, 23877; pyrogallic acid, 25510; pyrogallol, 25511; pyrogallol, 25512; pyrogallol, 25513; Pyrogallol, 25514;
201-771-8: sorbose, 29129;
201-774-4: FR-705, 12358; pentabromotoluene, 23131, 23132;
201-778-6: Dowicide 7, 8931; pentachlorophenol, 23135; pentachlorophenol, 23136; Santophen 20, 27417;
201-781-2: meat-sugar, 19196; muscle sugar, 20526; phaseomannite, 23579;
201-782-8: ACL 85, 90 Plus, 355; CDB 90, 5524; Fi-Clor, 11766; Queschlor, 25658; trichloroisocyanuric acid, 32218;
201-793-8: Ayrtol, 3514; chloroxylenol, 6159; Nipacide® MX, 21257;
201-794-3: Mesidine, 19398;
201-795-9: 2,4,6-trichlorophenol, 32222;
201-800-4: PVP, 25429;
201-803-0: 2-Furoic acid, 12497;
201-808-8: *o*-toluenesulfonamide, 31948;
201-814-0: Anti-oxidant 425, 2576; Cyanox® 425, 7485;
201-828-7: Ortholate, 22332;
201-841-8: Dibutylhydroquinone, 8368; Eastman® DTBHQ, 9466;
201-861-7: Dynamyte, 9361; Premerge, 24888; Tubotox, 32551;
201-865-9: Hager's reagent, 13535; Lyddite, 18068; Pertite, 23437; picric acid, 23821; reflorit, 26003;
201-877-4: Jayflex® TOTM, 15559;
201-928-0: Erythorbic acid, 10775; isoascorbic acid, 15338;
201-939-0: Fancol Menthol, 11449; menthol terpine hydrate, 19321;
201-943-2: pulegone, 25379;
201-944-8: thyme camphor, 31775; thymol, 31778;
201-945-3: resacetophenone, 26152;
201-958-4: phlorol, 23675;
201-961-0: salicylaldehyde, 27269;
201-962-6: mercufenol chloride, 19345;
201-963-1: *o*-anisidine, 2473;
201-972-0: Rose Ester, 26964;
201-983-0: Additin 30, 653; Akrochem® Antioxidant PANA, 1161; Antioxidant PAN, 2590; Naugard® PANA, 20798; Neozone A, 20982; Neozone D, 20985;
201-993-5: Dowicide 1, 8926; Nipacide® OPP, 21258; *o*-phenylphenol, 23648; Preventol® O Extra, 24931;
202-013-9: Actiron NX 3, 492;
202-016-5: GDL, 12705;
202-027-5: ketone base, 16140; Michler's ketone, 19646;
202-033-8: Benzhydrol, 3965;
202-041-1: dimethoxyphenol, 8500;
202-045-3: veratrole, 33568;
202-046-9: Decalin®, 7840;
202-049-5: naphthalene, 20706;
202-051-6: Leucol, 17024;
202-075-7: ethoxyquin, 11049; Polyflex, 24435; Santoquin, 27420;
202-077-8: Isatin, 15331;
202-078-3: β-methylnaphthalene, 19575;
202-085-1: quinaldine, 25678;
202-110-6: DAB, 7639;
202-143-6: Salt, Amido-R, 27295;
202-147-8: pyronine Y, 25523;
202-153-0: Cassella's acid, 5402;
202-154-6: Ebert and Merz's α-acid, 9495;
202-156-7: 2,3-naphthalenediol, 20710;
202-163-5: Biphenyl, 4216; diphenyl, 8592;
202-181-3: diphenyloxazole, 8602;
202-194-4: xanthene, 34589;
202-196-5: phenothiazine, 23618; PTZ® Phenothiazine Purified, 25378;
202-209-4: Schaeffer's acid, 27558;
202-213-6: jara jara, 15518; nerolin, 21007;
202-223-0: methyl eugenol, 19534;
202-230-9: Isothan Q-75, 15427;
202-249-2: Agerite® White, 1024;
202-252-9: creosol, 6990;
202-258-1: styrolyl alcohol, 29878;
202-259-7: methyl benzoate, 19515;
202-264-4: Astix, 3248; Duplosan New System CMPP, 9128; Duplosan® CMPP, 9130;
202-268-6: Vulkacit® 1000, 34124;
202-271-2: Kuron, 16496;
202-273-3: Trioxone, 32357;
202-280-1: Ablumide SDE, 81; Alkamide® DS-280/S, 1642; Alkamide® HTDE, 1643; Amldex SE, 2204; Karamide ST-DEA, 15752; Monamid® 718, 20117; Nopcogen 14-S, 21492;
202-281-7: Active #18, 503; Alkamide® DO-280, 1641; Alkamide® WRS 1-66, 1653; Amidex O, 2099; Crillon ODE, 7051; Emid® 6545, 10194; Hartamide 9137, 13659; Incromide OD, 14993; Laurel SD-400, 16840; Mackamide MO, 18160; Marlamid® D 1885, 18739; Mazamide® O 20, 19113; Ninol® 201, 21233; Norfox® F-221, 21549; Rewomid® DO 280, 26322; Rewomid® DO 280 SE, 26323; Schercomid ODA, 27653; Schercomid SO-A, 27661; Varamide® A-7, 33380;
202-307-7: Lexgard P, 17145; Nipasol M, 21282; propylparaben, 25231;
202-309-8: sodium *para*-aminohippurate, 28804;
202-310-3: Cadox® TDP, 4915; Cadox® TS-50S, 4916;
202-311-9: Nipabenzyl, 21248;
202-318-7: Butylparaben, 4797; Lexgard B, 17143; Nipabutyl, 21249;
202-319-2: Kodaflex® TEG-EH, 16279;
202-327-6: Abcure S-40-25, 35; AcetOxyl 2.5 and 5, 305; Aztec® Benzoyl Peroxide-70-77, 3553; Aztec® Benzoyl Peroxide-Dry, 3554; Benzoyl peroxide, 3983; Cadet® BPO-70W, 4891; Cadox® 40E, 4906; Cadox® BPO-W40, 4908; Cadox® BS, 4909; Cadox® BTA, 4910; Cadox® BTW-50, 4911; Clear by Design, 6385; Dermoxyl®, 8132; Florox, 12025; Lucidol, 17758; Lucidol 75FP, 17759; Lucidol GS, 17760; Lucidol RM, 17761; Lucidol-78, 17762; Lucilite, 17763; Lucipal, 17764; Luperco A, 17831; Luperco AC, 17832; Luperco AFR, 17833; Luperco AFR-250, 17834; Luperco AST, 17835; Luzidol, 18054; Nericur Gel 5, 21003; Novadeiox, 21657; Periygel®, 23275; Topex, 32043;
202-330-2: Chalkone, 5849;
202-345-4: safrol, 27229; shikimole, 28174;
202-360-6: Agrichem MCPA-25, 50, 1048; Agricorn 500, 1050; Agritox 50, 1077; Agroxone 50, 1086; Albar-M, 1351; Atlas MCPA, 3321; BH MCPA 75, 4119; Campbell's MCPA 25, 50, 5064; Campbell's MCPA 25, 50, 5065; Chafer MCPA 675, 5846; Empal, 10278; Farmon MCPA 50, 11487; FBC MCPA, 11516; MCPA, 19183; Mecpa, 19205; MSS MCPA 50, 20432; Phenoxylene 50, 23628; Phenoxylene Plus, 23629; Power MCPA, 24828; Quad MCPA 50%, 25566; Star MCPA, 29525;
202-361-1: Albar-40, 1349; Albar-Super, 1352; Atlas D, 3304; Campbell's Dioweed 50, 5058; Dicotox Extra, 8402; Dormone, 8810; Farmon 2,4-D, 11482; Fernimine, 11639; Fernoxone, 11642; For-Ester, 12244; MSS 2,4-D Amine, 20421; MSS 2,4-D Ester, 20422; Shell D 50, 28152; Syford, 30477; Verdone CDA, 33580;
202-365-3: Belmac Straight, 3913; Fisons 18-15, MCPB, 11850; MCPB, 19184, 19185; Tropotox, 32468;
202-366-9: 2,4-DB, 7810; Butoxone, 4752; Campbell's DB Straight, 5056; Embutox, 10039;
202-374-2: Keromet MD 60, MD 80, 16092; Metal Deactivator S, 19428;
202-377-9: Ethohexadiol, 10968; ethyl hexanediol, 11073;
202-387-3: Monosulfiram, 20257;
202-394-1: ADK STAB LA-32, 729; benzotriazole1H-benzotriazole, 3982; Preventol® CI 8, 24930;
202-397-8: Atlasol KAD, 3348; Crodazoline O, 7161; Crodazoline S, 7163; Hodag C-100-S, 14169; Miramine® GS, 19869; Schercozoline S, 27716; stearyl Hydroxyethyl imidazoline, 29606;
202-409-1: BBTS, 3863; Delac NS, 8001; Perkacit® TBBS, 23289; Santocure NS, 27402; Vanax® NS, 33292;
202-410-7: Morfax, 20314;
202-411-2: CBTS, 5484; cyclohexyl benzothiazole sulfenamide, 7523; Delac S, 8002; Durax®, 9181; Furbac, 12490; Perkacit® CBS, 23277; Santocure, 27400; Vancure D.A.A, 33308; Vulkacit® CZ/EGC, DZ/EGC, 34129;
202-414-9: Crodazoline O, 7162; Marlowet® 5440, 18928; Schercozoline O, 27715; Sovatex IM17H, 29160;
202-420-1: dimethylglyoxime, 8523;
202-422-2: *o*-xylene, 34623;
202-423-8: *o*-cresol, 6998;
202-425-9: Chloroben, 6117; Dizene, 8744;
202-426-4: *o*-chloroaniline, 6114;
202-430-6: *o*-phenylenediamine, 23640; Octolite 544, 22005;
202-443-7: toluhydroquinone, 31951;
202-462-0: 4-chlororesorcinol, 6145;
202-464-1: oxalic ether, 22416;
202-465-7: Durene, 9195;
202-467-8: 2,4,5-trichlorophenol, 32221; Dowicide 2, 8927;
202-473-0: Ageflex AMA, 966; Allyl methacrylate, 1784; Sipomer® AM, 28532; SR-201, 29360;
202-480-9: 2,3-dibromo-1-propanol, 8360; dibromo

propanol, 8361;
202-488-2: AB, 19;
202-490-3: diethyl ketone, 8436;
202-494-5: Dihydroxyacetone, 8459;
202-495-0: Thiovanol®, 31723;
202-496-6: methyl ethyl ketoxime, 19532;
202-500-6: methyl acrylate, 19509;
202-505-3: chlorothiophen2-chlorothiophene, 6152;
202-506-9: Akroform® ETU-22 PM, 1191; ethylene thiourea, 11132; ETU, 11158; Perkacit® ETU, 23283;
202-509-5: Agrisynth BLO, 1074; Agsol Ex BLO, 1090; BLO®, 4350; Butyrolactone, 4803; Dynasolve 699, 9389; GBL, 12704;
202-510-0: Texacar® EC, 31424;
202-525-2: Lowinox® 44S36, 17649; Rutenol, 27147; Ultranox® 236, 32846;
202-539-9: Telepaque, 31057;
202-567-1: Algafen, 1583; Algofen, 1602; Anthiphen, 2520; Dicestal, 8376; Dichlorophen, 8392; Ecco MP® 2004, 9513; Fungo®, 12483; Mascot Moss Killer, 19004; Nuophene, 21783; Panacide, 22652; Preventol®, 24926; Super Moss Killer & Lawn Fungicide, 30182; Super Mosstox, 30183; Super Mosstox, 30184;
202-577-6: Accelerator DT, 201; Akrochem® DOTG, 1167; D.O.T.G, 8813; Ditolyguanidine, 8732; Perkacit® DOTG, 23280; Vanax® DOTG, 33289;
202-592-8: Allantoin, 1739; Fancol TOIN, 11453;
202-597-5: ethyl methacrylate, 11077;
202-598-0: Solactol, 28883;
202-605-7: Ancazide IS, 2420; Monex, 20210; Mono Thiurad, 20216; Perkacit® TMTM, 23293; Thionex, 31703; TMTM, 31906; Unads®, 32907; Vulcaid 222, 34079;
202-607-8: Akrochem® TETD, 1180; Ancazide ET, 2419; Disulfiram, 8725; ethyl tuads®, 11085; ethyl tuex, 11086; ethylthiurad, 11144; Perkacit® TETD, 23291;
202-608-3: Crodasinic L, 7143; Hamposyl® L, 13600; lauroyl sarcosine, 16861; Vanseal® LS, 33357;
202-615-1: Butyl methacrylate, 4774;
202-617-2: Ageflex EGDMA, 969; ethylene glycol dimethacrylate, 11125; Perkalink® 401, 23315;
202-620-9: Borethyl, 4420;
202-625-6: QO® Tetrahydrofurfuryl Alcohol (THFA®), 25559; tetrahydrofurfuryl alcohol, 31371; THFA, 31660;
202-626-1: furfuryl alcohol, 12493; QO® Furfuryl Alcohol (FA®), 25555;
202-627-7: furfural, 12491; QO® Furfural, 25554;
202-637-1: Benzenesulfonamide, 3962;
202-640-8: CP0280, 6932; phenyltrichlorosilane, 23650;
202-647-6: Butyl-*m*-xylene, 4784; Lowinox® TBMX, 17675;
202-675-9: Butyl toluene, 4782; Lowinox® PTBT, 17673; 4-*t*-butyl phenol, 4800;
202-679-0: Lowinox® 070, 17650; Terbutol, 31228;
202-680-6: Lily of valley, artificial, 17258;
202-698-4: Accelerator 2P, 187; Akrochem® P.P.D, 1173; Vanax® 552, 33284;
202-700-3: Ajidew A-100, 1142; Pidolidone®, 23832;
202-704-5: cumene, 7352;
202-705-0: α-methylstyrene, 19579;
202-708-7: acetophenone, 300;
202-716-0: mirbane oil, 19935; nitrobenzene, 21348;
202-723-9: *m*-toluic acid, 31953;
202-724-4: *m*-Aminobenzoic Acid, 2167;
202-739-6: Mycose, 20551;
202-741-7: Gallicin, 12611;
202-746-4: Allisan, 1747; dicloran, 8395; Fumite Dicloran, 12459;
202-751-1: 3,5-dinitrobenzoic acid, 21349; Dinitrobenzoic acid, 8541;
202-759-5: carvone, 5363;
202-760-0: protocatechuic acid, 25289;
202-773-1: 1,3-diisopropyl benzene, 8467;
202-778-9: Carboxymethyl mercaptosuccinic acid, 5256; Evanacid® 3CS, 11251;
202-785-7: Lexgard M, 17144; methylparaben, 19577; Nigagin, 21171; Nipagin M, 21266;
202-787-8: Ethiodan, 10954; Iophendylate, 15234; Myodil, 20569; Pantopaque, 22695;
202-802-8: *p*-hydroxyacetophenone, 14658;
202-803-3: *p*-toluic acid, 31955;
202-810-1: *p*-nitroaniline, 21345;
202-811-7: *p*-nitrophenol, 21368; P.N.P, 22493;
202-819-0: dimethylaminobenzaldehyde, 8515;
202-826-9: 1,4-diisopropylbenzene, 8468;
202-831-6: Wurster's blue, 34576;
202-845-2: 2-diethylaminoethanol, 8440; Pennad 150, 23113;
202-851-5: styrol, 29875;
202-854-1: benzylamine, 3991; Sumine® 2005, 30110;
202-855-7: Benzonitrile, 3976;
202-859-9: Bentalol, 3946; benzyl alcohol, 3986;
202-860-4: Benzaldehyde, 3957; Ethereal Oil of Bitter Almonds, 10944;
202-864-6: Ronicol, 26948;
202-885-0: N-ethyl morpholine, 11140; Texacat® NEM, 31431; Toyocat® -NEM, 32106; Dabco® NCM, NEM, NMM, 7671;
202-898-1: Cyclamic acid, 7509; hexamic Acid, 14007;
202-905-8: Grasselerator 102, 13289; hexamethylamine, 14000; hexamethylene tetramine, 14006; hexamine, 14008;
202-908-4: Doverphos® 10, 8838; Doverphos® 10-HR, 8839; Lankromark® LE65, 16700; Syn-O-Ad® P-399, 30564; triphenyl phosphite, 32393; Weston® EGTPP, Weston® TPP, 34312;
202-910-5: Anilazine, 2465; Dairene®, 7698; Dyrene®, 9437;
202-915-2: Bidisin, 4132;
202-924-1: Nipaguard® TCC, 21270; TCC, 30856;
202-925-7: Atlas CIPC 40, 3303; Campbell's CIPC 40%, 5055; Chlorpropam, 6162; Mirvale, 19939; MSS CIPC, 20428; Residuren, 26166; Spud-Nic®, 29350; Warefog, 34215;
202-935-1: Baker P-8, 3596; Flexricin® P-8, 11961; glyceryl triacetyl ricinoleate, 13089; Naturechem® GTR, 20772;
202-936-7: Perkalink® 300, 23312; Perkalink® 301, 23313; triallyl cyanurate, 32191;
202-940-9: Mephaneine, 19323;
202-943-5: Ageflex CHMA, 968;
202-945-6: Farmon Mini Slug Pellets, 11489; Gastratox 6G Slug Pellets, 12693; Helarion, 13759; Meta, 19403; metaldehyde, 19431; Mifaslug, 19726; Mini Slugit Pellets, 19830; Mota, 20360; PBI Slug Pellets, 22936; Quad Mini Slug Pellets, 25567; Slug Destroyer, 28615; Slug Pellets, 28616; Slugit Liquid, 28619; Slugoids, 28620; Tripart® Mini Slug Pellets, 32381; Unicrop 6% Mini Slug Pellets, 32930; Vassgro Mini Slug Pellets, 33471;
202-951-9: UBOB®, 32699;
202-959-2: Michler's Base, 19644;
202-962-9: Variamine, 33387;
202-965-5: Agerite® Stalite, 1022; Agerite® Stalite® S, 1023; Naugalube® 438, 20783; Vanox® 12, 33335; Vanox® ODP, 33345;
202-966-0: Desmodur® VKS-2, VKS-4, VKS-18, 8160; Isonaphthol, 15377; Isonate® 125M, 15378; Isonate® 2125M, 15379; MDI, 19188; Rubinate® LF-168, 27108;
202-969-7: Flexzone 3C, 11967; N-Isopropyl-N'-phenyl-*p*-phenylene diamine, 15413; Permanax IPPD, 23364; Vanox® 3C, 33338; Vulkanox® 4010 NA, 34147;
202-974-4: Tonox® 22, Tonox® R, 32014;
202-981-2: diphenyl oxide, 8598; geranium crystals, 12887; phenyl ether, 23633;
202-984-9: Vanox® 6H, 33339;
202-988-0: Suconox-4®, 29923;
202-992-2: Kerobit® BPD, 16083; Naugalube® 403, 20782; Tenamine 2, 31091; Topanol® M, 32025;
202-995-9: carbanilic ether, 5179; Phenylurethane, 23652;
203-002-1: Akrochem® DPG, 1168; D.P.G, 7624; Dynamine, 9355; melaniline, 19259; Perkacit® DPG, 23281; Phenaldine, 23585; Vanax® DPG, 33290; Vulcaid DPG, 34081; Vulcogene ND, 34115; Vulkacit® D/C, 34130;
203-003-7: carbanilide, 5180; Zeonet U, 34722;
203-004-2: A-1, A-1 Thiocarbanilide, 5; Activit, 505; Akrochem® Thio No. 1, 1181; Anchoracel, 2426; diphenylthiourea, 8604; dithizone, 8731; Stabilizer C, 29407;
203-005-8: Phenol Carbonate, 23604;
203-007-9: Iphaneine, 15246;
203-041-4: Mazeen® 173, 19120; Neutrol® TE, 21036; Quadrol®, 25582; tetra(2-hydroxypropyl) ethylenediamine, 31356;
203-049-8: triethanolamine, 32240;
203-051-9: Kodaflex® Triacetin, 16281; triacetin, 32182;
203-052-4: Accelerator MF, 208; Amax® , Amax No 1, 1950; Delac MOR, 8000; OBTS, 21970; OMTS, 22161; Perkacit® MBS, 23284;
203-061-3: Syn-O-Ad® P-312, 30561; tributyl phosphite, 32208;
203-063-4: Armeen® 3-12, 2931;
203-065-5: cinnamoyl chloride, 6314;
203-066-0: Emperor alloy, 10289; Huntsman 201,

14469; Huntsman 240, 14470; Huntsman 312, 14471; Huntsman 331, 14472; Huntsman 351, 14473; Huntsman 474, 14474; Huntsman 765, 14475; Replay RP 2177, 26145; Replay RP 2236, 26146; Retain PS-4000, 26249; Rigipore, 26759;
203-079-1: 2-ethylhexyl acetate, 11138;
203-080-7: octyl acrylate, 22037;
203-090-1: Adimoll® DO, 693; dioctyl adipate, 8548; Good-rite® GP-223, 13201; Jayflex® DOA, 15554; Kodaflex® DOA, 16272; Monoplex® DOA, 20247; Palatinol® DOA, 22590; Plasthall® DOA, 24002; Plastomoll® DOA, 24058; Polycizer DOA, 24391; PX-238, 25440; Uniflex® DOA, 32945; Wickenol® 158, 34393;
203-091-7: Kodaflex® DOZ, 16275; Plasthall® DOZ, 24007; Priplast 3018, 25026;
203-098-5: toluylene, 31960;
203-103-0: Akrochem® Accelerator R, 1154; Naugex SD-1, 20801; Sulfasan, 29970; Vanax® A, 33287;
203-109-3: cinnamein, 6312;
203-111-4: Rapi-Cure EHVE, 25882;
203-117-7: Accelerator DBA, 200; Vulcaid 28, 34074;
203-139-7: phenetole, 23590;
203-148-6: phenylacetic acid, 23637; α-toluic acid, 31952;
203-149-1: Dabco® B-16, 7646; Pentamin BDMA etc, 23163;
203-154-9: Antisepsin, 2614;
203-157-5: Valadol, 33238;
203-166-4: *p*-methoxyphenylacetic acid, 19505;
203-180-0: Eltesol® TSX, 9876; Manro PTSA, 18600; Mod Acid, 20013; PTSA 70, 25377; toluene sulfonic acid, 31950;
203-192-6: Chlorphenesin, 6161;
203-205-5: Arizole® Anethole Extra., 2885;
203-212-3: cinnamyl alcohol, 6315; Peruvin, 23443; styrone, 29888;
203-213-9: cinnamaldehyde, 6311;
203-218-6: phenylmercuric oleate, 23645;
203-232-2: Dehyquart C Crystals, 7982; laurylpyridinium chloride, 16867; Ledmin LPC, 16960;
203-233-8: Armeen® L8D, 2952;
203-234-3: Aerofroth 88, 859; 2-ethylhexanol, 11136;
203-246-9: PTAL, 25375; *p*-tolyl aldehyde, 31961;
203-254-2: *p*-anisidine, 2474;
203-288-8: methyl cyanoacetate, 19524;
203-308-5: Pennzone E 0686, 23121;
203-310-6: Acetal, 280; Acetron®, Acetron® GP, Acetron® NS, 308; AT-20GF, 3282; Cadco® Acetal, 4884; Delrin® 100, 500, 8033; Delrin® 100AF, 500AF, 8034; Delrin® 100ST, 500T, 8035; Delrin® 107, 507, 8036; Delrin® 150A, 550SA, 8037; Delrin® 570, 8038; Delrin® 900, 8039; Delrin® AF Blend, 8040; Electrafil® J-80/CF/10/TF/10, 9739; Thermocomp® KB-1008, 31616;
203-313-2: Nipro (i), 21303;
203-315-3: Emalex PG-di-O, 9983;
203-326-3: Alperox-F, 1816; Laurox®, 16859; Laurydol, 16862;
203-327-3: QO® Furan, 25552;
203-327-9: Bisomer DBF, 4278; dibutyl fumarate, 8364;
203-328-4: Bisomer DBM, 4279; dibutyl maleate, 8365; Octomer DBM, 22008; PX-504, 25443;
203-341-5: Meraneine, 19330;
203-347-8: Emeressence® 1150, 10077; ethylene brassylate, 11121;
203-349-9: Uniflex® DCA, Unimate® DCA, 32943;
203-350-4: Adimoll® DB, 690; Cetiol® B, 5797;
203-351-5: Catinal CB-50, 5456; Dehyquart LDB, 7986; Retarder N, 26256; Vantoc CL, 33372;
203-363-5: Alkamuls® SDG, 1683; PEG-2 stearate, 23043;
203-364-0: PEG-2 oleate, 23042;
203-366-1: Ceroxin GL, 5758;
203-368-2: Alkamide® R-280, 1649; Rewomid® R 280, 26329;
203-369-8: Flexricin® 15, 11952; glycol ricinoleate, 13104;
203-375-0: Cephrol, 5695; citronellol, 6348;
203-376-6: citronellal, 6347;
203-377-1: Geraniol, 12885; Meranol, 19331;
203-378-7: Vernol®, 33600;
203-396-5: *p*-xylene, 34624;
203-398-6: *p*-cresol, 6999;
203-400-5: *p*-dichlorobenzene, 8386; Kaydox, 15811; Paradow, 22738; Santochlor, 27399;
203-401-0: *p*-chloroaniline, 6115, 6116;
203-404-7: Aminogen II, 2177; *p*-phenylenediamine, 23641;
203-412-0: Texacat® DMP, 31429;
203-419-9: dimethyl succinate, 8509;
203-439-8: Epichlorhydrin, 10632;
203-440-3: glycidyl acrylate, 13091;
203-441-9: glycidyl methacrylate 2,3-epoxypropyl methacrylate, 13094; SR-379, 29382;
203-442-4: Ageflex AGE, 965; Allyl glycidyl ether, 1783; Sipomer® AGE, 28531;
203-448-7: A-17, 7; Bu-Gas, 4660; Butane, 4734;
203-455-5: n-propyl mercaptan, 25227, 25228, 25229;
203-457-6: Barchlor, 3626;
203-458-1: ethylene dichloride, 11122;
203-466-5: Acrylonitrile, 436;
203-468-6: ethylenediamine, 11133;
203-470-7: propenol, 25181;
203-473-3: ethylene glycol, 11123; glycol, 13099; Ilexan E, 14883;
203-474-9: diformyl, 8453; glyoxal 40%, 13146; Protectol® GL 40, 25262;
203-483-8: taurine, 30845;
203-489-0: hexylene glycol, 14041, 14042; Isol, 15366;
203-490-6: Oxyneurine, 22463;
203-492-7: CH7310, 5841; H7310, 13530; Hexamethyldisiloxane, 14004;
203-497-4: CO9816, 6509; O9816, 21946;
203-500-9: Surfynol® 61, 30373;
203-508-2: Adogen® TA-100, 779; Adogen® TA-101, 780; Adogen® TA-101, 781; Arosurf® TA-100, 3081; Arquad® 218-100, 3111; Arquad® 218-75, 3112; Blandofen CT, 4318; Dehyquart DAM, 7985; Distearyldimonium chloride, 8714; Genamin DSAC, 12771; Prepagen WK, 24894; Varisoft® TA-100, 33428;
203-509-2: Sumquat® 6045, 30123;
203-514-5: Aztec® *t*-butyl peracetate-50 OMS, 60 OMS, 75 OMS, 3542; Esperox® 12MD, 10845; Lupersol 70, 17844; Trigonox® F-C50, 32300;
203-515-0: CA0900, 4854;
203-528-1: methyl propyl ketone, 19557;
203-538-6: methyl-glycocoll, 19570; sarcosine, 27463;
203-539-1: Arcosolv® PM, 2814; Dowanol® PM, 8896; Icinol PM, 14816; Poly-Solv® MPM, 24558; propylene glycol monomethyl ether, 25220; Solvenon® PM, Solvent PM, 29050;
203-542-8: Alkanolamine, 1701; Dabco® DMEA, 7669; deanol, 7829; dimethylethanolamine, 8520; Tegoamin® DMEA, 30995; Texacat® DME, 31428; Toyocat® -DMA, 32100;
203-544-9: 1-nitropropane, 21373; NiPar S-10, 21279;
203-545-4: Plyamul® 40305-00, 24266; Unocal 76 Res 6206, 33080; Unocal 76 Res S-55, 33086; vinyl acetate, 33853; Vinyl Acetate, 33854; Vinyl Acetate Monomer, 33855;
203-551-7: methyl amyl alcohol, 19512;
203-558-5: diisopropylamine, 8469;
203-563-2: isopropyl chloroformate, 15407;
203-564-8: acetic anhydride, 293;
203-570-0: succinic anhydride, 29916;
203-571-6: maleic anhydride, 18476;
203-572-1: Arconate® Propylene Carbonate, 2811; Solvenon® PC, 29049; Texacar® PC, 31425;
203-576-3: *m*-xylene, 34622;
203-577-9: *m*-cresol, 6997;
203-581-0: *m*-chloroaniline, 6113;
203-584-7: *m*-phenylenediamine, 23639;
203-585-2: Cohedur® RK, 6576; resorcin, 26228; resorcinol, 26230;
203-587-3: 2,6-Lutidine, 17995;
203-603-9: Arcosolv® PMA, 2815; Dowanol® PMA, 8897; Ektasolve® PM Acetate, 9672;
203-611-2: phloroglucinol, 23674;
203-613-3: Collidine, 6625;
203-618-0: cyanuric acid, 7494; isocyanuric acid, 15357; Zeonet A, 34720;
203-620-1: diisobutyl ketone, 8461; valerone, 33243;
203-624-3: methyl cyclohexane, 19525;
203-625-9: Dracyl, 8953; methyl benzene, 19514; toluene, 31946;
203-626-4: γ-picoline, 23817;
203-628-5: Abluton T30, 139; chlorobenzene, 6119;
203-629-0: cyclohexylamine, 7525;
203-630-6: Adronal, 792; cyclohexanol, 7521; Hexalin, 13998; Hydralin, 14536;
203-631-1: cyclohexanone, 7522; Sextone, 28118;
203-632-7: Carbolic Acid, 5206; phenic acid, 23592;
203-635-3: thiophenol, 31708;
203-636-9: β-picoline, 23816;
203-640-0: *p*-methyl morpholine, 19573; Texacat® NMM, 31432;
203-643-7: α-picoline, 23815;
203-646-3: 2-chloropyridine, 6144;
203-650-5: Aztec® *t*-Butyl Peroxyisobutyrate-75 OMS, 3546; Lupersol 80, 17845;

Trigonox® 41-C75, 32271;
203-661-5: Chemidex O, 5965; Incromine OPB, 15000; Lexamine O-13, 17099; Mackine 501, 18221; oleamidopropyl dimethylamine, 22109; Schercodine O, 27640;
203-664-1: Priplast 3013, 25025;
203-668-3: Butoxyethyl stearate, 4755; KP-23, 16376;
203-672-5: Kodaflex® DBS, 16265; Plasthall® DBS, 23994; Reomol DBS, 26129; Uniflex® DBS, 32942;
203-674-6: Pennzone B 0685, 23120;
203-677-2: n-valeric acid, 33242;
203-686-1: propyl acetate, 25203;
203-692-4: pentanen-pentane, 23164;
203-699-2: Amine C4, 2152;
203-700-6: Butyronitrile, 4805; nitrile C_4, 21338;
203-713-7: ethylene glycol methyl ether, 11127; methyl cellosolve®, 19521; methyl icinol, 19546; Poly-Solv® EM, 24557;
203-714-2: methylal, 19564;
203-716-3: diethylamine, 8439;
203-724-7: pyrrhol, 25533; Pyrrole, 25535;
203-726-8: Dynasolve 150, 9387; QO® Tetrahydrofuran (THF), 25558; tetrahydrofuran, 31370;
203-728-9: Pennodorant® 1013, 23117; tetrahydrothiophene, 31374;
203-729-4: Hopkin's lactic acid reagent, 14299; thiofurfuran, 31687; thiophene, 31707;
203-733-6: Aztec® Di-*t*-Butyl Peroxoide, 3557; di-*t*-butyl peroxide, 8369; Trigonox® B, 32295;
203-740-4: acid of amber, 334; sal succini, 27256; salt of amber, 27283; Succinellite, 29914; succinic acid, 29915;
203-742-5: maleic acid, 18475;
203-743-0: Fumaric acid, 12453;
203-744-6: Propamine D, 25170;
203-745-1: isobutyl acetate, 15345;
203-749-3: Crodasinic O, 7148; Hamposyl® O, 13604; Oleoyl sarcosine, 22123; Vanseal® OS, 33360;
203-750-9: M.B.A, 18103; methylene bisacrylamide, 19567;
203-751-4: Crodamol 1PM, 7116; Emerest® 2314, 10085; Estergel, 10883; isopropyl myristate, 15409; Kessco® IPM, 16115; Lexol IPM-NF, 17150; Liponate IPM, 17378; Promyr, 25152; Radia® 7190, 25729; Tegosoft® M, 31024; Unimate® IPM, 32971; Wickenol® 101, 105, 34378;
203-755-6: Abluwax EBS, 140; Acrawax® C, 383; Advawax® 290, 820; Alkamide® STEDA, 1652; Armowax EBS, 2985; Glycowax® 765, 13137; Kemamide® W-39, 15904; Kemwax, 15993; Nopcowax 22-DS, 21513; Radia® 7506, 25745; Uniwax 1760, 33069;
203-756-1: Advawax® 240, 819; Kemamide® W-20, 15903;
203-757-7: Adimoll® DH, 691;
203-759-8: Bumyr, 4670; Butyl myristate, 4775; Wickenol® 141, 34384;
203-766-6: methyl caprate, 19519;
203-767-1: methyl n-amyl ketone, 19552;
203-768-7: sorbic acid, 29105; Sorbistat, 29107;
203-775-5: 1,4-dibromobutane, 8357;
203-777-6: Exxsol® Hexane, 11361; n-hexane, 14009;
203-782-3: putrescine, 25423;
203-786-5: 1,4-butanediol, 4735; Dabco® BDO, 7647;
203-788-6: but-2-yne-1,4-diol, 4725;
203-794-9: Ansol E-121, 2494; monoglyme, 20227;
203-804-1: EE Solvent, 9636; Ethoxyethanol, 11045; ethyl icinol, 11074; Poly-Solv® EE, 24556;
203-806-2: cyclohexane, 7520;
203-808-3: Antepan, 2518; piperazine, 23873; Tasnon®, 30841; Upixon, 33094;
203-809-9: pyridine, 25481;
203-813-0: piperidine, 23874;
203-815-1: morpholine, 20346;
203-820-9: diisopropanolamine, 8466;
203-821-4: Dipropylene glycol, 8611;
203-825-6: Cosbiol®, 6865; Dermane, 8103; Robane®, 26870; Romane, 26928; squalane, 29357;
203-826-1: squalene, 29358; Supraene, 30301; Supraene®, 30302;
203-827-7: Ablunol GMO, 102; Aldo® MO, 1541; Capmul® GMO, 5112; glyceryl monooleate, 13084; Kemester® 2000, 15940; Kessco® GMO, 16111; Mazol® 300 K, 19134; Mazol® GMO, 19136; Monomuls® 90-O18, 20240; Radiasurf® 7150, 25801; Witconol 2421, 34500;
203-828-2: Alkamide® OIP, 1648; Mackamide OP, 18162; Rewomid® IPE 280, 26325; Schercomid OMI, 27655; Witcamide® 61, 34441;
203-838-7: heptanoic acid, 13791;
203-839-2: EE Acetate, 9635; Ethoxyethanol acetate, 11046; Sensolve EEA, 27954;
203-841-3: 3,3'-thiodipropionic acid, 31684;
203-845-5: ipomic acid, 15247; sebacic acid, 27795;
203-852-3: 1-hexanol, 14011; Epal® 6, 10615; Exxal® 6, 11334; hexyl alcohol, 14039;
203-856-5: glutaraldehyde, 13062; Protectol® GDA, GT 50, 25261; Relugan GT, 26046; Sonacide, 29073; Ucarcide® 225, 32711;
203-860-7: Shostakovsky Balsam, 28187;
203-865-4: D.E.H. 20, 7610; D.E.H. 52, 7612; diethylenetriamine, 8446;
203-867-5: Aminoethylethanolamine, 2172;
203-868-0: Dabco® DEOA-LF, 7667; Diethanolamine, 8434;
203-872-2: diethylene glycol, 8442; Her, 13798;
203-874-3: thiodiglycol, 31682;
203-881-1: ethylene glycol diacetate, 11124;
203-883-2: Ablumide SME, 82; Alkamide® S-280, 1650; Amidex SME, 2103; Mackamide SMA, 18167; Mazamide® SMEA, 19114; Monamid® S, 20122; Rewomid® S 280, 26330; Witcamide® 70, 34442;
203-884-8: Schercomid OME, 27654;
203-886-9: Ablunol EGMS, 100; Alkamuls® EGMS/C, 1664; Alkamuls® SEG, 1684; Emalex EGS-A, 9952; Monthyle, 20298; Schercemol EGMS, 27600;
203-887-4: Radia® 7185, 25727;
203-893-7: Neodene® 8, 20890; octene-1, 21991;
203-897-9: 1-heptanol, 13792; heptyl alcohol, 13797;
203-905-0: 2-butoxyethanol, 4753; Butyl Cellosolve®, 4767; Butyl Icinol, 4771; Dowanol® EB, 8894; Ektasolve® EB, 9668;
203-906-6: Dowanol® DM, 8891; Ektasolve® DM, 9666; methyl di-Icinol, 19526; Poly-Solv® DM, 24554;
203-911-3: methyl laurate, 19545;
203-916-0: Amine 8 D, 2125;
203-917-6: 1-octanol, 21988; capryl alcohol, 5157; Emery® 3322, 10166; Emery® 3324, 10168; Epal® 8, 10616; Lorol C8, 17616;
203-919-7: Carbitol®, 5194; Diethoxol, 8435; Dioxitol, 8576; Ektasolve® DE, DE-HG, 9664; Ethoxydiglycol, 11044; ethyl di-Icinol, 11068; Transcutol, 32137;
203-924-4: diglyme, 8455;
203-927-0: Adogen® 412, 767; Arquad® 12-37W, 3105; Arquad® 12-50, 3106; Chemquat 12-50, 6024; Dehyquart LT, 7987; Empigen® 5089, 10355; Octosol 562, 22020; Varisoft® LAC, 33423;
203-928-6: Ammonyx® CETAC, CETAC-30, 2261; Arquad® 16-29, 3107; Arquad® 16-50, 3108; Barquat® CT-29, 3667; Carsoquat® CT-429, 5349; Cetrimonium chloride, 5814; cetyl trimethyl ammonium chloride, 5826; Chemquat 16-50, 6025; Dehyquart A, 7980; Genamin CTAC, 12770; Incroquat CTC-30, 15037; Quatrex CTAC, 25624; Querton 16Cl-29, 25650; Radiaquat 6444, 25792; Rhodaquat® M242C/29, 26644; Varisoft® 250, 300, 355, 33407; Varisoft® 300, 355, 33408;
203-929-1: Arquad® 18-50, 3109; Octosol 474, 22019; Tomah Q-ST-50, 32000; Varisoft® ST-50, TSC, 33427; Varisoft® TSC, 33431;
203-930-7: CO9750, 6506;
203-931-2: Emery® 1202, 10138; Emery's L-114, 10187; Pelargonic acid, 23078;
203-933-3: 2-butoxyethanol acetate, 4754; Ektasolve® EB Acetate, 9669;
203-934-9: Tegosoft® S, 31027; Wickenol® 127, 34380;
203-938-0: Decanoyl chloride, 7844;
203-941-7: lauroyl chloride, 16860;
203-943-8: Armeen® DM12D, 2942; dimethyl lauramine, 8507; N,N-dimethyldodecylamine, 8519; Onamine 12, 22164;
203-950-6: Texacure EA-24, 31439; Texlin® 300, 31507; triethylenetetramine, 32245;
203-955-3: 1-Bromodecane, 4578; decyl bromide, 7868;
203-956-9: Epal® 10, 10617; Exxal® 10, 11338; Lorol C10, 17610;
203-961-6: Butyl Carbitol®, 4766; Butyl Di-icinol, 4769; Dowanol® DB, 8890; Ektasolve® DB, 9662;
203-962-1: Poly-Solv® TM, 24559;
203-965-8: undecylenic acid, 32916;
203-966-3: Emery® 2216, 10145; methyl palmitate, 19555; Radia® 7120, 25723;
203-968-4: dodecene-1, 8778; Neodene® 12, 20895;
203-970-5: Neodol® 1, 20908; Neoflex® 11,

20921;
203-982-0: 1-dodecanol, 8777; Emery® 3326, 10169; Emery® 3332, 10170; Emery® 3357, 10174; Epal® 12, 10618; Exxal® 12, 11339; lauryl alcohol, 16864; Lipocol L, 17353; Lorol C12, 17611; Lorol C12-C14, 17612; Lorol C8-C10 Special, 17617; Lorol Special, 17621; Philcohol 1200, 23664;
203-986-2: tetraethylenepentamine, 31367; Texlin® 400, 31508;
203-989-9: PEG-4, 23049;
203-990-4: Emery® 2218, 10146; Kemester® 4516, 15944; methyl stearate, 19560;
203-992-5: Emerest® 2301, 10081; Emery® 2219, 10147; Emery® 2301, 10155; Kemester® 104, 15935; methyl oleate, 19554; Priolube 1400, 25013; Witconol 2301, 34496;
203-997-2: Adma® 16, 743; Armeen® DM16D, 2943; Crodamine 3.A16D, 7106; N,N-dimethylhexadecylamine, 8524;
203-998-8: tridecyl alcohol, 32233;
204-000-3: Cachalot® M-43 NF, 4879; Dehydag Wax 14, 7943; Emery® 3334, 10171; Epal® 14, 10619; Lanette 14, 16661; Lanette 14, 16662; Lorol C14, 17613; myristyl alcohol, 20577; Philcohol 1400, 23666;
204-002-4: Adma® 14, 742; N,N-dimethyltetradecylamine, 8527; Empigen® AH, 10361; Onamine 14, 22165;
204-007-1: Distoline, 8724; Emersol® 210, 10120; Emersol® 6333 NF, 10123; Emersol® 7021, 10125; Industrene® 104, 15100; oleic Acid, 22112; Pamolyn 100 FGK, 22644; Pamolyn® 125, 22648; Priolene 6900, 25012; red oil, 25964;
204-009-2: Armid® E, 2961; Crodamide E, ER, 7096; erucamide, 10770; Kemamide® E, 15895; Petrac® Eramide®, 23464; Polydis® TR 131, 24412; Unislip 1753, 33044;
204-010-8: Hystrene® 5522, 14771; Hystrene® 9022, 14765; Prifrac 2989, 24956;
204-011-3: erucic acid, 10771; Prifrac 2990, 24957;
204-012-9: Neodene® 18, 20899;
204-015-5: Amine OL, 2162; Armeen® OL, 2956; Armeen® OLD, 2957; Jet Amine PO, 15595; Kemamine® P-989D, 15921; Radiamine 6172, 25757;
204-017-6: Adol® 62 NF, 783; Cachalot® S-56, 4881; Crodacol S70, 7071; Crodacol S95NF, 7072; Dehydag Wax 18, 7945; Emery® 3343, 10173; Epal® 18NF, 10627; Fancol SA, 11451; Lanette 18 DEO, 16664; Lipocol S, 17357; Lorol C18, 17615; Loxiol VPG 1354, 17684; Philcohol 1800, 23669; Rita SA, 26793; Stearal, 29591; stearyl alcohol, 29604; Varonic® BG, 33440;
204-020-2: Armeen® 2-18, 2927;
204-038-0: Liquapen, 17452; Penicillin G potassium, 23109; Penicillin G potassium, 23110;
204-043-8: Blattanex®, 4329; Blattanex® 20, 4330; Fumite Propoxur, 12465; Propoxur, 25198; Propoxur, 25199; Propoxur, 25200; propoxur, 25201; Suncide®, 30138; Unden®, Undene®, 32917;
204-060-0: metyridine, 19609; Promintic, 25143;
204-062-1: propylene, 25212;
204-065-8: dimethyl ether, 8505; Dymel®, 9331; methyl ether, 19529;
204-070-5: methyl butynol, 19518;
204-072-6: CE6350, 5530; ethyltrichlorosilane, 11147;
204-079-4: thiodan, 31680; Thionex, 31702;
204-082-0: Acarin, 183; Dicofol, 8398; Fumite Dicofol, 12460; Kelthane, 15881; Mitigan, 19959;
204-087-8: Bromcresol purple, 4567;
204-090-4: Lotusate, 17635;
204-101-2: AEPD®, 832; AEPD® 85, 833; 2-Amino-2-ethyl-1,3-propanediol, 2170;
204-104-9: Hercules® Improved Tech. PE, 13859; Hercules® Mono-PE, 13861; pentaerythritol, 23138, 28139; Pentaerythritol, 23139; Pentek, 23176;
204-110-1: Liponate PS-4, 17386; pentaerythrityl tetrastearate, 23142, 23143; Radia® 7176, Radiasurf® 7175, 25726;
204-112-2: Altal, 1869; Disflamoll® TP, 8642; Kronitex® TPP, 16437; triphenyl phosphate, 32391;
204-114-3: Dasanit®, 7787; Terracur® P, 31329;
204-116-4: Phanteine, 23569;
204-118-5: Genomoll P, 12838;
204-126-9: Electrafil® TR-1900/EC, 9757; Emralon 304, 10449; Emralon 8301-01, 10450; Fluon® AD1, AD1L, AD1H, 12054; Fluon® CDI, 12055; Fluon® G170, 12056; Fluorocomp® FC-101, 12094; Fluorocomp® FC-144, 12095; Fluorocomp® FC-174, 12096; Fluorocomp® FC-182, 12097; Fluorolene®, 12103; Fluoromelt®, 12106; Fluorosint® 500, 12110; Hostaflon® TF 1101, 14339; Hostaflon® TF 1620, 14340; Hostaflon® TF 2071, 14341; Hostaflon® TF 5032, 14342; Hostaflon® TF 5537, 14343; McLube 1777, 19180; MS-122, 20413; Polymist® F-5, 24496; polytetrafluoroethylene, 24640; polytetrafluoroethylene, 24641; Polytetrafluoroethylene, 24642; RT/Duroid® M, 27053; SLA 1611, 1612, 28581; SLA 1612, 28582;
204-132-1: Dantoin® MDMH, 7734;
204-134-2: Tedion V-18, 30919; tetradifon, 31366;
204-171-4: Neo Heliopan® OS, 20870; Tetrathal, 31393;
204-178-2: Ashlade TCNB, 3199; Bygran S, 4821; Fumite TCNB, 12467; Fumite TCNB Smoke, 12468; Hytec, 14775; Nebulin, 20839; New Hickstor 6, 21065; New Hystor, 21066; Quad Store, 25568; Quad-Keep, 25572; Tecgran, 30886; tecnazene, 30898; Tripart® Arena Granules, 32362; Tubodust, Tubostore, 32549;
204-187-1: quercetin, 25647; Ritacetin, 26799;
204-210-5: Dichlone, 8381;
204-211-0: Bis(2-ethylhexy) Phthalate, 4228; Bisoflex 81, 4249; Bisoflex 82, 4252; dioctyl phthalate, 8549; Kodaflex® DOP, 16273; Morflex 310, 20322; Morflex 410, 20325; NLA-20, 21396; Octoil, 22002; Palatinol® DOP, 22591; Plasthall® DOP, 24004; Plasticizer 28P, 24030; Polycizer DOP, 24390; PX-138, 25438; Reomol DCP, 26130; Vestinol AH, 33666; Witcizer 312, 34454;
204-212-6: Kodaflex® DMEP, 16270; Reomol P, 26132;
204-213-1: Dibutoxyethyl phthalate, 8363; Plasthall® 200, 23979;
204-258-7: 1,3-Dichloro-5,5-dimethyl hydantoin, 8389; Dantoin® DCDMH, 7732; Hydan, 14528;
204-259-2: phenyl salicylate, 23634;
204-260-8: Filtrosol A, 11800; Heliophan, 13766; Homosalate, 14292; Kemester® HMS, 15963;
204-263-4: Dermoblock OS, 8116; Escalol® 587, 10790; Uvinul® O-18, 33202;
204-265-5: ethyl salicylate, 11081; Sal-ethyl, 27264;
204-271-8: Veltol, 33544;
204-273-9: Julin's chloride, 15632;
204-274-4: chloranil, 6087; Vulklor, 34159;
204-275-6: anthrarufin, 2533;
204-277-0: uramil, 33113;
204-278-2: Fusarex, 12513; Hickstor, 14062; Hystor 10, 14754; Tripart® Arena 6, 32360;
204-278-6: 2,4,6-tribromophenol, 32202; Bromol, 4582; FR-613, 12357; Great Lakes PH-73, 13310;
204-284-9: *o*-toluic acid, 31954;
204-287-5: AA, 11; anthranilic acid, 2529;
204-288-0: *o*-hydroxyacetophenone, 14657;
204-289-6: Trotyl, 32474;
204-294-3: Jayflex® DTDP, 15555; PX-126, 25437;
204-295-9: Good-rite® GP-265, 13205; Morflex 125, 20316; Morflex 175, 20318; octyldecyl phthalate, 22042;
204-311-4: Cleve's ω-Acid or J-acid, 6412; Delta acid, 8044;
204-317-7: Gaultheria oil, 12701; Gaultheric acid, 12702; Teaberry Oil, 30872;
204-327-1: Anti-oxidant 2246, 2580; Cyanox® 2246, 7489; Lowinox® 22M46, 17652; Lowinox® 22M46, 17653; Ralox® 46, 25842; Ultranox® 246, 32847; Vulkanox® BKF, 34149;
204-335-5: Michler's hydrol, 19645;
204-337-6: ADK STAB 1413, 716;
204-340-2: tetrahydronaphthalene, 31373;
204-351-2: Cleve's β acid, 6410;
204-371-1: Sterilite Hop Defoliant, 29734;
204-375-3: β-naphthalene sulfonic acid, 20707; Dehscofix 918, 7926;
204-380-0: betapal Concentrate, 4063;
204-385-8: Nipacide® BCP, 21252;
204-390-5: Campbell's Redipon, 5070; Campbell's Redipon Extra, 5071; MSS 2,4-DP, 20424;
204-393-1: Ablumide LDE, 79; Alkamide® 327, 1634; Alkamide® LE, 1646; Amidex L-9, 2094; Amidex LD, 2095; Carsamide® SAL-7, 5323; Comperlan LD, 6676; Crillon LDE, 7049; Emalex

NN-7, 9974; Emid® 6519, 10193; Empilan® LDE, 10408; Hartamide LDA, 13661; Incromide L-90, 14987; Incromide LLT, 14990; Incromide LM-70, 14991; lauramide DEA, 16828; Mackamide L10, 18156; Mazamide® 1214, 19107; Mazamide® L-298, 19110; Monamid® 150-LMWC, 20115; Ninol® 30-LL, 21234; Schercomid 1214, 27646; Schercomid SL-Extra, 27659; Standamid® KDS, 29456; Varamide® ML-1, 33384; Witcamide® 5195, 34440;

204-399-4: ethyl *p*-hydroxybenzoate, 11061; Nipagin A, 21263;

204-402-9: Benzyl benzoate, 3987; Peruscabin, 23440;

204-403-4: Dibenzo GMF, 8353;

204-406-0: DPTT, 8952; Perkacit® DPTT, 23282; Sulfads®, 29960;

204-407-6: Benzoflex 2-45, 3970;

204-424-9: Altax®, 1872; M B T S, 18100; MBTS, 19172; Perkacit® MBTS, 23286; Vulkacit® DM/C, 34131;

204-427-5: kachin, 15651; pyrocatechin, 25497; pyrocatechol, 25498;

204-428-0: 1,2,4-trichlorobenzene, 32214; Hipochem GM, 14110;

204-450-0: D.N.T, 7622;

204-458-4: Amprol, 2364; Amprolium hydrochloride, 2365; Pancoxin, 22668;

204-464-7: Ethavan, 10937; ethyl vanillin, 11087; Rhodiarome, 26678; Vanbeenol, 33294;

204-465-2: vanilla, 33326; vanillin, 33327; Zimco, 34763;

204-473-6: metanilic acid, 19445;

204-479-9: benzethonium chloride, 3963; Hyamine® 10X, 14484; Hyamine® 1622 50%, 14485; Phemerol Chloride, 23581;

204-482-5: sulfanilic acid, 29966;

204-490-9: 2-amino-5-nitrothiazole, 2184; Entramin, 10595;

204-497-7: Fyfanon, 12533; malathion, 18460; Malathion 60, 18461; Prioderm, 25011;

204-498-2: Progallin P, 25111; propyl gallate, 25204; Sustane® PG, 30436;

204-514-8: Melilot, 19279;

204-516-9: Cumal, 7349; cuminaldehyde, 7354;

204-524-2: Dicofen, 8397; Fenitex, 11595; Fenitrothion EC, 11596; Folithion®, 12207; Micromite, 19683; Novathion, 21708;

204-526-3: Sumquat® 6050, 30124;

204-527-9: Ablumine 280, 85; Ammonyx® 4, 4B, 485, 4002, 2260; Amyx A-25-S 0040, 2391; Carsoquat® SDQ-25, 5350; Catinal OB-80E, 5458; Cycloton® SCS, 7542; Emcol® 4, 10044; Hetquat S-20, 13955; Incroquat SDQ-25, 15044; Mackernium SDC-25, 18210; Miracare® SCS, 19859; M-Quat® B-25, 20404; octadecylbenzenemethanaminium chloride, 21985; Quatrex STC-25, 25626; Rhodaquat® M270C/18, 26645; stearalkonium chloride, 29592; Stedbac®, 29609; Sumquat® 6210, 30126; Varisoft® SDAC-W, 33426;

204-528-4: triisopropanolamine, 32313;

204-534-7: Emerest® 2423, 10099; Kemester® 1000, 15938; Priolube 1435, 25019; Radia® 7363, 25739; triolein, 32351;

204-535-2: Boroflow, 4433; Gesatop, 12923; Herbazin 50, 13801; Simadex, 28459; Simanex, 28460; Simapron, 28461; simazine, 28464; Simflow, 28471; Sinazine, 28493; Syngran, 30543; Weedex S2, 34259;

204-539-4: OA-505, 21954;

204-542-0: MSS IPC 50, 20431; Propham, 25185, 25186;

204-546-2: *m*- cresyl acetate, 7011; Kresatin, 16402;

204-552-5: phosphorus ether, 23742; triethyl phosphite, 32244;

204-555-1: benzylidene acetone, 3992;

204-556-7: phenoxyacetic acid, 23626;

204-558-8: Monoplex® DOS, 20248; Octoil S, 22003; Plasthall® DOS, 24005; Reomol DOS, 26131; Uniflex® DOS, 32946;

204-566-1: styracin, 29860;

204-580-8: Glyconyl, 13124;

204-589-7: Emeressence® 1160 Rose Ether, 10079; Igepal® Cephene Distilled, 14850; Igepal® OD-410, 14873; Phenoxetol, 23622; 2-Phenoxyethanol, 23627; Prox-onic PH-01, 25334; Rewopal® MPG 10, 26344;

204-593-9: Acetoquat CPC, 302; cetyl pyridinium chloride, 5823;

204-602-6: *p*-anisaldehyde, 2470;

204-608-9: Butyrone, 4804;

204-614-1: Argus DLTDP, 2861; Carstab® DLTDP, 5356; Cyanox® LTDP, 7491; dilauryl thiodipropionate, 8476; Evanstab® 12, 11255; Lankromark® DLTDP, 16689; Lowinox® DLTDP, 17666; Wytox® LT, 34582;

204-616-2: *p*-aminophenol, 2185; Rodinal, 26900; Takatol, 30725; Unal, 32908;

204-617-8: Black and White Bleaching Cream, 4302; Eldopaque, 9723; hydroquinone, 14632; quinol, 25685; Tecquinol® Tech. Grade, 30910;

204-619-9: Bos MH, 4456; Burtolin, 4722; Chiltern Fazor, 6067; Malazide, 18464; maleic hydrazide, 18477; Mazide 25, 19129; MSS MH18, 20433; Regulox K, 26020; Royal Slo-Gro, 27034; Sucker Stuff, 29922;

204-623-0: propanal, 25172;

204-626-7: Diacetone alcohol, 8249; Pyraton, 25472;

204-629-3: Girard's reagent T, 12950;

204-634-0: acetylacetone, 314;

204-635-6: Lubrizol® 2153, 17745; succinimide, 29917;

204-639-8: paraldehyde, 22759;

204-646-6: Butyraldehyde, 4801;

204-649-2: levulinic acid, 17084;

204-650-8: Azodicarbonamide, 3524; Celogen® AZ 120, 130, 150, 180, 199, 5642; Ficel® AC2, 11762; Genitron, 12834; Kempore® 60/14FF, 15987; Porofor® ADC/E, 24698; Santechem 21-21, 27382;

204-653-4: glycol dimercaptoacetate, 13101;

204-659-7: Agallol, 954; Atiran, 3286; Baytan®, 3851; Ceresan, 5745;

204-661-8: Dioxane, 8574;

204-662-3: isoamyl acetate, 15336; Jargonelle pear essence, 15521; pear oil, 22971;

204-663-9: thiodiglycolic acid, 31683;

204-666-5: ADK STAB LS-8, 732; butyl stearate, 4781; Emerest® 2325, 10088; Kemester® 5510, 15948; Kessco® BS, 16103; Lexolube® BS-Tech, 17163; Oleo-Coll LP, 22118; Priolube 1451, 25020; Radia® 7051, 25718; Uniflex® BYS-Tech, Unimate® BYS, 32941; Witconol 2326, 34497;

204-669-1: Azelaic acid, 3519; Emerox® 1110, 10115; Emery's L-110, 10186;

204-672-8: Bretol®, 4506; Sumquat® 6020, 30121;

204-673-3: Adipic acid, 698; Adi-pure®, 706;

204-677-5: caprylic acid, 5158; Emery® 657, 10136; Prifrac 2901, 24950;

204-680-1: Emery® 2214, 10144; methyl myristate, 19550;

204-685-9: diethylene glycol butyl ether acetate, 8443;

204-690-6: Amine 10, 2126; Amine 12, 2127; Amine 12-98D, 2128; Armeen® 12, 2933; Armeen® 12D, 2934; Radiamine 6164, 25754;

204-693-2: Crodamide S, SR, 7098; Kemamide® S, 15900; Petrac® Vyn-Eze®, 23469; stearamide, 29593; Uniwax 1750, 33068;

204-694-8: Adma® 18, 744; Adogen® MA-108 SF, 776; Amine 2M18D, 2139; Armeen® DM18D, 2944; Crodamine 3.A18D, 7107; Dymanthine, 9327; Kemamine® T-9902, 15933; Onamine 18, 22167;

204-695-3: Amine 18-90, 2131; Armeen® 18, 2938; Armeen® 18D, 2939; Armid® HTD, 2963; Crodamine 1.18D, 7102; Kemamine® P-990, P-990D, 15922; Steamfilm FG, 29589;

204-696-9: carbon dioxide, 5222; Cardice, 5259; Dricold, 9026; Drikold, 9032; dry ice, 9057;

204-704-0: chlorobromoform, 6121;

204-709-8: 2-amino-2-methyl-1 propanol, 2182; 2-amino-2-methyl-1,3-propanediol, 2183; AMP, 2300; AMP-95, 2302; AMPD, 2304; AVT-75, 3504;

204-710-3: CV-4772, 7448;

204-718-7: Ledercort, 16954;

204-766-9: Bromochloro dimethyl hydantoin, 4574; Dantoin® GSD-550, 7733; Halobrom, 13560; Quesbrom, 25657;

204-769-5: S.S.T® Sump Saver Tablets, 27180; Tris Nitro®, 32408;

204-772-1: Soa, 28682; sucrose octa-acetate, 29935;

204-781-0: neopentyl glycol, 20937; NPG® Glycol, 21751;

204-782-6: Abrodil, 158; Methiodal sodium, 19485; Skiodan Sodium, 28570;

204-786-8: itrol, 15433; silver citrate, 28422;

204-794-1: Dipentaerythritol, 8581;

204-799-9: tris (2,3-dibromopropyl) phosphate, 32402;

204-800-2: Antifoam T, 2551; Kronitex® TBP, 16434; Phos-Ad 100, 23686; Syn-O-Ad® 8412, 30553; tributyl

phosphate, 32207;
204-803-9: CB2405, 5474;
204-809-1: Surfynol® 104, 30376; Surfynol® 104A, 30377; Surfynol® 104BC, 30378; Surfynol® 104E, 30379; Surfynol® 104H, 30380; Surfynol® 104PA, 30381; Surfynol® 104PG, 30382; Surfynol® 104S, 30383; Surfynol® DF-110D, DF-110L, 30392; Surfynol® PC, 30403; Surfynol® PG-50, 30404; Surfynol® TG, 30407; Surfynol® TG-E, 30408;
204-812-8: Empicol® 0585/A, 10302; Niaproof® Anionic Surfactant 08, 21092; Rhodapon® BOS, 26632; Sulfetal 4105, 29981;
204-818-0: Baypren® 110, 3828; Baypren® 110 VSC, 3829; Baypren® 216, 3830; Baypren® 310, 3831; Baypren® AT-H, AT-M, AT-S, 3832; Baypren® EM1, 3833; Baypren® Latex KA 8348, 3834; Baypren® Latex L 200A, 3835; Baypren® M1, 3836; chloroprene, 6136; Daubond DC-9300, 7794; Neoprene Latex 115, 20952; Neoprene NPG 6856, 20955; polychloroprene, 24386;
204-822-2: potassium acetate, 24733, 24734; sal diureticum, 27246;
204-823-8: Niacet Sodium Acetate Anhyd. Tech, 21089; sodium acetate, 28726;
204-824-3: pyruvic acid, 25537, 25538;
204-825-9: Acetosol, 304; Distillex DS4, 8720; Dowper, 8942; Nema, 20861; perchlorethylene, 23220; perchloroethylene, 23222; Perclene, 23225; Perclene TG, 23226; Perklone, 23321;
204-828-5: Basfapon®, 3705; Dowpon, 8943;
204-832-7: Abalyn, 24;
204-839-5: Aerosol® IB-45, 891; Aerosol® IB-45, 892; Bevaloid 6423, 4083; Geropon® CYA/DEP, 12906; Monawet MB-45, 20178; Monawet MB-45, 20179; Octosol IB-45, 22026; Rewopol® SBDB 45, 26371;
204-840-0: xanthophyll, 34591;
204-847-9: Chloramine B, 6085;
204-854-7: Antibacterin, 2537; Chloramine T, 6086; Chlorozone, 6160; Ketjensept, 16133; Tochlorine, 31916;
204-857-3: Ludigol F, 17772; sodium *m*-nitrobenzene sulfonate, 28795;
204-872-5: β-l-pinene, 23858; β-pinene, 23859;
204-875-1: Aquatreat KM, 2760;
204-876-7: Aquatreat SDM, 2761; methyl namate®, 19551; Octopol SDM-40, 22014; Perkacit® SDMC, 23288; sodium dimethyldithiocarbamate, 28758; Thiostop N, 31717;
204-878-8: N-chlorosuccinimide, 6148;
204-881-4: Anti-Oxydant Bayer, 2595; BHT, 4122; Butylated hydroxytoluene, 4788; DBPC, 7813; Deenax, 7874; Ionol, 15226; Lowinox® BHT, 17664; Nipanox® BHT, 21273; Oxyguard, 22457; Ralox® BHT food grade, 25845; Spectratech® CM 11340, KM 11264, 29240; Tenamine 3, 31092; Vanox® PCX, 33346; Vulkanox® KB, 34152;
204-884-0: 2,6-di-*t*-butylphenol, 4799; 2,6-dibutylphenol, 8370; Isonox® 103, 15382; Lowinox® 001, 17646;
204-886-1: saccharin sodium, 27196;
204-889-8: Surfak, 30362;
204-933-6: mercurochrome, 19354;
204-936-2: Tandearil, 30795;
204-949-3: Naganol, 20660;
204-956-1: acetrizoate sodium, 307;
204-976-0: Aerosol® NS, 894;
204-979-7: D.T.S., 7626;
205-011-6: dimethyl phthalate, 8508; Fermine, 11631; Kemester® DMP, 15956; Kodaflex® DMP, 16271; Palatinol® M, 22593; Unimoll® DM, 32979; Uniplex 110, 33009;
205-012-1: menadiol sodium diphosphate, 19319; Synkavit, 30548; Synkayvite, 30549;
205-014-2: DCP, 7818; Dicapryl phthalate, 8375;
205-016-3: Dapon 35, 7737; Daponite Sheet, 7739; diallyl phthalate, 8283; Nonflammable Decobest DA, 21448; RX® 1-501N, 27154; RX® 3-1-530, 27156;
205-024-7: Angiovist 282, 2457; Hypaque Meglumine, 14705; MD 60, 19187; Reno-M, 26113;
205-026-8: Cyasorb® UV 24, 7496; Spectra-Sorb UV 24, 29230;
205-027-3: Benzophenone-6, 3978; Uvinul® D-49, 33194;
205-028-9: Benzophenone-2, 3977; Uvinul® D-50, 33195;
205-029-4: Benzoresorcinol, 3981; Uvinul® 400, 33191;
205-031-5: Cyasorb® UV 9, 7495; Escalol® 567, 10789; Lankromark® LE296, 16698; Lankromark® LE296, 16699; Neo Heliopan® BB, 20867; Rhodialux A, 26675; Ritaphenone 3, 26823; Spectra-Sorb UV9, 29229; Syntase® 62, 30621; Uvinul® M-40, 33197;
205-032-0: 2-methylbenzophenone, 19565;
205-037-8: Dipicrylamine, 8607; Hexil, 14031;
205-038-3: Explosive D, 11314;
205-055-6: Dowicide A, 8932;
205-067-1: atoquinol, 3380; cinchophen, 6307;
205-079-7: naphthoresorcin, 20726;
205-087-0: Captan Granular, 5164; Captan, Captan-Col, Captan-50, Captan-83P, Captan Granular, 5165; Hormone Rooting Powder, 14310; Merpan, 19369; Orthocide, 22328; PP Captan 83, 24839; Vancide® 89, 33302;
205-088-6: Folpan, 12209; Folpet, 12210; Phaltan, 23567;
205-097-5: Hippuran-:ss131:ksl, 14125; Hipputope, 14128; Nephroflow, 20989;
205-101-5: Chryzoplus, Chryzopon, Chryzosan, Chryzotek, 6258; 4-indol-3-ylbutyric acid, 15086; Seradix, 28005;
205-105-7: paratartaric acid, 22803; sal tartar, 27257; tartaric acid, 30832;
205-107-8: Akrochem® Peptizer PTP, 1170; Renacit® 7, 26103;
205-109-9: Nipacide® DX, 21255; Nipacide® PX, 21260;
205-123-5: Naptol, 20739;
205-124-0: Diosal, 8572;
205-126-1: Cevalin, 5827; Liqui-Cee, 17460; sodium ascorbate, 28734; vitamin C sodium salt, 33939;
205-129-8: Neo Heliopan® MA, 20869;
205-132-4: methyl anthranilate, 19513;
205-137-1: Cryptonol, 7282; Quinosol, 25687;
205-138-7: Aminogen I, 2176; α-naphthylamine, 20729;
205-145-5: Monacrin, 20097;
205-146-0: Cleve's γ-Acid, 6411;
205-149-7: diethyl toluamide, 8438;
205-182-7: Antioxygene BN, 2600; hydronaphthol, 14623;
205-183-2: cupferron, 7365;
205-202-4: Odylen®, 22046;
205-223-9: Antioxidant PBN, 2591; Antioxygene MC, 2603; P.B.N, 22488; Stabilator A.R, 29400;
205-232-8: Accelerator BZ Powder, 198; Butasan Vulcanization Accelerator, 4741; Butazate, 4742; Butazate 50D, 4743; Butyl Zimate®, 4783; Octocure ZDB-50, 21995; Perkacit® ZDBC, 23295;
205-234-9: Amidex CP, 2091; Monamid® 150-CW, 20114; Standamid® CD, 29454;
205-238-0: Butyl Namate®, 4776; Octopol NB-47, 22012;
205-239-6: Dowicide B, 8933;
205-241-7: Eastman® Inhibitor RMB, 9470; Resorcinol benzoate, 26231;
205-251-1: Octoate Z, 21992;
205-252-7: Butyl benzoate, 4763; Hipochem B-3-M, 14106; Marvanol® Carrier BB, 18980; n-Butyl benzoate, 4790;
205-265-8: Retrocure® G, 26279; Vulkalent® TM, 34141;
205-271-0: Hodag C-100-L, 14167; lauryl hydroxyethyl imidazoline, 16865; Mackazoline L, 18200; Schercozoline L, 27714;
205-278-9: Pantholin, 22693; Ritapan CAP, 26817; vitamin B_5 calcium salt, 33936;
205-281-5: Closyl LA 3584, 6443; Crodasinic LS30, 7144; Crodasinic LS35, 7145; Gardol®, 12668; Hamposyl® L-30, 13601; Maprosyl® 30, 18671; Medialan LD, 19214; Sarkosyl NL30, 27465; Secosyl, 27823; sodium lauroyl sarcosinate, 28779; Tensyl 30, 31205; Vanseal® NALS-30, 33359; Zoharsyl L-30, 34876;
205-283-6: phlorone, 23676;
205-285-7: Adinol OT, 697; Arkopon T, 2898; Fenopon T-33 and T-43, 11608; Fenopon T-51, 11609; Fenopon T-77, 11610; Geropon® T-22/A, 12915; Geropon® T-33, 12916; Hostapon SO, 14395; Hostapon T Powd, 14397;
205-286-2: Agrichem Flowable Thiram, 1047; Akrochem® TMTD, 1182; Ancazide ME, 2421; Fernide, 11638; FS Thiram 15% Dust, 12427; Hortag Thiram, 14317; methyl tuads®, 19562; Perkacit® TMTD, 23292; Pomarsol®, 24677; Rezifilm, 26521; Spotret 75 WDG, 29329; Thiram, 31726; Thiurad, 31730; Tuads®, 32542; Tuex, 32553;
205-287-8: Akrochem® Cu.D.D, 1165; copper dimethyldithiocarbamate, 6779;

Cumate®, 7351; Perkacit® CDMC, 23278;
205-288-3: AAprotect, 16; Accelerator MZ Powder, 209; Ancazate ME, 2416; Methasan, 19478; Methazate, 19480; methyl zimate®, 19563; Octocure ZDM-50, 21997; Perkacit® ZDMC, 23297; Ultra Zinc DMC, 32782; Vancide® MZ-96, 33303; Zimate®, 34762;
205-290-4: Luprosil® Sodium Salt, 17922; sodium propionate, 28812; Spac, 29182;
205-293-0: metam-sodium, 19444; Sistan®, 28550; Unifume, 32954;
205-302-8: Lida-Mantle, 17213; lidocaine, 17216;
205-309-6: Accelerator A22, 192;
205-316-4: Butyl lactate, 4773;
205-324-8: Cetats®, 5786;
205-341-0: Achilles Dipentene, 323; Dipentene, 8584;
205-343-1: nitro base, 21341; Vulcaniline, 34095;
205-345-2: betazole hydrochloride, 4072;
205-348-9: sodium cyclamate, 28754;
205-352-0: BTC® 2565, 4639; BTC® 824, 4636; Catigene DC/100, 5445; Catigene® DC 100, 5455; Exameen 824 3724, 11284; FMB 65-15, 65-28 , 12135; JAQ Powdered Quaternary, 15517;
205-355-7: Hampshire® NTA Acid, 13611; nitrilotriacetic acid, 21339; TG Buffer, 31547;
205-358-3: disodium EDTA, 8644; edetate disodium, 9613; Endrate, 10553; Hamp-Ene® Na2, 13588; Questal DI 0770, 25661; Sequestrene® NA2, 27997; Sequestrene® NA2 Edetate USP, 27998; sodium versenate, 28831; Trilon® BD, 32325;
205-359-9: propazine, 25179; Prozinex, 25350;
205-360-4: Hampshire® DEG, 13608; sodium dihydroxyethyl glycinate, 28757;
205-364-6: Rheocin, 26557;
205-380-3: Niaproof® Anionic Surfactant 4, 21093;
205-381-9: Cheelox® 120, 5866; Cheelox® HE-24, 5868; Chel DM-41, 5872; Emkasene 800, 10259; Hamp-Ol® 120, 13594; Hamp-Ol® Crystals, 13596; Kalex OH, 15672; Questal FEC 0800, 25663; Trilon® D Liq, 32329;
205-388-7: Akyposal TLS42, 1322; Avirol® T 40, 3485; Carsonol® TLS, 5340; DeSonol T, 8191; Elfan® 240T and 240T/S, 9815; Empicol® TL40, 10342; Lorol TA and TAR, 17622; Maprofix TLS, 18666; Marlinat® DFL 40, 18852; Naturechem® MAR, 20773; Neopon LT, 20950; Norfox® TLS, 21555; Nutrapon TLS-500, 21827; Perlankrol® ATL40, 23324; Rewopol® TLS 40, 26383; Rhodapon® LT-6, 26638; Standapol® T, 29481; Stepanol® SPT, 29697; Sulfetal KT 400, 29985; Sulfochem TLS, 30016; TEA-lauryl sulfate, 30873; Tensopol LT, 31193; Texapon T 42, 31480; Ufarol TA-40, 32744; Witcolate TLS-500, 34479; Zoharpon LAT, 34869;
205-391-3: Cheelox® 80, 5864; Chel DTPA-41, 5874; Hamp-Ex® 80, 13591; Kalex Penta, 15673; Pentaquest Extra 0685, 23169; Pentaquest OPNA 0256, 23171; Pentasodium pentetate, 23172; Polyquest 80, 24533; Trilon® C Liq, 32328;
205-392-9: Flexricin® P-4, 11959; Methyl acetyl ricinoleate, 19508;
205-393-4: B.A.R, 3561;
205-398-1: cinnamic acid, 6313;
205-399-7: benzyl acetate, 3985; Plastolin I, 24051;
205-411-0: Aminoethylpiperazine, 2173;
205-413-1: Narceol, 20742;
205-414-7: Aminitrazole, 2164;
205-419-4: Bayer 5072, 3767;
205-427-8: Estragole, 10897;
205-428-3: Acetoquat CPB, 301;
205-443-5: Aero 343 Xanthate, 842;
205-450-3: diisobutyl adipate, 8460; Plasthall® DIBA, 23996;
205-455-0: Aldo® MR, 1542; Flexricin® 13, 11951; glyceryl monoricinoleate, 13085; Hodag GMR, 14183; Mazol® GMR, 19137; Softigen® 701, 28854;
205-468-1: Glaurin, 12978; PEG-2 laurate, 23041;
205-469-7: Avistin® PN, 3489; Chemical 39 Base, 5959; Marlamid® A 18, 18738;
205-470-2: Flexricin® 100, 11954; P® -10 Acid, 22499; ricinoleic acid, 26734;
205-471-8: Cenwax® ME, 5693; Naturechem® MHS, 20774; Paricin® 1, 22833;
205-472-3: Flexricin® P-1, 11957; methyl ricinoleate, 19558;
205-480-7: n-Butyl Acrylate, 4786;
205-483-3: Colamine, 6593; ethanolamine, 10930;
205-488-0: Formax, 12268; sodium formate, 28767;
205-489-6: CP0800, 6935; n-propyltrichlorosilane, 25232;
205-491-7: CD3780, 5501; D3780, 7636;
205-500-4: ethyl acetate, 11063;
205-503-0: malonic acid, 18491;
205-507-2: 2,6-diaminopyridine, 8305; diaminopyridine, 8306;
205-513-5: Hexetidine, 14030; Oraldene, 22263; Sterisil, 29739;
205-520-3: 1-hydroxy-2-pyridine, 14669;
205-525-0: calcium oleate, 4961;
205-526-6: Ablunol GML, 101; Aldo® MLD, 1540; glyceryl monolaurate, 13083; Grindtek ML 90, 13410; Hodag GML, 14180; Imwitor® 312, 14941; Kessco® GML, 16110; Monomuls® 90-L12, 20239;
205-530-8: Acetamide MEA, 283; Amidex AME, 2086; Carsamide® AMEA, 5320; Incromectant AMEA-100, 14976; Lipamide MEAA, 17326; Mackamide AME-75, AME-100, 18152; Schercomid AME, 27648;
205-535-5: Rhodapon® OLS, 26639;
205-538-1: MSG, 20416;
205-539-7: Hamposyl® S, 13605;
205-541-8: Alkamide® LIPA/C, 1647; Amidex LIPA, 2096; Empilan® LIS, 10409; Monamid® LIPA, 20119; Rewomid® IPL 203, 26326;
205-542-3: Emalex PGML, 9985; Schercemol PGML, 27620;
205-547-0: Campbell's Nabam Soil Fungicide, 5066;
205-550-7: caproic acid, 5127;
205-553-3: cupric acetate, 7372; verdigris, 33571;
205-554-9: magnesium acetate, 18357;
205-559-6: Butyl oleate, 4778; Butyl Oleate C-914, 4779; Emerest® 2328, 10089; Kemester® 4000, 15943; Plasthall® 503, 23989; Priolube 1405, 25014; Uniflex® BYO, 32940; Witcizer 100, 34453;
205-560-1: Ablumide LME, 80; Alkamide® L-203, 1644; Amidex LMMEA, 2097; Crillon LME, 7050; Empilan® LME, 10410; Hartamide LMEA, 13662; Incromide LCL, 14989; Mackamide LMM, 18158; Monamid® LMA, 20120; Ninol® LMP, 21237; Rewomid® L 203, 26328;
205-563-8: Exxsol® Heptane, 11360; n-heptane, 13790;
205-568-5: Atlasol 103, 3341; Empicol® 0758, 10303;
205-569-8: Amine 16D, 2130;
205-570-6: Ageflex FM-12, 986;
205-571-1: Emerest® 2316, 10086; isopropyl palmitate, 15410; Kessco® IPP, 16116; Lexol IPP, 17151; Liponate IPP, 17379; Propal, 25168; Radia® 7200, 25730; Tegosoft® P, 31026; Unimate® IPP, 32972; Wickenol® 111, 34379;
205-577-4: DEA-lauryl sulfate, 7827;
205-582-1: Emery® 650, 10134; Hystrene® 9512, 14766; lauric acid, 16856; Philacid 1200, 23660; Prifrac 2920, 24952;
205-590-5: Emkapol PO-18, 10248; Marley Cement Waterproofer, 18829; Octosol 449, 22018; potassium oleate, 24763, 24764;
205-591-0: Eunatrol, 11194; sodium oleate, 28802;
205-594-7: Ansol E-181, 2495;
205-596-8: Armeen® 16, 2935; Armeen® 16D, 2936, 2937; Crodamine 1.16D, 7101; Kemamine® P-880D, 15919;
205-597-3: Adol® 85, 786; Cachalot® O-15, 4880; Dermaffine®, 8093; Emery® 3312, 10162; Emery® 3317, 10163; Fancol OA-95, 11450; HD-Echelon 90/95, 13718; HD-Eutanol, 13719; Lipocol O, 17355; Novol, 21730; oleyl alcohol, 22131;
205-599-4: Cymag, 7577; sodium cyanide, 28753;
205-609-7: phenol red, 23605; phenolsulfonphthalein, 23611;
205-619-1: TMPD® Glycol, 31905;
205-620-7: arrhenal, 3128; Arsinette, 3145; Methar 30, 19477;
205-624-9: methyl selenac, 19559;
205-633-8: Baking soda, 3600; Bufferight, 4656; Col-Evac, 6601; saleratus, 27262; sodium bicarbonate, 28736; Tronacarb Sodium Bicarbonate, 32461;
205-634-3: oxalic acid, 22413; oxalic acid dihydrate, 22414;
205-643-2: Astryl, 3267;
205-696-1: mesotartaric acid, 19400; *meso*-tartaric acid, 30831;
205-702-2: proline, 25129;
205-706-4: Zoamix, 34844;
205-710-6: ethyl namate®, 11078; Octopol SDE-25, 22013; Thiostop E, N, 31716;
205-711-1: 8-hydroxyquinoline, 14670;
205-712-7: chromotrope acid, 6244;

205-724-2: R-Acid, 25710;
205-725-8: Hymush, 14689; Nemacin, 20862; Storite, 29796; Tecto, Tecto 60%, 30913; thiabendazole, 31661; Thibenzole, 31668; Tubazole, 32545;
205-736-8: Accelerator Mercapto, 207; Captax, 5166; MBT, 19170; MBT®, 19171; mercaptobenzothiazole, 19335; Perkacit® MBT, 23285; Rokon, 26918; Rotax®, 26999; Thiotax, 31719; Vulkacit® M, 34133; Vulkacit® M, Vulkacit® Merkapto/C, 34134;
205-737-3: lichen sugar, 17204;
205-738-9: Hostapon STT Paste, 14396; sodium methyl stearoyl taurate, 28792;
205-739-4: Hyraldite C Ext, 14730; sodium formaldehydesulfoxylate, 28766;
205-743-6: 2-ethylhexoic acid, 11137;
205-746-2: CM8930, 6461;
205-749-9: Gallic acid, 12610;
205-753-0: Pabagel, 22533; Pabanol, 22534; RVPaba Lipstick, 27153;
205-756-7: DLPA 375, 8748;
205-758-8: edetate trisodium, 9614; Hamp-Ene® Na3 Liq, 13589; Sequestrene® NA3, 28000;
205-769-8: Eastman® HQMME, 9467; hydroquinone monomethyl ether, 14633;
205-775-0: Cephreine, 5694;
205-776-6: phytol, 23780;
205-777-1: Vulcaid 27, 34073;
205-782-9: Lucofen S A, 17768;
205-788-1: Akyposal NLS, 1314; Alscoap LN-40, LN-90, 1853; Calfoam SLS-30, 4986; Carsonol® SLS Paste B, 5339; Chemsalan NLS 30, 6029; Dreft, 8976; Drene, 8977; Empicol® 0303, 10301; Empicol® LS30, 10327; Empicol® LXV, 10329; Empicol® LY28/S, 10330; Empicol® LZG 30, 10332; Empicol® LZP, 10333; Empimin® LR28, 10425; Hartenol LAS-30, 13669; Marlinat® DFK 30, 18851; Naxolate WA-97, 20818; Norfox® SLS, 21552; Nutrapon DL 3891, 21817; Nutrapon W 1367, 21829; Nutrapon WAQE 2364, 21830; Octosol SLS, 22027; Orvus WA, 22346; Perlankrol® DSA, 23326; Polystep® B-3, 24596; Rewopol® NLS 15 L, 26368; Rhodapon® 101-10, 26631; Rhodapon® LCP, 26636; sodium lauryl sulfate, 28781; Standapol® WAQ-LC, 29482; Stepanol® ME Dry, 29695; Stepanol® WA Extra, 29698; Sulfetal C 38, 29982; Sulfochem SLN, 30012; Sulfochem SLP-95, 30013; Sulfochem SLS, 30014; Sulfochem SLX, 30015; Sulfopon 101, 101 Special, 30023; Sulfopon 101/POL, 30024; Sulfopon P-40, 30027; Sulfotex LCX, 30032; Supralated ME, 30311; Texapon K-12, K-1296, L-100, 31461; Texapon K-1296, 31462; Texapon LS Highly Conc, 31464; Texapon OT Highly Conc. Needles, 31472; Texapon VHC Needles, ZHC Needles, 31481; Texapon ZHC Needles, 31482; Texapon ZHC Needles, 31483; Ufarol Na-30, 32743; Witcolate 6400, 34469;
205-790-2: Panogen M, 22682;
205-792-3: potassium cyanide, 24751;
205-797-0: betamethasone 21 phosphate, 4060;
205-801-0: Schradan, 27731; Sytam, 30667;
205-804-7: Kelfizina, 15838;
205-808-9: madder, 18289; rubianic acid, 27103;
205-814-1: rutin trihydrate, 27150;
205-827-2: Lanvis, 16771;
205-840-3: Octocure ZMBT-50, 21998; Oxaf, 22408; Perkacit® ZMBT, 23298; Vulkacit® Merkapto/MGC, 34135; Zetax®, 34741; ZMBT, 34842;
205-845-0: rhapontin, 26530;
205-855-5: Phenetidine, 23589;
205-857-6: sodium butyrate, 28742;
205-861-8: lime nitrogen, 17268;
205-912-4: Idryl, 14841;
205-923-4: Chrysene, 6252;
205-999-9: Dabco® 33-LV, 7640; Dabco® Crystalline, 7652; TEDA-L33, 30916; Tegoamin® 33, 30994; Texacat® TD-33, 31433;
206-019-2: Imidazole, 14895;
206-022-9: pyrrodiazole, 25534;
206-026-0: miazine, 19625; pyrimidine, 25484;
206-050-1: Folidol® M, 12200; Metacide, 19409; methyl parathion, 19556; Parathion-methyl, 22808;
206-052-2: Terrathion, 31343;
206-054-3: Disyston® FE-10, 8726; Solvigran, 29063;
206-055-9: EP-1, 10611;
206-059-0: K-Lyte, 16233;
206-066-9: Oxsoralen-Ultra, 22437;
206-067-4: Naxonate® 4KT, 20823; Nitro BT, 21342;
206-074-2: Gluconal® K, 13043; Jaon, 15507; Katorin, 15798; potassium d-gluconate, 24752; Potassium gluconate, 24755;
206-075-8: Gluconal® CA A, CAM, 13038; Neo-Calglucon, 20877;
206-076-3: Gluconal® FE, 13042;
206-082-6: Korlan, 16351; Nankor, 20677; ronnel, 26951; Trolene, 32443;
206-094-1: muscarine, 20524;
206-098-3: Bromex, 4569; Dibrom, 8356;
206-101-8: Synpro® 303, 30592;
206-103-9: Armid® O, 2964; Crodamide O, OR, 7097; Kemamide® O, 15898; oleamide, 22107; Petrac® Slip-Eze, 23468; Polydis® TR 121, 24411; Unislip 1759, 33045;
206-104-4: Goulard powder, 13216; lead acetate, 16902; Ledac, 16948; sal Saturni, 27253; salt of Saturn, 27287; sugar of lead, 29955;
206-108-6: Metacure® T-9, 19416;
206-110-7: Metasystox R, 19461; Metasystox® R, 19463;
206-114-9: Amerzine®, 2047; hydrazine, 14545; Scav-Ox® 35%, 27552; Scav-Ox® II, 27553; Ultra Pure, 32780; Zerox, 34738;
206-155-2: MPI DMSA Kidney Reagent, 20393;
206-156-8: Rochelle salt, 26882; sal Rupellensis, 27252;
206-162-0: Leukeran Tablets, 17029;
206-190-3: Dichlorotrifluoroethane, 8393; Genetron® 123, 12823;
206-191-9: Flutec PP9, 12124;
206-200-6: Fluorad® FX-8, 12067;
206-215-8: Aldrin, 1558; Aldrin Dust, 1559; Murald, 20502;
206-221-0: paraoxon, 22783;
206-245-1: bromacil, 4564; Hyvar X, 14793; Hyvar® X, 14794; Uragan, 33105;
206-289-1: Haloxon, 13567; Loxon, 17687;
206-330-3: α-ketoglutaric acid, 16137;
206-354-4: Direx® 4L, 8620; Diurex, 8734; diuron, 8736; Diuron Bayer, 8737; Karmex, 15771; Karmex®, 15772;
206-356-5: Afalon, 918; Ashlade Linuron, 3195; Atlas Linuron, 3318; Atlas Linuron, 3319; Campbell's Linuron 45%, 5061; Du Pont Linuron 50, 4L, 9073; Linex® 4L, 17294; Linurex, 17316; Linuron, 17317; Linuron 15, 17318; Linuron 450 FL, 17319; Norunil, 21592; Rotalin, 26997; Rotalin, 26998;
206-373-8: Agridin 60, 1053; Antlak, 2629; Diazinon Liquid, 8344; Diazitol, Diazitol Liquid, 8345; Diazol, 8347; Diziktol, 8745; Neocidol Veterinary Powder, 20880; Root Guard, 26957;
206-376-4: n-Capric acid, 5126; Prifrac 2906, 24951;
206-386-9: Flutec PP3, 12123;
206-397-9: Fluorad® FC-26, 12060;
206-423-9: isoflupredone acetate, 15362; Predef, 24870;
206-426-5: Lycanol, 18061;
206-537-9: BCF, 3864;
206-557-8: Genetron® HFC 125, 12828;
206-573-5: Flutec PP2, 12122;
206-585-0: Flutec PP1, 12121;
206-597-6: Lidex, 17215;
206-625-7: Advex 91025, 821; Forex®, 12245; Geon® 83457, 12857; Geon® 83718, 12858; Geon® 86100, 86101, and 86103, 12859; Geon® 8700A, 12853; Geon® 8720, 12854; Geon® 87239, 87241, 12860; Geon® 87396, 12861; Geon® 87420, 12862; Geon® 8812, 8813, 12855; Geon® 8896, 12856; Geon® HTX 92190, 12863; Geon® W015, 12865; Georgia Gulf 3131 Clear 02, 12866; Georgia Gulf 5006, 12867; Georgia Gulf 5006 General, 12868; Georgia Gulf 9105, 12869; Georgia Gulf 9151, 12870; Georgia Gulf 9175J, 12871; Georgia Gulf 9202, 12872; Georgia Gulf CL-7049, 12873; Georgia Gulf EH-71L, 12874; Georgia Gulf EX-240, 12875; Georgia Gulf HH-1900, 12876; Georgia Gulf HM-7054, 12877; Georgia Gulf SP-7107, 12878; Georgia Gulf UV-7160, 12879; Klegecell R30, 16216; Korogel, 16354; Koroplate, 16357; Lutofan®, 17996; Norvinyl DX 550, 21596; Novablend® 501, 21642; Novablend® 5555, 21643; polyvinyl chloride, 24668; Saran, 27450; Saran 313, 27451; Saran 510, 27452; Saran F-239, F-278, F-310, 27453; Saran Wrap, 27454; Saran® Filament, 27455; Ultraprene®, 32860; Vestolit®, 33679; Vestolit® B 7021, 33680; Vestolit® E 6003, E 6503, E 7003, 33682; Vestolit® HI, 33683; Vestolit® LF HI-SP 5735, 33684; Vestolit® M 5867, 33685;

Vestolit® P 1330 K, 33686; Vestolit® S 6058, 33687; Vinoflex®, 33839;
206-736-0: Fluorophenol, 12109;
206-761-7: nickel acetate, 21118;
206-835-9: Lingraine, 17296; Migraine Dolviran®, 19744;
206-857-9: Kerasol, 16058;
206-991-8: Carbogran, Carbogran E, Carbogran UF, 5201; Carbomant, 5210; Carborex, 5237; Carsilon, 5326; Exolon, 11302; Lonsicar, 17585; Meccarb, 19202; Resilon, 26172; silicon carbide, 28329; silundum, 28406; Simax, 28463;
207-069-8: Diacetyl, 8250;
207-108-9: Liquamar, 17450; Phenprocoumon, 23630;
207-114-1: gentisin, 12847;
207-125-1: Jodomiron, 15615; Uromiro(n), 33147;
207-136-1: Satric, 27491;
207-312-8: Dicyandiamide, 8411;
207-321-7: Fluorobenzene, 12093;
207-329-0: Cadaverine, 4883;
207-334-8: linolenic acid, 17306;
207-341-6: thiophosgene, 31711;
207-348-4: Bromopicrin, 4585;
207-387-7: neopine, 20939;
207-432-0: Birlane, 4220; Birlane, 4221; Chlorfenvinphos, 6098; Sapecron, 27436; Sedanox, 27831;
207-437-8: propiolic acid, 25189;
207-439-9: Aeromatt, 863; Albacar, 1342; Amical® 101, 2060; Amical® 85, 2059; Amical® SC, 2062; Atomite, 3378; Calofil, 5016; Calofort®, 5017; Calopake®, 5020; Camel-CAL®, Camel-CARB®, Camel-FIL, Camel-FINE, 5047; Camel-TEX®, CamelWITE®, 5049; Carbital®, 5191; Carbital® 35, 5192; Carbital® 50, 5193; CC-103, 5485; CP Filler, 6928; Drikalite®, 9031; Duramite®, 9161; Hallcote® 573, 13554; Hi-Pflex® 100, 14103; Hubercarb® Q 6-20, 14453; Hubercarb® W 2, 14454; Iceland spar, 14813; Kalvan, 15701; Kotamite®, 16370; Macromite, 18281; Marbledust, 18691; Marblemite, 18693; Micro-White® 07 Slurry, 19712; Micro-White® 10 Codex, 19713; Micro-White® 100, 19717; Micro-White® 15, 19714; Micro-White® 25, 19715; Micro-White® 40, 19716; mild lime, 19758; No. 1 White, 21410; No. 3 White, 21411; Opacicoat, 22195; Opacimite, 22196; sandscale, 27362; SC-53, 27535; Slab Dip AC699, 28585; Snowflake P. E., 28675; Snowflake White, 28676; Socal, 28703; statuary marble, 29575; Sturcal®, 29846; Super-Pflex® 100, 30261; T-Carb, 30855; Trucal, 32486; Trucarb, 32487; Ultramite, 32841; Ultra-Pflex®, 32856;
207-440-4: Cacodyl, 4882;
207-462-4: Dichloramine T, 8382;
207-495-4: naphthazarin, 20713;
207-521-4: rhein, 26534;
207-531-9: Tetryl, 31423;
207-567-5: Juglone, 15631;
207-581-1: imperatorin, 14912; peucedanin, 23541;
207-586-9: Indigo, 15074;
207-597-9: retene, 26271;
207-612-9: Morocide, 20340;
207-616-0: ulexine, 32761;
207-618-1: ninhydrin, 21230;
207-632-8: Daphnetin, 7736;
207-790-8: Vulcazol, 34112;
207-806-3: Hippuric acid, 14126;
207-830-4: Butyroin, 4802;
207-838-8: Consal, 6732; mild alkali, 19757; Oxyper, 22465; sal soda, 27255; salt of soda, 27288; scotch soda, 27758; soda ash, 28710; sodium carbonate, 28744; sodium carbonate monohydrate, 28745; Solvay® Soda, 29041; Trona Soda Ash, 32459; Tronalight Light Soda Ash, 32462;
207-840-9: oxazolidine, 22419;
207-877-0: equisetic acid, 10735;
207-889-6: carvacrol, 5362; oxycymol, 22445;
207-899-0: coumalic acid, 6901;
207-914-0: Ciba 1906, 6291; thiambutosine, 31663;
207-921-9: Cardolite® NC-507, 5267;
207-924-5: hydrocinnamic acid, 14577;
207-925-0: benzyl chloroformate, 3988;
207-926-6: Tolan, 31933;
207-929-2: chavicol, 5861;
207-938-1: ϵ-caprolactone monomer, 5138;
207-965-9: Perstoff, 23430;
207-984-2: orcinol, 22280;
207-987-9: Avitrol, 3494;
208-010-9: suberic acid, 29903;
208-046-5: dimethylamine hydrochloride, 8513;
208-051-2: cyanogen bromide, 7479;
208-058-0: baker's salt, 3598; sal volatile, 27258; salt of Hartshorn, 27284;
208-060-1: Guanidine nitrate, 13441;
208-080-0: DL-Borneol, 4428;
208-127-5: vitamin D+74, 33942;
208-146-9: raffinose, 25835;
208-156-3: pental, 23147;
208-167-3: Barium carbonate, 3636; Durex white, 9208;
208-174-1: acetyl methyl carbinol, 312;
208-178-3: abietic acid, 45; sylvic acid, 30503;
208-187-2: Canthaxanthin, 5095;
208-205-9: Camilol, 5053; Hydagen® B, 14525;
208-236-8: Nitroform, 21359; tetranitromethane, 31386;
208-245-7: Alkanet, 1695;
208-253-0: Fluor-Amps, 12069; Fluorescein disodium salt, 12081; Fluorescite, 12086; Fluor-I-Strip, 12091; uranine, 33115;
208-268-2: maclurin, 18235;
208-275-0: tritane, 32420;
208-287-6: kaempferol, 15654;
208-293-9: Dehydroacetic acid, 7964;
208-330-9: Dequalinium Chloride, 8077; Labosept, 16533;
208-349-2: parsley camphor, 22865;
208-355-5: pavlin, 22928;
208-383-8: flavone, 11903;
208-401-4: Gluconic Acid, 13048;
208-407-7: Asahi Aji®, 3174; Gluconal® NA, 13046; Naglusol, 20662; sodium gluconate, 28769;
208-408-2: copper gluconate, 6781; Gluconal® CU, 13041;
208-432-3: trimellitic acid, 32335;
208-456-4: Iodanisol, 15204;
208-497-8: Tetralol, 31381;
208-525-9: Acaprin®, 181; 1,3-di-6-quinolylurea, 8618;
208-531-1: chloroacetophenone, 6112;
208-534-8: sodium benzoate, 28735;
208-545-8: Chrysoidine, 6256;
208-576-7: Amerstat® 233, 2039; Dazomet, 7806; N-521® Biocide, 20640;
208-580-9: S-Carb, 27549; sodium sesquicarbonate, 28816;
208-586-1: iohydrin, 15223;
208-590-3: silver carbonate, 28420;
208-591-9: cesium carbonate, 5772;
208-594-5: Sylvan, 30497;
208-596-6: lysidine, 18087;
208-601-1: 4,6-dinitrocresol, 8542; Dekryll, 7999;
208-622-6: Crimidine, 7053;
208-651-4: *m*-anisidine, 2472;
208-673-4: Oxynone, 22464;
208-686-5: Captex® 8000, 5170; Miglyol® 808, 19736; Radiamuls® MCT 2108, 25776; tricaprylin, 32211;
208-687-0: Dynasan® 112, 9375; Trilaurin, 32317;
208-704-1: dicyclohexyl carbodiimide, 8412;
208-736-6: cetyl palmitate, 5822; Crodamol CP, 7119; Cutina® CP, 7437; Emalex CC-16, 9934; Kemester® CP, 15955; Kessco® 653, 16100; Precifac ATO, 24865; Radia® 7500, 25742; Rewowax CG, 26434; Schercemol CP, 27588; Starfol® CP, 29535; Waxenol® 815, 34238;
208-750-2: Acetylene dichloride, 320; Dioform, 8553;
208-760-7: *tert*-Butyl Acetate, 4785;
208-764-9: CD3770, 5500; D3770, 7635;
208-765-4: CH7260, 5839;
208-795-8: Exaltone, 11282; muscone, 20527; musk, 20531;
208-797-9: ADA, 615; β-ketoglutaric acid, 16138;
208-806-6: cyclohexyl chloride, 7524;
208-818-1: isobutyl formate, 15346;
208-826-5: Telone, 31069;
208-831-2: ethyl isothiocyanate, 11075; mustard oil, 20539;
208-849-0: Barium acetate, 3634;
208-863-7: calcium formate, 4948; Latibon®, 16817;
208-865-8: cupric formate, 7380;
208-875-2: Emery® 654, 10135; Hystrene® 9014, 14764; Hystrene® 9514, 14767; Myristic acid, 20574; Philacid 1400, 23662; Prifrac 2940, 24953;
208-904-9: CO9817, 6510;
208-912-2: tanacetone, 30775;
208-913-8: Lithostat, 17522;
208-915-9: Elastocarb Tech Light, Heavy, 9691; Elastocarb UF, 9692; Magocarb-33, 18404;
208-938-4: Loretine, 17606;
208-953-6: gentian violet, 12845; Pyoctanin, 25455;
208-958-3: Irigenin, 15301;
208-964-6: Laroxyl, 16801; Lentizol, 16992;
208-987-1: diazoresorcin, 8351;
209-001-2: Dimetridazole, 8528; Emtryl, 10482;
209-007-5: Iosol, 15237;

209-025-3: *o*-nitrobenzaldehyde, 21346;
209-027-4: Salysal, 27306;
209-032-1: Thonzide, 31746;
209-062-5: Liskonum Tablets, 17492; Lithane, 17500;
209-069-3: Iodozol, 15221;
209-077-7: trimesic acid, 32337;
209-084-5: *p*-nitrobenzaldehyde, 21347;
209-090-8: Aliso, 1617; aluminum isopropoxide, 1909;
209-096-0: Kloben®, 16225; Neburex, 20840; neburon, 20841; Noruben, 21591;
209-097-6: Dynasan® 118, 9378; Kemester® 5500, 15947; Kemester® 6000, 15952; Kemester® GMS (Powd.), 15962; Neobee® 62, 20874;
209-098-1: Dynasan® 116, 9377;
209-099-7: Dynasan® 114, 9376;
209-127-8: Fungisterol, 12479; N-glycylglycine, 13139;
209-132-5: methyl isothiocyanate, 19543;
209-136-7: CO9810, 6508; O9810, 21945;
209-141-4: Prenol, 24892;
209-150-3: Afco-Chem MGS, 923; magnesium stearate, 18367; Petrac® MG-20 NF, 23466; Synpro® 90, 30591;
209-151-9: Afco-Chem ZNS, 924; Antidust 2, 2541; DLG-10, 20, 8747; Hallcote® ZS 5050, 13556; Petrac® ZN-41, 23470; Synpro® 8, 30589; zinc stearate, 34793;
209-155-0: zinc undecenoate, 34798;
209-242-3: Ethion, 10955;
209-254-9: silver acetate, 28414;
209-321-2: pararosaniline, 22798;
209-323-3: Longifene, 17581;
209-386-7: 1,6-naphthalenediol, 20709;
209-391-4: Tonophosphan, 32013;
209-406-4: Aerosol OT, 877; Aerosol® GPG, 882; Aerosol® OT-70 PG, OT-S, 896; Aerosol® OT-75%, 897; Aerosol® OT-MSO, 898; Agrilan® AEC266, 1056; Alcopol O, 1499; Anonaid TH, 2481; Arowet SC-75, 3091; Arylene M40, 3165; Astrowet 0-70-PG, 3265; Astrowet 0-75, 3266; Bevaloid 1299, 4082; Chemax DOSS/70, 5895; Complemix, 6681; Coprol, 6796; Cropol 60, 7210; Dioctyl, 8547; dioctyl sodium sulfosuccinate, 8550; Discol DFW, 8631; Disponil SUS IC 8, 8709; Drewfax® S-700, 8991; Drewfex® 0007, 8992; Elfanol® 883, 9828; Emcol® 4500, 10051; Emcol® DOSS, 10058; Empimin® OP70, 10434; Empimin® OT, 10435; Geropon® CYA/60, 12905; Geropon® DOS, 12907; Hipochem EK-18, 14109; Hodag DOSS-70, 14178; Lankropol® KO2, 16715; Lankropol® KPH70, 16716; Leonil OS, 17002; Mackanate DOS-40, 18186; Manoxol OT, OT/P and OT/B, 18581; Manoxol OT60, 18582; Manoxolot, 18583; Marlinat® DF 8, 18850; Mazawet® DOSS 70, 19118; Modane Soft, 20015; Monawet MO, 20182; Monawet MO-65-150, 20183; Monawet MO-70, 20184; Monawet MO-84R2W, 20185; Ninate® DS 70, 21229; Norval, 21594; Octowet 40, 22035; Pentasol, 23174; Rewopol® SBDO 75, 26374; Rexowet ASG-81, 26505; Schercopol DOS-70, 27666; Schercowet DOS-70, 27711; Secosol DOS/70, 27810; Secosol® DOS 70, 27814; Servoxyl VLA 2170, 28071; Tensuccin D8, 31201; Thorowet G-40 3230, 31755; Triton® GR-5M, 32428; Triumphnetzer ZSG, 32435; Warcowet O, 34212;
209-433-1: Adamsite, 624;
209-439-4: Entamide, 10593;
209-441-5: Benzoin, 3975; Persian balsam, 23417;
209-470-3: Lauth's violet, 16870;
209-471-9: 2,6-naphthalenedisulfonic acid, 20708; Ebert and Merz's β-acid, 9496;
209-478-7: 2,7-naphthalenediol, 20711;
209-481-3: potassium benzoate, 24738;
209-496-3: Lawsone, 16891;
209-502-6: Anti-Oxidant MB, 2589; 2-mercaptobenzimidazole, 19334;
209-523-0: methionine hydroxy analog, 19486; MHA, 19624;
209-529-3: Montreal potash, 20302; pearl ash, 22977; Pearl Dust®, 22978; potassium carbonate, 24743; potassium carbonate, 24744; potassium carbonate, 24745; sal absinthii, 27242; salt of wormwood, 27292;
209-530-9: rubidium carbonate, 27104;
209-542-4: allethrin, 1744;
209-544-5: Scuranate, 27776; toluene diisocyanate, 31947; Voranate T-80, Type I, Type II, 34042;
209-545-0: anserine, 2490;
209-603-5: Dionosil, 8569; Propyliodone, 25225, 25226;
209-621-3: Eccobrite RB, 9524; stilbene, 29764;
209-661-1: ethylbutylcarbinol, 11119;
209-697-8: Grossmann reagent, 13426;
209-719-6: iodobenzene, 15210;
209-740-0: calcium cyanide, 4945; Cyanolime, 7480; powdered hydrocyanic acid, 24815;
209-753-1: 1-hexene, 14029; Neodene® 6, 20886;
209-786-1: potassium stearate, 24783, 24784;
209-800-6: Saytex® VBR, 27525;
209-812-1: Guanidine thiocyanate, 13442;
209-893-3: NEPD, 20988;
209-917-2: vinaconic acid, 33766;
209-940-8: dimethylethylamine, 8521; N,N-dimethylethylamine, 8522;
209-942-9: manganese white, 18551;
209-943-4: lead subcarbonate, 16918; Ledca, 16950;
209-945-5: Ultra-DMC, 32791;
209-954-4: DL-lactic acid, 16573;
209-967-5: 3,3'-diaminodiphenylsulfone, 8303;
209-968-0: cumyl phenol, 7355;
210-036-0: triphenyl phosphine, 32392;
210-165-2: galactitol, 12592; Melampyrite, 19258;
210-277-1: mandelic acid, 18518;
210-283-4: Zentralin, 34715;
210-355-5: naphthalol, 20712;
210-379-6: Mollit B, 20068; Plastic A, 24019; Uniplex 260, 33015;
210-382-2: *t*-butyl perbenzoate, 4798; Aztec® *t*-Butyl Perbenzoate, 3543; Esperox® 10, 10844; Polyvel CR-5T, 24659; Trigonox® 93, 32276; Trigonox® C, 32297;
210-410-3: 2-indanone, 15063;
210-436-5: 2,4-dibromophenol, 8359; FR-612, 12356;
210-483-1: pyrrolidone2-pyrrolidone, 25536; Soluphor® P, 29023;
210-514-9: apple acid, 2667;
210-535-3: CT2523, 7320;
210-568-3: 5-nitroisophthalic acid, 21361;
210-615-8: 4-(dimethylamino)benzoic acid, 8514;
210-623-1: Fraissite, 12374;
210-668-7: Preventol® CI 5, 24928;
210-702-0: Dynasan® 110, 9374;
210-721-4: Stabilizer 2013-P®, 29406;
210-736-6: H.A. Solvent, 13522;
210-827-0: Emalex EG-di-L, 9949; glycol dilaurate, 13100; Kemester® EGDL, 15958;
210-868-4: methyl ethyl sulfide, 19533;
210-871-0: Sulfa-Hitech® 0382, 29961;
210-903-3: nitric ether, 21333;
210-966-8: α-methylnaphthalene, 19574;
211-011-3: Kemester® EGDS, 15959;
211-014-3: Alkamuls® EGDS, 1663; Emalex EG-di-S, 9951; Emerest® 2355, 10091; ethylene glycol distearate VA, 11126; Genapol® PMs, 12795; glycol distearate, 13102; Kessco® EGDS, 16105; Lexemul® EGDS, 17136; Lipo EGDS, 17338; Mapeg® EGDS, 18641; McAlester EGDS, 19176; Pegosperse® 50 DS, 23065; Rewopal® PG 280, 26348; Rita EDGS, 26789; Secoster® DMS, 27817;
211-020-6: dimethyl adipate, 8502;
211-047-3: Amyl acetate, 2380;
211-057-8: Howard's silver, 14431;
211-074-0: Hexamethylene Glycol, 14005;
211-103-7: Acelan A, 268;
211-105-8: Neodene® 16, 20898;
211-107-9: Neodol® 5, 20910;
211-138-8: Lutate, 17966;
211-162-9: Ammonium acetate, 2226;
211-183-3: rose bengal, 26963;
211-185-4: Great Lakes PHT4, 13312; Saytex® RB-49, 27524; tetrabromophthalic anhydride, 31361;
211-241-8: Resorufin, 26232;
211-279-5: aluminum stearate, 1929; Duroseal, 9258; HiGel, 14081; LoGel, 17556; Med Gel, 19208; Metasap® 537, 19449; Novogel® ST, 21729; Synpro® 404, 30593;
211-284-2: triethanolamine hydrochloride, 32241;
211-322-8: Zisnet F-PT, 34838;
211-334-3: Manal, 18509; manganese acetate, 18536;
211-362-6: Ekatin, 9657; thiometon, 31700;
211-445-7: *cis*-stilbene, 29765;
211-466-1: Emerest® 2324, 10087; Estol 1476, 10893; Kemester® 5415, 15946; Kessco® IBS, 16113; Uniflex® IBYS, 32948;
211-479-2: Varitox, 33433;
211-519-9: Enisyl, 10583; L-(+)-lysine hydrochloride, 18089;
211-522-5: Eltesol® ST 40, 9872; Manro STS, 18606; Naxonat® 4ST, 20821; Reworyl® T, 26410; sodium *p*-toluenesulfonate, 28827;
211-546-6: Behenyl alcohol, 3899; Dehydag Wax 22 (Lanette), 7946; Emery® 3304, 10160; Loxiol VPG 1451, 17685;
211-566-5: Agaricic, 956;

211-569-1: Humectant SD-35, 14464; Panthenylethyl ether, 22691;
211-581-7: 2-phenyl imidazole, 23643;
211-656-4: Dynasil® M, 9385;
211-659-0: CT2090, 7316;
211-669-5: Dehyton® PAB-30, 7993; Zohartaine AB, 34877;
211-687-3: Esperox® 570, 10854; Lupersol 575, 17867; Trigonox® 121, 32281;
211-694-1: ethyl lactate, 11076;
211-702-3: Olminat, 22143;
211-748-4: Alkateric® PB, 1721; Mackam CET, 18141; Mirataine® BET-P-30, 19925;
211-750-5: Alkanox® 240-3T, 1704; Argus DSTDP, 2863; Carstab® DSTDP, 5357; Cyanox® STDP, 7493; Distearyl thiodipropionate, 8713; Evanstab® 18, 11258; Lankromark® DSTDP, 16691; Lowinox® DSTDP, 17667;
211-765-7: 2-methyl imidazole, 19572;
211-776-7: Diaminocyclohexane, 8300; 1,2-diaminocyclohexane, 8301;
211-797-1: dimethoxy Tetrahydrofuran, 8499; Protectol® DMT, 25260;
211-914-6: Propanil, 25174; Sorpur®, 29148; Surcopur, 30331;
211-952-3: Fluorfolpet, 12087;
211-986-9: Euparen® M, 11198;
212-004-1: diatrizoate sodium, 8337; MD 50, 19186; Urovist, Urovist Sodium 300, 33150;
212-052-3: dimethylmethyl phosphonate, 8525; Fyrol® DMMP, 12539;
212-073-8: Eptam 6E, 10728; EPTC, 10729;
212-094-2: Perkadox® SE-8, 23310;
212-104-5: CA0570, 4848;
212-112-9: Ektapro® EEP Solvent, 9659; ethyl 3-ethoxypropionate, 11062;
212-193-0: CP0160, 6931;
212-222-7: 1-Phenoxy-2-propanol, 23625;
212-293-4: Infusorial earth, 15125;
212-305-8: CP0320, 6933;
212-308-4: dimethyl anthranilate, 8503;
212-324-1: Trithion, 32422;
212-344-0: Akrochem® Antiozonant PD-2, 1164;
212-358-7: nitrovin, 21382;
212-371-8: lead formate, 16908; Ledfo, 16959;
212-391-7: calcium citrate, 4944;
212-411-4: cupric oxalate, 7385;
212-414-0: tin(II) oxalate, 31816;
212-440-2: Phospholipon® SC, 23730;
212-490-5: sodium stearate, 28820;
212-579-9: dimethoxane, 8498; Givgard DXN, 12951;
212-634-7: Ametrex, 2050; ametryn, 2051;
212-668-2: Sudan I, 29938;
212-702-6: Sipalin AOC, 28512;
212-707-3: Vesulong, 33700;
212-728-8: Indigo Carmine, 15075;
212-755-5: potassium citrate, 24750;
212-769-1: potassium bitartrate, 24740; soluble cream of tartar, 28979;
212-773-3: sodium tartrate, 28825;
212-782-2: Bisomer 2HEMA, 4273; hydroxyethyl methacrylate, 14662; Sipomer® HEM-D, 28536;
212-800-9: Rongalit® C, 26945;
212-806-1: Chembetaine OL-30, 5935; Incronam OD-50, 15022; Mackam OB-30, 18148; Mafo® OB, 18303; oleyl betaine, 22132;
212-819-2: Neodene® 10, 20891; Nipasol M Potassium, 21283;
212-828-1: Agsol Ex1, 1091; Micropure® Ultra, 19694; N-methyl-2-pyrrolidone, 19578; NMP, 21406; PartsPrep™ Degreaser, 22872;
212-855-9: dihydroisophorone, 8457;
212-864-8: *p*-toluoyl chloride, 31957;
212-895-7: Celogen® TSH, 5645;
212-950-5: Clarosan 1FG, 6372; Prebane 500, 24861; Terbutrex, 31229; terbutryn, 31230; Cestarsol, 5779;
213-033-2: cobaltous acetate tetrahydrate, 6535;
213-048-4: Aktisil AM, 1201; γ-aminopropyltriethoxysilane, 2186; CA0750, 4852; Dynasylan® AMEO, Dynasylan® AMEO-P, 9396; Prosil® 220, 25237; Union Carbide® A-1100, 32990;
213-052-6: Azotoz 580, 3538; :demeton-S-methyl, 8055; :Demetox, 8056; :Metasystox 55, 19460; :Metasystox® I, 19462; :Mifatox, 19727; :Power DSM, 24822; :Quad DSM, 25565; :Vassgro DSM, 33469;
213-067-8: soluble tartar, 28987;
213-085-6: Aerosol® AY-65, 889;
213-100-4: lead, 16900;
213-103-2: NM-AMD, 21404;
213-147-2: Aztec® *t*-Butyl Peroxypivalate-75 OMS, 3548; Esperox® 31M, 10848; Lupersol 11, 17843; Trigonox® 25-C75, 32268;
213-170-8: Emalex EG-di-O, 9950;
213-178-1: CH7332, 5842;
213-195-4: Diglycolamine® Agent (DGA®), 8454;
213-207-8: Kemester® 9022, 15953;
213-234-5: 2-ethyl-4-methyl imidazole, 11139; EMI-24, 10190;
213-271-7: 2,3,6-trichlorophenol, 32220;
213-371-0: CP0110, 6929;
213-408-0: Dyfonate, 9316;
213-427-4: CD6150, 5516;
213-447-3: Cytro-Lane, 7606; mephosfolan, 19326;
213-449-4: methidathion, 19483; Supracide, 30299; Suprathion, 30318;
213-482-4: Enide, 10581;
213-546-1: Acrex, 384; Dinobuton, 8543;
213-584-9: Rhodamine G and G Extra, 26617;
213-615-6: CT2520, 7319;
213-626-6: Butyl bisbutyl peroxy valerate, 4765; Luperco 230-XL, 17824; Lupersol 230, 17855;
213-637-6: CD4450, 5504;
213-650-7: CT2500, 7317;
213-658-0: diallyl maleate, 8282;
213-666-4: ABM Chlormequat 40, 72.5, 142; Ashlade 4-60 CCC, 700 CCC, 3185; Atlas Chlormequat 46, 700, 3302; CCC 700, 5497; Chlormequat chloride, 6110; Cleanacres PDR 675, 6382; Clifton Chlormequat 46, 6415; Cycogan, 7545; Fargro Chlormequat, 11471; Farmacel, 11478; Farmacel, Farmacel 645, 11477; Headland Swift, 13735; Hyquat 70, 75, 14729; Mandops Barleyquat B, 18520; Mandops Bettaquat B, 18521; Mandops Spring Poquaternary, 18525; MSS Chlormequat 40, 46, 60, 70, 20427; Portman Chlormequat 400, 460, 600, 700, 24711; Power 64, 640, 700, 24817; Quadrangle Chlormequat 700, 25575; Standup, 29487; Star Chlormequat, 29523; Terpal® CC, M, 31310; Titan, 31859; Tripart® Brevis, 32366; Tripart® Chlormequat 460, 32367;
213-668-5: Dynasylan® HMDS, 9409; H7301, 13529; hexamethyldisilazane, 14003; Prosil® HMDS, 25242;
213-680-0: Propyl Zithate®, 25206; Zithate®, 34839;
213-695-2: Ammonium stearate, 2253;
213-722-6: Flotox, 12027; flour of sulfur, 12028;
213-773-6: CH7250, 5838; H7250, 13528; Hexamethylcyclotrisilazane, 14001;
213-800-1: desmetryn, 8142; Semeron, 27936;
213-834-7: triallyl isocyanurate, 32192;
213-890-2: bunamidine hydrochloride, 4680; Scolaban, 27745;
213-909-4: chromium acetate (ic), 6229;
213-911-5: ABC Trieb, 34; Ammonium bicarbonate, 2231;
213-912-0: CD5470, 5507;
213-926-7: CP0810, 6936;
213-928-8: Metacure® T-1, 19414;
213-934-0: Aktisil VM, 1205; CV-5000, 7453; Dynasylan® VTMOEO, 9422; Union Carbide® A-172, 32985;
213-944-5: Aztec® 2,5-Tri, 3552; Esperal® 230, 10838; Lupersol 130, 17847; Polyvel CR-L10, 24660; Trigonox® 145, 32286;
213-990-6: CC3555, 5495; 3-cyanopropyltrichlorosilane, 7482;
213-997-4: Glyphogan, 13147; glyphosate, 13148; Roundup, 27011; Spasor, 29213; Sting, 29768;
214-005-2: Hal-Lub-D, 13558; Hal-Lub-N, 13559; Haro® Chem P28G, 13642; lead stearate, 16917; P-289, 22504; P-51, 22518;
214-060-2: 4-hydroxycoumarin, 14661;
214-073-3: Uniplex 108, 33008;
214-115-0: Nipaheptyl, 21272;
214-118-7: Dichlofluanid, 8379; Elvaron®, 9893; Euparen®, 11197;
214-122-9: Dapon M, 7738;
214-134-4: Lysivane, 18091;
214-183-1: cuprous thiocyanate, 7407;
214-189-4: CD5605, 5509;
214-197-8: Folimat®, 12201; Omethoate, 22155;
214-291-9: myrtrimonium bromide, 20589; Mytab®, 20595; Querton 14Br-40, 25649; Rhodaquat® M214B/99, 26642; Sumquat® 6110, 30125;
214-294-5: Zeonet B, 34721;
214-304-8: Kemester® 2050, 15941;
214-306-9: Neodene® 14, 20897;
214-312-1: Armeen® 2-10, 2926; Radiamine 6310, 25761;
214-355-6: 2-Acetyl pyridine, 321;
214-434-5: dihydrojasmone, 8458;
214-538-0: anthranol, 2530;
214-565-8: Oragrafin Calcium, 22260; Solu-Biloptin, 28977;
214-588-3: Tolinase, 31936;
214-604-9: Decabromodiphenyl oxide, 7837; FR-1210, 12365; Great Lakes DE-83R, 13304; Octoguard FR-01, 21999; Saytex® 102E, 27514; Thermoguard® 505, 31632;
214-620-6: Progallin LA, 25110;

214-684-5: tris (hydroxymethyl) aminomethane hydrochloride, 32404;
214-685-0: CM9100, 6465; Dynasylan® MTMS, 9415; Union Carbide® A-163, 32983;
214-724-1: cetyl stearate, 5825; Schercemol CS, 27589;
214-776-5: 3-quinuclidinone hydrochloride, 25693;
214-787-5: BH Prefix D, 4120; Casoron, 5394; Casoron G, 5395; Casoron G4, 5396; Fydulan, Fydumas, Fydusit, 12532; Prefix D, 24876;
214-818-2: Amikapron, 2121;
214-912-3: Dibenzosuberone, 8354;
214-945-3: Bilivist, 4144; Oragrafin Sodium, 22261;
214-955-8: Angio-Conray, 2456;
214-966-8: Curatin, 7415;
214-987-2: diphenyl octyl phosphate, 8597;
215-023-3: carminic acid, 5294;
215-089-3: xylenol, 34625;
215-090-9: Eltesol® SX 30, 9873; Esi-Terge SXS, 10831; Hartotrope SXS 40, Powd, 13694; Manro SXS, 18607; Naxonate® 4L, 20824; Naxonate® SX, 20828; Nutrol SXS 5418, 21849; Pilot SXS-40, 23843; Reworyl® NXS 40, 26409; sodium xylene sulfonate, 28832; Stepanate® SXS, 29674; Sulfotex SXS-40, 30036; Witconate SXS 40%, 34492;
215-100-1: Amerfloc® 2, 2007; Dynaflock, 9344; Dynagrout, 9348; Manfloc, 18528; sodium aluminate, 28731;
215-108-5: Ben-A-Gel®, 3929; Bentonite, 3954; BentoPharm, 3955; Brebent, 4496; Bregel, 4501; Fulbent, 12434; Fulbond, 12435; Gadorgel, 12563; K 129-H, 15640; Korthix, 16362; Korthix H-NF, 16363; MicroForm B, 19672; MicroForm BCS, 19673; Montigel, 20299; Polargel® HV, 24295; Polargel® NF, 24296; Wilkinite, 34409; Yellowstone, 34652;
215-116-9: arsenic pentoxide, 3140;
215-117-4: orpiment, 22317;
215-125-8: Boric anhydride, 4423;
215-127-9: Barium oxide, 3641; baryta, 3675;
215-128-4: Barium peroxide, 3642;
215-133-1: Berylla, 4053; Glucina, 13035;
215-134-7: Bismuth oxide, 4240; Bismuth trioxide, 4245; flowers of bismuth, 12032;
215-136-8: Bismuth subnitrate, 4243; flake white, 11871; magister of bismuth, 18320;
215-137-3: calcium hydroxide, 4952; Edelwit, 9601; lime water, 17270; Red Hot Pellets, 25959; slaked lime, 28591; Trulime, 32493;
215-138-9: calcium oxide, 4962; Caloxol CP2, 5023; Dynacal, 9337; KM Pebble Lime, 16240; lime, 17265;
215-139-4: calcium peroxide, 4964; Fertilox, 11707; Oxy-Gro, 22456;
215-142-0: Photophor®, 23763;
215-145-7: Alveograf, 1939; Periograf, 23274;
215-146-2: cadmium oxide, 4901;
215-154-6: cobaltous oxide, 6540; prepared cobalt oxide, 24897;
215-156-7: cobaltic oxide, 6532;
215-158-8: Guignet's Green, 13456;
215-160-9: Accrox, 258; chromic oxide, 6223; chromium oxide (ic), 6236; M100, 18107;
215-168-2: Black 103, 4301; Blackox, 4307; Crocus Martius, 7062; Deanox, 7830; Disperfin, 8666; ferric oxide, 11651; Ferroxide, 11694; Foundrox, 12344; Hydroferrox®, 14586; Indian Ocher, 15067; Jeweller's rouge, 15611; Kroma Red, 16418; Magnox, 18389; Mapico, 18654; No Vein Compound, 21409; ocher, 21976; Octotint 103, 22030; OSO® 1905, 22364; OSO® 440, 22363; Pferrico, 23555; Pferrisperse, 23556; Pferrox, 23560; Rainbow Custom Colored Mortars, 25836; Red 139, 25948; rouge, 27009; sal mineral, 27248; sienna, 28249; specularite, 29256; Transoxide, 32143; trip, 32358; Two Cubed Eight, 32638;
215-170-3: FR-20, 12352; Gilumag, 12944; Magnaspheres®, 18348; Magnifin® H10, 18378; Magnifin® H10C, 18379; Magnifin® H7A, 18377; Magoh-S, 18405; MGH-93, 19622; Mylanta, 20564; Neutramag®, 21031; Versamag® DC, 33607; Zerogen® 10, Zerogen® 60, 34733;
215-171-9: Burnt magnesia, 4714; Corox, 6846; Dynamag, 9350; Dynatherm, 9425; Elastomag® 100, 9704; Encapsulated MgO, 10538; Flamarret®, 11872; Insulmag®, 15152; Ken-Mag®, 16007; MagChem® 1060, 18312; MagChem® 10B, 18310; MagChem® 20, 18311; Maglite Y, 18323; Maglite® D, 18324; magnesia, 18351; Magotex, 18406; Magox® 98HR, 18407; Magox® Super Premium, 18408; Magrods, 18410; Plastomag® 170, 24052; Rhenomag, 26551; Scorchex, 27753; Scorchguard O, 27754; Scorchguard-bound, 27755; Sermag®, 28044; Tanbase®, 30794;
215-174-5: lead peroxide, 16913; Lepro, 17005; LP-100, 17691;
215-175-0: antimony trioxide, 2574; Crystic® Prefil F, 7304; Dechlorane A-O, 7849; FireShield® H, 11837; FireShield® HPM, 11838; FireShield® L, 11839; flowers of antimony, 12030; Octoguard FR-10, 22000; Petcat R-9, 23452; Thermoguard® L, 31636; UltraFine® II, 32799;
215-181-3: potassium hydroxide, 24756; vegetable alkali, 33512;
215-185-5: Pels, 23090; soda lye, 28714; sodium hydroxide, 28772;
215-191-8: turpeth mineral, 32612;
215-193-9: indium oxide, 15081;
215-197-0: hepar sulfur, 13787;
215-199-1: Double White, 8817; Kasil, 15778; potassium silicate, 24779; Potkem, 24795; soluble potash glass, 28983;
215-200-5: lanthanum oxide, 16766;
215-202-6: glass-maker's soap, 12972; KM Manganese Dioxide, 16238; manganese dioxide, 18541; Mangoxe, 18563;
215-204-7: molybdenum trioxide, 20081;
215-209-4: Oxone, 22435; oxygen powder, 22453; Solozone, 28962;
215-211-5: Hesthsulphid, 13886; sodium sulfide, 28822;
215-213-6: niobium oxide, 21246;
215-215-7: nickel oxide, 21130; Nico, 21147;
215-218-3: palladium oxide, 22615;
215-219-9: strontia, 29826;
215-222-5: Activox, 507; Activox B, 508; Canfelzo, 5087; Decelox, 7847; Denzox, 8068; Electrox, 9798; Entrox, 10596; Extrox, 11329; Felzodox, 11592; Finex-25, 11809; Finex-25-020, 11810; Fotofax, 12339; Garozinc, 12680; Ken-Zinc®, 16050; K-Zinc, 16527; Microx, 19718; Octocure 553, 21994; philosopher's wool, 23670; Photozinc, 23764; Rubox, 27113; Vita Zinc, 33922; zinc oxide, 34786; Zinc Oxide No. 185, 34787; Zinc Oxide No. 318, 34788; Zinc Oxide Transparent, 34789; Zinkoxyd Activ®, 34812; Zinox, 34815;
215-225-1: thorium dioxide, 31749;
215-227-2: Dynazirkon, 9429; Kontrastin, 16333; S945, S975, S987, S992, S994, 27184; SM 945, 28621; SM 975, 28622; SM 987, 28623; SM 992, 28624; SM 994, 28625; Zedox, 34698; zirconium oxide, 34824;
215-231-4: tungsten trioxide, 32591;
215-233-5: yttrium oxide, 34684;
215-234-0: ytterbium oxide, 34681;
215-235-6: minium tego, 19832; red lead, 25960;
215-236-1: phosphorus pentoxide, 23745; phosphorus pentoxide, 23746; Superphosphate, 30262;
215-237-7: Nyacol® A-1530, 21859; Thermoguard® FR, 31635; Timonox, 31806; TPL, 32112;
215-238-2: tantalum oxide, 30815;
215-239-8: KM Vanadium Pentoxide, 16246; vanadium pentoxide, 33275;
215-251-3: Sachtolith®, 27201; Spalerite, 29184; zinc sulfide, 34796;
215-252-9: stannic sulfide, 29496;
215-255-5: antimony pentasulfide, 2570; golden antimony sulfide, 13177; sulfur gold, 30045;
215-263-9: Moldag 200, 20049; molybdenum disulfide, 20077; Molydag, 20083; Molydag 204, 20084; Molydag 206, 20085; Molydag 208, 20086; Molydag 210, 20087; Molydag 211, 20088; Molydag 214, 20089; Molykote®, 20090; SLA 1208, 1261, 28576; SLA 1261, 28577; SLA 1286, 28580; SLA 2208, 28583;
215-264-4: manganite, 18555;
215-267-0: lead monoxide, 16910; Litharge, 17501; Vulcaid, 34072;
215-268-6: white pyrites, 34371;
215-269-1: cupric oxide, 7386; UP 13600, 33091;
215-270-7: Brown Copp., 4618; Cupridan, 7389; cuprous oxide, 7405; Cuprox, 7409; Lolotint 97, 17562; Nordox, 21540; Perenox, 23240; Purple Copp, 25417; Purplecopp 97N Premium, 25418; ruby ore, 27120; violet copper, 33875;
215-279-6: Atomite®, 3379; Calbux, 4929; calcium carbonate, 4940; calcium carbonate, 4941; calcium carbonate, 4942; Garden Lime, 12663; limestone, 17273; Portland stone, 24709; Supercoat®,

30228; Supermite®, 30256;
215-283-8: zeolites, 34718;
215-288-5: Fulcat Catalysts, 12436; Fulmont Activated Bleaching Earths, 12448; K 129, 15639; Lubrigel, 17736; Montmorillonite, 20300;
215-289-0: Sepigel A, 27985;
215-290-6: Ceruse, 5769; Halcarb 20, 13545; Kremser White, 16399; white lead, 34366;
215-293-2: cresylic acid, 7013; Medo, 19221; Productol, 25084; tricresol, 32229;
215-311-9: M.N.T, 18105;
215-321-3: Nibren wax, 21097;
215-329-7: Arosolve MN-LF, 3072;
215-343-3: Aerosol® OS, 895; Dehscofix 916, 7924; Rhodacal® IN, 26585; Supragil® WP, 30309;
215-347-5: Witconate DS, 34485;
215-350-1: Ceraphyl® 50, 5716; myristyl lactate, 20578; Wickenol® 506, 34400;
215-353-8: castor oil, 5417; Surfactol® 13, 30339;
215-354-3: Aldo® PGHMS KFG, 1544; Cerasynt® M, PA, 5736; CPH-52-SE, 6942; Emalex PGMS, 9986; Emerest® 2380, 10092; Grindtek PGMS 90, 13414; Hodag PGMS, 14215; Hodag PGS, 14224; Kessco® PGML, 16125; Kessco® PGMS, 16126; Lipo PGMS, 17343; Mazol® PGMS, 19140; Pegosperse® PMS CG, 23067; Promodan SP, 25144; Radiamuls® PG 2201, 25779; Schercemol PGMS, 27621; Witconol 2380, 34498;
215-355-9: Naturechem® GMHS, 20770; Paricin® 13, 22834;
215-356-4: Dinonyl phenol, 8544;
215-359-0: Kessco® GDS 386F, 16108;
215-474-6: Cooksons, 6753; Thermoguard® UF, 31637; Timonox, 31807; Timonox Blue Star, 31808;
215-477-2: Aloxicoll, 1813; aluminum chlorohydrate, 1903; Dow Corning® ACH-303, 8867; Locron, 17543; Micro-Dry®, 19669; Reach® 101, 201, 501, 25919; Ritachlor 50%, 26801;
215-478-8: Van Gel®, 33268;
215-481-4: arsenic trioxide, 3143; poison flour, 24286;
215-535-7: xylene, 34620;
215-540-4: antipyoninum, 2612; Dehybor, 7942; FB 48, 11513; sodium borate anhydrous, 28739; Solubor, 28990;
215-542-5: Ageflex FA-10, 979; Sipomer® IDA, 28539;
215-547-2: Bisomer D10M, 4275; Octomer DIOM, 22010;
215-548-8: Syn-O-Ad® 8484, 30558; TCP, 30857; tricresyl phosphate, 32230;
215-549-3: Emalex PGO, 9987; Prolein, 25126; Propylene glycol oleate, 25221; Radiamuls® PG 2206, 25780;
215-551-4: Country Fresh Disinfectant, 6905;
215-553-5: Monoplex® DIOA, 20246; Plasthall® DIOA, 24000;
215-559-8: Ablusol DBM, 125; Ammonium dodecylbenzene sulfonate, 2238; Hetsulf 50A, 13978; Nansa® AS 40, 20684; Novozone®, 21745;
215-566-6: zinc borate, 34769;
215-568-7: cupric sulfate, basic, 7388; Cusatrib, 7429;
215-587-0: Eltesol® PSA 65, 9868; Phenosulfonic acid, 23617;
215-607-8: chrome bronze, 6204; chromous trioxide, 6247; red oxide of chromium, 25965;
215-609-9: acetylene black, 319; Addipast, 652; Black Pearls® 1100, 4305; carbon black, 5220; Continex® LH-10, 6741; Continex® LH-10, N-351, 6742; Continex® N351, 6743; Derussole, 8135; Diablack® A, 8245; Efweko, 9641; Elftex® 675, 9832; Flex Carbon, 11918; Furnex, 12496; FW 18, 12527; FW 200 Beads and Powd, 12528; Ketjenblack® EC-310 NW, 16129;
215-630-3: subacetate of lead, 29902;
215-641-3: Halowax 1014, 13564;
215-648-1: Masil® SF 1,000,000, 19017; Masil® SF 5, 19018; Masil® SF 500, 19020; Masil® SF 500,000, 19021; Sentry Dimethicone, 27959;
215-654-4: Morflex MSC, 20331;
215-657-0: copper naphthenate, 6783;
215-661-2: 2-butanone peroxide, 4738; Butanox, 4739; Cadox® HBO-50, 4913; Cadox® L-30, 4914; Hi-Point® 90, 14120; Lupersol DDM-9, 17871; Lupersol DSW, 17873; methyl ethyl ketone peroxide, 19531; Quickset® Extra, 25672; Sprayset® MEKP, 29338;
215-662-8: Agenap HMW-H, 1011; naphthenic acid, 20714;
215-663-3: Ablunol S-20, 108; Alkamuls® S-20, 1676; Alkamuls® SML, 1685; Arlacel® 20, 2899; Atmer® 100, 3356; Crill 1, 7028; Dehymuls SML, 7974; Disponil SML 100 F1, 8695; Emsorb® 2515, 10461; Ethylan® GL20, 11109; Glycomul® L, 13111; Hetan SL, 13888; Hodag SML, 14242; Kemester® S20, 15966; Liposorb L, 17404; Montane 20, 20265; Prote-sorb SML, 25278; S-Maz® 20, 28640; Sorbax SML, 29098; Sorbon S-20, 29117; Span® 20, 29187;
215-664-9: Ablunol S-60, 110; Alkamuls® S-60, 1677; Alkamuls® SMS, 1687; Arlacel® 60, 2901; Capmul® S, 5119; Crill 3, 7030; Dehymuls SMS, 7976; Disponil SMS 100 F1, 8702; Drewsorb® 60K, 9019; Durtan® 60, 9275; Emalex SPE-100S, 9992; Emsorb® 2505, 10458; Ethylan® GS60, 11112; Famodan MS Kosher, 11423; Glycomul® S FG, 13114; Grindtek SMS, 13415; Hetan SS, 13890; Hodag SMS, 14245; Kemester® S60, 15968; Liposorb S, 17412; Montane 60, 20267; Polycon S-60 K, 24399; Prote-sorb SMS, 25281; Radiamuls® Sorb 2145, Sorb 2161, Sorb 2166, 25783; S-Maz® 60K, 28642; Sorbax SMS, 29101; Sorbon S-60, 29119; Span® 60, 60K, 29189;
215-665-4: Ablunol S-80, 111; Alkamuls® S-80, 1681; Alkamuls® SMO, 1686; Arlacel® 80, 2902; Atmer® 105, 3358; Capmul® O, 5115; Crill 4, 7031; Dehymuls SMO, 7975; DeSotan® SMO, 8194; Disponil SMO 100 F1, 8698; Emalex SPO-100, 9996; Emsorb® 2500, 10455; Ethylan® GO80, 11110; Glycomul® O, 13112; Hetan SO, 13889; Hodag SMO, 14243; Kemester® S80, 15970; Liposorb O, 17407; Montane 481, 20270; Montane 80, 20269; Polycon S-80 K, 24400; Prote-sorb SMO, 25279; Radiasurf® 7155, 25802; S-Maz® 80, 28644; Sorbax SMO, 29099; Sorbon S-80, 29121; Span® 80, 29191; Witconol 2500, 34501;
215-667-5: Uniplex 600, 33019; Uniplex 680, 33020;
215-676-4: Ammonium bifluoride, 2232; fluoram, 12068; Matt salt, 19067;
215-681-1: Celkate T-21, 5592; magnesium silicate, 18366;
215-683-2: Akrochem® Rubbersil RS-150/RS-200, 1177; Flexin, 11935;
215-684-8: Ketjensil® SM 405, 16134; Valfor® 100, 33244; Wessalith, 34284;
215-687-4: Acsil, 454; Crystal, 7284; Gasbinda, 12688; Marley Cement Dustproofer, 18827; N., 20633; sodium silicate, 28817;
215-691-6: A-2, 6; Abradux, 150; Abramant, 151; Abramax, 152; Abrarex, 153; Abrasit, 154; Activated alumina, 495; Adamant, 621; Alcan AA-100, 1411; Alcan C-70, C-71, C-72, C-73, C-75, 1414; Alcan FRF 5, 10, 20, 30, 40, 60, 80, 85, 1415; Alcan FRF LV2, LV4, LV5, LV6, LV7, LV8, LV9, 1416; Alcan H-10, 1429; Alexite, 1568; Alfrax B301, 1580; Aloxite, 1814; Alpha-Ruvite, 1832; Alufrit, 1880; alumina, 1891; aluminum oxide, 1919; aluminum oxide C, 1920; Alumite, 1933; Alundum®, 1934; Bikorit, 4139; Borite, 4426; Borolon, 4441; Boxite, 4468; Brasivol, 4483; C-1, 4839; Carbo Alumina, 5195; Carbo-corundum, 5196; Compalox, 6672; Corolox, Corotox, Corowalt, 6842; Corubin, 6862; Corundite, 6863; D-201, 7632; Dirubin, 8622; Dural, 9149; Dycron, 9309; Flame guard, 11874; H-30, 13527; Haltex 300, 13573; Italcor, 15431; Maftec®, 18304; Martipol, 18969; Martisorb, 18970; Martoxin, 18971; Purdox, 25404; Realox®, 25931; Rewagit, 26293; Rex, 26435; S-201, 27182; Saffil®, 27221; SB-136, 27527; SB-336, 27528; SB-632, 27529; Selexsorb® COS, 27866; T-1061, 30674; T-64, 30673; Versal 150, 33605;
215-696-3: mercuric sulfide, red and black, 19350;
215-709-2: Orthorix, 22337;
215-710-8: calcium silicate, 4969; Vansil®, 33364;
215-713-4: antimony trisulfide, 2575;
215-715-5: Lithopone, 17520;
215-721-8: glassite, 12971; glossite, 13013;
215-724-4: Carmine, 5289; Carmine 40, 5290;
215-739-6: Litmus, 17523;
215-744-3: chitin, 6082; Kytamer® PC, 16522;
215-750-6: Archil, 2808;
215-753-2: quebracho, 25636; tannic acid, 30807;
215-754-8: tylosin, 32663;

215-763-7: Vanobid, 33332;
215-800-7: Amplex, 2362; chlorophyll, 6134;
215-808-0: Veragel® 200, 33566;
215-811-7: Bayluscide®, 3805; clonitrilide, 6432;
215-899-7: Amine D, 2155;
215-911-0: CH7280, 5840; Hexamethyldisilane, 14002;
215-951-9: Morflex 1129, 20330; Uniplex 270, 33016;
215-968-1: Mirataine® HC-Acid, 19929;
215-969-7: dodecylthioethanol, 8781; DV-1936, 9296;
216-078-6: Maretin, 18704; Rametin, 25856;
216-087-5: Fluorad® FC-24, 12059; trifluoromethanesulfonic acid, 32251;
216-208-1: Morflex 560, 20329;
216-245-3: Eastobrite® OB-1, 9475; Ethenediyl diphenylene bisbenzoxazole, 10940;
216-260-5: Diapen, 8327;
216-319-5: CC3275, 5488; chloromethyldichloromethylsilane, 6128;
216-341-5: Geropon® MLS/A, 12909; sodium methallyl sulfonate, 28790;
216-343-6: Witconate NIS, 34487;
216-353-0: Carbodan, 5197; Curaterr®, 7414; Nex, 21079; Rampart, 25860; Sipcam UK Carbosip 5G, 28515; Tripart® Nex, 32382; Yaltox®, 34636;
216-428-8: Agriphlan 24, 1070; Ornamental Weeder, 22308; Treflan, 32162; Trifluralin, 32252; Trigard, 32256; Trilin® 10G, 32319; Trimaran, 32333; Tripart® Trifluralin 48 EC, 32387; Tristar, 32415;
216-472-8: Afco-Chem CS, 921; calcium stearate, 4970; Hallcote® CSD, 13555; Nopcote C-104, 21510; Petrac® CP-11, 23463; Synpro® 15F, 30590;
216-485-9: Alar, 1334; B-Nine, 4365; daminozide, 7723; Dazide, 7805;
216-491-1: mercuric acetate, 19346;
216-643-7: strontium carbonate, 29828;
216-652-6: Akrochem® TBUT, 1178; tetrabutylthiuram disulfide, 31363;
216-653-1: Driveron, 9046;
216-696-6: Franocide, 12378;
216-699-2: TBAB, 30851; tetrabutyl ammonium bromide, 31362;
216-700-6: Ablumox LO, 91; Ammonyx® LO, 2264; Empigen® OB, 10375; lauramine oxide, 16832; Rhodamox® LO, 26620; Schercamox DML, 27575;
216-715-8: azaperone, 3518; Stresnil, 29813;
216-751-4: Weston® TLTTP, 34322;
216-853-9: SR-272, 29374;
216-881-1: Actrilawn 10, 566; Totril, 32081;
216-882-7: Bromoxynil, 4590; Buctril, 4650;
216-884-8: nitroxynil, 21384; Trodax, 32439;
216-920-2: Alicep®, 1610; Atlas Silver, 3330; Atlas Silver, 3331; Better Flowable, 4076; Chiltern Pyrazol, 6071; chloridazon, 6103; Gladiator, 12953; Paramin DF, 22774; Pyramin®, 25466; Starter Flowable, 29545; Tripart® Gladiator, 32372; Trojan SC, 32442; Weedmaster, 34261;
216-935-4: Agrichem, 1045; Agricorn D, 1051; BH 2,4-D Ester 40, 4115; Campbell's Destox, 5057; CDA Dicotox Extra, 5519; Clopyralid, 6436; Dow Shield, 8887; Format, 12267; Hadranol, 13533; Lontrel, 17586; Shield, 28173; silvapron, 28409;
216-940-1: Texacat® ZR-70, 31437;
216-948-5: Sofibex, 28835;
216-996-7: Alugan, 1881; Bromocyclene, 4576; Bromodan, 4577;
217-006-6: CC3270, 5487; chloromethyldimethylchlorosilane, 6129;
217-007-1: CV-4720, 7447;
217-060-0: dimethyl azelate, 8504; Emery® 2914, 10158;
217-064-2: cresol red, 6996;
217-129-5: Aresin, 2839; Arresin, 3127;
217-157-8: formammidinesulfinic acid, 12262; Manofast, 18571; thiourea dioxide, 31721;
217-164-6: Aminoethylaminopropyltrimethoxy silane, 2171; CA0700, 4850; Dow Corning® Z-6020, 8878; Dynasylan® DAMO, 9404; Dynasylan® DAMO-P, 9405; Dynasylan® DAMO-T, 9406; Prosil® 3128, 25239; Union Carbide® A-1120, 32993;
217-175-6: Ammonium thiocyanate, 2256;
217-210-5: 2,4-dichlorobenzyl alcohol, 8387; Myacide® SP, 20546;
217-234-6: Aspartic acid, 3234;
217-291-7: Sinapoline, 28491;
217-316-1: Syn-O-Ad® P-316, 30562;
217-325-0: Varisoft® 432-100, 33409;
217-370-6: CT2970, 7326; EXP-51, 11306;
217-420-7: Lowinox® CA 22, 17665; Topanol® CA, 32023;
217-421-2: Cyasorb® UV 531, 7497; Hostavin® ARO 8, 14418; Lankromark® LE285, 16697; Lowilite® 22, 17642; Spectra-Sorb UV 531, 29232; Uvinul® 408, 33192;
217-430-1: Rhodapon® OS, 26640; Sulfopon O 680, 30026; Supralated LS, 30310;
217-450-0: 4,4'-(2-Ethyl-2-nitrotrimethylene)-dimorpholine, 11141; Bioban® P-1487®, 4160;
217-464-7: Dacthal, 7685; dimethyl tetrachloroterephthalate, 8512; Vegetable Turf and Ornamental Weeder, 33526;
217-465-2: Benfluralin, 3940;
217-488-8: tetrazolium chloride, 31395;
217-568-2: Perkadox® 30, 23305;
217-588-1: Bombardier, 4386; Bravo 500, 4490; Chiltern Ole, 6070; Chlorothalonil, 6150; Contact 75, 6738; Daconil Turf, 7681; Groutcide 75, 13432; Jupital, 15633; Nopcocide® N-40-D, 21487; Nopcocide® N-96, 21488; Nopcocide® N-96-S, 21489; NuoCide®, 21773; Power Chlorothalonil 50, 24818; Repulse, 26151; Siclor, 28204; Sipcam UK Rover 500, 28516; Tripart® Faber, 32371; Tripart® Ultrafaber, 32388;
217-615-7: Gramoxone, 13254; Gramoxone, 13255; Gramoxone X, 13256; Scythe, 27779; Speedway, 29264;
217-617-8: Ashlade 4% At Gran, 3184; Ashlade Atrazine 50 FL, 3188; Atraflow, 3387; Atranex, 3393; atrazine, 3394; Borocil A, 4430; Boroflow A, 4434; Boroflow A/ATA, 4435; Chlorea, 6094; Gesaprim 500FW, 12922; Herbazin Total, 13805; Mascot Gauntlet, 19002; Primatol AA, 24969;
217-618-3: trietazine, 32238;
217-635-6: Dicamba, 8374; Tracker, 32130;
217-636-1: Tordon, 32056; Tordon 22K, 32057;
217-638-2: Albras Propachlor, 1384; Albrass, 1385; Atlas Orange, 3322; Bexton, 4096; Croptex Amber, 7212; Portman propachlor 50FL, 24713; Prolex, 25128; Propachlor, 25163; Propachlor, 25164; Ramrod, 25861; Tripart® Granular, 32373; Tripart® Sentinel, 32385;
217-650-8: Pristane, 25037;
217-680-1: Planotox, 23914;
217-691-1: Esperox® 41-25A, 10850;
217-699-5: tartrazine, 30835;
217-752-2: *t*-Butylhydroquinone, 4793; Eastman® MTBHQ, 9471; Sustane® TBHQ, 30437;
217-830-6: Nitrile 10 D, 21335;
217-843-7: chloroxuron, 6158; Tenoran, 31119;
217-847-9: bioMeT TBTF, 4171;
217-850-5: sodium caprylate, 28743;
217-901-1: Locorten, 17541;
217-950-9: Amine 14D, 2129;
217-956-1: Acetamin 24, 284;
217-980-2: Clofazimine, 6427;
217-983-9: CM9050, 6464; Dynasylan® MTES, 9413; Union Carbide® A-162, 32982;
217-991-2: Borderland Black, 4416; Club, 6452; Draza®, 8973; Mesurol®, 19401; methiocarb, 19484;
218-058-2: Empicol® HL25, 10322;
218-134-5: Dichloroditane, 8390;
218-142-9: Brunol, 4625;
218-163-3: Asparagine, 3232;
218-216-0: Anox® PP 18, 2486; Lowinox® PO35, 17671; Naugard® 76, 20789; Ralox® 530, 25843; Ultranox® 276, 32849;
218-217-6: Weston® TSP, 34324;
218-222-3: CD5400, 5505;
218-254-8: V-40, 33216;
218-320-6: Abil® AV 8853, 20-1000, 47; Emalex MTS-30E, 9971; phenyl dimethicone, 23632; Phenyl trimethicone, 23635;
218-332-1: Tre-Hold, 32165;
218-429-9: pyronine B, 25522;
218-454-5: Trigonox® 111-B40, 32280;
218-463-4: Ageflex FA-12, 980; lauryl acrylate, 16863; SR-335, 29378;
218-495-9: Drexar 530, 9023; MSMA, 20417; Neoarsycodyl, 20871; Weed Hoe, 34257;
218-499-0: lenacil, 16984; Venzar®, 33562; Vizor, 34001, 34002;
218-500-4: Cottonex, 6894; Meturon® 4L, 19607;
218-507-2: Trigonox® D-E50, 32298;
218-583-7: Acetamin 86, 285; Acetamin T, 288;
218-645-3: DY 023, 9300; Heloxy® 62, 13771;
218-715-3: Apagallin, 2640; keraphen, 16055;
218-724-2: Nitron, 21367;
218-743-6: Synpro® Cadmium Stearate, 30597;
218-747-8: CV-5050, 7454;
218-748-3: Nitrobutylmorpholine, 21351;
218-763-5: dienochlor, 8429; Pentac Aquaflow, 23133;
218-791-8: Unihib® 314, 32959;
218-793-9: Akyposal ALS 33, 1302; Ammonium

lauryl sulfate, 2242; Avirol® A, 3477; Carsonol® ALS-R, 5332; DeSonol A, 8186; Empicol® AL30, 10306; Lorol NH, 17620; Maprofix NH and NHL, 18664; Marlinat® DFN 30, 18853; Neopon LAM, 20945; Nutrapon HA 3841, 21821; Nutrapon PP 3563, 21824; Octosol ALS-28, 22024; Perlankrol® DAF25, 23325; Polystep® B-7, 24597; Rhodapon® L-22, L-22/C, 26635; Sermul EA129, 28048; Standapol® A, 29467; Stepanol® AM, 29693; Sulfochem ALS, 29996; Tensopol N, 31194; Texapon ALS, 31453; Ufarol Am 30, 32742; Witcolate NH, 34478; Zoharpon LAA, 34864;
218-828-8: ACL 56, 59, 60, 66, 354; CDB Clearon, 5525; Clearon, 6392; dichloro-1,3,5-triazinetrione, potassium salt, 8384; potassium dichloroisocyanurate, 24753;
218-904-0: Oxyclozanide, 22444; Zanil, 34692;
218-960-6: cresol purple, 6995;
218-961-1: Avadex, 3449;
218-962-7: Avadex® BW, 3450; Far-Go/Avadex BW, 11470; triallate, 32190;
218-988-9: Croptex Bronze, 7213; pentanochlor, 23165; Solan, 28887;
218-996-2: phosalone, 23691; Zolone, 34888;
219-015-0: Sudafed Nasal Spray, 29937;
219-031-8: Fluorescein, 12080; Fluoresceine, 12082;
219-058-5: CC3285, 5489; chloromethyltrimethyl silane, 6130;
219-099-9: Ageflex DEGDMA, 962; diethylene glycol dimethacrylate, 8444;
219-143-7: Trigonox® BPIC, 32296;
219-258-2: potassium sorbate, 24781;
219-266-6: Tinaderm, 31818;
219-314-6: *p*-Butylcresol, 4792; Cao, 5098; Lowinox® MBPC, 17669;
219-351-8: Desmodur® RE, 8157;
219-352-3: quindoxin, 25682;
219-363-3: captafol, 5163; Difolatan, 8452; Merpafol, 19368; Sanspor, 27379;
219-371-7: Heloxy® 67, 13774;
219-376-4: Ageflex BGE, 967; Ageflex TBGE, 998; Epirez 501, 10654; Heloxy® 61, 13770;
219-378-5: Ageflex FA-2, 976;
219-440-1: Nitrile 12, 21336;
219-451-1: bufexamac, 4655; Parfenac, 22830;
219-455-3: Morestan, 20311; Morestan®, 20312; quinomethionate, 25686;
219-459-5: Dodine FL, WP, 8785; Radspor FT, 65WP, 25833;
219-460-0: Adame, 622; Ageflex FA-1, 974; Ageflex FA-1, 975;
219-470-5: Lowilite® 55, 17643; Topanex 100BT, 32019; Uvazol P, 33184;
219-529-5: Ageflex THFMA, 999; tetrahydrofurfuryl methacrylate, 31372;
219-644-0: Thiate E, 31666;
219-660-8: Nacap®, 20643; Nuodex 84, 21779;
219-671-8: Ageflex FM-25, 987;
219-673-9: Melcril 4085, 19267;
219-698-5: Ageflex FA-6, 977; Ageflex n-HA, 996;
219-700-4: Jasmacyclene, 15529;
219-754-9: MP-2, 20388;
219-784-2: Aktisil EM, 1202; CG6720, 5836; Dow Corning® Z-6040, 8881; Dynasylan® GLYMO, 9408; Prosil® 5136, 25240; Union Carbide® A-187, 32988;
219-785-8: CM8550, 6456; Dow Corning® Z-6030, 8879; Dynasylan® MEMO, 9412; Prosil® 248, 25238; Union Carbide® A-174, 32986;
219-787-9: CC3300, 5493; chloropropyltrimethoxysilane, 6138; 3-chloropropyltrimethoxysilane, 6142, 6143; Dow Corning® Z-6076, 8883; Dynasylan® CPTMO, 9403;
219-844-8: CC3291, 5491; chloropropyltrichlorosilane, 6139;
219-854-2: Esaflon, 10782; sulfur hexafluoride, 30046;
219-855-8: CA0567, 4847;
219-860-5: CD6000, 5514;
219-919-5: Admox® 18-85, 759; Amyx SO 3734, 2395; Barlox® 18S, 3652; Chemoxide ST, 6000; Incromine Oxide S, 15011; Mackamine SO, 18177; Mazox® SDA, 19155; Ninox® SO, 21244; Rewominox S 300, 26335; Rewominoxid S 300, 26338; Schercamox DMS, 27577;
219-976-6: V-601, 33219;
219-989-7: 1-hydroxybenzotriazole, 14659;
219-991-8: Aaterra WP, 17;
220-001-1: Schercemol CM, 27586;
220-021-0: Photine, 23756;
220-099-6: CD6210, 5517;
220-114-6: Etrofolan®, 11156; isoprocarb, 15404; isopropyl phenylmethyl carbamate, 15411; Mipcin®, 19846;
220-120-9: Nipacide® BIT, 21253;
220-147-6: Azinphos-ethyl, 3520; Cotnion-Ethyl, 6882; Gusathion® A, 13502;
220-153-9: Phospholipon® PC, 23729;
220-219-7: Aerosol® TR-70, 899; Monawet MT Series, 20186;
220-222-3: Chloroneb 65W Fungicide, 6131; Chloroneb Systemic Flowable Fungicide, 6132; Terraneb SP Turf Fungicide, 31337;
220-250-6: Agsol Ex 2, 1087; NEP, 20987;
220-264-2: Sipomer® TATM, 28541;
220-281-5: Vikane, 33754;
220-336-3: Emersol® 871, 10122; isostearic acid, 15421; Prisorine 3508, 25036; Proto-Lan IP, 25295;
220-404-2: CT3254, 7328;
220-449-8: CV-4917, 7452; Dynasylan® VTMO, 9421; Union Carbide® A-171, 32984;
220-476-5: Alkamuls® SS, 1688; Emalex CC-18, 9936; Hetester 412, 13892; Lexol SS, 17155; Liponate SS, 17388; Radia® 7501, 25743; Ritachol® SS, 26804;
220-548-6: Ektasolve® EP, 9671; ethylene glycol propyl ether, 11130;
220-552-8: Briquest® ADPA-60A, 4536; Fostex P, 12338; Unihib® 106, 32957;
220-666-8: Vestamin® TMD, 33656;
220-672-0: CT2928, 7324;
220-688-8: Ageflex FM-1, 983; Sipomer® 2M1M, 28528;
220-767-7: dichloro-1,3,5-triazinetrione, sodium salt, 8385;
220-780-8: CG6710, 5835; glycidoxypropyl methyl diethoxysilane, 13090; GP-137, 13221;
220-835-6: Octomer DOM, 22011; PX-538, 25444;
220-836-1: Wickenol® 159, 34394;
220-864-4: Chlorpyrifos, 6163; Crossfire, 7239; Dowco 179, 8902; Dursban®, 9272; Dursban® 14G, 9273; Dursban® 2E, 9274; Lorsban, 17623; Pyrinex, 25486; Spannit, 29198; suSCon® Blue, 30425; suSCon® Green, 30426; Talon, 30751;
220-941-2: CO9835, 6513; Dynasylan® OCTEO, 9416; Prosil® 9202, 25241; Union Carbide® A-137, 32980;
221-066-9: CP0330, 6934; phenyltrimethoxysilane, 23651;
221-109-1: Aerosol® MA-80, 893; Empimin® MA, 10427; Octosol HA-80, 22025;
221-110-7: Aztec® *t*-Butyl Peroctoate, 3544; Aztec® *t*-Butyl Peroctoate-50 OMS, 3545; Esperox® 28, 10847; Trigonox® 21, 32265;
221-111-2: Aztec® 1,1-Bis(*t*-Butylperoxy)Cyclohexane-80 BBP, 3550; Luperco 331-XL, 17827; Lupersol 331-80B, 17860; Trigonox® 22-BB80, 32266; USP® -400P, 33168;
221-112-8: HVA-2, 14480; Vanax® MBM, 33291;
221-188-2: Avirol® SA 4113, 3483; Rhodapon® TDS, 26641;
221-221-0: Ogtac 85 V, 22052;
221-231-5: Perkadox® 20, 23303;
221-266-6: murexide, 20506;
221-279-7: Akyporox RLM 22, 1291;
221-281-8: Mulsifan RT 23, 20458; Oxetal VD 20, 22425;
221-291-2: Weston® PNPG, 34316;
221-301-5: Metobromuron, 19583; Patoran, 22916; Patoran® FL, 22917; Pattonex, 22923;
221-304-1: Kemamide® B, 15894; Uniwax 1747, 33067;
221-310-4: Amerstat® 294, 2044; bistrichloromethylsulfone, 4285; Stauffer N-1386®, 29580;
221-328-2: CP0156, 6930;
221-334-5: CM8620, 6457;
221-336-6: CA0699, 4849;
221-356-5: Doverphos® 53, 8842; Doverphos® TLP, 8852; Weston® TLP, 34321;
221-416-0: Akyposal RLM 56 S, 1319; sodium laureth sulfate, 28778;
221-441-7: pyrolignite of iron, 25515;
221-450-6: Akyposal MGLS, 1309; Empicol® EGB, EGC, 10315; magnesium laureth sulfate, 18363; magnesium lauryl sulfate, 18364; Rhodapon® LM, 26637; Standapol® MG, 29475; Texapon MG, 31465; Zoharpon MgES, 34871;
221-453-2: Heloxy® 65, 13773;
221-486-2: Akypo NP 70, 1226;
221-490-4: Sudan II, 29939;
221-498-8: Benzophenone-9, 3979; Uvinul® DS-49, 33196;
221-499-3: Ageflex MEA, 994;
221-515-9: Dichan 100, 8377; dicyclohexylamine nitrite, 8414;
221-554-1: d-camphorsulfonic acid, 5076;
221-573-5: Cyasorb® UV 5411, 7503; Spectra-Sorb UV 5411, 29233; Uvazol 311, 33183;
221-658-7: Trigonox® ACS-M28, 32292;
221-660-8: CA0742, 4851;
221-661-3: Chemidex L, 5963; lauramidopropyl dimethylamine, 16831; Lexamine L-13, 17098; Mackine 801, 18224;

Schercodine L, 27638;
221-662-9: Cerasynt® 303, 5733;
221-695-9: Saytex® HBCD-LM, 27522;
221-787-9: Alkamuls® MM/M, 1670; Ceraphyl® 424, 5727; Cetiol® MM, 5801; Crodamol MM, 7122; Kemester® MM, 15965; Liponate MM, 17380; Schercemol MM, 27609; Waxenol® 810, 34237;
221-816-5: acid fuchsine, 333;
221-838-5: cupric nitrate, 7384;
221-890-9: Nifuraldezone, 21169;
221-906-4: CT2030, 7314;
221-913-2: FR-913, 12359; Great Lakes PHE-65, 13311; tribromophenyl allyl ether, 32203;
221-950-4: Ageflex TM 402, 403, 404, 410, 421, 423, 451, 461, 462, 1000; Ageflex TMPTMA, 1002; Perkalink® 400, 23314; SR-350, 29380;
221-967-7: dibromoneopentyl glycol, 8358; FR-522, 12355;
222-020-0: Kodaflex® TOTM, 16280; Palatinol® TOTM, 22595; Plasthall® TOTM, 24016; PX-338, 25441; Syn-O-Ad® P-374, 30563;
222-036-8: Saytex® BCL-462, 27518;
222-048-3: (3-Chloro-2-hydroxy)o-propyltrimethyl ammonium chloride, 6126; 3-Chloro-2-hydroxypropyl trimethylammonium chloride (50% aq solution), 6127; Catiomaster-C, 5459; CHPTA 65%, 6176; Quab 188, 25562; Reagens-CF2, 25928;
222-059-3: Admox® 14-85, 758; Barlox® 14, 3650; Empigen® OH25, 10377; Incromine Oxide M, 15007; Lilaminox M4, 17252; Mazox® MDA, 19153; Ninox® M, 21243; Schercamox DMM, 27576;
222-068-2: nickel carbonate, 21123;
222-077-1: Asulam, 3277; Asulox, 3278, 3279;
222-098-6: Delan-Col, 8006; Dithianone, 8729;
222-101-0: Nicfo, 21101;
222-191-1: Abate 1-SG, 2-CG, 4-E, 5CG, 26;
222-217-1: CE6250, 5528; Union Carbide® A-186, 32987;
222-226-0: Morillol®, 20335;
222-264-8: sodium undecylenate, 28830;
222-321-7: TAHP-80, 30719; Trigonox® TAHP-W85, 32309;
222-374-6: Neodene® 20, 20900;
222-376-7: Nonanol, 21436;
222-389-8: Butyl cumyl peroxide, 4768; Luperco 801-XL, 17829; Trigonox® T, 32308;
222-392-4: seminose, 27939;
222-402-7: amicarbalide, 2063;
222-468-7: Clelands Reagent, 6402;
222-477-6: Akrochem® 9930 Zinc Oxide Transparent, 1152; zinc carbonate, 34771;
222-512-5: Versapen, 33609;
222-534-5: Chlolografin Meglumine, 6172;
222-540-8: pentaerythrityl triacrylate, 23144; SR-444, 29384;
222-793-4: rhamnose, 26529;
222-823-6: Butylbenzene sulfonamide, 4789; Dellatol® BBS, 8026; Plasthall® BSA, 23993; Plastomoll® BMB, 24056;
222-848-2: Almora, 1790; magnesium gluconate, 18362;
222-884-9: Jayflex® DUP, 15556; PX-111, 25435;
222-899-0: Deriphat 160, 8090; disodium lauryliminodipropionate, 8650; Miranol® H2C-HA, 19897; Monateric 1188M, 20149;
222-904-6: Syn-O-Ad® P-310, 30560;
222-980-4: Cetiol®, 5793; Schercemol OLO, 27616; Starfol® OO, 29537; Wickenol® 143, 34386;
222-981-6: Ceraphyl® 140, 5722; Cetiol® V, 5806; decyl oleate, 7869; Schercemol DO, 27598; Tegosoft® DO, 31019;
222-995-2: Bladafum®, 4308;
223-003-0: Chlorophacinone, 6133; Drat, 8972; Karate, 15758; Ridene, 26746; Sakarat Special, 27239; Skaterpax, 28564;
223-024-5: Omadine® MDS, 22152;
223-055-4: Pennstop® 1866, 23118;
223-077-4: Crotonic Acid, 7260;
223-095-2: Bitrex®, 4291;
223-200-1: CB2100, 5473;
223-228-4: Ageflex FM-4, 984;
223-276-6: Distearyl pentaerythritol diphosphite, 8712; Doverphos® S-680, 8850; Weston® 618, Weston® 619, 34304;
223-289-7: potassium chlorate, 24746;
223-296-5: Sodium Omadine® 40% Aq. Sol'n, 28803; sodium pyrithione, 28813;
223-297-0: benazolin, 3935;
223-320-4: Fluorad® FC-118, 12058;
223-356-0: Trigonox® 36-C75, 32270; USP® -355M, 33167;
223-383-8: Uvazol 237, 33182;
223-404-0: fuberidazole, 12428;
223-445-4: hydroxybutylmethylphenylchlorobenzo triazole, 14660; Lowitite® 26, 17677; Uvazol 236, 33181;
223-459-0: 4,5,8-trimethylpsoralen, 32340;
223-470-0: 2-butyloctanol, 4795; Michel XO-150-12, 19640;
223-498-3: sodium monochloracetate, 28797; Somon, 29072;
223-534-8: Solvenon® IPP, 29047;
223-545-8: sinigrin, 28500;
223-635-7: Sipomer® MEM, 28540;
223-662-4: Nibiol, 21095;
223-772-2: Rhodialux S, 26676; Spectra-Sorb UV 284, 29231; Sungard, 30143; Syntase® 230, 30622; Uval, 33177; Uvinul® MS-40, 33199;
223-781-1: Pruv, 25354;
223-795-8: Amasil® P, 1947; calcium propionate, 4968; Luprosil® Salt, 17921;
223-805-0: Dowicil® 75, 8937;
223-819-7: Amine M218, 2160; Radiamine 6346, 25763;
223-861-6: Vestanat® IPDI, 33657;
223-943-1: CV-4800, 7449; Dow Corning® Z-6075, 8882;
224-033-7: Axiquel, 3509;
224-079-8: Uniplex 552, 33018;
224-160-8: glycol palmitate, 13103; Lanol P, 16737;
224-221-9: CM8980, 6462;
224-292-6: lauramidopropyl betaine, 16830; Lexaine® LM, 17094;
224-318-6: Ancaris, 2411; Canopar, 5093; thenium closylate, 31568;
224-320-7: Uvi-Nox 1494, 33189;
224-506-8: Permethyl 101A, 23387;
224-583-8: V-65, 33220;
224-588-5: Aktisil MM, 1203; CM8500, 6455; Dynasylan® MTMO, 9414; Prosil® 196, 25236; Union Carbide® A-189, 32989;
224-736-9: Gluconal® ZN, 13047; Zinc gluconate, 34776;
224-769-9: CD6220, 5518;
224-772-5: Afco-Chem LIS, 922; lithium stearate, 17512;
224-831-5: Esperox® 5100, 10859; Trigonox® 127, 32284;
224-970-1: dimethylaminoethyl chloride hydrochloride, 8516;
225-004-1: Farnesol, 11492;
225-101-9: Brasoran 50 WP, 4484;
225-141-7: Dextrone X, 8239; paraquat, 22796; Paraquat + Plus, 22797;
225-146-4: disodium inosinate, 8645;
225-190-4: Sulfochem K, 30007; Tensopol SPK, 31195;
225-202-8: diphenyl phosphite, 8599; Doverphos® 213, 8843; Doverphos® DPP, 8849; Weston® DPP, 34309;
225-208-0: Nipacide® BK, 21254;
225-214-3: Akyposal MLS 30, 1311; Empicol® LQ33/T, 10326;
225-219-0: nitralin, 21319; Planavin, 23909;
225-258-3: DMP, 8757;
225-268-8: Ebal, 9488; ethyl-benzaldehyde, 11118;
225-305-8: CT1750, 7311; T1750, 30675;
225-373-9: Kalidone®, 15678;
225-469-0: Hampamide B, 13584;
225-714-1: Nipagin M Sodium, 21268;
225-733-5: Ageflex FM-1Q80MC, 982; Madquat Q-6, 18293;
225-768-6: Cheelox® NTA-Na3, 5869; Hampshire® NTA 150, 13610; Hampshire® NTA Na3, 13612; NTA, 21759; Trilon® A, A-92, 32323;
225-805-6: 3-chloropropyltriethoxysilane, 6140, 6141; CC3292, 5492; Dynasylan® CPTEO, 9402;
225-875-8: Plondrel, 24160;
226-020-1: pyrimithate, 25485;
226-021-7: dimethirimol, 8495; Milcurb, 19756;
226-029-0: Uvinul® N-35, 33200;
226-031-1: carboxin, 5255;
226-066-2: oxycarboxin, 22441; Plantvax, 23921; Plantvax 20, 23922; Plantvax 75, 23923; Ringmaster, 26771;
226-097-1: Akyporox RLM 40, 1292; Hetoxol L-4, 13941; Mulsifan CPA, 20453;
226-112-1: CO9830, 6512;
226-159-8: Arlasolve® DMI, 2913;
226-182-3: H-K Mastitis, 14150; Versapen K, 33610;
226-191-2: Solricin® 435, 28967; Soricinol 40, 29143;
226-218-8: sulfamic acid, 29963;
226-242-9: Eutanol G, 11239; Exxal® 20, 11343; Michel XO-150-20, 19643; octyl dodecanol, 22038;
226-300-9: Lonzest® 143-S, 17595; Schercemol MP, 27610;
226-333-3: diaveridine, 8338;
226-342-2: CV-5100, 7455;
226-362-1: Mexitil, 19615;
226-394-6: citral (*cis* and *trans*), 6336; Neral, 20996;
226-420-6: Tomah PA-12EH, 31980;
226-546-1: Lactamide MEA, 16568; Mackamide

LME, 18157; Parapel® LAM-100, 22786; Schercomid LME, 27652;
226-602-5: Schercemol 65, 27579;
226-713-9: ethyl selenac, 11082; Selazate, 27853;
226-775-7: Escalol® 557, 10788; Neo Heliopan® AV, 20866; octyl methoxycinnamate, 22039; Parsol® MCX, 22867;
226-886-0: Vanceril, 33295;
226-889-7: pentagastrin, 23146; Peptavlon, 23201;
226-956-0: CM8645, 6458;
226-969-1: Iopydone, 15235;
227-011-5: Chlorpyrifos-methyl, 6164; Cooper Graincote, 6759; Reldan 50, 26037;
227-031-4: Lederspan, 16958;
227-197-8: drazoxolan, 8974; Ganocide, 12648; Mil-Col, 19755; Saisan, 27235;
227-264-1: Gluconal® CA M B, 13039;
227-335-7: Crolactil CSL, 7184; Pationic CSL, 22906; Radiamuls® CSL 2980, 25772; Verv®, 33644;
227-369-2: Trigonox® TMPH, 32310;
227-424-0: Racumin®, 25713;
227-561-6: Ageflex IBOA, 991; Sipomer® IBOA, 28537;
227-567-9: Bismuth subcarbonate, 4242;
227-595-1: Sinbar®, 28494; terbacil, 31220;
227-637-9: Terbuthylazine, 31227; Tyllanex, 32658;
227-813-5: Limonene, 17280;
227-847-0: Bis(1-methylamyl) Sodium Sulfosuccinate, 4227; Lankropol® KMA, 16713;
228-126-3: Ageflex FM-1620, 989; Ageflex FM-68, 988;
228-227-2: Schercotaine IAB, 27697;
228-229-3: Emalex PG-di-S, 9984;
228-250-8: Neo Heliopan® 303, 20865; Uvinul® N-539, 33201;
228-464-1: Emcol® E-607L, 10059; Lapyrium Chloride, 16782;
228-491-9: DV-2301, 9297;
228-504-8: Ceraphyl® 31, 5713;
228-506-9: Meglumine, 19243;
228-652-3: Amerstat® 282, 2043; N-948® Biocide, 20641; Tolcide MBT, 31934;
228-973-9: Eribate, 10753;
228-973-9: Neo-Cebitate®, 20878; sodium erythorbate, 28761, 28762;
229-146-5: Aminotrimethylene phosphonic acid, 2191; Chemphonate AMP, 6004; Fostex AMP, 12337; Unihib® 305-LC, 32958;
229-176-9: Kodaflex® DOTP, 16274;
229-222-8: DMDM hydantoin, 8754; Glydant®, 13140; Mackstat® DM, 18234; Nipaguard® DMDMH, 21269;
229-347-8: Ammonium nitrate, 2245; Ansax, 2488; Herco-Prills, 13835; Nitram, 21322; Nitrammite, 21323; Old Plantation, 22106;
229-457-6: Zoldine® ZT-55, 34887;
229-567-4: proglumide, 25116;
229-680-9: Afcolene, 926;
229-713-7: Diazabicycloundecene, 8341;
229-722-6: Anox® 20, 2483; Lowinox® PP35, 17672; Naugard® 10, 20788; Ralox® 630, 25844;
229-782-3: Aztec®, 3539; Aztec® 1,1-Bis(*t*-Butylperoxy)-3,3,5-Trimethyl Cyclohexane, 3549; Luperco 231-XL, 17825; Lupersol 231, 17856; Peroximon® S-164/40P, 23410; Trigonox® 29, 32269;
229-879-0: carbadox, 5177; Mecadox, 19200;
229-887-4: rhodinol, 26685;
229-912-9: Crystamet 1020, 7293; Drymet® 59, 9065; Lasilso, 16806; Metso, 19604; sodium metasilicate, 28788; sodium metasilicate pentahydrate, 28789;
229-919-7: CHP, 6173;
229-929-1: CD6010, 5515;
229-934-9: Kodaflex® TXIB, 16282;
229-966-3: Barium stearate, 3643; Synpro® Barium Stearate, 30596;
229-985-7: diethylene glycol propyl ether, 8445; Ektasolve® DP, 9667;
229-995-1: Ageflex FM-1Q80DMS, 981;
230-022-8: hydroxysuccinic acid, 14671; Malic Acid, 18483;
230-042-7: Monocron, 20223;
230-253-4: Famid, 11422;
230-279-6: 2-dimethylamino-2-methyl-1-propanol, 8517; DMAMP-80, 8753;
230-351-7: oleandocyn, 22111;
230-386-8: Chafer CMPP Super, 5845; Cleanacres CMPP, 6381; Clenecorn, 6403; Clifton CMPP Amine 60, 6416; Clovotox, 6449; Compitox Extra, 6680; FBC CMPP, 11514; Headland Charge, 13727; Hymec, 14685; Iso-Cornox, 15353; Mascot Cloverkiller, 19000; Mecoprop, 19204;
230-429-0: Ammonyx® CO, 2262; Amyx CO 3764, 2393; Barlox® 16S, 3651; Mazox® CDA, 19150;
230-525-2: Arquad® 210-50, 3110; BTC® 1010-80, 4638; BTC® 99, 4633; Catigene® 1011, 5449; FMB 210-8, 210-15, 12131; Radiaquat 6410, 6412, 25789; Radiaquat 6412, 25790; Rewoquat B 10, 26389;
230-528-9: Diamine OL, 8298; Duomeen® OL, 9114; Jet Amine DO, 15589; Kemamine® D-989, 15915;
230-597-5: Nerolidol, 21006;
230-636-6: carotene, 5308; Lucarotin® 10%, 17753; Solatene, 28900;
230-653-9: Bilopaque, 4145;
230-698-4: Catigene BR 80 B, 5444;
230-711-3: Cotton-Pro®, 6895; Gesagard, 12921; Prometrex, 25139; prometryn, 25140, 25141;
230-785-7: potassium pyrophosphate, 24777; Tetrakal, 31375;
230-896-0: Emalex O.T.G, 9976;
230-902-1: xylazine, 34619;
230-939-3, 214-302-7, 203-943-8: Adma® 8, 10, 12, 741;
230-990-1: Amine M210D, 2159; Armeen® M2-10D, 2953; Dama® 1010, 7718; didecyl methylamine, 8415;
230-993-8: Ageflex mDMDAC, 993; Ageflex NB-50, 995; Agefloc PC20HV, 1008; Agequat C505, 1013; Agestat 41, 1028;
231-025-7: Didronel, 8421;
231-072-3: aluminum, 1897;
231-096-4: iron, 15307; malleable iron, 18485;
231-099-0: lanthanum, 16763;
231-100-4: Haro® Mix CE-701, 13645; Haro® Mix CK-711, 13646; Haro® Mix MH-204, 13647; lead, 16901;
231-102-5: lithium, 17508;
231-104-6: magnesium, 18356;
231-105-1: Dienol, 8430; manganese, 18535; Tripart® Liquid Manganese, 32377; Tronamang Electrolytic Manganese Metal, 32463;
231-106-7: mercury, 19355;
231-107-2: molybdenum, 20075;
231-111-4: malleable nickel, 18486; nickel, 21106; Nickel 200, 21107; Nickel 201, 21108; Nickel 204, 21109; Nickel 205, 21110; Nickel 211, 21111; Nickel 212, 21112; Nickel 213, 21113; Nickel 222, 21114; Nickel 223, 21115; Nickel 229, 21116; Nickel 270, 21117; Raney nickel, 25867;
231-113-5: niobium, 21245;
231-114-0: osmium, 22356;
231-115-6: palladium, 22611; palladium black, 22613;
231-116-1: Platinum, 24099; platinum black, 24100;
231-119-8: potassium, 24732;
231-125-0: rhodium, 26686;
231-128-7: samarium, 27307;
231-129-2: scandium, 27545;
231-130-8: silicon, 28326;
231-131-3: Collosol Argentum, 6635; silver, 28413;
231-132-9: Metallic Sodium, 19436; Nametal, 20674; natrium, 20756; sodium, 28725;
231-133-4: strontium, 29827;
231-135-5: tantalum, 30813;
231-137-6: terbium, 31222;
231-138-1: thallium, 31552;
231-141-8: Haro® Mix ZT-514, 13650; tin, 31809;
231-142-3: titanium, 31865;
231-143-9: tungsten, 32583;
231-146-5: antimony, 2567; grey antimony, 13352; stibium, 29761; Thermoguard® CPA, 31634;
231-148-6: arsenic, 3136;
231-149-1: Barium, 3633;
231-150-7: Beryllium, 4054;
231-151-2: Boron, 4442; Trona Elemental Boron, 32454;
231-152-8: cadmium, 4893;
231-153-3: Bachite, 3575; Bastanet, 3735; Batchite, 3738; Calgon® Type SGL, 4988; Carbon, 5213; Carbon, 5214; Carbon, 5215; Carbon, 5216; Carbon, 5217; carbon, activated, 5225; Carboraffin, 5236; Cecarbon, 5535; Darco, 7768; Degusorb, 7916; Hydraffin, 14533; Hydrodarco, 14585; Metasil DA, 19454; Metasil W/2, 19459; Microsperse, 19699; Modulex, 20032; Mogul® L, 20036; Monarch® 1100, 20136; Neospectra, 20969; Neotex, 20975; Norit, C, PK, R, RO, 21560; Norithene, 21561; Novacarb, 21646; Printex, 24998; Printex 25 Beads and Powd, 24999; Printex P, 25000; Printex U Beads and Powd, 25001; Raven, 25903; Regal® 400R, 26008; S160 Beads and Powder, 27181; Sevacarb, 28109; Sorbonorit, 29128; Special Black 100 Powd, 29216; Special Black 4, 4A Beads and Powd, 29217; Statex, 29552; Sterling, 29743; sulfite carbon, 29991; Superba, 30219; Superjet,

30246; Tack, 30689; Thermax® Floform N-990, 31581; Thermax® Stainless, 31582; Thermax® Stainless Floform N-907, 31583; United, 33055; Velvetex, 33552; Vulcan® XC72R, 34093;
231-154-9: cerium, 5754;
231-155-4: cesium, 5770;
231-157-5: chromium, 6228;
231-158-0: cobalt, 6524;
231-159-6: casting copper, 5413; copper, 6777;
231-160-1: Erbium, 10746;
231-163-8: Gallium, 12615;
231-164-3: Germanium, 12897;
231-165-9: Gold, 13171;
231-169-0: holmium, 14278;
231-171-1: vanadium, 33270;
231-172-7: xenon, 34599;
231-173-2: ytterbium, 34680;
231-174-8: yttrium, 34683;
231-175-3: Delaville, 8010; zinc, 34765;
231-177-4: Bismuth, 4235;
231-179-5: calcium, 4935;
231-180-0: indium, 15080;
231-193-1: tellurium oxide, 31064;
231-194-7: selenium dioxide, 27862;
231-195-2: sulfur dioxide, 30044;
231-198-1: rubidium iodide, 27106;
231-198-9: Haro® Chem PTS-E, 13644;
231-208-1: aluminum chloride, anhydrous, 1902; Praestol® K2001, 24854;
231-210-2: Coclor, 6560; Coppertrace, 6793; cupric chloride dihydrate, 7379;
231-211-8: K. Tab, 15642; Kaon-Cl, 15729; Kay Ciel, 15807; K-Contin, 15818; kelp salt, 15859; K-Lor, 16226; K-Lyte/C1, 16234; KM Potassium Chloride, 16242; Leo K, 16996; Micro-K, 19676; muriate of potash, 20513; Pfiklor, 23564; potassium chloride, 24747; Potassium chloride, 24748; potassium chloride, 24749; Sando-K, 27333; Selora, 27931; Slow-K, 28609; Trona Muriate of Potash, 32455; Trona Potassium Chloride, 32456;
231-298-2: magnesium sulfate, 18368; sal amarum, 27243; sel d'Angleterre, 27850;
231-299-8: mercuric chloride, 19347;
231-301-7: rubidium sulfate, 27107;
231-306-4: Schercemol DIS, 27595; Unimate® DIPS, 32969;
231-308-5: Dialose, 8284;
231-314-8: Kricinol 35, 16411; Potassium ricinoleate, 24778; Solricin® 135, 28964;
231-389-7: Acrisorcin, 392;
231-403-1: Ageflex IBOMA, 992; Sipomer® IBOMA, 28538;
231-426-7: Emalex NN-15, 9973; Monamid® 150-MW, 20116;
231-442-4: Iodine, 15208;
231-448-7: Adjunct B, 708; disodium phosphate (anhydrous), 8652; disodium phosphate, dihydrate, 8653; salt perlate, 27293; tasteless salts, 30843;
231-449-2: Aspon, 3238; Monosorb, 20256; MSP, 20420; Oilfos, 22089; Recresal, 25944; sodium phosphate, 28810;
231-493-2: cyclodextrins, 7516;
231-509-8: Avgard, 3464; trisodium phosphate, anhydrous, 32412; trisodium phosphate, dodecahydrate, 32413; TSP-12, 32538; TSP-O, 32539;
231-511-9: KM Sodium Perchlorate, 16245;
231-512-4: Fraude's reagent, 12381; perchloric acid, 23221;
231-530-2: Tomah PA-14, 31982;
231-544-9: Herapath's salt, 13799;
231-545-4: Aerosil®, 875; Aerosil® 130, 150, 873; Aerosil® 200, 874; Cab-O-Sil® L-90, 4868; Cab-O-Sil® TS-530, 4869; Celatom, 5575; Celite®, 5586; Celite® 110, 5587; Celite® 209, 5588; Celite® 270, 5589; Celite® HSC, 5611; Celite® R-625, 5590; Celite® Snow Floss, 5591; Clarcel, 6357; diatomaceous earth, 8335; flint, 11971; Flocsil, 11991; Flo-Guard, 12034; Glaucosil, 12976; Hi-Sil, 14131; Hyflo Super-Cel, 14674; J Slip NS-77, 15466; Krystallos, 16469; Micropil, 19689; Milowite, 19799; Neosyl®, 20972; rock crystal, 26885; Santocel, 27398; Silantox, 28266; silica, 28313; Silica FK 320 DS, 28315; Silver Bond 30, 28418; Tamsil 8, 30770; Tamsil Gold Bond, 30771; Wessalon, 50, 50S, S, 34285;
231-548-0: Abisol, 70; Amersite® 2, 2035; Amersite® 2, 2036; Leucogen, 17023; sodium acid sulfite, 28728; sodium bisulfite, 28738;
231-551-7: sodium Molybdate, dihydrate, 28796;
231-552-2: sodium monofluorophosphate, 28798;
231-554-3: Chili saltpeter, 6063; Saliter, 27273; sodium nitrate, 28800;
231-555-9: sodium nitrite, 28801;
231-556-4: perborax, 23213; perborin, 23214; peroxydol, 23412; sodium perborate, 28806;
231-569-5: Boron trifluoride, 4446; Leecure B Series, 16963;
231-588-9: Fascat® 4400, 11497; Libavius' fuming spirit, 17185; opacite, 22197; salt of tin, 27291; stannic chloride, 29494;
231-589-4: cobalt chloride (ous), 6528; cobaltous chloride, 6537; Cobatope-60, 6543;
231-592-0: zinc chloride, 34772; Zinctrace, 34806;
231-594-1: tartarline, 30833;
231-595-7: hydrochloric acid, 14575, 14576; marine acid, 18718; soldering acid, 28909; spirit of salt, 29298;
231-596-2: palladium(II) chloride, 22617;
231-598-3: Alberger® Natural Flake, 1361; Betrox, 4075; Hopper salt, 14303; marine salt, 18722; muriate of soda, 20514; Natritope Chloride, 20755; rock salt, 26887; sal commune, 27244; sal Culinaris, 27245; Saltex, 27296; Slow-Sodium, 28610; sodium chloride, 28748; Superior® Granulated, 30244; Tru-Flake® Salt, 32490;
231-599-9: sodium bromide, 28741;
231-600-2: cesium chloride, 5773;
231-601-8: antimony pentachloride, 2568; Adogen® S-18 V, 778;
231-609-1: Chemidex S, 5968; Incromine SB, 15014; Lexamine S-13, 17100; Lipamine SPA, 17328; Mackine 301, 18217; Mazeen® S-13, 19124; Miramine® SODI, 19872; Schercodine S, 27641; TEGO Amid S18, 30957;
231-633-2: Marphos, 18951; NFB, 21080; phosphoric acid, 23738, 23739; Solklean™ 101, 28950;
231-634-8: hydrogen Fluoride, 14589;
231-635-3: ammonia, 2219; Morpan BC, 20344; Quadrilan® BC, 25580; Rewoquat B 50, 26390; Roccal, 26880; spirit of Hartshorn, 29296; Variquat® 50MC, 33397; Zephiran Chloride, 34728; Zoharquat 50, 34873;
231-639-5: fuming sulfuric acid, 12457; spirit of alum, 29295; spirit of vitriol, 29299; sulfuric acid, 30055;
231-659-4: potassium iodide, 24758, 24759;
231-667-8: Fluorinse, 12090; Fluorl, 12092; fluorol, 12101; Gel II, 12716; sodium fluoride, 28764;
231-668-3: Adeka Hypote, 664; Chlorasol®, 6089; Chloros, 6146; Chlorosoda, 6147; Dakin's solution, 7704; Domestos, 8795; HyPure N, 14728; Merclor D, 19342; oxychloride, 22442; salunol, 27302; sodium hypochlorite, 28773;
231-672-5: sodium iodate, 28774;
231-673-0: metabisulfite, 19404; sodium metabisulfite, 28783;
231-674-6: cuprous iodide, 7404;
231-679-3: sodium iodide, 28775;
231-694-5: sodium tripolyphosphate, 28828;
231-711-6: Killgerm® Py-Kill W, 16182; tetramethrin, 31384;
231-714-2: dynamite acid, 9357; fuming nitric acid, 12456; Nitraline, 21320; nitric acid, 21332;
231-718-4: zinc bromide, 34770;
231-721-0: Schercoteric CY-2, 27705;
231-722-6: Biosulphur Powder, 4210; Crystex®, Crystex® Regular, 7294; Gofrativ, 13165; Gofravik, 13166; Green Sulphur, 13324; Greggio, 13348; Hortag Aquasulf, 14313; Kumulus® DF, FL, 16482; Kumulus® S, 16483; Manox, 18577; mother of pearl sulfur, 20361; Octocure 456, 21993; roll sulfur, 26920; Rubbermakers Sulfur, 27094; Solfa, 28928; Solfa, 28929; Stoller Flowable Sulphur, 29782; Struktol® SU 109, 29837; Suffa, 29949; sulfur, 30043; Sulfur-F, 30054; Thiovit, 31724; Tripart® Imber, 32374; Vassgro Flowable Sulphur, 33470; Yellow sulphur, 34650;
231-727-3: zirconium hydride, 34823;
231-729-4: ferric chloride, 11648;
231-743-0: nickel chloride, 21125;
231-749-3: phosphorus trichloride, 23747;
231-753-5: green vitriol, 13337; Green-up Mossfree, 13346; iron sulfate (ous), 15321;
231-760-3: potassium permanganate, 24769, 24770; sin red, 28490;
231-764-5: Ammonium phosphate monobasic, 2251; Firesaife, 11836; Mono Ammonium Phosphate (Agricultural Grade), 20215; Phosal, 23687;
231-765-0: dioxogen, 8577; Genoxide®, 12840; hydrogen peroxide, 14590; hydrozone, 14672; Oxzone, 22478; Percarbamid, 23219; Perhydrol, 23268; Perone, 23399; Peroxal, 23403; Peroxol, 23411; Peroxyl, 23413; Proxy, 25348;

Truzone, 32496;
231-766-6: Caro's acid, 5306;
231-767-1: Nutrifos, 21833; sodium pyrophosphate, 28814; Tetron, 31396; TSPP, 32540;
231-768-7: violet phosphorus, 33876;
231-778-1: Bromine, 4572;
231-781-8: Anthion, 2519; potassium persulfate, 24771;
231-784-4: Albaryt, 1353; Barium sulfate, 3644; Blanc Fixe Micro, 4313; Citobaryum, 6332; Esophotrast, 10835; Ewo, 11279; fixed white, 11858; Huberbrite 1, 14452; Micropaque, 19686; Oratrast, 22271; Radiopaque, 25830; Raybar, 25909; Roebaryt, 26902; Sachtoperse® HU, 27202; Sacon®, 27206; Sparrnite, 29209; terra ponderosa, 31325;
231-786-5: Ammonium persulfate, 2249; thioxydant lumire, 31725;
231-791-2: Tritiotope, 32423;
231-793-3: zinc vitriol, 34799;
231-801-5: chromic acid, 6220;
231-810-4: Amine CS-1246, 2154; Oxaban®-E, 22405;
231-818-8: Petre, 23474; potassium nitrate, 24762; prismatic niter, 25035; sal niter, 27250; sal Prunella, 27251; saltpeter, saltpeter flour, 27298;
231-820-9: Crimidesa, 7052; sal Mirabil, 27249; salt cake, 27281; sodium sulfate, 28821; Trona Anhydrous Sodium Sulphate, 32451; Trona Salt Cake, 32458;
231-821-4: Santosite, 27424; sodium sulfite, 28823;
231-826-1: Calbrite, 4928; calcium hydrogen phosphate, 4951; Caliment, 5006; Calipharm, 5008; Dentplus Special, 8067; Lucaphos® 40, 48, 17752; monetite, 20209;
231-829-8: Bromox, 4589; potassium bromate, 24741;
231-830-3: potassium bromide, 24742;
231-834-5: potassium phosphate, dibasic, 24773;
231-835-0: Antelope, 2517; B.P. Pyro®, 3564; Donut Pyro®, 8800; Perfection®, 23243; Sapp #4, 27446; sodium acid pyrophosphate, 28727; Taterfos®, 30844; Victor Cream®, 33727;
231-836-6: Adox 3125, 787; C-2, 4842; Chloritane, 6109; sodium chlorite, 28749;
231-837-1: acid calcium phosphate, 330; calcium phosphate (monobasic), 4966; Cefkaphos®, 5549; H.T, 13524; Ibex, 14805; Py-Ran, 25468; Regent® 12XX, 26014; V-90®, 33222;
231-838-7: Curafos® STP, 7411; Flav-R-Keep FP-51, 11906; Hysorb, 14749; Lem-O-Fos®, 16979; Polygon, 24460; Rhodia-Phos, 26677; Sea-Gard® Formula FP-91, 27786;
231-840-8: calcium phosphate (tribasic), 4967; Ephos, 10628; precipitated phosphate, 24866;
231-842-9: Cuproid, 7394; cuprous chloride, 7403;
231-843-4: Ferrofloc, 11672; ferrous chloride, 11689;
231-845-5: lead chloride, 16905; Leclo, 16944;
231-846-0: Chrome yellow, 6212; mock vermilion, 20012;
231-847-6: cupric sulfate, 7387; Kocide® Copper Sulfate Pentahydrate Crystals, 16262; Mackechnie, 18202; Sulfato de Cobre Valles, 29974;
231-853-9: lunar caustic, 17816; silver nitrate, 28431;
231-867-5: Ametox, 2049; green mordant, 13321; Lycopon, 18065; sodium thiosulfate, 28826;
231-868-0: Fascat® 2004, 11496; single muriate of tin, 28497; stannous chloride, 29502; tin salts, 31814;
231-869-6: Mangatrace, 18559;
231-871-7: Amcide, 1986; Ammonium sulfamate, 2254; Root-Out, 26959;
231-887-4: Arpal Non Selex, 3095; Atlacide, 3290; Centex, 5678; Defol, 7889; Granular Weedkiller, 13270; KM Sodium Chlorate, 16244; sodium chlorate, 28747;
231-889-5: sodium chromate, 28750;
231-890-0: Blankit®, 4322; Geblitol, 12708; Hybaite, 14486; Hybrite®, 14492; Hydrolin, 14610; Hydronyx, 14625; Hydros, 14635; Hydros® F, 14637; hydrosulfite, 14643; Hydrosulfite AWC, 14645; Hydrosulfite BASF, 14646; Hydrosulphit®, 14648; Luredox® BP, PO, 17937; Reductone, 25983; sodium hydrosulfite, 28771;
231-891-6: sodium metaborate, 28784, 28785;
231-892-1: sodium persulfate, 28809;
231-895-8: lachiol, 14801; silver fluoride, 28423; tachiol, 30686;
231-900-3: Anhydrite, 2458; calcium sulfate (anhydrous), 4971; Clay Breaker, 6376; drierite, 9027; pearl-hardening, 22985;
231-901-9: Aresenid, 2838; arsenic acid, 3137;
231-902-4: Fatsco, 11506;
231-904-5: Calcars, 4930; calcium arsenate, 4938;
231-906-6: potassium bichromate, 24739; potassium dichromate, 24754;
231-907-1: potassium phosphate, tribasic, 24774;
231-908-7: calcium hypochlorite, 4953; chloride of lime, 6104; HTH, 14443; Hyporit, 14720; Induclor, 15089; Perchloron, 23223; Pittabs, 23888; Pittclor, 23890; Prestochlor, 24919; Pulsar, 25384; Repak, 26141; Stellos, 29641; Swim clear, 30460;
231-909-2: HyPure K, 14726; Potassium hypochlorite, 24757;
231-912-9: KM Potassium Perchlorate, 16243; potassium perchlorate, 24767;
231-913-4: Beycostat 148 K, 4099; Beycostat 273 P, 4102; potassium phosphate, monobasic, 24772;
231-915-5: glazier's salt, 12982; potassium sulfate, 24785; Trona Potassium Sulphate, 32457; Trona Sulphate of Potash, 32460;
231-943-8: zinc nitrate, 34782;
231-944-3: Delaphos, 8008; Diroval, 8621; zinc phosphate, 34790;
231-954-8: Fluorine, 12088;
231-955-3: Aquadag, 2690; Dag 137, 7686; Dag 154, 7687; Dag 197, 7689; graphite, 13277; Lonza KS, 17589; mineral carbon, 19820; Prodag, 25067; Rollit, 26921; SLA 1262, 1275, 28578; SLA 1275, 28579; SLA 2239, 28584;
231-956-9: oxygen, 22451;
231-957-4: selenium, 27861; Vandex®, 33311;
231-959-5: chlorine, 6107;
231-960-0: manganese sulfate, 18549; Mansu, 18612;
231-961-6: monogermane, 20225;
231-964-2: nitrosylsulfuric acid, 21379;
231-975-2: tungstic acid, 32592;
231-977-3: sulfuretted hydrogen, 30053;
231-981-5: telluretted hydrogen, 31060;
231-982-0: Ammonium thiosulfate, 2257;
231-984-1: Ammonium sulfate, 2255; Nipro (ii), 21304;
231-987-8: Ammonium phosphate dibasic, 2250;
231-992-5: mercuric sulfate, 19349;
231-995-1: magnesium fluoride, 18361; Magtran, 18412;
232-001-9: zinc fluoride, 34774;
232-015-5: silicon tetrafluoride, 28334;
232-018-1: zirconium tetrafluoride, 34827;
232-021-8: antimony pentafluoride, 2569;
232-029-1: tungsten hexafluoride, 32587;
232-033-3: silver chloride, 28421;
232-038-0: silver iodide, 28428;
232-051-1: Alcan Aluminum Fluoride, 1412;
232-055-3: Ammonium alum, 2229; Curb, 7416; Guardsman, 13450; Mandops Narsty, 18523;
232-056-9: aluminum phosphate, 1922;
232-059-5: arsenic trichloride, 3141;
232-060-0: arsenic trifluoride, 3142;
232-065-8: Macquer's salt, 18277; potassium arsenate, 24737;
232-066-3: arsine, 3144;
232-070-5: Harle's solution, 13637;
232-077-3: α-pinene, 23857;
232-090-4: Pan-O-Lite, 22683; sodium aluminum phosphate, 28733; Stabil-9, 29399;
232-094-6: magnesium chloride, 18360;
232-104-9: nickel sulfate, 21135;
232-122-7: Biju®, 4136; Bismica 46, 4234; Bismuth chloride oxide, 4237; Bismuth oxychloride, 4241; Brown 208, 4617; Mearlite® GBU, 19192; Pearl I, II, III, 22979; Pearl Supreme UVS, 22981; pearl white, 22982; Pearl-Glo®, 22984; Perlex, 23332; U Pink 113, U Violet 109, 32696;
232-130-0: cesium bromide, 5771;
232-131-6: cuprous bromide, 7402;
232-134-2: 2-chloroethylmethyldichlorosilane, 6122; CC3005, 5486;
232-137-9: chromic fluoride, 6222; fluorchrome, 12072;
232-141-0: Plessy's Green, 24113;
232-143-1: Ammonium dichromate, 2237;
232-145-2: cesium iodide, 5776;
232-146-8: cesium nitrate, 5777;
232-156-2: Fremy's salt, 12394;
232-160-4: Dyetone®, 9312;
232-166-7: Mancobride Mancanese, 18515;
232-167-2: cupric bromide, 7377;
232-188-7: calcium fluoride, 4947;
232-189-2: calcium hydride, 4949; Hydrolete, 14609; hydrolith, 14611;
232-191-3: calcium iodate, 4954;
232-192-9: calcium molybdate, 4958;
232-196-0: potassium periodate, 24768;

232-197-6: sodium metaperiodate, 28786;
232-218-9: lithium nitrate, 17511;
232-222-0: cadmium fluoride, 4897;
232-223-6: cadmium iodide, 4899;
232-232-5: Hypochlorous acid, 14718; HyPure A, 14724;
232-233-0: Chloric acid, 6102; HyPure C, 14725;
232-235-1: KM Ammonium Perchlorate, 16236;
232-240-9: rubidium chloride, 27105;
232-241-4: thallium chloride, 31554;
232-259-2: oxyammonia, 22439;
232-261-3: KM Ammonium Metavanadate, 16235;
232-263-4: silane, 28265; Silicane, 28318;
232-271-8: Egg oil, 9645; EmCon E-5, 10065;
232-272-3: tung oil, 32580;
232-273-9: Lipovol SUN, 17439; sunflower seed oil, 30141; Super Refined Sunflower Oil, 30199;
232-274-4: Lipovol SOY, 17438; soya bean oil, 29169; Soybean oil, 29173; Super Refined Soybean, 30198;
232-276-5: Lipovol SAF, 17435; Neobee® 18, 20872; safflower oil, 27222; saffron oil, 27225; Super Refined Safflower USP, 30195;
232-277-0: Lucca oil, 17754; olive oil, 22138; Super Refined Olive , 30193;
232-278-6: Linol, 17304; Linseed oil, 17309;
232-281-2: corn oil, 6834;
232-282-8: coconut butter, 6566; Coconut Oils® 76, 92, 110, 6567; Esi-Terge 40% Coconut Oil Soap, 10825; Konut, 16334; Super Refined Coconut Oil, 30188;
232-287-5: Creosote, 6991;
232-288-0: Jesuit's balsam, 15580;
232-289-6: Atlas Tanked Cod Oil, 3336; cod liver oil, 6569; Mainstay®, 18417; Super D, 30174;
232-290-1: ceresin, 5746;
232-292-2: hydrogenated castor oil, 14592;
232-293-8: AA Standard, AA USP, 12; Cosmetol® X, 6871; Crystal® O, Crystal® Crown, 7289;
232-294-3: Bone oil, 4402;
232-296-4: Katchung oil, 15788; peanut oil, 22969; Super Refined Peanut , 30194;
232-299-0: Actralube 7142, 538; rapeseed oil, 25874;
232-300-4: rosin oil, 26978;
232-302-5: cetyl esters, 5819; Crodamol SS, 7126; Dermalcare® SPS, 8101; Liponate SPS, 17387; Ritaceti, 26798; Ross Spermaceti Wax Substitute 573, 26992; spermaceti, 29276; Starfol® Wax CG, 29539; W.G.S. Synaceti 116 NF/USP, 34189;
232-304-6: tall oil, 30737;
232-306-7: Actrasol C-50, C-75, C-85, 549; Chemax SCO, 5926; Chemsulf SCO/75, 6046; Eureka 102, 11207; Laurel R-50, 16837; Marvanol® SCO (50%), 18989; NMP, NMP Conc, 21405; Nopcocastor, 21485; Nopcosulf CA-60, -70, 21504; Nopcosurf CA, 21507; Standapol® SCO, 29478;
232-307-2: Actiflo® 68, 70, 476; Alcolec® 439-C, 1481; Alcolec® 439-C, 1482; Alcolec® 440-WD, 1483; Alcolec® 495, 1484; Alcolec® BS, 1485; Alcolec® F-100, 1486; Alcolec® S, 1487; Aqualipid 95, 2701; Blendmax 322, 4346; Centrocap® 162SS, 162US, 5683; Centrol® CA, 5684; Centrolex® C, 5686; Centromix® CPS, 5687; Centrophase® C, 5688; Centrophase® HR, 5689; Centrophase® HR2B, HR2U, 5690; Centrophil® K, 5691; Crolec 4135, 7189; Dermasome® MT, 8107; Dermasome® TRF, 8112; Kelecin®, 15835; lecithin, 16939; Lecithin L-Range, 16940; Lecithin Water Dispersible CLR, 16941; Lipotin 100, 100J, SB, 17422; Phosal, 23688; PhosPho E-100, 23707; PhosPho F-97, 23708; PhosPho LCN-TS, 23711; PhosPho S-85, 23713; PhosPho T-20, 23714;
232-310-9: Nutimalt® Range, 21805;
232-311-4: Super Refined Menhaden , 30191;
232-313-5: Sicolub® E, 28219;
232-314-0: neatsfoot oil, 20836, 20837;
232-315-6: Glycolube® VL, 13109; Microlube A, 19679; Octowax 321, 22033; Paracol® 1886, 22722; Paracol® 800N, 22725; Paracol® 810N, 22726; Paracol® M161, 22727; paraffin, 22742; Paraffin, 22743; Paratulle®, 22812; petroleum wax, 23501; Ross Wax #100, 26993; Ross Wax #145, 165, 26994; Sasolwaks, 27476; Sasolwaks M3, 27477; Synwax, 30659; Vybar® 103, 34165;
232-316-1: Lipovol PAL, 17434; palm oil, 22625, 22626;
232-347-0: candelilla wax, 5084;
232-348-6: Anhydrous Lanolin HP-2050, 2461; Anhydrous Lanolin P80, 2462; Anhydrous Lanolin P9SRA, 2463; Corona, 6843; Coronet, 6844; Cosmetic Lanolin USP, 6870; Fluilan, 12046; lanolin, 16740; Lantrol® 1673, 16767; Lantrol® HP-2073, 16769; Ultra Anhydrous Lanolin HP-2060, 32779; Vigilan, 33750;
232-350-7: turpentine, 32610;
232-354-7: Ligroin, 17244; petroleum spirit, 23500;
232-355-4: Cardolite® NC-511, 5268;
232-360-1: Arlacel® 83, 2903; Crill 43, 7034; Dehymuis SSO, 7968; Disponil SSO 100 F1, 8704; Emalex SPO-150, 9997; Emsorb® 2502, 10456; Glycomul® SOC, 13115; Hodag SSO, 14246; Liposorb SQO, 17415; S-Maz® 83R, 28645;
232-366-4: Pearl, 22975;
232-370-6: Gingelly oil, 12946; Lipovol SES, 17436; Sesame oil, 28075; Super Refined Sesame, 30196;
232-373-2: petrolatum, 23495;
232-384-2: Balneol, 3609; Emulzome, 10532; Kaydol, 15810;
232-387-9: yeast extract, 34640;
232-388-4: Solricin® 235, 28965;
232-391-0: ADK CIZER O-130P, 710; Drapex® 6.8, 8969; Edenol D82, 9607; Epoxol 7-4, 10710; Hoe S 3680, 14260; Lankroflex® GE, 16680; Paraplex® G.62, 22790; Paraplex® G-60, 22793; Plasthall® ESO, 24008; PX-800, 25445;
232-393-1: Invert Sugar, 15200;
232-399-4: Carnauba, 5296; Carnauba wax, 5297; Emulsion C-340, 10516;
232-405-5: Actran Extra, 546;
232-406-0: oiticia, 22094;
232-410-2: Akorex, 1149; Code 321, 6571; Diamond D 31, 8310; Famous, 11426; hydrogenated soybean oil, 14596; Lipovol HS, 17428; S-Flakes, 28131; Witarix® 440, 34431;
232-417-0: polyphosphoric acid, 24520;
232-418-6: Degras Special, 7904;
232-425-4: palm kernel oil, 22624;
232-428-0: Avocado Oil, 3497, 3498; Avocado Oil CLR, 3499; Cropure Avocado, 7218; Lipovol A, 17425;
232-430-1: Argobase 125, 2851; Argowax Standard, 2859; Ceralan®, 5702; Fancol LA, 11446; Hartolan, 13676; Hartolite, 13677; Ivarlan 3310, 15444; Lanolin alcohol, 16741; Ritawax, 26835; Solulan, 28998; Super Hartolan, 30178;
232-436-4: corn syrup, 6838;
232-438-5: Myverol® 18-85K, 20627;
232-439-0: Perichthol, 23270;
232-440-6: Alcolec® Z-3, 1488; Centrolene® A, S, 5685; Lipotin H, 17424; PhosPho 642, 23706;
232-443-2: Aromatic Solvent 150, 3043; Hi-Sol® 10,15,70, 14134; naphtha, 20705; petroleum naphtha, 23498; Texsolve V, 31532;
232-444-8: East Indian Balsam of Copaiba, 9461;
232-447-4: Arquad® T-27W, 3121; Arquad® T-50, 3122; Jet Quat T-50, 15606; tallow trimonium chloride, 30747;
232-452-1: Fancol HL, 11441; Ivarlan HL, 15451; Lipolan, 17366; Satulan, 27492; Super-Sat, 30270;
232-455-8: Drakeol 10, 8962; Drakeol 19, 8963; Drakeol 5, 8960; Drakeol 7, 8961; Paratherm NF, 22806; Parol 70, 22859; Peneteck, 23103;
232-468-9: Pancrease, 22671; Pancrex V, 22675;
232-475-7: rosin, 26976;
232-479-9: Oulutac 90 D, 22391; Uni-Tac® R99, 33053; Zonester® 65, 34896;
232-483-0: Empilan® 2502, 10384;
232-490-9: Bitumen, 4294; petroleum pitch, 23499;
232-491-4: Emery® 2895 Foamaster Soap L, 10156; Foamaster Soap L, 12148; sodium tallowate, 28824;
232-501-7: valerian, 33241;
232-514-8: Precirol ATO, 24867; Precirol WL 2155, 24868;
232-519-5: gum arabic, 13480; picked turkey gum, 23811;
232-523-7: gum benjamin, 13482;
232-524-2: carrageenan, 5315; Clarifloc, 6363; Stamere®, 29437; Wakal® K Range, 34197;
232-528-4: Dammar Resin, 7724;
232-536-8: guar gum, 13445; Jaguar® C, 15495; Jaguar® Guar Gum, 15500; Prinza® Range, 25010; Progacyl® CP-7, 25108; Supercol® Guar Gum, 30229;
232-539-4: karaya gum, 15760;
232-541-5: Locust bean gum, 17545; Wakal® J

Range, 34196;
232-545-4: Imsil® A-10, 14936; Siogel, 28508; Sipernat® 22S, 28520; Sipernat® 283LS, 28523; Sipernat® 50, 28522; Sipernat® D17, 28524;
232-552-5: Dragon gum, 8956; gum tragacanth, 13490;
232-553-0: pectin, 23017; vegetable jelly, 33521;
232-554-6: Byco A, C, O, 4818; Crodyne BY19, 7182; gelatin, 12722; Hydrocoll G-40, 14581; Pharmagel, 23572; Pronel Capsules, 25158; Rousselot Gelatine, 27012; SPG Gelatine, 29281;
232-555-1: kalzose, 15703;
232-567-7: Arozyme TD, 3092; Clarase® 5,000, 40,000, 6356; diastase, 8331; Taka-Therm® L-340, 30724; Tenase® 1200, L-340, L-1200, 31094; Termamyl®, 31299; Termamyl® 120L, 31300;
232-577-1: Catalase L, 5429; Microcatalase®, 19661;
232-580-8: Chymopapain, 6284; Discase, 8624;
232-582-9: Collagenase, 6615;
232-588-1: Nervanzse, 21011;
232-615-7: sucrase, 29929;
232-619-9: Lipase, 17329; Pancreatic Lipase 250, 22672;
232-627-2: papain, 22698; Scintillase, 27740;
232-628-8: Penicillinase, 23112;
232-629-3: pepsin, 23200;
232-647-1: Kabikinase, 15650;
232-650-8: Parenzyme, 22827; trypsin, 32530;
232-656-0: urease, 33122;
232-658-1: Agar (Agar-agar), 955; vegetable gelatin, 33518;
232-659-7: Cortymol® LP, 6861;
232-664-4: Thytropar, 31782;
232-668-6: peroxidase, 23404;
232-671-2: Alpha Chymar, 1820; α-Chymotrypsins, 6286; Zolyse, 34889;
232-674-9: Celluflow C-25, 5616; cellulose, 5626; CF 31,000 C Coarse, 5830; CF 42,500T Medium, 5831; CF 70,000WDK, Ex. Superfine, 5832; Elcema® F150, G250, P100, 9721; Emcocel® 90M, 10042; Fibra-Cel®, 11748; Solka-Floc® BW-40, BW-100, BW-200, BW-2030, UF-900-FCC & NF, 28949;
232-678-0: Cromoist HYA, 7200;
232-679-6: Ceraphyl® 847, 5729; Control 1-100, 6749; corn starch, 6836; corn starch, 6837; D12F, 7629; Pure-Dent® B700, 25405; Purity® 21, 25411; Supercore® S13F, 30230;
232-680-1: alginic acid, 1593; Kelacid®, 15824; Saltialgine H8, 27297; Satialgine H8, 27481;
232-681-7: Heparin, Sodium Salt, 13788; Liquemin, 17453;
232-683-8: Liver sugar, 17527;
232-689-0: DPNR, 8949; Dynatex GTZ, 9424; ENR 25, 10585; Hyflo NS, S, 14673; PA-57, 22531; PA-80, 22532; Polyisoprene, 24471; rubber, 27089; Unitex, 33056;
232-692-7: Sol-U-Tein EA, 29028;
232-696-9: Cromoist CS, 7199;
232-697-4: Actigen C, 478; Cationic Collagen Polypeptides, 5460; Clearcol, 6390; collagen, 6610; Collagen CLR, 6612; Collagen Native Extra 1%, 6614; Collasol, 6619; Desamidocollagen, 8136; Polygeline+sn, 24451; Polymoist Mask, 24497; Ritacollagen BA-1, 26805; Secolan S-1, BA-1, BA-1G, 27803; Solu-Coll, 28992;
232-734-4: Cellulase, 5619; Cellulase, 5620; Cellulase 4000, 5621; Cellulase AC, 5622; Celluzyme® 2400 T, 5638;
232-752-2: AFP 2000, 946; protease, 25256;
232-796-2: rennet, 26110;
232-864-1: galactosidase, 12593;
232-877-2: Agidex, 1030; Aldomax GA-100, 1548; Diazyme® L-200, 8352; Distillase® L-200, 8716; Fermenzyme® L-200, 11628;
232-885-6: Clarex® L, 6359; Pearex-L®, 22972; pectinase, 23018; Pectinase AT, 23019; Spark-L® HPG, 29203;
232-917-9: urokinase, 33145;
232-936-2: Bovinal-20, 4465;
232-940-4: Lycadex®, 18059; Maltodextrin, 18500; Maltrin® M040, 18502; Maltrin® M510, 18504; Microduct®, 19670; Mor-Rex® I-920, 20348; Wickenol® 550, 34403;
233-000-6: Tachigaren 70, 30685;
233-020-5: Barium nitrate, 3639;
233-032-0: Nitral, 21318;
233-038-3: Chrometrace, 6218; chromic chloride, 6221;
233-042-5: Dynasylan® TCS, 9417; trichlorosilane, 32223;
233-046-7: phosphorus oxychloride, 23743;
233-047-2: antimony trichloride, 2573;
233-054-0: CT1800, 7312; silicon tetrachloride, 28333;
233-058-2: zirconium tetrachloride, 34826;
233-060-3: phosphorus pentachloride, 23744;
233-062-4: cobaltic fluoride, 6531; cobaltic trifluoride, 6534;
233-072-9: Elliott's Lawn Sand, 9848; Elliott's Moss Killer, 9849; ferric sulfate, 11652; ferrous sulfate, 11690; Greenmaster Autumn, 13342; Greenmaster Mosskiller, 13344; Hart Lawn Sand, 13657; Hart Moss Killer, 13658; iron sulfate (ic), 15320; Iron vitriol, 15322; Maxicrop Moss Killer & Conditioner, 19078; sal Martis, 27247; salt of steel, 27290; Shoemaker's black, 28184; Taylors Lawn Sand, 30849; Vitax Micro Gran, Vitax Turf Tonic, 33958; Vitax Turf Tonic, 33960; Walkover Moss Killer, 34198;
233-108-3: Anhydrone, 2460; Dehydrite®, 7963;
233-109-9: Hydriodic acid, 14555;
233-110-4: Hydrazine sulfate, 14546;
233-113-0: Hydrobromic acid, 14564;
233-131-9: strontium nitrate, 29830; Strotope, 29832;
233-135-0: Alcan Aluminum Sulfate Liquid, 1413; Alferric, 1576; Aluminoferric, 1894; aluminum sulfate, 1930; Clar+Ion® A410P, 6366; Lapotan, 16781; mountain butter, 20367; paperhanger's alum, 22701; pearl alum, 22976; pickle alum, 23813;
233-136-6: Boron nitride, 4443;
233-139-2: Ant Flip, 2497; Boric Acid, 4422; Homberg's salt, 14285; sal sedativus, 27254;
233-140-8: Cal Plus, 4923; calcium chloride, 4943; Dowflake, 8921; Homberg's phosphorus, 14284; Huppert's reagent, 14476; Ice Melt, 14808; Jarcal, 15519; Liquical, 17459; Liquidow, 17478; Marley Cement Accelerator, 18825; Snomelt, 28673; Sure-Step, 30334;
233-141-3: alum, 1883; aluminum potassium sulfate, 1923; potash alum, 24724;
233-146-0: thoron, 31752;
233-151-8: Mohr's salt, 20039;
233-153-3: Octojet 104, 22004; Purac, 25400;
233-160-7: Rohrbach's solution, 26915;
233-162-8: Anthium Dioxcide, 2521; chlorine dioxide, 6108;
233-163-3: chromous chloride, 6245;
233-165-4: rhodium chloride, 26687;
233-215-5: Naugard® 445, 20790;
233-226-5: Chemstat® HTSA#3, 6041; HTSA #3, 14446; Kemamide® E-180, 15896; Kemamide® E-221, 15897; N-octadecyl-13-docosenamide, 21986; stearyl erucamide, 29605;
233-232-8: Monox, 20259;
233-237-5: lanthanum chloride, 16764;
233-238-0: lanthanum nitrate, 16765;
233-245-9: lead nitrate, 16911; Ledni, 16961;
233-246-4: BSWL 202, 4634; lead silicate, 16915;
233-250-6: Cecasol, 5536; Extrusil, 11331; Keical-Ace, 15822; Microcal, 19657; Microcal ET, 19658; Micro-Cel® A, 19662; Micro-Cel® T-38, 19663; Paratemp, 22805; Silasorb, 28268; Silene, 28300;
233-252-7: zirconium silicate, 34825; Zircosil, 34828;
233-265-8: palladium(II) nitrate, 22618;
233-267-9: sodium selenite, 28815;
233-272-6: nitrous gas or air, 21381;
233-277-3: Kasal®, 15777; porous alum, 24702; sodium alum, 28730;
233-292-5: propylene glycol monolaurate, 25219;
233-296-7: Caddy, 4890; cadmium chloride, 4895;
233-307-5: Calomel, 5018;
233-321-1: Stahl's sulfur salt, 29425;
233-322-7: Acerdol, 275; calcium permanganate, 4963; Monol, 20229;
233-332-1: calcium nitrate, 4960; lime nitrate, 17267; nitrocalcite, 21352; saltpeter rot, 27299;
233-334-2: cobaltous sulfate, 6541;
233-335-8: white precipitate, 34370;
233-343-1: Kalex HMP, 15670; Limex G, 17276; Sodaphos, 28721; sodium hexametaphosphate, 28770; Vitrafos®, 33982; Water Brite, 34223;
233-401-6: chromic potassium sulfate, 6224;
233-402-1: cobaltous nitrate, 6539;
233-432-5: Finaplix, 11803; Trenbolone acetate, 32171;
233-433-0: Albaphos Dental Na 211, 1348;
233-459-2: lead molybdate, 16909;
233-469-7: Ammonium bisulfite, 2233;
233-474-4: Perkadox® 58, 23306;
233-517-7: vanadium chloride (ous), 33274;
233-520-3: bishydroxyethyl-N,N-Bis(2-hydroxyethyl) stearyl amine, 4231; Chemeen 18-2, 5943; Chemstat® 273-E, 6037; Ethomeen® 18/12, 10972; PEG

stearamine series, 23026; Varonic® S202, 33448;
233-539-7: Amerstat® 300, 2045; Biosperse® 240, 4207;
233-596-8: katarsit, 15786;
233-606-0: methamidophos, 19474; Monitor, 20213; Nitofol®, 21314; Prodex, 25069; Tamaron®, 30760;
233-634-3: Speedcure EDB, 29261;
233-638-5: V-30, 33215;
233-644-8: cupric arsenite, 7376; pickle green, 23814; Scheele's green, 27562;
233-653-7: silver sulfate, 28434;
233-657-9: Boron tribromide, 4444; Trona Boron Tribromide, 32452;
233-658-4: Boron trichloride, 4445; Trona Boron Trichloride, 32453;
233-660-5: Barium chromate, 3638;
233-662-6: cesium sulfate, 5778;
233-713-2: D-lactic acid, 16572;
233-758-8: Ludiomil, 17774;
233-782-9: sodium metaphosphate, 28787;
233-788-1: Barium chloride, 3637;
233-792-3: Bismuth nitrate, 4239;
233-822-5: Tenephrol, 31103;
233-826-7: magnesium nitrate, 18365; Magnisal, 18380;
233-864-4: cetyl ricinoleate, 5824; Liponate CRM, 17374; Naturchem® CR, 20767;
233-892-7: Bistrimethylsilyl acetamide, 4286; CB2500, 5479; Dynasylan® BSA, 9399;
233-899-5: ferric nitrate, 11650;
233-937-0: periodic acid, 23273;
233-940-7: resmethrin, 26218;
233-971-6: Stronscan-85, 29825; strontium chloride, 29829;
234-077-9: Lanamine®, 16646; Mixed isopropanolamines myristate, 19984;
234-081-0: Schercemol NGDL, 27613;
234-139-5: nitrothal-isopropyl, 21380;
234-140-0: Barium bromide, 3635;
234-190-3: sodium dichromate, 28756;
234-206-9: Argus DTDTDP, 2864; Cyanox® 711, 7486; Evanstab® 13, 11256;
234-232-0: Battal, 3740; Carbate Flowable, 5182; Carbate Flowable, 5183; Carbendazim, 5187; Carbendazim, 5188; Hinge, 14095; Konker®, 16327; Mascot Systemic Turf Fungicide, 19007; Maxim, 19082; Stempor DG, 29648; Tripart® Defensor FL, 32370; Turfclear, 32604;
234-316-7: Caprol® 10G100, 5131; Hodag PGO-1010 (formerly Hodag SVO-10107), 14217; Polyaldo® DGDO KFG, 24353; Polygliceryl-10 decaoleate, 24453;
234-319-3: Alcamizer 1, 1409; Alcamizer 2, 1410;
234-325-6: Ablunol GMS, 103; Aldo® HMS KFG, 1539; Aldo® MS FG, 1543; GMS Base, 13162;
234-336-6: phosphomolybdic acid, 23731;
234-347-6: M50, 18109;
234-364-9: Ammonium tungstate, 2258;
234-394-2: Dariloid® 100, 7774; Dricoid® 200, 9025; Kelflo®, 15839; Kelgum®, 15847; Kel-Lite+sn CM, 15853; Keltrol®, 15886; Keltrol® F, 15889; Kelzan®, 15890; Kelzan® AR, 15891; Kelzan® D, M, XC Polymer, 15892; Kelzan® S, 15893; Rhodicare XC, 26682; Rhodigel® EZ, 26683; Rhodigel®, Rhodigel® EZ, 26684; Rhodopol® 23, XGD, 26696; Rhodopol® XGD, 26697; Xanflood®, 34587;
234-401-9: Rewoteric® AM DM-35L, 26423;
234-409-2: zinc naphthenate, 34781;
234-448-5: satin white, 27483;
234-511-7: Meyerhofferite, 19617;
234-522-7: Borax Glass, 4412;
234-614-7: cobaltic oxide monohydrate, 6533;
234-634-6: chrome black, 6203;
234-722-4: Ammonium molybdate, 2244;
234-796-8: Roydazide, 27045;
234-851-6: thulium oxide, 31769;
234-856-3: terbium oxide, 31223;
234-963-5: zirconium boride, 34822;
234-975-0: Barium titanate, 3646;
234-988-1: calcium titanate, 4974;
235-015-3: holmium oxide, 14279;
235-030-5: preparing salt, 24898; sodium stannate, 28819;
235-042-0: scandia, 27544; scandium oxide, 27546;
235-043-6: samarium oxide, 27308;
235-044-1: strontium titanate, 29831;
235-045-7: Erbium oxide, 10747;
235-087-6: phosphotungstic acid, 23749, 23750, 23751;
235-088-1: Hartotrope STS-40, Powd, 13693;
235-106-8: marcasite, 18700;
235-111-5: Tetrabor®, 31357;
235-113-6: cupric carbonate, 7378; malachite, 18458; mountain blue, 20366; verditer blue, 33574;
235-123-0: Camite, 5054;
235-134-0: Antracol®, 2634; Fruvit®, 12423;
235-169-1: Ytterbium Yb-169 DTPA, 34682;
235-180-1: Unicrop Zineb, 32932; Zidanit, 34758; Zineb, 34807, 34808;
235-183-8: Ammonium bromide, 2234; FR-11, 12351;
235-185-9: Ammonium Fluoride, 2239;
235-186-4: Ammonium chloride, 2235; Katapone VV-328, 15785;
235-252-2: Halphos, 13569;
235-253-8: Aciculite, 326; Alphatex®, 1836; Altowhite LL, 1875; aluminum silicate, 1924; Aluminum Silicate P820, 1925; Alusil, 1937; Alusil ET, 1938; Burgess 2211, 4703; Burgess 30-P, 4702; Burgess KE, 4704; Dynamullit, 9360; Iceberg®, 14811; Icecap® K, 14812; Kaopolite® 1152, 15730; Kaopolite® AB, 15731; Kaopolite® SF, 15732; Kaylene, 15813; Kaylene-ol, 15814; Multex, 20461; Optiwhite®, 22247; Pasilex, 22876; pipeclay, 23872; Sipernat® 44, 28521; Tisyn®, 31857;
235-255-9: potassium stannate, 24782;
235-340-0: Hectabrite® AW, 13749; Hectalite® 200, 13750; Hectorite, 13752;
235-380-9: Halbase 10, 13543;
235-446-7: siccative, 28200;
235-654-8: Campbell's X-Spor, 5075; Headland Spirit, 13734; Maneb, 18526; Manguard®, 18564; Manzate®, 18624; Septal, 27988; Trimangol, 32331;
235-697-2: Sipomer® DCPA, 28533;
235-714-3: cobaltous carbonate, 6536;
235-715-9: nickel carbonate, basic, 21124;
235-741-0: Marlophor® IH-Acid, 18895; Rhodafac® PEH, 26601;
235-759-9: Molybdate Red, 20074;
235-799-7: Crafol AP-201, 6955; Tryfac® 5573, 32514;
235-802-1: Adinol CT, 696;
235-907-2: Rewoteric® AM V, 26428; Zoharteric LF, 34882;
235-921-9: Ageflex HDDA, 990;
235-935-5: Trigonox® 141, 32285; USP® -245, 33166;
235-963-8: Terbufos, 31226;
236-025-0: stearyl/lauryl thiodipropionate, 29608;
236-029-2: Ageflex FA-1Q80DMS, 972;
236-049-1: cyhexatin, 7568; Plictran, 24129, 24130;
236-050-7: Trigonox® 42, 32272; Trigonox® 42 PR, 32273; Trigonox® 42S, 32274;
236-062-2: Berol 302, 4026; Chemstat® 172, 6034;
236-068-5: nickel nitrate, 21128;
236-112-3: CD4153, 5502; Di-*t*-butoxydiacetoxysilane, 8362; Dynasylan® BDAC, 9398;
236-149-5: Setacin F Spezial Paste, 28082; disodium lauryl sulfosuccinate, 8649;
236-152-1: Mocap 10G, 20006;
236-164-7: Rewoteric® AM HC, 26425;
236-221-6: Thiofluor+sn, 31686;
236-239-4: Diak No. 4, 8278;
236-407-7: Torque, 32077;
236-411-9: Maloran, 18493;
236-487-3: cesium fluoride, 5774;
236-503-9: Nifurpirinol, 21170;
236-533-2: Ducobee-Hy, 9082; Neo-Cytamen, 20885;
236-565-7: CT3250, 7327;
236-623-1: gold trichloride, 13175;
236-645-1: salt of Lemery, 27285;
236-656-1: Afugan, 952; Missile, 19944;
236-671-3: Head and Shoulders, 13726; Zinc Omadine, 34783; Zinc Omadine®, 48% Fine Particle Disp, 34784; Zinc Omadine®, Powd, 34785; zinc pyrithione, 34791; Zink Pyrion, 34809;
236-675-5: Covermark, 6921; Diox DR 22, 8573; Finntitan, 11817; Hitox®, 14148; Hombitan®, 14288; Kronos®, 16439; Kronos®, 16440; Kronos® 1000 , 16441; Kronos® 2020, 16442; Kronos® 2073 , 16443; Kronos® 3020, 16444; Mearlin®, 19191; Octotint 138, 22031; Rayox, 25912; Runa, 27131; Rutiox, 27151; Siccolam, 28202; Spectra-Pearl®, 29228; Synthetic Rutile, 30643; Tiona®, 31839; Tiona® HSS, 31840; Tiona® RCL-4, 31841; Tiona® RCL-535, 31842; Tiona® RCS-P, 31843; Tioxide, 31846; Ti-Pure® R-103, 31850; Titan, 31858; Titan Design, 31861; titanium dioxide, 31867; Titanium Dioxide P25, 31869; Titanox, 31872; Titanox Design, 31873; Titanox RA-39, 31874; Titanweiss C, Extra T, Standard T, Standard A, 31875; Tronox Titanium Dioxide Pigments, Chloride Process, 32465; Tytanpol, 32690; Zopaque, 34904;
236-709-9: Barium nitrite, 3640; Stickstoffoxydbaryt, 29762;

236-710-4: Desicchlora, 8139;
236-740-8: V-59, 33217;
236-743-4: sodium tungstate, 28829;
236-751-8: aluminum nitrate, 1917;
236-753-9: Weston® PTP, 34317;
236-757-0: Permethyl 99A, 23390;
236-813-4: Telloy®, 31058; tellurium, 31062;
236-845-9: Naples yellow, 20733;
236-871-0: CT2015, 7313;
236-878-9: zinc chromate, 34773; zinc yellow, 34800;
236-942-6: Pationic 138, 22905;
236-948-9: Deccox, 7846; Dechlorane® Plus 25, Plus 35, Plus 515, 7850;
236-977-7: lithium molybdate, 17510;
237-003-3: Reinecke's salt, 26033;
237-029-5: cerium sulfate, 5755;
237-031-6: quinalphos, 25679; Savall, 27497, 27498;
237-057-8: Tornusil, 32073;
237-066-7: phosphorus acid, 23741;
237-081-9: sodium ferrocyanide, 28763; yellow prussiate of soda, 34648;
237-093-4: Locoid, 17540;
237-159-2: Fyrol® FR-2, 12540;
237-192-2: cuprous potassium cyanide, 7406;
237-199-0: betanal E, 4061; betanal Tandem, 4062; Goliath, 13188; Gusto, 13504; Headland Dephend, 13728; phenmedipham, 23595; Phenmedipham, 23596; phenmedipham, 23597; phenmedipham, 23598; Pistol, 23883; Pistol 400, 23884; Protrum K, 25299; Suplex, 30292; Tripart® Beta, 32364; Tripart® Beta 2, 32365; Vangard, 33323; Vanguard, 33324; Vanguard, 33325;
237-235-5: Antifungin, 2554;
237-272-7: Northovan, 21585;
237-323-3: yellow prussiate of potash, 34647;
237-340-6: Pyricit, 25479; sodium fluoroborate, 28765;
237-388-8: K-Van, 16499;
237-396-1: nickel sulfamate, 21134; Nimate, 21201;
237-410-6: Synkrolith, 30550;
237-412-7: Hyflux M, 14676;
237-414-8: Schlippe's salt, 27723;
237-435-2: Fischer's salt, 11848;
237-486-0: lead fluoborate, 16907;
237-511-5: Aminopropyltrimethoxysilane, 2187; CA0880, 4853; Dynasylan® AMMO, 9397; Union Carbide® A-1110, 32992;
237-558-1: HyPure L, 14727; lithium hypochlorite, 17509;
237-574-9: potassium tripolyphosphate, 24787;
237-696-2: Naugard® NBC, 20797; NBC, 20831; Perkacit® NDBC, 23287; Vanox® NBC, 33344;
237-732-7: CSC 2-aminobutane, 7308;
237-772-5: Nyad® , 21860; Nyad® Wollastonite, 21861; Nycor® R, 21884; Vansil® W-10, 33365; Wollastocoat®, 34545; Wollastokup®, 34546; wollastonite, 34547;
237-796-6: methyl benzoquate, 19516; Statyl, 29578;
238-102-4: Octomer DIBM, 22009;
238-107-1: hydrargaphen, 14539;
238-113-4: Cadmate®, 4892; cadmium diethyldithiocarbamate, 4896; ethyl cadmate, 11065;
238-177-3: Krystallazurin, 16468;
238-270-9: Accelerator EZ Powder, 204; Ancazate ET, 2415; Ethazate, 10938; ethyl zimate®, 11088; Octocure ZDE-50, 21996; Perkacit® ZDEC, 23296;
238-305-8: Dehyton® PLG, 7996;
238-306-3: Rewoteric® AM 2L-40, 26416;
238-310-5: Cerasynt® D, 5735; Witcamide® MAS, 34444;
238-311-0: Ammonyx® OAO, 2265; Chemoxide O, 5997; Incromine Oxide OD-50, 15009; Mackamine O2, 18174; Mazox® ODA, 19154; Standamox 01, 29459;
238-430-8: Afilan PP, 933; Emerest® 2485, 10101;
238-479-5: Aerosol® 18, 878; Empimin® MKK, 10431; Octosol A-18, 22022;
238-687-6: zinc ammonium chloride, 34766;
238-692-3: Ageflex ZDA, 1004; Akrochem® ZDA Powd, 1184;
238-778-0: Akrochem® Z.B.E.D, 1183; Arazate®, 2787; Perkacit® ZBEC, 23294;
238-781-7: Jet Amine DMOD, 15586;
238-836-5: Ketrax, 16149;
238-877-9: ABT-2500®, 166; Altalc 200 USP, 1870; Artic Mist, 3148; Ceramitalc, 5707; Cimfix 606, Cimpact 699, 710, 6306; I T Talc, 14800; J-13, 15468; Jetfill 700C, 15607; Lubestine, 17712; Luzenac B170, 18052; Microbloc®, 19656; MicroPflex 1200, 19687; Microtuff 1000, 19711; Mistron CB, 19951; Mistron Vapor-RE, 19952; Mistron ZSC, 19953; Nicron 325, 21161; Nicron 665, 21162; Nicron JS 422, 21163; Nytal® 100, 21932; Olympic, 22146; PolyTalc 445, 24624; Select-A-Sorb, 27856; Silverline 200, 28438; Silverline 665, 28439; Special Extender, 29218; Stellar 500, 29638; SteriLine 200, 29732; SteriLine 665, 29733; Super Lubestine, 30180; Supra EF, 30295; Suprafino A, 30303; talc, 30731; Talc MS, 30732; Vantalc®, 33368; Vertal 200, 33633; Vertal 92, 200, 33632;
238-878-4: quartz, 25596; rose quartz, 26967;
238-887-3: Sebacil®, 27796; Volathion®, 34011;
238-920-1: methyl green, 19540;
238-921-7: CD5635, 5511;
238-953-1: Bismuth subsalicylate, 4244;
239-032-7: Mirataine® H2C-HA, 19928;
239-138-3: Eosin, 10606;
239-148-8: cryolite, 7279; Kryalith, 16448;
239-263-3: Vicure® 55, 33735;
239-333-3: Devrinol, 8216; Devrinol T, 8217;
239-341-7: Q-1301, 25542;
239-352-7: benodanil, 3942; Calirus, 5009; Mascot Clearing, 18999;
239-354-8: Isobutyl Niclate®, 15347;
239-499-7: aluminum chlorate, 1901; Mallebrein, 18487;
239-518-9: Hematin, 13780;
239-556-6: Dytek® A, 9439;
239-557-1: Perkadox® 16, 23301; Perkadox® 16-W40-GB5, 23302;
239-560-8: methyl niclate®, 19553;
239-580-7: Thioset® M, 31714;
239-592-2: Dicurane 500 FW, 8409; Ludorum, 17778; Talisman, 30735; Tolurex, 31958; Tolurgan, 31959; Toro, 32074; Tripart® Ludorum 700, 32378;
239-593-8: V-70, 33221;
239-635-5: TechneScan PYP, 30890;
239-701-3: Ageflex TMPTA, 1001; Sipomer® TMPTA, 28543; SR-351, 29381;
239-707-6: sodium percarbonate, 28807;
239-716-5: Weston® EHDPP, 34313;
239-741-1: USP® -90MD, 33164;
239-802-2: sodium iron edetate, 28776;
239-934-0: Hahnmann's mercury, 13537;
240-110-8: Alachlor, 1327; Alagan, 1329; Alanex, 1331; Alazine, 1340;
240-171-0: CT3610, 7330;
240-173-1: Cholebrine, 6167;
240-178-9: zinc lactate, 34780;
240-211-7: Lupersol 221, 17851;
240-263-0: 1,4-Dithiothreitol, 8730;
240-282-4: Espercarb® 840, 10841; Lupersol 223, 17852; Trigonox® EHP, 32299;
240-286-6: Carbetamex, 5189;
240-354-5: CB2409.5, 5476;
240-367-6: Chemstat® HTSA#1, 6040; HTSA #1, 14445; Kemamide® P-181, 15899; oleyl palmitamide, 22133;
240-458-0: geraniol acetate, 12886; Geranyl acetate, 12889;
240-474-8: erythrosin, 10778; Iodesin, 15205; iodoeosin, 15212;
240-613-2: Argus DMTDP, 2862; Cyanox® MTDP, 7492; Evanstab® 14, 11257;
240-638-9: Trigonox® 169-OP50, 32289;
240-654-6: Tramisol®, 32134;
240-718-3: Cerone, 5756; Ethrel, 11050; Ethrel C, 11051; Ethrel-E, 11052; Ethrel-R, 11053; Power Ethephon 48, 24823;
240-736-1: Crodasinic LT40, 7146;
240-778-0: sodium bisulfide, 28737;
240-795-3: potassium metabisulfite, 24761;
240-815-0: Lannate®, 16727; methomyl, 19502; Sorex Golden Fly Bait, 29137;
240-851-7: Madecassol, 18291;
240-865-3: Kemamine® BQ-2802C, 15906;
240-924-3: Chemical Base 6532, 5960; Lexamine 22, 17096; stearamidoethyl diethylamine, 29594;
240-934-8: malladrite, 18484; salufer, 27301; sodium silicofluoride, 28818; SSF, 29390; Uba, 32698;
240-969-9: Aflammit TI, 939;
240-985-6: Aflammit ZR, 941;
240-991-9: calcium sulfate (dihydrate), 4972; Compactrol®, 6670;
241-004-4: sodium borohydride, 28740;
241-034-8: hydrofluosilicic acid, 14587; keramyl, 16054; montanine, 20271; sand acid, 27321;
241-053-1: Vanax® PML, 33293;
241-154-0: monensin, 20208; Romensin, 26930;
241-178-1: Edifenphos, 9618; Hinosan®, 14097;
241-189-1: Flosol, 12026;
241-234-5: Caustic baryta, 5468;
241-270-1: Danol, 7728;
241-327-0: Behentrimonium chloride, 3898; Genamin KDM-F, 12772; Varisoft® BT-85, 33417;
241-364-2: Dehyquart D, 7984;
241-533-0: Isonox® 132, 15384; Vanox® 1320, 33337;
241-640-2: Hest MS, 13885; Hetester MS, 13899;

Kemester® 1418, 15939; Schercemol MS, 27611;
241-646-5: Starfol® BB, 29534;
241-654-9: Cetiol® J600, 5799; Dynacerin®, 9339; Dynacerin® 660, 9340;
241-677-4: CE6345, 5529; Dynasylan® ETAC, 9407; ethyltriacetoxysilane, 11146;
241-774-1: Santogard PVI, 27411;
241-775-7: Benomyl, 3944;
241-789-3: Ageflex T4EGDA, 970;
241-869-8: 2-cyanoethyltriethoxysilane, 7478; CC3433, 5494;
241-950-8: 5-phenyltetrazole, 23649;
242-040-3: CT3600, 7329;
242-042-4: CB2790, 5481;
242-044-5: CO9819, 6511;
242-056-0: 3-chloropropylmethyldimethoxysilane, 6137; CC3290, 5490;
242-069-1: Elocril, 9852; Iodofenphos, 15213;
242-085-9: Phospholipon® MC, 23728;
242-159-0: flowers of tin, 12033; Superlite, 30252; tin ash, 31811; tin(IV) oxide, 31817;
242-201-8: Crodamol BE, 7117; Kemester® BE, 15954; Schercemol BE, 27585;
242-272-5: CI7810, 6288; Dynasylan® IBTMO, 9410; Prosil® 178, 25235;
242-287-7: Dynasil® CA, 9383;
242-332-0: Ethoquad® O/12, 11001;
242-339-9: chrome red, 6209;
242-348-8: Cutlass, 7442;
242-354-0: Hibiclens, 14049; Hibitane, 14054; Phisomed, 23671; Rotersept, 27002; Uniscrub®, 33039; Unisept® Solution, 33040;
242-471-7: Syncrowax HR-C, 30534;
242-472-2: CD5636, 5512;
242-480-6: Renovue-65, Renovue-DIP, 26115;
242-505-0: methabenzthiazuron, 19472; Tribunil®, 32204;
242-637-9: molybdenum dioxide, 20076;
242-640-5: Jasmopyrane, 15533;
242-670-9: Fixin, 11860;
242-748-2: Ledate®, 16949; methyl ledate, 19547; Super Sulfur No. 2, 30206;
242-777-0: oryzalin, 22347; Weed Stopper, 34258;
242-784-9: Weston® DSP, 34310;
242-960-5: Edenor PTO, 9610; Liponate PO-4, 17385; Mazol® PETO, 19139; Pentaerythrityl tetraoleate, 23141;
243-116-9: Variotin, 33396;
243-215-7: oxadiazon, 22406; Ronstar 2G, 26954;
243-424-3: Espercarb® 438M-60, 10840; Lupersol 225, 17854; Trigonox® ADC, 32293; Trigonox® SBP, 32307;
243-433-2: Dosaflo, 8812;
243-468-3: Amical® 48, 2058;
243-734-9: Arcosolv® TPM, 2816;
243-746-4: ferric hydroxide, 11649; ferrugo, 11700;
243-761-6: methazole, 19481; Probe, 25048;
243-814-3: zinc hydroxide, 34778;
243-815-9: Chiltern Kocide 101, 6069; Comac Parasol, 6659; cupric hydroxide, 7381, 7382, 7383; Schweitzer's reagent, 27738;
243-885-0: Great Lakes PHT4-Diol, 13313; tetrabromo phthalatediol, 31358;
243-957-1: silver(I) oxide , 28432;
244-058-7: osmic acid, 22355; osmium tetroxide, 22357;
244-068-1: Lanoxicaps, 16748; Lanoxin, 16749; Lanoxine-PG, 16750;
244-088-0: Phostoxin, 23754; Power Phosphine Pellets, 24830; Talunex, 30757;
244-121-9: Akrochem® TDEC, 1179; ethyl tellurac®, 11084; Perkacit® TDEC, 23290;
244-168-5: cadmium hydroxide, 4898;
244-209-7: Lexone®, 17165; metribuzin, 19595, 19596; Sencor®, 27947; Sencorex® WG, 27948;
244-238-5: Sulfetal CJOT 38, 29983;
244-288-8: Spectraban, 29225;
244-289-3: Escalol® 507, 10787; Padimate O, 22549; Solarchem® O, 28898;
244-299-8: Bismate®, 4232; Bismet, 4233; Bismuth dimethyldithiocarbamate, 4238;
244-325-8: dilauryl phosphite, 8475; Weston® DLP, 34306;
244-344-1: cesium hydroxide, 5775;
244-438-2: silver sulfide, 28435;
244-492-7: Alugel, 1882; Alumina hydrate, 1892; aluminum hydroxide, 1905; Colugel, 6655; Dialume, 8287; F-1000 Dried Gel, 11380; F-1000®, 11381; F-2000, 11382; F-2000 Dried Gel, 11383; F-2100 Dried Gel, 11386; F-2200 Dried Gel, 11388; F-500, -3600, etc, 11379; Hydral® 710, 14535; Hydroxal, 14656; Martifin, 18960; Martinal, 18962; Martinal® OL-111 LE, 18963; Martinal® ON-4608, 18964; Martinal® OS, 18965; Onyx Classica , 22185; Polarite® 880E(W), 24302; Rehydragel® Compressed Gel, 26025; SpaceRite S-11, 29183;
244-501-4: Miramine® O, 19871;
244-544-9: cyanazine, 7472; Fortrol, 12323; Match, 19055;
244-556-4: Vansil, 33363;
244-588-9: Ageflex FA-2Q50DMS, 973;
244-654-7: Kankerex, 15713; mercuric oxide, red and yellow, 19348; red oxide of mercury, 25966;
244-726-8: Jet Amine DE-13, 15584;
244-751-4: Briquest® 543-45AS, 4535;
244-754-0: Afilan EHS, 931; Cetiol® 868, 5794; Lexolube® T-110, 17164; octyl stearate, 22041; Wickenol® 156, 34392;
244-793-3: Palite, 22610;
244-838-7: Laraflex®, 16785;
244-848-1: Nemacur®, 20863;
244-906-6: Esperox® 497M, 10851; Trigonox® 97-C75, 32277;
245-039-6: Vicure® 10, 33734;
245-148-9: Ranide, 25868;
245-178-2: Vitride®, 33994;
245-204-2: Starfol® OS, 29538;
245-216-8: Bendiocarb, 3937; Ficam, 11760; Garvox 3G, 12685; Seedox SC, 27840;
245-217-3: Emalex PG-di-L, 9982;
245-228-3: O.O.D, 21944;
245-261-3: Aspartame, 3233; Equal, 10731; NutraSweet, 21832;
245-289-6: Ceraphyl® IPL, 5732;
245-366-4: CM9220, 6467;
245-396-8: Lofepramin Hydrochloride, 17552;
245-423-3: Kamillosan, 15707;
245-430-1: Abol, 146; Aphox, 2654; FBC Pirimicarb 50, 11517; Phantom, 23571; pirimicarb, 23879; Pirimor, 23880; Power Demo, 24819; Rapid, 25884;
245-445-3: Blade, 4310; Vydate®, 34171;
245-477-8: Butachlor, 4727; Butanex, 4736; Machete, 18124;
245-629-3: Aerosol® A-196-85, 887; Octosol TH-40, 22028; Rewopol® SBDC 40, 26372;
245-679-6: Esperox® 13M, 10846;
245-704-0: Fernex, 11636, 11637; Primicid, 24983; Primotec, 24986;
245-737-0: heptenophos, 13795; Hostaquick, 14407;
245-740-7: CDA Mildothane, 5520; Cercobin, 5738; Mildothane, 19759; Mildothane Turf Liquid, 19760; Sys Tec® 1998, 30661; thiophanate methyl, 31706;
245-741-2: Nemafax, 20864;
245-764-8: Canesten®, 5086; Gyne-Lotrimin, 13513; Lotrimin, 17631; Mono-Baycuten®, 20217;
245-876-7: CT2507, 7318; Union Carbide® A-1160, 32994;
245-949-3: Milgo, 19765; Milstem, 19800;
245-951-4: Campbell's Rapier, 5069; Kerb 50W, 16064; Kerb 50W, 16065; Kerb Propyzamide 50, 16066; Rapier, 25889;
245-986-5: Hostathion, 14413; Methoxone, 19503; MSS CMPP, 20429; triazophos, 32196;
246-140-8: aluminum nitride, 1918;
246-147-6: mepiquat chloride, 19327; Pix® ULV, 23898;
246-240-1: Di-*t*-butyl dicarbonate, 8367;
246-331-6: CM9160, 6466;
246-347-3: Calixin®, 5011;
246-376-1: potassium sorbate, 24780;
246-386-6: Photocure 51, 23757;
246-466-0: tripropylene glycol, 32400;
246-467-6: CI7840, 6289;
246-515-6: Decrolin®, 7866; Parolite, 22861; Reduxol Z, 25986; zinc formosul, 34775;
246-563-8: BHA, 4121; Butylated hydroxyanisole, 4787; Nipantiox, 21276; Nipantiox 1-F, 21277; Sustane® 1-F, 30432; Sustane® BHA, 30435;
246-584-2: Lexaine® O, 17095; Mackam HV, 18143; Mirataine® BET-O-30, 19924; Oleamidopropyl betaine, 22108;
246-585-8: Bentazon, 3947;
246-598-9: Chemoxide SAO, 5999; Mackamine SAO, 18176;
246-613-9: isooctyl thioglycolate, 15386;
246-614-4: triisooctyl phosphite, 32312; Weston® TIOP, 34320;
246-672-0: nonyl phenol, 21460;
246-675-7: Diaparene, 8326;
246-677-8: Coalite N.T.P, 6518; Kronitex® TXP, 16438; Syn-O-Ad® 8475M, 30554;
246-678-3: Bisbutyl peroxy diisopropyl benzene, 4230; Luperox 802, 17841; Perkadox® 14, 23300; Retilox® F 40 MG, 26274; Vul-Cup 40KE and R, 34118; Vul-Cup, Vul-Cup 40KE, Vul-Cup R, 34119;
246-680-4: DeSonate 50-S, 8176; DeSonate 60-S, 8177; Nacconol® 40G, 20647; Nansa® HS80S, 20687; Rhodacal® DDB-40, 26582; sodium dodecylbenzene

sulfonate, 28760; Sul-fon-ate AA-10, 30020;
246-684-6: Mackamine OAO, 18175; oleamidopropylamine oxide, 22110;
246-688-8: Aerosol® DPOS-45, 881;
246-698-2: Halso® 99, 13570;
246-705-9: Crodesta DKS F110, 7167; Crodesta F-160, 7169; Grilloten PSE141G, 13378; Ryoto Sugar Ester S-1170, 27169; Sucro Ester 15, 29932; Sucro Ester 7, 29931;
246-771-9: Texanol® Ester-Alcohol, 31451; 2,2,4-trimethyl-1,3-pentanediol monoisobutyrate, 32339;
246-784-2: Ablusol DBD, 124;
246-814-1: Amaze®, 1951; Amidocid®, 2106; Oftanol®, 22049; Pryfon, 25355;
246-839-8: Eltesol® XA, 9877; Manro XSA, 18610; Reworyl® X, 26412; XSA 80, 34611; Xylene sulfonic acid, 34621;
246-868-6: Afilan ICS, 932; Kemester® 5822, 15951; Kessco® ICS, 16114; Schercemol ICS, 27603;
246-873-3: Grilloten LSE87, 13376; Ryoto Sugar Ester LWA-1570, 27166;
246-929-7: Artodan SP 55 Kosher, 3150; Crolactil SSL, 7186; Emplex, 10443; Emulsilac S, 10514; Grindtek FAL 1, 13409; Pationic SSL, 22908; Radiamuls® SSL 2990, 25788;
246-933-9: Kantrex, 15718; Kantrim, 15720; Klebcil, 16209;
246-939-7: Swedex SSL-5AC, 30450;
246-941-2: Diacetin, 8248; Hallco® C-491, 13548;
246-960-6: Dehscofix 917, 7925; Supragil® NK, 30308;
246-985-2: Rhodapex® 674/C, 26621; Rhodapex® EST-30, 26629;
246-998-3: Doverphos® 6, 8835; Weston® TDP, 34318;
247-036-5: LABS 100/H.V., 16544; Nansa® TDB, 20696; Nusolv ABP-62, 21802; tridecylbenzene sulfonic acid, 32236;
247-038-6: Tegin® O, 30946;
247-098-3: Doverphos® 7, 8836; Weston® PDDP, 34315;
247-125-9: Filex, 11776; Propamocarb hydrochloride, 25171;
247-144-2: Emerest® 2419, 10097; glyceryl dioleate, 13080; Priolube 1409, 25016;
247-148-4: FR-1206, 12363; Great Lakes CD-75P, 13300; Hexabromocyclododecane, 13988;
247-280-2: Lopid, 17599;
247-310-4: disodium lauramido MEA-sulfosuccinate, 8647; Geropon® SBL-203, 12912; Incrosul LMS, 15055; Mackanate LM-40, 18189; Rewopol® SBL 203, 26377; Varsulf® SBL-203, 33461;
247-363-3: Krenite, 16400;
247-404-5: phenothrin, 23619; Sumithrin, 30117;
247-459-5: Equiben, 10733; Novazole, 21711;
247-464-2: Nipagin M Potassium, 21267;
247-481-5: morantel tartrate, 20307; Paratect Bolus, 22804;
247-525-3: ethofumesate, 10967; Nortron, 21586;
247-552-0: Deriphat 160C, 8091;
247-556-2: Arylan® PWS, 3158; Kowet 3300, 16374; Rhodacal® 330, 26577; Rhodacal® IPAM, 26586;
247-557-8: Ablusol DBC, 123; Agrilan® X98, 1063; Kowet 12, 16373;
247-561-6: Rhodacal® RM/210, 26590; Supragil® MNS/90, 30307;
247-564-6: Dehscofix 923, 7927;
247-568-8: Ablunol S-40, 109; Arlacel® 40, 2900; Crill 2, 7029; Disponil SMP 100 F1, 8700; Emsorb® 2510, 10460; Glycomul® P, 13113; Hodag SMP, 14244; Kemester® S40, 15967; Liposorb P, 17410; Montane 40, 20266; Prote-sorb SMP, 25280; Radiasurf® 7135, 25798; S-Maz® 40, 28641; Sorbax SMP, 29100; Sorbon S-40, 29118; Span® 40, 29188;
247-569-3: Ablunol S-85, 112; Alkamuls® S-85, 1682; Alkamuls® STO, 1689; Arlacel® 85, 2904; Atmer® 106, 3359; Crill 45, 7035; Disponil STO 100 F1, 8705; Emsorb® 2503, 10457; Ethylan® GT85, 11113; Glycomul® TO, 13116; Hodag STO, 14247; Kemester® S85, 15971; Liposorb TO, 17416; Prote-sorb STO, 25282; Radia® 7355, 25738; S-Maz® 85, 28646; Sorbax STO, 29102; Span® 85, 29192; Witconol 2503, 34502;
247-574-0: Abitol® E, 71;
247-658-7: diphenyl isooctyl phosphite, 8595; Doverphos® DPIOP, 8848; Weston® ODPP, 34314;
247-668-1: glyceryl caprylate, 13077; Imwitor® 308, 14940; Imwitor® 742, 14945; Witafrol® 7420, 34421;
247-669-7: Flexricin® 9, 11950; Naturechem® PGR, 20777; Propylene glycol ricinoleate, 25222;
247-693-8: diphenylcresyl phosphate, 8601; Disflamoll® DPK, TPK, 8641;
247-704-6: Hallco® C-918, 13549; monacetin, 20095;
247-706-7: Ryoto Sugar Ester P-1570, P-1670, 27168;
247-710-9: Ammonium xylene sulfonate, 2259; Eltesol® AX 40, 9865; Hartotrope AXS, 13691; Naxonate® 4AX, 20822; Stepanate® AXS, 29672;
247-730-8: Behenamine oxide, 3897;
247-759-6: Doverphos® 4-HR, 8845; Doverphos®4, 8834; Lankromark® LE109, 16695; Lowinox® TNPP, 17676; Trisnonylphenyl phosphite, 32411; Weston® 399, 34297; Weston® TNPP, 34323; Wytox® 312, 34578;
247-761-7: Kathon®, 15792; Kathon® 893, 15794; Micro-Chek® 11, 19664; Pancil T, 22666; Skane® M-8, 28563;
247-777-4: diphenyl isodecyl phosphite, 8594; Doverphos® 8, 8837; Weston® DPDP, 34308;
247-779-5: Weston® 600, 34303;
247-784-2: Marlopon® ADS 50, 18902;
247-872-0: Fairy Ring Destroyer, 11416; Funginex, 12478; Saprol, 27447; triforine, 32254;
247-891-4: Alkamuls® S-65, 1678; Alkamuls® STS, 1690; Crill 35, 7033; Disponil STS 100 F1, 8707; Emsorb® 2507, 10459; Famodan TS Kosher, 11424; Glycomul® TS KFG, 13117; Grindtek STS, 13416; Hodag STS, 14248; Kemester® S65, 15969; Liposorb TS, 17418; Montane 65, 20268; Prote-sorb STS, 25283; Radia® 7345, 25737; Radiamuls® Sorb 2344, Sorb 2345, 25786; Radiamuls® Sorb 2345, 25787; S-Maz® 65K, 28643; Sorbax STS, 29103; sorbitan tristearate, 29111; Span® 65, 29190;
247-897-7: Isoflurane, 15363;
247-899-8: Milldride® OSA, 19789;
247-952-5: Alkanox® P-24, 1705; Blendex® 340, 4342; Ultranox® 626, 32850; Ultranox® 627A, 32851;
247-954-6: Esperox® 750M, 10857;
247-955-1: Aztec® *t*-Butyl Peroxyneodecanoate-50 OMS, 75 OMS, 3547; Esperox® 33M, 10849; Lupersol 10, 17842; Trigonox® 23, 32267;
247-956-7: Esperox® 939M, 10858; Lupersol 188-M75, 17848; Lupersol 288-M75, 17859; Trigonox® 99-B75, 32278;
247-979-2: glycidyl decanoate, 13092; Glydexx N-10, 13141;
248-003-8: Larocin, 16787; Larotid, 16800; Qualamox, 25583;
248-012-7: Lozol, 17688;
248-052-5: Dantocol® DHE, 7730; DEDM hydantoin, 7872;
248-111-5: Timolate, 31805;
248-133-5: Exxal® 8, 11336; isooctyl alcohol, 15385;
248-289-4: dodecylbenzene sulfonic acid, 8779; Pentine Acid 5431, 23182; Rhodacal® ABSA, 26579; Witco® 1298, 34458;
248-296-2: Polystep® A-15-30K, 24587;
248-299-9: Ceraphyl® 230, 5724; Kodaflex® DIDA, 16268; Monoplex® DDA, 20245; Plasthall® DIDA, 23998; Schercemol DIA, 27592; Unimate® DIPA, 32968;
248-317-5: Crodesta DKS F10, 7166; Crodesta F-10, 7168; Ryoto Sugar Ester S-570, S-770, 27171;
248-351-0: Hetester HCA, 13895; Naturechem® GTH, 20771;
248-363-6: Exchem GO-1, 11294;
248-470-8: Adol® 66, 785; Diadol 18G, 8264; Michel XO-150-1620, 19642;
248-502-0: Neo Heliopan® Hydro, 20868;
248-523-5: Morflex 100, 20315; Palatinol® D10, 22586; Witcizer 313, 34455;
248-565-4: CT3795, 7331;
248-586-9: Capmul® GDL, 5111; Emulsynt GDL, 10525; glyceryl dilaurate, 13079; Kemester® GDL, 15961; Kessco® GDL, 16107; Lexemul® GDL, 17138;
248-587-4: Kemester® EE, 15957; Schercemol EE, 27599;
248-595-8: CO9745, 6505;
248-597-9: Anox® IC-14, 2485; Vanox® GT, 33342;
248-655-3: VA-044, 33225;
248-659-5: Marignac's salt, 18717;
248-666-3: Bisomer 2HPMA, 4274; hydroxypropyl methacrylate, 14667;
248-688-3: Schercemol NGDC, 27612;
248-710-1: Alkamide® CP-1255, 1638;
248-762-5: Akyporox NP 15, 1276;
248-811-0: Armeen® DMOD, 2947; Crodamine 3.AOD, 7109;
248-848-2: Syn-O-Ad® 8480, 30557;

249-014-0: Reslin, 26216;
249-060-1: Schercemol NGDO, 27614;
249-063-8: Calfax DB-45, 4983; Chemcogen AC, 5941; Rhodacal® DSB, 26584;
249-110-2: Duplosan® DP, 9131;
249-112-3: Cumene sulfonic acid, 7353; Eltesol® CA 65, 9866; Reworyl® C, 26403;
249-205-9: bromadiolone, 4565; Rentokil Deadline, 26116; Slaymor, 28597;
249-271-9: Lodosin, 17549;
249-277-1: Ajidew N-50, 1143; Nalidone®, 20672; Ritamectant PCA, 26816;
249-419-2: Cobex, 6544;
249-528-5: Actellic, 458; Actellifog, 459; Blex, 4349; Fumite Pirimiphos Methyl Smoke, 12463, 12464; Sybol, 30475;
249-530-6: Amilperoxy pivalate, 2122; Aztec® *t*-Amyl peroxypivalate-75 OM, 3540; Esperox® 551M, 10853; Lupersol 554-M50, 554-M75, 17865; Trigonox® 125-C75, 32283;
249-554-7: Tomah PA-13i, 31981;
249-596-6: Preventol® CI7-100, 24929; tolyltriazole, 31962;
249-707-8: Ageflex FA-8, 978; SR-440, 29383;
249-720-9: Seenox 412S, 27843;
249-793-7: Dermol OO, 8129; Schercemol OHS, 27615; Wickenol® 171, 34398;
249-828-6: Syn-O-Ad® 8479, 30556;
249-862-1: Bernel® Ester EHP, 4008; Ceraphyl® 368, 5725; Kessco® Octyl Palmitate, 16118; Lexol EHP, 17148; octyl palmitate, 22040; Schercemol OP, 27617; Tegosoft® OP, 31025; Unimate® EHP, 32970; Wickenol® 155, 34391;
249-951-5: Dowanol® DPnB, 8893;
249-978-2: Ageflex FM-10, 985; isodecyl methacrylate, 15359;
249-981-9: Arylmate®, 3166; Croneton®, 7206;
250-001-7: Bronidox L, 4592;
250-113-6: Oxivent, 22433;
250-135-6: Crodateric S, 7156;
250-151-3: Hamposyl® M-30, 13603;
250-188-5: Perlankrol® PA Conc, 23329;
250-228-1: Hartotrope KTS 44, 13692;
250-241-2: Orthene, 22322;
250-275-8: lauryl PCA, 16866;
250-426-8: CM8450, 6454; Colortrend, 6648;
250-475-5: Amipaque, 2197;
250-478-1: Endobil, 10548;
250-517-2: Propetamphos, 25183;
250-624-4: Flubendazole, 12039;
250-635-4: Equivurm Plus, 10736; Mebatreat, 19197; Mebendazole, 19198; Mebenvet, 19199; Ovitelmin, 22402; Telmin, 31066; Vermox, 33591;
250-651-1: Schercemol 318, 27582;
250-696-7: Emerest® 2308, 10083; Kemester® 5721, 15950; Lexolube® B-109, 17162; Liponate TDS, 17389;
250-705-4: Alkamuls® GMS/C, 1667; Alkamuls® GMS/C, 1668; Capmul® GMS, 5113; Cerasynt® WM, 5737; CPH-250-SE, 6944; Cutina® GMS, Cutina® KD16, Cutina® MD-A, 7440; Geleol, 12733; glyceryl stearate, 13087; Imwitor® 191, 14939; Radiamuls® MG 2141, MG 2142, MG 2600, MG 2900, 25777; Tegin® 4011, 30945; Tego® Care 150, Care 300, 30967; Witconol 2400, Witconol MST, Witconol RHT, 34499; Zohar GLST, 34850;
250-709-6: Alkanox® 240, 1703; Hostanox® PAR 24, 14375; Lowinox® 242, 17655; Naugard® 524, 20791;
250-778-2: dodemorph-acetate, 8782; F-238, 11376; Meltatox®, 19313;
250-801-6: Evangard® 18MP, 11252; octadecyl mercaptopropionate, 21983;
250-913-5: Eltesol® SC 93, 9871; Naxonate® 45SC, 20826; Naxonate® SC, 20827; Reworyl® NCS, 26406; sodium cumene sulfonate, 28752; Stepanate® SCS, 29673; Witconate SCS 45%, 34490;
251-013-5: octadecyl methacrylate, 21984; stearyl methacrylate, 29607;
251-035-5: BTC® 818, 4635;
251-084-2: FR-1205, 12362; Fyrol® PBR, 12541; Great Lakes DE-71, 13302; pentabromodiphenyl oxide, 23129, 23130;
251-087-9: FR-1208, 12364; Great Lakes DE-79, 13303; Octabromodiphenyl oxide, 21981; Saytex® 111, 27515;
251-118-6: Saytex® BT-93®, 27520;
251-136-4: Millithix® 925, 19794;
251-306-8: Schercotaine PAB, 27699;
251-375-4: Amitraz, 2202; Mitaban, 19954; Mitac 20, 19955; Taktic, 30729;
251-381-7: Croptex Fungex, 7215;
251-622-6: Sunett, 30140;
251-646-7: Adimoll® DN, 692; Jayflex® DINA, 15551; Plastomoll® NA, 24059; PX-209, 25439;
251-732-4: Naturechem® EGHS, 20769; Paricin® 15, 22835;
251-734-5: Naturechem® PGHS, 20776; Paricin® 9, 22842;
251-793-7: Bushwacker, 4724;
251-835-4: Arelon, 2830; Chiltern IPU, 6068; Hytane, 14774; Portman Isotop, 24712; Power Swing, 24834; Protugan, 25300; Sabre, 27191; Tolkan, 31937; Tolkan, 31938; Tolkan 500, 31939;
251-844-3: Vanox® SKT, 33347;
251-879-4: Pradone Plus, 24850;
251-932-1: Cetiol® A, 5796;
252-011-7: Caprol® 10G40, 5130; Drewpol® 10-4-O, 9014; Hodag PGO-104 (formerly Hodag SVO-1047), 14220; Hodag SVO-1047, 14249; Mazol® PGO-104, 19141;
252-029-5: Esperox® C-496, 10860;
252-091-3: Vanax® 808 HP, 33285;
252-104-2: Arcosolv® DPM, 2812; Icinol DPM, 14815; Poly-Solv® DPM, 24555;
252-125-7: Tokuthion®, 31932;
252-182-8: Texacat® DD, 31426;
252-287-4: diisooctyl phosphite, 8465; Doverphos® DIOP, 8846;
252-302-9: Dermol 108, 8119;
252-390-9: CT2910, 7322; Dynasylan® TRIAMO, 9418;
252-470-3: Comodor, Comodor 600, 6667; tebutam, 30879;
252-478-7: Ceraphyl® 28, 5712; cetyl lactate, 5820; Liponate CL, 17373;
252-487-6: Nipagin A Sodium, 21265;
252-488-1: Nipasol M Sodium, 21284;
252-512-0: Uniplex 225, 33013;
252-529-3: Diflubenzuron, 8450; Dimilin, 8529;
252-545-0: Bolstar, 4384; Helothion, 13768;
252-615-0: Enilconazole, 10582; Fungaflor, 12471; Imaverol, 14891; imazalil, 14892; Magnate, 18349;
252-626-0: Alfacron, 1571; Alfacron 10WP, 1572; azamethiphos, 3517;
252-816-3: Isonox® 129, 15383; Vanox® 1290, 33336;
252-873-4: Weston® DOPI, 34307;
252-964-9: Ceraphyl® ICA, 5731; Eutanol G16, 11240; Michel XO-150-16, 19641;
252-979-0: Lederfen, 16955;
252-981-1: vegetable rouge, 33523;
253-019-3: Empicol® SLL, 10338; Geropon® LSS, 12908;
253-034-5: Weston® DTDP, 34311;
253-040-8: Calfax 10L-45, 4982; Dowfax 3B2, 8911; Poly-Tergent® 3B2, 24630;
253-048-1: Nipagin A Potassium, 21264;
253-049-7: Nipabutyl Sodium, 21251;
253-057-0: FR-513, 12354; tribromoneopentyl alcohol, 32201;
253-084-8: Sorbon S-66, 29120;
253-138-0: Morflex 530, 20328;
253-149-0: Adol® 52 NF, 782; Cachalot® C-50 NF, 4878; Cetaffine®, 5781; Cetal, 5783; cetyl alcohol, 5815; cetyl alcohol, 5816; cetyl alcohol, 5817; Crodacol C70, 7068; Crodacol C95NF, 7069; Dehydag Wax 16, 7944; Emery® 3336, 10172; Epal® 16NF, 10626; ethal, 10910; Exxal® 16, 11341; Fancol CA, 11432; Lanette 16, 16663; Lipocol C, 17351; Lorol C16, 17614; Loxiol VPG 1743, 17686; Philcohol 1600, 23667; Rita CA, 26788;
253-178-9: CDA Roval, 5521; Roval Dust, 27021; Roval Flo, 27022; Roval Green, 27023; Roval WP, 27024; Turbair Roval, 32598;
253-211-7: Weston® 430, 34298;
253-332-5: Stabilizer, 29404;
253-349-8: Chymex, 6283;
253-363-4: Ammonyx® KP, 2263; Incroquat O-50, 15041; Mackernium KP, 18208; M-Quat® JO-50, 20407;
253-384-9: USP® -240, 33165;
253-407-2: Emerest® 2421, 10098;
253-452-8: Aerosol® A-268, 888;
253-458-0: Radiasurf® 7423, 25814;
253-466-4: Avoparcin, 3502;
253-483-7: Lopresoretic, 17601;
253-519-1: ACS 60, 453; Ammonium curmene sulfonate, 2236; Eltesol® AC60, 9863; Reworyl® ACS, 26402;
253-521-2: Caprylic imidazoline, 5159; Crodazoline Cy, 7160; Mackazoline CY, 18199; Monazoline CY, 20194;
253-703-1: Cropotex®, 7211;
253-718-3: Lancer, 16653; Perfluidone, 23248;
253-855-9: etrimfos, 11155; Satisfar, 27488;
253-874-2: Loniten, 17584;
253-946-3: stannic chromate, 29495;
253-980-9: Acylglutamate HS-11, 603;
253-981-4: Acylglutamate MS-11, 607;
254-009-1: Nipabutyl Potassium, 21250;
254-037-4: Alkanox® 24-44, 1702; Sandostab P-EPQ, 27353;
254-056-8: Rodeo, 26899;
254-187-0: Aerosol® 22, 879;

254-320-2: Aliette, 1613;
254-372-6: Abiol, 69; Germall® 115, 12893; Imidazolidinyl urea, 14896;
254-384-1: Manzanate, 18623;
254-408-0: Karathane, 15759;
254-445-2: Marlophor® HS-Acid, 18894;
254-485-0: fenpropathrin, 11615; Meothrin, 19322;
254-495-5: Caprol® 10G10S, 5128; Hodag PGS-1010, 14226;
254-585-4: Chemidex P, 5966; Palmitamidopropyl dimethylamine, 22630;
254-896-5: Aktisil PF 216, 1204; CB2494, 5478; Si 69, 28190;
254-938-2: pendimethalin, 23102; Stomp, 29784; Stomp H, 29785;
255-051-3: Amidex RC, 2101; Mackamide R, 18165; Rodea, 26898;
255-062-3: Aerosol® A-102, 885; Schercopol LPS, 27668; Setacin 103 Spezial, 28081; Surfagene S 30, 30361;
255-209-1: azocyclotin, 3523; Clermait®, 6408; Peropal®, 23401;
255-248-4: Neoscan, 20963;
255-349-3: Countdown, 6904; Goltix®, 13191; Metamitron, 19443;
255-350-9: DPPG, 8950; Emerest® 2388, 10094; Lexol PG 900, 17154; Schercemol PGDP, 27619;
255-391-2: Nimrod, 21220;
255-392-8: Anox® 70, 2484;
255-464-9: Loprox, 17604;
255-485-3: Iso Isostearyle WL 3196, 15334; isostearyl isostearate, 15422; Schercemol 1818, 27584; Starfol® IS, 29536;
255-490-0: Dipalmitoyl hydroxy proline, 8578; Lipacide DPHP, 17321;
255-574-7: Osyrol, 22373;
255-663-0: carnidazole, 5298; Spartrix, 29212;
255-674-0: Patlac® IL, 22909;
255-894-7: bifenox, 4135;
255-950-0: Laurel SBT, 16838; sulfated butyl tallate, 29972;
256-103-8: Bayleton® 5, 3800; Monterey Bayleton, 20287; Triadimefon, 32184;
256-120-0: disodium oleamido MIPA sulfosuccinate, 8651; Emcol® 416L, 10045; Emcol® K8300, 10062; Mackanate OP, 18194; Sole Terge 8, 28914;
256-145-7: fenbendazole, 11594; Panacur, 22655;
256-152-5: Avenge 2, 3460;
256-176-6: Ageflex FA-1Q75MC, 971;
256-214-1: Chemidex M, 5964; Schercodine M, 27639;
256-360-6: Ageflex PEA, 997; Laromer® POEA, 16791; Melcril 4087, 19268; SR-339, 29379;
256-599-6: Fumite Ronilan, 12466; Mascot Contact Turf Fungicide, 19001; Power Drive, 24821; Ronilan® DF, FL, 26949; Ronilon, 26950; vinclozolin, 33811;
256-634-5: acifluorfen, 343;
256-779-4: Glass H, 12967; Hexaphos, 14013; Metagon, 19425; Polyphos, 24519;
256-873-5: CT2902, 7321;
257-033-0: Sipomer® DCPM, 28534;
257-048-2: Amine CS-1135®, 2153; Canguard® 327, 5088; Oxaban®-A, 22404;
257-060-8: metolachlor, 19585;
257-074-4: hexazinone, 14020; hexazinone, 14021; Velpar,, 33536; Velpar®, 33537;
257-132-9: Scheibler's reagent, 27567;
257-141-8: Hoegrass, 14271;
257-326-3: fenvalerate, 11620; Sumicidin, 30080;
257-440-3: Ceraphyl® 60, 5718;
257-842-9: Cymbush, 7580; Cymperator, 7583; Cyperkill, 7587; cypermethrin, 7588; Cypersect, 7589; Cypertox, 7590;
257-843-4: Amihope LL-11, 2120;
257-974-7: cupric acetate, basic, 7373;
257-976-8: Atrinal, 3395;
258-004-5: Cops 1, 6797;
258-007-1: Hamposyl® M, 13602;
258-067-9: Ambush, 1982; Cooper Coopex, 6758; Darmycel Agarifume Smoke, 7780; Elimite®, 9838; Fumite Permethrin, 12462; Kafil, 15655; Nippon Ant Killer Powder, 21300; Nippon Ready For Use Ant and Crawling Insect Killer, 21302; Nix Creme Rinse, 21387; Nix Dermal Cream, 21388; Permasect, 23377; permethrin, 23384; Permethrin, 23385; permethrin, 23386; Picket, 23812; Quamilin, 25589; Turbair Permethrin, 32597;
258-154-1: flamprop-isopropyl, 11887; Flamprop-M-isopropyl, 11888; Gunner, 13493; Power Flame, 24824; Power Flamprop, 24825;
258-193-4: Mackamide ISA, 18155;
258-207-9: Bistetramethyl piperidinyl sebacate, 4284; Lowilite® 77, 17645; Tinuvin® 770, 31836; Uvaseb 770, 33178;
258-250-3: Saytex® BN-451, 27519;
258-256-6: Thripstick®, 31766;
258-290-1: Posistac, 24718; Salinomycin, 27271;
258-377-8: Schercoquat BAS, 27673;
258-436-4: Perkadox® 26-fl, 23304;
258-476-2: Laponite® D, 1677 Laponite® XLG, 16780; 8; sodium magnesium silicate, 28782;
258-636-1: Acylglutamate LT-12, 606;
258-659-7: Accelerase, 185;
258-714-5: oxfendazole, 22426; Synanthic, 30520; Systamex, 30662;
258-767-4: Echo, 9577; Embark, 10017; mefluidide, 19230; Mowchem, 20379;
258-847-9: Jayflex® TINTM, 15558; Plasthall® TIOTM, 24015; PX-339, 25442;
258-904-8: Vanox® MTI, 33343;
259-210-8: Herborane, 13807;
259-423-6: Arova 16, 3088;
259-513-5: Bitertanol, 4288;
259-537-6: Spinnaker, 29292; Summit, 30118; triadimenol, 32185;
259-559-6: Droncit®, 9049;
259-565-9: carbosulfan, 5245; Marshal 10G, 18955; Marshal/suSCon, 18956;
259-582-1: neomycin palmitate, 20929;
259-597-3: Timbrel, 31800;
259-715-3: acesulfame potassium, 277;
259-733-1: Clout, 6446;
259-817-8: Arpocox, 3096;
259-837-7: Emcol® 3780, 10047; Hetamine 5L-25, 13887; Incromate SDL, 14973; Lexamine S-13 Lactate, 17101; stearamidopropyl dimethylamine lactate, 29595;
259-978-4: difenacoum, 8449; Killgerm® Ratak Cut Wheat Rat Bait, 16183; Neosorexa, 20967; Ratak, 25896;
259-980-5: Matikus, 19061; Mouser, 20376; Talon, 30750;
260-070-5: Ceraphyl® 45, 5715;
260-081-5: Argidone®, 2848; arginine PCA, 2849;
260-143-1: Anlonyx® 12S, 2477; Mackanate OD-35, 18192; Monamate OPA-100, 20109; Standapol® SH-100, 29479; Texapon SH 100, 31479;
260-148-9: Lopressor, 17602; Lopressor SR, 17603;
260-391-0: Syn-O-Ad® 8478, 30555; Syn-O-Ad® 8485, 30559;
260-410-2: Schercomid SLE, 27658;
260-503-8: Katadolon, 15780;
260-515-3: Basilex, 3709; Risolex, 26782;
260-875-1: Fongarid, 12225; Furalaxyl, 12487;
260-929-4: Benazalox, 3934;
260-967-1: Closantel, 6441; Flukiver, 12051;
260-976-0: Ethoxy carbonyl phenylmethyl phenylformamidine, 11043; Givsorb® UV-1, 12952;
260-979-7: metalaxyl, 19430;
261-043-0: cymoxanil, 7582;
261-093-3: Dynasylan® IMEO, 9411;
261-145-5: Union Carbide® A-1106, 32991;
261-205-0: Rintal®, 26774;
261-222-3: Octosol A-1, 22021;
261-282-0: CB2805, 5482;
261-430-4: Chemstat® 106G/90, 6032;
261-451-9: Patafol, 22881;
261-521-9: Ceraphyl® 375, 5726; Dermol 185, 8120; Schercemol 185, 27581;
261-526-6: Saytex® 120, 27516;
261-619-1: Bernel® Ester CO, 4004; Cetearyl octanoate, 5791; cetyl octanoate, 5821; Crodamol CAP, 7118; Emalex CC-168, 9935; Lanol 1688, 16736; Luvitol EHO, 18036; Schercemol 1688, 27583; Schercemol CO, 27587; Tegosoft® CO, 31017; Tegosoft® Liquid, 31022;
261-673-6: Ceraphyl® 140-A, 5723; Mackester IDO, 18212; Schercemol IDO, 27604; Wickenol® 144, 34387;
261-675-7: Wickenol® 153, 34390;
261-684-6: Schercotaine MAB, 27698;
261-767-7: Actimer FR-1025M, 484;
261-819-9: Bernel® OPG, 4011; Schercemol OPG, 27618; Wickenol® 160, 34395;
261-931-8: Aludone®, 1878; aluminum PCA, 1921;
262-093-6: Iopamidol, 15233;
262-104-4: Bumper, 4669; Power Propiconazole, 24832; Powerspire, 24837; Radar, Radar Propiconazole, 25714;
262-108-6: Dermol 105, 8118;
262-134-8: Behenamidopropyl dimethylamine, 3896; Chemidex B, 5961; Incromine BB, 14997; Mackine 601, 18222; Schercodine B, 27635;
262-373-8: quartz glass, 25597; vitreosil, 33993;
262-706-6: Magnum 240, 18392; Magnum 2610, 18396; Magnum 275, 18393; Magnum 3661, 18397; Magnum 4420, 18398; Magnum 445 HQ, 18394; Magnum 788HP, 18395; Magnum 9450P, 18399; Magnum FG960, 18400; Multibase ABS 3075, 20463; Multibase

ABS 3525 CL, 20464; Multibase ABS 3959, 20465;
262-737-6: Actimer FR-803, 483;
262-854-2: Tiox, 31845;
262-872-0: Vanox® ZMTI, 33349;
262-895-6: Liponate PB-4, 17382;
262-912-7: Jasmacyclat, 15528;
262-975-0: Akrochem® Antioxidant 16, 1158; Alkylated phenol, 1734;
262-976-6: Amine HBG, 2156; Amine HBGD, 2157; Armeen® HT, 2950; Armeen® HTD, 2951; Crodamine 1.HT, 7103; Jet Amine PHT, 15594; Kemamine® P-970, 15920; Radiamine 6140, 25750; Radiamine 6141, 25751;
262-977-1: Amine KK, 2158; Armeen® C, 2940; Armeen® CD, 2941; Jet Amine PC, 15592; Kemamine® P-650D, 15918; Radiamine 6160, 25752; Radiamine 6161, 25753;
262-978-7: Hystrene® 1835, 14755;
262-979-2: Acelan L, 269; Acetadeps, Acelan A, 279; Acylan, 595; Fancol Acel, 11428; Modulan®, 20031; Ritacetyl®, 26800;
262-980-8: Hetlan AC, 13906;
262-988-1: Emery® 2253, 10152; methyl coconate, 19523; Radia® 7117, 25722;
262-990-2: Armeen® M2C, 2954; Jet Amine M2C, 15591;Kemamine® T-6501, 15929; Radiamine 6360, 25764;
262-991-8: Amine M2HBG, 2161; Armeen® M2HT, 2955; Kemamine® T-9701, 15931; Radiamine 6343, 25762;
263-016-9: Barlox® 12, 3649; Chemoxide WC, 6003; Cocamine oxide, 6557; Empigen® 5083, 10354; Genaminox CS, 12778; Genaminox KC, 12779; Karox AO-30, 15774; Mackamine CO, 18169; Naxide 1230, 20817; Schercamox DMC, 27574;
263-017-4: Armeen® DMSD, 2948; Jet Amine DMSD, 15587;
263-020-0: Amine 2M1218D, 2134; Amine 2MKKD, 2143; Armeen® DMCD, 2945; Jet Amine DMCD, 15585; Kemamine® T-6502D, 15930;
263-022-1: Amine 2MHBGD, 2142; Armeen® DMHTD, 2946;
263-027-9: glyceryl cocoate, 13078; Imwitor® 928, 14948;
263-030-5: Myverol® 18-50K, 20626;
263-032-6: Alphadim® 90LC, 1825; Myverol® 18-40, 20625; Tegomuls® 19, 31009;
263-035-2: Myverol® 18-30, 20623;
263-038-9: Adogen® 461, 771; Arquad® C-33W, 3114; Arquad® C-50, 3115; Chemquat C/33W, 6026; Coco trimethyl ammonium chloride, 6562; Jet Quat C-50, 15603; Marlazin® KC 30/50, 18753; Varisoft® 461, 33411;
263-049-9: Norfox® 1101, 21544;
263-058-8: Amphotensid B4 F, 2347; Cocamidopropyl betaine, 6551; Dehyton® PK, 7995; Incronam 30, 15016; Rewoteric® AM B-13, 26417; Schercotaine CAB, 27695; Tego® -Betaine C, 30961; Tego® -Betaine E, 30962; Tego® -Betaine L-7, 30964;
263-078-7: Ethoquad® CB/12, 11000; PEG-2 cocobenzonium chloride, 23038;
263-080-8: Alkaquat® DMB-451-50, DMB-451-80, 1707; Arquad® DMCB-80, 3116; Mariazin® KC 21/50, 18715; Querton KKBCl-50, 25656;
263-081-3: Arquad® DMHTB-75, 3117; Kemamine® BQ-9702C, 15909; Querton 441-BC, 25653;
263-082-9: Arquad® M2HTB-80, 3119; Kemamine® BQ-9701C, 15908;
263-085-5: Kemamine® BQ-9742C, 15910;
263-086-0: Armeen® 2C, 2928; Radiamine 6260, 25759;
263-087-6: Accoquat 2C-75, 2C-75H, 242; Adogen® 462, 772; Arquad® 2C-75, 3102; Dye Retarder #1, 9310; Jet Quat 2C-75, 15602; Kemamine® Q-6502C, 15924; M-Quat® 2475, 20403; Radiaquat 6462, 25793; Tomah Q-2C, 31989; Varisoft® 462, 33412;
263-089-7: Amine 2HBG, 2132; Amine2 VT, 2163; Armeen® 2HT, 2929; Kemamine® S-970, 15927;
263-090-2: Adogen® 442, 769; Adogen® 442-P100, 770; Arquad® 2HT-75, 3103; Kemamine® Q-9702C, 15925; M-Quat® 257, 20400; Querton 442, 25654; Radiaquat 6442, 25791; Radiaquat 6475, 6480, 25796;
263-093-9: calcium sulfonate, 4973; Lubrizol® 2152, 17744; TLA-256, 31891;
263-099-1: tallow, 30739;
263-107-3: Acintol® 736, 2122, D25LR, D30E, DFA, EPG, FA-1, FA-2, R Type3A, LO-3A, SB, SM4, 348; Acofor, 364; Oulu 102, 22378; Pamolyn, 22642; Xtol, 34614;
263-112-0: Jet Amine PS, 15596;
263-123-0: Armid® HT, 2962;
263-125-1: Adogen® 170, 765; Amine BG, 2151; Armeen® T, 2958; Armeen® TD, 2959; Crodamine 1.T, 7104; Genamin TA Grades, 12776; Jet Amine PT, 15597; Radiamine 6170, 25755; Radiamine 6171, 25756; tallow amine, 30740; tallowamine, 30748;
263-127-2: Eureka 392, 11208;
263-129-3: Industrene® 143, 15102; tallow fatty acid, 30743;
263-130-9: Glycon® S-65, 13120; hydrogenated tallow acid, 14598; Petrac® PHTA, 23467; Prifac 9428, 24949;
263-134-0: Adogen® 417, 768; Arquad® S-50, 3120; Jet Quat S-50, 15605; Tomah Q-S, 31999;
263-142-4: Arizona DRS-40, 2890; Arizona DRS-50, 2892; Diprosin K-80, 8613;
263-144-5: Arizona DRS-43, 2891; Diprosin N-70, 8614;
263-147-1: Acetamin C, 286; Amine Acetate KK, 2149; Radiamac 6169, 25749;
263-149-2: Acetamin HT, 287; Amine Acetate HBG, 2148; Radiamac 6149, 25748;
263-153-4: Marlamid® DF 1218, 18740;
263-155-5: Akypogene FP 35 T, 1255;
263-163-9: Berol 307, 4027; Chemstat® 273-C, 6036; Cocamide DEA, 6547; Ethokem C/12, 10970; Norfox® DC, 21548; Purton CFD, 25419;
263-164-4: Rewoteric® AM 2C NM, 26414;
263-170-7: Chemzoline C-22, 6051; Mackazoline C, 18198; Monazoline C, 20192; Schercozoline C, 27712; Sovatex EP 5288, 29157; Sovatex IM12H, 29158; Unamine® C, 32911; Varine C, 33392;
263-171-2: Chemzoline T-44, 6053; Hodag C-100-T, 14170; Miramine® HPS-B, 19870; Miramine® TO, 19873; Monazoline T, 20197; Textamine T-1, 31534; Varine T, 33395;
263-173-3: Fenopon TK32, 11612; Geropon® TK-32, 12918; sodium methyl tall oil acid taurate, 28793, 28794;
263-174-9: Geropon® TC-42, 12917; sodium methyl cocoyl taurate, 28791;
263-177-5: Ablumox T-15, 92; Accomeen T2, T5, T15, 226; Agrisorb, 1072; Berol 455, 4032; Berol 456, 4033; Chemeen T-2, 5949; Clifton Glyphosate Additive, 6417; Ethomeen® T/12, 10985; Ethox TAM-2, 11039; Ethylan® TT-15, 11117; Eumulgin PA 10, 11179; Exell, 11295; Frigate, 12408; Genamin T-020, 12775; Hetoxamine T-2, 13918; Hyspray, 14750; Icomeen® T-15, 14818; Icomeen® T-2, 14817; Jogral, 15616; Lo-Dose, 17548; Marlazin® T 10, 18757; Mazeen® T-2, 19128; Power Spray Save, 24833; Prox-onic MT-02, 25327; Team, 30875; Teric 17M2, 31262; Tomah E-T-2, 31978; Topup, 32053; Toximul® TA-2, 32098; Trymeen® 6606, 32522; Trymeen® TAM-15, 32528; Varonic® T202, T220, 33449; Varonic® T-220, 33450; Wayfarer, 34246; Witcamine® 6606, 34447; Zusomin TG 102, 34917;
263-179-6: Chemoxide T, 6001;
263-180-1: Aromox® C/12-W, 3047; Schercamox CMA, 27573;
263-186-4: Duomeen® TDO, 9117;
263-189-0: Diamine BG, 8295; Duomeen® T, 9116; Jet Amine DT, 15590; tallow diamine, 30742;
263-190-6: Deriphat 154, 8089; disodium tallowiminodipropionate, 8655; Mirataine® T2C-30, 19931; Monateric TDB-35, 20175;
263-191-1: Jet Amine TRT, 15600; tallow triamine, 30746; Triameen T, 32193;
263-193-2: Closyl 30 2089, 6442; Hampfoam 35, 13593; Hamposyl® C-30, 13599; Medialan KA, 19212; Vanseal® 35, 33355; Vanseal® NACS-30, 33358;
263-195-3: Cocodiamine, 6564; Diamine KKP, 8297; Duomeen® C, 9112; Duomeen® CD, 9113; Jet Amine DC, 15582; Radiamine 6560, 25766;
263-218-7: Mackamine LAO, 18172;
263-312-8: Miranol® HM-SF Conc, 19901;
263-560-7: Blazer®, 4333;
263-718-5: Resazurin, 26161;
264-016-1: Geropon® WS-25, WS-25-I, 12919; sodium dinonyl sulfosuccinate, 28759;
264-038-1: Be Square® 185, 3866; Forbest MW 23, 12238; Fortex®, 12298; Mekon® White, 19248; Microcrystalline wax, 19666; Multiwax® 180-M, 20492; Multiwax® HS, 20493; Paracol® 404C, 22724; Polymekon®, 24485; Starwax® 100, 29546; Ultraflex®, 32801; Victory®, 33731;
264-041-8: EHIDA Kit, 9649;

264-071-1: nuarimol, 21762; Triminol, 32344;
264-118-6: Neodol® 25, 20913;
264-119-1: Amerlate® P, 2018; Fancor IPL, 11456; Lanesta S, 16658; Ritasol, 26832;
264-150-0: Chlorcosane, 6092; Chlorinated paraffin, 6105; Electrofine® S-70, 9782; Granuform, 13269;
264-189-3: Crodateric Cy, 7154;
264-626-8: Dioleyl hydrogen phosphite, 8554;
264-846-4: Acticarbone, 461;
264-980-3: Alsystin, 1868; Starycide, 29547; triflumuron, 32248;
265-004-9: Maxahibit TT-50, 19074;
265-082-4: Viplex 895-BL, 33888;
265-103-7: Viplex 680-P, 33886;
265-105-8: HAN® 857, 13613;
265-114-7: Jayflex® 215, 15545; Penreco 2251 Oil, 23123;
265-149-8: Exxsol® D-40, D-60, D-80, D-110, D-130, 11359; Isopar® M, 15392;
265-150-3: Isopar® C, 15388; Isopar® E, 15389; Isopar® G, 15390; Isopar® L, 15391;
265-156-6: Jayflex® 210, 15544;
265-207-2: Neodene® 1420, 20905; Neodene® 1624, 20906; Neodene® 2024, 20907;
265-233-4: Norpar® 12, 21573;
265-307-6: Lidarral, 17214;
265-724-3: caprylic/capric acid triglyceride, 5161; Captex® 300, 5168; Emalex K.T.G, 9967; Labrafac Lipophile WL 1349, 16535; Lexol GT 855, GT 865, 17149; Liponate GC, 17377; Mazol® 1400, 19135; Miglyol® 812, 19737; Myritol 318, 20580; Neobee® M-5, 20875; Tegosoft® CT, 31018;
265-839-9: Waxenol® 801, 34236;
265-932-4: Ethoxy carbonyl phenylethyl phenylformamidine, 11042;
265-995-8: CF 1500, 5829;
266-043-4: Portland cement, 24708;
266-065-4: Crodamol W, 7127; Tegosoft® SH, 31028;
266-096-3: Monceren®, 20198; pencycuron, 23097, 23098, 23099; Trotis, 32473;
266-124-4: Imwitor® 780 K, 14946; Schercemol GMIS, 27601;
266-164-2: Omnipaque, 22159;
266-231-6: Crafol AP-53, 6959;
266-275-6: Topas 100, 32026;
266-362-9: Cheetah R, 5870;
266-367-6: Neodol® 91, 20917;
266-368-1: Amphoteen 24, 2343; Empigen® BB, 10363;
266-464-3: Vaderm, 33237;
266-492-6: Ambush C, 1983; Ashlade Halt, 3194; Chemtech Cypermethrin, 6048; Chiltern Cyperkill 10, 6066; FAL Cypermethrin 10, 11419; Quadrangle Cyper, 25577; Topclip Parasol, 32037; Toppel, 32049; Toppel, 32050;
266-533-8: Crolactil SISL, 7185; Pationic ISL, 22907;
266-583-0: Butisan, 4746; Butisan® S, 4747; metazachlor, 19465; Pree®, 24873; Track, 32129;
266-639-4: Corbel®, 6802; Mistral, 19949; Power Task, 24835;
266-777-5: Incroquat I-85, 15039;
266-778-0: Foamquat IAES, 12180; Monaquat ISIES, 20134; M-Quat® 522, 20401; Schercoquat IAS, 27679; Schercoquat IIS, 27683;
266-922-2: Amine 2M12D, 2135; Empigen® AB, 10358;
266-924-3: Armeen® 3-16, 2932;
266-944-2: Softenol® 3100, 28847;
266-945-8: Neustrene® 059, 21022;
266-951-0: Alphadim® 90NLK, 1826;
266-952-6: Alphadim® 90AB, 1824;
266-994-5: Mirage, 19864; Octave, 21990; Prelude, 24885; prochloraz, 25057; Sporak, 29316; Sporgon, 29317; Sportak, 29319; Sportak Delta, 29321;
267-013-3: caprylic/capric acid, 5160; Industrene® 365, 15105; Philacid 0810, 23658;
267-015-4: Radia® 7060, 25719;
267-017-5: Emery® 2209, 10143; methyl caprylate-caprate, 19520;
267-019-6: Exxal® L1315, 11345;
267-051-0: Marlican®, 18845;
267-052-6: Ethoquad® T/12, 11002; PEG-2 tallowmonium chloride, 23044;
267-057-3: Myverol® 18-04K, 20620;
267-058-9: Miramine® C, 19867;
267-101-1: Chemidex SI, 5970; Mackine 401, 18219; Mazeen® DAPI, 19122; Schercodine I, 27637;
267-360-0: Schercoquat SAS, 27686;
267-569-7: Schercoteric O-AA, 27710;
267-821-6: Emalex DISG-2, 9943;
268-040-3: Miranate® B, 19874;
268-062-3: Sul-fon-ate OA-5R, 30021;
268-072-8: Varisoft® DHT, 33422;
268-074-9: Radiaquat 6471, 25795;
268-084-3: Neustrene® 064, 21023;
268-085-9: Neustrene® 045, 21021;
268-093-2: Radia® 7370, 25740; Radia® 7371, 25741;
268-130-2: Ammonium lauroyl sarcosinate, 2241; Hamposyl® AL-30, 13597;
268-213-3: Chemstat® PS-101, 6043;
268-215-4: Amine B11, 2150;
268-217-5: Amine 2M16D, 2138;
268-221-7: Radiamine 6572, 25768;
268-365-0: Empicol® ML 26/F, 10336;
268-372-9: Actrasol SP, 557;
268-450-2: Karate, 15757;
268-452-3: Emulamid TO-21, 10484;
268-531-2: Varisoft® TIMS, 33430;
268-577-3: Octosol A-18-A, 22023;
268-581-5: Dipentaerythrityl hexacaprylate/hexacaprate, 8582; Liponate DPC-6, 17375;
268-589-9: Akypomine® BC 50, 1264;
268-616-4: Maltrin® M200, 18503; Maltrin® QD M600, 18505;
268-665-1: Weston® 494, 34302;
268-761-3: Amonyl 675 SB, 2298; Chembetaine CAS, 5932; Cocamidopropyl hydroxysultaine, 6555; Crosultaine C-50, 7241; Lexaine® CSB-50, 17092; Lonzaine® CS, 17593; Mafo® CSB, 18301; Rewoteric® AM CAS, 26421; Rewoteric® AM CAS-15, 26422; Sandobet SC, 27327; Schercotaine SCAB, 27700;
268-770-2: Ablumide CME, 78; Adeka Sole YA, 672; Alkamide® C-212, 1635; Amidex CME, 2090; Amidex KME, 2093; Carsamide® CMEA, 5322; Cocamide MEA, 6548; Cocamide MEA (1:1), 6549; Comperlan P 100, 6677; Emid® 6500, 10191; Empilan® CME, 10391; Foamole M, 12178; Incromide CM, 14986; Mackamide CMA, 18154; Marlamid® M 1218, 18744; Mazamide® CFAM, 19109; Monamid® CMA, 20118; Monamide, 20123; Ninol® CMP, 21235; Ninol® CNR, 21236; Rewomid® C 212, 26320; Varamide® C-212, 33381; Zoramide CM, 34905;
268-771-8: Chemidex C, 5962; Chemidex WC, 5973; Cocamidopropyl dimethylamine, 6552; Incromine CB, 14998; Lexamine C-13, 17097; Mackine 101, 18215; Mazeen® DAPL, 19123; Mazeen® SHCFA, 19127; Miramine® CODI, 19868; Schercodine C, 27636;
268-772-3: Amidex TD, 2104;
268-820-3: Alkaterge® E, 1717;
268-877-4: Radiaquat 6470, 25794;
268-891-0: Marlamid® M 1618, 18745;
268-910-2: Radiasurf® 7125, 25797;
268-938-5: Ablumox CAPO, 90; Aminoxid WS 35, 2196; Amyx CDO 3599, 2392; Barlox® C, 3653; Chemoxide CAW, 5994; Cocamidopropylamine oxide, 6556; Empigen® OS/A, 10378; Incromine Oxide C, 15003; Mackamine CAO, 18168; Mazox® CAPA, 19149; Monalux CAO, 20103; Ninox® FCA, 21241; Patogen AO-30, 22914; Rewominox B 204, 26333; Rhodamox® CAPO, 26619; Schercamox C-AA, 27572; Standamox CAW, 29460; Standamox PCAW, 29461; Varox® 1770, 33454;
268-949-5: Monamine T-100, 20130; Schercomid SO-T, 27662; Schercomid TO-2, 27664;
269-023-3: Wickenol® 13 1, 34381;
269-027-5: Emerest® 2384, 10093; Hydrophilol ISO, 14629; Propylene glycol isostearate, 25218;
269-084-6: Acylglutamate CT-12, 599;
269-087-2: Acylglutamate CS-11, 597;
269-123-7: oleite, 22114; Türkischrotöl 100%, 32541;
269-220-4: Albalan, 1344; Fancor Lanwax, 11457; Lanfrax® 1776, 16673; Lanocerin®, 16728;
269-409-1: Crotein HKP, 7251; Keramino 25, 16053; Kera-Tein AA, 16060;
269-410-7: Emsorb® 2518, 10463; Sorbitan diisostearate, 29109;
269-597-5: Esperox® 545M, 10852; Lupersol 546-M75, 17863; Trigonox® 123-C75, 32282;
269-598-0: Rhodapon® CAV, 26633;
269-637-1: calcium montanate, 4959; Hostalub® VP Ca W 2, 14365;
269-646-0: shale oil, 28136;
269-663-3: Larostat® 88, 16798; M-Quat® 1033, 20402;
269-682-7: Bacote®, 3578;
269-730-7: Duoquad® O-50, 9120;
269-789-9: Phos-chek, 23694;
269-793-0: Cocamide MIPA, 6550;
269-804-9: C-Flakes, 5834; Duratex, 9180;

Duromel, 9246; Duromel B108, 9247; Emvelop®, 10533; hydrogenated cottonseed oil, 14593; Lubritab®, 17742; Sta-Nut EE, 29508;
269-819-0: Empigen® CDR40, 10369; Schercoteric MS, 27707;
269-820-6: Aratex, 2783; BBS, 3862; Cirol, 6324; Creamtex, 6974; Durkex 500, 9228; Durko, 9230; Durlite F, 9232; hydrogenated vegetable oil, 14601; Hydrokote® 95, 14604; K.L.X., 15644; Kaokote F, 15723; Kaola, 15724; Kaomel, 15728; Kaoprem-E, 15733; Kaorich Beads, 15734; Kaorich Gold, 15735; Lipo SS, 17345; Lipodan SET Kosher, 17364; Magna A, 18326; Optima 23B, 22238; Sterotex®, 29753; Sterotex® HM, 29754; Wecobee® M, 34253;
269-855-7: Baytroid®, 3856; Bulldock, 4663; cyfluthrin, 7563; Solfac, 28930;
269-919-4: Arquard® B-100, 3123;
269-922-0: Querton 280, 25652;
269-926-2: Kemamine® T-9742D, 15932;
270-095-3: Neodene® 1112, 20904;
270-108-2: Vanax® 833, 33286;
270-109-8: Vanox® AT, 33341;
270-156-4: Cocoyl sarcosine, 6568; Hamposyl® C, 13598; Vanseal® CS, 33356;
270-302-7: Amerlate® LFA, 2017; Amerlate® WFA, 2019; Fancor LFA, 11458; Ritalafa®, 26808; Skliro Distilled, 28572;
270-310-0: Fancol Karite Extract, 11445; Shea butter extract, 28143;
270-311-6: Cetiol® SB45, 5804; Fancol Karite Butter, 11444; Lipex 102, 17331; Shea Butter, 28141; Shebu, Refined, 28145;
270-315-8: hydroxylated lanolin, 14663; Ivarlan OH, 15453; OHlan®, 22055; Ritahydrox, 26807;
270-329-4: Accobetaine CL, 222; Amphoteen BCM-30, 2345; Chembetaine BW, 5930; Chembetaine CB, 5933; Coco betaine, 6561; Incronam CD-30, 15020; Lonzaine® 12C, 17591; Mackam CB-35, 18140; Mafo® CB 40, 18300; Velvetex® AB-45, 33553;
270-331-5: FMB 302-8, 12132; Querton 210Cl-50, 25651;
270-351-4: Philcohol 1214, 23665;
270-355-6: Alkamide® DIN-295/S, 1640; Alkamide® SDO, 1651; Amidex S, 2102; Mackamide S, 18166; Marlamid® DF 1818, 18741; Purton SFD, 25420; Schercomid SLS, 27660; Stamid LS 5487, 29441;
270-356-1: Chemidex T, 5972;
270-407-8: Bio-Terge® AS-40, 4212; Carsonol® AOS, 5333; Nansa® LSS38/A, 20689; Norfox® ALPHA XL, 21546; Polystep® A-18, 24588; Rhodacal® 301-10, 26576; Witconate AOK, 34483;
270-414-6: Amine 2M14D, 2137;
270-416-7: Kemamine® D-974, 15914; Radiamine 6570, 25767;
270-431-9: Empilan® CIS, 10390;
270-664-6: Hoechst Wax S, 14270; Montan acid wax, 20262;
270-679-8: glyceryl montanate, 13086; Hostalub® WE4, 14366; molasses, 20043;
270-719-4: Viplex 885, 33887;
270-864-3: Schercopol OMS-Na, 27669;
270-939-0: Jet Amine PE 1214, 15593;
271-089-3: Jayflex® UDP, 15560; undecyl dodecyl phthalate, 32915;
271-090-9: diisononyl phthalate, 8463; Jayflex® DINP, 15552; Jayflex® DIOP, 15553; Palatinol® DN, 22589; Palatinol® N, 22594; PX-109, 25433;
271-091-4: Bisoflex BP9, 4255; diisodecyl phthalate, 8462; Jayflex® DIDP, 15550; Kodaflex® DIDP, 16269; NLA-30, 21397; Palatinol® DIDP, 22588; Palatinol® Z, 22596; PX-120, 25436;
271-102-2: Mackanate CP, 18183; Monamate C-1142, 20104;
271-130-5: Propomeen C/12, 25194;
271-233-5: Exxal® 9, 11337;
271-234-0: decyl alcohol, 7867;
271-235-6: Exxal® 13, 11340;
271-250-8: Neoflex® 9, 20922;
271-401-8: Propoquad® 2HT/11, 25196;
271-528-9: Nansa® 1042, 20681; Nansa® SSA, 20695;
271-531-5: Nansa® YS94, 20698;
271-532-0: Nansa® TS 50, 20697;
271-548-8: Stasoft J, 29549;
271-556-1: Empicol® 0031/T, 10298; Empicol® DA, 10310;
271-606-2: MB 450, 19168;
271-657-0: Active #2, 501; Adeka Sole CO, 671; Alkamide® 101 CG, 1631; Alkamide® CDE, 1637; Aminol KDE, 2179; Schercomid CDA, 27649; Schercomid SCE, 27657;
271-694-2: Radia® 7171, 25725;
271-696-6: Diamine HBG, 8296; Kemamine® D-970, 15913; Radiamine 6540, 25765;
271-704-5: Miranol® C2M-SF 70%, 19888; Miranol® FBS, 19896;
271-705-0: Miranol® CS Conc, 19891; Sandoteric CFL, 27354; Schercoteric MS-EP, 27709;
271-762-1: Duoquad® T-50, 9121; Jet Quat DT-50, 15604;
271-789-9: Miranol® SM Conc, 19911;
271-790-4: Miranol® DM Conc. 45%, 19892;
271-792-5: Miranol® J2M Conc, 19902; Miranol® JB, 19905;
271-793-0: Ampholak XCO-30, 2321; Miranol® CM Conc. NP, 19889; Miranol® FA-NP, 19894;
271-794-6: Empigen® CDL60, 10367; Miranol® H2M Conc, 19898; Miranol® HM Conc, 19900;
271-807-5: Nansa® SBA, 20692;
271-855-7: Tomah PA-19, 31986;
271-862-5: Miranol® OS-D, 19909; Sandoteric TFL Conc, 27355;
271-863-0: Miranol® JS Conc, 19907;
271-864-6: Miranol® H2M-SF Conc, 19899;
271-867-2: Akrochem® Antioxidant 12, 1157; Ultranox® 257, 32848;
271-870-9: Weston® 474, 34300;
271-929-9: Monateric ISA-35, 20168; Schercoteric I-AA, 27706;
271-957-1: Miranol® 2CIB, 19884;
271-972-3: Chemoxide TAO, 6002;
272-000-0: Ceraphyl® GA, 5730;
272-008-4: Alkylate 215, 1732;
272-043-5: Dehyton® G, 7990; Dehyton® PG, 7994; disodium cocoamphodiacetate, 8643; Miranol® C2M Conc. NP-PG, 19887; Miranol® FB-NP, 19895; Schercoteric MS-2, 27708;
272-047-7: Schercodine T, 27642;
272-103-0: Duomeen® OTM, 9115;
272-191-0: Armeen® 2T, 2930; Radiamine 6270, 25760;
272-192-6: Duomeen® TTM, 9118;
272-207-6: Adogen® 470, 773; Arquad® 2T-75, 3104;
272-219-1: Schercopol CMS-Na, 27665;
272-332-6: Telopar, 31070;
272-339-4: Amine 2MBGD-M, 2141; Amine 2MOLD, 2144; Armeen® DMTD, 2949; Jet Amine DMTD, 15588;
272-349-9: Actrasol SP 175K, 558;
272-363-5: Neutral Degras, 21025;
272-374-5: Miranol® S2M-SF Conc, 19910;
272-383-4: Amphoterge® KJ-2, 2352; Miranol® J2M-SF Conc, 19903; Miranol® JAS-50, 19904; Miranol® JBS, 19906; Monateric 1000, 20147; Monateric 811, 20143; Zoharteric LF-SF, 34883;
272-441-9: Hismanal, 14132;
272-647-9: Ageflex TPGDA, 1003; Laromer® TPGDA, 16792;
272-675-1: Empicol® TA40, 10340;
272-746-7: Varisoft® 2 TD, 33402;
272-778-1: Epal+20+m2, 10613;
272-787-0: Amine 740, 2145; Jet Amine TET, 15598; tallow tetramine, 30745;
272-897-9: Ampholak YCO-40, 2328; Miranol® C2M Anhyd. Acid, 19886;
272-964-2: Ceraphyl® 70, 5720;
273-049-0: Glucate® SS, 13033; methyl glucose sesquistearate, 19539;
273-066-3: Fyrquel® EHC, 12544; Kronitex® 100, 16429; Kronitex® 1840, 16431; Kronitex® 200, 16430; Kronitex® 25, 16427; Kronitex® 50, 16428; triaryl phosphate, 32194; triaryl phosphate, 32195;
273-118-5: hydrogenated stripped coconut acid, 14597; Hystrene® 5012, 14759; Industrene® 223, 15103;
273-160-4: Propomeen T/12, 25195;
273-168-8: Kronitex® TCP, 16435;
273-181-9: Akwilox 133, 1213;
273-187-1: Foamole B, 12177;
273-219-4: Tixogel VP, 31883;
273-222-0: Ceraphyl® 65, 5719;
273-234-6: Nansa® EVM50, 20686;
273-257-1: Empicol® LM, 10324; Empicol® LMV/T, 10325; Empicol® LZ, 10331; Empicol® LZV, 10334;
273-258-7: Empicol® TAS30, 10341;
273-299-0: Kessco® 887, 16102;
273-313-5: vegetable oil, 33522;
273-328-7: Akyposal DS 28, 1305; Akyposal DS 56, 1306;
273-429-6: Monazoline IS, 20195; Schercozoline I, 27713;
273-544-1: Benzalkonium chloride, 3958; Empigen® BCB50, 10364;
273-575-0: Axol® C 62, 3511;
273-576-6: hydrogenated tallow glyceride lactate, 14600; Lamegin® GLP 20, 16619;
273-604-7: Actrasol SS, 561;

273-612-0: Acetylated hydrogenated tallow glyceride, 316; Lamegin® EE, 16618;
273-613-6: Acidan N 12, 338; hydrogenated tallow glyceride citrate, 14599; Lamegin® ZE 30, 60, 16620;
273-616-2: Super Refined Shark, 30197;
273-627-2: Paramount B, 22778; Witarix® 212, 34429;
273-683-8: Nusolv ABP-103, 21801;
273-688-5: Halthal, 13574;
273-968-7: Amphisol, 2308; Surfagene FGD 600, 30354;
274-001-1: Crotein ASK, 7248; Crotein K, WKP, 7255; Hydrokeratin AL-30, 14603; Kera-Tein 1000, 16059; Nutrilan® Keratin W, 21836;
274-022-6: Querton GCI-50, 25655;
274-125-6: Fluazifop-butyl, 12038; Fusilade, 12519; Gallant, 12607;
274-267-9: Ampholak XOO-30P, 2324;
274-308-0: Soy-Tein NL, 29176;
274-322-7: Tomahawk, 32001;
274-357-8: Suttocide® A, 30440;
274-536-0: Eqvalan, 10737; Ivermectin, 15454;
274-572-7: Naugard® XL-1, 20799;
274-581-6: Parsol® 1789, 22866;
274-695-6: Hostapon KTW New, 14394; sodium lauroyl taurate, 28780;
274-764-0: Liponate NPGC-2, 17381;
274-834-0: Parapel® LIS, 22787;
274-846-6: Ethoquad® C/12, 10997; Variquat® 638, 33399;
274-923-4: Amphoteen BCA-30, 2344; Mirataine® BET-C-30, 19923;
274-925-5: Mirataine® BSC, 19926; Mirataine® CBS, CBS Mod, 19927;
275-117-5: Gellan gum, 12736; Gelrite®, 12746; Kelco-Gel® Gellan Gum, 15827;
275-226-8: Vanax® CPA, 33288;
275-521-1: Uniplex 310, 33017;
275-532-1: Arquad® 2C-70 Nitrite, 3101;
275-637-2: Dermol 89, 8117; Kessco® Octyl Isononanoate, 16117;
276-158-1: Jayflex® 77, 15543;
276-171-2: Crill 6, 7032; Emalex SPIS-100, 9994; Emsorb® 2516, 10462; Sorbitan isostearate, 29110;
276-432-0: Tomah DA-14, 31966;
277-298-6: Standapol® 1610, 29464;
277-328-8: Hinochloa®, 14096; mefenacet, 19229;
277-385-9: Ultravist, 32877;
277-682-3: Checkmate, 5863; Poast®, 24278; Sethoxydim, 28099;
277-728-2: Apollo 50C, 2661;
277-746-0: Alfaprostol, 1574;
278-306-0: Neodol® 23, 20911;
278-477-1: Marlophor® CS-Acid, 18891;
278-877-6: Choletec, 6170;
278-901-5: Trigonox® ADC-NS60, 32294;
278-928-2: Diazolidinyl urea, 8348; Germall® II, 12894;
279-498-9: Doverphos® 11, 8840; Doverphos® DPGDP, 8847; Weston® THOP, 34319;
279-499-4: Doverphos® 12, 8841; Weston® DHOP, 34305;
279-500-8: Weston® 491, 34301;
279-752-9: fluroxypyr 1-methylheptyl ester, 12118; Starane, 29528;
279-917-5: Amiter LGOD, 2198;
280-518-3: Phospholipid PTD, 23723;
281-991-9: Behenyl betaine, 3900; Incronam B-40, 15018;
282-968-6: Empimin® SDS, 10436; sodium decyl sulfate, 28755;
283-078-0: Radia® 7514, 25747;
283-291-9: Pristene R20, 25039;
283-390-7: Schercemol MEL-3, 27606;
284-219-9: Ampholyte KKDP-60, 2334; Ampholyte KKE-70, 2335; Mackam 151C, 18137;
284-863-0: Radia® 7040, 25717;
284-868-8: Priolube 1414, 25017; Radia® 7230, 25732;
284-980-7: Argonol 40, 2855;
284-987-5: Marlophen® 81N, 18882;
285-203-4: Priolube 1429, 25018; Propylene glycol dioleate, 25217; Radia® 7204, 25731;
285-206-0: Radia® 7187, 25728;
285-207-6: Radia® 7331, 25736;
285-540-7: Radia® 7231, 25733;
285-547-5: Radiasurf® 7175, 25805;
285-550-1: Radiasurf® 7410, 25811;
286-344-4: ADK STAB NA-11, 733;
286-490-9: Radiasurf® 7600, 25820;
287-011-6: C12-14 alkyl dimethylamine oxide, 4840; Lilaminox M24, 17251;
287-075-5: Radia® 7108, 25720;
287-488-0: L.A.S, 16528; Labrasol, 16543;
287-489-6: Corn oil PEG-6 esters, 6835; Labrafil M 2125 CS, 16540;
287-494-3: Marlon® AS3, 18878;
287-824-6: Radia® 7110, 25721;
288-048-0: Lilamin LSP 33, 17250;
288-305-7: Radiasurf® 7156, 25803;
288-459-5: Radiasurf® 7402, 25808; Radiasurf® 7403, 25809; Radiasurf® 7404, 25810; Radiasurf® 7443, 25815; Radiasurf® 7444, 25816;
288-668-1: Radia® 7241, 25734;
289-151-3: Incrosoft S-75, 15050; Quaternium-27, 25607; Varisoft® 475, 33415;
289-181-7: Mackam RA, 18149; Rewoteric® AM R40, 26426;
289-256-4: Schercemol SE, 27622;
289-325-9: Ceraphyl® 85, 5721;
290-580-3: Radia® 7505, 25744;
290-850-0: Rewopol® B 1003, 26358;
291-394-5: cetyl diethanolamine phosphate, 5818; Crodafos CDP, 7077;
291-990-5: Schercoquat SOAS, 27688;
292-564-1: Diamine B11, 8294;
292-932-1: Radia® 7266, 25735;
292-951-5: Priolube 1458, 25021; Radia® 7131, 25724;
292-960-4: Radia® 7510, 25746;
293-208-8: Radiasurf® 7900, 25821;
293-391-2: Schercemol MEP-3, 27608;
293-509-4: Exsyproteines 2%, 11318;
294-015-5: Crodamine 1.0, 1.0D, 7100;
294-352-4: H+72old EP-1, 13526; Tears Plus®, 30876; Vinisil, 33827;
294-538-5: Dynacet®, 9341;
294-563-1: Rewoquat W 75 H, 26395;
294-600-1: glyceryl stearate citrate, 13088; Imwitor® 369, 14942; Imwitor® 370, 14943; Monoglyceride citric ester, 20226;
295-264-9: Ampholak YCA/P, 2326;
296-473-8: Anhydrol, 2459; ASP®, 3230; Bilt-Cote®, 4147; Bilt-Plates®, 4148; Buca, 4646; Catalpo, 5432; Continental® Clay, 6740; copper oxychloride, 6784; Dixie Clay®, 8743; Fiberfrax® 6000 RPS, 11726; Fiberkal, 11736; Huber 40C, 14449; Huber 65A, 14450; Huber 95, 14451; kaolin, 15725; Kaowool®, 15736; Kayphobe-ABO, 15816; Langford Clay, 16674; McNamee Clay®, 19182; Par Clay®, 22713; Peerless®, 23023; Peerless® No. 1, 23024; Pharmolin, 23576; Polyplate 90, 24523; Satintone, 27486; SP-33, 29177; SPS, 29349; takizolit, 30726; Translink®, 32139; Whitetex, 34375;
297-364-8: Radiasurf® 7400, 25807;
297-495-0: Rewoquat DQ 35, 26393;
297-627-7: Permethyl 102A, 23388;
298-632-7: Rewoteric® AMKSF 40, 26430;
302-442-2: Incromide BEM, 14984;
305-318-6: Ampholak YJH-40, 2329;
305-488-1: Schercoquat ROAS, 27684;
306-235-8: Crosilk 10,000, 7233; Ritasilk, 26830; Silk Pro-Tein, 28362; Solu-Silk Protein, 29025;
306-522-8: Radiasurf® 7270, 25806; Radiasurf® 7414, 25812; Radiasurf® 7417, 25813; Radiasurf® 7453, 25817; Radiasurf® 7454, 25818; Radiasurf® 7473, 25819;
306-621-6: Ceraphyl® 791, 5728;Hetester HSS, 13897; isocetyl stearoyl stearate, 15349;
306-817-1: glyceryl lanolate, 13081; Lanesta G, 16657;
306-998-7: Ampholak 7TY, 2317;
307-030-6: Apricot kernel oil PEG-6 esters, 2675; Labrafil M 1944 CS, 16537;
307-332-8: Dimodan PVP Kosher, 8534; Dimodan S, 8535;
307-334-9: Acidan, 337;
307-455-7: Ampholak YCE, 2327;
307-456-2: Ampholak XCE, 2320;
307-458-3: Ampholak 7TX, 2313; Ampholak 7TX/C, 2314; Ampholak 7TX-SD 55, 2315; Ampholak 7TX-T, 2316; Ampholak XO7, 2323;
307-710-2: T.A.M, 30672;
307-919-9: Amine 780, 2147;
308-441-3: Marlowet® 5311, 18925;
309-148-3: Actigen E, 479;
309-206-8: Rewoteric® QAM 50, 26432;
310-296-6: Collagen Hydrolyzate Cosmetic 55, 6613; Collamino 25, 6616; Cropepsol, Cropeptone, 7208; Crotein A, C, O, 7244; Crotein CAA, 7249; Crotein CAA/SF, 7250; Crotein O, 7256; Crotein SPA, 7258; Extiat®, 11323; Hydrocoll AL-50, AL-55, EN-40, EN-55, EN-55-X, EN-SD, EN-SD-1M, EN-SD-10M, 14580; Lexein® X-250HP, 17131; Nutrilan® FPK, H, M, 21834; Nutrilan® I-50, 21835; Nutrilan® L, 21837; Parenamine, 22823; Peptein® 2000®, 23202; Polypeptide 10, 24516; Polypeptide 37, 24517; protogest, 25290;
371-792-5: Ampholak XJO, 2322;
401-990-0: Chimassorb® 119FL, 6072;
402-140-1: CM8650, 6459;
402-780-1: Facet®, 11405;
403-320-2: Lupersol 533-M75, 17862;
403-640-2: Folicur, 12198; Horizon/Horizont,

14308; Raxil, 25908; Silvacur, 28408;
403-730-1: Agsol Ex 12, 1089;
406-060-8: Mikolite, 19749; vermicu lite, 33589;
541-253-3: Lewisite, 17089;

PART IV

DIRECTORY

Manufacturers and Suppliers

Manufacturers and Suppliers

3M Co./Industrial Chemicals,
3M Center Bldg. 223-6S-04,
St. Paul, MN, 55144-1000, USA
612-736-1394, 800-541-6752
www.3m.com

3M Co./Specialty Chem,
3M Center Bldg. 223-6S-04,
St. Paul,MN, 55144-1000, USA
612-733-3064,
Fax 612-737-7635, www.3m.com

3M Pharmaceuticals,
Bldg 275-2E-13,
PO Box 33275,
St Paul, MN, 55133-3275, USA
612-736-4030
Fax 612-733-6068, www.3m.com

3M United Kingdom plc,
Commercial Chemicals Div,
3M House, PO Box 1,
Bracknell, Berks, RG12 1JU, UK
www.3m.com

3V Inc.
9140 Arrowpoint Blvd., Suite 120,
Charlotte, NC, 28273-8120, USA
704-523-5252
Fax 704-522-1763

3V Sigma SpA,
T. Tasso 58, PO Box 219,
I-24100 Bergamo, Italy
39-11-35-212274
Fax 39-11-35-239569

AAA Molybdenum Products,
7233 W. 116th Place,
Broomfield, CO 88020 USA
303-460-0844, 800-443-6812
Fax 303-460-0851

Abatron, Inc.,
5501 95th Ave.,
Kenosha, WI 53144,USA.
414-653-2000, 800-445-1754
Fax 414-653-2019,
www.abatron.com

Abbott Laboratories,
1400 Sheridan Rd.,
N. Chicago, IL 60064-4000,USA
708-937-8800, 800-240-1043
Fax 708-937-6676,
www.abbott.com

Abbott Laboratories,
Queenborough, Kent, ME11 5EL,
UK
44 1795-580-099
Fax 44 1795-580-404,
www.abbott.com

Ablestik,
Station Rd.,
Linton, Cambridge
Cambridgeshire, CB1 6NW, UK
44-1-223-893-771
Fax 44-1-223-893-546,
www.ablestik.com

ABM Chemicals Ltd.,
Poleacre Lane,
Woodley, Stockport,
Cheshire, SK6 1PQ, UK
44 161-430-4391
Fax 44 161-430-4364

Abril Industrial Waxes Ltd,
Unit B4, Waterslade House,
Thame Road, Haddenham,
Bucks, HP17 8NT, UK
44 1844-299099
Fax 44 1844-299098,
www.abril.co.uk

Accro-Seal,
316 W. Briggs St.,
PO Box 210,
Vicksburg, MI 49097-0210,USA
616-649-1014, 800-225-7138
Fax 616-649-1067,
www.acroseal.com

Aceto Corporation,
1 Hollow Lane, Suite 201,
Lake Success, NY
11042-1215,USA
516-627-6000
Fax 516-627-6093,
www.aceto.com

Acheson A.N.Z. Pty.,
PO Box 98,
Revesby, N.S.W., 2212, Australia
61-2-755-3099
www.acheson.com

Acheson Colloids (Canada),
PO Box 665,
Shaver St.,
Brantford, Ont., N3T 5P9,
Canada.
519-752-5461,
www.acheson.com

Acheson Colloids Co.,
1600 Washington Ave.,
PO Box 611747,
Port Huron, MI 48061-1747,
USA.
810-984-5581, 800-255-1908
Fax 810-984-1446,
www.acheson.com

Acheson do Brasil,
Rua Howard A. Acheson Jr., 279,
Cotia-SP CEP 06700, Brazil.
55-11-492-4000,
www.acheson.com

Acheson Industries Europe,
Sun Life House,
85 Queens Rd.,
Reading, Berks., RG1 4PT, UK
44-1734-588844
Fax 44-1734-574897,
www.acheson.com

Active Organics, Inc,
1097 Yates St.,
Lewisville, TX 75057, USA.
972-221-7500
Fax 972-221-3324,
www.activeorganics.com

Adasco Inc.,
2029 South Broadway, Building
Geneva, OH 44041, USA
216-466-2114

Addagrip Surface Treatment,
Bird-in-Eye Hill,
Uckfield,East Sussex, TN22 5HA,
UK
44-1825-761333
Fax 44-1825-768566

ADM Tronics Unlimited,
224-S Pegasus Ave,
Northvale, NJ 07647, USA
201-767-6040
Fax 201-784-0620,
www.admtronics.com

Adria Laboratories Inc.,
PO Box 16529,
582 W Goodale Blvd,
Columbus, OH 43216-6529, USA
614-764-8100

Adshead Ratcliffe,
Derby Road, Belper,
Derby, Derbyshire, DE5 1WJ, UK
44 1773 826661
Fax 44 1773-821215

Advance Web Products,
529 5th Ave,
New York, NY 10017, USA
401-946-8629, 401-943-2757

Advanced Elastomer Belgium,
Ave. de Tervuren 270-272,
B-1150 Brussels, Belgium
32-2-774-0411
Fax 32-2-774-0410,
www.aestpe.com

Advanced Elastomer Canada,
Box 787, Streetsville Postal S,
Mississauga, Ont., L5M
2G4,Canada
416-826-9575, www.aestpe.com

Advanced Elastomer Systems,
260 Springside Dr., PO Box 558,
Akron, OH 44334-0584, USA
216-668-3600, 216-668-8242
www.aestpe.com

Advanced Polymer Coatings,
PO Box 127,
West Point, PA 19486, USA
215-794-5466, Fax 215-794-5468

AEI Compounds, Power,
Gravesend, Kent, DA11 9AF, UK
44-1474-564466, Fax 44-
1474-564386

Agan Chemical Manufacturing,
PO Box 262, Ashdod,
77102 Israel.
972-7-629-6611, Fax 972-7-628-
0304
www.agan.co.il

Agrichem (International),
Industrial Estate, Station Road,
Whittlesey,Cambridgeshire, PE7
2EY, UK
44-1733-204019
Fax 44-1733-204162

Agri-Technics Ltd.,
Muston Gorse, Redmile,
Nottingham, Notts, NG13 0GN,
UK
44-1949-42255
Fax 44-1949-43407.

Aicello Chemical Co.Ltd.,
45 Koshikawa,
Ishimaki-honmach,
Toyohashi City, Aichi Pref, 441
11, Japan
81-532-88-0611
Fax 81-532-88-5102.

Air Prods. Nederland BV,
Kanaalweg 15, PO Box 3193,
3502 GD Utrecht, The
Netherlands
31-30-857100
Fax 31-30-857111,
www.airproducts.com

Air Prods./Polyurethane,
7201 Hamilton Blvd.,
Allentown, PA 18195-1501, USA
610-481-6799, 800-345-3148
Fax 610-481-4381,
www.airproducts.com

Air Products and Chemicals,
7201 Hamilton Blvd.,
Allentown, PA 18195-1501, USA
610-481-4911, 800-345-3148
Fax 610-481-5900,
www.airproducts.com

Air Products and Chemicals Germany,
Postfach 5108, Robert-Koch-Str.
27,
D-22821 Norderstedt, Germany.
49-40-529009-0
Fax 49-40-52900999,
www.airproducts.com

Air Products and Chemicals Mexico,
Rio Guadiana 23, Piso 5,
Colonia Cuauhtemoc,
Mexico D.F., 06500, Mexico
525-591-0800,
Fax 525-592-3018,
www.airproducts.com

Manufacturers and Suppliers

Air Products and Chemicals/Polymers,
7201 Hamilton Blvd.,
Allentown, PA 18195-1501,USA
610-481-6799, 800-345-3148,
Fax 610-481-4381,
www.airproducts.com

Ajinomoto Co., Inc.,
15-1, Kyobashi 1-chome,
Chuo-ku,
Tokyo 104, Japan
81-3-5250-8152
Fax 81-3-5250-8259
www.ajinomoto.com

Ajinomoto Europe Sal,
Stubbenhuk 3,
20459 Hamburg, Germany
40-3749-3650
Fax 40-372087
www.ajinomoto.com

Ajinomoto USA Inc.,
Glenpointe Centre West,
500 Frank W. Burr Blvd.,
Teaneck, NJ 07666-6894,USA
201-907-3250
Fax 201-907-3252
www.ajinomoto.com

Akrochem Chemical Co,
255 Fountain St.,
Akron, OH 44304, USA
216-535-2108, 800-321-2260
Fax 216-535-8947

Akros Chemicals Ltd (Azko Nobel),
PO Box 1, Eccles,
Manchester M3O 0BH, UK
44-161-7851111
Fax 44-161-7887886,
www.akzonobel.com/

Akzo Nobel Base Chemicals,
4 Stationsplien/3818 LE,
PO Box 247,
3800 AE Amersfoort, The Netherlands
31-33-4676270
Fax 31-33-4676110,
www.akzonobel.com/hc/home.htm

Akzo Nobel Chemicals,
300 S. Riverside Plaza,
Chicago, IL 60606-6697,USA
312-906-7500, 800-227-7070
Fax 312-906-7811,
www.akzonobel.com/

Akzo Nobel Chemicals,
1 City Center Dr., Suite 320,
Mississauga,Ontario, L5B IM2, Canada
905-273-5959
Fax 905-273-7339,
www.akzonobel.com/

Akzo Nobel Chemicals,
1-5, Queens Road, Hersham,
Walton-on-Thames, Surrey, KT12 5NL, UK
44-1932-247891
Fax 44-1932-231204,
www.akzonobel.com/

Akzo Nobel Chemicals bv,
4 Stationsplein/3818,
PO Box 247,
3800 AE Amersfoort, The Netherlands
31-33-4676536
Fax 31-33-4676110,
www.akzonobel.com/

Akzo Nobel Chemicals/BU
Base Chemicals,
6 Grand Ave., PO Box 80,
Camellia, NSW 2142, Australia
61-2-638 4555
Fax 61-2-638 4681,
www.akzonobel.com/

Akzo Nobel Chemie GmbH,
Postfach 100132,
52301 Duren, Germany
49-2421-492261
Fax 49-2421-595380,
www.akzonobel.com/

Akzo Nobel Engineering B.V,
Postbus 9300,Velperweg 76,
NL-6800 SB Arnhem, The Netherlands
31-85-662714
Fax 31-85-665140,
www.akzonobel.com/

Akzo Nobel Resins,
1 Synthesebaan/4612 RB, PO Box 79
4600 AB Bergen op Zoom, The Netherlands
31-164-276200
Fax 31-164-276259,
www.akzonobel.com/

Akzo Nobel Salt Europe,
PO Box 247,
3800 AE Amersfoort, The Netherlands
31-33-676767
Fax 31-33-676132,
www.akzonobel.com/

Akzo PQ Silica VoF,
PO Box 247,
3800 AE Amesfoort, The Netherlands
31-33-46786282
Fax 31-33-4676169,
www.akzonobel.com/

Alba International,
508 Clearwater Dr.,
N. Aurora, IL, 60542 USA
708-897-4200, 800-669-9333
Fax 708-377-5330

Alban Muller Int'l.,
212, rue de Rosny,
93102 Montreuil, France
33-1-48-58-30-25
Fax 33-1-48-58-03-71

Albion Group Ltd.
113 Station Road,
Hampton, Middlesex, TW12 2DY,UK

Albright & Wilson (Australia), Ltd.
PO Box 20,
Yarraville,
Yarraville, VIC 3013, Australia
61-3-9688-7777
Fax 61-3-9688-7788,
www.albright-wilson.com

Albright & Wilson Americas,
PO Box 4439,
Glen Allen, VA 23058-4439 USA
804-550-4300, 800-446-3700
Fax 804-550-4385
www.albright-wilson.com

Albright & Wilson Ltd.,
No. 2 Okamotoya Bldg. 6 Fl.,
1-24, Toranomon 1-chome,
Minato, Tokyo 105, Japan
81-3-3508-9461
Fax 81-3-3591-0733,
www.albright-wilson.com

Albright & Wilson UK,
P O Box 3, 210-222 Hagley Road West,
Oldbur, Warley, W. Midlands, B68 0NN, UK
44-121-420-5297
Fax 44-121-420-5462,
www.albright-wilson.com

Alcan Chemicals,
3690 Orange Place, Suite 400,
Cleveland, OH 44122-4438,USA
216-765-2550, 800-321-3864
Fax 216-765-2570

Alcan Chemicals Europe,
Ditton Rd.,
Widnes, Ches., WA8 0PH, UK
44-1592-411000
Fax 44-151-802 2999

Alcan Chemicals Ltd.,
Chalfont Park, Gerrards Cross,
Bucks, SL9 0QB, UK
44-1753-887373
Fax 44-1753-881556

Alchemie Ltd.,
Brookhampton Lane,
Kineton, Warwickshire, CV35 0JA, UK
44 1926 641600
Fax 44 1926-641698

Alco Chemical Corp.,
909 Mueller Dr.,
PO Box 5401,
Chattanooga, TN 37406-0401,USA
423-629-1405, 800-251-1080
Fax 423-698-9367

Alcoa,
201 Isabella St./7th. St. Bridge
Pittsburgh, PA 15212-5858
412-553-4545
www.alcoa.com

Alcoa Industrial Chemical,
4701 Alcoa Rd.,
PO Box 300,
Bauxite, AR, 72011, USA
501-776-4717, 800-643-8771
Fax 501-776-4904,
www.alcoa.com

Alcoa Industrial Chemical/Asia,
2 Havelock RD,
#07-5 Apollo Center, 0105, Singapore
65-538-0070
Fax 65-538-3237,
www.alcoa.com

Alcoa Industrial Chemical/Europe,
im Atzelnest 3,
D-6380 Bad Homburg, Germany
49-06172-4068-0
Fax 49-06172-4068-13,
www.alcoa.com

Alcoa Inter-America,
396 Alhambra Circle, Suite 200,
Coral Gables, FL 33114, USA
305-445-8544
Fax 305-444-8924,
www.alcoa.com

Alcoa Kasei Ltd.,
Toranomom 4-chrome,
Minato-ku, Tokyo 105, Japan
81-3-5472-3201
Fax 81-3-5472-3209,
www.alcoa.com

Alcon Laboratories,
6201 South Freeway,
Ft Worth, TX 76134, USA
817-293-0450

Alconox Inc.,
9 E. 40th Street, Ste. 200",
New York, NY 10016, USA
212-532-4040
Fax 212-532-4301

Aldo Products Co. Inc.,
1604 N Main Street,
Kannapolis, NC 28081, USA
704-932-3054
Fax 704-932-3041

Manufacturers and Suppliers

Allchem Industries,
4001 Newberry Rd, Suite E-3,
Gainesville, FL 32607, USA
904-378-9696
Fax 904-338-0400

Allergan Inc.,
P. O. Box 19534
Irvine, CA 92623-9534
714-246-4500 800-347-4500
Fax 714-246-6987
www.allergan.com

Allied Colloids Ltd.,
P. O. Box 38, Cleckheaton Rd.,
Low Moor, Bradford, BD12 0JZ
U.K.
44-1274-41-70-00
Fax 44-1274-60-64-99

Allied Signal Inc.,
101 Columbia Rd.,
PO Box 2245,
Morristown, NJ 07962, USA
201-455-2000, 973-455-5445
800-522-8001, 800-810-4340,
www.alliedsignal.com

Allied Signal Engineering,
PO Box 2332,
Morristown, NJ 07962-2332, USA
201-455-5010,
www.alliedsignal.com

Allied Signal Europe,
Haasrode Research Park,
Grauwmeer 1,
B-3001 Heverlee
Leuvein, Belgium
32-16-39-12-33
Fax 32-16-40-03-77
www.alliedsignal.com

Allied Signal Europe N.V.,
Haasrode Research Park,
Grauwmeer 1,
B-3001 Heverlee
Leuvein, Belgium
32-16-39-12-11
Fax 32-16-400-039,
www.alliedsignal.com

Allied Signal Inc.,
PO Box 1053,
101 Columbia Rd.,
Morristown, NJ 07960, USA
201-455-2000, 800-526-0717
Fax 201-707-4555,
www.alliedsignal.com

Allied Signal Inc.,
P.O. Box 1057,
Morritown, NJ 07962, USA
201-455-2000 800-707-4555
Fax 201-455-3198
www.alliedsignal.com

Allied Signal Inc./Performance Additives,
PO Box 1039,
101 Columbia Rd.,
Morristown, NJ 07962-1039, USA
201-455-2145, 800-222-0094
Fax 201-455-6154,
www.alliedsignal.com

Alox Corp.,
3943 Buffalo Ave.,
PO Box 517,
Niagara Falls, NY 14302, USA
716-282-1295
Fax 716-282-2289

Altex Chemical Co. Ltd.,
Clayfield Works, Slaithwaite,
Huddersfield, West Yorkshire,
BD12 0JZ, UK

Alzo Inc.,
650 Jernee Mill Rd,
Sayreville, NJ O8879, USA
908-254-1901
Fax 908-254-4423

AMC SPREA S.p.A.,
Sede Amministrativa e Comercic,
Via Pasubio 37,
21040 Venegono S,
Milan, Italy
39-2-331-827500
Fax 39-2-331-827492

Amerchol Corp.,
PO Box 4051,
136 Talmadge Rd.,
Edison, NJ USA
732-248-6000, 800-367-3534
Fax 732-287-4186

Amerchol Europe,
Havenstraat 86,
B-1800 Vilvoorde, Belgium
32-2-252-4012
Fax 32-2-252-4909

Amerchol, Ikeda Corp.,
New Tokyo Bldg., No. 3-1,
Marunouchi 3-Chome,
Chiyoda-Ku, Tokyo,100, Japan
81-3-3212-8791
Fax 813-3215-5069

American Chemet Corp.,
400 Lake Cook Rd.,
Deerfield, IL 60015, USA
708-948-0800
Fax 708-948-0811

American Chemical Services,
PO Box 190,
Griffith, IN 46319, USA
219-924-4370
Fax 219-924-5298

American Colloid Co.,
1500 W. Shure Dr.,
Arlington Hts., IL 60004-1434,
USA
708-394-8730
Fax 708-506-6199

American Colloid Co.,
Highway 212 West,
PO Box 160,
Belle Fourche, SD 57717, USA
605-892-2591, 800-535-1935
Fax 605-892-4880

American Cyanamid Co.,
#1 Campus Dr.,
Parsippany, NJ 7054, USA
973-683-2000
Fax 973-683-4041

American Emulsions Co.,
PO Box 3787,
Dalton, GA 30721, USA
404-226-7028
Fax 404-278-5183

American Ingredients,
3947 Broadway,
Kansas City, MO 64111, USA
816-561-9050, 800-669-4092
Fax 816-561-1132

American Ingredients,
3947 Broadway,
Kansas City, MO 64111, USA
816-561-9050, 800-669-2250
Fax 816-561-0422

American Laboratories,
4410 South 102 St.,
Omaha, NE 68127, USA
402-339-2494, 800-445-5989
Fax 402-339-0801

American Lecithin Co,
115 Hurley Rd., Unit 2B,
Oxford, CT 06478, USA
203-262-7100
Fax 203-262-7101

American McGaw,
2525 McGaw Ave,
Irvine, CA 92714-5895, USA

American Optical,
Soft Contact Lens Business,
14 Mechanic St,
Southbridge, MA 01550, USA
508-765-9711

Ameripol Synpol Corp.,
PO Box 667,
1215 Main St.,
Port Neches, TX 77651, USA
409-722-8321, 800-547-0622
Fax 409-724-8713,
www.ameripol.com

Ameripol Synpol Corp.,
146 S. High St., 7th floor,
Akron, OH 44308-1493, USA
216-762-4422, 800-321-9001
Fax 216-762-2549,
www.ameripol.com

Amersham Corp.,
2636 S Clearbrook Dr.,
Arlington Heights, IL 60005, USA
708-593-6300

Amersham International,
Amersham Place, Little Chalfont,
Amersham, Bucks, HP7 9NA, UK
44 1494 544000
Fax 44 1494 542266

Amoco Chemical (Europe),
15, Rue Rothschild,
1211 Geneva 21, Switzerland
41-22-715-0701,
www.amoco.com

Amoco Chemical Asia,
16th Floor, Great Eagle Centre,
23 Harbour Rd.
Hong Kong
852-2586-8899, www.amoco.com

Amoco Chemicals,
200 East Randolph Dr.,
Mail code 7802,
Chicago, IL 60601, USA
312-856-4729, 800-621-4567
Fax 312-856-6225,
www.amoco.com

Amoco Performance Products,
10th Floor, Tonichi Bldg.,
2-31 Roppongi 6-Chome,
Mianto Ku, Tokyo, 106, Japan
www.amoco.com

Ampacet Corp.,
660 White Plains Rd.,
Tarrytown, NY 10591-5130, USA
914-631-6600
Fax 914-631-7197,
www.ampacet.com

Ampacet Europe S.A.,
Rue D'Ampacet 1,
6780 Messancy,
Belgium
32-63- 371490
Fax 32-63-371499,
www.ampacet.com

Anaquest,
2005 W Beltline Highway,
Madison, WI 53713, USA
608-273-0019

Anchor Group plc,
777 Canterbury Road,
Westlake, OH 44145, USA

Manufacturers and Suppliers

Andeno BV,
Grubbenvorsterweg 8,
5928 NX, Venlo-Holland, The Netherlands
31-77-899-555
Fax 31-77-299300

Anderson Development,
1415 E. Michigan St.,
Adrian, MI 49221, USA
517-263-2121
Fax 517-263-1000

Andrea Aromatics,
PO Box 3091,
Princeton, NJ 08543-3091, USA
609-695-7710
Fax 609-392-8914

Anedco, Inc.,
10429 Koenig Rd.,
Houston, TX 77034, USA
713-484-3900
Fax 713-484-3931

Angus Chemie GmbH
Zeppelinstrasse 30
49479 Ibbenbüren, Germany
49-5459-560
Fax 49-5459-56-267
www.angus.de

Ansul Co.,
1 Stanton St,
Marinette, WI 54143, USA
715-735-7411

Antec International,
Windham Road,
Chilton Industrial Estate,
Sudbury, Suffolk, CO10 6XD,UK
44-1787-77305
Fax 44-1787-310846

Anzon Ltd.,
Cookson House, Willington Quay,
Wallsend, Tyne & Wear,
NE28 6UQ, UK

Aqualon Canada Inc.,
5407 Eglinton Ave. West,
Etobicoke, Ont., M9C 5K6, Canada
416-620-5400

Aqualon Co.,
1313 North Market St.,
PO Box 8740,
Wilmington, DE 19899-8740, USA
302-594-6600, 800-345-8104,
Fax 302-594-6660

Aqualon France,
3 Rue Eugene & Armand Peugeot,
92500 Rueil-Malmaison, France
33-1-4751-2919
Fax 33-1-4777-0614

Aquarium Systems Inc,
8141 Tyler Boulevard,
Mentor, OH 44060-4889, USA
216-255-1997, 800-822-1100
Fax 216-255-8994,

Arcmann-Denmark A/S,
Strandparken 15,
DK 8000 Aarhus C, Denmark
45-6-12 65 44

ARCO Chemical Co.,
3801 West Chester Pike
Newtown Square, PA 19073-2387
610-359-200 Fax 610-359-2722
www.arcochem.com

ARCO Chemical Europe Inc.,
ARCO Chemical House, Bridge Ave.,
Maidenhead, Berks, SL6 1YP, UK
44-1628-77-5000 Fax 44-1628-77-5180
www.arcochem.com

ARCO Chemical Asia Pacific Ltd.,
41st. Floor, The Lee Gardens
33 Hysan Ave., Causeway Bay
Hong Kong
852-2-822-2668
Fax 852-2-840-1690
www.arcochem.com

Arco Chemical Europe,
Weenahuis,Weena 141,
NL-30 13 CK Rotterdam, The Netherlands
31-10-401 0400
Fax 31-10-411 4849,
www.arco.com

Argus Division
www.argus.com

Arista Industries, Inc.,
1082 Post Rd.,
Darien, CT 06820, USA
203-655-0881, 800-255-6457
Fax 203-656-0328

Aristech Chemical Co.,
600 Grant St., Room 1170,
Pittsburgh, PA 15230-0250, USA
412-433-7800
Fax 412-433-7721

Arizona Chemical Co.,
1001 E. Business Hwy. 98,
Panama City, FL 32401-3633, USA
904-785-6700, 800-526-5294
Fax 904-785-2203

Armour Pharmaceuticals,
500 Arcola Road,
P.O. Box 1200,
Collegeville, PA 19422, USA
215-454-8000, 800-72-RORER
Fax 215-454-8940

Arol Chemical Products,
649 Ferry St.,
Newark, NJ 07105, USA
201-344-1510
Fax 201-344-7127

Asahi Chemical Industries,
Hibiya Mitsui Bldg., 1-2, Yura,
Chiyoda-ku, Tokyo 100, Japan
81-3-3507-2730
Fax 81-3-3507-2495,
www.asahi.com

Asahi Denka Kogyo Kogyo,
Furukawa Bldg. 2-8,
Nihonbashi Muro-machi 2-chome Ch,
Tokyo 103, Japan,
81-3-5255-9002
Fax 81-3-3270-2463,
www.asahi.com

Ashland Chemical Co.,
PO Box 2219,
Columbus, OH 43216, USA
614-790-3333, 800-526-4032
Fax 614-889-3465,
www.ashchem.com/

Ashland Chemical Co.,
2463 Royal Windsor Drive,
Mississauga, Ont., L5J 1K9, Canada
905-823-7975
www.ashland.com/

Ashland-Chemie,
Reisholzstrasse 16,
40721 Hilden, Germany
49-21-711030
Fax 49-21-711-0335,
www.ashland.com

Ashley Polymers,
5114 Fort Hamilton Parkway,
Brooklyn, NY 11219, USA
718-851-8111
Fax 718-972-3256,
www.ashleypoly.com

Associated Lead Mfg,
Crescent House, P O Box 19E,
Newcastle-upon-Tyne,
Tyne & Wear, NE99 1GE, UK
44-1632-610161

Astor Products Inc./Blue Arrow Division,
P.O. Box 2366,
5244 Edgewood Court,
Jacksonville, FL 32203, USA
904-783-5000

Astra Chemicals GmbH,
Postfach 249,
Tinsdaler Weg 18,
D-22876 Wedel, Germany
49-4103-70 80
Fax 49-4103-70-82-93,
www.astra.com

Astra Pharmaceutical,
50 Otis St.,
Westboro, MA, 01581-4500, USA
508-366-1100, 800-225-6333
Fax 508-366-7406,
www.astra.com

Astro Industries Inc.,
PO Box 2559,
114 Industrial Blvd.,
Morganton, NC 28655, USA
704-584-3800, 800-872-7876
Fax 704-584-3885,
www.astro.com

Atlas Refinery, Inc.
142 Lockwood St.,
Newark, NJ 07105, USA
973-589-2002
Fax 973-589-7377,
www.atlasrefinery.com

Atomergic Chemetals,
222 Sherwood Ave.,
Farmingdale, NY 11735-1718, USA
516-694-9090
Fax 516-694-9177

Atrachem L.P.,
7201 W 65th Street,
Bradford Park, IL 60638, USA
708-458-8450
Fax 708-458-0286

Ausimont UK Ltd.,
93-99 Upper Richmond Road,
Putney, London, SW15 2TJ, UK
44-181-780-0399, 44181 780-4000
Fax 44-181-780-2871,
www.ausimont.com

Ausimont USA Inc.,
Crown Pt. Rd. & Leonards Lane,
PO Box 26,
Thorofare, NJ 08086, USA
609-853-8119, 800-323-2874
Fax 609-853-6405,
www.ausimont.com

Australian Bakels PTY,
PO Box 6100,
Silverwater NSW 2128, Australia
61-2-9739-9300
Fax 61-2-9739-9464

Australian Synthetic,
Maidstone Street, PO Box 33,
Altona, 3018, Australia

Manufacturers and Suppliers

Avitrol Corp.,
7644 E 46th Street,,
Tulsa, OK 74145, USA
918-622-7763, 800-633-5069
Fax 918-622-2527

Avon Packers Ltd.,
Salisbury Road,
Downton,Wilts., SP5 3JJ, UK
44-1725-22822
Fax 44-1725-22840

Axel Plastics Research Laboratories Inc,
Box 770 855,
Woodside, NY 11377, USA
718-672-8300
Fax 718-565-7447,
www.axelplast.com

Ayrton Saunders plc,
34 Hanover Street,
Liverpool, Merseyside, L1 4LN, UK

B.C. Ames,
131 Loxington Street,
Box 70,
Waltham, MA USA

Bacon Industries Inc,
192 Pleasant St.,
Watertown, MA 02172, USA
617-926-2550
Fax 617-926-2022

Bakelite GmbH,
Postfach 7154,
Gennaer Strasse 2-4,
58642 Iserlohn, Germany
49-2374-510
Fax 49-2374-51409

Bakelite Polymers (UK),
Syer House, Stafford Court,
Stafford Park, Telford,
Shropshire, TF3 3BD, UK

Baker Performance Chemicals,
3920 Essex Lane,
PO Box 27714,
Houston, TX 77227-7714, USA
713-599-7400, 800-231-3606
Fax 713-552-1937

Baker Performance Chemicals,
P.O. Box 11192,
Bakersfield, CA 93389, USA
805-763-5137, 800-231-3606
Fax 805-765-6046

Baker Rubber Inc.,
700 West Chippewa,
PO Box 2438,
South Bend, IN 46680, USA
219-291-5101
Fax 219-291-9152

Baker Sillavan Ltd.,
Weaver Valley Rd.,
Wharton,Winsford,
Cheshire, CW7 3BU, UK
44-1606-553151
Fax 44-1606-593008

Barnes-Hind Inc.,
333 Howard Ave,
Des Plaines, IL 60018-1907, USA

BASF AG,
Carl-Bosch Str. 38,
67056 Ludwigshafen,
ZOA/KI-C-100, Germany
49-621-60-49223
Fax 49-621-60-72866,
www.basf.com

BASF Australia Ltd.,
500 Princes Highway,
Nobel Park,Vic., 3174, Australia
61-39-2-12-1500
Fax 61-39-2-12-1511,
www.basf.com

BASF Belgium S.A./N.,
Ave. Hamoir-laan 14,
1180 Brussels, Belgium
32-2-373-2111
Fax 32-2-375-1042,
www.basf.com

BASF Canada Inc.,
345 Carlingview Dr.,
Toronto, Ont., M9W 5N9, Canada
416-675-3611
Fax 416-674-2588,
www.basf.com

BASF Corp.,
3000 Continental Dr. North,
Mount Olive, NJ 07828-1234, USA
201-426-2600, 800-367-9861
Fax 201-426-4999,
www.basf.com

BASF Corp./Chemicals,
3000 Continental Dr. North,
Mount Olive, NJ 07828-1234, USA
201-426-2800, 800-669-2273
Fax 201-426-2610,
www.basf.com

BASF Corp./Coatings,
3000 Continental Dr. North,
Mount Olive, NJ 07828-1234, USA
201-426-2800, 800-669-2273
Fax 201-426-2610,
www.basf.com

BASF Corp./Fibers ,
4330 Chesapeake Dr.,
Charlotte, NC 28266, USA
704-392-4313, 800-247-0557
Fax 704-393-3649,
www.basf.com

BASF Corp./Plastic Materials,
3000 Continental Dr. North,
Mount Olive, NJ 07828-1234, USA
201-426-2600, 800-669-2273
Fax 201-426-3912,
www.basf.com

BASF Espanola S.A.,
Paseo de Gracia, 99
E-08008, Barcelona, Spain
34-3-488-1010
Fax 34-3-488-2020,
www.basf.com

BASF Mexicana, S.A.,
Insurgentes Sur 975,
Col. Ciud,Delegacion Benito Juarez,
03710 Mexico, D.F., Mexico
52-5-325-2600
Fax 52-5-325-2777,
www.basf.com

BASF plc,
PO Box 4, Earl Road,
Cheadle Hulme, Cheshire, SK8 6QG, UK
44-161-485-6222
Fax 44-161-486-0891,
www.basf.com

Bausch & Lomb, Inc.,
One Baush & Lomb Place,
Rochester, NY 14604, USA
716-338-6000
Fax 716-338 6007

Bausch & Lomb, Inc.,
265 Baush & Lomb Dr.,
Oakland, MD 21550, USA
301-334-9933
Fax 301-334-8775

Baxenden Chemicals Ltd.,
Union Lane,
Droitwich, WR9 9BB, UK
44-1905-794-795
Fax 44-1905-794-002

Baxter Health Care,
One Baxter Pkwy,
Deerfield, IL 60015, USA
708-948-2000

Bay Resins, Inc.,
PO Box 630, Route 313,
Millington, MD 21651, USA
410-928-3083
Fax 410-928-5412

Bayer AG,
Bayerwerk,
D-51368 Leverkusen, Germany
49-214-30-1
Fax 49-214-30-65136,
www.bayer.com

Bayer Corp./Agriculture,
77 Belfield Rd.,
Etobicoke, Ont., M9W 1G6, Canada
416-248-0771
Fax 416-614-1058,
www.bayer.com

Bayer Corp./Fibers,
Bayer Rd.,
Pittsburgh, PA 15205-9741, USA
412-777-2000, 800-662-2927
Fax 412-777-7840,
www.bayer.com

Bayer Corp./Fibers,
2603 W. Market St.,
Akron, OH 44313, USA
216-836-0451, 800-321-0997
Fax 216-838-0200,
www.bayer.com

Bayer Corp./Industrial Chemicals,
100 Bayer Rd.,
Pittsburgh, PA 15205-9741, USA
412-777-2000, 800-662-2927
www.bayer.com

Bayer Corp./Polymers,
Bayer Rd.,
Pittsburgh, PA 15205-9741, USA
412-777-2000, 800-662-2927
www.bayer.com

Bayer Hispania Indus,
Pablo Claris 196,
E-80037 Barcelona, Spain
34-3-218 45 50
Fax 34-3-217 41 49,
www.bayer.com

Bayer plc,
Stoke Court, Stoke Poges,
Slough, Berks., SL2 4LY, UK
44-1753-645151
www.bayer.com

Bayer UK Ltd., Agric.,
Eastern Way,
Bury St Edmunds, Suffolk, IP32 7AH, UK
44-1284-763200
Fax 44 1284-702810,
www.bayer.com

BDH Chemicals Ltd.,
Broom Road, Parkstone,,
Poole, Dorset, BH12 4NN, UK
44-1202-745520
Fax 44-1202-738299

BDH Inc., Pigments Div.,
350 Evans Ave.,
Toronto, Ont., M8Z 1K5, Canada
416-255-8521

Beacon Chemical Co.,
PO Box 2500,
125 Macquiesten Pkwy.,
Mt. Vernon, NY 10550, USA
914-699-3400
Fax 914-699-2783

Belzak Corp.,
850 Bloomfield Ave.,
Clifton, NJ 07012, USA
201-773-0602
Fax 201-773-0602

Benzsay & Harrison Inc.,
Railroad Ave,
PO Box 459,
Delanson, NY 12053, USA
518-895-2311
Fax 518-895-8475

Bergvik Kemi,
Box 66,
S-82022 Sandarne, Sweden
46-270-62500
Fax 46-270- 60100

Berk Chemicals Ltd.,
PO Box 56,
Priestley Rd.,
Basingstoke, Hants, RG24 9QB, UK
44-1256-29292
Fax 44-1256-64711

Berk Pharmaceuticals,
St Leonards House, St Leonards,
Eastbourne, Sussex, BN21 3YG, UK
44-1323-501111

Berlex Laboratories,
300 Fairfield Road,
Wayne, NJ 07470, USA
201-694-4100, 800-221-1756
Fax 201-305-5365

Bernel Chemical Co.,
174 Grand Ave.,
Englewood, NJ 07631, USA
201-569-8934
Fax 201-569-1741

Berol Nobel AB.,
S-44485 Stenungsund,
Sweden
46-303-85000
Fax 46-303-84659

Berol Nobel Ltd.,
23 Grosvenor Road,
St. Albans, Herts, AL1 3AW, UK
44-1727-841421
Fax 44-1727-841529

BF Goodrich Chemical,
Goerlitzer Strasse 1,
41460 Neuss 1, Germany
49-2131-18050
Fax 49-2131-180-530,
www.BFGSolutions.com

BF Goodrich Chemical,
742 Rue de Verdun,
B-1130 Brussels, Belgium
32-2-247-1911
Fax 32-2-247-1991,
www.BFGSolutions.com

BF Goodrich Co./Geon
www.BFGSolutions.com

BF Goodrich Specialty,
9911 Brecksville Rd.,
Brecksville, OH 44141-3247, USA
216-447-5000, 800-331-1144
Fax 216-447-5770,
www.BFGSolutions.com

Bioglan Laboratories,
1 The Cam Centre, Wilbury Way,
Hitchin, Herts, S94 OTW, UK
44-1462-438444
Fax 44-1462-421242,
www.BFGSolutions.com

BIP Chemicals Ltd.,
PO Box 6, Popes Lane,
Oldbury, Warley,West Midlands,
B69 4PD, UK
44-121-552-1551
Fax 44-121-552-4267

Blew Chemical Co.,
PO Box 501,
Palos Heights, IL 60463, USA
708-448-5780
Fax 708-448-5781

BMC,
278 E 7th St,
St Paul, MN 55101, USA
651-228-6400
Fax 651-228-6572

BOC Group plc,
Chertsey Road,
Windlesham, Surrey, GU20 6HJ, UK
44-1276-77222
Fax 44-1276-71333,
www.boc.com

Boehringer Ingelheim, GmbH
Binger Strasse 173,
D-55216 Ingelheim, Germany
49-6132-773-666
Fax 49-6132-773-755

Boehringer Ingelheim,
Ellesfield Avenue,
Bracknell, Berks., RG12 8YS, UK
44-1344-424-600
Fax 44-1344-424-600

Boehringer Ingelheim Pharmacuetical s,
900 Ridgebury Road,
P.O. Box 368,
Ridgefield, CT 06877-0368, USA

Boliden Intertrade,
3379 Peachtree Rd. NE, Suite 3,
Atlanta, GA 30326, USA
404-239-6700, 800-241-1912
Fax 404-239-6701

The Boots Co. plc,
Head Office,
Nottingham, Notts, NG2 3AA, UK
44-115-950-6111
Fax 44-115-959-2727,
www.boots.co.uk

Borax Europe Ltd.,
170 Priestley Rd.,
Guildford, Surrey, GU2 5RQ, UK
44-1483-242035
Fax 44-1483-242097,
www.borax.com

Borax Ltd.,
Gorsey Lane,
Widnes,Cheshire, WA8 0RP, UK
44-1483-734000
Fax 44-1483-457676,
www.borax.com

Borden (UK) Ltd.,
North Baddesley,
Southampton, Hants, SO52 9ZB, UK
44-1703-732-131
Fax 44-1703-738-656

Borden Chemical Div.,
180 E. Broad St.,
Columbus, OH 43215, USA
614-225-4000, 800-225-8044
Fax 614-225-3476

Borregaard LignoTech,
PO Box 162,
N-1701 Sarpsborg, Norway
46-9-11 80 00
Fax 46-9-11 87 70

Bos Chemicals Ltd.,
Paget Hall, Tydd St Giles",
Wisbech, Cambridgeshire, PE13 5FL, UK
44-1945-870-118
Fax 44-1945-870-264

Bostik Div./Emhart,
Boston St.,
Middleton, MA 01949, USA
508-777-0100, 800-726-7845
Fax 508-774-7376

Bostik Inc.,
211 Boston St.,
Middleton, MA 01949, USA
987-777-0100, 800-726-7845
Fax 987-750-7212

Bostik Ltd.,
Ulverscroft Road,
Leicester, Leics, LE4 6BW, UK
44-116-2689-353
Fax 44-1162689-299

BP Chemicals Inc.,
4440 Warrensville Center Rd.,
Warrensville Hts., OH 44128, USA
216-586-6455, 800-272-4367
Fax 216-586-5588, www.bp.com

BP Chemicals Ltd.,
Britannic House, 6th Floor,
1 Finsbury Circus, London,
EC2M 7BA, UK
44-171-496-4867
Fax 44-171-496-4898,
www.bp.com

BP Oil UK Ltd.,
BP House, Breakspear Way,
Hemel Hempstead, Herts, HP2 4UL, UK
44-1442-232-323
Fax 44-1442-225-225,
www.bp.com

BP Performance Polymers,
60 Walnut Ave., Suite 100,
Clark, NJ 07066, USA
908-815-7843
Fax 908-815-7844, www.bp.com

Bracco Industria Chimicas,
Casella Postale 12064,
Via E Folli 50,
I-20134 Milan, Italy,
39-2-2 1771
Fax 39-2-21773

Brent Chemicals International,
Ridgeway,
Ivor, Bucks, SL0 9JJ, UK
44-1753-651-812
Fax 44-1753-652-460

Brian Jones & Associates,
Fluorocarbon Building,
Caxton Hill, Hertford, Herts,
SG13 7NH, UK
44-1992-553-065
Fax 44-1992-551-873

Bristol Laboratories,
PO Box 4500755,
Princeton, NJ 08543-4500, USA
609-897-2000
Fax 315-432-4804

Bristol-Myers Co. Ltd.,
Swakeleys House, Milton Rd.,
Ickenham, Uxbridge, Middlesex,
UB10 8NS, UK
44-1895-639-911
Fax 44-1895-636-975,
www.bristolmyers.com

Bristol-Myers Squibb Co.,
345 Park Avenue,
New York, NY 10154-0037, USA
212-546-4000
Fax 212-546-5664,
www.bristolmyers.com

Bristol-Myers Squibb Pharmaceuticals,
BMS House,
141-149 Staines Road,
Hounslow, Middlesex, TW3 3JA, UK
44-181-572-7422
Fax 44-181-577-1756,
www.bristolmyers.com

British Chrome & Chemicals,
Urlay Nook, Eaglescliffe,
Stockton-on-Tees, Cleveland,
TS16 0QG, UK
44-1642-787-755
Fax 44-1642-781-935

British Nova Works Ltd.,
57-61 Lea Road,
Southall, Middlesex, UB2 5QB, UK
44-181-574-6531
Fax 44-181-571-7572

Brocades Pharma,
Brocades House, Pyrford Road,
West Byfleet, Surrey, KT14 6RA, UK
44-1932-355-535
Fax 44-1932-353-458

Broemmel Pharmaceutical,
19901 Nordhoff St,
Northridge, CA 91324, USA

Brooks Industries Inc.,
70 Tyler Place,
South Plainfield, NJ 07080, USA
908-561-5200
Fax 908-561-9174

Burgess Pigment Co.,
PO Box 349,
Sandersville, GA 31082, USA
912-552-2544, 800-841-8999
Fax 912-552-1772

Burlington Chemical,
PO Box 111,
Burlington, NC 27216, USA
910-584-0111, 800-672-5888
Fax 910-584-3548,
www.burco.com

Burmah-Castrol Chemicals,
Burmah Castrol House,
Pipers W,
Swindon,Wilts, SN3 1RE, UK
44-1793-511-521
Fax 44-1793-612-524

Burroughs Wellcome Co,
3030 Cornwallis Rd,
Research Triangle Park, NC 27709, USA
919-248-3000, 800-722-9292
Fax 919-248-8375

Bush Beach Ltd.,
Paul Ungerer House, Earl Rd, S
Handforth, Wilmslow, Cheshire,
SK9 3RL, UK
44-161-485-8231
Fax 44-161-485-6445

Bush Boake Allen Ltd,
Blackhorse Lane,
Walthamstow, London, E17 5QP, UK
44-181-531-4211
Fax 44-181-527-2360

C.P. Hall Co.,
7300 South Central Ave.,
Chicago, IL 60638-0428, USA
708-594-6000, 800-321-8242
Fax 708-458-0428,
www.cphall.com

Cabot Australasia Ltd,
PO Box 19,
300 Millers Road,
Altona, Vic., 3018, Australia
61-3-391-1622
Fax 61-3-391-9370,
www.cabot-corp.com

Cabot Canada Ltd.,
350 Wilton St.,
Sarnia, Ont., N6T 7N4, Canada
519-336-2261
Fax 519-336-8501,
www.cabot-corp.com

Cabot Carbon Ltd.
Barry Site, Sully Moors Rd.,
Sully, S. Glamorgan, CF6 2XP, UK
44-1446-736999
Fax 44-1446-737123,
www.cabot-corp.com

Cabot Carbon Ltd.
Lees Lane,
Stanlow-Ellesmere Port,
S. Wirral, Cheshire, L65 4HT, UK
44-151-355-3677
Fax 44-151-356-0712,
www.cabot-corp.com

Cabot Corp./Cab-O-Si,
PO Box 188,
Tuscola, IL 61953-0188, USA
217-253-3370, 800-222-6745
Fax 217-253-4334
www.cabot-corp.com

Cabot Europe Ltd./Special,
Le Nobel 4B,
2 rue Marcel Monge,
F-92158 Suresnes Cedex, France
33-1-46-97-5800
Fax 33-1-47-72-6647,
www.cabot-corp.com

Cabot Plastics Belgium,
Rue E Vandervelde 131,
B-4431 Loncin, Belgium
32-41-46-8211
Fax 32-41-46-5499
www.cabot-corp.com

Cabot Plastics International,
Interleuvenlaan 5,
B-3001 Leuven, Belgium
32-1-639-0111
Fax 32-1-640-1253
www.cabot-corp.com

Cabot Plastics Ltd.,
Gate St.,
Dunkinfield, SK16 4RU, UK
44-161-330-5051
Fax 44-161-308-2641,
www.cabot-corp.com

Cadillac Plastic & Chemicals,
2855 Coolidge Highwat, #300
Troy, MI 48084, USA
313-583-1200, 800-274-100

Cadillac Plastic (Canada),
91 Kelfield St.,
Rexdale, Toronto, Ont., Canada
416-249-8311
Fax 416-249-0148

Cadillac Plastic GmbH,
Alfred-Nobel-Strasse 17,
D-68517 Viernheim, Germany
49-6204-70960
Fax 49-6204-75899

Cadillac Plastic Ltd,
Rivermead Drive, Westlea,
Swindon,Wilts, SN5 7YT, UK
44-1793-514-949
Fax 44-1793-511-762

Cadillac Plastics Australia,
PO Box 145,
Ermington, Sydney, NSW 2115, Australia
61-2-9714-1166
Fax 61-2-9648-5487

Caffaro SpA,
Via Friuli, 55,
20031 Cesano Maderno, Italy
39-362-51-4266
Fax 39-362-51-4405

Cal Polymers, Inc.,
2115 Gaylord St.,
Long Beach, CA 90813, USA
213-436-7372

Calbiochem Corp.,
PO Box 12087,
San Diego, CA 92119, USA
619-450-9600

Calbiochem-Novabiochem,
10933 North Torrey Pines Road,
La Jolla, CA 92037, USA
619-450-9600
Fax (619)453-3552

Calder Colours (Ashby),
Nottingham Road,
Ashby-de-la-Zouch,
Leicestershire, LE6 5DR, UK
44-1530-412885
Fax 44-1530-417315

Calgene Chemical Inc,
7247 North Central Park Ave.,
Skokie, IL 60076-4093, USA
708-675-3950, 800-432-7187
Fax 708-675-3013

Calgon Carbon Canada,
130 Royal Crest Ct,
Markum, Ont., L3R 0A1, Canada
905-477-9242, 905-477-4511
Fax 905-673-8883,
www.calgoncarbon.com

Calgon Carbon Corp.,
PO Box 717,
Pittsburgh, PA 15230-0717, USA
412-787-4519, 800-422-7266
Fax 412-787-6676,
www.calgoncarbon.com

Calgon Australia,
RMB L 303
Ballarat, Vic., 3353, Australia
1-800-999-988
Fax 61-353-447-308,
www.calgon.com

Callaway Chemical Co.,
6003 Hamilton Rd.,
PO Box 2335,
Columbus, GA 31993-3599, USA
706-576-2000
Fax 706-576-6455, www.vul.com

Calumet Lubricants Co.,
2780 Waterfront Pkwy. E.
Drive,Suite 200,
Indianapolis, IN 46214, USA
317-328-5660
Fax 317-328-5668

Calumet Lubricants Co.,
14000 Mackinaw Ave.,
Chicago, IL 60633, USA
708-862-9100

Camtex Fabrics Ltd.,
Blackwood Road, Lilyhall,
Workington, Cumbria, CA14 4JJ, UK
44-1900-602646
Fax 44-1900-66827

Canadian Harvest USA
1001 S. Cleveland St.,
PO Box 272,
Cambridge, MN 55008, USA
612-689-5800
Fax 612-689-5949

Manufacturers and Suppliers

Cancarb Ltd.,
1702 Brier Park Crescent N.W.,
PO Box 310,
Medicine Hat, Alberta, T1A 7G1, Canada
403-527-1121
Fax 403-529-6093

Caramba Chemie GmbH,
Wanheimer Strasse 334/6,
D-47055 Duisburg, Germany

Carborundum Co.,
168 Creekside Dr,
Amherst, NY 14228, USA
716-691-2051

Carborundum Resistance,
Trafford Park Road, Trafford Park,
PO Box 55, Manchester, M17 1HP, UK
44-161-872-2381
Fax 44-1744-882941

Cardolite Corp.,
500 Doremus Ave.,
Newark, NJ 07105-4805, USA
201-344-5015, 800-322-7365
Fax 201-344-1197

Carl Freudenberg,
Postfach 1369,
D-69469 Weinheim, Germany

Carless Refining & Marketing,
St. James House,
Eastern Road,
Romford, Essex, RM1 3NL, UK
44-1708-755557
Fax 44-1708-753890

Carlton Laboratories,
4 Manor Parade,
Salvington Road, Durrington,
Worthing,West Sussex, BN13 2JP, UK

Carpenter Technology,
PO Box 14662,
101 W. Bern St.,
Reading, PA 19612-4662, USA
215-371-2000

Carter-Wallace Ltd.,
Wear Bay Road,
Folkestone, Kent, CT19 6PG, UK
44-1303-850661

CasChem Inc.,
40 Ave. A,
Bayonne, NJ 07002, USA
201-858-7900, 800-CASCHEM
Fax 201-437-2728

Cashew Co., Ltd.,
407-1, Yoshino-cho 1-chome,
Ohmiya-shi, Saitama, 330, Japan
81-48-663-2461
Fax 81-48-666-0219

Cassella AG,
Hanauer Landstrasse 526,
W-60386 Frankfurt/M, Germany
49-69-410901
Fax 49-69-41092100

Catalysis Div., Urban Industry,
Hans Neumann-Parcela
13-Valenc,
Estado Carabobo, Venezuela

Catawba-Charlab, Inc,
5046 Old Pineville Rd.,
PO Box 240497,
Charlotte, NC 28224, USA
704-523-4242
Fax 704-522-8142

Catomance Ltd.,
96 Bridge Rd East,
Welwyn Garden City, Herts, AL7 1JW, UK
44-1707-324373
Fax 44-1707-372191

Ceca S.A.,
22, place de l'Iris,
Cedex 54, 92062 Paris-La Defense, France
33-1-47-96-9090
Fax 33-1-47-96-9234

Celite (UK) Ltd.,
Livingston Rd.,
Hessle, North Humberside, HU13 OEG, UK
44-1482-64-5265
Fax 44-1482-64-1176

Celite Corp.,
PO Box 519,
Lompoc, CA 93438-0519, USA
805-735-7791, 800-342-8667
Fax 805-735-5699

Celite Corp. (Canada),
295 The West Mall,
Etobicoke, Ont., M9C 4Z7, Canada
416-626-8175
Fax 416-626-8235

Celite France,
9 rue du Colonel-de-Rochebrune,B.P. 240,
92504 Rueil-Malmaison, France
33-47-49-0560
Fax 33-47-08-3025

Celite Mexico S.A. d'Alejandro Dumas
No. 103, 3er, Col. Polanco, C.P. 11560,
Mexico
52-5-203-5611
Fax 52-5-255-1835

Celite Pacific,
Suite 284, Sui On Centre,
8 Harbor Road, Hong Kong
852 582 5609
Fax 852 827 9392

Central Soya Aarhus,
Skansevej 2,
PO Box 380,
DK-8100 Aarhus C, Denmark
45-89-31-2111
Fax 45-89-31-2112

Central Soya Co.,
PO Box 2507,
1946 West Cook Rd.,
Fort Wayne, IN 46801-2507, USA
219-425-5432, 800-348-0960
Fax 219-425-5485

Certified Laboratories,
PO Box 70, Oldbury,
Warley, West Midlands, B69 4AD, UK
44-121-525-6678
Fax 44-121-500-5386

Chemax International,
155 N. Main St.,
New City, NY 10956, USA
914-634-0451
Fax 914-634-0937,
www.chemax.com

Chemax, Inc.,
PO Box 6067,
Greenville, SC 29606, USA
864-277-7000, 800-334-6234
Fax 864-277-7807,
www.chemax.com

Chemcentral Corp.,
7050 W. 71 St.,
PO Box 730,
Bedford Park, IL 60499-0730, USA
708-594-7000, 800-331-6174
Fax 708-594-6328

Chemetall GmbH,
Reuterweg 14,
D-60271 Frankfurt/M, Germany
49-69-159-2554
Fax 49-69-159-2053

Chemfax Inc.,
3 Rivers Rd.,
PO Box 2389,
Gulfport, MS 39505, USA
601-863-6511
Fax 601-868-3669

Chemical Combine,
6005 Stara Zagora, SHK
Bulgaria

Chemische Fabrik GmbH,
Postfach 1260,
Hauptstrasse 4,
67271 Kleinkarlbach, Germany
49-6359-8010
Fax 49-6359-801209

Chemische Fabriken O,
Postfach 1328,
IM Schleeke 77,
D-38642 Goslar, Germany
49-5321-7510
Fax 49-5321-751 192

Chemlease Int'l.,
PO Box 951628,
Lake Mary, FL 32795-1628, USA

Chemoxy International,
All Saints Refinery, Cargo Fleet Road,
Middlesbrough, Cleveland, TS3 6AF, UK
44-1642-248555
Fax 44-1642-244340

Chemron Corp.,
PO Box 2299,
Paso Robles, CA 93447, USA
805-239-1550, 800-423-1148
Fax 805-239-8551

Chemsal Chemicals,
Kupferstrasse 1,
PO Box 100262,
D-46422 Emmerich, Germany
49-2822 711-0
Fax 49-2822 18294

Chemsearch (UK) Ltd.,
Landchard House, Victoria Street,
West Bromwich, West Midlands, B70 8ER, UK
44-121-525-1666
Fax 44-121-500-5386

Chemseco,
4800 Blue Parkway,
Kansas City, MO 64130,USA

Chemson GmbH,
Post Box 12,
A-9601 Arnoldstein, Austria
43-4255-222-6324
Fax 43-4255-222-6350

Chemson Ltd.,
Hayhole Works, Willington Quay,
Wallsend, Tyne & Wear, NE28 0PB, UK
44-191-258-5892
Fax 44-191-258-1549

Chemson Polymer Additives,
Postfach 43, A 9601 Arnoldstein, Austria
43-4255-222-6324
Fax 43-4255-222- 6350

Chemtech (Crop Protection),
The Arable Centre,Winterbourne Monkton,
Swindon, Wilts, SN4 9NW, UK
44-1672-3591

Chem-Trend A/S,
Smedeland 14,
Postboks 1384,
DK-2600 Glostrup, Denmark
45-42-45-6711
Fax 45-43-63-03-50,

Chem-Trend Inc.,
1445 W. McPherson Park Dr.,
PO Box 860,
Howell, MI 48844-0860, USA
517-546-4520, 800-727-7730
Fax 517-546-6875

Chemtronics Inc.,
8125 Cobb Center Dr.,
Kennesaw, GA 30144, USA
404-424-4888, 800-645-5244
Fax 404-423-0748

Chevron Chemical Co.,
1301 McKinney,
PO Box 3766,
Houston, TX 77253-3766, USA
713-754-2000, 800-231-3260
www.chevron.com

Chevron Chemical Co.,
6001 Bollinger Canyon Rd.,
San Ramon, CA 94583, USA
925-842-5764
Fax 925-842-0378,
www.chevron.com

Chevron Chemical Co.,
PO Box 3766,
Houston, TX 77253, USA
713-754-2000, 800-231-3260
www.chevron.com

Chevron International,
PO Box 7146,
San Francisco, CA 94120-7146,
USA
415-894-5341, 800-344-5650
Fax 415-894-1083,
www.chevron.com

Chiltern Farm Chemicals,
11 High Street,
Thornborough, Buckingham,
Bucks, MK18 2DF,UK
44-1280-822400
Fax 44-1280-822082

Chipman Ltd.,
Portland Building, Portland St.,
Staple Hill, Bristol, Avon, BS16
4PS, UK
44-1179-574574
Fax 44-1272-563461

Church & Dwight Co.,
Box CN5297,
469 N. Harrison St.,
Princeton, NJ 08543-5297, USA
609-497-7113, 800-221-0453
Fax 609-497-7176

Ciba Agrochemicals,
Whittlesford,
Cambridge, Cambridgeshire,
CB2 4QT, UK
44-1223-833621
Fax 44-1223-835211,
www.ciba.com/

Ciba Ltd.,
CH-4002, Basel, Switzerland
41-061- 696-4534
Fax 41-061-697-1111,
www.ciba.com/

Ciba Pharmaceuticals,
Wimblehurst Road,
Horsham, West Sussex, RH12
4AB, UK
44-1403-50101
Fax 44-1403- 56643,
www.ciba.com/

Ciba Pharmaceuticals Co.,
556 Morris Ave.,
Summit, NJ 07901, USA
800-742-2422
www.ciba.com/

Ciba Pigments,
Ashton New Road,
Clayton, Manchester, M11 4AR,
UK
44-161-223-1341
Fax 44-161-231-7422,
www.ciba.com/

Ciba Plastics UK,
Duxford,
Cambridge, Cambridgeshire,
CB2 4QA, UK
44-1223-832121
Fax 44-1223-838404,
www.ciba.com/

Ciba Specialty Chemicals Water Treatments Ltd.,
PO Box 38,
Bradford, West Yorkshire, BD12
0JZ, UK
44-1274-417-000
Fax 44-1274-606-499,
www.ciba.com/

Ciba Specialty Chemicals Water Treatments, Inc.,
900 Route 9 North,
Woodbridge, NJ 07095-1015,
USA
www.ciba.com/

Ciba-Geigy (Japan) L,
10-66, Miyuki-cho,
Takarazuka-shi, Hyogo, 665,
Japan
81-797-74-2472
Fax 81-797-74-2515,
www.ciba.com/

Ciba-Geigy Corp.,
540 White Plains Rd.,
Tarrytown, NY 101591, USA
914-785-2000, 800-431-1874
www.ciba.com/

Ciba-Geigy Corp./Additives,
540 White Plains Rd.,
PO Box 2005,
Tarrytown, NY 10591-9005, USA
914-785-4461, 914-785-2000
800-431-2360, 800-431-1900,
www.ciba.com/

Ciba-Geigy Corp./Chelates,
PO Box 18300,
Greensboro, NC 27419, USA
800-334-9481
www.ciba.com/

Ciba-Geigy Corp./Chemicals,
410 Swing Rd.,
Greensboro, NC 27409-2080,
USA
910-632-6000, 800-334-9481
Fax 910-632-2861,
www.ciba.com/

Ciba-Geigy Corp./Dye,
PO Box 18300,410 Swing Rd.,
Greensboro, NC 27419, USA
910-632-2011, 800-334-9481
Fax 910-632-7008,
www.ciba.com/

Ciba-Geigy Corp./Furane,
5121 San Fernando Road West,
Los Angeles, CA 90039-1071,
USA
818-247-6210
Fax 818-507-0167,
www.ciba.com/

Ciba-Geigy Corp./Pigments,
335 Water Street,
Newport, DE 19804-2434, USA
302-992-5600, 302-633-2000
800-431-2777, www.ciba.com/

Ciba-Geigy Corp./Specialty Chemicals,
281 Fields Lane,
Brewster, NY 10509, USA
914-785-3000, 800-222-1906
Fax 914-785-3476,
www.ciba.com/

Ciba-Geigy Corp./Textiles,
www.ciba.com/

Ciba-Geigy Marienber,
Postfach 1253,
D-64623 Bensheim, Germany
49-6254-790
Fax 49-6254-79505,
www.ciba.com/

Cilag AG,
Hochstrasse 201-209,
CH-8201 Schaffhausen,
Switzerland
41-53-82 91 11
Fax 41-53-82 94 43

Clark Chemical Industries,
103 Walnut Grove Rd.,
Cartersville, GA 30120, USA
770-386-6397
Fax 770-386-6393

Clark Colors,
155 Helen St., PO Box 705,
South Plainfield, NJ 07080, USA
201-757-4500
Fax 201-757-3170

Clayton Aniline Co.,
PO Box 2, Ashton New Road,
Clayton, Manchester, M11 4AP,
UK
44-161-223-1391
Fax 44-161-223-4315

Cleanacres Ltd.,
Adoversford,
Cheltenham, Gloucestershire,
GL54 4LZ, UK
44-1242-820481
Fax 44-1242-820807

Climax Molybdenum B.V.,
Postbus 1130,
NL-3180 AC Rozenburg, The
Netherlands
31-15933

Climax Molybdenum Co,
1370 Washington Pike,
Bridgeville, PA 15017-2839, USA
412-257-5181
Fax 412-257-5191

Climax Performance Materials,
PO Box 22015,
Tempe, AZ 85285-2015, USA

CNC International, Ltd.,
PO Box 3000,
20 Privilege St.,
Woonsocket, RI 02895, USA
401-769-6100
Fax 401-769-4509

Coal Products Ltd.,
PO Box 16, Mill Lane,
Wingerworth,
Chesterfield, Derbyshire, S42
6JJ, UK
44-1246-277001
Fax 44-1246-212212

Coalite Chemicals Division,
PO Box 152,
Buttermilk Lane,
Bolsover, Chesterfield,
Derbyshire, S44 6AZ, UK
44-1246-826816
Fax 44-1246-240309

Manufacturers and Suppliers

Coates Coatings International,
Station Lane,
Witney, Oxfordshire, OX8 6XZ, UK
44-1993-702969
Fax 44-1993-775579

Collinda Ltd.,
Collinda House,
25 Ottways Lane,
Ashstead, Surrey, KT21 2PZ, UK
44-1372-278-416
Fax 44-1372-278-559

Colonial Chemical, Inc.,
PO Box 11525,
Chattanooga, TN 37401-2525, USA
615-267-8947
Fax 615-266-0770

Colonial Metals, Inc.,
Triumph Industrial Complex,
PO Box 726,
Elkton, MD 21921, USA
410-398-7200, 800-962-1537
Fax 410-398-2918

Colorcon,
415 Moyer Blvd.,
PO Box 24,
West Point, PA 19486-0024, USA
215-699-7733
Fax 215-661-2605

Colorcon GmbH,
Hauptstrasse 5,
61462 Königstein, Germany
49-6174-93890
Fax 49-6174-23698

Colorcon Ltd.,
Murray Road, St Paul's Cray,
Orpington, Kent, BR5 3QY, UK
44-1689- 838301
Fax 44-1689-878342

Colorcon P.R., Inc.,
PO Box 979,
Punta Santiago, PR 00661, USA
809-852-3815
Fax 809-852-0030

Colorcon s.a.r.l.,
Bureaux de la Jonchure,
62-70 rue Yvan Tourgueneff,
78380 Bougival, France
33-1-30-82-1582
Fax 33-1-30-82-7879

Colorcon S.r.l.,
via Disciplini 18,
20123 Milano, Italy
39-2-805-2222
Fax 39-2-861-229

Colour-Chem Ltd.,
Ravindra Annexe, Dinshaw Vachh,
194 Churchgate Reclamation,
Bombay, 400 020, India

Colourex Ltd.,
Plot 7 Wimbledon Avenue,
Brandon, Suffolk, IP27 0NZ, UK
44-1842-811693

Columbia Nitrogen Co,
PO Box 1483 (13)
Augusta, GA,30913,USA

Compounding Ingredients,
Unit 217, Walton Summit Centre,
Bamber Bridge, Preston, Lancs, PR5 8AL, UK
44-1772-322888
Fax 44-1772-315853

Compounding Technologies,
13435 Estelle St.,
Corona, CA 91720, USA
714-371-7701, 800-325-1564
Fax 714-371-7724

Conap, Inc.,
1405 Buffalo St.,
Olean, NY 14760, USA
716-372-9650
Fax 716-372-1594

Consolidated Chemica,
Abbey Road, The Industrial Estate,
Wrexham, Clwyd, LL13 9PW, Wales UK
44-1978-661-351
Fax 44-1978-661-673

Continental Polymers,
2225 East Del Amo Blvd.,
Compton, CA 90220, USA
310-637-2103, 800-441-3943
Fax 310-637-2415

Contract Chemicals,
Penrhyn Road,
Knowsley Industrial Park South,
Prescot, Merseyside, L34 9HY, UK
44-151-548-8840
Fax 44-151-548-6548

Cool-Amp Conducto-Lube,
15834 Upper Boones Ferry Rd,
Lake Oswego, OR 97035, USA
503-624-6426
Fax 503-624-6436

Cooper Chem Co.,
40 Parker Rd.,
Long Valley, NJ 07853, USA
908-876-3231
Fax 908-876-3857

The Cooper Co. Inc.,
PO Box 726,
Gulf Breeze, FL 32562-0726, USA
904-932-5005
Fax 904-932-1923

Coopers Creek Chem Co.,
884 River Rd.,
West Conshohocken, PA 19428, USA
610-828-0375
Fax 610-828-9720

Copolymer Rubber,
Scenic Highway, PO Box 2591,
Baton Rouge, LA 70821, USA
504-355-5655, 800-535-9960
Fax 504-355-8056

Coulter Electronics,
Northwell Drive,
Luton, Bedfordshire, LU3 3RH, UK
44-1582-491414
Fax 44-1582-490390

Courtaulds Advanced,
72 Lockhurst Lane,
Coventry, West Midlands, CV6 5RS, UK
44-1203-582000
Fax 44-1203-687328,
www.courtaulds.com

Courtaulds Aerospace Ltd.,
72 Lockhurst Lane,
Coventry, West Midlands, CV6 5RS, UK
44-1203-583067
Fax 44-1203-583826,
www.courtaulds.com

Courtaulds Chemicals,
Nelson Acetate Works,
Caton Rd.,
Lancaster, Lancs., LA1 3PF, UK
44-1524-66111
Fax 44-1524 846384,
www.courtaulds.com

Courtaulds Chemicals,
Macclesfield Road,
Leek, Staffordshire, ST13 8UZ, UK
www.courtaulds.com

Courtaulds Engineering,
PO Box 11, Foleshill Rd,
Coventry, West Midlands, CV6 5AB, UK
44-1203-688771
Fax 44-1203-687925,
www.courtaulds.com

Courtaulds Fine Chemicals,
PO Box 5,
Spondon, Derbyshire, DE2 7BP, UK
44-1332-661422
Fax 44-1332 280610,
www.courtaulds.com

Courtaulds plc, Patents,
PO Box 111,
72 Lockhurst Lane,
Coventry, West Midlands, CV6 5RS, UK
44-1203-688771
Fax 44-1203-583837,
www.courtaulds.com

Courtaulds Water Soluble,
PO Box 5,
Spondon, Derbyshire, DE21 7BP, UK
44-1332- 661422
Fax 44-1332-661078,
www.courtaulds.com

Coventry Chemicals Ltd.,
Woodhams Road, Siskin Drive,
Coventry, West Midlands, CV3 4FX, UK
44-1203-639739
Fax 44-1203-639717

Cox Chemicals Ltd.,
Overley Hill,Wellington,
Telford, Shropshire, TF6 5HD, UK
44-1952-86333
Fax 44-1952-86207

Cox Pharmaceuticals,
Whiddon Valley,
Barnstaple, Devon, EX32 8NS, UK
44-1271-75001
Fax 44-1271-46106

CP International Chemicals Ltd.,
Northgate House, 21 Northgate,
Bishops Stortford, Herts, CM23 2ET, UK
44-1279-506330
Fax 44-1279-755873

CPS Kemi Aps,
Hejreskovv 22,
3490 Kvistagaard, Denmark
45-42-890533
Fax 45-42-238077

Cray Valley,
Cedex 101,
F92970 Paris la Defense, France
33-1-41-35-6810
Fax 33-1-41- 35-6143

Cri-Tech, Inc.,
85 Winter St.,
Hanover, MA 02339, USA
617-826-5600
Fax 617-826-5770

Croda Inc.,
7 Century Dr,
Parsippany, NJ 07054-4698, USA
973-644-4900
www.croda.com

Manufacturers and Suppliers

Croda Adhesives Ltd.,
Winthorpe Rd,
Newark, Notts, NG24 2AL, UK
44-1636 646711
Fax 44-1636-605187,
www.croda.com

Croda Bakery Service,
Falcon St.,
Oldham, Lancs, OL8 1JU, UK
www.croda.com

Croda Canada Ltd.,
78 Tisdale Ave.,
Toronto, Ont., M4A 1Y7, Canada
416-751-3571
Fax 416-751-9611,
www.croda.com

Croda Chemicals Ltd.,
Cowick Hall, Snaith,
Goole, North Humberside, DN14 9AA, UK
44-1405-860551
Fax 44-1405-862253,
www.croda.com

Croda Colloids Ltd.,
Foundry Lane,
Ditton, Widnes, Cheshire, WA8 8UB, UK
44-151-423-3441
Fax 44-151-423-3205,
www.croda.com

Croda do Brazil Ltda,
Rua Croda 230 Distrito Industr,
CEP 13.053,Campinas/SP-C.P. 1098, Brazil
www.croda.com

Croda Food Services,
Falcon St.,
Oldham, Lancs., OL8 1JU, UK
44-161-652-6311
Fax 44-161-627-2346,
www.croda.com

Croda Food Services,
Falcon St.,
Oldham, Lancs., OL8 13U, UK
44-161-652-6311
Fax 44-161-627-2346,
www.croda.com

Croda Inc.,
7 Century Dr.,
Parsippany, NJ 07054-4698, USA
201-644-4900, 973-644-4900,
Fax 201-644-9222,
www.croda.com

Croda International,
Cowick Hall, Snaith,
Goole, North Humberside, DN14 9AA, UK
44-1405-860551
Fax 44-1405-860205,
www.croda.com

Croda Italiana Srl,
via Grocco, N917 27036,
Mortara (PV), Italy
www.croda.com

Croda Japan K.K.,
Aceman Bldg.,
1-10, Tokui-cho 1-chome,
Chuo-ku,
Osaka, 540, Japan
81-6-942-1791
Fax 81-6-942-1790,
www.croda.com

Croda Oleochemicals Personal Care,
Cowick Hall, Snaith,
Goole, East Yorkshire, DN14 9AA, UK
44-1405-860551
Fax 44-1405-860205,
www.croda.com

Croda Resins Ltd.,
Crabtree Manorway South,
Belvedere, Kent, DA17 6BA, UK
44-181-311-9109
Fax 44-181-310-9878,
www.croda.com

Croda Singapore Pte.,
20 Chia Ping Rd.,
S-2261, Singapore
65-261-3008
Fax 65-261-2825,
www.croda.com

Croda Surfactants Ltd.,
Cowick Hall, Snaith,
Goole, North Humberside, DN14 9AA, UK
44-1405-860551
Fax 44-1405-860205,
www.croda.com

Croda Universal Ltd.,
Oak Rd.,Clough Rd.,
Hull, North Humberside, HU6 7PH, UK
44-1482-443181
Fax 44-1483-341792,
www.croda.com

Croda Universal, Inc,
4014 Walnut Pond Dr.,
Houston, TX 77059, USA
713-282-0022
Fax 713-282-0024,
www.croda.com

Crompton & Knowles Co.,
PO Box 33157,
Charlotte, NC 28233, USA
704-372-5890, 800-323-4383
Fax 704-332-8785,
www.crompton-knowles.com

Crompton & Knowles Co.,
1595 MacArthur Blvd.,
Mahwah, NJ 07430, USA
201-818-1200, 800-343-4860
Fax 201-818-2173,
www.crompton-knowles.com

Crookes Laboratories,
1 Thane Road,
Nottingham,Notts, NG2 3AA, UK
44-1159-592-584
Fax 44-1602-595-508

Crosfield B.V.,
Ir. Rocourstraat 28,
6245 ZG Eijsden, The Netherlands
31-4409-9333
Fax 31-4409-3995

Crosfield Brazil,
Ave Maria Coehlo de Aguiar 215,
Bloco C 3 andar,CEP 05804
Sao Paulo, Brazil
55-11-545-1481
Fax 55-11-545-2226

Crosfield Co.,
101 Ingalls Ave.,
Joliet, IL 60435, USA
815-727-3651, 800-727-3651
Fax 815-727-5312

Crosfield Group,
PO Box 26,
Warrington, Cheshire, WA5 1AB, UK
44-1925-416100
Fax 44-1925-59828

Crosfield SpA,
via Dei Cipressi, 10,
37033, Montorio Verona, Italy
39-45-557159
Fax 39-45-884 0099

Crowley Chemical Co.,
261 Madison Ave.,
New York, NY 10016, USA
212-682-1200
Fax 212-953-3487

Crowley Tar Products,
261 Madison Ave.,
New York, NY 10016, USA
212-682-1200

Crown Metro Inc.,
PO Box 5857,
Greenville, SC 29606, USA
803-299-1331, 800-368-1331
Fax 803-299-1678

Crown Technology, Inc.,
7513 E. 96 St.,
PO Box 50426,
Indianapolis, IN 46250-0426, USA
317-845-0045, 800-432-0045
Fax 317-845-9086

Croxton & Garry Ltd.,
Curtis Rd. Industrial Estate,
Dorking, Surrey, RH4 1XA, UK
44-1306-886688
Fax 44-1306-887780

Crucible Chemical Co.,
PO Box 6786, Donaldson Center,
Greenville, SC, 29606 USA
803-277-1284, 800-845-8873
Fax 803-299-1192

Custom Coating & Lamination,
715 Plantation Street,
Worcester, MA, 01605 USA

Custom Fibers Europe,
13 Rassau Industrial Estate,
Ebbw Vale, Gwent, Wales UK
44-1495-350655

Custom Fibers International,
PO Box 940,
2045 Lebec Rd.,
Lebec, CA 93243, USA
800-321-5324, Fax 805-248-1123

Cutter Laboratories,
4th & Parker,
PO Box 1986,
Berkeley, CA, 94701, USA

Cuyahoga Plastics,
1265 Babbitt Rd.,
Cleveland, OH 44132-2798, USA
216-261-2744
Fax 216-261-3537

Cyanamid Aerospace,
Abenbury Way,
Wrexham Industrial Estate,
Wrexham, Clwyd, LL13 9UF, Wales UK

Cyanamid Agriculture UK Ltd.,
154 Fareham Road
Gosport, Hants., PO13 0AS, UK
44-1239-224-142 Fax 44-1329-224-335
www.agrocentre.co.uk

Cyanamid Canada Inc.,
88 McNabb St.,
Markham, Ont., L3R 6E6, Canada
416-470-3600
Fax 416-470-3852

Cyanamid Fothergill,
Abenbury Way,
Wrexham Industrial Estate,
Wrexham, Clwyd, LL13 9UF, Wales UK

D.J. Enterprises, Inc.,
PO Box 31366,
Cleveland, OH 44131-0366, USA
216-524-3879
Fax 216-524-3879

D.L. Forster Ltd.,
12 The Ongar Road Trading Estate,
Great Dunmow, Essex, CM6 1EU, UK
44-1371-865201
Fax 44-1371-876541

Dainippon Pharmaceuticals,
2-6-8, Dosho-machi,
Chuo-ku, Osaka, 541, Japan
81-6-203-5321
Fax 81-6-203-6581

Dales Pharmaceutical,
Snaygill Industrial Estate,
Keighley Road, Skipton,
North Yorkshire, BD23 2RW, UK

Dampney Co. Inc.,
85 Paris Street,
Everett, MA 02149, USA
617-389-2805

Daniel Products Co.,
400 Claremont Ave.,
Jersey City, NJ 07304, USA
201-432-0800
Fax 201-432-0266

Darmex Corp.,
71 Jane Street,
Roslyn, NY, 11577, USA

Dart Industries Belgium,
Pierre Corneliskaai 35,
B-9300 Aalst, Belgium
32-53-70 14 11
Fax 32-53-78 27 26

Daubert Chemical Co.,
4700 S. Central Ave.,
Chicago, IL 60638, USA
708-496-7350

Davis & Geck,
RD2 Km 47 Hm 4,
Manati, PR 00701, USA

Dax Products Ltd.,
PO Box 119,
Nottingham, Notts., NG3 5ED, UK
44-1159-609996
Fax 44-1602-604620

DCS Color & Supply Co,
2011 S. Allis St.,
Milwaukee, WI 53207, USA
414-769-2580
Fax 414-769-2598,

DDSA Pharmaceuticals Ltd.,
310 Old Brompton Road,
London, SW5 9JQ, UK
44-171-373-7884
Fax 44-171-370-4321

Dead Sea Bromine/Ameribrom Inc.,
52 Vanderbilt Avenue
New York NY 10017
212-286-4000
Fax 212-286-4475
www.deadseabromine.com

Dearborn Chemicals Ltd.,
Foundry Lane,
Widnes, Cheshire, WA8 8UD, UK

Degussa AG,
Postfach 110533,
Weossfrauenstrasse 9,
D-60287 Frankfurt/M, Germany
49-69-218-01
Fax 49-69-218-3218,
www.degussa.com

Degussa Corp.,
65 Challenger Rd.,
Ridgefield Park, NJ 07660, USA
201-807-3224, 800-237-6745
Fax 201-807-3111,
www.degussa.com

Degussa Corp. Rubbers & Pigments Div.,
65 Challenger Rd.,
Ridgefield Park, NJ 07660, USA
201-641-6100
Fax 201-641-2270,
www.degussa.com

Degussa Ltd.,
Earl Rd, Stanley Green,
Handforth, Wilmslow, Cheshire, SK9 3RL, UK
44-161-486-6211
Fax 44-161-485-6445,
www.degussa.com

Degussa Metal Group,
3900 S. Clinton Ave.,
S. Plainfield, NJ 07080, USA
201-561-1100
Fax 201-769-9456,
www.degussa.com

Delaware Chemical Co.,
PO Box 126,
Daleville, IN 47334, USA

Denoon CDS,
The Campbell House,
Bilbrough,York, North Yorkshire, YO2 3PN, UK
44-1937-835515

Dermal Laboratories,
Tatmore Place, Gosmore,
Hitchin, Herts, SG4 7QR, UK
44-1462-458866

Dermik Laboratories,
500 Arcola Road,
P.O. Box 1200,
Collegeville, PA 19426-0107, USA
312-687-7440

Dexine Rubber Co. Ltd.,
Spotland Bridge Works,
Rochdale, Manchester, OL12 6AU, UK
44-1706-40011
Fax 44-1706-527714

Dexter Aerospace Materials,
P.O.Box 312,
2850 Willow Pass Rd.,
Pittsburg, CA 94565-0031, USA
510-458-8000
Fax 510-458-8030

Dexter Chemical Corp,
845 Edgewater Rd.,
Bronx, NY 10474, USA
718-542-7700
Fax 718-991-7684

Dexter Corp./Frekote,
One Dexter Drive,
Seabrook, NH 03874, USA
603-474-5541
Fax 603-474-5545

Dexter Distributor Products,
One Dexter Drive,
Seabrook, NH 03874, USA
603-474-5541
Fax 603-474-5545

Dexter Hysol
Dexter Distribution Programs
One Dexter Drive
Seabrook NH 03874
603-474-5541 Fax 474-5545

Dexter Specialty Company
One East Water St.,
Waukegan, IL 60085, USA
708-623-4200

Diacot Ltd.,
Kirkhaw Lane,
Pontefract, W. Yorkshire, WF11 8RD, UK
44-1977-675270

Dimex Ltd.,
46 Peckover St,
Bradford, W. Yorkshire, BD1 5BD, UK
44-1274 308052
Fax 44-1274-737058

Dista Products Co.,
Lilly Research Laboratoies,
Indianapolis, IN 46285, USA
317-276-4000

Dista Products Ltd.,
Fleming Road, Speke,
Liverpool, Merseyside, L24 9LN, UK
44-151-486-3939
Fax 44-151-486-8750

Distillex Ltd.,
Unit 117 & 120, Clydesdale Place,
Leyland, Preston, Lancs., PR5 3QS, UK
44-1772-454129
Fax 44-1772-622258

Doff Portland Ltd.,
Bolsover Street, Hucknall,
Nottingham, Notts, NG15 7TY, UK
44-1159-632842
Fax 44-1602-638657

Dorsey Laboratories,
59 Route 10,
East Hanover, NJ 07936, USA
201-503-7500
Fax 201-503-8265

Dorsey Pharmaceuticals,
4000 Monroe Rd,
Charlotte, NC 28205, USA
704-331-7000
Fax 704-377-1064

Dover Chemical Corp.,
3676 Davis Rd. N.W., PO Box 40
Dover, OH 44622, USA
330-343-7711, 800-321-8805
Fax 330-364-1579

Dow Ahlen,
Theodor-Schwarte-Str. 39,
D-59227 Ahlen, Germany
49-2382-891-0
Fax 49-2382-5151

Dow Chemical Canada,
PO Box 1012,
1086 Modeland Rd.,
Sarnia, Ont., N7T 7K7, Canada
519-339-3131, 800-441-4369,
www.dow.com/

Dow Chemical Co. Ltd,
2 Heathrow Boulevard,
284 Bath Road,
West Drayton, Middlesex, UB7 0DQ, UK
44-181-917-5000
Fax 44-181-917-5400,
www.dow.com/

Dow Chemical Europe S.A.,
Bachtobelstrasse 3,
CH-8810 Horgen, Switzerland
41-1-728-2095
Fax 41-1-728-3081,
www.dow.com/

Dow Chemical France,
BP 203, European Health Care
300 route des Cretes,
F-06904 Sophia Antipolis, France
33-92-94-40-00
Fax 33-78-62-78-98,
www.dow.com/

Manufacturers and Suppliers

Dow Chemical Kabushik,
507-1, Kishi Yamakita-cho,
Ashigarakami-gun, Kanagawa
258-01, Japan
81-465-76-3108
Fax 81-465-75-1064,
www.dow.com/

Dow Chemical North America,
2040 Willard H. Dow Center,
Midland, MI 48674, USA
517-636-1000, 800-441-4DOW
Fax 517-636-9752,
www.dow.com/

Dow Chemical Pacific,
39th Floor, Sun Hung Kai Center,
30 Harbour Rd., Wanchai, PO
Box,
Hong Kong
www.dow.com/

Dow Chemical Plastics,
2040 W.H. Dow Center,
Midland, MI 48674, USA
800-441-4DOW
Fax 517-638-9942,
www.dow.com/

Dow Chemical Quimica Mexicana,
Av. Paseo de las Palmas 555-3,
Lomas de Chapultepec,
Mexico, 11000 D.F., Mexico
52-5-227-1900
www.dow.com/

Dow Corning Belgium,
Rue General de Gaulle, 62,
B-1310 La Hulpe, Belgium
32-2-655-21-11
Fax 32-2-655-20-01,
www.dowcorning.com

Dow Corning Corp.,
PO Box 0994,
Midland, MI 48686-0994, USA
517-496-4000, 800-248-2481
Fax 517-496-4586,
www.dowcorning.com

Dow Corning Ltd.,
Kings Court, 185 Kings Rd,
Reading, Berks, RG1 4EX, UK
44-1734-507251
Fax 44-1734-575051,
www.dowcorning.com

Dow Corning Ophthalmics,
Midland, MI 48640, USA
www.dowcorning.com

Dow Corning STI,
47799 Halyard Dr., Suite 99,
Plymouth, MI 48170, USA
313-459-7792
Fax 313-459-0204,
www.dowcorning.com

DowElanco,
9330 Zionsville Rd.,
Indianapolis, IN 46268-1054,
USA
317-337-7364, 800-258-3033
Fax 317-337-7374

DowElanco Ltd.,
Latchmore Court, Brand Street,
Hitchin, Herts, SG5 1HZ, UK
44-1462-457272
Fax 44-1462-453906

Doyle Specialties,
9800 Cozycroft Avenue,
Chatsworth, CA 91311, USA

Dr. Madis Laboratories,
375 Huyler St., PO Box 2247,
South Hackensack, NJ 07606,
USA
201-440-5000
Fax 201-342-8000

Drew Ameroid Nederland,
Triathlonstraat 33,
3078 HX Rotterdam, The
Netherlands
31-10-479-0144
Fax 31-10-479-6525

Drew Ameroid Pte. Ltd.,
27 Tanjong Penjuru,
Jurong, 2260, Singapore
65-261-6544
Fax 65-265-0959

Drew Chemical Ltd.,
Drew Court,
525 Finley Ave.,
Ajax, Ont., L1S 2E5, Canada
416-683-0150
Fax 416-427-0688

Drew Industrial Division,
One Drew Plaza,
Boonton, NJ 07005, USA
201-263-7800, 800-526-1015
x7800
Fax 201-263-4483

Drexel Chemical Co.,
1700 Channel Ave.,
Memphis, TN 38106-1412, USA
901-774-4370
Fax 901-774-4666

DSM Andeno B.V.,
Noorderpoort 9,
PO Box 81, 5900 AB Venlo, The
Netherlands
31-77-3-899555
Fax 31-77-3-899300,
www.dsm.nl

DSM Chemicals & Fertilizer,
P.O. Box 43,
6130 AA Sittard, The Netherlands
31-46-773612
Fax 31-46-773912, www.dsm.nl

DSM China,
Shen Xin Mansion Suite
1007-1108,
200 Ninghai Dong Lu,
2000021 Shanghai, China
86-21-63741910
Fax 86-21-741909, www.dsm.nl

DSM Deutschland GmbH,
Tersteegenstrasse 77,
40474 Dusseldorf, Germany
49-211-454940
Fax 49-211-4370917,
www.dsm.nl

DSM Engineering Plastics,
233 Arvin Ave.,
Stoney Creek, Ont., L8E 2L9,
Canada
416-662-1866
Fax 416-662-3493, www.dsm.nl

DSM Engineering Plastics,
PO Box 3333,
2267 West Mill Road,
Evansville, IN 47732-3333, USA
812-435-7500, 800-333-4237
Fax 812-435-7702, www.dsm.nl

DSM Fine Chemicals,
Park 80 West, Plaza Two,
Saddlebrook, NJ 07663-5817,
USA
201-845-4404
Fax 201-845-4406, www.dsm.nl

DSM France SA,
Tour Atlantique,
9 Place de la Pyramide,
La Défense 9,
92911 Paris La Défense Cedex,
France
33-141-250505
Fax 33-147-760100, www.dsm.nl

DSM Japan K.K.,
Hanai House 7th Floor,
1-2-9 Shiba Koen,
Minato-Ku, Tokyo 105, Japan
81-3-34-377670
Fax 81-3-34377680, www.dsm.nl

DSM Kunstharze GmbH,
Postfach 1351,
Am Kreisforst 1,
D-49703 Meppen, Germany
49-5931-15160
Fa
x 49-5931-156101, www.dsm.nl

DSM Melamine America,
PO Box 327,
Addis, LA 70710-0327, USA
www.dsm.nl

DSM España, S.A.
c/ Frederic Mompou, 5
08960 Sant Just Desvern, Spain
34-93-473-1100
Fax 34-93-473-6373,
www.dsm.nl

DSM Resins UK Ltd.,
PO Box 8, Ellesmere Port,
South Wirral, L65 0HB, UK
44-151-355-6170
Fax 44-151-357-1282,
www.dsm.nl

DSM Thermoplastic Elastomers,
Postbus 43,
6130 AA Sittard, The Netherlands
31-46-763037
Fax 31-46-760040, www.dsm.nl

DSM Thermoplastic Elastomers,
29 Fuller St.,
Leominster, MA 01453-9895,
USA
508-534-1010, 800-524-0120,
Fax 508-534-1005, www.dsm.nl

DSM United Kingdom,
DSM House,
Papermill Dr,
Redditich, Worcs., B98 8QJ, UK
44-1527-590590
Fax 44-1527-590555,
www.dsm.nl

Duncan Flockhart & Co.,
Stockley Park West,
Uxbridge, Middlesex, UB11 1B7,
UK
44-181-990-9333
Fax 44-181-990-4325

Duncan Flockhart Ltd.,
Cobden Street,
Montrose,
Angus, DD10 8FB, Scotland UK,

Dunlop Adhesives,
Chester Road,
Birmingham,West Midlands, B35
7AL, UK
44-121-373-8101
Fax 44-121-384-2826,
www.dunlop.com

Dunlop Rubber,
Silvertown House,
Vincent Square, London, SW1P
2PL, UK
44-171-834-3848
Fax 44-171-834-3879,
www.dunlop.com

DuPont (Australia) Ltd.,
Northside Gardens,
168 Walker St.,
PO Box 930, N. Sydney,
NSW,2060, Australia
61-2 923-6111
www.dupont.com

DuPont (Germany),
DuPont Strasse 1,
D-61343 Bad Hamburg V,
Germany
49 61-72-87-1313
Fax 49-61-72-87-1314,
www.dupont.com

DuPont (UK) Ltd.,
Wedgewood Way,
Stevenage, Herts, SG1 4QN, UK
44-1438-734026
Fax 44-1438-734379,
www.dupont.com

DuPont (UK) Ltd./Agriculture,
Wedgewood Way,
Stevenage, Herts, SG1 4QN, UK
44-1438-734000
Fax 44-1438-734154,
www.dupont.com

DuPont (UK) Ltd./Polymers,
Maylands Ave.,
Hemel Hempstead, Herts, HP2 7DP, UK
44-1442-61251
www.dupont.com

DuPont Adhesive Polymers,
1007 Market St.,Suite D-5070-3,
Wilmington, DE 19898, USA
800-441-7111
Fax 302-773-2128,
www.dupont.com

DuPont Argentina S.A,
Av. Eduardo Madero 1020,
Buenos Aires, Argentina
54-1-312-2011, 54-1-319-4342
www.dupont.com

DuPont Asia Pacific,
1122 New World Office Bldg.,
East Wing,
Salisbury Rd.,
Kowloon, Hong Kong
852-734-5345
Fax 852-724-4458,
www.dupont.com

DuPont Automotive Products,
950 Stephenson Hwy.,
Troy, MI 48006-7013, USA
313-583-8000
www.dupont.com

DuPont Canada Inc.,
PO Box 2200,
Streetsville Postal Station,
Mississauga, Ont., L5M 2H3, Canada
416-821-3300, 800-668-6942
www.dupont.com

DuPont Chemicals,
1007 Market St.,
Wilmington, DE 19898, USA
800-441-7515
www.dupont.com

DuPont China Ltd.,
1122 New World Office Bldg., E,
Salisbury Rd.,
Kowloon, Hong Kong
852-734-5345
www.dupont.com

DuPont Co./Agriculture,
PO Box 80038,
Walker's Mill, Barley Mill Plaza,
Wilmington, DE 19880-0038, USA
800-441-7515
Fax 302-992-2276,
www.dupont.com

DuPont Elastomers,
505 Blue Ball Rd.,
PO Box 306,
Elkton, MD 21922-0306, USA
410-392-2532, 800-452-1454,
Fax 410-392-2540,
www.dupont.com

DuPont France,
137 rue de L'Université
F-75334 Paris Cedex 0, France
33-45-50-63-32
Fax 33-47-53-09-65,
www.dupont.com

DuPont Medical Products,
549-4 Albany St.,
Boston, MA 02118, USA
617-482-9595, 800-323-8903,
Fax 617-542-8463,
www.dupont.com

DuPont Merck Pharmaceuticals,
Barley Mill Plaza,
Wilmington, DE 19880-0025, USA
302-992-5000, 800-441-9861,
Fax 302-892-8530,
www.dupont.com

DuPont Mitsui Polychems,
Kasumigaseki Bldg., 24th floor,
2-5, Kasumigaseki 3-chome,
Chiyo, Tokyo 100, Japan
81-3-3580-5531
www.dupont.com

DuPont Nylon,
Barley Mill Plaza,
PO Box 80025,
Wilmington, DE 19880-0025, USA
302-999-3213, 800-231-0998
Fax 302-999-3441,
www.dupont.com

DuPont S.A. de C.V.,
Apartado Postal 5819,
06500 Mexico D.F., Mexico
52-5-250-9033
Fax 52-5-250-9033,
www.dupont.com

DuPont Singapore Pte,
Maritime Square,
#07-01World Trade Center,
Singapore 0409
65-273-2244
www.dupont.com

DuPont/Petrochemical,
Wilmington, DE 19898, USA
302-999-5053, 800-231-0998,
www.dupont.com

DuPont/Polymer Products,
Kirk Mill Bldg.,
Wilmington, DE 19898, USA
302-992-3010, 800-262-2745,
www.dupont.com

DuPont-NEN Medical Products,
1007 Market St,
Wilmington, DE 19898, USA
302-774-1000
Fax 302-774-7321,
www.dupont.com

DuPont-Showa Denko Co.,
Denpa Bldg.,
1-11-15, Higashi,
Shinagawa-ku, Tokyo 141, Japan
81-3-3444-5161
Fax 81-3-3444-5140,
www.dupont.com

DuPont-Toray Co., Ltd.,
Chuo Bldg., No. 10,
1-5-6, Nih, Chuo-ku,Tokyo 103, Japan
81-3-3245-5081
Fax 81-3-3231-1604,
www.dupont.com

Dwight Products, Inc,
10 Stuyvesant Ave.,
PO Box 909,
Lyndhurst, NJ 07071, USA
201-438-3388
Fax 201-438-0594

Dylon Industries, Inc.,
7700 Clinton Rd.,
Cleveland, OH 44144-1045, USA
216-651-1300, 800-237-8246
Fax 216-651-1777

Dylon International Ltd.,
Worsley Bridge Rd.,
Sydenham, London, SE26 5HD, UK
44-181-650-4801
Fax 44-181-658-8735

Dymax Corp.,
51 Greenwoods Road,
Torrington, CT 06791, USA
860-482-1010
Fax 860-496-0608,
www.dymax.com

Dymax Europe GmbH,
Trakehner Strasse 3,
D-60487 Frankfurt/M., Germany
49-69-7165-3568
Fax 49-69-7165-3830,
www.dymax.com

Dynaloy, Inc.,
7 Great Meadow Lane,
Hanover, NJ 07936, USA
201-887-9270
Fax 201-887-3678

Dynamit Nobel AG,
Postfach 1261, Kaiserstrasse,
D-53840 Troisdorf, Germany
49-2241-890
Fax 49-2241-89 15 40

Dynamold Inc.,
2905 Shamrock Avenue,
PO Box 9617,
Fort Worth, TX 76107, USA

Dynochem UK Ltd.,
Duxford,
Cambridge, Cambridgeshire,
CB2 4QB, UK
44-1223-837370
Fax 44-1223-832386

E. Merck,
Postfach 4119,
Frankfurter Strasse 250,
D-64293 Darmstadt, Germany
49-6151-72-0
Fax 49-6151-72-2000,
www.merck.com

E/M Corp.,
PO Box 2400,
2801 Kent Ave.,
W. Lafayette, IN 47906, USA
317-497-6346, 800-428-7802,
Fax 317-497-6348, www.em-corporation.com

Eagle-Picher Industries,
PO Box 550,
C & Porter Sts.,
Joplin, MO 64801, USA
417-623-8000
Fax 417-782-1923

E-A-R Specialty Comp,
7911 Zionsville Rd.,
Indianapolis, IN 46268, USA
317-692-1111
Fax 317-692-3111

Eastern Color & Chem,
35 Livingston St.,
PO Box 6161,
Providence, RI 02904, USA
401-331-9000
Fax 401-331-2155

Eastman Chemical Brazil,
Rua George Eastman,
213, Caixa Postal 225,
Sao Paulo, Brazil
55-11-543-5122
www.eastman.com

Eastman Chemical Co.,
Office Unit 704, 7th Fl., Lido,
Ji Chang Rd., Jiang Tai Rd.,
Beijing, 100004, China
86-10-436-7376
Fax 86-10-436-7380,
www.eastman.com

Eastman Chemical International,
Hertizentrum 6,
3263 Zug, Switzerland
41-42-23-25-25
Fax 41-42-21-12-52,
www.eastman.com

Eastman Chemical International,
11 Spring St.,
Chatswood, NSW 2067, Australia
61-2-411-3399
Fax 61-2-411-6430,
www.eastman.com

Eastman Chemical Products,
PO Box 431,
Kingsport, TN 37662-5280, USA
423-229-2000, 800-EASTMAN,
Fax 423-229-1196,
www.eastman.com/

Ebnother AG,
CH-6203 Sempach-Stati,
Switzerland
41-98-91-11
Fax 41-98-22-46

ECC International,
5775 Peachtree-Dunwoody Rd.
NE,Suite 200G,
Atlanta, GA 30342, USA
404-303-4415, 800-843-3222
Fax 404-303-4384, www.ecc.com

ECC International Ltd,
John Keay House,
St. Austell, Cornwall, PL25 4DJ,
UK
44-1726-74482
Fax 44-1726-623019,
www.ecc.com

ECC International S.A.,
2 rue du Canal,
B-4551 Lixhe, Belgium
32-41-79-9811
Fax 32-41-79-8279,
www.ecc.com

ECC International/Calcium Products,
100 Mansell Ct. East,Suite 300,
Roswell, GA 30076, USA
770-594-0660, 770-645-3384,
800-251-6327
www.ecc.com

ECC Japan Ltd.,
1-5-11, Shiba,
Minato-ku,Tokyo 105, Japan
81-3-5443-3144
Fax 81-3-5443-3137,
www.ecc.com

Efkay Chemicals Ltd.,
Banderway House,
156-162 Kilburn High Rd,
London, NW6 4JD, UK
44-171-625-4445
Fax 44-171-328-9101

Eka Nobel (Australia),
22 Commercial Dr.,
Dandenong,Vic., 3175, Australia
61-3-706-4488
www.ekanobel.com

Eka Nobel (NZ) Inc.,
Totara St.,
PO Box 5136,
Mount Maanganui, 3030, New
Zealand
64-7-5757089
Fax 64-7-5752 484,
www.ekanobel.com

Eka Nobel AB,
S-445 80 Bohus, Sweden
46-31-58-7000
Fax 46-31-98-1774,
www.ekanobel.com

Eka Nobel Inc.,
2622 Nashville Ferry Rd. East,
PO Box 2167,
Columbus, MS 39701, USA
770-956-2520, 800-821-9486,
www.ekanobel.com

Eka Nobel Ltd.,
Unit 304 Worle Pkwy.,
Summer Lane, Worle,
Weston-Super-Mare, Avon, BS22
OWA, UK
44-1934-522244
Fax 44-1934-522577,
www.ekanobel.com

Eka Nobel Paper Chemicals,
PO Box 3045,
6202 NA Maastricht, The
Netherlands
31-43-689595
Fax 31-43-634755,
www.ekanobel.com

Elementis Specialties,
Birtley,
Chester-le-Street, Co. Durham,
DH3 1QX, UK
44-191-410-2361
Fax 44-191-410-6005

Elf Atochem Canada,
PO Box 278,
Oakville, Ont., L6J 5A3, Canada
905-827-9841
Fax 905-827-7913,
www.elf-atochem.com

Elf Atochem North Am,
2000 Market St.,
Philadelphia, PA 19103-3222,
USA
215-419-7000, 800-628-4453
Fax 215-419-7875,
www.elf-atochem.com

Elf Atochem S.A.,
4, Cours Michelet,
La Défense 10,
F-92091 Paris Cedex 4, France
33-49-00-8080
Fax 33-49-00-83-96,
www.elf-atochem.com

Elf Atochem UK Ltd.,
Colthrop Lane, Thatcham,
Newbury, Berks, RG13 4LW, UK
44-1635-870-000
Fax 44-1635-861212,
www.elf-atochem.com

Eli Lilly & Co.,
Lilly Corporate Center,
Indianapolis, IN 46285, USA
317-276-2000
Fax 317-276-6876, www.lilly.com

Ellis & Everard plc,
Pine Street,
South Bank Road, Cargo Fleet,
Middlesbrough,TS3 8BD, UK
01642-227388
Fax 01642-242609

EM Industries, Inc.,
5 Skyline Dr.,
Hawthorne, NY 10532, USA
914-592-4660
Fax 914-592-9469

Emco Services Inc.,
PO Box 2191,
Taunton, MA 02780, USA
508-823-8852
Fax 508-824-6735

Emerson & Cuming Composites,
59 Walpole St.,
Canton, MA 02021-1838, USA
617-821-4250
Fax 617-828-0124

Emkay Chemical Co.,
319-325 Second St.,
PO Box 6537,
Elizabeth, NJ 07206, USA
908-352-7053, 888-365-2924
Fax 908-352-6398

EMS-American Grilon,
Corporate Way Road,
PO Box 1717,
Sumter, SC 29151-1717, USA
803-481-9173
Fax 803-481-6121,
www.emschem.com

EMS-Chemie AG,
Selnaustrasse 16,
CH-8039 Zurich, Switzerland
41-081-632-6434
Fax 41-081-632-7408,
www.emschem.com

EMS-Dottikon AG
CH-5605 Dottikon, Switzerland
41-56-616-81-11
Fax 41-56-616-81-20, www.ems-dottikon.ch

EMS-Grilon (UK) Ltd.,
Drummond Road,
Astonfields Industrial Estate,
Stafford, Staffse, ST16 3EL, UK
44-1785-59121
Fax 44-1785-213068,
www.emschem.com

EMS-Japan Corp.,
Kikuchi Bldg.,
4-6 Nihonbashi,
3-Chome, Chuo-ku, Tokyo 103,
Japan
www.emschem.com

Endo Laboratories Inc.,
1000 Stewart Ave,
Garden City, NY 11530, USA

Engelhard (Hong Kong),
Block B2, 6/F,
Eldex Industrial Estate,
21 Ma Tau Wei Road, Hunghom,
Kowloon, Hong Kong
852-2365-0302
Fax 852-2765-6406,
www.Engelhard.com

Engelhard Australia,
10-12 Prospect St.,
Box Hill 3128, Vic., Australia
61-3-899-6330
Fax 61-3-899-6360,
www.Engelhard.com

Engelhard Canada Ltd.,
195 Riviera Dr.,
Markham, Ont., L3R 5J6, Canada
905-940-4020
Fax 905-940-4470,
www.Engelhard.com

Engelhard Corp.,
101 Wood Ave. South,
PO Box 770,
Iselin, NJ 08830-0770, USA
732-205-5000, 800-631-9505
Fax 908-906-0337,
www.Engelhard.com

Engelhard GmbH,
Lise Meitner Strasse 7,
63303 Dreieich, Germany
49-6103-9345-0
Fax 49-6103-34787,
www.Engelhard.com

Engelhard Ltd.,
Chancery House,
St. Nicholas Way,
Sutton, Surrey, SM1 1JB, UK
44-181-643-8080
Fax 44-181-643-6063,
www.engelhard.com

Engelhard s.r.l.,
via Ronchi, 17,
20134 Milan, Italy
39-2-264-251
Fax 39-2-215-4602,
www.Engelhard.com

Engelhard Technologies,
www.Engelhard.com

Eni SpA
Piazzale Enrico
Mattei 1
00144 Rome, Italy
39-6-59821

EniChem America, Inc.,
2000 West Loop South,
Suite 2010,
Houston, TX 77027, USA
713-940-0700, 800-441-3646
Fax 713-940-0761
www.eni.it

EniChem Elastomeri Srl,
Strada 3, Palazzo B1, Milanofiori,
I-20090 Assago, Milan, Italy
39-2-5201
Fax 39-2-52026077
www.eni.it

EniChem Elastomers Am.,
2000 West Loop South,
Suite 2010,
Houston, TX 77027, USA
713-940-0700, 800-882-7080
Fax 713-940-0761

EniChem UK Ltd.,
Cadland Road,
Hardley, Hythe,
Southampton,Hants, SO45 3YY,
UK
44-1703-883-244
Fax 44-1703-883309

Enso-Gutzeit OY,
PO Box 309,
SF-00101 Helsinki, Finland
358-16291
Fax 358-1629471

Epoleon Corp.,
D.S. Bldg. 1-39
Ikenohata 2-Ch,
Tokyo, 110, Japan
81-3-823-1111

Ernst Jager GmbH,
Postfach 130380,
Oerschbachstrasse 35-39,
D-40589 Düsseldorf, Germany
49-211-792271
Fax 49-211-793144

Essex Specialty Products,
1250 Harmon Rd.,
Auburn, MI 48326, USA
810-391-6300
Fax 810-391-6417

Ethicon Inc.,
Rt 22, PO Box 151,
Somerville, NJ 08876, USA
908-218-0707

Ethox Chemicals, Inc.,
PO Box 5094, Sta. B,
Donaldson Center,
Greenville, SC 29606, USA
803-277-1620
Fax 803-277-8981

Ethyl Corpn.,
330 S. Fourth St.,
Richmond, VA 23217
(804) 788-5000

European Colour (Pigments),
Bankfield Street,
Stockport, Cheshire, SK5 7PB,
UK
44-161-480-3891
Fax 44-161-480-9852

European Vinyls Corp,
Emil-von-Behring-Strasse 2,
D-60439 Frankfurt/M-5, Germany
49-69-580-101
Fax 49-69-580-1640

Eval Co. of America,
1001 Warrenville Rd., Suite 20,
Lisle, IL 60532, USA
708-719-4610, 800-423-9762
Fax 708-719-4622

Evans Clay Co.,
PO Box 6,
Summit, NJ 07902-0006, USA
908-273-2500
Fax 908-273-8718

Evans Vanodine International,
Brierley Rd,
Walton Summit Centre,
Bamber Bridge, Preston, Lancs.,
PR5 8AH, UK
44-1772-322200
Fax 44-1772-626000

Evode Ltd.,
Common Road,
Stafford, Staffs., ST16 3EH, UK
44-1785-57755
Fax 44-1785-41818

Evode Plastics Ltd.,
Wanlip Road, Syston,
Leicester, Leicestershire, LE7
8PD, UK
44-1162-696752
Fax 44-1162 693209

Evode-Tanner Industries,
PO Box 1967,
Greenville, SC 29602, USA
803-232-3893
Fax 803-232-3094

Exchem plc,
Great Oakley Works, Great
Oakley,
Harwich, Essex, CO12 5JW, UK
44-1255-880239
Fax 44-1255-880429

Expancel,
PO Box 13000,
S-85013 Sundsvall, Sweden
46-60-134000
Fax 46-60-569518

Expancel, Inc.,
2150-H Northmont Pkwy.,
Duluth, GA 30136, USA
404-813-9126
Fax 404-813-8639

Exsymol,
4 Ave. Prince Hereditaire Albe,
MC 98000, Monaco
377-92-05-66-77
Fax 377-92-05-25-02

Exxon Chemical Belgium,
Boulevard du Souverain 280,
B-1160 Brussels, Belgium
32-2-674 41 11
Fax 32-2-674 41 29,
www.exxonn.com/exxonchemical

Exxon Chemical Co.,
PO Box 3272,
Houston, TX 77253-3272, USA
713-870-6000, 800-526-0749
Fax 713-870-6661,
www.exxon.com/exxonchemical

Exxon Chemical Europe,
Blvd du Souverain 280,
B-1160 Bruxelles, Belgium
32-2-769-31
Fax 32-2-769-3225,
www.exxonn.com/exxonchemical

Exxon Chemical Geopo,
PO Box 122,
4600 Parkway, Solent Business
Park,
Whiteley, Fareham, Hants, PO15
7AZ, UK
44-1489 884400
Fax 44-1489-884403,
www.exxonn.com/exxonchemical

Exxon Chemical International
33rd Floor, Shui on Centre,
8 Harbour Road,
Wanchai, Hong Kong
852-582-0888
www.exxonn.com/exxonchemical

Exxon Chemical Japan,
TBS Kaikan Bldg.,3-3,
Akasaka 5-Chome,
Minato-Ku,Tokyo 107, Japan
81-3-582-9243
www.exxonn.com/exxonchemical

Exxon Chemical Mediterrane,
via Paleocapa 7,
I-20121 Milan, Italy
39-2-88031
Fax 39-2 8803231,
www.exxonn.com/exxonchemical

FAIR Laboratories,
Claremont Way Industrial Estate,
London, NW2 1AL, UK
44-181-905-5111
Fax 44-181-9055222

Fairfax Biological Laboratories,
PO Box 300,
Electronic Road,
Clinton Corners, NY 12514, USA
914-266-3705
Fax 914-266-4892

Fairmount Chemical Co.,
117 Blanchard St.,
Newark, NJ 07105, USA
201-344-5790, 800-872-9999
Fax 201-690-5298

The Fanning Corp.,
2450 W. Hubbard St.,
Chicago, IL 60612-1408, USA
312-563-1234
Fax 312-563-0087

Farbwerke Hoechst,
Brueningstrasse 50, Hoechst,
49-65-929 Frankfurt/M, Germany
49-69-3050
Fax 49-69 303666

Fargro Ltd.,
Toddington Lane,
Littlehampton,West Sussex,
BN17 7PP, UK
44-1903-721591
Fax 44-1903-730737

Farmers Crop Chemical,
Thorn Farm, Evesham Road,
Inkberrow, Worcestershire, WR7
4LJ, UK
44-01386-793401
Fax 44-01386-793184

Farmitalia (Farmaceutic),
Viale E Bezzi 24,
I-20146 Milano, Italy,

Feb Ltd.,
Albany House, Swinton Hall
Road, Swinton,
Manchester, M27 1DT, UK
44-161-794-7411
Fax 44-161-793 4529

Ferguson & Menzies Ltd.,
312 Broomloan Road,
Glasgow, G51 2JW, UK
44-141-445-3555
Fax 44-141-425-1079

Ferguson & Timpson Ltd.,
5 Atholl Avenue,
Glasgow, G52 4UA, UK
44-141-882-4691
Fax 44-141-810-3402

Fernox Manufacturing Co. Ltd.,
Britannia Works, Clavering,
Saffron Walden, Essex, CB11
4QZ, UK
44-1799-550811
Fax 44-1799-550853

Ferro Corp./Bedford,
7050 Krick Rd.,
Bedford, OH 44146, USA
216-641-8580, 800-321-9946
Fax 216-439-7686,
www.ferro.com

Ferro Corp./Color,
4150 E. 56th St., PO Box 6550,
Cleveland, OH 44101, USA
216-641-8580
Fax 216-641-8831,
www.ferro.com

Ferro Corp./Electron,
4150 E. 56th St., P.O. Box 6550,
Cleveland, OH 44101, USA
216-641-8580
Fax 216-641-8857,
www.ferro.com

Ferro Corp./Engineer,
7500 East Pleasant Valley Rd.,
Independence, OH 44131, USA
216-641-8580
Fax 216-524-0493,
www.ferro.com

Ferro Corp./Filled & Reinforced,
5001 O'Hara Dr.,
Evansville, IN 47711, USA
812-423-5218
Fax 812-423-1864,
www.ferro.com

Ferro Corp./Filtros,
603 W. Commercial St.,
East Rochester, NY 14445, USA
716-586-8770, 800-633-2143
Fax 716-586-7154,
www.ferro.com

Ferro Corp./Frit Div.,
4150 E. 56th St., P.O. Box 6550,
Cleveland, OH 44101, USA
216-641-8580
Fax 216-641-2049,
www.ferro.com

Ferro Corp./Grant Chemicals,
111 W. Irene Rd.,
Zachary, LA 70791-9738, USA
www.ferro.com

Ferro Corp./International,
Alexander House, Crown Gate,
Runcorn, Cheshire, WA7 2UP,
UK
44-1928-71-9737
www.ferro.com

Ferro Corp./Keil Chemicals,
3000 Sheffield Ave.,
Hammond, IN 46320, USA
219-931-2630, 800-628-9079
Fax 219-931-0895,
www.ferro.com

Ferro Corp./Liquid Co.,
54 Kellogg Ct.,
Edison, NJ 08817-2509, USA
908-287-1930
Fax 908-287-8966,
www.ferro.com

Ferro Corp./Liquid Co.,
1301 N. Flora St.,
Plymouth, IN 46563, USA
219-935-5131, 800-424-2009
Fax 219-935-5278,
www.ferro.com

Ferro Corp./Plastic,
103 Railroad Ave.,
Stryker, OH 43557, USA
419-682-3311
Fax 419-682-4924,
www.ferro.com

Ferro Corp./Polymer,
7050 Krick Rd.,
Walton Hills, OH 44146-4494,
USA
www.ferro.co

Ferro Corp./Powder Coatings,
4150 E. 56th St., P.O. Box 6550,
Cleveland, OH 44101, USA
216-641-8580, 800-626-7890
Fax 216-641-8596,
www.ferro.com

Ferro Corp./Refractories,
661 Willet Rd.,
Buffalo, NY 14218, USA
716-825-7900
Fax 716-825-0421,
www.ferro.com

Ferro Corp./Specialties,
603 W. Commercial St., P.O. Box
389,
East Rochester, NY 14445, USA
716-586-8770, 800-633-2143
Fax 716-586-7154,
www.ferro.com

Ferro Corp./Transelco,
Box 217, 1789 Transelco Dr.,
Penn Yan, NY 14527, USA
315-536-3357
Fax 315-536-8091,
www.ferro.com

Ferro Corp./World Headquarters,
1000 Lakeside Ave.,
Cleveland, OH 44114-1183, USA
216-641-8580
Fax 216-696-6958,
www.ferro.com

Ferro Corporation/Polymer Additives Division,
1000 Wayside Rd.,
Cleveland, OH 44110,USA
216-641-8580x6755,
800-321-4236
Fax 216-486-6638,
www.ferro.com

Ferro Enamel Espanol,
A.C. 232, Castellon,
12080 Spain
34-52-2211
Fax 34-53-4051, www.ferro.com

Ferro-Plast Srl,
via Modigliani 2,
Milano, Italy
39-2-2137295 Fax 39-2-2139576,
www.ferro.com

Filtec Ltd.,
Suite A, Constance House,
Constance Ind. Est., Waterloo
Road,
Widnes, Cheshire, WA8 0QR, UK
44-151-495-1988
Fax 44-151-420-1407

Fina Chemicals,
52 Rue de l'Industrie,
B-1040 Brussels, Belgium
32-2-288 9132
Fax 32-2-288-3322

Fina plc,
Fina House, 1 Ashley Ave,
Epsom, Surrey, KT18 5AD, UK
44-1372-726226
Fax 44-1372-744520

Fine Agrochemicals,
3 The Bull Ring,
Worcester, WR2 5AA, UK
44-1905-748444
Fax 44-1905-748440

Fine Organics Ltd.,
Seal Sands,
Middlesbrough, Cleveland, TS2
1UB, UK
44-1642 546666
Fax 44-1642-546823

Finetex Inc.,
418 Falmouth Ave., PO Box 216,
Elmwood Park, NJ 07407, USA
201-797-4686
Fax 201-797-6558

Firestone Synthetic,
PO Box 26611,
381 W. Wilbeth Rd.,
Akron, OH 44319-0006, USA
216-379-7727, 800-282-0222
Fax 216-379-7875,
www.firesyn.com

Firestone Textiles Co.,
PO Box 486,
Woodstock, Ont., N4S 7Y9,
Canada
800-265-2237

Fisons Corp.,
755 Jefferson Rd,
Rochester, NY 14623, USA
716-475-9000, 800-334-6433

Fisons plc, Pharmaceuticals,
Weyside Park, Catteshall Lane,
Godalming, Surrey, GU7 1XE,
UK
44-1483-410210
Fax 44-1483-410220

Flexshield Chemical Mfg.,
128 E Campbell,
Chandler, AZ 85225, USA

Floridin Co.,
PO Box 510,
1101 N. Madison St.,
Quincy, FL 32351-0510, USA
904-627-7688, 800-228-1131
Fax 904-875-1757

Fluka,
www.sigma-aldrich.com

FMC Asia Pacific,
Sanbancho KS Bldg., 7th Fl.,
2 Sanbancho, Chiyoda-ku,
Tokyo 102, Japan
81-3-5210-9670
Fax 81-3-5210-9671,
www.fmc.com

FMC Corp. (UK) Ltd./P,
Tenax Rd.,Trafford Park,
Manchester, Lancs, M17 1WT,
UK
44-161-872-2323
Fax 44-161-873-3177,
www.fmc.com

FMC Corp. Canada,
625 Howe St., Suite 1200,
Vancouver, B.C., V6C 2T6,
Canada
604-685-6508
Fax 604-689-5917, www.fmc.com

FMC Corp./Ag Chem Group,
1735 Market St.,
Philadelphia, PA 19103, USA
215-299-6000
Fax 215-299-5999, www.fmc.com

FMC Corp./Chemical Products Group,
1735 Market St.,
Philadelphia, PA 19103, USA
215-299-6000, 800-346-5101
Fax 215-299-5999, www.fmc.com

FMC Corp./Food Ingredients,
1735 Market St.,
Philadelphia, PA 19103, USA
215-299-6000, 800-346-5101
Fax 215-299-6291, www.fmc.com

FMC Corp./Food Phosphates,
1735 Market St.,
Philadelphia, PA 19103, USA
215-299-6884
Fax 215-299-6887, www.fmc.com

FMC Corp./Lithium,
449 N. Cox Road, PO Box 3925,
Gastonia, NC 28054-0020, USA
704-868-5300, 800-362-2548
Fax 704-868-5370, www.fmc.com

FMC Corp./Marine Colloids,
1735 Market St.,
Philadelphia, PA 19103, USA
215-299-6242, 800-526-3649
Fax 215-299-6291, www.fmc.com

FMC Corp./Pharmaceuticals,
1735 Market St.,
Philadelphia, PA 19103, USA
215-299-6000, 800-362-3773
Fax 215-299-6821, www.fmc.com

FMC Corp./Process Additives,
1735 Market St.,
Philadelphia, PA 19103, USA
215-299-6121, 800-545-6532
www.fmc.com

FMC de Mexico S.A.
d,Arquimedes 130 piso 7,
Colonia Polanco, 11560 Mexico
DF, Mexico
525-280-7666
Fax 525-281-4576, www.fmc.com

FMC Europe SA,
Ave. Louise 480-B9,
1050 Brussels, Belgium
32-2-645-9211
Fax 32-2-646-4454,
www.fmc.com

Ford Smith & Co. Ltd,
Lyndean Industrial Estate,
Felixstowe Road, Abbey Wood,
London, SE2 9SG, UK
44-181-310-8127
Fax 44-181-310-9563

Fordath Engineering,
West Bromwich,W. Midlands, UK,

Foseco (FS) Ltd.,
Tamworth, Staffs, B78 3TL, UK
44-1827-289999
Fax 44-1827-250806,
www.foseco.com

Foseco Ltd./Metallurgy,
Tamworth, Staffs, B78 3TL, UK
www.foseco.com

Fothergill & Harvey,
PO Box 3, Church Street,
Littleborough, Lancs., OL15 8HG,
UK
44-1706-7015
Fax 44-1706-70576

Fothergill Cables Ltd.,
PO Box 3, Church Street,
Littleborough, Lancs.,OL15 8HG,
UK
44-1706-7015
Fax 44-1706-70576

Fothergill Composite,
Dunball Park, Dunball,
Bridgwater, Somerset, TA6 4TP,
UK

Fothergill Engineered Fibers,
PO Box 1,
Littleborough, Lancs., OL15 9QP,
UK
44-1706-72414

Fothergill Engineered Surfactants,
Long Causeway,
Leeds, LS9 0NY, UK

Fothergill Tygaflor Ltd.,
PO Box 2, Summit,
Littleborough, Lancs., OL5 0LT,
UK
44-1706-378837
Fax 44-1706-370203

Frank B. Ross Co.,
22 Halladay St.,
PO Box 4085,
Jersey City, NJ 07304-0085, USA
201-433-4512
Fax 201-332-3555

Franklin Industrial,
612 Tenth Ave. North,
Nashville, TN 37203, USA
615-259-4222, 800-872-3740
Fax 615-726-2693

Franklin Industrial,
821 Tilton Bridge Rd., S.E.,
Dalton, GA 30721, USA
404-277-3740
Fax 404-277-9827

Franklin Mineral Products,
PO Drawer 390,
Hartwell, GA 30643, USA
706-376-3174
Fax 706-376-3044

Franklin Oil Corp.,
Box 46030,
Cleveland, OH 44146-0030, USA
216-232-3000

Freeman Industries,
100 Marbledale Rd.,
P.O. Box 415,
Tuckahoe, NY 10707-0415, USA
914-961-2100
Fax 914-961-5793

Fudow Company, Ltd.,
4-11-26 Nishi Rokugo 4-chome,
Ohta-ku,Tokyo 144, Japan
81-3-3737-0611
Fax 81-3-3738-0554

Furon Co.,
386 Metacom Ave.,
Bristol, RI 02809, USA
401-253-2000, 800-336-3534
Fax 401-253-8211

Furon/CHR Div.,
407 East St.,
New Haven, CT 06509-9988,
USA
203-777-3631, 800-525-2523

Galderma Laboratories,
3000 Altamesa Blvd., Suite 300,
Fort Worth, TX 76133, USA
817-551-8664

Galen Ltd.,
19 Lower Sagoe Industrial
Estate,
Portadown, Craigavon,
Armagh, BT63 5QA, N. Ireland
01762-334974
Fax 01762-350206

Gard Corp.,
2727 Roe Lane,
Kansas City, KS 66103, USA
913-236-5000
Fax 913-432-8309

Garvey Chemical Corp,
PO Box 127,
1300 N 7th,
St Joseph, MO 64502, USA

Gattefosse Corp.,
372 Kinderkamack Rd.,
Westwood, NJ 07675, USA
201-358-1700
Fax 201-358-4050

Gattefosse S.A.,
36 Chemin de Genas,
BP 603,
69804 Saint Priest, C, France
33-78- 90-6311
Fax 33-78-90-4567

Gaylord Chemical Co.,
PO Box 1209,
106 Galeria Blvd.,
Slidell, LA 70459-1209, USA
504-639-5633, 800-426-6620
Fax 504-649-0068

Geeco,
Gore Road Industrial Estate,
New Milton, Hants, BH25 6SE,
UK
44-1425 614600
Fax 44-1425-619463

Geistlich Sohne AG,
CH-6110 Wolhusen, Switzerland
41-71-0333

General Chemical Canada,
201 City Centre Dr.,
Mississauga, Ont., L5B 3A3,
Canada
416-896-9595, 800-668-0433
Fax 416-276-6594,
www.generalchem.com

General Chemical Chemicals,
90 East Halsey Rd.,
Parsippany, NJ 07054-0373,
USA
201-515-0900, 800-631-8050
Fax 201-515-2468,
www.generalchem.com

General Color & Chemicals,
Valley St., P.O. Box 7,
Minerva, OH 44657, USA
216-868-4161
Fax 216-868-5880

General Electric Co.,
21800 Tungsten Rd,
Cleveland, OH 44117, USA
216-266-2451
Fax 216-266-3372, www.ge.com

General Electric Co.,
One Plastics Ave.,
Pittsfield, MA 01201, USA
413-448-4808, 800-845-0600
www.ge.com

General Electric Co.,
260 Hudson River Rd.,
Waterford, NY 12188, USA
518-237-3330, 800-255-8886
Fax 518-233-3931, www.ge.com

General Electric Europe,
Cyprusweg 2,
1044-AA Amsterdam, The
Netherlands
31-20-5806911
www.ge.com

General Electric Hong Kong,
15/F Convention Center,
No. 1 Harbor Rd.,
Wanchai, Hong Kong
852-5-8105616,
www.ge.com

General Electric Plastics,
Plasticslaan 1, PO Box 117,
4600 AC Bergen op Zoo, The
Netherlands
31-1640-32911
Fax 31-1640-43949,
www.ge.com

General Electric Plastics ABS Ltd.,
Bo'ness Road,
Grangemouth Central, FK3 9XF, Scotland UK
44-324-483490
Fax 44-324-471570,
www.ge.com

General Electric Plastics Japan Ltd.,
Shionogi Honcho Kyodo Bldg., 3-7-2, Nihonbashi-Honcho, Chuo-K,Tokyo 103, Japan
81-3-5695-4877
Fax 81-3-5695-4860,
www.ge.com

General Electric Plastics UK,
Old Hall Rd.,
Sale, Manchester, M33 2HG, UK
44-161-905-5000,
www.ge.com

General Electric Plastics-Canada,
2300 Meadowvale Blvd.,
Mississauga, Ont., L5N 5P9, Canada
905-858-5700, 800-668-4646
www.ge.com

General Electric, Components,
1975 Noble St.,
Cleveland, OH 44112, USA
216-266-2451
Fax 216-266-3372, www.ge.com

General Electric-Huntsman,
One Noryl Ave.,
Selkirk, NY 12158, USA
518-475-5734
Fax 518-475-5583, www.ge.com

Genesee Polymers Corp,
G-5251 Fenton Rd.,
PO Box 7047,
Flint MI, 48507-0047, USA
810-238-4966
Fax 810-767-3016,

Geoliquids, Inc.,
15 E. Palatine Rd., Suite 109,
Prospect Heights, IL 60070, USA
708-215-0938, 800-827-2411
Fax 708-215-9821

George Mann & Co., Inc.,
PO Box 9066,
Harborside Blvd.,
Providence, RI 02940-9066, USA
401-781-5600
Fax 401-941-0830

Georgia Gulf Corp.,
PO Box 105197,
400 Perimeter Center Terrace,
Atlanta, GA 30348, USA
504-389-2500

Georgia Gulf Corp./PVC,
PO Box 629,
Plaquemine, LA 70765-0629, USA
504-685-1200, 800-241-2673
Fax 504-685-1270

Georgia Gulf Sulfur,
P.O. Box 1165,
Valdosta, GA 31603, USA
912-244-0000
Fax 912-245-1664

Giulini Adolfomer Inc.,
Rua Ferreira Viana, 656,
Sao Paulo, 04761-010, Brazil
55-11-523-4877
Fax 55-11 247-0648

Giulini Chemie GmbH,
Giulini-strasse 2,
67029 Ludwigshafen, Germany
49-621 570 901

Givaudan Iberica S.A,Pl
d'en Batelle, Sant Celoni,
E-08470 Barcelona, Spain
34-3-867-0600
Fax 34-3-867-0319

Givaudan-Roure Corp.,
100 Delawanna Ave.,
Clifton, NJ 07014, USA
201-365-8277
Fax 201-777-9304

Glaxo Laboratories,
Glaxo House, Berkeley Avenue,
Greenford, Middlesex, UB6 0NN, UK
44-171-493-4060
Fax 44-181-966-8330

Glaxo Welcome plc,
Glaxo Wellcome House,
Berkeley Ave,
Greenford, Middlesex, UB6 0NN, UK
0171-493-4060
Fax 0181-966-8330,
www.glaxowellcome.co.uk

Glaxo Wellcome Inc.,
P.O. Box 13398,
Research Triangle Park, NC 27709, USA
919-248-2100
www.glaxowellcome.com

Glaxo Wellcome Inc. Canada,
7333 Mississauga Rd North,
Mississauga, Ont., L5N 6L4, Canada
905-819-3000
Fax 905-819-3099,
www.glaxowellcome.ca/

Glenwood Inc.,
83 N. Summit St,
PO Box 518,
Tenafly, NJ 07670, USA
201-569-0050

Goldschmidt AG,
Goldschmidtstrasse 100,
Postfach 101461,
45127 Essen, Germany
49-201-173-01
Fax 49-201-173-3000,
www.goldschmidt.com

Goldschmidt AG,
PO Box 1222, Randburg 2125,
Johannesburg/Transvaal,
Rep. of S. Africa
27-787-5622
Fax 27-787 5065,
www.goldschmidt.com

Goldschmidt AG,
Goldschmidtstrasse 100,
D-45014 Essen, Germany
49-201-173-2850
Fax 49-201-173-1990,
www.goldschmidt.com

Goldschmidt Canada,
2150 Winston Park Dr.,Unit 201,
Oakville, Ont., L6H 5V1, Canada
905-829-2233, 800-799-2483
Fax 905-829-2575,
www.goldschmidt.com

Goldschmidt Chemical,
914 E. Randolph Rd.,
PO Box 1299,
Hopewell, VA 23860, USA
804-541-8658, 800-446-1809
Fax 804-541-2783,
www.goldschmidt.com

Goldschmidt Industries,
941 Robinson Hwy.,
PO Box 279,
McDonald, PA 15057-0279, USA
412-796-1511, 800-426-7273
Fax 412-922-6657,
www.goldschmidt.com

Goldschmidt Industries,
Caixa Postal 106,
07111-970 Guarulhos S, Brazil
55-602-0888
Fax 55-602-1383,
www.goldschmidt.com

Goldschmidt Japan K.,
Rm. 1113, Shuwa Kioi-cho TBR B,
5-chome, Koji-machi,
Chiyoda-ku,Tokyo 102, Japan
81-3-234-2831
Fax 81-3-234-3019,
www.goldschmidt.com

Goldschmidt Ltd.,
The,Tego House, Chippenham Dr.,
Kingston, Milton Keynes, Bucks, MK10 0AE, UK
44-1908-582250
Fax 44-1908-582254,
www.goldschmidt.com

Goldschmidt S.A., N.,
Kapucijnenlaan 1,
1030 Brussels, Belgium
32-2-41-8750
Fax 32-2-41-5452,
www.goldschmidt.com

Goldsmith & Eggleton,
300 First St.,
Wadsworth, OH 44281-2084, USA
216-336-6616, 800-321-0954
Fax 216-334-4709

Goodyear Canada Inc.,
45 Raynes Ave.,
Bowmanville,Ont., Canada,
800-848-8266
www.goodyear.com

Goodyear Chemicals Europe,
14 Ave. des Tropiques,
Z.A. de Courtaboeuf 2,
91955 Les Ulis Cedex, France
33-1-69-29-27-00
Fax 33-1-69-29-27-01,
www.goodyear.com

Goodyear Int'l. Corp.,
Sankaido Bldg.,
1-9-13 Akasaka,
Minato-Ku,Tokyo 107, Japan
81-3-582-0926
Fax 81-3-582-1877,
www.goodyear.com

Goodyear Tire & Rubber,
1144-East Market St.,
Akron, OH 44316, USA
216-796-3845, 800-321-2385
www.goodyear.com

Goodyear Tire & Rubber,
1485 E. Archwood Ave.,
Akron, OH 44306-3299, USA
216-796-6400, 800-548-8107
Fax 216-796-2617,
www.goodyear.com

Grace Dearborn Ltd.,
Waterside Lane, Ditton,
Widnes, Cheshire, WA8 8UD, UK
44-151-424-5351
Fax 44-151-423-2722

Grain Processing Corp.,
1600 Oregon St.,
Muscatine, IA 52761, USA
319-264-4265
Fax 319-264-4289

Great Lakes Chemical,
PO Box 44,
Oil Sites Road,
Ellesmere Port,
South Wirral, L65 4GD, UK
44-151-356-8489
Fax 44-151-356-8490,
www.greatlakeschem.com

Great Lakes Chemical,
PO Box 2200,
One Great Lakes Blvd.,
W. Lafayette, IN 47906-0200, USA
317-497-6100, 800-621-9521,
Fax 317-497-6123,
www.greatlakeschem.com

Great Lakes Chemical,
LaVie Sakuragicho Bldg.,
5-26-3, Sakuragicho, Nishi-Ku,
Yokohama 220, Japan
81-45-212-9541
Fax 81-45-212-9539,
www.greatlakeschem.com

Great Lakes Europe,
5 rue de la Grand Ourse,
Cergy St.-Christophe,
F-95801 Cergy Pontois, France
33-1-34-41-6000
Fax 33-1-30-75-93-00,
www.greatlakeschem.com

Great Lakes/Asia Pacific,
1004 Printing House,
6 Duddell St.,
Central Hong Kong, China
852-2-537-4238
Fax 8522-973-0029,
www.greatlakeschem.com

Great Lakes-QO Chemical,
via Montecatini, 15,
20144-Milano, Italy
39-2-423-9471
Fax 39-2-483-00036,
www.greatlakeschem.com

Great Lakes-QO Chemical,
Industrieweg 12,
Haven 391,
B-2030 Antwerp, Belgium
32-3-541-2165
Fax 32-3-541-6503,
www.greatlakeschem.com

Greene, Tweed & Co.,
Box 305,
Kulpsville, PA 19443, USA
215-256-9521
Fax 215-256-0189

Griffin Corp.,
PO Box 1847, Rock Ford Rd.,
Valdosta, GA 31603-1847, USA
912-242-8635, 800-237-1854
Fax 912-244-5978

Griffin Corp. Asia Pacific,
Ste. 206, Hawaii Kai,
333 Keahole St.,
Honolulu, HI 96825, USA
808-395-5669
Fax 808-395-4121

Griffin Europe S.A.,
49 Ave. de La Gare,
200 Neuchatel, Switzerland
41-382-58600
Fax 41-382-47128,

Griffith Laboratories,
1 Griffith St.,
Scoresby,Vic., 3179, Australia
61-3-763-5300
Fax 61-3-763-4887

Griffith Laboratories,
Cotes Park,
Somercotes, Derbyshire, DE5 4NN, UK
44-1773-832171
Fax 44-1773-835294

Grindsted Products Denmark,
Edwin Rahrs Vej 38,
DK-8220 Brabrand, Denmark
45-86-25-3366
Fax 45-6-25-1077

Grünau Illertissen GmbH,
Postfach 1063,
Robert-Hansen-Strasse 1,
D-89251 Illertissen, Germany
49-7303-13-706
Fax 49-7303-13-203,
www.gruenau-illertissen.de

Guardian Chemical,
230 Marcus Blvd,
PO Box 2500,
Smithtown, NY 11787, USA
516-273-0900, 800-645-5566
Fax 516-273-0858

Guthrie Latex, Inc.,
7400 N. Oracle, Suite 330,
Tucson, AZ 85712, USA
520-742-3087
Fax 520-575-0511

H.A. Smith & Sons Ltd.,
Torrington Ave, Till Hill,
Coventry, West Midlands, CV4 9GX, UK
44-1203-461111
Fax 44-1203-465325

H.B. Fuller Co.,
3530 Lexington Ave. North,
St. Paul, MN 55126-8076, USA
612-481-1588, 800-468-6358
Fax 612-481-1828

H.B. Fuller Japan Co,
700, Matsushima-cho,
Hamamatsu,
Shizuoka 430, Japan
81-53-425-0751
Fax 81-53-426-1672

H.B. Fuller UK Ltd.,
Amber Business Centre,Greenhill Lane,
Leabrooks, Derbyshire, DE55 4BR, UK
44-1773-608877
Fax 44-1773-602673

Haarmann & Reimer Australia,
9 Garling Rd.,
Marayong, NSW, 2148, Australia
61-2-671-3444
Fax 61-2-621 8086

Haarmann & Reimer Co.,
PO Box 175,
70 Diamond Road,
Springfield, NJ 07081, USA
201-467-5600,800-422-1559
Fax 201-467-3514

Haarmann & Reimer Co.,
1127 Myrtle St., PO Box 932,
Elkhart, IN 46515, USA
219-262-7874, 800-348-7414
Fax 219-262-6747

Haarmann & Reimer GmbH,
Postfach 1253,
Rumohrtalstrasse 1,
37603 Holzminden, Germany
49-5531-900
Fax 49-5531-901649

Haarmann & Reimer Ltd.,
Fieldhouse Lane,
Marlow, Bucks., SL7 1NA, UK
44-1628-472051
Fax 44-1628 890795

Haarmann & Reimer Ltd.,
Denison Rd,
Selby, North Yorkshire, YO8 8EF, UK
44-1757-703691
Fax 44-1757-701468

Haifa Chemicals Ltd.,
PO Box 1809,
31018 Haifa, Israel
972-4-469611
Fax 972-4-457849,

Hallam Polymer Engineering,
Callywhite Lane, Dronfield,
Sheffield, South Yorkshire, S18 6XR, UK
44-1246-415511
Fax 44-1246-414818

Halstab,
3100 Michigan St.,
Hammond, IN 46323, USA
219-844-3980
Fax 219-844-7287,
www.halstab.com

Haltermann GmbH,
Ferdinand Strasse 55-57,
D-20095 Hamburg, Germany
49-40-333-8280
Fax 49-40-333-8214

Haltermann Ltd.,
16717 Jacintopart Blvd.,
Houston, TX 77015, USA
713-452-5951
Fax 713-457-1128

Haltermann N.V.,
Kentenislaan 3,
B-9130 Kallo, Belgium
32-3-750-0211
Fax 32-3-775-0261

Harcros Chemicals Inc.,
PO Box 2930,
5200 Speaker Rd.,
Kansas City, KS 66110-2930, USA
913-621-7749
Fax 913-621-7746,
www.harcoschem.com

Harcros Durham Chemicals,
Birtley,
Chester-le-Street, Co. Durham, DH3 1QX, UK
44-1914-102361
Fax 44-1914-106005,
www.harcoschem.com

Harcros Pigments Inc,
11 Executive Dr., Suite 1,
Fairview Heights, IL 62208, USA
618-628-2300, 800-323-7796
Fax 618-628-1029,
www.harcospigments.com

Hardman Inc.,
600 Cortlandt St.,
Belleville, NJ 07109, USA
201-751-3000
Fax 201-751-8407

Harlow Chemical Co.,
Central Road, Templefields,
Harlow, Essex, CM20 2AH,UK
44-1279-436211
Fax 44-1279-444025

Harshaw Chemical Co.,
1945 E. 97th St.,
Cleveland, OH 44106, USA

Hart Chemicals Ltd.,
256 Victoria Rd. South,
Guelph, Ont., N1H 6K8, Canada
519-824-3280
Fax 519-824-0755

Hart Metals Inc.,
P.O. Box 428,
Route 209 N,
Tamaqua, PA 18252, USA
717-668-0001
Fax 717-668-6526

Hart Products Corp.,
173 Sussex St.,
Jersey City, NJ 07302, USA
201-433-6632
Fax 201-435-7268

Harwick Chemical Corp.,
60 S. Seiberling St.,
PO Box 9360,
Akron, OH 44305-0360, USA
216-798-9300
Fax 216-798-0214

Manufacturers and Suppliers

Haynes Int'l. Inc.,
1020 West Park Ave.,
PO Box 9013,
Kokomo, IN 46904-9013, USA
800-342-9637, 800-354-0806,
Fax 317-456-6905

Haynes Int'l. Ltd.,
PO Box 10, Parkhouse St.,
Openshaw, Manchester, M11
2ER, UK
44-161-230-7777
Fax 44-161-223-2412

Haysite Reinforced Plastics,
Sales & Marketing Division,
5599 New Perry Highway,
Erie, PA 16509, USA
814-868-3691
Fax 814-864-7803

Henkel (Ireland) Ltd.,
Western Industrial Estate,
Naas Road,
Dublin 12, Ireland
353-1-450-5622
Fax 353-1-450-3649,
www.henkel.com

Henkel Argentina S.A,
Carabelas 2398,
1870 Avellaned,
Casilla de Correo 3496,
AR-1000 Buenos Aires,
Argentina
54-1-204-2056
Fax 54-1-205-3360,
www.henkel.com

Henkel Australia Pty,
1 Clyde St.,
Silverwater, NSW, 2141,
Australia
61-2-748-4355
Fax 61-2-748-3863,
www.henkel.com

Henkel Canada Ltd.,
2290 Argentia Rd.,
Mississauga, Ont., L5N 6H9,
Canada
416-542-7588, 800-668-6023
Fax 416-542-7566,
www.henkel.com

Henkel Chimica SpA,
via Scalabrini 24,
22073 Fino Mornasco, Italy
39-31-88-42-01
Fax 39-31-88-43-60,
www.henkel.com

Henkel Coordination,
16 Avenue de Port,
1080 Bruxelles Belgium
32-423-17-11
Fax 32-428-34-67,
www.henkel.com

Henkel Corp./Coating,
300 Brookside Ave.,
Ambler, PA 19002-3498, USA
215-628-1000, 800-445-2207,
www.henkel.com

Henkel Corp./Cospha,
300 Brookside Ave.,
Ambler, PA 19002, USA
215-628-1476, 800-955-1456,
Fax 215-628-1450,
www.henkel.com

Henkel Corp./Emery,
5051 Estecreek Rd.,
Cincinnati, OH 45232-1446, USA
513-530-7300, 800-543-7370
Fax 513-530-7581,
www.henkel.com

Henkel Corp./Emery Japan,
PO Box 191, World Trade
Center,
2-4-1 Hamamatsu-cho,
Minato-Ku,Tokyo 105, Japan
83-43-55-6112
www.henkel.com

Henkel Corp./Emery OPG,
3300 Westinghouse Blvd.,
Charlotte, NC 28217, USA
800-634-2436,
www.henkel.com

Henkel Corp./Extraction Tech.,
2430 N. Huachuca Dr.,
Tucson, AZ 85745-1273, USA
520-622-8891, 800-328-6198
Fax 520-624-0912,
www.henkel.com

Henkel Corp./Fine Chemicals,
5325 South 9th Ave.,
La Grange, IL 60525-3602, USA
708-579-6150, 800-328-6199
Fax 708-579-6152,
www.henkel.com

Henkel Corp./Functional Products,
300 Brookside Ave.,
Ambler, PA 19002, USA
215-628-1583, 800-654-7588
Fax 215-628-1155,
www.henkel.com

Henkel Corp./Organic,
300 Brookside Ave.,
Ambler, PA 19002, USA
215-628-1441, 800-634-2436
Fax 215-628-1200,
www.henkel.com

Henkel Corp./Process & Polymers,
300 Brookside Ave.,
Ambler, PA 19002, USA
215-628-1000, 800-654-7588
Fax 215-628-1457,
www.henkel.com

Henkel Corp./Textiles,
11709 Fruehauf Dr.,
Charlotte, NC 28241, USA
704-587-3807, 800-634-2436
Fax 704-587-3838,
www.henkel.com

Henkel France S.A.,
BP 309, 150 rue Gallieni,
F-92102 Boulogne Bill, France
33-46-84-90-00
Fax 33-46-84-90-90,
www.henkel.com

Henkel Hakusui Corp.,
Nissho Iwai Bldg.,
5-8, Imabashi 2-chome, Chuo-ku,
Osaka 541, Japan
81-6-202-7347
Fax 81-6-223-1576,
www.henkel.com

Henkel Iberica s.a.
Pasaje Mariner no. 9,
08025 Barcelona, Spain
34-3-290-4763
Fax 34-3-290-4879,
www.henkel.com

Henkel Iberica s.a.
Sector E C/42, Zona Franca,
08040 Barcelona, Spain
34-3-290-4850
Fax 34-3-290-4878,
www.henkel.com

Henkel KGaA/COK-Coatings,
D-40191 Düsseldorf, Germany
49-211-797-2022
www.henkel.com

Henkel KgaA/Cospha,
Postfach 101100,
D-40191 Düsseldorf, Germany
49-211-797-0
Fax 49-211-798-7696,
www.henkel.com

Henkel KgaA/Dehydag,
Postfach 101100,
D-40191 Düsseldorf, Germany
49-211-797-4561
Fax 49-211-798-8558,
www.henkel.com

Henkel Ltd.,
Henkel House, 292-308
Southbury Road,
Enfield, Middlesex, EN1 1TS, UK
44-181-804-3343
Fax 44-181-443-2777,
www.henkel.com

Henkel Ltd./Adhesive,
292-308 Southbury Road,
Enfield, Middlesex, EN1 1TS, UK
44-181-804-3343
Fax 44-181-443-2777,
www.henkel.com

Henkel Mexicana S.A.,
Calz. de la Viga S/N,Fracc.
Los Laureles en Tulpetlac,
Ecatepec de Morelos, Mexico
52-5-787-1899
Fax 52-5-729-9804,
www.henkel.com

Henkel Nopco A/S,
Postboks 2040,
Stromse,
3003 Drammen, Norway
47-3220-2200
Fax 47-3288-0701,
www.henkel.com

Henkel Organics,
Henkel House, 292-308
Southbury Rd.,
Enfield, Middlesex, EN1 1TS, UK
44-181-804-3343
Fax 44-181-443-4392,
www.henkel.com

Henkel S.A. Industries,
Avenida das Nacoes Unidas
10.989,
CEP 04578
Sao Paulo, Brazil
55-11-828-2340
Fax 55-11-828-2326,
www.henkel.com

Henkel South Africa,
PO Box 3933,
Johannesburg,
2000, Rep. of S. Africa
27-11-864-4950
Fax 27-11-864-7888,
www.henkel.com

Herbert Laboratories,
2525 Dupont Drive,
Irvine, CA 92713, USA
714-252-4500

Hercules B.V.,
8 Veraartlaan,
NL-2288GM Rijswijk, The
Netherlands
31-70-150-000
Fax 31-70-398-9893,
www.herc.com

Hercules B.V./Aqualo,
Postbus 5832,
NL-2280 HV Rijswijk, The
Netherlands
31-70-315-0226
Fax 31-70-390-7560,
www.herc.com

Hercules Europe S.A.,
Avenue de Tervuren 300,
B-1150 Brussels, Belgium
32-2-761-5511
www.herc.com

Hercules Food Ingredients,
1313 North Market St.,
PO Box 8740,
Wilmington, DE 19899-8740,
USA
302-594-5000, 800-654-6529
Fax 302-594-6660,
www.herc.com

Hercules Inc.,
Hercules Plaza-6205SW,
Wilmington, DE 19894-0001,
USA
302-594-5000, 800-247-4372
Fax 302-594-5400,
www.herc.com

Hercules Inc./Aqualon,
1313 North Market St.,
Wilmington, DE 19894-0001,
USA
302-594-5000, 800-345-8104
Fax 302-594-6660,
www.herc.com

Hercules Inc./Carbon,
PO Box 98,
Magna, UT 84044, USA
www.herc.com

Hercules Inc./Paper,
500 Hercules Rd.,CSD Bldg.
8145,
Wilmington, DE 18908-1599,
USA
302-995-4584
Fax 302-995-4077,
www.herc.com

Hercules Inc./PFW Ar,
Hercules Plaza,
1313 North Market St,
Wilmington, DE 19894-001, USA
302/594-5000
www.herc.com

Hercules Ltd.,
31 London Road,
Reigate, Surrey, RH2 9YA, UK
44-1737-242434
Fax 44-1737-224288,
www.herc.com

Heresite Protective,
PO Box 249,
822 S. 14th St.,
Manitowoc, WI 54220, USA
414-684-6646

Hermetite Products Ltd.,
Tavistock Road,
West Drayton, UB7 7RA, UK

Heterene Chemical Co,
PO Box 247,
295 Vreeland Ave.,
Paterson, NJ 07543, USA
201-278-2000
Fax 201-278-7512

Hexcel Corp./Chemicals,
215 N. Centennial St.,
Zeeland, MI 49464, USA
616-772-2193
Fax 616-772-7344

Hexcel Corp./Resins,
20701 Nordhoff St.,
Chatsworth, CA 91311, USA
213-322-8050, 800-423-5451
Fax 818-709-0399

Hexcel Corp./Trevarn,
5794 W. Las Positas Blvd.,
Pleasanton, CA 94588, USA
510-847-9500, 800-444-3923
Fax 510-734-9688

Hi Temp Lubricants Inc.,
7019 Corporate Way,
Dayton, OH 45459, USA

Hickson & Welch Ltd.,
Wheldon Road,
Castleford, West Yorkshire,
WF10 2JT, UK
44-1977-556565
Fax 44-1977-518058

Hickson Danchem Corp,
1975 Richmond Blvd.,
PO Box 400,
Danville, VA 24540, USA
804-797-8105, 800-797-8100
Fax 804-799-2814

Hickson Manro Ltd.,
Bridge St.,
Stalybridge, Cheshire, SK15
1PH, UK
44-161-338-5511
Fax 44-161-303-2991

High Point Chemical,
PO Box 2316,
243 Woodbine St.,
High Point, NC 27261, USA
910-884-2214, 800-727-2214
Fax 910-884-5039

Himont U.S.A., Inc.,
Three Little Falls Centre,
2801 Centerville Rd., POB 154,
Wilmington, DE 19850-5439,
USA
302-996-6000, 800-545-7719
Fax 302-996-5587

Hitox Corp. of America,
PO Box 2544,
Corpus Christi, TX 78403, USA
512-882-5175
Fax 512-882-6948

Hodgson Chemicals Ltd.,
Chantry Lane, PO Box 7,
Beverley,North Humberside,
HU17 0NN, UK
44-1482-881133
Fax 44-1482-871888

Hoechst (UK) Ltd./Ag,
East Winch Hall, East Winch,
King's Lynn, Norfolk, PE32 1HN,
UK
44-1553-841581
Fax 44-1553-841090,
www.hcc.com/

Hoechst (UK) Ltd./Films,
Hoechst House, Salisbury Rd.,
Hounslow, Middlesex, TW4 6JH,
UK
44-181-570-7712
Fax 44-181-572-4854,
www.hcc.com/

Hoechst (UK) Ltd./Polymers,
Walton Manor, Walton,
Milton Keynes, Bucks, MK7 7A3,
UK
44-1908-665050
Fax 44-1908-680516,
www.hcc.com/

Hoechst AG,
Postfach 800320,
D-65926 Frankfurt/M, Germany
49-69-305-5753
Fax 49-69-316700,
www.hcc.com/

Hoechst AG
Entwicklung TH 1,
D-65926 Frankfurt/M, Germany
49-69-305-2298
Fax 49-69-318435,
www.hcc.com/

Hoechst Canada Inc.,
800 Blvd Rene Levesque W,
23rd Floor,
Montreal, PQ, H3B 1Z1, Canada
514-871-5511
www.hcc.com/

Hoechst Chemical Group,
1601 West LBJ Freeway,
Dallas, TX 75234, USA
214-277-4000, 800-235-2637
Fax 214-277-4920,
www.hcc.com/

Hoechst Chemicals (UK),
Hoechst House, Salisbury Rd.,
Hounslow, Middlesex, TW4 6JH,
UK
44-181-570-7712
Fax 44-181-577-1854,
www.hcc.com/

Hoechst Far East Ltd.,
801 Hong Kong Club Bldg.,
3A Chater Rd.,
Central Hong Kong, China
www.hcc.com/

Hoechst Japan Ltd.,
New Hoechst Bldg.,
10-16, Akas,Minato-ku,
Tokyo 107, Japan
81-3-3479-5118
Fax 81-3-3479-6715,
www.hcc.com/

Hoechst Mitsubishi Kase,
Hoechst Bldg.,
10-33, Akasaka, Minato-ku,
Tokyo 107, Japan
81-3-3582-8452
Fax 81-3-3582-2375,
www.hcc.com/

Hoechst Schering Agr.
P.O. Box 27 06 54,
D-13476 Berlin, Germany
49-30-4390-8-0
Fax 49-30-4390-8222,
www.hcc.com/

Hoechst/Advanced Materials,
90 Morris Ave.,
Summit, NJ 07901, USA
908-598-4000, 800-526-4960
Fax 908-598-4330,
www.hcc.com/

Hoechst/Bulk Pharmacueticals,
1601 West LBJ Freeway,
PO Box 819005,
Dallas, TX 75381-9005, USA
214-277-4783
Fax 214-277-3858,
www.hcc.com/

Hoechst/Engineering,
90 Morris Ave.,
Summit, NJ 07901, USA
201-635-2600
Fax 201-635-4300,
www.hcc.com/

Hoechst/Fine Chemicals,
5200 77 Center Dr.,
Charlotte, NC 28217, USA
704-559-6000, 800-242-6222
x6183
Fax 704-559-6153,
www.hcc.com/

Hoechst/Int'l. Headquarters,
PO Box 2500, Rt. 202-206
North Somerville, NJ
08876-1258, USA
908-231-2000, 800-235-2637
www.hcc.com/

Hoechst/Polymer Additives,
5200-77 Center Dr.,
PO Box 1026,
Charlotte, NC 28201-1026, USA
704-559-6038
Fax 704-559-6780,
www.hcc.com/

Hoechst/Specialty Chemicals,
5200 77 Center Dr.,
PO Box 1026,
Charlotte, NC 28217, USA
704-559-6000, 800-365-2436
Fax 704-559-6342,
www.hcc.com/

Hoechst/Surfactants,
5200 77 Center Dr.,
Charlotte, NC 28217, USA
704-599-4000, 800-255-6189
Fax 704-559-6323,
www.hcc.com/

Hoechst/Waxes, Lubricants,
Route 202-206,
PO Box 2500,
Somerville, NJ 08876-1258, USA
908-704-7043
Fax 908-704-7059,
www.hcc.com/

Hoechst-Roussel Pharmaceuticals,
Route 202-206 North,
P.O. Box 2500,
Somerville, NJ 08876-1258, USA
201-231-2000, 800-451-4455
www.hcc.com/

Hoeganaes Corp.,
Sub Interlake,
River Rd. & Taylors Lane,
Riverton, NJ 08077, USA
609-829-2220
Fax 609-786-2574

Hoffmann Mineral,
Münchenerstrasse 75, PO Box 1460,
D-86633 Neuburg, Germany
49-84-31-53-0
Fax 49-84-31-53-330,

Hoffmann-LaRoche Inc,
340 Kingsland St.,
Nutley, NJ 07110, USA
201-909-8332, 800-526-0189
Fax 201-909-8414

Hoffmann-LaRoche S.A,
Grenzacherstrasse 124,
CH-4002 Basle, Switzerland
41-61-688 1111
Fax 41-61-688 6590

Holliday Dyes & Chem,
PO Box B22, Leeds Road,
Huddersfield, West Yorkshire,
HD2 1UH, UK
44-1484-421841
Fax 44-1484-515328

Holt Lloyd Corp.,
4647 Hugh Howell Road,
Tucker, GA 30084, USA

Holt Lloyd Ltd.,
Lloyds House, Alderley Rd,
Wilmslow, Cheshire, SK9 1QT,
UK
44-1625-526838
Fax 44-1625-526962

Horlicks,
New Horizon Court,
Brentford, Middlesex, TW8 9EP,
UK
44-181 975 2000

Hormel Foods Corp.,
1 Hormel Place,
Austin, MN 55912-3680, USA
507-437-5676
Fax 507-437-5120

Horn's Crop Service,
P O Box 326,
Bellevue, OH 44811, USA

Hortichem Ltd.,
1 Edison Road,
Churchfields Industrial Estate,
Salisbury,Wilts., SP2 7NU, UK
44-1722-20133
Fax 44-1722-26799

Hough, Hoseason & Co,
20-22 Chapel Street,
Levenshulme,
Manchester, M19 3PT, UK
44-161-224 3271
Fax 44-161-257-2076

Howlett Adhesives Ltd.,
Horsley Road, Off Kingsthorpe,
Northampton, Northants, NN2 6LL, UK
44-1604-712977
Fax 44-1604-791471

Hubron Ltd.,
Albion St. Works, Failsworth,
Manchester, M35 0FP, UK
44-161-681-2691
Fax 44-161-683-4658

Hüls AG,
Postfach 1320,
D-45764 Marl, Germany
49-2365-49-1
Fax 49-2365-49-2000,
www.huls.com

Hüls AG/Troisdorf,
PO Box 1347,
D-53839 Troisdorf, Germany
49-2241-85-4321
Fax 49-2241-85-4319,
www.huls.com

Hüls America Inc.,
220 Davidson Ave.,
Somerset, NJ 08873, USA
908-560-6345, 800-631-5275
www.huls.com

Hüls America Inc.,
80 Centennial Ave.,
PO Box 456,
Piscataway, NJ 08855-0456,
USA
908-980-6984
Fax 908-980-6970,
www.huls.com

Hüls America Inc.,
235 Orenda Rd.,
Brampton, Ont., L6T 1E6,
Canada
905-451-3810
Fax 905-451-4469,
www.huls.com

Huntington Laboratories,
970 East Tipton St.,
Huntington, IN 46750, USA
219-356-8100, 800-537-5724
Fax 219-356-6485

Huntsman Corp.,
3040 Post Oak Blvd.,
Houston, TX 77056, USA
713-235-6000, 800-231-3107
Fax 800-831-1782,
www.huntsman.com

Huntsman Corp. Belgium,
Woluwe Office Garden,
Woluwedal 26,
B-1932 Zaventem, Belgium
32-2-718-01-20
Fax 32-2-718-02-11,
www.huntsman.com

Huntsman de Brasil,
R. Heloisa Pamplona 628,
Sao Caetano Do Sul,
SP 09520, Sao Paulo, Brazil
55-11-442-9264
Fax 55-11-441-7216,
www.huntsman.com

Huntsman Expandable,
5100 Bainbridge Blvd.,
Chesapeake, VA 23320, USA
804-494-2500
Fax 804-494-2770,
www.huntsman.com

Huntsman International,
350 Orchard Rd.,
#11-07/10 Shaw House,
Singapore 0922,
65-730-0288
Fax 65-730-0222,
www.huntsman.com

Huntsman International Trading,
Baumwall 5,
D-20459 Hamburg, Germany
49-40-37670-0
Fax 49-40-37282-9,
www.huntsman.com

Hyland Therapeutics,
444-W Gelnoaks Blvd,
Glendale, CA 91202, USA

Hynson Westcott & Dunning,
Charles & Chase Sts,
Baltimore, MD 21201, USA

Hyperlast,
Station Road, Birch Vale,
Stockport, Cheshire, SK12 5BR,
UK
44-1663-746518
Fax 44-1663-746605

ICI Acrylics Inc.,
305 Water St.,
Newport, DE 19804-2410, USA
302-999-6200
Fax 302-999-6232, www.ici.com

ICI Acrylics, Europe,
Lincoln House,
137-143 Hammersmith Rd.,
London, W14 0QL, UK
44-171-331-7100
Fax 44-171-331-7110,
www.ici.com

ICI Agrochemicals,
Fernhurst,
Haslemere, Surrey, GU27 3JE,
UK
44-1428-644061
Fax 44-1428-652922,
www.ici.com

ICI Americas, Inc.,
PO Box 15391,
Concord Plaza,
3411 Silverside R,
Wilmington, DE 19850, USA
302-887-3000, 800-441-7780
Fax 302-887-2972, www.ici.com

ICI Atkemix Inc.,
PO Box 1085,
70 Market St.,
Brantford, Ont., N3T 5T2,
Canada
519-756-6181
Fax 519-758-8140, www.ici.com

ICI Australia Ltd.,
1 Nicholson St.,
Melbourne, Vic., 3000, Australia
61-3-9665-7111
Fax 61-3-9665-7937,
www.ici.com

ICI Belgium nv/sa,
Everslaan 45,
B-3078 Everberg, Belgium
32-2-758 92 11
Fax 32-2-759 77 22, www.ici.com

Manufacturers and Suppliers

ICI Canada Inc.,
90 Sheppard Ave. East,
PO Box 200, Station A,
North York, Ont., M2N 6H2,
Canada
416-229-7000
www.ici.com

ICI Chemicals & Polymers,
475 Creamery Way,
Exton, PA 19341, USA
610-363-4737, 800-ICI-PTFE,
Fax 610-363-4748, www.ici.com

ICI Chemicals & Polymers,
PO Box 14, The Heath,
Runcorn, Cheshire, WA7 4QF,
UK
44-1928-514444
Fax 44-1928-515555,
www.ici.com

ICI China Ltd.,
33/F Dorset House,
Taikoo Place, 979 King's Rd.,
Quarry Bay, Hong Kong, China
852-2968-2828
Fax 852-2968-1001, www.ici.com

ICI Chlor-Chemicals,
PO Box 14, The Heath,
Runcorn, Cheshire, WA7 4QG,
UK
44-1928-514444
Fax 44-1928-580504,
www.ici.com

ICI Deutsche GmbH,
Postfach 500728,
Emil-von-Behring-Strasse 2,
60439 Frankfurt/M, Germany
49-69-5801-00
Fax 49-69-5801-234,
www.ici.com

ICI Films,
Cedar Creek Rd.,
PO Box 630,
Fayetteville, NC 28302, USA
910-433-8200
Fax 910-323-5012, www.ici.com

ICI Films (Belgium),
Everslaan 45,
B-3078 Everberg, Belgium
32-2-758-92-11
Fax 32-2-758- 91-84,
www.ici.com

ICI Fluorochemicals, ICI Klea,
PO Box 13, The Heath,
Runcorn, Cheshire, WA7 4QG,
UK
44-1928-514444
Fax 44-1928-511418,
www.ici.com

ICI Fluoropolymers UK,
Hillhouse International,
PO Box 4, Thornton, Cleveleys,
Blackpool, Lancs., FY5 4QD, UK
44-1253-861444
Fax 44-1253-861950,
www.ici.com

ICI Forest Products,
31st Floor, 630 René Levesque,
Montreal,PQ, H3B 1S6, Canada
514-397-6100
Fax 514-397-6109, www.ici.com

ICI France SA,
196 Rue Houdan,
92330 Sceaux, France
33-41-13-32-32
Fax 33-41-13-32-90, www.ici.com

ICI Garden Products,
Fernhurst,
Haslemere, Surrey, GU27 3JE,
UK
44-1428-645454
Fax 44-1428-657222,
www.ici.com

ICI Group Headquarters,
Imperial Chemical House,
9 Millbank,
London, SW1P 3JF, UK
44-171-834-4444
Fax 44-171-834-2042,
www.ici.com

ICI Holland BV,
Rozenburg Works,
Merseyweg 10,
3197 KG Rotterdam/Bot, The
Netherlands
31-181-299111
Fax 31-181-293900, www.ici.com

ICI International Ltd.,
Ul. Usacheva 35 Korpus 1,
119048 Moscow, Russia
7-95-2455970
Fax 95-2455017, www.ici.com

ICI Japan Ltd.,
13th Fl., NYK Tennoz Bldg.,
2-20 Higashi-Shinagawa,
2-chome, Tokyo 140, Japan
81-3-5462-8415
Fax 81-3-5462-8647,
www.ici.com

ICI Katalco,
PO Box 1,
Billingham,
Cleveland, TS23 1LB, UK
44-1642-553601
Fax 44-1642-522542,
www.ici.com

ICI Katalco,
2 Transam Plaza Dr., Suite 230,
Oak Brook Terrace, IL 60181,
USA
708-268-6300
Fax 708-268-9797, www.ici.com

ICI Norden AB,
Box 184,
S-401 23 Gøteborg, Sweden
46-31-773-7000
Fax 46-31-773-7075,
www.ici.com

ICI Nutrition,
Alexander House, Crown Gate,
Runcorn, Cheshire, WA7 2UP,
UK
44-1928-793090
Fax 44-1928-716997,
www.ici.com

ICI Paints,
Wexham Rd.,
Slough, Berks., SL2 5DS, UK
44-1753-550000
Fax 44-1753-578218,
www.ici.com

ICI Paints,
The Glidden, 925 Euclid Ave.,
Cleveland, OH 44115-1487, USA
216-344-8000
Fax 216-344-8900, www.ici.com

ICI Petrochemicals & Fertilizers,
PO Box 90, Wilton,
Middlesbrough, Cleveland, TS90
8JE, UK
44-1642-454144
Fax 44-1642-432444,
www.ici.com

ICI Polyester,
PO Box 90, Wilton,
Middlesbrough, Cleveland, TS90
8JE, UK
44-1642-454144
Fax 44-1642-432444,
www.ici.com

ICI Polymer Additives,
PO Box 751,
Wilmington, DE 19897, USA
302-886-3564, 800-456-3669 x
3564
Fax 302-886-5267, www.ici.com

ICI Polyurethanes Belgium,
Everslaan 45,
B-3078 Everberg, Belgium
32-2-758-92-11
Fax 32-2-758-97-23, www.ici.com

ICI Polyurethanes Group,
286 Mantua Grove Rd.,
West Deptford, NJ 08066-1732,
USA
609-423-8300, 800-257-5547
Fax 609-423-8601, www.ici.com

ICI Polyurethanes UK,
Hitchen Lane,
Shepton Mallet, BA4 5TZ, UK
44-01749-343061
Fax 44-01749-346283,
www.ici.com

ICI Polyurethanes/Canada,
2795 Slough St.,
Mississauga, Ont., L4T 1G2,
Canada
905-678-9150
Fax 905-678-9350, www.ici.com

ICI Resins US.
730 Main St.,
Wilmington, MA, 01887-0677,
USA
800-225-0947
Fax 508-657-7978, www.ici.com

ICI Specialty Chemicals,
Concord Pike & New Murphy Rd.,
Wilmington, DE 19897, USA
302-886-3000, 800-822-8215
Fax 302-886-2972, www.ici.com

ICI Surfactants (Australia),
Newsom St.,
Ascot Vale, Vic., 3032, Australia
61-3-9272-5355
Fax 61-3-9272-5353,
www.ici.com

ICI Surfactants (Belgium),
Everslaan 45,
B-3078 Everberg, Belgium
32-2-758-92 11
Fax 32-2-758-96 86, www.ici.com

ICI Surfactants America,
Delaware Corporate Center 1,
1 Righter Pkwy.,
Wilmington, DE 19803, USA
302-887-3000, 800-822-8215
Fax 302-887-3525, www.ici.com

ICN Nutritional Biochemicals,
26201 Miles Road,
Cleveland, OH 44128, USA

ICN Pharmaceuticals,
ICN Plaza,
3300 Hyland Ave,
Costa Mesa, CA 92626, USA
714-545-0100, 800-556-1937

Ideal Manufacturing,
Atlas House, Burton Road,
Finedon, Wellingborough,
Northants., NN9 5HX, UK
44-1933-681616
Fax 44-1933-681042

IGI,
POB 383,
85 Old Eagle School Rd.,
Wayne, PA 19087, USA
610-687-9030
Fax 610-254-8548
www.igiwax.com

IGI Boler, Inc.,
85 Old Eagle School Rd.,
Wayne, PA 19087, USA
215-687-9030, 800-852-6537
www.igiwax.com

Manufacturers and Suppliers

Imperial Chemical Ind. Plc,
PO Box 6, Shire Park,
Bessemer Rd.,
Welwyn Garden City, Herts, AL7 1HD, UK
44-1707-323400
Fax 44-1707-337332,
www.ici.com

Imperial Oil, Chemical,
111 St. Clair Ave. West,
Toronto, Ont., M5W 1K3, Canada
416-968-4046

Incitec Ltd.,
PO Box 140, Morningside,
Queensland, 4170, Australia
61-7-867-9300
Fax 61-7-867-9310

Inco Alloys Int'l. Inc,
3200 Riverside Dr.,
Huntington, WV 25720-1771, USA
304-526-5100, 800-344-INCO
Fax 304-526-5643

Inco Europe Ltd.,
1-3 Grosvenor Place,
London, SW1X 7EA, UK
44-171-235-2040
Fax 44-171-235-4358

Indspec Chemical Corp.,
411 Seventh Ave., Suite 300,
Pittsburgh, PA 15219, USA
412-765-1200
Fax 412-765-0439

Indspec Chemical Corp.,
Gebouw de Goudsesingel, Th. Kipstraat 8-10,
3011 RT Rotterdam, The Netherlands
31-10-412-0122
Fax 31-10-414-3035

Industrial Adhesives,
130 N. Campbell Ave.,
Chicago, IL 60612, USA
312-666-2686, 800-223-2686
Fax 312-666-1824

Industrial Adhesives,
Moor Road,
Chesham, Bucks., HP5 1SB, UK
44-1494-784444
Fax 44-1494-791903

Industrial Fibers, Inc.,
2889 N. Nagel Court,
Lake Bluff, IL 60044, USA
708-295-0046
Fax 708-295-0520

Industrial Quimica Del Espania,
Avda De Galicia 31,
33005 Iviedo, Spain
34-524-0694
Fax 34-8-525 8466

Industrias Quimicas,
Avenida Rafael de Casanova 81,
Mollet del Vallés,
E-08100 Barcelona, Spain
34-3-570-56-96
Fax 34-3-593-80-11

Inolex Chemical Co.,
Jackson & Swanson Sts.,
Philadelphia, PA 19148-3497, USA
215-271-0800, 800-521-9891
Fax 215-289-9065

International Dioxcide,
544-Ten Rod Road,
North Kingstown, RI 02852-4220, USA
908-499-9660, 800-477-6071
Fax 908-388-3648

Intervet Inc.,
405 State Street, P O Box 318,
Millsboro, DE 19966, USA
302-934-8051
Fax 302-934-6087

Invequimica & CIA SC,
PO Box 3227,
Medellin,
Columbia,

Iolab Corp.,
500 Iolab Drive,
Claremont, CA 91711, USA
714-624-2020, 800-423-1871
Fax 714-399-1646

ISC Chemicals Ltd.,
St Andrews Road,
Avonmouth, Bristol, Avon, BS11 9HP, UK
44-1179-823631
Fax 44-1272-822688

ISP (Österreich) GmbH,
Belvederegasse 18/1,
A-1040 Vienna, Austria
43-1-504-76-21
Fax 43-1-505-89-44,
www.ispcorp.com

ISP (Australasia) Pty,
73-75 Derby St., Silverwater,
Sydney N.S.W., 2141, Australia
61-2-648-5177
Fax 61-2-647-1608,
www.ispcorp.com

ISP (Canada) Inc.,
1075 The Queensway East,
Box 1740, Station B,
Mississauga, Ont., L4Y 4C1, Canada
905-277-0381
Fax 905-272-0552,
www.ispcorp.com

ISP Asia Pacific Pte,
200 Cantonment Rd.,
Hex 06-07 Southpoint,
0208 Singapore
65-224-9406
Fax 65-226-0853,
www.ispcorp.com

ISP Europe,
40 Alan Turing Rd.,
Surrey Research Park,
Guildford, Surrey, GU2 5YF, UK
44-1483-301757
Fax 44-1483-302175,
www.ispcorp.com

ISP Van Dyk, Inc.,
11 William St.,
Belleville, NJ 07109, USA
201-450-7722
Fax 201-751-2047,
www.ispcorp.com

ISP, International SP,
1361 Alps Rd.,
Wayne, NJ 07470-3688, USA
201-628-4000, 800-522-4423
Fax 201-628-4117,
www.ispcorp.com

ITW Devcon,
30 Endicott St.,
Danvers, MA 01923, USA
508-777-1100, 800-933-8266
Fax 800-765-4329

Ivax Industries Inc.,
1880 Langston St.,
Rock Hill, SC 29730, USA
803-327-8868, 800-343-7872
Fax 803-366-7256

J.B. Roerig,
235 East 42nd Street,
New York, NY 10017, USA
212-573-2323, 800-533-4535

J.C. Bottomley,
Brookfoot, Brighouse,
West Yorkshire, WDG 2QZ, UK
44-1484-714574
Fax 44-1484-717718

J.M. Huber Corp,
333 Thornall St,
Edison, NJ 8818, USA
908-549-8600
908-549-2239, www.huber.com/

J.M. Huber Corp.,
PO Box 310,
701 Fontain Street,
Havre de Grace, MD 21078, USA
410-939-3500
Fax 410-939-7301
www.huber.com/

J.M. Huber Corp./Carbon,
PO Box 2831,
Borger, TX 79008-2831, USA
806-274-6331, 800-631-6331
www.huber.com/

J.M. Huber Corp./Clay,
One Huber Rd.,
Macon, GA 31298, USA
912-745-4751
Fax 912-745-1116,
www.huber.com/

J.M. Huber Corp./Engineered Materials,
1100 Penn Ave.,
PO Box 2831,
Borger, TX 79008-2831, USA
806-274-6331
www.huber.com/

J.M. Huber Corp./Engineered Materials,
One Huber Rd.,
Macon, GA 31298, USA
912-745-4751, 800-TRY-HUBER,
Fax 912-745-1116,
www.huber.com/

J.M. Huber Corp./Engineered Materials,
4940 Peachtree Industrial Blvd, Suite 340,
Norcross, GA 30071, USA
404-441-1301
Fax 404-368-9908,
www.huber.com/

J.R. Technology Ltd.,
81 North End, Meldreth,
Royston, Herts., SG8 6NU, UK
44-1763-260721
Fax 44-1763-262002

James Briggs Ltd.,
Salmon Fields, Royton,
Oldham, Lancs., OL2 6HZ, UK
44-161-627-0101
Fax 44-161-627-0971

James Robinson Ltd.,
PO Box B3, Hillhouse Lane,
Huddersfield,West Yorkshire, HD1 6BU, UK
44-1484-435577
Fax 44-1484-435580

Jan Dekker B.V.,
Postbus 10,
NL-1520 AA Wormerveer, The Netherlands
31-75-27-8278
Fax 31-75-21-3883

Janssen Chimica,
Janssen Pharmaceuticalaan 3,
B-2440 Geel, Belgium
32-14-60-420
Fax 32-14-60-4220

Janssen Pharmaceuticals,
1125 Trenton-Harbourton Road,
P.O. Box 200,
Titusville, NJ 08560-0200, USA
609-730-2000, 800-253-3682
Fax 609-730-3044

Japan Synthetic Rubber,
11-24, Tsukiji
2-chome,Chuo-Ku,Tokyo104,
Japan
81-3-5565-6500
Fax 81-3-5565-6636

Jetco Chemicals, Inc.,
PO Box 1898,
Corsicana, TX 75110, USA
903-872-3011, 800-477-5353
Fax 903-872-4216

Jet-Lube, Inc.,
4849 Homestead Rd. 77028,
PO Box 21258,
Houston, TX 77226-1258, USA
713-674-7617, 800-JET-LUBE
Fax 713-678-4604

Johnson & Johnson-Merck,
Camp Hill Rd.,
Ft. Washington, PA 19034, USA
215-233-7700
Fax 215-233-8315, www.jnj-merck.com

Johnson Matthey Inc.,
17370 N. Laurel Park Dr.,
Suite 400 East,
Livonia, MI 48152, USA
313-591-4031,
Fax 313-591-4032

Johnson Matthey plc,
York Way,
Royston, Herts., SG8 5HJ, UK
44-1763-253200
Fax 44-1763-253492

Johnson Matthey plc,
Elton House, North
Powell Street,
Birmingham B1 3DD, UK
44-121-693 3555
Fax 44-121-236-3351

Johnson Matthey GmbH Alpha
Postfach 6540,
Zeppelinstrasse 7
D76185 Karlsruhe, Germany
49-6196-7038-21
Fax 49-6196-7038-012

Johnsons of Hendon Ltd.,
Hempstalls Lane,
Newcastle, Staffs., ST5 0SW, UK
44-1782-717100
Fax 44-1782-717707

K & K Greeff Chemicals,
Suffolk House, George Streeet,
Croydon, Greater London, CR9
3QL, UK
44-181-686-0544
Fax 44-181-686-4792

Kabi Pharmacia,
800 Centennial Ave,
P.O. Box 1327,
Piscataway, NJ 08855-1327,
USA
908-457-8000, 800-526-3619
Fax 908-457-8283

KabiVitrum AB,
Lindhagensgaten 133,
S-112 87 Stockholm, Sweden
46-8-13-8000
Fax 46-8-54-8020

Kao Corp.,
14-10, Nihonbashi,
Kayabacho 1,
Chuo-ku,Tokyo 103, Japan
81-3-3660-7111

Kao Corp. S.A.,
Puig dels Tudons, 10,
08210 Barbera Del Valles,
Barcelona, Spain
34-3-729-0000
Fax 34-3-719 0534,

Kao Corp./Edible Fat & Oil,
14-10 Nihonbashi,
Kayabacho 1,
Chuo-ku, Tokyo 103, Japan
81-3-3660-7860

Kaopolite, Inc.,
2444
Morris Ave.,
Union, NJ 07083, USA
908-789-0609
Fax 908-851-2974

Karlshamns AB.
S 37482 Karlshamn, Sweden
46-454-82000
Fax 46-454-18453

Kelco,
8355 Aero Drive,
PO Box 23576,
San Diego, CA 92123-1718, USA
619-292-4900, 800-535-2656
Fax 619-467-6520
www.monsanto.com

Kelco International,
Neuer Wall 63,
D-20354 Hamburg 36, Germany
49-40-37-35-91
Fax 49-40-36-57-47

Kelco International,
Tadworth Surrey,
Waterfield, KT20 5HQ, UK
44-1737-377000
Fax 44-1737-377100

Kelco International,
Les Mercuriales,
40 Rue Jean Jaures,
93176 Bagnolet Cedex, France
33-1-49-72-2800
Fax 33-1-43-62-8038

Kemira Ince Ltd.,
Ince,
Chester, Cheshire, CH2 4LB, UK
44-151-357-2777
Fax 44-151-357-2144

Kemtron International,
P O Box 2508,
289 Sherman Avenue,
Newark, NJ 07114, USA
201-623-7787

Kenrich Petrochemicals,
140 E. 22nd St.,
PO Box 32,
Bayonne, NJ 07002-0032, USA
201-823-9000, 800-LICA KPI
Fax 201-823-0691

Kensol Corp.,
PO Box 3179,
Allentown, PA 18106, USA
800-776-0612

Kerr-McGee Chemical
PO Box 25861,
Oklahoma City, OK 73125
405-270-1313, 800654-3911
Fax 405-270-3123,
www.kerr-mcgee.com

Kerr-McGee Chemical Europe,
Hohegrabenweg 87,
40667 Meerbusch-Buderich,
Düsseldorf, Germany
www.kerr-mcgee.com

Killgerm Chemicals Ltd.,
Denholme Drive,
Ossett, West Yorkshire, WF5
9BW, UK
44-1924-277631
Fax 44-1924-265033

Kincaid Enterprises,
PO Box 549, Plant Rd.,
Nitro, WV 25143, USA
304-755-3377
Fax 304-755-4547

King Industries, Inc.,
Science Rd.,
Norwalk, CT 06852, USA
203-866-5551, 800-431-7900
Fax 203-866-1268

King Industries, Inc.,
Kattensingel 7,
2801 CA Gouda, The
Netherlands
31-1820-28577
Fax 31-1820-29249

Knoll Pharmaceutical,
30 N Jefferson Rd,
Whippany, NJ 07981, USA
201-887-8300, 800-526-0710

Koch Chemical Co.,
PO Box 2608,
Corpus Christi, TX 78403, USA
512-242-8362
Fax 512-242-8353

Koch Chemical Co./Muskegon,
1725 Warner St.,
Whitehall, MI 49461, USA
616-894-4018
Fax 616-893-4141

Koch Chemicals Ltd.,
2 Marshgate Drive,
Hertford, Herts., SG13 7JY, UK
44-1992-553781
Fax 44-1992-586961

Koppert (UK) Ltd.,
1 Wadhurst Business Park,
Faircriouch Lane,
Wadhurst, East Sussex, TN5
6PT, UK
44-1892-884411
Fax 44-1892-882469

Kronos Canada, Inc.,
4 Place Ville-Marie, Suite 500,
Montreal, PQ, H3B 4M5, Canada
514-397-3501
Fax 514-393-1186

Kronos Ltd.,
Barons Court, Manchester Rd.,
Wilmslow, Cheshire, SK9 1BQ,
UK
44-1625-529511
Fax 44-1625-533123

Kronos, Inc.,
PO Box 4272,
Houston, TX 77210, USA
713-987-6300, 800-866-5600
Fax 713-987-6358

Kureha Chemical Industries,
9-11, Nihonbashi
Horidome-cho,
Chuo-ku,Tokyo 103, Japan
81-3-3249-4666
Fax 81-3-3661-1277

Kyowa America Corp.,
385 Clinton,
Costa Mesa, CA 92626, USA
714-641-0411
Fax 714-540-5849

Kyowa Chemical Industries,
305, Yashima-Nishimachi,
Takamatsu-shi,
Kagawa, 761-01, Japan
81-877-47-2500
Fax 81-877-47-4208

Kyowa Hakko Kogyo Co.,
Ohtemachi Bldg.,6-1,
Ohtemachi 1-chome,
Chiyoda-ku, Tokyo 100, Japan
81-3-3201-7211
Fax 81-3-3284-1968

Kyowa Hakko USA Inc.,
599 Lexington Ave., Suite 2780,
New York, NY 10022, USA
212-319-5353
Fax 212-421-1283

L W Vass (Agriculture),
Springfield Farm, Silsoe Road,
Maulden, Beds, MK45 2AX, UK
44-1525-403041
Fax 44-1525-402282

Lagap Pharmaceutical,
Woolmer Way,
Bordon, Hants., GU35 9QE, UK
44-1420-478301
Fax 44-1420-476726

Lakeland Laboratories,
Peel Lane, Astley Green,
Tyldesley, Manchester, M29 7FE,
UK
44-1942-873555
Fax 44-1942-884409

Lambson Ltd.,
Aire & Calder Works,
Cinder Lane,
Castleford, West Yorkshire,
WF10 1LU, UK
44-1977-510511
Fax 44-1977 603049

Lancashire Chemical,
High Street West,
Glossop, Derbyshire, SK13 8ES,
UK
44-1457-860006
Fax 44-1457-868394

Langley Smith & Co.,
36 Spital Square,
London, E1 6DY, UK
44-171-247-7473
Fax 44-171-375-1470

Lanstar Ltd.,
Liverpool Road, Cadishead,
Manchester, M30 5DT, UK
44-161-775-2644
Fax 44-161-776-1077

Laporte (North America),
c/o Royale Pigments & Chemical,
12 Rt. 17 North, Suite 309,
Paramus, NJ 07652, USA
201-845-4666
Fax 201-845-0719

Laporte Absorbents,
PO Box 2,
Moorfield Rd,
Widnes, Cheshire, WA8 0JU, UK
44-151-495-2222
Fax 44-151-420-4088

Laporte Inc.,
1212 Church St.,
PO Box 44,
Gonzales, TX 78629, USA
210-672-2891, 800-324-2891
Fax 210-672-3650

Laporte Industries SEA,
171 Chin Swee Rd.,
#05-05 San Centre,
Singapore
65-532-0676
Fax 65-532-0502

Laporte plc,
Nations House,
103 Wigmore Street,
London, W1H9AB, UK
44-171-399-2473
Fax 44-171-399-2471

Larkhall Laboratories,
225 Putney Bridge Road,
London, SW15 2PY, UK
44-181-874-1130
Fax 44-181-871-0066

LaRoche Chemicals Inc.,
PO Box 1031,
Airline Hwy.,
Baton Rouge, LA 70821-1031,
USA
504-356-8406, 800-548-6336
Fax 504-356-8405,
www.larocheind.com

LaRoche Industries Inc.,
1100 Johnson Ferry Rd. NE,
Atlanta, GA 30342, USA
404-851-0300
Fax 404-851-0476,
www.larocheind.com

Laserson & Sabetay,
BP57, Avenue des Grenots,
F-91151 Etampes Cedex, France
33-64-94-31-24
Fax 33-64-94-98-97

Laserson S.A.,
BP 57, Zone Industrielle,
91151 Etampes Cedex, France
33-64-94-31-24
Fax 33-64-94-98-97

Laur Silicone Rubber,
4930 S. M-18, Box 509,
Beaverton, MI 48612-0509, USA
517-435-7706
Fax 517-535-7707

Laurel Industries,
30195 Chagrin Blvd.,
Cleveland, OH 44124-5794, USA
216-831-5747, 800-711-1304
Fax 216-831-8479

Lawn & Garden Products,
P O Box 5317,
Fresno, CA 93755, USA
209 225 4770
Fax 209 225 1319

Lederle Laboratories,
One Cyanamid Plaza,
Wayne, NJ 07470, USA
914-735-2815
www.ahp.com

Lederle Laboratories,
Professional Services Dept.,
Pearl River, NY 10965, USA
914-735-2815
www.ahp.com

Leo Laboratories,
Longwick Road, Princes Risboro,
Aylesbury, Bucks., HP17 9RR,
UK
44-184 44-7333
Fax 44-184-44-2278

Lever Industrial Ltd.,
P O Box 20, Cressex Industrial,
High Wycombe, Bucks., HP12
3TL, UK
44-1494-461234
Fax 44-1494-462565

Lever Industriel,
103 Rue DeParis,
9300 Bobigny, France

Leverton-Clarke Ltd.,
Unit 16, Sherrington Way,
Lister Rd Industrial Estate,
Basingstoke, Hants., RG22 4DQ,
UK
44-1256-810393
Fax 44-1256-479324

LiphaTech, Inc.,
3600 W. Elm St.,
Milwaukee, WI 53209, USA
414-351-1476, 800-558-1003
Fax 414-351-1847

Lipo Chemicals Inc.,
207 19th Ave.,
Paterson, NJ 07504, USA
201-345-8600
Fax 201-345-8365,
www.lipochemicals.com/

Lipo do Brasil Ltda.,
Rua Roque Petrella, 376,
04581 Sao Paulo, Brazil
55-11-533-2354
Fax 55-11-533-8997,
www.lipochemicals.com/

Liquid Plastics Ltd.,
PO Box 7, London Road,
Preston, Lancs., PR1 4AJ, UK
44-1772-59781
Fax 44-1772-202627

Llewellyn Ryland Ltd.,
Balsall Heath Works,
Haden Street,
Birmingham, West Midlands, B12
9DB, UK
44-121-440-2284
Fax 44-121-440-0281

Lloyd Laboratories Inc,
PO Box 573,
Peabody, MA 01960-7573, USA
508-531-0053, 800-842-2605
Fax 508-532-6381

Loes Enterprises Inc.,
1457 Iglehart Avenue,
St Paul, MN 55104, USA
612-646-1385
Fax 612-646-3057

Lonza France SARL,
55, rue Aristide Briand,
F-92309 Levallois-Per, France
33-1-40-89-9925
Fax 33-1-40-89-9921,
www.lonza.com

Lonza Inc.,
17-17 Route 208,
Fair Lawn, NJ USA
201-794-2400, 800-777-1875
Fax 201-794-2695,
www.lonza.com

Lonza Japan Ltd.,
Kyowa Shinkawa Bldg., 8th Fl.,
20-8, Shinkawa 2-chome,
Chuo-ku,Tokyo 104, Japan
81-3-5566-0612
Fax 81-3-5566-0619,
www.lonza.com

Lonza Ltd./Fine Chemicals,
Münchensteinerstrasse 38,
CH-4002 Basle, Switzerland
49-41-61-316-8111
Fax 49-41-61-316-8301,
www.lonza.com

Lonza SpA,
via Vittor Pisani, 31,
I-20124 Milan, Italy
39-2-66-9991
Fax 39-2-66-98-7630,
www.lonza.com

Lonza UK Ltd.,
Imperial House,
Lypiatt Road,
Cheltenham, Gloucestershire,
GL50 2QJ, UK
44-1242-513211
Fax 44-1242-222294,
www.lonza.com

Lord Corp (UK) Ltd.,
Stretford Motorway Industrial,
Barton Dock Road, Stretford,
Manchester, M32 0ZH, UK
44-161-865-8048
Fax 44-161-865-0096,
www.lordcorp.com

Lord Corp./Chemical,
2000 West Grandview Blvd.,
Erie,PA,16514-0038, USA
814-868-3611, 800-243-6565,
Fax 814-864-3452,
www.lordcorp.com

The Lubrizol Corp.,
29400 Lakeland Blvd.,
Wickliffe, OH 44092-2298, USA
216-943-4200
Fax 216-943-5337,
www.lubrizol.com

Lubrizol France S.A.,
25 quai de France,
F-76100 Rouen, France
33-35-72-04-09
www.lubrizol.com

Lubrizol Japan Ltd.,
3-5-1, Toranomon, Minato-ku,
Tokyo 105, Japan
81-3-5041-4170
Fax 81-3-5401-4178,
www.lubrizol.com

Lucas Meyer (UK) Ltd,
Unit 46, Deeside Ind. Park,
First Ave., Deeside,
Clwyd, CH5 2NU, Wales UK
44-1244-281169
Fax 44-1244-281167,
www.lucas-meyer.com

Lucas Meyer GmbH & Co.,
Ausschläger Elbdeich 62,
D-20539 Hamburg, Germany
49-40-789-55-0
Fax 49-40-789-83-29,
www.lucas-meyer.com

Lucas Meyer Inc.,
PO Box 3218
Decatur, IL 62524-3218, USA
217-875-3660, 800-769-3660
Fax 217-877-5046,
www.lucas-meyer.com

Luzenac America, Inc,
9000 E. Nichols Ave., Suite 20,
Englewood, CO 80112, USA
303-643-0400, 800-525-8252
Fax 303-643-0444,
www.luzenac.com

Luzenac, Inc.,
1075 North Service Rd. W.,
Oakville, Ont., L6M 2G2, Canada
416-825-3930
Fax 416-825-3932,
www.luzenac.com

M. Michel & Co., Inc,
90 Broad St.,
New York, NY 10004, USA
212-344-3878
Fax 212-344-3880

Mach-1 Compounding,
775 E. Highland Rd.,
Macedonia, OH 44056, USA
330-467-8108
Fax 330-467-3570
www.mach1.com

Mackenzie Corp.,
78015 Chemical Rd.,
Bush, LA 70431, USA
504-886-2000
Fax 504-886-2174

Magnesium Elektron,
500 Point Breeze Rd.,
Flemington, NJ, 08822, USA
908-782-5800, 800-366-9596
Fax 908-782-7768

Magnolia Plastics, Inc.,
5547 Peachtree Industrial Blvd,
Chamblee, GA 30341, USA
404-451-2777
Fax 404-451-5376

Makhteshim Chemical,
PO Box 60,
Beer-Sheva,
84100, Israel,
972-7-296611
Fax 972-7-280304

Mallinckrodt,
Brakesplan Road South, Harefie,
Uxbridge, Middlesex, UB9 7LS,
UK
www.mallchem.com

Mallinckrodt Canada,
7500 Trans Canada Hwy.,
Pointe Claire, PQ, H9R 5H8,
Canada
514-695-1220
www.mallchem.com

Mallinckrodt Laboratory Chemicals,
Postfach 1268,
Industriestrasse 19-21,
D-64802 Dieburg, Germany
49-6071-20040
Fax 49-6071-200444,
www.mallchem.com

Mallinckrodt Laboratory Chemicals,
2443 Warrenville Rd,
Lisle, IL 60532, USA
708-955-4555
www.mallchem.com

Mallinckrodt Laboratory Chemicals,
4100 North Elston Ave,
Chicago, IL 60618, USA
312-478-1118
www.mallchem.com

Mallinckrodt Laboratory Chemicals,
222 Red School Street,
Phillipsburg, NJ O8865, USA
908-859-6916, 800-354-2050
Fax 908-859-6916,
www.mallchem.com

Mallinckrodt Veterinary,
PO 5840,
St Louis, MO 63134, USA
314-654-2000,1-888-744-1414
Fax 314-654-8410,
www.mallchem.com

Mallinckrodt, Inc.,
Mallinckrodt & 2nd Street,
PO Box 5439,
St Louis, MO 63147, USA
314-895-2000, 800-325-8888
Fax 314-539-1251,
www.mallchem.com

Manox Ltd.,
Manox House,
Coleshill Street,
Miles Platting,Manchester, M10 7AA, UK

Marfleet Refining Co,
Hedon Road, Marfleet,
Hull, North Humberside, HU9 5NJ, UK

Marion Merrell Dow Inc.,
9300 Ward Parkway,
P.O. Box 8480,
Kansas City, MO 64114-0480, USA
800-552-3656

A.H. Marks & Co. Ltd,
Wyke Lane, Wyke,
Bradford, West Yorkshire, BD12 9EJ, UK
44-1274-691234
Fax 44-1274-691176,
www.ahmarks.com

Marley Floors Ltd.,
Lenham,
Maidstone, Kent, ME17 2DE, UK
44-1622-858877
Fax 44-1622-858944

Marlowe-Van Loan Corp.,
PO Box 1851,
High Point, NC 27261, USA
910-886-7126, 800-422-4MVL
Fax 910-889-6663

Martin Marietta Magnesia,
PO Box 15470,
2323 Eastern Blvd., Bldg. E,
Baltimore, MD 21220-0470, USA
410-780-5500, 800-648-7400
Fax 410-780-5777

Martinswerk GmbH,
PO Box 1209,
Kölner Strasse 110,
50102 Bergheim, Germany
49-2271-9020
Fax 49-2271-902557

Master Bond Inc.,
154 Hobart St.,
Hackensack, NJ 07601, USA
201-343-8983

May & Baker Ltd.,
Rainham Road South,
Dagenham, Essex, RM10 7XS, UK
44-118-592-3060

Maybrook Inc.,
570 Broadway,
PO Box 68,
Lawrence, MA 01841, USA
508-682-1853
Fax 508-682-2544

Mayco Oil & Chemical,
775 Louis Dr.,
PO Box 2809,
Warminster, PA 18974-0357, USA
215-672-6600, 800-523-3903,
Fax 215-443-7094

Mazzucchelli Celluloide,
via S & P Mazzucchelli 7,
Castiglione Olona,
I-21043 Varese, Italy
39-826-211
Fax 39-826-213

McGhan NuSil Corp.,
1150 Mark Ave.,
Carpinteria, CA 93013, USA
805-684-8780
Fax 805-684-2365

McIntyre Group Ltd.,
24601 Governors Hwy.,
University Park, IL 60466-4127, USA
708-534-6200, 800-645-6457
Fax 708-534-6216,
www.mcintyregroup.com

McIntyre Chemicals Ltd.,
Blk 513, Bishan Town Centre St., 13
#01-500 Singapore 570513
65-354-3547, 65-354-0353
Fax 65-353-9068,
www.mcintyregroup.com

McKechnie Chemicals,
PO Box 4, Tanhouse Lane,
Widnes, Cheshire, WA8 0PG, UK
44-151-424-2611
Fax 44-151-424-4221

McLaughlin Gormley King,
8810 10th Ave. N.,
Minneapolis, MN 55427, USA
612-544-0341
Fax 612-544-6437

McLube,
9 Crozerville Rd.,
Aston, PA 19014, USA
215-459-1890, 800-2-MCLUBE
Fax 215-459-9538

McNeil Consumer Products,
Camp Hill Rd,
Fort Washington, PA 19304, USA
215-233-7000

McNeil Pharmaceuticals,
Spring House, PA 19477-0776, USA
215-628-5000

Mead Johnson & Co.,
P.O. Box 4500,
Princeton, NJ 08543-4500, USA
609-897-2000, 800-321-1335

Medi-Physics Inc.,
2636 Clearbrook Drive,
Arlington Heights, IL 60005, USA
708-593-6300

Meer Corp.,
PO Box 9006,
9500 Railroad Ave.,
N. Bergen, NJ 07047-1206, USA
201-861-9500
Fax 201-861-9267

Mendell Oy,
Maitotie 4,
15560 Nastola, Finland
358-18-6112290

Menley & James Laboratories,
1500 Spring Garden St,
Philadelphia, PA 19101, USA

Mercian Corp.,
5-8, Kyobashi 1-chome,
Chuo-ku,Tokyo 104, Japan
81-3-3231-3917
Fax 81-3-3276-0151

Merck & Co., Inc.,
PO Box 4,
West Point, PA 19486-0004, USA
800-672-6372
www.merck.com

Merck & Co., Inc.,
PO Box 2000,
Rahway, NJ 07065-0900, USA
908-594-4000
Fax 908-594-5431,
www.merck.com

Merck & Co., Inc./AgVet,
P.O. Box 2000,
Rahway, NJ 07065, USA
908-855-4277
Fax 908-855-9366,
www.merck.com

Merck Japan Ltd.,
Arco Tower, 8-1,
Shimomeguro 1,
Meguro-ku, Tokyo 153, Japan
81-3-5434-4700
Fax 81-3-5434-4705,
www.merck.com

Merck KgaA,
Postfach 4119,
Frankfurter Strasse 250,
D-64293 Darmstadt, Germany
49-6151-72-0
Fax 49-6151-72-2000,
www.merck.com

Merck Ltd.,
Merck House,
Poole, Dorset, BH15 1TD, UK
44-1202-669700
Fax 44-1202- 665599,
www.merck.com

Merck Pty. Ltd.,
207 Colchester Rd.,
Kilsyth, Vic., 3137, Australia
61-3-728-5855
Fax 61-3-728-1351,
www.merck.com

Merck Sharp & Dohme/Iso,
4545 Oleatha Ave.,
St. Louis, MO 63116, USA
800-325-9034
Fax 314-353-3754,
www.merck.com

Merix Chemical Co.,
2234 E. 75th St.,
Chicago, IL 60649, USA
312-221-8242
Fax 312-221-3047

Merquinsa,
Gran Vial 17,
E-08160 Montmeló
Barcelona, Spain
34-3- 572-1100
Fax 34-3-572-0934

Messer UK Limited,
Cedar House, 39 London Road,
Reigate, Surrey, RH2 9QE, UK
44-173-724-1133, 44-173-724-1133,
Fax 44-173-724-1842

Metalcrete Mfg Co.,
10330 Brecksville Road,
Cleveland, OH 44141, USA
216-526-5600
Fax 216-526-5601

Microfluidics Corp.,
90 Oak St.,
PO Box 9101,
Newton, MA 02164-9101, USA
617-969-5452, 800-370-5452
Fax 617-965-1213

Midland Resources Inc,
3211 Clinton Parkway CT #1,
Lawrence, KS 66047-2627, USA
913-842-7424, 800-879-6353
Fax 913-842-3150

Midwest Elastomers,
700 Industrial Dr., PO Box 412,
Wapakoneta, OH 45895, USA
419-738-9634
Fax 419-738-4504

Midwest Film Corp.,
4848 S. Hoyne,
Chicago, IL 60609, USA
312-254-5959, 800-538-9400
Fax 312-254-0443

Midwest Grain Producers,
1300 Main St.,
PO Box 130,
Atchison, KS 66002, USA
913-367-1480, 800-255-0302
Fax 913-367-0192

Midwest Rubber Reclaimation,
PO Box 2349,
E. St. Louis, IL 62202-2349, USA
618-337-6400

Midwest Zinc,
1001 W. Weed St.,
Chicago, IL 60622, USA
312-944-1505
Fax 312-944-5943

Miles Inc.
Bayer Inc., Mobay Rd.,
Pittsburgh, PA 15205-9741, USA
412-777-2000, 800-662-2927
Fax 412-777-2608

Miles Inc./Pharmaceuticals
Bayer Inc., 400 Morgan Lane,
West Haven, CT 06516-4175, USA
203-937-2000, 800-468-0894
Fax 412-394-5578

Miles Inc./Polysar,
2603 W. Market St.,
Akron, OH 44313, USA
216-836-0451, 800-321-0997
Fax 216-836-0200

Miles Inc./Polyurethane,
Mobay Rd.,
Pittsburgh, PA 15205-9741, USA
412-777-2000, 800-662-2927

Miles Laboratories Inc.,
PO Box 390,
Shawnee, KS 66201, USA

Miles Ltd./Biotechnology,
PO Box 37, Stoke Court,
Stoke Poges, Slough, Berks.,
SL2 4LY, UK
44-1281-45151
Fax 44-1281-43893

Milliken Chemical,
PO Box 1927, M-400,
Spartanburg, SC 29304-1927, USA
864-503-6188, 800-345-0372
Fax 803-503-2430

Milliken Chemical N.V.,
18-24 Ham,
B-9000 Gent, Belgium
32-9-265-1082
Fax 32-9-265-1195

Minas de Gádor S.A.,
General Zabala 24,
E-28002 Madrid, Spain
34-1411-0355
Fax 34-1562-2830

Mineral Research & Development,
One Woodlawn Green,
Charlotte, NC 28217, USA
704-525-2771, 800-334-0417
Fax 704-527-8232

Mirfield Sales Services,
Moorend House, Moorend Lane,
Dewsbury, West Yorkshire,
WF13 4QQ, UK
44-1484-842851
Fax 44-1484-847066

Mission Pharmacal Co.,
1325 East Durango,
San Antonio, TX 78210, USA
512-533-7118

Mitchell Cotts Chemical Ltd.,
PO Box No 6, Steanard Lane,
Mirfield, West Yorkshire, WF14 8QB, UK
44-1924-493861
Fax 44-1924-490972

Mitsubishi Chemical,
Mitsubishi Bldg.,
5-2, Marunou,
Chiyoda-ku,Tokyo 100, Japan
81-3-3283-6531
Fax 81-3-3283-6658

Mitsubishi Chemical,
Niederkasseler Lohweg 8,
D-40547 Düsseldorf, Germany
49-211-52392-0
Fax 49-211-591272

Mitsubishi Gas Chemical,
Mitsubishi Bldg.,
5-2, Marunouchi, 2-chome
Chiyoda-ku,Tokyo 100, Japan
81-3-3283-5000
Fax 81-3283-5120

Mitsubishi Gas Chemical,
520 Madison Ave., 17th Floor,
New York, NY 10022, USA
212-752-4620
Fax 212-758-4012,
www.mgc-a.com

Mitsubishi International,
520 Madison Ave.,
New York, NY 10022-4223, USA
212-605-2193, 800-442-6266
Fax 212-605-1704

Mitsubishi Kasei Corp.,
Mitsubishi Bldg.,
5-2, Marunou,
Chiyoda-ku,Tokyo 100, Japan
81-3-3283-6254

Mitsubishi Kasei Poly.,
Mitsubishi Bldg.,
5-2, Marunou,
Chiyoda-ku,Tokyo 100, Japan
81-3-3283-4405
Fax 81-3-3283-4480

Mitsubishi Materials,
1-5-1, Ohte-machi,
Chiyoda-ku,Tokyo 100, Japan
81-3-5252-5200
Fax 81-3-5252-5270/1

Mitsubishi Oil Co.,
Sanyu Bldg.,
2-4, Toranomon 1-,
Minato-ku, Tokyo 105, Japan
81-3-3595-7663
Fax 81-3-3508-2521

Mitsubishi Petrochemical,
Mitsubishi Bldg.,
5-2, Marunou,
Chiyoda-ku,Tokyo 100, Japan
81-3-3283-5700
Fax 81-3-3283-5472

Mitsubishi Rayon Co.,
3-19, Kyobashi 2-chome,
Chuo-ku,Tokyo 104, Japan
81-3-3272-4321
Fax 81-3-3245-8781

Mitsubishi Yuka Badisch,
1000, Kawajiri-cho,
Yokkaichi,Mie., 510, Japan
81-593-45-7230
Fax 81-593-45-7246

Mitsui Co. Ltd.,
Enterprise Plaza,
5600 North May Ave,
Oklahoma City, OK 73112, USA
405-842-2233
Fax 405-842-9901

Mitsui Petrochemical,
250 Park Ave., Suite 950,
New York, NY 10017, USA
212-682-2366
Fax 212-490-6694

Mitsui Petrochemical,
Kasumigaseki Bldg.,
2-5, Kasum,
Chiyoda-ku,Tokyo 100, Japan
81-3-3580-3616
Fax 81-3-3593-0028

Mitsui Toatsu Chemicals,
Kasumigaseki Bldg.,
2-5, Kasum,
Chiyoda-ku,Tokyo 100, Japan
81-3-3592-4111
Fax 81-3-3592-4267

Mitsui Toatsu Chemicals,
2500 Westchester Ave., Suite 1,
Purchase, NY 10577, USA
914-253-0777
Fax 914-253-0790

Mitsui Toatsu Liquid Cataysts,
Otsuka Bldg., 23-4,
Shinbashi,
Minato-ku,Tokyo 105, Japan
81-3-3431-9131
Fax 81-3-3433-0402

Mobil Chemical Co.,
PO Box 3029,
Edison, NJ 08818-3029, USA
908-321-6000
www.mobil.com

Mobil Chemical Co./Films,
1150 Pittsford-Victor Rd.,
Pittsford, NY 14534, USA
800-654-3436 (NY),
800-828-6381
www.mobil.com

Mobil Chemical Co./Petrochemicals,
15600 J.F. Kennedy Blvd., Suite 800,
Houston, TX 77032, USA
713-590-7700
Fax 713-590-7908,
www.mobil.com

Mobil Chemical Co./Polystyrene,
Rt. 27 & Vinyard Rd., Box 3029,
Edison, NJ 08818-3029, USA
908-321-3500, 800-922-0380
Fax 908-321-3501,
www.mobil.com

Mobil Mining & Minerals,
P.O. Box 26683,
Richmond, VA 23261, USA
804-798-4291
www.mobil.com

Mobil Oil Corp., Spec. Products,
3225 Gallows Rd.,7W004,
Fairfax, VA 22037-0001, USA
703-849-3609, 800-662-4525
Fax 703-849-6637,
www.mobil.com

Mobil Plastics Europe,
Zoning Industriel de Latour,
B-6761 Virton, Belgium
32-63-21 32 11
Fax 32-63-21 34 24
www.mobil.com

Momar Industrial Services,
Barracks Road,
Sand Lane Industrial Estate,
Stourport on Severn,Worcestershire, DY13 9QB, UK
44-1299-827232
Fax 44-1299-827608

Mona Industries Inc.,
PO Box 425,76 E. 24th St.,
Paterson, NJ 07544, USA
201-345-8220, 800-553-6662
Fax 201-345-3527,
www.monaweb.com

Monomer-Polymer & Dajac Laboratories Inc,
1675 Bustleton Pike,
Feasterville, PA 19053, USA
215-364-1155
Fax 215-364-1583

Monsanto Australia Ltd.,
600 St. Kilda Rd.,12th Floor,
Melbourne,Vic.a, 3004, Australia
61-3-522-7122
Fax 61-3-525-2253,
www.monsanto.com

Monsanto Canada Inc.,
2330 Argentia Rd.,Box 787,
Streetsville Postal Station,
Mississauga, Ont., L5M 2G4, Canada
905-826-9222
Fax 905-826-3119,
www.monsanto.com

Monsanto Chemical Co.,
800 N. Lindbergh Blvd.,
St. Louis, MO 63167, USA
314-694-1000, 800-325-4330
Fax 314-694-7625,
www.monsanto.com

Monsanto Commercial,
Bosques de Durazno 61,
3R Piso,Bosques de las Lomas,
Mexico City, DF 11700, Mexico
52-5-251-2715
Fax 52-5-251-7923,
www.monsanto.com

Monsanto Plastics,
800 N. Lindbergh Blvd.,
St. Louis, MO 63167, USA
800-325-4330
www.monsanto.com

Monsanto plc, Agriculture,
Thames Tower, Burleys Way,
Leicester, Leicestershire, LE1 3TP, UK
44-1162 620864
Fax 44-1162 530320,
www.monsanto.com

Monsanto, Detergents,
Rue Laid Burniat,
1348 Louvain la Neuve, France,
www.monsanto.com

Montana Talc Co.,
28769 Sappington Rd.,
Three Forks, MT 59752, USA
406-285-3286
Fax 406-285-3530

Montedison Belgio S.A.,
8 rue de l'Industrie,
B-1400 Nivelles Belgium
32-67-21 86 82
Fax 32-67-21 86 82

Montedison Deutschland,
Kölner Strasse 3A,
Postfach 5648,
D-65731 Eschborn/TS.1, Germany
49-6196-92201
Fax 49-6196-482389

Montedison do Brazil,
Avenida Brig. Faria, Lima 888-,
CEP 01452 Jardim Paulistano,
Sao Paulo SP, Brazil
55-11-210-3325
Fax 55-11-813-1700

Montedison SpA,
Casella Postale 10528,
Foro Buonoparte 31,
I-20121 Milano, Italy
39-2-633 31/627 01
Fax 39-2-805 0969

Mooney Chemicals Inc.,
2301 Scranton Rd.,
Cleveland, OH 44113, USA
216-781-8383, 800-321-9696

Mooney Plastics Ltd.,
Braintree Road,
Ruislip, Middlesex, HA4 0XX, UK
44-181 841 4211

Morflex, Inc.,
2110 High Point Rd.,
Greensboro, NC 27403, USA
910-292-1781
Fax 910-854-4058

Morton International,
Greville House, Hibernia Road,
Hounslow, Middlesex, TW3 3RX, UK
44-181-570-7766
Fax 44-181-570-6943,
www.mortonintl.com

Morton International,
100 North Riverside Plaza,
Chicago, IL 60606-1598, USA
312-807-2000
Fax 312-807-3150,
www.mortonintl.com

Morton International,
Westward House,155-157 Staines Rd.,
Hounslow, Middlesex, TW3 3JB, UK
44-181-570-7766
Fax 44-181-570-6943,
www.mortonintl.com

Morton International,
7900-A Taschereau Blvd.,
Brossard, PQ, J4X 1C2, Canada
514-466-7764
Fax 514-466-7771,
www.mortonintl.com

Morton International,
2700 E. 170 St.,
Lansing, IL 60438, USA
708-474-7000, 800-323-3224
Fax 708-868-7490,
www.mortonintl.com

Morton International,
10 S. Electric St.,
West Alexandria, OH 45381, USA
513-839-4612, 800-348-8846
Fax 513-839-5615,
www.mortonintl.com

Morton International,
150 Andover St.,
Danvers, MA 01923-1480, USA
508-774-3100
Fax 508-750-9512,
www.mortonintl.com

Morton International,
100 North Riverside Plaza,
Chicago, IL 60606, USA
312-807-2000
www.mortonintl.com

Morton International,
2000 West St.,
Cincinnati, OH 45215, USA
513-733-2100
Fax 513-733-2133,
www.mortonintl.com

Morton International,
PO Box 3089,
130 Mountain Creek Church Rd.,
Greenville, SC 29602, USA
803-292-5700, 800-845-6810
Fax 803-292-5713,
www.mortonintl.com

Morton International,
Oppauer Str 43,
D-68305 Mannheim, Germany
49-621-76260
www.mortonintl.com

Morton Salt,
100 North Riverside Plaza,
Chicago, IL 60606-1597, USA
312-807-2562
Fax 312-807-2228,
www.mortonintl.com

MSD Agvet,
Hertford Road,
Hoddesdon, Herts., EN11 9BU, UK
44-1992-467272
Fax 44-1992-467270

MTM Agrochemicals Ltd.,
18 Liverpool Road, Great Sankey,
Warrington, Cheshire, WA5 1QR, UK
44-1925-33232
Fax 44-1925-52679

Multibase Inc.,
3835 Copley Rd.,
Copley, OH 44321, USA
216-867-5124, 800-343-5626
Fax 216-666-7419

Multicrom S.A.,
Almirante Brown 778,
1704 Ramos Mejia, Argentina
54-1-658-8091-94
Fax 54-1-656-3905

MultiTherm Corp.,
125 S. Front St.,
Colwyn, PA 19023, USA
610-461-6442, 800-225-7440
Fax 610-461-0160

Munzing Chemie GmbH,
Salzstrasse 174,
Postfach 27 62,
D-74076 Heilbronn, Germany
49-7131-987-0
Fax 49-7131-987-125

Murlin Chemical Inc.,
Balligo Road,
West Conshohocken, PA 19428, USA
610-825-1165
Fax 610-825-8659

Murphy Chemical Co.,
Paper Mill Lane, Bramford,
Ipswich, Suffolk, IP8 4BZ, UK
44-1473-830492
Fax 44-1473-830046

Murphy Chemical Ltd.,
Latchmore Court, Brand Street,
Hitchin, Herts., SG5 1HZ, UK

Mydrin Ltd.,
Albion Road,
Carlton Industrial Estate,
Barnsley, South Yorkshire, S71 3PL, UK
44-1226-723661
Fax 44-1226-728298

Mykroy/Mycalex Ceramics,
125 Clifton Blvd.,
Clifton, NJ 07011, USA
201-779-8866
Fax 201-779-2013

Nalco Chemical B.V.,
Postbus 5131,
NL-5004 EC Tilburg, The Netherlands
31-13-63 55 55
Fax 31-13-67 45 45,
www.nalco.com

Nalco Chemical Co.,
One Nalco Center,
Naperville, IL 60563-1198, USA
708-305-1041, 800-527-7753
Fax 708-305-2998,
www.nalco.com

Nalco Fuel Tech.,
P.O. Box 3031,
Naperville, IL 60566-7031, USA
708-983-3242, www.nalco.com

Nalco/Exxon Energy C,
7705 Hwy. 90A,
Sugar Land, TX 77478, USA
800-333-3714
Fax 713-263-7149,
www.nalco.com

Nalco/Process Chemicals,
6216 W. 66th Place,
Chicago, IL 60638-5299, USA
708-496-5041, 800-435-0861
Fax 708-496-5290,
www.nalco.com

Napp Laboratories Ltd.,
Cambridge Science Park, Milton,
Cambridge, Cambridgeshire,
CB4 4BH, UK

National Starch & Chemical,
107 Neythal Rd.,
Jurong,
S-2262, Singapore
65-261-5528
Fax 65-264-1870,
www.nationalstarch.com

National Starch & Chemical,
Box 6500, 10 Finderne Ave.,
Bridgewater, NJ 08807-3300, USA
908-685-5000, 800-797-4992
Fax 908-417-5696,
www.nationalstarch.com

National Starch & Chemical,
Prestbury Court, Greencourts B,
333 Styal Rd., Manchester, M22 5LW, UK
44-161-435-3200
Fax 44-161-435-3300,
www.nationalstarch.com

National Starch & Chemical,
Galvin Rd, Slough Trading Estate,
Slough, Berks., SL1 4DF, UK
44-1753-533494
Fax 44-1753-539338,
www.nationalstarch.com

Neville Chemical Co.,
2800 Neville Rd.,
Pittsburgh, PA 15225-1496, USA
412-331-4200
Fax 412-777-4234

Niacet Corp.,
PO Box 258, 400 47th St.,
Niagara Falls, NY 14304, USA
716-285-1474, 800-828-1207
Fax 716-285-1497

Nicholas Laboratories,
225 Bath Road,
Slough, Berks., SL1 4AU, UK
44-1753-570340
Fax 44-1753-523971

Nickerson Chemicals,
Mill St East,
Dewsbury, West Yorkshire,
WF12 9BQ, UK
44-1924-453886
Fax 44-1924 458995

Nickerson Seeds Ltd.,
JNRC, Rothwell,
Lincolnshire, LN7 6DT, UK
44-1472-89471
Fax 44-1472-89602

Nihon Emulsion Co. Ltd.,
5-32-7, Koenji Minami,
Suginami-ku,Tokyo 166, Japan
81-3-3314-3211
Fax 81-3-3312-7207

Nihon Kagaku Sangyo,
20-5, Shitaya 2-chome,
Taito-ku,Tokyo 110, Japan
81-3-3876-3131
Fax 81-3-3876-3278

Nihon Nohyaku Co., Ltd.,
2-5, Nihonbashi 1-Chome,
Chuo-ku,Tokyo 103, Japan
81-3-3278-0461
Fax 81-3-3281-2443

Nihon Schering K.K.,
6-64 Nishimiyahara 2-chome,
Yodogawa-ku,Osaka 532, Japan
81-6-396-2300
Fax 81-6-398-2215

Nihon Surfactants,
3-24-3 Hasune,
Itabashi-Ku, Tokyo, Japan
81-3-966-7331

Nipa Hardwicke Inc.,
2411 Silverside Rd.,
104 Hagley Bldg.,
Wilmington, DE 19810, USA
302-478-1522
Fax 302-478-4097,
www.nipa.com

Nipa Hardwicke Inc./Germany,
Hans-Boeckler-Ring 9,
Norderstedt, D-22851, Germany
49-40-529-54-60
Fax 49-40-529-54-630,
www.nipa.com

Nipa Hardwicke Inc./UK,
Llantwit Farde, Pontypridd,
Mid Glamorgan, CF38 2SN,
Wales UK
44-443-205311
Fax 44-443-207746,
www.nipa.com

Nipa Laboratories Ltd.,
Llanwit Fardre, Pontypridd,
Mid Glamorgan, CF38
2SN,Wales UK
44-443-205311
Fax 44-443-207746,
www.nipa.com

Nisshin Oil Mills, Ltd.,
23-1 1-Chome, Shinkawa,
Chuo-Ku, Tokyo 104, Japan
81-3-3206-5113
Fax 81-3-3206-6456

Nitto Electric Industries,
1-2 Shimohozumi 1-chome,
Ibaraki, Osaka 567, Japan

Norit Americas Inc.,
1050 Crown Pointe Pkwy.,
Atlanta, GA 30338, USA
404-512-4610, 800-641-9245
Fax 404-512-4622

Norit N.V.,
Postbus 105,
NL-3800 AC Amersfoort, The Netherlands
31-64-8911
Fax 31-61-7429

Norit UK Ltd.,
Clydesmill Place,
Cambuslang Industrial Estate,
Glasgow, G32 8RF, UK
44-141-641-8841
Fax 44-141-641- 0742

Norma Products Ltd.,
Arnham Road,
Newbury, Berks., RG14 5RU, UK
44-1635 521 880
Fax 44-1635 523 418

Norman, Fox & Co.,
5511 S. Boyle Ave.,
PO Box 58727,
Vernon, CA 90058, USA
323-583-0016, 800-632-1777
Fax 323-583-9769,
www.norfoxx.com

Norplex,
PO Box 1448,
La Crosse, WI 54601, USA

Norsk Hydro AS,
Bygdoyalle 2,
N 0240 Oslo 2, Norway
47-2-243-2100
Fax 47-2-243-2725

Norsk Hydro Plast Oy,
Sabiansgatan 12 G49, PL 299,
358-00131 Helsinki, Finland
358-0-660446

Norwich Eaton Pharmaceuticals,
PO Box 191,
27 Eaton Ave,
Norwich, NY 13815-1799, USA
607-335-2565, 800-448-4878,
Fax 607-335-2098

Nova Polymers, Inc.,
2650 Eastside Park Dr.,
Evansville, IN 47716-8466, USA
812-476-0339
Fax 812-476-0592

NovaChem Corporation,
PO Box 6379,
201 Old Thomasville Rd,
High Point, NC 27262, USA
336-885-0041,
www.novachem.com

Novacor Chemicals.,
2550 Busha Hwy.,
Marysville, MI 48040, USA
313-364-5555, 800-627-1221
Fax 313-364-4670

Novatec Plastics & Chemicals,
PO Box 597,
275 Industrial Way,
Eatontown, NJ 07724, USA
908-542-6600, 800-PVC-NOVA
Fax 908-389-0431

Novo Nordisk A/S,
Bi, Novo Allé,
DK-2880 Bagsvaerd, Denmark
45-44-44-8888
Fax 45-44-44-60 88,
www.novo.dk

Novo Nordisk Bioindustries,
33 Turner Rd.,
Danbury, CT 06813-1907, USA
800-251-6686
Fax 203-790-2748, www.novo.dk

Novo Nordisk Bioindustries,
Makuhari Techno Garden CB-6,
1-3, Nakase 1-chome,
Chiba 261-01, Japan
81-43-296-6767
Fax 81-43-296-6767,
www.novo.dk

Novon International,
181 Cooper Ave.,
Tonawanda, NY 14150, USA

NYCO Minerals, Inc.,
124 Mountain View Dr.,
PO Box 368,
Willsboro, NY 12996-0368, USA
518-963-4262
Fax 518-963-1110

NYCO Minerals, Inc.,
Ordrupvej 24,
PO Box 88,
DK-2920 Charlottenlun, Denmark
45-39-64-3370
Fax 45-39-64-3710

Nylon Corp. of America,
333 Sundial Ave.,
Manchester, NH 03103, USA
603-627-5150, 800-851-2001
Fax 603-627-5154

Oak International Inc.,
1160 White Street, P O Box 837,
Sturgis, MI 49091, USA
616 651 9790
Fax 616 651 7849

Oakite Products, Inc,
50 Valley Rd.,
Berkeley Hts., NJ 07922, USA
908-464-6900, 800-526-4473
Fax 908-464-6031

Occidental Chemical,
Occidental Tower, 5005 LBJ Freeway,
Dallas, TX 75244, USA
214-404-3800, 800-752-5151
Fax 214-404-3669,
www.oxychem.com/

Occidental Chemical,
Toranomon 34 Mori Bldg., 9th Floor,
25-5 Toranomon 1-Chome,
Minato-ku, Tokyo 105, Japan
81-3-3502-4651
Fax 81-3-3502-4640,
www.oxychem.com

Occidental Chemical,
PO Box 344,
Niagara Falls, NY 14302-0344, USA
800-733-1165
Fax 716-278-7297,
www.oxychem.com

Occidental Chemical,
Suite 2, 16th Floor,
275 Alfred St.,
N. Sidney, NSW, 2060, Australia
61-2-9-957-6722
Fax 61-2-9-957-1548,
www.oxychem.com

Occidental Chemical,
Tampa, FL
USA
813-286-3800
www.oxychem.com

Occidental Chemical,
PO Box 27702,
Houston, TX 77227-7702, USA
713-623-7615, 800-800-4373
www.oxychem.com

Occidental Chemical,
Occidental Tower,
5005 LBJ Freeway,
Dallas, TX 75244, USA
214-404-3300
Fax 214-404-4815,
www.oxychem.com

Occidental Chemical,
PO Box 809050,
5005 LBJ Freeway,
Dallas, TX 75380-9050, USA
214-404-4932, 800-733-3339
Fax 214-404-4981,
www.oxychem.com

Occidental Chemical,
Holidaystraat 5,
B-1831 Diegem, Belgium
32-725-4450
Fax 32-725-4676,
www.oxychem.com

Octavius Hunt Ltd.,
5 Dove Lane, Redfield,
Bristol, Avon, BS5 9NQ, UK
44-1179-555304
Fax 44-1272-557875

Octel Chemicals Ltd.,
Halebank,
Widnes, Cheshire, WA8 8NS, UK
44-151-424-3671
Fax 44-151-420-1301

Olin Australia Ltd.,
PO Box 141, Level 2, Suite 1,
601 Pacific Hwy.,
St. Leonards, N.S.W., 2065, Australia
61-2-906-4455
Fax 61-2-439-4198,
www.olin.com

Olin Brasil Ltd.,
Nacoes Unidas, 11.857, 12th Floor,
Sao Paulo, 04578-000, Brazil
55-11-505-0382
Fax 55-11 505-1950,
www.olin.com

Olin Chemicals,
501 Merritt 7,
PO Box 4500,
Norwalk, CT 06856-4500, USA
203-750-3429, 800-462-6546
Fax 203-270-3752, www.olin.com

Olin Chemicals USA
PO Box 586,
Cheshire, CT 6410, USA
800-344-9168
Fax 203-271-4060, www.olin.com

Olin Chlor-Alkali Products,
650 25th St., Suite 300,
Cleveland, OH 37311, USA
615-336-4850
www.olin.com

Olin Corp./Agriculture,
PO Box 991,
Little Rock, AR 72203, USA
www.olin.com

Olin Far East Ltd.,
80-6 Soosong-dong,
Suktan Bldg.,
Chongro-ku, Seoul, 110-140, Korea
82-2-737-2840
Fax 82-2-730-7387,
www.olin.com

Olin Far East Ltd.,
7F-2 No. 137,
Fu Shing S. Road Sec. 1,
Taipei, Taiwan, R.O.C.,
886-2-752-4413
Fax 886-2-741-2113,
www.olin.com

Olin Industrial H.K.,
111 Peninsula Centre,
67 Mody Rd.,Tsim Sha Tsui-East,
Kowloon, Hong Kong
852-366-8303
Fax 852-367-1309, www.olin.com

Olin Japan Inc.,
Shiozaki Bldg.,
7-1 Hirakawa-Cho 2-Chome,
Chiyod,
Tokyo,102, Japan
81-3-3263-4615
www.olin.com

Olin Pte., Ltd.,
501 Orchard Rd.,
08-03 Lane Crawford Rd.,
Singapore 0923
65-735-1268
Fax 65-735-1298, www.olin.com

Olin Pty. Ltd.,
PO Box 114,
Bergvlei, 2012,
Rep. of S. Africa
27-11-444-2244
Fax 27-11-444-2241,
www.olin.com

Olin Quimica, S.A.
Campos Eliseos No. 385,
Piso 9, Col. Polanco, Delg.
Miguel Hidalgo,
11560, Mexico
52-5-281-2045
Fax 52-5-281-2037,
www.olin.com

Olin Research Center,
350 Knotter Dr.,
PO Box 586,
Cheshire, CT 06410-0586, USA
203-271-4000, 800-654-6763
Fax 203-271-4060, www.olin.com

Olin, Electronic Materials,
2873 North Nevada St.,
Chandles, AZ 85225, USA
602-926-2020, 800-367-4868
Fax 602-926-9332, www.olin.com

Omex Agriculture Ltd,
Bardney Airfield, Tupholme,
Lincoln, Lincs, LN3 5TP, UK
44-1526-398661
Fax 44-1526-398434

Oral-B Laboratories,
Gatehouse Road,
Aylesbury,Bucks, HP19 3ED, UK
44-1296 432601
Fax 44-1296-434283

Organon Inc.,
375 Mt Pleasant Ave,
West Orange, NJ 07052, USA
201-325 4500

Ortho Pharmaceutical,
PO Box 300,
Raritan, NJ 08869, USA
908-218-6000

Owens-Corning Fiberglass,
Fiberglass Tower,
Toledo, OH 43659, USA
419-248-7185, 800-462-3435
Fax 419-248-6712,
www.owenscorning.com

Owens-Illinois Inc.,
One Sea Gate,
Toledo, OH 43666, USA

Oxychem, Petrochemicals,
5 Greenway Plaza, Suite 2400,
Houston, TX 77046, USA
713-623-2246, 800-448-2246
Fax 713-968-6318,
www.oxychem.com

Pacific Anchor Chemical,
1224 Mindon Road,
Cumberland, RI 02864, USA
401-333-4100
Fax 401-333-4630
www.airproducts.com

Pacific Anchor Chemical,
3305 E. 26th St.,
Los Angeles, CA 90023, USA
213-264-0311
www.airproducts.com

Pacific Dispersions,
4615 Ardine Street,
Cudahy, CA 90201, USA

Pan Britannica Industries,
Britannica House,
Waltham Cross, Herts., EN8
7DY, UK
44-1992 23691
Fax 44-1992-26452

Parke-Davis,
Mitchell House, Southampton
Row,
Eastleigh, Hants., SO5 5RY, UK
www.parke-davis.com

Parke-Davis,
188 Howard Ave,
Holland, MI 49424, USA
616-392-2375
Fax 616-392-8914,
www.parke-davis.com

Parke-Davis,
201 Tabor Rd,
Morris Plains, NJ 07950, USA
201-540-2000, 800-223-0423
Fax 201-540-2248,
www.parke-davis.com

Pasminco Europe Ltd.,
Alloys House, PO Box 36,
Willenhall Lane,
Bloxwich, Walsall, West
Midlands, WS3 2XW, UK
44-1922-408444
Fax 44-1922-710043

PCR, Inc.,
PO Box 1466,
Gainesville, FL 32602-1466, USA
904-376-8246, 352-376-8246,
800-331-6313
Fax 904-371-6246,
352-371-6246,

Pecora Corp.,
165 Wambold Road,
Harleysville, PA 19438, USA
215-723 6051
Fax 215-721 0286

Penick Corp.,
158 Mt Olivet Ave,
Newark, NJ 07114, USA
201-621 2802
Fax 201-621 2816

Pennwalt Italia SpA,
via del Porto,
I-28040 Marano-Ticino, Italy
321-9791
Fax 321-979246

Penreco,
138 Petrolia St.,
Karns City, PA 16041, USA
412-756-0110, 800-245-3952
Fax 412-756-1050

Penta Manufacturing,
PO Box 1448,
Fairfield, NJ 07007-1448, USA
201-740-2300
Fax 201-740-1839

Pentagon Chemicals Ltd,
Northside,
Workington, Cumbria, CA14 1JJ,
UK
44-1900-604371
Fax 44-1900-66943

Pentagon Urethanes Ltd,
Northside,
Workington, Cumbria, CA14 1JJ,
UK

Penwest Pharmaceuticals Co.,
Church House,
48 Church St.,
Reigate, Surrey, RH1 6YS, UK
44-1737-222 323

Penwest Pharmaceuticals Co.,
2981 Rt. 22,
Patterson, NY 12563-9970, USA
914-878-3414, 800-431-2457
Fax 914-878-3484

Penwest Pharmaceuticals Co.,
Postfach 1207,
25430 Uetersen, Germany
49-6135-951410
Fax 49-6135-951223

Peridot Chemicals,
90 East Halsey Rd, Suite 204A,
Parsippany, NJ 07054-3709,
USA
800-222-0121

Peridot Chemicals,
PO Box 15487,
Augusta, GA 30919, USA
706-737-0661, 800-222-0121

Perma-Flex Mold Co.,
1919 East Livingston Ave.,
Columbus, OH 43209, USA
614-242-8034, 800-736-6653
Fax 614-252-8572

Permutit Co. Ltd., T,
632-652 London Road,
Isleworth, Middlesex, TW7 4EZ,
UK
44-181-560-6431
Fax 44-181-568-9772

Person & Covey Inc.,
PO Box 25018,
616 Allen Ave,
Glendale, CA 91221-5018, USA
818-240-1030, 800-423-2341

Perstorp Ferguson Ltd,
Aycliffe Industrial Estate,
Newton Aycliffe, Co. Durham,
DL5 6UE, UK
44-1325-300666
Fax 44-1325-300385

Petrolite Corp./Head,
6910 E. 14th St.,
Tulsa, OK 74112, USA
918-836-1601, 800-331-5516
Fax 918-834-9718

Petrolite Corp./Industries,
369 Marshall Ave,
St. Louis, MO 63119, USA

Petrolite GmbH,
PO Box 2031,
Kaiser-Friedrich Promenade 59,
61290 Bad Homburg, Germany

Petrolite Ltd./EuroC,
Kirkby Bank Rd,
Knowsley Industrial Estate,
Liverpool, Merseyside, L33 7SY,
UK
44-151-546-2855
Fax 44-151-549-1858

Petrolite Pacific Pt,
2 Tanjong Penjuru Crescent,
Jurong, Singapore 2260
65-268-0517
Fax 65-262-0344

Pfizer Asia/Australia,
PO Box 57,
West Ryde, NSW, 2114,
Australia
61-2-858-9500,
www.pfizer.com

Pfizer Canada,
PO Box 800,
Point Claire/Dorval,
Montreal,PQ, H9R 4V2, Canada
514-695-0500
www.pfizer.com

Pfizer K.K.,
3-22 Toranomon 2-chome,
Minato-ku,Tokyo 105, Japan
81-3-3503-0441
Fax (03) 3503-0447,
www.pfizer.com

Pfizer S.A.,
Principe de Vergara 109,
E-28002 Madrid, Spain
34-1-262 11 00,
www.pfizer.com

Pfizer/Dairy & Brewery,
4215 N. Port Washington Rd.,
Milwaukee, WI 53212, USA
414-332-3545, 800-231-1590
www.pfizer.com

Pharmaceutical Basic,
8755 West Higgins Road/810,
Chicago, IL 60631, USA
312-380 0080

Pharmax Ltd., Bourne Road,
Bexley, Kent, DA5 1NX, UK
44-1322 550550
Fax 44-1322-555469

Phillips Chemical Co,
PO Box 968,
Borger, TX 79008, USA
806-274-5236, 800-858-4327
Fax 806-274-5230

Phillips Chemical Co,
101 ARB Plastics Technical
Center,
Bartlesville, OK 74004, USA
918-661-9845
Fax 918-662-2929

Phillips Petroleum Chemical,
Steenweg op Brussels 355,
B-3090 Overijse, Belgium
32-2-689-1211
Fax 32-2-689-1472

Phillips Petroleum Co.,
Philips Quadrant, 35 Guildford,
Woking, Surrey, GU22 7QT, UK
44-1483-756666
Fax 44-1483-752371

Phillips Petroleum Inc.,
Shin-Tokyo Bldg.,
3-1, Marunou,
Chioyda-ku,Tokyo 100,Japan
81-3-3216-6951
Fax 81-3-3216-6960

Phillips Yeast Products,
Park Royal Road,
London, NW10 7JX, UK

Piccadilly Products,
199 Piccadilly,
London, W1V 9LE, UK

Pickering Laboratories,
1951 Colony Street, Suite S,
Mountain View, CA 94043, USA
415-968-9502
Fax 415-968-0749

Pierce & Stevens Corp.,
PO Box 1092,
Buffalo, NY 14240-1092, USA
716-856-4910, 800-888-4910
Fax 716-856-7530

Pilot Chemical Co.,
11756 Burke St.,
Santa Fe Springs, CA 90670,
USA
310-723-0036, 800-707-4568
Fax 310-945-1877,
www.pilotchemical.com

Plascoat Systems Ltd.,
Trading Estate,
Farnham, Surrey, GU9 9NY, UK
44-1252-733-3777
Fax 44-1252-725-719

Plastic Coatings Ltd.,
Woodbridge Industrial Estate,
Guildford, Surrey, GU1 1BG, UK
44-1483-31155
Fax 44-1483 33534

Plasticolors, Inc.,
2600 Michigan Ave.,
PO Box 816,
Ashtabula, OH 44004, USA
216-997-5137
Fax 216-992-3613

PMC Specialities International,
65B Wigmore Street,
London, W1H 9LG, UK
44-171-935-4058
Fax 44-171-935 9895

PMC Specialties Group,
501 Murray Rd.,
Cincinnati, OH 45217, USA
513-242-3300, 800-543-2466
Fax 513-482-7353

PMC Specialties Group,
20525 Center Ridge Road,
Rocky River, OH 44116, USA
216-356-0700, 800-543-2466
Fax 216-356-2787

Pointing Ltd.,
Princess Way,
Prudhoe, Northumberland, NE42
6NJ, UK
44-1661 832621
Fax 44-1661-835650

Polimex, Inc.,
204 North Dooley St.,
Grapevine, TX 76051, USA
817-481-3547
Fax 817-488-4816

Polymer Composites,
PO Box 30010,
4610 Theurer Blvd.,
Winona, MN 55987, USA
507-454-4150, 800-526-4960
Fax 507-457-4040

The Polymer Corp.,
2120 Fairmount Ave, PO Box
14235,
Reading, PA 19612-4235, USA
215-320 6600, 800-729-0101
Fax 800-366-0301

Polymeric Systems Inc.,
Wheatland & Mason Streets,
Phoenixville, PA 19460, USA

Polysciences Inc.,
400 Valley Road,
Warrington, PA 18976-2590,
USA
215-343-6484, 800-523-2575
Fax 800-343-3291

Polysciences, Europe,
Postfach 1130,
D-69208 Eppelheim, Germany
49-6221-765767
Fax 49-6221-764620

Polyurethane Corp. of America,
624 Schuyler Ave.,
Lyndhurst, NJ 07071, USA
201-438-2325
Fax 201-507-1367

Polyurethane Special,
624 Schuyler Ave.,
Lyndhurst, NJ 07071, USA
201-438-2325
Fax 201-507-1367

Polyvinyl Chemical Industries,
730 Main St.,
Wilmington, MA 01887, USA

Portman Agrochemical,
Apex House, Grand Arcade,
Tally-Ho Corner, North Finchley,
London, N12 0EH, UK
44-181-446-8383
Fax 44-181-445-6045

Potters Industries Inc.,
Southpoint Corporate
Headquarters,
PO Box 840,
Valley Forge, PA 19482-0840,
USA
610-651-4700
Fax 610-408-9723

Poythress Laboratories,
16 N 22nd St,
PO Box 26946,
Richmond, VA 23261, USA

PPG Industries Inc.,
Urb. Industrial Mario Julia,
A&B Street Caparra Heights,
San Juan, 920, Puerto Rico
www.ppg.com

**PPG Industrial do Brasil
Limitada,**
Rua Sampaio Viana,
277,13o Andar-Conj.132,
Sao Paulo-SP, Paraiso,
CEP 04004-000, Brazil
55-11-887-6522
Fax 551-887-2224,
www.ppg.com

PPG Industries (UK),
Carrington Business Park,
Carrington,Urmston,
Manchester, M31 4DD, UK
44-161-777-9203
Fax 44-161-777-9064,
www.ppg.com

PPG Industries Asia/Pacific,
LTD,Takanawa Court, 5th Floor,
13-1, Takanawa 3 chome,
Minato-ku-Tokyo 108, Japan
03-3280-2861
Fax 03-3280-2920,
www.ppg.com

PPG Industries Sales, Inc.,
Immeuble Scor,
1 Avenue du President Wilson,
Paris LaDéfense Cedex, 92704,
France
33-1-4698-8100
Fax 33-1-4698-8263,
www.ppg.com

PPG Industries Taiwan,
Suite 601, Worldwide House, No.
131,
Ming East Rd., Sec. 3,
Taipei 105, Taiwan, R.O.C.
886-2-514-8052
Fax 886-2-514-7957,
www.ppg.com

PPG Industries, Inc.,
One PPG Place, 34 North,
Pittsburgh, PA 15272, USA
412-434-3131, 800-CHEM-PPG
Fax 412-434-2891,
www.ppg.com

PPG Industries, Inc.,
961 Division St.,
Adrian, MI 49221, USA
517-263-7831
Fax 517-263-2552,
www.ppg.com

PPG Industries, Inc.,
PO Box 98,
100 Station Ave.,
Stockertown, PA 18083-0098, USA
215-759-3690
Fax 215-759-3692,
www.ppg.com

PPG Industries, Inc.,
3938 Porett Dr.,
Gurnee, IL 60031, USA
847-244-3410, 800-323-0856
Fax 847-244-9633,
www.ppg.com

PPG Industries-Asia,
Takanawa Court, 5th floor,
13-1 Takanawa 3-Chome,
Minato-Ku, Tokyo 108, Japan
81-3-3280-2911
Fax 81-3-3280-2920,
www.ppg.com

PPG Mexico de Mexico DA de CV,
Av. Presidente Juarez No. 1978,
Col. San Jeronimo Tepetlacalco,
Tlalnepantla, Edo., C.P. 54090,
Mexico
52-5-397-8011
Fax 52-5-398-5133,
www.ppg.com

PPG Ouvrie S.A.,
64, rue Faldherbe,B.P. 127,
59811 Lesquin Cedex, France
33-2087-0510
Fax 33-2087-5631,
www.ppg.com

PQ Australia Pty. Ltd.,
Melbourne, Vic., Australia
61-3-791-4066
Fax 61-3-791-6511

PQ Corp.,
PO Box 840,
Valley Forge, PA 19482-0840, USA
610-651-4200, 800-944-7411
Fax 610-251-9118

Presperse Inc.,
141 Ethel Rd W,
Piscataway, NJ 08854-5928, USA
732-819-8009
Fax 732-819-7155

Procter & Gamble Co.,
Rm C2N09A, 11530 Reed Hartman Hwy.,
Cincinnati, OH 45241, USA
513-626-3701, 800-477-8899
Fax 513-626-3145,
www.pg.com/chemicals

Procter & Gamble Inc.,
4711 Yonge St.,
PO Box 355, Station A,
Toronto, Ont., M5W 1C5, Canada
416-730-4064
Fax 416-730-4122,
www.pg.com/chemicals

Procter & Gamble Inc.,
1-17 Koyocho-naka 1-chome,
Higashinaka-ku,
Kobe 658, Japan
81-78-845-5328
Fax 81-78-845-6912,
www.pg.com/chemicals

Procter & Gamble Ltd.,
PO Box 9, Hayes Gate House,
27 Uxbridge Rd.,
Hayes, Middlesex, UB4 0JD, UK
44-181-242-2303
Fax 44-181-242-2294,
www.pg.com/chemicals

Production Chemicals,
Dalton Way,
Middlewich Motorway Estate,
Middlewich, Cheshire, CW10 0HS, UK
44-160-684-5557
Fax 44-160-684-6408

Protex Chemicals Ltd.,
Astley Lane Industrial Estate,
Astley Way, Swillington,
Leeds, L228 8XT, UK
44-113-2876002
Fax 44-113-2875003

Protex Chemie Basel,
Thannerstrasse 72,
CH-4054 Basel, Switzerland
41-61-302 0066
Fax 41-61-302 0076

Protex Extrosa,
Postfach 1415,
79504 Lorrach, Germany
49-7621-84772
Fax 49-7621-12429

Protex Korea Co. Ltd.,
PO Box 1193,
Seoul 1351010, Korea
82-2-548-6993
Fax 82-2-548-6564

Protex Nederland,
Broekhovenseweg 130 R,
502 LJ Tilburg, The Netherlands
31-13-36-7477
Fax 31-13-36-6592

Protex S.A.,
B.P. 177, 6 rue Barbès,
92305 Levallois-Paris, France
33-1-41-34-1400
Fax 33-1-41-34-1416

Puma Chemical Co. Inc.,
1601 109th St,
Grand Prairie, TX 75050, USA

Punati Chemical Corp,
615 S Eton,
Birmingham, MI 48009, USA

QO Chemicals (Australia),
16 Princess St.,
Kew, Vic., 3101, Australia
61-3-853-6464
Fax 61-3-853-5929

QO Chemicals Europe,
8, rue Bellini,
75782 Paris Cedex 16, France
33-1-4503-1450
Fax 33-1-4704-3604

QO Chemicals, Inc.,
Industrieweg 12,
Haven 391,
2030 Antwerp, Belgium
32-3-541-21-65
Fax 32-3541-65-03

Quantum Chemical Co.,
11500 Northlake Dr.,
PO Box 429550,
Cincinnati, OH 45249, USA
513-530-6500, 800-323-4905
Fax 513-530-6119

Quantum Chemical Co.,
1 Pierce Place, Suite 250C,
Itasca, IL 60143, USA
708-285-1111, 800-323-7659
Fax 708-285-3316

Quantum Chemical Europe,
Lange Bunder 7,
4854 MB Bavel, The Netherlands
31-1613-6600
Fax 31-1613-3500

Quantum Composites,
4702 James Savage Rd.,
Midland, MI 48642-8642, USA
517-496-2884
Fax 517-496-2333

Ques Industries Inc.,
5420 W.140th St.,
Cleveland, OH 44142, USA
216-267-8989, 800-340-8989
Fax 216-267-8998

Quest International,
Postbus 2,
1400 CA Bussum, The Netherlands
31-2159-99111
Fax 31- 2159-46067

Quest International,
5115 Sedge Blvd.,
Hoffman Estates, IL 60192, USA
800-621-4710,
Fax 847-645-7061

Quest International,
Bromborough Port,
Wirral, Merseyside, L62 4SU, UK
44-151-645-2060
Fax 44-151-645-6975

Quest International,
12 Britton St.,
Smithfield, NSW, 2164, Australia
61-2 827-4000
Fax 61-2-604-7926

Quest International,
Postfach 650170,
Poppenbütteler Chaussee 36,
D-22361 Hamburg 65, Germany
49-40-607970
Fax 49-40-6079710

Quest International,
10 Painters Mill Rd.,
Owings Mills, MD 21117-3686, USA
410-363-2550
Fax 410-363-7514

Quest International,
400 International Dr.,
Mt. Olive, NJ 07828, USA
201-691-7100, 800-598-5986
Fax 201-691-7479

Quest International,
PO Box 630, Woods Corners,
Norwich, NY 13815, USA
607-334-9951
Fax 607-334-5022

Quest International,
Kennington Road,
Ashford, Kent, TN24 0LT, UK
44-1233 644444
Fax 44-1233-644146

Quinoderm Ltd.,
Manchester Road, Hollinwood,
Oldham, Lancs, OL8 4PB, UK
44-161-624-9307
Fax 44-161-627-0928

R.H. Carlson Co.,
41 Chestnut St.,
Greenwich, CT 06830, USA
203-531-5500, 800-243-5404
Fax 203-531-0032

R.T. Vanderbilt Co.,
30 Winfield St,
PO Box 5150,
Norwalk, CT 06856-5150, USA
203-853-1400, 800-243-6064
Fax 203-853-1452,
www.rtvanderbilt.com

Manufacturers and Suppliers

R.W. Greeff & Co. Inc.,
777 West Putnam Ave,
Greenwich, CT 06830, USA
203-532-2900
Fax 203-532-2980

Radilon Inc.,
PO Box 18367,
Spartanburg, SC 29318, USA
803-579-2729

Ralco Industries, Inc.,
1112 River St., PO Box 509,
Woonsocket, RI 02895, USA
401-767-2700, 800-343-4166
Fax 401-767-2823

Raschig AG,
Mundenheimer Strasse 100,
D-67061 Ludwigshafen, Germany
49-621-56180
Fax 49-621-5618661

Raschig Corp.,
5000 Old Osborne Tpke.,
PO Box 7656,
Richmond, VA 23231, USA
804-222-9516
Fax 804-226-1569

Raschig France S.A.R,
49, ave. de Versailles,
F-75016 Paris, France
33-1-45-24- 0636
Fax 33-1-45-20- 8408

Raschig UK Ltd.,
Dock Office, Trafford Rd.,
Salford Quays,
Salford, Lancs., M5 2XB, UK
44-161-877-3933
Fax 44-161-877-3944

Reckitt's Colours Ltd.,
Morley Street,
Kingston-upon-Hull,
North Humberside, HU8 8DN, UK
44-11482-329875
Fax 44-14182-223114

Reed Corp.,
233 W. Pkwy.,
Pompton Plains, NJ 07444, USA
201-831-0636
Fax 201-831-0791

Regal Chemical Co.,
PO Box 900,
Alpharetta, GA 30201, USA
404-475-4837
Fax 404-475-1254

Reheis Inc.,
PO Box 609,
235 Snyder Ave.,
Berkeley Heights, NJ 07922,
USA
908-464-1500
Fax 908-464-8094,
www.reheis.com

Reheis Ireland,
Kilbarrack Rd.,
Dublin 5, Ireland,
353-1-8322621
Fax 353-1-8392205,
www.reheis.com

Reichhold Chemicals,
2400 Ellis Rd.,
Durham, NC 27703-5543, USA
919-990-7500, 800-448-3482
Fax 919-990-7711,
www.reichhold.com

Reichhold Chemicals/Polymers,
PO Box 1433,
Pensacola, FL 32596-1433, USA
904-433-7621, 800-874-0868
Fax 904-433-3655,
www.reichhold.com

Reichhold GmbH,
Postfach 10,
Breitenleer-strasse 97-99,
A-1222 Vienna, Austria
43-1-233551
Fax 43-1-233519232,
www.reichhold.com

Reichhold/Emulsion Polymers,
PO Box 13582,
Research Triangle Park, NC
27709-3582, USA
919-990-7500, 800-441-6461,
919-990-7746
www.reichhold.com

Reichhold/Reactive Polymers,
8540 Baycenter Rd., PO Box
191,
Jacksonville, FL 32245, USA
904-739-2170
www.reichhold.com

Reilly Industries Inc.,
1500 S. Tibbs Ave.,
Indianapolis, IN 46241, USA
317-248-6411, 800-777-3536
Fax 317-248-6402,
www.reillyind.com

Releasomers, Inc.,
PO Box 82,
Bradfordwoods, PA 15015, USA
412-452-4474
Fax 412-452-1965

Rentokil Ltd.,
Felcourt,
East Grinstead, West Sussex,
RH19 2JY, UK
44-1342-833022
Fax 44-1342-326229

Resinas Sintéticas SA,
Aribau, 185-6a planta,
E-08021 Barcelona, Spain
34-3-200-6911
Fax 34-3-209-3900

Resinoid Engineering,
7557 St. Louis Ave.,
Skokie, IL 60076, USA
708-673-1050
Fax 708-673-2160

Resinous Chemicals Ltd.,
Cross Lane,
Dunston, Tyne & Wear, NE11
9HQ, UK
44-191-493-2525
Fax 44-191-460-6270

Revertex Ltd.,
Templefields, Harlow,
Essex, CM20 2BH, UK
44-1279-429555
Fax 44-1279-412984

Rexene Products Co.,
5005 LBJ Freeway, Occidental
Tower,
Dallas, TX 75244, USA
214-450-9000, 800-233-1159
Fax 214-450-9028

Reynolds Metal Co.,
PO Box 27003,
6603 W. Broad St.,
Richmond, VA 23261-7003, USA
804-281-2000, 800-253-2060

Reynolds Metal Co./Chemicals,
PO Box 97,
Bauxite, AR 72011, USA
501-776-4104, 800-253-2060
Fax 501-776-4103

Rhein Chemie Corp.,
1008 Whitehead Road Ext.,
Trenton, NJ 08638, USA
609-771-9100
Fax 609-771-0232

Rhein-Chemie Rheinau GmbH,
Postfach 810409,
Müulheimer Strasse 24-28,
D-68219 Mannheim, Germany
49-621-8-9070

Rheox Europe N.V./S.A.,
Rue de l'Hopital ,31
Gasthuisstraat,
B1000-Brussels 08520, Belgium
32-2-549-0068
Fax 32-2-513-2425,
www.rheox.com

Rheox Inc.,
PO Box 700,
Wyckoffs Mill Road,
Hightstown, NJ 08520, USA
609-443-2500, 800-866-6800
Fax 609-443-2422,
www.rheox.com

Rhône-Poulenc ABM,
Poleacre Lane, Woodley,
Stockport, Cheshire, SK6 1PQ,
UK
44-161-430-4391
Fax 44-161-430-8523,
www.rhone-poulenc.com

Rhône-Poulenc Ag Co.,
PO Box 12014, 2 TW Alexander
Dr.,
Research Triangle Park, NC
27709, USA
919-549-2000, 800-334-9745
Fax 919-549-3924,
www.rhone-poulenc.com

Rhône-Poulenc Agrochemicals,
14-20 rue Pierre Baizet, B.P.
9163,
69263 Lyon Cedex 09, France
33-7-229-2525
Fax 33-7-229-2885,
www.rhone-poulenc.com

Rhône-Poulenc Basic,
One Corporate Dr., Box 881,
Shelton, CT 06484, USA
800-642-4200
Fax 203-925-3627,
www.rhone-poulenc.com

Rhône-Poulenc Chemicals,
Woodley,
Stockport, Cheshire, SK6 1PQ,
UK
44-161-430-4391
Fax 44-161-430-4364,
www.rhone-poulenc.com

Rhône-Poulenc Chemicals,
Staveley,
Chesterfield, Derbyshire, S43
2PB, UK
44-1246-277251
Fax 44-1246-280090,
www.rhone-poulenc.com

Rhône-Poulenc Chemicals,
Oak House,Reeds Crescent,
Watford, Herts, WD1 1QH, UK
44-181 984 3342
Fax 44-181 984 1701,
www.rhone-poulenc.com

Rhône-Poulenc Chemicals,
25 Quai Paul Doumer,
F-92408 Courbevoie Cedex 29,
France
33-47-68-1234
Fax 33-47-68-23-00,
www.rhone-poulenc.com

Rhône-Poulenc Chemicals,
Ketenislaan 1,
B-9130 Kallo, Belgium
32-37-551759
Fax 32-37-756995,
www.rhone-poulenc.com

Rhône-Poulenc Food Ingredients,
CN 7500, Prospect Plains Rd.,
Cranbury, NJ 08512, USA
609-860-4600, 800-253-5052
www.rhone-poulenc.com

Rhône-Poulenc Gerona,
via Milano 78,
Ospiate Di Bollate,
I-20021 Milano, Italy
39-2-38 33 41
Fax 39-2-3833 43 01,
www.rhone-poulenc.com

Rhône-Poulenc N.V.,
Kuhlmannkaai,
B-9020 Ghent, Belgium
32-91 44-8891
www.rhone-poulenc.com

Rhône-Poulenc North America,
One Corporate Dr.,
Shelton, CT 06484, USA
203-925-8164
Fax 203-925-3670,
www.rhone-poulenc.com

Rhône-Poulenc Rorer,
Rainham Rd South,
Dagenham, Essex, RM10 7XS, UK
44-181 592 3060
Fax 44-181-593 2140,
www.rhone-poulenc.com

Rhône-Poulenc Rorer,
500 Arcola Road,
Collegeville, PA 19426-0107, USA
610-454-8000
www.rhone-poulenc.com

Rhône-Poulenc Silico,
Cranbury, NJ 08512, USA
800-288-1175
www.rhone-poulenc.com

Rhône-Poulenc Surfactant,
Les Miroirs-Défense 3,
18 ave. d'Alsace, Cedex 29,
92097 Paris, LaDéfense, France
(33-1) 4768 1234
Fax (33-1) 4768 0900,
www.rhone-poulenc.com

Rhône-Poulenc Surfactant,
3265 Wolfdale Rd.,
Mississauga, Ont., L5C 1V8, Canada
905-270-5534
Fax 905-270-5816,
www.rhoône-poulenc.com

Rhône-Poulenc Surfactants,CN 7500,
Prospect Plains Rd.,
Cranberry, NJ 08512-7500, USA
609-860-4000, 800-922-2189
Fax 609-860-0459,
www.rhone-poulenc.com

Rhône-Poulenc Silico,
27 06/07 The Concourse,
300 Beach Road,
Singapore 0719
65-291-1921
Fax 65-296-6044,
www.rhone-poulenc.com

Rhône-Poulenc, Inc.,
CN7500, Prospect Plain Rd.,
Cranbury, NJ 08512-7500, USA
609-860-4000, 800-922-2189
Fax 609-860-0269,
www.rhone-poulenc.com

Rhône-Poulenc, Inc./Specialty Chemicals,
PO Box 769,
Marietta, GA 30061, USA
404-422-1250, 800-677-7412
Fax 404-427-0874,
www.rhone-poulenc.com

Rhône-Poulenc/Coatings & Paint,
CN 7500, Prospect Plains Rd.,
Cranbury, NJ 08512, USA
609-860-4600, 800-253-5052
www.rhone-poulenc.com

Rhône-Poulenc/Detergents,
CN 7500, Prospect Plains Rd.,
Cranbury, NJ 08512, USA
609-860-4600, 800-253-5052
www.rhone-poulenc.com

Rhône-Poulenc/Film Div.,
2754 West Park Dr.,
Holcomb, NY 14469, USA
716-657-5800
Fax 716-657-5838,
www.rhone-poulenc.com

Rhône-Poulenc/Perf. Resins,
1525 Church St. Ext.,
Marietta, GA 30060, USA
404-422-1250
Fax 404-427-0874,
www.rhone-poulenc.com

Rhône-Poulenc/Textiles,
1000 Hurricane Shoals Rd.,
Lawrenceville, GA 30243, USA
www.rhone-poulenc.com

Rhône-Poulenc/Tire & Rubber,
CN 7500,
Cranbury, NJ 08512-7500, USA
609-860-3068
Fax 609-395-1632,
www.rhone-poulenc.com

Rhône-Poulenc/Water Solvents,
CN 7500,
Cranberry, NJ 08512-7500, USA
609-860-4000, 800-922-2189
Fax 609-860-0459,
www.rhone-poulenc.com

Rhône-Poulenc/Water Treatment,
One Gatehall Dr.,
Parsippany, NJ 07054, USA
201-292-2900, 800-848-7659
Fax 201-292-5295,
www.rhone-poulenc.com

Richardson-Vicks Inc,
P.O. Box 5516,
Cincinnati, OH 45201, USA
800-358-8707

Ricon Resins, Inc.,
569 24 1/4 Road,
Grand Junction, CO 81505, USA
303-245-8148
Fax 303-245-4348

Ridge Technologies,
117 Lyons Rd.,
Basking Ridge, NJ 07920, USA
908-766-1915
Fax 908-204-0312

Rigby Taylor Ltd.,
The Riverway Estate,
Portsmouth Road, Peasmarsh,
Guildford, Surrey, GU3 1LZ, UK
44-1483-35657
Fax 44-1483-34058

Rit-Chem Co. Inc.,
109 Wheeler Ave.,
PO Box 435,
Pleasantville, NY 10570-0435, USA
914-769-9110
Fax 914-769-1408

Robeco Chemicals Inc.,
99 Park Ave.,
New York, NY 10016, USA
212-986-6410
Fax 212-986-6419

Robinson Brothers Ltd.,
Phoenix Street,
West Bromwich, West Midlands,
B70 0AH, UK
44-121-553-2451
Fax 44-121-500-5183

Roche Laboratories,
340 Kingsland St,
Nutley, NJ 07110, USA
201-235-3381
Fax 201-235-8023

Roche Vitamins,
340 Kingsland St.,
Nutley, NJ 07110-1199, USA
201-235-5000
Fax 201-535-7606

Rocol Ltd.,
Rocol House, Swillington,
Leeds, West Yorkshire, LS26 8BS, UK
44-1132-866511
Fax 44-1132 872159

Rogers Corp.,
One Technology Drive,
Rogers, CT 06263, USA
203-774-9605
Fax 203-774-9630

Rohm and Haas (Australia),
969 Burke Rd., PO Box 11,
Camberwell, Vic. 3124, Australia
www.rohmhaas.com

Rohm and Haas (UK) Ltd.,
Lennig House,
2 Mason's Avenue, Croydon,
Greater London, CR9 3NB, UK
44-181-686-8844
Fax 44-181-686 8329,
www.rohmhaas.com

Rohm and Haas Canada,
2 Manse Rd.,
West Hill, Ont., M1E 3T9, Canada
416-284-4711, 800-268-4201
Fax 416-284-2982,
www.rohmhaas.com

Rohm and Haas Co.,
100 Independence Mall West,
Philadelphia, PA 19106-2399, USA
215-592-3000, 800-323-4165
Fax 215-592-6909,
www.rohmhaas.com

Rohm and Haas Co. Europe,
Chesterfield House, Bloomsbury,
London, WC1A 2TP, UK
44-171-242-4455
Fax-44-171-404-4126,
www.rohmhaas.com

Ronacrete Ltd.,Ronac House,
Selinas Lane,
Dagenham, Essex, RM8 1QL, UK
44-1815-937621
Fax 44-181-595 6969

Roquette (UK) Ltd.,
Pantiles House,
2 Nevill St.,
Tunbridge Wells, Kent, TN2 5TT, UK
44-1892-500200
Fax 44-1892-510872

Ross Laboratories,
625 Cleveland Ave,
Columbus, OH 43216, USA
614-624-7677

Rostone Corp.,
PO Box 7497,
2450 Sagamore Pkwy. South,
Lafayette, IN 47903, USA
317-474-2421, 800-637-4851
Fax 317-474-8785

Roussel Laboratories,
Broadwater Park, North Orbital,
Uxbridge, Middlesex, UB9 5HP,
UK
44-1895- 834343
Fax 44-1895-834578

Roussel Uclaf, Fine Chemicals,
Tour Roussel Hoechst,
F-92080 Paris la Défense,
France
33-40-81-4884
Fax 33-40-90-0032

Rowell Laboratories,
Baudette, MN 56623, USA

Ruco Polymer Corp.,
New South Rd.,
Hicksville, NY 11802, USA
516-931-8100
Fax 516-931-8179

Ruetgers-Nease Chemicals,
201 Struble Rd.,
State College, PA 16801, USA
814-238-2424
Fax 814-238-1567

Rutgers Organics GmbH,
Postfach 310160,
Sanhofer Str. 95,
D-68261 Mannheim, Germany
49-621-7654-286
Fax 49-621-7654-413

Rystan Co. Inc.,
PO Box 214,
47 Center Avenue,
Little Falls, NJ 07424, USA
201-256-3737

S & D Chemicals Ltd.,
Cunningham House, Westfield
Lane,
Harrow, Middlesex, HA3 9ED, UK
44-181-907-8822
Fax 44-181-907-1798

S F C,
3 Rue des Carrires,
93800 Epinay s/Seine, France

Sachtleben Chemie GmbH,
Huntingdon House, Princess St.,
Bolton, BL1 1EJ, UK
44-1204-363634
Fax 44-1204-36-1144

Sandoz Inc,
608 5th Avenue,
New York, NY 10020, USA
212-307 1122

Sandoz AG,
Lichtstrasse 35,
CH-4002 Basel, Switzerland
61-24-1111
Fax 61-24-8081

Sandoz Pharmaceuticals,
Route 10,
East Hanover, NJ 07936, USA
201-503-7500

Sandoz Products Ltd.,
Frimley Business Park, Frimley,
Camberley, Surrey, GU16 5SG,
UK
44-1276-692255
Fax 44-1276-692508

Sanofi Bio-Industries,
620 Progress Ave.,
PO Box 1609,
Waukesha, WI 53187, USA
414-547-5531

Sanofi Bio-Industries,
8 Neshaminy Interplex,Suite 213,
Trevose, PA 19053, USA
215-638-7801
Fax 215-638-8168

Sanofi Winthrop Ltd.,
One Onslow Street,
Guildford, Surrey, GU1 4YS, UK
44-1483-505515
Fax 44-1483-35432

Sartomer Company Inc/America,
Oaklands Corp. Center,
502 Thomas Jones Way,
Exton, PA 19341, USA
610-363-4117, 800-SARTOMER
Fax 610-363-6849,
www.sartomer.com

Sartomer Company Inc/Asia,
331 Northbridge Rd,
#23-06 Odeon Towers,
188720, Singapore
65-334-2645, 65-334-2645
Fax 65-334-2647,
www.sartomer.com

Sartomer Company Inc/Europe,
Le Diamant B,
92970 Paris La Défense-Cedex,
France
33-1-4135-68-21
Fax 33-1-4135-62-70
www.sartomer.com

Sasolchem,
2 Sturdee Avenue, Rosebank,
Johannesburg, 2196, Rep. of S.
Africa
27-11-880-1322
Fax 27-11-880-1362

Savage Laboratories,
60 Baylis Road,
PO Box 2006,
Melville, NY 11747, USA

Schaefer Technologies,
3000 Carrollton Road,
Saginaw, MI 48604, USA
517-753-1877, 800-444-9034
Fax 517-753-3346

Schenectady Chemical,
319 Comstock Rd.,
Scarborough, Ont., M1L 2H3,
Canada

Schenectady Chemical,
PO Box 1046,
Schenectady, NY 12301, USA
518-370-4200
Fax 518-346-3111

Schenectady-Midland,
Four Ashes, Wolverhampton,
West Midlands, WV10 7BT, UK
44-1902 790555
Fax 44-1902 791640

Scher Chemicals, Inc,
Industrial West & Styertowne R,
PO Box 4317,
Clifton, NJ 07012, USA
201-471-1300
Fax 201-471-3783

Schering AG,
Postfach 650311,
Müllerstrasse 170-178,
D-13303 Berlin 65, Germany
30-4680
Fax 30-4685305,
www.schering.com

Schering Agrochemical,
Hauxton, Cambridge,
Cambridgeshire, CB2 5HU, UK
44-1223 870312
Fax 44-1223 872142,
www.schering.com

Schering Agrochemical,
Chesterford Park Industrial
Research Station,
Saffron Walden, Essex, CB10
1XL, UK
44-1799-30123
Fax 44-1799-30991,
www.schering.com

Schering Corp.,
Galloping Hill Rd,
Kenilworth, NJ 07033, USA
908-298-4000, 800-526-4099
www.schering.com

Schering Health Care,
The Brow, Burgess Hills,
West Sussex, RH15 9NE, UK
44-1444-232323
Fax 44-1444-246613,
www.schering.com

Schering Industrial,
Gorsey Lane, Widnes,
Cheshire, WA8 0HE, UK
44-151-495-1989
Fax 44-151-495-2003,
www.schering.com

Schweizerhall Inc.,
10 Corporate Place South,
Piscataway, NJ 08854, USA
908-981-8200, 800-243-6564
Fax 908-981-8282

Scientific Hospital,
100 Wavertree Blvd,
Wavertree Technology Park,
Liverpool, Merseyside, L7 9PQ,
UK
44-151 708-8008

Scott Bader Co. Ltd.,
Wollaston, Wellingborough,
Northants., NN8 7RL, UK
44-01933-663100
Fax 44-1933-665650

Scott Bader S.A.,
65 rue Sully,
F-80044-Amiens, France
33-22 44-42 00
Fax 33-22 44-06 24

Scottish Adhesives Co.,
9-23 Farnell Street,
Glasgow, KA30 8PR, UK
44-141-332-1736
Fax 44-141-332-3736

The Scotts Company Ltd
Professional Products,
Paper Mill Lane, Bramford,
Ipswich, Suffolk, IP8 LBZ, UK
44-1473-830492
Fax 44-1473-830386

Seal Ltd.,
46 Chesterfield Road,
Leicester, Leicestershire, LE5
5LP, UK
44-1162-739501

Seedcote Systems Ltd.,
Telford Way, Thetford,
Norfolk, IP24 1HU, UK
44-1842-66261
Fax 44-1842-66263

Selectokil Ltd.,
The Highlands,
A249 Detling, Maidstone,
Kent, ME14 3HT, UK
44-1622-734-214
Fax 44-1622-735-106

Seppic,
75 Quai d'Orsay,
75321 Paris Cedex 07, France
33-1-40-62-5555
Fax 33-1-40-62-5253

Seppic Inc.,
30 Two Bridges Rd., Suite 225,
Fairfield, NJ 07004, USA
201-882-5597
Fax 201-882-5178

Service Chemicals Ltd.,
17 Lanchester Way,
Royal Oak Industrial Estate,
Daventry, Northants., NN11 4PH, UK
44-1327-704444
Fax 44-1327-711154

Servier Laboratories,
Fulmer Hall, Windmill Road,
Fulmer, Slough, Berks., SL3 6HH, UK

Seton Healthcare Group,
Turbiton House,
Oldham,Lancs., OL1 3HS, UK
44-161-652-2222
Fax 44-161-626-9090

Sevalco Ltd.,
Severn Rd, Avonmouth,
Bristol, Avon, BS11 0YL, UK
44-1179-822611
Fax 44-1272 35444

Seven Seas Ltd.,
Marfleet, Hull,
Humberside, HU9 5NJ, UK
44-1482-275234
Fax 44-1482-74345

Shell Canada Chemical,
PO Box 100, Station M,
Calgary, Alberta, T2P 2H5, Canada
800-567-8728
Fax 800-567-8862,
www.shell.com

Shell Chemical Co.,
910 Louisiana Street,
Houston, TX 77252, USA
800-USA-SHELL
Fax 713-241-6367,
www.shell.com

Shell Chemical Co.,
One Shell Plaza, PO Box 2463,
Houston, TX 77252, USA
713-241-6161, 800-872-7435
Fax 713-241-4043,
www.shell.com

Shell Chemicals UK Ltd.,
Heronbridge House, Chester Business Park,
Wrexham Rd, Chester, Cheshire, CH4 9QA, UK
44-1244-685678
Fax 44-1244-685010,
www.shell.com

Sherex Chemical Co.,
PO Box 646,
5777 Frantz Rd.,
Dublin, OH 43017, USA
614-764-6500, 800-848-7370
Fax 614-764-6650,
www.witco.com

Siber Hegner (UK) Ltd.,
221-241 Beckenham Rd.,Makenzie House,
Beckenham, Kent, BR3 4UF, UK

Siber Hegner Ltd.,
County House,
221-241 Beckenham Road,
Beckenham,
Kent, BR3 4UF, UK
44-181-659-2345
Fax 44-181-659-1292

Sigma Coatings,
Fina House, Ashley Avenue,
Epsom, Surrey, KT18 5AD, UK
44-1372-726226
Fax 44-1372-744520,
www.sigmaaldrich.com

Sigma-Aldrich Chemical Australia,
Unit 2, 10 Anella Ave.,
Castle Hill, NSW, 2154, Australia
61-2-899-9977
Fax 61-2-899-9742,
www.sigmaaldrich.com

Sigma-Aldrich Chemical Corp.,
3050 Spruce Street,
St. Louis, MO 63103, USA
314-771-5765
www.sigmaaldrich.com

SIGV Division Agricultural,
Av. Rafael Casanova, 81,
08100 Mollet Del Valles-Barcelona, Spain
34-93-570-56-96
Fax 34-93-593-80-11

Sihi Pumps (UK) Ltd.,
Broadheath, Altrincham,
Cheshire, WA14 INB, UK
44-161-928-6371
Fax 44-161-928-3022

Silberline Ltd.,
Banbeath Rd.,
Leven, Fife, KY8 5HD, UK
44-1333-24734
Fax 44-1333-21369

Silberline Manufacturing,
Lincoln Drive, PO Box B,
Tamaqua, PA 18252-0420, USA
717-668-6050, 800-348-4824
Fax 717-668-0197

Siltech Inc.,
4437 Park Dr., Suite E,
Norcross, GA 30093, USA
770-279-8601, 800-849-8227
Fax 770-279-8535

SKW BioSystems Ltd,
SKW House, Kelvin Road,
Newbury, Berks., RG14 2DB, UK
44-1635-38343
Fax 44-1635-37896

Smith & Nephew plc,
2 Temple Place, Victoria Embankment,
London, WC2R 3BP, UK
44-171-836-7922
Fax 44-171-240-7088

SmithKline Beecham plc,
Clarendon Road,
Worthing, West Sussex, BN14 8QH, UK

Smooth-On, Inc.,
1000 Valley Rd.,
Gillette, NJ 07933, USA
201-647-5800
Fax 201-604-2224

Solec/Solar Energy Co.,
Box 3065,
Princeton, NJ 08543-3065, USA
609-883-7700
Fax 609-497-0182

Solochart Ltd.,
Brookhampton Lane,
Kineton,Warwickshire, CV35 0JA, UK

Solrec Ltd.,
Middleton Road, Heysham,
Morecambe, Lancs., LA3 3JW, UK
44-1524 853053
Fax 44-1524-851284

Solutia Australia Limited,
1/437 Canterbury Road,
Surrey Hills, Vic., 3127, Australia
61-3-98884539
www.solutia.com

Solutia Beijing,
C612 Bejing Lufthasan Center,
Beijing 100016, China
011-8610-463-8046
www.solutia.com

Solutia Canada,
2330 Argentina Rd.,
Mississauga, Ont., L5M 2G4, Canada
905-826-9222
www.solutia.com

Solutia Europe SA,
Parc Scientifique-Fleming,
Rue Laid Burniate, 3,
B-1348 Louvain-la-Neuve, Belgium
32-10-48-13-21
Fax 32-10-48-12-24,
www.solutia.com

Solutia Japan Limited,
Nihonbashi Daini Building, 41-12,
Nihonbashi Hakozaki-cho,
Chuo-ku, Tokyo 103, Japan
81-3-56441638
www.solutia.com

Solutia Mexico,
Bosques de Duranzo 61,3er Piso,
Bosques de las Lomas,
D. F. 11700 MX,Mexico
www.solutia.com

Solutia Moscow,
Volkov Lane 19,
Moscow, 123242, Russia
7-502-221-0000,
www.solutia.com

Solutia Singapore,
101 Thompson Rd,
#19-00 United Square,
Singapore 307591, Singapore
65-733-1611
www.solutia.com

Solutia USA
1460 Broadway,
New York, NY 10036,
212-382-9600
www.solutia.com

Solutia Venezuela,
Avenida Francisco de Miranda,
Edificio Parque Cristal,
Torre Este Piso 8 Ofc. 8-12,
Caracas, Los Palos Grandes,1062, Venezulea
58-2-285-0944
www.solutia.com

Solvay Enzymes GmbH,
Postfach 690307,
Hans Böckler Allee 20,
30612 Hannover, Germany
49-511-8570
Fax 49-511-8572371,
www.solvay.com

Solvay Enzymes, Inc.,
PO Box 4859,
1230 Randolph St.,
Elkhart, IN 46514-0859, USA
219-523-3700, 800-342-2097
Fax 219-523-3800,
www.solvay.com

Solvay Interox Ltd,
PO Box 7, Warrington,
Cheshire, WA4 6HB, UK
44-1925-651277
Fax 44-1925-655856,
www.solvay.com

Solvay Polymers Inc.,
3333 Richmond Ave.,
PO Box 27328,
Houston, TX 77227-7328, USA
713-525-4000, 800-231-6313
Fax 713-522-7890,
www.solvay.com

Solvay S.A.,
33 rue du Prince Albert,
B-1050 Brussels, Belgium
32-509-6111
Fax 32-509-6617,
www.solvay.com

Manufacturers and Suppliers

Sorex Ltd.,
St Michael's Industrial Estate,
Hale Road, Widnes,
Cheshire, WA8 8TJ, UK
44-151-420-7151
Fax 44-151-495-1163

Specialty Chem Products,
Two Stanton St.,
Marinette, WI 54143, USA
715-735-9033
Fax 715-735-5304

Specialty Chemical Chemicals,
PO Box 2606,
Cleveland, TN 37311, USA
615-479-9664
Fax 615-472-6158,
www.bayer.com

Specialty Products Ltd.,
227 Norseman St.,
Toronto, Ont., M8Z 2R5, Canada
416-239-6541
Fax 416-231-7264,
www.bayer.com

Specialty Products,
75 Montgomery St.,
PO Box 306,
Jersey City, NJ 07303-0306, USA
201-434-4700, 800-321-8506
Fax 201-434-6052,
www.bayer.com

Spectra Brands plc,
Factory 475, Treloggan Industrial Estate,
Newquay, Cornwall, TR7 2SX, UK
44-1637 871171
Fax 44-1637 878627

Sphere Laboratories,
The Yews, Main Street,
Chilton, Oxfordshire, OX11 0RZ, UK
44-1235-833896

Stafford-Miller Contine,
Nijverheidsstraat 9,
B-2431 Oevel, Belgium
31-14-58-9481
Fax 31-14-58-9485

Standard Chem. & Pharmaceuticals,
P.O. Box 74,
Hsin-Ying 730,
Taiwan
886-6-6362916-244
Fax 886-6-6362515

Standard Oil Chemicals,
200 Public Square,
Cleveland, OH 44114-2375, USA

Standard Tar Products,
2456 W Cornell Street,
Milwaukee, WI 53209-6294, USA
414-873-7650, 800-825-7650
Fax 414-873-7737,
www.standardtar.com

Star Agrochem Ltd.,
Odder Farm, Saxilby Road,
Burton, Lincoln, Lincs., LN1 2BB, UK
44-1522 03777

Steetley Chemicals Ltd.,
Berk House, PO Box 56,
Basing, Basingstoke, Hants.,
RG21 2EG, UK

Steetley Magnesia Products,
PO Box 8, Hartlepool Works,
Hartlepools, Cleveland, TS24 0BY, UK
44-1429-267071
Fax 44-1429-266600

Steetley Quarry Products,
PO Box 128,
Woodville, OH 43469, USA
419-849-2321, 800-445-3930
Fax 419-849-3589

Stella-Meta Filters,
Laverstoke Mill,
Whitchurch, Hants., RG28 7NR, UK
44-1256-895959
Fax 44-1256-892074

Stepan Canada,
PO Box 307,
Orillia, Ont., L3V 6J6, Canada
705-326-7329
Fax 705-326-4523,
www.stepan.com

Stepan Co.,
22 West Frontage Rd.,
Northfield, IL 60093, USA
708-446-7500, 800-745-7837
Fax 708-501-2443,
www.stepan.com

Stepan Europe,
BP127,
38340 Voreppe, France
33-76-50-51-00
Fax 33-7656-7165,
www.stepan.com

Stepan/PVO.
100 West Hunter Ave.,
Maywood, NJ 07607, USA
201-845-3030
Fax 201-845-6754,
www.stepan.com

Stephenson Thompson Textiles,
P O Box 305, Listerhills Road,
Bradford, West Yorkshire, BD7 IHY, UK
44-1274-723811
Fax 44-1274-370108

Sterling Research Laboratories,
Sterling-Winthrop House,Onslow Street,
Guildford, Surrey, GU1 4YS, UK
44-1483-505515
Fax 44-1483-35432

Sterwin Chemicals Inc.,
PO Box 537,
Rt 113-Dupont Highway,
Millsborough, DE 19966, USA

Stockhausen, Inc.,
2401 Doyle St.,
Greensboro, NC 27406, USA
910-333-3500
Fax 910-333-3545

Stoller Chemicals Ltd.,
53 Bradley Hall Trading Estate,
Bradley Lane, Standish, Lancs,
WN6 0XQ, UK
44-1257-427722
Fax 44-1257-427888,

Stoner Inc.,
1070 Robert Fulton Hwy.,
PO Box 65,
Quarryville, PA 17566, USA
717-786-7355, 800-227-5538
Fax 800-515-5150

Stowlin Ltd.,
Radnor Rd, South Wigston,
Leicester, Leicestershire, LE18 4XY, UK
44-1162-785373
Fax 44-1162-772616

Struktol Co.,
201 E. Steels Corner Rd., PO B,
Stow, OH 44224-0649, USA
216-928-5188, 800-327-8649
Fax 216-928-8726

Struktol Co., Ltd.,
60 Venture Dr., Unit 23,
Scarborough, Ont., M1B 3S4, Canada
416-286-4040
Fax 416-286-4043

Stull Chemical Company,
PO Box 47907,
San Antonio, TX 78265, USA

Suchema AG,
Haupstrasse 15,
CH-8251 Kaltenbach,
Switzerland
49-41-54-41-1265
Fax 49-41-54-41-3234

Surfachem Ltd.,
Wellington Park House, Thirsk,
Leeds, West Yorkshire, LS1 4DP, UK
44-1132-342636
Fax 44-1132 445910

Surmak Products Ltd.,
99 Mabgate,
Leeds, West Yorkshire, LS9 7DR, UK
44-1132-450371
Fax 44-1132 428701

Sutton Laboratories,
116 Summit Ave., PO Box 837,
Chatham, NJ 07928-0837, USA
201-635-1551
Fax 201-635-4964

Swale Coatings & Ink Ltd.,
Taylor Rd, Trafford Park,
Urmston, Manchester, M31 2TE, UK
44-161-748-7340
Fax 44-161-748-7685

Sybron Chemicals Canada,
666 Appleby Line,
Burlington, Ont., L7L5Y3,
Canada
905-637-3337, 909-637-6198,
www.sybronchemicals.com

Sybron Chemicals Inc,
PO Box 66, Birmingham Rd,
Birmingham, NJ 8011, USA
609-893-1100, 800-678-0020
www.sybronchemicals.com

Sybron Chemie Nederlands,
Postbus 46,
NL-6710 BA Ede, The Netherlands
31-8380-70911
Fax 31-8380-30236,
www.sybronchemicals.com

Synchemicals Ltd.,
Owen Street,
Coalville, Leicestershire, LE6 2DE, UK
44-1530-510060
Fax 44-1530-510299

Synthetic Chemicals,
Four Ashes, Wolverhampton,
West Midlands, WV10 7BP, UK
44-1902-794000
Fax 44-1902-794300

Synthetic Rubber Technology,
3898 Shawnee St. N.W., PO Box 639,
Uniontown, OH 44685, USA
216-699-1256
Fax 216-699-1404

Synthite Ltd.,
Alyn Works, Denbigh Rd,
Mold, Clwyd, CH7 1B7, Wales UK
44-1352-752521
Fax 44-1352-700182

TACC International,
Air Station Industrial Park,
Rockland, MA 02370, USA
617-878-7015
Fax 617-871-6727

Taiwan Surfactant Co,
8 Fl., No 11, Sec. 1,
Chung Shan North Road,
Taipei, Taiwan, R.O.C.
886-2-2541-1122
Fax 886-2-2542-3773,
www.taiwansurfactant.com.tw

Tego Chemie Service,
Gerlingstr. 64,
D-45139 Essen, Germany
49-201-173-06
Fax 49-201-173-1939

Tego Chemie Service,
PO Box 1299,
914 E. Randolph Rd.,
Hopewell, VA 23860, USA
804-541-8658, 800-446-1809
Fax 804-541-2783

Teknor Apex Co.,
505 Central Ave.,
Pawtucket, RI 02861, USA
401-725-8000, 800-554-9893
Fax 401-724-6250

Tennant-KVK Ltd.,
69 Grosvenor St,
London, W1X 0BP, UK
44-171-493- 5451
Fax 44-171-491-4922

Tenneco Malros Ltd.,
Rockingham Works, Avonmouth,
Bristol, Avon, BS11 0YT, UK

Testworth Laboratories,
401 S Main Street, PO Box 91,
Columbia City, IN 46725, USA
219 244-5137
Fax 219 244-5138

Texaco Chemical Co.,
PO Box 27707,
Houston, TX 77227-7707, USA
713-961-3711
Fax 713-235-6437,
www.texaco.com

Texaco Chemical Co./Oxides & Specialties,
4800 Fournace Place, PO Box 430,
Bellaire, TX 77401, USA
www.texaco.com

Texaco France S.A.,
5, rue Bellini, Tour Arago,
F-92806 Puteaux Cedex, France
33-1-47-17-2602
Fax 33-1-47-76-3050,
www.texaco.com

Texapol Corp.,
177 Mikron Rd.,
Bethlehem, PA 18017, USA
215-759-8222, 800-523-9242
Fax 215-759-9433

Thermal Ceramics,
2102 Old Savannah Rd.,
Augusta, GA 30906, USA
706-796-4200, 800-KAOWOOL
Fax 706-796-4328

Thermedics Inc.,
470 Wildwood Street, PO Box 29,
Woburn, MA 01888-1799, USA

Thermoset Plastics Inc.,
5101 East 65th St., PO Box 20902,
Indianapolis, IN 46220, USA
317-259-4161
Fax 317-252-8402

Thibaut & Walker Co.,
PO Box 296,
49 Rutherford St.,
Newark, NJ 07101, USA
201-589-3331
Fax 201-589-7231

Thomas Swan & Co. Ltd.,
Crookhall, Consett,
Co Durham, DH8 7ND, UK
44-1207-505131
Fax 44-1207-590467

Thor Chemicals Ltd. (UK),
Earl Rd., Cheadle Hulme,
Cheshire, SK8 6QP, UK
44-161-486-2028
Fax 44-161-488-4155

Thor Chemicals, Inc.,
Brook House, 37 North Ave.,
Norwalk, CT 06851, USA
203-846-8613
Fax 203-846-4810

Tiarco Chemical Div.,
1300 Tiarco Drive,
Dalton, GA 30720, USA
706-277-1300
Fax 706-277-9039

Tilcon Ltd.,
Conyngham Hall,
Knaresborough,
North Yorkshire, HG5 9AY, UK
44-1423-862841
Fax 44-1423-864555

Tillots Laboratories,
Unit 24, Henlow Trading Estate,
Henlow, Beds., SG16 6DS, UK

Tioxide Group Ltd.,
Lincoln House,
137-143 Hammersmith Road,
London, W14 0QL, UK
44-171-331-7777
Fax 44-171-331-7778

Tosoh Canada Ltd.,
1200 Sheppard Ave. East, Suite 511, Willowdale, Ont., M2K 2S5,
Canada
416-756-2226
Fax 416-756-2750,
www.tosoh.com

Tosoh Corp.,
7-7 Akasaka 1-chome,
Minato-ku,Tokyo 107, Japan
81-3-3582-8120, www.tosoh.com

Tosoh Europe B.V.,
World Trade Centre Amsterdam,
Strawinskylaan 1351,
1077 XX Amsterdam, The Netherlands
31-20-644026, 31-20-623412
www.tosoh.com

Tosoh USA Inc.,
1100 Circle 75 Pkwy., Suite 60,
Atlanta, GA 30339, USA
770-956-1100
Fax 770-956-7368,
www.tosoh.com

Tra-Con, Inc.,
45 Wiggins Ave.,
Bedford, MA 1730, USA
617-275-6363, 800-872-2661
Fax 617-275-9249

Travenol Laboratories Inc.,
One Baxter Parkway,
Deerfield, IL 60015, USA

Tremco Ltd./Adhesive,
86-88 Bestobell Road,
Slough, Berks., SL1 4SZ, UK
44-1753-691696
Fax 44-1753-822640

Tri-K Industries,
PO Box 128,
Northvale, NJ 07647-0128, USA
201-261-2800, 800-526-0372
Fax 201-261-1432

Tri-Star Chem. Co.,
PO Box 38627,
Dallas, TX 75238, USA
214-341-0054

Troy Chemical Co.,
Uiverlaan 12e, PO Box 132,
3140 AC Maassluis, The Netherlands
31-10-5927494
Fax 31-10-592-8877,
www.troy.com

Troy Chemical Co. UK,
Zenith House, Northolme Rd.,
Louth, Lincs., LN11 0HQ, UK
44-1507-609606
Fax 44-1507-607107,
www.troy.com

Troy Chemical Co., Ltd.,
157 Overture Rd.,
Scarborough, Ont., M1E 2W5,
Canada
416-287-9116
Fax 416-287-9779,
www.troy.com

Troy Chemie GmbH,
Uerdingerstrasse 541,
47800 Krefeld 1, Germany
49-2151-59-03-38
Fax 49-2151-59-81-45,
www.troy.com

Troy Corp.,
PO Box 955, 8 Vreeland Rd.,
Florham Park, NJ 07932-0955,
USA
201-443-0003
Fax 201-443-0257,
www.troy.com

Truchem Ltd.,
Brook House, 30 Larwood Grove,
Sherwood, Nottingham, Notts.,
HG5 3JD,UK
44-1159-260762
Fax 44-1602-671153

Tufnol Ltd.,
PO Box 376, Perry Barr,
Birmingham, West Midlands, B42 2TB, UK
44-121-356-9351
Fax 44-121-331-4235

Typharm Ltd.,
16 Parkstone Road,
Poole, Dorset, BH15 2PG, UK
44-1202-666626
Fax 44-1202-666309

U.S. Biochemical (UK),
25 Signet Court, Newmarket Rd,
Cambridge, Cambridgeshire,
CB5 8LA,UK
44-1223-467064
Fax 44-1223-60732

U.S. Biochemical Corp.,
PO Box 22400,
Cleveland, OH 44122, USA
216-765-5000, 800-321-9322
Fax 216-464-5075

U.S. Borax Inc.,
26877 Tourney Rd.,
Valencia, CA 91355-1847, USA
805-287-5400, 800-US BORAX
Fax 800-626-4872

U.S. Cosmetics,
313 Lake Rd., PO Box 859,
Dayville, CT 06241, USA
203-779-3990, 800-752-0490
Fax 203-779-3994

UCB (Chem) Ltd.,
Star House, 69 Clarendon Road,
Watford, Herts., WD1 1DJ, UK
44-1923-248011
Fax 44-1923-250225

UCB Chemical Corp. Malaysia,
PT 12701, Tuanku Jaafar
Industrial Park,
71450 Seremban, Negeri
Sembilan D.K.,
Malaysia
60-6-675-1112
Fax 60-6-675-1115

UCB Chemicals Corp.,
2000 Lake Park Dr.,
Smyrna, GA 30080, USA
770-434-6188, 800-433-2873
Fax 770-434-8314

UCB N.V. Filmsektor,
Ottergemsesteenweg 801, PO
Box 369,
B-9000 Ghent, Belgium
091-40-32-11
Fax 091-40-88-00

UCB S.A./Chemical,
33 rue d'Anderlecht,
B-1620 Drogenbos, Belgium
32-2-3714923
Fax 32-2-3714924

UCB S.A./Chemical,
60, Allée de la Recherche
B-1070 Brussels, Belgium
32-2-641-1411
Fax 32-2-640-9860, www.ucb.be

UCIB/Usines Chimique,
Route D'Oulins,
28260 Anet, France
33-37-62-8200
Fax 33-37-41-9132

Ultrachem Inc.,
The Galleria, 2 Bridge St,
Red Bank, NJ 7701, USA
732-224-0200
732-224-0017,
www.ultrachem.com

ULTRAFINE Chemicals,
Synergy House, Guildhall Close,
Manchester Science Park,
Manchester, M15 6SY, UK
44-161-226-8774
Fax 44-161-227-9758,
www.u-net/~ultrfine

Unger Fabrikker AS,
PO Boks 254,
N-1601 Fredrikstad, Norway
47-69-32-0020
Fax 47-69-32-3775,
www.unger.no/

Unichem,
11211 FM 2920,
Tomball, TX 77375, USA
713-357-2700
Fax 713-357-2701,
www.unichem.com

Unichem plc,
Unichem House, Cox Lane,
Chessington, Surrey, KT9 1SH,
UK
44-181-391-2323
Fax 44-181-974-1707,
www.unichem.com

Unichem, Inc.,
P.O. Box 612,
916 Main St.,
Haw River, NC 27258, USA
910-578-5476
Fax 910-578-1871,
www.unichem.com

Unichema Australia,
164 Ingles St.,
Port Melbourne,Vic., 3207,
Australia
61-3-647-9311
Fax 61-3-645-3001

Unichema Chemicals Ltd.,
Bebington,
Wirral, Merseyside, L62 4UF, UK
44-151-645-2020
Fax 44-151-645-9197

Unichema Chemie B.V.,
Postbus 2,
2800 AA Gouda, The
Netherlands
31-1820-42911
Fax 31-1820-42250

Unichema Chemie GmbH,
Postfach 100963,
D-46249 Emmerich, Germany
49-2822-720
Fax 49-2822-72276

Unichema France S.A.,
148 Boulevard Haussemann,
75008 Paris, France
33-1-44-95-0840
Fax 33-1- 42-563188

Unichema International,
Postbus 2,
2800 AA Gouda, The
Netherlands
31-1820-42911
Fax 31-1820-42250

Unichema North America,
4650 S. Racine Ave.,
Chicago, IL 60609, USA
312-376-9000, 800-833-2864
Fax 312-376-0095

Unigreg Ltd.,
Enterprise House, 181-189 Garth
Road,
Morden, Surrey, SM4 4LL, UK
44-181-330-1421
Fax 44-181-330-6812

Unilever plc,
PO Box 68, Unilever House,
Blackfriars,
London, EC4P 4BQ, UK
44-171- 822-5252
Fax 44-171-822-5951,
www.unilever.com

Unimin Canada Ltd.,
RR #4, PO Box 2000,
Havelock, Ont., K0L 1Z0, Canada
705-877-2210, 800-363-4140
Fax 705-877-3343

Unimin Corp.,
258 Elm St.,
New Canaan, CT 06840, USA
203-966-8880, 800-243-9004
Fax 203-966-3453

Union Camp Chemicals,
Vigo Lane,
Chester-le-Street, Co. Durham,
DH3 2RB, UK
44-91-410-2631
Fax 44-91-410-9391,
www.unioncamp.com

Union Camp Corp./Chemicals,
1600 Valley Rd.,
Wayne, NJ 07470, USA
201-628-2290, 800-733-1374
Fax 201-628-2840
www.unioncamp.com

Union Camp Corp./Chemicals,
PO Box 2668,
Savannah, GA 31402, USA
912-238-6000
www.unioncamp.com

Union Carbide (Europe),
7, rue du Pre-Bouvier,
CH-1217 Meyrin (Geneve),
Switzerland
41-22-989 66 53
Fax 41-22-989 6545,
www.unioncarbide.com

Union Carbide (UK),
93-95 High Street,
Rickmansworth, Herts., WD3
1RB, UK
44-1923 720 366
Fax 44-1923-896721,
www.unioncarbide.com

Union Carbide Canada,
7400 Blvd des Galleries,
d'Anjou, PQ, H1M 3M2, Canada
514-493-2610
Fax 514-493-2619,
www.unioncarbide.com

Union Carbide Corp.,
39 Old Ridgebury Rd.,
Danbury,CT,06817-0001, USA
203-794-2000, 800-335-8550
Fax 203-794-3170,
www.unioncarbide.com

Union Carbide Corp.,
PO Box 12014,
Research Triangle Park, NC
27709, USA
www.unioncarbide.com

Uniroyal Chemical Co.,
199 Benson Road,
Middlebury, CT 06749, USA
203-573-2269, 800-322-3243
Fax 203-573-2165,
www.uniroyal.com

Uniroyal Chemical Ltd.,
Kennet House, 4 Langley Quay,
Slough, Berks., SL3 6EH, UK
44-1753-603000
Fax 44-1753-603078,
www.uniroyal.com

United Catalysts Inc.,
PO Box 32370,
Louisville, KY 40232, USA
502-634-7200
Fax 502-637-3732

United Coconut Chemicals,
UCPB Bldg., 17th Fl.,
Makati Ave., Makati,
Metro Manila, Philippines
632 815-4104
Fax 632 817-2251

United Composites, Inc.,
PO Box 180601,
Arlington, TX 76096-0601, USA
817-468-2929
Fax 817-468-3122

Unitex Chemical Corp.,
PO Box 16344,
520 Broome Rd.,
Greensboro, NC 27406, USA
910-378-0965
Fax 910-272-4312

Unitex Ltd.,
Halfpenny Lane, Knaresborough,
North Yorkshire, HG5 0PP, UK
44-1423-862677
Fax 44-1423-868340

Universal Chemicals & Coatings,
1975 Fox Lane,
Elgin, IL 60123, USA
708-931-1700
Fax 708-931-1799

Universal-Matthey Products,
Jeffreys Road, Brimsdown,
Enfield, Middlesex, EN3 7PN, UK

Manufacturers and Suppliers

Unocal Hydrocarbon Sales,
1701 Golf Rd., Suite 1-1101,
Rolling Meadows, IL 60008-4295, USA
708-734-7622, 800-967-7601
Fax 708-734-7677,
www.unocal.com

Upjohn Co./Fine Chemicals,
7000 Portage Rd.,
Kalamazoo, MI 49001, USA
616-323-5505, 800-253-8600
Fax 616-329-3604,
www.upjohn.com

Upjohn Ltd.,
PO Box 8, Fleming Way,
Crawley, West Sussex, RH10 2NJ, UK
44-1293-531133
Fax 44-1293-548850,
www.upjohn.com

Van Den Bergh Foods,
2200 Cabot Dr.,
Lisle, IL 60532, USA
708-505-5300, 800-949-7344
Fax 708-955-5497

Vedag GmbH,
Flinschstrasse 10-16,
60388 Frankfurt/M, Germany

Veitsiluoto Oy/Fores,
PO Box 196,
SF-90101 Oulu 10, Finland
358-81-316 3111
Fax 358-81-378 5755

Vencel Resil Ltd.,
Arndale House,
18-20 Spital Street,
Dartford, Kent, DA1 2HT, UK
44-1322-626600
Fax 44-1322-626610

Venture Chemical Co.,
P.O. Box 88,
Adelphia, NJ 07710, USA
908-780-7171
Fax 908-462-3644

Vianova Resins GmbH,
Rheingaustrabe 190-196,
65203 Wiesbaden, Germany
49-611-962-5261
Fax 49-611-962-9343

Vinamul Ltd.,
Mill Lane, Carshalton,
Surrey, SM5 2JU, UK
44-181-669-4422
Fax 44-181-669-3189

Virkler Chemical Co.,
12345 Steele Creek Road,
Charlotte, NC 28273, USA

Vistakon Inc.,
4500 S Salisbury Road,
Jacksonville, FL 32216, USA
904-443-1000

Vitax Ltd.,
Owen Street,
Coalville, Leicestershire,LE67 3DE, UK
44-1530-510060
Fax 44-1530-510299

Vulcan Plastics Ltd.,
Hosey Hill, Westerham,
Kent, TN16 1TB, UK
44-1959-562304

W.A. Cleary Chemical,
1049 Route 27,
Somerset, NJ 08875-0010, USA
908-247-800 Fax 908-247-6977

W.A. Cleary Chemical,
Southview Industrial Park,
178 Ridge Rd., Suite A,
Dayton, NJ 08810, USA
908-329-8399, 800-524-1662
Fax 908-274-0894

W.J. Rendell Ltd.,
Ickleford Manor,
Hitchin, Herts., SG5 3XE, UK
44-1462-432596

W.J. Ruscoe Co.,
483 Kenmore Blvd.,
Akron, OH 44301, USA
216-253-8148
Fax 216-253-2933

W.R. Grace/Davison Co.,
PO Box 2117,
Baltimore, MD 21203-2117, USA
301-659-9000
Fax 410-659-9213,
www.grace.com

Wacker Chemicals (US),
3301 Sutton Rd.,
Adrian, MI 49221, USA
517-264-8131, 800-485-3686
Fax 517-264-8795

Wacker Chemicals Ltd.,
The Clock Tower,
Mount Felix,Bridge Street,
Walton-on-Thames, Surrey, KT12 1AS, UK
44-1932-246111
Fax 44-1932-240141

Wacker Silicones Corp.,
3301 Sutton Rd.,
Adrian, MI 49221-9397, USA
517-264-8500, 800-248-0063
Fax 517-264-8246,
www.wacker.silicones.com

Wacker-Chemie GmbH,
Hanns-Seidel-Platz 4,
D-81737 München, Germany
49-89-6279-01
Fax 49-89-6279-1770,
www.wacker.de

Wako Chemicals GmbH,
Nissanstrasse 2,
D-41468 Neuss, Germany
49-2131-3110
Fax 49-2131-31-1100,
www.wako-chem.co.ip

Wako Chemicals USA
1600 Bellwood Rd.,
Richmond, VA 23237, USA
804-271-7677
Fax 804-271-7791,
www.wako-chem.co.ip

Wako Pure Chemical Industries Ltd.,
1,2-Doshomachi 3-Chome,
Chuo-ku, Osaka 541, Japan
81-6-203-3741
Fax 81-6-222-1203,
www.wako-chem.co.ip

Wallace Laboratories,
Cranbury, NJ 08512, USA
609-665-6000

Warwick International Ltd.,
Wortley Moor Road,
Leeds, West Yorkshire, LS12 4JE, UK
44-1132-637331
Fax 44-1132-794795

Weatherguard/Marble,
2515 Newbold Avenue,
Bronx, NY 10462, USA

Werner G. Smith, Inc,
1730 Train Ave.,
Cleveland, OH 44113, USA
216-861-3676, 800-535-8343
Fax 216-861-3680

Westbrook Lanolin Co.,
Argonaut Works, Laisterdyke,
Bradford, West Yorkshire, BD4 8AU, UK
44-1274-663331
Fax 44-1274-667665

Westvaco Corp., Chemicals,
PO Box 70848,
Charleston Hts., SC 29415-0848, USA
803-740-2300
Fax 803-740-2329

Westvaco Corp., Custom Chemicals,
PO Box 237, Hwy. 60 East,
Mulberry, FL 33860, USA
813-425-3043

Westwood Chemical Co.,
46 Tower Drive,
Middletown, NY 10940, USA
914-692-6721
Fax 914-695-1906

Weyl America,
201 Struble Rd.,
State College, PA 16801, USA
814-231-9261, 800-458-3434
Fax 814-238-4235

Whitehall Laboratories,
22-24 Torrington Place,
London, WC1E 7ET, UK
44-171-636-8080
Fax 44-171-580-6037

Whitfield Plastics,
Unit Ten, Chrystleton Court,
Manor Park,
Runcorn, Cheshire, WA7 1SU, UK
44-1928-571000
Fax 44-1928-571010

Whitney & Oettler Inc.,
PO Box 8024,
Savannah, GA 31412, USA
912-232-7166

William F. Nye, Inc.,
PO Box 8927,
New Bedford, MA 02742-8927, USA
508-996-6721
Fax 508-997-5285

William H. Rorer Inc.,
500 Virginia Drive,
Fort Washington, PA 19034, USA

William Pearson Ltd.,
Clough Road,
Hull, North Humberside, HU6 7QA, UK
44-1482-443151
Fax 44-1482-440444

Winchem Ltd.,
7-11 Claremont Rd,
West Byfleet, Weybridge,
Surrey, KT14 6DY, UK
44-1932-340597
Fax 44-1932-340459

Windsor Healthcare Ltd.,
Ellesfield Avenue,
Bracknell, Berks., RG12 4YS, UK

Witco (Europe) S.A.,
7, rue du Pre-Bouvier,
CH-1217 Meyrin, Geneva, Switzerland
41-22-989-2111
Fax 41-22-989-2391,
www.witco.com

Witco Asia Pacific Pte Ltd,
12 Science Park Drive,
#03-04 The Mendel,
Singapore Science Park, 118225, Singapore
65-774-4800
Fax 65-770-5148,
www.witco.com

Witco Canada Ltd.,
565 Coronation Drive,
Westhill, Ont., M1E2K3, Canada
416-497-991
Fax 416-284-8141,
www.witco.com

Witco Chemical Ltd.,
Paragon Works, Baxenden,
Near Accrington,
Lancs., BB5 2SL, UK
44-125-439-8616
Fax 44-125-439-8586,
www.witco.com

Witco Corp.,
One American Lane,
Greenwich, CT 06831-2559, USA
203-552-2000, 800-779-4826
x6400
Fax 203-552-2010,
www.witco.com

Witco Corp./Allied-Kelite,
17050 Lathrop Ave.,
Harvey, IL 60426, USA
800-323-9784
www.witco.com

Witco Corp./Concarb.,
10500 Richmond, Suite 116,
PO Box 42817,
Houston, TX 77242-2817, USA
713-978-5745, 800-231-4591
Fax 713-978-5728,
www.witco.com

Witco Corp./Lubricants,
10100 Santa Monica Blvd., Suite
1470,
Los Angeles, CA 90067-4183,
USA
310-277-4511, 800-429-4826
Fax 310-201-0383,
www.witco.com

Witco Corp./Organics,
One American Lane,
Greenwich, CT 06831-2559, USA
203-552-2000
Fax 203-552-2010,
www.witco.com

Witco do Brasil Ltda,
Rua Verbo Divino 16661 cj 64,
Sao Paulo, 04719-002, Brasil
55-11-5181-2799
Fax 55-11-5181-7972,
www.witco.com

Witco Surfactants UK,
Paragon Works,
Baxenden, Nr. Accrington,
Lancs., BB5 2SL, UK
44-1254-39-8616
Fax 44-1254-39-8586,
www.witco.com

Witco UK,
Union Lane, Droitwich,
Worcestershire, WR9 9BB, UK
44-190 579 4795
Fax 44-190 579 4002,
www.witco.com

WWH (Witco Corp.),
One American Lane,
Greenwich, CT 06831-2559, USA
203-552-2000, 800-779-4826
Fax 203-552-2010,
www.witco.com

Wyeth Laboratories,
Huntercombe Lane South,
Taplow,
Maidenhead, Berks., SL6 0PH,
UK
44-1628-604377
Fax 44-1628-666368,
www.ahp.com

Wyeth Laboratories Inc.,
P.O. Box 8299,
Philadelphia, PA 19101, USA
215-688-4400
www.ahp.com

Yokkaichi Chemical Chemicals,
2-1, Miyahigashi-cho 2-chome,
Yokkaichi-shi, Mie, 510, Japan
81-593-45-1161
Fax 81-593-45-1168

Yorkshire Chemicals,
Kirkstall Rd,
Leeds, West Yorkshire, LS3 1LL,
UK
44-1132-443111
Fax 44-1132-421670

Yorkshire Pat-Chem Inc.,
11 Worley Rd., PO Box 1926,
Greenville, SC 29602, USA
803-233-3941, 800-443-9358
Fax 803-232-3542

Zeelan Industries Inc.,
3M Center, Bldg. 220-8E-04,
St. Paul, MN 55144-1000, USA
617-737-1751
Fax 612-737-1764

Zeeland Chemicals, Inc.,
215 N. Centennial St.,
Zeeland, MI 49464, USA
616-772-2193, 800-223-0453
Fax 616-772-7344,
www.cambrex.com

Zeon Chemicals, Inc.,
4111 Bells Lane,
Louisville, KY 40211, USA
800-735-3388

Zeon Europe GmbH,
Am Seester 18 (Euro Center),
D-40547 Düsseldorf 11, Germany
49-211-52670
Fax 49-211-5267-160

Zohar Detergent Factory,
19239 Kibbutz Daila, Israel
972-4-9897-234
Fax 972-4-9897-200,

Zschimmer & Schwarz,
Postfach 2179,
Max-Schwarz-Str. 3-5,
D-56112 Lahnstein/Rhein,
Germany
49-2621-12-276
Fax 49-2621-12407

Zschimmer & Schwarz France,
10 rue Saint-Marc,
F-75002 Paris, France,
33-42-33-1033
Fax 33-40-26-2381,

Zschimmer & Schwarz Italy,
via Vercelli 81,
13038 Tricerro, Italy
39-161-821421

Zyma S.A.,
Route de l'Etraz,
CH-1260 Nyon, Switzerland
41-22-63-3111
Fax 41-22-62-1744,